THE AUTHORITY SINCE 1868

THE WORLD ALMANAC®
AND BOOK OF FACTS
2000

WORLD ALMANAC BOOKS

THE WORLD ALMANAC®
AND BOOK OF FACTS
2000

Editorial Director: Robert Famighetti
Deputy Editor: William A. McGeveran, Jr.
Senior Editor: Lori P. Wiesenfeld
Associate Editors: Beth R. Ellis, Kevin Seabrooke
Desktop Publishing Associate: Elizabeth J. Lazzara
Contributing Editors: Jacqueline Laks Gorman, Geoffrey M. Horn, Eileen O'Reilly, Donald Young
Cover: Bill Smith Studio

PRIMEDIA REFERENCE INC.
General Manager: Alfred De Seta
Director of Editorial Production: Andrea J. Pitluk
Director–Purchasing and Production: Edward A. Thomas
Director–Information Technology: Bill C. Chehade
Managing Editor: Gail S. Schultz

Associate Editor:	**Production Editor:**	**Publishing Systems Associate:**
Ileana Parvulescu	Donna J. Schindler	Christy A. Gera

Desktop Publishing Assistant: Hana Shaki
Director of Indexing Services: Marjorie B. Bank
Index Editor: Walter Kronenberg

WORLD ALMANAC BOOKS
Vice President–Sales and Marketing:	**Marketing Manager:**
James R. Keenley	Jacqueline J. Ogle

The editors acknowledge with thanks the many letters of helpful comment and criticism from readers of THE WORLD ALMANAC. Because of the volume of mail directed to the editorial offices, it is not possible to reply to each letter writer. However, every communication is read by the editors and all comments and suggestions receive careful attention. THE WORLD ALMANAC's e-mail address is Walmanac@aol.com.

THE WORLD ALMANAC does not decide wagers.

The first edition of THE WORLD ALMANAC, a 120-page volume with 12 pages of advertising, was published by the New York World in 1868. Annual publication was suspended in 1876. Joseph Pulitzer, publisher of the New York World, revived THE WORLD ALMANAC in 1886 with the goal of making it a "compendium of universal knowledge." It has been published annually since then.

WORLD ALMANAC BOOKS
An Imprint of PRIMEDIA Reference Inc.
One International Boulevard, Suite 630
Mahwah, New Jersey 07495-0017

CONTENTS

GENERAL INDEX

Note: Page numbers in **boldface** indicate key references. Page numbers in *italics* indicate photos.

The World Almanac
and Book of Facts
2000

What day marks the beginning of the new millennium? Is it Jan. 1, 2000—or Jan. 1, 2001? What events were being planned to mark the transition? What new developments can we look forward to in the new century and millennium? What were the most dramatic events, outstanding achievements, and incredible changes of the past 1,000 years? In this special section and in special features throughout the book, *The World Almanac and Book of Facts 2000* addresses these and other millennial questions.

First of all, when does the new millennium begin? The calendar used in most of the Western world, and parts of Asia and Africa, takes the year following Christ's birth—or what was once thought to be the year—as its starting date. (The birth of Christ was originally figured as occurring in 1 BC, but modern scholars place it about 4 BC or earlier.) This calendar starts with the year AD 1 (there was no year 0). A millennium is a period of 1,000 years. Counting from AD 1, then, the 2d millennium ends on Dec. 31, 2000, and the 3d millennium begins on Jan. 1, 2001. The year 2001 has been officially adopted by the Greenwich Observatory in England as the start of the new millennium. But for most people, the night to celebrate will be Dec. 31, 1999, regardless of the "official" ruling.

In the following pages, former U.S. Secretary of Labor Robert B. Reich describes his vision of a new century and millennium in which the opposing forces of technology and tribalism will compete to shape the course of events. In addition, experts in various fields give their own "ten greatest" lists of the most significant people, events, and accomplishments of the second millennium as it passes into history. The contributors are: Pulitzer-Prize-winning historian and writer Arthur M. Schlesinger Jr., Pulitzer-Prize-winning historian and author David Herbert Donald, military history editor, author, and television commentator Rod Paschall, art historian Anthony F. Janson, Pulitzer-Prize-winning author John Updike, television science guru Mr. Wizard (Don Herbert), and the late Nobel-Prize-winning chemist Glenn T. Seaborg.

Other articles in this section provide a Millennium Calendar, with the latest information on numerous millennium events around the world, describe the Y2K (year 2000) problem, or "millennium bug," threatening computers everywhere at the turn of the millennium, list some notable expert predictions that were proven false in the 20th century, offer a selection of memorable quotes from the 20th century, and look back on some major events 50 or 100 years ago.

In addition, throughout the book you will find Millennium Fact Boxes, highlighting intriguing facts and dramatic changes in the century or millennium just ending, or the new era about to begin. For example, there are lists of the ten most significant buildings of the 20th century, according to noted architect Richard Meier (Buildings, Bridges, Tunnels, and Dams chapter) or the 100 top news stories of the century, according to a survey by the Newseum in Arlington, VA (Arts and Media chapter). You can look back 100 years at the kings and queens of Europe in 1900, or the countries of the world at that time, with basic facts about each, as originally printed in *The World Almanac 1900* (Nations of the World chapter). Or see what the biggest U.S. cities were (Cities chapter) or what major events took place (World History chapter) as originally reported in *The World Almanac* 100 years ago. Other Millennium Fact Boxes feature information ranging from changes in people's diets during the 20th century (Health chapter) to the make-up of baseball's All-Century team (Sports chapter) to a description of the International Space Station (Aerospace chapter) being built for the century to come.

The Two Great Forces of the Future

By Robert B. Reich

Robert B. Reich, best-selling author of The Work of Nations *(1991),* Locked in the Cabinet *(1997), and many other books and articles, served as U.S. Secretary of Labor from 1993 to 1997. He is now University Professor of social and economic policy at Brandeis University.*

Beyond the obvious, specific hazards ahead (such as global warming, excessive population growth, and nuclear proliferation), a more universal drama will play itself out in the coming century. Two great, opposing forces are likely to grow stronger, and the contest between them may well determine the fate of humankind. The first force is technology. The second is tribalism.

Technology and Tribalism

Today we know technology as the Internet, satellite television, supersonic jets, and wonder drugs. Tomorrow it will be small, personal digital devices linking us immediately to everything and everyone we could possibly want to be connected with (and even some things to

which we'd rather not be connected). And it will be genetic manipulations allowing us to live decades longer, and to commit the ultimate narcissistic act of self-cloning. Regardless of its precise form, technology is based on knowledge, rationality, and invention. It is premised upon insights about how the physical world around us actually works—its atoms, electrons, molecules, and genetic material.

Tribalism is a fundamentally different force. Today we witness this fierce group loyalty in the conflicts between Albanians and Serbs in the Balkans, and Protestants and Roman Catholics in Northern Ireland, and in the obstacles in varying degrees faced by ethnic minorities almost everywhere—the Chinese in Indonesia,

Kurds in Turkey, Palestinians in the West Bank and Gaza Strip, Francophones in Canada, African-Americans in the United States. In sharp contrast with technology, tribalism is based on passion, ethnicity, and myth. It is premised on feelings of solidarity among certain people sharing the same history, language, religion, race, customs, or homeland.

Do not assume that technology will be unambiguously positive, and tribalism necessarily negative. Technology can improve our lives, to be sure, but it also can generate weapons capable of ever more devastating destruction, and it almost certainly will shrink the already scant cloak of privacy we human beings still possess. Tribalism can be expected to motivate unspeakable atrocities, as it has too often in the past, but the loyalty to a group, its traditions, and its aspirations may inspire great works of art, enhance the human spirit, and deepen people's sense of community and mutual responsibility. Even democracy must rely on the existence of some tribal-like bond, which we commonly call "citizenship." Democracy can be enriched when political responsibility is devolved to sub-national groups, as it has been recently in Scotland and Wales.

An Interconnected World

Technology and tribalism are likely to come into ever-greater conflict with one another in the century ahead. Consider, first, that within a few decades all human beings on the globe will have immediate connection to one another through digital voice and imaging and through ever faster and cheaper travel. But rather than weaken the lure of tribalism, such wondrously efficient connections are likely to reinforce tribal ties. Geographically dispersed ethnic or religious minorities will find it easier to stay in contact with one another and thus preserve their ways of life, notwithstanding local pressures to assimilate. The very impersonality and ubiquity of global technology may serve to deepen tribal ties. Superficially connected with everyone and everything, people everywhere may feel greater need for authentic connection and identification with others who share common beliefs or customs. For these reasons, despite greater interconnectedness, the next century may well feature frequent outbursts of ethnic pride, chauvinism, and patriotism.

As money, goods, services, and even people move across borders with ever greater speed, the modern nation-state—largely a product of the 18th and 19th centuries—will become less relevant, and political sovereignty of less practical significance. Yet people will, quite naturally, still yearn to control their destinies. Every economic crisis brought on by the bursting of a speculative bubble and the resulting rush of capital or people from one region to another will ignite demands to reassert political control over borders. Hence we can expect periodic outbreaks of economic tribalism—protectionist tariffs, capital controls, economic isolationism, and restrictions on immigration.

As the world becomes more interconnected and intercommunicative, English will become the universal language. Global culture will grow more homogenized and ubiquitous—dominated by Western music, film, and video—and delivered ever more efficiently to the farthest corners of the world via satellite and the Internet. Sad to say, the lure of profits is likely to drive much entertainment toward the titillating, violent, and prurient. This will surely threaten tribal customs and norms. Expect eruptions of cultural conservatism and tribal orthodoxy—tight restrictions on what can be said, worn, viewed, or displayed. In response to ever more-intrusive technology, we may also witness an upsurge in spirituality—an interest in the "inner life" which technology cannot reach. Expect a greater role for religion, for the occult, and for tribal ritual. Some of these manifestations will be ennobling and will enrich the human spirit. Some may be murderously intolerant.

Conflicts in the New Century

The wars of the 21st century will pit technology against tribalism. The world's most technologically advanced people are likely to exercise power through the use of complex weapons systems. Technologically impoverished groups, meanwhile, are more likely to resort to "ethnic cleansing," religious persecution, and terrorism. Electronic weapons will cause more direct physical damage and greater loss of life in the short run. But tribal warfare will be more enduring and, if allowed to fester, cause greater misery overall. The world's major powers will almost certainly spend ever-larger sums seeking to police or quell tribal conflict. In this, they will find their "smart" bombs virtually useless.

The clash between technology and tribalism also is likely to be reflected in the world's emerging social structure. Increasingly, the best-educated in every region of the globe—including China, the Asian subcontinent, and Africa—will converge into a global, technological elite whose members have more in common with one another than with the ethnic, religious, or national groups of which they are nominally a part. The elite will communicate incessantly around the globe—doing business with one another, learning from one another, visiting and vacationing together. The advancement and spread of technology will make members of this elite far wealthier than the populations surrounding them. They can be expected to oppose tribalism in all its forms—rejecting economic isolationism, cultural conservatism, and narrow religious orthodoxy.

The world's less educated are more likely be attracted to resurgent tribalism. Although they will enjoy many of the advantages of global technology, they also will be threatened by it. Their jobs will be among the first to be jeopardized by global flows of financial capital, goods, and people. Their communities, families, and traditions will be sharply disrupted. They are the most likely to find the new global culture to be offensive. Hence they will be most readily persuaded to embrace economic isolation, tribal solidarity, and cultural purity. While many of the less educated will be better off materially than they are today, they will not be nearly as wealthy as the technological elite. As the economic gap between them and the elite widens, their resentments toward the elite may build.

The Central Dilemma

The most enduring dramas of the 21st century will be acted out on the fields where these two emerging classes meet, and where they will determine—by brute force, by negotiation, or by gradual accommodation—the appropriate balance between technology and tribalism. The central dilemma of the new century will be how to gain the material advantages of technology while maintaining the human bonds and stable communities that give life enduring meaning.

This dilemma will be faced within every nation, in every region of the globe. It will give rise to violence, cultural transformations, and spiritual movements. It will never be solved definitively, for there is no single solution. All we can do now, at the brink of the new century and millennium, is to hope that the process of seeking a balance will be as respectful of humanity as is humanly possible.

The Ten Most Influential People
of the Second Millennium

By Arthur M. Schlesinger Jr.

Arthur M. Schlesinger Jr., the Pulitzer-Prize-winning historian and writer, served as special assistant to the president in the Kennedy administration.

Names are listed in order of importance.

1. William Shakespeare, 1564-1616
2. Isaac Newton, 1642-1727
3. Charles Darwin, 1809-82
4. Nicolaus Copernicus, 1473-1543
5. Galileo Galilei, 1564-1642
6. Albert Einstein, 1879-1955
7. Christopher Columbus, 1451-1506
8. Abraham Lincoln, 1809-65
9. Johann Gutenberg, c. 1397-1468
10. William Harvey, 1578-1657

The Ten Most Significant Events
of the Second Millennium

By David Herbert Donald

David Herbert Donald, noted historian and author, is a professor emeritus at Harvard University. His writings include Charles Sumner and the Coming of the Civil War *(1960) and* Look Homeward: A Life of Thomas Wolfe *(1987), for both of which he won Pulitzer Prizes.*

Items are listed chronologically.

1. Invention of **gunpowder** (in the West), early 1300s
2. The **Black Death** devastates Europe, 1347-51
3. Johann Gutenberg uses movable type to **print** early Bibles, c. 1455
4. Christopher **Columbus** reaches America, 1492
5. James Watt perfects his **steam engine**, 1775
6. The **American Revolution**, 1775-83
7. Charles **Darwin** publishes *Origin of Species*, 1859
8. Henry Ford begins commercial development of the **automobile**, 1903
9. **First World War**, 1914-18
10. Dropping of the **atomic bomb** on Japan, 1945

The Ten Most Significant Battles
of the Second Millennium

By Rod Paschall

Rod Paschall is the editor of MHQ: The Quarterly Journal of Military History, *author of* The Defeat of Imperial Germany *and other books, a military history television commentator, and former director of the U.S. Army Military History Institute.*

Items are listed chronologically.

1. **The Battle of Hastings**, Oct. 14, 1066. Norman rule is established in England.
2. **The Battle of Ain Jalut**, Sept. 3, 1260. The Mamluks halt Mongol progress through the Middle East.
3. **The Battle of Hakata Bay**, June 21-Aug. 15, 1281. The Mongol invasion of Japan is defeated by Japanese warriors and a storm ("The Divine Wind").
4. **The Battle for Constantinople**, Apr.-May 1453. The Turks conquer Constantinople in a successful drive for dominance of Byzantium.
5. **The Battle of Yorktown**, Sept.-Oct. 1781. The United States wins independence from Britain.
6. **The Battle of Waterloo**, June 18, 1815. Britain destroys Napoleon's ambitions.
7. **The Battle of Tsushima Strait**, May 27-28, 1905. Japan defeats the Russian fleet and emerges as a world power.
8. **The Battle of the Atlantic**, 1917-18. The Allies overcome the German submarine menace, moving toward the defeat of Imperial Germany.
9. **The Battle of Moscow**, Nov. 15-Dec. 6, 1941. The Red Army halts Hitler and bogs down Germany.
10. **The Battle of Normandy**, June-July 1944. The Allied victory paves the way for the Western democracies to control Western Europe by the end of World War II.

The Ten Most Significant Works of Art
of the Second Millennium

By Anthony F. Janson

Anthony F. Janson, revising author of the widely read History of Art *(originally written by his father, H.W. Janson), has been an art history professor at the University of North Carolina in Wilmington since 1994. He was also chief curator of the Indianapolis Museum of Art, the John and Mable Ringling Museum of Art, and the North Carolina Museum of Art.*

In Professor Janson's words, **"These works of art are ranked simply by how much they have moved me when I saw them in person. They provided me with my greatest artistic experiences to date."**

1. Chartres Cathedral, 12th-13th centuries
2. *Isenheim Altarpiece*, Matthias Grünewald, c. 1510-15
3. Sistine Chapel ceiling, Michelangelo, 1508-12
4. *Apollo and Daphne*, Gianlorenzo Bernini, 1622-24
5. *Mary Magdalen*, Donatello, c. 1455
6. *The Calling of St. Matthew*, Caravaggio, c. 1599-1602
7. *Pierrot*, Antoine Watteau, c. 1720
8. *Mérode Altarpiece*, Robert Campin, c. 1425-30
9. *Moses Well*, Claus Sluter, 1395-1406
10. *Guernica*, Pablo Picasso, 1937

The Ten Greatest Works of Literature
of the Second Millennium

By John Updike

John Updike, the acclaimed American novelist, short-story writer, poet, and literary critic, is perhaps best known for his "Rabbit" novels, beginning with Rabbit, Run *(1960), and ending with* Rabbit at Rest *(1990), one of two novels for which he won a Pulitzer Prize. His literary criticism is displayed in such works as* More Matter *(1999).*

Items are listed chronologically.

1. Thomas Aquinas, *Summa Theologica*, written c. 1265-73.
2. Dante Alighieri, *The Divine Comedy*, written c. 1307-21.
3. Miguel de Cervantes Saavedra, *Don Quixote*, Part I, 1605; Part II, 1615.
4. William Shakespeare, Comedies, Histories, and Tragedies, written c. 1590-1613, published 1623.
5. Voltaire, *Candide*, 1759.
6. Edward Gibbon, *The History of the Decline and Fall of the Roman Empire*, 1776-88.
7. Leo Tolstoy, *War and Peace*, 1865-69.
8. Fyodor Dostoyevsky, *The Possessed*, 1871-72.
9. Marcel Proust, *Remembrance of Things Past*, 1913-27.
10. James Joyce, *Ulysses*, 1922.

The Ten Most Significant Inventions
of the Second Millennium

By Mr. Wizard (Don Herbert)

Don Herbert, better known as Mr. Wizard, was host of the weekly science television program Watch Mr. Wizard *from 1951 to 1965. Since then, he has written, produced, and hosted numerous educational programs on science and technology; he is currently host of the cable TV program* Teacher to Teacher With Mr. Wizard.

Items are listed chronologically.

1. **Magnifying glass**, 1250. A precursor to all optical instruments.
2. **Printing press**, c. 1450. Led to general knowledge.
3. **Electric battery**, 1800. Beginning of battery-operated devices.
4. **Refrigerator**, c. 1850. A way of keeping food edible.
5. **Gasoline engine**, 1885. Freed humans from the horse.
6. **Airplane**, 1903. Beginning of high-speed transportation.
7. **Frozen food**, 1923. Made long-term storage of perishables possible.
8. **Transistor**, 1948. Made modern communication and computation possible.
9. **Artificial satellite**, 1957. Made global communication practical.
10. **Minicomputer**, 1960. Computing for every desk.

The Ten Greatest Scientists
of the Second Millennium

By Glenn T. Seaborg

Glenn T. Seaborg, who shared the 1951 Nobel Prize in chemistry, was the co-discoverer of plutonium and a number of other elements. He was chairman of the U.S. Atomic Energy Commission from 1961 to 1971. He taught chemistry at the University of California at Berkeley and was associate director of the Lawrence Berkeley National Laboratory, until his death in 1999.

Names are listed chronologically.

1. Leonardo da Vinci, 1452-1519
2. Isaac Newton, 1642-1727
3. Jöns Jakob Berzelius, 1779-1848
4. Charles Darwin, 1809-82
5. Dmitri Mendeleyev, 1834-1907
6. Ernest Rutherford, 1871-1937
7. Albert Einstein, 1879-1955
8. Niels Bohr, 1885-1962
9. Werner Heisenberg, 1901-76
10. Enrico Fermi, 1901-54

Millennium Calendar

(Dates and details subject to change.)

1999

After 150 years under Portuguese rule, Macau becomes a special administrative region of China, **Dec. 20**.

The U.S. transfers control over the Panama Canal to Panama, **Dec. 31**.

Ceremonies mark the opening of the huge Millennium Dome at Greenwich, England, **Dec. 31**.

Worldwide on **Dec. 31,** tens of millions of people gather together to celebrate what, for most people, is the dawn of the new millennium. A 24-hour television special, involving networks in more than 50 countries, is expected to reach more than 3 billion people via video, radio, and the Internet.

2000

As revelers ring in the New Year, **Jan. 1**, programmers anxiously check whether their computers have survived the switch from 1999 to 2000—the dreaded Y2K "millennium bug."

Some 1,000 groups in more than 130 countries celebrate **Jan. 1** as World Peace Day.

An estimated 4 million pilgrims are expected to visit Bethlehem, Nazareth, Jerusalem, and other sacred places in Israel and the West Bank for Holy Land 2000, **Jan. 1-Dec.**

31. Another 40 million pilgrims and tourists are expected to visit Rome and its environs.

Odyssey 2000, a round-the-world bicycle trek, has enlisted more than 250 participants to cycle 20,000 miles across four dozen countries during the year 2000.

The Year of the Dragon, year 4698 on the Chinese lunar calendar, begins **Feb. 5**. To celebrate the new year, 10,000 young Chinese will build a 10,000-ft-long dragon along the Great Wall near Beijing.

New Zealand hosts the America's Cup 2000 championship yacht races, **Feb. 19-Mar. 4**.

The first centesimal leap year since 1600 adds a day to the calendar, **Feb. 29**, designated by the U.S. and many other countries as International Children's Day.

The U.S. conducts its decennial census **Apr. 1.**

Apr. 6 marks the 1st full day of the year 1421 on the Islamic calendar.

Organizers hope to enlist more than 300 million people to participate **Apr. 22** in the largest Earth Day ever.

The Expo 2000 world's fair is held in Hannover, Germany, **June 1-Oct. 31**.

Participation by more than 30 million people is the goal for the global March for Jesus, **June 10**, with prayer and worship processions in over 2,000 cities.

More than 40,000 spectator vessels are expected to join a flotilla of tall ships from at least 60 countries when OpSail 2000 reaches New York harbor, **July 3-9**.

The Holy Shroud of Turin, Italy, is on public display, **Aug. 26-Oct. 22**.

The XXVII Olympiad is held in Sydney, Australia, **Sept. 15-Oct. 1**. Sydney also hosts the Games of the XI Paralympiad for athletes with disabilities, **Oct. 18-29**, and "Harbour of Life," an Olympic arts festival.

On **Sept. 30**, Rosh Hashanah, a Jewish holy day, marks the 1st full day of the year 5761 on the Hebrew calendar.

The U.S. elects its 1st president of the 3d millennium, **Nov. 7**.

New Millennium's Eve, **Dec. 31.** Most authorities say the 3d millennium technically begins **Jan. 1, 2001**.

Some Websites Relating to the Millennium

America's Cup 2000 . . . . . . http://www.americascup2000.org	Millennium Around the
Canada and the Millennium http://www.millennium.gc.ca	World . . . . http://www.millenniumworld.org
Countdown 2000 . . . . http://www.countdown2000.com	Odyssey 2000. . . . . http://www.kneeland.com
Everything 2000 . . . . http://www.everything2000.com	OpSail 2000 . . . . http://www.opsail.org
Expo 2000 . . . . http://www.expo2000.de	Sydney 2000. . . . . http://www.sydney.olympic.org
Greenwich 2000. . . . . http://millennium.greenwich2000.com	White House Millennium http://www.whitehouse.gov/
Holy See. . . . . http://www.vatican.va/phome_en.htm	Program . . . . Initiatives/Millennium
Millennium Photo Project . . http://www.millenniumphoto.com	Year 2000 Conversion http://www.y2k.gov

---

The Year 2000 Problem

By Richard Hantula

Richard Hantula is a freelance writer and editor who often writes on science and technology.

Businesses and governments have been spending billions of dollars in the 1990s to fix and upgrade their computer systems, so as to minimize the potentially disruptive effects of a glitch known as the Year 2000 (or "Y2K") Problem—popularly dubbed the "Millennium Bug."

Legacy of Bygone Days

The Y2K problem—the malfunctioning of computer systems because they are unable to properly identify dates in the year 2000 and thereafter—is not really a bug in the strict sense ("an error in coding or logic" or a "recurring physical problem"). It originated in a computer design practice that was not only intentional but at one time saved large sums of money.

In bygone days when even the simplest computers were enormously expensive and difficult to operate, computer memory was in short supply. Software programmers and hardware designers took extraordinary pains to make efficient use of this scarce resource. If you could economize on memory space by dropping the first 2 digits of a year (as we do in such date expressions as 11/13/59, where the 59 stands for 1959), it made very good sense to do so. In the overwhelming majority of cases, after all, references were to years in the 20th century, so there was no problem with abbreviating 1959 as 59.

As the years passed, more powerful computers were developed, but the habit of abbreviating dates remained, although experts were aware that trouble was in the wind. Practical difficulties with the year 2000 were encountered as long ago as 1970, when mortgage companies needed to calculate 30-year mortgages. As the 21st century drew closer, more and more public attention was paid to the potential hazards lurking in the misunderstanding of 00 as 1900 instead of 2000.

By 1999, almost all current computer products accommodated 4-digit years, but enormous numbers of systems were still in use that either were manufactured years before or were designed in accordance with the old conventions. The problem potentially affected both software—programs and databases—and hardware, including many electronic devices that use microprocessors (especially those of older vintage), such as telephone switchboards, building security systems, VCRs, time-lock safes, and credit card scanners.

Readiness and Repair

In the U.S., federal government Y2K efforts were coordinated by the President's Council on Year 2000 Conversion, which also offered pointers to individuals. People were advised to "stay informed" regarding Y2K measures taken by businesses and governments and to take common sense steps to prepare for possible temporary disruptions in basic services—as they would if, say, a major storm were approaching. The Federal Trade Commission suggested that people owning equipment using microchips, such as a PC, fax, camcorder, monitored security system, or Global Positioning system unit, check with the manufacturer to see if there might be a problem. The commission also concluded that, at least for the transition period, it would be a good idea for people to keep paper records of activity regarding loan payments and financial accounts and to obtain an up-to-date copy of their credit reports.

The Federal Reserve said that nearly all insured financial institutions were prepared and that ATM systems were generally in good shape. The Fed did not recommend that people withdraw extra money (beyond ordinary holiday needs) but said additional currency would be printed as a precaution to meet possible increased demand.

Small companies were advised by the U.S. Small Business Administration on what they should do to check, fix, and test their computer systems for Y2K readiness. The agency stood ready to help with modernization loans. Consumers also needed to check out the Y2K status of their computer systems. In many cases this was a relatively simple matter, since a number of hardware and software makers placed Y2K-compliance tests and corrective software on the World Wide Web for downloading free of charge. Commercial products were also available that could be of help.

Many businesses and governments, however, use customized hardware and software. Merely finding all the places where computer code or microchips might need repair turned into a major ordeal. In some cases the original source code (the code as written by a programmer, prior to conversion to the machine code that a computer could understand) was no longer available. Or the source code was written in languages that many of today's programmers were not familiar with. Much of the software was written years ago in COBOL, an idiosyncratic programming language introduced in 1960. And this software was often written before modern standards of logical software analysis and design were adopted, which meant the Y2K problem might pop up in unexpected spots.

Government Action

To minimize a predicted deluge of lawsuits over the costs associated with Y2K damage, the U.S. Congress enacted the Y2K Act in July 1999. It applies to Y2K failures that occur before Jan. 1, 2003, and does not affect personal injury and wrongful death claims. Among its provisions: a 30- to 90-day cooling-off period before the filing of Y2K lawsuits, so companies have a chance to fix problems; a $250,000 ceiling on punitive damages for many small businesses; and "proportional liability" guidelines to ensure that defendants pay only for the damages they cause.

Other federal measures included a 1998 law, the Year 2000 Information and Readiness Disclosure Act, which encouraged the dissemination and sharing of information about Y2K-related issues. Among measures targeted at specific industries, the U.S. Securities and Exchange Commission required brokerage firms and nonbank transfer agents to bring their computer systems into Y2K

compliance by Dec. 1, 1999. Meanwhile, the federal government planned to set up a crisis management system, the Y2K Information Coordination Center, late in 1999 near the White House in Washington, DC, slated to operate until mid-2000.

In a June 1999 report, the White House Office of Management and Budget put the federal government's spending on fixing the Y2K problem at $8 billion, with the Defense Department accounting for nearly half that sum. A subsequent report by the agency said that, as of mid-August, 97% of the federal government's mission-critical systems were Y2K compliant. The Social Security system had already been declared to be fully Y2K ready in late 1998.

A September 1999 U.S. Senate committee report gave particularly high marks to the financial services industry for Y2K readiness; the utilities and telecommunications industries also received generally favorable evaluations, although isolated disruptions could not be ruled out. Concerns were voiced about the health care industry, however. Although federal computer systems were believed to be by and large in good shape, the smooth operation of programs such as Medicare and Medicaid, as well as private health plans, depended also on a myriad of systems belonging to doctors, health care facilities, and local government agencies—systems that were still not known to be entirely Y2K ready.

Many countries lacked the financial or technical resources to adopt as proactive a stance as the U.S. appeared to have done. In a mid-1999 analysis, the U.S. State Department looked at more than 160 countries and found that half of them faced a medium to high risk of a Y2K-related disruption of their telecommunications, energy, or transportation industries.

Predictions That Failed

Predicting the future is, of course, a highly risky enterprise. Here is a brief sampling of seemingly informed predictions and observations from the past that, in the 20th century (or before), were clearly proved false.

"Everything that can be invented has been invented."
Attributed to *Charles H. Duell*, U.S. commissioner of patents (1899).

"Heavier-than-air flying machines are impossible."
William Thomson, 1st Baron Kelvin, mathematician, physicist, and president of the Royal Society, London (late 19th cent.).

"The actual building of roads devoted to motor cars is not for the near future, in spite of many rumors to that effect."
Harper's Weekly, Aug. 2, 1902.

"[The incandescent light bulb is] good enough for our transatlantic friends...but unworthy of the attention of practical or scientific men."
Committee of the British Parliament, 1878.

"The abolishment of pain in surgery is a chimera.... Knife and pain are two words in surgery that must forever be associated in the consciousness of the patient."
Surgeon *Alfred Velpeau*, 1839.

"All mankind has heard much of M. Lesseps and his Suez Canal.... I have a very strong opinion that such a canal will not and cannot be made."
Anthony Trollope, *The West Indies and the Spanish Main* (1860; canal opened, 1869).

"It seems certain that before 1985 mankind will enter a genuine age of scarcity in which many things...will be in short supply."
Biologist *Paul R. Ehrlich*, *The End of Affluence* (1974).

"For the second time in our history, a British prime minister has returned from Germany bringing peace with honor. I believe it is peace for our time."
Prime Min. *Neville Chamberlain* describing the Munich Pact (1938) with Adolf Hitler.

"I think there is a world market for about five computers."
Reportedly said by IBM chief *Thomas J. Watson* in 1943, a year before the introduction of the Mark I, IBM's first large-scale calculating computer.

"People will soon get tired of staring at a plywood box every night."
Darryl F. Zanuck, film producer, on the future of television (1946).

"Whether you like it or not, history is on our side. We will bury you."
Reportedly said by Soviet leader *Nikita Khrushchev*, at a reception in Moscow (1956).

"You won't have Nixon to kick around anymore, because, gentlemen, this is my last press conference."
Richard Nixon addressing reporters after losing the California gubernatorial election (1962).

"As for the case at hand, if properly managed by the District Court, it appears to us highly unlikely to occupy any substantial amount of petitioner's time."
The U.S. Supreme Court (*Clinton* v. *Jones*, 1997), in turning down Pres. Bill Clinton's petition to postpone a sexual harassment lawsuit until he left the White House.

Notable Quotes of the 20th Century

"That's one small step for man, one giant leap for mankind."
Neil Armstrong, the first person to set foot on the moon (July 1969).

"May the Force be with you!"
George Lucas, screenplay for *Star Wars* (1977).

"Speak softly and carry a big stick."
Adage cited by *Pres. Theodore Roosevelt* (Sept. 1901).

"The world must be made safe for democracy."
Pres. Woodrow Wilson (Apr. 1917), calling on Congress to authorize U.S. entry into World War I.

"The only thing we have to fear is fear itself."
Pres. Franklin D. Roosevelt, first inaugural address (Mar. 1933).

"Nonviolence and truth are inseparable and presuppose one another. There is no god higher than truth."
Mohandas K. Gandhi, liberator of India, saying published in 1939.

"We shall defend our island, whatever the cost may be, we shall fight on the beaches, we shall fight on the landing grounds, we shall fight in the fields and in the streets, we shall fight in the hills; we shall never surrender."
British Prime Min. Winston Churchill (June 1940).

"Yesterday, December 7, 1941—a day which will live in infamy...."
Pres. Franklin D. Roosevelt, in his war message to Congress, describing the Japanese attack on Pearl Harbor.

"I shall return."
Gen. Douglas MacArthur, on leaving the Philippines to Japanese invaders (Mar. 1942).

"In spite of everything, I still believe that people are really good at heart."
Diary entry in July 1944 by *Anne Frank,* a Holocaust victim.

"E = mc2"
*Albert Einstein'*s statement of the relationship between energy (E), mass (m), and the speed of light (c), the formula that unlocked the power of the atom.

"The buck stops here."
Sign on the desk of *Pres. Harry S. Truman.*

"Every Communist must grasp the truth: 'Political power grows out of the barrel of a gun.'"
Chinese Communist leader *Mao Zedong, Selected Works* (1965).

"From Stettin in the Baltic to Trieste in the Adriatic an iron curtain has descended across the Continent."
Winston Churchill (Mar. 1946), at the dawn of the cold war.

"Big Brother is watching you."
George Orwell, in the novel *1984* (1948).

"And so, my fellow Americans, ask not what your country can do for you; ask what you can do for your country."
Pres. John F. Kennedy, inaugural address (Jan. 1961).

"I have a dream that one day this nation will rise up and live out the true meaning of its creed.... I have a dream that my four little children will one day live in a nation where they will not be judged by the color of their skin but by the content of their character."
Civil rights leader *Martin Luther King Jr.,* speech at the March on Washington (Aug. 1963).

"We shall overcome."
Anthem of the civil rights movement.

"Extremism in the defense of liberty is no vice! And ... moderation in the pursuit of justice is no virtue!"
Pres. nominee *Barry Goldwater* (July 1964).

"We will seek no wider war."
Pres. Lyndon B. Johnson (Aug. 1964), after ordering bombing raids on North Vietnam, in what proved to be an early stage of U.S. involvement in Vietnam.

"People have got to know whether or not their president is a crook. Well, I'm not a crook."
Pres. Richard Nixon (Nov. 1973) during the Watergate scandal, which led to his resignation in Aug. 1974.

"Read my lips: no new taxes."
Pres. George Bush, campaign pledge (Aug. 1988), two years before he signed a bill raising taxes.

"I did not have sexual relations with that woman, Miss Lewinsky. I never told anybody to lie, not a single time—never."
Pres. Bill Clinton (Jan. 1998) at the start of the Monica Lewinsky scandal, which led to his impeachment 11 months later.

"Three passions, simple but overwhelmingly strong, have governed my life: the longing for love, the search for knowledge, and unbearable pity for the suffering of mankind."
Philosopher *Bertrand Russell, Autobiography* (1967).

"God, give us grace to accept with serenity the things that cannot be changed, courage to change the things which should be changed, and the wisdom to distinguish the one from the other."
Reinhold Niebuhr, The Serenity Prayer (1943).

"I have found it impossible to carry the heavy burden of responsibility and to discharge my duties as King as I would wish to do without the help and support of the woman I love."
Edward VIII of England, abdication speech, 1936.

"One is not born a woman, one becomes one."
French author *Simone de Beauvoir, The Second Sex* (1949).

"The great question ... which I have not been able to answer, despite my thirty years of research into the feminine soul, is 'What does a woman want?'"
Statement attributed to *Sigmund Freud.*

"No woman can call herself free who does not own and control her body."
Margaret Sanger, birth-control crusader (Dec. 1963)

"There is less in this than meets the eye."
Tallulah Bankhead, actress (Jan. 1922).

"Genius is one percent inspiration and ninety-nine percent perspiration."
Thomas Edison, Life (1932).

"All I know is just what I read in the papers."
Remark frequently made by humorist *Will Rogers.*

"Nice guys finish last."
Remark by *Leo Durocher* (undated).

"Winning isn't everything. It's the only thing."
Maxim attributed to *Vince Lombardi,* who coached the Green Bay Packers to victory in the first two Super Bowls.

"The medium is the message."
Marshall McLuhan, Understanding Media (1964).

"Rose is a rose is a rose is a rose, is a rose."
Gertrude Stein, Sacred Emily (1913).

"Things fall apart, the center cannot hold;
Mere anarchy is loosed upon the world."
William Butler Yeats, from the poem *The Second Coming* (1921).

"I saw the best minds of my generation destroyed by madness, starving hysterical naked."
Allen Ginsberg, from the poem *Howl* (1956).

"Turn On, Tune In, Drop Out."
Timothy Leary, psychologist, lecture title in 1960s.

"This is the way the world ends
Not with a bang but a whimper."
T. S. Eliot, from the poem *The Hollow Men* (1925).

"In the future everyone will be world-famous for fifteen minutes."
Pop artist *Andy Warhol* (1968).

"Hindsight is always twenty-twenty."
Billy Wilder, film director, quoted in 1979.

"Don't look back. Something may be gaining on you."
Baseball pitcher *Satchel Paige, How to Keep Young* (1953).

"It ain't over till it's over."
Attributed to baseball catcher *Yogi Berra.*

Historical Anniversaries
1900 — 100 Years Ago

The U.S. and Britain sign Hay-Pauncefote Treaty, **Feb. 5,** giving the U.S. the right to build a canal in Central America.

Congress adopts the Currency Act, **Mar. 14,** making gold the standard of U.S. currency.

Acts of Congress establish Puerto Rico, **Apr. 12,** and Hawaii, **Apr. 30,** as U.S. territories.

In Boer War, British garrison at Mafeking is relieved, **May 17,** after a 7-month siege. British capture Johannesburg, **May 31,** and Pretoria, **June 5.**

Russia annexes Manchuria, **May 21.**

International Ladies' Garment Workers Union founded in New York City, **June 3.**

Boxer Rebellion begins in China, **June 20,** when young rebels attack foreigners in Beijing. The U.S. announces, **June 26,** that its troops will join those of other nations in thwarting the rebels. Multinational forces restore order, **Aug. 14.**

Count Ferdinand von Zeppelin launches the 1st rigid airship at Friedrichshafen, Germany, **July 2.**

An anarchist assassinates Italy's King Umberto I, **July 29;** he is succeeded by his son, Victor Emmanuel III.

A hurricane in Galveston, TX, in late **Aug.**-early **Sept.** kills 6,000 and causes $20 million in property damage.

More than 100,000 coal miners go on strike, **Sept. 17;** strike ends, **Oct. 25,** with miners receiving a pay increase.

Republican William McKinley is reelected president, **Nov. 6,** defeating William Jennings Bryan; Theodore Roosevelt is elected vice president.

Art. Paul Cézanne's *Still Life With Onions,* Pablo Picasso's *Le Moulin de la Galette,* Albert Pinkham Ryder's *Toilers of the Sea,* Henri de Toulouse-Lautrec's *La Modiste.*

Literature. *The Wonderful Wizard of Oz* by L. Frank Baum, *Lord Jim* by Joseph Conrad, *Sister Carrie* by Theodore Dreiser, *The Tales of Peter Rabbit* by Beatrix Potter, *Monsieur Beaucaire* by Booth Tarkington.

Nonfiction. *The Interpretation of Dreams* by Sigmund Freud.

Music. The operas *Louise* by Gustave Charpentier, *Tosca* by Giacomo Puccini, and *The Tale of Tsar Saltan* (with "The Flight of the Bumblebee") by Nikolai Rimsky-Korsakov.

Musicals. *Fiddle-dee-dee* with Lillian Russell, Fay Templeton, Joe Weber, Lew Fields, music by John Stromberg, lyrics by Edgar Smith.

Popular Songs. "A Bird in a Gilded Cage" by Harry Von Tilzer, lyrics by Arthur J. Lamb; "Goodbye, Dolly Gray" by Paul Barnes, lyrics by Will D. Cobb.

Science and Technology. German physicist Max Planck formulates quantum theory of light. F. E. Dorn discovers radon. Scientists rediscover and make public Gregor Mendel's laws of heredity.

Sports. Baseball's American League is formed, **Jan. 29.** The 2d modern-day Olympic Games are held in Paris, **May 20-Oct. 28,** with France winning the most gold medals. The U.S. beats Great Britain in the 1st-ever Davis Cup tennis match, **Aug. 8-10.**

Miscellaneous. Temperance activist Carry Nation begins attacking saloons with an ax. English archaeologist Arthur John Evans unearths remains of the ancient palace of Knossos on Crete. The Automobile Club of America sponsors the 1st auto show. Eastman Kodak introduces the Brownie box camera for $1.

1950 — 50 Years Ago

Masked bandits rob $2.8 million from the Boston office of Brink's Inc., **Jan. 17.** (Finally, in 1956, 10 suspects were arrested. Two died before trial; 8 were convicted and jailed.)

Alger Hiss is convicted of 2 counts of perjury, **Jan. 21,** for denying charges of espionage; he is sentenced to 2 concurrent 5-year prison terms, **Jan. 25.**

India becomes independent democratic republic, **Jan. 26.**

Pres. Harry S. Truman authorizes U.S. production of the H-bomb, **Jan. 31.**

The U.S. recognizes the Bao Dai regime in Vietnam, **Feb. 7,** later supplying arms and military advisers.

Sen. Joseph R. McCarthy (R, WI) launches his anti-Communist crusade, **Feb. 9.**

China and the Soviet Union sign treaty of alliance and friendship, **Feb. 14.**

Chiang Kai-shek is proclaimed president of nationalist China, **Mar. 1.**

Korean War begins, **June 25,** when North Korean forces cross 38th parallel into South Korea. Responding to UN call, Pres. Truman orders Air Force, Navy personnel to region, **June 27;** approves use of ground forces, air strikes, **June 30.** Seoul falls to North Korea, **June 28.** Gen. Douglas MacArthur appointed commander of UN forces, **July 8.**

To prevent a general strike, the U.S. Army seizes all U.S. railroads on Truman's order, **Aug. 27;** they are returned to their owners in 1952.

U.S. forces land at Inchon, **Sept. 15;** they recapture Seoul, **Sept. 26.** South Korean troops reach 38th parallel, **Sept. 29.**

Congress passes Internal Security Act over Truman's veto, **Sept. 23,** setting up Subversive Activities Control Board to monitor Communists' activities.

U.S. troops cross the 38th parallel into North Korea, **Oct. 7.** UN forces take Pyongyang, **Oct. 20.**

Chinese Communist forces invade Tibet, **Oct. 21.**

Pres. Truman escapes assassination attempt by 2 Puerto Rican nationalists outside Washington's Blair-Lee House, **Nov. 1;** a White House policeman and 1 attacker are killed.

UN forces reach the Chinese border, **Nov. 20.** The Chinese launch a counterattack, **Nov. 26,** halting offensive.

Pres. Truman declares a state of national emergency, **Dec. 16,** as a result of Korean War setbacks and threat of Communist aggression.

Art. Willem de Kooning's *Excavation,* Jackson Pollock's *Lavender Mist* and *One* (paintings); Alberto Giacometti's *Seven Figures and a Head* (sculpture).

Drama. *The Cocktail Party* by T. S. Eliot, *Come Back, Little Sheba* by William Inge, *The Country Girl* by Clifford Odets.

Literature. *The Martian Chronicles* by Ray Bradbury, *The Wall* by John Hersey, *The Cardinal* by Henry Morton Robinson, *World Enough and Time* by Robert Penn Warren.

Movies. *All About Eve* with Bette Davis, Anne Baxter; *The Asphalt Jungle* with Sterling Hayden, Louis Calhern, Jean Hagen; *Born Yesterday* with Judy Holliday, William Holden, Broderick Crawford; Disney's animated *Cinderella; Father of the Bride* with Spencer Tracy, Elizabeth Taylor; *Harvey* with James Stewart; *Sunset Boulevard* with Gloria Swanson, William Holden.

Music. The opera *The Consul* by Gian-Carlo Menotti.

Musicals. *Call Me Madam* with Ethel Merman, music and lyrics by Irving Berlin; *Guys and Dolls* with Robert Alda, Vivian Blaine, music and lyrics by Frank Loesser.

Nonfiction. *Dianetics* by L. Ron Hubbard, *The Lonely Crowd* by David Riesman, Reuel Denney, Nathan Glazer.

Popular Songs. "Good Night, Irene" by Huddie Ledbetter, John Lomax; "If I Knew You Were Comin' I'd 'Ave Baked a Cake" by Al Hoffman, Bob Merrill, Clem Watts; "Music! Music! Music!" by Stephen Weiss, and Bernie Baum; "Silver Bells" by Jay Livingston, Ray Evans.

Science and Technology. The element californium discovered by scientists at University of California-Berkeley. E. I. du Pont introduces Orlon. Haloid Corp. of Rochester, NY, introduces Xerox copying machine.

Sports. Ezzard Charles defends world heavyweight title, **Aug. 15, Sept. 27,** and **Dec. 5.**

Television. *Your Show of Shows* with Sid Caesar, Imogene Coca; the quiz show *You Bet Your Life* with Groucho Marx.

Miscellaneous. Charles Schulz's *Peanuts* comic strip 1st appears. The Brooklyn-Battery Tunnel opens and the UN Secretariat Building is completed in New York City.

Clinton Impeached and Acquitted

By Geoffrey M. Horn

Geoffrey M. Horn is a freelance writer and editor who frequently writes on political and cultural topics.

The impeachment of Pres. William Jefferson Clinton by the U.S. House of Representatives (in Dec. 1998), his Senate trial, and his acquittal by the Senate on Feb. 12, 1999, gave the nation a lesson in constitutional history, the laws of perjury and conspiracy, and the arcane rules of congressional procedure. For only the 2d time in U.S. history, a president was impeached, and also acquitted (Andrew Johnson was impeached and acquitted in 1868). The dramatic votes by the House, and the somber verdict of the Senate, where the outcome was never really in doubt, also offered a primer on the institutional differences between the 2 chambers and the inescapable power of political arithmetic.

The Path to Impeachment

The events that led to Clinton's impeachment by the House had played out against the backdrop of the midterm congressional elections. Day after day, from late Jan. to early Oct. 1998, voters had been inundated with seamy revelations concerning the president's liaison with Monica Lewinsky, a former White House intern, and his efforts to conceal the nature of their relationship from lawyers for Paula Corbin Jones, who was pursuing a sexual harassment lawsuit against him, and from his family, his aides, and the public at large. Ultimately, independent counsel Kenneth Starr submitted a report to Congress, promptly made public, citing what it called "substantial and credible information" that Clinton had lied under oath and obstructed justice in his cover-up of the liaison.

When the House voted on Oct. 8 to launch proceedings leading to possible impeachment, Republicans held a 228-seat majority and had every hope of substantially increasing their margin. But polls showed a majority of the public supporting the president and skeptical of Republicans' motives. The Republicans' loss of 5 House seats in the Nov. 3 election was a bitter setback for the GOP.

The electoral setback, which some observers thought might slow down the impeachment steamroller, had precisely the opposite effect. Republican members, who in Oct. had refused to impose a time limit on the impeachment inquiry, now were determined to bring the matter to a vote before the 105th Congress adjourned and their narrow majority was further eroded. Democrats, many of whom had been wary of defending Clinton, increasingly rallied to the president's support—in part because of a belief that the public leaned toward Clinton's side.

Articles of Impeachment

During Nov. and early Dec. 1998, the House Judiciary Committee, consisting of 21 Republicans and 16 Democrats, heard public presentations from independent counsel Kenneth Starr, from experts on the consequences of perjury and the constitutional standards for impeachment, and from counsel for the president and the committee—but not from any witnesses with direct involvement in the case. The committee chairman, Rep. Henry J. Hyde (R, IL), submitted a series of 81 questions to Clinton, who in his written responses acknowledged the "hard truth" that his conduct had been "wrong," but adamantly denied that he had lied under oath or urged anyone else to do so.

After rancorous debate, the committee on Dec. 11-12 approved 4 articles of impeachment. The articles charged that Clinton had (I) given "perjurious, false and misleading testimony" to a federal grand jury on Aug. 17, 1998; (II) lied under oath on Dec. 23, 1997, and Jan. 17, 1998, in the Paula Corbin Jones civil case; (III) obstructed justice by influencing others, including Lewinsky and Clinton's personal secretary, Betty Currie, to conceal evidence and testify falsely; and (IV) abused the powers of his office by lying to Congress about the actions specified in the first three articles. Articles I, III, and IV were approved 21-16, by straight party-line vote. One Republican broke ranks

on Article II, which passed by a vote of 20-17 despite the fact that the Jones case had recently been settled out of court, with the president agreeing Nov. 13 to pay her $850,000. A motion to censure rather than impeach the president, proposed by committee Democrats and endorsed by the White House, failed by a margin of 22-14. The White House lost further ground over the next few days, as a number of moderate Republicans, many of whom represented districts that had supported Clinton in 1992 and 1996, announced they would vote to impeach. Threats of primary challenges in 2000 from anti-Clinton conservatives likely played some part in the moderates' decisions.

The impeachment votes by the whole House on Dec. 19 climaxed a tumultuous week in which the U.S. and Britain carried out intensive air strikes against Iraq, and the House speaker-designate, Bob Livingston (R, LA), withdrew his candidacy after acknowledging past extramarital affairs of his own. Article I, charging Clinton with grand jury perjury, passed on a vote of 228 to 206, with only 5 Republicans voting against and 5 Democrats in favor. Twelve Republicans voted no and the same 5 Democrats voted yes on Article III, alleging obstruction of justice, which passed by a margin of 221-212. Articles II and IV were rejected by votes of 205-229 and 148-285, respectively. House leaders did not allow a censure resolution to reach the floor.

Verdict by the Numbers

The Senate proceedings, only the 2d presidential impeachment trial in U.S. history, opened Jan. 7, 1999. Chief Justice William H. Rehnquist presided, wearing a black robe with four gold stripes on each sleeve.

The ceremonies had pageantry aplenty, but little real drama. Led by Hyde, the 13 "managers"—all Republicans—appointed by the House faced virtually insurmountable obstacles in prosecuting their case. Although Republicans in the Senate had the numbers to carry every procedural vote, for which only a simple majority was needed, conviction of the president required a two-thirds majority. Even if the managers could hold all 55 Senate Republicans, they would still need to convince at least 12 of the 45 Democrats both that Clinton had committed at least one of the offenses charged and that, as a result, he should be removed from office. They never came close.

Seeking to avoid the partisan acrimony that had tainted proceedings in the House, Senate leaders agreed on a truncated trial in which the managers and the White House legal team, headed by Charles F. C. Ruff, were allowed to state—and restate—their case. No witnesses were called. The lone concession to the managers' fervent plea for live witnesses was the introduction of videotaped depositions and from only 3 witnesses: from Lewinsky; from the president's friend Vernon Jordan, who helped her get a job while the Jones lawyers were seeking her testimony; and from presidential aide Sidney Blumenthal, who had a shadowy role in what the managers alleged was a White House campaign to smear Lewinsky and vilify Starr after her relationship with Clinton was disclosed. Little emerged that was new. In her deposition, Lewinsky continued to deny she had been asked by anyone to deny a sexual relationship with Clinton in her affidavit in the Jones case.

The trial ended Feb. 12 with verdicts of not guilty on both articles of impeachment passed by the House. The first article, alleging grand jury perjury, failed by a vote of 45-55, and the second, charging obstruction, fell 17 votes short on a tally of 50-50. Not a single Democrat voted for either article. Rehnquist pronounced the Senate verdict of acquittal at 12:39 PM, bringing to an abrupt resolution the scandal that had dominated the national agenda for more than 12 months.

Conflict in Kosovo

By Geoffrey M. Horn

Geoffrey M. Horn is a freelance writer and editor who frequently writes on political and cultural topics.

Not quite 2 weeks short of its 50th birthday, the North Atlantic Treaty Organization on Mar. 24, 1999, mounted its first-ever military operation against a sovereign country, the Federal Republic of Yugoslavia (FRY). The 78-day war, which pitted the world's most powerful military alliance against a nation no bigger than Kentucky, was hardly the kind of conflict NATO's founders had anticipated.

When the foreign ministers of NATO's 12 original members signed the North Atlantic Treaty on Apr. 4, 1949, the organization was expected to stand as a bulwark against the Soviet Union and its allies. By 1999, however, the Soviet empire was in shambles, and most Eastern European countries were gravitating toward the West. Even as NATO on Mar. 12 expanded its membership to 19, adding 3 former Warsaw Pact nations—Hungary, Poland, and the Czech Republic—some commentators questioned whether the organization had a clear mission in the post cold-war world.

Milosevic vs. the Kosovars

What were NATO's strategic interests in Yugoslavia? The country, consisting of the republics of Serbia and Montenegro, had few economic assets. Its military posed no direct threat to the West. But its president, Slobodan Milosevic, had a long history of stirring up tensions in the Balkans.

This time, the focus of contention between NATO and Milosevic was Kosovo, a rocky region covering some 4,200 sq mi of southern Serbia, with a population of about 2,000,000. Serbs have claimed Kosovo for centuries, and still commemorate a defeat there by the Ottoman Turks in 1389 as a national calamity.

Within the FRY more than 60% of the people were Serbs, but in Kosovo at least 90% were ethnic Albanians. These Albanians, known as Kosovars, are mainly Muslim, while the Serbs are predominantly Orthodox Christian. Differences in language and culture also divide the 2 peoples.

From the end of World War II through the 1970s, while Tito governed Yugoslavia, the Kosovars exercised substantial autonomy. In 1989, however, Kosovo's autonomy was revoked by Milosevic, who had seized on Serb nationalism as his path to power. In the early 1990s he supported Serb secessionists in civil wars in Croatia and Bosnia. All sides were guilty of atrocities, but the Serbs were especially ruthless in their reliance on "ethnic cleansing"—the systematic use of violence, including murder and rape—to purge millions of Croats and Bosnian Muslims from Serb-held lands.

Milosevic became president of the FRY in July 1997. Hostilities intensified in Kosovo during the next 19 months, as the Kosovo Liberation Army (KLA), a guerrilla group, attacked Serb targets and Belgrade launched brutal counteroffensives. Western leaders warned that Milosevic was planning a massive campaign of ethnic cleansing in Kosovo.

A refugee crisis in Kosovo would not only be a humanitarian disaster; if unanswered by the West, it might also have serious political consequences. A massive influx of Kosovar refugees could destabilize governments in Albania and Macedonia, triggering a wider war in the Balkans. Nor were large numbers of Kosovar refugees likely to receive a warm welcome in NATO countries like Germany and Italy, which already had unemployment rates above 10%.

Increasingly, NATO staked its prestige on forcing Milosevic to back down. An ultimatum in Oct. 1998 threatened air strikes if Serbian security forces remained in Kosovo; later that month, when Yugoslav troops began leaving the province, the bombing was indefinitely postponed. A new cycle of violence in late 1998 and early 1999 prompted further negotiations, but Milosevic rejected NATO's key demand that a 28,000-member peacekeeping force, led by NATO, be allowed to patrol Kosovo for up to 3 years.

What Went Right—and Wrong

During NATO's 11-week military mission, called Operation Allied Force, the U.S. and, to a lesser extent, Great Britain carried the burden of the air war. From the commencement of hostilities until the formal suspension of bombing on June 10, more than 35,000 sorties were flown, an average of about 450 per day.

Eventually the bombing took a catastrophic toll on Yugoslavia's infrastructure, destroying bridges and roads, demolishing factories, and disrupting power and communications. The Economist Intelligence Unit concluded in August that the war would end up costing Yugoslavia more than $60 billion and make it the poorest country in Europe. As of June 1999, FRY reports of Yugoslav casualties included 462 soldiers, 114 police, and about 2,000 civilians; some Western sources put the overall death toll closer to 5,000. Remarkably, NATO suffered no combat fatalities.

What the high-altitude aerial attack could not do, at least in the short run, was to halt Serb ethnic cleansing in Kosovo. Within weeks, hundreds of thousands of Kosovars were streaming toward the borders, victims of an accelerating terror campaign that killed large numbers of men, women, and children, looted homes, burned villages, and turned Pristina, the provincial capital, into a ghost town. By the end of April more than 380,000 Kosovars had fled to Albania, and nearly 175,000 were in Macedonia.

The bombing campaign got off to a slow start, in part because of bad weather, in part because of concern in Washington and other NATO capitals that any allied casualties, however minimal, would jeopardize public support for the war. While some criticized the air campaign as too limited and cautious, a number of errors were made that added to the civilian death toll. In one case, on May 7, bombs from a U.S. aircraft hit the Chinese embassy in Belgrade, killing 3 Chinese citizens. The attack, which the U.S. insisted was unintentional, soured relations between Beijing and Washington for months.

The Kosovo conflict also raised tensions with Russia, a traditional ally of the Serbs and a member of the NATO Partnership for Peace program. (Russia suspended formal cooperation with NATO during the war.) Anxious to reach a settlement before a scheduled G-8 summit conference in mid-June, Russian Pres. Boris Yeltsin authorized his envoy to the Balkans, Viktor Chernomyrdin, to join with Finnish Pres. Martti Ahtisaari in persuading Milosevic to come to terms.

An Uncertain Future

The air war—and perhaps the eventual rumored threat of a ground invasion—induced the Serbs, on June 9, to agree to a 50,000-member, multinational Kosovo force, known as KFOR, which included 12,000 troops from Britain and 7,000 from the U.S., and to allow the repatriation of Kosovar refugees.

Russia was eager to play an important role, and on June 12, unilaterally took control of Pristina's airport. Ordered by NATO's supreme European commander, U.S. Gen. Wesley Clark, to force them out, the British commander, Lt. Gen. Sir Michael Jackson, refused, observing, "It's not worth starting World War III." An agreement providing for the redeployment of Russian troops under NATO command was reached a week later.

NATO faced the challenge of maintaining order in Kosovo, but it was doubtful whether Serbs and Kosovars were truly ready to live side-by-side. Many Serbs now were fleeing Kosovo, and those who remained risked revenge attacks from Kosovars. Complicating NATO's mission and Kosovo's future was the fact that Milosevic retained power, despite his indictment in May by a UN war crimes tribunal for atrocities committed against the Kosovars.

THE TOP TEN NEWS STORIES OF 1999

Bill Clinton became only the 2d president in U.S. history to be impeached by the House of Representatives and tried in the Senate. He had been impeached by the House, Dec. 19, 1998, on charges of giving false and misleading testimony to a grand jury and obstruction of justice. Both charges grew out of an investigation into Clinton's cover-up of a sexual liaison with former White House intern Monica Lewinsky. The trial started in the U.S. Senate on Jan. 7, 1999, and ended with his acquittal on Feb. 12. On the first article of impeachment, 55 Senators voted for acquittal; the vote on the second was 50-50, with a two-thirds majority needed for conviction.

After more than a year of fighting between Serbian forces and ethnic Albanians in the Kosovo region of Yugoslavia, NATO began air strikes against Yugoslavia on Mar. 24 to induce a Serbian pullback from the area. As the air strikes proceeded, Serb attacks on the Albanians increased, and a tide of refugees poured out of Kosovo into neighboring areas. On June 9, Yugoslav Pres. Slobodan Milosevic signed an agreement to end the fighting, withdraw Serb soldiers from Kosovo, and allow international peacekeeping forces to enter. During the bombing, an indictment had been issued against Milosevic by a Yugoslav war crimes tribunal for "crimes against humanity."

On Apr. 20, in Littleton, CO, 2 teenagers, aged 18 and 17, fatally shot 12 students and a teacher, at Columbine High School, before ultimately killing themselves. The incident was the most dramatic of several tragedies involving guns and loss of life, which fueled a national debate over gun control and the alleged harmful influence of violence in the media.

On Feb. 7, King Hussein of Jordan, 63, died of non-Hodgkin's lymphoma. Hussein was regarded as a leading Middle East peacemaker and had attended the Wye summit in October 1998 to press for peace, an action seen as helping to turn the tide toward a new accord. A new Israeli prime minister, Ehud Barak, elected in May, promised to pursue a final settlement with the Palestinians and to open peace talks with Syria.

On Aug. 17, an earthquake measuring 7.4 on the Richter scale struck a heavily populated area of northwest Turkey, killing at least 16,000 people and causing massive injury and damage. Many of the dead were trapped and killed in collapsed buildings said to have been poorly constructed. The quake was Turkey's deadliest in 60 years.

A small plane piloted by John F. Kennedy Jr., and carrying his wife, Carolyn Bessette Kennedy, and her sister Lauren Bessette, crashed off Martha's Vineyard, MA, on July 16, killing all on board. The accident set off a worldwide outpouring of sorrow and evoked memories of the long line of tragedies suffered by the Kennedy family. Kennedy, the only surviving son of assassinated Pres. John F. Kennedy and Jacqueline Kennedy Onassis, was the publisher of *George*, a political magazine.

A large majority of voters in the territory of East Timor voted for independence from Indonesia in a UN-sponsored referendum held on Aug. 30. Prior to the vote, anti-separatist militias reportedly had engaged in a terror campaign, which escalated after the results were published. Hundreds were massacred, and the infrastructure of Dili, the provincial capital, was severely damaged. After much delay, the Indonesian government, Sept. 12, agreed to accept an international peacekeeping force in the area.

The summer of 1999 saw high temperatures and a severe drought in the eastern U.S. By July, West Virginia and portions of 5 other states hit by drought had been declared disaster areas. Hardest hit were farmers, who saw their crops literally withering away. The drought was followed Sept. 16-17 by Hurricane Floyd, a major storm that came ashore in the U.S. near Wilmington, NC, leaving large portions of that state flooded with 2 feet of rain. The storm then progressed up the east coast, causing extensive flooding and power outages all the way into New England.

The 2000 presidential election was news in 1999, as prospective candidates in both major parties jockeyed for position. Texas Gov. George W. Bush, son of former Pres. George Bush, who formally launched his campaign June 12 in Iowa, was by then the apparent Republican front-runner. Vice Pres. Al Gore, regarded as the Democratic front-runner, officially declared his candidacy June 16 in Tennessee. Other leading candidates ranged from Arizona Sen. John McCain, a former Vietnam POW, on the GOP side to Bill Bradley, former pro-basketball star and U.S. senator from New Jersey, on the Democratic side. Outside presidential politics, First Lady Hillary Rodham Clinton was taking steps toward seeking a U.S. Senate seat in New York.

People around the world prepared for Jan. 1, 2000, and the dawn of a new century and millennium (officially the 20th century does not end until Dec. 31, 2000, but this was not a major concern to most people). Celebrations were planned, along with many retrospectives of the century and millennium gone by. As Jan. 1 approached, opinions were still divided over the impact on today's high-tech society of the so-called Y2K computer problem.

CHRONOLOGY OF THE YEAR'S EVENTS

Reported Month by Month, Nov. 1, 1998, to Oct. 15, 1999

NOVEMBER 1998

National

Democrats Make Small Gains in Election—The Democratic Party, which had once feared disaster because of the cloud of scandal over the Clinton White House, did surprisingly well in the **Nov. 3** elections. Democrats gained 5 seats in the House of Representatives—the first time since 1934 that the party holding the presidency had picked up House seats in a nonpresidential year. The Republicans, however, still held a majority, 223-211, with 1 independent, the closest division between the parties since 1953. In the Senate, each party took 3 seats from the other side, and the GOP thus kept its 55-45 margin. The Republicans retained a lopsided margin in governorships, 31-17, with 1 Independent and 1 Reform Party candidate.

With the impeachment process gathering momentum, the Republicans ran TV ads in some districts that alluded to presidential scandals. But Pres. Bill Clinton's job approval rating remained high among the general public, and blacks, who were among his strongest supporters, turned out in

larger numbers than expected. Many voters seemed tired of hearing about his affair with a young intern, Monica Lewinsky, and the subsequent cover-up.

In the most expensive senatorial contest, Sen. Alfonse D'Amato (R, NY), chair of the Senate Banking Committee, was upset by Rep. Charles E. Schumer (D). In North Carolina, Democrat John Edwards defeated incumbent Lauch Faircloth (R), a strong critic of Clinton. Evan Bayh won the open seat in Indiana for a 3d Democratic pickup. Carol Moseley-Braun (IL), the Senate's only black member, who had been plagued by ethical controversies, was the only Democratic senator to lose; she was defeated by State Sen. Peter G. Fitzgerald. Jim Bunning (the Hall of Fame baseball pitcher) and Gov. George V. Voinovich won open seats in Kentucky and Ohio, respectively, for 2 more Republican gains.

Republican governors, stressing sound management and lower taxes, were reelected in many large states, including New York, Wisconsin, Michigan, and Pennsylvania. George W. Bush was reelected governor of Texas in a landslide, and his brother Jeb Bush won a closer race for governor of Florida. Republicans George H. Ryan and Bob Taft won open contests in Illinois and Ohio, respectively. In California, however, Lt. Gov. Gray Davis, a Democrat, took the statehouse away from Republicans. And in Minnesota, the Reform Party candidate, former professional wrestler Jesse ("The Body") Ventura, won the governorship.

Gingrich Announces Resignation—Newt Gingrich (R, GA) announced **Nov. 6** that he would not seek reelection as House Speaker and would leave Congress at the end of his term. Gingrich, a conservative, had led the GOP to a majority in the House in 1994 for the first time in 40 years, and was rewarded with election as Speaker. However, he did not fare well against Pres. Clinton in legislative confrontations, and in 1997 he was reprimanded by the House for misuse of tax-exempt donations. He was seriously weakened by the poor Republican showing in the 1998 elections.

Glenn Completes 2d Trip Through Space—On **Nov. 7**, Sen. John Glenn, 77, the oldest person ever to travel through space, landed at Cape Canaveral, FL, with 6 other crew members after a 10-day mission aboard the shuttle *Discovery*. The shuttle was launched **Oct. 29**, 36 years after Glenn became the first American to orbit the earth. Experiments were made to study effects of space travel on an older person, and the crew deployed satellites, **Oct. 30** and **Nov. 1**.

Clinton Settles Paula Jones Suit—Pres. Clinton, **Nov. 13**, settled a lawsuit brought by Paula Corbin Jones in 1994, by agreeing to pay her $850,000, but with no apology or admission of guilt. She alleged that Clinton, while governor of Arkansas in 1991, had made an unwanted sexual advance to her in a Little Rock hotel. Clinton's relationship with former White House intern Monica Lewinsky had been discovered while Jones's lawyers were seeking to establish a pattern of sexual misconduct. Although Federal District Judge Susan Webber Wright dismissed Jones's lawsuit in April 1998, its revival by a U.S. appeals court was thought possible after Clinton acknowledged that he had withheld information about his relationship with Lewinsky during his January 1998 deposition in the Jones case.

Starr Indicts Hubbell Again—Webster Hubbell, a friend of the Clintons, was indicted **Nov. 13** for the 3d time by the Whitewater grand jury of independent counsel Kenneth Starr. Hubbell, associate attorney general in 1993-94, had later been jailed for embezzlement. The latest indictment alleged that Hubbell lied about work he had performed for the Castle Grande real estate project, whose failure helped bring down the savings and loan company owned by the Clintons' Whitewater business partners. A judge had dismissed a 2d indictment for tax evasion.

Fed Cuts Interest Rates—On **Nov. 17**, for the 3d time in less than 2 months, the Federal Reserve Board announced cuts in short-term interest rates—from 5% to 4.75% for the rate banks charge each other on overnight loans, and from 4.75% to 4.5% for the Fed's rate on loans to commercial banks. Many banks reduced their prime rate from 8% to 7.75%. On **Nov. 23**, for the 29th time in 1998, the Dow Jones industrial average closed at a record high—9374.27—which also was the high for the year.

GOP Backs Livingston for Speaker—At their party caucus on **Nov. 18**, Republicans who would serve in the new House unanimously supported Bob Livingston (LA) for Speaker. Earlier, on **Nov. 16**, House Democrats again chose Richard A. Gephardt (MO) as minority leader. David E. Bonior was again named minority whip. In contested elections, Democrats chose Martin Frost (TX) and Robert Menendez (NJ) to be chair and vice chair of the Democratic caucus.

Livingston, whose declared candidacy for Speaker came just hours before Speaker Newt Gingrich announced his retirement, had another opponent briefly when Christopher Cox (CA) announced, **Nov. 6**, that he would also run for Speaker. Cox dropped out **Nov. 9**. House Majority Leader Dick Armey (TX) had a more difficult time in the GOP caucus, requiring 3 ballots to win. Majority whip Tom DeLay

(TX) was chosen again without opposition. J.C. Watts (OK), the only black Republican in the House, ousted John A. Boehner (OH) for chair of the Republican Conference; Tillie K. Fowler (FL) was chosen as vice chair.

Starr Testifies at Impeachment Hearing—Independent counsel Kenneth Starr restated his case for Clinton's impeachment in testimony before the House Judiciary Committee **Nov. 19**. In a 12-hour appearance he summarized his detailed report and argued that Clinton was guilty of perjury and obstruction of justice in relation to the Lewinsky affair. He acknowledged that thus far he had no evidence implicating the president in illegalities in other areas of inquiry. Committee Democrats, generally ignoring the specific factual allegations, argued that they did not warrant impeachment and denounced Starr's report as partisan and needlessly graphic.

Earlier, on **Nov. 9**, 19 historians and constitutional scholars testifying before the committee provided no consensus as to whether Clinton's alleged offenses would have been deemed impeachable by authors of the Constitution. That same day the Supreme Court refused to hear appeals of lower-court rulings that failed to shield the Secret Service and some of the president's lawyers from testifying.

A lawyer for former White House intern Monica Lewinsky said, **Nov. 16**, that Lewinsky had signed a contract for a book that would tell her story of the affair. Samuel Dash, Starr's ethics adviser, resigned **Nov. 20**, claiming Starr had violated the 1978 independent counsel law by becoming an advocate for prosecution in testimony before the committee. Clinton, **Nov. 27**, replied in writing to 81 questions submitted to him by the committee; he contended that while he had misled others, he never lied under oath. Many Republicans objected to the substance and dismissive tone of Clinton's responses.

Kevorkian Charged in Death Shown on TV—A videotape of a man receiving a fatal injection was shown on the TV program *60 Minutes* **Nov. 22**. The man, Thomas Youk, suffered from amyotrophic lateral sclerosis, a fatal neurological disorder also known as Lou Gehrig's disease. He had turned to Dr. Jack Kevorkian, who had assisted in more than 130 suicides. This was the first time, however, Kevorkian directly killed the patient. Kevorkian taped the procedure, including administering a lethal injection of potassium chloride. On **Nov. 25**, Michigan prosecutors charged him with first-degree murder.

FBI Report Shows Crime Decline—The FBI reported **Nov. 22** that the U.S. murder rate in 1997 was at its lowest since 1967—6.8 per 100,000 residents. The number of murders was 18,209, a decline of about one-fourth since 1993. An overall decline in serious crimes continued in 1997; some 13.2 million were reported, off 11% from 1991.

States Settle With Tobacco Companies—The country's 4 largest tobacco companies signed a settlement, **Nov. 23**, with 46 states, plus the District of Columbia and 4 territories, in which the companies agreed to pay $206 billion over 25 years to cover public health costs related to smoking. Eight states negotiated the plan. Four other states had already settled with the industry. The new agreement did not shield the companies from private or class-action lawsuits seeking damages. The companies would not be allowed to sell merchandise with tobacco-brand logos or advertise on billboards or mass-transit vehicles. An increase in the price of cigarettes was not required, but was likely to occur. The settlement did not impose penalties based on teen smoking rates.

Gore Won't Face Probe—Attorney Gen. Janet Reno announced **Nov. 24** that she would not appoint an independent counsel to investigate Vice Pres. Al Gore in connection with phone calls he had made to raise money for the 1996 presidential campaign. He had denied making calls from the White House to raise hard money, although he did attend a 1995 White House meeting at which depositing funds into hard money accounts was mentioned.

U.S. Economy at a Glance November 1998	
Unemployment rate	4.4%
Consumer prices (change over Oct.)	+0.2%
Producer prices (change over Oct.)	−0.2%
Trade deficit	$14.66 bil
Dow Jones high (Nov. 23)	9374.27
Dow Jones low (Nov. 2 and 3)	8706.15
Index of leading economic indicators (change over Oct.)	+0.6%

International

Malaysians Protest Trial of Ex-Official—The trial of former Deputy Prime Min. Anwar bin Ibrahim of Malaysia began in Kuala Lumpur, **Nov. 2.** Prime Min. Datuk Seri Mahathir bin Mohamad had removed Anwar in September, and 2 weeks later he was charged with corruption and sodomy. Anwar pleaded not guilty. The accusations prompted weeks of street protests. At the annual economic summit of Pacific Rim nations in Kuala Lumpur, **Nov. 16**, Vice Pres. Al Gore, representing the United States, asserted that Malaysia was suppressing freedom. Malaysian officials expressed anger at Gore's remarks.

Saudi Millionaire Indicted in Bombings—Osama bin Laden, a Saudi millionaire allegedly linked to a number of terrorist acts, was indicted **Nov. 4** by a U.S. grand jury in New York in connection with the August bombings at U.S. embassies in Kenya and Tanzania. The indictment said that bin Laden had headed a terrorist conspiracy since 1989, utilizing assistance from other terrorist groups and several governments. He was accused of training Somalis who killed 18 U.S. soldiers in Somalia in 1993.

Clinton Lifts Sanctions on India, Pakistan—Pres. Clinton made it known **Nov. 6** that he was lifting some sanctions that—as required by U.S. law—he had imposed on India and Pakistan after they conducted nuclear weapons tests in May. The prime ministers of the 2 countries had said in September that they would impose a moratorium on tests, and in October the longtime rivals held talks. In ending some sanctions, Clinton allowed international trade and investment agencies to resume activities in the 2 countries and allowed loans from the World Bank and the International Monetary Fund.

Cambodia Rivals Form Government—Two longtime rivals, Hun Sen and Prince Norodom Ranariddh, agreed **Nov. 13** to set up a coalition government in Cambodia. Hun Sen's party had won the July election but lacked enough seats to form a government. Under a **Nov. 23** agreement, Hun Sen's party received 12 cabinet posts and Ranariddh's got 11. Hun Sen would be premier and Ranariddh president of the assembly. A new body, the Senate, would be headed by a member of Hun Sen's party. The assembly approved the new government **Nov. 30.**

Rescue of Brazil's Economy Approved—A $41.5 billion aid package designed to rescue Brazil's ailing economy was announced **Nov. 13.** The International Monetary Fund would provide $18 billion and the United States $5 billion; the rest would come from the World Bank, the Inter-American Development Bank, and 19 other governments. To qualify for the money, Brazil would have to cut spending, increase taxes, and bring its federal budget into a surplus.

Bombers on Way to Iraq Are Called Back—A U.S. aerial assault on Iraq was stopped at the last minute, **Nov. 15**, when Pres. Clinton, having accepted Iraqi concessions, called back bombers carrying cruise missiles. On **Oct. 31** Iraq ended its cooperation with the UN Special Commission on Iraq (UNSCOM), which monitored Iraq's destruction of its missiles and chemical and biological weapons. On **Nov. 5**, the UN Security Council unanimously condemned Iraq for its action. The last UN weapons inspectors and most other UN personnel left Iraq **Nov. 11.** That same day Clinton warned Iraqi Pres. Saddam Hussein that an attack was imminent and U.S. Defense Sec. William Cohen ordered more aircraft and troops to the region.

On **Nov. 14**, Iraqi Deputy Prem. Tariq Aziz sent a letter to UN Sec. Gen. Kofi Annan, stating that Iraq would allow inspections to resume. Iraq's UN representative Nizar Hamdoon sent a note the next day saying that the letter was an "unconditional decision" to cooperate with UNSCOM. Clinton then called back the bombers. Later on **Nov. 15**, Clinton appeared to call for the overthrow of Saddam Hussein. He stressed that Iraq had to allow full UNSCOM access to sites and provide requested documents. UN aid workers began to return **Nov. 16**, followed by arms inspectors **Nov. 17.** Earlier on **Nov. 1,** Denis Halliday resigned as coordinator of the UN's oil-for-food program. He blamed economic sanctions for the deaths of Iraqi children.

3 Convicted for Crimes Against Serbs—Two Bosnian Muslims and a Bosnian Croat were convicted and sentenced to prison **Nov. 16**, for war crimes against Serb prisoners. Sentences for murder, torture, and rape were imposed in the Netherlands by the International Criminal Tribunal for the former Yugoslavia. Fresh reports of atrocities in the Serbian province of Kosovo prompted the UN Security Council, **Nov. 17**, to ask Pres. Slobodan Milosevic of Yugoslavia to allow international investigation of possible war crimes.

Clinton Urges Japan to Act on Economy—Pres. Clinton traveled to Japan and South Korea. In Japan **Nov. 19-20**, he urged his hosts to deal more forcefully with their economic problems, and to take in more imports from countries suffering economic downturns. In South Korea, Clinton expressed concern over reports that North Korea was stepping up missile production, but at a press conference **Nov. 21** he and Pres. Kim Dae Jung defended their policy of engagement with that country.

China Arrests Supporters of New Party—Five advocates of a new China Democracy Party were arrested by Chinese authorities **Nov. 30.** Since first attempting to register the party in June, supporters had been harassed.

General

Countries Cope With Hurricane's Aftermath—Hurricane Mitch, which had plowed through Central America in late October, left a legacy of devastation. There were nearly 11,000 deaths, including 6,400 in Honduras and 4,000 in Nicaragua; 25% and 20% of the population were left homeless in Honduras and Nicaragua, respectively. Mayor Cesar Castellanos of Tegucigalpa, Honduras, died in a helicopter crash **Nov. 1** as he surveyed the damage. In Nicaragua, where one volcano collapsed and another erupted, damage was put at $1.5 billion. In Honduras, damage was estimated at $4 billion, and 90% of the banana plants were destroyed. El Salvador and Guatemala also were hit hard.

DNA Links Jefferson to Son by Slave—The British journal *Nature* reported in its **Nov. 5** issue that a DNA analysis showed it was likely that Pres. Thomas Jefferson fathered a son by his slave Sally Hemings. Blood samples had been taken from living descendants of Jefferson and Hemings. An unusual Y chromosome, found in both families, appeared to link Jefferson to Hemings's son, Eston, bolstering rumors that Jefferson had had a relationship with Hemings.

DECEMBER 1998

National

Exxon and Mobil Announce Plans to Merge—Exxon Corp. announced **Dec. 1** that it planned to buy Mobil Corp. and thus create the world's largest company as measured by revenue. The 2 oil-industry giants had combined 1997 incomes of $203.1 billion and total net profits of $11.8 billion. Their combined market value, $239 billion, would rank 3d, behind that of Microsoft and General Electric. The merger, if approved by federal regulators, would be accomplished by an $80 billion stock transaction in which Mobil shareholders would receive 1.32 Exxon shares for each of their Mobil shares. Some 9,000 employees were expected to lose their jobs when the corporations merged.

Former Agriculture Secretary Found Not Guilty—Mike Espy, who had served as Pres. Clinton's secretary of agriculture, was found not guilty of corruption charges by a federal jury in Washington, DC, **Dec. 2**. Espy's indictment had been obtained by an independent counsel, Donald Smaltz, who charged that Espy had accepted $34,000 in illegal gifts from companies that the Dept. of Agriculture regulated. The Smaltz investigation, which cost $17 million, led to 15 convictions and the levying of $11 million in fines, mostly from companies accused of illegal gifts to Espy.

Hoffa Wins Teamsters Presidency—James P. Hoffa, a Michigan labor lawyer, won the presidency of the International Brotherhood of Teamsters on **Dec. 5**, when his principal opponent, Tom Leedham, conceded defeat in the union's mail-in election. The election was to fill the last 3 years of the 5-year term to which Ron Carey was elected in 1996. (Federal overseers had removed Carey, after finding he had "closed his eyes" to a scheme that fed union funds into his reelection campaign.) Hoffa is the son of James R. Hoffa, who headed the Teamsters from 1957 to 1971. In the December election, Hoffa won 55% of the vote in a 3-way contest, Leedham won 39%, and John Metz, 6%.

Clinton Spared Fund-Raising Investigation—Attorney Gen. Janet Reno said, **Dec. 7**, that she would not appoint an independent counsel to look into whether Pres. Clinton broke the law in spending Democratic Party funds on television ads for the 1996 campaign. In November she had also turned down an outside inquiry into the use of party funds by Vice Pres. Al Gore. Clinton had used money intended for general party-building to, in effect, promote his reelection. Reno concluded that Clinton and Gore had acted on the basis of advice from lawyers and had "lacked the criminal intent to violate the law."

Earlier, on **Dec. 1**, Federal Election Commission auditors concluded that both major parties had broken the law by using party funds that had no restrictions on spending to bankroll advertising for their presidential nominees. The ads did not explicitly urge the election or defeat of any candidate. The FEC auditors said that the Clinton–Gore campaign should return $7 million of the public funds that it had received and that the campaign of Robert Dole, the Republican presidential nominee, should return $17.7 million. The FEC board, however, rejected these recommendations, 6–0, on **Dec. 10**, expressing doubts about the auditors' conclusions.

In Los Angeles **Dec. 14**, U.S. District Judge Manuel L. Real sentenced Democratic Party donor Johnny Chung to 5 years of probation and 3,000 hours of community service on charges, to which he had pleaded guilty, that included $20,000 in illegal gifts to the Clinton–Gore campaign. The Democratic Party had returned nearly $400,000 in gifts from Chung that were of dubious legality.

Puerto Rico Says No to Statehood—In a nonbinding plebiscite held **Dec. 13**, Puerto Ricans had an opportunity to express a preference as to the future political status of Puerto Rico. Three of the 5 options offered—independence, commonwealth status, and free association with the United States—got little support. Statehood was favored by 46.5%. However, the "none of the above" option was supported by 50.2%, indicating that most wished to retain Puerto Rico's current status. Although Puerto Rico's current status with the U.S. was as a commonwealth, some people objected to the wording of the commonwealth option which they felt made Puerto Rico sound like a colony.

House of Representatives Impeaches Clinton—The U.S. House of Representatives gave its approval, **Dec. 19**, to 2 articles of impeachment charging Pres. Clinton with perjury and obstruction of justice. The allegations, approved by the House Judiciary Committee earlier in the month,

involved actions that the president had taken to conceal his relationship with a former White House intern, Monica Lewinsky.

The White House, **Dec. 8**, presented a 184-page legal brief defending the president. With respect to obstruction of justice allegations, it asserted that Clinton had sought to find a job for Lewinsky long before she became a witness in a sexual harassment suit filed by Paula Corbin Jones against Clinton. It also stated that Clinton's conversation with his secretary, Betty Currie, in which he appeared to be encouraging her to agree to a version of events favorable to himself, had taken place before she was actually called as a witness in that case. Charles Ruff, the White House counsel, testified **Dec. 9** that although the president's conduct was reprehensible, the House should not seek to overturn "the mandate of the American people." Five former prosecutors testified **Dec. 9** that it would be difficult to prove that Clinton had intentionally lied—a requirement for a perjury conviction. None of the principal figures in the case were called as witnesses by the committee, whose final debates consisted mainly of restatements of familiar positions.

Committee Republicans, **Dec. 9**, presented 4 proposed articles of impeachment, based on recommendations from the independent counsel, Kenneth Starr, in his September report to Congress. The committee approved the 4 articles—the first 3 on **Dec. 11** and the 4th on **Dec. 12**. Articles I, III, and IV were approved on straight party-line votes, 21–16, with all Republicans in support and all Democrats opposed. Article II was approved 20–17, with 1 Republican, Lindsey Graham (SC), voting no.

Article I charged that Clinton had "willfully corrupted and manipulated the judicial process of the United States" by providing false information to a federal grand jury on Aug. 17, 1998, concerning, among other matters, the "nature and details" of his relationship with Lewinsky. Article II declared that Clinton had lied about his relationship with Lewinsky in a Jan. 17, 1998, deposition in the Jones case. Article III asserted that a number of actions taken by the president constituted obstruction of justice. Although individuals were not named, it was clear that the actions included his alleged encouragement to Lewinsky to provide a false affidavit in the Jones case, efforts to find her a job, efforts to conceal evidence (subpoenaed gifts exchanged with Lewinsky), and coaching of Currie. The article also asserted that he had allowed his attorney Robert Bennett to make false statements at the Jones deposition. Article IV said that Clinton had made "perjurious, false and misleading sworn statements" in his written response to questions submitted to him by the committee in November. All 4 articles concluded with the declaration that Clinton warranted "removal from office and disqualification to hold and enjoy any office of honor, trust or profit under the United States."

Clinton, **Dec. 11**, again apologized for his actions. The next day the Judiciary Committee voted down, 22–14, a motion to censure him as an alternative to impeachment. Rep. Bob Livingston (LA), whom Republicans had picked to be Speaker of the House beginning in January 1999, said he would not schedule a vote on censure because it was not part of the constitutional impeachment process.

The House debated the articles on **Dec. 18** and **19**. Rep. Henry J. Hyde (R, IL), chairman of the Judiciary Committee, argued that "lying under oath . . . is a public act, not a private act" and constitutes "willful, premeditated, deliberate corruption of the nation's system of justice." Rep. John Conyers Jr. (D, MI), defending Clinton, argued that "impeachment was designed to rid this nation of traitors and tyrants, not [of] attempts to cover up extramarital affairs."

After 14 hours of emotional argument, the House, **Dec. 19**, approved Article I, 228–206, with 5 Democrats in support and 5 Republicans opposed. Article II failed, 229–205 (5 Democrats in support, 28 Republicans opposed). Article III was approved, 221–212 (5 Democrats in support, 12 Republicans opposed). Article IV was defeated, 285–148 (1

Democrat in support, 81 Republicans opposed). The lone independent voted against all 4 articles. One Democrat was absent for all 4 votes and another for the last 2.

The president, appearing on the White House lawn with First Lady Hillary Rodham Clinton and about 100 House Democrats after the voting, appealed to the Senate to find a resolution of the matter that would punish him without removing him from office. Former Presidents Gerald Ford and Jimmy Carter, **Dec. 21**, urged the Senate to support a censure resolution rather than bring on a trial. After Trent Lott (R, MS), the Senate majority leader, **Dec. 29** indicated his support for a short trial with no witnesses, Hyde wrote him, **Dec. 30**, warning the Senate not to act in haste without hearing all of the evidence.

Speaker-Designate Announces He Will Leave House— Rep. Bob Livingston (LA), chosen by his Republican colleagues in November to be the next Speaker of the House, announced **Dec. 19** that he would not be a candidate and that he would soon resign his House seat.

Two days earlier, on **Dec. 17**, Livingston admitted that he himself had had extramarital affairs "on occasion." Larry Flynt, the publisher of *Hustler* magazine, disclosed **Dec. 18** that the magazine was, in fact, investigating Livingston's personal life. Flynt had offered up to $1 million to anyone who could document an affair with a member of Congress.

Stock Market Rose Again in 1998—On Wall Street, the Dow Jones industrial average closed **Dec. 31** at 9181.43, 16.1% higher than on the last trading day of 1997. Although 1998 was the 4th consecutive year of dramatic advance in stock values, most of the gain had been registered in the first 6 months of the year; prices had fallen sharply and then revived during the last 6 months. The Nasdaq index of over-the-counter stocks, consisting of smaller companies, soared by nearly 40% during 1998.

U.S. Economy at a Glance December 1998	
Unemployment rate	4.3%
Consumer prices (change over Nov.)	+0.1%
Producer prices (change over Nov.)	+0.5%
Trade deficit	$14.24 bil
Dow Jones high (Dec. 29)	9320.98
Dow Jones low (Dec. 14)	8695.60
Index of leading economic indicators (change over Nov.)	+0.2%
4th-quarter GDP (change over 3d quarter)	+5.9%

Calendar Year 1998	
Unemployment rate	4.5%
Consumer prices (change over 1997)	+1.6%
Producer prices (change over 1997)	−0.1%
Trade deficit	$164.28 bil
Dow Jones high (Nov. 23)	9374.27
Dow Jones low (Aug. 31)	7539.07
GDP (change over 1997)	+3.9%

International

Bosnian Serb General Arrested for Genocide—Maj. Gen. Radislav Krstic was arrested **Dec. 2** in Bosnia–Herzegovina by U.S. and NATO troops. The international Yugoslav war crimes tribunal had drawn up a sealed indictment of Krstic, who was the highest-ranking official to be charged in connection with the Bosnian conflict. The indictment asserted that he was personally involved in the apparent killing of Muslim men from Srebrenica, 7,000 of whom had disappeared after the town was captured by Serb forces in 1995. Krstic pleaded not guilty **Dec. 7**.

U.S. Adds Component to Space Station—Assembly of the international space station began in earnest after the U.S. shuttle *Endeavour* lifted off from the Kennedy Space Center in Florida on **Dec. 4**. The shuttle carried *Unity*, the 2d component of the station, which was to be attached to *Zarya*, the Russian-built module that had been in orbit for 2 weeks. The

shuttle caught up with *Zarya* **Dec. 6**. One astronaut grabbed *Zarya* with a robot arm, and another fired thruster rockets to bring the 2 vehicles together. The united craft, 77 feet long, weighed 70,000 pounds. During a 7-hour spacewalk, **Dec. 7**, 2 astronauts connected 40 cables between the components. Crew members spent some time in both components, then returned to the shuttle, which separated from the space station **Dec. 13**. The shuttle landed in Florida **Dec. 15**.

Last Guerrilla Troops Surrender in Cambodia—The Khmer Rouge movement in Cambodia apparently died out **Dec. 5**, when its last 500 to 1,000 troops surrendered to the government. Millions of Cambodians lost their lives in the 1970s by execution and enforced hardships during the brief but brutal rule of the Khmer Rouge. The movement's long-time leader, Pol Pot, had died in April 1998.

Clinton Visits Middle East—Pres. Clinton began a trip to the Middle East **Dec. 12**. After meeting with Israeli Prime Min. Benjamin Netanyahu in Israel, Clinton landed by helicopter at the new Gaza International Airport in Palestinian territory on **Dec. 14**. With Clinton present, the Palestine National Council reaffirmed cancellation of provisions in the Palestinian charter that called for destruction or nonacceptance of Israel. Clinton, Netanyahu, and Palestinian leader Yasir Arafat met together at a border point **Dec. 15**, but reached little agreement on further steps toward peace.

U.S., Britain Bomb Iraqi Military Sites—Beginning on **Dec. 16**, and over a period of 4 nights, the U.S. and Great Britain bombed some 100 Iraqi military sites from the air. The action, long threatened, came after the allies concluded Iraq would not cooperate with UN arms inspectors.

In November, Pres. Clinton had called back planes already in the air when Iraq appeared to make concessions. On **Dec. 8**, inspectors with the UN Special Commission on Iraq (UNSCOM) began surprise searches of Iraqi sites; the next day they were denied access to offices of the ruling Baath Party. UNSCOM Chairman Richard Butler reported to the UN **Dec. 15** that Iraq had failed to provide promised cooperation, was imposing new restrictions, and had removed materials from some sites before inspection. The inspectors left Iraq on **Dec. 16**.

Cruise missile attacks from jets based on land and sea began that night. Targets included military command posts, air defense sites, and facilities used by the Republican Guard, Iraq's elite soldiers. One oil refinery was targeted. Clinton said the strikes were intended to damage Iraq's ability to produce weapons of mass destruction. Russia and China objected to the lack of consultation with the UN Security Council. The attack on Iraq began as the House of Representatives was moving toward an impeachment vote; Sen. Trent Lott (R, MS), the Senate majority leader, questioned "both the timing and policy."

Iraq declared, **Dec. 19**, that UN inspectors would not be allowed back into the country. In a televised address **Dec. 19**, Clinton claimed the attacks had weakened Iraq militarily and damaged its weapons program. Iraqi Pres. Saddam Hussein declared **Dec. 20** that the confrontation had been a victory for Iraq over the "enemies of God and humanity." Deputy Prem. Tariq Aziz said that 62 soldiers and many civilians had been killed.

On **Dec. 28** and **30**, U.S. jets fired back at ground sites in the "no-flight" zones (areas Western nations had declared off-limits to Iraqi aircraft), after being fired at.

5 More Men Indicted in Embassy Bombing—U.S. prosecutors indicted 5 more men, **Dec. 16**, in the August 1998 bombing of the U.S. Embassy in Tanzania. According to the indictment, the 5, who were charged with conspiracy and murder, had bought supplies, including a truck, for the attack. None were in custody.

Israeli Parliament Topples Government—The shaky coalition of Israeli Prime Min. Benjamin Netanyahu collapsed, **Dec. 21**, when the Knesset (parliament) voted 81–30 to dissolve the government. Netanyahu's Likud Party dominated the coalition. But it also included Orthodox religious parties and nationalists opposed to the Wye Memorandum (agreed to by Netanyahu), which contemplated more Israeli military withdrawals from the West Bank. Earlier on **Dec. 21**, the Knesset, 56–48, had rejected Netanyahu's peace plan. After the vote, resignations from Likud gave promise that the next election, scheduled for May 1999, would be a free-for-all.

General

Octuplets Born in Texas—A woman in Houston gave birth in December to 8 babies—the only known set of octuplets to be born alive in the U.S. The parents were U.S. citizens who had been born in Nigeria. The mother, Nkem Chukwu, 27, who had taken fertility drugs, gave birth to 1 infant on **Dec. 8** and to 7 more on **Dec. 20**. The 7 babies born together, by caesarean section, were 10 weeks premature. The 8 babies ranged in weight from 10.3 to 26 ounces. The smallest baby died **Dec. 27**.

Olympic Committee Looks Into Bribery Allegations—The International Olympic Committee **Dec. 11** began an internal investigation of rumors that bribes had been offered by cities seeking to be chosen as sites for Olympic Games. The IOC's Marc Hodler of Switzerland said **Dec. 12** that he knew that committee members had taken bribes. Although initial attention was focused on bribes in connection with the choice of Salt Lake City, UT, to host the 2002 Winter Games, Hodler said that the host cities for the 1996, 1998, and 2000 Games also were implicated.

1998 Was Warmest Year on Record—The UN's World Meteorological Organization said **Dec. 17** that 1998 was the warmest year ever recorded. Temperatures had been running above normal for 20 years, and 7 of the 10 warmest years ever recorded had occurred in the 1990s. In 1998 the average surface temperature was 58°F. Some scientists attributed the warming to the release of industrial gases in the atmosphere. Others said that El Niño, the recurring weather phenomenon that alters weather patterns, was responsible.

JANUARY 1999

National

Presidential Race Begins to Shape Up—After the congressional and state elections of November 1998, potential candidates for president in the 2000 election began to examine their options seriously. Sen. John McCain (R, AZ) formed an exploratory committee to consider a presidential campaign, on **Dec. 30**. Vice Pres. Al Gore officially entered the contest on **Dec. 31**, when he filed a notice with the Federal Election Commission, though he made no public statement at the time. On **Jan. 4**, Sen. Robert Smith (R, NH) announced the formation of a campaign committee. Sen. John Ashcroft (R, MO) said **Jan. 5** that he would not run for president but would instead focus on getting reelected to the Senate. On **Jan. 8**, former Gov. Lamar Alexander (R, TN) formed an exploratory committee. Sen. Paul David Wellstone (D, MN) said, **Jan. 9**, that he would not run for president. On **Jan. 12**, former Sen. Bill Bradley (D, NJ) filed a notice of candidacy with the FEC; Rep. John R. Kasich (R, OH) filed papers on **Jan. 13**.

New Congress Opens Session—The first session of the 106th Congress convened on **Jan. 6**.

During earlier deliberations, Republicans had settled on J. Dennis Hastert (IL) as their new candidate for House Speaker and on **Jan. 6** he was elected over Minority Leader Richard A. Gephardt (D, MO), 220–205. Hastert had served in the Illinois legislature; he was first elected to the House in 1986. Known as a conservative, he also was seen as a conciliator who would be less ideological and confrontational than Gingrich had been. In his acceptance speech, **Jan. 6**, the new Speaker said that his goals were saving Social Security and Medicare, lowering taxes, providing more federal

support for education, and establishing greater military readiness.

Senate Begins Clinton's Impeachment Trial—The trial of Pres. Clinton on charges that he had committed misdeeds worthy of removal from office began in the U.S. Senate **Jan. 7**. The White House contended, **Jan. 11**, that the charges against Clinton, even if true, did not measure up to impeachable offenses.

The 13 House "managers" made their presentation **Jan. 14-16**, contending that perjury and obstruction of justice were impeachable offenses. In presenting their defense, **Jan. 19-21**, Clinton's attorneys dismissed the case as consisting of circumstantial evidence and unsupported testimony, all interpreted in the way most unfavorable to the president. On **Jan. 24**, 3 House managers informally interviewed former White House intern Monica Lewinsky. After days of wrangling over witnesses—the House managers wanted to call 15—the Senate by a vote of 56–44, on **Jan. 27**, approved a greatly slimmed-down list of only 3. Lewinsky was to be one, along with Clinton friend and adviser Vernon Jordan and Sidney Blumenthal, a White House adviser.

Inflation at 12-Year Low—The government said **Jan. 14** that consumer prices had risen 1.6% in 1998. This was slightly below the 1997 rate of 1.7%, and represented the smallest increase in prices since 1986. On Wall Street, stocks continued their strong advance, with the Dow Jones industrial average closing at all-time highs on **Jan. 6** and **8**, the latter being 9643.32.

Clinton Reports on "State of the Union"—Pres. Clinton gave his annual State of the Union address to a joint session of Congress and a national television audience, **Jan. 19**, while he was still on trial at the Capitol and faced the potential of being removed from office. Democrats gave Clinton a rousing welcome and frequent applause, while public opinion polls taken after the speech showed his job-approval ratings at or near an all-time high, despite the controversy.

Clinton urged that 62% of the $4.4 trillion in projected budget surpluses over the next 15 years be applied to saving Social Security and that up to $700 billion of the budget surplus be applied to shoring up Medicare funds. He also proposed that $700 billion in Social Security funds be invested in the stock market.

Reps. Jennifer Dunn (WA) and Steve Largent (OK) gave the Republican response to Clinton. In light of the projected budget surpluses, they advocated lowering income taxes. Largent urged Congress to "scrap the Internal Revenue tax code" and to ban so-called partial-birth abortions.

Alan Greenspan, chairman of the Federal Reserve Board, on **Jan. 20**, supported the plan to shore up Social Security using projected federal budget surpluses but said that no investment of Social Security funds by the government could be insulated from political control. In general, Republicans opposed government investment of funds, but favored an approach allowing workers to invest some contributions in private funds.

U.S. Economy at a Glance January 1999	
Unemployment rate	4.3%
Consumer prices (change over Dec.)	+0.1%
Producer prices (change over Dec.)	+0.3%
Trade deficit	$16.15 bil
Dow Jones high (Jan. 8)	9643.32
Dow Jones low (Jan. 22)	9120.67
Index of leading economic indicators (change over Dec.)	+0.5%

International

New European Currency Becomes Legal Tender—The European Union's new currency, the euro, became legal tender on **Jan. 1** in 11 participating nations. The currency could be utilized for personal checks, travelers' checks, and credit cards, and in the trading of stocks and bonds. However, bills and coins bearing the denomination would not appear until 2002, and national currencies would continue to exist until then. Trading in the currency opened **Jan. 4**, with the euro pegged at U.S. $1.1668. At the end of the day the euro stood at U.S. $1.1827.

Currency Upheaval Continues in Brazil—The Brazilian currency, the real, continued to be a matter of international concern. On **Jan. 6**, Gov. Itamar Franco of the Brazilian state of Minas Gerais suspended payments on the state's $15.4 billion debt to the national government, saying that there was no money to pay it. Brazilian Pres. Fernando Henrique Cardoso struggled to make fiscal changes so as to eliminate the national budget deficit and make Brazil eligible for $42 billion in aid from the International Monetary Fund. On **Jan. 13**, the national government devalued the real, triggering declines in markets around the world and fears of economic instability in Latin America at least. The government, **Jan. 15**, allowed the real to float freely on international markets. Immediately, markets in Brazil and elsewhere surged, as investors snapped up Brazilian stocks at bargain prices. The real continued to decline against the U.S. dollar.

U.S. Spied on Iraq Through Inspectors—U.S. officials acknowledged **Jan. 6** that the U.S. had spied on Iraq through the United Nations Special Commission (UNSCOM) conducting weapons inspections in Iraq. Spies posing as inspectors had installed devices to monitor communications of Iraqi security forces, including those guarding Pres. Saddam Hussein. Britons and Israelis were said to have helped interpret the data collected. Earlier, on **Jan. 5**, the U.S. fired missiles at Iraqi jets. On **Jan. 7**, the U.S. fired a missile at an antiaircraft battery. The U.S. State Dept. said **Jan. 11** it had reports that Iraq, in the previous 2 months, had executed 500 distrusted military officers and dissidents in its Shiite-dominated territory.

U.S. Complains to Japan About Steel Exports—The U.S., **Jan. 7**, asked Japan to voluntarily cut back on its steel exports to the U.S. From July to October 1998, imports into the U.S. had increased by some 50% compared with the same period in 1997. This led to layoffs, bankruptcies, and the closing of some U.S. mills. U.S. companies and workers complained that the appeal to Japan was too weak, and on **Jan. 12** the Clinton administration said it would impose trade penalties on Japan unless it cut back its steel exports. Japan said **Jan. 13** that it had no plans to do so.

Bosnian Serb War Crimes Suspect Killed—French NATO forces killed a suspected war criminal in Bosnia, **Jan. 9**, while attempting to arrest him. Dragan Gagovic had been indicted by the Yugoslavia war crimes tribunal in 1996 on charges related to the rape and torture of Muslim women during a Serb offensive in eastern Bosnia in 1992-93. The French troops fired on Gagovic's car in the town of Foca after he drove through a roadblock, killing him and causing injury to several passengers.

Serbs Kill 45 Civilians in Kosovo—In one incident that achieved wide publicity, Serb forces killed 45 ethnic Albanian civilians in Kosovo, **Jan. 15**. NATO, **Jan. 19**, warned Pres. Slobodan Milosevic of Yugoslavia that he must honor the 1998 cease-fire negotiated with the rebels in Kosovo or face air strikes. Since early 1998, the fighting had claimed 1,000 lives and driven up to 280,000 people from their homes. The U.S. carrier *Enterprise* was sent by NATO to the Adriatic Sea, **Jan. 20**. On **Jan. 29,** Serb forces killed 24 men, mostly civilians, in a stronghold of the Kosovo Liberation Army. NATO ambassadors **Jan. 30** gave NATO the authority to attack military targets in Serbia if Milosevic failed to agree to negotiate within a week.

Peace Agreement Collapses in Angola—UN Sec. Gen. Kofi Annan said, **Jan. 17**, that the peace process had collapsed in Angola and that "the country is now in a state of war." The government and the National Union for the Total Independence of Angola (UNITA) had signed an accord in 1994, but fighting had resumed. On **Jan. 21**, the UN Security Council voted unanimously to continue its mission in Angola. The council blamed the UNITA rebels for the renewed fighting and demanded that it give up territory it controlled to the government.

Brother of Ex-Mexican President Convicted—Raúl Salinas de Gortari, brother of former Pres. Carlos Salinas de Gortari of Mexico, was convicted **Jan. 21** of masterminding the 1994 murder of José Francisco Ruiz Massieu. The victim, who was shot to death, was the secretary general of the ruling Institutional Revolutionary Party. Salinas was sentenced to 50 years in prison without parole. He remained under investigation in connection with other crimes.

Pope Visits Mexico, U.S.—Pope John Paul II arrived in Mexico City **Jan. 22** on a visit to Mexico and the U.S. The pontiff, who was thought to be suffering from Parkinson's disease, appeared frail during his trip. He met with Pres. Ernesto Zedillo Ponce de León and separately with Cuauhtemoc Cardenas Solorzano, leader of a major opposition party, and said mass for 100,000 people at Mexico City's Aztec Stadium **Jan. 25**.

Pres. Clinton welcomed the pope to St. Louis on **Jan. 26.** That day, at a sports arena, John Paul appealed to 20,000 teenagers to forgo drugs and premarital sex. On **Jan. 27**, the pope said mass at the Trans World Dome in St. Louis, then left for Rome.

Jordan's Hussein Picks New Successor—On **Jan. 24**, King Hussein of Jordan named his son Abdullah as crown prince with the right of succession to the throne. Abdullah replaced the king's younger brother, Hassan, in that position. Hussein, who was seriously ill with non-Hodgkin's lymphoma, had undergone treatment for 6 months at the Mayo Clinic in Rochester, MN, returning on **Jan. 19**. On **Jan. 26,** in a letter whose contents were broadcast on Jordanian television, the king outlined complaints against his younger brother; these included assertions that Hassan had meddled in the army and foreign ministry. Hussein said Hassan had not agreed, once king himself, to appoint one of Hussein's sons as crown prince. Abdullah was sworn in as crown prince, **Jan. 26**.

General

Tennessee Wins College Football Title—The University of Tennessee Volunteers defeated the Florida State Seminoles in the Fiesta Bowl, 23–16, on **Jan. 4** and were acclaimed in both press association polls as the undisputed division 1-A national champions. The flexibility built into the Bowl Championship Series had permitted the matching of the 2 teams that had been ranked a consensus 1st and 2d in season-end polls and computer calculations.

Agreement Ends Lockout of NBA Players—An agreement reached **Jan. 6** brought an end to a 6-month player lockout by owners of National Basketball Association teams. The labor dispute had threatened to wipe out the entire 1998-99 season. Money had been the problem: Owners were concerned about rising player salaries, which now averaged $2.6 million a season. The agreement, covering 6 years, set no limit for the first 3 years on how much the players could receive in league revenue. Maximum salaries for players were put at $9 million to $14 million, depending on years played. Minimum salaries were pegged at $287,500 for rookies and $1 million for 10-year veterans. On **Feb. 5**, the NBA got its regular season underway.

Michael Jordan Announces Retirement—Michael Jordan, regarded by many as the greatest player ever, announced his retirement **Jan. 13**. The 35-year-old guard for the Chicago Bulls, who had led his team to 6 championships—most recently in the 1997-98 season—had scored 29,277 points in 11 seasons, had led the league in scoring 10 times, and had the highest NBA career scoring average (31.5 points per game). He had retired once before after the 1992-93 season to play minor league baseball, but he returned to basketball in 1995. He had been named most valuable player during 5 regular seasons and for 6 playoffs.

Olympic Committee Members Expelled—As the International Olympic Committee scandal continued to unfold, the IOC on **Jan. 24** recommended the expulsion of 6 IOC members—from Chile, Republic of the Congo, Ecuador, Kenya, Mali, and Sudan. Earlier in January, several people involved with the Salt Lake City, UT, committee—which had successfully won the 2002 Winter Olympics—resigned. One of them said, **Jan. 8**, that members of the committee had made payments up to $70,000 and provided health and real estate services to IOC members and their families. On **Jan. 22**, John Coates, president of the Australian Olympic Committee, said he had offered $35,000 each to 2 IOC members before they voted on Sydney's successful bid for the 2000 Summer Games.

The IOC panel investigating alleged misconduct by Olympic officials released its findings **Jan. 24**, by which time 3 IOC members—from Finland, Libya, and Swaziland—already had resigned. IOC Pres. Juan Antonio Samaranch denied that he had known in advance of the offers made to the IOC members. He said that the 2000 and 2002 Games would not be moved. The 6 IOC members were officially expelled for corruption **Mar. 17** following a vote of the IOC.

AIDS Virus Traced to African Chimpanzee Species—A longtime suspicion about the origin of HIV, the virus that causes acquired immune deficiency syndrome, received confirmation in a report released **Jan. 31**. A team of international scientists, led by Beatrice Hahn of the University of Alabama-Birmingham, reported that the predominant strain of the virus had been traced to a chimpanzee subspecies that lived in Cameroon, the Congo Republic, Equatorial Guinea, and Gabon. The virus did not appear to affect the health of the chimpanzees themselves. Hahn speculated that human exposure to the animals' blood may have occurred when the animals bit people or when their meat was being prepared for food.

Denver Broncos Repeat as Super Bowl Champs—The Denver Broncos, led by quarterback John Elway, retained their National Football League championship, **Jan. 31**, defeating the Atlanta Falcons, 34–19, in Super Bowl XXXIII in Miami.

Denver, **Jan. 17**, had defeated the New York Jets, 23–10, to win the American Football Conference title. Atlanta, **Jan. 17**, had upset the Minnesota Vikings, 30–27, in overtime to take the National Football Conference title. The winning points were from a 38-yard field goal kicked by Morten Andersen with 3:08 left in the extra period.

In the Super Bowl, Elway completed 18 of 29 pass attempts for 336 yards, including 1 touchdown of 80 yards, and he ran for another.

FEBRUARY 1999

National

Clinton Submits Budget With $117 Billion Surplus—Pres. Clinton, **Feb. 1**, sent to Congress a $1.77 trillion budget for fiscal year 2000 that included a surplus of $117 billion. The administration foresaw surpluses rising to nearly $400 billion per year over the next decade. The projections were based in part on assumptions that the growth in the gross domestic product would be at or near a modest 2% per year for the next few years. Major goals of the Clinton budget were reducing the national debt (now $5.6 trillion) and rescuing the Social Security system from potential collapse. The budget also provided for targeted tax cuts aimed at middle-class families, as well as increases in discretionary spending on defense, education, and health care. On **Feb. 2**, Rep. John R. Kasich (R, OH), chairman of the House Budget Committee, countered with a proposal for a 10% tax cut for everyone.

Quayle In, Gephardt Out of Presidential Race—Rep. Richard A. Gephardt (MO) announced, **Feb. 3**, that he would not seek the Democratic nomination for president in

2000. Gephardt, the House minority leader, said he would focus on winning a Democratic majority in the House in 2000, which would open the way to his election as Speaker. Former Vice Pres. Dan Quayle said, **Feb. 3**, that he was forming an exploratory committee to consider running for the Republican nomination; his speech stressed themes of lower taxes and family values. On **Feb. 18**, Sen. Robert Smith (NH), a conservative who opposed abortion and gun control, formally announced his candidacy for the GOP nomination. It was reported, **Feb. 22**, that former Gov. Pete Wilson of California would not seek the Republican nomination. Sen. John F. Kerry (D, MA) took himself out of consideration for 2000, **Feb. 26**.

Police Fire at Unarmed Man—An unarmed African immigrant who had no criminal record was killed **Feb. 4** when struck by 19 of 41 shots fired at him by 4 New York City police officers. The 4 had been looking for a rape suspect when they encountered the man, Amadou Diallo, 22, in the vestibule of his Bronx apartment building. Lawyers for the officers said they fired at Diallo because they thought he was reaching for a gun. The 4 officers were charged with second-degree murder **Mar. 25**.

Clinton Acquitted on Impeachment Charges—The Senate, **Feb. 12**, concluded the impeachment trial of Pres. Clinton, finding him not guilty of both counts against him. In both cases, the votes in favor of conviction and removal from office, for perjury and obstruction of justice, fell far short of the two-thirds required.

The formal round of questioning of witnesses by the 13 House Republican impeachment "managers" began **Feb. 1**, with the closed-door questioning of former White House intern Monica Lewinsky, with whom the president had had an affair. Clinton's lawyers declined to cross-examine her, but one of them read a one-sentence note of apology to her from the president for her difficult experiences of the past year. Vernon Jordan, a friend of Clinton's, was questioned on **Feb. 2** and Sidney Blumenthal, a White House aide, on **Feb. 3**.

A vote cast by the senators on **Feb. 4** appeared to foreshadow a verdict of not guilty. The Senate, 70–30, rejected a proposal supported by the House managers to have Lewinsky testify in person before the Senate itself. The managers did not seek to have Jordan or Blumenthal called; instead, on **Feb. 4**, the Senate agreed, 62–38, to let the managers show their videotaped depositions to the Senate.

A transcript of Lewinsky's testimony, released **Feb. 5**, showed she still denied that anyone had asked her to lie or submit a false affidavit in the Paula Corbin Jones sexual harassment case against Clinton. She said she now had "mixed feelings" toward Clinton. On **Feb. 6**, House managers played excerpts from the testimony of all 3 witnesses to the Senate, and Clinton's lawyers played excerpts from what Lewinsky and Jordan had said.

All 13 House managers participated, **Feb. 8**, in the closing arguments. Rep. Henry Hyde (R, IL), chairman of the House Judiciary Committee, admonished the senators that if they concluded that the president had committed the misdeeds charged against him yet voted not guilty, they would be "saying a perjurer and obstructer of justice can be president." In his closing remarks, **Feb. 8**, White House Counsel Charles Ruff maintained that Clinton had not committed perjury or obstruction.

The Senate debated the president's fate, behind closed doors, **Feb. 9-12**, and the final votes took place on **Feb. 12**. On Article I, the perjury charge, the tally was 55–45 against conviction. On Article II, the obstruction of justice charge, the Senate divided equally, 50–50; a two-thirds majority, or 67 votes, was required. All 45 Democrats voted not guilty on both counts. Forty-five Republicans voted guilty twice. Five Republicans—Slade Gorton (WA), Richard Shelby (AL), Ted Stevens (AK), Fred Thompson (TN), and John W. Warner (VA)—voted guilty only on the obstruction charge, while 5 other Republicans—John H.

Chafee (RI), Susan M. Collins (ME), Jim Jeffords (VT), Olympia J. Snowe (ME), and Arlen Specter (PA)—voted not guilty on both counts.

Following the impeachment votes, a bipartisan effort, led by Sens. Dianne Feinstein (D, CA) and Robert F. Bennett (R, UT), to censure Clinton failed as well. A 56–43 majority obtained in a procedural vote fell short of the two-thirds needed to overcome parliamentary maneuvering by Sen. Phil Gramm (R, TX), who opposed censure as unconstitutional.

In a brief statement, **Feb. 12**, Clinton said he was "profoundly sorry" for his actions, and urged the country to move on—a sentiment echoed by political leaders in both parties.

Racist Convicted in Dragging Death of Black Man—John William King, who described himself as a white supremacist, was convicted **Feb. 23** in Jasper County, TX, of the brutal murder of a black man, James Byrd Jr., that had shocked the country. In June 1998, Byrd had been seized and chained to the back of a truck and dragged 3 miles on rural roads east of Jasper, suffering fatal injuries. Two more men faced charges in connection with the killing. The county jury that convicted King sentenced him, **Feb. 25**, to death by lethal injection.

Daley Wins Another Term as Chicago Mayor—Mayor Richard M. Daley of Chicago was reelected **Feb. 23** with 73% of the vote. Daley, the son of the late Mayor Richard J. Daley, defeated U.S. Rep. Bobby L. Rush, a fellow Democrat. Rush, who is black, contended that minority neighborhoods had missed out on much of the city's recent prosperity, but Daley was endorsed by a number of black leaders.

Republican Elected to Gingrich's Seat—A Republican, Johnny Isakson, was elected, **Feb. 23**, to the U.S. House seat in Georgia's 6th District left vacant by the resignation of former Speaker Newt Gingrich. Isakson, a former state legislator, won 65% of the vote in a nonpartisan primary against 5 other candidates.

"Jane Doe" Alleges Assault by Clinton—In a television interview broadcast **Feb. 24**, an Arkansas woman claimed that in 1978, Bill Clinton, then the state attorney general, had assaulted her in a Little Rock hotel room. She said he came to her hotel room, and then held her down, bit her lip, and assaulted her sexually. In 1998, Juanita Broaddrick had sworn in an affidavit that the incident never happened. In that case she was identified as Jane Doe No. 5. Later that year, however, in response to questions from the office of the independent counsel, Kenneth Starr, she stated that the assault had indeed occurred. David Kendall, Pres. Clinton's attorney, denounced Broaddrick's allegation as "absolutely false."

U.S. Economy at a Glance February 1999	
Unemployment rate	4.4%
Consumer prices (change over Jan.)	+0.1%
Producer prices (change over Jan.)	−0.5%
Trade deficit	$18.52 bil
Dow Jones high (Feb. 22)	9552.68
Dow Jones low (Feb. 9)	9133.03
Index of leading economic indicators (change over Jan.)	+0.3%

International

Brazil, IMF Reach New Agreement—Brazil and the International Monetary Fund announced, **Feb. 4**, that they had reached a modified agreement on economic goals that would trigger the next payment in the IMF's $42 billion rescue package. Previous calculations had been upset after the currency, the real, had been devalued in January. Brazilian officials said they would increase their goal for budget surpluses and step up the process of privatization of industry.

Ethiopia, Eritrea Fight Over Location of Border—A dispute over a contested border area in the Badme region led to the resumption **Feb. 6** of armed conflict between Ethiopia and Eritrea. Ethiopia had agreed to negotiations, and the UN Security Council, **Feb. 10**, called on Eritrea to withdraw from the region at issue, which it had seized in 1998, and begin to negotiate. On **Feb. 27**, Eritrea accepted the peace plan, which had originally been put forth by the Organization of African Unity. Ethiopia claimed, **Feb. 28**, that its army had defeated Eritrean forces in the Badme region.

King Hussein of Jordan Dies—The 46-year reign of King Hussein of Jordan ended **Feb. 7** when he died in Amman. Hussein, who had been seriously ill with cancer since the summer of 1998, was immediately succeeded by his eldest son, who assumed the throne as Abdullah II.

On **Feb. 6** the Jordanian cabinet had officially declared he was unable to govern and in accordance with his wishes, named Abdullah regent.

Immediately upon assuming the throne, Abdullah named his half-brother Hamzah, 18, the son of Queen Noor, as his crown prince and successor. Pres. Clinton and 3 former U.S. presidents—Gerald Ford, Jimmy Carter, and George Bush—attended Hussein's funeral in Amman, **Feb. 8**. Other world leaders present included Prince Charles and Prime Min. Tony Blair of Great Britain, Pres. Jacques Chirac of France, Chancellor Gerhard Schroeder of Germany, Pres. Boris Yeltsin of Russia, and Prime Min. Benjamin Netanyahu of Israel.

Turkey's Seizure of Kurd Leader Stirs Protests—Turkish agents **Feb. 15** captured Abdullah Ocalan, head of the Kurdistan Workers' Party, a major group fighting for Kurdish independence in Turkey, Iraq, Iran, and Syria. Ocalan was seized in Nairobi, Kenya, after he left the Greek embassy there, where he had sought refuge. Ocalan was flown to Turkey and, on **Feb. 23**, charged with treason. Turks saw him as responsible for the deaths of thousands during the Kurds' armed struggle.

Kurd protestors **Feb. 16** took hostages at Greek and Kenyan diplomatic facilities; all were freed unharmed. Israel also was suspected of having assisted in the capture, and Kurds stormed the Israeli consulate in Berlin, **Feb. 17**. Israeli troops killed 4 Kurds. A possible Greek role in the capture led Greek Prem. Costas Simitis, **Feb. 18**, to remove 3 cabinet members, including the foreign and interior ministers. American officials said **Feb. 19** that the U.S. had helped Turkey capture Ocalan.

Northern Ireland Government Structure Approved—Northern Ireland's legislature, **Feb. 16**, approved, 77–29, the structure for a new executive government for the strife-torn province—a major step in the implementation of a peace agreement reached in 1998. The structure included a Northern Ireland Executive consisting of 10 departments whose heads would constitute a cabinet. These departments would be responsible for taxation, law enforcement, and other local matters. Sinn Fein, the political arm of the Provisional Irish Republican Army, would get 2 of the cabinet slots.

Fund Aids Forced Laborers for Nazis—Chancellor Gerhard Schroeder of Germany said **Feb. 16** that a fund, calculated at $1.7 billion, would be set aside to compensate Holocaust victims, including those who were forced by the Nazis into slave labor for German companies before and during World War II. German companies would finance the fund. A second fund would educate people about the Holocaust. Several class-action lawsuits filed in the U.S. sought damages from German companies that had used slave labor.

UN Votes to Pull Peacekeepers From Angola—As fighting between Angolan rebels and the government continued, Pres. José Eduardo dos Santos said, **Feb. 17**, that he wanted UN peacekeepers to leave his country. The UN Security Council, reversing a January decision, voted, **Feb. 26**, to pull its peacekeeping force out of Angola.

Rebels in Congo Launch Offensive—Ernest Wamba dia Wamba, leader of the rebel forces in Congo (formerly Zaire), said **Feb. 17** that he was resuming the offensive against the government of Pres. Laurent Kabila. He complained that Kabila had not negotiated directly with the rebels. It was reported that the rebels, supported by troops from neighboring Rwanda, would seek to gain control of the diamond-rich province that provided the government's principal source of income.

Prime Ministers of India, Pakistan Meet—Prime Mins. Atal Bihari Vajpayee of India and Nawaz Sharif of Pakistan met in Pakistan, **Feb. 20** and **21,** to discuss the divisions between their countries. Both nations had tested nuclear weapons in 1998. The leaders pledged to reduce the danger of a nuclear conflict and to notify each other of any future tests. They also agreed that a resolution of the dispute over Kashmir was essential to future peace.

Kosovo Peace Talks Reach Tentative Agreement—Negotiators for the Serbs and ethnic Albanians—including delegates from the Kosovo Liberation Army—reached a tentative agreement **Feb. 23** on the future of the province of Kosovo. The KLA wanted independence for the province, populated mostly by ethnic Albanians, which Serbia's parent government of Yugoslavia opposed.

The talks opened near Paris on **Feb. 6**. Pres. Clinton announced, **Feb. 13,** that U.S. troops would participate in a NATO mission in Kosovo, and on **Feb. 17** the U.S. ordered 51 more planes to Europe for use in possible air strikes. NATO Supreme Commander Gen. Wesley Clark, an American, sought, **Feb. 22,** to persuade the ethnic Albanians to sign the draft plan even though it did not permit a referendum on independence. The plan conditionally accepted by both sides on **Feb. 23** granted some autonomy to Kosovo for 3 years, after which NATO would assess the situation. Yugoslavia was resisting another element of the proposed agreement—that 30,000 NATO troops be deployed in the province.

Nigerians Elect a President—Nigeria's transition to civilian rule was nearly completed **Feb. 27** with the election of Olusegun Obasanjo as president. He had been a military leader during the civil war with Biafra, 1967–70, and became president in 1976 after his predecessor was assassinated while seeking to implement civilian rule. Obasanjo had completed the transition, handing over the presidency to an elected civilian in 1979. The military later seized power again, and Obasanjo was imprisoned from 1995 until 1998, when Nigeria's ruler Gen. Sani Abacha died. Obasanjo received 63% of the vote **Feb. 27;** but his opponent, Olue Falae, claimed there had been vote fraud.

General

2d Report Issued on Olympic Scandal—A panel of independent investigators working for the Salt Lake Organizing Committee issued its report **Feb. 9** on the Olympic bribery scandal. The report put most blame on 3 members of the Salt Lake City, UT, committee that had successfully sought to obtain the 2002 Winter Olympics. The 3 members of the committee had resigned. The report said that committee members gave members of the International Olympic Committee cash and other gifts without notifying other Salt Lake committee members. The report said that nobody was suspected of actual illegal actions.

MARCH 1999

National

Republicans Enter Presidential Race—Pat Buchanan, a conservative writer and television commentator, launched his 3d run for president on **Mar. 2.** He announced in New Hampshire, site of his upset primary victory in 1996, and reiterated his opposition to trade laws that sent U.S. jobs overseas. Gov. George W. Bush (TX) said, **Mar. 2,** that he would set up a committee to explore a run for president. Public opinion polls showed him with a wide lead among Republicans for the 2000 presidential nomination.

Former Gov. Lamar Alexander (TN), a 1996 aspirant, announced his candidacy for the GOP nomination, **Mar. 9.**

Elizabeth Dole, former head of the Red Cross and wife of former presidential candidate Bob Dole, said, **Mar. 10,** that she was creating an exploratory committee to consider a 2000 run. Magazine publisher Steve Forbes announced his 2d bid for the GOP nomination, **Mar. 16.** He endorsed the flat tax, opposed abortion, and advocated giving parents vouchers to put their children in private schools. On **Mar. 24,** Rev. Jesse Jackson, the civil rights leader, said he would not make a 3d try for the Democratic nomination.

Monica Lewinsky in Media Spotlight—Former White House intern Monica Lewinsky discussed her relationship with Pres. Clinton in a taped television interview with Barbara Walters that was shown on ABC, **Mar. 3,** to a TV audience estimated at 70 million people. An immunity agreement with Kenneth Starr, the independent counsel, prevented Lewinsky from discussing how she had been treated by his office. Lewinsky apologized to the nation for her role in the affair. She said Clinton had indicated to her he might leave his wife and said she now had mixed feelings about the president. Lewinsky also revealed for the first time that as a result of a relationship with a Pentagon co-worker during a gap in her affair with the president, she had become pregnant and had had an abortion.

On **Mar. 4,** her book, *Monica's Story,* written by Andrew Morton, went on sale. Proceeds from the book and from TV appearances outside the United States (she was not paid by ABC) would help pay her substantial legal bills. In the book—not subject to any restrictions by Starr—she claimed that on the January 1998 day when she was first confronted by Starr's deputies, they had threatened her with 27 years in prison and resisted her plea to call her mother (whom they said they might prosecute) or a lawyer.

U.S. Fires Scientist at Nuclear Lab—In the midst of a burgeoning investigation of the possible theft of U.S. nuclear secrets by China, the Dept. of Energy, **Mar. 8,** dismissed a computer scientist who worked at the Los Alamos (NM) National Laboratory. The employee, Wen Ho Lee, was officially ousted for having failed to "properly safeguard classified material" and for not telling his employers of his contacts with a "sensitive country." The investigation, conducted by the Energy Dept. and the FBI, sought to determine if secrets obtained from the U.S. had permitted China to make smaller nuclear warheads that could be carried in a cluster on a single missile. U.S. national security adviser, Samuel R. Berger, said **Mar. 14** that China had benefited from information obtained from Los Alamos. The CIA, **Mar. 15,** began an inquiry into how U.S. national security might have been damaged.

Dr. Kevorkian Guilty of Murder—Dr. Jack Kevorkian, who claimed that he had helped 130 people take their own lives, was convicted of 2d-degree murder **Mar. 26** in Oakland County (MI) Circuit Court. In 4 previous trials, Kevorkian had never been convicted in connection with a death; in those instances the patient had self-administered lethal drugs. In the current case, Kevorkian was convicted of injecting a lethal drug into Thomas Youk, who was fatally ill with amyotrophic lateral sclerosis (Lou Gehrig's disease). A tape of Youk's death had been played on national television. Long an advocate of the legalization of assisted suicide, Kevorkian had acted as his own attorney in the trial. On **Apr. 13,** he was sentenced to 10 to 25 years in prison.

Dow Jones Stock Average Tops 10,000—On **Mar. 29,** the Dow Jones industrial average closed above 10,000 for the first time. The final charge saw the Dow jump 184.54 points to rise slightly above the mythical barrier at 10006.78. The average of 30 large companies had been advancing steeply over the past several years, although it had skidded badly, almost 20%, at one point in 1998 because of fears about the Asian financial crisis. The rebound left many smaller companies behind; they were still showing declines in 1999.

Jury Gives $81 Million to Smoker's Family—The family of a cigarette smoker who died of lung cancer was awarded $81 million, **Mar. 30,** by a jury in Multnomah County (OR) Circuit Court. Attorneys for the plaintiffs presented documents from cigarette maker Philip Morris showing an apparent effort by the company to conceal the health threats and addictiveness of smoking. The jury's award, including $79.5 million in punitive damages, was the highest ever against a tobacco company. Philip Morris said the award would be appealed. On **May 13,** a state judge reduced the punitive portion to $32 million.

U.S. Economy at a Glance March 1999	
Unemployment rate	4.2%
Consumer prices (change over Feb.)	+0.2%
Producer prices (change over Feb.)	+0.3%
Trade deficit	$19.31 bil
Dow Jones high (Mar. 29).	10006.78
Dow Jones low (Mar. 3)	9275.88
Index of leading economic indicators (change over Feb.)	Unchanged
1st-quarter GDP (change over 4th quarter of 1998)	+3.7%

International

8 Tourists Murdered in Ugandan Park—Eight tourists, a Ugandan game warden, and 3 rangers were murdered in a national forest in Uganda on **Mar. 1.** The killers were Rwandan Hutus who apparently sought to destabilize the Ugandan government, which relies heavily on tourist revenue. The slain tourists were from Britain (4), New Zealand (2), and the United States (2). The Ugandan army reported, **Mar. 4** and **8,** that Ugandan and Rwandan forces had killed 25 rebels, some of whom were wearing the tourists' clothing.

Pilot Acquitted of Manslaughter in Ski-Lift Deaths in Italy—The U.S. Marine pilot of a plane that had snapped a ski-lift cable above a valley in Italy in 1998, causing 20 people to fall to their deaths, was acquitted **Mar. 4** of charges of involuntary homicide and manslaughter. Lawyers for Capt. Richard Ashby argued at his court-martial at Camp Lejeune, NC, that poor communications, improper training, faulty safety equipment, and unmarked maps were factors to be considered in his defense. The Italian government and relatives of victims denounced the verdict. Capt. Joseph Schweitzer, the navigator on the plane, pleaded guilty, **Mar. 29,** to obstruction and conspiracy for destroying a videotape that had recorded part of the flight; he was sentenced to dismissal from the Marine Corps, **Apr. 2.** On **May 7,** Ashby was also found guilty of obstruction and conspiracy; he was sentenced **May 10** to 6 months in prison and dismissed from the Marines.

El Salvador Elects New President—A member of the ruling party in El Salvador was elected president on **Mar. 7.** Francisco Flores of the Arena Party got 51% of the vote; his nearest rival received 29%. This was the 2d presidential election since the UN-brokered peace agreement ended 12 years of civil war in 1992. Flores said he would continue the policies of his predecessor, Armando Calderon Sol, who had pushed for privatization of the economy.

Clinton Tours Countries Hit by Hurricane—Pres. Clinton, **Mar. 8,** began a tour of 4 Central American countries, 2 of which had suffered heavy damage when Hurricane Mitch struck in October 1998. He met with the presidents of each country. Clinton, in Nicaragua on **Mar. 8,** announced $120 million in hurricane-recovery aid for that country. On **Mar. 9,** he visited Honduras, where about 6,000 were known to have been killed by the hurricane. On **Mar. 10** Clinton became the first U.S. president in 30 years to visit Guatemala, where he apologized for U.S. support for former right-wing governments that had committed widespread human-rights abuses.

3 East European Nations Join NATO—Three ex-members of the Cold War era alliance led by the Soviet Union formally joined the North Atlantic Treaty Organization on **Mar. 12.** The foreign ministers of the 3—the Czech Repub-

lic, Hungary, and Poland—attended a ceremony with U.S. Sec. of State Madeleine Albright in Independence, MO, home of Harry S. Truman, who was president when NATO was founded. NATO planned to commit considerable resources to upgrading the military forces of the 3 countries.

20 European Union Leaders Quit Under Fire—The 20 members of the European Commission, the executive wing of the European Union, resigned **Mar. 16.** In a report issued **Mar. 15,** investigators appointed by the European Parliament found that the commissioners had mismanaged programs for which they were responsible and had hired friends who were unqualified for their positions. Six commissioners, including former Prem. Edith Cresson of France, came under sharpest criticism. It was concluded that she knew millions of euros were being wasted on her programs.

Paraguay Leader Quits After Assassination—Vice Pres. Luis Maria Argaña of Paraguay was assassinated **Mar. 23** in Asunción, the capital, and Pres. Raúl Cubas Grau resigned 5 days later. Argaña's faction of the ruling party had called for Cubas's impeachment for his defiance of a Supreme Court order to return to prison Gen. Lino Cesar Oviedo, who had been sentenced for planning a coup against a former president. Four men killed Argaña and his driver in a grenade and gun attack on their jeep. The lower house of Congress voted, **Mar. 24,** to begin impeachment proceedings against Cubas immediately. Six people were killed during riots calling for his ouster. On **Mar. 29,** Cubas went to Brazil, where he was granted political asylum, and Luis Angel González Macchi, president of the Senate, was sworn in as his successor.

Kosovo Talks Fail, NATO Bombs Serbia—On **Mar. 24,** NATO launched attacks on targets in Yugoslavia after the Serbs refused to sign a peace agreement worked out for the future of the rebellious province of Kosovo.

Representatives of ethnic Albanians, who constituted most of Kosovo's population, had signed the accord, worked out in February, on **Mar. 18.** Pres. Slobodan Milosevic of Yugoslavia, the umbrella government for Serbia and Montenegro, had rejected the agreement because it provided for NATO forces in Kosovo. The UN High Commissioner for Refugees estimated, **Mar. 19,** that 240,000 people had fled from their homes in Kosovo because of the conflict. Talks between Milosevic and U.S. envoy Richard C. Holbrooke, **Mar. 22** and **23,** reached no agreement.

NATO, which had warned Serbia that failure of the peace talks would result in air strikes, launched attacks on the evening of **Mar. 24.** Targets in Montenegro and Kosovo were also hit. Cruise missiles were fired from U.S. and British ships and submarines in the Adriatic Sea and from U.S. B-52 bombers; other NATO warplanes attacked from bases in Italy. The participation of 4 German jets marked the first combat by German aircraft since the end of World War II. In 50 years, NATO had never attacked a sovereign nation. The U.S. reported that 3 Yugoslav MiG jets were shot down during the first night.

Pres. Clinton said in an address **Mar. 24** that NATO sought to stop the Serbian offensive against civilians in Kosovo and to damage the Serbian military. Targets included radar and missile sites and command and communication centers. Pres. Boris Yeltsin of Russia denounced the attacks, **Mar. 24,** and both Russia and China criticized the NATO offensive at the UN Security Council the next day.

Attacks, which continued nightly, expanded to include military headquarters, barracks, arms factories, and supply lines. NATO planes downed 2 MiGs over Bosnia-Herzegovina, **Mar. 26.** A U.S. F-117 was shot down **Mar. 27;** the pilot was rescued.

None of this appeared to slow the Serb push in Kosovo. Reports of massacres proliferated. Growing numbers of refugees poured into Albania, Macedonia, and Montenegro. They said that the Serbs were terrorizing and forcibly expelling large numbers of refugees. A number of the 19 NATO countries pledged financial aid to the refugees.

Prem. Yevgeny Primakov of Russia met with Milosevic, **Mar. 30,** in an effort to find a diplomatic solution. Three U.S. soldiers, members of a peacekeeping force in Macedonia that was in the process of withdrawing, were captured by Serb forces on **Mar. 31** near the Yugoslav-Macedonia border.

Pinochet's Arrest Ruled Lawful—An appellate committee of Britain's Law Lords held, **Mar. 24,** that the 1998 arrest of Gen. Augusto Pinochet of Chile in England had been legal. The majority agreed that the international convention against torture applies to ex-heads of state. Extradition to Spain remained unresolved. Spain, **Mar. 27,** added 33 new criminal charges against Pinochet.

European Union Backs Palestinian State—The European Union, **Mar. 26,** threw its support behind the "unqualified" right of Palestinians to have an independent state—although not immediately. The 15-member EU said that right was not subject to an Israeli veto. The U.S. argued that the future of the West Bank, Gaza, and Jerusalem should be decided by talks between Israel and the Palestinians.

Ford Purchases Volvo—The Ford Motor Co. completed the purchase of Volvo Cars, a unit of Sweden's Volvo A.B., for $6.45 billion on **Mar. 31.** They had announced the agreement **Jan. 28**. The Volvo Cars division specialized in medium-priced luxury cars and station wagons.

General

3d Inquiry Criticizes Olympic Committee—The 3d of 5 investigations into the International Olympic Committee was concluded **Mar. 1** with the release of a final report. The independent panel, headed by former Sen. George Mitchell (D, ME), concluded that the bribery related to the effort by Salt Lake City, UT, leaders to win the 2002 Winter Games was only part of a broader corruption of the IOC. On **Mar. 17,** the IOC voted to expel 6 of its members implicated in the bribery scandal. Five other IOC members had already resigned.

Baseball Great Joe DiMaggio Dies—Fans nationwide mourned the death of Joe DiMaggio, **Mar. 8,** at the age of 84, from pneumonia and other complications following surgery for lung cancer in 1998. Dubbed the "Yankee Clipper," he was a star center fielder for the New York Yankees in a 13-season career that began in 1936, interrupted only by 3 years of service during World War II. He amassed a lifetime batting average of .325, a record-setting 56-game hitting streak, and a home run total of 361. He appeared in 11 All-Star games, played in 10 World Series, and was named the American League's most valuable player 3 times. He was briefly married to Hollywood star Marilyn Monroe.

Championship Boxing Match Investigated—A fight for the heavyweight boxing championship of the world ended in a draw, **Mar. 13,** and in controversy. The opponents were Evander Holyfield, an American, and Lennox Lewis, from Britain, who between them held the 3 heavyweight titles recognized by leading organizations. The fight, in New York City's Madison Square Garden, was declared a draw though most fans and boxing officials thought Lewis had clearly won. Influenced by the controversial decision, the NY state legislature approved legislation **June 15** requiring open scoring in boxing.

2 in Balloon Circumnavigate Globe Nonstop—Two balloonists completed the first nonstop circumnavigation of the globe by balloon, **Mar. 20,** after flying about 29,000 miles. Bertrand Piccard of Switzerland, the captain, and Brian Jones of Britain took off from Switzerland on **Mar. 1** in the *Breitling Orbiter 3,* a 180-foot tall helium-powered balloon. They soon picked up a high-altitude jet stream and traveled east across Africa, Asia, the Pacific Ocean, Central America, the Atlantic Ocean, and back to Africa, where they landed southwest of Cairo, Egypt, on **Mar. 21**.

"Shakespeare in Love" Named Best Picture—A romantic comedy about William Shakespeare was the surprise winner of the best-picture Oscar awarded **Mar. 21** by the Academy of Motion Picture Arts and Sciences. In all,

Shakespeare in Love won 7 awards. *Saving Private Ryan,* a graphic film depicting the Allied invasion of France during World War II, missed the top prize but won 5 awards, including 1 for best director, Steven Spielberg. Gwyneth Paltrow, who starred in *Shakespeare,* was named best actress. Roberto Benigni, an Italian, was named best actor for *Life Is Beautiful,* which was also named best foreign-language film.

Connecticut Wins NCAA Basketball Title—The Connecticut Huskies won their first NCAA men's basketball championship **Mar. 29,** defeating the Duke Blue Devils, 77–74. Duke, the heavy favorite, carried a 32-game winning streak into the title game. MVP Richard Hamilton scored 27 points for the Huskies. A day earlier, on **Mar. 28,** Purdue won its first women's title, also defeating Duke, 62–45. Purdue's coach, Carolyn Peck, was the first black woman to coach the women's championship team.

APRIL 1999
National

Man Pleads Guilty in Death of Gay Student—Russell Henderson, 21, pleaded guilty **Apr. 5** in the October 1998 beating death of Matthew Shepard, a homosexual student at the Univ. of Wyoming. Henderson and Aaron McKinney had been charged with kidnapping and first-degree murder. Henderson said McKinney, who was awaiting trial, had beaten Shepard to death. Judge Jeffrey Donnell sentenced Henderson to 2 consecutive life terms; if convicted by a jury, he could have faced the death penalty.

Clinton Held in Contempt of Court—U.S. District Judge Susan Webber Wright, in Little Rock, AR, **Apr. 12,** held Pres. Clinton in civil contempt of court for testifying falsely in a January 1998 deposition. Wright cited 2 "intentionally false" statements by Clinton—that he had not had sexual relations with Monica Lewinsky and that he did not recall having been alone with her. The deposition was taken in the sexual-harassment lawsuit brought against Clinton by Paula Corbin Jones. Wright fined Clinton $1,202, the cost of her trip to Washington to preside at the deposition, and said he was liable for some expenses incurred by Jones. Clinton was the first sitting president ever to be held in contempt of court.

Clintons' Whitewater Partner Acquitted of Obstruction—Susan McDougal, a partner of then-Gov. Bill Clinton and Hillary Rodham Clinton in the Whitewater Development Corp., was found not guilty **Apr. 12** of obstruction of justice. Kenneth Starr, the independent counsel, had indicted McDougal for refusing to testify about the role of the Clintons in Whitewater and other matters. McDougal and her late husband, James, had been convicted of fraud at an earlier trial. In the current trial, McDougal told the Little Rock, AR, federal jury that she knew of no wrongdoing by the president. She said she had refused to testify because she believed Starr was interested primarily in damaging Clinton politically. The jury deadlocked on 2 counts of criminal contempt against McDougal, and Judge George Howard Jr. declared a mistrial on these counts.

Independent Counsel Opposes Renewal of Law—Kenneth Starr, the independent counsel investigating the Whitewater affair and related matters, advised Congress **Apr. 14** not to renew the 1978 law that had been the basis of his authority. Testifying before the Senate Governmental Affairs Committee, Starr defended his own investigation. But he said the independent counsel act did not give the attorney general enough flexibility in deciding when to request appointment of an outside counsel to investigate alleged wrongdoing by high-level officials. The Justice Dept. also opposed renewing the law, which now had little support in Congress. Many critics believed the law had allowed investigations of public officials to become excessively expensive, wide-ranging, and protracted.

The independent counsel law expired **June 30**.

Quayle, Bauer Seek GOP Nomination—Former Vice Pres. Dan Quayle announced, **Apr. 14**, that he would seek the Republican presidential nomination. Quayle, who had been defeated in 1992 with Pres. George Bush in their bid for a second term, launched his bid for the top spot in his home town, Huntington, IN. He urged a 30% tax cut for all and stressed his foreign-policy experience.

Conservative Christian activist Gary Bauer announced his candidacy, **Apr. 21**, at his high school in Newport, KY. Bauer blamed the Littleton, CO, shooting tragedy the day before on a "culture of death" apparent in contemporary American values and in the advocacy of abortion rights.

2 Students Kill 13 at High School—Two teenage boys killed 12 of their fellow students and a teacher **Apr. 20** at Columbine High School in Littleton, CO, south of Denver. The 2 gunmen then shot themselves fatally. More than 30 people were wounded, some seriously or critically. Police identified the killers as Eric Harris, 18, and Dylan Klebold, 17. The tragedy was by far the worst of a recent series of shootings by boys at their schools.

Harris and Klebold, armed with a semiautomatic rifle, 2 sawed-off shotguns, and a semiautomatic pistol, and wearing ski masks, entered the school about 11:20 AM and began their assault, shooting off weapons and throwing bombs. Many of those killed were school athletes.

Police received emergency 911 alerts beginning at 11:26 AM, and a makeshift SWAT team entered the school within 20 minutes; other SWAT teams entered later. The FBI, the Colorado National Guard, and the federal Bureau of Alcohol, Tobacco, and Firearms also participated in the rescue. Students were evacuated in small groups as the school was cleared room by room. The dead gunmen and 10 of those they killed were found in the library. Because of the need to search the premises carefully, the bodies of the dead were not removed until the next day. Police found several weapons and many bombs. One gasoline-filled container, found **Apr. 22**, also contained shrapnel. A bomb had exploded in the library and another in a car in the parking lot.

Both assailants had been in minor trouble with authorities. On his Internet site, Harris had provided information on building a bomb. Both gunmen were members of the so-called Trench Coat Mafia—a dozen or so students who, it was said, appeared aloof from other students, and were often targets of disparaging remarks. Some of these students seemed interested in Nazi lore. The attack came on the birthday of Adolf Hitler, and the size of the boys' arsenal suggested it had been long planned. Entries in Harris's diary, made public beginning **Apr. 24**, indicated that the plan was to kill as many as 500 people. Vice Pres. Al Gore was among 70,000 people who attended an **Apr. 25** memorial service for the victims. Authorities, **Apr. 27**, put the total number of bombs found around the school and at Harris's home at 51.

The Colorado shootings prompted so-called copycat threats throughout the U.S. On **Apr. 28**, a 14-year-old boy shot 2 students, one fatally, at a high school in Taber, Alberta, Canada. On **Apr. 29**, Klebold's 18-year-old girlfriend admitted that she had bought the 2 shotguns and the rifle used in the attack at gun shows.

Transfer of Technology to China Investigated—The investigation of the transfer of U.S. nuclear secrets to China continued. Energy Sec. Bill Richardson confirmed, **Apr. 28**, that Wen Ho Lee, a scientist dismissed in March from the Los Alamos National Laboratory, had transferred large quantities of data from high-security computers to an accessible network. Earlier, on **Apr. 21**, intelligence officials gave an assessment, mostly classified, to the president and Congress. They confirmed China had obtained secret data on nuclear weapons, and that it was likely China engaged in more spying.

School Tragedy Revives Gun-Control Debate—The shootings at Columbine High School revived the gun-control issue. On **Apr. 27**, a week after the tragedy, Pres. Clin-

ton submitted several gun-control proposals to Congress. Clinton asked Congress to mandate background checks for those wanting to buy explosives or to buy weapons at gun shows. He also proposed raising from 18 to 21 the minimum age at which someone could own a handgun.

Prior to the shootings, on **Apr. 6**, in the first state referendum of its kind, voters in Missouri, 52% to 48%, had defeated a proposal to allow the carrying of concealed weapons. The National Rifle Assn. had reportedly spent $4 million in support of the referendum.

It so happened that the NRA scheduled a 3-day convention in Denver for the week following the Columbine shootings. The NRA, **April 21**, said it would cut its program to one day; Denver Mayor Wellington Webb urged instead that the convention be canceled, but it was not. At the convention on **April 30**, NRA leaders reaffirmed their support for the constitutional right to bear arms. Outside, several thousand people demonstrated against the NRA.

U.S. Economy at a Glance April 1999	
Unemployment rate	4.3%
Consumer prices (change over Mar.).	+0.7%
Producer prices (change over Mar.)	+0.5%
Trade deficit	$18.79 bil
Dow Jones high (Apr. 29)	10878.38
Dow Jones low (Apr. 1)	9832.51
Index of leading economic indicators (change over Mar.).	+0.1%

International

NATO Bombing of Yugoslavia Continues—NATO forces stepped up bombing of Yugoslavia from the air, in an effort to persuade Serbia to reach an agreement on Kosovo. The Serbian news agency said **Apr. 1** that 3 captured U.S. soldiers would be tried by a military court. A report the same day concluded that the Serbs had killed 800 ethnic Albanians in the past week. In its first attack on Belgrade, the Yugoslav capital, **Apr. 3**, NATO struck the Yugoslav and Serbian Interior Ministry buildings. Serbian troops and armored vehicles also became targets, beginning **Apr. 4**. By **Apr. 5**, 4 bridges over the Danube River had been destroyed. On **Apr. 4** Serbia reversed its position and said it would not try the 3 captured soldiers, but would free them when the bombing stopped. At the same time, several NATO countries agreed to take in refugees forced out of Kosovo. Pres. Slobodan Milosevic of Yugoslavia declared a unilateral cease-fire in Kosovo, **Apr. 6**, but NATO said that he had to withdraw Serb troops and allow refugees to return under international protection. NATO reported, **Apr. 7**, that the Serbs had burned 50 Kosovo villages since **Apr. 4**. NATO said **Apr. 9** it had received disturbing reports of rapes of ethnic Albanian women by Serbian soldiers. On **Apr. 11**, NATO released aerial photographs of newly turned earth that appeared to indicate mass graves, and there were reports from refugees describing mass executions.

Pres. Boris Yeltsin of Russia warned NATO, **Apr. 9**, not to push his country into military support of Serbia. NATO missiles on **Apr. 12** struck a train on a bridge, killing 10 civilians. Serbian troops briefly seized an Albanian border post, **Apr. 13**. Yugoslavia claimed **Apr. 14** that at least 64 refugees were killed that day on a roadway by NATO aircraft; NATO confirmed the next day that civilians had been "mistakenly" killed in the belief that they were part of a military convoy.

NATO, **Apr. 20**, destroyed the headquarters of Milosevic's political party, and on **Apr. 22** one of his residences, in Belgrade, was bombed. The state television headquarters was hit **Apr. 22**, killing at least 10 people.

U.S. Gen. Wesley Clark, the NATO commander, estimated, **Apr. 27**, that Serbia had driven 700,000 refugees out of Kosovo since the bombing had begun. Pres. Clinton, **Apr. 27**, ordered the call-up of 33,102 reservists to assist the mission in Yugoslavia. On **Apr. 29**, NATO bombed the Yugoslav army headquarters and Defense Ministry in Belgrade. The U.S. House of Representatives, on a 213–213 tie vote, failed **Apr. 28** to pass a resolution supporting the air attacks.

Merger Creates Giant Oil and Gas Producer—A merger announced **Apr. 1** would create a company with annual revenues approaching $100 billion. BP Amoco PLC, just formed from a merger of British Petroleum PLC and the Amoco Corp., said it now planned to acquire Atlantic Richfield Co.—like Amoco, a U.S. firm. The merger was expected to cost 2,000 employees, mostly in the U.S., their jobs. The new company would rank first in oil and natural gas production in Britain and the U.S.

Canada Creates a Territory for Inuits—Canada, **Apr. 1,** established a new territory, Nunavut, as a means of providing autonomy for the Inuit people. Nunavut had been the eastern part of the Northwest Territories. Only 25,000 people, mostly Inuits, lived in the sparsely settled region, which covered about 820,000 square miles. In return for the autonomy granted them, the Inuits had dropped their land claims against Canada.

Libya Hands Over Plane-Bombing Suspects—Libya, **Apr. 5,** released to a UN official for trial 2 suspects in the 1988 bombing of Pan Am Flight 103. The explosion of the plane over Scotland killed all 259 on board and 11 others who were struck by the wreckage. The 2 men were to be tried in the Netherlands under Scottish law. By turning over the suspects, Libya gained automatic release from UN, though not from U.S., sanctions.

Chinese Premier Visits U.S., Canada—During a U.S. visit that began **Apr. 6**, Chinese Prem. Zhu Rongji was unable to reach a comprehensive trade agreement with the Clinton administration. In concessions **Apr. 7**, China lifted an import ban on wheat from the Pacific Northwest and on USDA-inspected meat and poultry products. Zhu met with Pres. Clinton **Apr. 8** and with congressional leaders **Apr. 9**. He traveled to other cities as well.

During Zhu's visit to St. John's, Newfoundland, **Apr. 14**, the mayor flew the flag of Tibet, which China had occupied since 1950. China, on **Apr. 15**, agreed to buy Can$100 million worth of Canadian wheat. Zhu and Prime Min. Jean Chrétien announced **Apr. 16** that China would cut tariffs on some Canadian imports.

Iraqi Confrontation with UN Continues—On **Apr. 8**, Iraq rejected a proposal from the United Nations that would permit resumption of UN arms inspections inside Iraq. The UN plan included relaxed sanctions and a more generous food-for-oil arrangement under which Iraq sold oil to raise funds for basic human needs. On **Apr. 2**, after a hiatus of more than 2 weeks, British and U.S. planes had resumed bombing Iraq.

President of Niger Assassinated—Pres. Ibrahim Bare Mainassara of Niger was shot to death **Apr. 9** at the airport in Niamey, the capital. A U.S. State Dept. spokesman said he had been killed by members of the presidential guard. Tensions had run high since a February election for local and regional offices. The Supreme Court, **Apr. 6**, had annulled some of the results. After the assassination, a military junta led by Daouda Malam Wanke, the commander of the presidential guard, took power.

India, Pakistan Test Ballistic Missiles—India, **Apr. 11,** tested an intermediate-range ballistic missile able to carry a nuclear warhead. Pakistan, **Apr. 14** and **15**, tested 2 similar missiles. In accord with a February agreement, each country told the other in advance of the tests. Indian Prime Min. Atal Behari Vajpayee, who had supported his country's 1998 entry into the nuclear "club," lost a parliamentary vote of confidence, 270–269, **Apr. 17**, and resigned. His party was hurt by its advocacy of India as a predominantly Hindu state. India's president dissolved Parliament on **Apr. 26** and called for a general election within 6 months.

Malaysia's Ex-Deputy Premier Sentenced—The former deputy prime minister of Malaysia was found guilty, **Apr. 14**, of 4 charges of corruption and was sentenced to 6 years in prison. Anwar bin Ibrahim, who had been charged with trying to prevent an investigation into his alleged adultery and sodomy, blamed the verdict on the regime of Prime Min. Datuk Seri Mahathir bin Mohamad. Ibrahim's case had attracted international attention and prompted widespread antigovernment demonstrations in Malaysia.

Algeria Elects President After Rivals Quit—Abdelaziz Bouteflika, a former foreign minister, was elected president of Algeria on **Apr. 15**, getting more than 70% of the vote according to official returns. On **Apr. 14**, his 6 rivals had pulled out of the contest, saying that the election was a fraud. Bouteflika was the choice of many leaders of the country's military, political, and business establishment. The election occurred against a backdrop of an apparently waning military effort by Islamic fundamentalists to gain power.

Ex-Pakistani Leader Convicted of Corruption—On **Apr. 15** in Rawalpindi, former Prime Min. Benazir Bhutto of Pakistan, who had been forced from power in 1996, was convicted of corruption, along with her husband, Asif Ali Zardari. The 2 judges found that the pair had taken kickbacks from a Swiss company that Bhutto had chosen to monitor collection of import duties. They were sentenced to 5 years in prison and fined $8.6 million. Bhutto, who was in London at the time, said she would not return to Pakistan until her appeal was heard.

NATO Celebrates 50th Anniversary—The North Atlantic Treaty Organization celebrated its 50th anniversary in Washington, DC, **Apr. 23-25**, but the occasion fell awkwardly in the middle of NATO's bombing campaign against Yugoslavia. The fete drew presidents and prime ministers from the 19 member nations. NATO leaders, **Apr. 23**, issued a communique renewing their insistence on the withdrawal of Serbian forces from Kosovo and on the safe return of refugees under international protection. On **Apr. 25**, NATO leaders met with leaders of 7 non-NATO nations surrounding Yugoslavia and pledged to protect them from Yugoslav aggression.

Palestinians Back Off on Statehood—The Palestinians decided **Apr. 29** not to declare statehood on May 4. Palestinian leader Yasir Arafat returned, **Apr. 13**, from consultations in 2 dozen nations on the proper course of action. The U.S., **Apr. 14,** criticized Israel's ongoing construction of new settlements on the West Bank. Arafat, **Apr. 26**, received a letter from Pres. Clinton urging him not to act unilaterally on statehood but also giving a signal of support for eventual statehood. At the urging of Arafat, Palestinian leaders on **Apr. 29** abandoned the plan to declare statehood, thus cooling off the issue prior to the Israeli election in May.

General

A Solar System With 3 Planets Discovered—Astronomers announced **Apr. 15** that they had discovered evidence of another solar system—the only one known aside from our own—in the constellation Andromeda. Up to that time evidence had been found of the existence of 18 new planets, each orbiting a different star, or sun. Never before had more than 1 planet been connected to a single star. None of the 3 planets in the new solar system was seen directly; their presence was inferred from gravitational tugs on their star.

The 3 planets—all of them large gaseous objects similar to Jupiter—were circling Upsilon Andromedae, about 44 light-years distant. The largest and most distant from its star was over 4 times as massive as Jupiter and required 4 years to circle the star. The discovery was announced by 2 teams of astronomers, one at San Francisco State Univ. and the other from the Harvard–Smithsonian Center for Astrophysics and the National Center for Atmospheric Research.

Hockey's Wayne Gretzky Retires—Wayne Gretzky, 38, who held or shared 61 National Hockey League records, announced his retirement **Apr. 16** after a 21-season professional career. His last game with the New York Rangers was **Apr. 18**. Gretzky gained his greatest fame leading the Edmonton Oilers to 4 Stanley Cup championships. During his career, "The Great One," acclaimed by many as the game's greatest player ever, scored a record 894 goals and chalked up a record 1,963 assists. On **June 23**, Gretzky was officially selected to be inducted into the Hockey Hall of Fame, after the usual 3-year waiting period was waived.

MAY 1999

National

6 Students Wounded by Teen Gunman—By **May 1**, in schools in every state except Vermont, authorities had dealt with threats of "copycat" situations following the bloodbath at Columbine High School in Littleton, CO. In an echo of that incident, a 15-year-old student at Heritage High School in Conyers, GA, allegedly shot 6 fellow students, with a .22-caliber rifle **May 20**. None of those wounded suffered life-threatening injuries. The boy, Thomas Solomon, gave himself up to school authorities. On **May 20**, Pres. Clinton and First Lady Hillary Rodham Clinton met with students, teachers, and victims' families in Littleton.

Treasury Sec. Rubin Resigns—Sec. of the Treasury Robert Rubin, a leading figure in the Clinton cabinet whose policies had been given much of the credit for the roaring U.S. economy, announced **May 12** that he would soon resign. Pres. Clinton announced the same day that he would nominate Deputy Treasury Sec. Lawrence H. Summers, a former chief economist for the World Bank, to succeed Rubin; Summers was sworn in **July 2**.

Senate Struggles With Gun Control—The Senate passed a bill **May 20** that sought to deal with youth violence. It almost foundered, however, over the inclusion of gun-control measures. On **May 10**, Pres. Clinton met with representatives of gun manufacturers and the Internet as well as leaders in the television, motion picture, and video game industries to discuss youth violence. The Senate, **May 12**, defeated, 51–47, an amendment to the youth violence bill that would have required background checks for all firearms purchases at gun shows. (Under existing law only licensed gun dealers were required to conduct such checks.) Republicans, who had mostly opposed the amendment, then supported a mandatory check that Democrats opposed because of alleged loopholes; this version passed, 48–47, **May 14**.

On **May 18**, the Senate approved, 78–20, an amendment requiring licensed handgun dealers to include child-safety devices on their guns. The background-check controversy came up again, **May 20**, in a Democratic-sponsored amendment that closed some loopholes previously allowed. The newest version passed, 51–50, on Vice Pres. Al Gore's tie-breaking vote. The Senate gave final approval to the bill on **May 20**, 73–25. Other provisions banned the sale of semiautomatic weapons to persons under 18, allowed youths as young as 14 to be charged as adults, forbade anyone convicted of a violent crime as a juvenile to own firearms, and authorized $5 billion to deal with juvenile crime problems.

Fund-Raiser for Democrats Pleads Guilty—Yah Lin (Charlie) Trie, a friend of Pres. Clinton, pleaded guilty **May 21** in the ongoing investigation of political fund-raising abuses. Earlier, on **May 11**, Johnny Chung, another fund-raiser for the Democrats, testified before the House Government Reform Committee. He said that a Chinese general, Ji Shengde, who ran China's military intelligence agency, had sent him $300,000 in 1996 after asking him to give it to the campaign to reelect Clinton. Chung testified that he had not given any of the money to the Democrats and had, in fact, used it for personal and business expenses. Contributions to U.S. campaigns by foreign nationals is illegal.

Trie pleaded guilty to a felony charge of causing false statements to be made to the Federal Election Commission and to a misdemeanor count. After doubts arose about its legality, Democrats had returned $600,000 Trie raised for them.

"Emergency" Spending Bill Pays for Kosovo—Pres. Clinton, **May 21**, signed a so-called emergency spending bill that covered many of the costs associated with the bombing campaign in Yugoslavia. Because most of the $14.5 billion allocated by the measure was designated as emergency spending, the sums were not subject to budget caps imposed by Congress on federal outlays. Some $10.9 billion was approved for the U.S. share of NATO's air war in defense of the Yugoslav province of Kosovo. More than $1 billion went for the assistance of Kosovar refugees. The bill also authorized $687 million for hurricane relief in Central America, $900 million for tornado victims in Kansas and Oklahoma, and $574 million in aid for U.S. farmers.

House Report Details Chinese Espionage—In a largely unclassified version of its report on Chinese espionage, released **May 25**, a U.S. House select committee asserted that China had stolen information on the most advanced U.S. nuclear weapons. Earlier, at a hearing **May 5**, members of a Senate committee deplored the failure of investigators to examine, for 3 years after he came under suspicion, the computer of Wen Ho Lee at the Los Alamos National Laboratory. They also objected that scientists at the laboratories could still download classified data and take it from the laboratories. Through his lawyer, **May 6**, Lee denied allegations of espionage.

The **May 25** House committee report said that Chinese data thefts had continued from the 1970s and had included information about every U.S. nuclear warhead currently employed. The report also said that Hughes Space and Communications International and Loral Space & Communications Ltd. had transferred "missile design information" to China without obtaining required licenses.

Independent Counsel Drops 2 Prosecutions—On **May 25**, Kenneth Starr, the independent counsel investigating the Whitewater affair and other allegations, dropped pending criminal contempt charges against Susan McDougal, a former partner of the Clintons in Whitewater Development Corp. A jury deadlocked on those charges, producing a mistrial. He also dropped charges of obstructing justice and making false statements against Julie Hiatt Steele. Steele had confirmed then contradicted the story of Kathleen Willey, who said Clinton had made an unwanted advance to her in the White House in 1993. On **May 7** a federal judge had declared a mistrial in the Steele case after a jury could not reach a verdict.

Republican Wins Louisiana House Seat—In a **May 29** runoff election, the Republicans held the Louisiana House seat vacated when Rep. Robert Livingston resigned amid reports of marital infidelity. Two Republicans—state Rep. David Vitter and David Conner Treen, a former governor and U.S. representative—had survived the first round of voting, **May 1**. Vitter, a conservative, won the runoff election, 51% to 49%.

U.S. Economy at a Glance May 1999	
Unemployment rate	4.2%
Consumer prices (change over Apr.)	Unchanged
Producer prices (change over Apr.)	+0.2%
Trade deficit	$21.39 bil
Dow Jones high (May 13)	11107.19
Dow Jones low (May 27)	10466.93
Index of leading economic indicators (change over Apr.)	+0.3%

International

Bombing of Yugoslavia Continues—NATO air attacks on Yugoslavia continued through May, but without persuading Serbia to withdraw its forces from Kosovo. On **May 1**, NATO bombed a bridge in Serbia as a bus was crossing it, causing over 30 deaths.

The Rev. Jesse Jackson, who met **May 1** with Yugoslav Pres. Slobodan Milosevic, was instrumental in getting the release, **May 2**, of 3 U.S. soldiers who had been captured near the Serbian-Macedonian border **Mar. 31** by Serbian forces. Two other U.S. soldiers were killed **May 5** when their Apache helicopter crashed in Albania during a training flight. Macedonia, **May 5,** closed its borders to refugees. The UN high commissioner for refugees said, **May 5**, that Kosovo was being "brutally and methodically" emptied. The foreign ministers of the leading industrialized nations, including Russia, reached an agreement **May 6** on principles for a solution in Kosovo. Russia agreed that there should be an international presence in Kosovo after Serb forces pulled out.

On **May 7**, a U.S. stealth bomber bombed the Chinese embassy in Belgrade, killing 3 Chinese citizens and wounding 27 people. The incident was attributed to error resulting from use of an outdated map. U.S. and NATO apologies failed to prevent massive Chinese protests, as a result of which U.S. Ambassador Jim Sasser was unable to leave the U.S. embassy in Beijing until **May 12**. On **May 13**, in another accident, NATO bombed Korisa, a village in Kosovo, killing at least 79 ethnic Albanians. In an effort begun **May 21**, NATO bombing disrupted electrical power to 80% of Serbia. NATO **May 25** approved a plan to put 48,000 peacekeeping troops into Kosovo.

Louise Arbour, chief prosecutor of the international war crimes tribunal investigating the Yugoslav conflict, announced **May 27** that Milosevic had been indicted for murder and other war crimes. Four other Serbian leaders were indicted on similar charges. None were in custody.

Panama Elects Woman President—Panama, **May 2**, elected its first woman president, Mireya Elisa Moscoso, who won 45% of the vote in a 3-way contest. The widow of a former president, Arnulfo Arias, she had lived with him in exile in the U.S for many years. In Panama, she had led his political party since 1991.

Yeltsin Fires Premier, Avoids Impeachment—On **May 12,** Pres. Boris Yeltsin of Russia removed Prem. Yevgeny Primakov for what Yeltsin called his failure to get the economy moving. Primakov was the 3d premier Yeltsin had replaced in 14 months. Yeltsin nominated Interior Min. Sergei Stepashin to replace him. On **May 14**, the Duma (parliament), 243–20, called on Yeltsin to resign. Meanwhile, the Duma was debating 5 articles of impeachment against him. All failed on votes taken **May 15**, with the charge that he had started the war against the Republic of Chechnya in 1994 coming closest to getting the required two-thirds vote. The Duma, **May 19**, confirmed Stepashin as premier.

Netanyahu Loses Reelection Bid in Israel—Israeli voters chose a new prime minister, Ehud Barak, in the national election **May 17**. Barak, head of the center-left One Israel coalition, defeated incumbent Prime Min. Benjamin Netanyahu, winning 1,791,020 votes (56%) to 1,402,474 (44%). Barak was head of the Labor Party, the principal component of the coalition. Three other candidates for prime minister had withdrawn from the contest on **May 15** and **16**. Barak, a former army chief of staff, had promised to continue the policies of slain former Prime Min. Yitzhak Rabin. He stressed the importance of reaching a final settlement with the Palestinians.

Voters **May 17** also chose the 120 members of the Knesset (parliament). Both major parties—Labor (now absorbed into the coalition) and Likud—lost ground, while Shas, an ultra-religious party supported by Israelis of Middle Eastern and North African origin, gained sharply. One Israel ended up with 26 seats, Likud with 19, and Shas with 17. In all, 15 parties ended up with seats. Netanyahu announced he would resign as head of Likud.

India Attacks Islamic Force in Kashmir—New fighting flared **May 26** in Kashmir, which was claimed by both India and Pakistan. Some 500 armed Islamic militants had occupied positions on the Indian side of the cease-fire line. India said that Pakistan was backing the militants. From early May, clashes between the militants and the Indian army had left 300 people dead. On **May 26**, India began attacking the

militants from the air. On **May 27**, Pakistan shot down 2 Indian jets that, India said, had crossed into Pakistani air space unintentionally.

Civilian President Takes Over in Nigeria—On **May 29**, for the first time in 15 years, Nigeria had a democratically chosen leader. Olusegun Obasanjo, who had been elected in February, was sworn in as president. On **May 5** his predecessor, Abdulsalami Abubakar, signed a new constitution providing a framework for the new government and granting more power and money to states and localities. Fighting between 2 ethnic groups in an oil-producing region in the Niger River delta flared **May 30**, and within days had taken 200 lives.

General

Bid for Racing's Triple Crown Ends in Injury—Charismatic, a 31-1 long shot, won the 125th Kentucky Derby in Louisville on **May 1**. Gamblers with a $2 ticket on Charismatic to win got a $64.60 payoff, 3d highest in Derby history. Charismatic ran the 1.25-mile race in 2 minutes, 3.29 seconds, finishing a neck ahead of Menifee.

Charismatic also won the Preakness Stakes, the 2d leg in horse racing's Triple Crown, in Baltimore, **May 15**. Charismatic's attempt to win the 3d race in the Triple Crown series, the Belmont Stakes in Elmont Park, NY, on **June 5**, ended in near tragedy, when the horse suffered 2 broken bones in his left front leg as he crossed the finish line in 3d place. The winner was another long shot (29-1), Lemon Drop Kid. Charismatic underwent successful surgery on **June 6** but would never race again.

Super Bowl Hero John Elway Retires—John Elway of the Denver Broncos, who in his last 2 seasons quarterbacked his team to Super Bowl titles, announced his retirement **May 2**. Over his career, Elway was the NFL's all-time winningest starting quarterback (148-82-1; .643) and ranked 2d (behind Dan Marino) in passing yards (51,475) and pass completions (4,123). He was 3d in touchdown passes (300).

76 Tornadoes Claim Lives in the Plains—Tornadoes tore across the northern and southern plains states **May 3**, causing about 50 deaths, injuring more than 700, and destroying or damaging more than 2,000 houses. In all, 76 twisters were tallied; one of them, reportedly a mile wide, plowed 20 miles through the Oklahoma City metropolitan area. Oklahoma, hardest hit, endured 45 tornadoes and reported more than 40 deaths. Five died in Kansas. Texas, Nebraska, and South Dakota were also hard hit. Damages in Oklahoma and Kansas were put at $200 million and $140 million, respectively.

JUNE 1999

National

NYC Officers Guilty in Torture Case—The case of 5 New York City police officers accused of the 1997 torture of Abner Louima, a Haitian immigrant, came to a close **June 8,** when Officer Charles Schwarz was convicted by a U.S. District Court jury in Brooklyn on 1 assault charge. On **May 25,** the main defendant, Officer Justin Volpe, had pleaded guilty to 6 federal charges connected with torturing and sodomizing Louima with a stick in a station-house restroom, following Louima's arrest in a nightclub brawl. The other 3 officers were acquitted. The case focused national attention on the issue of police brutality.

Bush, Gore Formally Enter 2000 Race—To no one's surprise, Gov. George W. Bush of Texas and Vice Pres. Al Gore formally declared their candidacies for president in June. Bush opened his campaign in Iowa, the state that would hold the first voting in caucuses in early 2000. Polls showed him as the top choice of Republican voters for the nomination. He led his GOP rivals in raising campaign money and in endorsements by fellow Republican leaders. In the town of Amana, IA, **June 12**, Bush said flatly for the first time that he was running, and he embraced a philosophy of "compassionate conservatism." He promised to cut taxes and urged Americans to recognize that they were responsible for their actions. In New Hampshire, **June 14**, he said he would not require that his nominees for federal judgeships oppose abortion.

Gore announced his candidacy **June 16** in Carthage, TN, near his family farm. He said he knew how to keep the economic boom rolling, and he reaffirmed his support for abortion rights. Observers thought his stress on "moral leadership" was an effort to separate himself from the scandals of the Clinton administration. In an interview **June 16**, he said that Pres. Clinton's behavior in the Monica Lewinsky matter was "inexcusable."

Sen. Orrin G. Hatch (UT) said **June 22** that he would seek the Republican nomination. A conservative member of the Senate since 1977, he was known for negotiating agreements with the Democrats. Bush reported **June 30** that his campaign had raised $36.25 million, by far the most ever for a presidential candidate this early.

Gun Legislation Defeated in House—A month after the Senate had struggled to pass legislation imposing new restrictions on the sale of guns, the House, **June 18**, rejected a similar bill. The Senate bill had put the gun regulations in a juvenile-crime bill, but the House split the 2 subjects. The juvenile-crime bill, similar to what the Senate had approved, passed **June 17**. It would provide $1.5 billion more to states to deal with crime by minors. Anyone 14 years old could be tried as an adult. A hotly debated amendment was approved that would allow the Ten Commandments to be displayed in schools and government buildings.

On **June 17**, an amendment to the gun bill, approved 218–211, would allow only a 24-hour background check when someone bought a gun at a gun show. As a result, many in the House who wanted a tougher bill, as well as those who opposed any gun legislation, joined in voting against final passage on **June 18**, and the bill failed, 280–147.

Former Fund-Raiser for Democrats Arraigned—John Huang, a former fund-raiser for the Democrats, was arraigned **June 21**, on a felony count—conspiracy to defraud the Federal Election Commission—in connection with $7,500 he had raised for the California party between 1992 and 1994. He was not charged in connection with the $3.4 million he had raised for the Democrats for the 1996 campaign. The party had returned nearly half of that money as potentially tainted. Huang, in accepting the felony count, agreed to cooperate with the ongoing investigation. Huang pleaded guilty **Aug. 12**. U.S. District Judge Richard A. Paez fined him $10,000, put him on probation for a year, and ordered him to perform 500 hours of community service.

Report Issued on Breast Implants and Disease—In a report released **June 21**, independent investigators from the National Academy of Sciences' Institute of Medicine concluded that silicone breast implants did not cause systemic diseases. The implants had been linked to cancer, rheumatoid arthritis, and other illnesses. Thousands of lawsuits had led to a $4 billion settlement with 3 manufacturers, and on **June 1** a $3.2 billion settlement with one company, Dow Corning, had been reported. The report did conclude that the implants could cause local conditions, including injuries to tissue and infections of the breast.

War Over, Capital Turns to Domestic Issues—With the military conflict in Kosovo apparently ended, national U.S. political leaders again gave a priority to domestic legislation. Congressional Republicans were pressing Pres. Clinton to support a big tax cut. Many also called for building a missile defense system and reducing the role of the federal government in education. On **June 25**, Clinton urged Congress to reform Medicare and Social Security to ensure the programs' future solvency. He also urged Congress to reform campaign financing, pass a "bill of rights" for members of health maintenance organizations (HMOs), and increase the minimum wage. Clinton announced **June 28** that new pro-

jections showed federal budget surpluses totaling nearly $3 trillion through 2009.

Analysts feared that Medicare would be insolvent by 2015. Clinton, **June 29**, proposed to take $794 billion from projected budget surpluses over the next 15 years to protect Medicare until 2027. He also proposed a new benefit, payment through Medicare for prescription drugs used outside the hospital (drugs used in the hospital already were covered). As effective new drugs were coming on the market, their prices were climbing. Clinton put the cost of the drug payment at $118 billion over 10 years.

Key Interest Rate Is Increased—After leaving a key interest rate alone since November 1998, the Federal Reserve Board, **June 30**, announced an increase in the federal-funds rate—the rate banks charged each other on overnight loans—from 4.75% to 5%. The increase was an attempt to head off possible inflationary pressures. The discount rate—the rate at which commercial banks borrowed from the Fed—remained at 4.5%.

Hubbell Pleads Guilty in Arkansas Land Deal—On **June 30**, Webster Hubbell, a former associate U.S. attorney general and friend and political associate of the Clintons, pleaded guilty to reduced charges in the investigation being conducted by independent counsel Kenneth Starr. In his negotiated plea with Starr, Hubbell acknowledged that he had concealed from investigators legal work that he and Hillary Rodham Clinton had done in connection with a land deal described by prosecutors as a sham. He also pleaded guilty to a misdemeanor count of failing to pay federal income taxes. Hubbell would serve no time in prison and would pay nominal fines. He had spent 18 months in prison on charges previously brought by Starr. Hubbell told reporters that Mrs. Clinton had done nothing illegal.

U.S. Economy at a Glance June 1999	
Unemployment rate	4.3%
Consumer prices (change over May)	Unchanged
Producer prices (change over May)	−0.1%
Trade deficit	$24.60 bil
Dow Jones high (June 30)	10970.80
Dow Jones low (June 11)	10490.51
Index of leading economic indicators (change over May)	+0.3%
2d-quarter GDP (change over 1st quarter of 1999)	+1.9%

International

New President Inaugurated in South Africa—On **June 2**, in a national parliamentary election, South African voters kept the African National Congress in power, thereby assuring that its leader, Thabo Mbeki, would succeed Nelson Mandela, who was retiring as the nation's president. The ANC received about two-thirds of the vote. Election issues included political corruption and high crime and unemployment rates.

Mbeki received a masters degree in economics from Sussex Univ. in England. He had been a leader in the ANC and in the Communist Party (until he left the latter in 1990). Appointed deputy president by the elderly Mandela in 1994, Mbeki was viewed as the day-to-day leader of the government. The National Assembly elected Mbeki as president **June 14**. At his inauguration, **June 16**, he said he would seek to lift millions of South Africans from poverty.

South Koreans Sink North Korean Boat—A series of confrontations in disputed waters led **June 15** to the sinking of a North Korean torpedo boat by fire from South Korean ships. Twenty North Koreans were killed, including the 17 crew members. North Korean military and fishing boats had been crossing a UN-drawn demarcation line in the Yellow Sea, which North Korea did not recognize.

Leaders of Top Industrial Nations Meet—The leaders of the world's major industrialized nations met in Cologne, Germany, on **June 18**, just as the conflict in Kosovo was winding down. The heads of government of the so-called Group of Seven nations supported a plan to cut the debt obligations of the poorest nations.

When Pres. Boris Yeltsin of Russia joined the meeting, it became the Group of Eight. He met with Pres. Clinton, **June 20**, and they agreed to resume arms-control negotiations. The G-7 leaders, **June 20**, pledged to provide substantial aid for the reconstruction of Kosovo and the rest of the Balkans, but made it clear that Serbia, in order to share in the assistance, must commit to political and economic reforms.

Kosovo Conflict Ends as Serb Forces Pull Out—NATO ended its bombing campaign against Yugoslavia, **June 20**, as Serb forces completed their withdrawal from Kosovo.

On **June 2**, Pres. Martti Ahtisaari of Finland and the Russian envoy, Viktor Chernomyrdin, had presented a proposal supported by major Western nations to Pres. Slobodan Milosevic of Yugoslavia in Belgrade. The revised plan required that the Serbs leave Kosovo, and it provided for protection by an international force for ethnic Albanian refugees returning to their homes. The plan dropped NATO's previous support for a referendum by Kosovars on their future, and it restricted the international force to Kosovo. On **June 3**, Milosevic and Serbia's parliament, by a vote of 136–74, accepted the plan. NATO and U.S. estimates **June 4** put the number of Serb soldiers killed and wounded in the bombing at 10,000 to 15,000.

NATO and Yugoslav military officials began meeting **June 5** at Kosovo's border, and on **June 9**, Yugoslavia signed an agreement pledging that Serbian forces would complete a withdrawal within 11 days. Enforcement was to be overseen by an international authority. The NATO force in Kosovo (the Kosovo Force, or KFOR) was to be commanded by Lt. Gen. Michael Jackson of Britain. KFOR, operating under the UN flag, would seek to keep the peace, assist returning refugees, and disarm the Kosovo Liberation Army (KLA). Some 30 countries were expected to contribute a total of 50,000 troops to KFOR. The bombing campaign was suspended **June 10**. Foreign ministers of leading Western nations agreed the same day on a series of stabilizing measures for the Balkans aimed at promoting democracy, human rights, and economic recovery.

The first NATO troops entered Kosovo **June 12**, though to their surprise Russian troops, a few hours earlier, had already entered Pristina, Kosovo's capital. The role for the Russians had not been agreed to, and the Russians' appearance created international tension. Meanwhile, NATO soldiers were discovering widespread looting, vandalism, and burning, most of it apparently carried out by the Serbs. Ethnic Serb civilians began to flee Kosovo for Serbia. Two Serbs were killed **June 13** after exchanging gunfire with NATO troops; 2 German journalists were killed by gunmen the same day. British Foreign Min. Geoff Hoon said **June 17** that 10,000 ethnic Albanians might have been killed by Serbs. On **June 18**, investigators for the UN International Criminal Tribunal for the Former Yugoslavia arrived in Kosovo.

In talks ending **June 18**, Russian and NATO representatives agreed that Russia would have areas of control within the French, German, and U.S. sectors. After NATO confirmed, **June 20**, that the Serb withdrawal was complete, the bombing was formally ended. The first NATO fatalities occurred **June 21** when 2 British soldiers and 2 civilians died while clearing an explosive device. The KLA agreed, **June 21**, to disarm within 90 days.

Pres. Clinton visited Slovenia **June 21** and Macedonia **June 22**, where he met with refugees. By then, ethnic Albanian refugees were streaming back to Kosovo, and violent incidents between Albanians and Serbs were being reported. U.S. Marines killed a Serb gunman **June 23**. On **June 24** the U.S. offered a $5 million reward for information leading to the capture of Milosevic or any of the other 4 government leaders indicted in May for war crimes.

Israelis and Muslim Guerrillas Clash—Israeli forces battled the Shiite Muslim guerrilla group Hezbollah in June,

in the harshest exchange along the Lebanon-Israeli border in 3 years. Hezbollah fired mortars into northern Israel **June 24**, and Hezbollah fighters battled with the South Lebanese Army, which was an ally of Israel. Prime Min. Benjamin Netanyahu ordered air strikes on Hezbollah **June 24**. In fighting that continued **June 25**, 10 civilians—8 Lebanese and 2 Israelis—were killed, and many people were injured.

Turks Sentence Kurdish Leader to Death—Abdullah Ocalan, the leader of the Kurdish rebellion in Turkey, was convicted of treason by a Turkish court, **June 29**, and sentenced to death. He had been captured in February. Ocalan had told the court that if he was spared the death penalty he would use his influence among the Kurds to end the conflict with the government. Mary Robinson, UN high commissioner for human rights, criticized the trial for a lack of due process.

General

Dallas Wins Stanley Cup in 4-Hour Game—The Dallas Stars won the National Hockey League championship, **June 19-20**, defeating the Buffalo Sabres, 2–1, in a game that lasted 3 overtimes and more than 4 hours (including breaks). The Stars prevailed in the championship series, 4 games to 2, to gain the franchise's first title since it was founded as the Minnesota North Stars in 1967. Brett Hull scored the winning goal, which was allowed although he had a skate in the goaltender's crease area when he shot. The Stars' center, Joe Nieuwendyk, was named MVP in the playoffs.

San Antonio Wins First NBA Title—The San Antonio Spurs, broke through **June 25** to capture their first National Basketball Assn. title. They defeated the New York Knicks, 78–77, to win the championship series, 4 games to 1. The Spurs were led by center David Robinson and forward Tim Duncan, who was named MVP in the final series. He averaged 27.4 points and 14 rebounds per game in the finals. The Spurs, coached by Gregg Popovich, won the decisive game in New York when the Knicks' Latrell Sprewell, who had scored 35 points, missed a last-second shot. The Knicks, the Eastern Conference champions, had been seeded 8th (and last); they had barely made the playoffs. Their top player, Patrick Ewing, had been knocked out of the playoffs with an injury during the 3d-round series against Indiana.

JULY 1999

National

First Lady Eyes U.S. Senate Seat From New York—After months of speculation about her plans, First Lady Hillary Rodham Clinton, **July 6**, took a first formal step toward becoming a candidate in 2000 for the U.S. Senate seat in New York being vacated by Daniel Patrick Moynihan (D). She notified the Federal Election Commission that she was forming an exploratory committee to consider running. No first lady had ever before sought public office. The first lady had lived in Illinois before joining her husband in Arkansas, and had never been a resident of New York. She said that she cared deeply about issues important to New Yorkers, and from **July 7** to **10**, she toured upstate New York in the first of a series of "listening tours."

Class-Action Plaintiffs Win Tobacco Suit—A 6-member Miami-Dade County jury in Miami, **July 7**, held leading tobacco companies liable for various illnesses in smokers. The class-action lawsuit, filed in 1994 against the tobacco companies on behalf of Florida smokers, was the first of its type that actually came to trial. After 8 months of testimony, the jury concluded that cigarettes were addictive and could cause at least 20 diseases, and that the companies had sold a defective product and conspired to mislead the public about the dangers of smoking. The decision was subject to appeal and represented only the first phase of the trial; damages and claims of individual Florida residents were to be taken up in a 2d and 3d phase, respectively.

Race for 2000 Proceeds—A year before the nominating conventions, presidential aspirants were campaigning nearly full time. Vice Pres. Al Gore, the front-running Democratic candidate, outlined his anti-crime package, **July 12**; it emphasized gun-control measures: a ban on cheap handguns, a limit of 1 gun purchase per person per month, a gun-safety test for prospective gun owners, and a nationwide system of licensing for gun owners. His rival for the Democratic nomination, former senator Bill Bradley, had called for similar measures, and also for the registration of handguns.

On the Republican side, Sen. Robert Smith (NH) withdrew from the nomination contest, **July 13**, and from the party as well, changing his party designation to Independent. Smith's defection cut the Republican majority in the Senate to 54–45. Rep. John R. Kasich (OH) dropped out of the GOP presidential race, **July 14**, and also announced that he would not seek reelection to the House in 2000. Gov. George W. Bush (TX), the leader in polls for the Republican nomination, said, **July 15**, that he would not accept federal matching funds for his nomination bid, thus freeing him from spending caps under the law. Bush had already raised more money than any previous candidate for a presidential nomination. Steve Forbes, whose own campaign for the GOP nomination was largely self-financed, had also declined matching funds.

The Reform Party convention opened in Dearborn, MI, **July 23**, and heard from its highest elected official, Gov. Jesse Ventura (MN), who said he would not be a candidate for president in 2000. On **July 25** the delegates chose as chairman Jack Gargan, who had Ventura's support. The vote was a rebuff to party founder Ross Perot.

Senate Approves "Bill of Rights" for Patients—The Senate **July 15** approved a so-called patients' bill of rights. The 53–47 vote, largely along party lines (no Democrat supported it and Pres. Clinton vowed a veto), was just one step down a trail toward ultimate reform. The bill aimed to protect patients enrolled in health insurance plans, including health maintenance organizations (HMOs). Patients in employer-sponsored health plans would be allowed to appeal denials of coverage for treatment. Democrats contended that too few Americans were protected under the bill and that too few protections were offered.

JFK Jr., Wife, Sister-in-law Die in Plane Crash—John F. Kennedy Jr., son of the former president, died in a plane crash **July 16** along with his wife, Carolyn Bessette Kennedy, and his sister-in-law, Lauren Bessette. A toddler during his father's White House years, Kennedy, with his sister, Caroline, had been shielded from public view during their childhoods. As an adult he sought to carry on the family tradition in his own way and, while pursued by the media, avoided the limelight where possible. At his death, Kennedy was editor of *George* magazine.

On the evening of **July 16** the 3 took off from Fairfield, NJ, with Kennedy piloting his Piper Saratoga II HP, a small single-engine plane. They were bound first for Martha's Vineyard, to drop off Bessette, and then for Hyannisport, MA, where the Kennedys were to attend a cousin's wedding. A report that the plane was overdue prompted a massive sea search. On **July 17**, debris from the plane washed ashore on Martha's Vineyard. A large portion of the plane was found **July 20**, about 7 miles offshore. Radar data released that day indicated that the plane had made an apparently unnecessary right turn, then 2 more right turns, then begun a rapid descent to the sea. Aviation experts said that the data suggested Kennedy may have lost his bearings and then lost control of the plane.

The bodies of the 3 victims were recovered **July 21**. The state medical examiner's office, reporting the results of the autopsies, **July 22**, stated that the 3 had died instantaneously of multiple traumatic injuries. In a private ceremony attended by family members on the U.S. Navy destroyer *Briscoe*, the ashes of the 3 were buried at sea off the coast of Massachusetts, **July 22**. On **July 30**, the National Transportation Safety Board reported it had found no evidence of problems with the plane itself. The NTSB's probe was expected to last another 6-9 months.

Republican in House Joins Democrats—The narrow Republican margin in the U.S. House of Representatives shrank a bit more **July 17** when Rep. Michael P. Forbes, from Long Island, NY, switched from the GOP to the Democratic Party. Forbes said that the Republicans had been taken over by extremists who were unable to govern. There were now 222 Republicans in the House, 211 Democrats, and 1 Independent. One seat was vacant as a result of the death, **July 15**, of longtime Rep. George E. Brown, Jr. (D, CA).

Heat Wave, Drought Cause Devastation—The eastern U.S. experienced record heat **July 19-Aug. 1**, resulting in at least 200 deaths, with at least 80 in Illinois. High temperatures, accompanied by little rainfall, contributed to severe drought conditions. On **Aug. 2**, Pres. Clinton declared 6 states (Kentucky, Maryland, Ohio, Pennsylvania, Virginia, and West Virginia) agricultural disaster areas, entitling farmers to low-interest federal loans to cover crop losses. The U.S. Dept. of Agriculture estimated the drought could cost farmers as much as $1.1 billion in lost income.

House GOP Pushes Huge Tax Cut—House Republicans spearheaded a drive in July for a $792 billion reduction in taxes for individuals and businesses. Pres. Clinton promised a veto. Voting mostly along party lines, the House on **July 22** approved the bill, 223–208. It provided a 10% tax cut over 10 years for everyone, reduced taxes for married couples who paid more taxes than if they were single, cut the capital gains tax, and gave tax breaks to businesses. Federal budget surpluses had been forecast to total nearly $3 trillion over the next 10 years. Democrats favored putting most of that money into reducing the national debt and shoring up Social Security and Medicare. Federal Reserve Board Chairman Alan Greenspan said, **July 22**, that "the timing is not right" for a tax cut as large as the House had approved. He urged that Congress reduce the debt instead and save tax cuts for a time of economic recession. Greenspan repeated his points **July 28**.

On **July 30**, the Senate, 57–43, also approved a $792 billion tax cut, but with provisions that differed from the House bill. A Senate–House conference had the task of reconciling the 2 bills.

Woman Commands Space Shuttle Flight—On **July 23**, with the launch of the U.S. space shuttle *Columbia*, Air Force Col. Eileen M. Collins became the first woman to command a shuttle flight. Collins had previously piloted 2 shuttle flights. On **July 23** the crew launched into orbit the Chandra X-ray Observatory, which was intended to detect radiation from black holes, quasars, exploding stars, and gas clouds. *Columbia* landed at Cape Canaveral on **July 27**.

Court Fines Clinton for Lying About Lewinsky—U.S. District Court Judge Susan Webber Wright, **July 29**, ordered Pres. Clinton to pay $89,000 in legal expenses because he had deliberately given false testimony about his relationship with Monica Lewinsky in his deposition in the Paula Corbin Jones sexual-harassment case. Wright, in April, had held Clinton in contempt of court. Most of the money went to lawyers representing Jones, and about $9,480 went to the Rutherford Institute, which had helped pay Jones's legal expenses. Wright wrote that she took "no pleasure in imposing contempt sanctions against this nation's president."

Heavy Loser in Stock Trading Kills 12—Mark Barton, a securities trader who had suffered heavy losses, shot 9 people to death and wounded 13 others in Atlanta on **July 29** before taking his own life. The bodies of his wife and 2 children were also found, along with a note from Barton confessing to their murders, which apparently had occurred a few days earlier. Barton was a day-trader who bought and sold stocks quickly, a practice considered very risky. Armed with a handgun and a pistol, he entered the offices of 2 day-trading firms, Momentum Securities and All-Tech Investment Group, where he shot all of his victims. Several hours later, as police closed in on his van outside Atlanta, Barton stopped at a service station and shot himself to death. Barton had been the only suspect in the 1993 murder of his first wife and her mother.

Tripp Indicted for Taping Lewinsky—Linda Tripp, who had secretly taped conversations between herself and her then-friend, Monica Lewinsky, was indicted by a Maryland grand jury **July 30**. The tapes were used in evidence against Pres. Clinton when he was impeached by the U.S. House in 1998. Tripp was charged specifically with illegally taping a conversation in December 1997 and also for disclosing the contents of that conversation to *Newsweek* magazine.

U.S. Economy at a Glance July 1999	
Unemployment rate	4.3%
Consumer prices (change over June)	+0.3%
Producer prices (change over June)	+0.2%
Trade deficit	$24.89 bil
Dow Jones high (July 16)	11209.84
Dow Jones low (July 30)	10655.15
Index of leading economic indicators (change over June)	+0.3%

International

New Head of UN Mission Arrives in Kosovo—On **July 2**, UN Sec. Gen. Kofi Annan announced that French Health Min. Bernard Kouchner would head the UN Mission in Kosovo responsible for reestablishing civil government as a prelude to a return to full self-government. On **July 3**, British soldiers shot and killed 2 Kosovo Albanians. NATO representatives and Russian officials, during talks in Moscow **July 4** and **5**, worked out the final details for the deployment of 3,600 Russian soldiers in Kosovo; the Russians entered the province **July 6**.

Kouchner arrived in the province **July 15**. That day, his predecessor, Sergio Vieira de Mello, warned that since the Kosovars had returned, Serbs, Gypsies, and other minorities were being killed and subjected to other human-rights abuses. In a vehicular accident, **July 18**, 2 U.S. soldiers became the first Americans killed since the entry into Kosovo. In the worst incidence of violence so far since the return of the Kosovars, 14 Serb farmers were shot to death in Gracko on **July 23**.

On **July 28**, nearly 100 countries and international organizations and agencies pledged $2.1 billion ($500 million from the U.S.) to assist Kosovo. On **July 30**, the U.S. agreed to pay $4.5 million in compensation to the injured victims and families of those killed in the U.S. bombing of the Chinese embassy in Belgrade in May.

Islamic Militants Reportedly Leave Kashmir—The latest conflict between India and Pakistan showed signs of quieting down with the reported departure of Islamic militants who had occupied the disputed region of Kashmir. Prime Min. Nawaz Sharif of Pakistan promised Pres. Clinton in Washington, **July 4**, to restore a cease-fire line in Kashmir, which the militants had crossed. His ability to impose this commitment on his own military remained uncertain, and on **July 7** leaders of the militants vowed to continue fighting to establish an autonomous Islamic state in Kashmir. At a border meeting **July 11**, military commanders agreed to a settlement, with a full withdrawal of the militants set by **July 16**. Clashes continued however. Muslims killed 20 Hindus **July 20** in 3 attacks in Kashmir; 16 were reported dead in battles on **July 21**. Indian military officials said **July 26** that the militants had been pushed out of Kashmir, ending the crisis at least for the time being.

Barak Takes Oath as Israeli Prime Minister, Launches Diplomatic Offensive—Ehud Barak, who had been elected in May and then put together a governing coalition, took the oath as prime minister of Israel, **July 6**. His government, embracing 7 parties from the political center and left, had control of 75 of the 120 seats in the Knesset. Barak pledged

to act boldly to achieve a final settlement with the Palestinians (but without giving up any of Jerusalem), and he said he also would seek peace with Syria. Barak decided to serve as defense minister; as foreign minister he named David Levy, who had held that position for a time under the previous prime minister, Benjamin Netanyahu.

Barak met with key Middle Eastern leaders, including Pres. Hosni Mubarak of Egypt in Alexandria **July 9**, Palestinian leader Yasir Arafat at the Israeli–Gaza border **July 11**, and King Abdullah II of Jordan in Aqaba **July 13**. Arafat renewed his vow to thwart terrorist activity, and Barak restated that Israel would build no more Jewish settlements in the West Bank. Meeting with Pres. Clinton in Washington and Camp David **July 15-18**, Barak set a 15-month deadline for achieving peace in the Middle East. He said he was prepared to make "painful compromises" with Syria over the Golan Heights. Barak and Arafat met again **July 27**.

Sierra Leone Rivals Sign Peace Pact—An agreement signed **July 7** by Pres. Ahmad Tejan Kabbah of Sierra Leone and Foday Sankoh, head of the Revolutionary United Front, granted amnesty to the rebels and 4 cabinet seats in a new government. Whether the agreement would end the 8-year civil war remained to be seen; the parties had signed a previous peace agreement that later collapsed. Sankoh had been imprisoned since 1997 and was under a death sentence. Fighting in Freetown, the capital, in January 1999, had claimed thousands of lives, and the overall death toll during the civil war was now put at 14,000.

Students in Iran Clash With Police—A court in Iran, **July 7**, banned publication of a leading moderate newspaper, and university students launched protests beginning **July 8**. That night, members of an Islamic vigilante group attacked demonstrating students in Tehran, and police stormed a dormitory. Many students were beaten, and at least 1 was killed. Within days the student protests had spread to 18 cities. Clashes between protesters and authorities disrupted Tehran, where most businesses were closed on **July 13**. Supporters of the regime staged their own rally in Tehran on **July 14**. Pres. Mohammed Khatami, a moderate, condemned the student demonstrations, perhaps fearing a crackdown by hardline Islamic authorities. Students halted their demonstrations **July 17**.

6 Nations Sign Congo Cease-Fire Agreement—The Democratic Republic of the Congo and 5 other nations—all of whom had troops in the Congo—signed a cease-fire agreement **July 10** that sought to end the civil war in the Congo. Signatories included Angola, Namibia, and Zimbabwe, all supporters of the government of the Congo, and Rwanda and Uganda, countries that supported the rebels. The cease-fire called for the integration of government and rebel soldiers into the army and for a national dialogue on the future of the Congo. The agreement also provided for the disarming of the Hutu militias. The 2 major rebel factions that failed to sign the accord in July signed **Aug. 31**.

Ban on British Beef Exports Lifted—In a vote by the European Commission, the European Union, **July 14**, ended its 3-year ban on British beef exports. In 1996, British scientists had reported a possible link between bovine spongiform encephalopathy, or "mad cow disease," which afflicted some British cattle, and Creutzfeldt–Jakob disease, a fatal brain disorder in humans. Britain had since slaughtered thousands of cattle and barred the use of animal meat and bone in the feed given to its cattle.

China Detains Thousands in Spiritual Sect—China on **July 22** declared illegal the Falun Gong group, which had claimed millions of adherents in China. Falun Gong, or Buddhist Law, emphasizes spirituality, moral values, and good health practices, including exercise. The group, which had no apparent political agenda, had been pressing the government for protection of its rights. On **July 23**, Chinese authorities began detaining thousands of its members and seizing its publications. Among those picked up were 1,200 government officials who were members. On **July 29** the government issued a warrant for the arrest of the group's leader, Li Hongzhi, who was living in the U.S.

King Hassan of Morocco Dies—King Hassan II of Morocco, an influential leader in the Arab world, died **July 23** at age 70. A supporter of Palestinian independence, he had also maintained diplomatic ties with Israel and was on good terms with both his fellow Arabs and the West. He had been king since 1961, 5 years after Morocco achieved independence. Hassan's son on **July 23** ascended the throne as Mohammed VI. He had studied law and international relations in France. Hassan's funeral, **July 25**, was attended by many world leaders; the U.S. contingent was headed by Pres. Clinton, the first lady, and former Pres. George Bush.

General

Sampras, Davenport Win Wimbledon Titles—Top-seeded Pete Sampras won his 6th Wimbledon men's singles title **July 4**, defeating fellow American Andre Agassi, 6–3, 6–4, 7–5. Also on **July 4**, another American, Lindsay Davenport, the 3d seed, won the women's singles title, her first, defeating Steffi Graf, 6–4, 7–5. Graf, a 7-time singles winner from Germany announced later, on **Aug. 13**, that she was retiring from tennis, ending a 17-year career that included 22 Grand Slam victories.

U.S. Women Defeat China for Soccer Title—The U.S. team won the Women's World Cup in soccer, **July 10**, defeating China in the final on penalty kicks. The Americans had won once before, in 1991, in the quadrennial competition. The 1999 tournament was held in the U.S., with the final played before 90,185 spectators at the Rose Bowl in Pasadena, CA. Some 40 million Americans watched at least part of the game on television. The Americans' best player, Michelle Akers, had to leave the game after regulation because of heat exhaustion and a concussion. Regulation time ended without any scoring, as did the overtime period. The game was then decided on penalty kicks. The Americans scored all 5 penalty kick attempts, but U.S. goalkeeper Briana Scurry blocked 1 of the 5 Chinese kicks. American Kristine Lilly was named Most Valuable Player.

Overcoming Cancer, Cyclist Wins Tour de France—Lance Armstrong, **July 25**, became the first American on an American team to win the Tour de France. (American Greg LeMond won in 1986, 1989, and 1990, but he had ridden for French teams.) Over a 3-week period, the 27-year-old Armstrong completed the 2,290-mile course in 91 hours, 32 minutes, and 16 seconds. In 1996, Armstrong had gone through a life-threatening struggle with testicular cancer, which spread to his lungs, abdomen, and brain. After surgery and chemotherapy, he was declared free of the cancer in 1997, and he resumed competition in 1998.

AUGUST 1999

National

First Lady Discusses President's Behavior—First Lady Hillary Rodham Clinton said in an interview that Pres. Bill Clinton had lied at first about his relationship with Monica Lewinsky to protect his wife. Her comments were published in the September issue of the new *Talk* magazine, available **Aug. 2**. She said that her husband had suffered abuse as a child deriving from a "terrible conflict" between his mother and grandmother. Both the president and the first lady subsequently denied that they meant to offer any excuses for his bad behavior.

Congress Passes Big GOP Tax Cut—Congress completed action on a Republican-backed $792 billion tax-cut bill. Pres. Clinton had vowed to veto it. A Senate and House conference committee approved the final draft of the bill **Aug. 3**. All income tax rates would be cut by 1%. The top rate on capital gains would drop from 20% to 18%. The inheritance tax on large estates would be phased out, and many married couples would get a tax break. The House gave its final approval to the measure **Aug. 5**, 221–206.

With 4 Republicans defecting, the Senate approved the bill **Aug. 5** by just 50–49. Clinton had opposed the measure as too extravagant; he said he could support only a much smaller tax cut.

Holbrooke Confirmed as UN Ambassador—The Senate **Aug. 5** confirmed Richard C. Holbrooke as ambassador to the UN by an 81–16 vote. The confirmation of Holbrooke—a former ambassador to Germany, assistant secretary of state, and diplomatic envoy—had been delayed 14 months because of ethics allegations against him and moves by some Senate Republicans. He was sworn in **Aug. 25**.

Gunman Attacks Los Angeles Jewish Center—A white supremacist gunman, Buford Furrow Jr., opened fire in the lobby of a Los Angeles Jewish community center **Aug. 10**, wounding 3 young children and 2 staff members. After leaving the North Valley Jewish Community Center, Furrow killed a Filipino-American letter carrier a few miles away. He surrendered to authorities in Las Vegas, NV, **Aug. 11**, saying he had committed the shootings because he was "concerned about the decline of the white race." On **Aug. 19** Furrow was indicted on federal charges in the death of the postal worker, Joseph Ileto, as well as for using a firearm in a murder and being a felon in illegal possession of a firearm. He also faced state counts of murder and attempted murder.

Evolution Dropped From Kansas Curriculum—The Kansas Board of Education voted 6–4, **Aug. 11**, to drop the theory of evolution from the public-school science curriculum. The state board also approved omitting from the curriculum references to the Big Bang theory of the origin of the universe. Kansas Gov. Bill Graves (R) and leading educators in Kansas criticized the board's action.

Clinton Offers Clemency to 16 Puerto Rican Terrorists—Pres. Clinton, **Aug. 11**, offered to commute the sentences of 16 Puerto Ricans who belonged to an organization that used violent means to promote independence for the island. Between 1974 and 1983, some 130 bombings were blamed on the Armed Forces of National Liberation (FALN). Six people were killed by the explosions and many wounded, but Clinton said those to whom he offered clemency had not been directly involved in attacks that harmed others. Most of the 16 had served at least 19 years in prison. Those offered clemency had to agree to renounce violence and comply with parole requirements.

The plan was supported by Hispanic leaders and human rights activists but opposed by law-enforcement officials and many members of Congress in both parties. The first lady issued a statement **Sept. 4** opposing the clemency on the grounds that the prisoners had not come forward to renounce violence. The White House denied charges that Clinton's offer was a political move, designed to bolster support among the large Puerto Rican population in New York, where Hillary Rodham Clinton appeared ready to run for a U.S. Senate seat. The White House announced **Sept. 7** that 12 of the activists had accepted the clemency offer with the conditions attached to it.

Bush Places First in Iowa Poll—Gov. George W. Bush of Texas placed first in a presidential straw poll by Iowa Republicans, **Aug. 14**. Nine of the 10 GOP aspirants competed. At the fund-raiser for the state Republican Party, anyone buying a $25 ticket could vote. Bush received 31% of the 23,685 votes cast. Steve Forbes was 2d (21%) and Elizabeth Dole 3d (14%). Although no national convention delegates were chosen, some candidates spent lavishly to turn out votes. On **Aug. 16**, Lamar Alexander, who got only 6% of the votes and finished 6th, withdrew from the race, saying he did not have enough support or financial resources to continue.

Within days, Bush was being asked by the press if he had ever used cocaine. All other major-party presidential candidates had responded to the question by saying they had not. Bush acknowledged he was once "young and irresponsible," but refused to be specific. He eventually said that he had not used illegal drugs in at least the last 25 years. Previously, he had acknowledged a drinking problem until age 40; he was now 53. In Texas, Bush had signed a law increasing penalties for possession of cocaine.

Fed Raises Interest Rates—The Federal Reserve Board announced **Aug. 24** that it was raising 2 short-term interest rates—the federal-funds rate (the rate banks charge each other on overnight loans), from 5% to 5.25%, and the discount rate (the rate the Fed charges on loans made to commercial banks), from 4.5% to 4.75%. The former rate had also been increased in June. The Fed said the increases would "markedly" reduce the risk of future inflation.

FBI Admission Revives Waco Cult Story—Six years after some 80 cult members perished during a fire and assault by federal agents near Waco, TX, the FBI admitted, **Aug. 25,** that agents had fired pyrotechnic tear gas canisters at the compound, which bounced harmlessly away. Previously, the agency had denied firing any incendiary devices. The FBI still maintained that the subsequent fire was ignited by members of the cult. Attorney Gen. Janet Reno said, **Aug. 26,** that at the time of the Waco incident she had forbidden the use of incendiary devices. On **Sept. 1**, she ordered a new investigation into the assault. On **Sept. 8,** she named former Sen. John Danforth (R, MO) to head the inquiry.

U.S. Economy at a Glance August 1999	
Unemployment rate	4.2%
Consumer prices (change over July)	+0.3%
Producer prices (change over July)	+0.5%
Trade deficit	$24.10 bil
Dow Jones high (Aug. 25)	11326.04
Dow Jones low (Aug. 2)	10645.96
Index of leading economic indicators (change over July)	−0.1%

International

Results of Indonesian Election Announced—Almost 2 months after the votes were cast, Indonesian Pres. B. J. Habibie announced the result of parliamentary elections, **Aug. 3**. The opposition Indonesian Democratic Party of Struggle polled 34% and the Golkar party, Habibie's party, received 22.4%. In all, 48 parties competed in the election, the nation's first free election in 44 years. Complaints of fraud had delayed the announcement of the vote.

Smaller Yugoslav Republic Seeks Autonomy—Montenegro, the smaller of Yugoslavia's 2 remaining republics, proposed **Aug. 5** that it be granted equal status with Serbia, the nation's larger and dominant republic. Montenegro's government opposed the policies of the federal president, Slobodan Milosevic, toward the breakaway province of Kosovo. Under the Montenegran proposal, the 2 republics would no longer be a federation; instead, they would be a commonwealth of states on equal footing.

Yeltsin Dismisses Another Premier—Pres. Boris Yeltsin of Russia **Aug. 9** dismissed Prem. Sergei Stepashin, who had held office for 3 months. Stepashin was the 4th premier in 17 months to be ousted. Yeltsin apparently was dissatisfied with Stepashin's lack of progress in dealing with the economy and fighting crime and corruption. He chose as the new premier Vladimir Putin, who had served for 15 years in the KGB, the Soviet Union's secret service agency. Yeltsin, who in 1998 had appointed Putin director of the Federal Security Service, also indicated that he favored him as his successor in the 2000 presidential election. On **Aug. 16**, after Putin stressed the need to establish law and order in Russia, the State Duma, the lower house of parliament, approved his nomination.

Russia Ends Islamic Revolt in Dagestan—Islamic militants in the region of Dagestan in southern Russia declared Dagestan to be an independent state, **Aug. 10**. Chechnya, Dagestan's neighbor, had been a center of violent rebellion. On **Aug. 14**, Russia said it had bombed suspected Dagestan rebel sites in Chechnya. Fighting on the ground was heavy. Russia said **Aug. 25** that the militants had been driven out of

Dagestan. Russian and Chechen officials said 59 and 38 combatants on their respective sides had been killed; the actual death toll may have been higher.

Downed Plane Adds to India–Pakistani Tension—An Indian military plane shot down a Pakistani naval plane **Aug. 10**, killing all 16 aboard. Relations between the countries already were strained by their recent nuclear tests and by another confrontation over Kashmir. India said the plane was 6 miles inside Indian airspace. Pakistan denied this and said the unarmed plane was on a routine training mission. On **Aug. 11**, Pakistan fired a missile at 2 Indian planes and 3 helicopters that were flying journalists to the crash site.

2 Mergers Planned in Aluminum Industry—Aluminum companies based in 3 countries announced **Aug. 11** that they planned a merger that would create the largest company in the industry. The 3 companies were Alcan Aluminum Ltd. (Canada), Alusuisse Lonza Group AG (Switzerland), and Pechiney SA (France); their total revenue in 1998 was about $25 billion.

Aluminum Co. of America (Alcoa) announced **Aug. 19** that it would buy Reynolds Metals Co., a U.S.-based company. If both mergers went through, Alcoa—heretofore the world's largest aluminum company—would lose that claim to the Alcan-led company. However, the Alcoa–Reynolds combination would rank as the leading aluminum producer in the world since it would have fewer non-aluminum businesses than the Alcan-led firm.

Earthquake Takes Heavy Toll in Turkey—A severe earthquake rocked Turkey **Aug. 17** and caused the deaths of at least 16,000 people. Over 20,000 were injured, and many others were missing and presumed dead. The quake measured 7.4 on the Richter scale, an intensity seldom surpassed. The epicenter was near Izmit, a city about 55 miles east of Istanbul. In Izmit alone, some 76,000 buildings may have collapsed, and most of the victims there lived in large apartment buildings. A frantic rescue effort continued for several days, with many countries providing aid. Much of the blame for the destruction of so many buildings was placed on the shoddy workmanship allegedly allowed by a government that failed to enforce construction codes.

Berlin Once Again Capital of Germany—For the first time since World War II, Berlin became the capital of a united Germany on **Aug. 23**, when Chancellor Gerhard Schroeder moved to the city. Berlin had been the heart of imperial Germany and the focal point for the Nazi regime of Adolf Hitler. After the fall of Hitler's regime, Berlin was divided between sectors occupied by the Western Allies (which became part of the German Federal Republic) and the Russian sector, which became part of East Germany. Bonn was the capital of the Federal Republic (West Germany) and, after the fall of the Berlin Wall in 1989, of a united Germany. Mayor Eberhard Diepgen welcomed Schroeder and the rest of the government, including the federal parliament, to Berlin.

Final Crew Leaves Russian Space Station—The Russian space station *Mir* was vacated **Aug. 27** by its last crew after the Russian government decided to abandon the aging station because of funding problems. The 2 Russian cosmonauts and 1 French astronaut on board landed in Kazakhstan **Aug. 28**. In space a record 13 years, *Mir* was expected to fall from orbit in 2000 and burn up in the atmosphere. The Soviet Union had launched the space station in 1986, and it was utilized for some 17,000 scientific experiments. In 1995, Valery Polyakov completed a record 439 consecutive days in space aboard *Mir*. A U.S. astronaut, Shannon Lucid, set the woman's endurance record (and the record for any American) on *Mir*, 188 days, in 1996.

East Timor Votes on Leaving Indonesia—Almost all eligible voters in East Timor, a province of Indonesia, turned out **Aug. 30** to vote in a UN-sponsored referendum on autonomy or independence. Indonesia annexed East Timor, a former Portuguese colony, in 1975. Many in the province had resisted their subsequent mistreatment at the hands of the Indonesians. After the fall of Pres. Suharto in 1998, Pres. B. J. Habibie proposed autonomy for East Timor, with Indonesia continuing to control finance, defense, and foreign affairs. He had said that if autonomy was rejected, he would just set East Timor free. In the past year, pro-Indonesia militias inside East Timor had killed hundreds of East Timorese, and they resumed their attacks **Aug. 31** while the result of the voting was awaited.

General

Water Found First Time in "Extraterrestrial"—Scientists reported in the **Aug. 27** issue of *Science* that, for the first time ever, liquid water had been found in an object from outer space. In 1998, a meteorite, once part of an asteroid, had fallen to earth in Monahans, TX. Scientists put its age at 4.5 billion years. Halite crystals, or salt crystals, similar to those formed on earth by evaporating water, were visible; scientists found tiny pockets of briny water inside the crystals.

SEPTEMBER 1999

National

Clintons Buy New York House—After Pres. Clinton and First Lady Hillary Rodham Clinton completed weeks of house-hunting, the White House announced, **Sept. 2,** that they had decided to buy a 5-bedroom house in Chappaqua, NY, a prosperous suburb north of New York City, for $1.7 million. (The Clintons had no home of their own while living in the Arkansas governor's residence and the White House.) Mrs. Clinton was expected to announce soon that she would run for the U.S. Senate from New York. In order to become a senator from New York, she had to establish residency there by Election Day 2000. A wealthy friend and Democratic fund-raiser, Terrence McAuliffe, deposited $1.35 million of his own money in an account as collateral for the mortgage. After this arrangement attracted controversy, the Clintons instead obtained a loan from PNC Mortgage, a subsidiary of PNC Bank Corp.

Cable Network Conglomerate to Buy CBS—Viacom, the world's largest cable network company, announced **Sept. 7** that it planned to buy CBS Corp., which owned CBS Television. The acquisition would be accomplished in a $41 billion stock transaction, and Viacom would assume a $1.4 billion debt. Sumner Redstone, the Viacom chairman, was to head the combined firm, which would retain Viacom as a name. Viacom owns Paramount film and television studios as well as Nickelodeon, MTV, other cable outlets, and other properties.

Ex-Cabinet Secretary Pleads Guilty—Henry Cisneros, Pres. Clinton's first secretary of Housing and Urban Development, pleaded guilty, **Sept. 7**, to lying to the FBI. He acknowledged that payments he had made to an ex-mistress were much higher than the $60,000 figure he had given to the FBI in 1992 and 1993, during background checks related to his nomination to the Cabinet. The guilty plea, to a misdemeanor, was part of an agreement with an independent counsel, David Barrett; Cisneros would pay a $10,000 fine but would not go to jail. The plea closed an investigation that cost more than $9 million.

Bradley, McCain Formally Enter 2000 Race—Bill Bradley returned to his hometown of Crystal City, MO, **Sept. 8** to declare that he was a candidate for the Democratic nomination for president. A basketball player at Princeton who then starred with the New York Knicks, Bradley had gone on to serve 18 years in the U.S. Senate from New Jersey. In announcing his candidacy, he called for campaign finance reform and greater government aid for health care and poor children.

Sen. John McCain (R, AZ) formally declared his candidacy for the Republican nomination Sept. 27. A U.S. Naval Academy graduate, McCain spent 23 years in the Navy, 5 1/2 of them as a North Vietnamese prisoner. Retiring as a cap-

tain, he served in the U.S. House, then the Senate. McCain, announcing in Nashua, NH, called on Americans to "take our government back from the power brokers and special interests."

On the same day, former Vice Pres. Dan Quayle withdrew from the race for the GOP nomination. He had finished 8th in the Iowa straw poll, and thereafter had difficulty raising money.

On **Sept. 29**, Gore announced that he was moving his campaign headquarters from Washington, DC, to Nashville, TN, in what appeared to some as an attempt to rejuvenate his campaign (Gore remained the apparent front-runner, but Bradley was rising in the polls). Gore also was seeking to cut costs by reducing rents and by easing out some of his many high-salaried advisers.

Gunman Kills 7 in Baptist Church—Armed with 2 semi-automatic handguns, a local resident, Larry Ashbrook, shot 7 people to death in a Baptist church in Fort Worth, TX, **Sept. 15**. At the time, some 150 worshippers were attending a service for young people; 3 of those killed were teenagers. Ashbrook also wounded 7 others. He then shot himself to death.

Future Presidents to Get $400,000 Salary—Congress **Sept. 16** completed action on the appropriations bill for the Dept. of the Treasury, the Office of the President, and some other agencies. The bill included a salary increase for U.S. presidents, beginning in 2001, from the current $200,000 to $400,000. The president's pay had not been increased since 1969.

2d Man Convicted in Dragging Death—A 2d white man, Lawrence Brewer, was convicted, **Sept. 20**, by a jury in Bryan, TX, of the 1998 murder of a black man, James Byrd Jr., who died after being dragged while chained to a pickup truck. Another man, John King, had previously been convicted of the crime, and a 3d, Shawn Berry, was awaiting trial. The jury, **Sept. 23**, sentenced Brewer to death.

U.S. Sues Tobacco Companies—The U.S. Justice Dept. **Sept. 22** sued 5 major American tobacco companies and 2 defunct lobbying groups, charging that they had colluded to defraud the public, misleading smokers about the addictiveness and dangers of tobacco products. In 1998 the companies had settled lawsuits by the states for $246 billion. The U.S. suit sought to recover expenditures made by Medicare and by veterans' and federal employees' health plans.

Mission to Mars Fails—NASA reported **Sept. 23** that it had lost communication with the *Mars Climate Orbiter*, a spacecraft scheduled to enter orbit around Mars that day. The craft had apparently broken up after coming too close to the planet because of a navigational error. Launched in December 1998, the $125 million orbiter was supposed to gather data for 2 years on the Martian atmosphere. A 2d spacecraft, the *Mars Polar Lander*, was scheduled to land on the planet in December.

Clinton Vetoes $792 Billion Tax Cut—As promised, Pres. Clinton **Sept. 23** vetoed the $792 billion tax cut approved by the Republican-controlled Congress. He argued that a tax cut of that magnitude would prevent a reduction in the national debt and jeopardize the future of Social Security and Medicare.

U.S. Economy at a Glance September 1999	
Unemployment rate	4.2%
Consumer prices (change over Aug.)	+0.4%
Producer prices (change over Aug.)	+1.1%
Dow Jones high (Sept. 9)	11079.40
Dow Jones low (Sept. 29)	10213.48
3d-quarter GDP (change over 2d quarter of 1999)	+4.8%

International

Terror Bombs in Russia Kill Hundreds—A series of bombings in Russia resulted in heavy loss of life; the government blamed the bombings on Islamic terrorists from the breakaway republic of Chechnya. An explosion in a Moscow shopping center, **Aug. 31**, had killed one person and wounded about 40. Then a car bomb near an apartment complex in Dagestan, **Sept. 4**, killed over 60 and wounded more than 100. A bomb in a Moscow apartment building, **Sept. 9**, took more than 90 lives. On **Sept. 13**, at least 118 died in another huge explosion at an apartment building in Moscow. At least 18 died and more than 200 were injured **Sept. 16** in an explosion in an apartment building in Volgodonsk.

Russian planes, **Sept. 23-25**, bombed an airport, oil refinery, arms depot, radar installation, and other industrial sites in Chechnya. Russia **Sept. 30** sent ground troops into Chechnya.

East Timor Votes for Independence—The UN reported **Sept. 4** that East Timor had voted by a large majority—78.5%—for independence, rather than for the option of remaining within Indonesia with greater autonomy. Indonesian Pres. B. J. Habibie had said he would free East Timor if the autonomy option was defeated. Voter turnout was put at 98.6%. In reaction to the vote, pro-Indonesia militias went on a rampage **Sept. 4-9,** destroying property and killing people in Dili, the East Timor capital, and elsewhere. Bishop Carlos Belo, co-winner of the 1996 Nobel Peace Prize, fled to Australia **Sept. 7** as the militias seized 7,000 refugees at or near his home and drove them away in trucks. Thousands fled to the Indonesian province of West Timor. The new wave of killings added to the toll of more than 100,000 East Timorese already thought to have been slain in conflicts with Indonesia over 2 decades. Indonesian military forces reportedly joined the assaults in some cases.

After some delay, Habibie agreed **Sept. 12** to allow foreign peacekeeping troops to enter East Timor. Threats of economic sanctions had apparently persuaded him to cooperate. The UN Security Council voted unanimously **Sept. 15** to deploy a multinational force in an attempt to restore peace. Australia, a neighbor of Indonesia, was to ultimately supply about half of the projected 8,000 troops. Some 200 U.S. military personnel were to provide logistical support.

Some 2,500 foreign troops in the newly constituted INTERFET (International Force for East Timor) arrived in Dili **Sept. 20**. By then, relief officials estimated that 400,000 people had either fled to West Timor or into the hills. On **Sept. 21** Habibie urged the legislature to ratify East Timor's independence. INTERFET **Sept. 23** raided a militia headquarters and arrested 6 militia members. The UN High Commission on Human Rights, **Sept. 23**, documented that widespread human rights abuses had occurred in East Timor.

Charges Dropped in Death of Princess Diana—Charges filed against 9 photographers and a motorcycle driver, in connection with the 1997 crash that killed Princess Diana and 2 others, were dropped **Sept. 3**. Two French magistrates concluded that no action by those charged, who had been pursuing Diana's car, could be definitely linked to the accident. The report blamed the accident essentially on the intoxication of Henri Paul, the driver. Paul and Diana's companion, Emad Mohamed (Dodi) al-Fayed, were also killed in the accident; Trevor Rees-Jones, a bodyguard, was seriously injured.

Middle East Peace Process Back on Track—Palestinian leader Yasir Arafat and Ehud Barak, Israel's new prime minister, signed an agreement **Sept. 5** that gave new impetus to the regional peace process. The leaders agreed to further Israeli withdrawals from the West Bank that would leave 40% of the area in full or partial Palestinian control by January 2000. The Palestinians in effect promised not to declare independence unilaterally, while the Israelis agreed not to allow more settlements on the West Bank or Gaza Strip. More Palestinian prisoners would be released, but not any prisoner who had killed an Israeli. The leaders set a September 2000 deadline for a final peace agreement. The Israeli Knesset (Parliament), **Sept. 8**, approved the accord, 54-23.

Israel freed 199 Palestinians, **Sept. 9**, with 151 more prisoners scheduled for release in October. On **Sept. 10**, Israel transferred 7% of the West Bank to partial Palestinian con-

trol. Representatives of the 2 sides began talks, **Sept. 13**, on a permanent settlement.

Clinton, Chinese President Confer—Pres. Clinton and Pres. Jiang Zemin of China met privately **Sept. 11** during the Asia–Pacific Economic Conference in Auckland, New Zealand. The 2 presidents concluded that recent stresses between their countries were behind them, and agreed to work toward Chinese membership in the World Trade Organization. Clinton urged China not to use force against Taiwan, which had said in July it would deal with the mainland on a state-to-state basis.

North Korea Halts Testing, Sanctions Eased—North Korea **Sept. 12** agreed to stop test launches of its long-range ballistic missiles. In response, Pres. Clinton **Sept. 17** ordered the relaxation of some sanctions against the Communist state. North Korea could now import consumer goods from the U.S., and U.S. companies could invest in raw-material industries in North Korea. People in the U.S. could send money to North Koreans, and U.S. carriers could now transport people and goods between the 2 countries.

Nuclear Accident Alarms Japan—An accident at a nuclear fuel plant 70 miles northeast of Tokyo, **Sept. 30**, caused high levels of radiation to be released into the air in Japan's worst nuclear accident. Workers had poured 35 pounds of uranium into a purification tank containing nitric acid, almost 7 times the proper amount. More than 50 people apparently were exposed to the radiation, and 300,000 people in the vicinity were told to remain indoors. Trains heading for the region were stopped, and highways were closed. Japan had a history of nuclear-power accidents.

General

Agassi, Serena Williams Win Tennis Titles—Americans won the singles titles at the U.S. Open tennis tournament in New York City.

The women's final nearly became a match between sisters. In a semi-final match on **Sept. 10**, the top seed, Martina Hingis of Switzerland, defeated the American Venus Williams 6–1, 4–6, 6–3. Hingis then lost the final to Serena Williams, 17, the younger sister of Venus Williams, 6–3, 7–6, on **Sept. 11**. Williams was the 2d black woman, after Althea Gibson in 1957 and 1958, to win the U.S. Open championship.

In the men's competition, top-seed Pete Sampras and defending champion Patrick Rafter dropped out because of injuries. Andre Agassi won the final, **Sept. 12**, over Todd Martin, 6–4, 6–7, 6–7, 6–3, 6–2.

Hurricane Causes Widespread Damage—Hurricane Floyd—packing 140 mph winds—battered the Bahamas, **Sept. 14**, then caused major damage in the U.S. Some 3 million Americans evacuated their homes, the most ever in response to a hurricane threat. After coming ashore near Wilmington, NC, **Sept. 16**, Floyd moved up the coast, leaving New England **Sept. 17**. Many counties were declared disaster areas, and damage in North Carolina alone was put at $6 billion. In North Carolina, more than 40 people died as a result of the storm. At least 28 died in other states. Widespread flooding added to the distress in several states.

Taiwan Earthquake Kills 2,300—An earthquake measuring 7.6 on the Richter scale struck Taiwan in the early morning of **Sept. 21**, killing at least 2,300 people, injuring 8,700, and leaving 100,000 homeless. Many people were asleep at the time and were trapped in high-rise apartment buildings that toppled to the ground. The epicenter was in the mountainous central region of the island; Taichung and Nantou were among the cities hardest hit.

OCTOBER 1-15, 1999

National

MCI Seeks to Buy Sprint in $129 Billion Deal—In a deal approved by the boards of both companies **Oct. 4**, MCI WorldCom, Inc. agreed to acquire the Sprint Corp. The union, to be accomplished in a $129 billion stock swap, would constitute the largest corporate acquisition ever. The

2 phone companies, if joined, would still rank below AT&T in long-distance market share. The deal, announced **Oct. 5**, still needed approval by the Federal Communications Commission and the Antitrust Division of the Justice Dept.

Company Settles Diet Pill Case for $3.75 Billion— Some 6 million users of a diet pill combination called fen-phen were expected to share in a $3.75 billion settlement agreed to by the product's manufacturer, American Home Products, **Oct. 7**. The pills had been sold to overweight people as a weight-loss alternative to exercise and dieting. The company withdrew the pills in 1997 at the request of the U.S. Food and Drug Administration, after studies linked them to heart valve defects. Later, some 6,500 lawsuits were filed against American Home Products. Monetary rewards were expected to range from as little as $30 to those who had filled prescriptions to $1.5 million to users who suffered heart valve injuries.

House Passes Tough Bill on Patients' Rights—The U.S. House of Representatives **Oct. 7** approved, 275–151, a bill that would set uniform national standards for health insurance and allow patients to sue health insurance plans that provided poor treatment or harmed patients by denying care. The Senate had passed a weaker bill that covered 48 million Americans; the House bill covered 161 million. House Speaker J. Dennis Hastert (R, IL) warned that the bill, if enacted, would bring on a flood of lawsuits. However, 68 Republicans deserted their party leaders and supported the bill, which would have to be reconciled with the Senate version in conference committee.

Stock Prices Slump—The stock market fell off sharply during the week ending **Oct. 15**. The Dow Jones industrial average declined 630.05 points, or 5.9%, the largest weekly point decrease ever and the worst weekly percentage decline since the week ending Oct. 13, 1989. The Dow fell 266.90 points on Friday alone, and closed at 10019.71, some 1,300 points below its August high of 11326.04. Analysts cited fears that interest rates might be raised again as part of an effort to control inflation. On **Oct. 15**, the government reported that wholesale prices in September had risen 1.1%, the highest one-month rise in 9 years.

Senate Rejects Nuclear Test Ban Treaty—The Senate, **Oct. 13**, rejected a treaty signed by the U.S. that banned all underground nuclear testing. With a two-thirds approval required, only 48 senators voted for the treaty, while 51 voted nay. Opponents said the treaty was unverifiable and unenforceable and "would not stop other nations from testing or developing nuclear weapons." Pres. Clinton, who had signed the treaty in 1996, excoriated the Senate, warning of a return to isolationism. The Senate had not rejected a treaty since twice voting down the Treaty of Versailles and the League of Nations (in 1919 and 1920). Clinton pledged **Oct. 15** that he would abide by the provisions of the treaty. It could not become effective until all 44 nations with nuclear capability ratified it. So far, 26 had done so.

AFL–CIO Endorses Gore for President—Vice Pres. Al Gore, a candidate for the Democratic presidential nomination, received the endorsement of the AFL–CIO **Oct. 13**. It was a major endorsement, but lacked the support of important unions within the AFL–CIO, including the Teamsters and the United Automobile Workers. Labor had generally been slow to back Gore because of his support for free-trade treaties that many U.S. workers believed caused a drain of jobs to other countries. Gore said that, as president, he would not support any agreement that put U.S. workers at risk.

International

Russians Push Into Chechnya—Following a series of explosions in September for which Chechen Islamic terrorists were blamed, the conflict between Russia and the rebellious Russian republic of Chechnya heated up again. On **Oct. 1** the Kremlin declared a small group of pro-Moscow Chechens as the "sole legitimate authority" in the republic. Chechnya's president, Aslan Maskhadov, said the

next day that a Russian invasion was underway. On **Oct. 4** he claimed that Russian air raids in 6 regions had killed 400 civilians. That same day the Chechens reportedly downed a Russian plane, killing the pilot. Russian Prime Min. Vladimir Putin said, **Oct. 5**, that his goal was "full destruction of the terrorists." Russia claimed **Oct. 5** that federal forces controlled one-third of the republic. More than 100,000 civilians had fled into neighboring Ingushetia. A Russian general, Valery Manilov, said **Oct. 8** that 20 to 30 Russian soldiers had been killed in the fighting.

UN Leader Sees Long Presence in East Timor—UN Sec. Gen. Kofi Annan proposed **Oct. 5** that the United Nations assume responsibility for guiding East Timor to statehood. He said that basic institutions were collapsing or nonexistent. Anti-independence militias had killed many people, and driven thousands into exile. UN peacekeepers from Australia, **Oct. 6**, killed 2 anti-independence militiamen who had attempted to ambush them. The U.S. military arrived in force, **Oct. 8**, with the appearance of the *Belleau Wood*, an amphibious assault ship, in the harbor of Dili, the capital. Only about 50 U.S. military personnel actually had duty on shore, and some of them stayed on the ship at night. Maj. Gen. Peter Cosgrove, the Australian officer in charge of the peacekeeping mission, charged **Oct. 11** that the Indonesian army was backing the militias.

Voters in India Back Hindu Nationalists—Parliamentary voting completed in India on **Oct. 7** strengthened the majority for the government of Prime Min. Atal Behari Vajpayee, head of the Hindu nationalist Bharatiya Janata Party. The ruling coalition increased its number of seats in Parliament, suggesting that political instability in India was over for the present. The Congress Party, led by Sonia Gandhi, widow and daughter-in-law of 2 former prime ministers, trailed badly. Vajpayee promised to aim at getting peace talks restarted between India and Pakistan.

Pakistani Army Ousts Civilian Government—The army of Pakistan ousted the elected government of Prime Min. Nawaz Sharif, **Oct. 12**, just hours after Sharif had dismissed the head of the army, Gen. Pervez Musharraf. The latter said, **Oct. 13**, that the military had acted to prevent further destabilizing of the country. Sharif, whose regime was viewed by many Pakistanis as corrupt and incompetent, had apparently been taking steps to strengthen his control over the institutions of government, including the army. The coup, apparently bloodless, was completed within a few hours. Sharif and several other leaders were put under arrest. Musharraf, the de facto ruler, imposed martial law **Oct. 15**, suspended the constitution, and dismissed Parliament.

International Doctors' Group Wins Nobel Prize—The Nobel Peace Prize was awarded **Oct. 15** to Doctors Without Borders (Médecins Sans Frontières), an organization that brought medical care to hungry, sick, and wounded people in many countries, and also was outspoken against human rights abuses and other injustices. The group currently had 2,000 personnel in 80 countries. One-fourth of the volunteer workers were French, and the rest were from 45 countries. Outspoken members of the organization had been expelled from several countries.

General

Grand Jury Indicts No One in Death of Girl—Prosecutors said **Oct. 13** that they had insufficient evidence to charge anyone in the murder of 6-year-old JonBenet Ramsey, whose body was found Dec. 26, 1996, beaten and strangled, in the basement of her family's Boulder, CO, home. The young girl, who was named Little Miss Colorado of 1995 after winning a beauty-queen competition, had been reported missing by her parents, John and Patsy Ramsey, who also produced a purported ransom note. Some experts said the note could have been written by her mother. Many observers believed authorities had botched the case, allowing evidence to be tainted. The grand jury was dismissed without having heard from the Ramseys.

Notable Supreme Court Decisions, 1998-99

The 1998-99 term of the U.S. Supreme Court began Oct. 5, 1998, and ended June 23, 1999, the earliest finish in 30 years.

In his 28th year on the bench and his 13th year as chief justice, William H. Rehnquist took time out from his Court duties to preside over only the 2d presidential impeachment trial in U.S. history, Jan. 7-Feb. 12.

Signed opinions were issued in 75 cases, 16 fewer than in 1997-98. The full Court ruled unanimously in 34 cases and reached 5-4 verdicts in 16. As usual, Associate Justices Anthony M. Kennedy and Sandra Day O'Connor defined the Court's ideological center; each dissented only 8 times during the 1998-99 term. One of the Court's more liberal members, Associate Justice John Paul Stevens, cast the most dissenting votes (20).

The opening session attracted nearly 1,000 civil rights activists protesting lack of minority representation among the Court's law clerks. The National Association for the Advancement of Colored People (NAACP), which backed the protest, said that of the 428 clerks hired since 1972 by the 9 currently sitting justices, fewer than 2% were black and only 1% were Latino. The 35 law clerks hired for the 1999-2000 term included 5 minority group members; men outnumbered women, 23 to 12.

Former Associate Justice Harry A. Blackmun, 90, died in Arlington, VA, on Mar. 4, 1999. Blackmun, who joined the Court as a moderate conservative in 1970 and departed as a liberal in 1994, wrote the majority opinion in *Roe* v. *Wade* (1973), which legalized abortion.

Census 2000. In a ruling that dealt a blow to the Clinton administration and congressional Democrats, the Court held, 5-4, that statistical sampling methods may not be used to determine the official census count employed in apportioning House seats to each state (*Dept. of Commerce* v. *U.S. House of Representatives*; Jan. 25). The decision did not preclude the use of such methods for other purposes, such as allocation of federal grants.

Criminal Law. By a 6-3 vote, the justices struck down an anti-gang ordinance under which Chicago police had arrested more than 42,000 people for loitering in public places between 1992 and 1995 (*City of Chicago* v. *Morales*; June 10). Writing for the majority, Justice Stevens argued that the law failed to distinguish whether suspected gang members and others might be loitering near Wrigley Field "to rob an unsuspecting fan or just to get a glimpse of Sammy Sosa leaving the ballpark."

The Court ruled unanimously that when police stop a motorist for a minor traffic violation, they are not authorized to conduct a full search of the car and driver (*Knowles* v. *Iowa*; Dec. 8). In a 6-3 decision, however, the Court broadened police authority in traffic stops by ruling that an officer may search a passenger's belongings for contraband when there is probable cause that the driver has committed a crime (*Wyoming* v. *Houghton*; Apr. 5). A unanimous Court curbed the practice of "ride-alongs,"

holding that when reporters and photographers accompany police on raids of private homes, the suspects' 4th Amendment protections against unreasonable searches and seizures are violated (*Wilson* v. *Layne*; May 24).

Disabilities. By a 7-2 margin, the justices took a narrow view of the Americans with Disabilities Act (1990), holding that Congress did not intend the law to cover people whose impairments are remedied by medication, eyeglasses, contact lenses, or other devices (*Sutton* v. *United Air Lines*; June 22). In other disability cases, the Court held, 6-3, that states had an obligation, wherever possible, to treat people in small, community-based facilities rather than large institutions (*Olmstead* v. *L.C.*; June 22) and ruled by a 7-2 margin that a federal statute known since 1990 as the Individuals with Disabilities Education Act required schools to pay for the special nonmedical needs of disabled pupils (*Cedar Rapids* v. *Garret F.*; Mar. 3).

Federal-State Relations. In 3 rulings, all issued on June 23 and all decided by 5-4 votes, the Court relied on the doctrine of "sovereign immunity" to shield states from suits brought under federal law. In *Alden* v. *Maine* the Court upheld the dismissal of a suit brought by state employees in state court under the federal Fair Labor Standards Act (1938). Related decisions in *Florida Prepaid* v. *College Savings Bank* and *College Savings Bank* v. *Florida Prepaid* overturned 2 federal laws, both enacted in 1992, that authorized suits against states in patent infringement and trademark cases, respectively.

Gifts to Public Officials. Rejecting the theory under which independent counsel Donald Smaltz had prosecuted former Agriculture Secretary Mike Espy and others, the justices ruled unanimously that providing free meals and other gifts to public officials is not illegal unless there is a demonstrated connection between the gratuity and the specific action it was intended to reward (*U.S.* v. *Sun-Diamond Growers of California*; Apr. 27).

Sexual Harassment. The Court continued to break new ground in this area, holding that school districts may, under certain circumstances, be sued for damages under federal law if they fail to stop harassment of one student by another (*Davis* v. *Monroe County Board of Education*; May 24). Writing for a 5-4 majority, Justice O'Connor cautioned that districts may properly be held liable "only where they are deliberately indifferent to sexual harassment, of which they have actual knowledge, that is so severe, pervasive, and objectively offensive that it can be said to deprive the victims of access to the educational opportunities or benefits provided by the school."

Welfare. A 7-2 decision in *Saenz* v. *Roe* (May 17) struck down as a violation of the 14th Amendment a California law that limited new state residents, during their 1st year of residency, to the welfare benefits they would have received had they remained in their home states. Fifteen states had established such "two-tiered" systems since Congress revamped the welfare system in 1996.

The 1999 Nobel Prizes

The 1999 Nobel Prize winners were announced Sept. 30-Oct. 15. Each prize consisted of a large solid gold medal and a cash award worth 7.9 million Swedish kronor (about $960,000).

Chemistry: The Egyptian-born Ahmed H. Zewail won for his development of femtosecond spectroscopy, a technique employing ultrashort laser flashes to take "snapshots" of how atoms move in a chemical reaction.

Memorial Prize in Economic Science: The Canadian-born Robert A. Mundell was cited for demonstrating how exchange rates affect monetary and fiscal policy; his analysis of "optimum currency areas" paved the way for the euro, the common European currency.

Literature: The German novelist and political activist Günter Grass won for *The Tin Drum, Cat and Mouse, Dog*

Years, The Flounder, and other "frolicsome black fables [that] portray the forgotten face of history."

Peace: The international humanitarian organization Doctors Without Borders (Médecins Sans Frontières), founded in 1971, was honored for its emergency relief efforts, which in 1999 included aid to Kosovo and East Timor.

Physics: Two Dutch researchers, Gerardus 't Hooft and Martinus J. G. Veltman, shared the physics prize for strengthening the mathematical foundations of particle physics and developing calculation methods that led to confirmation of the existence of the "top quark" in 1995.

Physiology or Medicine: The German-born biologist Günter Blobel won for showing how intrinsic signaling mechanisms, analogous to ZIP codes or airline baggage tags, regulate the movement of proteins within a cell.

Notable Quotes in 1999

"Hear ye! Hear ye! Hear ye! All persons are commanded to keep silent on pain of imprisonment."
Senate sergeant-at-arms *James W. Ziglar*, using a proclamation from the 1868 impeachment trial of Pres. Andrew Johnson at the opening of the impeachment trial of Pres. Bill Clinton, Jan. 7, 1999.

"Now that the Senate has fulfilled its constitutional responsibility, bringing this process to a conclusion, I want to say again to the American people how profoundly sorry I am for what I said and did to trigger these events and the great burden they have imposed on the Congress and on the American people."
Pres. Bill Clinton, following his acquittal in his impeachment trial.

"Mommy made a big mistake."
Former White House intern *Monica Lewinsky*, when asked what she would tell her future children about her relationship with Pres. Bill Clinton.

"The school's in a panic, and I'm in the library. I've got students down."
Columbine High School teacher, in a frantic call to the police during the school shooting in Littleton, CO.

"Tell my girls I love them."
Last words of Columbine High School teacher *Dave Sanders*, who died in the Littleton, CO, shooting.

"John was a shining light in all our lives, and in the lives of the nation and the world that first came to know him as a little boy."
Sen. Edward M. Kennedy, in a statement released after the death of his nephew John F. Kennedy Jr.

"As I stepped out, I felt like putting a jacket over my head—so embarrassed."
King Abdullah II, the new king of Jordan, on having to take a 6-car motorcade, instead of a taxi, to see a movie during a visit to Washington, DC.

"It didn't collapse, it blew away."
Moore, OK, resident *Clifford L. Dodson*, after his house was destroyed by a tornado.

"Why am I here, and where are my parents?"
6-year-old boy *Chang Ching-Hung*, who was rescued after being trapped for 87 hours after the Sept. 21 earthquake in Taiwan.

"They killed my 10 brothers, and I am alone."
Naxhije Zymi, Kosovar refugee.

"I'd like to name the baby America, but I have to talk to my wife and parents first."
Kosovar refugee *Naim Karaliju*, after the birth of his son in the U.S. less than 24 hours after his mother's arrival from a Macedonian refugee camp.

"Who is this coming? Oh, a famous actor. Too bad. What I need is a doctor."
Hatixhe Ajeti, a refugee, on seeing Richard Gere visiting the camp she lived in.

"And you had the audacity to go on national television, show the world what you did, and dare the legal system to stop you. Well, sir, consider yourself stopped."
Circuit Court Judge Jessica Cooper, to Dr. Jack Kevorkian before she sentenced him to prison for his videotaped mercy-killing of a patient suffering from Lou Gehrig's disease.

"First you're painted into a corner, then you're hung out to dry, and finally you're framed."
Former Sec. of State Warren Christopher, describing similarities between public life and sitting for a portrait.

"When I tell people I'm from Chappaqua, now they don't confuse it with Chappaquiddick."
Chappaqua, NY, resident *Robby Bernstein,* on the benefit of having the Clintons buy a house in his town.

"As my daughter said, 'Hey, Dad, you're not as cool as they think you are.'"
Texas *Gov. George W. Bush*, acknowledging the high expectations for his run for the presidency.

"We're moving it from K Street to the aisles of Kmart."
Vice Pres. Al Gore, on trying to boost his presidential campaign by moving the headquarters to Tennessee.

"Speed and quickness are more important than bulk."
Bill Bradley, a Democratic presidential candidate, who hoped to overtake Gore.

"It's an abuse of office, but I promised him."
GOP candidate, *Sen. John McCain* (AZ), explaining that he told his 11-year-old son he would let him go to the front of the line at Disney World if he became president.

"I think she is the checkout person at the local market."
Attorney Gen. Janet Reno, telling what she heard a couple say when they thought she looked familiar.

"During my service in the U.S. Congress, I took the initiative in creating the Internet."
Vice Pres. Al Gore, taking credit for the system created by the Defense Dept.

"I was constantly a candidate for 20 years. It kept me young."
German novelist *Günter Grass*, who won the 1999 Nobel Prize for Literature.

"With grief in our soul, we're abandoning a piece of Russia, abandoning something we constructed in space, and it's unclear what we'll build next."
Cosmonaut *Viktor Afanasyev*, on leaving the *Mir* space station.

"Organized religion is a sham and a crutch for weak-minded people who need strength in numbers."
Minnesota *Gov. Jesse Ventura,* interviewed for the November issue of *Playboy* magazine, where he appeared to be disparaging religion.

"The first thought I had was, 'Oh, good!' The second thought I had was 'Oh, no!'"
British author *David Chuter*, on learning he had won the Bulwer-Lytton Fiction Contest for bad writing.

"This is a terrible mistake because I used up all my English... Grazie all'Italia. Grazie all'America."
Italian actor/director *Roberto Benigni*, after winning his 2d Academy Award for the movie *Life Is Beautiful*.

"I truly never believed that this would happen."
Soap opera star *Susan Lucci*, on winning her 1st Daytime Emmy, after being nominated and losing for 18 years straight.

"I had to set my priorities straight."
Movie fan *Vance Rego*, on quitting his job to be 1st on line for tickets for *Star Wars: Episode I—The Phantom Menace*.

"I can assure you, it was a difficult weekend at our house."
Florida *Gov. Jeb Bush*, after his wife was fined for reporting to Customs a much lower than accurate total for clothes she bought on a trip to Paris.

"Crooks have calendars too."
Tony Aveling, of the Australian Bankers Assn., telling customers not to take a lot of money out of the bank in fear of Y2K problems because criminals might take the occasion to steal it.

"Honey, I was around for Y1K."
96-year-old *Sen. Strom Thurmond* (SC), speaking to a woman at a year 2000 ceremony.

"After landing, actually having been somewhat surprised at the fact that we were able to make a successful touchdown, I realized I was going to have to say something."
Astronaut *Neil Armstrong*, on his famous quote the day he stepped on the Moon.

"I thought, 'Should I scream, should I yell or should I cry?' And I guess I ended up doing them all."
Tennis player *Serena Williams*, on her upset victory in the 1999 U.S. Open.

"I'm devastated I will no longer be a hockey player. I will miss every part of the game, because I loved every part of the game."
Hockey great *Wayne Gretzky*, following his retirement.

"I've had a great time."
Basketball great *Michael Jordan*, on his retirement from the sport.

Offbeat News Stories, 1999

Message in a Bottle: A letter written in 1914 by Pvt. Thomas Hughes to his beloved wife in England was finally delivered in May 1999—to his daughter, Emily Crowhurst, now living in New Zealand. Hughes, who wrote the note on his way to the front during the early days of World War I, had slipped it into a ginger beer bottle and dropped it into the English Channel only 2 days before he died. The letter, still dry after 85 years, was found by a fisherman at the mouth of the Thames.

Old Gold: A fish named Tish, listed by the *Guinness Book of Records* as the world's oldest captive goldfish, was found dead at the bottom of his tank in early Aug. 1999. The silver-colored specimen, who lived with Hilda and Gordon Hand of Yorkshire, England, was said to be at least 43 years old, having been won at a fair by their son Peter in 1956.

Ig Nobel Calling: On Sept. 30, 1999, Sanders Theater at Harvard—yes, Harvard—University hosted the Ig Nobel Prizes, honoring researchers and inventors whose accomplishments "cannot or should not be reproduced." Among the honorees were Takeshi Makino, president of a Japanese detective agency, for helping to develop an "infidelity detection spray that wives can apply to their husband's underwear," and Dr. Arvid Vatle of Norway, for his painstaking study of the kinds of household containers in which his patients submitted urine samples. Four real-life Nobel laureates demonstrated the Ig Nobel environmental protection award winner, a scratch-and-sniff "self-perfuming business suit."

"Dog" Days: On Apr. 17, 1999, while NATO was bombing Belgrade, Yugoslavia's film academy awarded top honors to *Wag the Dog*, a Hollywood satire in which White House aides distract the nation from a presidential sex scandal by fabricating a crisis in the Balkans. Belgrade TV stations broadcast the movie 2 days after the bombing began. On May 10 a Yugoslav cabinet minister, Goran Matic, claimed to reporters that what the Western media portrayed as throngs of terrified Kosovars fleeing their homeland really weren't refugees at all, but a few thousand Albanians hired for $5.50 apiece by the CIA to look like refugees.

Missed Spelling: When Lee Williams wanted a tattoo recently, he went to Eternal Tattoos. He should have brought a dictionary with him, since neither he nor any of the tattoo workers apparently knew how to correctly spell the word "villain." Emblazoned on his right forearm for all to see was "villian" instead. When a friend pointed out the error, Williams, 23, pointed his finger at the tattoo parlor and in Feb. 1999 filed suit for $25,000 in damages. In the meantime, he had plastic surgery to correct the misteak—um, mistake.

The System Worked: A newly hired dispatcher was nabbed by Connecticut state police after he typed in his own name and birth date while being trained on a computer system designed to let police know whether a suspect has any outstanding arrest warrants. The trainee, Gregory Zeoli, 23, was taken into custody after the computer turned up a warrant against him for passing bad checks.

Wanted: Exotic Dancer: An ad ran in the Palm Beach Post in mid-April 1999 seeking exotic dancers willing to move to the town of Stuart, FL, and dance at a nightclub. Not so unusual in itself—but the address to send a resume to was certainly unexpected: Dept. of Labor/Bureau of Workforce Program Support. The reason the Labor Dept. placed the ad is that the club wanted to hire a dancer from outside the U.S., and federal law mandates that the department first see whether any Americans want the job.

Mayo Diet: Eiichi Urata, who set out May 6, 1999, with enough rations for a one-night hike, survived for another 2 weeks on 2 large squeeze-tubes of mayonnaise, after he got lost on frigid Mt. Iwasuge, near Nagano, Japan. Urata, 59, mixed the mayo with snow, making a snow cone that he described as "quite good and tasty."

Miscellaneous Facts, 1999

GPS and 9999: A key event on the high-tech calendar came at 7 p.m. EST on Aug. 21, when the clocks of 24 satellites of the Global Positioning System were reset; outages vexed users of older GPS units, especially in Japan, where thousands of automobile navigation systems failed. On a happier note, computers emerged unscathed Sept. 9, the 9th day of the 9th month of 1999. Some experts had feared that older systems might misinterpret the 9999 date as a software code to shut down.

Worldwide Weapons: The U.S. was the world's leading arms seller in 1998, says a report issued in Aug. 1999 by the Congressional Research Service. New U.S. arms deals totaled $7.1 billion, followed by Germany ($5.5 billion) and France ($3 billion). The leading purchasers were Saudi Arabia ($2.7 billion), United Arab Emirates ($2.5 billion), and Malaysia ($2.1 billion). Developing nations bought more than two-thirds of all weapons sold during 1991-98.

Alms and the Man: Estimates issued in May 1999 by the American Association of Fund-Raising Counsel indicate that U.S. charitable giving rose 11% to a record $175 billion in 1998. Microsoft Chairman Bill Gates (whose stock holdings surpassed $100 billion in 1999) and his wife, Melinda, emerged as the nation's leading philanthropists, establishing a new charitable foundation valued at $17.1 billion in Aug. 1999. The 2 announced a 20-year, $1 billion scholarship program for minority students on Sept. 16.

Health Bulletins: In a comprehensive study of foodborne illnesses, the Centers for Disease Control and Prevention (CDC) concluded in Sept. 1999 that food poisoning afflicts some 76 million Americans a year and causes about 5,000 deaths annually. Earlier CDC reports indicated that measles had been virtually eradicated in the U.S. and that death rates from cardiovascular disease had dropped 60% since 1950. A study released in Aug. by Harvard University researchers reported that women who walk briskly for 3 hours a week can cut their risk of heart disease by up to 40%. Research made public Sept. 7 by the *Journal of the American Medical Association* warned of the dangers of head injuries to young athletes, especially football and soccer players; one report estimated that high school varsity athletes suffered more than 62,000 cases of mild traumatic brain injury each year, with football accounting for 63% of the total.

Press Under Fire: In a survey of 1,001 U.S. adults sponsored by the First Amendment Center of Vanderbilt University, Feb. 26-March 24, 1999, 50% of respondents chose freedom of speech as "most important to American society," while only 6% named freedom of the press. The poll found that 28% of Americans think the First Amendment "goes too far," and 53% believe the nation's press has too much freedom, up from 38% two years earlier.

Pace Setters: In Apr. 1999 a Chilean immigrant, Maria Grasso, working in Boston, struck it rich with a Big Game jackpot worth $197 million, the largest lottery prize ever won by a single individual in the U.S. The baseball Mark McGwire hit for his 70th homer in 1998 sold at auction for over $3 million on Jan. 12, 1999, and a U.S. silver dollar minted in 1834 went for a record $4.14 million on Aug. 30. Gusts from a tornado that whipped through the Oklahoma City area on May 3 were clocked at 318 mph, the fastest wind speed ever recorded on earth. The *Wall Street Journal* reported on July 12 that Billy Mitchell, 33, of Fort Lauderdale, FL, had played the first certified perfect game in Pac-Man history—earning a score of 3,333,360 using only one man.

OBITUARIES

A

Adams, Alice, 72, author of elegant short stories and novels (*Superior Women,* 1984); San Francisco, CA, May 27, 1999.

Agronsky, Martin, 84, broadcast correspondent and developer of the TV program *Agronsky and Company* (1969-87); Washington, DC, July 25, 1999.

B

Bart, Lionel, 68, British songwriter who created the musical *Oliver!* (1960); London, England, Apr. 3, 1999.

Bates, Clayton (Peg Leg), 91, famed one-legged tap dancer; Fountain Inn, SC, Dec. 6, 1998.

Beriosova, Svetlana, 66, Lithuanian-born dancer with the Royal Ballet from the 50s to the 70s; London, England, Nov. 10, 1998.

Bird, Vere, 89, led Antigua to independence and served as prime minister (1981-94); St. John's, Antigua, June 28, 1999.

Bishop, Hazel, 92, chemist/entrepreneur; developed the first long-wearing lipstick; Rye, NY, Dec. 5, 1998.

Blackmun, Harry A., 90, liberal Supreme Court justice (1970-94); authored the 1973 *Roe* v. *Wade* abortion decision; Arlington, VA, Mar. 4, 1999.

Bogarde, (Sir) Dirk, 78, British matinee idol later known for weighty roles in such films as *The Servant* (1963) and *Death in Venice* (1971); London, England, May 8, 1999.

Boxcar Willie (Lecil Travis Martin), 67, country singer/songwriter who styled himself as a hobo; Branson, MO, Apr. 12, 1999.

Bradley, Marion Zimmer, 69, science fiction writer of *The Mists of Avalon* (1983) and the Darkover series; Berkeley, CA, Sept. 25, 1999.

C

Calhoun, Rory, 76, star of numerous westerns and TV's *The Texan* (1958-60); Burbank, CA, Apr. 28, 1999.

Callahan, Harry, 86, photographer known for his sophisticated, complex pictures; Atlanta, GA, Mar. 15, 1999.

Carmichael, Stokely (Kwame Ture), 57, militant civil rights leader; coined the phrase "black power" in the 60s; Conakry, Guinea, Nov. 15, 1998.

Cass, Peggy, 74, comic actress best-known as the unwed pregnant secretary in *Auntie Mame*; New York, NY, Mar. 8, 1999.

Castelli, Leo, 91, influential art dealer; New York, NY, Aug. 21, 1999.

Chamberlain, Wilt, 63, intimidating center (1959-73) perhaps the most dominant player in basketball history; only one to score 100 points in a game; holds the all-time NBA rebounding record; Los Angeles, CA, Oct. 12, 1999.

Chaudhuri, Nirad, 101, Indian-born author of *The Autobiography of an Unknown Indian* (1951) and other provocative works; Oxford, England, Aug. 1, 1999.

Chiles, Lawton, 68, Democratic governor of Florida since 1991; also served in the Senate (1971-89); Tallahassee, FL, Dec. 12, 1998.

Cockerell, Sir Christopher, 88, British engineer who invented the Hovercraft. Southhampton, England, June 1, 1999.

Cody, Iron Eyes, 94, Native American actor famous for shedding a tear in a 1971 "Keep America Beautiful" commercial; Los Angeles, CA, Jan. 4, 1999.

Conrad, Pete, 69, *Apollo 12* commander who was the 3d man to walk on the moon; also flew on 3 other missions; in a motorcycle accident; Ojai, CA, July 8, 1999.

Corby, Ellen, 87, actress who played the grandmother on TV's *The Waltons* in the 70s; Woodland Hills, CA, Apr. 14, 1999.

Corio, Ann, 84, star burlesque queen who toured for 30 years in the show *"This Was Burlesque"*; Englewood, NJ, Mar. 1, 1999.

D

Darrow, Whitney, Jr., 89, witty cartoonist for *The New Yorker* magazine for 50 years; Burlington, VT, Aug. 10, 1999.

Delany, Sarah (Sadie), 109, coauthor of the best-seller *Having Our Say: The Delany Sisters' First 100 Years* (1993); Mount Vernon, NY, Jan. 25, 1999.

DiMaggio, Joe, 84, flawless NY Yankees center fielder (1936-42, 1946-51), one of the greatest players ever; had record 56-game hitting streak, 1941; Hollywood, FL, Mar. 8, 1999.

Dmytryk, Edward, 90, director of films such as *Crossfire* (1947) and *Murder, My Sweet* (1944); later blacklisted; Encino, CA, July 1, 1999.

Drapeau, Jean, 83, mayor of Montreal for nearly 30 years; Montreal, Quebec, Canada, Aug. 12, 1999.

Dubus, Andre, 62, short-story writer whose finest work came after he was disabled in a 1986 car accident; Haverhill, MA, Feb. 24, 1999.

E

Edison, Harry (Sweets), 83, jazz trumpeter with the Count Basie Band; Columbus, OH, July 27, 1999.

Ehrlichman, John, 73, adviser to Richard Nixon and a central figure in the Watergate cover-up; Atlanta, GA, Feb. 14, 1999.

Elion, Gertrude, 81, chemist who shared a 1988 Nobel Prize for work that led to important drugs; Chapel Hill, NC, Feb. 21, 1999.

Ewbank, Weeb, 91, Hall of Fame coach who led the Baltimore Colts (1959) and NY Jets (1969) to league championships; Oxford, OH, Nov. 17, 1998.

Exner, Judith Campbell, 65, reputed mistress of both Pres. John Kennedy and mobster Sam Giancana; Duarte, CA, Sept. 24, 1999.

F

Fadiman, Clifton, 95, essayist and anthologist; helped found the Book-of-the-Month Club and starred on radio's *Information Please;* Sanibel Island, FL, June 20, 1999.

Falk, Lee, 87, creator of the comic strips "The Phantom" and "Mandrake the Magician"; New York, NY, Mar. 13, 1999.

Farmer, Art, 71, be-bop musician who was a master on the trumpet and fluegelhorn; New York, NY, Oct. 4, 1999.

Farmer, James, 79, leading civil rights activist; founded the Congress of Racial Equality in 1942; Fredericksburg, VA, July 9, 1999.

Fell, Norman, 74, actor who played Mr. Roper on TV's *Three's Company* in the 70s; Woodland Hills, CA, Dec. 14, 1998.

Forrest, Helen, 82, big-band singer with Benny Goodman, Harry James, and Artie Shaw; Los Angeles, CA, July 11, 1999.

Funt, Allen, 84, creator and host of TV's long-running *Candid Camera*; Pebble Beach, FL, Sept. 5, 1999.

G

Gaddis, William, 75, author of complex novels, including *The Recognitions* (1955) and *JR* (1975); East Hampton, NY, Dec. 16, 1998.

Galindo, Héctor Alejandro, 93, Mexican film director; Mexico, Feb. 1, 1999.

Garrity, W. Arthur, Jr., 79, federal judge who issued a controversial 1974 order mandating busing to desegregate Boston schools; Wellesley, MA, Sept. 16, 1999.

Godden, Rumer, 90, British writer of novels (*Black Narcissus,* 1939), plays, poems, and children's books; Dumfriesshire, Scotland, Nov. 8, 1998.

Gold, Ernest, 77, composer who won an Oscar for the score of *Exodus* (1960); Santa Monica, CA, Mar. 17, 1999.

Goldwater, John L., 83, creator (with illustrator Bob Montana) of comic book teen Archie Andrews; New York, NY, Feb. 26, 1999.

Gorbachev, Raisa, 67, stylish, high-profile wife of former Soviet Pres. Mikhail Gorbachev; Münster, Germany, Sept. 20, 1999.

Gore, Albert, Sr., 90, powerful Tennessee Democrat in the House (1939-53) and Senate (1953-71); father of Vice Pres. Al Gore; Carthage, TN, Dec. 5, 1998.

Gould, Sandra, 73, character actress who played Gladys Kravitz on TV's *Bewitched* (1966-72); and was the voice of Betty Rubble on *The Flintstones*; Burbank, CA, July 20, 1999.

Greenfield, Meg, 68, influential columnist; longtime editorial page editor of *The Washington Post*; Washington, DC, May 13, 1999.

Grotowski, Jerry, 65, visionary Polish stage director; Pontedera, Italy, Jan. 14, 1999.

H

Hall, Huntz, 78, one of Hollywood's "Dead End Kids"; starred in scores of movies beginning in the 30s; Los Angeles, CA, Jan. 30, 1999.

Hart, Owen, 33, World Wrestling Federation star; died after a fall from an aerial device into the ring; Kansas City, MO, May 23, 1999.

Hassan II, King, 70, ruler of Morocco for 38 years; known as a peacemaker; Rabat, Morocco, July 23, 1999.

Herlihy, Ed, 89, announcer known as the voice of Kraft Foods; narrator of World War II newsreels; emceed *Horn & Hardart's Children's Hour;* New York, NY, Feb. 2, 1999.

Herzberg, Gerhard, 94, German-born chemist who won a 1971 Nobel for work in molecular spectroscopy; Ottawa, Ontario, Mar. 3, 1999.

Hess, Leon, 85, millionaire who built the Amerada Hess oil company and owned the NY Jets; New York, NY, May 7, 1999.

Hirt, Al, 76, popular Dixieland jazz trumpeter; New Orleans, LA, Apr. 27, 1999.

Hobson, Valerie, 81, British actress (*Kind Hearts and Coronets*, 1949); married to disgraced politician John Profumo; London, England, Nov. 13, 1998.

Hodgkin, Sir Alan Lloyd, 84, English neurophysiologist; shared a 1963 Nobel for work on nerve transmission; Cambridge, England, Dec. 20, 1998.

Hume, Cardinal Basil, 76, leader of the Roman Catholic Church in England and Wales since 1976; London, England, June 17, 1999.

Hunter, Jim "Catfish," 53, Hall of Fame pitcher who led the A's and Yankees to 5 World Series titles in the 70s; Hertford, NC, Sept. 9, 1999.

Hussein, King, 63, ruler of Jordan since 1952; credited with creating stability at home and seeking peace with Israel; Amman, Jordan, Feb. 7, 1999.

I

Ivan, Tommy, 88, Hall of Fame hockey coach and executive with the Detroit Red Wings and Chicago Blackhawks; Chicago, IL, June 26, 1999.

J

Johnson, Frank, Jr., 80, federal judge whose rulings helped end Southern segregation; Montgomery, AL, July 23, 1999.

Jones, Henry, 86, award-winning character actor; Los Angeles, CA, May 17, 1999.

K

Kane, Bob, 83, cartoonist who created Batman; Los Angeles, CA, Nov. 3, 1998.

Kanin, Garson, 86, prolific writer and director of plays (*Born Yesterday*, 1946) and films (*Adam's Rib*, 1949); New York, NY, Mar. 13, 1999.

Kelley, DeForest, 79, actor who as Dr. McCoy on *Star Trek* treated the voyagers of the starship *Enterprise;* Woodland Hills, CA, June 11, 1999.

Kendall, Henry W., 72, physicist who shared a 1990 Nobel Prize for confirming the existence of quarks; near Tallahassee, FL, Feb. 15, 1999.

Kennedy, John F., Jr., 38, charismatic son of Pres. Kennedy; an American icon seeking to carve out his own place; founder of *George* magazine; killed with his wife and sister-in-law when his plane crashed off Martha's Vineyard, July 16, 1999.

Khalifa, Sheik Isa bin Salman al-, 65, emir of Bahrain since 1961; Manama, Bahrain, Mar. 6, 1999.

Kiley, Richard, 76, distinguished actor who starred as Don Quixote in the musical *Man of La Mancha* (1965); Middletown, NY, Mar. 5, 1999.

Killanin, Lord, 84, Irish peer who was president of the International Olympic Committee (1972-80); Dublin, Ireland, Apr. 25, 1999.

Kimbro, Henry, 87, star outfielder in the Negro leagues (1937-53); Nashville, TN, July 11, 1999.

Kirkland, Lane, 77, longtime AFL-CIO president (1979-95); Washington, DC, Aug. 14, 1999.

Kraus, Alfredo, 71, Spanish lyric tenor who specialized in the bel canto repertory; Madrid, Spain, Sept. 10, 1999.

Kubrick, Stanley, 70, acclaimed director of such films as *Dr. Strangelove* (1964) and *2001: A Space Odyssey* (1968); Hertfordshire, England, Mar. 7, 1999.

L

Leontief, Wassily, 93, Russian-born economist who won a 1993 Nobel Prize; New York, NY, Feb. 5, 1999.

Lewis, Janet, 99, poet and novelist (*The Wife of Martin Guerre*, 1941); Los Altos, CA, Dec. 1, 1998.

Lowe, Alex, 40, tenacious climber regarded as the world's best mountaineer; swept away by an avalanche; Shisha Pangma, Tibet, Oct. 5, 1999.

M

Mainassara, Ibrahim Bare, 49, president of Niger since 1996; assassinated in Naimey, Niger, Apr. 9, 1999.

Marais, Jean, 84, French star of over 70 films, including Jean Cocteau's *Beauty and the Beast* (1946); Cannes, France, Nov. 8, 1998.

Mars, Forrest, 95, candy magnate who created M&Ms and helped develop Milky Way; Miami, FL, July 1, 1999.

Mature, Victor, 86, brawny star of films such as *Samson and Deliliah* (1949), and *The Robe* (1953); San Diego County, CA, Aug. 4, 1999.

McCann, Donal, 56, leading Irish stage actor; also starred in the film *The Dead* (1987); Dublin, Ireland, July 17, 1999.

Mellon, Paul, 91, philanthropist and arts patron; championed the National Gallery of Art; Upperville, VA, Feb. 1, 1999.

Menuhin, Sir Yehudi, 82, legendary violinist and conductor who began his career as a child prodigy in the 20s; Berlin, Germany, Mar. 12, 1999.

Mizell, Wilmer (Vinegar Bend), 68, country boy who was a major league pitcher (1952-62), later North Carolina congressman; Kerrville, TN, Feb. 21, 1999.

Monsoon, Gorilla (Robert Marella), 62, villainous pro wrestling star from the 60s to the 80s; Willingboro, NJ, Oct. 6, 1999.

Moore, Archie, 84, light-heavyweight boxing champion (1952-62); San Diego, CA, Dec. 9, 1998.

Moore, Brian, 77, Irish-born writer best-known for his novel *The Lonely Passion of Judith Hearne* (1956); Malibu, CA, Jan. 11, 1999.

Morita, Akio, 78, engineer and co-founder of Sony Corp.; spearheaded the internationalization of Japanese business; Tokyo, Japan, Oct. 3, 1999.

Morris, Willie, 64, writer who explored his Southern roots in books like *North Toward Home;* also edited *Harper's Magazine;* Jackson, MS, Aug. 2, 1999.

Motley, Marion, 79, powerful Cleveland Browns fullback (1946-53); among the NFL's first black players; Cleveland, OH, June 27, 1999.

Murdoch, Iris, 79, British writer and philosopher whose novels included *The Sea, the Sea* (1978); Oxford, England, Feb. 8, 1999.

Murray, Kathryn, 92, ballroom dancer who hosted TV's *Arthur Murray Party* with her husband in the 50s; Honolulu, HI, Aug. 6, 1999.

N

Newhouser, Hal, 77, Hall of Fame pitcher (1939-55), mostly with the Detroit Tigers; only pitcher to win 2 consecutive MVP awards; Southfield, MI, Nov. 10, 1998.

Newley, Anthony, 67, multitalented British entertainer who co-wrote, directed, and starred in the musical *Stop the World—I Want to Get Off* (1961); Jensen Beach, FL, Apr. 14, 1999.

Nkomo, Joshua, 82, African nationalist who led Zimbabwe's fight for independence; Harare, Zimbabwe, July 1, 1999.

Norvo, Red, 91, jazz vibraphonist and bandleader; Santa Monica, CA, Apr. 6, 1999.

Nyerere, Julius, 77, founding father of Tanzania and prominent African nationalist; served as his country's first president, 1962-85; London, England, Oct. 14, 1999.

O

Ogilvy, David, 88, British-born founder of the ad agency Ogilvy & Mather; created campaigns for Hathaway shirts and Schweppes; Bonnes, France, July 21, 1999.

P

Pakula, Alan, 70, director of such films as *Klute* (1971), *All the President's Men* (1976), and *Sophie's Choice* (1982); in a highway accident; Suffolk County, NY, Nov. 19, 1998.

Papadopoulos, George, 80, leader of a 1967 Greek military coup; later convicted of treason and insurrection; Athens, Greece, June 27, 1999.

Paterson, Jennifer, 71, eccentric British costar of the TV cooking show *Two Fat Ladies;* London, England, Aug. 10, 1999.

Perrot, Kim, 32, star guard with the WNBA-champion Houston Comets; Houston, TX, Aug, 19, 1999.

Plato, Dana, 34, actress who starred in TV's *Diff'rent Strokes* (1978-84); of an apparent drug overdose; Moore, OK, May 8, 1999.

Pritzker, Jay, 76, billionaire who built the Hyatt hotel chain; Chicago, IL, Jan. 23, 1999.

Puzo, Mario, 78, author of *The Godfather* (1969) and its sequels and Oscar-winning film adaptations; Bay Shore, NY, July 2, 1999.

Q

Quarry, Jerry, 53, heavyweight contender of the 60s and 70s; later suffered from boxing-induced brain damage; Templeton, CA, Jan. 3, 1999.

Quintero, José, 74, Panamanian-born director celebrated for staging the plays of Eugene O'Neill; New York, NY, Feb. 26, 1999.

R

Reed, Oliver, 61, British actor who starred in such films as *Oliver!* (1968) and *Women in Love* (1969); Valetta, Malta, May 2, 1999.

Reese, Pee Wee, 81, Hall of Fame shortstop for the Brooklyn Dodgers who eased Jackie Robinson's entry into the majors; Louisville, KY, Aug. 14, 1999.

Ripken, Cal, Sr., 63, longtime Baltimore Orioles coach; father of star player Cal Ripken Jr.; Baltimore, MD, Mar. 25, 1999.

Rodbell, Martin, 73, biochemist who shared a 1994 Nobel Prize for work on cellular communication; Chapel Hill, NC, Dec. 7, 1998.

Rodrigo, Joaquín, 97, major composer of Spanish classical music, including "Concierto de Aranjuez"; Madrid, Spain, July 6, 1999.

Rodríguez, Claudio, 65, Spanish writer of lyrical poems infused with symbolism; Madrid, Spain, July 22, 1999.

Rogers, Buddy, 94, star of the classic film *Wings* (1927); married to actress Mary Pickford; Rancho Mirage, CA, Apr. 21, 1999.

Rolle, Esther, 78, actress who played Florida Evans on TV's *Maude* and *Good Times* in the 70s; Los Angeles, CA, Nov. 17, 1998.

Roman, Ruth, 75, glamorous star of the films *Champion, Strangers on a Train,* and *Colt .45;* Laguna Beach, CA, Sept. 9, 1999.

Rountree, Martha, 87, co-creator and 1st moderator of *Meet the Press;* Washington, DC, Aug. 23, 1999.

S

Sabines Gutiérrez, Jaime, 72, popular Mexican poet of lyrical, vivid works; Mexico City, Mexico, Mar. 19, 1999.

St. Cyr, Lili, 80, striptease artist known for her onstage bubble baths; Hollywood, CA, Jan. 29, 1999.

Sarazen, Gene, 97, top golfer; first to win all 4 major championships; Naples, FL, May 13, 1999.

Schawlow, Arthur, 77, physicist who shared a 1981 Nobel Prize for developing the laser; Palo Alto, CA, Apr. 28, 1999.

Scott, George C., 71, intense award-winning actor known for daring performances in such films as *The Hustler* (1961), *Dr. Strangelove* (1964), and *Patton* (1970); Westlake Village, CA, Sept. 22, 1999.

Seaborg, Glenn T., 86, Nobel Prize-winning scientist who led the 1941 team that created plutonium; Lafayette, CA, Feb. 25, 1999.

Semon, Waldo, 100, B.F. Goodrich chemist who invented vinyl and guided the development of synthetic rubber; Hudson, OH, May 26, 1999.

Señor Wences, (Wenceslao Moreño), 103, Spanish-born ventriloquist who was a regular on TV variety shows; New York, NY, Apr. 20, 1999.

Shaw, Robert, 82, choral conductor and music director of the Atlanta Symphony (1967-88); New Haven, CT, Jan. 25, 1999.

Sidney, Sylvia, 88, actress who starred in such films as *Dead End* (1937) and *Summer Wishes, Winter Dreams* (1973); New York, NY, July 1, 1999.

Silverstein, Shel, 66, writer/cartoonist best known for children's poetry books (*Where the Sidewalk Ends,* 1974; *A Light in the Attic,* 1981); Key West, FL, found dead May 9, 1999.

Siskel, Gene, 53, influential film critic paired with Roger Ebert in a popular movie-review team on TV; Evanston, IL, Feb. 20, 1999.

Springfield, Dusty, 59, British pop singer whose 60s hits included "I Only Want to Be With You"; Henley-on-Thames, England, Mar. 2, 1999.

Stanky, Eddie, 82, aggressive infielder who played on 3 pennant-winning teams in the 40s and 50s; Fairhope, AL, June 6, 1999.

Stein, Herbert, 83, free-market economist and prominent Nixon adviser; Washington, DC, Sept. 8, 1999.

Steinberg, Saul, 84, Romanian-born artist whose whimsical drawings appeared in *The New Yorker* since 1941; New York, NY, May 12, 1999.

Stone, Jesse, 97, influential record producer; wrote the 1954 hit "Shake, Rattle and Roll"; Altamonte Springs, FL, Apr. 1, 1999.

Strasberg, Susan, 60, stage and film actress lauded at 17 for playing Anne Frank on Broadway; New York, NY, Jan. 21, 1999.

T

Talbert, Bill, 80, tennis champ who won 8 doubles titles at the U.S. Open in the 40s; New York, NY, Feb. 28, 1999.

Tilberis, Liz, 51, British-born editor of *Harper's Bazaar* who wrote about her battle with ovarian cancer; New York, NY, Apr. 21, 1999.

Torme, Mel, 73, velvet-voiced pop-jazz vocalist; also a prolific songwriter; Los Angeles, CA, June 5, 1999.

Trow, Bob, 72, actor who played Bob Dog and other characters on TV's *Mister Rogers' Neighborhood;* New Alexandria, PA, Nov. 2, 1998.

Turner, Clyde (Bulldog), 79, football Hall of Fame center and linebacker; Galveston, TX, Oct. 30, 1998.

U

Udall, Morris K., 76, Arizona congressman (1961-91) who was a leading liberal and environmentalist; Washington, DC, Dec. 12, 1998.

V

Vander Pyl, Jean, 79, voice of Wilma Flintstone and other cartoon characters; Dana Point, CA, Apr. 10, 1999.

Vaughan, Frankie, 71, popular English singer known as Mr. Moonlight; Buckinghamshire, England, Sept. 17, 1999.

W

Webster, Katie, 63, blues singer and pianist known as the Swamp Boogie Queen; League City, TX, Sept. 5, 1999.

West, Morris, 83, Australian writer of the popular novels *The Devil's Advocate* (1959) and *The Shoes of the Fisherman* (1963); Sydney, Australia, Oct. 9, 1999.

Whitney, Ruth, 71, feminist editor of *Glamour* magazine (1967-98); Irvington, NY, June 4, 1999.

Whyte, William H., 81, author of the 1956 best-seller *The Organization Man*; New York, NY, Jan. 12, 1999.

Williams, Joe, 80, legendary jazz singer known for "Every Day (I Have the Blues)"; Las Vegas, NV, Mar. 29, 1999.

Wilson, Flip, 64, comedian famous for his outrageous characters; first black entertainer to host a network TV variety show; Malibu, CA, Nov. 25, 1998.

Wisdom, John Minor, 93, pioneering judge whose opinions helped end segregation in the 60s; New Orleans, LA, May 15, 1999.

Wong, Martin, 53, New York painter known for his visionary realism; San Francisco, CA, Aug. 12, 1999.

Wynn, Early, 79, Hall of Fame pitcher who won 300 games in a 23-year career; Venice, FL, Apr. 4, 1999.

Z

Zaslow, Michael, 54, longtime soap opera star on *Guiding Light* and *One Life to Live;* New York, NY, Dec. 6, 1998.

U.S. PRESIDENTIAL CAMPAIGN 2000

Source: Democratic National Committee; Republican National Committee; Federal Election Commission; as of Oct. 20, 1999

The 2 leading U.S. political parties are the Democrats, who won 49.25% of the popular vote for president in 1996, and the Republicans, who won 40.73%. Presidential nominees for 2000 will be selected at the Republican National Convention, held at the First Union Center in Philadelphia, July 29-Aug. 4, 2000, and at the Democratic National Convention, held at the Staples Center in Los Angeles, Aug. 14-17, 2000. The general election takes place on Nov. 7, 2000.

Debates: The following presidential candidate forums, town meetings, and debates have been tentatively scheduled for the 2000 primary season; many will be televised nationally, most often by C-SPAN and CNN. Republicans: Jan. 6, Durham, NH; Jan. 7, SC (site to be selected); Jan. 10, Cedar Rapids, IA; Jan. 15, Johnston, IA; Jan. 26, Manchester, NH; Feb. 15, Columbia, SC; Mar. 2, Los Angeles, CA. Democrats: Jan. 5, Durham, NH; Jan. 8, Johnston, IA; Jan. 12, Cedar Rapids, IA; Jan. 27, Manchester, NH; Mar. 1, Los Angeles, CA.

How Delegates Are Chosen: Each party allocates convention delegates among the states according to a complex formula that takes into account the state's population, voting history, and other factors. The District of Columbia, Puerto Rico, American Samoa, Guam, and the U.S. Virgin Islands (and, for the Democrats, a voting delegation of Democrats Abroad) are also represented. Preliminary allocations for the Democratic National Convention called for 4,367 individual delegates to cast a total of 4,336 delegate votes (some delegates cast fractional votes). The Republican National Convention was expected to have 2,066 delegates. Most delegates are chosen through caucuses (meetings of voters or officials) or in primary elections. (However, some primaries or caucuses do not result in a choice of delegates bound to a particular candidate.) Since 1992, many states have moved their presidential primaries or caucuses to dates earlier in the year. As a result, in the year 2000 about 70% of the delegates to the national conventions will be selected by the end of March.

How to Read This Table: The table lists by date the presidential primaries and caucuses for the 2000 presidential election, along with the number of delegates chosen in each state or territory. Dates and requirements for most primaries are set by state legislatures; in a few cases, however, the party runs its own primary. Delegate counts (abbreviated "Dels.") reflect the total voting strength at the national convention; this total may exceed, often by a substantial margin, the number of any delegates chosen on the date cited. All dates and delegate totals are subject to change.

DATE	State/Terr.	Democrats Dels.	Type	Republicans Dels.	Type
1/15	LA			28	Caucus
1/24	AK			23	Caucus/Straw Poll
	IA	56	Caucus	25	Caucus
2/1	NH	29	Primary	17	Primary
2/5	DE	22	Primary		
2/8	DE			12	Primary
2/19	SC			37	Primary
2/22	AZ			30	Primary
	MI			58	Primary
2/26	Am. Samoa	4	Caucus/Exec. Comm.		
	Guam	4	Caucus/Conv.		
	Virgin Is.	4	Caucus/Conv.		
2/27	PR	14	Primary		
2/29	ND	19	Caucus		
	VA	55	Primary		
	WA	94	Primary	37	Primary
3/7	CA	434	Primary	162	Primary
	CT	67	Primary	25	Primary
	GA	92	Primary	54	Primary
	HI	33	Caucus		
	ID	23	Caucus		
	ME	32	Primary	14	Primary
	MD	92	Primary	31	Primary
	MA	118	Primary	37	Primary
	MN	91	Caucus		
	MO	92	Primary	35	Primary
	NY	294	Primary	101	Primary
	ND	22	Caucus		
	OH	170	Primary	69	Primary
	RI	32	Primary	14	Primary
	VT	22	Primary	12	Primary
	Am. Samoa	6	Caucus/Conv.		
3/9	SC	52	Primary		
3/10	CO	61	Primary	40	Primary
	UT	29	Primary	29	Primary
	WY	18	Caucus	22	Primary

DATE	State/Terr.	Democrats Dels.	Type	Republicans Dels.	Type
3/11	AZ	55	Primary		
	MI	157	Caucus		
3/12	NV	30	Caucus		
3/14	FL	185	Primary	80	Primary
	LA	74	Primary		
	MS	46	Primary	33	Primary
	OK	52	Primary	38	Primary
	TN	81	Primary	37	Primary
	TX	231	Primary Caucus	124	Primary
3/21	IL	189	Primary	74	Primary
3/25	AK	19	Caucus		
3/26	PR	59	Primary		
4/1	Virgin Is.	6	Caucus		
4/4	KS	42	Primary	35	Primary
	WI	92	Primary	37	Primary
4/15, 4/17	VA	99	Caucus		
4/25	MN			34	Caucus/Conv.
	PA	191	Primary	80	Primary
5/2	DC	33	Primary	15	Primary
	IN	89	Primary	55	Primary
	NC	103	Primary	62	Primary
5/6	Guam	6	Caucus/Conv.		
5/9	NE	32	Primary	30	Primary
	WV	42	Primary	18	Primary
5/16	OR	58	Primary	24	Primary
5/19	HI			14	Caucus/Conv.
5/23	AR	48	Primary	24	Primary
	ID			28	Primary
	KY	58	Primary	31	Primary
5/25	NV			17	Caucus/Conv.
6/6	AL	63	Primary	44	Primary
	MT	24	Primary	23	Primary
	NJ	124	Primary	54	Primary
	NM	35	Primary	21	Primary
	SD	22	Primary	22	Primary

Other U.S. Political Parties

(Dates and details subject to change.)

Constitution Party. Formerly U.S. Taxpayers Party. Share of 1996 pres. vote: 0.19%. Convention held Sept. 1-6, 1999, St. Louis, MO; nominated Howard Phillips (pres.) and Joseph Sobran (vice pres.).

Green Party. Share of 1996 pres. vote: 0.68%. Association of State Green Parties convention planned for June 24-25, 2000, Denver, CO.

Libertarian Party. Share of 1996 pres. vote: 0.50%. Convention planned for June 30-July 4, 2000, Anaheim, CA.

Natural Law Party. Share of 1996 pres. vote: 0.12%. Convention planned for July 2000, Washington, DC.

Reform Party. Share of 1996 pres. vote: 8.40%. Convention planned for Aug. 10-13, 2000, Long Beach, CA.

UNITED STATES GOVERNMENT

EXECUTIVE BRANCH	LEGISLATIVE BRANCH	JUDICIAL BRANCH
PRESIDENT **Vice President** **Executive Office of the President** White House Office Office of the Vice President Council of Economic Advisers Council on Environmental Quality National Security Council Office of Administration Office of Management and Budget Office of National Drug Control Policy Office of Policy Development Office of Science and Technology Policy Office of the U.S. Trade Representative	**CONGRESS** **Senate House** Architect of the Capitol U.S. Botanic Garden General Accounting Office Government Printing Office Library of Congress Congressional Budget Office Tax Court	**Supreme Court of the United States** Courts of Appeals District Courts Territorial Courts Court of International Trade Court of Federal Claims Court of Appeals for the Armed Forces Court of Veterans Appeals Administrative Office of the Courts Federal Judicial Center Sentencing Commission

The Clinton Administration

As of Oct. 1999; mailing addresses are for Washington, DC.

Terms of office of the president and vice president: Jan. 20, 1997, to Jan. 20, 2001.

President — Bill Clinton receives an annual salary of $200,000 (taxable), and an annual expense allowance of $50,000 (nontaxable) for costs resulting from official duties. In addition, up to $100,000 a year may be spent on travel expenses and $19,000 on official entertainment (both nontaxable), available for allocation within the Executive Office of the President.
Website: http://www.whitehouse.gov/WH/EOP/html/ OP_Home.html
E-mail: president@whitehouse.gov

Vice President — Al Gore receives an annual salary of $181,400 (as of Jan. 1, 2000), plus $10,000 for expenses, all taxable.
Website: http://www.whitehouse.gov/WH/EOP/OVP/ index.html
E-mail: vice.president@whitehouse.gov

The Cabinet Department Heads

(Salary: $157,000 per year, as of Jan. 1, 2000)

Secretary of State — Madeleine K. Albright
Secretary of the Treasury — Lawrence H. Summers
Secretary of Defense — William S. Cohen
Attorney General — Janet Reno
Secretary of the Interior — Bruce Babbitt
Secretary of Agriculture — Dan Glickman
Secretary of Commerce — William M. Daley
Secretary of Labor — Alexis M. Herman
Secretary of Health and Human Services — Donna E. Shalala
Secretary of Housing and Urban Development — Andrew M. Cuomo
Secretary of Transportation — Rodney E. Slater
Secretary of Energy — Bill Richardson
Secretary of Education — Richard W. Riley
Secretary of Veterans Affairs — Togo D. West Jr.

The White House Staff

1600 Pennsylvania Ave. NW 20500
Website: http://www.whitehouse.gov

Chief of Staff to the President — John Podesta
Asst. to the President & Deputy Chief of Staff — Maria Echaveste
Asst. to the President & Deputy Chief of Staff — Stephen J. Ricchetti
Senior Adviser on Policy & Strategy — Douglas B. Sosnik
Assistants to the President:
　Counsel to the President — Beth Nolan
　Deputy Counsel to the President — Bruce Lindsey
　Domestic Policy Council — Bruce Reed
　　Office of National AIDS Policy — Sandy Thurman, dir.
　Presidential Personnel — Bob Nash
　Press Secretary — Joseph Lockhart
　Legislative Affairs — Lawrence Stein
　Communications — Loretta Ucelli/Sidney Blumenthal
　National Economic Policy — Gene Sperling

　Intergovernmental Affairs — Mickey Ibarra
　National Security — Samuel R. Berger
　Staff Secretary — Sean Maloney
　Political Affairs — Minyon Moore
　Public Liaison — Mary Beth Cahill
　Management & Administration — Mark F. Lindsay
　Counselor to the President — Ann Lewis
　Cabinet Secretary — Thurgood Marshall Jr.
　Director of Presidential Scheduling — Stephanie Streett
　Director of Speechwriting — J. Terry Edmonds
　Chief of Staff to the First Lady — Melanne Verveer
　　E-mail: first.lady@whitehouse.gov
　Senior Advisor for Policy & Communications — Joel Johnson
　Director of Advance — Dan Rosenthal
　Director of the President's Initiative for One America — R. Ben Johnson
　Special Envoy for the Americas — Kenneth (Buddy) MacKay
　Counselor to the Chief of Staff — Karen Tramontano

Executive Agencies

Council of Economic Advisers — Martin Baily, chair
Website: http://www.whitehouse.gov/WH/EOP/CEA/ html/index.html
Office of Administration — Michael J. Lyle, act. dir.
Website: http://www.whitehouse.gov/WH/EOP/html/ other/OA.html
Office of Science & Technology Policy — Neal F. Lane
Website: http://www.whitehouse.gov/WH/EOP/OSTP/ html/OSTP_Home.html
Office of Nat. Drug Control Policy — Barry R. McCaffrey
Website: http://www.whitehousedrugpolicy.gov
Office of Management and Budget — Jacob J. Lew, dir.
Website: http://www.whitehouse.gov/omb
U.S. Trade Representative — Charlene Barshefsky
Website: http://www.ustr.gov
Council on Environ. Quality — George Frampton, chair
Website: http://www.whitehouse.gov/CEQ

Department of State

2201 C St. NW 20520
Website: http://www.state.gov

Secretary of State — Madeleine K. Albright
Deputy Secretary — Strobe Talbott
Chief of Staff — Elaine K. Shocas
U.S. Ambassador to the United Nations — Richard C. Holbrooke
Under Sec. for Political Affairs — Thomas R. Pickering
Under Sec. for Management — Bonnie R. Cohen
Under Sec. for Global Affairs — Frank E. Loy
Under Sec. for Economic, Business, & Agricultural Affairs — vacant
Under Sec. for Arms Control & International Security Affairs — John D. Holum
Policy Planning Director — Morton Halperin
Chief of Protocol — Mary Mel French

Inspector General — Jacqueline L. Williams-Bridgers
Legal Adviser — David R. Andrews
Director General of the Foreign Service & Director of Personnel — Edward W. Grehm Jr.
Assistant Secretaries for:
  Administration — Patrick F. Kennedy
  African Affairs — Susan E. Rice
  Consular Affairs — Mary A. Ryan
  Democracy, Human Rights, & Labor — Harold Koh
  Diplomatic Security — David Carpenter
  East Asian & Pacific Affairs — Stanley Roth
  Economic & Business Affairs — Alan Larson
  European & Canadian Affairs — Marc Grossman
  Intelligence & Research — Donald W. Keyser, act.
  Inter-American Affairs — Peter Romero, act.
  International Narcotics & Law — Rand Beers
  International Organization Affairs — David Welch
  Legislative Affairs — Barbara Larkin
  Near Eastern Affairs — Martin S. Indyk
  Oceans, International Environmental, & Scientific Affairs — Melinda Kimble, act.
  Politico-Military Affairs — Eric D. Newson
  Population, Refugees, & Migration — Julia V. Taft
  Public Affairs — James P. Rubin
  South Asian Affairs — Karl Inderfurth

Department of the Treasury

1500 Pennsylvania Ave. NW 20220
Website: http://www.ustreas.gov

Secretary of the Treasury — Lawrence H. Summers
Deputy Sec. of the Treasury — Stu Eizenstat
Under Sec. for Domestic Finance — Gary Gensler
Under Sec. for International Affairs — Tim Geithner
Under Sec. for Enforcement — James Johnson
General Counsel — Neal Wolin, act.
Inspector General — Jeffrey Rush
Inspector General for Tax Administration — David Williams
Assistant Secretaries for:
  Economic Policy — David Wilcox
  Enforcement — Elizabeth Bresee
  Financial Institutions — Richard Carnell
  Fiscal Affairs — Donald Hammond
  International Affairs — Ted Truman
  Legislative Affairs — Linda Robertson
  Management — Nancy Killefer
  Public Affairs — vacant
  Tax Policy — vacant
Treasurer of the U.S. — Mary Ellen Withrow
Bureaus:
  Alcohol, Tobacco, & Firearms — John W. Magaw, dir.
  Comptroller of the Currency — John Hawke, comm.
  Customs — Raymond W. Kelly, comm.
  Engraving & Printing — Tom Ferguson, dir.
  Federal Law Enforcement Training Center — W. Ralph Basham, dir.
  Financial Management Service — Richard Gregg, comm.
  Internal Revenue Service — Charles Rossotti, comm.
  Mint — Philip N. Diehl, dir.
  Office of Thrift Supervision — Ellen S. Seidman
  Public Debt — Van Zeck, comm.
  U.S. Secret Service — Brian L. Stafford, dir.

Department of Defense

The Pentagon 20301
Website: http://www.defenselink.mil

Secretary of Defense — William S. Cohen
Deputy Secretary — John J. Hamre
Under Sec. for Acquis. and Technol. — Jacques S. Gansler
Under Sec. for Personnel & Readiness — Rudy de Leon
Under Sec. for Policy — Walter B. Slocombe
Assistant Secretaries for:
  Command, Control, Communications, & Intelligence — Arthur Money
  Force Management — Frank Rush, act.
  Health Affairs — Dr. Sue Bailey
  International Security Affairs — Franklin D. Kramer
  Legislative Affairs — John K. Veroneau

Public Affairs — Kenneth H. Bacon
Reserve Affairs — Charles Cragin, act.
Spec. Operations & Low-Intensity Conflict — Brian Sheridan
Strategy & Threat Reduction — Edward L. Warner III
Program Analysis & Evaluation — Robert Soule, dir.
Inspector General — Donald Mancuso, act.
Comptroller — William J. Lynn III
General Counsel — Judith A. Miller
Intelligence Oversight — George Lotz
Operational Test & Evaluation — Phillip E. Coyle III, dir.
Chairman, Joint Chiefs of Staff — Gen. Henry H. Shelton
Secretary of the Army — Louis Caldera
Secretary of the Navy — Richard Danzig
Commandant of the Marine Corps — James Jones
Secretary of the Air Force — F. Whitten Peters

Department of Justice

Constitution Ave. & 10th St. NW 20530
Website: http://www.usdoj.gov

Attorney General — Janet Reno
Deputy Attorney General — Eric H. Holder Jr.
Associate Attorney General — Raymond C. Fisher
Office of Dispute Resolution — Peter R. Steenland Jr.
Solicitor General — Seth P. Waxman
Office of Inspector General — Robert L. Ashbaugh, act.
Assistants:
  Antitrust Division — Joel I. Klein
  Civil Division — David W. Ogden, act.
  Civil Rights Division — Bill Lann Lee, act.
  Criminal Division — James K. Robinson
  Environ. & Nat. Resources Division — Lois J. Schiffer
  Justice Programs — Laurie Robinson
  Legal Counsel — Randolph D. Moss, act.
  Policy Development — Eleanor D. Acheson
  Legislative Affairs — Jon P. Jennings, act.
  Administration — Stephen R. Colgate
  Tax Division — Loretta C. Argrett
Executive Secretariat — Anna-Marie Kilmade Gatons, dir.
Office of Investigative Agency Policies — vacant
Office of Public Affairs — Myron Marlin, dir.
Office of Information & Privacy — Richard L. Huff/Daniel J. Metcalfe
Community Oriented Policing Services — Mary Lou Leary, act. dir.
Federal Bureau of Investigation — Louis J. Freeh, dir.
Exec. Off. for Immigration Review — Kevin D. Rooney, dir.
Bureau of Prisons — Kathleen Hawk Sawyer, dir.
Community Relations Service — Rose M. Ochi, dir.
Drug Enforcement Admin. — Donnie R. Marshall, act.
Office of Intelligence Policy & Review — Fran Fragos Townsend, counsel
Office of Professional Responsibility — H. Marshall Jarrett, counsel
Exec. Off. for U.S. Trustees — Joseph Patchan, dir.
Foreign Claims Settlement Comm. — vacant
Exec. Office for U.S. Attorneys — Mary H. Murguia, dir.
Immigration & Naturalization Service — Doris Meissner, comm.
Pardon Attorney — Roger C. Adams
U.S. Parole Commission — Michael J. Gaines, chair
U.S. Marshals Service — George R. Havens, act. dir.
U.S. Natl. Cen. Bureau of INTERPOL — John J. Imhoff, chief
Office of Intergovernmental Affairs — Brian de Vallance, dir.
Office of Tribal Justice — Mark VanNorman, dir.
Violence Against Women Act — Bonnie Campbell, dir.
National Drug Intelligence Center — Michael T. Horn, dir.

Department of the Interior

1849 C St. NW 20240
Website: http://www.doi.gov

Secretary of the Interior — Bruce Babbitt
Deputy Secretary — David Hayes, act.
Assistant Secretaries for:
  Fish, Wildlife, & Parks — Donald Barry
  Indian Affairs — Kevin Gover
  Intergovernmental Affairs — Grace Garcia

Land & Minerals — Sylvia Baca, act.
Policy, Management, & Budget — M. John Berry
Water & Science — Patricia J. Beneke
Bureau of Land Management — Patrick Shea, dir.
Bureau of Reclamation — Eluid L. Martinez, comm.
Fish & Wildlife Service — Jamie Rappaport Clark, dir.
Geological Survey — Thomas Casadevall, act. dir.
Mineral Management Service — Walt Rosenbusch, dir.
National Park Service — Robert G. Stanton, dir.
Surf. Mining Reclam. & Enforcement — Kathy Karpan, dir.
Communications — Michael Gauldin, dir.
Off. of Congressional & Legislative Affairs — Lenna Aoki
Solicitor — John D. Leshy
External Affairs — Jana Prewitt
Exec. Secretariat & Regulatory Affairs — Julie Faulkner

Department of Agriculture

1400 Independence Ave. SW 20250
Website: http://www.usda.gov

Secretary of Agriculture — Dan Glickman
Deputy Secretary — Richard Rominger
Under Secretaries for:
  Farm & Foreign Agric. Services — Gus Schumacher Jr.
  Food, Nutrition, & Consumer Services — Shirley R. Watkins
  Food Safety — Catherine Woteki
  Marketing & Regulatory Programs — Michael Dunn
  Natural Resources & Environment — Jim Lyons
  Research, Education, & Economics — Miley Gonzalez
  Rural Development — Jill Long Thompson
Assistant Secretaries for:
  Administration — Sally Thompson, act.
  Congressional Relations — Andrew C. Fish
General Counsel — Charlie Rawls
Inspector General — Roger C. Viadero
Chief Financial Officer — Sally Thompson
Chief Information Officer — Anne F. Thomson Reed
Chief Economist — Keith Collins
Communications — Sedelta Verble, dir.
Press Secretary — Andrew Solomon

Department of Commerce

14th St. between Constitution &
Pennsylvania Ave. NW 20230
Website: http://www.doc.gov

Secretary of Commerce — William M. Daley
Deputy Secretary — Robert Mallett
Chief of Staff — David Lane
General Counsel — Andrew Pincus
Assistant Secretaries:
  Chief Financial Officer & Asst. Secretary for Administration — vacant
  Economic Development Admin. — Phillip Singerman
  Export Admin. — Chester Straub Jr., act.
  Export Enforcement — F. Amanda Debusk
  Import Administration — Robert LaRussa
  Legislative Affairs — Deborah Kilmer
  Market Access & Compliance — Patrick Mulloy
  National Telecomm. Information Administration — Clarence Irving Jr.
  Oceans & Atmosphere — Terry Garcia
  Patent & Trademark Office & Comm. — Q. Todd Dickinson, act.
  Trade Development — Michael Copps
  U.S. & Foreign Commercial Service — Awilda Marquez
Bureau of the Census — Kenneth Prewitt, dir.
Under Sec. for Oceans & Atmosphere — D. James Baker
Under Sec. for Export Admin. — William Reinsch
Under Sec. for International Trade — David Aaron
Under Sec. for Econ. Affairs — Robert Shapiro
Under Sec. for Technology — Gary Bachula, act.
Natl. Institute for Standards & Tech. — Raymond Kammer, dir.
Minority Business Dev. Agency — Courtland Cox, dir.
Public Affairs/Press Secretary — Maurice Goodman

Department of Labor

200 Constitution Ave. NW 20210
Website: http://www.dol.gov

Secretary of Labor — Alexis M. Herman
Deputy Secretary — Edward Montgomery, act.
Chief of Staff — Lee Satterfield
Assistant Secretaries for:
  Admin. & Management — Patricia W. Lattimore
  Congressional & Intergov. Affairs — Geri Palast
  Employment & Training — Ray Bramucci
  Employment Standards — Bernard E. Anderson
  Occupational Safety & Health — Charles Jeffress
  Mine Safety & Health — Davitt McAteer
  Pension & Welfare Benefits — Rick McGahey
  Policy — Susan Green, act.
  Public Affairs — Howard Waddell, act.
  Veterans Employment & Training — Al Borrego
Solicitor of Labor — Henry Solano
Bureau of International Affairs — Andrew Samet, act.
Women's Bureau — Dolores Crockett, act.
Inspector General — Charles C. Masten
Bureau of Labor Statistics — Katharine G. Abraham

Department of Health and Human Services

200 Independence Ave. SW 20201
Website: http://www.os.dhhs.gov

Secretary of Health & Human Services — Donna E. Shalala
Deputy Secretary — Kevin L. Thurm
Chief of Staff — Mary Beth Donahue
Assistant Secretaries for:
  Health — David Satcher
  Legislation — Richard J. Tarplin
  Management & Budget — John J. Callahan
  Planning & Evaluation — Margaret Ann Hamburg
  Public Affairs — Melissa Skolfield
  Aging — Jeanette C. Takamura
  Children & Families — Olivia Golden
General Counsel — Harriet S. Rabb
Inspector General — June Gibbs Brown
Office of Civil Rights — Thomas E. Perez, dir.
Surgeon General — David Satcher
Health Care Financing Admin. — Nancy-Ann DeParle

Department of Housing and Urban Development

451 7th St. SW 20410
Website: http://www.hud.gov

Secretary of Housing & Urban Development — Andrew M. Cuomo
Deputy Secretary — Saul Ramirez
Assistant Secretaries for:
  Administration — Joseph Smith
  Community Planning & Development — Cardell Cooper
  Fair Housing & Equal Opportunity — Eva Plaza
  Housing & Federal Housing Comm. — William Apgar
  Cong. & Intergov. Relations — Halbert C. DeCell III
  Policy Development & Research — Xavier Briggs
  Public Affairs — Ginny Terzano, act.
  Public & Indian Housing — Harold Lucas
General Counsel — Gail Laster
Inspector General — Susan M. Gaffney
Chief Financial Officer — vacant
Government National Mortgage Assn. — George Anderson
Off. of Federal Housing Enterprise Oversight — vacant

Department of Transportation

400 7th St. SW 20590
Website: http://www.dot.gov

Secretary of Transportation — Rodney E. Slater
Deputy Secretary — Mortimer L. Downey
Assistant Secretaries for:
  Administration — Melissa Allen
  Budget & Programs — Jack Basso
  Governmental Affairs — Michael Frazier, act.
  Aviation & International Affairs — Bradley Mims, act.
  Transportation — Eugene Conti

Public Affairs — William Schulz
U.S. Coast Guard Commandant — Adm. James M. Loy
Federal Aviation Admin. — Jane Garvey
Federal Highway Admin. — Kenneth Wykle
Federal Railroad Admin. — Jolene Molitoris
Maritime Admin. — Clyde Hart
Natl. Highway Traffic Safety Admin. — Ricardo Martinez
Federal Transit Admin. — Gordon J. Linton
Research & Special Programs Admin. — Kelley Coyner
St. Lawrence Seaway Devel. Corp. — Albert Jacques

Department of Energy
1000 Independence Ave. SW 20585
Website: http://home.doe.gov

Secretary of Energy — Bill Richardson
Deputy Secretary — T. J. Glauthier
Under Secretary — Ernest I. Moniz
Chief of Staff — Gary Falle
Deputy Chief of Staff for Intl. Policy — Rebecca Gaghen
Deputy Chief of Staff for Administration & Domestic Policy — LeeAnn Inadomi
General Counsel — Maryann Sullivan
Inspector General — John C. Layton
Assistant Secretaries for:
  Congressional & Intergov. Affairs — John Angell
  Energy Efficiency & Renewable Energy — Dan Reicher
  Defense Programs — Thomas F. Gioconda, act.
  Policy — Mark Mazur, dir.
  International Affairs — David Goldwyn
  Environmental Restoration & Waste Management — Carolyn Huntoon
  Administration & Human Resource Management — Richard Farrell
  Environment, Safety, & Health — Dr. David Michaels
  Fossil Energy — Robert Gee
Nuclear Energy — Bill Magwood, dir.
Energy Information Admin. — Jay E. Hakes, adm.
Economic Impact & Diversity — Sarah Summerville, dir.
Hearings & Appeals — George Breznay, dir.
Energy Research — Martha Krebs, dir.
Civilian Radioactive Waste Management — Lake H. Barrett, act. dir.
Nonproliferation & National Security — Rose Gottemoller
Chief Financial Officer — Mike Telson
Energy Advisory Board — vacant
Office of Public Affairs — Brooke Anderson, dir.

Department of Education
400 Maryland Ave., SW 20202
Website: http://www.ed.gov

Secretary of Education — Richard W. Riley
Deputy Secretary — Marshall S. Smith, act.
Chief of Staff — Leslie T. Thornton
Inspector General — Lorraine P. Lewis
General Counsel — Judith Winston
Assistant Secretaries for:
  Adult & Vocational Education — Patricia McNeil
  Civil Rights — Norma V. Cantu
  Educational Research & Improvement — Kent McGuire
  Elementary & Secondary Educ. — Judith Johnson, act.
  Intergov. & Interagency Affairs — Mario Moreno
  Legislative & Congressional Affairs — Scott Fleming
  Postsecondary Education — Claudio Prieto, act.
  Special Educ. & Rehab. Services — Judith Heumann
Bilingual Education & Minority Language Affairs — Art Love, act. dir.
Rehab. Services Admin. — Frederic K. Schroeder, comm.
Education Statistics — Gary Phillips, act. comm.

Department of Veterans Affairs
810 Vermont Ave. NW 20420
Website: http://www.va.gov

Secretary of Veterans Affairs — Togo D. West Jr.
Deputy Secretary — Hershel W. Gober
Assistant Secretaries for:
  Congressional Affairs — Sheila C. McCready, act.
  Management — Edward A. Powell
  Human Resources & Admin. — Eugene Brickhouse
  Policy & Planning — Dennis Duffy
  Public & Intergovernmental Affairs — John Hanson
Inspector General — Richard J. Griffin
Under Sec. for Benefits — Joseph Thompson
Under Sec. for Health — Thomas L. Garthwaite, M.D., act.
National Cemetery System — Roger R. Rapp, act.
General Counsel — Leigh Bradley
Board of Veterans Appeals — Eligah Dane Clark, chair
Board of Contract Appeals — Guy H. McMichael III, chair
Small & Disadvantaged Business Utilization — Scott S. Denniston, dir.
Veterans Service Organization Liaison — Allen F. Kent

Notable U.S. Government Agencies
Source: *The U.S. Government Manual*; National Archives and Records Administration; World Almanac research
All addresses are Washington, DC, unless otherwise noted; as of Oct. 1999;
* = independent agency

Bureau of Alcohol, Tobacco, and Firearms — John W. Magaw, dir. (Dept. of Treas., 650 Mass. Ave NW, 20226).
Website: http://www.atf.treas.gov
Bureau of the Census — Kenneth Prewitt, dir. (Dept. of Commerce, 4700 Silver Hill Rd., Suitland, MD 20746).
Website: http://www.census.gov
Bureau of Economic Analysis — J. Steven Landerfeld, dir. (Dept. of Commerce, 1441 L St. NW, 20230).
Website: http://www.bea.doc.gov
Bureau of Indian Affairs — Kevin Gover, asst. sec. (Dept. of the Interior, 1849 C St. NW, 20240).
Website: http://www.doi.gov/bureau-indian-affairs.html
Bureau of Prisons — Kathleen Hawk Sawyer, dir. (Dept. of Justice, 320 First St. NW, 20534).
Website: http://www.bop.gov
Centers for Disease Control & Prevention — Jeffrey P. Koplan, dir. (Dept. of HHS, 1600 Clifton Rd. NE, Mailstop D14, Atlanta, GA 30333).
Website: http://www.cdc.gov
*Central Intelligence Agency — George J. Tenet, dir. (Wash., DC 20505).
Website: http://www.odci.gov/cia
*Commission on Civil Rights — Mary Frances Berry, chair (624 9th St. NW, 20425).
Website: http://www.usccr.gov
*Commodity Futures Trading Commission — William J. Rainer, chair (3 Lafayette Center, 1155 21st St. NW, 20581).
Website: http://www.cftc.gov

*Consumer Product Safety Commission — Ann Brown, chair (East West Towers, 4330 East West Hwy., Bethesda, MD 20814).
Website: http://www.cpsc.gov
*Environmental Protection Agency — Carol M. Browner, adm. (401 M St. SW, 20460).
Website: http://www.epa.gov
*Equal Employment Opportunity Commission — Ida L. Castro, chair (1801 L St. NW, 20507).
Website: http://www.eeoc.gov
*Export-Import Bank of the United States — James A. Harmon, pres. and chair (811 Vermont Ave. NW, 20571).
Website: http://www.exim.gov
*Farm Credit Administration — Marsha P. Martin, chair, Farm Credit Administration Board (1501 Farm Credit Drive, McLean, VA 22102).
Website: http://www.fca.gov
Federal Aviation Administration — Jane Garvey, adm. (Dept. of Trans., 800 Independence Ave. SW, 20591).
Website: http://www.faa.gov
Federal Bureau of Investigation — Louis J. Freeh, dir. (Dept. of Justice, 935 Pennsylvania Ave. NW, 20535).
Website: http://www.fbi.gov
*Federal Communications Commission — William E. Kennard, chair (445 12th St. SW, 20554).
Website: http://www.fcc.gov

*Federal Deposit Insurance Corporation — Donna Tanoue, chair (550 17th St. NW, 20429).
Website: http://www.fdic.gov

*Federal Election Commission — Scott Thomas, act. chair (999 E St. NW, 20463).
Website: http://www.fec.gov

*Federal Emergency Management Agency — James Lee Witt, dir. (500 C St. SW, 20472).
Website: http://www.fema.gov

*Federal Energy Regulatory Commission — James J. Hoecker, chair (888 1st St. NE, 20426).
Website: http://www.ferc.fed.us

Federal Highway Administration — Kenneth Wykle, adm. (Dept. of Trans., 400 7th St. SW, 20590).
Website: http://www.fhwa.dot.gov

*Federal Maritime Commission — Harold J. Creel Jr., chair (800 N. Capitol St. NW, 20573).
Website: http://www.fmc.gov

*Federal Mine Safety & Health Review Commission — Mary Lu Jordan, chair (1730 K St. NW, 20006).
Website: http://www.fmshrc.gov

*Federal Reserve System — Alan Greenspan, chair, Board of Governors (20th St. & C St. NW, 20551).
Website: http://www.federalreserve.gov

*Federal Trade Commission — Robert Pitofsky, chair (600 Pennsylvania Ave. NW, 20580).
Website: http://www.ftc.gov

Fish & Wildlife Service — Jamie Rappaport Clark, dir. (Dept. of the Interior, 1849 C St. NW, 20240).
Website: http://www.fws.gov

Food and Drug Administration — Jane E. Henney, MD, comm. (5600 Fishers Lane, Rockville, MD 20857).
Website: http://www.fda.gov

Forest Service — Mike Dombeck, chief (Dept. of Agriculture, 201 14th St. SW, 20250).
Website: http://www.fs.fed.us

General Accounting Office — (cong. agency) David Michael Walker, comptroller gen. (441 G St. NW, 20548).
Website: http://www.gao.gov

*General Services Administration — David J. Barram, adm. (1800 F St. NW, 20405).
Website: http://www.gsa.gov

Government Printing Office — (cong. agency) Michael F. DiMario, public printer (732 N. Capitol St. NW, 20401).
Website: http://www.gpo.gov

Immigration & Naturalization Service — Doris Meissner, comm. (Dept. of Justice, 425 I St. NW, 20536).
Website: http://www.ins.usdoj.gov

*Inter-American Foundation — Maria Otero, chair (901 N Stuart St., 10th floor, Arlington, VA 22203).
Website: http://www.iaf.gov

Internal Revenue Service — Charles Rossotti, comm. (Dept. of Treas., 1111 Constitution Ave. NW, 20224).
Website: http://www.irs.gov

Library of Congress — (cong. agency) Dr. James H. Billington, Librarian of Congress (101 Indep. Ave. SE, 20540).
Website: http://www.loc.gov

*National Aeronautics and Space Administration — Daniel S. Goldin, adm. (300 E St. SW, 20546).
Website: http://www.nasa.gov

*National Archives & Records Administration — John W. Carlin, archivist (8601 Adelphi Rd., College Park, MD 20740).
Website: http://www.nara.gov

*National Endowment for the Arts — William J. Ivey, chair (1100 Pennsylvania Ave. NW, 20506).
Website: http://arts.endow.gov

*National Endowment for the Humanities — William Ferris, chair (1100 Pennsylvania Ave. NW, 20506).
Website: http://www.neh.fed.us

National Institutes of Health — Harold E. Varmus, dir. (9000 Rockville Pike, Bethesda, MD 20892).
Website: http://www.nih.gov

*National Labor Relations Board — John C. Truesdale, chair (1099 14th St. NW, 20570).
Website: http://www.nlrb.gov

National Oceanic and Atmospheric Administration — D. James Baker, undersec. (Dept. of Commerce, 14th & Constitution Ave. NW, 20230).
Website: http://www.noaa.gov

National Park Service — Robert G. Stanton, dir. (Dept. of the Interior, 1849 C St. NW, 20240).
Website: http://www.nps.gov

*National Railroad Passenger Corp. (Amtrak) — George Warrington, Pres. & CEO (60 Mass. Ave. NE, 20002).
Website: http://www.amtrak.com

*National Science Foundation — Eamon Kelly, chair, National Science Board (4201 Wilson Blvd., Arlington, VA 22230).
Website: http://www.nsf.gov

*National Transportation Safety Board — Jim Hall, chair (490 L'Enfant Plaza SW, 20594).
Website: http://www.ntsb.gov

*Nuclear Regulatory Commission — Greta Joy Dicus, chair (11555 Rockville Pike, Rockville, MD 20852).
Website: http://www.nrc.gov

Occupational Safety & Health Administration — Charles Jeffress, asst. sec. (Dept. of Labor, 200 Constitution Ave. NW, 20210).
Website: http://www.osha.gov

*Occupational Safety & Health Review Commission — Thomasina V. Rogers, chair (1120 20th St. NW, 9th Floor, 20036).
Website: http://www.oshrc.gov

*Office of Government Ethics — Stephen D. Potts, dir. (1201 New York Ave. NW, Suite 500, 20005).
Website: http://www.usoge.gov

*Office of Personnel Management — Janice Lachance, dir. (1900 E St. NW, 20415-0001).
Website: http://www.opm.gov

*Office of Special Counsel — Elaine D. Kaplan, special counsel (1730 M St. NW, Suite 216, 20036).
Website: http://www.osc.gov

*Peace Corps — Chuck Caquet, act. dir. (1111 20th St., NW, 20526).
Website: http://www.peacecorps.gov/home.html

*Postal Rate Commission — Edward J. Gleiman, chair (1333 H St. NW, Suite 300, 20268).
Website: http://www.prc.gov

*Securities and Exchange Commission — Arthur Levitt, chair (450 5th St. NW, 20549).
Website: http://www.sec.gov

*Selective Service System — Gil Coronado, dir. (National Headquarters, 1515 Wilson Blvd., Arlington, VA 22209-2425).
Website: http://www.sss.gov

*Small Business Administration — Aida Alvarez, adm. (409 Third St. SW, 20416).
Website: http://www.sba.gov

Smithsonian Institution — (quasi-official agency) I. Michael Heyman[1], sec. (1000 Jefferson Dr. SW, Rm. 354, 20560-0033).
Website: http://www.si.edu

*Social Security Administration — Kenneth S. Apfel, comm. (6401 Security Blvd., Baltimore, MD 21235).
Website: http://www.ssa.gov

Surgeon General — David Satcher (Public Health Service, HHS, Parklawn Bldg., 5600 Fishers Lane, Rm. 18-67, Rockville, MD 20857).
Website: http://www.surgeongeneral.gov

*Tennessee Valley Authority — Craven Crowell, chair, Board of Directors (400 W. Summit Hill Dr., Knoxville, TN 37902, and One Mass. Ave. NW, Suite 300, 20444).
Website: http://www.tva.gov

*Trade and Development Agency — J. Joseph Grandmaison, dir. (1621 N. Kent St., Suite 300, Arlington, VA 22209).
Website: http://www.tda.gov

United States Coast Guard — Adm. James M. Loy, commandant (Dept. of Trans., 2100 2d St. SW, 20593).
Website: http://www.uscg.mil

United States Customs Service — Raymond W. Kelly, comm. (1300 Pennsylvania Ave. NW, 20229).
Website: http://www.customs.treas.gov

*United States International Trade Commission — Lynn M. Bragg, chair (500 E St. SW, 20436).
Website: http://www.usitc.gov

United States Mint — Philip N. Diehl, dir. (Dept. of Treas., 633 3d St. NW, 20220).
Website: http://www.usmint.gov

*United States Postal Service — William J. Henderson, Postmaster General (475 L'Enfant Plaza SW, 20260).
Website: http://www.usps.gov

United States Secret Service — Brian L. Stafford, dir. (Dept. of Treas., 950 H St. NW, Ste. 8000, 20001).
Website: http://www.ustreas.gov/usss

(1) Lawrence Small was expected to replace I. Michael Heyman in Jan. 2000.

CONGRESS

The One Hundred and Sixth Congress
With Official 1998 Election Results

Source: Voter News Service; World Almanac research
The 106th Congress convened on Jan. 6, 1999.

The Senate

Rep., 53; Dem., 45; Ind., 1; Vacant, 1; Total, 100. *Incumbent. Boldface denotes the 1998 election winner. Data as of Oct. 1999.

Terms are for 6 years and end Jan. 3 of the year preceding the senator's name in the following table. Annual salary (as of Jan. 1, 2000), $141,300; President Pro Tempore, Majority Leader, and Minority Leader, $157,000. To be eligible for the Senate, one must be at least 30 years old, a U.S. citizen for at least 9 years, and a resident of the state from which chosen. Congress must meet annually on Jan. 3, unless it has, by law, appointed a different day.

The ZIP code of the Senate is 20510; the telephone number is 202-224-3121; the website is http://www.senate.gov. Senate officials in 1999 were: President Pro Tempore, Strom Thurmond; Majority Leader, Trent Lott; Majority Whip, Don Nickles; Minority Leader, Tom Daschle; Minority Whip, Harry Reid.

D-Democrat; R-Republican; ACP-A Connecticut Party; C-Conservative; I-Independent; IN-Independence; L-Liberal; RL-Right to Life

Term ends	Senator (Party)/Service from[1]	1998 Election	Term ends	Senator (Party)/Service from[1]	1998 Election
	Alabama			**Indiana**	
2003	Jeff Sessions (R)/1/7/97		2001	Richard G. Lugar (R)/1977	
2005	**Richard Shelby*** (R)/1/6/87	817,973	2005	**Evan Bayh** (D)/1/6/99	1,012,244
	Clayton Suddith (D)	474,568		Paul Helmke (R)	552,732
	Alaska			**Iowa**	
2003	Ted Stevens (R)/12/24/68		2003	Tom Harkin (D)/1985	
2005	**Frank H. Murkowski*** (R)/1981	165,227	2005	**Chuck Grassley*** (R)/1981	648,480
	Joseph A. "Joe" Sonneman (D)	43,743		David Osterberg (D)	289,049
	Arizona			**Kansas**	
2001	Jon Kyl (R)/1/4/95		2003	Pat Roberts (R)/1/7/97	
2005	**John McCain*** (R)/1/6/87	696,577	2005	**Sam Brownback*** (R)/1/7/97	474,639
	Ed Ranger (D)	275,224		Paul Feleciano, Jr. (D)	229,718
	Arkansas			**Kentucky**	
2003	Tim Hutchinson (R)/1/7/97		2003	Mitch McConnell (R)/1985	
2005	**Blanche Lambert Lincoln** (D)/1/6/99	385,878	2005	**Jim Bunning** (R)/1/6/99	569,817
	Fay Boozman (R)	295,870		Scotty Baesler (D)	563,051
	California			**Louisiana**	
2001	Dianne Feinstein (D)/11/10/92		2003	Mary L. Landrieu (D)/1/7/97	
2005	**Barbara Boxer*** (D)/1993	4,411,705	2005	**John B. Breaux*** (D)/1/6/87	620,502
	Matt Fong (R)	3,576,351		"Jim" Donelon (R)	306,616
	Colorado			**Maine**	
2003	Wayne Allard (R)/1/7/97		2001	Olympia J. Snowe (R)/1/4/95	
2005	**Ben Nighthorse Campbell*** (R)/1993	829,370	2003	Susan M. Collins (R)/1/7/97	
	Dottie Lamm (D)	464,754		**Maryland**	
	Connecticut			2001	Paul S. Sarbanes (D)/1977	
2001	Joe Lieberman (D,ACP)/1989		2005	**Barbara Ann Mikulski*** (D)/1/6/87	1,062,810
2005	**Christopher J. Dodd*** (D)/1981	628,306		Ross Z. Pierpont (R)	444,637
	Gary A. Franks (R)	312,177		**Massachusetts**	
	Delaware			2001	Edward M. Kennedy (D)/11/7/62	
2001	William V. Roth, Jr. (R)/1/1/71		2003	John F. Kerry (D)/1/2/85	
2003	Joseph R. Biden, Jr. (D)/1973			**Michigan**	
	Florida			2001	Spencer Abraham (R)/1/4/95	
2001	Connie Mack (R)/1989		2003	Carl Levin (D)/1979	
2005	**Bob Graham*** (D)/1/6/87	2,436,407		**Minnesota**	
	Charlie Crist (R)	1,463,755	2001	Rod Grams (R)/1/4/95	
	Georgia		2003	Paul David Wellstone (D)/1991	
2003	Max Cleland (D)/1/7/97			**Mississippi**	
2005	**Paul Douglas Coverdell*** (R)/1993	918,540	2001	Trent Lott (R)/1989	
	Michael J. Coles (D)	791,904	2003	Thad Cochran (R)/12/27/78	
	Hawaii			**Missouri**	
2001	Daniel K. Akaka (D)/4/28/90		2001	John Ashcroft (R)/1/4/95	
2005	**Daniel K. Inouye*** (D)/1963	315,252	2005	**Christopher (Kit) Bond*** (R)/1/6/87	830,625
	Crystal Young (R)	70,964		Jeremiah W. (Jay) Nixon (D)	690,208
	Idaho			**Montana**	
2003	Larry E. Craig (R)/1991		2001	Conrad Burns (R)/1989	
2005	**Mike Crapo** (R)/1/6/99	262,966	2003	Max Baucus (D)/12/15/78	
	Bill Mauk (D)	107,375		**Nebraska**	
	Illinois			2001	Bob Kerrey (D)/1989	
			2003	Chuck Hagel (R)/1/7/97	
2003	Richard J. Durbin (D)/1/7/97			**Nevada**	
2005	**Peter G. Fitzgerald** (R)/1/6/99	1,709,041	2001	Richard H. Bryan (D)/1989	
	Carol Moseley-Braun* (D)/1993	1,610,496	2005	**Harry Reid*** (D)/1/6/87	208,650
				John Ensign (R)	208,222

Term ends	Senator (Party)/Service from[1]	1998 Election	Term ends	Senator (Party)/Service from[1]	1998 Election
	New Hampshire			**South Carolina**	
2003	Robert Smith (I)[2]/12/7/90		2003	Strom Thurmond (R)/11/7/56	
2005	**Judd Gregg*** (R)/1993	213,477	2005	**Ernest Hollings*** (D)/11/9/66	562,791
	George Condodemetra (D)	88,883		Bob Inglis (R)	488,132
	New Jersey			**South Dakota**	
2001	Frank R. Lautenberg (D)/12/27/82		2003	Tim Johnson (D)/1/7/97	
2003	Robert G. Torricelli (D)/1/7/97		2005	**Tom Daschle*** (D)/1/6/87	162,884
	New Mexico			Ron Schmidt (R)	95,431
2001	Jeff Bingaman (D)/1983			**Tennessee**	
2003	Pete V. Domenici (R)/1973		2001	Bill Frist (R)/1/4/95	
	New York		2003	Fred Thompson (R)/12/9/94	
2001	Daniel Patrick Moynihan (D,L)/1977			**Texas**	
2005	**Charles E. Schumer** (D,IN,L)/1/6/99	2,551,065	2001	Kay Bailey Hutchison (R)/6/5/93	
	Alfonse D'Amato* (R,C,RL)/1981	1,954,423	2003	Phil Gramm (R)/1985	
	North Carolina			**Utah**	
2003	Jesse Helms (R)/1973		2001	Orrin G. Hatch (R)/1977	
2005	**John Edwards** (D)/1/6/99	1,029,237	2005	**Robert F. Bennett*** (R)/1993	316,652
	Lauch Faircloth* (R)/1993	945,943		Scott Leckman (D)	163,172
	North Dakota			**Vermont**	
2001	Kent Conrad (D)/1/6/87		2001	Jim Jeffords (R)/1989	
2005	**Byron L. Dorgan*** (D)/12/14/92	134,747	2005	**Patrick Leahy*** (D)/1975	154,567
	Donna Nalewaja (R)	75,013		Fred H. Tuttle (R)	48,051
	Ohio			**Virginia**	
2001	Mike Dewine (R) /1/4/95		2001	Charles S. Robb (D)/1989	
2005	**George V. Voinovich** (R)/1/6/99	1,922,087	2003	John W. Warner (R)/1/2/79	
	Mary O. Boyle (D)	1,482,054		**Washington**	
	Oklahoma		2001	Slade Gorton (R)/1989	
2003	James M. Inhofe (R)/11/21/94		2005	**Patty Murray*** (D)/1993	1,103,184
2005	**Don Nickles*** (R)/1981	570,682		Linda Smith (R)	785,377
	Don E. Carroll (D)	268,898		**West Virginia**	
	Oregon		2001	Robert C. Byrd (D)/1959	
2003	Gordon Smith (R)/1/7/97		2003	John D. Rockefeller IV (D)/1/15/85	
2005	**Ron Wyden*** (D)/2/6/96	682,425		**Wisconsin**	
	John Lim (R)	377,739	2001	Herbert H. Kohl (D)/1989	
	Pennsylvania		2005	**Russ Feingold*** (D)/1993	890,059
2001	Rick Santorum (R)/1/4/95			Mark W. Neumann (R)	852,272
2005	**Arlen Specter*** (R)/1981	1,814,180		**Wyoming**	
	Bill Lloyd (D)	1,028,839	2001	Craig Thomas (R)/1/4/95	
	Rhode Island		2003	Michael B. Enzi (R)/1/7/97	
2001	Vacant[3]				
2003	John F. Reed (D)/1/7/97				

(1) Jan. 3, unless otherwise noted. (2) Republican Sen. Robert Smith announced July 13, 1999, that he had changed his party designation to Independent. (3) Seat fell vacant after the death of John H. Chafee on Oct. 24, 1999. Governor was to appoint successor.

The House of Representatives

Rep., 222; Dem., 211; Ind., 1; Vacant, 1; Total, 435.
*Incumbent. Boldface denotes the 1998 election winner. Data as of Oct. 1999.

Members' terms to Jan. 3, 2001. Annual salary (as of Jan. 1, 2000), $141,300; Speaker of the House, $181,400; Majority Leader and Minority Leader, $157,000. To be eligible for membership, a person must be at least 25 years of age, a U.S. citizen for at least 7 years, and a resident of the state from which he or she is chosen. The ZIP code of the House is 20515; the telephone number is 202-225-3121. The website is http://www.house.gov.

House officials in 1999 were: Speaker, J. Dennis Hastert; Majority Leader, Dick Armey; Majority Whip, Tom DeLay; Minority Leader, Richard A. Gephardt; Minority Whip, David E. Bonior.

D-Democrat; R-Republican; AI-American Independent; C-Conservative; GR-Green;
I-Independent; IA- Independent American; IN-Independence; L-Liberal;
LB-Libertarian; NL-Natural Law; RF-Reform; RL-Right to Life; TX-Taxpayers.

Dist.	Representative (Party)	1998 Election	Dist.	Representative (Party)	1998 Election
	Alabama		5.	**Bud Cramer*** (D)	134,819
1.	**H. L. Sonny Callahan*** (R)	Unopposed		Gil Aust (R)	58,536
2.	**Terry Everett*** (R)	131,428	6.	**Spencer Bachus*** (R)	154,761
	Joe Fondren (D)	58,136		Donna Wesson Smalley (D)	60,657
3.	**Bob Riley*** (R)	101,731	7.	**Earl F. Hillard*** (D)	Unopposed
	Joe Turnham (D)	73,357		**Alaska**	
4.	**Robert Aderholt*** (R)	106,297		**Don Young*** (R)	139,676
	Don Bevill (D)	82,065		Jim Duncan (D)	77,232

Dist.	Representative (Party)	1998 Election
	Arizona	
1.	**Matt Salmon*** (R)	98,840
	David Mendoza (D)	54,108
2.	**Ed Pastor*** (D)	57,178
	Ed Barron (R)	23,628
3.	**Bob Stump*** (R)	137,618
	Stuart Marc Starky (D)	66,979
4.	**John Shadegg*** (R)	102,722
	Eric Ehst (D)	49,538
5.	**Jim Kolbe*** (R)	103,952
	Tom Volgy (D)	91,030
6.	**J.D. Hayworth*** (R)	106,891
	Steve Owens (D)	88,001
	Arkansas	
1.	**Marion Berry*** (D)	Unopposed
2.	**Vic Snyder*** (D)	100,334
	Phil Wyrick (R)	72,737
3.	**Asa Hutchinson*** (R)	154,780
	Ralph Forbes (RF)	36,917
4.	**Jay Dickey*** (R)	92,346
	Judy Smith (D)	68,194
	California	
1.	**Mike Thompson** (D)	121,713
	Mark C. Luce (R)	64,622
2.	**Wally Herger*** (R)	128,372
	Roberts "Rob" Braden (D)	70,837
3.	**Doug Ose** (R)	100,621
	Sandie Dunn (D)	86,471
4.	**John T. Doolittle*** (R)	155,306
	David Shapiro (D)	85,394
5.	**Robert T. Matsui*** (D)	130,715
	Robert S. Dinsmore (R)	47,307
6.	**Lynn Woolsey*** (D)	158,446
	Ken McAuliffe (R)	69,295
7.	**George Miller*** (D)	125,842
	Norman H. Reece (R)	38,290
8.	**Nancy Pelosi*** (D)	148,027
	David J. Martz (R)	20,781
9.	**Barbara Lee*** (D)	140,722
	Claiborne "Clay" Sanders (R)	22,431
10.	**Ellen O. Tauscher*** (D)	127,134
	Charles Ball (R)	103,299
11.	**Richard W. Pombo*** (R)	95,496
	Robert L. Figueroa (D)	56,345
12.	**Tom Lantos*** (D)	128,135
	Robert H. Evans, Jr. (R)	36,562
13.	**Fortney Pete Stark*** (D)	101,671
	James R. Goetz (R)	38,050
14.	**Anna G. Eshoo*** (D)	129,663
	John C. "Chris" Haugen (R)	53,719
15.	**Tom Campbell*** (R)	111,876
	Dick Lane (D)	70,059
16.	**Zoe Lofgren*** (D)	85,503
	Horace Eugene Thayn (R)	27,494
17.	**Sam Farr*** (D)	103,719
	Bill McCampbell (R)	52,470
18.	**Gary A. Condit*** (D)	118,842
	Linda M. Degroat (LB)	18,089
19.	**George Radanovich*** (R)	131,105
	Jonathan Richter (LB)	34,044
20.	**Cal Dooley*** (D)	60,599
	Cliff Unruh (R)	39,183
21.	**Bill Thomas*** (R)	115,989
	John Evans (RF)	30,994
22.	**Lois Capps*** (D)	111,388
	Thomas J. Bordonaro, Jr. (R)	86,921
23.	**Elton W. Gallegly*** (R)	96,362
	Daniel "Dan" Gonzalez (D)	64,068
24.	**Brad Sherman*** (D)	103,491
	Randy Hoffman (R)	69,501
25.	**Howard "Buck" McKeon*** (R)	114,013
	Bruce R. Acker (LB)	38,669
26.	**Howard L. Berman*** (D)	69,000
	Juan Carlos Ros (LB)	6,556
27.	**James E. Rogan** (R)	80,702
	Barry A. Gordon (D)	73,875
28.	**David Dreier*** (R)	90,607
	Janice M. Nelson (D)	61,721

Dist.	Representative (Party)	1998 Election
29.	**Henry A. Waxman*** (D)	131,561
	Mike Gottlieb (R)	40,282
30.	**Xavier Becerra*** (D)	58,230
	Patricia Parker (R)	13,441
31.	**Matthew G. Martinez*** (D)	61,173
	Frank C. Moreno (R)	19,786
32.	**Julian C. Dixon*** (D)	112,253
	Laurence Ardito (R)	14,622
33.	**Lucille Roybal-Allard*** (D)	43,310
	Wayne Miller (R)	6,364
34.	**Grace Flores Napolitano** (D)	76,471
	Ed Perez (R)	32,321
35.	**Maxine Waters*** (D)	78,732
	Gordon Michael Mego (AI)	9,413
36.	**Steven T. Kuykendall** (R)	88,843
	Janice Hahn (D)	84,624
37.	**Juanita Millender-McDonald*** (D)	70,026
	Saul E. Lankster (R)	12,301
38.	**Steve Horn*** (R)	71,386
	Peter Mathews (D)	59,767
39.	**Ed Royce*** (R)	97,366
	A. "Cecy" R. Groom (D)	52,815
40.	**Jerry Lewis*** (R)	97,406
	Robert "Bob" Conaway (D)	47,897
41.	**Gary G. Miller** (R)	68,310
	Eileen R. Ansari (D)	52,264
42.	**Vacant**[1]	
43.	**Ken Calvert*** (R)	83,012
	Mike Rayburn (D)	56,373
44.	**Mary Bono*** (R)	97,013
	Ralph Waite (D)	57,697
45.	**Dana Rohrabacher*** (R)	94,296
	Patricia W. Neal (D)	60,022
46.	**Loretta Sanchez*** (D)	47,964
	Robert Kenneth "Bob" Dornan (R)	33,388
47.	**Christopher Cox*** (R)	132,711
	Christina Avalos (D)	57,938
48.	**Ron Packard*** (R)	138,948
	Sharon K. Miles (NL)	23,262
49.	**Brian P. Bilbray*** (R)	90,516
	Christine T. Kehoe (D)	86,400
50.	**Bob Filner*** (D)	Unopposed
51.	**Randy "Duke" Cunningham*** (R)	126,229
	Dan Kripke (D)	71,706
52.	**Duncan Hunter*** (R)	116,251
	Lynn Badler (LB)	21,933
	Colorado	
1.	**Diana DeGette*** (D)	116,628
	Nancy McClanahan (R)	52,452
2.	**Mark Udall** (D)	113,946
	Bob Greenlee (R)	108,385
3.	**Scott McInnis*** (R)	156,501
	Reed Kelley (R)	74,479
4.	**Bob Schaffer*** (R)	131,318
	Susan Kirkpatrick (D)	89,973
5.	**Joel Hefley*** (R)	155,790
	Ken Alford (D)	55,609
6.	**Tom Tancredo** (R)	111,374
	Henry L. Strauss (D)	82,662
	Connecticut	
1.	**John B. Larson** (D)	97,681
	Kevin O'Connor (R)	69,668
2.	**Sam Gejdenson*** (D)	99,567
	Gary M. Koval (R)	57,860
3.	**Rosa L. DeLauro*** (D)	109,726
	Martin T. Reust (R)	42,090
4.	**Christopher Shays*** (R)	94,767
	Jonathan Kantrowitz (D)	40,988
5.	**Jim Maloney*** (D)	78,394
	Mark Nielsen (R)	76,051
6.	**Nancy L. Johnson*** (R)	101,630
	Charlotte Koskoff (D)	69,201
	Delaware	
	Michael N. Castle* (R)	119,811
	Dennis E. Williams (D)	57,446

Dist.	Representative (Party)	1998 Election
	Florida	
1.	**Joe Scarborough*** (R)	**Unopposed**
2.	**Allen Boyd*** (D)	**Unopposed**
3.	**Corrine Brown*** (D)	**66,621**
	Bill Randall (R)	53,530
4.	**Tillie K. Fowler*** (R)	**Unopposed**
5.	**Karen L. Thurman*** (D)	**132,005**
	Jack Gargan (RF)	67,147
6.	**Clifford (Cliff) B. Stearns*** (R)	**Unopposed**
7.	**John L. Mica*** (R)	**Unopposed**
8.	**Bill McCollum*** (R)	**104,298**
	Al Krulick (D)	54,245
9.	**Michael Bilirakis*** (R)	**Unopposed**
10.	**C.W. Bill Young*** (R)	**Unopposed**
11.	**Jim Davis*** (D)	**85,262**
	Joe Chillura (R)	46,176
12.	**Charles T. Canady*** (R)	**Unopposed**
13.	**Dan Miller*** (R)	**Unopposed**
14.	**Porter Goss*** (R)	**Unopposed**
15.	**Dave Weldon*** (R)	**129,278**
	David R. Golding (D)	75,654
16.	**Mark Foley*** (R)	**Unopposed**
17.	**Carrie P. Meek*** (D)	**Unopposed**
18.	**Ileana Ros-Lehtinen*** (R)	**Unopposed**
19.	**Robert Wexler*** (D)	**Unopposed**
20.	**Peter Deutsch*** (D)	**Unopposed**
21.	**Lincoln Diaz-Balart*** (R)	**84,018**
	Patrick Cusack (D)	28,378
22.	**Clay Shaw*** (R)	**Unopposed**
23.	**Alcee L. Hastings*** (D)	**Unopposed**
	Georgia	
1.	**Jack Kingston*** (R)	**Unopposed**
2.	**Sanford Dixon Bishop, Jr.*** (D)	**77,953**
	Joseph Francis McCormick, Jr. (R)	59,305
3.	**Michael A. (Mac) Collins*** (R)	**Unopposed**
4.	**Cynthia McKinney*** (D)	**100,622**
	Sunny J. Warren (R)	64,146
5.	**John Lewis*** (D)	**109,177**
	John H. Lewis, Sr. (R)	29,877
6.	**Johnny Isakson** (R)[2]	**51,548**
	Christina Jeffrey (R)	20,115
7.	**Bob Barr*** (R)	**85,982**
	James F. "Jim" Williams (D)	69,293
8.	**Saxby Chambliss*** (R)	**87,993**
	Ronald L. Cain (D)	53,079
9.	**Nathan Deal*** (R)	**Unopposed**
10.	**Charlie Norwood*** (R)	**88,527**
	Marion Denise Spencer Freeman (D)	60,004
11.	**John Linder*** (R)	**120,909**
	Vincent Littman (D)	53,510
	Hawaii	
1.	**Neil Abercrombie*** (D)	**116,693**
	Gene Ward (R)	68,905
2.	**Patsy Takemoto Mink*** (D)	**144,254**
	Carol J. Douglass (R)	50,423
	Idaho	
1.	**Helen Chenoweth*** (R)	**113,231**
	Dan Williams (D)	91,653
2.	**Mike Simpson** (R)	**91,337**
	Richard H. Stallings (D)	77,736
	Illinois	
1.	**Bobby L. Rush*** (D)	**151,890**
	Marlene White Ahimaz (R)	18,429
2.	**Jesse L. Jackson, Jr.*** (D)	**148,985**
	Robert Gordon III (R)	16,075
3.	**William O. Lipinski*** (D)	**115,887**
	Robert Marshall (R)	44,012
4.	**Luis V. Gutierrez*** (D)	**54,244**
	John Birch (R)	10,529
5.	**Rod R. Blagojevich*** (D)	**95,738**
	Alan Spitz (R)	33,687
6.	**Henry J. Hyde*** (R)	**111,603**
	Thomas A. Cramer (D)	49,906
7.	**Danny K. Davis*** (D)	**130,984**
	Dorn E. Van Cleave III (LB)	9,984

Dist.	Representative (Party)	1998 Election
8.	**Philip M. Crane*** (R)	**104,242**
	Mike Rothman (D)	47,614
9.	**Janice D. (Jan) Schakowsky** (D)	**107,878**
	Herbert Sohn (R)	33,448
10.	**John E. Porter*** (R)	**Unopposed**
11.	**Gerald C. Weller*** (R)	**100,597**
	Gary S. Mueller (D)	70,458
12.	**Jerry F. Costello*** (D)	**99,605**
	William Melvin Price (R)	65,409
13.	**Judy Biggert** (R)	**121,929**
	Susan W. Hynes (D)	77,909
14.	**J. Dennis Hastert*** (R)	**117,304**
	Robert A. Cozzi, Jr. (D)	50,844
15.	**Thomas W. Ewing*** (R)	**104,255**
	Laurel Lunt Prussing (D)	65,054
16.	**Donald Manzullo*** (R)	**Unopposed**
17.	**Lane A. Evans*** (D)	**100,128**
	Mark Baker (R)	94,072
18.	**Ray LaHood*** (R)	**Unopposed**
19.	**David D. Phelps** (D)	**122,430**
	Brent Winters (R)	87,614
20.	**John M. Shimkus*** (R)	**121,103**
	Rick Verticchio (D)	76,475
	Indiana	
1.	**Peter J. Visclosky*** (D)	**92,634**
	Michael Petyo (R)	33,503
2.	**David M. McIntosh*** (R)	**99,608**
	Sherman A. Boles (D)	62,452
3.	**Tim Roemer*** (D)	**84,625**
	Daniel A. Holtz (R)	61,041
4.	**Mark E. Souder*** (R)	**93,671**
	Mark J. Wehrle (D)	54,286
5.	**Steve Buyer*** (R)	**101,567**
	David Steele (D)	58,504
6.	**Dan Burton*** (R)	**135,250**
	Bob Kern (D)	31,472
7.	**Edward A. Pease*** (R)	**109,712**
	Samuel (Dutch) Hillenburg (D)	44,823
8.	**John N. Hostettler*** (R)	**92,785**
	Gail Riecken (D)	81,871
9.	**Baron Hill** (D)	**92,973**
	Jean Leising (R)	87,797
10.	**Julia M. Carson*** (D)	**69,682**
	Gary A. Hofmeister (R)	47,017
	Iowa	
1.	**Jim Leach*** (R)	**106,419**
	Bob Rush (D)	79,529
2.	**Jim Nussle*** (R)	**104,613**
	Bob Tully (D)	83,405
3.	**Leonard L. Boswell*** (D)	**107,947**
	Larry McKibben (R)	78,063
4.	**Greg Ganske*** (R)	**129,942**
	Jon Dvorak (D)	67,550
5.	**Tom Latham*** (R)	**Unopposed**
	Kansas	
1.	**Jerry Moran*** (R)	**152,775**
	Jim Phillips (D)	36,618
2.	**Jim Ryun*** (R)	**108,527**
	Jim Clark (D)	69,521
3.	**Dennis Moore** (D)	**103,376**
	Vince Snowbarger* (R)	93,938
4.	**Todd Tiahrt*** (R)	**94,785**
	Jim Lawing (D)	62,737
	Kentucky	
1.	**Edward Whitfield*** (R)	**95,308**
	Tom Barlow (D)	77,402
2.	**Ron Lewis*** (R)	**113,285**
	Bob Evans (D)	62,848
3.	**Anne Meagher Northup*** (R)	**100,690**
	Chris Gorman (D)	92,865
4.	**Ken Lucas** (D)	**93,485**
	Gex "Jay" Williams (R)	81,547
5.	**Harold "Hal" Rogers*** (R)	**142,215**
	Sidney Jane Bailey-Bamer (D)	39,585
6.	**Ernest Fletcher** (R)	**104,046**
	Ernie Scorsone (D)	90,033

Dist.	Representative (Party)	1998 Election
	Louisiana	
1.	**David Vitter** (R)[3]	**61,661**
	David Conner Treen (R)	59,849
2.	**William J. Jefferson*** (D)	**102,247**
	David Reed (D)	10,803
3.	**W.J. "Billy" Tauzin*** (R)	**Unopposed**
4.	**"Jim" McCrery*** (R)	**Unopposed**
5.	**John Cooksey*** (R)	**Unopposed**
6.	**Richard Baker*** (R)	**97,044**
	Marjorie McKeithen (D)	94,201
7.	**Chris John*** (D)	**Unopposed**

In Louisiana, all candidates of all parties ran against each other on Nov. 3, 1998, in an open primary, unless they were unopposed incumbents, in which case they were declared elected. Candidates who received more than 50% of the primary vote were also declared elected. All districts had candidates that were declared elected after the Nov. 3 election, and therefore a Dec. runoff was not needed.

Dist.	Representative (Party)	1998 Election
	Maine	
1.	**Thomas H. Allen*** (D)	**134,335**
	Ross J. Connelly (R)	79,160
2.	**John E. Baldacci*** (D)	**146,202**
	Johnathan Reisman (R)	45,674
	Maryland	
1.	**Wayne T. Gilchrest*** (R)	**135,771**
	Irving Pinder (D)	60,450
2.	**Robert L. Ehrlich Jr.*** (R)	**145,711**
	Kenneth T. Bosley (D)	64,474
3.	**Benjamin L. Cardin*** (D)	**137,501**
	Colin Felix Harby (R)	39,667
4.	**Albert R. Wynn*** (D)	**129,139**
	John B. Kimble (R)	21,518
5.	**Steny H. Hoyer*** (D)	**126,792**
	Robert B. Ostrom (R)	67,176
6.	**Roscoe Bartlett*** (R)	**127,802**
	Timothy D. McCown (D)	73,728
7.	**Elijah E. Cummings*** (D)	**112,699**
	Kenneth Kondner (R)	18,742
8.	**Constance A. Morella*** (R)	**133,145**
	Ralph G. Neas (D)	87,497
	Massachusetts	
1.	**John W. Olver*** (D)	**121,863**
	Gregory L. Morgan (R)	48,055
2.	**Richard E. Neal*** (D)	**Unopposed**
3.	**James P. McGovern*** (D)	**108,613**
	Matthew J. Amorello (R)	79,174
4.	**Barney Frank*** (D)	**Unopposed**
5.	**Martin T. Meehan*** (D)	**127,418**
	David E. Coleman (R)	52,725
6.	**John F. Tierney*** (D)	**117,132**
	Peter G. Torkildsen (R)	90,986
7.	**Edward J. Markey*** (D)	**137,178**
	Patricia H. Long (R)	56,977
8.	**Michael E. Capuano** (D)	**99,603**
	Philip Hyde III (R)	14,125
9.	**John Joseph Moakley*** (D)	**Unopposed**
10.	**William D. Delahunt*** (D)	**164,917**
	Eric V. Bleicken (R)	70,466
	Michigan	
1.	**Bart Stupak*** (D)	**130,129**
	Michelle A. McManus (R)	87,630
2.	**Peter Hoekstra*** (R)	**146,854**
	Bob Shrauger (D)	63,573
3.	**Vernon Ehlers*** (R)	**146,364**
	John Ferguson Jr. (D)	49,489
4.	**Dave Camp*** (R)	**155,343**
	Dan Marsh (LB)	10,404
5.	**James A. Barcia*** (D)	**135,254**
	Donald W. Brewster (R)	51,442

Dist.	Representative (Party)	1998 Election
6.	**Fred Upton*** (R)	**113,292**
	Clarence J. Annen (D)	45,358
7.	**Nick Smith*** (R)	**104,656**
	Jim Berryman (D)	72,998
8.	**Debbie Stabenow*** (D)	**125,169**
	Susan Munsell (R)	84,254
9.	**Dale E. Kildee*** (D)	**105,457**
	Tom McMillin (R)	79,062
10.	**David E. Bonior*** (D)	**108,770**
	Brian Palmer (R)	94,027
11.	**Joe Knollenberg*** (R)	**144,264**
	Travis Reeds (D)	76,107
12.	**Sander Levin*** (D)	**105,824**
	Leslie Touma (R)	79,619
13.	**Lynn Nancy Rivers*** (D)	**99,935**
	Tom Hickey (R)	68,328
14.	**John Conyers, Jr.*** (D)	**126,321**
	Vendella M. Collins (R)	16,140
15.	**Carolyn Cheeks Kilpatrick*** (D)	**108,582**
	Chrysanthea D. Boyd-Fields (R)	12,887
16.	**John D. Dingell*** (D)	**116,145**
	William Morse (R)	54,121
	Minnesota	
1.	**Gil Gutknecht*** (R)	**131,233**
	Tracy L. Beckman (D)	108,420
2.	**David Minge*** (D)	**148,933**
	Craig Duehring (R)	99,490
3.	**Jim Ramstad*** (R)	**203,731**
	Stan J. Leino (D)	66,505
4.	**Bruce F. Vento*** (D)	**128,726**
	Dennis Newinski (R)	95,388
5.	**Martin Olav Sabo*** (D)	**145,535**
	Frank Taylor (R)	60,035
6.	**Bill Luther*** (D)	**148,728**
	John Kline (R)	136,866
7.	**Collin C. Peterson*** (D)	**169,907**
	Aleta Edin (R)	66,562
8.	**James L. Oberstar*** (D)	**173,734**
	Jerry Shuster (R)	69,667
	Mississippi	
1.	**Roger F. Wicker*** (R)	**66,738**
	Rex N. Weathers (D)	30,438
2.	**Bennie G. Thompson*** (D)	**80,507**
	Will Chipman (LB)	32,533
3.	**Charles W. "Chip" Pickering, Jr.*** (R)	**84,785**
	C.T. Scarborough (LB)	15,465
4.	**Ronnie Shows** (D)	**73,252**
	Delbert Hosemann (R)	61,551
5.	**Gene Taylor*** (D)	**78,661**
	Randy McDonnell (R)	19,341
	Missouri	
1.	**William (Bill) Clay, Sr.*** (D)	**90,840**
	Richmond A. Soluade, Sr. (R)	30,635
2.	**James M. Talent*** (R)	**142,313**
	John Ross (D)	57,565
3.	**Richard A. Gephardt*** (D)	**98,287**
	William J. Federer (R)	74,005
4.	**Ike Skelton*** (D)	**133,173**
	Cecilia D. Noland (R)	51,005
5.	**Karen McCarthy*** (D)	**101,313**
	Penny D. Bennett (R)	47,582
6.	**Pat (Patsy Ann) Danner*** (D)	**136,774**
	Jeff Bailey (R)	51,679
7.	**Roy Blunt*** (R)	**129,746**
	Marc Perkel (R)	43,416
8.	**Jo Ann Emerson*** (R)	**104,271**
	Anthony J. (Tony) Heckemeyer (D)	59,426
9.	**Kenny Hulshof*** (R)	**117,196**
	Linda Vogt (D)	66,861
	Montana	
	Rick Hill* (R)	**175,748**
	Dusty Deschamps (D)	147,073

Dist.	Representative (Party)	1998 Election
	Nebraska	
1.	**Doug Bereuter*** (R)	136,058
	Don Eret (D)	48,826
2.	**Lee Terry** (R)	106,782
	Michael Scott (D)	55,722
3.	**Bill Barrett*** (R)	149,896
	Jerry Hickman (LB)	27,278
	Nevada	3
1.	**Shelley Berkley** (D)	79,315
	Don Chairez (R)	73,540
2.	**Jim Gibbons*** (R)	201,623
	Christopher Horne (IA)	20,738
	New Hampshire	
1.	**John E. Sununu*** (R)	104,430
	Peter Flood (D)	51,783
2.	**Charles Bass*** (R)	85,746
	Mary Rauh (D)	72,321
	New Jersey	
1.	**Robert E. Andrews*** (D)	90,279
	Ronald L. Richards (R)	27,855
2.	**Frank A. LoBiondo*** (R)	93,248
	Derek Hunsberger (D)	43,563
3.	**Jim Saxton*** (R)	97,508
	Steven J. Polansky (D)	55,248
4.	**Christopher H. Smith*** (R)	92,991
	Larry Schneider (D)	52,281
5.	**Marge Roukema*** (R)	106,304
	Mike Schneider (D)	55,487
6.	**Frank Pallone Jr.*** (D)	78,102
	Michael Ferguson (R)	55,180
7.	**Bob Franks*** (R)	77,751
	Maryanne S. Connelly (D)	65,776
8.	**Bill J. Pascrell Jr.*** (D)	81,068
	Mathew J. Kirnan (R)	46,289
9.	**Steven R. Rothman*** (D)	91,330
	Steve Lonegan (R)	47,817
10.	**Donald M. Payne*** (D)	82,244
	William Stanley Wnuck (R)	10,678
11.	**Rodney P. Frelinghuysen*** (R)	100,910
	John P. Scollo (D)	44,160
12.	**Rush Holt** (D)	92,528
	Mike Pappas* (R)	87,221
13.	**Robert Menendez*** (D)	70,308
	Theresa de Leon (R)	14,615
	New Mexico	
1.	**Heather A. Wilson*** (R)	86,784
	Phillip J. Maloof (D)	75,040
2.	**Joe Skeen*** (R)	85,077
	E. Shirley Baca (D)	61,796
3.	**Tom Udall** (D)	91,248
	Bill Redmond* (R)	74,266
	New York	
1.	**Michael P. Forbes*** (D)[4]	99,460
	William G. Holst (D,I)	55,630
2.	**Rick A. Lazio*** (R,C)	85,089
	John C. Bace (D)	37,949
3.	**Peter T. King*** (R,C,RL)	117,258
	Kevin N. Langberg (D)	63,628
4.	**Carolyn McCarthy*** (D,IN)	90,256
	Gregory R. Becker (R,C,RL)	79,984
5.	**Gary L. Ackerman*** (D,IN,L)	97,404
	David C. Pinzon (R,C)	49,586
6.	**Gregory W. Meeks*** (D,IN,L)	Unopposed
7.	**Joseph Crowley*** (D)	50,924
	James J. Dillon (R)	18,896
8.	**Jerrold L. Nadler*** (D,L)	112,948
	Theodore Howard (R)	18,383
9.	**Anthony Weiner** (D,IN)	69,439
	Louis Telano (R)	24,486

Dist.	Representative (Party)	1998 Election
10.	**Edolphus Towns*** (D,L)	83,528
	Ernestine M. Brown (R)	5,577
11.	**Major R. Owens*** (D,L)	75,773
	David Greene (R,C)	7,284
12.	**Nydia M. Velazquez*** (D)	53,269
	Rosemarie Markgraf (R)	7,405
13.	**Vito J. Fossella*** (R,C,RL)	76,138
	Eugene V. Prisco (D,L)	40,167
14.	**Carolyn B. Maloney*** (D,IN,L)	111,072
	Stephanie E. Kupferman (R)	32,458
15.	**Charles B. Rangel*** (D,L)	90,424
	David E. Cunningham (R)	5,633
16.	**Jose E. Serrano*** (D,L)	67,367
	Thomas W. Bayley, Jr. (R)	2,457
17.	**Eliot L. Engel*** (D,L)	80,947
	Peter Fiumefreddo (R,C,IN)	11,037
18.	**Nita M. Lowey*** (D)	91,623
	Daniel McMahon (D)	12,594
19.	**Sue W. Kelly*** (R,C)	104,467
	Dick Collins (D)	56,378
20.	**Benjamin A. Gilman*** (R)	98,546
	Paul J. Feiner (D,IN,L)	65,589
21.	**Michael R. McNulty*** (D,C,IN)	146,729
	Lauren Ayers (D)	50,931
22.	**John E. Sweeney** (R,C,IN)	106,919
	Jean P. Bordewich (D)	81,296
23.	**Sherwood L. Boehlert*** (R)	111,242
	David Vickers (C,RL)	26,493
24.	**John M. McHugh*** (R,C)	116,682
	Neil P. Tallon (D)	31,011
25.	**James T. Walsh*** (R,C)	121,204
	Yvonne Rothenberg (D,L)	53,461
26.	**Maurice D. Hinchey*** (D,IN,L)	108,204
	Wm. H. Bud Walker (R,C)	54,776
27.	**Thomas M. Reynolds*** (R,C)	102,042
	Bill Cook (D,IN,RL)	75,978
28.	**Louise M. Slaughter*** (D)	118,856
	Richard A. Kaplan (R,IN)	56,443
29.	**John J. La Falce*** (D,IN,L)	97,335
	Chris Collins (R,C)	69,481
30.	**Jack Quinn*** (R,C,IN)	115,993
	Crystal D. Peoples (D)	55,199
31.	**Amo Houghton*** (R,C)	107,615
	Caleb Rossiter (D)	40,091
	North Carolina	
1.	**Eva M. Clayton*** (D)	85,125
	Ted Tyler (R)	50,578
2.	**Bob Etheridge*** (D)	100,550
	Dan Page (R)	72,997
3.	**Walter B. Jones*** (R)	83,529
	Jon Williams (D)	50,041
4.	**David Price*** (D)	129,157
	Tom Roberg (R)	93,469
5.	**Richard Burr*** (R)	119,103
	Mike Robinson (D)	55,806
6.	**Howard Coble*** (R)	112,740
	Jeffrey D. Bentley (LB)	14,454
7.	**Mike McIntyre*** (D)	124,366
	Paul Meadows (LB)	11,924
8.	**Robert C. "Robin" Hayes** (R)	67,505
	Mike Taylor (D)	64,127
9.	**Sue Myrick*** (R)	120,570
	Rory Blake (D)	51,345
10.	**T. Cass Ballenger*** (R)	118,541
	Deborah Garrett Eddins (LB)	19,970
11.	**Charles H. Taylor*** (R)	112,908
	David Young (D)	84,256
12.	**Mel Watt*** (D)	82,305
	John "Scott" Keadle (R)	62,070
	North Dakota	
	Earl Pomeroy* (D)	119,668
	Kevin Cramer (R)	87,511

Dist.	Representative (Party)	1998 Election
	Ohio	
1.	**Steve Chabot*** (R)	**92,421**
	Roxanne Qualls (D)	82,003
2.	**Rob Portman*** (R)	**154,344**
	Charles W. Sanders (D)	49,293
3.	**Tony P. Hall*** (D)	**114,198**
	John S. Shondel (R)	50,544
4.	**Michael G. Oxley*** (R)	**112,011**
	Paul McClain (D)	63,529
5.	**Paul E. Gillmor*** (R)	**123,979**
	Susan Davenport Darrow (D)	61,926
6.	**Ted Strickland*** (D)	**102,852**
	Nancy P. Hollister (R)	77,711
7.	**Dave Hobson*** (R)	**120,765**
	Donald E. Minor, Jr. (D)	49,780
8.	**John A. Boehner*** (R)	**127,979**
	John W. Griffin (D)	52,912
9.	**Marcy Kaptur*** (D)	**130,793**
	Edward S. Emery (R)	30,312
10.	**Dennis J. Kucinich*** (D)	**110,552**
	Joe Slovenec (R)	55,015
11.	**Stephanie Tubbs Jones** (D)	**115,226**
	James D. Hereford (R)	18,592
12.	**John R. Kasich*** (R)	**124,197**
	Edward S. Brown (D)	60,694
13.	**Sherrod Brown*** (D)	**116,309**
	Grace L. Drake (R)	72,666
14.	**Thomas C. Sawyer*** (D)	**106,046**
	Tom Watkins (R)	63,027
15.	**Deborah Pryce*** (R)	**113,846**
	Adam Clay Miller (D)	49,334
16.	**Ralph Regula*** (R)	**117,426**
	Peter D. Ferguson (D)	66,047
17.	**James A. Traficant, Jr.*** (D)	**123,718**
	Paul H. Alberty (R)	57,703
18.	**Bob Ney*** (R)	**113,119**
	Robert L. Burch (D)	74,571
19.	**Steven C. LaTourette*** (R)	**126,786**
	Elizabeth Kelley (D)	64,090
	Oklahoma	
1.	**Steve Largent*** (R)	**91,031**
	Howard Plowman (D)	56,309
2.	**Tom A. Coburn*** (R)	**85,581**
	Kent Pharaoh (D)	59,042
3.	**Wes Watkins*** (R)	**89,832**
	Walt Roberts (D)	55,163
4.	**J.C. Watts, Jr.*** (R)	**83,272**
	Ben Odom (D)	52,107
5.	**Ernest Istook*** (R)	**103,217**
	M. C. Smothermon (D)	48,182
6.	**Frank D. Lucas*** (R)	**85,261**
	Paul M. Barby (D)	43,555
	Oregon	
1.	**David Wu** (D)	**119,993**
	Molly Bordonaro (R)	112,827
2.	**Greg Walden** (R)	**132,316**
	Kevin M. Campbell (D)	74,924
3.	**Earl Blumenauer*** (D)	**153,889**
	Bruce Alexander Knight (LB)	16,930
4.	**Peter A. DeFazio*** (D)	**157,524**
	Steve J. Webb (R)	64,143
5.	**Darlene Hooley*** (D)	**124,916**
	Marylin Shannon (R)	92,215
	Pennsylvania	
1.	**Robert A. Brady*** (D)	**77,788**
	William M. Harrison (R)	15,898
2.	**Chaka Fattah*** (D)	**102,763**
	Anne Marie Mulligan (R)	16,001
3.	**Robert A. Borski*** (D)	**66,270**
	Charles F. Dougherty (R)	45,390

Dist.	Representative (Party)	1998 Election
4.	**Ron Klink*** (D)	**103,183**
	Mike Turzai (R)	58,485
5.	**John E. Peterson*** (R)	**99,502**
	William M. Belitskus (GR)	17,734
6.	**Tim Holden*** (D)	**85,374**
	John Meckley (R)	54,579
7.	**Curt Weldon*** (R)	**119,491**
	Martin J. D'Urso (D)	46,920
8.	**Jim Greenwood*** (R)	**93,697**
	Bill Tuthill (D)	48,320
9.	**Bud Shuster*** (R)	**Unopposed**
10.	**Don Sherwood** (R)	**84,275**
	Patrick Casey (D)	83,760
11.	**Paul E. Kanjorski*** (D)	**88,933**
	Stephen A. Urban (R)	44,123
12.	**John P. Murtha*** (D)	**100,528**
	Timothy E. Holloway (R)	46,239
13.	**Joseph M. Hoeffel*** (D)	**95,105**
	Jon D. Fox* (R)	85,915
14.	**William J. Coyne*** (D)	**83,355**
	Bill Ravotti (R)	52,745
15.	**Pat Toomey** (R)	**81,755**
	Roy C. Afflerbach (D)	66,930
16.	**Joseph R. Pitts*** (R)	**95,979**
	Robert S. Yorczyk (D)	40,092
17.	**George W. Gekas*** (R)	**Unopposed**
18.	**Mike Doyle*** (D)	**98,363**
	Dick Walker (R)	46,945
19.	**Bill Goodling*** (R)	**96,284**
	Linda G. Ropp (D)	40,674
20.	**Frank Mascara*** (D)	**Unopposed**
21.	**Phil English*** (R)	**94,518**
	Larry Klemens (D)	54,591
	Rhode Island	
1.	**Patrick J. Kennedy*** (D)	**92,788**
	Ronald G. Santa (R)	38,460
2.	**Robert A. Weygand*** (D)	**110,917**
	John O. Matson (R)	38,169
	South Carolina	
1.	**Mark Sanford*** (R)	**118,414**
	Joe Innella (NL)	11,586
2.	**Floyd D. Spence*** (R)	**119,583**
	Jane Frederick (D)	84,864
3.	**Lindsey Graham*** (R)	**Unopposed**
4.	**Jim DeMint** (R)	**105,264**
	D. Glenn Reese (D)	73,314
5.	**John Spratt*** (D)	**95,105**
	Mike Burkhold (R)	66,299
6.	**James E. Clyburn*** (D)	**116,507**
	Gary McLeod (R)	41,421
	South Dakota	
	John R. Thune* (R)	**194,157**
	Jeff Moser (D)	64,433
	Tennessee	
1.	**William L. Jenkins*** (R)	**68,904**
	Kay C. White (D)	30,710
2.	**John J. Duncan*** (R)	**90,860**
	Robert O. Watson (I)	4,372
3.	**Zach Wamp*** (R)	**75,100**
	James M. Lewis, Jr. (D)	37,144
4.	**William V. Hilleary*** (R)	**62,829**
	Jerry W. Cooper (D)	42,627
5.	**Bob Clement*** (D)	**74,611**
	William M. Lancaster (I)	6,162
6.	**Bart Gordon*** (D)	**75,055**
	Walt Massey (R)	62,277
7.	**Ed Bryant*** (R)	**Unopposed**
8.	**John S. Tanner*** (D)	**Unopposed**
9.	**Harold E. Ford, Jr.*** (D)	**75,428**
	Claude Burdikoff (R)	18,078

Dist.	Representative (Party)	1998 Election	Dist.	Representative (Party)	1998 Election
	Texas			**Vermont**	
1.	**Max Sandlin*** (D)	**80,788**		**Bernie Sanders*** (I)	**136,403**
	Dennis Boerner (R)	55,191		Mark Candon (R)	70,740
2.	**Jim Turner*** (D)	**81,556**		**Virginia**	
	Brian Babin (R)	56,891	1.	**Herbert H. "Herb" Bateman*** (R)	**76,474**
3.	**Sam Johnson*** (R)	**106,690**		Bradford L. Phillips (I)	13,235
	Ken Ashby (LB)	10,288	2.	**Owen B. Pickett*** (D)	**Unopposed**
4.	**Ralph M. Hall*** (D)	**82,989**	3.	**Robert C. "Bobby" Scott*** (D)	**48,129**
	Jim Lohmeyer (R)	58,954		Robert S. "Bob" Barnett (I)	14,453
5.	**Pete Sessions*** (R)	**61,714**	4.	**Norman Sisisky*** (D)	**Unopposed**
	Victor Morales (D)	48,073	5.	**Virgil H. Goode, Jr.*** (D)	**Unopposed**
6.	**Joe Barton*** (R)	**112,957**	6.	**Robert W. "Bob" Goodlatte*** (R)	**89,177**
	Ben B. Boothe (D)	40,112		David A. Bowers (D)	39,487
7.	**Bill Archer*** (R)	**111,010**	7.	**Thomas J. "Tom" Bliley, Jr.*** (R)	**77,044**
	Drew Parks (LB)	7,889		Bradley E. Evans (I)	20,293
8.	**Kevin Brady*** (R)	**123,372**	8.	**James P. Moran, Jr.*** (D)	**97,545**
	Don L. Richards (LB)	9,576		Demaris H. Miller (R)	48,352
9.	**Nick Lampson*** (D)	**86,055**	9.	**Frederick C. "Rick" Boucher*** (D)	**87,163**
	Tom Cottar (R)	49,101		J. A. "Joe" Barta (R)	55,918
10.	**Lloyd Doggett*** (D)	**116,127**	10.	**Frank R. Wolf*** (R)	**103,648**
	Vincent J. May (LB)	20,155		Cornell W. Brooks (D)	36,476
11.	**Chet Edwards*** (D)	**71,142**	11.	**Thomas M. Davis III*** (R)	**91,603**
	Vince Hanke (LB)	15,161		C. W. "Levi" Levy (I)	18,807
12.	**Kay Granger*** (R)	**66,740**		**Washington**	
	Tom Hall (D)	39,084	1.	**Jay Inslee** (D)	**112,726**
13.	**Mac Thornberry*** (R)	**81,141**		Rick White* (R)	99,910
	Mark Harmon (D)	37,027	2.	**Jack Metcalf*** (R)	**124,125**
14.	**Ron Paul*** (R)	**84,459**		Grethe Cammermeyer (D)	100,776
	Loy Sneary (D)	68,014	3.	**Brian Baird** (D)	**120,364**
15.	**Ruben Hinojosa*** (D)	**47,957**		Don Benton (R)	99,855
	Tom Haughey (R)	34,221	4.	**Doc Hastings*** (R)	**121,684**
16.	**Silvestre Reyes*** (D)	**67,486**		Gordon Allen Pross (D)	43,043
	Stu Nance (LB)	5,329	5.	**George Nethercutt*** (R)	**110,040**
17.	**Charlie Stenholm*** (D)	**75,367**		Brad Lyons (D)	73,545
	Rudy Izzard (R)	63,700	6.	**Norm Dicks*** (D)	**143,308**
18.	**Sheila Jackson Lee*** (D)	**82,091**		Bob Lawrence (R)	66,291
	James Galvan (LB)	9,176	7.	**Jim McDermott*** (D)	**183,076**
19.	**Larry Combest*** (R)	**108,266**		Stan Lippmann (RF)	19,545
	Sidney Blankenship (D)	21,162	8.	**Jennifer Dunn*** (R)	**135,539**
20.	**Charlie Gonzalez** (D)	**50,356**		Heidi Behrens-Benedict (D)	91,371
	James Walker (R)	28,347	9.	**Adam Smith*** (D)	**111,948**
21.	**Lamar Smith*** (R)	**165,047**		Ron Taber (R)	61,108
	Jeffrey Charles Blunt (LB)	15,561		**West Virginia**	
22.	**Tom DeLay*** (R)	**87,840**	1.	**Alan B. Mollohan*** (D)	**105,101**
	Hill Kemp (D)	45,386		Richard Kerr (LB)	19,013
23.	**Henry Bonilla*** (R)	**73,177**	2.	**Bob Wise*** (D)	**99,357**
	Charlie Urbina Jones (D)	40,281		Sally Anne Kay (R)	29,136
24.	**Martin Frost*** (D)	**56,321**	3.	**Nick Joe Rahall, II*** (D)	**78,814**
	Shawn Terry (R)	40,105		Joe Whelan (LB)	12,196
25.	**Ken Bentsen*** (D)	**58,591**		**Wisconsin**	
	John M. Sanchez (R)	41,848	1.	**Paul Ryan** (R)	**108,475**
26.	**Dick Armey*** (R)	**120,332**		Lydia Carol Spottswood (D)	81,164
	Joe Turner (LB)	16,182	2.	**Tammy Baldwin** (D)	**116,377**
27.	**Solomon P. Ortiz*** (D)	**61,638**		Josephine W. Musser (R)	103,528
	Erol A. Stone (R)	34,284	3.	**Ron Kind*** (D)	**128,256**
28.	**Ciro D. Rodriguez*** (D)	**71,849**		Troy A. Brechler (R)	51,001
	Edward Elmer (LB)	7,504	4.	**Jerry Kleczka*** (D)	**105,841**
29.	**Gene Green*** (D)	**44,179**		Tom Reynolds (R)	76,666
	Lea Sherman (I)	2,013	5.	**Tom Barrett*** (D)	**121,129**
30.	**Eddie Bernice Johnson*** (D)	**57,603**		Jack Melvin (R)	33,506
	Carrie Kelleher (R)	21,338	6.	**Thomas E. Petri*** (R)	**144,144**
	Utah			Timothy J. Farness (TX)	11,267
1.	**James V. Hansen*** (R)	**109,708**	7.	**David R. Obey*** (D)	**115,613**
	Steve Beierlein (D)	49,307		Scott West (R)	75,049
2.	**Merrill Cook*** (R)	**93,718**	8.	**Mark Green*** (R)	**112,418**
	Lily Eskelsen (D)	77,198		Jay Johnson* (D)	93,441
3.	**Chris Cannon*** (R)	**100,830**	9.	**F. James Sensenbrenner, Jr.*** (R)	**175,533**
	Will Christensen (IA)	20,720		Jeffrey M. Gonyo (I)	16,419
				Wyoming	
				Barbara Cubin* (R)	**100,687**
				Scott Farris (D)	67,399

The following members of Congress are nonvoting: Carlos A. Romero-Barceló (D), resident commissioner, Puerto Rico; Eleanor Holmes Norton (D), District of Columbia; Robert A. Underwood (D), Guam; Eni F. H. Faleomavaega (D), American Samoa; Donna M. Christian-Christensen (D), Virgin Islands. (1) Seat fell vacant after the death of George E. Brown, Jr. on July 15, 1999. A special election to fill the seat was to be held Nov. 16, 1999. (2) Johnny Isakson won a special election held Feb. 23, 1999, to fill the seat left vacant by the resignation of Newt Gingrich. (3) David Vitter won a special election held May 29, 1999, to fill the seat left vacant by the resignation of Bob Livingston. (4) Michael P. Forbes changed his party designation to Democrat in 1999.

Congressional Committees

Senate Standing Committees
(as of Oct. 1999)

Agriculture, Nutrition, and Forestry
Chairman: Richard G. Lugar, IN
Ranking Dem.: Tom Harkin, IA

Appropriations
Chairman: Ted Stevens, AK
Ranking Dem.: Robert C. Byrd, WV

Armed Services
Chairman: John W. Warner, VA
Ranking Dem.: Carl Levin, MI

Banking, Housing, and Urban Affairs
Chairman: Phil Gramm, TX
Ranking Dem.: Paul S. Sarbanes, MD

Budget
Chairman: Pete V. Domenici, NM
Ranking Dem.: Frank R. Lautenberg, NJ

Commerce, Science, and Transportation
Chairman: John McCain, AZ
Ranking Dem.: Ernest Hollings, SC

Energy and Natural Resources
Chairman: Frank H. Murkowski, AK
Ranking Dem.: Jeff Bingaman, NM

Environment and Public Works
Chairman: Vacant[1]
Ranking Dem.: Max Baucus, MT

Finance
Chairman: William V. Roth, Jr., DE
Ranking Dem.: Daniel Patrick Moynihan, NY

Foreign Relations
Chairman: Jesse Helms, NC
Ranking Dem.: Joseph R. Biden, Jr., DE

Governmental Affairs
Chairman: Fred Thompson, TN
Ranking Dem.: Joe Lieberman, CT

Health, Education, Labor, and Pensions
Chairman: Jim Jeffords, VT
Ranking Dem.: Edward M. Kennedy, MA

Indian Affairs
Chairman: Ben Nighthorse Campbell, CO
Ranking Dem.: Daniel K. Inouye, HI

Judiciary
Chairman: Orrin G. Hatch, UT
Ranking Dem.: Patrick Leahy, VT

Rules and Administration
Chairman: Mitch McConnell, KY
Ranking Dem.: Christopher J. Dodd, CT

Small Business
Chairman: Christopher (Kit) Bond, MO
Ranking Dem.: John F. Kerry, MA

Veterans' Affairs
Chairman: Arlen Specter, PA
Ranking Dem.: John D. Rockefeller IV, WV

(1)Vacancy due to the death of Sen. John H. Chafee (RI) on Oct. 24, 1999.

Senate Special Committees
(as of Oct. 1999)

Aging
Chairman: Charles Grassley, IA
Ranking Dem.: John B. Breaux, LA

Year 2000 Technology Problem
Chairman: Robert F. Bennett, UT
Ranking Dem.: Christopher J. Dodd, CT

Senate Select Committees
(as of Oct. 15, 1999)

Ethics
Chairman: Pat Roberts, KS
Ranking Dem.: Harry Reid, NV

Intelligence
Chairman: Richard Shelby, AL
Ranking Dem.: Bob Kerrey, NE

House Select Committee
(as of Oct. 1999)

Intelligence
Chairman: Porter Goss, FL
Ranking Dem.: Julian C. Dixon, CA

Joint Committees of Congress
(as of Oct. 1999)

Economic
Chairman: Sen. Connie Mack, FL
V. Chairman: Rep. Jim Saxton, NJ

Library
Chairman: Sen. Ted Stevens, AK
V. Chairman: Rep. Bill Thomas, CA

Printing
V. Chairman: Sen. Mitch McConnell, KY
Chairman: Rep. Bill Thomas, CA

Taxation
Chairman: Rep. Bill Archer, TX
V. Chairman: Sen. William V. Roth, Jr., DE

House Standing Committees
(as of Oct. 1999)

Agriculture
Chairman: Larry Combest, TX
Ranking Dem.: Charlie Stenholm, TX

Appropriations
Chairman: C. W. Bill Young, FL
Ranking Dem.: David R. Obey, WI

Armed Services
Chairman: Floyd D. Spence, SC
Ranking Dem.: Ike Skelton, MO

Banking and Financial Services
Chairman: Jim Leach, IA
Ranking Dem.: John J. La Falce, NY

Budget
Chairman: John R. Kasich, OH
Ranking Dem.: John Spratt, SC

Commerce
Chairman: Thomas J. "Tom" Bliley, Jr., VA
Ranking Dem.: John D. Dingell, MI

Education and the Workforce
Chairman: Bill Goodling, PA
Ranking Dem.: William (Bill) Clay, Sr., MO

Government Reform
Chairman: Dan Burton, IN
Ranking Dem.: Henry A. Waxman, CA

House Administration
Chairman: Bill Thomas, CA
Ranking Dem.: Steny H. Hoyer, MD

International Relations
Chairman: Benjamin A. Gilman, NY
Ranking Dem.: Sam Gejdenson, CT

Judiciary
Chairman: Henry J. Hyde, IL
Ranking Dem.: John Conyers, Jr., MI

Resources
Chairman: Don Young, AK
Ranking Dem.: George Miller, CA

Rules
Chairman: David Dreier, CA
Ranking Dem.: John Joseph Moakley, MA

Science
Chairman: F. James Sensenbrenner, Jr., WI
Ranking Dem.: Ralph M. Hall, TX

Small Business
Chairman: James M. Talent, MO
Ranking Dem.: Nydia M. Velazquez, NY

Standards of Official Conduct
Chairman: Lamar S. Smith, TX
Ranking Dem.: Howard L. Berman, CA

Transportation and Infrastructure
Chairman: Bud Shuster, PA
Ranking Dem.: James L. Oberstar, MN

Veterans' Affairs
Chairman: Bob Stump, AZ
Ranking Dem.: Lane A. Evans, IL

Ways and Means
Chairman: Bill Archer, TX
Ranking Dem.: Charles B. Rangel, NY

Speakers of the House of Representatives

(as of Oct. 15, 1999)

Party designations: A, American; D, Democratic; DR, Democratic-Republican; F, Federalist; R, Republican; W, Whig

Name	Party	State	Tenure	Name	Party	State	Tenure
Frederick Muhlenberg	F	PA	1789-1791	James G. Blaine	R	ME	1869-1875
Jonathan Trumbull	F	CT	1791-1793	Michael C. Kerr	D	IN	1875-1876
Frederick Muhlenberg	F	PA	1793-1795	Samuel J. Randall	D	PA	1876-1881
Jonathan Dayton	F	NJ	1795-1799	Joseph W. Keifer	R	OH	1881-1883
Theodore Sedgwick	F	MA	1799-1801	John G. Carlisle	D	KY	1883-1889
Nathaniel Macon	DR	NC	1801-1807	Thomas B. Reed	R	ME	1889-1891
Joseph B. Varnum	DR	MA	1807-1811	Charles F. Crisp	D	GA	1891-1895
Henry Clay	DR	KY	1811-1814	Thomas B. Reed	R	ME	1895-1899
Langdon Cheves	DR	SC	1814-1815	David B. Henderson	R	IA	1899-1903
Henry Clay	DR	KY	1815-1820	Joseph G. Cannon	R	IL	1903-1911
John W. Taylor	DR	NY	1820-1821	Champ Clark	D	MO	1911-1919
Philip P. Barbour	DR	VA	1821-1823	Frederick H. Gillett	R	MA	1919-1925
Henry Clay	DR	KY	1823-1825	Nicholas Longworth	R	OH	1925-1931
John W. Taylor	D	NY	1825-1827	John N. Garner	D	TX	1931-1933
Andrew Stevenson	D	VA	1827-1834	Henry T. Rainey	D	IL	1933-1935
John Bell	D	TN	1834-1835	Joseph W. Byrns	D	TN	1935-1936
James K. Polk	D	TN	1835-1839	William B. Bankhead	D	AL	1936-1940
Robert M. T. Hunter	D	VA	1839-1841	Sam Rayburn	D	TX	1940-1947
John White	W	KY	1841-1843	Joseph W. Martin Jr.	R	MA	1947-1949
John W. Jones	D	VA	1843-1845	Sam Rayburn	D	TX	1949-1953
John W. Davis	D	IN	1845-1847	Joseph W. Martin Jr.	R	MA	1953-1955
Robert C. Winthrop	W	MA	1847-1849	Sam Rayburn	D	TX	1955-1961
Howell Cobb	D	GA	1849-1851	John W. McCormack	D	MA	1962-1971
Linn Boyd	D	KY	1851-1855	Carl Albert	D	OK	1971-1977
Nathaniel P. Banks	A	MA	1856-1857	Thomas P. O'Neill Jr.	D	MA	1977-1987
James L. Orr	D	SC	1857-1859	James Wright	D	TX	1987-1989
William Pennington	R	NJ	1860-1861	Thomas S. Foley	D	WA	1989-1995
Galusha A. Grow	R	PA	1861-1863	Newt Gingrich	R	GA	1995-1999
Schuyler Colfax	R	IN	1863-1869	J. Dennis Hastert	R	IL	1999-
Theodore M. Pomeroy	R	NY	1869				

Floor Leaders in the U.S. Senate Since the 1920s

Majority Leaders				**Minority Leaders**			
Name	Party	State	Tenure	Name	Party	State	Tenure
Charles Curtis[1]	R	KS	1925-1929	Oscar W. Underwood[2]	D	AL	1920-1923
James E. Watson	R	IN	1929-1933	Joseph T. Robinson	D	AR	1923-1933
Joseph T. Robinson	D	AR	1933-1937	Charles L. McNary	R	OR	1933-1944
Alben W. Barkley	D	KY	1937-1947	Wallace H. White	R	ME	1944-1947
Wallace H. White	R	ME	1947-1949	Alben W. Barkley	D.	KY	1947-1949
Scott W. Lucas	D	IL	1949-1951	Kenneth S. Wherry	R	NE	1949-1951
Ernest W. McFarland	D	AZ	1951-1953	Henry Styles Bridges	R	NH	1951-1953
Robert A. Taft	R	OH	1953	Lyndon B. Johnson	D	TX	1953-1955
William F. Knowland	R	CA	1953-1955	William F. Knowland	R	CA	1955-1959
Lyndon B. Johnson	D	TX	1955-1961	Everett M. Dirksen	R	IL	1959-1969
Mike Mansfield	D	MT	1961-1977	Hugh D. Scott	R	PA	1969-1977
Robert C. Byrd	D	WV	1977-1981	Howard H. Baker Jr.	R	TN	1977-1981
Howard H. Baker Jr.	R	TN	1981-1985	Robert C. Byrd	D	WV	1981-1987
Robert J. Dole	R	KS	1985-1987	Robert J. Dole	R	KS	1987-1995
Robert C. Byrd	D	WV	1987-1989	Thomas A. Daschle	D	SD	1995-
George J. Mitchell	D	ME	1989-1995				
Robert J. Dole	R	KS	1995-1996				
Trent Lott	R	MS	1996-				

Note: Majority and Minority Leaders as of Oct. 15, 1999. (1) First Republican to be designated floor leader. (2) First Democrat to be designated floor leader.

Political Divisions of the U.S. Senate and House of Representatives, 1901-99

Source: *1995-1996 Congressional Directory*; Senate Library; all figures reflect immediate post-election party breakdown

		SENATE					HOUSE OF REPRESENTATIVES				
Congress	Years	No. of Sen.	Demo-crats	Repub-licans	Other parties	Vacant	No. of Rep.	Demo-crats	Repub-licans	Other parties	Vacant
57th	1901-03	90	29	56	3	2	357	153	198	5	1
58th	1903-05	90	32	58			386	178	207		1
59th	1905-07	90	32	58			386	136	250		
60th	1907-09	92	29	61		2	386	164	222		
61st	1909-11	92	32	59		1	391	172	219		
62d	1911-13	92	42	49		1	391	228	162	1	
63d	1913-15	96	51	44	1		435	290	127	18	
64th	1915-17	96	56	39	1		435	231	193	8	3
65th	1917-19	96	53	42	1		435	210[1]	216	9	
66th	1919-21	96	47	48	1		435	191	237	7	
67th	1921-23	96	37	59			435	132	300	1	2
68th	1923-25	96	43	51	2		435	207	225	3	
69th	1925-27	96	40	54	1	1	435	183	247	5	
70th	1927-29	96	47	48	1		435	195	237	3	
71st	1929-31	96	39	56	1		435	163	267	1	4
72d	1931-33	96	47	48	1		435	216[2]	218	1	
73d	1933-35	96	59	36	1		435	313	117	5	
74th	1935-37	96	69	25	2		435	322	103	10	

Congress	Years	SENATE No. of Sen.	Demo-crats	Repub-licans	Other parties	Vacant	HOUSE OF REPRESENTATIVES No. of Rep.	Demo-crats	Repub-licans	Other parties	Vacant
75th	1937-39	96	75	17	4		435	333	89	13	
76th	1939-41	96	69	23	4		435	262	169	4	
77th	1941-43	96	66	28	2		435	267	162	6	
78th	1943-45	96	57	38	1		435	222	209	4	
79th	1945-47	96	57	38	1		435	243	190	2	
80th	1947-49	96	45	51			435	188	246	1	
81st	1949-51	96	54	42			435	263	171	1	
82d	1951-53	96	48	47	1		435	234	199	2	
83d	1953-55	96	46	48	2		435	213	221	1	
84th	1955-57	96	48	47	1		435	232	203		
85th	1957-59	96	49	47			435	234	201		
86th	1959-61	98	64	34			436[3]	283	153		
87th	1961-63	100	64	36			437[4]	262	175		
88th	1963-65	100	67	33			435	258	176		1
89th	1965-67	100	68	32			435	295	140		
90th	1967-69	100	64	36			435	248	187		
91st	1969-71	100	58	42			435	243	192		
92d	1971-73	100	54	44	2		435	255	180		
93d	1973-75	100	56	42	2		435	242	192	1	
94th	1975-77	100	60	37	2		435	291	144	1	
95th	1977-79	100	61	38	1		435	292	143		
96th	1979-81	100	58	41	1		435	277	158		
97th	1981-83	100	46	53	1		435	242	192	1	
98th	1983-85	100	46	54			435	269	166		
99th	1985-87	100	47	53			435	253	182		
100th	1987-89	100	55	45			435	258	177		
101st	1989-91	100	55	45			435	260	175		
102d	1991-93	100	56	44			435	267	167	1	
103d	1993-95	100	57	43			435	258	176	1	
104th	1995-97	100	48	52			435	204	230	1	
105th	1997-99	100	45	55			435	207	227	1	
106th	1999-2001	100	45[5]	55[5]			435	211[5]	223[5]	1[5]	

(1) Democrats organized House with help of other parties. (2) Democrats organized House because of Republican deaths. (3) Proclamation declaring Alaska a state issued Jan. 3, 1959. (4) Proclamation declaring Hawaii a state issued Aug. 21, 1959. (5) As of Oct. 25, 1999, there were 45 Democrats, 53 Republicans, 1 Independent, and 1 vacant seat in the Senate, and 211 Democrats, 222 Republicans, 1 Independent, and 1 vacant seat in the House of Representatives.

Congressional Bills Vetoed, 1789-1999

Source: Senate Library

	Regular vetoes	Pocket vetoes	Total vetoes	Vetoes overridden		Regular vetoes	Pocket vetoes	Total vetoes	Vetoes overridden
Washington	2	—	2	—	Benjamin Harrison	19	25	44	1
John Adams	—	—	—	—	Cleveland	42	128	170	5
Jefferson	—	—	—	—	McKinley	6	36	42	—
Madison	5	2	7	—	Theodore Roosevelt	42	40	82	1
Monroe	1	—	1	—	Taft	30	9	39	1
John Q. Adams	—	—	—	—	Wilson	33	11	44	6
Jackson	5	7	12	—	Harding	5	1	6	—
Van Buren	—	1	1	—	Coolidge	20	30	50	4
William Harrison	—	—	—	—	Hoover	21	16	37	3
Tyler	6	4	10	1	Franklin Roosevelt	372	263	635	9
Polk	2	1	3	—	Truman	180	70	250	12
Taylor	—	—	—	—	Eisenhower	73	108	181	2
Fillmore	—	—	—	—	Kennedy	12	9	21	—
Pierce	9	—	9	5	Lyndon Johnson	16	14	30	—
Buchanan	4	3	7	—	Nixon	26	17	43	7
Lincoln	2	4	6	—	Ford	48	18	66	12
Andrew Johnson	21	8	29	15	Carter	13	18	31	2
Grant	45	48	93	4	Reagan	39	39	78	9
Hayes	12	1	13	1	Bush[1]	29	15	44	1
Garfield	—	—	—	—	Clinton[2,3]	27	—	27	4
Arthur	4	8	12	1	Total[1,3]	1,475	1,064	2,539	108
Cleveland	304	110	414	2					

(1) Excluded from the figures are 2 additional bills, which Pres. Bush claimed to be vetoed but Congress considered enacted into law because the president failed to return them to Congress during a recess period. (2) As of Oct. 1, 1999. (3) Does not include line-item veto, which was ruled unconstitutional by the Supreme Court on June 25, 1998.

Librarians of Congress

Librarian	Served	Appointed by President	Librarian	Served	Appointed by President
John J. Beckley	1802-1807	Jefferson	Herbert Putnam	1899-1939	McKinley
Patrick Magruder	1807-1815	Jefferson	Archibald MacLeish	1939-1944	F. D. Roosevelt
George Watterston	1815-1829	Madison	Luther H. Evans	1945-1953	Truman
John Silva Meehan	1829-1861	Jackson	L. Quincy Mumford	1954-1974	Eisenhower
John G. Stephenson	1861-1864	Lincoln	Daniel J. Boorstin	1975-1987	Ford
Ainsworth Rand Spofford	1864-1897	Lincoln	James H. Billington	1987-	Reagan
John Russell Young	1897-1899	McKinley			

U.S. JUDICIARY

(data as of Oct. 1999)

Justices of the United States Supreme Court

The Supreme Court comprises the chief justice of the U.S. and 8 associate justices, all appointed by the president with advice and consent of the Senate. Salaries (as of Jan. 1, 2000): chief justice, $181,400 annually; associate justice, $173,600 annually. The Supreme Court is at the U.S. Supreme Court Bldg., 1 First St. NE, Washington, DC 20543. Internet sites containing the full text of Supreme Court decisions include: http://supct.law.cornell.edu/supct

Members of the Supreme Court at the start of the 1999–2000 term (Oct. 4, 1999): Chief justice: William H. Rehnquist; associate justices: Stephen G. Breyer, Ruth Bader Ginsburg, Anthony M. Kennedy, Sandra Day O'Connor, Antonin Scalia, David H. Souter, John Paul Stevens, Clarence Thomas.

Name,[1] apptd. from	Service Term	Yrs	Born	Died
John Jay, NY	1789-1795	5	1745	1829
John Rutledge, SC	1789-1791	1	1739	1800
William Cushing, MA	1789-1810	20	1732	1810
James Wilson, PA	1789-1798	8	1742	1798
John Blair, VA	1789-1796	6	1732	1800
James Iredell, NC	1790-1799	9	1751	1799
Thomas Johnson, MD	1791-1793	1	1732	1819
William Paterson, NJ	1793-1806	13	1745	1806
John Rutledge[2], SC	1795	—	1739	1800
Samuel Chase, MD	1796-1811	15	1741	1811
Oliver Ellsworth, CT	1796-1800	4	1745	1807
Bushrod Washington, VA	1798-1829	31	1762	1829
Alfred Moore, NC	1799-1804	4	1755	1810
John Marshall, VA	1801-1835	34	1755	1835
William Johnson, SC	1804-1834	30	1771	1834
Henry B. Livingston, NY	1806-1823	16	1757	1823
Thomas Todd, KY	1807-1826	18	1765	1826
Joseph Story, MA	1811-1845	33	1779	1845
Gabriel Duval, MD	1811-1835	22	1752	1844
Smith Thompson, NY	1823-1843	20	1768	1843
Robert Trimble, KY	1826-1828	2	1777	1828
John McLean, OH	1829-1861	32	1785	1861
Henry Baldwin, PA	1830-1844	14	1780	1844
James M. Wayne, GA	1835-1867	32	1790	1867
Roger B. Taney, MD	1836-1864	28	1777	1864
Philip P. Barbour, VA	1836-1841	4	1783	1841
John Catron, TN	1837-1865	28	1786	1865
John McKinley, AL	1837-1852	15	1780	1852
Peter V. Daniel, VA	1841-1860	19	1784	1860
Samuel Nelson, NY	1845-1872	27	1792	1873
Levi Woodbury, NH	1845-1851	5	1789	1851
Robert C. Grier, PA	1846-1870	23	1794	1870
Benjamin R. Curtis, MA	1851-1857	6	1809	1874
John A. Campbell, AL	1853-1861	8	1811	1889
Nathan Clifford, ME	1858-1881	23	1803	1881
Noah H. Swayne, OH	1862-1881	18	1804	1884
Samuel F. Miller, IA	1862-1890	28	1816	1890
David Davis, IL	1862-1877	14	1815	1886
Stephen J. Field, CA	1863-1897	34	1816	1899
Salmon P. Chase, OH	1864-1873	8	1808	1873
William Strong, PA	1870-1880	10	1808	1895
Joseph P. Bradley, NJ	1870-1892	21	1813	1892
Ward Hunt, NY	1872-1882	9	1810	1886
Morrison R. Waite, OH	1874-1888	14	1816	1888
John M. Harlan, KY	1877-1911	34	1833	1911
William B. Woods, GA	1880-1887	6	1824	1887
Stanley Matthews, OH	1881-1889	7	1824	1889
Horace Gray, MA	1881-1902	20	1828	1902
Samuel Blatchford, NY	1882-1893	11	1820	1893
Lucius Q.C. Lamar, MS	1888-1893	5	1825	1893
Melville W. Fuller, IL	1888-1910	21	1833	1910
David J. Brewer, KS	1889-1910	20	1837	1910
Henry B. Brown, MI	1890-1906	15	1836	1913
George Shiras Jr., PA	1892-1903	10	1832	1924
Howell E. Jackson, TN	1893-1895	2	1832	1895
Edward D. White, LA	1894-1910	16	1845	1921
Rufus W. Peckham, NY	1895-1909	13	1838	1909

Name,[1] apptd. from	Service Term	Yrs	Born	Died
Joseph McKenna, CA	1898-1925	26	1843	1926
Oliver W. Holmes, MA	1902-1932	29	1841	1935
William R. Day, OH	1903-1922	19	1849	1923
William H. Moody, MA	1906-1910	3	1853	1917
Horace H. Lurton, TN	1909-1914	4	1844	1914
Charles E. Hughes, NY	1910-1916	5	1862	1948
Willis Van Devanter, WY	1910-1937	26	1859	1941
Joseph R. Lamar, GA	1910-1916	5	1857	1916
Edward D. White, LA	1910-1921	10	1845	1921
Mahlon Pitney, NJ	1912-1922	10	1858	1924
James C. McReynolds, TN	1914-1941	26	1862	1946
Louis D. Brandeis, MA	1916-1939	22	1856	1941
John H. Clarke, OH	1916-1922	5	1857	1945
William H. Taft, CT	1921-1930	8	1857	1930
George Sutherland, UT	1922-1938	15	1862	1942
Pierce Butler, MN	1922-1939	16	1866	1939
Edward T. Sanford, TN	1923-1930	7	1865	1930
Harlan F. Stone, NY	1925-1941	16	1872	1946
Charles E. Hughes, NY	1930-1941	11	1862	1948
Owen J. Roberts, PA	1930-1945	15	1875	1955
Benjamin N. Cardozo, NY	1932-1938	6	1870	1938
Hugo L. Black, AL	1937-1971	34	1886	1971
Stanley F. Reed, KY	1938-1957	19	1884	1980
Felix Frankfurter, MA	1939-1962	23	1882	1965
William O. Douglas, CT	1939-1975	36	1898	1980
Frank Murphy, MI	1940-1949	9	1890	1949
Harlan F. Stone, NY	1941-1946	5	1872	1946
James F. Byrnes, SC	1941-1942	1	1879	1972
Robert H. Jackson, NY	1941-1954	12	1892	1954
Wiley B. Rutledge, IA	1943-1949	6	1894	1949
Harold H. Burton, OH	1945-1958	13	1888	1964
Fred M. Vinson, KY	1946-1953	7	1890	1953
Tom C. Clark, TX	1949-1967	18	1899	1977
Sherman Minton, IN	1949-1956	7	1890	1965
Earl Warren, CA	1953-1969	16	1891	1974
John Marshall Harlan, NY	1955-1971	16	1899	1971
William J. Brennan Jr., NJ	1956-1990	33	1906	1997
Charles E. Whittaker, MO	1957-1962	5	1901	1973
Potter Stewart, OH	1958-1981	23	1915	1985
Byron R. White, CO	1962-1993	31	1917	
Arthur J. Goldberg, IL	1962-1965	3	1908	1990
Abe Fortas, TN	1965-1969	4	1910	1982
Thurgood Marshall, NY	1967-1991	24	1908	1993
Warren E. Burger, VA	1969-1986	17	1907	1995
Harry A. Blackmun, MN	1970-1994	24	1908	1999
Lewis F. Powell Jr., VA	1971-1987	16	1907	1998
William H. Rehnquist, AZ	1971-1986	15	1924	
John Paul Stevens, IL	1975-		1920	
Sandra Day O'Connor, AZ	1981-		1930	
William H. Rehnquist, AZ	1986-		1924	
Antonin Scalia, VA	1986-		1936	
Anthony M. Kennedy, CA	1988-		1936	
David H. Souter, NH	1990-		1939	
Clarence Thomas, VA	1991-		1948	
Ruth Bader Ginsburg, DC	1993-		1933	
Stephen Breyer, MA	1994-		1938	

(1) Chief justices in italics. (2) Rejected Dec. 15, 1795.

U.S. Courts of Appeals

(Salaries, $150,000, as of Jan. 1, 2000. CJ means Chief Judge)

Federal Circuit — Haldane Robert Mayer, CJ; Giles S. Rich, Pauline Newman, Paul R. Michel, S. Jay Plager, Alan D. Lourie, Raymond C. Clevenger III, Randall R. Rader, Alvin A. Schall, William C. Bryson, Arthur J. Gajarsa; Clerk's Office, Washington, DC 20439.

District of Columbia Circuit — Harry T. Edwards, CJ; Patricia M. Wald, Laurence H. Silberman, Stephen F. Williams, Douglas Ginsburg, David B. Sentelle, Karen LeCraft Henderson, A. Raymond Randolph, Judith W. Rogers, David S. Tatel, Merrick B. Garland; Clerk's Office, Washington, DC 20001.

First Circuit (ME, MA, NH, RI, Puerto Rico) — Juan R. Torruella, CJ; Bruce M. Selya, Michael Boudin, Norman H. Stahl, Sandra Lynch, Kermit Lipez; Clerk's Office, Boston, MA 02210.

Second Circuit (CT, NY, VT) — Ralph K. Winter, CJ; Wilfred Feinberg, James L. Oakes, Ellsworth Van Graafeiland, Thomas J. Meskill, Amalya L. Kearse, Jon O. Newman, Richard J. Cardamone, Roger J. Miner, John M. Walker, Joseph M. McLaughlin, Dennis Jacobs, Pierre N. Leval, Guido Calabresi, José A. Cabranes, Fred I. Parker, Chester J. Straub, Rosemary S. Pooler, Robert D. Sack, Sonia Sotomayor, Robert A. Katzman; Clerk's Office, New York, NY 10007.

Third Circuit (DE, NJ, PA, Virgin Islands) — Edward R. Becker, CJ; Dolores K. Sloviter, Carol Los Mansmann, Morton I. Greenberg, Anthony J. Scirica, Richard L. Nygaard, Samuel A. Alito Jr., Jane R. Roth, Theodore A. McKee, Marjorie O. Rendell; Clerk's Office, Philadelphia, PA 19106.

Fourth Circuit (MD, NC, SC, VA, WV) — J. Harvie Wilkinson III, CJ; H. Emory Widener Jr., Francis D. Murnaghan Jr., William W. Wilkins Jr., Paul V. Niemeyer, Clyde H. Hamilton, J. Michael Luttig, Karen J. Williams, M. Blane Michael, Diana Gribbon Motz, William B. Traxler Jr., Robert B. King; Clerk's Office, Richmond, VA 23219.

Fifth Circuit (LA, MS, TX) — Carolyn Dineen King, CJ; E. Grady Jolly, Patrick E. Higginbotham, W. Eugene Davis, Edith H. Jones, Jerry E. Smith, Jacques L. Wiener Jr., Rhesa H. Barksdale, Emilio M. Garza, Harold R. DeMoss Jr., Fortunato P. Benavides, Carl E. Stewart, Robert M. Parker, James L. Dennis; Clerk's Office, New Orleans, LA 70130.

Sixth Circuit (KY, MI, OH, TN) — Boyce F. Martin Jr., CJ; Gilbert S. Merritt, David A. Nelson, James L. Ryan, Danny J. Boggs, Alan E. Norris, Richard F. Suhrheinrich, Eugene E. Siler Jr., Alice M. Batchelder, Martha Craig Daughtrey, Karen Nelson Moore, R. Guy Cole Jr., Eric L. Clay, Ronald Lee Gilman; Clerk's Office, Cincinnati, OH 45202.

Seventh Circuit (IL, IN, WI) — Richard A. Posner, CJ; Thomas E. Fairchild, Wilbur F. Pell Jr., William J. Bauer, Harlington Wood Jr., Richard D. Cudahy, Jesse E. Eschbach, John L. Coffey, Joel M. Flaum, Frank H. Easterbrook, Kenneth F. Ripple, Daniel A. Manion, Michael S. Kanne, Ilana D. Rovner, Diane P. Wood, Terence T. Evans; Clerk's Office, Chicago, IL 60604.

Eighth Circuit (AR, IA, MN, MO, NE, ND, SD) — Roger L. Wollman, CJ; Pasco M. Bowman, Theodore McMillian, Richard S. Arnold, George G. Fagg, C. Arlen Beam, James B. Loken, David R. Hansen, Morris S. Arnold, Diana E. Murphy; Clerk's Office, St. Louis, MO 63101.

Ninth Circuit (AK, AZ, CA, HI, ID, MT, NV, OR, WA, Guam, N. Mariana Islands) — Procter Hug Jr., CJ; James R. Browning, Melvin Brunetti, Ferdinand F. Fernandez, William A. Fletcher, Susan P. Graber, Michael Daly Hawkins, Andrew J. Kleinfeld, Alex Kozinski, M. Margaret McKeown, Thomas G. Nelson, Diarmuid F. O'Scannlain, Harry Pregerson, Stephen Reinhardt, Pamela Ann Rymer, Mary M. Schroeder, Barry G. Silverman, A. Wallace Tashima, Sidney R. Thomas, David R. Thompson, Stephen S. Trott, Kim M. Wardlaw; Clerk's Office, San Francisco, CA 94119.

Tenth Circuit (CO, KS, NM, OK, UT, WY) — Stephanie K. Seymour, CJ; John C. Porfilio, Stephen H. Anderson, Deanell R. Tacha, Bobby R. Baldock, Wade Brorby, David M. Ebel, Paul J. Kelly, Robert H. Henry, Mary Beck Briscoe, Carlos Lucero, Michael R. Murphy; Clerk's Office, Denver, CO 80257.

Eleventh Circuit (AL, FL, GA) — R. Lanier Anderson III, CJ; Gerald B. Tjoflat, J. L. Edmondson, Emmett R. Cox, Stanley F. Birch Jr., Joel F. Dubina, Susan H. Black, Ed Carnes, Rosemary Barkett, Frank M. Hull, Stanley Marcus, Charles R. Wilson; Clerk's Office, Atlanta GA 30303.

U.S. District Courts

(Salaries, $141,300, as of Jan. 1, 2000. CJ means Chief Judge)

Alabama — Northern: U. W. Clemon, CJ; Edwin Nelson, Inge Johnson, H. Dean Buttram III, Sharon L. Blackburn, C. Lynwood Smith Jr., Sam C. Pointer Jr., J. Foy Guin, Seybourn Lynne, William Acker, James Hancock, Robert Propst; Clerk's Office, Birmingham 35203. **Middle:** W. Harold Albritton, CJ; Myron H. Thompson, Ira DeMent; Clerk's Office, Montgomery 36101. **Southern:** Charles R. Butler Jr., CJ; A.T. Howard, R.W. Vollmer, W.B. Hand, T.V. Pittman; Clerk's Office, Mobile 36602.

Alaska — James K. Singleton, CJ; H. Russel Holland, John W. Sedwick; Clerk's Office, Anchorage 99513.

Arizona — Stephen M. McNamee, CJ; William D. Browning, Paul G. Rosenblatt, Roger G. Strand, John M. Roll, Roslyn Silver, Frank R. Zappata, Alfredo Marquez, Raner C. Collins; Clerk's Office, Phoenix 85025.

Arkansas — **Eastern:** Susan Webber Wright, CJ; Stephen M. Reasoner, George Howard Jr., William R. Wilson Jr., James M. Moody; Clerk's Office, Little Rock 72201-3325. **Western:** Jimm Larry Hendren, CJ; H. Franklin Waters, Robert T. Dawson, Harry F. Barnes; Clerk's Office, Fort Smith 72902-1547.

California — **Northern:** Marilyn H. Patel, CJ; Samuel Conti, Spencer Williams, William H. Orrick Jr., William A. Ingram, William W Schwarzer, Thelton E. Henderson, D. Lowell Jensen, Charles A. Legge, Fern M. Smith, Vaughn R. Walker, James Ware, Saundra Brown Armstrong, Ronald M. Whyte, Claudia Wilken, Maxine M. Chesney, Susan Illston, Charles R. Breyer, Martin J. Jenkins, Jeremy Fogel, William Alsup; Clerk's Office, San Francisco 94102. **Eastern:** William B. Shubb, CJ; Lawrence K. Karlton, CJ Emeritus; David F. Levi, Garland E. Burrell, Anthony W. Ishii, Frank C. Damrell Jr., Milton L. Schwartz, Edward J. Garcia; Clerk's Office, Sacramento 95814. **Central:** Terry J. Hatter Jr., CJ; Manuel L. Real, Consuelo B. Marshall, Alicemarie H. Stotler, William D. Keller, Stephen V. Wilson, J. Spencer Letts, Dickran Tevrizian, Ronald S. W. Lew, Gary L. Taylor, Lourdes G. Baird, Audrey B. Collins, Richard A. Paez, Robert J. Timlin, George H. King, Dean D. Pregerson, Christina A. Snyder, Carlos R. Moreno, Margaret M. Morrow, A. Howard Matz, David O. Carter, Nora M. Manella, Gary A. Feess; Clerk's Office, Los Angeles 90012. **Southern:** Marilyn L. Huff, CJ; Judith N. Keep, Irma E. Gonzalez, Napoleon A. Jones Jr., Barry T. Moskowitz, Jeffrey T. Miller, Thomas J. Whelan, Edward J. Schwartz, Howard B. Turrentine, Gordon Thompson Jr., Leland C. Nielsen, William B. Enright, John S. Rhoades, Rudi M. Brewster, Earl B. Gilliam; Clerk's Office, San Diego 92101-8900.

Colorado — Richard P. Matsch, CJ; Lewis T. Babcock, Edward W. Nottingham, Daniel B. Sparr, Wiley Y. Daniel, Walker D. Miller; Clerk's Office, Denver 80294.

Connecticut — Alfred V. Covello, CJ; Robert N. Chatigny, Dominic J. Squatrito, Alvin W. Thompson, Janet Bond Arterton, Janet C. Hall, Christopher F. Droney, Stefan Underhill; Clerk's Office, Bridgeport 06604, Hartford 06103, New Haven 06510.

Delaware — Joseph J. Farnan Jr., CJ; Sue L. Robinson, Roderick R. McKelvie, Gregory M. Sleet; Clerk's Office, Wilmington 19801.

District of Columbia — Norma Holloway Johnson, CJ; Thomas P. Jackson, Thomas F. Hogan, Royce C. Lamberth, Gladys Kessler, Paul L. Friedman, Ricardo M. Urbina, Emmet G. Sullivan, James Robertson, Colleen Kollar-Kotelly, Henry H. Kennedy Jr., Richard W. Roberts, Clerk's Office, Washington DC 20001.

Florida — Northern: C. Roger Vinson, CJ; Lacey A. Collier, Robert L. Hinkle, Stephan Mickle; Clerk's Office, Tallahassee 32301. **Middle:** Elizabeth A. Kovachevich, CJ; G. Kendall Sharp, Patricia C. Fawsett, Harvey E. Schlesinger, Ralph W. Nimmons Jr., Anne C. Conway, Steven D. Merryday, Susan C. Bucklew, Henry Lee Adams Jr., Richard A. Lazzara; Clerk's Offices, Tampa 33602. **Southern:** Edward B. Davis, CJ; William J. Zloch, Kenneth L. Ryskamp, Federico A. Moreno, Shelby Highsmith, Donald L. Graham, K. Michael Moore, Ursula Ungaro-Benages, Wilkie D. Ferguson Jr., Daniel T. K. Hurley, Joan A. Lenard, Donald M. Middlebrooks, Alan S. Gold, William P. Dimitrouleas, Patricia A. Seitz; Clerk's Office, Miami 33128.

Georgia — Northern: Orinda D. Evans, CJ; Harold L. Murphy, G. Ernest Tidwell, J. Owen Forrester, Jack T. Camp, Julie E. Carnes, Clarence Cooper, Willis B. Hunt Jr., Thomas W. Thrash Jr., Richard W. Story; Clerk's Office, Atlanta 30303. **Middle:** Duross Fitzpatrick, CJ; J. Robert Elliott, W. Louis Sands, Hugh Lawson; Clerk's Office, Macon 31202. **Southern:** Dudley H. Bowen Jr., CJ; B. Avant Edenfield, William T. Moore Jr., Anthony A. Alaimo; Clerk's Office, Savannah 31412.

Hawaii — David Alan Ezra, CJ; Alan C. Kay, Helen Gillmor, Susan Oki Mollway; Clerk's Office, Honolulu 96850.

Idaho — Edward J. Lodge, CJ; B. Lynn Winmill; Clerk's Office, Boise 83724.

Illinois — Northern: Marvin E. Aspen, CJ; Charles P. Kocoras, Charles R. Norgle Sr, James F. Holderman, Ann C. Williams, Harry D. Leinenweber, James B. Zagel, Suzanne B. Conlon, George M. Marovich, George W. Lindberg, Wayne R. Andersen, Philip G. Reinhard, Ruben Castillo, Blanche M. Manning, David H. Coar, Robert W. Gettleman, Elaine E. Bucklo, Joan B. Gottschall; Clerk's Office, Chicago 60604. **Central:** Joe Billy McDade, CJ; Michael M. Mihm, Michael P. McCuskey, Jeanne E. Scott; Clerk's Office, Springfield 62701. **Southern:** J. Phil Gilbert, CJ; Paul E. Riley, G. Patrick Murphy, David R. Herndon; Clerk's Office, East St. Louis 62202.

Indiana — Northern: William C. Lee, CJ; Allen Sharp, James T. Moody, Robert L. Miller Jr., Rudy Lozano; Clerk's Office, South Bend 46601. **Southern:** Sarah E. Barker, CJ; S. Hugh Dillin, Larry J. McKinney, John Daniel Tinder, David F. Hamilton, Richard L. Young; Clerk's Office, Indianapolis 46204.

Iowa — Northern: Michael J. Melloy, CJ; Mark W. Bennett; Clerk's Office, Cedar Rapids 52401. **Southern:** Ronald E. Longstaff, CJ; Charles R. Wolle, Robert W. Pratt; Clerk's Office, Des Moines 50306-9344.

Kansas — G. Thomas Van Bebber, CJ; John W. Lungstrum, Monti L. Belot, Kathryn H. Vratil, J. Thomas Marten, Carlos Murguia; Clerk's Office, Kansas City 66101.

Kentucky — Eastern: Henry R. Wilhoit Jr., CJ; William O. Bertelsman, Karl S. Forester, Joseph M. Hood, Jennifer B. Coffman; Clerk's Office, Lexington 40588-3074. **Western:** Charles R. Simpson III, CJ; John G. Heyburn II, Jennifer B. Coffman, Thomas B. Russell, Joseph H. McKinley Jr.; Clerk's Office, Louisville 40202.

Louisiana — Eastern: A. J. McNamara, CJ; Morley L. Sear, Martin L. C. Feldman, Edith Brown Clement, Ginger Berrigan, Stanwood R. Duval Jr., Eldon E. Fallon, Sarah S. Vance, Mary Ann Viel Lemmo, G. Thomas Porteous Jr., Ivan L. R. Lemelle, Carl J. Barbier; Clerk's Office, New Orleans 70130. **Middle:** Frank J. Polozola, CJ; Ralph E. Tyson; Clerk's Office, Baton Rouge 70801. **Western:** F. A. Little Jr., CJ; Rebecca F. Doherty, Richard T. Haik Sr, James T. Trimble Jr., Donald E. Walter, Tucker L. Melançon, Robert G. James; Clerk's Office, Shreveport 71101.

Maine — D. Brock Hornby, CJ; Gene Carter, Morton A. Brody; Clerk's Office, Portland 04101.

Maryland — J. Frederick Motz, CJ; Frederic N. Smalkin, William M. Nickerson, Marvin J. Garbis, Benson Everett Legg, Catherine C. Blake, Andre M. Davis, Deborah K. Chasanow, Peter J. Messitte, Alexander Williams Jr.; Clerk's Office, Baltimore 21201.

Massachusetts — William G. Young, CJ; Joseph L. Tauro, Robert E. Keeton, Mark L. Wolf, Douglas P. Woodlock, Edward F. Harrington, Nathaniel M. Gorton, Richard G. Stearns, Reginald C. Lindsay, Patti B. Saris, Nancy Gert-

ner, George A. O'Toole, Rya W. Zobel, Michael A. Ponsor; Clerk's Office, Boston 02210.

Michigan — Eastern: Lawrence P. Zatkoff, CJ; Avern Cohn, Patrick J. Duggan, Bernard A. Friedman, Paul V. Gadola, Gerald E. Rosen, Robert H. Cleland, Nancy G. Edmunds, Denise Page Hood, Paul D. Borman, John Corbett O'Meara, Arthur J. Tarnow, Victoria A. Roberts, George C. Steeh; Clerk's Office, Detroit 48226. **Western:** Richard A. Enslen, CJ; Robert H. Bell, David W. McKeague, Gordon J. Quist; Clerk's Office, Grand Rapids 49503.

Minnesota — Paul A. Magnuson, CJ; Richard H. Kyle, Donovan W. Frank, James M. Rosenbaum, Michael J. Davis, John R. Tunheim, Ann D. Montgomery; Clerk's Offices, Minneapolis 55415, St. Paul 55101.

Mississippi — Northern: Neal Biggers, CJ; Glen H. Davidson, W. Allen Pepper Jr.; Clerk's Office, Oxford 38655. **Southern:** Tom S. Lee, CJ; William H. Barbour Jr., Henry T. Wingate, Walter J. Gex III, Charles W. Pickering Sr, David Bramlette; Clerk's Office, Jackson 39201.

Missouri — Eastern: Jean C. Hamilton, CJ; Donald J. Stohr, Carol E. Jackson, Charles A. Shaw, Catherine D. Perry, E. Richard Webber, Rodney W. Sippel; Clerk's Office, St. Louis 63101. **Western:** D. Brook Bartlett, CJ; Dean Whipple, Fernando J. Gaitan Jr., Ortrie D. Smith, Gary A. Fenner, Nanette Laughrey; Clerk's Office, Kansas City 64106.

Montana — Jack D. Shanstrom, CJ; Charles C. Lovell, Donald W. Molloy, Lou Aleksich Jr.; Clerk's Office, Billings 59101.

Nebraska — William G. Cambridge, CJ; Richard G. Kopf, Thomas M. Shanahan, Joseph F. Batillion; Clerk's Office, Omaha 68101.

Nevada — Howard D. McKibben, CJ; Philip M. Pro, David W. Hagen, Johnnie B. Rawlinson; Clerk's Office, Las Vegas 89101, Reno 89501.

New Hampshire — Paul J. Barbadoro, CJ; Joseph A. DiClerico, Steven J. McAuliffe; Clerk's Office, Concord 03301.

New Jersey — Anne E. Thompson, CJ; John W. Bissell, Maryanne Trump Barry, Garrett E. Brown Jr., Alfred J. Lechner Jr., Nicholas H. Politan, Alfred M. Wolin, John C. Lifland, William G. Bassler, Mary L. Cooper, Joseph E. Irenas, Jerome B. Simandle, William H. Walls, Stephen M. Orlofsky, Joseph A. Greenaway Jr., Katharine S. Hayden; Clerk's Office, Newark 07101.

New Mexico — John Edwards Conway, CJ; James A. Parker, C. Leroy Hansen, Martha Vazquez, Bruce D. Black; Clerk's Office, Albuquerque 87102.

New York — Northern: Thomas J. McAvoy, CJ; Frederick J. Scullin Jr., Lawrence E. Kahn, David N. Hurd; Clerk's Office, Syracuse 13261-7367. **Eastern:** Charles P. Sifton, CJ; Thomas C. Platt Jr., Raymond J. Dearie, Edward R. Korman, Reena Raggi, Arthur D. Spatt, Carol Bagley Amon, Sterling Johnson Jr., Denis R. Hurley, David G. Trager, Joanna Seybert, Allyne R. Ross, John Gleeson, Nina Gershon, Fredric Block, Jacob Mishler Sr., Eugene H. Nickerson, Jack D. Weinstein Sr., Leonard D. Wexler; Clerk's Office, Brooklyn 11201. **Southern:** Thomas P. Griesa, CJ; Charles L. Brieant, John E. Sprizzo, Lewis A. Kaplan, Michael B. Mukasey, Kimba Wood, Robert P. Patterson Jr., Lawrence McKenna, John S. Martin Jr., Loretta A. Preska, Harold Baer Jr., Deborah A. Batts, Denny Chin, Denise L. Cote, John Koeltl, Allen G. Schwartz, Barrington D. Parker Jr., Shira A. Scheindlin, Sidney H. Stein, Jed S. Rakoff, Barbara S. Jones, Richard C. Casey; Clerk's Office New York City 10007. **Western:** David G. Larimer, CJ; Richard J. Arcara, William M. Skretny, Charles J. Siragusa, John T. Curtin, John T. Elfvin, Michael A. Telesca; Clerk's Office, Buffalo 14202, Rochester 14614.

North Carolina — Eastern: Terrence W. Boyle, CJ; James C. Fox, Malcolm J. Howard; Clerk's Office, Raleigh 27611. **Middle:** N. Carlton Tilley Jr., CJ; Frank W. Bullock, William L. Osteen Sr, James A. Beaty Jr.; Clerk's Office, Greensboro 27402. **Western:** Graham C. Mullen, CJ; Richard L. Voorhees, Lacy H. Thornburg; Clerk's Office, Asheville 28801, Charlotte 28202, Statesville 28687.

North Dakota — Rodney S. Webb, CJ; Patrick A. Conmy; Clerk's Office, Bismarck 58502.

Ohio — **Northern:** Paul R. Matia, CJ; Lesley Brooks Wells, James G. Carr, Solomon Oliver Jr., David A. Katz, Kathleen McDonald O'Malley, Peter C. Economus, Donald C. Nugent, Patricia A. Gaughan, James S. Gwin, Dan Aaron Polster; Clerk's Office, Cleveland 44114. **Southern:** Walter Herbert Rice, CJ; John D. Holschuh, Herman J. Weber, James L. Graham, George C. Smith, S. Arthur Spiegel, Sandra S. Beckwith, Edmund A. Sargus Jr., Susan J. Dlott, Joseph P. Kinneary, Algenon L. Marbley; Clerk's Office, Columbus 43215.

Oklahoma — **Northern:** Terry C. Kern, CJ; Sven Erik Holmes, Michael Burrage; Clerk's Office, Tulsa 74103. **Eastern:** Michael Burrage, CJ; Frank H. Seay; Clerk's Office, Muskogee 74401. **Western:** David L. Russell, CJ; Ralph G. Thompson, Robin J. Cauthron, Tim Leonard, Vicki Miles-LaGrange; Clerk's Office, Oklahoma City 73102.

Oregon — Michael R. Hogan, CJ; Garr M. King, Robert E. Jones, Ann L. Aiken, Ancer L. Haggerty; Clerk's Office, Portland 97204.

Pennsylvania — **Eastern:** James T. Giles, CJ; Robert F. Kelly, Franklin S. Van Antwerpen, Jan E. Dubois, Herbert J. Hutton, Jay C. Waldman, Ronald L. Buckwalter, William H. Yohn Jr., Harvey Bartle III, Stewart Dalzell, John R. Padova, J. Curtis Joyner, Eduardo C. Robreno, Anita B. Brody, Bruce W. Kauffman; Clerk's Office, Philadelphia 19106-1797. **Middle:** Thomas I. Vanaskie, CJ; Sylvia H. Rambo, James F. McClure Jr., A. Richard Caputo, James M. Munley, Yvette Kane; Clerk's Office, Scranton 18501. **Western:** Donald E. Ziegler, CJ; William L. Standish, D. Brooks Smith, Donald J. Lee, Donetta W. Ambrose, Gary L. Lancaster, Robert J. Cindrich, Sean J. McLaughlin; Clerk's Office, Pittsburgh 15230.

Rhode Island — Ronald R. Lagueux, CJ; Ernest C. Torres, Mary M. Lisi; Clerk's Office, Providence 02903.

South Carolina — C. Weston Houck, CJ; G. Ross Anderson Jr., Joseph F. Anderson Jr., David C. Norton, Dennis W. Shedd, Henry M. Herlong Jr., Cameron McGowan Currie, Patrick Michael Duffy, Margaret B. Seymour; Clerk's Office, Columbia 29201.

South Dakota — Lawrence L. Piersol, CJ; Charles B. Kornmann, Karen E. Schreier; Clerk's Office, Rapid City 57701.

Tennessee — **Eastern:** R. Allan Edgar, CJ; James H. Jarvis, Thomas G. Hull, R. Leon Jordan, Curtis L. Collier; Clerk's Office, Knoxville 37902. **Middle:** Robert L. Echols, CJ; Todd J. Campbell, Aleta A. Trauger; Clerk's Office, Nashville 37203. **Western:** Julia S. Gibbons, CJ; James D. Todd, Jerome Turner, Jon Phipps McCalla, Bernice B. Donald; Clerk's Office, Memphis 38103.

Texas — **Northern:** Jerry Buchmeyer, CJ; Mary Lou Robinson, A. Joe Fish, Robert B. Maloney, Sidney A. Fitzwater, Samuel R. Cummings, John H. McBryde, Jorge A. Solis, Terry Means, Joe Kendall, Sam A. Lindsay; Clerk's Office, Dallas 75242. **Southern:** George P. Kazen, CJ; Filemon B. Vela, Hayden W. Head Jr., Ricardo H. Hinojosa, Lynn N. Hughes, David Hittner, Kenneth M. Hoyt, Sim Lake, Melinda Harmon, John D. Rainey, Samuel B. Kent, Ewing Werlein Jr., Lee H. Rosenthal, Janis Graham Jack, Vanessa D. Gilmore, Nancy F. Atlas, Hilda G. Taglia, Keith P. Ellison; Clerk's Office, Houston 77002. **Eastern:** Richard A. Schell, CJ; Howell Cobb, Paul N. Brown, John Hannah Jr., David Folsom, Thad Heartfield, T. John Ward, Joe J. Fisher, William M. Steger; Clerk's Office, Tyler 75702. **Western:** Harry Lee Hudspeth, CJ; David Briones, Hipolito F. Garcia, Edward C. Prado, Fred Biery, Orlando L. Garcia, James R. Nowlin, Sam Sparks, Walter S. Smith Jr., W. Royal Furgeson; Clerk's Office, San Antonio 78206.

Utah — Dee Benson, CJ; Tena Campbell, Dale A. Kimball, J. Thomas Greene, Bruce Jenkins, David K. Winder, Brian Theodore "Ted" Stewart; Clerk's Office, Salt Lake City 84101.

Vermont — J. Garvan Murtha, CJ, William K. Sessions III; Clerk's Office, Burlington 05402.

Virginia — **Eastern:** Claude M. Hilton, CJ; James R. Spencer, Thomas S. Ellis III, Rebecca Beach Smith, Henry Coke Morgan Jr., Robert E. Payne, Raymond A. Jackson, Leonie M. Brinkema, Jerome B. Friedman, Gerald Bruce Lee; Clerk's Office, Alexandria 22314. **Western:** Samuel G. Wilson, CJ; James C. Turk, James P. Jones, Norman K. Moon; Clerk's Office, Roanoke 24006.

Washington — **Eastern:** Wm. Fremming Nielsen, CJ; Fred Van Sickle, Robert H. Whaley, Edward F. Shea, Allan A. McDonald; Clerk's Office, Spokane 99210. **Western:** John C. Coughenour, CJ; Barbara Jacobs Rothstein, Robert J. Bryan, Thomas S. Zilly, Franklin D. Burgess, Robert S. Lasnik; Clerk's Office, Seattle 98104, Tacoma 98402.

West Virginia — **Northern:** Frederick P. Stamp Jr., CJ; Irene M. Keeley, W. Craig Broadwater, Robert E. Maxwell; Clerk's Office, Wheeling 26003. **Southern:** Charles H. Haden II, CJ; John T. Copenhaver Jr., David A. Faber, Joseph R. Goodwin, Robert C. Chambers; Clerk's Office, Charleston 25329.

Wisconsin — **Eastern:** J. P. Stadtmueller, CJ; Rudolph T. Randa, Charles N. Clevert, Lynn S. Adelman; Clerk's Office, Milwaukee 53202. **Western:** John C. Shabaz; CJ; Barbara B. Crabb, Clerk's Office, Madison 53701.

Wyoming — William F. Downes, CJ; Alan B. Johnson, Clarence A. Brimmer; Clerk's Office, Cheyenne 82001.

U.S. Territorial District Courts

Guam — John S. Unpingco, CJ; Clerk's Office, Agana 96910.

Northern Mariana Islands — Alex R. Munson, CJ; Clerk's Office, Saipan MP 96950.

Puerto Rico — Hector M. Laffitte, CJ; Juan M. Perez-Gimenez, Jose Antonio Fuste, Salvador E. Casellas, Daniel R. Dominguez; Clerk's Office, Hato Rey 00918.

Virgin Islands — Raymond L. Finch, CJ; Thomas K. Moore; Clerk's Office, St. Croix 00820.

U.S. Court of International Trade

New York, NY 10278-0001

(Salaries, $141,300, as of Jan. 1, 2000)

Chief Judge — Gregory W. Carman

Judges — Jane A. Restani, Thomas J. Aquilino Jr., Richard W. Goldberg, Donald C. Pogue, Evan J. Wallach, Judith M. Barzilay, Delissa A. Ridgway.

U.S. Court of Federal Claims

Washington, DC 20005 (Salaries, $141,300, as of Jan. 1, 2000)

Chief Judge — Loren A. Smith

Judges — John P. Wiese, Christine Odell Cook Miller, Marian Blank Horn, Eric G. Bruggink, Bohdan A. Futey, Roger B. Andewelt, James T. Turner, Robert H. Hodges, Diane Gilbert Weinstein.

U.S. Tax Court

Washington, DC 20217 (Salaries, $141,300 as of Jan. 1, 2000)

Chief Judge — Mary Ann Cohen

Judges — Renato Beghe, Herbert L. Chabot, John O. Colvin, Joel Gerber, Julian I. Jacobs, Carolyn Miller Parr, Robert P. Ruwe, James S. Halpern, Carolyn P. Chiechi, David Laro, Stephen J. Swift, Thomas B. Wells, Laurence J. Whalen, Maurice B. Foley, Juan F. Vasquez, Joseph H. Gale, L. Paige Marvel, Michael B. Thornton.

U.S. Court of Appeals for Veterans Claims

Washington, DC 20004 (Salaries, $141,300, as of Jan. 1, 2000)

Chief Judge — Frank Q. Nebeker

Judges — Kenneth B. Kramer, John J. Farley 3d, Ronald M. Holdaway, Donald L. Ivers, Jonathan R. Steinberg, William P. Greene Jr.

STATE AND LOCAL GOVERNMENT

Mayors of Selected U.S. Cities

As of Oct. 1999

D, Democrat; R, Republican; N-P, Non-Partisan; I, Independent; Prog. Coal., Progressive Coalition

City	Name	Next Election
Abilene, TX	Grady Barr, N-P	2002, May
Akron, OH	Donald L. Plusquellic, D	1999, Nov.
Alameda, CA	Ralph J. Appezzato, N-P	2002, Nov.
Albany, GA	Thomas Coleman, D	1999, Nov.
Albany, NY	Gerald D. Jennings, D	2001, Nov.
Albuquerque, NM	Jim Baca, D	2001, Nov.
Alexandria, LA	Edward G. Randolph Jr., D	2002, Nov.
Alexandria, VA	Kerry J. Donley, D	2000, May
Alhambra, CA	Talmage Burke, N-P	(1)
Allentown, PA	William Heydt, R	2001, Nov.
Amarillo, TX	Kel Seliger, N-P	2001, May
Ames, IA	Ted Tedesco, N-P	2001, Nov.
Anaheim, CA	Tom Daly, D	2002, Nov.
Anchorage, AK	Rick Mystrom, R	2000, Apr.
Anderson, IN	J. Mark Lawler, D	1999, Nov.
Anderson, SC	Richard A. Shirley, N-P	2000, June
Ann Arbor, MI	Ingrid B. Sheldon, R	2000, Nov.
Annapolis, MD	Dean L. Johnson, R	2001, Nov.
Appleton, WI	Timothy M. Hanna, N-P	2000, Apr.
Arcadia, CA	Roger Chandler, N-P	2000, Apr.
Arlington, MA	John W. Hurd, N-P	2000, Mar.
Arlington, TX	Elzie Odom, N-P	2001, May
Arlington Hts., IL	Arlene J. Mulder, N-P	2001, Apr.
Arvada, CO	Robert G. Frie, N-P	1999, Nov.
Asheville, NC	Leni Sitnick, N-P	2001, Nov.
Athens, GA	Doc Eldridge, D	2002, Nov.
Atlanta, GA	Bill Campbell, D	2001, Nov.
Atlantic City, NJ	James Whelan, N-P	2002, May
Augusta, GA	Bob Young, N-P	2002, Nov.
Augusta, ME	William E. Dowling, N-P	2000, Nov.
Aurora, CO	Paul E. Tauer, N-P	1999, Nov.
Aurora, IL	David L. Stover, N-P	2001, Apr.
Austin, TX	Kirk P. Watson, N-P	2000, May
Bakersfield, CA	Bob Price, N-P	2000, Mar.
Baldwin Park, CA	Manuel Lozano, N-P	2001, Nov.
Baltimore, MD	Kurt L. Schmoke, D	1999, Nov.
Baton Rouge, LA	Tom Edward McHugh, D	2000, Nov.
Battle Creek, MI	Ted Dearing, N-P	1999, Nov.
Bayonne, NJ	Joseph V. Doria Jr., N-P	2002, May
Baytown, TX	Pete C. Alfaro, N-P	2001, May
Beaumont, TX	David W. Moore, N-P	2000, May
Belleville, IL	Mark Kern, N-P	2001, Apr.
Bellevue, WA	Mike Creighton, N-P	1999, Nov.
Bellflower, CA	Joe Cvetko, N-P	2000, Mar.
Bellingham, WA	Mark Asmundson, N-P	1999, Nov.
Berkeley, CA	Shirley Dean, N-P	2002, Nov.
Bethlehem, PA	Donald T. Cunningham Jr., D	2001, Nov.
Beverly Hills, CA	Thomas S. Levy, N-P	2000, Apr.
Billings, MT	Charles F. Tooley, N-P	2001, Nov.
Biloxi, MS	A. J. Holloway, R	2001, June
Binghamton, NY	Richard A. Bucci, R	2001, Nov.
Birmingham, AL	William Bell, D	(2)
Bismarck, ND	Bill Sorensen, R	2002, June
Bloomfield, NJ	John Bukowski Jr., R	2001, Nov.
Bloomington, IL	Judy Markowitz, N-P	2001, Apr.
Bloomington, IN	John Fernandez, D	1999, Nov.
Bloomington, MN	Coral Houle, N-P	1999, Nov.
Boca Raton, FL	Carol G. Hanson, N-P	2001, Mar.
Boise, ID	H. Brent Coles, N-P	2001, Nov.
Bossier City, LA	George Dement, N-P	2001, Apr.
Boston, MA	Thomas M. Menino, D	2001, Nov.
Boulder, CO	Will Toor, N-P	2001, Nov.
Bridgeport, CT	Joseph P. Ganim, D	1999, Nov.
Bristol, CT	Frank N. Nicastro Sr., D	1999, Nov.
Brockton, MA	John T. Yunits Jr., D	1999, Nov.
Broken Arrow, OK	Jim Reynolds, N-P	2000, Apr.
Brooklyn Park, MN	Grace Arbogast, N-P	2002, Nov.
Brownsville, TX	Blanca S. Vela, N-P	2003, May
Bryan, TX	Lonnie Stabler, N-P	2001, May
Buena Park, CA	Jack Mauller, N-P	1999, Nov.
Buffalo, NY	Anthony M. Masiello, D	2001, Nov.
Burbank, CA	Stacey Murphy, N-P	2000, May
Burlington, VT	Peter A. Clavelle, Prog. Coal.	2001, Mar.
Calumet City, IL	Gerome P. Genova, I	2001, Apr.
Camarillo, CA	Kevin B. Kildee, N-P	1999, Nov.
Cambridge, MA	Francis H. Duehay, D	2000, Jan.
Camden, NJ	Milton Milan, D	2001, May
Canton, OH	Richard D. Watkins, R	1999, Nov.
Cape Coral, FL	Roger G. Butler, N-P	2000, Nov.
Carlsbad, CA	Claude A. Lewis, N-P	2002, Nov.
Carson, CA	Peter D. Fajardo, N-P	2001, Mar.
Carson City, NV	Ray Masayko, N-P	2000, Nov.
Casper, WY	James W. "Tim" Monroe, N-P	1999, Nov.
Cedar Rapids, IA	Lee R. Clancey, N-P	1999, Nov.
Champaign, IL	Gerald Schweighart, N-P	2003, Apr.
Chandler, AZ	Jay Tibshraeny, N-P	2000, Jan.
Charleston, SC	Joseph P. Riley Jr., D	1999, Nov.
Charleston, WV	Jay Goldman, D	2003, May
Charlotte, NC	Patrick McCrory, R	1999, Nov.
Charlottesville, VA	Virginia Daugherty, D	2000, May
Chattanooga, TN	Jon Kinsey, N-P	2001, Mar.
Chesapeake, VA	William E. Ward, N-P	2000, May
Chester, PA	Dominic F. Pileggi, R	1999, Nov.
Cheyenne, WY	Leo A. Pando, N-P	2000, Nov.
Chicago, IL	Richard M. Daley, D	2003, Feb.
Chicopee, MA	Richard J. Kos, N-P	1999, Nov.
Chino, CA	Eunice Ulloa, R	2000, Nov.
Chula Vista, CA	Shirley A. Horton, N-P	2002, June
Cicero, IL	Betty Loren-Maltese, R	2001, Apr.
Cincinnati, OH	Roxanne Qualls, D	1999, Nov.
Clarksville, TN	Johnny Piper, N-P	2002, Nov.
Clearwater, FL	Brian Aungst, N-P	2002, Mar.
Cleveland, OH	Michael R. White, D	2001, Nov.
Cleveland Hts., OH	Edward J. Kelley, N-P	2000, Jan.
Clinton, IA	La Metta Wynn, N-P	1999, Nov.
Clifton, NJ	James A. Anzaldi, R	2002, May
Colorado Spgs., CO	Mary Lou Makepeace, R	2003, Apr.
Columbia, MO	Darwin Hindman, N-P	2001, Apr.
Columbia, SC	Robert D. Coble, N-P	2002, Apr.
Columbus, GA	Bobby G. Peters, D	2002, Nov.
Columbus, OH	Gregory S. Lashutka, R	1999, Nov.
Compton, CA	Omar Bradley, N-P	2001, June
Concord, CA	Michael Pastrick, N-P	1999, Nov.
Concord, NH	William J. Veroneau, N-P	1999, Nov.
Coon Rapids, MN	Lonni McCauley, N-P	2000, Nov.
Coral Gables, FL	Raul Valdes-Fauli, N-P	2001, Apr.
Coral Springs, FL	John Sommerer, N-P	2000, Mar.
Corona, CA	Janice Rudman, N-P	1999, Nov.
Corpus Christi, TX	Samuel Loyd Neal, N-P	2001, Apr.
Costa Mesa, CA	Gary Monahan, N-P	(3)
Council Bluffs, IA	Tom Hanafan, N-P	2001, Nov.
Covington, KY	Denny Bowman, D	1999, Nov.
Cranston, RI	John O'Leary, D	2002, Nov.
Cuyahoga Falls, OH	Donald L. Robart, R	2001, Nov.
Dallas, TX	Ronald Kirk, N-P	2003, May
Daly City, CA	Adrienne J. Tissier, N-P	1999, Nov.
Danbury, CT	Gene F. Eriquez, D	1999, Nov.
Danville, VA	Ruby B. Archie, N-P	2000, May
Davenport, IA	Phillip Yerington, N-P	1999, Nov.
Davis, CA	Julie Partansky, N-P	2000, June
Dayton, OH	Michael R. Turner, R	2001, Nov.
Daytona Beach, FL	Baron H. Asher, N-P	1999, Nov.
Dearborn, MI	Michael A. Guido, N-P	2001, Nov.
Dearborn Hts., MI	Ruth A. Canfield, N-P	2001, Nov.
Decatur, IL	Terry M. Howley, N-P	2003, Apr.
Delray Beach, FL	Jay Alperin, N-P	2000, Mar.
Denton, TX	Jack Miller, N-P	2000, May
Denver, CO	Wellington E. Webb, N-P	2003, May
Des Moines, IA	Preston A. Daniels, N-P	1999, Nov.
Des Plaines, IL	Paul W. Jung, N-P	2001, Apr.
Detroit, MI	Dennis W. Archer, D	2001, Nov.
Dothan, AL	Chester L. Sowell, N-P	2001, July
Dover, DE	James L. Hutchinson, N-P	2000, May
Downey, CA	Gary P. McCaughan, N-P	(4)
Dubuque, IA	Terrance M. Duggan, N-P	2001, Nov.
Duluth, MN	Gary L. Doty, N-P	1999, Nov.
Durham, NC	Nicholas J. Tennyson, N-P	1999, Nov.
East Hartford, CT	Timothy Larson, D	1999, Nov.
East Lansing, MI	Mark S. Meadows, N-P	1999, Nov.
East Orange, NJ	Robert L. Bowser, D	2001, Nov.
Edison, NJ	George Spadoro, D	2001, Nov.
Edmond, OK	Robert Rudkin, N-P	2001, May
El Cajon, CA	Mark Lewis, N-P	2002, Nov.
Elgin, IL	Ed Schock, N-P	2003, Apr.
Elizabeth, NJ	J. Christian Bollwage, D	2000, Nov.
Elkhart, IN	James P. Perron, D	1999, Nov.
El Monte, CA	L. Rachel Montes, N-P	2001, Mar.
El Paso, TX	Carlos Moises Ramirez, N-P	2001, May
Elyria, OH	Michael B. Keys, D	1999, Nov.
Enfield, CT	Mary Lou Strom, N-P	1999, Nov.
Enid, OK	J. Doug Frantz, N-P	2001, Apr.
Erie, PA	Joyce Savocchio, D	2001, Nov.
Escondido, CA	Lori Holt Pfieler, N-P	2000, Nov.
Euclid, OH	Paul Oyaski, D	1999, Nov.
Eugene, OR	James D. Torrey, N-P	2000, Nov.
Evanston, IL	Lorraine Morton, N-P	2001, Apr.

City	Name	Next Election
Evansville, IN	Frank F. McDonald II, D	1999, Nov.
Everett, WA	Edward D. Hansen, N-P	2001, Nov.
Fairbanks, AK.	James C. Hayes, N-P	2001, Oct.
Fairfield, CA	George Pettygrove, N-P	2001, Nov.
Fairfield, CT	Kenneth A. Flatto, D	1999, Nov.
Fall River, MA	Edward M. Lambert Jr., D	1999, Nov.
Fargo, ND	Bruce W. Furness, N-P	2002, Apr.
Farmington Hills, MI	Aldo Vagnozzi, N-P	1999, Nov.
Fayetteville, NC	J. L. Dawkins, N-P	1999, Nov.
Fitchburg, MA	Mary Whitney, N-P	1999, Nov.
Flagstaff, AZ	Christopher Bavasi, N-P	2000, Apr.
Flint, MI	Woodrow Stanley, D	1999, Nov.
Florissant, MO	James J. Eagan, N-P	2003, Apr.
Fontana, CA	David Eshleman, D	2002, Nov.
Ft. Collins, CO	Ray Martinez, N-P	2001, Apr.
Ft. Lauderdale, FL	Jim Naugle, N-P	2000, Mar.
Ft. Smith, AR	C. Raymond Baker, N-P	2002, Nov.
Ft. Wayne, IN	Paul Helmke, R	1999, Nov.
Ft. Worth, TX	Kenneth L. Barr, N-P	2001, May
Fountain Valley, CA	John Collins, N-P	1999, Nov.
Frankfort, KY	William I. May Jr., N-P	1999, Nov.
Fremont, CA.	Gus Morrison, N-P	2000, Nov.
Fresno, CA	Jim Patterson, N-P	2001, Jan.
Fullerton, CA	Jan Flory, N-P	1999, Nov.
Gadsden, AL	Steven A. Means, N-P	2002, Nov.
Gainesville, FL	Paula M. DeLaney, N-P	2001, Nov
Galveston, TX	Roger R. "Bo" Quiorga, N-P	2000, May
Gardena, CA	Donald L. Dear, N-P	2001, Mar.
Garden Grove, CA	Bruce A. Broadwater, N-P	2000, Nov.
Garland, TX	Jim Spence, N-P	2001, May
Gary, IN	Scott L. King, D	1999, Nov.
Gastonia, NC	Porter L. McAteer, N-P	1999, Nov.
Glendale, AZ	Elaine M. Scruggs, N-P	2000, May
Glendale, CA	Ginger Bremberg, N-P	2000, Apr.
Grand Forks, ND	Patricia A. Owens, N-P	2000, June
Grand Prairie, TX	Charles V. England, N-P	2000, May
Grand Rapids, MI	John H. Logie, N-P	1999, Nov.
Greeley, CO	LaVern C. Nelson, N-P	1999, Nov.
Green Bay, WI	Paul F. Jadin, N-P	2003, Apr.
Greensboro, NC	Carolyn S. Allen, N-P	1999, Nov.
Greenville, SC	Knox H. White, R	1999, Nov.
Greenwich, CT	Tom R. Ragland, R	1999, Nov.
Groton, CT	Jane Dauphinais, N-P	1999, Nov.
Gulfport, MS.	Robert C. Short, R	2001, June
Hamden, CT	Barbara DeNicola, R	1999, Nov.
Hamilton, OH	Thomas Nye, N-P	1999, Nov.
Hammond, IN	Duane W. Dedelow Jr., R	1999, Nov.
Hampton, VA	Joseph H. Spencer II, N-P	2000, May
Harrisburg, PA	Stephen R. Reed, D	2001, Nov.
Hartford, CT	Michael P. Peters, N-P	1999, Nov.
Haverhill, MA	James A. Rurak, N-P	1999, Nov.
Hawthorne, CA	Larry Guidi, N-P	1999, Nov.
Hayward, CA	Roberta Cooper, N-P	2002, Nov.
Helena, MT.	Colleen McCarthy, N-P	2001, Nov.
Henderson, NV.	James B. Gibson, N-P	2001, June
Hialeah, FL	Raul L. Martinez, R	2001, Nov.
High Point, NC	Rebecca R. Smothers, N-P	1999, Nov.
Hoboken, NJ	Anthony Russo, N-P	2001, May
Hollywood, FL	Mara Giulianti, N-P	2000, Mar.
Holyoke, MA.	Daniel J. Szostkiewicz, D	1999, Nov.
Honolulu, HI	Jeremy Harris, N-P	2000, Nov.
Houston, TX	Lee P. Brown, N-P	1999, Nov.
Huntington, WV	Jean Dean, R	2000, Nov.
Huntington Beach, CA	Peter Green, N-P	1999, Nov.
Huntington Park, CA	Rosario Marin, N-P	2000, Apr.
Huntsville, AL	Loretta Spencer, N-P	2000, Aug.
Idaho Falls, ID	Linda Milam, N-P	2001, Nov.
Independence, MO.	Ron Stewart, N-P	2002, Apr.
Indianapolis, IN	Stephen Goldsmith, R	1999, Nov.
Inglewood, CA	Roosevelt F. Dorn, N-P	2002, Nov.
Iowa City, IA	Ernest W. Lehman, N-P	1999, Nov.
Irvine, CA	Christina Shea, N-P	2000, Nov.
Irving, TX	Joe Putnam, N-P	2002, May
Irvington, NJ	Sara B. Bost, D	2002, May
Jackson, MS.	Harvey Johnson, D	2001, June
Jacksonville, FL	John A. Delaney, R	2003, May
Janesville, WI	Thomas J. Stehura, N-P	(5)
Jefferson City, MO	Thomas P. Rackers, R	2003, Apr.
Jersey City, NJ	Bret Schundler, R	2001, May
Johnson City, TN	Vance W. Cheek Jr., N-P	2001, May
Joliet, IL	Arthur Schultz, N-P	2003, Apr.
Juneau, AK.	Dennis Egan, D	2000, Oct.
Kalamazoo, MI	Robert B. Jones, N-P	1999, Nov.
Kansas City, KS	Carol S. Marinovich, N-P	2001, Apr.
Kansas City, MO.	Kay Barnes, N-P	2003, Mar.
Kenner, LA	Louis J. Congemi, R	2002, Apr.
Kenosha, WI	John Antaramian, D	2000, Apr.
Kettering, OH	Marilou W. Smith, N-P	2001, Nov.
Killeen, TX	Fred L. Latham, N-P	2002, May
Knoxville, TN	Victor H. Ashe, R	1999, Nov.
Kokomo, IN	James Trobaugh, R	1999, Nov.
LaCrosse, WI.	John D. Medinger, N-P	2001, Apr.
Lafayette, IN	Dave Heath, R	1999, Nov.
La Habra, CA.	Juan Garcia, N-P	1999, Nov.
Lake Charles, LA.	Willie L. Mount, D	2001, May
Lakeland, FL	Ralph L. Fletcher, N-P	2000, Nov.
Lakewood, CA	Joseph Esquivel, N-P	2000, Mar.
Lakewood, CO.	Linda Morton, N-P	1999, Nov.
Lakewood, OH.	Madeline Cain, D	1999, Nov.
La Mesa, CA	Arthur Madrid, N-P	2000, Nov.
La Mirada, CA	Bob Chotiner, N-P	2000, Mar.
Lancaster, CA	Frank C. Roberts, N-P	2000, Apr.
Lancaster, PA	Charlie Smithgall, R	2002, Nov.
Lansing, MI	David C. Hollister, N-P	2001, Nov.
Laredo, TX.	Elizabeth G. "Betty" Flores, N-P	2002, May
Largo, FL	Thomas D. Feaster, N-P	2000, Mar
Las Cruces, NM.	Ruben A. Smith, D	1999, Nov.
Las Vegas, NV.	Oscar B. Goodman, D	2003, June
Lawrence, KS	Ervin E. Hodges, N-P.	2000, Apr.
Lawrence, MA	Patricia Dowling, N-P	2001, Nov.
Lawton, OK	Cecil E. Powell, D	2001, Mar.
Lexington, KY	Pam Miller, N-P	2002, Nov.
Lima, OH	David J. Berger, N-P	2002, Nov.
Lincoln, NE	Don Wesely, N-P	2003, May
Little Rock, AR.	Jim Dailey, N-P	2002, Nov.
Livermore, CA	Cathie Brown, N-P	1999, Nov.
Livonia, MI	Jack E. Kirksey, N-P	1999, Nov.
Lodi, CA.	Keith Laud, N-P	1999, Nov.
Long Beach, CA	Beverly O'Neill, N-P	2002, Apr.
Longmont, CO	Leona Stoecker, N-P	1999, Nov.
Longview, TX	David L. McWhorter, N-P	2000, May
Lorain, OH	Joe Koziura, D	1999, Nov.
Los Angeles, CA	Richard Riordan, N-P	2001, June
Louisville, KY	David Armstrong, D	2002, Nov.
Lowell, MA	Eileen M. Donoghue, D	2000, Jan.
Lubbock, TX	Windy Sitton, R	2000, May
Lynchburg, VA	D.L. "Pete" Warren, N-P	2000, June
Lynn, MA	Patrick J. McManus, D	1999, Nov.
Lynwood, CA.	Richard Sanchez, N-P	1999, Nov.
Macon, GA	Jim Marshall, D	1999, Nov.
Madison, WI	Susan J.M. Bauman, N-P	2003, Apr.
Malden, MA	Richard Howard, D	1999, Nov.
Manchester, CT.	Stephen T. Cassano, N-P	1999, Nov.
Manchester, NH.	Raymond J. Wieczorek, R	1999, Nov.
Mansfield, OH	Lydia J. Reid, D	1999, Nov.
Marietta, GA	Ansley L. Meaders, D	1999, Nov.
McAllen, TX.	Leo Montalvo, R	2001, May
Medford, MA	Michael J. McGlynn, D	1999, Nov.
Medford, OR	Lindsay D. Berryman, N-P	2000, Nov.
Melbourne, FL	John Buckley, N-P	2000, Nov.
Memphis, TN	Willie W. Herenton, D	2003, Oct.
Mentor, OH	Richard Hennig, N-P	2002, Nov.
Merced, CA	MaryJo Knudsen, N-P	1999, Nov.
Meriden, CT.	Joseph Marinan Jr., N-P	1999, Nov.
Meridian, MS	John Robert Smith, R	2001, June
Mesa, AZ.	Wayne Brown, N-P.	2000, Mar.
Mesquite, TX	Mike Anderson, N-P.	2001, May
Miami, FL.	Joe Carollo, N-P.	2001, Nov.
Miami Beach, FL	Neisen O. Kasdin, N-P.	2001, Nov.
Midland, TX.	Robert E. Burns, N-P.	2001, May
Midwest City, OK	Eddie O. Reed, N-P.	2002, Apr.
Milford, CT	Frederick L. Lisman, R.	1999, Nov.
Milpitas, CA.	Henry C. Manayan, N-P	2000, Nov.
Milwaukee, WI.	John O. Norquist, D	2000, Apr.
Minneapolis, MN	Sharon Sayles Belton, D	2001, Nov.
Minnetonka, MN	Karen J. Anderson, N-P	2001, Nov.
Mobile, AL	Michael C. Dow, R	2001, Aug.
Modesto, CA	Richard A. Lang, N-P	1999, Nov.
Monroe, LA	Abe E. Pierce III, D	2000, Mar.
Montclair, NJ	William N. Farlie Jr., N-P	2000, May
Montebello, CA	Kathy Salazar, N-P	1999, Nov.
Monterey Park, CA	Judy Chu, N-P	2000, Mar.
Montgomery, AL	Emory Folmar, R	(2)
Montpelier, VT.	Charles Karparis, N-P	(5)
Moreno Valley, CA	Frank West, N-P	1999, Nov.
Mt. Prospect, IL	Gerald L. "Skip" Farley, N-P	2001, Apr.
Mt. Vernon, NY	Ernest D. Davis, D	1999, Nov.
Mountain View, CA	Mary Lou Zoglin, N-P	1999, Nov.
Muncie, IN	Dan Cannan, R	1999, Nov.
Muskogee, OK.	Jim Bushnell, N-P	2000, Nov.
Napa, CA.	Ed Henderson, N-P	2001, Mar.
Naperville, IL	George A. Pradel, N-P	2003, Apr.
Nashua, NH.	Donald C. Davidson, N-P	1999, Nov.

City	Name	Next Election	City	Name	Next Election
Nashville, TN	Bill Purcell, N-P	2003, Aug.	Rock Hill, SC	Doug Echols, N-P	2001, Oct.
National City, CA	George H. Waters, R	2002, Nov.	Rock Island, IL	Mark W. Schwiebert, N-P	2001, Apr.
Newark, NJ	Sharpe James, D	2002, May	Rockford, IL	Charles E. Box, D	2001, Apr.
New Bedford, MA	Frederick M. Kalisz Jr., N-P	1999, Nov.	Rockville, MD	Rose G. Krasnow, N-P	1999, Nov.
New Britain, CT	Lucian J. Pawlak, D	1999, Nov.	Rome, NY	Joseph A. Griffo, R	1999, Nov.
New Haven, CT	John DeStefano Jr., D	1999, Nov.	Rosemead, CA	Joel Vasquez, N-P	2000, Apr.
New Orleans, LA	Marc H. Morial, D	2002, Feb.	Roseville, MI	Gerald K. Alsip, N-P	2001, Nov.
Newport Beach, CA	Dennis O'Neil	1999, Nov.	Roswell, NM	Bill B. Owen, N-P	2002, Mar.
Newport News, VA	Joe S. Frank, N-P	2002, May	Royal Oak, MI	Dennis G. Cowan, N-P	1999, Nov.
New Rochelle, NY	Timothy Idoni, D	1999, Nov.	Sacramento, CA	Joseph Serna Jr., N-P	2000, Nov.
Newton, MA	David B. Cohen, N-P	2001, Nov.	Saginaw, MI	Gary L. Loster, N-P	2001, Nov.
New York, NY	Rudolph W. Giuliani, R	2001, Nov.	St. Charles, MO	Patricia M. York, N-P	2003, Apr.
Niagara Falls, NY	James Galie, D	1999, Nov.	St. Clair Shores, MI	Curtis L. Dumas, N-P	1999, Nov.
Norfolk, VA	Paul D. Fraim, N-P	2000, June	St. Cloud, MN	Larry Meyer, N-P	2001, Nov.
Norman, OK	Bob Thompson, N-P	2001, Apr.	St. Joseph, MO	Larry R. Stobbs, N-P	2002, Apr.
N. Charleston, SC	R. Keith Summey, R	2003, June	St. Louis, MO	Clarence Harmon, D	2001, Apr.
N. Little Rock, AR	Patrick Henry Hays, N-P	2001, Nov.	St. Louis Park, MN	Gail Dorfman, N-P	1999, Nov.
Norwalk, CA	Michael A. Mendez, N-P	2000, Mar.	St. Paul, MN	Norm Coleman, N-P	2001, Nov.
Norwalk, CT	Frank J. Esposito, R	1999, Nov.	St. Petersburg, FL	David J. Fischer, N-P	2001, Mar.
Novato, CA	Michael DiGiorgio, N-P	1999, Nov.	Salem, OR	Michael Swaim, N-P	2000, Nov.
Oakland, CA	Jerry Brown, D	2002, Nov.	Salinas, CA	Anna M. Caballero, N-P	2000, Nov.
Oak Park, IL	Barbara Furlong, N-P	2001, Apr.	Salt Lake City, UT	Deedee Corradini, D	1999, Nov.
Oceanside, CA	Dick Lyon, N-P	2000, Nov.	San Angelo, TX	Johnny Fender, N-P	2001, May
Odessa, TX	Mike Atkins, N-P	2000, May	San Antonio, TX	Howard W. Peak, N-P	2001, May
Ogden, UT	Glenn J. Mecham, N-P	1999, Nov.	San Bernardino, CA	Judith Valles, D	2001, Nov.
Oklahoma City, OK	Kirk Humphreys, N-P	2002, Apr.	San Diego, CA	Susan Golding, R	2000, Nov.
Olympia, WA	Bob Jacobs, N-P	1999, Nov.	Sandy City, UT	Thomas M. Dolan, N-P	2001, Nov.
Omaha, NE	Hal J. Daub, R	2001, June	San Francisco, CA	Willie L. Brown Jr., N-P	1999, Nov.
Ontario, CA	Gary C. Ovitt, N-P	2002, Nov.	San Jose, CA	Ron Gonzales, D	2002, Nov.
Orange, CA	Joanne Coontz, N-P	2000, Nov.	San Leandro, CA	Shelia Young, N-P	2000, June
Orlando, FL	Glenda E. Hood, N-P	2000, Sept.	San Mateo, CA	Claire Mack, N-P	1999, Nov.
Oshkosh, WI	Melanie L. Bloechl, N-P	2000, Nov.	San Rafael, CA	Albert J. Boro, N-P	1999, Nov.
Overland Park, KS	Ed Eilert, R	2001, Apr.	Santa Ana, CA	Miguel Pulido, N-P	2000, Nov.
Owensboro, KY	Waymond Morris, N-P	1999, Nov.	Santa Barbara, CA	Harriet Miller, N-P	2001, Nov.
Oxnard, CA	Manuel M. Lopez, N-P	2000, Nov.	Santa Clara, CA	Judy Nadler, N-P	2002, Nov.
Palm Springs, CA	William G. Kleindienst, N-P	1999, Nov.	Santa Clarita, CA	JoAnne Darcy, N-P	1999, Nov.
Palo Alto, CA	Gary Fazzino, N-P	2000, Nov.	Santa Cruz, CA	Katherine Beiers	1999, Nov.
Parma, OH	Gerald M. Boldt, D	1999, Nov.	Santa Fe, NM	Larry Delgado, N-P	2002, Mar.
Pasadena, CA	Bill Bogaard, N-P	2003, May	Santa Maria, CA	Don Lahr, N-P	2000, Nov.
Pasadena, TX	Johnny Isbell, N-P	2001, May	Santa Monica, CA	Pam O'Connor, N-P	2000, Nov.
Passaic, NJ	Margie Semler, N-P	2001, May	Santa Rosa, CA	Janet Condvon, N-P	2000, Nov.
Paterson, NJ	Martin G. Barnes, R	2002, May	Sarasota, FL	Mollie C. Cardamone, N-P	2003, Apr.
Pawtucket, RI	James E. Doyle, D	1999, Nov.	Savannah, GA	Floyd Adams Jr., N-P	1999, Nov.
Peabody, MA	Peter Torigian, D	1999, Nov.	Schaumburg, IL	Al Larson, R	2003, Apr.
Pembroke Pines, FL	Alex G. Fekete, N-P	2000, Mar.	Schenectady, NY	Albert P. Jurczynski, R	1999, Nov.
Pensacola, FL	John R. Fogg, N-P	2001, June	Scottsdale, AZ	Sam Kathryn Campana, R	2000, Mar.
Peoria, IL	Lowell G. Grieves, N-P	2001, Apr.	Scranton, PA	James P. Connors, D	2001, Nov.
Philadelphia, PA	Edward Rendell, D	1999, Nov.	Seattle, WA	Paul Schell, N-P	1999, Nov.
Phoenix, AZ	Skip Rimsza, N-P	2001, Sept.	Sheboygan, WI	James R. Schramm, N-P	2001, Apr.
Pico Rivera, CA	Carlos A. Garcia, N-P	2000, Mar.	Shreveport, LA	Keith Hightower, D	2002, Nov.
Pierre, SD	Gary Drewes, N-P	2000, Apr.	Simi Valley, CA	Bill Davis, N-P	2000, Nov.
Pine Bluff, AR	Jerry Taylor, N-P	2000, Nov.	Sioux City, IA	Thomas Padgett, N-P	2000, Nov.
Pittsburgh, PA	Tom J. Murphy, D	2001, Nov.	Sioux Falls, SD	Gary Hanson, N-P	2002, Apr.
Pittsfield, MA	Gerald S. Doyle Jr., N-P	1999, Nov.	Skokie, IL	George Van Dusen, N-P	2001, Apr.
Plainfield, NJ	Albert McWilliams, N-P	2002, Nov.	Somerville, MA	Dorothy A. KellyGay, N-P	2001, Apr.
Plano, TX	John Longstreet, N-P	2000, May	South Bend, IN	Stephen J. Luecke, D	1999, Nov.
Plantation, FL	Rae Carole Armstrong, D	2003, Mar.	South Gate, CA	Henry Gonzalez, N-P	2000, Apr.
Pocatello, ID	Gregory R. Anderson, N-P	2001, Nov.	Southfield, MI	Donald F. Fracassi, R	2001, Nov.
Pomona, CA	Edward S. Cortez, N-P	2000, Nov.	Sparks, NV	Tony Armstrong, N-P	2003, June
Pompano Beach, FL	William F. Griffin, N-P	2000, Mar.	Spartanburg, SC	James E. Talley, N-P	2001, Nov.
Pontiac, MI	Walter Moore, N-P	2001, Nov.	Spokane, WA	John Talbott, N-P	2001, Nov.
Port Arthur, TX	Oscar Ortiz, D	2002, May	Springfield, IL	Karen Hasara, N-P	2003, May
Portland, ME	Nicholas Mavodones Jr., N-P	2000, June	Springfield, MA	Michael J. Albano, D	1999, Nov.
Portland, OR	Vera Katz, N-P	2000, Nov.	Springfield, MO	Leland L. Gannaway, N-P	2001, Apr.
Portsmouth, VA	James W. Holley III, N-P	2000, May	Springfield, OH	Warren Copeland, N-P	2000, Jan.
Providence, RI	Vincent A. Cianci Jr., I	2002, Nov.	Stamford, CT	Dannel P. Malloy, D	2001, Nov.
Provo, UT	Lewis K. Billings, N-P	2001, Nov.	Sterling Hts., MI	Richard J. Notte, N-P	1999, Nov.
Quincy, IL	Charles W. Scholz, D	2001, Apr.	Stockton, CA	Gary Podesto, N-P	2000, Nov.
Quincy, MA	James A. Sheets, D	1999, Nov.	Stratford, CT	Debbie Rose, N-P	1999, Nov.
Racine, WI	James M. Smith, N-P	2003, Apr.	Suffolk, VA	E. Dana Dickens III, N-P	2002, July
Raleigh, NC	Tom Fetzer, N-P	(2)	Sunnyvale, CA	Manuel Valerio, N-P	1999, Nov.
Rancho Cucamonga, CA	William Alexander, N-P	2000, Nov.	Sunrise, FL	Steven B. Feren, N-P	2001, Mar.
Rapid City, SD	Jim Shaw, N-P	2001, May	Syracuse, NY	Roy A. Bernardi, R	2001, Nov.
Reading, PA	Paul Angstadt, R	1999, Nov.	Tacoma, WA	Brian Ebersole, N-P	1999, Nov.
Redding, CA	Bob Anderson, N-P	2000, Apr.	Tallahassee, FL	Scott Maddox, N-P	2001, Feb.
Redondo Beach, CA	Gregory C. Hill, N-P	2001, Mar.	Tampa, FL	Dick A. Greco, N-P	2003, Mar.
Redwood City, CA	Dianne Howard, N-P	1999, Nov.	Taunton, MA	Robert G. Nunes, D	1999, Nov.
Reno, NV	Jeff Griffin, N-P	2002, Nov.	Taylor, MI	Gregory E. Pitoniak, D	2001, Nov.
Rialto, CA	Ray Farmer, N-P	2000, Nov.	Tempe, AZ	Neil G. Giuliano, N-P	2000, May
Richardson, TX	Gary Slagel, N-P	2001, May	Temple, TX	Keifer Marshall Jr., N-P	2000, May
Richmond, CA	Rosemary M. Corbin, D	2001, Nov.	Terre Haute, IN	James Jenkins, D	1999, Nov.
Richmond, VA	Timothy M. Kaine, N-P	2000, July	Thornton, CO	Margaret W. Carpenter, N-P	1999, Nov.
Riverside, CA	Ronald O. Loveridge, N-P	2001, Nov.	Thousand Oaks, CA	Linda Parks, N-P	1999, Nov.
Roanoke, VA	David A. Bowers, D	2000, May	Titusville, FL	Larry D. Bartley, N-P	2000, Nov.
Rochester, MN	Charles J. Canfield, N-P	2002, Nov.	Toledo, OH	Carty Finkbeiner, N-P	2001, Nov.
Rochester, NY	William A. Johnson Jr., D	1999, Nov.	Topeka, KS	Joan Wagnon, N-P	2001, Apr.
Rochester Hills, MI	Kenneth D. Snell, N-P	1999, Nov.	Torrance, CA	Dee Hardison, N-P	2002, Mar.
			Trenton, NJ	Douglas H. Palmer, N-P	2002, May

City	Name	Next Election	City	Name	Next Election
Troy, MI	Jeanne M. Stine, N-P	2001, Apr.	W. Allis, WI	Jeannette Bell, N-P	2000, Mar.
Troy, NY	Mark Pattison, D	1999, Nov.	W. Covina, CA	Kathy Howard, N-P	2000, Mar.
Tucson, AZ	George Miller, D	1999, Nov.	W. Hartford, CT	Robert Bouvier, R	1999, Nov.
Tulsa, OK	M. Susan Savage, D	2002, Mar.	W. Haven, CT	H. Richard Borer Jr., D.	1999, Nov.
Tuscaloosa, AL	Alvin DuPont, D	2001, Aug.	W. Palm Beach, FL	Joel T. Daves, N-P	2003, Mar.
Tyler, TX	Kevin Eltife, N-P	2000, May	Westland, MI	Robert J. Thomas, D	2001, Nov.
Union City, NJ	Raul "Rudy" Garcia, D	2000, May	Westminster, CA	Frank G. Fry, N-P	2000, Nov.
Upland, CA	Robert R. Nolan, N-P	2000, Nov.	Westminster, CO	Nancy Heil, N-P	1999, Nov.
Utica, NY	Edward Hanna, I	1999, Nov.	Wheaton, IL	C. James Carr, N-P	2003, Apr.
Vacaville, CA	David A. Fleming, N-P	2002, Nov.	White Plains, NY	Joseph Delfino, R	2001, Nov.
Vallejo, CA	Gloria Exlin, D	1999, Nov.	Whittier, CA	Greg Nordbak, N-P	2003, Apr.
Vancouver, WA	Royce E. Pollard, N-P	1999, Nov.	Wichita, KS	Bob Knight, N-P	2003, Apr.
Vineland, NJ	Anthony Campanella, R	2000, June	Wichita Falls, TX	Kay Yeager, N-P	2000, May
Virginia Beach, VA	Meyera E. Oberndorf, I	2000, May	Wilkes-Barre, PA	Thomas McGroarty, D	1999, Nov.
Visalia, CA	Wally Gregory, N-P	1999, Nov.	Wilmington, DE	James H. Sills Jr., D	2000, Nov.
Vista, CA	Gloria E. McClellan, R	2002, Nov.	Wilmington, NC	Hamilton E. Hicks Jr., N-P	1999, Nov.
Waco, TX	Michael D. Morrison, N-P	2000, May	Winston-Salem, NC	Jack Cavanagh Jr., N-P	2001, Nov.
Walnut Creek, CA.	Gwen Regalia, N-P	1999, Nov.	Woodbridge, NJ	James E. McGreevey, D	1999, Nov.
Waltham, MA	William F. Stanley, D	1999, Nov.	Woonsocket, RI	Susan D. Menard, N-P	1999, Nov.
Warren, MI	Mark A. Steenbergh, N-P	1999, Nov.	Worcester, MA	Raymond V. Mariano, N-P	1999, Nov.
Warren, OH	Henry Angelo, D	1999, Nov.	Wyandotte, MI	Lawrence S. Stec, N-P	2001, Apr.
Warwick, RI	Lincoln D. Chafee, R	2000, Nov.	Wyoming, MI	Douglas L. Hoekstra Jr., N-P	2001, Nov.
Washington, DC	Anthony A. Williams, D	2002, Nov.	Yakima, WA	John Puccinelli, N-P	2000, Jan.
Waterbury, CT	Philip A. Giordano, R	1999, Nov.	Yonkers, NY	John Spencer, R	1999, Nov.
Waterloo, IA	John R. Rooff III, R	1999, Nov.	York, PA	Charles Robertson, D	2001, Nov.
Waukegan, IL	William F. Durkin, D	2001, Apr.	Youngstown, OH	George M. McKelvey, D	2001, Nov.
Waukesha, WI	Carol Lombardi, N-P	2002, Apr.	Yuma, AZ	Marilyn R. Young, N-P	2001, Nov.
Wauwatosa, WI	Maricolette Walsh, N-P	2000, Apr.			

(1) Mayoralty rotated among city council members every 9 mos. (2) Runoff election to be held Nov. 2, 1999. (3) Mayor appointed by city council. (4) Mayoralty rotated among city council members every 12 mos. (5) City manager; hired, not elected.

Governors of States and Puerto Rico
As of Oct. 1999

State	Capital, ZIP Code	Governor	Party	Term years	Term expires	Annual salary[1]
Alabama	Montgomery 36130	Don Siegelman	Dem.	4	Jan. 2003	$94,655
Alaska	Juneau 99811	Tony Knowles	Dem.	4	Dec. 2002	81,648
Arizona	Phoenix 85007	Jane Dee Hull	Rep.	4	Jan. 2003	95,000
Arkansas	Little Rock 72201	Mike Huckabee	Rep.	4	Jan. 2003	60,000
California	Sacramento 95814	Gray Davis	Dem.	4	Jan. 2003	165,000
Colorado	Denver 80203	Bill Owens	Rep.	4	Jan. 2003	90,000
Connecticut	Hartford 06106	John G. Rowland	Rep.	4	Jan. 2003	78,000
Delaware	Dover 19901	Thomas R. Carper	Dem.	4	Jan. 2001	107,000
Florida	Tallahassee 32399	Jeb Bush	Rep.	4	Jan. 2003	112,304
Georgia	Atlanta 30334	Roy E. Barnes	Dem.	4	Jan. 2003	115,939
Hawaii	Honolulu 96813	Ben Cayetano	Dem.	4	Dec. 2002	94,780
Idaho	Boise 83720	Dirk Kempthorne	Rep.	4	Jan. 2003	92,500
Illinois	Springfield 62706	George H. Ryan	Rep.	4	Jan. 2003	140,132
Indiana	Indianapolis 46204	Frank O'Bannon	Dem.	4	Jan. 2001	77,200
Iowa	Des Moines 50319	Tom Vilsack	Dem.	4	Jan. 2003	104,352
Kansas	Topeka 66612	Bill Graves	Rep.	4	Jan. 2003	91,742
Kentucky	Frankfort 40601	Paul Patton	Dem.	4	Dec. 1999	97,067
Louisiana	Baton Rouge 70804	M. J. "Mike" Foster Jr.	Rep.	4	Jan. 2004	95,000
Maine	Augusta 04333	Angus S. King Jr.	Ind.	4	Jan. 2003	70,000
Maryland	Annapolis 21401	Parris N. Glendening	Dem.	4	Jan. 2003	120,000
Massachusetts	Boston 02133	Paul Cellucci	Rep.	4	Jan. 2003	90,000
Michigan	Lansing 48909	John Engler	Rep.	4	Jan. 2003	127,300
Minnesota	St. Paul 55155	Jesse Ventura	RF[2]	4	Jan. 2003	120,303
Mississippi	Jackson 39205	Kirk Fordice	Rep.	4	Jan. 2000	83,160
Missouri	Jefferson City 65102	Mel Carnahan	Dem.	4	Jan. 2001	112,755
Montana	Helena 59620	Marc Racicot	Rep.	4	Jan. 2001	83,672
Nebraska	Lincoln 68509	Mike Johanns	Rep.	4	Jan. 2003	65,000
Nevada	Carson City 89710	Kenny C. Guinn	Rep.	4	Jan. 2003	117,000
New Hampshire	Concord 03301	Jeanne Shaheen	Dem.	2	Jan. 2001	90,547
New Jersey	Trenton 08625	Christine Todd Whitman	Rep.	4	Jan. 2002	85,000
New Mexico	Santa Fe 87503	Gary E. Johnson	Rep.	4	Jan. 2003	90,000
New York	Albany 12224	George E. Pataki	Rep.	4	Jan. 2003	179,000
North Carolina	Raleigh 27603	James B. Hunt Jr.	Dem.	4	Jan. 2001	113,656
North Dakota	Bismarck 58505	Edward T. Schafer	Rep.	4	Jan. 2001	76,879
Ohio	Columbus 43266	Bob Taft	Rep.	4	Jan. 2003	119,235
Oklahoma	Oklahoma City 73105	Frank Keating	Rep.	4	Jan. 2003	101,140
Oregon	Salem 97310	John Kitzhaber	Dem.	4	Jan. 2003	88,300
Pennsylvania	Harrisburg 17120	Tom Ridge	Rep.	4	Jan. 2003	105,035
Rhode Island	Providence 02903	Lincoln C. Almond	Rep.	4	Jan. 2003	95,000
South Carolina	Columbia 29211	Jim Hodges	Dem.	4	Jan. 2003	106,078
South Dakota	Pierre 57501	William J. Janklow	Rep.	4	Jan. 2003	87,276
Tennessee	Nashville 37243	Don Sundquist	Rep.	4	Jan. 2003	85,000
Texas	Austin 78711	George W. Bush	Rep.	4	Jan. 2003	115,345
Utah	Salt Lake City 84114	Michael O. Leavitt	Rep.	4	Jan. 2001	93,000
Vermont	Montpelier 05609	Howard Dean	Dem.	2	Jan. 2001	115,763
Virginia	Richmond 23219	James S. Gilmore III	Rep.	4	Jan. 2002	124,855
Washington	Olympia 98504	Gary Locke	Dem.	4	Jan. 2001	132,000
West Virginia	Charleston 25305	Cecil H. Underwood	Rep.	4	Jan. 2001	90,000
Wisconsin	Madison 53707	Tommy G. Thompson	Rep.	4	Jan. 2003	115,699
Wyoming	Cheyenne 82002	Jim Geringer	Rep.	4	Jan. 2003	95,000
Puerto Rico	San Juan 00936	Pedro J. Rossello	NPP[3]	4	Jan. 2001	70,000

(1) Salary in effect in 1999. (2) Reform Party. (3) New Progressive Party.

Races for Governor, 1998

Source: Voter News Service

State	Democrat	Vote	Republican	Vote	Other	Vote
AL....	**DON SIEGELMAN**	**760,155**	Fob James, Jr.*	554,746		
AK ...	**TONY KNOWLES***	**112,879**	John Lindauer	39,331		
AZ...	Paul Johnson...............	361,552	**JANE DEE HULL***	**620,188**		
AR ...	Bill Bristow................	272,923	**MIKE HUCKABEE***	**421,989**		
CA ...	**GRAY DAVIS**............	**4,860,702**	Dan Lungren	3,218,030		
CO ...	Gail Schoettler............	639,905	**BILL OWENS**	**648,202**		
CT ...	Barbara B. Kennelly.......	354,187	**JOHN G. ROWLAND***	**628,707**		
FL	Buddy MacKay	1,773,054	**JEB BUSH**.............	**2,192,105**		
GA ...	**ROY E. BARNES**	**941,076**	Guy Millner...............	790,201		
HI ...	**BEN CAYETANO***	**204,206**	Linda Lingle	198,952		
ID ...	Robert C. Huntley	110,815	**DIRK KEMPTHORNE**	**258,095**		
IL	Glenn W. Poshard.........	1,594,191	**GEORGE H. RYAN**	**1,714,094**		
IA	**TOM VILSACK**	**500,231**	Jim Ross Lightfoot.........	444,787		
KS ...	Tom Sawyer	168,243	**BILL GRAVES***	**544,882**		
ME ...	Thomas J. Connolly........	50,506	James B. Longley, Jr........	79,716	**ANGUS S. KING JR.*** (I)..	**246,772**
MD ...	**PARRIS N. GLENDENING***...	**846,972**	Ellen R. Sauerbrey.........	688,357		
MA ...	Scott Harshbarger	901,843	**PAUL CELLUCCI***	**967,160**		
MI ...	Geoffrey Fieger...........	1,143,574	**JOHN ENGLER***	**1,883,005**		
MN ...	Hubert H. "Skip" Humphrey III.	587,528	Norm Coleman............	717,350	**JESSE VENTURA** (RF)..	**773,713**
NE ...	Bill Hoppner.............	250,678	**MIKE JOHANNS**	**293,910**		
NV ...	Jan Laverty Jones	182,281	**KENNY GUINN**	**223,892**		
NH ...	**JEANNE SHAHEEN***	**210,769**	Jay Lucas................	98,473		
NM ...	Martin J. Chavez..........	226,755	**GARY E. JOHNSON***	**271,948**		
NY ...	Peter F. Vallone	1,570,317	**GEORGE E. PATAKI***	**2,571,991**	Tom Golisano (IP).....	364,056
OH ...	Lee Fisher	1,498,956	**BOB TAFT**.............	**1,678,721**		
OK ...	Laura Boyd	357,552	**FRANK KEATING***	**505,498**		
OR ...	**JOHN KITZHABER***	**717,061**	Bill Sizemore.............	334,001	Peg Luksik (CP)	315,761
PA ...	Ivan Itkin	938,745	**TOM RIDGE***	**1,736,844**		
RI ...	Myrth York	129,105	**LINCOLN C. ALMOND***	**156,180**		
SC ...	**JIM HODGES**	**570,070**	David Beasley*............	484,088		
SD ...	Bernie Hunhoff	85,473	**WILLIAM J. JANKLOW***	**166,621**		
TN ...	John J. Hooker	287,750	**DON SUNDQUIST***	**669,973**		
TX ...	Garry Mauro	1,165,444	**GEORGE W. BUSH***	**2,551,454**		
VT....	**HOWARD DEAN***	**121,425**	Ruth Dwyer	89,726		
WI ...	Ed Garvey..............	679,553	**TOMMY G. THOMPSON*** ...	**1,047,716**		
WY ...	John P. Vinich	70,754	**JIM GERINGER***	**97,235**		

* = Incumbent. Uppercase denotes winner. (CP)=Constitutional Party, (I)=Independent, (IP)=Independence Party, (RF)=Reform Party

State Officials, Salaries, Party Membership

As of Oct. 1999; I=independent

Alabama
Governor — Don Siegelman, D, $94,655
Lt. Gov. — Steve Windom, R, $12 per day, plus $50 per day expenses, plus $3,780 per mo expenses
Sec. of State — Jim Bennett, R, $66,722
Atty. Gen. — William Pryor, R, $124,951
Treasurer — Lucy Baxley, D, $66,722
Legislature: meets annually at Montgomery the 3d Tues. in Apr., 1st year of term of office; 1st Tues. in Feb., 2d and 3d yr; 2d Tues. in Jan., 4th yr. Members receive $10 per day salary, plus $50 per day expenses, plus $2,280 per mo expenses.
Senate — Dem., 23; Rep., 12. Total, 35
House — Dem., 69; Rep., 36. Total, 105

Alaska
Governor — Tony Knowles, D, $81,648
Lt. Gov — Fran Ulmer, D, $76,188
Atty. General — Bruce Botelho, D, $86,808
Legislature: meets annually in Jan. at Juneau for 120 days with a 10-day extension possible upon 2/3 vote. First session in odd years. Members receive $24,012 annually, plus $173 per diem.
Senate — Dem., 5; Rep., 15. Total, 20
House — Dem., 14; Rep., 26. Total, 40

Arizona
Governor — Jane Dee Hull, R, $95,000
Sec. of State — Betsey Bayless, R, $70,000
Atty. Gen. — Janet Napolitano, D, $90,000
Treasurer — Carol Springer, R, $70,000
Legislature: meets annually in Jan. at Phoenix. Each member receives an annual salary of $24,000.
Senate — Dem., 14; Rep., 16. Total, 30
House — Dem., 20; Rep., 40. Total, 60

Arkansas
Governor — Mike Huckabee, R, $60,000
Lt. Gov. — Winthrop P. Rockefeller, R, $29,000
Sec. of State — Sharon Priest, D, $37,500
Atty. Gen. — Mark Pryor, D, $50,000
Treasurer — Jimmie Lou Fisher, D, $37,500
Auditor — Gus Wingfield, D, $37,500
General Assembly: meets odd years in Jan. at Little Rock. Members receive $12,500 annually.
Senate — Dem., 28; Rep., 7. Total, 35
House — Dem., 85; Rep., 14; 1 vacancy. Total, 100

California
Governor — Gray Davis, D, $165,000
Lt. Gov. — Cruz Bustamante, D, $123,750
Sec. of State — Bill Jones, R, $123,750
Controller — Kathleen Connell, D, $132,000
Atty. Gen. — Bill Lockyer, D, $140,250
Legislature: meets at Sacramento on the 1st Mon. in Dec. of even-numbered years; each session lasts 2 years. Members receive $99,000 annually, plus $121 per diem.
Senate — Dem., 25; Rep., 15. Total, 40
Assembly — Dem., 47; Rep., 32; Green, 1. Total, 80

Colorado
Governor — Bill Owens, R, $90,000
Lt. Gov. — Joe Rogers, R, $68,500
Sec. of State — Donetta Davidson, R, $68,500
Atty. Gen. — Ken Salazar, D, $80,000
Treasurer — Mike Coffman, R, $68,500
General Assembly: meets annually in Jan. at Denver. Members receive $30,000 annually.
Senate — Dem., 15; Rep., 20. Total, 35
House — Dem., 24; Rep., 41. Total, 65

Connecticut
Governor — John G. Rowland, R, $78,000
Lt. Gov. — M. Jodi Rell, R, $71,500
Sec. of State — Susan Bysiewicz, D, $65,000
Treasurer — Denise Nappier, D, $70,000
Comptroller — Nancy S. Wyman, D, $65,000
Atty. Gen. — Richard Blumenthal, D, $75,000
General Assembly: meets annually odd years in Jan. and even years in Feb., at Hartford. Members receive $21,788 annually, plus $5,500 (senator), $4,500 (representative) per year for expenses.
Senate — Dem., 19; Rep., 17. Total, 36
House — Dem., 96; Rep., 54; 1 vacancy. Total, 151

Delaware

Governor — Thomas R. Carper, D, $107,000
Lt. Gov. — Ruth Ann Minner, D, $47,900
Sec. of State — Edward J. Freel, D, $95,500
Atty. Gen. — M. Jane Brady, R, $105,200
Treasurer — Jack Markell, D, $84,800
General Assembly: meets annually the 2d Tues. in Jan. and continues until June 30, at Dover. Members receive $30,700 annually.
Senate — Dem., 13; Rep., 8. Total, 21
House — Dem., 15; Rep., 26. Total, 41

Florida

Governor — Jeb Bush, R, $117,240
Lt. Gov. — Kenneth "Buddy" McKay, D, $112,304
Sec. of State — Sandra Mortham, R, $116,056
Comptroller — Robert R. Milligan, R, $116,056
Atty. Gen. — Robert Butterworth, D, $116,056
Treasurer — Bill Nelson, D, $116,056
Legislature: meets annually at Tallahassee. Members receive $26,388 annually, plus expense allowance.
Senate — Dem., 15; Rep., 25. Total, 40
House — Dem., 46; Rep., 71; 3 vacancies. Total, 120

Georgia

Governor — Roy E. Barnes, D, $115,939
Lt. Gov. — Mark Taylor, D, $75,725
Sec. of State — Lewis Massey, D, $93,120
Atty. Gen. — Thurbert Baker, D, $106,300
General Assembly: meets annually in Atlanta. Members receive $11,348 annually ($75 per diem and $4,800 expense reimbursement).
Senate — Dem., 34; Rep., 22. Total, 56
House — Dem., 102; Rep., 78. Total, 180

Hawaii

Governor — Ben Cayetano, D, $94,780
Lt. Gov. — Mazie K. Hirono, D, $90,041
Atty. Gen. — Earl I. Anzai, $85,302
Comptroller — Raymond H. Sato, $85,302
Dir. of Budget & Finance — Neal H. Miyahira, $85,302
Legislature: meets annually on 3d Wed. in Jan. at Honolulu. Members receive $32,000 annually, plus expenses.
Senate — Dem., 23; Rep., 2. Total, 25
House — Dem., 39; Rep., 12. Total, 51

Idaho

Governor — Dirk Kempthorne, R, $92,500
Lt. Gov. — C. L. "Butch" Otter, R, $24,500
Sec. of State — Pete T. Cenarrusa, R, $75,000
Treasurer — Ron Crane, R, $75,000
Atty. Gen. — Alan Lance, R, $82,500
Legislature: meets annually the Mon. on or nearest Jan. 9 at Boise. Members receive $14,760 annually, plus $75 per day during session if required to maintain a 2d residence, $40 if no 2d residence; plus $50 per day when engaged in legislative business when legislature is not in session.
Senate — Dem., 4; Rep., 31. Total, 35
House — Dem., 12; Rep., 58. Total, 70

Illinois

Governor — George H. Ryan, R, $140,132
Lt. Gov. — Corrine Wood, R, $107,160
Sec. of State — Jesse White, D, $123,646
Comptroller — Daniel Hynes, D, $107,160
Atty. Gen. — James Ryan, R, $123,646
Treasurer — Judy Baar Topinka, R, $107,160
General Assembly: meets annually in Nov. and Jan. at Springfield. Members receive $53,581 annually.
Senate — Dem., 27; Rep., 32. Total, 59
House — Dem., 62; Rep., 56. Total, 118

Indiana

Governor — Frank O'Bannon, D, $77,200
Lt. Gov. — Joseph E. Kernan, D, $64,000
Sec. of State — Sue Anne Gilroy, R, $66,000
Atty. Gen. — Jeffrey A. Modesitt, D, $79,400
Treasurer — Tim Berry, R, $66,000
Auditor — Connie Nass, R, $66,000
General Assembly: meets annually on the Tues. after the 2d Mon. in Jan. at Indianapolis. Members receive $11,600 annually, plus $112 per day while in session, $25 per day while not in session.
Senate — Dem., 19; Rep., 31. Total, 50
House — Dem., 53; Rep., 47. Total, 100

Iowa

Governor — Tom Vilsack, D, $104,352
Lt. Gov. — Sally Pederson, D, $73,046
Sec. of State — Chester J. Culver D, $82,939
Atty. Gen. — Tom Miller, D, $102,361
Treasurer — Michael L. Fitzgerald, D, $85,429
Auditor — Richard D. Johnson, R, $85,429
Sec. of Agriculture — Patty Judge, D, $85,429
General Assembly: meets annually in Jan. at Des Moines. Members receive $21,381 annually, plus expense allowance.
Senate — Dem., 20; Rep., 30. Total, 50
House — Dem., 44; Rep., 56 Total, 100

Kansas

Governor — Bill Graves, R, $91,742
Lt. Gov. — Gary Sherrer, R, $25,072
Sec. of State — Ron Thornburgh, R, $71,269
Atty. Gen. — Carla Stovall, R, $81,958
Treasurer — Tim Shallenburger, R, $71,269
Insurance Commissioner — Kathleen Sebelius, D, $71,269
Legislature: meets annually on the 2d Mon. of Jan. at Topeka. Members receive $74.58 per day salary, plus $80 per day expenses while in session, $5,400 total allowance while not in session.
Senate — Dem., 13; Rep., 27. Total, 40
House — Dem., 48; Rep., 77. Total, 125

Kentucky

Governor — Paul Patton, D, $97,067
Lt. Gov. — Steve Henry, D, $82,521
Sec. of State — John Y. Brown III, D, $82,521
Atty. Gen. — A. B. Chandler III, D, $82,521
Treasurer — John Kennedy Hamilton, D, $82,521
Auditor — Ed Hatchett, D, $82,521
Sec. of Economic Dev. — Gene Strong, $154,922
General Assembly: meets even years in Jan. at Frankfort. Members receive $151 per day, plus $88 per day expenses during session and $1,435 per month for expenses for interim.
Senate — Dem., 18; Rep., 20. Total, 38
House — Dem., 65; Rep., 35. Total, 100

Louisiana

Governor — M. J. "Mike" Foster Jr., R, $95,000
Lt. Gov. — Kathleen Babineaux Blanco, D, $85,000
Sec. of State — W. Fox McKeithen, R, $85,000
Atty. Gen. — Richard Ieyoub, D, $85,000
Treasurer — Ken Duncan, D, $85,000
Legislature: meets in odd-numbered years at Baton Rouge starting last Mon. in Mar., for 60 legislative days of 85 calendar days; meets in even-numbered years on last Mon. in Apr. for 30 days of 45 calendar days. Members receive $16,800 annually, plus $97 per day expenses while in session and $500 per month as an unvouchered expense allowance.
Senate — Dem., 25; Rep., 14. Total, 39
House — Dem., 77; Rep., 26, 2 vacancies. Total, 105

Maine

Governor — Angus S. King Jr., I, $70,000
Sec. of State — Dan A. Gwadosky, D, $61,526
Atty. Gen. — Andrew Ketterer, D, $90,272
Treasurer — Dale McCormick, D, $61,526
State Auditor — Gail M. Chase, D, $72,405
Legislature: meets in odd-numbered years at Augusta on first Wed. in Dec.; meets in even-numbered years on Wed. after first Tues. in Jan. Members receive $10,500 for first regular session, $7,500 for 2d, plus a daily expense allowance.
Senate — Dem., 20; Rep., 14; 1 ind. Total, 35
House — Dem., 79; Rep.,71; 1 ind. Total, 151

Maryland

Governor — Parris N. Glendening, D, $120,000
Lt. Gov. — Kathleen Kennedy Townsend, D, $100,000
Comptroller — William Donald Schaefer, D, $100,000
Atty. Gen. — J. Joseph Curran Jr., D, $100,000
Sec. of State — John Willis, D, $70,000
Treasurer — Richard N. Dixon, D, $100,000
General Assembly: meets 90 consecutive days annually beginning on 2d Wed. in Jan. at Annapolis. Members receive $29,700 annually, plus expenses.
Senate — Dem., 32; Rep., 15. Total, 47
House — Dem., 106; Rep., 35. Total, 141

Massachusetts

Governor — Paul Cellucci, R, $90,000
Lt. Gov. — Jane Swift, R, $60,000
Sec. of State — William Francis Galvin, D, $75,000
Atty. Gen. — Thomas F. Reilly, D, $80,000
Treasurer — Shannon P. O'Brien, D, $75,000
Auditor — A. Joseph DeNucci, D, $75,000
General Court (legislature): meets Jan. biennially in Boston. Members receive $46,410 annually.
Senate — Dem., 33; Rep., 7. Total, 40
House — Dem., 130; Rep., 27; 1 ind.; 2 vacancies. Total, 160

Michigan
Governor — John Engler, R, $127,300
Lt. Gov. — Dick Posthumus, R, $93,978
Sec. of State — Candice S. Miller, R, $124,900
Atty. Gen. — Jennifer M. Granholm, D, $124,900
Treasurer — Mark A. Murray (appointed), $111,000
Legislature: meets annually in Jan. at Lansing. Members receive $56,981 annually.
Senate — Dem., 15; Rep., 23. Total, 38
House — Dem., 52; Rep., 58. Total, 110

Minnesota
(RP=Reform Party; DFL=Democratic-Farmer-Labor Party)
Governor — Jesse Ventura, RP, $120,303
Lt. Gov. — Mae Schunk, RP, $66,168
Sec. of State — Mary Kiffmeyer, R, $66,168
Atty. Gen. — Michael Hatch, DFL, $93,983
Treasurer — Carol Johnson, DFL, $66,168
Auditor — Judith H. Dutcher, R, $72,187
Legislature: meets for a total of 120 days within every 2 years, at St. Paul. Members receive $31,140 annually, plus expense allowance during session.
Senate — DFL, 42; R, 24; 1 ind. Total, 67
House — DFL, 63; R, 71. Total, 134

Mississippi
Governor — Kirk Fordice, R, $83,160
Lt. Gov. — Ronnie Musgrove, D, $40,800
Sec. of State — Eric Clark, D, $75,000
Atty. Gen. — Mike Moore, D, $90,800
Treasurer — Marshall Bennett, D, $75,000
Auditor — Phil Bryant, R, $75,000
Legislature: meets annually in Jan. at Jackson. Members receive $10,000 per regular session, plus travel allowance, and $1,500 per month when not in session.
Senate — Dem., 34; Rep., 18. Total, 52
House — Dem., 84; Rep., 34; 2 ind., 2 vacancies. Total, 122

Missouri
Governor — Mel Carnahan, D, $112,755
Lt. Gov. — Roger B. Wilson, D, $68,195
Sec. of State — Rebecca McDowell Cook, D, $90,471
Atty. Gen. — Jeremiah W. Nixon, D, $97,899
Treasurer — Bob Holden, D, $90,471
State Auditor — Claire McCaskill, D, $90,471
General Assembly: meets annually at Jefferson City beginning 1st Wed. after 1st Mon. in Jan. Members receive $29,080 annually.
Senate — Dem., 18; Rep., 16. Total, 34
House — Dem., 86; Rep., 76; 1 ind. Total, 163

Montana
Governor — Marc Racicot, R, $83,672
Lt. Gov. — Judy Martz, R, $58,961
Sec. of State — Mike Cooney, D, $63,571
Atty. Gen. — Joe Mazurek, D, $71,638
Legislative Assembly: meets odd years in Jan. at Helena. Members receive $58.50 per legislative day, plus $70 per day for expenses while in session.
Senate — Dem., 16; Rep., 34. Total, 50
House — Dem., 35; Rep., 65. Total, 100

Nebraska
Governor — Mike Johanns, R, $65,000
Lt. Gov. — Dave Maurstad, R, $47,000
Sec. of State — Scott Moore, R, $52,000
Atty. Gen. — Don Stenberg, R, $64,500
Treasurer — David Heineman, R, $49,500
Legislature: Unicameral body composed of 49 members who are elected on a nonpartisan ballot and are called senators; meets annually in Jan. at Lincoln. Members receive $12,000 annually, plus expenses.

Nevada
Governor — Kenny C. Guinn, R, $117,000
Lt. Gov. — Lorraine Hunt, R, $50,000
Sec. of State — Dean Heller, R, $80,000
Controller — Kathy Augustine, R, $80,000
Atty. Gen. — Frankie Sue Del Papa, D, $110,000
Treasurer — Brian Krolicki, R, $80,000
Legislature: meets at Carson City odd years starting on 1st Mon. in Feb. for 120 days. Members receive $130 per day salary, plus $80 per day expenses, while in session.
Senate — Dem., 9; Rep., 12. Total, 21
Assembly — Dem., 28; Rep., 14. Total, 42

New Hampshire
Governor — Jeanne Shaheen, D, $90,547
Sec. of State — William M. Gardner, D, $72,206
Atty. Gen. — Philip T. McLaughlin, D, $80,832
Treasurer — Georgie A. Thomas, R, $72,206

General Court (Legislature): meets every year in Jan. at Concord. Members receive $200, presiding officers $250, biannually.
Senate — Dem., 13; Rep., 11. Total, 24
House — Rep., 242; Dem., 154; 1 ind.; 3 vacancies. Total, 400

New Jersey
Governor — Christine Todd Whitman, R, $85,000
Sec. of State — DeForest B. Soaries, R, $115,000
Atty. Gen. — John J. Farmer Jr, R, $115,000
Treasurer — Roland H. Machold (acting), R, $115,000
Legislature: meets throughout the year at Trenton. Members receive $35,000 annually, except president of Senate and speaker of Assembly, who receive 1/3 more.
Senate — Dem., 16; Rep., 24. Total, 40
Assembly — Dem., 32; Rep., 48. Total, 80

New Mexico
Governor — Gary E. Johnson, R, $90,000
Lt. Gov. — Walter Bradley, R, $65,000
Sec. of State — Rebecca Vigil-Giron, D, $65,000
Atty. Gen. — Patricia Madrid, D, $72,500
Treasurer — Michael A. Montoya, D, $65,000
Legislature: meets starting on the 3d Tues. in Jan. at Santa Fe; odd years for 60 days, even years for 30 days. Members receive $75 per day while in session.
Senate — Dem., 25; Rep., 17. Total, 42
House — Dem., 40; Rep., 30. Total, 70

New York
Governor — George E. Pataki, R, $179,000
Lt. Gov. — Mary O. Donohue, R, $151,500
Sec. of State — Alexander F. Treadwell, R, $120,800
Comptroller — H. Carl McCall, D, $151,500
Atty. Gen. — Eliot Spitzer, D, $151,500
Legislature: meets annually in Jan. at Albany. Members receive $79,500 annually, plus $130 per day expenses.
Senate — Dem., 24; Rep., 36; 1 vacancy. Total, 61
Assembly — Dem., 98; Rep., 52. Total, 150

North Carolina
Governor — James B. Hunt Jr., D, $113,656
Lt. Gov. — Dennis Wicker, D, $94,552, plus expenses
Sec. of State — Elaine F. Marshall, D, $94,552
Atty. Gen. — Michael Easley, D, $94,552
Treasurer — Harlan E. Boyles, D, $94,552
General Assembly: meets odd years in Jan. at Raleigh. Members receive $13,951 annually and an expense allowance of $559 per month, plus subsistence and travel allowance while in session. Also meets in even years for a short session (about 6-8 weeks), usually in May.
Senate — Dem., 35; Rep., 15. Total, 50
House — Dem., 66; Rep., 54. Total, 120

North Dakota
Governor — Edward T. Schafer, R, $76,879
Lt. Gov. — Rosemarie Myrdal, R, $63,183
Sec. of State — Alvin A. Jaeger, R, $58,262
Atty. Gen. — Heidi Heitkamp, D, $65,753
Treasurer — Kathi Gilmore, D, $58,262
Legislative Assembly: meets odd years in Jan. at Bismarck. Members receive $250 per month salary, plus $111 per calendar day salary during session and $42 per day expenses plus any additional state or local taxes on lodging, with a limit of $650 per month.
Senate — Dem., 18; Rep., 31. Total, 49
House — Dem., 34; Rep., 64. Total, 98

Ohio
Governor — Robert Taft, R, $119,235
Lt. Gov. — Maureen O'Connor, R, $62,500
Sec. of State — J. Kenneth Blackwell, R, $88,082
Atty. Gen. — Betty D. Montgomery, R, $88,082
Treasurer — Joseph T. Deters, R, $88,082
Auditor — Jim Petro, R, $88,082
General Assembly: begins odd years at Columbus starting on 1st Mon. in Jan. Members receive $42,426 annually.
Senate — Dem., 12; Rep., 21. Total, 33
House — Dem., 40; Rep., 59. Total, 99

Oklahoma
Governor — Frank Keating, R, $101,140
Lt. Gov. — Mary Fallin, R, $75,530
Sec. of State — Mike Hunter, R, $65,000
Atty. Gen. — Drew Edmondson, D, $94,349
Treasurer — Robert Butkin, D, $82,004
Auditor — Clifton Scott, D, $82,004

Legislature: meets annually at noon the first Mon. in Feb. at Oklahoma City. In odd-numbered years, the session includes one day (1st Tuesday after 1st Monday) in Jan. Members receive $38,400 annually.
Senate — Dem., 33; Rep., 15. Total, 48
House — Dem., 61; Rep., 40. Total, 101

Oregon

Governor — John Kitzhaber, D, $88,300
Sec. of State — Phil Keisling, D, $67,900
Atty. Gen. — Hardy Myers, D, $72,800
Treasurer — Jim Hill, D, $67,900
Legislative Assembly: meets odd years in Jan. at Salem. Members receive $1,208 monthly, $86 expenses per day during session and when attending meetings during the interim, plus $400 expense account during interim.
Senate — Dem., 13; Rep., 17. Total, 30
House — Dem., 25; Rep., 35. Total, 60

Pennsylvania

Governor — Tom Ridge, R, $105,035
Lt. Gov. — Mark Schweiker, R, $83,027
Sec. of the Commonwealth — Kim Pizzingrilli, R, $95,346
Atty. Gen. — D. Michael Fisher, R, $104,000
Treasurer — Barbara Hafer, R, $104,000
General Assembly: convenes annually in Jan. at Harrisburg. Members receive $57,367 annually, plus expenses.
Senate — Dem., 20; Rep., 30. Total, 50
House — Dem., 99; Rep., 104. Total, 203

Rhode Island

Governor — Lincoln C. Almond, R, $95,000
Lt. Gov. — Charles J. Fogarty, D, $80,000
Sec. of State — James R. Langevin, D, $80,000
Atty. Gen. — Sheldon Whitehouse, D, $85,000
Treasurer — Paul J. Tavares, D, $80,000
General Assembly: meets annually in Jan. at Providence. Members receive $10,000 annually.
Senate — Dem., 42; Rep., 8. Total, 50
House — Dem., 86; Rep., 13; 1 ind. Total, 100

South Carolina

Governor — Jim Hodges, D, $106,078
Lt. Gov. — Robert L. Peeler, R, $46,545
Sec. of State — Jim Miles, R, $92,007
Comptroller Gen. — James A. Lander, D, $92,007
Atty. Gen. — Charles M. Condon, R, $92,007
Treasurer — Grady L. Patterson Jr., D, $92,007
General Assembly: meets annually in Jan. at Columbia. Members receive $10,400 annually, plus $88 per day for expenses.
Senate — Dem., 24; Rep., 22. Total, 46
House — Dem., 57; Rep., 66; 1 vacancy. Total, 124

South Dakota

Governor — William J. Janklow, R, $87,276
Lt. Gov. — Carole Hillard, R, $63,377
Sec. of State — Joyce Hazeltine, R, $59,300
Treasurer — Dick Butler, D, $59,300
Atty. Gen. — Mark Barnett, R, $74,131
Auditor — Vernon Larson, R, $59,300
Legislature: meets annually in Jan. at Pierre. Members receive $6,000 for 40-day session in odd-numbered years, and $6,000 for 35-day session in even-numbered years, plus $95 per legislative day.
Senate — Dem., 13; Rep., 22. Total, 35
House — Dem., 19; Rep., 51. Total, 70

Tennessee

Governor — Don Sundquist, R, $85,000
Lt. Gov. — John S. Wilder, D, $49,500
Sec. of State — Riley C. Darnell, D, $120,000
Comptroller — William Snodgrass, D, $120,000
Atty. Gen. — John Knox Walkup, D, $112,068
General Assembly: meets annually in Jan. at Nashville. Members receive $16,500 annual salary, plus $114 per day expenses while in session.
Senate — Dem., 18; Rep., 15. Total, 33
House — Dem., 59; Rep., 40. Total, 99

Texas

Governor — George W. Bush, R, $115,345
Lt. Gov. — Rick Perry, R, $7,200
Sec. of State — Elton Bomer, R, $76,966
Comptroller — Carole Keeton Rylander, R, $92,217
Atty. Gen. — John Cornyn, R, $92,217
Railroad Commissioners — Tony Garza, R, Chair; Michael Williams, R; Charles R. Matthews, R; $92,217

Legislature: meets odd years in Jan. at Austin. Members receive $7,200 annually, plus $95 per day expenses while in session.
Senate — Dem., 15; Rep., 16. Total, 31
House — Dem., 78; Rep., 72. Total, 150

Utah

Governor — Michael O. Leavitt, R, $93,000
Lt. Gov. — Olene S. Walker, R, $72,300
Atty. Gen. — Jan Graham, D, $78,200
Auditor — Auston G. Johnson, R, $74,600
Treasurer — Edward T. Alter, R, $72,300
Legislature: convenes for 45 days on 3d Mon. in Jan. each year at Salt Lake City. Members receive $100 per day, plus $38 a day expenses.
Senate — Dem., 11; Rep., 18. Total, 29
House — Dem., 21; Rep., 54. Total, 75

Vermont

Governor — Howard Dean, D, $115,763
Lt. Gov. — Douglas A. Racine, D, $48,258
Sec. of State — Deborah L. Markowitz, D, $72,845
Atty. Gen. — William H. Sorrell, D, $87,507
Treasurer — James H. Douglas, R, $72,845
Auditor — Edward Flanagan, D, $72,845
General Assembly: meets in Jan. at Montpelier (annual and biennial session). Members receive $536 per week while in session plus $105 per day for special session, plus expenses.
Senate — Dem., 17; Rep., 13. Total, 30
House — Dem., 77; Rep., 66; Prog. Coalition, 4; Libertarian/Rep., 1; 2 ind. Total, 150

Virginia

Governor — James S. Gilmore III, R, $124,855
Lt. Gov. — John H. Hager, R, $36,321
Atty. Gen. — Mark L. Earley, R, $110,667
Sec. of the Commonwealth — Anne P. Petera, R, $124,435
Treasurer — Mary G. Morris, R, $106,026
General Assembly: meets annually in Jan. at Richmond. Members receive $18,000 (senate), $17,640 (assembly) annually, plus expense and mileage allowances.
Senate — Dem., 19; Rep., 21. Total, 40
House — Dem., 50; Rep., 49; 1 ind. Total, 100

Washington

Governor — Gary Locke, D, $132,000
Lt. Gov. — Brad Owen, D, $69,000
Sec. of State — Ralph Munro, R, $75,900
Atty. Gen. — Christine Gregoire, D, $120,000
Treasurer — Mike Murphy, D, $92,500
Legislature: meets annually in Jan. at Olympia. Members receive $31,130 annually, plus $82 per diem while in session, and $82 per diem for attending meetings during interim.
Senate — Dem., 27; Rep., 22. Total, 49
House — Dem., 49; Rep., 49. Total, 98

West Virginia

Governor — Cecil H. Underwood, R, $90,000
Sec. of State — Ken Hechler, D, $65,000
Atty. Gen. — Darrell McGraw, D, $75,000
Treasurer — John D. Perdue, D, $70,000
Comm. of Agric. — Gus Douglass, D, $70,000
Auditor — Glen B. Gainer 3d, D, $70,000
Legislature: meets annually in Jan. at Charleston, except after gubernatorial elections, when the legislature meets in Feb. Members receive $15,000 annually.
Senate — Dem., 29; Rep., 5. Total, 34
House — Dem., 75; Rep., 25. Total, 100

Wisconsin

Governor — Tommy G. Thompson, R, $115,699
Lt. Gov. — Scott McCallum, R, $60,183
Sec. of State — Douglas La Follette, D, $54,610
Treasurer — Jack Voight, R, $54,610
Atty. Gen. — James E. Doyle, D, $112,274
Legislature: meets in Jan. at Madison. Members receive $41,809 annually, plus $75 per day expenses.
Senate — Dem., 17; Rep., 16. Total, 33
Assembly — Dem., 45; Rep., 54. Total, 99

Wyoming

Governor — Jim Geringer, R, $95,000
Sec. of State — Joseph B. Meyer, R, $72,500
Atty. Gen. — Gay Woodhouse, R, $72,500
Treasurer — Cynthia Lummis, R, $72,500
Dir., Dept. of Audit — Michael Geesey, R, $72,500
Legislature: meets odd years in Jan., even years in Feb., at Cheyenne. Members receive $125 per day while in session, plus $80 per day for expenses.
Senate — Dem., 10; Rep., 20. Total, 30
House — Dem., 17; Rep., 43. Total, 60

CABINETS OF THE U.S.

The U.S. Cabinet and Its Role

The heads of major executive departments of government constitute the Cabinet. This institution, not provided for in the U.S. Constitution, developed as an advisory body out of the desire of presidents to consult on policy matters. Aside from its advisory role, the Cabinet as a body has no function and wields no executive authority. The president may or may not consult it and is not bound by its advice. Most presidents also confer with numerous advisers outside the Cabinet. A group of regular informal advisers to the president has been known in American history as a "kitchen cabinet." The formal Cabinet (which may include other officials besides department heads, as designated by the president) meets at times set by the president. Members of Pres. Bill Clinton's Cabinet listed here are as of Oct. 15, 1999.

Secretaries of State

The Department of Foreign Affairs was created by act of Congress on July 27, 1789, and the name changed to Department of State on Sept. 15.

President	Secretary	Home	Apptd.	President	Secretary	Home	Apptd.
Washington	Thomas Jefferson	VA	1789	Harrison, B.	James G. Blaine	ME	1889
"	Edmund Randolph	VA	1794	"	John W. Foster	IN	1892
"	Timothy Pickering	PA	1795	Cleveland	Walter Q. Gresham	IN	1893
Adams, J.	Timothy Pickering	PA	1797	"	Richard Olney	MA	1895
"	John Marshall	VA	1800	McKinley	Richard Olney	MA	1897
Jefferson	James Madison	VA	1801	"	John Sherman	OH	1897
Madison	Robert Smith	MD	1809	"	William R. Day	OH	1898
"	James Monroe	VA	1811	"	John Hay	DC	1898
Monroe	John Quincy Adams	MA	1817	Roosevelt, T.	John Hay	DC	1901
Adams, J.Q.	Henry Clay	KY	1825	"	Elihu Root	NY	1905
Jackson	Martin Van Buren	NY	1829	"	Robert Bacon	NY	1909
"	Edward Livingston	LA	1831	Taft	Robert Bacon	NY	1909
"	Louis McLane	DE	1833	"	Philander C. Knox	PA	1909
"	John Forsyth	GA	1834	Wilson	Philander C. Knox	PA	1913
Van Buren	John Forsyth	GA	1837	"	William J. Bryan	NE	1913
Harrison, W.H.	Daniel Webster	MA	1841	"	Robert Lansing	NY	1915
Tyler	Daniel Webster	MA	1841	"	Bainbridge Colby	NY	1920
"	Abel P. Upshur	VA	1843	Harding	Charles E. Hughes	NY	1921
"	John C. Calhoun	SC	1844	Coolidge	Charles E. Hughes	NY	1923
Polk	John C. Calhoun	SC	1845	"	Frank B. Kellogg	MN	1925
"	James Buchanan	PA	1845	Hoover	Frank B. Kellogg	MN	1929
Taylor	James Buchanan	PA	1849	"	Henry L. Stimson	NY	1929
"	John M. Clayton	DE	1849	Roosevelt, F.D.	Cordell Hull	TN	1933
Fillmore	John M. Clayton	DE	1850	"	E.R. Stettinius Jr.	VA	1944
"	Daniel Webster	MA	1850	Truman	E.R. Stettinius Jr.	VA	1945
"	Edward Everett	MA	1852	"	James F. Byrnes	SC	1945
Pierce	William L. Marcy	NY	1853	"	George C. Marshall	PA	1947
Buchanan	William L. Marcy	NY	1857	"	Dean G. Acheson	CT	1949
"	Lewis Cass	MI	1857	Eisenhower	John Foster Dulles	NY	1953
"	Jeremiah S. Black	PA	1860	"	Christian A. Herter	MA	1959
Lincoln	Jeremiah S. Black	PA	1861	Kennedy	Dean Rusk	NY	1961
"	William H. Seward	NY	1861	Johnson, L.B.	Dean Rusk	NY	1963
Johnson, A.	William H. Seward	NY	1865	Nixon	William P. Rogers	NY	1969
Grant	Elihu B. Washburne	IL	1869	"	Henry A. Kissinger	DC	1973
"	Hamilton Fish	NY	1869	Ford	Henry A. Kissinger	DC	1974
Hayes	Hamilton Fish	NY	1877	Carter	Cyrus R. Vance	NY	1977
"	William M. Evarts	NY	1877	"	Edmund S. Muskie	ME	1980
Garfield	William M. Evarts	NY	1881	Reagan	Alexander M. Haig Jr.	CT	1981
"	James G. Blaine	ME	1881	"	George P. Shultz	CA	1982
Arthur	James G. Blaine	ME	1881	Bush	James A. Baker 3d	TX	1989
"	F.T. Frelinghuysen	NJ	1881	"	Lawrence S. Eagleburger	MI	1992
Cleveland	F.T. Frelinghuysen	NJ	1885	Clinton	Warren M. Christopher	CA	1993
"	Thomas F. Bayard	DE	1885	"	Madeleine K. Albright	DC	1997
Harrison, B.	Thomas F. Bayard	DE	1889				

Secretaries of the Treasury

The Treasury Department was organized by act of Congress on Sept. 2, 1789.

President	Secretary	Home	Apptd.	President	Secretary	Home	Apptd.
Washington	Alexander Hamilton	NY	1789	Tyler	Walter Forward	PA	1841
"	Oliver Wolcott	CT	1795	"	John C. Spencer	NY	1843
Adams, J.	Oliver Wolcott	CT	1797	"	George M. Bibb	KY	1844
"	Samuel Dexter	MA	1801	Polk	Robert J. Walker	MS	1845
Jefferson	Samuel Dexter	MA	1801	Taylor	William M. Meredith	PA	1849
"	Albert Gallatin	PA	1801	Fillmore	Thomas Corwin	OH	1850
Madison	Albert Gallatin	PA	1809	Pierce	James Guthrie	KY	1853
"	George W. Campbell	TN	1814	Buchanan	Howell Cobb	GA	1857
"	Alexander J. Dallas	PA	1814	"	Phillip F. Thomas	MD	1860
"	William H. Crawford	GA	1816	"	John A. Dix	NY	1861
Monroe	William H. Crawford	GA	1817	Lincoln	Salmon P. Chase	OH	1861
Adams, J.Q.	Richard Rush	PA	1825	"	William P. Fessenden	ME	1864
Jackson	Samuel D. Ingham	PA	1829	"	Hugh McCulloch	IN	1865
"	Louis McLane	DE	1831	Johnson, A.	Hugh McCulloch	IN	1865
"	William J. Duane	PA	1833	Grant	George S. Boutwell	MA	1869
"	Roger B. Taney	MD	1833	"	William A. Richardson	MA	1873
"	Levi Woodbury	NH	1834	"	Benjamin H. Bristow	KY	1874
Van Buren	Levi Woodbury	NH	1837	"	Lot M. Morrill	ME	1876
Harrison, W.H.	Thomas Ewing	OH	1841	Hayes	John Sherman	OH	1877
Tyler	Thomas Ewing	OH	1841	Garfield	William Windom	MN	1881

President	Secretary	Home	Apptd.	President	Secretary	Home	Apptd.
Arthur	Charles J. Folger	NY	1881	Truman	Fred M. Vinson	KY	1945
"	Walter Q. Gresham	IN	1884	"	John W. Snyder	MO	1946
"	Hugh McCulloch	IN	1884	Eisenhower	George M. Humphrey	OH	1953
Cleveland	Daniel Manning	NY	1885	"	Robert B. Anderson	CT	1957
"	Charles S. Fairchild	NY	1887	Kennedy	C. Douglas	NJ	1961
Harrison, B.	William Windom	MN	1889	Johnson, L.B.	C. Douglas Dillon	NJ	1963
"	Charles Foster	OH	1891	"	Henry H. Fowler	VA	1965
Cleveland	John G. Carlisle	KY	1893	"	Joseph W. Barr	IN	1968
McKinley	Lyman J. Gage	IL	1897	Nixon	David M. Kennedy	IL	1969
Roosevelt, T.	Lyman J. Gage	IL	1901	"	John B. Connally	TX	1971
"	Leslie M. Shaw	IA	1902	"	George P. Shultz	IL	1972
"	George B. Cortelyou	NY	1907	"	William E. Simon	NJ	1974
Taft	Franklin MacVeagh	IL	1909	Ford	William E. Simon	NJ	1974
Wilson	William G. McAdoo	NY	1913	Carter	W. Michael Blumenthal	MI	1977
"	Carter Glass	VA	1918	"	G. William Miller	RI	1979
"	David F. Houston	MO	1920	Reagan	Donald T. Regan	NY	1981
Harding	Andrew W. Mellon	PA	1921	"	James A. Baker 3d	TX	1985
Coolidge	Andrew W. Mellon	PA	1923	"	Nicholas F. Brady	NJ	1988
Hoover	Andrew W. Mellon	PA	1929	Bush	Nicholas F. Brady	NJ	1989
"	Ogden L. Mills	NY	1932	Clinton	Lloyd Bentsen	TX	1993
Roosevelt, F.D.	William H. Woodin	NY	1933	"	Robert E. Rubin	NY	1995
"	Henry Morgenthau, Jr.	NY	1934	"	Lawrence H. Summers	CT	1999

Secretaries of Defense

The Department of Defense, originally designated the National Military Establishment, was created on Sept. 18, 1947. It is headed by the secretary of defense, who is a member of the president's Cabinet. The departments of the army, of the navy, and of the air force function within the Defense Department, and since 1947 the secretaries of these departments have not been members of the president's Cabinet.

President	Secretary	Home	Apptd.	President	Secretary	Home	Apptd.
Truman	James V. Forrestal	NY	1947	Nixon	Elliot L. Richardson	MA	1973
"	Louis A. Johnson	WV	1949	"	James R. Schlesinger	VA	1973
"	George C. Marshall	PA	1950	Ford	James R. Schlesinger	VA	1974
"	Robert A. Lovett	NY	1951	"	Donald H. Rumsfeld	IL	1975
Eisenhower	Charles E. Wilson	MI	1953	Carter	Harold Brown	CA	1977
"	Neil H. McElroy	OH	1957	Reagan	Caspar W. Weinberger	CA	1981
"	Thomas S. Gates Jr.	PA	1959	"	Frank C. Carlucci	PA	1987
Kennedy	Robert S. McNamara	MI	1961	Bush	Richard B. Cheney	WY	1989
Johnson, L.B.	Robert S. McNamara	MI	1963	Clinton	Les Aspin	WI	1993
"	Clark M. Clifford	MD	1968	"	William J. Perry	CA	1994
Nixon	Melvin R. Laird	WI	1969	"	William S. Cohen	ME	1997

Secretaries of War

The War Department (which included jurisdiction over the navy until 1798) was created by act of Congress on Aug. 7, 1789, and Gen. Henry Knox was commissioned secretary of war under that act on Sept. 12, 1789.

President	Secretary	Home	Apptd.	President	Secretary	Home	Apptd.
Washington	Henry Knox	MA	1789	Grant	John A. Rawlins	IL	1869
"	Timothy Pickering	PA	1795	"	William T. Sherman	OH	1869
"	James McHenry	MD	1796	"	William W. Belknap	IA	1869
Adams, J.	James McHenry	MD	1797	"	Alphonso Taft	OH	1876
"	Samuel Dexter	MA	1800	"	James D. Cameron	PA	1876
Jefferson	Henry Dearborn	MA	1801	Hayes	George W. McCrary	IA	1877
Madison	William Eustis	MA	1809	"	Alexander Ramsey	MN	1879
"	John Armstrong	NY	1813	Garfield	Robert T. Lincoln	IL	1881
"	James Monroe	VA	1814	Arthur	Robert T. Lincoln	IL	1881
"	William H. Crawford	GA	1815	Cleveland	William C. Endicott	MA	1885
Monroe	John C. Calhoun	SC	1817	Harrison, B.	Redfield Proctor	VT	1889
Adams, J.Q.	James Barbour	VA	1825	"	Stephen B. Elkins	WV	1891
"	Peter B. Porter	NY	1828	Cleveland	Daniel S. Lamont	NY	1893
Jackson	John H. Eaton	TN	1829	McKinley	Russel A. Alger	MI	1897
"	Lewis Cass	MI	1831	"	Elihu Root	NY	1899
"	Benjamin F. Butler	NY	1837	Roosevelt, T.	Elihu Root	NY	1901
Van Buren	Joel R. Poinsett	SC	1837	"	William H. Taft	OH	1904
Harrison, W.H.	John Bell	TN	1841	"	Luke E. Wright	TN	1908
Tyler	John Bell	TN	1841	Taft	Jacob M. Dickinson	TN	1909
"	John C. Spencer	NY	1841	"	Henry L. Stimson	NY	1911
"	James M. Porter	PA	1843	Wilson	Lindley M. Garrison	NJ	1913
"	William Wilkins	PA	1844	"	Newton D. Baker	OH	1916
Polk	William L. Marcy	NY	1845	Harding	John W. Weeks	MA	1921
Taylor	George W. Crawford	GA	1849	Coolidge	John W. Weeks	MA	1923
Fillmore	Charles M. Conrad	LA	1850	"	Dwight F. Davis	MO	1925
Pierce	Jefferson Davis	MS	1853	Hoover	James W. Good	IL	1929
Buchanan	John B. Floyd	VA	1857	"	Patrick J. Hurley	OK	1929
"	Joseph Holt	KY	1861	Roosevelt, F.D.	George H. Dern	UT	1933
Lincoln	Simon Cameron	PA	1861	"	Harry H. Woodring	KS	1937
"	Edwin M. Stanton	PA	1862	"	Henry L. Stimson	NY	1940
Johnson, A.	Edwin M. Stanton	PA	1865	"	Robert P. Patterson	NY	1945
"	John M. Schofield	IL	1868	Truman	Kenneth C. Royall[1]	NC	1947

(1) Last member of the Cabinet with this title. The War Department became the Department of the Army and became a branch of the Department of Defense in 1947.

Secretaries of the Navy

The Navy Department was created by act of Congress on Apr. 30, 1798.

President	Secretary	Home	Apptd.
Adams, J.	Benjamin Stoddert	MD	1798
Jefferson	Benjamin Stoddert	MD	1801
"	Robert Smith	MD	1801
Madison	Paul Hamilton	SC	1809
"	William Jones	PA	1813
"	Benjamin W. Crowninshield	MA	1814
Monroe	Benjamin W. Crowninshield	MA	1817
"	Smith Thompson	NY	1818
"	Samuel L. Southard	NJ	1823
Adams, J.Q.	Samuel L. Southard	NJ	1825
Jackson	John Branch	NC	1829
"	Levi Woodbury	NH	1831
"	Mahlon Dickerson	NJ	1834
Van Buren	Mahlon Dickerson	NJ	1837
"	James K. Paulding	NY	1838
Harrison, W.H.	George E. Badger	NC	1841
Tyler	George E. Badger	NC	1841
"	Abel P. Upshur	VA	1841
"	David Henshaw	MA	1843
"	Thomas W. Gilmer	VA	1844
"	John Y. Mason	VA	1844
Polk	George Bancroft	MA	1845
"	John Y. Mason	VA	1846
Taylor	William B. Preston	VA	1849
Fillmore	William A. Graham	NC	1850
"	John P. Kennedy	MD	1852
Pierce	James C. Dobbin	NC	1853
Buchanan	Isaac Toucey	CT	1857
Lincoln	Gideon Welles	CT	1861
Johnson, A.	Gideon Welles	CT	1865
Grant	Adolph E. Borie	PA	1869
"	George M. Robeson	NJ	1869
Hayes	Richard W. Thompson	IN	1877
"	Nathan Goff Jr.	WV	1881
Garfield	William H. Hunt	LA	1881
Arthur	William E. Chandler	NH	1882
Cleveland	William C. Whitney	NY	1885
Harrison, B.	Benjamin F. Tracy	NY	1889
Cleveland	Hilary A. Herbert	AL	1893
McKinley	John D. Long	MA	1897
Roosevelt, T.	John D. Long	MA	1901
"	William H. Moody	MA	1902
"	Paul Morton	IL	1904
"	Charles J. Bonaparte	MD	1905
"	Victor H. Metcalf	CA	1906
"	Truman H. Newberry	MI	1908
Taft	George von L. Meyer	MA	1909
Wilson	Josephus Daniels	NC	1913
Harding	Edwin Denby	MI	1921
Coolidge	Edwin Denby	MI	1923
"	Curtis D. Wilbur	CA	1924
Hoover	Charles Francis Adams	MA	1929
Roosevelt, F.D.	Claude A. Swanson	VA	1933
"	Charles Edison	NJ	1940
"	Frank Knox	IL	1940
"	James V. Forrestal	NY	1944
Truman	James V. Forrestal[1]	NY	1945

(1) Last member of Cabinet with this title. The Navy Department became a branch of the Department of Defense when the latter was created on Sept. 18, 1947.

Attorneys General

The office of attorney general was established by act of Congress on Sept. 24, 1789. It officially reached Cabinet rank in Mar. 1792, when the first attorney general, Edmund Randolph, attended his initial Cabinet meeting. The Department of Justice, headed by the attorney general, was created June 22, 1870.

President	Attorney General	Home	Apptd.
Washington	Edmund Randolph	VA	1789
"	William Bradford	PA	1794
"	Charles Lee	VA	1795
Adams, J.	Charles Lee	VA	1797
Jefferson	Levi Lincoln	MA	1801
"	John Breckenridge	KY	1805
"	Caesar A. Rodney	DE	1807
Madison	Caesar A. Rodney	DE	1807
"	William Pinkney	MD	1811
"	Richard Rush	PA	1814
Monroe	Richard Rush	PA	1817
"	William Wirt	VA	1817
Adams, J.Q.	William Wirt	VA	1825
Jackson	John M. Berrien	GA	1829
"	Roger B. Taney	MD	1831
"	Benjamin F. Butler	NY	1833
Van Buren	Benjamin F. Butler	NY	1837
"	Felix Grundy	TN	1838
"	Henry D. Gilpin	PA	1840
Harrison, W.H.	John J. Crittenden	KY	1841
Tyler	John J. Crittenden	KY	1841
"	Hugh S. Legare	SC	1841
"	John Nelson	MD	1843
Polk	John Y. Mason	VA	1845
"	Nathan Clifford	ME	1846
"	Isaac Toucey	CT	1848
Taylor	Reverdy Johnson	MD	1849
Fillmore	John J. Crittenden	KY	1850
Pierce	Caleb Cushing	MA	1853
Buchanan	Jeremiah S. Black	PA	1857
"	Edwin M. Stanton	PA	1860
Lincoln	Edward Bates	MO	1861
"	James Speed	KY	1864
Johnson, A.	James Speed	KY	1865
"	Henry Stanbery	OH	1866
"	William M. Evarts	NY	1868
Grant	Ebenezer R. Hoar	MA	1869
"	Amos T. Akerman	GA	1870
"	George H. Williams	OR	1871
"	Edwards Pierrepont	NY	1875
"	Alphonso Taft	OH	1876
Hayes	Charles Devens	MA	1877
Garfield	Wayne MacVeagh	PA	1881
Arthur	Benjamin H. Brewster	PA	1882
Cleveland	Augustus Garland	AR	1885
Harrison, B.	William H. H. Miller	IN	1889
Cleveland	Richard Olney	MA	1893
"	Judson Harmon	OH	1895
McKinley	Joseph McKenna	CA	1897
"	John W. Griggs	NJ	1898
"	Philander C. Knox	PA	1901
Roosevelt, T.	Philander C. Knox	PA	1901
"	William H. Moody	MA	1904
"	Charles J. Bonaparte	MD	1906
Taft	George W. Wickersham	NY	1909
Wilson	J.C. McReynolds	TN	1913
"	Thomas W. Gregory	TX	1914
"	A. Mitchell Palmer	PA	1919
Harding	Harry M. Daugherty	OH	1921
Coolidge	Harry M. Daugherty	OH	1923
"	Harlan F. Stone	NY	1924
"	John G. Sargent	VT	1925
Hoover	William D. Mitchell	MN	1929
Roosevelt, F.D.	Homer S. Cummings	CT	1933
"	Frank Murphy	MI	1939
"	Robert H. Jackson	NY	1940
"	Francis Biddle	PA	1941
Truman	Thomas C. Clark	TX	1945
"	J. Howard McGrath	RI	1949
"	J.P. McGranery	PA	1952
Eisenhower	Herbert Brownell Jr.	NY	1953
"	William P. Rogers	MD	1957
Kennedy	Robert F. Kennedy	MA	1961
Johnson, L.B.	Robert F. Kennedy	MA	1963
"	N. de B. Katzenbach	IL	1964
"	Ramsey Clark	TX	1967
Nixon	John N. Mitchell	NY	1969
"	Richard G. Kleindienst	AZ	1972
"	Elliot L. Richardson	MA	1973
"	William B. Saxbe	OH	1974
Ford	William B. Saxbe	OH	1974
"	Edward H. Levi	IL	1975
Carter	Griffin B. Bell	GA	1977
"	Benjamin R. Civiletti	MD	1979
Reagan	William French Smith	CA	1981
"	Edwin Meese 3d	CA	1985
"	Richard Thornburgh	PA	1988
Bush	Richard Thornburgh	PA	1989
"	William P. Barr	NY	1991
Clinton	Janet Reno	FL	1993

Secretaries of the Interior

The Department of the Interior was created by act of Congress on Mar. 3, 1849.

President	Secretary	Home	Apptd.
Taylor	Thomas Ewing	OH	1849
Fillmore	Thomas M. T. McKennan	PA	1850
"	Alex H. H. Stuart	VA	1850
Pierce	Robert McClelland	MI	1853
Buchanan	Jacob Thompson	MS	1857
Lincoln	Caleb B. Smith	IN	1861
"	John P. Usher	IN	1863
Johnson, A.	John P. Usher	IN	1865
"	James Harlan	IA	1865
"	Orville H. Browning	IL	1866
Grant	Jacob D. Cox	OH	1869
"	Columbus Delano	OH	1870
"	Zachariah Chandler	MI	1875
Hayes	Carl Schurz	MO	1877
Garfield	Samuel J. Kirkwood	IA	1881
Arthur	Henry M. Teller	CO	1882
Cleveland	Lucius Q.C. Lamar	MS	1885
"	William F. Vilas	WI	1888
Harrison, B.	John W. Noble	MO	1889
Cleveland	Hoke Smith	GA	1893
"	David R. Francis	MO	1896
McKinley	Cornelius N. Bliss	NY	1897
"	Ethan A. Hitchcock	MO	1898
Roosevelt, T.	Ethan A. Hitchcock	MO	1901
"	James R. Garfield	OH	1907
Taft	Richard A. Ballinger	WA	1909
"	Walter L. Fisher	IL	1911
Wilson	Franklin K. Lane	CA	1913
"	John B. Payne	IL	1920
Harding	Albert B. Fall	NM	1921
"	Hubert Work	CO	1923
Coolidge	Hubert Work	CO	1923
"	Roy O. West	IL	1929
Hoover	Ray Lyman Wilbur	CA	1929
Roosevelt, F.D.	Harold L. Ickes	IL	1933
Truman	Harold L. Ickes	IL	1945
"	Julius A. Krug	WI	1946
"	Oscar L. Chapman	CO	1949
Eisenhower	Douglas McKay	OR	1953
"	Fred A. Seaton	NE	1956
Kennedy	Stewart L. Udall	AZ	1961
Johnson, L.B.	Stewart L. Udall	AZ	1963
Nixon	Walter J. Hickel	AK	1969
"	Rogers C.B. Morton	MD	1971
Ford	Rogers C.B. Morton	MD	1971
"	Stanley K. Hathaway	WY	1975
"	Thomas S. Kleppe	ND	1975
Carter	Cecil D. Andrus	ID	1977
Reagan	James G. Watt	CO	1981
"	William P. Clark	CA	1983
"	Donald P. Hodel	OR	1985
Bush	Manuel Lujan	NM	1989
Clinton	Bruce Babbitt	AZ	1993

Secretaries of Agriculture

The Department of Agriculture was created by act of Congress on May 15, 1862. On Feb. 8, 1889, its commissioner was renamed secretary of agriculture and became a member of the Cabinet.

President	Secretary	Home	Apptd.
Cleveland	Norman J. Colman	MO	1889
Harrison, B.	Jeremiah M. Rusk	WI	1889
Cleveland	J. Sterling Morton	NE	1893
McKinley	James Wilson	IA	1897
Roosevelt, T.	James Wilson	IA	1901
Taft	James Wilson	IA	1909
Wilson	David F. Houston	MO	1913
"	Edwin T. Meredith	IA	1920
Harding	Henry C. Wallace	IA	1921
Coolidge	Henry C. Wallace	IA	1923
"	Howard M. Gore	WV	1924
"	William M. Jardine	KS	1925
Hoover	Arthur M. Hyde	MO	1929
Roosevelt, F.D.	Henry A. Wallace	IA	1933
"	Claude R. Wickard	IN	1940
Truman	Clinton P. Anderson	NM	1945
Truman	Charles F. Brannan	CO	1948
Eisenhower	Ezra Taft Benson	UT	1953
Kennedy	Orville L. Freeman	MN	1961
Johnson, L.B.	Orville L. Freeman	MN	1963
Nixon	Clifford M. Hardin	IN	1969
"	Earl L. Butz	IN	1971
Ford	Earl L. Butz	IN	1974
"	John A. Knebel	VA	1976
Carter	Bob Bergland	MN	1977
Reagan	John R. Block	IL	1981
"	Richard E. Lyng	CA	1986
Bush	Clayton K. Yeutter	NE	1989
"	Edward Madigan	IL	1991
Clinton	Mike Espy	MS	1993
"	Dan Glickman	KS	1995

Secretaries of Commerce and Labor

The Department of Commerce and Labor, created by Congress on Feb. 14, 1903, was divided by Congress Mar. 4, 1913, into separate departments of Commerce and Labor. The secretary of each was made a Cabinet member.

Secretaries of Commerce and Labor

President	Secretary	Home	Apptd.
Roosevelt, T.	George B. Cortelyou	NY	1903
"	Victor H. Metcalf	CA	1904
"	Oscar S. Straus	NY	1906
Taft	Charles Nagel	MO	1909

Secretaries of Labor

President	Secretary	Home	Apptd.
Wilson	William B. Wilson	PA	1913
Harding	James J. Davis	PA	1921
Coolidge	James J. Davis	PA	1923
Hoover	James J. Davis	PA	1929
"	William N. Doak	VA	1930
Roosevelt, F.D.	Frances Perkins	NY	1933
Truman	L.B. Schwellenbach	WA	1945
"	Maurice J. Tobin	MA	1949
Eisenhower	Martin P. Durkin	IL	1953
"	James P. Mitchell	NJ	1953
Kennedy	Arthur J. Goldberg	IL	1961
"	W. Willard Wirtz	IL	1962
Johnson, L.B.	W. Willard Wirtz	IL	1963
Nixon	George P. Shultz	IL	1969
"	James D. Hodgson	CA	1970
"	Peter J. Brennan	NY	1973
Ford	Peter J. Brennan	NY	1974
"	John T. Dunlop	CA	1975
"	W.J. Usery Jr.	GA	1976
Carter	F. Ray Marshall	TX	1977
Reagan	Raymond J. Donovan	NJ	1981
"	William E. Brock	TN	1985
"	Ann D. McLaughlin	DC	1987
Bush	Elizabeth Hanford Dole	NC	1989
"	Lynn Martin	IL	1991
Clinton	Robert B. Reich	MA	1993
"	Alexis M. Herman	AL	1997

Secretaries of Commerce

President	Secretary	Home	Apptd.
Wilson	William C. Redfield	NY	1913
"	Joshua W. Alexander	MO	1919
Harding	Herbert C. Hoover	CA	1921
Coolidge	Herbert C. Hoover	CA	1923
"	William F. Whiting	MA	1928
Hoover	Robert P. Lamont	IL	1929
"	Roy D. Chapin	MI	1932
Roosevelt, F.D.	Daniel C. Roper	SC	1933
"	Harry L. Hopkins	NY	1939
"	Jesse Jones	TX	1940
"	Henry A. Wallace	IA	1945
Truman	Henry A. Wallace	IA	1945
"	W. Averell Harriman	NY	1947
"	Charles Sawyer	OH	1948
Eisenhower	Sinclair Weeks	MA	1953
"	Lewis L. Strauss	NY	1958
"	Frederick H. Mueller	MI	1959
Kennedy	Luther H. Hodges	NC	1961

President	Secretary	Home	Apptd.	President	Secretary	Home	Apptd.
Johnson, L.B.	Luther H. Hodges	NC	1963	Carter	Juanita M. Kreps	NC	1977
"	John T. Connor	NJ	1965	"	Philip M. Klutznick	IL	1979
"	Alex B. Trowbridge	NJ	1967	Reagan	Malcolm Baldrige	CT	1981
"	Cyrus R. Smith	NY	1968	"	C. William Verity Jr.	OH	1987
Nixon	Maurice H. Stans	MN	1969	Bush	Robert A. Mosbacher	TX	1989
"	Peter G. Peterson	IL	1972	"	Barbara H. Franklin	PA	1992
"	Frederick B. Dent	SC	1973	Clinton	Ronald H. Brown	DC	1993
Ford	Frederick B. Dent	SC	1974	"	Mickey Kantor	CA	1996
"	Rogers C.B. Morton	MD	1975	"	William M. Daley	IL	1997
"	Elliot L. Richardson	MA	1975				

Secretaries of Housing and Urban Development

The Department of Housing and Urban Development was created by act of Congress on Sept. 9, 1965.

President	Secretary	Home	Apptd.	President	Secretary	Home	Apptd.
Johnson, L.B.	Robert C. Weaver	WA	1966	Carter	Patricia Roberts Harris	DC	1977
"	Robert C. Wood	MA	1969	"	Moon Landrieu	LA	1979
Nixon	George W. Romney	MI	1969	Reagan	Samuel R. Pierce Jr.	NY	1981
"	James T. Lynn	OH	1973	Bush	Jack F. Kemp	NY	1989
Ford	James T. Lynn	OH	1974	Clinton	Henry G. Cisneros	TX	1993
"	Carla Anderson Hills	CA	1975	"	Andrew M. Cuomo	NY	1997

Secretaries of Transportation

The Department of Transportation was created by act of Congress on Oct. 15, 1966.

President	Secretary	Home	Apptd.	President	Secretary	Home	Apptd.
Johnson, L.B.	Alan S. Boyd	FL	1966	Reagan	Andrew L. Lewis Jr.	PA	1981
Nixon	John A. Volpe	MA	1969	"	Elizabeth Hanford Dole	NC	1983
"	Claude S. Brinegar	CA	1973	"	James H. Burnley	NC	1987
Ford	Claude S. Brinegar	CA	1974	Bush	Samuel K. Skinner	IL	1989
"	William T. Coleman Jr.	PA	1975	"	Andrew H. Card Jr.	MA	1992
Carter	Brock Adams	WA	1977	Clinton	Federico F. Peña	CO	1993
"	Neil E. Goldschmidt	OR	1979	"	Rodney E. Slater	AR	1997

Secretaries of Energy

The Department of Energy was created by federal law on Aug. 4, 1977.

President	Secretary	Home	Apptd.	President	Secretary	Home	Apptd.
Carter	James R. Schlesinger	VA	1977	Bush	James D. Watkins	CA	1989
"	Charles Duncan Jr.	WY	1979	Clinton	Hazel R. O'Leary	MN	1993
Reagan	James B. Edwards	SC	1981	"	Federico F. Peña	CO	1997
"	Donald P. Hodel	OR	1982	"	Bill Richardson	NM	1998
"	John S. Herrington	CA	1985				

Secretaries of Health, Education, and Welfare

The Department of Health, Education, and Welfare was created by Congress on Apr. 11, 1953. On Sept. 27, 1979, it was divided by Congress into separate departments of Education and of Health and Human Services, with the secretary of each being a Cabinet member.

President	Secretary	Home	Apptd.	President	Secretary	Home	Apptd.
Eisenhower	Oveta Culp Hobby	TX	1953	Nixon	Robert H. Finch	CA	1969
"	Marion B. Folsom	NY	1955	"	Elliot L. Richardson	MA	1970
"	Arthur S. Flemming	OH	1958	"	Caspar W. Weinberger	CA	1973
Kennedy	Abraham A. Ribicoff	CT	1961	Ford	Caspar W. Weinberger	CA	1974
"	Anthony J. Celebrezze	OH	1962	"	Forrest D. Mathews	AL	1975
Johnson, L.B.	Anthony J. Celebrezze	OH	1963	Carter	Joseph A. Califano Jr.	DC	1977
"	John W. Gardner	NY	1965	"	Patricia Roberts Harris	DC	1979
"	Wilbur J. Cohen	MI	1968				

Secretaries of Health and Human Services

President	Secretary	Home	Apptd.	President	Secretary	Home	Apptd.
Carter	Patricia Roberts Harris	DC	1979	Reagan	Otis R. Bowen	IN	1985
Reagan	Richard S. Scyhweiker	PA	1981	Bush	Louis W. Sullivan	GA	1989
"	Margaret M. Heckler	MA	1983	Clinton	Donna E. Shalala	WI	1993

Secretaries of Education

President	Secretary	Home	Apptd.	President	Secretary	Home	Apptd.
Carter	Shirley Hufstedler	CA	1979	Bush	Lauro F. Cavazos	TX	1989
Reagan	Terrel Bell	UT	1981	"	Lamar Alexander	TN	1991
"	William J. Bennett	NY	1985	Clinton	Richard W. Riley	SC	1993
"	Lauro F. Cavazos	TX	1988				

Secretaries of Veterans Affairs

The Department of Veterans Affairs was created on Oct. 25, 1988, when Pres. Ronald Reagan signed a bill that made the Veterans Administration into a Cabinet department, effective Mar. 15, 1989.

President	Secretary	Home	Apptd.	President	Secretary	Home	Apptd.
Bush	Edward J. Derwinski	IL	1989	Clinton	Togo D. West Jr.	NC	1998
Clinton	Jesse Brown	IL	1993				

ECONOMICS

U.S. Budget Receipts and Outlays, 1996-99

Source: Financial Management Service, U.S. Dept. of the Treasury

For the fiscal year 1999 the federal budget showed a surplus of $122.7 bil, or 1.4% of GDP. The surplus, nearly twice the size of the previous year's, was the largest ever in dollar terms and the largest as a percentage of GDP since 1951. It was also the first time since 1956-1957 there had been 2 consecutive years of surpluses.

(in millions of dollars; many figures do not add to totals because of independent rounding or omitted subcategories, including some subcategories with negative values.)

	Fiscal 1996[1]	Fiscal 1997[1]	Fiscal 1998[1]	Fiscal 1999[1]
NET RECEIPTS				
Individual income taxes	$656,417	$737,466	$828,597	$879,480
Corporation income taxes	171,824	182,294	188,677	184,680
Social insurance taxes and contributions:				
Federal old-age and survivors insurance	311,869	336,728	358,784	383,559
Federal disability insurance	55,623	55,261	57,016	60,910
Federal hospital insurance	104,998	110,710	119,863	132,268
Railroad retirement fund	3,872	4,051	4,353	4,143
Total employment taxes and contributions	476,362	506,750	540,015R	580,880
Other insurance and retirement:				
Unemployment	28,584	28,202	27,484	26,480
Federal employees retirement	4,389	4,344	4,261	4,399
Non-federal employees	80	74	74	73
Total social insurance taxes and contributions	**509,415**	**539,371**	**571,835**	**611,832**
Excise taxes	54,015	56,926	57,669	70,399
Estate and gift taxes	17,189	19,845	24,076	27,782
Customs duties	18,671	17,927	18,297	18,336
Deposits of earnings by Federal Reserve Banks	20,477	19,636	24,540	25,917
All other miscellaneous receipts	4,755	5,491	5,027	5,108
Net Budget Receipts	**1,452,763**	**1,578,955**	**1,721,421**	**1,827,285**
NET OUTLAYS				
Legislative Branch	2,272	2,362	2,600R	2,621
The Judiciary	3,061	3,259	3,463	3,793
Executive Office of the President:				
The White House Office	39	39	46	51
Office of Management and Budget	55	56	56R	59
Total Executive Office	**202**	**219**	**236R**	**416**
International Assistance Program:				
International security assistance	4,254	4,403	4,950R	5,395
Multilateral assistance	2,077	2,141	1,850	1,857
Agency for International Development	3,059	2,814	2,435R	2,346
International Development Assistance	5,229	2,902	2,494R	2,418
Total International Assistance Program	**9,716**	**10,128**	**8,980**	**10,059**
Agriculture Department:				
Food stamp program	25,359	22,857	20,141	19,051
Farm Service Agency	8,350	7,417	10,421	19,508
Forest Service	3,411	3,209	3,399	3,423
Total Agriculture Department	**54,338**	**52,549**	**53,950**	**62,885**
Commerce Department:				
Bureau of the Census	260	282	542	1,131
Total Commerce Department	**3,703**	**3,780**	**4,047**	**5,036**
Defense Department—Military:				
Military personnel	66,669	69,722	68,976	69,503
Operation and maintenance	88,629	92,465	93,473R	96,420
Procurement	48,912	47,691	48,207R	48,824
Research, development, test, evaluation	36,561	37,026	37,421R	37,362
Military construction	6,684	6,188	6,046	5,519
Total Defense Department—Military	**253,258**	**258,330**	**256,124R**	**261,379**
Defense Department—Civil	32,535	30,282	31,216R	32,008
Education Department	29,900	30,014	31,498R	33,521
Energy Department	16,199	14,470	14,444	16,079
Health and Human Services Department:				
Public Health Service	21,405	21,755	23,680R	25,547
Health Care Financing Adm	354,898	369,714	379,950	390,181
Food and Drug Administration	865	873	838	951
National Institutes of Health	10,217	11,199	12,501	13,815
Total Health and Human Services Dept.	**319,802**	**339,541**	**350,571R**	**359,700**
Housing and Urban Development Department	25,512	27,525	30,224	32,736
Interior Department	6,720	6,722	7,232R	7,773
Justice Department:				
Federal Bureau of Investigation	2,305	2,700	2,949	3,040
Drug Enforcement Agency	746	969	1,099	1,203
Immigration and Naturalization Service	2,246	2,770	3,593	3,775
Federal Prison System	3,013	2,939	2,682	3,204
Total Justice Department	**11,951**	**14,315**	**16,169R**	**18,318**
Labor Department:				
Unemployment Trust Fund	26,146	24,299	23,408	24,870
Total Labor Department	**32,496**	**30,461**	**30,002**	**32,459**
State Department	4,955	5,245	5,373R	6,464
Transportation Department:				
Federal Aviation Administration	8,925	8,815	9,242	9,507
Total Transportation Department	**38,776**	**39,835**	**39,467R**	**41,819**
Treasury Department:				
Internal Revenue Service	28,595	31,386	33,153	31,386
Interest on the public debt	343,955	355,796	363,824	353,511
Total Treasury Department	**365,336**	**379,345**	**390,094R**	**387,280**
Veterans Affairs Department	36,915	39,277	41,776	43,169
Environmental Protection Agency	6,046	6,167	6,288R	6,752

	Fiscal 1996[1]	Fiscal 1997[1]	Fiscal 1998[1]	Fiscal 1999[1]
General Services Administration	$625	$1,083	$1,095R	$–46
National Aeronautics and Space Administration	13,882	14,358	14,206	13,664
Office of Personnel Management	42,872	45,404	46,307	47,515
Small Business Administration	872	334	–78	58
Social Security Administration	375,232	393,309	408,202	419,790
Other independent agencies:				
Corporation for Natl. and Community Service	477	564	591R	609
Corporation for Public Broadcasting	275	260	250	281
District of Columbia	712	717	818	–2,691
Equal Employment Opportunity Commission	224	231	244	255
Export-Import Bank of the U.S.	–560	–114	–208	–168
Federal Communications Commission	978	1,001	1,769	3,293
Federal Deposit Insurance Corporation	–8,732	–14,181	–4,122	–4,702
Legal Services Corporation	282	282	285	298
National Archives & Records Adm.	199	198	210	225
National Foundation on the Arts and Humanities	285	230	207	217
National Labor Relations Board	166	175	177	182
National Science Foundation	3,012	3,131	3,188	3,285
Nuclear Regulatory Commission	57	51	38	37
Railroad Retirement Board	5,007	4,870	4,837	4,830
Securities and Exchange Commission	42	–20	–231	–255
Smithsonian Institution	432	491	488	486
Tennessee Valley Authority	757	–337	–784	2
Total other independent agencies	8,577	–2,489	10,653R	6,865
Undistributed offsetting receipts	–135,649	–154,970	–161,036R	–159,080
NET BUDGET OUTLAYS	**$1,560,094**	**$1,600,911**	**$1,652,224R**	**$1,704,545**
Less net receipts	1,452,763	1,578,955	1,721,421	1,827,285
DEFICIT (-) OR SURPLUS (+)	**$–107,331**	**$–21,957**	**$+70,039**	**$+122,740**

(1) Fiscal year ends Sept. 30. R=revised.

Summary of Receipts, Outlays, and Surpluses or Deficits, 1936-95

Source: Financial Management Service, U.S. Dept. of the Treasury
(millions of dollars)

Fiscal Year[1]	Receipts	Outlays	Surplus or Deficit (–)[2]	Fiscal Year[1]	Receipts	Outlays	Surplus or Deficit (–)[2]
1936	$3,923	$8,228	$–4,304	1967	$148,822	$157,464	$–8,643
1937	5,387	7,580	–2,193	1968	152,973	178,134	–25,161
1938	6,751	6,840	–89	1969	186,882	183,640	3,242
1939	6,295	9,141	–2,846	1970	192,807	195,649	–2,842
1940	6,548	9,468	–2,920	1971	187,139	210,172	–23,033
1941	8,712	13,653	–4,941	1972	207,309	230,681	–23,373
1942	14,634	35,137	–20,503	1973	230,799	245,707	–14,908
1943	24,001	78,555	–54,554	1974	263,224	269,359	–6,135
1944	43,747	91,304	–47,557	1975	279,090	332,332	–53,242
1945	45,159	92,712	–47,553	1976	298,060	371,779	–73,719
1946	39,296	55,232	–15,936	Transition quarter[3]	81,232	95,973	–14,741
1947	38,514	34,496	4,018	1977	355,559	409,203	–53,644
1948	41,560	29,764	11,796	1978	399,561	458,729	–59,168
1949	39,415	38,835	580	1979	463,302	503,464	–40,162
1950	39,443	42,562	–3,119	1980	517,112	590,920	–73,808
1951	51,616	45,514	6,102	1981	599,272	678,209	–78,936
1952	66,167	67,686	–1,519	1982	617,766	745,706	–127,940
1953	69,608	76,101	–6,493	1983	600,562	808,327	–207,764
1954	69,701	70,855	–1,154	1984	666,457	851,781	–185,324
1955	65,451	68,444	–2,993	1985	734,057	946,316	–212,260
1956	74,587	70,640	3,947	1986	769,091	990,231	–221,140
1957	79,990	76,578	3,412	1987	854,143	1,003,804	–149,661
1958	79,636	82,405	–2,769	1988	908,166	1,063,318	–155,151
1959	79,249	92,098	–12,849	1989	990,701	1,144,020	–153,319
1960	92,492	92,191	301	1990	1,031,308	1,251,776	–220,469
1961	94,388	97,723	–3,335	1991	1,054,265	1,323,757	–269,492
1962	99,676	106,821	–7,146	1992	1,090,453	1,380,794	–290,340
1963	106,560	111,316	–4,756	1993	1,153,226	1,408,532	–255,306
1964	112,613	118,528	–5,915	1994	1,257,451	1,460,553	–203,102
1965	116,817	118,228	–1,411	1995	1,351,495	1,515,412	–163,917
1966	130,835	134,532	–3,698				

(1) Fiscal years 1936 to 1976 end June 30; after 1976, fiscal years end Sept. 30. (2) May not equal difference between figures shown, because of rounding. (3) Transition quarter covers July 1, 1976-Sept. 30, 1976.

Budget Receipts and Outlays, 1789-1935

Source: U.S. Dept. of the Treasury; annual statements for years ending June 30 unless otherwise noted
(thousands of dollars)

Yearly Average	Receipts	Outlays	Yearly Average	Receipts	Outlays	Yearly Average	Receipts	Outlays
1789-1800[1]	$5,717	$5,776	1866-1870	$447,301	$377,642	1901-1905	$559,481	$535,559
1801-1810[2]	13,056	9,086	1871-1875	336,830	287,460	1906-1910	628,507	639,178
1811-1820[2]	21,032	23,943	1876-1880	288,124	255,598	1911-1915	710,227	720,252
1821-1830[2]	21,928	16,162	1881-1885	366,961	257,691	1916-1920	3,483,652	8,065,333
1831-1840[2]	30,461	24,495	1886-1890	375,448	279,134	1921-1925	4,306,673	3,578,989
1841-1850[2]	28,545	34,097	1891-1895	352,891	363,599	1926-1930	4,069,138	3,182,807
1851-1860	60,237	60,163	1896-1900	434,877	457,451	1931-1935	2,770,973	5,214,874
1861-1865	160,907	683,785						

(1) Average for period March 4, 1789, to Dec. 31, 1800. (2) Years from 1801 to 1842 end Dec. 31; average for 1841-1850 is for the period Jan. 1, 1841, to June 30, 1850.

Public Debt of the U.S.

Source: Bureau of Public Debt, U.S. Dept. of the Treasury

Fiscal year	Debt (billions)	Debt per cap. (dollars)	Interest paid (billions)	% of federal outlays	Fiscal year	Debt (billions)	Debt per cap. (dollars)	Interest paid (billions)	% of federal outlays
1870	$2.4	$61.06	—	—	1981	$997.9	$4,338	$95.6	14.1
1880	2.0	41.60	—	—	1982	1,142.0	4,913	117.4	15.7
1890	1.1	17.80	—	—	1983	1,377.2	5,870	128.8	15.9
1900	1.2	16.60	—	—	1984	1,572.3	6,640	153.8	18.1
1910	1.1	12.41	—	—	1985	1,823.1	7,598	178.9	18.9
1920	24.2	228	—	—	1986	2,125.3	8,774	190.2	19.2
1930	16.1	131	—	—	1987	2,350.3	9,615	195.4	19.5
1940	43.0	325	$1.0	10.5	1988	2,602.3	10,534	214.1	20.1
1950	256.1	1,688	5.7	13.4	1989	2,857.4	11,545	240.9	21.0
1955	272.8	1,651	6.4	9.4	1990	3,233.3	13,000	264.8	21.1
1960	284.1	1,572	9.2	10.0	1991	3,665.3	14,436	285.5	21.6
1965	313.8	1,613	11.3	9.6	1992	4,064.6	15,846	292.3	21.2
1970	370.1	1,814	19.3	9.9	1993	4,411.5	17,105	292.5	20.8
1975	533.2	2,475	32.7	9.8	1994	4,692.8	18,025	296.3	20.3
1976	620.4	2,852	37.1	10.0	1995	4,974.0	18,930	332.4	22.0
1977	698.8	3,170	41.9	10.2	1996	5,224.8	19,805	344.0	22.0
1978	771.5	3,463	48.7	10.6	1997	5,413.1	20,026	355.8	22.2
1979	826.5	3,669	59.8	11.9	1998	5,526.2	20,443	363.8	22.0
1980	907.7	3,985	74.9	12.7					

Note: Through 1976 the fiscal year ended June 30. From 1977 on, the fiscal year ends Sept. 30.

Consumer Price Index

The Consumer Price Index (CPI) is a measure of the average change in prices over time of one or more kinds of basic consumer goods and services.

From Jan. 1978, the Bureau of Labor Statistics began publishing CPIs for 2 population groups: (1) a CPI for all urban consumers (CPI-U), which covers about 87% of the total population; and (2) a CPI for urban wage earners and clerical workers (CPI-W), which covers about 32% of the total population. The CPI-U includes, in addition to wage earners and clerical workers, groups such as professional, managerial, and technical workers, the self-employed, short-term workers, the unemployed, retirees, and others not in the labor force.

The CPI is based on prices of food, clothing, shelter, and fuels; transportation fares; charges for doctors' and dentists' services; drug prices; and prices of other goods and services bought for day-to-day living. The index currently measures price changes from a designated reference period, 1982-84, which equals 100.0. Use of this reference period began in Jan. 1988.

U.S. Consumer Price Indexes, First Half 1999

Source: Bureau of Labor Statistics, U.S. Dept. of Labor

(Data are semiannual averages of monthly figures)

(1982–84=100)	CPI-U (all urban consumers)		CPI-W (urban wage-earners/clerical)	
	1st half 1999	% change 2d half 1998 to 1st half 1999	1st half 1999	% change 2d half 1998 to 1st half 1999
ALL ITEMS	165.4	1.0	162.0	1.1
Food, beverages	163.9	1.2	163.1	1.2
Housing	162.8	0.9	159.0	0.8
Apparel	131.8	−0.7	130.5	−0.7
Transportation	142.1	0.6	140.8	0.4
Medical care	248.6	1.8	247.8	1.8
Recreation	102.0	0.8	101.4	0.5
Other goods, services	255.3	5.8	258.4	7.4
Services	187.5	1.1	184.1	1.0
SPECIAL INDEXES				
All items less food	165.8	1.0	161.7	1.0
Commodities less food	132.8	0.8	133.1	0.9
Nondurables	149.6	1.6	149.5	1.8
Energy	101.9	−0.2	101.1	−0.1
All items less energy	173.7	1.2	170.3	1.1

Consumer Price Indexes (CPI-U),[1] Annual Percent Change, 1987-98

Source: Bureau of Labor Statistics, U.S. Dept. of Labor

	1987	1988	1989	1990	1991	1992	1993	1994	1995	1996	1997	1998
ALL ITEMS	3.6	4.1	4.8	5.4	4.2	3.0	3.0	2.6	2.8	3.0	2.3	1.6
Food	4.1	4.1	5.8	5.8	2.9	1.2	2.2	2.4	2.8	3.3	2.6	2.2
Shelter	4.7	4.8	4.5	5.4	4.5	3.3	3.0	3.1	3.2	3.2	3.1	3.3
Rent, residential	4.1	3.8	3.9	5.6	6.1	2.5	2.3	2.5	2.5	2.7	2.9	3.2
Fuel and other utilities	−1.1	−1.4	3.3	3.5	3.3	2.2	3.0	1.0	0.7	3.1	2.6	−1.8
Apparel and upkeep	4.4	4.3	2.8	4.6	3.7	2.5	1.4	−0.2	−1.0	−0.2	0.9	0.1
Private transportation	−3.0	3.3	4.9	5.2	2.6	2.2	2.3	3.1	3.7	2.7	0.7	−2.2
New cars	3.6	2.0	2.0	1.8	3.8	2.5	2.4	3.4	2.2	1.7	0.2	−0.6
Gasoline	−4.0	0.9	9.5	14.1	−1.8	−0.2	−1.3	0.5	1.6	6.1	−0.1	−13.4
Public transportation	3.5	1.8	5.0	10.1	4.4	1.7	10.3	3.0	2.3	3.4	2.6	1.9
Medical care	6.6	6.5	7.7	9.0	8.7	7.4	5.9	4.8	4.5	3.5	2.8	3.2
Entertainment	3.3	4.3	5.2	4.7	4.5	2.8	2.5	2.9	2.5	3.4	2.1	1.5
Commodities	3.2	3.5	4.7	5.2	4.2	2.0	1.9	1.7	1.9	2.6	1.4	0.1

(1) The Consumer Price Index CPI-U measures the average change in prices of goods and services purchased by all urban consumers.

U.S. Consumer Price Indexes for Selected Items and Groups, 1970-98

Source: Bureau of Labor Statistics, U.S. Dept. of Labor

(1982-84 = 100, unless otherwise noted. Annual averages of monthly figures. For all urban consumers.)

	1970	1975	1980	1985	1990	1995	1997	1998
ALL ITEMS	38.8	53.8	82.4	107.6	130.7	152.4	160.5	163.0
Food and beverages	40.1	60.2	86.7	105.6	132.1	148.9	157.7	161.1
Food	39.2	59.8	86.8	105.6	132.4	148.4	157.3	160.7
Food at home	39.9	61.8	88.4	104.3	132.3	148.8	158.1	161.1
Cereals and bakery products	37.1	62.9	83.9	107.9	140.0	167.5	177.6	181.1
Meats, poultry, fish, and eggs	44.6	67.0	92.0	100.1	130.0	138.8	148.5	147.3
Dairy products	44.7	62.6	90.9	103.2	126.5	132.8	145.5	150.8
Fruits and vegetables	37.8	56.9	82.1	106.4	149.0	177.7	187.5	198.2
Sugar and sweets	30.5	65.3	90.5	105.8	124.7	137.5	147.3	150.2
Fats and oils	39.2	73.5	89.3	106.9	126.3	137.3	141.7	146.9
Nonalcoholic beverages	27.1	41.3	91.4	104.3	113.5	131.7	133.4	133.0
Other foods	39.6	58.9	83.6	106.4	131.2	151.1	161.2	165.5
Food away from home	37.5	54.5	83.4	108.3	133.4	149.0	157.0	161.1
Alcoholic beverages	52.1	65.9	86.4	106.4	129.3	153.9	162.8	165.7
Housing	36.4	50.7	81.1	107.7	128.5	148.5	156.8	160.4
Shelter	35.5	48.8	81.0	109.8	140.0	165.7	176.3	182.1
Rent of primary residence[1]	46.5	58.0	80.9	111.8	138.4	157.8	166.7	172.1
Fuel and other utilities[1]	29.1	45.4	75.4	106.5	111.6	123.7	130.8	128.5
Energy services	31.8	50.0	75.8	106.9	117.4	119.2	125.1	121.2
Household furnishings and operation	46.8	63.4	86.3	103.8	113.3	123.0	125.4	126.6
House furnishings	55.5	69.8	88.5	101.7	106.7	111.2	125.4	126.6
Apparel	59.2	72.5	90.9	105.0	124.1	132.0	132.9	133.0
Men's and boys'	62.2	75.5	89.4	105.0	120.4	126.2	130.1	131.8
Women's and girls'	71.8	85.5	96.0	104.9	122.6	126.9	126.1	126.0
Footwear	56.8	69.6	91.8	102.3	117.4	125.4	127.6	128.0
Transportation	37.5	50.1	83.1	106.4	120.5	139.1	144.3	141.6
Private	37.5	50.6	84.2	106.2	118.8	136.3	141.0	137.9
New vehicles	53.0	62.9	88.4	106.1	121.4	139.0	144.3	143.4
Used cars and trucks	31.2	43.8	62.3	113.7	117.6	156.5	151.1	150.6
Gasoline	27.9	45.1	97.5	98.6	101.0	99.8	105.8	91.6
Public	35.2	43.5	69.0	110.5	142.6	175.9	186.7	190.3
Medical care	34.0	47.5	74.9	113.5	162.8	220.5	234.6	242.1
Entertainment	47.5	62.0	83.6	107.9	132.4	153.9	162.5	—[2]
Other goods and services	40.9	53.9	75.2	114.5	159.0	206.9	224.8	237.7
Tobacco products	43.1	54.7	72.0	116.7	181.5	225.7	243.7	274.8
Personal care	43.5	57.9	81.9	106.3	130.4	147.1	152.7	156.7
Personal care products	42.7	58.0	79.6	107.6	128.2	143.1	144.2	148.3
Personal care services	44.2	57.7	83.7	108.9	132.8	151.5	162.4	166.0

(1) Dec. 1982 = 100. (2) The BLS stopped tracking this category after 1997, and began tracking a category classified as Recreation. The Recreation index for 1998 is 99.6; for 1997, 97.4; for 1995, 94.5.

Consumer Price Indexes by Region and Selected Cities, 1998-99

Source: Bureau of Labor Statistics, U.S. Dept. of Labor

(1982-84 = 100)	CPI-U Indexes[1]			% change	CPI-W Indexes[2]			% change
	Aug. 1998	July 1999	Aug. 1999	Aug. 1998-Aug. 1999	Aug. 1998	July 1999	Aug. 1999	Aug. 1998-Aug. 1999
U.S. CITY AVERAGE	163.4	166.7	167.1	2.3	160.0	163.3	163.8	2.4
Northeast urban	170.5	173.4	174.1	2.1	167.1	170.2	170.9	2.3
Size A—More than 1,500,000	171.4	174.5	175.1	2.2	167.1	170.3	171.0	2.3
Size B/C—50,000 to 1,500,000	102.2	103.9	104.3	2.1	101.7	103.4	103.8	2.1
Midwest urban	159.5	162.9	163.2	2.3	155.6	159.1	159.4	2.4
Size A—More than 1,500,000	161.0	164.6	164.8	2.4	156.4	159.9	160.2	2.4
Size B/C—50,000 to 1,500,000	102.0	103.9	104.2	2.2	101.7	103.8	104.0	2.6
Size D—Nonmetro. (less than 50,000)	153.3	157.2	157.7	2.9	151.4	155.4	156.1	3.1
South urban	159.5	162.2	162.6	1.9	157.5	160.1	160.6	2.0
Size A—More than 1,500,000	158.9	161.4	161.9	1.9	156.3	158.9	159.5	2.0
Size B/C—50,000 to 1,500,000	102.5	104.3	104.4	1.9	102.1	103.9	104.0	1.9
Size D—Nonmetro. (less than 50,000)	160.2	162.6	163.7	2.2	160.6	163.0	164.1	2.2
West urban	164.8	168.9	169.5	2.9	160.7	164.7	165.3	2.9
Size A—More than 1,500,000	165.6	169.9	170.5	3.0	159.7	164.0	164.7	3.1
Size B/C—50,000 to 1,500,000	102.5	104.9	105.2	2.6	102.3	104.7	105.1	2.7
SELECTED AREAS								
Chicago–Gary–Kenosha, IL–IN–WI	165.4	169.4	169.3	2.4	159.6	163.4	163.5	2.4
L.A.–Riverside–Orange County, CA	162.6	165.8	166.3	2.3	156.1	159.2	159.8	2.4
New York, NY–Northern NJ–Long Island, NY–NJ–CT–PA	174.2	177.2	177.6	2.0	169.7	172.5	173.2	2.1
Boston–Brockton–Nashua, MA–NH–ME–CT	—	175.3	—	—	—	173.3	—	—
Cleveland–Akron, OH	—	162.8	—	—	—	154.9	—	—
Dallas–Fort Worth, TX	—	158.3	—	—	—	158.0	—	—
Washington–Baltimore, DC–MD–VA–WV	—	104.6	—	—	—	104.3	—	—
Atlanta, GA	161.9	—	165.9	2.5	159.1	—	163.2	2.6
Detroit–Ann Arbor–Flint, MI	160.5	—	164.2	2.3	155.1	—	158.7	2.3
Houston–Galveston–Brazoria, TX	147.4	—	148.9	1.0	146.1	—	147.9	1.2
Miami–Fort Lauderdale, FL	160.8	—	162.3	0.9	158.0	—	160.0	1.3
Philadelphia–Wilmington–Atlantic City, PA–DE–NJ–MD	168.6	—	173.1	2.7	167.9	—	172.6	2.8
San Francisco–Oakland–San Jose, CA	166.6	—	173.5	4.1	162.7	—	170.0	4.5
Seattle–Tacoma–Bremerton, WA	168.5	—	173.4	2.9	163.8	—	168.8	3.1

(1) For all urban consumers. (2) For urban wage-earners and clerical workers. — = not available.

Consumer Price Index, 1915-1999

Source: Bureau of Labor Statistics, U.S. Dept. of Labor
(1967 = 100. Annual averages of monthly figures, specified for all urban consumers.)

Prices as measured by the U.S. Consumer Price Index have risen steadily since World War II. What cost $1.00 in 1967 (the reference year) cost about 30 cents in 1915, 54 cents in 1945, and $4.96 by 1999.

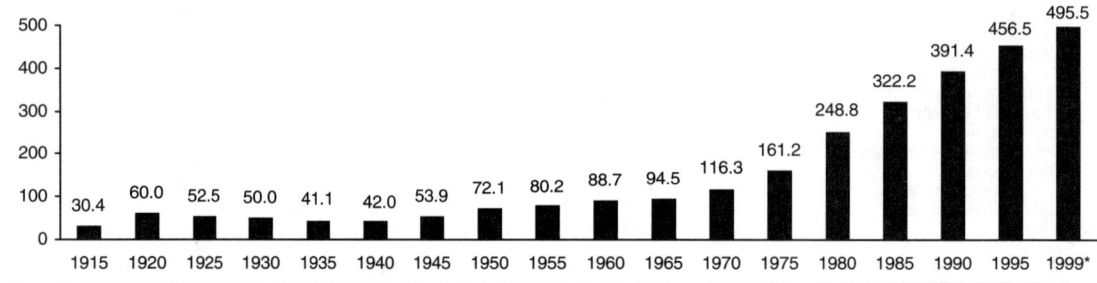

*Average for 1st half 1999.

Percentage Change in Consumer Prices in Selected Countries

Source: International Monetary Fund
(annual averages)

COUNTRY	1975-1980	1980-1985	1991-1992	1992-1993	1993-1994	1994-1995	1995-1996	1996-1997	1997-1998
Canada...................	8.7	7.4	1.5	1.8	0.2	2.2	1.6	1.6	1.0
France	10.5	9.6	2.4	2.1	1.7	1.8	2.0	1.2	0.7
Germany	4.1	3.9	4.0	4.1	3.0	1.8	1.5	1.8	1.0
Italy......................	16.3	13.7	5.1	4.5	4.0	5.2	4.0	2.0	2.0
Japan.....................	6.5	2.7	1.7	1.3	0.7	−0.1	0.1	1.7	0.6
Spain.....................	18.6	12.2	5.9	4.6	4.7	4.7	3.6	2.0	1.8
Sweden	10.5	9.0	2.3	4.6	2.2	2.5	0.5	0.5	−0.1
Switzerland...............	2.3	4.3	4.1	3.3	0.8	1.8	0.8	0.5	0.1
United Kingdom	14.4	7.2	3.7	1.6	2.5	3.4	2.4	3.1	3.4
United States	8.9	5.5	3.0	3.0	2.6	2.8	3.0	2.3	1.6

Index of Leading Economic Indicators

Source: The Conference Board

The index of leading economic indicators is used to project the U.S. economy's performance. The index is made up of 10 measurements of economic activity that tend to change direction in advance of the overall economy. The index has predicted economic downturns from 8 to 20 months in advance and recoveries from 1 to 10 months in advance; however, it can be inconsistent, and has occasionally shown "false signals" of recessions.

Components

Average weekly hours of production workers in manufacturing

Average weekly initial claims for unemployment insurance, state programs

Manufacturers' new orders for consumer goods and materials, adjusted for inflation

Vendor performance (slower deliveries diffusion index)

Manufacturers' new orders, nondefense capital goods industries, adjusted for inflation

New private housing units authorized by local building permits

Stock prices, 500 common stocks

Money supply: M-2, adjusted for inflation

Interest rate spread, 10-yr Treasury bonds less federal funds

Consumer expectations (researched by Univ. of Michigan)

U.S. Gross Domestic Product, Gross National Product, Net National Product, National Income, and Personal Income

Source: Bureau of Economic Analysis, U.S. Dept. of Commerce
(billions of dollars)

	1960	1970	1980	1990	1997	1998
GROSS DOMESTIC PRODUCT	—	—	—	$5,546.1	$8,110.9	$8,511.0
GROSS NATIONAL PRODUCT	$515.3	$1,015.5	$2,732.0	5,567.8	8,102.9	8,490.5
Less: Consumption of fixed capital.....................	46.4	88.8	303.8	602.7	871.8	908.0
Equals: Net national product	468.9	926.6	2,428.1	4,965.1	7,231.1	7,582.5
Less: Indirect business tax and nontax liability	45.3	94.0	213.3	444.0	627.2	655.3
Business transfer payments	2.0	4.1	12.1	26.8	35.1	36.1
Statistical discrepancy	−2.8	−1.1	4.9	7.8	−55.8	−76.5
Plus: Subsidies less current surplus of government enterprises	0.4	2.9	5.7	4.5	21.9	27.1
Equals: National income	424.9	832.6	2,203.5	4,491.0	6,646.5	6,994.7
Less: Corporate profits with inventory valuation and capital consumption adjustments........................	49.5	74.7	177.2	380.6	817.9	824.6
Net interest.................................	11.3	41.2	200.9	463.7	432.0	449.3
Contributions for social insurance...................	21.9	62.2	216.5	503.1	727.0	767.5
Wage accruals less disbursements	0.0	0.0	0.0	0.1	3.7	4.0
Plus: Personal interest income.......................	27.5	81.8	312.6	666.3	747.3	764.8
Personal dividend income.......................	24.9	69.3	271.9	698.2	260.3	263.1
Government transfer payments to persons...........	12.9	22.2	52.9	144.4	1,083.3	1,120.8
Business transfer payments	2.0	4.1	12.1	21.3	27.2	28.2
Equals: PERSONAL INCOME	409.4	831.8	2,258.5	4,673.8	6,784.0	7,126.1

U.S. Gross Domestic Product

Source: Bureau of Economic Analysis, U.S. Dept. of Commerce
(billions of dollars)

	1998	First Quarter 1999[1]	Second Quarter 1999[1]		1998	First Quarter 1999[1]	Second Quarter 1999[1]
Gross domestic product	$8,511.0	$8,808.7	$8,893.3	**Net exports of goods and services**	$-151.2	$-196.9	$-225.7
Personal consumption expenditures	5,807.9	6,050.6	6,148.3	Exports	959.0	962.7	972.9
Durable goods	724.7	771.2	777.6	Goods	680.8	677.7	683.1
Nondurable goods	1,662.4	1,736.0	1,771.3	Services.	278.2	285.0	289.8
Services	3,420.8	3,543.4	3,599.4	Imports	1,110.2	1,159.6	1,198.6
Gross private domestic investment	1,367.1	1,417.4	1,426.7	Goods	932.4	975.2	1,009.6
Fixed investment	1,307.8	1,377.9	1,407.1	Services.	177.8	184.5	189.0
Nonresidential	938.2	972.6	994.0	**Government consumption expenditures and gross investment**	1,487.1	1,537.5	1,544.1
Structures.	246.9	255.0	255.7	Federal	520.6	536.6	533.3
Producers' durable equipment	691.3	717.6	738.3	National defense	340.4	345.5	343.5
Residential.	369.6	405.3	413.1	Nondefense	180.2	191.1	189.8
Change in business inventories	59.3	39.5	19.6	State and local.	966.5	1,000.9	1,010.8

(1) Seasonally adjusted at annual rates.

Countries With Highest Gross Domestic Product and Per Capita GDP[1]

Source: Central Intelligence Agency, *The World Factbook 1998*

Gross Domestic Product (billions of dollars; 1997 estimates unless otherwise noted)				Per Capita Gross Domestic Product[3] (dollars; 1997 estimates unless otherwise noted)			
1. United States	$8,083.0	21. Pakistan	$344.0	1. Luxembourg	$33,700	21. Germany	$20,800
2. China	4,250.0[2]	22. Netherlands	343.9	2. United States	30,200	22. Finland	20,000
3. Japan	3,080.0	23. Taiwan	308.0	3. Norway	27,400	San Marino	20,000
4. Germany	1,740.0	24. Poland	280.7	4. Monaco	25,000[4]	24. Sweden	19,700
5. India	1,534.0	25. South Africa	270.0	5. Singapore	24,600	25. Bahamas	19,400
6. France	1,320.0	26. Egypt	267.1	6. Japan	24,500	26. Ireland	18,600
7. United Kingdom	1,242.0	27. Philippines	244.0	7. U. Arab Emirates	24,000	27. Andorra	18,000[4]
8. Italy	1,240.0	28. Belgium	236.3	8. Switzerland	23,800	Brunei	18,000
9. Brazil	1,040.0	29. Colombia	231.1	9. Belgium	23,200	29. New Zealand	17,700
10. Indonesia	960.0	30. Malaysia	227.0	Denmark	23,200	30. Israel	17,500
11. Mexico	694.3	31. Saudi Arabia	206.5	11. Liechtenstein	23,000[5]	31. Qatar	16,700
12. Russia	692.0	32. Venezuela	185.0	12. France	22,700	32. Spain	16,400
13. Canada	658.0	33. Sweden	176.2	13. Kuwait	22,300	33. Portugal	15,200
14. Spain	642.4	34. Austria	174.1	14. Netherlands	22,000	34. Cyprus	15,000[6]
15. South Korea	631.2	35. Switzerland	172.4	15. Canada	21,700	35. Taiwan	14,200
16. Thailand	525.0	36. Chile	168.5	16. Italy	21,500	36. South Korea	13,700
17. Australia	394.0	37. Bangladesh	167.0	17. Australia	21,400	Bahrain	13,700
18. Turkey	388.3	38. Portugal	149.5	Austria	21,400	38. Greece	13,000
19. Iran	371.2	39. Greece	137.4	19. United Kingdom	21,200	39. Malta	12,900
20. Argentina	348.2	40. Nigeria	132.7	20. Iceland	21,000	40. Malaysia	11,100

(1) U.S. data from *The World Factbook* may differ from data elsewhere from the U.S. Bureau of Economic Analysis. International GDP estimates are derived from purchasing power parity calculations, which involve the use of intl. dollar price weights applied to the quantities of goods and services produced in a given economy. (2) 1997 estimate as extrapolated from World Bank estimate with use of official Chinese growth figures for 1996-97; may overstate the GDP by as much as 25%. Hong Kong, a special administrative region of China since July 1, 1997, had a GDP of $175.2 billion and a per capita GDP of $26,800 in 1997. (3) These territories had large per capita GDPs: Bermuda (UK, 1996) $29,000, Cayman Islands (UK, 1996) $23,800, Aruba (Neth., 1996) $21,000, Guam (U.S., 1996) $19,000, Faroe Islands (Den., 1996) $16,300, Greenland (Den.) $16,100, Macao (Port.) $15,600. (4) 1995 estimate. (5) 1996 estimate. (6) Does not include Turkish-held area.

Chapter 11

Chapter 11 refers to the provisions in the Federal Bankruptcy Code for court-supervised reorganization of debtor companies. A company files for Chapter 11 protection when it can no longer pay its creditors or when it expects future liabilities it cannot hope to pay, such as product liability damage awards. In 1991, the U.S. Supreme Court ruled that the provision of federal bankruptcy law that permits corporations to reorganize while continuing to operate was also available for use by individuals. The Bankruptcy Reform Act of 1994 further amended Chapter 11.

Process

1. Bankruptcy filing imposes an automatic stay.
• Creditors generally cannot file or continue suits for repayment.
• Debts are frozen and creditors generally must stop collection actions. This is called the "automatic stay."
• Debtor's day-to-day operations continue.
• Spending, borrowing, and asset sales outside of the debtor's normal course of business must be approved by the court.
• Secured creditors can ask the court for exemption from the automatic stay to undertake or continue to recover the collateral that secures their claim.

2. Unsecured creditors form a committee.
• The U.S. trustee appoints the committee, which ordinarily consists of the 7 largest unsecured creditors who are willing to serve on the panel.
• The U.S. trustee can appoint additional committees to represent other creditors and shareholders.

• The committee chooses representatives to deal with the debtor company.
• The committee and U.S. trustee oversee the debtor's business operations.
• Creditors and the U.S. trustee can ask the court to appoint an examiner to investigate possible fraud or mismanagement.
• Creditors and the U.S. trustee can ask the court to order the appointment of a case trustee to run the debtor company.
• If the court orders the appointment, the U.S. trustee selects the case trustee unless a party asks that creditors be allowed to elect the case trustee.

3. The committee, other creditors, and the debtor company negotiate a reorganization plan.
• Parties negotiate a plan for the reorganization of the debtor's business and repayment of frozen debts. This step can take months or years.
• Only the debtor can file a reorganization plan with the court for the first 120 days of the bankruptcy case. The court

can extend the duration of the so-called "exclusivity" period and often does so.
• If the debtor does not file a plan during the exclusivity period, if the debtor's plan is not approved by the court, or if a trustee is appointed, any party can file a plan.
• The proponent of the plan prepares a disclosure statement, which must be approved by the court at a separate hearing.

4. Creditors and shareholders vote on the plan.
• Only creditors and shareholders whose claims and interests are impaired or affected by the plan vote on it.
• A class of creditors accepts the plan if the plan is approved by creditors who hold more than half of the claims in the class by number and at least two-thirds of the claims by amount.
• A class of shareholders accepts the plan if the plan is approved by shareholders who hold at least two-thirds of the equity interest in the class by amount.

5. Judge considers the plan.
• The bankruptcy judge approves the plan if it complies with the Bankruptcy Code and all impaired classes approve.
• If at least one of the impaired classes approves the plan and it meets certain statutory tests, the judge can confirm the plan in a "cramdown," even if not all impaired classes approve.

6. Reorganized company emerges.
• Generally, the debtor's debts are discharged.

• The debtor and creditors must comply with the confirmed plan.
• The automatic stay ends and a permanent injunction goes into effect against any effort to collect prepetition debts other than as provided in the plan.
• The reorganized debtor operates like a normal company.
• Only about 17% of the debtors who file Chapter 11 cases get their plans confirmed.

Expedited Procedure for Small Businesses
• The Bankruptcy Reform Act of 1994 included an expedited confirmation process to be used in Chapter 11 cases filed by small businesses.
• The debtor can elect to use the new process if it has less than $2 million in debts and its primary business is not owning or operating real estate.
• The court can order that a creditors' committee not be appointed.
• Unless the court orders otherwise, the debtor's exclusivity period for filing a plan is shortened to 100 days and all plans must be filed within 160 days.
• The court may conditionally approve the disclosure statement. This saves time by combining the court hearing on the disclosure statement with the hearing on confirmation of the plan.

State Finances: Revenue, Expenditures, Debt, and Taxes

Source: Census Bureau, U.S. Dept. of Commerce
(fiscal year 1997)

STATE	Revenue (millions)	Expenditures (millions)	Debt (millions)	Per capita[1] debt	Per capita[1] taxes	Per capita[1] expenditures
Alabama	$14,008	$12,945	$3,780	$875	$1,270	$2,997
Alaska	9,439	5,722	3,291	5,403	2,659	9,396
Arizona	13,692	12,419	2,742	602	1,500	2,726
Arkansas	8,844	7,685	2,248	891	1,497	3,046
California	131,349	117,643	45,337	1,405	1,911	3,646
Colorado	12,780	10,861	3,402	874	1,359	2,790
Connecticut	14,520	13,826	17,051	5,214	2,491	4,228
Delaware	4,211	3,404	3,434	4,692	2,381	4,650
Florida	41,432	37,464	16,022	1,093	1,439	2,557
Georgia	24,028	21,975	6,186	826	1,456	2,936
Hawaii	6,701	6,093	5,253	4,425	2,601	5,133
Idaho	4,289	3,674	1,598	1,321	1,620	3,037
Illinois	39,038	35,302	23,801	2,001	1,559	2,968
Indiana	17,537	16,370	6,140	1,047	1,552	2,792
Iowa	9,509	9,348	2,014	706	1,643	3,278
Kansas	7,950	7,496	1,211	467	1,630	2,889
Kentucky	15,033	12,949	7,120	1,822	1,745	3,313
Louisiana	15,929	14,286	7,030	1,615	1,297	3,283
Maine	5,215	4,441	3,203	2,579	1,626	3,576
Maryland	20,128	16,200	9,873	1,938	1,689	3,180
Massachusetts	26,538	25,791	29,386	4,803	2,175	4,216
Michigan	45,509	36,092	14,431	1,477	2,032	3,693
Minnesota	22,882	18,443	4,862	1,038	2,395	3,936
Mississippi	9,400	9,006	2,455	899	1,471	3,298
Missouri	16,601	14,230	7,579	1,403	1,447	2,634
Montana	3,524	3,204	2,056	2,339	1,489	3,645
Nebraska	5,537	4,802	1,494	902	1,538	2,898
Nevada	6,494	5,130	2,769	1,651	1,809	3,059
New Hampshire	3,561	3,324	5,848	4,986	780	2,833
New Jersey	36,087	29,430	26,591	3,302	1,790	3,654
New Mexico	8,188	7,059	2,458	1,421	1,920	4,080
New York	95,442	83,243	74,078	4,084	1,922	4,590
North Carolina	25,527	22,864	5,677	765	1,708	3,079
North Dakota	2,818	2,426	900	1,404	1,660	3,784
Ohio	45,250	37,407	13,437	1,201	1,468	3,344
Oklahoma	11,328	9,593	3,795	1,144	1,526	2,892
Oregon	15,004	12,388	5,841	1,801	1,525	3,820
Pennsylvania	49,318	39,296	15,368	1,279	1,612	3,269
Rhode Island	4,229	4,002	5,302	5,372	1,666	4,054
South Carolina	13,805	12,847	5,350	1,423	1,431	3,417
South Dakota	2,316	2,070	1,841	2,494	1,041	2,806
Tennessee	15,696	14,284	3,315	618	1,233	2,661
Texas	63,864	48,887	12,462	641	1,184	2,515
Utah	7,724	6,818	2,451	1,190	1,462	3,311
Vermont	2,370	2,123	2,037	3,459	1,527	3,605
Virginia	24,322	19,287	9,941	1,476	1,430	2,864
Washington	26,841	22,207	9,493	1,692	1,997	3,958
West Virginia	7,467	7,145	3,040	1,674	1,600	3,935
Wisconsin	23,592	18,200	9,832	1,902	1,970	3,520
Wyoming	2,559	2,127	872	1,817	1,380	4,431
ALL STATES[2]	**1,039,423**	**893,827**	**455,697**	**1,706**	**1,660**	**3,346**

(1) Per capita amounts are based on population figures of the resident U.S. population (excluding the District of Columbia) as of July 1, 1996. (2) Totals in this line may not add because of rounding.

State and Local Government Receipts and Current Expenditures

Source: Bureau of Economic Analysis, U.S. Dept. of Commerce
(billions of dollars)

	1997R	1998	First Quarter 1999[1]		1997R	1998	First Quarter 1999[1]
Receipts	$1,094.3	$1,148.1	$1,192.9	Net interest paid	$−77.4	$−83.0	$−87.0
Personal tax and nontax receipts .	219.9	240.3	252.8	Interest paid.	63.3	63.9	64.3
Income taxes	164.3	180.7	190.5	Less: Interest received by			
Nontaxes	32.0	34.5	36.2	government	140.6	146.9	151.4
Other	23.6	25.0	26.1	Less: Dividends received by			
Corporate profits tax accruals . . .	36.0	35.2	36.9	government	14.8	16.1	16.7
Indirect business tax and nontax				Subsidies less current surplus of			
accruals	533.4	559.4	577.9	government enterprises.	−10.6	−9.5	−9.0
Sales taxes.	261.5	271.6	283.8	Subsidies.	0.4	0.4	0.4
Property taxes	209.1	217.4	223.9	Less: Current surplus of			
Other	62.8	70.4	70.1	government enterprises	10.9	9.9	9.4
Contributions for social insurance	79.9	82.1	84.2	Less: Wage accruals less			
Federal grants-in-aid	225.0	231.1	241.1	disbursements.	0.0	0.0	0.0
Current expenditures.	**960.1**	**997.9**	**1,023.2**	**Surplus or deficit (−),**			
Consumption expenditures.	758.8	789.1	810.2	**national income and**			
Transfer payments to persons . . .	304.1	317.4	325.7	**product accounts**	134.1	150.2	169.7

(R) Revised figures. (1) Seasonally adjusted at annual rates.

State and Local Government Current Expenditures and Gross Investment, by Function

Source: Bureau of Economic Analysis, U.S. Dept. of Commerce
(millions of dollars)

	1996			1997		
	Total[1]	Current Expends.	Gross Investment[2]	Total[1]	Current Expends.	Gross Investment[2]
TOTAL	$1,084,663	$922,571	$162,092	$1,135,758	$960,147	$175,611
Central executive, legislative, and judicial						
activities	66,708	63,465	3,243	71,725	68,150	3,575
Administrative, legislative, and judicial activities .	36,807	34,688	2,119	38,921	36,593	2,328
Tax collection and financial management	29,901	28,777	1,124	32,804	31,557	1,247
Civilian safety.	112,194	104,597	7,597	118,689	110,537	8,152
Police	50,382	47,749	2,633	53,218	50,390	2,828
Fire.	19,496	17,975	1,521	20,297	18,613	1,684
Correction.	42,316	38,873	3,443	45,174	41,534	3,640
Education	388,080	352,079	36,001	407,721	367,955	39,766
Elementary and secondary	296,980	271,668	25,312	312,962	284,993	27,969
Higher.	66,887	57,247	9,640	70,195	59,693	10,502
Libraries	5,454	4,836	618	5,747	5,104	643
Other	18,759	18,328	431	18,817	18,165	652
Health and hospitals	27,652	22,237	5,415	27,450	21,827	5,623
Health.	25,476	23,649	1,827	26,408	24,406	2,002
Hospitals	2,176	−1,412	3,588	1,042	−2,579	3,621
Income support, social security, and welfare . . .	245,655	244,892	763	255,974	255,216	758
Govt. employees retirement and disability	−2,387	−2,387	—	1,768	1,768	—
Workers' compensation and temporary disability						
insurance	9,811	9,811	—	10,021	10,021	—
Medical care.	163,610	163,610	—	169,123	169,123	—
Welfare and social services	74,621	73,858	763	75,062	74,304	758
Veterans' benefits and services	269	148	21	277	260	17
Housing and community services	28,640	4,282	24,358	30,877	5,525	25,352
Housing, comm. dev., urban renewal	5,319	1,579	3,740	6,852	2,899	3,953
Water	5,770	−3,265	9,035	6,308	−3,428	9,736
Sewerage.	10,423	455	9,968	10,612	597	10,015
Sanitation	7,128	5,513	1,615	7,105	5,457	1,648
Recreational and cultural activities	16,648	12,234	4,414	17,142	12,388	4,754
Energy	−3,573	−7,518	3,945	−3,250	−7,688	4,438
Gas utilities.	−918	−1,224	306	−1,139	−1,404	265
Electric utilities	−2,655	−6,294	3,639	−2,111	−6,284	4,173
Agriculture	4,504	4,247	257	4,643	4,379	264
Natural resources.	11,481	9,066	2,415	11,897	9,331	2,566
Transportation	126,383	65,238	61,145	134,408	67,866	66,542
Highways	100,371	52,453	47,918	106,923	54,610	52,313
Water	1,572	63	1,509	1,705	71	1,634
Air.	2,459	−1,199	3,658	2,741	−1,268	4,009
Transit and railroad.	21,981	13,921	8,060	23,039	14,453	8,586
Economic development, regulation, and						
services	8,090	7,734	356	8,488	8,103	385
Labor training and services	5,484	5,360	124	5,474	5,345	129
Commercial activities	−12,572	−12,840	268	−13,919	−14,224	305
Publicly owned liquor store systems	−624	−635	11	−648	−658	10
Govt.-administered lotteries, parimutuels.	−12,239	−12,239	—	−13,527	−13,527	—
Other	291	34	257	256	−39	295
Net interest paid[2]	−2,117	−2,117	—	−6,452	−6,452	—
Other and unallocable	61,137	49,367	11,770	64,614	51,629	12,985

(1) Sum of current expenditures and gross investment. (2) Excludes interest received by social insurance funds, which is netted against expenditures for the appropriate functions.

Banks in the U.S.—Number, Deposits

Source: Federal Deposit Insurance Corp. (as of Dec. 31, 1998)

Comprises all FDIC-insured commercial and savings banks, including savings and loan institutions (S&Ls).

| | TOTAL NUMBER OF BANKS | | | | | TOTAL DEPOSITS (millions of dollars) | | | |
| | | Commercial banks[1] | | | | | Commercial banks[1] | | |
Year	ALL BANKS	Natl.	State	Non-members	All savings	ALL DEPOSITS	Natl.	State	Non-members	All savings
1935.....	15,295	5,386	1,001	7,735	1,173	$ 45,102[2]	$24,802	$13,653	$5,669	$978[2]
1940.....	15,772	5,144	1,342	6,956	2,330	67,494	35,787	20,642	7,040	4,025
1945.....	15,969	5,017	1,864	6,421	2,667	151,524	77,778	41,865	16,307	15,574
1950.....	16,500	4,958	1,912	6,576	3,054	171,963	84,941	41,602	19,726	25,694
1955.....	17,001	4,692	1,847	6,698	3,764	235,211	102,796	55,739	26,198	50,478
1960.....	17,549	4,530	1,641	6,955	4,423	310,262	120,242	65,487	34,369	90,164
1965.....	18,384	4,815	1,405	7,327	4,837	467,633	185,334	78,327	51,982	151,990
1970.....	18,205	4,621	1,147	7,743	4,694	686,901	285,436	101,512	95,566	204,367
1975.....	18,792	4,744	1,046	8,595	4,407	1,157,648	450,308	143,409	187,031	376,900
1980.....	18,763	4,425	997	9,013	4,328	1,832,716	656,752	191,183	344,311	640,470
1985.....	18,033	4,959	1,070	8,378	3,626	3,140,827	1,241,875	354,585	521,628	1,022,739
1990.....	15,158	3,979	1,009	7,355	2,815	3,637,292	1,558,915	397,797	693,438	987,142
1993.....	13,220	3,304	969	6,685	2,262	3,528,487	1,576,725	476,093	701,512	774,157
1994.....	12,603	3,075	976	6,400	2,152	3,611,618	1,630,171	533,261	711,006	737,180
1995.....	11,970	2,858	1,042	6,040	2,030	3,769,477	1,695,817	614,924	716,829	741,907
1996.....	11,670	2,763	1,024	5,902	1,981	3,788,905	1,795,110	567,809	698,497	727,489
1997.....	10,922	2,597	992	5,554	1,779	4,125,811	2,004,855	729,009	687,832	704,115
1998.....	10,461	2,458	994	5,322	1,687	5,090,585	2,137,855	810,511	733,157	704,531

(1) "Nonmembers" are banks that are not members of the Federal Reserve System; "National" and "State" institutions are members.
(2) Figures for 1935 do not include data for S&Ls (not available).

50 Largest U.S. Bank Holding Companies

Source: *American Banker* (as of Dec. 31, 1998)

Bank Holding Company	Assets (millions)	Bank Holding Company	Assets (millions)
Citigroup Inc., New York, NY	$667,400	BB&T Corp., Winston-Salem, NC	$34,427
BankAmerica Corp., San Francisco, CA	617,679	HSBC Americas Inc., Buffalo, NY	33,944
Chase Manhattan Corp., New York, NY	365,875	Summit Bancorp, Princeton, NJ	33,101
Bank One Corp., Chicago, IL	261,496	Union Planters Corp., Memphis, TN	31,692
J.P. Morgan & Co., New York, NY	261,067	UnionBanCal Corp., San Francisco, CA	31,407
First Union Corp., Charlotte, NC	237,363	LaSalle National Corp., Chicago, IL	28,983
Wells Fargo Bank, San Francisco, CA	202,475	Fifth Third Bancorp, Cincinnati, OH	28,922
Bankers Trust Corp., New York, NY	133,115	Huntington Bancshares Inc., Columbus, OH	28,293
Fleet Financial Group Inc., Boston, MA	104,382	Northern Trust Corp., Chicago, IL	27,870
SunTrust Banks Inc., Atlanta, GA	93,170	MBNA Corp., Wilmington, DE	25,806
National City Corp., Cleveland, OH	88,246	Harris Bankcorp, Chicago, IL	25,112
KeyCorp., Cleveland, OH	80,020	Popular Inc., San Juan, PR	23,205
PNC Bank Corp., Pittsburgh, PA	77,207	First Security Corp., Salt Lake City, UT	21,768
U.S. Bancorp Inc., Minneapolis, MN	76,438	Marshall & Ilsley Corp., Milwaukee, WI	21,566
BankBoston Corp.	73,513	First American Corp., Nashville, TN	20,732
Wachovia Corp., Winston-Salem, NC	64,123	M&T Bank Corp., Buffalo, NY	20,584
Bank of New York, NY	63,579	AmSouth Bancorp, Birmingham, AL	19,902
Mellon Bank Corp., Pittsburgh, PA	50,777	First Tennessee National Corp., Memphis, TN	18,734
Republic New York Corp., New York, NY	50,424	Citizens Financial Group, Inc., Providence, RI	18,430
State Street Corp., Boston, MA	47,082	First Maryland Bancorp, Baltimore, MD	18,295
Firstar Corp., Milwaukee, WI	38,476	Compass Bancshares Inc., Birmingham, AL	17,289
SouthTrust Corp., Birmingham, AL	38,134	Zions Bancorp., Salt Lake City, UT	16,649
Regions Financial Corp., Birmingham, AL	36,832	Old Kent Financial Corp., Grand Rapids, MI	16,589
Comerica Inc., Detroit, MI	36,601	BancWest Corp., Honolulu, HI	15,050
Mercantile Bancorp, Inc., St. Louis, MO	34,571	Pacific Century Financial Corp., Honolulu, HI	15,017

U.S. Bank Failures

Source: Federal Deposit Insurance Corp.

Year	Closed or assisted	Year	Closed or assisted	Year	Closed or assisted	Year	Closed or assisted	Year	Closed or assisted
1934........	61	1960	2	1971	6	1982	42	1991	127
1935........	32	1961	9	1972	3	1983	48	1992	122
1936........	72	1963	2	1973	6	1984	80	1993	41
1937........	84	1964	8	1975	14	1985	120	1994	13
1938........	81	1965	9	1976	17	1986	145	1995	6
1939........	72	1966	8	1978	7	1987	203	1996	5
1940........	48	1967	4	1979	10	1988	221	1997	1
1955........	5	1969	9	1980	11	1989	207	1998	3
1959........	3	1970	8	1981	10	1990	169		

World's 50 Largest Banking Companies[1]

Source: *American Banker* (Aug. 1999)

Banks	Assets (millions)	Banks	Assets (millions)
Deutsche Bank AG, Frankfurt, Germany	$735,171.0	Bank of Tokyo-Mitsubishi Ltd., Tokyo, Japan	$579,791.7
UBS Group, Zurich, Switzerland	687,379.7	HypoVereinsbank AG, Munich, Germany	540,851.1
Citigroup, New York, United States	668,641.0	ABN-AMRO Bank NV, Amsterdam, Netherlands	507,216.7
Bank of America Corp., Charlotte, N.C., United States	617,679.0	HSBC Holdings PLC, London, United Kingdom	482,920.9
		Credit Suisse Group, Zurich, Switzerland	475,017.8

Banks	Assets (millions)	Banks	Assets (millions)
ING Group, Amsterdam, Netherlands	$463,597.6	Lloyds TSB Group, London, United Kingdom	$279,343.2
Credit Agricole Group, Paris, France	457,037.0	Tokai Bank Ltd., Nagoya, Japan	264,449.4
DaHchi Kangyo Bank Ltd., Tokyo, Japan	455,900.7	DG Bank, Frankfurt, Germany	261,722.1
Societe Generale, Paris, France	450,224.8	Bank One Corp., Chicago, Illinois, United States	261,496.0
Dresdner Bank AG, Frankfurt, Germany	429,027.3	JP Morgan & Co. Inc., New York, United States	261,067.0
Sumitomo Bank Ltd., Osaka, Japan	428,000.8	Credit Lyonnals, Paris, France	245,199.4
Sanwa Bank Ltd., Osaka, Japan	418,373.0	Halifax PLC, Leeds, United Kingdom	240,395.7
Westdeutsche Landesbank Girozentrale, Duesseldorf, Germany	415,954.0	Asahi Bank Ltd., Tokyo, Japan	237,851.9
Norinchukin Bank, Tokyo, Japan	407,624.0	First Union Corp., Charlotte, N.C., United States	237,363.0
Sakura Bank, Ltd., Tokyo, Japan	392,098.9	Dexia Group, Brussels, Belgium	234,001.7
Fuji Bank Ltd., Tokyo, Japan	385,252.8	Bankgesellschaft Berlin AG, Berlin, Germany	221,558.3
Commerzbank AG, Frankfurt, Germany	382,935.6	Wells Fargo & Co., San Francisco, United States	202,475.0
Banque Nationale de Paris, France	381,316.5	Istituto Bancario San Paolo IMI, Turin, Italy	185,403.0
Industrial Bank of Japan Ltd., Tokyo, Japan	370,394.0	Canadian Imperial Bank of Commerce, Toronto, Canada	183,043.9
Chase Manhattan Corp., New York, United States	365,875.0	Grupo Santander, Santander, Spain	182,014.9
Barclays PLC, London, United Kingdom	365,128.3	Banca Intesa SPA Group, Milan, Italy	179,654.2
Fortis Bank, Brussels, Belgium	323,567.0	Royal Bank of Canada, Montreal, Canada	178,470.9
Parlbas, Paris, France	309,364.0	Norddeutsche Landesbank Girozentrale, Hanover, Germany	178,337.4
NatWest Group, London, United Kingdom	309,266.7	KBC Bank and Insurance, Brussels, Belgium	173,413.1
Abbey National Plc, London, United Kingdom	295,608.6	**Totals for the Top 50 in Assets**	**$17,886,854.9**
Rabobank Group, Utrecht, Netherlands	293,141.5		
Bayerische Landesbank Girozentrale, Munich, Germany	285,237.9		

(1) Includes bank holding companies and commercial and savings banks. **NOTE:** Data for U.S. companies listed include assets not included in "50 Largest U.S. Bank Holding Companies" table.

Federal Deposit Insurance Corporation (FDIC)

The Federal Deposit Insurance Corporation (FDIC) is the independent deposit insurance agency created by Congress to maintain stability and public confidence in the nation's banking system. In its unique role as deposit insurer of banks and savings associations, and in cooperation with other federal and state regulatory agencies, the FDIC seeks to promote the safety and soundness of insured depository institutions in the U.S. financial system by identifying, monitoring, and addressing risks to the deposit insurance funds. The FDIC aims at promoting public understanding and sound public policies by providing financial and economic information and analyses. It seeks to minimize disruptive effects from the failure of banks and savings associations. It seeks to ensure fairness in the sale of financial products and the provision of financial services.

The FDIC's income consists of assessments on insured banks and income from investments; it receives no appropriations from Congress. The Corporation may borrow from the U.S. Treasury, not to exceed $30 billion outstanding, but the agency has made no such borrowings since it was organized in 1933. The FDIC's Bank Insurance Fund was $29.8 billion (unaudited) and the Savings Association Insurance Fund stood at $9.1 billion (unaudited), as of June 30, 1999.

Federal Reserve Board Discount Rate

The discount rate is the rate of interest set by the Federal Reserve that member banks are charged when borrowing money through the Federal Reserve System. Includes any changes through Oct. 1999.

Effective date	Rate	Effective date	Rate	Effective date	Rate	Effective date	Rate	Effective date	Rate
1980:		**1982:**		**1985:**		**1989:**		**1994:**	
Feb. 15	13	July 20	11½	May 20	7½	Feb. 24	7	May 17	3½
May 30	12	Aug. 2	11	**1986:**		**1990:**		Aug. 16	4
June 13	11	Aug. 16	10½	March 7	7	Dec. 18	6½	Nov. 15	4¾
July 28	10	Aug. 27	10	April 21	6½	**1991:**		**1995:**	
Sept. 26	11	Oct. 12	9½	July 11	6	Apr. 30	5½	Feb. 1	5¼
Nov. 17	12	Nov. 22	9	Aug. 21	5½	Sept. 13	5	**1996:**	
Dec. 5	13	Dec. 15	8½	**1987:**		Nov. 6	4½	Jan. 31	5
1981:		**1984:**		Sept. 4	6	Dec. 20	3½	**1998:**	
May 5	14	April 9	9	**1988:**		**1992:**		Oct. 15	4¾
Nov. 2	13	Nov. 21	8½	Aug. 9	6½	July 3	3	Nov. 17	4½
Dec. 4	12	Dec. 24	8					**1999:**	
								Aug. 24	4¾

Federal Reserve System

The Federal Reserve System is the central bank for the U.S. The system was established on Dec. 23, 1913, originally to give the country an elastic currency, to provide facilities for discounting commercial paper, and to improve the supervision of banking. Since then, the system's responsibilities have been broadened. Over the years, stability and growth of the economy, a high level of employment, stability in the purchasing power of the dollar, and reasonable balance in transactions with other countries have come to be recognized as primary objectives of governmental economic policy.

The Federal Reserve System consists of the Board of Governors, the 12 District Reserve Banks and their branch offices, and the Federal Open Market Committee. Several advisory councils help the board meet its varied responsibilities.

The hub of the system is the 7-member Board of Governors in Washington. The members of the board are appointed by the president and confirmed by the Senate, to serve 14-year terms. The president also appoints the chairman and vice chairman of the board from among the board members for 4-year terms that may be renewed. As of Oct. 1999 the board members were: Alan Greenspan, Chair; Roger W. Ferguson Jr., Vice Chair; Edward W. Kelley Jr.; and Edward M. Gramlich; there were two vacancies.

The board is the policy-making body. In addition to those responsibilities, it supervises the budget and operations of the Reserve Banks, approves the appointments of their presidents, and appoints 3 of each District Bank's directors, including the chairman and vice chairman of each Reserve Bank's board.

The 12 Reserve Banks and their branch offices serve as the decentralized portion of the system, carrying out day-to-day operations such as circulating currency and coin and providing fiscal agency functions and payments mechanism services. The District Banks are in Boston, New York, Philadelphia, Cleveland, Richmond, Atlanta, Chicago, St. Louis, Minneapolis, Kansas City, Dallas, and San Francisco.

The system's principal function is monetary policy, which it controls using 3 tools: reserve requirements, the discount rate, and open market operations. Uniform reserve requirements, set by the board, are applied to the transaction accounts and nonpersonal time deposits of all depository institutions. Responsibility for setting the discount rate (the interest rate at which depository institutions can borrow money from the Reserve Banks) is shared by the Board of Governors and the Reserve Banks. Changes in the discount rate are recommended by the individual boards of directors of the Reserve Banks and are subject to approval by the Board of Governors.

The most important tool of monetary policy is open market operations (the purchase and sale of government securities). Responsibility for influencing the cost and availability of money and credit through the purchase and sale of government securities lies with the Federal Open Market Committee (FOMC), which is composed of the 7 members of the Board of Governors, the president of the Federal Reserve Bank of New York, and 4 other Federal Reserve Bank presidents, who each serve one-year terms on a rotating basis. The committee bases its decisions on economic and financial developments and outlook, setting yearly growth objectives for key measures of money supply and credit. The decisions of the committee are carried out by the Domestic Trading Desk of the Federal Reserve Bank of New York.

The Federal Reserve Act prescribes a Federal Advisory Council, consisting of 1 member from each Federal Reserve District, who is elected annually by the Board of Directors of each of the 12 Federal Reserve Banks. The council meets with the Federal Reserve Board 4 times a year to discuss business and financial conditions, as well as to make advisory recommendations.

The Consumer Advisory Council is a statutory body, including both consumer and creditor representatives, which advises the Board of Governors on its implementation of consumer regulations and other consumer-related matters.

Following the congressional passage of the Monetary Control Act of 1980, the Federal Reserve System's Board of Governors established the Thrift Institutions Advisory Council to provide information and perspectives on the special needs and problems of thrift institutions. This group is composed of representatives of mutual savings banks, savings and loan associations, and credit unions.

United States Mint

Source: United States Mint, U.S. Dept. of the Treasury

The United States Mint was created on Apr. 2, 1792, by an act of Congress, which established the U.S. national coinage system. Supervision of the mint was a function of the secretary of state, but in 1799 the mint became an independent agency reporting directly to the president. The mint was made a statutory bureau of the Treasury Department in 1873, with a director appointed by the president to oversee its operations.

The mint manufactures and ships all U.S. coins for circulation to Federal Reserve banks and branches, which in turn issue coins to the public and business community through depository institutions. The mint also safeguards the Treasury Department's stored gold and silver, as well as other monetary assets.

The composition of dimes, quarters, and half dollars, traditionally produced from silver, was changed by the Coinage Act of 1965, which mandated that these coins from here on in be minted from a cupronickel-clad alloy and reduced the silver content of the half dollar to 40%. In 1970, legislative action mandated that the half dollar and a dollar coin be minted from the same alloy.

The Eisenhower dollar was minted from 1971 through 1978, when legislation called for the minting of the smaller Susan B. Anthony dollar coin. The Anthony dollar, which was minted from 1979 through 1981, marked the first time that a woman, other than a mythical figure, appeared on a U.S. coin produced for general circulation. Authorized by the U.S. Dollar Coin Act of 1997 to replace the Susan B. Anthony dollar in 2000, is the Golden Dollar Coin. Golden in color, with a smooth edge and wide border, the obverse side depicts Sacagawea (a Shoshone woman who helped guide Lewis and Clark) and her infant son. The reverse shows an American eagle and 17 stars, one for each of the states at the time of the Lewis and Clark expedition.

Mint headquarters are in Washington, DC. Mint production facilities are in Philadelphia, Denver, San Francisco, and West Point, NY. In addition, the mint is responsible for the U.S. Bullion Depository at Fort Knox, KY.

Proof coin sets, silver proof coin sets, and uncirculated coin sets are available from the mint. The mint also produces ongoing series of national and historic medals in honor of significant persons, events, and sites.

Since 1982, the mint has produced the following congressionally authorized commemorative coins: 1982 George Washington half dollar; 1984 U.S. Olympic coins; 1986 U.S. Statue of Liberty coins; 1987 Bicentennial of the U.S. Constitution coins; 1989 U.S. Congressional coins; 1990 Eisenhower Centennial coin; 1991 United Services Organization 59th Anniversary coin; 1991 Korean War Memorial coin; 1991 Mount Rushmore Anniversary coins; 1992 U.S. Olympic coins; 1992 White House 200th Anniversary coin; 1992 Christopher Columbus Quincentenary coins; 1993 Bill of Rights coins; 1993 World War II 50th Anniversary coins; 1994 World Cup USA coins; Thomas Jefferson 250th Anniversary coin; U.S. Veterans coins (featuring the Prisoner of War coin, Vietnam Veterans Memorial coin, and Women in Military Service for America coin); Bicentennial of the U.S. Capitol Commemorative Silver Dollar; 1995 Civil War Battlefield coins; 1995/1996 U.S. Olympic Games of the Atlanta Centennial Games; 1997 U.S. Botanic Garden Silver Dollar; 1997 Franklin Delano Roosevelt Gold coin; 1997 Gold and Silver Jackie Robinson Commemorative coins; 1997 National Law Enforcement Memorial Silver Dollar; Black Revolutionary War Patriots Silver Dollar; Robert F. Kennedy Silver Dollar; and the National Law Enforcement Officers Memorial Silver Dollar; 1999 Yellowstone National Park Silver Dollar; 1999 George Washington five-dollar gold coin; and the Dolley Madison Silver Dollar.

The congressionally authorized American Eagle gold, platinum, and silver bullion coins are available through dealers worldwide. The gold and platinum eagles are sold in one-ounce, half-ounce, quarter-ounce, and one-tenth-ounce sizes. The American eagle silver bullion coin contains one troy ounce of .999 fine silver and is priced according to the daily market value of silver. These coins also are available directly from the mint in proof condition, separately priced.

The mint offers free public tours and operates sales centers at the U.S. mints in Denver and Philadelphia; it also operates a sales center at Union Station, in Washington, DC.

Further information is available from the U.S. Mint, Customer Service Center, 10003 Derekwood Ln., Lanham, MD 20706. Telephone: (202) 283-COIN. Website: http://www. usmint.gov

Portraits on U.S. Treasury Bills, Bonds, Notes, and Savings Bonds

Denomination	Savings bonds	Treasury bills*	Treasury bonds*	Treasury notes*
$50	Washington		Jefferson	
75	Adams			
100	Jefferson		Jackson	
200	Madison			
500	Hamilton		Washington	
1,000	Franklin	H. McCulloch	Lincoln	Lincoln
5,000	Revere	J. G. Carlisle	Monroe	Monroe
10,000	J. Wilson	J. Sherman	Cleveland	Cleveland
50,000		C. Glass		
100,000		A. Gallatin	Grant	Grant
1,000,000		O. Wolcott	T. Roosevelt	T. Roosevelt
100,000,000				Madison
500,000,000				McKinley

*The U.S. Treasury discontinued issuing treasury bill, bond, and note certificates in 1986. Since then, all issues of marketable treasury securities have been available only in book-entry form, although some certificates remain in circulation.

M I L L E N N I U M   F A C T   B O X

Spare Change

Source: The United States Mint, U.S. Department of Treasury

The penny is nearly as old as the United States. The U.S. Mint has produced different coins at different times. The dollar, half-dollar, quarter, and ten-cent coin denominations were originally produced from gold and silver. Reeded edges were eventually incorporated into the design of these coins so as to deter counterfeiting and fraudulent use, such as filing down the edges of a coin in an attempt to recover precious metals. (The dime has 118 reeds, a quarter has 119, a half-dollar has 150, the Eisenhower dollar has 189 reeds, and the Susan B. Anthony dollar has 133.)

Pennies and nickels are considered "minor" coins of the United States and have never contained precious metals.

Currently, no coins produced for circulation contain precious metals. However, the continued use of reeded edges on current circulating coinage of larger denominations is useful to the visually impaired. For example, the ten-cent and one-cent coins are similar in size; the reeding of the ten-cent coin makes it easily identifiable by touch.

- Before the creation of the U.S. Mint, currency in common use included foreign and colonial currency, livestock, produce, and wampum.
- The Mint's first delivery of coins occurred in March 1793 and consisted of 11,178 copper cents.
- The two-cent coin (1864) was the first coin to bear the motto "In God We Trust."
- The first U.S. commemorative coin was produced in 1892 and featured Christopher Columbus.
- In 1893, Queen Isabella of Spain became the first woman to be featured on a U.S. commemorative coin.
- The first coin to feature an African-American was the Booker T. Washington Memorial Half Dollar.
- Calvin Coolidge was the first president to have his portrait appear on a coin struck during his lifetime.
- In 1943, the content of cent coins was changed to zinc-coated steel because of a wartime copper shortage.
- The Lincoln cent is the only circulating coin currently produced in which the portrait faces to the right.
- The approximate average life span of a coin is 25 years.

DENOMINATION	FIRST ISSUED	LAST ISSUED
HALF CENT	1793	1857
CENT	**1793**	**still issued**
TWO-CENT PIECE	1864	1873
SILVER THREE-CENT PIECE	1851	1873
NICKEL THREE-CENT PIECE	1865	1889
NICKEL FIVE-CENT PIECE	**1866**	**still issued**
HALF DIME	1794	1873
DIME	**1796**	**still issued**
TWENTY-CENT PIECE	1875	1878
QUARTER DOLLAR	**1796**	**still issued**
HALF DOLLAR	**1794**	**still issued**
DOLLAR (Includes Eisenhower, 1971-1978, and Susan B. Anthony, 1979-1981)	**1794**	**still issued**

(Denominations in bold type are still authorized.) **Note:** This list does not include regular issue gold coins.

New Commemorative State Quarters, 1999-2008

Source: United States Mint, U.S. Dept. of the Treasury

Beginning in Jan. 1999, a series of five quarter dollars with new reverses are being issued each year through 2008, celebrating each of the 50 states. To make room on the reverse of the commemorative quarters for each state's design, certain design elements have been moved, thereby creating a new obverse design as well. The coins are being issued in the sequence the states became part of the Union (date each state entered the union is shown below).

1999
Delaware
  Dec. 7, 1787
Pennsylvania
  Dec. 12, 1787
New Jersey
  Dec. 18, 1787
Georgia
  Jan. 2, 1788
Connecticut
  Jan. 9, 1788

2000
Massachusetts
  Feb. 6, 1788
Maryland
  Apr. 28, 1788
South Carolina
  May 23, 1788
New Hampshire
  June 21, 1788
Virginia
  June 25, 1788

2001
New York
  July 26, 1788
North Carolina
  Nov. 21, 1789
Rhode Island
  May 29, 1790
Vermont
  Mar. 4, 1791
Kentucky
  June 1, 1792

2002
Tennessee
  June 1, 1796
Ohio
  Mar. 1, 1803
Louisiana
  Apr. 30, 1812
Indiana
  Dec. 11, 1816
Mississippi
  Dec. 10, 1817

2003
Illinois
  Dec. 3, 1818
Alabama
  Dec. 14, 1819
Maine
  Mar. 15, 1820
Missouri
  Aug. 10, 1821
Arkansas
  June 15, 1836

2004
Michigan
  Jan. 26, 1837
Florida
  Mar. 3, 1845
Texas
  Dec. 29, 1845
Iowa
  Dec. 28, 1846
Wisconsin
  May 29, 1848

2005
California
  Sept. 9, 1850
Minnesota
  May 11, 1858
Oregon
  Feb. 14, 1859
Kansas
  Jan. 29, 1861
West Virginia
  June 20, 1863

2006
Nevada
  Oct. 31, 1864
Nebraska
  Mar. 1, 1867
Colorado
  Aug. 1, 1876
North Dakota
  Nov. 2, 1889
South Dakota
  Nov. 2, 1889

2007
Montana
  Nov. 8, 1889
Washington
  Nov. 11, 1889
Idaho
  July 3, 1890
Wyoming
  July 10, 1890
Utah
  Jan. 4, 1896

2008
Oklahoma
  Nov. 16, 1907
New Mexico
  Jan. 6, 1912
Arizona
  Feb. 14, 1912
Alaska
  Jan. 3, 1959
Hawaii
  Aug. 21, 1959

Denominations of U.S. Currency

Since 1969 the largest denomination of U.S. currency that has been issued is the $100 bill. As larger-denomination bills reach the Federal Reserve Bank, they are removed from circulation. Because some discontinued currency is expected to be in the hands of holders for many years, the description of the various denominations below is continued.

Amt.	Portrait	Embellishment on back	Amt.	Portrait	Embellishment on back
$1	Washington	Great Seal of U.S.	$100	Franklin	Independence Hall
2	Jefferson	Signers of Declaration	500	McKinley	Ornate denominational marking
5	Lincoln	Lincoln Memorial	1,000	Cleveland	Ornate denominational marking
10	Hamilton	U.S. Treasury	5,000	Madison	Ornate denominational marking
20	Jackson	White House	10,000	Chase	Ornate denominational marking
50	Grant	U.S. Capitol	100,000*	W. Wilson	Ornate denominational marking

*For use only in transactions between Federal Reserve System and Treasury Department.

U.S. Currency and Coin

Source: Financial Management Service, U.S. Dept. of the Treasury (Mar. 31, 1999)

Amounts Outstanding and in Circulation

Currency	Total currency and coin	Total currency	Federal Reserve notes[1]	U.S. notes	Currency no longer issued
Amounts outstanding	$692,580,332,727	$666,465,772,829	$665,942,414,837	$268,715,616	$254,642,376
Less amounts held by:					
Treasury	95,245,940	17,299,530	17,085,787	20,739	193,004
Federal Reserve banks . .	174,665,754,026	174,227,836,433	174,227,831,988	400	4,045
Amounts in circulation	$517,829,332,761	$492,220,636,866	$491,697,497,062	$268,694,477	$254,445,327

Coins[2]	Total	Dollars[3]	Fractional coins
Amounts outstanding	$26,114,559,898	$2,024,703,898	$24,089,856,000
Less amounts held by:			
Treasury	77,946,410	5,878,410	72,068,000
Federal Reserve banks	427,917,593	36,369,121	391,548,472
Amounts in circulation	$25,608,695,895	$1,982,456,367	$23,626,239,528

Currency in Circulation by Denominations

Denomination	Total currency in circulation	Federal Reserve notes[1]	U.S. notes	Currency no longer issued
$1	$6,723,236,225	$6,576,489,748	$143,481	$146,602,996
$2	1,169,217,536	1,036,748,694	132,456,266	12,576
$5	7,733,317,335	7,591,783,160	110,210,110	31,324,065
$10	13,534,938,060	13,512,558,260	5,950	22,373,850
$20	87,037,399,660	87,017,295,200	3,380	20,101,080
$50	49,917,140,150	49,905,648,700	—	11,491,450
$100	325,789,401,800	325,741,535,800	25,875,200	21,990,800
$500	143,757,500	143,569,500	—	188,000
$1,000	167,033,000	166,828,000	—	205,000
$5,000	1,755,000	1,700,000	—	55,000
$10,000	3,440,000	3,340,000	—	100,000
Fractional parts	485	—	—	485
Partial notes[4]	115	—	90	25
TOTAL CURRENCY	**$492,220,636,866**	**$491,697,497,062**	**$268,694,477**	**$254,445,327**

Comparative Totals of Money in Circulation — Selected Dates

Date	Dollars (in millions)	Per capita[5]	Date	Dollars (in millions)	Per capita[5]
Mar. 31, 1999	$517,829.0	$1,902.21	June 30, 1960	$32,064.6	$177.47
Mar. 31, 1998	474,979.0	1,762.42	June 30, 1955	30,229.3	182.90
Mar. 31, 1997	444,534.0	1,664.58	June 30, 1950	27,156.3	179.03
Mar. 31, 1996	416,280.0	1,573.15	June 30, 1945	26,746.4	191.14
Mar. 31, 1995	401,610.0	1,531.39	June 30, 1940	7,847.5	59.40
Mar. 31, 1990	257,664.4	1,028.71	June 30, 1935	5,567.1	43.75
June 30, 1985	185,890.7	778.58	June 30, 1930	4,522.0	36.74
June 30, 1980	127,097.2	558.28	June 30, 1925	4,815.2	41.56
June 30, 1975	81,196.4	380.08	June 30, 1920	5,467.6	51.36
June 30, 1970	54,351.0	265.39	June 30, 1915	3,319.6	33.01
June 30, 1965	39,719.8	204.14	June 30, 1910	3,148.7	34.07

(1) Issued on or after July 1, 1929. (2) Excludes coins sold to collectors at premium prices. (3) Includes $481,781,898 in standard silver dollars. (4) Represents the value of certain partial denominations not presented for redemption. (5) Based on Bureau of the Census estimates of population. The requirement for a gold reserve against U.S. notes was repealed by Public Law 90-269, approved Mar. 18, 1968. Silver certificates issued on and after July 1, 1929, became redeemable from the general fund on June 24, 1968. The amount of security after those dates has been reduced accordingly.

New U.S. Currency Designs

On Mar. 25, 1996, the U.S. Treasury issued a redesigned $100 note incorporating many new and modified anticounterfeiting features. The note was the first of the U.S. currency series to be redesigned. A new $50 note was issued Oct. 27, 1997, and the new $20 bill was released into circulation Sept. 24, 1998. New $10 and $5 notes were expected to be unveiled by the end of 1999 and to be in circulation by summer 2000; a new $1 note with a more modest redesign will follow. Old notes are being removed from circulation as they are returned to the Federal Reserve.

The new $100 bill has: a larger portrait, moved off-center; a watermark (seen only when held up to the light) to the right of the portrait, depicting the same person; a security thread that glows red when exposed to ultraviolet light in a dark environment; color-shifting ink that changes from green to black when viewed at different angles, to appear in the numeral on the lower, front right-hand corner of the bill; microprinting in the numeral in the note's lower, front left-hand corner and on the portrait; and other features for security, machine authentication, and processing of the currency. The redesigned $20 and $50 bills incorporate the same features as the $100 bill, with the notable addition of a low-vision feature, a large (14-mm high, as compared to 7.8-mm on the old design), dark numeral on a light background on the back of the note. (The security thread in the $50 glows yellow; in the $20 it glows green.) The low-vision feature will appear on subsequent redesigned notes in the series. More new currency information is available on the U.S. Treasury's website: http://www.ustreas.gov

Consumer Credit Outstanding, 1996-98

Source: Federal Reserve System
(billions of dollars)
Estimated amounts of credit outstanding as of end of year. Not seasonally adjusted.

	1996	1997	1998		1996	1997	1998
TOTAL	**$1,211.6**	**$1,264.1**	**$1,331.7**	Credit unions	17.8	19.6	19.9
Major holders				Savings institutions	10.3	11.4	12.5
Commercial banks	526.8	512.6	508.9	Nonfinancial business	44.9	45.0	39.2
Finance companies	152.4	160.0	168.5	Pools of securitized assets[1]	188.7	221.5	272.3
Credit unions	144.1	152.4	155.4	Nonrevolving	688.7	708.2	745.2
Savings institutions	44.7	47.2	51.8	Commercial banks	298.2	292.7	298.6
Nonfinancial business	77.7	78.9	74.9	Finance companies	119.9	121.4	136.2
Pools of securitized assets[1]	265.8	313.1	372.4	Credit unions	126.3	132.8	135.5
Major Types of Credit[2]				Savings institutions	34.4	35.7	39.2
Revolving	522.9	555.9	586.5	Nonfinancial business	32.8	34.0	35.7
Commercial banks	228.6	219.8	210.3	Pools of securitized assets[1]	77.1	91.6	100.1
Finance companies	32.5	38.6	32.3				

(1) Outstanding balances of pools upon which securities have been issued; these balances are no longer carried on the balance sheets of the loan originators. (2) Includes estimates for holders that do not separately report consumer credit holding by type.

Leading U.S. Businesses in 1998

Source: FORTUNE Magazine
(millions of dollars in revenues)

Aerospace
Boeing	$56,154
Lockheed Martin	26,266
United Technologies	25,715
AlliedSignal	15,128
Textron	11,549
Northrop Grumman	8,902
General Dynamics	4,970
B.F. Goodrich	3,951

Airlines
AMR	$19,205
UAL	17,561
Delta Air Lines	14,138
NWA	9,045
US Airways Group	8,688
Continental Airlines	7,951
Southwest Airlines	4,164
Trans World Airlines	3,259
America West Holdings	2,023
Alaska Air Group	1,898

Apparel
Nike	$9,553
VF	5,479
Reebok International	3,225
Liz Claiborne	2,535
Fruit of the Loom	2,170
Warnaco Group	1,950
Nine West Group	1,917
Kellwood	1,782
Jones Apparel Group	1,685

Beverages
Pepsico	$22,348
Coca-Cola	18,813
Coca-Cola Enterprises	13,414
Anheuser-Busch	11,246
Adolph Coors	1,900
Whitman	1,845

Building Materials, Glass
Owens-Illinois	$5,450
Owens-Corning	5,009
Corning	3,689
USG	3,130
Armstrong World Inds.	2,746
Johns Manville	1,781
Vulcan Materials	1,776

Chemicals
E. I. du Pont de Nemours	$39,130
Dow Chemical	18,441
Monsanto	8,648
PPG Industries	7,510
Union Carbide	5,659
Sherwin-Williams	4,934
Praxair	4,833
Eastman Chemical	4,481
FMC	4,378

Commercial Banks
BankAmerica Corp.	$50,777
Chase Manhattan Corp.	32,379
Bank One Corp.	25,595
First Union Corp.	21,543
Wells Fargo	20,482
J.P. Morgan & Co.	18,425
Bankers Trust Corp.	12,048
Fleet Finl. Group	10,002

Computer and Data Services
Electronic Data Systems	$16,891
Unisys	7,208
Computer Sciences	6,601
First Data	5,118
Automatic Data Proc.	4,798
Comdisco	3,243
America Online	2,600
Micro Warehouse	2,220
Dun & Bradstreet	2,042
Wang Laboratories	1,887

Computer Peripherals
Seagate Technology	$6,819
Quantum	5,805
EMC	3,974
Western Digital	3,542
Lexmark International	3,021
Storage Technology	2,258
Imation	2,047
Iomega	1,694

Computer Software
Microsoft	$14,484
Oracle	7,144
Computer Assoc. Intl.	4,719

Computers, Office Equipment
IBM	$81,667
Hewlett-Packard	47,061
Compaq Computer	31,169
Xerox	20,019
Dell Computer	18,243
Sun Microsystems	9,791
Gateway 2000	7,468
NCR	6,505
Apple Computer	5,941

Diversified Financials
Citigroup	$76,431
Fannie Mae	31,499
American Express	19,132
Freddie Mac	18,048
Household International	8,708
Marsh & McLennan	7,190
AON	6,493

Electronics, Electrical Equip.
General Electric	$100,469
Lucent Technologies	30,147
Motorola	29,398
Raytheon	19,530
Emerson Electric	13,447
Whirlpool	10,323
Honeywell	8,427
Rockwell International	8,025

Entertainment
Walt Disney	$22,976
Time Warner	14,582
Viacom	12,096
CBS	9,061

Food
ConAgra	$23,841
Sara Lee	20,011
RJR Nabisco Holdings	17,037
Archer Daniels Midland	16,109
IBP	12,849
H. J. Heinz	9,209
Farmland Industries	8,775
BestFoods	8,374
Campbell Soup	7,505
Tyson Foods	7,414

Food and Drug Stores
Kroger	$28,203
Safeway	24,484
American Stores	19,867
Albertson's	16,005
Walgreen	15,307
CVS	15,274
Fred Meyer	14,879
Winn-Dixie Stores	13,617
Publix	12,067
Rite Aid	11,375

Food Services
McDonald's	$12,421
Tricon Global Restaurants	8,468
Nebco Evans	7,421
Darden Restaurants	3,287
Advantica	1,962
Wendy's International	1,948

Forest and Paper Products
International Paper	$19,500
Georgia-Pacific	13,223
Kimberly-Clark	12,298
Weyerhaeuser	10,766
Fort James	7,301
Boise Cascade	6,162
Champion International	5,653
Mead	4,579
Union Camp	4,503

Furniture
Leggett & Platt	$3,370
Steelcase	2,760
Lifestyle Furns. Intl.	2,002
Furniture Brands Intl.	1,960

General Merchandisers
Wal-Mart Stores	$139,208
Sears Roebuck	41,322
Kmart	33,674
Dayton Hudson	30,951
J. C. Penney	30,678
Federated Dept. Stores	15,833
May Department Stores	13,413

Health Care
Cigna	$21,437
Aetna	20,604
Columbia/HCA Healthcare	18,681
United Healthcare	17,355

Hotels, Casinos, Resorts
Marriott International	$7,968
Starwood Hotels	4,832
Hilton Hotels	4,064
Host Marriott	3,519

Industrial and Farm Equip.
Caterpillar	$20,977
Deere	13,822
Ingersoll-Rand	8,292
ITT Industries	7,523
American Standard	6,654
Baker Hughes	6,312
Cummins Engine	6,266
Case	6,149

Insurance (Life and Health)
TIAA-CREF[2]	$35,889
Prudential of America[1]	34,427
Metropolitan Life[1]	26,735
New York Life (Mutual)	19,849
Northwestern Mut. Life (Mut.)	14,645

Insurance (Property and Casualty)
State Farm Ins. (Mutual)	$48,114
American Intl. Group (Stock)	33,296
Allstate (Stock)	25,879
Loews (Stock)	20,713

Hartford Fin'l. Svces. (Stock)	$15,022
Liberty Mutual Group (Mutual)	13,166
Nationwide Ins. Enterprise[1]	13,105

Mail, Pkg., Freight Delivery
United Parcel Svce.	$24,788
Fdx	15,873

Metal Products
Gillette	$10,056
Crown Cork & Seal	8,300
ITW	5,648
Fortune Brands	4,797
Masco	4,345

Metals
Alcoa	$15,489
Reynolds Metals	5,859
Bethlehem Steel	4,478
LTV	4,273
Nucor	4,151
Ryerson Tull	4,093
Allegheny Teledyne	3,923

Motor Vehicles and Parts
General Motors	$161,315
Ford Motor	144,416
Dana	12,839
Johnson Controls	12,587
TRW	11,886

Petroleum Refining
Exxon	$100,697
Mobil	47,678
Texaco	31,707
Chevron	26,801
USX	24,754

Pharmaceuticals
Merck	$26,898
Johnson & Johnson	23,657
Bristol-Myers Squibb	18,284
Pfizer	14,704
American Home Products	13,463
Abbott Laboratories	12,478
Warner-Lambert	10,214
Eli Lilly	10,051
Schering-Plough	8,077
Pharmacia & Upjohn	6,893

Publishing & Printing
R.R. Donnelley & Sons	$5,900
Gannett	5,121
McGraw-Hill	3,729
Times Mirror	3,291
Knight-Ridder	3,100
Tribune	2,981
New York Times	2,937
Reader's Digest Assn.	2,634

Railroads
Union Pacific	$10,553
CSX	9,898
Burlington Northern Santa Fe	8,941
Norfolk Southern	4,428

Rubber and Plastic Prods.
Goodyear Tire	$12,649
Rubbermaid	2,554
Sealed Air	2,507
M.A. Hanna	2,286

Scientific, Photo., and Control Equip.
Minnesota Mining & Mfg.	$15,021
Eastman Kodak	13,406
Applied Materials	4,042

Securities
Merrill Lynch	$35,853
Morgan Stanley	31,131
Lehman Bros. Holdings	19,894

Bear Stearns	$7,980
Paine Webber Group	7,250

Semiconductors
Intel	$26,273
Texas Instruments	8,460
Advanced Micro Devices	2,542
National Semiconductor	2,537

Soaps, Cosmetics
Procter & Gamble	$37,154
Colgate-Palmolive	8,972
Avon Products	5,213
Estée Lauder	3,618
Clorox	2,741
Revlon	2,252

Specialty Retailers
Home Depot	$30,219
Costco Cos.	24,270
Lowe's	12,245
Toys "R" Us	11,200
Limited	9,347
Gap	9,054
Office Depot	8,998
Circuit City Group	8,871
Best Buy	8,378

Telecommunications
AT&T	$53,588
Bell Atlantic	31,566
SBC Communications	28,777
GTE	25,473
BellSouth	23,123
MCI WorldCom	17,678
Ameritech	17,154
Sprint	17,134
US West	12,378

Temporary Help
Manpower	$8,814
Olsten	4,603
Kelly Services	4,092

Textiles
Shaw Industries	$3,542
Mohawk Industries	2,639
Springs Industries	2,180
Burlington Industries	2,010
Westpoint Stevens	1,779

Tobacco
Philip Morris	$57,813
Universal	4,287
Dimon	2,563

Toys, Sporting Goods
Mattel	$4,782
Hasbro	3,304

Transportation Equipment
Brunswick	$3,945
Trinity Industries	2,473
Harley-Davidson	2,064

Utilities, Gas and Electric
PG&E Corp.	$19,942
Duke Energy	17,610
Texas Utilities	14,736
Utilicorp United	12,563
Entergy	11,495
Houston Industries	11,488
Southern	11,403

Wholesalers
Ingram Micro	$22,034
McKesson HBOC	20,857
Supervalu	17,201
Cardinal Health	15,918
Sysco	15,328
Fleming	15,069

(1) Not a stock company, but reported financial data according to Generally Accepted Accounting Principle. (2) Not a mutual company, but reported financial data based on statutory accounting.

Largest Corporate Mergers or Acquisitions in U.S.

Source: Securities Data Co.

(as of Oct. 15, 1999; an * denotes an announced merger or acquisition not yet complete; year = year effective or announced)

Company	Acquirer	Dollars	Year	Company	Acquirer	Dollars	Year
Sprint*	MCI WorldCom	$129.0 bil	1999	Amoco Corp.	British Petroleum Co. PLC	$55.0 bil	1998
Mobil Corp.*	Exxon Corp.	86.4 bil	1998	US WEST*	Qwest Communication	48.5 bil	1999
Citicorp	Travelers Group Inc.	72.6 bil	1998	MCI Communications	WorldCom Inc.	43.4 bil	1998
Ameritech Corp.*	SBC Communications Inc.	72.4 bil	1998	CBS Corp.*	Viacom	40.9 bil	1999
GTE Corp.*	Bell Atlantic Corp.	70.9 bil	1998	Chrysler Corp.	Daimler-Benz AG	40.5 bil	1998
Tele-Communications	AT&T	69.9 bil	1998	Wells Fargo & Co.	Norwest Corp.	34.4 bil	1998
AirTouch Communications	Vodafone Group PLC	65.9 bil	1999	ARCO*	BP Amoco PLC	33.7 bil	1999
MediaOne Group*	AT&T	63.1 bil	1999	US West Media Group	shareholders	31.7 bil	1998
BankAmerica Corp.	NationsBank Corp.	61.6 bil	1998	NYNEX	Bell Atlantic	30.8 bil	1997
				Electronic Data Sys.	shareholders	29.7 bil	1996

Company	Acquirer	Dollars	Year	Company	Acquirer	Dollars	Year
First Chicago NBD	BANC ONE Corp.	$29.6 bil	1998	Vodafone AirTouch-Wireless Ops*	Bell Atlantic–Wireless Ops	$15.0 bil	1999
RJR Nabisco	Kohlberg Kravis Roberts	29.4 bil	1989	MFS Communications	WorldCom	14.9 bil	1996
Associates First Capital	shareholders	26.6 bil	1998	Barnett Banks	NationsBank Corp.	14.8 bil	1998
Lucent Technologies	shareholders	24.1 bil	1996	HF Ahmanson & Co.	Washington Mutual	14.7 bil	1998
Pacific Telesis Group	SBC Communications	22.4 bil	1997	HBO & Co.	McKesson Corp.	13.8 bil	1998
General Re Corp.	Berkshire Hathaway Inc.	22.3 bil	1998	Kraft	Philip Morris	13.7 bil	1988
Ascend Communications	Lucent Technologies	21.1 bil	1999	ITT Corp.	Starwood Lodging	13.5 bil	1998
Waste Management	USA Waste Services	20.0 bil	1998	Gulf Oil	Standard Oil of CA	13.4 bil	1984
Capital Cities/ABC	Walt Disney	18.3 bil	1996	Fred Meyer	Kroger Co.	13.0 bil	1998
SunAmerica Inc.	American Int'l. Group.	18.1 bil	1998	Conrail	Investor group	12.8 bil	1997
CoreStates Financial	First Union Corp.	17.1 bil	1998	PacifiCorp*	Scottish Power PLC	12.6 bil	1998
McCaw Cellular Communications	AT&T	16.7 bil	1994	HFS	CUC International	12.4 bil	1997
BankBoston Corp.*	Fleet Finl. Group	15.9 bil	1999	Central & South West*	American Electric Power	12.3 bil	1997
McDonnell Douglas	Boeing	15.8 bil	1997	Squibb	Bristol-Myers	11.9 bil	1989
Honeywell*	Allied Signal	15.5 bil	1999	American Stores Co.*	Albertson's Inc.	11.8 bil	1998
Warner Comm.	Time	15.1 bil	1990	Allstate	shareholders	11.8 bil	1995

U.S. Corporations With Largest Revenues in 1998

Source: FORTUNE Magazine

(millions of dollars)

Company, headquarters	Revenues	Company, headquarters	Revenues	Company, headquarters	Revenues
General Motors, Detroit, MI ..	$161,315	BankAmerica Corp., Charlotte, NC............	$50,777	TIAA-CREF, New York, NY ...	$35,889
Ford Motor, Dearborn, MI....	144,416	State Farm Insurance Cos., Bloomington, IL.........	48,114	Merrill Lynch, New York, NY ..	35,853
Wal-Mart Stores, Bentonville, AR.........	139,208	Mobil, Fairfax, VA...........	47,678	Prudential Insurance Co. of America, Newark, NJ	34,427
Exxon, Irving, TX	100,697	Hewlett-Packard, Palo Alto, CA...........	47,061	Kmart, Troy, MI	33,674
General Electric, Fairfield, CT	100,469	Sears Roebuck, Hoffman Estates, IL.......	41,322	American Intl. Group, New York, NY	33,296
IBM, Armonk, NY	81,667	E. I. Du Pont de Nemours, Wilmington, DE	39,130	Chase Manhattan Corp., New York, NY	32,379
Citigroup, New York, NY ..	76,431	Procter & Gamble, Cincinnati, OH	37,154	Texaco, White Plains, NY	31,707
Philip Morris, New York, NY ..	57,813			Bell Atlantic, New York, NY ...	31,566
Boeing, Seattle, WA	56,154				
AT&T, New York, NY........	53,588				

Fastest-Growing U.S. Franchises in 1998[1]

Source: Entrepreneur Magazine, Jan. 1999

Company	Business	Minimum start-up cost[2]	Company	Business	Minimum start-up cost[2]
Yogen Früz Worldwide	frozen yogurt	$25,000	TCBY Treats	frozen yogurt	$113,000
McDonald's	hamburgers, chicken, salads	413,100	Mail Boxes Etc.	postal & business services	115,000
Jani-King	commercial cleaning	8,400	Re/Max Int'l. Inc.	real estate services	20,000
7-Eleven Convenience Stores	convenience stores	12,500	Miracle Ear Hearing Systems	misc. health-care businesses	82,800
Jackson Hewitt Tax Service	tax services	54,000	CD Warehouse Inc.	miscellaneous retail businesses	123,300
KFC	chicken	1,100,000			
Subway	submarine sandwiches	61,900	Great Clips Inc.	hair care	86,900
Century Small Business Solutions	accounting & tax services	20,400	The Quizno's Corp.	submarine sandwiches	162,200
Wendy's Int'l. Inc.	hamburgers, salads, chicken	845,500	Super 8 Motels Inc.	hotels & motels	240,600
			Jazzercise Inc.	fitness centers	1,500
CleanNet USA Inc.	commercial cleaning	3,900	Tim Hortons	donuts	396,200
Taco Bell Corp.	Mexican fast food	236,400	Christmas Decor Inc.	misc. decorative products & services	17,600
Coldwell Banker Real Estate Corp.	real estate services	21,000	Futurekids Inc.	computer training	70,000
Blimpie Int'l. Inc.	submarine sandwiches	103,000	PostNet Postal & Business Centers	postal & business services	84,200
Curves for Women	fitness centers	20,400	Schlotzsky's Deli	misc. sandwiches	1,300,000
Papa John's Pizza	pizza	152,000	Orion Food Systems Inc.	misc. fast food	15,900

(1) Based on the number of new franchise units added. (2) Not including franchise fee, which varies.

U.S. Capital Gains Tax

Source: George W. Smith III, CPA, Nationally Sydicated Tax Author and Columnist; as of Oct. 1999

The following shows how the maximum tax rate on net long-term capital gains for individuals has changed since 1960.

Year	Max %	Year	Max %	Year	Max %	Year	Max %
1960.............	25.0	1972	35.0[1]	1987................	28.0	1997	20.0[4]
1970.............	29.5	1978	28.0	1988................	33.0[2]	1999	20.0[5]
1971.............	32.5	1981	20.0	1990................	28.0[3]	2000	20/18[6]

(1) From 1972 to 1976, the interplay of minimum tax and maximum tax resulted in a marginal rate of 49.125%. (2) Statutory maximum of 28%, but "phase-out" notch increased marginal rate to 33%; interplay of all "phase-outs" could have increased the effective marginal rate to 49.5%. (3) The Budget Act of 1990 increased the statutory rate to 31% and capped the marginal rate at 28%; however, some taxpayers faced effective marginal rates of more than 34% because of the phase-out of personal exemptions and itemized deductions. (4) New rate is for those who, after July 28, 1997, sell capital assets held for more than 18 mos (12 mos for sales after Dec. 31, 1997). A 10% capital gains rate applies to individuals in the 15% income tax bracket. (Those who, after July 28, 1997, but before Jan. 1, 1998, sell capital assets held between 12 and 18 mos will be taxed at the old top rate of 28%. Those who sold capital assets after May 6, 1997, but before July 29, 1997, will be taxed at the 20% rate, so long as such assets were held for at least a year.) (5) The IRS Restructuring and Reform Act of 1998 repealed the more-than-18-month holding period for sales after Dec. 31, 1997. Beginning Jan. 1, 1998, capital assets need only be held 12 months to have the 20%/10% capital gains rates apply. (6) For capital assets sold after 2000 and held for at least 5 years, a top rate of 18% may apply (8% for those in the 15% income tax bracket).

1999 Federal Corporate Tax Rates

Taxable Income Amount	Tax Rate	Taxable Income Amount	Tax Rate
Not more than $50,000	15%	$335,001 to $10,000,000	34%
$50,001 to $75,000	25%	$10,000,001 to $15,000,000	35%
$75,001 to $100,000	34%	$15,000,001 to $18,333,333	38%
$100,001 to $335,000	39%	More than $18,333,333	35%

Personal service corporations (used by incorporated professionals such as attorneys and doctors) pay a flat rate of 35%.

Global Stock Markets

Source: The Conference Board; not seasonally adjusted

Stock price indexes (1990=100):	1997	1998	1999 Jan.	Feb.	Mar.	Apr.	May	June
United States	263.3	327.0	384.6	372.2	336.7	401.3	391.3	412.6
Japan	63.5	53.0	50.3	49.8	54.9	57.9	55.9	60.8
Germany	178.7	299.1	305.1	290.4	288.8	318.9	299.8	318.0
France	152.8	206.0	233.9	225.2	231.0	242.4	239.4	249.6
United Kingdom	206.6	243.7	249.0	261.0	267.4	279.8	266.9	272.2
Italy	134.3	214.9	231.7	234.1	245.4	245.9	237.1	237.5
Canada	188.8	197.5	196.7	184.5	192.9	205.0	200.0	204.9

Foreign Exchange Rates, 1970-98

Source: International Monetary Fund

(National currency units per dollar except as indicated; data are annual averages)

Year	Australia[1] (dollar)	Austria (schilling)	Belgium (franc)	Canada (dollar)	Denmark (krone)	France (franc)	Germany[2] (deutsche mark)	Greece (drachma)
1970	1.1136	25.880	49.680	1.0103	7.489	5.5200	3.6480	30.00
1975	1.3077	17.443	36.799	1.0175	5.748	4.2876	2.4613	32.29
1980	1.1400	12.945	29.237	1.1693	5.634	4.2250	1.8175	42.62
1985	0.7003	20.690	59.378	1.3655	10.596	8.9852	2.9440	138.12
1990	0.7813	11.370	33.418	1.1668	6.189	5.4453	1.6157	158.51
1991	0.7791	11.676	34.148	1.1457	6.396	5.6421	1.6595	182.27
1992	0.7353	10.989	32.150	1.2087	6.036	5.2938	1.5617	190.62
1993	0.6801	11.632	34.597	1.2901	6.484	5.6632	1.6533	229.25
1994	0.7317	11.422	33.456	1.3656	6.361	5.5520	1.6228	242.60
1995	0.7415	10.081	29.480	1.3724	5.602	4.9915	1.4331	231.66
1996	0.7829	10.587	30.962	1.3635	5.799	5.1155	1.5048	240.71
1997	0.7441	12.204	35.774	1.3846	6.604	5.8367	1.7341	273.06
1998	0.6294	12.379	36.299	1.4835	6.701	5.8995	1.7597	295.53

Year	India (rupee)	Ireland[1] (pound)	Italy (lira)	Japan (yen)	Malaysia (ringgit)	Mexico (new peso)	Netherlands (guilder)	Norway (krone)
1970	7.576	2.3959	623	357.60	3.0900	—	3.5970	7.1400
1975	8.409	2.2216	653	296.78	2.4030	—	2.5293	5.2282
1980	7.887	2.0577	856	226.63	2.1767	—	1.9875	4.9381
1985	12.369	1.0656	1,909	238.54	2.4830	—	3.3214	8.5972
1990	17.504	1.6585	1,198	144.79	2.7049	2.8126	1.8209	6.2597
1991	22.742	1.6155	1,241	134.71	2.7501	3.0184	1.8697	6.4829
1992	25.918	1.7053	1,232	126.65	2.5474	3.0949	1.7585	6.2145
1993	30.493	1.4671	1,574	111.20	2.5741	3.1156	1.8573	7.0941
1994	31.374	1.4978	1,612	102.21	2.6243	3.3751	1.8200	7.0576
1995	32.427	1.6038	1,629	94.06	2.5044	6.4194	1.6057	6.3352
1996	35.433	1.6006	1,543	108.78	2.5159	7.6009	1.6859	6.4498
1997	36.313	1.5180	1,703	120.99	2.8132	7.9141	1.9513	7.0734
1998	41.259	1.4257	1,736	130.91	3.9244	9.1360	1.9837	7.5451

Year	Portugal (escudo)	Singapore (dollar)	South Korea (won)	Spain (peseta)	Sweden (krona)	Switzerland (franc)	Thailand (baht)	United Kingdom[1] (pound)
1970	28.75	3.0800	310.57	69.72	5.1700	4.3160	21.000	2.3959
1975	25.51	2.3713	484.00	57.43	4.1530	2.5839	20.379	2.2216
1980	50.08	2.1412	607.43	71.76	4.2309	1.6772	20.476	2.3243
1985	170.39	2.2002	870.02	170.04	8.6039	2.4571	27.159	1.2963
1990	142.55	1.8125	707.76	101.93	5.9188	1.3892	25.585	1.7847
1991	144.48	1.7276	733.35	103.91	6.0475	1.4340	25.517	1.7694
1992	135.00	1.6290	780.65	102.38	5.8238	1.4062	25.400	1.7655
1993	160.80	1.6158	802.67	127.26	7.7834	1.4776	25.319	1.5020
1994	165.99	1.5274	803.45	133.96	7.7160	1.3677	25.150	1.5316
1995	151.11	1.4174	771.27	124.69	7.1333	1.1825	24.915	1.5785
1996	154.24	1.4100	804.45	126.66	6.7060	1.2360	25.343	1.5617
1997	175.31	1.4848	951.29	146.41	7.6349	1.4513	31.364	1.6377
1998	180.10	1.6736	1,401.44	149.40	7.9499	1.4498	41.359	1.6564

(1) Value of one unit of foreign currency in dollars. (2) West Germany prior to 1991.

Foreign Direct Investment[1] in the U.S. by Selected Countries and Territories

Source: Bureau of Economic Analysis; U.S. Dept. of Commerce

(millions of dollars)

	1997	1998		1997	1998
ALL COUNTRIES[2].	$681,651	$811,756	Mexico.	$1,723	$4,029
Canada	64,022	74,840	Panama.	6,645	7,025
Europe[3].	425,220	539,906	Other Western Hemisphere[3].	25,652	20,294
Austria	1,831	4,872	Bahamas.	1,986	2,141
Belgium	6,771	9,577	Bermuda.	3,423	2,674
Denmark.	3,025	3,229	Netherlands Antilles	7,701	4,727
Finland	3,089	4,321	UK islands, Caribbean region	11,954	10,395
France	47,088	62,167	Africa[3].	1,608	884
Germany	69,701	95,045	Middle East[3].	6,882	7,831
Ireland	10,514	13,227	Israel	2,292	2,459
Italy.	3,318	3,830	Kuwait	2,881	NA
Luxembourg	6,218	20,214	Saudi Arabia	1,573	NA
Netherlands	84,862	96,904	Asia and Pacific[3].	148,218	156,085
Norway.	3,971	3,616	Australia.	16,229	14,755
Spain	2,643	2,292	Hong Kong.	1,757	2,097
Sweden	13,147	14,564	Japan.	123,514	132,569
Switzerland.	38,574	54,011	Malaysia	465	89
United Kingdom	129,551	151,335	Singapore	2,776	1,813
South and Central America[3]	10,049	11,916	South Korea.	−327	285
Brazil	698	609	Taiwan.	2,778	3,120

(1) The book value of foreign direct investors' equity in, and net outstanding loans to, their U.S. affiliates. A U.S. affiliate is a U.S. business enterprise in which a single foreign direct investor owns at least 10% of the voting securities or the equivalent. (2) Total includes sources not reflected in regional subtotals. (3) Totals include countries or territories not shown.

U.S. Direct Investment[1] Abroad in Selected Countries and Territories

Source: Bureau of Economic Analysis, U.S. Dept. of Commerce

(millions of dollars)

	1990	1997	1998		1990	1997	1998
ALL COUNTRIES[2].	$424,086	$860,723	$980,565	Mexico.	$9,398	$25,395	$25,877
Canada	67,033	99,859	103,908	Panama.	7,409	20,958	26,957
Europe.	211,194	420,934	489,539	Other Western Hemisphere[3].	30,113	56,489	66,978
Austria	889	2,621	3,838	Bahamas.	3,309	1,515	287
Belgium	9,050	17,403	18,920	Barbados.	NA	801	1,077
Denmark.	1,597	2,576	2,628	Bermuda.	21,737	33,092	41,076
Finland	551	1,338	1,700	Dominican Republic	NA	476	535
France	18,874	34,615	39,188	Jamaica.	604	1,687	2,105
Germany	27,259	43,931	42,853	Netherlands Antilles	−2,229	5,393	4,472
Greece	288	638	660	Trinidad and Tobago	508	602	1,054
Ireland	6,880	14,476	15,936	UK islands, Caribbean			
Italy.	13,117	17,749	14,638	region	4,800	12,143	15,713
Luxembourg	1,390	9,796	14,930	Africa[3].	4,861	10,253	13,491
Netherlands	22,658	64,648	79,386	Egypt.	1,465	1,570	1,955
Norway.	3,815	6,262	7,609	Nigeria.	161	1,465	1,925
Portugal	598	1,498	1,474	South Africa.	956	2,347	2,363
Spain	7,704	11,642	12,807	Middle East[3].	3,973	8,959	10,599
Sweden	1,600	7,299	6,053	Israel	756	2,286	3,067
Switzerland.	25,199	35,203	37,616	Saudi Arabia	1,981	3,079	4,209
Turkey.	494	1,076	1,069	United Arab Emirates	519	682	710
United Kingdom	68,224	138,765	178,648	Asia and Pacific[3].	61,869	142,704	161,797
Other	NA	7,743	9,588	Australia.	14,846	26,125	33,676
South America[3]	23,760	67,112	73,290	China.	NA	5,013	6,348
Argentina	2,956	9,766	11,489	Hong Kong.	6,187	19,065	20,802
Brazil	14,918	35,727	37,802	India.	513	1,684	1,480
Chile.	1,368	7,767	9,132	Indonesia.	3,226	7,395	6,932
Colombia	1,728	3,727	4,317	Japan.	20,997	35,569	38,153
Ecuador	387	1,175	952	Malaysia	1,384	5,623	6,193
Peru	410	2,595	2,587	New Zealand	3,131	5,191	6,136
Venezuela	1,490	5,176	5,697	Philippines.	1,629	3,403	3,192
Central America[3]	17,719	48,881	56,387	Singapore	3,385	17,514	19,783
Costa Rica	NA	1,580	2,126	South Korea.	2,178	6,528	7,365
Guatemala	NA	357	429	Taiwan.	2,014	4,944	4,937
Honduras	NA	183	186	Thailand.	1,585	3,537	5,721

(1) The book value of U.S. direct investors' equity in, and net outstanding loans to, their foreign affiliates. A foreign affiliate is a foreign business enterprise in which a single U.S. investor owns at least 10% of the voting securities or the equivalent. (2) Total includes countries not reflected in regional totals. (3) Total includes countries not shown. NA = not available.

U.S. Holdings of Foreign Stocks

Source: Bureau of Economic Analysis, U.S. Dept. Of Commerce

(billions of dollars)

	1996[R]	1997[R]	1998		1996[R]	1997[R]	1998
Western Europe	$468.8	$714.1	$959.8	Canada.	$67.0	$70.7	$61.6
Of which: Switzerland.	33.9	61.3	72.2	Japan.	126.4	134.8	123.1
Netherlands	64.8	106.2	135.0	Latin America.	76.8	88.9	51.6
France	42.8	84.2	110.1	Of which: Mexico.	22.1	34.8	21.7
Germany	40.5	64.5	82.3	Other countries and territories	137.8	192.5	211.0
Sweden	34.2	38.3	43.2	Of which: Hong Kong.	37.3	27.9	26.7
United Kingdom	185.4	217.4	289.5	Australia	26.1	31.1	35.9
Spain	22.8	24.0	35.3	TOTAL HOLDINGS.	876.8	1,201.0	1,407.1

(R) Revised figures.

U.S. International Transactions

Source: Bureau of Economic Analysis, U.S. Dept. of Commerce; revised as of July 1997

(millions of dollars)

	1965	1970	1975	1980	1985	1990	1995	1997[R]	1998
Exports of goods, services, and income[1]	$42,722	$68,387	$157,936	$344,440	$382,749	$700,455	$991,490	$1,197,206	$1,192,231
Merchandise adjusted, excluding military[2]	26,461	42,469	107,088	224,250	215,915	389,307	575,871	679,715	670,246
Services	8,824	14,171	25,497	47,584	73,155	147,824	218,739	258,828	263,661
Income receipts on U.S. assets abroad	7,437	11,748	25,351	72,606	93,679	163,324	196,880	258,663	258,324
Imports of goods, services, and income	−32,708	−59,901	−132,745	−333,774	−484,037	−757,758	−1,086,539	−1,298,705	−1,368,718
Merchandise adjusted, excluding military[2]	−21,510	−39,866	−98,185	−249,750	−338,088	−498,337	−749,431	−876,366	−917,178
Services	−9,111	−14,520	−21,996	−41,491	−72,862	−120,019	−147,036	−166,907	−181,011
Income payments on foreign assets in the U.S.	−2,088	−5,515	−12,564	−42,532	−73,087	−139,402	−190,072	−255,432	−270,529
Unilateral transfers, net	−4,583	−6,156	−7,075	−8,349	−22,700	−34,588	−34,046	−41,966	−44,075
Capital account transactions, net	N/A	N/A	N/A	N/A	N/A	N/A	N/A	292	617
U.S. assets abroad, net (increase/ capital outflow [–])	−5,716	−9,337	−39,703	−86,967	−39,889	−74,011	−307,207	−465,296	−292,818
U.S. official reserve assets, net	1,225	2,481	−849	−8,155	−3,858	−2,158	−9,742	−1,010	−6,784
U.S. government assets, other than official reserve assets, net	−1,605	−1,589	−3,474	−5,162	−2,821	2,307	−549	68	−429
U.S. private assets, net	−5,336	−10,229	−35,380	−73,651	−33,211	−74,160	−296,916	−464,354	−285,605
Foreign assets in the U.S., net (increase/capital inflow [+])	742	6,359	17,170	62,612	146,383	140,992	451,234	751,661	502,637
Statistical discrepancy (sum of above items with sign reversed)	−457	−219	4,417	20,886	17,494	24,911	−14,931	−143,192	10,126
Memorandum:									
Balance on current account	5,431	2,331	18,116	2,317	−123,987	−91,892	−129,095	−143,465	−220,562

R = Revised. (1) Excludes transfers of goods and services under U.S. military grant programs. (2) Excludes exports of goods under U.S. military agency sales contracts identified in Census export documents, excludes imports of goods under direct defense expenditures identified in Census import documents, and reflects various other adjustments.

Gold Reserves of Central Banks and Governments

Source: International Financial Statistics, IMF; million fine troy ounces

Year end	All countries[1]	United States	Belgium	Canada	France	Germany[2]	Italy	Japan	Nether- lands	Switzer- land	United Kingdom
1975	1,018.71	274.71	42.17	21.95	100.93	117.61	82.48	21.11	54.33	83.20	21.03
1980	952.99	264.32	34.18	20.98	81.85	95.18	66.67	24.23	43.94	83.28	18.84
1985	949.39	262.65	34.18	20.11	81.85	95.18	66.67	24.33	43.94	83.28	19.03
1986	949.11	262.04	34.18	19.72	81.85	95.18	66.67	24.23	43.94	83.28	19.01
1987	944.49	262.38	33.63	18.52	81.85	95.18	66.67	24.23	43.94	83.28	19.01
1988	946.65	261.87	33.67	17.14	81.85	95.18	66.67	24.23	43.94	83.28	19.00
1989	941.04	261.93	30.23	16.10	81.85	95.18	66.67	24.23	43.94	83.28	18.99
1990	939.01	261.91	30.23	14.76	81.85	95.18	66.67	24.23	43.94	83.28	18.94
1991	938.01	261.91	30.23	12.96	81.85	95.18	66.67	24.23	43.94	83.28	18.89
1992	927.55	261.84	25.04	9.94	81.85	95.18	66.67	24.23	43.94	83.28	18.61
1993	922.02	261.79	25.04	6.05	81.85	95.18	66.67	24.23	35.05	83.28	18.45
1994	917.98	261.73	25.04	3.89	81.85	95.18	66.67	24.23	34.77	83.28	18.44
1995	908.79	261.70	20.54	3.41	81.85	95.18	66.67	24.23	34.77	83.28	18.43
1996	906.10	261.66	15.32	3.09	81.85	95.18	66.67	24.23	34.77	83.28	18.43
1997	890.57	261.64	15.32	3.09	81.89	95.18	66.67	24.23	27.07	83.28	18.42
1998	966.26	261.61	9.52	2.49	102.37	118.98	83.36	24.23	33.83	83.28	23.00

(1) Covers IMF members with reported gold holdings. For countries not listed above, see *International Financial Statistics.* (2) West Germany prior to 1991.

U.S. National Income by Industry[1]

Source: Bureau of Economic Analysis, U.S. Dept. of Commerce

(billions of dollars)

	1960	1970	1980	1990	1996	1997	1998
National income without capital consumption adjustment	$428.6	$835.1	$2,263.9	$4,513.6	$6,212.7	$6,598.0	$6,928.6
Domestic industries	425.1	827.8	2,216.3	4,492.0	6,200.3	6,606.0	6,949.3
Private industries	371.6	695.4	1,894.5	3,830.2	5,351.8	5,728.5	6,043.0
Agriculture, forestry, fisheries	17.8	25.9	61.4	98.0	106.4	106.0	104.2
Mining	5.6	8.4	43.8	36.8	47.9	52.5	50.6
Construction	22.5	47.4	126.6	222.0	289.2	305.1	331.1
Manufacturing	125.3	215.6	532.1	859.5	1,085.9	1,151.0	1,168.7
Durable goods	73.4	127.7	313.7	483.1	617.9	659.4	684.2
Nondurable goods	52.0	87.9	218.4	376.3	468.0	491.6	484.4
Transportation, public utilities	35.8	64.4	177.3	326.3	464.7	480.9	500.8
Transportation	18.5	31.5	85.8	139.2	195.0	208.0	216.2
Communications	8.2	17.6	48.1	91.6	137.0	139.3	149.3
Electric, gas, sanitary services	9.1	86.8	43.4	95.5	132.7	133.6	135.3
Wholesale trade	25.0	47.5	143.3	261.7	350.9	384.2	409.2
Retail trade	41.3	79.9	189.4	392.3	509.6	543.2	580.0
Finance, insurance, real estate	51.3	96.4	279.5	684.2	1,089.2	1,192.0	1,273.5
Services	46.9	109.8	341.0	949.4	1,407.9	1,513.6	1,624.9
Government	53.5	132.4	321.8	661.1	848.5	877.5	906.3

(1) Figures may not add because of rounding. Total national income also includes income from outside the U.S.

U.S. National Income by Type of Income[1]
Source: Bureau of Economic Analysis, U.S. Dept. of Commerce
(billions of dollars)

	1960	1970	1980	1990	1996	1997	1998
NATIONAL INCOME[2]	$424.9	$832.6	$2,203.5	$4,491.0	$6,256.0	$6,646.5	$6,994.7
Compensation of employees	296.7	618.3	1,638.2	3,297.6	4,409.0	4,687.2	4,981.0
Wages and salaries	272.8	551.5	1,372.0	2,745.0	3,640.4	3,893.6	4,153.9
Government	49.2	117.1	260.1	516.0	640.9	664.2	689.3
Other	223.7	434.3	1,111.8	2,229.0	2,999.5	3,229.4	3,464.6
Supplements to wages and salaries	23.8	66.8	266.3	552.5	768.6	793.7	827.1
Employer contrib. for social ins.	12.6	34.3	127.9	278.3	381.7	400.7	420.1
Other labor income	11.2	32.5	138.4	274.3	387.0	392.9	406.9
Proprietors' income with adjustments	52.1	80.2	180.7	363.3	527.7	551.2	577.2
Farm	11.6	14.7	20.5	41.9	38.9	35.5	28.7
Nonfarm	40.5	65.4	160.1	321.4	488.8	515.8	548.5
Rental income of persons, with capital consumption adjustment	15.3	18.2	6.6	−14.2	150.2	158.2	162.6
Corp. profits with inventory adjustment	49.8	69.5	194.0	354.7	679.0	741.2	732.3
Corp. profits before tax	49.9	76.0	237.1	365.7	680.2	734.4	717.8
Corp. profits tax liability	22.7	34.4	84.8	138.7	226.1	246.1	240.1
Corp. profits after tax	27.2	41.7	152.3	227.1	454.1	488.3	477.7
Dividends	12.9	22.5	54.7	153.5	261.9	275.1	279.2
Undistributed profits	14.3	19.2	97.6	73.6	192.3	213.2	198.5
Inventory valuation adjustment	−0.2	−6.6	−43.1	−11.0	−1.2	6.9	14.5
Net interest	11.3	41.2	200.9	463.7	418.6	432.0	449.3

(1) Figures do not add, because of rounding and incomplete enumeration. (2) National income is the aggregate of labor and property earnings that arises in the production of goods and services. It is the sum of employee compensation, proprietors' income, rental income, adjusted corporate profits, and net interest. It measures the total factor costs of goods and services produced by the economy. Income is measured before deduction of taxes. Total national income figures include adjustments not itemized.

Distribution of U.S. Total Personal Income[1]
Source: Bureau of Economic Analysis, U.S. Dept. of Commerce
(billions of dollars)

Year	Personal income	Personal taxes	Disposable personal income	Personal outlays	Personal Savings Amount	Personal Savings As pct. of disposable income
1960	$411.7	$48.7	$362.9	$339.6	$23.3	6.4%
1965	555.8	61.9	493.9	456.2	37.8	7.6
1970	836.1	109.0	727.1	666.1	61.0	8.4
1975	1,315.6	156.4	1,159.2	1,054.8	104.4	9.0
1980	2,285.7	312.4	1,973.3	1,811.5	161.8	8.2
1985	3,439.6	437.7	3,002.0	2,795.8	206.2	6.9
1990	4,791.6	624.8	4,166.8	3,958.1	208.7	5.0
1991	4,968.5	624.8	4,343.7	4,097.4	246.4	5.7
1992	5,264.2	650.5	4,613.7	4,341.0	272.6	5.9
1993	5,480.1	689.9	4,790.2	4,575.8	214.4	4.5
1994	5,757.9	739.1	5,018.9	4,842.1	176.8	3.5
1995	6,072.1	795.0	5,277.0	5,097.2	179.8	3.4
1996	6,425.2	890.5	5,534.7	5,376.2	158.5	2.9
1997	6,784.0	989.0	5,795.1	5,674.1	121.0	2.1
1998	7,126.1	1,098.3	6,027.9	6,000.2	27.7	0.5

(1) Personal income minus taxes=disposable income minus outlays=savings. Figures may not add because of rounding.

Record One-Day Gains and Losses on the Dow Jones Industrial Average
Source: Dow Jones & Co., Inc.; as of Oct. 15, 1999

GREATEST POINT GAINS

Rank	Date	Close	Net Chg	% Chg
1.	9/8/98	8,020.78	380.53	4.98
2.	10/28/97	7,498.32	337.17	4.71
3.	10/16/98	8,299.36	330.58	4.15
4.	9/1/98	7,827.43	288.36	3.82
5.	3/5/99	9,736.08	268.68	2.84
6.	9/2/97	7,879.78	257.36	3.38
7.	9/23/98	8,154.41	257.21	3.26
8.	9/3/99	11,078.45	235.24	2.17
9.	1/6/99	9,544.97	233.78	2.51
10.	11/3/97	7,674.39	232.31	3.12

GREATEST % GAINS

Rank	Date	Close	Net Chg	% Chg
1.	10/6/31	99.34	12.86	14.87
2.	10/30/29	258.47	28.40	12.34
3.	9/21/32	75.16	7.67	11.36
4.	10/21/87	2,027.85	186.84	10.15
5.	8/3/32	58.22	5.06	9.52
6.	2/11/32	78.60	6.80	9.47
7.	11/14/29	217.28	18.59	9.36
8.	12/18/31	80.69	6.90	9.35
9.	2/13/32	85.82	7.22	9.19
10.	5/6/32	59.01	4.91	9.08

GREATEST POINT LOSSES

Rank	Date	Close	Net Chg	% Chg
1.	10/27/97	7,161.15	−554.26	−7.19
2.	8/31/98	7,539.07	−512.61	−6.37
3.	10/19/87	1,738.74	−508.00	−22.61
4.	8/27/98	8,165.99	−357.36	−4.19
5.	8/4/98	8,487.31	−299.43	−3.40
6.	9/10/98	7,615.54	−249.48	−3.17
7.	8/15/97	7,694.66	−247.37	−3.11
8.	9/30/98	7,842.62	−237.90	−2.94
9.	5/27/99	10,466.93	−235.23	−2.20
10.	1/14/99	9,120.93	−228.63	−2.45

GREATEST % LOSSES

Rank	Date	Close	Net Chg	% Chg
1.	10/19/87	1,738.74	−508.00	−22.61
2.	10/28/29	260.54	−38.33	−12.82
3.	10/29/29	230.07	−30.57	−11.73
4.	11/6/29	232.13	−25.55	−9.92
5.	12/18/1899	58.27	−5.57	−8.72
6.	8/12/32	63.11	−5.79	−8.40
7.	3/14/07	76.23	−6.89	−8.29
8.	10/25/87	1,793.93	−156.83	−8.04
9.	7/21/33	88.71	−7.55	−7.84
10.	10/18/37	125.73	−10.57	−7.75

Dow Jones Industrial Average Since 1962

High		YEAR	Low		High		YEAR	Low	
Jan. 3	726.01	**1962**	June 26	535.76	Apr. 27	1024.05	**1981**	Sept. 25	824.01
Dec. 18	767.21	**1963**	Jan. 2	646.79	Dec. 27	1070.55	**1982**	Aug. 12	776.92
Nov. 18	891.71	**1964**	Jan. 2	766.08	Nov. 29	1287.20	**1983**	Jan. 3	1027.04
Dec. 31	969.26	**1965**	June 28	840.59	Jan. 6	1286.64	**1984**	July 24	1086.57
Feb. 9	995.15	**1966**	Oct. 7	744.32	Dec. 16	1553.10	**1985**	Jan. 4	1184.96
Sept. 25	943.08	**1967**	Jan. 3	786.41	Dec. 2	1955.57	**1986**	Jan. 22	1502.29
Dec. 3	985.21	**1968**	Mar. 21	825.13	Aug. 25	2722.42	**1987**	Oct. 19	1738.74
May 14	968.85	**1969**	Dec. 17	769.93	Oct. 21	2183.50	**1988**	Jan. 20	1879.14
Dec. 29	842.00	**1970**	May 6	631.16	Oct. 9	2791.41	**1989**	Jan. 3	2144.64
Apr. 28	950.82	**1971**	Nov. 23	797.97	July 16	2999.75	**1990**	Oct. 11	2365.10
Dec. 11	1036.27	**1972**	Jan. 26	889.15	Dec. 31	3168.83	**1991**	Jan. 9	2470.30
Jan. 11	1051.70	**1973**	Dec. 5	788.31	June 1	3413.21	**1992**	Oct. 9	3136.58
Mar. 13	891.66	**1974**	Dec. 6	577.60	Dec. 29	3794.33	**1993**	Jan. 20	3241.95
July 15	881.81	**1975**	Jan. 2	632.04	Jan. 31	3978.36	**1994**	Apr. 4	3593.35
Sept. 21	1014.79	**1976**	Jan. 2	858.71	Dec. 13	5216.47	**1995**	Jan. 30	3832.08
Jan. 3	999.75	**1977**	Nov. 2	800.85	Dec. 27	6560.91	**1996**	Jan. 10	5032.94
Sept. 8	907.74	**1978**	Feb. 28	742.12	Aug. 6	8259.31	**1997**	Apr. 11	6391.69
Oct. 5	897.61	**1979**	Nov. 7	796.67	Nov. 23	9374.27	**1998**	Aug. 31	7539.07
Nov. 20	1000.17	**1980**	Apr. 21	759.13	Aug. 25	11326.04	**1999***	Jan. 22	9120.67

*As of Oct.15.

Milestones of the Dow Jones Industrial Average

(as of Oct. 15, 1999)

First close over...

100	Jan. 12, 1906
500	Mar. 12, 1956
1,000	Nov. 14, 1972
1,500	Dec. 11, 1985
2,000	Jan. 8, 1987
2,500	July 17, 1987
3,000	April 17, 1991
3,500	May 19, 1993
4,000	Feb. 23, 1995
4,500	June 16, 1995
5,000	Nov. 21, 1995
5,500	Feb. 8, 1996
6,000	Oct. 14, 1996
6,500	Nov. 25, 1996
7,000	Feb. 13, 1997
7,500	June 10, 1997

First close over...

8,000	July 16, 1997
8,100	July 24, 1997
8,200	July 30, 1997
8,300	Feb. 12, 1998
8,400	Feb. 18, 1998
8,500	Feb. 27, 1998
8,600	Mar. 10, 1998
8,700	Mar. 16, 1998
8,800	Mar. 19, 1998
8,900	Mar. 20, 1998
9,000	Apr. 6, 1998
9,100	Apr. 14, 1998
9,200	May 13, 1998
9,300	July 16, 1998
9,500	Jan. 6, 1999*
9,600	Jan. 8, 1999

First close over...

9,700	Mar. 5, 1999
9,800	Mar. 11, 1999
9,900	Mar. 15, 1999
10,000	Mar. 29, 1999
10,100	Apr. 8, 1999
10,300	Apr. 12, 1999*
10,400	Apr. 14, 1999
10,500	Apr. 21, 1999
10,700	Apr. 22, 1999*
10,800	Apr. 27, 1999
11,000	May 3, 1999*
11,100	May 13, 1999
11,200	July 12, 1999
11,300	Aug. 25, 1999

*9,400, 10,200, 10,600, and 10,900 are not listed because the Dow had risen another 100 points or more by the time the market closed for the day.

Components of the Dow Jones Averages

(as of Nov. 1, 1999)

Dow Jones Industrial Average

AlliedSignal	Exxon	Merck
Aluminum Co. of America (Alcoa)	General Electric	Microsoft*
American Express	General Motors	Minnesota Mining & Manufacturing
AT&T	Hewlett-Packard	Morgan (J.P.)
Boeing	Home Depot*	Philip Morris
Caterpillar	IBM	Procter & Gamble
Citigroup	Intel*	SBC Communications*
Coca-Cola	International Paper	United Technologies
DuPont	Johnson & Johnson	Wal-Mart
Eastman Kodak	McDonald's	Walt Disney

*These companies became component stocks of the DJIA Nov. 1, 1999, replacing Chevron; Goodyear Tire & Rubber; Sears, Roebuck; and Union Carbide. The inclusion of Intel and Microsoft, both traded on the Nasdaq stock market, marks the first time a DJIA component has not been listed on the NYSE since the Dow's inception in 1896.

Dow Jones Transportation Average

Airborne Freight	FDX	Southwest Air Lines
Alexander & Baldwin	GATX	UAL (United Air Lines)
AMR (American Airlines)	J.B. Hunt Transportation	Union Pacific
Burlington Northern Santa Fe	Norfolk Southern	US Airways
CNF Transportation	Northwest Airlines	USFreightways
CSX	Roadway Express	Yellow Corp.
Delta Air Lines	Ryder System	

Dow Jones Utility Average

American Electric Power	Edison International	Public Service Enterprise Group
Columbia Energy Group	Enron	Southern Co.
Consolidated Edison	Reliant Energy	Texas Utilities
Consolidated Natural Gas	PECO	Unicom
Duke Energy	PG&E	Williams Cos.

Most Active Common Stocks in 1998

New York Exchange	Volume (millions of shares)	American Exchange	Volume (millions of shares)	NASDAQ	Volume (millions of shares)
Compaq Computer Corp.	2,868.4	Nabors Industries, Inc.	248.9	Dell Computer Corp.	8,524.6
Philip Morris Cos., Inc.	1,457.4	Viacom Inc. Class B	233.2	Intel Corp.	4,411.1
Cendant Corp.	1,436.1	Trans World Airlines	226.0	Cisco Systems, Inc.	4,072.6
Lucent Technologies Inc.	1,167.1	Grey Wolf, Inc.	179.9	Microsoft Corp.	3,837.6
AT&T Corp.	1,165.7	First Australia Prime Income, Inc.	175.2	MCI Worldcom, Inc.	2,893.3
America Online.	1,159.2	Echo Bay Mines Ltd.	165.6	Amazon.com, Inc.	2,638.2
General Electric Company	1,031.8	Royal Oak Mines Inc.	165.0	Oracle Corp.	2,305.9
Micron Technology	911.0	Hasbro, Inc.	127.5	Applied Materials, Inc.	1,939.9
IBM.	905.9	Keane, Inc.	120.7	3Com Corporation	1,904.0
Boeing Co.	866.5	Harken Energy Corporation	114.0	Yahoo! Inc.	1,818.0
PepsiCo, Inc.	852.0	Berma Gold Corporation	91.0	Ascend Communications, Inc.	1,620.8
Travelers Group Inc.	842.2	Forest Laboratories, Inc.	89.6	Sun Microsystems, Inc.	1,567.8
Pfizer Inc.	834.4	PLC Systems, Inc.	80.4	LM Ericsson Telephone Company	1,432.0
Telecomm Brasil Telebras	828.8	IVAX Corporation	78.0	HBO & Company.	1,363.6
Chase Manhattan Corp.	793.2	Metromedia International Group, Inc.	77.7	Netscape Communications Corp.	1,324.9

Average Yields of Long-Term Treasury, Corporate, and Municipal Bonds

Source: Office of Market Finance, U.S. Dept. of the Treasury

Period	Treasury 30-year bonds	New Aa corporate bonds[1]	New Aa municipal bonds[2]	Period	Treasury 30-year bonds	New Aa corporate bonds[1]	New Aa municipal bonds[2]
1986				**1995**			
June	7.57	9.39	7.75	June	6.57	7.42	5.61
Dec.	7.37	8.87	6.70	Dec.	6.06	7.02	5.46
1987				**1996**			
June	8.57	9.64	7.69	June	7.06	8.00	5.82
Dec.	9.12	10.22	7.83	Dec.	6.55	7.45	5.47
1988				**1997**			
June	9.00	10.08	7.67	June	6.77	7.71	5.39
Dec.	9.01	10.05	7.40	Dec.	5.99	6.68	5.07
1989				**1998**			
June	8.27	9.24	6.94	Jan.	5.81	6.62	4.93
Dec.	7.90	9.23	6.76	Feb.	5.89	6.66	4.96
1990				Mar.	5.95	6.63	5.10
June	8.46	9.69	6.98	Apr	5.92	6.59	5.10
Dec.	8.24	9.55	6.85	May	5.93	6.63	5.17
1991				Jun	5.70	6.43	5.01
June	8.47	9.37	6.90	July	5.68	6.36	5.04
Dec.	7.70	8.55	6.43	Aug.	5.54	6.34	5.06
1992				Sept.	5.20	6.26	4.94
June	7.84	8.45	6.32	Oct	5.01	6.21	4.82
Dec.	7.44	8.12	6.02	Nov.	5.25	6.42	4.93
1993				Dec.	5.06	6.13	4.90
June	6.81	7.48	5.54	Oct	5.01	6.21	4.82
Dec.	6.25	7.22	5.27	**1999**			
1994				Jan.	5.16	6.14	4.94
June	7.40	8.16	5.96	Feb.	5.37	6.33	4.89
Dec.	7.87	8.66	6.63	Mar.	5.58	6.52	5.02

(1) Treasury series based on 3-week moving average of reoffering yields of new corporate bonds rated Aa by Moody's Investors Service with an original maturity of at least 20 years. (2) Index of new reoffering yields on 20-year general obligations rated Aa by Moody's Investors Service.

Performance of Mutual Funds by Type

Source: CDA/Wiesenberger, Rockville, MD, 800-232-2285

(data for period ending Sept. 30, 1999)

FUND TYPE	Fund objective	AVERAGE RETURN 1-Year	AVERAGE RETURN 5-Year	FUND TYPE	Fund objective	AVERAGE RETURN 1-Year	AVERAGE RETURN 5-Year
Stock	Natural resources	23.22%	8.26%	Bond	Corporate-Investment grade	-1.10%	7.08%
	Equity income	13.24%	16.16%		Corporate high yield	3.79%	8.08%
	Financial services	8.70%	19.86%		Convertible	20.34%	12.69%
	Precious metals	8.21%	-11.39%		U.S. Treasury	-2.07%	7.24%
	Growth and income	19.67%	18.17%		U.S. Gov't./agency	-1.60%	6.53%
	Health/biotechnology	10.81%	18.19%		U.S. Gov't./ short & intermediate	0.32%	5.89%
	Aggressive growth	40.35%	16.49%		U.S. Gov't./long	-4.48%	7.43%
	Small caps	26.57%	13.93%		General Bond–short & intermediate	1.02%	6.29%
	Real estate	-3.43%	7.67%		General Bond–long	-1.70%	7.59%
	S&P 500 Index	27.13%	24.46%		General Bond– Investment grade	-0.46%	6.66%
	Technology/comm.	97.69%	31.12%		General mortgage	1.98%	6.33%
	Utilities	14.65%	15.58%		Loan participation	6.33%	7.01%
	Mid-caps	31.64%	16.87%		Multi-sector	0.88%	7.45%
	Domestic growth	28.71%	19.56%	Municipal bond	National	-1.84%	5.42%
International stock	Emerging equity	41.60%	-5.02%		California	-2.39%	5.80%
	Global income	1.12%	6.39%		New York	-2.81%	5.56%
	Global equity	27.76%	12.37%		Single state	-2.50%	5.62%
	Non-U.S. equity	37.93%	6.96%		High yield	-1.84%	6.20%
	Emerging income	23.89%	6.75%		Insured	-2.68%	5.63%
Hybrid	Asset allocation-Domestic	14.29%	14.07%				
	Asset allocation-Global	13.06%	9.26%				
	Balanced-domestic	12.14%	14.04%				
	Balanced-global	14.62%	9.95%				

Chicago Board of Trade, Contracts Traded 1989, 1998

	1989	1998	% change 1989-98		1989	1998	% change 1989-98
FUTURES GROUP				Metals	8,346	154	−98.2
Agricultural	31,280,793	47,561,302	52.0	PCS insurance	—	7,753	—
Financial	79,355,659	165,960,763	109.1	**Total options**	26,255,904	64,050,508	144.0
Stock index	1,100,495	3,567,512	224.2	**COMBINED FUTURES AND OPTIONS**			
Energy	—	272	—	Agricultural	35,473,981	68,749,036	65.6
Metals	358,466	49,079	−86.3	Financial	101,410,029	218,570,232	115.5
Total futures	**112,095,413**	**217,138,938**	**93.7**	Stock index	1,100,495	3,812,910	246.5
OPTIONS GROUP				Energy	—	272	—
Agricultural	4,193,188	11,187,734	166.8	Metals	366,812	49,233	−86.6
Financial	22,054,370	52,609,469	138.5	Insurance	—	—	—
Stock index	—	245,398	—	PCS insurance	—	7,753	—
Energy	—	0	—	**GRAND TOTAL**	**138,351,317**	**281,189,436**	**103.2**

Selected Personal Consumption Expenditures in the U.S., 1991-97

Source: Bureau of Economic Analysis, U.S. Dept. of Commerce
(billions of dollars)

	1991	1992	1993R	1994R	1995R	1996R	1997
Food & tobacco	$693.8	$709.5	$733.4	$761.7	$783.8	$805.2	$832.3
Food purchased for off-premise consumption	419.1	423.3	435.6	451.6	462.2	477.0	494.2
Purchased meals and beverages	223.1	228.6	243.0	254.3	264.1	268.8	277.2
Tobacco products	43.8	49.6	46.6	47.3	48.7	50.2	51.4
Clothing, accessories, jewelry	265.7	283.5	298.1	312.7	323.4	338.0	353.3
Shoes	31.9	33.6	34.4	36.0	36.8	38.5	39.8
Clothing and accessories less shoes	179.3	191.7	201.8	211.6	217.7	226.9	237.9
Jewelry and watches	31.4	33.2	35.6	37.7	39.3	41.4	43.1
Personal care	59.1	63.1	65.1	68.4	71.9	75.0	79.4
Toilet articles, preparations	39.4	41.4	43.1	45.3	47.2	49.7	52.6
Barber shops, beauty parlors, health clubs	19.7	21.8	22.0	23.0	24.7	25.3	26.8
Housing	616.5	646.8	672.8	712.7	750.3	787.4	829.8
Owner-occupied nonfarm dwellings—space rent	434.1	457.8	480.9	507.0	532.2	559.1	590.3
Tenant-occupied nonfarm dwellings—rent	155.8	160.5	162.1	174.0	184.6	193.2	203.2
Rental value of farm dwellings	5.2	5.3	5.5	5.8	5.9	6.1	6.3
Household operation	448.4	470.6	504.1	535.0	562.8	592.8	620.7
Furniture, including bedding	38.4	39.8	42.7	45.9	48.0	50.6	54.8
Kitchen and other household appliances	21.6	22.2	24.0	25.6	27.2	28.5	29.7
China, glassware, tableware, utensils	18.9	20.7	22.0	24.0	25.3	27.0	28.6
Other durable house furnishings	42.3	45.5	48.2	52.3	54.5	57.9	61.8
Semidurable house furnishings	21.6	23.2	25.0	27.2	28.9	30.7	32.8
Household utilities	145.4	148.6	160.3	163.8	168.5	176.6	178.5
Telephone, telegraph	63.5	70.3	74.5	82.6	90.2	97.1	104.2
Medical care	668.7	733.2	785.5	826.1	871.6	912.4	957.3
Drug preparations, sundries	70.9	75.0	78.1	81.6	85.7	91.1	98.1
Physicians	152.1	167.2	172.5	180.0	191.4	198.2	205.2
Dentists	34.7	38.5	40.8	43.9	47.6	49.5	52.6
Hospitals and nursing homes	293.4	320.0	341.1	357.0	375.9	389.8	408.1
Health insurance	37.3	42.7	53.6	55.0	53.6	57.4	58.0
Personal business	318.9	341.7	357.4	370.4	389.1	416.2	459.1
Brokerage charges, investment counseling	25.3	30.4	35.7	36.2	38.8	46.6	54.4
Bank service charges, trust services, safe deposit box	25.7	28.0	30.7	31.6	33.9	37.3	41.5
Legal services	42.9	46.5	47.9	48.8	49.1	53.0	55.9
Funeral, burial expenses	9.4	10.1	10.8	11.1	12.2	13.3	13.8
Transportation	436.8	471.5	504.0	542.2	572.3	611.6	636.4
User-operated transportation	401.4	435.7	465.5	502.6	530.1	567.3	588.3
New autos	75.3	82.1	86.4	91.2	87.1	85.8	86.2
Used autos	32.0	35.5	40.2	44.1	52.4	55.8	57.3
Repair, greasing, washing, parking, storage, rental, leasing	85.2	94.4	102.4	116.4	128.7	143.6	154.9
Gasoline and oil	103.9	106.6	107.6	109.4	114.4	124.5	126.5
Tolls	2.1	2.3	2.5	2.6	2.8	2.8	3.0
Insurance premiums less claims paid	22.6	25.5	26.8	27.5	29.4	31.5	34.4
Purchased local transportation	7.9	8.0	8.4	8.9	9.2	10.0	10.4
Mass transit systems	5.3	5.4	5.6	5.9	6.0	6.5	6.8
Taxicab	2.6	2.6	2.8	3.0	3.2	3.5	3.6
Purchased intercity transportation	27.5	27.9	30.1	30.7	33.0	34.3	37.7
Railway (excl. commutation)	0.8	0.8	0.8	0.7	0.8	0.8	0.8
Bus	1.1	1.1	1.0	1.1	1.3	1.1	1.2
Airline	23.0	23.3	25.4	25.8	27.7	28.5	31.5
Recreation	292.0	310.8	340.2	370.2	402.5	432.3	462.9
Books, maps	16.9	17.7	19.0	20.6	22.1	24.2	25.2
Magazines, newspapers, sheet music	21.9	21.6	22.7	24.5	25.5	27.6	29.1
Nondurable toys and sport supplies	32.8	34.2	36.6	39.7	42.2	45.1	47.8
Wheel goods, sports and photographic equipment, boats, pleasure aircraft	29.5	29.9	32.6	35.6	39.1	42.3	48.1
Video & audio prods., computers, musical instruments	57.3	61.2	68.1	78.5	85.2	92.0	96.5
Flowers, seeds, potted plants	11.3	12.3	12.7	13.4	13.9	14.8	15.9
Admissions to specified spectator amusements	15.7	16.6	18.1	19.0	20.2	21.9	23.3
Motion picture theaters	5.3	5.0	5.2	5.6	6.0	6.2	6.6
Legitimate theater, opera	6.0	6.8	7.8	8.2	8.7	9.3	10.0
Spectator sports	4.5	4.8	5.1	5.2	5.5	6.4	6.7
Clubs, fraternal organizations	9.6	10.3	11.2	11.8	12.7	13.0	13.8
Commercial participant amusements	23.8	27.2	31.5	36.2	41.5	44.7	49.1
Education and research	86.1	93.1	98.5	104.7	112.2	119.7	129.4
Higher education	48.0	52.0	55.5	59.0	62.2	65.7	69.6
Nursery, elementary, and secondary schools	18.0	19.3	20.1	21.4	22.8	23.5	25.7
Religious and welfare activities	104.1	115.6	121.3	131.2	139.8	151.1	157.6
TOTAL personal consumption expenditures	$3,975.1	$4,219.8	$4,459.2	$4,717.0	$4,957.7	$5,215.7	$5,493.7

R = revised figures.

Minerals

Source: U.S. Geological Survey, U.S. Dept. of the Interior; as of mid-1999

Aluminum: the second most abundant metallic element in the earth's crust. Bauxite is the main source of aluminum; convert to aluminum equivalent by multiplying by 0.232. Guinea, Brazil, and Australia have 58% of the world's reserves. Aluminum is used in the U.S. principally in transportation (35%), packaging (25%), and building (15%).

Chromium: about 3/4 of the world's production of chromite, the chief source of chromium, is in India, Kazakhstan, Turkey, and South Africa. The chemical and metallurgical industries use about 90% of the chromite consumed in the world.

Cobalt: used in superalloys for jet engines, chemicals (paint driers, glass and ceramics, catalysts, magnetic coatings, and rechargeable batteries), permanent magnets, and cemented carbides for cutting tools. Australia, Canada, Congo (formerly Zaire), Finland, Norway, Russia, and Zambia account for most of the world cobalt refinery production.

Columbium (niobium): used mostly as an additive in steel-making and in superalloys. Brazil and Canada are the world's leading columbium raw materials (feedstock) producers. There is no U.S. columbium mining industry.

Copper: main uses of copper in the U.S. are in building construction (42%), electrical and electronic products (25%), transportation (13%), industrial machinery and equipment (11%), and consumer and general products (9%). The leading producer is Chile, followed by the U.S., Indonesia, Canada, Australia, Peru, Russia, China, Poland, Mexico, Kazakhstan, and Zambia. Principal mining States are Arizona, Utah, and New Mexico.

Gold: used in the U.S. in jewelry and the arts (55%), electronics and other industries (42%), and dentistry (3%). South Africa has about half of the world's resources; significant quantities also are present in the U.S., Australia, Russia, Uzbekistan, Canada, and Brazil. Gold is mined in nearly all the Western U.S. States and in Alaska.

Iron ore: the source of primary iron for the world's iron and steel industries. Major iron ore producers include Australia, Brazil, China, and the former Soviet Union.

Lead: China, Australia, the U.S., Peru, and Canada are the world's largest producers of lead. Transportation accounts for the major end use in the U.S., with 90% used in batteries, bearings, casting metals, and solders. Other uses include emergency power supply batteries, construction sheeting, sporting ammunition, and power cable coverings. The U.S. produces and consumes about 25% of the world's lead metal, including primary and recycled material.

Manganese: essential to iron and steel production. The U.S., Japan, and Western Europe have exhausted nearly all of their economically minable manganese. South Africa and the former Soviet Union have over 85% of the world's identified resources.

Nickel: vital to the stainless steel industry; used to make superalloys for the chemical and aerospace industries. Leading producers include Russia, Canada, Australia, New Caledonia, and Indonesia.

Platinum-Group Metals: the platinum group consists of 6 related metals: platinum, palladium, rhodium, ruthenium, iridium, and osmium. They commonly occur together in nature and are among the scarcest of the metallic elements. They are consumed in the U.S. by the following industries: automotive, electrical and electronic, chemical, and dental and medical. The automotive, chemical, and petroleum-refining industries use platinum-group metals mainly as catalysts. Russia and South Africa have most of the world's reserves.

Silver: used in the following U.S. industries: photography, electrical and electronic products, sterlingware, electroplated ware, and jewelry. Silver is mined in more than 60 countries. Nevada produces more than 40% of U.S. silver, Idaho 16%.

Tantalum: a refractory metal with unique electrical, chemical, and physical properties; used in the U.S. mostly to produce electronic components, mainly tantalum capacitors. Australia, Brazil, and Canada are the world's leading tantalum raw materials (feedstock) producers. There is no U.S. tantalum mining industry.

Titanium: approximately 95% of consumption is in the form of titanium dioxide, a white pigment in paint, paper, and plastics. As a metal, titanium is used primarily in commercial and military aerospace. Major mining operations are in Australia, Canada, Norway, and South Africa. U.S. mine production is in Florida and Virginia.

Vanadium: used as an alloying element in steel and aerospace titanium alloys, as a catalyst in the production of maleic and phthalic anhydride, and in the production of sulfuric acid. South Africa, Russia, and China are the world's largest producers of vanadium-bearing ores and concentrates.

Zinc: used as a protective coating on steel, as diecastings, as an alloying metal with copper to make brass, and as a component of chemical compounds in rubber and paints. It is mined in 46 countries. China is the leading producer, followed by Canada, Australia, Peru, the U.S., and Mexico. In the U.S., mine production comes mostly from Alaska, Tennessee, New York, and Missouri.

World Mineral Reserve Base

Source: U.S. Geological Survey, U.S. Dept. of the Interior; as of mid-1999

Mineral	Reserve Base[1]
Aluminum	34,000 mil metric tons[2]
Chromium	7,600 mil metric tons
Cobalt	9.5 mil metric tons
Columbium	5.6 mil metric tons
Copper	650 mil metric tons
Gold	77,000 metric tons[3]
Iron ore	300,000 mil metric tons
Lead	140 mil metric tons

Mineral	Reserve Base[1]
Manganese	5,000 mil metric tons
Nickel	140 mil metric tons
Platinum-Group Metals	78 mil kilograms
Silver	420,000 metric tons
Tantalum	24,000 metric tons
Titanium	630 mil metric tons[4]
Vanadium	27 mil metric tons
Zinc	430 mil metric tons

(1) Includes demonstrated reserves that are currently economic or marginally economic, plus some that are currently subeconomic. (2) Bauxite. (3) Excludes China and some other countries for which reliable data were not available. (4) Titanium dioxide (TiO_2) content of ilmenite and rutile.

U.S. Nonfuel Mineral Production—10 Leading States in 1998

Source: U.S. Geological Survey, U.S. Dept. of the Interior

Rank/State	Value (millions)	Percent of U.S. total	Principal minerals, in order of value
1. Nevada	3,100	7.72	Gold, silver, copper, sand & gravel (construction), lime
2. California	2,970	7.41	Sand & gravel (construction), cement, boron minerals, stone (crushed), gold
3. Arizona	2,820	7.04	Copper, sand & gravel (construction), cement, molybdenum, stone (crushed)
4. Georgia	2,140	5.33	Clays, stone (crushed), cement, sand & gravel (construction)
5. Florida	1,960	4.90	Phosphate rock, stone (crushed), cement, sand & gravel (construction), zirconium concentrates
6. Texas	1,920	4.80	Cement, stone (crushed), sand & gravel (construction), magnesium metal, lime
7. Michigan	1,660	4.15	Cement, iron ore, sand & gravel (construction), magnesium compounds, stone (crushed)
8. Minnesota	1,560	3.88	Iron ore, sand & gravel (construction), stone (crushed), stone (dimension), sand & gravel (industrial)
9. Missouri	1,360	3.38	Stone (crushed), cement, lead, lime, zinc
10. Utah	1,300	3.25	Copper, sand & gravel (construction), magnesium metal, gold, cement

U.S. Nonfuel Minerals Production

Source: U.S. Geological Survey, U.S. Dept. of the Interior

Production as measured by mine shipments, sales, or marketable production (including consumption by producers).

		1993	1994	1995	1996	1997	1998
Beryllium (metal equivalent)	metric tons	198	173	202	211	231	243
Copper (recoverable content of ores, etc.)	thousand metric tons	1,800	1,850	1,850	1,920	1,940	1,860
Gold (recoverable content of ores, etc.)	metric tons	331.0	326.2	317.0	326.0	360.0	363.0
Iron ore, usable (includes byproduct material)	million metric tons	55.7	58.5	62.5	62.1	63.0	62.9
Lead (in concentrate)	thousand metric tons	355	363	386	426	448	481
Magnesium metal (primary)	thousand metric tons	132	128	142	133	125	106
Molybdenum (content of ore and concentrate)	metric tons	36,803	46,810	58,000	56,000	58,900	53,300
Nickel (content of ore and concentrate)	metric tons	2,464	—	1,557	1,333	—	—
Silver (recoverable content of ores, etc.)	metric tons	1,640	1,490	1,560	1,570	2,150	2,040E
Zinc (recoverable content of ores, etc.)	thousand metric tons	488	570	614	598	605	742
Asbestos	thousand metric tons	14	10	9	10	7	6
Barite	thousand metric tons	315	583	543	662	692	476
Boron minerals	thousand metric tons	574	550	728	581	604	580
Bromine	million kilograms	177	195	218	227	247	230
Cement (portland, masonry, etc.)	thousand metric tons	73,807	77,948	76,906	79,266	82,582	85,800E
Clays	thousand metric tons	40,700	42,000	43,100	43,100	42,000	43,000
Diatomite	thousand metric tons	599	646	722	729	773	725
Feldspar	thousand metric tons	770	765	880	890	900	820E
Fluorspar	thousand metric tons	56	49	51	8	—	—
Garnet (industrial)	metric tons	55,800	44,700	46,300	60,900	64,900	74,000
Gemstones	million dollars	57.7	50.5	48.7	43.6	25.0	23.0
Gypsum	thousand metric tons	15,800	17,200	16,600	17,500	18,500	19,000
Helium (extracted from natural gas)	million cubic meters	99.3	112.0	101.0	103.0	116.0	118.0
Helium (Grade A sold)	million cubic meters	95.6	100.4	96.1	94.7	107.0	108.0
Iodine	thousand kilograms	1,935	1,630	1,220	1,270	1,320	1,490
Lime	thousand metric tons	16,932	17,393	18,530	19,225	19,678	20,132
Mica (scrap & flake)	thousand metric tons	88	110	108	97	114	87
Peat	thousand metric tons	616	574	648	549	661	676
Perlite (sold and used by producers)	thousand metric tons	569	644	700	684	706	685
Phosphate rock (marketable product)	thousand metric tons	35,494	41,115	43,500	45,400	45,900	44,200
Potash (K2O equivalent)	thousand metric tons	1,506	1,400	1,480	1,390	1,400	1,400
Pumice and pumicite	thousand metric tons	469	490	529	612	577	583
Salt	thousand metric tons	38,200	39,700	40,800	42,900	40,600	40,700
Sand and gravel (construction)	thousand metric tons	869,000	891,000	907,000	914,000	961,000	1,020,000E
Sand and gravel (industrial)	thousand metric tons	26,220	27,300	28,200	27,800	28,500	28,200
Soda ash (sodium carbonate)	thousand metric tons	8,959	9,321	10,100	10,200	10,700	10,100
Sodium sulfate (natural)	thousand metric tons	322	298	327	306	318	290
Stone (crushed)	million metric tons	1,120	1,230	1,260	1,330	1,420	1,500E
Stone (dimension)	thousand metric tons	1,280	1,190	1,160	1,150	1,180	1,080E
Sulfur (in all forms)	thousand metric tons	10,959	11,500	11,800	12,000	12,000	11,300E
Talc	thousand metric tons	968	935	1,060	994	1,050	1,060
Vermiculite	thousand metric tons	190	177	171	W	W	W

(E) Estimated. (W) Withheld to avoid disclosing company proprietary data. (—) No production.

U.S. Reliance on Foreign Supplies of Minerals

Source: U.S. Geological Survey, U.S. Dept. of the Interior

Mineral	Percent imported in 1998	Major sources (1994-1997)	Major uses
Arsenic	100	China, Hong Kong, Japan	Wood preservatives, herbicides, nonferrous alloys
Bauxite & alumina	100	Australia, Guinea, Jamaica, Brazil	Aluminum production, abrasives, chemicals, refractories
Bismuth	100	Belgium, Mexico, United Kingdom, China	Pharmaceuticals, chemicals, alloys, metallurgical additives
Columbium (niobium)	100	Brazil, Canada, Germany	Steelmaking, superalloys
Fluorspar	100	China, South Africa, Mexico	Hydrofluoric acid, aluminum fluoride, steelmaking
Graphite (natural)	100	Mexico, China, Canada, Madagascar	Refractories, brake linings, pencils
Manganese	100	South Africa, Gabon, Australia, France	Steelmaking, batteries, agricultural chemicals
Mica, sheet (natural)	100	India, Belgium, Germany, China	Electronic & electrical equipment
Strontium	100	Mexico, Germany	Television picture tubes, ferrite magnets, pyrotechnics
Thallium	100	Mexico, Belgium, Canada, Germany	Superconductor materials, electronics, alloys, glass
Thorium	100	Australia, France	Ceramics, welding electrodes, catalysts
Yttrium	100	China, France, United Kingdom, Belgium	Television phosphors, fluorescent lights, oxygen sensors, ceramics
Gemstones	99	Israel, Belgium, India	Jewelry, carvings, gem & mineral collections
Platinum	94	South Africa, United Kingdom, Germany, Russia	Catalysts, jewelry, dental & medical alloys
Palladium	88	Russia, South Africa, Belgium, United Kingdom	Catalysts, capacitors
Tin	85	Brazil, Indonesia, Bolivia, China	Solder, tinplate, chemicals, alloys
Antimony	84	China, Mexico, Bolivia, South Africa	Flame retardants, batteries, chemicals, ceramics & glass
Barite	80	China, India, Mexico, Morocco	Oil & gas well drilling fluids, chemicals
Potash	80	Canada, Russia, Belarus	Fertilizers, chemicals
Tantalum	80	Australia, Thailand, China, Brazil	Capacitors, superalloys, cemented carbide tools
Chromium	79	South Africa, Russia, Turkey, Zimbabwe, Kazakhstan	Steel, chemicals, refractories
Tungsten	78	China, Russia, Germany, Bolivia	Cemented carbides, electrical & electronic components, tool steels, alloys
Cobalt	77	Norway, Finland, Zambia, Canada	Superalloys, cemented carbides, magnetic alloys, catalysts
Iodine	72	Chile, Japan	Sanitation, pharmaceuticals, heat stabilizers, catalysts, animal feed
Zinc	70	Canada, Mexico, Spain, Peru	Galvanizing, alloys
Stone (dimension)	69	Italy, India, Brazil, Canada	Construction, monuments
Nickel	65	Canada, Norway, Russia, Australia	Stainless steel, alloys, plating
Peat	52	Canada	Horticulture, agriculture
Diamond (dust, grit, and powder)	51	Ireland, China, Russia	Grinding, drilling, cutting, polishing

U.S. Copper, Lead, and Zinc Production, 1950-98

Source: U.S. Geological Survey, U.S. Dept. of the Interior

	Copper		Lead		Zinc			Copper		Lead		Zinc	
Year	Quantity (metric tons)	Value ($1,000)	Quantity (metric tons)	Value ($1,000)	Quantity (metric tons)	Value ($1,000)	Year	Quantity (metric tons)	Value ($1,000)	Quantity (metric tons)	Value ($1,000)	Quantity (metric tons)	Value ($1,000)
1950	827	379,122	390,839	113,078	565,516	167,000	1992	1,760	4,179,000	397,076	307,337	523,430	673,800
1960	1,037	733,706	223,774	57,722	395,013	112,365	1993	1,800	3,635,000	355,185	248,540	488,283	496,795
1970	1,560	1,984,484	518,698	178,609	484,560	163,650	1994	1,850	4,430,000	363,000	298,000	570,000	619,000
1975	1,282	1,814,763	563,783	267,230	425,792	366,097	1995	1,850	5,640,000	386,000	359,000	614,000	756,000
1980	1,181	2,666,931	550,366	515,189	317,103	261,671	1996	1,920	4,610,000	426,000	459,000	598,000	674,000
1985	1,105	1,631,000	413,955	174,008	226,545	201,607	1997	1,940	4,570,000	448,000	460,000	605,000	860,000
1990	1,586	4,310,000	483,704	490,750	515,355	847,485	1998	1,860	3,235,000	481,000	480,000	742,000	841,000
1991	1,630	3,931,000	465,931	343,907	517,804	602,426							

U.S. Pig Iron and Raw Steel Output, 1940-98

Source: American Iron and Steel Institute

(net tons)

Year	Total pig iron	Raw steel	Year	Total pig iron	Raw steel	Year	Total pig iron	Raw steel
1940	46,071,666	66,982,686	1975	79,923,000	116,642,000	1993	53,082,000	97,877,000
1945	53,223,169	79,701,648	1980	68,721,000	111,835,000	1994	54,426,000	100,579,000
1950	64,586,907	96,836,075	1985	50,446,000	88,259,000	1995	56,097,000	104,930,000
1955	76,857,417	117,036,085	1990	54,750,000	98,906,000	1996	54,485,000	105,309,478
1960	66,480,648	99,281,601	1991	48,637,000	87,896,000	1997	54,679,000	108,561,182
1965	88,184,901	131,461,601	1992	52,224,000	92,949,000	1998	53,164,000	108,752,334
1970	91,435,000	131,514,000						

Steel figures include only that portion of the capacity and production of steel for castings used by foundries operated by companies producing steel ingots.

World Gold Production, 1975-98

Source: U.S. Geological Survey, U.S. Dept. of the Interior

(troy ounces)

		Africa			North and South America				Other			
Year	World prod.	South Africa	Ghana	Congo Dem. Rep	United States	Canada	Mexico	Colombia	Australia	China	Philippines	USSR[1]
1975	38,476,371	22,937,820	523,889	115,743	1,052,252	1,653,611	144,710	308,864	526,821	NA	502,577	NA
1980	39,197,315	21,669,468	353,000	96,452	969,782	1,627,477	195,991	510,439	547,591	NA	753,452	8,425,000
1985	49,283,691	21,565,230	299,363	257,206	2,427,232	2,815,118	265,693	1,142,385	1,881,491	1,950,000	1,062,997	8,700,000
1986	51,534,056	20,513,665	287,127	257,206	3,739,015	3,364,700	250,615	1,285,878	2,413,842	2,100,000	1,296,400	8,850,000
1987	53,033,614	19,176,500	327,598	385,809	4,947,040	3,724,000	256,822	853,600	3,558,954	2,300,000	1,048,081	8,850,000
1988	60,308,973	19,965,611	355,620	401,884	6,459,534	4,334,338	292,508	932,822	5,046,059	2,507,758	980,019	8,925,046
1989	65,335,998	19,530,290	429,470	340,798	8,543,449	5,127,850	276,914	948,640	6,544,702	2,893,567	964,265	9,773,826
1990	70,206,932	19,454,414	541,419	299,002	9,458,395	5,446,722	311,283	943,689	7,849,186	3,215,074	790,619	9,709,524
1991	70,422,599	19,326,133	845,918	282,927	9,454,311	5,676,278	326,073	1,120,260	7,530,283	3,858,089	833,219	8,359,193
1992	73,529,583	19,742,838	997,702	225,055	10,616,561	5,189,194	318,003	1,032,618	7,825,491	4,501,104	729,886	8,231,554
1993	73,300,000	19,907,772	1,250,000	280,000	10,642,314	4,916,781	356,873	883,149	7,947,535	5,144,119	508,818	8,228,179
1994	72,500,000	16,650,000	1,400,000	357,000	10,500,000	4,710,000	446,895	668,000	8,236,634	4,240,000	870,000	8,172,719*
1995	71,000,000	16,800,000	1,710,000	38,000	10,200,000	4,890,000	652,000	193,000	8,150,000	4,500,000	873,000	4,250,000*
1996	72,700,000	16,000,000	1,580,000	40,000	10,500,000	5,350,000	787,000	48,000	9,310,000	4,660,000	1,020,000	3,950,000*
1997	77,200,000	15,800,000	1,750,000	13,000	11,600,000	5,510,000	836,000	16,000	10,000,000	5,630,000	1,090,000	3,700,000*
1998	79,500,000	15,200,000	2,360,000	3,000	11,700,000	5,340,000	817,000	6,000	10,000,000	5,720,000	800,000	3,330,000*

NA=not available. (1) USSR as constituted prior to Dec. 1991. * Russia only.

U.S. and World Silver Production, 1930-98

Source: U.S. Geological Survey, U.S. Dept. of the Interior

(metric tons)

Year[1]	United States	World	Year[1]	United States	World	Year[1]	United States	World
1930	1,578.	7,736	1965	1,238.	8,007	1992	1,800	14,600
1935	1,428.	6,865	1970	1,400.	9,670	1993	1,640	14,300
1940	2,164.	8,565	1975	1,087.	9,428	1994	1,490	14,000
1945	904.	5,039	1980	1,006.	10,556	1995	1,560	15,100
1950	1,347.	6,323	1985	1,227.	13,051	1996	1,570	15,200
1955	1,134.	9,967	1990	2,120.	16,600	1997	2,150	16,400
1960	1,120.	7,505	1991	1,860.	15,600	1998	2,040E	16,140E

(1) Largest production of silver in the United States was in 1915—2,332 metric tons. E = Estimated.

Aluminum Summary, 1980-98

Source: U.S. Geological Survey, U.S. Dept. of the Interior

Item	Unit	1980	1985	1990	1992	1993	1994	1995	1996[4]	1997[4]	1998[4]
U.S. production	1,000 metric tons	6,231	5,262	6,441	6,798	6,639	6,385	6,563	6,860	7,150	7,150
Primary aluminum	1,000 metric tons	4,654	3,500	4,048	4,042	3,695	3,299	3,375	3,577	3,603	3,713
Secondary aluminum[1]	1,000 metric tons	1,577	1,762	2,393	2,756	2,944	3,086	3,188	3,310	3,550	3,440
Primary aluminum value	Billion dollars	7.8	3.8	6.6	5.1	4.3	5.2	6.4	5.6	6.1	5.4
Price (Primary aluminum)[2]	Cents/pound	76.1	48.8	74.0	57.5	53.3	71.2	85.9	71.3	77.1	65.5
Imports for consumption[3]	1,000 metric tons	647	1,420	1,514	1,725	2,544	3,382	2,975	2,810	3,080	3,550
Exports[3]	1,000 metric tons	1,346	908	1,659	1,453	1,207	1,365	1,610	1,500	1,570	1,590
World production	1,000 metric tons	15,383	15,398	19,299	19,500	19,800	19,200	19,700	20,800	21,500	22,100

(1) Recoverable metal content from purchased scrap, old and new. (2) Average prices for primary aluminum, quoted by *Metals Week*. (3) Crude and semicrude (incl. metal and alloys, plates, bars, etc., and scrap). (4) All data in metric tons, except primary production, have been rounded to 3 significant figures.

Economic and Financial Glossary

Source: Reviewed by William M. Gentry, Graduate School of Business, Columbia University

Annuity contract: An investment vehicle sold by insurance companies. Annuity buyers can elect to receive periodic payments for the rest of their lives. Annuities provide insurance against outliving one's wealth.

Arbitrage: A form of hedged investment meant to capture slight differences in the prices of 2 related securities— for example, buying gold in London and selling it at a higher price in New York.

Balanced budget: A budget is balanced when receipts equal expenditures. When receipts exceed expenditures, there is a **surplus;** when they fall short of expenditures, there is a **deficit.**

Balance of payments: The difference between all payments, for some categories of transactions, made to and from foreign countries over a set period of time. A *favorable* balance of payments exists when more payments are coming in than going out; an *unfavorable* balance of payments obtains when the reverse is true. Payments may include gold, the cost of merchandise and services, interest and dividend payments, money spent by travelers, and repayment of principal on loans.

Balance of trade (trade gap): The difference between exports and imports, in both actual funds and credit. A nation's balance of trade is *favorable* when exports exceed imports and *unfavorable* when the reverse is true.

Bear market: A market in which prices are falling.

Bearer bond: A bond issued in bearer form rather than being registered in a specific owner's name. Ownership is determined by possession.

Bond: A written promise, or IOU, by the issuer to repay a fixed amount of borrowed money on a specified date and generally to pay interest at regular intervals in the interim.

Bull market: A market in which prices are on the rise.

Capital gain (loss): An increase (decrease) in the market value of an asset over some period of time. For tax purposes, capital gains are typically calculated from when an asset is bought to when it is sold.

Commercial paper: An extremely short-term corporate IOU, generally due in 270 days or less.

Convertible bond: A corporate bond (see below) that may be converted into a stated number of shares of common stock. Its price tends to fluctuate along with fluctuations in the price of the stock and with changes in interest rates.

Consumer price index (CPI): A statistical measure of the change in the price of consumer goods.

Corporate bond: A bond issued by a corporation. The bond normally has a stated life and pays a fixed rate of interest. Considered safer than the common or preferred stock of the same company.

Cost of living: The cost of maintaining a standard of living measured in terms of purchased goods and services. Inflation typically measures changes in the cost of living.

Cost-of-living adjustments: Changes in promised payments, such as retirement benefits, to account for changes in the cost of living.

Credit crunch (liquidity crisis): A situation in which cash for lending is in short supply.

Debenture: An unsecured bond backed only by the general credit of the issuing corporation.

Deficit spending: Government spending in excess of revenues, generally financed with the sale of bonds. A deficit increases the government debt.

Deflation: A decrease in the level of prices.

Depression: A long period of economic decline when prices are low, unemployment is high, and there are many business failures.

Derivatives: Financial contracts, such as options, whose values are based on, or *derived* from, the price of an underlying financial asset or indicator such as a stock or an interest rate.

Devaluation: The official lowering of a nation's currency, decreasing its value in relation to foreign currencies.

Discount rate: The rate of interest set by the Federal Reserve that member banks are charged when borrowing money through the Federal Reserve System.

Disposable income: Income after taxes that is available to persons for spending and saving.

Diversification: Investing in more than one asset in order to reduce the riskiness of the overall asset portfolio. By holding more than one asset, losses on some assets may be offset by gains realized on other assets.

Dividend: Discretionary payment by a corporation to its shareholders, usually in the form of cash or stock shares.

Dow Jones Industrial Average: An index of stock market prices, based on the prices of 30 leading companies on the New York Stock Exchange.

Econometrics: The use of statistical methods to study economic and financial data.

Federal Deposit Insurance Corporation (FDIC): A U.S. government-sponsored corporation that insures accounts in national banks and other qualified institutions against bank failures.

Federal Reserve System: The entire banking system of the U.S., incorporating 12 Federal Reserve banks (one in each of 12 Federal Reserve districts), 24 Federal Reserve branch banks, all national banks, and state-chartered commercial banks and trust companies that have been admitted to its membership. The governors of the system greatly influence the nation's monetary and credit policies.

Full employment: The economy is said to be at full employment when everyone who wishes to work at the going wage-rate for his or her type of labor is employed, save only for the small amount of unemployment due to the time it takes to switch from one job to another.

Futures: A futures contract is an agreement to buy or sell a specific amount of a commodity or financial instrument at a particular price at a set date in the future. For example, futures based on a stock index (such as the Dow Jones Industrial Average) are bets on the future price of that group of stocks.

Golden parachute: Provisions in contracts of some high-level executives guaranteeing substantial severance benefits if they lose their position in a corporate takeover.

Government bond: A bond issued by the U.S. Treasury, considered a safe investment. Government bonds are divided into 2 categories—those that are not marketable and those that are. *Savings bonds* cannot be bought and sold once the original purchase is made. Marketable bonds fall into several categories. *Treasury bills* are short-term U.S. obligations, maturing in 3, 6, or 12 months. *Treasury notes* mature in up to 10 years. *Treasury bonds* mature in 10 to 30 years. *Indexed bonds* are adjusted for inflation.

Greenmail: A company buying back its own shares for more than the going market price to avoid a threatened hostile takeover.

Gross domestic product (GDP): The market value of all goods and services that have been bought for final use during a period of time. It became the official measure of the size of the U.S. economy in 1991, replacing *gross national product (GNP),* in use since 1941. GDP covers workers and capital employed within the nation's borders. GNP covers production by U.S. residents regardless of where it takes place. The switch aligned U.S. terminology with that of most other industrialized countries.

Hedge fund: A flexible investment fund for a limited number of large investors (the minimum investment is typically $1 million). Hedge funds use a variety of investment techniques, including those forbidden to mutual funds, such as short-selling and heavy leveraging.

Hedging: Taking 2 positions whose gains and losses will offset each other if prices change, in order to limit risk.

Individual retirement account (IRA): A self-funded tax-advantaged retirement plan that allows employed individuals to contribute up to a maximum yearly sum. With a *traditional* IRA, individuals contribute pre-tax earnings and defer income taxes until retirement. With a *Roth* IRA, individuals contribute after-tax earnings but do not pay taxes on future withdrawals (the interest is never taxed). *401(k) plans* are employer-sponsored plans similar to traditional IRAs, but having higher contribution limits.

Inflation: An increase in the level of prices.

Insider information: Important facts about the condition or plans of a corporation that have not been released to the general public.

Interest: The cost of borrowing money.

Investment bank: A financial institution that arranges the initial issuance of stocks and bonds and offers companies advice about acquisitions and divestitures.

Junk bonds: Bonds issued by companies with low credit ratings. They typically pay relatively high interest rates because of the fear of default.

Leading indicators: A series of 11 indicators from different segments of the economy used by the U.S. Commerce Department to predict when changes in the level of economic activity will occur.

Leverage: The extent to which a purchase was paid for with borrowed money. Amplifies the potential gain or loss for the purchaser.

Leveraged buyout (LBO): An acquisition of a company in which much of the purchase price is borrowed, with the debt to be repaid from future profits or by subsequently selling off company assets. A leveraged buyout is typically carried out by a small group of investors, often including incumbent management.

Liquid assets: Assets consisting of cash and/or items that are easily converted into cash.

Margin account: A brokerage account that allows a person to trade securities on credit. A **margin call** is a demand for more collateral on the account.

Money supply: The currency held by the public, plus checking accounts in commercial banks and savings institutions.

Mortgage-backed securities: Created when a bank, builder, or government agency gathers together a group of mortgages and then sells bonds to other institutions and the public. The investors receive their proportionate share of the interest payments on the loans as well as the principal payments. Usually, the mortgages in question are guaranteed by the government.

Municipal bond: Issued by governmental units such as states, cities, local taxing authorities, and other agencies. Interest is exempt from U.S.—and sometimes state and local—income tax. *Municipal bond unit investment trusts* offer a portfolio of many different municipal bonds chosen by professionals. The income is exempt from federal income taxes.

Mutual fund: A portfolio of professionally bought and managed financial assets in which you pool your money along with that of many other people. A share price is based on net asset value, or the value of all the investments owned by the funds, less any debt, and divided by the total number of shares. The major advantage, relative to investing individually in only a small number of stocks, is less risk—the holdings are spread out over many assets and if one or two do badly the remainder may shield you from the losses. *Bond funds* are mutual funds that deal in the bond market exclusively. *Money market mutual funds* buy in the so-called money market—institutions that need to borrow large sums of money for short terms. These funds often offer special checking account advantages.

National debt: The debt of the national government, as distinguished from the debts of political subdivisions of the nation and of private business and individuals.

National debt ceiling: Total borrowing limit set by Congress beyond which the U.S. national debt cannot rise. This limit is periodically raised by congressional vote.

Option: A type of contractual agreement between a buyer and a seller to buy or sell shares of a security. A **call** option contract gives the right to purchase shares of a specific stock at a stated price within a given period of time. A **put** option contract gives the buyer the right to sell shares of a specific stock at a stated price within a given period of time.

Per capita income: The total income of a group divided by the number of people in the group.

Prime interest rate: The rate charged by banks on short-term loans to large commercial customers with the highest credit rating.

Producer price index: A statistical measure of the change in the price of wholesale goods. It is reported for 3 different stages of the production chain: crude, intermediate, and finished goods.

Program trading: Trading techniques involving large numbers and large blocks of stocks, usually used in conjunction with computer programs. Techniques include *index arbitrage,* in which traders profit from price differences between stocks and futures contracts on stock indexes, and *portfolio insurance,* which is the use of stock-index futures to protect stock investors from potentially large losses when the market drops.

Public debt: The total of a nation's debts owed by state, local, and national government. Increases in this sum, reflected in public-sector deficits, indicate how much of the nation's spending is being financed by borrowing rather than by taxation.

Recession: A mild decrease in economic activity marked by a decline in real (inflation-adjusted) GDP, employment, and trade, usually lasting from 6 months to a year, and marked by widespread decline in many sectors of the economy.

Savings Association Insurance Fund (SAIF): Created in 1989 to insure accounts in savings and loan associations up to $100,000.

Seasonal adjustment: Statistical changes made to compensate for regular fluctuations in data that are so great they tend to distort the statistics and make comparisons meaningless. For instance, seasonal adjustments are made for a slowdown in housing construction in midwinter and for the rise in farm income in the fall after summer crops are harvested.

Short-selling: Borrowing shares of stock from a brokerage firm and selling them, hoping to buy the shares back at a lower price, return them, and realize a profit from the decline in prices.

Stagnation: Economic slowdown in which there is little growth in the GDP, capital investment, and real income.

Stock: *Common stocks* are shares of ownership in a corporation. For publicly held firms, the stock typically trades on an exchange, such as the New York Stock Exchange; for closely held firms, the founders and managers own most of the stock. There can be wide swings in the prices of this kind of stock. *Preferred stock* is a type of stock on which a fixed dividend must be paid before holders of common stock are issued their share of the issuing corporation's earnings. Preferred stock is less risky than common stock. *Convertible preferred stock* can be converted into the common stock of the company that issued the preferred. *Over-the-counter stock* is not traded on the major or regional exchanges, but rather through dealers from whom you buy directly. *Blue chip* stocks are so called because they have been leading stocks for a long time. *Growth* stocks are from companies that reinvest their earnings, rather than pay dividends, with the expectation of future stock price appreciation.

Supply-side economics: A school of thinking about economic policy holding that lowering income tax rates will inevitably lead to enhanced economic growth and general revitalization of the economy.

Takeover: Acquisition of one company by another company or group by sale or merger. A *friendly takeover* occurs when the acquired company's management is agreeable to the merger; when management is opposed to the merger, it is a *hostile* takeover.

Tender offer: A public offer to buy a company's stock; usually priced at a premium above the market.

Zero coupon bond: A corporate or government bond that is issued at a deep discount from the maturity value and pays no interest during the life of the bond. It is redeemable at face value.

AGRICULTURE
U.S. Farms—Number and Acreage by State, 1997 and 1998
Source: National Agricultural Statistics Service, U.S. Dept. of Agriculture

STATE	Farms (1,000) 1997	1998	Acreage (mil) 1997	1998	Acreage per farm 1997	1998	STATE	Farms (1,000) 1997	1998	Acreage (mil) 1997	1998	Acreage per farm 1997	1998
Alabama	49.0	49.0	9.6	9.5	196	194	Nebraska	55.0	55.0	46.4	46.4	844	844
Alaska	0.6	0.6	0.9	0.9	1,625	1,625	Nevada	3.0	3.0	6.9	6.9	2,300	2,300
Arizona	7.9	7.9	28.3	28.3	3,582	3,582	New Hampshire	3.0	3.1	0.4	0.4	140	135
Arkansas	49.0	49.5	14.8	14.8	302	298	New Jersey	9.6	9.6	0.8	0.8	86	86
California	87.0	89.0	28.7	28.5	330	320	New Mexico	15.5	16.0	45.3	45.3	2,923	2,831
Colorado	29.5	29.5	32.5	32.2	1,102	1,092	New York	38.0	38.0	7.8	7.8	205	205
Connecticut	4.1	4.1	0.4	0.4	93	93	N. Carolina	59.0	58.0	9.4	9.4	161	162
Delaware	2.8	2.7	0.6	0.6	209	215	N. Dakota	31.5	31.0	39.5	39.5	1,260	1,274
Florida	45.0	45.0	10.6	10.6	236	236	Ohio	79.0	80.0	14.9	14.9	189	186
Georgia	49.0	50.0	11.3	11.3	231	226	Oklahoma	83.0	83.0	34.0	34.0	410	410
Hawaii	5.5	5.5	1.4	1.4	262	262	Oregon	39.0	39.5	17.2	17.2	449	435
Idaho	24.5	24.5	12.0	12.0	490	490	Pennsylvania	60.0	60.0	7.7	7.7	128	128
Illinois	79.0	79.0	27.8	27.8	352	352	Rhode Island	0.8	0.8	0.7	0.7	87	87
Indiana	66.0	66.0	15.6	15.6	236	236	S. Carolina	25.0	25.0	5.0	4.9	200	196
Iowa	98.0	97.0	33.0	33.0	337	340	S. Dakota	32.5	32.5	44.0	44.0	1,354	1,354
Kansas	65.0	65.0	47.5	47.5	731	731	Tennessee	91.0	91.0	12.0	11.9	132	131
Kentucky	91.0	90.0	13.9	13.9	153	154	Texas	225.0	226.0	131.5	131.5	584	582
Louisiana	30.0	30.0	8.2	8.2	273	273	Utah	15.0	15.0	11.6	11.6	773	773
Maine	7.0	6.9	1.3	1.3	183	186	Vermont	6.6	6.7	1.3	1.3	202	200
Maryland	13.0	12.5	2.2	2.1	169	168	Virginia	49.0	49.0	8.8	8.8	180	180
Massachusetts	6.0	6.0	0.6	0.6	95	95	Washington	39.0	40.0	15.7	15.7	403	393
Michigan	53.0	52.0	10.4	10.4	196	200	W. Virginia	21.0	21.0	3.7	3.7	176	176
Minnesota	81.0	80.0	29.1	28.9	359	361	Wisconsin	79.0	78.0	16.5	16.4	209	210
Mississippi	42.0	42.0	11.7	11.6	279	276	Wyoming	9.2	9.2	34.6	34.6	3,761	3,761
Missouri	110.0	110.0	30.1	30.1	274	274							
Montana	27.0	27.5	57.8	57.5	2,141	2,091	**UNITED STATES**	**2,190**	**2,192**	**956**	**954**	**436**	**435**

Decline in Farm Workers, 1820-1994*
Source: U.S. Dept. of Agriculture, Economic Research Service

Of the approximately 2.9 mil workers in the U.S. in 1820, 71.8%, or about 2.1 mil, were employed in farm occupations. The percentage of U.S. workers in farm occupations had declined drastically by the turn of the century, and by 1994 only 2.5% of all U.S. workers were employed in farm occupations.

(percent of total U.S. workers in farm occupations)

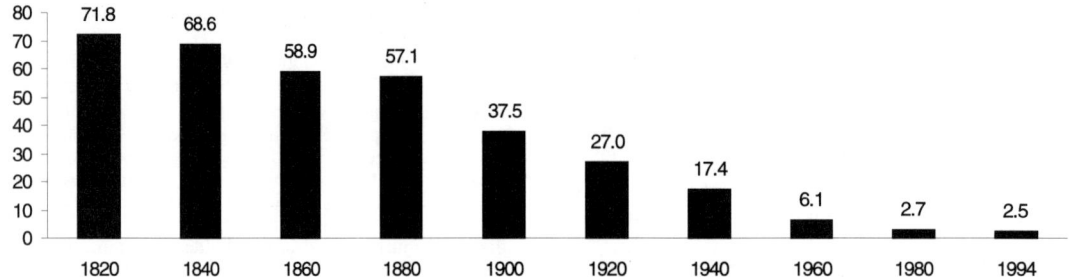

* Figures not compiled for years after 1994. Total workers for 1994 are employed workers age 15 and older; total workers for 1980 are members of the experienced civilian labor force ages 16 and older; total workers for 1900 to 1960 are members of the experienced civilian labor force 14 and older; total workers for 1820 to 1880 are gainfully employed workers 10 and older.

U.S. Farms, 1940-98
Source: National Agricultural Statistics Service, U.S. Dept. of Agriculture

Since 1940, the number of farms in the U.S. has sharply declined, while the size of the average farm has grown.

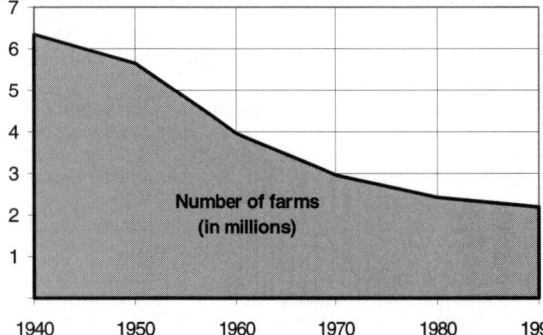

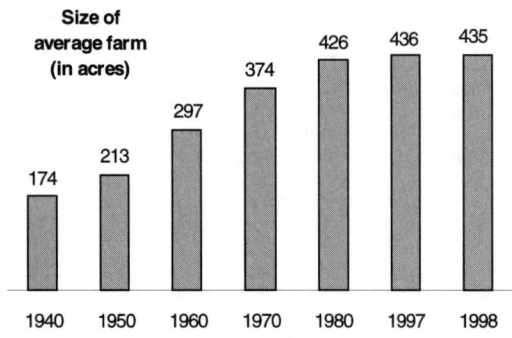

Eggs: U.S. Production, Price, and Value, 1997-98[1]

Source: National Agricultural Statistics Service, U.S. Dept. of Agriculture

STATE	Eggs produced 1997 (mil)	1998	Price per dozen[2] 1997 (dollars)	1998	Value of production 1997 (1,000 dollars)	1998	STATE	Eggs produced 1997 (mil)	1998	Price per dozen[2] 1997 (dollars)	1998	Value of production 1997 (1,000 dollars)	1998
AL...	2,499	2,518	1.060	1.030	220,745	216,128	NH...	46	42	0.825	0.750	3,163	2,613
AR..	3,215	3,233	1.030	0.978	275,954	263,490	NJ...	463	488	0.623	0.570	24,037	23,180
CA..	6,663	6,608	0.621	0.561	344,810	308,924	NM...	302	299	0.593	0.550	14,924	13,704
CO..	857	945	0.720	0.671	51,420	52,841	NY...	931	986	0.666	0.620	51,671	50,943
CT..	917	839	0.598	0.589	45,697	41,181	NC..	2,794	2,555	0.875	0.890	203,729	189,496
DE..	112	92	1.150	1.380	10,733	10,580	ND...	49	60	0.530	0.450	2,164	2,250
FL...	2,499	2,528	0.594	0.535	123,701	112,707	OH...	6,976	7,395	0.614	0.570	356,939	351,263
GA..	4,867	5,126	0.885	0.880	358,941	375,907	OK...	901	927	0.873	0.780	65,548	60,255
HI...	172	156	0.906	0.871	12,986	11,323	OR...	783	758	0.644	0.590	42,021	37,268
ID...	236	248	0.707	0.696	13,904	14,384	PA...	5,900	5,983	0.653	0.610	321,058	304,136
IL...	841	838	0.622	0.618	43,592	43,157	RI...	24	22	0.589	0.639	1,178	1,172
IN...	5,652	5,831	0.637	0.588	300,027	285,709	SC...	1,228	1,189	0.768	0.696	78,592	68,962
IA...	5,527	5,969	0.526	0.452	242,267	224,832	SD...	542	618	0.500	0.430	22,583	22,145
KS...	323	392	0.566	0.468	15,225	15,269	TN...	255	299	0.931	0.926	19,784	23,073
KY..	710	863	0.744	0.725	44,020	52,140	TX...	4,186	4,257	0.768	0.715	267,904	253,646
LA...	460	475	0.908	0.848	34,807	33,567	UT...	483	478	0.576	0.520	23,184	20,713
ME..	1,434	1,373	0.694	0.630	82,933	72,083	VT...	56	55	0.608	0.628	2,827	2,852
MD..	882	867	0.732	0.697	53,802	50,358	VA...	806	860	0.951	0.933	63,876	66,865
MA..	156	139	0.610	0.642	7,930	7,431	WA...	1,379	1,394	0.653	0.594	75,024	69,023
MI...	1,327	1,395	0.560	0.500	61,927	58,104	WV...	245	241	1.210	1.250	24,704	25,104
MN...	2,957	3,152	0.559	0.480	137,747	126,080	WI...	998	1,031	0.564	0.514	46,906	44,161
MS..	1,547	1,578	1.130	1.210	145,676	159,115	WY...	3.5	3.6	0.630	0.570	184	171
MO..	1,719	1,732	0.592	0.530	84,804	76,497	Other[3]	61	86	0.610	0.545	3,111	3,904
MT..	88	84	0.570	0.550	4,180	3,850							
NE..	2,469	2,706	0.520	0.430	106,990	96,965	**U.S.[4]**	**77,532**	**79,717**	**0.703**	**0.655**	**4,539,929**	**4,349,521**

(1) Estimates cover the 12-month period from Dec. 1 of the previous year through Nov. 30. (2) Average of all eggs sold by producers, including hatching eggs. (3) AK, AZ, and NV combined to avoid disclosure of individual operations; totals listed under "other." AK price estimates discontinued in 1996. (4) Total states may not equal U.S. total because of rounding.

Livestock on Farms in the U.S., 1900-99

Source: National Agricultural Statistics Service, U.S. Dept. of Agriculture

(in thousands)

Year (On Jan. 1)	All cattle[1]	Milk cows	Sheep and lambs	Hogs and pigs[2]	Year (On Jan. 1)	All cattle[1]	Milk cows	Sheep and lambs	Hogs and pigs[2]
1900..	59,739	16,544	48,105	51,055	1985..	109,582	10,777	10,716	54,073
1910..	58,993	19,450	50,239	48,072	1990..	95,816	10,015	11,358	53,788
1920..	70,400	21,455	40,743	60,159	1991..	96,393	9,966	11,174	54,416
1930..	61,003	23,032	51,565	55,705	1992..	97,556	9,688	10,797	57,649
1940..	68,309	24,940	52,107	61,165	1993..	99,176	9,581	10,906	58,795
1950..	77,963	23,853	29,826	58,937	1994..	100,974	9,494	9,836	60,847
1955..	96,592	23,462	31,582	50,474	1995..	102,785	9,466	8,989	59,329
1960..	96,236	19,527	33,170	59,026	1996..	103,548	9,372	8,465	56,038
1965..	109,000	16,981	25,127	56,106	1997..	101,656	9,252	8,024	57,366
1970..	112,369	12,091	20,423	57,046	1998..	99,744	9,199	7,825	62,213
1975..	132,028	11,220	14,515	54,693	1999[3].	98,522	9,143	7,238	60,536
1980..	111,242	10,758	12,699	67,318					

(1) From 1966, includes milk cows and heifers that have calved. (2) 1900-95, as of Dec. 1 of preceding year; 1996-98 as of June 1 of same year. (3) Total estimated value on farms, as of Jan. 1, 1999 (Dec. 1, 1998 for hogs and pigs), was (avg. value per head in parentheses): cattle, $58,560,000 ($594); sheep and lambs, $640,428,000 ($88); hogs and pigs, $2,831,847,000 ($46).

U.S. Meat Production and Consumption, 1940-98

Source: Economic Research Service, U.S. Dept. of Agriculture

(in millions of pounds)

Year	Beef Production	Beef Consumption[2]	Veal Production	Veal Consumption[2]	Lamb and mutton Production	Lamb and mutton Consumption[2]	Pork Production	Pork Consumption[2]	All red meats[1] Production	All red meats[1] Consumption[2]	All Poultry Production	All Poultry Consumption[2]
1940.....	7,175	7,257	981	981	876	873	10,044	9,701	19,076	18,812	NA	NA
1950.....	9,534	9,529	1,230	1,206	597	596	10,714	10,390	22,075	21,721	3,174	3,097
1960.....	14,728	15,465	1,109	1,118	769	857	13,905	14,057	30,511	31,497	6,310	6,168
1970.....	21,684	23,451	588	613	551	669	14,699	14,957	37,522	39,689	10,193	9,981
1980.....	21,643	23,560	400	420	318	351	16,617	16,838	38,978	41,701	14,173	13,525
1990.....	22,743	24,031	327	325	363	397	15,354	16,031	38,787	40,784	23,468	22,151
1991.....	22,917	24,113	306	305	363	396	15,999	16,392	39,585	41,207	24,700	23,270
1992.....	23,086	24,261	310	312	348	388	17,233	17,461	40,977	42,422	26,201	24,394
1993.....	23,049	24,006	285	286	337	381	17,088	17,408	40,759	42,081	27,329	25,097
1994.....	24,386	25,125	293	291	308	345	17,696	17,811	42,683	43,571	29,113	25,754
1995.....	25,222	25,533	319	319	287	348	17,849	17,768	43,677	43,968	30,393	25,944
1996.....	25,525	25,863	378	378	268	334	17,117	16,795	43,288	43,370	32,015	26,760
1997.....	25,490	25,609	334	333	260	333	17,274	16,821	43,358	43,096	32,964	27,261
1998.....	25,760	26,303	262	265	251	359	19,011	18,308	45,284	45,235	33,352	27,821

(1) Meats may not add to total because of rounding. (2) Consumption (also called total disappearance) is estimated as: production plus beginning stocks, plus imports, minus exports, minus ending stocks. NA = not available.

U.S. Government Agricultural Payments by State, 1998[1]

Source: Economic Research Service, U.S. Dept. of Agriculture
(in thousands of dollars)

STATE	Feed Grains[2]	Wheat[2]	Rice[2]	Cotton[2]	Wool Act[2]	Conservation[3]	Total[4]
Alabama	$-8	$-2	$0	$-17	$0	$20,832	$96,832
Alaska	0	0	0	0	0	1,070	1,404
Arizona	-19	-3	0	-63	0	287	78,987
Arkansas	-8	-36	-17	-95	0	9,037	466,712
California	-27	8	-50	-155	-1	7,730	352,710
Colorado	-115	-36	0	0	0	61,681	257,887
Connecticut	-2	0	0	0	0	79	2,433
Delaware	0	0	0	0	0	349	10,770
Florida	0	-0	0	0	0	4,393	24,918
Georgia	-32	-39	0	-131	0	16,236	178,283
Hawaii	0	0	0	0	0	73	231
Idaho	-12	-82	0	0	0	32,168	196,340
Illinois	-373	-6	0	0	0	63,184	934,043
Indiana	-191	-2	0	0	0	28,402	463,999
Iowa	-409	0	0	0	0	134,951	1,146,046
Kansas	-163	-86	0	0	0	115,623	872,524
Kentucky	-55	-5	0	0	0	18,908	140,107
Louisiana	-10	0	-30	-72	0	7,923	257,694
Maine	0	0	0	0	0	2,343	6,483
Maryland	-6	-0	0	0	0	2,746	38,037
Massachusetts	0	0	0	0	0	137	1,717
Michigan	-115	-4	0	0	0	17,488	208,077
Minnesota	-278	-39	0	0	-7	60,790	762,449
Mississippi	-6	-1	-6	-76	0	33,311	281,899
Missouri	-89	-10	0	-4	0	92,446	424,018
Montana	-29	-80	0	0	0	104,763	357,904
Nebraska	-470	-11	0	0	0	58,692	797,382
Nevada	0	0	0	0	0	366	2,673
New Hampshire	0	0	0	0	0	255	1,859
New Jersey	0	0	0	0	0	241	5,492
New Mexico	-6	-8	0	-6	0	20,932	60,441
New York	-35	0	0	0	0	6,612	59,675
North Carolina	-26	-1	0	-13	0	5,105	129,375
North Dakota	-70	-120	0	0	0	125,246	601,846
Ohio	-90	-3	0	0	0	26,486	313,123
Oklahoma	-4	-112	0	-6	0	37,580	302,236
Oregon	-3	-14	0	0	0	18,294	99,952
Pennsylvania	-27	-2	0	0	0	4,707	45,356
Rhode Island	0	0	0	0	0	25	167
South Carolina	-19	-4	0	-7	0	8,405	62,366
South Dakota	-98	-6	0	0	0	81,100	429,613
Tennessee	-22	-18	0	-10	0	15,400	127,962
Texas	-334	-56	24	-148	280	133,726	998,457
Utah	0	-13	0	0	0	6,142	24,981
Vermont	-3	0	0	0	0	575	4,475
Virginia	-4	-3	0	0	0	3,201	45,603
Washington	-37	-41	0	0	0	35,637	257,165
West Virginia	0	0	0	0	0	424	5,280
Wisconsin	-255	-5	0	0	0	40,570	252,787
Wyoming	-5	-3	0	0	-2	8,491	28,690
UNITED STATES	**$-3,457**	**$-844**	**$21**	**$-804**	**$270**	**$1,475,161**	**$12,219,559**

(1) Includes both cash payments and payment-in-kind (PIK) for fiscal year. (2) Negatives indicate that the current year's Advanced Deficiency Payments were less than refunds from producers to government because advances paid in the previous year were too high. (3) Includes amount paid under agriculture and conservation programs (Conservation Reserve, Agriculture Conservation, Emergency Conservation, and Great Plains Program). (4) Total government payments include various other categories not shown, including production flexibility contracts and loan deficiency payments.

U.S. Federal Food Assistance Programs, 1989-98[1]

Source: Food and Nutrition Service, U.S. Dept. of Agriculture
(in millions of dollars)

	1989	1990	1991	1992	1993	1994	1995	1996	1997	1998
Food stamps[2]	$12,932	$15,491	$18,769	$22,462	$23,653	$24,493	$24,620	$24,325	$21,485	$18,916
Puerto Rico nutrition asst.[3]	908	937	963	1,002	1,040	1,079	1,131	1,143	1,174	1,204
Natl. school lunch[4]	3,769	3,834	4,224	4,564	4,750	5,016	5,160	5,355	5,554	5,828
School breakfast[5]	513	596	685	787	869	959	1,048	1,119	1,214	1,271
WIC[6]	1,911	2,122	2,301	2,597	2,825	3,169	3,440	3,695	3,844	3,890
Summer food service[7]	146	164	182	204	220	230	237	250	244	262
Child/adult care[7]	697	813	945	1,094	1,225	1,354	1,464	1,534	1,571	1,552
Special milk	18	19	20	20	19	18	17	17	17	17
Nutrition for the elderly[4]	146	142	144	151	153	152	148	145	145	141
Food distrib. to Indian reserv.[7]	65	66	65	62	63	65	65	70	71	72
Commodity supp. food prog.[7]	73	85	93	105	113	107	99	100	99	94
Food dist.—charitable inst.[8]	136	104	93	116	91	105	64	11	6	9
Emergency food assistance[9]	310	334	301	272	271	264	135	80	192	235
Other costs[10]	72	75	82	91	111	124	144	146	139	139
TOTAL[11]	**$21,696**	**$24,781**	**$28,867**	**$33,527**	**$35,403**	**$37,135**	**$37,772**	**$37,990**	**$35,755**	**$33,630**

(1) All data are for fiscal (not calendar) years. (2) Includes federal share of state administrative expenses and other federal costs. (3) Puerto Rico participated in the Food Stamp Program from FY 1975 until July 1982, when it initiated a separate grant program. (4) Includes cash payments and commodity costs (entitlement, bonus, and cash in lieu). (5) Excludes startup costs. (6) Includes the WIC Farmers Market Nutrition Program, program studies and special grants. (7) Includes commodity costs and administrative expenditures. (8) Includes summer camps. (9) Includes the Emergency Food Assistance Program (TEFAP) for all years, and the Soup Kitchens/Food Banks Program (1989-96). (10) Includes certain miscellaneous child nutrition costs. (11) Excludes Food Program Administration (federal) costs. Totals may not add because of rounding.

U.S. Farm Marketings by State, 1997-98

Source: Economic Research Service, U.S. Dept. of Agriculture

(in thousands of dollars)

STATE/RANK[1]	1997 FARM MARKETINGS Total	Crops	Livestock and products	1998 FARM MARKETINGS Total	Crops	Livestock and products
Alabama (25)	$3,227,434	$796,157	$2,431,277	$3,283,129	$696,391	$2,586,728
Alaska (50)	32,426	26,152	6,274	47,037	19,860	27,177
Arizona (29)	2,151,560	1,263,148	888,412	2,368,073	1,425,152	942,921
Arkansas (12)	5,862,203	2,446,386	3,415,817	5,421,870	2,171,726	3,250,144
California (1)	24,849,653	18,555,414	6,294,239	24,616,242	17,771,171	6,845,071
Colorado (17)	4,399,249	1,387,670	3,011,579	4,309,508	1,452,853	2,856,655
Connecticut (43)	496,459	278,719	217,740	508,594	280,854	227,740
Delaware (40)	747,653	174,181	573,472	773,791	164,339	609,452
Florida (9)	6,243,366	4,978,525	1,264,841	6,761,965	5,354,616	1,407,349
Georgia (11)	5,885,210	2,443,316	3,441,894	5,454,249	2,046,648	3,407,601
Hawaii (42)	482,599	414,602	67,997	509,951	417,970	91,981
Idaho (24)	3,315,014	1,926,430	1,388,584	3,320,352	1,735,389	1,584,963
Illinois (6)	9,276,040	7,339,362	1,936,678	7,742,280	6,167,021	1,575,259
Indiana (15)	5,506,236	3,610,023	1,896,213	4,884,568	3,245,248	1,639,320
Iowa (3)	12,840,692	7,310,947	5,529,745	10,994,252	6,216,645	4,777,607
Kansas (5)	9,001,475	3,984,532	5,016,943	7,784,013	3,246,911	4,537,102
Kentucky (19)	3,632,928	1,654,874	1,978,054	3,920,208	1,786,527	2,133,681
Louisiana (33)	2,139,969	1,480,538	659,431	1,890,624	1,245,131	645,493
Maine (45)	486,119	227,767	258,352	505,645	223,586	282,059
Maryland (35)	1,538,383	623,088	915,295	1,520,218	571,292	948,926
Massachusetts (44)	532,276	430,352	101,924	507,340	394,873	112,467
Michigan (22)	3,587,753	2,235,720	1,352,033	3,480,343	2,157,557	1,322,786
Minnesota (7)	8,155,006	4,100,778	4,054,228	7,679,914	3,925,101	3,754,813
Mississippi (23)	3,476,300	1,469,913	2,006,387	3,454,358	1,285,208	2,169,150
Missouri (16)	5,563,617	2,768,474	2,795,143	4,681,843	2,261,666	2,420,177
Montana (34)	2,063,342	1,071,848	991,494	1,799,189	934,272	864,917
Nebraska (4)	10,092,232	4,550,182	5,542,050	8,848,014	3,724,514	5,123,500
Nevada (47)	310,008	129,884	180,124	336,613	142,748	193,865
New Hampshire (48)	166,070	96,733	69,337	151,147	82,167	68,980
New Jersey (39)	776,311	596,189	180,122	828,318	650,058	178,260
New Mexico (32)	1,913,484	559,783	1,353,701	1,950,222	513,221	1,437,001
New York (26)	2,895,699	1,036,861	1,858,838	3,145,799	1,054,005	2,091,794
North Carolina (8)	8,302,454	3,608,816	4,693,638	7,163,967	3,246,807	3,917,160
North Dakota (28)	3,312,959	2,702,197	610,762	3,003,856	2,454,587	549,269
Ohio (14)	5,344,658	3,475,782	1,868,876	4,972,519	3,124,412	1,848,107
Oklahoma (20)	4,362,549	1,301,498	3,061,051	3,900,273	1,062,193	2,838,080
Oregon (27)	3,111,750	2,372,207	739,543	3,091,992	2,329,995	761,997
Pennsylvania (18)	4,127,690	1,338,876	2,788,814	4,174,592	1,260,720	2,913,872
Rhode Island (49)	82,849	73,616	9,233	64,866	55,770	9,096
South Carolina (36)	1,694,361	897,383	796,978	1,511,115	747,911	763,204
South Dakota (21)	4,237,050	2,417,031	1,820,019	3,508,037	1,951,260	1,556,777
Tennessee (31)	2,291,672	1,286,707	1,004,965	2,215,587	1,177,114	1,038,473
Texas (2)	13,460,835	5,276,900	8,183,935	13,206,203	4,986,047	8,220,156
Utah (37)	952,959	238,069	714,890	980,866	244,728	736,138
Vermont (41)	512,612	96,830	415,782	556,765	84,425	472,340
Virginia (30)	2,393,848	856,183	1,537,665	2,328,428	767,810	1,560,618
Washington (13)	5,382,015	3,778,247	1,603,768	5,154,635	3,424,474	1,730,161
West Virginia (46)	394,104	70,587	323,517	404,745	69,180	335,565
Wisconsin (10)	5,756,477	1,686,063	4,070,414	6,193,000	1,700,577	4,492,423
Wyoming (38)	844,675	198,796	645,879	850,295	169,656	680,639
UNITED STATES	$208,212,283	$111,644,336	$96,567,947	$199,761,410	$102,222,386	$94,539,024

(1) States ranked by 1998 total.

Value of U.S. Agricultural Exports and Imports, 1977-98[1]

Source: Economic Research Service, U.S. Dept. of Agriculture

(in billions of dollars, except percent)

Year	Trade balance	Agric. exports	Percentage of all exports	Agric. imports	Percentage of all imports	Year	Trade balance	Agric. exports	Percentage of all exports	Agric. imports	Percentage of all imports
1977	10.6	24.0	20	13.4	9	1988	14.3	35.3	12	21.0	5
1978	13.4	27.3	21	13.9	8	1989	18.1	39.7	12	21.6	5
1979	15.8	32.0	19	16.2	8	1990	17.7	40.4	11	22.7	5
1980	23.2	40.5	19	17.3	7	1991	15.1	37.8	10	22.7	5
1981	26.4	43.8	19	17.3	7	1992	18.2	42.6	10	24.5	5
1982	23.6	39.1	18	15.5	6	1993	18.3	42.9	10	24.6	4
1983	18.5	34.8	18	16.3	7	1994	17.4	44.0	9	26.6	4
1984	19.1	38.0	18	18.9	6	1995	24.9	54.7	10	29.9	4
1985	11.5	31.2	15	19.7	6	1996	27.3	59.9	10	32.6	4
1986	5.4	26.3	13	20.9	6	1997	21.6	57.4	9	35.8	4
1987	7.2	27.9	12	20.7	5	1998	16.7	53.7	8	37.0	4

(1) Fiscal year (Oct.-Sept.).

Farm Business Real Estate Debt Outstanding, by Lender Groups,[1] 1960-98

Source: Economic Research Service, U.S. Dept. of Agriculture

(in thousands of dollars)

Dec. 31	Total farm real estate debt[2]	AMOUNTS HELD BY PRINCIPAL LENDER GROUPS				
		Farm Credit System[2]	Farm Services Agency[3]	Life insurance companies[4]	All operating banks	Other[5]
1960.....	$11,309,593	$2,222,301	$623,895	$2,651,587	$1,355,733	$4,456,068
1970.....	27,505,932	6,420,357	2,179,873	5,122,291	3,328,876	10,454,540
1980.....	89,692,429	33,224,684	7,435,059	11,997,922	7,765,058	29,269,705
1985.....	100,076,120	42,168,554	9,820,913	11,272,689	10,731,881	26,082,096
1988.....	77,832,498	28,445,452	8,979,749	9,039,395	14,433,688	16,934,218
1989.....	75,978,245	26,895,927	8,203,215	9,113,109	15,685,485	16,080,503
1990.....	74,731,876	25,924,490	7,639,490	9,703,958	16,288,128	15,169,299
1991.....	74,943,893	25,305,300	7,040,851	9,545,804	17,416,527	15,631,629
1992.....	75,421,255	25,407,547	6,394,446	8,765,021	18,756,851	16,095,415
1993.....	76,036,358	24,899,573	5,837,377	8,985,489	19,594,554	16,719,356
1994.....	77,679,838	24,596,715	5,465,063	9,025,132	21,079,145	17,513,779
1995.....	79,286,920	24,851,298	5,055,018	9,091,957	22,276,503	18,012,138
1996.....	81,657,044	25,729,867	4,701,970	9,468,069	23,275,938	18,481,196
1997.....	85,359,385	27,097,928	4,372,663	9,698,796	25,239,726	18,950,271
1998.....	89,615,293	28,887,735	4,073,399	10,723,206	27,168,314	18,762,640

(1) Exclude operator households. (2) Includes data for joint stock land banks and real estate loans by Agricultural Credit Assn. (3) Includes loans made directly by Farm Services Agency for farm ownership, soil and water loans to individuals, Native American tribe land acquisition, grazing associations, and half of economic emergency loans. Also includes loans for rural housing on farm tracts and labor housing. (4) American Council of Life Insurance. (5) Estimated by ERS, USDA. Includes Commodity Credit Corporation storage and drying facility loans.

Grain, Hay, Potato, Cotton, Soybean, Tobacco Production, by State, 1998

Source: National Agricultural Statistics Service, U.S. Dept. of Agriculture

STATE	Barley (1,000 bu)	Corn, grain (1,000 bu)	Cotton (Upland) (1,000 b)	All hay (1,000 t)	Oats (1,000 bu)	Potatoes (1,000 cwt)	Soybeans (1,000 bu)	Tobacco (1,000 lb)	All wheat (1,000 bu)
Alabama.........	—	12,600	570.0	1,575	816	780	7,040	—	3,570
Alaska	—	—	—	—	—	—	—	—	—
Arizona..........	6,160	5,250	580.0	1,740	—	2,284	—	—	15,840
Arkansas	—	21,500	1,220.0	2,250	720	—	85,000	—	45,900
California	7,500	41,600	1,150.0	8,115	2,250	13,703	—	—	38,550
Colorado.........	9,430	155,150	—	4,602	1,750	28,230	—	—	103,710
Connecticut	—	NE	—	128	—	—	—	4,573	—
Delaware	1,800	15,500	—	55	—	1,012	7,128	—	3,723
Florida	—	3,410	68.0	575	—	8,798	690	17,102	559
Georgia	—	22,525	1,550.0	1,495	1,325	—	4,620	92,400	10,320
Hawaii...........	—	—	—	—	—	—	—	—	—
Idaho	59,280	7,800	—	5,549	2,250	139,650	—	—	102,410
Illinois..........	—	1,473,450	—	3,395	3,920	1,421	468,600	—	57,600
Indiana..........	—	760,350	—	2,690	1,500	1,600	235,200	17,000	35,750
Iowa	—	1,769,000	—	5,332	10,915	306	501,600	—	1,280
Kansas	280	418,950	13.4	8,020	2,700	—	75,000	—	494,900
Kentucky.........	441	135,700	—	5,705	—	—	36,000	460,910	24,750
Louisiana	—	43,740	645.0	726	—	—	22,470	—	3,960
Maine	—	NE	—	280	1,752	18,060	—	—	—
Maryland	3,456	43,600	—	632	350	1,081	14,260	9,100	10,750
Massachusetts....	—	NE	—	202	—	660	—	1,764	—
Michigan.........	1,300	227,550	—	3,565	4,830	14,725	73,710	—	30,780
Minnesota........	22,825	1,032,750	—	7,110	19,530	21,170	285,600	—	80,444
Mississippi	—	43,000	1,450.0	1,738	—	—	48,000	—	6,750
Missouri	—	285,000	350.0	7,703	611	1,892	170,000	6,300	57,500
Montana.........	57,600	2,070	—	5,020	3,240	3,180	—	—	168,790
Nebraska	400	1,239,750	—	7,680	5,320	9,781	165,000	—	82,800
Nevada	400	—	—	1,556	—	2,726	—	—	1,240
New Hampshire ...	—	NE	—	110	—	—	—	—	—
New Jersey.......	232	9,016	—	237	—	702	3,164	—	2,288
New Mexico	—	14,025	90.0	1,548	—	3,204	—	—	7,950
New York	—	66,120	—	3,110	6,510	7,290	3,977	—	7,020
North Carolina	1,140	53,900	1,005.0	1,486	1,160	3,430	38,205	566,890	27,880
North Dakota	106,150	88,275	—	4,190	26,040	28,670	48,800	—	310,650
Ohio	—	470,940	—	3,875	6,500	1,200	193,160	17,934	74,240
Oklahoma........	235	28,600	140.0	3,380	1,025	—	6,120	—	198,900
Oregon..........	8,060	6,270	—	3,374	3,850	26,229	—	—	57,490
Pennsylvania	5,025	116,550	—	3,915	8,480	3,360	15,800	15,720	9,690
Rhode Island	—	NE	—	22	—	147	—	—	—
South Carolina	141	11,000	350.0	640	1,125	—	10,500	96,750	7,680
South Dakota	4,560	429,550	—	8,160	20,100	1,248	133,380	—	120,884
Tennessee	—	59,520	545.0	3,969	—	—	35,090	117,969	15,170
Texas	215	185,000	3,500.0	6,870	6,890	4,867	5,940	—	136,500
Utah	7,055	3,384	—	2,778	630	728	—	—	8,834
Vermont	—	NE	—	504	—	—	—	—	—
Virginia	4,270	25,200	139.8	2,604	—	1,380	11,040	98,625	11,025
Washington	33,800	19,000	—	3,156	1,125	93,225	—	—	157,425
West Virginia	—	2,720	—	1,157	200	—	—	2,380	456
Wisconsin........	3,380	404,150	—	6,370	18,300	30,895	51,700	4,230	7,635
Wyoming	7,310	7,620	—	2,445	1,408	120	—	—	6,790
UNITED STATES..	**352,445**	**9,761,085**	**13,366.2**	**151,338**	**167,122**	**477,754**	**2,756,794**	**1,529,647**	**2,550,383**

NE = Not estimated, bu = bushels, b = bales (480-lbs), t = tons, cwt = hundredweight.

Production of Principal U.S. Crops, 1988-98

Source: National Agricultural Statistics Service, U.S. Dept. of Agriculture

Year	Corn for grain (1,000 bu)	Oats (1,000 bu)	Barley (1,000 bu)	Sorghum for grain (1,000 bu)	All wheat (1,000 bu)	Rye (1,000 bu)	Flax-seed (1,000 bu)	Upland Cotton (1,000 b)	Cotton-seed (1,000 t)
1988......	4,928,681	217,375	289,994	576,686	1,812,201	14,689	1,615	15,412.5	6,061.8
1989......	7,531,953	373,587	404,203	615,420	2,036,618	13,647	1,215	12,196.6	4,677.4
1990......	7,934,028	357,654	422,196	573,303	2,729,778	10,176	3,812	15,505.4	5,968.5
1991......	7,474,765	243,851	464,326	584,860	1,980,139	9,734	6,200	17,614.3	6,925.5
1992......	9,476,698	294,229	455,090	875,022	2,466,798	11,440	3,288	16,219.5	6,230.1
1993......	6,336,470	206,770	398,041	534,172	2,396,440	10,340	3,480	16,134.6	6,343.2
1994......	10,102,735	229,008	374,862	649,206	2,320,981	11,341	2,922	19,662.0	7,603.9
1995......	7,373,876	162,027	359,562	460,373	2,182,591	10,064	2,211	17,532.2	6,848.7
1996......	9,293,435	155,273	395,751	802,974	2,285,133	9,016	1,602	18,413.5	7,143.5
1997[1].....	9,206,832	167,246	359,878	633,545	2,481,466	8,132	2,420	18,245.0	6,934.6
1998[2].....	9,761,085	167,122	352,445	519,933	2,550,383	11,795	6,708	13,366.2	5,181.6

Year	Tobacco (1,000 lb)	All hay (1,000 t)	Beans, dry edible (1,000 cwt)	Peas, dry edible (1,000 cwt)	Peanuts[3] (1,000 lb)	Soy-beans[4] (1,000 bu)	Potatoes (1,000 cwt)	Sweet potatoes (1,000 cwt)
1988......	1,369,500	125,736	19,253	3,868	3,980,917	1,548,841	356,438	10,945
1989......	1,367,188	144,706	23,729	3,883	3,989,995	1,923,666	370,444	11,358
1990......	1,626,380	146,212	32,379	2,372	3,602,770	1,925,947	402,110	12,594
1991......	1,664,372	152,073	33,765	3,715	4,926,570	1,986,539	417,622	11,203
1992......	1,721,671	146,903	22,615	2,535	4,284,416	2,190,354	425,367	12,005
1993......	1,613,319	146,799	21,913	3,292	3,392,415	1,870,958	428,693	11,053
1994......	1,582,896	150,060	29,028	2,255	4,247,455	2,516,694	467,054	13,395
1995......	1,268,538	154,166	30,812	4,765	4,247,455	2,176,814	443,606	12,906
1996......	1,517,334	149,457	27,960	2,671	3,661,205	2,382,364	498,633	13,456
1997[1].....	1,787,399	152,536	29,370	5,752	3,539,380	2,688,750	467,091	13,327
1998[2].....	1,529,647	151,338	30,828	5,934	3,931,275	2,756,794	477,754	11,887

Year	Rice (1,000 cwt)	Sugar-cane (1,000 t)	Sugar beets (1,000 t)	Pecans[5] (1,000 lb)	Apples (1,000 t)	Grapes (1,000 t)	Peaches (1,000 t)	Oranges[6] (1,000 bx)	Grape-fruit[6] (1,000 bx)
1988......	159,897	29,904	24,810	308,200	4,560.0	6,033.7	1,311.1	200,250	68,700
1989......	154,487	29,426	25,131	250,500	4,958.4	5,930.9	1,181.5	209,050	69,500
1990......	156,088	28,136	27,513	205,000	4,828.4	5,659.9	1,121.1	184,415	49,300
1991......	159,367	30,252	28,203	299,000	4,853.4	5,555.9	1,347.8	178,950	55,500
1992......	179,658	30,363	29,143	166,000	5,284.3	6,052.1	1,336.0	209,610	55,265
1993......	156,110	31,101	26,249	365,000	5,342.4	6,023.2	1,330.1	255,760	68,375
1994......	197,779	30,929	31,853	199,000	5,667.8	5,870.6	1,253.3	240,450	65,100
1995......	173,871	30,944	27,954	268,000	5,292.5	5,922.3	1,150.8	263,605	71,050
1996......	171,321	29,462	26,680	221,500	5,196.0	5,554.3	1,058.2	263,890	66,200
1997[1].....	182,992	31,709	29,886	335,000	5,161.9	7,290.9	1,312.3	292,620	70,200
1998[2].....	188,051	33,717	32,660	146,400	5,693.7	5,903.0	1,214.7	315,525	64,150

(1) Revised. (2) Preliminary. (3) Harvested for nuts. (4) Harvested for beans. (5) Utilized production only. (6) Crop year ending in year cited.

Principal U.S. Crops: Area Planted and Harvested, 1996-98

Source: National Agricultural Statistics Service, U.S. Dept. of Agriculture

(in thousand acres)

STATE	Area planted[1] 1996	1997	1998	Area harvested[1] 1996	1997	1998	STATE	Area planted[1] 1996	1997	1998	Area harvested[1] 1996	1997	1998
AL.....	2,228	2,310	2,252	2,142	2,139	2,092	NE....	18,801	19,142	18,960	18,222	18,693	18,565
AZ.....	835	814	775	831	806	769	NV....	515	523	513	512	521	510
AR....	8,692	8,497	8,550	8,550	8,354	8,265	NH....	84	79	71	82	78	70
CA....	5,202	5,193	4,944	4,760	4,664	4,420	NJ....	427	439	450	394	416	408
CO....	6,276	6,489	6,277	5,530	6,055	5,927	NM....	1,318	1,278	1,225	933	1,124	943
CT....	119	113	101	114	108	96	NY....	3,011	3,046	2,994	2,934	2,987	2,934
DE....	505	535	519	497	522	505	NC....	4,881	5,073	5,017	4,650	4,828	4,786
FL.....	1,109	1,120	1,124	1,084	1,089	1,028	ND....	22,501	22,273	20,801	22,087	21,152	20,131
GA....	4,306	4,333	4,067	3,960	3,957	3,442	OH....	10,273	10,748	10,651	10,092	10,532	10,520
HI.....	46	34	34	46	34	34	OK....	11,111	10,850	10,607	8,924	9,229	8,597
ID.....	4,517	4,473	4,509	4,393	4,317	4,361	OR....	2,404	2,329	2,235	2,324	2,248	2,158
IL.....	23,801	23,600	23,751	23,063	23,386	23,552	PA....	4,140	4,304	4,347	4,035	4,195	4,247
IN.....	12,648	12,764	13,029	12,395	12,560	12,696	RI.....	12	12	14	12	12	14
IA.....	24,189	24,709	24,891	23,982	24,467	24,688	SC....	1,971	1,990	1,902	1,892	1,910	1,757
KS....	24,151	23,324	23,065	20,879	22,526	22,143	SD....	16,910	16,860	16,545	16,235	15,986	16,113
KY....	5,844	5,531	5,869	5,640	5,268	5,637	TN....	4,899	4,799	4,836	4,603	4,547	4,574
LA.....	4,035	4,095	4,055	3,994	4,019	3,752	TX....	24,343	23,475	23,784	18,202	20,137	16,814
ME....	325	295	285	314	288	279	UT....	1,133	1,131	1,105	1,064	1,079	1,047
MD....	1,575	1,555	1,470	1,521	1,506	1,415	VT....	347	369	357	332	361	352
MA....	126	124	132	120	119	129	VA....	2,925	2,842	2,931	2,781	2,705	2,768
MI.....	6,953	6,871	6,790	6,694	6,740	6,662	WA....	4,461	4,353	4,382	4,378	4,215	4,251
MN....	20,051	20,175	20,310	19,722	19,749	19,990	WV....	657	661	659	646	654	652
MS....	4,880	4,740	4,810	4,787	4,666	4,717	WI....	8,170	8,191	8,082	7,859	7,836	7,792
MO....	13,360	13,387	13,629	12,879	13,210	13,330	WY....	1,831	1,886	1,779	1,780	1,819	1,692
MT....	10,734	10,283	9,787	10,292	9,799	9,184	**U.S.**[2]..	**333,682**	**332,072**	**329,323**	**313,202**	**317,662**	**310,847**

(1) Crops included in area planted are corn, sorghum, oats, barley, winter wheat, rye, durum wheat, other spring wheat, rice, soybeans, peanuts, sunflower, cotton, dry edible beans, potatoes, and sugar beets. Harvested acreage is used for all hay, tobacco, and sugarcane in computing total area planted. Includes double-cropped acres and unharvested small grains planted as cover crops. (2) State figures do not add to U.S. totals because of sunflower and sugar-beet unallocated acreage.

Average Prices Received by U.S. Farmers, 1940-98

Source: National Agricultural Statistics Service, U.S. Dept. of Agriculture

Figures below represent dollars per 100 lb for hogs, beef cattle, veal calves, sheep, lamb, and milk (wholesale); dollars per head for milk cows; cents per lb for chickens, broilers, turkeys, and wool; cents per dozen for eggs; weighted calendar year prices for livestock and livestock products other than wool. For 1943-63, wool prices are weighted on marketing year basis. The marketing year was changed in 1964 from a calendar year to a Dec.-Nov. basis for hogs, chickens, broilers, and eggs.

Year	Hogs	Cattle (beef)	Calves (veal)	Sheep	Lambs	Milk cows	Milk	Chickens (excl. broilers)	Broilers	Turkeys	Eggs	Wool
1940...	5.39	7.56	8.83	3.95	8.10	61	1.82	13.0	17.3	15.2	18.0	28.4
1950...	18.00	23.30	26.30	11.60	25.10	198	3.89	22.2	27.4	32.8	36.3	62.1
1960...	15.30	20.40	22.90	5.61	17.90	223	4.21	12.2	16.9	25.4	36.1	42.0
1970...	22.70	27.10	34.50	7.51	26.40	332	5.71	9.1	13.6	22.6	39.1	35.4
1975...	46.10	32.20	27.20	11.30	42.10	412	8.75	9.9	26.3	34.8	54.5	44.8
1980...	38.00	62.40	76.80	21.30	63.60	1,190	13.05	11.0	27.7	41.3	56.3	88.1
1985...	44.00	53.70	62.10	23.90	67.70	860	12.76	14.8	30.1	49.1	57.1	63.3
1986...	49.30	52.60	61.10	25.60	69.00	820	12.51	12.5	34.5	47.1	61.6	66.8
1987...	51.20	61.10	78.50	29.50	77.60	920	12.54	11.0	28.7	34.8	54.9	91.7
1988...	42.30	66.60	89.20	25.60	69.10	990	12.26	9.2	33.1	38.6	52.8	138.0
1989...	42.50	69.50	90.80	24.40	66.10	1,030	13.56	14.9	36.6	40.9	68.9	124.0
1990...	53.70	74.60	95.60	23.20	55.50	1,160	13.74	9.3	32.6	39.4	70.9	80.0
1991...	49.10	72.70	98.00	19.70	52.20	1,100	12.27	7.1	30.8	38.4	67.8	55.0
1992...	41.60	71.30	89.00	25.80	59.50	1,130	13.15	8.6	31.8	37.7	57.6	74.0
1993...	45.20	72.60	91.20	28.60	64.40	1,160	12.84	10.0	34.0	39.0	63.4	51.0
1994...	39.90	66.70	87.20	30.90	65.60	1,170	13.01	7.6	35.0	40.4	61.4	78.0
1995...	40.50	61.80	73.10	28.00	78.20	1,130	12.78	6.5	34.4	41.6	62.4	104.0
1996...	51.90	58.70	58.40	29.90	82.20	1,090	14.75	6.6	38.1	43.3	74.9	70.0
1997[1]..	52.90	63.10	78.90	37.90	90.30	1,100	13.36	7.7	37.7	39.9	70.3	84.0
1998[2]..	34.40	59.60	78.80	30.60	72.30	1,120	15.41	8.0	39.3	38.0	65.5	60.0

Figures below represent cents per lb for cotton, apples, and peanuts; dollars per bushel for oats, wheat, corn, barley, and soybeans; dollars per 100 lb for rice, sorghum, and potatoes; dollars per ton for cottonseed and baled hay; weighted crop year prices. The marketing year is described as follows: apples, June-May; wheat, oats, barley, hay, and potatoes, July-June; cotton, rice, peanuts, and cottonseed, Aug.-July; soybeans, Sept.-Aug.; and corn and sorghum grain, Oct.-Sept.

Year	Corn	Wheat	Upland cotton*	Oats	Barley	Rice	Soybeans	Sorghum	Peanuts	Cottonseed	Hay	Potatoes	Apples
1940...	0.62	0.67	9.8	0.30	0.39	1.80	0.89	0.87	3.7	21.70	9.78	0.85	NA
1950...	1.52	2.00	39.9	0.79	1.19	5.09	2.47	1.88	10.9	86.60	21.10	1.50	NA
1960...	1.00	1.74	30.1	0.60	0.84	4.55	2.13	1.49	10.0	42.50	21.70	2.00	2.7
1970...	1.33	1.33	21.9	0.62	0.97	5.17	2.85	2.04	12.8	56.40	26.10	2.21	6.5
1975...	2.54	3.55	51.1	1.45	2.42	8.35	4.92	4.21	19.0	97.00	52.10	4.48	8.8
1980...	3.11	3.91	74.4	1.79	2.86	12.80	7.57	5.25	25.1	129.00	71.00	6.55	12.1
1985...	2.23	3.08	56.8	1.23	1.98	6.53	5.05	3.45	24.4	66.00	67.60	3.92	17.3
1986...	1.50	2.42	51.5	1.21	1.61	3.75	4.78	2.45	29.2	80.00	59.70	5.03	19.1
1987...	1.94	2.57	63.7	1.56	1.81	7.27	5.88	3.04	28.0	82.50	65.00	4.38	12.7
1988...	2.54	3.72	55.6	2.61	2.80	6.83	7.42	4.05	28.0	118.00	85.20	6.02	17.4
1989...	2.36	3.72	63.6	1.49	2.42	7.35	5.69	3.75	28.0	105.00	85.40	7.36	13.9
1990...	2.28	2.61	67.1	1.14	2.14	6.68	5.74	3.79	34.7	121.00	80.60	6.08	20.9
1991...	2.37	3.00	56.8	1.21	2.10	7.58	5.58	4.01	28.3	71.00	71.20	4.96	25.1
1992...	2.07	3.24	53.7	1.32	2.04	5.89	5.56	3.38	30.0	97.50	74.30	5.52	19.5
1993...	2.50	3.26	58.1	1.36	1.99	7.98	6.40	4.13	30.4	113.00	84.70	6.18	18.4
1994...	2.26	3.45	72.0	1.22	2.03	6.78	5.48	3.80	28.9	101.00	86.70	5.58	18.6
1995...	3.24	4.55	75.4	1.67	2.89	9.15	6.72	5.69	29.3	106.00	82.20	6.77	24.0
1996...	2.71	4.30	69.3	1.96	2.74	9.96	7.35	4.17	28.1	126.00	95.80	4.93	20.8
1997[1]..	2.43	3.38	65.2	1.60	2.38	9.70	6.47	3.95	28.3	121.00	100.00	5.62	22.1
1998[2]..	1.90	2.65	64.2	1.10	1.98	8.50	5.35	3.10	25.7	129.00	84.60	5.24	17.1

*Beginning in 1964, 480-lb net weight bales. NA = Not available. (1) Revised. (2) Preliminary.

U.S. Grain Storage Capacity, by Region

Source: National Agricultural Statistics Service, U.S. Dept. of Agriculture

(in thousand acres)

Region	1993-94	1994-95	1995-96	1996-97	1997-98	1998-99
Northeast[1].............................	443	431	399	394	377	378
Southeast[2]	988	936	915	883	865	841
Delta[3]	574	564	559	540	531	528
Eastern Corn Belt[4]	5,183	5,128	5,115	5,025	4,988	4,985
Western Corn Belt[5]	5,282	5,144	5,062	5,003	4,891	4,896
Southern Plains[6]........................	1,504	1,439	1,368	1,319	1,177	1,098
Central Plains[7]	3,392	3,214	3,267	3,196	3,134	3,102
Northern Plains[8]	2,190	2,112	2,091	2,033	2,033	1,996
Pacific Northwest[9]......................	681	661	645	652	636	633
West[10]................................	162	152	142	140	139	140
Unallocated............................	356	331	311	281	271	291
U.S. TOTAL...........................	**20,775**	**20,112**	**19,874**	**19,466**	**19,042**	**18,888**

(1) ME, MA, NH, VT, CT, RI, NY, NJ, PA, MD, DE, WV. (2) VA, KY, TN, NC, SC, GA, AL, FL. (3) LA, AR, MS. (4) WI, IL, IN, MI, OH. (5) MN, IA, MO. (6) TX, NM, OK. (7) CO, KS, NE. (8) MT, WY, ND, SD. (9) WA, OR, ID. (10) CA, NV, UT, AZ.

World Wheat, Rice, and Corn Production, 1998

Source: UN Food and Agriculture Organization

(in thousands of metric tons)

COUNTRY	Wheat	Rice[1]	Corn	COUNTRY	Wheat	Rice[1]	Corn
Afghanistan	2,834	450	240	Madagascar	10F	2,447	152
Argentina	10,000*	1,036	19,100	Malaysia	—	1,940*	50*
Australia	21,855	1,335	340	Mexico	3,241	483	18,411
Austria	1,342	—	1,573F	Moldova	1,047	—	1,229
Bangladesh	1,803	28,292	2	Morocco	4,388	20	200
Belgium-Lux.	1,803	—	251	Myanmar	92	16,651	308
Brazil	2,222	7,795	29,296	Nepal	1,030	3,640F	1,476
Bulgaria	3,295*	12F	1,340*	Netherlands	1,040	—	77*
Cambodia	—	3,515	61	New Zealand	256	—	176
Canada	24,400	—	8,912	Nigeria	98*	3,268F	5,858*
Chile	1,400	104	943	Pakistan	18,694	6,587	1,260F
China	110,000*	192,971*	125,395*	Peru	146	1,548	932
Colombia	25*	1,850F	1,200*	Philippines	—	8,554	3,823
Croatia	1,020	—	1,982	Poland	9,537	—	496
Cuba	—	388F	125F	Portugal	143	167*	850
Czech Rep.	3,931	—	173	Romania	4,804	5	8,623
Denmark	4,933	—	—	Russia	26,900	410	800
Ecuador	15*	1,071F	687F	Slovakia	1,789	—	637
Egypt	6,093	5,585F	5,330F	South Africa	1,469	3F	7,574
Ethiopia	1,143	—	2,344	Spain	5,347	754	4,154
Finland	491*	—	—	Sri Lanka	—	2,692	33
France	39,862	113	14,426	Sweden	2,287	—	—
Germany	20,187	—	2,781	Switzerland	616	—	185
Greece	2,058	209	1,816	Syria	4,070	—	303F
Hungary	4,974	7F	6,500	Thailand	—	23,240	4,986
India	66,000*	122,244*	10,000*	Turkey	21,000	317	2,300
Indonesia	—	48,472	10,058	Turkmenistan	600F	55F	50F
Iran	12,000*	2,600F	1,000	Ukraine	14,937	73*	2,300
Iraq	1,100F	250F	125*	United Kingdom	15,449	—	—
Ireland	673	—	—	United States	69,410	8,529	247,943
Italy	8,250	1,400F	9,030	Uruguay	500*	865	225
Japan	569*	11,200*	—	Uzbekistan	3,094	340*	200*
Kazakhstan	4,746	236	166	Venezuela	—	672	938
Kenya	250F	55F	2,450	Vietnam	—	29,141	1,612
Korea, North	165*	2,063	1,765	Yugoslavia	2,937*	—	5,600*
Korea, South	7F	7,312F	74F	Zimbabwe	280*	—	1,418
Laos	—	1,674	106	**WORLD, TOTAL**	**588,842**	**563,188**	**604,012**

* Unofficial figure. F=Food and Agriculture Organization (FAO) estimate. Where production is small or nonexistent, — is indicated. Because not all countries are reported on this table, country totals do not add to world totals. (1) Rice paddy.

Wheat, Rice, and Corn—Exports and Imports of 10 Leading Countries

Source: UN Food and Agriculture Organization

(in thousands of metric tons; ranked for 1997)

LEADING EXPORTERS	EXPORTS[1] Wheat 1995	1996	1997	LEADING IMPORTERS	IMPORTS[1] Wheat 1995	1996	1997
U.S.	32,420	31,150	25,768	Italy	5,078	6,261	6,976
Australia	7,818	14,567	19,377	Egypt	5,069	6,008*	6,902*
Canada	16,960	16,520	18,857	Japan	5,965	5,927	6,315
France	16,310	14,550	14,600	Iran	3,100*	3,874*	6,017*
Argentina	6,913	3,532	8,766	Brazil	6,135	7,663	4,850
Germany	3,681	4,199	3,861	Indonesia	4,054	4,116	3,611
United Kingdom	2,669	3,674	3,645	Algeria	3,504	1,971	3,508
Kazakhstan	2,485	1,909	2,792	Korea (South)	2,341	2,222	3,325
Denmark	1,540	823	1,059	Spain	2,757	2,103	2,973
Hungary	2,764	299	970	Belgium-Lux.	2,719	2,776	2,854

	Rice 1995	1996	1997		Rice 1995	1996	1997
Thailand	6,197	5,454	3,240	Iran	1,633*	1,150	973
Vietnam	1,988	3,003	3,000F	Brazil	870	792	816
U.S.	3,083	2,640	2,296	Nigeria	300*	345*	731*
India	4,913	2,511	2,133*	Philippines	263	866	722
Pakistan	1,852	1,600	1,767	Iraq	225	214	684
China	235	356	1,009	Saudi Arabia	522	721	665
Australia	541	566	654	Malaysia	427	577	639
Uruguay	462	603	648	South Africa	466	482	591
Italy	523	608	632	Japan	28	444	568
Argentina	390	259	537	Côte d'Ivoire	404	313	470

	Corn 1995	1996	1997		Corn 1995	1996	1997
U.S.	60,240	52,410	41,791	Japan	16,580	16,003	16,097
Argentina	6,000	6,424	10,965	Korea (South)	9,035	8,678	8,312
France	6,474	6,651	7,340	China	11,702	6,428	5,786
China	112	158	6,617	Egypt	2,425	2,471*	3,059*
South Africa	1,508	1,948	1,690	Malaysia	2,383	2,227	2,744
Hungary	600	129	1,192	Mexico	2,686	5,842	2,518
Zimbabwe	287	234	402	Spain	2,912	2,031	2,503
Brazil	11	351	358	Netherlands	1,589	1,453	1,769
Germany	244	276	353	Colombia	1,153	1,700	1,734
Canada	443	512	263	United Kingdom	1,501	1,373	1,472

* Unofficial figure. F=Food and Agriculture Organization (FAO) estimate. (1) By marketing years.

World Commercial Catch of Fish, Crustaceans, and Mollusks[1], by Major Fishing Areas, 1992-97[2]

Source: U.S. Dept. of Commerce, Natl. Oceanic and Atmospheric Admin., Natl. Marine Fisheries Service

(in thousands of metric tons; live weight)

Area	1992	1993	1994	1995	1996	1997
Marine						
Pacific Ocean	54,378	55,691	61,906	61,917	63,562	62,518
Atlantic Ocean	24,343	23,687	23,675	24,777	24,764	25,903
Indian Ocean	7,363	7,892	7,758	8,045	8,232	8,358
TOTAL	86,084	87,270	93,339	94,739	97,558	96,780
Inland Waters						
N. America	584	540	524	531	554	598
S. America	357	375	403	443	445	484
Europe	504	498	509	529	513	519
Former USSR	667	536	471	417	414	401
Asia	11,668	13,211	15,077	17,321	19,430	21,267
Africa	1,839	1,903	1,843	2,039	2,005	2,065
Oceania	25	23	21	23	22	23
TOTAL	15,644	17,087	18,850	21,303	23,384	25,358
GRAND TOTAL	101,728	104,356	112,189	116,042	119,942	122,138

(1) Does not include marine mammals and aquatic plants. (2) Revised back to 1992.

Commercial Catch of Fish, Crustaceans, and Mollusks[1], by Selected Country, 1992-97[2]

Source: U.S. Dept. of Commerce, Natl. Oceanic and Atmospheric Admin., Natl. Marine Fisheries Service

(in thousands of metric tons; live weight)

COUNTRY	1992	1993	1994	1995	1996	1997
China	16,579	19,708	23,834	28,418	31,937	35,038
Peru	7,508	9,009	12,005	8,943	9,522	7,877
Japan	8,502	8,081	7,398	6,787	6,765	6,689
Chile	6,502	6,036	7,839	7,591	6,909	6,084
United States[3]	5,604	5,941	5,926	5,638	5,395	5,448
India	4,233	4,546	4,738	4,906	5,258	5,378
Russia	5,611	4,461	3,781	4,374	4,730	4,715
Indonesia	3,439	3,685	3,913	4,139	4,291	4,404
Thailand	3,246	3,385	3,522	3,573	3,515	3,488
Norway	2,561	2,588	2,570	2,803	2,960	3,223

(1) Does not include marine mammals and aquatic plants. (2) Revised back to 1992. (3) Includes weight of clam, oyster, scallop, and other mollusk shells. This weight is not included in U.S. landings statistics shown elsewhere.

U.S. Commercial Landings of Fish and Shellfish, 1986-98[1]

Source: U.S. Dept. of Commerce, Natl. Oceanic and Atmospheric Admin., Natl. Marine Fisheries Service

Year	Landings for human food		Landings for industrial purposes[2]		TOTAL	
	mil lb	mil dollars	mil lb	mil dollars	mil lb	mil dollars
1986	3,393	$2,641	2,638	$122	6,031	$2,763
1987	3,946	2,979	2,950	136	6,896	3,115
1988	4,588	3,362	2,604	158	7,192	3,520
1989	6,204	3,111	2,259	127	8,463	3,238
1990	7,041	3,366	2,363	156	9,404	3,522
1991	7,031	3,169	2,453	139	9,484	3,308
1992	7,618	3,531	2,019	147	9,637	3,678
1993	8,214	3,317	2,253	154	10,467	3,471
1994	7,936	3,751	2,525	95	10,461	3,846
1995	7,667	3,625	2,121	145	9,788	3,770
1996	7,474	3,355	2,091	132	9,565	3,487
1997	7,244	3,285	2,598	163	9,842	3,448
1998	7,173	3,009	2,021	119	9,194	3,128

Note: Data does not include landings outside the 50 states or products of aquaculture, except oysters and clams. (1) Statistics on landings are shown in round weight for all items except univalve and bivalve mollusks such as clams, oysters, and scallops, which are shown in weight of meats (excluding the shell). All data are preliminary. (2) Processed into meal, oil, solubles, and shell products or used as bait or animal food.

U.S. Domestic Landings, by Regions, 1997-98[1]

Source: U.S. Dept. of Commerce, Natl. Oceanic and Atmospheric Admin., Natl. Marine Fisheries Service

Region	1997		1998	
	1,000 lb	1,000 dollars	1,000 lb	1,000 dollars
New England	640,621	$555,913	595,611	$537,442
Middle Atlantic	236,881	199,912	261,686	181,177
Chesapeake	688,142	169,319	653,365	179,869
South Atlantic	298,685	213,386	239,912	197,010
Gulf	1,790,310	758,681	1,536,583	718,925
Pacific Coast and Alaska	6,125,787	1,464,962	5,843,268	1,235,709
Great Lakes	26,256	16,797	27,116	16,274
Hawaii	36,568	68,693	36,426	62,065
TOTAL	9,843,250	3,447,663	9,193,967	3,128,471

(1) Landings reported in round (live) weight items except for univalve and bivalve mollusks (e.g., clams, oysters, scallops), which are reported in weight of meats (excluding shell). Landings for Mississippi River Drainage Area states not included (not available).

EMPLOYMENT

Employment and Unemployment in the U.S., 1950-98

Source: Bureau of Labor Statistics, U.S. Dept. of Labor

(civilian labor force, persons 16 years of age and older; annual averages; in thousands)

Year[1]	Employed	Unemployed	Unemployment rate	Year[1]	Employed	Unemployed	Unemployment rate
1950	58,918	3,288	5.0%	1990[2]	118,793	7,047	5.6%
1960	65,778	3,852	5.5	1991	117,718	8,628	6.8
1970	78,678	4,093	4.9	1992	118,482	9,613	7.5
1980	99,303	7,637	7.1	1993	120,259	8,940	6.9
1985	107,150	8,312	7.2	1994[3]	123,060	7,996	6.1
1986	109,597	8,237	7.0	1995	124,900	7,404	5.6
1987	112,440	7,425	6.2	1996	126,708	7,236	5.4
1988	114,988	6,701	5.5	1997[4]	129,558	6,739	4.9
1989	117,342	6,528	5.3	1998[4]	131,463	6,210	4.5

(1) **Early unemployment rates:** 1915, 9.7; 1916, 4.8; 1917, 4.8; 1918, 1.4; 1919, 2.3; 1920, 4.0; 1921, 11.9; 1922, 7.6; 1923, 3.2; 1924, 5.5; 1925, 4.0; 1926, 1.9; 1927, 4.1; 1928, 4.4; 1929, 3.2; 1930, 8.7; 1931, 15.9; 1932, 23.6; 1933, 24.9; 1934, 21.7; 1935, 20.1; 1936, 16.9; 1937, 14.3; 1938, 19.0; 1939, 17.2; 1940 (Persons 14 years of age and older), 14.6. (2) Beginning in 1990, data incorporate 1990 census-based population controls, adjusted for the estimated undercount. (3) Beginning in 1994, not strictly comparable with prior years, because of a major redesign of the survey used. (4) 1997 and 1998 not strictly comparable with 1994-96 because of revisions in population controls used in the household survey.

Unemployment Insurance Data, by State, 1998

Source: Employment and Training Admin., U.S. Dept. of Labor; state programs only

STATE	Monetarily eligible claimants	First payments	Final payments	Initial claims	Benefits paid	Average weekly benefit	Employers subject to state law
AL.	169,742	145,482	27,920	328,699	$191,465,234	$152.29	87,092
AK	47,865	43,750	17,706	100,310	102,257,513	176.00	15,995
AZ.	93,787	68,329	19,313	163,565	133,868,133	148.75	98,462
AR	115,754	85,547	24,192	199,151	167,267,262	185.54	57,884
CA	1,434,799	1,074,824	401,353	2,831,990	2,438,277,739	154.37	855,931
CO	94,074	57,354	21,516	112,583	151,645,140	224.88	122,963
CT	122,647	108,712	27,195	196,101	325,065,928	214.17	94,654
DE	28,695	24,695	5,021	45,037	62,322,009	197.20	23,429
DC	23,191	18,278	9,842	27,974	58,759,586	231.42	24,799
FL.	311,169	240,236	93,990	415,851	638,478,441	205.37	366,576
GA	245,842	174,750	47,404	359,339	239,754,196	180.42	176,528
HI	45,865	36,819	11,833	95,506	141,649,876	269.09	27,398
ID	56,024	46,198	12,688	107,794	91,110,651	195.35	36,663
IL	340,259	299,676	95,300	600,164	1,042,859,964	226.87	272,705
IN	168,093	127,747	34,719	253,328	223,109,534	201.31	124,302
IA	92,678	72,383	12,922	139,641	144,970,197	214.16	67,964
KS	61,651	49,164	12,517	99,334	117,277,126	214.17	65,379
KY	138,614	109,645	18,928	241,694	211,129,913	185.64	82,617
LA.	98,451	67,825	17,902	156,007	142,278,850	148.40	92,661
ME	47,314	40,498	21,572	79,237	81,306,988	148.61	37,030
MD	140,273	101,399	31,832	210,914	283,963,091	202.32	125,620
MA	224,551	183,120	55,438	355,102	694,415,091	261.00	163,512
MI	548,266	408,121	87,039	777,438	953,148,501	234.63	210,596
MN	130,041	106,529	28,320	200,334	315,604,973	256.95	123,611
MS	89,752	60,436	14,197	174,036	100,327,276	146.01	51,409
MO	201,953	140,357	35,133	367,849	259,912,808	163.68	126,036
MT	34,102	26,956	7,605	57,059	52,837,849	173.20	30,102
NE	39,667	27,706	7,447	57,000	41,841,173	164.09	43,488
NV	88,663	63,179	19,775	121,539	171,009,647	207.51	40,707
NH	24,254	15,574	653	34,276	23,690,012	183.12	36,957
NJ.	309,475	265,724	112,321	494,940	1,030,756,305	265.66	218,631
NM	37,816	32,512	9,886	60,461	79,554,023	169.38	40,613
NY	515,433	471,413	228,307	994,673	1,491,962,919	205.78	447,759
NC	348,228	222,716	37,978	750,471	378,426,836	206.65	163,349
ND	19,095	12,278	4,195	30,035	33,247,447	190.42	18,650
OH	336,054	263,215	46,265	555,738	656,907,708	215.45	230,942
OK	63,412	46,857	11,562	103,792	88,755,212	188.71	73,445
OR	174,617	147,539	39,209	366,571	399,048,287	215.12	97,413
PA.	517,245	418,539	104,690	1,107,439	1,316,049,001	237.77	243,342
PR	152,520	145,984	66,331	274,235	247,006,890	99.35	47,181
RI	52,792	46,742	13,611	101,286	133,823,829	226.84	31,592
SC	156,855	101,526	18,758	306,750	157,752,812	174.24	83,988
SD	11,392	8,427	848	19,629	15,230,083	161.70	21,607
TN	219,081	165,387	47,896	408,255	300,943,630	174.42	108,194
TX.	633,705	337,812	154,526	709,123	888,602,229	207.79	378,870
UT	52,069	37,301	10,078	63,617	81,062,786	194.98	48,686
VT.	23,109	19,243	2,917	36,745	39,512,240	180.52	19,965
VI	1,965	1,764	654	2,715	4,112,273	154.31	NA
VA.	158,338	101,245	22,136	275,830	180,907,809	182.86	155,188
WA	279,962	177,642	59,380	517,811	726,706,041	259.61	177,463
WV	62,778	51,885	10,482	88,429	114,555,007	186.97	38,237
WI.	259,242	219,771	36,907	485,025	443,860,128	199.97	119,539
WY	18,693	11,079	2,690	23,336	22,904,392	189.49	18,167
U.S.	9,661,912	7,331,890	2,262,899	16,685,758	$18,433,292,591	$199.91	6,465,885

NA = Not available.

Unemployment Rates, by Selected Country, 1975-98

Source: Bureau of Labor Statistics, U.S. Dept. of Labor; civilian labor force, seasonally adjusted; Oct. 1999

Time Period	U.S.	Australia	Canada	France	Germany[1]	Italy[2]	Japan	Sweden	UK
1975	8.5	4.9	6.9	4.2	3.4	3.4	1.9	1.6	4.6
1980	7.1	6.1	7.5	6.5	2.8	4.4	2.0	2.0	7.0
1981	7.6	5.8	7.6	7.6	4.0	4.9	2.2	2.5	10.5
1982	9.7	7.2	11.0	8.3	5.6	5.4	2.4	3.1	11.3
1983	9.6	10.0	11.9	8.6	6.9[3]	5.9	2.7	3.5	11.8
1984	7.5	9.0	11.3	10.0	7.1	5.9	2.8	3.1	11.7
1985	7.2	8.3	10.5	10.5	7.2	6.0	2.6	2.8	11.2
1986	7.0	8.1	9.6	10.6	6.6	7.5[3]	2.8	2.6	11.2
1987	6.2	8.1	8.9	10.8	6.3	7.9	2.9	2.2[3]	10.3
1988	5.5	7.2	7.8	10.3	6.3	7.9	2.5	1.9	8.6
1989	5.3	6.2	7.5	9.6	5.7	7.8	2.3	1.6	7.2
1990	5.6[3]	6.9	8.1	9.1	5.0	7.0	2.1	1.8	6.9
1991	6.8	9.6	10.4	9.6	4.3	6.9[3]	2.1	3.1	8.8
1992	7.5	10.8	11.3	10.4[3]	4.6	7.3	2.2	5.6	10.1
1993	6.9	10.9	11.2	11.8	5.7	10.2[3]	2.5	9.3	10.5
1994	6.1[3]	9.7	10.4	12.3	6.5	11.3	2.9	9.6	9.7
1995	5.6	8.5	9.5	11.8	9.4	12.0[P]	3.2	9.1	8.7
1996	5.4	8.6	9.7	12.5	10.4	12.1[P]	3.4	9.9	8.2[P]
1997	4.9	8.6	9.2	12.4	7.8	12.3	3.4	10.1	7.0
1998	4.5	8.0	8.3	11.8	7.5	12.3	4.1	8.4	6.3
1st quarter.....	4.6	8.1	8.6	12.0	7.7	12.2	3.7	8.7	6.4
2d quarter.....	4.4	8.1	8.4	11.8	7.6	12.3	4.2	8.6	6.2
3d quarter.....	4.5	8.2	8.3	11.7	7.4	12.4	4.3	8.4	6.3
4th quarter	4.4	7.7	8.0	11.5	7.3	12.4	4.4	7.6	6.2

P=Preliminary. **NOTE:** For the sake of comparisons, U.S. unemployment rate concepts were applied to unemployment data for other countries. Quarterly and monthly figures for France were calculated by applying annual adjustment factors to current published data and are less precise indicators of unemployment under U.S. concepts than the annual figures. (1) Former West Germany only, through 1994; from 1995 on figures are for unified Germany and not adjusted by BLS. (2) Quarterly rates are for first month of quarter. (3) As a result of revisions in survey methodology, there are breaks in the data series for the U.S. (1990, 1994), France (1992), Germany (1983), Italy (1986, 1991, 1993), and Sweden (1987); data prior to a survey change are not fully comparable to data after a survey change. (—)Dashes indicate unavailable data.

Employed Persons in the U.S., by Occupation and Sex, 1997-98

Source: Bureau of Labor Statistics, U.S. Dept. of Labor

(in thousands)

	TOTAL 16 years and older		MEN 16 years and older		WOMEN 16 years and older	
	1997	1998	1997	1998	1997	1998
TOTAL	**129,558**	**131,463**	**69,685**	**70,693**	**59,873**	**60,771**
Managerial and professional specialty..................	37,686	38,937	19,249	19,867	18,437	19,070
Executive, administrative, and managerial	18,440	19,054	10,271	10,585	9,170	8,169
Officials and administrators, public administration	694	719	372	389	322	330
Other executive, administrative, and managerial	13,143	13,635	7,951	8,181	5,191	5,454
Management-related occupations	4,604	4,700	1,948	2,015	2,655	2,685
Professional specialty	19,245	19,889	8,978	9,282	10,267	10,602
Engineers	2,036	2,052	1,841	1,824	195	228
Mathematical and computer scientists..............	1,494	1,747	1,040	1,243	454	505
Natural scientists	529	519	385	359	164	161
Health diagnosing occupations	1,027	1,083	769	798	259	285
Health assessment and treating occupations	2,886	2,898	391	428	2,495	2,470
Teachers, college and university	869	919	498	530	371	389
Teachers, except college and university	4,798	4,962	1,166	1,225	3,632	3,737
Lawyers and judges	925	951	678	679	247	272
Other professional specialty occupations	4,681	4,750	2,231	2,196	2,450	2,555
Technical, sales, and administrative support	36,309	38,521	13,760	13,792	24,549	24,728
Technicians and related support	4,214	4,261	2,028	1,976	2,186	2,285
Sales occupations	15,734	15,850	7,840	7,875	7,894	7,975
Administrative support, including clerical	18,361	18,410	3,892	3,941	14,469	14,469
Service occupations...............................	17,537	17,836	7,122	7,222	10,416	10,614
Precision production, craft, and repair	14,124	14,411	12,868	13,208	1,256	1,203
Mechanics and repairers	4,675	4,786	4,494	4,592	181	194
Construction trades	5,378	5,594	5,251	5,485	127	109
Other precision production, craft, and repair	4,071	4,031	3,123	3,131	948	900
Operators, fabricators, and laborers	18,389	18,258	1,358	13,769	4,540	4,487
Machine operators, assemblers, and inspectors	7,962	7,791	4,962	4,882	3,000	2,909
Transportation and material moving occupations	5,389	5,353	4,872	4,818	518	545
Motor vehicle operators	4,089	4,069	3,629	3,601	461	468
Other transportation and material moving occupations ..	1,300	1,294	1,243	1,217	57	77
Handlers, equipment cleaners, helpers, and laborers.....	5,048	5,102	4,025	4,069	1,023	1,033
Construction laborers	811	821	773	784	37	37
Other handlers, equipment cleaners, etc.	2,307	4,282	1,794	3,285	514	996
Farming, forestry, and fishing	3,503	3,502	2,828	2,835	675	668

NOTE: Totals may not add because of independent rounding.

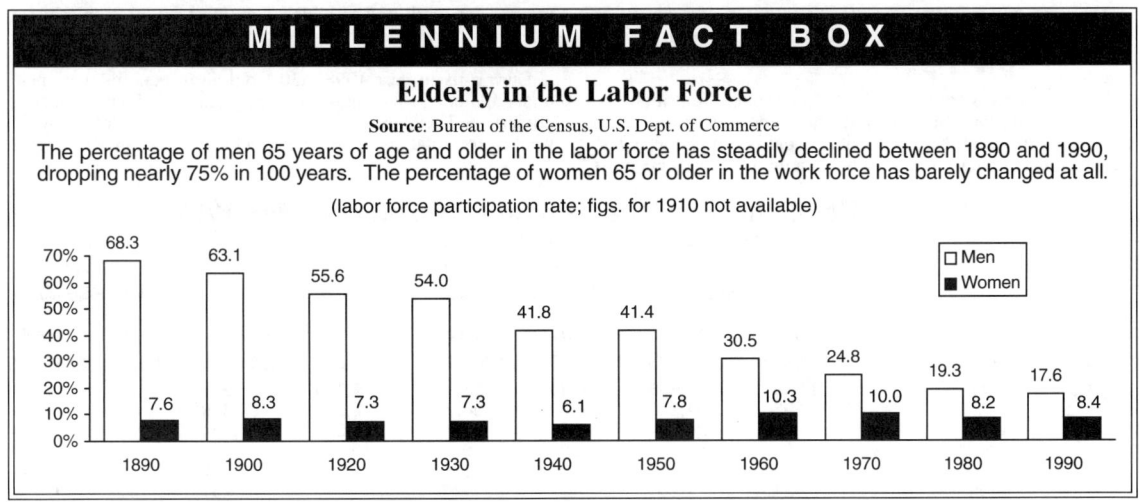

MILLENNIUM FACT BOX

Elderly in the Labor Force

Source: Bureau of the Census, U.S. Dept. of Commerce

The percentage of men 65 years of age and older in the labor force has steadily declined between 1890 and 1990, dropping nearly 75% in 100 years. The percentage of women 65 or older in the work force has barely changed at all.

(labor force participation rate; figs. for 1910 not available)

U.S. Unemployment Rates by Selected Characteristics, 1996-99[1]

Source: Bureau of Labor Statistics, U.S. Dept. of Labor; seasonally adjusted, quarterly averages

	1996				1997				1998				1999	
	I	II	III	IV	I	II	III	IV	I	II	III	IV	I	II
TOTAL (all civilian workers)	5.6	5.5	5.3	5.3	5.2	5.0	4.9	4.7	4.6	4.4	4.5	4.4	4.3	4.3
Men, 20 years and older	4.9	4.8	4.4	4.4	4.4	4.2	4.0	4.0	3.8	3.6	3.8	3.6	3.4	3.5
Women, 20 years and older	4.9	4.8	4.7	4.9	4.6	4.5	4.3	4.1	4.2	4.0	4.0	4.0	3.8	3.9
Both sexes, 16 to 19 years	17.3	16.6	15.4	16.6	16.9	16.1	16.3	14.8	14.6	14.2	14.7	14.9	14.6	13.4
White	4.9	4.8	4.5	4.6	4.5	4.2	4.2	4.0	4.0	3.8	3.9	3.8	3.7	3.8
Black	10.6	10.4	10.6	10.7	10.7	10.3	9.5	9.7	9.3	8.8	9.2	8.4	8.0	7.5
Black and other	9.5	9.3	9.2	9.3	9.3	9.1	8.3	8.4	8.2	7.6	7.8	7.4	7.2	6.7
Hispanic origin	9.6	9.3	8.6	8.0	8.2	7.8	7.5	7.4	7.0	7.0	7.3	7.4	5.4	6.8
Married men, spouse present	3.1	3.0	3.0	3.0	2.8	2.7	2.6	2.5	2.5	2.3	2.3	2.3	2.2	2.3
Married women, spouse present	3.7	3.7	3.4	3.6	3.3	3.2	3.1	2.9	3.1	2.8	2.9	2.8	2.8	2.7
Women who maintain families	7.8	8.0	8.7	8.6	8.7	7.9	7.9	7.8	7.6	7.4	7.1	6.7	5.4	6.6
OCCUPATION														
Managerial and professional specialty . .	2.4	2.4	2.3	2.3	2.1	2.0	2.0	1.8	1.9	1.8	1.8	1.8	1.9	2.0
Technical, sales, and administrative														
support	4.5	4.4	4.5	4.6	4.3	4.1	4.1	4.0	4.0	3.8	3.8	3.8	3.8	3.6
Precision production, craft, and repair. . .	5.7	5.5	5.3	5.4	5.0	4.9	4.7	4.8	4.4	4.2	4.4	3.7	3.8	4.3
Operators, fabricators, and laborers	8.3	8.2	7.6	7.7	8.0	7.4	7.5	7.0	6.5	6.6	6.8	6.7	5.9	6.3
Farming, forestry, and fishing	7.9	8.3	6.8	7.3	7.3	7.2	6.7	7.1	6.7	6.3	6.5	6.4	7.5	7.6
INDUSTRY														
Nonagricultural private wage and salary														
workers	5.7	5.6	5.4	5.4	5.3	5.1	5.0	4.7	4.7	4.5	4.7	4.5	4.3	4.3
Goods-producing industries	6.3	6.2	5.7	6.0	5.7	5.3	5.2	4.9	4.9	4.6	4.9	4.6	4.5	4.5
Mining	6.0	4.6	4.4	5.3	4.3	3.0	4.0	3.6	3.3	2.8	3.5	3.0	6.8	6.7
Construction	10.7	10.2	9.3	9.7	9.5	8.8	8.7	8.5	8.2	7.5	7.6	6.7	7.2	7.4
Manufacturing	5.0	4.9	4.5	4.8	4.5	4.3	4.1	3.8	3.8	3.7	4.1	3.9	3.5	3.5
Durable goods	4.8	4.7	4.0	4.6	4.0	3.6	3.3	3.2	3.3	3.1	3.8	3.2	3.2	3.3
Nondurable goods	5.3	5.3	5.2	5.1	5.2	5.3	5.3	4.5	4.6	4.6	4.6	4.9	4.1	3.8
Service-producing industries	5.5	5.4	5.3	5.2	5.2	4.9	4.9	4.7	4.6	4.5	4.6	4.5	4.2	4.3
Transportation and public utilities . . .	4.0	4.4	4.0	4.1	4.0	3.2	3.6	3.2	3.4	3.3	3.4	3.3	2.9	3.0
Wholesale and retail trade.	6.6	6.5	6.3	6.2	6.4	6.3	6.1	6.0	5.7	5.4	5.6	5.4	5.3	5.4
Finance, insurance, and real estate. .	2.6	2.6	2.8	2.9	3.2	3.1	3.1	2.5	2.5	2.2	2.4	2.7	2.2	2.5
Services.	5.6	5.5	5.3	5.1	4.9	4.6	4.5	4.4	4.6	4.6	4.6	4.5	4.1	4.1
Government workers	2.8	3.0	2.9	3.0	2.9	2.6	2.6	2.3	2.5	2.2	2.3	2.1	2.2	2.5
Agricultural wage/salary workers.	10.6	10.4	9.7	10.0	8.8	9.2	9.0	9.3	9.3	8.2	7.8	7.5	10.1	10.0

(1) Some figures from 1996 through the 2d quarter of 1998 have been revised. **NOTE:** Beginning in Jan. 1997, data reflect revised population controls used in the household survey. Beginning in Jan. 1998, data reflect new composite estimation procedures and revised population controls used in the household survey. Beginning in Jan. 1999, data reflect new population controls used in the household survey.

Unemployment Insurance

Source: Unemployment Insurance Service, U.S. Dept. of Labor

Unlike old-age and survivors insurance, which is entirely a federal program, unemployment insurance in the U.S. is a federal-state system that provides insured wage earners partial replacement for lost wages during a period of involuntary unemployment. The program protects most wage and salary workers. During fiscal year 1998, an estimated 122 million workers in commerce, industry, agriculture, and government were covered under the federal-state system.

Each state, as well as the District of Columbia, Puerto Rico, and the Virgin Islands, has its own law and operates its own program. The amount and duration of the weekly benefits are determined by state laws and are based on prior wages and length of employment. States are required to extend the dura-

tion of benefits when unemployment in the state rises to and remains above specified levels; costs of extended benefits are shared by the state and federal governments.

Under the Federal Unemployment Tax Act, the federal tax rate is 6.2% on the first $7,000 paid to each employee of employers with one or more employees in 20 weeks of the year or with a quarterly payroll of $1,500 or more. A credit of up to 5.4% is allowed for taxes paid under state unemployment insurance laws that meet certain criteria, for a net federal rate of 0.8%; subject employers also pay a state unemployment tax. Governmental agencies and certain nonprofit organizations are not subject to the federal tax; these employers reimburse states for benefits paid to former employees.

The secretary of labor certifies states for administrative grants to operate the program (under the Social Security Act) and for employer tax credit (under the Federal Unemployment Tax Act).

Benefits are financed solely by employer contributions, except in Alaska, New Jersey, and Pennsylvania, where employees also contribute. Benefits are paid through the states' public employment offices, at which unemployed workers must register for work and to which they must report regularly for referral to a possible job during the time when they are drawing weekly benefit payments.

During fiscal year 1998, $20.1 billion in benefits were paid under all unemployment insurance programs to 7.4 million beneficiaries. They received an average payment of $190.51 weekly for total unemployment, which lasted an average of 14.3 weeks.

Civilian Employment of the Federal Government, May 1999

Source: Statistical Analysis and Services Division, U.S. Office of Personnel Management
(payroll in thousands of dollars)

	ALL AREAS Employment	Payroll	UNITED STATES Employment	Payroll	WASH., D.C., MSA Employment	Payroll	OVERSEAS Employment	Payroll
TOTAL, all agencies[1]	2,768,817	$9,843,173	2,666,008	$9,505,058	323,872	$1,505,952	102,809	$338,115
Legislative Branch	30,394	123,965	30,385	123,895	29,093	117,909	9	70
Congress	16,930	64,349	16,930	64,349	16,930	64,349	—	—
U.S. Senate	6,399	24,054	6,399	24,054	6,399	24,054	—	—
House of Representatives	10,513	40,223	10,513	40,223	10,513	40,223	—	—
Architect of the Capitol	1,863	6,309	1,863	6,309	1,863	6,309	—	—
Congressional Budget Ofc	215	1,251	215	1,251	215	1,251	—	—
General Accounting Ofc	3,312	18,966	3,310	18,951	2,374	13,835	2	15
Government Printing Ofc	3,265	12,507	3,265	12,507	2,940	11,748	—	—
Library of Congress	4,368	18,446	4,361	18,391	4,341	18,324	7	55
U.S. Tax Court	262	1,320	262	1,320	257	1,297	—	—
Judicial Branch	31,873	133,416	31,489	131,972	1,616	8,205	384	1,444
Supreme Court	390	1,427	390	1,427	390	1,427	—	—
U.S. Courts	31,483	131,989	31,099	130,545	1,226	6,778	384	1,444
Executive Branch	2,706,550	9,585,792	2,604,134	9,249,191	293,163	1,379,838	102,416	336,601
Exec Ofc of the President	1,594	9,866	1,585	9,802	1,585	9,802	9	64
White House Office	385	2,069	385	2,069	385	2,069	—	—
Ofc of Vice President	20	153	20	153	20	153	—	—
Ofc of Mgmt & Budget	502	3,397	502	3,397	502	3,397	—	—
Ofc of Administration	178	916	178	916	178	916	—	—
Council Economic Advisors	27	159	27	159	27	159	—	—
Ofc of Policy Development	30	211	30	211	30	211	—	—
National Security Council	42	283	42	283	42	283	—	—
Ofc of Natl Drug Control	104	706	104	706	104	706	—	—
Ofc of U.S. Trade Rep	167	1,132	158	1,068	158	1,068	9	64
Executive Departments	1,627,703	6,097,393	1,544,781	5,817,645	216,210	1,033,746	82,922	279,748
State	25,067	117,843	8,940	45,261	7,699	37,797	16,127	72,582
Treasury	156,873	579,876	155,694	574,126	21,806	110,151	1,179	5,750
Defense, Total	699,865	2,508,919	643,124	2,340,180	65,094	288,596	56,741	168,739
Defense, Mil Funct Total	675,038	2,432,410	618,392	2,263,952	64,247	285,056	56,646	168,458
Defense, Civ Funct Total	24,827	76,509	24,732	76,228	847	3,540	95	281
Dept of the Army	233,841	677,541	209,884	609,199	19,532	45,612	23,957	68,342
Army, Mil Funct Total	209,015	601,033	185,153	532,972	18,685	42,072	23,862	68,061
Army, Civil Funct Total	24,826	76,508	24,731	76,227	847	3,540	95	281
Corps of Engineers	24,784	76,170	24,689	75,889	805	3,202	95	281
Dept of the Navy	191,001	896,034	184,643	862,254	25,395	136,287	9,358	33,780
Dept of the Air Force	162,389	530,526	155,047	514,653	5,514	36,633	7,342	15,873
Defense Log Agcy	40,078	151,392	38,889	147,105	2,541	11,754	1,189	4,287
Other Def Act (excl DLA)	72,556	253,426	57,661	206,969	12,112	58,310	14,895	46,457
Justice	123,779	526,403	121,248	515,285	21,987	115,588	2,531	11,118
Interior	71,872	243,568	71,520	242,432	7,909	33,736	352	1,136
Agriculture	99,293	332,374	97,998	328,685	11,239	50,474	1,295	3,689
Commerce	48,571	165,166	47,785	161,803	20,616	94,050	786	3,363
Labor	15,884	65,943	15,853	65,823	5,398	24,875	31	120
Health and Human Services	60,592	265,603	60,378	264,515	26,752	128,329	214	1,088
Housing & Urban Dev	9,573	46,677	9,495	46,303	2,747	15,053	78	374
Transportation	64,114	367,793	63,585	364,902	9,863	56,295	529	2,891
Energy	15,937	88,221	15,932	88,176	5,422	34,573	5	45
Education	4,676	21,926	4,672	21,910	3,224	15,702	4	16
Veterans Affairs	231,607	767,081	228,557	758,244	6,484	28,527	3,050	8,837
Independent Agencies	1,077,253	3,478,533	1,057,768	3,421,744	75,368	336,290	19,485	56,789
Bd of Gov, Fed Rsrv Sys	1,670	8,812	1,670	8,812	1,670	8,812	—	—
Environmtl Protect Agcy	18,743	89,354	18,694	89,129	6,267	32,817	49	225
Equal Employ Opp Comm	2,642	11,377	2,642	11,377	678	3,350	—	—
Federal Communic Comm	1,947	10,143	1,945	10,129	1,600	8,620	2	14
Federal Deposit Ins Corp	7,472	40,568	7,461	40,507	2,562	15,617	11	61
Fed. Emerg. Mgmt Agcy	5,342	19,050	4,843	18,105	2,279	8,690	499	945
General Svcs Admin	14,200	59,207	14,109	58,922	4,812	22,442	91	285
Natl Aero & Space Admin	18,554	96,736	18,536	96,634	3,987	21,934	18	102
Natl Fnd Arts & Humanities	359	1,708	359	1,708	359	1,708	—	—
Peace Corps	1,088	5,840	654	3,573	524	3,231	434	2,267
Securities & Exchange Comm	2,836	14,151	2,836	14,151	1,787	8,923	—	—
Smithsonian Inst., Total	5,288	18,502	5,118	18,011	4,701	16,446	170	491
Social Security Admin	64,469	242,112	63,963	240,281	1,714	7,396	506	1,831
U.S. Postal Service	872,814	2,606,891	868,621	2,592,271	23,167	78,481	4,193	14,620

(1) Totals include agencies not listed.

U.S. Occupational Illnesses, by Industry and Type of Illness, 1997

Source: Bureau of Labor Statistics, U.S. Dept. of Labor

(percent distribution)

Occupational illness TOTAL [1,833,380 cases]	All private sector[1]	GOODS PRODUCING				SERVICE PRODUCING				
		Agri-culture[2]	Mining[3]	Con-struc-tion	Manu-facturing	Trans. and pub. utilities	Whole-sale	Retail	Finance[4]	Service
	100.0	100.0	100.0	100.0	100.0	100.0	100.0	100.0	100.0	100.0
Nature of injury, illness:										
Sprains, strains.........	43.6	32.5	36.3	36.9	38.2	52.2	46.6	41.4	38.0	50.3
Bruises, contusions	9.0	8.7	12.2	8.0	9.0	10.5	8.1	10.1	6.3	8.4
Cuts, lacerations	7.3	12.0	6.3	9.6	8.4	4.0	6.6	10.9	6.1	4.1
Fractures	6.5	8.0	17.1	10.3	6.6	5.4	7.1	5.8	7.0	5.0
Carpal tunnel syndrome..	1.6	0.6	0.4	0.6	2.7	1.0	1.3	1.0	5.7	1.4
Tendinitis	1.0	0.4	—	0.4	1.8	0.4	0.6	1.0	1.3	0.9
Chemical burns	0.7	0.5	0.8	0.5	0.9	0.3	0.8	0.8	0.3	0.6
Amputations...........	0.6	0.8	1.4	0.6	1.2	0.3	0.6	0.4	0.1	0.2
Multiple traumatic injuries.	3.3	2.9	3.7	3.1	2.9	4.2	3.4	3.0	4.4	3.2
Source of injury, illness:										
Chemicals/chem. products	1.8	1.6	6.1	1.5	2.3	1.1	1.3	1.6	1.9	1.8
Containers	14.6	8.2	5.3	5.8	13.9	24.7	22.7	20.3	11.0	8.5
Furniture, fixtures.......	3.5	0.9	0.2	1.8	2.7	1.7	2.3	5.9	6.3	4.8
Machinery	6.7	6.3	13.5	5.9	11.6	2.2	7.2	7.1	4.9	3.7
Parts and materials	11.3	8.8	22.4	24.2	18.1	7.2	12.3	7.0	6.2	3.6
Worker motion or position	15.1	12.1	4.9	12.5	18.1	14.8	14.9	13.0	21.9	15.1
Floor, ground surface....	15.7	16.5	15.1	17.4	10.1	16.4	13.8	19.1	22.2	17.8
Tools, instruments, equip..	5.9	8.5	9.0	10.6	6.6	3.5	3.9	5.6	4.0	5.0
Vehicles	8.1	8.5	6.5	5.5	4.9	17.8	12.2	7.4	6.5	6.8
Health care patient......	4.7	—	—	—	(4)	0.5	0.1	—	0.5	19.9
Event or exposure:										
Contact with object/equip.	27.0	34.4	472	34.8	33.8	20.9	27.0	28.2	17.89	17.8
Struck by object	13.1	15.5	25.6	19.2	14.2	10.3	12.8	14.5	9.8	9.1
Struck against object	7.0	9.6	8.4	7.6	7.9	5.9	7.1	8.7	5.0	5.1
Caught in object........	4.4	5.2	12.3	4.0	8.3	2.8	4.7	3.2	1.6	1.9
Fall to lower level	5.42	8.4	9.7	11.6	3.2	7.1	5.9	42	7.6	4.3
Fall to same level	10.8	8.8	6.5	7.0	7.6	9.4	8.0	15.5	15.7	13.9
Slips, trips (without fall) ..	3.1	2.2	2.0	3.0	2.5	3.3	3.2	3.5	3.9	3.5
Overexertion...........	27.7	17.0	21.8	23.3	26.3	30.3	31.4	24.8	22.2	32.5
Overexertion in lifting	16.2	9.3	8.4	13.4	14.5	17.1	20.6	16.8	14.2	17.1
Repetitive motion	4.1	1.8	0.6	1.5	8.0	2.3	3.2	2.9	10.2	30.
Exposed to harmful substance...........	4.6	6.2	3.7	3.7	5.5	3.0	2.9	5.6	4.1	4.5
Transportation accidents .	4.0	4.0	3.2	3.0	1.9	9.3	5.8	3.1	5.2	4.1
Fires, explosions	0.2	—	0.5	0.3	0.2	0.2	0.1	0.3	0.3	0.1
Assault, by person	1.2	—	(4)	0.1	0.1	0.4	0.2	0.3	0.9	3.8

NOTE: Dashes (—) indicate data are not available or do not meet publication guidelines. Because of rounding and classifications not shown, percentages may not add to 100. All injuries and illnesses reported involved days away from work. (1) Private sector includes all industries except government, but excludes farms with fewer than 11 employees. (2) Agriculture includes forestry and fishing, but excludes farms with fewer than 11 employees. (3) Data conforming to OSHA definition for mining operators in coal, metal, and nonmetal mining and for employers in railroad transportation are provided to the Bureau of Labor Statistics by the Mine Safety and Health Administration, U.S. Dept. of Labor; and by the Federal Railroad Administration, U.S. Dept. of Transportation. Independent mining contractors are excluded from the coal, metal, and nonmetal industries. (4) Finance includes insurance and real estate.

Fatal Occupational Injuries, 1998

Source: Bureau of Labor Statistics, U.S. Dept. of Labor

	FATALITIES			FATALITIES	
	Number	Percentage		Number	Percentage
TRANSPORTATION INCIDENTS	**2,630**	**44**	**CONTACT WITH OBJECTS AND EQUIPMENT**......................	**941**	**16**
Highway	1,431	24	Struck by object......................	517	9
Collision between vehicles, mobile equip. .	701	12	Struck by falling object...............	317	5
Vehicle struck stationary object	306	5	Struck by flying object	58	1
Noncollision	373	6	Caught in or compressed by equipment or objects.......................	266	4
Nonhighway (farm, industrial premises)...	384	6	Caught in or crushed in collapsing materials	140	2
Aircraft	223	4			
Worker struck by a vehicle.............	413	7	**EXPOSURE TO HARMFUL SUBSTANCE OR ENVIRONMENTS** ...	**572**	**9**
Water vehicle	112	2	Contact with electric current	334	6
Railway...........................	60	1	Contact with temperature extremes.......	46	1
ASSAULTS AND VIOLENT ACTS	**960**	**16**	Exposure to caustic, noxious, or allergenic substances	104	2
Homicide	709	12	Oxygen deficiency....................	87	1
Shooting...........................	569	9			
Stabbing...........................	61	1	**FIRES AND EXPLOSIONS**.............	**205**	**3**
Self-inflicted injury	223	4	**OTHER EVENTS OR EXPOSURES**......	**16**	**—**
FALLS	**702**	**12**	**TOTAL**...........................	**6,026**	**100**
Fall to lower level	623	10			
Fall on same level....................	51	1			

NOTE: Totals for categories may include subcategories not shown separately. Percentages, based on incidence rate per total fatalities, may not add to totals because of rounding. Dashes (—) indicate less than 0.5% or unavailable data.

U.S. Wage and Salary Workers Paid Hourly Rates, Second Quarter 1999

Source: Bureau of Labor Statistics, U.S. Dept. of Labor; unpublished tabulations from Current Population Survey

(in thousands)

SEX AND AGE	Total hourly workers	$5.15[1] or less	Less than $10.00	$10.00 or more
Total, 16 years and older..........................	72,142	3,382	38,624	33,518
16 to 24 years	16,841	1,673	14,285	2,556
20 to 24 years......................	10,205	642	7,947	2,258
25 years and older..................	55,301	1,708	24,340	30,961
25 to 54 years	47,851	1,396	20,574	27,277
25 to 34 years	16,906	577	8,290	8,616
35 to 44 years	18,147	485	7,395	10,752
45 to 54 years	12,798	333	4,890	7,908
55 years and older..............	7,450	313	3,766	3,684
55 to 64 years	5,753	160	2,598	3,155
65 years and older..............	1,697	153	1,168	529
Men, 16 years and older.............	36,193	1,306	16,102	20,091
16 to 24 years	8,657	734	6,953	1,704
20 to 24 years................	5,262	286	3,788	1,474
25 years and older...............	27,536	572	9,149	18,387
Women, 16 years and older............	35,949	2,076	22,522	13,427
16 to 24 years	8,183	939	7,331	852
20 to 24 years................	4,943	356	4,159	784
25 years and older...............	27,765	1,136	15,191	12,574
RACE AND HISPANIC ORIGIN				
White				
Total, 16 years and older	58,862	2,669	30,847	28,015
Men........................	29,999	1,049	13,016	16,983
Women	28,863	1,620	17,831	11,032
Black				
Total, 16 years and older	10,151	569	6,125	4,026
Men........................	4,702	189	2,450	2,252
Women	5,450	380	3,676	1,774
Hispanic origin				
Total, 16 years and older	9,139	455	6,233	2,906
Men........................	5,391	219	3,346	2,045
Women	3,748	235	2,886	862
FULL- AND PART-TIME STATUS				
Full-time workers				
Total, 16 years and older..............	54,883	1,308	25,127	29,756
Men........................	30,603	544	11,436	19,167
Women	24,280	764	13,691	10,589
Part-time workers				
Total, 16 years and older	17,129	2,067	13,442	3,687
Men........................	5,517	762	4,631	886
Women	11,612	1,305	8,811	2,801

NOTE: Data refer to the sole or principal job, exclude the self-employed, and are not seasonally adjusted. Totals may not add because of independent rounding or because all subcategories are not listed. Full- or part-time status on the principal job is not identifiable for some multiple jobholders. Data for "other races" are not presented, and Hispanics are included in both white and black population groups. The data are from unpublished work tables and should not be considered as if part of an official BLS news release. (1) $5.15 = minimum wage starting Sept. 1, 1997.

Federal Minimum Hourly Wage Rates Since 1950

Source: Bureau of Labor Statistics, U.S. Dept. of Labor

The Fair Labor Standards Act of 1938 and subsequent amendments provide for minimum wage-coverage applicable to nonprofessional workers in specified nonsupervisory employment categories.

EFFECTIVE DATE	NONFARM WORKERS			FARM WORKERS[4]	EFFECTIVE DATE	NONFARM WORKERS			FARM WORKERS[4]
	Under laws prior to 1966[1]	Percent of avg. earnings[2]	Under 1966 and later provis.[3]			Under laws prior to 1966[1]	Percent of avg. earnings[2]	Under 1966 and later provis.[3]	
Jan. 25, 1950 ...	$0.75	54	NA	NA	Jan. 1, 1976....	$2.30	46	$2.20	2.00
Mar. 1, 1956....	1.00	52	NA	NA	Jan. 1, 1977....	(5)	(5)	2.30	2.20
Sept. 3, 1961 ...	1.15	50	NA	NA	Jan. 1, 1978....	2.65	44	2.65	2.65
Sept. 3, 1963 ...	1.25	51	NA	NA	Jan. 1, 1979....	2.90	45	2.90	2.90
Feb. 1, 1967....	1.40	50	$1.00	$1.00	Jan. 1, 1980....	3.10	43	3.10	3.10
Feb. 1, 1968....	1.60	54	1.15	1.15	Jan. 1, 1981....	3.35	42	3.35	3.35
Feb. 1, 1969....	(5)	(5)	1.30	1.30	Apr. 1, 1990....	3.80[6]	35	3.80	3.80[6]
Feb. 1, 1970....	(5)	(5)	1.45	(5)	Apr. 1, 1991....	4.25[6]	38	4.25	4.25[6]
Feb. 1, 1971....	(5)	(5)	1.60	(5)	Oct. 1, 1996....	4.75[7]	37	4.75	4.75[7]
May 1, 1974....	2.00	46	1.90	1.60	Sept. 1, 1997...	5.15	NA	5.15	5.15
Jan. 1, 1975....	2.10	45	2.00	1.80					

NA = not applicable. (1) Applies to workers covered prior to 1961 Amendments and, after Sept. 1965, to workers covered by 1961 Amendments. Rates set by 1961 Amendments were: Sept. 1961, $1.00; Sept. 1964, $1.15; and Sept. 1965, $1.25. (2) Percent of gross average hourly earnings of production workers in manufacturing. (3) Applies to workers newly covered by Amendments of 1966, 1974, and 1977, and Title IX of Education Amendments of 1972. (4) Included in coverage as of 1966, 1974, and 1977 Amendments. (5) No change in rate. (6) Training wage for workers age 16-19 in first 6 months of first job: Apr. 1, 1990, $3.35; Apr. 1, 1991, $3.62. The training wage expired Mar. 31, 1993. (7) Under 1996 legislation, a subminimum training wage of $4.25 an hour was established for employees under 20 years of age during their first 90 consecutive calendar days of employment with an employer. For workers receiving gratuities, the minimum wage remained $2.13 per hour.

Hourly Compensation Costs, by Selected Country, 1975-97

Source: Bureau of Labor Statistics, U.S. Dept. of Labor

(in U.S. dollars, compensation for production workers in manufacturing)

Country/Territory	1975	1985	1990	1997	Country/Territory	1975	1985	1990	1997
Australia	$5.62	$8.20	$13.07	$16.00	Luxembourg	$6.50R	$7.81R	$16.74R	$22.55[2]
Austria	4.51	7.58	17.75	21.92	Mexico	1.47	1.59	1.58R	1.75
Belgium	6.41	8.97	19.17R	22.82	Netherlands	6.58	8.75	18.06R	20.61
Canada	5.96	10.94	15.84R	16.55	New Zealand	3.21	4.47	8.33	11.02
Denmark	6.28	8.13	18.02R	22.02	Norway	6.77	10.37	21.47	23.72
Finland	4.61	8.16	21.03	21.44	Portugal	1.58	1.53	3.77	5.29
France	4.52	7.52	15.98R	17.97	Singapore	0.84	2.47	3.78	8.24
Germany[1]	6.35	9.60	22.03R	28.28	Spain	2.53	4.66	11.38R	12.16
Greece	1.69	3.66	6.76R	9.63[2]	Sri Lanka	0.28	0.28	0.35	0.48[2]
Hong Kong[3]	0.76	1.73	3.20	5.42	Sweden	7.18	9.66	20.93	22.24
Ireland	3.03	5.92	11.66R	13.57	Switzerland	6.09	9.66	20.86	24.19
Israel	2.25	4.06	8.55	12.05	Taiwan	0.40	1.50	3.93R	5.89
Italy	4.67	7.63	17.45R	16.74	United Kingdom	3.37	6.27	12.70R	15.47
Japan	3.00	6.34	12.80	19.37	United States	6.36	13.01	14.91	18.24
Korea, South	0.32	.23	3.71	7.22					

(1) Data are for area covered by the former West Germany. (2) 1996. (3) Now part of China. (R) Revised.

Top 15 U.S. Metropolitan Areas, by Average Annual Salary, 1997

Source: Bureau of Labor Statistics, U.S. Dept. of Labor

Rank	Metropolitan area	Average annual salary[1]	Rank	Metropolitan area	Average annual salary[1]
1.	San Jose, CA	$48,702	8.	Bergen–Passaic, NJ	$38,509
2.	New York, NY	47,281	9.	Washington, D.C.-MD-VA-WV	38,487
3.	San Francisco, CA	42,583	10.	Jersey City, NJ	38,455
4.	New Haven–Bridgeport–Stamford– Danbury–Waterbury, CT	42,485	11.	Detroit, MI	37,126
5.	Middlesex–Somerset– Hunterdon, NJ	41,796	12.	Hartford, CT	36,643
			13.	Oakland, CA	36,377
6.	Newark NJ	40,411	14.	Seattle-Bellevue-Everett, WA	36,311
7.	Trenton, NJ	39,835	15.	Boston-Worcester-Lawrence-Lowell- Brockton, MA-NH	36,210

NOTE: Jacksonville, NC, recorded the **lowest average annual pay** among U.S. metropolitan areas in 1997—$18,960—followed by Yuma, AZ ($19,467), Myrtle Beach, SC ($19,527), McAllen–Edinburg–Mission, TX ($19,779), and Brownsville–Harlingen–San Benito, TX ($20,041). The average annual salary in the 5 bottom-ranked metropolitan areas averaged 37-40% below the nationwide metropolitan average of $31,717. (1) Data are preliminary and include workers covered by Unemployment Insurance and Unemployment Compensation for Federal Employees programs.

Average Hours and Earnings of U.S. Production Workers, 1969-98[1]

Source: Bureau of Labor Statistics, U.S. Dept. of Labor

(annual averages)

	Weekly hours	Hourly earnings	Weekly earnings		Weekly hours	Hourly earnings	Weekly earnings
1969	37.7	$3.04	$114.61	1984	35.2	$8.32	$292.86
1970	37.1	3.23	119.83	1985	34.9	8.57	299.09
1971	36.9	3.45	127.31	1986	34.8	8.76	304.85
1972	37.0	3.70	136.90	1987	34.8	8.98	312.50
1973	36.9	3.94	145.39	1988	34.7	9.28	322.02
1974	36.5	4.24	154.76	1989	34.6	9.66	334.24
1975	36.1	4.53	163.53	1990	34.5	10.01	345.35
1976	36.1	4.86	175.45	1991	34.3	10.32	353.98
1977	36.0	5.25	189.00	1992	34.4	10.57	363.61
1978	35.8	5.69	203.70	1993	34.5	10.83	373.64
1979	35.7	6.16	219.91	1994	34.7	11.12	385.86
1980	35.3	6.66	235.10	1995	34.5	11.43	394.34
1981	35.2	7.25	255.20	1996R	34.4	11.82	406.61
1982	34.8	7.68	267.26	1997R	34.6	12.28	424.89
1983	35.0	8.02	280.70	1998	34.6	12.77	441.84

(1) Private-industry production workers in mining and manufacturing; construction workers; nonsupervisory workers in services, transportation, and public utilities; wholesale or retail trade; finance, insurance, or real estate. R=revised figures.

Median Income, by Sex, Race, Age, and Education, 1997-98

Source: Bureau of the Census, U.S. Dept. of Commerce

	1997	1998		1997	1998
MALE	**$35,248**	**$36,252**	**FEMALE**	**$26,029**	**$26,855**
Race			**Race**		
White	36,118	37,196	White	26,470	27,304
Black	26,897	27,472	Black	22,764	23,864
Hispanic origin[1]	21,799	22,505	Hispanic origin[1]	19,676	19,817
Age			**Age**		
Under 65 years	35,126	36,170	Under 65 years	25,978	26,838
65 and over	45,648	43,157	65 and over	30,358	28,326
Educational attainment			**Educational attainment**		
Less than 9th grade	19,291	19,380	Less than 9th grade	14,161	14,467
9th-12th grade (no diploma)	24,726	23,958	9th-12th grade (no diploma)	16,697	16,482
High school graduate	31,215	31,477	High school graduate	22,067	22,780
Some college, no degree	35,945	36,934	Some college, no degree	26,335	27,420
Associate degree	38,022	40,274	Associate degree	28,812	29,924
Bachelor's degree or more	53,450	56,524	Bachelor's degree or more	38,038	39,786

NOTE: Includes only full-time, year-round workers, 15 years old and over as of Mar. of the following year. (1) May be of any race.

Median Weekly Earnings of Wage and Salary Workers in the U.S. by Age, Sex, and Union Affiliation, 1997-98

Source: Bureau of Labor Statistics, U.S. Dept. of Labor

	1997				1998			
SEX AND AGE	**TOTAL**	**Members of unions[1]**	**Repre-sented by unions[2]**	**Non-union**	**TOTAL**	**Members of unions[1]**	**Repre-sented by unions[2]**	**Non-union**
Total, 16 years and older ...	$503	$640	$632	$478	$523	$659	$653	$499
16 to 24 years	306	385	384	302	319	415	410	315
25 years and older	540	655	648	511	572	673	667	537
25 to 34 years	481	579	572	466	502	595	591	489
35 to 44 years	579	675	666	548	597	683	678	576
45 to 54 years	607	704	697	578	620	716	712	592
55 to 64 years	558	661	657	512	592	697	692	560
65 years and older ...	393	614	609	374	405	610	597	383
Men, 16 years and older ...	579	683	679	539	598	699	696	573
16 to 24 years	317	402	404	313	334	430	424	326
25 years and older	615	697	693	595	639	712	709	617
25 to 34 years	515	607	603	503	544	513	615	524
35 to 44 years	651	712	708	630	677	722	719	680
45 to 54 years	713	744	741	698	732	755	755	719
55 to 64 years	669	702	701	649	699	738	737	674
65 years and older .	452	677	672	415	482	657	659	445
Women, 16 years and older .	431	577	568	411	456	596	593	430
16 to 24 years	292	353	351	289	305	389	382	301
25 years and older	462	587	581	437	485	605	602	483
25 to 34 years	427	521	514	416	451	542	542	439
35 to 44 years	482	592	585	461	498	605	605	479
45 to 54 years	495	627	620	465	516	651	645	488
55 to 64 years	433	582	575	408	476	602	596	448
65 years and older ...	348	—	586	324	350	548[3]	522	329

Note: Data refer to the sole or principal job of full-time workers. Excluded are self-employed workers whose businesses are incorporated, although they technically qualify as wage and salary workers. (1) Including members of an employee association similar to a union. (2) Including members of a labor union or employee association similar to a union, and others whose jobs are covered by a union or an employee-association contract. (3) Data not shown where base is less than 50,000.

Work Stoppages (Strikes and Lockouts) in the U.S., 1960-98

Source: Bureau of Labor Statistics, U.S. Dept. of Labor; involving 1,000 workers or more

Year	Number of stoppages[1]	Workers involved[1] (thousands)	Work days idle[1] (thousands)	Year	Number of stoppages[1]	Workers involved[1] (thousands)	Work days idle[1] (thousands)
1960	222	896	13,260	1984	62	376	8,499
1965	268	999	15,140	1985	54	324	7,079
1970	381	2,468	52,761	1986	69	533	11,861
1971	298	2,516	35,538	1987	46	174	4,481
1972	250	975	16,764	1988	40	118	4,381
1973	317	1,400	16,260	1989	51	452	16,996
1974	424	1,796	31,809	1990	44	185	5,926
1975	235	965	17,563	1991	40	392	4,584
1976	231	1,519	23,962	1992	35	364	3,989
1977	298	1,212	21,258	1993	35	182	3,981
1978	219	1,006	23,774	1994	45	322	5,020
1979	235	1,021	20,409	1995	31	192	5,771
1980	187	795	20,844	1996	37	273	4,889[2]
1981	145	729	16,908	1997	29	339	4,497
1982	96	656	9,061	1998	34	387	5,116
1983	81	909	17,461				

(1) Numbers cover stoppages that began in the year indicated. Days of idleness include all stoppages in effect. (2) Revised.

Work Stoppages Involving 5,000 Workers or More Beginning in 1998

Source: Bureau of Labor Statistics, U.S. Dept. of Labor

EMPLOYER, LOCATION, UNION	Began	Ended	Workers involved[1]	Estimated days idle in 1998[1]
Kaiser Permanente, Northern CA; California Nurses Assn.....	1/28	1/29	8,000[2]	16,000
Kaiser Permanente, Northern CA; California Nurses Assn.....	2/24	2/24	7,100	7,100
General Motors Corp., Interstate; Automobile Workers (UAW)	6/5	7/29	152,200[2]	3,313,000
Bell Atlantic Corporation, Interstate; Communication Workers	8/9	8/11	73,000	146,000
Consolidated Rail Company, Interstate;Brotherhood of Maintenance of Way Employees......................	8/14	8/14	3,500	23,500
US West Corporation, Interstate; Communications Workers.	8/16	8/30	34,000	340,000
Southern New England, Telecommunications Company, Interstate; Communications Workers......................	8/23	9/17	6,300	113,400
Northwest Airlines, Interstate; Air Line Pilots Assn.	8/29	9/10	33,700	214,600

(1) Workers and days idle are rounded to the nearest 100. (2) Excludes workers in Canada and Mexico.

Labor Union Directory

Source: Bureau of Labor Statistics, U.S. Dept. of Labor; AFL-CIO; World Almanac research, as of Oct. 1999.

(*) Independent union; all others affiliated with AFL-CIO.

Actors and Artistes of America, Associated (AAAA), 165 W 46th St., Suite 500, New York, NY 10036; founded 1919; Theodore Bikel, Pres.; no individual members, 7 National Performing Arts Unions are affiliates; approx. 100,000 combined membership.

Actors' Equity Association, 165 W 46th St., New York, NY 10036; founded 1913; Ron Silver, Pres. (since 1991); 40,000 active members.

Air Line Pilots Association, 1625 Massachusetts Ave. NW, Washington, DC 20036; founded 1931; Capt. Duane Woerth, Pres. (since 1999); 55,000 members, 51 airlines.

American Federation of Labor & Congress of Industrial Organizations (AFL-CIO), 815 16th St. NW, Washington, DC 20006; founded 1955; John J. Sweeney, Pres. (since 1995); 13 mil. members.

Automobile, Aerospace & Agricultural Implement Workers of America, International Union, United (UAW), 8000 E Jefferson Ave., Detroit, MI 48214; founded 1935; Stephen P. Yokich, Pres. (since 1995); 760,000 active (500,000 ret.) members, 1,000+ locals.

Bakery, Confectionery, Tobacco Workers and Grain Millers International Union (BCTGM), 10401 Connecticut Ave., Kensington, MD 20895; founded 1886; Frank Hurt, Pres. (since 1992); 125,000 members.

Boilermakers, Iron Ship Builders, Blacksmiths, Forgers and Helpers, International Brotherhood of (IBBISB/ BF&H), 753 State Ave., Suite 565, Kansas City, KS 66101; founded 1880; Charles W. Jones, Int'l Pres. (since 1983); 80,000 members, 368 locals.

Bricklayers and Allied Craftworkers, International Union of, 815 15th St. NW, Washington, DC 20005; founded 1865; John J. Flynn, Pres. (since 6/1/99); 100,000 members, 300 locals.

Carpenters and Joiners of America, United Brotherhood of, 50 F St, NW, Washington, DC 20001; founded 1881; Douglas J. McCarron, Gen. Pres. (since 1995); 509,000 members, 1,000 locals.

Communications Workers of America (CWA), 501 3d St. NW, Washington, DC 20001; founded 1938; Morton Bahr, Pres. (since 1985); 630,000+ members, 1,400 locals.

*****Education Association, National,** 1201 16th St. NW, Washington, DC 20036; founded 1857; Bob Chase, Pres. (since 1996); 2.4 mil. members, 13,500 affiliates.

Electrical Workers, International Brotherhood of (IBEW), 1125 15th St. NW, Washington, DC 20005; founded 1891; John J. Barry, Pres. (since 1986); 750,000 members, 1,090 locals.

Electronic, Electrical, Salaried, Machine and Furniture Workers, International Union of (IUE), 1126 16th St. NW, Washington, DC 20036; founded 1949; Edward L. Fire, Pres. (since 1997); 120,000 members, 380 locals.

Engineers, International Union of Operating (IUOE), 1125 17th St. NW, Washington, DC 20036; founded 1896; Frank Hanley, Pres.; 360,000 members, 175 locals.

Farm Workers of America, United (UFW), 29700 Woodford-Tehachapi Rd., PO Box 62, Keene, CA 93531; founded 1962; Arturo S. Rodríguez, Pres. (since 1993); 50,000 members.

*****Federal Employees, Federal District 1, National Federation of (NFFE FD1, IAMAW, AFL-CIO),** 1016 16th St. NW, Suite 300, Washington, DC 20036; founded 1917; Richard Brown, Pres. (1998); 120,000 members, 300 locals.

Fire Fighters, International Association of, 1750 New York Ave. NW, Washington, DC 20006; founded 1918; Alfred K. Whitehead, Pres. (since 1988); 250,000 members, 2,508 locals.

Firemen and Oilers, National Conference of, 1900 L St., NW, Suite 502, Washington, DC 20036; founded 1898; George J. Francisco, Jr., Pres.; 26,000 members, 147 locals.

Flight Attendants, Association of, 1275 K St. NW, Washington, DC 20005; founded 1945; Patricia A. Friend, Int'l Pres.; 46,000 members, 26 carriers.

Food and Commercial Workers International Union, United (UFCW), 1775 K St. NW, Washington, DC 20006- 1598; founded 1979 following merger; Douglas H. Dority, Natl. Pres. (since 1994); 1.4 mil. members, 997 locals.

Glass, Molders, Pottery, Plastics & Allied Workers Intl. Union (GMP), 608 E Baltimore Pike, PO Box 607, Media, PA 19063; founded 1842; James Rankin, Pres. (since 1997); 65,000 members, 370 locals.

Government Employees, American Federation of (AFGE), 80 F St. NW, Washington, DC 20001; founded 1932; Bobby L. Harnage Sr., Pres.; 180,000 members, 1,100 locals.

Graphic Communications International Union (GCIU), 1900 L St. NW, Washington, DC 20036; founded 1983; James J. Norton, Pres. (since 1985); 160,000 members, 330 locals.

Hotel Employees and Restaurant Employees International Union, 1219 28th St. NW, Washington, DC 20007; John W. Wilhelm, Gen. Pres. (since 1998); 350,000 members, 140 locals.

Iron Workers, International Association of Bridge, Structural, Ornamental and Reinforcing, 1750 New York Ave. NW, Suite 400, Washington, DC 20006; founded 1896; Jake West, Gen. Pres. (since 1989); 120,000 members, 242 locals.

Laborers' International Union of North America (LIUNA), 905 16th St. NW, Washington, DC 20006-1765; founded 1903; Arthur A. Coia, Pres. (since 1993); 820,000 members, 680 locals.

Leather Goods, Plastics Novelty, and Service Workers' Union, International, 265 W 14th St., Suite 711, New York, NY 10011; Rosemary Berman, Gen. Pres. (since 1997); 5,500 members, 80 locals.

Letter Carriers, National Association of (NALC), 100 Indiana Ave. NW, Washington, DC 20001-2144; founded 1889; Vincent R. Sombrotto, Pres. (since 1978); 315,000 members, 2,800 locals.

Locomotive Engineers, Brotherhood of (BLE), The Standard Bldg. Mezzanine, 1370 Ontario St., Cleveland, OH 44113-1702; founded 1863; Edward Dubroski, Pres. (since 1999); 56,000 members, 600+ divisions.

Longshore & Warehouse Union, International (ILWU), 1188 Franklin St., San Francisco, CA 94109-6800; founded 1937; Brian T. McWilliams, Pres. (since 1994); 60,000 members, 74 locals.

Longshoremen's Association, International (ILA), 17 Battery Pl., Suite 1530, New York, NY 10004; John Bowers, Pres. (since 1987); 65,000 members, 345 locals.

Machinists and Aerospace Workers, International Association of (IAMAW), 9000 Machinists Pl., Upper Marlboro, MD 20772-2687; founded 1888; R. Thomas Buffenbarger, Pres. (since 1997); 780,000 members, 1,194 locals.

Maintenance of Way Employes, Brotherhood of (BMWE), 26555 Evergreen Rd., Suite 200, Southfield, MI 48076; founded 1887; M. A. "Mac" Fleming, Pres. (since 1990); 50,000 members, 790 locals.

Marine Engineers' Beneficial Assn. (MEBA), 444 N Capitol St. NW, Suite 800, Washington, DC 20001; founded 1875; Lawrence O'Toole, Pres. (since 1998).

Maritime Union, National (NMU), 1150 17th St. NW, Washington, DC 20036; Rene Lioeanjie, Pres. (since 1997); 6,000 members. (Affiliated with Seafarers Int'l. Union of N. America.)

Mine Workers of America, United (UMWA), 900 15th St. NW, Washington, DC 20005; founded 1890; Cecil E. Roberts, Pres. (since 1995); 130,000 members, 600 locals.

Musicians of the United States and Canada, American Federation of (AFM), 1501 Broadway, Suite 600, New York, NY 10036; founded 1896; Steve Young, Pres. (since 1995); 125,000 members, 275 locals.

Needletrades, Industrial, and Textile Employees, Union of (UNITE), 1710 Broadway, New York, NY 10019; founded 1995; Jay Mazur, Pres. (since 1995); 250,000 members, 900 locals.

Newspaper Guild-Communications Workers of America (CWA) The, 501 3d St. NW, Suite 250, Washington, DC 20001-2797; founded 1933; Linda Foley, Pres. (since 1995); 34,000 members, 90 locals.

*Nurses Association, American (ANA), 600 Maryland Ave. SW, Suite 100-W, Washington, DC 20024-2571; founded 1897; Beverly L. Malone, Pres. (since 1996); 177,000 members, 53 constituent state & territorial assns.

Office and Professional Employees International Union (OPEIU), 265 W 14th St., Suite 610, New York, NY 10011; founded 1945 (AFL Charter); Michael Goodwin, Pres. (since 1994); 130,000 members, 200 locals.

PACE International Union, AFL-CIO, CLC (PACE), 3340 Perimeter Hill Dr., PO Box 1475, Nashville, TN 37202; founded 1884; Boyd D. Young, Pres. (since 1999); 320,000 members, 1,600 locals.

Painters and Allied Trades, International Brotherhood of (IBPAT), 1750 New York Ave. NW, Washington, DC 20006; founded 1887; Michael E. Monroe, Gen. Pres.; 130,000 members, 440 locals.

*Plant Guard Workers of America, International Union, United (UPGWA), 25510 Kelly Rd., Roseville, MI 48066; founded 1948; Gene McConville, Pres. (since 1990); 12,000 members, 180 locals.

Plasterers' and Cement Masons' International Association of the United States and Canada, Operative, 14405 Laurel Pl., Suite 300, Laurel, MD 20707; founded 1864; John J. Dougherty, Pres.; 40,000 members, 100 locals.

Plumbing and Pipe Fitting Industry of the United States and Canada, United Association of Journeymen and Apprentices of the, 901 Massachusetts Ave. NW, PO Box 37800, Washington, DC 20013; founded 1889; Martin J. Maddaloni, Gen. Pres. (since 1997); 304,000 members, 335 locals.

*Police, National Fraternal Order of, 1410 Donelson Pike, A-17, Nashville, TN 37217; Gilbert G. Gallegos, Natl. Pres. (since 1995); 280,000 members, 1,992 affiliates.

Police Associations, International Union of, 1421 Prince St., Suite 330, Alexandria, VA 22314; Samuel A. Cabral, Pres. (since 1995); 50,000 members, 400 locals.

*Postal Supervisors, National Association of, 1727 King St., Suite 400, Alexandria, VA 22314-2753; Vincent Palladino, Pres. (since 1992); 38,000 members, 400 locals.

Postal Workers Union, American (APWU), 1300 L St. NW, Washington, DC 20005; founded 1971; Moe Biller, Pres. (since 1980); 350,000 members, 1,600+ locals.

Roofers, Waterproofers & Allied Workers, United Union of, 1660 L St. NW, Suite 800, Washington, DC 20036; founded 1906; Earl Kruse, Pres. (since 1985); 25,000 members, 110 locals.

*Rural Letter Carriers' Association, National, 1630 Duke St., 4th Fl., Alexandria, VA 22314; founded 1903; Steven Smith, Pres. (since 1997); 98,000 members; 50 state org.

Seafarers International Union of North America (SIU), 5201 Auth Way and Britannia Way, Camp Springs, MD 20746; founded 1938; Michael Sacco, Pres. (since 1988); 85,000 members, 18 affiliates.

Service Employees International Union (SEIU), 1313 L St. NW, Washington, DC 20005; founded 1921; Andrew L. Stern, Pres. (since 1996); 1.3 million members, 424 locals.

Sheet Metal Workers' International Association (SMWIA), 1750 New York Ave. NW, Washington, DC 20006; founded 1888; Michael J. Sullivan, Pres. (since 1999); 150,000 members, 203 locals.

State, County, and Municipal Employees, American Federation of (AFSCME), 1625 L St. NW, Washington, DC 20036; Gerald W. McEntee, Pres. (since 1981); 1.3 mil. members, 3,617 locals.

Steelworkers of America, United (USWA), 5 Gateway Center, Pittsburgh, PA 15222; founded 1936; George F. Becker, Pres. (since 1994); 700,000 members, 2,600 locals.

Teachers, American Federation of (AFT), 555 New Jersey Ave. NW, Washington, DC 20001; founded 1916; Sandra Feldman, Pres. (since 1997); 984,000 members, 2,267 locals.

Teamsters, International Brotherhood of (IBT), 25 Louisiana Ave. NW, Washington, DC 20001; founded 1903; James P. Hoffa, Gen. Pres. (since 1999); 1.4 mil. members, 569 locals.

Television and Radio Artists, American Federation of, 260 Madison Ave., 7th fl., New York, NY 10016; founded 1937; Shelby Scott, Natl. Pres. (since 1993); 75,000 members, 35 locals.

Theatrical Stage Employees, Moving Picture Technicians, Artists and Allied Crafts of the United States, Its Territories, and Canada, International Alliance of (IATSE), 1515 Broadway, Suite 601, New York, NY 10036; founded 1893; Thomas C. Short, Pres. (since 1994); 95,000 members, 555+ locals.

Transit Union, Amalgamated (ATU), 5025 Wisconsin Ave. NW, 3rd Fl., Washington, DC 20016; founded 1892; James La Sala, Pres. (since 1986); 165,000 members, 275 locals.

Transportation-Communications International Union (TCU), 3 Research Place, Rockville, MD 20850; founded 1899; Robert A. Scardelletti, Pres. (since 1991); 110,000 members, 412 locals.

Transportation Union, United (UTU), 14600 Detroit Ave., Cleveland, OH 44107; founded 1969; Charles L. Little, Pres. (since 1995); 140,000 members, 680 locals.

Transport Workers Union of America, 80 West End Ave., 5th Fl., New York, NY 10023; founded 1934; Sonny Hall, Int'l. Pres. (since 1993); 125,000+ members, 92 locals.

*Treasury Employees Union, National (NTEU), 901 E St. NW, Suite 600, Washington, DC 20004; founded 1938; Colleen M. Kelley, Natl. Pres. (since 1999); 155,000 represented, 285 chapters.

*University Professors, American Association of (AAUP), 1012 14th St. NW, Suite 500, Washington, DC 20005; founded 1915; James T. Richardson, Pres.; 44,000 members, 850 chapters.

Utility Workers Union of America (UWUA), 815 16th St. NW, Washington, DC 20006; founded 1945; Donald Wightman, Pres. (since 1996); 50,000 members, 225 locals.

U.S. Union Membership, 1930-98

Source: Bureau of Labor Statistics, U.S. Dept. of Labor

Year	Labor force[1] (thousands)	Union members[2] (thousands)	Percentage of labor force	Year	Labor force[1] (thousands)	Union members[2] (thousands)	Percentage of labor force
1930....	29,424	3,401	11.6	1987	99,303	16,913	17.0
1935....	27,053	3,584	13.2	1988	101,407	17,002	16.8
1940....	32,376	8,717	26.9	1989	103,480	16,960	16.4
1945....	40,394	14,322	35.5	1990	103,905	16,740	16.1
1950....	45,222	14,267	31.5	1991	102,786	16,568	16.1
1955....	50,675	16,802	33.2	1992	103,688	16,390	15.8
1960....	54,234	17,049	31.4	1993	105,067	16,598	15.8
1965....	60,815	17,299	28.4	1994	107,989	16,748	15.5
1970....	70,920	19,381	27.3	1995	110,038	16,360	14.9
1975....	76,945	19,611	25.5	1996	111,960	16,269	14.5
1980....	90,564	19,843	21.9	1997	114,533	16,110	14.1
1985....	94,521	16,996	18.0	1998	116,730	16,211	13.9
1986....	96,903	16,975	17.5				

(1) Does not include agricultural employment; from 1985, does not include self-employed or unemployed persons. (2) From 1930 to 1980, includes dues-paying members of traditional trade unions, regardless of employment status; from 1985, includes members of employee associations that engage in collective bargaining with employers.

TAXES

Federal Income Tax

Source: George W. Smith III, CPA, Nationally Syndicated Tax Author and Columnist

As national and state elections come closer, it can be expected that a variety of tax-related legislative proposals will be introduced by both the House and Senate. Ultimately, the taxpayer will have to contend with a multitude of new tax laws, along with the multitude that already exists.

Current Highlights

Smoking Withdrawal. The IRS reversed a 20-year-old position in 1999 by allowing taxpayers to deduct two types of treatments for cigarette smoking as a medical expense: (1) the cost of participation in a smoking-cessation program, and (2) the purchase of drugs that require a physician's prescription to alleviate the effects of nicotine withdrawal. However, individuals cannot deduct over-the-counter medications such as nonprescription nicotine patches and gum.

Garage Sales. In most situations, revenues received from garage sales do not result in taxable income. It's a good probability that the product sold cost more than the revenue received for it. By the same token, losses attributable to a garage sale are personal and therefore not deductible. The reasoning for this is that the individual doing the selling is not in a trade or business on a regular basis.

Electioneering. Campaign expenses of a candidate running for election or reelection to office are not deductible. Expenses include travel, qualification and registration fees, posters and novelty items.

Gambling Winnings and Losses. Total lottery and other gambling winnings must be reported on page 1 of a taxpayer's return, Form 1040. All gambling expenses are reported on Schedule A. Such expenses are deductible only up to the amount of winnings reported on page 1. No winnings, no deduction. A recent 1999 U.S. Tax Court decision reaffirmed that professional gamblers engaged in wagering as a trade or business cannot deduct their business losses in excess of their winnings.

Educational Assistance. Up to $5,250 of benefits provided by an employer for educational assistance is excludable from an employee's gross income. This provision has been extended for undergraduate courses beginning before June 1, 2000.

Educational expenses that lead to a new job position, a substantial advancement, or an advanced degree may be deductible on Schedule A if the expense is required to retain salary status or employment. However, these expenses are not deductible if they qualify the individual for a new trade or business. For more information on what expenses do and do not qualify call the IRS at 1-800-829-3676 and ask for your free Publication 508, Educational Expenses.

An expense deduction is allowed for the cost of obtaining a master's degree of law (LLM), provided the attorney has already passed the bar and is working in this particular field.

Parking Tickets. Penalties and fines paid to a governmental unit or department are not deductible. This includes penalties and fines for parking and speeding tickets. Penalties for late filing of an individual's tax return also are not deductible, nor are expenses incurred in a criminal case resulting in a conviction.

Work Commuting. An employee who is required to haul tools, instruments, or other items in his or her automobile to and from work can deduct the additional cost of hauling these items. This would include rent paid on a trailer to carry these items.

Truck/Bus Drivers. The deduction percentage for meals consumed while away from home by individuals subject to hours of service limitations of the Department of Transportation, such as interstate truck and bus drivers, certain railroad employees, and merchant marines, is 55% in 1999. The deduction gradually increases to 80% by 2008.

Investment Expenses. Investors can deduct as a miscellaneous deduction on Schedule A investment fees, custodial fees, trust administration fees, newspapers reporting investment advice, financial reports, and other expenses paid for managing their investment portfolio that produce taxable income.

Investors cannot deduct any portion of expenses for attending a convention, seminar, or similar meeting for investment purposes.

Fitness Expense. Health spa expenses are not deductible even if there is a job requirement to stay in good physical condition. For instance, the costs for a law enforcement officer to belong to a fitness center to stay physically fit are not deductible even if this is a job-related requirement.

Funeral Expenses. Funeral expenses are not deductible on an individual's income tax return. They are deductible on the deceased's federal estate tax return (Form 706).

Excess Estate Deduction. If the total deductions in the estate's last tax year are more than the estate's gross income for that year, the beneficiaries succeeding to the estate's property can claim such excess as a Schedule A miscellaneous deduction.

Divorce Tax Expense. Taxpayers can deduct the portion of their legal fees incident to a divorce if the attorney's invoice actually specifies how much is for tax advice. Also deductible are legal fees paid to collect taxable alimony.

Tax Filing Memo. Taxpayers should make their checks payable to the United States Treasury, rather than to the Internal Revenue Service. To speed processing the IRS asks taxpayers not to staple their checks to the tax return. To protect the taxpayer's privacy, social security numbers no longer appear on mailing labels.

Recent Legislation

Sale of Homestead. Married couples who lived in their principal residence for at least 2 years during a 5-year period ending on the date of sale and filing a 1999 joint income tax return may exclude up to $500,000 in gain from the sale of their residence. This deduction is reusable every 2 years.

Homeowners who have resided in their home fewer than 2 years and have to sell because of a change in the place of their employment or for health reasons or because of certain unforeseen circumstances may prorate the exclusion based on the amount of time in the home.

For single taxpayers the excludable amount of gain is up to $250,000. Married couples who do not share a principal residence with their spouse but still continue to file a joint tax return also may claim the $250,000 exclusion for a qualifying sale or exchange of each spouse's principal residence.

Mileage. The standard mileage tax rate deduction for business use of an auto was decreased for 1999 to 31 cents a mile. However, the IRS postponed the effective date to April 1, 1999. Accordingly, the higher 1998 mileage rate of 32.5 cents per mile continues to apply to business mileage before April 1, 1999. The standard mileage rate is now allowed for lease cars.

Charitable Travel. In 1998 Congress increased the mileage deduction to 14 cents per mile for an individual who uses his or her automobile in volunteer work for qualified charities. The medical and moving mileage rates remain at 10 cents per mile.

Damage Awards. Congress recently enacted legislation that limits tax-free treatment for damage awards to physical injury or sickness or to actual medical expenses.

Innocent Spouse Relief. The IRS Reform Act provides a separate liability section for taxpayers who are divorced, legally separated, or living apart for at least 12 months. In effect, this section prevents a divorced or separated spouse from being held liable for the other spouse's tax liability.

$500 Child Credit. The Taxpayer Relief Act provides a $500 credit that is deducted from a taxpayer's 1999 federal tax for each child under the age of 17 who can be taken as a dependent. Dependents must be either the taxpayer's son or daughter, a descendant of the son or daughter, a stepchild, or an eligible foster child. Previously, the credit was $400.

The allowable portion of the child credit is reduced for higher-income taxpayers. Lower-income families may use the credit to offset Social Security taxes as well as income taxes.

Student Loan Interest. The maximum deduction for interest paid on qualified education loans for 1999 increases from $1,000 to $1,500. There is a 60-month time limit. If interest payments were made before 1999, the months in which those payments were required count against the 60-month time limit.

The education loan must have been for the taxpayer, spouse, or any dependents. However, an individual who is a dependent of another taxpayer may not deduct any interest he or she paid. Married individuals must file a joint return in order to claim the deduction. The deduction begins to phase out when adjusted gross income (AGI) on a joint return reaches $60,000, and is completely gone when it reaches $75,000. For single individuals the amounts are $40,000 to $55,000.

The maximum interest deduction is $1,500 for 1999, $2,000 for the year 2000, and $2,500 thereafter. This is a page 1, Form 1040 deduction, allowed whether or not the taxpayer itemizes deductions on Schedule A.

Savings Bonds. Interest earned on U.S. Series EE bonds issued after 1989 may be exempt from federal income tax if the bonds are used to pay college tuition and fees for a taxpayer, spouse, or dependent. There is a phaseout rule if income is too high. Married taxpayers filing separately are not eligible.

Capital Gains. The long-term capital gains tax rate for individual taxpayers is 20% for qualified investments held more than 12 months. For taxpayers in the 15% tax bracket, the maximum rate is 10%.

Five-Year Rule. Starting Jan. 1, 2001, Congress lowered the capital gains rate for investments held more than 5 years to 18%, 8% if in the 15% tax bracket. Further, if the taxpayer is in a bracket above 15%, the 5-year holding period would apply only to investments acquired after Dec. 31, 2000. For individuals in the 15% tax bracket, investments do not have to be acquired after the year 2000 to have the 5-year period begin.

Hobbies. Capital gains on collectibles such as art, antiques, jewelry, stamps, and coins are taxed at a maximum 28%. Certain newly minted gold and silver coins issued by the federal government are eligible for the lower rates.

Real Estate. Under the new capital gains rules, gain attributable to depreciation that was deducted from real estate will be recaptured at a maximum 25% tax rate. Any excess qualifying gain will be taxed at a maximum rate of 20%.

Self-Employed. The deduction for medical insurance premiums by self-employed individuals increases from 45% in 1998 to 60% in 1999. This is a page 1, Form 1040 deduction from total income. The percentage is slated to increase annually to 100% in the year 2003.

Home Office. Congress enacted a new law starting in 1999 for taxpayers who set up an office at home to take care of the administrative or management side of their business. As a result, taxpayers will not be barred from taking a home office deduction because the home will be considered a principal place of business. Prior rules governing the deduction still must be followed. Further, the advantages of the deduction should be weighed against the 2-year qualification rules for the $250,000/$500,000 exclusion of gain on sale of a principal residence.

Children's Income. Parents may elect to include on their income tax return the unearned income of a dependent child under age 14 whose unearned income is more than $700 but less than $7,000. The income must consist solely of interest and dividends. Form 8814, *Parent's Election to Report Child's Interest and Dividends,* must be attached to the parents' tax return. This election is not available if estimated tax payments were made during the year in the child's name.

An individual may not claim a dependency exemption in 1999 for a child who qualifies as a full-time student and is over age 23 at the end of the year unless the child's gross income is less than $2,750. ($2,800 for the year 2000.)

If a dependent child with taxable income cannot file an income tax return, the parent, guardian or other legally responsible person must file a return for the child. Otherwise, that individual may be held responsible for not filing the child's tax return and for the payment of any taxes owed.

Estate and Gift Tax. The unified estate and gift tax exemption increases in annual increments to $650,000 in 1999 and $675,000 for the years 2000 and 2001. The exemption amount eventually maximizes to $1 million in 2006.

Confidentiality. The recently legislated IRS Reform Act extends attorney-client confidentiality to certified public accountants (CPAs) and enrolled agents (EAs) in connection with noncriminal tax proceedings before the IRS and courts.

Death Benefits. Qualified accelerated death benefits paid under a life insurance contract to terminally ill persons (certified as expected to die within 24 months) now are excludable from gross income. A similar exclusion applies to the sale or assignment of death benefits under a life insurance contract to another person. Accelerated death benefits paid to a chronically ill person under a long-term care rider are tax-free up to $175 per day.

Domestic Workers. The annual threshold dollar amount for reporting and paying Social Security and federal unemployment taxes on domestic employees, including nannies and housekeepers, is $1,100. Household workers under 18 are exempt unless working in a household is their principal occupation. Household employers must apply for an employer federal identification number and issue W-2 wage statements.

Foreign Income. The Taxpayer Relief Act increased the foreign earned income exclusion to $74,000 in 1999, $76,000 in the year 2000, $78,000 for 2001, and $80,000 thereafter. The maximum dollar amount starting in 2002 will be adjusted for any cost-of-living increase.

Employment Tax Rates. The maximum wage base for withholding Social Security tax for 1999 increased to $72,600. The tax rate remains at 6.2%. The Medicare tax rate is 1.45%. There is no maximum wage base for Medicare; all wages are subject to the Medicare tax. Both employer and employee (each) pay these tax rates. Self-employed individuals pay both parts, 12.4% and 2.9%, for a total of 15.3% on net earnings.

Elective Withholding. Taxpayers receiving Social Security benefits (and certain other federal payments) may elect to have federal tax withheld at a rate of 7%, 15%, 28%, or 31%. States also must permit elective federal withholding from unemployment compensation at a 15% rate.

Underpayment. Taxpayers will not have to pay a penalty for underpayment of estimated federal income tax unless the amount of the underpayment in 1999 is $1,000 or more. For the tax filing year 2000, taxpayers with Adjusted Gross Income (AGI) over $150,000 can avoid an underpayment penalty by prepaying 106% of the preceding year's (1999) income tax liability.

Estimated Taxes. Due dates for individual federal quarterly estimated tax payments for the income tax year of 2000 are: 1st quarter, Mon., Apr. 17, 2000; 2d quarter, Thurs., June 15; 3d quarter, Fri., Sept. 15; with the 4th and final quarterly payment due Tues., Jan. 16, 2001. Different quarterly filing dates may apply for state and city estimated tax payments.

Installment Payments. Depending upon the amount owed for federal income tax on April 17, 2000, taxpayers may apply for monthly installment payments by attaching Form 9465, *Installment Agreement Request,* to their tax return. Penalty and interest will continue to accrue on any balance owed. There is a $43 IRS filing fee if the request is approved.

Postmark. The IRS must accept the postmark of qualified couriers such as UPS and FedEx as proof of timely mailing.

Electronic Tax Refunds. Taxpayers may have their refund electronically deposited directly into their checking or savings account. The direct deposit of tax refunds continues to grow in popularity with approximately 22 million taxpayers using direct deposit in 1999, as compared to 19 million for the previous year. According to the IRS, qualifying refunds to taxpayers filing their returns electronically are issued within 21 days.

IRS Services. Federal tax forms, tax legislation, relevant court decisions, and other information and resources are available from the IRS via the following:

Internet website: www.irs.ustreas.gov

Telnet: iris.irs.ustreas.gov

File Transfer Protocol: ftp.irs.ustreas.gov

Direct Dial (by modem): (703)-321-8020

Fax: (703)-368-9694

Forms/Publications: 1-800-829-3676

English/Spanish. The Internal Revenue Service provides videotaped instructions both in English and Spanish at participating libraries. Many IRS publications and tax forms including instructions are also printed in Spanish. For more information, call 1-800-TAX-FORM and ask for the free IRS Publication 1SP, *Derechos del Con-tribuyente.*

Hearing Impaired. The IRS telephone service for hearing-impaired persons is available for taxpayers that have access to TDD equipment. The toll-free number is 1-800-829-4059.

Who Must File a Tax Return?

Most U.S. citizens and resident aliens will have to file a 1999 income tax return if gross income for the year is at least as much as the amount shown in the following table:

Filing Status	1999 Gross Income
Single	
Under 65	$7,050
65 or older	8,100
Married filing jointly	
Both spouses under 65	12,700
One spouse 65 or older	13,550
Both spouses 65 or older	14,400
Married filing separately	2,750
Head of Household	
Under 65	9,100
65 or older	10,150
Qualifying widow(er)	
Under 65	9,950
65 or older	10,800

A tax return must also be filed if:

• Taxpayer had net earnings of $400 or more from self-employment for the year.
• Taxpayer received advance earned income credit payments during the year from an employer or is entitled to receive a refundable earned income credit.
• Taxpayer paid estimated income tax payments during 1999 or expects an income tax refund.
• Taxpayer has losses to be carried back or forward.
• Additional taxes are owed for:
— Social Security tax on unreported tips.
— Alternative minimum tax.
— Recapture of investment credit.
— Tax attributable to qualified retirement distributions including IRAs, annuities, and modified endowment contracts.

When to File

U.S. individual income tax returns for 1999 are required to be filed with the IRS no later than Monday, April 17, 2000. An individual who cannot file on time should file Form 4868, *Application for Automatic Extension of Time to File U.S. Individual Income Tax Return.* This form gives the taxpayer an automatic 4-month extension of time to file until Tuesday, August 15, 2000. This is not an extension of time to pay the tax. The taxpayer will still owe interest and may be charged a penalty on any federal income tax owed and not paid to the IRS by April 17, 2000.

Which Tax Return to File?

Most U.S. citizens can use one of the following income tax forms: Form 1040, 1040A, or 1040EZ. Forms 1040A and 1040EZ are shorter and simpler than Form 1040.

You may be able to use the shortest, Form 1040EZ, if:
• You are single or married filing jointly and do not claim any dependents.
• You are not 65 or older or blind.
• You have income only from wages, salaries, tips, taxable scholarships or fellowships, unemployment compensation, or Alaska Permanent Fund dividends.
• Your taxable income is less than $50,000, and you do not have over $400 of taxable interest income.
• You do not claim a student loan interest deduction or an education credit.
• You do not itemize deductions, claim any adjustments to income, or have tax credits other than the earned income credit.
• You received no advance earned income credit payments.
• You did not make any estimated tax payments.

You may be able to use Form 1040A if:
• You have income only from wages, salaries, tips, taxable scholarships or fellowships, interest and dividends, IRA distributions, pensions, annuities, unemployment compensation, and/or taxable Social Security or railroad retirement benefits.
• Your taxable income is less than $50,000.
• You do not itemize deductions.
• You claim a deduction for qualified IRA contributions.
• You claim a credit for child and dependent care expenses, credit for the elderly or the disabled, the earned income credit, the adoption credit, child tax credit, or education credits.

• You report employment taxes on wages paid to household employees on Schedule H.
• You take the education exclusion for interest income earned from Series EE U.S. Savings Bonds.
• You received advance earned income credit payments.
• You owe alternative minimum tax.
• You have made estimated tax payments.

You will have to file Form 1040 if any of these apply:
• Your taxable income is $50,000 or more. (However, you may also use Form 1040 for lower amounts.)
• You plan to itemize deductions.
• You receive any nontaxable dividends or capital gain distributions.
• You have foreign bank accounts and/or foreign trusts.
• You have taxable refunds from state or local income taxes.
• You have business, farm, or rental income or losses.
• You sold or exchanged capital assets or business property.
• You have miscellaneous income such as alimony that is not allowed on Form 1040A or 1040EZ.
• You have additional adjustments to income such as payments for alimony or moving expenses.
• You are allowed a foreign tax credit or certain other credits to which you are entitled.
• You have other taxes to pay, such as self-employment tax or Social Security tax on tips.
• You have losses that are to be carried back or forward.
• You are required to file additional forms such as **Form 2106,** Employee Business Expenses; **Form 2555,** Foreign Earned Income; **Form 3903,** Moving Expenses; **Form 4972,** Tax on Lump-Sum Distributions.

1999 Individual Income Tax Rates

Single

Tax Rate	Taxable Income
15%	$0 to $25,750
28%	$25,751 to $62,450
31%	$62,451 to $130,250
36%	$130,251 to $283,150
39.6%	More than $283,150

Married Filing Jointly or Qualifying Widow(er)

Tax Rate	Taxable Income
15%	$0 to $43,050
28%	$43,051 to $104,050
31%	$104,051 to $158,550
36%	$158,551 to $283,150
39.6%	More than $283,150

Married Filing Separately

Tax Rate	Taxable Income
15%	$0 to $21,525
28%	$21,526 to $52,025
31%	$52,026 to $79,275
36%	$79,276 to $141,575
39.6%	More than $141,575

Head of Household

Tax Rate	Taxable Income
15%	$0 to $34,550
28%	$34,551 to $89,150
31%	$89,151 to $144,400
36%	$144,401 to $283,150
39.6%	More than $283,150

Estates and Trusts

Tax Rate	Taxable Income
15%	$0 to $1,750
28%	$1,751 to $4,050
31%	$4,051 to $6,200
36%	$6,201 to $8,450
39.6%	More than $8,450

The AMT Tax. The purpose of the alternative minimum tax (AMT) is to increase an individual's tax if certain benefits result in a regular income tax that is lower than the tax that would apply if these benefits were added back to taxable income. If certain tax deductions do reduce an individual's regular income tax liability below the amount that would have to be paid under the AMT, the taxpayer must pay the difference in the form of an alternative minimum tax liability.

The AMT rate for noncorporate taxpayers is 26% for alternative minimum taxable income less the exemption amount up to $175,000 ($87,500 for married individuals filing separately). Above that dollar level, a 28% rate applies.

Dependent Exemptions

The deductible exemption amount for each individual taxpayer or dependent for 1999 has been adjusted for cost of living to $2,750. For the year 2000, the amount increases another $50 to $2,800.

Exemption Phaseout. The deduction for each exemption is reduced by 2% for each $2,500 ($1,250 for married filing separately) or fraction thereof by which adjusted gross income for 1999 exceeds the following amounts:

Married filing jointly	$189,950
Qualifying widow(er)	$189,950
Head of household	$158,300
Single	$126,600
Married filing separately	$94,975

The exemption amount is fully phased out when adjusted gross income is more than $122,500 ($61,250 for married filing separately) over the threshold amount.

Standard Deduction

The standard deduction is a flat dollar amount that is subtracted from the adjusted gross income (AGI) of taxpayers who do not itemize deductions. The amount depends on filing status and is adjusted annually for inflation.

1999 Basic Standard Deduction

Single	$4,300
Married filing jointly or qualifying widow(er)	$7,200
Married filing separately	$3,600
Head of household	$6,350

These figures are not applicable if an individual can be claimed as a dependent on another person's tax return. *Caution:* Taxpayers with itemized deductions totaling more than the above amounts usually should itemize their deductions on Schedule A, Form 1040.

Dependent's Standard Deduction. An individual claimed as a dependent on another person's income tax return generally may claim on his or her own tax return only the larger of $700 or earned income amount plus $250, not to exceed $4,300 (a blind dependent may add $1,050 to this amount). Earned income includes wages, salaries, commissions, and tips. It also includes net profit from self-employment. Any part of a scholarship or fellowship grant that must be included in gross income also is considered earned income.

Taxpayers in certain categories such as the elderly or blind may claim an additional standard deduction.

1999 Additional Standard Deduction

Single or head of household, 65 or older OR blind: . . . $1,050
Single or head of household, 65 or older AND blind: . . $2,100
Married filing jointly or qualifying widow(er), 65 or older
   OR blind (per person): . . . . $ 850
Married filing jointly or qualifying widow(er), 65 or older
   AND blind (per person): $1,700
Married filing separately, 65 or older OR blind: . . . . $ 850
Married filing separately, 65 or older AND blind: $1,700

Persons who claim a deduction because of blindness must attach a doctor's statement to their income tax return.

Tax Tip: For tax purposes, an individual is considered 65 years of age on the day preceding his or her 65th birthday. Thus, a taxpayer whose 65th birthday falls on Jan. 1, 2000, may take an additional standard deduction for 1999.

Adjustments to Income

IRA Deduction. The maximum tax-deferred Individual Retirement Arrangement (IRA) contribution for a married couple filing a joint return was increased to $4,000 per year, but not to exceed the total combined earned income if less than $4,000. Each spouse can contribute up to $2,000 annually even if a spouse had little or no income. There are income limitations.

Taxpayers may contribute to their IRAs even if they are covered by an employer-sponsored qualified retirement plan. Married taxpayers filing jointly in 1999 with adjusted gross income (AGI) less than $51,000 may take the maximum IRA deduction allowed regardless of whether either spouse is an active participant in a qualified retirement plan. Single taxpayers in a qualified plan may deduct up to the maximum IRA contribution provided their AGI is less than $31,000. The IRA deduction begins to phase out over the next $10,000 of AGI if a taxpayer is an active participant in a qualified retirement plan.

Congress went further and recently provided that an individual is not considered an active participant in an employer-sponsored plan merely because the individual's spouse is an active participant for any part of a plan year. However, the maximum deductible IRA contribution for an individual who is not an active participant but whose spouse is phased out at a new AGI dollar level between $150,000 and $160,000.

IRA Withdrawals. There is a 10% early withdrawal penalty for distribution before age 59½. However, a distribution before the IRA owner reaches age 59½ is not subject to the 10% penalty if the IRA distribution meets one of the following exceptions:

• Paid to the beneficiary after the death of the owner.
• Made on account of the disability of the owner.
• Part of a series of substantially equal periodic payments.
• Made to an employee following separation from service after age 55. This exception does not apply if a qualified distribution from a pension plan is rolled into an IRA.
• Used to pay certain unreimbursed medical expenses during the year.
• Used to pay certain qualifying higher education expenses.
• Used to pay certain qualified first time home buyer acquisition costs.

Reading Material. For more information on IRAs call the IRS at 1-800 829-3676 for a free copy of Publication 590, *Individual Retirement Arrangements (IRA).*

The Roth IRA. Although contributions paid into a Roth IRA are not deductible, distributions of funds, including investment earnings held in the account for 5 years or longer and paid after age 59½, are free both of income tax and the early withdrawal penalty at the time of distribution. However, there are limitations on what amounts can be paid into a Roth IRA.

Any funds paid from the Roth IRA after the 5-year exclusion period to an estate or beneficiary on or after an individual's death, including funds paid to an individual who is disabled, are tax and penalty free regardless of age. This includes withdrawals up to $10,000 if used for a first-time home purchase.

Funds paid for "qualified higher education expenses" of the taxpayer, spouse, or any child or grandchild of the taxpayer or spouse are taxable but not subject to the early withdrawal penalty if held for 5 years. Withdrawals from a Roth IRA conversion held less than 5 years are subject both to income tax and the withdrawal penalty. This holds true regardless of age at the time of distribution.

Starting in 1999, all traditional IRA income resulting from the rollover will be taxed in the year of distribution. If the funds are left in the Roth IRA, they will not be taxed upon distribution as long as all requirements are met. There is no penalty on the rollover as long as the funds remain in the Roth IRA.

Education IRA. A trust or educational custodial account can now be set up for paying qualified higher education expenses of the account holder. Annual contributions must be paid by Dec. 31 and are limited to $500 a year per child. The age limit is 18. Earnings on the contributions are distributed tax-free to the IRA beneficiary if used to pay for qualified education expenses. Contributions are limited depending on the creator's adjusted gross income (AGI) and they are nondeductible.

Moving Expenses. Taxpayers who change jobs or are transferred during the year usually can deduct part of their moving expenses. These expenses include travel and the cost of moving household goods to their new home. The cost of meals is no longer deductible. The standard mileage rate for automobiles remains at 10 cents per mile, plus parking and tolls.

In order to take a moving expense deduction, the new job must be at least 50 miles farther from the former home than the old job. Employees must work full time for at least 39 weeks during the first 12 months after they arrive in the general area of their new job. Moving expenses are now an adjustment to income and are reported on page 1, Form 1040. Moves within the U.S. are reported on Form 3903, *Moving Expenses.*

Itemized Deductions

If the total amount of itemized deductions is more than the standard deduction, taxpayers generally should itemize their deductions. The following examples are but a few of the many deductions that may be reported on Schedule A, Form 1040.
- Long-term care insurance premiums are now deductible up to certain annual limits based on age. Any long-term care benefits received under a qualifying policy will be tax-free, subject to per-diem restrictions.
- Most mortgage interest paid on a taxpayer's primary residence and on a taxpayer's second home is fully deductible. However, there are limitations.
- Interest paid on home equity loans is deductible, but only on the first $100,000 of equity debt.
- Borrowers can generally deduct points paid on their principal home mortgage loan on Schedule A. The buyer can also deduct "seller-paid points" on the purchase of a principal residence.
- Investment interest expense is deductible only to the extent of net investment income. Any investment interest expense not currently deducted is carried over to future years.
- State and local income taxes, real estate taxes, and personal property taxes are fully deductible. Sales taxes are no longer deductible.
- Cosmetic surgery for congenital abnormality, personal injury resulting from an accident or trauma, or a disfiguring disease is allowed as a medical deduction. Only the total amount of medical expenses that exceeds 7.5% of the taxpayer's adjusted gross income is deductible.
- Casualty and theft losses are deductible subject to the $100 limitation rule for each occurrence. Only the remaining excess amount over 10% of AGI is deductible.
- Taxpayers deducting individual charitable contributions of $250 or more must obtain written substantiation from the charity. If the amount is $75 or more, the charity must include a breakdown of the payment indicating how much was a (deductible) contribution and what (if any) was the (nondeductible) value of goods, meals or services received.
- Miscellaneous items including union and professional dues, tax preparation fees, safe-deposit box rental expense, and employee business expenses are deductible insofar as they exceed 2% of AGI.
- Amounts spent for tools and supplies used at work are deductible if they wear out within 1 year of the date of purchase. Tools expected to last more than a year will have to be depreciated. These expenses also are subject to the 2% rule.
- Armed forces reservists can deduct the unreimbursed cost of their uniforms if regulations restrict the individual from wearing them except while on duty as a reservist.
- Unreimbursed employee business expenses including travel, automobile, telephone, and gifts are deductible on Schedule A as miscellaneous itemized deductions. Only 50% of the cost of customer meals and entertainment is deductible. These expenses are further subject to the 2% rule.
- Employment fees paid to agencies, resume costs, postage, travel, and other related expenses to look for a new job in your present occupation are deductible even if you do not get a new job.

More Limitations. Many itemized deductions otherwise allowed are further reduced by the smaller of these two figures: 3% of a taxpayer's AGI in excess of the 1999 threshold amount of $126,600 ($63,300 for married taxpayers filing separately) OR 80% of the amount of these itemized deductions otherwise allowable for the year. This provision does not apply to medical expenses, investment interest expense, casualty losses, or gambling losses to the extent of gambling winnings.

Business Expenses

Intangible Write-offs. Recent legislation provides that patents, trademarks, and certain other intangible assets are amortized over a 15-year period. The cost of goodwill and customer/patient name lists is included in this provision.

Employee Meals. Meals provided on the premises for the convenience of the employer will not be taxable to the employee. The cost of the meals will be fully deductible by the employer if more than half of all employees receiving them are given the meals for the employer's convenience.

Payroll Deposit. The employer's federal payroll tax threshold for deposit requirement has increased from $500 to $1,000 per quarter.

Corporate AMT Tax. The corporate alternative minimum tax (AMT) is repealed for small businesses, provided that a 3-year average of $5 million or less annual gross receipts test is met. A corporation will continue to be exempt from AMT as long as its average gross receipts for the prior 3 years does not exceed $7.5 million.

Business Meals. The deduction for qualified business meals and entertainment expenses is limited to 50% of their cost. A receipt is required for business meals, entertainment, and transportation costs above $75. Adequate records must be kept substantiating the time, place, date, and purpose of the expense.

Travel. Travel expenses paid for other individuals (including a spouse) traveling with the taxpayer on a business trip are not deductible unless the individual (1) is an employee, (2) has a bona fide business purpose for the travel, and (3) would otherwise be allowed to deduct the travel expense. Expenses paid for business assignments away from home in a single location that last for more than one year are no longer considered temporary or deductible.

Association Dues. Dues paid to business, social, athletic, luncheon, sporting, and country clubs, including airport and hotel clubs, are no longer deductible. However, dues paid to the Chamber of Commerce and business economic clubs remain tax deductible.

Empowerment Zones. A 20% employment tax credit is available to most employers for qualified wages paid to each full or part-time employee who is a resident of a federal empowerment zone. These zones are distressed areas designated for economic revitalization by the U.S. government.

The 20% credit applies to the first $15,000 of wages paid for each employee. To qualify, an employee must perform substantially all employment services within the zone and in the employer's trade or business.

Corporate Eligibility. The maximum number of eligible shareholders of an S corporation was increased from 35 to 75. S corporations can now own 80% or more of a C corporation.

Business Equipment. The election to expense currently instead of depreciating the cost of certain business machinery and other assets over a period of years is called a "Section 179 Expense Election." The maximum amount deductible for 1999 is $19,000, $500 more than 1998. For the year 2000 the maximum increases to $20,000, and it increases annually thereafter until it reaches $25,000 for 2003.

Tax Credits

Adoption Credit. An adoption expense credit is available for up to $5,000 of qualified expenses for each eligible adopted person. The credit limit is per person, not per year. The adoption credit increases to $6,000 for an eligible person with special needs. Adoption expenses paid through a nondiscriminatory employee adoption-assistance program may be excluded from gross income. Both the adoption credit and exclusion begin to phase out when AGI reaches $75,000.

Earned Income Credit. Lower income workers who have dependent children and maintain a household may be eligible for a refundable earned income credit. The credit is based on total earned income such as wages, commissions, and tips.

The maximum 1999 earned income credit for an individual with one qualifying child is $2,312. The credit is phased out as earned income increases, and is completely phased out once adjusted gross income reaches $26,928. For an individual with 2 or more qualifying children, the maximum credit is $3,816 and is fully phased out once AGI reaches $30,580.

The credit has been extended to include persons who do not have a qualifying child. For individuals without children the maximum credit is $347. To qualify: (1) Earned income and adjusted gross income (AGI) must be less than $10,200, (2) the individual or spouse must be at least 25 years old and less than 65 years old, and (3) the individual cannot be claimed as a dependent on another person's return.

The Welfare Reform Act added several more restrictions to the earned income credit. (1) The credit cannot be taken by individuals who are not authorized to be employed in the U.S.; (2) individuals must include their Social Security number and, if married, their spouse's Social Security number on the return claiming the credit; (3) the individual's "disqualified" income cannot exceed $2,350. Disqualified income includes interest, dividends, and if greater than zero, net rent, royalty income, and capital-gains net income.

The IRS will assist individuals filing for the credit if they need assistance. Individuals may qualify for the credit even if they are not otherwise required to file a return. However, a tax return *must be filed* to receive the refund.

Education Credits. Many individuals now will qualify for an educational credit, the *Hope Scholarship Credit.* The amount of the credit is up to a maximum of $1,500 per student. This credit is deducted from an individual's federal income tax liability. The Hope Scholarship credit applies to qualified tuition and related expenses for the student's first 2 years of postsecondary education in a degree or certificate program at an eligible educational institution. The credit does not apply to room and board or cost of books, nor is the credit refundable.

Another new educational credit, the *Lifetime Learning Credit,* is available for taxpayers whose postsecondary education expenses are not eligible for the Hope credit. This credit is equal to 20% of qualified tuition and fees paid for by the taxpayer, spouse, or dependents. Taxpayers may claim this credit up to $1,000 ($5,000 of expenses x 20%) for total qualified tuition and related expenses paid for all eligible students in 1999 who are enrolled in an eligible educational institution. After the year 2003, the credit increases to $2,000 ($10,000 x 20%). Allowable credits are deducted from the individual's federal income tax and reported on Form 8863, *Education Credits (Hope and Lifetime Learning Credits).*

The Lifetime Learning Credit is allowed only for years in which the Hope credit is not used. The credit may not be taken in any year in which funds are withdrawn from an Educational IRA. Any excess of the credit not used is nonrefundable.

The credit begins to phase out when modified adjusted gross income (AGI) exceeds $40,000 for singles and $80,000 on a joint return, with full phaseout at $50,000 for singles and $100,000 on joint returns.

Taxable Social Security Benefits

Earnings Limitations. *Age 62 to 64.* For 1999, individuals in this age group will lose $1 of their Social Security benefits for every $2 of earned income exceeding $9,600. Earned income consists of wages, salaries, commissions, tips, and other such compensation.

Age 65 to 69. For each $3 of earned income above $15,500 individuals lose $1 of benefits.

Age 70 or Over. Individuals age 70 or over will not lose any benefits regardless of the amount of their earnings.

Taxable Benefits. Up to 50% of Social Security benefits may be taxable income if the person's total income is:
• more than $25,000 but less than $34,000 for a single individual, a head of household, a qualifying widow(er), or a married person who is filing separately *if the spouses lived apart all year.*
• over $32,000 but less than $44,000 for married individuals filing jointly.

For people with incomes exceeding these maximum amounts, 85% of Social Security benefits become taxable. If the taxpayer is married and filing separately, and lived with a spouse at any time during the year, the percentage amounts are reduced to zero.

Most Social Security benefits will not be taxable if they are the only income received during 1999.

Retirement Planning

Retirement Planning. Previous legislation eliminated the 5-year averaging for lump sum distributions from qualified retirement plans beginning after 1999. However, prior rules to individuals who were age 50 before 1986 still apply.

Age 70½ Plus. The owner of a traditional IRA must begin receiving distributions from the IRA by Apr. 1 of the calendar year following the year in which he or she reaches age 70½, even if the individual is not retired. However, any employee who works beyond age 70½ and is not a 5% or more owner of the business can continue to defer profit sharing and pension retirement plan distributions.

The SIMPLE Plan. A simplified retirement plan titled Savings Incentive Match Plan for Employees (SIMPLE) is available for businesses with 100 or fewer employees, including self-employed individuals. This retirement plan is generally easier to implement and more cost effective to administrate than a traditional 401(k). For 1999, employees can defer up to $6,000 in compensation, and, of course, defer the tax liability on these amounts to a future date. A SIMPLE retirement plan can operate either as an IRA or as a 401(k).

Retired and Moved. States may not impose an income tax on retirement income if the person is no longer a resident.

IRS Tax Audit

The IRS received more than 125 million individual income tax returns in 1999. Only about 1 out of every 100 of those returns will be audited . . . good news unless that one happens to be yours. The IRS is very good at selecting returns that will yield additional taxes.

If after an audit, the IRS concludes you have additional tax liability, you can meet with the examiner's supervisor to discuss your case further. If you still do not agree, you can appeal the findings through a separate Appeals Office. You can also appeal to the U.S. Tax Court.

If you would like further information, call the IRS TeleTax at 1-800-829-4477 for recorded information on topic number 151, *Your Appeal Rights.* The information is also available at http://www.irs.ustreas.gov

Your Rights as a Taxpayer

Under the *Taxpayer Bill of Rights 1,* the IRS now must explain, in easy-to-understand language, any actions it proposes to take against a taxpayer. The law also requires the IRS to modify some of its audit and collection procedures.

When Congress enacted the *Taxpayer Bill of Rights 2,* it created an Office of the Taxpayer Advocate within the IRS with authority to order IRS personnel to issue refund checks and meet deadlines for resolving disputes. Also, the agency must pay legal fees if the taxpayer wins the case and the IRS cannot show it was "substantially justified" in pursuing the matter.

More recently, Congress created a 9-member independent oversight board to watch over the IRS management and the administration of its duties.

The legislation shifts the burden of proof to the IRS under certain circumstances in disputes dealing with income, estate, and gift taxes. Further, it establishes formal procedures designed to ensure due process when the IRS seeks to collect taxes by levy, including seizure.

For more information see IRS Publication 1, *Your Rights as a Taxpayer;* call 1-800-TAX-FORM for a free copy.

State Government Individual Income Taxes

Source: Reproduced with permission from *CCH State Tax Guide,* published and copyrighted by CCH Inc., 2700 Lake Cook Road, Riverwoods, IL 60015

Below are basic state tax rates on taxable income, for 1999 unless otherwise indicated. Alaska, Florida, Nevada, South Dakota, Texas, Washington, and Wyoming did not have state income taxes and are thus not listed. For further details, see notes which follow.

Alabama
1st	$1,000	2%
Next	$5,000	4%
Over	$6,000	5%

Arizona*
1st	$20,000	2.87%
Next	$30,000	3.2%
Next	$50,000	3.74%
Next	$200,000	4.72%
$300,001 and over		5.04%

Arkansas
1st	$2,999	1%
Next	$3,000	2.5%
Next	$3,000	3.5%
Next	$6,000	4.5%
Next	$10,000	6%
$25,000 or over		7%

California*
$0 to $10,528		1%
$10,529 to $24,954		2%
$24,955 to $39,384		4%
$39,385 to $54,674		6%
$54,675 to $69,096		8%
Over $69,096		9.3%

Colorado
4.75% of federal taxable income

Connecticut
1st	$20,000	3%
Over	$20,000	4.5%

Delaware
$2,001 to $5,000		3.1%
Next	$5,000	4.85%
Next	$10,000	5.8%
Next	$5,000	6.15%
Next	$5,000	6.45%
Over	$30,000	6.9%

District of Columbia
1st	$10,000	6%
2d	$10,000	8%
Over	$20,000	9.5%

Georgia
1st	$1,000	1%
Next	$2,000	2%
Next	$2,000	3%
Next	$2,000	4%
Next	$3,000	5%
Over	$10,000	6%

Hawaii
1st	$4,000	1.6%
Next	$4,000	3.9%
Next	$8,000	6.8%
Next	$8,000	7.2%
Next	$8,000	7.5%
Next	$8,000	7.8%
Next	$20,000	8.2%
Next	$20,000	8.5%
Over	$80,000	8.75%

Idaho*
1st	$1,000	2%
2d	$1,000	4%
3d	$1,000	4.5%
4th	$1,000	5.5%
5th	$1,000	6.5%
Next	$2,500	7.5%
Next	$12,500	7.8%
Next	$20,000	8.2%

Illinois
3% of taxable net income

Indiana
3.4% of adj. gross income

Iowa
$0 to $1,148	0.36%
$1,149 to $2,296	0.72%
$2,297 to $4,592	2.43%
$4,593 to $10,332	4.5%
$10,333 to $17,220	6.12%
$17,221 to $22,960	6.48%
$22,961 to $34,440	6.8%
$34,441 to $51,660	7.92%
Over $51,660	8.98%

Kansas
1st	$30,000	3.5%
Next	$30,000	6.25%
Over	$60,000	6.45%

Kentucky
1st	$3,000	2%
Next	$1,000	3%
Next	$1,000	4%
Next	$3,000	5%
Over	$8,000	6%

Louisiana*
1st	$10,000	2%
Next	$40,000	4%
Over	$50,000	6%

Maine
Less than $4,150	2%
$4,150 to $8,249	4.5%
$8,250 to $16,499	7%
$16,500 or more	8.5%

Maryland
1st	$1,000	2%
2d	$1,000	3%
3d	$1,000	4%
Over	$3,000	4.85%

Massachusetts
Short term cap. gains		12%
5 classes of cap. gain income		0-5%
All other income		5.95%

Michigan
4.4% of taxable income

Minnesota
$0 to $25,220	5.5%
$25,221 to $100,200	7.25%
Over $100,200	8%

Mississippi
1st	$5,000	3%
Next	$5,000	4%
Over	$10,000	5%

Missouri
1st	$1,000	1.5%
2d	$1,000	2%
3rd	$1,000	2.5%
4th	$1,000	3%
5th	$1,000	3.5%
6th	$1,000	4%

7th	$1,000	4.5%
8th	$1,000	5%
9th	$1,000	5.5%
Over	$9,000	6%

Montana
$0 to $1,999	2%
$2,000 to $3,999 less $20	3%
$4,000 to $7,999 less $60	4%
$8,000 to $12,099 less $140	5%
$12,100 to $16,099 less $251	6%
$16,100 to $20,099 less $422	7%
$20,100 to $28,199 less $623	8%
$28,200 to $40,199 less $905	9%
$40,200 to $70,399 less $1,307	10%
$70,400 and over less $2,011	11%

Nebraska
1st	$4,000	2.62%
Next	$26,000	3.65%
Next	$16,750	5.24%
Over	$46,750	6.99%

New Hampshire
5% of interest and dividends

New Jersey
1st	$20,000	1.4%
Next	$30,000	1.75%
Next	$20,000	2.45%
Next	$10,000	3.5%
Next	$70,000	5.525%
Over	$150,000	6.37%

New Mexico*
Not over $8,000	1.7%
$8,001 to $16,000	3.2%
$16,001 to $24,000	4.7%
$24,001 to $40,000	6%
$40,001 to $64,000	7.1%
$64,001 to $100,000	7.9%
Over $100,000	8.2%

New York
1st	$16,000	4%
Next	$6,000	4.5%
Next	$4,000	5.25%
Next	$14,000	5.9%
Over	$40,000	6.85%

North Carolina
Up to $21,250		6%
Next	$78,750	7%
Over	$100,000	7.75%

North Dakota
1st	$3,000	2.67%
Next	$2,000	4%
Next	$3,000	5.33%
Next	$7,000	6.67%
Next	$10,000	8%

Next	$10,000	9.33%
Next	$15,000	10.67%
Over	$50,000	12%

Ohio
1st	$5,000	0.673%
Next	$5,000	1.347%
Next	$5,000	2.694%
Next	$5,000	3.368%
Next	$20,000	4.040%
Next	$40,000	4.715%
Next	$20,000	5.388%
Next	$100,000	6.255%
Over	$200,000	6.799%

Oklahoma
1st	$2,000	0.5%
Next	$3,000	1%
Next	$2,500	2%
Next	$2,300	3%
Next	$2,400	4%
Next	$2,800	5%
Next	$6,000	6%
Remainder		6.75%

Oregon
1st	$2,350	5%
Next	$3,500	7%
Over	$5,850	9%

Pennsylvania 2.8%

Rhode Island
26.5% of federal liability

South Carolina
1st	$2,340	2.5%
Next	$2,340	3%
Next	$2,340	4%
Next	$2,340	5%
Next	$2,340	6%
$11,701 and over		7%

Tennessee
6% of interest and dividends

Utah
1st	$1,500	2.3%
Next	$1,500	3.3%
Next	$1,500	4.2%
Next	$1,500	5.2%
Next	$1,500	6%
Over	$7,500	7%

Vermont
25% of federal income tax

Virginia
1st	$3,000	2%
Next	$2,000	3%
Next	$12,000	5%
Over	$17,000	5.75%

West Virginia
1st	$10,000	3%
Next	$15,000	4%
Next	$15,000	4.5%
Next	$20,000	6%
Over	$60,000	6.5%

Wisconsin*
$0 to $10,000	4.77%
$10,001 to $20,000	6.37%
$20,001 and over	6.77%

* = Community property state in which, in general, one-half of the community income is taxable to each spouse.

Alabama: Rates shown are for married persons filing jointly. Single persons, heads of families, married persons filing separately, and estates or trusts are taxed at 2% of the first $500 of taxable income, 4% on the next $2,500, and 5% on the rest.

Arizona: Rates shown are for married persons filing jointly and heads of households. For single taxpayers and married taxpayers filing separately, rates range from 2.87% of the first $10,000 of taxable income to 5.04% of taxable income over $150,000.

California: Rates shown are the 1999 inflation-adjusted rates for residents who are joint taxpayers or surviving spouses with dependents. For single taxpayers, married persons filing separately, and fiduciaries, rates range from 1% on the first $5,264 of taxable income to 9.3% on taxable income of $34,548 and over. For unmarried heads of households, rates range from 1% on the first $10,531 of taxable income to 9.3% on taxable income of $47,025 and over. A 7% alternative minimum tax is imposed.

Colorado: Alternative minimum tax imposed. Qualified taxpayers may pay alternative tax of 0.5% of gross receipts from sales.

Connecticut: Rates shown are for married individuals filing jointly or persons filing as a surviving spouse. For: (1) unmarried individuals and married individuals filing separately, rates are 3% on the first $10,000 of Conn. taxable income and $300 plus 4.5% of the excess over $10,000; (2) for heads of households, rates are 3% of the first $16,000 of taxable income and $480 plus 4.5% of the excess over $16,000; and (3) for trusts or estates, rates are 4.5% of taxable income. Although the rate reductions are effective as noted above, employer withholding will continue per the July 1, 1996, withholding tables. The Commissioner of Revenue services issued new withholding tables effective July 1, 1998. Resident estates and trusts are subject to the 4.5% income tax rate on all of their income. Additional state minimum tax imposed on resident individuals, trusts, and estates is equal to the amount by which the Conn. minimum tax exceeds the Conn. basic income tax [the lesser of (a) 19% of adjusted federal tentative minimum tax, or (b) 5% of adjusted federal alternative minimum taxable income]. Separate provisions apply for non- and part-year resident individuals, trusts, and estates.

District of Columbia: The tax on unincorporated business is 9.975%. Minimum tax, $100.

Georgia: Rates shown are for married persons filing jointly and heads of households. Single persons pay at rates ranging from 1% on taxable net income not over $750 to 6% on taxable net income over $7,000. Married persons filing separately pay at rates ranging from 1% on taxable net income not over $500 to 6% on taxable net income over $5,000.

Hawaii: Rates shown are for taxpayers filing jointly and surviving spouses. For heads of households, rates range from 1.6% of taxable income up to $3,000 to 8.75% of taxable income of $60,000 and over. For unmarried individuals (other than a surviving spouse or head of household), married individuals filing separately, and estates and trusts, the rates range from 1.6% of taxable income up to $2,000 to 8.75% of taxable income over $40,000.

Idaho: Each person (joint returns deemed one person) filing return pays additional $10.

Illinois: Additional personal property replacement tax of 1.5% of net income is imposed on partnerships, trusts, and S corporations.

Iowa: An alternative minimum tax is imposed equal to 75% of the maximum state individual income tax rate for the tax year of the state alternative minimum taxable income.

Kansas: Rates shown are for married individuals filing joint returns. For tax year 1999 and all tax years thereafter, for single individuals and married individuals filing separate returns, the rate is 3.5% of the first $15,000 of Kansas taxable income; for taxable income that is more than $15,000 but not over $30,000, the rate is $525 plus 6.25% of the excess over $15,000. For taxable income that is more than $30,000 the rate is $1,462.50 plus 6.45% of the excess over $30,000.

Louisiana: These are the maximum tax rates for individuals. For joint returns, the tax is determined as if net income and personal exemption credits were reduced by one-half. Actual tax is determined from tax tables.

Maine: Rates shown are for single individuals and married persons filing separately. For unmarried or legally separated individuals who qualify as heads of household, tax rates range from 2% if taxable income is less than $6,200 to 8.5% if taxable income is $24,750 or more. For married individuals filing jointly and widows or widowers permitted to file a joint federal return, tax rates range from 2% if taxable income is less than $8,250 to 8.5% if taxable income is $33,000 or more. Additional state minimum tax is imposed equal to the amount by which the state minimum tax (27% of adjusted federal tentative minimum tax) exceeds Maine income tax liability, other than withholding tax liability.

Maryland: For a tax year beginning after 2000 but before 2002, income over $3,000 will be taxed at a rate of 4.8%. For a tax year beginning after 2001, income over $3,000 will be taxed at a rate of 4.75%.

Michigan: For the year 2000, the personal income tax rate decreases to 4.3%. Persons with business activity in Michigan are also subject to a single business tax on an adjusted tax base.

Minnesota: Rates shown are 1999 amounts for married individuals filing jointly and surviving spouses. For single individuals, the tax is 5.5% on the first $17,250 of taxable income, 7.25% on income over $17,250 but not over $56,680, and 8% on income over $56,680; for married individuals filing separately, tax is 2.2% on the first $12,610 of income, 7.25% on income over $12,610 but not over $50,100, and 8% on income over $50,100; for unmarried heads of households, tax is 5.5% on the first $21,240 of taxable income, 7.25% on income over $21,240 but not over $85,350, and 8% on income over $85,350. A 7% alternative minimum tax is imposed.

Montana: Rates shown are 1998 amounts, as indexed for inflation. Minimum tax, $1.

Nebraska: Rates shown are for married couples filing jointly and qualified surviving spouses. Rates for married couples filing separately range from 2.62% of the first $2,000 to 6.99% of taxable income over $23,375. Rates for heads of households range from 2.62% of the first $3,800 to 6.99% of taxable income over $35,000. Rates for single individuals range from 2.62% of the first $2,400 to 6.99% of taxable income over $26,500. Rates for estates range from 2.62% of the first $500 to 6.99% for taxable income over $15,150.

New Jersey: Rates shown are for married persons filing jointly, heads of households, and surviving spouses. Rates for married persons filing separately, unmarried individuals, and estates and trusts range from 1.4% of the first $20,000 of taxable income to 6.37% of taxable income over $75,000.

New Mexico: Rates shown are for married persons filing jointly and surviving spouses. For married persons filing separately, rates range from 1.7% on the first $4,000 of taxable income to 8.2% on taxable income over $50,000. For heads of households, rates range from 1.7% on the first $7,000 of taxable income to 8.2% on taxable income over $83,000. For single individuals, estates, and trusts, rates range from 1.7% of the first $5,500 of taxable income to 8.2% of taxable income over $65,000. Qualified taxpayers may pay alternative tax of 0.75% of gross receipts from New Mexico sales.

New York: Rates shown are for married individuals filing jointly and surviving spouses. Separate schedules are set out for heads of households (ranging from 4% on the first $11,000 of taxable income to 6.85% on taxable income over $30,000) and for unmarried individuals, married individuals filing separately, and estates and trusts (ranging from 4% of the first $8,000 of taxable income to 6.85% of taxable income over $20,000). In addition, individuals, estates, and trusts are subject to a 6% tax on minimum taxable income. A tax table benefit recapture supplemental tax is imposed on some individuals.

North Carolina: Rates shown are for married persons filing jointly. For heads of households rates are 6% on the first $17,000, 7% of next $63,000, 7.75% of excess over $80,000. For unmarried individuals (other than surviving spouses and heads of households), the rate is 6% of first $12,750, 7% of next $47,250, 7.75% of excess over $60,000. For married filing separately the rate is 6% of first $10,625, 7% of next $39,375, 7.75% of excess over $50,000.

North Dakota: Individuals, estates, and trusts are allowed an optional method of computing the tax. The optional tax is 14% of the taxpayer's adjusted federal income tax liability for the tax year.

Oklahoma: Rates shown are for heads of households, married persons filing jointly, and a surviving spouse not deducting federal income taxes. Single persons, married persons filing separately, and estates and trusts not deducting federal income taxes pay at rates ranging from 0.5% on the first $1,000 of taxable income to 6.75% on taxable income over $10,000. Optional rates (ranging from 0.5% to 10%) are enacted for taxpayers who deduct federal income taxes.

Oregon: Rates shown are for single or married filing separately. Rates for joint filers, heads of households, and qualifying widow(er)s are 5% of the first $4,700, 7% for $4,701 to $11,700, and 9% over $11,700.

Rhode Island: This is the 1999 rate. For 2000, the rate is 26% of federal liability.

Utah: Rates shown are for married persons filing jointly, heads of households, or qualifying widow(er). Married taxpayers filing separately, single taxpayers, and estates and trusts pay at rates ranging from 2.3% on taxable income not over $750 to 7% on taxable income over $3,750.

West Virginia: Rates shown are for single taxpayers, married persons filing jointly, heads of household, surviving spouses, and estates and trusts. For married taxpayers filing separately, rates range from 3% of the first $5,000 of taxable income to 6.5% of income over $30,000. A minimum tax is also imposed, equal to the excess by which an amount equal to 25% of any federal minimum tax or alternative minimum tax for the tax year exceeds the total tax due for the tax year.

Wisconsin: These are the 1998 rates. Rates shown are for married persons filing jointly. Rates for married persons filing separately range from 4.77% of the first $5,000 of taxable income to 6.77% of taxable income over $10,000. The rates for fiduciaries and single individuals range from 4.77% of the first $7,500 of taxable income to 6.77% of taxable income over $15,000. Alternative minimum tax is imposed. For tax years after 1998, a temporary recycling surcharge is imposed on individuals, estates, partnerships, trusts, and exempt trusts, except those entities engaged only in farming, at the rate of the greater of $25 or 0.2173% of net business income. The maximum surcharge is $9,800. An individual, estate, trust, exempt trust, or partnership engaged in farming is subject to a surcharge of $25.

ENERGY

U.S. Energy Summary, 1998

Source: Energy Information Administration, U.S. Dept. of Energy, *Annual Energy Review 1998*

Energy consumption in the U.S. in 1998 declined slightly to 94.2 quadrillion British thermal units (Btu), the first decline since 1991, according to preliminary data. Continued warm winter weather was the major cause. Decreases in the consumption of natural gas and conventional hydroelectric power were largely offset by increases in the consumption of nuclear electric power, petroleum, and coal. Natural gas consumption decreased by 3.1% to 21.3 trillion cubic feet. The consumption of coal was at an all-time high of 1.04 billion short tons in 1998, up 0.8% from the 1997 level. The energy intensity of the economy, measured in terms of energy consumption per dollar of gross domestic product, fell from 1997, continuing a long-term trend. About 12,500 Btu of energy were consumed for each 1992 dollar in 1998, compared with about 17,800 Btu in 1978 and about 14,200 Btu in 1988.

U.S. total energy production in 1998 grew by less than 1%, although reaching an all-time high of 72.9 quadrillion Btu. The increase that occurred resulted from a rise in the production of coal, nuclear electric power, natural gas, and biomass; production of conventional hydroelectric power declined by 8.4% while crude oil production fell by 3.2%. Coal production increased 2.6% to a record level of 1.12 billion short tons, while generation of nuclear electric power increased by 7.2% to 674 billion kilowatt hours. Crude oil (including lease condensate) production continued to drop, and at 13.2 quadrillion Btu was at its lowest level in 47 years.

U.S. net imports of energy (imports minus exports) rose to an all-time high of 21.8 quadrillion Btu in 1998, an increase of 4.5% from the 1997 level. Most of the increase was accounted for by petroleum net imports, which increased by 3.1% to a record level of 20.3 quadrillion Btu. U.S. net imports of petroleum totaled 9.5 million barrels per day in 1998. Members of OPEC supplied 4.8 million barrels per day, or 50.7% of the total. Although coal remained the primary U.S. energy export, coal exports fell 7.5% to 77.2 million short tons in 1998.

U.S. Energy Overview, 1960-98

Source: Energy Information Administration, U.S. Dept. of Energy, *Annual Energy Review 1998*

(in quadrillion Btu)

	1960	1965	1970	1975	1980	1985	1990[1]	1995	1997[P]	1998[P]
Production	41.49	49.34	62.07	59.86	64.76	64.87	70.76[R]	71.04[R]	72.51[R]	72.90
Fossil fuels	39.87	47.23	59.19	54.73	59.01	57.54	58.56	57.41	58.76[R]	58.92
Coal	10.82	13.06	14.61	14.99	18.60	19.33	22.46	21.98	23.21[R]	23.82
Natural gas (dry)	12.66	15.78	21.67	19.64	19.91	16.98	18.36	19.10	19.39[R]	19.47
Crude oil[2]	14.93	16.52	20.40	17.73	18.25	18.99	15.57	13.89	13.66[R]	13.22
Natural gas plant liquids	1.46	1.88	2.51	2.37	2.25	2.24	2.17	2.44	2.50[R]	2.41
Nuclear electric power	0.01	0.04	0.24	1.90	2.74	4.15	6.16	7.18	6.68[R]	7.16
Hydroelectric pumped storage[3]	(4)	(4)	(4)	(4)	(4)	(4)	−0.04	−0.03	−0.04	−0.05
Renewable energy	1.61	2.07	2.65	3.23	3.01	3.18	6.07[R]	6.48[R]	7.12[R]	6.87
Conventional hydroelectric power[5]	1.61	2.06	2.63	3.15	2.90	2.97	3.01	3.21	3.70[R]	3.39
Geothermal energy	(*)	(*)	0.01	0.07	0.11	0.20	0.33[R]	0.32[R]	0.32[R]	0.31
Biofuels[6]	(*)	(*)	(*)	(*)	(*)	0.01	2.63	2.85[R]	2.98[R]	3.05
Solar energy	0	0	0	0	0	0	0.07	0.07	0.07[R]	0.07
Wind energy	0	0	0	0	0	(*)	0.02	0.03	0.04	0.04
Imports	4.23	5.92	8.39	14.11	15.97	12.10	18.99	22.48	25.53[R]	26.15
Coal	0.01	(*)	(*)	0.02	0.03	0.05	0.07	0.18	0.19	0.22
Natural gas	0.16	0.47	0.85	0.98	1.01	0.95	1.55	2.90	3.06[R]	3.21
All crude oil and petroleum pdcts.[7]	4.00	5.40	7.47	12.95	14.66	10.61	17.12	18.86	21.75[R]	22.21
Other[8]	0.06	0.04	0.07	0.16	0.28	0.49	0.26	0.54	0.53[R]	0.51
Exports	1.48	1.85	2.66	2.36	3.72	4.23	4.91	4.58	4.63[R]	4.32
Coal	1.02	1.38	1.94	1.76	2.42	2.44	2.77	2.32	2.19	2.03
Natural gas	0.01	0.03	0.07	0.07	0.05	0.06	0.09	0.16	0.16	0.16
All crude oil and petroleum pdcts.[7]	0.48	0.39	0.55	0.44	1.16	1.66	1.82	1.99	2.10	1.95
Other[8]	0.02	0.06	0.11	0.08	0.09	0.08	0.23	0.11	0.18[R]	0.18
Adjustments[9]	−0.43	−0.72	−1.37	−1.07	−1.05	1.24	−0.75[R]	1.93[R]	0.97[R]	-0.50
Consumption[10]	43.80	52.68	66.43	70.55	75.96	73.98	84.09[R]	90.86[R]	94.37[R]	94.23
Fossil fuels	42.14	50.58	63.52	65.35	69.98	66.22	71.95[R]	76.94	80.39[R]	80.06
Coal	9.84	11.58	12.26	12.66	15.42	17.48	19.10	20.09[R]	21.44[R]	21.62
Coal coke net imports	−0.01	−0.02	−0.06	0.01	−0.04	−0.01	(*)	0.03	0.02	0.03
Natural gas[11]	12.39	15.77	21.79	19.95	20.39	17.83	19.30	22.16	22.54[R]	21.84
Petroleum[12]	19.92	23.25	29.52	32.73	34.20	30.92	33.55	34.66	36.38[R]	36.57
Nuclear electric power	0.01	0.04	0.24	1.90	2.74	4.15	6.16	7.18	6.68[R]	7.16
Hydroelectric pumped storage[3]	(4)	(4)	(4)	(4)	(4)	(4)	−0.04	−0.03	−0.04	−0.05
Renewable energy	1.66	2.06	2.67	3.29	3.23	3.61	6.17[R]	6.76[R]	7.33[R]	7.07
Conventional hydroelectric power[5,13]	1.66	2.06	2.65	3.22	3.12	3.40	3.10	3.47	3.92[R]	3.60
Geothermal energy[14]	(*)	(*)	0.01	0.07	0.11	0.20	0.35[R]	0.34[R]	0.32[R]	0.32
Biofuels[6]	(*)	(*)	(*)	(*)	(*)	0.01	2.63	2.85[R]	2.98[R]	3.05
Solar energy	0	0	0	0	0	0	0.07	0.07	0.07[R]	0.07
Wind energy	0	0	0	0	0	(*)	0.02	0.03	0.04	0.04

(1) Starting in 1990, expanded coverage of nonelectric utility use of renewable energy resulted in an increase in total energy production and consumption figures. (2) Includes lease condensate. (3) Total pumped storage facility production minus energy used for pumping. (4) Before 1990, pumped storage is included in conventional hydroelectric power. (5) Starting in 1990, pumped storage is removed and expanded coverage of industrial use of hydroelectric power is included. (6) These include wood, wood waste, peat, wood liquors, railroad ties, pitch, wood sludge, municipal solid waste, agricultural waste, straw, tires, landfill gases, fish oils, and/or other waste. (7) Includes imports of crude oil for the Strategic Petroleum Reserve, which began in 1977. (8) Coal coke and small amts. of electricity transmitted across U.S. borders with Canada and Mexico. (9) A balancing item. (10) Starting in 1990, "Consumption" includes the part of net imports of electricity derived from nonrenewable energy sources. (11) Includes supplemental gaseous fuels. (12) Petroleum products supplied, including natural gas plant liquids and crude oil burned as fuel. (13) Starting in 1990, includes only the part of net imports of electricity derived from hydroelectric power. (14) Includes electricity imports from Mexico derived from geothermal energy. R=revised data. P=preliminary data. (*)=Less than 0.005 quadrillion Btu. **NOTE:** Totals may not equal sum of components as a result of independent rounding.

World Energy Consumption and Production Trends, 1997

Source: Energy Information Administration, U.S. Dept. of Energy, International Energy Database, July 1999

The world's consumption of primary energy—petroleum, natural gas, coal, net hydroelectric, nuclear, geothermal, solar, wind electric power, and biomass (primarily for the United States)—increased from 376 quadrillion Btu (British thermal units) in 1996 to 380 quadrillion Btu in 1997. The 29 countries of the Organization for Economic Cooperation and Development (OECD), which includes some of the world's largest economies (the United States, Japan, and Germany), continued to dominate global energy use. OECD nations accounted for more than 58% of the world's primary energy consumption in 1997. World production of primary energy increased from 375 quadrillion Btu in 1996 to 382 quadrillion Btu in 1997. World production of petroleum in 1997 was over 72 million barrels per day, or 151 quadrillion Btu; petroleum remained the most heavily used source of energy.

In 1997, 3 countries—the United States, Russia, and China—retained their position as the world's leading producers (39%) and consumers (42%) of energy. Russia and the United States alone supplied 30% of the world total. The U.S. alone accounted for 25% of the world's total energy consumption. The U.S. consumed 30% more energy than it produced—an imbalance of 21.9 quadrillion Btu.

World's Major Producers of Primary Energy, 1997

Source: Energy Information Administration, International Energy Database; quadrillion Btu

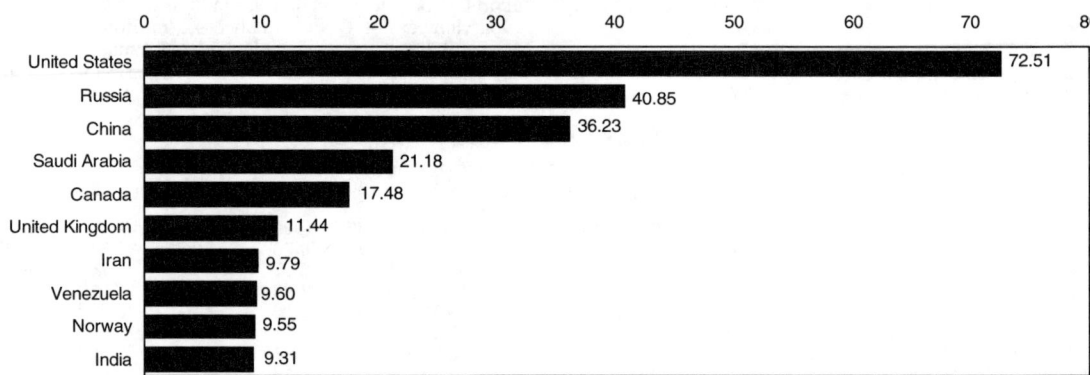

World's Major Consumers of Primary Energy, 1997

Source: Energy Information Administration, International Energy Database; quadrillion Btu

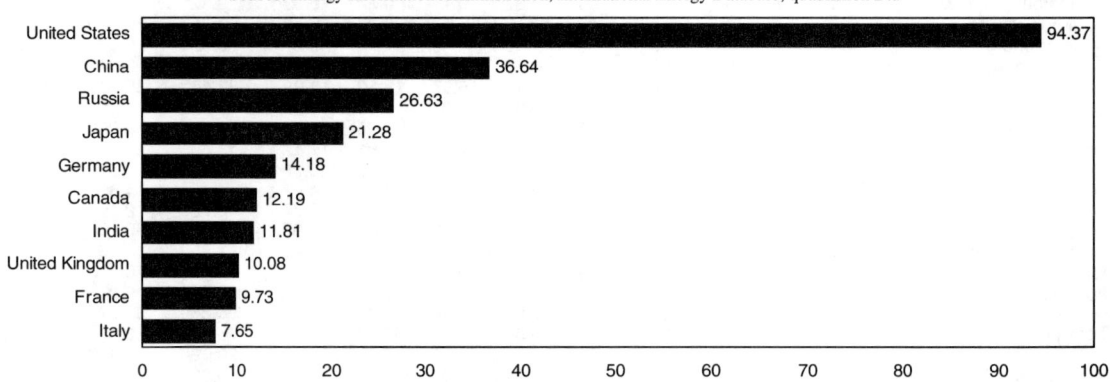

U.S. Petroleum Trade, 1975-98

Source: Energy Information Administration, U.S. Dept. of Energy, *Monthly Energy Review,* June 1998
(in thousands of barrels per day; average for the year)

Year	Imports from Persian Gulf[1]	Total imports	Total exports	Net imports[2]	Petroleum products supplied	Year	Imports from Persian Gulf[1]	Total imports	Total exports	Net imports[2]	Petroleum products supplied
1975	1,165	6,056	209	5,846	16,322	1987	1,077	6,678	764	5,914	16,665
1976	1,840	7,313	223	7,090	17,461	1988	1,541	7,402	815	6,587	17,283
1977	2,448	8,807	243	8,565	18,431	1989	1,861	8,061	859	7,202	17,325
1978	2,219	8,363	362	8,002	18,847	1990	1,966	8,018	857	7,161	16,988
1979	2,069	8,456	471	7,985	18,513	1991	1,845	7,627	1,001	6,626	16,714
1980	1,519	6,909	544	6,365	17,056	1992	1,778	7,888	950	6,938	17,033
1981	1,219	5,996	595	5,401	16,058	1993	1,782	8,620	1,003	7,618	17,237
1982	696	5,113	815	4,298	15,296	1994	1,728	8,996	942	8,054	17,718
1983	442	5,051	739	4,312	15,231	1995	1,573	8,835	949	7,886	17,725
1984	506	5,437	722	4,715	15,726	1996	1,604	9,399	981	8,419	18,234
1985	311	5,067	781	4,286	15,726	1997	1,755	10,162	1,003	9,158	18,620
1986	912	6,224	785	5,439	16,281	1998	2,091	10,708	945	9,764	18,680

(1) Bahrain, Iran, Iraq, Kuwait, Qatar, Saudi Arabia, and the United Arab Emirates. (2) Net imports are total imports minus total exports. **Notes:** Beginning in Oct. 1977, imports for the Strategic Petroleum Reserves are included. U.S. geographic coverage includes the 50 states and the District of Columbia. U.S. exports include shipments to U.S. territories, and imports include receipts from U.S. territories. Figures in this table may not add, because of independent rounding.

Appliance Use in U.S. Households, 1980-98

Source: Energy Information Administration, U.S. Dept. of Energy, *Annual Energy Review 1998*

(percentage of households)

Appliance	1980	1982	1984	1987	1990	1993	1998	Change 1980-98
Total households (millions)	82	84	86	90	94	97	101	+20
Type of appliances								
Electric appliances								
Television set (color)	82	85	88	93	96	98	99	+17
Clothes washer	74	72	74	76	76	77	77	+3
Range (stove-top burner)	54	53	54	57	58	61	60	+7
Oven, regular or microwave	59	59	63	79	88	91	91	+32
Oven, microwave	14	21	34	61	79	84	83	+69
Clothes dryer	47	45	46	51	53	57	55	+8
Dishwasher	37	36	38	43	45	45	50	+13
Dehumidifier	9	9	9	10	12	9	NA	NA
Swimming-pool pump[2]	3	3	NA	NA	5	5	5	+2
Gas appliances[3]								
Range (stove-top or burner)	46	47	45	43	42	38	39	-7
Oven	42	42	42	41	41	36	37	-5
Clothes dryer	14	15	16	15	16	15	16	+2
Outdoor gas grill	9	11	13	20	26	29	NA	NA
Refrigerators[4]								
One	86	86	88	86	84	85	85	-1
Two or more	14	13	12	14	15	15	15	+1
Air conditioning								
Central[5]	27	28	30	36	39	44	47	+20
Individual room units[5]	30	30	30	30	29	25	25	-5
Portable kerosene heaters	([1])	3	6	6	5	2	2	+2

(1) Less than 0.5%. (2) All reported swimming pools were assumed to have an electric pump for filtering and circulating the water, except for 1993, when a filtering system was made explicit. (3) Includes natural gas or liquefied petroleum gases. (4) Fewer than 0.5% of the households did not have a refrigerator. (5) Households with both central and individual room units are counted only under "Central." NA= not available. **NOTE:** Percentages may not add because of independent rounding.

Energy Consumption, Total and Per Capita, by State, 1997

Source: Energy Information Administration, U.S. Dept. of Energy, State Energy Data Report 1997 (released Sept. 1998)

	TOTAL CONSUMPTION					CONSUMPTION PER CAPITA				
Rank	State	Trillion Btu	Rank	State	Trillion Btu	Rank	State	Million Btu	Rank	State	Million Btu
1.	Texas	11,396.1	27.	Iowa	1,136.4	1.	Alaska	1,143.5	27.	Oregon	349.1
2.	California	7,727.5	28.	Colorado	1,133.4	2.	Louisiana	940.0	28.	Nevada	348.0
3.	Ohio	4,144.3	29.	Oregon	1,132.9	3.	Wyoming	892.2	29.	Georgia	345.4
4.	New York	4,093.2	30.	Mississippi	1,123.7	4.	Texas	587.6	30.	Utah	334.6
5.	Louisiana	4,093.0	31.	Kansas	1,033.1	5.	North Dakota	554.9	31.	Michigan	333.1
6.	Pennsylvania	3,900.7	32.	Arkansas	1,030.2	6.	Kentucky	462.6	32.	District of Columbia	333.1
7.	Illinois	3,900.2	33.	West Virginia	809.2	7.	Indiana	457.5	33.	South Dakota	327.7
8.	Florida	3,614.7	34.	Connecticut	795.8	8.	Alabama	457.3	34.	North Carolina	326.2
9.	Michigan	3,259.1	35.	Alaska	697.3	9.	West Virginia	445.6	35.	Illinois	325.2
10.	Indiana	2,683.6	36.	Utah	691.2	10.	Maine	445.3	36.	Pennsylvania	324.6
11.	Georgia	2,588.4	37.	New Mexico	647.1	11.	Montana	429.4	37.	Missouri	323.2
12.	New Jersey	2,585.4	38.	Nebraska	617.1	12.	Oklahoma	422.9	38.	New Jersey	320.7
13.	North Carolina	2,425.2	39.	Nevada	584.4	13.	Idaho	411.6	39.	Virginia	315.4
14.	Washington	2,164.2	40.	Maine	553.4	14.	Mississippi	411.2	40.	Colorado	291.1
15.	Virginia	2,126.4	41.	Idaho	497.7	15.	Arkansas	408.1	41.	Vermont	283.5
16.	Tennessee	2,084.2	42.	Wyoming	428.3	16.	Iowa	397.9	42.	Maryland	266.8
17.	Alabama	1,977.5	43.	Montana	377.5	17.	Kansas	397.0	43.	New Hampshire	259.0
18.	Wisconsin	1,835.4	44.	North Dakota	355.8	18.	South Carolina	389.0	44.	Arizona	252.9
19.	Kentucky	1,809.6	45.	New Hampshire	303.9	19.	Tennessee	387.8	45.	Massachusetts	250.6
20.	Missouri	1,748.9	46.	Delaware	267.2	20.	Washington	385.3	46.	Florida	246.2
21.	Minnesota	1,685.8	47.	South Dakota	241.9	21.	New Mexico	375.2	47.	Connecticut	243.3
22.	Massachusetts	1,534.1	48.	Hawaii	239.5	22.	Nebraska	372.3	48.	California	240.0
23.	South Carolina	1,474.2	49.	Rhode Island	235.1	23.	Ohio	370.1	49.	Rhode Island	237.9
24.	Oklahoma	1,405.2	50.	District of Columbia	176.6	24.	Delaware	363.2	50.	New York	225.3
25.	Maryland	1,360.0	51.	Vermont	167.1	25.	Minnesota	359.5	51.	Hawaii	201.0
26.	Arizona	1,152.4		**TOTAL U.S.**	**94,063.6**	26.	Wisconsin	352.8		**TOTAL U.S.**	**351.2**

Gasoline Retail Prices, U.S. City Average, 1974-99

Source: Energy Information Administration, U.S. Dept. of Energy, *Monthly Energy Review,* Aug. 1999

(cents per gallon, including taxes)

AVERAGE	Leaded regular	Unleaded regular	Unleaded premium	All types[1]	AVERAGE	Leaded regular	Unleaded regular	Unleaded premium	All types[1]
1974	53.2	NA	NA	NA	1987	89.7	94.8	109.3	95.7
1975	56.7	NA	NA	NA	1988	89.9	94.6	110.7	96.3
1976	59.0	61.4	NA	NA	1989	99.8	102.1	119.7	106.0
1977	62.2	65.6	NA	NA	1990	114.9	116.4	134.9	121.7
1978	62.6	67.0	NA	65.2	1991	NA	114.0	132.1	119.6
1979	85.7	90.3	NA	88.2	1992	NA	112.7	131.6	119.0
1980	119.1	124.5	NA	122.1	1993	NA	110.8	130.2	117.3
1981[2]	131.1	137.8	147.0[3]	135.3	1994	NA	111.2	130.5	117.4
1982	122.2	129.6	141.5	128.1	1995	NA	114.7	133.6	120.5
1983	115.7	124.1	138.3	122.5	1996	NA	123.1	141.3	128.8
1984	112.9	121.2	136.6	119.8	1997	NA	123.4	141.6	129.1
1985	111.5	120.2	134.0	119.6	1998	NA	105.9	125.0	111.5
1986	85.7	92.7	108.5	93.1	1999 (Jan.-June)	NA	107.0	126.5	112.7

(1) Also includes types of motor gasoline not shown separately. (2) In Sept. 1981, the Bureau of Labor Statistics changed the weights used in the calculation of average motor gasoline prices. Starting in Sept. 1981, gasohol is included in the average for all types, and unleaded premium is weighted more heavily. (3) Based on Sept. through Dec. data only. **NOTE:** Geographic coverage for 1974-77 is 56 urban areas; for 1978 and later, 85 urban areas. NA = not available.

World Crude Oil and Natural Gas Reserves, Jan. 1, 1998

Sources: Energy Information Administration, U.S. Dept. of Energy, *Annual Energy Review 1998;*
Oil and Gas Journal (OGJ), Dec. 1997; *World Oil (WO)*, Aug. 1998

Region and country	Crude oil (billion barrels) OGJ	WO	Natural gas (trillion cubic feet) OGJ	WO	Region and country	Crude oil (billion barrels) OGJ	WO	Natural gas (trillion cubic feet) OGJ	WO
North America	**67.4**	**68.8**	**295.1**	**298.2**	Iraq	112.5	99.7	109.8	112.6
Canada	4.8	5.5	65.0	67.5	Kuwait	96.5	93.5	52.9	56.7
Mexico	40.0	40.8	63.9	63.5	Oman	5.2	3.8	27.5	21.3
United States	22.5	22.5	167.2	167.2	Qatar	3.7	4.2	300.0	270.0
Central and South					Saudi Arabia	261.5	263.8	190.5	208.0
America	**86.2**	**63.0**	**222.3**	**221.5**	Syria	2.5	2.3	8.3	8.4
Argentina	2.6	2.6	24.3	24.1	United Arab Emirates	97.8	64.2	204.9	208.0
Bolivia	0.1	0.1	4.6	4.2	Yemen	4.0	3.1	16.9	17.0
Brazil	4.8	7.1	5.6	8.0	Other	0.0	0.0	0.2	0.6
Colombia	2.8	2.6	14.2	8.0	**Africa**	**70.1**	**76.7**	**348.6**	**355.2**
Ecuador	2.1	2.8	3.7	3.6	Algeria	9.2	13.8	130.6	139.5
Peru	0.8	0.8	7.0	7.0	Angola	5.4	3.9	1.7	1.7
Trinidad and Tobago	0.6	0.5	15.9	18.3	Cameroon	0.4	0.6	3.9	3.8
Venezuela	71.7	45.5	143.1	145.5	Congo Republic	1.5	1.6	3.2	4.3
Other	0.7	0.9	3.9	2.8	Egypt	3.8	3.7	27.6	28.8
Western Europe	**18.3**	**19.7**	**173.1**	**159.6**	Libya	29.5	26.9	46.3	45.5
Denmark	0.9	1.0	4.0	3.7	Nigeria	16.8	21.2	114.9	109.2
Germany	0.4	0.4	12.1	12.3	Tunisia	0.3	0.3	2.5	2.8
Italy	0.7	0.6	10.5	8.1	Other	3.1	4.7	17.9	19.7
Netherlands	0.1	0.1	61.3	63.1	**Far East and Oceania**	**42.3**	**58.0**	**320.6**	**436.3**
Norway	10.4	11.7	52.3	41.4	Australia	1.8	2.3	19.4	51.9
United Kingdom	5.0	5.2	26.8	27.0	Brunei	1.4	1.1	14.1	13.3
Other	0.7	0.8	6.1	4.3	China	24.0	34.0	41.0	42.4
Eastern Europe and					India	4.3	3.5	17.4	13.5
Former USSR	**59.0**	**64.3**	**2,000.4**	**1,903.5**	Indonesia	5.0	9.1	72.3	137.8
Hungary	0.1	0.1	3.2	1.6	Malaysia	3.9	5.0	79.8	87.0
Romania	1.6	0.9	14.0	4.3	New Zealand	0.1	0.2	2.4	2.0
Former USSR	48.6	54.8	1,700.0	1,705.0	Pakistan	0.2	0.2	21.0	23.3
Other[1]	8.7	8.4	283.2	192.6	Papua New Guinea	0.3	0.3	9.0	6.0
Middle East	**677.0**	**624.4**	**1,726.1**	**1,720.7**	Thailand	0.3	0.3	7.0	12.5
Bahrain	0.2	0.2	5.1	5.0	Other	0.9	2.0	37.2	46.8
Iran	93.0	89.7	810.0	812.2	**WORLD**	**1,020.1**	**975.0**	**5,087.2**	**5,095.2**

(1) Albania, Azerbaijan, Belarus, Bulgaria, Croatia, Czech Republic, Georgia, Hungary, Kazakhstan, Kyrgyzstan, Lithuania, Poland, Romania, Serbia, Slovakia, Tajikistan, Turkmenistan, Ukraine, Uzbekistan. **NOTES:** Data for Kuwait and Saudi Arabia include one-half of the reserves in the Neutral Zone between Kuwait and Saudi Arabia. All reserve figures except those for the former USSR and natural gas reserves in Canada are *proved reserves* recoverable with present technology and prices. Former USSR figures and natural gas figures for Canada are *explored reserves*, which include proved, probable, and some partially possible. Totals may not equal sum of components as a result of independent rounding.

Nuclear Electricity Generation by Selected Country, Mar. 1999

Source: Energy Information Administration, U.S. Dept. of Energy, *Monthly Energy Review*, June 1999

(billion kilowatt-hours; E = estimate)

Argentina	0.7	France	34.3	Lithuania	1.0	Sweden	7.5E
Belgium	4.4	Germany	14.2	Mexico	0.9	Switzerland	2.3
Brazil	0.4	Hungary	1.1	Netherlands	0.4	Taiwan	2.9
Bulgaria	1.9E	India	1.1	Russia	11.7	Ukraine	8.0
Canada	7.2	Japan	27.7	South Africa	1.4	United Kingdom	9.3
Finland	2.1	Korea, South	7.9	Spain	4.2	United States	60.9E

Nations Most Reliant on Nuclear Energy, 1998

Source: International Atomic Energy Agency, May 1999

(Nuclear electricity generation as % of total)

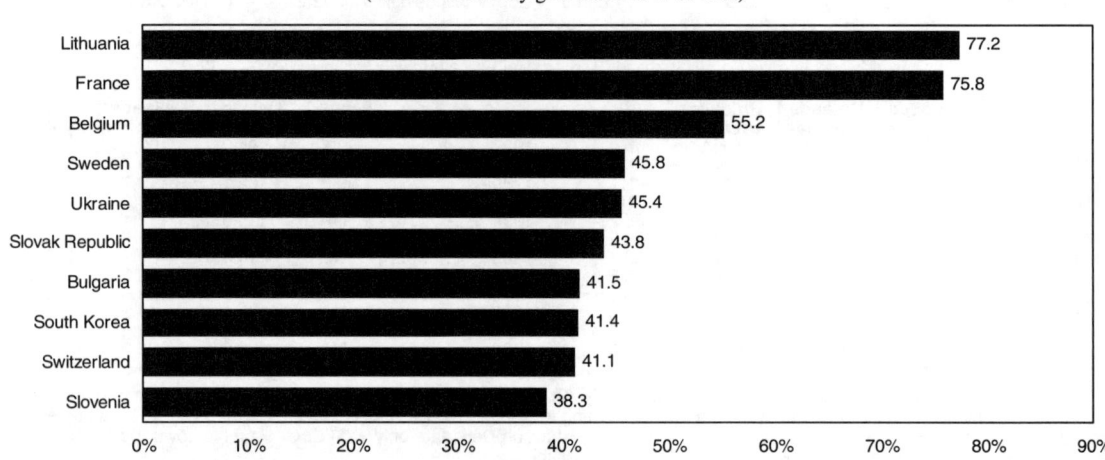

World Nuclear Power Summary, 1998

Source: International Atomic Energy Agency, May 1999

Country	Reactors in operation — No. of units	Reactors in operation — Total MW(e)	Reactors under construction — No. of units	Reactors under construction — Total MW(e)	Nuclear electricity supplied in 1998 — TW(e).h[1]	Nuclear electricity supplied in 1998 — % of total	Total operating experience to Dec. 31, 1998 — Years	Total operating experience to Dec. 31, 1998 — Months
Argentina	2	935	1	692	6.93	10.04	40	7
Armenia	1	376	—	—	1.42	24.69	31	3
Belgium	7	5,712	—	—	43.89	55.16	156	7
Brazil	1	626	1	1,229	3.27	1.08	16	9
Bulgaria	6	3,538	—	—	15.49	41.50	101	1
Canada.	14	9,998	—	—	67.50	12.44	405	2
China	3	2,167	6	4,420	13.46	1.16	17	5
Czech Republic . . .	4	1,648	2	1,824	12.35	20.50	50	8
Finland	4	2,658	—	—	20.98	27.44	79	4
France	58	61,653	1	1,450	368.40	75.77	1052	1
Germany	20	22,282	—	—	145.20	28.29	570	7
Hungary	4	1,729	—	—	13.12	35.62	54	2
India	10	1,695	4	808	10.15	2.51	159	1
Iran.	—	—	2	2,111	—	—	—	—
Japan	53	43,691	2	1863	306.94	35.86	863	5
Kazakhstan	1	70	—	—	0.09	0.18	25	6
Korea, South	15	12,340	3	2,550	85.19	41.39	137	5
Lithuania.	2	2,370	—	—	12.29	77.21	26	6
Mexico	2	1,308	—	—	8.83	5.41	13	11
Netherlands	1	449	—	—	3.59	4.13	54	—
Pakistan	1	125	1	300	0.34	0.65	27	3
Romania.	1	650	1	650	4.90	10.35	2	6
Russia	29	19,843	4	3,375	95.38	13.08	613	6
Slovakia	5	2,020	3	1164	11.39	43.80	73	11
Slovenia	1	632	—	—	4.79	38.33	17	3
South Africa	2	1,842	—	—	13.58	7.25	28	3
Spain	9	7,377	—	—	56.68	31.66	174	2
Sweden	12	10,040	—	—	70.00	45.75	255	2
Switzerland.	5	3,079	—	—	24.37	41.07	118	10
Taiwan	6	4,884	1	1,300	35.41	24.77	104	1
Ukraine.	16	13,765	4	3,800	70.64	45.42	222	1
United Kingdom . . .	35	12,968	—	—	91.14	27.09	1,168	4
United States	104	96,423	—	—	673.70	18.69	2,351	8
TOTAL	**434**	**348,891**	**36**	**27,536**	**2,291.41**	**—**	**9,012**	**6**

(1) 1 terawatt-hour [TW(e).h] = 10^6 megawatt-hour [MW(e).h]. For an average power plant, 1 TW(e).h = 0.39 megatonnes of coal equivalent (input) and 0.23 megatonnes of oil equivalent (input).

U.S. Nuclear Reactor Units and Power Plant Operations

Source: Energy Information Administration, U.S. Dept. of Energy, *Monthly Energy Review,* June 1999

	Number of reactor units — Licensed for operation — Operable	Number of reactor units — Licensed for operation — In startup	Number of reactor units — Construction permits — Granted	Number of reactor units — Construction permits — Pending	Number of reactor units — On order	Number of reactor units — Announced	Number of reactor units — Total	Total design capacity (million KWs)	Nuclear-based electricity generation (million net KW-hours)	Nuclear portion of domestic electricity generation (percent)
1977	65	2	78	49	13	2	209	203	250,883	11.8
1978	70	0	88	32	5	0	195	191	276,403	12.5
1979	68	0	90	24	3	0	185	180	255,155	11.4
1980	70	1	82	12	3	0	168	162	251,116	11.0
1981	74	0	76	11	2	0	163	157	272,674	11.9
1982	77	2	60	3	2	0	144	134	282,773	12.6
1983	80	3	53	0	2	0	138	129	293,677	12.7
1984	86	6	38	0	2	0	132	123	327,634	13.6
1985	95	3	30	0	2	0	130	121	383,691	15.5
1986	100	7	19	0	2	0	128	119	414,038	16.6
1987	107	4	14	0	2	0	127	119	455,270	17.7
1988	108	3	12	0	0	0	123	115	526,973	19.5
1989	110	1	10	0	0	0	121	113	529,355	19.0
1990	111	0	8	0	0	0	119	111	576,862	20.5
1991	111	0	8	0	0	0	119	111	612,565	21.7
1992	109	0	8	0	0	0	117	111	618,776	22.1
1993	109	0	7	0	0	0	116	110	610,291	21.2
1994	109	0	7	0	0	0	116	110	640,440	22.0
1995	109	1	6	0	0	0	116	110	673,402	22.5
1996	110	0	6	0	0	0	116	110	674,729	21.9
1997	107	0	3	0	0	0	110	102	628,644	20.1
1998	104	0	3	0	0	0	107	99	673,702	21.0

ENVIRONMENT

Greenhouse Effect and Global Warming

Source: U.S. Environmental Protection Agency

The Earth naturally absorbs incoming solar radiation and emits thermal radiation back into space. Some of the thermal radiation is trapped by certain so-called greenhouse gases in the atmosphere, which increases warming of the Earth's surface and atmosphere. In recent years, carbon dioxide (CO_2), a naturally occurring greenhouse gas, has been building up in the atmosphere as the result of human activities such as the burning of fossil fuels (coal, oil, and natural gas) and deforestation. Water vapor, methane (CH_4), nitrous oxide (N_2O), and ozone (O_3) are also naturally occurring greenhouse gases. Greenhouse gases that are mostly human-made include chlorofluorocarbons (CFCs), hydrochlorofluorocarbons (HCFCs), hydrofluorocarbons (HFCs), perfluorocarbons (PFCs), and sulfur hexafluoride (SF_6). In addition, several non-greenhouse gases (carbon monoxide [CO], oxides of nitrogen [NOx], and nonmethane volatile organic compounds [NMVOCs]) contribute indirectly to the greenhouse effect by producing greenhouse gases during chemical transformations or by influencing the atmospheric lifetimes of greenhouse gases.

Since 1800, atmospheric concentrations of CO_2, CH_4, and N_2O have increased by 30%, 145%, and 15%, respectively. This increasing buildup is believed by many scientists to be the major cause of higher than normal average global temperatures in the 1990s; 1998 was the hottest year on record (global average temp. 58.1°F). The 2d-hottest year was 1997, and 7 of the 10 hottest years on record were in the 1990s. Over the past century, the Earth's average temperature has risen by approximately 1°F, and some scientists believe that it could rise by 2° to 6°F over the next century. This global warming could speed the melting of the polar ice caps, inundate coastal lowlands, and bring about major changes in crop production and in natural habitat. The United States is the world's leading producer of CO_2, followed by China, Russia, Japan, India, and Germany.

In Dec. 1997, a United Nations summit on global warming was held in Kyoto, Japan. Delegates from over 150 nations adopted an international treaty to set some limits on emissions of CO_2, CH_4, N_2O, HFCs, PFCs, and SF_6. The accord, known as the Kyoto Protocol, called for an overall reduction in emissions of 5.2% below 1990 levels by the year 2012, significantly short of the 15% reduction proposed by the European Union. Under the accord, the 15 EU nations agreed to reductions of 8%, the U.S. to 7%, and Japan to 6%. Developing nations were permitted to limit their emissions voluntarily. The accord allowed high-emissions nations to meet their targets by purchasing pollution rights from nations that exceed their goal, although the mechanism for doing so was left unsettled. No penalties for noncompliance were specified.

At a follow-up conference in Buenos Aires in Nov. 1998, it was agreed to resolve unsettled issues by the year 2000. The U.S. signed the treaty on Nov. 12, 1998; U.S. ratification required Senate approval.

U.S. Greenhouse Gas Emissions From Human Activities, 1990-97

Source: U.S. Environmental Protection Agency

GAS AND SOURCE	1990	1993	1994	1995	1996	1997
Carbon dioxide (CO_2)	**1,344.3**	**1,379.2**	**1,403.5**	**1,419.2**	**1,469.3**	**1,487.9**
Fossil fuel combustion	1,327.2	1,360.6	1,383.9	1,397.8	1,447.7	1,466.0
Methane (CH_4)	**169.9**	**172.0**	**175.5**	**178.6**	**178.3**	**179.6**
Coal Mining	24.0	19.2	19.4	20.3	18.9	18.8
Natural gas systems	32.9	34.1	33.5	33.2	33.7	33.5
Enteric fermentation	32.7	33.6	34.5	34.9	34.5	34.1
Nitrous oxide (N_2O)	**95.7**	**100.4**	**108.3**	**105.4**	**108.2**	**109.0**
Agricultural soil management	65.3	67.0	73.4	70.2	72.0	74.1
Hydrofluorocarbons (HFCs), perfluorocarbons (PFCs), and sulfur hexafluoride (SF_6)[1]	**22.2**	**23.4**	**25.9**	**30.8**	**34.7**	**37.1**
TOTAL U.S. EMISSIONS	**1,632.1**	**1,675.0**	**1,713.2**	**1,733.9**	**1,790.5**	**1,813.6**
NET U.S. EMISSIONS[2]	**1,320.6**	**1,466.5**	**1,504.7**	**1,525.4**	**1,582.0**	**1,605.0**

Note: Emissions are given in millions of metric tons of carbon equivalent (MMTCE), a measurement used by the Intergovernmental Panel on Climate Change (IPCC) to compare greenhouse gases. Totals may not equal sum of individual source categories due to rounding. Subcategories (indented) are not all-inclusive. (1) These gases have extremely high global warming potential, and PFCs and SF_6 have long atmospheric lifetimes. (2) Total emissions minus carbon dioxide absorbed by forests or other means.

U.S. Greenhouse Gas Emissions, 1997

Source: U.S. Environmental Protection Agency

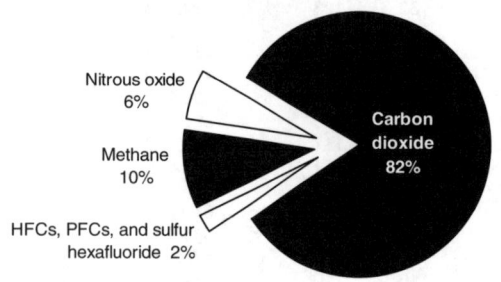

World Carbon Dioxide Emissions from the Use of Fossil Fuels, 1997

Source: Energy Information Administration, 1999

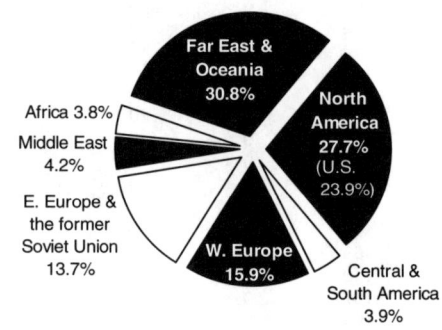

Toxics Release Inventory, 1996-97

Source: U.S. Environmental Protection Agency

Reported industrial releases of toxic chemicals into the environment in the U.S. by major manufacturing facilities (excluding power plants and mining facilities) increased 2.2% from the 1996 figure and decreased 42.8% from the figure for 1988, the baseline year. Totals below may not add because of rounding.

Pollutant releases	1997 mil lb	1996 mil lb	Top industries, total releases	1997 mil lb	1996 mil lb
Air releases	1,332	1,459	Chemicals	797	792
Surface water releases	218	179	Primary metals	695	620
Underground injection	220	204	Paper	234	229
On-site land releases	347	332	Plastics	108	117
TOTAL	**2,116**	**2,174**	Transportation equipment	102	108
Pollutant transfers			**Top carcinogens, air/water/land releases**		
To recycling	2,381	2,151	Dichloromethane	48	54
To energy recovery	508	477	Styrene	45	42
To treatment	259	290	Formaldehyde	22	21
To publicly owned treatment works	267	236	Trichloroethylene	18	21
Other transfers	0	3	Acetaldehyde	13	13
TOTAL	**3,415**	**3,157**	Chloroform	7	10

Top 10 States, Total Releases, 1995-97

Source: U.S. Environmental Protection Agency

State	1997 mil lb	1996 mil lb	1995 mil lb	State	1997 mil lb	1996 mil lb	1995 mil lb
Texas	262	265	304	Indiana	123	113	115
Louisiana	186	186	178	Tennessee	107	107	111
Ohio	159	153	157	Utah	104	87	79
Pennsylvania	143	119	128	Alabama	95	103	115
Illinois	128	114	119	Florida	95	86	81

Emissions of Principal Air Pollutants in the U.S., 1988-1997

Source: U.S. Environmental Protection Agency, Office of Air Quality Planning and Standards

(in thousand short tons; estimated)

Source	1988	1989	1990	1991	1992	1993	1994	1995	1996	1997
Carbon monoxide[1]	116,081	103,480	95,794	97,790	94,400	94,526	98,854	89,151	90,611	87,451
Lead	7.1	5.5	5.0	4.2	3.8	3.9	4.0	3.9	3.9	3.9
Nitrogen oxides[2]	23,718	23,414	23,436	23,520	23,789	24,046	24,345	23,768	23,391	23.582
Volatile organic compounds[2]	24,027	22,274	20,935	21,063	20,642	20,830	21,465	20,558	19,306	19,214
Particulate matter[3]	3,527	3,507	3,333	3,252	3,288	3,170	3,134	3,161	3,157	3,112
Sulfur dioxide	23,154	23,308	23,678	23,057	22,819	22,478	21,880	19,189	19,836	20,371
TOTAL[4]	**190,514**	**175,989**	**167,181**	**168,686**	**164,942**	**165,054**	**169,682**	**155,831**	**156,305**	**153,734**

(1) The observed increase in carbon monoxide emissions between 1993 and 1994 is attributed to 2 sources: transportation emissions (up 2%) and wildfire emissions (up 160%). (2) Ozone, a major air pollutant and the primary constituent of smog, is not emitted directly to the air but is formed by sunlight acting on emissions of nitrogen oxides and volatile organic compounds. (3) Does not include natural sources. (4) Totals are rounded, as are components of totals.

Carbon Monoxide Emission Estimates, 1988-97

Source: U.S. Environmental Protection Agency, Office of Air Quality Planning and Standards; in thousand short tons

Source	1988	1989	1990	1991	1992	1993	1994	1995	1996	1997
Fuel combustion	7,373	7,443	5,510	5,856	6,154	5,585	5,519	5,934	5,980	4,817
Industrial processes	7,034	7,013	5,852	5,740	5,683	5,898	5,839	5,791	5,816	6,052
Transportation	85,779	80,870	73,224	77,443	75,511	76,030	77,883	70,377	69,353	67,014
Miscellaneous	15,895	8,154	11,208	8,751	7,052	7,013	9,613	7,049	9,462	9,568
TOTAL[1]	**116,081**	**103,480**	**95,794**	**97,790**	**94,400**	**94,526**	**98,854**	**89,151**	**90,611**	**87,451**

(1) Totals may not add because of rounding.

Lead Emission Estimates, 1988-97

Source: U.S. Environmental Protection Agency, Office of Air Quality Planning and Standards; in short tons

Source	1988	1989	1990	1991	1992	1993	1994	1995	1996	1997
Fuel combustion	511	505	500	495	490	495	494	488	493	496
Industrial processes	3,090	3,161	3,278	3,081	2,734	2,869	3,005	2,873	2,892	2,897
Transportation	3,452	1,802	1,197	592	584	547	544	563	525	522
TOTAL[1]	**7,053**	**5,468**	**4,975**	**4,168**	**3,808**	**3,911**	**4,043**	**3,924**	**3,910**	**3,915**

(1) Totals may not add because of rounding.

Nitrogen Oxides Emission Estimates, 1988-97

Source: U.S. Environmental Protection Agency, Office of Air Quality Planning and Standards; in thousand short tons

Source	1988	1989	1990	1991	1992	1993	1994	1995	1996	1997
Fuel combustion	10,472	10,538	10,895	10,779	10,928	11,111	11,015	10,828	10,519	10,724
Industrial processes	860	852	892	816	857	861	878	873	879	917
Transportation	11,659	11,731	11,278	11,639	11,750	11,849	12,069	11,830	11,650	11,595
Miscellaneous	727	293	371	286	254	225	383	237	343	346
TOTAL[1]	**23,718**	**23,414**	**23,436**	**23,520**	**23,789**	**24,046**	**24,345**	**23,768**	**23,391**	**23,582**

(1) Totals may not add because of rounding.

Air Quality of Selected U.S. Metropolitan Areas[1], 1988-97

Source: U.S. Environmental Protection Agency, Office of Air Quality Planning and Standards

Metropolitan statistical area	1988	1989	1990	1991	1992	1993	1994	1995	1996	1997
Atlanta, GA.	44	17	52	24	19	42	13	43	22	26
Bakersfield, CA	126	114	97	109	100	97	97	104	109	55
Baltimore, MD	60	28	29	50	23	48	41	36	28	30
Boston, MA–NH	28	12	7	13	9	6	10	8	2	8
Chicago, IL	40	16	5	21	4	3	8	21	6	9
Dallas, TX.	37	18	24	2	11	11	15	36	12	15
Denver, CO.	35	16	11	7	8	3	2	2	0	0
Detroit, MI.	35	18	11	28	8	5	13	14	13	12
El Paso, TX	15	26	22	7	11	9	8	5	7	3
Fresno, CA.	110	91	62	83	61	59	55	60	65	50
Hartford, CT	39	19	13	23	15	14	18	15	5	16
Houston, TX.	72	43	54	37	32	28	45	65	28	47
Las Vegas, NV–AZ.	30	45	22	10	6	9	14	4	5	0
Los Angeles–Long Beach, CA	239	222	174	174	178	137	144	109	94	63
Miami, FL	8	6	1	1	3	6	1	2	1	3
Minneapolis–St. Paul, MN–WI	11	8	4	2	3	0	2	7	2	0
New Haven–Meriden, CT	26	11	15	30	10	12	13	14	8	19
New York, NY.	57	30	37	50	11	19	21	19	15	23
Orange County, CA	56	58	46	35	38	25	15	9	9	3
Philadelphia, PA–NJ.	53	44	39	48	24	50	26	30	22	32
Phoenix–Mesa, AZ.	29	34	13	11	15	17	11	25	17	15
Pittsburgh, PA.	43	21	19	22	9	13	19	25	11	20
Riverside–San Bernardino, CA	185	190	158	159	175	167	148	125	118	106
Sacramento, CA.	88	71	66	69	48	22	37	34	33	2
St. Louis, MO–IL	44	29	24	33	16	9	32	35	20	15
Salt Lake City–Ogden, UT	16	22	5	20	9	2	4	4	9	1
San Diego, CA	123	128	97	67	66	57	45	47	31	14
San Francisco, CA	1	0	0	0	0	0	0	2	0	0
Seattle–Bellevue–Everett, WA	20	7	10	5	3	0	3	0	6	1
Ventura, CA	108	93	70	89	55	44	64	66	62	44
Washington, DC–MD–VA–WV	56	27	26	49	15	47	21	30	18	28

(1) Data indicate the number of days metropolitan statistical areas failed to meet acceptable air-quality standards at trend sites (Pollutant Standards Index rating over 100). All figures were revised based on new standards set in 1998.

Hazardous Waste Sites in the U.S., 1999

Source: U.S. Environmental Protection Agency, *National Priorities List*, July 1999

STATE	Final Gen	Final Fed	Proposed Gen	Proposed Fed	Total	STATE	Final Gen	Final Fed	Proposed Gen	Proposed Fed	Total
Alabama	10	3	0	0	13	Nevada	1	0	0	0	1
Alaska	1	6	0	0	7	New Hampshire	17	1	0	0	18
Arizona	7	3	0	0	10	New Jersey	103	7	3	1	114
Arkansas	12	0	0	0	12	New Mexico	9	1	1	0	11
California	69	24	3	0	96	New York	81	4	1	0	86
Colorado	13	3	2	0	18	North Carolina	23	2	1	0	26
Connecticut	13	1	0	0	14	North Dakota	0	0	0	0	0
Delaware	16	1	0	0	17	Ohio	29	3	2	2	36
District of Columbia	0	1	0	0	1	Oklahoma	11	1	1	0	13
Florida	45	6	1	0	52	Oregon	8	2	1	0	11
Georgia	12	2	1	0	15	Pennsylvania	92	6	3	0	101
Hawaii	1	3	0	0	4	Rhode Island	10	2	0	0	12
Idaho	5	2	2	0	9	South Carolina	23	2	0	0	25
Illinois	35	4	4	0	43	South Dakota	0	1	0	0	1
Indiana	29	0	1	0	30	Tennessee	10	3	1	1	15
Iowa	15	1	1	0	17	Texas	29	4	3	0	36
Kansas	9	1	0	1	11	Utah	8	4	6	0	18
Kentucky	15	1	0	0	16	Vermont	9	0	0	0	9
Louisiana	14	1	2	0	17	Virginia	20	9	0	0	29
Maine	10	3	0	0	13	Washington	33	14	1	0	48
Maryland	10	8	1	0	19	West Virginia	5	2	1	0	8
Massachusetts	22	8	1	0	31	Wisconsin	39	0	1	0	40
Michigan	69	0	1	1	71	Wyoming	2	1	0	0	3
Minnesota	24	2	0	0	26	American Samoa	0	0	0	0	0
Mississippi	1	0	2	0	3	Guam	1	1	0	0	2
Missouri	20	3	2	0	25	Puerto Rico	9	0	0	0	9
Montana	8	0	3	0	11	Virgin Islands	2	0	0	0	2
Nebraska	9	1	0	0	10	TOTALS	1,068	158	53	6	1,285

Note: Gen = general superfund sites; Fed = federal facility sites.

Watersheds in the U.S.

Source: U.S. Environmental Protection Agency

A watershed is a water drainage area, or land areas bounded by ridges that catch rain and snow and drain to rivers, lakes, and groundwater within the drainage area. In a comprehensive assessment of watersheds in the continental U.S. released in Sept. 1999, the Environmental Protection Agency (EPA) concluded that 15% of the 2,262 watersheds had good water quality, 36% had moderate water quality, and 23% had more serious problems; there was insufficient information to fully characterize the remaining 26%. The data indicate that polluted runoff from urban and rural areas is a major contributor to water quality problems, threatening water quality even in currently healthy watersheds.

The EPA categorized the watersheds by combining nationally available data from 15 individual databases, from both public and private sources, into a single Index of Watershed Indicators. The indicators include 7 used to assess watershed conditions (quality) and 8 used to assess vulnerability to degradation from pollution. You can find information about your own watershed on the Internet by going to the following website: http://www.epa.gov/surf3/index.html

U.S. List of Endangered and Threatened Species

Source: Fish and Wildlife Service, U.S. Dept. of Interior; as of Sept. 30, 1999

Group	ENDANGERED		THREATENED		Total listed species	Species with recovery plans
	U.S. only	Foreign only	U.S. only	Foreign only		
Mammals	61	248	8	16	333	48
Birds	74	178	15	6	273	76
Reptiles	14	65	22	14	115	30
Amphibians	9	8	8	1	26	12
Fishes	69	11	42	0	122	91
Snails	18	1	10	0	29	20
Clams	61	2	8	0	71	45
Crustaceans	17	0	3	0	20	12
Insects	28	4	9	0	41	27
Arachnids	5	0	0	0	5	5
Animals, subtotal	**356**	**517**	**125**	**37**	**1,035**	**366**
Flowering plants	550	1	135	0	686	530
Conifers	2	0	1	2	5	2
Ferns and others	26	0	2	0	28	28
Plants, subtotal	**578**	**1**	**138**	**2**	**719**	**560**
GRAND TOTAL	**934**	**518**	**263**	**39**	**1,754[1]**	**886[2]**

(1) When separate populations of a species are listed as endangered and as threatened, those species are tallied twice. Those species are the argali, bull trout, chimpanzee, chinook salmon, gray wolf, green sea turtle, leopard, olive ridley sea turtle, piping plover, roseate tern, saltwater crocodile, sockeye salmon, steelhead, and Steller sea lion. (2) There are 525 approved recovery plans. Some recovery plans cover more than one species, and a few species have separate plans covering different parts of their ranges. Recovery plans are drawn up only for listed species that occur in the U.S.

Some Endangered Animal Species

Source: Fish and Wildlife Service, U.S. Dept. of the Interior

Common name	Scientific name	Range
Armadillo, giant	Pridontes maximus	Venezuela, Guyana to Argentina
Bat, gray	Myotis grisescens	Central, southeastern U.S.
Bear, brown	Ursus arctos arctos	Italy
Bison, wood	Bison bison athabascae	Canada, northwestern U.S.
Bobcat	Felis rufus escuinapae	Central Mexico
Camel, Bactrian	Camelus bactrianus	Mongolia, China
Caribou, woodland	Rangifer tarandus caribou	U.S., Canada
Cheetah	Acinonyx jubatus	Africa to India
Chimpanzee, pygmy	Pan paniscus	Congo (formerly Zaire)
Chinchilla	Chinchilla brevicaudata boliviana	Bolivia
Condor, California	Gymnogyps californianus	U.S. (AZ, CA, OR), Mexico (Baja California)
Crane, hooded	Grus monacha	Japan, Russia
Crane, whooping	Grus americana	Canada, Mexico, U.S. (Rocky Mts. to Carolinas)
Crocodile, American	Crocodylus acutus	U.S. (FL), Mexico, Caribbean Sea, Central and S America
Deer, Columbian white-tailed	Odocoileus virginianus leucurus	U.S. (OR, WA)
Dolphin, Chinese river	Lipotes vexillifer	China
Elephant, Asian	Elephas maximus	S central and southeastern Asia
Fox, northern swift	Vulpes velox hebes	U.S., Canada
Gorilla	Gorilla gorilla	Central and W Africa
Hawk, Hawaiian	Buteo solitarius	U.S. (HI)
Hyena, brown	Hyaena brunnea	Southern Africa
Kangaroo, Tasmanian forester	Macropus giganteus tasmaniensis	Australia (Tasmania)
Leopard	Panthera pardus	Africa and Asia
Lion, Asiatic	Panthera leo persica	Turkey to India
Manatee, West Indian	Trichechus manatus	Southeastern U.S., Caribbean Sea, S America
Monkey, spider	Ateles geoffroyi frontatus	Costa Rica, Nicaragua
Ocelot	Felis pardalis	U.S. (AZ, TX) to Central and S America
Orangutan	Pongo pygmaeus	Borneo, Sumatra
Ostrich, West African	Struthio camelus spatzi	W Sahara
Otter, marine	Lutra felina	Peru south to Straits of Magellan
Panda, giant	Ailuropoda melanoleuca	China
Panther, Florida	Felis concolor coryi	U.S. (LA, AR east to SC, FL)
Parakeet, golden	Aratinga guarouba	Brazil
Parrot, imperial	Amazona imperialis	West Indies (Dominica)
Penguin, Galapagos	Spheniscus mendiculus	Ecuador (Galapagos Islands)
Puma, eastern	Puma concolor couguar	Eastern N America
Python, Indian	Python molurus molurus	Sri Lanka, India
Rhinoceros, black	Diceros bicornis	Sub-Saharan Africa
Rhinoceros, northern white	Ceratotherium simum cottoni	Congo (formerly Zaire), Sudan, Uganda, Central African Republic
Salamander, Chinese giant	Andrias davidianus davidianus	Western China
Squirrel, Carolina northern flying	Glaucomys sabrinus coloratus	U.S. (NC, TN)
Stork, oriental white	Ciconia ciconia boyciana	China, Japan, Korea, Russia
Tiger	Panthera tigris	Asia
Tortoise, Galapagos	Geochelone elephantopus	Ecuador (Galapagos Islands)
Turtle, leatherback sea	Dermochelys coriacea	Tropical, temperate, and subpolar seas
Turtle, Plymouth red-bellied	Pseudemys rubriventris bangsi	U.S. (MA)
Whale, gray	Eschrichtius robustus	N Pacific Ocean
Whale, humpback	Megaptera novaeangliae	Oceania
Wolf, red	Canis rufus	Southeastern U.S. to central TX
Woodpecker, ivory-billed	Campephilus principalis	S central and southeastern U.S., Cuba
Yak, wild	Bos grunniens mutus	China (Tibet), India
Zebra, mountain	Equus zebra zebra	South Africa

Classification

Source: *Funk & Wagnalls New Encyclopedia*

In biology, classification is the identification, naming, and grouping of organisms into a formal system. The 2 fields that are most directly concerned with classification are taxonomy and systematics. Although the 2 disciplines overlap considerably, taxonomy is more concerned with nomenclature (naming) and with constructing hierarchical systems, and systematics with uncovering evolutionary relationships. Two kingdoms of living forms, Plantae and Animalia, have been recognized since Aristotle established the first taxonomy in the 4th century BC. In addition, there are the following 3 kingdoms: Protista (one-celled organisms), Monera (bacteria and blue-green algae; also known as the kingdom Procaryotae), and Fungi. The 7 basic categories of classification (from most general to most specific) are: kingdom, phylum (or division), class, order, family, genus, and species. Below are 2 examples:

ZOOLOGICAL HIERARCHY

Kingdom	Phylum	Class	Order	Family	Genus	Species name	Common name
Animalia	Chordata	Mammalia	Primates	Hominidae	Homo	Homo sapiens	Human

BOTANICAL HIERARCHY

Kingdom	Division*	Class	Order	Family	Genus	Species name	Common name
Plantae	Magnoliophyta	Magnoliopsida	Magnoliales	Magnoliaceae	Magnolia	M. virginiana	Sweet Bay

* In botany, the division is generally used in place of the phylum.

Gestation, Longevity, and Incubation of Animals

Information reviewed and updated by Ronald M. Nowak, author *Walker's Mammals of the World* (6th ed., Johns Hopkins University Press, 1999). Average longevity figures supplied by Ronald T. Reuther. These apply to animals in captivity; the potential life span of animals is rarely attained in nature. Figures on gestation and incubation are averages based on estimates made by leading authorities.

ANIMAL	Gestation (days)	Average longevity (years)	Maximum longevity (yr-mo)
Ass	365	12	47
Baboon	187	20	45
Bear: Black	219	18	36-10
Grizzly	225	25	50
Polar	240	20	45
Beaver	105	5	50
Bison	285	15	40
Camel	406	12	50
Cat (domestic)	63	12	28
Chimpanzee	230	20	60
Chipmunk	31	6	10
Cow	284	15	30
Deer (white-tailed)	201	8	20
Dog (domestic)	61	12	20
Elephant (African)	660	35	70
Elephant (Asian)	645	40	77
Elk	250	15	26-8
Fox (red)	52	7	14
Giraffe	457	10	36-2
Goat (domestic)	151	8	18
Gorilla	258	20	54
Guinea pig	68	4	8
Hippopotamus	238	41	61
Horse	330	20	50
Kangaroo (gray)	36	7	24

ANIMAL	Gestation (days)	Average longevity (years)	Maximum longevity (yr-mo)
Leopard	98	12	23
Lion	100	15	30
Monkey (rhesus)	166	15	37
Moose	240	12	27
Mouse (meadow)	21	3	4
Mouse (dom. white)	19	3	6
Opossum (American)	13	1	5
Pig (domestic)	112	10	27
Puma	90	12	20
Rabbit (domestic)	31	5	13
Rhinoceros (black)	450	15	45-10
Rhinoceros (white)	480	20	50
Sea lion (California)	350	12	34
Sheep (domestic)	154	12	20
Squirrel (gray)	44	10	23-6
Tiger	105	16	26-3
Wolf (maned)	63	5	15-8
Zebra (Grant's)	365	15	50

Incubation time (days)

Chicken	21
Duck	30
Goose	30
Pigeon	18
Turkey	26

Speeds of Animals

Source: *Natural History* magazine. Copyright © The American Museum of Natural History, 1974

ANIMAL	mph	ANIMAL	mph	ANIMAL	mph
Cheetah	70	Mongolian wild ass	40	Elephant	25
Pronghorn antelope	61	Greyhound	39.35	Black mamba snake	20
Wildebeest	50	Whippet	35.50	Six-lined race runner	18
Lion	50	Rabbit (domestic)	35	Wild turkey	15
Thomson's gazelle	50	Mule deer	35	Squirrel	12
Quarterhorse	47.5	Jackal	35	Pig (domestic)	11
Elk	45	Reindeer	32	Chicken	9
Cape hunting dog	45	Giraffe	32	Spider (Tegenaria atrica)	1.17
Coyote	43	White-tailed deer	30	Giant tortoise	0.17
Gray fox	42	Wart hog	30	Three-toed sloth	0.15
Hyena	40	Grizzly bear	30	Garden snail	0.03
Zebra	40	Cat (domestic)	30		

Most of these measurements are for maximum speeds over approximate quarter-mile distances. Exceptions are the lion and elephant, whose speeds were clocked in the act of charging; the whippet, which was timed over a 200-yd course; the cheetah, timed over a 100-yd distance; and the black mamba, six-lined race runner, spider, giant tortoise, three-toed sloth, and garden snail, which were measured over various small distances.

Major Venomous Animals

Snakes

Asian pit viper — from 2 ft to 5 ft long; throughout Asia; reactions and mortality vary, but most bites cause tissue damage, and mortality is generally low.

Australian brown snake — 4 ft to 7 ft long; very slow onset of cardiac or respiratory distress; moderate mortality, but because death can be sudden and unexpected, it is the most dangerous of the Australian snakes; antivenom.

Barba Amarilla or fer-de-lance — up to 7 ft long; from tropical Mexico to Brazil; severe tissue damage common; moderate mortality; antivenom.

Black mamba — up to 14 ft long, fast-moving; S and C Africa; rapid onset of dizziness, difficulty breathing, erratic heartbeat; mortality high, nears 100% without antivenom.

Boomslang — less than 6 ft long; in African savannahs; rapid onset of nausea and dizziness, often followed by slight recovery and then sudden death from internal hemorrhaging; bites rare, mortality high; antivenom.

Bushmaster — up to 12 ft long; wet tropical forests of C and S America; few bites occur, but mortality rate is high.

Common or Asian cobra — 4 ft to 8 ft long; throughout southern Asia; considerable tissue damage, sometimes paralysis; mortality probably not more than 10%; antivenom.

Copperhead — less than 4 ft long; from New England to Texas; pain and swelling; very seldom fatal; antivenom seldom needed.

Coral snake — 2 ft to 5 ft long; in Americas south of Canada; bite may be painless; slow onset of paralysis, impaired breathing; mortalities rare, but high without antivenom and mechanical respiration.

Cottonmouth water moccasin — up to 5 ft long; wetlands of southern U.S. from Virginia to Texas. Rapid onset of severe pain, swelling; mortality low, but tissue destruction can be extensive; antivenom.

Death adder — less than 3 ft long; Australia; rapid onset of faintness, cardiac and respiratory distress; at least 50% mortality without antivenom.

Desert horned viper — in dry areas of Africa and western Asia; swelling and tissue damage; low mortality; antivenom.

European viper — 1 ft to 3 ft long; bleeding and tissue damage; mortality low; antivenom.

Gaboon viper — more than 6 ft long; fat; 2-in. fangs; south of the Sahara; massive tissue damage, internal bleeding; few recorded bites.

King cobra — up to 16 ft long; throughout southern Asia; rapid swelling, dizziness, loss of consciousness, difficulty breathing, erratic heartbeat; mortality varies sharply with amount of venom involved, but most bites involve nonfatal amounts; antivenom.

Krait — up to 5 ft long; in SE Asia; rapid onset of sleepiness; numbness; as much as 50% mortality even with use of antivenom.

Puff adder — up to 5 ft long; fat; south of the Sahara and throughout the Middle East; rapid large swelling, great pain, dizziness; moderate mortality, often from internal bleeding; antivenom.

Rattlesnake — 2 ft to 6 ft long; throughout W Hemisphere; rapid onset of severe pain, swelling; mortality low, but amputation of affected digits is sometimes necessary; antivenom. Mojave rattler may produce temporary paralysis.

Ringhals, or spitting, cobra — 5 ft to 7 ft long; southern Africa; squirts venom through holes in front of fangs as a defense; venom is severely irritating, can cause blindness.

Russell's viper or tic-polonga — more than 5 ft long; throughout Asia; internal bleeding; bite reports common; moderate mortality rate; antivenom.

Saw-scaled or carpet viper — as much as 2 ft long; in dry areas from India to Africa; severe bleeding, fever; high mortality, causes more human fatalities than any other snake; antivenom.

Sea snakes — throughout Pacific, Indian oceans except NE Pacific; almost painless bite, variety of muscle pain, paralysis; mortality rate low, many bites not envenomed; some antivenoms.

Sharp-nosed pit viper or one hundred pace snake — up to 5 ft long; in S Vietnam, Taiwan, and China; the most toxic of Asian pit vipers; very rapid onset of swelling and tissue damage, internal bleeding; moderate mortality; antivenom.

Taipan — up to 11 ft long; in Australia and New Guinea; rapid paralysis with severe breathing difficulty; mortality nears 100% without antivenom.

Tiger snake — 2 ft to 6 ft long; S Australia; pain, numbness, mental disturbances with rapid onset of paralysis; may be the most deadly of all land snakes, although antivenom is quite effective.

Yellow or Cape cobra — 7 ft long; in S Africa; most toxic venom of any cobra; rapid onset of swelling, breathing and cardiac difficulties; mortality is high without treatment; antivenom.

Note: Not all bites by venomous snakes are actually envenomed. Any animal bite, however, carries the danger of tetanus, and anyone suffering a venomous snake bite should seek medical attention. Antivenoms do not cure; they are only an aid in the treatment of bites. Mortality rates above are for envenomed bites; low mortality, c. 2% or less; moderate, 2%-5%; high, 5%-15%.

Lizards

Gila monster — as much as 24 in. long, with heavy body and tail; in high desert in SW U.S. and N Mexico; immediate severe pain and transient low blood pressure; no recent mortality.

Mexican beaded lizard — similar to Gila monster, Mexican west coast; reaction and mortality rate similar to Gila monster.

Insects

Ants, bees, wasps, hornets, etc. Global distribution. Usual reaction is piercing pain in area of sting. Not directly fatal, except in cases of massive multiple stings. However, many people suffer allergic reactions — swelling and rashes — and a few may die within minutes from severe sensitivity to the venom (anaphylactic shock).

Spiders, Scorpions

Atrax spider — also known as funnel whip spider; several varieties, often large; in Australia; slow onset of breathing, circulation difficulties; low mortality; antivenom.

Black widow — small, round-bodied with red hourglass marking; the widow and its relatives are found in tropical and temperate zones; severe musculoskeletal pain, weakness, breathing difficulty, convulsions; may be more serious in small children; low mortality; antivenom. The **redback** spider of Australia has the hourglass marking on its back, rather than on its front, but is otherwise identical to the black widow.

Brown recluse, or fiddleback, spider — small, oblong body; throughout U.S.; pain with later ulceration at place of bite; in severe cases fever, nausea, and stomach cramps; ulceration may last months; very low mortality.

Scorpion — crablike body with stinger in tail, various sizes, many varieties throughout tropical and subtropical areas; various symptoms may include severe pain spreading from the wound, numbness, severe agitation, cramps; severe reaction may include respiratory failure; low mortality, usually in children; antivenoms.

Tarantula — large, hairy spider found around the world; the American tarantula, and probably all other tarantulas, are harmless to humans, though their bite may cause some pain and swelling.

Sea Life

Cone-shell — mollusk in small, beautiful shell; in the S Pacific and Indian oceans; shoots barbs into victims; paralysis; low mortality.

Octopus — global distribution, usually in warm waters; all varieties produce venom, but only a few can cause death; rapid onset of paralysis with breathing difficulty.

Portuguese man-of-war — jellyfishlike, with tentacles up to 70 ft long; in most warm water areas; immediate severe pain; not directly fatal, though shock may cause death in rare cases.

Sea wasp — jellyfish, with tentacles up to 30 ft long, in the S Pacific; very rapid onset of circulatory problems; high mortality because of speed of toxic reaction; antivenom.

Stingray — several varieties of differing sizes; found in tropical and temperate seas and some fresh water; severe pain, rapid onset of nausea, vomiting, breathing difficulties; wound area may ulcerate, gangrene may appear; seldom fatal.

Stonefish — brownish fish that lies motionless as a rock on bottom in shallow water; throughout S Pacific and Indian oceans; extraordinary pain, rapid paralysis; low mortality; antivenom available, amount determined by number of puncture wounds; warm water relieves pain.

Major U.S. Public Zoological Parks

Source: *World Almanac* questionnaire, 1999; budget and attendance in millions

Zoo	Budget	Atten-dance	Acres	Species	Some major attractions
Albuquerque (NM) Biological Park	$13.5	1.2	63	300	Polar Bears, Tropical America, Mexican Wolves, Koalas *for further information: (505) 764-6200.*
Arizona-Sonora Desert Museum (Tucson, AZ)	NA	0.6	100	300+	Desert Loop Trail, Desert Waters, Pollinator Gardens *for further information: (520) 883-2702.*
Audubon Zoo (New Orleans)	10.0	0.9	58	400	Louisiana Swamp, Jaguar Jungle, white tigers *for further information: (800) 774-7394.*
Baltimore Zoo	8.4	0.6	80	283	Children's zoo, Bog Turtle Exhibit, ZooLights *for further information: (410) 366-LION.*
Bronx Zoo (N.Y.C.)	37.1	2.2	265	600+	Congo Gorilla Forest, Wild Asia, Jungle World *for further information: (718) 367-1010.*
Brookfield Zoo (Chicago area)	41.0	2.0	216	400	Living Coast, The Swamp, Habitat Africa, 7 Seas Panorama *for further information: (708) 485-0263.*
Buffalo (NY) Zoological Gardens	4.3	0.4	23.5	175	Diversity of Life, Reptile House, World of Wildlife *for further information: (716) 837-3900.*
Cincinnati Zoo and Botanical Garden	18.0	1.3	72	700+	Manatee Springs, Gorilla World, Jungle Trails *for further information: (800) 94-HIPPO.*
Cleveland Metroparks Zoo	9.8	1.2	165	617	Rainforest, Wolf Wilderness, Australian Adventure *for further information: (216) 661-6500.*
Columbus Zoo and Aquarium (Powell, OH)	15.0	1.3	404	700	Manatee Coast, African Forest, songbird aviary *for further information: (614) 645-3550.*
Dallas Zoo	9.0	0.8	95	700	Wilds of Africa, Chimpanzee Forest, Endangered Tiger Habitat *for further information: (214) 670-2525.*
Denver Zoo	14.0	1.7	80	715	Tropical Discovery, Primate Panorama, Dragons of Komodo *for further information: (303) 376-4800.*
Detroit Zoological Park (Royal Oak, MI)	16.5	1.4	125	280	Penguinarium, indoor Butterfly and Hummingbird Garden *for further information: (248) 398-0900.*
Houston Zoological Gardens	NA	1.5	55	800	Tropical Bird House, Koala Crossing, Primate Exhibit *for further information: (713) 523-5888.*
Lincoln Park Zoological Gardens (Chicago)	17.0	3.0	35	290	Great Ape House, Farm-in-the-Zoo, Kovler Sea Lion Pool *for further information: (312) 742-2000.*
Los Angeles Zoo	17.2	1.3	80	350	Chimpanzees of Mahale Mountains, Red Ape Rain Forest *for further information: (323) 644-6400.*
Louisville (KY) Zoo	8.0	0.7	75	133	BOMA African Petting Zoo, Islands Exhibit *for further information: (502) 459-2181.*
Memphis (TN) Zoo	7.6	0.7	70+	400	Cat Country, Once Upon a Farm, Primate Canyon *for further information: (901) 276-WILD.*
Miami Metrozoo	7.2	0.5	300	195	white tiger, koalas, Komodo dragons, condor exhibits *for further information: (305) 251-0401.*
Milwaukee County Zoo	11.4	1.3	194	347	Jurassic Park/Lost World, Butterfly Exhibit *for further information: (414) 771-3040.*
Minnesota Zoo (Apple Valley)	17.8	1.2	500	488	Discovery Bay, Southeast tropical rain forest, World of Birds *for further information: (800) 366-7811.*
National Zoo (Washington, DC)	20.0	3.0	163	435	Amazonia, Panda Exhibit, Tiger Tracks *for further information: (202) 673-4666.*
Oklahoma City Zoological Park & Botanical Garden	9.9	0.7	110	2,200	Aquaticus, Cat Forest, Lion Overlook, Great EscApe *for further information: (405) 424-3344.*
Omaha's Henry Doorly Zoo	9.8	1.1	130	640	Indoor rain forest, cat complex, aquarium, bird aviary *for further information: (402) 733-8401.*
Oregon Zoo (Portland)	30.0	1.0	64	232	Penguinarium, Great Northwest, Alaska Tundra *for further information: (503) 226-1561.*
Philadelphia Zoo	20.0	1.1	42	350	Primate Reserve, Amphibian and Reptile House *for further information: (215) 243-1100.*
Phoenix (AZ) Zoo	13.2	1.2	125	300	Arizona Trail, Discovery Trail, Africa Trail, Tropics Trail *for further information: (602) 273-1341.*
Point Defiance Zoo & Aquarium (Tacoma, WA)	6.8	0.5	27	300	Rocky Shores, Arctic Tundra, Penguin Point, Elephants! *for further information: (253) 591-5337.*
Riverbanks Zoo & Garden (Columbia, SC)	5.0	0.9	170	400	bird pavilion, walled botanical garden *for further information: (803) 779-8717.*
St. Louis Zoo	28.0	2.9	90	705	Big Cat Country, Jungle of the Apes, Children's Zoo *for further information: (314) 781-0900.*
San Diego Wild Animal Park	40.0	1.5	2,200	400	Heart of Africa walking safari, Wgasa Bush Line Railway *for further information: (619) 234-6541.*
San Diego Zoo	80.0	3.5	100	800	Giant Pandas, Polar Bear Plunge, Gorilla Tropics, Skyfari *for further information: (619) 234-3153.*
San Francisco Zoo	15.0	0.9	125	300	Gorilla World, Koala Crossing, Lion House, Penguin Island *for further information: (415) 753-7080.*
Toledo (OH) Zoo	11.0	0.9	60	659	Hippoquarium, Primate Forest, Kingdom of the Apes *for further information: (419) 385-5721.*
Tulsa (OK) Zoo and Living Museum	7.0	0.7	70	500	Tropical American Rain Forest, North American Living Museum *for further information: (918) 669-6600.*
Woodland Park Zoo (Seattle)	15.0	1.0	92	280+	Tropical Rain Forest, Elephant Forest, African Savanna *for further information: (206) 684-4800.*
Zoo Atlanta	11.5	0.9	48	220	Gorillas of the Ford African Rain Forest, giant pandas *for further information: (404) 624-5600.*

Note: NA = Not available.

Major Canadian Public Zoological Parks

Source: *World Almanac* questionnaire, 1999; budget in millions of dollars (Canadian), attendance in millions

Zoo	Budget	Atten-dance	Acres	Species	Some major attractions
Assiniboine Park Zoo (Winnipeg)...	$2.4	0.4	50	279	Bird of Prey exhibition, Boo at the Zoo, Lights of the Wild *for further information: (204) 986-6921.*
Calgary Zoo	NA	0.9	136	252	Canadian Wilds, Primate Complex, African Bushveld *for further information: (800) 588-9993.*
Granby Zoo (Quebec)	6.3	0.4	85	225	AFRIKA Pavilion, AMAZOO water park, Reptile Pavilion *for further information: (877) GRANBYZOO.*
Toronto Zoo	22.9	1.2	710	469	African Savanna, Zoo Geography, Safari Camp *for further information: (416) 392-5900.*

Note: NA = Not available.

Top 50 American Kennel Club Registrations

Source: American Kennel Club, New York, NY; covers (new) dogs registered during calendar year shown

Breed	Rank 1998	Number registered 1998	Rank 1997	Number registered 1997	Breed	Rank 1998	Number registered 1998	Rank 1997	Number registered 1997
Labrador Retriever	1	157,936	1	158,366	Pekingese	26	11,734	27	12,871
Golden Retriever	2	65,681	4	70,158	English Springer Spaniel.	27	11,578	25	13,796
German Shepherd Dog ..	3	65,326	3	75,177	Great Dane	28	10,686	29	11,878
Rottweiler	4	55,009	2	75,489	Lhasa Apso	29	10,037	31	11,223
Dachshund.............	5	53,896	7	51,904	Dalmatian	30	9,722	17	22,726
Beagle	6	53,322	6	54,470	Collie	31	9,474	30	11,724
Poodle	7	51,935	5	54,773	Brittany	32	9,252	33	10,710
Chihuahua	8	43,468	12	38,926	W. Highland White Terrier	33	9,210	36	8,805
Yorkshire Terrier	9	42,900	9	41,283	Pembroke Welsh Corgi ..	34	8,932	37	8,281
Pomeranian	10	38,540	10	39,357	Chinese Shar-Pei	35	8,614	32	10,772
Shih Tzu..............	11	38,468	11	39,075	Akita	36	8,237	34	10,124
Boxer	12	36,345	13	38,047	Weimaraner...........	37	8,119	39	7,701
Cocker Spaniel.........	13	34,632	8	41,439	Saint Bernard	38	7,153	38	7,849
Miniature Schnauzer	14	31,063	14	32,351	Chow Chow...........	39	6,241	35	9,536
Shetland Sheepdog	15	27,978	15	32,086	Australian Shepherd	40	5,668	40	6,140
Miniature Pinscher......	16	22,675	18	22,297	Mastiff	41	5,148	41	5,270
Pug..................	17	21,487	19	20,082	Scottish Terrier	42	4,938	43	5,151
Siberian Husky..........	18	21,078	16	24,432	Chesapeake Bay Retriever	43	4,685	42	5,204
Boston Terrier..........	19	18,308	20	18,185	Cairn Terrier...........	44	4,632	45	4,423
Maltese...............	20	18,013	21	17,428	Great Pyrenees	45	4,085	44	4,709
Basset Hound..........	21	15,726	23	17,152	Alaskan Malamute......	46	3,699	46	4,409
Doberman Pinscher.....	22	15,367	22	17,385	Papillon	47	3,205	52	2,914
Bulldog	23	13,836	26	13,673	Vizsla................	48	2,902	50	2,992
Germ. Shorthaired Pointer	24	12,927	24	14,380	Italian Greyhound	49	2,896	54	2,788
Bichon Frise	25	12,806	28	12,587	Airedale Terrier	50	2,891	47	3,225

Cat Breeds

Source: The Cat Fanciers' Association, Manasquan, NJ

Only a small percentage of house cats in the U.S. are pedigreed or registered with one of the official registering bodies. The largest is the Cat Fanciers' Assn., Inc., with 671 member clubs. The Cat Fanciers' Assn. recognized 37 breeds as of Dec. 31, 1998 (in order of registration totals): Persian, Maine Coon, Siamese, Exotic, Abyssinian, Oriental, Scottish Fold, American Shorthair, Birman, Tonkinese, Burmese, Cornish Rex, Ocicat, Devon Rex, Russian Blue, Norwegian Forest Cat, Colorpoint Shorthair, British Shorthair, Somali, Manx, Ragdoll, Egyptian Mau, Sphynx, Turkish Angora, Japanese Bobtail, American Curl, Chartreux, Balinese, Singapura, Javanese, Selkirk Rex, Korat, Bombay, Havana Brown, American Wirehair, Turkish Van, and European Burmese.

Trees of the U.S.

Source: American Forests, Washington, DC, 1999

Approximately 825 native and naturalized species of trees are grown in the U.S. The oldest living tree is believed to be a bristlecone pine tree in California named Methuselah, estimated to be 4,700 years old. The world's largest known living tree, the General Sherman giant sequoia in California, weighs more than 6,167 tons—as much as 41 blue whales or 740 elephants.

American Forests recognizes and lists the "National Champion" (largest known, by total mass) of each U.S. tree species. Anyone can nominate candidates for the 2000-01 *National Register of Big Trees*; for information, write to American Forests, PO Box 2000, Washington, DC 20013, or check their website: http://www.amfor.org

Listed here, in alphabetical order, are ten National Champion trees selected by American Forests as worthy of note.

Selected National Champion Trees

Tree Type	Girth at 4.5 ft. (in.)	Height (ft.)	Crown Spread (ft.)	Total Points	Location
American Beech[1]	279	115	138	429	Lothian, MD
Black Willow.............	400	76	92	499	Grand Traverse County, MI
Coast Douglas-Fir	438	329	60	782	Coos County, OR
Coast Redwood[2]	867	313	101	1,205	Prairie Creek Redwoods State Park, CA
Giant Sequoia	998	275	107	1,300	Sequoia National Park, CA
Loblolly Pine	188	148	83	357	Warren, AR
Pinyon Pine	213	69	52	295	Cuba, NM
Sugar Maple.............	274	65	54	353	Kitzmiller, MD
Sugar Pine	442	232	29	681	Dorrington, CA
White Oak	382	96	119	508	Wye Mills State Park, MD

(1) Replaces American Elm, which was damaged by fire. (2) Remeasurement in 1997.

ARTS AND MEDIA

Some Notable Movies, Sept. 1998 – Aug. 1999

Movies	Stars	Director
Affliction	Nick Nolte, James Coburn, Sissy Spacek, Willem Dafoe	Paul Schrader
American History X	Edward Norton, Edward Furlong	Tony Kaye
American Pie	Jason Biggs, Chris Klein, Natasha Lyonne, Thomas Ian Nichols	Paul Weitz
Analyze This	Robert DeNiro, Billy Crystal, Lisa Kudrow	Harold Ramis
Antz	Woody Allen, Sylvester Stallone, Sharon Stone, Gene Hackman, Jennifer Lopez, Christopher Walken, Grant Shaud	Eric Darnell, Tim Johnson
Austin Powers: The Spy Who Shagged Me	Mike Myers, Heather Graham, Michael York, Robert Wagner, Seth Green, Rob Lowe, Verne J. Troyer, Mindy Sterling	Jay Roach
Big Daddy	Adam Sandler, Joey Lauren Adams, Jon Stewart, Rob Schneider	Dennis Dugan
Blair Witch Project, The	Heather Donohue, Michael Williams, Joshua Leonard	Daniel Myrick, Eduardo Sanchez
Bowfinger	Steve Martin, Eddie Murphy, Christine Baranski, Heather Graham	Frank Oz
Bug's Life, A	Kevin Spacey, Dave Foley, Julia Louis-Dreyfus, Madeline Kahn	John Lasseter, Andrew Stanton
Central Station	Fernanda Montenegro, Marilia Pera, Vinicius de Oliveira, Soia Lira	Walter Salles
Civil Action, A	John Travolta, Robert Duvall	Steven Zaillian
EDtv	Matthew McConaughey, Jenna Elfman, Woody Harrelson, Ellen DeGeneres, Sally Kirkland, Martin Landau, Rob Reiner	Ron Howard
Election	Matthew Broderick, Reese Witherspoon, Jessica Campbell	Alexander Payne
Elizabeth	Cate Blanchett, Geoffrey Rush, Christopher Eccleston, Joseph Fiennes, John Gielgud, Richard Attenborough	Shekhar Kapur
Enemy of the State	Will Smith, Gene Hackman	Tony Scott
Eyes Wide Shut	Tom Cruise, Nicole Kidman	Stanley Kubrick
Gods and Monsters	Ian McKellen, Lynn Redgrave, Brendan Fraser	Bill Condon
Hilary and Jackie	Emily Watson, Rachel Griffiths	Anand Tucker
Iron Giant, The	Harry Connick Jr., Eli Marienthal, Christopher McDonald	Brad Bird
Life Is Beautiful	Roberto Benigni, Nicoletta Braschi	Roberto Benigni
Little Voice	Michael Caine, Brenda Blethyn, Jim Broadbent, Jane Horrocks	Mark Herman
Matrix, The	Keanu Reeves, Laurence Fishburne	The Wachowski Brothers
Notting Hill	Julia Roberts, Hugh Grant	Roger Michell
October Sky	Jake Gyllenhaal, Chris Cooper, Chris Owen, Laura Dern	Joe Johnston
One True Thing	Meryl Streep, Renee Zellweger, William Hurt	Carl Franklin
Patch Adams	Robin Williams, Monica Potter, Philip Seymour Hoffman	Tom Shadyac
Pleasantville	Tobey Maguire, Jeff Daniels, Joan Allen, Reese Witherspoon, William H. Macy	Gary Ross
Prince of Egypt, The	Val Kilmer, Ralph Fiennes, Michelle Pfeiffer, Danny Glover, Sandra Bullock, Jeff Goldblum, Patrick Stewart	Brenda Chapman, Steve Hickner, Simon Wells
Rugrats Movie, The	Whoopi Goldberg, Tim Curry, David Spade, Busta Rhymes	Norton Virgien, Igor Kovalyov
Runaway Bride	Julia Roberts, Richard Gere, Joan Cusack, Hector Elizondo	Garry Marshall
Rushmore	Jason Schwartzman, Olivia Williams, Bill Murray	Wes Anderson
Shakespeare in Love	Gwyneth Paltrow, Joseph Fiennes, Geoffrey Rush, Colin Firth, Ben Affleck, Judi Dench, Rupert Everett	John Madden
Simple Plan, A	Bill Paxton, Billy Bob Thornton, Bridget Fonda	Sam Raimi
Sixth Sense, The	Bruce Willis, Haley Joel Osment, Toni Collette, Olivia Williams	M. Night Shyamalan
Star Trek Insurrection	Patrick Stewart, Jonathan Frakes, Brent Spiner, LeVar Burton	Jonathan Frakes
Star Wars: Episode I- The Phantom Menace	Liam Neeson, Ewan McGregor, Natalie Portman, Jake Lloyd	George Lucas
Stepmom	Julia Roberts, Susan Sarandon, Ed Harris	Chris Columbus
Summer of Sam	John Leguizamo, Adrien Brody, Mora Sorvino, Jennifer Esposito	Spike Lee
Tarzan	Minnie Driver, Tony Goldwyn, Rosie O'Donnell, Phil Collins	Kevin Lima, Chris Buck
Tea With Mussolini	Cher, Judi Dench, Joan Plowright, Maggie Smith, Lily Tomlin	Franco Zeffirelli
Thin Red Line, The	Sean Penn, Nick Nolte, Adrien Brody, Jim Caviezel, Elias Koteas	Terrence Malick
True Crime	Clint Eastwood, Isaiah Washington, Denis Leary, James Woods	Clint Eastwood
200 Cigarettes	Martha Plimpton, Paul Rudd, Christina Ricci, Casey Affleck, Janeane Garofalo, Kate Hudson, Ben Affleck	Risa Bramon Garcia
Waking Ned Devine	Ian Bannen, David Kelly, Fionnula Flanagan	Kirk Jones
Waterboy, The	Adam Sandler, Kathy Bates, Fairuza Balk	Frank Coraci
What Dreams May Come	Robin Williams, Cuba Gooding Jr., Annabella Sciorra	Vincent Ward
Wild Wild West	Will Smith, Kevin Kline, Kenneth Branagh, Salma Hayak	Barry Sonnenfeld
Winslow Boy, The	Nigel Hawthorne, Rebecca Pidgeon, Guy Edwards	David Mamet
You've Got Mail	Tom Hanks, Meg Ryan, Jean Stapleton, Greg Kinnear	Nora Ephron

Top 50 Movies, 1998

Source: *Variety*, Feb. 1-Feb. 7, 1999; box-office grosses in the U.S. and Canada during calendar year 1998

Rank	Title	Gross (millions)	Rank	Title	Gross (millions)	Rank	Title	Gross (millions)
1.	Titanic	$488.2	18.	Antz	$88.7	35.	Snake Eyes	$55.6
2.	Armageddon	201.6	19.	The Rugrats Movie	86.4	36.	What Dreams May Come	55.4
3.	Saving Private Ryan	190.8	20.	The X-Files	83.9	37.	Small Soldiers	55.1
4.	There's Something About Mary	174.4	21.	The Wedding Singer	80.2	38.	The Prince of Egypt	55.1
5.	The Waterboy	147.9	22.	City of Angels	78.9	39.	Halloween: H2O	55.0
6.	Dr. Doolittle	144.2	23.	The Horse Whisperer	75.4	40.	Star Trek: Insurrection	53.9
7.	Deep Impact	140.5	24.	Six Days, Seven Nights	74.3	41.	Tomorrow Never Dies	52.0
8.	Godzilla	136.3	25.	Blade	70.1	42.	Patch Adams	46.4
9.	Rush Hour	136.1	26.	Lost in Space	69.1	43.	Practical Magic	46.4
10.	Good Will Hunting	134.1	27.	A Perfect Murder	67.7	44.	The Negotiator	44.7
11.	Lethal Weapon 4	130.4	28.	The Parent Trap	66.3	45.	Meet Joe Black	43.2
12.	A Bug's Life	127.6	29.	Ever After	65.7	46.	Wag the Dog	42.9
13.	The Truman Show	125.6	30.	You've Got Mail	63.8	47.	Ronin	41.6
14.	As Good As It Gets	124.1	31.	Hope Floats	60.1	48.	The Siege	39.9
15.	Mulan	120.6	32.	U.S. Marshals	57.8	49.	Primary Colors	39.3
16.	The Mask of Zorro	93.7	33.	The Man in the Iron Mask	57.0	50.	I Still Know What You Did Last Summer	38.9
17.	Enemy of the State	92.0	34.	Everest	55.7			

National Film Registry, 1989-98

Source: National Film Registry, Library of Congress

"Culturally, historically, or esthetically significant" films placed on the registry. * = selected in 1998.

Adam's Rib (1949)
The Adventures of Robin Hood (1938)
The African Queen (1951)
All About Eve (1950)
All That Heaven Allows (1955)
All Quiet on the Western Front (1930)
An American in Paris (1951)
American Graffiti (1973)
A Movie (1958)
Annie Hall (1977)
The Apartment (1960)
The Awful Truth (1937)
Badlands (1973)
The Band Wagon (1953)
The Bank Dick (1940)
The Battle of San Pietro (1945)
Ben-Hur (1926)
The Best Years of Our Lives (1946)
Big Business (1929)
The Big Parade (1925)
The Big Sleep (1946)
The Birth of a Nation (1915)
The Black Pirate (1926)
Blacksmith Scene (1893)
Blade Runner (1982)
The Blood of Jesus (1941)
Bonnie and Clyde (1967)
Bride of Frankenstein (1935)*
The Bridge on the River Kwai (1957)
Bringing Up Baby (1938)
Broken Blossoms (1919)
Cabaret (1972)
Carmen Jones (1954)
Casablanca (1942)
Castro Street (1966)
Cat People (1942)
Chan Is Missing (1982)
The Cheat (1915)
Chinatown (1974)
Chulas Fronteras (1976)
Citizen Kane (1941)
The City (1939)*
City Lights (1931)
The Conversation (1974)
The Cool World (1963)
Cops (1922)
A Corner in Wheat (1909)
The Crowd (1928)
Czechoslovakia 1968 (1968)
David Holzman's Diary (1968)
The Day the Earth Stood Still (1951)
Dead Birds (1964)*
The Deer Hunter (1978)
Destry Rides Again (1939)
Detour (1946)
Dodsworth (1936)
Dog Star Man (1964)
Don't Look Back (1967)*
Double Indemnity (1944)
Dr. Strangelove (or, How I Learned to Stop Worrying and Love the Bomb) (1964)
Duck Soup (1933)
Easy Rider (1969)*
Eaux D'Artifice (1953)
El Norte (1983)

E.T.: The Extra-Terrestrial (1982)
The Exploits of Elaine (1914)
Fantasia (1940)
Fatty's Tintype Tangle (1915)
Flash Gordon serial (1936)
Footlight Parade (1933)
Force of Evil (1948)
The Forgotten Frontier (1931)
42nd Street (1933)*
The Four Horsemen of the Apocalypse (1921)
Frankenstein (1931)
Frank Film (1973)
Freaks (1932)
The Freshman (1925)
From the Manger to the Cross (1912)*
Fury (1936)
The General (1927)
Gerald McBoing Boing (1951)
Gertie the Dinosaur (1914)
Gigi (1958)
The Godfather (1972)
The Godfather, Part II (1974)
The Gold Rush (1925)
Gone With the Wind (1939)
The Graduate (1967)
The Grapes of Wrath (1940)
Grass (1925)
The Great Dictator (1940)
The Great Train Robbery (1903)
Greed (1924)
Gun Crazy (1949)*
Harlan County, U.S.A. (1976)
Harold and Maude (1972)
The Heiress (1949)
Hell's Hinges (1916)
High Noon (1952)
High School (1968)
Hindenburg Disaster Newsreel Footage (1937)
His Girl Friday (1940)
The Hitch-Hiker (1953)*
Hospital (1970)
The Hospital (1971)
How Green Was My Valley (1941)
How the West Was Won (1962)
The Hustler (1961)
I Am a Fugitive From a Chain Gang (1932)
The Immigrant (1917)*
Intolerance (1916)
Invasion of the Body Snatchers (1956)
It Happened One Night (1934)
It's a Wonderful Life (1946)
The Italian (1915)
Jammin' the Blues (1944)
The Jazz Singer (1927)
Killer of Sheep (1977)
King Kong (1933)
Knute Rockne, All American (1940)
The Lady Eve (1941)
Lassie Come Home (1943)
The Last of the Mohicans (1920)
The Last Picture Show (1972)*

Lawrence of Arabia (1962)
The Learning Tree (1969)
Letter From an Unknown Woman (1948)
The Life and Death of 9413— A Hollywood Extra (1928)
Life and Times of Rosie the Riveter (1980)
The Little Fugitive (1953)
Little Miss Marker (1934)*
The Lost World (1925)*
Louisiana Story (1948)
Love Me Tonight (1932)
Magical Maestro (1952)
The Magnificent Ambersons (1942)
The Maltese Falcon (1941)
The Manchurian Candidate (1962)
Manhattan (1921)
March of Time: Inside Nazi Germany—1938 (1938)
Marty (1955)
M*A*S*H (1970)
Mean Streets (1973)
Meet Me in St. Louis (1944)
Meshes of the Afternoon (1943)
Midnight Cowboy (1969)
Mildred Pierce (1945)
Modern Times (1936)
Modesta (1956)*
Morocco (1930)
Motion Painting No. 1 (1947)
Mr. Smith Goes to Washington (1939)
The Music Box (1932)
My Darling Clementine (1946)
The Naked Spur (1953)
Nanook of the North (1922)
Nashville (1975)
A Night at the Opera (1935)
The Night of the Hunter (1955)
Ninotchka (1939)
North by Northwest (1959)
Nothing but a Man (1964)
One Flew Over the Cuckoo's Nest (1975)
On the Waterfront (1954)
The Outlaw Josey Wales (1976)
Out of the Past (1947)
The Ox-Bow Incident (1943)*
Pass the Gravy (1928)*
Paths of Glory (1957)
Phantom of the Opera (1925)*
The Philadelphia Story (1940)
Pinocchio (1940)
A Place in the Sun (1951)
Point of Order (1964)
The Poor Little Rich Girl (1917)
Powers of Ten (1978)*
Primary (1960)
The Prisoner of Zenda (1937)
The Producers (1968)
Psycho (1960)
The Public Enemy (1931)*
Pull My Daisy (1959)
Raging Bull (1980)
Rear Window (1954)
Rebel Without a Cause (1955)
Red River (1948)

Republic Steel Strike Riots Newsreel Footage (1937)
Return of the Secaucus 7 (1980)
Ride the High Country (1962)
Rip Van Winkle (1896)
The River (1937)
Road to Morocco (1942)
Safety Last (1923)
Salesman (1969)
Salt of the Earth (1954)
Scarface (1932)
The Searchers (1956)
Seventh Heaven (1927)
Shadow of a Doubt (1943)
Shadows (1959)
Shane (1953)
She Done Him Wrong (1933)
Sherlock, Jr. (1924)
Shock Corridor (1963)
Show Boat (1936)
Singin' in the Rain (1952)
Sky High (1922)*
Snow White (1933)
Snow White and the Seven Dwarfs (1937)
Some Like It Hot (1959)
Stagecoach (1939)
Star Wars (1977)
Steamboat Willie (1928)*
Sullivan's Travels (1941)
Sunrise (1927)
Sunset Boulevard (1950)
Sweet Smell of Success (1957)
Tabu (1933)
Tacoma Narrows Bridge Collapse (1940)*
Taxi Driver (1976)
Tevye (1939)
The Thief of Bagdad (1924)
The Thin Man (1934)
To Be or Not To Be (1942)
To Fly (1976)
To Kill a Mockingbird (1962)
Tootsie (1982)*
Topaz (1943-45)
Top Hat (1935)
Touch of Evil (1958)
The Treasure of the Sierra Madre (1948)
Trouble in Paradise (1932)
Tulips Shall Grow (1942)
Twelve O'Clock High (1949)*
2001: A Space Odyssey (1968)
Verbena Tragica (1939)
Vertigo (1958)
Westinghouse Works 1904 (1904)*
West Side Story (1961)
What's Opera, Doc? (1957)
Where Are My Children? (1916)
The Wind (1928)
Wings (1927)
Within Our Gates (1920)
The Wizard of Oz (1939)
A Woman Under the Influence (1974)
Woodstock (1970)
Yankee Doodle Dandy (1942)
Zapruder Film (1963)

Most Popular Movie Videos

Source: Alexander & Associates/Video Flash, New York, NY

Top 10 Rentals, 1998	All-Time Top 10 Rentals[1]	Top 10 Sales, 1998	All-Time Top 10 Sales[2]
1. Titanic (1997)	1. Top Gun	1. Titanic	1. The Lion King
2. Air Force One	2. Pretty Woman	2. Hercules	2. Aladdin
3. Con Air	3. The Little Mermaid	3. The Little Mermaid	3. Beauty and the Beast
4. Face/Off	4. Home Alone	4. Lion King 2: Simba's Pride	4. Snow White and the Seven Dwarfs
5. As Good As It Gets	5. Ghost	5. Peter Pan[3]	5. Forrest Gump
6. My Best Friend's Wedding	6. Cinderella	6. Lady and the Tramp[3]	6. Toy Story
7. Men in Black	7. Beauty and the Beast	7. Flubber	7. The Little Mermaid
8. The Devil's Advocate	8. Lion King	8. Men in Black	8. 101 Dalmatians (animated)
9. Armageddon	9. Terminator 2: Judgment Day	9. Anastasia	9. Jurassic Park
10. G.I. Jane	10. Aladdin	10. Air Force One	10. Pocahontas

(1) Rented Mar. 1, 1987-Dec. 29, 1998. (2) Sold Feb. 16, 1988-Dec. 30, 1998. (3) Re-release

100 Best American Movies of All Time
Source: American Film Institute

Compiled in 1998 based on ballots sent to 1,500 figures, mostly from the film world. Criteria for judging included historical significance, critical recognition and awards, and popularity. The year each film was first released is in parentheses.

1. Citizen Kane (1941)
2. Casablanca (1942)
3. The Godfather (1972)
4. Gone With the Wind (1939)
5. Lawrence of Arabia (1962)
6. The Wizard of Oz (1939)
7. The Graduate (1967)
8. On the Waterfront (1954)
9. Schindler's List (1993)
10. Singin' in the Rain (1952)
11. It's a Wonderful Life (1946)
12. Sunset Boulevard (1950)
13. The Bridge on the River Kwai (1957)
14. Some Like It Hot (1959)
15. Star Wars (1977)
16. All About Eve (1950)
17. The African Queen (1951)
18. Psycho (1960)
19. Chinatown (1974)
20. One Flew Over the Cuckoo's Nest (1975)
21. The Grapes of Wrath (1940)
22. 2001: A Space Odyssey (1968)
23. The Maltese Falcon (1941)
24. Raging Bull (1980)
25. E.T.: The Extra-Terrestrial (1982)
26. Dr. Strangelove (1964)
27. Bonnie and Clyde (1967)
28. Apocalypse Now (1979)
29. Mr. Smith Goes to Washington (1939)
30. Treasure of the Sierra Madre (1948)
31. Annie Hall (1977)
32. The Godfather Part II (1974)
33. High Noon (1952)
34. To Kill a Mockingbird (1962)
35. It Happened One Night (1934)
36. Midnight Cowboy (1969)
37. The Best Years of Our Lives (1946)
38. Double Indemnity (1944)
39. Doctor Zhivago (1965)
40. North by Northwest (1959)
41. West Side Story (1961)
42. Rear Window (1954)
43. King Kong (1933)
44. The Birth of a Nation (1915)
45. A Streetcar Named Desire (1951)
46. A Clockwork Orange (1971)
47. Taxi Driver (1976)
48. Jaws (1975)
49. Snow White and the Seven Dwarfs (1937)
50. Butch Cassidy and the Sundance Kid (1969)
51. The Philadelphia Story (1940)
52. From Here to Eternity (1953)
53. Amadeus (1984)
54. All Quiet on the Western Front (1930)
55. The Sound of Music (1965)
56. M*A*S*H (1970)
57. The Third Man (1949)
58. Fantasia (1940)
59. Rebel Without a Cause (1955)
60. Raiders of the Lost Ark (1981)
61. Vertigo (1958)
62. Tootsie (1982)
63. Stagecoach (1939)
64. Close Encounters of the Third Kind (1977)
65. The Silence of the Lambs (1991)
66. Network (1976)
67. The Manchurian Candidate (1962)
68. An American in Paris (1951)
69. Shane (1953)
70. The French Connection (1971)
71. Forrest Gump (1994)
72. Ben-Hur (1959)
73. Wuthering Heights (1939)
74. The Gold Rush (1925)
75. Dances With Wolves (1990)
76. City Lights (1931)
77. American Graffiti (1973)
78. Rocky (1976)
79. The Deer Hunter (1978)
80. The Wild Bunch (1969)
81. Modern Times (1936)
82. Giant (1956)
83. Platoon (1986)
84. Fargo (1996)
85. Duck Soup (1933)
86. Mutiny on the Bounty (1935)
87. Frankenstein (1931)
88. Easy Rider (1969)
89. Patton (1970)
90. The Jazz Singer (1927)
91. My Fair Lady (1964)
92. A Place in the Sun (1951)
93. The Apartment (1960)
94. Goodfellas (1990)
95. Pulp Fiction (1994)
96. The Searchers (1956)
97. Bringing Up Baby (1938)
98. Unforgiven (1992)
99. Guess Who's Coming to Dinner (1967)
100. Yankee Doodle Dandy (1942)

All-Time Top 50 American Movies Through 1998
Source: *Variety* magazine

Rank	Title/Date	Gross[1]	Rank	Title/Date	Gross[1]	Rank	Title/Date	Gross[1]
1.	Titanic (1997)	$600.8	18.	The Lost World: Jurassic Park (1997)	$229.1	35.	Pretty Woman (1990)	$178.4
2.	Star Wars (1977)	461.0	19.	Mrs. Doubtfire (1993)	219.2	36.	Tootsie (1982)	177.2
3.	E.T.: The Extra-Terrestrial (1982)	399.8	20.	Ghost (1990)	217.6	37.	Top Gun (1986)	176.8
4.	Jurassic Park (1993)	357.1	21.	Aladdin (1992)	217.4	38.	Snow White and the Seven Dwarfs (1937)	175.3
5.	Forrest Gump (1994)	329.7	22.	Back to the Future (1985)	208.2	39.	There's Something About Mary (1998)	175.1
6.	The Lion King (1994)	312.9	23.	Terminator 2 (1991)	204.8	40.	Crocodile Dundee (1986)	174.8
7.	Return of the Jedi (1983)	309.2	24.	Armageddon (1998)	201.6	40.	Home Alone 2 (1992)	173.6
8.	Independence Day (1996)	306.2	25.	Indiana Jones and the Last Crusade (1989)	197.2	42.	Rain Man (1988)	172.8
9.	The Empire Strikes Back (1980)	290.3	26.	Gone with the Wind (1939)	191.9	43.	Air Force One (1997)	172.4
10.	Home Alone (1990)	285.8	27.	Toy Story (1995)	191.8	44.	Apollo 13 (1995)	172.1
11.	Jaws (1975)	260.0	28.	Saving Private Ryan (1998)	190.9	45.	Three Men and a Baby (1987)	167.8
12.	Batman (1989)	251.2	29.	Dances With Wolves (1990)	184.2	46.	Robin Hood: Prince of Thieves (1991)	165.5
13.	Men in Black (1997)	250.0	30.	Batman Forever (1995)	184.0	47.	The Exorcist (1973)	165.0
14.	Raiders of the Lost Ark (1981)	242.4	31.	The Fugitive (1993)	183.9	48.	Batman Returns (1992)	162.8
15.	Twister (1996)	241.7	32.	Mission: Impossible (1996)	181.0	49.	The Sound of Music (1965)	160.5
16.	Ghostbusters (1984)	238.6	33.	Liar, Liar (1997)	179.5	50.	The Firm (1993)	158.3
17.	Beverly Hills Cop (1984)	234.8	34.	Indiana Jones and the Temple of Doom (1984)	179.9			

(1) Gross is in millions of absolute dollars based on box office sales in the U.S. and Canada. Ticket prices favor recent films, but older films have the advantage of reissues.

Top 50 Record Long-Run Broadway Plays[1]
Source: The League of American Theatres and Producers, Inc., New York, NY

Title	Performances	Title	Performances	Title	Performances
*Cats	6,949	*Beauty and the Beast	2,136	Evita	1,567
A Chorus Line	6,137	Pippin	1,944	The Voice of the Turtle	1,557
Oh! Calcutta! (revival)	5,962	South Pacific	1,925	Barefoot in the Park	1,530
*Les Miserables	5,031	Magic Show	1,920	Dreamgirls	1,521
*The Phantom of the Opera	4,734	Gemini	1,819	Mame	1,508
42nd Street	3,485	Deathtrap	1,793	Grease (revival)	1,505
*Miss Saigon	3,396	Harvey	1,775	Same Time, Next Year	1,453
Grease (original)	3,388	Dancin'	1,774	Arsenic and Old Lace	1,444
Fiddler on the Roof	3,242	*Smokey Joe's Cafe	1,773	The Sound of Music (orig.)	1,443
Life With Father	3,224	La Cage aux Folles	1,761	How to Succeed in Business	1,417
Tobacco Road	3,182	Hair	1,750	Me and My Girl	1,409
Hello Dolly	2,844	The Wiz	1,672	Hellzapoppin	1,404
My Fair Lady	2,717	Born Yesterday	1,642	The Music Man	1,375
Annie	2,377	Crazy for You	1,638	Funny Girl	1,348
Man of La Mancha	2,329	Ain't Misbehavin'	1,604	Mummenchanz	1,326
Abie's Irish Rose	2,327	The Best Little Whorehouse	1,584	Oh! Calcutta! (orig.)	1,314
Oklahoma!	2,212	Mary, Mary	1,572		

* Still running May 30, 1999. (1) Number of performances through May 30, 1999.

Broadway Season Statistics, 1959-98

Source: The League of American Theatres and Producers, Inc., New York, NY

Season	Gross (mil $)	Attendance (mil)	Playing Weeks	New Productions	Season	Gross (mil $)	Attendance (mil)	Playing Weeks	New Productions
1959-1960	46	7.9	1,156	58	1979-1980	146	9.6	1,540	61
1960-1961	44	7.7	1,210	48	1980-1981	197	11.0	1,544	60
1961-1962	44	6.8	1,166	53	1981-1982	223	10.1	1,455	48
1962-1963	44	7.4	1,134	54	1982-1983	209	8.4	1,258	50
1963-1964	40	6.8	1,107	63	1983-1984	227	7.9	1,097	36
1964-1965	50	8.2	1,250	67	1984-1985	208	7.3	1,075	33
1965-1966	54	9.6	1,295	68	1985-1986	190	6.5	1,045	33
1966-1967	55	9.3	1,269	69	1986-1987	206	7.0	1,038	41
1967-1968	59	9.5	1,259	74	1987-1988	252	8.1	1,116	32
1968-1969	58	8.6	1,209	67	1988-1989	265	8.0	1,097	30
1969-1970	53	7.1	1,047	62	1989-1990	282	8.0	1,061	35
1970-1971	55	7.4	1,107	49	1990-1991	267	7.3	970	28
1971-1972	52	6.5	1,157	55	1991-1992	293	7.4	903	37
1972-1973	45	5.4	889	55	1992-1993	328	7.9	1,019	33
1973-1974	46	5.7	907	43	1993-1994	356	8.1	1,061	37
1974-1975	57	6.6	1,101	54	1994-1995	406	9.0	1,118	29
1975-1976	71	7.3	1,136	55	1995-1996	436	9.5	1,146	38
1976-1977	93	8.8	1,349	54	1996-1997	499	10.6	1,347	37
1977-1978	114	9.6	1,433	42	1997-1998	558	11.5	1,442	34
1978-1979	134	9.6	1,542	50	1998-1999	588	11.7	1,441	39

Some Notable Broadway Theater Openings, 1998-99 Season

Amy's View. Drama about how a strong-willed, famous actress deals with her professional life and conflicts with her adult daughter. By David Hare. Directed by Richard Eyre. With Judi Dench, Samantha Bond, Tate Donovan, Ronald Pickup, and Anne Pitoniak.

Annie Get Your Gun. Revival of Irving Berlin's musical about Annie Oakley and Frank Butler. Book by Herbert and Dorothy Fields, revised by Peter Stone. Directed by Graciela Daniele. With Bernadette Peters and Tom Wopat.

Death of a Salesman. A revival of the Arthur Miller drama. Directed by Robert Falls. With Brian Dennehy, Elizabeth Franz, Kevin Anderson, Ted Koch, and Howard Witt.

Electra. Sophocles's drama about a woman's obsessive desire to avenge her father's murder. Adapted by Frank McGuinness. Directed by David Leveaux. With Zoë Wanamaker.

Fosse. A musical retrospective of director and choreographer Bob Fosse's work. Directed by Richard Maltby Jr. and Ann Reinking. Choreography by Bob Fosse, and Ann Reinking, recreated by Chet Walker. With Valarie Pettiford and Jane Lanier.

The Iceman Cometh. A revival of the Eugene O'Neill drama. Directed by Howard Davies. With Kevin Spacey, Tony Danza, Tim Pigott-Smith, James Hazeldine, and Robert Sean Leonard.

It Ain't Nothing But the Blues. A musical based on the history of the blues, from its inception to fusion with other musical genres. Book by Charles Bevel, Lita Gaithers, Randal Myler, Ron Taylor, and Dan Wheetman. Directed by Randal Myler. With "Mississippi" Charles Bevel, Gretha Boston, Carter Calvert, Eloise Laws, Ken Page, Gregory Porter, and Dan Wheetman.

Side Man. A drama about jazz and the sidemen's passion for it, from the music's heyday through the eighties. By Warren Leight. Directed by Michael Mayer. With Robert Sella, Frank Wood, Angelica Torn, Joseph Lyle Taylor, Kevin Geer, Michael Mastro, and Wendy Makkena.

The Weir. Drama set in a rural Irish pub, about three men and a woman who exchange ghost stories. By Conor McPherson. Directed by Ian Rickson. With Kieran Ahern, Brendan Coyle, Dermot Crowley, Michelle Fairley, and Jim Norton.

Some Notable Nonprofit Professional Theater Companies in the U.S.

Source: Theatre Communications Group, Inc.

Theater Company	City	State	Theater Company	City	State
Actors Theatre of Louisville	Louisville	KY	Huntington Theatre Company	Boston	MA
Alabama Shakespeare Festival	Montgomery	AL	La Jolla Playhouse	La Jolla	CA
Alley Theater	Houston	TX	Lincoln Center Theater	New York	NY
Alliance Theatre Company	Atlanta	GA	Long Wharf Theatre	New Haven	CT
American Conservatory Theatre	San Francisco	CA	Manhattan Theatre Club	New York	NY
American Repertory Theatre	Cambridge	MA	Mark Taper Forum	Los Angeles	CA
Arena Stage	Washington	DC	McCarter Theatre	New Brunswick	NJ
Berkeley Repertory Theatre	Berkeley	CA	Milwaukee Repertory Theater	Milwaukee	WI
Center Stage	Baltimore	MD	Oregon Shakespeare Festival	Ashland	OR
Children's Theatre Company, The	Minneapolis	MN	Roundabout Theatre Company	New York	NY
Cincinnati Playhouse in the Park	Cincinnati	OH	San Jose Repertory Theatre	San Jose	CA
Cleveland Play House, The	Cleveland	OH	Seattle Repertory Theatre	Seattle	WA
Denver Center Theatre Company	Denver	CO	South Coast Repertory	Costa Mesa	CA
Goodman Theatre	Chicago	IL	Steppenwolf Theatre Company	Chicago	IL
Guthrie Theater, The	Minneapolis	MN	Trinity Repertory Company	Providence	RI
Hartford Stage Company	Hartford	CT			

U.S. Symphony Orchestras[1]

Source: American Symphony Orchestra League, 33 West 60th St., New York, NY 10023; data as of August 1999

Symphony Orchestra[2]	Music Director[3]	Symphony Orchestra[2]	Music Director[3]
Alabama Symphony (AL)	Richard Westerfield	Chicago (IL)	Daniel Barenboim
American (NY)	Leon Botstein	Cincinnati (OH)	Jesus Lopez-Cobos
Atlanta (GA)	Yoel Levi	Cleveland (OH)	Christoph von Dohnányi
Austin (TX)	Peter Bay	Colorado (Denver)	Marin Alsop
Baltimore (MD)	Yuri Temirkanov	Colorado Springs (CO)	Yaacov Bergman
Boston (MA)	Seiji Ozawa	Columbus (OH)	Alessandro Siciliani
Brooklyn Philharmonic (NY)	Robert Spano	Dallas (TX)	Andrew Litton
Buffalo Philharmonic (NY)	JoAnne Falletta	Dayton Philharmonic (OH)	Neal Gittleman
Charlotte (NC)	Peter McCoppin	Delaware (Wilmington)	—

Symphony Orchestra[2]	Music Director[3]	Symphony Orchestra[2]	Music Director[3]
Detroit (MI)	Neeme Jarvi	New Mexico (Albuquerque)	David Lockington
Florida Orchestra (Tampa)	Jahja Ling	New York Philharmonic (NYC)	Kurt Masur
Florida Philharmonic	James Judd	North Carolina Symphony (Raleigh)	Gerhardt Zimmermann
(Ft. Lauderdale)		Oklahoma City Philharmonic (OK)	Joel A. Levine
Florida Symphonic Pops	Crafton Beck	Omaha Symphony (NE)	Victor Yampolsky
(Boca Raton)		Oregon Symphony (Portland)	James DePreist
Fort Wayne Philharmonic (IN)	Edvard Tchivzhel	Pacific Symphony (Santa Ana, CA)	Carl St. Clair
Fort Worth (TX)	John Giordano	Philadelphia (PA)	Wolfgang Sawallisch
Grand Rapids (MI)	—	Philharmonia Baroque	
Grant Park (Chicago, IL)	—	(San Francisco, CA)	Nicholas McGegan
Hartford (CT)	Michael Lankeste	Phoenix (AZ)	Hermann Michael
Houston (TX)	Christoph Eschenbach	Pittsburgh (PA)	Mariss Jansons
Honolulu (HI)	Samuel Wong	Portland (ME)	Toshiyuki Shimada
Indianapolis (IN)	Raymond Leppard	Richmond Symphony (VA)	Gerardo Edelstein
Jacksonville (FL)		Rochester Philharmonic Orch. (NY)	Christopher Seaman
Kansas City (MO)	Anne Manson	St. Louis (MO)	Hans Vonk
Knoxville (TN)	Kirk Trevor	St. Paul Chamber Orch. (MN)	Hugh Wolff
Long Beach (CA)	JoAnn Falletta	San Antonio (TX)	Christopher Wilkins
Los Angeles Philharmonic (CA)	Esa-Pekka Salonen	San Francisco (CA)	Michael Tilson Thomas
Louisiana Philharmonic	Klauspeter Seibel	San Jose (CA)	Leonid Grin
(New Orleans)		Savannah (GA)	Philip B. Greenberg
Louisville Orchestra (KY)	Uriel Segal	Seattle (WA)	Gerard Schwarz
Memphis (TN)	David Loebel	Spokane (WA)	Fabio Mechetti
Milwaukee (WI)	Andreas Delfs	Syracuse (NY)	—
Minnesota (Minneapolis)	Eiji Oue	Toledo (OH)	Andrew Massey
Naples Philharmonic (FL)	Christopher Seaman	Tucson (AZ)	George Hanson
Nashville Symphony (TN)	Kenneth D. Schermerhorn	Tulsa Philharmonic (OK)	Kenneth Jean
National (Washington, DC)	Leonard Slatkin	Utah (Salt Lake City)	Keith Lockhart
New Haven (CT)	Jung-Ho Pak	Virginia Symphony (Norfolk)	JoAnn Falletta
New Jersey (Newark)	Zdenek Macal	West Virginia (Charleston)	Thomas B. Conlin

(1) Includes only orchestras with annual expenses $2 mil or greater. (2) Orchestra name = place name + Symphony Orchestra, unless otherwise indicated. (3) General title; listed is highest-ranking member of conducting personnel. — indicates vacancy.

U.S. Opera Companies With Budgets of $1 Million or More
Source: OPERA America, 1156 15th Street NW, Washington, DC 20005-1704; July 1999

Academy of Vocal Arts Opera Theatre (Philadelphia, PA); K. James McDowell, dir.
American Musical Theatre of San Jose (CA); Dianna Shuster, art. dir.
Arizona Opera (Tucson); David Speers, gen. dir.
Aspen Opera Theater Center (CO); Robert Harth, pres./ceo
Atlanta Opera (GA); Alfred Kennedy, exec. dir.
Austin Lyric Opera (TX); Joseph McClain, gen. dir.
Baltimore Opera Company (MD); Michael Harrison, gen. dir.
Boston Lyric Opera Company (MA); Janice Mancini Del Sesto, gen. dir.
Brooklyn Academy of Music (NY); Karen Brooks Hopkins, pres.
Central City Opera (Denver, CO); Pelham Pearce, gen. dir.
Cincinnati Opera (OH); Patricia Beggs, mng. dir.
Civic Light Opera (Pittsburgh, PA); Van Kaplan, exec. prod.
Cleveland Opera (OH); David Bamberger, gen. dir.
Connecticut Opera (Hartford); Willie Anthony Waters, gen./art. dir.
Dallas Opera (TX); Plato Karayanis, gen. dir.
Dayton Opera (OH); Ardith Hamilton, gen. mng.
Des Moines Metro Opera, Inc. (IA); Jerilee Mace, exec. dir.
Florentine Opera Company (Milwaukee, WI); Dennis Hanthorn, gen. dir.
Florida Grand Opera (Miami, FL); Robert Heuer, gen. mng./ceo
Fort Worth Opera (TX); William Walker, gen. dir.
Fullerton Civic Light Opera (CA); Griff Duncan, gen. mng.
Glimmerglass Opera (Cooperstown, NY); Esther Nelson, gen. dir.
Goodspeed Opera House (East Haddam, CT); Michael Price, exec. dir.
Hawaii Opera Theatre (Honolulu); Henry Akina, gen./art. dir.
Houston Grand Opera (TX); David Gockley, gen. dir.
Indianapolis Opera (IN); John C. Pickett, exec. dir.
Kentucky Opera (Louisville); Deborah S. Sandler, gen. dir.
Knoxville Opera Company (TN); Susan Arp, dev./admin. dir.
Los Angeles Opera (CA); Peter Hemmings, gen. dir.
Lyric Opera of Chicago (IL); William Mason, gen. dir.
Lyric Opera of Kansas City (MO); Evan R. Luskin, gen. dir.
Metro Lyric Opera (Allenhurst, NJ); Era M. Tognoli, gen./art. dir.
Metropolitan Opera (New York, NY); Joseph Volpe, gen. mng.
Michigan Opera Theatre (Detroit); David Dichiera, gen. dir.
Minnesota Opera (Minneapolis); Kevin Smith, gen. dir.
Nashville Opera Association (TN); Carol Penterman, ceo/exec. dir.
New Jersey State Opera (Newark); Alfredo Silipigni, art. dir.
New Orleans Opera Association (LA); Robert Lyall, gen. dir.
New York City Opera (NY); Paul Kellogg, gen. dir.
Ohio Light Opera (Wooster); James Stuart, art. dir.
Opera Carolina (Charlotte, NC); TBA
Opera Colorado (Denver); Stephen Seifert, exec. dir.
Opera/Columbus (OH); William F. Russell, gen. dir.
Opera Company of Philadelphia (PA); Robert B. Driver, gen. dir.
OperaDelaware (Wilmington); Leland P. Kimball III, gen. dir.
Opera Festival of New Jersey (Princeton); Karen Tiller, gen. dir.
Opera Grand Rapids (MI); Robert Lyall, gen. dir.
Opera Memphis (TN); Michael Ching, gen./art. dir.
Opera Omaha (NE); Jane Hill, exec. dir.
Opera Pacific (Irvine, CA); Martin G. Hubbard, exec. dir.
Opera Theatre of Saint Louis (MO); Charles MacKay, gen. dir.
Opera San José (CA); Irene Dalis, gen. dir.
Orlando Opera (FL); Robert Swedberg, gen. dir.
Palm Beach Opera (FL); Herbert P. Benn, gen. dir.
Pittsburgh Opera (PA); Mark Weinstein, exec. dir.
Portland Opera (OR); Robert Bailey, gen. dir.
San Diego Civic Light Opera Association (CA); Dave Twomey, pres./bd. dirs.
San Diego Opera (CA); Ian D. Campbell, gen. dir.
San Francisco Opera (CA); Lotfi Mansouri, gen. dir.
Santa Barbara Civic Light Opera (CA); Paul Iannaccone, exec. prod.
Santa Fe Opera (NM); John O. Crosby, gen. dir.
Sarasota Opera (FL); Susan T. Danis, exec. dir.
Seattle Opera (WA); Speight Jenkins, gen. dir.
Skylight Opera Theatre (Milwaukee, WI); Joan Lounsbery, mng. dir.
Tulsa Opera (OK); Carol I. Crawford, gen. dir.
Utah Festival Opera Company (Logan); Michael Ballam, gen. dir.
Utah Opera (Salt Lake City); Anne Ewers, gen. dir.
Virginia Opera (Norfolk); Peter Mark, gen./art. dir.
Washington Opera (DC); Patricia L. Mossel, exec. dir.
West Virginia Symphony Orchestra (Charleston); Paul A. Helfrich, exec. dir.

Some Notable U.S. Dance Companies
Source: DanceUSA

Organization	City	State	Organization	City	State
African-American Dance Ensemble	Durham	NC	Ballet Omaha	Omaha	NE
Alabama Ballet	Birmingham	AL	Ballet West	Salt Lake City	UT
Alvin Ailey American Dance Theater	New York	NY	BalletMet Columbus	Columbus	OH
American Ballet Theatre	New York	NY	Betty Salamun's DANCECIRCUS	Milwaukee	WI
American Repertory Ballet Company	New Brunswick	NJ	Bill T. Jones/Arnie Zane Dance Company	New York	NY
The Atlanta Ballet, Inc.	Atlanta	GA	Boston Ballet	Boston	MA
Ballet Arizona	Phoenix	AZ	Caribbean Dance Company of the Virgin		
Ballet Austin	Austin	TX	Islands	St. Croix	USVI
Ballet Concierto de Puerto Rico	Santurce	PR	Carolina Ballet	Raleigh	NC
Ballet Hispanico of New York	New York	NY	Carolyn Dorfman Dance Company	Union	NJ
Ballet Internationale, Inc.	Indianapolis	IN	Charleston Ballet Theatre	Charleston	SC
Ballet Memphis	Memphis	TN	Chen & Dancers	New York	NY

Organization	City	State	Organization	City	State
Cincinnati Ballet	Cincinnati	OH	Lucinda Childs Dance Company	New York	NY
Collage Dance Theatre	Los Angeles	CA	Malashock Dance & Company	San Diego	CA
Contemporary Dance/Fort Worth	Forth Worth	TX	Margaret Jenkins Dance Company	San Francisco	CA
Cunningham Dance Foundation	New York	NY	Maria Benitez Teatro Flamenco	Santa Fe	NM
Dallas Black Dance Theatre	Dallas	TX	Mark Morris Dance Group	New York	NY
Dance Alloy	Pittsburgh	PA	Meredith Monk/The House Foundation	New York	NY
Dance Theatre of Harlem	New York	NY	Milwaukee Ballet	Milwaukee	WI
DanceBrazil	New York	NY	Monte/Brown Dance	New York	NY
David Gordon/Pick Up Co.	New York	NY	Montgomery Ballet	Montgomery	AL
Dayton Ballet	Dayton	OH	Muntu Dance Theatre	Chicago	IL
Dayton Contemporary Dance Co.	Dayton	OH	Nai-Ni Chen Dance Company	Fort Lee	NJ
Demetrius Klein Dance Company	Lake Worth	FL	Nashville Ballet	Nashville	TN
Denishawn Repertory Dancers	New York	NY	New York City Ballet	New York	NY
Diavolo Dance Theater	Santa Monica	CA	North Carolina Dance Theatre	Charlotte	NC
Donald Byrd/The Group	Brooklyn	NY	Ohio Ballet	Akron	OH
Doug Varone & Dancers/DOVA, Inc.	New York	NY	Oregon Ballet Theatre	Portland	OR
EIKO & KOMA	New York	NY	Pacific Northwest Ballet	Seattle	WA
Flamenco Vivo Carlota Santana	New York	NY	The Parsons Dance Company	New York	NY
Fort Worth Dallas Ballet	Forth Worth	TX	Pat Graney Company	Seattle	WA
Garth Fagan Dance	Rochester	NY	Paul Taylor Dance Foundation	New York	NY
Hartford Ballet	Hartford	CT	Paula Josa-Jones/Performance		
Houston Ballet Foundation	Houston	TX	Works	Chilmark	MA
Hubbard Street Dance Chicago	Chicago	IL	Pittsburgh Ballet Theatre	Pittsburgh	PA
James Sewell Ballet	Minneapolis	MN	Rhythm In Shoes	Dayton	OH
Jazz Tap Ensemble	Los Angeles	CA	Richmond Ballet	Richmond	VA
Joe Goode Performance Group	San Francisco	CA	River North Dance Company	Chicago	IL
The Joffrey Ballet of Chicago	Chicago	IL	San Francisco Ballet	San Francisco	CA
June Watanabe in Company	San Rafael	CA	San Jose Cleveland Ballet	San Jose	CA
Karen Bamonte Dance Works	Philadelphia	PA	Smuin Ballets/SF	San Francisco	CA
Kim Robards Dance	Denver	CO	The Solomons Company/Dance	New York	NY
Ko-Thi Dance Company	Milwaukee	WI	State Ballet of Missouri	Kansas City	MO
Lar Lubovitch Dance Company	New York	NY	Stephen Petronio Dance Company	New York	NY
Laura Dean Musicians and Dancers	Bahama	NC	Stuart Pimsler Dance & Theater	Columbus	OH
Lily Cai Chinese Dance Company	San Francisco	CA	Tennessee Children's Dance Ensemble	Knoxville	TN
Limón Dance Company	New York	NY	Trinity Irish Dance Company	Chicago	IL
LINES Contemporary Ballet	San Francisco	CA	Trisha Brown Company	New York	NY
Liz Lerman Dance Exchange	Takoma Park	MD	Tulsa Ballet Theatre	Tulsa	OK
Louisville Ballet	Louisville	KY	The Washington Ballet	Washington	DC

Some Notable Museums

This unofficial list of the largest museums in the U.S. by budget was compiled with the assistance of the American Association of Museums, a national association representing the concerns of the museum community. Association members also include zoos, aquariums, arboretums, botanical gardens, and planetariums, but these are not included in *The World Almanac* listing. See also Major U.S. Public Zoological Parks and Major Canadian Public Zoological Parks in the Environment chapter.

Museum	City	State	Museum	City	State
American Museum of Natural History	New York	NY	Museum of Contemporary Art	Los Angeles	CA
Amon Carter Museum of Western Art	Ft. Worth	TX	Museum of Fine Arts	Boston	MA
The Art Institute of Chicago	Chicago	IL	Museum of Fine Arts	Houston	TX
Autry Museum of Western Heritage	Los Angeles	CA	Museum of Modern Art	New York	NY
Brooklyn Museum of Art	Brooklyn	NY	Museum of New Mexico	Santa Fe	NM
Busch-Reisinger Museum	Cambridge	MA	Museum of Science	Boston	MA
California Academy of Science	San Francisco	CA	The Museum of Television & Radio	Beverly Hills	CA
Carnegie Museums of Pittsburgh	Pittsburgh	PA	Mystic Seaport Museum	Mystic	CT
Chicago Historical Society	Chicago	IL	National Air & Space Museum	Washington	DC
Children's Museum of Indianapolis	Indianapolis	IN	National Baseball Hall of Fame and		
Cincinnati Art Museum	Cincinnati	OH	Museum, Inc.	Cooperstown	NY
Cincinnati Museum Center	Cincinnati	OH	National Gallery of Art	Washington	DC
Cleveland Museum of Art	Cleveland	OH	National Museum of American History-		
Colonial Williamsburg	Williamsburg	VA	Smithsonian Inst.	Washington	DC
Corning Museum of Glass	Corning	NY	National Museum of Natural History	Washington	DC
Dallas Museum of Art	Dallas	TX	Nelson-Atkins Museum of Art	Kansas City	MO
Denver Art Museum	Denver	CO	New York Historical Society	New York	NY
Denver Museum of Natural History	Denver	CO	New York State Museum	Albany	NY
Detroit Institute of Arts	Detroit	MI	The Newseum	Arlington	VA
Exploratorium	San Francisco	CA	Ohio Historical Society	Columbus	OH
The Field Museum of Natural History	Chicago	IL	Peabody Essex Museum	Salem	MA
Fine Arts Museum of San Francisco	San Francisco	CA	Pennsylvania Historical & Museum		
Franklin Institute	Philadelphia	PA	Commission	Harrisburg	PA
The Frick Collection	New York	NY	Philadelphia Museum of Art	Philadelphia	PA
Harvard University Art Museum	Cambridge	MA	Public Museum of Grand Rapids	Grand Rapids	MI
Henry F. Dupont Winterthur Museum	Winterthur	DE	Rock & Roll Hall of Fame and Museum Inc.	Cleveland	OH
Henry Ford Museum/Greenfield Village	Dearborn	MI	San Diego Museum of Art	San Diego	CA
High Museum of Art	Atlanta	GA	San Francisco Museum of Modern Art	San Francisco	CA
Houston Museum of Natural Science	Houston	TX	Science Museum of Minnesota	Saint Paul	MN
Jamestown-Yorktown Foundation	Williamsburg	VA	Scottsdale Museum of Contemp. Art	Scottsdale	AZ
Jewish Museum	New York	NY	St. Louis Science Center	St. Louis	MO
L.A. County Museum of Art	Los Angeles	CA	Toledo Museum of Art	Toledo	OH
Liberty Science Center, Liberty State Park	Jersey City	NJ	U.S. Holocaust Memorial Museum	Washington	DC
Maryland Academy of Sciences	Baltimore	MD	Univ. of Pennsylvania Museum, University		
Maryland Science Center	Baltimore	MD	of Pennsylvania	Philadelphia	PA
Metropolitan Museum of Art	New York	NY	Virginia Museum of Fine Arts	Richmond	VA
Milwaukee Public Museum	Milwaukee	WI	Wadsworth Atheneum	Hartford	CT
Minneapolis Institute of Art	Minneapolis	MN	Walker Art Center	Minneapolis	MN
Museum of African American History	Detroit	MI	Whitney Museum of American Art	New York	NY

100 Best-Selling U.S. Magazines

Source: Audit Bureau of Circulations, Schaumburg, IL

General magazines, exclusive of groups and comics; also excluding magazines that failed to file reports to ABC by press time. Based on total average paid circulation during the 6 months ending Dec. 31, 1998.

Magazine	Circulation	Magazine	Circulation	Magazine	Circulation
1. Modern Maturity	20,534,357	33. Star	1,821,209	68. Self	1,141,145
2. NRTA/AARP Bulletin	20,357,541	34. Field & Stream	1,763,741	69. Sesame Street Magazine	1,133,817
3. Reader's Digest	13,767,575	35. Ebony	1,750,027	70. The Family Handyman	1,108,882
4. TV Guide	12,579,912	36. Parents	1,742,824	71. Us	1,105,241
5. National Geographic Magazine	8,612,102	37. Country Living	1,682,404	72. Soap Opera Digest	1,100,069
6. Better Homes and Gardens	7,613,249	38. Life	1,626,547	73. Bon Appetit	1,086,997
		39. Men's Health	1,624,242	74. Family Fun	1,077,707
7. Family Circle	5,004,902	40. First for Women	1,604,363	75. Vanity Fair	1,076,150
8. Good Housekeeping	4,584,879	41. Woman's World	1,593,385	76. Scouting	1,061,074
9. Ladies' Home Journal	4,575,996	42. Popular Science	1,563,778	77. Health	1,059,079
10. Woman's Day	4,242,097	43. Golf Digest	1,554,134	78. Country Home	1,056,780
11. McCall's	4,202,809	44. Sunset	1,458,702	79. PC/Computing	1,044,252
12. Time	4,060,074	45. Entertainment Weekly	1,452,973	80. Kiplinger's Personal Finance Magazine	1,029,386
13. People Weekly	3,635,146	46. Popular Mechanics	1,432,301		
14. Playboy	3,336,213	47. Cooking Light	1,425,107	81. Michigan Living	1,015,754
15. Sports Illustrated	3,264,345	48. Golf Magazine	1,401,671	82. Home	1,013,141
16. Newsweek	3,153,281	49. Outdoor Life	1,356,813	83. American Health for Women	1,007,972
17. Prevention	3,143,783	50. Boys' Life	1,291,380		
18. Redbook	2,867,951	51. In Style	1,264,529	84. American Homestyle & Gardening	1,001,149
19. Cosmopolitan	2,768,251	52. American Rifleman	1,258,252		
20. The American Legion Magazine	2,711,501	53. Scholastic Parent & Child	1,257,337	85. Essence	1,000,623
		54. Parenting Magazine	1,257,124	86. Penthouse	997,934
21. Southern Living	2,518,732	55. Consumers Digest	1,255,309	87. Travel & Leisure	990,668
22. Via Magazine	2,508,341	56. Rolling Stone	1,251,010	88. Elle	974,819
23. Seventeen	2,415,727	57. Car and Driver	1,249,939	89. Victoria	974,645
24. Martha Stewart Living	2,354,284	58. Discover	1,241,488	90. Fitness	955,871
25. National Enquirer	2,244,213	59. Vogue	1,211,771	91. Today's Homeowner	953,983
26. YM	2,186,706	60. Motor Trend	1,197,118	92. The American Hunter	929,002
27. U.S. News & World Report	2,181,402	61. Mademoiselle	1,191,719	93. Jet	926,675
		62. PC Magazine	1,182,181	94. Business Week	908,953
28. Glamour	2,163,640	63. New Woman	1,179,184	95. Gourmet	891,797
29. 'Teen	2,077,653	64. Endless Vacation	1,151,193	96. House Beautiful	887,976
30. Smithsonian	2,041,134	65. PC World	1,147,034	97. Food & Wine	872,822
31. Money	1,905,158	66. Shape	1,143,409	98. Nation's Business	858,718
32. V.F.W. Magazine	1,882,847	67. Weight Watchers Magazine	1,143,045	99. Child	850,159
				100. Allure	845,861

Some Notable New Books, 1998

Source: List published by American Library Association, Chicago, IL, 1999, for books published in 1998

Fiction

Triage, Scott Anderson
Flanders, Patricia Anthony
The Voyage of the Narwhal, Andrea Barrett
Collected Fictions, tr. by Andrew Hurley, Jorge Luis Borge
The Coast of Good Intentions, Michael Byers
The Farming of Bones, Edwidge Danticat
About a Boy, Nick Hornby
Charming Billy, Alice McDermott
Birds of America, Lorrie Moore
I Married a Communist, Philip Roth
Beach Boy, Ardashir Vakil

Poetry

Sweet Machine, Mark Doty
Without, Donald Hall
Tales from Ovid, Ted Hughes

Nonfiction

The Life of Thomas More, Peter Ackroyd
Lindbergh, A. Scott Berg
Pillar of Fire: America in the King Years, 1963-65, Taylor Branch
Titan: The Life of John D. Rockefeller, Sr., Ron Chernow
The Road to Ubar: Finding the Atlantis of the Sands, Nicholas Clapp
Articles of Faith: A Frontline History of the Abortion Wars, Cynthia Gorney
King Leopold's Ghost: A Story of Greed, Terror, and Heroism in Colonial Africa, Adam Hochschild
Ship of God in the Deep Blue Sea, Gary Kinder
A Beautiful Mind, Sylvia Nasar
A Hope in the Unseen: An American Odyssey from the Inner City to the Ivy League, Ron Suskind

Young Adults

Nonfiction

The Amazing True Story of a Teenage Single Mom, Katherine Arnoldi
33 Things Every Girl Should Know: Stories, Songs, Poems and Smart Talk by 33 Extraordinary Women, Tonya Bolden
Corpses, Coffins, and Crypts: A History of Burial, Penny Colman
Invisible Enemies: Stories of Infectious Disease, Jeanette Farrell
Flash! The Associated Press Covers the World, Vincent Alabisco, Ed.
Martha Graham: A Dancer's Life, Russell Freedman
Modoc: The Story of the Greatest Elephant That Ever Lived, Ralph Helfer
I, Too, Sing America: Three Centuries of African American Poetry, Catherine Clinton, Ed.
There's a Hair in My Dirt: A Worm's Story, Gary Larson
No Pretty Pictures: A Child of War, Anita Lobel
Commander-in-Chief Abraham Lincoln and the Civil War, Albert Marrin
The Shared Heart, Adam Mastoon
Young, Black, and Determined: A Biography of Lorraine Hansberry, Patricia C. and Frederick L. McKissack
No More Strangers Now: Young Voices from a New South Africa; interviews, Tim McKee; photos, Anne Blackshaw
The Space between Our Footsteps: Poems and Paintings from the Middle East, Naomi Shihab Nye, Ed
Knots in My Yo-Yo String: The Autobiography of a Kid, Jerry Spinelli
Behind the Mask: The Life of Queen Elizabeth I, Jane Resh Thomas

Lest We Forget: The Passage from Africa to Slavery and Emancipation, Velma Maia Thomas
A Lion's Hunger: Poems of First Love, Ann Marshall Turner
War and the Pity of War, Neil Philip, Ed.

Fiction

Go and Come Back, Joan Abelove
Rules of the Road, Joan Bauer
Life in the Fat Lane, Cherie Bennett
The Shakespeare Stealer, Gary L. Blackwood
Smack, Melvin Burgess
Heroes, Robert Cormier
Someone like You, Sarah Dessen
Love Among the Walnuts, Jean Ferris
The Skin I'm In, Sharon G. Flake
Whirligig, Paul Fleischman
Shadow Spinner, Susan Fletcher
The Other Shepards, Adele Griffin
Among the Hidden, Margaret Haddix
Sunshine Rider: The First Vegetarian Western, Ric Lynden Hardman
Help Wanted: Short Stories about Young People Working, Anita Silvey, Ed.
Kissing Doorknobs, Terry Spencer Hesser
A Life for a Life, Ernest Hill
The Maze, Will Hobbs
My Louisiana Sky, Kimberly Willis Holt
The Circuit: Stories from the Life of a Migrant Child, Francisco Jiménez
Heaven, Angela Johnson
The Falcon, Jackie French Koller
The Wreckers, Iain Lawrence
Dust Devils, Robert Laxalt
The Sacrifice, Diane Matcheck
The Pirate's Son, Geraldine McCaughrean
Petey, Ben Mikaelson
Sirena, Donna Jo Napoli
The Dark Light, Mette Newth

Tribes of Palos Verdes, Joy Nicholson
Soldier's Heart, Gary Paulsen
A Long Way From Chicago, Richard Peck
Strays Like Us, Richard Peck
Treasures in the Dust, Tracey Porter
Zebra and Other Stories, Chaim Potok
A Door Near Here, Heather Quarles
Choosing up Sides, John H. Ritter
Hero, S.L. Rottman

Harry Potter and the Sorcerer's Stone, J.K. Rowling
Holes, Louis Sachar
Jungle Dogs, Graham Salisbury
I Am Mordred: A Tale from Camelot, Nancy Springer
The Spirit Window, Joyce Sweeney
Making Up Megaboy, Virginia Walter
Hard Ball, Will Weaver

The Killer's Cousin, Nancy Werlin
To Say Nothing of the Dog: or, How We Found the Bishop's Bird Stump at Last, Connie Willis
I Rode a Horse of Milk White Jade, Diane Lee Wilson
If You Come Softly, Jacqueline Woodson
Armageddon Summer, Jane Yolen and Bruce Coville

Some Notable New Books for Children, 1998

Source: List published by American Library Association, Chicago, IL, 1999, for books published in 1998

Younger Readers

Arlene Alda's 1 2 3, Arlene Alda
Ouch! Natalie Babbitt
And If the Moon Could Talk, Kate Banks
I Lost My Bear, Jules Feiffer
Mama Cat Has Three Kittens, Denise Fleming
Cowboy Baby, Sue Heap
Zoom City, Thacher Hurd
How Santa Got His Job, Stephen Krensky
Zelda and Ivy, Laura McGee Kvasnosky
10 Minutes till Bedtime, Peggy Rathmann
Cendrillon: A Caribbean Cinderella, Robert D. San Souci
No, David! David Shannon
Snow, Uri Shulevitz
Fire Truck, Peter Sís
Pete's a Pizza, William Steig
Elizabeti's Doll, Stephanie Stuve-Bodeen
My Name Is Georgia, Jeanette Winter

Middle Readers

Boss of the Plains: The Hat That Won the West, Laurie Carlson
Bodies from the Bog, James M. Deem
The Number Devil: A Mathematical Adventure, Hans Magnus Enzensberger
Bandit's Moon, Sid Fleischman
Joey Pigza Swallowed the Key, Jack Gantos
The Wild Boy, Mordicai Gerstein
Cool Melons Turn to Frogs! Matthew Gollub

Chuck Close, Up Close, Jan Greenberg and Sandra Jordan
Snowflake Bentley, Jacqueline Briggs Martin
Beautiful Warrior: The Legend of the Nun's Kung Fu, Emily Arnold McCully
Secret Letters from 0 to 10, Susie Hoch Morgenstern
Duke Ellington: The Piano Prince and His Orchestra, Andrea Davis Pinkney
Joan of Arc, Josephine Poole
Harry Potter and the Sorcerer's Stone, J.K. Rowling
Home to Medicine Mountain, Chiori Santiago
G is for Googol: A Math Alphabet Book, David M. Schwartz
Joan of Arc, Diane Stanley
I Have Heard of a Land, Joyce Carol Thomas

Older Readers-JHS

Go and Come Back, Joan Ablelove
Shipwreck at the Bottom of the World: The Extraordinary True Story of Shackleton and the Endurance, Jennifer Armstrong
Rules of the Road, Joan Bauer
The Shakespeare Stealer, Gary Blackwood
Shadow Spinner, Susan Fletcher
Martha Graham: A Dancer's Life, Russell Freedman
The Other Shepards, Adele Griffin
My Louisiana Sky, Kimberly Willis Holt

No Pretty Pictures: A Child of War, Anita Lobel
The Pirate's Son, Geraldine McCaughrean
No More Strangers Now: Young Voices from a New South Africa, Tim McKee
Gone a-Whaling: The Lure of the Sea and the Hunt for the Great Whale, Jim Murphy
Restless Spirit: The Life and Work of Dorothea Lange, Elizabeth Partridge
A Long Way from Chicago, Richard Peck
Thanks to My Mother, Schoschana Rabinovici
Holes, Louis Sachar
Making Up Megaboy, Virginia Walter
Bat 6, Virginia Euwer Wolff

All Ages

Voices in the Park, Anthony Browne
Home Run: The Story of Babe Ruth, Robert Burleigh
Insectlopedia, Douglas Florian
This Land Is Your Land, Woody Guthrie
I See the Rhythm, Igus Toyomi
A Caldecott Celebration: Six Artists Share Their Paths to the Caldecott Medal, Leonard Marcus
Tibet: Through the Red Box, Peter Sís
With a Whoop and a Holler: A Bushel of Lore from Way Down South, Nancy Van Lann
You Can't Take a Balloon in the Metropolitan Museum, Jacqueline Preiss Weitzman
Walter Wick's Optical Tricks, Walter Wick

Best-Selling Books, 1998

Source: *Publishers Weekly*, Mar. 29, 1999

Rankings are based on copies "shipped and billed" in 1998, minus returns through early 1999.

Fiction

1. *The Street Lawyer*, John Grisham
2. *Rainbow Six*, Tom Clancy
3. *Bag of Bones*, Stephen King
4. *Man in Full*, Tom Wolfe
5. *Mirror Image*, Danielle Steel
6. *The Long Road Home*, Danielle Steel
7. *The Klone and I*, Danielle Steel
8. *Point of Origin*, Patricia Cornwell
9. *Paradise*, Toni Morrison
10. *All Through the Night*, Mary Higgins Clark
11. *I Know This Much Is True*, Wally Lamb
12. *Tell Me Your Dreams*, Sidney Sheldon
13. *The Vampire Armand*, Anne Rice
14. *The Loop*, Nicholas Evans
15. *You Belong to Me*, Mary Higgins Clark

Nonfiction

1. *The 9 Steps to Financial Freedom*, Suze Orman
2. *The Greatest Generation*, Tom Brokaw
3. *Sugar Busters!*, H. Leighton Steward, Morrison C. Bethea, Sam S. Andrews and Luis A. Balart
4. *Tuesdays with Morrie*, Mitch Albom
5. *The Guinness Book of Records 1999*
6. *Talking to Heaven*, James Van Praagh
7. *Something More: Excavating Your Authentic Self*, Sarah Ban Breathnach

8. *In the Meantime*, Iyanla Vanzant
9. *A Pirate Looks at Fifty*, Jimmy Buffett
10. *If Life Is a Game These Are the Rules*, Cherie Carter-Scott
11. *Angela's Ashes*, Frank McCourt
12. *For the Love of the Game: My Story*, Michael Jordan
13. *The Day Diana Died*, Christopher Andersen
14. *The Century*, Peter Jennings and Todd Brewster
15. *Eat Right 4 Your Type*, Peter J. D'Adam

Trade Paperbacks

1. *Don't Sweat the Small Stuff... And It's All Small Stuff*, Richard Carlson, Ph.D.
2. *Divine Secrets of the Ya-Ya Sisterhood*, Rebecca Wells
3. *Chicken Soup for the Teenage Soul*, Jack Canfield, Mark Victor Hansen et al. (dual edition)
4. *Chicken Soup for the Teenage Soul*, Jack Canfield, Mark Victor Hansen et al. (dual edition)
5. *Don't Sweat the Small Stuff with Your Family*, Richard Carlson, Ph.D.
6. *Chicken Soup for the Kid's Soul*, Jack Canfield, Mark Victor Hansen et al.
7. *Chicken Soup for the Pet Lover's Soul*, Jack Canfield, Mark Victor Hansen et al.
8. *Chicken Soup for the Mother's Soul*, Jack Canfield, Mark Victor Hansen et al
9. *A 2nd Helping of Chicken Soup for the Woman's Soul*, Jack Canfield, Mark Victor Hansen et al.

10. *Here on Earth,* Alice Hoffman
11. *Prescription for Nutritional Healing: A Practical A-Z Reference to Drug-Free Remedies Using Vitamins, Minerals, Herbs & Food Supplements,* James F. Balch, M.D., and Phyllis A. Balch, C.N.C.
12. *Midwives,* Chris Bohjalian
13. *Cold Mountain,* Charles Frazier
14. *James Cameron's Titanic,* Ed W. Marsh
15. *A 5th Portion of Chicken Soup for the Soul,* Jack Canfield and Mark Victor Hansen

Almanacs, Atlases, & Annuals

1. *The World Almanac and Book of Facts 1999,* Robert Famighetti
2. *The Ernst & Young Tax Guide 1998*
3. *The World Almanac and Book of Facts 1998,* Robert Famighetti
4. *The Old Farmer's Almanac 1999,* Robert B. Thomas
5. *The Best American Short Stories 1998,* Edited by Garrison Keillor

Mass Market

1. *The Partner,* John Grisham
2. *The Ghost,* Danielle Steel
3. *The Ranch,* Danielle Steel
4. *Special Delivery,* Danielle Steel
5. *Unnatural Exposure,* Patricia Cornwell
6. *Pretend You Don't See Her,* Mary Higgins Clark
7. *Power Plays: ruthless.com,* Tom Clancy
8. *Rising Tides,* Nora Roberts
9. *Wizard and Glass,* Stephen King
10. *Dr. Atkins' New Diet Revolution,* Robert C. Atkins, M.D.
11. *Into Thin Air,* Jon Krakauer
12. *Tom Clancy's Op-Center V.,* Created by Tom Clancy and Steve Pieczenik
13. *The Notebook,* Nicholas Sparks
14. *Fear Nothing,* Dean Koontz
15. *Sanctuary,* Nora Roberts

Leading U.S. Daily Newspapers, 1998

Source: 1999 *Editor & Publisher International Yearbook*
(Circulation as of Sept. 30, 1998; m = morning, e = evening)

As of Feb. 1, 1999, the number of U.S. daily newspapers had dropped to 1,489, for a net loss of 20 since Feb. 1, 1998. Most of the change was the result of mergers and conversions. The number of cities with more than 1 daily paper continued its downward trend, going from 55 to 49. Average daily circulation for the 6 months ending Sept. 30, 1998, was 56,182,092, down 545,810 from the same period in 1997, for a decrease of less than 1%. Although Sunday editions increased by 11, the overall number of Sunday papers decreased by 5 to 898. Average Sunday circulation for the 6 months ending Sept. 30, 1998, fell 420,571, to 60,065,892.

Newspaper		Circulation	Newspaper		Circulation
1. New York (NY) *Wall Street Journal*	(m)	1,740,450	51. San Antonio (TX) *Express-News*	(m)	218,661
2. Arlington (VA) *USA Today*	(m)	1,653,428	52. Hartford (CT) *Courant*	(m)	211,041
3. Los Angeles (CA) *Times*	(m)	1,067,540	53. Richmond (VA) *Times-Dispatch*	(m)	207,175
4. New York (NY) *Times*	(m)	1,066,658	54. Oklahoma City (OK) *Daily Oklahoman*	(m)	204,963
5. Washington (DC) *Post*	(m)	759,122	55. Los Angeles (CA) *Daily News*	(m)	201,107
6. New York (NY) *Daily News*	(m)	723,143	56. St. Paul (MN) *Pioneer Press*	(m)	199,119
7. Chicago (IL) *Tribune*	(m)	673,508	57. Norfolk (VA) *Virginian-Pilot*	(m)	197,773
8. Long Island (NY) *Newsday*	(m)	572,444	58. Seattle (WA) *Post-Intelligencer*	(m)	196,271
9. Houston (TX) *Chronicle*	(m)	550,763	59. Cincinnati (OH) *Enquirer*	(m)	196,181
10. Chicago (IL) *Sun-Times*	(m)	485,666	60. Nashville (TN) *Tennessean*	(m)	184,979
11. Dallas (TX) *Morning News*	(m)	479,863	61. Austin (TX) *American-Statesman*	(m)	183,319
12. San Francisco (CA) *Chronicle*	(m)	475,324	62. Philadelphia (PA) *Daily News*	(m)	175,448
13. Boston (MA) *Globe*	(m)	470,825	63. Rochester (NY) *Democrat and Chronicle*	(m)	174,579
14. New York (NY) *Post*	(m)	437,467	64. Little Rock (AR) *Democrat-Gazette*	(m)	173,316
15. Phoenix (AZ) *Arizona Republic*	(m)	435,330	65. West Palm Beach (FL) *Palm Beach Post*	(m)	173,074
16. Philadelphia (PA) *Inquirer*	(m)	428,895	66. Jacksonville (FL) *Times-Union*	(m)	172,511
17. Newark (NJ) *Star-Ledger*	(m)	407,026	67. Providence (RI) *Journal*	(m)	167,381
18. Cleveland (OH) *Plain Dealer*	(m)	382,933	68. Memphis (TN) *Commercial Appeal*	(m)	163,603
19. San Diego (CA) *Union-Tribune*	(all day)	378,112	69. Des Moines (IA) *Registers*	(m)	163,292
20. Orange County (CA) *Registers*	(m)	356,953	70. Tulsa (OK) *World*	(m)	162,186
21. Miami (FL) *Herald*	(m)	349,114	71. Riverside (CA) *Press-Enterprise*	(m)	161,612
22. Portland (OR) *Oregonian*	(all day)	346,593	72. Neptune (NJ) *Asbury Park Press*	(all day)	159,472
23. St. Petersburg (FL) *Times*	(m)	344,784	73. Raleigh (NC) *News & Observer*	(m)	157,634
24. Denver (CO) *Post*	(m)	341,554	74. Fresno (CA) *Bee*	(m)	155,931
25. Minneapolis (MN) *Star Tribune*	(m)	334,751	75. Dayton (OH) *Daily News*	(m)	152,308
26. Denver (CO) *Rocky Mountain News*	(m)	331,978	76. White Plains (NY) *Journal News.*	(m)	151,695
27. St. Louis (MO) *Post-Dispatch*	(m)	329,582	77. Las Vegas (NV) *Review-Journal*	(m)	151,162
28. Baltimore (MD) *Sun*	(m)	314,033	78. Birmingham (AL) *News*	(m)	148,835
29. Atlanta (GA) *Constitution.*	(m)	303,698	79. Toledo (OH) *Blade*	(m)	146,138
30. San Jose (CA) *Mercury News*	(m)	290,885	80. Akron (OH) *Beacon Journal*	(m)	143,199
31. Milwaukee (WI) *Journal Sentinel*	(m)	285,776	81. Bergen County (NJ) *Record*	(m)	141,368
32. Sacramento (CA) *Bee*	(m)	283,598	82. Arlington Heights (IL) *Daily Herald*	(m)	141,331
33. Kansas City (MO) *Star*	(m)	281,596	83. Grand Rapids (MI) *Press*	(e)	139,703
34. Detroit (MI) *Free Press*	(m)	278,286	84. Salt Lake City (UT) *Tribune*	(m)	129,612
35. Boston (MA) *Herald*	(m)	241,425	85. Allentown (PA) *Morning Call*	(m)	129,522
36. New Orleans (LA) *Times-Picayune*	(m)	259,317	86. Tacoma (WA) *News Tribune*	(m)	129,247
37. Fort Lauderdale (FL) *Sun-Sentinel*	(m)	258,959	87. Wilmington (DE) *News Journal.*	(all day)	125,401
38. Orlando (FL) *Sentinel*	(all day)	258,726	88. Columbia (SC) *State.*	(m)	120,433
39. Los Angeles (CA) *Investors Business*	(m)	251,124	89. Knoxville (TN) *News-Sentinel*	(m)	115,248
40. Columbus (OH) *Dispatch*	(m)	246,528	90. Spokane (WA) *Spokesman-Review*	(m)	114,475
41. Detroit (MI) *News.*	(e)	245,351	91. San Francisco (CA) *Examiner*	(e)	113,198
42. Charlotte (NC) *Observer*	(m)	243,818	92. Lexington (KY) *Herald-Leader*	(m)	113,036
43. Pittsburgh (PA) *Post-Gazette*	(m)	243,453	93. Albuquerque (NM) *Journal*	(m)	112,751
44. Buffalo (NY) *News*	(all day)	237,229	94. Sarasota (FL) *Herald-Tribune*	(m)	109,438
45. Tampa (FL) *Tribune*	(m)	235,786	95. Charleston (SC) *Post & Courier*	(m)	109,341
46. Fort Worth (TX) *Star-Telegram*	(m)	232,112	96. Atlanta (GA) *Journal*	(e)	106,272
47. Indianapolis (IN) *Star*	(m)	230,223	97. Worcester (MA) *Telegram & Gazette*	(m)	105,896
48. Louisville (KY) *Courier-Journal*	(m)	228,144	98. Jackson (MS) *Clarion-Ledger*	(m)	105,382
49. Seattle (WA) *Times*	(e)	227,715	99. Long Beach (CA) *Press-Telegram*	(m)	105,167
50. Omaha (NE) *World-Herald*	(all day)	219,891	100. Honolulu (HI) *Advertiser*	(m)	102,358

MILLENNIUM FACT BOX

Newseum Top 100 News Stories of the Century

Source: © The Freedom Forum Newseum Inc.

The Newseum in Arlington, VA, asked journalists and historians to select the top 100 news stories of the 20th century. The following list is the result. Stories are ranked in order of importance as determined by survey respondents.

1. 1945 United States drops atomic bombs on Hiroshima, Nagasaki: Japan surrenders to end World War II.
2. 1969 American astronaut Neil Armstrong becomes the first human to walk on the moon.
3. 1941 Japan bombs Pearl Harbor: United States enters World War II.
4. 1903 Wilbur and Orville Wright fly the first powered airplane.
5. 1920 Women win the vote.
6. 1963 President John F. Kennedy assassinated in Dallas.
7. 1945 Horrors of Nazi Holocaust, concentration camps exposed.
8. 1914 World War I begins in Europe.
9. 1954 *Brown* v. *Board of Education* ends "separate but equal" school segregation.
10. 1929 U.S. stock market crashes: The Great Depression sets in.
11. 1928 Alexander Fleming discovers the first antibiotic, penicillin.
12. 1953 Structure of DNA discovered.
13. 1991 Soviet Union dissolves, Mikhail Gorbachev resigns: Boris Yeltsin takes over.
14. 1974 President Richard M. Nixon resigns after Watergate scandal.
15. 1939 Germany invades Poland: World War II begins in Europe.
16. 1917 Russian Revolution ends: Communists take over.
17. 1913 Henry Ford organizes the first major U.S. assembly line to produce Model T cars.
18. 1957 Soviets launch Sputnik, first space satellite: space race begins.
19. 1905 Albert Einstein presents special theory of relativity: general relativity theory to follow.
20. 1960 FDA approves birth-control pill.
21. 1953 Dr. Jonas Salk's polio vaccine proven effective in University of Pittsburgh tests.
22. 1933 Adolf Hitler named chancellor of Germany: Nazi Party begins to seize power.
23. 1968 Civil rights leader Martin Luther King assassinated in Memphis, TN.
24. 1944 D-Day invasion marks the beginning of the end of World War II in Europe.
25. 1981 Deadly AIDS disease identified.
26. 1964 Congress passes landmark Civil Rights Act outlawing segregation.
27. 1989 Berlin Wall falls as East Germany lifts travel restrictions.
28. 1939 Television debuts in America at New York World's Fair.
29. 1949 Mao Zedung establishes People's Republic of China: Nationalists flee to Formosa (Taiwan).
30. 1927 Charles Lindbergh crosses the Atlantic in first solo flight.
31. 1977 First mass market personal computers launched.
32. 1989 World Wide Web revolutionizes the Internet.
33. 1948 Scientists at Bell Labs invent the transistor.
34. 1933 FDR launches "New Deal:" sweeping federal economic, public works legislation to combat depression.
35. 1962 Cuban Missile Crisis threatens World War III.
36. 1912 "Unsinkable" *Titanic*, largest man-made structure, sinks.
37. 1945 Germany surrenders: V.E. Day celebrated.
38. 1973 *Roe* v. *Wade* decision legalizes abortion.
39. 1918 World War I ends with Germany's defeat.
40. 1909 First regular radio broadcasts begin in America.
41. 1918 Worldwide flu epidemic kills 20 million.
42. 1946 ENIAC accelerates digital computing.
43. 1941 Regular TV broadcasting begins in the United States.
44. 1947 Jackie Robinson breaks baseball's color barrier.
45. 1948 Israel achieves statehood.
46. 1909 Plastic invented: revolutionizes products, packaging.
47. 1955 Montgomery, AL, bus boycott begins after Rosa Parks refuses to give up her seat to a white person.
48. 1945 Atomic bomb tested in New Mexico.
49. 1993 Apartheid ends in South Africa: law to treat races equally.
50. 1963 Civil rights march converges on Washington, DC: Martin Luther King gives "I Have A Dream" speech.
51. 1959 American scientists patent the computer chip.
52. 1901 Marconi transmits radio signal across the Atlantic.
53. 1998 White House sex scandal leads to impeachment of President William Jefferson Clinton.

54. 1947 Sec. of State George Marshall proposes European recovery program (The Marshall Plan).
55. 1968 Presidential candidate Robert F. Kennedy assassinated in California.
56. 1920 U.S. Senate rejects Versailles Treaty: dooms League of Nations.
57. 1962 Rachel Carson's *Silent Spring* stimulates environmental protection movement.
58. 1962 British rock group The Beatles takes the United States by storm after debut on the Ed Sullivan show.
59. 1965 Congress passes Voting Rights Act, outlawing measures used to suppress minority votes.
60. 1961 Yuri Gagarin becomes first man in space.
61. 1939 First jet airplane takes flight.
62. 1965 U.S. combat troops arrive in South Vietnam: U.S. planes bomb North Vietnam.
63. 1975 North Vietnamese forces take over Saigon.
64. 1942 Manhattan Project begins secret work on atomic bomb: Fermi triggers first atomic chain reaction.
65. 1945 Congress strengthens "GI Bill of Rights" to help veterans.
66. 1961 Alan Shepard becomes first American in space.
67. 1973 Watergate scandal engulfs Nixon administration.
68. 1906 Earthquake hits San Francisco: "Paris of the West" burns.
69. 1945 United Nations is officially established.
70. 1961 Communists build wall to divide East and West Berlin.
71. 1920 Mohandas Gandhi begins leading nonviolent reform movement in India.
72. 1911 Standard Oil loses Supreme Court antitrust suit: monopolies suffer blow.
73. 1973 United States withdraws last ground troops from Vietnam.
74. 1949 North Atlantic Treaty Organization (NATO) established.
75. 1928 Joseph Stalin begins forced modernization of the Soviet Union: resulting famines claim 25 million.
76. 1932 Democrat Franklin D. Roosevelt beats incumbent President Herbert Hoover.
77. 1985 Mikhail Gorbachev becomes Soviet premier: begins era of "Glasnost."
78. 1900 Max Planck proposes quantum theory of energy.
79. 1997 Scientists clone sheep, dubbed Dolly, in Scotland.
80. 1956 Congress passes interstate highway bill.
81. 1914 Panama Canal opens, linking the Atlantic and Pacific oceans.
82. 1963 Betty Friedan's *The Feminine Mystique* inaugurates modern women's rights movement.
83. 1986 The Space Shuttle *Challenger* explodes killing crew including school teacher Christa McAuliffe.
84. 1950 United States sends troops to defend South Korea.
85. 1968 Violence erupts at Democratic National Convention in Chicago.
86. 1900 Sigmund Freud publishes *The Interpretation of Dreams*.
87. 1958 China begins "Great Leap Forward" modernization program: estimated 20 million die in ensuing famine.
88. 1917 United States enters World War I.
89. 1927 Babe Ruth hits 60 home runs—a single-season record that would last for 34 years.
90. 1962 John Glenn becomes first American to orbit the earth.
91. 1964 North Vietnamese boats reportedly attack U.S. ships: Congress passes Gulf of Tonkin resolution.
92. 1997 *Pathfinder* lands on Mars, sending back astonishing photos.
93. 1938 Hitler launches "Kristallnacht," ordering Nazis to commit acts of violence against German Jews.
94. 1940 Winston Churchill designated prime minister of Great Britain.
95. 1978 Louise Brown, first "test-tube baby," born healthy.
96. 1948 Soviets blockade West Berlin: Western allies respond with massive airlift.
97. 1975 Bill Gates and Paul Allen start Microsoft Corp. to develop software for Altair computer.
98. 1986 Chernobyl nuclear plant leak results in eventual deaths of an estimated 7,000.
99. 1925 Teacher John Scopes' trial pits creation against evolution in Tennessee.
100. 1964 The U.S. surgeon general warns about smoking-related health hazards.

Leading Canadian Daily Newspapers, 1998

Source: 1999 Editor & Publisher International Yearbook
(Circulation as of Sept. 30, 1998; m = morning)

For the year ending Feb. 1, 1999, Canadian daily circulation increased for the second year in a row. Total circulation jumped by 334,130 to 5,081,738, up from 4,747,608 from the same period a year ago. Three Sunday editions began, and overall Sunday circulation increased by 144,685 to 3,307,832. During the same period, Sunday circulation increased by 131,332, to 3,163,147.

Newspaper	Circulation	Newspaper	Circulation
Toronto (ON) *Star*..................(m)	460,654	Vancouver (BC) *Sun* ....(m)	189,823
Toronto (ON) *National Post* ....(m)	325,000	Montreal (QC) *La Presse* ....(m)	168,881
Toronto (ON) *Globe and Mail*........(m)	309,045	Vancouver (BC) *Province* ....(m)	156,688
Montreal (QC) *Le Journal*...........(m)	254,957	Montreal (QC) *Gazette* ....(m)	141,595
Toronto (ON) *Sun*..................(m)	240,164	Edmonton (AB) *Journal*.............(m)	141,583

Top 15 News/Information/Entertainment Websites

Source: Media Matrix, Inc.

Rank		Visitors[1]	Rank		Visitors[1]
1.	www.aol.com/mynews/home.adp	13,951	8.	www.weather.com.........................	5,573
2.	www.aol.com/webcenters/entertainment/ home.adp	11,599	9.	www.disney.go.com.......................	5,372
			10.	www.digitalcity.com.......................	5,198
3.	www.about.com.........................	8,569	11.	www.ivillage.com.........................	4,507
4.	www.aol.com/webcenters/sports/home.adp	8,464	12.	www.sony.com	4,398
5.	www.aol.com/webcenters/computing/home.adp .	8,220	13.	www.pathfinder.com......................	4,395
6.	www.zdnet.com.........................	7,418	14.	www.cnn.com	4,163
7.	www.msnbc.com	6,273	15.	www.espn.go.com........................	4,133

(1) Number of visitors in thousands who visited website at least once in Aug. 1999.

U.S. Commercial Radio Stations, by Format, 1993-99

Source: M Street Corporation, Nashville, TN © 1999; counts are for Aug. of each year

Stations, by primary format	1993	1994	1995	1996	1997	1998	1999
1. Country	2,612	2,642	2,608	2,558	2,502	2,393	2,321
2. Adult Contemporary (AC) ...	1,895	1,784	1,661	1,592	1,521	1,562	1,576
3. News, Talk, Business, Sports	841	1,028	1,165	1,262	1,313	1,356	1,396
4. Oldies, Classic Hits, R&B Oldies....	734	714	718	725	753	975	1,109
5. Religion (Teaching and Music)	915	926	970	996	1,054	1,075	1,088
6. Rock (Album, Modern, Classic, Alternative)	643	721	808	868	942	782	803
7. Spanish and Ethnic	421	470	492	515	549	565	613
8. Adult Standards.................	421	435	469	474	536	563	572
9. Top-40	441	358	324	314	351	379	391
10. Urban and Urban AC.............	321	328	342	348	358	347	278
11. Jazz........................	45	43	58	54	50	90	77
12. Pre-Teen	13	19	26	30	35	32	47
13. Easy Listening.................	116	106	85	91	87	40	46
14. Variety	68	63	63	65	51	51	40
15. Classical, Fine Arts	45	44	39	41	46	42	38
16. Comedy.....................	0	1	0	0	0	0	0
Off Air	345	369	323	298	162	114	100
Changing formats/not available	14	6	9	6	3	14	11
TOTAL STATIONS	**9,890**	**10,057**	**10,160**	**10,237**	**10,313**	**10,380**	**10,506**

Top-Grossing North American Concert Tours, 1985-98

Source: Pollstar, Fresno, CA

Artist (Year)	Total gross[1]	Cities/ Shows	Artist (Year)	Total gross[1]	Cities/ Shows
1. The Rolling Stones (1994)...........	$121.2	43/60	11. The Grateful Dead (1994)...........	52.4	29/84
2. Pink Floyd (1994)	103.5	39/59	12. Elton John/Billy Joel (1994)	47.7	14/21
3. The Rolling Stones (1989)..........	98.0	33/60	13. Elton John (1998)...................	46.2	52/63
4. The Rolling Stones (1997)..........	89.3	26/33	14. The Grateful Dead (1993)...........	45.6	29/81
5. U2 (1997)	79.9	37/46	15. Kiss (1996)......................	43.6	75/92
6. The Eagles (1994).................	79.4	32/54	16. Boyz II Men (1995).................	43.2	133/134
7. The New Kids on the Block (1990)	74.1	122/152	17. Billy Joel (1990)..................	43.0	53/95
8. U2 (1992)	67.0	61/73	18. The Who (1989)..................	41.7	27/39
9. The Eagles (1995)................	63.3	46/58	19. Dave Mathews Band (1998)..........	40.1	76/85
10. Barbra Streisand (1994)	58.9	6/22	20. Bruce Springsteen and the E Street Band (1985)	39.1	21/40

(1) In mils. Not adjusted for inflation.

Sales of Recorded Music and Music Videos, by Genre and Format, 1994-98

Source: Recording Industry Assn. of America, Washington, DC
Breakdown is by percentage of all recorded music sold.

GENRE	1994	1995	1996	1997	1998	GENRE	1994	1995	1996	1997	1998
Rock..................	35.1	33.5	32.6	32.5	25.7	Oldies	0.8	1.0	0.8	0.8	0.7
Country	16.3	16.7	14.7	14.4	14.1	New Age	1.0	0.7	0.7	0.8	0.6
R&B	9.6	11.3	12.1	11.2	12.8	Children's.............	0.4	0.5	0.7	0.9	0.4
Rap..................	7.9	6.7	8.9	9.4	10.0						
Pop..................	10.3	10.1	9.3	10.1	9.7	**FORMAT**	**1994**	**1995**	**1996**	**1997**	**1998**
Other	5.3	7.0	5.2	5.7	7.9	Compact disc (CD)	58.4	65.0	68.4	70.2	74.8
Gospel	3.3	3.1	4.3	4.5	6.3	Cassette	32.1	25.1	19.3	18.2	14.8
Classical..............	3.7	2.9	3.4	2.8	3.3	Singles (all types)	7.4	7.5	9.3	9.3	6.8
Jazz	3.0	3.0	3.3	2.8	1.9	Music video	0.8	0.9	1.0	0.6	1.0
Soundtracks	1.0	0.9	0.8	1.2	1.7	LP	0.8	0.5	0.6	0.7	0.7

Note: Totals may not equal 100% because of "Don't know/no answer" responses to survey.

Sales of Recorded Music and Music Videos, by Units Shipped and Value, 1990-98

Source: Recording Industry Assn. of America, Washington, DC

(in millions, net after returns)

FORMAT	1990	1991	1992	1993	1994	1995	1996	1997	1998	% change 1997-98
Compact disc (CD)										
Units shipped	286.5	333.3	407.5	495.4	662.1	722.9	778.9	753.1	847.0	12.5
Dollar value	3,451.6	4,337.7	5,326.5	6,511.4	8,464.5	9,377.4	9,934.7	9,915.1	11,416.0	15.1
CD single										
Units shipped	1.1	5.7	7.3	7.8	9.3	21.5	43.2	66.7	56.0	-16.0
Dollar value	6.0	35.1	45.1	45.8	56.1	110.9	184.1	272.7	213.2	-21.8
Cassette										
Units shipped	442.2	360.1	366.4	339.5	345.4	272.6	225.3	172.6	158.5	-8.2
Dollar value	3,472.4	3,019.6	3,116.3	2,915.8	2,976.4	2,303.6	1,905.3	1,522.7	1,419.9	-6.8
Cassette single										
Units shipped	87.4	69.0	84.6	85.6	81.1	70.7	59.9	42.2	26.4	-37.4
Dollar value	257.9	230.4	298.8	298.5	274.9	236.3	189.3	133.5	94.4	-29.3
LP/EP										
Units shipped	11.7	4.8	2.3	1.2	1.9	2.2	2.9	2.7	3.4	25.9
Dollar value	86.5	29.4	13.5	10.6	17.8	25.1	36.8	33.3	34.0	2.1
Vinyl single										
Units shipped	27.6	22.0	19.8	15.1	11.7	10.2	10.1	7.5	5.4	-28.0
Dollar value	94.4	63.9	66.4	51.2	47.2	46.7	47.5	35.6	25.7	-27.8
Music video										
Units shipped	9.2	6.1	7.6	11.0	11.2	12.6	16.9	18.6	27.2	46.2
Dollar value	172.3	118.1	157.4	213.3	231.1	220.3	236.1	323.9	508.0	56.8
DVD										
Units shipped	—	—	—	—	—	—	—	—	0.5	—
Dollar value	—	—	—	—	—	—	—	—	12.2	
TOTAL UNITS	865.7	801.0	895.5	955.6	1,122.7	1,112.7	1,137.2	1,063.4	1,124.3	5.7
TOTAL VALUE	7,541.1	7,834.2	9,024.0	10,046.6	12,068.0	12,320.3	12,533.8	12,236.8	13,723.5	12.1

Multi-Platinum and Platinum Awards for Recorded Music and Music Videos, 1998

Source: Recording Industry Assn. of America, Washington, DC

To achieve platinum status, an album must reach a minimum sale of 1 mil units in LPs, tapes, and CDs, with a manufacturer's dollar volume of at least $2 mil based on one-third of the suggested retail list price for each record, tape, or CD sold. To achieve multi-platinum status, an album must reach a minimum sale of at least 2 mil units in LPs, tapes, and CDs, with a manufacturer's dollar volume of at least $4 mil based on one-third of the list price.

Singles must sell 1 mil units to achieve a platinum award and 2 mil to achieve a multi-platinum award. EP singles count as 2

units. Double-CD sets count as 2 units. Music videos (long form) must sell 100,000 units to qualify for a platinum award and must sell more than 200,000 units for a multi-platinum award. Video singles, which must have a maximum running time of 15 minutes and no more than 2 songs per title, must sell 50,000 units to qualify for a platinum award and at least 100,000 units to qualify for a multi-platinum award.

Awards listed were for albums and singles released in 1998 and for music videos released at any time. No multi-platinum or platinum video singles were awarded in 1998.

Albums, Multi-Platinum
(numbers in parentheses = millions sold)
'N Sync, *'N Sync* (4)
Alabama, *For the Record–41 Number One Hits* (2)
Barenaked Ladies, *Stunt* (3)
Beastie Boys, *Hello Nasty* (3)
Brandy, *Never Say Never* (3)
Garth Brooks, *Double Live* (12)
Mariah Carey, *Ones* (2)
Celine Dion, *These Are Special Times* (3)
Dixie Chicks, *Wide Open Spaces* (3)
Eightball, *Lost* (2)
Lauryn Hill, *The Miseducation of Lauryn Hill* (3)
Natalie Imbruglia, *Left of the Middle* (2)
Jay-Z, *Hard Knock Life, Volume 2* (3)
Jewel, *Spirit* (2)
Madonna, *Ray of Light* (3)
Master P, *MP Da Last Don* (4)
Dave Matthews Band, *Before These Crowded Streets* (2)
Alanis Morissette, *Supposed Former Infatuation Junkie* (3)
R. Kelly, *R.* (3)
Snoop Doggy Dogg, *Da Game is to Be Sold, Not to be Told* (2)
Soundtrack, *Hope Floats* (2)
Soundtrack, *City of Angels* (4)
Soundtrack, *Doctor Doolittle: The Album* (2)
Soundtrack, *Armageddon* (3)
U2, *Best of 1980-1990/The B-Sides* (2)

Albums, Platinum
'N Sync, *Home for Christmas*
Aerosmith, *A Little South of Sanity*
All Saints, *All Saints*
Big Punisher, *Capital Punishment*
Brooks & Dunn, *If You See Her*
Eric Clapton, *Pilgrim*
Sheryl Crow, *The Globe Sessions*
Jermaine Dupri, *Jermaine Dupri Presents–Life in 1472*
Eve 6, *Eve 6*
Fastball, *All the Pain Money Can Buy*
Goo Goo Dolls, *Dizzy Up the Girl*
Hanson, *Three-Car Garage*
Dru Hill, *Enter the Dru*

Hole, *Celebrity Skin*
Hootie & The Blowfish, *Musical Chairs*
Whitney Houston, *My Love is Your Love*
Alan Jackson, *High Mileage*
Korn, *Follow the Leader*
Master P, *I Got the Hook Up (Soundtrack)*
Reba McEntire, *If You See Him*
Natalie Merchant, *Ophelia*
Method Man, *Tical 2000*
George Michael, *Ladies and Gentlemen*
Monica, *The Boy is Mine*
Mya, *Mya*
Offspring, *Americana*
Outkast, *Aquemini*
Pearl Jam, *Yield*
LeAnn Rimes, *Sittin' On Top of the World*
Scarface, *My Homies*
Jerry Seinfeld, *I'm Telling You For the Last Time*
Brian Setzer Orchestra, *The Dirty Boogie*
Vonda Shepard, *Songs from Ally McBeal (Soundtrack)*
Silkk the Shocker, *Charge it 2 Da Game*
Smashing Pumpkins, *Adore*
Soundtrack, *The Wedding Singer*
Soundtrack, *Player's Club*
Soundtrack, *Godzilla*
Soundtrack, *Back to Titanic*
Soundtrack, *Touched By an Angel*
Soundtrack, *South Park–Chef Aid–The South Park Album*
Soundtrack, *The Prince of Egypt*
Bruce Springsteen, *Tracks*
George Strait, *One Step at a Time*
Keith Sweat, *Still in the Game*
Various, *Jock Jams, Volume 4*
Rob Zombie, *Hellbilly Deluxe*

Singles, Multi-Platinum
(numbers in parentheses = millions sold)
Brandy and Monica, *The Boy is Mine* (2)

Singles, Platinum
98 Degrees, *Because of You*
Backstreet Boys, *Everybody (Backstreet's Back)*

Mariah Carey, *My All*
Deborah Cox, *Nobody's Supposed to be Here*
Divine, *Lately*
Faith Hill, *This Kiss*
Wyclef Jean (Fugee AllStars), *Gone Till November*
Jon B, *Are U Still Down?*
Montell Jordan, *Let's Ride*
Lord Tariq & Peter Gunz, *Deja Vu (Uptown Baby)*
Master P, *Make 'em Say Uhh!*
Monica, *The First Night*
Next, *Too Close*
Public Announcement, *Body Bumpin Yippie-Yi-Yo*
Puff Daddy (featuring Jimmy Page), *Come With Me*
R. Kelly & Celine Dion, *I'm Your Angel*
Shania Twain, *You're Still the One*
Usher, *Nice and Slow*
Usher, *My Way*

Music Videos, Multi-Platinum
(numbers in parentheses = millions sold)
Hanson, *Tulsa, Tokyo, and the Middle of Nowhere* (5)
'N Sync, *'N the Mix* (3)

Backstreet Boys, *All Access Video* (4)
Backstreet Boys, *A Night Out with the Backstreet Boys* (2)
Metallica, *Binge & Purge* (12)
Michael Flatley, *Lord of the Dance* (12)
Various, *Old Friends* (2)
Veggie Tales, *Josh and the Big Wall* (5)
Veggie Tales, *Madame Blueberry* (4)
Yanni, *Live at the Acropolis* (6)
Yanni, *Tribute* (2)

Music Videos, Platinum
Fleetwood Mac, *The Dance*
Hanson, *Road to Albertane*
Jay-Z, *Streets is Watching*
Rage Against the Machine, *Rage Against the Machine*
Various, *Homecoming Texas Style*
Various, *Moments to Remember*
Various, *All Day Singin' and Dinner on the Ground*
Various, *Sing Your Blues Away*
Various, *Joy in the Camp*
Stevie Ray Vaughan and Double Trouble, *Live from Austin, Texas*

Top-Selling Video Games, 1998

Source: The NPD TRSTS Video Game Tracking Service, The NPD Group, Inc., Port Washington, NY; ranked by units sold

Title
1. Nintendo 64 GoldenEye 007
2. Nintendo 64 Zelda: Ocarina Time
3. Sony Playstation Gran Rurismo Racing
4. Ninendo 64 Banjo Kazooie
5. Nintendo 64 Super Mario 64
6. Sony Playstation Resident Evil 2
7. Nintendo 64 WCW/NWO Revenge
8. Sony Playstation WWF Warzone

Title
9. Sony Playstation Crash Bandicot 2
10. Sony Playstation Madden 99
11. Nintendo 64 Mario Kart
12. Sony Playstation Tekken 3
13. Sony Playstation Crash Bandicoot Warp
14. Sony Playstation NFL Gameday 99
15. Sony Playstation Metla Gear Solid

U.S. Television Set Owners

Source: Nielsen Media Research; January 1, 1999

Of the 98.0 million homes (98% of U.S. households) that owned at least one TV set in 1998:

99% had color televisions	40% had 3 or more TV sets	67% received basic cable
34% had 2 TV sets	85% had a VCR	39% received premium cable

Some Television Addresses, Phone Numbers, Internet Sites

ABC–American Broadcasting Co.
77 W 66th St.
New York, NY 10023 (212) 456-7777
Website: http://www.abc.com
CBS–Columbia Broadcasting System
51 W 52nd St.
New York, NY 10019 (212) 975-4321
Website: http://www.cbs.com
NBC–National Broadcasting Co.
30 Rockefeller Plaza
New York, NY 10112 (212) 664-4444
Website: http://www.nbc.com
Fox Television
205 E 67th St.
New York, NY 10021 (212) 452-5555
Website: http://www.fox.com
PBS–Public Broadcasting Service
1320 Braddock Place
Alexandria, VA 22314 (703) 739-5000
Website: http://www.pbs.org

CABLE
A&E–Arts & Entertainment Network
235 E 45th St.
New York, NY 10017 (212) 210-1400
Website: http://www.aande.com
AMC–American Movie Classics
Rainbow Media Holdings, Inc.
111 Stewart Avenue
Bethpage, NY 11714 (516) 396-3000
Website: http://www.amctv.com

BET–Black Entertainment Television
1 BET Plaza, 1900 W Place, NE
Washington, DC 20018 (202) 608-2000
Website: http://www.msbet.com
CNBC–Consumer News and Business Channel
2200 Fletcher Ave.
Fort Lee, NJ 07024 (201) 585-2622
Website: http://www.cnbc.com
CNN–Cable News Network
One CNN Center, Box 105366
Atlanta, GA 30348-5366 (404) 827-1500
Website: http://www.cnn.com
C-SPAN–Cable-Satellite Public Affairs Network
400 N Capitol St. NW, Suite 650
Washington, DC 20001 (202) 737-3220
Website: http://www.c-span.org
DIS–The Disney Channel
3800 W Alameda Ave.
Burbank, CA 91505 (818) 569-7500
Website: http://www.disneychannel.com
ESPN–ESPN, Inc.
ESPN Plaza, 935 Middle St.
Bristol, CT 06010 (860) 585-2000
Website: http://espn.com
LIF–Lifetime
309 W 49th St.
New York, NY 10019 (212) 424-7000
Website: http://www.lifetimetv.com

MSNBC
1 MSNBC Plaza
Secaucus, NJ 07094 (201) 583-5000
Website: http://www.msnbc.com
MTV–Music Television
MTV Networks, Inc.
1515 Broadway
New York, NY 10036 (212) 258-8000
Website: http://www.mtv.com
NICK–Nickelodeon/Nick at Nite
MTV Networks, Inc.
1515 Broadway
New York, NY 10036 (212) 258-8000
Websites: http://www.nick.com
http://www.nick-at-nite.com
TBS–Turner Broadcasting System
Turner Entertainment Group
One CNN Center, Box 105366
Atlanta, GA 30348-5366
(404) 827-1700
Website: http://www.turner.com
TDC–The Discovery Channel
Discovery Communications
7700 Wisconsin Ave., Suite 700
Bethesda, MD 20814 (301) 986-0444
Website: http://www.discovery.com
USA–USA Network
USA Networks
1230 Ave. of the Americas
New York, NY 10020 (212) 408-9100
Website: http://www.usanetwork.com

Number of Cable TV Systems, 1975-99

Source: *Television and Cable Factbook*, Warren Publishing, Inc., Washington, DC; estimates as of Jan. 1

Year	Systems	Year	Systems	Year	Systems	Year	Systems
1975	3,506	1982	4,825	1988	8,500	1994	11,214
1976	3,681	1983	5,600	1989	9,050	1995	11,215
1977	3,832	1984	6,200	1990	9,575	1996	11,220
1978	3,875	1985	6,600	1991	10,704	1997	10,943
1979	4,150	1986	7,500	1992	11,073	1998	10,845
1980	4,225	1987	7,900	1993	11,108	1999	10,700
1981	4,375						

Top 20 Cable TV Networks, 1999

Source: *Cable Television Developments,* Natl. Cable Television Assn., Jan.-Apr. 1999; ranked by number of subscribers

Network[1]	Affiliates	Subscribers (mil)	Network[1]	Affiliates	Subscribers (mil)
1. TBS (1976)	11,668	77.0	12. The Weather Channel (1982)	7,000	72.0
2. The Discovery Channel (1985)	NA	76.4	13. QVC Network (1986)	7,297	70.1
3. ESPN (1979)	NA	76.2	14. The Learning Channel (1980)	NA	70.0
4. USA Network (1980)	NA	75.8	15. MTV: Music Television (1981)	9,176	69.4
5. C-SPAN (1979)	7,047	75.7	16. AMC (American Movie Classics)		
6. TNT (Turner Network Television) (1988)	10,637	75.6	(1984)	NA	69.0
7. FOX Family Channel (1998[3])	13,818	74.0	17. CNBC (1989)	5,822	68.0
8. TNN (The Nashville Network) (1983)	NA	74.0	18. Nickelodeon (1979)/Nick at Nite		
9. Life Time Television (1984)	11,000	73.4	(1985)	11,788	67.0
10. A&E Television Networks (1984)	12,000[2]	73.0	19. VH1 (Music First) (1985)	6,481	65.6
11. CNN (1980)	11,528	73.0	20. ESPN2 (1993)	NA	64.5

NA = Not available. **Note:** Data include noncable affiliates. (1) Date in parentheses is year service began. (2) U.S. and Canada. (3) Began 1977 as the Family Channel; relaunched as FOX Family Channel, 1998.

U.S. Households With Cable Television, 1977-98

Source: Nielsen Media Research

Year	Basic cable subscribers	As % of households with TVs	Year	Basic cable subscribers	As % of households with TVs
1977	12,168,450	16.6	1988	48,636,520	53.8
1978	13,391,910	17.9	1989	52,564,470	57.1
1979	14,814,380	19.4	1990	54,871,330	59.0
1980	17,671,490	22.6	1991	55,786,390	60.6
1981	23,219,200	28.3	1992	57,211,600	61.5
1982	29,340,570	35.0	1993	58,834,440	62.5
1983	34,113,790	40.5	1994	60,483,600	63.4
1984	37,290,870	43.7	1995	62,956,470	65.7
1985	39,872,520	46.2	1996	64,654,160	66.7
1986	42,237,140	48.1	1997	65,929,420	67.3
1987	44,970,880	50.5	1998	67,011,180	67.4

Average Television Viewing Time, May 1999

Source: Nielsen Media Research (hours: minutes per week)

Group	Age	Mon.-Fri. 10 AM-4:30 PM	Mon.-Fri. 4:30 PM-7:30 PM	Mon.-Sun. 8-11 PM	Sat. 7 AM-1 PM	Mon.-Fri. 11:30 PM-1 AM
Women	18+	5:26	3:47	9:01	0:42	1:32
	18-24	4:04	2:34	5:54	0:33	1:20
	25-54	4:27	3:03	8:23	0:40	1:33
	55+	7:50	5:40	11:28	0:48	1:35
Men	18+	3:19	2:47	8:11	0:35	1:26
	18-24	2:26	1:51	4:58	0:26	1:14
	25-54	2:42	2:18	7:51	0:35	1:30
	55+	5:10	4:26	10:35	0:42	1:24
Teens	12-17	1:59	2:56	5:51	0:42	0:50
Children	2-5	5:35	2:40	4:02	1:01	0:26
	6-11	1:47	2:34	4:44	1:04	0:26
TOTAL		**4:02**	**3:10**	**7:44**	**0:43**	**1:16**

TV Viewing Shares, Broadcast Years 1989-1998[1]

Source: *Cable TV Facts,* Cable Advertising Bureau, New York, NY

	All Television Households[2] '89 '90 '91 '92 '93 '94 '95 '96 '97 '98	All Cable Households[2] '89 '90 '91 '92 '93 '94 '95 '96 '97 '98	Pay Cable Households[2] '89 '90 '91 '92 '93 '94 '95 '96 '97 '98
Network			
Affiliates	58 55 53 54 53 52 48 46 43 41	49 46 46 47 46 44 41 40 38 36	45 43 41 43 42 42 38 36 35 34
Indep. TV			
Stations[3]	20 20 21 20 21 21 22 21 20 20	16 16 17 16 17 17 17 17 17 16	16 16 16 16 16 17 17 18 17 17
Public TV			
Stations	3 3 3 4 4 3 3 3 3	3 3 2 3 3 3 3 3 3	16 16 16 16 16 17 17 18 17 2
Basic Cable	17 21 24 24 25 26 30 33 36 40	28 32 35 35 36 37 42 43 46 49	2 2 2 2 2 3 2 3 2 49
Pay Cable	7 6 6 6 5 5 6 6 13 7	11 10 9 8 8 8 8 8 9	27 30 34 33 35 36 41 43 46 12

(1) Broadcast year (season) ends in May of the year shown, began the previous Sept. (2) Share figures refer to percentage of the viewing audience for all television viewing, 24 hours/day. As a result of multiset use and rounding of numbers, share figures add to more than 100. (3) Independent shares include those for Fox.

Favorite Syndicated Programs, 1998-99

Source: Nielsen Media Research, Aug. 31, 1998-May 16, 1999

Average audience percentages, or ratings, are estimates of the percentage of TV-owning households watching a program.

Rank	Program	Avg. audience (%)	Rank	Program	Avg. audience (%)
1.	Wheel of Fortune-Syn (Mon.-Fri.)	11.5	11.	Frasier-Syn (Mon.-Fri.)	5.5
2.	Jeopardy (Mon.-Fri.)	9.6	12.	Home Improvement-Syn (Mon.-Fri.)	5.4
3.	ESPN NFL Regular Season	6.9	13.	X-Files-Syn (various)	5.3
4.	Jerry Springer (Mon.-Fri.)	6.8	14.	Wheel of Fortune (various)	5.2
5.	Judge Judy (Mon.-Fri.)	6.8	15.	Nat'l Geo. on Assignment (various weeks)	4.4
6.	Friends-Syn (Mon.-Fri)	6.6	16.	ER-Syn (various weeks)	4.3
7.	Oprah Winfrey Show (Mon.-Fri.)	6.4	16.	Warner Bros Volume 30 (various weeks)	4.3
7.	Seinfeld (Mon.-Fri.)	6.4	18.	Seinfeld (various)	4.2
9.	Entertainment Tonight	5.9	19.	Hollywood Squares (Mon.-Fri.)	4.1
10.	Buena Vista I (various weeks)	5.8	20.	Star Trek: Deep Space Nine (various)	4.1

TV Parental Guidelines

On Dec. 19, 1996, representatives of the television industry announced the creation of TV Parental Guidelines, a rating system intended to give parents advance information about the content of programs. The guidelines, modeled after the Motion Picture Ratings System and developed by a broad spectrum of industry representatives, began to appear on broadcast and cable television programs in Jan. 1997. On July 10, 1997, most of the television industry, after negotiations with advocacy groups, agreed to add the labels D, L, S, and V to the existing ratings. The added labels, which went into effect by Oct. 1, provide more specific information about the degree of violence, coarse language, and sexually suggestive content. Some of the networks that did not add the labels Oct. 1 began to add their own parental advisories to shows.

There are two categories of ratings, one for children's programs and one for programs not specifically designed for children. The ratings are as follows:

The following categories apply to programs designed solely for children:

All Children. *This program is designed to be appropriate for all children.* Whether animated or live action, the themes and elements in this program are specifically designed for a very young audience, including children ages 2-6. This program is not expected to frighten younger children.

Directed to Older Children. *This program is designed for children age 7 and above.* It may be more appropriate for children who have acquired the developmental skills needed to distinguish between make-believe and reality. Themes and elements in this program may include mild fantasy or comedic violence, or may frighten children under the age of 7. Therefore, parents may wish to consider the suitability of this program for their very young children. Programs containing fantasy violence that may be more intense or more combative than other programs in this category are designated as **TV-Y7-FV.**

The following categories apply to programs designed for the entire audience:

General Audience. *Most parents would find this program suitable for all ages.* Although this rating does not signify a program designed specifically for children, most parents may let younger children watch this program unattended. It contains little or no violence, no strong language, and little or no sexual dialogue or situations.

Parental Guidance Suggested. *This program contains material that parents may find unsuitable for younger children.* Many parents may want to watch it with their younger children. The theme itself may call for parental guidance and/or the program contains one or more of the following: moderate violence (V), some sexual situations (S), infrequent coarse language (L), or some suggestive dialogue (D).

Parents Strongly Cautioned. *This program contains some material that many parents would find unsuitable for children under 14 years of age.* Parents are strongly urged to exercise greater care in monitoring this program and are cautioned against letting children under the age of 14 watch unattended. This program contains one or more of the following: intense violence (V), intense sexual situations (S), strong coarse language (L), or intensely suggestive dialogue (D).

Mature Audience Only. *This program is specifically designed to be viewed by adults and therefore may be unsuitable for children under 17.* This program contains one or more of the following: graphic violence (V), explicit sexual activity (S), or crude, indecent language (L).

When a program is broadcast, the appropriate icon should appear in the upper left corner of the picture frame for the first 15 seconds. If the program is longer than 1 hour, the icon should be repeated at the beginning of the 2d hour. Guidelines are also displayed in TV listings in a number of newspapers and magazines.

Favorite Prime-Time Television Programs, 1998-99

Source: Nielsen Media Research

Data are for regularly scheduled network programs (Sept. 1, 1998–May 26, 1999); ranked by average audience percentage. Average audience percentages, or ratings, are estimates of the percentage of all TV-owning households that are watching a particular program. Audience share percentages are estimates of the percentage of those watching TV that are tuned into a particular program. The top 51 programs are listed (there are 6 programs tied for 46th place).

Rank	Program	Average audience	Audience share	Rank	Program	Average audience	Audience share
1.	E.R.	17.8	29	26.	X-Files	9.4	14
2.	Friends	15.7	26	27.	Dharma & Greg	9.3	16
3.	Frasier	15.6	24	28.	Walker, Texas Ranger	9.2	17
4.	NFL Monday Night Football	13.9	22		Spin City	9.2	14
5.	Jesse	13.7	22	30.	NBC Sunday Night Movie	9.1	14
	Veronica's Closet	13.7	21	31.	Diagnosis Murder	9.0	14
7.	60 Minutes	13.2	22		60 Minutes II	9.0	14
8.	Touched by an Angel	13.1	20	33.	Practice, The	8.9	15
9.	CBS Sunday Movie	12.1	19		Dateline-Wednesday	8.9	15
10.	20/20-Wednesday	11.2	19	35.	Will & Grace	8.8	14
11.	Home Improvement	11.0	18		Just Shoot Me	8.7	14
12.	Everybody Loves Raymond	10.6	16		NFL Monday Blast	8.7	15
13.	NYPD Blue	10.5	18	38.	Whose Line Is It Anyway?	8.6	13
14.	Law and Order	10.1	17		Cosby	8.6	14
15.	Drew Carey Show	9.9	16		King of Queens	8.6	13
	20/20-Friday	9.9	18	41.	Hughleys, The	8.5	13
17.	JAG	9.8	16	42.	20/20-Monday	8.3	13
	NFL Monday Showcase	9.8	16	43.	Nash Bridges	8.2	15
	Providence	9.8	18		20/20-Sunday	8.2	12
	Dateline-Friday	9.8	17		Secret Lives Of Men	8.2	13
21.	Ally McBeal	9.7	14	46.	Simpsons	8.1	13
	Becker	9.7	14		Norm Show, The	8.1	13
	CBS Tuesday Movie	9.7	16		Promised Land	8.1	13
	Dateline-Monday	9.7	16		World's Most Amazing Videos	8.1	13
	Dateline-Tueday	9.7	16		Dateline-Friday	8.1	15
					48 Hours	8.1	13

All-Time Top Television Programs

Source: Nielsen Media Research, Jan. 1961-Feb. 1, 1999

Estimates exclude unsponsored or joint network telecasts or programs under 30 minutes long. Ranked by rating (percentage of TV-owning households tuned in to the program).

Rank	Program	Telecast date	Network	Rating (%)	Avg. audience (in thousands)
1.	M*A*S*H (last episode)	2/28/83	CBS	60.2	50,150
2.	Dallas (Who Shot J.R.?)	11/21/80	CBS	53.3	41,470
3.	Roots-Pt. 8	1/30/77	ABC	51.1	36,380
4.	Super Bowl XVI	1/24/82	CBS	49.1	40,020
5.	Super Bowl XVII	1/30/83	NBC	48.6	40,480
6.	XVII Winter Olympics - 2d Wed.	2/23/94	CBS	48.5	45,690
7.	Super Bowl XX	1/26/86	NBC	48.3	41,490
8.	Gone With the Wind-Pt. 1	11/7/76	NBC	47.7	33,960
9.	Gone With the Wind-Pt. 2	11/8/76	NBC	47.4	33,750
10.	Super Bowl XII	1/15/78	CBS	47.2	34,410
11.	Super Bowl XIII	1/21/79	NBC	47.1	35,090
12.	Bob Hope Christmas Show	1/15/70	NBC	46.6	27,260
13.	Super Bowl XVIII	1/22/84	CBS	46.4	38,800
	Super Bowl XIX	1/20/85	ABC	46.4	39,390
15.	Super Bowl XIV	1/20/80	CBS	46.3	35,330
16.	Super Bowl XXX	1/28/96	NBC	46.0	44,150
	ABC Theater (The Day After)	11/20/83	ABC	46.0	38,550
18.	Roots-Pt. 6	1/28/77	ABC	45.9	32,680
	The Fugitive	8/29/67	ABC	45.9	25,700
20.	Super Bowl XXI	1/25/87	CBS	45.8	40,030
21.	Roots-Pt. 5	1/27/77	ABC	45.7	32,540
22.	Super Bowl XXVIII	1/30/94	NBC	45.5	42,860
	Cheers (last episode)	5/20/93	NBC	45.5	42,360
24.	Ed Sullivan	2/9/64	CBS	45.3	23,240
25.	Super Bowl XXVII	1/31/93	NBC	45.1	41,990
26.	Bob Hope Christmas Show	1/14/71	NBC	45.0	27,050
27.	Roots-Pt. 3	1/25/77	ABC	44.8	31,900
28.	Super Bowl XXXII	1/25/98	NBC	44.5	43,630
29.	Super Bowl XI	1/9/77	NBC	44.4	31,610
	Super Bowl XV	1/25/81	NBC	44.4	34,540
31.	Super Bowl VI	1/16/72	CBS	44.2	27,450
32.	XVII Winter Olympics - 2d Fri.	2/25/94	CBS	44.1	41,540
	Roots-Pt. 2	1/24/77	ABC	44.1	31,400
34.	Beverly Hillbillies	1/8/64	CBS	44.0	22,570
35.	Roots-Pt. 4	1/26/77	ABC	43.8	31,190
	Ed Sullivan	2/16/64	CBS	43.8	22,445
37.	Super Bowl XXIII	1/22/89	NBC	43.5	39,320
38.	Academy Awards	4/7/70	ABC	43.4	25,390
39.	Super Bowl XXXI	1/26/97	FOX	43.3	42,000
40.	Thorn Birds-Pt. 3	3/29/83	ABC	43.2	35,990
41.	Thorn Birds-Pt. 4	3/30/83	ABC	43.1	35,900
42.	CBS NFC Championship	1/10/82	CBS	42.9	34,960
43.	Beverly Hillbillies	1/15/64	CBS	42.8	21,960
44.	Super Bowl VII	1/14/73	NBC	42.7	27,670
45.	Thorn Birds-Pt. 2	3/28/83	ABC	42.5	35,400

Top-Rated TV Shows of Each Season, 1950-51 to 1998-99

Source: Nielsen Media Research; regular series programs, Sept.-May season

Season	Program	Rating[1]	TV-owning households (in thousands)	Season	Program	Rating[1]	TV-owning households (in thousands)
1950-51	Texaco Star Theatre	61.6	10,320	1975-76	All in the Family	30.1	69,600
1951-52	Godfrey's Talent Scouts	53.8	15,300	1976-77	Happy Days	31.5	71,200
1952-53	I Love Lucy	67.3	20,400	1977-78	Laverne & Shirley	31.6	72,900
1953-54	I Love Lucy	58.8	26,000	1978-79	Laverne & Shirley	30.5	74,500
1954-55	I Love Lucy	49.3	30,700	1979-80	60 Minutes	28.2	76,300
1955-56	$64,000 Question	47.5	34,900	1980-81	Dallas	31.2	79,900
1956-57	I Love Lucy	43.7	38,900	1981-82	Dallas	28.4	81,500
1957-58	Gunsmoke	43.1	41,920	1982-83	60 Minutes	25.5	83,300
1958-59	Gunsmoke	39.6	43,950	1983-84	Dallas	25.7	83,800
1959-60	Gunsmoke	40.3	45,750	1984-85	Dynasty	25.0	84,900
1960-61	Gunsmoke	37.3	47,200	1985-86	Cosby Show	33.8	85,900
1961-62	Wagon Train	32.1	48,555	1986-87	Cosby Show	34.9	87,400
1962-63	Beverly Hillbillies	36.0	50,300	1987-88	Cosby Show	27.8	88,600
1963-64	Beverly Hillbillies	39.1	51,600	1988-89	Roseanne	25.5	90,400
1964-65	Bonanza	36.3	52,700	1989-90	Roseanne	23.4	92,100
1965-66	Bonanza	31.8	53,850	1990-91	Cheers	21.6	93,100
1966-67	Bonanza	29.1	55,130	1991-92	60 Minutes	21.7	92,100
1967-68	Andy Griffith	27.6	56,670	1992-93	60 Minutes	21.6	93,100
1968-69	Rowan & Martin Laugh-In	31.8	58,250	1993-94	Home Improvement	21.9	94,200
1969-70	Rowan & Martin Laugh-In	26.3	58,500	1994-95	Seinfeld	20.5	95,400
1970-71	Marcus Welby, MD	29.6	60,100	1995-96	E.R.	22.0	95,900
1971-72	All in the Family	34.0	62,100	1996-97	E.R.	21.2	97,000
1972-73	All in the Family	33.3	64,800	1997-98	Seinfeld	22.0	98,000
1973-74	All in the Family	31.2	66,200	1998-99	E.R.	17.8	99,400
1974-75	All in the Family	30.2	68,500				

(1) Rating is percent of TV-owning households tuned in to the program. Data prior to 1988-89 exclude Alaska and Hawaii.

100 Leading U.S. Advertisers, 1998

Source: Competitive Media Reporting and Publishers Information Bureau, New York, © copyright 1999
(in thousands of dollars)

Rank	Advertiser	Ad spending	Rank	Advertiser	Ad spending	Rank	Advertiser	Ad spending
1.	General Motors	$2,121,040.9	35.	JC Penney Co	$371,971.5	69.	Toyota Motor Corp Dlr Assn	$217,265.0
2.	Procter & Gamble	1,724,259.7	36.	Sprint	359,082.7	70.	S.C. Johnson & Son	217,038.6
3.	DaimlerChrysler Ag	1,410,748.7	37.	Toyota Motor Corp*	350,798.8	71.	Dillard	211,227.2
4.	Philip Morris Cos	1,264,353.2	38.	US Govt	348,095.1	72.	SBC Communications	211,194.8
5.	Ford Motor Co	1,147,589.2	39.	Anheuser-Busch Cos	340,663.8	73.	Gap	206,936.7
6.	Time Warner	829,185.0	40.	Pepsico	339,378.8	74.	Best Buy Co	206,694.0
7.	Walt Disney Co	809,878.7	41.	RJR Nabisco Holdings	325,734.2	75.	First Union	203,856.4
8.	Sears Roebuck & Co	720,543.9	42.	Seagram Co Ltd	325,127.7	76.	Mazda Motor	199,614.6
9.	Unilever	691,203.5	43.	Kellogg Co	324,288.1	77.	Nike	198,113.5
10.	Diageo Plc	659,093.6	44.	Bristol-Myers Squibb Co	323,275.6	78.	Nissan Motor Co Ltd*	196,656.3
11.	Johnson & Johnson	658,775.8	45.	American Home Pdts	321,519.4	79.	Wendys Intl	190,340.0
12.	MCI Worldcom	658,771.8	46.	Coca-Cola Co	315,793.4	80.	Ralston Purina Co.	189,305.4
13.	Toyota Motor	651,975.4	47.	K Mart	314,308.5	81.	Hershey Foods	186,218.3
14.	Ford Motor Co*	633,245.2	48.	IBM	306,465.7	82.	Glaxo Wellcome Plc	183,830.5
15.	News Corp Ltd	611,388.7	49.	American Express Co	295,966.7	83.	Gateway	183,221.3
16.	McDonalds	571,747.9	50.	Schering-Plough	295,361.2	84.	Quaker Oats Co	180,024.7
17.	Sony	567,142.0	51.	Valassis Communications	287,237.4	85.	Kimberly-Clark	175,621.9
18.	Federated Dept. Stores	563,893.9	52.	Mars	276,535.5	86.	Mitsubishi Motors	174,711.9
19.	AT&T	556,763.9	53.	Nestle SA	273,768.1	87.	MacAndrews & Forbes Holdings	172,578.5
20.	National Amusements	497,662.4	54.	DaimlerChrysler Ag Dlr Assn	265,771.8	88.	Honda Motor Co Ltd*	167,068.6
21.	Tricon Global Restaurants	496,616.0	55.	Hasbro	264,220.2	89.	Gillette Co	163,673.6
22.	May Dept Stores Co.	488,237.4	56.	Volkswagen Ag	254,441.3	90.	Morgan Stanley Dean Witter Dscvr. & Co	160,536.4
23.	Honda Motor Co Ltd.	484,888.1	57.	Mattel	245,213.3	91.	Credit Lyonnais SA	160,483.0
24.	General Motors *	473,864.3	58.	Campbell Soup Co	244,848.6	92.	Mastercard Intl	157,462.8
25.	General Motors Corp Dlr Assn	473,178.0	59.	Wal-Mart Stores	240,702.0	93.	Philips Electronics NV.	156,277.2
26.	Circuit City Stores	472,740.0	60.	SmithKline Beecham Plc	238,954.5	94.	Montgomery Ward & Co	154,994.6
27.	Ford Motor Co Dlr Assn	468,300.3	61.	Clorox Co	237,433.2	95.	Cendant	153,278.3
28.	Political Adv	467,277.9	62.	Microsoft	236,072.4	96.	Eastman Kodak Co	151,098.8
29.	Nissan Motor Co Ltd	461,267.2	63.	Bayer AG Group	235,129.8	97.	Sara Lee	150,148.2
30.	General Mills	430,509.4	64.	Pfizer	234,258.6	98.	Merck & Co	147,557.4
31.	Loreal SA	403,149.1	65.	Visa USA	226,100.7	99.	Compaq Computer	144,711.2
32.	Warner-Lambert Co.	391,841.7	66.	Bell Atlantic	225,258.0	100.	Darden Restaurants	139,552.5
33.	DaimlerChrysler Ag*	385,946.1	67.	Novartis AG	220,794.1			
34.	Dayton Hudson	377,647.1	68.	Home Depot	217,866.1			

* local dealers

U.S. Ad Spending by Selected Categories, 1998

Source: Competitive Media Reporting and Publishers Information Bureau, New York, © copyright 1999
(in thousands of dollars, Jan.-Dec. 1998)

Category	Total	Magazines	Sunday Magazines	News-papers	Network Television	Sport Television	Syndicated Television	Cable TV Networks	Network Radio
Automotive	$14,073,45	$1,702,738	$35,034	$4,566,184	$2,401,722	$3,833,976	$172,073	$664,239	$27,722
Retail	11,572,471	525,989	124,822	5,556,887	1,131,526	2,729,666	113,856	375,801	105,847
Media/adv't.	4,121,887	616,174	16,345	874,256	870,243	676,566	148,595	322,663	81,986
Drugs/remedies.	3,916,790	825,057	90,967	66,031	1,646,645	260,373	358,013	510,493	102,745
Toiletries/cosmet.	3,833,432	1,199,578	20,923	7,051	1,500,599	251,746	321,510	434,515	57,952
Food	3,335,298	646,292	45,117	20,189	947,716	803,833	258,890	462,523	31,251
Travel/hotels	3,045,165	619,947	46,095	1,041,460	149,500	435,684	19,987	172,837	31,167
Candy, snacks/soft drinks	2,463,843	294,044	12,680	12,645	1,084,540	316,729	270,141	337,484	60,890
Computers/office equip.	2,421,515	949,383	9,344	131,012	516,128	137,141	36,682	269,802	33,064
Direct Response .	2,340,072	1,232,772	381,961	158,970	80,259	80,100	47,811	232,086	29,571
Ins./real estate...	2,073,165	223,952	15,600	692,045	255,357	422,961	18,437	126,083	47,025
Apparel, footwear	1,435,274	835,627	28,286	5,254	308,970	55,972	30,916	123,492	1,933
Sporting goods/toys.	1,363,551	280,330	1,315	15,568	404,746	146,425	108,698	378,309	298
Audio/video equip.	1,017,507	238,889	30,674	26,348	292,807	111,367	68,705	184,259	19,957
Hshld equip.	1,017,133	233,746	24,101	10,189	403,870	107,906	67,866	134,788	21,557
Beer/Wine	896,291	56,513	1,525	7,528	468,255	117,117	16,461	128,533	12,141
Soaps/cleansers .	766,281	127,665	9,101	2,377	339,092	73,566	74,001	117,575	14,908
Cigarettes, tobacco	550,737	363,683	10,631	10,849	3,194	823	1,421	2,024	0
Building materials	480,070	164,000	8,235	28,780	66,088	87,533	10,172	94,494	1,338
Mfg. materials/freight	423,037	133,701	114	11,172	161,044	33,642	2,446	39,455	318
Misc. merch.	405,439	128,795	3,773	7,862	142,324	15,067	4,917	83,163	3,888
Hhld. furn.	388,518	273,416	9,111	6,590	42,301	20,078	10,024	20,256	1,706
Pets, foods/supplies	361,798	85,905	10,340	1,083	134,082	38,442	27,576	59,992	1,529
Gasoline, lubricants	331,240	48,831	1,878	2,317	81,969	81,416	12,918	52,885	1,742
Jewelry/watches .	295,108	206,114	2,982	7,615	23,842	18,898	8,429	12,669	100
Liquor	291,790	217,256	5,519	6,194	8	1,169	0	2,691	0
Horticul. farming .	209,492	12,343	5,039	26,565	39,271	40,631	14,552	29,509	1,029
TOTAL	**79,312,048**	**13,780,249**	**1,029,447**	**16,130,928**	**16,271,972**	**15,486,766**	**2,691,648**	**6,671,978**	**824,007**

The **20TH**
CENTURY
In PICTURES

CLOCKWISE FROM UPPER LEFT:
Charles Lindbergh; Winston
Churchill; Babe Ruth;
Mother Teresa; Martin
Luther King Jr.; Elvis
Presley; Albert Einstein

TEDDY ROOSEVELT

With typical exuberance, Pres. Theodore Roosevelt campaigns for reelection in 1904. A champion of reform, conservation, and national expansion, he captured the mood of the nation at the turn of the century.

WOODROW WILSON

Pres. Woodrow Wilson on his cross-country trip in 1919 to promote the League of Nations. After the devastation of World War I, he regarded the League as vital to world peace.

PROHIBITION

Police destroy barrels of beer during the 1920s era of bootleggers, speakeasies, corruption, and crime, typified by the infamous Chicago gangster Al Capone.

Al Capone

THE GREAT DEPRESSION

Homeless men line up outside a public bathhouse during the Great Depression, when unemployment reached more than 12 million. In "fireside chats" to the nation, Pres. Franklin D. Roosevelt broadcast hope for recovery under his "New Deal."

FDR

SUBURBIA

Rows of look-alike houses in Levittown, Long Island, in 1954. Levittown's low-cost, middle-class housing, built primarily for World War II veterans and their families, was widely imitated.

THE WOMEN'S MOVEMENT

Supporters of the National Organization for Women march in New York City. In the 1960s and later, NOW and similar organizations sought equality for women, with some success.

EQUALITY
THE TIME IS NOW

DON'T BE A CLOWN TAKE WOMEN SERIOUSLY

CORBIS/BETTMANN

CORBIS/BETTMANN

KENNEDY ASSASSINATION

Pres. John Kennedy, in a Dallas motorcade shortly before his assassination, Nov. 22, 1963. Above, Jacqueline Kennedy stands by as Lyndon Johnson is sworn in as president, on a flight back to Washington.

THE 1963 MARCH ON WASHINGTON

More than 200,000 people gather in Washington, DC, in a peaceful demonstration calling for action on civil rights. They were addressed by the Rev. Martin Luther King Jr.

AP/WIDE WORLD PHOTOS

THE TURBULENT '60S

In a decade known for youthful rebellion and social activism, police disperse anti-Vietnam War protesters outside the 1968 Democratic National Convention in Chicago.

AP/WIDE WORLD PHOTOS

WATERGATE

Following his resignation from the presidency, Aug. 9, 1974, under threat of impeachment in the Watergate scandal, Richard Nixon waves the victory sign and leaves the White House by helicopter.

THE REAGAN ERA

After easily winning reelection in 1984, popular conservative Pres. Ronald Reagan is sworn in for a second term as Nancy Reagan looks on.

THE SIMPSON TRIAL

Following a sensational trial, former football star O.J. Simpson hears a jury find him not guilty of the murders of his ex-wife and a friend of hers.

THE OKLAHOMA CITY BOMBING

Rescue workers and volunteers attend a memorial service in front of the bombed-out Alfred P. Murrah Federal Building in Oklahoma City, after the search for victims had ended. The 1995 terrorist attack left 168 dead.

WORLD EVENTS

CORBIS

WORLD WAR I

Troops take up positions in trenches near Verdun, France, during World War I (1914–18). More than 600,000 soldiers were killed in the long and bloody Battle of Verdun, and over 8 million lives in all were lost in this "war to end all wars."

CORBIS

Adolf Hitler

CORBIS/HULTON-DEUTSCH COLLECTION

WORLD WAR II

Adolf Hitler declares war on Poland (left), Sept. 1, 1939; as his troops march in, Britain and France declare war on Germany. When Japan bombs Pearl Harbor, Dec. 7, 1941, the U.S. enters a war that involves 61 countries and directly kills an estimated 55 million soldiers and civilians. In addition, Nazis round up and kill some 5–6 million Jews and millions of others, in the Holocaust. On June 6, 1944, D-Day, Allied invasion troops hit the beaches of Normandy, in occupied France. In February 1945, anticipating victory, Allied leaders meet at Yalta to plan future steps. Germany surrenders in May; Japan surrenders shortly after the U.S., in August, drops atom bombs on Hiroshima and Nagasaki.

CORBIS

The Holocaust

Pearl Harbor

THE RUSSIAN REVOLUTION

Demonstrators in St. Petersburg flee from government troops firing into the crowd, in 1917 during the revolution in Russia that led to establishment of the communist Soviet Union.

CORBIS

D-Day

CORBIS/BETTMANN

AP/WIDE WORLD PHOTOS

Atom Bomb, Nagasaki

Churchill, Roosevelt, Stalin at Yalta

CORBIS/BETTMANN

CHAIRMAN MAO

Mao Zedong, chairman of the Chinese Communist Party, led the revolution that in 1949 established China as a communist state.

MAHATMA GANDHI

Mahatma Gandhi inspired a historic campaign of nonviolence and civil disobedience leading to India's independence from Britain in 1947.

THE KOREAN WAR

Casualties are evacuated near Chosin, Korea, 1950. The Korean conflict, an offshoot of the cold war, began when the communist North invaded the South in 1950, and grew into a wider war involving the U.S., China, and many other nations.

CORBIS/HULTON-DEUTSCH COLLECTION

THE CUBAN MISSILE CRISIS

The specter of nuclear war arises in 1962 after photos confirm the presence of Soviet missiles in Cuba, backing Cuban leader Fidel Castro. The crisis ends when the Soviet Union yields and removes them.

AP/WIDE WORLD PHOTOS

MISSILE TRANSPORTERS

12 PROBABLE GUIDELINE MISSILES

HEAVY EQUIPMENT

5 MISSILE DOLLIES

20 LONG CYLINDRICAL TANKS

MISSILE TRANSPORTER

COTE FIDEL CASTRO

OPEN STORAGE

THE VIETNAM WAR

In a famous photo, children flee a burning Vietnamese village after a napalm attack on suspected Communist guerrillas there in 1972. As the Vietnam War dragged on, with heavy casualties on all sides, U.S. participation became increasingly controversial at home.

AP/WIDE WORLD PHOTOS

THE HOSTAGE CRISIS

One of 63 American hostages is exhibited outside the U.S. embassy in Tehran, Iran. Militant followers of Ayatollah Khomeini held the embassy and hostages from November 1979 until January 1981.

CORBIS/BETTMANN

Ayatollah Ruhollah Khomeini

THE CAMP DAVID ACCORDS

Egyptian Pres. Anwar al-Sadat, U.S. Pres. Jimmy Carter, and Israeli Prime Min. Menachem Begin sign a 1978 peace accord. It was reached between Israel and Egypt in negotiations hosted by Carter at his Camp David retreat.

CORBIS

Mikhail Gorbachev

Fall of the Berlin Wall

END OF THE COLD WAR

The new openness of Soviet leader Mikhail Gorbachev led to the collapse of Soviet-dominated regimes in Eastern Europe and breakup of the Soviet Union, ending the cold war with the West. At right, Berliners celebrate the 1989 fall of the Berlin Wall, long a symbol of cold war divisions.

Nelson Mandela

SOUTH AFRICA

Black South Africans cheer Nelson Mandela's 1990 release after 27 years imprisonment by the white minority government. In 1994, following the dismantling of apartheid and the first free elections, the black nationalist leader became president of South Africa.

CHARLES AND DIANA

Britain's Prince Charles and his bride, Diana, kiss on the balcony of Buckingham Palace after their 1981 wedding. The world watched as their marriage unraveled and, in 1997, mourned the death of the popular princess in a Paris car crash.

THE SAN FRANCISCO EARTHQUAKE

In 1906 an earthquake of 8.3 magnitude rocks San Francisco and causes a fire that rages for three days. Most of the city is destroyed, and some 700 lives are lost.

SINKING OF THE *TITANIC*

The *New York Times* headlines one of the worst maritime disasters in history: On Apr. 14–15, 1912, the British luxury liner *Titanic* hit an iceberg in the North Atlantic and sank on its maiden voyage. More than 1,500 people perished.

CHERNOBYL

Man guards empty town after explosions at a nuclear reactor at Chernobyl, Ukraine, in 1986 spread radiation across much of Europe and led to evacuations of some areas. Thirty-one persons were killed at the time; thousands of later deaths were attributed to radiation from the accident—the worst ever at a nuclear power plant.

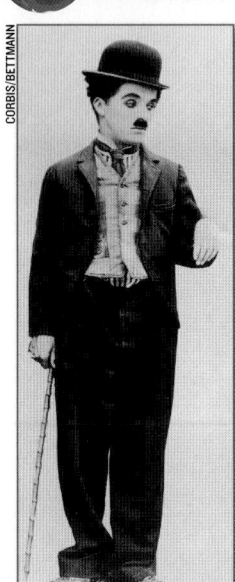

CHARLIE CHAPLIN

British actor Charlie Chaplin, a master of pantomime in silent films, won renown for his portrayal of the "little tramp" who triumphs over adversity.

CASABLANCA

While Dooley Wilson plays piano and sings "As Time Goes By," Humphrey Bogart romances Ingrid Bergman in the 1942 American film classic.

MARILYN MONROE

America's all-time pinup girl, movie actress Marilyn Monroe appears in a scene from *The Seven Year Itch* (1955) that was immortalized in thousands of posters.

PABLO PICASSO

The Spanish painter and sculptor Pablo Picasso, shown here beside one of his paintings, had an innovative, distinctly modern style; he is generally considered the greatest artist of the century.

THE BEATLES

Beatlemania hit the U.S. in the 1960s when the British rock group appeared on TV's *Ed Sullivan Show*. The innovative Beatles were the most influential band of the rock music era.

FRANK SINATRA

Master showman Frank Sinatra sings at Carnegie Hall in 1986. A heartthrob of bobby-soxers in the 1940s, he became the first pop superstar in a career spanning more than 50 years.

I LOVE LUCY

I Love Lucy, the zany hit TV sitcom starring (left to right) Desi Arnaz, Lucille Ball, Vivian Vance, and William Frawley, ran originally from 1951 to 1957; it may live on forever in reruns.

SESAME STREET

Cookie Monster and Ernie, two of Jim Henson's lovable Muppets, join Bob McGrath, an original cast member, in 1998 to celebrate the 30th anniversary of the critically acclaimed PBS children's series Sesame Street.

JESSE OWENS
One of the greatest track-and-field athletes of all time, Jesse Owens leaps to his 4th gold medal at the 1936 Olympics in Berlin, setting a world record in the running broad jump. Adolf Hitler refused to acknowledge the champion because of his race.

BABE DIDRIKSON ZAHARIAS
Often called the greatest woman athlete of the first half of the 20th century, Babe Didrikson Zaharias excelled in many sports, especially golf, in which she won every major women's championship between 1936 and 1954.

JACKIE ROBINSON
The first black to break the color barrier in major league baseball, Jackie Robinson was a daring base runner highly popular with fans. He batted .311 in his 10 seasons with the Brooklyn Dodgers (1947–56) and was elected to the Hall of Fame in 1962.

MUHAMMAD ALI

Muhammad Ali, dubbed "The Greatest," became world heavyweight boxing champ in 1964, and regained the title 2 more times; he is shown here (right) defeating Joe Frazier in a 1975 bout.

BILLIE JEAN KING

Billie Jean King, on the Wimbledon court. A leading activist for women's equality in tennis, King herself won Wimbledon 6 times and was the U.S. Open champion 4 times in the 1970s.

MICHAEL JORDAN

With his crowd-thrilling play, Michael Jordan leads the Chicago Bulls to their sixth NBA championship in 1998 (left). One of the greatest players ever, he retired in 1999 having been named MVP 6 times in the finals and 5 in the regular season.

JOE MONTANA

San Francisco 49er quarterback Joe Montana drops back for another pass. Famous for making the clutch play, and often considered the best quarterback in pro football history, he was named Super Bowl MVP a record 3 times.

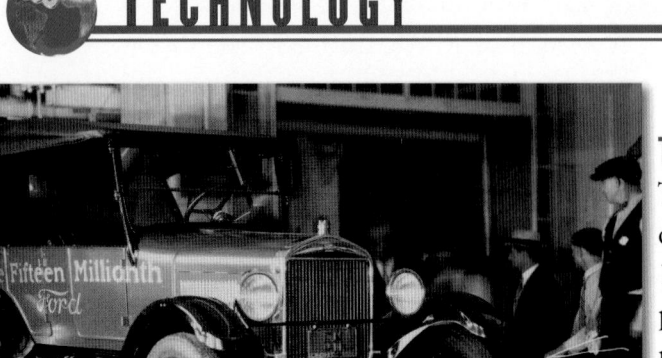

CORBIS/BETTMANN

THE MODEL T FORD

The 15-millionth Model T rolls off the Ford assembly line in 1927. Introduced in 1908, the practical, low-cost "Tin Lizzie" put the common man into the driver's seat and ushered in the age of mass automobile travel.

THOMAS EDISON

Prolific inventor Thomas Edison, shown here in his lab in 1901, held more than 1,000 patents. His practical light bulb, electric generating system, sound-recording device, and movie projector were among the inventions that shaped the 20th century.

CORBIS/BETTMANN

NASA

MAN ON THE MOON

In July 1969, astronauts Neil Armstrong and Buzz Aldrin (the latter shown at left) became the first humans to set foot on the moon.

THE NEW COMPUTER AGE

In the last decades of the century, the computer has transformed industry and society. The personal computer, with software and Internet connection, has become a fundamental tool in offices, homes, and schools. At right, Microsoft founder Bill Gates, a pioneer in PC operating systems and applications, sits beside the ubiquitous PC.

AP/WIDE WORLD PHOTOS

NATIONAL DEFENSE

Data as of Oct. 15, 1999.

Chief Commanding Officers of the U.S. Military

Chairman, Joint Chiefs of Staff
Gen. Henry Hugh Shelton

Vice Chairman
Gen. Joseph W. Ralston

The Joint Chiefs of Staff consists of the Chairman and Vice Chairman of the Joint Chiefs of Staff; the Chief of Staff, U.S. Army; the Chief of Naval Operations; the Chief of Staff, U.S. Air Force; and the Commandant of the Marine Corps.

Army

Chief of Staff	Date of Rank
Eric K. Shinseki	Aug. 5, 1997

Other Generals

	Date of Rank
Abrams, John N.	Sept. 14, 1998
Bramlett, David A.	Sept. 1, 1996
Clark, Wesley K.	June 21, 1996
Coburn, John G.	May 14, 1999
Keane, John M.	Jan. 22, 1999
Meigs, Montgomery C.	Nov. 10, 1998
Schoomaker, Peter J.	Oct. 24, 1997
Schwartz, Thomas A.	Aug. 31, 1998
Shelton, Henry H.	Mar. 1, 1996
Shinseki, Eric K.	Aug. 5, 1997
Tilelli, John H., Jr.	July 19, 1994
Wilson, Johnnie E.	May 1, 1996

Navy

Chief of Naval Operations	Date of Rank
Jay L. Johnson (aviator)	Apr. 1, 1996

Other Admirals

	Date of Rank
Abbot, Charles S. (aviator)	Sept. 1, 1998
Blair, Dennis C. (surface warfare)	May 1, 1999
Bowman, Frank L. (submariner)	Oct. 1, 1996
Clark, Vernon E. (surface warfare)	Aug. 6, 1999
Clemins, Archie R. (submariner)	Jan. 1, 1997
Ellis, Jr., James O. (aviator)	Jan. 1, 1999
Gehman, Harold W., Jr. (surface warfare)	Oct. 1, 1996
Lopez, Thomas J. (surface warfare)	July 31, 1996
Mies, Richard W. (submariner)	Aug. 1, 1998
Pilling, Donald L. (surface warfare)	Oct. 30, 1997
Prueher, Joseph W. (aviator)	June 1, 1995

Air Force

Chief of Staff	Date of Rank
Michael E. Ryan	Apr. 4, 1996

Other Generals

	Date of Rank
Babbitt, George T., Jr.	June 1, 1997
Eberhart, Ralph E.	Aug. 1, 1997
Gamble, Patrick K.	Oct. 1, 1998
Gordon, John A.	Oct. 31, 1997
Jumper, John P.	Nov. 17, 1997
Myers, Richard B.	Sept. 1, 1997
Newton, Lloyd W.	Apr. 1, 1997
Ralston, Joseph W.	July 1, 1995
Robertson, Charles T., Jr.	Sept. 1, 1998

Marine Corps

Commandant of the Marine Corps (CMC)	
Gen. James L. Jones	July 1, 1999

Other Generals

Dake, Terrence R.	Sept. 5, 1998
Wilhelm, Charles E.	Sept. 25, 1997
Zinn, Anthony C.	Aug. 8, 1997

Coast Guard

Commandant, with rank of Admiral

James M. Loy	May 29, 1998

Vice Commandant, with rank of Vice Admiral

James C. Card	May 23, 1997

Unified Defense Commands Commanders in Chief

U.S. European Command, Stuttgart-Vaihingen, Germany — Gen. Wesley K. Clark (USA) (concurrently NATO Supreme Allied Commander, Europe)

U.S. Pacific Command, Honolulu, HI — Adm. Dennis C. Blair (USN)

U.S. Atlantic Command, Norfolk, VA — Adm. Harold W. Gehman, Jr. (USN) (concurrently NATO Supreme Allied Commander, Atlantic)

U.S. Special Operations Command, MacDill AFB, Florida — Gen. Peter J. Schoomaker (USA)

U.S. Transportation Command, Scott AFB, Illinois — Gen. Tony Robertson (USAF)

U.S. Central Command, MacDill AFB, Florida — Gen. Anthony C. Zinni (USMC)

U.S. Southern Command, Miami, FL — Gen. Charles E. Wilhelm (USMC)

U.S. Space Command, Peterson AFB, Colorado — Gen. Richard B. Myers (USAF)

U.S. Strategic Command, Offutt AFB, Nebraska — Adm. Richard W. Mies (USN)

North Atlantic Treaty Organization International Commands

NATO Headquarters:
Chairman, NATO Military Committee — Adm. Guido Venturoni (Italian Navy)

Strategic Command:
Allied Command Europe (ACE) — Gen. Wesley K. Clark (USA), Supreme Allied Commander Europe

Subordinate Command:
Allied Forces South Europe (AFSOUTH) — Adm. James O. Ellis Jr. (USN), Commander-in-Chief, South
Allied Forces North Europe (AFNORTH) — Gen. Joachim Spiering (GEA), Commander-in-Chief, North

Strategic Command:
Allied Command Atlantic (ACLANT) — Adm. Harold W. Gehman, Jr. (USN), Supreme Allied Commander, Atlantic

Subordinate Commands:
Western Atlantic (WESTLANT) — Adm. Vernon E. Clark (USN), Commander-in-Chief, Western Atlantic
Southern Atlantic (SOUTHLANT) — Vice Adm. Alexandre Reis Rodrigues (PON), Commander-in-Chief, Southern Atlantic
Eastern Atlantic (EASTLANT) — Adm. Nigel Essenhigh, (Royal Navy), Commander-in-Chief, Eastern Atlantic

Principal U.S. Military Training Centers

Army

Name, PO address	ZIP	Nearest city	Name, PO address	ZIP	Nearest city
Aberdeen Proving Ground, MD	21005	Aberdeen	Fort Lee, VA.	23801	Petersburg
Carlisle Barracks, PA	17013	Carlisle	Fort McClellan, AL.	36205	Anniston
Fort Benning, GA	31905	Columbus	Fort Rucker, AL	36362	Dothan
Fort Bliss, TX	79916	El Paso	Fort Sill, OK.	73503	Lawton
Fort Bragg, NC	28307	Fayetteville	Fort Leonard Wood, MO	65473	Rolla
Fort Gordon, GA.	30905	Augusta	Joint Readiness Training Center,		
Fort Huachuca, AZ	85613	Sierra Vista	Ft. Polk, LA	71459	Leesville
Fort Jackson, SC	29207	Columbia	National Training Center, Ft. Irwin, CA	92311	Barstow, CA
Fort Knox, KY.	40121	Radcliff	The Judge Advocate General		
Fort Leavenworth, KS	66027	Leavenworth	School, VA.	22901	Charlottesville

Navy

Name, PO address	ZIP	Nearest city	Name, PO address	ZIP	Nearest city
Naval Education & Training Ctr.	32508	Pensacola, FL	Naval Post Graduate School	93943	Monterey, CA
Naval Air Training Center	78419	Corpus Christi,TX	Naval Submarine School	06349	Groton, CT
			Naval Training Ctr., Great Lakes	60088	N. Chicago, IL
Training Command Fleet	23511	Norfolk, VA	Naval War College	02841	Newport, RI
Training Command Fleet	92113	San Diego, CA	Naval Air Tech. Training Ctr.	32508	Pensacola, FL
Naval Aviation Schools Command	32508	Pensacola, FL	Fleet Antisubmarine Warfare	92147	San Diego, CA
Naval Education & Training Ctr.	02841	Newport, RI			

Marine Corps

Name, PO address	ZIP	Nearest city	Name, PO address	ZIP	Nearest city
MCB Camp Lejeune, NC	28542	Jacksonville	MCAS Cherry Point, NC	28533	Havelock
MCB Camp Pendleton, CA.	92055	Oceanside	MCAS Miramar, CA.	92145	San Diego
MCB Kaneohe Bay, HI	96863	Kailua	MCAS New River, NC	28545	Jacksonville
MCAGCC Twentynine Palms, CA.	92278	Palm Springs	MCAS Beaufort, SC	29904	Beaufort
MCCDC Quantico, VA	22134	Quantico	MCAS Yuma, AZ	85369	Yuma
MCRD Parris Island, SC.	29905	Beaufort	MCMWTC Bridgeport, CA.	93517	Bridgeport
MCRD San Diego, CA	92140	San Diego			

MCB = Marine Corps Base. MCCDC = Marine Corps Combat Development Command. MCAS = Marine Corps Air Station. MCRD = Marine Corps Recruit Depot. MCAGCC = Marine Corps Air-Ground Combat Center. MCMWTC = Marine Corps Mountain Warfare Training Center.

Air Force

Name, PO address	ZIP	Nearest city	Name, PO address	ZIP	Nearest city
Goodfellow AFB, TX.	76908	San Angelo	Maxwell AFB, AL	36112	Montgomery
Keesler AFB, MS	39534	Biloxi	Sheppard AFB, TX	76311	Wichita Falls
Lackland AFB, TX	78236	San Antonio			

All are Air Education and Training Command Bases.

Personal Salutes and Honors, U.S.

The U.S. national salute, 21 guns, is also the salute to a national flag. U.S. independence is commemorated by the salute to the Union—one gun for each state—fired at noon July 4, at all military posts provided with suitable artillery.

A 21-gun salute on arrival and departure, with 4 ruffles and flourishes, is rendered to the president of the United States, to an ex-president, and to a president-elect. The national anthem or "Hail to the Chief," as appropriate, is played for the president, and the national anthem for the others. A 21-gun salute on arrival and departure, with 4 ruffles and flourishes, also is rendered to the sovereign or chief of state of a foreign country or a member of a reigning royal family, and the national anthem of his or her country is played. The music is considered an inseparable part of the salute and immediately follows the ruffles and flourishes without pause. For the Honors March, generals receive the "General's March," admirals receive the "Admiral's March," and all others receive the 32-bar medley of "The Stars and Stripes Forever."

GRADE, TITLE, OR OFFICE	SALUTE (IN GUNS) Arriving	Leaving	Ruffles and flourishes	Music
Vice president of United States	19		4	Hail, Columbia
Speaker of the House	19		4	Honors March
U.S. or foreign ambassador	19		4	Nat. anthem of official
Premier or prime minister	19		4	Nat. anthem of official
Secretary of Defense, Army, Navy, or Air Force	19	19	4	Honors March
Other cabinet members, Senate president pro tempore, governor, or chief justice of U.S.	19		4	Honors March
Chairman, Joint Chiefs of Staff	19	19	4	
Army chief of staff, chief of naval operations, Air Force chief of staff, Marine commandant	19	19	4	Honors March
General of the Army, general of the Air Force, fleet admiral	19	19	4	
Generals, admirals	17	17	4	
Assistant secretaries of Defense, Army, Navy, or Air Force	17	17	4	Honors March
Chair of a committee of Congress	17		4	Honors March

OTHER SALUTES (on arrival only) include: 15 guns, with 3 ruffles and flourishes, for U.S. envoys or ministers and foreign envoys or ministers accredited to the U.S.; 15 guns, for a lieutenant general or vice admiral; 13 guns, with 2 ruffles and flourishes, for a major general or rear admiral (upper half) and for U.S. ministers resident and ministers resident accredited to the U.S.; 11 guns, with 1 ruffle and flourish, for a brigadier general or rear admiral (lower half) and for U.S. charges d'affaires and like officials accredited to the U.S.; 11 guns, no ruffles and flourishes, for consuls general accredited to the U.S.

Military Units, U.S. Army and Air Force

ARMY UNITS. Squad: In infantry usually 10 enlisted personnel under a staff sergeant. Platoon: In infantry 4 squads under a lieutenant. Company: Headquarters section and 3 platoons under a captain. (Company-size unit in the artillery is a battery; in the cavalry, a troop.) Battalion: Hdqts. and 4 or more companies under a lieutenant colonel. (Battalion-size unit in the cavalry is a squadron.) Brigade: Hdqts. and 3 or more battalions under a colonel. Division: Hdqts. and 3 brigades with artillery, combat support, and combat service support units under a major general. Army Corps: Two or more divisions with corps troops under a lieutenant general. Field Army: Hdqts. and 2 or more corps with field Army troops under a general.

AIR FORCE UNITS. Flight: Numerically designated flights are the lowest level unit in the Air Force. They are used primarily where there is a need for small mission elements to be incorporated into an organized unit. Squadron: A squadron is the basic unit in the Air Force. It is used to designate the mission units in operational commands. Group: The group is a flexible unit composed of 2 or more squadrons whose functions may be operational, support, or administrative in nature. Wing: An operational wing normally has 2 or more assigned mission squadrons in an area such as combat, flying training, or airlift. Numbered Air Forces: Normally an operationally oriented agency, the numbered air force is designed for the control of 2 or more wings with the same mission and/or geographical location. Major Command: A major subdivision of the Air Force that is assigned a major segment of the USAF mission.

The Federal Service Academies

U.S. Military Academy, West Point, NY. Founded 1802. Awards BS degree and Army commission for a 5-year service obligation. For admissions information, write Admissions Office, Bldg. 606, USMA, West Point, NY 10996.

U.S. Naval Academy, Annapolis, MD. Founded 1845. Awards BS degree and Navy or Marine Corps commission for a 5-year service obligation. For admissions information, write Dean of Admissions, Naval Academy, Annapolis, MD 21402.

U.S. Air Force Academy, Colorado Springs, CO. Founded 1954. Awards BS degree and Air Force commission for a 6-year service obligation. For admissions information, write Registrar, U.S. Air Force Academy, CO 80840-5025.

U.S. Coast Guard Academy, New London, CT. Founded 1876. Awards BS degree and Coast Guard commission for a 5-year service obligation. For admissions information, write Director of Admissions, Coast Guard Academy, New London, CT 06320.

U.S. Merchant Marine Academy, Kings Point, NY. Founded 1943. Awards BS degree, a license as a deck, engineer, or dual officer, and a U.S. Naval Reserve commission. Service obligations vary according to options taken by the graduate. For admissions information, write Admission Office, U.S. Merchant Marine Academy, Kings Point, NY 11024.

U.S. Army, Navy, Air Force, Marine Corps, and Coast Guard Insignia

Source: Dept. of the Army, Dept. of the Navy, Dept. of the Air Force, U.S. Dept. of Defense

Army

General of the Armies — Gen. John J. Pershing, the only person to have held this rank, in life, was authorized to prescribe his own insignia, but never wore in excess of four stars. The rank originally was established posthumously by Congress for George Washington in 1799, and he was promoted to the rank by joint resolution of Congress, approved by Pres. Gerald Ford, Oct. 19, 1976.

General of the Army — Five silver stars fastened together in a circle and the coat of arms of the United States in gold color metal with shield and crest enameled.

General	Four silver stars
Lieutenant General	Three silver stars
Major General	Two silver stars
Brigadier General	One silver star
Colonel	Silver eagle
Lieutenant Colonel	Silver oak leaf
Major	Gold oak leaf
Captain	Two silver bars
First Lieutenant	One silver bar
Second Lieutenant	One gold bar

Warrant Officers

Grade Five — Silver bar with 4 enamel silver squares
Grade Four — Silver bar with 4 enamel black squares
Grade Three — Silver bar with 3 enamel black squares
Grade Two — Silver bar with 2 enamel black squares
Grade One — Silver bar with 1 enamel black squares

Noncommissioned Officers

Sergeant Major of the Army (E-9) — Three chevrons above 3 arcs, with an American Eagle centered on the chevrons, flanked by 2 stars—one star on each side of the eagle. Also wears distinctive red and white shield collar insignia.

Command Sergeant Major (E-9) — Three chevrons above 3 arcs with a 5-pointed star with a wreath around the star between the chevrons and arcs.

Sergeant Major (E-9) — Three chevrons above 3 arcs with a 5-pointed star between the chevrons and arcs.

First Sergeant (E-8) — Three chevrons above 3 arcs with a lozenge between the chevrons and arcs.

Master Sergeant (E-8) — Three chevrons above 3 arcs.
Sergeant First Class (E-7) — Three chevrons above 2 arcs.
Staff Sergeant (E-6) — Three chevrons above 1 arc.
Sergeant (E-5) — Three chevrons.
Corporal (E-4) — Two chevrons.

Specialists

Specialist (E-4) — Eagle device only.

Other enlisted

Private First Class (E-3) — One chevron above one arc.
Private (E-2) — One chevron.
Private (E-1) — None.

Air Force

Insignia for Air Force officers are identical to those of the Army. Insignia for enlisted personnel are worn on both sleeves and consist of a star and an appropriate number of rockers. Chevrons appear above 5 rockers for the top 3 noncommissioned officer ranks, as follows (in ascending order): Master Sergeant, 1 chevron; Senior Master Sergeant, 2 chevrons; and Chief Master Sergeant, 3 chevrons. The insignia of the Chief Master Sergeant of the Air Force has 3 chevrons and a wreath around the star design.

Navy

The following are worn on the lower sleeves of the Service Dress Blue uniform. They are of gold embroidery.

Rank	Insignia
Fleet Admiral*	1 two inch with 4 one-half inch
Admiral	1 two inch with 3 one-half inch
Vice Admiral	1 two inch with 2 one-half inch
Rear Admiral (upper half)	1 two inch with 1 one-half inch
Rear Admiral (lower half)	1 two inch
Captain	4 one-half inch
Commander	3 one-half inch
Lieutenant Commander	2 one-half inch with 1 one-quarter inch between
Lieutenant	2 one-half inch
Lieutenant (j.g.)	1 one-half inch with one-quarter inch above
Ensign	1 one-half inch

Warrant Officer-W-4— $1/_2$″ stripe with 1 break
Warrant Officer W-3 — $1/_2$″ stripe with 2 breaks, 2″ apart
Warrant Officer W-2— $1/_2$″ stripe with 3 breaks, 2″ apart
Warrant Officer W-1— $1/_4$″ stripe with 3 breaks, 2″ apart

Enlisted personnel (noncommissioned petty officers)—A rating badge worn on the upper left sleeve, consisting of a spread eagle, appropriate number of chevrons, and centered specialty mark.

*The rank of Fleet Admiral is reserved for wartime use only.

Marine Corps

Marine Corps' distinctive cap and collar ornament is the Marine Corps Emblem—a combination of the American eagle, a globe, and an anchor. Marine Corps and Army officer insignia are similar. Marine Corps enlisted insignia, although basically similar to Army's, feature crossed rifles beneath the chevrons. Marine Corps enlisted rank insignia are as follows:

Sergeant Major of the Marine Corps (E-9) — Same as Sergeant Major (below) but with Marine Corps emblem in the center with a 5-pointed star on both sides of the emblem.

Sergeant Major (E-9) — Three chevrons above 4 rockers with a 5-pointed star in the center.

Master Gunnery Sergeant (E-9) — Three chevrons above 4 rockers with a bursting bomb insignia in the center.

First Sergeant (E-8) — Three chevrons above 3 rockers with a diamond in the middle.

Master Sergeant (E-8) — Three chevrons above 3 rockers with crossed rifles in the middle.

Gunnery Sergeant (E-7) — Three chevrons above 2 rockers with crossed rifles in the middle.

Staff Sergeant (E-6) —Three chevrons above 1 rocker with crossed rifles in the middle.

Sergeant (E-5) — Three chevrons above crossed rifles.
Corporal (E-4) — Two chevrons above crossed rifles.
Lance Corporal (E-3) — One chevron above crossed rifles.
Private First Class (E-2) — One chevron.
Private (E-1) — None.

Coast Guard

Coast Guard insignia follow Navy custom, with certain minor changes such as the officer cap insignia. The Coast Guard shield is worn on both sleeves of officers and on the right sleeve of all enlisted personnel.

U.S. Army Personnel on Active Duty[1]

Source: Dept. of the Army, U.S. Dept. of Defense

Date[2]	Total strength	Commissioned officers Total	Male	Female[3]	Warrant officers Male[4]	Female	Enlisted personnel Total	Male	Female
1940	267,767	17,563	16,624	939	763	—	249,441	249,441	—
1942	3,074,184	203,137	190,662	12,475	3,285	—	2,867,762	2,867,762	—
1943	6,993,102	557,657	521,435	36,222	21,919	0	6,413,526	6,358,200	55,325
1944	7,992,868	740,077	692,351	47,726	36,893	10	7,215,888	7,144,601	71,287
1945	8,266,373	835,403	772,511	62,892	56,216	44	7,374,710	7,283,930	90,780
1946	1,889,690	257,300	240,643	16,657	9,826	18	1,622,546	1,605,847	16,699
1950	591,487	67,784	63,375	4,409	4,760	22	518,921	512,370	6,551
1955	1,107,606	111,347	106,173	5,174	10,552	48	985,659	977,943	7,716
1960	871,348	91,056	86,832	4,224	10,141	39	770,112	761,833	8,279
1965	967,049	101,812	98,029	3,783	10,285	23	854,929	846,409	8,520
1970	1,319,735	143,704	138,469	5,235	23,005	13	1,153,011	1,141,537	11,476
1975	781,316	89,756	85,184	4,572	13,214	22	678,324	640,621	37,703
1980 (Sept. 30)	772,661	85,339	77,843	7,496	13,265	113	673,944	612,593	61,351
1985 (Sept. 30)	776,244	94,103	83,563	10,540	15,296	288	666,557	598,639	67,918
1990 (Mar. 31)	746,220	91,330	79,520	11,810	15,177	470	639,713	567,015	72,698
1994	553,627	74,956	64,281	10,675	12,448	535	465,688	405,664	60,024
1995	521,036	72,646	62,250	10,396	12,053	599	435,807	377,832	57,975
1996 (May 31)	493,330	68,850	58,875	9,975	11,456	660	408,511	351,669	56,842
1997 (May 31)	487,297	67,986	58,270	9,716	11,021	719	403,072	342,817	60,255
1998	491,707	67,048	56,650	10,398	10,989	661	402,000	345,149	56,851
1999	479,100	66,613	56,952	9,661	10,767	757	388,211	329,803	58,408

(1) Represents strength of the active Army, including Philippine Scouts, retired Regular Army personnel on extended active duty, and National Guard and Reserve personnel on extended active duty; excludes U.S. Military Academy cadets, contract surgeons, and National Guard and Reserve personnel not on extended active duty. (2) June 30, unless otherwise noted; data for 1940 to 1946 include personnel in the Army Air Forces and its predecessors (Air Service and Air Corps). (3) Includes women doctors, dentists, and Medical Service Corps officers for 1946 and subsequent years, women in the Army Nurse Corps for all years, and the Women's Army Corps and Women's Medical Specialists Corps (dietitians, physical therapists, and occupational specialists) for 1943 and subsequent years. (4) Act of Congress approved Apr. 27, 1926, directed the appointment as warrant officers of field clerks still in active service. Includes flight officers as follows: 1943, 5,700; 1944, 13,615; 1945, 31,117; 1946, 2,580.

U.S. Navy Personnel on Active Duty

Source: Dept. of the Navy, U.S. Dept. of Defense

Date	Officers	Nurses	Enlisted	Officer Candidates	Total
1940 (June)	13,162	442	144,824	2,569	160,997
1945 (June)	320,293	11,086	2,988,207	61,231	3,380,817
1950 (June)	42,687	1,964	331,860	5,037	381,538
1960 (June)	67,456	2,103	544,040	4,385	617,984
1970 (June)	78,488	2,273	605,899	6,000	692,660
1980 (June)[1]	63,100	—	464,100	—	527,200
1990 (Sept.)	74,429	—	530,133	—	604,562
1993 (Mar.)	66,787	—	445,409	—	512,196
1994 (Apr.)	64,430	—	418,378	—	482,808
1995 (May)	61,075	—	402,626	—	463,701
1996 (June)	60,013	—	376,595	—	436,608
1997 (June)	57,341	—	340,616	—	397,957
1998 (Sept.)	55,007	—	326,196	—	381,203
1999 (June)	55,726	—	322,372	—	378,098

(1) Starting in 1980, "Nurses" are included with "Officers," and "Officer Candidates" are included with "Enlisted."

U.S. Marine Corps Personnel on Active Duty

Source: Dept. of the Marines, U.S. Dept. of Defense
(midyear personnel figures)

Year	Officers	Enlisted	Total	Year	Officers	Enlisted	Total	Year	Officers	Enlisted	Total
1940	1,800	26,545	28,345	1990	19,958	176,694	196,652	1995	18,017	153,929	171,946
1945	37,067	437,613	474,680	1991	19,753	174,287	194,040	1996	18,146	154,141	172,287
1950	7,254	67,025	74,279	1992	19,132	165,397	184,529	1997	18,089	154,240	172,329
1960	16,203	154,418	170,621	1993	18,878	161,205	180,083	1998	17,984	154,648	172,632
1970	24,941	234,796	259,737	1994	18,430	159,949	178,379	1999	17,892	155,250	173,142
1980	18,198	170,271	188,469								

U.S. Air Force Personnel on Active Duty

Source: Air Force Dept., U.S. Dept. of Defense
(as of May 1)

Year[1]	Strength	Year[1]	Strength	Year[1]	Strength	Year[1]	Strength
1918	195,023	1943	2,197,114	1980	557,969	1994	426,327
1920	9,050	1944	2,372,292	1986	608,200	1995	400,051
1930	13,531	1945	2,282,259	1990	535,233	1996	389,400
1940	51,165	1950	411,277	1991	510,432	1997	378,681
1941	152,125	1960	814,213	1992	470,315	1998	363,479
1942	764,415	1970	791,078	1993	444,351	1999	357,929

(1) Prior to 1947, data are for U.S. Army Air Corps and Air Service of the Signal Corps.

U.S. Coast Guard Personnel on Active Duty

Source: U.S. Coast Guard, U.S. Dept. of Defense
(midyear personnel figures)

Year	Total	Officers	Cadets	Enlisted	Year	Total	Officers	Cadets	Enlisted
1970	37,689	5,512	653	31,524	1990	37,308	6,475	820	29,860
1975	36,788	5,630	1,177	29,981	1991	38,280	7,095	900	30,285
1980	39,381	6,463	877	32,041	1992	39,185	7,348	919	30,918
1981	39,760	6,519	981	32,260	1993	38,832	7,724	691	30,417
1983	39,708	6,535	811	32,362	1994	37,284	7,401	881	29,002
1984	38,705	6,790	759	31,156	1995	36,731	7,489	841	28,401
1985	38,595	6,775	733	31,087	1996	35,229	7,270	830	27,129
1986	37,284	6,577	754	29,953	1997	34,717	7,079	868	26,770
1987	38,576	6,644	859	31,073	1998	34,890	7,140	805	26,945
1988	37,723	6,530	887	30,306	1999	35,267	7,135	880	27,251
1989	37,453	6,614	867	29,972					

Chairmen of the Joint Chiefs of Staff, 1949-99

Gen. of the Army Omar N. Bradley, USA . . .	8/16/49 –8/14/53	Gen. George S. Brown, USAF.	7/1/74 – 6/20/78
Adm. Arthur W. Radford, USN	8/15/53 – 8/14/57	Gen. David C. Jones, USAF	6/21/78 – 6/18/82
Gen. Nathan F. Twining, USAF.	8/15/57 – 9/30/60	Gen. John W. Vessey Jr., USA	6/18/82 – 9/30/85
Gen. Lyman L. Lemnitzer, USA	10/1/60 – 10/30/62	Adm. William J. Crowe, Jr., USN	10/1/85 – 9/30/89
Gen. Maxwell D. Taylor, USA	10/1/62 – 7/3/64	Gen. Colin L. Powell, USA.	10/1/89 – 9/30/93
Gen. Earle G. Wheeler, USA	7/3/64 – 7/2/70	Gen. John M. Shalikashvili, USA.	10/1/93 – 9/30/97
Adm. Thomas H. Moorer, USN	7/3/70 – 6/30/74	Gen. Henry H. Shelton, USA.	10/1/97 –

Women in the U.S. Armed Forces

Source: U.S. Dept. of Defense

Women in the Army, Navy, Air Force, Marines, and Coast Guard are fully integrated with male personnel. Expansion of military women's programs began in the Department of Defense in fiscal year 1973.

Admission of women to the service academies began in the fall of 1976.

Under rules instituted in 1993, women were allowed to fly combat aircraft and serve aboard warships. Women remained restricted from service in ground combat units.

Between Apr. 1993 and July 1994, almost 260,000 positions in the armed forces were opened to women. By 1994, 80.2% of all jobs and 92% of all career fields in the military had been opened to women. As of June 30, 1999, women made up 14.1% of the armed forces.

Women Active Duty Troops in 1999

Service	% Women
Army	15.0
Navy	13.3
Marines	5.6
Air Force	18.0
Coast Guard	11.0

Women on Active Duty, All Services*: 1973-99

Year	% Women	Year	% Women
1973	2.5	1987	10.2
1975	4.6	1993	11.6
1981	8.9	1999	14.1

*Not including the Coast Guard, which is a part of the Dept. of Transportation.

For Further Information on the U.S. Armed Forces

Army — Office of the Chief of Public Affairs, Attention: Media Relations Division—MRD, Army 1500 Wash., DC 20310-1500. **Website:** http://www.army.mil

Navy — Chief of Information, 1200 Navy Pentagon, Wash., DC 20350-1200. **Website:** http://www.navy.mil

Air Force — Office of Public Affairs, 1690 Air Force, Pentagon, Wash., DC 20330-1690. **Website:** http://www.af.mil

Marine Corps — Commandant of the Marine Corps (Code PA), Headquarters, U.S. Marine Corps, Wash. DC 20380-1775. **Website:** http://www.usmc.mil

Coast Guard — Commandant (G-IP), U.S. Coast Guard, 2100 Second St. SW, Wash., DC 20593-0001. **Website:** http://www.uscg.mil

Additional information on all the U.S. Armed Forces branches, as well as many other related organizations, can be accessed through DefenseLINK, the official Internet site of the Dept. of Defense: http://www.defenselink.mil

African American Service in U.S. Wars

American Revolution. About 5,000 African Americans served in the Continental Army, mostly in integrated units, some in all-black combat units.

Civil War. Some 200,000 African Americans served in the Union Army; 38,000 were killed, and 22 won the Medal of Honor (the nation's highest award).

World War I. About 367,000 African Americans served in the armed forces, 100,000 in France.

World War II. Over 1 mil African Americans served in the armed forces; all-black fighter and bomber AAF units and infantry divisions gave distinguished service. (By 1954, armed forces were completely desegregated.)

Korean War. Approximately 3,100 African Americans lost their lives in the Korean combat.

Vietnam War. 274,937 African Americans served in the armed forces (1965-74); 5,681 were killed in combat.

Persian Gulf War. About 104,000 African Americans served in the Kuwaiti theater—20% of all U.S. troops, compared with 8.7% of all troops for World War II and 9.8% for Vietnam.

Defense Contracts

Source: U.S. Dept. of Defense
(in thousands of dollars)

The 50 companies (including their subsidiaries) or organizations receiving the largest dollar volume of prime contract awards from the U.S. Department of Defense during fiscal year 1998.

Lockheed Martin.	$12,341,236	Allied Signal.	$655,994	Worldcorp	$316,215
Boeing	10,865,899	Computer Sciences	646,655	Highmark.	314,923
Raytheon	5,661,161	Foundation Health Systems.	592,990	Booz Allen & Hamilton	311,704
General Dynamics	3,679,867	CBS	567,485	Johns Hopkins Univ.	305,842
Northrop Grumman	2,690,742	DynCorp	537,330	Philipp Holzmann	
United Technologies.	1,983,147	Standard Missile	475,088	Aktiengesells.	304,869
Textron	1,837,815	IT Group	435,920	New Energy Ventures	297,075
Litton Industries	1,644,465	Rockwell International	435,264	FDX.	288,999
Newport News Shipbuilding .	1,546,634	Triwest Healthcare Alliance .	419,973	Halliburton.	285,623
TRW.	1,347,903	Avondale Industries	398,821	Unisys Corporation	281,197
Carlyle Group.	1,328,834	Mitre.	394,233	Ocean Shipholdings	278,784
Science Applications Intl. . . .	1,223,978	Mass. Inst. of Technology . .	371,883	Government of the U.S..	265,291
General Electric	1,161,392	Texas Instruments	349,120	Electronic Data Systems . . .	260,804
Humana	867,453	Rolls-Royce PLC	345,054	Nassco Holdings	257,650
GTE	787,073	Aerospace	339,055	Sverdrup	255,834
ITT Industries	780,794	Longbow LLC.	331,067	Shell Oil.	254,625
The General Elect. Co. PLC.	732,057	Alliant Techsystems	316,569	Nichols Research	243,722

U.S. Veteran Population

Source: U.S. Dept. of Veterans Affairs; as of July 1999

(in thousands)

TOTAL VETERANS IN CIVILIAN LIFE[1]	**24,803**
Total wartime veterans[2]	**18,968**
Total Persian Gulf War	2,223
Persian Gulf War with service in Vietnam era	308
Persian Gulf War with no prior wartime service	1,915
Total Vietnam era	8,113
Vietnam era with service in Korean conflict	460
Vietnam era with no prior wartime service	7,653
Total Korean conflict	4,064
Korean conflict with service in WWII	608
Korean conflict with no prior wartime service	3,456
World War II	5,940
World War I	3
Total peacetime veterans	**5,835**
Total post-Vietnam era	3,010
Service between Korean conflict and Vietnam era only	2,695
Other peacetime	130

NOTE: Details may not add to total shown because of rounding. (1) There are an indeterminate number of Mexican Border period veterans, 14 of whom were receiving benefits in July 1999. (2) The total for "wartime veterans" consists only of veterans from each listed war that had no prior wartime service. The data refer only to veterans living in the U.S. and Puerto Rico; data on veterans living elsewhere are not available.

Veterans Compensation and Pension Case Payments

Source: 1900-1980: Dept. of Veterans Affairs; 1990-1998: Natl. Center for Veteran Analysis and Statistics

Fiscal year	Living veteran cases	Deceased veteran cases	Total cases	Total expenditures (dollars)	Fiscal year	Living veteran cases	Deceased veteran cases	Total cases	Total expenditures (dollars)
1900	752,510	241,019	993,529	$138,462,130	1970	3,127,338	1,487,176	4,614,514	$5,253,839,611
1910	602,622	318,461	921,083	159,974,056	1980	3,195,395	1,450,785	4,646,180	11,046,637,368
1920	419,627	349,916	769,543	316,418,030	1990	2,746,329	837,596	3,583,925	14,674,411,000
1930	542,610	298,223	840,833	418,432,809	1995	2,668,576	661,679	3,330,255	17,765,045,000
1940	610,122	239,176	849,298	429,138,465	1996	2,671,026	637,232	3,308,258	17,055,809,000
1950	2,368,238	658,123	3,026,361	2,009,462,298	1997	2,666,785	613,976	3,280,761	19,284,287,000
1960	3,008,935	950,802	3,959,737	3,314,761,383	1998	2,668,030	594,782	3,262,812	20,168,906,000

Active Duty U.S. Military Personnel Strengths, Worldwide

Source: U.S. Dept. of Defense

(as of Sept. 30, 1998)

U.S. Territories & Special Locations

U.S., 48 contiguous states	960,817
Alaska	16,351
Hawaii	34,643
Guam	3,935
Johnston Atoll	220
Puerto Rico	1,970
Transients	30,368
Afloat	98,650
TOTAL[1]	**1,146,959**

Europe

Belgium	1,645
Bosnia and Herzegovina	6,912
Croatia	119
Germany	69,663
Greece	441
Greenland	130
Hungary	1,379
Iceland	1,651
Italy	11,519
Macedonia, F.Y.R. of	442
Netherlands	685
Portugal	1,033
Spain	3,219
Sweden	3,219
Turkey	2,518
United Kingdom	10,156
Afloat	4,314
TOTAL[1]	**116,444**

East Asia & Pacific

Australia	322
Japan	40,364
Korea, South	36,890
Singapore	152
Thailand	124
Afloat	17,627
TOTAL[1]	**95,680**

Sub-Saharan Africa

Somalia	231
TOTAL[1]	**515**

North Africa, Middle East & South Asia

Bahrain	916
Diego Garcia	715
Egypt	1,041
Kuwait	3,921
Saudi Arabia	4,873
United Arab Emirates	313
Afloat	15,781
TOTAL[1]	**27,869**

Other Western Hemisphere

Canada	156
Cuba (Guantánamo)	1,348
Ecuador	367
Haiti	356
Honduras	594
Panama	4,694
Afloat	2,920
TOTAL[1]	**10,786**

TOTAL WORLDWIDE[2] **1,406,830**

(1) Countries and areas with fewer than 100 assigned U.S. military members not listed; regional totals include personnel stationed in those countries and areas not shown. (2) Total worldwide also includes U.S. military personnel stationed in the former Soviet Union (89), as well as undistributed personnel (8,488 total—8,216 ashore, 272 afloat).

The Medal of Honor

The Medal of Honor is the highest military award for bravery that can be given to any individual in the United States. The first Army Medals were awarded on Mar. 25, 1863, and the first Navy Medals went to sailors and Marines on Apr. 3, 1863.

On Dec. 21, 1861, Pres. Abraham Lincoln signed into law a bill to create the Navy Medal of Honor. Lincoln later (July 14, 1862) approved a resolution providing for the presentation of Medals of Honor to enlisted men of the Army and Voluntary Forces, making it a law. The law was amended on March 3, 1863 to extend its provisions to include officers as well as enlisted men.

The Medal of Honor is awarded in the name of Congress to a person who, while a member of the armed forces, distinguishes himself or herself conspicuously by gallantry and intrepidity at the risk of life above and beyond the call of duty while engaged in an action against any enemy of the United States; while engaged in military operations involving conflict with an opposing foreign force; or while serving with friendly foreign forces engaged in an armed conflict against an opposing armed force in which the United States is not a belligerent party. The deed performed must have been one of personal bravery or self-sacrifice so conspicuous as to clearly distinguish the individual above his or her comrades and must have involved risk of life. Incontestable proof of the performance of service is required, and each recommendation for award of this decoration is considered on the standard of extraordinary merit.

Prior to World War I, the 2,625 Army Medal of Honor awards up to that time were reviewed to determine which past awards met new stringent criteria. The Army removed 911 names from the list, most of them former members of a volunteer infantry group during the Civil War who had been induced to extend their enlistments when they were promised the medal. However, in 1977 a medal was restored to

Dr. Mary Walker, and in 1989 medals were restored to Buffalo Bill Cody and 7 other Indian scouts.

Since that review, Medals of Honor have been awarded in the following numbers:

World War I	124	Korean War	131
Peacetime (1920-40)	18	Vietnam War	240
World War II	441	Somalia	2

The figure for World War II includes 7 African-American soldiers who were awarded Medals of Honor (6 of them posthumously) in Jan. 1997. Previously, no black soldier had received the medal for World War II service; an Army inquiry begun in 1993 concluded that the prevailing political climate and Army practices of the time had prevented proper recognition of heroism on the part of black soldiers in that war.

Nations With Largest Armed Forces, by Active-Duty Troop Strength[1]

Source: *The Military Balance* (International Institute for Strategic Studies, published by Oxford University Press, UK)

	Troop strength Active troops (thousands)	Reserve troops (thousands)	Defense expend. ($ bil)	Tanks (MBT) (army only)	Navy Cruisers/ Frigates/ Destroyers	Sub-marines	Combat aircraft FGA (air force only)	fighters (air force only)
1. CHINA	2,840.0	1,200.0	36.6	8,500	36F/18D	61	400+	2,748 est.
2. UNITED STATES	1,447.6	1,711.7	273.0[2]	7,836	30C/44F/57D**	95	52 tactic. ftr. sqn	
3. RUSSIA	1,240.0	2,400.0	64.0	15,500	22C/18F/19D**	128	725	415
4. India*	1,145.0	528.4	12.8	3,314	18F/6D**	17	17 sqn	20 sqn
5. N. Korea	1,055.0	4,700.0	5.4	3,000	3F	26	525 total FGA/ftr.	
6. S. Korea	672.0	4,500.0	14.7	2,130	33F/7D	6	255	130
7. Turkey	639.0	378.7	8.1	4,205	16F/5D	15	11 sqn	7 sqn
8. Pakistan*	587.0	513.0	3.5	2,120#	8F/3D	9	123	242
9. Iran	518.0	350.0	4.7	1,390	3F/1D	3	150	114
10. Vietnam	492.0	3,000.0	1.0	1,315	7F	—	71	124
11. Indonesia	284.0	400.0	4.8	355#	17F	2	54	12
12. Egypt	450.0	254.0	2.7	3,700	8F/1D	8	135	338
13. Myanmar	429.0	NA	2.2	130	—	—	24	36
14. Iraq	387.5	650.0	1.3	2,700	2F	—	130	180
15. Ukraine	387.4	1,000.0	1.3	4,063	4 total	3	200	424
16. FRANCE	380.8	292.5	41.5	768#	1C/35F/4D**	14	7 sqn	5 sqn
17. Taiwan	376.0	1,657.5	13.7	719	18F/18D	4	344 total FGA/ftr.	
18. Germany	347.1	315.0	33.4	3,248	12F/3D	16	8 sqn	8 sqn
19. Italy	325.2	484.0	21.8	1,325	1C/26F/4D**	8	8 sqn	7 sqn
20. Syria	320.0	500.0	2.2	4,600	4F	3	154	310
21. Brazil	314.7	1,115.0	13.9	287#	18F/3D**	6	78	16
22. Thailand	266.0	200.0	3.2	277#	14F**	—	47	41
23. Poland	241.8	406.0	3.1	1,729	1F/1D	3	109	231
24. Japan	235.6	46.7	40.9	1,110	48F/10D	16	70	228
25. Romania	227.0	427.0	0.8	1,255	6F/1D	1	88	203
26. UNITED KINGDOM	213.8	320.8	35.7	541	23F/12D**	15	11 sqn	6 sqn
27. Spain	197.5	431.9	7.7	776	17F**	8	3 sqn	9 sqn
28. Morocco	196.3	150.0	1.4	524	1F	—	47	15
29. Israel	175.0	430.0	11.1	4,300	—	3	412 total FGA/ftr.	
30. Mexico	175.0	300.0	3.7	—	4F/3D	—	—	10

Nations with known strategic nuclear capability in all capital letters. *India and Pakistan tested nuclear devices in 1998. MBT=main battle tank. FGA=fighter, ground attack; Sqn= squadron (12-24 aircraft). # =light tanks only. **Denotes navies with aircraft carriers, as follows: U.S. 12, UK 3, France 2, India 1, Italy 1, Russian 1, Brazil 1, Spain 1, Thailand 1. (1) All figures are for 1997.

Directors of the Central Intelligence Agency

In 1942, Pres. Franklin D. Roosevelt established the Office of Strategic Services (OSS); it was disbanded in 1945. In 1946, Pres. Harry Truman established the Central Intelligence Group (CIG) to operate under the National Intelligence Authority (NIA). A 1997 law replaced the NIA with the National Security Council and the CIG with the Central Intelligence Agency.

Director	Served	Appointed by President
Adm. Sidney W. Souers	1946	Truman
Gen. Hoyt S. Vandenberg	1946 -1947	Truman
Adm. Roscoe H. Hillenkoetter	1947-1950	Truman
Gen. Walter Bedell Smith	1950-1953	Truman
Allen W. Dulles	1953 -1961	Eisenhower
John A. McCone	1961-1965	Kennedy
Adm. William F. Raborn Jr.	1965-1966	Johnson
Richard Helms	1966 -1973	Johnson
James R. Schlesinger	1973	Nixon

Director	Served	Appointed by President
William E. Colby	1973 -1976	Nixon
George Bush	1976 -1977	Ford
Adm. Stansfield Turner	1977-1981	Carter
William J. Casey	1981-1987	Reagan
William H. Webster	1987-1991	Reagan
Robert M. Gates	1991-1993	Bush
R. James Woolsey	1993 -1995	Clinton
John M. Deutch	1995 -1997	Clinton
George J. Tenet	1997-	Clinton

Nuclear Arms Treaties and Negotiations: A Historical Overview

Aug. 5, 1963—Limited Test Ban Treaty signed in Moscow by U.S., USSR, and Britain; prohibited testing of nuclear weapons in space, above ground, and under water.

Jan. 27, 1967—Outer Space Treaty banned the introduction of nuclear weapons and other weapons of mass destruction into orbit around the earth, their installation on the moon or other celestial body, or their station in space.

July 1, 1968—Nuclear Nonproliferation Treaty, with U.S., USSR, and Great Britain as major signers, limited spread of nuclear material for military purposes by agreement not to assist nonnuclear nations in getting or making nuclear weapons. Extended indefinitely, May 11, 1995.

May 26, 1972—Strategic Arms Limitation Treaty (SALT I) signed in Moscow by U.S. and USSR. This short-term agreement imposed a 5-year freeze on both testing and deployment of intercontinental ballistic missiles (ICBMs) as well as submarine-launched ballistic missiles (SLBMs). In the area of defensive nuclear weapons, the separate **ABM Treaty** limited antiballistic missiles to 2 sites of 100 antiballistic missile launchers in each country (amended in 1974 to one site in each country). ABM Treaty amended Sept. 1997 to allow flexibility in development of shorter-range nuclear weapons.

July 3, 1974—ABM Treaty Revision (protocol on antiballistic missile systems) and **Threshold Test Ban Treaty** on limiting underground testing of nuclear weapons to 150 kilotons were signed by U.S. and USSR in Moscow.

Sept. 1977—U.S. and USSR agreed to continue to abide by SALT I, despite its expiration date.

June 18, 1979—SALT II signed in Vienna by the U.S. and USSR, constrained offensive nuclear weapons, limiting each side to 2,400 missile launchers and heavy bombers; ceiling to apply until Jan. 1, 1985. Treaty also set a subceiling of 1,320 ICBMs and SLBMs with multiple warheads on each side. SALT II never reached the Senate floor for ratification because Pres. Jimmy Carter withdrew support following Dec. 1979 Soviet invasion of Afghanistan.

Dec. 8, 1987—Intermediate-Range Nuclear Forces (INF) Treaty signed in Washington, D.C., by USSR leader Mikhail Gorbachev and U.S. Pres. Ronald Reagan, eliminating all U.S. and Soviet intermediate- and shorter-range nuclear missiles from Europe and Asia. Ratified, with conditions, by U.S. Senate on May 27, 1988; by USSR on June 1, 1988. Entered into force June 1, 1988.

July 31, 1991—Strategic Arms Reduction Treaty (START I) signed in Moscow by USSR and U.S. to reduce strategic offensive arms by about 30% in 3 phases over 7 years. START I was the first treaty to mandate reductions by the superpowers. Treaty was approved by U.S. Senate Oct. 1, 1992. With the Soviet Union breakup in Dec. 1991, 4 former Soviet republics became independent nations with strategic nuclear weapons—Russia, Ukraine, Kazakhstan, and Belarus. The last 3 agreed in principle in 1992 to transfer their nuclear weapons to Russia and ratify START I. The Russian Supreme Soviet voted to ratify, Nov. 4, 1992, but Russia decided not to provide instruments of ratification until the other 3 republics ratified START I and acceded to the Nuclear Nonproliferation Treaty (NPT) as nonnuclear nations. By late 1994, all 3 nations had done so, and NPT entered into force on Dec. 5, 1994. In Dec. 1996, Belarus was the last of the 3 to give up its nuclear weapons.

Jan. 3, 1993—START II signed in Moscow by U.S. and Russia. Potentially the broadest disarmament pact in history, it called for both sides to reduce their long-range nuclear arsenals to about one-third of their then-current levels within a decade and disable and dismantle launching systems. The U.S. ratified START II on Jan. 26, 1996. On Sept. 26, 1997, the U.S. and Russia signed an agreement that would delay the dismantling of launching systems under START II to the end of 2007 (they would still be disabled by 2003). The accord was expected to facilitate Russian ratification of START II. Russia and the U.S. also agreed in writing to work toward further strategic arms cuts in a 3d round of START negotiations.

Sept. 24, 1996—Comprehensive Test Ban Treaty (CTBT) signed by U.S. and Russia. The CTBT bans all nuclear weapon tests and other nuclear explosions. It is intended to help prevent the nuclear powers from developing more advanced weapons, while limiting the ability of other states to acquire such devices. As of Oct. 1999, the CTBT had been signed by 154 nations, including China, Russia, the U.S., the U.K, and France. It had been ratified by 26, including France and the U.K., but not the U.S. or China.

Monthly Military Pay Scale[1]

Source: U.S. Dept. of Defense; effective Jan. 1, 1999

Rank/Grade	Years of Service						
	2	4	8	12	16	20	26
General—0-10 (2)	$8,114.40	$8,114.40	$8,425.80	$8,892.60	$9,528.90	$10,167.00	$10,800.00
Lt. General—0-9.................	7,129.20	7,129.20	7,466.10	7,776.90	8,425.80	8,892.60	9,528.90
Major General—0-8	6,481.20	6,481.20	7,129.20	7,466.10	7,776.90	8,425.80	8,425.80
Brig. General—0-7...............	5,583.90	5,583.90	5,583.90	5,583.90	7,129.20	7,129.20	7,129.20
Colonel—0-6	4,257.30	4,257.30	4,257.30	4,257.30	5,432.40	5,834.40	6,694.20
Lt. Colonel—0-5.................	3,639.30	3,639.30	3,639.30	4,224.30	4,845.00	5,277.50	5,277.50
Major—0-4.....................	3,181.20	3,181.20	3,608.70	4,071.90	4,444.80	4,444.80	4,444.80
Captain—0-3	2,714.10	3,210.60	3,484.80	3,855.30	3,855.30	3,855.30	3,855.30
1st Lt.—0-2....................	2,312.10	2,871.30	2,871.30	2,871.30	2,871.30	2,871.30	2,871.30
2d Lt.—0-1	1,913.10	1,913.10	1,913.10	1,913.10	1,913.10	1,913.10	2,231.70
Chief Warrant—W-4.............	2,653.80	2,714.10	2,962.80	3,303.00	3,577.80	3,792.00	4,224.30
Warrant Officer—W-1............	1,880.70	2,037.90	2,221.50	2,407.20	2,591.70	2,777.70	2,777.70
Sgt. Major—E-9 (3)	(3)	(3)	(3)	2,942.10	3,078.00	3,207.50	3,704.70
Master Sgt.—E-8	0.00	0.00	2,412.60	2,547.30	2,682.90	2,811.30	3,308.40
Sgt. 1st class—E-7..............	1,818.90	1,952.10	2,082.90	2,216.70	2,382.60	2,480.40	2,976.60
Staff Sgt.—E-6	1,579.80	1,715.40	1,844.10	2,010.00	2,140.20	2,140.20	2,140.20
Sergeant—E-5	1,384.20	1,514.70	1,680.30	1,811.10	1,811.10	2,140.20	2,140.20
Corporal—E-4	1,252.80	1,426.60	1,426.60	1,426.60	1,426.60	1,426.60	1,426.60
Pvt. 1st class—E-3..............	1,179.00	1,274.70	1,274.70	1,274.70	1,274.70	1,274.70	1,274.70
Private—E-2	0.00	0.00	0.00	0.00	0.00	0.00	0.00
Recruit—E-1...................	0.00	0.00	0.00	0.00	0.00	0.00	0.00

(1) The basic pay shown in this table is without a cap. The actual amount of pay received is limited to $9,225.00 per month. (2) While serving as Chairman or Vice Chairman of the Joint Chiefs of Staff, Chief of Staff of the Army or Air Force, Chief of Naval Operations, Commandant of the Marine Corps or Coast Guard, the amount of basic pay is $11,502.60, regardless of years of service; however, the amount received is limited to $9,225.00 per month. (3) While serving as Sergeant Major of the Army, Master Chief Petty Officer of the Navy or Coast Guard, Chief Master Sergeant of the Air Force, or Sergeant Major of the Marine Corps, basic pay is $4,346.40 per month.

Casualties in Principal Wars of the U.S.

Source: U.S. Dept. of Defense, U.S. Coast Guard

Data prior to World War I are based on incomplete records in many cases. Casualty data are confined to dead and wounded personnel and, therefore, exclude personnel captured or missing in action who were subsequently returned to military control. Dash (—) indicates information is not available. Off. = officers.

War	Branch of service	Number serving	CASUALTIES			
			Battle deaths	Other deaths	Wounds not mortal[7]	Total
Revolutionary War	Total	—	4,435	—	6,188	—
1775-83	Army	184,000	4,044	—	6,004	—
	Navy	to	342	—	114	—
	Marines	250,000	49	—	70	—
War of 1812	Total	286,730[8]	2,260	—	4,505	6,765
1812-15	Army	—	1,950	—	4,000	5,950
	Navy	—	265	—	439	704
	Marines	—	45	—	66	111
Mexican War	Total	78,789[8]	1,733	11,550	4,152	17,435
1846-48	Army	—	1,721	11,550	4,102	17,373
	Navy	—	1	—	3	4
	Marines	—	11	—	47	58
	Coast Guard[12]	71 off.	—	—	—	—
Civil War	Total	2,213,582[8]	140,415	224,097	281,881	646,392
Union forces	Army	2,128,948	138,154	221,374	280,040	639,568
1861-65	Navy	—	2,112	2,411	1,710	6,233
	Marines	84,415	148	312	131	591
Confederate forces	Total	—	74,524	59,297	—	133,821
(estimate)[1]	Army	600,000	—	—	—	—
1863-66	Navy	to	—	—	—	—
	Marines	1,500,000	—	—	—	—
	Coast Guard[12]	219 off.	1	—	—	—
Spanish-American War	Total	307,420	385	2,061	1,662	4,108
1898	Army[3]	280,564	369	2,061	1,594	4,024
	Navy	22,875	10	0	47	57
	Marines	3,321	6	0	21	27
	Coast Guard[12]	660	0	—	—	—
World War I	Total	4,743,826	53,513	63,195	204,002	320,710
April 6, 1917-	Army[4]	4,057,101	50,510	55,868	193,663	300,041
Nov. 11, 1918	Navy	599,051	431	6,856	819	8,106
	Marines	78,839	2,461	390	9,520	12,371
	Coast Guard	8,835	111	81	—	192
World War II	Total	16,353,659	292,131	115,185	671,846	1,079,162
Dec. 7, 1941-	Army[5]	11,260,000	234,874	83,400	565,861	884,135
Dec. 31, 1946[2]	Navy[6]	4,183,466	36,950	25,664	37,778	100,392
	Marines	669,100	19,733	4,778	68,207	91,718
	Coast Guard	241,093	574	1,343	—	1,917
Korean War[9]	Total	5,764,143	33,667	3,249	103,284	140,200
June 25, 1950-	Army	2,834,000	27,709	2,452	77,596	107,757
July 27, 1953	Navy	1,177,000	493	160	1,576	2,226
	Marines	424,000	4,267	339	23,744	28,353
	Air Force	1,285,000	1,198	298	368	1,864
	Coast Guard	44,143	—	—	—	—
Vietnam War[10]	Total	8,752,000	47,393	10,800	153,363	211,556
Aug. 4, 1964-	Army	4,368,000	30,929	7,272	96,802	135,003
Jan. 27, 1973	Navy	1,842,000	1,631	931	4,178	6,740
	Marines	794,000	13,085	1,753	51,392	66,230
	Air Force	1,740,000	1,741	842	931	3,514
	Coast Guard	8,000	7	2	60	69
Persian Gulf War	Total	467,939[11]	148	151	467	766
1991	Army	246,682	98	105	—	—
	Navy	98,852	6	14	—	—
	Marines	71,254	24	26	—	—
	Air Force	50,751	20	6	—	—
	Coast Guard	400	—	—	—	—

(1) Authoritative statistics for the Confederate forces are not available. An estimated 26,000-31,000 Confederate personnel died in Union prisons. (2) Data are for Dec. 1, 1941, through Dec. 31, 1946, when hostilities were officially terminated by Presidential Proclamation; few battle deaths or wounds not mortal were incurred after Japanese acceptance of Allied peace terms on Aug. 14,1945. Numbers serving Dec. 1, 1941-Aug. 31, 1945, were: Total—14,903,213; Army—10,420,000; Navy—3,883,520; Marine Corps—599,693. (3) Number serving covers the period April 21-Aug. 13, 1898, while dead and wounded data are for the period May 1-Aug. 31, 1898. Active hostilities ceased on Aug. 13, 1898, but ratifications of the treaty of peace were not exchanged between the United States and Spain until April 11, 1899. (4) Includes Army Air Forces battle deaths and wounds not mortal, as well as casualties suffered by American forces in northern Russia to Aug. 25, 1919, and in Siberia to April 1, 1920. Other deaths covered the period April 1, 1917-Dec. 31, 1918. (5) Includes Army Air Forces. (6) Battle deaths and wounds not mortal include casualties incurred in Oct. 1941 due to hostile action. (7) Marine Corps data for World War II, the Spanish-American War, and prior wars represent the number of individuals wounded, whereas all other data in this column represent the total number (incidence) of wounds. (8) As reported by the Commissioner of Pensions in his Annual Report for Fiscal Year 1903. (9) As a result of an ongoing Dept. of Defense review of available Korean War casualty record information, updates to previously reported figures for battle deaths and other deaths are reflected in this table. (10) Number serving covers the period Aug. 4, 1964-Jan. 27, 1973 (date of ceasefire). Includes casualties incurred in Mayaguez Incident. Wounds not mortal exclude 150,332 persons not requiring hospital care. (11) Estimated, because deployment figures changed continually. (12) Actually the U.S. Revenue Cutter Services, predecessor to the U.S. Coast Guard.

AEROSPACE

Memorable Moments in Human Spaceflight

Sources: National Aeronautics and Space Administration; Congressional Research Service; World Almanac research

Note: U.S. space missions are in **boldface**. Other missions were sponsored by the Soviet Union or, later, the Commonwealth of Independent States. All dates are Eastern standard time. EVA = extravehicular activity. ASTP = Apollo-Soyuz Test Project. Number of total flights by each crew member is given in parentheses when flight listed is not the first.

Dates	Mission[1]	Crew (no. of flights)	Duration (hr:min)	Remarks
4/12/61	Vostok 1	Yuri A. Gagarin	1:48	1st human orbital flight
5/5/61	Mercury-Redstone 3	**Alan B. Shepard Jr.**	**0:15**	**1st American in space**
7/21/61	Mercury-Redstone 4	**Virgil I. Grissom**	**0:15**	**Spacecraft sank, Grissom rescued**
8/6/61-8/7/61	Vostok 2	Gherman S. Titov	25:18	1st spaceflight of more than 24 hrs
2/20/62	Mercury-Atlas 6	**John H. Glenn Jr.**	**4:55**	**1st American in orbit; 3 orbits**
5/24/62	Mercury-Atlas 7	**M. Scott Carpenter**	**4:56**	**Manual retrofire error caused 250-mi landing overshoot**
8/11/62-8/15/62	Vostok 3	Andrian G. Nikolayev	94:22	Vostok 3 and 4 made 1st group flight
8/12/62-8/15/62	Vostok 4	Pavel R. Popovich	70:57	On 1st orbit it came within 3 mi of Vostok 3
10/3/62	Mercury-Atlas 8	**Walter M. Schirra Jr.**	**9:13**	**Landed 5 mi from target**
5/15/63-5/16/63	Mercury-Atlas 9	**L. Gordon Cooper**	**34:19**	**1st U.S. evaluation of effects of one day in space on a person; 22 orbits**
6/14/63-6/19/63	Vostok 5	Valery F. Bykovsky	119:06	Vostok 5 and 6 made 2d group flight
6/16/63-6/19/63	Vostok 6	Valentina V. Tereshkova	70:50	1st woman in space; passes within 3 mi of Vostok 5
10/12/64-10/13/64	Voskhod 1	Vladimir M. Komarov, Konstantin P. Feoktistov, Boris B. Yegorov	24:17	1st 3-person orbital flight; 1st without space suits
3/18/65-3/19/65	Voskhod 2	Pavel I. Belyayev, Aleksei A. Leonov	26:02	Leonov made 1st "space walk" (10 min)
3/23/65	Gemini-Titan 3	**Grissom (2), John W. Young**	**4:53**	**1st piloted spacecraft to change its orbital path**
6/3/65-6/7/65	Gemini-Titan 4	**James A. McDivitt, Edward H. White 2d**	**97:56**	**White was 1st American to "walk in space" (36 min)**
8/21/65-8/29/65	Gemini-Titan 5	**Cooper (2), Charles Conrad Jr.**	**190:55**	**Longest-duration human flight to date**
12/15/65-12/16/65	Gemini-Titan 6A	**Schirra (2), Thomas P. Stafford**	**25:51**	**Completed 1st U.S. space rendezvous, with Gemini 7**
12/4/65-12/18/65	Gemini-Titan 7	**Frank Borman, James A. Lovell**	**330:35**	**Longest-duration Gemini flight**
3/16/66	Gemini-Titan 8	**Neil A. Armstrong, David R. Scott**	**10:41**	**1st docking of one space vehicle with another; mission aborted, control malfunction; 1st Pacific landing**
6/3/66-6/6/66	Gemini-Titan 9A	**Stafford (2), Eugene A. Cernan**	**72:21**	**Performed rendezvous maneuvers, including simulation of lunar module rendezvous**
7/18/66-7/21/66	Gemini-Titan 10	**Young (2), Michael Collins**	**70:47**	**1st use of Agena target vehicle's propulsion systems; 1st orbital docking**
9/12/66-9/15/66	Gemini-Titan 11	**Conrad (2), Richard F. Gordon Jr.**	**71:17**	**1st tethered flight; highest Earth-orbit altitude (850 mi)**
11/11/66-11/15/66	Gemini-Titan 12	**Lovell (2), Edwin W. "Buzz" Aldrin Jr.**	**94:34**	**Final Gemini mission; 5 1/2 hr EVA**
4/23/67-4/24/67	Soyuz 1	Komarov (2)	26:40	Crashed on reentry, killing Komarov
10/11/68-10/22/68	Apollo-Saturn 7	**Schirra (3), Donn F. Eisele, R. Walter Cunningham**	**260:09**	**1st piloted flight of Apollo spacecraft command-service module only; live TV footage of crew**
12/21/68-12/27/68	Apollo-Saturn 8	**Borman (2), Lovell (3), William A. Anders**	**147:00**	**1st lunar orbit and piloted lunar return reentry (command-service module only); views of lunar surface televised to Earth**
1/14/69-1/17/69	Soyuz 4	Vladimir A. Shatalov	71:21	Docked with Soyuz 5
1/15/69-1/18/69	Soyuz 5	Boris V. Volyanov, Aleksei S. Yeliseyev, Yevgeny V. Khrunov	72:54	Docked with 4; Yeliseyev and Khrunov transferred to Soyuz 4 via a spacewalk
3/3/69-3/13/69	Apollo-Saturn 9	**McDivitt (2), D. Scott (2), Russell L. Schweickart**	**241:00**	**1st piloted flight of lunar module**
5/18/69-5/26/69	Apollo-Saturn 10	**Stafford (3), Young (3), Cernan (2)**	**192:03**	**1st lunar module orbit of Moon, 50,000 ft from Moon surface**
7/16/69-7/24/69	Apollo-Saturn 11	**Armstrong (2), Collins (2), Aldrin (2)**	**195:18**	**1st lunar landing made by Armstrong and Aldrin (720); collected 48.5 lb of soil, rock samples; lunar stay time 21:36:21**
10/11/69-10/16/69	Soyuz 6	Georgi S. Shonin, Valery N. Kubasov	118:43	1st welding of metals in space
10/12/69-10/17/69	Soyuz 7	Anatoly V. Flipchenko, Vladislav N. Volkov, Viktor V. Gorbatko	118:40	Space lab construction test made; Soyuz 6, 7, and 8: 1st time 3 spacecraft, 7 crew members orbited the Earth at once
10/13/69[2]	Soyuz 8	Shatalov (2), Yeliseyev (2)	118:51	Part of space lab construction team
11/14/69-11/24/69	Apollo-Saturn 12	**Conrad (3), Richard F. Gordon Jr. (2), Alan L. Bean**	**244:36**	**Conrad and Bean made 2d Moon landing (11/18); collected 74.7 lb of samples, lunar stay time 31:31**
4/11/70-4/17/70	Apollo-Saturn 13	**Lovell (4), Fred W. Haise Jr., John L. Swigart Jr.**	**142:54**	**Aborted after service module oxygen tank ruptured; crew returned safely using lunar module**

Dates	Mission[1]	Crew (no. of flights)	Duration (hr:min)	Remarks
6/1/70-6/19/70	Soyuz 9	Nikolayev (2), Vitaliy I. Sevastyanov	424:59	Longest human spaceflight to date
1/31/71-2/9/71	**Apollo-Saturn 14**	**A. Shepard (2), Stuart A. Roosa, Edgar D. Mitchell**	**216:01**	**Shepard and Mitchell made 3d Moon landing (2/3); collected 96 lb of lunar samples; lunar stay 33:31**
4/19/71[2]	Salyut 1[3]	(Occupied by Soyuz 11 crew)		1st space station
4/22/71[2]	Soyuz 10	Shatalov (3), Yeliseyev (3), Nikolay N. Rukavishnikov	47:46	1st successful docking with a space station; failed to enter space station
6/6/71-6/30/71	Soyuz 11	Georgi T. Dobrovolskiy, V. Volkov (2), Viktor I. Patsayev	570:22	Docked and entered Salyut 1 space station; orbited in Salyut 1 for 23 days, crew died during reentry from loss of pressurization
7/26/71-8/7/71	**Apollo-Saturn 15**	**D. Scott (3), James B. Irwin, Alfred M. Worden**	**295:12**	**Scott and Irwin made 4th Moon landing (7/30); 1st lunar rover use; 1st deep space walk; 170 lb of samples; 66:55 stay**
4/16/72-4/27/72	**Apollo-Saturn 16**	**Young (4), Charles M. Duke Jr., Thomas K. Mattingly 2d**	**265:51**	**Young and Duke made 5th Moon landing (4/20); collected 213 lb of lunar samples; lunar stay 71:2**
12/7/72-12/19/72	**Apollo-Saturn 17**	**Cernan (3), Ronald E. Evans, Harrison H. Schmitt**	**301:51**	**Cernan and Schmitt made 6th lunar landing (12/11); collected 243 lb of samples; record lunar stay of more than 75 hr**
5/14/73[2]	**Skylab 1[4]**	**(Occupied by Skylab 2, 3, and 4 crews)**		**1st U.S. space station**
5/25/73-6/22/73	**Skylab 2**	**Conrad (4), Joseph P. Kerwin, Paul J. Weitz**	**672:49**	**1st Amer. piloted orbiting space station; crew repaired damage caused during boost**
7/28/73-9/25/73	**Skylab 3**	**Bean (2), Owen K. Garriott, Jack R. Lousma**	**1,427:09**	**Crew systems and operational tests; exceeded pre-mission plans for scientific activities; 13 hr EVA 13:44**
11/16/73-2/8/74	**Skylab 4**	**Gerald P. Carr, Edward G. Gibson, William Pogue**	**2,017:15**	**Final Skylab mission**
7/15/75-7/21/75	Soyuz 19 (ASTP)	Leonov (2), Kubasov (2)	143:31	U.S.-USSR joint flight; crews linked up in space (7/17), conducted experiments, shared meals, and held a joint news conference
7/15/75-7/24/75	**Apollo (ASTP)**	**Vance Brand, Stafford (4), Donald K. Slayton**	**217:28**	**Joint flight with Soyuz 19**
12/10/77[2]	Soyuz 26	Yuri V. Romanenko, Georgiy M. Grechko (2)	2,314:00	1st multiple docking to a space station (Soyuz 26 and 27 docked at Salyut 6)
1/10/78[2]	Soyuz 27	Vladimir A. Dzhanibekov	142:59	*See Soyuz 26*
3/2/78[2]	Soyuz 28	Aleksei A. Gubarev (2), Vladimir Remek	190:16	1st international crew launch; Remek was 1st Czech in space
4/12/81-4/14/81	**Columbia (STS-1)**	**Young (5), Robert L. Crippen**	**54:21**	**1st space shuttle to fly into Earth's orbit**
11/12/81-11/14/81	**Columbia (STS-2)**	**Joe H. Engle, Richard H. Truly**	**54:13**	**1st scientific payload; 1st reuse of space shuttle**
11/11/82-11/16/82	**Columbia (STS-5)**	**Brand (2), Robert Overmyer, William Lenoir, Joseph Allen**	**122:14**	**1st 4-person crew**
6/18/83-6/24/83	**Challenger (STS-7)**	**Crippen (2), Frederick Hauck, Sally K. Ride, John M. Fabian, Norman Thagard**	**146:24**	**Ride was 1st U.S. woman in space; 1st 5-person crew**
6/27/83[2]	Soyuz T-9	Vladimir A. Lyakhov (2), Aleksandr Pavlovich Aleksandrov	3,585:46	Docked at Salyut 7; 1st construction in space
8/30/83-9/5/83	**Challenger (STS-8)**	**Truly (2), Daniel Brandenstein, William Thornton, Guion Bluford, Dale Gardner**	**145:09**	**Bluford was 1st U.S. black in space**
11/28/83-12/8/83	**Columbia (STS-9)**	**Young (6), Brewster Shaw Jr., Robert Parker, Garriott (2), Byron Lichtenberg, Ulf Merbold**	**247:47**	**1st 6-person crew; 1st Spacelab mission**
2/3/84-2/11/84	**Challenger (41-B)**	**Brand (3), Robert Gibson, Ronald McNair, Bruce McCandless, Robert Stewart**	**191:16**	**1st untethered EVA**
2/8/84-4/11/84	Soyuz T-10B	Leonid Kizim, Vladimir Solovyov, Oleg Atkov	1,510:43	Docked with Salyut 7; crew set space duration record of 237 days
4/3/84-10/2/84	Soyuz T-11	Yury Malyshev (2), Gennady Strekalov (3), Rakesh Sharma	4,365:48	Docked with Salyut 7; Sharma 1st Indian in space
4/6/84-4/13/84	**Challenger (41-C)**	**Crippen (3), Francis R. Scobee, George D. Nelson, Terry J. Hart, James D. van Hoften**	**167:40**	**1st in-orbit satellite repair**
7/17/84[2]	Soyuz T-12	Dzhanibekov (4), Svetlana Y. Savitskaya (2), Igor P. Volk	283:14	Docked at Salyut 7; Savitskaya was 1st woman to perform EVA
8/30/84-9/5/84	**Discovery (41-D)**	**Henry W. Hartsfield (2), Michael L. Coats, Richard M. Mullane, Steven A. Hawley, Judith A. Resnik, Charles D. Walker**	**144:56**	**1st flight of U.S. nonastronaut (Walker)**
10/5/84-10/13/84	**Challenger (41-G)**	**Crippen (4), Jon A. McBride, Kathryn D. Sullivan, Ride (2), Marc Garneau, David C. Leestma, Paul D. Scully-Power**	**197:24**	**1st 7-person crew**
11/8/84-11/16/84	**Discovery (51-A)**	**Hauck (2); David M. Walker, Dr. Anna L. Fisher, J. Allen (2), D. Gardner (2)**	**191:45**	**1st satellite retrieval/repair**
4/12/85-4/19/85	**Discovery (51-D)**	**Karol J. Bobko, Donald E. Williams, Jake Garn, Walker (2), Jeffrey A. Hoffman, S. David Griggs, M. Rhea Seddon**	**167:55**	**Garn was 1st U.S. senator in space**

Dates	Mission[1]	Crew (no. of flights)	Duration (hr:min)	Remarks
6/17/85- 6/24/85	Discovery (51-G)	Brandenstein (2), John O. Creighton, Shannon W. Lucid, Steven R. Nagel, Fabian (2), Prince Sultan Salman al-Saud, Patrick Baudry	169:39	Launched 3 satellites; Salman al-Saud was 1st Arab in space; Baudry was 1st French person on U.S. mission
10/3/85- 10/7/85	Atlantis (51-J)	Bobko (3), Ronald J. Grabe, David C. Hilmers, Stewart (2), William A. Pailes	97:47	1st Atlantis flight
10/30/85- 11/6/85	Challenger (61-A)	Hartsfield (3), Nagel (2), Buchli (2), Bluford (2), Bonnie J. Dunbar, Wubbo J. Ockels, Richard Furrer, Ernst Messerschmid	168:45	1st 8-person crew; 1st German Spacelab mission
11/26/85- 12/3/85	Atlantis (61-B)	Shaw (2), Bryan D. O'Connor, Sherwood C. Spring, Mary L. Cleave, Jerry L. Ross, C. Walker (3), Rodolfo Neri	165:05	Space structures assembly test; Neri was 1st Mexican in space
1/12/86- 1/18/86	Columbia (61-C)	R. Gibson (2), Charles F. Bolden Jr., Hawley (2), G. Nelson (2), Franklin R. Chang-Diaz, Robert J. Cenker, Bill Nelson	146:04	B. Nelson was 1st U.S. Representative in space; material and astronomy experiments conducted
1/28/86	Challenger (51-L)	Scobee (2), Michael J. Smith, Resnik (2), Ellison S. Onizuka (2), Ronald E. McNair, Gregory B. Jarvis, Christa McAuliffe	—	Exploded 73 sec after liftoff; all were killed
2/20/86[2]	Mir[3]	—	—	Space station with 6 docking ports launched
3/13/86[2]	Soyuz T-15	Kizim (3), Solovyov (2)	3,000:01	Ferry between stations; docked at Mir
2/5/87- 12/29/87	Soyuz TM-2	Romanenko (3), Aleksandr I. Laveikin	7,835:38	Romanenko set endurance record, since broken
7/22/87- 12/29/87	Soyuz TM-3	Aleksandr Viktorenko, Aleksandr Pavlovich Aleksandrov (2), Mohammed Faris	3,847:16	Docked with Mir; Faris 1st Syrian in space
12/21/87- 12/21/88	Soyuz TM-4	V. Titov (2), Muso Manarov, Anatoly Levchenko	8,782:39	Docked with Mir
6/7/88- 6/17/88	Soyuz TM-5	Viktor Savinykh (3), Anatoly Solovyev, Aleksandr Panayotov Aleksandrov	236:13	Docked with Mir; Aleksandrov 1st Bulgarian in space
9/29/88- 10/3/88	Discovery (STS-26)	Hauck (3), Richard O. Covey (2), Hilmers (2), G. Nelson (2), John M. Lounge (2)	97:00	Redesigned shuttle makes 1st flight
5/4/89- 5/8/89	Atlantis (STS-30)	D. Walker (2), Grabe (2), Thagard (2), Cleave (2), Mark C. Lee	96:56	Launched Venus orbiter Magellan
10/18/89- 10/23/89	Atlantis (STS-34)	Williams (2), Michael J. McCulley, Lucid (2), Chang-Diaz (2), Ellen S. Baker	119:39	Launched Jupiter probe and orbiter Galileo
4/24/90- 4/29/90	Discovery (STS-31)	McCandless (2), Sullivan (2), Loren J. Shriver (2), Bolden (2), Hawley (3)	121:16	Launched Hubble Space Telescope
10/6/90- 10/10/90	Discovery (STS-41)	Richard N. Richards (2), Robert D. Cabana, Bruce E. Melnick, William M. Shepherd (2), Thomas D. Akers	98:10	Launched Ulysses spacecraft to investigate interstellar space and the Sun
4/5/91- 4/11/91	Atlantis (STS-37)	Nagel (3), Kenneth D. Cameron, Linda Godwin, Ross (3), Jay Apt	144:32	Launched Gamma Ray Observatory to measure celestial gamma rays
5/18/91- 10/10/91	Soyuz TM-12	Anatoly Artsebarskiy, Sergei Krikalev (2) (to Mir), Helen Sharman	3,471:22	Docked with Mir; Sharman 1st from United Kingdom in space
3/17/92- 3/25/92	Soyuz TM-14	Viktorenko (3) (to Mir), Alexandr Kaleri (to Mir), Klaus-Dietrich Flade, Aleksandr Volkov (3) (from Mir), Krikalev (2) (from Mir)	3,495:11	First human CIS space mission; docked with Mir 3/19; Viktorenko and Kaleri to Mir; Volkov and Krikalev from Mir; Krikalev was in space 313 days
5/7/92- 5/16/92	Endeavour (STS-49)	Brandenstein (4), Kevin C. Chilton, Melnick (2), Pierre J. Thuot (2), Richard J. Hieb (2), Kathryn Thornton (2), Akers (2)	213:30	1st 3-person EVA; satellite recovery and redeployment
9/12/92- 9/21/92	Endeavour (STS-47)	R. Gibson (4), Curtis L. Brown Jr., Lee (2), Apt (2), N. Jan Davis, Mae Carol Jemison, Mamoru Mohri	190:30	Jemison was 1st black woman in space; Lee and Davis were 1st married couple to travel together in space; 1st Japanese Spacelab
10/22/92- 11/1/92	Columbia (STS-52)	James D. Wetherbee (2), Michael A. Baker (2), Shepherd (3), Tamara E. Jernigan (2), Charles L. Veach (2), Steven G. MacLean	236:57	Studied influence of gravity on basic fluid and solidification processes
4/8/93- 4/17/93	Discovery (STS-56)	Cameron (2), Stephen S. Oswald (2), C. Michael Foale (2), Ellen Ochoa, Kenneth D. Cockrell	222:08	2d atmospheric mission; Ochoa was 1st Hispanic woman in space
6/21/93- 7/1/93	Endeavour (STS-57)	Grabe (4), Brian J. Duffy (2), G. David Low (3), Nancy J. Sherlock, Peter J. K. Wisoff (2), Janice E. Voss	239:46	Carried Spacelab commercial payload module
10/18/93- 11/1/93	Columbia (STS-58)	John E. Blaha (4), Richard A. Searfoss, Lucid (4), David A. Wolf, William A. McArthur, Martin J. Fettman	336:29	Studied effects of microgravity
12/2/93- 12/13/93	Endeavour (STS-61)	Covey (3), Kenneth D. Bowersox (2), Claude Nicollier (2), Story Musgrave (5), Akers (3), K. Thornton (3), Hoffman (4)	259:58	Hubble Space Telescope repaired; Akers set new U.S. EVA duration record (29 hr, 40 min)

Dates	Mission[1]	Crew (no. of flights)	Duration (hr:min)	Remarks
2/3/94- 2/11/94	Discovery (STS-60)	Bolden (3), Kenneth S. Reightier Jr. (2), Davis (2), Chang-Diaz (3), Ronald M. Sega, Krikalev (3)	199:10	Krikalev was 1st Russian on U.S. shuttle
4/9/94- 4/20/94	Endeavour (STS-59)	Sidney M. Gutierrez (2), Chilton (2), Apt (3), Michael R. Clifford (2), Godwin (2), Thomas D. Jones	269:50	Gathered data about Earth and the effects humans have on its carbon, water, and energy cycles
7/1/94- 11/4/94	Soyuz TM-19	Yuri I. Malenchenko, Talgat A. Musabayev, Merbold (2) (from *Mir*)	3,022:53	Docked with *Mir*; Merbold from *Mir*
9/9/94- 9/20/94	Discovery (STS-64)	Richards (4), L. Blaine Hammond Jr. (2), Jerry M. Linenger, Susan J. Helms (2), Carl J. Meade (3), Lee (3)	262:50	Performed atmospheric research; 1st untethered EVA in over 10 years
2/3/95- 2/11/95	Discovery (STS-63)	Wetherbee (3), Eileen M. Collins, Bernard A. Harris (2), Foale (3), Voss (2), V. Titov (4)	198:29	*Discovery* and Russian space station rendezvous
3/2/95- 3/18/95	Endeavour (STS-67)	Oswald (3), William G. Gregory, Samuel T. Durrance (2), Ronald Parise (2), Wendy B. Lawrence, Jernigan (3), John M. Grunsfeld	399:09	Shuttle data made available on the Internet; astronomy research conducted
3/14/95- 3/22/95	Soyuz TM-21	Thagard (2), Vladimir Dezhurov, Strekalov (5)	2,688[5]	Docked with *Mir* 3/16/95; Thagard was 1st Amer. on the Russ. spacecraft; Valery Polyakov returned to Earth, 3/22/95, after record stay in space (439 days)
6/27/95- 7/7/95	Atlantis (STS-71)	R. Gibson (5), Charles J. Precourt (2), E. Baker (3), Gregory J. Harbaugh (3), Dunbar (4), Solovyev (4) (to *Mir*), Nikolai M. Budarin (to *Mir*), Thagard (5) (from *Mir*), Strekalov (from *Mir*), Dezhurov (from *Mir*)	269:47	1st *Mir* docking; exchanged crew members with *Mir*; Thagard, with his stay on *Mir*, had spent 115 days in space
10/20/95- 11/5/95	Columbia (STS-73)	Bowersox (3), Kent Rominger, K. Thornton (4), Catherine Coleman, Michael Lopez-Alegria, Fred Leslie, Albert Sacco	381:52	Most ever first-time space flyers; near-weightlessness experiments conducted in microgravity laboratory
11/8/95- 11/20/95	Atlantis (STS-74)	Cameron (3), James D. Halsell Jr. (2), Chris Hadfield, Ross (5), McArthur (2)	196:30	2d *Mir* docking (11/15-11/18); erected a 15-ft permanent docking tunnel to *Mir* for future use by U.S. orbiters
1/11/96- 1/20/96	Endeavour (STS-72)	Duffy (3), Brent W. Jett Jr., Winston E. Scott, Leroy Chiao (2), Daniel T. Barry, Koichi Wakata	214:01	Released NASA space probe; retrieved Japanese satellite; 13 hr EVA
2/22/96- 3/9/96	Columbia (STS-75)	Andrew M. Allen (3), Scott J. Horowitz, Chang-Diaz (5), Umberto Guidoni, Hoffman (5), Maurizio Cheli, Nicollier (3)	377:40	Lost an Italian satellite when its tether was severed; microgravity experiments performed; singe marks found on 2 O-rings
3/22/96- 3/31/96	Atlantis (STS-76)	Chilton (3), Searfoss (2), Sega (2), Clifford (3) Godwin (3), Lucid (5)(to *Mir*)	221:15	3d *Mir* docking (5 days); Lucid to *Mir*; 2-person EVA
6/20/96- 7/7/96	Columbia (STS-78)	Terence T. Henricks (4), Kevin R. Kregel (2), Helms (3), Richard M. Linnehan, Charles E. Brady, Jean-Jacques Favier, Robert Brent Thirsk	405:48	Studied weightlessness with the Life/Microgravity Spacelab on board
9/16/96- 9/26/96	Atlantis (STS-79)	Apt (4), Terry Wilcutt (2), William Readdy (3), Akers (4), Carl E. Walz (3), Lucid (5) (from *Mir*), Blaha (5) (to *Mir*)	243:19	Docked with *Mir* 9/18/96; exchanged crew members, including Lucid, who set U.S. and women's individual duration in space record (188 days)
11/19/96- 12/7/96	Columbia (STS-80)	Cockrell (3), Rominger (2), Jernigan (4), Jones (3), Musgrave (6)	423:53	Longest-duration shuttle flight; Musgrave was oldest person to fly in space; 2 science satellites deployed and retrieved
1/12/97- 1/22/97	Atlantis (STS-81)	M. Baker (4), Jett (2), Wisoff (3), Grunsfeld (2), Marsha Ivins (4), Linenger (2) (to *Mir*), Blaha (5) (from *Mir*)	243:30	Docked with *Mir* 1/14-1/19/97; Linenger to *Mir*; Blaha from *Mir*, spent 128 days in space
2/11/97- 2/21/97	Discovery (STS-82)	Bowersox (4), Horowitz (2), Joe Tanner (2), Hawley (4), Harbaugh (4), Lee (4), Steve Smith (2)	238:47	Increased capabilities of Hubble Space Telescope; 5 EVAs used to service it
5/15/97- 5/24/97	Atlantis (STS-84)	Precourt (3), E. Collins (2), Jean-François Clervoy (2), Carlos Noriega (2), Ed Lu, Elena Kondakova, Foale (4) (to *Mir*), Linenger (2) (from *Mir*)	221:20	Docked with *Mir* 5/16-5/21/97; Foale to *Mir*; Linenger from *Mir*, 132 days in space, 2d longest time for an American; stay on *Mir* marked by troubles incl. fire 2/23
7/1/97- 7/17/97	Columbia (STS-94)	Halsell (4), Susan L. Still (2), Janice Voss (4), Donald A. Thomas (4), Michael Gernhardt (3), Roger Crouch (2), Greg Linteris (2)	376:46	Reflight of Microgravity Science Laboratory-1 mission (STS-83) that was aborted 4/8/97 because of problem with fuel cell
8/5/97- 2/19/98	Soyuz TM-26	Solovyev (5), Pavel Vinogradov	4,743:35	Docked with *Mir* 8/7/97; repaired damaged space station
8/7/97- 8/19/97	Discovery (STS-85)	Brown (4), Rominger (3), Davis (3), Robert L. Curbeam Jr., Stephen K. Robinson, Bjarni V. Tryggvason	284:27	Deployed and retrieved satellite designed to study Earth's middle atmosphere; demonstrated robotic arm
9/25/97- 10/6/97	Atlantis (STS-86)	Wetherbee (4), Michael J. Bloomfield, V. Titov (4), Scott Parazynski (2), Jean-Loup Chrétien (3), Lawrence (2), Wolf (2) (to *Mir*), Foale (4) (from *Mir*)	236:24	Docked with *Mir* 9/27-10/3/97; delivered new computer to *Mir*; Wolf to *Mir*; Foale from *Mir*; stay on *Mir* marked by collision with cargo ship 6/25, worst such collision ever

Dates	Mission[1]	Crew (no. of flights)	Duration (hr:min)	Remarks
1/22/98-1/31/98	Endeavour (STS-89)	Wilcutt (3), Joe F. Edwards Jr., Dunbar (5), Michael P. Anderson, James F. Reilly II, Salizhan Sharipov, Andrew Thomas (2) (to Mir), Wolf (2) (from Mir)	211:48	Docked with Mir 1/24-1/29/98; delivered water and cargo; Thomas to Mir; Wolf from Mir, 128 days in space
1/29/98-8/25/98	Soyuz TM-27	Musabayev (2), Budarin (2), Leopold Eyharts	4,923:36	Docked with Mir 1/31/98
4/17/98-5/3/98	Columbia (STS-90)	Searfoss (3), Scott D. Altman, Linnehan (2), Dafydd Rhys Williams, Kathryn P. Hire, Jay C. Buckey, James A. Pawelczyk	381:50	Studied effects of microgravity on the nervous systems of the crew and over 2,000 live animals; 1st surgery in space on animals meant to survive
6/2/98-6/12/98	Discovery (STS-91)	Precourt (4), Dominic L. Gorie, Lawrence (3), Chang-Diaz (6), Janet L. Kavandi, Valery Ryumin (4), A. Thomas (2) (from Mir)	235:53	Final docking mission with Mir; Thomas from Mir, 141 days in space
10/29/98-11/7/98	Discovery (STS-95)	Brown (5), Steven W. Lindsey (2), Parazynski (3), Robinson (2), Pedro Duque, Chiaki Mukai (2), Glenn (2)	213:44	Sen. John Glenn, 77, was oldest person to fly in space; Duque was 1st Spaniard in space; series of experiments to study aging process performed on Glenn; Spartan 201 satellite, which studied the Sun, deployed and retrieved
12/4/98-12/15/98	Endeavour (STS-88)	Cabana (4), Frederick W. Sturckow, Nancy J. Currie (3), Ross (6), James H. Newman (3), Krivalev (4)	283:18	1st assembly of International Space Station; attached U.S.-built Unity connecting module with already-deployed Russian-built Zarya control module; 1st crew to enter ISS
5/27/99-6/6/99	Discovery (STS-96)	Rominger (4), Rick D. Husband, Ochoa (3), Jernigan (5), Barry (2), Julie Payette, Valery Ivanovich Tokarev	235:13	Transferred nearly 2 tons of supplies to International Space Station; small satellite STARSHINE deployed and observed by students on Earth
7/23/99-7/27/99	Columbia (STS-93)	Collins (3), Jeffrey S. Ashby, Hawley (5), Coleman (2), Michel Tognini (2)	118:50	Collins was 1st woman to command a space shuttle; deployed Chandra X-ray Observatory, a telescope designed to study the universe

Note: As of Oct. 15, 1999, there have been 95 space shuttle flights, 70 since the 1986 Challenger explosion. Active shuttles include the Columbia (26 flights), the Discovery (26), the Atlantis (20), and the Endeavour (13). (The Challenger completed 9 missions.)

Four Soviets died in spaceflights: Komarov was killed on Soyuz 1 (1967) when the parachute lines tangled during descent; the 3-person Soyuz 11 crew (1971) was asphyxiated. Seven Americans died in the Challenger explosion, and 3 astronauts—Virgil I. Grissom, Edward H. White, and Roger B. Chaffee—died in the Jan. 27, 1967, Apollo 1 fire on the ground at Cape Kennedy, FL.

(1) For space shuttle flights, mission name is in parentheses following the name of the orbiter. (2) Launch date. (3) Space stations, such as the Salyuts and Mir, were used to house crews starting in 1971. (4) Skylab 1 deteriorated and fell from orbit without burning up upon entering the atmosphere. Pieces fell on Australia and into the Indian Ocean; no one was injured. (5) The approximate crew duration for Thagard's stay. Crew did not return together.

Individuals Who Have Flown in Space, 1961-99

Source: Congressional Research Service; World Almanac research; as of Oct. 15, 1999

Country	No. of individs.	Country	No. of individs.	Country	No. of individs.	Country	No. of individs.
United States	257	Cuba	1	Japan	5	Slovakia	1
Russia/CIS	94	Czechoslovakia	1	Mexico	1	Spain	2
Afghanistan	1	France	8	Mongolia	1	Switzerland	1
Austria	1	Germany	10	Netherlands	1	Syria	1
Belgium	1	Hungary	1	Poland	1	United Kingdom	1
Bulgaria	2	India	1	Romania	1	Vietnam	1
Canada	9	Italy	3	Saudi Arabia	1	TOTAL	408

Note: All individuals flew on either a Russian/CIS-sponsored mission or on a U.S.-sponsored mission. All cosmonauts who were citizens of the USSR at the time of launch are included under "Russia/CIS." "Germany" includes former E and W Germany.

M I L L E N N I U M   F A C T   B O X

International Space Station

The International Space Station (ISS) being built for the new millennium will be the largest cooperative scientific project in history.

16 cooperating nations: U.S., Russia, Canada, Belgium, Denmark, France, Germany, Italy, Netherlands, Norway, Spain, Sweden, Switzerland, United Kingdom, Japan, and Brazil

The station when completed:
- mass of 1,040,000 lb
- 356' x 290', with almost an acre of solar panels
- 6 laboratories; living space for up to 7 people

Assembly:
- 11/20/98: Russian-built Zarya control module launched by rocket—1st step in assembly of the station
- 12/4/98: U.S.-built Unity connecting module launched on space shuttle Endeavour; shuttle crew attached Unity and Zarya
- 5/27/99: space shuttle Discovery launched, bringing supplies; 1st docking with ISS
- in March 2000, 1st crew to live aboard ISS
- to be completed by 2004, after 46 total missions

Examples of research planned:
- growing living cells for research in an environment free of gravity
- studying the effects on humans of long-term exposure to reduced gravity
- studying large-scale long-term changes in Earth's environment by observing Earth from orbit

Summary of Worldwide Successful Announced Payloads, 1957-98

Source: National Aeronautics and Space Administration

(A *payload* is something carried into space by a rocket.)

Year	Total[1]	Russia[2]	United States	Japan	European Space Agency	China	France	India	United Kingdom	Germany	Canada
1957-59	24	6	18	—	—	—	—	—	—	—	—
1960-69	1,035	399	614	—	2	—	4	—	1	—	—
1970-79	1,366	1,028	247	18	5	8	14	1	6	3	4
1980-89	1,431	1,132	191	26	14	16	5	9	4	7	5
1990	159	96	31	7	1	5	2	1	5	1	0
1991	157	101	30	2	4	1	6	1	2	1	2
1992	128	77	27	3	1	2	3	2	0	1	1
1993	104	59	29	1	2	1	2	1	0	0	0
1994	109	64	27	4	1	5	0	2	0	2	0
1995	87	45	24	2	2	1	3	1	0	1	1
1996	69	23	32	1	10	2	0	1	0	0	0
1997	85	27	37	2	12	6	0	1	0	0	0
1998	77	24	34	1	12	6	0	0	0	0	0
TOTAL	**4,831**	**3,081**	**1,341**	**67**	**66**	**53**	**39**	**20**	**18**	**16**	**13**

(1) Includes launches sponsored by countries not shown. (2) Figures for 1957-91 are for the Soviet Union; 1992-96 figures are for the Commonwealth of Independent States.

Notable U.S. Planetary Science Missions

Source: National Aeronautics and Space Administration

Spacecraft	Launch date (Coordinated Universal Time)	Mission	Remarks
Mariner 2	Aug. 27, 1962	Venus	Passed within 22,000 mi of Venus 12/14/62; contact lost 1/3/63 at 54 million mi
Ranger 7	July 28, 1964	Moon	Yielded over 4,000 photos of lunar surface
Mariner 4	Nov. 28, 1964	Mars	Passed behind Mars 7/14/65; took 22 photos from 6,000 mi
Ranger 8	Feb. 17, 1965	Moon	Yielded over 7,000 photos of lunar surface
Surveyor 3	Apr. 17, 1967	Moon	Scooped and tested lunar soil
Mariner 5	June 14, 1967	Venus	In solar orbit; closest Venus flyby 10/19/67
Mariner 6	Feb. 24, 1969	Mars	Came within 2,000 mi of Mars 7/31/69; collected data, photos
Mariner 7	Mar. 27, 1969	Mars	Came within 2,000 mi of Mars 8/5/69
Mariner 9	May 30, 1971	Mars	First craft to orbit Mars 11/13/71; sent back over 7,000 photos
Pioneer 10	Mar. 2, 1972	Jupiter	Passed Jupiter 12/4/73; exited the planetary system 6/13/83; transmission ended 3/31/97 at 6.39 billion mi
Pioneer 11	Apr. 5, 1973	Jupiter, Saturn	Passed Jupiter 12/3/74; Saturn 9/1/79; discovered an additional ring and 2 moons around Saturn; operating in outer solar system; transmission ended 9/95
Mariner 10	Nov. 3, 1973	Venus, Mercury	Passed Venus 2/5/74; arrived Mercury 3/29/74. 1st time gravity of 1 planet (Venus) used to whip spacecraft toward another (Mercury)
Viking 1	Aug. 20, 1975	Mars	Landed on Mars 7/20/76; did scientific research, sent photos; functioned 6$^1/_2$ years
Viking 2	Sept. 9, 1975	Mars	Landed on Mars 9/3/76; functioned 3$^1/_2$ years
Voyager 1	Sept. 5, 1977	Jupiter, Saturn	Encountered Jupiter 3/5/79; provided evidence of Jupiter ring; passed near Saturn 11/12/80
Voyager 2	Aug. 20, 1977	Jupiter, Saturn, Uranus, Neptune	Encountered Jupiter 7/9/79; Saturn 8/25/81; Uranus 1/24/86; Neptune 8/25/89
Pioneer Venus 1	May 20, 1978	Venus	Entered Venus orbit 12/4/78; spent 14 years studying planet; ceased operating 10/19/92
Pioneer Venus 2	Aug. 8, 1978	Venus	Encountered Venus 12/9/78; probes impacted on surface
Magellan	May 4, 1989	Venus	Landed on Venus 8/10/90; orbited and mapped Venus; monitored geological activity on surface; ceased operating 10/11/94
Galileo	Oct. 18, 1989	Jupiter	Used Earth's gravity to propel it toward Jupiter; encountered Venus Feb. 1990; encountered Jupiter 12/7/95; released probe to Jovian surface; encountered moons Ganymede, Europa, Io, and Callisto
Mars Observer	Sept. 25, 1992	Mars	Communication was lost 8/21/93
Near Earth Asteroid Rendezvous (NEAR)	Feb. 17, 1996	Asteroid Eros	Expected rendezvous with Eros, early 2000; to orbit and study the asteroid for about 1 year
Mars Global Surveyor	Nov. 7, 1996	Mars	Began orbiting Mars 9/11/97; began 2-year mapping survey of entire Martian surface 3/9/99; discovered magnetism on planet; observed Martian moon Phobos
Mars Pathfinder	Dec. 4, 1996	Mars	Landed on Mars 7/4/97; rover Sojourner made measurements of the Martian climate and soil composition, sending thousands of surface images; ceased operating 9/27/97
Cassini	Oct. 15, 1997	Saturn	Scheduled to reach Saturn in 2004; 4-year mission to study planet's atmosphere, rings, and moons; probe will land on moon Titan
Lunar Prospector	Jan. 6, 1998	Moon	Began orbiting Moon 1/11/98; mapped abundance of 11 elements on Moon's surface; discovered evidence of water-ice at both Lunar poles; made 1st precise gravity map of entire lunar surface; crashed into crater near Moon's south pole 7/31/99 to end mission
Mars Climate Orbiter	Dec. 11, 1998	Mars	Communication was lost 9/23/99
Mars Polar Lander	Jan. 3, 1999	Mars	Scheduled to land on Mars 12/3/99; 3-month mission to study soil and search for near-surface ice; will release 2 microprobes to search for water-ice 3 ft below Martian surface
Stardust	Feb. 7, 1999	Comet Wild-2	Scheduled to reach comet in 2004; to gather dust samples and return them to Earth in 2006

Notable Proposed U.S. Space Missions

Source: National Aeronautics and Space Administration

Planned Launch date	Mission	Purpose
2000	High Energy Transient Explorer II (HETE II)	Study gamma ray bursts
July 4, 2000	High Energy Solar Spectroscopic Imager (HESSI)	Explore the physics of particle acceleration and energy release in solar flares
Jan. 2001	Genesis	Collect solar wind (pieces of the sun) and return samples to Earth
Dec. 2001	Space InfraRed Telescope Facility (SIRTF)	Make high-sensitivity observations of celestial sources

Note: All spacecraft to be launched by expendable rockets.

Passenger Traffic at World Airports, 1998

Source: Airports Council International-North America

AIRPORT	Passenger Arrivals and Departures	AIRPORT	Passenger Arrivals and Departures
London, UK (Heathrow)	60,659,593	Tokyo, Japan (Narita)	24,441,365
Tokyo/Haneda, Japan (Tokyo Intl.)	51,240,704	Singapore (Changi)	23,803,180
Frankfurt, Germany (Rhein/Main)	42,716,270	Sidney, Australia (Kingsford Smith)	21,152,001
Paris, France (Charles De Gaulle)	38,628,926	Munich, Germany (Munich)	19,321,355
Amsterdam, Netherlands (Schiphol)	34,420,143	Zurich, Switzerland (Zurich)	19,276,782
Seoul, South Korea (Kimpo Intl.)	29,429,044	Osaka, Japan (Kansai Intl.)	19,161,035
London, UK (Gatwick)	29,173,196	Mexico City, Mexico (Mexico City)	18,946,440
Hong Kong, China (Hong Kong Intl.)	27,919,935	Brussels, Belgium (Brussels Intl.)	18,481,897
Toronto, Ontario (Lester B. Pearson Intl.)	26,744,530	Palma De Mallorca, Spain (Palma de Mallorca)	17,660,402
Bangkok, Thailand (Bangkok Intl.)	25,623,901	Fukuoka, Japan (Fukuoka Intl.)	17,574,001
Madrid, Spain (Barajas)	25,466,612	Manchester, UK (Manchester)	17,556,839
Rome, Italy (Fiumicino)	25,337,365	Beijing, China (Beijing Capital Intl.)	17,318,999
Paris, France (Orly)	24,951,984		

Note: Excludes U.S. airports. Includes only airports participating in the Airports Council International Annual Airport Traffic Statistics collection.

Passenger Traffic at U.S. Airports, 1998

Source: Airports Council International-North America

AIRPORT	Passenger Arrivals and Departures	AIRPORT	Passenger Arrivals and Departures
Atlanta (Hartsfield Intl.—ATL)	73,474,298	New York (J. F. Kennedy Intl.—JFK)	31,436,478
Chicago (O'Hare—ORD)	72,485,228	Houston (George Bush Intercontinental—IAH)	31,026,369
Los Angeles (LAX)	61,215,712	Minneapolis/St. Paul (MSP)	30,347,920
Dallas/Ft. Worth (DFW)	60,482,700	Las Vegas (McCarran Intl.—LAS)	30,227,287
San Francisco (SFO)	40,060,326	St. Louis (Lambert-St. Louis Intl.—STL)	28,700,622
Denver (DEN)	36,831,400	Orlando (MCO)	27,748,571
Miami (MIA)	33,935,491	Boston (Logan Intl.—BOS)	26,526,708
Newark (EWR)	32,512,106	Seattle-Tacoma (SEA)	25,863,132
Phoenix (Sky Harbor Intl.—PHX)	31,769,113	Philadelphia (PHL)	24,230,374
Detroit (DTW)	31,544,426	Charlotte (Charlotte/Douglas Intl.—CLT)	22,951,636

U.S. Scheduled Airline Traffic, 1996-98

Source: Air Transport Association of America

	1996	1997	1998
Revenue passengers enplaned (000)	**581,200**	**599,100**	**614,200**
Revenue passenger miles (000)	578,663,000	605,574,000	619,456,000
Available seat miles (000)	835,071,000	860,803,000	874,170,000
Revenue passenger load factor (%)	69.3	70.3	70.9
Cargo traffic	**17,775,000**	**20,514,000**	**20,476,000**
Revenue freight and express (ton miles)	15,301,000	17,959,000	18,116,000
Revenue U.S. Mail (ton miles)	2,454,000	2,555,000	2,360,000
Financial			
Passenger revenue ($000)	$75,286,000	$79,471,000	$80,986,000
Net profit ($000)	$2,804,000	$5,170,000	$4,894,000
Employees	**564,425**	**586,509**	**621,058**

Leading U.S. Passenger Airlines, 1998

Source: Air Transport Association of America

(in thousands)

Airline	Passengers	Airline	Passengers	Airline	Passengers
Delta	105,213	Alaska	13,029	American Trans Air	4,274
United	86,799	American Eagle	10,284	Mesaba	4,120
American	81,431	Continental Express	5,683	Atlantic Southeast	4,027
Southwest	59,053	AirTran	5,464	Air Wisconsin	2,757
US Airways	57,990	Aloha	5,144	Trans States	2,450
Northwest	50,490	Reno	5,099	Continental Micronesia	2,282
Continental	41,613	Hawaiian	5,000	Midwest Express	1,881
TWA	23,893	Horizon Air	4,389	Midway	1,875
America West	17,769				

National Aviation Hall of Fame

The National Aviation Hall of Fame at Dayton, OH, is dedicated to honoring the outstanding pioneers of air and space. 166 aviation and space leaders have been inducted since it was established in 1962. For more information, write to National Aviation Hall of Fame, P.O. Box 31096, Dayton, OH 45437 or call (937) 256-0944. The website is http://www.nationalaviation.org

Airline On-Time Arrivals, 1995-99

Source: Office of General Counsel, U.S. Dept. of Transportation

(percent of arrivals within 15 min. of scheduled time, for leading airlines)

AIRLINE	2d quarter 1999	1st quarter 1999 (rank)	1998 (rank)	1997 (rank)	1996 (rank)	1995 (rank)
1. Northwest	79.3	75.2 (5)	70.6 (9)	74.7 (9)	76.6 (3)[1]	80.7 (2)
2. Southwest	78.2	80.2 (1)	80.8 (1)	81.9 (1)	81.8 (1)	82.3 (1)
3. TWA..................	77.0	75.8 (4)	78.3 (5)	80.2 (2)	68.5 (10)	74.3 (10)
4. Delta	76.9	77.0 (3)	79.6 (3)	74.1 (10)	71.2 (7)	76.2 (9)
5. Continental	74.2	78.3 (2)	77.3 (6)	78.2 (5)	76.6 (2)[1]	79.5 (4)
6. US Airways	72.8	68.5 (10)	78.9 (4)	80.1 (3)	75.7 (4)	79.8 (3)
7. Alaska	72.5	69.6 (9)	72.0 (8)	74.9 (8)	68.6 (9)	76.7 (8)
8. America West	72.3	74.1 (7)	68.5 (10)	77.5 (6)	70.8 (8)	77.6 (6)
9. United	71.3	74.6 (6)	73.8 (7)	75.9 (7)	73.8 (5)	77.7 (5)
10. American.............	66.5	70.4 (8)	80.1 (2)	79.1 (4)	72.2 (6)	77.5 (7)
AVERAGE for all 10 airlines	**74.3**	**74.8**	**77.2**	**77.7**	**74.5**	**78.7**

Note: All domestic scheduled-service passenger flights, including those with mechanical delays, are included. A canceled flight is counted as a delay. The on-time performance database tracks only these 10 leading airlines, which account for more than 90% of domestic operating revenues. (1) When figures are carried out to several decimal places, Continental had the better on-time performance of these two carriers.

U.S. Airline Safety, Scheduled Commercial Carriers, 1981-98

Source: Air Transport Association of America

	Departures (millions)	Fatal accidents	Fatalities	Fatal accidents per 100,000 departures		Departures (millions)	Fatal accidents	Fatalities	Fatal accidents per 100,000 departures
1981 ...	5.2	4	4	0.077	1990 ...	6.9	6	39	0.087
1982[1] ..	5.0	4	234	0.060	1991 ...	6.8	4	62	0.059
1983 ...	5.0	4	15	0.079	1992 ...	7.1	4	33	0.057
1984 ...	5.4	1	4	0.018	1993 ...	7.2	1	1	0.014
1985 ...	5.8	4	197	0.069	1994 ...	7.5	4	239	0.053
1986[1] ..	6.4	2	5	0.016	1995 ...	8.1	2	166	0.025
1987[1] ..	6.6	4	231	0.046	1996 ...	8.2	3	342	0.036
1988[1] ..	6.7	3	285	0.030	1997 ...	8.2	3	3	0.037
1989 ...	6.6	8	131	0.121	1998 ...	8.3	1	1[2]	0.012

(1) Sabotage-caused accidents are included in the number of fatal accidents and fatalities, but not in the calculation of accident rates.
(2) On-ground employee fatality.

Aircraft Operating Statistics, 1998

Source: Air Transport Association of America; figures are averages for most commonly used models

	No. of seats	Speed airborne (mph)	Flight length (mi)	Fuel (gal per hr)	Operating cost per hr		No. of seats	Speed airborne (mph)	Flight length (mi)	Fuel (gal per hr)	Operating cost per hr
B747-100	458	515	2,350	3,742	$6,284	B737-800......	154	363	363	774	$2,221
B747-400	383	539	4,899	3,364	6,787	MD-90	150	445	693	840	1,689
B747-200/300..	370	523	3,376	3,660	7,632	B727-200......	150	436	713	1,311	2,505
B747-F	0	511	2,231	3,666	6,791	B727-F........	0	468	603	1,296	4,753
L-1011-100/200	315	495	1,352	2,502	4,253	A320-100/200 ..	147	461	1,179	807	2,147
B-777	290	522	2,994	2,124	3,810	B737-400......	142	413	656	785	1,933
DC-10-10	289	499	1,525	2,271	5,157	MD-80	140	432	795	940	2,010
DC-10-40	285	508	2,108	2,689	5,685	B737-300......	132	416	605	783	1,879
DC-10-30	262	523	2,706	2,683	5,627	DC-9-50.......	124	373	334	922	1,962
MD-11	249	524	3,068	2,468	5,928	B737-100/200 ..	111	390	460	830	1,864
A300-600	244	476	1,262	1,700	5,196	B737-500......	110	417	640	752	1,828
L-1011-500....	251	492	1,356	2,253	3,857	DC-9-40.......	108	386	490	844	1,693
B767-300ER...	213	498	2,273	1,606	3,348	DC-9-30.......	98	382	480	839	2,022
B757-200	186	467	1,240	1,064	2,614	F-100.........	97	386	513	656	1,907
B767-200ER...	180	490	2,307	1,480	3,130	DC-9-10.......	88	324	476	805	1,265

Some Notable Aviation Firsts[1]

1903 — On Dec. 17, near Kitty Hawk, NC, brothers Wilbur and Orville Wright made the first human-carrying, powered flight. Each made 2 flights; the longest, about 852 ft, lasted 59 sec.

1907 — U.S. airplane manufacturing company formed by Glenn H. Curtiss.

1908 — 1st airplane passenger, Lt. Frank P. Lahm, rode with Wilbur Wright in a brief (6 min, 24 sec) flight.

1911 — 1st transportation of mail by airplane officially approved by the U.S. Postal Service began on Sept. 23. It lasted one week. In 1918, limited scheduled air mail service began. By 1921, scheduled transcontinental airmail service began between New York City and San Francisco.

1914 — 1st scheduled passenger airline service began. It operated between St. Petersburg and Tampa, FL.

1919 — 1st airline food, a basket lunch, was served as part of a commercial airline service.

1930 — Ellen Church became 1st flight attendant.

(1)Excludes notable around-the-world and international trips.

1939 — On Aug. 27, the German Heinkel He 178 made the first successful flight powered by a jet engine.

1947 — Mach 1, the sound barrier, was broken by Amer. Charles E. ("Chuck") Yeager in a Bell X-1 rocket-powered aircraft.

1947 — Largest airplane ever flown, Howard Hughes's "Spruce Goose," flew 1 mi at an altitude of 80 ft.

1953 — Jacqueline Cochran became 1st woman to fly faster than sound.

1960 — Convair B-58, 1st supersonic bomber, was introduced.

1968 — The supersonic speed of Mach 2 was accomplished for 1st time, in a Tupolev Tu-144. The plane had an approximate maximum speed of 1,200 mph.

1970 — The Tupolev Tu-144, during commercial transport, exceeded Mach 2. It reached about 1,335 mph at 53,475 ft.

1976 — The Concorde began 1st scheduled supersonic commercial service.

Some Notable Around-the-World and Intercontinental Trips

	From/To	Miles	Time	Date
Nellie Bly	New York/New York		72d 06h 11m	1889
George Francis Train	New York/New York		67d 12h 03m	1890
Charles Fitzmorris	Chicago/Chicago		60d 13h 29m	1901
J. W. Willis Sayre	Seattle/Seattle		54d 09h 42m	1903
J. Alcock-A.W. Brown [1]	Newfoundland/Ireland	1,960	16h 12m	June 14-15, 1919
2 U.S. Army airplanes	Seattle/Seattle	26,103	35d 01h 11m	1924
Richard E. Byrd, Floyd Bennett [2]	Spitsbergen (Nor.)/N. Pole	1,545	15h 30m	May 9, 1926
Amundsen-Ellsworth-Nobile Polar Expedition (in a dirigible)	Spitsbergen (Nor.)/over N. Pole to Teller, Alaska		80h	May 11-14,1926
E.S. Evans and L. Wells (New York World)	New York/New York	18,410[3]	28d 14h 36m 05s	June 16-July 14, 1926
Charles Lindbergh [4]	New York/Paris	3,610	33h 29m 30s	May 20-21, 1927
Amelia Earhart, W. Stultz, L. Gordon	Newfoundland/Wales		20h 40m	June 17-18, 1928
Graf Zeppelin	Friedrichshafen, Ger./Lakehurst, NJ	6,630	4d 15h 46m	Oct. 11-15, 1928
Graf Zeppelin	Friedrichshafen, Ger./Lakehurst, NJ	21,700	20d 04h	Aug. 14-Sept. 4, 1929
Wiley Post and Harold Gatty (Monoplane Winnie Mae)	New York/New York	15,474	8d 15h 51m	July 1, 1931
C. Pangborn-H. Herndon Jr. [5]	Misawa, Japan/Wenatchee, Wash.	4,458	41h 34m	Oct. 3-5, 1931
Amelia Earhart [6]	Newfoundland/Ireland	2,026	14h 56m	May 20-21, 1932
Wiley Post (Monoplane Winnie Mae)[7]	New York/New York	15,596	115h 36m 30s	July 15-22, 1933
Hindenburg Zeppelin	Lakehurst, NJ/Frankfort, Ger.		42h 53m	Aug. 9-11, 1936
H. R. Ekins (Scripps-Howard Newspapers in race) (Zeppelin Hindenburg to Germany, airplanes from Frankfurt)	Lakehurst, NJ/Lakehurst, NJ	25,654	18d 11h 14m 33s	Oct. 19, 1936
Howard Hughes and 4 assistants	New York/New York	14,824	3d 19h 08m 10s	July 10-13, 1938
Douglas Corrigan	New York/Dublin		28h 13m	July 17-18, 1938
Mrs. Clara Adams (Pan American Clipper)	Port Washington, NY/ Newark, NJ		16d 19h 04m	June 28- July 15, 1939
Globester, U.S. Air Transport Command	Washington, DC/Washington, DC	23,279	149h 44m	Oct. 4, 1945
Capt. William P. Odom (A-26 Reynolds Bombshell)	New York/New York	20,000	78h 55m 12s	Apr. 12-16, 1947
America, Pan American 4-engine Lockheed Constellation[8]	New York/New York	22,219	101h 32m	June 17-30, 1947
Col. Edward Eagan	New York/New York	20,559	147h 15m	Dec. 13, 1948
USAF B-50 Lucky Lady II (Capt. James Gallagher) [9]	Ft. Worth, TX/Ft. Worth, TX	23,452	94h 01m	Mar. 2, 1949
Col. D. Schilling, USAF [10]	England/Limestone, ME	3,300	10h 01m	Sept. 22, 1950
C.F. Blair Jr.	Norway/Alaska	3,300	10h 29m	May 29, 1951
2 U.S. S-55	Massachusetts/Scotland	3,410	42h 30m	July 15-31, 1952
Canberra Bomber [11]	N. Ireland/Newfoundland	2073	04h 34m	Aug. 26, 1952
	Newfoundland/N. Ireland	2073	03h 25m	Aug. 26, 1952
3 USAF B-52 Strato-fortresses [12]	Merced, CA/CA	24,325	45h 19m	Jan. 15-18, 1957
Max Conrad	Chicago/Rome	5,000	34h 03m	Mar. 5-6, 1959
USSR TU-114 [13]	Moscow/New York	5,092	11h 06m	June 28, 1959
Boeing 707-320	New York/Moscow	c.5,090	08h 54m	July 23, 1959
Peter Gluckmann (solo)	San Francisco/San Francisco	22,800	29d	Aug. 22-Sept. 20, 1959
Sue Snyder	Chicago/Chicago	21,219	62h 59m	June 22-24, 1960
Max Conrad	Miami/Miami	25,946	8d 18h 35m 57s	Feb. 28-Mar. 8, 1961
Sam Miller & Louis Fodor	New York/New York	23,129	46h 28m	Aug. 3-4, 1963
Robert & Joan Wallick	Manila/Manila	23,129	5d 06h 17m 10s	June 2-7, 1966
Arthur Godfrey, Richard Merrill Fred Austin, Karl Keller	New York/New York	23,333	86h 9m 01s	June 4-7, 1966
Trevor K. Brougham	Darwin, Australia/Darwin	24,800	5d 05h 57m	Aug. 5-10, 1972
Walter H. Mullikin, Albert Frink, Lyman Watt, Frank Cassaniti, Edward Shields	New York/New York	23,137	1d 22h 50s	May 1-3, 1976
Arnold Palmer	Denver/Denver	22,985	57h 7m 12s	May 17-19, 1976
Boeing 747[14]	San Francisco/San Francisco	26,382	57h 25m 42s	Oct. 28-31, 1977
Richard Rutan & Jeana Yeager[15]	Edwards AFB, CA	24,986	09d 03m 44s	Dec. 14-23, 1986
Concorde	New York/New York	1,114 mph	31h 27m 49s	Aug. 15-16, 1995
Col. Douglas L. Raaberg and crew, B1 bomber[16]	Dyess AFB, Abilene, TX/Dyess AFB	6,250	36h 13m 36s	June 3, 1995
Linda Finch[17]	Oakland, CA/Oakland, CA	26,000	73d	Mar. 17-May 28, 1997
Bertrand Piccard, Brian Jones[18]	Switzerland/Egypt	29,054.6	19d 21h 55m	Mar. 1-21, 1999

(1) Nonstop transatlantic flight. (2) Claim of reaching N. Pole in dispute; if claim is untrue, then Amundsen-Ellsworth-Nobile were the first to fly over N. Pole. (3) Includes mileage by train and auto, 4,110; by plane, 6,300; by steamship, 8,000. (4) Solo transatlantic flight in the Ryan monoplane "Spirit of St. Louis." (5) Nonstop transpacific flight. (6) First woman's transoceanic solo flight. (7) First to fly solo around N circumference of the world and first to fly twice around the world. (8) Inception of regular commercial global air service. (9) First nonstop round-the-world flight, refueled 4 times in flight. (10) Nonstop jet transatlantic flight. (11) Transatlantic round trip on same day. (12) First nonstop global flight by jet planes; refueled in flight by KC-97 aerial tankers; average speed approx. 525 mph. (13) Nonstop between Moscow and New York. (14) Speed record around the world over both Earth's poles. (15) Circled Earth nonstop without refueling. (16) Refueled in flight 6 times. Tested B-1B bomber by bombing 3 pre-arranged target sites on 3 continents. (17) Followed the intended around-the-world flight route (1937) of Amelia Earhart. (18) First to circumnavigate the globe nonstop in a balloon.

METEOROLOGY
National Weather Service Watches and Warnings
Source: National Weather Service, NOAA, U.S. Dept. of Commerce; *Glossary of Meteorology,* American Meteorological Society

National Weather Service forecasters issue a *Severe Thunderstorm* or *Tornado Watch* for a specific area when a severe convective storm that usually covers a relatively small geographic area or moves in a narrow path is sufficiently intense to threaten life and/or property. Examples include thunderstorms with large hail, damaging winds, and/or tornadoes. Excessive localized convective rains are not classified as severe storms but are often the product of severe local storms. Such rainfall may result in phenomena that threaten life and property, such as flash floods. Although cloud-to-ground lightning is not a criterion for severe local storms, it is acknowledged to be a leading cause of storm deaths and injuries.

A *Watch* alerts people that threatening weather is likely. Under a Watch, they should remain alert for approaching storms, activate a plan for action, and monitor ongoing events closely. A *Warning* means that severe weather is occurring or has been indicated by radar; immediate action should be taken by people in the storm's path.

Severe Thunderstorm—a thunderstorm that produces a tornado, winds of at least 50 knots (58 mph), and/or hail at least 3/4 inch in diameter. A thunderstorm with winds of at least 35 knots (40 mph) and/or hail at least 1/2 inch in diameter is defined as approaching severe. A *Severe Thunderstorm Watch* is issued for a specific area where such storms are most likely to develop. A *Severe Thunderstorm Warning* indicates that a severe thunderstorm has been sighted or indicated by radar.

Tornado—a violent rotating column of air (winds over 200 mph), usually pendant to a cumulonimbus cloud, with circulation reaching the ground. A tornado nearly always starts as a funnel cloud and may be accompanied by a loud roaring noise. On a local scale, it is the most destructive of all atmospheric phenomena. Tornado paths have varied in length from a few feet to more than 100 miles (avg. 5 mi); in diameter from a few feet to more than a mile (avg. 220 yd); average forward speed, 30 mph.

Cyclone—an atmospheric circulation of winds rotating counterclockwise in the northern hemisphere and clockwise in the southern hemisphere. Tornadoes, hurricanes, and the lows shown on weather maps are all examples of cyclones of various size and intensity. Cyclones are usually accompanied by precipitation or stormy weather.

Subtropical Storm—an atmospheric circulation of one-minute sustained surface winds, 34 knots (39 mph) or more. Depending on its characteristics and intensity, it can develop into a tropical storm or a hurricane.

Tropical Storm—an atmospheric circulation of one-minute sustained surface winds within a range of 34 to 63 knots (39 to 73 mph). A *Tropical Storm Watch* is an announcement that a tropical storm or tropical storm conditions may pose a threat to coastal areas generally within 36 hours. A *Tropical Storm Warning* is an announcement that tropical storm conditions pose a threat along a specified segment of coastline within 24 hours.

Hurricane—a severe cyclone originating over tropical ocean waters and having one-minute sustained surface winds 64 knots (73 mph) or higher. (West of the international date line, in the western Pacific, such storms are known as *typhoons.*) The area of hurricane-force winds forms a circle or an oval, sometimes as wide as 300 mi in diameter. In the lower latitudes, hurricanes usually move west or northwest at 10 to 15 mph. When the center approaches 25° to 30° North Latitude, the direction of motion often changes to northeast, with increased forward speed.

Blizzard—a severe weather condition characterized by strong winds bearing a great amount of snow. The National Weather Service specifies winds of 35 mph or higher and sufficient falling and/or blowing snow to frequently reduce visibility to less than 1/4 mi. for at least 3 hours.

Flood—Flooding takes many forms. *River Flooding:* This natural process occurs when rains, sometimes coupled with melting snow, fill river basins with too much water too quickly; torrential rains from decaying hurricanes or tropical systems can also be a major cause of river flooding. *Coastal Flooding:* Winds from tropical storms and hurricanes or intense offshore low pressure systems can drive ocean water inland and cause significant flooding. Coastal floods can also be produced by sea waves called *tsunamis,* sometimes referred to as tidal waves; these waves are produced by earthquakes or volcanic activity. *Flash Flooding:* Usually due to copious amounts of rain falling in a short time, flash flooding typically occurs within 6 hours of the rain event. Flash floods account for the majority of flood deaths in the U.S. *Urban Flooding:* Urbanization significantly increases runoff over what would occur on natural terrain, making flash flooding in these areas extremely dangerous. Streets can become swift-moving rivers, and basements can become death traps as they fill with water. *Ice Jam Flooding:* Ice can accumulate at natural or artificial obstructions and stop the flow of water. As the water flow is stopped, water builds up and flooding can occur upstream. If the jam suddenly gives way, the gush of ice and water can cause serious downstream flash flooding.

Flash Flood or Flood Watch: Flash flooding or flooding is possible within a designated area.

Flash Flood or Flood Warning: Flash flooding or flooding has been reported or is imminent; all necessary precautions should be taken immediately.

Urban and Small Stream Advisory: Small streams, streets, and low-lying areas such as railroad underpasses and urban storm drains are flooding.

National Weather Service Marine Warnings and Advisories

Small Craft Advisory alerts mariners to sustained (exceeding 2 hours) weather and/or sea conditions, either present or forecast, potentially hazardous to small boats. Although "small craft" is not defined, hazardous conditions generally include winds of 18 to 33 knots and/or dangerous wave conditions. It is the responsibility of the mariner, based on experience and on the location and size or type of boat, to determine whether conditions are hazardous to the boat. Upon receiving word of a Small Craft Advisory, the mariner should immediately obtain the latest marine forecast to determine the reason for the advisory.

Gale Warning indicates that winds within the range 34 to 47 knots, not directly associated with a tropical storm, are forecast for the area.

Tropical Storm Warning indicates that winds within the range of 34 to 63 knots are forecast in a specified coastal area to occur within 24 hours or less. Issued only for winds of tropical weather systems.

Storm Warning indicates that winds 48 knots or above, not directly associated with a tropical storm, are forecast for the area.

Hurricane Warning indicates that winds 64 knots or greater are forecast for the area within 24 hours. Issued only for winds produced by tropical weather systems.

Special Marine Warning indicates potentially hazardous weather conditions, usually of short duration (2 hours or less) and producing wind speeds of 34 knots or more, not adequately covered by existing marine warnings.

Primary sources of dissemination are commercial radio, TV, U.S. Coast Guard radio stations, and NOAA VHF-FM broadcasts. These NOAA broadcasts on 162.40 to 162.55 MHz can usually be received 20-40 mi from the transmitting antenna site, depending on terrain and quality of the receiver used. Where transmitting antennas are on high ground, the range may be somewhat greater, reaching 60 mi or more.

Monthly Normal Temperatures, Precipitation

Source: National Climatic Data Center, NESDIS, NOAA, U.S. Dept. of Commerce

The temperatures given here are based on records for the 30-year period 1961-90. For stations that did not have continuous records from the same site for the entire 30 years, the means have been adjusted to the record at the present site.

Figures are for airport stations unless otherwise indicated. * = city station. T = temperature in Fahrenheit; P = precipitation in inches; L = less than 0.05 inch.

	Jan.		Feb.		Mar.		Apr.		May		June		July		Aug.		Sept.		Oct.		Nov.		Dec.	
Station	T	P	T	P	T	P	T	P	T	P	T	P	T	P	T	P	T	P	T	P	T	P	T	P
Albany, NY	21	2.4	24	2.3	34	2.9	46	3.0	58	3.4	67	3.6	72	3.2	70	3.5	61	3.0	50	2.8	40	3.2	27	2.9
Albuquerque, NM	34	0.4	40	0.5	47	0.5	55	0.5	64	0.5	74	0.6	79	1.4	76	1.6	69	1.0	57	0.9	44	0.4	35	0.5
Anchorage, AK	15	0.8	19	0.8	26	0.7	36	0.7	47	0.7	54	1.1	58	1.7	56	2.4	48	2.7	35	2.0	21	1.1	16	1.1
Asheville, NC	36	3.3	39	3.9	47	4.6	55	3.4	63	4.4	69	4.2	73	4.5	72	4.7	66	3.9	56	3.6	48	3.6	40	3.5
Atlanta, GA	41	4.8	45	4.8	54	5.8	62	4.3	69	4.3	76	3.6	79	5.0	78	3.7	73	3.4	62	3.1	53	3.9	45	4.3
Atlantic City, NJ	31	3.5	33	3.1	42	3.6	50	3.6	60	3.3	69	2.6	75	3.8	73	4.1	66	2.9	55	2.8	46	3.6	36	3.3
Baltimore, MD	32	3.1	35	3.1	44	3.4	53	3.1	63	3.7	73	3.7	77	3.7	76	3.9	69	3.4	57	3.0	47	3.3	37	3.4
Barrow, AK	–13	0.2	–18	0.2	–15	0.2	–2	0.2	19	0.2	34	0.3	39	0.9	38	1.0	31	0.6	14	0.5	–2	0.3	–11	0.2
Birmingham, AL	42	5.1	46	4.7	54	6.2	62	5.0	69	4.9	76	3.7	80	5.3	79	3.6	73	3.9	63	2.8	53	4.3	45	5.1
Bismarck, ND	9	0.5	16	0.4	28	0.8	43	1.7	55	2.2	64	2.7	71	2.1	68	1.7	57	1.5	46	0.9	29	0.5	14	0.5
Boise, ID	29	1.5	36	1.2	43	1.3	49	1.2	58	1.1	67	1.8	74	0.4	73	0.4	63	0.8	52	0.8	40	1.5	30	1.4
Boston, MA	29	3.6	30	3.6	39	3.7	48	3.6	58	3.3	68	3.1	74	2.8	72	3.2	65	3.1	55	3.3	45	4.2	34	4.0
Buffalo, NY	24	2.7	25	2.3	34	2.7	45	2.9	57	3.1	66	3.6	71	3.1	69	4.2	62	3.5	51	3.1	41	3.8	29	3.7
Burlington, VT	16	1.8	18	1.6	31	2.2	44	2.8	56	3.1	65	3.5	71	3.7	68	4.1	59	3.3	48	2.9	37	3.1	23	2.4
Caribou, ME	9	2.4	12	1.9	25	2.4	38	2.5	51	3.1	61	2.9	66	4.0	63	4.1	54	3.5	43	3.1	31	3.6	15	3.2
Charleston, SC	48	3.5	51	3.3	58	4.3	65	2.7	73	4.0	78	6.4	82	6.8	81	7.2	76	4.7	67	2.9	58	2.5	51	3.2
Chicago, IL	21	1.5	25	1.4	37	2.7	49	3.6	59	3.3	69	3.8	73	3.7	72	4.2	64	3.8	53	2.4	40	2.9	27	2.5
Cleveland, OH	25	2.0	27	2.2	37	2.9	48	3.1	58	3.5	68	3.7	72	3.5	70	3.4	64	3.4	53	2.5	43	3.2	31	3.1
Columbus, OH	26	2.2	30	2.2	41	3.3	51	3.2	61	3.9	69	4.0	73	4.3	72	3.7	66	3.0	54	2.2	43	3.2	32	2.9
Dallas-Ft. Worth, TX	43	1.8	48	2.2	57	2.8	66	3.5	73	4.9	81	3.0	85	2.3	85	2.2	77	3.4	67	3.5	56	2.3	47	1.8
Denver, CO	30	0.5	33	0.6	39	1.3	48	1.7	57	2.4	67	1.8	74	1.9	71	1.5	62	1.2	51	1.0	39	0.9	31	0.6
Des Moines, IA	19	1.0	25	1.1	37	2.3	51	3.4	62	3.7	72	4.5	77	3.8	74	4.2	65	3.5	54	2.6	39	1.8	24	1.3
Detroit, MI	23	1.8	25	1.7	36	2.6	47	3.0	58	2.9	68	3.6	72	3.2	71	3.4	63	2.9	51	2.1	40	2.7	28	2.8
Dodge City, KS	30	0.5	35	0.6	43	1.6	55	2.0	64	3.0	74	3.1	80	3.2	78	2.7	69	1.9	57	1.3	43	0.8	32	0.6
Duluth, MN	7	1.2	12	0.8	24	1.9	39	2.3	51	3.0	60	3.8	66	3.6	64	4.0	54	3.8	44	2.5	28	1.8	13	1.2
Fairbanks, AK	–10	0.5	–4	0.4	11	0.4	31	0.3	49	0.6	60	1.4	63	1.9	57	2.0	46	1.0	25	0.9	3	0.8	–7	0.9
Fresno, CA	46	2.0	51	1.8	55	1.9	61	1.0	69	0.3	77	0.1	82	L	80	L	75	0.2	65	0.5	54	1.4	45	1.4
Galveston, TX*	53	3.3	55	2.3	62	2.2	69	2.4	76	3.6	81	4.4	83	4.0	84	4.5	80	5.9	73	2.8	64	3.4	56	3.5
Grand Junction, CO	25	0.6	34	0.5	43	0.9	52	0.7	62	0.9	72	0.5	79	0.6	76	0.8	67	0.8	55	1.0	40	0.7	29	0.6
Grand Rapids, MI	22	1.8	24	1.4	34	2.6	46	3.4	58	3.1	67	3.7	72	3.2	70	3.6	61	4.2	50	2.8	38	3.3	27	2.9
Hartford, CT	25	3.4	28	3.2	38	3.6	49	3.9	60	4.1	69	3.8	74	3.2	72	3.7	63	3.8	52	3.6	42	4.0	30	3.9
Helena, MT	20	0.6	26	0.4	34	0.7	43	1.0	53	1.8	62	1.9	69	1.1	67	1.3	55	1.2	45	0.6	32	0.5	21	0.6
Honolulu, HI	73	3.6	73	2.2	74	2.2	76	1.5	78	1.1	79	0.5	81	0.6	81	0.4	81	0.8	80	2.3	77	3.0	74	3.8
Houston, TX	50	3.2	54	3.3	61	2.7	68	4.2	75	4.7	80	4.0	83	3.3	82	3.7	78	4.9	70	3.7	61	3.4	54	3.7
Huron, SD	13	0.4	19	0.8	32	1.2	46	2.0	58	2.7	68	3.3	74	2.3	72	2.0	61	1.4	49	1.4	32	0.7	18	0.5
Indianapolis, IN	26	2.3	30	2.5	41	3.8	52	3.7	63	4.0	72	3.5	75	4.5	73	3.6	67	2.9	55	2.6	43	3.2	31	3.3
Jackson, MS	44	5.2	48	4.7	57	5.8	65	5.6	72	5.1	79	3.2	82	4.5	81	3.8	76	3.6	65	3.3	56	4.8	48	5.9
Jacksonville, FL	52	3.3	55	3.9	61	3.7	67	2.8	73	3.6	79	5.7	82	5.6	81	7.9	78	7.0	70	2.9	62	2.1	55	2.7
Juneau, AK	24	4.5	28	3.7	33	3.3	40	2.8	47	3.4	53	3.1	56	4.2	55	5.3	49	6.7	42	7.8	32	4.9	27	4.4
Kansas City, MO	26	1.1	31	1.1	43	2.5	55	3.1	64	5.0	73	4.7	79	4.4	76	4.0	68	4.9	57	3.3	43	1.9	30	1.6
Knoxville, TN	36	4.2	40	4.1	49	5.1	58	3.7	65	4.1	73	4.0	77	4.7	76	3.1	70	3.1	58	2.8	49	3.8	40	4.5
Lander, WY	20	0.5	25	0.6	34	1.2	43	2.1	53	2.3	63	1.5	71	0.8	69	0.5	58	1.1	47	1.1	31	0.8	21	0.6
Lexington, KY	31	2.9	35	3.2	45	4.4	55	3.9	64	4.5	72	3.7	76	5.0	75	3.9	68	3.2	57	2.6	46	3.4	36	4.0
Little Rock, AR	39	3.9	44	4.4	53	5.3	62	6.2	70	7.0	78	7.8	82	8.2	81	8.1	74	7.4	63	6.3	52	5.2	43	4.3
Los Angeles, CA*	58	2.9	60	3.1	61	2.6	63	1.0	66	0.2	70	L	74	L	75	0.1	74	0.5	70	0.3	63	2.0	58	2.0
Louisville, KY	32	2.9	36	3.3	46	4.7	56	4.2	65	4.6	73	3.5	77	4.5	76	3.5	70	3.2	58	2.7	47	3.7	37	3.6
Marquette, MI*	12	2.2	14	1.7	24	2.8	37	2.6	50	3.0	59	3.5	65	2.9	63	3.4	54	4.1	44	3.6	30	2.9	17	2.6
Memphis, TN	40	3.7	44	4.4	53	5.4	63	5.5	71	5.0	79	3.6	83	3.8	81	3.4	74	3.5	63	3.0	53	5.1	44	5.7
Miami, FL	67	2.0	69	2.1	72	2.4	75	2.9	79	6.2	81	9.3	83	5.7	83	7.6	82	7.6	78	5.6	74	2.7	69	1.8
Milwaukee, WI	19	1.6	23	1.5	33	2.7	44	3.5	55	2.8	65	3.2	71	3.5	69	3.5	62	3.4	50	2.4	38	2.5	24	2.3
Minneapolis, MN	12	1.0	18	0.9	31	1.9	46	2.4	59	3.4	68	4.1	74	3.5	71	3.6	61	2.7	49	2.2	33	1.6	18	1.1
Mobile, AL	50	4.8	53	5.5	61	6.4	68	4.5	75	5.7	80	5.0	82	6.9	82	7.0	78	5.9	68	2.9	60	4.1	53	5.3
Moline, IL	20	1.5	25	1.2	37	3.0	50	3.9	61	4.3	71	4.3	75	5.0	73	4.2	65	4.0	53	2.9	40	2.5	25	2.2
Nashville, TN	36	3.6	40	3.8	50	4.9	59	4.4	68	4.9	76	3.6	79	4.0	78	3.5	72	3.5	60	2.6	50	4.1	41	4.6
Newark, NJ	31	3.4	33	3.0	42	3.9	52	3.8	63	4.1	73	3.2	78	4.5	76	3.9	69	3.7	58	3.1	47	3.9	36	3.5
New Orleans, LA	51	5.1	54	6.0	62	4.9	69	4.5	75	4.6	80	5.8	82	6.1	82	6.2	78	5.5	69	3.1	61	4.4	55	5.8
New York, NY*	32	3.4	34	3.3	42	4.1	53	4.2	63	4.4	72	3.7	77	4.4	76	4.0	68	3.9	58	3.6	48	4.5	37	3.9
Norfolk, VA	39	3.8	41	3.5	49	3.7	57	3.1	66	3.8	74	3.8	78	5.1	77	4.8	72	3.9	61	3.2	53	2.9	44	3.2
Oklahoma City, OK	36	1.1	41	1.6	50	2.7	60	2.8	68	5.2	77	4.3	82	2.6	81	2.6	73	3.8	62	3.2	50	2.0	39	1.4
Omaha, NE	21	0.7	27	0.8	39	2.0	52	2.7	62	4.5	72	3.9	77	3.5	74	3.2	65	3.7	53	2.3	39	1.5	25	1.0
Philadelphia, PA	30	3.2	33	2.8	42	3.5	52	3.6	63	3.8	72	3.7	77	4.3	76	3.8	68	3.4	56	2.6	46	3.3	36	3.4
Phoenix, AZ	54	0.7	58	0.7	62	0.9	70	0.2	79	0.1	88	0.1	94	0.8	92	1.0	86	0.9	75	0.7	62	0.7	54	1.0
Pittsburgh, PA	26	2.5	29	2.4	39	3.4	50	3.2	60	3.6	68	3.7	72	3.8	71	3.2	64	3.0	52	2.4	42	2.9	32	2.9
Portland, ME	21	3.5	23	3.3	33	3.7	43	4.1	53	3.6	62	3.4	69	3.1	67	2.9	59	3.1	49	3.9	39	5.2	27	4.6
Portland, OR	40	5.4	44	3.9	47	3.6	51	2.4	57	2.1	64	1.5	68	0.6	69	1.1	63	1.8	55	2.7	46	5.3	40	6.1
Providence, RI	28	4.1	30	3.7	37	4.3	47	4.0	57	3.5	67	2.8	73	3.0	71	4.0	64	3.5	54	3.8	44	4.2	33	4.5
Raleigh, NC	39	3.6	42	3.4	50	3.7	59	2.9	67	3.7	74	3.7	78	4.4	77	4.4	71	3.3	60	2.7	51	2.9	43	3.1
Rapid City, SD	22	0.4	27	0.5	34	1.0	45	1.9	55	2.7	65	3.1	72	2.0	71	1.7	60	1.2	49	1.1	35	0.6	24	0.5
Reno, NV	33	1.1	38	1.0	43	0.7	49	0.4	57	0.7	65	0.5	72	0.3	70	0.3	60	0.4	51	0.4	40	0.9	33	1.0
Richmond, VA	37	3.2	39	3.2	48	3.6	57	3.0	66	3.8	74	3.6	78	5.0	77	4.4	70	3.3	59	3.5	50	3.2	40	3.3
St. Louis, MO	29	1.8	34	2.1	45	3.6	57	3.5	66	4.0	75	3.7	80	3.9	78	2.9	70	3.1	58	2.7	46	3.3	34	3.0
Salt Lake City, UT	28	1.1	34	1.2	42	1.9	50	2.1	59	1.8	69	0.9	78	0.8	76	0.9	65	1.3	53	1.4	41	1.3	30	1.4
San Antonio, TX	49	1.7	54	1.8	62	1.5	69	2.5	76	4.2	82	3.8	85	2.2	85	2.5	79	3.4	70	3.2	60	2.6	52	1.5
San Diego, CA	57	1.8	59	1.5	60	1.8	62	0.8	64	0.2	67	0.1	71	L	73	0.1	71	0.2	68	0.4	62	1.5	57	1.6
San Francisco, CA	49	4.4	52	3.2	53	3.1	56	1.4	58	0.2	62	0.1	63	L	64	0.1	65	0.2	61	1.2	55	2.9	49	3.1
San Juan, PR	77	2.8	77	2.1	78	2.3	79	3.8	81	5.9	82	4.0	83	4.4	83	5.3	82	5.3	82	5.7	80	5.9	78	4.7
Sault Ste. Marie, MI*	13	2.4	14	1.7	24	2.3	38	2.4	51	2.7	58	3.1	64	2.7	63	3.6	55	3.7	45	3.2	33	3.5	19	2.9
Savannah, GA	49	3.6	52	3.2	59	3.8	66	3.0	74	4.1	79	5.7	82	6.4	81	7.4	77	4.5	67	2.4	59	2.2	52	3.0
Scottsbluff, NE	25	0.5	30	0.5	36	1.1	47	1.6	56	2.8	67	2.6	74	2.1	72	1.1	61	1.1	50	0.8	36	0.6	26	0.6
Seattle, WA	41	5.4	44	4.0	47	3.8	50	2.5	56	1.8	61	1.6	65	0.9	66	1.2	61	1.9	54	3.3	46	5.7	42	6.0
Spokane, WA	27	2.0	33	1.5	39	1.5	46	1.2	54	1.4	62	1.3	69	0.7	68	0.7	59	0.7	47	1.0	35	2.2	28	2.4
Springfield, MO	31	1.8	36	2.2	46	3.9	56	4.2	65	4.4	73	5.1	78	2.9	77	3.5	69	4.6	58	3.6	46	3.8	35	3.2
Syracuse, NY	22	2.3	24	2.2	34	2.8	46	3.3	57	3.3	65	3.8	70	3.8	68	3.5	62	3.8	51	3.2	41	3.7	28	3.2
Tampa, FL	60	2.0	62	3.1	67	3.0	71	1.2	77	3.1	81	5.5	82	6.6	82	7.6	81	6.0	75	2.0	68	1.8	62	2.2
Washington, DC	31	2.7	34	2.8	43	3.2	53	3.1	62	4.0	71	3.9	76	3.5	74	3.9	67	3.4	55	3.2	45	3.3	35	3.2
Wilmington, DE	31	3.0	33	2.9	43	3.4	52	3.4	63	3.8	72	3.6	76	4.2	75	3.4	68	3.4	56	2.9	46	3.3	36	3.5

Normal High and Low Temperatures, Precipitation

Source: National Climatic Data Center, NESDIS, NOAA, U.S. Dept. of Commerce

The normal temperatures given here are based on records for the 30-year period 1961-90. The extreme temperatures (through 1990) are listed for the stations shown and may not agree with the state records shown on page 231.

Figures are for airport stations unless otherwise indicated. * = city station. Temperatures are Fahrenheit.

State	Station	NORMAL TEMPERATURE January Max.	Min.	July Max.	Min.	EXTREME TEMPERATURE Highest	Lowest	ANNUAL PRECIPITATION (inches)
Alabama	Mobile	60	40	91	73	104	3	63.96
Alaska	Anchorage	21	8	65	52	85	−34	15.91
Alaska	Barrow	−7	−19	45	34	79	−56	4.49
Arizona	Phoenix	66	41	106	81	122	17	7.66
Arkansas	Little Rock	49	29	92	72	112	−5	72.10
California	Los Angeles*	68	49	84	65	112	28	14.77
California	San Diego	66	49	76	66	111	29	9.9
California	San Francisco	56	42	72	54	106	20	19.70
Colorado	Denver	43	16	88	59	104	−30	15.40
Connecticut	Hartford	33	16	85	62	102	−26	44.14
Delaware	Wilmington	39	22	86	67	102	−14	40.84
District of Columbia	Washington-National	42	27	89	71	104	−5	38.63
Florida	Jacksonville	64	41	91	72	105	7	51.32
Florida	Miami	75	59	89	76	98	30	55.91
Georgia	Atlanta	50	32	88	70	105	−8	50.77
Georgia	Savannah	60	38	91	72	105	3	49.22
Hawaii	Honolulu	80	66	88	74	94	53	22.02
Idaho	Boise	36	22	90	58	111	−25	12.11
Illinois	Chicago	29	13	84	63	104	−27	35.82
Illinois	Moline	28	11	86	65	106	-27	39.08
Indiana	Indianapolis	34	17	86	65	104	−23	39.94
Iowa	Des Moines	28	11	87	67	108	−24	33.12
Kentucky	Lexington	39	22	86	66	103	−21	44.55
Kentucky	Louisville	40	23	87	67	105	−20	44.39
Louisiana	New Orleans	61	42	91	73	102	11	61.88
Maine	Caribou	19	−2	77	55	96	−41	36.60
Maine	Portland	30	11	79	58	103	−39	44.34
Maryland	Baltimore	40	23	87	67	105	−7	40.76
Massachusetts	Boston	36	22	82	65	102	−12	41.51
Michigan	Detroit	30	16	83	61	104	−21	32.62
Michigan	Sault Ste. Marie*	21	5	76	51	98	−36	34.23
Minnesota	Duluth	16	−2	77	55	97	−39	30.00
Minnesota	Minneapolis-St. Paul	21	3	84	63	105	−34	28.32
Mississippi	Jackson	56	33	92	71	106	2	55.37
Missouri	Kansas City	35	17	89	68	109	−23	37.62
Missouri	St. Louis	38	21	89	70	107	−18	37.51
Montana	Helena	30	10	85	53	105	−42	11.60
Nebraska	Omaha	31	11	88	66	114	−23	29.86
Nebraska	Scottsbluff	38	12	90	59	109	−42	15.27
Nevada	Reno	45	21	92	51	105	−16	7.53
New Jersey	Atlantic City	40	21	85	65	106	−11	40.29
New Mexico	Albuquerque	47	22	93	64	105	−17	8.88
New York	Albany	30	11	84	60	100	−28	36.17
New York	Buffalo	30	17	80	62	99	−20	38.58
New York	New York-La Guardia	37	26	84	69	107	−3	42.12
North Carolina	Asheville	47	25	83	62	100	−16	47.59
North Carolina	Raleigh	49	29	88	68	105	−9	41.43
North Dakota	Bismarck	20	−2	84	56	109	−44	15.47
Ohio	Cleveland	32	18	82	61	104	−19	36.63
Ohio	Columbus	34	19	84	63	102	−19	38.09
Oregon	Portland	45	34	80	57	107	−3	36.30
Pennsylvania	Philadelphia	38	23	86	67	104	−7	41.41
Pennsylvania	Pittsburgh	34	19	83	62	103	−18	36.85
Rhode Island	Providence	37	19	82	63	104	−13	45.53
South Carolina	Charleston	58	38	90	73	104	6	51.53
South Dakota	Huron	24	2	87	62	112	−39	20.08
South Dakota	Rapid City	34	11	86	58	110	−30	16.64
Tennessee	Memphis	49	31	92	73	108	−13	52.10
Tennessee	Nashville	46	27	90	69	107	−17	47.30
Texas	Galveston*	58	47	87	79	101	8	42.28
Texas	Houston	61	40	93	72	107	7	46.07
Utah	Salt Lake City	36	19	92	64	107	−30	16.18
Vermont	Burlington	25	8	81	60	101	−30	34.47
Virginia	Norfolk	47	31	86	70	104	−3	44.64
Virginia	Richmond	46	26	88	68	105	−12	43.16
Washington	Seattle-Tacoma	45	35	75	55	99	0	37.19
Washington	Spokane	33	21	83	54	108	−25	16.49
Wisconsin	Milwaukee	26	12	80	62	103	−26	32.93
Wyoming	Lander	31	8	86	56	101	−37	13.01

Mean Annual Snowfall (inches) based on record through 1990: Boston, MA, 42; Sault Ste. Marie, MI, 113; Albany, NY, 65.2; Burlington, VT, 78.6; Lander, WY, 66; Juneau, AK, 105.8.

Wettest Spot: Mount Waialeale, HI, on the island of Kauai, is the rainiest place in the United States and in the world, according to the National Geographic Society; it has an average annual rainfall of 460 inches.

Below are the official temperature extremes through mid-1998. There are many unofficial claims. To qualify as official meteorological data, readings must be taken on approved instruments in a sheltered and ventilated location.

Highest Temperature: A temperature of 136° F observed at Azizia (Al Aziziyah), near Tripoli, Libya, on Sept. 13, 1922, is generally accepted as the world's highest temperature recorded under standard conditions.

The record high in the United States was 134° F in Death Valley, CA, July 10, 1913.

Lowest Temperature: A record low temperature of −129° F was recorded at the Soviet Antarctica station of Vostok on July 21, 1983.

The record low in the United States was −80° F at Prospect Creek, AK, Jan. 23, 1971.

The lowest official temperature on the North American continent was recorded at −81° F in Feb. 1947, at an airport in the Yukon called Snag.

Annual Climatological Data, 1998

Source: National Climatic Data Center, NESDIS, NOAA, U.S. Dept. of Commerce

Station	Elev. (ft.)	Temperature °F Highest	Date	Lowest	Date	Precipitation Total (in.)	Greatest in 24 hours	Date	Sleet or snow Total (in.)	Greatest in 24 hours	Date	Fastest wind MPH	Date	No. of days Prec. .01 in. or more	Snow, sleet 1 in. or more
Albany, NY	275	89	9/6	−7	1/1	38.94	2.22	10/10	28.8	4.8	1/23	55	5/31	142	10
Albuquerque, NM	5,311	102	6/29	11	12/8	9.83	1.45	3/15	6.5	2.7	12/6-7	45	2/24	57	2
Anchorage, AK	114	72	8/1	−11	1/5	11.65	0.66	6/6-7	58.0	13.9	12/4	31	8/16	101	2
Asheville, NC	2,140	92	9/5	11	1/1	48.02	4.67	1/7-8	13.5	11.9	1/27	36	1/28	130	1
Atlanta, GA	1,010	95	7/19	19	3/12	46.16	3.72	3/7-8	—	—	—	38	6/19	51	0
Atlantic City, NJ	64	98	7/22	8	12/26	39.92	2.52	2/23-24	—	—	—	40	2/25	52	—
Baltimore, MD	148	96	8/25	9	12/26	34.37	2.05	2-4/5	5.8	3.0	12/23	34	2/25	114	2
Barrow, AK	31	67	6/29	−37	1/26	4.70	0.35	7/15	33.7	2.5	4/18-19	48	10/24	101	10
Birmingham, AL	620	99	7/8	19	3/12	67.27	3.34	1/7	—	—	—	—	—	118	—
Bismarck, ND	1,647	97	7/19	−24	1/13	23.70	4.64	8/21-22	41.3	7.1	11/9	39	6/1	93	11
Boise, ID	2,838	108	7/17	−2	12/21	16.71	1.13	5/25-26	15.5	6.1	12/25	46	4/23	112	5
Boston, MA	15	93	7/22	7	2/15	53.67	5.69	6/13	—	—	—	47	9/15	119	—
Buffalo, NY	705	89	8/8	8	12/30	34.70	2.21	1/7-8	52.5	8.7	3/21	46	11/11	146	12
Burlington, VT	332	89	7/22	−5	1/1	50.42	3.62	8/11	64.0	11.3	1/15-16	34	1/13	165	15
Caribou, ME	624	90	8/10	−20	1/23	36.02	2.23	7/9-10	102.7	9.5	1/23-24	36	1/4	163	32
Charleston, SC	40	100	6/30	22	3/13	67.72	10.52	9/21-22	—	—	—	41	9/3	115	—
Chicago, IL	658	95	6/27	−1	1/13	37.53	2.93	10/17-18	20.4	6.5	3/9	41	11/10	117	3
Cleveland, OH	777	94	6/25	7	12/23	32.83	1.80	6/11-12	21.7	3.7	3/10	41	7/21	135	6
Columbus, OH	813	97	9/6	11	12/23	37.57	2.56	12/21-22	9.0	1.7	12/31	43	3/28	135	2
Dallas-Ft. Worth, TX	551	110	7/12	20	12/22	34.24	3.54	10/2	—	—	—	43	5/8	81	—
Denver, CO	5,282	100	7/20	−19	12/22	15.93	2.79	7/24-25	—	—	—	46	7/14	86	—
Des Moines, IA	938	98	7/20	−10	12/31	37.70	2.45	6/18	—	—	—	53	8/10	121	—
Detroit, MI	637	96	6/25	2	12/30	34.13	4.34	7/7	13.9	3.2	1/23	53	7/21	122	6
Duluth, MN	1,428	90	7/13	−21	12/30	31.52	2.48	11/9-10	84.5	7.2	11/10	47	3/31	112	26
Fairbanks, AK	436	87	7/5	−48	1/4	10.74	0.83	7/6-7	25.3	3.5	12/5	30	5/22	109	7
Fresno, CA	328	109	8/5	21	12/23	17.65	1.80	6/6	—	—	—	36	2/3	71	—
Grand Rapids, MI	793	94	6/25	1	1/14	30.37	1.58	3/8-9	42.3	8.2	3/9	49	11/10	128	12
Hartford, CT	169	93	8/24	5	1/1	45.11	2.70	6/30	—	—	—	41	4/23	126	—
Helena, MT	3,828	97	7/17	−29	1/12	12.57	1.23	7/28	—	—	—	46	9/15	95	—
Honolulu, HI	7	89	10/2	53	1/31	4.52	0.84	12/31	—	—	—	34	12/4	74	—
Houston, TX	96	104	8/2	27	12/26	54.84	6.33	11/12-13	0.0	0.0	—	36	7/14	99	0
Huron, SD	1,281	100	9/10	−25	1/13	23.32	1.74	10/4	53.1	12.8	3/31	59	8/24	99	13
Indianapolis, IN	795	94	9/6	4	12/31	46.98	3.15	6/14-15	5.4	3.1	12/30-31	49	7/20	135	2
Jackson, MS	291	100	8/29	23	3/12	51.55	4.34	1/21-22	T	T	12/24	55	9/28	104	0
Jacksonville, FL	26	103	6/19	27	1/1	56.72	4.32	2/16-17	0	0	—	57	7/2	105	0
Kansas City, MO	979	98	7/21	−7	3/12	49.53	2.91	10/4-5	8.6	4.8	3/8-9	47	6/25	106	3
Knoxville, TN	979	97	9/13	12	1/1	53.90	3.69	4/18-19	0.5	0.4	1/27	—	—	140	1
Lander, WY	5,557	100	7/18	−27	12/21	17.53	1.41	6/17	91.2	11.6	3.29	55	12/28	83	21
Lexington, KY	966	97	9/6	9	3/12	49.60	5.04	6/29	—	—	—	44	6/29	140	—
Los Angeles, CA	97	95	9/1	37	12/22	27.06	3.10	2/3	—	—	—	38	3/28	59	—
Louisville, KY	477	97	9/6	13	3/12	46.33	1.87	7/7	23.3	9.2	2/4	54	6/10	136	4
Marquette, MI	1,415	93	7/14	−14	3/16	28.34	1.57	8/22-23	135.7	9.9	1/8-9	—	—	144	41
Memphis, TN	258	102	7/6	18	3/12	51.81	2.53	4/29	3.8	2.0	1/16	—	—	110	2
Miami, FL	7	98	7/3	49	3/13	70.23	5.85	11/4-5	T	T	5/6	55	2/2	143	0
Milwaukee, WI	679	95	6/25	−4	1/13	34.64	3.97	8/5-6	31.6	12.4	1/8-9	52	11/10	126	8
Minn.-St. Paul, MN	834	94	7/13	−23	1/13	33.39	2.60	6/26-27	36.3	4.0	1/25	49	5/15	123	16
Mobile, AL	211	99	7/9	27	3/12	86.52	10.06	9/27-28	—	—	—	51	9/28	115	—
Moline, IL	592	98	7/20	−4	12/31	48.37	4.14	10/17	18.1	3.3	1/21	49	11/10	128	6
Nashville, TN	590	99	9/6	15	3/12	52.01	5.24	6/4-5	—	—	—	40	4/8	139	—
Newark, NJ	7	98	7/22	14	1/1	43.47	2.70	8/17-18	6.5	3.1	3/22	44	9/7	114	8
New Orleans, LA	4	99	8/28	34	3/12	79.04	6.03	9/10-11	—	—	—	48	1/7	104	—
New York, NY	132	93	7/22	14	1/1	48.69	2.64	1/23-24	7.5	5.0	3/22	34	3/21	110	2
Norfolk, VA	24	98	7/22	24	12/31	54.76	4.78	2/3-4	—	—	—	46	8/27	119	—
North Little Rock, AR	N/A	104	7/7	16	3/12	40.98	3.06	2/10	T	T	12/23	—	—	105	0
Oklahoma City, OK	1,285	107	9/4	10	12/22	35.35	2.77	4/26	1.0	1.0	12/21	41	3/27	71	1
Philadelphia, PA	5	95	6/26	16	1/1	31.65	2.33	1/23-24	2.6	1.5	12/23	51	6/1	116	1
Phoenix, AZ	1,109	117	7/15	32	12/24	10.53	1.02	7/22	T	T	12/6	43	3/6	42	0
Pittsburgh, PA	1,137	91	8/7	6	1/1	34.22	2.37	6/27	10.9	2.4	3/11	53	6/30	143	3
Portland, ME	43	92	7/22	−6	12/31	54.77	6.51	10/9-10	39.4	8.6	1/15-16	57	8/24	114	9
Portland, OR	21	101	7/28	11	12/23	46.02	2.41	11/25-26	—	—	—	43	11/23	183	—
Providence, RI	51	93	7/22	9	1/1	52.70	3.29	6/13	—	—	—	38	11/11	116	—
Raleigh, NC	416	99	7/22	18	1/1	52.68	3.36	3/18-19	2.0	2.0	1/19	38	9/2	130	1
Rapid City, SD	3,162	104	7/8	−22	12/21	21.89	2.46	10/16-17	—	—	—	59	2/25	104	—
Reno, NV	4,404	104	7/17	−3	12/23	12.03	1.02	3/23-24	—	—	—	52	11/23	80	—
Richmond, VA	164	102	7/22	16	1/1	46.75	2.53	3/8-9	—	—	—	44	6/16	129	—
St. Louis, MO	535	98	8/24	4	3/12	43.62	3.23	7/29-30	14.1	2.3	1/8	41	11/10	120	7
Salt Lake City, UT	4,221	102	7/19	1	12/21	23.81	1.76	2/24-25	65.6	18.0	2/24-25	47	5/26	112	19
San Antonio, TX	788	107	6/15	23	12/26	42.05	13.35	10/17-18	—	—	—	38	8/13	82	—
San Diego, CA	13	92	9/3	39	12/25	16.05	1.57	2/3-4	T	T	3/28	38	2/23	65	0
San Francisco, CA	8	91	8/3	31	12/24	32.63	3.41	2/2-3	—	—	—	43	12/5	108	—
San Juan, PR	13	94	9/10	68	2/13	72.55	4.87	9/21-22	0.0	0.0	—	79	9/21	203	0
Sault Ste. Marie, MI	718	89	7/14	−8	1/30	28.30	1.84	5/30-31	—	—	—	34	11/24	151	—
Savannah, GA	46	102	6/30	24	3/13	49.47	3.74	1/22-23	—	—	—	43	6/25	113	—
Scottsbluff, NE	3,943	104	7/18	−27	12/22	17.25	1.67	7/7	33.8	7.4	11.2	48	2/25	98	—
Seattle, WA	400	97	7/28	15	12/20	44.06	3.12	11/25-26	—	—	—	37	11/24	152	—
Spokane, WA	2,356	103	7/27	−4	12/20	17.77	0.95	11/20	23.9	5.2	12/25	41	12/28	122	9
Springfield, MO	1,278	99	9/5	4	12/20	48.47	2.43	10/5	10.1	4.2	1/8	44	6/21	118	3
Syracuse, NY	410	90	8/9	−9	1/1	37.06	2.26	8/24-25	79.9	13.9	2/24-25	59	9/7	161	23
Tampa, FL	19	98	6/21	39	3/13	55.35	5.21	7/7-8	T	T	7/5	36	2/4	100	0
Washington, DC	10	97	8/25	18	1/1	35.94	2.13	2/4-5	0.5	0.4	12/23	40	7/21	117	0
Wilmington, DE	74	93	6/26	13	12/26	36.46	2.53	1/23-24	—	—	—	43	3/21	115	—

(T) Trace. (—) Data not available or incomplete. (1) Where one date is shown, it is the starting date of the storm. (2) Sustained for at least 2 minutes, not peak gust.

Record Temperatures by State Through 1997

Source: National Climatic Data Center, NESDIS, NOAA, U.S. Dept. of Commerce

State	Lowest °F	Highest °F	Latest date	Station	Approx. elevation in feet
Alabama	-27		Jan. 30, 1966	New Market	760
		112	Sept. 5, 1925	Centerville	345
Alaska	-80		Jan. 23, 1971	Prospect Creek	1,100
		100	June 27, 1915	Fort Yukon	420
Arizona	-40		Jan. 7, 1971	Hawley Lake	8,180
		128	June 29, 1994[1]	Lake Havasu City	505
Arkansas	-29		Feb. 13, 1905	Pond	1,250
		120	Aug. 10, 1936	Ozark	396
California	-45		Jan. 20, 1937	Boca	5,532
		134	July 10, 1913	Greenland Ranch	-178
Colorado	-61		Feb. 1, 1985	Maybell	5,920
		118	July 11, 1888	Bennett	5,484
Connecticut	-32		Feb. 16, 1943	Falls Village	585
		106	July 15, 1995	Danbury	450
Delaware	-17		Jan. 17, 1893	Millsboro	20
		110	July 21, 1930	Millsboro	20
Florida	-2		Feb. 13, 1899	Tallahassee	193
		109	June 29, 1931	Monticello	207
Georgia	-17		Jan. 27, 1940	CCC Camp F-16	1,000
		112	July 24, 1952	Louisville	132
Hawaii	12		May 17, 1979	Mauna Kea Obs. 111.2	13,770
		100	Apr. 27, 1931	Pahala	850
Idaho	-60		Jan. 18, 1943	Island Park Dam	6,285
		118	July 28, 1934	Orofino	1,027
Illinois	-35		Feb. 3, 1996[1]	Mount Carroll	817
		117	July 14, 1954	East St. Louis	410
Indiana	-36		Jan. 19, 1994	New Whiteland	785
		116	July 14, 1936	Collegeville	672
Iowa	-47		Feb. 3, 1996[1]	Elkader	770
		118	July 20, 1934	Keokuk	614
Kansas	-40		Feb. 13, 1905	Lebanon	1,812
		121	July 24, 1936[1]	Alton (near)	1,651
Kentucky	-37		Jan. 19, 1994	Shelbyville	730
		114	July 28, 1930	Greensburg	581
Louisiana	-16		Feb. 13, 1899	Minden	194
		114	Aug. 10, 1936	Plain Dealing	268
Maine	-48		Jan. 19, 1925	Van Buren	510
		105	July 10, 1911[1]	North Bridgton	450
Maryland	-40		Jan. 13, 1912	Oakland	2,461
		109	July 10, 1936[1]	Cumberland; Frederick	623; 325
Massachusetts	-35		Jan. 12, 1981	Chester	640
		107	Aug. 2, 1975	Chester; New Bedford	640; 120
Michigan	-51		Feb. 9, 1934	Vanderbilt	785
		112	July 13, 1936	Mio	963
Minnesota	-60		Feb. 2, 1996	Tower	1,430
		114	July 6, 1936[1]	Moorhead	904
Mississippi	-19		Jan. 30, 1966	Corinth	420
		115	July 29, 1930	Holly Springs	600
Missouri	-40		Feb. 13, 1905	Warsaw	700
		118	July 14, 1954[1]	Warsaw; Union	700; 560
Montana	-70		Jan. 20, 1954	Rogers Pass	5,470
		117	July 5, 1937	Medicine Lake	1,950
Nebraska	-47		Feb. 12, 1899	Camp Clarke	3,700
		118	July 24, 1936[1]	Minden	2,169
Nevada	-50		Jan. 8, 1937	San Jacinto	5,200
		125	June 29, 1994[1]	Laughlin	605
New Hampshire	-46		Jan. 28 1925	Pittsburg	1,575
		106	July 4, 1911	Nashua	125
New Jersey	-34		Jan. 5, 1904	River Vale	70
		110	July 10, 1936	Runyon	18
New Mexico	-50		Feb. 1, 1951	Gavilan	7,350
		122	June 27, 1994	Waste Isolat. Pilot Plt.	3,418
New York	-52		Feb. 18, 1979[1]	Old Forge	1,720
		108	July 22, 1926	Troy	35
North Carolina	-34		Jan. 21, 1985	Mt. Mitchell	6,525
		110	Aug. 21, 1983	Fayetteville	213
North Dakota	-60		Feb. 15, 1936	Parshall	1,929
		121	July 6, 1936	Steele	1,857
Ohio	-39		Feb. 10, 1899	Milligan	800
		113	July 21, 1934[1]	Gallipolis (near)	673
Oklahoma	-27		Jan. 18, 1930	Watts	958
		120	June 27, 1994[1]	Tipton	1,350
Oregon	-54		Feb. 10, 1933[1]	Seneca	4,700
		119	Aug. 10, 1898	Pendleton	1,074
Pennsylvania	-42		Jan. 5, 1904	Smethport	1,500
		111	July 10, 1936[1]	Phoenixville	100
Rhode Island	-25		Feb. 5, 1996	Greene	425
		104	Aug. 2, 1975	Providence	51
South Carolina	-19		Jan. 21, 1985	Caesars Head	3,115
		111	June 28, 1954[1]	Camden	170
South Dakota	-58		Feb. 17, 1936	McIntosh	2,277
		120	July 5, 1936	Gannvalley	1,750
Tennessee	-32		Dec. 30, 1917	Mountain City	2,471
		113	Aug. 9, 1930[1]	Perryville	377
Texas	-23		Feb. 8, 1933[1]	Seminole	3,275
		120	Aug. 12, 1936	Seymour	1,291

State	Lowest °F	Highest °F	Latest date	Station	Approx. elevation in feet
Utah.	−69		Feb. 1, 1985	Peter's Sink	8,092
		117	Jul. 5, 1985	Saint George	2,880
Vermont.	−50		Dec. 30, 1933	Bloomfield	915
		105	July 4, 1911	Vernon.	310
Virginia	−30		Jan. 22, 1985	Mountain Lake Bio. Station	3,870
		110	July 15, 1954	Balcony Falls	725
Washington.	−48		Dec. 30, 1968	Mazama; Winthrop	2,120; 1,755
		118	Aug. 5, 1961[1]	Ice Harbor Dam.	475
West Virginia.	−37		Dec. 30, 1917	Lewisburg	2,200
		112	July 10, 1936[1]	Martinsburg	435
Wisconsin	−54		Jan. 24, 1922	Danbury.	908
		114	July 13, 1936	Wisconsin Dells	900
Wyoming	−66		Feb. 9, 1933	Riverside R.S.	6,650
		114	July 12, 1900	Basin	3,500

* Estimated. (1) Also on earlier dates at the same or other places.

World Temperature and Precipitation

Source: World Meteorological Organization

Average daily maximum and minimum temperatures and annual precipitation are based on records for the 30-year period 1961-90. The length of record of extreme temperatures includes all available years of data for a given location and is usually for a longer period; record temperatures may have been measured at a different location within the city. Surface elevations are supplied by the WMO and may differ from city elevation figures in other sections of *The World Almanac*. NA = not available.

Station	Surface elevation (feet)	July Max.	July Min.	January Max.	January Min.	EXTREME Max.	EXTREME Min.	Average annual precipitation (inches)
Algiers, Algeria	82	61.7	42.6	87.1	65.3	NA	NA	27.0
Athens, Greece	49	56.1	44.6	88.9	73.0	NA	NA	14.6
Auckland, New Zealand	20	74.8	61.2	58.5	46.4	NA	NA	49.4
Bangkok, Thailand	66	89.6	69.8	90.9	77.0	104	51	59.0
Berlin, Germany	190	35.2	26.8	73.6	55.2	107	−4	23.3
Bogotá, Colombia	8,357	67.3	41.7	64.6	45.5	75	21	32.4
Bombay (Mumbai), India.	36	85.3	66.7	86.2	77.5	110	46	85.4
Bucharest, Romania	298	34.7	22.1	83.8	60.1	105	−18	23.4
Budapest, Hungary	456	34.2	24.8	79.7	59.7	103	−10	20.3
Buenos Aires, Argentina	82	85.8	67.3	59.7	45.7	104	22	45.2
Cairo, Egypt.	243	65.8	48.2	93.9	71.1	118	34	1.0
Cape Town, South Africa.	138	79.0	60.3	63.3	44.6	105	28	20.5
Caracas, Venezuela	2,739	79.9	60.8	81.3	66.0	96	45	36.1
Casablanca, Morocco	203	62.8	47.1	77.7	66.7	NA	NA	16.8
Copenhagen, Denmark.	16	35.6	28.4	68.9	55.0	NA	NA	NA
Damascus, Syria	2,004	54.3	32.9	97.2	61.9	NA	NA	5.6
Dublin, Ireland.	279	45.7	36.5	66.0	52.5	86	8	28.8
Geneva, Switzerland	1,364	38.3	27.9	76.3	53.2	101	−3	35.6
Havana, Cuba	164	78.4	65.5	88.3	74.8	NA	NA	46.9
Hong Kong, China.	203	65.5	56.5	88.7	79.9	97	32	87.2
Istanbul, Turkey	108	47.8	37.2	82.8	65.3	105	7	27.4
Jerusalem, Israel.	2,483	53.4	39.4	83.8	63.0	107	26	23.2
Lagos, Nigeria	125	90.0	72.3	82.8	72.1	NA	NA	59.3
Lima, Peru.	43	79.0	66.9	66.4	59.4	NA	NA	0.2
London, England.	203	44.1	32.7	71.1	52.3	99	2	29.7
Manila, Philippines	79	85.8	74.8	89.1	76.8	NA	NA	49.6
Mexico City, Mexico.	7,570	70.3	43.7	73.8	53.2	NA	NA	33.4
Montreal, Canada	118	21.6	5.2	79.2	59.7	100	−36	37.0
Nairobi, Kenya.	5,897	77.9	50.9	71.6	48.6	NA	NA	41.9
Paris, France.	213	42.8	33.6	75.2	55.2	105	−1	25.6
Prague, Czech Republic	1,197	32.7	22.5	73.9	53.2	98	−16	20.7
Reykjavik, Iceland	200	35.4	26.6	55.9	46.9	76	−3	31.5
Rome, Italy	79	53.8	35.4	88.2	62.1	NA	NA	33.0
San Salvador, El Salvador	2,037	86.5	61.3	86.2	66.4	105	45	68.3
São Paulo, Brazil.	2,598	81.1	65.7	71.2	53.1	NA	NA	57.4
Shanghai, China.	23	45.9	32.9	88.9	76.6	104	10	43.8
Singapore	52	85.8	73.6	87.4	75.6	NA	NA	84.6
Stockholm, Sweden	171	30.7	23.0	71.4	56.1	97	−26	21.2
Sydney, Australia.	10	79.5	65.5	62.4	43.9	114	32	46.4
Tehran, Iran.	3,906	45.0	30.0	98.2	75.2	109	−5	9.1
Tokyo, Japan	118	49.1	34.2	83.8	72.1	NA	NA	55.4
Toronto, Canada	567	27.5	12.0	80.2	57.6	105	−26	30.8

Hurricane and Tornado Classifications

Source: National Weather Service, NOAA, U.S. Dept. of Commerce

The Saffir-Simpson Hurricane Scale is a 1-5 rating based on a hurricane's intensity. The scale is used to give an estimate of the potential property damage and flooding expected along the coast from a hurricane landfall. Wind speed is the determining factor in the scale. The Fujita (or F) Scale, created by T. Theodore Fujita, is used to classify tornadoes. The F Scale uses rating numbers from 0 to 5, based on the amount and type of wind damage.

Saffir-Simpson Scale (Hurricanes)

Category	Wind Speed	Severity	Storm Surge[1]
1	74-95 MPH	Weak	4-5 feet
2	96-110 MPH	Moderate	6-8 feet
3	111-130 MPH	Strong	9-12 feet
4	131-155 MPH	Very Strong	13-18 feet
5	more than 155 MPH	Devastating	more than 18 feet

(1) Above normal tides.

Fujita Scale (Tornadoes)

Rank	Wind Speed	Damage	Strength
F-0	40-72 MPH	Light	Weak
F-1	73-112 MPH	Moderate	Weak
F-2	113-157 MPH	Considerable	Strong
F-3	158-206 MPH	Severe	Strong
F-4	207-260 MPH	Devastating	Violent
F-5	more than 261 MPH	Incredible	Violent

Hurricane Names in 2000

Source: National Weather Service, NOAA, U.S. Dept. of Commerce

Atlantic hurricanes — Alberto, Beryl, Chris, Debby, Ernesto, Florence, Gordon, Helene, Isaac, Joyce, Keith, Leslie, Michael, Nadine, Oscar, Patty, Rafael, Sandy, Tony, Valerie, William.

Eastern Pacific hurricanes — Aletta, Bud, Carlotta, Daniel, Emilia, Fabio, Gilma, Hector, Ileana, John, Kristy, Lane, Miriam, Norman, Olivia, Paul, Rosa, Sergio, Tara, Vicente, Willa, Xavier, Yolanda, Zeke.

Tides and Their Causes

Source: U.S. Dept. of Commerce, Natl. Oceanic & Atmospheric Admin. (NOAA), Natl. Ocean Service (NOS)

The tides are a natural phenomenon involving the alternating rise and fall in the large fluid bodies of the earth caused by the combined gravitational attraction of the sun and moon. The combination of these two variable influences produces the complex recurrent cycle of the tides. Tides may occur in both oceans and seas, to a limited extent in large lakes, in the atmosphere, and, to a very minute degree, in the earth itself. The length of time between succeeding tides varies as the result of many factors.

The tide-generating force represents the difference between (1) the centrifugal force produced by the revolution of the earth around the common center-of-gravity of the earth-moon system and (2) the gravitational attraction of the moon acting upon the earth's overlying waters. Since, on the average, the moon is only 238,856 miles from the earth compared with the sun's much greater distance of 92,980,000 miles, this closer distance outranks the much smaller mass of the moon compared with that of the sun, and the moon's tide-raising force is, accordingly, 2.5 times that of the sun.

The effect of the tide-generating forces of the moon and sun acting tangentially to the earth's surface (the so-called "tractive force") tends to cause a maximum accumulation of the waters of the oceans at two diametrically opposite positions on the surface of the earth and to withdraw compensating amounts of water from all points 90° degrees removed from the positions of these tidal bulges. As the earth rotates beneath the maxima and minima of these tide-generating forces, a sequence of two high tides, separated by two low tides, ideally is produced each day (semidiurnal tide).

Twice in each lunar month, when the sun, moon, and earth are directly aligned, with the moon between the earth and the sun (at new moon) or on the opposite side of the earth from the sun (at full moon), the sun and the moon exert their gravitational force in a mutual or additive fashion. The highest high tides and lowest low tides are produced at these times. These are called *spring* tides. At two positions 90° degrees in between, the gravitational forces of the moon and sun—imposed at right angles—tend to counteract each other to the greatest extent, and the range

between high and low tides is reduced. These are called *neap* tides. This semi-monthly variation between the spring and neap tides is called the *phase inequality.*

The inclination to the equator of the moon's monthly orbit and the inclination of the sun to the equator during the earth's yearly orbit produce a difference in the height of succeeding high tides and in the extent of depression of succeeding low tides that is known as the *diurnal inequality*. In most cases, this produces a so-called *mixed tide*. In extreme cases, these phenomena may result in only one high tide and one low tide each (*diurnal tide*). There are other monthly and yearly variations in the tide because of the elliptical shape of the orbits themselves.

The datum for Charting and Predictions is Mean Lower Low Water (MLLW). This became effective Nov. 1980 according to the convention of 1980, which prescribed that data on all United States coastlines would be the same; namely, Mean Higher High Water (MHHW), Mean High Water (MHW), Mean Tide Level (MTL), Mean Sea Level (MSL), Mean Low Water (MLW), Mean Lower Low Water (MLLW). Diurnal range of tide is the difference in height between MHHW and MLLW. Mean range of tide is the difference in height between MHW and MLW.

The actual range of tide in the open ocean is less than in the shoreline regions. However, as the ocean tide approaches shoal waters and its effects are augmented, the tidal range may be greatly increased. In Nova Scotia along the narrow channel of the Bay of Fundy, the range of tides, or difference between high and low waters, may reach 43-1/2 feet or more (under spring tide conditions) as a result of resonant amplification.

At New Orleans, the periodic rise and fall of the diurnal tide is affected by the seasonal stages of the Mississippi River, being about 10 inches at low stage and zero at high. The Canadian Tide Tables for 1972 gave a maximum range of nearly 50 feet at Leaf Basin, Ungava Bay, Quebec.

In every case, actual high or low tide can vary considerably from the average, as a result of weather conditions such as strong winds, abrupt barometric pressure changes, or prolonged periods of extreme high or low pressure.

The Average Rise and Fall of Tides [1]

Places	Ft.	In.	Places	Ft.	In.	Places	Ft.	In.
Baltimore, MD	1	8	Hampton Roads, VA	2	10	St. John's, Nfld.	2	7[2]
Boston, MA	10	4	Key West, FL	1	10	St. Petersburg, FL	2	3
Charleston, SC	5	10	Mobile, AL	1	6	San Diego, CA	5	9
Cristobal, Panama	1	1	New London, CT	3	1	Sandy Hook, NJ	5	2
Eastport, ME	19	4	Newport, RI	3	11	San Francisco, CA	5	10
Ft. Pulaski, GA	7	6	New York, NY	5	1	Seattle, WA	11	4
Galveston, TX	1	5	Philadelphia, PA	6	9	Vancouver, B.C.	10	6
Halifax, N.S.	4	5[2]	Portland, ME	9	11	Washington, DC	3	2
						St. Petersburg, FL	2	3

(1) Diurnal range. (2) Mean range.

Speed of Winds in the U.S.

Source: National Climatic Data Center, NESDIS, NOAA, U.S. Dept. of Commerce

Miles per hour — average high through 1997. Wind velocities in true values.

Station	Avg.	High	Station	Avg.	High	Station	Avg.	High
Albuquerque, NM	8.9	52	Helena, MT	7.7	73	Mt. Washington, NH	35.4	231
Anchorage, AK	7.1	75	Honolulu, HI	11.3	46	New Orleans, LA	8.1	69
Atlanta, GA	9.1	60	Houston, TX	7.8	51	New York, NY(b)	9.3	40
Baltimore, MD	9.0	80	Indianapolis, IN	9.6	47	Omaha, NE	10.5	58
Bismarck, ND	10.2	54	Jacksonville, FL	7.9	46	Philadelphia, PA	9.5	73
Boston, MA	12.5	54	Kansas City, MO	10.7	48	Phoenix, AZ	6.2	43
Buffalo, NY	11.9	91	Las Vegas, NV	9.3	53	Pittsburgh, PA	9.1	58
Cape Hatteras, NC	11.0	60	Lexington, KY	9.1	46	Portland, OR	7.9	88
Casper, WY	12.8	81	Little Rock, AR	7.8	65	Rochester, NY	9.7	59
Chicago, IL	10.4	58	Los Angeles, CA	6.2	49	St. Louis, MO	9.7	52
Cleveland, OH	10.5	53	Louisville, KY	8.3	46	Salt Lake City, UT	8.8	71
Dallas-Ft. Worth, TX	10.7	73	Memphis, TN	8.8	51	San Diego, CA	7.0	56
Denver, CO	8.6	46	Miami, FL	9.2	(a)86	San Francisco, CA	8.7	47
Des Moines, IA	10.7	76	Milwaukee, WI	11.5	54	Seattle, WA	9.0	66
Detroit, MI	10.3	51	Minn.-St. Paul, MN	10.5	51	Spokane, WA	8.9	59
Hartford, CT	8.4	46	Mobile, AL	8.9	63	Washington, DC	9.4	46

(a) Highest velocity ever recorded in Miami area was 132 mph, at former station in Miami Beach in Sept. 1926. (b) Data for Central Park; Battery Place data through 1960, avg. 14.5, high 113.

El Niño

Source: National Weather Service, NOAA

El Niño is a naturally occurring climate phenomenon characterized by warmer-than-normal ocean temperatures in the equatorial eastern Pacific and along the tropical western coasts of Central and South America. The term *El Niño*, Spanish for "the Christ Child," was originally used by Ecuadorian and Peruvian fishermen to refer to a warm ocean current typically appearing around Christmastime and lasting for several months. Fish are less abundant during these warm intervals, so fishermen often take a break to repair equipment and spend time with their families. In some years, however, the water remains especially warm into May or even June. Over the years, the term has come to be reserved for those exceptionally strong, warm intervals that not only disrupt fishermen's lives but also bring heavy rains.

The first known record of El Niño is attributed to Francisco Pizarro, a Spaniard who in 1525 described unusual desert rainfall in northern Peru and its El Niño association. El Niño episodes occur generally every 2 to 6 years and typically last 12 to 18 months. Recent episodes include 1972-73, 1977-78, 1982-83, 1986-87, 1991-92, and 1997-98.

The intensity of El Niño events varies—some are strong, such as the 1982-83 and 1997-98 events; others are considerably weaker, based on intensity and area encompassed by the abnormally warm ocean temperatures. The eastward extent of the warmer than normal water varies from episode to episode. Both of these characteristics affect the patterns of temperature and precipitation variations associated with El Niño in the U.S. and elsewhere.

The 1997-98 El Niño, one of the most powerful climate events of the century, strongly impacted global weather patterns. The extremely warm temperatures in the equatorial Pacific, combined with shifts in trade winds across the tropics, contributed to wildfires in Indonesia; significant crop loss in Argentina and New Zealand; devastating floods in Chile, Peru, southern Brazil, and northern Argentina; mudslides in California; and record rains in the southeastern U.S.

El Niño has a significant influence on weather and climate patterns around the globe, and its impacts are most clearly seen in the wintertime. During El Niño years, winter temperatures in the continental U.S. tend to be warmer than normal in the northern and west coast states and cooler than normal in the Southeast. Conditions tend to be wetter than normal over central and southern California and the southwest U.S. and across much of the southern third of the contiguous 48 states, particularly along the Gulf Coast, and drier than normal over the northern portions of the Rocky Mountains and in the Ohio valley region. Globally, El Niño brings wetter than normal conditions to Peru and Chile and drier than normal conditions to Australia and Indonesia. It should be noted that El Niño is only one of a number of factors influencing seasonal variations of climate.

The opposite of El Niño is La Niña, with colder than normal sea surface temperatures in the tropical Pacific. La Niña typically brings wetter than normal conditions to the Pacific Northwest and warmer than normal temperatures to much of the southern U.S. during winter months.

El Niño and La Niña episodes are detected and monitored by observing systems, including satellites, moored buoys, and drifting buoys released by volunteer ships crossing the Pacific Ocean. Highly sophisticated numerical computer models of the global ocean and atmosphere use data from the observing systems to predict the onset and evolution of El Niño and its associated impacts. Numerous other models at research institutions worldwide also use the data from the observing systems to increase the understanding of El Niño and improve forecasting techniques.

Wind Chill Table

Source: National Weather Service, NOAA, U.S. Dept. of Commerce

Temperature and wind combine to cause heat loss from body surfaces. The following table shows that, for example, a temperature of 20 degrees Fahrenheit, plus a wind of 20 miles per hour, causes a body heat loss equal to that in minus 10 degrees temperature with no wind. In other words, a 20-mph wind makes 20 degrees feel like minus 10.

The top line of figures shows temperatures in degrees Fahrenheit. The column at far left shows wind speeds up to 45 mph. (Wind speeds greater than 45 mph have little additional chilling effect.)

MPH	35	30	25	20	15	10	5	0	−5	−10	−15	−20	−25	−30	−35	−40	−45
5	33	27	21	16	12	7	0	−5	−10	−15	−21	−26	−31	−36	−42	−47	−52
10	22	16	10	3	−3	−9	−15	−22	−27	−34	−40	−46	−52	−58	−64	−71	−77
15	16	9	2	−5	−11	−18	−25	−31	−38	−45	−51	−58	−65	−72	−78	−85	−92
20	12	4	−3	−10	−17	−24	−31	−39	−46	−53	−60	−67	−74	−81	−88	−95	−103
25	8	1	−7	−15	−22	−29	−36	−44	−51	−59	−66	−74	−81	−88	−96	−103	−110
30	6	−2	−10	−18	−25	−33	−41	−49	−56	−64	−71	−79	−86	−93	−101	−109	−116
35	4	−4	−12	−20	−27	−35	−43	−52	−58	−67	−74	−82	−89	−97	−105	−113	−120
40	3	−5	−13	−21	−29	−37	−45	−53	−60	−69	−76	−84	−92	−100	−107	−115	−123
45	2	−6	−14	−22	−30	−38	−46	−54	−62	−70	−78	−85	−93	−102	−109	−117	−125

Heat Index

The heat index is a measure of the contribution high humidity makes, in combination with abnormally high temperatures, to reducing the body's ability to cool itself. For example, the index shows that an air temperature of 100 degrees Fahrenheit with a relative humidity of 50% has the same effect on the human body as a temperature of 120 degrees. Sunstroke and heat exhaustion are likely when the heat index reaches 105. This index is a measure of what hot weather "feels like" to the average person for various temperatures and relative humidities.

| Relative Humidity | Air Temperature* | | | | | | | | | | |
| | 70 | 75 | 80 | 85 | 90 | 95 | 100 | 105 | 110 | 115 | 120 |
	Apparent Temperature*										
0%	64	69	73	78	83	87	91	95	99	103	107
10%	65	70	75	80	85	90	95	100	105	111	116
20%	66	72	77	82	87	93	99	105	112	120	130
30%	67	73	78	84	90	96	104	113	123	135	148
40%	68	74	79	86	93	101	110	123	137	151	
50%	69	75	81	88	96	107	120	135	150		
60%	70	76	82	90	100	114	132	149			
70%	70	77	85	93	106	124	144				
80%	71	78	86	97	113	136					
90%	71	79	88	102	122						
100%	72	80	91	108							

*Degrees Fahrenheit

Ultraviolet (UV) Index Forecast

Source: National Weather Service, NOAA, U.S. Dept. of Commerce

The National Weather Service (NWS), Environmental Protection Agency (EPA), and Centers for Disease Control and Prevention (CDC) developed and began offering a UV index on June 28, 1994, in response to increasing incidence of skin cancer, cataracts, and other effects from exposure to the sun's harmful rays. The UV Index is now a regular element of NWS atmospheric forecasts.

UV Index number and forecast. The UV Index number, ranging from 0 to 10+, is an indication of the amount of UV radiation reaching the earth's surface over the one-hour period around noon. The lower the number, the less the radiation. The UV Index forecast is produced for 58 cities by the NWS Climate Prediction Center. The index number is based on several factors: latitude, day of year, time of day, total atmospheric ozone, elevation, and predicted cloud conditions. The index is valid for a radius of about 30 miles around a listed city; however, adjustments should be made for a number of factors.

Ozone. Ozone is measured by a NOAA polar orbiting satellite. The more ozone, the lower the UV radiation at the surface.

Cloudiness. Increased cloudiness lowers the Index number.

Reflectivity. Reflective surfaces intensify UV exposure. As an example, grass reflects 2.5% to 3% of UV radiation reaching the surface; sand, 20% to 30%; snow and ice, 80% to 90%; water, up to 100% (depending on reflection angle).

Elevation. At higher elevations, UV radiation travels a shorter distance to reach the surface so there is less atmosphere to absorb the rays. For every 4,000 ft. one travels above sea level, the UV Index increases by 1 unit. Snow and lack of pollutants intensify UV exposure at higher altitudes.

Latitude. The closer to the equator, the higher the UV radiation level.

Accuracy. After gathering data from 20 UV sensors (during June-Oct. 1994), the NWS determined that 32% of UV Index forecasts for that period were correct, 76% were within ±1 UV Index unit, and about 90% were within ±2 units. Unpredictable cloudiness, haze, and pollution contribute to forecast error.

SPF number. The UV Index is not linked in any way to the SPF number on suntan lotions and sunscreens. For an explanation of the SPF factor, contact the product's manufacturer or the Food and Drug Administration.

Further information. For precautions to take after learning the UV Index number, call the U.S. EPA hotline (800-296-1996) or your doctor. For questions on scientific aspects, call the NWS at 301-713-0622.

Global Measured Extremes of Temperature and Precipitation

Source: National Climatic Data Center; based on records through Mar. 1998

Highest Temperature Extremes

Continent	Highest Temp. (deg F)	Place	Elevation (Feet)	Date
Africa	136	El Azizia, Libya	367	Sept. 13, 1922
North America	134	Death Valley, CA (Greenland Ranch)	−178	July 10, 1913
Asia	129	Tirat Tsvi, Israel	−722	June 21, 1942
Australia	128	Cloncurry, Queensland	622	Jan. 16, 1889
Europe	122	Seville, Spain	26	Aug. 4, 1881
South America	120	Rivadavia, Argentina	676	Dec. 11, 1905
Oceania	108	Tuguegarao, Philippines	72	Apr. 29, 1912
Antarctica	59	Vanda Station, Scott Coast	49	Jan. 5, 1974

Lowest Temperature Extremes

Continent	Lowest Temp. (deg F)	Place	Elevation (Feet)	Date
Antarctica	−129.0	Vostok	11,220	July 21, 1983
Asia	−90.0	Oimekon, Russia	2,625	Feb. 6, 1933
Asia	−90.0	Verkhoyansk, Russia	350	Feb. 7, 1892
Greenland	−87.0	Northice	7,687	Jan. 9, 1954
North America	−81.4	Snag, Yukon, Canada	2,120	Feb. 3, 1947
Europe	−67.0	Ust'Shchugor, Russia	279	Jan.*
South America	−27.0	Sarmiento, Argentina	879	June 1, 1907
Africa	−11.0	Ifrane, Morocco	5,364	Feb. 11, 1935
Australia	−9.4	Charlotte Pass, NSW	5,758	June 29, 1994
Oceania	14.0	Haleakala Summit, Maui, HI	9,750	Jan. 2, 1961

*Exact day and year unknown, lowest in 15-year period.

Highest Average Annual Precipitation Extremes

Continent	Highest Avg. (Inches)	Place	Elevation (Feet)	Years of Record
South America	523.6[1,2]	Lloro, Colombia	520[3]	29
Asia	467.4[1]	Mawsynram, India	4,597	38
Oceania	460.0[1]	Mt. Waialeale, Kauai, HI	5,148	30
Africa	405.0	Debundscha, Cameroon	30	32
South America	354.0[2]	Quibdo, Colombia	120	16
Australia	340.0	Bellenden Ker, Queensland	5,102	9
North America	256.0	Henderson Lake, British Columbia	12	14
Europe	183.0	Crkvica, Bosnia-Herzegovina	3,337	22

(1) The value given is continent's highest and possibly the world's depending on measurement practices, procedures, and period of record variations. (2) The official greatest average annual precipitation for South America is 354 inches at Quibdo, Colombia. The 523.6 inches average at Lloro, Colombia (14 miles SE and at a higher elevation than Quibdo) is an estimated amount. (3) Approximate elevation.

Lowest Average Annual Precipitation Extremes

Continent	Lowest Avg. (Inches)	Place	Elevation (Feet)	Years of Record
South America	0.03	Arica, Chile	95	59
Africa	<0.1	Wadi Halfa, Sudan	410	39
Antarctica	0.8[1]	Amundsen-Scott South Pole Station	9,186	10
North America	1.2	Batagues, Mexico	16	14
Asia	1.8	Aden, Yemen	22	50
Australia	4.05	Mulka (Troudaninna), South Australia	160[2]	42
Europe	6.4	Astrakhan, Russia	45	25
Oceania	8.93	Puako, Hawaii	5	13

(1) The value given is the average amount of solid snow accumulating in one year as indicated by snow markers. The liquid content of the snow is undetermined. (2) Approximate elevation.

DISASTERS

As of Oct. 15, 1999. Listings are selective and generally do not include disasters with relatively low fatalities.

Some Notable Shipwrecks Since 1854

(Figures indicate estimated lives lost. Does not include most military disasters.)

1854, Mar.—City of Glasgow; Brit. steamer missing in N Atlantic; 480.

1854, Sept. 27—Arctic; U.S. (Collins Line) steamer sunk in collision with French steamer *Vesta* near Cape Race; 285-351.

1856, Jan. 23—Pacific; U.S. (Collins Line) steamer missing in N Atlantic; 186-286.

1858, Sept. 23—Austria; German steamer destroyed by fire in N Atlantic; 471.

1863, Apr. 27—Anglo-Saxon; Brit. steamer wrecked at Cape Race; 238.

1865, Apr. 27—Sultana; Mississippi River steamer blew up near Memphis, TN; 1,450.

1869, Oct. 27—Stonewall; steamer burned on Mississippi River below Cairo, IL; 200.

1870, Jan. 25—City of Boston; Brit. (Inman Line) steamer vanished between New York and Liverpool; 177.

1870, Oct. 19—Cambria; Brit. steamer wrecked off N Ireland; 196.

1872, Nov. 7—Mary Celeste; U.S. half-brig sailed from New York for Genoa; found abandoned; loss of life unknown.

1873, Jan. 22—Northfleet; Brit. steamer foundered off Dungeness, England; 300.

1873, Apr. 1—Atlantic; Brit. (White Star) steamer wrecked off Nova Scotia; 585.

1873, Nov. 23—Ville du Havre; French steamer sank after collision with Brit. sailing ship *Loch Earn*; 226.

1875, May 7—Schiller; German steamer wrecked off Scilly Isles; 312.

1875, Nov. 4—Pacific; U.S. steamer sank after collision off Cape Flattery; 236.

1878, Sept. 3—Princess Alice; Brit. steamer sank after collision in Thames River; 700.

1878, Dec. 18—Byzantin; French steamer sank after collision in Dardanelles; 210.

1881, May 24—Victoria; steamer capsized in Thames River, Canada; 200.

1883, Jan. 19—Cimbria; German steamer sank in collision with Brit. steamer *Sultan* in North Sea; 389.

1887, Nov. 15—Wah Yeung; Brit. steamer burned at sea; 400.

1890, Feb. 17—Duburg; Brit. steamer wrecked, China Sea; 400.

1890, Sept. 19—Ertogrul; Turkish frigate foundered off Japan; 540.

1891, Mar. 17—Utopia; Brit. steamer sank in collision with Brit. ironclad *Anson* off Gibraltar; 562.

1895, Jan. 30—Elbe; German steamer sank in collision with Brit. steamer *Craithie* in North Sea; 332.

1895, Mar. 11—Reina Regenta; Spanish cruiser foundered near Gibraltar; 400.

1898, Feb. 15—Maine; U.S. battleship blown up in Havana Harbor; 260.

1898, July 4—La Bourgogne; French steamer sank in collision with Brit. sailing ship *Cromartyshire* off Nova Scotia; 549.

1898, Nov. 26—Portland; U.S. steamer wrecked off Cape Cod; 157.

1904, June 15—General Slocum; excursion steamer burned in East River, New York City; 1,030.

1904, June 28—Norge; Danish steamer wrecked on Rockall Island, Scotland; 620.

1906, Aug. 4—Sirio; Italian steamer wrecked off Cape Palos, Spain; 350.

1908, Mar. 23—Matsu Maru; Japanese steamer sank in collision near Hakodate, Japan; 300.

1909, Aug. 1—Waratah; Brit. steamer, Sydney to London, vanished; 300.

1910, Feb. 9—General Chanzy; French steamer wrecked off Minorca, Spain; 200.

1911, Sept. 25—Liberté; French battleship exploded at Toulon; 285.

1912, Mar. 5—Principe de Asturias; Spanish steamer wrecked off Spain; 500.

1912, Apr. 14-15—Titanic; Brit. (White Star) steamer hit iceberg in N Atlantic; 1,503.

1912, Sept. 28—Kichemaru; Japanese steamer sank off Japanese coast; 1,000.

1914, May 29—Empress of Ireland; Brit. (Canadian Pacific) steamer sunk in collision with Norwegian collier in St. Lawrence River; 1,014.

1915, May 7—Lusitania; Brit. (Cunard Line) steamer torpedoed and sunk by German submarine off Ireland; 1,198.

1915, July 24—Eastland; excursion steamer capsized in Chicago River; 812.

1916, Feb. 26—Provence; French cruiser sank in Mediterranean; 3,100.

1916, Mar. 3—Principe de Asturias; Spanish steamer wrecked near Santos, Brazil; 558.

1916, Aug. 29—Hsin Yu; Chinese steamer sank off Chinese coast; 1,000.

1917, Dec. 6—Mont Blanc, Imo; French ammunition ship and Belgian steamer collided in Halifax Harbor; 1,600.

1918, Apr. 25—Kiang-Kwan; Chinese steamer sank in collision off Hankow; 500.

1918, July 12—Kawachi; Japanese battleship blew up in Tokayama Bay; 500.

1918, Oct. 25—Princess Sophia; Canadian steamer sank off Alaskan coast; 398.

1919, Jan. 17—Chaonia; French steamer lost in Straits of Messina, Italy; 460.

1919, Sept. 9—Valbanera; Spanish steamer lost off Florida coast; 500.

1921, Mar. 18—Hong Kong; steamer wrecked in South China Sea; 1,000.

1922, Aug. 26—Niitaka; Japanese cruiser sank in storm off Kamchatka, USSR; 300.

1927, Oct. 25—Principessa Mafalda; Italian steamer blew up, sank off Porto Seguro, Brazil; 314.

1928, Nov. 12—Vestris; Brit. steamer sank off Virginia; 113.

1934, Sept. 8—Morro Castle; U.S. steamer, Havana to New York, burned off Asbury Park, NJ; 134.

1939, May 23—Squalus; U.S. submarine sank off Portsmouth, NH; 26.

1939, June 1—Thetis; submarine sank, Liverpool Bay; 99.

1942, Feb. 18—Truxtun and Pollux; U.S. destroyer and cargo ship ran aground, sank off Newfoundland; 204.

1942, Oct. 2—Curacoa; Brit. cruiser sank after collision with liner Queen Mary; 338.

1944, Dec. 17-18—3 U.S. Third Fleet destroyers sank during typhoon in Philippine Sea; 790.

1947, Jan. 19—Himera; Greek steamer hit a mine off Athens; 392.

1947, Apr. 16—Grandcamp; French freighter exploded in Texas City, TX, harbor, starting fires; 510.

1948, Nov.—Chinese army evacuation ship exploded and sank off S Manchuria; 6,000.

1948, Dec. 3—Kiangya; Chinese refugee ship wrecked in explosion S of Shanghai; 1,100+.

1949, Sept. 17—Noronic; Canadian Great Lakes Cruiser burned at Toronto dock; 130.

1952, Apr. 26—Hobson and Wasp; U.S. destroyer and aircraft carrier collided in Atlantic; 176.

1954, May 26—Pennington; sank off Rhode Island; 103.

1954, Sept. 26—Toya Maru; Japanese ferry sank in Tsugaru Strait, Japan; 1,172.

1956, July 26—Andrea Doria and **Stockholm;** Italian liner and Swedish liner collided off Nantucket; 51.

1957, July 14—Eshghabad; Soviet ship ran aground in Caspian Sea; 270.

1960, Dec. 19—Constellation; U.S. aircraft carrier caught fire in Brooklyn Navy Yard, NY; 49.

1961, Apr. 8—Dara; Brit. liner exploded in Persian Gulf; 236.

1961, July 8—Save; Portuguese ship ran aground off Mozambique; 259.

1963, Apr. 10—Thresher; U.S. Navy atomic submarine sank in N Atlantic; 129.

1964, Feb. 10— Australian destroyer *Voyager* sank after collision with aircraft carrier *Melbourne* off New South Wales; 82.

1965, Nov. 13—Yarmouth Castle; Panamanian registered cruise ship burned and sank off Nassau; 89.

1967, July 29—Forrestal; U.S. aircraft carrier caught fire off N Vietnam; 134.

1968, Jan. 25—Dakar; Israeli submarine vanished in Mediterranean Sea; 69.

1968, late May—Scorpion; U.S. nuclear submarine sank in Atlantic near Azores; 99 (located Oct. 31).

1969, June 2—Evans; U.S. destroyer cut in half by Australian carrier *Melbourne*, S China Sea; 74.

1970, Mar. 4—Eurydice; French submarine sank in Mediterranean near Toulon; 57.

1970, Dec. 15—Namyong-Ho; South Korean ferry sank in Korea Strait; 308.

1974, May 1—Motor launch capsized off Bangladesh; 250.

1974, Sept. 26—Soviet destroyer sank in Black Sea; 200+.

1975, Nov. 10—Edmund Fitzgerald; U.S. cargo ship sank during storm on Lake Superior; 29.

1976, Oct. 20—George Prince and **Frosta;** ferryboat and Norwegian tanker collided on Mississippi R. at Luling, LA; 77.

1976, Dec. 25—Patria; Egyptian liner caught fire and sank in the Red Sea; 100.

1979, Aug. 14—23 yachts competing in Fastnet yacht race sank or abandoned during storm in S Irish Sea; 18.

1981, Jan. 27—Tamponas II; Indonesian passenger ship caught fire and sank in Java Sea; 580.

1983, Feb. 12—Marine Electric; coal freighter sank during storm off Chincoteague, VA; 33.

1983, May 25—10th of Ramadan; Nile steamer caught fire and sank in Lake Nasser; 357.

1986, Apr. 20—ferry sank near Barisal, Bangladesh; 262.

1986, Aug. 31—Soviet passenger ship *Admiral Nakhimov* and Soviet freighter *Pyotr Vasev* collided in Black Sea; 398.

1987, Mar. 6—British ferry capsized off Zeebrugge, Belgium; 189.

1987, Dec. 20—Philippine ferry *Dona Paz* and oil tanker *Victor* collided in Tablas Strait; 4,341.

1988, Aug. 6—Indian ferry capsized on Ganges R.; 400+.

1989, Aug. 20—Brit. barge *Bowbelle* struck Brit. pleasure cruiser *Marchioness* on Thames R. in central London; 56.

1989, Sept. 10—Romanian pleasure boat and Bulgarian barge collided on Danube R.; 161.

1991, Apr. 10—Auto ferry and oil tanker collided outside Livorno Harbor, Italy; 140.

1991, Dec. 14—Salem Express; ferry rammed coral reef near Safaga, Egypt; 462.

1993, Feb. 17—Neptune; ferry capsized off Port-au-Prince, Haiti; 500+.

1993, Oct. 10—West Sea Ferry; capsized in Yellow Sea near W South Korea during storm; 285.

1994, Sept. 28—Estonia; ferry sank in Baltic Sea; 1,049.

1996, May 21—Bukoba; ferry sank in Lake Victoria (Africa); 500.

1997, Mar. 28—Albanian refugee boat sank in Adriatic Sea after being rammed by Italian navy warship *Sibilla*; 83.

1997, Sept. 8—Pride of la Gonâve; Haitian ferry sank off Montrouis, Haiti; 200+.

1998, Apr. 4—passenger boat capsized off coast near Ibaka beach, Nigeria; 280.

1998, Sept. 2—2 passenger boats capsized on Lake Kivu, near Bukavu, Congo; 200+.

1998, Sept. 18—ferry sank S of Manila; 97.

1999, Feb. 6—Harta Rimba; cargo ship sank off Indonesia; 280+.

1999, Mar. 26—passenger boat overturned off coast, Sierra Leone; 150+.

1999, Apr. 2—passenger ferry sank off coast of Nigeria; 100+.

1999, May 1—amphibious excursion boat sank in Lake Hamilton, AR; 13.

1999, May 8—passenger ferry capsized off Bangladesh; 200+.

Some Notable Aircraft Disasters Since 1937

Date	Aircraft	Site of accident	Deaths
1937, May 6	German zeppelin Hindenburg	Burned at mooring, Lakehurst, NJ	36*
1944, Aug. 23	U.S. Air Force B-24 Liberator bomber	Hit school, Freckleton, England	61*
1945, July 28	U.S. Army B-25	Hit Empire State Building, New York, NY.	14*
1952, Dec. 20	U.S. Air Force C-124	Fell, burned, Moses Lake, WA	87
1953, Mar. 3	Canadian Pacific Comet Jet	Karachi, Pakistan	11[1]
1953, June 18	U.S. Air Force C-124	Crashed, burned near Tokyo	129
1955, Oct. 6	United Airlines DC-4	Crashed in Medicine Bow Peak, WY	66
1955, Nov. 1	United Airlines DC-6B	Exploded, crashed near Longmont, CO	44[2]
1956, June 20	Venezuelan Super-Constellation	Crashed in Atlantic off Asbury Park, NJ	74
1956, June 30	TWA Super-Const., United DC-7	Collided over Grand Canyon, AZ	128
1960, Dec. 16	United DC-8 jet, TWA Super-Const.	Collided over New York City	134[3]
1962, Mar. 16	Flying Tiger Super-Constellation	Vanished in W Pacific	107
1962, June 3	Air France Boeing 707	Crashed on takeoff from Paris	130
1962, June 22	Air France Boeing 707 jet	Crashed in storm, Guadeloupe, W.I.	113
1963, June 3	Chartered Northwest Airlines DC-7	Crashed in Pacific off British Columbia.	101
1963, Nov. 29	Trans-Canada Airlines DC-8F	Crashed after takeoff from Montreal.	118
1965, May 20	Pakistani Boeing 720-B	Crashed at Cairo, Egypt, airport.	121
1966, Jan. 24	Air India Boeing 707 jetliner	Crashed on Mont Blanc, France-Italy.	117
1966, Feb. 4	All-Nippon Boeing 727	Plunged into Tokyo Bay	133
1966, Mar. 5	BOAC Boeing 707 jetliner	Crashed on Mount Fuji, Japan.	124
1966, Dec. 24	U.S. military-chartered CL-44.	Crashed into village in South Vietnam	129*
1967, Apr. 20	Swiss Britannia turboprop	Crashed at Nicosia, Cyprus	126
1967, July 19	Piedmont Boeing 727, Cessna 310	Collided in air, Hendersonville, NC.	82
1968, Apr. 20	S. African Airways Boeing 707	Crashed on takeoff, Windhoek, South-West Africa.	122
1968, May 3	Braniff International Electra	Crashed in storm near Dawson, TX	85
1969, Mar. 16	Venezuelan DC-9	Crashed after takeoff from Maracaibo, Venezuela	155[4]
1969, Dec. 8	Olympic Airways DC-6B	Crashed near Athens in storm	93
1970, Feb. 15	Dominican DC-9	Crashed into sea on takeoff from Santo Domingo	102
1970, July 3	British chartered jetliner	Crashed near Barcelona, Spain	112
1970, July 5	Air Canada DC-8	Crashed near Toronto International Airport	108
1970, Aug. 9	Peruvian turbojet	Crashed after takeoff from Cuzco, Peru	101*
1970, Nov. 14	Southern Airways DC-9	Crashed in mountains near Huntington, WV	75[5]
1971, July 30	All-Nippon Boeing 727 and Japanese Air Force F-86	Collided over Morioka, Japan	162[6]
1971, Sept. 4	Alaska Airlines Boeing 727	Crashed into mountain near Juneau, AK	111
1972, Aug. 14	East German Ilyushin-62	Crashed on takeoff, East Berlin	156
1972, Oct. 13	Aeroflot Ilyushin-62	Crashed near Moscow	176
1972, Dec. 3	Chartered Spanish airliner	Crashed on takeoff, Canary Islands	155
1972, Dec. 29	Eastern Airlines Lockheed Tristar.	Crashed on approach to Miami Intl. Airport	101
1973, Jan. 22	Chartered Boeing 707	Burst into flames during landing, Kano Airport, Nigeria	176
1973, Feb. 21	Libyan jetliner	Shot down by Israeli fighter planes over Sinai	108
1973, Apr. 10	British Vanguard turboprop	Crashed during snowstorm at Basel, Switzerland	104
1973, June 3	Soviet Supersonic TU-144	Crashed near Goussainville, France.	14[7]
1973, July 11	Brazilian Boeing 707	Crashed on approach to Orly Airport, Paris	122
1973, July 31	Delta Airlines jetliner	Crashed, landing in fog at Logan Airport, Boston.	89
1973, Dec. 23	French Caravelle jet	Crashed in Morocco	106
1974, Mar. 3	Turkish DC-10 jet	Crashed at Ermenonville near Paris	346
1974, Apr. 23	Pan American 707 jet	Crashed in Bali, Indonesia	107
1974, Dec. 1	TWA-727	Crashed in storm, Upperville, VA	92
1974, Dec. 4	Dutch-chartered DC-8	Crashed in storm near Colombo, Sri Lanka	191
1975, Apr. 4	Air Force Galaxy C-5A	Crashed near Saigon, S Viet., after takeoff (carrying orphans)	172
1975, June 24	Eastern Airlines 727 jet	Crashed in storm, JFK Airport, NY	113
1975, Aug. 3	Chartered 707	Hit mountainside, Agadir, Morocco.	188
1976, Sept. 10	British Airways Trident, Yugoslav DC-9	Collided near Zagreb, Yugoslavia	176
1976, Sept. 19	Turkish 727.	Hit mountain, S Turkey	155
1976, Oct. 13	Bolivian 707 cargo jet	Crashed in Santa Cruz, Bolivia.	100[8]
1977, Mar. 27	KLM 747, Pan American 747	Collided on runway, Tenerife, Canary Islands.	582[9]
1977, Nov. 19	TAP Boeing 727	Crashed on Madeira	130
1977, Dec. 4	Malaysian Boeing 737	Hijacked, then exploded in mid-air over Straits of Johore.	100
1977, Dec. 13	U.S. DC-3	Crashed after takeoff at Evansville, IN	29[10]
1978, Jan. 1	Air India 747	Exploded, crashed into sea off Bombay	213
1978, Sept. 25	Boeing 727, Cessna 172	Collided in air, San Diego, CA	150
1978, Nov. 15	Chartered DC-8	Crashed near Colombo, Sri Lanka	183
1979, May 25	American Airlines DC-10	Crashed after takeoff at O'Hare Intl. Airport, Chicago	275[11]

Date	Aircraft	Site of accident	Deaths
1979, Aug. 17	Two Soviet Aeroflot jetliners	Collided over Ukraine	173
1979, Nov. 26	Pakistani Boeing 707	Crashed near Jidda, Saudi Arabia	156
1979, Nov. 28	New Zealand DC-10	Crashed into mountain in Antarctica	257
1980, Mar. 14	Polish Ilyushin 62	Crashed making emergency landing, Warsaw	87[12]
1980, Aug. 19	Saudi Arabian Tristar	Burned after emergency landing, Riyadh	301
1981, Dec. 1	Yugoslavian DC-9	Crashed into mountain in Corsica	178
1982, Jan. 13	Air Florida Boeing 737	Crashed into Potomac R. after takeoff	78
1982, July 9	Pan Am Boeing 727	Crashed after takeoff in Kenner, LA	153[13]
1983, Sept. 1	S. Korean Boeing 747	Shot down after violating Soviet airspace	269
1983, Nov. 27	Colombian Boeing 747	Crashed near Barajas Airport, Madrid	183
1985, Feb. 19	Spanish Boeing 727	Crashed into Mt. Oiz, Spain	148
1985, June 23	Air-India Boeing 747	Crashed into Atlantic Ocean S of Ireland	329
1985, Aug. 2	Delta Air Lines L-1011	Crashed at Dallas-Ft. Worth Intl. Airport	137
1985, Aug. 12	Japan Air Lines Boeing 747	Crashed into Mt. Ogura, Japan	520[14]
1985, Dec. 12	Arrow Air DC-8	Crashed after takeoff in Gander, Newfoundland	256[15]
1986, Mar. 31	Mexican Boeing 727	Crashed NW of Mexico City	166
1986, Aug. 31	Aeromexico DC-9	Collided with Piper PA-28 over Cerritos, CA	82[16]
1987, May 9	Polish Ilyushin 62M	Crashed after takeoff in Warsaw, Poland	183
1987, Aug. 16	Northwest Airlines MD-82	Crashed after takeoff in Romulus, MI	156
1987, Nov. 28	S. African Boeing 747	Crashed into Indian Ocean near Mauritius	159
1987, Nov. 29	S. Korean Boeing 707	Exploded over Thai-Burmese border	155
1988, Mar. 17	Colombian Boeing 707	Crashed into mountainside near Venezuela border	137
1988, July 3	Iranian A300 Airbus	Shot down by U.S. Navy warship Vincennes over Persian Gulf	290
1988, Dec. 21	Pan Am Boeing 747	Exploded and crashed in Lockerbie, Scotland	270[17]
1989, Feb. 8	U.S. Boeing 707	Crashed into mountain in Azores Islands off Portugal	144
1989, June 7	Suriname DC-8	Crashed near Paramaribo Airport, Suriname	168
1989, July 19	United Airlines DC-10	Crashed while landing in Sioux City, IA	111
1989, Sept. 19	French DC-10	Exploded in air over Niger	171
1990, Oct. 2	Chinese airline Boeing 737	Hijacked; upon landing in Guangzhou, crashed on ground	132
1991, May 26	Lauda-Air Boeing 767-300	Exploded over rural Thailand	223
1991, July 11	Nigerian DC-8	Crashed while landing at Jidda, Saudi Arabia	261
1991, Oct. 5	Indonesian military transport	Crashed after takeoff from Jakarta	137*
1992, July 31	Thai Airbus A-300-310F	Crashed into mountain S. of Kathmandu, Nepal	113
1992, Oct. 4	El Al Boeing 747-200F	Crashed into 2 apartment bldgs., Amsterdam, Netherlands	120*
1994, Jan. 3	Aeroflot TU-154	Crashed and exploded after takeoff in Irkhutsk, Russia	125[18]
1994, Apr. 26	China Airlines Airbus A-300-600R	Crashed at Japan's Nagoya Airport	264
1994, June 16	China Northwest Airlines TU-154	Crashed 10 min. after takeoff	160
1994, Sept. 8	USAir Boeing 737-300	Crashed in Aliquippa, PA, near Pittsburgh Intl. Airport	132
1994, Oct. 31	American Eagle ATR-72-210	Crashed in field near Roselawn, IN	68
1995, Aug. 11	Aviateca Boeing 737	Crashed into Chichontepec volcano, El Salvador	65
1995, Dec. 20	American Airlines Boeing 757	Crashed into mountain 50 mi N of Cali, Colombia	160
1996, Jan. 8	Antonova 32 cargo jet	Crashed into central market, Kinshasa, Zaire	350+*
1996, Feb. 6	Turkish Boeing 757	Crashed into Atlantic Ocean, off Dominican Republic	189
1996, Apr. 25	T-43, a military version of a Boeing 737	Crashed into mountain near Dubrovnik, Croatia	35[19]
1996, May 11	ValuJet DC-9	Crashed into the Florida Everglades after takeoff	110
1996, July 17	Trans World Airlines Boeing 747	Exploded and crashed in Atlantic Ocean, off Long Isl., NY	230
1996, Aug. 29	Vnukovo TU-154	Crashed into mountain on Arctic island of Spitsbergen	141
1996, Oct. 2	Aeroperu Boeing 757	Crashed in Pacific after takeoff from Lima, Peru	70
1996, Oct. 31	Brazilian TAM Fokker-100	Crashed into houses in São Paulo, Brazil	98[20]
1996, Nov. 7	Nigerian Boeing 727	Crashed into a lagoon 40 mi SE of Lagos, Nigeria	143
1996, Nov. 12	Saudi Arabian Boeing 747, Kazakh Ilyushin-76 cargo plane	Collided in midair near New Delhi, India	349[21]
1996, Nov. 23	Ethiopian Boeing 767	Hijacked, then crashed in Indian Ocean off the Comoros	127
1997, Jan. 9	Comair Embraer 120	Crashed on approach into Detroit Metro. Airport	29
1997, Feb. 4	2 Sikorsky CH-53 transport helicopters	Collided in midair over northern Galilee, Israel	73
1997, May 8	China Southern Airlines Boeing 737	Crashed on approach into Shenzhen's Huangtian Airport	35
1997, July 11	Cubana de Aviación Antonov-24	Crashed into the Caribbean off SE Cuba	44
1997, Aug. 6	Korean Air Boeing 747-300	Crashed into jungle on Guam on approach into airport	228
1997, Sept. 3	Vietnamese Airlines Tupolev TU-134	Crashed on approach into Phnom Penh airport	64
1997, Sept. 14	U.S. C-141 cargo plane, German TU-154	Collided in midair off SW Africa	33
1997, Sept. 26	Indonesian Airbus A-300	Crashed near Medan, Indonesia, airport	234
1997, Oct. 10	Austral Airlines DC-9-32	Crashed and exploded near Neuvo Berlin, Uruguay	74
1997, Dec. 6	Russian AN-124 transport cargo plane	Crashed into apartment complex near Irkutsk, Siberia	67*
1997, Dec. 15	Chartered TU-154 from Tajikistan	Crashed in desert near Sharja, U.A.E., airport	85
1997, Dec. 17	Chartered Yakovlev-42 from Ukraine	Crashed in mountains near Katerini, Greece	70
1997, Dec. 19	SilkAir Boeing 737-300	Crashed in Musi River, Sumatra, Indonesia	104
1998, Jan. 14	Afghan cargo plane	Crashed into mountain, SW Pakistan	50+
1998, Feb. 2	Cebu Pacific Air DC-9-32	Crashed into mountain near Cagayan de Oro, Philippines	104
1998, Feb. 16	China Airlines Airbus 300-622R	Crashed on approach to airport, Taipei, Taiwan	203[22]
1998, Apr. 20	Air France Boeing 727-200	Crashed into mountain after takeoff from Bogotá, Colombia	53
1998, Sept. 2	Swissair MD-11	Crashed into Atlantic Ocean off Halifax, Nova Scotia	229
1998, Sept. 25	Pauknair BAE146	Crashed into hillside in Morocco	38
1998, Oct. 11	Congo Oar Lines Boeing 727	Shot down by rebels in Kindu, Congo	40
1998, Dec. 11	Thai Airways Airbus A310-200	Crashed short of runway at Surat Thani airport, southern Thailand	101
1999, Feb. 3	Chartered Antonov plane	Crashed in residential area of Luanda, Angola	28
1999, Feb. 8	Chartered Air Karibu plane	Crashed near Menkao, Congo, 60 mi E of Kinshasa	8
1999, Feb. 24	China Southwest Airlines TU-154	Crashed on approach to Wenzhou airport, eastern China	61
1999, Sept. 1	LAPA Boeing 737-200	Crashed on takeoff from Jorge Newbery Airport, Buenos Aires	74[23]
1999, Sept. 1	Chartered Northern Air Cessna 404	Crashed into Mount Meru, Arusha, Tanzania	12

*Including those on ground and in buildings. (1) First fatal crash of commercial jet plane. (2) Caused by bomb planted by John G. Graham in insurance plot to kill his mother, a passenger. (3) Incl. all 128 aboard planes and 6 on ground. (4) Killed 84 on the plane and 71 on the ground. (5) Incl. 43 Marshall Univ. football players and coaches. (6) Airliner-fighter crash; pilot of fighter parachuted to safety, was arrested for negligence. (7) First supersonic plane crash; killed 6 crewmen and 8 on ground; there were no passengers. (8) Crew of 3 killed; 97, mostly children, killed on the ground. (9) World's worst airline disaster. (10) Incl. Univ. of Evansville basketball team. (11) Incl. 2 on the ground. Highest death toll in U.S. aviation history. (12) Incl. 22 members of U.S. boxing team. (13) Incl. 8 on the ground. (14) Worst single-plane disaster. (15) Incl. 248 members of U.S. 101st Airborne Division. (16) Incl. 15 on the ground. (17) Incl. 11 on the ground. (18) Incl. 1 on the ground. (19) Incl. U.S. Sec. of Commerce Ronald Brown. (20) Incl. 2 on the ground. (21) World's worst midair collision. (22) Incl. 6 on the ground. (23) Incl. 10 on the ground.

Some Notable Railroad Disasters

Date	Location	Deaths	Date	Location	Deaths
1876, Dec. 29	Ashtabula, OH	92	1944, Dec. 31	Bagley, UT	50
1880, Aug. 11	Mays Landing, NJ	40	1945, Aug. 9	Michigan, ND	34
1887, Aug. 10	Chatsworth, IL	81	1946, Mar. 20	Aracaju, Mexico	185
1888, Oct. 10	Mud Run, PA	55	1946, Apr. 25	Naperville, IL	45
1891, June 14	Nr. Basel, Switzerland	100	1947, Feb. 18	Gallitzin, PA	24
1896, July 30	Atlantic City, NJ	60	1949, Oct. 22+	Nr. Dwor, Poland	200
1903, Dec. 23	Laurel Run, PA	53	1950, Feb. 17	Rockville Centre, NY	31
1904, Aug. 7	Eden, CO	96	1950, Sept. 11	Coshocton, OH	33
1904, Sept. 24	New Market, TN	56	1950, Nov. 22	Richmond Hill, NY	79
1906, Mar. 16	Florence, CO	35	1951, Feb. 6	Woodbridge, NJ	84
1906, Oct. 28	Atlantic City, NJ	40	1952, Mar. 4	Nr. Rio de Janeiro, Brazil	119
1906, Dec. 30	Washington, DC	53	1952, July 9	Rzepin, Poland	160
1907, Jan. 2	Volland, KS	33	1952, Oct. 8	Harrow, England	112
1907, Jan. 19	Fowler, IN	29	1953, Mar. 27	Conneaut, OH	21
1907, Feb. 16	New York, NY	22	1955, Apr. 3	Guadalajara, Mexico	300
1907, Feb. 23	Colton, CA	26	1956, Jan. 22	Los Angeles, CA	30
1907, May 11	Lompoc, CA	36	1957, Sept. 1	Kendal, Jamaica	178
1907, July 20	Salem, MI	33	1957, Sept. 29	Montgomery, W Pakistan	250
1910, Mar. 1	Wellington, WA	96	1957, Dec. 4	London, England	90
1910, Mar. 21	Green Mountain, IA	55	1958, May 8	Rio de Janeiro, Brazil	128
1911, Aug. 25	Manchester, NY	29	1958, Sept. 15	Elizabethport, NJ	48
1912, July 4	East Corning, NY	39	1960, Nov. 14	Pardubice, Czech.	110
1912, July 5	Ligonier, PA	23	1962, Jan. 8	Woerden, Netherlands	91
1914, Aug. 5	Tipton Ford, MO	43	1962, May 3	Tokyo, Japan	163
1914, Sept. 15	Lebanon, MO	28	1964, July 26	Porto, Portugal	94
1915, May 22	Nr. Gretna, Scotland	227	1970, Feb. 1	Buenos Aires, Argentina	236
1916, Mar. 29	Amherst, OH	27	1972, June 16	Vierzy, France	107
1917, Sept. 28	Kellyville, OK.	23	1972, July 21	Seville, Spain	76
1917, Dec. 12[1]	Modane, France	543	1972, Oct. 6	Saltillo, Mexico	208
1917, Dec. 20	Shepherdsville, KY	46	1972, Oct. 30	Chicago, IL	45
1918, June 22	Ivanhoe, IN	68	1974, Aug. 30	Zagreb, Yugoslavia	153
1918, July 9	Nashville, TN	101	1975, Feb. 28	London subway train	41
1918, Nov. 1	Brooklyn, NY	97	1977, Jan. 18	Granville, Australia	83
1919, Jan. 12	South Byron, NY	22	1981, June 6	Bihar, India	700+
1919, Dec. 20	Onawa, ME	23	1982, Jan. 27	El Asnam, Algeria	130
1921, Feb. 27	Porter, IN	37	1982, July 11	Tepic, Mexico	120
1921, Dec. 5	Woodmont, PA	27	1983, Feb. 19	Empalme, Mexico	100
1922, Aug. 5	Sulphur Spring, MO	34	1988, Dec. 12	London, England	115
1922, Dec. 13	Humble, TX.	22	1989, Jan. 15+	Maizdi Khan, Bangladesh	110
1923, Sept. 27	Lockett, WV	31	1990, Jan. 4+	Sindh Prov., Pakistan	210
1925, June 16	Hackettstown, NJ	50	1991, May 14	Shigaraki, Japan	42
1926, Sept. 5	Waco, CO	30	1993, Sept. 22	Big Bayou Conot, AL	47
1937, July 16	Nr. Patna, India	107	1994, Mar. 8	Nr. Durban, South Africa	63
1938, June 19	Saugus, MT	47	1994, Sept. 22	Tolunda, Angola	300
1939, Aug. 12	Harney, NV	24	1995, Aug. 20	Firozabad, India	358
1939, Dec. 22	Near Magdeburg, Germany	132	1997, Mar. 3	Punjab State, Pakistan	125
1939, Dec. 22	Near Friedrichshafen, Germany	99	1997, Mar. 31	Huarte Arakil, Spain	21
1940, Apr. 19	Little Falls, NY	31	1997, Apr. 29	Hunan, China	58
1940, July 31	Cuyahoga Falls, OH	43	1997, May 4+	Rwanda	100
1943, Aug. 29	Wayland, NY	27	1997, Sept. 14	Central India	77
1943, Sept. 6	Frankford Junction, Philadelphia, PA	79	1998, June 3	Eschede, Germany	102
1943, Dec. 16	Between Rennert and Buie, NC	72	1998, Feb. 19+	Yaounde, Cameroon	100
1944, Jan. 16	Leon Prov., Spain	500	1999, Mar. 15	Bourbonnais, IL	11
1944, Mar. 2	Salerno, Italy	521	1999, Mar. 24	Nairobi, Kenya	32+
1944, July 6	High Bluff, TN	35	1999, Aug. 2	Gauhati, India	285+
1944, Aug. 4	Near Stockton, GA	47	1999, Oct. 5	London, England	30+
1944, Sept. 14	Dewey, IN	29			

(1) World's worst train wreck; passenger train derailed.

Principal U.S. Mine Disasters Since 1900

Source: Bureau of Mines, U.S. Dept. of the Interior; Mine Safety and Health Admin., U.S. Dept. of Labor

(All are bituminous-coal mines unless otherwise noted.)

Date	Location	Deaths	Date	Location	Deaths
1900, May 1	Scofield, UT	200	1917, June 8	Butte, MT[1]	163
1902, May 19	Coal Creek, TN	184	1919, June 5	Wilkes-Barre, PA[2]	92
1902, July 10	Johnstown, PA	112	1922, Nov. 6	Spangler, PA	77
1903, June 30	Hanna, WY	169	1922, Nov. 22	Dolomite, AL	90
1904, Jan. 25	Cheswick, PA	179	1923, Feb. 8	Dawson, NM	120
1905, Feb. 26	Virginia City, AL	112	1923, Aug. 14	Kemmerer, WY	99
1907, Jan. 29	Stuart, WV	84	1924, Mar. 8	Castle Gate, UT	171
1907, Dec. 6	Monongah, WV	361	1924, Apr. 28	Benwood, WV	119
1907, Dec. 19	Jacobs Creek, PA	239	1926, Jan. 13	Wilburton, OK	91
1908, Nov. 28	Marianna, PA	154	1927, Apr. 30	Everettville, WV	97
1909, Nov. 13	Cherry, IL	259	1928, May 19	Mather, PA	195
1910, Jan. 31	Primero, CO	75	1930, Nov. 5	Millfield, OH	82
1910, May 5	Palos, AL	90	1940, Jan. 10	Bartley, WV	91
1910, Nov.8	Delagua, CO	79	1947, Mar. 25	Centralia, IL	111
1911, Apr. 8	Littleton, AL	128	1951, Dec. 21	West Frankfort, IL	119
1911, Dec. 9	Briceville, TN	84	1968, Nov. 20	Farmington, WV	78
1912, Mar. 26	Jed, WV	83	1972, May 2	Kellogg, ID[1]	91
1913, Apr. 23	Finleyville, PA	96	1976, Mar. 9	Oven Fork, KY	15
1913, Oct. 22	Dawson, NM	263	1981, Apr. 15	Redstone, CO	15
1914, Apr. 28	Eccles, WV	181	1981, Dec. 8	Whitwell, TN	13
1915, Mar. 2	Layland, WV	112	1984, Dec. 19	Huntington, UT	27
1917, Apr. 27	Hastings, CO	121	1989, Sept. 13	Sturgis, KY	10

Note: World's worst mine disaster killed 1,549 workers in Honkeiko Colliery in Manchuria, Apr. 25, 1942. (1) Metal mine. (2) Anthracite mine.

Some Notable U.S. Tornadoes Since 1925

Date	Location	Deaths	Date	Location	Deaths
1925, Mar. 18	MO, IL, IN	689	1968, May 15	Midwest	71
1927, Apr. 12	Rock Springs, TX	74	1969, Jan. 23	MS	32
1927, May 9	AR, Poplar Bluff, MO	92	1971, Feb. 21	Mississippi delta	110
1927, Sept. 29	St. Louis, MO	90	1973, May 26-27	South, Midwest (series)	47
1930, May 6	Hill, Navarro, Ellis Co., TX	41	1974, Apr. 3-4	AL, GA, TN, KY, OH	315
1932, Mar. 21	AL (series of tornadoes)	268	1977, Apr. 4	AL, MS, GA	22
1936, Apr. 5	MS, GA	455	1979, Apr. 10	TX, OK	60
1936, Apr. 6	Gainesville, GA	203	1984, Mar. 28	NC, SC	57
1938, Sept. 29	Charleston, SC	32	1985, May 31	NY, PA, OH, Ont. (series)	75
1942, Mar. 16	Central to NE Mississippi	75	1987, May 22	Saragosa, TX	29
1942, Apr. 27	Rogers and Mayes Co., OK	52	1989, Nov. 15	Huntsville, AL	18
1944, June 23	OH, PA, WV, MD	150	1990, Aug. 28	Northern IL	25
1945, Apr. 12	OK-AR	102	1991, Apr. 26	KS, OK	23
1947, Apr. 9	TX, OK, KS	169	1992, Nov. 21-23	South, Midwest	26
1948, Mar. 19	Bunker Hill and Gillespie, IL	33	1994, Mar. 27-28	AL, TN, GA, NC, SC	
1949, Jan. 3	LA and AR	58		(series)	52
1952, Mar. 21	AR, MO, TN (series)	208	1995, May 6-7	Southern OK, northern TX	23
1953, May 11	Waco, TX	114	1997, Mar. 1	Central AR	26
1953, June 8	MI, OH	142	1997, May 27	Jarrell, TX	27
1953, June 9	Worcester and vicinity, MA	90	1998, Feb. 22-23	Central FL	42
1953, Dec. 5	Vicksburg, MS	38	1998, Mar. 20	Northeast GA	12
1955, May 25	KS, MO, OK, TX	115	1998, Mar. 24	Eastern India	145
1957, May 20	KS, MO	48	1998, Apr. 8	AL, GA, MS	39
1958, June, 4	NW Wisconsin	30	1998, Apr. 16	AK, KY, TN	10
1959, Feb. 10	St. Louis, MO	21	1998, May 30	Spencer, SD	6
1960, May 5, 6	Southeastern OK, AR	30	1999, Jan. 17	Western TN	8
1965, Apr. 11	IN, IL, OH, MI, WI	271	1999, Jan. 21	AK, TN	8
1966, Mar. 3	Jackson, MS	57	1999, Apr. 3	Northwestern LA	6
1966, Mar. 3	MS, AL	61	1999, Apr. 9	OH, IL, IN, MO	6
1967, Apr. 21	IL, MI	33	1999, May 3	OK, KS	49

Some Notable Hurricanes, Typhoons, Blizzards, Other Storms

H.—hurricane; T.—typhoon

Date	Location	Deaths	Date	Location	Deaths
1888, Mar. 11-14	Blizzard, eastern U.S.	400	1970, Oct. 15	T. *Titang,* Philippines	526
1900, Aug.-Sept.	H., Galveston, TX	6,000	1970, Nov. 13	Cyclone, Bangladesh	300,000
1906, Sept. 19-24	H., LA, MS.	350	1971, Aug. 1	T. *Rose,* Hong Kong	130
1906, Sept. 18	Typhoon, Hong Kong	10,000	1972, June 19-29	H. *Agnes,* FL to NY	118
1915, Sept. 29	H., LA	500	1972, Dec. 3	T. *Theresa,* Philippines	169
1926, Sept. 11-22	H., FL, AL	243	1973, June-Aug.	Monsoon rains, India	1,217
1926, Oct. 20	H., Cuba	600	1974, June 11	Storm Dinah, Luzon Isl., Phil.	71
1928, Sept. 6-20	H., southern FL	1,836	1974, July 11	T. *Gilda,* Japan, S. Korea	108
1930, Sept. 3	H., Dominican Republic	2,000	1974, Sept. 19-20	H. *Fifi,* Honduras	2,000
1935, Aug. 29- Sept. 10	H., Caribbean, southeastern U.S.	400+	1974, Dec. 25	Cyclone leveled Darwin, Austral.	50
1938, Sept. 21	H., Long Island, NY; New England	600	1975, Sept. 13-27	H. *Eloise,* Caribbean, NE U.S.	71
			1976, May 20	T. *Olga,* floods, Philippines	215
1940, Nov. 11-12	Blizzard, NE, Midwest U.S.	144	1977, July 25, 31	T. *Thelma,* T. *Vera,* Taiwan	39
1942, Oct. 15-16	H., Bengal, India	40,000	1978, Oct. 27	T. *Rita,* Philippines	c. 400
1944, Sept. 9-16	H., NC to New England	46	1979, Aug. 30 - Sept. 7	H. *David,* Caribbean, E U.S.	1,100
1947, Dec. 26	Blizzard, NYC, N Atlantic states	55	1980, Aug. 4-11	H. *Allen,* Caribbean, TX	272
1952, Oct. 22	Typhoon, Philippines	440	1981, Nov. 25	T. *Irma,* Luzon Isl., Phil.	176
1954, Aug. 30	H. *Carol,* northeastern U.S.	68	1983, June	Monsoon, India	900
1954, Oct. 5-18	H. *Hazel,* E Canada, U.S.; Haiti.	347	1984, Sept. 2	T. *Ike,* S Philippines	1,363
1955, Aug. 12-13	H. *Connie,* NC, SC, VA, MD	43	1985, May 25	Cyclone, Bangladesh	10,000
1955, Aug. 7-21	H. *Diane,* eastern U.S.	400	1985, Oct. 26-Nov. 6	H. *Juan,* SE U.S.	97
1955, Sept. 19	H. *Hilda,* Mexico	200	1987, Nov. 25	T. *Nina,* Philippines	650
1955, Sept. 22-28	H. *Janet,* Caribbean	500	1988, Sept. 10-17	H. *Gilbert,* Caribbean, Gulf of Mex.	260
1956, Feb. 1-29	Blizzard, W Europe	1,000	1989, Sept. 16-22	H. *Hugo,* Caribbean, SE U.S.	504
1957, June 25-30	H. *Audrey,* TX to AL	390	1990, May 6-11	Cyclones, SE India	450
1958, Feb. 15-16	Blizzard, northeastern U.S.	171	1991, Apr. 30	Cyclone, Bangladesh	139,000
1959, Sept. 17-19	T. *Sarah,* Japan, S. Korea	2,000	1991, Nov. 5	Tropical storm, Philippines	7,000+
1959, Sept. 26-27	T. *Vera,* Honshu, Japan	4,466	1992, Aug. 24-26	H. *Andrew,* southern FL, LA	14
1960, Sept. 4-12	H. *Donna,* Caribbean, E U.S.	148	1993, Mar. 13-14	Blizzard, eastern U.S.	200
1961, Sept. 11-14	H. *Carla,* TX.	46	1993, June	Monsoon, Bangladesh	2,000
1961, Oct. 31	H. *Hattie,* Br. Honduras	400	1994, Nov. 8-18	Storm Gordon, Caribbean, FL	830
1963, May 28-29	Windstorm, Bangladesh	22,000	1995, Sept. 4-6	H. *Luis,* Caribbean	14
1963, Oct. 4-8	H. *Flora,* Caribbean	6,000	1995, Sept. 13-22	H. *Marilyn,* Virgin Isls., Carib.	13
1964, Oct. 4-7	H. *Hilda,* LA, MS, GA.	38	1995, Oct. 2-4	H. *Opal,* S Mexico, FL, AL	59
1964, June 30	T. *Winnie,* N Philippines	107	1995, Nov. 2-3	T. *Angela,* Philippines	600+
1964, Sept. 5	T. *Ruby,* Hong Kong and China	735	1996, Jan. 7-8	Blizzard, northeastern U.S.	100
1965, May 11-12	Windstorm, Bangladesh	17,000	1996, July 8-13	H. *Bertha,* Carib., eastern U.S.	15
1965, June 1-2	Windstorm, Bangladesh	30,000	1996, Aug. 22	Blizzard, Himalayas, N India	239
1965, Sept. 7-12	H. *Betsy,* FL, MS, LA	74	1996, Aug. 29- Sept. 6	H. *Fran,* Carib., NC, VA, WV.	28
1965, Dec. 15	Windstorm, Bangladesh	10,000	1996, Sept. 9-10	H. *Hortense,* Caribbean	24
1966, June 4-10	H. *Alma,* Honduras, SE U.S.	51	1996, Sept. 9	T. *Sally,* S China.	114
1966, Sept. 24-30	H. *Inez,* Carib., FL, Mexico	293	1996, Nov. 6	Cyclone, Andhra Pradesh, India.	1,000+
1967, July 9	T. *Billie,* SW Japan	347	1996, Nov. 24-25	Ice storms, TX to MO	26
1967, Sept. 5-23	H. *Beulah,* Carib., Mex., TX.	54	1996, Dec. 25	Tropical storm, E Malaysia	100+
1967, Dec. 12-20	Blizzard, Southwest U.S.	51	1997, May 19	Cyclone, Bangladesh	108
1968, Nov. 18-28	T. *Nina,* Philippines	63	1997, May 26	Rain storm, Philippines	29
1969, Aug. 17-18	H. *Camille,* MS, LA	256	1997, July 2	Storms, southeastern MI	16
1970, July 30- Aug. 5	H. *Celia,* Cuba, FL, TX	31	1997, Aug. 18	Typhoon, Taiwan	24
1970, Aug. 20-21	H. *Dorothy,* Martinique	42	1997, Sept. 27	Cyclone, S Bangladesh	c. 35
1970, Sept. 15	T. *Georgia,* Philippines	300			
1970, Oct. 14	T. *Sening,* Philippines	583			

Date	Location	Deaths	Date	Location	Deaths
1997, Oct. 8-10	H. *Pauline*, SW Mexico	230	1998, Sept. 21-23	H. *Georges*, Caribbean, FL	
1997, Oct. 13	Cyclone, Tongi, Bangladesh	15+		Keys, U.S. Gulf Coast	600+
1998, Feb. 4-6	Blizzard, KY, WV	10+	1998, Oct. 27-29	H. *Mitch*, Honduras, Nicaragua,	
1998, June 9	Cyclone, Gujarat, India	1,320		Guatemala, El Salvador	10,866+
1998, Aug.	Monsoon, Bangladesh	326	1999, Sept. 4-17	H. *Floyd,* Bahamas, eastern	
				seaboard, U.S.	69+

Some Notable Floods, Tidal Waves

Date	Location	Deaths	Date	Location	Deaths
1228	Holland	100,000	1976, June 5	Teton Dam collapse, ID	11
1642	China	300,000	1976, July 31	Big Thompson Canyon, CO	139
1883, Aug. 27	Indonesia	36,000	1976, Nov. 17	East Java, Indonesia	136
1887	Huang He River, China	900,000	1977, July 19-20	Johnstown, PA	68
1889, May 31	Johnstown, PA	2,209	1977, Nov. 6	Toccoa, GA	39
1900, Sept. 8	Galveston, TX	5,000	1978, June-Sept.	N India	1,200
1903, June 15	Heppner, OR	325	1979, Jan.-Feb.	Brazil	204
1911	Chang Jiang River, China	100,000	1979, July 17	Lomblem Isl., Indonesia	539
1913, Mar. 25-27	OH, IN	732	1979, Aug. 11	Morvi, India	15,000
1915, Aug. 17	Galveston, TX	275	1980, Feb. 13-22	Southern CA, AZ	26
1928, Mar. 13	Dam collapse, Saugus, CA	450	1981, Apr.	N China	550
1928, Sept. 13	Lake Okeechobee, FL	2,000	1981, July	Sichuan, Hubei Prov., China	1,300
1931, Aug.	Huang He River, China	3,700,000	1982, Jan. 23	Nr. Lima, Peru	600
1937, Jan. 22	OH, MS Valleys	250	1982, May 12	Guangdong, China	430
1939	N China	200,000	1982, Sept. 17-21	El Salvador, Guatemala	1,300+
1946, Apr. 1	HI, AK	159	1984, Aug-Sept.	South Korea	200+
1947, Sept. 20	Honshu Island, Japan	1,900	1985, July 19	Dam collapse, N Italy	361
1951, Aug.	Manchuria	1,800	1987, Aug.-Sept.	N Bangladesh	1,000+
1953, Jan. 31	W Europe	2,000	1988, Sept.	N India	1,000+
1954, Aug. 17	Farahzad, Iran	2,000	1990, June 14	Shadyside, OH	23
1955, Oct. 7-12	India, Pakistan	1,700	1991, Dec. 18-26	TX	18
1959, Nov. 1	W Mexico	2,000	1993, July-Aug.	Midwest	48
1959, Dec. 2	Frejus, France	412	1994, July	GA, AL	32
1960, Oct. 10	Bangladesh	6,000	1995, Jan. 30-Feb. 9	NW Europe	40
1960, Oct. 31	Bangladesh	4,000	1995, July	Hunan Province, China	1,200
1962, Feb. 17	North Sea coast, Germany	343	1995, Aug. 19	SW Morocco	136
1962, Sept. 27	Barcelona, Spain	445	1995, Dec. 25	KwaZulu Natal, South Africa	166
1963, Oct. 9	Dam collapse, Vaiont, Italy	1,800	1996, Jan.	Northeastern U.S.	15+
1966, Nov. 3-4	Florence, Venice, Italy	113	1996, Feb. 17	Biak Isl., Indonesia	105
1967, Jan. 18-24	E Brazil	894	1996, April	Afghanistan	100+
1967, Mar. 19	Rio de Janeiro, Brazil	436	1996, June-July	S China	950+
1967, Nov. 26	Lisbon, Portugal	464	1996, Aug. 7	Pyrenees Mts., Spain	71
1968, Aug. 7-14	Gujarat State, India	1,000	1996, Dec.-1997, Jan.	Northwestern U.S.	29
1968, Oct. 7	NE India	780	1997, Mar.	Ohio R. Valley	35
1969, Jan. 18-26	Southern CA	100	1997, July	Poland, Czech Republic	98
1969, Mar. 17	Mundau Valley, Alagoas, Brazil	218	1997, Oct.	Israel, Egypt, Jordan	19
1969, Aug. 20-22	Western VA	189	1997, Nov.	Spanish-Portuguese border	31+
1969, Sept. 15	South Korea	250	1997, Nov.	Bardera, Somalia	1,300+
1969, Oct. 1-8	Tunisia	500	1998, Jan.	Kenya	86
1970, May 20	Central Romania	160	1998, Feb.	California to Tijuana, Mexico	30+
1970, July 22	Himalayas, India	500	1998, Mar.	SW Pakistan	300+
1971, Feb. 26	Rio de Janeiro, Brazil	130	1998, July-Aug.	China	4,150
1972, Feb. 26	Buffalo Creek, WV	118	1998, July-Sept.	Bangladesh	1,441
1972, June 9	Rapid City, SD	236	1998, July 17	Papua New Guinea	3,000
1972, Aug. 7	Luzon Isl., Philippines	454	1998, Aug. 24	S Texas, Mexico	16
1972, Aug. 19-31	Pakistan	1,500	1999, Aug. 1-4	S. Korea, Philippines, Vietnam,	
1974, Mar. 29	Tubaro, Brazil	1,000		Thailand	188+
1974, Aug. 12	Monty-Long, Bangladesh	2,500	1999, Sept.-Oct.	NE Mexico	350+

Some Major Earthquakes

Source: Global Volcanism Network, Smithsonian Institution; U.S. Geological Survey, Dept. of the Interior; World Almanac research

Magnitude of earthquakes (Mag.) is measured on the Richter scale; each higher number represents a tenfold increase in energy. Adopted in 1935, the scale is applied to earthquakes as far back as reliable seismograms are available.

Date	Location	Deaths	Mag.	Date	Location	Deaths	Mag.
526, May 20	Antioch, Syria	250,000	NA	1896, June 15	Japan, sea wave	27,120	NA
856	Corinth, Greece	45,000	"	1905, Apr. 4	Kangra, India	19,000	8.6
1057	Chihli, China	25,000	"	1906, Apr. 18-19	San Francisco, CA	503[2]	8.3
1169, Feb. 11	Near Mt. Etna, Sicily	15,000[1]	"	1906, Aug. 17	Valparaiso, Chile	20,000	8.6
1268	Cilicia, Asia Minor	60,000	"	1907, Oct. 21	Central Asia	12,000	8.1
1290, Sept. 27	Chihli, China	100,000	"	1908, Dec. 28	Messina, Italy	83,000	7.5
1293, May 20	Kamakura, Japan	30,000	"	1915, Jan. 13	Avezzano, Italy	29,980	7.5
1531, Jan. 26	Lisbon, Portugal	30,000	"	1918, Oct. 11	Mona Passage, P.R.	116	7.5
1556, Jan. 24	Shaanxi, China	830,000	"	1920, Dec. 16	Gansu, China	200,000	8.6
1667, Nov.	Shemaka, Caucasia	80,000	"	1923, Sept. 1	Yokohama, Japan	143,000	8.3
1693, Jan. 11	Catania, Italy	60,000	"	1925, Mar. 16	Yunnan, China	5,000	7.1
1730, Dec. 30	Hokkaido, Japan	137,000	"	1927, May 22	Nan-Shan, China	200,000	8.3
1737, Oct. 11	India, Calcutta	300,000	"	1932, Dec. 25	Gansu, China	70,000	7.6
1755, June 7	N Persia	40,000	"	1933, Mar. 2	Japan	2,990	8.9
1755, Nov. 1	Lisbon, Portugal	60,000	8.75*	1933, Mar. 10	Long Beach, CA	115	6.2
1783, Feb. 4	Calabria, Italy	30,000	NA	1934, Jan. 15	India, Bihar-Nepal	10,700	8.4
1797, Feb. 4	Quito, Ecuador	41,000	"	1935, Apr. 21	Taiwan (Formosa)	3,276	7.4
1811-12	New Madrid, MO (series)	NA	8.7*	1935, May 30	Quetta, India	50,000	7.5
1822, Sept. 5	Asia Minor, Aleppo	22,000	NA	1939, Jan. 25	Chillan, Chile	28,000	8.3
1828, Dec. 28	Echigo, Japan	30,000	"	1939, Dec. 26	Erzincan, Turkey	30,000	8.0
1868, Aug. 13-15	Peru, Ecuador	40,000	"	1946, Dec. 20	Honshu, Japan	1,330	8.4
1875, May 16	Venezuela, Colombia	16,000	"	1948, June 28	Fukui, Japan	5,390	7.3
1886, Aug. 31	Charleston, SC	60	6.6	1949, Aug. 5	Pelileo, Ecuador	6,000	6.8

Date	Location	Deaths	Mag.	Date	Location	Deaths	Mag.
1950, Aug. 15	Assam, India	1,530	8.7	1990, May 30	N Peru	115	6.3
1953, Mar. 18	NW Turkey	1,200	7.2	1990, June 20	W Iran	40,000+	7.7
1956, June 10-17	N Afghanistan	2,000	7.7	1990, July 16	Luzon, Philippines	1,621	7.8
1957, July 2	N Iran	1,200	7.4	1991, Feb. 1	Pakistan, Afgh. border	1,200	6.8
1957, Dec. 13	W Iran	1,130	7.3	1991, Oct. 19	N India	2,000	7.0
1960, Feb. 29	Agadir, Morocco	12,000	5.9	1992, Mar. 13, 15	E Turkey	4,000	6.2/6.0
1960, May 21-30	S Chile	5,000	9.5	1992, June 28	S California	1	7.5/6.6
1962, Sept. 1	NW Iran	12,230	7.3	1992, Dec. 12	Flores Isl., Indonesia	2,500	7.5
1963, July 26	Skopje, Yugoslavia	1,100	6.0	1993, July 12	off Hokkaido, Japan	200+	7.7
1964, Mar. 27	Alaska	131	9.2	1992, Sept. 1	SW Nicaragua	116	7.0
1966, Aug. 19	E Turkey	2,520	7.1	1992, Oct. 12	Cairo, Egypt	450	5.9
1968, Aug. 31	NE Iran	12,000	7.3	1993, Sept. 29	Maharashtra, S India	9,748[3]	6.3
1970, Jan. 4	Yunnan Prov., China	10,000	7.5	1994, Jan. 17	Northridge, CA	61	6.8
1970, Mar. 28	W Turkey	1,100	7.3	1994, Feb. 15	S Sumatra, Indon.	215	7.0
1970, May 31	N Peru	66,000	7.8	1994, June 6	Cauca, SW Colombia	1,000	6.8
1971, Feb. 9	San Fernando Val., CA	65	6.6	1994, Aug. 19	N Algeria	164	6.0
1972, Apr. 10	S Iran	5,054	7.1	1995, Jan. 16	Kobe, Japan	5,502	6.9
1972, Dec. 23	Managua, Nicaragua	5,000	6.2	1995, May 27	Sakhalin Isl., Russia	1,989	7.5
1974, Dec. 28	Pakistan (9 towns)	5,200	6.3	1995, Oct. 1	SW Turkey	73	6.0
1975, Sept. 6	Turkey (Lice, etc.)	2,300	6.7	1995, Oct. 9	W coast, Mexico	c. 40+	7.6
1976, Feb. 4	Guatemala	23,000	7.5	1996, Feb. 3	SW China	200+	7.0
1976, May 6	NE Italy	1,000	6.5	1996, Feb. 17	Irian Jaya, Indonesia	53	7.5
1976, June 25	Irian Jaya, New Guinea	422	7.1	1997, Feb. 4	Turkmen.-Iran border	79	6.9
1976, July 27	Tangshan, China	255,000	8.0	1997, Feb. 27	W Pakistan	100+	7.3
1976, Aug. 16	Mindanao, Philippines	8,000	7.8	1997, Feb. 28	NW Iran	1,000+	6.1
1976, Nov. 24	NW Iran-USSR border	5,000	7.3	1997, May 10	N Iran	1,560	7.5
1977, Mar. 4	Romania	1,500	7.2	1997, May 21	Madhya Pradesh, India	40+	6.1
1977, Aug. 19	Indonesia	200	8.0	1997, July 9	NE Venezuela	82	6.9
1977, Nov. 23	NW Argentina	100	8.2	1997, Sept. 26	Central Italy	11	5.5/5.7
1978, Sept. 16	NE Iran	15,000	7.8	1997, Sept. 28	Sulawesi, Indonesia	17+	5.9
1979, Sept. 12	Indonesia	100	8.1	1997, Oct. 15	Illapel, Chile	8.12	6.8
1979, Dec. 12	Colombia, Ecuador	800	7.9	1998, Jan. 10	Zhangbei, China	50	6.2
1980, Oct. 10	NW Algeria	3,500	7.7	1998, Feb. 4, 8	Takhar province, NE Afghanistan	2,323	6.1
1980, Nov. 23	S Italy	3,000	7.2	1998, May 22	Central Bolivia	105	6.5
1981, June 11	S Iran	3,000	6.9	1998, May 30	NE Afghanistan	4,700+	6.9
1981, July 28	S Iran	1,500	7.3	1998, June 27	Adana, Turkey	144	6.3
1982, Dec. 13	W Arabian Peninsula	2,800	6.0	1998, July 9	Azores, Portugal	10	5.8
1983, May 26	N Honshu, Japan	81	7.7	1998, Nov. 29	East Indonesia	34	7.8
1983, Oct. 30	E Turkey	1,342	6.9	1999, Jan. 25	Armenia, Colombia	1,185+	6.0
1985, Mar. 3	Chile	146	7.8	1999, Feb. 11	Central Afghanistan	60	6.0
1985, Sept. 19	Michoacan, Mexico	9,500	8.1	1999, Mar. 28	Uttar Pradesh, India	87	6.8
1986, Oct. 10	El Salvador	1,000+	5.5	1999, May 7	Southern Iran	26+	6.2
1987, Mar. 6	Colombia-Ecuador	4,000+	7.0	1999, June 16	Puebla, Mexico	16	6.7
1988, Aug. 20	India-Nepal border	1,450	6.6	1999, Aug. 17	Western Turkey	16,695+[4]	7.4
1988, Nov. 6	China-Burma border	1,000	7.3	1999, Sept. 7	Athens, Greece	143	5.9
1988, Dec. 7	Soviet Armenia	55,000	7.0	1999, Sept. 21	Taichung, Taiwan	2,321	7.6
1989, Oct. 17	San Francisco Bay area	62	7.1	1999, Sept. 30	Oaxaca, Mexico	20	7.5

(*) estimated from earthquake intensity. NA=not available. (1) Once thought to have been a volcanic eruption; evidence indicates a destructive earthquake and tsunami occurred on this date. (2) With subsequent fires, death toll rose to 700. (3) Official death toll as released by Indian government. Other sources reported estimates of about 30,000 deaths. (4) Confirmed dead as of Sept. 20, 1999.

Other Recent Earthquakes

Source: Global Volcanism Network, Smithsonian Institution; dates are Greenwich Mean Time

Date	Location	Magnitude	Date	Location	Magnitude
1998, Mar. 25	Balleny Islands	8.2	1999, Jan. 19	Papua New Guinea	7.0
Apr. 1	Sumatra, Indonesia	7.0	Jan. 28	Aleutian Islands	6.6
Apr. 27	Irian Jaya, Indonesia	7.4	Feb. 6	Santa Cruz Islands	7.3
May 3	Ryukyu Island, SE of Taiwan	7.4	Apr. 5	New Britain, Papua New Guinea	7.4
June 1	Kamchatka, Russia	6.5	Apr. 8	E Russia, NE China	7.1
Nov. 9	Banda Sea	6.6/7.0	May 10, 16	New Britain, Papua New Guinea	7.1
Dec. 29	Fiji Islands	6.9	Oct. 16	S. California	7.0

Some Notable Fires Since 1835

(See also Some Notable Explosions Since 1910.)

Date	Location	Deaths	Date	Location	Deaths
1835, Dec. 16	New York, NY, 500 bldgs. destroyed	—	1911, Mar. 25	Triangle Shirtwaist factory, NY, NY	146
1845, May	Canton, China, theater	1,670	1913, Oct. 14	Mid Glamorgan, Wales, colliery	439
1871, Oct. 8	Chicago, $196 million loss; 17,000 bldgs. destroyed	250	1918, Apr. 13	Norman, OK, state hospital	38
1871, Oct. 8	Peshtigo, WI, forest fire	1,182	1918, Oct. 12	Cloquet, MN, forest fire	400
1872, Nov. 9	Boston, 800 bldgs. destroyed	—	1919, June 20	Mayagüez Theater, San Juan, P.R.	150
1876, Dec. 5	Brooklyn, NY, theater	295	1923, May 17	Camden, SC, school	76
1877, June 20	St. John, New Brunswick	100	1924, Dec. 24	Babb's Switch, OK, school	35
1881, Dec. 8	Ring Theater, Vienna	850	1929, May 15	Cleveland, OH, clinic	125
1887, May 25	Opera Comique, Paris	200	1930, Apr. 21	Columbus, OH, penitentiary	320
1887, Sept. 4	Exeter, England, theater	200	1931, July 24	Pittsburgh, PA, home for aged	48
1894, Sept. 1	MN, forest fire	413	1934, Dec. 11	Hotel Kerns, Lansing, MI	34
1897, May 4	Paris, charity bazaar	150	1938, May 16	Atlanta, GA, Terminal Hotel	35
1900, June 30	Hoboken, NJ, docks	326	1940, Apr. 23	Natchez, MS, dance hall	198
1902, Sept. 20	Birmingham, AL, church	115	1942, Nov. 28	Cocoanut Grove, Boston	491
1903, Dec. 30	Iroquois Theater, Chicago	602	1942, Dec. 12	St. John's, Nfld., hostel	100
1908, Jan. 13	Rhoads Theater, Boyertown, PA	170	1943, Sept. 7	Gulf Hotel, Houston, TX	55
1908, Mar. 4	Collinwood, OH, school	176	1944, July 6	Ringling Circus, Hartford, CT	168
			1946, June 5	LaSalle Hotel, Chicago	61

Date	Location	Deaths	Date	Location	Deaths
1946, Dec. 7	Winecoff Hotel, Atlanta	119	1977, May 28	Southgate, KY, nightclub	164
1946, Dec. 12	NY, NY, ice plant, tenement	37	1977, June 9	Abidjan, Ivory Coast, nightclub	41
1949, Apr. 5	Effingham, IL, hospital	77	1977, June 26	Columbia, TN, jail	42
1950, Jan. 7	Davenport, IA, Mercy Hospital	41	1977, Nov. 14	Manila, Philippines, hotel	47
1953, Mar. 29	Largo, FL, nursing home	35	1978, Jan. 28	Kansas City, Coates House Hotel	16
1953, Apr. 16	Chicago, metalworking plant	35	1978, Aug. 19	Abadan, Iran, movie theater	425+
1957, Feb. 17	Warrenton, MO, home for aged	72	1979, July 14	Saragossa, Spain, hotel	80
1958, Mar. 19	New York, NY, loft building	24	1979, Dec. 31	Chapais, Quebec, social club	42
1958, Dec. 1	Chicago, parochial school	95	1980, May 20	Kingston, Jamaica, nursing home	157
1958, Dec. 16	Bogotá, Colombia, store	83	1980, Nov. 21	MGM Grand Hotel, Las Vegas	84
1959, June 23	Stalheim, Norway, resort hotel	34	1980, Dec. 4	Stouffer Inn, Harrison, NY	26
1960, Mar. 12	Pusan, Korea, chemical plant	68	1981, Jan. 9	Keansburg, NJ, boarding home	30
1960, July 14	Guatemala City, mental hospital	225	1981, Feb. 10	Las Vegas Hilton	8
1960, Nov. 13	Amude, Syria, movie theater	152	1981, Feb. 14	Dublin, Ireland, discotheque	44
1961, Jan. 6	Thomas Hotel, San Francisco	20	1982, Sept. 4	Los Angeles, apartment house	24
1961, Dec. 8	Hartford, CT, hospital	16	1982, Nov. 8	Biloxi, MS, county jail	29
1961, Dec. 17	Niteroi, Brazil, circus	323	1983, Feb. 13	Turin, Italy, movie theater	64
1963, May 4	Diourbel, Senegal, theater	64	1983, Dec. 17	Madrid, Spain, discotheque	83
1963, Nov. 18	Surfside Hotel, Atlantic City, NJ	25	1984, May 11	Great Adventure Amusement Pk., NJ	8
1963, Nov. 23	Fitchville, OH, rest home	63	1985, Apr. 21	Tabaco, Phil., movie theater	44
1963, Dec. 29	Roosevelt Hotel, Jacksonville, FL	22	1985, Apr. 26	Buenos Aires, Argentina, hospital	79
1964, May 8	Manila, apartment bldg.	30	1985, May 11	Bradford, England, soccer stadium	53
1964, Dec. 18	Fountaintown, IN, nursing home	20	1986, Dec. 31	Puerto Rico, Dupont Plaza Hotel	96
1965, Mar. 1	LaSalle, Quebec, apartment	28	1987, May 6-June 2	N China, forest fire	193
1965, Aug. 11-16	Watts riot fires, CA	30+	1987, Nov. 17	London, England, subway	30
1966, Mar. 11	Numata, Japan, 2 ski resorts	31	1988, Mar. 20	Lashio, Burma, 2,000 buildings	134
1966, Aug. 13	Melbourne, Australia, hotel	29	1990, Mar. 25	Bronx, NY, social club	87
1966, Oct. 17	New York, NY, bldg. (firefighters)	12	1991, Mar. 3	Addis Ababa, Ethiopia, munitions	
1966, Dec. 7	Erzurum, Turkey, barracks	68		dump	260+
1967, Feb. 7	Montgomery, AL, restaurant	25	1991, Sept. 3	Hamlet, NC, processing plant	25
1967, May 22	Brussels, Belgium, store	322	1991, Oct. 20-21	Oakland, Berkeley, CA, wildfire	24
1967, July 16	Jay, FL, state prison	37	1993, Apr. 19	Waco, TX, cult compound	72
1968, Feb. 26	Shrewsbury, England, hospital	22	1994, May 10	Bangkok, Thailand, toy factory	213
1968, May 11	Vijayawada, India, wedding hall	58	1994, July 4-10	Glenwood Springs, CO (firefighters)	14
1968, Nov. 18	Glasgow, Scotland, factory	24	1994, Dec. 10	Karamay, China, theater	300
1969, Dec. 2	Notre Dame, Can., nursing home	54	1994, Nov. 2	Durunka, Egypt, burning fuel flood	500
1970, Jan. 9	Marietta, OH, nursing home	27	1995, Oct. 28	Baku, Azerbaijan, subway train	300
1970, Nov. 1	Grenoble, France, dance hall	145	1995, Dec. 23	Mandi Dabwali, India, school	500+
1970, Dec. 20	Tucson, AZ, hotel	28	1996, Mar. 19	Quezon City, Philippines,	
1971, Mar. 6	Burghoezli, Switzerland,			nightclub	150+
	psychiatric clinic	28	1996, Mar. 28	Bogor, Indonesia, shopping mall	78
1971, Apr., 20	Bangkok, Thailand, hotel	24	1996, Apr. 11	Düsseldorf, Germany, airport	16
1971, Dec., 25	Seoul, South Korea, hotel	162	1996, Oct. 22	Caracas, Venezuela, jail	25
1972, May 13	Osaka, Japan, nightclub	116	1996, Nov. 20	Hong Kong, building	39
1972, July 5	Sherborne, England, hospital	30	1997, Feb. 23	Baripada, India, worship site	164
1973, Feb. 6	Paris, France, school	21	1997, Apr. 15	Mina, Saudi Arabia, encampment	343
1973, June 24	New Orleans, LA, bar	32	1997, June 7	Thanjavur, India, temple	60+
1973, Nov. 6	Fukui, Japan, train	28	1997, June 13	New Delhi, India, movie theater	60
1973, Nov. 29	Kumamoto, Japan, dept. store	107	1997, July 11	Pattaya, Thailand, hotel	90
1973, Dec. 2	Seoul, South Korea, theater	50	1997, Sept. 29	Home for retarded children, near	
1974, Feb. 1	São Paulo, Brazil, bank building	189		Colina, Chile	30
1974, June 30	Port Chester, NY, discotheque	24	1998, Mar. 26	School dorm., Mazeras, India	22
1974, Nov. 3	Seoul, S. Korea, hotel, disco	88	1998, Dec. 3	Orphanage, Manila, Philippines	28
1975, Dec. 12	Mina, Saudi Arabia, tent city	138	1999, Feb. 10	Police hdqtrs., Samara, Russia	23
1976, Oct. 24	Bronx, NY, social club	25	1999, Mar. 24	Mont Blanc tunnel, France and Italy	40
1977, Feb. 25	Moscow, Russia, Rossiya hotel	45	1999, June 30	Camp dormitory, Hwasung, S. Korea	23

Some Notable Explosions Since 1910

(See also Principal U.S. Mine Disasters Since 1900.)

Date	Location	Deaths	Date	Location	Deaths
1910, Oct. 1	Los Angeles Times Bldg.	21	1958, Apr. 18	Sunken munitions ship,	
1913, Mar. 7	Dynamite, Baltimore harbor	55		Okinawa, Japan	40
1915, Sept. 27	Gasoline tank car, Ardmore, OK	47	1958, May 22	Nike missiles, Leonardo, NJ	10
1917, Apr. 10	Munitions plant, Eddystone, PA	133	1959, Apr. 10	World War II bomb, Philippines	38
1917, Dec. 6	Halifax Harbor, Canada	1,654	1959, June 28	Rail tank cars, Meldrin, GA	25
1918, May 18	Chemical plant, Oakdale, PA	193	1959, Aug. 7	Dynamite truck, Roseburg, OR	13
1918, July 2	Explosives, Split Rock, NY	50	1959, Nov. 2	Jamuri Bazar, India, explosives	46
1918, Oct. 4	Shell plant, Morgan Station, NJ	64	1959, Dec. 13	2 apt. bldgs., Dortmund, Ger.	26
1919, May 22	Food plant, Cedar Rapids, IA	44	1960, Mar. 4	Belgian munitions ship, Havana, Cuba	100
1920, Sept. 16	Wall Street, NY, NY, bomb	30	1960, Oct. 25	Gas, Windsor, Ont., store	11
1921, Sept. 21	Chem. storage facility, Oppau, Ger.	561	1962, Jan. 16	Gas pipeline, Edson, Alberta	8
1924, Jan. 3	Food plant, Pekin, IL	42	1962, Oct. 3	Telephone Co. office, NY, NY	23
1927, May 18	Bath school, Lansing, MI	38	1963, Jan. 2	Packing plant, Terre Haute, IN	16
1928, April 13	Dance hall, West Plains, MO	40	1963, Mar. 9	Dynamite plant, S. Africa	45
1937, Mar. 18	New London, TX, school	311	1963, Aug. 13	Explosives dump, Gauhaiti, India	32
1940, Sept. 12	Hercules Powder, Kenvil, NJ	55	1963, Oct. 31	State Fair Coliseum, Indianapolis, IN	73
1942, June 5	Ordnance plant, Elwood, IL	49	1964, July 23	Bone, Algeria, harbor munitions	100
1944, Apr. 14	Bombay, India, harbor	700	1965, Mar. 4	Gas pipeline, Natchitoches, LA	17
1944, July 17	Port Chicago, CA, pier	322	1965, Aug. 9	Missile silo, Searcy, AR	53
1944, Oct. 21	Liquid gas tank, Cleveland	135	1965, Oct. 21	Bridge, Tila Bund, Pakistan	80
1947, Apr. 16	Texas City, TX, pier	576	1965, Oct. 30	Cartagena, Colombia	48
1948, July 28	Farben works, Ludwigshafen, Ger.	184	1965, Nov. 24	Armory, Keokuk, IA	20
1950, May 19	Munitions barges, S. Amboy, NJ	30	1967, Dec. 25	Apartment bldg., Moscow, USSR	20
1956, Aug. 7	Dynamite trucks, Cali, Colombia	1,100	1968, Apr. 6	Sports store, Richmond, IN	43

Date	Location	Deaths
1970, Apr. 8	Subway construction, Osaka, Japan...	73
1971, June 24	Tunnel, Sylmar, CA	17
1971, June 28	School, fireworks, Puebla, Mexico	13
1971, Oct. 21	Shopping center, Glasgow, Scotland	20
1973, Feb., 10	Liquid gas tank, Staten Island, NY	40
1975, Dec. 27	Coal mine, Chasnala, India	431
1976, Apr. 13	Lapua, Finland, munitions works	40
1977, Nov. 11	Freight train, Iri, South Korea	57
1977, Dec. 22	Grain elevator, Westwego, LA	35
1978, Feb. 24	Derailed tank car, Waverly, TN	12
1978, July 11	Propylene tank truck, Spanish coastal campsite	150
1980, Oct. 23	School, Ortuella, Spain	64
1982, Apr. 25	Antiques exhibition, Todi, Italy	33
1982, Nov. 2	Salang Tunnel, Afghanistan	1,000-3,000
1984, Feb. 25	Oil pipeline, Cubatao, Brazil	508
1984, June 21	Naval supply depot, Severomorsk, USSR	200+
1984, Nov. 19	Gas storage area, NE Mexico City	334
1984, Dec. 3	Chemical plant, Bhopal, India	3,849
1984, Dec. 5	Coal mine, Taipei, Taiwan	94
1985, June 25	Fireworks factory, Hallett, OK	21
1988, Apr. 10	Pakistani army ammunitions dump near Rawalpindi and Islamabad	100
1988, July 6	Oil rig, North Sea	167
1989, June 3	Gas pipeline, between Ufa, Asha, USSR	650+
1992, Mar. 3	Coal mine, Kozlu, Turkey	270+
1992, Apr. 22	Sewer, Guadalajara, Mexico	190
1992, May 9	Coal mine, Plymouth, Nova Scotia	26
1993, Feb. 26	World Trade Center, NY, NY	6
1994, July 18	Jewish community center, Buenos Aires, Argentina	100
1995, Apr. 19	Fed'l. office building, Oklahoma City...	168[1]
1995, Apr. 29	Subway construction, South Korea	110
1995, Nov. 13	Military facility, Riyadh, Saudi Arabia	7
1996, Jan. 31	Bank, Colombo, Sri Lanka	53
1996, Feb. 25	Jerusalem and Ashkelon, Israel	27
1996, Mar. 3-4	Jerusalem and Tel Aviv, Israel	33
1996, June 25	U.S. military housing complex, near Dhahran, Saudi Arabia	19

Date	Location	Deaths
1996, July 24	Train, Colombo, Sri Lanka	86
1996, Nov. 10	Cemetery, Moscow, Russia	13
1996, Nov. 16	Russian military apartment, Dagestan region, Russia	68
1996, Nov. 21	Building, San Juan, Puerto Rico	29
1996, Nov. 27	Coal mine, Shanxi province, China	91+
1996, Dec. 2	Train, Haryana, India	12
1996, Dec. 30	Train, Assam, India	59+
1997, Jan. 18	Near courthouse, Lahore, Pakistan	25
1997, Mar. 19	Ammunition depot, Jalalabad, Afghanistan	16
1997, July 8	Train, Punjab, India	36
1997, July 9	Military airfield, S Romania	16
1997, July 30	Market, Jerusalem	15
1997, Nov. 19	Car, Hyderabad, India	23
1997, Dec. 2	Coal mine, Novokuznetsk, Siberia	68
1997, Dec. 6	Trains, southern India	10+
1998, Jan. 17	Coal mine, Sokobanja, Serbia	29
1998, Feb. 14	Oil tankers (2), Yaounde, Cameroon	120
1998, Feb. 14	17 bombs, Coimbatore, India	50
1998, Feb. 23	Train, near El Affroune, Algiers	18
1998, Mar. 5	Bus, Colombo, Sri Lanka	32
1998, Mar. 9	Train, Lahore, Pakistan	10
1998, Apr. 4	Coal mine, Donetsk, Ukraine	63
1998, Aug. 7	Bomb, U.S. Embassy, Nairobi, Kenya	213
	Bomb, U.S. Embassy, Dar-es-Salaam, Tanzania	11
1998, Aug. 15	Car bomb, Omagh, Ireland	29
1998, Aug. 16	Coal mine, Luhansk, Ukraine	24
1998, Aug. 31	Marketplace in Algiers	17
1998, Sept. 8	Two buses, Sao Paulo, Brazil	59
1998, Oct. 17	Oil pipeline, Jesse, Nigeria	700+
1999, May 16	Fuel truck, Punjab province, Pakistan	75
1999, July 27	Truck, Chongqing, China	15
1999, July 29	Gold mine, Carletonville, S. Africa	17
1999, Sept. 10	Apartment building, Moscow	94
1999, Sept. 13	Apartment building, Moscow	118
1999, Sept. 16	Apartment building, Moscow	18
1999, Sept. 26	Fireworks factory, Celaya, Mexico	56

(1) Includes a rescue worker who died during the rescue effort.

Notable Nuclear Accidents

Oct. 7, 1957 — A fire in the Windscale plutonium production reactor N of Liverpool, England, released radioactive material; later blamed for 39 cancer deaths.

Jan. 3, 1961 — A reactor at a federal installation near Idaho Falls, ID, killed 3 workers. Radiation contained.

Oct. 5, 1966 — A sodium cooling system malfunction caused a partial core meltdown at the Enrico Fermi demonstration breeder reactor, near Detroit, MI. Radiation contained.

Jan. 21, 1969 — A coolant malfunction from an experimental underground reactor at Lucens Vad, Switzerland, released a large amount of radiation into a cavern, which was then sealed.

Mar. 22, 1975 — Fire at the Brown's Ferry reactor in Decatur, AL, caused dangerous lowering of cooling water levels.

Mar. 28, 1979 — The worst commercial nuclear accident in the U.S. occurred as equipment failures and human mistakes led to a loss of coolant and a partial core meltdown at the Three Mile Island reactor in Middletown, PA.

Feb. 11, 1981 — Eight workers were contaminated when over 100,000 gallons of radioactive coolant leaked into the containment building of TVA's Sequoyah 1 plant in Tennessee.

Apr. 25, 1981 — Some 100 workers were exposed to radiation during repairs of a nuclear plant at Tsuruga, Japan.

Jan. 6, 1986 — A cylinder of nuclear material burst after being improperly heated at a Kerr-McGee plant at Gore, OK. One worker died; 100 were hospitalized.

Apr. 26, 1986 — In the worst accident in the history of nuclear power, fires and explosions resulting from an unauthorized experiment at the Chernobyl nuclear power plant near Kiev, USSR (now in Ukraine), left at least 31 dead in the immediate aftermath and spread radioactive material over much of Europe. An estimated 135,000 people were evacuated from areas around Chernobyl, some of which were uninhabitable for years. As a result of the radiation released into the atmosphere, tens of thousands of excess cancer deaths (as well as increased birth defects) were expected.

Sept. 30, 1999 — Japan's worst nuclear accident ever occurred at a uranium-reprocessing facility in Tokaimura, NE of Tokyo, when workers accidentally overloaded a container with uranium, thereby exposing workers and area residents to extremely high radiation levels.

Record Oil Spills

The number of tons can be multiplied by 7 to estimate roughly the number of barrels spilled; the exact number of barrels in a ton varies with the type of oil. Each barrel contains 42 gallons.

Name, place	Date	Cause	Tons
Ixtoc I oil well, S Gulf of Mexico	June 3, 1979	Blowout	600,000
Nowruz oil field, Persian Gulf	Feb. 1983	Blowout	600,000 (est.)
Atlantic Empress & *Aegean Captain*, off Trinidad and Tobago	July 19, 1979	Collision	300,000
Castillo de Bellver, off Cape Town, South Africa	Aug. 6, 1983	Fire	250,000
Amoco Cadiz, near Portsall, France	Mar. 16, 1978	Grounding	223,000
Torrey Canyon, off Land's End, England	Mar. 18, 1967	Grounding	119,000
Sea Star, Gulf of Oman	Dec. 19, 1972	Collision	115,000
Urquiola, La Coruna, Spain	May 12, 1976	Grounding	100,000
Hawaiian Patriot, N Pacific	Feb. 25, 1977	Fire	99,000
Othello, Tralhavet Bay, Sweden	Mar. 20, 1970	Collision	60,000-100,000

Other Notable Oil Spills

Name, place	Date	Cause	Gallons
Persian Gulf	began Jan. 23, 1991	Spillage by Iraq	130,000,000[1]
Braer, off Shetland Islands	Jan. 5, 1993	Grounding	26,000,000
Aegean Sea, off N Spain	Dec. 3, 1992	Unknown	21,500,000
Sea Empress, off SW Wales	Feb. 15, 1996	Grounding	18,000,000
World Glory, off South Africa	June 13, 1968	Hull failure	13,524,000
Exxon Valdez, Prince William Sound, AK	Mar. 24, 1989	Grounding	10,080,000
Keo, off MA	Nov. 5, 1969	Hull failure	8,820,000
Storage tank, Sewaren, NJ	Nov. 4, 1969	Tank rupture	8,400,000
Ekofisk oil field, North Sea	Apr. 22, 1977	Well blowout	8,200,000
Storage tank, Monongahela River	Jan. 2, 1988	Tank rupture	3,800,000 (est.)[2]

(1) Est. by Saudi Arabia. Some estimates as low as 25 mil gal. (2) Other estimates are as high as 84.6 mil gal.

Historic Assassinations Since 1865

1865—Apr. 14. U.S. Pres. Abraham Lincoln shot by John Wilkes Booth in Washington, DC; died Apr. 15.

1881—Mar. 13. Alexander II, of Russia.—July 2. U.S. Pres. James A. Garfield shot by Charles J. Guiteau, Washington, DC; died Sept. 19.

1894—June 24. Pres. Sadi Carnot of France, by Italian anarchist, Sante Caserio, in Lyon.

1898—Sept. 10. Empress Elizabeth of Austria, stabbed by Italian anarchist Luigi Luccheni.

1900—July 29. Umberto I, king of Italy.

1901—Sept. 6. U.S. Pres. William McKinley in Buffalo, NY; died Sept. 14. Leon Czolgosz executed for the crime.

1908—Feb. 1. King Carlos I of Portugal and his son Luis Felipe, in Lisbon.

1913—Feb. 23. Mex. Pres. Francisco I. Madero and Vice Pres. Jose Pino Suarez.—Mar. 18. George, king of Greece.

1914—June 28. Archduke Francis Ferdinand of Austria-Hungary and his wife in Sarajevo, Bosnia (later part of Bosnia and Herzegovina), by Gavrilo Princip.

1916—Dec. 30. Grigori Rasputin, politically powerful Russian monk.

1918—July 12. Grand Duke Michael of Russia, at Perm.—July 16. Nicholas II, abdicated as czar of Russia; his wife, the Czarina Alexandra; their son, Czarevitch Alexis; their daughters, Grand Duchesses Olga, Tatiana, Marie, Anastasia; and 4 members of their household, executed by Bolsheviks at Ekaterinburg.

1920—May 20. Mexican Pres. Gen. Venustiano Carranza in Tlaxcalantongo.

1922—Aug. 22. Michael Collins, Irish revolutionary.—Dec. 16. Polish Pres. Gabriel Narutowicz in Warsaw.

1923—July 20. Gen. Francisco "Pancho" Villa, ex-rebel leader, in Parral, Mexico.

1928—July 17. Gen. Alvaro Obregon, president-elect of Mexico, in San Angel, Mexico.

1932—May 6. Pres. Paul Doumer of France shot by Russian émigré, Pavel Gorgulov, in Paris.

1934—July 25. In Vienna, Austrian Chancellor Engelbert Dollfuss by Nazis.

1935—Sept. 8. U.S. Sen. Huey P. Long shot in Baton Rouge, LA, by Dr. Carl Austin Weiss, who was slain by Long's bodyguards; Long died Sept. 10.

1940—Aug. 20. Leon Trotsky (Lev Bronstein), 63, exiled Russian war minister, near Mexico City, by Ramon Mercador del Rio, a Spaniard.

1948—Jan. 30. Mohandas K. Gandhi, 78, shot in New Delhi, India, by Nathuram Vinayak Godse.—Sept. 17. Count Folke Bernadotte, UN mediator for Palestine, by Jewish extremists in Jerusalem.

1951—July 20. King Abdullah ibn Hussein of Jordan.—Oct. 16. Prime Min. Liaquat Ali Khan of Pakistan shot in Rawalpindi.

1956—Sept. 21. Pres. Anastasio Somoza of Nicaragua, shot in Leon; died Sept. 29.

1957—July 26. Pres. Carlos Castillo Armas of Guatemala, in Guatemala City by one of his own guards.

1958—July 14. King Faisal of Iraq; his uncle, Crown Prince Abdullah; and July 15, Prem. Nuri as-Said, in Baghdad.

1959—Sept. 25. Prime Min. Solomon Bandaranaike of Ceylon, by Buddhist monk in Colombo.

1961—Jan. 17. Ex-Prem. Patrice Lumumba of the Congo, in Katanga Province.—May 30. Dominican dictator Rafael Leonidas Trujillo Molina, near Ciudad Trujillo.

1963—June 12. Medgar W. Evers, NAACP's Mississippi field secretary, by Byron De La Beckwith in Jackson, MS.—Nov. 2. Pres. Ngo Dinh Diem of South Vietnam and his brother, Ngo Dinh Nhu, in a military coup.—Nov. 22. U.S. Pres. John F. Kennedy shot in Dallas, TX; accused gunman Lee Harvey Oswald was murdered by Jack Ruby while awaiting trial.

1965—Jan. 21. Iranian Prem. Hassan Ali Mansour in Tehran; 4 executed.—Feb. 21. Malcolm X, black nationalist, shot in New York City.

1966—Sept. 6. Prime Min. Hendrik F. Verwoerd of South Africa stabbed to death in parliament at Cape Town.

1968—Apr. 4. Rev. Dr. Martin Luther King Jr. fatally shot in Memphis, TN; James Earl Ray convicted of crime.—June 5. Sen. Robert F. Kennedy (D, NY) shot in Los Angeles; Sirhan Sirhan, convicted of crime.

1971—Nov. 28. Prime Min. Wasfi Tal of Jordan, in Cairo, by Palestinian guerrillas.

1973—Mar. 2. U.S. Amb. Cleo A. Noel Jr., U.S. Charge d'Affaires George C. Moore, and Belgian Charge d'Affaires Guy Eid killed by Palestinian guerrillas in Khartoum, Sudan.

1974—Aug. 19. U.S. Amb. to Cyprus, Rodger P. Davies, killed by sniper's bullet in Nicosia.

1975—Feb. 11. Pres. Richard Ratsimandrava, of Madagascar, shot in Tananarive.—Mar. 25. King Faisal of Saudi Arabia shot by nephew Prince Musad Abdel Aziz, in royal palace, Riyadh.—Aug. 15. Bangladesh Pres. Sheik Mujibur Rahman killed in coup.

1976—Feb. 13. Nigerian head of state, Gen. Murtala Ramat Mohammed, by self-styled "young revolutionaries."

1977—Mar. 16. Kamal Jumblat, Lebanese Druse chieftain, shot near Beirut.—Mar. 18. Congo Pres. Marien Ngouabi shot in Brazzaville.

1978—July 9. Former Iraqi Prem. Abdul Razak Al-Naif shot in London.

1979—Feb. 14. U.S. Amb. Adolph Dubs shot by Afghan Muslim extremists in Kabul.—Aug. 27. Lord Mountbatten, World War II hero, and 2 others killed when a bomb exploded on his fishing boat off the coast of Co. Sligo, Ire. IRA claimed responsibility.—Oct. 26. South Korean Pres. Park Chung Hee and 6 bodyguards fatally shot by Kim Jae Kyu, head of South Korean CIA, and 5 aides in Seoul.

1980—Apr. 12. Liberian Pres. William R. Tolbert slain in military coup.—Sept. 17. Former Nicaraguan Pres. Anastasio Somoza Debayle shot in Paraguay.

1981—Oct. 6. Egyptian Pres. Anwar al-Sadat shot by commandos while reviewing military parade in Cairo.

1982—Sept. 14. Lebanese Pres.-elect Bashir Gemayel killed by bomb in east Beirut.

1983—Aug. 21. Philippine opposition leader Benigno Aquino Jr. shot by gunman at Manila International Airport.

1984—Oct. 31. Indian Prime Min. Indira Gandhi shot and killed by 2 Sikh bodyguards, in New Delhi.

1986—Feb. 28. Swedish Prem. Olof Palme shot by gunman on Stockholm street.

1987—June 1. Lebanese Prem. Rashid Karami killed when bomb exploded aboard a helicopter.

1988—Apr. 16. PLO military chief Khalil Wazir (Abu Jihad) gunned down by Israeli commandos in Tunisia.

1989—Aug. 18. Colombian presidential candidate Luis Carlos Galan killed by Medellín cartel drug traffickers at campaign rally in Bogotá.—Nov. 22. Lebanese Pres. Rene Moawad killed when bomb exploded next to his motorcade.

1990—Mar. 22. Presidential candidate Bernando Jamamillo Ossa shot by gunman at an airport in Bogotá.

1991—May 21. Rajiv Gandhi, former prime min. of India, killed by bomb during election rally in Madras.

1992—June 29. Mohammed Boudiaf, pres. of Algeria, shot by gunman in Annaba.

1993—May 1. Ranasinghe Premadasa, pres. of Sri Lanka, killed by bomb in Colombo.

1994—Mar. 23. Luis Donaldo Colosio Murrieta, Mexican presidential candidate, shot by gunman Mario Aburto Martinez. —Apr. 6. Burundian Pres. Cyprien Ntaryamira and Rwandan Pres. Juvenal Habyarimana killed, with 8 others, when their plane was apparently shot down.

1995—Nov. 4. Yitzhak Rabin, prime min. of Israel, shot by gunman Yigal Amir at peace rally in Tel Aviv.

1996—Oct. 2. Andrei Lukanov, former Bulgarian prime minister, shot outside his home by an unidentified gunman.

1998—Feb. 6. Claude Erigmac, prefect of Corsica, shot in the back while walking to a concert, by two unidentified gunmen.

1999—Mar. 23. Paraguayan Vice-Pres. Luis Maria Argaña, ambushed and shot to death, along with his driver, by four unidentified assailants.—Apr. 9. Niger's Pres. Ibrahim Bare Mainassara, ambushed and killed by dissident soldiers.

Assassination Attempts

1912—Oct. 14. Former U.S. Pres. Theodore Roosevelt shot and wounded by demented man in Milwaukee, WI.

1933—Feb. 15. In Miami, FL, Joseph Zangara, anarchist, shot at Pres.-elect Franklin D. Roosevelt, but a woman seized his arm, and the bullet fatally wounded Mayor Anton J. Cermak, of Chicago, who died Mar. 6.

1944—July 20. Adolf Hitler was injured when a bomb, planted by a German officer, exploded in Hitler's headquarters. One aide was killed and 12 were injured in the explosion.

1950—Nov. 1. In an attempt to assassinate Pres. Harry Truman, 2 members of a Puerto Rican nationalist movement—Griselio Torresola and Oscar Collazo—tried to shoot their way into Blair House. Torresola was killed, and a White House policeman, Pvt. Leslie Coffelt, was fatally shot.

1970—Nov. 27. Pope Paul VI unharmed by knife-wielding assailant who attempted to attack him in Manila airport.

1972—May 15. Alabama Gov. George Wallace shot in Laurel, MD, by Arthur Bremer; seriously crippled.

1975—Sept. 5. Pres. Gerald R. Ford unharmed when a Secret Service agent grabbed a pistol aimed at him by Lynette (Squeaky) Fromme, a Charles Manson follower, in Sacramento.—Sept. 22. Pres. Ford again unharmed when Sara Jane Moore fired a revolver at him.

1980—May 29. Civil rights leader Vernon E. Jordan Jr. shot and wounded in Ft. Wayne, IN.

1981—Jan. 16. Irish political activist Bernadette Devlin McAliskey and her husband shot and seriously wounded by 3 members of a Protestant paramilitary group in Co. Tyrone, Ire.—

Mar. 30. Pres. Ronald Reagan, along with Press Sec. James Brady, Secret Service agent Timothy J. McCarthy, and Washington, DC, policeman Thomas Delahanty shot and seriously wounded by John W. Hinckley Jr. in Washington, DC.—May 13. Pope John Paul II and 2 bystanders shot and wounded by Mehmet Ali Agca, an escaped Turkish murderer, in St. Peter's Square, Rome.

1982—May 12. Pope John Paul II unharmed after guards overpowered a man with a knife, in Fatima, Portugal.

1984—Oct. 12. British Prime Min. Margaret Thatcher narrowly escaped injury when a bomb, said to have been planted by the IRA, exploded at the Grand Hotel in Brighton, England, during a Conservative Party conference. Four died, including a member of Parliament.

1986—Sept. 7. Chilean Pres. Gen. Augusto Pinochet Ugarte escaped unharmed when his motorcade was attacked by rebels using rockets, bazookas, grenades, and rifles.

1995—June 26. Egyptian Pres. Hosni Mubarak unharmed when gunmen fired on his motorcade in Addis Ababa, Ethiopia. Four died, including 2 Ethiopian police officers.

1997—Feb. 12. Colombian Pres. Ernesto Samper Pizano unharmed when a bomb exploded on a runway in Barranquilla as his plane was preparing to land.—Apr. 30. Tajik Pres. Imamali Rakhmanov injured when a grenade was thrown at him. 2 others were killed.

1998—Feb. 9. Georgian Pres. Eduard A. Shevardnadze unharmed when gunmen fired on his motorcade in Tbilisi, Georgia. Three died, including 2 bodyguards and 1 assailant.

Notable U.S. Kidnappings Since 1924

Robert Franks, 13, in Chicago, **May 22, 1924,** by 2 youths, Richard Loeb and Nathan Leopold, who killed boy. Demand for $10,000 ignored. Loeb died in prison; Leopold paroled 1958.

Charles A. Lindbergh Jr., 20 mos. old, in Hopewell, NJ, **Mar. 1, 1932;** found dead **May 12.** Ransom of $50,000 paid to man identified as Bruno Richard Hauptmann, 35, paroled German convict who entered U.S. illegally. Hauptmann was convicted after spectacular trial at Flemington, and electrocuted in Trenton, NJ, prison, **Apr. 3, 1936.**

William A. Hamm Jr., 39, in St. Paul, **June 15, 1933.** $100,000 paid. Alvin Karpis given life, paroled in 1969.

Charles F. Urschel, in Oklahoma City, **July 22, 1933.** Released **July 31** after $200,000 paid. George "Machine Gun" Kelly and 5 others sentenced to life.

Brooke L. Hart, 22, in San Jose, CA. Thomas Thurmond and John Holmes arrested after demanding $40,000 ransom. When Hart's body was found in San Francisco Bay, **Nov. 26, 1933,** a mob attacked the jail and lynched the 2 kidnappers.

George Weyerhaeuser, 9, in Tacoma, WA, **May 24, 1935.** Returned home **June 1** after $200,000 paid. Kidnappers given 20 to 60 years.

Charles Mattson, 10, in Tacoma, WA, **Dec. 27, 1936.** Found dead **Jan. 11, 1937.** Kidnapper asked $28,000, but failed to contact for delivery.

Arthur Fried, in White Plains, NY, **Dec. 4, 1937.** Body not found. Two kidnappers executed.

Robert C. Greenlease, 6, taken from Kansas City, MO, school **Sept. 28, 1953,** held for $600,000. Body was found Oct. 7. Bonnie Brown Heady and Carl A. Hall pleaded guilty and were executed.

Peter Weinberger, 32 days old, Westbury, NY, **July 4, 1956,** for $2,000 ransom, not paid. Child found dead. Angelo John LaMarca, 31, convicted, executed.

Lee Crary, 8, in Everett, WA, **Sept. 22, 1957;** $10,000 ransom, not paid. He escaped after 3 days, led police to George E. Collins, who was convicted.

Frank Sinatra Jr., 19, from hotel room in Lake Tahoe, CA, **Dec. 8, 1963.** Released **Dec. 11** after his father paid $240,000 ransom. Three men sentenced to prison.

Barbara Jane Mackle, 20, abducted **Dec. 17, 1968,** from Atlanta, GA, motel; found unharmed 3 days later, buried in a coffin-like box 18 inches underground, after her father had paid $500,000 ransom; Gary Steven Krist sentenced to life, Ruth Eisenmann-Schier to 7 years.

Mrs. Roy Fuchs, 35, and 3 children held hostage 2 hours, **May 14, 1969,** in Long Island, NY, released after her husband, a bank manager, paid kidnappers $129,000 in bank funds; 4 men arrested, ransom recovered.

Virginia Piper, 49, abducted **July 27, 1972,** from her home in suburban Minneapolis; found unharmed near Duluth 2 days later after husband paid $1 million ransom.

Patricia "Patty" Hearst, 19, taken from her Berkeley, CA, apartment **Feb. 4, 1974.** "Symbionese Liberation Army" captors demanded her father, publisher Randolph Hearst, give millions to the area's poor. Implicated in a San Francisco bank holdup, **Apr. 15.** The FBI, **Sept. 18, 1975,** captured her and others; they were indicted on various charges. Patricia Hearst convicted of bank robbery, **Mar. 20, 1976;** released from prison under executive clemency, **Feb. 1, 1979.** In 1978, William and Emily Harris were sentenced to 10 years to life for the kidnapping; both were paroled in 1983.

J. Reginald Murphy, 40, an editor of *Atlanta* (GA) *Constitution,* kidnapped **Feb. 20, 1974;** freed **Feb. 22** after newspaper paid $700,000 ransom. William A. H. Williams arrested; most of the money recovered.

E. B. Reville, Hepzibah, GA, banker, and wife, Jean, kidnapped **Sept. 30, 1974.** Ransom of $30,000 paid. He was found alive; Jean Reville was found dead **Oct. 2.**

Jack Teich, Kings Point, NY, steel executive, seized **Nov. 12, 1974;** released **Nov. 19** after payment of $750,000.

Adam Walsh, 6, abducted from a Hollywood, FL, department store, **July 27, 1981.** Although his severed head was found 2 weeks later, his body was never recovered. John Walsh, Adam's father, became active in raising awareness about missing children.

Sidney J. Reso, oil company executive, seized **Apr. 29, 1992;** died **May 3;** Arthur D. Seale and wife, Irene, arrested **June 19.** Arthur Seale pleaded guilty, sentenced to life in prison; Irene Seale sentenced to 20-year prison term.

Polly Klaas, 12, Petaluma, CA, abducted at knife point, **Oct. 1, 1993,** during a slumber party at her home. Police arrested Richard Allen Davis on **Nov. 30;** he led them to her body, found **Dec. 4** in wooded area of Cloverdale, CA. Davis found guilty **June 18, 1996,** and sentenced to death **Sept. 26.**

Marshall I. Wais, 79, owner of 2 San Francisco steel companies, kidnapped **Nov. 19, 1996,** from his San Francisco home. Released unharmed the same day after $500,000 ransom paid; Thomas William Taylor and Michael K. Robinson arrested the same day.

EDUCATION

Historical Overview of U.S. Public Elementary and Secondary Schools

Source: National Center for Education Statistics, U.S. Dept. of Education

	1979-80	1989-90	1990-91	1991-92	1992-93	1993-94	1994-95	1995-96	1996-97
Population statistics (thousands)									
Total U.S. population[1]	224,567	246,819	249,440	252,137	255,028	257,783	260,292	262,761	265,179
Population 5-17 years of age	48,041	44,947	45,312	45,918	46,662	47,419	48,110	48,911	49,698
Percentage 5-17 years of age	21.4	18.2	18.2	18.2	18.3	18.4	18.5	18.6	18.7
Enrollment (thousands)									
Elementary and secondary	41,651	40,543	41,217	42,047	42,816	43,465	44,111	44,840	45,611
Kindergarten & grades 1-8	28,034	29,152	29,878	30,506	31,081	31,504	31,898	32,341	32,764
Grades 9-12	13,616	11,390	11,338	11,541	11,735	11,961	12,213	12,500	12,847
Percentage pop. 5-17 enrolled	86.7	90.2	91.0	91.6	91.8	91.7	91.7	91.7	91.8
Percentage in high schools	32.7	28.1	27.5	27.4	27.4	27.5	27.7	27.9	28.2
High school graduates (thousands)	2,748	2,320	2,235	2,212	2,233	2,221	2,274	2,273	2,341
School term; staff									
Average school term (in days)	178.5	*	179.8	*	*	*	*	*	*
Total instructional staff (thousands)	2,406	2,986	3,051	3,104	3,140	3,209	3,281	3,352	3,448
Teachers, librarians, and other non-supervisory instructional staff (thousands)	2,300	2,860	2,924	2,975	3,017	3,088	3,161	3,231	3,324
Revenue and expenditures (millions)									
Total revenue	$96,881	$208,548	$223,341	$234,486	$247,626	$260,142	$273,149	$287,703	$305,055
Total expenditures	95,962	212,770	229,430	241,567	252,935	265,285	279,000	293,646	313,131
Current elem. and secondary	86,984	188,229	202,038	211,216	220,948	231,543	243,878	255,107	270,152
Capital outlay	6,506	17,781	19,771	20,797	22,172	23,747	24,456	27,556	31,434
Others	598	2,983	3,296	4,392	4,379	4,682	5,149	4,725	4,647
Interest on school debt	1,874	3,776	4,325	5,162	5,437	5,335	5,518	6,259	6,899
Salaries and pupil cost									
Avg. annual salary of instruct. staff[2]	$16,715	$32,638	$34,401	$35,550	$36,454	$37,383	$38,441	$39,465	40,562
Expenditure per capita total pop.	427	862	920	958	992	1,029	1,072	1,118	1,181
Current expenditure per pupil ADA[3]	2,272	4,980	5,258	5,421	5,584	5,767	5,989	6,147	6,392

NOTE: Because of rounding, details may not add to totals. * = Data not collected. (1) As of July 1; excludes armed forces overseas. (2) Includes supervisors, principals, teachers, and nonsupervisory instructional staff. (3) ADA means average daily attendance.

Programs for the Disabled, 1989-98[1]

Source: Office of Special Education and Rehabilitative Services, U.S. Dept. of Education

(Number of children up to 21 years old served annually in educational programs for the disabled; in thousands)

Type of Disability	1989-90	1990-91	1991-92	1992-93	1993-94	1994-95	1995-96	1996-97	1997-98
ALL DISABILITIES	4,641	4,762	4,949	5,176	5,365	5,378	5,573	5,729	5,904
Learning disabilities	2,050	2,130	2,234	2,351	2,408	2,489	2,579	2,649	2,726
Speech impairments	973	985	997	994	1,014	1,015	1,022	1,043	1,059
Mental retardation	548	534	538	518	536	555	570	579	589
Serious emotional disturbance	381	390	399	400	414	427	438	445	453
Hearing impairments	57	58	60	60	64	64	67	68	69
Orthopedic impairments	48	49	51	52	56	60	63	66	67
Other health impairments	52	55	58	65	82	106	133	160	190
Visual impairments	22	23	24	23	24	24	25	25	25
Multiple disabilities	86	96	97	102	108	88	93	98	106
Deafness/blindness	2	1	1	1	1	1	1	1	1
Autism and other disabilities	*	*	5	19	24	29	39	44	54
Preschool disabilities[2]	422	441	484	590	634	519	544	552	564

NOTE: Counts are based on reports from the 50 states and the District of Columbia. Details may not add to totals because of rounding. * = Data not collected. (1) Includes students served under Chapter I and Individuals With Disabilities Education Act (IDEA). (2) Includes preschool children 3-5 years and 0-5 years served under Chapter I and IDEA, respectively.

Technology in U.S. Public Schools, 1996-99

Source: Quality Education Data, Inc., Denver, CO

	Number of Schools				Percentage of Schools			
	1996	1997	1998	1999	1996	1997	1998	1999
Schools with computers[1]	82,675	84,080	85,903	86,197	97.8	98.1	98.5	98.4
Elementary[2]	50,997	51,974	52,187	52,276	98.6	99.6	98.8	98.6
Junior high[3]	13,556	13,760	14,189	14,419	97.8	97.7	98.5	99.1
Senior high[4]	16,434	16,609	17,294	17,307	97.1	96.4	98.4	98.1
Schools with networks[1]	29,875	32,299	49,178	51,099	35.3	37.7	56.4	58.3
Elementary[2]	14,868	16,441	26,422	27,661	28.8	31.5	50.0	52.2
Junior high[3]	5,590	6,035	9,003	9,572	40.3	42.8	62.5	65.8
Senior high[4]	9,166	9,565	12,853	13,064	54.2	55.5	73.1	74.1
Schools with CD-ROMs[1]	43,499	46,388	64,200	65,769	51.5	54.1	73.6	75.1
Elementary[2]	24,353	26,377	37,908	38,718	47.1	50.5	71.7	73.0
Junior high[3]	7,952	8,410	11,023	11,502	57.3	59.7	76.5	79.0
Senior high[4]	10,756	11,140	13,985	14,367	63.5	64.6	79.5	81.5
Schools with Internet access[1]	14,211	35,762	60,224	61,093	16.8	41.7	69.1	69.7
Elementary[2]	7,608	21,026	34,195	34,812	14.7	40.3	64.7	65.6
Junior high[3]	2,707	5,752	10,888	11,127	19.5	40.8	75.5	76.4
Senior high[4]	3,736	8,984	13,829	13,900	22.1	52.1	78.6	78.8

(1) Includes schools for special and adult education, not shown separately. (2) Includes preschool and schools with grade spans of K-3, K-5, K-6, K-8, and K-12. (3) Includes schools with grade spans of 4-8, 7-8, and 7-9. (4) Includes vocational, technical, and alternative high schools and schools with grade spans of 7-12, 9-12, and 10-12.

Students per Computer in U.S. Public Schools

Source: *QED's Technology in Public Schools, 1997-1998*

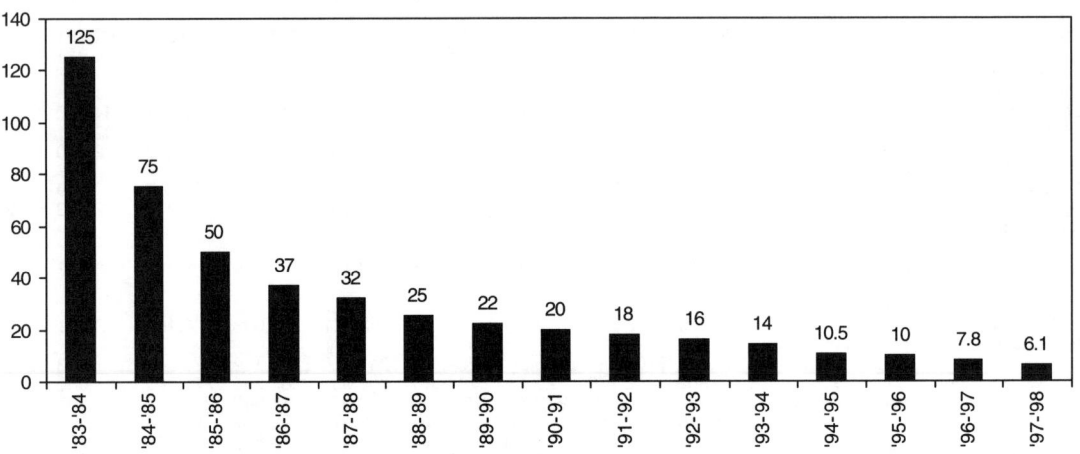

Overview of U.S. Public Schools, Fall 1997*

Source: National Center for Education Statistics, U.S. Dept. of Education; National Education Association

	Local school districts	Elementary schools	Secondary schools	Classroom teachers	Total enrollment	Pupils per teacher	Teacher's avg. pay[1]	Expend. per pupil[2]
AL......	127	881	310	45,973	749,187	16.3	$32,818	$4,903
AK	53	202	90	7,625	132,123	17.3	51,738	9,097
AZ	329	981	333	41,129	814,113	19.8	33,850	4,940
AR	311	687	417	26,932	456,497	16.9	30,578	4,840
CA	994	6,010	1,907	268,581	5,803,734	21.6	43,725	5,414
CO	176	1,096	359	37,840	687,167	18.2	37,052	5,728
CT	166	807	207	37,658	535,164	14.2	50,730	8,901
DE	19	118	45	6,850	111,960	16.3	42,439	7,804
DC	1	121	35	4,399	77,111	17.5	46,350	9,019
FL......	67	2,031	464	124,473	2,294,077	18.4	34,475	5,986
GA	180	1,445	306	85,005	1,375,980	16.2	37,378	5,708
HI......	1	190	51	10,653	189,887	17.8	38,377	6,144
ID	112	403	213	13,207	244,403	18.5	32,775	4,732
IL	929	3,092	1,017	118,734	1,998,289	16.8	43,873	6,557
IN	295	1,393	450	57,371	987,483	17.2	39,682	6,605
IA	377	1,073	442	32,717	501,054	15.3	34,040	6,047
KS	304	1,019	424	31,527	469,687	14.9	36,811	6,158
KY	176	982	346	40,488	669,322	16.5	34,525	5,929
LA	66	1,016	316	48,599	776,813	16.0	29,650	5,201
ME	284	545	162	15,700	212,526	13.5	34,349	6,774
MD	24	1,056	217	48,318	830,744	17.2	41,739	7,543
MA	351	1,487	336	67,170	949,006	14.1	43,930	7,818
MI......	674	2,614	873	90,529	1,702,672	18.8	49,277	7,568
MN	380	1,221	704	51,998	853,621	16.4	39,106	6,371
MS	153	569	307	29,441	504,792	17.1	29,547	4,312
MO	525	1,504	630	60,869	910,654	15.0	33,975	5,823
MT	461	524	365	10,228	162,335	15.9	30,617	6,112
NE	640	980	352	20,139	292,681	14.5	32,668	6,472
NV	17	339	100	16,053	296,621	18.5	37,093	5,541
NH	179	414	99	12,931	201,629	15.6	36,640	6,236
NJ......	608	1,792	432	89,671	1,250,276	13.9	50,442	10,211
NM	89	542	188	19,647	331,673	16.9	30,152	4,674
NY	705	3,014	942	190,874	2,861,823	15.0	49,034	9,658
NC	117	1,609	381	77,785	1,236,083	15.9	33,315	5,315
ND	233	349	218	8,070	118,572	14.7	28,230	5,198
OH	661	2,781	972	110,757	1,847,035	16.7	38,977	6,517
OK	547	1,218	599	40,215	623,681	15.5	30,606	5,150
OR	198	937	259	26,935	541,346	20.1	42,150	6,792
PA......	501	2,348	783	108,014	1,815,151	16.8	47,650	7,686
RI......	36	255	56	10,598	153,321	14.5	44,300	8,307
SC	90	798	277	42,336	659,256	15.6	33,608	5,371
SD	176	502	309	9,282	142,443	15.3	27,341	4,924
TN	139	1,127	366	54,142	893,020	16.5	35,340	5,011
TX......	1,042	4,729	1,888	254,557	3,891,877	15.3	33,648	5,736
UT	40	489	244	21,115	482,957	22.9	32,950	4,045
VT......	286	265	68	7,909	105,984	13.4	36,299	7,171
VA......	141	1,414	391	75,524	1,110,815	14.7	36,654	5,677
WA	296	1,348	565	49,074	991,235	20.2	38,788	6,182
WV	55	610	207	20,947	301,419	14.4	33,398	6,519
WI......	426	1,528	540	57,227	881,780	15.4	39,899	7,398
WY	48	284	120	6,677	97,115	14.5	32,022	6,448
TOTAL U.S.	**14,805**	**62,739**	**21,682**	**2,744,493**	**46,127,194**	**16.8**	**$39,385**	**$6,392**

* Full-time elementary and secondary day schools only. (1) National Education Association estimate. (2) Fall 1996.

Mathematics, Reading, and Science Achievement of U.S. Students

Source: National Assessment of Educational Progress, National Center for Education Statistics, U.S. Dept. of Education

Percent of students who scored at or above proficient level in national tests.

STATE[1]	GRADE 4 Math. 1992	GRADE 4 Math. 1996	GRADE 4 Reading 1994	GRADE 4 Reading 1998	GRADE 8 Math. 1996	GRADE 8 Reading 1998	GRADE 8 Science 1996	STATE[1]	GRADE 4 Math. 1992	GRADE 4 Math. 1996	GRADE 4 Reading 1994	GRADE 4 Reading 1998	GRADE 8 Math. 1996	GRADE 8 Reading 1998	GRADE 8 Science 1996
AL	43	48	23	24	45	21	18	MT	NA	71	35	37	75	38	41
AK	NA	65	NA	NA	68	NA	31	NE	67	70	34	NA	76	NA	35
AZ	53	57	24	22	57	28	23	NH	NA	NA	NA	38	NA	NA	NA
AR	47	54	24	23	52	23	22	NV	NA	57	36	21	NA	24	NA
CA	46	46	18	20	51	22	20	NJ	68	68	33	NA	NA	NA	NA
CO	61	67	28	34	67	30	32	NM	50	51	21	22	51	24	19
CT	67	75	38	46	70	42	36	NY	57	64	27	29	61	34	27
DE	55	54	23	25	55	25	21	NC	50	64	30	28	56	31	24
DC	23	20	NA	10	20	12	5	ND	72	75	38	NA	77	NA	41
FL	52	55	23	23	54	23	21	OK	NA	NA	NA	30	NA	29	NA
GA	53	53	26	24	51	25	21	OR	NA	65	NA	28	67	33	32
HI	52	53	19	17	51	19	15	PA	65	68	30	NA	NA	NA	NA
IN	60	72	33	NA	68	NA	30	RI	54	61	32	32	60	30	26
IA	72	74	35	35	78	NA	36	SC	48	48	20	22	48	22	17
KS	NA	NA	NA	34	NA	35	NA	TN	47	58	27	25	53	26	22
KY	51	60	26	29	56	29	23	TX	56	69	26	29	59	28	23
LA	39	44	15	19	38	18	13	UT	66	69	30	28	70	31	32
ME	75	75	41	36	77	42	41	VT	NA	67	NA	NA	72	NA	34
MD	55	59	26	29	57	31	25	VA	59	62	26	30	58	33	27
MA	68	71	36	37	68	36	37	WA	NA	67	27	29	67	32	27
MI	61	68	NA	28	67	NA	32	WV	52	63	26	29	54	27	21
MN	71	76	33	36	75	37	37	WI	71	74	35	34	75	33	39
MS	36	42	18	18	36	19	12	WY	69	64	32	30	68	29	34
MO	62	66	31	29	64	29	28	U.S. AVG.	57	62	28	29	61	31	27

NA = Not administered. (1) Only participating states are included.

Revenues[1] for Public Elementary and Secondary Schools, by State, 1997-98

Source: National Education Association; estimated; in thousands

STATE	Total	Federal Amount	Federal %	State Amount	State %	Local and intermediate Amount	Local and intermediate %
Alabama	$4,030,356	$379,707	9.4	$2,601,160	64.5	$1,049,489	26.0
Alaska	1,183,024*	148,479*	12.6*	751,882*	63.6*	282,663*	23.9*
Arizona	4,388,915*	344,440*	7.8*	2,108,912*	48.1*	1,935,563*	44.1*
Arkansas	2,322,451	193,028	8.3	1,535,550	66.1	593,873	25.6
California	35,054,650*	3,108,332*	8.9*	19,920,649*	56.8*	12,025,669*	34.3*
Colorado	4,183,998*	235,150*	5.6*	1,860,095*	44.5*	2,088,753*	49.9*
Connecticut	5,112,950	235,980	4.6	2,064,560	40.4	2,812,410	55.0
Delaware	966,422	68,906	7.1	650,099	67.3	247,417	25.6
District of Columbia	452,084*	67,102*	14.8*	0	0	384,982*	85.2*
Florida	14,583,108	1,058,483	7.3	7,067,649	48.5	6,456,976	44.3
Georgia	8,579,628	563,428	6.6	4,485,220	52.3	3,530,980	41.2
Hawaii	1,364,412	101,842	7.5	1,231,799	90.3	30,771	2.3
Idaho	1,344,671*	87,524*	6.5*	870,387*	64.7*	386,760*	28.8*
Illinois	13,649,628*	954,160*	7.0*	3,651,254*	26.7*	9,044,214*	66.3*
Indiana	7,006,752*	346,390*	4.9*	3,758,925*	53.6*	2,901,437*	41.4*
Iowa	3,189,269	117,484	3.7	1,726,779	54.1	1,345,006	42.2
Kansas	3,090,829	165,692	5.4	1,814,830	58.7	1,110,307	35.9
Kentucky	3,881,816*	290,277*	7.5*	2,562,918*	66.0*	1,028,621*	26.5*
Louisiana	4,251,305*	510,950*	12.0*	2,117,148*	49.8*	1,623,207*	38.2*
Maine	1,520,325	95,181	6.3	698,005	45.9	727,139	47.8
Maryland	6,267,608	358,393	5.7	2,524,148	40.3	3,385,067	54.0
Massachusetts	7,533,212*	385,065*	5.1*	2,719,766*	36.1*	4,428,381*	58.8*
Michigan	13,579,423*	896,321*	6.6*	11,102,410*	81.8*	1,580,692*	11.6*
Minnesota	6,504,296*	252,395*	3.9*	3,725,754*	57.3*	2,526,147*	38.8*
Mississippi	2,502,975*	325,505	13.0*	1,422,642*	56.8*	754,828*	30.2*
Missouri	5,841,090*	344,710*	5.9*	2,336,931*	40.0*	3,159,449*	54.1*
Montana	989,202	97,500	9.9	481,142	48.6	410,560	41.5
Nebraska	1,688,662*	65,586*	3.9*	627,379*	37.2*	995,697*	59.0*
Nevada	1,754,717	74,068	4.2	550,787	31.4	1,129,862	64.4
New Hampshire	1,365,391*	42,742*	3.1*	83,529*	6.1*	1,239,120*	90.8*
New Jersey	12,555,896*	390,079*	3.1*	4,737,699*	37.7*	7,428,118*	59.2*
New Mexico	2,328,142	204,965	8.8	1,638,853	70.4	484,324	20.8
New York	27,690,556	1,807,000	6.5	11,165,000	40.3	14,718,556	53.2
North Carolina	7,127,549	506,055*	7.1	4,689,505	65.8	1,931,989	27.1
North Dakota	668,941	77,790	11.6	279,109	41.7	312,042	46.6
Ohio	12,694,407	857,850	6.8	5,495,244	43.3	6,341,313	50.0
Oklahoma	3,119,028*	302,007*	9.7*	1,950,627*	62.5*	866,394*	27.8*
Oregon	3,525,000	250,000	7.1	2,175,000	61.7	1,100,000	31.2
Pennsylvania	15,327,396*	843,587*	5.5*	6,315,350*	41.2*	8,168,459*	53.3*
Rhode Island	1,271,975	66,103	5.2	538,797	42.4	667,075	52.4
South Carolina	4,156,500	312,000	7.5	2,176,100	52.4	1,668,400	40.1
South Dakota	787,412*	73,441*	9.3*	249,994*	31.7*	463,977*	58.9*
Tennessee	4,491,405*	355,640*	7.9*	2,314,163*	51.5*	1,821,602*	40.6*
Texas	23,920,057	1,832,912	7.7	10,282,362	43.0	11,804,783	49.4
Utah	2,248,932	141,412	6.3	1,408,832	62.6	698,688	31.1
Vermont	814,567*	39,447*	4.8*	228,020*	28.0*	547,100*	67.2*
Virginia	6,661,612	361,680	5.4	2,468,859	37.1	3,831,073	57.5*
Washington	6,722,916	465,411	6.9	4,588,341	68.2	1,669,164	24.8
West Virginia	2,176,238	186,962	8.6	1,367,620	62.8	621,656	28.6
Wisconsin	7,054,119	304,000	4.3	3,881,491	55.0	2,868,628	40.7
Wyoming	661,472	44,654	6.8	317,818	48.0	299,000	45.2
TOTAL U.S.	$314,187,289	$21,337,815	6.8	$155,321,093	49.4	$137,528,381	43.8

* Indicates NEA estimate. (1) Included as revenue receipts are all appropriations from general funds of federal, state, county, and local governments; receipts from taxes levied for school purposes; income from permanent school funds and endowments; and income from leases of school lands and miscellaneous sources (interest on bank deposits, tuition, gifts, school lunch charges, etc.).

M I L L E N N I U M   F A C T   B O X

Public and Private Schools

Source: National Center for Education Statistics, U.S. Dept. of Education

During the 20th century, enrollment in both public and private schools in the U.S. rose along with the population, though there were dips, especially in the 1970s and early 1980s. The proportion of students attending private, as opposed to public, schools generally stayed around 10%.

School year[1]	Public school enrollment[2]	Private school enrollment[2]	% Private	School year[1]	Public school enrollment[2]	Private school enrollment[2]	% Private
1899-1900 ...	15,503	1,352	8.7	1959-60	35,182	5,675	16.1
1909-10	17,814	1,558	8.7	1969-70	45,550	5,500[3]	12.1
1919-20	21,578	1,699	7.9	1979-80	41,651	5,000[3]	12.0
1929-30	25,678	2,651	10.3	1989-90	40,543	5,355[3]	13.2
1939-40	25,434	2,611	10.3	1996-97[4]	45,592	5,783[3]	12.7
1949-50	25,111	3,380	13.5	2006-07[5]	48,368	6,088	12.6

(1) Fall enrollment. (2) In thousands. (3) Estimated. (4) Preliminary. (5) Projected.

Public High School Graduation Rates, 1996-97

Source: National Center for Education Statistics, U.S. Dept. of Education

	Rate (%)[1]	Rank		Rate (%)[1]	Rank		Rate (%)[1]	Rank
Alabama.........	57.3	44	Louisiana........	55.0	49	Ohio...........	69.7	29
Alaska	63.8	36	Maine...........	71.1	24	Oklahoma.......	77.1	12
Arizona.........	62.1	39	Maryland	71.2	22 T	Oregon	67.4	33
Arkansas	69.8	28	Massachusetts....	75.8	15	Pennsylvania.....	75.7	16
California	66.2	35	Michigan	68.9	30	Rhode Island.....	70.9	25
Colorado.........	72.3	21	Minnesota	77.3	11	South Carolina ...	52.4	51
Connecticut	74.0	20	Mississippi.......	56.1	46	South Dakota	80.9	7
Delaware	63.0	37	Missouri.........	70.6	26	Tennessee.......	55.9	48
District of Columbia	57.0	45	Montana.........	81.0	6	Texas...........	58.9	42
Florida	56.0	47	Nebraska	82.2	4	Utah............	77.8	10
Georgia	53.2	50	Nevada	66.4	34	Vermont.........	81.1	5
Hawaii...........	62.6	38	New Hampshire...	68.2	31	Virginia	75.5	17
Idaho...........	78.7	9	New Jersey	85.8	2	Washington......	71.2	22 T
Illinois...........	76.5	14	New Mexico......	57.9	43	West Virginia.....	74.4	18
Indiana..........	70.4	27	New York	60.9	41	Wisconsin	79.5	8
Iowa............	84.7	3	North Carolina	61.3	40	Wyoming........	77.0	13
Kansas..........	74.1	19	North Dakota	86.9	1	**TOTAL U.S.......**	**67.1**	
Kentucky.........	67.8	32						

T=Tied in rank with one or more states. **NOTE:** Data exclude ungraded pupils and have not been adjusted for interstate migration. (1) Graduates as percentage of fall 1993 9th-grade enrollment.

Institutions of Higher Education—Charges, 1969-70 to 1999-2000

Source: National Center for Education Statistics, U.S. Dept. of Education; The College Board

Figures for 1969-70 are average charges for full-time resident degree-credit students; figures for later years are average charges per full-time equivalent student. Room and board are based on full-time students. These figures are enrollment-weighted, according to the number of full-time-equivalent undergraduates, and thus vary from averages given elsewhere.

	TUITION AND FEES			BOARD RATES (7-day basis)[1]			DORMITORY CHARGES		
	All institutions	2-yr	4-yr	All institutions	2-yr	4-yr	All institutions	2-yr	4-yr
PUBLIC (in-state)									
1969-70	$323	$178	$427	$511	$465	$540	$369	$308	$395
1979-80	583	355	840	867	894	898	715	572	749
1989-90	1,356	756	2,035	1,635	1,581	1,728	1,513	962	1,561
1990-91	1,454	824	2,159	1,691	1,594	1,767	1,612	1,050	1,658
1991-92	1,624	937	2,410	1,780	1,612	1,852	1,731	1,074	1,789
1992-93	1,782	1,025	2,349	1,841	1,668	1,854	1,756	1,106	1,816
1993-94	1,942	1,125	2,537	1,880	1,681	1,895	1,873	1,190	1,934
1994-95	2,057	1,192	2,681	1,949	1,712	1,967	1,959	1,232	2,023
1995-96[2].........	NA	1,330	2,811	NA	—[3]	3,932[4]	NA	—[4]	—[4]
1996-97[2].........	NA	1,465	2,975	NA	—[3]	4,167[4]	NA	—[4]	—[4]
1997-98[2].........	NA	1,567	3,111	NA	—[3]	4,358[4]	NA	—[4]	—[4]
1998-99[2].........	NA	1,554	3,247	NA	—[3]	4,522[4]	NA	—[4]	—[4]
1999-2000[2].......	NA	1,627	3,356	NA	—[3]	4,730	NA	—[4]	—[4]
PRIVATE									
1969-70	1,533	1,034	1,809	561	546	608	436	413	503
1979-80	3,130	2,062	3,811	955	924	1,078	827	769	999
1989-90	8,147	5,196	10,348	1,948	1,811	2,339	1,923	1,663	2,411
1990-91	8,772	5,570	11,379	2,074	1,989	2,470	2,063	1,744	2,654
1991-92	9,434	5,752	12,192	2,252	2,090	2,727	2,221	1,789	2,860
1992-93	9,942	6,059	10,294	2,344	1,875	2,354	2,348	1,970	2,362
1993-94	10,572	6,370	10,952	2,434	1,970	2,445	2,490	2,067	2,506
1994-95	11,111	6,914	11,481	2,509	2,023	2,520	2,587	2,233	2,601
1995-96[2].........	NA	6,339	12,216	NA	4,063[4]	5,166[4]	NA	—[4]	—[4]
1996-97[2].........	NA	6,613	12,994	NA	4,346[4]	5,363[4]	NA	—[4]	—[4]
1997-98[2].........	NA	7,079	13,785	NA	4,442[4]	5,575[4]	NA	—[4]	—[4]
1998-99[2].........	NA	6,940	14,709	NA	4,373[4]	5,754[4]	NA	—[4]	—[4]
1999-2000[2].......	NA	7,182	15,380	NA	4,583	5,959	NA	—[4]	—[4]

NA = not available. (1) Data for 1989-90 to 1993-94 reflect 20 meals per week rather than 7 days per week. (2) 1995-96 through 1999-2000 figures supplied by the College Board; earlier figures from National Center for Education Statistics. (3) Sample too small to provide meaningful information. (4) Board and dormitory figures for 1995-96 through 1999-2000 are combined.

Top 20 Colleges and Universities in Endowment Assets[1]

Source: National Association of College and University Business Officers (NACUBO)

College/University	Endowment assets[2]	College/University	Endowment assets[2]
1. Harvard University	$13,019,736	11. Columbia University	$3,425,992
2. University of Texas	7,647,309	12. University of Pennsylvania	3,059,401
3. Yale University	6,624,449	13. Rice University	2,790,627
4. Princeton University	5,582,800	14. Cornell University	2,527,871
5. Emory University	5,104,801	15. Northwestern University	2,397,715
6. Stanford University	4,559,066	16. University of Chicago	2,359,358
7. University of California	3,787,884	17. University of Michigan	2,303,054
8. Massachusetts Institute of Technology	3,678,127	18. University of Notre Dame	1,766,176
9. Texas A&M University	3,531,517	19. Vanderbilt University	1,539,242
10. Washington University	3,445,743	20. Dartmouth College	1,519,708

NOTE: Figures are for market value of endowment assets, excluding pledges and working capital. (1) As of June 30, 1998. (2) In thousands.

U.S. Higher Education Trends: Bachelor's Degrees Conferred

Source: National Center for Education Statistics, U.S. Dept. of Education

Figures for 1999-2000 are projected.

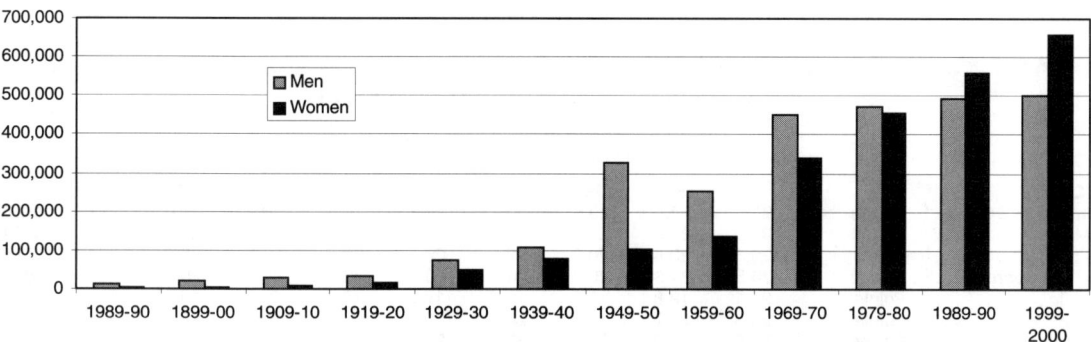

Financial Aid for College and Other Postsecondary Education

Reviewed by National Assoc. of Student Financial Aid Administrators

The cost of postsecondary education in the U.S. has increased in recent years, but financial aid, which may be in the form of grants (no repayment needed), loans, and/or work-study programs, is widely available to help families meet these expenses. Most aid is limited to family financial need as determined by standard formulas. Students interested in receiving aid are advised to apply, without making prior assumptions. Financial aid personnel at each school can provide information about programs available to students, steps to apply for them, and deadlines, all of which may vary.

First-time applicants for federal aid must file a Free Application for Federal Student Aid (FAFSA), generally as soon as possible after Jan. 1 for the academic year starting the following September. Figures provided must agree with federal income tax forms filed for the previous year. Other possible sources of aid include state governments, employers and unions, civic organizations, and the institutions themselves. There are also special federal programs that pay for postsecondary education in return for service: AmeriCorps (phone: 1-800-942-2677) and ROTC (phone: 1-800-USA-ROTC). Additional forms and certain fees may be required if a student is to be considered for institutional aid. Aid must be reapplied for annually.

A federal formula, based on information provided on the FAFSA, takes into account such factors as family after-tax income in the preceding calendar year, parental assets (excluding the parents' home) and length of time to retirement, and unusual expenses (such as very high medical expenses). The resulting Expected Family Contribution, or EFC (which is divided among the family members in college), is subtracted from the total cost of attendance for each person (including room and board or allowance for living costs) to determine financial need, and thus the maximum federal aid for which the family may be eligible. (Some institutions use a separate formula for needs-based institu-

tional aid.) Some schools guarantee to meet full financial need of each admitted student; others try to do so but may fall short, depending on availability of funds. Outside scholarships (even if non-needs-based) are taken into account in determining need.

The aid package offered by each school may include one or more of the following resources: Federal Pell Grants, for those with relatively great financial need; Federal Supplementary Educational Opportunity Grants, for those with greatest financial need; grants from the school; federal work-study or other work programs; low-interest Perkins loans; and subsidized and unsubsidized Stafford loans. Unsubsidized Stafford loans are available without need, as are all PLUS loans to parents. Loans have varying interest rates and other requirements. Repayment of Perkins and Stafford loans does not begin until after graduation; deferments are available under certain circumstances. For PLUS loans, parents must pass a credit check and begin repayment of both principal and interest while the student is still in school. Legislation passed by Congress in 1998 reduced interest rates on student loans and increased maximum Pell Grants.

Certain federal income tax credits—dollar for dollar reductions of the amount of tax due—are available to families who meet income and other requirements; see the chapter on Taxes.

Rules for financial aid are complex and changeable. *The Student Guide*, a comprehensive resource on financial aid from the U.S. Dept. of Education, can be found at the website http://www.ed.gov/prog_info/SFA/StudentGuide

Further information and FAFSA forms are available from the school or from the Federal Student Aid Information Center, PO Box 84, Washington, DC 20044; phone: 1-800-4-FED-AID, Mon.-Fri., 8 AM - 8 PM Eastern Time. The Information Center also has a free booklet called *The EFC Formula Book*. FAFSA forms can be obtained online at http://www.fafsa.ed.gov

Salaries of College Professors, 1998-99

Source: American Association of University Professors

TEACHING LEVEL	MEN Type of institution			WOMEN Type of institution		
	Public	Private/ Independent	Church-related	Public	Private/ Independent	Church-related
Doctoral level						
Professor	$80,379	$99,979	$84,796	$72,885	$90,611	$77,972
Associate	57,653	65,843	60,059	54,322	61,956	56,180
Assistant.	48,647	57,296	50,009	45,203	52,521	46,427
Master's level						
Professor	64,414	70,643	66,151	61,711	65,593	60,588
Associate	51,812	54,260	52,634	49,615	51,273	48,189
Assistant.	42,673	44,511	42,317	41,189	43,002	40,312
General 4-year						
Professor	58,432	68,145	52,945	57,045	64,089	49,678
Associate	48,643	51,044	43,412	46,808	49,202	41,791
Assistant.	40,625	41,551	36,534	39,245	40,634	36,017
2-year						
Professor	57,067	45,099	36,422	52,461	40,252	35,496
Associate	48,321	40,515	36,359	44,835	35,513	34,609
Assistant.	41,515	35,715	30,342	39,561	34,219	29,774

ACT (formerly American College Testing) Mean Scores and Characteristics of College-Bound Students, 1990-99

Source: ACT, Inc.

(for school year ending in year shown)

SCORES[1]	Unit[1]	1990[2]	1991[2]	1992[2]	1993[2]	1994[2]	1995[2]	1996[2]	1997[2]	1998[2]	1999[2]
Composite Scores . . .	**Points**	20.6	20.6	20.6	20.7	20.8	20.8	20.9	21.0	21.0	21.0
Male	Points	21.0	20.9	20.9	21.0	20.9	21.0	21.0	21.1	21.2	21.1
Female	Points	20.3	20.4	20.5	20.5	20.7	20.7	20.8	20.8	20.9	20.9
English Score	**Points**	20.5	20.3	20.2	20.3	20.3	20.2	20.3	20.3	20.4	20.5
Male	Points	20.1	19.8	19.8	19.8	19.8	19.8	19.8	19.9	19.9	20.0
Female	Points	20.9	20.7	20.6	20.6	20.7	20.6	20.7	20.7	20.8	20.9
Math Score	**Points**	19.9	20.0	20.0	20.1	20.2	20.2	20.2	20.6	20.8	20.7
Male	Points	20.7	20.6	20.7	20.8	20.8	20.9	20.9	21.3	21.5	21.4
Female	Points	19.3	19.4	19.5	19.6	19.6	19.7	19.7	20.1	20.2	20.2
PARTICIPANTS											
Total Number	**1,000**	817	796	832	875	892	945	925	959	995	1,019
Male	Percent	46	45	45	45	45	44	44	44	43	43
White	Percent	79	79	79	79	79	80	79	74	76	72
Black	Percent	9	9	9	9	9	9	9	10	11	10
Composite Scores											
27 or above	Percent	12	11	12	12	13	13	13	14	14	14
18 or below.	Percent	35	35	35	35	34	34	34	33	33	33

(1) Minimum point score, 1; maximum score, 36. Test scores and characteristics of college-bound students are based on the performance of all ACT-tested students who graduated in the spring of a given school year and who took the ACT Assessment during junior or senior year of high school. (2) Beginning with the Oct. 1989 test (1990 scores), an entirely new ACT Assessment was introduced. The Enhanced ACT Assessment increases the emphasis on rhetorical skills in the measurement of writing proficiency, increases the number of advanced math items, and includes a new reading test that features inferential and reasoning skills and a test designed to measure science reasoning. The Enhanced ACT also provides subscores in English, mathematics, and reading. The composite scores for 1989 have been converted to provide a basis of comparison; all 1990-99 scores are for the Enhanced ACT. It is not possible to compare directly these data and data from earlier years.

ACT Average Composite Scores by State, 1998-99

Source: ACT, Inc.

STATE	Avg. Composite Score	% Grads Taking ACT[1]	STATE	Avg. Composite Score	% Grads Taking ACT[1]	STATE	Avg. Composite Score	% Grads Taking ACT[1]	STATE	Avg. Composite Score	% Grads Taking ACT[1]
AL.	20.2	65	IL	21.4	67	MT	21.8	54	RI	22.7	3
AK	21.1	35	IN	21.2	19	NE	21.7	73	SC.	19.1	18
AZ.	21.4	28	IA	22.0	66	NV	21.5	41	SD.	21.2	70
AR	20.3	69	KS	21.5	75	NH	22.2	5	TN	19.9	77
CA	21.3	12	KY	20.1	68	NJ	20.7	4	TX	20.3	31
CO	21.5	62	LA	19.6	76	NM.	20.1	64	UT.	21.4	68
CT	21.6	3	ME	22.1	4	NY	22.0	14	VT.	21.9	9
DE	20.5	3	MD	20.9	10	NC	19.4	12	VA	20.6	7
DC	18.6	13	MA	22.0	6	ND	21.4	79	WA	22.6	18
FL.	20.6	39	MI.	21.3	69	OH	21.4	59	WV	20.2	58
GA	20.0	16	MN	22.1	64	OK	20.6	69	WI	22.3	67
HI	21.6	18	MS	18.7	82	OR.	22.6	11	WY	21.4	66
ID	21.4	60	MO	21.6	67	PA	21.4	7	**U.S. AVG.**	**21.0**	**36**

(1) Based on number of high school graduates in 1998, as projected by the Western Interstate Commission for Higher Education, and number of students in the class of 1998 who took the ACT.

SAT Mean Verbal and Math Scores of College-Bound Seniors, 1989-99

Source: The College Board

(recentered scale; for school year ending in year shown)

	Unit	1989	1990	1991	1992	1993	1994	1995	1996	1997	1998	1999
Verbal Scores	**Points**	**504**	**500**	**499**	**500**	**500**	**499**	**504**	**505**	**505**	**505**	**505**
Male	Points	510	505	503	504	504	501	505	507	507	509	509
Female	Points	498	496	495	496	497	497	502	503	503	502	502
Math Scores	**Points**	**502**	**501**	**500**	**501**	**503**	**504**	**506**	**508**	**511**	**512**	**511**
Male	Points	523	521	520	521	524	523	525	527	530	531	531
Female	Points	482	483	482	484	484	487	490	492	494	496	495

Note: In 1995, the College Board recentered the scoring scale for the SAT by reestablishing the original mean score of 500 on the 200-800 scale. For 1997-99, individual student scores were converted to the recentered score. For 1996, scores are based on recentered scores plus scores converted from the original to the new scale. Scores for 1989-95 were recomputed after individual scores were converted from the original to the new scale.

SAT Mean Scores by State, 1989 and 1996-99

Source: The College Board

(recentered scale; for school year ending in year shown)

	1989		1996		1997		1998		1999		% Grads
STATE	V	M	V	M	V	M	V	M	V	M	Taking SAT[1]
Alabama	556	539	565	558	561	555	562	558	561	555	9
Alaska	519	505	521	513	520	517	521	520	516	514	50
Arizona	528	523	525	521	523	522	525	528	524	525	34
Arkansas	547	536	556	550	567	558	568	555	563	556	6
California	498	509	495	511	496	514	497	516	497	514	49
Colorado	534	530	536	538	536	539	537	542	536	540	32
Connecticut	512	498	507	504	509	507	510	509	510	509	80
Delaware	512	494	508	495	505	498	501	493	503	497	67
District of Columbia	481	466	489	473	490	475	488	476	494	478	77
Florida	497	494	498	496	499	499	500	501	499	498	53
Georgia	479	475	484	477	486	481	486	482	487	482	63
Hawaii	482	507	485	510	485	510	483	513	482	513	52
Idaho	541	523	543	536	544	539	545	544	542	540	16
Illinois	537	539	564	575	562	578	564	581	569	585	12
Indiana	490	487	494	494	494	497	497	500	496	498	60
Iowa	585	585	590	600	589	601	593	601	594	598	5
Kansas	569	561	579	571	578	575	582	585	578	576	9
Kentucky	552	539	549	544	548	546	547	550	547	547	12
Louisiana	549	534	559	550	560	553	562	558	561	558	8
Maine	508	493	504	498	507	504	504	501	507	503	68
Maryland	510	505	507	504	507	507	506	508	507	507	65
Massachusetts	509	499	507	504	508	508	508	508	511	511	78
Michigan	534	534	557	565	557	566	558	569	557	565	11
Minnesota	550	550	582	593	582	592	585	598	586	598	9
Mississippi	547	536	569	557	567	551	562	549	563	548	4
Missouri	546	538	570	569	567	568	570	573	572	572	8
Montana	545	542	546	547	545	548	543	546	545	546	21
Nebraska	562	560	567	568	562	564	565	571	568	571	8
Nevada	516	512	508	507	508	509	510	513	512	517	34
New Hampshire	524	510	520	514	521	518	523	520	520	518	72
New Jersey	500	497	498	500	497	508	497	508	498	510	80
New Mexico	558	550	554	548	554	545	554	551	549	542	12
New York	495	496	497	499	495	502	495	503	495	502	76
North Carolina	474	469	490	486	490	488	490	492	493	493	61
North Dakota	574	581	596	599	588	595	590	599	594	605	5
Ohio	528	520	536	535	535	536	536	540	534	568	25
Oklahoma	554	542	566	557	568	560	568	564	567	560	8
Oregon	519	509	523	521	525	524	528	528	525	525	53
Pennsylvania	501	490	498	492	498	495	497	495	498	495	70
Rhode Island	506	492	501	491	499	493	501	495	504	499	70
South Carolina	476	469	480	474	479	474	478	473	479	475	61
South Dakota	573	560	574	566	574	570	584	581	585	588	4
Tennessee	561	542	563	552	564	556	564	557	559	553	13
Texas	492	490	495	500	494	501	494	501	494	499	50
Utah	572	555	583	575	576	570	572	570	570	568	5
Vermont	512	497	506	500	508	502	508	504	514	506	70
Virginia	507	498	507	496	506	497	507	499	508	499	65
Washington	524	515	519	519	523	523	524	526	525	526	52
West Virginia	525	515	526	506	524	508	525	513	527	512	18
Wisconsin	553	554	577	586	579	590	581	594	584	595	7
Wyoming	538	537	544	544	543	543	548	546	546	551	10
NATIONAL AVERAGE	**504**	**502**	**505**	**508**	**505**	**511**	**505**	**512**	**505**	**511**	**43**

NOTE: In 1995, the College Board recentered the scoring scale for the SAT by reestablishing the original mean score of 500 on the 200-800 scale. The College Board states that comparing states or ranking them on the basis of SAT scores alone is invalid, and the College Board discourages doing so. (1) Based on number of high school graduates in 1999, as projected by the Western Interstate Commission for Higher Education, and number of students in the class of 1999 who took the SAT.

Top 50 Public Libraries in the U.S. and Canada

Source: Public Library Data Service, Statistical Report 1999, Public Library Association

Ranked at end of the 1998 fiscal year by population served.

Population served	Library name and location	No. of branches[1]	No. of holdings	Circulation	Annual acquisition expenditures
3,722,500	Los Angeles Public Library (CA)	67	5,739,540	12,846,569	$9,629,463
3,397,460	Los Angeles Public Library, County of (CA)	85	6,883,210	14,989,561	6,230,378
3,070,302	New York Public Library, The Branch Libraries (NY)	85	10,483,628	12,652,619	12,096,818
2,783,726	Chicago Public Library (IL)	78	9,222,449	8,177,472	11,046,825
2,385,421	Toronto Public Library (Ontario)	97	8,718,431	24,918,622	8,621,443
2,300,664	Brooklyn Public Library (NY)	58	5,129,468	10,316,731	9,434,066
1,975,676	Queens Borough Public Library (NY)	62	9,510,814	16,882,963	11,483,476
1,840,436	Houston Public Library (TX)	36	4,913,944	5,507,535	5,503,474
1,737,811	Miami-Dade Public Library System (FL)	30	3,882,275	4,769,816	3,569,459
1,585,577	Free Library of Philadelphia (PA)	52	7,968,978	6,066,954	7,218,460
1,430,677	Broward County Library System (FL)	34	2,075,118	6,781,202	5,178,480
1,363,600	San Antonio Public Library (TX)	18	1,802,614	3,977,212	2,277,476
1,336,449	Carnegie Library of Pittsburgh (PA)	19	6,632,303	2,888,361	1,800,024
1,220,710	Phoenix Public Library (AZ)	12	1,849,473	6,685,104	3,011,189
1,201,900	San Diego Public Library (CA)	33	2,798,189	6,494,616	2,983,815
1,171,600	Hawaii State Public Library System (HI)	49	3,127,664	7,781,129	229,549
1,110,504	Sacramento Public Library (CA)	22	1,788,730	3,747,409	2,540,190
1,104,815	Harris County Public Library (TX)	25	2,066,557	5,034,035	1,852,237
1,087,393	King County Library System (WA)	44	3,717,247	12,966,258	5,459,057
1,052,300	Dallas Public Library (TX)	22	3,098,091	3,807,124	2,564,651
1,027,974	Detroit Public Library (MI)	23	2,832,169	1,513,185	2,636,293
1,016,376	Montréal, Bibliothèque de (Quebec)	24	2,507,470	5,360,796	1,699,245
1,001,383	Providence Public Library (RI)	9	2,545,174	832,975	749,494
998,200	San Bernardino County Library (CA)	27	1,100,500	2,698,800	962,500
985,500	Las Vegas-Clark County Library District (NV)	24	2,275,920	5,257,083	3,770,000
968,532	Buffalo & Erie County Public Library (NY)	52	3,524,297	8,734,854	3,672,354
948,800	Fairfax County Public Library (VA)	20	2,208,342	9,696,887	4,251,702
931,324	San Diego County Library (CA)	32	1,080,758	2,941,685	1,408,415
915,900	Tampa-Hillsborough County Public Library (FL)	19	1,959,190	3,963,350	3,200,106
893,969	San Jose Public Library System (CA)	17	1,563,000	6,477,930	3,493,370
857,616	Cincinnati & Hamilton County, Public Library of (OH)	41	9,271,068	12,805,278	8,208,848
846,584	Memphis & Shelby County Public Libraries (TN)	21	1,848,643	3,701,931	1,964,374
843,000	St. Louis Public Library (MO)	18	3,138,143	8,327,623	4,139,448
837,000	Montgomery County Dept. of Public Libraries (MD)	21	2,509,343	9,836,900	5,560,781
819,334	Calgary Public Library (Alberta)	15	2,078,206	11,009,451	2,670,012
817,851	Tucson-Pima County (AZ)	18	1,173,549	5,110,542	2,140,930
794,794	Atlanta-Fulton Public Library (GA)	34	2,101,661	2,534,636	4,167,429
789,600	San Francisco Public Library (CA)	26	2,383,391	5,462,172	3,373,224
787,900	Contra Costa County Library (Pleasant Hill, CA)	22	1,304,550	3,669,522	1,509,998
779,534	Jacksonville Public Libraries (FL)	14	2,773,082	3,577,969	2,069,884
773,810	Prince George's County Memorial Library System (MD)	18	2,113,130	4,062,748	2,549,700
770,684	Indianapolis-Marion County Public Library (IN)	21	1,793,719	9,506,576	5,405,126
768,850	Fresno County Library (CA)	34	915,982	1,553,603	482,786
762,235	Columbus Metropolitan Library (OH)	20	2,520,257	11,171,752	6,721,308
749,227	Orange County Library System (FL)	12	1,650,496	4,080,833	2,346,699
724,060	Baltimore County Public Library (MD)	16	1,876,088	9,944,704	4,429,868
720,895	Hennepin County Library (MN)	26	1,764,379	10,539,250	3,652,683
718,000	Enoch Pratt Free Library (Baltimore, MD)	26	2,866,633	1,324,013	2,955,220
717,400	Macomb County Library (MI)	0	139,631	184,676	327,026
713,968	Rochester Public Library (NY)	10	1,033,437	1,632,263	1,399,700

(1) Main branch not included.

Number of Public Libraries and Operating Income, by State

Source: Public Libraries Survey, National Center for Education Statistics, U.S. Dept. of Education

(data for fiscal year 1996 unless otherwise indicated; operating income in thousands)

STATE	No. of libraries[1]	Operating income[2]	STATE	No. of libraries[1]	Operating income[2]	STATE	No. of libraries[1]	Operating income[2]
Alabama	269	$51,915	Kentucky	186	$54,942	Ohio	688	$484,821
Alaska	103	20,366	Louisiana	324	77,036	Oklahoma	202	39,844
Arizona	164	77,558	Maine	275	20,666[3]	Oregon	198	77,022
Arkansas	202	27,821	Maryland	182	138,029	Pennsylvania	650	201,961[3]
California	1,042	619,364	Massachusetts	488	158,733	Rhode Island	73	25,584
Colorado	235	108,732	Michigan	661	209,905[3]	South Carolina	183	52,851
Connecticut	245	107,809	Minnesota	361	117,693	South Dakota	134	11,226
Delaware	30	11,242	Mississippi	242	27,316	Tennessee	285	60,512
District of Columbia	27	21,657	Missouri	348	113,055	Texas	762	215,643[3]
			Montana	110	12,852	Utah	100	39,099
Florida	441	269,520	Nebraska	245	29,331[3]	Vermont	201	10,020[3]
Georgia	371	103,353	Nevada	79	34,185	Virginia	313	143,550
Hawaii	49	21,631	New Hampshire	238	24,458	Washington	312	172,626
Idaho	141	18,582	New Jersey	453	258,632	West Virginia	174	20,488
Illinois	778	383,737[3]	New Mexico	92	24,076	Wisconsin	451	125,666
Indiana	425	179,793	New York	1,068	663,735	Wyoming	74	11,732
Iowa	556	54,198	North Carolina	357	110,012			
Kansas	374	53,344	North Dakota	86	7,047	**U.S. TOTAL**	**16,047**	**$5,904,967**

(1) Includes central libraries and branches. (2) Some totals may be underestimated because of nonresponse. (3) These libraries reported data for fiscal year 1994 or 1995.

American Colleges and Universities
General Information for the 1998–99 Academic Year
Source: Peterson's, a Thomson Learning Company, Copyright 1999

These listings include only **accredited undergraduate degree-granting institutions** in the United States and the U.S. territories that have a total institutional **enrollment of 1,000 or more**. Four-year colleges (those that award a bachelor's degree as their highest undergraduate degree) are listed first, followed by two-year colleges (those that award an associate as their highest or primary undergraduate degree). Data are not reported for certain institutions that did not provide updated information on Peterson's Annual Survey of Undergraduate Institutions for the 1998–99 academic year.

All institutions are coeducational except those where the ZIP code is followed directly by: (1)–men only, (2)–primarily men, (3)–women only, (4)–primarily women.

Year is that of founding.

The **Tuition & Fees** column shows annual tuition and required fees for full-time students, or (where indicated) tuition and fees per credit hour (cr. hr.) for part-time students. Where tuition costs vary according to residence, the figure is given for the most local residents and is coded: (A)–area residents, (S)–state residents; all other tuition figures apply to all students regardless of residence. Where annual expenses are expressed as a lump sum (including full-time tuition, mandatory fees, room and board), the figure is coded: (C) for comprehensive fee. **Room & Board** is average cost for one academic year.

Control: 1–independent (nonprofit), 2–independent-religious, 3–proprietary (profit-making), 4–federal, 5–state, 6–commonweath (Puerto Rico), 7–territory (U.S. territories), 8–county, 9–district, 10–city, 11–state and local, 12–state-related. **Degree** means the highest degree offered (B–bachelor's, M–master's, F–first professional, D–doctorate).

Enrollment is the total number of matriculated undergraduate and (if applicable) graduate students.

Faculty is the total number of faculty members teaching undergraduate courses and (if available) graduate courses.

Any data not available or not applicable are indicated as NA.

Four-Year Colleges

Name, address	Year	Tuition & Fees	Room & Board	Control, Degree	Enroll-ment	Faculty
Abilene Christian U, Abilene, TX 79699-9100	1906	$9,180	$4,010	2-D	4,630	303
Academy of Art Coll, San Francisco, CA 94105-3410	1929	$14,910	NA	3-M	4,964	493
Adams State Coll, Alamosa, CO 81102	1921	$1,958 (S)	$4,630	5-M	2,512	134
Adelphi U, Garden City, NY 11530	1896	$14,720	$6,630	1-D	5,806	333
Adrian Coll, Adrian, MI 49221-2575	1859	$12,830	$4,320	2-B	1,055	113
Alabama Ag & Mech U, Normal, AL 35762-1357	1875	$2,168 (S)	$2,678	5-D	5,128	298
Alabama State U, Montgomery, AL 36101-0271	1867	$2,030 (S)	$3,500	5-M	5,552	357
Albany St U, Albany, GA 31705-2717	1903	$2,124 (S)	$3,226	5-M	3,200	173
Albertus Magnus Coll, New Haven, CT 06511-1189	1925	$17,262	$6,324	2-M	1,669	68
Albion Coll, Albion, MI 49224-1831	1835	$16,806	$5,072	2-B	1,490	117
Albright Coll, Reading, PA 19612-5234	1856	$18,310	$5,612	2-B	1,471	120
Alcorn State U, Lorman, MS 39096-9402	1871	$2,429 (S)	$2,370	5-M	2,860	222
Alfred U, Alfred, NY 14802-1205	1836	$8,602 (S)	$6,790	1-D	2,393	197
Allegheny Coll, Meadville, PA 16335	1815	$19,360	$4,840	2-B	1,898	162
Allentown Coll of St. Francis de Sales, Center Valley, PA 18034-9568	1964	$11,750	$5,690	2-M	2,250	111
Alma Coll, Alma, MI 48801-1599	1886	$13,823	$5,250	2-B	1,442	144
Alvernia Coll, Reading, PA 19607-1799	1958	$11,320	$5,200	2-B	1,294	136
Alverno Coll, Milwaukee, WI 53234-3922 (3)	1887	$9,722	$4,250	2-M	2,018	198
Amber U, Garland, TX 75041-5595	1971	$4,025	NA	2-M	1,584	65
American Intl Coll, Springfield, MA 01109-3189	1885	$11,244	$6,034	1-D	2,000	122
American Military U, Manassas Park, VA 20111	1991	$4,835	NA	3-M	1,689	130
American U, Washington, DC 20016-8001	1893	$18,555	$7,652	2-D	10,611	1,086
American U of Puerto Rico, Bayamón, PR 00960-2037	1963	$3,070	NA	1-B	3,515	206
Amherst Coll, Amherst, MA 01002-5000	1821	$23,027	$6,361	1-B	1,651	201
Anderson Coll, Anderson, SC 29621-4035	1911	$9,475	$4,228	2-B	1,096	86
Anderson U, Anderson, IN 46012-3495	1917	$13,360	$4,330	2-D	2,231	222
Andrews U, Berrien Springs, MI 49104	1874	$11,577	$3,630	2-D	2,901	324
Angelo State U, San Angelo, TX 76909	1928	$2,242 (S)	$3,908	5-M	6,315	266
Anna Maria Coll, Paxton, MA 01612	1946	$12,240	$5,560	2-M	1,470	84
Appalachian State U, Boone, NC 28608	1899	$1,840 (S)	$3,190	5-D	12,386	828
Aquinas Coll, Grand Rapids, MI 49506-1799	1886	$12,950	$4,432	2-M	2,483	198
Arizona State U, Tempe, AZ 85287	1885	$2,059 (S)	$4,825	5-D	43,732	1,976
Arizona State U East, Mesa, AZ 85212	1995	$2,059 (S)	NA	5-M	1,095	79
Arizona State U West, Phoenix, AZ 85069-7100	1984	$2,059 (S)	NA	5-M	4,878	296
Arkansas State U, State University, AR 72467	1909	$2,280 (S)	$2,990	5-D	10,364	522
Arkansas Tech U, Russellville, AR 72801-2222	1909	$2,126 (S)	$2,924	5-M	4,458	294
Armstrong Atlantic State U, Savannah, GA 31419-1997	1935	$1,836 (S)	$4,284	5-M	5,570	407
Art Ctr Coll of Design, Pasadena, CA 91103-1999	1930	$17,180	NA	1-M	1,395	385
Asbury Coll, Wilmore, KY 40390-1198	1890	$12,020	$3,390	2-B	1,280	155
Ashland U, Ashland, OH 44805-3702	1878	$13,601	$5,300	2-D	5,768	234
Assumption Coll, Worcester, MA 01615-0005	1904	$15,595	$6,400	2-M	2,598	211
Athens State U, Athens, AL 35611-1902	1822	$1,845 (S)	NA	5-B	2,739	135
Auburn U, Auburn University, AL 36849-0002	1856	$2,610 (S)	NA	5-D	21,775	1,262
Auburn U Montgomery, Montgomery, AL 36124-4023	1967	$2,289 (S)	NA	5-D	5,408	348
Audrey Cohen Coll, New York, NY 10013-1919	1964	$8,860	NA	1-M	1,108	127
Augsburg Coll, Minneapolis, MN 55454-1351	1869	$14,616	$5,134	2-M	2,907	281
Augustana Coll, Rock Island, IL 61201-2296	1860	$15,300	$4,842	2-B	2,301	198
Augustana Coll, Sioux Falls, SD 57197	1860	$13,112	$4,056	2-M	1,698	167
Augusta State U, Augusta, GA 30904-2200	1925	$1,926 (S)	NA	5-M	5,295	274
Aurora U, Aurora, IL 60506-4892	1893	$11,310	$4,491	1-M	2,121	251
Austin Coll, Sherman, TX 75090-4400	1849	$14,205	$5,393	2-M	1,250	106
Austin Peay State U, Clarksville, TN 37040	1927	$2,280 (S)	$3,270	5-M	7,508	477
Averett Coll, Danville, VA 24541-3692	1859	$12,500	$4,200	2-M	2,220	253
Avila Coll, Kansas City, MO 64145-1698	1916	$10,860	$4,600	2-M	1,270	159
Azusa Pacific U, Azusa, CA 91702-7000	1899	$13,947	$4,512	2-D	5,368	540
Babson Coll, Babson Park, MA 02457-0310	1919	$20,365	$8,100	1-M	3,401	206
Baker Coll of Flint, Flint, MI 48507-5508	1911	$6,300	NA	1-M	4,581	154
Baker Coll of Jackson, Jackson, MI 49202	1994	$6,300	NA	1-B	1,010	63

Name, address	Year	Tuition & Fees	Room & Board	Control, Degree	Enroll- ment	Faculty
Baker Coll of Muskegon, Muskegon, MI 49442-3497	1888	$6,480	NA	1-B	2,370	115
Baker Coll of Owosso, Owosso, MI 48867-4400	1984	$5,850	NA	1-B	1,850	109
Baker Coll of Port Huron, Port Huron, MI 48060-2597	1990	$6,300	NA	1-B	1,036	85
Baker U, Baldwin City, KS 66006-0065	1858	$10,900	$4,650	2-M	2,400	105
Baldwin-Wallace Coll, Berea, OH 44017-2088	1845	$13,275	$5,060	2-M	4,528	334
Ball State U, Muncie, IN 47306-1099	1918	$3,414 (S)	$4,316	5-D	18,924	1,050
Bard Coll, Annandale-on-Hudson, NY 12504	1860	$22,220	$7,016	1-D	1,423	160
Barnard Coll, New York, NY 10027-6598 (3)	1889	$20,976	$8,898	1-B	2,270	283
Barry U, Miami Shores, FL 33161-6695	1940	$13,550	$6,090	2-D	7,426	552
Barton Coll, Wilson, NC 27893-7000	1902	$10,150	$3,778	2-B	1,271	77
Baruch Coll of the City U of New York, New York, NY 10010-5585	1919	$3,330 (S)	NA	11-D	14,936	828
Bates Coll, Lewiston, ME 04240-6028	1855	$28,650 (C)	NA	1-B	1,713	176
Bayamón Central U, Bayamón, PR 00960-1725	1970	$4,025	NA	2-M	3,188	199
Bayamón Technological U Coll, Bayamón, PR 00959-1919	1971	$1,090 (S)	NA	6-B	5,826	247
Baylor U, Waco, TX 76798	1845	$10,266	$4,615	2-D	12,987	715
Beaver Coll, Glenside, PA 19038-3295	1853	$15,840	$6,900	2-D	2,746	301
Becker Coll, Worcester, MA 01615-0071	1784	$10,130	$5,280	1-B	1,113	127
Belhaven Coll, Jackson, MS 39202-1789	1883	$9,370	$3,660	2-M	1,407	129
Bellarmine Coll, Louisville, KY 40205-0671	1950	$10,970	$3,840	2-M	2,917	189
Bellevue U, Bellevue, NE 68005-3098	1965	$3,650	NA	1-M	2,929	93
Belmont U, Nashville, TN 37212-3757	1951	$10,300	$4,086	2-M	2,963	393
Beloit Coll, Beloit, WI 53511-5596	1846	$19,050	$4,326	1-B	1,258	119
Bemidji State U, Bemidji, MN 56601-2699	1919	$3,118 (S)	$3,084	5-M	4,350	151
Benedict Coll, Columbia, SC 29204	1870	$7,502	$4,182	2-B	2,208	142
Benedictine Coll, Atchison, KS 66002-1499	1859	$11,750	$4,620	2-M	1,284	72
Benedictine U, Lisle, IL 60532-0900	1887	$12,330	$4,610	2-D	2,700	263
Bentley Coll, Waltham, MA 02452-4705	1917	$16,495	$7,160	1-M	5,775	367
Berea Coll, Berea, KY 40404	1855	$195	$3,510	1-B	1,517	151
Berklee Coll of Music, Boston, MA 02215-3693	1945	$15,100	$7,890	1-B	2,953	377
Berry Coll, Mount Berry, GA 30149-0159	1902	$10,210	$4,750	2-M	2,104	126
Bethel Coll, Mishawaka, IN 46545-5591	1947	$11,500	$3,700	2-M	1,627	140
Bethel Coll, St. Paul, MN 55112-6999	1871	$13,840	$5,160	2-M	2,769	226
Bethune-Cookman Coll, Daytona Beach, FL 32114-3099	1904	$8,047	$4,984	2-B	2,481	215
Biola U, La Mirada, CA 90639-0001	1908	$14,286	$5,140	2-D	3,621	283
Birmingham-Southern Coll, Birmingham, AL 35254	1856	$13,960	$5,365	2-M	1,540	148
Black Hills State U, Spearfish, SD 57799-9500	1883	$2,878 (S)	$2,690	5-M	3,623	135
Bloomfield Coll, Bloomfield, NJ 07003-9981	1868	$9,650	$5,000	2-B	1,958	193
Bloomsburg U of Pennsylvania, Bloomsburg, PA 17815-1905	1839	$4,278 (S)	$3,550	5-M	7,647	411
Bluefield Coll, Bluefield, VA 24605-1799	1922	$9,100	$4,610	2-B	1,035	161
Bluefield State Coll, Bluefield, WV 24701-2198	1895	$2,044 (S)	NA	5-B	2,405	158
Bluffton Coll, Bluffton, OH 45817-1196	1899	$12,375	$5,121	2-M	1,000	94
Boise State U, Boise, ID 83725-0399	1932	$2,294 (S)	$3,488	5-D	15,696	823
Boricua Coll, New York, NY 10032-1560	1974	$6,300	NA	1-M	1,190	116
Boston Coll, Chestnut Hill, MA 02467-3800	1863	$20,292	$8,020	2-D	13,765	1,127
Boston U, Boston, MA 02215	1839	$23,148	$7,870	1-D	29,131	3,133
Bowdoin Coll, Brunswick, ME 04011-2546	1794	$22,905	$6,285	1-B	1,583	170
Bowie State U, Bowie, MD 20715-9465	1865	$3,357 (S)	$4,230	5-M	5,024	295
Bowling Green State U, Bowling Green, OH 43403	1910	$4,422 (S)	$4,392	5-D	17,735	784
Bradley U, Peoria, IL 61625-0002	1897	$12,690	$5,170	1-M	5,813	476
Brandeis U, Waltham, MA 02454-9110	1948	$22,851	$6,970	1-D	4,405	532
Brewton-Parker Coll, Mt. Vernon, GA 30445-0197	1904	$5,760	$2,850	2-B	1,416	183
Briar Cliff Coll, Sioux City, IA 51104-2100	1930	$11,880	$4,194	2-B	1,001	71
Bridgewater Coll, Bridgewater, VA 22812-1599	1880	$13,270	$5,970	2-B	1,124	89
Bridgewater State Coll, Bridgewater, MA 02325-0001	1840	$3,324 (S)	$4,502	5-M	9,161	489
Brigham Young U, Provo, UT 84602-1001	1875	$2,630	$4,150	2-D	32,202	1,797
Brigham Young U–Hawaii Campus, Laie, HI 96762-1294	1955	$2,665	$5,075	2-B	2,301	174
Brooklyn Coll of the City U of New York, Brooklyn, NY 11210-2889	1930	$3,413 (S)	NA	11-M	14,973	884
Brown U, Providence, RI 02912	1764	$23,124	$6,898	1-D	7,782	687
Bryant Coll, Smithfield, RI 02917-1284	1863	$14,800	$6,700	1-M	3,365	193
Bryn Mawr Coll, Bryn Mawr, PA 19010-2899 (3)	1885	$21,430	$7,870	1-D	1,796	122
Bucknell U, Lewisburg, PA 17837	1846	$21,210	$5,355	1-M	3,626	296
Buena Vista U, Storm Lake, IA 50588	1891	$14,848	$4,507	2-M	2,753	98
Butler U, Indianapolis, IN 46208-3485	1855	$15,690	$5,570	1-F	4,126	442
Cabrini Coll, Radnor, PA 19087-3698	1957	$13,200	$6,900	2-M	2,043	195
Caldwell Coll, Caldwell, NJ 07006-6195	1939	$10,800	$5,600	2-M	1,964	156
California Baptist U, Riverside, CA 92504-3206	1950	$8,236	$4,326	2-M	2,094	159
California Coll of Arts & Crafts, San Francisco, CA 94107	1907	$15,052	NA	1-M	1,143	239
California Inst of Technology, Pasadena, CA 91125-0001	1891	$18,816	$5,883	1-D	1,858	353
California Inst of the Arts, Valencia, CA 91355-2340	1961	$18,185	NA	1-M	1,192	279
California Lutheran U, Thousand Oaks, CA 91360-2787	1959	$15,415	$5,985	2-M	2,753	209
California Polytechnic State U, San Luis Obispo, San Luis Obispo, CA 93407	1901	$2,231 (S)	$5,355	5-M	16,296	951
California State Polytechnic U, Pomona, Pomona, CA 91768-2557	1938	$1,923 (S)	$5,477	5-M	17,577	1,056
California State U, Bakersfield, Bakersfield, CA 93311-1099	1970	$1,965 (S)	$4,230	5-M	5,594	357
California State U, Chico, Chico, CA 95929-0722	1887	$2,075 (S)	$5,493	5-M	14,983	893
California State U, Dominguez Hills, Carson, CA 90747-0001	1960	$1,821 (S)	NA	5-M	12,054	721
California State U, Fresno, Fresno, CA 93740-0057	1911	$1,806 (S)	$5,203	5-D	18,101	1,074
California State U, Fullerton, Fullerton, CA 92834-9480	1957	$1,947 (S)	NA	5-M	25,675	1,457
California State U, Hayward, Hayward, CA 94542-3000	1957	$1,827 (S)	NA	5-M	12,888	703
California State U, Long Beach, Long Beach, CA 90840-0119	1949	$1,846 (S)	$5,200	5-M	28,637	1,610
California State U, Los Angeles, Los Angeles, CA 90032-8530	1947	$1,757 (S)	NA	5-M	19,732	1,213
California State U, Northridge, Northridge, CA 91330	1958	$1,980 (S)	NA	5-M	27,203	1,564
California State U, Sacramento, Sacramento, CA 95819-6048	1947	$1,982 (S)	$5,117	5-M	23,676	1,414
California State U, San Bernardino, San Bernardino, CA 92407-2397	1965	$1,896 (S)	$4,338	5-M	13,280	571
California State U, San Marcos, San Marcos, CA 92096-0001	1990	$1,720 (S)	NA	5-M	5,025	330
California State U, Stanislaus, Turlock, CA 95382	1957	$1,915 (S)	$6,100	5-M	6,351	368
California U of Pennsylvania, California, PA 15419-1394	1852	$4,475 (S)	$4,268	5-M	5,800	364
Calvin Coll, Grand Rapids, MI 49546-4388	1876	$12,250	$4,500	2-M	4,121	321
Cambridge Coll, Cambridge, MA 02138-5304	1971	$7,620	NA	1-M	2,087	150

Name, address	Year	Tuition & Fees	Room & Board	Control, Degree	Enroll- ment	Faculty
Cameron U, Lawton, OK 73505-6377	1908	$2,180 (S)	$2,720	5-M	5,136	382
Campbellsville U, Campbellsville, KY 42718-2799	1906	$7,302	$3,730	2-M	1,646	117
Campbell U, Buies Creek, NC 27506	1887	$10,003	$3,738	2-D	3,281	445
Canisius Coll, Buffalo, NY 14208-1098	1870	$13,882	$5,940	2-M	4,949	411
Capital U, Columbus, OH 43209-2394	1830	$14,760	$4,550	2-F	4,047	430
Cardinal Stritch U, Milwaukee, WI 53217-3985	1937	$10,130	$4,410	2-D	5,165	572
Carleton Coll, Northfield, MN 55057-4001	1866	$21,885	$4,584	1-B	1,881	206
Carlow Coll, Pittsburgh, PA 15213-3165 (4)	1929	$11,708	$4,880	2-M	2,378	213
Carnegie Mellon U, Pittsburgh, PA 15213-3891	1900	$20,375	$6,555	1-D	8,185	934
Carroll Coll, Helena, MT 59625-0002	1909	$11,490	$4,540	2-B	1,247	121
Carroll Coll, Waukesha, WI 53186-5593	1846	$14,420	$4,440	2-M	2,621	220
Carson-Newman Coll, Jefferson City, TN 37760	1851	$10,610	$3,830	2-M	2,336	183
Carthage Coll, Kenosha, WI 53140-1994	1847	$15,365	$4,630	2-M	2,191	131
Case Western Reserve U, Cleveland, OH 44106	1826	$17,940	$5,240	1-D	9,486	2,030
Castleton State Coll, Castleton, VT 05735	1787	$4,506 (S)	$5,206	5-M	1,766	158
Catawba Coll, Salisbury, NC 28144-2488	1851	$11,352	$4,650	2-M	1,255	117
The Catholic U of America, Washington, DC 20064	1887	$17,110	$7,360	2-D	5,486	660
Cedar Crest Coll, Allentown, PA 18104-6196 (3)	1867	$15,820	$5,975	2-B	1,451	136
Cedarville Coll, Cedarville, OH 45314-0601	1887	$9,312	$4,716	2-B	2,664	204
Centenary Coll of Louisiana, Shreveport, LA 71134-1188	1825	$11,400	$4,050	2-M	1,064	114
Center for Creative Studies–Coll of Art & Design, Detroit, MI 48202-4034	1926	$14,496	NA	1-B	1,056	197
Central Coll, Pella, IA 50219-1999	1853	$12,802	$4,592	2-B	1,208	114
Central Connecticut State U, New Britain, CT 06050-4010	1849	$3,614 (S)	$5,472	5-M	11,686	816
Central Methodist Coll, Fayette, MO 65248-1198	1854	$10,710	$4,150	2-M	1,267	109
Central Michigan U, Mount Pleasant, MI 48859	1892	$3,546 (S)	$4,480	5-D	25,595	898
Central Missouri State U, Warrensburg, MO 64093	1871	$2,640 (S)	$3,970	5-M	10,763	499
Central State U, Wilberforce, OH 45384	1887	$3,318 (S)	$4,695	5-M	1,026	102
Central Washington U, Ellensburg, WA 98926	1891	$2,826 (S)	$4,788	5-M	8,355	434
Centre Coll, Danville, KY 40422-1394	1819	$14,600	$5,050	2-B	1,052	101
Chadron State Coll, Chadron, NE 69337	1911	$2,148 (S)	$3,250	5-M	2,809	166
Chaminade U of Honolulu, Honolulu, HI 96816-1578	1955	$10,900	$5,410	2-M	2,690	268
Champlain Coll, Burlington, VT 05402-0670	1878	$9,785	$7,165	1-B	2,274	156
Chapman U, Orange, CA 92866	1861	$18,750	$7,044	2-F	3,822	397
Charleston Southern U, Charleston, SC 29423-8087	1964	$9,248	$3,776	2-M	2,445	133
Chestnut Hill Coll, Philadelphia, PA 19118-2693 (3)	1924	$14,265	$6,510	2-D	1,536	214
Cheyney U of Pennsylvania, Cheyney, PA 19319	1837	$4,023 (S)	$4,646	5-M	1,743	132
Chicago State U, Chicago, IL 60628	1867	$2,496 (S)	$5,700	5-M	8,416	482
Christian Brothers U, Memphis, TN 38104-5581	1871	$11,930	$3,950	2-M	1,964	181
Christopher Newport U, Newport News, VA 23606-2998	1960	$3,466 (S)	$4,950	5-M	5,004	309
The Citadel, The Military Coll of South Carolina, Charleston, SC 29409 (2)	1842	$3,499 (S)	$2,953	5-M	4,015	191
City Coll of the City U of New York, New York, NY 10031-9198	1847	$3,309 (S)	NA	11-D	11,355	981
City U, Bellevue, WA 98004-6442	1973	$6,000	NA	1-M	13,038	1,100
Claflin Coll, Orangeburg, SC 29115	1869	$5,580	$3,314	2-B	1,161	84
Claremont McKenna Coll, Claremont, CA 91711	1946	$19,020	$6,890	1-B	1,024	130
Clarion U of Pennsylvania, Clarion, PA 16214	1867	$4,419 (S)	$3,480	5-M	5,866	369
Clark Atlanta U, Atlanta, GA 30314	1865	$9,348	$5,870	2-D	5,410	482
Clarke Coll, Dubuque, IA 52001-3198	1843	$12,439	$4,886	2-M	1,279	145
Clarkson U, Potsdam, NY 13699	1896	$18,593	$6,712	1-D	2,749	174
Clark U, Worcester, MA 01610-1477	1887	$20,940	$4,150	1-D	3,042	309
Clayton Coll & State U, Morrow, GA 30260-0285	1969	$2,168 (S)	NA	5-B	4,274	239
Clemson U, Clemson, SC 29634	1889	$3,392 (S)	$4,434	5-D	16,685	1,302
Cleveland State U, Cleveland, OH 44115-2440	1964	$3,528 (S)	$4,852	5-D	16,326	850
Clinch Valley Coll of the U of Virginia, Wise, VA 24293	1954	$3,348 (S)	$4,614	5-B	1,483	98
Coastal Carolina U, Conway, SC 29528-6054	1954	$3,100 (S)	$4,800	5-M	4,556	275
Coe Coll, Cedar Rapids, IA 52402-5070	1851	$16,320	$4,760	2-M	1,263	127
Colby Coll, Waterville, ME 04901-8840	1813	$29,190 (C)	NA	1-B	1,802	162
Coleman Coll, La Mesa, CA 91942-1532	1963	$9,000	NA	1-M	1,060	101
Colgate U, Hamilton, NY 13346-1386	1819	$22,770	$6,235	1-M	2,776	256
Coll Misericordia, Dallas, PA 18612-1098	1924	$13,830	$6,210	2-M	1,697	158
Coll of Aeronautics, Flushing, NY 11371 (2)	1932	$7,600	NA	1-B	1,272	57
Coll of Charleston, Charleston, SC 29424-0001	1770	$3,290 (S)	$3,900	5-B	9,484	671
Coll of Mount St. Joseph, Cincinnati, OH 45233-1670	1920	$11,950	$5,760	2-M	2,307	218
Coll of Mount Saint Vincent, Riverdale, NY 10471-1093	1911	$13,580	$6,750	1-M	1,615	157
The Coll of New Jersey, Ewing, NJ 08628	1855	$4,843 (S)	$6,161	5-M	6,711	626
The Coll of New Rochelle, New Rochelle, NY 10805-2308 (4)	1904	$11,100	$5,850	1-M	7,214	219
Coll of Notre Dame, Belmont, CA 94002-1997	1851	$14,976	$6,650	2-M	1,764	194
Coll of Notre Dame of Maryland, Baltimore, MD 21210-2476 (3)	1873	$14,086	$6,130	2-M	3,165	90
Coll of Our Lady of the Elms, Chicopee, MA 01013-2839	1928	$12,950	$5,000	2-M	1,001	94
Coll of Saint Benedict, Saint Joseph, MN 56374-2091 (3)	1887	$14,758	$4,861	2-B	1,977	144
Coll of St. Catherine, St. Paul, MN 55105-1789 (3)	1905	$14,258	$4,290	2-M	3,152	218
Coll of Saint Elizabeth, Morristown, NJ 07960-6989 (4)	1899	$13,060	$6,200	2-M	1,793	172
Coll of Saint Mary, Omaha, NE 68124-2377 (3)	1923	$11,814	$4,420	2-B	1,035	148
The Coll of Saint Rose, Albany, NY 12203-1419	1920	$11,719	$6,024	1-M	4,039	330
Coll of St. Scholastica, Duluth, MN 55811-4199	1912	$13,995	$4,134	2-M	2,084	165
Coll of Santa Fe, Santa Fe, NM 87505-7634	1947	$13,240	$4,796	1-M	1,535	163
Coll of Staten Island of the City U of New York, Staten Island, NY 10314	1955	$3,326 (S)	NA	11-M	11,980	742
Coll of the Holy Cross, Worcester, MA 01610-2395	1843	$21,080	$7,100	2-B	2,789	268
Coll of the Ozarks, Point Lookout, MO 65726	1906	$150	$2,500	2-B	1,488	107
The Coll of West Virginia, Beckley, WV 25802-2830	1933	$3,600	$3,960	1-M	1,979	143
Coll of William & Mary, Williamsburg, VA 23187-8795	1693	$5,032 (S)	$4,673	5-D	7,590	699
The Coll of Wooster, Wooster, OH 44691-2363	1866	$19,300	$5,260	2-B	1,747	146
Colorado Christian U, Lakewood, CO 80226-7499	1914	$10,010	$5,160	2-M	1,978	156
The Colorado Coll, Colorado Springs, CO 80903-3294	1874	$19,980	$5,328	1-M	1,991	210
The Colorado Inst of Art, Denver, CO 80203-2903	1952	$10,260	$5,760	3-B	1,750	130
Colorado School of Mines, Golden, CO 80401-1887	1874	$5,069 (S)	$4,985	5-D	3,192	317
Colorado State U, Fort Collins, CO 80523-0015	1870	$3,083 (S)	$5,184	5-D	22,556	982
Colorado Tech U, Colorado Springs, CO 80907-3896	1965	$6,838	NA	3-D	1,178	92
Columbia Coll, Chicago, IL 60605-1996	1890	$8,618	NA	1-M	8,843	1,124

Name, address	Year	Tuition & Fees	Room & Board	Control, Degree	Enrollment	Faculty
Columbia Coll, Columbia, MO 65216-0002	1851	$9,244	$4,270	2-M	7,959	81
Columbia Coll, New York, NY 10027	1754	$22,650	$7,492	1-B	3,841	NA
Columbia Coll, Caguas, PR 00726	1966	$2,400	NA	3-B	1,100	62
Columbia Coll, Columbia, SC 29203-5998 (3)	1854	$12,150	$4,500	2-M	1,388	145
Columbia Union Coll, Takoma Park, MD 20912-7796	1904	$11,790	$4,290	2-B	1,126	43
Columbia U, School of General Studies, New York, NY 10027-6939	1754	$19,623	$7,492	1-B	1,079	520
Columbia U, The Fu Foundation School of Engineering & Applied Sci, New York, NY 10027	1864	$22,650	$7,492	1-D	NA	NA
Columbus Coll of Art & Design, Columbus, OH 43215-1758	1879	$11,880	$5,900	1-B	1,542	154
Columbus State U, Columbus, GA 31907-5645	1958	$2,463 (S)	$3,560	5-M	5,122	343
Concord Coll, Athens, WV 24712-1000	1872	$2,310 (S)	$3,850	5-B	2,662	167
Concordia Coll, Moorhead, MN 56562	1891	$12,655	$3,645	2-B	2,969	278
Concordia U, Irvine, CA 92612-3299	1972	$14,550	$5,380	2-M	1,074	95
Concordia U, River Forest, IL 60305-1499	1864	$11,987	$5,024	2-M	1,959	183
Concordia U, Seward, NE 68434-1599	1894	$11,310	$3,786	2-M	1,241	125
Concordia U at St. Paul, St. Paul, MN 55104-5494	1893	$11,980	$4,726	2-M	1,492	117
Concordia U Wisconsin, Mequon, WI 53097-2402	1881	$10,760	$3,770	2-M	4,541	195
Connecticut Coll, New London, CT 06320-4196	1911	$28,475 (C)	NA	1-M	1,800	170
Converse Coll, Spartanburg, SC 29302-0006 (3)	1889	$14,445	$4,490	1-M	1,463	87
Coppin State Coll, Baltimore, MD 21216-3698	1900	$3,624 (S)	$5,200	5-M	3,765	202
Cornell Coll, Mount Vernon, IA 52314-1098	1853	$17,840	$4,995	2-B	1,024	133
Cornell U, Ithaca, NY 14853-0001	1865	$9,374 (S)	$7,555	1-D	18,649	1,532
Cornerstone U, Grand Rapids, MI 49525-5897	1941	$10,026	$4,392	2-B	1,226	104
Covenant Coll, Lookout Mountain, GA 30750	1955	$12,900	$4,315	2-M	1,031	60
Creighton U, Omaha, NE 68178-0001	1878	$12,756	$5,190	2-D	6,226	863
The Culinary Inst of America, Hyde Park, NY 12538-1499	1946	$13,990	NA	1-B	2,120	124
Cumberland Coll, Williamsburg, KY 40769-1372	1889	$8,430	$3,976	2-M	1,739	102
Cumberland U, Lebanon, TN 37087-3554	1842	$8,190	$3,500	1-M	1,206	100
Curry Coll, Milton, MA 02186-9984	1879	$15,700	$6,340	1-M	2,175	210
Daemen Coll, Amherst, NY 14226-3592	1947	$10,980	$5,800	1-M	1,905	183
Dakota State U, Madison, SD 57042-1799	1881	$3,027 (S)	$2,724	5-B	1,844	87
Dallas Baptist U, Dallas, TX 75211-9299	1965	$7,800	$3,510	2-M	3,721	260
Daniel Webster Coll, Nashua, NH 03063-1300	1965	$13,845	$5,953	1-B	1,091	75
Dartmouth Coll, Hanover, NH 03755	1769	$23,012	$6,147	1-D	5,258	528
Davenport Coll of Business, Grand Rapids, MI 49503	1866	$8,508	NA	1-M	2,453	136
Davenport Coll of Business, Kalamazoo Campus, Kalamazoo, MI 49006-2791	1866	$8,418	NA	1-B	1,284	78
Davenport Coll of Business, Lansing Campus, Lansing, MI 48933-2197	1979	$8,418	NA	1-B	1,130	105
David Lipscomb U, Nashville, TN 37204-3951	1891	$8,470	$4,140	2-F	2,381	204
David N. Myers Coll, Cleveland, OH 44115-1096	1848	$7,800	NA	1-B	1,181	85
Davidson Coll, Davidson, NC 28036-1719	1837	$20,595	$6,126	2-B	1,639	153
Delaware State U, Dover, DE 19901-2277	1891	$2,970 (S)	$5,130	5-M	3,155	174
Delaware Valley Coll, Doylestown, PA 18901-2697	1896	$14,929	$5,826	1-M	2,090	141
Delta State U, Cleveland, MS 38733-0001	1924	$2,354 (S)	$2,600	5-D	3,979	285
Denison U, Granville, OH 43023	1831	$20,250	$5,590	1-B	2,156	174
DePaul U, Chicago, IL 60604-2287	1898	$13,490	$5,841	2-D	18,565	1,354
DePauw U, Greencastle, IN 46135-1772	1837	$17,050	$5,840	2-B	2,250	221
Detroit Coll of Business, Dearborn, MI 48126-3799	1962	$6,264	NA	1-M	3,261	279
Detroit Coll of Business–Flint, Flint, MI 48504-1700	1974	$4,644	NA	1-B	1,029	121
Detroit Coll of Business, Warren Campus, Warren, MI 48092-5209	1962	$6,264	NA	1-M	2,107	175
DeVry Inst of Technology, Phoenix, AZ 85021-2995	1967	$7,308	NA	3-B	3,747	108
DeVry Inst of Technology, Long Beach, CA 90806	1984	$7,308	NA	3-B	2,398	102
DeVry Inst of Technology, Pomona, CA 91768-2642	1983	$7,308	NA	3-B	3,562	142
DeVry Inst of Technology, Decatur, GA 30030-2198	1969	$7,308	NA	3-B	3,000	139
DeVry Inst of Technology, Addison, IL 60101-6106	1982	$7,308	NA	3-B	3,944	149
DeVry Inst of Technology, Chicago, IL 60618-5994	1931	$7,308	NA	3-B	3,819	109
DeVry Inst of Technology, Kansas City, MO 64131-3698	1931	$7,308	NA	3-B	2,587	100
DeVry Inst of Technology, Columbus, OH 43209-2705	1952	$7,308	NA	3-B	3,276	116
DeVry Inst of Technology, Irving, TX 75063-2439	1969	$7,308	NA	3-B	2,970	164
Dickinson Coll, Carlisle, PA 17013-2896	1773	$21,600	$6,030	1-B	1,844	186
Dickinson State U, Dickinson, ND 58601-4896	1918	$2,131 (S)	$2,618	5-B	1,800	108
Dillard U, New Orleans, LA 70122-3097	1869	$8,000	$4,464	2-B	1,722	124
Doane Coll, Crete, NE 68333-2430	1872	$11,450	$3,550	2-M	1,982	134
Dominican Coll of Blauvelt, Orangeburg, NY 10962-1210	1952	$11,620	$6,950	1-M	1,711	178
Dominican Coll of San Rafael, San Rafael, CA 94901-2298	1890	$15,424	$7,246	2-M	1,437	196
Dominican U, River Forest, IL 60305-1099	1901	$13,700	$4,880	2-M	2,068	180
Dordt Coll, Sioux Center, IA 51250-1697	1955	$11,450	$3,260	2-M	1,420	95
Dowling Coll, Oakdale, NY 11769-1999	1955	$12,630	NA	1-D	5,935	424
Drake U, Des Moines, IA 50311-4516	1881	$15,200	$4,970	1-D	5,115	258
Drew U, Madison, NJ 07940-1493	1867	$21,396	$6,392	2-D	2,368	223
Drexel U, Philadelphia, PA 19104-2875	1891	$15,048	$7,518	1-D	11,617	832
Drury Coll, Springfield, MO 65802-3791	1873	$10,060	$4,050	1-M	2,018	122
Duke U, Durham, NC 27708-0586	1838	$22,173	$7,283	2-D	11,581	2,100
Duquesne U, Pittsburgh, PA 15282-0001	1878	$14,066	$6,158	2-D	9,552	766
D'Youville Coll, Buffalo, NY 14201-1084	1908	$10,040	$5,120	1-M	1,893	165
Earlham Coll, Richmond, IN 47374-4095	1847	$18,618	$4,674	2-B	1,075	127
East Carolina U, Greenville, NC 27858-4353	1907	$1,848 (S)	$3,860	5-D	18,263	1,351
East Central U, Ada, OK 74820-6899	1909	$1,812 (S)	$2,066	5-M	4,062	225
Eastern Coll, St. Davids, PA 19087-3696	1952	$13,200	$5,654	2-M	2,675	282
Eastern Connecticut State U, Willimantic, CT 06226-2295	1889	$3,838 (S)	$5,346	5-M	4,724	345
Eastern Illinois U, Charleston, IL 61920-3099	1895	$3,112 (S)	$3,919	5-M	11,735	679
Eastern Kentucky U, Richmond, KY 40475-3101	1906	$2,060 (S)	$2,144	5-M	15,402	NA
Eastern Mennonite U, Harrisonburg, VA 22802-2462	1917	$12,600	$4,700	2-F	1,340	135
Eastern Michigan U, Ypsilanti, MI 48197	1849	$3,529 (S)	$4,660	5-D	22,463	1,312
Eastern Nazarene Coll, Quincy, MA 02170-2999	1918	$11,440	$4,176	2-M	1,620	67
Eastern New Mexico U, Portales, NM 88130	1934	$1,716 (S)	$3,224	5-M	3,492	241
Eastern Oregon U, La Grande, OR 97850-2899	1929	$3,231	$4,425	5-M	2,235	103
Eastern Washington U, Cheney, WA 99004-2431	1882	$2,622 (S)	$4,294	5-M	7,688	348
East Stroudsburg U of Pennsylvania, East Stroudsburg, PA 18301-2999	1893	$4,322 (S)	$3,780	5-M	5,790	295

Name, address	Year	Tuition & Fees	Room & Board	Control, Degree	Enrollment	Faculty
East Tennessee State U, Johnson City, TN 37614-0734	1911	$2,100 (S)	$2,870	5-D	11,730	935
East Texas Baptist U, Marshall, TX 75670-1498	1912	$6,750	$3,098	2-B	1,221	102
Eckerd Coll, St. Petersburg, FL 33711	1958	$17,130	$4,810	2-B	1,504	126
Edgewood Coll, Madison, WI 53711-1997	1927	$10,280	$4,600	2-M	1,965	162
Edinboro U of Pennsylvania, Edinboro, PA 16444	1857	$4,193 (S)	$3,674	5-M	7,108	414
Elizabeth City State U, Elizabeth City, NC 27909-7806	1891	$1,720 (S)	$3,472	5-B	1,932	149
Elizabethtown Coll, Elizabethtown, PA 17022-2298	1899	$16,930	$5,200	2-B	1,730	158
Elmhurst Coll, Elmhurst, IL 60126-3296	1871	$11,900	$5,104	2-M	2,787	174
Elmira Coll, Elmira, NY 14901	1855	$20,276	$6,690	1-M	1,977	74
Elon Coll, Elon College, NC 27244	1889	$11,542	$4,376	2-M	3,845	274
Embry-Riddle Aeronautical U, Prescott, AZ 86301-3720	1978	$9,920	$4,950	1-B	1,526	92
Embry-Riddle Aeronautical U, Daytona Beach, FL 32114-3900	1926	$9,890	$4,600	1-M	4,699	250
Embry-Riddle Aeronautical U, Extended Campus, Daytona Beach, FL 32114	1970	$1,590	NA	1-M	7,276	2,513
Emerson Coll, Boston, MA 02116-1511	1880	$17,826	$8,480	1-D	3,865	287
Emmanuel Coll, Boston, MA 02115 (3)	1919	$14,550	$7,025	2-M	1,499	108
Emory U, Atlanta, GA 30322-1100	1836	$21,120	$6,972	2-D	11,353	2,486
Emporia State U, Emporia, KS 66801-5087	1863	$1,982 (S)	$3,560	5-D	5,419	313
Endicott Coll, Beverly, MA 01915-2096	1939	$13,508	$7,160	1-M	1,367	176
Evangel U, Springfield, MO 65802-2191	1955	$8,850	$3,550	2-M	1,631	122
The Evergreen State Coll, Olympia, WA 98505	1967	$2,742 (S)	$4,530	5-M	4,194	173
Fairfield U, Fairfield, CT 06430-5195	1942	$18,310	$7,234	2-M	5,208	386
Fairleigh Dickinson U, Florham–Madison Campus, Madison, NJ 07940-1099	1942	$14,122	$6,274	1-M	3,411	243
Fairleigh Dickinson U, Teaneck–Hackensack Campus, Teaneck, NJ 07666	1942	$14,122	$6,404	1-D	3,765	797
Fairmont State Coll, Fairmont, WV 26554	1865	$2,040 (S)	$3,696	5-B	6,712	423
Fashion Inst of Technology, New York, NY 10001-5992	1944	$2,710 (S)	$5,600	11-M	11,196	884
Faulkner U, Montgomery, AL 36109-3398	1942	$6,980	$3,800	2-F	2,491	115
Fayetteville State U, Fayetteville, NC 28301-4298	1867	$1,662 (S)	$3,600	5-D	NA	234
Felician Coll, Lodi, NJ 07644-2198	1942	$10,012	$5,400	2-M	1,203	121
Ferris State U, Big Rapids, MI 49307-2742	1884	$3,908 (S)	$4,966	5-F	9,651	512
Finch U of Health Scis/The Chicago Medical School, North Chicago, IL 60064	1912	$11,342	NA	1-D	1,382	390
Fitchburg State Coll, Fitchburg, MA 01420-2697	1894	$3,346 (S)	$4,440	5-M	6,416	305
Flagler Coll, St. Augustine, FL 32085-1027	1968	$5,950	$3,680	1-B	1,669	152
Florida Ag & Mech U, Tallahassee, FL 32307-3200	1887	$2,105 (S)	$3,698	5-D	11,828	732
Florida Atlantic U, Boca Raton, FL 33431-0991	1961	$2,022 (S)	$4,774	5-D	19,562	1,264
Florida Gulf Coast U, Fort Myers, FL 33965-6565	1991	$1,656 (S)	$4,792	5-M	3,014	279
Florida Inst of Technology, Melbourne, FL 32901-6975	1958	$15,550	$4,870	1-D	4,267	498
Florida Intl U, Miami, FL 33199	1965	$2,035 (S)	$5,070	5-D	30,527	1,281
Florida State U, Tallahassee, FL 32306	1857	$1,988 (S)	$4,706	5-D	31,071	1496
Fontbonne Coll, St. Louis, MO 63105-3098	1917	$10,150	$4,614	2-M	1,984	175
Fordham U, New York, NY 10458	1841	$17,014	$7,667	2-D	13,623	1,183
Fort Hays State U, Hays, KS 67601-4099	1902	$1,992 (S)	$3,551	5-M	5,401	302
Fort Lewis Coll, Durango, CO 81301-3999	1911	$2,084 (S)	$4,380	5-B	4,314	276
Fort Valley State U, Fort Valley, GA 31030-3298	1895	$2,157 (S)	$3,330	5-D	2,689	174
Framingham State Coll, Framingham, MA 01701-9101	1839	$3,150 (S)	$3,999	5-M	5,685	280
Franciscan U of Steubenville, Steubenville, OH 43952-1763	1946	$11,370	$4,870	2-M	2,022	144
Francis Marion U, Florence, SC 29501-0547	1970	$3,390 (S)	$3,550	5-M	3,947	182
Franklin & Marshall Coll, Lancaster, PA 17604-3003	1787	$22,664	$5,400	1-B	1,862	222
Franklin Pierce Coll, Rindge, NH 03461-0060	1962	$16,170	$5,400	1-M	1,495	130
Franklin U, Columbus, OH 43215-5399	1902	$5,314	NA	1-M	4,302	236
Freed-Hardeman U, Henderson, TN 38340-2399	1869	$7,524	$4,120	2-M	1,733	115
Fresno Pacific U, Fresno, CA 93702-4709	1944	$11,936	$4,130	2-M	1,742	169
Friends U, Wichita, KS 67213	1898	$9,975	NA	1-M	3,027	188
Frostburg State U, Frostburg, MD 21532-1099	1898	$3,544 (S)	$4,846	5-M	5,260	334
Furman U, Greenville, SC 29613	1826	$16,419	$4,608	1-M	2,993	200
Gallaudet U, Washington, DC 20002-3625	1864	$6,283	$6,922	1-D	1,701	297
Gannon U, Erie, PA 16541-0001	1925	$12,994	$5,050	2-D	3,250	269
Gardner-Webb U, Boiling Springs, NC 28017	1905	$9,620	$4,780	2-M	2,920	174
Geneva Coll, Beaver Falls, PA 15010-3599	1848	$11,534	$4,850	2-M	1,986	147
George Fox U, Newberg, OR 97132-2697	1891	$15,520	$5,120	2-D	2,338	199
George Mason U, Fairfax, VA 22030-4444	1957	$4,296 (S)	$5,190	5-D	24,010	1,449
Georgetown Coll, Georgetown, KY 40324-1696	1829	$10,190	$4,280	2-M	1,664	133
Georgetown U, Washington, DC 20057	1789	$21,405	$8,416	2-D	12,433	1,844
The George Washington U, Washington, DC 20052	1821	$21,360	$8,140	1-D	19,481	2,468
Georgia Coll & State U, Milledgeville, GA 31061	1889	$2,064 (S)	$4,070	5-M	5,168	345
Georgia Inst of Technology, Atlanta, GA 30332-0001	1885	$2,901 (S)	$4,884	5-D	13,954	696
Georgian Court Coll, Lakewood, NJ 08701-2697 (4)	1908	$11,116	$3,750	2-M	2,405	194
Georgia Southern U, Statesboro, GA 30460-1000	1906	$2,256 (S)	$3,318	5-D	13,904	711
Georgia Southwestern State U, Americus, GA 31709-4693	1906	$2,145 (S)	$3,330	5-M	2,581	147
Georgia State U, Atlanta, GA 30303-3083	1913	$2,673 (S)	NA	5-D	22,686	1,431
Gettysburg Coll, Gettysburg, PA 17325-1411	1832	$22,430	$5,346	2-B	2,123	221
Glenville State Coll, Glenville, WV 26351-1200	1872	$1,956 (S)	$3,600	5-B	2,235	168
Golden Gate U, San Francisco, CA 94105-2968	1853	$8,472	NA	1-D	5,418	523
Goldey-Beacom Coll, Wilmington, DE 19808-1999	1886	$7,200	NA	1-M	1,650	73
Gonzaga U, Spokane, WA 99258	1887	$15,487	$5,170	2-D	4,059	273
Gordon Coll, Wenham, MA 01984-1899	1889	$15,760	$4,950	2-M	1,519	110
Goshen Coll, Goshen, IN 46526-4794	1894	$11,450	$4,160	2-B	1,045	109
Goucher Coll, Baltimore, MD 21204-2794	1885	$18,525	$7,130	1-M	1,461	152
Governors State U, University Park, IL 60466	1969	$2,278 (S)	NA	5-M	6,180	282
Graceland Coll, Lamoni, IA 50140	1895	$10,860	$4,210	2-M	4,167	92
Grambling State U, Grambling, LA 71245	1901	$2,088 (S)	$2,636	5-D	5,070	224
Grand Canyon U, Phoenix, AZ 85017-3030	1949	$8,946	$3,740	2-M	2,486	175
Grand Valley State U, Allendale, MI 49401-9403	1960	$3,408 (S)	$4,840	5-M	16,751	888
Grand View Coll, Des Moines, IA 50316-1599	1896	$11,410	$3,890	2-B	1,371	133
Grantham Coll of Engineering, Slidell, LA 70460-6815	1951	$3,800	NA	3-B	1,200	13
Greenville Coll, Greenville, IL 62246-0159	1892	$12,586	$4,850	2-M	1,077	88
Grinnell Coll, Grinnell, IA 50112-0805	1846	$17,568	$5,414	1-B	1,345	154
Grove City Coll, Grove City, PA 16127-2104	1876	$6,576	$3,912	2-B	2,338	145
Guilford Coll, Greensboro, NC 27410-4173	1837	$14,750	$5,400	2-B	1,398	116

Name, address	Year	Tuition & Fees	Room & Board	Control, Degree	Enroll-ment	Faculty
Gustavus Adolphus Coll, St. Peter, MN 56082-1498	1862	$16,120	$4,150	2-B	2,527	224
Gwynedd-Mercy Coll, Gwynedd Valley, PA 19437-0901	1948	$12,280	$6,000	2-M	1,776	187
Hamilton Coll, Clinton, NY 13323-1296	1812	$22,700	$5,850	1-B	1,733	217
Hamline U, St. Paul, MN 55104-1284	1854	$14,850	$4,968	2-D	2,919	325
Hampshire Coll, Amherst, MA 01002	1965	$23,780	$6,435	1-B	1,160	94
Hampton U, Hampton, VA 23668	1868	$9,596	$4,442	1-D	5,704	378
Hannibal-LaGrange Coll, Hannibal, MO 63401-1999	1858	$7,590	$2,927	2-B	1,026	85
Hanover Coll, Hanover, IN 47243-0108	1827	$10,085	$4,440	2-B	1,087	112
Harding U, Searcy, AR 72149-0001	1924	$7,712	$4,148	2-M	3,904	247
Hardin-Simmons U, Abilene, TX 79698-0001	1891	$8,130	$3,441	2-F	2,317	179
Harris-Stowe State Coll, St. Louis, MO 63103-2136	1857	$2,528 (S)	NA	5-B	1,735	130
Hartwick Coll, Oneonta, NY 13820-4020	1797	$22,235	$6,080	1-B	1,488	164
Harvard U, Cambridge, MA 02138	1636	$22,802	$7,487	1-D	17,675	2,220
Hastings Coll, Hastings, NE 68901-7696	1882	$11,368	$3,870	2-M	1,138	108
Haverford Coll, Haverford, PA 19041-1392	1833	$21,740	$7,370	1-B	1,147	109
Hawaii Pacific U, Honolulu, HI 96813-2785	1965	$7,500	$7,350	1-M	8,505	657
Heidelberg Coll, Tiffin, OH 44883-2462	1850	$16,260	$5,135	2-M	1,517	107
Henderson State U, Arkadelphia, AR 71999-0001	1890	$2,166 (S)	$2,936	5-M	3,685	229
Hendrix Coll, Conway, AR 72032-3080	1876	$10,408	$4,160	2-B	1,047	99
Heritage Coll, Toppenish, WA 98948-9599	1982	$6,450	NA	1-M	1,152	168
High Point U, High Point, NC 27262-3598	1924	$10,420	$5,300	2-M	3,030	206
Hillsdale Coll, Hillsdale, MI 49242-1298	1844	$12,680	$5,550	1-B	1,190	123
Hiram Coll, Hiram, OH 44234-0067	1850	$16,514	$5,414	2-B	1,129	93
Hobart & William Smith Colls, Geneva, NY 14456-3397	1822	$22,380	$6,795	1-B	1,786	171
Hofstra U, Hempstead, NY 11549	1935	$13,544	$6,880	1-D	12,807	1,114
Hollins U, Roanoke, VA 24020-1688 (3)	1842	$15,320	$6,125	1-M	1,063	95
Holy Family Coll, Philadelphia, PA 19114-2094	1954	$10,620	NA	2-M	2,615	225
Hood Coll, Frederick, MD 21701-8575 (4)	1893	$16,418	$6,700	2-M	1,748	86
Hope Coll, Holland, MI 49422-9000	1866	$14,878	$4,884	2-B	2,920	283
Houghton Coll, Houghton, NY 14744	1883	$12,765	$4,840	2-B	1,355	98
Houston Baptist U, Houston, TX 77074-3298	1960	$8,535	$3,300	2-M	2,306	170
Howard Payne U, Brownwood, TX 76801-2715	1889	$7,620	$3,630	2-B	1,488	133
Howard U, Washington, DC 20059-0002	1867	$8,985	$5,714	1-D	10,211	1,242
Humboldt State U, Arcata, CA 95521-8299	1913	$1,926 (S)	$5,360	5-M	7,475	543
Hunter Coll of the City U of New York, New York, NY 10021-5085	1870	$3,329 (S)	NA	11-M	19,611	1,133
Huntingdon Coll, Montgomery, AL 36106-2148	1854	$11,230	$5,000	2-B	NA	67
Husson Coll, Bangor, ME 04401-2999	1898	$8,800	$4,880	1-M	1,280	96
Idaho State U, Pocatello, ID 83209	1901	$1,726 (S)	$3,730	5-D	12,257	725
Illinois Inst of Technology, Chicago, IL 60616-3793	1890	$16,460	$5,005	1-D	5,906	493
Illinois State U, Normal, IL 61790-2200	1857	$4,004 (S)	$3,975	5-D	20,394	1,019
Illinois Wesleyan U, Bloomington, IL 61702-2900	1850	$18,376	$4,824	1-B	2,022	172
Immaculata Coll, Immaculata, PA 19345-0500 (4)	1920	$12,115	$6,200	2-D	2,586	215
Indiana Inst of Technology, Fort Wayne, IN 46803-1297	1930	$11,500	$4,430	1-M	1,645	173
Indiana State U, Terre Haute, IN 47809-1401	1865	$3,196 (S)	$4,304	5-D	10,970	697
Indiana U Bloomington, Bloomington, IN 47405	1820	$3,929 (S)	$5,000	5-D	35,600	1,829
Indiana U East, Richmond, IN 47374-1289	1971	$2,849 (S)	NA	5-B	2,280	163
Indiana U Kokomo, Kokomo, IN 46904-9003	1945	$2,890 (S)	NA	5-M	2,796	182
Indiana U Northwest, Gary, IN 46408-1197	1959	$2,895 (S)	NA	5-M	4,792	323
Indiana U of Pennsylvania, Indiana, PA 15705-1087	1875	$4,204 (S)	$3,662	5-D	13,790	845
Indiana U–Purdue U Fort Wayne, Fort Wayne, IN 46805-1499	1917	$3,321 (S)	NA	5-M	10,653	629
Indiana U–Purdue U Indianapolis, Indianapolis, IN 46202-2896	1969	$3,441 (S)	$3,050	5-D	27,821	2,497
Indiana U South Bend, South Bend, IN 46634-7111	1922	$2,985 (S)	NA	5-M	7,387	506
Indiana U Southeast, New Albany, IN 47150-6405	1941	$2,847 (S)	NA	5-M	5,813	373
Indiana Wesleyan U, Marion, IN 46953-4999	1920	$11,204	$4,310	2-M	1,912	138
Inter American U of Puerto Rico, Arecibo Campus, Arecibo, PR 00614-4050	1957	$2,620	NA	1-M	4,115	238
Inter American U of Puerto Rico, Bayamón Campus, Bayamón, PR 00957	1912	$2,800	NA	1-B	5,047	278
Inter American U of Puerto Rico, Metro Campus, San Juan, PR 00919-1293	1960	NA	NA	1-M	12,000	NA
Inter American U of Puerto Rico, San Germán Campus, San Germán, PR 00683-5008	1912	$3,354	$2,200	1-M	5,735	348
International Acad of Merchandising & Design, Ltd., Chicago, IL 60602-9736	1977	$9,900	NA	3-B	1,048	90
Iona Coll, New Rochelle, NY 10801-1890	1940	$13,420	$7,950	2-M	4,645	370
Iowa State U of Sci & Technology, Ames, IA 50011	1858	$2,766 (S)	$3,958	5-D	25,585	1,587
Ithaca Coll, Ithaca, NY 14850-7020	1892	$16,900	$7,652	1-M	5,895	542
Jackson State U, Jackson, MS 39217	1877	$2,380 (S)	$3,458	5-D	6,292	382
Jacksonville State U, Jacksonville, AL 36265-1602	1883	$2,060 (S)	$2,770	5-M	7,618	352
Jacksonville U, Jacksonville, FL 32211-3394	1934	$13,900	$5,060	1-M	2,117	213
James Madison U, Harrisonburg, VA 22807	1908	$4,148 (S)	$5,008	5-D	14,996	861
Jamestown Coll, Jamestown, ND 58405	1883	$8,770	$3,180	2-B	1,141	85
John Brown U, Siloam Springs, AR 72761-2121	1919	$9,802	$4,478	2-M	1,454	96
John Carroll U, University Heights, OH 44118-4581	1886	$14,620	$5,804	2-M	4,473	414
John F. Kennedy U, Orinda, CA 94563-2689	1964	$9,252	NA	1-D	1,749	792
John Jay Coll of Criminal Justice, the City U of New York, New York, NY 10019	1964	$3,309 (S)	NA	11-M	9,706	689
Johns Hopkins U, Baltimore, MD 21218-2699	1876	$21,675	$7,675	1-D	5,127	429
Johnson & Wales U, North Miami, FL 33181	1992	$15,855	NA	1-B	1,156	46
Johnson & Wales U, Providence, RI 02903-3703	1914	$12,807	$5,550	1-D	8,381	332
Johnson & Wales U, Charleston, SC 29403	1984	$13,689	NA	1-B	1,450	49
Johnson C. Smith U, Charlotte, NC 28216-5398	1867	$8,469	$3,846	1-B	1,443	98
Johnson State Coll, Johnson, VT 05656-9405	1828	$4,641 (S)	$5,206	5-M	1,627	126
The Juilliard School, New York, NY 10023-6588	1905	$15,000	$6,700	1-D	1,081	253
Juniata Coll, Huntingdon, PA 16652-2119	1876	$17,580	$5,205	2-B	1,244	108
Kalamazoo Coll, Kalamazoo, MI 49006-3295	1833	$17,976	$5,454	2-B	1,357	103
Kansas State U, Manhattan, KS 66506	1863	$2,467 (S)	$3,780	5-D	20,885	1,047
Kean U, Union, NJ 07083	1855	$3,669 (S)	NA	5-M	11,338	825
Keene State Coll, Keene, NH 03435	1909	$4,340 (S)	$4,824	5-M	4,354	346
Kennesaw State U, Kennesaw, GA 30144-5591	1963	$2,013 (S)	NA	5-M	12,861	600
Kent State U, Kent, OH 44242-0001	1910	$4,460 (S)	$4,314	5-D	20,947	1,384
Kentucky State U, Frankfort, KY 40601	1886	$2,050 (S)	$3,276	12-M	2,303	157
Kenyon Coll, Gambier, OH 43022-9623	1824	$22,850	$4,110	1-B	1,568	139

Name, address	Year	Tuition & Fees	Room & Board	Control, Degree	Enroll-ment	Faculty
Kettering U, Flint, MI 48504-4898	1919	$14,232	$3,863	1-M	3,256	156
King's Coll, Wilkes-Barre, PA 18711-0801	1946	$14,000	$6,360	2-M	2,256	168
Knox Coll, Galesburg, IL 61401	1837	$19,074	$5,076	1-B	1,194	126
Kutztown U of Pennsylvania, Kutztown, PA 19530-0730	1866	$4,219 (S)	$3,820	5-M	7,903	400
Lafayette Coll, Easton, PA 18042-1798	1826	$21,202	$6,841	2-B	2,244	232
LaGrange Coll, LaGrange, GA 30240-2999	1831	$9,726	$4,440	2-M	1,002	76
Lake Forest Coll, Lake Forest, IL 60045-2399	1857	$19,560	$4,720	1-M	1,210	126
Lakeland Coll, Sheboygan, WI 53082-0359	1862	$11,230	$4,525	2-M	3,281	64
Lake Superior State U, Sault Sainte Marie, MI 49783-1626	1946	$3,642 (S)	$4,738	5-M	3,427	191
Lamar U, Beaumont, TX 77710	1923	$1,868 (S)	$3,466	5-D	10,139	554
Lander U, Greenwood, SC 29649-2099	1872	$3,600 (S)	$3,800	5-M	2,600	174
La Roche Coll, Pittsburgh, PA 15237-5898	1963	$10,400	$5,624	2-M	1,527	166
La Salle U, Philadelphia, PA 19141-1199	1863	$14,850	$6,510	2-M	5,381	404
La Sierra U, Riverside, CA 92515-8247	1922	$14,025	$4,137	2-D	1,282	99
Lawrence Technological U, Southfield, MI 48075-1058	1932	$9,340	NA	1-M	4,290	328
Lawrence U, Appleton, WI 54912-0599	1847	$19,620	$4,710	1-B	1,235	157
Lebanon Valley Coll, Annville, PA 17003-0501	1866	$15,980	$5,300	2-M	1,984	110
Lee U, Cleveland, TN 37320-3450	1918	$5,638	$3,970	2-M	3,088	215
Lehigh U, Bethlehem, PA 18015-3094	1865	$21,350	$6,410	1-D	6,363	485
Lehman Coll of the City U of New York, Bronx, NY 10468-1589	1931	$3,320 (S)	NA	11-M	9,009	694
Le Moyne Coll, Syracuse, NY 13214-1399	1946	$13,450	$5,930	2-M	3,122	230
Lenoir-Rhyne Coll, Hickory, NC 28603	1891	$12,386	$4,500	2-M	1,603	137
Lesley Coll, Cambridge, MA 02138-2790 (3)	1909	$14,606	$6,950	1-D	5,953	886
LeTourneau U, Longview, TX 75607-7001	1946	$10,744	$5,068	2-M	2,506	153
Lewis & Clark Coll, Portland, OR 97219-7899	1867	$18,530	$6,036	1-F	2,938	168
Lewis-Clark State Coll, Lewiston, ID 83501-2698	1893	$1,868 (S)	$3,258	5-B	3,073	307
Lewis U, Romeoville, IL 60446	1932	$13,024	$5,500	2-M	4,072	366
Liberty U, Lynchburg, VA 24502	1971	$7,680	$4,800	2-D	6,767	246
Lincoln Memorial U, Harrogate, TN 37752-1901	1897	$8,000	$3,500	1-M	1,750	116
Lincoln U, Jefferson City, MO 65102	1866	$2,076 (S)	$3,396	5-M	3,214	202
Lincoln U, Lincoln University, PA 19352	1854	$4,180 (S)	$4,720	12-M	2,084	187
Lindenwood U, St. Charles, MO 63301-1695	1827	$10,150	$5,450	2-M	5,184	278
Lindsey Wilson Coll, Columbia, KY 42728-1298	1903	$8,760	$4,400	2-M	1,425	80
Linfield Coll, McMinnville, OR 97128-6894	1849	$16,960	$5,180	2-B	2,760	129
Lock Haven U of Pennsylvania, Lock Haven, PA 17745-2390	1870	$4,062 (S)	$3,976	5-M	3,718	225
Loma Linda U, Loma Linda, CA 92350	1905	$10,530	NA	2-D	3,506	1,162
Long Island U, Brooklyn Campus, Brooklyn, NY 11201-8423	1926	$14,496	$4,900	1-D	7,971	923
Long Island U, C.W. Post Campus, Brookville, NY 11548-1300	1954	$14,530	$6,150	1-D	9,281	1,101
Long Island U, Southampton Coll, Southampton, NY 11968-4198	1963	$14,600	$7,400	1-M	2,801	229
Longwood Coll, Farmville, VA 23909-1800	1839	$4,416 (S)	$4,456	5-M	3,340	208
Loras Coll, Dubuque, IA 52004-0178	1839	$13,750	$5,005	2-M	1,683	136
Louisiana Coll, Pineville, LA 71359-0001	1906	$6,763	$3,100	2-B	1,047	96
Louisiana State U & Ag & Mech Coll, Baton Rouge, LA 70803	1860	$2,711 (S)	$3,820	5-D	29,868	1,345
Louisiana State U in Shreveport, Shreveport, LA 71115-2399	1965	$2,230 (S)	NA	5-M	4,410	201
Louisiana Tech U, Ruston, LA 71272	1894	$2,567 (S)	$2,850	5-D	9,656	452
Lourdes Coll, Sylvania, OH 43560-2898	1958	$8,100	NA	2-B	1,298	120
Loyola Coll in Maryland, Baltimore, MD 21210-2699	1852	$16,560	$7,450	2-D	6,181	432
Loyola Marymount U, Los Angeles, CA 90045-8350	1911	$16,495	$7,004	2-F	6,998	518
Loyola U Chicago, Chicago, IL 60611-2196	1870	$16,054	$6,700	2-D	13,811	887
Loyola U New Orleans, New Orleans, LA 70118-6195	1912	$13,354	$6,040	2-F	4,968	383
Lubbock Christian U, Lubbock, TX 79407-2099	1957	$8,578	$3,220	2-M	1,353	104
Luther Coll, Decorah, IA 52101-1045	1861	$15,630	$3,750	2-B	2,472	231
Lycoming Coll, Williamsport, PA 17701-5192	1812	$16,160	$4,700	2-B	1,475	107
Lynchburg Coll, Lynchburg, VA 24501-3199	1903	$16,415	$4,400	2-M	1,986	202
Lyndon State Coll, Lyndonville, VT 05851-0919	1911	$4,516 (S)	$5,206	5-M	1,160	108
Lynn U, Boca Raton, FL 33431-5598	1962	$17,200	$6,250	1-D	1,806	171
Macalester Coll, St. Paul, MN 55105-1899	1874	$18,758	$5,593	2-B	1,791	203
Macon State Coll, Macon, GA 31206-5144	1968	$1,268 (S)	NA	5-B	3,568	164
Madonna U, Livonia, MI 48150-1173	1947	$6,040	$4,384	2-M	3,924	289
Maharishi U of Mgmt, Fairfield, IA 52557	1971	$14,670	$4,960	1-D	1,155	112
Malone Coll, Canton, OH 44709-3897	1892	$11,280	$4,925	2-M	2,235	183
Manchester Coll, North Manchester, IN 46962-1225	1889	$12,660	$4,770	2-M	1,037	94
Manhattan Coll, Riverdale, NY 10471	1853	$14,555	$7,350	2-M	3,029	244
Manhattanville Coll, Purchase, NY 10577-2132	1841	$17,300	$8,000	1-M	2,094	250
Mansfield U of Pennsylvania, Mansfield, PA 16933	1857	$4,404 (S)	$3,770	5-M	2,979	202
Marian Coll, Indianapolis, IN 46222-1997	1851	$13,406	$4,644	2-B	1,339	134
Marian Coll of Fond du Lac, Fond du Lac, WI 54935-4699	1936	$11,370	$5,916	2-M	2,244	140
Marietta Coll, Marietta, OH 45750-4000	1835	$16,150	$4,774	1-M	1,288	109
Marist Coll, Poughkeepsie, NY 12601-1387	1929	$13,098	$7,494	1-M	4,747	444
Marquette U, Milwaukee, WI 53201-1881	1881	$15,384	$5,754	2-D	10,754	984
Marshall U, Huntington, WV 25755-2020	1837	$2,184 (S)	$4,576	5-D	13,606	929
Mars Hill Coll, Mars Hill, NC 28754	1856	$8,900	$3,800	2-B	1,265	123
Mary Baldwin Coll, Staunton, VA 24401 (4)	1842	$14,415	$7,000	2-M	1,566	143
Marygrove Coll, Detroit, MI 48221-2599	1905	$9,056	NA	2-M	3,603	60
Maryland Inst, Coll of Art, Baltimore, MD 21217-4192	1826	$16,760	NA	1-M	1,194	180
Marylhurst U, Marylhurst, OR 97036-0261	1893	$9,960	$5,928	2-M	1,085	230
Marymount Manhattan Coll, New York, NY 10021-4597	1936	$12,290	NA	1-B	2,319	323
Marymount U, Arlington, VA 22207-4299	1950	$12,770	$5,980	2-M	3,427	311
Maryville U of Saint Louis, St. Louis, MO 63141-7299	1872	$10,910	$5,200	1-M	3,057	267
Mary Washington Coll, Fredericksburg, VA 22401-5358	1908	$3,556 (S)	$5,258	5-M	3,806	242
Marywood U, Scranton, PA 18509-1598	1915	$14,003	$6,200	2-D	2,885	247
Massachusetts Coll of Art, Boston, MA 02115-5882	1873	$3,964 (S)	$6,386	5-M	2,304	182
Massachusetts Coll of Liberal Arts, North Adams, MA 01247-4100	1894	$3,437 (S)	$4,993	5-M	1,627	115
Massachusetts Coll of Pharmacy & Health Scis, Boston, MA 02115-5896	1823	$14,508	$7,800	1-D	1,662	109
Massachusetts Inst of Technology, Cambridge, MA 02139-4307	1861	$23,100	$6,750	1-D	9,885	923
The Master's Coll & Seminary, Santa Clarita, CA 91321-1200	1927	$12,180	$4,950	2-F	1,295	131
McKendree Coll, Lebanon, IL 62254-1299	1828	$10,400	$4,170	2-B	1,883	164
McMurry U, Abilene, TX 79697	1923	$9,075	$4,017	2-B	1,366	134

Name, address	Year	Tuition & Fees	Room & Board	Control, Degree	Enrollment	Faculty
McNeese State U, Lake Charles, LA 70609-2495	1939	$2,012 (S)	$2,328	5-M	8,117	300
MCP Hahnemann U, Philadelphia, PA 19102-1192	1848	$9,660	$8,550	1-D	3,008	1,446
Medaille Coll, Buffalo, NY 14214-2695	1875	$10,470	$5,100	1-M	1,179	103
Medgar Evers Coll of the City U of New York, Brooklyn, NY 11225-2298	1969	$3,282 (S)	NA	11-B	4,720	359
Medical Coll of Georgia, Augusta, GA 30912	1828	$2,526 (S)	NA	5-D	2,053	733
Mercer U, Macon, GA 31207-0003	1833	$14,656	$5,080	2-D	6,935	339
Mercy Coll, Dobbs Ferry, NY 10522-1189	1951	$7,600	$6,600	1-M	8,460	665
Mercyhurst Coll, Erie, PA 16546	1926	$12,750	$4,884	2-M	2,821	190
Meredith Coll, Raleigh, NC 27607-5298 (3)	1891	$8,490	$3,900	2-M	2,612	261
Merrimack Coll, North Andover, MA 01845-5800	1947	$14,530	$7,230	2-M	3,010	240
Mesa State Coll, Grand Junction, CO 81502-2647	1925	$1,986 (S)	$4,830	5-M	4,966	295
Messiah Coll, Grantham, PA 17027	1909	$12,990	$5,500	2-B	2,697	229
Methodist Coll, Fayetteville, NC 28311-1420	1956	$11,900	$4,580	2-B	1,851	178
Metropolitan State Coll of Denver, Denver, CO 80217-3362	1963	$1,976 (S)	NA	5-B	17,273	976
Metropolitan State U, St. Paul, MN 55106-5000	1971	NA	NA	5-M	4,894	497
Miami U, Oxford, OH 45056	1809	$5,512 (S)	$5,070	12-D	16,340	1,032
Michigan State U, East Lansing, MI 48824-1020	1855	$4,789 (S)	$4,172	5-D	43,189	3,372
Michigan Technological U, Houghton, MI 49931-1295	1885	$4,062 (S)	$4,590	5-D	6,257	412
MidAmerica Nazarene U, Olathe, KS 66062-1899	1966	$10,022	$4,810	2-M	1,428	120
Middlebury Coll, Middlebury, VT 05753-6002	1800	$29,340 (C)	NA	1-D	2,280	227
Middle Tennessee State U, Murfreesboro, TN 37132	1911	$2,196 (S)	$3,382	5-D	18,432	920
Midland Lutheran Coll, Fremont, NE 68025-4200	1883	$12,250	$3,450	2-B	1,034	70
Midwestern State U, Wichita Falls, TX 76308-2096	1922	$2,091 (S)	$3,638	5-M	5,687	262
Miles Coll, Birmingham, AL 35208	1905	$4,150	$2,852	2-B	1,390	56
Millersville U of Pennsylvania, Millersville, PA 17551-0302	1855	$4,400 (S)	$4,650	5-M	7,466	448
Millikin U, Decatur, IL 62522-2084	1901	$14,079	$5,271	2-B	2,173	251
Millsaps Coll, Jackson, MS 39210-0001	1890	$13,612	$6,276	2-M	1,362	99
Mills Coll, Oakland, CA 94613-1000 (3)	1852	$16,522	$7,084	1-M	1,108	167
Milwaukee School of Engineering, Milwaukee, WI 53202-3109 (2)	1903	$14,325	$4,020	1-M	2,904	235
Minnesota State U, Mankato, Mankato, MN 56002-8400	1868	$2,983 (S)	$5,079	5-M	11,193	574
Minot State U, Minot, ND 58707-0002	1913	$2,139 (S)	$2,531	5-M	3,156	218
Mississippi Coll, Clinton, MS 39058	1826	$8,364	$3,630	2-F	3,449	258
Mississippi State U, Mississippi State, MS 39762	1878	$2,731 (S)	$3,510	5-D	15,718	851
Mississippi U for Women, Columbus, MS 39701-9998	1884	$2,284 (S)	$2,557	5-M	3,314	206
Mississippi Valley State U, Itta Bena, MS 38941-1400	1946	$2,353 (S)	$2,444	5-M	2,447	155
Missouri Baptist Coll, St. Louis, MO 63141-8698	1964	$8,820	$4,230	2-B	2,716	113
Missouri Southern State Coll, Joplin, MO 64801-1595	1937	$2,384 (S)	$3,370	5-B	5,547	283
Missouri Valley Coll, Marshall, MO 65340-3197	1889	$10,500	$5,000	2-B	1,404	68
Missouri Western State Coll, St. Joseph, MO 64507-2294	1915	$2,534 (S)	$3,458	5-B	5,182	320
Molloy Coll, Rockville Centre, NY 11571-5002	1955	$10,400	NA	1-M	2,250	289
Monmouth Coll, Monmouth, IL 61462-1998	1853	$14,630	$4,410	2-B	1,077	104
Monmouth U, West Long Branch, NJ 07764-1898	1933	$14,442	$6,793	1-M	5,311	398
Montana State U–Billings, Billings, MT 59101-9984	1927	$2,517 (S)	$3,240	5-M	4,256	248
Montana State U–Bozeman, Bozeman, MT 59717-2640	1893	$2,677 (S)	$4,275	5-D	11,688	684
Montana State U–Northern, Havre, MT 59501-7751	1929	$2,504 (S)	$3,650	5-M	1,704	118
Montana Tech of The U of Montana, Butte, MT 59701-8997	1895	$2,542 (S)	$3,900	5-M	1,809	119
Montclair State U, Upper Montclair, NJ 07043-1624	1908	$3,694 (S)	$5,952	5-D	12,757	774
Montreat Coll, Montreat, NC 28757-1267	1916	$10,042	$4,098	2-M	1,054	51
Moody Bible Inst, Chicago, IL 60610-3284	1886	$830	$4,670	2-M	1,403	138
Moorhead State U, Moorhead, MN 56563-0002	1885	$2,908 (S)	$3,256	5-M	6,666	313
Moravian Coll, Bethlehem, PA 18018-6650	1742	$17,276	$5,720	2-M	1,867	165
Morehead State U, Morehead, KY 40351	1922	$2,150 (S)	$3,405	5-M	8,254	458
Morehouse Coll, Atlanta, GA 30314 (1)	1867	$9,724	$6,582	1-B	2,925	252
Morgan State U, Baltimore, MD 21251	1867	$3,412 (S)	$5,296	5-D	6,299	340
Morningside Coll, Sioux City, IA 51106-1751	1894	$12,306	$4,390	2-M	1,210	121
Morris Brown Coll, Atlanta, GA 30314-4140	1881	$8,210	$4,750	2-B	2,069	165
Mount Aloysius Coll, Cresson, PA 16630-1999	1939	$9,520	$4,580	2-B	1,221	125
Mount Holyoke Coll, South Hadley, MA 01075 (3)	1837	$22,340	$6,820	1-B	1,891	204
Mount Marty Coll, Yankton, SD 57078-3724	1936	$9,168	$3,826	2-M	1,004	83
Mount Mary Coll, Milwaukee, WI 53222-4597 (3)	1913	$10,740	$3,970	2-M	1,324	152
Mount Mercy Coll, Cedar Rapids, IA 52402-4797	1928	$11,860	$4,120	2-B	1,257	106
Mount Olive Coll, Mount Olive, NC 28365	1951	$8,700	$3,750	2-B	1,553	128
Mount Saint Mary Coll, Newburgh, NY 12550-3494	1960	$10,200	$5,450	1-M	2,082	167
Mount St. Mary's Coll, Los Angeles, CA 90049-1599 (4)	1925	$15,216	$7,808	2-M	2,022	211
Mount Saint Mary's Coll & Seminary, Emmitsburg, MD 21727-7799	1808	$15,650	$6,450	2-F	1,719	149
Mount Senario Coll, Ladysmith, WI 54848-2128	1962	$9,500	NA	1-B	1,145	66
Mount Union Coll, Alliance, OH 44601-3993	1846	$14,290	$4,170	2-B	2,069	136
Mount Vernon Nazarene Coll, Mount Vernon, OH 43050-9500	1964	$9,977	$3,933	2-M	1,876	139
Muhlenberg Coll, Allentown, PA 18104-5586	1848	$18,660	$5,200	2-B	2,460	210
Murray State U, Murray, KY 42071-0009	1922	$2,300 (S)	$3,560	5-M	8,903	481
Muskingum Coll, New Concord, OH 43762	1837	$10,785	$4,750	2-M	1,738	87
National-Louis U, Evanston, IL 60201-1796	1886	NA	NA	1-D	7,577	NA
National U, La Jolla, CA 92037-1011	1971	$6,975	NA	1-M	14,062	1,500
Nazareth Coll of Rochester, Rochester, NY 14618-3790	1924	$12,985	$6,375	1-M	2,820	173
Nebraska Wesleyan U, Lincoln, NE 68504-2796	1887	$11,220	$3,723	2-B	1,741	158
Neumann Coll, Aston, PA 19014-1298	1965	$12,940	$6,580	2-M	1,416	159
New Coll of California, San Francisco, CA 94102-5206	1971	$8,376	NA	1-M	1,000	90
New Hampshire Coll, Manchester, NH 03106-1045	1932	$13,570	$5,980	1-D	5,653	201
New Jersey City U, Jersey City, NJ 07305-1597	1927	$3,828 (S)	$5,200	5-M	8,603	503
New Jersey Inst of Technology, Newark, NJ 07102-1982	1881	$5,466 (S)	$6,502	5-D	8,191	591
Newman U, Wichita, KS 67213-2097	1933	$9,000	$3,400	2-M	1,903	305
New Mexico Highlands U, Las Vegas, NM 87701	1893	$1,662 (S)	$2,706	5-M	2,510	169
New Mexico Inst of Mining & Technology, Socorro, NM 87801	1889	$2,073 (S)	$3,584	5-D	1,445	138
New Mexico State U, Las Cruces, NM 88003-8001	1888	$2,196 (S)	$3,316	5-D	15,409	658
New Orleans Baptist Theological Seminary, New Orleans, LA 70126-4858	1917	$1,900	NA	2-D	1,736	28
New School Bachelor of Arts, New School U, New York, NY 10011-8603	1919	$13,420	NA	1-D	1,202	618
New York Inst of Technology, Old Westbury, NY 11568-8000	1955	$10,630	$6,830	1-F	9,133	673
New York U, New York, NY 10012-1019	1831	$21,730	$8,414	1-D	36,719	5,371

Name, address	Year	Tuition & Fees	Room & Board	Control, Degree	Enrollment	Faculty
Niagara U, Niagara University, NY 14109	1856	$12,890	$6,078	2-M	2,888	249
Nicholls State U, Thibodaux, LA 70310	1948	$2,507 (S)	$2,820	5-M	7,402	305
Nichols Coll, Dudley, MA 01571-5000	1815	$11,325	$6,710	1-M	1,427	76
Norfolk State U, Norfolk, VA 23504-3907	1935	$3,000 (S)	$4,992	5-D	7,115	545
North Carolina Ag & Tech State U, Greensboro, NC 27411	1891	$1,622 (S)	$3,860	5-D	7,465	533
North Carolina Central U, Durham, NC 27707-3129	1910	$1,944 (S)	$3,475	5-F	5,743	413
North Carolina State U, Raleigh, NC 27695	1887	$2,200 (S)	$4,560	5-D	27,960	1,593
North Carolina Wesleyan Coll, Rocky Mount, NC 27804-8677	1956	$8,144	$4,952	2-B	1,660	65
North Central Coll, Naperville, IL 60566-7063	1861	$13,845	$5,097	2-M	2,567	205
North Central U, Minneapolis, MN 55404-1322	1930	$7,480	$3,730	2-B	1,070	64
North Dakota State U, Fargo, ND 58105	1890	$2,566 (S)	$3,246	5-D	9,533	534
Northeastern Illinois U, Chicago, IL 60625-4699	1961	$2,470 (S)	NA	5-M	10,545	476
Northeastern State U, Tahlequah, OK 74464-2300	1846	$1,740 (S)	$2,610	5-D	8,503	449
Northeastern U, Boston, MA 02115-5096	1898	$16,511	$8,520	1-D	24,027	1,960
Northeast Louisiana U, Monroe, LA 71209-0001	1931	$1,952 (S)	$2,560	5-D	10,527	588
Northern Arizona U, Flagstaff, AZ 86011	1899	$2,080 (S)	NA	5-D	19,940	1,356
Northern Illinois U, De Kalb, IL 60115-2854	1895	$3,837 (S)	$4,310	5-D	22,473	1,192
Northern Kentucky U, Highland Heights, KY 41099	1968	$2,120 (S)	$3,316	5-F	11,795	789
Northern Michigan U, Marquette, MI 49855-5301	1899	$2,986 (S)	$4,468	5-M	7,779	354
Northern State U, Aberdeen, SD 57401-7198	1901	$2,535 (S)	$2,848	5-M	2,815	119
North Georgia Coll & State U, Dahlonega, GA 30597-1001	1873	$2,052 (S)	$3,374	5-M	3,313	193
North Greenville Coll, Tigerville, SC 29688-1892	1892	$7,400	$4,280	2-B	1,081	102
North Park U, Chicago, IL 60625-4895	1891	$14,690	$5,030	2-D	2,154	102
Northwestern Coll, Orange City, IA 51041-1996	1882	$11,300	$3,400	2-B	1,190	104
Northwestern Coll, St. Paul, MN 55113-1598	1902	$13,920	$4,176	2-B	1,814	192
Northwestern Oklahoma State U, Alva, OK 73717-2799	1897	$1,802 (S)	$2,316	5-M	1,970	115
Northwestern State U of Louisiana, Natchitoches, LA 71497	1884	$2,177 (S)	$2,606	5-D	8,572	299
Northwestern U, Evanston, IL 60208	1851	$22,458	$6,675	1-D	17,428	2,464
Northwest Missouri State U, Maryville, MO 64468-6001	1905	$2,813 (S)	$3,890	5-M	6,294	259
Northwest Nazarene U, Nampa, ID 83686-5897	1913	$12,456	$3,519	2-M	1,693	96
Northwood U, Midland, MI 48640-2398	1959	$10,889	$5,058	1-M	2,640	62
Norwich U, Northfield, VT 05663	1819	$14,950	$5,718	1-M	2,785	274
Notre Dame Coll, Manchester, NH 03104-2299	1950	$12,396	$5,713	2-M	1,220	99
Nova Southeastern U, Fort Lauderdale, FL 33314-7721	1964	$10,570	$5,930	1-D	16,050	1,306
Nyack Coll, Nyack, NY 10960-3698	1882	$11,100	$5,340	2-F	1,594	119
Oakland U, Rochester, MI 48309-4401	1957	$3,734 (S)	$4,555	5-D	14,289	693
Oakwood Coll, Huntsville, AL 35896	1896	$7,628	$4,680	2-B	1,805	163
Oberlin Coll, Oberlin, OH 44074-1090	1833	$22,438	$6,238	1-M	2,947	323
Occidental Coll, Los Angeles, CA 90041-3392	1887	$19,957	$6,180	1-M	1,597	183
Oglethorpe U, Atlanta, GA 30319-2797	1835	$15,920	$5,190	1-M	1,181	123
Ohio Dominican Coll, Columbus, OH 43219-2099	1911	$9,350	$4,840	2-B	1,977	115
Ohio Northern U, Ada, OH 45810-1599	1871	$19,815	$4,875	2-F	2,987	253
The Ohio State U, Columbus, OH 43210	1870	$3,660 (S)	$5,289	5-D	48,511	3,331
Ohio U, Athens, OH 45701-2979	1804	$4,275 (S)	$4,914	5-D	19,564	1,186
Ohio U–Chillicothe, Chillicothe, OH 45601-0629	1946	$3,102 (S)	NA	5-B	1,596	118
Ohio U–Zanesville, Zanesville, OH 43701-2695	1946	$3,117 (S)	NA	5-M	1,222	70
Ohio Wesleyan U, Delaware, OH 43015	1842	$20,040	$6,370	2-B	1,873	174
Oklahoma Baptist U, Shawnee, OK 74804	1910	$8,336	$3,250	2-M	2,171	171
Oklahoma Christian U of Sci & Arts, Oklahoma City, OK 73136-1100	1950	$8,278	$3,840	2-M	1,690	133
Oklahoma City U, Oklahoma City, OK 73106-1402	1904	$8,512	$3,890	2-F	4,407	341
Oklahoma Panhandle State U, Goodwell, OK 73939-0430	1909	$1,510 (S)	$2,192	5-B	1,100	80
Oklahoma State U, Stillwater, OK 74078	1890	$2,357 (S)	$4,536	5-D	20,466	1,199
Old Dominion U, Norfolk, VA 23529	1930	$3,976 (S)	$5,000	5-D	18,552	971
Olivet Nazarene U, Kankakee, IL 60901-0592	1907	$10,838	$4,696	2-M	2,386	145
Oral Roberts U, Tulsa, OK 74171-0001	1963	$10,460	$4,728	2-D	3,508	272
Oregon Inst of Technology, Klamath Falls, OR 97601-8801	1947	$3,309 (S)	$3,910	5-M	2,679	156
Oregon State U, Corvallis, OR 97331	1868	$3,540 (S)	$5,064	5-D	15,176	1,335
Otterbein Coll, Westerville, OH 43081	1847	$14,997	$4,944	2-M	2,726	160
Ouachita Baptist U, Arkadelphia, AR 71998-0001	1886	$8,090	$3,100	2-B	1,536	144
Our Lady of Holy Cross Coll, New Orleans, LA 70131-7399	1916	$5,580	NA	2-M	1,280	113
Our Lady of the Lake U of San Antonio, San Antonio, TX 78207-4689	1895	$10,872	$4,206	2-D	3,689	259
Pace U, New York, NY 10038	1906	$13,820	$6,120	1-D	13,151	1,012
Pacific Lutheran U, Tacoma, WA 98447	1890	$15,680	$4,890	2-M	3,684	324
Pacific Union Coll, Angwin, CA 94508-9707	1882	$13,530	$4,305	2-M	1,558	113
Pacific U, Forest Grove, OR 97116-1797	1849	$16,695	$4,564	1-D	1,864	204
Palm Beach Atlantic Coll, West Palm Beach, FL 33416-4708	1968	$9,900	$4,130	2-M	2,065	144
Palmer Coll of Chiropractic, Davenport, IA 52803-5287	1897	$14,520	NA	1-F	1,753	113
Park Coll, Parkville, MO 64152-4358	1875	$4,410	$4,600	2-M	8,591	100
Parsons School of Design, New School U, New York, NY 10011-8878	1896	$18,540	$8,857	1-M	2,708	403
Peirce Coll, Philadelphia, PA 19102-4699	1865	$7,112	NA	1-B	2,474	257
Pennsylvania State U Abington Coll, Abington, PA 19001-3918	1950	$5,682 (S)	NA	12-B	3,243	225
Pennsylvania State U Altoona Coll, Altoona, PA 16601-3760	1939	$5,682 (S)	$4,840	12-B	3,873	237
Pennsylvania State U at Erie, The Behrend Coll, Erie, PA 16563	1948	$5,832 (S)	$4,840	12-M	3,470	247
Pennsylvania State U Berks Campus of the Berks–Lehigh Valley Coll, Reading, PA 19610-6009	1924	$5,682 (S)	$4,840	12-B	1,939	137
Pennsylvania State U Harrisburg Campus of the Capital Coll, Middletown, PA 17057-4898	1966	$5,832 (S)	$4,840	12-D	3,456	235
Pennsylvania State U Schuylkill Campus of the Capital Coll, Schuylkill Haven, PA 17972-2208	1934	$5,654 (S)	$4,840	12-B	1,070	71
Pennsylvania State U U Park Campus, University Park, PA 16802-1503	1855	$5,832 (S)	$4,840	12-D	41,114	2,302
Pepperdine U, Malibu, CA 90263-0002	1937	$20,210	$6,840	2-D	7,948	323
Peru State Coll, Peru, NE 68421	1867	$2,085 (S)	$3,120	5-M	1,695	97
Pfeiffer U, Misenheimer, NC 28109-0960	1885	$9,816	$4,130	2-M	1,682	52
Philadelphia Coll of Bible, Langhorne, PA 19047-2990	1913	$9,520	$4,890	2-M	1,375	128
Philadelphia Coll of Textiles & Sci, Philadelphia, PA 19144-5497	1884	$13,466	$6,354	1-M	3,371	412
Piedmont Coll, Demorest, GA 30535-0010	1897	$8,200	$3,930	2-M	1,564	169
Pittsburg State U, Pittsburg, KS 66762-5880	1903	$2,016 (S)	$3,544	5-M	6,268	313
Plattsburgh State U of New York, Plattsburgh, NY 12901-2681	1889	$3,845 (S)	$4,476	5-M	5,937	378

Name, address	Year	Tuition & Fees	Room & Board	Control, Degree	Enrollment	Faculty
Plymouth State Coll, Plymouth, NH 03264-1595	1871	$4,342 (S)	$4,706	5-M	3,990	314
Point Loma Nazarene U, San Diego, CA 92106-2899	1902	$12,464	$5,220	2-M	2,659	251
Point Park Coll, Pittsburgh, PA 15222-1984	1960	$11,406	$5,252	1-M	2,424	219
Polytechnic U, Brooklyn Campus, Brooklyn, NY 11201-2990	1854	$19,150	$4,600	1-D	3,181	364
Polytechnic U, Farmingdale Campus, Farmingdale, NY 11735-3995	1854	$19,150	NA	1-D	3,181	364
Polytechnic U of Puerto Rico, Hato Rey, PR 00919	1966	$4,560	NA	1-M	4,996	249
Pomona Coll, Claremont, CA 91711	1887	$20,680	$8,270	1-B	1,605	156
Pontifical Catholic U of Puerto Rico, Ponce, PR 00717-0777	1948	$3,910	$2,660	2-F	7,256	654
Portland State U, Portland, OR 97207-0751	1946	$3,180 (S)	$6,150	5-D	17,186	746
Prairie View A&M U, Prairie View, TX 77446-0188	1878	$2,364 (S)	$3,475	5-M	5,996	355
Pratt Inst, Brooklyn, NY 11205-3899	1887	$17,151	$7,530	1-M	3,819	627
Presbyterian Coll, Clinton, SC 29325	1880	$14,806	$4,447	2-B	1,081	118
Princeton U, Princeton, NJ 08544-1019	1746	$23,820	$6,711	1-D	6,514	892
Providence Coll, Providence, RI 02918	1917	$16,655	$7,125	2-M	5,534	391
Purchase Coll, State U of New York, Purchase, NY 10577-1400	1967	$3,879 (S)	$5,584	5-M	3,626	252
Purdue U, West Lafayette, IN 47907	1869	$3,368 (S)	$5,260	5-D	36,878	2,268
Purdue U Calumet, Hammond, IN 46323-2094	1951	$3,088 (S)	NA	5-M	9,974	457
Purdue U North Central, Westville, IN 46391-9528	1967	$2,979 (S)	NA	5-M	3,373	230
Queens Coll, Charlotte, NC 28274-0002	1857	$9,410	$5,830	2-M	1,715	115
Queens Coll of the City U of New York, Flushing, NY 11367-1597	1937	$3,393 (S)	NA	11-M	16,195	1,129
Quincy U, Quincy, IL 62301-2699	1860	$12,410	$4,390	2-M	1,139	106
Quinnipiac Coll, Hamden, CT 06518-1940	1929	$14,880	$7,590	1-F	5,929	400
Radford U, Radford, VA 24142	1910	$3,180 (S)	$4,636	5-M	8,368	495
Ramapo Coll of New Jersey, Mahwah, NJ 07430-1680	1969	$4,206 (S)	$6,117	5-M	4,812	292
Randolph-Macon Coll, Ashland, VA 23005-5505	1830	$16,240	$4,345	2-B	1,114	151
Reed Coll, Portland, OR 97202-8199	1908	$22,340	$6,400	1-M	1,357	121
Regis Coll, Weston, MA 02493 (3)	1927	$15,250	$7,200	2-M	1,304	152
Regis U, Denver, CO 80221-1099	1877	$14,970	$6,200	2-M	NA	109
Reinhardt Coll, Waleska, GA 30183-0128	1883	$6,210	$4,470	2-B	1,046	105
Rensselaer Polytechnic Inst, Troy, NY 12180-3590	1824	$20,604	$7,456	1-D	6,509	357
Rhode Island Coll, Providence, RI 02908-1924	1854	$3,076 (S)	$5,500	5-D	8,683	640
Rhode Island School of Design, Providence, RI 02903-2784	1877	$19,670	$6,490	1-F	2,001	315
Rhodes Coll, Memphis, TN 38112-1690	1848	$17,518	$5,332	2-M	1,466	150
Rice U, Houston, TX 77251-1892	1912	$14,306	$6,400	1-D	4,318	594
The Richard Stockton Coll of New Jersey, Pomona, NJ 08240-0195	1969	$3,776 (S)	$4,980	5-M	6,157	340
Rider U, Lawrenceville, NJ 08648-3001	1865	$15,410	$6,510	1-M	5,327	408
Rivier Coll, Nashua, NH 03060-5086	1933	$13,190	$5,525	2-M	2,737	215
Roanoke Coll, Salem, VA 24153-3794	1842	$16,410	$5,250	2-B	1,733	174
Robert Morris Coll, Chicago, IL 60605	1913	$10,500	NA	1-B	3,728	323
Robert Morris Coll, Moon Township, PA 15108-1189	1921	$8,339	$5,512	1-M	4,737	296
Roberts Wesleyan Coll, Rochester, NY 14624-1997	1866	$12,400	$4,394	2-M	1,459	129
Rochester Inst of Technology, Rochester, NY 14623-5604	1829	$16,359	$6,645	1-D	12,602	1,046
Rockford Coll, Rockford, IL 61108-2393	1847	$14,750	$4,900	1-M	1,243	132
Rockhurst U, Kansas City, MO 64110-2561	1910	$11,790	$4,920	2-M	2,862	191
Roger Williams U, Bristol, RI 02809	1956	$15,840	$7,460	1-F	3,695	325
Rollins Coll, Winter Park, FL 32789-4499	1885	$20,010	$6,426	1-M	2,234	258
Roosevelt U, Chicago, IL 60605-1394	1945	$11,030	$5,500	1-D	6,837	530
Rose-Hulman Inst of Technology, Terre Haute, IN 47803-3920	1874	$18,105	$5,300	1-M	1,722	130
Rosemont Coll, Rosemont, PA 19010-1699 (3)	1921	$13,340	$6,760	2-M	1,077	157
Rowan U, Glassboro, NJ 08028-1701	1923	$4,241 (S)	$5,492	5-D	9,480	386
Russell Sage Coll, Troy, NY 12180-4115 (3)	1916	$14,230	$5,770	1-B	1,000	146
Rutgers, The State U of New Jersey, Camden Coll of Arts & Scis, Camden, NJ 08102	1927	$5,190 (S)	$5,606	5-B	2,352	173
Rutgers, The State U of New Jersey, Coll of Engineering, Piscataway, NJ 08854-8058	1864	$5,836 (S)	$6,115	5-B	2,192	141
Rutgers, The State U of New Jersey, Coll of Nursing, Newark, NJ 07102-1803	1956	$5,130 (S)	$6,026	5-D	NA	34
Rutgers, The State U of New Jersey, Coll of Pharmacy, Piscataway, NJ 08855	1927	$5,836 (S)	$6,115	5-D	1,124	63
Rutgers, The State U of New Jersey, Cook Coll, New Brunswick, NJ 08903	1921	$5,817 (S)	$5,922	5-B	3,309	117
Rutgers, The State U of New Jersey, Douglass Coll, New Brunswick, NJ 08901-1414 (3)	1918	$5,349 (S)	$6,123	5-B	3,064	941
Rutgers, The State U of New Jersey, Livingston Coll, Piscataway, NJ 08854	1969	$5,382 (S)	$6,125	5-B	3,336	941
Rutgers, The State U of New Jersey, Newark Coll of Arts & Scis, Newark, NJ 07102-1896	1946	$5,151 (S)	$6,026	5-B	3,615	349
Rutgers, The State U of New Jersey, Rutgers Coll, New Brunswick, NJ 08901	1766	$5,386 (S)	$6,113	5-B	10,737	941
Rutgers, The State U of New Jersey, U Coll–Newark, Newark, NJ 07102-1896	1934	$138/cr. hr. (S)	NA	5-B	1,695	349
Rutgers, The State U of New Jersey, U Coll–New Brunswick, New Brunswick, NJ 08903	1934	$138/cr. hr. (S)	NA	5-B	3,018	941
Sacred Heart U, Fairfield, CT 06432-1000	1963	$13,475	$6,570	2-M	5,605	410
Saginaw Valley State U, University Center, MI 48710	1963	$3,448 (S)	$4,590	5-M	8,010	466
St. Ambrose U, Davenport, IA 52803-2898	1882	$12,850	$4,810	2-D	2,819	233
Saint Anselm Coll, Manchester, NH 03102-1310	1889	$16,670	$6,160	2-B	2,064	171
Saint Augustine's Coll, Raleigh, NC 27610-2298	1867	$6,560	$4,292	2-B	1,598	108
St. Bonaventure U, St. Bonaventure, NY 14778-2284	1858	$13,100	$5,400	2-M	2,857	173
St. Cloud State U, St. Cloud, MN 56301-4498	1869	$3,082 (S)	$3,600	5-D	13,627	633
St. Edward's U, Austin, TX 78704-6489	1885	$10,730	$4,710	2-M	3,422	200
St. Francis Coll, Brooklyn Heights, NY 11201-4398	1884	$7,680	NA	2-B	2,448	172
Saint Francis Coll, Loretto, PA 15940-0600	1847	$14,342	$6,250	2-M	1,916	174
St. John Fisher Coll, Rochester, NY 14618-3597	1948	$12,500	$5,800	2-M	2,474	187
Saint John's U, Collegeville, MN 56321 (1)	1857	$14,758	$4,758	2-F	1,854	182
St. John's U, Jamaica, NY 11439	1870	$12,230	NA	2-D	18,336	1,096
Saint Joseph Coll, West Hartford, CT 06117-2700 (3)	1932	$14,490	$6,070	2-M	1,820	84
Saint Joseph's Coll, Standish, ME 04084-5263	1912	$11,710	$5,770	2-M	1,308	97
St. Joseph's Coll, New York, NY 11205-3688	1916	$8,326	NA	1-B	1,292	135
St. Joseph's Coll, Suffolk Campus, Patchogue, NY 11772-2399	1916	$8,917	NA	1-M	2,912	173
Saint Joseph's U, Philadelphia, PA 19131-1395	1851	$16,165	$7,070	2-D	6,484	430
St. Lawrence U, Canton, NY 13617-1455	1856	$21,435	$6,575	1-M	1,937	182
Saint Leo U, Saint Leo, FL 33574-2008	1889	$10,996	$5,420	2-M	1,668	116
Saint Louis U, St. Louis, MO 63103-2097	1818	$15,050	$5,702	2-D	14,253	3,020

Name, address	Year	Tuition & Fees	Room & Board	Control, Degree	Enrollment	Faculty
Saint Martin's Coll, Lacey, WA 98503-7500	1895	$13,120	$4,590	2-M	1,477	68
Saint Mary-of-the-Woods Coll, Saint Mary-of-the-Woods, IN 47876 (3)	1840	$12,975	$5,010	2-M	1,272	56
Saint Mary's Coll, Notre Dame, IN 46556 (3)	1844	$15,652	$5,632	2-B	1,355	174
Saint Mary's Coll of California, Moraga, CA 94556	1863	$15,998	$7,000	2-D	4,346	245
St. Mary's Coll of Maryland, St. Mary's City, MD 20686	1840	$6,875 (S)	$5,645	5-B	1,682	173
Saint Mary's U of Minnesota, Winona, MN 55987-1399	1912	$12,495	$4,270	2-D	5,575	301
St. Mary's U of San Antonio, San Antonio, TX 78228-8507	1852	$10,608	$4,908	2-D	4,189	297
Saint Michael's Coll, Colchester, VT 05439	1904	$15,900	$7,253	2-M	2,773	194
St. Norbert Coll, De Pere, WI 54115-2099	1898	$14,434	$5,162	2-M	2,031	168
St. Olaf Coll, Northfield, MN 55057-1098	1874	$16,500	$4,180	2-B	2,981	404
Saint Peter's Coll, Jersey City, NJ 07306-5997	1872	$14,366	$5,060	2-M	3,512	380
St. Thomas Aquinas Coll, Sparkill, NY 10976	1952	$10,700	$6,910	1-M	2,215	150
St. Thomas U, Miami, FL 33054-6459	1961	$11,840	$4,200	2-F	2,264	132
Saint Vincent Coll, Latrobe, PA 15650-2690	1846	$13,461	$4,782	2-B	1,155	109
Saint Xavier U, Chicago, IL 60655-3105	1847	$12,560	$5,430	2-M	3,854	279
Salem Coll, Winston-Salem, NC 27108-0548 (4)	1772	$12,415	$7,610	2-M	1,029	81
Salem State Coll, Salem, MA 01970-5353	1854	NA	$4,200	5-M	8,736	409
Salisbury State U, Salisbury, MD 21801-6837	1925	$3,842 (S)	$5,390	5-M	6,080	371
Salve Regina U, Newport, RI 02840-4192	1934	$16,300	$7,250	2-D	2,217	189
Samford U, Birmingham, AL 35229-0002	1841	$9,432	$4,406	2-D	4,473	403
Sam Houston State U, Huntsville, TX 77341	1879	$1,586 (S)	$3,290	5-D	12,205	519
San Diego State U, San Diego, CA 92182	1897	$1,854 (S)	$5,344	5-D	31,453	2,407
San Francisco State U, San Francisco, CA 94132-1722	1899	$1,982 (S)	$6,280	5-M	27,446	1,557
San Jose State U, San Jose, CA 95192-0001	1857	$2,017 (S)	$5,736	5-M	26,628	1,689
Santa Clara U, Santa Clara, CA 95053-0001	1851	$16,635	$7,323	2-D	7,707	579
Sarah Lawrence Coll, Bronxville, NY 10708	1926	$23,076	$7,612	1-M	1,408	238
Savannah Coll of Art & Design, Savannah, GA 31402-3146	1978	$13,500	$6,375	1-M	3,965	210
Savannah State U, Savannah, GA 31404	1890	$2,226 (S)	$3,650	5-M	2,745	155
School of the Art Inst of Chicago, Chicago, IL 60603-3103	1866	$17,160	NA	1-M	2,297	444
School of the Museum of Fine Arts, Boston, MA 02115	1876	$15,890	NA	1-M	1,235	142
School of Visual Arts, New York, NY 10010-3994	1947	$13,890	NA	3-M	5,323	852
Seattle Pacific U, Seattle, WA 98119-1997	1891	$14,541	$5,574	2-D	3,394	213
Seattle U, Seattle, WA 98122	1891	$14,805	$5,334	2-D	5,667	414
Seton Hall U, South Orange, NJ 07079-2697	1856	$13,600	$7,230	2-D	10,359	872
Seton Hill Coll, Greensburg, PA 15601 (4)	1883	$12,640	$4,800	2-M	1,076	121
Shawnee State U, Portsmouth, OH 45662-4344	1986	$3,063 (S)	$4,096	5-B	3,438	242
Shaw U, Raleigh, NC 27601-2399	1865	$6,304	$4,174	2-M	2,569	270
Shenandoah U, Winchester, VA 22601-5195	1875	$14,400	$4,950	2-D	2,174	275
Shepherd Coll, Shepherdstown, WV 25443-3210	1871	$2,228 (S)	$4,139	5-B	4,055	292
Shippensburg U of Pennsylvania, Shippensburg, PA 17257-2299	1871	$4,344 (S)	$4,012	5-M	6,741	330
Shorter Coll, Rome, GA 30165-4298	1873	$8,260	$4,450	2-M	1,849	178
Siena Coll, Loudonville, NY 12211-1462	1937	$12,710	$5,835	2-M	2,981	234
Siena Heights U, Adrian, MI 49221-1796	1919	$10,700	$4,330	2-M	1,994	192
Simmons Coll, Boston, MA 02115 (3)	1899	$18,564	$7,590	1-D	3,401	371
Simpson Coll, Indianola, IA 50125-1297	1860	$13,095	$4,570	2-B	1,992	108
Simpson Coll & Graduate School, Redding, CA 96003-8606	1921	$9,110	$4,400	2-M	1,259	81
Skidmore Coll, Saratoga Springs, NY 12866-1632	1903	$21,988	$6,652	1-M	2,573	164
Slippery Rock U of Pennsylvania, Slippery Rock, PA 16057	1889	$4,302 (S)	$3,722	5-M	6,923	391
Smith Coll, Northampton, MA 01063 (3)	1871	$21,512	$7,560	1-D	3,212	276
Sonoma State U, Rohnert Park, CA 94928-3609	1960	$2,130 (S)	$6,049	5-M	7,003	520
South Carolina State U, Orangeburg, SC 29117-0001	1896	$2,974 (S)	$2,616	5-D	4,742	229
South Dakota School of Mines & Technology, Rapid City, SD 57701-3995	1885	$3,378 (S)	$3,026	5-D	2,211	142
South Dakota State U, Brookings, SD 57007	1881	$2,912 (S)	$2,728	5-D	8,630	514
Southeastern Coll of the Assemblies of God, Lakeland, FL 33801-6099	1935	$4,999	$3,382	2-B	1,078	87
Southeastern Louisiana U, Hammond, LA 70402	1925	$2,155 (S)	$2,400	5-M	15,308	645
Southeastern Oklahoma State U, Durant, OK 74701-0609	1909	$1,879 (S)	$2,689	5-M	3,855	211
Southeast Missouri State U, Cape Girardeau, MO 63701-4799	1873	$3,000 (S)	$4,080	5-M	8,487	378
Southern Adventist U, Collegedale, TN 37315-0370	1892	$9,736	$3,678	2-M	1,724	101
Southern Arkansas U–Magnolia, Magnolia, AR 71753	1909	$1,896 (S)	$2,690	5-M	2,712	162
Southern Connecticut State U, New Haven, CT 06515-1355	1893	$3,568 (S)	$5,782	5-M	11,264	754
Southern Illinois U Carbondale, Carbondale, IL 62901-6806	1869	$3,420 (S)	$3,777	5-D	22,251	1,024
Southern Illinois U Edwardsville, Edwardsville, IL 62026-0001	1957	$2,276 (S)	$4,066	5-F	11,520	781
Southern Methodist U, Dallas, TX 75275	1911	$16,790	$6,579	2-D	10,038	710
Southern Nazarene U, Bethany, OK 73008-2694	1899	$8,222	$4,168	2-M	1,950	137
Southern Oregon U, Ashland, OR 97520	1926	$3,204 (S)	$4,557	5-M	5,457	221
Southern Polytechnic State U, Marietta, GA 30060-2896	1948	$1,998 (S)	$3,450	5-M	3,678	133
Southern U & Ag & Mech Coll, Baton Rouge, LA 70813	1880	$2,068 (S)	$3,320	5-D	9,567	539
Southern U at New Orleans, New Orleans, LA 70126-1009	1959	NA	NA	5-M	NA	235
Southern Utah U, Cedar City, UT 84720-2498	1897	$1,854 (S)	$2,432	5-M	5,539	217
Southern Wesleyan U, Central, SC 29630-1020	1906	$10,180	$3,706	2-M	1,337	46
Southwest Baptist U, Bolivar, MO 65613-2597	1878	$8,347	$2,905	2-M	3,708	244
Southwestern Adventist U, Keene, TX 76059	1894	$8,500	$4,248	2-M	1,163	102
Southwestern Oklahoma State U, Weatherford, OK 73096-3098	1901	$1,798 (S)	$2,400	5-F	4,657	237
Southwestern U, Georgetown, TX 78626	1840	$14,000	$5,190	2-B	1,255	149
Southwest Missouri State U, Springfield, MO 65804-0094	1905	$3,214 (S)	$3,594	5-M	16,794	892
Southwest State U, Marshall, MN 56258-1598	1963	$3,056 (S)	$3,326	5-M	3,483	131
Southwest Texas State U, San Marcos, TX 78666	1899	$2,214 (S)	$3,992	5-D	21,481	948
Spalding U, Louisville, KY 40203-2188	1814	$10,396	$2,754	2-D	1,575	149
Spelman Coll, Atlanta, GA 30314-4399 (3)	1881	$10,095	$6,560	1-B	NA	209
Spring Arbor Coll, Spring Arbor, MI 49283-9799	1873	$10,686	$4,190	2-M	2,384	103
Springfield Coll, Springfield, MA 01109-3797	1885	$14,825	$5,348	1-D	2,490	232
Spring Hill Coll, Mobile, AL 36608-1791	1830	$13,860	$5,250	2-M	1,373	93
Stanford U, Stanford, CA 94305-9991	1891	$21,389	$8,427	1-D	17,154	1,534
State U of New York at Albany, Albany, NY 12222-0001	1844	$4,173 (S)	$5,472	5-D	16,867	946
State U of New York at Binghamton, Binghamton, NY 13902-6000	1946	$4,110 (S)	$5,308	5-D	12,259	739
State U of New York at Buffalo, Buffalo, NY 14260	1846	$4,340 (S)	$5,604	5-D	23,370	1,690
State U of New York at Farmingdale, Farmingdale, NY 11735	1912	$3,950 (S)	$5,780	5-B	5,492	337
State U of New York at New Paltz, New Paltz, NY 12561	1828	$3,885 (S)	$5,020	5-M	6,402	534

Name, address	Year	Tuition & Fees	Room & Board	Control, Degree	Enrollment	Faculty
State U of New York at Oswego, Oswego, NY 13126	1861	$3,945 (S)	$5,888	5-M	7,718	416
State U of New York at Stony Brook, Stony Brook, NY 11794	1957	$3,932 (S)	$6,030	5-D	18,628	1,682
State U of New York Coll at Brockport, Brockport, NY 14420-2997	1867	$3,940 (S)	$5,150	5-M	8,581	435
State U of New York Coll at Buffalo, Buffalo, NY 14222-1095.	1867	$3,791 (S)	$4,560	5-M	11,030	636
State U of New York Coll at Cortland, Cortland, NY 13045	1868	$3,974 (S)	$5,300	5-M	6,511	439
State U of New York Coll at Fredonia, Fredonia, NY 14063	1826	$4,075 (S)	$5,200	5-M	4,809	365
State U of New York Coll at Geneseo, Geneseo, NY 14454-1401	1871	$4,016 (S)	$4,820	5-M	5,497	334
State U of New York Coll at Old Westbury, Old Westbury, NY 11568-0210 . .	1965	$3,731 (S)	$5,345	5-B	3,360	222
State U of New York Coll at Oneonta, Oneonta, NY 13820-4015.	1889	$3,908 (S)	$5,790	5-M	5,380	354
State U of New York Coll at Potsdam, Potsdam, NY 13676	1816	$3,899 (S)	$5,440	5-M	4,029	302
State U of New York Coll of Environmental Sci & Forestry, Syracuse, NY 13210	1911	$3,413 (S)	$7,370	5-D	1,729	136
State U of New York Empire State Coll, Saratoga Springs, NY 12866-4391 .	1971	$3,545 (S)	NA	5-M	7,542	329
State U of New York Health Sci Ctr at Syracuse, Syracuse, NY 13210-2334.	1950	$3,710 (S)	$6,650	5-D	1,142	45
State U of New York Inst of Technology at Utica/Rome, Utica, NY 13504-3050	1966	$3,939 (S)	$5,880	5-M	2,533	120
State U of West Georgia, Carrollton, GA 30118	1933	$2,088 (S)	$3,532	5-M	8,665	392
Stephen F. Austin State U, Nacogdoches, TX 75962	1923	$1,513 (S)	$4,118	5-D	12,132	638
Stetson U, DeLand, FL 32720-3781	1883	$15,765	$5,730	1-F	2,967	230
Stevens Inst of Technology, Hoboken, NJ 07030	1870	$19,360	$6,690	1-D	3,467	213
Stillman Coll, Tuscaloosa, AL 35403-9990	1876	$5,460	$3,764	2-B	1,017	73
Stonehill Coll, Easton, MA 02357	1948	$15,730	$7,450	2-M	2,658	260
Strayer U, Washington, DC 20005-2603.	1892	$8,100	NA	3-M	10,449	398
Suffolk U, Boston, MA 02108-2770.	1906	$12,920	$8,956	1-D	6,445	515
Sullivan Coll, Louisville, KY 40205	1864	$8,904	NA	3-M	2,975	88
Sul Ross State U, Alpine, TX 79832	1920	$1,680 (S)	$3,480	5-M	3,101	77
Susquehanna U, Selinsgrove, PA 17870-1001	1858	$18,350	$5,390	2-B	1,765	168
Swarthmore Coll, Swarthmore, PA 19081-1397	1864	$22,000	$7,500	1-B	1,388	181
Syracuse U, Syracuse, NY 13244-0003	1870	$18,056	$8,040	1-D	14,561	1,397
Tarleton State U, Stephenville, TX 76402	1899	$2,464 (S)	$3,192	5-M	6,333	332
Taylor U, Upland, IN 46989-1001	1846	$13,484	$4,544	2-B	1,880	144
Teikyo Post U, Waterbury, CT 06723-2540	1890	$12,260	$5,600	1-B	1,340	24
Temple U, Philadelphia, PA 19122-6096.	1884	$6,150 (S)	$5,924	12-D	27,539	2,596
Tennessee State U, Nashville, TN 37209-1561	1912	$3,069 (S)	$3,060	5-D	8,750	369
Tennessee Technological U, Cookeville, TN 38505	1915	$2,116 (S)	$3,220	5-D	8,215	460
Texas A&M Intl U, Laredo, TX 78041-1900.	1969	$2,101 (S)	NA	5-M	2,986	155
Texas A&M U, College Station, TX 77843.	1876	$2,777 (S)	$3,357	5-D	43,389	2,327
Texas A&M U at Galveston, Galveston, TX 77553-1675	1962	$2,834 (S)	$3,653	5-B	1,168	125
Texas A&M U–Commerce, Commerce, TX 75429-3011.	1889	$2,288 (S)	$3,912	5-D	7,793	382
Texas A&M U–Corpus Christi, Corpus Christi, TX 78412-5503	1947	$1,954 (S)	NA	5-D	6,335	405
Texas A&M U–Kingsville, Kingsville, TX 78363.	1925	$2,180 (S)	NA	5-D	5,940	403
Texas A&M U–Texarkana, Texarkana, TX 75505-5518.	1971	$1,586 (S)	NA	5-M	1,156	61
Texas Christian U, Fort Worth, TX 76129-0002	1873	$11,090	$4,000	2-D	7,395	550
Texas Lutheran U, Seguin, TX 78155-5999	1891	$10,370	$3,866	2-B	1,512	122
Texas Southern U, Houston, TX 77004-4584	1947	$2,064 (S)	$4,000	5-D	6,316	387
Texas Tech U, Lubbock, TX 79409	1923	$2,607 (S)	$4,539	5-D	24,158	1,059
Texas Wesleyan U, Fort Worth, TX 76105-1536.	1890	$7,950	$3,850	2-F	3,086	242
Texas Woman's U, Denton, TX 76204 (4).	1901	$1,980 (S)	$3,783	5-D	9,352	814
Thomas Jefferson U, Philadelphia, PA 19107.	1824	$16,140	NA	1-M	2,168	203
Thomas More Coll, Crestview Hills, KY 41017-3495	1921	$11,250	$4,500	2-M	1,550	174
Tiffin U, Tiffin, OH 44883-2161	1888	$9,210	$4,650	1-M	1,439	80
Toccoa Falls Coll, Toccoa Falls, GA 30598-1000	1907	$7,525	$3,782	2-M	1,012	76
Touro Coll, New York, NY 10010.	1971	$8,980	NA	1-F	8,500	810
Towson U, Towson, MD 21252-0001.	1866	$4,120 (S)	$5,450	5-M	15,923	1,067
Transylvania U, Lexington, KY 40508-1797	1780	$13,260	$5,150	2-B	1,073	102
Trevecca Nazarene U, Nashville, TN 37210-2877	1901	$9,090	$4,236	2-M	1,582	141
Trinity Coll, Hartford, CT 06106-3100	1823	$22,470	$6,630	1-M	2,258	223
Trinity Coll, Washington, DC 20017-1094 (3)	1897	$12,490	$6,310	2-M	1,576	141
Trinity Coll of Vermont, Burlington, VT 05401-1470 (4).	1925	$13,420	$6,700	2-M	1,020	132
Trinity Intl U, Deerfield, IL 60015-1284	1897	$12,630	$4,800	2-D	2,571	161
Trinity U, San Antonio, TX 78212-7200.	1869	$14,724	$5,970	2-M	2,512	258
Tri-State U, Angola, IN 46703-1764	1884	$11,900	$4,850	1-B	1,081	83
Troy State U, Troy, AL 36082.	1887	$2,250 (S)	$3,660	5-M	6,630	357
Troy State U Dothan, Dothan, AL 36304-0368	1961	$2,229 (S)	NA	5-M	2,215	136
Troy State U Montgomery, Montgomery, AL 36103-4419	1965	$2,085 (S)	NA	5-M	3,455	185
Truman State U, Kirksville, MO 63501-4221.	1867	$3,274 (S)	$4,192	5-M	6,439	396
Tufts U, Medford, MA 02155	1852	$22,811	$7,108	1-D	8,876	1,002
Tulane U, New Orleans, LA 70118-5669.	1834	$22,066	$6,710	1-D	11,109	775
Tusculum Coll, Greeneville, TN 37743-9997.	1794	$11,800	$3,900	2-M	1,544	162
Tuskegee U, Tuskegee, AL 36088.	1881	$8,662	$4,710	1-F	3,080	306
Union Coll, Schenectady, NY 12308-2311	1795	$22,135	$6,417	1-M	2,484	208
The Union Inst, Cincinnati, OH 45206-1925	1969	$7,296	NA	1-D	2,019	109
Union U, Jackson, TN 38305-3697.	1823	$8,180	$3,360	2-M	2,223	187
United States Air Force Acad, USAF Academy, CO 80840-5025 (2)	1954	$0 (C)	NA	4-B	4,072	577
United States Intl U, San Diego, CA 92131-1799	1952	$11,745	$5,040	1-D	1,319	111
United States Military Acad, West Point, NY 10996 (2)	1802	$0 (C)	NA	4-B	4,209	577
United States Naval Acad, Annapolis, MD 21402-5000	1845	$0 (C)	NA	4-B	4,020	600
Universidad del Turabo, Gurabo, PR 00778-3030.	1972	$2,860	NA	1-M	NA	NA
Universidad Metroa, Río Piedras, PR 00928-1150.	1980	$3,392	NA	1-M	NA	NA
The U of Akron, Akron, OH 44325-0001.	1870	$3,660 (S)	$4,835	5-D	23,506	1,648
The U of Alabama, Tuscaloosa, AL 35487	1831	$2,594 (S)	$3,810	5-D	18,426	965
The U of Alabama at Birmingham, Birmingham, AL 35294	1969	$2,850 (S)	NA	5-D	15,056	1,803
The U of Alabama in Huntsville, Huntsville, AL 35899	1950	$2,832 (S)	$3,700	5-D	6,998	430
U of Alaska Anchorage, Anchorage, AK 99508-8060.	1954	$2,466 (S)	$6,591	5-M	14,765	982
U of Alaska Fairbanks, Fairbanks, AK 99775-7480.	1917	$2,410 (S)	$4,150	5-D	6,895	655
U of Alaska Southeast, Juneau, AK 99801-8625	1972	$2,164 (S)	NA	5-M	NA	134
The U of Arizona, Tucson, AZ 85721	1885	$2,058 (S)	$5,042	5-D	34,327	1,551
U of Arkansas, Fayetteville, AR 72701-1201.	1871	$2,661 (S)	$4,030	5-D	14,612	957
U of Arkansas at Little Rock, Little Rock, AR 72204-1099	1927	$3,026 (S)	NA	5-D	10,541	701
U of Arkansas at Pine Bluff, Pine Bluff, AR 71601-2799.	1873	$2,046 (S)	$3,724	5-M	3,069	213

Name, address	Year	Tuition & Fees	Room & Board	Control, Degree	Enrollment	Faculty
U of Baltimore, Baltimore, MD 21201-5779	1925	$3,804 (S)	NA	5-F	4,624	249
U of Bridgeport, Bridgeport, CT 06601	1927	$13,644	$6,810	1-D	2,503	311
U of California, Berkeley, Berkeley, CA 94720-1500	1868	$4,355 (S)	$7,788	5-D	31,011	1,787
U of California, Davis, Davis, CA 95616	1905	$4,332 (S)	$6,831	5-D	24,865	1,602
U of California, Irvine, Irvine, CA 92697	1965	$4,050 (S)	$5,733	5-D	18,142	997
U of California, Los Angeles, Los Angeles, CA 90095	1919	$4,050 (S)	$6,545	5-D	35,796	3,435
U of California, Riverside, Riverside, CA 92521-0102	1954	$4,126 (S)	$6,579	5-D	10,602	620
U of California, San Diego, La Jolla, CA 92093-5003	1959	$4,200 (S)	$6,897	5-D	19,370	1,598
U of California, Santa Barbara, Santa Barbara, CA 93106	1909	$4,098 (S)	$6,899	5-D	19,360	938
U of California, Santa Cruz, Santa Cruz, CA 95064	1965	$4,181 (S)	$6,900	5-D	10,981	591
U of Central Arkansas, Conway, AR 72035-0001	1907	$2,692 (S)	$3,000	5-D	8,695	450
U of Central Florida, Orlando, FL 32816	1963	$2,025 (S)	$3,735	5-D	30,206	NA
U of Central Oklahoma, Edmond, OK 73034-5209	1890	$1,806 (S)	$2,690	5-M	13,927	701
U of Central Texas, Killeen, TX 76540-1416	1973	$3,224	$2,742	1-M	1,157	56
U of Charleston, Charleston, WV 25304-1099	1888	$11,600	$4,358	1-M	1,337	113
U of Chicago, Chicago, IL 60637-1513	1891	$22,476	$7,604	1-D	12,140	1,779
U of Cincinnati, Cincinnati, OH 45221-0091	1819	$4,359 (S)	$5,958	5-D	28,162	1,195
U of Colorado at Boulder, Boulder, CO 80309	1876	$2,939 (S)	$4,908	5-D	25,126	1,827
U of Colorado at Colorado Springs, Colorado Springs, CO 80933	1965	$2,520 (S)	$4,790	5-D	6,596	261
U of Colorado at Denver, Denver, CO 80217-3364	1912	$2,204 (S)	NA	5-D	13,772	668
U of Colorado Health Scis Ctr, Denver, CO 80262	1883	$3,079 (S)	NA	5-D	2,530	1,701
U of Connecticut, Storrs, CT 06269	1881	$5,096 (S)	$5,544	5-D	18,140	1,123
U of Dallas, Irving, TX 75062-4799	1955	$12,144	$5,324	2-D	3,086	232
U of Dayton, Dayton, OH 45469-1300	1850	$14,670	$4,670	2-D	10,185	797
U of Delaware, Newark, DE 19716	1743	$4,574 (S)	$4,952	12-D	18,574	1,016
U of Denver, Denver, CO 80208	1864	$17,886	$5,877	1-D	8,870	425
U of Detroit Mercy, Detroit, MI 48219-0900	1877	$12,986	$5,210	2-D	6,626	511
U of Dubuque, Dubuque, IA 52001-5099	1852	$12,640	$4,440	2-F	1,032	73
U of Evansville, Evansville, IN 47722-0002	1854	$13,880	$4,910	2-M	2,917	193
The U of Findlay, Findlay, OH 45840-3653	1882	$13,878	$5,510	2-M	4,105	300
U of Florida, Gainesville, FL 32611	1853	$1,930 (S)	$4,960	5-D	42,336	1,447
U of Georgia, Athens, GA 30602	1785	$2,838 (S)	$4,390	5-D	30,009	3,075
U of Great Falls, Great Falls, MT 59405	1932	$8,100	$3,520	2-M	1,122	101
U of Guam, Mangilao, GU 96923	1952	NA	NA	7-M	NA	230
U of Hartford, West Hartford, CT 06117-1599	1877	$17,320	$7,230	1-D	6,892	625
U of Hawaii at Hilo, Hilo, HI 96720-4091	1970	$2,186 (S)	$4,849	5-B	2,462	281
U of Hawaii at Manoa, Honolulu, HI 96822	1907	$2,950 (S)	$3,926	5-D	17,004	1,396
U of Houston, Houston, TX 77004	1927	$1,993 (S)	$4,405	5-D	32,296	856
U of Houston–Clear Lake, Houston, TX 77058-1098	1974	$2,106 (S)	NA	5-M	6,806	411
U of Houston–Downtown, Houston, TX 77002-1001	1974	$2,046 (S)	NA	5-B	8,393	453
U of Houston–Victoria, Victoria, TX 77901-4450	1973	$1,776 (S)	NA	5-M	1,512	84
U of Idaho, Moscow, ID 83844-4140	1889	$1,942 (S)	$3,824	5-D	11,437	638
U of Illinois at Chicago, Chicago, IL 60607-7128	1946	$3,898 (S)	$5,690	5-D	24,652	2,719
U of Illinois at Springfield, Springfield, IL 62794-9243	1969	$2,789 (S)	NA	5-M	4,334	269
U of Illinois at Urbana–Champaign, Urbana, IL 61801	1867	$4,120 (S)	$4,978	5-D	36,303	1,834
U of Indianapolis, Indianapolis, IN 46227-3697	1902	$12,990	$4,690	2-D	3,286	322
The U of Iowa, Iowa City, IA 52242-1316	1847	$2,760 (S)	$4,162	5-D	28,705	1,702
U of Kansas, Lawrence, KS 66045	1866	$2,385 (S)	$3,832	5-D	27,625	2,056
U of Kentucky, Lexington, KY 40506-0032	1865	$2,736 (S)	$3,470	5-D	23,707	2,270
U of La Verne, La Verne, CA 91750-4443	1891	$15,160	$4,820	1-D	2,367	NA
U of Louisville, Louisville, KY 40292-0001	1798	$2,630 (S)	$3,360	5-D	20,195	1,785
U of Maine, Orono, ME 04469	1865	$4,344 (S)	$5,084	5-D	9,126	634
U of Maine at Augusta, Augusta, ME 04330-9410	1965	$3,255 (S)	NA	5-B	5,248	207
U of Maine at Farmington, Farmington, ME 04938-1990	1863	$3,520 (S)	$4,522	5-B	2,306	153
U of Maine at Presque Isle, Presque Isle, ME 04769-2888	1903	$3,210 (S)	$3,970	5-B	1,344	93
U of Mary, Bismarck, ND 58504-9652	1959	$8,055	$3,150	2-M	2,148	155
U of Mary Hardin-Baylor, Belton, TX 76513	1845	$6,944	$3,290	2-M	2,479	182
U of Maryland, Baltimore County, Baltimore, MD 21250-5398	1963	$4,570 (S)	$5,194	5-D	10,122	797
U of Maryland, Coll Park, College Park, MD 20742-5045	1856	$4,460 (S)	$5,988	5-D	32,925	1,836
U of Maryland Eastern Shore, Princess Anne, MD 21853-1299	1886	$3,240 (S)	$4,530	5-D	3,204	280
U of Maryland U Coll, College Park, MD 20742-1600	1947	$5,490 (S)	NA	5-M	14,142	655
U of Massachusetts Amherst, Amherst, MA 01003	1863	$5,572 (S)	$4,520	5-D	24,545	1,265
U of Massachusetts Boston, Boston, MA 02125-3393	1964	$4,297 (S)	NA	5-D	13,481	828
U of Massachusetts Dartmouth, North Dartmouth, MA 02747-2300	1895	$1,744 (S)	$4,828	5-D	6,599	418
U of Massachusetts Lowell, Lowell, MA 01854-2881	1894	$4,422 (S)	$4,677	5-D	12,141	598
The U of Memphis, Memphis, TN 38152	1912	$2,412 (S)	NA	5-D	20,100	1,205
U of Miami, Coral Gables, FL 33124	1925	$19,512	$7,606	1-D	13,422	2,309
U of Michigan, Ann Arbor, MI 48109	1817	$5,710 (S)	$5,488	5-D	37,197	3,710
U of Michigan–Dearborn, Dearborn, MI 48128-1491	1959	$4,850 (S)	NA	5-M	8,112	411
U of Michigan–Flint, Flint, MI 48502-1950	1956	$3,559 (S)	NA	5-M	6,656	242
U of Minnesota, Crookston, Crookston, MN 56716-5001	1966	$4,568 (S)	$3,315	5-B	2,492	85
U of Minnesota, Duluth, Duluth, MN 55812-2496	1947	$4,316 (S)	$4,011	5-M	8,765	462
U of Minnesota, Morris, Morris, MN 56267-2134	1959	$4,554 (S)	$3,879	5-B	1,959	120
U of Minnesota, Twin Cities Campus, Minneapolis, MN 55455-0213	1851	$4,450 (S)	$4,311	5-D	46,973	2,722
U of Mississippi, University, MS 38677-9702	1844	$2,631 (S)	$2,710	5-D	11,337	509
U of Mississippi Medical Ctr, Jackson, MS 39216-4505	1955	$2,106 (S)	NA	5-D	1,862	644
U of Missouri–Columbia, Columbia, MO 65211	1839	$4,280 (S)	$4,454	5-D	22,780	1,675
U of Missouri–Kansas City, Kansas City, MO 64110-2499	1929	$4,278 (S)	$4,450	5-D	10,610	867
U of Missouri–Rolla, Rolla, MO 65409-0910	1870	$4,194 (S)	$4,363	5-D	4,918	401
U of Missouri–St. Louis, St. Louis, MO 63121-4499	1963	$4,396 (S)	$4,603	5-D	15,880	1,035
U of Mobile, Mobile, AL 36663-0220	1961	$7,260	$4,080	2-M	1,918	161
The U of Montana–Missoula, Missoula, MT 59812-0002	1893	$2,630 (S)	$4,212	5-D	12,138	649
U of Montevallo, Montevallo, AL 35115	1896	$3,180 (S)	$3,332	5-M	3,177	209
U of Nebraska at Kearney, Kearney, NE 68849-0001	1903	$2,269 (S)	$3,180	5-M	6,849	430
U of Nebraska at Omaha, Omaha, NE 68182	1908	$2,356 (S)	NA	5-D	13,274	784
U of Nebraska–Lincoln, Lincoln, NE 68588	1869	$2,829 (S)	$3,865	5-D	22,408	1,531
U of Nebraska Medical Ctr, Omaha, NE 68198	1869	$2,732 (S)	NA	5-D	2,599	740
U of Nevada, Las Vegas, Las Vegas, NV 89154-9900	1957	$1,642 (S)	$5,300	5-D	21,001	1,168

Name, address	Year	Tuition & Fees	Room & Board	Control, Degree	Enroll-ment	Faculty
U of Nevada, Reno, Reno, NV 89557	1874	$2,109 (S)	$5,060	5-D	12,303	651
U of New England, Biddeford, ME 04005-9526	1831	$14,085	$5,820	1-F	2,606	237
U of New Hampshire, Durham, NH 03824	1866	$5,889 (S)	$4,636	5-D	13,408	729
U of New Haven, West Haven, CT 06516-1916	1920	$13,100	$6,220	1-D	4,953	617
U of New Mexico, Albuquerque, NM 87131-2039	1889	$2,165 (S)	$4,382	5-D	23,852	2,164
U of New Orleans, New Orleans, LA 70148	1958	$2,512 (S)	$3,150	5-D	15,629	730
U of North Alabama, Florence, AL 35632-0001	1830	$2,184 (S)	$3,352	5-M	5,773	286
U of North Carolina at Asheville, Asheville, NC 28804-3299	1927	$1,834 (S)	$3,870	5-M	3,175	287
The U of North Carolina at Chapel Hill, Chapel Hill, NC 27599	1789	$2,165 (S)	$4,830	5-D	24,255	2,640
U of North Carolina at Charlotte, Charlotte, NC 28223-0001	1946	$1,718 (S)	$3,670	5-D	16,861	941
U of North Carolina at Greensboro, Greensboro, NC 27402-6170	1891	$2,021 (S)	$3,981	5-D	12,700	758
U of North Carolina at Pembroke, Pembroke, NC 28372-1510	1887	$1,510 (S)	$3,202	5-M	2,998	209
U of North Carolina at Wilmington, Wilmington, NC 28403-3201	1947	$1,796 (S)	$4,420	5-M	9,643	543
U of North Dakota, Grand Forks, ND 58202	1883	$2,946 (S)	$3,243	5-D	10,369	732
U of Northern Colorado, Greeley, CO 80639	1890	$2,578 (S)	$4,570	5-D	11,860	563
U of Northern Iowa, Cedar Falls, IA 50614	1876	$2,752 (S)	$3,636	5-D	13,545	820
U of North Florida, Jacksonville, FL 32224-2645	1965	$2,006 (S)	$4,634	5-D	11,537	645
U of North Texas, Denton, TX 76203	1890	$2,187 (S)	$3,938	5-D	25,514	990
U of Notre Dame, Notre Dame, IN 46556	1842	$19,947	$5,366	2-D	10,301	NA
U of Oklahoma, Norman, OK 73019-0390	1890	$2,311 (S)	$4,314	5-D	23,862	1,923
U of Oregon, Eugene, OR 97403	1872	$3,408 (S)	$5,100	5-D	17,318	1,154
U of Osteopathic Medicine & Health Scis, Des Moines, IA 50312	1898	$12,070	NA	1-F	1,155	108
U of Pennsylvania, Philadelphia, PA 19104	1740	$22,250	$7,700	1-D	NA	3,696
U of Phoenix, Phoenix, AZ 85072-2069	1976	$5,352	NA	3-M	41,467	4,621
U of Pittsburgh, Pittsburgh, PA 15260	1787	$6,164 (S)	$5,598	12-D	25,873	3,468
U of Pittsburgh at Greensburg, Greensburg, PA 15601-5860	1963	$6,074 (S)	$4,380	12-B	1,536	95
U of Pittsburgh at Johnstown, Johnstown, PA 15904-2990	1927	$6,154 (S)	$5,060	12-B	3,170	188
U of Portland, Portland, OR 97203-5798	1901	$15,520	$4,990	2-M	2,770	256
U of Puerto Rico, Aguadilla U Coll, Aguadilla, PR 00604-0160	1972	$1,638 (S)	NA	6-B	3,312	133
U of Puerto Rico at Arecibo, Arecibo, PR 00614-4010	1967	$1,559 (S)	NA	6-B	4,580	268
U of Puerto Rico at Ponce, Ponce, PR 00732-7186	1970	$1,090 (S)	NA	6-B	4,202	186
U of Puerto Rico, Cayey U Coll, Cayey, PR 00737	1967	$970 (S)	NA	6-B	3,944	246
U of Puerto Rico, Humacao U Coll, Humacao, PR 00791	1962	$1,020 (S)	NA	6-B	4,508	286
U of Puerto Rico, Mayagüez Campus, Mayagüez, PR 00681-5000	1911	NA	$4,800	6-D	12,883	801
U of Puerto Rico, Medical Scis Campus, San Juan, PR 00936-5067 (4)	1950	$1,200 (S)	NA	6-D	2,822	750
U of Puerto Rico, Río Piedras, San Juan, PR 00931	1903	NA	NA	6-D	21,385	1,293
U of Puget Sound, Tacoma, WA 98416-0005	1888	$18,940	$5,070	1-M	2,978	259
U of Redlands, Redlands, CA 92373-0999	1907	$18,545	$7,224	1-M	1,579	581
U of Rhode Island, Kingston, RI 02881	1892	$4,592 (S)	$5,962	5-D	14,319	626
U of Richmond, U of Richmond, VA 23173	1830	$17,570	$4,143	1-F	3,683	339
U of Rio Grande, Rio Grande, OH 45674	1876	$3,589 (A)	$4,800	1-M	2,000	111
U of Rochester, Rochester, NY 14627-0250	1850	$21,020	$7,289	1-D	8,100	1,335
U of St. Francis, Joliet, IL 60435-6169	1920	$11,950	$4,960	2-M	2,556	139
U of Saint Francis, Fort Wayne, IN 46808-3994	1890	$10,710	$4,540	2-M	1,726	175
U of St. Thomas, St. Paul, MN 55105-1096	1885	$14,660	$5,100	2-D	10,790	719
U of St. Thomas, Houston, TX 77006-4696	1947	$10,550	$4,730	2-D	2,696	215
U of San Diego, San Diego, CA 92110-2492	1949	$15,780	$7,600	2-D	6,753	543
U of San Francisco, San Francisco, CA 94117-1080	1855	$15,950	$7,610	2-D	7,990	886
U of Sci & Arts of Oklahoma, Chickasha, OK 73018-0001	1908	$1,368 (S)	$2,070	5-B	1,498	67
The U of Scranton, Scranton, PA 18510	1888	$15,880	$7,406	2-M	4,711	383
U of Sioux Falls, Sioux Falls, SD 57105-1699	1883	$10,750	$3,500	2-M	1,039	71
U of South Alabama, Mobile, AL 36688-0002	1963	$2,838 (S)	$2,882	5-D	11,445	946
U of South Carolina, Columbia, SC 29208	1801	$3,534 (S)	$3,966	5-D	25,250	1,449
U of South Carolina Aiken, Aiken, SC 29801-6309	1961	$3,014 (S)	$3,890	5-M	3,179	219
U of South Carolina Spartanburg, Spartanburg, SC 29303-4999	1967	$3,014 (S)	$3,550	5-M	3,729	233
U of South Dakota, Vermillion, SD 57069-2390	1862	$3,012 (S)	$2,988	5-D	7,293	472
U of Southern California, Los Angeles, CA 90089	1880	$20,480	$6,978	1-D	28,739	2,583
U of Southern Colorado, Pueblo, CO 81001-4901	1933	$2,191 (S)	$4,656	5-M	5,296	337
U of Southern Indiana, Evansville, IN 47712-3590	1965	$2,705 (S)	NA	5-M	8,415	453
U of Southern Maine, Portland, ME 04104-9300	1878	$3,938 (S)	$4,987	5-F	10,462	596
U of Southern Mississippi, Hattiesburg, MS 39406-5167	1910	$2,590 (S)	$2,976	5-D	14,537	726
U of South Florida, Tampa, FL 33620-9951	1956	$2,086 (S)	$4,734	5-D	33,654	1,607
U of Southwestern Louisiana, Lafayette, LA 70504	1898	$1,947 (S)	$2,592	5-D	16,933	522
The U of Tampa, Tampa, FL 33606-1490	1931	$14,652	$4,850	1-M	2,990	230
U of Tennessee at Chattanooga, Chattanooga, TN 37403-2598	1886	$2,200 (S)	NA	5-M	8,682	595
The U of Tennessee at Martin, Martin, TN 38238-1000	1900	$2,240 (S)	$3,396	5-M	5,837	230
U of Tennessee, Knoxville, TN 37996	1794	$2,576 (S)	$3,916	5-D	26,784	1,257
The U of Texas at Arlington, Arlington, TX 76013	1895	$2,088 (S)	NA	5-D	18,662	1,141
The U of Texas at Austin, Austin, TX 78712-1111	1883	$2,866 (S)	$4,537	5-D	48,906	2,509
The U of Texas at Brownsville, Brownsville, TX 78520-4991	1973	NA	NA	5-M	NA	399
The U of Texas at Dallas, Richardson, TX 75083-0688	1969	$2,414 (S)	NA	5-D	9,517	463
The U of Texas at El Paso, El Paso, TX 79968-0001	1913	$2,266 (S)	NA	5-D	14,677	795
The U of Texas at San Antonio, San Antonio, TX 78249-0617	1969	$2,744 (S)	NA	5-D	18,397	858
The U of Texas at Tyler, Tyler, TX 75799-0001	1971	$2,084 (S)	NA	5-M	3,393	219
The U of Texas–Houston Health Sci Ctr, Houston, TX 77225-0036	1972	$2,455 (S)	NA	5-D	3,089	1,046
The U of Texas Medical Branch at Galveston, Galveston, TX 77555	1891	$1,674 (S)	NA	5-D	1,987	1,202
The U of Texas of the Permian Basin, Odessa, TX 79762-0001	1969	NA	NA	5-M	NA	128
The U of Texas–Pan American, Edinburg, TX 78539-2999	1927	$1,973 (S)	$2,663	5-D	12,373	633
The U of the Arts, Philadelphia, PA 19102-4944	1870	$15,070	NA	1-M	1,806	323
U of the District of Columbia, Washington, DC 20008-1175	1976	$2,360 (S)	NA	9-M	5,247	615
U of the Incarnate Word, San Antonio, TX 78209-6397	1881	$10,840	$4,430	2-D	3,583	332
U of the Pacific, Stockton, CA 95211-0197	1851	$19,365	$5,770	1-D	5,554	592
U of the Sacred Heart, San Juan, PR 00914-0383	1935	$4,095	NA	2-M	5,149	325
U of the Scis in Philadelphia, Philadelphia, PA 19104-4495	1821	$13,290	$6,540	1-D	2,225	205
U of the South, Sewanee, TN 37383-1000	1857	$17,730	$5,030	2-D	1,507	158
U of the Virgin Islands, Charlotte Amalie, VI 00802-9990	1962	$2,676 (S)	$5,466	7-M	2,638	240
U of Toledo, Toledo, OH 43606-3398	1872	$3,952 (S)	$4,340	5-D	20,411	1,210
U of Tulsa, Tulsa, OK 74104-3189	1894	$12,930	$4,540	2-D	4,246	302

Name, address	Year	Tuition & Fees	Room & Board	Control, Degree	Enrollment	Faculty
U of Utah, Salt Lake City, UT 84112-1107	1850	$2,601 (S)	$4,801	5-D	25,213	1,360
U of Vermont, Burlington, VT 05405-0160	1791	$7,550 (S)	$5,440	5-D	10,337	1,055
U of Virginia, Charlottesville, VA 22903	1819	$4,786 (S)	$4,421	5-D	22,099	1,871
U of Washington, Seattle, WA 98195	1861	$3,366 (S)	$4,779	5-D	35,108	3,892
The U of West Alabama, Livingston, AL 35470	1835	$2,568 (S)	$2,952	5-M	2,147	130
U of West Florida, Pensacola, FL 32514-5750	1963	$1,985 (S)	NA	5-D	8,003	213
U of Wisconsin–Eau Claire, Eau Claire, WI 54702-4004	1916	$2,872 (S)	$3,133	5-M	10,674	513
U of Wisconsin–Green Bay, Green Bay, WI 54311-7001	1968	$2,738 (S)	NA	5-M	5,601	285
U of Wisconsin–La Crosse, La Crosse, WI 54601-3742	1909	$2,859 (S)	$3,250	5-M	9,318	478
U of Wisconsin–Madison, Madison, WI 53706-1380	1848	$3,242 (S)	$4,880	5-D	NA	2,545
U of Wisconsin–Milwaukee, Milwaukee, WI 53201-0413	1956	$3,327 (S)	$3,696	5-D	21,525	1,381
U of Wisconsin–Oshkosh, Oshkosh, WI 54901-8602	1871	$2,609 (S)	$2,938	5-M	10,960	538
U of Wisconsin–Parkside, Kenosha, WI 53141-2000	1968	$2,705 (S)	$4,200	5-M	4,582	262
U of Wisconsin–Platteville, Platteville, WI 53818-3099	1866	NA	$3,334	5-M	5,135	280
U of Wisconsin–River Falls, River Falls, WI 54022-5001	1874	$2,750 (S)	$3,036	5-M	5,617	318
U of Wisconsin–Stevens Point, Stevens Point, WI 54481-3897	1894	$2,790 (S)	$3,266	5-M	8,514	427
U of Wisconsin–Stout, Menomonie, WI 54751	1891	$2,619 (S)	$3,254	5-M	7,604	392
U of Wisconsin–Superior, Superior, WI 54880-4500	1893	$2,652 (S)	$3,298	5-M	2,593	160
U of Wisconsin–Whitewater, Whitewater, WI 53190-1790	1868	$2,772 (S)	$3,790	5-M	10,632	609
U of Wyoming, Laramie, WY 82071	1886	$2,330 (S)	$4,278	5-D	11,124	666
U System Coll for Lifelong Learning, Concord, NH 03301	1972	$4,535 (S)	NA	11-B	2,149	350
Upper Iowa U, Fayette, IA 52142-1857	1857	$9,750	$3,958	1-M	4,311	182
Urbana U, Urbana, OH 43078-2091	1850	$10,530	$4,500	2-M	1,144	79
Ursinus Coll, Collegeville, PA 19426-1000	1869	$17,380	$5,880	2-B	1,229	143
Ursuline Coll, Pepper Pike, OH 44124-4398 (4)	1871	$12,128	$4,460	2-M	1,306	133
Utah State U, Logan, UT 84322	1888	$2,175 (S)	NA	5-D	19,322	685
Utica Coll of Syracuse U, Utica, NY 13502-4892	1946	$14,912	$5,932	1-M	1,973	183
Valdosta State U, Valdosta, GA 31698	1906	$1,974 (S)	$3,780	5-D	9,386	530
Valparaiso U, Valparaiso, IN 46383-6493	1859	$15,060	$4,130	2-F	3,720	399
Vanderbilt U, Nashville, TN 37240-1001	1873	$21,478	$7,738	1-D	9,996	2,133
Vanguard U of Southern California, Costa Mesa, CA 92626-6597	1920	$11,848	$4,980	2-M	1,315	158
Vassar Coll, Poughkeepsie, NY 12604	1861	$22,090	$6,620	1-M	2,396	258
Villa Julie Coll, Stevenson, MD 21153	1952	$9,240	NA	1-M	2,086	176
Villanova U, Villanova, PA 19085-1699	1842	$19,133	$7,940	2-D	9,951	870
Virginia Commonwealth U, Richmond, VA 23284-9005	1838	$4,111 (S)	$4,624	5-D	23,125	2,560
Virginia Military Inst, Lexington, VA 24450	1839	$6,380 (S)	$4,210	5-B	1,328	144
Virginia Polytechnic Inst & State U, Blacksburg, VA 24061	1872	$4,147 (S)	$3,648	5-D	25,608	1,578
Virginia State U, Petersburg, VA 23806-0001	1882	$3,307 (S)	$5,006	5-M	4,341	241
Virginia Union U, Richmond, VA 23220-1170	1865	$8,980	$4,120	2-D	1,700	104
Virginia Wesleyan Coll, Norfolk, VA 23502-5599	1961	$13,400	$5,550	2-B	1,420	114
Viterbo Coll, La Crosse, WI 54601-4797	1890	$11,150	$4,250	2-M	2,574	227
Wagner Coll, Staten Island, NY 10301-4495	1883	$16,000	$6,200	1-M	2,050	172
Wake Forest U, Winston-Salem, NC 27109	1834	$19,450	$5,500	2-D	6,082	1,655
Walla Walla Coll, College Place, WA 99324-1198	1892	$12,693	$3,447	2-M	1,675	184
Walsh Coll of Accountancy & Business Administration, Troy, MI 48007-7006	1922	$4,758	NA	1-M	3,112	127
Walsh U, North Canton, OH 44720-3396	1958	$10,680	$5,110	2-M	1,555	122
Wartburg Coll, Waverly, IA 50677-1003	1852	$13,610	$4,100	2-B	1,541	142
Washburn U of Topeka, Topeka, KS 66621	1865	$3,150 (S)	$3,300	10-F	6,065	440
Washington & Jefferson Coll, Washington, PA 15301-4801	1781	$18,000	$4,350	1-B	1,225	101
Washington & Lee U, Lexington, VA 24450-0303	1749	$16,195	$5,800	1-F	2,066	254
Washington Coll, Chestertown, MD 21620-1197	1782	$18,250	$5,740	1-M	1,144	103
Washington State U, Pullman, WA 99164	1890	$3,270 (S)	$4,540	5-D	20,641	1,261
Washington U in St. Louis, St. Louis, MO 63130-4899	1853	$22,422	$6,778	1-D	12,035	2,515
Wayland Baptist U, Plainview, TX 79072-6998	1908	$6,200	$3,140	2-M	4,295	286
Waynesburg Coll, Waynesburg, PA 15370-1222	1849	$10,550	$4,430	2-M	1,318	105
Wayne State Coll, Wayne, NE 68787	1910	$2,140 (S)	$3,170	5-M	3,835	231
Wayne State U, Detroit, MI 48202	1868	$3,399 (S)	NA	5-D	31,203	2,762
Weber State U, Ogden, UT 84408-1001	1889	$1,935 (S)	$3,605	5-M	13,900	464
Webster U, St. Louis, MO 63119-3194	1915	$10,910	$5,332	1-D	11,853	662
Wellesley Coll, Wellesley, MA 02481 (3)	1870	$21,660	$6,990	1-B	2,287	332
Wentworth Inst of Technology, Boston, MA 02115-5998 (2)	1904	$11,500	$6,400	1-B	3,076	166
Wesleyan U, Middletown, CT 06459-0260	1831	$22,980	$6,380	1-D	3,204	303
Wesley Coll, Dover, DE 19901-3875	1873	$11,709	$5,019	2-M	1,298	77
West Chester U of Pennsylvania, West Chester, PA 19383	1871	$4,162 (S)	$4,460	5-M	11,578	734
Western Carolina U, Cullowhee, NC 28723	1889	$1,839 (S)	$3,228	5-D	6,286	486
Western Connecticut State U, Danbury, CT 06810-6885	1903	$2,062 (S)	$5,170	5-M	5,372	357
Western Illinois U, Macomb, IL 61455-1390	1899	$3,037 (S)	$4,188	5-M	12,610	644
Western Intl U, Phoenix, AZ 85021-2718	1978	$5,280	NA	3-M	2,506	100
Western Kentucky U, Bowling Green, KY 42101-3576	1906	$2,140 (S)	$2,800	5-M	14,866	913
Western Maryland Coll, Westminster, MD 21157-4390	1867	$17,730	$5,350	1-M	3,234	216
Western Michigan U, Kalamazoo, MI 49008	1903	$3,655 (S)	$4,591	5-D	26,575	1,093
Western Montana Coll of The U of Montana, Dillon, MT 59725-3598	1893	$2,036 (S)	$3,574	5-B	1,053	75
Western New England Coll, Springfield, MA 01119-2654	1919	$11,448	$6,544	1-F	4,941	314
Western New Mexico U, Silver City, NM 88062-0680	1893	$855 (S)	$2,786	5-M	2,580	145
Western Oregon U, Monmouth, OR 97361-1394	1856	$3,153 (S)	$4,410	5-M	4,283	239
Western State Coll of Colorado, Gunnison, CO 81231	1901	$2,152 (S)	$4,744	5-B	2,432	143
Western Washington U, Bellingham, WA 98225-5996	1893	$2,772 (S)	$4,728	5-M	11,655	584
Westfield State Coll, Westfield, MA 01086	1838	$3,094 (S)	$4,554	5-M	NA	NA
West Liberty State Coll, West Liberty, WV 26074	1837	$2,200 (S)	$3,200	5-B	2,475	137
Westminster Coll, New Wilmington, PA 16172-0001	1852	$15,430	$4,315	2-M	1,577	122
Westminster Coll, Salt Lake City, UT 84105-3697	1875	$11,246	$4,358	1-M	2,217	212
Westmont Coll, Santa Barbara, CA 93108-1099	1937	$17,998	$6,382	2-B	1,316	139
West Texas A&M U, Canyon, TX 79016-0001	1909	$1,744 (S)	$3,195	5-M	6,348	362
West Virginia State Coll, Institute, WV 25112-1000	1891	$2,184 (S)	$3,550	5-B	4,817	276
West Virginia U, Morgantown, WV 26506	1867	$2,336 (S)	$4,832	5-D	23,233	1,575
West Virginia U Inst of Technology, Montgomery, WV 25136	1895	$2,370 (S)	$4,048	5-M	2,508	179
West Virginia Wesleyan Coll, Buckhannon, WV 26201	1890	$16,750	$4,100	2-M	1,642	149
Wheaton Coll, Wheaton, IL 60187-5593	1860	$13,780	$4,910	2-D	2,760	253

Name, address	Year	Tuition & Fees	Room & Board	Control, Degree	Enroll- ment	Faculty
Wheaton Coll, Norton, MA 02766	1834	$20,820	$6,620	1-B	1,451	137
Wheeling Jesuit U, Wheeling, WV 26003-6295	1954	$14,200	$5,280	2-M	1,530	102
Wheelock Coll, Boston, MA 02215 (4)	1888	$15,520	$6,300	1-M	1,213	115
Whitman Coll, Walla Walla, WA 99362-2083	1859	$19,756	$5,750	1-B	1,387	175
Whittier Coll, Whittier, CA 90608-0634	1887	$18,634	$6,594	1-F	2,164	128
Whitworth Coll, Spokane, WA 99251-0001	1890	$15,593	$5,300	2-M	2,076	102
Wichita State U, Wichita, KS 67260	1895	$2,489 (S)	$3,865	5-D	14,350	491
Widener U, Chester, PA 19013-5792	1821	$14,380	$6,850	1-D	7,252	441
Wilkes U, Wilkes-Barre, PA 18766-0002	1933	$15,091	$6,830	1-D	2,200	225
Willamette U, Salem, OR 97301-3931	1842	$20,290	$5,530	2-F	2,446	260
William Carey Coll, Hattiesburg, MS 39401-5499	1906	$6,624	$1,790	2-M	NA	172
William Jewell Coll, Liberty, MO 64068-1843	1849	$11,850	$3,900	2-B	1,160	168
William Paterson U of New Jersey, Wayne, NJ 07470-8420	1855	$3,786 (S)	$5,320	5-M	9,384	335
William Penn Coll, Oskaloosa, IA 52577-1799	1873	$11,490	$3,840	2-B	1,096	63
Williams Coll, Williamstown, MA 01267	1793	$22,990	$6,520	1-M	2,108	250
William Woods U, Fulton, MO 65251-1098	1870	$12,300	$5,200	2-M	1,509	86
Wilmington Coll, New Castle, DE 19720-6491	1967	$5,750	NA	1-D	4,607	520
Wilmington Coll, Wilmington, OH 45177	1870	$12,500	$4,680	2-B	1,093	108
Wingate U, Wingate, NC 28174-0159	1896	$11,690	$4,300	2-M	1,230	105
Winona State U, Winona, MN 55987-5838	1858	$3,019 (S)	$3,150	5-M	6,753	350
Winston-Salem State U, Winston-Salem, NC 27110-0003	1892	$1,574 (S)	$3,403	5-B	2,778	250
Winthrop U, Rock Hill, SC 29733	1886	$3,938 (S)	$3,904	5-M	5,591	444
Wittenberg U, Springfield, OH 45501-0720	1845	$19,140	$5,006	2-B	2,088	166
Wofford Coll, Spartanburg, SC 29303-3663	1854	$15,390	$4,605	2-B	1,081	85
Woodbury U, Burbank, CA 91510	1884	$15,170	$5,728	1-M	1,123	209
Worcester Polytechnic Inst, Worcester, MA 01609-2280	1865	$18,910	$6,578	1-D	3,821	273
Worcester State Coll, Worcester, MA 01602-2597	1874	$2,615 (S)	$4,140	5-M	5,323	247
Wright State U, Dayton, OH 45435	1964	$3,708 (S)	$4,916	5-D	17,000	950
Xavier U, Cincinnati, OH 45207-2111	1831	$14,520	$5,900	2-D	6,407	530
Xavier U of Louisiana, New Orleans, LA 70125-1098	1925	$8,215	$4,900	2-F	3,655	250
Yale U, New Haven, CT 06520	1701	$23,100	$7,050	1-D	11,032	2,986
Yeshiva U, New York, NY 10033-3201	1886	$14,590	$4,770	1-D	5,481	1,092
York Coll of Pennsylvania, York, PA 17405-7199	1787	$6,100	$4,550	1-M	5,105	476
York Coll of the City U of New York, Jamaica, NY 11451-0001	1967	$3,292 (S)	NA	11-B	5,649	451
Youngstown State U, Youngstown, OH 44555-0001	1908	$3,558 (S)	$4,560	5-D	12,533	801

Two-Year Colleges

Unless otherwise indicated, the highest undergraduate degree offered by two-year colleges is the associate degree. Figures for Room & Board are given where applicable.

Name, address	Year	Tuition & Fees	Room & Board	Control, Degree	Enroll- ment	Faculty
Abraham Baldwin Ag Coll, Tifton, GA 31794-2601	1933	$1,490 (S)	$3,000	5	2,631	118
Adirondack Comm Coll, Queensbury, NY 12804	1960	$2,268 (S)		11	3,379	244
Aiken Tech Coll, Aiken, SC 29802-0696	1972	$1,018 (S)		11	2,463	101
Aims Comm Coll, Greeley, CO 80632-0069	1967	$1,329 (A)		9	6,637	370
Alabama Southern Comm Coll, Monroeville, AL 36461	1965	$1,792 (S)		5	1,600	107
Alamance Comm Coll, Graham, NC 27253-8000	1959	$591 (S)		5	3,504	240
Albuquerque Tech Voc Inst, Albuquerque, NM 87106-4096	1965	$741 (S)		5	16,239	919
Alexandria Tech Coll, Alexandria, MN 56308-3707	1961	$2,355 (S)		5	1,825	89
Allan Hancock Coll, Santa Maria, CA 93454-6399	1920	$390 (S)		11	8,851	497
Allegany Coll of Maryland, Cumberland, MD 21502-2596	1961	$2,465 (A)		11	2,687	210
Allen County Comm Coll, Iola, KS 66749-1607	1923	$1,504 (S)	$2,900	11	1,950	140
Allentown Business School, Allentown, PA 18103-3880	1869	$7,305		3	1,050	50
Alpena Comm Coll, Alpena, MI 49707-1495	1952	$1,790 (A)		11	2,115	89
Alvin Comm Coll, Alvin, TX 77511-4898	1949	$742 (A)		11	2,505	228
Amarillo Coll, Amarillo, TX 79178-0001	1929	$510 (A)		11	7,558	424
American River Coll, Sacramento, CA 95841-4286	1955	$390 (S)		9	21,373	790
Angelina Coll, Lufkin, TX 75902-1768	1968	$657 (A)		11	3,970	201
Anne Arundel Comm Coll, Arnold, MD 21012-1895	1961	$1,873 (A)		11	12,113	570
Anoka-Ramsey Comm Coll, Coon Rapids, MN 55433-3470	1965	$2,338 (S)		5	4,387	170
Anson Comm Coll, Polkton, NC 28135-0126	1962	$581 (S)		5	1,676	108
Antelope Valley Coll, Lancaster, CA 93536-5426	1929	$390 (S)		11	9,578	400
Arapahoe Comm Coll, Littleton, CO 80160-9002	1965	$1,784 (S)		5	7,495	298
Arizona Western Coll, Yuma, AZ 85366-0929	1962	$840 (S)	$2,850	11	5,889	411
Arkansas State U–Beebe, Beebe, AR 72012-1000	1927	$1,128 (S)	$2,180	5	3,000	120
The Art Inst of Atlanta, Atlanta, GA 30328	1949	$9,984		3-B	1,721	91
The Art Inst of Dallas, Dallas, TX 75231-9959	1978	$10,800		3	1,300	84
The Art Inst of Fort Lauderdale, Fort Lauderdale, FL 33316-3000	1968	$9,945		3-B	2,654	150
The Art Inst of Houston, Houston, TX 77056-4115	1978	$10,890		3	1,685	78
The Art Inst of Philadelphia, Philadelphia, PA 19103-5198	1966	$10,827		3	2,374	168
The Art Inst of Pittsburgh, Pittsburgh, PA 15222-3269	1921	$10,305	$5,925	3	2,601	107
The Art Inst of Seattle, Seattle, WA 98121-1642	1982	$10,260		3	2,720	200
Asheville-Buncombe Tech Comm Coll, Asheville, NC 28801-4897	1959	$586 (S)		5	4,538	435
Ashland Comm Coll, Ashland, KY 41101-3683	1937	$1,100 (S)		5	2,140	144
Asnuntuck Community-Technical Coll, Enfield, CT 06082-3800	1972	$1,814 (S)		5	1,913	106
Athens Area Tech Inst, Athens, GA 30601-1500	1958	$849 (S)		5	1,747	100
Atlanta Metro Coll, Atlanta, GA 30310-4498	1974	$1,300 (S)		5	1,924	96
Atlantic Cape Comm Coll, Mays Landing, NJ 08330-2699	1966	$2,070 (A)		8	4,881	260
Augusta Tech Inst, Augusta, GA 30906	1961	$822 (S)		5	2,815	189
Austin Comm Coll, Austin, TX 78752-4390	1972	$1,110 (A)		9	29,000	1,406
Bainbridge Coll, Bainbridge, GA 31717	1972	$1,256 (S)		5	1,050	64
Bakersfield Coll, Bakersfield, CA 93305-1299	1913	$392 (S)		11	12,951	496
Baltimore City Comm Coll, Baltimore, MD 21215-7893	1947	$2,050 (S)		5	5,974	429
Barstow Coll, Barstow, CA 92311-6699	1959	$390 (S)		11	3,420	113
Barton County Comm Coll, Great Bend, KS 67530-9283	1969	$1,408 (S)	$2,904	11	5,318	292
Bay de Noc Comm Coll, Escanaba, MI 49829-2511	1963	$1,754 (A)		8	2,240	143

Name, address	Year	Tuition & Fees	Room & Board	Control, Degree	Enroll-ment	Faculty
Beaufort County Comm Coll, Washington, NC 27889-1069	1967	$578 (S)		5	1,562	110
Belleville Area Coll, Belleville, IL 62221-5899	1946	$1,044 (A)		9	13,878	1,117
Bellevue Comm Coll, Bellevue, WA 98007-6484	1966	$1,446 (S)		5	10,321	572
Belmont Tech Coll, St. Clairsville, OH 43950-9735	1971	$2,223 (S)		5	1,636	101
Bergen Comm Coll, Paramus, NJ 07652-1595	1965	$2,555 (A)		8	11,812	610
Berkeley Coll, West Paterson, NJ 07424-3353	1931	$11,835	$7,485	3	2,035	127
Berkeley Coll, New York, NY 10017-4604	1936	$11,835		3-B	1,521	106
Berkshire Comm Coll, Pittsfield, MA 01201-5786	1960	$2,112 (S)		5	2,328	180
Bessemer State Tech Coll, Bessemer, AL 35021-0308	1966	$1,560 (S)		5	1,417	105
Bevill State Comm Coll, Sumiton, AL 35148	1969	$1,275 (S)		5	3,616	63
Big Bend Comm Coll, Moses Lake, WA 98837-3299	1962	$1,515 (S)	$3,700	5	1,888	170
Bishop State Comm Coll, Mobile, AL 36603-5898	1965	$1,437 (S)		5	3,660	217
Bismarck State Coll, Bismarck, ND 58506-5587	1939	$1,809 (S)	$2,970	5	2,594	118
Black Hawk Coll, Moline, IL 61265-5899	1946	$1,696 (A)		11	5,500	534
Blackhawk Tech Coll, Janesville, WI 53547-5009	1968	$1,798 (S)		9	3,381	293
Black River Tech Coll, Pocahontas, AR 72455	1972	$1,178 (A)		5	1,156	70
Blinn Coll, Brenham, TX 77833-4049	1883	$1,030 (A)	$2,950	11	10,481	410
Blue Mountain Comm Coll, Pendleton, OR 97801-1000	1962	$1,602 (S)		11	1,734	256
Blue Ridge Comm Coll, Flat Rock, NC 28731-9624	1969	$590 (S)		11	1,739	207
Blue Ridge Comm Coll, Weyers Cave, VA 24486-0080	1967	$1,475 (S)		5	2,772	143
Bossier Parish Comm Coll, Bossier City, LA 71111-5801	1967	$1,110 (S)		5	3,920	158
Bowling Green State U–Firelands Coll, Huron, OH 44839-9791	1968	$3,258 (S)		5	1,228	82
Brazosport Coll, Lake Jackson, TX 77566-3199	1968	$386 (A)		11	3,502	178
Brevard Comm Coll, Cocoa, FL 32922-6597	1960	$936 (S)		5	14,344	1,335
Briarcliffe Coll, Bethpage, NY 11714	1966	$8,320		3-B	1,293	84
Bristol Comm Coll, Fall River, MA 02720-7395	1965	$2,340 (S)		5	5,517	276
Bronx Comm Coll of the City U of New York, Bronx, NY 10453	1959	$2,610 (S)		11	7,653	390
Brookdale Comm Coll, Lincroft, NJ 07738-1597	1967	$2,046 (A)		8	11,423	570
Brooks Coll, Long Beach, CA 90804-3291	1971	$9,210	$5,080	3	1,100	100
Broome Comm Coll, Binghamton, NY 13902-1017	1946	$2,391 (S)		11	5,422	332
Broward Comm Coll, Fort Lauderdale, FL 33301-2298	1960	$1,170 (S)		5	30,333	775
Brown Inst, Mendota Heights, MN 55120	1946	$8,550		3	1,800	96
Bucks County Comm Coll, Newtown, PA 18940-1525	1964	$2,130 (A)		8	8,659	533
Bunker Hill Comm Coll, Boston, MA 02129	1973	$2,294 (S)		5	6,417	126
Burlington County Coll, Pemberton, NJ 08068-1599	1966	$1,786 (A)		8	5,735	292
Butler County Comm Coll, El Dorado, KS 67042-3280	1927	$1,318 (S)	$3,040	11	7,150	573
Butler County Comm Coll, Butler, PA 16003-1203	1965	$1,650 (A)		8	3,096	220
Cabrillo Coll, Aptos, CA 95003-3194	1959	$420 (S)		9	13,147	552
Caldwell Comm Coll & Tech Inst, Hudson, NC 28638-2397	1964	$578 (S)		5	2,983	369
Calhoun Comm Coll, Decatur, AL 35609-2216	1965	$1,350 (S)		5	7,313	347
Camden County Coll, Blackwood, NJ 08012-0200	1967	$1,920 (A)		11	11,785	627
Cañada Coll, Redwood City, CA 94061-1099	1968	$422 (S)		11	5,500	231
Cape Cod Comm Coll, West Barnstable, MA 02668-1599	1961	$2,453 (S)		5	3,494	258
Cape Fear Comm Coll, Wilmington, NC 28401-3993	1959	$581 (S)		5	4,590	211
Capital Community–Technical Coll, Hartford, CT 06105-2354	1946	$1,814 (S)		5	2,911	210
Carl Albert State Coll, Poteau, OK 74953-5208	1934	$1,260 (S)	$2,272	5	2,169	160
Carl Sandburg Coll, Galesburg, IL 61401-9576	1967	$1,792 (A)		11	2,635	208
Carroll Comm Coll, Westminster, MD 21157	1993	$2,271 (A)		11	2,435	164
Carroll Tech Inst, Carrollton, GA 30116	1968	$1,473 (S)		5	1,775	101
Carteret Comm Coll, Morehead City, NC 28557-2989	1963	$587 (S)		5	1,415	93
Casper Coll, Casper, WY 82601-4699	1945	$1,090 (S)	$2,550	9	3,899	185
Cayuga County Comm Coll, Auburn, NY 13021-3099	1953	$2,618 (S)		11	2,820	155
Cecil Comm Coll, North East, MD 21901-1999	1968	$1,920 (A)		8	1,339	162
Cedar Valley Coll, Lancaster, TX 75134-3799	1977	$530 (A)		5	2,700	120
Central Alabama Comm Coll, Alexander City, AL 35011-0699	1965	$1,437 (S)		5	1,609	140
Central Arizona Coll, Coolidge, AZ 85228-9779	1961	$794 (S)	$3,280	8	3,897	440
Central Carolina Comm Coll, Sanford, NC 27330-9000	1962	$587 (S)		11	3,303	192
Central Carolina Tech Coll, Sumter, SC 29150-2499	1963	$848 (A)		5	2,356	184
Central Comm Coll–Grand Island Campus, Grand Island, NE 68802-4903	1976	$1,320 (S)		11	2,926	112
Central Comm Coll–Hastings Campus, Hastings, NE 68902-1024	1966	$1,325 (S)	$2,560	11	2,413	114
Central Comm Coll–Platte Campus, Columbus, NE 68602-1027	1968	$1,325 (S)	$2,560	11	2,543	84
Central Florida Comm Coll, Ocala, FL 34478-1388	1957	$1,312 (S)		11	5,778	212
Centralia Coll, Centralia, WA 98531-4099	1925	$1,416 (S)		5	3,600	186
Central Lakes Coll, Brainerd, MN 56401-3904	1938	$2,209 (S)		5	2,857	140
Central Maine Tech Coll, Auburn, ME 04210-6498	1964	$2,340 (S)	$3,700	5	1,163	77
Central Ohio Tech Coll, Newark, OH 43055-1767	1971	$2,484 (S)		5	1,716	158
Central Oregon Comm Coll, Bend, OR 97701-5998	1949	$1,802 (A)	$4,675	9	3,789	238
Central Piedmont Comm Coll, Charlotte, NC 28235-5009	1963	$560 (S)		11	14,878	1,235
Central Texas Coll, Killeen, TX 76540-1800	1967	$672 (S)	$2,742	11	15,238	328
Central Virginia Comm Coll, Lynchburg, VA 24502-2498	1966	$1,477 (S)		5	4,021	182
Central Wyoming Coll, Riverton, WY 82501-2273	1966	$1,290 (S)	$2,618	11	1,681	168
Century Comm & Tech Coll, White Bear Lake, MN 55110	1970	$2,172 (S)		5	6,141	247
Cerro Coso Comm Coll, Ridgecrest, CA 93555-9571	1973	$390 (S)		5	10,474	286
Chabot Coll, Hayward, CA 94545-5001	1961	$390 (S)		5	12,925	933
Chaffey Coll, Rancho Cucamonga, CA 91737-3002	1883	$410 (S)		9	14,800	540
Chandler-Gilbert Comm Coll, Chandler, AZ 85225-2479	1985	$1,120 (A)		11	3,910	195
Charles County Comm Coll, La Plata, MD 20646-0910	1958	$2,418 (A)		11	5,897	338
Charles Stewart Mott Comm Coll, Flint, MI 48503-2089	1923	$1,836 (A)		9	9,040	420
Chattahoochee Tech Inst, Marietta, GA 30060	1961	$906 (S)		5	2,066	76
Chattahoochee Valley State Comm Coll, Phenix City, AL 36869-7928	1974	$1,425 (S)		5	1,681	86
Chattanooga State Tech Comm Coll, Chattanooga, TN 37406-1097	1965	$1,147 (S)		5	8,439	626
Chemeketa Comm Coll, Salem, OR 97309-7070	1955	$1,530 (S)		11	8,936	651
Chesapeake Coll, Wye Mills, MD 21679-0008	1965	$2,117 (A)		11	2,178	141
Chesterfield-Marlboro Tech Coll, Cheraw, SC 29520-1007	1967	$1,000 (A)		11	1,112	65
Chipola Jr Coll, Marianna, FL 32446-3065	1947	$1,290 (S)		5	2,038	142
Chippewa Valley Tech Coll, Eau Claire, WI 54701-6162	1912	$1,710 (S)		9	3,756	400

Name, address	Year	Tuition & Fees	Room & Board	Control, Degree	Enrollment	Faculty
Cincinnati State Tech & Comm Coll, Cincinnati, OH 45223-2690	1966	$3,218 (S)		5	6,145	347
Cisco Jr Coll, Cisco, TX 76437-9321	1940	$920 (A)	$2,600	11	2,607	98
Citrus Coll, Glendora, CA 91741-1899	1915	$414 (S)		11	11,015	392
City Coll of San Francisco, San Francisco, CA 94112-1821	1935	$410 (S)		11	90,000	1,117
City Colls of Chicago, Harold Washington Coll, Chicago, IL 60601-2449	1962	$1,475 (A)		11	8,434	232
City Colls of Chicago, Harry S Truman Coll, Chicago, IL 60640-5616	1956	$1,400 (A)		11	5,600	160
City Colls of Chicago, Kennedy-King Coll, Chicago, IL 60621-3733	1935	$1,400 (A)		11	7,117	100
City Colls of Chicago, Malcolm X Coll, Chicago, IL 60612-3145	1911	$1,400 (A)		11	8,791	119
City Colls of Chicago, Olive-Harvey Coll, Chicago, IL 60628-1645	1970	$1,400 (A)		11	3,631	133
City Colls of Chicago, Richard J. Daley Coll, Chicago, IL 60652-1242	1960	$1,475 (A)		11	11,007	140
City Colls of Chicago, Wilbur Wright Coll, Chicago, IL 60634-1591	1934	$1,475 (A)		11	12,472	230
Clackamas Comm Coll, Oregon City, OR 97045-7998	1966	$1,296 (S)		9	6,496	519
Clark Coll, Vancouver, WA 98663-3598	1933	$1,568 (S)		5	7,039	526
Clark State Comm Coll, Springfield, OH 45501-0570	1962	$2,430 (S)		5	2,437	161
Clatsop Comm Coll, Astoria, OR 97103-3698	1958	$1,665 (S)		8	1,155	165
Cleveland Comm Coll, Shelby, NC 28152	1965	$588 (S)		5	2,131	91
Cleveland Inst of Electronics, Cleveland, OH 44114-3636	1934	$1,335		3	4,823	9
Cleveland State Comm Coll, Cleveland, TN 37320-3570	1967	$1,162 (S)		5	3,330	191
Clinton Comm Coll, Clinton, IA 52732-6299	1946	$1,872 (S)		11	1,107	75
Clinton Comm Coll, Plattsburgh, NY 12901-9573	1969	$2,446 (S)		11	1,696	179
Clovis Comm Coll, Clovis, NM 88101-8381	1971	$548 (A)		5	3,810	180
Coahoma Comm Coll, Clarksdale, MS 38614-9799	1949	$885 (S)	$2,288	11	1,141	85
Coastal Bend Coll, Beeville, TX 78102-2197	1965	$520 (S)	$2,375	8	2,988	140
Coastal Carolina Comm Coll, Jacksonville, NC 28546-6899	1964	$756 (S)		11	3,636	223
Coastal Georgia Comm Coll, Brunswick, GA 31520-3644	1961	$1,316 (S)		5	1,875	94
Coastline Comm Coll, Fountain Valley, CA 92708-2597	1976	$402 (S)		11	11,665	350
Cochise Coll, Douglas, AZ 85607-9724	1962	$810 (S)	$3,006	11	1,343	88
Cochise Coll, Sierra Vista, AZ 85635-2317	1977	$810 (S)		11	3,026	253
Coconino Comm Coll, Flagstaff, AZ 86003	1991	$880 (S)		5	2,985	205
Coffeyville Comm Coll, Coffeyville, KS 67337-5063	1923	$1,280 (S)	$2,925	11	1,933	61
Colby Comm Coll, Colby, KS 67701-4099	1964	$1,152 (S)	$3,176	11	2,110	63
Coll of Alameda, Alameda, CA 94501-2109	1970	$394 (S)		11	4,681	166
Coll of DuPage, Glen Ellyn, IL 60137-6599	1967	$1,440 (A)		11	28,968	1,532
Coll of Eastern Utah, Price, UT 84501-2699	1937	$1,312 (S)	$3,140	5	2,617	134
Coll of Lake County, Grayslake, IL 60030-1198	1967	$1,530 (A)		9	13,733	965
Coll of Marin, Kentfield, CA 94904	1926	$412 (S)		11	8,589	464
Coll of St. Catherine–Minneapolis, Minneapolis, MN 55454-1494 (4).	1964	$10,500		2-M	1,047	136
Coll of San Mateo, San Mateo, CA 94402-3784	1922	$410 (S)		11	11,681	476
Coll of The Albemarle, Elizabeth City, NC 27906-2327	1960	$588 (S)		5	2,071	122
Coll of the Canyons, Santa Clarita, CA 91355-1899	1969	$322 (S)		11	9,000	400
Coll of the Desert, Palm Desert, CA 92260-9305	1959	$418 (S)		11	10,420	320
Coll of the Mainland, Texas City, TX 77591-2499	1967	$439 (A)		11	3,307	180
Coll of the Redwoods, Eureka, CA 95501-9300	1964	$410 (S)	$4,857	11	6,927	376
Coll of the Sequoias, Visalia, CA 93277-2234	1925	$414 (S)		11	9,969	445
Coll of the Siskiyous, Weed, CA 96094-2899	1957	$410 (S)	$4,122	11	3,205	151
Collin County Comm Coll District, Plano, TX 75093-8309	1985	$724 (A)		11	11,251	686
Colorado Mountain Coll, Alpine Campus, Steamboat Springs, CO 80487	1965	$1,270 (A)	$4,800	9	1,293	71
Colorado Mountain Coll, Timberline Campus, Leadville, CO 80461	1965	$1,270 (A)	$4,800	9	1,090	65
Colorado Northwestern Comm Coll, Rangely, CO 81648-3598	1962	$470 (A)	$3,950	5	1,919	187
Columbia Basin Coll, Pasco, WA 99301-3397	1955	$1,635 (S)		5	6,056	428
Columbia Coll, Sonora, CA 95370	1968	$390 (S)		11	2,353	121
Columbia-Greene Comm Coll, Hudson, NY 12534-0327	1969	$2,358 (S)		11	1,602	114
Columbia State Comm Coll, Columbia, TN 38402-1315	1966	$1,156 (S)		5	4,366	248
Columbus State Comm Coll, Columbus, OH 43216-1609	1963	$2,124 (S)		5	16,600	911
Comm Coll of Allegheny County, Pittsburgh, PA 15233-1894	1966	$2,201 (A)		8	16,191	3,073
Comm Coll of Aurora, Aurora, CO 80011-9036.	1983	$1,746 (S)		5	4,300	215
The Comm Coll of Baltimore County–Catonsville Campus, Catonsville, MD 21228-5381	1957	$2,056 (A)		8	8,857	478
The Comm Coll of Baltimore County–Dundalk Campus, Baltimore, MD 21222	1970	$1,882 (A)		8	2,630	178
The Comm Coll of Baltimore County–Essex Campus, Baltimore, MD 21237	1957	$1,988 (A)		11	7,842	385
Comm Coll of Beaver County, Monaca, PA 15061-2588	1966	$2,080 (A)		5	2,094	134
Comm Coll of Denver, Denver, CO 80217-3363	1970	$1,956 (S)		5	6,112	596
Comm Coll of Philadelphia, Philadelphia, PA 19130-3991	1964	$2,310 (A)		11	42,000	1,182
Comm Coll of Rhode Island, Warwick, RI 02886-1807	1964	$1,746 (S)		5	15,366	680
Comm Coll of Southern Nevada, North Las Vegas, NV 89030-4296	1971	$1,185 (S)		5	28,854	1,125
Comm Coll of the Air Force, Maxwell Air Force Base, AL 36112-6613	1972	$0 (C)		4	384,887	6,500
Comm Coll of Vermont, Waterbury, VT 05676-0120	1970	$3,204 (S)		5	4,447	518
Compton Comm Coll, Compton, CA 90221-5393	1927	$414 (S)		11	7,003	347
Connors State Coll, Warner, OK 74469-9700	1908	$1,223 (S)	$1,874	5	1,977	137
Contra Costa Coll, San Pablo, CA 94806-3195	1948	$390 (S)		11	7,074	222
Corning Comm Coll, Corning, NY 14830-3297	1956	$2,666 (S)		11	5,019	192
Cosumnes River Coll, Sacramento, CA 95823-5799	1970	$390 (S)		9	15,002	425
County Coll of Morris, Randolph, NJ 07869-2086	1966	$2,387 (A)		8	7,557	474
Cowley County Comm Coll & Area Voc–Technical School, Arkansas City, KS 67005-1147	1922	$1,240 (S)	$2,850	11	3,756	184
Crafton Hills Coll, Yucaipa, CA 92399-1799	1972	$358 (S)		11	5,200	184
Craven Comm Coll, New Bern, NC 28562-4984	1965	$588 (S)		5	2,265	202
Crowder Coll, Neosho, MO 64850-9160	1963	$1,440 (A)	$3,000	11	1,741	173
Cuesta Coll, San Luis Obispo, CA 93403-8106	1964	$412 (S)		9	9,156	361
Cumberland County Coll, Vineland, NJ 08362-0517	1963	$2,696 (A)		11	2,490	167
Cuyahoga Comm Coll, Eastern Campus, Highland Hills, OH 44122-6104	1971	$1,871 (A)		11	4,673	201
Cuyahoga Comm Coll, Metro Campus, Cleveland, OH 44115-3123	1963	$1,871 (A)		11	18,415	320
Cuyahoga Comm Coll, Western Campus, Parma, OH 44130-5199	1966	$1,871 (A)		11	10,810	580
Cypress Coll, Cypress, CA 90630-5897	1966	$390 (S)		11	14,048	425
Dabney S. Lancaster Comm Coll, Clifton Forge, VA 24422	1964	$1,646 (S)		5	1,489	160
Dalton State Coll, Dalton, GA 30720-3797	1963	$1,218 (S)		5-B	2,967	112
Danville Comm Coll, Danville, VA 24541-4088	1967	$1,535 (S)		5	3,686	163
Darton Coll, Albany, GA 31707-3098	1965	$1,330 (S)		5	2,635	171

Name, address	Year	Tuition & Fees	Room & Board	Control, Degree	Enroll- ment	Faculty
Davidson County Comm Coll, Lexington, NC 27293-1287	1958	$579 (S)		11	2,366	346
Daytona Beach Comm Coll, Daytona Beach, FL 32120-2811	1958	$1,245 (S)		5	10,672	961
Dean Coll, Franklin, MA 02038-1994	1865	$13,275	$6,680	1	1,303	70
De Anza Coll, Cupertino, CA 95014-5793	1967	$474 (S)		11	26,451	775
DeKalb Tech Inst, Clarkston, GA 30021-2397	1961	$1,152 (S)		5	3,111	443
Delaware County Comm Coll, Media, PA 19063-1094	1967	$2,070 (A)		11	9,040	519
Delaware Tech & Comm Coll, Jack F. Owens Campus, Georgetown, DE 19947	1967	$1,386 (S)		5	3,221	200
Delaware Tech & Comm Coll, Stanton/Wilmington Campus, Newark, DE 19713	1968	$1,386 (S)		5	7,193	463
Delaware Tech & Comm Coll, Terry Campus, Dover, DE 19901	1972	$1,386 (S)		5	1,994	159
Delgado Comm Coll, New Orleans, LA 70119-4399	1921	$1,256 (S)		5	13,355	794
Del Mar Coll, Corpus Christi, TX 78404-3897	1935	$877 (A)		11	9,958	535
Delta Coll, University Center, MI 48710	1961	$1,776 (A)		9	9,472	515
Denmark Tech Coll, Denmark, SC 29042-0327	1948	$1,080 (S)	$2,862	5	1,189	44
Des Moines Area Comm Coll, Ankeny, IA 50021-8995	1966	$1,837 (S)		11	10,306	616
DeVry Inst, North Brunswick, NJ 08902-3362	1969	$7,308		3-B	3,365	135
Diné Coll, Tsaile, AZ 86556	1968	$620	$2,940	4	1,989	152
Dixie Coll, St. George, UT 84770-3876	1911	$1,372 (S)		5	5,215	233
Dodge City Comm Coll, Dodge City, KS 67801-2399	1935	$1,342 (S)	$3,240	11	2,259	163
Doña Ana Branch Comm Coll, Las Cruces, NM 88003-8001	1973	$768 (A)	$3,229	11	4,299	265
Durham Tech Comm Coll, Durham, NC 27703-5023	1961	$584 (S)		5	5,302	428
Dutchess Comm Coll, Poughkeepsie, NY 12601-1595	1957	$2,395 (S)		11	6,263	406
Dyersburg State Comm Coll, Dyersburg, TN 38024	1969	$1,152 (S)		5	2,265	153
East Arkansas Comm Coll, Forrest City, AR 72335-2204	1974	$792 (S)		5	1,193	101
East Central Coll, Union, MO 63084-0529	1968	$1,392 (A)		9	3,239	181
East Central Comm Coll, Decatur, MS 39327-0129	1928	$1,000 (S)	$1,840	11	2,077	124
Eastern Arizona Coll, Thatcher, AZ 85552-0769	1888	$652 (S)	$3,250	11	5,823	371
Eastern Maine Tech Coll, Bangor, ME 04401-4206	1966	$2,664 (S)	$3,600	5	1,248	133
Eastern New Mexico U–Roswell, Roswell, NM 88202-6000	1958	$678 (A)	$3,364	5	2,858	150
Eastern Oklahoma State Coll, Wilburton, OK 74578-4999	1908	$1,288 (S)	$1,920	5	2,217	54
Eastern Wyoming Coll, Torrington, WY 82240-1699	1948	$1,450 (S)	$2,600	11	1,413	228
Eastfield Coll, Mesquite, TX 75150-2099	1970	$530 (A)		11	7,859	91
East Georgia Coll, Swainsboro, GA 30401-2699	1973	$1,232 (S)		5	1,073	41
East Mississippi Comm Coll, Scooba, MS 39358-0158	1927	$1,040 (S)	$1,850	11	1,923	85
ECPI Coll of Technology, Hampton, VA 23666	1966	$6,210		3	2,155	25
ECPI Coll of Technology, Virginia Beach, VA 23462	1966	$6,210		3	2,155	127
Edgecombe Comm Coll, Tarboro, NC 27886-9399	1968	$581 (S)		11	1,784	162
Edison Comm Coll, Fort Myers, FL 33906-6210	1962	$1,250 (S)		11	9,021	345
Edison State Comm Coll, Piqua, OH 45356-9253	1973	$2,325 (S)		5	2,509	295
Edmonds Comm Coll, Lynnwood, WA 98036-5999	1967	$1,401 (S)	$4,020	11	8,033	439
Education America–Tampa Tech Inst Campus, Tampa, FL 33612-8410	1948	$8,100		3-B	1,450	48
Elaine P. Nunez Comm Coll, Chalmette, LA 70043-1249	1992	$1,110 (S)		5	2,107	113
El Camino Coll, Torrance, CA 90506-0001	1947	$410 (S)		5	23,985	533
El Centro Coll, Dallas, TX 75202-3604	1966	$650 (A)		8	4,003	363
Elgin Comm Coll, Elgin, IL 60123-7193	1949	$1,290 (A)		11	9,549	467
Elizabethtown Comm Coll, Elizabethtown, KY 42701-3081	1964	$1,100 (S)		5	3,451	181
El Paso Comm Coll, El Paso, TX 79998-0500	1969	$990 (S)		8	20,744	1,444
Enterprise State Jr Coll, Enterprise, AL 36331-1300	1965	$1,487 (S)		5	1,617	125
Erie Comm Coll, City Campus, Buffalo, NY 14203-2698	1971	$2,600 (S)		11	2,275	254
Erie Comm Coll, North Campus, Williamsville, NY 14221-7095	1946	$2,600 (S)		11	5,063	418
Erie Comm Coll, South Campus, Orchard Park, NY 14127-2199	1974	$2,600 (S)		11	3,141	402
Essex County Coll, Newark, NJ 07102-1798	1966	$2,186 (A)		8	8,353	397
Eugenio María de Hostos Comm Coll of the City U of New York, Bronx, NY 10451	1968	$2,550 (S)		11	4,177	429
Everett Comm Coll, Everett, WA 98201-1327	1941	$1,481 (S)		5	6,399	290
Evergreen Valley Coll, San Jose, CA 95135-1598	1975	$420 (S)		11	10,067	300
Fairleigh Dickinson U, New Coll of General & Continuing Studies, Hackensack, NJ 07601-6112	1964	$13,776	$6,404	1-M	NA	NA
Fashion Inst of Design & Merchandising, Los Angeles Campus, Los Angeles, CA 90015-1421	1969	$13,100		3	2,121	125
Fayetteville Tech Comm Coll, Fayetteville, NC 28303-0236	1961	$569 (S)		5	6,930	851
Feather River Comm Coll District, Quincy, CA 95971	1968	$412 (S)		11	1,200	95
Fergus Falls Comm Coll, Fergus Falls, MN 56537-1009	1960	$2,331 (S)		5	1,567	90
Finger Lakes Comm Coll, Canandaigua, NY 14424-8395	1965	$2,504 (S)		11	4,370	234
Fiorello H. LaGuardia Comm Coll of the City U of New York, Long Island City, NY 11101-3071	1970	$2,610 (A)		11	9,283	666
Fisher Coll, Boston, MA 02116-1500	1903	$12,300	$6,800	1-B	2,638	36
Flathead Valley Comm Coll, Kalispell, MT 59901-2622	1967	$1,626 (A)		11	1,769	111
Florence-Darlington Tech Coll, Florence, SC 29501-0548	1963	$1,120 (A)		5	3,453	209
Florida Comm Coll at Jacksonville, Jacksonville, FL 32202-4030	1963	$1,245 (S)		5	20,640	1,458
Florida Keys Comm Coll, Key West, FL 33040-4397	1965	$1,274 (S)		5	1,756	166
Florida Natl Coll, Hialeah, FL 33012	1982	$4,000		3	1,141	70
Floyd Coll, Rome, GA 30162-1864	1970	$1,864 (S)		5	2,198	80
Foothill Coll, Los Altos Hills, CA 94022-4599	1958	$488 (S)		11	16,039	524
Forsyth Tech Comm Coll, Winston-Salem, NC 27103-5197	1964	$574 (S)		5	5,231	535
Fort Scott Comm Coll, Fort Scott, KS 66701	1919	$1,260 (S)	$2,700	11	1,555	173
Fox Valley Tech Coll, Appleton, WI 54913-2277	1967	$2,118 (S)		11	7,255	1,241
Frank Phillips Coll, Borger, TX 79008-5118	1948	$780 (A)	$2,236	11	1,101	96
Frederick Comm Coll, Frederick, MD 21702-2097	1957	$2,330 (A)		11	4,294	272
Fresno City Coll, Fresno, CA 93741-0002	1910	$412 (S)		9	17,816	832
Front Range Comm Coll, Westminster, CO 80030-2105	1968	$1,580 (S)		5	11,455	703
Fullerton Coll, Fullerton, CA 92832-2095	1913	$410 (S)		11	21,091	678
Fulton-Montgomery Comm Coll, Johnstown, NY 12095-3790	1964	$2,484 (S)		11	1,850	129
Gadsden State Comm Coll, Gadsden, AL 35902-0227	1985	$1,437 (S)	$2,250	5	5,198	288
Gainesville Coll, Gainesville, GA 30503-1358	1964	$1,221 (S)		5	2,847	111
Galveston Coll, Galveston, TX 77550-7496	1967	$744 (S)		11	2,159	105
Garden City Comm Coll, Garden City, KS 67846-6399	1919	$1,216 (S)	$3,250	8	2,084	148
Garland County Comm Coll, Hot Springs, AR 71913	1973	$908 (A)		11	2,029	103
Gaston Coll, Dallas, NC 28034-1499	1963	$593 (S)		11	4,030	370
Gateway Comm Coll, Phoenix, AZ 85034-1795	1968	$1,120 (A)		11	7,201	300

Name, address	Year	Tuition & Fees	Room & Board	Control, Degree	Enroll-ment	Faculty
Gateway Community-Technical Coll, New Haven, CT 06511-5918	1992	$1,794 (S)		5	3,981	291
Gateway Tech Coll, Kenosha, WI 53144-1690	1911	$1,782 (S)		11	8,147	488
Gavilan Coll, Gilroy, CA 95020-9599	1919	$332 (S)		11	5,000	164
Genesee Comm Coll, Batavia, NY 14020-9704	1966	$2,790 (S)		11	4,066	219
George Corley Wallace State Comm Coll, Selma, AL 36702-1049	1966	$1,451 (S)		5	1,823	78
George C. Wallace Comm Coll, Dothan, AL 36303-9234	1949	$1,464 (S)		5	3,083	180
Georgia Military Coll, Milledgeville, GA 31061-3398	1879	$7,084	$4,050	11	4,238	179
Georgia Perimeter Coll, Decatur, GA 30034-3897	1964	$1,522 (S)		5	14,085	1,170
Germanna Comm Coll, Locust Grove, VA 22508-2102	1970	$1,490 (S)		5	3,720	135
Glendale Comm Coll, Glendale, AZ 85302-3090	1965	$1,098 (A)		11	17,809	784
Glendale Comm Coll, Glendale, CA 91208-2894	1927	$436 (S)		11	14,786	807
Glen Oaks Comm Coll, Centreville, MI 49032-9719	1965	$1,590 (A)		11	1,350	96
Gloucester County Coll, Sewell, NJ 08080	1967	$2,015 (A)		8	4,597	218
Gogebic Comm Coll, Ironwood, MI 49938	1932	$1,431 (A)		11	1,211	114
Golden West Coll, Huntington Beach, CA 92647-2748	1966	$426 (S)		11	11,742	501
Gordon Coll, Barnesville, GA 30204-1762	1852	$1,300 (S)	$2,534	5	2,476	99
Grand Rapids Comm Coll, Grand Rapids, MI 49503-3201	1914	$1,614 (A)		9	13,098	511
Grays Harbor Coll, Aberdeen, WA 98520-7599	1930	$1,493 (S)		5	3,104	108
Grayson County Coll, Denison, TX 75020-8299	1964	$960 (A)	$2,200	11	3,267	176
Great Basin Coll, Elko, NV 89801-3348	1967	$1,140 (S)		5	2,438	291
Great Lakes Coll, Midland, MI 48642 (4)	1907	$5,933		1	1,518	144
Greenfield Comm Coll, Greenfield, MA 01301-9739	1962	$2,387 (S)		5	2,240	152
Green River Comm Coll, Auburn, WA 98092-3699	1965	$1,658 (S)		5	7,227	381
Greenville Tech Coll, Greenville, SC 29606-5616	1962	$1,080 (A)		5	8,906	539
Griffin Tech Inst, Griffin, GA 30223	1965	$1,137 (S)		5	1,551	89
Grossmont Coll, El Cajon, CA 92020-1799	1961	$410 (S)		11	15,947	653
Guam Comm Coll, Guam Main Facility, GU 96921-3069	1977	$944 (S)		7	2,010	199
Guilford Tech Comm Coll, Jamestown, NC 27282-0309	1958	$560 (S)		11	6,764	385
Gulf Coast Comm Coll, Panama City, FL 32401-1058	1957	$1,259 (S)		5	7,052	501
Gwinnett Tech Inst, Lawrenceville, GA 30046-1505	1984	$936 (S)		5	3,500	84
Hagerstown Comm Coll, Hagerstown, MD 21742-6590	1946	$2,430 (A)		8	2,802	203
Halifax Comm Coll, Weldon, NC 27890-0809	1967	$572 (S)		11	1,449	68
Harford Comm Coll, Bel Air, MD 21015-1698	1957	$2,004 (A)		11	4,690	340
Harrisburg Area Comm Coll, Harrisburg, PA 17110-2999	1964	$1,973 (A)		11	10,476	649
Harry M. Ayers State Tech Coll, Anniston, AL 36202-1647	1966	$1,200 (S)		5	1,200	39
Hartnell Coll, Salinas, CA 93901-1697	1920	$398 (S)		9	8,352	363
Hawkeye Comm Coll, Waterloo, IA 50704-8015	1967	$2,175 (S)		11	4,158	238
Haywood Comm Coll, Clyde, NC 28721-9453	1964	$578 (S)		11	1,674	118
Hazard Comm Coll, Hazard, KY 41701-2403	1968	$1,100 (S)		5	2,263	156
Heald Coll, Schools of Business & Technology, Hayward, CA 94545-1557	1863	$6,300		1	1,070	75
Heald Coll, Schools of Business & Technology, Milpitas, CA 95035	1863	$8,400		1	1,100	30
Heald Coll, Schools of Business & Technology, San Francisco, CA 94105-2206	1863	$6,660		1	12,000	37
Heald Coll, Schools of Business & Technology, Honolulu, HI 96814-3797	1863	$2,220		1	1,110	42
Henderson Comm Coll, Henderson, KY 42420-4623	1963	$1,100 (S)		5	1,150	96
Henry Ford Comm Coll, Dearborn, MI 48128-1495	1938	$1,710 (A)		9	12,984	994
Herkimer County Comm Coll, Herkimer, NY 13350	1966	$2,430 (S)		11	2,542	143
Hesser Coll, Manchester, NH 03103-7245	1900	$8,220	$4,540	3-B	2,660	99
Hibbing Comm Coll, Hibbing, MN 55746-3300	1916	$2,271 (S)		5	2,619	100
Highland Comm Coll, Freeport, IL 61032-9341	1962	$1,373 (A)		11	2,504	169
Highland Comm Coll, Highland, KS 66035	1858	$1,519 (S)	$2,704	11	2,600	195
Highline Comm Coll, Des Moines, WA 98198-9800	1961	$1,515 (S)		5	9,319	223
High-Tech Inst, Phoenix, AZ 85014-4901	NA	$9,325		3	1,622	72
Hill Coll of the Hill Jr Coll District, Hillsboro, TX 76645-0619	1923	$810 (A)	$2,660	9	2,421	157
Hillsborough Comm Coll, Tampa, FL 33631-3127	1968	$1,245 (S)		5	16,695	654
Hinds Comm Coll, Raymond, MS 39154-9799	1917	$1,070 (S)	$1,850	11	11,599	890
Hocking Comm Coll, Nelsonville, OH 45764-9588	1968	$2,157 (S)		5	4,684	238
Holmes Comm Coll, Goodman, MS 39079-0369	1928	$1,054 (S)	$1,570	11	2,705	125
Holyoke Comm Coll, Holyoke, MA 01040-1099	1946	$2,476 (S)		5	5,468	235
Hopkinsville Comm Coll, Hopkinsville, KY 42241-2100	1965	$1,100 (S)		5	2,752	158
Horry-Georgetown Tech Coll, Conway, SC 29528-6066	1965	$1,115 (S)		11	3,500	240
Housatonic Community-Technical Coll, Bridgeport, CT 06604-4704	1965	$1,814 (S)		5	3,551	177
Houston Comm Coll System, Houston, TX 77270-7849	1971	$900 (A)		11	37,616	2,327
Howard Coll, Big Spring, TX 79720-3702	1945	$748 (A)	$2,604	11	2,058	161
Howard Comm Coll, Columbia, MD 21044-3197	1966	$2,607 (A)		11	5,100	318
Hudson County Comm Coll, Jersey City, NJ 07306	1974	$2,918 (A)		11	4,174	322
Hudson Valley Comm Coll, Troy, NY 12180-6096	1953	$2,465 (A)		11	9,519	511
Hutchinson Comm Coll & Area Voc School, Hutchinson, KS 67501-5894	1928	$1,312 (S)	$2,830	11	3,770	302
Illinois Central Coll, East Peoria, IL 61635-0001	1967	$1,344 (A)		11	12,341	658
Illinois Eastern Comm Colls, Frontier Comm Coll, Fairfield, IL 62837-2601	1976	$1,056 (A)		11	1,652	143
Illinois Eastern Comm Colls, Lincoln Trail Coll, Robinson, IL 62454	1969	$1,056 (A)		11	1,279	76
Illinois Eastern Comm Colls, Olney Central Coll, Olney, IL 62450	1962	$1,146 (A)		11	1,381	88
Illinois Eastern Comm Colls, Wabash Valley Coll, Mount Carmel, IL 62863	1960	$1,056 (A)		11	2,698	97
Illinois Valley Comm Coll, Oglesby, IL 61348-9692	1924	$1,315 (A)		9	4,202	189
Imperial Valley Coll, Imperial, CA 92251-0158	1922	$312 (S)		11	7,697	298
Independence Comm Coll, Independence, KS 67301-0708	1925	$1,332 (S)		5	1,602	153
Indiana Business Coll, Indianapolis, IN 46204-1108	1902	$5,985		3	2,236	70
Indian Hills Comm Coll, Ottumwa, IA 52501-1398	1966	$1,650 (S)		11	3,375	136
Indian River Comm Coll, Fort Pierce, FL 34981-5596	1960	$1,260 (S)		5	11,500	869
Instituto Comercial de Puerto Rico Jr Coll, San Juan, PR 00919-0304	1946	$3,172		3	1,790	97
Interboro Inst, New York, NY 10019-3602	1888	$6,100		3	1,131	41
Inver Hills Comm Coll, Inver Grove Heights, MN 55076-3224	1969	$2,168 (S)		5	4,793	220
Iowa Lakes Comm Coll, Estherville, IA 51334-2295	1967	$2,210 (S)	$2,860	11	2,608	107
Iowa Western Comm Coll, Council Bluffs, IA 51502	1966	$2,160 (S)	$3,000	9	4,000	218
Irvine Valley Coll, Irvine, CA 92620-4399	1979	$412 (S)		11	10,511	344
Isothermal Comm Coll, Spindale, NC 28160-0804	1965	$588 (S)		5	1,891	92
Itasca Comm Coll, Grand Rapids, MN 55744	1922	$2,422 (S)		5	1,019	72
Itawamba Comm Coll, Fulton, MS 38843	1947	$900 (S)		11	3,500	102
Ivy Tech State Coll–Central Indiana, Indianapolis, IN 46206-1763	1963	$1,937 (S)		5	5,116	360
Ivy Tech State Coll–Columbus, Columbus, IN 47203-1868	1963	$1,937 (S)		5	2,734	222

Name, address	Year	Tuition & Fees	Room & Board	Control, Degree	Enrollment	Faculty
Ivy Tech State Coll–Eastcentral, Muncie, IN 47302-9448	1968	$1,937 (S)		5	2,808	217
Ivy Tech State Coll–Kokomo, Kokomo, IN 46903-1373	1968	$1,937 (S)		5	1,800	132
Ivy Tech State Coll–Lafayette, Lafayette, IN 47905-5266	1968	$1,937 (S)		5	2,394	184
Ivy Tech State Coll–Northcentral, South Bend, IN 46619-3837	1968	$1,937 (S)		5	2,512	177
Ivy Tech State Coll–Northeast, Fort Wayne, IN 46805-1430	1969	$1,937 (S)		5	3,265	283
Ivy Tech State Coll–Northwest, Gary, IN 46409-1499	1963	$1,937 (S)		5	3,707	280
Ivy Tech State Coll–Southcentral, Sellersburg, IN 47172-1829	1968	$1,937 (S)		5	1,634	125
Ivy Tech State Coll–Southeast, Madison, IN 47250-1883	1963	$1,937 (S)		5	1,089	89
Ivy Tech State Coll–Southwest, Evansville, IN 47710-3398	1963	$1,937 (S)		5	2,587	213
Ivy Tech State Coll–Wabash Valley, Terre Haute, IN 47802	1966	$1,937 (S)		5	2,561	178
Ivy Tech State Coll–Whitewater, Richmond, IN 47374-1220	1963	$1,937 (S)		5	1,142	109
Jackson Comm Coll, Jackson, MI 49201-8399	1928	$1,613 (A)		8	6,643	435
Jackson State Comm Coll, Jackson, TN 38301-3797	1967	$1,218 (S)		5	3,628	222
James H. Faulkner State Comm Coll, Bay Minette, AL 36507	1965	$1,680 (S)	$2,325	5	2,777	153
James Sprunt Comm Coll, Kenansville, NC 28349-0398	1964	$600 (S)		5	1,122	90
Jefferson Coll, Hillsboro, MO 63050-2441	1963	$1,408 (A)		11	3,963	197
Jefferson Comm Coll, Watertown, NY 13601	1961	$2,494 (S)		11	3,377	168
Jefferson Comm Coll, Steubenville, OH 43952-3598	1966	$1,770 (A)		11	1,335	160
Jefferson State Comm Coll, Birmingham, AL 35215-3098	1965	$1,395 (A)		5	5,137	292
John A. Logan Coll, Carterville, IL 62918-9900	1967	$1,080 (A)		11	5,224	246
Johnson County Comm Coll, Overland Park, KS 66210-1299	1967	$1,472 (S)		11	15,572	686
Johnston Comm Coll, Smithfield, NC 27577-2350	1969	$588 (S)		5	2,718	257
John Tyler Comm Coll, Chester, VA 23831-5316	1967	$1,480 (S)		5	5,055	233
John Wood Comm Coll, Quincy, IL 62301-9147	1974	$1,664 (A)		9	2,187	156
Joliet Jr Coll, Joliet, IL 60431-8938	1901	$1,380 (A)		11	11,137	537
Jones County Jr Coll, Ellisville, MS 39437-3901	1928	$792 (S)	$1,800	11	4,430	175
J. Sargeant Reynolds Comm Coll, Richmond, VA 23285-5622	1972	$1,457 (S)		5	10,069	560
Kalamazoo Valley Comm Coll, Kalamazoo, MI 49003-4070	1966	$1,271 (A)		11	8,793	414
Kankakee Comm Coll, Kankakee, IL 60901-0888	1966	$1,216 (A)		11	3,379	145
Kansas City Kansas Comm Coll, Kansas City, KS 66112-3003	1923	$1,140 (S)		11	5,417	350
Kaskaskia Coll, Centralia, IL 62801-7878	1966	$1,280 (A)		11	2,993	157
Keiser Coll, Daytona Beach, FL 32114	1995	$10,500		3	2,434	19
Keiser Coll, Fort Lauderdale, FL 33309	1977	$10,500		3	2,434	56
Keiser Coll, Melbourne, FL 32901-1461	1989	$10,500		3	2,434	24
Keiser Coll, Sarasota, FL 34236	1995	$10,500		3	2,434	17
Keiser Coll, Tallahassee, FL 32308	1992	$10,500		3	2,434	27
Kellogg Comm Coll, Battle Creek, MI 49017-3397	1956	$1,470 (A)		11	4,870	312
Kent State U, Stark Campus, Canton, OH 44720-7599	1967	$3,056 (S)		5-B	2,756	170
Kent State U, Trumbull Campus, Warren, OH 44483-1998	1954	$3,056 (S)		5	2,268	139
Kent State U, Tuscarawas Campus, New Philadelphia, OH 44663-9403	1962	$3,056 (S)		5	1,572	130
Kilgore Coll, Kilgore, TX 75662-3299	1935	$832 (A)		11	4,155	212
Kingsborough Comm Coll of the City U of New York, Brooklyn, NY 11235	1963	$2,600 (S)		11	15,175	961
Kingwood Coll, Kingwood, TX 77339-3801	1984	$864 (A)		11	3,895	224
Kirkwood Comm Coll, Cedar Rapids, IA 52406-2068	1966	$1,767 (S)		11	11,164	582
Kirtland Comm Coll, Roscommon, MI 48653-9699	1966	$1,580 (A)		9	1,222	95
Kishwaukee Coll, Malta, IL 60150	1967	$1,400 (A)		11	3,164	201
Labette Comm Coll, Parsons, KS 67357-4299	1923	$1,209 (S)	$2,460	11	1,408	234
Lackawanna Jr Coll, Scranton, PA 18509	1894	$7,310		1	1,071	46
Lake Area Tech Inst, Watertown, SD 57201	1964	$2,700		5	1,130	65
Lake City Comm Coll, Lake City, FL 32025-8703	1962	$1,140 (S)	$3,700	5	2,355	209
Lake Land Coll, Mattoon, IL 61938-9366	1966	$1,424 (A)		11	5,388	335
Lakeland Comm Coll, Kirtland, OH 44094-5198	1967	$1,963 (A)		11	8,267	631
Lake Michigan Coll, Benton Harbor, MI 49022-1899	1946	$1,530 (A)		9	3,295	240
Lakeshore Tech Coll, Cleveland, WI 53015-1414	1967	$1,734 (S)		11	2,709	635
Lake-Sumter Comm Coll, Leesburg, FL 34788-8751	1962	$1,236 (S)		11	2,543	141
Lake Tahoe Comm Coll, South Lake Tahoe, CA 96150-4524	1975	$417 (S)		11	3,400	177
Lake Washington Tech Coll, Kirkland, WA 98034-8506	1949	$1,401		9	4,670	307
Lamar Comm Coll, Lamar, CO 81052-3999	1937	$2,062 (S)	$4,040	5	1,248	51
Lamar U–Orange, Orange, TX 77630-5899	1969	$1,818 (S)		5	1,579	80
Lamar U–Port Arthur, Port Arthur, TX 77641-0310	1909	$1,770 (S)		5	2,398	137
Lane Comm Coll, Eugene, OR 97405-0640	1964	$1,640 (S)		11	11,437	503
Laney Coll, Oakland, CA 94607-4893	1953	$394 (S)		11	11,120	315
Lansing Comm Coll, Lansing, MI 48901-7210	1957	$1,207 (A)		11	15,690	919
Laramie County Comm Coll, Cheyenne, WY 82007-3299	1968	$1,070 (S)	$3,488	8	4,087	255
Laredo Comm Coll, Laredo, TX 78040-4395	1946	$700 (A)		11	7,446	359
Lawson State Comm Coll, Birmingham, AL 35221-1798	1965	$1,404 (S)		5	1,643	105
Lee Coll, Baytown, TX 77522-0818	1934	$422 (A)		9	6,074	358
Lehigh Carbon Comm Coll, Schnecksville, PA 18078-2598	1967	$2,130 (A)		11	3,798	260
Lenoir Comm Coll, Kinston, NC 28502-0188	1960	$578 (S)		5	2,205	163
Lewis & Clark Comm Coll, Godfrey, IL 62035-2466	1970	$1,368 (A)		9	5,992	347
Lima Tech Coll, Lima, OH 45804-3597	1971	$2,368 (A)		5	2,470	167
Lincoln Land Comm Coll, Springfield, IL 62794-9256	1967	$1,356 (A)		9	11,251	400
Linn-Benton Comm Coll, Albany, OR 97321	1966	$1,620 (S)		11	5,624	540
Long Beach City Coll, Long Beach, CA 90808-1780	1927	$410 (S)		5	26,176	946
Longview Comm Coll, Lee's Summit, MO 64081-2105	1969	$1,457 (A)		11	8,091	394
Lorain County Comm Coll, Elyria, OH 44035	1963	$2,291 (A)		11	6,798	299
Lord Fairfax Comm Coll, Middletown, VA 22645-0047	1969	$1,446 (S)		5	3,610	161
Los Angeles City Coll, Los Angeles, CA 90029-3590	1929	$390 (S)		9	15,000	625
Los Angeles Harbor Coll, Wilmington, CA 90744-2397	1949	$407 (S)		11	7,503	117
Los Angeles Mission Coll, Sylmar, CA 91342-3245	1974	$384 (S)		11	6,569	115
Los Angeles Pierce Coll, Woodland Hills, CA 91371-0001	1947	$436 (S)		11	13,999	510
Los Angeles Valley Coll, Valley Glen, CA 91401-4096	1949	$412 (S)		11	15,655	482
Los Medanos Coll, Pittsburg, CA 94565-5197	1974	$312 (S)		9	7,152	244
Louisiana State U at Alexandria, Alexandria, LA 71302-9121	1960	$1,132 (S)		5	2,362	89
Louisiana State U at Eunice, Eunice, LA 70535-1129	1967	$1,056 (S)		5	2,633	114
Lower Columbia Coll, Longview, WA 98632-0310	1934	$1,464 (S)		5	3,361	160
Luzerne County Comm Coll, Nanticoke, PA 18634-9804	1966	$1,800 (A)		8	6,123	435
Macomb Comm Coll, Warren, MI 48093-3896	1954	$1,628 (A)		9	33,309	793
Madison Area Tech Coll, Madison, WI 53704-2599	1911	$1,934 (S)		9	13,479	1,881

Name, address	Year	Tuition & Fees	Room & Board	Control, Degree	Enroll- ment	Faculty
Madisonville Comm Coll, Madisonville, KY 42431-9185	1968	$1,100 (S)		5	1,821	167
Manatee Comm Coll, Bradenton, FL 34206-7046	1957	$1,207 (S)		5	7,262	267
Manchester Community-Technical Coll, Manchester, CT 06045-1046	1963	$1,814 (S)		5	5,252	205
Maple Woods Comm Coll, Kansas City, MO 64156-1299	1969	$1,457 (A)		11	5,042	394
Marion Tech Coll, Marion, OH 43302-5694	1971	$2,421 (S)		12	1,666	105
Marshalltown Comm Coll, Marshalltown, IA 50158-4760	1927	$1,856 (S)		9	1,152	107
Massasoit Comm Coll, Brockton, MA 02302-3996	1966	$2,310 (S)		5	6,472	357
Maysville Comm Coll, Maysville, KY 41056	1967	$1,100 (S)		5	1,246	111
McDowell Tech Comm Coll, Marion, NC 28752-9724	1964	$577 (S)		5	1,049	58
McHenry County Coll, Crystal Lake, IL 60012-2761	1967	$1,214 (A)		11	5,050	277
McLennan Comm Coll, Waco, TX 76708-1499	1965	$1,140 (A)		8	5,630	297
Mendocino Coll, Ukiah, CA 95482-0300	1973	$412 (S)		11	4,600	193
Merced Coll, Merced, CA 95348-2898	1962	$412 (S)		11	7,305	421
Mercer County Comm Coll, Trenton, NJ 08690-1004	1966	$2,130 (A)		11	7,974	361
Meridian Comm Coll, Meridian, MS 39307	1937	$1,000 (S)		11	3,007	255
Merritt Coll, Oakland, CA 94619-3196	1953	$364 (S)		11	4,852	201
Mesabi Range Comm & Tech Coll, Virginia, MN 55792-3448	1918	$2,347 (S)		5	1,333	71
Mesa Comm Coll, Mesa, AZ 85202-4866	1965	$972 (A)		11	24,000	852
Metropolitan Comm Coll, Omaha, NE 68103-0777	1974	$1,238 (S)		11	11,583	608
Miami-Dade Comm Coll, Miami, FL 33132-2296	1960	$1,344 (S)		11	44,686	2,070
Miami U–Hamilton Campus, Hamilton, OH 45011-3399	1968	$3,118 (S)		5-B	2,665	170
Miami U–Middletown Campus, Middletown, OH 45042-3497	1966	$3,118 (S)		5-B	2,892	158
Middle Georgia Coll, Cochran, GA 31014-1599	1884	$1,550 (S)	$3,200	5	1,953	150
Middlesex Comm Coll, Bedford, MA 01730-1655	1970	$2,114 (A)		5	6,664	377
Middlesex Community–Technical Coll, Middletown, CT 06457-4889	1966	$1,814 (S)		5	2,273	120
Middlesex County Coll, Edison, NJ 08818-3050	1964	$2,578 (A)		8	10,397	552
Midland Coll, Midland, TX 79705-6399	1969	$900 (A)		11	4,602	194
Midlands Tech Coll, Columbia, SC 29202-2408	1974	$1,100 (A)		11	9,798	661
Mid Michigan Comm Coll, Harrison, MI 48625-9447	1965	$1,610 (A)		11	2,331	227
Mid-Plains Comm Coll, North Platte, NE 69191	1965	$1,302 (S)		9	1,726	108
Mid-South Comm Coll, West Memphis, AR 72301	1993	$822 (A)		5	1,451	81
Mid-State Tech Coll, Wisconsin Rapids, WI 54494-5599	1917	$1,821 (S)		11	1,927	96
Mineral Area Coll, Park Hills, MO 63601-1000	1922	$1,178 (A)		9	2,581	130
Minneapolis Comm & Tech Coll, Minneapolis, MN 55403-1779	1965	$2,206 (S)		5	5,728	363
Minnesota State Coll–Southeast Tech, Winona, MN 55987	1992	$2,206 (S)		5	1,514	84
MiraCosta Coll, Oceanside, CA 92056-3899	1934	$414 (S)		5	9,121	388
Mission Coll, Santa Clara, CA 95054-1897	1977	$392 (S)		11	8,594	376
Mississippi County Comm Coll, Blytheville, AR 72316-1109	1975	$874 (A)		5	2,154	79
Mississippi Delta Comm Coll, Moorhead, MS 38761-0668	1926	$920 (S)	$2,370	9	2,456	130
Mississippi Gulf Coast Comm Coll, Perkinston, MS 39573-0548	1911	$890 (S)	$1,706	9	10,609	753
Mitchell Comm Coll, Statesville, NC 28677-5293	1852	$584 (S)		5	1,637	100
Moberly Area Comm Coll, Moberly, MO 65270-1304	1927	$1,115 (A)		11	2,293	146
Modesto Jr Coll, Modesto, CA 95350-5800	1921	$334 (S)		11	16,058	505
Mohave Comm Coll, Kingman, AZ 86401	1971	$720 (S)		5	5,967	349
Mohawk Valley Comm Coll, Utica, NY 13501-5394	1946	$2,716 (S)	$4,378	11	5,225	274
Monroe Coll, Bronx, NY 10468-5407	1933	$9,140		3-B	2,915	115
Monroe Comm Coll, Rochester, NY 14623-5780	1961	$2,654 (S)		11	13,628	1,236
Monroe County Comm Coll, Monroe, MI 48161-9047	1964	$1,144 (A)		8	3,629	201
Montana State U Coll of Technology–Great Falls, Great Falls, MT 59405	1969	$2,064 (S)		5	1,102	108
Montcalm Comm Coll, Sidney, MI 48885-0300	1965	$1,629 (A)		11	2,036	125
Monterey Peninsula Coll, Monterey, CA 93940-4799	1947	$430 (S)		5	15,475	388
Montgomery Coll, Conroe, TX 77384	1995	$830 (A)		11	4,644	179
Montgomery Coll–Germantown Campus, Germantown, MD 20876	1975	$2,402 (A)		11	3,787	163
Montgomery Coll–Rockville Campus, Rockville, MD 20850-1196	1965	$2,402 (A)		11	12,206	647
Montgomery Coll–Takoma Park Campus, Takoma Park, MD 20912	1946	$2,402 (A)		11	4,367	231
Montgomery County Comm Coll, Blue Bell, PA 19422-0796	1964	$2,220 (A)		8	8,852	693
Moorpark Coll, Moorpark, CA 93021-1695	1967	$442 (S)		8	12,846	450
Moraine Park Tech Coll, Fond du Lac, WI 54936-1940	1967	$1,814 (S)		11	6,816	402
Moraine Valley Comm Coll, Palos Hills, IL 60465-0937	1967	$1,457 (A)		11	13,417	584
Morgan Comm Coll, Fort Morgan, CO 80701-4399	1967	$1,792 (S)		5	1,200	153
Morton Coll, Cicero, IL 60804-4398	1924	$1,488 (A)		11	4,661	214
Motlow State Comm Coll, Tullahoma, TN 37388-8100	1969	$1,210 (S)		5	3,365	194
Mountain Empire Comm Coll, Big Stone Gap, VA 24219-0700	1972	$1,575 (S)		5	2,800	150
Mountain View Coll, Dallas, TX 75211-6599	1970	$530 (A)		11	5,128	264
Mt. Hood Comm Coll, Gresham, OR 97030-3300	1966	$1,620 (S)		11	7,655	547
Mount Ida Coll, Newton Centre, MA 02459-3310	1899	$12,562	$8,950	1-B	1,800	222
Mt. San Antonio Coll, Walnut, CA 91789-1399	1946	$332 (S)		9	24,003	805
Mt. San Jacinto Coll, San Jacinto, CA 92583-2399	1963	$408 (S)		11	8,508	357
Mount Wachusett Comm Coll, Gardner, MA 01440-1000	1963	$2,910 (S)		5	3,167	187
Murray State Coll, Tishomingo, OK 73460-3130	1908	$1,446 (S)	$2,218	5	1,700	73
Muscatine Comm Coll, Muscatine, IA 52761-5396	1929	$1,872 (S)		5	1,245	95
Muskegon Comm Coll, Muskegon, MI 49442-1493	1926	$1,487 (A)		11	4,346	150
Muskingum Area Tech Coll, Zanesville, OH 43701-2626	1969	$2,910 (S)		11	2,151	104
Napa Valley Coll, Napa, CA 94558-6236	1942	$290 (S)		11	5,998	304
Nash Comm Coll, Rocky Mount, NC 27804-0488	1967	$588 (S)		5	1,997	106
Nashville Auto Diesel Coll, Nashville, TN 37206-3298 (2)	1919	$15,372	$2,640	3	1,250	48
Nashville State Tech Inst, Nashville, TN 37209-4515	1970	$1,202 (S)		5	7,271	294
Nassau Comm Coll, Garden City, NY 11530-6793	1959	$2,210 (S)		11	20,337	1,639
Naugatuck Valley Community–Technical Coll, Waterbury, CT 06708-3000	1992	$1,814 (S)		5	4,736	299
Navarro Coll, Corsicana, TX 75110-4899	1946	$979 (A)		11	3,467	175
Neosho County Comm Coll, Chanute, KS 66720-2699	1936	$1,271 (S)	$2,900	11	1,500	96
Newbury Coll, Brookline, MA 02445-5796	1962	$12,330	$7,000	1-B	3,394	92
New England Coll of Finance, Boston, MA 02111-2645	1909	$6,040		1-B	1,126	135
New England Inst of Technology, Warwick, RI 02886-2244	1940	$10,350		1-B	2,537	208
New England Inst of Technology at Palm Beach, West Palm Beach, FL 33407	1983	$6,900		3	1,196	68
New Hampshire Comm Tech Coll, Berlin/Laconia, Berlin, NH 03570-3717	1966	$3,136 (S)		5	1,433	81
New Hampshire Comm Tech Coll, Manchester/Stratham, Manchester, NH 03102-8518	1945	$3,164 (S)		5	2,007	200
New Hampshire Comm Tech Coll, Nashua/Claremont, Nashua, NH 03063	1967	$3,164 (S)		5	1,282	118

Name, address	Year	Tuition & Fees	Room & Board	Control, Degree	Enroll- ment	Faculty
New Hampshire Tech Inst, Concord, NH 03301-7412	1964	$3,388 (S)	$4,515	5	2,435	146
New Mexico Jr Coll, Hobbs, NM 88240-9123	1965	$332 (A)	$3,200	11	2,622	129
New Mexico State U–Alamogordo, Alamogordo, NM 88311-0477	1958	$768 (A)		5	2,006	124
New River Comm Coll, Dublin, VA 24084-1127	1969	$1,501 (S)		5	3,595	181
New York City Tech Coll of the City U of New York, Brooklyn, NY 11201-2983	1946	$3,289 (S)		11-B	11,180	893
Niagara County Comm Coll, Sanborn, NY 14132-9460	1962	$2,610 (S)		11	4,828	274
Nicolet Area Tech Coll, Rhinelander, WI 54501-0518	1968	$2,338 (S)		11	1,384	84
Normandale Comm Coll, Bloomington, MN 55431-4399	1968	$2,228 (S)		5	6,729	216
Northampton County Area Comm Coll, Bethlehem, PA 18020-7599	1967	$2,190 (A)	$4,200	11	5,505	519
North Arkansas Coll, Harrison, AR 72601	1974	$936 (A)		11	1,825	112
North Central Michigan Coll, Petoskey, MI 49770-8717	1958	$1,515 (A)		8	2,184	128
North Central Missouri Coll, Trenton, MO 64683-1824	1925	$1,425 (A)	$3,000	9	1,375	76
North Central Tech Coll, Mansfield, OH 44901-0698	1961	$2,668 (S)		5	2,714	182
Northcentral Tech Coll, Wausau, WI 54401-1899	1912	$1,933 (S)		9	2,290	228
North Central Texas Coll, Gainesville, TX 76240-4699	1924	$600 (A)		8	4,054	195
North Country Comm Coll, Saranac Lake, NY 12983-0089	1967	$2,360 (S)		11	1,027	118
North Dakota State Coll of Sci, Wahpeton, ND 58076	1903	$1,769 (S)		5	2,469	159
Northeast Alabama Comm Coll, Rainsville, AL 35986-0159	1963	$1,197 (S)		5	1,617	51
Northeast Comm Coll, Norfolk, NE 68702-0469	1973	$1,344 (S)		11	4,754	220
Northeastern Jr Coll, Sterling, CO 80751-2399	1941	$1,876 (S)	$4,440	5	2,038	75
Northeast Iowa Comm Coll, Calmar Campus, Calmar, IA 52132-0480	1966	$2,338 (S)		11	1,139	137
Northeast Iowa Comm Coll, Peosta Campus, Peosta, IA 52068-9776	1970	$2,328 (S)		11	1,843	66
Northeast Mississippi Comm Coll, Booneville, MS 38829	1948	$950 (S)		5	2,962	126
Northeast State Tech Comm Coll, Blountville, TN 37617-0246	1966	$1,150 (S)		5	3,961	224
Northeast Texas Comm Coll, Mount Pleasant, TX 75456-1307	1985	$950 (A)		11	2,052	114
Northern Essex Comm Coll, Haverhill, MA 01830	1960	$1,992 (S)		5	5,809	404
Northern New Mexico Comm Coll, Española, NM 87532	1909	$538 (S)	$2,640	5	2,250	160
Northern Oklahoma Coll, Tonkawa, OK 74653-0310	1901	$1,103 (S)	$1,880	5	2,530	80
Northern Virginia Comm Coll, Annandale, VA 22003-3796	1965	$1,441 (S)		5	37,411	1,334
North Harris Coll, Houston, TX 77073-3499	1972	$830 (A)		11	9,726	196
North Hennepin Comm Coll, Minneapolis, MN 55445-2231	1966	$2,496 (S)		5	4,700	200
North Idaho Coll, Coeur d'Alene, ID 83814-2199	1933	$1,128 (A)		11	3,520	280
North Iowa Area Comm Coll, Mason City, IA 50401-7299	1918	$1,975 (S)	$3,124	11	2,722	186
North Lake Coll, Irving, TX 75038-3899	1977	$530 (A)		8	6,233	342
Northland Comm & Tech Coll, Thief River Falls, MN 56701	1965	$2,522 (S)		5	1,804	111
Northland Pioneer Coll, Holbrook, AZ 86025-0610	1974	$720 (S)		11	4,205	275
North Seattle Comm Coll, Seattle, WA 98103-3599	1970	$1,497 (S)		5	9,118	379
North Shore Comm Coll, Danvers, MA 01923-4093	1965	$2,404 (S)		5	5,529	378
NorthWest Arkansas Comm Coll, Bentonville, AR 72712	1989	$912 (A)		11	3,517	201
Northwest Coll, Powell, WY 82435-1898	1946	$1,326 (S)	$2,932	11	1,688	181
Northwestern Coll, Lima, OH 45805-1498	1920	$6,804		1	1,978	58
Northwestern Connecticut Community-Technical Coll, Winsted, CT 06098-1798	1965	$1,814 (S)		5	1,848	94
Northwestern Michigan Coll, Traverse City, MI 49686-3061	1951	$1,767 (A)	$4,400	11	4,090	93
Northwestern Tech Inst, Rock Springs, GA 30739	1966	$816		5	1,320	62
Northwest Mississippi Comm Coll, Senatobia, MS 38668-1701	1927	$1,000 (S)	$1,810	11	4,850	200
Northwest-Shoals Comm Coll, Muscle Shoals, AL 35662	1961	$1,485 (S)		5	3,235	211
Northwest State Comm Coll, Archbold, OH 43502-9542	1968	$2,346 (S)		5	2,388	151
Norwalk Community-Technical Coll, Norwalk, CT 06854-1655	1961	$1,814 (S)		5	4,974	347
Oakland Comm Coll, Bloomfield Hills, MI 48304-2266	1964	$1,496 (A)		11	24,237	855
Oakton Comm Coll, Des Plaines, IL 60016-1268	1969	$1,193 (A)		9	9,785	617
Ocean County Coll, Toms River, NJ 08754-2001	1964	$2,164 (A)		8	7,195	322
Odessa Coll, Odessa, TX 79764-7127	1946	$856 (A)	$1,949	11	4,591	208
Ohlone Coll, Fremont, CA 94539-5884	1967	$390 (S)		11	10,000	434
Okaloosa-Walton Comm Coll, Niceville, FL 32578-1295	1963	$1,080 (S)		11	9,318	274
Oklahoma City Comm Coll, Oklahoma City, OK 73159-4419	1969	$1,308 (S)		5	9,637	377
Oklahoma State U, Oklahoma City, OK 73107-6120	1961	$1,667 (S)		5	3,877	223
Oklahoma State U, Okmulgee, Okmulgee, OK 74447-3901	1946	$2,273 (S)		5	2,320	138
Olympic Coll, Bremerton, WA 98337-1699	1946	$1,503 (S)		5	6,694	310
Onondaga Comm Coll, Syracuse, NY 13215-2099	1962	$2,626 (A)		11	7,000	488
Orangeburg-Calhoun Tech Coll, Orangeburg, SC 29118-8299	1968	$1,008 (A)		11	1,930	121
Orange Coast Coll, Costa Mesa, CA 92628-5005	1947	$352 (S)		11	24,999	763
Orange County Comm Coll, Middletown, NY 10940-6437	1950	$2,285 (S)		11	5,565	440
Otero Jr Coll, La Junta, CO 81050-3415	1941	$1,446 (S)	$3,600	5	1,086	69
Owensboro Comm Coll, Owensboro, KY 42303-1899	1986	$1,100 (S)		5	2,212	131
Owens Comm Coll, Findlay, OH 45840	1983	$1,916 (S)		5	1,807	158
Owens Comm Coll, Toledo, OH 43699-1947	1966	$1,916 (S)		5	14,071	726
Oxnard Coll, Oxnard, CA 93033-6699	1975	$410 (S)		5	6,823	288
Ozarks Tech Comm Coll, Springfield, MO 65802	1990	$1,364 (A)		9	5,317	150
Paducah Comm Coll, Paducah, KY 42002-7380	1932	$1,100 (A)		5	2,794	143
Palm Beach Comm Coll, Lake Worth, FL 33461-4796	1933	$1,245 (S)		5	16,587	737
Palo Alto Coll, San Antonio, TX 78224-2499	1987	$878 (A)		11	6,536	347
Palomar Coll, San Marcos, CA 92069-1487	1946	$412 (S)		11	25,235	1,152
Palo Verde Coll, Blythe, CA 92225-1118	1947	$390 (S)		11	2,278	69
Panola Coll, Carthage, TX 75633-2397	1947	$796 (A)	$2,764	11	1,504	104
Paradise Valley Comm Coll, Phoenix, AZ 85032-1200	1985	$1,120 (A)		11	6,300	309
Paris Jr Coll, Paris, TX 75460-6298	1924	$974 (A)		11	3,100	117
Parkland Coll, Champaign, IL 61821-1899	1967	$1,410 (A)		9	7,937	435
Pasadena City Coll, Pasadena, CA 91106-2041	1924	$412 (S)		11	23,006	824
Pasco-Hernando Comm Coll, New Port Richey, FL 34654-5199	1972	$1,266 (S)		5	5,073	275
Passaic County Comm Coll, Paterson, NJ 07505-1179	1968	$2,460 (S)		8	4,017	281
Patrick Henry Comm Coll, Martinsville, VA 24115-5311	1962	$1,410 (S)		5	2,689	141
Paul D. Camp Comm Coll, Franklin, VA 23851-0737	1971	$1,430 (S)		5	1,550	59
Pearl River Comm Coll, Poplarville, MS 39470	1909	$980 (S)	$1,706	11	2,720	167
Pellissippi State Tech Comm Coll, Knoxville, TN 37933-0990	1974	$1,172 (S)		5	8,170	412
Pennsylvania Coll of Technology, Williamsport, PA 17701-5778	1965	$6,540 (S)	$4,085	12-B	5,298	401
Pennsylvania State U Delaware County Campus of the Commonwealth Coll, Media, PA 19063-5596	1966	$5,654 (S)		12-B	1,671	117
Pennsylvania State U DuBois Campus of the Commonwealth Coll, DuBois, PA 15801-3199	1935	$5,654 (S)		12	1,152	82

Name, address	Year	Tuition & Fees	Room & Board	Control, Degree	Enrollment	Faculty
Pennsylvania State U Hazleton Campus of the Commonwealth Coll, Hazleton, PA 18201-1291	1934	$5,654 (S)	$4,840	12	1,321	88
Pennsylvania State U Mont Alto Campus of the Commonwealth Coll, Mont Alto, PA 17237-9703.	1929	$5,654 (S)	$4,840	12	1,165	96
Pennsylvania State U Shenango Campus of the Commonwealth Coll, Sharon, PA 16146-1537	1965	$5,654 (S)		12-B	1,024	94
Pennsylvania State U Worthington Scranton Campus of the Commonwealth Coll, Dunmore, PA 18512-1699.	1923	$5,654 (S)		12	1,575	108
Pennsylvania State U York Campus of the Commonwealth Coll, York, PA 17403-3298.	1926	$5,654 (S)		12	2,086	116
Penn Valley Comm Coll, Kansas City, MO 64111	1969	$1,457 (A)		11	4,495	288
Pensacola Jr Coll, Pensacola, FL 32504-8998	1948	$1,304 (S)		5	15,000	819
Petit Jean Coll, Morrilton, AR 72110	1961	$990 (S)		5	1,059	62
Phillips Comm Coll of the U of Arkansas, Helena, AR 72342-0785	1965	$888 (A)		11	2,493	70
Phoenix Coll, Phoenix, AZ 85013-4234	1920	$1,098 (A)		11	10,900	540
Piedmont Comm Coll, Roxboro, NC 27573-1197	1970	$587 (S)		5	1,600	95
Piedmont Tech Coll, Greenwood, SC 29648-1467.	1966	$1,590 (S)		5	3,715	225
Piedmont Virginia Comm Coll, Charlottesville, VA 22902-7589	1972	$1,572 (S)		5	4,059	287
Pierce Coll, Lakewood, WA 98498-1999.	1967	$1,650 (S)		5	10,109	579
Pikes Peak Comm Coll, Colorado Springs, CO 80906-5498	1968	$1,740 (S)		5	9,556	476
Pine Tech Coll, Pine City, MN 55063	1965	$2,286 (S)		5	1,026	42
Pitt Comm Coll, Greenville, NC 27835-7007.	1961	$587 (S)		11	4,793	265
Pittsburgh Tech Inst, Pittsburgh, PA 15222-2560.	1946	$13,080	$5,850	3	1,321	85
Porterville Coll, Porterville, CA 93257-6058	1927	$430 (S)		5	3,589	140
Portland Comm Coll, Portland, OR 97280-0990	1961	$1,665 (S)		11	38,245	1,385
Potomac State Coll of West Virginia U, Keyser, WV 26726-2698	1901	$1,926 (S)	$3,934	5	1,121	83
Prairie State Coll, Chicago Heights, IL 60411-8226	1958	$1,680 (A)		11	5,275	319
Pratt Comm Coll & Area Voc School, Pratt, KS 67124-8317	1938	$1,280 (S)	$2,836	11	1,421	54
Prestonsburg Comm Coll, Prestonsburg, KY 41653-1815	1964	$1,100 (A)		5	2,371	129
Prince George's Comm Coll, Largo, MD 20774-2199.	1958	$2,830 (A)		8	12,435	708
Pueblo Comm Coll, Pueblo, CO 81004-1499	1933	$1,791 (S)		5	4,216	310
Pulaski Tech Coll, North Little Rock, AR 72118	1945	$1,056 (S)		5	3,356	145
Queensborough Comm Coll of the City U of New York, Bayside, NY 11364	1958	$2,602 (S)		11	10,305	719
Quincy Coll, Quincy, MA 02169-4522	1958	$2,580		10	4,706	69
Quinebaug Valley Community-Technical Coll, Danielson, CT 06239-1440.	1971	$1,814 (S)		5	1,214	82
Quinsigamond Comm Coll, Worcester, MA 01606-2092.	1963	$2,250 (S)		5	5,178	371
Randolph Comm Coll, Asheboro, NC 27204-1009	1962	$582 (S)		5	1,662	108
Ranken Tech Coll, St. Louis, MO 63113	1907	$6,300		1	1,423	93
Rappahannock Comm Coll, Glenns, VA 23149-2616	1970	$1,461 (S)		12	1,560	141
Raritan Valley Comm Coll, Somerville, NJ 08876-1265	1965	$2,100 (A)		8	5,614	289
Reading Area Comm Coll, Reading, PA 19603-1706	1971	$2,040 (A)		8	2,857	249
Redlands Comm Coll, El Reno, OK 73036-5304	1938	$1,416 (S)		5	2,062	124
Red Rocks Comm Coll, Lakewood, CO 80228-1255	1969	$1,842 (S)		5	7,957	276
Reedley Coll, Reedley, CA 93654-2099	1926	$390 (S)	$3,560	11	7,804	221
Rend Lake Coll, Ina, IL 62846-9801	1967	$1,152 (A)		5	3,902	180
Renton Tech Coll, Renton, WA 98056	1942	$1,710		5	5,792	422
Richard Bland Coll of the Coll of William & Mary, Petersburg, VA 23805-7100.	1961	$2,040 (S)		5	1,322	57
Richland Comm Coll, Decatur, IL 62521-8513	1971	$1,295 (A)		9	3,267	210
Richmond Comm Coll, Hamlet, NC 28345-1189.	1964	$584 (S)		5	1,324	100
Ricks Coll, Rexburg, ID 83460-4107.	1888	$1,950	$2,627	2	8,551	389
Ridgewater Coll, Willmar, MN 56201-1097	1961	$2,330 (S)		5	3,625	244
Rio Hondo Coll, Whittier, CA 90601-1699.	1960	$454 (S)		11	15,000	710
Rio Salado Coll, Tempe, AZ 85281-6950	1978	$1,120 (A)		11	9,457	597
Riverland Comm Coll, Austin, MN 55912	1940	$2,287 (S)		5	2,615	152
Roane State Comm Coll, Harriman, TN 37748-5011	1971	$1,066 (S)		5	5,366	337
Robeson Comm Coll, Lumberton, NC 28359-1420	1965	$588 (S)		5	1,640	114
Rochester Comm & Tech Coll, Rochester, MN 55904-4999	1915	$2,206 (S)		5	3,900	225
Rockingham Comm Coll, Wentworth, NC 27375-0038.	1964	$587 (S)		5	1,900	114
Rockland Comm Coll, Suffern, NY 10901-3699	1959	$2,400 (S)		11	6,080	395
Rock Valley Coll, Rockford, IL 61114-5699.	1964	$1,390 (A)		9	8,600	260
Rogers State U, Claremore, OK 74017-3252	1909	$1,349 (S)	$1,401	5	3,248	270
Rogue Comm Coll, Grants Pass, OR 97527-9298	1970	$1,620 (S)		11	3,715	459
Rose State Coll, Midwest City, OK 73110-2799	1968	$1,008 (S)		11	7,575	412
Rowan-Cabarrus Comm Coll, Salisbury, NC 28145-1595	1963	$589 (S)		5	3,961	229
Roxbury Comm Coll, Roxbury Crossing, MA 02120-3400	1973	$2,400 (S)		5	2,334	172
Sacramento City Coll, Sacramento, CA 95822-1386	1916	$312 (S)		11	16,583	726
Saint Augustine Coll, Chicago, IL 60640-3501	1980	$6,420		1-B	1,134	140
Saint Charles County Comm Coll, St. Peters, MO 63376-0975	1986	$1,440 (A)		5	5,416	253
St. Clair County Comm Coll, Port Huron, MI 48061-5015.	1923	$1,885 (A)		11	3,862	273
St. Cloud Tech Coll, St. Cloud, MN 56303-1240	1948	$2,158 (S)		5	2,440	121
St. Johns River Comm Coll, Palatka, FL 32177-3897.	1958	$1,167 (S)		5	4,192	157
St. Louis Comm Coll at Florissant Valley, St. Louis, MO 63135-1499.	1963	$1,344 (A)		9	7,365	350
St. Louis Comm Coll at Forest Park, St. Louis, MO 63110-1316	1962	$1,344 (A)		9	5,700	366
St. Louis Comm Coll at Meramec, Kirkwood, MO 63122-5720.	1963	$1,408 (A)		9	12,713	570
St. Paul Tech Coll, St. Paul, MN 55102-1800.	1919	$2,161 (S)		12	4,544	565
St. Petersburg Jr Coll, St. Petersburg, FL 33731-3489.	1927	$1,660 (S)		11	20,560	826
St. Philip's Coll, San Antonio, TX 78203-2098	1898	$878 (A)		9	7,848	454
Salem Comm Coll, Carneys Point, NJ 08069-2799	1972	$1,780 (A)		8	1,127	96
Salish Kootenai Coll, Pablo, MT 59855-0117	1977	$2,442		1-B	1,016	55
Salt Lake Comm Coll, Salt Lake City, UT 84130-0808	1948	$1,542 (S)		5	18,691	1,322
Sampson Comm Coll, Clinton, NC 28329-0318	1965	$591 (S)		11	1,305	102
Sandhills Comm Coll, Pinehurst, NC 28374-8299	1963	$588 (S)		11	2,628	148
San Diego Mesa Coll, San Diego, CA 92111-4998.	1964	$412 (S)		11	23,294	676
San Diego Miramar Coll, San Diego, CA 92126-2999	1969	$395 (S)		11	6,659	231
San Jacinto Coll–North Campus, Houston, TX 77049-4599	1974	$644 (A)		11	3,961	240
San Jacinto Coll–South Campus, Houston, TX 77089-6099	1979	$684 (A)		11	4,608	215
San Joaquin Delta Coll, Stockton, CA 95207-6370.	1935	$390 (S)		9	18,526	572
San Juan Coll, Farmington, NM 87402-4699	1958	$360 (S)		8	3,870	284
Santa Barbara City Coll, Santa Barbara, CA 93109-2394	1908	$407 (S)		11	11,288	578

Name, address	Year	Tuition & Fees	Room & Board	Control, Degree	Enroll- ment	Faculty
Santa Fe Comm Coll, Gainesville, FL 32606-6200	1966	$1,494 (S)		11	12,473	625
Santa Fe Comm Coll, Santa Fe, NM 87505-4887	1983	$506 (A)		11	3,280	285
Santa Monica Coll, Santa Monica, CA 90405-1628	1929	$446 (S)		11	25,824	1,075
Sauk Valley Comm Coll, Dixon, IL 61021	1965	$1,408 (A)		9	2,452	152
Savannah Tech Inst, Savannah, GA 31405	1929	$891 (S)		5	1,753	112
Schenectady County Comm Coll, Schenectady, NY 12305-2294	1968	$2,455 (S)		11	3,484	194
Schoolcraft Coll, Livonia, MI 48152-2696	1961	$1,640 (A)		9	9,292	434
Scott Comm Coll, Bettendorf, IA 52722-6804	1966	$1,755 (S)		11	3,807	235
Scottsdale Comm Coll, Scottsdale, AZ 85250-2699	1969	$1,120 (A)		11	9,418	455
Seattle Central Comm Coll, Seattle, WA 98122-2400	1966	$1,527 (S)		5	10,303	388
Seminole Comm Coll, Sanford, FL 32773-6199	1966	$1,328 (S)		11	8,124	641
Seminole State Coll, Seminole, OK 74818-0351	1931	$1,335 (S)	$2,200	5	1,633	77
Seward County Comm Coll, Liberal, KS 67905-1137	1969	$1,408 (S)	$3,100	11	2,325	214
Shawnee Comm Coll, Ullin, IL 62992-9725	1967	$1,216 (A)		11	1,961	200
Shelby State Comm Coll, Memphis, TN 38174-0568	1970	$1,152 (S)		5	4,542	265
Sheridan Coll, Sheridan, WY 82801-1500	1948	$1,296 (S)	$3,290	11	2,681	213
Shoreline Comm Coll, Seattle, WA 98133-5696	1964	$1,449 (S)		5	8,539	419
Sierra Coll, Rocklin, CA 95677-3397	1936	$410 (S)		5	18,000	518
Sinclair Comm Coll, Dayton, OH 45402-1460	1887	$1,457 (A)		11	17,325	975
Skagit Valley Coll, Mount Vernon, WA 98273-5899	1926	$1,458 (S)		5	5,862	374
Skyline Coll, San Bruno, CA 94066-1698	1969	$334 (S)		11	8,514	268
Snead State Comm Coll, Boaz, AL 35957-0734	1898	$1,380 (S)	$1,766	5	1,815	99
Snow Coll, Ephraim, UT 84627-1203	1888	$1,254 (S)	$2,800	5	2,643	126
Solano Comm Coll, Suisun City, CA 94585-3197	1945	$407 (S)		11	10,076	374
Somerset Comm Coll, Somerset, KY 42501-2973	1965	$1,140 (S)		5	2,482	153
South Arkansas Comm Coll, El Dorado, AR 71731-7010	1975	$922 (A)		5	1,203	63
Southeast Arkansas Coll, Pine Bluff, AR 71603	1991	$910 (S)		5	2,138	89
Southeast Comm Coll, Cumberland, KY 40823-1099	1960	$1,140 (S)		5	2,152	150
Southeast Comm Coll, Lincoln Campus, Lincoln, NE 68520-1299	1973	$1,356 (S)		9	5,205	580
Southeastern Baptist Theological Seminary, Wake Forest, NC 27588-1889	1950	$4,020		2-D	1,326	53
Southeastern Comm Coll, Whiteville, NC 28472-0151	1964	$581 (S)		5	1,610	145
Southeastern Comm Coll, North Campus, West Burlington, IA 52655-0180	1968	$1,740 (S)	$2,700	11	1,918	94
Southeastern Illinois Coll, Harrisburg, IL 62946-4925	1960	$1,056 (A)		5	3,477	184
Southeast Tech Inst, Sioux Falls, SD 57107-1301	1968	$2,374		5	2,203	113
Southern Maine Tech Coll, South Portland, ME 04106	1946	$2,585 (S)	$4,100	5	2,308	148
Southern State Comm Coll, Hillsboro, OH 45133-9487	1975	$2,619 (S)		5	1,582	96
Southern Union State Comm Coll, Wadley, AL 36276	1922	$1,437 (S)	$2,250	5	4,500	217
Southern U at Shreveport, Shreveport, LA 71107	1964	$1,110 (S)		5	1,399	112
South Florida Comm Coll, Avon Park, FL 33825-9356	1965	$1,260 (S)		5	2,500	213
South Georgia Coll, Douglas, GA 31533-5098	1906	$1,312 (S)	$2,490	5	1,148	54
South Mountain Comm Coll, Phoenix, AZ 85040	1979	$898 (A)		11	2,571	184
South Plains Coll, Levelland, TX 79336-6595	1958	$786 (A)	$2,400	11	6,705	353
South Puget Sound Comm Coll, Olympia, WA 98512-6292	1970	$1,615 (S)		5	5,511	247
South Seattle Comm Coll, Seattle, WA 98106-1499	1970	$1,482 (S)		5	6,582	194
Southside Virginia Comm Coll, Alberta, VA 23821-9719	1970	$1,547 (S)		5	3,273	182
South Suburban Coll, South Holland, IL 60473-1270	1927	$1,332 (A)		11	7,810	384
Southwestern Coll, Chula Vista, CA 91910-7299	1961	$435 (S)		11	16,139	760
Southwestern Comm Coll, Creston, IA 50801	1966	$2,046 (S)		5	1,093	75
Southwestern Comm Coll, Sylva, NC 28779	1964	$577 (S)		5	1,651	180
Southwestern Michigan Coll, Dowagiac, MI 49047-9793	1964	$1,674 (A)		11	3,019	209
Southwestern Oregon Comm Coll, Coos Bay, OR 97420-2912	1961	$1,734	$4,905	11	2,907	251
Southwest Mississippi Comm Coll, Summit, MS 39666	1918	$850 (S)		11	1,656	89
Southwest Missouri State U—West Plains, West Plains, MO 65775	1963	$2,130 (S)	$3,700	5	1,369	NA
Southwest Texas Jr Coll, Uvalde, TX 78801-6297	1946	$702 (A)	$2,120	11	3,452	166
Southwest Virginia Comm Coll, Richlands, VA 24641-1510	1968	$1,430 (S)		5	3,938	235
Southwest Wisconsin Tech Coll, Fennimore, WI 53809-9778	1967	$1,799 (S)		11	2,074	101
Spokane Comm Coll, Spokane, WA 99217-5399	1963	$1,452 (S)		5	6,294	367
Spokane Falls Comm Coll, Spokane, WA 99224-5288	1967	$1,401 (S)		5	8,842	346
Spoon River Coll, Canton, IL 61520-9801	1959	$1,536 (A)		5	1,813	189
Springfield Tech Comm Coll, Springfield, MA 01105-1296	1967	$2,510 (S)		5	6,509	247
Stanly Comm Coll, Albemarle, NC 28001-7458	1971	$588 (S)		5	1,483	76
Stark State Coll of Technology, Canton, OH 44720-7299	1970	$3,114 (S)		11	4,413	210
State Tech Inst at Memphis, Memphis, TN 38134-7693	1967	$1,156 (S)		5	8,835	763
State U of New York Coll of Ag & Tech at Cobleskill, Cobleskill, NY 12043	1916	$3,771 (S)	$5,730	5-B	2,320	145
State U of New York Coll of Ag & Tech at Morrisville, Morrisville, NY 13408	1908	$3,765 (S)	$5,440	5-B	2,767	171
State U of New York Coll of Technology at Alfred, Alfred, NY 14802	1908	$3,757 (S)	$5,258	5-B	2,840	141
State U of New York Coll of Technology at Canton, Canton, NY 13617	1906	$3,500 (S)	$5,230	5-B	2,078	124
State U of New York Coll of Technology at Delhi, Delhi, NY 13753	1913	$3,687 (S)	$5,770	5-B	2,002	141
Sullivan County Comm Coll, Loch Sheldrake, NY 12759	1962	$2,656 (S)		11	1,677	106
Surry Comm Coll, Dobson, NC 27017-0304	1965	$623 (S)		5	2,852	140
Sussex County Comm Coll, Newton, NJ 07860	1981	$2,310 (A)		11	2,250	188
Tacoma Comm Coll, Tacoma, WA 98466	1965	$1,521 (S)		5	5,122	391
Taft Coll, Taft, CA 93268-2317	1922	$420 (S)	$2,720	11	1,054	66
Tallahassee Comm Coll, Tallahassee, FL 32304-2895	1966	$1,205 (S)		11	10,563	388
Tarrant County Coll, Fort Worth, TX 76102-6599	1967	$884 (A)		8	25,570	1,189
Technical Career Insts, New York, NY 10001-2705	1909	$6,245		3	3,500	192
Technical Coll of the Lowcountry, Beaufort, SC 29901-1288	1972	$1,000 (S)		5	1,853	69
Temple Coll, Temple, TX 76504-7435	1926	$930 (A)	$2,830	9	3,147	125
Terra State Comm Coll, Fremont, OH 43420-9670	1968	$2,268 (S)		5	2,630	136
Texarkana Coll, Texarkana, TX 75599-0001	1927	$690 (A)		11	4,003	196
Texas State Tech Coll, Sweetwater, TX 79556-4108	1970	$1,620 (S)	$4,340	5	1,157	135
Texas State Tech Coll–Harlingen, Harlingen, TX 78550-3697	1967	$1,293 (S)	$3,255	5	3,311	172
Texas State Tech Coll–Waco/Marshall Campus, Waco, TX 76705-1695	1965	$2,031 (S)		5	1,067	285
Thomas Nelson Comm Coll, Hampton, VA 23670-0407	1968	$1,546 (S)		5	7,058	320
Three Rivers Comm Coll, Poplar Bluff, MO 63901-2393	1966	$1,054 (A)		11	2,300	67
Tidewater Comm Coll, Norfolk, VA 23510	1968	$1,696 (S)		5	17,907	760
Tomball Coll, Tomball, TX 77375-4036	1988	$884 (A)		11	4,240	230
Tompkins Cortland Comm Coll, Dryden, NY 13053-9533	1968	$2,708 (S)		11	2,554	196
Treasure Valley Comm Coll, Ontario, OR 97914-3423	1962	$1,680 (S)	$3,589	11	3,516	113

Name, address	Year	Tuition & Fees	Room & Board	Control, Degree	Enroll-ment	Faculty
Tri-County Comm Coll, Murphy, NC 28906-7919	1964	$581 (S)		5	1,035	55
Trident Tech Coll, Charleston, SC 29423-8067	1964	$1,064 (A)		11	8,730	523
Trinidad State Jr Coll, Trinidad, CO 81082-2396	1925	$1,962 (S)	$3,282	5	2,543	176
Trinity Valley Comm Coll, Athens, TX 75751-2765	1946	$608 (A)	$2,908	11	4,623	224
Triton Coll, River Grove, IL 60171-9983	1964	$1,518 (A)		5	17,815	642
Truckee Meadows Comm Coll, Reno, NV 89512-3901	1971	$948 (S)		5	9,954	536
Truett-McConnell Coll, Cleveland, GA 30528	1946	$5,550	$3,000	2	1,955	169
Tulsa Comm Coll, Tulsa, OK 74135-6198	1968	$1,140 (S)		5	26,000	1,200
Tunxis Comm Tech Coll, Farmington, CT 06032-3026	1969	$1,814 (S)		5	3,335	175
Tyler Jr Coll, Tyler, TX 75711-9020	1926	$826 (A)		11	7,955	364
Ulster County Comm Coll, Stone Ridge, NY 12484	1961	$2,596 (S)		11	2,678	185
Umpqua Comm Coll, Roseburg, OR 97470-0226	1964	$1,610 (S)		11	1,789	148
Union County Coll, Cranford, NJ 07016-1528	1933	$2,579 (A)		11	8,900	391
The U of Akron–Wayne Coll, Orrville, OH 44667-9192	1972	$3,443 (S)		5	1,385	113
U of Alaska Anchorage, Kenai Peninsula Coll, Soldotna, AK 99669	1964	$2,229 (S)		5	1,688	107
U of Alaska Anchorage, Matanuska-Susitna Coll, Palmer, AK 99645	1958	$1,772 (S)		5	1,182	111
U of Alaska Southeast, Sitka Campus, Sitka, AK 99835-9418	1962	$1,896 (S)		5	1,290	104
U of Arkansas Comm Coll at Hope, Hope, AR 71801-0140	1966	$860 (A)		5	1,327	68
U of Cincinnati Clermont Coll, Batavia, OH 45103-1785	1972	$3,051 (S)		5	1,950	163
U of Cincinnati Raymond Walters Coll, Cincinnati, OH 45236-1007	1967	$3,573 (S)		5	3,483	265
U of Hawaii–Hawaii Comm Coll, Hilo, HI 96720-4091	1954	$1,034 (S)	$3,813	5	2,285	149
U of Hawaii–Honolulu Comm Coll, Honolulu, HI 96817-4598	1920	$956 (S)		5	4,124	185
U of Hawaii–Kapiolani Comm Coll, Honolulu, HI 96816-4421	1957	$956 (S)		5	7,236	317
U of Hawaii–Leeward Comm Coll, Pearl City, HI 96782-3393	1968	$999 (S)		5	6,000	236
U of Hawaii–Maui Comm Coll, Kahului, HI 96732	1967	$1,002 (S)		5	2,797	136
U of Kentucky, Lexington Comm Coll, Lexington, KY 40506-0235	1965	$2,036 (S)		5	6,111	355
U of New Mexico–Gallup, Gallup, NM 87301-5603	1968	$720 (S)		5-B	2,612	159
U of New Mexico–Los Alamos Branch, Los Alamos, NM 87544-2233	1980	$752 (S)		5	1,000	96
U of New Mexico–Valencia Campus, Los Lunas, NM 87031-7633	1981	$720 (S)		5	1,507	93
U of North Dakota–Lake Region, Devils Lake, ND 58301-1598	1941	$1,878 (S)	$2,612	5	1,134	54
U of Puerto Rico, Colegio Regional de la Montaña, Utuado, PR 00641	1979	$1,486 (S)		6-B	1,330	76
U of South Carolina Beaufort, Beaufort, SC 29902-4601	1959	$1,988 (S)		5	1,039	80
U of South Carolina Lancaster, Lancaster, SC 29721-0889	1959	$1,988 (S)		5	1,064	52
U of South Carolina Sumter, Sumter, SC 29150-2498	1966	$1,988 (S)		5	1,233	72
U of Wisconsin–Fox Valley, Menasha, WI 54952-8002	1933	$2,081 (S)		5	1,328	58
U of Wisconsin–Waukesha, Waukesha, WI 53188-2799	1966	$2,098 (S)		5	1,635	77
Utah Valley State Coll, Orem, UT 84058-5999	1941	$1,519 (S)		5-B	21,647	967
Valencia Comm Coll, Orlando, FL 32802-3028	1967	$1,275 (S)		5	24,655	954
Vance-Granville Comm Coll, Henderson, NC 27536-0917	1969	$588 (S)		5	3,197	284
Ventura Coll, Ventura, CA 93003-3899	1925	$412 (S)		11	11,791	574
Vermont Tech Coll, Randolph Center, VT 05061-0500	1866	$5,224 (S)	$5,206	5-B	1,003	107
Vernon Regional Jr Coll, Vernon, TX 76384-4092	1970	$801 (A)	$2,153	11	1,928	98
Victoria Coll, Victoria, TX 77901-4494	1925	$784 (A)		8	3,733	120
Victor Valley Coll, Victorville, CA 92392-5849	1961	$308 (S)		5	10,218	325
Vincennes U, Vincennes, IN 47591-5202	1801	$2,541 (S)	$4,194	5	5,788	383
Virginia Highlands Comm Coll, Abingdon, VA 24212-0828	1967	$1,525 (S)		5	3,443	129
Virginia Western Comm Coll, Roanoke, VA 24038	1966	$1,434 (S)		5	7,092	290
Vista Comm Coll, Berkeley, CA 94704-5102	1974	$360 (S)		11	3,843	146
Volunteer State Comm Coll, Gallatin, TN 37066-3188	1970	$1,146 (S)		5	6,718	404
Wake Tech Comm Coll, Raleigh, NC 27603-5696	1958	$568 (S)		11	8,186	546
Wallace State Comm Coll, Hanceville, AL 35077-2000	1966	$1,170 (S)		5	4,659	341
Walla Walla Comm Coll, Walla Walla, WA 99362-9267	1967	$1,545 (S)		5	5,287	416
Washington State Comm Coll, Marietta, OH 45750-9225	1971	$2,565 (S)		5	2,019	132
Washtenaw Comm Coll, Ann Arbor, MI 48106	1965	$1,726 (A)		11	10,441	599
Waubonsee Comm Coll, Sugar Grove, IL 60554-9799	1966	$1,340 (A)		9	7,021	558
Wayne Comm Coll, Goldsboro, NC 27533-8002	1957	$588 (S)		11	2,853	214
Wayne County Comm Coll District, Detroit, MI 48226-3010	1967	$1,770 (A)		11	12,000	400
Weatherford Coll, Weatherford, TX 76086-5699	1869	$858 (A)	$2,599	11	2,576	146
Wenatchee Valley Coll, Wenatchee, WA 98801-1799	1939	$1,461 (S)	$3,780	11	3,074	203
Westchester Business Inst, White Plains, NY 10602	1915	$10,890		3	1,086	48
Westchester Comm Coll, Valhalla, NY 10595-1698	1946	$2,583 (S)		11	10,603	791
Western Iowa Tech Comm Coll, Sioux City, IA 51102-5199	1966	$2,040 (S)		5	4,096	238
Western Nebraska Comm Coll, Scottsbluff, NE 69361	1926	$1,020 (S)		11	1,704	196
Western Nevada Comm Coll, Carson City, NV 89703-7316	1971	$1,095 (S)		5	5,100	354
Western Oklahoma State Coll, Altus, OK 73521-1397	1926	$1,196 (S)	$1,800	5	1,701	93
Western Piedmont Comm Coll, Morganton, NC 28655-4511	1964	$579 (S)		5	2,400	132
Western Texas Coll, Snyder, TX 79549-9502	1969	$907 (A)		11	1,180	55
Western Wisconsin Tech Coll, La Crosse, WI 54602-0908	1911	$1,843 (S)		9	4,760	187
Western Wyoming Comm Coll, Rock Springs, WY 82902-0428	1959	$1,098 (S)	$2,610	11	2,624	232
West Hills Comm Coll, Coalinga, CA 93210-1399	1932	$390 (S)		5	3,463	170
Westmoreland County Comm Coll, Youngwood, PA 15697	1970	$1,440 (A)		8	5,561	390
West Shore Comm Coll, Scottville, MI 49454-0277	1967	$1,234 (A)		9	1,338	81
West Valley Coll, Saratoga, CA 95070-5698	1963	$444 (S)		11	11,000	560
West Virginia Northern Comm Coll, Wheeling, WV 26003-3699	1972	$1,438 (S)		5	2,670	166
West Virginia U at Parkersburg, Parkersburg, WV 26101-9577	1971	$1,240 (S)		5-B	3,500	176
Wharton County Jr Coll, Wharton, TX 77488-3298	1946	$884 (A)		11	4,208	224
Whatcom Comm Coll, Bellingham, WA 98226-8003	1970	$1,440 (S)		5	3,801	200
Wilkes Comm Coll, Wilkesboro, NC 28697	1965	$588 (S)		5	1,948	196
William Rainey Harper Coll, Palatine, IL 60067-7398	1965	$1,438 (A)		11	14,986	1,018
Wilson Tech Comm Coll, Wilson, NC 27893-3310	1958	$581 (S)		5	1,444	100
Wisconsin Indianhead Tech Coll, New Richmond Campus, New Richmond, WI 54017-1738	1972	$1,824 (S)		9	1,234	65
Wisconsin Indianhead Tech Coll, Rice Lake Campus, Rice Lake, WI 54868	1941	$1,824 (S)		9	1,335	79
Wor-Wic Comm Coll, Salisbury, MD 21804	1976	$1,584 (A)		11	2,057	103
Wytheville Comm Coll, Wytheville, VA 24382-3308	1967	$1,445 (S)		5	2,328	142
Yakima Valley Comm Coll, Yakima, WA 98907-2520	1928	$1,431 (S)		5	3,946	292
Yavapai Coll, Prescott, AZ 86301-3297	1966	$744 (S)	$3,240	11	4,834	487
York Tech Coll, Rock Hill, SC 29730-3395	1961	$948 (A)		5	3,476	230

ASSOCIATIONS AND SOCIETIES

Source: World Almanac questionnaire; World Almanac research

Selected list, by first key word in each title. (Listed by acronym when that is the official name.) Founding year in parentheses; last figure after ZIP code = membership as reported. Information, especially website addresses, subject to change. Where there is no punctuation at end of line in a website address, do not add punctuation or space. For other organizations, see Directory of Sports Organizations, under Sports; Where to Get Help directory, under Health; Labor Union Directory, under Employment; lists of religious groups' headquarters, under Religious Information; international organizations, under Nations of the World.

AACSB-The Intl. Association for Management Education, (1916), 600 Emerson Rd., Ste. 300, St. Louis, MO 63141; 900 organizations; http://www.aacsb.edu

Aaron Burr Accord (1985), P.O. Box 4644, Seattle, WA 98104; 384.

Accountants, American Institute of Certified Public (1887), 1211 Ave. of the Americas, New York, NY 10036; 330,000; http://www.aicpa.org

Accountants, Natl. Assn. of Enrolled Federal Tax (1960), P.O. Box 59-009, Chicago, IL 60659.

Acoustical Society of America (1929), 500 Sunnyside Blvd., Woodbury, NY 11797; 6,800; http://asa.aip.org/index.html

Actuaries, Society of (1949), 475 N. Martingale Rd., Ste. 800, Schaumburg, IL 60173; 16,500; http://www.soa.org

Advertisers, Assn. of Natl. (1910), 708 Third Ave., New York, NY 10017; 225 cos.; http://www.ana.net

Advertising Agencies, American Assn. of (1917), 405 Lexington Ave., New York, NY 10174; 600 agencies.

Aeronautic Assn., Natl. (1904), 1815 N. Fort Myer Dr., Arlington, VA 22209; 6,000; http://www.naa.ycg.org

Aerospace Industries Assn. of America (1919), 1250 Eye St. NW, Wash., DC 20005; 50 cos.; http://www.aia-aerospace.org

Aerospace Medical Assn. (1929), 320 S. Henry St., Alexandria, VA 22314; 3,400; http://www.asma.org

African Violet Soc. of America Inc. (1946), 2375 North St., Beaumont, TX 77702; 9,000; http://avsa.org

Afro-American Life and History, Assn. for the Study of (1915), 7961 Eastern Ave., Ste. 301, Silver Spring, MD 20910; 1,400; http://www.artnoir.com/asalh

AFS-USA (1947), 198 Madison Ave., 8th Floor, New York, NY 10016; 475,000.

Agricultural Economics Assn., American (1910), 415 S. Duff Ave., Ste. C, Ames, IA 50010; 3,500; http://www.aaea.org

Agricultural Engineers, American Soc. of (ASAE) (1907), 2950 Niles Rd., St. Joseph, MI 49085; 8,000; http://www.asae.org

Agricultural History Society (1919), 1800 M St. NW, Rm. 2103, Wash., DC 20036; 1,400.

Agronomy, American Society of (1907), 677 S. Segoe Rd., Madison, WI 53711; 11,466; http://www.agronomy.org

Aircraft Owners and Pilots Assn. (1939), 421 Aviation Way, Frederick, MD 21701; 350,000+; http://www.aopa.org

Air Force Assn. (1946), 1501 Lee Hwy., Arlington, VA 22209; 150,000; http://www.afa.org

Air Force Gunners Assn. (1986), 453 Plaza Circle, Bossier City, LA 71111; 1,700.

Air & Waste Management Assn. (1907), One Gateway Center, 3d Fl., Pittsburgh, PA 15222; 12,000; http://www.awma.org

Al-Anon Family Groups, Inc. (1951), 1600 Corporate Landing Pkwy., Virginia Beach, VA 23454; 31,000 groups; http://www.al-anon.alateen.org

Alcoholics Anonymous (1935), P.O. Box 459, Grand Central Station, New York, NY 10163; 2 mil+; http://www.alcoholics-anonymous.org

Alcoholism and Drug Dependence, Inc., Natl. Council on (1944), 12 W. 21st St., New York, NY 10010; http://www.ncadd.org

Alexander Graham Bell Assn. for the Deaf (1890), 3417 Volta Pl. NW, Wash., DC 20007; 5,000; http://www.agbell.org

Allergy, Asthma, and Immunology, American Academy of (1943), 611 E. Wells St., Milwaukee, WI 53202; 5,882; http://www.aaaai.org

Alpha Delta Kappa (1947), 1615 West 92d St., Kansas City, MO 64114; 56,000; http://www.alphadeltakappa.org

Alpha Lambda Delta, Natl. (1924), P.O. Box 4403, Macon, GA 31208; 575,000; http://www.mercer.edu/ald

Alpine Club, American (1902), 710 Tenth St., Ste. 100, Golden, CO 80401; 3,750; http://www.americanalpineclub.org

Alzheimer's Assn. (1980), 919 Michigan Ave., Chicago, IL 60611; http://www.alz.org

Amateur Radio Union, Intl. (IARU) (1925), P.O. Box 310905, Newington, CT 06131; 148 org.; http://www.iaru.org

AMBUCS, Inc., Natl. (1921), P.O. Box 5127, High Point, NC 27262; 6,500+; http://www.ambucs.com

American Bar Association, 750 N. Lake Shore Dr., Chicago, IL 60611; http://www.abanet.org

American Indian Affairs, Inc., Assn. on (1922), P.O. Box 268, Sisseton, SD 57262; 25,000.

American Indians, Natl. Congress of (1944), 2010 Massachusetts Ave. NW, Wash., DC 20036; 3,000; http://www.ncai.org

American Legion (1919), P.O. Box 1055, 700 N. Pennsylvania St., Indianapolis, IN 46204; 2.9 mil.; http://www.legion.org

American Legion Auxiliary (1920), 777 N. Meridian St., 3d Fl., Indianapolis, IN 46204; 900,000; http://www.legion-aux.org

Americares Foundation (1982), 161 Cherry St., New Canaan, CT 06840; http://www.americares.org

AMIDEAST (formerly American Mideast Educational & Training Services) (1951), 1730 M St. NW, Ste. 1100, Wash., DC 20036; http://www.amideast.org

Amnesty Intl. USA (1961), 322 8th Ave., New York, NY 10001; http://rights.amnesty.org

Amputation Foundation, Inc., Natl. (1919), 38-40 Church St., Malverne, NY 11565; 2,500.

AMVETS (American Veterans) (1947); **AMVETS Natl. Auxiliary** (1946), 4647 Forbes Blvd., Lanham, MD 20706; 250,000; http://www.amvets.org

Amusement Parks and Attractions, Intl. Assn. of (1918), 1448 Duke St., Alexandria, VA 22314; 5,600; http://www.iaapa.org

Animals, American Society for Prevention of Cruelty to (ASPCA) (1866), 424 E. 92d St., New York, NY 10128; 299,000; http://www.aspca.org

Animal Protection Institute (1968), 2831 Fruitridge Rd., Sacramento, CA 95820; 65,000; http://www.api4animals.org

Animal Welfare Institute (1951), P.O. Box 3650, Wash., DC 20007; 5,500; other: 13,160; http://www.animalwelfare.com

Anthropological Assn., American (1902), 4350 N. Fairfax Dr., Ste. 640, Arlington, VA 22203; 10,000; http://www.aaanet.org

Anti-Vivisection Society, American (AAVS) (1883), 801 Old York Rd., Ste. 204, Jenkintown, PA 19046; 10,000; http://www.aays.org

Antiquarian Society, American (1812), 185 Salisbury St., Worcester, MA 01609; 655.

Appalachian Mountain Club (1876), 5 Joy St., Boston, MA 02108; 76,000; http://www.outdoors.org

Appalachian Trail Conference (1925), Washington & Jackson Sts., Harpers Ferry, WV 25425; 23,000; http://www.atconf.org

Appraisers, American Society of (1936), 555 Herndon Pkwy., Herndon, VA 22070; 5,900; http://www.appraisers.org

Arab Americans, Natl. Assn. of (1972), 1212 New York Ave. NW, Wash., DC 20005; http://www.naaa.net/index.html

Arbitration Assn., American (1926), 140 W. 51st St., New York, NY 10020; 10,000; http://www.adr.org

Arc of the United States, The (1950), 500 E. Border St., Ste. 300, Arlington, TX 76010; 140,000; http://www.thearc.org

Archaeological Institute of America (1879), 656 Beacon St., Boston, MA 02215; 10,000; http://www.archaeological.org

Archery Assn., Natl. (1879), One Olympic Plaza, Colorado Springs, CO 80909; 6,000; http://www.USArchery.org

Architects, American Institute of (1857), 1735 New York Ave. NW, Wash., DC 20006; 55,000; http://www.aia.online.org

ARMA Intl. (formerly Assn. of Records Managers & Administrators) (1975), 4200 Somerset Dr., Ste. 215, Prairie Village, KS 66208; 10,300; http://www.arma.org/hq

AFCEA (Armed Forces Communications and Electronics Assn.) (1946), 4400 Fair Lakes Ct., Fairfax, VA 22033; 40,000; http://www.afcea.org

Army, Assn. of the United States (1950), 2425 Wilson Blvd., Arlington, VA 22201; 117,000; http://www.ausa.org

Arthritis Foundation (1948), 1330 W. Peachtree St., Atlanta, GA 30309; http://www.arthritis.org

Arts, American Council for the (1960), One E. 53d St., New York, NY 10022; 1,500; http://www.artsusa.org

Arts, American Federation of (1909), 41 E. 65th St., New York, NY 10021; 520+ museums/institutions.

Arts, Americans for the (1996), 1000 Vermont Ave. NW, 12th Fl., Wash., DC 20005; 1,000; http://www.artsusa.org

Arts and Letters, American Academy of (1898), 633 W. 155 St., New York, NY 10032; 250.

Arts and Letters, Natl. Society of (1944), 4227 46th St., NW, Wash., DC 20016; 1,450; http://www.arts-nsal.org

Arts and Sciences, American Academy of (1780), Norton's Woods, 136 Irving St., Cambridge, MA 02138; 3,600 fellows; http://www.amacad.org

Associated Press (1848), 50 Rockefeller Plaza, New York, NY 10020; 1,700 newspapers, 5,000 U.S. broadcast stations, 8,500 intl. subscribers; http://www.ap.org

Association Executives, American Society of (1920), 1575 I St. NW, Wash., DC 20005; 25,000; http://www.asaenet.org

Astrologers, Inc., American Federation of (1938), P.O. Box 22040, Tempe, AZ 85285; 3,500+.

Astronautical Society, American (1954), 6352 Rolling Mill Pl., Springfield, VA 22152; 1,500; http://www.astronautical.org

Astronomical Society, American (1899), 2000 Florida Ave. NW, Ste. 400, Wash., DC 20009; 1,600+; http://www.aas.org

Ataxia Foundation, Natl. (1957), 2600 Fernbrook Ln., Ste. 119, Minneapolis, MN 55447; 10,300; http://www.ataxia.org

Atheists, American (1963), P.O. Box 5733, Parsippany, NJ 07054; 2,000; http://www.atheists.org

Auctioneers Assn., Natl. (1949), 8880 Ballentine St., Overland Park, KS 66214; 5,826; http://www.auctioneers.org

Audubon Society, Natl. (1905), 700 Broadway, New York, NY 10003; 550,000; http://www.audubon.org

Authors Guild, The (1912), 330 W. 42d St., 29th Floor, New York, NY 10036; 7,200; http://www.authorsguild.org

Authors Registry Inc., The (1995), 330 W. 42d St., 29th Fl., New York, NY 10036; representing approx. 50,000 authors http://www.authorsregistry.org

Autism Society of America (1965), 7910 Woodmont Ave., Ste. 650, Bethesda, MD 20814; 23,000; http://www.autism-society.org

Autograph Collectors Club, Universal (1965), P.O. Box 6181, Wash., DC 20044; 1,800; http://www.uacc.org

Automobile Assn., American (1902), 1000 AAA Dr., Heathrow, FL 32746; 40 mil; http://www.aaa.com

Automobile Club of America, Antique (1935), 501 W. Governor Rd., Hershey, PA 17033; 53,000; http://www.aaca.org/frame.htm

Automobile Dealers Assn., Natl. (1917), 8400 Westpark Dr., McLean, VA 22102; 19,500; http://www.nadanet.com

Automobile License Plate Collectors Assn. (1954), 226 Ridgeway Dr., Bridgeport, WV 26330; 3,000; http://www.alpca.org

Automotive Hall of Fame (1936), 21400 Oakwood Blvd., Dearborn, MI 48124; 700 visiting members.

Badminton, USA (1936), One Olympic Plaza, Colorado Springs, CO 80909; 2,200; http://www.usabadminton.org

Baker Street Irregulars (1934), P.O. Box 2189, Easton, MD 21601; 300.

Bald-Headed Men of America (1972), 102 Bald Dr., Morehead City, NC 28557; approx. 36,000.

Ball Players of America, Assn. of Prof. (1924), 12062 Valley View St., Ste. 211, Garden Grove, CA 92845; 10,000.

Bankers Assn., American (1875), 1120 Connecticut Ave. NW, Wash., DC 20036; http://www.aba.com

Bar Assn., Federal (1920), 2215 M Street, Wash., DC 20037; 15,002; http://www.fedbar.org

Barber Shop Quartet Singing in America, Inc., Soc. for the Preservation & Encouragement of (1938), 6315 Third Ave., Kenosha, WI 53143; 34,000; http://www.spebsqsa.org

Baseball Congress, American Amateur (1935), 118 Redfield Plaza, Marshall, MI 49068; 14,500 teams; http://www.voyager.net/aabc

Baseball Congress, Natl. (1931), 300 S. Sycamore, Wichita, KS 67213; 7,500; http://www.wichitawranglers.com

Baseball Players of America, Assn. of Prof. (1924), 12062 Valley View St., Ste. 211, Garden Grove, CA 92845; 16,000+.

Baseball Research, Inc., Society for American (1971), 812 Huron Rd E #719, Cleveland, OH 44115; 6,100; http://www.sabr.org

Battleship Assn., American (1964), P.O. Box 711247, San Diego, CA 92171; 1,190.

Beer Can Collectors of America (1970), 747 Merus Ct., Fenton, MO 63026; 4,000; http://www.bcca.com

Beta Gamma Sigma, Inc. (1913), 11701 Borman Dr., Ste. 295, St. Louis, MO 63146; 415,000; http://www.betagammasigma.org

Beta Sigma Phi (1931), 1800 W. 91st Pl., Box 8500, Kansas City, MO 64114; 200,000; http://www.dmr1.com/bsp

Better Business Bureaus, Council of (1970), 4200 Wilson Blvd., Ste. 800, Arlington, VA 22203; 150 bureaus; http://www.bbb.org

Bible Society, American (1816), 1865 Broadway, New York, NY 10023; 280,000; http://www.americanbible.org

Biblical Literature, Society of (1880), 825 Houston Mill Rd., Ste. 350, Atlanta, GA 30329; 8,000; http://www.sbl-site.org

Bibliographical Society of America (1904), P.O. Box 1537, Lenox Hill Station, New York, NY 10021; 1,200; http://www.bibsocamer.org

Big Brothers/Big Sisters of America (1904), 230 N. 13th St., Philadelphia, PA 19107; 494 agencies; http://bbbsa.org

Biochemistry and Molecular Biology, American Society for (1906), 9650 Rockville Pike, Bethesda, MD 20814; 10,000; http://www.faseb.org/asbmb

Biological Sciences, American Institute of (1947), 730 11th St. NW, Wash., DC 20001; 5,000; http://www.aibs.org

Blind, American Council of the (1961), 1155 15th St. NW, Ste. 720, Wash., DC 20005; 45,000; http://www.acb.org

Blind, Natl. Federation of the (1940), 1800 Johnson St., Baltimore, MD 21230; 50,000; http://www.nfb.org

Blinded Veterans Assn. (1945), 477 H St. NW, Wash., DC 20001; 7,900.; http://bva.org

Blindness America, Prevent (1908), 500 E. Remington Rd., Schaumburg, IL 60173; http://www.preventblindness.org

B'nai B'rith Intl. (1843), 1640 Rhode Island Ave. NW, Wash., DC 20036; 250,000; http://www.bnaibrith.org

Boat Owners Assn. of the U.S. (1966), 880 S. Pickett St., Alexandria, VA 22304; 500,000; http://www.boatus.com

Booksellers Assn., American (1900), 828 S. Broadway, Tarrytown, NY 10591; 8,000; http://www.bookweb.org/aba

Boy Scouts of America (1910), 1325 Walnut Hill Lane, Irving, TX 75015; 4.7 mil; http://www.bsa.scouting.org

Boys & Girls Clubs of America (1906), 1230 W. Peachtree St. NW, Atlanta, GA 30309; 2.85 mil; http://www.bgca.org

Bread for the World, Inc. (1974), 1100 Wayne Ave., Ste. 1000, Silver Spring, MD 20910; 44,000; http://www.bread.org

Brith Sholom (1905), 3939 Conshohocken Ave., Philadelphia, PA 19131; 6,000.

Broadcasters, Natl. Assn. of (1922-23), 1771 N St. NW, Wash., DC 20036; http://www.nab.org

Burroughs Bibliophiles, The (1960), 454 Elaine Dr., Pittsburgh, PA 15236; 863.

Business Communicators, Intl. Assn. of (1970), One Hallidie Plaza, Ste. 600, San Francisco, CA 94102; 12,500; http://www.iabc.com/homepage.htm

Business Education Assn., Natl. (1946), 1914 Association Dr., Reston, VA 20191; 12,000; http://www.nbea.org/nbea.html

Business Women's Assn., American (1949), 9100 Ward Pkwy., Kansas City, MO 64114; 70,000; http://www.abwahq.org

Button Society, Natl. (1938), c/o Lois Pool, 2733 Juno Pl., Akron, OH 44333; 4,600+.

Byron Society of America, The (1973), c/o Prof. Charles E. Robinson, Dept. of English, Univ. of Delaware, Newark, DE 19716; 350.

Camp Fire Boys & Girls (1910), 4601 Madison Ave., Kansas City, MO 64112; 667,000; http://www.campfire.org

Camping Assn., American (1910), 5000 State Rd. 67 N., Martinsville, IN 46151; 5,000; http://www.acacamps.org

Cancer Society, American (1913), 1599 Clifton Rd. NE, Atlanta, GA 30329; over 2 mil.; http://www.cancer.org

Cartoonists Society, Natl. (1946), Columbus Circle Station, P.O. Box 20267, New York, NY 10023; 500+; http://www.unitedmedia.com/ncs/ncs.html

Cat Fanciers' Assn. (1906), 1805 Atlantic Ave., P.O. Box 1005, Manasquan, NJ 08736; 650 clubs; http://www.cfainc.org/cfa

Catholic Bishops, U.S. Natl. Conference of (1917), 3211 4th St. NE, Wash., DC 20017; 402 members, 350 staff.

Catholic Church Extension Society (1905), 150 S. Wacker Dr., 20th floor, Chicago, IL 60606; 90,000; http://www.knight.org/advent/cathen/14078a.htm

Catholic Daughters of the Americas (1903), 10 W. 71st St., New York, NY 10023; 117,815.

Catholic Educational Assn., Natl. (1904), 1077 30th St. NW, Ste. 100, Wash., DC 20007; 26,000; http://www.ncea.org

Catholic Historical Soc., American (1884), 263 S. Fourth St., Philadelphia, PA 19106; 625; http://www.AM/CHS.org

Catholic Library Association (1921), 100 North St., Ste. 224, Pittsfield, MA 01201; 1,000; http://www.cathla.org

Catholic War Veterans, USA, Inc. (1935), 441 N. Lee St., Alexandria, VA 22314; 25,000.

Cemetery and Funeral Assn., Intl. (1887), 1895 Preston White Dr., #220, Reston, VA 22091; 2,200; http://www.icfa.org

Ceramic Society, American (1898), 735 Ceramic Place, Westerville, OH 43081; 10,000; http://www.acers.org

Cerebral Palsy Assns., Inc., United (1949), 1660 L St. NW, Ste. 700, Wash., DC 20036; 150; http://www.ucpa.org

Certified Electronics Technicians, Intl. Society of (1970), 2708 W. Berry St., Ft. Worth, TX 76109; 1,717; http://www.iscet.org

Chamber of Commerce of the U.S.A. (1912), 1615 H St. NW, Wash., DC 20062; 215,000.

Chamber Music Players, Inc., Amateur (1947), 1123 Broadway, New York, NY 10010; 4,300; http://www.acmp.net

Chartered Life Underwriters, American Soc. of (1927), 270 Bryn Mawr Ave., Bryn Mawr, PA 19010; 30,000.

Checker Federation, American (1949), P.O.Box 241, Petal, MS 39465; 1,000.

Chemical Engineers, American Inst. of (1908), 345 E. 47th St., New York, NY 10017; 60,000; http://www.che.ufl.edu.aiche

Chemical Manufacturers Assn. (1872), 1300 Wilson Blvd., Arlington, VA 22209; 191 cos.; http://www.cmahq.com

Chemical Society, American (1876), 1155 16th St. NW, Wash., DC 20036; 151,000; http://www.acs.org

Chess Federation, U.S. (1949), 186 Rt. 9W, New Windsor, NY 12553; 84,327; http://www.uschess.org

Chess League of America, Correspondence (1897), P.O. Box 59625, Schaumburg, IL 60159; 1,000; http://www.newtan.loyola.edu/ccla

Chiefs of Police, Intl. Assn. of (1893), 515 N. Washington St., Alexandria, VA 22314; 17,000; http://www.theiacp.org

Childhood Education Intl., Assn. for (1892), 17904 Georgia Ave., Ste. 215, Olney, MD 20832; 12,000; http://www.udel.edu/bateman/acei

Children's Aid Intl. (1977), P.O. Box 83220, San Diego, CA 92138; http://www.childrensaid.org

Children's Book Council, The (1945), 568 Broadway, Ste. 404, New York, NY 10012; 85 publishers; http://www.cbcbooks.org.

Child Welfare League of America (1920), 440 First St. NW, Ste. 310, Wash., DC 20001; 1,000+ agencies; http://www.cwla.org

Chiropractic Assn., American (1963), 1701 Clarendon Blvd., Arlington, VA 22209; 19,000; http://www.amerchiro.org/aca

Chris-Craft Antique Boat Club, Inc. (1973), 217 S. Adams St., Tallahassee, FL 32301; 2,700; http://www.chris-craft.org

Christian Children's Fund, Inc. (1938), 2821 Emerywood Pkwy., Richmond, VA 23274; Employees: 125; http://www.christianchildrensfund.org

Christian Endeavor Union, World's (1895), 1221 E. Broad St., Columbus, OH 43205; 3 mil.

Christian Laity Counseling Board, Inc. (1970), 5901 Plainfield Dr., Charlotte, NC 28215; 52.5 mil.

Christians and Jews, Natl. Conference of (1928), 71 Fifth Ave., Ste. 1100, New York, NY 10003.

Church Women United (1941), 475 Riverside Dr., Ste. 500, New York, NY 10115; http://www.churchwomen.org

Cincinnati, Society of the (1783), 2118 Massachusetts Ave. NW, Wash., DC 20008; 3,300.

Cities, Natl. League of (1924), 1301 Pennsylvania Ave. NW, Wash., DC 20004; 1,550 cities; http://www.nlc.org

Civil Air Patrol (1947), 105 S. Hansell St., Maxwell AFB, AL 36112; 60,000; http://www.cap.af.mil

Civil Engineers, American Society of (1852), 1801 Alexander Bell Dr., Reston, VA 20191; 104,000; http://www.asce.org

Civil Liberties Union, American (ACLU) (1920), 125 Broad St., 18 Fl., New York, NY 10004; 250,000; http://www.aclu.org

Civitan International, Inc. (1920), P.O. Box 130744., Birmingham, AL 35213-0744; 30,000; http://www.civitan.org

Clean Energy Research Institute (1974), Univ. of Miami, Coral Gables, FL, 33124; 70 fellows.

Clinical Pathologists, American Society of (1922), 2100 W. Harrison St., Chicago, IL 60612; 79,000; http://www.ascp.org

Coaster Enthusiasts, American (1978), 5800 Foxridge Dr., Ste. 115, Mission, KS 66202; 5,800.

Coast Guard Combat Veterans Assn. (1985), 17728 Striley Dr., Ashton, MD 20861; 1,800.

Codependents Anonymous (1986), 5150 N. 16th St., Phoenix, AZ 85016; http://www.ourcoda.org

Collectors Association Inc., American (1939), P.O. Box 39106, Minneapolis, MN 55439; 3,586; http://www.collector.com

College Admission Counseling, Natl. Assn. for (1937), 1631 Prince St., Alexandria, VA 22314; 6,300; http://www.nacac.com

College Board, The (1900), 45 Columbus Ave., New York, NY 10023; 2,900 institutions; http://www.collegeboard.org

College English Assn. (1937), English Dept., Winthrop Univ., Rock Hill, SC 29733; 1,000.

College Music Society (1958), 202 W. Spruce St., Missoula, MT 59802; 7,000; http://www.music.org

Colleges and Employers, Natl. Assn. of (1956), 62 Highland Ave., Bethlehem, PA 18017; 3,375; http://www.jobweb.org

Colleges and Universities, Assn. of American (1915), 1818 R St. NW, Wash., DC 20009; 680 institutions; http://www.aacu-edu.org

Colonial Dames XVII Century, Natl. Society of (1915), 1300 New Hampshire Ave. NW, Wash., DC 20036; 14,000.

Commercial Collectors, Int'l. Assn. of (1970), 4040 W. 70th St., Minneapolis, MN 55435; 350; http://www.commercialcollector.com

Commercial Law League of America (1895), 150 N. Michigan Ave., # 600, Chicago, IL 60601; 4,400; http://www.clla.org

Commercial Travelers of America, The Order of United (1888), 632 N. Park St., Columbus, OH 43215; 170,000.

Common Cause (1970), 1250 Connecticut Ave., NW, Wash., DC 20036; 250,000; http://www.commoncause.org

Communication Assn., National (1914), 5105 Backlick Rd., Bldg. E, Annandale, VA 22003; 7,000; http://natcom.org

Community Cultural Center Assn., American (1992), 149 Cannongate III, Nashua, NH 03063; http://pw1.netcom.com/~mjanz/index.html

Community Colleges, American Assn. of (1920), One Dupont Circle NW, Ste. 410, Wash., DC 20036; 1,113 inst; http://www.aacc.nche.edu

Composers, Authors & Publishers, American Soc. of (ASCAP) (1914), One Lincoln Plaza, New York, NY 10023; 24,000; http://www.ascap.com

Composers/USA, Natl. Assn. of (1932), Box 49256, Barrington Station, Los Angeles, CA 90049; 700; http://www.thebook.com/nacusa

Computing Machinery, Assn. for (1947), 1515 Broadway, 17th Fl., New York, NY 10036; http://www.acm.org

Computing Professionals, Inst. for Certification of (1973), 2200 E. Devon Ave., Ste. 247, Des Plaines, IL 60018; 50,000+; http://www.iccp.org

Concerned Women for America (1979), 370 L'Enfant Promenade SW, Ste. 800, Wash., DC 20024; 500,000; http://www.cwfa.org

Concrete Institute, American (1904), 22400 W. Seven Mile Rd., Detroit, MI 48219; 20,000; http://www.aci-int.org

Congress of Racial Equality (CORE) (1942), 817 Broadway, 3d Fl., New York, NY 10003; 100,000; http://www.core-online.org

Conscientious Objectors, Central Committee for (1948), 1515 Cherry St., Philadelphia, PA 19102; 4,000; http://www.libertynet.org/ccco

Constantian Society, The (1970), 5505 Fifth Ave., Pittsburgh, PA 15232; 750; http://members.tripod.com/~constantian/index.html

Construction Industry Manufacturers Assn. (1911), 111 E. Wisconsin Ave., Milwaukee, WI 53202; 500 cos.; http://www.cimanet.com

Construction Specifications Institute (1948), 601 Madison St., Alexandria, VA 22314; 17,000; http://www.csinet.org

Consumer Credit Assn., Intl. (1912), 243 N. Lindbergh Blvd., St. Louis, MO 63141; 20,000.

Consumer Federation of America (1968), 1424 16th St. NW, Ste. 604, Wash., DC 20036; 250 organizations; http://www.stateandlocal.org

Consumer Information Center (1970), Pueblo, CO 81009; http://www.pueblo.gsa.gov

Consumer Interests, American Council on (ACCI) (1953), 240 Stanley Hall, Univ. of Missouri, Columbia, MO 65211; 1,200; http://acci.ps.missouri.edu

Consumers Union of the U.S. (1936), 101 Truman Ave., Yonkers, NY 10703; 405,990; http://www.consumersunion.org

Contract Bridge League, American (1938), 2990 Airways Blvd., Memphis, TN 38116; 170,000; http://www.acbl.org

Co-op America (1983), 1612 K St. NW, Ste. 600, Wash., DC 20006; 50,000 individuals, 2,000 businesses; http://www.coopamerica.org

Correctional Assn., American (1870), 4380 Forbes Blvd., Lanham, MD 20706; 20,000+; http://www.corrections.com/aca

Correctional Officers, Intl. Assn. of (1977), 8600 Glenarden Pkwy., Glenarden, MD 20706.

Cosmetology Assn., Natl. (1921), 3510 Olive St., St. Louis, MO 63103; 32,000; http://www.nca-now.com

Cotton Council of America, Natl. (1938), 1918 N. Pkwy., Memphis, TN 38112; http://www.cotton.org

Counseling Assn., American (1952), 5999 Stevenson Ave., Alexandria, VA 22304; 55,000; http://www.counseling.org

Count Dracula Society (1962), 334 W. 54th St., Los Angeles, CA 90037; 500.

Country Music Assn. (1958), One Music Circle S, Nashville, TN 37203; 6,700; http://www.CMAworld.com

Crafts & Creative Industries, Assn. of (ACCI) (1976), 1100-H Brandyvine Blvd., Zanesville, OH 43702; 6,327; http://www.creative-industries.com

Creative Children and Adults, Natl. Assn. for (1974), 8080 Springvalley Dr., Cincinnati, OH 45236; 6,000.

Credit Union Natl. Assn. & Affiliates (1934), 5710 Mineral Point Rd., Madison, WI 53705; 51 credit union leagues.

Cribbage Congress, American (1979), P.O. Box 10486, Napa, CA 94581; 6,800; http://www.cribbage.org

Crime and Delinquency, Natl. Council on (1907), 685 Market St., Ste. 620, San Francisco, CA 94105; 700+ members; http://www.nccd-crc.com

Criminology, American Society of (1941), 1314 Kinnear Rd., Ste. 212, Columbus, OH 43212; 2,600; http://www.asc41.com

Crop Protection Assn., American (1933), 1156 15th St. NW, Ste. 400, Wash., DC 20005; 80 cos.; http://www.acpa.org

Crop Science Society of America (1955), 677 S. Segoe Rd., Madison, WI 53711; 4,360; http://www.crops.org

Cryogenic Soc. of America, Inc. (1964), 1033 South Blvd., Ste. 13, Oak Park, IL 60302; http://www-csa.fnal.gov

Customs Brokers and Forwarders Assn. of America, Natl. (1897), 1200 18th St. NW, #901, Wash., DC 20036; 800 cos.; http://ncbfaa.org

Cystic Fibrosis Foundation (1955), 6931 Arlington Rd., Bethesda, MD 20814; 30,000; http://www.cff.org

Dairy Council, Natl. (1915), 6300 N. River Rd., Rosemont, IL 60018; http://www.nationaldairycouncil.org

Dairy Goat Assn., American (1904), 209 W. Main St., Spindale, NC 28160; 13,000; http://www.adga.org

Dark-Sky Association, Intl. (1987), 3225 N. First Ave., Tucson, AZ 85719; 3,189; http://www.darksky.org

Daughters of the American Revolution, Natl. Society (1890), 1776 D St. NW, Wash., DC 20006; 178,000; http://www.dar.org

Daughters of the British Empire, Natl. Society (1909), 800 Carrington Dr., Raleigh, NC 27615; 5,000; http://www.mind-spring.com/~dbesociety

Daughters of the Confederacy, United (1894), 328 North Blvd., Richmond, VA 23220; 20,000; http://www.hqudc.org

Deaf, Natl. Assn. of the (1880), 814 Thayer Ave., Silver Spring, MD 20910; 5,500; http://www.nad.org

Defenders of Wildlife (1947), 1101 Fourteenth St. NW, Ste. 1400, Wash., DC 20005; 300,000; http://www.defenders.org

Delta Kappa Gamma Society Intl. (1929), 416 W. 12th St., Austin, TX 78701; 165,000.

Delta Mu Delta Honor Society (1913), P.O. Box 46935, St. Louis, MO 63146; 100,000; http://www.deltamudelta.org

Democratic Natl. Committee (1848), 430 S. Capitol St. SE, Wash., DC 20003; 432 elected members; http://www.democrats.org/index.html

DeMolay International (1919), 10200 N. Executive Hills Blvd., Kansas City, MO 64153; 30,000; http://www.demolay.org

Dental Assn., American (1859), 211 E. Chicago Ave., Chicago, IL 60611; 141,000; http://www.ada.org

Descendants of Washington's Army at Valley Forge, Society of (1976), P.O. Box 915, Valley Forge, PA 19482; 982.

Desert Protective Council (1954), P.O. Box 2312, Valley Center, CA 92082; 400+.

Destroyer Escort Sailors Assn., Inc. (1975), P.O. Box 805, Vienna, VA 22183; 11,000; http://members.tripod.com/DESA-1mainNew.htm

Diabetes Assn., American (1940), 1660 Duke St., Alexandria, VA 22304; 380,000+; http://www.diabetes.org

Diabetes Institute, American (1976), 5901 Plainfield Dr., Charlotte, NC 28215.

Dialect Society, American (1889), c/o Allan Metcalf, English Dept., MacMurray College, Jacksonville, IL 62650; 550; http://www.americandialect.org

Digital Printing & Imaging Assn. (1992), 10015 Main St., Fairfax, VA 22031; 900 firms; http://www.dpia.org

Directors Guild of America (1936), 7920 Sunset Blvd., Los Angeles, CA 90046; 9,700; http://dga.org

Disabled American Veterans (1920), P.O. Box 14301, Cincinnati, OH 45250; 1,050,000; http://www.dav.org

Disabled American Veterans Auxiliary (1922), 3725 Alexandria Pike, Cold Spring, KY 41076; 127,000.

Disabled Sports USA (1967), 451 Hungerford Dr., Ste. 100, Rockville, MD 20850; 60,000+; http://www.dsusa.org/~dsusa/dsusa.html

Dogs on Stamps Study Unit (1979), 202A Newport Rd., Cranbury, NJ 08512; 400; http://www.dossu.org

Down Syndrome Society, Natl. (1979), 666 Broadway, Ste. 810, New York, NY 10012; 50,000; http://www.ndss.org

Dozenal Society of America (1944), Six Brancatelli, W. Islip, NY 11795; 144.

Ducks Unlimited (1937), One Waterfowl Way, Memphis, TN 38120; 620,000; http://www.ducks.org

Eaglehunters Intl. (1994), P.O. Box 1539, Hernando, FL 34442; 1,000; http://www.sharkhunters.com

Eagles, Fraternal Order of (1898), 12660 W. Capitol Dr., Brookfield, WI 53055; 1.1 mil.

Easter Seals (1919), 230 W. Monroe, Ste. 1800, Chicago, IL 60606; http://www.easter-seals.org

Eastern Star, General Grand Chapter, Order of the (1876), 1618 New Hampshire Ave. NW, Wash., DC 20009; 1.5 mil.; http://www.easternstar.org

Economic Assn., American (1885), 2014 Broadway, Ste. 305, Nashville, TN 37203; 20,000.

Edsel Club, The (1967), 1435 Larkspur Ave., Fort Myers, FL 33901.

Education, American Council on (1918), One Dupont Circle NW, #800, Wash., DC 20036; 1,800; http://www.acenet.edu

Education, Council for Advancement & Support of (1974), 11 Dupont Circle NW, Wash., DC 20036; 2,950 schools; http://www.case.org

Educational Research Assn., American (1916), 1230 17th St. NW, Wash., DC 20036; 23,000; http://www.aera.net

Education of Young Children, Natl. Assn. for the (1926), 1509 16th St. NW, Wash., DC 20036; 103,000; http://www.naeyc.org

Educators for World Peace, Intl. Assn. of (1969), P.O. Box 3282, Mastin Lake Station, Huntsville, AL 35810; 25,500.

8th Air Force Historical Society (1975), 711 S. Smith St., St. Paul, MN 55107; 18,000.

88th Infantry Division Assn., Inc. (1948), P.O. Box 8795, Lancaster, PA 17604; 5,166.

84th Infantry Div. Railsplitters Society, Inc. (1945), P.O. Box 827, Sioux Falls, SD 57101; 2,900.

82d Airborne Division Assn., Inc. (1946), NFCS, P.O. Box 9308, Fayetteville, NC 28311; 23,000+; http://www.fayettevillenc.com/airborne82dassn

Electrical and Electronics Engineers, Institute of (1884), 345 E. 47th St., New York, NY 10017; 320,000; http://www.ieee.org

Electrical Manufacturers Assn., Natl. (1926), 2101 L St. NW, Wash., DC 20037; 560 cos.; http://www.nema.org

Electrochemical Society, Inc. (1902), 65 South Main St., Pennington, NJ 08534; 7,000+; http://www.electrochem.org

Electronic Industries Assn. (1924), 2500 Wislon Blvd., Arlington, VA 22201; 1,058 cos.; http://www.eia.org

Elks, U.S.A., Benevolent and Protective Order of (1868), 2750 N. Lakeview Ave., Chicago, IL 60614; 1.2 mil.; http://www.elks.org

Energy Engineers, Assn. of (1977), 4025 Pleasantdale Rd., Ste. 420, Atlanta, GA 30340; 8,000; http://www.aeecenter.org

Engineering, Natl. Academy of (1964), 2101 Constitution Ave. NW, Wash., DC 20418; 1,893.

Engineering in Agricultural, Food, and Biological Systems, Society for (1907), 2950 Niles Rd., St. Joseph, MI 49085; 8,000.

Engineers, Natl. Society of Professional (1934), 1420 King St., Alexandria, VA 22314; 54,000; http://www.nspe.org

English, U.S. (1983), 1747 Pennsylvania Ave. NW, Ste. 1100, Wash., DC 20006; approx. 1 mil.

English-Speaking Union of the U.S. (1920), 16 E. 69th St., New York, NY 10021; 18,000; http://www.english-speakingunion.org

Entomological Society of America (1889), 9301 Annapolis Rd., Lanham, MD 20706; 6,500; http://www.entsoc.org

Environmental Health Assn., Natl. (1937), 720 S. Colorado Blvd., Ste. 970 South, Denver, CO 80222; 5,100.

Environmental Medicine, American Academy of (1965), P.O. Box CN 1001-8001, New Hope, PA 18938; 550; http://www.aaem.com

Esperanto League for North America Inc. (1954), P.O. Box 1129, El Cerrito, CA 94530; 800; http://www.esperanto-usa.org

Exchange Club, Natl. (1911), 3050 Central Ave., Toledo, OH 43606; 33,000; http://www.nationalexchangeclub.com

Experimental Aircraft Assn. (1953), 3000 Poberezny Road, Oshkosh, WI 54903; 170,000; http://www.eaa.org

Exploration Geophysicists, Society of (1959), P.O. Box 702740, Tulsa, OK 74170; http://www.seg.org.

Ex-Prisoners of War, American (1949), 3201 E. Pioneer Pkwy., Arlington, TX 76010; 31,000; http://www.ax-pow.org

Fairs & Expositions, Intl. Assn. of (1919), P.O. Box 985, Springfield, MO 65809; 2,500.

Family Campers and RVers (1949), 4804 Transit Rd., Bldg. 2, Depew, NY 14043; 56,000 families.

Family and Consumer Sciences, American Assn. of (1909), 1555 King St., Alexandria, VA 22314; 20,000.

Family Physicians, American Academy of (1947), 8880 Ward Pkwy., Kansas City, MO 64114; 85,000; http://www.aafp.org

Family Relations, Natl. Council on (1938), 3989 Central Ave. NE, Ste. 550, Minneapolis, MN 55421; 4,000; http://www.ncfr.org

Family Service America (1911), 11700 W. Lake Park Dr., Milwaukee, WI 53224; 250 agencies.

Farm Bureau Federation, American (1919), 225 Touhy Ave., Park Ridge, IL 60068; 4 mil; http://www.fb.com

Farmers of America Org., Natl. Future (1928), 5632 Mt. Vernon Memorial Hwy., Alexandria, VA 22309; 294,000 families; http://www.ffa.org

Farmers Union, Natl. (1902), 11900 E. Cornell Ave., Aurora, CO 80014; 300,000; http://www.nfu.org

Fat Acceptance, Inc., Natl. Assn. to Advance (NAAFA) (1969), P.O. Box 188620, Sacramento, CA 95818; 5,000; http://www.naafa.org

Fellowship of Reconciliation (1915), 521 N. Broadway, Nyack, NY 10960; 16,000; http://www.nonviolence.org/for

Feminists for Life of America (1972), 733 15th St. NW, Ste. 1100, Wash., DC 20005; http://www.feministsforlife.org

Financial Executives Institute (1938), 10 Madison Ave., Morristown, NJ 07962; http://www.fei.org

Financial Service Professionals, Soc. of (formerly American Society of CLU & ChFC (1928), 270 S. Bryn Mawr Ave., Bryn Mawr, PA 19010; 32,000; http://www.financialpro.org

Financial Women Intl. (1921 as the National Assoc. of Bank Women), 200 N. Glebe Rd., Ste. 820, Arlington, VA 22203; 8,000; http://www.fwi.org

Financiers, Inc., Intl. Society of (1979), P.O. Box 18508, Asheville, NC 28814; 300; http://insofin.com

Fire Chiefs, Intl. Assn. of (1873), 4025 Fair Ridge Dr., Fairfax, VA 22033; 12,000; http://www.iafc.org

Fire Protection Assn., Natl. (1896), One Batterymarch Park, Quincy, MA 02269; 68,000; http://www.nfpa.org

Fire Protection Engineers, Soc. of (1950), 7315 Wisconsin Ave., Ste. 1225W, Bethesda, MD 20814; 4,000; http://www.sfpe.org

First Amendment Studies, Inc., Institute for (1984), P.O. Box 589, Great Barrington, MA 01230; 10,000; http://www.ifas.org

Fisheries Soc., American (1870), 5410 Grosvenor Lane, Ste. 110, Bethesda, MD 20814; 10,000; http://www.fisheries.org

Fleet Reserve Assn. (1924), 125 N. West St., Alexandria, VA 22314; 158,000.

Fly Fishers, Fed. of (1965), 502 S. 19th, Ste. 1, Bozeman, MT 59715; 11,000; http://www.fedflyfishers.org/index2.shtml

Food Industry Suppliers (1919), 1451 Dolley Madison Blvd., McLean, VA 22101; http://www.iafis.org

Food Technologists, Institute of (1939), 221 N. LaSalle, Ste. 300, Chicago, IL 60601; 28,000; http://www.ift.org

Footwear Industries of America (1869), 1420 K St. NW, Ste. 600, Wash., DC 20005; 350; http://www.fia.org

Foreign Study, American Institute for (1964), 102 Greenwich Ave., Greenwich, CT 06830; 300,000; http://www.aifs.com

Foreign Trade Council, Inc., Natl. (1914), 1625 K St. NW, #1090, Wash., DC 20006; 550+ cos.; http://www.usaengage.org

Forensic Sciences, American Academy of (1948), P.O. Box 669, Colorado Springs, CO 80901; 5,100; http://www.aafs.org

Foresters, Society of American (1900), 5400 Grosvenor La., Bethesda, MD 20814; 17,500; http://www.safnet.org

Forest History Society (1946), 701 Wm. Vickers Ave., Durham, NC 27701; 933; http://www.lib.duke.edu/forest

Forest & Paper Assn., American (1993), 1111 19th St. NW, Wash., DC 20036; 400 cos.; http://www.afandpa.org

Forest Products Society (1947), 2801 Marshall Ct., Madison, WI 53705; 2,500.

Forests, American (1875), 910 17th St. NW, Ste. 600, Wash., DC 20006; 25,000; http://www.amfor.org

Foundrymen's Society, American (1896), 505 State St., Des Plaines, IL 60016; 13,000; http://www.afsinc.org

4-H Clubs (1914), 1400 Independence Ave., U.S. Dept of Agriculture, Wash., DC 20250; 5.5 mil.

458th Service Squadron Assn. (1991), 2114 W 29th St., Erie, PA 16508; 130.

Frederick A. Cook Society, (1940), P.O. Box 11421, Pittsburgh, PA 15238; 168; http://www.cookpolar.org

Freedom From Religion Foundation (1978), P.O. Box 750, Madison, WI 53701; 4,200; http://www.ffrf.org

Freedoms Foundation at Valley Forge (1949), 1601 Valley Forge Rd., P.O. Box 706, Valley Forge, PA 19482; http://www.ffvf.org

Free Men, Natl. Coalition of (1977), P.O. Box 129, Manhasset, NY 11030; 2,000; http://www.ncfm.org

French Institute/Alliance Française (1898), 22 E. 60th St., New York, NY 10022; 7,500; http://www.fiaf.org

Friendship and Good Will, Intl. Soc. of (1978), 412 Cherry Hills Dr., Bakersfield, CA 93309; 3,968.

Frozen Food Institute, American (1942), 2000 Corporate Ridge, Ste. 1000, McLean, VA 22102; 584; http://www.affi.com

Funeral & Memorial Societies of America (1963), P.O. Box 10, Hinesburg, VT 05461; 500,000; http://www.funerals.org/famsa

Funeral Consumers Alliance (FAMSA) (1963), P.O. Box 10, Hinesburg, VT 05461; 500,000; http://www.funerals.org/famsa

Future Business Leaders of America-/FBLA-PBL, Inc. Phi Beta Lambda, Inc. (1940), 1912 Association Drive, Reston, VA 20191; 246,827; http://www.fbla-pbl.org

Gamblers Anonymous (1957), P.O. Box 17173, Los Angeles, CA 90017; http://www.gamblersanonymous.org

Garden Club of America (1913), 14 E. 60th St., 3rd Floor, New York, NY 10022; 15,000; http://www.gcamerica.org

Garden Clubs, Natl. Council of State (1929), 4401 Magnolia Ave., St. Louis, MO 63110; 308,623; http://www.gardenclub.org

Gardeners of America Inc., The (1932), 5560 Merle Hay Rd., Johnston, IA 50131; 7,400.

Gas Appliance Manufacturers Assn. (1934), 1901 N. Moore St., Ste. 1100, Arlington, VA 22209; 244 cos.; http://www.gamanet.org

Gas Assn., American (1918), 400 North Capitol St. NW, Wash., DC 20001; 187 cos.; http://www.aga.org

Gay and Lesbian Task Force, Natl. (1973), 1700 Kalorama Rd. NW, Wash., DC 20009; 35,000; http://www.ngltf.org

General Contractors of America, The Associated (1918), 1957 E St. NW, Wash., DC 20006; 33,000; http://www.agc.org

Genealogical Society, Natl. (1903), 4527 17th St. NW, Arlington, VA 22207; 17,800; http://www.ngsgenealogy.org

Genetic Association, American (1904), P.O. Box 257, Buckeystown, MD 21717.

Geographers, Assn. of American (1904), 1710 16th St. NW, Wash., DC 20009; 6,900; http://www.aag.org

Geographic Education, Natl. Council for (1915), 16A Leonard Hall, IUP, Indiana, PA 15705; 2,300; http://www.ncge.org

Geographic Society, Natl. (1888), 1145 17th St. NW, Wash., DC 20036; 8.5 mil; http://www.nationalgeographic.com

Geographical Society, The American (1851), 120 Wall St., Ste. 100, New York, NY 10005; approx. 1,500.

Geological Society of America (1888), 3300 Penrose Pl., Boulder, CO 80301; 17,000; http://www.geosociety.org

Geriatrics Society, American (1942), 770 Lexington Ave., Ste. 300, New York, NY 10021; 6,500.

Gideons Intl. (1899), 2900 Lebanon Rd., Nashville, TN 37214; 131,000; http://www.gideons.org

Gifted Children, Natl. Assn. for (1954), 1707 L St. NW, Ste. 550, Wash., DC 20036; 8,000; http://www.nagc.org

Girl Scouts of the U.S.A. (1912), 420 5th Ave., New York, NY 10018; 3.5 mil; http://www.gsusa.org

Girls Incorporated (1945), 30 E. 33d St., New York, NY 10016; http://www.girlsinc.org

Glenn Miller Birthplace Society (1980), P.O. Box 61, Clarinda, IA 51632; 1,500; http://glennmiller.org

Gold Star Mothers, Inc., American (1928), 2128 Leroy Place NW, Wash., DC 20008; under 2,000.

Golden Key National Honor Society (1977), 1189 Ponce deLeon Ave., Atlanta, GA 30306; 800,000; http://gknhs.gsu.edu

Golf Assn., U.S. (1894), Golf House, P.O. Box 708, Far Hills, NJ 07931; 750,000; http://www.usga.org

Gospel Music Assn. (1964), 1205 Division St., Nashville, TN 37203; 5,500; http://www.gospelmusic.org

Government Finance Officers Assn. (1906), 180 N. Michigan Ave., Ste 800, Chicago, IL 60601; 13,600; http://www.financenet.gov/gfoa.htm

Governors' Assn., Natl. (1908), Hall of the States, 444 N. Capitol, Wash., DC 20001; 55 govs.; http://www.nga.org

Graduate Schools, Council of (1960), One Dupont Circle NW, #430 Wash., DC 20036; 415 instits.; http://www.cgsnet.org

Grandmother Clubs of America, Inc., Natl. Federation of (1938), 27 E. Monroe St., Rm. 519, Chicago, IL 60603; 4,200.

Grange of the Order of Patrons of Husbandry, Natl. (1867), 1616 H St. NW, Wash., DC 20006; 300,000+; http://www.nationalgrange.org

Graphic Arts, American Institute of (1914), 164 5th Ave., New York, NY 10010; 9,000; http://www.aiga.org

Gray Panthers (1970), 733 15th St., Ste 437, Wash., DC 20005; approx. 17,000; http://www.graypanthers.org.

Great Council of the U.S., Improved Order of Red Men (1847), 4521 Speight Ave., Waco, TX 76711; 26,000; http://www.members.xoom.com/redmen

Green Mountain Club (1910), 4711 Waterbury-Stowe Rd., Waterbury Ctr., VT 05677; 7,000; http://www.greenmountainclub.org

Grocers Assn., Nat'l. (1982), 1825 Samuel Morse Dr., Reston, VA 22190; 2,500.

Grocery Manufacturers of America (1908), 1010 Wisconsin Ave., Ste. 800, Wash., DC 20007; 140 cos.; http://www.gmabrands.com

Ground Water Assn., Natl. (1948), 601 Dempsey Rd., Westerville, OH 43081; 17,000; http://www.ngwa.org

Group Against Smokers' Pollution, Inc. (GASP) (1971), P.O. Box 632, College Park, MD 20741; 10,000+.

Guide Dog Foundation for the Blind, Inc. (1946), 371 E. Jericho Turnpike, Smithtown, NY 11787; 162,500; http://www.guidedog.org

Gyro Intl. (1912), 1096 Mentor Ave., Painesville, OH 44077; 4,600.

Hadassah, the Women's Zionist Organization of America (1912), 50 W. 58th St., New York, NY 10019; 385,000; http://www.hadassah.org

Handball Assn., U.S. (1951), 2333 N. Tucson Blvd., Tucson, AZ 85716; 8,200; http://www.ushandball.org

Health Council, Natl. (1920), 1730 M St. NW, Ste. 500, Wash., DC 20036; http://www.nhcouncil.org

Health Info. Management Assn., American (1928), 233 N. Michigan Ave., Ste. 2100, Chicago, IL 60601; 39,000; http://www.ahima.org

Hearing Society, Intl. (1951), 16880 Middlebelt Rd., Ste. 4, Livonia, MI 48152; 3,500; http://www.hearingihs.org

Hearing and Speech Action, Natl. Assn. for (1910), 10801 Rockville Pike, Rockville, MD 20852.

Heart Assn., American (1924), 7272 Greenville Ave., Dallas, TX 75231; 31,000; http://www.americanheart.org

Heating, Refrigerating & Air-Conditioning Engineers, Inc., American Soc. of (1894), 1791 Tullie Cir. NE, Atlanta, GA 30329; 50,000; http://www.ashrae.org

Helicopter Society, American (1943), 217 N. Washington St., Alexandria, VA 22314; 6,140; http://www.vtol.org

Hemispheric Affairs, Council on (1975), 1444 I St., NW, Ste. 211, Wash., DC 20005; 1,900; http://www.coha.org

Hibernians in America, Ancient Order of (1836), 1301 S.W. 26th Ave., Ft. Lauderdale, FL 33312; 200,000; http://www.aoh.com

Highpointers Club (1987), P.O. Box 1496, Golden, CO 80402; 1,700; http://highpointers.com

High School Assns., Natl. Federation of State (1920), P.O. Box 20626, Kansas City, MO 64195; 51 state assns.

High School Band Directors Hall of Fame, Natl. (1985), 519 N. Halifax Ave., Daytona Beach, FL 32118; http://www.bocanet.com/band/default.htm

High Twelve International (1921), 15456 Ivanhoe Dr., Visalia, CA 93292; 20,000.

Hiking Society, American (1976), P.O. Box 20160, Wash., DC 20041; 7,000; http://www.americanhiking.org

Historians, Organization of American (1907), 112 N. Bryan St., Bloomington, IN 47408; 12,000; http://www.indiana.edu/~oah

Historic Preservation, Natl. Trust for (1949), 1785 Massachusetts Ave. NW, Wash., DC 20036; 250,000; http://www.nationaltrust.org

Historical Assn., American (1884), 400 A St. SE, Wash., DC 20003; 16,000; http://www.theaha.org

Historical Society, United States (1971), 1st and Main Sts., Richmond, VA 23219; 250,000; http://www.ushsdolls.com

Hockey, U.S.A. (1936), 1775 Bob Johnson Dr., Colorado Springs, CO 80906; 540,000; http://usahockey.com

Home Builders, Natl. Assn. of (1942), 1201 15th St. NW, Wash., DC 20005; 157,000; http://www.nahb.com

Homeless, Natl. Coalition for the (1983), 1012 14th St., Ste. 600, Wash., DC 20005; 14,000; http://nch.ari.net

Homemakers of America, Future (1945), 1910 Association Dr., Reston, VA 22091; 281,000+; http://www.fhahero.org

Honor Society, Natl. (1921), 1904 Association Dr., Reston, VA 20191; 21,000; http://www.tassp.org/nhs

Horatio Alger Soc. (1965), P.O. Box 70361, Richmond, VA 23255; 250; http://www.ihot.com/~has

Horse Council, American (1969), 1700 K St. NW, #300, Wash., DC 20006; 1,800; http://www.horsecouncil.org

Horse Protection Assn., American (1966), 1000 29th St. NW, Ste. T-100, Wash., DC 20007; 8,000.

Hospital Assn., American (1899), 1 N. Franklin, Chicago, IL 60606; 5,100 hospitals; http://www.aha.org

Hostelling Intl., American Youth Hostels (1934), 733 15th Street NW, Ste. 840, Wash., DC 20005; 115,000; http://www.hiayh.org

Hotel & Motel Assn., American (1910), 1201 New York Ave. NW, Wash., DC 20005; 10,000+; http://www.ahma.com

Hot Rod Assn., Natl. (1951), 2035 Financial Way, Glendora, CA 91741; 85,000; http://www.nhraonline.com

Huguenot Society, Natl. (1951), 9033 Lyndale Ave. S, Ste. 108, Bloomington, MN 55420; 5,000; http://huguenot.netnation.com.

Humane Society of the U.S. (1954), 2100 L St. NW, Wash., DC 20037; 650,000; http://www.hsus.org

Human Resource Management, Society for (1948), 1800 Duke St., Alexandria, VA 22314; 115,000; http://www.shrm.org

Hybrid & Alternative Vehicle Society (1994), 3301 N. Belaire Dr., Altadena, CA 91001; 5,000.

Hydrogen Energy, Intl. Assn. for (1975), P.O. Box 248266, Coral Gables, FL 33124; 2,500; http://www.iahe.org

Identification, Intl. Assn. for (1915), 2535 Pilot Knob Road, Ste. 117, Mondota Heights, MN 55120; 4,600; http://theiai.org

Illuminating Engineering Society of N. America (1906), 120 Wall St., 17th Fl., New York, NY 10005; 10,000; http://www.iesna.org

Illustrators, Inc., Society of (1901), 128 E. 63d St., New York, NY 10021; 950; http://www.societyillustrators.org

Immigration Reform, Federation for American (1979), 1666 Connecticut Ave. NW, #400, Wash., DC; 70,000

Impotence Inst. of America (1983), 119 S. Ruth St., Maryville, TN 37803.

Independent Community Bankers of America, (1930), One Thomas Circle NW, #400, Wash., DC 20005; 5,400; http://www.icba.org

Industrial and Applied Mathematics, Society for (1952), 3600 Univ. City Science Ctr., Philadelphia, PA 19104; 9,000; http://www.iam.org

Industrial Designers Society of America (1965), 1142-E Walker Rd., Great Falls, VA 22066; 2,350; http://www.idsa.org

Industrial Security, American Soc. for (1955), 1625 Prince St., Alexandria, VA 22313; 30,000; http://www.asisonline.org

Information and Image Management, Assn. for (1943), 1100 Wayne Ave., Ste. 1100, Silver Spring, MD 20910; 11,000; http://www.aiim.org

Insurance Assn., American (1964), 1130 Connecticut Ave. NW, Ste. 1000, Wash., DC 20036; 250+ cos.; http://www.aiadc.org

Integrative and Comparative Biology, Society for (1890), 401 N. Michigan Ave. Chicago, IL 60611; 2,200. http://www.sicb.org

Intellectual Property Owners Assoc. (1972), 1255 23d St. NW, Ste. 200, Wash., DC 20037; 450; http://www.ipo.org

Intelligence Officers, Assn. of Former (1975), 6723 Whittier Ave., Ste. 303A, McLean, VA 22101; 2,700.

Intercollegiate Athletics, Natl. Assn. of (1937), 6120 S. Yale Ave., Ste. 1450, Tulsa, OK 74136; 352 schools; http://www.naia.org

Interior Designers, American Society of (1975), 608 Massachusetts Ave. NE, Wash., DC 20008; 30,000; http://www.asid.org

Intl. Colleges and Universities, Assn. of (1973), 1301 S. Noland Rd., Independence, MO 64055; 8,729 ind., 26 inst.

Intl. Education, Institute of (1919), 809 United Nations Plaza, New York, NY 10017; 650 U.S. colleges and universities; http://www.iie.org

Intl. Educational Exchange, Council on (1947), 205 E. 42d St., New York, NY 10017; 240 organizations; http://www.ciee.org

Intl. Educators, Assn. of (NAFSA) (1948), 1875 Connecticut Ave., Ste. 1000, Wash., DC 20009; 7,500; http://www.nafsa.org

Intl. Law, American Society of (1906), 2223 Massachusetts Ave. NW, Wash., DC 20008; 4,000; http://www.asil.org

Inventors, American Soc. of (1953), P.O. Box 58426, Philadelphia, PA 19102; 150.

Investigative Pathology, Inc., American Soc. for (1900), 9650 Rockville Pike, Bethesda, MD 20814; 1,723; http://www.asip.uthscsa.edu

Investment Management and Research, Assn. for (1990), 5 Boar's Head La., Charlottesville, VA 22901; 29,000.

Investors Corp., Natl. Assn. of (1951), 711 W. Thirteen Mile Rd., Madison Heights, MI 48701; 640,000; http://www.better-investing.org

IPC-Association Connecting Electronics Industries (formerly The Institute for Interconnecting & Packaging Electronic Circuits), (1957), 2215 Sanders Rd., Northbrook, IL 60062; 2,600; http://www.ipc.org

Irish American Cultural Inst. (1962), 1 Lackawanna Pl., Morristown, NJ 07960; 5,000; http://www.irishaci.com

Irish Historical Society, American (1897), 991 5th Ave., New York, NY 10028; 850; http://www.aihs.org

Iron and Steel Engineers, Assn. of (1907), Three Gateway Center, Ste. 1900, Pittsburgh, PA 15222; 11,650; http://www.aise.org

Iron and Steel Institute, American (1855), 1101 17th St. NW, Ste. 1300, Wash., DC 20036; 1,100; http://www.steel.org

Islamic Relations, Council on American- (1994), 1050 17th St. NW, Ste. 490, Wash., DC 20036; http://www.cair-net.org

Italian Historical Society of America (1949), 111 Columbia Heights, Brooklyn, NY 11201.

Jail Assn., American (1981), 2053 Day Rd., Ste. 100, Hagerstown, MD 21740; 4,600; http://www.corrections.com/aja

Jane Austen Society of North America (1979), 200 E. 57th St., #15B, New York, NY 10022; 3,800; http://www.jasna.org

Japanese American Citizens League (1929), 1765 Sutter St., San Francisco, CA 94115; 23,900; http://www.jacl.org

Jewish Book Council (1946), 15 E. 26th St., New York, NY 10010.

Jewish Committee, American (1906), 165 E. 56th St., New York, NY 10022; 50,000; http://www.ajc.org

Jewish Community Centers Assn. of North America (1917), 15 E. 26th St., New York, NY 10010; http://www.uja.org

Jewish Congress, American (1918), 15 E. 84th St., New York, NY 10028; 50,000; http://www.ajcongress.org/rght_col.htm

Jewish Federations, Council of (1932), 730 Broadway, New York, NY 10003; 200 agencies.

Jewish Historical Society, American (1892), 2 Thornton Rd., Waltham, MA 02453; approx. 3,000; http://www.ajhs.org

Jewish War Veterans of the U.S.A. (1896), 1811 R St. NW, Wash., DC 20009; 100,000; http://www.penfed.org/jwv/home.htm

Jewish Women, Natl. Council of (1893), 53 W. 23d St., 6th Fl., New York, NY 10010; 90,000; http://www.ncjw.org

Job's Daughters, Intl. Order of (1921), 233 W. 6th St., Papillion, NE 68046; 21,000; http://www.iojd.org

John Birch Society (1958), 770 Westhill Blvd, P.O. Box 8040, Appleton, WI 54914; nearly 100,000; http://www.jbs.org

Joint Action in Community Service (JACS), 5225 Wisconsin Ave. NW, Ste. 404, Wash., DC 20015.

Joseph Diseases Foundation, Inc., Intl. (1997), 4047 First St., Ste. 107, Livermore, CA 94550; 3,578.

Journalists, Society of Professional (1909), 16 S. Jackson St., Greencastle, IN 46135; 13,500; http://spj.org

Journalists and Authors, American Society of (1948), 1501 Broadway, Ste. 302, New York, NY 10036; 1,055; http://www.asja.org

Judaism, American Council for (1943), P.O. Box 9009, Alexandria, VA 22304.

Judicature Society, American (1913), 180 N. Michigan Ave., Ste. 600, Chicago, IL 60601; 10,000; http://www.ajs.org

Jugglers Assn., Intl. (1947), P.O. Box 218, Montague, MA 01351; 2,500; http://www.juggle.org

Junior Achievement (1919), One Education Way, Colorado Springs, CO 80906; 300,000; http://www.ja.org

Junior Auxiliaries, Natl. Assn. of (1941), P.O. Box 1873, Greenville, MS 38702; 12,500; http://www.tecinfo.com/~najanet

Junior Chamber of Commerce, U.S. (1915), P.O. Box 7, 4 W. 21st St., Tulsa, OK 74102; 200,000; http://www.usjaycees.org

Junior College Athletic Assn., Natl. (1938), P.O. Box 7305, Colorado Springs, CO 80933; 510; http://www.njcaa.org

Junior Leagues, Assn. of (1921), 660 First Ave., New York, NY 10016; 195,000.

Kappa Delta Epsilon, Inc. (1933), 2561 Rocky Ridge Road, Birmingham, AL 35243; about 40,591 -47 chapters.

Kidney Fund, The American (1971), 6110 Executive Blvd., Ste. 1010, Rockville, MD 20852; http://www.arbon.com/kidney

Kiwanis International (1915), 3636 Woodview Trace, Indianapolis, IN 46268; 303,000; http://www.kiwanis.org

Knights of Columbus (1882), One Columbus Plaza, New Haven, CT 06510; 1,560,633; http://www.kofe-supreme-council.org

Knights of Pythias (1864), 1495 Hancock St., Quincy, MA 02169; http://www.pythias.org

La Leche League Intl. (1956), 1400 N. Meacham Rd., Schaumburg, IL 60173; 50,000+; http://www.lalecheleague.org

Lady Bird Johnson Wildflower Center (1982), 4801 La Crosse Ave., Austin, TX 78739; 22,000; http://www.wildflower.org

Lambs Inc., The (1874), 3 W. 51st St., New York, NY 10019; 200.

Landscape Architects, American Society of (1899), 636 I St. NW, Wash., DC 20001; 12,000; http://www.asla.org

Law Libraries, American Assn. of (1906), 53 W. Jackson Blvd., #940, Chicago, IL 60604; 4,900; http://www.aallnet.org

Learned Societies, American Council of (1919), 228 E. 45th St., New York, NY 10017; 56 societies; http://www.acls.org

Lefthanders Intl. (1975), P.O. Box 8249, Topeka, KS 66608; 25,000.

Legal Administrators, Assn. of (1971), 175 E. Hawthorn Pkwy., Ste. 325, Vernon Hills, IL 60061; 8,200; http://www.alanet.org

Legal Secretaries, Natl. Assn. of (1929), 314 E 3rd St., Ste. 210, Tulsa, OK 74120; http://www.nals.org

Legion of Valor of the U.S.A. (1890), 92 Oak Leaf Lane, Chapel Hill, NC 27516; 753; http://members.aol.com/LValor1890

Leprosy Missions, American (1906), One Alm Way, Greenville, SC 29601; http://www.leprosy.org

Leukemia Society of America (1949), 600 Third Ave., New York, NY 10016; approx. 1 mil volunteers; http://www.leukemia.org

Lewis and Clark Trail Heritage Foundation, Inc. (1969), P.O. Box 3434, Great Falls, MT 59403; 2,500; http://www.lewisandclark.org

Lewis Carroll Society of North America (1974), 18 Fitzharding Pl., Owings Mill, MD 21117; 350; http://www.lewiscarroll.org/carroll.html

Libertarian Party (1971), 2600 Virginia Ave. NW, Ste. 100, Wash., DC 20037; 32,000; http://www.lp.org

Liberty Lobby (1955), 300 Independence Ave. SE, Wash., DC 20003; 20,000; http://www.spotlight.org

Libraries Assn., Special (1909), 1700 18th St. NW, Wash., DC 20009; 15,000; http://www.sla.org

Library Assn., American (1879), 50 E. Huron St., Chicago, IL 60611; 56,000; http://www.ala.org

Life Insurance of America, Federation (1913), 2335 S. 13th St., Milwaukee, WI 53215.

Lighter-Than-Air Society (1952), 1436 Triplett Blvd., Akron, OH 44306; 900.

Linguistic Society of America (1924), 1325 18th St. NW, Ste. 211, Wash., DC 20036; 7,000; http://www.lsadc.org

Lions Clubs Intl. (1917), 300 22d St., Oak Brook, IL 60523; 1,400,000; http://www.lionsclubs.org

Literacy Volunteers of America, Inc. (1962), 635 James St., Syracuse, NY 13203; 111,319; http://www.literacyvolunteers.org

Little League Baseball and Softball, Inc. (1939), P.O. Box 3485, Williamsport, PA 17701; approx. 4 mil; http://www.littleleague.org

Little People of America Inc. (1961), Box 745, Lubbock, TX 79408; 6,000; http://www.lpaonline.org

London Club (1975), 214 North 2100 Rd., Lecompton, KS 66050; 100+; http://www.londonclub.org

Lung Assn., American (1904), 1740 Broadway, New York, NY 10019; http://www.lungusa.org

Lutheran Education Assn. (1942), 7400 Augusta St., River Forest, IL 60305; 3,800.

Magazine Publishers of America (1919), 919 Third Ave., 22d Fl., New York, NY 10022; 275; http://www.magazine.org

Magicians, Intl. Brotherhood of (1922), 11155 S. Towne Sq., Ste. B2&C, St. Louis, MO 63123; 15,000; http://www.magician.org

Management Accountants, Institute of (1919), 10 Paragon Dr., Montvale, NJ 07645; 80,000; http://www.rutgers.edu/Accounting/raw/ima/ima.htm

Management Assn. Intl., American (1923), 1601 Broadway, New York, NY 10019; 70,000+; http://www.amanet.org

Management Consulting Firms, Assn. of (1929), 521 5th Ave., 35th Fl., New York, NY 10175; 50 firms.

Manufacturing Engineers, Soc. of (1932), One SME Dr., P.O. Box 930, Dearborn, MI 48121; 60,000+; http://www.sme.org

Manufacturers, Natl. Assn. of (1895), 1331 Pennsylvania Ave. NW, Ste. 1500 N. Tower, Wash., DC 20004; 14,000 cos.; http://www.nam.org

March of Dimes Birth Defects Foundation (1938), 1275 Mamaroneck Ave., White Plains, NY 10605; http://www.modimes.org

Marine Conservation, Ctr. for (1972), 1725 DeSales St. NW, #600, Wash., DC 20036; 120,000; http://www.CMC-ocean.org

Marine Corps League (1937), P.O. Box 3070, Merrifield, VA 22116; 42,000; http://www.mcleague.org

Market Technicians Assn. (1973), One World Trade Center, Ste. 4447, New York, NY 10048; 1,000; http://www.mta.org

Marketing Assn., American (1937), 250 S. Wacker Dr., Ste. 200, Chicago, IL 60606; 41,000; http://www.ama.org

Masonic Relief Assn. of U.S. and Canada (1889), 3827 Canal St., New Orleans, LA 70119.

Masons, Royal Arch, General Grand Chapter (1797), P.O. Box 489, Danville, KY 40423; 220,000.

Material & Process Engineering, Soc. for the Advancement of (1944), 1161 Parkview Dr., Covina, CA 91724; 6,000; http:/www.sampe.org

Mathematical Society, American (1888), 201 Charles St., Providence, RI 02904; 30,000; http://www.ams.org

Mathematical Statistics, Institute of (1935), 3401 Investment Blvd., Ste. 7, Hayward, CA 94545; 3,800; http://www.mstat.org

Mayflower Descendants, General Society of (1897), 4 Winslow St., P.O. Box 3297, Plymouth, MA 02361; 30,000+.

Mayors, U.S. Conference of (1932), 1620 Eye St. NW, Wash., DC 20006; http://www.usmayors.org

Mechanical Engineers, American Soc. of (1881), 345 E. 47th St., New York, NY 10017; 120,000; http://www.asme.org

Medical Assn., American (1847), 515 N. State St., Chicago, IL 60610; 300,000; http://www.ama-assn.org

Medical Assn., Natl. (1895), 1012 Tenth St. NW, Wash., DC 20001; 22,000.

Medical Library Assn. (1898), 6 N. Michigan Ave., Ste. 300, Chicago, IL 60602; 3,800 people, 1,200 instits; http://www.mlanet.org

Medieval Academy of America (1925), 1430 Massachusetts Ave., Ste. 313, Cambridge, MA 02138; 4,500; http://www.georgetown.edu/medievalacademy

MENC: The Natl. Assn. for Music Educators (formerly Music Educators Natl. Conference) (1907), 1806 Robert Fulton Dr., Reston, VA 20191; 85,000; http://www.menc.org

Mended Hearts (1955), 7272 Greenville Ave., Dallas, TX 75231; 25,000; http://www.mendedhearts.org

Mensa, Ltd., American (1960), 1229 Corporate Dr. W, Arlington, TX 76006; 44,825; http://www.us.mensa.org

Mental Health Assn., Natl. (1909), 1021 Prince St., Alexandria, VA 22314; 340 affiliates; http://www.nmha.org

Mental Health Program Directors, Natl. Assn. of State (1959), 66 Canal Ctr. Plaza, Ste. 302, Alexandria, VA 22314; 55.

Mentally Ill, Natl. Alliance for the (1980), 200 North Glebe Rd., Arlington, VA 22203; 170,000; http://www.nami.org

Merchant Marine Veterans of World War II, U.S. (1989), P.O. Box 629, San Pedro, CA 90733; 3,456; http://www.lanevictoryship.com

Merrill's Marauders Assn. (1979), 11244 N. 33rd St., Phoenix, AZ 85028; 1,777; http://www.marauder.org

Metal Powder Industries Federation (1944), 105 College Rd. E, Princeton, NJ 08540; approx. 3,000; http://www.mpif.org

Metals Intl. (ASM), American Society for (1913), 9639 Kinsman Rd., Materials Park, OH 44073; 43,000; http://www.asm-intl.org

Meteorological Society, American (1919), 45 Beacon St., Boston, MA 02108; 11,700; http://www.ametsoc.org/AMS

Metric Assn., Inc., U.S. (1916), 10245 Andasol Ave., Northridge, CA 91235; 1,200; http://lamar.colostate.edu/~hillger/

Microbiology, American Society for (1899), 1325 Massachusetts Ave. NW, Wash., DC 20005; 42,000; http://www.asmusa.org

Military Order of the Loyal Legion of the U.S. (1888), 1805 Pine St., Philadelphia, PA 19103; 1,000; http://suvcw.org/mollus.htm

Military Order of the Purple Heart of the USA (1958), 5413-B Backlick Rd., Springfield, VA 22151; 30,000; http://www.purpleheart.org

Military Order of the World Wars (1919), 435 N. Lee St., Alexandria, VA 22304; 12,000; http://moww.org

Military Surgeons of the U.S., Assn. of (1898), 9320 Old Georgetown Rd., Bethesda, MD 20814; 11,000; http://www.amsus.org

Mining Association, Natl. (1995), 1130 17th St. NW, Washington DC 20036; 340; http://www.nma.org

Mining Engineers, Society of (1871), 8307 Shaffer Pkwy., Littleton, CO 80127; 23,058.

Mining, Metallurgy and Exploration, Inc., Society for (1871), P.O. Box 625002, Littleton, CO 80162; 16,141; http://www.smenet.org

Mining, Metallurgical and Petroleum Engineers, American Institute of (1871), 3 Park Ave., New York, NY 10016; 90,000; http://www.idis.com/aime

Missing and Exploited Children, Natl. Center for (1984), The Charles B. Wang International Children's Building, Alexandria, VA 22314; http://www.missingkids.com

Model A Ford Club of America, Inc. (1957), 250 S Cypress St., La Habra, CA 90631; 15,500; http://www.mafca.com

Model Railroad Assn., Natl. (1935), 4121 Cromwell Rd., Chattanooga, TN 37421; 24,600; http://www.nmra.org

Modern Language Assn. of America (1883), 10 Astor Pl., New York, NY 10003; 32,000; http://www.mla.org

Moose Intl., Inc. (1888), Rt. 31, Mooseheart, IL 60539; 1.8 mil.

Mothers, Inc.®, American (1935), 301 Park Ave., New York, NY 10022; 6,000; http://www.americanmothers.org

Mothers of Twins Clubs, Natl. Organization of (1960), P.O. Box 23188, Albuquerque, NM 87192; 14,000; http://www.nomotc.org

Motion Picture Arts & Sciences, Academy of (1927), 8949 Wilshire Blvd., Beverly Hills, CA 90211; 6,300; http://www.oscars.org

Motion Picture & Television Engineers, Soc. of (1916), 595 W. Hartsdale Ave., White Plains, NY 10607; 10,000; http://www.smpte.org

Motorcyclist Assn., American (1924), 33 Collegeview Rd., Westerville, OH 43081; 225,000; http://www.ama-cycle.org

Motor Fire Apparatus in America, Soc. for the Preservation & Appreciation of Antique (1958), P.O. Box 2005, Syracuse, NY 13220; 3,000.

Motorists Association Natl. (1982), 402 W. 2nd St., Waunakee, WI 53597; 7,500; http://www.motorists.org

Multiple Sclerosis Society, Natl. (1946), 733 Third Ave., New York, NY 10017; 518,567; http://www.nmss.org

Muscular Dystrophy Assn., Inc. (1950), 3300 E. Sunrise Dr., Tucson, AZ 85718; 2 mil. volunteers; http://www.mdausa.org

Museums, American Assn. of (1906), 1575 Eye St. NW, Ste. 400, Wash., DC 20005; 15,800; http://www.aam-us.org

Music Center, American (1939), 30 W. 26th St., New York, NY 10010; 2,500; http://www.amc.net

Music Scholarship Assn., American (1956), 1030 Carew Tower, Cincinnati, OH 45202; 1,000; http://www.amsa_wpc.org

Music Teachers Natl. Assn. (1876), 441 Vine St., Ste. 505, Cincinnati, OH 45202; 24,000; http://www.mtna.org

Musicological Society, American (1934), 201 S. 34th St., Philadelphia, PA 19104; 3,500; http://www.musdra.ucdavis.edu/Documents/AMS/AMS.html

Muzzle Loading Rifle Assn., Natl. (1933), P.O. Box 67, Friendship, IN 47021; 21,000; http://nmlra@nmlra.org

Myasthenia Gravis Foundation of America (1952), 123 W. Madison, Ste. 800, Chicago, IL 60602; 30,000; http://www.myasthenia.org

Mystery Writers of America (1945), 17 E. 47th St., 6th Fl., New York, NY 10017; 2,200; http://www.mysterywriters.org

NA'AMAT USA (1925), 200 Madison Ave., New York, NY 10016; 50,000; U.S.; 900,000 worldwide; http://www.naamat.org

Name Society, American (1951), Dept. of Modern Languages, Box G-1224, Baruch College, 17 Lexington Ave., New York, NY 10010; 700.

Narcotics Anonymous World Services (1953), 19737 Nordhoff Pl., Chatsworth, CA 91311; 250,000-500,000; http://www.na.org

Natl. Assn. for the Advancement of Colored People (NAACP) (1909), 4805 Mt. Hope Dr., Baltimore, MD 21215; http://www.naacp.org

National Defense Industrial Assn. (1919), 2111 Wilson Blvd., # 400, Arlington, VA 22201; 28,000.

National Down Syndrome Society (1979), 666 Broadway, Ste 810,New York, NY 10012; 50,000; http://www.ndss.org.

National Guard Assn. of the U.S. (1878), One Massachusetts Ave. NW, Wash., DC 20001; 56,000; http://www.ngaus.org

National Party for America (1996), 10799 Sherman Grove Ave., # 18, Sunland, CA; 91040; 5,200.

National Press Club (1908), 529 14th St. NW, Wash., DC 20045; 4,800; http://npc.press.org

Nature Conservancy, The (1951), 4245 N. Fairfax Drive, Ste. 100, VA 22203; 900,000; http://www.tnc.org

Naturists, Inc., The (1980), P.O. Box 132, Oshkosh, WI 54902; 20,000; http://www.naturistsociety.com

Naval Architects & Marine Engineers, Society of (1893), 601 Pavonia Ave., Ste. 400, Jersey City, NJ 07306; 10,000.

Naval Institute, U.S. (1873), 118 Maryland Ave., Annapolis, MD 21402; http://www.usni.org

Naval Reserve Assn. (1954), 1619 King St., Alexandria, VA 22314; 23,000; http://www.navy-reserve.org/nra

Navigation, The Institute of (1945), 1800 Diagonal Rd., Ste. 480, Alexandria, VA 22314; 3,200; http://www.ion.org

Navy League of the U.S. (1902), 2300 Wilson Blvd., Arlington, VA 22201; 71,308.

Needlework Guild of America, Inc (NGA, Inc.) (1885), 1007-B Street Rd., Southampton, PA 18966; 100,000.

Negro College Fund, United (1944), 8260 Willow Oaks Corporate Dr., Fairfax, VA; 41 institutions; http://www.uncf.org

Neurofibromatosis Foundation, Natl. (1978), 95 Pine St., 16th Fl., New York, NY 10005; 40,000; http://www.nf.org

Newspaper Assn. of America (NAA) (1992), 1921 Gallows Rd., #4, Vienna, VA 22182; 1,800; http://www.naa.org

Ninety-Nines (Intl. Organization of Women Pilots) (1929), Box 965, Will Rogers Airport, Oklahoma City, OK 73159; 6,400; http://www.ninety-nines.org

Nobel Committee, American (1946), P.O. Box 20202, Sarasota, FL 34276; 180 & 500 CEOs.

Non-Commissioned Officers Assn. (1960), 10635 IH 35 North, San Antonio, TX 78233; 160,000; http://www.ncoausa.org

Northern Cross Society (1986), 214 North 2100 Road, Lecompton, KS 66050; 100+.

NOT-SAFE Nat'l Organization Taunting Safety and Fairness Everywhere (1983), P.O. Box 5743-WA, Montecito, CA 93150; 1,800.

Notaries, American Society of (1965), P.O. Box 5707, Tallahassee, FL 32314; 20,000; http://www.notaries.com

NSAC (Natl. Soc. of Accountants for Cooperatives) (1936), 6320 Augusta Dr., Ste. 800, Springfield, VA 22150; 1,900.

Nuclear Society, American (1954), 555 N. Kensington Ave., La Grange Park, IL 60525; 16,000; http://www.ans.org

Nude Recreation, American Assn. for (1931), 1703 N. Main St., Kissimmee, FL 34744; 45,000; http://www.aanr.com

Numismatic Assn., American (1891), 818 N. Cascade Ave., Colorado Springs, CO 80903; 30,000; http://www.money.org

Numismatic Society, The American (1858), Broadway at 155th St., New York, NY 10032; http://www.amnumsoc.org

Nursing, Natl. League for (1952), 350 Hudson St., New York, NY 10014; 16,000; http://www.nln.org

Nutritional Sciences, American Society for (1928), 9650 Rockville Pike, Bethesda, MD 20814; 3,400; http://www.arvo.org/asns

Odd Fellows, Independent Order of (1819), 422 Trade St., Winston-Salem, NC 27101; 380,000; .http://www.ioof.org

Old Crows, Assn. of (1964), 1000 N. Payne St., Alexandria, VA 22314; 25,000; http://www.aochq.org

Opthalmology, American Academy of (1979), P.O. Box 7424, San Francisco, CA 94120; 21,000; http://www.eyenet.org

Optical Society of America (1917), 2010 Massachusetts Ave. NW, Wash., DC 20036; 11,000; http://www.osa.org

Optimist Intl. (1919), 4494 Lindell Blvd., St. Louis, MO 63108; 140,000; http://www.optimist.org

Optometric Assn., American (1898), 243 N. Lindbergh Blvd., St. Louis, MO 63141; 33,000; http://www.aoanet.org

Organ Sharing United Network, for (1977), 1100 Boulders Parkway, Ste. 500, Richmond, VA 23225; 434; http://www.unos.org

Organists, American Guild of (1896), 475 Riverside Dr., Ste. 1260, New York, NY 10115; 20,200; http://www.agohq.org

Oriental Society, American (1842), Univ. of Michigan, Hatcher Graduate Library, 110D, Ann Arbor, MI 48109; 1,350; http://umich.edu/~aos

ORT Federation, American (Org. for Rehabilitation Through Training) (1924), 817 Broadway, 10th Fl., New York, NY 10003; 20,000.

Ornithologists' Union, American (1883), c/o Division of Birds, MRC-116, Smithsonian Institution, Wash., DC 20560; 4,500; http://www.pica.wru.umt.edu/AOU/AOU.html

Osteopathic Assn., American (1897), 142 E. Ontario, Chicago, IL 60611; 28,974; http://www.am-osteo-assn.org

Ostomy Assn., Inc., United (1962), 19772 MacArthur Blvd., Ste. 200, Irvine, CA 92612, 30,000; http://www.uoa.org

Outdoors Conservancy, The Great, (1998), 4311 Manatee Ave. W., Suite 210, Bradenton, FL 34209, 550; http://www.TheGreatOutdoors.org

Outlaw and Lawman History, Inc., Natl. Assn. for (NOLA) (1974), 1201 Holly Ct., Harker Heights, TX 76548; approx. 450; http://www.webdots.com/nola

Overeaters Anonymous (1960) 6075 Zenith Ct., NE, Rio Rancho, NM 87124; approx. 100,000; http://www.overeatersanonymous.org

Paralyzed Veterans of America (1947), 801 18th St. NW, Wash., DC 20006; 18,000; http://www.pva.org

Parametric Analysts, Intl. Soc. of (1978), P.O. Box 6402, Chesterfield, MO 63006; 275; http://ISPA-cost.org

Parapsychology Institute of America (1971), P.O. Box 5442, Babylon, NY 11707; 1,100.

Parents Without Partners (1957), 401 N. Michigan Ave., Chicago, IL 60611; 70,000; http://www.parentswithout partners.org

Parkinson's Disease Foundation, Inc. (1957), William Black Medical Bldg., Columbia-Presbyterian Medical Center, 710 W. 168th St., New York, NY 10032; 95,000; http://www.pdf.org

Parliamentarians, Natl. Assn. of (1930), 213 South Main St., Independence, MO 64050; 3,700; http://www.parliamentarians.org

Pasta Assn., Natl. (1904), 2101 Wilson Blvd., Ste. 920, Arlington, VA 22201; 74 cos.; http://www.ilovepasta.org

Patton Society (1970), 3116 Thorn St., San Diego, CA 92104; 250; http://members.aol.com/PattonsGH/homeghq.html.

Pearl Harbor History Associates, Inc. (1986), P.O. Box 1007, Stratford, CT 06615; approx. 300.

PEN American Center, Inc. (1922), 568 Broadway, Rm. 401, New York, NY 10012; 2,800; http://www.pen.org

Pen Friends, Intl. (1967), P.O. Box 290065, Brooklyn, NY 11229; 300,000; http://www.global-homebiz.com/ipf.html

Pension Plan, Committee for a Natl. (1979), P.O. Box 27851, Las Vegas, NV 89126; 325.

Pen Women, Natl. League of American (1897), 1300 17th St. NW, Wash., DC 20036; aprox. 4,500.

P.E.O. (Philanthropic Educational Organization) Sisterhood (1869), 3700 Grand Ave., Des Moines, IA 50312; 242,000.

People for the Ethical Treatment of Animals (PETA) (1980), 501 Front St., Norfolk, VA 23510; 600,000; http://www.peta-online.org

Performance Improvement, Intl. Society for (1962), 1300 L St. NW, #1250, Wash., DC 20005; 6,100; http://www.ispi.org

Petroleum Institute, American (1919), 1220 L St. NW, Wash., DC 20005; 400 companies; http://www.api.org

Pharmaceutical Assn., American (1852), 2215 Constitution Ave. NW, Wash., DC 20037; 50,000; http://www.aphanet.org

Phi Beta Kappa (1776), 1811 Q St. NW, Wash., DC 20009; approx. 500,000; http://www.pbk.org

Phi Delta Kappa (1911), 408 N. Union St., P.O. Box 789, Bloomington, IN 47402; 120,000+; http://www.pdkintl.org

Phi Kappa Phi (1897), P.O. Box 16000-Louisiana State University, Baton Rouge, LA 70893; 818,132; http://www.phikappaphi.org

Phi Theta Kappa (1918), Center for Excellence, 1625 Eastover Drive, Jackson, MS 39211; 1 mil.; http://www.ptk.org

Philatelic Society, American (1886), 100 Oakwood Ave., P.O. Box 8000, State College, PA 16803; 56,000; http://www.stamps.org

Philological Assn., American (1869), Dept. of Classics, College of the Holy Cross, Worcester, MA 01610; 3,300.

Philosophical Assn., American (1900), Univ. of Delaware, Newark, DE 19716; 10,500; http://www.udel.edu/apa

Philosophical Enquiry, Intl. Soc. For (1974), 60 Murifield Ct., St. Louis, MO 63141.

Philosophical Society, American (1743), 104 S. 5th St., Philadelphia, PA 19106; 690.

Photographers of America, Inc., Professional (1880), 229 Peachtree Street, NE, Atlanta, GA 30303; 14,000; http://www.ppa-world.org

Photographic Society of America, Inc. (1934), 3000 United Founders Blvd., Ste. 103, Oklahoma City, OK 73112; 6,200; http://www.psa-photo.org

Physical Therapy Assn., American (1921), 1111 N. Fairfax St., Alexandria, VA 22314; 70,000; http://www.apta.org

Physically Handicapped, Inc., Natl. Assn. of the (1858), NAPH Business Office, Scarlet Oaks, 440 Lafayette Ave., #GA4, Cincinnati, OH 45220; approx. 600.

Physics, American Inst. of (1931), One Physics Ellipse, College Park, MD 20740; 123,500; http://www.aip.org

Physiological Society, American (1887), 9650 Rockville Pike, Bethesda, MD 20814; 8,300; http://www.faseb.org/aps

Pilgrim Society (1820), 75 Court St., Plymouth, MA 02360; 900.

Pilot Intl. & Pilot Intl. Foundation (1921, 1975), P.O. Box 4844, 244 College St., Macon, GA 31208; 26,000.

Pi Mu Epsilon National Honorary Mathematics Society (1914), 2 Thornhill Drive, Pulaski, PA 16143; 110,000; http://www.PME-math.org

Planetary Society (1979), 65 N. Catalina Ave., Pasadena, CA 91106; 100,000.

Planned Parenthood Federation of America, Inc. (1916), 810 Seventh Ave., New York, NY 10019; http:/www.plannedparent hood.org

Plastic Modelers Society, Intl. (1963), P.O. Box 6138, Warner Robins, GA 31095; 5,111; http://www.ipmsusa.org

Plastics Engineers, Society of (1942), P.O. Box 403, Brookfield, CT 06804; 32,000+; http://www.4spe.org

Plastics Industry, Inc., Society of the (1937), 1801 K Street, NW, Ste. 600K, Wash., DC 20006; 2,000+ companies; http://www.socplas.org

Poetry Society of America (1910), 15 Gramercy Park, New York, NY 10003; approx. 3,000; http://www.poetrysociety.org

Poets, The Academy of American (1934), 584 Broadway, Ste. 1208, New York, NY 10012; 8,000; http://www.poets.org

Police Assn., Intl. (1950 in UK, 1962 in U.S.), P.O. Box 613-1822, South Miami, FL 33243; 276,000+; http://www.ipa-usa.org

Polish Army Veterans Assn. of America (1921), 119 E. 15th St., Ste. 1, New York, NY 10003; 3,000.

Polish Cultural Society of America, Inc. (1940), P.O. Box 31, Wall St., New York, NY 10005; 109,845.

Political Items Collectors, American (1945), P.O. Box 340339, San Antonio, TX 78234; 3,200; http://www.collectors.org

Political Science Assn., American (1903), 1527 New Hampshire Ave. NW, Wash., DC 20036; 16,200; http://www.apsanet.org

Political Science Assn., Southern (1928), Dept. of Political Science, University of Mississippi, University, MS 38677; 1,800; http://www.olemiss.edu/orgs/spsa

Political & Social Science, American Academy of (1889), 3937 Chestnut St., Philadelphia, PA 19104; 5,000.

Polo Assn., U.S. (1890), 4059 Iron Works Parkway, Ste. 1, Lexington, KY 40511; over 3,000; http://www.uspolo.org

Population Assn. of America (1931), 721 Ellsworth Dr., Ste. 303, Silver Spring, MD 20910; 3,000.

Portuguese American Federation, Inc., (1974), P.O. Box 694, Bristol, RI 02809; 250.

Portuguese Continental Union of the U.S.A. (1925), 899 Boylston St., Boston, MA 02115; 6,042; http://members.aol.com/UPCEUA

Postcard Dealers, Inc., International Federation of (1979), P.O. Box 1765, Manassas, VA 20108; 253.

Postmasters of the U.S., Natl. Assn. of (1898), 8 Herbert St., Arlington, VA 22305; 43,000.

Postmasters of the U.S., Natl. League of (1887), 1023 N. Royal St., Alexandria, VA 22314; 30,000; http://www.post-masters.org

Power Boat Assn., American (1903), P.O. Box 377, Eastpointe, MI 48021; 6,000.

Printing Industries of America, Inc. (1887), 100 Dangerfield Rd., Alexandria, VA 22314; 14,000; http://www.printing.org

Procrastinators Club of America (1956), P.O. Box 712, Bryn Athyn, PA 19006; 14,000+.

Production and Inventory Control Soc., American (1957), 500 W. Annandale Rd., Falls Church, VA 22046; 69,114.

Protection of Old Fishes, Soc. for the (1967), School of Fisheries, 357980, c/o Prof. G. Brown, Univ. of Washington, Seattle, WA 98195; 200.

Psoriasis Foundation, Natl. (1968), 6600 SW 92d Ave., Ste. 300, Portland, OR 97223; 40,000; http://www.psoriasis.org

Psychiatric Assn., American (1844), 1400 K St. NW, Wash., DC 20005; 40,453; http://www.psych.org

Psychical Research, American Society for (1885), 5 W. 73d St., New York, NY 10023; approx 5,000.

Psychoanalytic Assn., American (1911), 309 E. 49th St., New York, NY 10017; 3,400; http://apsa.org

Psychological Assn., American (1892), 750 1st St. NE, Wash., DC 20002; 159,000; http://www.apa.org

Psychological Assn. for Psychoanalysis, Inc., Natl. (1948), 150 W. 13th St., New York, NY 10011; 365; http://www.npap.org

PTA, Natl. (1897), 330 N. Wabash Ave., Ste. 2100, Chicago, IL 60611; 6.5 mil; http://www.pta.org

Public Administration, American Soc. for (1939), 1120 G St. NW, Wash., DC 20005; 11,000+; http://www.aspanet.org

Public Health Assn., American (1872), 1015 15th St. NW, Wash., DC 20005; http://www.apha.org

Public Relations Soc. of America, Inc. (1947), 33 Irving Pl., 3d Fl., New York, NY 10003; 17,383; http://www.prsa.org

Publishers, Assn. of American (1970), 71 5th Ave., New York, NY 10003; 200 cos.; http://www.publishers.org

Pulp and Paper Industries, Technical Assn. of the (TAPPI) (1915), 15 Technology Pkwy. S, Norcross, GA 30092; 34,000; http://www.tappi.org

Puppeteers of America (1936), 5 Cricklewood Path, Pasadena, CA 91107; 2,400; http://www.puppeteers.org

Quality Control, American Society for (ASQC) (1946), 611 E. Wisconsin Ave., Milwaukee, WI 53201; 140,000; http://www.asqc.org

Quota International, Inc. (1919), 1420 21st St. NW, Wash., DC 20036; 11,000+; http://www.quota.org

Rabbis, Central Conference of American (1889), 355 Lexington Ave., New York, NY 10017; 1,800; http://ccarnet.org

Racquetball Assn., U.S. (1968), 1685 W. Uintah, Colorado Springs, CO 80904; 25,000; http://www.racquetball.org

Radio Relay League, American (1914), 225 Main St., Newington, CT 06111; 172,000; http://www.arrl.org

Radio and Television Society Foundation, Intl. (1939), 420 Lexington Ave., Ste. 1714, New York, NY 10170; 1,787; http://www.irts.org

Railway Historical Society, Natl. (1935), P.O. Box 58547, Philadelphia, PA 19102; 14,402; http://www.rrhistorical.com/nrhs

Railway Progress Institute (1908), 700 N. Fairfax St., Ste. 601, Alexandria, VA 22314; 95 cos.; http://www.rpi.org

Reading Assn., Intl. (1956), 800 Barksdale Rd., P.O. Box 8139, Newark, DE 19714; 90,000; http://www.reading.org

Real Estate Institute, Intl. (1968), 1224 N. Nokomis, Alexandria, MN 56308; 4,806.

Rebekah Assemblies, Intl. Assn. of (1922), 422 Trade St., Winston-Salem, NC 27101; 114,651.

Recreation and Park Assn., Natl. (1965), 22377 Belmont Ridge Road, Ashburn, VA 20148; 21,957; http://www.nrpa.org

Recycling Coalition, Natl. (1979), 1727 King St., Ste. 105, Alexandria, VA 22514; 3,500.

Red Cross, American (1881), 8111 Gatehouse Rd., Falls Church, VA 22042; 1.44 mil volunteers; http://www.redcross.org

Refugee Committee, American (1978), 2344 Nicollet Ave. S., Ste. 350, Minneapolis, MN 55404; http://www.archq.org

Rehabilitation Assn., Natl. (1927), 633 S. Washington St., Alexandria, VA 22314; approx. 11,000; http://www.nationalrehab.org

Religion, American Academy of (1964), 1703 Clifton Rd. NE, Ste. G-5, Atlanta, GA 30329; 8,000; http://www.aar-site.org

Renaissance Society of America (1954), 24 W. 12th St., 3d Fl., New York, NY 10011; 3,900; http://www.r-s-a.org

Republican National Committee (1856), 310 1st St. SE, Wash., DC 20003; http://www.rnc.org

Reserve Officers Assn. of the U.S. (1922), One Constitution Ave. NE, Wash., DC 20003; 95,000; http://www.rua.org.

Restaurant Assn., Natl. (1919), 1200 17th St. NW, Wash., DC 20036; 33,000; http://www.restaurant.org

Retail Federation, Natl. (1908), 325 7th St. NW, Ste. 1100, Wash., DC 20004; 50,000; http://www.nrf.com

Retired Credit Union People, Natl. Assn. for (1978), P.O. Box 391, Madison, WI 53705; 81,180.

Retired Federal Employees, Natl. Assn. of (1921), 606 N. Washington St., Alexandria, VA 22314; 425,000 http://www.narfe.org

Retired Officers Assn. (1929), 201 N. Washington St., Alexandria, VA 22314; 396,000; http://www.troa.org

Retired Persons, American Assn. of (1958), 601 E St. NW, Wash., DC 20049; 32 mil.; http://www.aarp.org

Retired Teachers Assn., Natl. (1947), 601 E St., NW, Wash., DC 20049; 540,000.

Reye's Syndrome Foundation, Natl. (1974), 426 N. Lewis, St. P.O. Box 829, Bryan, OH 43506; 238; http://www.bright.net/~reyessyn

Richard III Society (1969), P.O. Box 13786, New Orleans, LA 70185; 700; http://www.r3.org

Rifle Assn., Natl. (1871), 11250 Waples Mill Rd., Fairfax, VA 22030; approx 3 mil; http://www.nra.org

Road & Transportation Builders Assn., American (1902), The ARTBA Building, 1010 Massachusetts Ave. NW, Wash., DC 20001; 5,000; http://www.artba.org

Roller Skating, U.S.A. (1937), 4730 South St., P.O. Box 6579, Lincoln, NE 68506; 30,000; http://www.usacrs.org

Rose Society, American (1892), 8877 Jefferson Page Rd, Shreveport, LA 72119; 20,000; http://www.ars.org

Rotary Intl. (1905), 1560 Sherman Ave., Evanston, IL 60201; 1,203,726; http://www.rotary.org

Running and Fitness Assn., American (1968), 4405 East West Highway, Ste. 405, Bethesda, MD 20814; approx. 16,500; http://www.arfa.org

Ruritan Natl., Inc. (1928), P.O. Box 487, Dublin, VA 24084; 33,447.

Safety Council, Natl. (1913), 1121 Spring Lake Dr., Itasca, IL 60143; 16,000; http://www.nsc.org

Safety Engineers, American Soc. of (1911), 1800 E. Oakton St., Des Plaines, IL 60018; 32,000; http://www.asse.org

Salt Institute (1914), 700 N. Fairfax St., Ste. 600, Alexandria, VA, 22314; 7 U.S., 30 Int'l.; http://www.saltinstitute.org

Sand Castle Builders, Intl. Society of (1988), 172 N. Pershing Ave., Akron, OH 44313; 200.

Save-the-Redwoods League (1918), 114 Sansome St., Ste. 605, San Francisco, CA 94104; 50,000; http://www.savetheredwoods.org

School Administrators, American Assn. of (1865), 1801 N. Moore St., Arlington, VA 22209; 14,000; http://www.aasa.org

School Boards Assn., Natl. (1940), 1680 Duke St., Alexandria, VA 22314; http://www.nsba.org

School Counselor Assn., American (1952), 801 N. Fairfax St., Ste. 310, Alexandria, VA 22314; 12,000; http://www.schoolcounselor.org

Schools of Art and Design, Natl. Assn. of (1944), 11250 Roger Bacon Dr., Reston, VA 20190; 553 institutions.

Science, American Assn. for the Advancement of (1848), 1200 New York Ave. NW, Wash., DC 20005; 144,000; http://www.aaas.org

Science Fiction Society, World (1939), P.O. Box 1270, Kendall Sq. Sta., Cambridge, MA 02142; 5,000.

Science Service Inc. (1921), 1719 N St. NW, Wash., DC 20036; http://www.sciserv.org

Sciences, Natl. Academy of (1863), 2101 Constitution Ave. NW, Wash., DC 20418; 4,000+; http://www.nas.edu

Science Teachers Assn., Natl. (1944), 1840 Wilson Blvd., Arlington, VA 22201; 53,000; http://www.nsta.org

Science Writers, Natl. Assn. of (1934), P.O. Box 294, Greenlawn, NY 11740; 1,801; http://www.nasw.org

Scrabble® Assn., Natl. (1980), P.O. Box 700, 120 Front St., Greenport, NY 11944; 10,000+; http://www.scrabble-assoc.com

Screen Actors Guild (1933), 5757 Wilshire Blvd., Los Angeles, CA 90036; 90,000; http://www.sag.com

Screenprinting & Graphic Imaging Assn. Intl. (1948), 10015 Main St., Fairfax, VA 22031; 3,800 ; http://www.sgia.org

Sculpture Soc., Natl. (1893), 1177 Ave. of the Americas, New York, NY 10036; 270.

Seamen's Service, United (1942), One World Trade Center, Ste. 2161, New York, NY 10048.

2d Air Division Assn. (1948), P.O. Box 484, Elkhorn, WI 53121; 7,000.

Secondary School Principals, Natl. Assn. of (1916), 1904 Association Dr., Reston, VA 20191; 41,000; http://www.nassp.org

Secretaries Intl.®, Professional/The Assn. for Office Professionals™ (1942), 10502 N.W. Ambassador Dr., Kansas City, MO 64152; 27,000; http://www.gvi.net/psi

Secular Humanism, Council for (1980), 1310 Sweet Home Rd., Amherst, NY 14226; 23,000; http://www.secularhumanism.org

Securities Industry Assn. (1972), 120 Broadway, New York, NY 10271; 715 firms.

Separation of Church & State, Americans United for (1947), 518 C St. NE, Wash., DC 20002; 60,000; http://www.au.org

Sertoma International (1912), 1912 E. Meyer Blvd., Kansas City, MO 64132; 24,992; http://www.sertoma.org

Sexuality Information & Education Council of the U.S. (SIECUS) (1964), 130 W. 42d St., Ste. 350, New York, NY 10036; http://www.siecus.org

Sharkhunters Int'l. (1983), P.O. Box 1539, Hernando, FL 34442; 4,000; http://www.sharkhunters.com

Shipbuilders Council of America (1921), 901 Washington St., Ste. 204, Alexandria, VA 22314; 50 organizations; http://www.shipbuilders.org

Ships in Bottles Assn. (1983), P.O. Box 180550, Coronado, CA 92178; 250.

Shrine of North America, The (1872), 2900 N. Rocky Point Dr., Tampa, FL 33607; approx 600,000; http://shrinershq.org

Sierra Club (1892), 85 2d St., 2d Fl., San Francisco, CA 94105; 550,000; http://www.sierraclub.org

Sigma Beta Delta (1994), P.O. Box 46935, St. Louis, MO 63146;12,000; http://www.sigmabetadelta.org

Skeet Shooting Assn., Natl. (1946), P.O. Box 680007, San Antonio, TX 78268; 15,800; http://nssa-nsca.com/nssa/index.html

Ski Team Foundation, U.S. (1964), 1500 Kearns Blvd., Park City, UT 84060; 60,000.

Small Business United, Natl. (1937), 1156 15th St. NW, Ste. 1100, Wash., DC 20005; 65,000+; http://www.nsbu.org

Social Sciences, Natl. Institute of (1865), 114 Clinton St., Ste. 3-A, Brooklyn Heights, NY 11201; 315.

Social Work Education, Council on (1952), 1600 Duke St., Alexandria, VA 22314; 4,000; http://www.cswe.org

Sociological Assn., American (1905), 1307 New York Avenue NW, Ste. 700, Wash., DC 20005; 13,500; http://www.asanet.org

Softball Assn./USA Softball, Amateur (1933), 2801 N.E. 50th St., Oklahoma City, OK 73111; 250,000; http://www.softball.org

Software and Information Industry Assn. (formerly Information Industry Assn.) (1999), 1730 M. St., Ste. 700, Wash., DC 20036; 1,400; http://www.siia.net

Soil Science Society of America (1936), 677 S. Segoe Rd., Madison, WI 53711; 5,714; http://www.soils.org

Soil & Water Conservation Society of America (1949), 7515 N.E. Ankeny Rd., Ankeny, IA 50021; 10,000.

Soldiers', Sailors', Marines' and Airmen's Club (1919), 283 Lexington Avenue, New York, NY 10016; 2,000; http://www.ssmaclub.org

Songwriters Guild of America (1931), 1500 Harbor Blvd., Weehawken, NJ 07087; 5,000+.

Sons of the American Legion (1932), Box 1055, Indianapolis, IN 46206; 161,376.

Sons of the American Revolution, Natl. Society of (1889), 1000 S. Fourth St., Louisville, KY 40203; 27,000; http://www.sar.org

Sons of Confederate Veterans (1896), 740 Mooresville Pike, Columbia, TN 38401; 26,000; http://www.scv.org

Sons of the Desert Laurel & Hardy Appreciation Society (1965), P.O. Box 8341, Universal City, CA 91608; 15,000; http://www.wayoutwest.org.

Sons of Italy in America, Order of (1905), 219 E St. NE, Wash. DC 20002; 500,000; http://www.osia.org

Sons of Norway (1895), 1455 W. Lake St., Minneapolis, MN 55408; 68,925; http://www.sofn.com

Soroptimist Intl. of the Americas (1921), Two Penn Center Plaza, Ste. 1000, Philadelphia, PA 19102; 50,000; http://www.soroptimist.org

Southern Christian Leadership Conference (1957), 334 Auburn Ave. NE, Atlanta, GA 30303; 1 mil.

Space Education Assn., U.S. (1973), P.O. Box 249, Rheems, PA 17570; 1,000.

Space Society, Natl. (1974), 600 Pennsylvania Ave SE, Ste. 201, Wash., DC 20003; 20,000; http://www.nss.org

Speech-Language-Hearing Assn., American (1925), 10801 Rockville Pike, Rockville, MD 20852; 93,000; http://www.asha.org

Speedskating Union of the U.S., Amateur (1927), 1033 Shady Lane, Glen Ellyn, IL 60137; 3,000.

Speleological Society, Natl. (1941), 2813 Cave Ave., Huntsville, AL 35810; 12,000; http://www.caves.org

Sports Car Club of America (1944), 9033 E. Eastern Pl., Englewood, CO 80112; 50,000+; http://www.scca.org

Sportscasters Assn., The American (1980), 5 Beekman St., New York, NY 10038; 500.

State & Local History, American Assn. for (1944), 1717 Church St., Nashville, TN 37203; 5,000; http://www.aaslh.org

State Governments, Council of (1933), 2760 Research Park Drive, P.O. Box 11910, Lexington, KY 40517; 50 states, 4 territories; http://www.csg.org

Statistical Assn., American (1839), 1429 Duke St., Alexandria, VA 22314; 18,000; http://www.amstat.org

Steamship Historical Society of America, Inc. (1935), 300 Ray Dr., Ste. 4, Providence, RI 02906; 3,500; http://www.sshsa.org

Stock Exchange, American (1911), 86 Trinity Pl., New York, NY 10006; 871; http://www.amex.com

Stock Exchange, New York (1792), 11 Wall St., New York, NY 10005; http://www.nyse.com

Stock Exchange, Philadelphia (1790), 1900 Market St., Philadelphia, PA 19103; 504; http://www.phlx.com

Student Councils, Natl. Assn. of (1931), 1904 Association Dr., Reston, VA 20191; 9,000 schools.

Stuttering Project, Natl. (1977), 5100 E. LaPalma Ave., #208, Anaheim Hills, CA 92807; 2,800; http://www.nspstutter.org

Submarine Veterans of WWII, U.S. (1955), 826 Chatham Ave., Elmhurst, IL 60126; 8,250.

Sudden Infant Death Syndrome Alliance (1987), 1314 Bedford Ave., Suite 210, Baltimore, MD 21208; http://www.sidsalliance.org

Supreme Council, 33°, A.A.S.R. (1801), 1733 16th St. NW, Wash., DC 20009; 400,000; http://www.srmason-sj.org.

Surgeons, American College of (1913), 633 N. Saint Clair St., Chicago, IL 60611; 57,000; http://www.facs.org

Symphony Orchestra League, American (1942), 1156 Fifteenth St. NW, Ste. 800, Wash., DC 20005; 850; http://www.symphony.org

Table Tennis Assn., U.S. (1933), One Olympic Plaza, Colorado Springs, CO 80909; 8,500; http://www.usatt.org

Tailhook Assn. (1956), 9696 Business Park Ave., P.O. Box 26700, San Diego, CA 92131; 11,800.

Tall Buildings and Urban Habitat, Council on (1969), Lehigh Univ., 11 E. Packer Ave., Bethlehem, PA 18038; 1,500; http://www.lehigh.edu/ctbuh/

Tau Beta Pi Association (1885), P.O. Box 2697, Knoxville, TN 37901; 387,000; http://www.tbp.org

Tax Administrators, Federation of (1937), 444 N. Capitol St. NW, Ste. 348, Wash., DC 20001; 60; http://www.taxadmin.org

Tax Foundation (1937), 1250 H St. NW, Ste. 750, Wash., DC 20005; 50 U.S. states; http://www.taxfoundation.org

Taxpayers Union, Natl. (1969), 108 N. Alfred St., Alexandria, VA 22314; 300,000+; http://www.ntu.org

Tea Assn. of the U.S.A., Inc. (1899), 420 Lexington Ave., Ste. 825, New York, NY 10170; 150 corps.

Teachers of English, Natl. Council of (1911), 1111 W. Kenyon Rd., Urbana, IL 61801; 68,000; http://www.ncte.org

Teachers of English to Speakers of Other Languages (1966), 1600 Cameron St., Ste. 300, Alexandria, VA 22314; 16,500; http://www.tesol.edu

Teachers of French, American Assn. of (1927), Mailcode 4510, Southern Illinois University, Carbondale, IL 62901; 9,550; http://aatf.utsa.edu

Teachers of German, Inc., American Assn. of (AATG) (1926), 112 Haddontowne Ct. #104, Cherry Hill, NJ 08034; 7,000; http://www.aatg.org

Teachers of Mathematics, Natl. Council of (1920), 1906 Association Dr., Reston, VA 20191; 120,000; http://www.nctm.org

Teachers of Singing, Natl. Assn. of (1944), 2800 Univ. Blvd. N., J.U. Sta., Jacksonville, FL 32211; 5,574; http://www.nats.org

Teachers of Spanish & Portuguese, American Assn. of (1917), Univ. of Northern Colorado, 210 Butler-Hancock, Greeley, CO 80639; 12,000; http://www.aatsp.org

Telephone Pioneer Assn., Independent (1920), 1401 H Street NW, Ste. 600, Wash., DC 20005.

Television Arts & Sciences, Natl. Academy of (1947), 111 W. 57th St., Ste. 1020, New York, NY 10019; 12,000; http://www.emmyonline.org

Testing & Materials, American Society for (1898), 100 Barr Harbor Dr., West Conshohocken, PA 19428; 35,000; http://www.astm.org

Tetra Society of North America (1992), Box 27, Ste. A-304, Plaza of Nations, 770 Pacific Blvd. S., Vancouver, BC V6B 5E7; 500.

Textile Manufacturers Institute, American (1949), 1130 Connecticut Ave. NW, Ste. 1200, Wash., DC 20036; 114 cos.; http://www.atmi.org

Theodore Roosevelt Assn. (1920), P.O. Box 719, Oyster Bay, NY 11771; 1,600.

Theological Library Assn., American (1946), 820 Church St., Ste. 300, Evanston, IL 60201; 250 inst.; http://www.atla.com

Theological Schools in the U.S. and Canada, The Assn. of (1918), 10 Summit Park Dr., Pittsburgh, PA 15275; 237; http://www.ats.edu

Theosophical Society in America (1911), 1926 N. Main St., Wheaton, IL 60189; 4,000; http://www.theosophical.org

Thoreau Society (1941), 44 Baker Farm, Lincoln, MA 01773; 2,000; http://www.walden.org

Thoroughbred Racing Assns. (1942), 420 Fair Hill Dr., Ste. 1, Elkton, MD 21921; 49 racing associations; http://www.traofna.com

Tin Can Sailors (1976), P.O. Box 100, Somerset, MA 02726; 18,000; http://www.destroyers.org

Titanic Historical Society, Inc. (1963), 208 Main St., P.O. Box 51053, Indian Orchard, MA 01151; 8,000; http://www.titanic1.org

Toastmasters Intl. (1924), 23182 Arroyo Vista, Rancho Santa Margarite, CA 92688; 173,000; http://www.toastmasters.org

Topical Assn., American (1949), P.O. Box 50820, Albuquerque, NM 87181; 6,000; http://home.prcn.org/~pauld/ata.

Toy Manufacturers of America (1916), 200 Fifth Ave., New York, NY 10010; 265; http://www.toy~tma.com

Totally Useless Skills, Institute of (1987), P.O. Box 181, Temple, NH 03084; 387; http://www.jlc.net/~useless

Trade Assn., Intl. (1990), 8383 E. Evans Rd., Scottsdale, AZ 85260; 1,000.

Trademark Assn., Intl. (1878), 1133 Avenue of the Americas, New York, NY 10036; 3,500; http://www.inta.org

Trail Assn., North Country (1980), 49 Monroe Center NW, Ste. 200B, Grand Rapids, MI 49503; 750; http://people. delphi.com/wesboyd/ncnst.htm

Training in Communication, Intl. (1938), 2519 Woodland Dr., Anaheim, CA 92801; 15,000; http://www.escape.ca~itcintl

Transit Assn., American Public (1974), 1201 New York Ave. NW, Wash., DC 20005; 1,100 organizations; http://www.apta.com

Translators Assn., American (1959), 225 Reinekers Lane, Ste. 590, Alexandria, VA 22314; 7,000; http://www.atanet.org

Transportation Alternatives (1973), 115 W. 30th St., #1203, New York, NY 10001; 3,500; http://www.transalt.org

Transportation Engineers, Inst. of (1930), 525 School St. SW, Ste. 410, Wash., DC 20024; 12,800; http://www.ite.org

Trapshooting Assn. of America, Amateur (1923), 601 W. National Road, Vandalia, OH 45377; 54,000; http://www.shootata.com

Travel Agents, American Society of (1931), 1101 King St., Ste. 200, Alexandria, VA 22314; 28,500; http://www.astanet.com

Travelers Protective Assn. of America (1890), 3755 Lindell Blvd., St. Louis, MO 63108; 120,065.

Trilateral Commission (1973), 345 E. 46th St., Ste. 711, New York, NY 10017; 365; http://www.trilateral.org

Truck Historical Soc., American (1971), 300 Office Park Dr., Ste. 120, Birmingham, AL 35223; 21,100; http://www.aths.org

Trucking Assns., American (1933), 2200 Mill Rd., Alexandria, VA 22314; 4,000; http://www.truckline.com

T. S. Eliot Society (1981), 5007 Waterman Blvd., St. Louis, MO 63108; 250.

U.F.O. Society of America (1997), 10799 Sherman Grove Ave, #18, Sunland, CA 91040; 550.

UFOs, Natl. Investigation Committee on (1967) 14617 Victory Blvd., Van Nuys, CA 91411; 1,500; http://www.NICUFO.org

UNICEF, U.S. Committee for (1937), 333 E. 38th St., New York, NY 10016; http://www.unicefusa.org

Underwriters, Natl. Assn. of Life (1890), 1922 F St. NW, Wash., DC 20006; 143,000.

Underwriters (CPCU), Soc. of Chartered Property and Casualty (1944), 720 Providence Rd., P.O. Box 3009, Malvern, pa 19355; 28,000; http://www.cpusociety.org

Uniformed Services, Natl. Assn. for (1968), 5535 Hempstead Way, Springfield, VA 22151; 160,000; http://www.naus.org

United Nations Assn. of the U.S.A. (1923, as League of Nations Assn.; 1945), 801 2nd Ave., New York, NY 10017; 25,000; http://www.unausa.org

United Order True Sisters, Inc., (1846), 100 State St., Albany, NY 12207; 3,200.

United Press Intl. (1907), 1510 H St., Ste. 600, Wash., DC 20005; http://www.upi.com/homepage.html

United Service Organizations (USO) (1941), Washington Navy Yard, 1008 Eberle Place, Wash., DC 20374; http://www.uso.org

United Way of America (1918), 701 N. Fairfax St., Alexandria, VA 22314; 1,353; http://www.unitedway.org

Universities, Assn. of American (1900), 1200 New York Ave., NW, Ste. 550, Wash., DC 20005; 62 institutions; http://www.tulane.edu/~aau

University Continuing Education Assn., (1915), One Dupont Circle, Ste. 615, Wash., DC 20036; 420 institutions; http://www.nucea.edu

University Foundation, Intl. (1973), 1301 S. Noland Rd., Independence, MO 64055; 62,311.

University Women, American Assn. of (1881), 1111 16th St. NW, Wash., DC 20036; 150,000; http://www.aauw.org

Urban League, Natl. (1910), 120 Wall St., New York, NY 10005; 50,000; http://www.nul.org

USENIX Association (1975), 2560 Ninth Street, Ste. 215, Berkeley, CA 94710; 7,024; http://www.usenix.org

USS *Forrestal* **CVA/CV/AVT-59 Assn., Inc.** (1991), 300 Cassady Ave., Virginia Beach, VA 23452; 1,600.

U.S.S. *Idaho* **Assn.** (1957), P.O. Box 711247, San Diego, CA 92171; 410.

Vampire Research Center (1972), P.O. Box 5442, Babylon, NY 11707; 700.

Variety Clubs Intl. (1928), 350 5th Ave., Ste. 1119, New York, NY 10018; 15,000.

Ventriloquists, North American Assn. of (1944), P.O. Box 420, Littleton, CO 80160; 1,756.

Veterans of Foreign Wars of the U.S. (1899), 406 W. 34th St., Kansas City, MO 64111; 1.9 mil.; http:/www.vfw.org

Veterans of Foreign Wars of the U.S., Ladies Auxiliary to the (1914), 406 W. 34th St., Kansas City, MO 64111; 727,921; http://www.ladiesauxvfw.com

Veterans of Underage Military Service (1991), 100 Village Lane, Philadelphia, PA 19154; 1,138.

Veterans of the Vietnam War, Inc. (1987), 760 Jumper Rd., Wilkes-Barre, PA 18702; 15,000; http://www.vvnw.org

Veterans of World War I of the USA, Inc. (1958), P.O. Box 8027, Alexandria, VA 22306; 7,900.

Veterinary Medical Assn., American (1863), 1931 N. Meacham Rd., Ste. 100, Schaumburg, IL 60173; 61,500; http://www.avma.org

Victorian Society in America (1966), 219 S. Sixth St., Philadelphia, PA 19106; 1,500; http://www.libertynet.org/vicsoc

Viewers for Quality Television, Inc. (1987), P.O. Box 195, Fairfax Station, VA 22039; 3,000; http://www.vqt.com

Virgil Fox Society (1977), 88 Chestnut St., Brooklyn, NY 11208; 400.

Volleyball, U.S.A. (1928), 3595 E. Fountain Blvd., Ste. I-2, Colorado Springs, CO 80910; 110,000.

War Mothers, American (1925), 2615 Woodley Pl. NW, Wash., DC 20008; under 1,000.

Watch & Clock Collectors, Natl. Assn. of (1943), 514 Poplar St., Columbia, PA 17512; 35,500; http://www.nawcc.org

Watercolor Society, American (1866), 47 5th Ave., New York, NY 10003; 500+; http://www.watercolor-online.com/aws

Water Environment Federation (1928), 601 Wythe St., Alexandria, VA 22314; 40,000; http://www.wef.org

Water Ski Assn., American (1939), 799 Overlook Dr. SE, Winter Haven, FL 33830; 20,000.

Water Works Assn., American (1881), 6666 W. Quincy Ave., Denver, CO 80235; 55,000; http://www.awwa.org

Welding Society, American (1919), 550 N.W. LeJeune Rd., Miami, FL 33126; 50,400; http://www.aws.org

Wheelchair Sports, USA (1956), 3595 E. Fountain Blvd., Ste. L-1, Colorado Springs, CO 80910; 4,600.

Widows, Society of Military (1968), 5535 Hempstead Way, Springfield, VA 22151; 2,000.

Wilderness Institute, Natl. (1989), P.O. Box 25766, Wash., DC 20007; http://www.nwi.org/

Wildlife Federation, Natl. (1936), 8925 Leesburg Pike, Vienna, VA 22184; 4.7 mil; http://www.nwf.org

Wildlife Management Institute (1911), 1101 14th St. NW, Ste. 801, Wash., DC 20005; 250; http://www.wildlifemgt.org/wmi

William Penn Assn. (1886), 709 Brighton Rd., Pittsburgh, PA 15233; 90,000.

Wireless Pioneers Inc., The Society of (1967), P.O. Box 86, Geyserville, CA 95441; 1,000; http://access.mountain.net/~carto/sowp001.htm

Wizard of Oz Club, Inc., The Intl. (1957), P.O. Box 266, Kalamazoo, MI 49004; 1,700; http://www.ozclub.org

Women, Natl. Organization for (NOW) (1966), 1000 16th St. NW, Ste. 700, Wash., DC 20036; 250,000; http://www.now.org

Women and Families, Natl. Partnership for (1971), 1875 Connecticut Ave. NW, Ste. 710, Wash., DC 20009; 2,500; http://www.nationalpartnership.org

Women Artists Inc., Natl. Assn. of (1889), 41 Union Sq. W, #906, New York, NY 10003; 800.

Women in Communications, Association for (1909 as Theta Sigma Phi), 1244 Ritchie Hwy., Ste. 6, Arnold, MD 21012; 7,500; http://www.womcom.org

Women Engineers, Society of (1950), 120 Wall St., 11th Fl., New York, NY 10005; 16,500; http://www.swe.org

Women in Radio and TV, Inc., American (1950), 1650 Tyson's Blvd., Ste. 200, McLean, VA 22102; 1,500.

Women of the U.S., Inc., Natl. Council of (1888), 777 UN Plaza, 7th Fl., New York, NY 10017; 500 members, 33 affiliate org.

Women Strike for Peace (1961), 110 Maryland Ave. NE, Ste. 102, Wash., DC 20002; 5,000.

Women Voters of the U.S., League of (1920), 1730 M St. NW, Wash., DC 20036; 100,000; http://www.lwv.org

Women World War Veterans (1919), 237 Madison Ave., New York, NY 10016; 35,000.

Women's Army Corps Veterans Assn. (1951), P.O. Box 5577, Ft. McClellan, AL 36205; 4,000.

Women's Christian Temperance Union, Natl. (1874), 1730 Chicago Ave., Evanston, IL 60201; 10,000; http://www.wctu.org

Women's Clubs, General Federation of (1890), 1734 N St. NW, Wash., DC, 20036; 300,000 U.S; http://www.gfwc.org

Women's Clubs, Natl. Fed. of Business & Prof. (1919), 2012 Massachusetts Ave. NW, Wash., DC 20036; 70,000.

Women's Intl. League for Peace & Freedom (1915), 1213 Race St., Philadelphia, PA 19107; 8,000.

Woodmen of America, Modern (1883), 1701 1st Ave., Rock Island, IL 61201; 375,000; http://www.modern-woodmen.com

Woodmen of the World Life Insurance Soc. (1890), 1700 Farnam St., Omaha, NE 68102; 850,000.

Workmen's Circle (1900), 45 E. 33d St., New York, NY 10016; 35,000; http://www.circle.org

World Council of Churches, U.S. Conference for the (1948), 475 Riverside Drive, New York, NY 10115; 317 denominations.

World Federalist Assn. (1975), P.O. Box 15250, Wash., DC 20003; 9,000; http://www.wfa.org

World Future Society (1966), 7910 Woodmont Ave., Ste. 450, Bethesda, MD 20814; 30,000; http://www.wfs.org

World Wildlife Fund (1961), 1250 24th St. NW, Wash., DC 20037; 1 mil+; http://www.worldwildlife.org

World's Fair Collectors Soc., Inc. (1968), P.O. Box 20806, Sarasota, FL 34276; 525; http://members.aol.com/bbqprod/wfcs.html

Writers Guild of America, West (1954), 7000 W. Third St., Los Angeles, CA 90048; 8,300; http://www.wga.org

Yachting Assn., Southern California (1921), 5489 E. Ocean Boulevard, Long Beach, CA 90803; 90 clubs & orgs., 21,500 families; http://www.scya.org

YMCA (Young Men's Christian Assns.) of the U.S.A. (1864 in IL, 1883 in NY,), 101 N. Wacker Dr., Chicago, IL 60606; 17 mil.; http://www.ymca.net

Young Women's Christian Assn. of the U.S.A. (1906), Empire State Bldg., Ste. 301, 350 Fifth Ave., New York, NY 10018; approx. 2 mil; http://www.ywca.org

Zero Population Growth (1968), 1400 16th St. NW, Ste. 320, Wash., DC 20036; 60,000+; http://www.zpg.org

Zionist Organization of America (1897), 4 E. 34th St., New York, NY 10016; 50,000+; http://www.zoa.org

Zoo and Aquarium Assn., American (1924), 8403 Colesville Rd., Ste. 710, Silverspring, MD 20907; 5,700; http://www.aza.org

ASTRONOMY AND CALENDAR

Edited by Dr. Lee T. Shapiro, Planetarium Director
Morehead Planetarium, University of North Carolina at Chapel Hill

Celestial Events Summary, 2000

2000 is a good year for total lunar eclipses. There are 2 of them. The first should be visible throughout N America the night of Jan. 20-21; the 2d, on July 16, centers over the eastern Pacific Ocean. There are also an unusual number of solar eclipses (4), but they are all partial.

At the start of the year, Jupiter, Saturn, and Mars are all visible in the early evening sky, with Venus noticeable in the morning sky. Jupiter has been slowly approaching Saturn, finally passing it at the end of May. Jupiter catches and passes Saturn approximately every 20 years. If conditions are right, there is sometimes a triple passing as a result of retrograde motion, but not in this cycle. After May, Jupiter and Saturn separate, but they begin to draw back together near the end of the year because of retrograde motion. In early May, the 5 planets visible with the naked eye slowly gather together in the sky although they are not particularly visible because of proximity to the Sun. Mercury, Venus, Mars, Jupiter, Saturn, the Moon, and the Sun are all within 2 hours of right ascension on both May 4 and May 5. The 6 objects other than the Moon are most tightly congregated in the sky on May 17, when they are within 1.4 hours of right ascension. The gravitational, tidal, and magnetic effects of this assembly are insignificant. During the 2d half of the year, Mars, Jupiter, and Saturn are mainly in the morning sky, with Venus in the evening sky. However, in late November, Jupiter and Saturn join Venus in the evening sky. Mercury, as always, switches back and forth between the evening and morning sky and is never obviously visible. The best opportunities for the northern hemisphere to see Mercury in the year 2000 are in mid-February and early June in the evening and mid-November in the morning.

From January through April there are pretty views of Venus with the waning crescent Moon in the morning and from August to December with the waxing crescent Moon in the early evening sky. On Apr. 6, look for the grouping of Jupiter, Saturn, Mars, and the crescent waxing Moon. Because of too much moonlight, most meteor showers will not be clearly visible.

Astronomical Positions Defined

Two celestial bodies are in **conjunction** when they are due N and S of each other, either in **right ascension** (with respect to the N celestial pole) or in **celestial longitude** (with respect to the N ecliptic pole). If the bodies are seen near each other, they will rise and set at nearly the same time. For the inner planets—Mercury and Venus—**inferior conjunction** occurs when the planets pass between Earth and the Sun, while **superior conjunction** occurs when either Mercury or Venus is on the far side of the Sun. They are in **opposition** when their Right Ascensions differ by exactly 12 hours, or when their Celestial Longitudes differ by 180°. One of the 2 objects in opposition will rise while the other is setting. **Quadrature** refers to the arrangement where the coordinates of 2 bodies differ by exactly 90°. These terms may refer to the relative positions of any 2 bodies as seen from Earth, but one of the bodies is so frequently the Sun that mention of the Sun is omitted in that case; otherwise, both bodies are named. When objects are in conjunction, the alignment is not perfect, and one is usually passing above or below the other. The geocentric angular separation between the Sun and an object is termed **elongation**. Elongation is limited only for Mercury and Venus; the greatest elongation for each of these bodies is noted in the appropriate table and is approximately the time for longest observation. **Perihelion** means the point in an orbit that is nearest to the Sun, and **aphelion**, the point farthest from the Sun. **Perigee** means the point in an orbit that is nearest Earth, **apogee** the point that is farthest from Earth. An **occultation** of a planet or a star is an **eclipse** of it by some other body, usually the Moon.

Astronomical Constants; Speed of Light

The following were adopted as part of the International Astronomical Union System of Astronomical Constants (1976): **Speed of light,** 299,792.458 km per sec., or about 186,282 statute mi per sec.; **solar parallax,** 8".794148; **Astronomical Unit,** 149,597,870 km, or 92,955,807 mi; **constant of nutation,** 9".2025; and **constant of aberration,** 20".49552.

Celestial Events Highlights, 2000

(Coordinated Universal Time, or UTC—the standard time of the prime meridian)

January

Mercury is moving closer to the Sun and remains hidden until the end of the month when it may be seen after sunset.

Venus, bright in the SE before sunrise, though somewhat lower by month's end, passes Antares on the 7th.

Mars, low in the SW as the sky darkens, holds that position through the month.

Jupiter, prominent in the S after sunset, triples with waxing quarter Moon and Saturn on the 14th and 15th.

Saturn, close to Jupiter in the S, makes it three planets easily seen in the early evening.

Moon occults Neptune on the 8th, Uranus on the 9th, Aldebaran on the 17th, passes Venus on the 3d3d, Jupiter on the 14th, and Saturn on the 15th, with a total lunar eclipse on the 21st visible throughout most of N America. Watch for the triple grouping of the Moon, Venus, and Antares an hour before sunrise on the 3d and the triple grouping of the Moon, Jupiter, and Saturn after sunset on the 14th.

Jan. 1—Uranus and Neptune, in Capricornus, stay there all year. Pluto, in Ophiuchus, stays there all year. Saturn is in Aries. Jupiter in Pisces. Mars in Aquarius. Venus in Libra. Mercury in Sagittarius.

Jan. 2—Mercury at aphelion. Venus enters Scorpius.

Jan. 3—Moon 3° N of Venus. Earth at perihelion closest to Sun, 91.4 mil mi.

Jan. 7—Venus 7° N of star Antares in Scorpius. Venus enters Ophiuchus.

Jan. 8—Moon 0.2° S of Neptune occults Neptune.

Jan. 9—Moon 0.4° S of Uranus occults Uranus.

Jan. 10—Moon 1.9° S of Mars.

Jan. 13—Saturn stationary, resumes direct motion.

Jan. 14—Moon 4° S of Jupiter.

Jan. 15—Moon 3° S of Saturn.

Jan. 16—Mercury at superior conjunction, passing beyond the Sun.

Jan. 17—Moon 1.2° N of star Aldebaran in Taurus occults Aldebaran.

Jan. 20—Sun enters Capricornus.

Jan. 21—Total lunar eclipse; see details under Eclipses. Venus enters Sagittarius.

Jan. 24—Neptune in conjunction with Sun.

February

Mercury, visible in the very early evening about half an hour after sunset very low in the WSW, best about mid-month.

Venus, still visible in the SE before sunrise, but getting lower.

Mars continues visible low in the SW in the early evening.

Jupiter, prominent in the S, again triples with the waxing quarter Moon and Saturn on the 11th and 12th.

Saturn, gradually being overtaken by Jupiter, makes the 4th naked-eye planet in the early evening.

Moon passes Venus on the 2d, Mercury on the 6th, Mars on the 8th, Jupiter on the 11th, Saturn on the 12th, and occults Aldebaran on the 14th.

Feb. 2—Moon 1.4° N of Venus; Mars enters Pisces.

Feb. 5—Partial solar eclipse; see details under Eclipses.

Feb. 6—Uranus in conjunction with Sun; Moon 1.8° S of Mercury.

Feb. 8—Moon 4° S of Mars.

Feb. 11—Moon 4° S of Jupiter.

Feb. 12—Moon 3° S of Saturn.

Feb. 14—Moon 1.2° N of Aldebaran occults Aldebaran.

Feb. 15—Mercury at greatest eastern elongation of 18° (E of Sun and setting after Sun); Mercury at perihelion.

Feb. 16—Jupiter enters Aries; Sun enters Aquarius.

Feb. 18—Venus enters Capricornus.

Feb. 21—Mars enters Cetus.

Feb. 22—Mercury stationary, begins retrograde motion; Mars enters Pisces.

March

Mercury, having left the evening sky, appears in the morning sky after the 1st week, passing Venus on the 15th.

Venus, low in the ESE, may be used to help find Mercury shortly before sunrise on the 15th.

Mars, now low in the W after sunset, is drawing closer to the giant planets.

Jupiter, still prominent, is now in the W after sunset.

Saturn continues to be approached by Jupiter, with waxing crescent Moon passing 3 bright planets in 3 days, 8th-10th.

Moon occults Neptune on the 3d, Uranus and Venus on the 4th, Neptune again on the 30th, Uranus again on the 31st, passes Mars on the 8th, Jupiter on the 9th, Saturn on the 10th, passing 6 planets in one week.

Mar. 1—Mercury at inferior conjunction passing between Earth and Sun.

Mar. 3—Moon 0.4° S of Neptune occults Neptune.

Mar. 4—Venus 0.7° S of Uranus; Moon 0.7° S of Uranus occults Uranus; Moon 0.6° S of Venus occults Venus.

Mar. 8—Moon 5° S of Mars.

Mar. 9—Moon 4° S of Jupiter.

Mar. 10—Moon 3° S of Saturn.

Mar. 11—Venus enters Aquarius; Sun enters Pisces.

Mar. 13—Mercury stationary, resumes direct motion.

Mar. 15—Mercury 2° N of Venus.

Mar. 16—Pluto stationary, begins retrograde motion

Mar. 20—Vernal Equinox at 2:35 AM EST (7:35 UTC), spring begins in the northern hemisphere, autumn in the southern hemisphere.

Mar. 22—Venus at aphelion; Mars enters Aries.

Mar. 28—Mercury at greatest western elongation of 28° (W of Sun and rising before the Sun).

Mar. 30—Moon 0.7° S of Neptune occults Neptune; Mercury at aphelion.

Mar. 31—Moon 1.0° S of Uranus occults Uranus.

April

Mercury continues low in the ESE, passes Venus on the 28th as it is lost from view.

Venus now very low in the E before sunrise.

Mars, now low in the W, is tripled with the two giant planets, passing Jupiter on the 6th and Saturn on the 16th.

Jupiter, low in the W and disappearing by mid-month, is part of an usual quadruple grouping of 3 planets and the waxing crescent Moon on the 6th.

Saturn also drops lower in the W after sunset and also is gone in the glare of sunset at mid-month.

Moon passes Mercury on the 2d, Venus on the 3d, Mars, Jupiter, and Saturn all on the 6th, Neptune on the 26th with an occultation, and Uranus on the 27th.

Apr. 2—Moon 1.6° S of Mercury; Venus enters Pisces.

Apr. 3—Moon 3° S of Venus.

Apr. 6—Moon 5° S of Mars, 4° S of Jupiter, and 3° S of Saturn; Mars 1.1° N of Jupiter.

Apr. 12—Venus enters Cetus.

Apr. 16—Mars 2° N of Saturn; Venus enters Pisces.

Apr. 18—Sun enters Aries.

Apr. 24—Mars enters Taurus.

Apr. 26—Moon 1.0° S of Neptune occults Neptune; Mars enters Taurus.

Apr. 27—Moon 1.3° S of Uranus.

Apr. 28—Mercury 0.3° S of Venus.

May

Mercury reemerges from the Sun's glow late in the month low in the WNW.

Venus disappears from view in the glow of sunrise.

Mars quickly disappears in the glow of sunset.

Jupiter begins to reappear in the early morning sky by; month's end, passing Saturn on the 31st.

Saturn also begins to reappear in the early morning sky by month's end.

Moon passes Mars on the 5th, Neptune on the 24th, and Uranus on the 25th.

May 1—Venus enters Aries.

May 5—Moon 5° S of Mars.

May 8—Jupiter in conjunction with Sun; Neptune stationary, begins retrograde motion.

May 9—Mercury at superior conjunction.

May 10—Saturn in conjunction with Sun.

May 13—Mercury at perihelion; Sun enters Taurus.

May 19—Mars 6° N of Aldebaran; Mercury 7° N of Aldebaran; Mercury 1.1° N of Mars.

May 24—Moon 1.2° S of Neptune; Venus enters Taurus.

May 25—Moon 1.5° S of Uranus; Uranus stationary, begins retrograde motion.

May 31—Jupiter 1.2° N of Saturn; Saturn and Jupiter enter Taurus.

June

Mercury still visible in the WNW after sunset, but disappearing after mid-month.

Venus remains hidden in the glare of the Sun.

Mars remains hidden in the glare of the Sun.

Jupiter is low in the ENE in the early morning.

Saturn, paired with Jupiter, is low in the ENE in the early morning.

Moon passes Saturn and Jupiter on the 1st, Mercury on the 4th, Neptune on the 20th, Uranus on the 21st, Saturn again on the 28th, and Jupiter again on the 29th.

June 1—Moon 3° S of Saturn and 4° S of Jupiter; Pluto at opposition; Jupiter enters Taurus.

June 4—Moon 4° S of Mercury.

June 9—Mercury at greatest eastern elongation of 24°.

June 11—Venus at superior conjunction.

June 18—Mars enters Gemini.

June 19—Venus enters Gemini.

June 20—Moon 1.3 ° S of Neptune; Northern Solstice at 9:48 PM EDT (1:48 UTC June 21).

June 21—Moon 1.6° S of Uranus; Sun enters Gemini.

June 22—Mercury stationary, begins retrograde motion.

June 26—Mercury at aphelion.

June 28—Moon 3° S of Saturn.

June 29—Moon 4° S of Jupiter.

July

Mercury returns to the morning sky after mid-month.

Venus reappears in the evening sky very low in the WNW after sunset near the end of the month.

Mars remains hidden in the glare of the Sun.

Jupiter, now prominent in the E before sunset, begins to slowly separate from Saturn.

Saturn rises higher in the E before sunset, with triple grouping of waning crescent Moon and Jupiter on the 26th.

Moon occults Mercury on the 29th, passes Neptune on the 17th, Uranus on the 18th, Saturn and Jupiter on the 26th, and undergoes a total lunar eclipse on the 16th (mostly not visible in N America).

July 1—Mars in conjunction with Sun. Partial solar eclipse; see details under Eclipses.

July 4—Earth at perihelion, furthest distance from the Sun, 94.5 mil mi.

July 6—Mercury at inferior conjunction.

July 11—Venus enters Cancer.

July 12—Venus at perihelion.

July 16—Total lunar eclipse; see details under Eclipses.

July 17—Mercury stationary, resumes direct motion; Moon 1.2° S of Neptune.

July 18—Moon 1.6° S of Uranus.

July 20—Sun enters Cancer.

July 26—Moon 2° S of Saturn and 3° S of Jupiter.

July 27—Mercury at greatest western elongation of 20°; Neptune at opposition; Venus enters Leo.

July 28—Mars enters Cancer.

July 29—Moon 0.8° N of Mercury occults Mercury.

July 31—Partial solar eclipse; see details under Eclipses.

August

Mercury in the morning sky is lost in the glare of the Sun after mid-month, passes Pollux on the 3d and Mars on the 10th.

Venus is low in the W after sunset; passes Regulus on the 6th.

Mars appears in the morning sky low in the ENE.

Jupiter, rising soon after midnight, is now prominent; in ESE before sunrise.

Saturn now visible nearly all of the 2d half of the night.

Moon occults Mars on the 28th, passes Neptune on the 13th, Uranus on the 14th, Saturn on the 22d, Jupiter on the 23d, and Venus on the 30th.

Aug. 3—Mercury 7° S of star Pollux in Gemini.

Aug. 6—Venus 1.1° N of star Regulus in Leo.

Aug. 9—Mercury at perihelion.

Aug. 10—Mercury 0.09° S of Mars; Sun enters Leo.

Aug. 11—Uranus at opposition.

Aug. 13—Moon 1.1° S of Neptune occults Neptune.

Aug. 14—Moon 1.4° S of Uranus.

Aug. 22—Mercury at superior conjunction; Moon 2° S of Saturn; Pluto stationary, resumes direct motion.

Aug. 23—Moon 3° S of Jupiter.

Aug. 25—Venus enters Virgo.

Aug. 28—Moon 0.9° N of Mars occults Mars; Mars enters Leo.

Aug. 30—Moon 4° N of Venus.

September

Mercury, very low in the W after sunset, passes Spica on the 23d.

Venus, low in the WSW after sunset, passes Spica on the 18th.

Mars, low in the E before sunrise, passes Regulus on the 16th.

Jupiter is visible throughout the 2d half of the night, passes Aldebaran on the 7th.

Saturn continues paired with Jupiter, and both are near the Hyades star cluster this month.

Moon passes Neptune on the 9th, Uranus on the 11th, Saturn and Jupiter on the 19th, Mars on the 25th, Mercury on the 29th, and Venus on the 30th.

Sept. 7—Jupiter 5° N of Aldebaran

Sept. 9—Moon 1.2° S of Neptune.

Sept. 11—Moon 1.4° S of Uranus.

Sept. 12—Saturn stationary, begins retrograde motion.

Sept. 16—Mars 0.8° N of Regulus; Sun enters Virgo.

Sept. 18—Venus 3° N of Spica in Virgo.

Sept. 19—Moon 1.8° S of Saturn and 2° S of Jupiter.

Sept. 22—Autumnal Equinox at 1:27 PM EDT (17:27 UTC), autumn begins in the northern hemisphere, spring begins in the southern hemisphere; Mercury at aphelion.

Sept. 23—Mercury 0.7° N of Spica.

Sept. 25—Moon 2° N of Mars.

Sept. 29—Moon 8° N of Mercury; Jupiter stationary, begins retrograde motion.

Sept. 30—Moon 5° N of Venus.

October

Mercury, very low in the SW after sunset, is lost in the glow of the Sun near the end of the month.

Venus, low in the SW after sunset, passes Antares on the 26th.

Mars is in the E in the early morning.

Jupiter, up more than half the night, is high in the W before sunrise and passes Aldebaran on the 21st.

Saturn is also high in the W before sunrise.

Moon passes Neptune on the 7th, Uranus on the 8th, Saturn on the 16th, Jupiter on the 17th, Mars on the 24th, and Venus on the 30th.

Oct. 1—Venus enters Libra.

Oct. 6—Mercury at greatest eastern elongation of 26°.

Oct. 7—Moon 1.3° S of Neptune.

Oct. 8—Moon 1.5° S of Uranus.

Oct. 15—Neptune stationary, resumes direct motion.

Oct. 16—Moon 1.6° S of Saturn.

Oct. 17—Moon 2° S of Jupiter.

Oct. 18—Mercury stationary, begins retrograde motion; Venus enters Scorpius.

Oct. 21—Jupiter 5° N of Aldebaran.

Oct. 24—Moon 3° N of Mars; Mars enters Virgo.

Oct. 25—Venus enters Ophiuchus.

Oct. 26—Uranus stationary, resumes direct motion; Venus 3° N of Antares.

Oct. 30—Mercury at inferior conjunction; Moon 4° N of Venus; Sun enters Libra.

November

Mercury appears in the morning sky after the 1st week.

Venus is low in the SW after sunset.

Mars, prominent in the SE before sunrise, is one of 4 naked-eye planets visible early in the morning at mid-month.

Jupiter is up the all night long, prominent in the W before sunrise.

Saturn is up the whole night long, visible low in the ENE after sunset.

Moon passes Neptune on the 3d, Uranus on the 4th, Saturn on the 12th, Jupiter on the 13th, Mars on the 21st, Mercury on the 24th, and Venus on the 29th.

Nov. 2—Venus and Mars at aphelion.
Nov. 3—Moon 1.6° S of Neptune.
Nov. 4—Moon 1.8° S of Uranus.
Nov. 5—Mercury at perihelion.
Nov. 7—Mercury stationary, resumes direct motion.
Nov. 10—Venus enters Sagittarius.
Nov. 12—Moon 1.6° S of Saturn.
Nov. 13—Moon 2° S of Jupiter.
Nov. 15—Mercury at greatest western elongation of 19°.
Nov. 19—Saturn at opposition.
Nov. 21—Moon 4° N of Mars.
Nov. 22—Sun enters Scorpius.
Nov. 24—Moon 3° N of Mercury.
Nov. 28—Jupiter at opposition.
Nov. 29—Moon 2° N of Venus; Sun enters Ophiuchus.

December

Mercury disappears in the glow of sunrise early in the month.

Venus, bright in the SW after sunset, passes Neptune on the 11th and Uranus on the 23d.

Mars, rising a couple of hours after midnight, is prominent in the SE before sunrise, passes Spica on the 11th.

Jupiter is bright in the E after sunset.

Saturn, higher than Jupiter, is prominent in the E after sunset.

Moon passes Uranus on the 2d, Saturn on the 9th, Jupiter on the 10th, Mars on the 20th, Neptune on the 28th, and Uranus and Venus on the 29th.

Dec. 2—Moon 2° S of Uranus.
Dec. 4—Pluto in conjunction with Sun.
Dec. 7—Venus enters Capricornus.
Dec. 9—Moon 1.8° S of Saturn.
Dec. 10—Moon 3° S of Jupiter.
Dec. 11—Mars 4° N of Spica; Venus 3° S of Neptune.
Dec. 17—Sun enters Sagittarius.
Dec. 19—Mercury at aphelion.
Dec. 20—Moon 4° N of Mars.
Dec. 21—Southern Solstice at 8:37 AM EST (13:37 UTC), winter begins in the northern hemisphere, summer begins in the southern hemisphere.
Dec. 23—Venus 1.3° S of Uranus.
Dec. 25—Partial solar eclipse; see details under Eclipses. Mercury at superior conjunction.
Dec. 28—Moon 2° S of Neptune.
Dec. 29—Moon 2° S of Uranus and 1.8° S of Venus.

Meteorites and Meteor Showers

When a chunk of material, ice or rock, plunges into Earth's atmosphere and burns up in a fiery display, the event is a **meteor**. While the chunk of material is still in space, it is a **meteoroid**. If a portion of the material survives passage through the atmosphere and reaches the ground, the remnant on the ground is a **meteorite**.

Meteorites found on Earth are classified into types, depending on their composition: **irons,** those composed chiefly of iron, a small percentage of nickel, and traces of other metals such as cobalt; **stones,** stony meteors consisting of silicates; and **stony irons,** containing varying proportions of both iron and stone.

The serious study of meteorites as non-earth objects began in the 20th century. Scientists now use sophisticated chemical analysis, X-rays, and mass spectrography in determining their origin and composition. In 1996, the results of a study of a Mars rock recovered 12 years earlier from the Allan Hills region of Antarctica suggested that life once existed on that planet. Although most meteorites are now believed to be fragments of asteroids or comets, geochemical studies have shown that a few Antarctic stones came from the moon or from Mars, from which they presumably were ejected by the explosive impact of asteroids.

The **largest** known meteorite, estimated to weigh about 55 metric tons, is situated at Hoba West near Grootfontein, Namibia. The largest known crater believed to have been produced by a meteorite was discovered in 1950 in northwestern Québec, Canada. It consists of a circular pit 4 km (2.5 mi) in diameter, containing a lake and surrounded by concentric piles of shattered granite.

Sporadic meteors, which enter the atmosphere throughout the year, seem to originate from the asteroid belt. Other meteors that come in groups and tend to occur at the same time each year create what are called **meteor showers**; these are the meteors associated with comets. As a comet orbits the Sun, the Sun slowly boils away some of the comet's material, and the comet leaves a trail of tiny particles which are dispersed along the comet's path. If Earth's orbit and this path intersect, then once a year, as Earth reaches that particular point in its orbit, there will be a meteor shower.

Meteor showers vary in strength, but usually the 3 best meteor showers of the year are the **Perseids**, which occur around Aug. 12, the **Orionids**, which occur around Oct. 21, and the **Geminids**, which occur around Dec. 13. These showers feature meteors at the rate of about 60 per hour. Best observing conditions occur with the absence of moonlight, usually when the Moon's phase is between waning crescent and waxing 1st quarter. Meteor showers are also usually better after the middle of the night.

For most meteor showers the cometary debris is relatively uniformly scattered along the comet's orbit. However, in the case of the **Leonid meteor shower,** which occurs every year around Nov. 17-18, the cometary debris, from Comet Temple-Tuttle, seems to be bunched up in one stretch. That means that most years when Earth crosses the orbit of this comet, the meteor shower produced is relatively weak. However, approximately every 33 years Earth encounters the bunched-up debris. Sometimes the storm is a disappointment, as it was in 1899 and 1933; at other times it is a roaring success, as in 1833 and 1866. In 1966 observers on the west coast of the United States were treated to an awesome display of meteors in the early morning as the rate peaked at 150,000 meteors per hour. The Leonid meteor storm reached an apparent peak in 1998 and 1999, but there was a possibility that this shower would also be strong in 2000, although a waning quarter Moon would interfere with observing.

Rising and Setting of Planets, 2000

Coordinated Universal Time (0 designates midnight)

Venus, 2000

Date	20° N Latitude Rise	Set	30° N Latitude Rise	Set	40° N Latitude Rise	Set	50° N Latitude Rise	Set	60° N Latitude Rise	Set
	h m	h m	h m	h m	h m	h m	h m	h m	h m	h m
Jan. 1	3 43	14 52	3 59	14 36	4 19	14 16	4 47	13 48	5 32	13 02
11	3 58	15 00	4 16	14 41	4 39	14 18	5 11	13 46	6 06	12 51
21	4 13	15 11	4 32	14 51	4 57	14 26	5 33	13 50	6 33	12 50
31	4 27	15 24	4 47	15 03	5 12	14 38	5 48	14 02	6 50	13 00
Feb. 10	4 39	15 39	4 59	15 19	5 23	14 55	5 57	14 21	6 55	13 23
20	4 49	15 55	5 06	15 37	5 28	15 15	5 59	14 45	6 49	13 55
Mar. 1	4 56	16 10	5 11	15 56	5 29	15 38	5 54	15 13	6 35	14 32
11	5 00	16 26	5 11	16 14	5 25	16 00	5 44	15 42	6 14	15 12
21	5 01	16 40	5 09	16 32	5 18	16 23	5 31	16 11	5 50	15 52
31	5 01	16 53	5 05	16 50	5 09	16 46	5 15	16 40	5 24	16 32
Apr. 10	5 00	17 07	5 00	17 07	4 59	17 08	4 58	17 09	4 56	17 12
20	4 59	17 20	4 54	17 24	4 49	17 30	4 41	17 39	4 28	17 52
30	4 59	17 33	4 50	17 42	4 39	17 53	4 24	18 09	4 01	18 33
May 10	5 00	17 48	4 47	18 00	4 31	18 17	4 09	18 39	3 34	19 15
20	5 03	18 03	4 47	18 19	4 26	18 40	3 58	19 09	3 11	19 57
30	5 09	18 19	4 50	18 39	4 26	19 03	3 51	19 38	2 52	20 38
June 9	5 19	18 36	4 57	18 57	4 30	19 24	3 51	20 04	2 41	21 14
19	5 31	18 51	5 08	19 14	4 40	19 42	3 59	20 23	2 43	21 40
29	5 46	19 05	5 24	19 27	4 55	19 55	4 15	20 35	3 00	21 50
July 9	6 02	19 16	5 41	19 36	5 15	20 02	4 38	20 39	3 31	21 45
19	6 18	19 24	6 00	19 42	5 37	20 04	5 05	20 36	4 11	21 29
29	6 34	19 28	6 19	19 43	6 01	20 01	5 35	20 26	4 53	21 07
Aug. 8	6 49	19 30	6 38	19 41	6 25	19 54	6 06	20 12	5 36	20 41
18	7 03	19 30	6 56	19 36	6 48	19 44	6 36	19 55	6 19	20 12
28	7 16	19 28	7 13	19 30	7 10	19 33	7 06	19 36	7 00	19 42
Sept. 7	7 28	19 25	7 30	19 23	7 33	19 20	7 36	19 17	7 41	19 12
17	7 41	19 23	7 48	19 17	7 56	19 09	8 06	18 58	8 22	18 41
27	7 55	19 23	8 06	19 12	8 19	18 58	8 37	18 40	9 05	18 12
Oct. 7	8 09	19 24	8 24	19 09	8 42	18 50	9 08	18 25	9 48	17 44
17	8 25	19 28	8 43	19 09	9 06	18 46	9 38	18 14	10 32	17 20
27	8 41	19 35	9 02	19 13	9 29	18 46	10 07	18 08	11 14	17 01
Nov. 6	8 57	19 45	9 20	19 22	9 50	18 52	10 31	18 10	11 49	16 53
16	9 12	19 58	9 36	19 34	10 05	19 04	10 49	18 21	12 09	17 00
26	9 24	20 12	9 47	19 49	10 15	19 21	10 57	18 39	12 12	17 24
Dec. 6	9 32	20 27	9 53	20 06	10 19	19 40	10 56	19 04	12 00	17 59
16	9 36	20 41	9 54	20 23	10 16	20 01	10 47	19 30	11 38	18 39
26	9 36	20 53	9 50	20 39	10 08	20 21	10 32	19 58	11 10	19 19

Mars, 2000

Date	20° N Latitude Rise	Set	30° N Latitude Rise	Set	40° N Latitude Rise	Set	50° N Latitude Rise	Set	60° N Latitude Rise	Set
	h m	h m	h m	h m	h m	h m	h m	h m	h m	h m
Jan. 1	9 37	21 03	9 49	20 52	10 03	20 38	10 21	20 19	10 51	19 50
11	9 23	20 57	9 32	20 49	9 42	20 38	9 57	20 24	10 19	20 02
21	9 08	20 51	9 14	20 45	9 21	20 38	9 31	20 28	9 46	20 14
31	8 52	20 45	8 56	20 41	9 00	20 38	9 05	20 32	9 13	20 25
Feb. 10	8 36	20 38	8 37	20 37	8 38	20 37	8 39	20 36	8 40	20 35
20	8 20	20 31	8 18	20 33	8 15	20 36	8 12	20 39	8 06	20 45
Mar. 1	8 04	20 23	7 59	20 28	7 53	20 34	7 45	20 42	7 33	20 55
11	7 48	20 16	7 40	20 23	7 31	20 32	7 19	20 45	7 00	21 05
21	7 32	20 09	7 22	20 18	7 10	20 31	6 53	20 48	6 27	21 14
31	7 16	20 01	7 04	20 13	6 49	20 29	6 28	20 50	5 55	21 24
Apr. 10	7 01	19 54	6 47	20 08	6 29	20 26	6 04	20 52	5 24	21 33
20	6 47	19 47	6 31	20 03	6 10	20 24	5 41	20 53	4 53	21 42
30	6 34	19 39	6 15	19 58	5 52	20 21	5 20	20 53	4 24	21 49
May 10	6 21	19 32	6 01	19 52	5 36	20 17	5 00	20 53	3 58	21 56
20	6 09	19 24	5 48	19 45	5 21	20 12	4 42	20 50	3 34	22 00
30	5 57	19 15	5 35	19 38	5 07	20 06	4 27	20 46	3 13	22 01
June 9	5 47	19 06	5 24	19 29	4 55	19 58	4 14	20 40	2 56	21 57
19	5 37	18 57	5 14	19 20	4 45	19 49	4 03	20 31	2 44	21 49
29	5 27	18 46	5 04	19 09	4 35	19 37	3 54	20 19	2 37	21 36
July 9	5 17	18 35	4 55	18 57	4 27	19 24	3 47	20 05	2 33	21 18
19	5 07	18 22	4 46	18 43	4 20	19 10	3 41	19 48	2 33	20 56
29	4 58	18 08	4 38	18 28	4 13	18 53	3 37	19 29	2 34	20 31
Aug. 8	4 48	17 53	4 29	18 12	4 06	18 35	3 33	19 07	2 37	20 03
18	4 37	17 38	4 20	17 54	3 59	18 15	3 30	18 45	2 40	19 34
28	4 27	17 21	4 11	17 36	3 53	17 54	3 26	18 20	2 43	19 03
Sept. 7	4 15	17 03	4 02	17 16	3 46	17 33	3 23	17 55	2 46	18 31
17	4 04	16 45	3 53	16 56	3 39	17 10	3 19	17 29	2 49	17 59
27	3 52	16 26	3 43	16 35	3 31	16 46	3 16	17 02	2 51	17 26
Oct. 7	3 39	16 07	3 32	16 14	3 24	16 22	3 12	16 34	2 53	16 53
17	3 27	15 47	3 22	15 52	3 16	15 58	3 07	16 06	2 54	16 19
27	3 14	15 27	3 11	15 30	3 07	15 33	3 02	15 38	2 54	15 45
Nov. 6	3 01	15 07	3 00	15 07	2 59	15 08	2 57	15 10	2 55	15 12
16	2 47	14 46	2 49	14 45	2 50	14 43	2 52	14 41	2 55	14 38
26	2 34	14 26	2 37	14 22	2 41	14 18	2 46	14 13	2 54	14 05
Dec. 6	2 20	14 05	2 26	14 00	2 32	13 53	2 41	13 45	2 54	13 31
16	2 07	13 45	2 14	13 38	2 23	13 29	2 35	13 17	2 53	12 58
26	1 53	13 25	2 02	13 16	2 13	13 05	2 28	12 49	2 52	12 26

Jupiter, 2000

Date	20° N Latitude Rise	20° N Latitude Set	30° N Latitude Rise	30° N Latitude Set	40° N Latitude Rise	40° N Latitude Set	50° N Latitude Rise	50° N Latitude Set	60° N Latitude Rise	60° N Latitude Set
	h m	h m	h m	h m	h m	h m	h m	h m	h m	h m
Jan. 1	12 38	1 10	12 31	1 18	12 21	1 27	12 08	1 40	11 48	2 01
11	12 01	0 33	11 53	0 41	11 43	0 51	11 30	1 04	11 09	1 25
21	11 24	23 54	11 16	0 06	11 06	0 16	10 52	0 30	10 30	0 52
31	10 49	23 20	10 40	23 29	10 29	23 39	10 15	23 54	9 52	0 20
Feb. 10	10 14	22 47	10 05	22 56	9 53	23 07	9 38	23 23	9 14	23 47
20	9 40	22 15	9 30	22 24	9 18	22 36	9 02	22 53	8 36	23 19
Mar. 1	9 06	21 43	8 56	21 54	8 43	22 06	8 26	22 24	7 58	22 52
11	8 33	21 13	8 23	21 24	8 09	21 37	7 51	21 56	7 21	22 25
21	8 01	20 43	7 50	20 54	7 35	21 08	7 16	21 28	6 44	22 00
31	7 29	20 13	7 17	20 25	7 02	20 40	6 41	21 01	6 08	21 35
Apr. 10	6 57	19 44	6 45	19 57	6 29	20 13	6 07	20 35	5 31	21 11
20	6 26	19 15	6 13	19 28	5 56	19 45	5 32	20 09	4 55	20 46
30	5 55	18 46	5 41	19 00	5 23	19 18	4 59	19 42	4 19	20 23
May 10	5 24	18 17	5 09	18 32	4 51	18 50	4 25	19 16	3 43	19 59
20	4 53	17 48	4 38	18 04	4 18	18 23	3 52	18 50	3 07	19 35
30	4 22	17 19	4 06	17 35	3 46	17 56	3 18	18 24	2 32	19 10
June 9	3 51	16 50	3 35	17 07	3 14	17 28	2 45	17 57	1 56	18 46
19	3 20	16 21	3 03	16 38	2 42	17 00	2 12	17 30	1 21	18 20
29	2 49	15 51	2 32	16 09	2 10	16 31	1 38	17 02	0 46	17 55
July 9	2 18	15 21	2 00	15 39	1 37	16 02	1 05	16 34	0 11	17 28
19	1 46	14 50	1 27	15 09	1 04	15 32	0 32	16 04	23 32	17 00
29	1 13	14 19	0 54	14 37	0 31	15 01	23 54	15 34	22 57	16 31
Aug. 8	0 40	13 46	0 21	14 05	23 54	14 29	23 20	15 03	22 22	16 01
18	0 06	13 13	23 43	13 32	23 19	13 57	22 45	14 31	21 46	15 30
28	23 28	12 39	23 08	12 58	22 44	13 23	22 09	13 57	21 09	14 57
Sept. 7	22 52	12 03	22 32	12 23	22 08	12 47	21 33	13 22	20 32	14 23
17	22 15	11 27	21 55	11 46	21 31	12 11	20 56	12 46	19 55	13 46
27	21 36	10 48	21 17	11 08	20 52	11 32	20 17	12 07	19 16	13 08
Oct. 7	20 56	10 08	20 37	10 28	20 12	10 53	19 37	11 28	18 37	12 28
17	20 15	9 27	19 56	9 47	19 31	10 11	18 57	10 46	17 56	11 46
27	19 33	8 45	19 14	9 04	18 49	9 28	18 15	10 03	17 15	11 03
Nov. 6	18 50	8 01	18 31	8 20	18 06	8 44	17 32	9 18	16 33	10 18
16	18 06	7 16	17 47	7 35	17 23	7 59	16 49	8 33	15 50	9 31
26	17 21	6 31	17 02	6 49	16 39	7 13	16 05	7 47	15 08	8 44
Dec. 6	16 36	5 45	16 18	6 04	15 54	6 27	15 21	7 00	14 25	7 57
16	15 52	5 00	15 34	5 19	15 11	5 42	14 38	6 14	13 42	7 10
26	15 09	4 16	14 50	4 34	14 28	4 57	13 55	5 30	13 00	6 24

Saturn, 2000

Date	20° N Latitude Rise	20° N Latitude Set	30° N Latitude Rise	30° N Latitude Set	40° N Latitude Rise	40° N Latitude Set	50° N Latitude Rise	50° N Latitude Set	60° N Latitude Rise	60° N Latitude Set
	h m	h m	h m	h m	h m	h m	h m	h m	h m	h m
Jan. 1	13 32	2 16	13 21	2 27	13 07	2 41	12 47	3 00	12 17	3 31
11	12 52	1 36	12 41	1 47	12 27	2 01	12 08	2 21	11 37	2 51
21	12 13	0 57	12 02	1 09	11 47	1 23	11 28	1 42	10 57	2 13
31	11 34	0 19	11 23	0 30	11 09	0 45	10 49	1 04	10 18	1 35
Feb. 10	10 56	23 38	10 45	23 50	10 30	0 08	10 11	0 27	9 39	0 59
20	10 19	23 01	10 07	23 13	9 53	23 28	9 32	23 48	9 00	0 24
Mar. 1	9 42	22 25	9 30	22 37	9 15	22 52	8 54	23 13	8 21	23 46
11	9 06	21 50	8 53	22 02	8 38	22 18	8 17	22 39	7 43	23 13
21	8 30	21 15	8 17	21 28	8 02	21 43	7 40	22 05	7 05	22 40
31	7 54	20 41	7 41	20 54	7 25	21 10	7 03	21 32	6 27	22 08
Apr. 10	7 19	20 06	7 06	20 20	6 49	20 36	6 26	20 59	5 49	21 36
20	6 44	19 33	6 30	19 46	6 13	20 03	5 50	20 27	5 12	21 05
30	6 09	18 59	5 55	19 13	5 38	19 30	5 14	19 54	4 34	20 34
May 10	5 34	18 25	5 20	18 39	5 02	18 57	4 37	19 22	3 57	20 03
20	5 00	17 52	4 45	18 06	4 27	18 24	4 01	18 50	3 20	19 31
30	4 25	17 18	4 10	17 33	3 51	17 51	3 25	18 17	2 43	19 00
June 9	3 50	16 44	3 35	16 59	3 16	17 18	2 49	17 45	2 06	18 28
19	3 15	16 10	3 00	16 25	2 40	16 45	2 13	17 12	1 29	17 56
29	2 40	15 35	2 24	15 51	2 04	16 11	1 37	16 38	0 52	17 24
July 9	2 04	15 01	1 48	15 17	1 28	15 37	1 01	16 04	0 15	16 51
19	1 29	14 25	1 12	14 41	0 52	15 02	0 24	15 30	23 34	16 17
29	0 52	13 50	0 36	14 06	0 16	14 26	23 44	14 55	22 56	15 42
Aug. 8	0 16	13 13	23 55	13 30	23 35	13 50	23 06	14 19	22 18	15 07
18	23 35	12 36	23 18	12 53	22 57	13 13	22 29	13 42	21 41	14 30
28	22 57	11 58	22 40	12 15	22 19	12 36	21 50	13 04	21 02	13 53
Sept. 7	22 18	11 20	22 01	11 36	21 41	11 57	21 12	12 26	20 24	13 14
17	21 39	10 41	21 22	10 57	21 02	11 18	20 33	11 46	19 45	12 35
27	20 59	10 00	20 42	10 17	20 22	10 37	19 53	11 06	19 05	11 54
Oct. 7	20 18	9 20	20 02	9 36	19 41	9 56	19 13	10 25	18 25	11 12
17	19 37	8 38	19 21	8 54	19 01	9 14	18 32	9 43	17 45	10 30
27	18 55	7 56	18 39	8 12	18 19	8 32	17 51	9 00	17 05	9 46
Nov. 6	18 13	7 13	17 57	7 29	17 37	7 49	17 10	8 17	16 24	9 03
16	17 31	6 30	17 15	6 46	16 56	7.06	16 28	7.33	15 43	8 18
26	16 49	5 47	16 33	6 03	16 14	6 22	15 46	6 50	15 02	7 34
Dec. 6	16 06	5 05	15 51	5 20	15 32	5 39	15 05	6 06	14 21	6 50
16	15 24	4 22	15 09	4 38	14 50	4 57	14 23	5 23	13 40	6 07
26	14 43	3 40	14 28	3 56	14 09	4 14	13 42	4 41	12 59	5 24

Star Tables

These tables include stars of visual magnitude 2.4 and brighter (the lower the number, the brighter the star). Stars of variable magnitude are designated by v. Coordinates are for mid-2000. If no parallax figures are given, the trigonometric parallax figure is smaller than the margin for error, and the distance given is obtained by indirect methods. Greek letters in the star names indicate perceived degree of relative brightness within the constellation, alpha being the brightest.

To find the time when the star is on the meridian, subtract Right Ascension of Mean Sun (see the table Greenwich Sidereal Time for 0ʰ UTC) from the star's Right Ascension, first adding 24h to the latter if necessary. Mark this result PM if less than 12h, but if greater than 12, subtract 12h and mark the remainder AM.

Star	Magni-tude	Paral-lax "	Light-yrs	Right ascen. h m	Decli-nation ° '	Star	Magni-tude	Paral-lax "	Light-yrs	Right ascen. h m	Decli-nation ° '
α Andromedae (Alpheratz)	2.06	0.02	90	0 08.4	+29 05	β Ursae Majoris (Merak)	2.37	0.04	78	11 01.8	+56 23
β Cassiopeiae (Caph)	2.27v	0.07	45	0 09.2	+59 09	α Ursae Majoris (Dubhe)	1.79	0.03	105	11 03.7	+61 45
α Phoenicia (Ankaa)	2.39	0.04	93	0 26.3	−42 18	β Leonis (Denebola)	2.14	0.08	43	11 49.1	+14 34
α Cassiopeiae (Schedar)	2.23	0.01	150	0 40.5	+56 32	α Crucis (Acrux)	1.58		370	12 26.6	−63 06
β Ceti (Deneb Kaitos)	2.04	0.06	57	0 43.6	−17 59	γ Crucis (Gacrux)	1.63		220	12 31.2	−57 07
β Andromedae (Mirach)	2.06	0.04	76	1 09.7	+35 37	γ Centauri	2.17		160	12 41.6	−48 58
α Eridani (Achernar)	0.46	0.02	118	1 37.7	−57 14	β Crucis (Becrux)	1.25v		490	12 47.7	−59 42
γ Andromedae (Almaak)	2.26		260	2 03.9	+42 20	ε Ursae Majoris (Alioth)	1.77v	0.01	68	12 54.0	+55 58
α Arietis (Hamal)	2.00	0.04	76	2 07.2	+23 28	ζ Ursae Majoris (Mizar)	2.27	0.04	88	13 23.9	+54 56
o Ceti (Mira)	2.00	0.01	103	2 19.2	−2 59	α Virginis (Spica)	0.97v	0.02	220	13 25.2	−11 10
α Ursae Minoris (Polaris)	2.02v		680	2 31.5	+89 16	ε Centauri	2.30v		570	13 39.9	−53 28
β Persei (Algol)	2.12v	0.03	105	3 08.2	+40 57	η Ursae Majoris (Alkaid)	1.86		100	13 47.6	+49 19
α Persei (Mirfak)	1.80	0.03	150	3 24.3	+49 51	β Centauri (Hadar)	0.61v	0.02	490	14 03.9	−60 23
α Tauri (Aldebaran)	0.85v	0.05	68	4 35.9	+16 30	η Centauri (Menkent)	2.06	0.06	55	14 06.7	−36 22
β Orionis (Rigel)	0.12v		900	5 14.5	−8 12	α Bootis (Arcturus)	−0.04	0.09	36	14 15.7	+19 11
α Aurigae (Capella)	0.08v	0.07	45	5 16.7	+46 00	ε Centauri	2.31v		390	14 35.5	−42 10
γ Orionis (Bellatrix)	1.64	0.03	470	5 25.1	+6 21	α Centauri (Rigel Kentaurus)	−0.01	0.75	4.3	14 39.6	−60 50
β Tauri (Elnath)	1.65	0.02	300	5 26.3	+28 36	α Lupi	2.30v		430	14 42.0	−47 23
δ Orionis (Mintaka)	2.23v		1500	5 32.0	−0 18	ε Bootis (Izar)	2.40	0.01	103	14 44.9	+27 04
ε Orionis (Alnilam)	1.70		1600	5 36.2	−1 12	β Ursae Minoris (Kochab)	2.08	0.03	105	14 50.7	+74 10
ζ Orionis (Alnitak)	2.05	0.02	1600	5 40.6	−1 56	α Coronae Borealis (Gemma)	2.23v	0.04	76	15 34.7	+26 43
κ Orionis (Saiph)	2.06	0.01	2100	5 47.7	−9 40	δ Scorpii (Dschubba)	2.32		590	16 00.4	−22 37
α Orionis (Betelgeuse)	0.50v		430	5 55.2	+7 24	α Scorpii (Antares)	0.96v	0.02	600	16 29.4	−26 26
β Aurigae (Menkalinan)	1.90v	0.04	88	5 59.5	+44 57	α Trianguli Australis (Atria)	1.92	0.02	82	16 48.7	−69 02
β Canis Majoris (Mirzam)	1.98	0.01	750	6 22.7	−17 57	ε Scorpii	2.29	0.05	66	16 50.2	−34 18
α Carinae (Canopus)	−0.72	0.02	310	6 23.9	−52 42	λ Scorpii (Shaula)	1.63v		310	17 33.6	−37 06
γ Geminorum (Alhena)	1.93	0.03	105	6 37.7	+16 24	α Ophiuchi (Rasalhague)	2.08	0.06	58	17 35.0	+12 34
α Canis Majoris (Sirius)	−1.46	0.38	8.7	6 45.1	−16 43	θ Scorpii	1.87	0.02	650	17 37.4	−43 00
α Canis Majoris (Adhara)	1.50		680	6 58.6	−28 58	γ Draconis (Eltanin)	2.23	0.02	108	17 56.6	+51 29
δ Canis Majoris (Wezen)	1.86		1790	7 08.4	−26 24	ε Sagittarii (Kaus Australis)	1.85	0.02	140	18 24.2	−34 23
α Geminorum (Castor)	1.99	0.07	45	7 34.65	+31 53	α Lyrae (Vega)	0.03	0.12	26.5	18 37.0	+38 47
α Canis Minoris (Procyon)	0.38	0.29	11.3	7 39.3	+5 13	o Sagittarii (Nunki)	2.02		300	18 55.3	−26 18
β Geminorum (Pollux)	1.14	0.09	35	7 45.3	+28 02	α Aquilae (Altair)	0.77	0.20	16.5	19 50.8	+8 52
δ Puppis (Naos)	2.25		2400	8 03.6	−40 00	γ Cygni (Sadr)	2.20		750	20 22.3	+40 15
γ Velorum (Al Suhail)	1.82		520	8 09.5	−47 20	α Pavonis (Peacock)	1.94		310	20 25.7	−56 44
ε Carinae (Avior)	1.86		340	8 22.5	−59 31	α Cygni (Deneb)	1.25		1600	20 41.5	+45 17
δ Velorum	1.96	0.04	76	8 44.7	−54 42	ε Pegasi (Enif)	2.39		780	21 44.2	+9 53
λ Velorum (Suhail)	2.21	0.02	750	9 08.0	−43 26	α Gruis (Al Nair)	1.74	0.05	64	22 08.3	−46 57
β Carinae (Miaplacidus)	1.68	0.04	86	9 13.2	−69 43	β Gruis	2.11v		280	22 42.7	−46 53
ι Carinae (Tureis)	2.25		750	9 17.1	−59 17	α Piscis Austrinis (Fomalhaut)	1.16	0.14	22.6	22 57.7	−29 37
α Hydrae (Alphard)	1.98	0.02	94	9 27.6	−8 40						
α Leonis (Regulus)	1.35	0.04	84	10 08.4	+11 58						

Morning and Evening Stars, 2000
(Coordinated Universal Time)

	Morning	Evening		Morning	Evening
Jan.	Mercury to Jan. 16	Mercury from Jan. 16		Neptune	Jupiter
	Venus	Mars		Pluto	Saturn
	Neptune from Jan. 24	Jupiter			Uranus to Feb. 6
	Pluto	Saturn	**Mar.**	Mercury from Mar. 1	Mercury to Mar. 1
		Uranus		Venus	Mars
		Neptune to Jan. 24		Uranus	Jupiter
Feb.	Venus	Mercury		Neptune	Saturn
	Uranus from Feb. 6	Mars		Pluto	

	Morning	Evening		Morning	Evening
Apr.	Mercury			Jupiter	Uranus from Aug. 11
	Venus	Mars		Saturn	Neptune
	Uranus	Jupiter		Uranus to Aug. 11	Pluto
	Neptune	Saturn	**Sept.**	Mars	Mercury
	Pluto			Jupiter	Venus
May	Mercury to May 9	Mercury from May 9		Saturn	Uranus
	Venus	Mars			Neptune
	Jupiter from May 8	Jupiter to May 8			Pluto
	Saturn from May 10	Saturn to May 10	**Oct.**	Mercury from Oct. 30	Mercury to Oct. 30
	Uranus			Mars	Venus
	Neptune			Jupiter	Uranus
	Pluto			Saturn	Neptune
June	Venus to June 11	Mercury			Pluto
	Jupiter	Venus from June 11	**Nov.**	Mercury	Venus
	Saturn	Mars		Mars	Jupiter from Nov. 28
	Uranus	Pluto from June 1		Jupiter to Nov. 28	Saturn from Nov. 19
	Neptune			Saturn to Nov. 19	Uranus
	Pluto to June 1				Neptune
July	Mercury from July 6	Mercury to July 6			Pluto
	Mars from July 1	Venus	**Dec.**	Mercury to Dec. 25	Mercury from Dec. 25
	Jupiter	Mars to July 1		Mars	Venus
	Saturn	Neptune from July 27			Jupiter
	Uranus	Pluto			Saturn
	Neptune to July 27				Uranus
Aug.	Mercury to Aug. 22	Mercury from Aug. 22			Neptune
	Mars	Venus			Pluto

Greenwich Sidereal Time for 0ʰ UTC (Coordinated Universal Time), 2000

(Add 12 hours to obtain Right Ascension of Mean Sun)

Date	d	h	m	Date	d	h	m	Date	d	h	m
Jan.	1	6	39.9	May	10	15	12.4	Sept.	7	23	05.5
	11	7	19.3		20	15	51.8		17	23	44.9
	21	7	58.7		30	16	31.3		27	0	24.4
	31	8	38.1	June	9	17	10.7	Oct.	7	1	03.8
Feb.	10	9	17.6		19	17	50.1		17	1	43.2
	20	9	57.0		29	18	29.5		27	2	22.6
Mar.	1	10	36.4	July	9	19	09.0	Nov.	6	3	02.1
	11	11	15.9		19	19	48.4		16	3	41.5
	21	11	55.3		29	20	27.8		26	4	20.9
	31	12	34.7	Aug.	8	21	07.2	Dec.	6	5	00.4
Apr.	10	13	14.1		18	21	46.7		16	5	39.8
	20	13	53.6		28	22	26.1		26	6	19.2
	30	14	33.0								

The Zodiac

The Sun's apparent yearly path among the stars is known as the **ecliptic**. The zone, 18° wide, 9° on each side of the ecliptic, is known as the **zodiac**. Inside this zone are the apparent paths of the Sun, Moon, Earth, and the other planets. Only Pluto regularly strays outside this band on the celestial sphere. The zodiac is used both astrologically and astronomically. Though the two had a common beginning, they are no longer the same.

Beginning at the point on the ecliptic that marks the position of the Sun at the vernal equinox and proceeding eastward, the astrological zodiac is divided into 12 signs of approximately 30° each. These signs are named from the 12 constellations of the zodiac with which the signs coincided in the time of the astronomer Hipparchus, about 2,000 years ago.

Owing to the precession of the equinoxes, that is to say, to the retrograde motion of the equinoxes along the ecliptic, each sign in the zodiac has, in the course of 2,000 years, moved backward about 30° into the constellation W of it; the sign Aries is now in the constellation Pisces, for example, and so on. The vernal equinox will move from Pisces into Aquarius about the middle of the 26th century.

The astronomical constellations of the zodiac, unlike the astrological signs, are not equal in size. The ecliptic actually moves through parts of 13, not 12, astronomical constellations, the 13th being Ophiuchus. Also, the constellation of the scorpion is called Scorpius, while the sign is called Scorpio. In actuality, the planets (other than Pluto) may appear in parts of 21 different constellations.

The signs of the zodiac, with their Latin and English names are given in the next column.

Spring	1.	♈	Aries	The Ram
	2.	♉	Taurus	The Bull
	3.	♊	Gemini	The Twins
Summer	4.	♋	Cancer	The Crab
	5.	♌	Leo	The Lion
	6.	♍	Virgo	The Virgin
Autumn	7.	♎	Libra	The Balance
	8.	♏	Scorpio	The Scorpion
	9.	♐	Sagittarius	The Archer
Winter	10.	♑	Capricorn	The Goat
	11.	♒	Aquarius	The Water Bearer
	12.	♓	Pisces	The Fishes

On Mar. 27, 2000, the disk of the Sun clips a corner of the constellation of Cetus. The constellations of the zodiac, with the approximate dates that the Sun is in each constellation in 2000, are as follows:

Jan.	1 -	Jan.	20	Sagittarius
Jan.	20 -	Feb.	16	Capricornus
Feb.	16 -	Mar.	11	Aquarius
Mar.	11 -	Apr.	18	Pisces
Apr.	18 -	May	13	Aries
May	13 -	June	21	Taurus
June	21 -	July	20	Gemini
July	20 -	Aug.	10	Cancer
Aug.	10 -	Sept.	16	Leo
Sept.	16 -	Oct.	30	Virgo
Oct.	30 -	Nov.	22	Libra
Nov.	22 -	Nov.	29	Scorpius
Nov.	29 -	Dec.	17	Ophiuchus
Dec.	17 -	Dec.	31	Sagittarius

Constellations

Culturally, constellations are imagined patterns among the stars that, in some cases, have been recognized through millennia. Knowledge of constellations was once necessary in order to function as an astronomer. For today's astronomers, constellations are simply areas on the entire sky in which interesting objects await observation and interpretation.

Because Western culture has prevailed in establishing modern science, equally viable and interesting constellations and celestial traditions of other cultures are not well known outside their regions of origin. Even the patterns with which we are most familiar today have undergone considerable change over the centuries, because the Western heritage embraces disparate in time as well as place.

Today, 88 constellations are recognized. Although many have antient origins, some are "modern," contrived out of unclaimed stars by astronomers a few centuries ago. Unclaimed stars were those too faint or inconveniently placed to be included in the more prominent constellations. Stars in a constellation are not necessarily near each other; they are just located in the same direction on the celestial sphere.

When astronomers began to travel to S Africa in the 16th and 17th cent., they found an unfamiliar sky that showed numerous brilliant stars. Thus, we find constellations in the southern hemisphere that depict technological marvels of the time, as well as some arguably traditional forms, such as the "fly."

Many of the commonly recognized constellations had their origins in ancient Asia Minor. These were adopted by the Greeks and Romans, who translated their names and stories into their own languages, modifying some details in the process. After the declines of these cultures, most such knowledge entered oral tradition or remained hidden in monastic libraries. From the 8th cent., the Muslim explosion spread through the Mediterranean world. Wherever possible, everything was translated into Arabic to be taught in the universities the Muslims established all over their new-found world.

In the 13th cent., Alfonso X of Castile, an avid student of astronomy, had Ptolemy's *Almagest,* translated into Latin. It thus became widely available to European scholars. In the process, the constellation names were translated, but the star names were retained in their Arabic forms. Transliterating Arabic into the Roman alphabet has never been an exact art, so many of the star names we use today only seem Arabic to those who are not scholars.

Until the 1920s, astronomers used curved boundaries for the constellation areas. As these were rather arbitrary at best, the International Astronomical Union adopted new constellation boundaries that ran due north-south and east-west, filling the sky much as the contiguous states fill up the area of the "lower 48" United States.

Names of stars often indicated what parts of the traditional figures they represented: Deneb, the tail of the swan; Betelgeuse, the armpit of the giant. Avoiding traditional names, astronomers may designate the brighter stars in a constellation with Greek letters, usually in order of brightness. Thus, the "alpha star" is often the brightest star of that constellation. The "of" implies possession, so the genitive (possessive) form of the constellation name is used, as in Alpha Orionis, the first star of Orion (Betelgeuse). Astronomers usually use a 3-letter form for the constellation name, as indicated here.

Within these boundaries, and occasionally crossing them, popular "asterisms" are recognized: the so-called Big Dipper is a small part of the constellation Ursa Major, the big bear; the Sickle is the traditional head and mane of Leo, the lion; one of the horn tips of Taurus, the bull, properly belongs to Auriga, the charioteer; the northeast star of the Great Square of Pegasus is Alpha Andromedae.

It is unlikely that further change will occur in the realm of the celestial constellations.

Name	Genitive Case	Abbr.	Meaning
Andromeda	Andromedae	And	Chained Maiden
Antlia	Antliae	Ant	Air Pump
Apus	Apodis	Aps	Bird of Paradise
Aquarius	Aquarii	Aqr	Water Bearer
Aquila	Aquilae	Aql	Eagle
Ara	Arae	Ara	Altar
Aries	Arietis	Ari	Ram
Auriga	Aurigae	Aur	Charioteer
Boötes	Boötis	Boo	Herdsmen
Caelum	Caeli	Cae	Chisel
Camelopardalus	Cameloparadalis	Cam	Giraffe
Cancer	Cancri	Cnc	Crab
Canes Venatici	Canum Venaticorum	CVn	Hunting Dogs
Canis Major	Canis Majoris	CMa	Great Dog
Canis Minor	Canis Minoris	CMi	Little Dog
Capricornus	Capricorni	Cap	Sea-goat
Carina	Carinae	Car	Keel
Cassiopeia	Cassiopeiae	Cas	Queen
Centaurus	Centauri	Cen	Centaur
Cepheus	Cephei	Cep	King
Cetus	Ceti	Cet	Whale
Chamaeleon	Chamaeleontis	Cha	Chameleon
Circinus	Circini	Cir	Compasses (art)
Columba	Columbae	Col	Dove
Coma Berenices	Comae Berenices	Com	Berenice's Hair
Corona Australis	Coronae Australis	CrA	Southern Crown
Corona Borealis	Coronae Borealis	CrB	Northern Crown
Corvus	Corvi	Crv	Crow
Crater	Crateris	Crt	Cup
Crux	Crucis	Cru	Cross (southern)
Cygnus	Cygni	Cyg	Swan
Delphinus	Delphini	Del	Dolphin
Dorado	Doradus	Dor	Goldfish
Draco	Draconis	Dra	Dragon
Equuleus	Equulei	Equ	Little Horse
Eridanus	Eridani	Eri	River
Fornax	Fornacis	For	Furnace
Gemini	Geminorum	Gem	Twins
Grus	Gruis	Gru	Crane (bird)
Hercules	Herculis	Her	Hercules
Horologium	Horologii	Hor	Clock
Hydra	Hydrae	Hya	Water Snake (female)
Hydrus	Hydri	Hyi	Water Snake (male)
Indus	Indi	Ind	Indian
Lacerta	Lacertae	Lac	Lizard
Leo	Leonis	Leo	Lion
Leo Minor	Leonis Minoris	LMi	Little Lion
Lepus	Leporis	Lep	Hare
Libra	Librae	Lib	Balance
Lupus	Lupi	Lup	Wolf
Lynx	Lyncis	Lyn	Lynx
Lyra	Lyrae	Lyr	Lyre
Mensa	Mensae	Men	Table Mountain
Microscopium	Microscopii	Mic	Microscope
Monoceros	Monocerotis	Mon	Unicorn
Musca	Muscae	Mus	Fly
Norma	Normae	Nor	Square (rule)
Octans	Octantis	Oct	Octant
Ophiuchus	Ophiuchi	Oph	Serpent Bearer
Orion	Orionis	Ori	Hunter
Pavo	Pavonis	Pav	Peacock
Pegasus	Pegasi	Peg	Flying Horse
Perseus	Persei	Per	Hero
Phoenix	Phoenicis	Phe	Phoenix
Pictor	Pictoris	Pic	Painter
Pisces	Piscium	Psc	Fishes
Piscis Austrinius	Piscis Austrini	PsA	Southern Fish
Puppis	Puppis	Pup	Stern (deck)
Pyxis	Pyxidis	Pyx	Compass (sea)
Reticulum	Reticuli	Ret	Reticle
Sagitta	Sagittae	Sge	Arrow
Sagittarius	Sagittarii	Sgr	Archer
Scorpius	Scorpii	Sco	Scorpion
Sculptor	Sculptoris	Scl	Sculptor
Scutum	Scuti	Sct	Shield
Serpens	Serpentis	Ser	Serpent
Sextans	Sextantis	Sex	Sextant
Taurus	Tauri	Tau	Bull
Telescopium	Telescopii	Tel	Telescope
Triangulum	Trianguli	Tri	Triangle
Triangulum Australe	Trianguli Australis	TrA	Southern Triangle
Tucana	Tucanae	Tuc	Toucan
Ursa Major	Ursae Majoris	UMa	Great Bear
Ursa Minor	Ursae Minoris	UMi	Little Bear
Vela	Velorum	Vel	Sail
Virgo	Virginis	Vir	Maiden
Volans	Volantis	Vol	Flying Fish
Vulpecula	Vulpeculae	Vul	Fox

Aurora Borealis and Aurora Australis

The **Aurora Borealis,** also called the **Northern Lights,** is a broad display of rather faint light in the northern skies at night. The **Aurora Australis,** a similar phenomenon, appears at the same time in southern skies. The aurora appears in a wide variety of forms. Sometimes it is seen as a quiet glow, almost foglike in character; sometimes as vertical streamers in which there may be considerable motion; sometimes as a series of luminous expanding arcs. There are many colors, with white, yellow, and red predominating.

The auroras are most vivid and most frequently seen at about 20 degrees from the magnetic poles, along the northern coast of the N American continent and the eastern part of the northern coast of Europe. The Aurora Borealis has been seen as far S as Key West, and the Aurora Australis has been seen as far N as Australia and New Zealand. Such occurrences are rare, however.

The Sun produces a stream of charged particles, called the solar wind. These particles, mainly electrons and protons, approach Earth at speeds on the order of 300 mi per second. Some of these particles are trapped by Earth's magnetic field, forming the Van Allen belts—two donut-shaped radiation bands around Earth. Excess amounts of these charged particles, often produced by solar flares, follow Earth's magnetic lines of force toward Earth's magnetic poles. High in the atmosphere, collisions between solar and terrestrial atoms result in the glow in the upper atmosphere called the aurora. The glow may be vivid where the lines of magnetic force converge near the magnetic poles.

The auroral displays appear at heights ranging from 50 to about 600 mi and have given us a means of estimating the extent of Earth's atmosphere.

The auroras are often accompanied by magnetic storms whose forces, also guided by the lines of force of Earth's magnetic field, disrupt electrical communication. Since the Sun is near the peak of the current solar cycle (#23), the increase in sunspots is expected to have an effect on both aurora and electrical communication.

Eclipses, 2000

(in Coordinated Universal Time, standard time of the prime meridian)

There are 6 eclipses in 2000: 4 partial eclipses of the Sun and 2 total eclipses of the Moon.

I. Total eclipse of the Moon, Jan. 21

The beginning of the umbral phase of this eclipse is visible in N America, Central America, S America, the Atlantic Ocean, Greenland, Europe, most of Africa, and western Asia. The end of the umbral phase is visible in Hawaii, most of the N Pacific Ocean, the eastern S Pacific Ocean, N America, Central America, S America, most of the Atlantic Ocean, most of Europe, and extreme western Africa.

Circumstances of the Eclipse

Event	Date	h	m
Moon enters penumbra	Jan. 21	2	2.9
Moon enters umbra	21	3	1.4
Moon enters totality	21	4	4.6
Middle of eclipse	21	4	43.5
Moon leaves totality	21	5	22.3
Moon leaves umbra	21	6	25.4
Moon leaves penumbra	21	7	24.1

Magnitude of eclipse: 1.33

II. Partial eclipse of the Sun, Feb. 5

This partial solar eclipse is visible from the S Pacific Ocean, S Atlantic Ocean, Antarctica, and the S Indian Ocean.

Circumstances of the Eclipse

Event	Date	h	m
Eclipse begins	Feb. 5	10	55.7
Greatest eclipse	5	12	49.3
Eclipse ends	5	14	43.1

III. Partial eclipse of the Sun, July 1

This partial solar eclipse is visible from the S Pacific Ocean and the southern tip of S America.

Circumstances of the Eclipse

Event	Date	h	m
Eclipse begins	July 1	18	7.1
Greatest eclipse	1	19	32.4
Eclipse ends	1	20	57.8

IV. Total eclipse of the Moon, July 16

The beginning of the umbral phase of this eclipse is visible in western N America, most of Mexico, Hawaii, the Pacific Ocean, Australia, New Zealand, Antarctica, and the east coast of Asia. The end of the umbral phase is visible in the western Pacific Ocean, Australia, New Zealand, most of Asia, the Indian Ocean, and the east coast of Africa.

Circumstances of the Eclipse

Event	Date	h	m
Moon enters penumbra	July 16	10	46.6
Moon enters umbra	16	11	57.2
Moon enters totality	16	13	2.0
Middle of eclipse	16	13	55.5
Moon leaves totality	16	14	49.0
Moon leaves umbra	16	15	53.8
Moon leaves penumbra	16	17	4.5

Magnitude of eclipse: 1.772

V. Partial eclipse of the Sun, July 31

This partial solar eclipse is visible from Russia, Siberia, Greenland, the Arctic Ocean, Canada, and the northwestern United States.

Circumstances of the Eclipse

Event	Date	h	m
Eclipse begins	July 31	0	37.4
Greatest eclipse	31	2	13.0
Eclipse ends	31	3	48.8

VI. Partial eclipse of the Sun, Dec. 25

This partial solar eclipse is visible from most of N America, Mexico, southern Greenland, and the western part of the N Atlantic Ocean.

Circumstances of the Eclipse

Event	Date	h	m
Eclipse begins	Dec. 25	15	26.6
Greatest eclipse	25	17	34.8
Eclipse ends	25	19	43.2

Eclipses in the U.S. in the 21st Century

During the 21st century Halley's Comet will return (2061-62), and there will be 8 total solar eclipses that are visible somewhere in the continental United States. The first comes after a long gap; the last one to be seen there was on Feb. 26, 1979, in the northwestern U.S.

Date	Path of Totality
Aug. 21, 2017	Oregon to South Carolina
Apr. 8, 2024	Mexico to Texas and up through Maine
Aug. 23, 2044	Montana to North Dakota
Aug. 12, 2045	N California to Florida

Date	Path of Totality
Mar. 30, 2052	Florida to Georgia
May 11, 2078	Louisiana to North Carolina
May 1, 2079	New Jersey to the lower edge of New England
Sept. 14, 2099	North Dakota to Virginia

Total Solar Eclipses, 1950-2020

Total solar eclipses actually take place nearly as often as total lunar eclipses; they occur at a rate of about 3 every 4 years, while total lunar eclipses come at a rate of about 5 every 6 years. However, total lunar eclipses are visible over at least half of the Earth, while total solar eclipses can be seen only along a very narrow path up to a few hundred miles wide and a few thousand miles long. Observing a total solar eclipse is thus a rarity for most people. Unlike lunar eclipses, solar eclipses can be dangerous to observe. This is not because the Sun emits more potent rays during a solar eclipse than at other times, but because the Sun is always dangerous to observe directly and people are particularly likely to stare at it during a solar eclipse.

Date	Duration[1] m	s	Width (mi)	Path of Totality
1950, Sept. 12	1	13	83	Arctic Ocean, Siberia, Pacific Ocean
1952, Feb. 25	3	9	85	Africa, Middle East, Soviet Union
1954, June 30	2	35	95	U.S., Canada, Iceland, Europe, Middle East
1955, June 20	7	7	157	SE Asia, Phillippines, Pacific Ocean
1956, June 8	4	44	266	S Pacific Ocean
1958, Oct. 12	5	10	129	Pacific Ocean, Chile, Argentina
1959, Oct. 2	3	1	75	New England, Atlantic Ocean, Africa
1961, Feb. 15	2	45	160	Europe, Soviet Union
1962, Feb. 5	4	8	91	Borneo, New Guinea, Pacific Ocean
1963, July 20	1	39	63	Pacific Ocean, Alaska, Canada, Maine
1965, May 30	5	15	123	New Zealand, Pacific Ocean
1966, Nov. 12	1	57	52	Pacific Ocean, S America, Atlantic Ocean
1968, Sept. 22	0	39	64	Soviet Union, China
1970, Mar. 7	3	27	95	Pacific Ocean, Mexico, Eastern U.S., Canada
1972, July 10	2	35	109	Siberia, Alaska, Canada
1973, June 30	7	3	159	Atlantic Ocean, Central Africa, Indian Ocean
1974, June 20	5	8	214	Indian Ocean, Australia
1976, Oct. 23	4	46	123	Africa, Indian Ocean, Australia
1977, Oct. 12	2	37	61	Pacific Ocean, Colombia, Venezuela
1979, Feb. 26	2	49	185	NW U.S., Canada, Greenland
1980, Feb. 16	4	8	92	Africa, Indian Ocean, India, Burma, China
1981, July 31	2	2	67	Soviet Union, Pacific Ocean
1983, June 11	5	10	123	Indian Ocean, Indonesia, New Guinea
1984, Nov. 22	1	59	53	New Guinea, Pacific Ocean
1985, Nov. 12	1	58	430	Antarctica
1986, Oct. 3[h]	0	1	1	N Atlantic Ocean
1987, Mar. 29[h]	0	7	3	S Atlantic Ocean, Africa
1988, Mar. 18	3	46	104	Sumatra, Borneo, Philippines, Pacific Ocean
1990, July 22	2	32	125	Finland, Soviet Union, Aleutian Islands
1991, July 11	6	53	160	Hawaii, Mexico, Central America, Colombia, Brazil
1992, June 30	5	20	182	S Atlantic Ocean
1994, Nov. 3	4	23	117	Peru, Bolivia, Paraguay, Brazil
1995, Oct. 24	2	9	48	Iran, India, SE Asia
1997, Mar. 9	2	50	221	Mongolia, Siberia
1998, Feb. 26	4	8	94	Galapagos Islands, Panama, Colombia, Venezuela
1999, Aug. 11	2	22	69	Europe, Middle East, India
2001, June 21	4	56	125	Atlantic Ocean, Africa, Madagascar
2002, Dec. 4	2	4	54	S Africa, Indian Ocean, Australia
2003, Nov. 23	1	57	338	Antarctica
2005, Apr. 8[h]	0	42	17	Pacific Ocean, NW S America
2006, Mar. 29	4	7	118	Atlantic Ocean, Africa, Asia
2008, Aug. 1	2	27	157	Arctic Ocean, Asia
2009, July 22	6	39	160	Asia, Pacific Ocean
2010, July 11	5	20	164	Pacific Ocean, southern S America
2012, Nov. 13	4	2	112	N Australia, Pacific Ocean
2013, Nov. 3[h]	1	40	36	Atlantic Ocean, Africa
2015, Mar. 20	2	47	304	N Atlantic Ocean, Arctic Ocean
2016, Mar. 9	4	10	96	Indonesia, Pacific Ocean
2017, Aug. 21	2	40	71	Pacific Ocean, U.S., Atlantic Ocean
2019, July 2	4	33	125	S Pacific Ocean, S America
2020, Dec. 14	2	10	56	S Pacific Ocean, S America, S Atlantic Ocean

h = indicates annular-total hybrid eclipse. (1) Duration refers to length of time at optimal viewing area.

Beginnings of the Universe

One of the dominating astronomical discoveries of the 20th century was the realization that the galaxies of the universe all seem to be moving away from us. It turned out that they are moving away not just from us but from one another—that is, the universe seems to be expanding. Hence, scientists conclude that the universe must once, very long ago, have been extremely compact and dense. Although there are alternatives to this theory and still many questions unresolved, much of the observational evidence currently available supports the idea that the universe we know began its existence between 8 and 20 bil years ago as an explosion of a super-dense, super-small concentration of matter.

This explosion of matter giving birth to the universe is called the **Big Bang**. On the subatomic level, according to this theory, there were vast changes of energy and matter and the way physical laws operated during the first 5 minutes. After those minutes the percentages of the basic matter of the universe—hydrogen, helium, and lithium—were set. Everything was so compact and so hot that radiation dominated the early universe and there were no stable, un-ionized atoms. At first, the universe was opaque, in the sense that any energy emitted was quickly absorbed and then re-emitted by free electrons. As the universe expanded, the density and the temperature continued to drop. A few hundred thousand years after the initial Big Bang, the temperature had dropped far enough that electrons and nuclei could combine to form stable atoms as the universe became transparent. Once that had occurred the radiation, which had been trapped, was free to escape.

In the 1940s, George Gamov and others predicted that astronomers should be able to see remnants of this escaped radiation. Astronomers continued to refine the theories and were preparing to build equipment to search for this background radiation when physicists Arno Penzias and Robert Wilson of the Bell Telephone Laboratories inadvertently beat them to the punch (the 2 were later awarded a Nobel Prize). Despite the Big Bang's success at predicting the existence of **cosmic background radiation,** there are still many unresolved questions, and astronomers are still working on modifications of the theory.

The Solar System

The planets of the solar system, in order of mean distance from the Sun, are Mercury, Venus, Earth, Mars, Jupiter, Saturn, Uranus, Neptune, and Pluto (Pluto sometimes nearer than Neptune). Both Uranus and Neptune are visible through good binoculars, but Pluto is so distant and so small that only large telescopes or long-exposure photographs can make it visible.

Because Mercury and Venus are nearer to the Sun than is Earth, their motions about the Sun are seen from Earth as wide swings first to one side of the Sun then to the other, though both planets are passing continuously around the Sun in almost circular orbits. When their passage takes them either between Earth and the Sun or beyond the Sun as seen from Earth, they are invisible to us. Because of geometry of the planetary orbits, Mercury and Venus require much less time to pass between Earth and the Sun than around the far side of the Sun; so their periods of visibility and invisibility are unequal.

The planets that lie farther from the Sun than does Earth may be seen for longer periods and are invisible only when so located in our sky that they rise and set at about the same time as the Sun—and thus become overwhelmed by the Sun's great brilliance. Although several of the giant planets seem to emit their own energy, they are observed from Earth as a result of sunlight reflecting from their surfaces or cloud layers. Mercury and Venus, because they are between Earth and the Sun, show phases very much as the Moon does. The planets farther from the Sun are always seen as full, although Mars does occasionally present a slightly gibbous phase—like the Moon when not quite full.

The planets appear to move rapidly among the stars because of being closer. The stars are also in motion, some of them at tremendous speeds, but they are so far away that their motion does not change their apparent positions in the heavens sufficiently to be perceived. The nearest star is about 7,000 times farther away than the most distant planet in our solar system.

Planets and the Sun, by Selected Characteristics

Sun and Planets	Semi- Diameter: at unit distance "	at mean least distance "	in mi mean s.d.	Volume 1	Mass[1]	Density[1]	Sidereal period d	h	m	s	Gravity at surface[1]	Reflecting power Pct°	Daytime surface temp. °F
Sun......	959.91	—	432,600	1,304,000	332,950	0.26	25	9	7		28.0	—	+9,941
Mercury ...	3.36	6.3	1,516	0.056	0.0553	0.98	58	15	36		0.38	0.11	725
Venus.....	8.35	30.7	3,761	0.857	0.815	0.94	244	R			0.91	0.65	865
Earth	—	—	3,960	1.000	1.000	1.00		23	56	4.2	1.00	0.37	68
Moon	2.40	986.6	1,080	0.0203	0.0123	0.61	27	7	44		0.17	0.12	266
Mars......	4.68	12.26	2,107	0.151	0.107	0.71		24	37	22	0.38	0.15	−24
Jupiter	96.41	24.4	43,450	1,321	317.83	0.24		9	55	30	2.36	0.52	−160
Saturn.....	80.31	10.03	36,191	764	95.16	0.12		10	30	0	0.92	0.47	−220
Uranus	34.98	2.02	15,763	63	14.54	0.24		17	14	R	0.89	0.51	−320
Neptune ...	33.96	1.18	15,304	58	17.15	0.30		16	7		1.12	0.41	−330
Pluto......	1.57	0.05	707	0.006	0.0021	0.37	6	9	18	R	0.07	0.30	−370

(1) Earth = 1. R = Retrograde rotation.

Planet Superlatives

Largest, most massive, planet . . . . .Jupiter	Smallest, least massive planet. . . . . .Pluto
Fastest orbiting planet . . . . .Mercury	Slowest orbiting planet . . . . .Pluto
Most eccentric orbit . . . . .Pluto	Most circular orbit. . . . . .Venus
Longest (synodic) day. . . . . .Mercury	Shortest (synodic) day . . . . .Jupiter
Coldest planet. . . . . .Pluto	Hottest planet. . . . . .Venus
Most moons . . . . .Saturn, Uranus (18)	No moons. . . . . .Mercury, Venus
Planet with largest moon . . . . .Jupiter	Planet with moon with most eccentric orbit .Neptune
Greatest average density . . . . .Earth	Lowest average density . . . . .Saturn
Tallest mountain .Mars	Deepest oceans . . . . .Jupiter

Largest Telescopes

Astronomers indicate the size of telescopes not by length or magnification, but by the diameter of the primary light-gathering component of the system—such as the lens or mirror. This measurement is a direct indication of the telescope's light-gathering power. The bigger the diameter, the fainter the objects you are enabled to see. For larger telescopes, the Earth's atmosphere limits the resolution of what you see. That is why the Hubble Space Telescope, which is outside the atmosphere, can have better resolution than larger telescopes on the Earth. Large mirror telescopes can be made less expensively than large lens telescopes, so all modern large optical telescopes are made with mirrors rather than lenses. Radio telescopes view at wavelengths not visible to optical telescopes, which are limited to the wavelengths detectable by the human eye. Radio telescopes have to be made larger than optical telescopes because resolving power requires larger diameters at longer wavelengths such as radio wavelengths.

Largest Refracting (lens) Optical Telescope:
Yerkes Observatory—1 m (40 in), at Williams Bay, WI

Largest Reflecting (mirror) Optical Telescope:
Keck—10 m (394 in), on Mauna Kea in Hawaii (segmented mirror)

Largest Space Telescope:
Hubble Space Telescope—2.4 m (94 in), in orbit around the Earth

Largest Single Radio Dish:
Arecibo Observatory—305 m (1,000 ft), in Puerto Rico

Largest Radio Interferometer:
10 telescopes of the Very Long Baseline Array (VLBA), scattered from Hawaii to the Virgin Islands with a resolution equal to a radio dish of 6,000 km (3,700 mi)

The Planets: Motion, Distance, and Brightness

Planet	Mean daily motion "	Orbital velocity mi per sec.	Sidereal revolution days	Synodic revolution days	Distance from Sun in millions of mi		Distance from Earth in millions of mi		Light at[1]	
					Max.	Min.	Max.	Min.	peri-helion	ap-helion
Mercury ..	14,732	29.75	88.0	115.9	43.4	28.6	138	48	10.56	4.59
Venus	5,768	21.77	224.7	583.9	67.7	66.8	162	24	1.94	1.89
Earth	3,548	18.51	365.3	—	94.5	91.4	—	—	1.03	0.97
Mars	1,887	15.00	687.0	779.9	154.9	128.4	249	34	0.52	0.36
Jupiter ...	299	8.12	4,332.6	398.9	507.1	460.3	602	366	0.041	0.034
Saturn ...	120	6.00	10,759.2	378.1	936.2	837.5	1,031	743	0.012	0.0099
Uranus ...	42	4.24	30,685.4	369.7	1,867.0	1,699.0	1,962	1,604	0.0030	0.0025
Neptune ..	21	3.41	60,189.0	367.5	2,822.0	2,770.0	2,916	2,676	0.0011	0.0011
Pluto	14	2.95	90,465.0	366.7	4,587.0	2,763.0	4,682	2,668	0.0011	0.00041

(1) Light at perihelion and aphelion is solar illumination in units of mean illumination at Earth.

Planets of the Solar System

Note: AU = astronomical unit (92.96 mil mi, mean distance of Earth from the Sun); d = 1 Earth synodic (solar) day (24 hrs); synodic day = rotation period of a planet measured with respect to the Sun (the "true" day, i.e. the time from midday to midday, or from sunrise to sunrise); sidereal day = the rotation period of a planet with respect to the stars

Mercury

```
Distance from Sun
  Perihelion . . . . . . . . . . . . . . . . . . . . . . . . . . . 28.6 mil mi
  Semi-major axis. . . . . . . . . . . . . . . . . . . . . . 0.387 AU
  Aphelion . . . . . . . . . . . . . . . . . . . . . . . . . . . . 43.4 mil mi
Period of revolution around Sun . . . . . . . . . . . . . . 87.97 d
Orbital eccentricity. . . . . . . . . . . . . . . . . . . . . . . 0.2056
Orbital inclination. . . . . . . . . . . . . . . . . . . . . . . . 7.00°
Synodic day (midday to midday) . . . . . . . . . . . . . . 175.97 d
Sidereal day. . . . . . . . . . . . . . . . . . . . . . . . . . . 58.65 d
Rotational inclination . . . . . . . . . . . . . . . . . . . . . . ~0.1°
Mass (Earth = 1) . . . . . . . . . . . . . . . . . . . . . . . 0.0553
Mean radius. . . . . . . . . . . . . . . . . . . . . . . . . . 1,516 mi
Mean density (Earth = 1). . . . . . . . . . . . . . . . . . 0.983
Natural satellites . . . . . . . . . . . . . . . . . . . . . . . . . . 0
Average surface temperature . . . . . . . . . . . . . . . 333° F
```

Venus

```
Distance from Sun
  Perihelion . . . . . . . . . . . . . . . . . . . . . . . . . . . 66.8 mil mi
  Semi-major axis . . . . . . . . . . . . . . . . . . . . . . . 0.723 AU
  Aphelion . . . . . . . . . . . . . . . . . . . . . . . . . . . . 67.7 mil mi
Period of revolution around Sun . . . . . . . . . . . . . . 224.70 d
Orbital eccentricity . . . . . . . . . . . . . . . . . . . . . . . 0.0068
Orbital inclination . . . . . . . . . . . . . . . . . . . . . . . . 3.39°
Synodic day (midday to midday) . . . . 116.75 d (retrograde)
Sidereal day . . . . . . . . . . . . . . . . . 243.02 d (retrograde)
Rotational inclination . . . . . . . . . . . . . . . . . . . . 177.3°
Mass (Earth = 1). . . . . . . . . . . . . . . . . . . . . . . . 0.815
Mean radius . . . . . . . . . . . . . . . . . . . . . . . . . . 3,761 mi
Mean density (Earth = 1) . . . . . . . . . . . . . . . . . . 0.943
Natural satellites . . . . . . . . . . . . . . . . . . . . . . . . . . 0
Average surface temperature . . . . . . . . . . . . . . . 865° F
```

Mercury, the nearest planet to the Sun, is the 2d-smallest of the 9 known planets. Its diameter is 3,032 mi; its mean distance from the Sun is 36,000,000 mi.

Mercury moves with great speed around the Sun, averaging about 30 mi per second to complete its circuit in about 88 Earth days. Mercury rotates upon its axis over a period of nearly 59 days, thus exposing all its surface periodically to the Sun. Because its orbital period is only about 50% longer than its sidereal rotation, the solar (synodic) day on Mercury, or the time from one sunrise to the next, is about 176 days, twice as long as a Mercurian year. It is believed that the surface passing before the Sun may reach a temperature of about 845° F, while the temperature on the nighttime side may fall as low as –300° F. Although Mercury is the closest planet to the Sun, it has by far the largest range of temperature change from day to night.

Uncertainty about conditions on Mercury and its motion arises from its short angular distance from the Sun as seen from Earth. Mercury is too much in line with the Sun to be observed against a dark sky, but is always seen during either morning or evening twilight.

Mariner 10 passed Mercury 3 times in 1974 and 1975. Less than half of the surface was photographed, revealing a degree of cratering similar to that of the Moon. The most imposing feature on Mercury, the Caloris Basin, is a huge impact crater more than 800 mi in diameter. Mercury also has a higher percentage of iron than any other planet. A very thin atmosphere of hydrogen and helium may be made up of gases of the solar wind temporarily concentrated by the presence of Mercury. The discovery of a weak but permanent magnetic field was a surprise to scientists. It has been held that both a fluid core and rapid rotation are necessary for the generation of a planetary magnetic field. Mercury may demonstrate the contrary; the field may reveal something about the history of Mercury. In 1992, radar mapping of Mercury with radio telescopes on Earth revealed evidence of possible water ice near its north and south poles.

Venus, slightly smaller than Earth, moves about the Sun at a mean distance of 67,000,000 mi in 225 Earth days. Its synodical revolution—its return to the same relationship with Earth and the Sun, which is a result of the combination of its own motion with that of Earth—is 584 days. As a result, every 19 months Venus is nearer to Earth than any other planet. Venus is covered with a dense, white, cloudy atmosphere that conceals whatever is below it. This same cloud reflects sunlight efficiently so that Venus is the 3d-brightest object in the sky, exceeded only by the Sun and the Moon.

Spectral analysis of sunlight reflected from Venus's cloud tops has shown features that can best be explained by identifying material of the clouds as sulfuric acid. In 1956, radio astronomers at the Naval Research Laboratories in Washington, DC, found a temperature for Venus of about 600° F. Subsequent data from the *Mariner 2* space probe in 1962 confirmed a high temperature. *Mariner 2* was unable to detect the existence of a magnetic field even as weak as 1/100,000 of Earth's magnetic field.

In 1967, a Soviet space probe, *Venera 4*, and the American *Mariner 5* arrived at Venus within a few hours of each other. *Venera 4* was designed to allow an instrument package to land gently on the surface, but it ceased to transmit information when its temperature reading went above 500° F, when it was still about 20 mi above the surface. The orbiting *Mariner 5*'s radio signals passed to Earth through Venus's atmosphere twice (once on the night side and once on the day side). The results were startling. Venus's atmosphere is nearly all carbon dioxide (96.5%), with 3.5% nitrogen and trace amounts of sulfur dioxide, argon, water, carbon dioxide, helium, and neon. It exerts a pressure at the planet's surface of as much as 90 times Earth's normal sea-level pressure of one atmosphere.

Because Earth and Venus are about the same size and were presumably formed at the same time by the same general process and from the same mixture of chemical elements, one is faced with the question: Why the difference?

Recent measurements indicate that Venus has a surface temperature of over 860° F as a result of a runaway greenhouse effect in the past. Because of the thick atmosphere, the temperature is essentially the same both day and night.

Radar astronomers determined the rotation period of Venus to be 243 days clockwise—in other words, contrary to the spin of the other planets and to its own motion around the Sun. If it were exactly 243.16 days, Venus would present the same face toward Earth at every inferior conjunction. This rate and sense of rotation allows a solar day (sunrise to sunrise) on Venus of 116.8 Earth days. Any part of Venus will receive sunlight on its clouds for more than 58 days and then return to darkness for 58 days.

Mariner 10 passed Venus before traveling on to Mercury in 1974. The carbon dioxide found in abundance in the atmosphere is rather opaque to certain ultraviolet wavelengths, enabling sensitive cameras to photograph the Venusian cloud cover. Photos radioed to Earth showed a spiral pattern in the clouds from the equator to the poles. Soviet spacecraft discovered that the clouds are confined in a 12-mi layer between 30 to 42 mi above the surface.

In 1978, two U.S. *Pioneer* probes confirmed expected high surface temperatures and high winds aloft. Winds of about 200 mi per hour there may account for the transfer of heat into the night side despite the low rotation speed of the planet. The probes detected 4 layers of clouds and more light on the surface than expected solely from sunlight. This light allowed Soviet scientists to obtain, in 1975 and later in 1982, 4 photos of rocks on the surface. Sulfur seems to play a large role in the chemistry of Venus, and reactions involving sulfur may be responsible for the glow. The *Pioneer* orbited confirmed the cloud pattern and its circulation shown by *Mariner 10*. Its radar produced maps of the entire planet showing large craters, continent-size highlands, and extensive dry lowlands.

The Venus orbiter *Magellan* launched in 1989 used sophisticated radar techniques to observe Venus and map 98% of the surface. The spacecraft observed over 1,600 volcanoes and volcanic features, enabling creation of a 3-dimensional map of the Venusian surface.

Craters more than 20 mi wide are believed to have been caused by impacting bodies. Theia Mons, a huge shield volcano, has a diameter of over 600 mi and a height of over 3.5 mi. (Compare this to the largest Hawaiian volcano, which is only about 125 mi in diameter, but with a height of nearly 5.5 mi from the ocean floor.) Many lava flows have been seen, and some old craters and plains seem to be filled with lava.

Most of the surface of Venus is believed to be younger than 1 bil to 400 mil years old. Modifications of previously existing surface features have been caused by weathering and by tectonic actions such as faulting.

Tectonic actions on Venus are distinctly different from such actions on Earth. No activity on Venus seems to be similar to Earth's moving tectonic plates, but local stretching and compressing may produce rift valleys and higher plains and mountains. Extensive sand dunes have been seen, and windblown deposits indicate stable wind patterns for very long periods of time. A channel about 4,200 mi long, due to lava flows, has been mapped. The orbit of *Magellan* was adjusted to a nearly circular shape about 300 mi from the planet's surface in 1993. In this mode, variation in *Magellan*'s orbital speed revealed information on irregularities in the gravitational field, presumably due to details in the internal structure of the planet. Although *Magellan* ceased operating in 1994, its data about the topography of Venus's surface were keeping teams of analysts and theoreticians busy in subsequent years.

Mars

Mars is the first planet beyond Earth, away from the Sun. Mars's diameter is about 4,213 mi. Although Mars's orbit is nearly circular, it is somewhat more eccentric than the orbits of many of the other planets, and Mars is more than 26 mil mi

Distance from Sun	
Perihelion	128.4 mil mi
Semi-major axis	1.524 AU
Aphelion	154.9 mil mi
Period of revolution around Sun	686.98 d (1.88 y)
Orbital eccentricity	0.0934
Orbital inclination	1.85°
Synodic day (midday to midday)	24h 41m 58s
Sidereal day	24h 37m 22s
Rotational inclination	25.19°
Mass (Earth = 1)	0.107
Mean radius	2,107 mi
Mean density (Earth = 1)	0.713
Natural satellites	2
Average surface temperature	−80°

farther from the Sun in some parts of its year than it is in others. Mars takes 687 Earth days to make one circuit of the Sun, traveling at about 15 mi a second. The planet rotates upon its axis in almost the same period of time as Earth—24 hours and 37 minutes. Mars's mean distance from the Sun is 141 mil mi, so its temperature would be lower than that on Earth even if its atmosphere were not so thin. *Mariner 4,* in 1965, reported that atmospheric pressure on Mars is between 1% and 2% of Earth's atmospheric pressure. As is the case with Venus, thin atmosphere appears to be composed largely of carbon dioxide. The planet is exposed to an influx of cosmic radiation about 100 times as intense as that on Earth.

Although early Earth telescopic observations led some to believe the colors they saw were indications of some sort of vegetation, the findings of *Mariner 4* of a complete lack of water and oxygen showed that such growth is not possible.

Mars's axis of rotation is inclined from a vertical to the plane of its orbit about the Sun by about 25°, and therefore Mars has seasons as does Earth. White caps form about the poles of Mars, growing in the winter and shrinking in the summer. These polar caps are now believed to be both water ice and carbon dioxide ice. It is the carbon dioxide that is seen to come and go with the seasons. The water ice is apparently in many layers with dust between them, indicating climatic cycles.

Mariners 6 and *7* in 1969 sent back many photographs of higher quality showing cratering similar to the earlier views, but also other types of terrain. Some regions seemed featureless over large areas; others were chaotic, showing high relief without apparent organization into mountain chains or craters. *Mariner 9,* the first spacecraft to orbit Mars (1971), transmitted photos and other data showing that Mars resembles no other planet we know, yet there were features clearly of volcanic origin. One of these is Olympus Mons, apparently a shield volcano whose caldera is more than 40 mi wide and whose outer slopes are 370 mi in diameter; it stands 15 mi above the surrounding plain—the tallest known mountain in the solar system. Some features may have been produced by cracking (faulting) and stretching of the surface. Valles Marineris, extending more than 2,500 mi, is an example on a colossal scale. Many craters seem to have been produced by impacting bodies that may have come from the nearby asteroid belt. Features near the S pole may have been produced by glaciers no longer present; a huge series of interrelated canyons stretches more than 3,000 mi.

Although the Russians landed a probe on the Martian surface in 1971, it transmitted for only 90 seconds. In 1976, the U.S. landed 2 *Viking* spacecraft on the Martian surface. The landers had devices aboard to perform chemical analyses of the soil in search of evidence of life; results were inconclusive. The 2 *Viking* orbiters returned the best pictures up to then of Martian topographic features. Scientists believe many of these features can be explained only if Mars once had large quantities of flowing water.

Two U.S. spacecraft—the Mars *Pathfinder* and the Mars *Global Surveyor*—were launched towards Mars in 1996. On July 4, 1997, using a unique array of balloons, *Pathfinder*, with its small movable robot named Sojourner, bounced to a safe landing on Mars. It actually bounded about 40 feet high after striking the ground at 40 mph and bounced 15 more times before coming to a halt. Sojourner spent 3 months examining rocks near *Pathfinder*. Geological results from the *Pathfinder* indicate that in its beginning stages Mars melted to a sufficient extent to separate into dense and lighter layers. It also appears that there was an era when the planet had large amounts of flooding waters on its surface. The *Surveyor,* which went into orbit around Mars in Sept. 1997, began an extensive mapping of the planet. *Surveyor* reported the presence of a very weak magnetic field that may have been stronger in the distant past. *Surveyor* results support the idea of an early active Martian surface, provide detailed mapping of the surface, indicate that water flowed from the southern hemisphere to the lower northern hemisphere, and detail the vast Hellas impact basin in the southern hemisphere, which is about 1,300 mi across.

In Dec. 1998, the *Mars Climate Orbiter* was launched to study atmospheric conditions on the planet, but scientists lost communication with it in Sept. 1999. The *Mars Polar Lander,* launched in Jan. 1999, was scheduled to land on Mars in Dec. 1999 to study the soil and search for water.

Mars's position in its orbit and its speed around that orbit in relation to Earth's position and speed bring the planet fairly close to Earth on occasions about 2 years apart and then move Mars and Earth too far apart for favorable observation. Every 15-17 years, the close approaches are especially favorable for observation.

Mars has 2 satellites, discovered in 1877 by Asaph Hall. The outer satellite, Deimos, revolves around the planet in about 31 hours. The inner satellite, Phobos, whips around Mars in a little more than 7 hours, making 3 trips around the planet each Martian day. Since it orbits Mars faster than the planet rotates, Phobos rises in the W and sets in the E, opposite to what other bodies appear to do in the Martian sky. *Mariner* and *Viking* photos show these satellites to be irregularly shaped and pitted with numerous craters. Phobos also exhibits a system of linear grooves, each about 1/3 mi across and roughly parallel. Phobos measures about 8 by 12 mi and Deimos about 5 by 7.5 mi.

Of the tens of thousands of meteorites found on Earth, approximately a dozen of them may have originated on Mars. In 1996, a NASA research team concluded that a meteorite found in 1984 on an Antarctic ice field not only might be a rock blasted from the surface of Mars but also might contain evidence that life existed on Mars more than 3.5 bil years ago. The meteorite has been age-dated to about 4.5 bil years. The scientists theorize that 3.5 bil years ago, Mars may have been warmer and wetter, and microscopic life may have formed and left evidence in the rock, including possible fossilized microscopic organisms. Then, 16 mil years ago, it is believed that a huge asteroid or comet struck Mars, blasting material, including this rock, into space. The rock may have entered Earth's atmosphere about 13,000 years ago, landing in Antarctica. The evidence is intriguing, but not conclusive, and even if the above conclusions are correct they indicate the presence only of microscopic life, at a time far in the past.

Jupiter

Jupiter, largest of the planets, has an equatorial diameter of nearly 89,000 mi, 11 times the diameter of Earth. Its polar diameter is almost 6,000 mi shorter. This noticeable oblateness is a result of the liquidity of the planet and its extremely rapid rate of rotation; a day is less than 10 Earth hours long. For a planet this size, this rotational speed is amazing. A point on Jupiter's equator moves at a speed of 22,000 mph, as compared with 1,000 mph for a point on Earth's equator.

Distance from Sun	
Perihelion	460.3 mil mi
Semi-major axis	5.203 AU
Aphelion	507.1 mil mi
Period of revolution around Sun	11.86 y
Orbital eccentricity	0.0484
Orbital inclination	1.305°
Synodic day (midday to midday)	9h 55m 33s
Sidereal day	9h 55m 30s
Rotational inclination	3.12°
Mass (Earth = 1)	317.8
Mean radius	43,450 mi
Mean density (Earth = 1)	0.24
Natural satellites	16
Average temperature*	–160° F
*i.e., temperature where atmosphere pressure equals 1 Earth atmosphere.	

Jupiter is at an average distance of 480 mil mi from the Sun and takes almost 12 Earth years to make one complete circuit of the Sun.

The major chemical constituents of Jupiter's atmosphere are molecular hydrogen (H_2) and helium (He). Minor constituents include methane (CH_4), ammonia (NH_3), and water (H_2O). The temperature at the tops of clouds may be about –280° F. The highest clouds are probably ammonia ice crystals, with a lower layer of ammonium hydrosulfide crystals, and perhaps a cloud layer of water ice crystals. The total atmosphere may be only a few hundred mi in depth, pulled down by the surface gravity (2.36 times Earth's gravity) to a relatively thin layer. The gases become denser with depth, until they may turn into a slush or slurry. There is no sharp interface between the gaseous atmosphere and the hydrogen ocean that accounts for most of Jupiter's volume. *Pioneer 10* and *11*, passing Jupiter in 1973 and 1974, provided evidence for considering Jupiter almost entirely liquid hydrogen. Thus Jupiter has a liquid hydrogen ocean more than 35,000 mi deep. It likely has a rocky core about the size of Earth, but 13 times more massive.

Jupiter's magnetic field is by far the strongest of any planet. Electrical activity caused by this field is so strong that it discharges billions of watts into Earth's magnetic field daily. At lower layers, under enormous pressure, the liquid hydrogen takes on the properties of a metal. It is likely that this liquid metallic hydrogen is the source for both Jupiter's persistent radio noise and its improbably strong magnetic field.

Fourteen of Jupiter's 16 known satellites were found through Earth-based observations. Four of the moons, Io, Europa, Ganymede, and Callisto—all discovered by Galileo in 1610—are large and bright, rivaling Earth's Moon and Mercury in diameter, and may be seen through binoculars. They move rapidly around Jupiter, and it is easy to observe their change of position from night to night. The other satellites are much smaller, in all but one instance much farther from Jupiter, and cannot be seen except through powerful telescopes. The 4 outermost satellites revolve around Jupiter clockwise as seen from the north, contrary to the motions of most satellites in the solar system and to the direction of revolution of planets around the Sun. These moons may be captured asteroids. Jupiter's mass is more than twice the mass of all the other planets, moons, and asteroids put together.

Photographs from *Pioneer 10* and *11* were far surpassed by those of *Voyager 1* and *2*, both of which rendezvoused with Jupiter in 1979. The Great Red Spot exhibited internal counterclockwise rotation. Much turbulence was seen in adjacent material passing N or S of it. The satellites Amalthea, Io, Europa, Ganymede, and Callisto were photographed, some in great detail. Io has active volcanoes that probably have ejected material into a doughnut-shaped ring enveloping its orbit about Jupiter. This is not to be confused with the thin, flat disklike ring closer to Jupiter's surface.

In 1994, 21 large fragments of Comet Shoemaker-Levy 9 collided with Jupiter in a dramatic barrage. Moving at 134,000 mph, stretched out like a 21-car freight train, the fragments impacted one after another against Jupiter. Massive plumes of gas erupted from the impact sites, forming brilliant fireballs and leaving dark blotches and smears behind. One of the largest chunks, labeled the G fragment, impacted with the force of 6 mil megatons of TNT, 100,000 times the power of the largest nuclear bomb ever detonated. It produced a plume 1,200-1,600 mi high and 5,000 mi wide and left a dark discoloration larger than Earth.

The *Galileo* spacecraft went into orbit around Jupiter and released an atmospheric probe into the Jovian atmosphere in Dec. 1995. The probe, traveling at a speed of over 100,000 mph, plunged into Jupiter's atmosphere relaying information about Jupiter's atmosphere for 57.6 minutes. The atmospheric probe revealed a relatively dry atmosphere for the planet, with the upper atmosphere being warmer and denser than expected. It also gave evidence of wind speeds of more than 400 mph and a relative absence of lightning. *Galileo* continues an extended mission to study the 4 large moons. *Galileo* observations show extensive high temperature volcanic eruptions on Io, that Europa may be slushy beneath its icy crust and possibly even warmer further down, that Ganymede has a magnetosphere and a thin oxygen atmosphere, and that Callisto may have a subsurface salty ocean.

Saturn

```
Distance from Sun
    Perihelion ........................... 837.5 mil mi
    Semi-major axis......................... 9.537 AU
    Aphelion ............................ 936.2 mil mi
Period of revolution around Sun ................29.46 y
Orbital eccentricity............................ 0.0542
Orbital inclination ............................. 2.484°
Synodic day (midday to midday) ........... 10h 30m 2s
Sidereal day............................. 10h 30m 0s
Rotational Inclination......................... 26.73°
Mass (Earth = 1) ............................. 95.16
Mean radius............................. 36,191 mi
Mean density (Earth = 1) ..................... 0.124
Natural satellites................................18
Average temperature* ...................... –220° F
*i.e., temperature where atmosphere pressure equals 1
    Earth atmosphere.
```

Saturn, last of the planets visible to the unaided eye, is almost twice as far from the Sun as Jupiter, almost 900 mil mi. It is 2d in size to Jupiter, but its mass is much smaller. Saturn's specific gravity is less than that of water. Its diameter is almost 75,000 mi at the equator; its rotational speed spins it completely around in a little more than 10 hours, and its atmosphere is much like that of Jupiter, except that the temperature at the top of its cloud layer is at least 500° F lower. At about 300° F below zero, the ammonia would be frozen out of Saturn's clouds. The theoretical construction of Saturn resembles that of Jupiter; it likely has a small dense center surrounded by a layer of liquid and a deep atmosphere.

Until *Pioneer 11* passed Saturn in 1979, only 10 satellites of the planet were known from ground-based observations. *Pioneer 11* discovered 2 more, and the other 6 were found in the *Voyager 1* and *2* flybys, which also yielded more information about Saturn's icy satellites.

Saturn's ring system begins about 7,000 mi above the visible disk of Saturn, lying above its equator and extending about 35,000 mi into space. The diameter of the ring system visible from Earth is about 170,000 mi; the rings are estimated to be no thicker than 10 mi. In 1973, radar observation showed the ring particles to be large chunks of material averaging a meter on a side.

Voyager 1 and *2* observations showed the rings to be considerably more complex than had been believed. To the untrained eye, the *Voyager* photographs could be mistaken for pictures of a colorful phonograph record. Launched in Oct. 1997, the *Cassini* spacecraft is scheduled to reach Saturn in 2004 to study the planet, its rings, and its satellites.

Uranus

```
Distance from Sun
    Perihelion ......................... 1,699 mil mi
    Semi-major axis .......................19.191 AU
    Aphelion ........................... 1,867 mil mi
Period of revolution around Sun ................ 84.01 y
Orbital eccentricity ........................... 0.0472
Orbital inclination ............................0.770°
Synodic day (midday to midday)..17h 14m 23s (retrograde)
Sidereal day .................17h 14m 24s (retrograde)
Rotational inclination ........................97.86°
Mass (Earth = 1) ............................ 14.54
Mean radius............................ 15,763 mi
Mean density (Earth = 1) ...................... 0.239
Natural satellites................................18
Average temperature*........................–323° F
*i.e., temperature where atmosphere pressure equals 1
    Earth atmosphere.
```

Voyager 2, after passing Saturn in 1981, headed for a rendezvous with Uranus, culminating in a flyby in 1986.

Uranus, discovered by Sir William Herschel on Mar. 13, 1781, lies 1.8 bil mi from the Sun, taking 84 years to make its circuit around our star. Uranus has a diameter of about 32,000 mi and spins once in some 17.4 hours, according to flyby magnetic data.

One of the most fascinating features of Uranus is how far over it is tipped. Its N pole lies 98° from being directly up and down to its orbit plane. Thus, its seasons are extreme. When the Sun rises at the N pole, it stays up for 42 Earth years; then it sets, and the N pole is in darkness (and winter) for 42 Earth years.

Uranus has 18 moons (the most recent was photographed by *Voyager 2* in 1986, but the images were not recognized until 1999), which have orbits lying in the plane of the planet's equator. (Of these, 5 are relatively large, while 11 are very small and more recently discovered.) In that plane there is also a complex of rings, 9 of which were discovered in 1978. Invisible from Earth, the 9 original rings were found by observers watching Uranus pass before a star. As they waited, they saw their photoelectric equipment register several short eclipses of the star; then the planet occulted the star as expected. After the star came out from behind Uranus, the star winked out several more times. Subsequent observations and analyses indicated the 9 narrow, nearly opaque rings circling Uranus. Evidence from the *Voyager 2* flyby showed the ring particles to be predominantly a yard or so in diameter.

In addition to photos of the 11 new, very small satellites, *Voyager 2* returned detailed photos of the 5 large satellites. As in the case of other satellites newly observed in the *Voyager* program, these bodies proved to be entirely different from one another and from any others. Miranda has grooved markings, reminiscent of Jupiter's Ganymede, but often arranged in a chevron pattern. Ariel shows rifts and channels. Umbriel is extremely dark, prompting some observers to regard its surface as among the oldest in the system. Titania has rifts and fractures, but not the evidence of flow found on Ariel. Oberon's main feature is its surface saturated with craters, unrelieved by other formations.

Uranus likely has a rocky core, surrounded by a thick, icy mantle or perhaps a liquid mantle of water, methane, and ammonia, on top of which is a slushy layer of hydrogen and helium that gradually becomes an atmosphere. In addition to its rotational tilt, Uranus's magnetic field axis is tipped an incredible 58.6° from its rotational axis and is displaced about 1/3 of its radius away from the planet's center.

Neptune

Neptune lies at an average distance of 2.8 bil mi. It was the last planet visited in *Voyager 2*'s epic 12-year trek (1977-89) from Earth.

As with other giant planets, Neptune may have no solid surface, or exact diameter. However, a mean value of 30,600 mi may be assigned to a diameter between atmosphere levels where the pressure is about the same as sea level on Earth.

```
Distance from Sun
  Perihelion ............................2,770 mil mi
  Semi-major axis.......................30.07 AU
  Aphelion ..............................2,822 mil mi
Period of revolution around Sun ...............164.79 y
Orbital eccentricity...........................0.0086
Orbital inclination.............................1.769°
Synodic day (midday to midday)............16h 6m 37s
Sidereal day.............................16h 6m 36s
Rotational inclination.........................28.32°
Mass (Earth = 1)...............................17.15
Mean radius.................................15,304 mi
Mean density (Earth = 1)......................0.297
Natural satellites ..................................8
Average temperature*......................–330° F
*i.e., temperature where atmosphere pressure equals 1
  Earth atmosphere.
```

```
Distance from Sun
  Perihelion ..........................2,763 mil mi
  Semi-major axis .......................39.48 AU
  Aphelion...........................4,587 mil mi
Period of revolution around Sun ...............247.68 y
Orbital eccentricity ..........................0.2488
Orbital inclination............................17.14°
Synodic day (midday to midday)....6d 9h 18m (retrograde)
Sidereal day ...................6d 9h 18m (retrograde)
Rotational inclination ........................122.46°
Mass (Earth = 1)..............................0.0021
Mean radius................................707 mi
Mean density (Earth = 1) ......................0.371
Natural satellites..................................1
Average surface temperature .................–369° F
```

Without a solid surface is it challenging to determine a "true" rotation rate for a giant planet. Astronomers use a determination of the rotation rate of the planet's magnetic field to indicate the internal rotation rate, which in the case of Neptune is 16.1 hours. Neptune orbits the Sun in 164.8 years in a nearly circular orbit. Neptune was discovered in 1846; not until 2010 will it have completed one full trip around the Sun since its discovery.

Voyager 2, which passed 3,000 mi from Neptune's N pole, found a magnetic field that is considerably asymmetric to the planet's structure, similar to, but not so extreme as, that found at Uranus.

Neptune's atmosphere was seen to be quite blue, with quickly changing white clouds often suspended high above an apparent surface. There is a Great Dark Spot, reminiscent of the Great Red Spot of Jupiter. Observations with the Hubble Space Telescope have shown that the Great Dark Spot originally seen by *Voyager* has apparently dissipated, but a new dark spot has since appeared. Atmospheric constituents are mostly hydrocarbon compounds. Although lightning and auroras have been found on other giant planets, only the aurora phenomenon has been seen on Neptune.

Six new satellites were definitively discerned around Neptune by *Voyager 2.* Five of these satellites orbit Neptune in a half day or less. Of the 8 satellites of Neptune in all, the largest, Triton, is in a retrograde orbit, suggesting that it was captured rather than being coeval with Neptune. Triton's large size, sufficient to raise significant tides on the planet, may one day, billions of years from now, cause Triton to come close enough to Neptune for it to be torn apart. Nereid was found in 1949 and has the highest orbital eccentricity (0.75) of any moon. Its long looping orbit suggests that it, too, was captured. Each of the satellites that has been photographed by the 2 *Voyagers* in the planetary encounters has been different from any of the other satellites, and certainly different from any of the planets. Only about half of Triton has been observed, but its terrain shows cratering and a strange regional feature described as resembling the skin of a cantaloupe. Triton has a tenuous atmosphere of nitrogen with a trace of hydrocarbons and evidence of active geysers injecting material into it. At –390° F, the wintertime parts of Triton are the coldest regions yet found in the solar system.

Voyager 2 also confirmed the existence of at least 3 rings composed of very fine particles. There may be some clumpiness in the rings' structure. It is not known whether Neptune's satellites influence the formation or maintenance of the rings.

As with the other giant planets, Neptune is emitting more energy than it receives from the Sun. *Voyager* found the excess to be 2.7 times the solar contribution. Cooling from internal heat sources and from the heat of formation of the planets is thought to be responsible.

Pluto

Although Pluto on the average stays about 3.6 bil mi from the Sun, its orbit is so eccentric that its minimum distance of 2.76 bil mi is less than Neptune's distance from the Sun.

Pluto is currently the most distant planet, but for about 20 years of its orbit, Pluto is closer to the Sun than Neptune. At its mean distance, Pluto takes 247.7 years to circumnavigate the Sun, a 3/2 resonance with Neptune. Until recently, this was about all that was known of Pluto.

About a century ago, a hypothetical planet was believed to lie beyond Neptune and Uranus because neither planet followed paths predicted by astronomers when all known gravitational influences were considered. In little more than a guess, a mass of 1 Earth was assigned to the mysterious body, and mathematical searches were begun. Amid some controversy about the validity of the predictive process, Pluto was discovered nearly where it had been predicted to lie, by Clyde Tombaugh at the Lowell Observatory in Flagstaff, AZ, in 1930.

At the U.S. Naval Observatory, also in Flagstaff, in 1978, James Christy obtained a photograph of Pluto that was distinctly elongated. Repeated observations of this shape and its variation were convincing evidence of the discovery of a satellite of Pluto, now named Charon. Subsequent observations show it to be 728 mi across, more than 12,000 mi from Pluto, and taking 6.4 days to move around Pluto. In this same length of time, Pluto and Charon both rotate once around their individual axes. The Pluto-Charon system thus appears to rotate as virtually a rigid body. Gravitational laws allow these interactions to give the mass of Pluto as 0.0021 of Earth. This mass, together with a new diameter for Pluto of 1,413 mi, make the density about twice that of water. Theorists predict that Pluto has a rocky core, surrounded by a thick mantle of ice.

It is now clear that Pluto, the body found by Tombaugh, could not have influenced Neptune and Uranus to go astray. Although a 10th planet might be out there somewhere, theorists no longer believe there are unexplained perturbations in the orbit of Uranus or Neptune that might be caused by it. Astronomers have found more than 2 dozen asteroid-size objects, somewhat beyond Pluto, in a region called the Kuiper Belt, where some comets are believed to originate.

Because the rotational axis of the system is tipped somewhat more then 120°, there is only a few year interval every 125 years when Pluto and Charon alternately eclipse each other. Both worlds are approximately spherical and have comparable densities. Charon's surface is identified as water ice; Pluto's surface is frozen methane. Large regions on Pluto are dark, others light; Pluto has spots and, perhaps, polar caps. Although extremely cold, Pluto appears to possess a thin nitrogen-methane atmosphere, at least while it is closer to the Sun. When Pluto occulted a star, the star's light faded in such a way as to have passed through a haze layer lying above the planet's surface, indicating an inversion of temperatures—110 K above and 50 K below—suggesting Pluto has primitive weather. A recent controversy raised the issue of Pluto's planet status. Pluto is clearly different from both the rocky terrestrial planets and the giant planets. Although some astronomers think Pluto most closely resembles the Kuiper Belt Objects and should be grouped with them, most astromers still classify Pluto as a planet.

MILLENNIUM FACT BOX

Searching for Planets

People have known of the existence of the planets in the Solar System that are closest to the Sun (Mercury, Venus, Mars, Jupiter and Saturn) since ancient times because they could be seen with the naked eye. However, the 3 farthest (Uranus, Neptune, and Pluto) were discovered only since the invention of the telescope. The first, Uranus, was discovered in 1781 by the English astronomer William Herschel. Next, Neptune's existence and location were predicted through its action upon Uranus, by both John Couch Adams of England and Urbain Jean Joseph Le Verrier of France in 1845, leading to its discovery the following year. Finally, Pluto was discovered in 1930 by the American astronomer Clyde Tombaugh.

During the last 10 years of the 20th century astronomers began to detect the presence of planets orbiting stars other than the Sun. As of yet, they are not actually seeing those objects, but merely sensing their existence by their effect on their parent star. Although the Sun is not an average star, it is typical. With over 200 bil stars in the Milky Way galaxy, it seems plausible that other stars might have planets.

Using the Doppler Effect to detect radial velocity changes in the motions of individual stars, astronomers are more likely to find high-mass planets in close orbits around stars, because that situation produces larger and more noticeable changes. About a dozen and a half star systems have been found that appear to have at least one planet less than 13 times the mass of Jupiter. Another dozen systems may have more massive planets. In 2 cases planets may have been detected in orbit around pulsars.

The star Upsilon Andromedae seems to have 3 planets, with masses 0.71, 2.11, and 4.61 times the mass of Jupiter, yet 2 of the planets are closer to their star than Earth is to the Sun. Astronomers are puzzled as to how planets the size of Jupiter or larger can exist so close to a star. In another case, the star Gliese 876 in Aquarius, which is only 15 light-years away, appears to have a planet about twice the mass of Jupiter. That planet takes just 61 days to go around its star and is closer to that star than Mercury is to our Sun.

The Sun

The Sun, the controlling body of Earth's solar system, is a star often described as average. Yet, the Sun's mass and luminosity are greater than that of 80% of the stars in Earth's galaxy. On the other hand, most of the stars that can be easily seen on any clear night are bigger and brighter than the Sun. It is the Sun's proximity to Earth that makes it appear tremendously large and bright. The Sun is 400,000 times as bright as the full moon and gives Earth 6 mil times as much light as do all the other stars put together. A series of nuclear fusion reactions where hydrogen nuclei are converted to helium nuclei produces the heat and light that make life possible on Earth.

The Sun has a diameter of 865,000 mi and, on average, is 92,956,000 mi from Earth. It is 1.41 times as dense as water. The light of the Sun reaches Earth in 499 seconds, or in slightly more than 8 minutes. The average solar surface temperature has been measured at a value of 5,778 K, or about 9,941° F. The interior temperature of the Sun is theorized to be about 28,000,000° F.

When sunlight is analyzed with a spectroscope, it is found to consist of a continuous spectrum composed of all the colors of the rainbow in order, crossed by many dark lines. The dark "absorption lines" are produced by gaseous materials in the outer layers of the Sun. More than 60 of the natural terrestrial elements have been identified in the Sun, all in gaseous form because of the Sun's intense heat.

Spheres and Corona

The radiating surface of the Sun is called the **photosphere;** just above it is the **chromosphere**. The chromosphere is visible to the naked eye only at total solar eclipses, appearing then to be a pinkish-violet layer with occasional great prominences projecting above its general level. With proper instruments, the chromosphere can be seen or photographed whenever the Sun is visible without waiting for a total eclipse. Above the chromosphere is the **corona,** also visible to the naked eye only at times of total eclipse. Instruments also permit the brighter portions of the corona to be studied whenever conditions are favorable. The pearly light of the corona surges mil of mi from the Sun. Iron, nickel, and calcium are believed to be principal contributors to the composition of the corona, all in a state of extreme attenuation and high ionization that indicates temperatures nearly 2 mil degrees Fahrenheit.

Sunspots

There is an intimate connection between sunspots and the corona. At times of low sunspot activity, the fine streamers of the corona are longer above the Sun's equator than over the polar regions of the Sun; during periods of high sunspot activity, the corona extends fairly evenly outward from all regions of the Sun, but to a much greater distance in space. Sunspots are dark, irregularly shaped regions whose diameters may reach tens of thousands of mi. The average life of a sunspot group is 2 months, but some sunspot groups have lasted for more than a year by being carried repeatedly around as the Sun rotated upon its axis.

Sunspots reach a low point, on average, every 11.3 years, with a peak of activity occurring irregularly between 2 successive minima. We are approaching the maximum of the current sunspot cycle, which is expected to be stronger than average. Launched in Dec. 1995, the SOHO spacecraft was designed to provide several years of study of the Sun from an orbit around the Sun. Observations from SOHO show that magnetic arches, called prominences, extending tens of thousands of mi. into the corona may release enormous amounts of energy heating the corona.

The Moon

The Moon completes a circuit around Earth in a period whose mean or average duration is 27 days, 7 hours, 43.2 minutes. This is the Moon's sidereal period. Because of the motion of the Moon in common with Earth around the Sun, the mean duration of the lunar month—the period from one New Moon to the next New Moon—is 29 days, 12 hours, 44.05 minutes. This is the Moon's **synodic period.**

The mean distance of the Moon from Earth is 238,906 mi. Because the orbit of the Moon about Earth is not circular but elliptical, however, the actual distance varies considerably. The maximum distance from Earth that the Moon may reach is 252,020 mi and the least distance is 225,792 mi. (All distances given here are from the center of one body to the center of the other.)

Distance from Earth	
Perigee	225,792 mi
Semi-major axis	238,906 mi
Apogee	252,020 mi
Period of revolution	27.322 d
Synodic orbital period (period of phases)	29.53 d
Orbital eccentricity	0.0549
Orbital inclination	5.145°
Sidereal day (rotation period	27.322 d
Rotational inclination	6.68°
Mass (Earth = 1)	0.0123
Mean radius	1,080 mi
Mean density (Earth = 1)	0.6051
Average surface temperature	−10° F

The Moon rotates on its axis in a period of time that is exactly equal to its sidereal revolution about Earth: 27.322 days. Thus the backside or farside of the Moon always faces away from Earth. This does not mean that the backside is always dark, since the Sun is the main source of light in the Solar System. The farside of the Moon gets just as much direct sunlight as the nearside. At New Moon phase, the far-side of the Moon is fully lit. With its long day and night, the daytime temperature can reach 266° F, while the coldest nighttime temperature may reach −292° F. This day-to-night temperature change is exceeded only by that on Mercury.

The Moon's revolution about Earth is irregular because of its elliptical orbit. The Moon's rotation, however, is regular, and this, together with the irregular revolution, produces what is called "libration in longitude," which permits the observer on Earth to see first farther around the E side and then farther around the W side of the Moon. The Moon's variation N or S of the ecliptic permits one to see farther over first one pole and then the other of the Moon; this is

called "libration in latitude." These two libration effects permit observers on Earth to see a total of about 60% of the Moon's surface over a period of time.

The hidden side of the Moon was first photographed in 1959 by the Soviet space vehicle *Lunik III*. The moon's far-side does appear noticeably different from the nearside, in that the farside has practically none of the large lava plains, called maria, so prominent on the nearside of the Moon.

From 1969 through 1972, 6 American spacecraft brought 12 astronauts to walk on the surface of the Moon. In 1998 NASA's *Lunar Prospector* spacecraft provided evidence for the presence of 300 million metric tons of water ice at the lunar poles. *Lunar Prospector* results also indicate that the Moon has a small core, supporting the idea that most of the mass of the Moon was ripped away from the early Earth when a Mars-size object collided with Earth.

Tides on Earth are caused mainly by the Moon, because of its proximity to Earth. The ratio of the tide-raising power of the Moon to that of the Sun is 11 to 5.

Harvest Moon and Hunter's Moon

The Harvest Moon, the full Moon nearest the autumnal equinox, ushers in a period of several successive days when the Moon rises soon after sunset. This phenomenon gives farmers in temperate latitudes extra hours of light in which to harvest their crops before frost and winter. The 2000 Harvest Moon falls on Sept. 13 UTC. Harvest Moon in the southern hemisphere temperate latitudes falls on Mar. 20.

The next full Moon after Harvest Moon is called the Hunter's Moon; it is accompanied by a similar but less marked phenomenon. In 2000, the Hunter's Moon occurs on Oct. 13, in the northern hemisphere and on Apr. 18 in the southern hemisphere.

Moon's Perigee and Apogee, 2000

(Coordinated Universal Time, standard time of the prime meridian)

	Perigee					Apogee			
Date	Hour		Date	Hour	Date	Hour		Date	Hour
Jan. 19	23		July 30	8	Jan. 4	12		July 15	16
Feb. 17	3		Aug. 27	14	Feb. 1	1		Aug. 11	12
Mar. 15	00		Sept. 24	8	Feb. 28	21		Sept. 8	13
Apr. 8	22		Oct. 19	22	Mar. 27	17		Oct. 6	7
May 6	9		Nov. 14	23	Apr. 24	12		Nov. 3	3
June 3	13		Dec. 12	22	May 22	4		Dec. 1	0
July 1	22				June 18	13		Dec. 28	15

Moon Phases, 2000

(Coordinated Universal Time, standard time of the prime meridian)

New Moon				Waxing Quarter				Full Moon				Waning Quarter			
Month	d	h	m	Month	d	h	m	Month	d	h	m	Month	d	h	m
Jan.	6	18	14	Jan.	14	13	34	Jan.	21	4	40	Jan.	28	7	57
Feb.	5	13	3	Feb.	12	23	21	Feb.	19	16	27	Feb.	27	3	53
Mar.	6	5	17	Mar.	13	6	59	Mar.	20	4	44	Mar.	28	0	21
Apr.	4	18	12	Apr.	11	13	30	Apr.	18	17	41	Apr.	26	19	30
May	4	4	12	May	10	20	0	May	18	7	34	May	26	11	55
June	2	12	14	June	9	3	29	June	16	22	27	June	25	1	0
July	1	19	20	July	8	12	53	July	16	13	55	July	24	11	2
July	31	2	25	Aug.	7	1	2	Aug.	15	5	13	Aug.	22	18	51
Aug.	29	10	19	Sept.	5	16	27	Sept.	13	19	37	Sept.	21	1	28
Sept.	27	19	53	Oct.	5	10	59	Oct.	13	8	53	Oct.	20	7	59
Oct.	27	7	58	Nov.	4	7	27	Nov.	11	21	15	Nov.	18	15	24
Nov.	25	23	11	Dec.	4	3	55	Dec.	11	9	3	Dec.	18	0	41
Dec.	25	17	22												

Earth: Size, Computation of Time, Seasons

Earth is the 5th-largest planet and the 3d from the Sun. Its mass is 6,580,000,000,000,000,000,000 tons. Earth's equatorial diameter is 7,928 mi while its polar radius is only 7,901 mi.

Size and Dimensions

Earth is considered a solid mass, yet it has a large, liquid iron, magnetic core with a radius of about 2,200 mi. Surpris-

ingly, it has a solid inner core that may be a large iron crystal, with a radius of 800 mi. Around the core is a thick shell or mantle of dense rock. This mantle is composed of materials rich in iron and magnesium. It is somewhat plastic-like and under slow steady pressure can flow like a liquid. The mantle, in turn, is covered by a thin crust forming the solid granite and basalt base of the continents and ocean basins.

Distance from the Sun	
Perihelion	91.4 mil mi
Semi-major axis.	1.0000 AU
Aphelion	94.5 mil mi
Period of revolution	365.256 d
Orbital eccentricity.	0.0167
Orbital inclination.	0.0°
Sidereal day (Rotation period).	23h 56m 4.2s
Synodic day (midday to midday)	24h 0m 0s
Rotational inclination	23.45°
Mass (Earth = 1)	1.00
Mean radius.	3,960 mi
Mean density (Earth = 1).	1.00
Natural satellites	1
Average surface temperature	45° F

Over broad areas of Earth's surface, the crust has a thin cover of sedimentary rock such as sandstone, shale, and limestone formed by weathering of Earth's surface and deposits of sands, clays, and plant and animal remains.

The temperature in Earth increases about 1°F with every 100 to 200 feet in depth, in the upper 100 km of Earth, and reaches nearly 8,500° F at the center. The heat is believed to be derived from radioactivity in the rocks, pressures developed within Earth, and the original heat of formation.

Atmosphere of Earth

Earth's atmosphere is a blanket composed of nitrogen, oxygen, and argon, in amounts of about 77%, 21%, and 1% by volume. Also present in minute quantities are carbon dioxide, hydrogen, neon, helium, krypton, and xenon. Water vapor displaces other gases and varies from nearly zero to about 4% by volume. The atmosphere rests on Earth's surface with the weight equivalent to a layer of water 34 ft deep. For about 300,000 ft upward, the gases remain in the proportions stated. Gravity holds the gases to Earth. The weight of the air compresses it at the bottom so that the greatest density is at Earth's surface. Pressure and density decrease as height increases because the weight pressing upon any layer is always less than that pressing upon the layers below.

The lowest layer of the atmosphere extending up about 7.5 mi is the **troposphere**, which contains 90% of the air and the tallest mountains. This is also where most weather phenomena occur. The temperature drops with increasing height throughout this layer. The atmosphere for about 23 mi above the troposphere is the **stratosphere,** where the temperature generally increases with height. The stratosphere contains ozone, which prevents ultraviolet rays from reaching Earth's surface. Since there is very little convection in the stratosphere, jets regularly cruise in the lower parts to provide a smoother ride for passengers.

Above the stratosphere is the **mesosphere,** where the temperature again decreases with height for another 19 mi. Extending above the mesosphere to the outer fringes of the atmosphere is the **thermosphere,** a region where temperature once more increases with height to a value measured in thousands of degrees Fahrenheit. The lower portion of this region, extending from 50 to about 400 mi in altitude, is characterized by a high ion density and is thus called the **ionosphere.** Most meteors are in the lower thermosphere or the mesosphere at the time they are observed.

Longitude, Latitude

Position on the globe is measured by meridians and parallels. Meridians, which are imaginary lines drawn around Earth through the poles, determine **longitude**. The meridian running through Greenwich, England, is the **prime meridian** of longitude, and all others are either E or W. Parallels, which are imaginary circles parallel with the equator, determine **latitude**. The length of a degree of longitude varies as the cosine of the latitude. At the equator a degree of longitude is 69.171 statute mi; this is gradually reduced toward the poles. Value of a longitude degree at the poles is zero.

Latitude is reckoned by the number of degrees N or S of the equator, an imaginary circle on Earth's surface everywhere equidistant between the two poles. According to the International Astronomical Union ellipsoid of 1964, the length of a degree of latitude is 68.708 statute mi at the equator and varies slightly N and S because of the oblate form of the globe; at the poles it is 69.403 statute mi.

Definitions of Time

Earth rotates on its axis and follows an elliptical orbit around the Sun. The rotation makes the Sun appear to move across the sky from E to W. This rotation determines day and night, and the complete rotation, in relation to the Sun, is called the **apparent** or **true solar day**. A sundial thus measures **apparent solar time**. This length of time varies, but an average determines the **mean solar day** of 24 hours.

The mean solar day and **mean solar time** are in universal use for civil purposes. Mean solar time may be obtained from apparent solar time by correcting observations of the Sun for the **equation of time**. Mean solar time may be as much as 16 minutes behind or 14 minutes ahead of apparent solar time.

Sidereal time is the measure of time defined by the diurnal motion of the vernal equinox and is determined from observation of the meridian transits of stars. One complete rotation of Earth relative to the equinox is called the **sidereal day**. The **mean sidereal day** is 23 hours, 56 minutes, 4.091 seconds of mean solar time.

The interval required for Earth to make one absolute revolution around the Sun is a **sidereal year;** it consisted of 365 days, 6 hours, 9 minutes, and 9.5 seconds of mean solar time (approximately 24 hours per day) in 1900 and has been increasing at the rate of 0.0001 second annually.

The **tropical year,** upon which our calendar is based, is the interval between 2 consecutive returns of the Sun to the vernal equinox. The tropical year consisted of 365 days, 5 hours, 48 minutes, and 46 seconds in 1900. It has been decreasing at the rate of 0.530 second per century. The **calendar year** begins at 12 o'clock midnight precisely, local clock time, on the night of Dec. 31-Jan. 1. The day and the calendar month also begin at midnight by the clock.

On Jan. 1, 1972, the Bureau International des Poids et Mesures in Paris introduced **International Atomic Time (TAI)** as the most precisely determined time scale for astronomical usage. The fundamental unit of TAI in the international system of units is **the second,** defined as the duration of 9,192,631,770 periods of the radiation corresponding to the transition between 2 hyperfine levels of the ground state of the cesium 133 atom. **Coordinated Universal Time (UTC),** which serves as the basis for civil timekeeping and is the standard time of the prime meridian, is officially defined by a formula which relates UTC to mean sidereal time in Greenwich, England. (UTC has replaced GMT as the basis for standard time for the world.)

The Zones and Seasons

The 5 zones of Earth's surface are the Torrid, lying between the Tropics of Cancer and Capricorn; the N Temperate, between Cancer and the Arctic Circle; the S Temperate, between Capricorn and the Antarctic Circle; and the 2 Frigid Zones, between the Polar Circles and the Poles.

The inclination or tilt of Earth's axis, 23° 27′ away from a perpendicular to the Earth's orbit of the Sun, determines the seasons. These are commonly marked in the N Temperate Zone, where spring begins at the vernal equinox, summer at the summer solstice, autumn at the autumnal equinox, and winter at the winter solstice.

In the S Temperate Zone, the seasons are reversed. Spring begins at the autumnal equinox, summer at the winter solstice, etc.

The points at which the Sun crosses the equator are the equinoxes, when day and night are most nearly equal. The points at which the Sun is at a maximum distance from the equator are the solstices. Days and nights are then most unequal. However, at the equator, day and night are equal throughout the year.

In June, the North Pole is tilted 23° 27′ toward the Sun, and the days in the northern hemisphere are longer than the nights, while the days in the southern hemisphere are shorter than the nights. In Dec., the North Pole is tilted 23° 27′ away from the Sun, and the situation is reversed.

The Seasons in 2000

In 2000 the 4 seasons begin in the northern hemisphere as shown. (Add one hour to Eastern Standard Time for Atlantic Time; subtract one hour for Central, 2 for Mountain, 3 for Pacific, 4 for Alaska, 5 for Hawaii-Aleutian. Also shown is Coordinated Universal Time.)

Seasons	Date	EST	UTC
Vernal Equinox (spring)	Mar. 20	2:35	7:35
Northern Solstice (summer)	June 21	20:48*	1:48
Autumnal Equinox (autumn)	Sept. 22	12:27	17:27
Southern Solstice (winter)	Dec. 21	8:37	13:37

*previous day

Poles of Earth

The geographic (rotation) poles, or points where Earth's axis of rotation cuts the surface, are not absolutely fixed in the body of Earth. The pole of rotation describes an irregular curve about its mean position.

Two periods have been detected in this motion: (1) an annual period due to seasonal changes in barometric pressure, to load of ice and snow on the surface, and to other phenomena of seasonal character; (2) a period of about 14 months due to the shape and constitution of Earth.

In addition, there are small but as yet unpredictable irregularities. The whole motion is so small that the actual pole at any time remains within a circle of 30 or 40 feet in radius centered at the mean position of the pole.

The pole of rotation for the time being is of course the pole having a latitude of 90° and an indeterminate longitude.

Magnetic Poles

Although Earth's magnetic field resembles that of an ordinary bar magnet, this magnetic field is probably produced by electric currents in the liquid currents of the Earth's outer core. The **north magnetic pole** of Earth is that region where the magnetic force is vertically downward, and the **south magnetic pole** is that region where the magnetic force is vertically upward. A compass placed at the magnetic poles experiences no directive force in azimuth.

There are slow changes in the distribution of Earth's magnetic field. This slow temporal change is referred to as the Secular change of the main magnetic field and the magnetic poles shift due this. The location of the N magnetic pole was first measured in 1831 at Cape Adelaide on the west coast of Boothia Peninsula in Canada's Northwest

Territories (about latitude 70° N and longitude 96° W). Since then it has moved over 500 miles. In 1994, the Geological Survey of Canada measured the N magnetic pole to be at latitude 78.3° N and longitude 104.0° W. Measurement for the past several decades by Canadian scientists indicate the NW motion of the pole continues, averaging about 6 mi per year.

The direction of the horizontal components of the magnetic field at any point is known as magnetic N at that point, and the angle by which it deviates E or W of true N is known as the magnetic declination.

A compass without error points in the direction of magnetic north. (In general, this is not the direction of the magnetic north pole.) If one follows the direction indicated by the N end of the compass, he or she will travel along a rather irregular curve that eventually reaches the north magnetic pole (though not usually by a great-circle route). However, the action of the compass should not be thought of as due to any influence of the distant pole, but simply as an indication of the distribution of Earth's magnetism at the place of observation.

Rotation of Earth

The speed of rotation of Earth about its axis has been found to be slightly variable. The variations may be classified as:

(A) Secular. Tidal friction acts as a brake on the rotation and causes a slow secular increase in the length of the day, about 1 millisecond per century.

(B) Irregular. The speed of rotation may increase for a number of years, about 5 to 10, and then start decreasing. The maximum difference from the mean in the length of the day during a century is about 5 milliseconds. The accumulated difference in time has amounted to approximately 44 seconds since 1900. The cause is probably motion in the interior of Earth.

(C) Periodic. Seasonal variations exist with periods of 1 year and 6 months. The cumulative effect is such that each year, Earth is late about 30 milliseconds near June 1 and is ahead about 30 milliseconds near Oct. 1. The maximum seasonal variation in the length of the day is about 0.5 millisecond. It is believed that the principal cause of the annual variation is the seasonal change in the wind patterns of the northern and southern hemispheres. The semiannual variation is due chiefly to tidal action of the Sun, which distorts the shape of Earth slightly.

The secular and irregular variations were discovered by comparing time based on the rotation of Earth with time based on the orbital motion of the Moon about Earth and of the planets about the Sun. The periodic variation was determined largely with the aid of quartz-crystal clocks. The introduction of the cesium-beam atomic clock in 1955 made it possible to determine in greater detail than before the nature of the irregular and periodic variations.

Chronological Eras

Era	Year	Begins in 2000	Era	Year	Begins in 2000
Byzantine	7509	Sept. 14	Grecian (Seleucidae)		Sept. 14
Jewish	5761	Sept. 29[1]		2312	or Oct. 14
Roman (Ab Urbe Condita)	2753	Jan. 14	Diocletian	1717	Sept. 11
Nabonassar (Babylonian)	2749	Apr. 23	Indian (Saka)	1922	Mar. 21
Japanese	2660	Jan. 1	Islamic/Muslim (Hijra)	1421	Apr. 5[1]

(1) Year begins at sunset.

Chronological Cycles, 2000

Dominical Letter	BA	Golden Number (Lunar Cycle)	VI	Roman Indiction	8
Epact	24	Solar Cycle	21	Julian Period (year of)	6713

Twilight

Twilight is that evening period of waning light from the time of sunset to dark, often termed dusk. Morning twilight, a time of increasing light, is called **dawn**. The source of this light is the Sun shining on the atmosphere above the observer. Twilight is a time of very slowly changing sky illumination with no abrupt variations. Nevertheless, there are 3 commonly accepted divisions in this smooth continuum defined by the distance the Sun lies below the astronomical horizon: civil twilight, nautical twilight, and astronomical twilight.

The **astronomical horizon** is that great circle lying 90° from the zenith, the point directly over the observer's head. Twilight ends in the evening or begins in the morning at a particular time. Nominally, evening events are repeated in reverse order in the morning.

Civil twilight is the time from the moment of sunset, when the Sun's apparent upper edge is just at the horizon, until the center of the Sun is 6° directly below the horizon. In many states, this is the time in the evening when automobile headlights must be turned on, not to see better, but to be seen by other drivers. After this time, a newspaper becomes increasingly difficult to read in the absence of artificial light. **Nautical twilight** ends when the Sun's center is 12° below the horizon. By this time in the evening, the bright stars used by navigators have appeared, and the horizon may still be seen. After this time, the horizon is more difficult to perceive, preventing navigators from sighting stars. **Astronomical twilight** ends in the evening when the Sun is 18° below the horizon and the sky is dark enough, at least away from the Sun's location, to allow astronomical work to proceed. Sunlight, however, is still shining on the higher levels of the atmosphere from the observer's zenith to the horizon toward the Sun. Although not named as a period of twilight, when the Sun is 24° below the horizon, no part of the observer's atmosphere, even toward the Sun, receives any sunlight.

In the tropics, the Sun moves nearly vertically, accomplishing its 6°, 12°, or 18° depression very quickly. In the polar regions, the Sun's diurnal motion may actually be nearly along the horizon, prolonging the twilight period or even not permitting darkness to fall at all. In mid-latitudes, civil twilight may last about a half hour; nautical twilight, an hour; and astronomers can go to work after approximately 90 minutes.

The twilight tables given in *The World Almanac* are for the beginning of morning twilight and the end of evening astronomical twilight. Although the instant of the Sun's horizontal depression may be calculated precisely, the phenomena associated with the event are sufficiently imprecise that the table need not be recalculated each year.

Astronomical Twilight—Meridian of Greenwich

Date	20° Morn. h m	20° Eve. h m	30° Morn. h m	30° Eve. h m	40° Morn. h m	40° Eve. h m	50° Morn. h m	50° Eve. h m	60° Morn. h m	60° Eve. h m
Jan. 1	5 16	18 50	5 30	18 36	5 45	18 22	6 00	18 07	6 18	17 49
11	5 20	18 56	5 33	18 43	5 45	18 30	5 59	18 17	6 15	18 01
21	5 21	19 02	5 32	18 50	5 43	18 40	5 53	18 29	6 05	18 18
Feb. 1	5 20	19 07	5 29	18 59	5 36	18 51	5 43	18 45	5 49	18 39
11	5 17	19 12	5 23	19 06	5 27	19 02	5 30	19 00	5 29	19 01
21	5 12	19 15	5 15	19 13	5 16	19 12	5 13	19 15	5 04	19 25
Mar. 1	5 06	19 18	5 06	19 19	5 03	19 22	4 56	19 30	4 39	19 47
11	4 59	19 22	4 55	19 25	4 48	19 33	4 34	19 47	4 07	20 15
21	4 50	19 25	4 43	19 32	4 31	19 44	4 10	20 05	3 32	20 45
Apr. 1	4 40	19 28	4 28	19 40	4 11	19 58	3 42	20 27	2 46	21 25
11	4 30	19 32	4 15	19 48	3 53	20 10	3 15	20 48	1 55	22 12
21	4 21	19 37	4 02	19 56	3 34	20 24	2 47	21 13	0 18	
May 1	4 13	19 42	3 50	20 05	3 16	20 39	2 17	21 40		
11	4 06	19 47	3 40	20 14	3 00	20 54	1 45	22 11		
21	4 01	19 53	3 31	20 23	2 46	21 08	1 09	22 48		
Jun. 1	3 57	19 59	3 25	20 31	2 34	21 22				
11	3 56	20 03	3 22	20 37	2 28	21 31				
21	3 57	20 06	3 23	20 41	2 28	21 36				
Jul. 1	4 00	20 07	3 26	20 41	2 33	21 35				
11	4 05	20 06	3 32	20 38	2 42	21 28		23 54		
21	4 10	20 02	3 40	20 32	2 54	21 17	1 14	22 55		
Aug. 1	4 16	19 56	3 49	20 22	3 10	21 02	1 54	22 16		
11	4 21	19 49	3 58	20 11	3 24	20 45	2 24	21 44		
21	4 25	19 40	4 06	19 59	3 38	20 27	2 50	21 14		23 33
Sep. 1	4 30	19 30	4 15	19 44	3 52	20 06	3 16	20 42	1 56	21 59
11	4 33	19 20	4 22	19 31	4 04	19 48	3 36	20 15	2 40	21 10
21	4 35	19 10	4 28	19 17	4 16	19 29	3 55	19 50	3 15	20 29
30	4 38	19 02	4 33	19 06	4 25	19 14	4 10	19 28	3 42	19 56
Oct. 1	4 38	19 01	4 34	19 05	4 26	19 12	4 12	19 26	3 44	19 53
11	4 40	18 53	4 40	18 53	4 36	18 56	4 28	19 04	4 11	19 21
21	4 43	18 46	4 46	18 43	4 47	18 42	4 44	18 44	4 35	18 53
Nov. 1	4 47	18 40	4 53	18 34	4 58	18 29	5 00	18 26	5 00	18 26
11	4 51	18 37	5 00	18 28	5 08	18 20	5 15	18 13	5 21	18 06
21	4 56	18 36	5 07	18 25	5 17	18 14	5 28	18 03	5 39	17 52
Dec. 1	5 01	18 37	5 14	18 24	5 27	18 12	5 40	17 58	5 56	17 42
11	5 07	18 40	5 21	18 26	5 35	18 12	5 50	17 57	6 08	17 38
21	5 12	18 45	5 26	18 30	5 41	18 16	5 56	18 00	6 16	17 41
31	5 16	18 50	5 30	18 36	5 44	18 22	6 00	18 07	6 18	17 48

Calculation of Rise Times

The charts contain rise and set times for the Sun and Moon for the Greenwich Meridian at N latitudes 20°, 30°, 40°, 50°, and 60°. From day to day, the values for the Sun at any particular latitude do not change very much. This means that whatever time the Sun rises or sets at the 0° meridian, it will rise or set at the same time at the Standard Time meridian of your time zone. Standard Time meridians occur every 15° of longitude (15° E and W, 30° E and W, etc.) The corrections necessary to observe that event from your location will be to account for your distance from the Standard Time meridian and for your latitude. Thus, if your latitude is about 45°, sunrise on Jan. 1, 2000, is roughly halfway between 7:22 and 7:59 AM on the Standard Time meridian for your time zone. If you are 7.5° west of your Standard Time meridian, sunrise will be about 1/2 hour later than this; if 7.5° east, about 1/2 hour earlier.

The Moon, however, moves its own diameter, about one-half degree, in an hour, or about 13.2° in one complete turn of Earth—one day. Most of this is eastward against the background stars of the sky, but some is also N or S movement. All this motion considerably affects the times of rise or set, as you can see from the adjacent entries in the table. Thus, it is necessary to take your longitude into account in addition to your latitude. If you have no need for total accuracy, simply note that the time will be between the 4 values (see example below) you find surrounding your location and the dates of interest.

The process of finding more accurate corrections is called interpolation. In the example, linear interpolation involving simple differences is used. In extreme cases, higher order interpolation should be used. If such cases are important to you, it is suggested that you plot the times, draw smooth curves through the plots, and interpolate by eye between the relevant curves. Some people find this exercise fun.

Let's find the times of the moonrise for the September Full Moon and sunset the same day at Dodge City, KS.

First, where is Dodge City, KS? Find Dodge City's latitude and longitude in the "Latitude, Longitude, and Altitude of U.S. and Canadian Cities" table found in the World Exploration and Geography section of *The World Almanac*. You must also know the time zone in which the city is located, which you can estimate from the "International Time Zones" map in the map section of *The World Almanac*.

I. Dodge City, KS: 37° 45′ 10″ N
                        100° 1′ 0″ W

IA. Convert these values to decimals:
10/60 = 0.17
45 + 0.17 = 45.17
45.17/60 = 0.75
37 + 0.75 = 37.75 N
0/60 = 0.00
1 + 0.00 = 1.00
1.00/60 = 0.02
100 + 0.02 = 100.02 W

IB. Fraction Dodge City lies between 30° and 40°:
37.75 − 30 = 7.75; 7.75/10 = 0.775

IC. Fraction world must turn between Greenwich and Dodge City:
100.02/360 = 0.278

ID. Dodge City is in the Central Standard Time zone and the CST meridian is 90°; thus 100.02 is 100.02 − 90 = 10.02° W of the Central Standard Meridian. In 24 hours, there are 24 x 60 = 1,440 minutes; 1,440/360 = 4 minutes for every degree around Earth. So events happen 4 x 10.02 = 40.1 minutes later in Dodge City than at the 90° meridian. (If the location is E of the Standard Meridian, events happen earlier.)

IE. The values IB and IC are interpolates for Dodge City; ID is the time correction from local to Standard time for Dodge City. These values need never be calculated again for Dodge City.

IIA. To find the time of moonrise we start from the table of Moon Phases, 2000. We see that September's Full Moon occurs on September 13. We need the Greenwich times for moonrise at latitudes 30° and 40°, and for September 13 and 14, the day of the Full Moon and the next day. These values are found in the Astronomy Daily Calendar 2000; we then compute the difference between the two latitudes.

	30°	Diff.	40°
Sept. 13	18:18	0:09	18:27
Sept. 14	18:52	0:03	18:55

IIB. We want IB and the September 13 time difference:
0.775 x 9 = 7.0

Add this to the September 13, 30° rise time:
18:18 + 7.0 = 18:25.0

And for September 14:
0.775 x 3 = 2.3

Add this to the September 14, 30° rise time:
18:52 + 2.3 = 18:54.3

These 2 times are for the latitude of Dodge City, but for the Greenwich meridian.

IIC. To get the time for Dodge City meridian, take the difference between these 2 times just determined,
18:54.3 − 18:25.0 = 29.3 minutes,

and find what fraction of this 24-hour change took place while Earth turned between Greenwich and Dodge City, 0.278 (See IC).
29.3 x 0.278 = 8.1 minutes after 18:25.0

Thus 18:25.0 + 8.1 = 18:33.1 is the time the Full Moon will rise in the local time of Dodge City.

IID. But this happens 40.1 minutes (See ID) later by CST clock time at Dodge City, thus
18:33.1 + 40.1 = 19:13.2 CST

But this is late summer, and daylight time is in effect;
19:13 + 1:00 = 20:13 CDT is the rise time for the Full Moon at Dodge City the evening of September 13, 2000.

IIIA. To find the time of sunset we need the Greenwich times for sunset at latitudes 30° and 40°. These values are found in the Astronomy Daily Calendar 2000; we then compute the difference between the two latitudes.

	30°	Diff.	40°
Sept. 13	18:08	0:04	18:12

IIIB. We want IB and the September 13 time difference:
0.775 x 4 = 3.1

Add this to the September 13, 30° set time:
18:08 + 3.1 = 18:11.1

This is the local time for the latitude of Dodge City.

IIIC. But this happens 40.1 minutes (See ID) later by CST clock time at Dodge City, thus
18:11.1 + 40.1 = 18:51.2 CST

But this is summer, and daylight time is in effect;
18:51 + 1:00 = 19:51 CDT is sunset at Dodge City on September 13, 2000.

JANUARY 2000

1st MONTH   **31 DAYS**

All times are Coordinated Universal Time (Greenwich Mean Time)

NOTE: Degrees are North Latitude. For each latitude (20°– 60°), numbers on first line indicate Sun.

Numbers on second line indicate Moon.

Moon Phases: FM = Full Moon: LQ = Last (Waning) Quarter: NM = New Moon, FQ = First (Waxing) Quarter

Sun's distance is in Astronomical Units

CAUTION: Must be converted to local time. For instructions see "Calculation of Rise Times."

Day of month, of week, of year	Sun on Meridian / Moon Phase (h m s) / Distance	Sun's Declination (° ') / Distance	20° RISE Sun / Moon (h m)	20° SET Sun / Moon (h m)	30° RISE Sun / Moon (h m)	30° SET Sun / Moon (h m)	40° RISE Sun / Moon (h m)	40° SET Sun / Moon (h m)	50° RISE Sun / Moon (h m)	50° SET Sun / Moon (h m)	60° RISE Sun / Moon (h m)	60° SET Sun / Moon (h m)
1 SA	12 03 17	−23 04	6 35	17 32	6 56	17 11	7 22	16 45	7 59	16 08	9 03	15 04
1		.9833	2 05	13 54	2 13	13 44	2 23	13 32	2 38	13 15	3 00	12 50
2 SU	12 03 45	−23 00	6 35	17 32	6 56	17 12	7 22	16 46	7 59	16 09	9 02	15 06
2		.9833	2 54	14 33	3 06	14 20	3 21	14 03	3 41	13 41	4 14	13 06
3 MO	12 04 13	−22 54	6 36	17 33	6 56	17 12	7 22	16 47	7 58	16 10	9 02	15 07
3		.9833	3 44	15 14	3 59	14 58	4 18	14 38	4 44	14 10	5 27	13 26
4 TU	12 04 41	−22 49	6 36	17 34	6 56	17 13	7 22	16 47	7 58	16 11	9 01	15 09
4		.9833	4 34	15 57	4 52	15 39	5 14	15 16	5 44	14 44	6 36	13 52
5 WE	12 05 08	−22 42	6 36	17 34	6 57	17 14	7 22	16 48	7 58	16 12	9 00	15 10
5		.9833	5 24	16 43	5 43	16 23	6 07	15 59	6 41	15 25	7 39	14 26
6 TH	12 05 35	−22 36	6 36	17 35	6 57	17 15	7 22	16 49	7 58	16 14	8 59	15 12
6	18 14 NM	.9833	6 14	17 31	6 33	17 11	6 58	16 47	7 33	16 12	8 33	15 12
7 FR	12 06 02	−22 29	6 37	17 36	6 57	17 15	7 22	16 50	7 57	16 15	8 58	15 14
7		.9834	7 02	18 21	7 21	18 02	7 45	17 39	8 19	17 05	9 17	16 09
8 SA	12 06 28	−22 21	6 37	17 36	6 57	17 16	7 22	16 51	7 57	16 16	8 58	15 16
8		.9834	7 48	19 13	8 06	18 56	8 28	18 34	8 59	18 05	9 51	17 15
9 SU	12 06 53	−22 14	6 37	17 37	6 57	17 17	7 22	16 52	7 57	16 17	8 56	15 18
9		.9834	8 32	20 05	8 48	19 50	9 07	19 33	9 33	19 08	10 16	18 27
10 MO	12 07 18	−22 05	6 37	17 37	6 57	17 18	7 22	16 53	7 56	16 19	8 55	15 20
10		.9834	9 14	20 57	9 27	20 46	9 42	20 32	10 03	20 14	10 36	19 44
11 TU	12 07 42	−21 56	6 37	17 38	6 57	17 19	7 21	16 54	7 56	16 20	8 54	15 22
11		.9834	9 55	21 50	10 04	21 42	10 14	21 34	10 29	21 22	10 52	21 02
12 WE	12 08 06	−21 47	6 38	17 39	6 57	17 19	7 21	16 55	7 55	16 21	8 53	15 24
12		.9835	10 34	22 43	10 39	22 40	10 45	22 36	10 53	22 31	11 06	22 23
13 TH	12 08 29	−21 38	6 38	17 39	6 57	17 20	7 21	16 56	7 55	16 23	8 52	15 26
13		.9835	11 13	23 37	11 14	23 38	11 15	23 40	11 17	23 42	11 19	23 45
14 FR	12 08 51	−21 28	6 38	17 40	6 57	17 21	7 21	16 57	7 54	16 24	8 50	15 28
14	13 34 FQ	.9836	11 53	none	11 50	none	11 46	none	11 40	none	11 32	none
15 SA	12 09 13	−21 17	6 38	17 41	6 57	17 22	7 20	16 58	7 53	16 26	8 49	15 30
15		.9836	12 36	0 33	12 28	0 38	12 19	0 45	12 06	0 55	11 47	1 10
16 SU	12 09 34	−21 06	6 38	17 41	6 57	17 23	7 20	17 00	7 52	16 27	8 47	15 33
16		.9836	13 22	1 31	13 10	1 41	12 55	1 54	12 35	2 11	12 05	2 38
17 MO	12 09 55	−20 55	6 38	17 42	6 56	17 24	7 20	17 01	7 52	16 29	8 46	15 35
17		.9837	14 12	2 32	13 56	2 46	13 37	3 04	13 11	3 28	12 29	4 08
18 TU	12 10 14	−20 43	6 38	17 43	6 56	17 24	7 19	17 02	7 51	16 30	8 44	15 37
18		.9838	15 07	3 36	14 49	3 53	14 26	4 15	13 55	4 45	13 02	5 36
19 WE	12 10 33	−20 31	6 38	17 43	6 56	17 25	7 19	17 03	7 50	16 32	8 42	15 40
19		.9838	16 08	4 41	15 48	5 00	15 23	5 24	14 49	5 58	13 50	6 57
20 TH	12 10 51	−20 19	6 38	17 44	6 56	17 26	7 18	17 04	7 49	16 33	8 40	15 42
20		.9839	17 12	5 44	16 52	6 04	16 28	6 29	15 54	7 03	14 56	8 02
21 FR	12 11 09	−20 06	6 38	17 45	6 56	17 27	7 17	17 05	7 48	16 35	8 39	15 44
21	4 40 FM	.9840	18 17	6 45	17 59	7 03	17 38	7 26	17 08	7 57	16 17	8 50
22 SA	12 11 26	−19 53	6 38	17 45	6 55	17 28	7 17	17 06	7 47	16 36	8 37	15 47
22		.9841	19 20	7 40	19 06	7 56	18 49	8 15	18 25	8 41	17 46	9 23
23 SU	12 11 42	−19 39	6 38	17 46	6 55	17 29	7 16	17 07	7 46	16 38	8 35	15 49
23		.9842	20 21	8 30	20 12	8 42	19 59	8 57	19 42	9 16	19 16	9 46
24 MO	12 11 57	−19 25	6 37	17 47	6 55	17 30	7 16	17 09	7 45	16 40	8 33	15 52
24		.9843	21 19	9 16	21 14	9 23	21 07	9 33	20 57	9 45	20 43	10 04
25 TU	12 12 11	−19 11	6 37	17 47	6 54	17 31	7 15	17 10	7 44	16 41	8 31	15 54
25		.9844	22 14	9 58	22 13	10 01	22 11	10 05	22 09	10 10	22 06	10 19
26 WE	12 12 25	−18 56	6 37	17 48	6 54	17 31	7 14	17 11	7 42	16 43	8 29	15 57
26		.9845	23 07	10 37	23 10	10 36	23 14	10 35	23 19	10 34	23 26	10 31
27 TH	12 12 38	−18 41	6 37	17 48	6 53	17 32	7 13	17 12	7 41	16 45	8 27	15 59
27		.9846	23 58	11 15	none	11 10	none	11 04	none	10 56	none	10 44
28 FR	12 12 50	−18 26	6 37	17 49	6 53	17 33	7 13	17 13	7 40	16 46	8 24	16 02
28	7 57 LQ	.9847	none	11 53	0 05	11 44	0 14	11 34	0 25	11 19	0 44	10 57
29 SA	12 13 01	−18 10	6 37	17 50	6 52	17 34	7 12	17 15	7 39	16 48	8 22	16 05
29		.9849	0 49	12 32	0 59	12 19	1 12	12 04	1 31	11 44	2 00	11 12
30 SU	12 13 11	−17 54	6 36	17 50	6 52	17 35	7 11	17 16	7 37	16 50	8 20	16 07
30		.9850	1 39	13 12	1 53	12 57	2 10	12 38	2 34	12 12	3 13	11 31
31 MO	12 13 21	−17 38	6 36	17 51	6 51	17 36	7 10	17 17	7 36	16 51	8 18	16 10
31		.9851	2 29	13 54	2 46	13 36	3 06	13 15	3 35	12 44	4 24	11 54

FEBRUARY 2000

2d MONTH   **29 DAYS**

All times are Coordinated Universal Time (Greenwich Mean Time)

NOTE: Degrees are North Latitude. For each latitude (20°– 60°), numbers on first line indicate Sun.

Numbers on second line indicate Moon.

Moon Phases: FM = Full Moon: LQ = Last (Waning) Quarter: NM = New Moon, FQ = First (Waxing) Quarter

Sun's distance is in Astronomical Units

CAUTION: Must be converted to local time. For instructions see "Calculation of Rise Times."

Day of month, of week, of year	Sun on Meridian / Moon Phase / Distance	Sun's Declination	20° RISE Sun / Moon	20° SET Sun / Moon	30° RISE Sun / Moon	30° SET Sun / Moon	40° RISE Sun / Moon	40° SET Sun / Moon	50° RISE Sun / Moon	50° SET Sun / Moon	60° RISE Sun / Moon	60° SET Sun / Moon
	h m s	° ′	h m	h m	h m	h m	h m	h m	h m	h m	h m	h m
1 TU	12 13 30	−17 21	6 36	17 51	6 51	17 37	7 09	17 18	7 35	16 53	8 15	16 12
32	.9853		3 19	14 39	3 38	14 20	4 01	13 56	4 34	13 22	5 29	12 26
2 WE	12 13 38	−17 04	6 35	17 52	6 50	17 37	7 08	17 19	7 33	16 55	8 13	16 15
33	.9854		4 09	15 26	4 28	15 06	4 53	14 42	5 28	14 07	6 27	13 07
3 TH	12 13 45	−16 47	6 35	17 53	6 50	17 38	7 07	17 21	7 32	16 56	8 11	16 18
34	.9856		4 57	16 16	5 17	15 56	5 41	15 32	6 16	14 58	7 15	14 00
4 FR	12 13 52	−16 29	6 35	17 53	6 49	17 39	7 06	17 22	7 30	16 58	8 08	16 20
35	.9857		5 45	17 07	6 03	16 49	6 26	16 27	6 58	15 56	7 52	15 03
5 SA	12 13 57	−16 12	6 34	17 54	6 48	17 40	7 05	17 23	7 29	17 00	8 06	16 23
36	13 03 NM .9859		6 30	18 00	6 46	17 44	7 07	17 25	7 35	16 59	8 21	16 15
6 SU	12 14 02	−15 54	6 34	17 54	6 48	17 41	7 04	17 24	7 27	17 02	8 03	16 26
37	.9860		7 13	18 53	7 27	18 40	7 43	18 25	8 06	18 05	8 42	17 31
7 MO	12 14 06	−15 35	6 34	17 55	6 47	17 42	7 03	17 25	7 26	17 03	8 01	16 28
38	.9862		7 55	19 46	8 05	19 37	8 17	19 27	8 34	19 13	9 00	18 50
8 TU	12 14 10	−15 17	6 33	17 55	6 46	17 42	7 02	17 27	7 24	17 05	7 58	16 31
39	.9863		8 35	20 39	8 41	20 35	8 48	20 30	8 59	20 22	9 14	20 11
9 WE	12 14 12	−14 58	6 33	17 56	6 45	17 43	7 01	17 28	7 22	17 07	7 56	16 33
40	.9865		9 14	21 33	9 16	21 33	9 19	21 33	9 22	21 33	9 27	21 33
10 TH	12 14 14	−14 38	6 32	17 56	6 45	17 44	7 00	17 29	7 21	17 09	7 53	16 36
41	.9867		9 54	22 28	9 52	22 32	9 49	22 38	9 45	22 45	9 40	22 56
11 FR	12 14 14	−14 19	6 32	17 57	6 44	17 45	6 59	17 30	7 19	17 10	7 51	16 39
42	.9869		10 35	23 24	10 28	23 33	10 20	23 44	10 10	23 59	9 54	none
12 SA	12 14 14	−13 59	6 31	17 57	6 43	17 46	6 58	17 31	7 17	17 12	7 48	16 41
43	23 21 FQ .9870		11 18	none	11 08	none	10 55	none	10 37	none	10 10	0 22
13 SU	12 14 14	−13 40	6 31	17 58	6 42	17 47	6 56	17 33	7 15	17 14	7 45	16 44
44	.9872		12 05	0 23	11 51	0 36	11 33	0 51	11 09	1 13	10 31	1 49
14 MO	12 14 12	−13 19	6 30	17 58	6 41	17 47	6 55	17 34	7 14	17 15	7 43	16 47
45	.9874		12 56	1 23	12 39	1 40	12 18	2 00	11 48	2 28	10 59	3 15
15 TU	12 14 10	−12 59	6 30	17 59	6 41	17 48	6 54	17 35	7 12	17 17	7 40	16 49
46	.9876		13 53	2 25	13 33	2 44	13 09	3 08	12 35	3 41	11 39	4 36
16 WE	12 14 07	−12 39	6 29	17 59	6 40	17 49	6 53	17 36	7 10	17 19	7 37	16 52
47	.9878		14 53	3 28	14 33	3 47	14 09	4 12	13 34	4 47	12 34	5 46
17 TH	12 14 03	−12 18	6 28	18 00	6 39	17 50	6 51	17 37	7 08	17 21	7 35	16 55
48	.9880		15 56	4 28	15 37	4 47	15 14	5 11	14 42	5 44	13 47	6 40
18 FR	12 13 59	−11 57	6 28	18 00	6 38	17 50	6 50	17 38	7 06	17 22	7 32	16 57
49	.9882		16 59	5 24	16 44	5 41	16 24	6 03	15 57	6 32	15 11	7 19
19 SA	12 13 53	−11 36	6 27	18 01	6 37	17 51	6 49	17 40	7 05	17 24	7 29	17 00
50	16 27 FM .9884		18 02	6 17	17 50	6 30	17 35	6 47	17 14	7 10	16 41	7 47
20 SU	12 13 48	−11 15	6 27	18 01	6 36	17 52	6 47	17 41	7 03	17 26	7 26	17 02
51	.9886		19 02	7 04	18 54	7 14	18 44	7 26	18 31	7 42	18 10	8 07
21 MO	12 13 41	−10 53	6 26	18 02	6 35	17 53	6 46	17 42	7 01	17 27	7 24	17 05
52	.9888		19 59	7 48	19 56	7 54	19 51	8 01	19 46	8 09	19 37	8 23
22 TU	12 13 34	−10 31	6 25	18 02	6 34	17 53	6 45	17 43	6 59	17 29	7 21	17 08
53	.9890		20 54	8 30	20 55	8 31	20 56	8 32	20 58	8 34	21 01	8 37
23 WE	12 13 26	−10 10	6 25	18 02	6 33	17 54	6 43	17 44	6 57	17 31	7 18	17 10
54	.9892		21 47	9 09	21 52	9 06	21 59	9 02	22 08	8 57	22 22	8 49
24 TH	12 13 18	−9 48	6 24	18 03	6 32	17 55	6 42	17 45	6 55	17 32	7 15	17 13
55	.9895		22 39	9 48	22 48	9 41	23 00	9 32	23 15	9 21	23 40	9 03
25 FR	12 13 09	−9 26	6 23	18 03	6 31	17 56	6 40	17 46	6 53	17 34	7 12	17 15
56	.9897		23 30	10 27	23 43	10 16	23 59	10 03	none	9 45	none	9 17
26 SA	12 12 59	−9 03	6 23	18 04	6 30	17 56	6 39	17 48	6 51	17 36	7 09	17 18
57	.9899		none	11 07	none	10 53	none	10 36	0 21	10 12	0 56	9 34
27 SU	12 12 49	−8 41	6 22	18 04	6 29	17 57	6 38	17 49	6 49	17 38	7 06	17 20
58	3 53 LQ .9903		0 21	11 49	0 37	11 32	0 57	11 11	1 24	10 43	2 09	9 55
28 MO	12 12 39	−8 18	6 21	18 04	6 28	17 58	6 36	17 50	6 47	17 39	7 04	17 23
59	.9904		1 12	12 33	1 30	12 14	1 52	11 51	2 24	11 18	3 18	10 24
29 TU	12 12 28	-7 56	6 20	18 05	6 27	17 59	6 35	17 51	6 45	17 41	7 01	17 25
60	.9907		2 02	13 19	2 21	13 00	2 46	12 35	3 20	12 00	4 19	11 01

MARCH 2000

3d MONTH   **31 DAYS**

All times are Coordinated Universal Time (Greenwich Mean Time)

NOTE: Degrees are North Latitude. For each latitude (20°– 60°), numbers on first line indicate Sun.

Numbers on second line indicate Moon.

Moon Phases: FM = Full Moon: LQ = Last (Waning) Quarter: NM = New Moon, FQ = First (Waxing) Quarter

Sun's distance is in Astronomical Units

CAUTION: Must be converted to local time. For instructions see "Calculation of Rise Times."

Day of month, of week, of year / Moon Phase	Sun on Meridian h m s / Distance	Sun's Declination ° ′	20° RISE Sun/Moon h m	20° SET Sun/Moon h m	30° RISE Sun/Moon h m	30° SET Sun/Moon h m	40° RISE Sun/Moon h m	40° SET Sun/Moon h m	50° RISE Sun/Moon h m	50° SET Sun/Moon h m	60° RISE Sun/Moon h m	60° SET Sun/Moon h m
1 WE	12 12 16	−7 33	6 20	18 05	6 26	17 59	6 33	17 52	6 43	17 43	6 58	17 28
61	.9909		2 51	14 08	3 11	13 48	3 35	13 24	4 10	12 49	5 10	11 49
2 TH	12 12 04	−7 10	6 19	18 06	6 25	18 00	6 32	17 53	6 41	17 44	6 55	17 31
62	.9912		3 39	14 59	3 58	14 40	4 21	14 17	4 55	13 44	5 52	12 49
3 FR	12 11 51	−6 47	6 18	18 06	6 24	18 01	6 30	17 54	6 39	17 46	6 52	17 33
63	.9914		4 25	15 51	4 42	15 35	5 04	15 14	5 34	14 45	6 23	13 58
4 SA	12 11 38	−6 24	6 17	18 06	6 22	18 01	6 29	17 55	6 37	17 47	6 49	17 36
64	.9917		5 09	16 44	5 23	16 31	5 42	16 14	6 07	15 51	6 47	15 13
5 SU	12 11 25	−6 01	6 17	18 07	6 21	18 02	6 27	17 56	6 35	17 49	6 46	17 38
65	.9919		5 51	17 38	6 03	17 28	6 17	17 16	6 36	16 59	7 06	16 33
6 MO	12 11 11	−5 38	6 16	18 07	6 20	18 03	6 25	17 58	6 32	17 51	6 43	17 41
66	5 17 NM .9922		6 32	18 32	6 40	18 26	6 49	18 19	7 02	18 10	7 21	17 54
7 TU	12 10 57	−5 15	6 15	18 07	6 19	18 03	6 24	17 59	6 30	17 52	6 40	17 43
67	.9924		7 12	19 27	7 16	19 26	7 20	19 24	7 26	19 21	7 35	19 18
8 WE	12 10 42	−4 51	6 14	18 08	6 18	18 04	6 22	18 00	6 28	17 54	6 37	17 46
68	.9927		7 53	20 23	7 52	20 26	7 51	20 29	7 50	20 35	7 48	20 42
9 TH	12 10 27	−4 28	6 13	18 08	6 17	18 05	6 21	18 01	6 26	17 56	6 34	17 48
69	.9929		8 34	21 19	8 29	21 27	8 22	21 36	8 14	21 49	8 01	22 09
10 FR	12 10 12	−4 04	6 12	18 08	6 16	18 05	6 19	18 02	6 24	17 57	6 31	17 51
70	.9932		9 17	22 18	9 08	22 30	8 56	22 44	8 40	23 04	8 17	23 36
11 SA	12 09 56	−3 41	6 12	18 08	6 14	18 06	6 18	18 03	6 22	17 59	6 28	17 53
71	.9935		10 03	23 18	9 50	23 33	9 33	23 53	9 11	none	8 35	none
12 SU	12 09 40	−3 17	6 11	18 09	6 13	18 07	6 16	18 04	6 20	18 01	6 25	17 56
72	.9937		10 53	none	10 36	none	10 15	none	9 47	0 19	9 00	1 03
13 MO	12 09 24	−2 53	6 10	18 09	6 12	18 07	6 14	18 05	6 18	18 02	6 22	17 58
73	6 59 FQ .9940		11 46	0 19	11 27	0 37	11 04	1 00	10 31	1 32	9 35	2 26
14 TU	12 09 07	−2 30	6 09	18 09	6 11	18 08	6 13	18 06	6 15	18 04	6 19	18 00
74	.9942		12 44	1 20	12 24	1 39	11 59	2 04	11 24	2 39	10 24	3 39
15 WE	12 08 50	−2 06	6 08	18 10	6 10	18 08	6 11	18 07	6 13	18 05	6 16	18 03
75	.9945		13 44	2 19	13 25	2 39	13 01	3 03	12 27	3 38	11 29	4 36
16 TH	12 08 33	−1 42	6 07	18 10	6 08	18 09	6 10	18 08	6 11	18 07	6 13	18 05
76	.9948		14 46	3 15	14 28	3 33	14 07	3 56	13 37	4 27	12 47	5 19
17 FR	12 08 16	−1 19	6 07	18 10	6 07	18 10	6 08	18 09	6 09	18 09	6 10	18 08
77	.9950		15 47	4 07	15 33	4 23	15 16	4 42	14 52	5 07	14 13	5 49
18 SA	12 07 58	−0 55	6 06	18 11	6 06	18 10	6 06	18 10	6 07	18 10	6 07	18 10
78	.9953		16 46	4 56	16 37	5 07	16 24	5 22	16 08	5 41	15 41	6 11
19 SU	12 07 40	−0 31	6 05	18 11	6 05	18 11	6 05	18 11	6 05	18 12	6 04	18 13
79	.9956		17 44	5 40	17 39	5 48	17 32	5 57	17 23	6 09	17 09	6 28
20 MO	12 07 22	−0 08	6 04	18 11	6 04	18 12	6 03	18 12	6 02	18 13	6 01	18 15
80	4 44 FM .9959		18 40	6 22	18 39	6 25	18 38	6 29	18 36	6 34	18 34	6 42
21 TU	12 07 04	+0 16	6 03	18 11	6 02	18 12	6 02	18 13	6 00	18 15	5 58	18 18
81	.9962		19 34	7 02	19 38	7 01	19 42	7 00	19 48	6 58	19 57	6 55
22 WE	12 06 46	+0 40	6 02	18 12	6 01	18 13	6 00	18 14	5 58	18 16	5 55	18 20
82	.9964		20 28	7 42	20 35	7 36	20 44	7 30	20 57	7 21	21 18	7 07
23 TH	12 06 28	+1 04	6 01	18 12	6 00	18 13	5 58	18 15	5 56	18 18	5 52	18 22
83	.9967		21 20	8 21	21 31	8 12	21 45	8 00	22 05	7 45	22 36	7 21
24 FR	12 06 10	+1 27	6 00	18 12	5 59	18 14	5 57	18 16	5 54	18 20	5 49	18 25
84	.9970		22 12	9 01	22 27	8 48	22 45	8 32	23 10	8 11	23 52	7 36
25 SA	12 05 52	+1 51	6 00	18 12	5 58	18 15	5 55	18 17	5 51	18 21	5 46	18 27
85	.9973		23 03	9 43	23 21	9 27	23 42	9 07	none	8 40	none	7 56
26 SU	12 05 34	+2 14	5 59	18 13	5 56	18 15	5 53	18 18	5 49	18 23	5 43	18 30
86	.9976		23 54	10 26	none	10 08	none	9 45	0 13	9 14	1 04	8 21
27 MO	12 05 16	+2 38	5 58	18 13	5 55	18 16	5 52	18 19	5 47	18 24	5 40	18 32
87	.9979		none	11 12	0 13	10 52	0 37	10 28	1 11	9 53	2 09	8 54
28 TU	12 04 57	+3 01	5 57	18 13	5 54	18 16	5 50	18 20	5 45	18 26	5 37	18 35
88	0 21 LQ .9982		0 43	12 00	1 03	11 40	1 28	11 14	2 04	10 39	3 05	9 38
29 WE	12 04 39	+3 25	5 56	18 14	5 53	18 17	5 49	18 21	5 43	18 28	5 34	18 37
89	.9985		1 32	12 49	1 51	12 30	2 16	12 06	2 51	11 32	3 50	10 33
30 TH	12 04 21	+3 48	5 55	18 14	5 52	18 18	5 47	18 22	5 41	18 29	5 31	18 40
90	.9988		2 18	13 41	2 36	13 23	2 59	13 01	3 31	12 30	4 25	11 38
31 FR	12 04 03	+4 11	5 54	18 14	5 50	18 18	5 45	18 23	5 38	18 31	5 28	18 42
91	.9990		3 02	14 33	3 19	14 18	3 39	14 00	4 06	13 34	4 51	12 51

APRIL 2000

4th MONTH   **30 DAYS**

All times are Coordinated Universal Time (Greenwich Mean Time)

NOTE: Degrees are North Latitude. For each latitude (20°– 60°), numbers on first line indicate Sun.

Numbers on second line indicate Moon.

Moon Phases: FM = Full Moon: LQ = Last (Waning) Quarter: NM = New Moon, FQ = First (Waxing) Quarter

Sun's distance is in Astronomical Units

CAUTION: Must be converted to local time. For instructions see "Calculation of Rise Times."

Day of month, of week, of year	Sun on Meridian / Moon Phase (h m s) / Distance	Sun's Declination (° ′)	20° RISE Sun/Moon	20° SET Sun/Moon	30° RISE Sun/Moon	30° SET Sun/Moon	40° RISE Sun/Moon	40° SET Sun/Moon	50° RISE Sun/Moon	50° SET Sun/Moon	60° RISE Sun/Moon	60° SET Sun/Moon
1 SA	12 03 46	+4 34	5 53	18 14	5 49	18 19	5 44	18 24	5 36	18 32	5 25	18 44
92	.9993		3 45	15 27	3 58	15 15	4 14	15 01	4 36	14 41	5 11	14 09
2 SU	12 03 28	+4 58	5 53	18 15	5 48	18 19	5 42	18 25	5 34	18 34	5 22	18 47
93	.9996		4 27	16 21	4 36	16 13	4 48	16 04	5 03	15 51	5 27	15 30
3 MO	12 03 10	+5 21	5 52	18 15	5 47	18 20	5 41	18 26	5 32	18 35	5 19	18 49
94	.9999		5 07	17 16	5 13	17 12	5 19	17 08	5 28	17 03	5 41	16 54
4 TU	12 02 53	+5 44	5 51	18 15	5 46	18 21	5 39	18 27	5 30	18 37	5 16	18 52
95	18 12 NM	.0002	5 48	18 12	5 49	18 13	5 50	18 15	5 52	18 17	5 54	18 20
5 WE	12 02 36	+6 06	5 50	18 15	5 44	18 21	5 37	18 29	5 28	18 39	5 13	18 54
96	.0005		6 29	19 10	6 26	19 16	6 22	19 23	6 16	19 33	6 07	19 48
6 TH	12 02 19	+6 29	5 49	18 16	5 43	18 22	5 36	18 30	5 26	18 40	5 10	18 57
97	.0008		7 13	20 09	7 05	20 20	6 55	20 32	6 42	20 50	6 22	21 18
7 FR	12 02 02	+6 52	5 48	18 16	5 42	18 22	5 34	18 31	5 23	18 42	5 07	18 59
98	.0011		7 59	21 10	7 46	21 25	7 31	21 43	7 11	22 08	6 39	22 49
8 SA	12 01 46	+7 14	5 48	18 16	5 41	18 23	5 33	18 32	5 21	18 43	5 04	19 02
99	.0013		8 48	22 13	8 32	22 30	8 13	22 52	7 46	23 23	7 02	none
9 SU	12 01 29	+7 37	5 47	18 17	5 40	18 24	5 31	18 33	5 19	18 45	5 01	19 04
100	.0016		9 42	23 15	9 23	23 34	9 00	23 59	8 27	none	7 33	0 16
10 MO	12 01 13	+7 59	5 46	18 17	5 39	18 24	5 30	18 34	5 17	18 46	4 58	19 06
101	.0019		10 39	none	10 19	none	9 53	none	9 18	0 34	8 18	1 33
11 TU	12 00 57	+8 21	5 45	18 17	5 37	18 25	5 28	18 35	5 15	18 48	4 55	19 09
102	13 30 FQ	.0022	11 38	0 15	11 18	0 35	10 53	1 00	10 18	1 36	9 18	2 36
12 WE	12 00 42	+8 43	5 44	18 17	5 36	18 26	5 26	18 36	5 13	18 49	4 52	19 11
103	.0025		12 39	1 12	12 21	1 31	11 58	1 54	11 26	2 27	10 33	3 23
13 TH	12 00 27	+9 05	5 43	18 18	5 35	18 26	5 25	18 37	5 11	18 51	4 49	19 14
104	.0027		13 39	2 04	13 24	2 21	13 05	2 41	12 39	3 09	11 56	3 55
14 FR	12 00 12	+9 26	5 43	18 18	5 34	18 27	5 23	18 38	5 09	18 53	4 46	19 16
105	.0030		14 38	2 53	14 27	3 06	14 12	3 22	13 53	3 44	13 22	4 18
15 SA	11 59 57	+9 48	5 42	18 18	5 33	18 27	5 22	18 39	5 07	18 54	4 43	19 19
106	.0033		15 35	3 37	15 28	3 46	15 19	3 57	15 07	4 12	14 48	4 35
16 SU	11 59 43	+10 09	5 41	18 19	5 32	18 28	5 20	18 40	5 05	18 56	4 40	19 21
107	.0036		16 30	4 19	16 28	4 24	16 24	4 30	16 20	4 38	16 12	4 50
17 MO	11 59 29	+10 30	5 40	18 19	5 31	18 29	5 19	18 41	5 03	18 57	4 37	19 24
108	.0038		17 24	4 59	17 26	4 59	17 28	5 00	17 31	5 01	17 35	5 02
18 TU	11 59 15	+10 51	5 40	18 19	5 30	18 29	5 18	18 42	5 01	18 59	4 34	19 26
109	17 41 FM	.0041	18 17	5 37	18 23	5 34	18 31	5 29	18 41	5 23	18 56	5 14
19 WE	11 59 02	+11 12	5 39	18 20	5 29	18 30	5 16	18 43	4 59	19 00	4 31	19 29
110	.0044		19 10	6 16	19 20	6 08	19 32	5 59	19 49	5 46	20 16	5 26
20 TH	11 58 50	+11 33	5 38	18 20	5 28	18 30	5 15	18 44	4 57	19 02	4 28	19 31
111	.0047		20 02	6 56	20 16	6 44	20 33	6 30	20 56	6 11	21 34	5 41
21 FR	11 58 37	+11 53	5 37	18 20	5 27	18 31	5 13	18 45	4 55	19 04	4 25	19 34
112	.0049		20 54	7 37	21 11	7 22	21 32	7 04	22 01	6 38	22 49	5 58
22 SA	11 58 26	+12 14	5 37	18 20	5 26	18 32	5 12	18 46	4 53	19 05	4 22	19 36
113	.0052		21 46	8 20	22 05	8 02	22 28	7 40	23 01	7 10	23 58	6 20
23 SU	11 58 14	+12 34	5 36	18 21	5 25	18 32	5 10	18 47	4 51	19 07	4 20	19 39
114	.0055		22 36	9 05	22 56	8 45	23 21	8 21	23 57	7 47	none	6 49
24 MO	11 58 03	+12 53	5 35	18 21	5 24	18 33	5 09	18 48	4 49	19 08	4 17	19 41
115	.0058		23 25	9 52	23 45	9 32	none	9 06	none	8 30	0 59	7 28
25 TU	11 57 53	+13 13	5 35	18 21	5 23	18 34	5 08	18 49	4 47	19 10	4 14	19 44
116	.0060		none	10 41	none	10 21	0 11	9 56	0 47	9 20	1 48	8 19
26 WE	11 57 43	+13 32	5 34	18 22	5 22	18 34	5 06	18 50	4 45	19 11	4 11	19 46
117	19 30 LQ	.0063	0 12	11 31	0 31	11 13	0 55	10 49	1 29	10 16	2 27	9 20
27 TH	11 57 34	+13 52	5 33	18 22	5 21	18 35	5 05	18 51	4 43	19 13	4 08	19 49
118	.0066		0 57	12 23	1 14	12 06	1 36	11 46	2 06	11 17	2 56	10 29
28 FR	11 57 25	+14 11	5 33	18 23	5 20	18 36	5 04	18 52	4 41	19 14	4 06	19 51
119	.0069		1 40	13 15	1 54	13 02	2 12	12 45	2 37	12 22	3 17	11 45
29 SA	11 57 17	+14 29	5 32	18 23	5 19	18 36	5 02	18 53	4 40	19 16	4 03	19 53
120	.0071		2 21	14 08	2 32	13 58	2 46	13 46	3 05	13 30	3 34	13 04
30 SU	11 57 09	+14 48	5 31	18 23	5 18	18 37	5 01	18 54	4 38	19 17	4 00	19 56
121	.0074		3 01	15 02	3 08	14 56	3 18	14 50	3 30	14 40	3 49	14 26

MAY 2000

5th MONTH   **31 DAYS**

All times are Coordinated Universal Time (Greenwich Mean Time)

NOTE: Degrees are North Latitude. For each latitude (20°– 60°), numbers on first line indicate Sun.

Numbers on second line indicate Moon.

Moon Phases: FM = Full Moon: LQ = Last (Waning) Quarter: NM = New Moon, FQ = First (Waxing) Quarter

Sun's distance is in Astronomical Units

CAUTION: Must be converted to local time. For instructions see "Calculation of Rise Times."

Day of month, of week, of year	Sun on Meridian / Moon Phase (h m s) / Distance	Sun's Declination (° ') / Distance	20° RISE Sun/Moon (h m)	20° SET Sun/Moon (h m)	30° RISE Sun/Moon (h m)	30° SET Sun/Moon (h m)	40° RISE Sun/Moon (h m)	40° SET Sun/Moon (h m)	50° RISE Sun/Moon (h m)	50° SET Sun/Moon (h m)	60° RISE Sun/Moon (h m)	60° SET Sun/Moon (h m)
1 MO	11 57 02	+15 06	5 31	18 24	5 17	18 37	5 00	18 55	4 36	19 19	3 57	19 58
122	.0076		3 41	15 57	3 44	15 56	3 48	15 55	3 53	15 53	4 01	15 50
2 TU	11 56 55	+15 24	5 30	18 24	5 16	18 38	4 59	18 56	4 34	19 21	3 55	20 01
123	.0079		4 22	16 54	4 20	16 58	4 19	17 02	4 17	17 09	4 14	17 18
3 WE	11 56 49	+15 42	5 30	18 24	5 15	18 39	4 57	18 57	4 33	19 22	3 52	20 03
124	.0081		5 04	17 54	4 58	18 02	4 51	18 12	4 42	18 27	4 27	18 49
4 TH	11 56 44	+15 59	5 29	18 25	5 14	18 39	4 56	18 58	4 31	19 24	3 49	20 06
125	4 12 NM .0084		5 49	18 55	5 39	19 08	5 26	19 24	5 09	19 46	4 43	20 22
5 FR	11 56 39	+16 17	5 28	18 25	5 14	18 40	4 55	18 59	4 29	19 25	3 47	20 08
126	.0086		6 39	19 59	6 24	20 16	6 06	20 37	5 42	21 06	5 03	21 54
6 SA	11 56 34	+16 34	5 28	18 25	5 13	18 41	4 54	19 00	4 28	19 27	3 44	20 11
127	.0089		7 32	21 04	7 14	21 23	6 52	21 47	6 21	22 22	5 30	23 20
7 SU	11 56 30	+16 50	5 27	18 26	5 12	18 41	4 53	19 01	4 26	19 28	3 41	20 13
128	.0091		8 30	22 07	8 10	22 28	7 45	22 53	7 10	23 29	6 10	none
8 MO	11 56 27	+17 07	5 27	18 26	5 11	18 42	4 52	19 02	4 24	19 30	3 39	20 16
129	.0094		9 30	23 07	9 10	23 27	8 44	23 51	8 08	none	7 06	0 32
9 TU	11 56 24	+17 23	5 26	18 27	5 11	18 43	4 51	19 03	4 23	19 31	3 36	20 18
130	.0096		10 32	none	10 13	none	9 49	none	9 16	0 26	8 18	1 25
10 WE	11 56 22	+17 39	5 26	18 27	5 10	18 43	4 50	19 04	4 21	19 33	3 34	20 21
131	20 00 FQ .0098		11 34	0 02	11 17	0 20	10 57	0 42	10 28	1 12	9 41	2 01
11 TH	11 56 21	+17 54	5 26	18 27	5 09	18 44	4 48	19 05	4 20	19 34	3 31	20 23
132	.0100		12 33	0 52	12 20	1 06	12 05	1 24	11 43	1 48	11 07	2 27
12 FR	11 56 20	+18 09	5 25	18 28	5 08	18 45	4 47	19 06	4 18	19 35	3 29	20 25
133	.0102		13 31	1 37	13 22	1 48	13 11	2 01	12 56	2 18	12 33	2 45
13 SA	11 56 19	+18 24	5 25	18 28	5 08	18 45	4 47	19 07	4 17	19 37	3 27	20 28
134	.0105		14 26	2 19	14 21	2 26	14 16	2 34	14 09	2 44	13 58	3 00
14 SU	11 56 19	+18 39	5 24	18 29	5 07	18 46	4 46	19 08	4 15	19 38	3 24	20 30
135	.0107		15 19	2 59	15 19	3 01	15 19	3 04	15 19	3 07	15 20	3 12
15 MO	11 56 20	+18 53	5 24	18 29	5 06	18 47	4 45	19 09	4 14	19 40	3 22	20 33
136	.0109		16 12	3 37	16 16	3 35	16 21	3 32	16 29	3 29	16 40	3 23
16 TU	11 56 21	+19 07	5 23	18 29	5 06	18 47	4 44	19 10	4 12	19 41	3 20	20 35
137	.0111		17 04	4 15	17 12	4 09	17 23	4 01	17 37	3 51	18 00	3 35
17 WE	11 56 22	+19 21	5 23	18 30	5 05	18 48	4 43	19 10	4 11	19 42	3 17	20 37
138	.0113		17 55	4 54	18 08	4 43	18 23	4 31	18 44	4 14	19 18	3 48
18 TH	11 56 25	+19 34	5 23	18 30	5 05	18 48	4 42	19 11	4 10	19 44	3 15	20 39
139	7 34 FM .0115		18 47	5 34	19 03	5 20	19 22	5 03	19 50	4 40	20 34	4 03
19 FR	11 56 28	+19 47	5 22	18 31	5 04	18 49	4 41	19 12	4 09	19 45	3 13	20 42
140	.0117		19 39	6 16	19 57	5 59	20 20	5 38	20 52	5 09	21 47	4 22
20 SA	11 56 31	+20 00	5 22	18 31	5 04	18 50	4 40	19 13	4 07	19 46	3 11	20 44
141	.0119		20 30	7 00	20 50	6 41	21 15	6 17	21 50	5 44	22 51	4 48
21 SU	11 56 35	+20 12	5 22	18 31	5 03	18 50	4 40	19 14	4 06	19 48	3 09	20 46
142	.0121		21 20	7 46	21 40	7 26	22 06	7 00	22 42	6 25	23 46	5 23
22 MO	11 56 39	+20 24	5 22	18 32	5 03	18 51	4 39	19 15	4 05	19 49	3 07	20 48
143	.0123		22 07	8 34	22 27	8 14	22 52	7 48	23 28	7 12	none	6 09
23 TU	11 56 44	+20 36	5 21	18 32	5 02	18 51	4 38	19 16	4 04	19 50	3 05	20 50
144	.0125		22 53	9 24	23 11	9 05	23 34	8 40	none	8 05	0 28	7 06
24 WE	11 56 50	+20 47	5 21	18 33	5 02	18 52	4 37	19 17	4 03	19 52	3 03	20 53
145	.0127		23 36	10 15	23 52	9 57	none	9 35	0 07	9 04	1 00	8 12
25 TH	11 56 56	+20 58	5 21	18 33	5 01	18 53	4 37	19 17	4 02	19 53	3 01	20 55
146	.0129		none	11 06	none	10 51	0 12	10 33	0 39	10 07	1 24	9 25
26 FR	11 57 02	+21 08	5 21	18 33	5 01	18 53	4 36	19 18	4 01	19 54	2 59	20 57
147	11 55 LQ .0131		0 17	11 58	0 30	11 46	0 46	11 32	1 08	11 13	1 42	10 41
27 SA	11 57 09	+21 19	5 21	18 34	5 01	18 54	4 36	19 19	4 00	19 55	2 57	20 59
148	.0133		0 57	12 50	1 06	12 43	1 17	12 33	1 33	12 21	1 57	12 00
28 SU	11 57 16	+21 28	5 20	18 34	5 00	18 54	4 35	19 20	3 59	19 56	2 55	21 01
149	.0134		1 36	13 43	1 41	13 40	1 47	13 36	1 56	13 31	2 09	13 22
29 MO	11 57 24	+21 38	5 20	18 35	5 00	18 55	4 35	19 21	3 58	19 57	2 54	21 03
150	.0136		2 15	14 38	2 16	14 40	2 17	14 41	2 19	14 43	2 21	14 47
30 TU	11 57 33	+21 47	5 20	18 35	5 00	18 55	4 34	19 21	3 57	19 58	2 52	21 04
151	.0138		2 55	15 36	2 52	15 42	2 47	15 49	2 42	15 59	2 33	16 15
31 WE	11 57 42	+21 56	5 20	18 35	5 00	18 56	4 34	19 22	3 56	20 00	2 50	21 06
152	.0139		3 38	16 36	3 30	16 46	3 20	17 00	3 07	17 18	2 47	17 46

JUNE 2000

6th MONTH　　　　　　　　　　　　　　　　　　　　　　　　**30 DAYS**

All times are Coordinated Universal Time (Greenwich Mean Time)

NOTE: Degrees are North Latitude. For each latitude (20°– 60°), numbers on first line indicate Sun.

Numbers on second line indicate Moon.

Moon Phases: FM = Full Moon: LQ = Last (Waning) Quarter: NM = New Moon, FQ = First (Waxing) Quarter

Sun's distance is in Astronomical Units

CAUTION: Must be converted to local time. For instructions see "Calculation of Rise Times."

Day of month, of week, of year	Sun on Meridian / Moon Phase / h m s	Sun's Decli-nation ° ′ / Distance	20° RISE Sun Moon h m	20° SET Sun Moon h m	30° RISE Sun Moon h m	30° SET Sun Moon h m	40° RISE Sun Moon h m	40° SET Sun Moon h m	50° RISE Sun Moon h m	50° SET Sun Moon h m	60° RISE Sun Moon h m	60° SET Sun Moon h m
1 TH	11 57 51	+22 04	5 20	18 36	4 59	18 57	4 33	19 23	3 56	20 01	2 49	21 08
153		.0141	4 25	17 39	4 13	17 54	3 57	18 13	3 37	18 38	3 04	19 21
2 FR	11 58 00	+22 12	5 20	18 36	4 59	18 57	4 33	19 24	3 55	20 02	2 48	21 10
154	12 14 NM	.0142	5 17	18 45	5 01	19 03	4 40	19 26	4 12	19 58	3 27	20 53
3 SA	11 58 10	+22 19	5 20	18 37	4 59	18 58	4 32	19 24	3 54	20 02	2 46	21 11
155		.0144	6 14	19 51	5 54	20 11	5 30	20 36	4 57	21 12	4 01	22 15
4 SU	11 58 21	+22 26	5 20	18 37	4 59	18 58	4 32	19 25	3 54	20 03	2 45	21 13
156		.0145	7 15	20 54	6 55	21 15	6 29	21 41	5 52	22 17	4 50	23 19
5 MO	11 58 32	+22 33	5 20	18 37	4 59	18 59	4 32	19 26	3 53	20 04	2 44	21 14
157		.0147	8 19	21 54	7 59	22 13	7 34	22 36	6 58	23 09	5 57	none
6 TU	11 58 42	+22 40	5 20	18 38	4 59	18 59	4 32	19 26	3 53	20 05	2 43	21 16
158		.0148	9 23	22 48	9 05	23 03	8 43	23 23	8 12	23 50	7 20	0 03
7 WE	11 58 54	+22 46	5 20	18 38	4 58	19 00	4 31	19 27	3 52	20 06	2 42	21 17
159		.0149	10 26	23 36	10 11	23 48	9 53	none	9 29	none	8 48	0 33
8 TH	11 59 05	+22 51	5 20	18 38	4 58	19 00	4 31	19 27	3 52	20 07	2 41	21 18
160		.0150	11 25	none	11 15	none	11 02	0 03	10 45	0 23	10 17	0 54
9 FR	11 59 17	+22 56	5 20	18 39	4 58	19 00	4 31	19 28	3 51	20 07	2 40	21 20
161	3 29 FQ	.0152	12 22	0 20	12 16	0 28	12 09	0 37	11 59	0 50	11 43	1 10
10 SA	11 59 29	+23 01	5 20	18 39	4 58	19 01	4 31	19 28	3 51	20 08	2 39	21 21
162		.0153	13 16	1 00	13 14	1 04	13 13	1 08	13 10	1 14	13 07	1 22
11 SU	11 59 41	+23 05	5 20	18 39	4 58	19 01	4 31	19 29	3 51	20 09	2 38	21 22
163		.0154	14 08	1 39	14 11	1 38	14 15	1 37	14 20	1 36	14 28	1 34
12 MO	11 59 53	+23 09	5 20	18 40	4 58	19 02	4 31	19 29	3 51	20 09	2 38	21 23
164		.0155	15 00	2 16	15 07	2 11	15 16	2 05	15 28	1 57	15 47	1 45
13 TU	12 00 06	+23 13	5 20	18 40	4 58	19 02	4 31	19 30	3 50	20 10	2 37	21 24
165		.0156	15 51	2 54	16 02	2 45	16 16	2 34	16 35	2 20	17 05	1 57
14 WE	12 00 18	+23 16	5 20	18 40	4 58	19 02	4 31	19 30	3 50	20 11	2 37	21 25
166		.0157	16 43	3 33	16 57	3 21	17 15	3 05	17 40	2 44	18 22	2 11
15 TH	12 00 31	+23 19	5 20	18 41	4 59	19 03	4 31	19 31	3 50	20 11	2 36	21 25
167		.0158	17 34	4 14	17 51	3 58	18 13	3 39	18 44	3 12	19 35	2 28
16 FR	12 00 44	+23 21	5 21	18 41	4 59	19 03	4 31	19 31	3 50	20 11	2 36	21 26
168	22 27 FM	.0159	18 25	4 57	18 45	4 39	19 09	4 16	19 44	3 44	20 43	2 51
17 SA	12 00 57	+23 23	5 21	18 41	4 59	19 03	4 31	19 31	3 50	20 12	2 36	21 26
169		.0159	19 15	5 42	19 36	5 22	20 02	4 57	20 38	4 22	21 42	3 22
18 SU	12 01 10	+23 24	5 21	18 41	4 59	19 03	4 31	19 32	3 50	20 12	2 36	21 27
170		.0160	20 04	6 30	20 24	6 09	20 50	5 44	21 26	5 07	22 29	4 03
19 MO	12 01 23	+23 25	5 21	18 42	4 59	19 04	4 31	19 32	3 50	20 12	2 36	21 27
171		.0161	20 50	7 19	21 10	6 59	21 34	6 34	22 07	5 58	23 04	4 56
20 TU	12 01 36	+23 26	5 21	18 42	4 59	19 04	4 31	19 32	3 51	20 13	2 36	21 28
172		.0162	21 34	8 10	21 51	7 51	22 13	7 28	22 42	6 55	23 30	6 00
21 WE	12 01 49	+23 26	5 22	18 42	4 59	19 04	4 31	19 32	3 51	20 13	2 36	21 28
173		.0162	22 16	9 01	22 30	8 45	22 48	8 25	23 12	7 57	23 50	7 10
22 TH	12 02 02	+23 26	5 22	18 42	5 00	19 04	4 32	19 32	3 51	20 13	2 36	21 28
174		.0163	22 55	9 52	23 06	9 39	23 19	9 23	23 37	9 01	none	8 25
23 FR	12 02 15	+23 26	5 22	18 42	5 00	19 05	4 32	19 33	3 51	20 13	2 36	21 28
175		.0164	23 34	10 43	23 40	10 34	23 49	10 23	none	10 07	0 05	9 42
24 SA	12 02 28	+23 25	5 22	18 43	5 00	19 05	4 32	19 33	3 52	20 13	2 37	21 28
176		.0164	none	11 35	none	11 30	none	11 23	0 00	11 15	0 18	11 01
25 SU	12 02 40	+23 23	5 23	18 43	5 00	19 05	4 32	19 33	3 52	20 13	2 37	21 28
177	1 00 LQ	.0165	0 11	12 27	0 14	12 27	0 18	12 26	0 22	12 24	0 30	12 22
26 MO	12 02 53	+23 21	5 23	18 43	5 01	19 05	4 33	19 33	3 52	20 13	2 38	21 27
178		.0165	0 50	13 22	0 48	13 25	0 47	13 30	0 44	13 36	0 41	13 46
27 TU	12 03 05	+23 19	5 23	18 43	5 01	19 05	4 33	19 33	3 53	20 13	2 39	21 27
179		.0166	1 30	14 19	1 24	14 27	1 17	14 37	1 08	14 51	0 53	15 14
28 WE	12 03 18	+23 16	5 23	18 43	5 01	19 05	4 34	19 33	3 53	20 13	2 40	21 27
180		.0166	2 14	15 19	2 04	15 32	1 51	15 48	1 34	16 09	1 08	16 45
29 TH	12 03 30	+23 13	5 24	18 43	5 02	19 05	4 34	19 33	3 54	20 13	2 40	21 26
181		.0167	3 02	16 22	2 48	16 39	2 30	17 00	2 05	17 29	1 27	18 17
30 FR	12 03 42	+23 10	5 24	18 43	5 02	19 05	4 34	19 33	3 55	20 13	2 41	21 25
182		.0167	3 55	17 28	3 38	17 48	3 15	18 12	2 44	18 47	1 54	19 45

JULY 2000

7th MONTH   **31 DAYS**

All times are Coordinated Universal Time (Greenwich Mean Time)

NOTE: Degrees are North Latitude. For each latitude (20°– 60°), numbers on first line indicate Sun.

Numbers on second line indicate Moon.

Moon Phases: FM = Full Moon: LQ = Last (Waning) Quarter: NM = New Moon, FQ = First (Waxing) Quarter

Sun's distance is in Astronomical Units

CAUTION: Must be converted to local time. For instructions see "Calculation of Rise Times."

Day of month, of week, of year	Sun on Meridian / Moon Phase (h m s)	Sun's Decli-nation ° ' / Distance	20° RISE Sun/Moon h m	20° SET Sun/Moon h m	30° RISE Sun/Moon h m	30° SET Sun/Moon h m	40° RISE Sun/Moon h m	40° SET Sun/Moon h m	50° RISE Sun/Moon h m	50° SET Sun/Moon h m	60° RISE Sun/Moon h m	60° SET Sun/Moon h m
1 SA	12 03 54	+23 06	5 24	18 43	5 03	19 05	4 35	19 33	3 55	20 12	2 42	21 25
183	19 20 NM	.0167	4 55	18 34	4 34	18 55	4 09	19 21	3 34	19 57	2 34	21 00
2 SU	12 04 05	+23 02	5 25	18 44	5 03	19 05	4 35	19 32	3 56	20 12	2 44	21 24
184		.0167	5 58	19 37	5 38	19 57	5 12	20 22	4 35	20 57	3 32	21 56
3 MO	12 04 16	+22 57	5 25	18 44	5 03	19 05	4 36	19 32	3 57	20 12	2 45	21 23
185		.0167	7 04	20 36	6 45	20 53	6 21	21 15	5 48	21 45	4 50	22 33
4 TU	12 04 27	+22 52	5 25	18 44	5 04	19 05	4 37	19 32	3 57	20 11	2 46	21 22
186		.0167	8 10	21 28	7 54	21 42	7 34	21 59	7 06	22 22	6 19	22 59
5 WE	12 04 37	+22 47	5 26	18 44	5 04	19 05	4 37	19 32	3 58	20 11	2 47	21 21
187		.0167	9 13	22 15	9 01	22 25	8 46	22 37	8 25	22 53	7 52	23 17
6 TH	12 04 47	+22 41	5 26	18 44	5 05	19 05	4 38	19 32	3 59	20 10	2 49	21 20
188		.0167	10 13	22 58	10 05	23 04	9 56	23 10	9 43	23 18	9 23	23 31
7 FR	12 04 57	+22 34	5 26	18 44	5 05	19 05	4 38	19 31	4 00	20 10	2 50	21 19
189		.0167	11 09	23 38	11 06	23 39	11 03	23 40	10 57	23 41	10 50	23 43
8 SA	12 05 06	+22 28	5 27	18 43	5 06	19 04	4 39	19 31	4 01	20 09	2 52	21 17
190	12 53 FQ	.0167	12 04	none	12 05	none	12 07	none	12 09	none	12 13	23 54
9 SU	12 05 15	+22 21	5 27	18 43	5 06	19 04	4 40	19 31	4 02	20 08	2 53	21 16
191		.0167	12 56	0 17	13 02	0 13	13 09	0 09	13 19	0 03	13 34	none
10 MO	12 05 24	+22 13	5 27	18 43	5 07	19 04	4 40	19 30	4 03	20 08	2 55	21 15
192		.0166	13 48	0 55	13 57	0 47	14 10	0 38	14 26	0 25	14 53	0 06
11 TU	12 05 32	+22 05	5 28	18 43	5 07	19 04	4 41	19 30	4 04	20 07	2 57	21 13
193		.0166	14 39	1 34	14 52	1 22	15 09	1 08	15 32	0 49	16 10	0 19
12 WE	12 05 39	+21 57	5 28	18 43	5 08	19 03	4 42	19 29	4 05	20 06	2 58	21 12
194		.0166	15 30	2 13	15 47	1 59	16 07	1 40	16 36	1 15	17 24	0 35
13 TH	12 05 46	+21 49	5 28	18 43	5 08	19 03	4 42	19 29	4 06	20 05	3 00	21 10
195		.0165	16 21	2 56	16 40	2 38	17 04	2 16	17 37	1 46	18 34	0 56
14 FR	12 05 53	+21 40	5 29	18 43	5 09	19 03	4 43	19 28	4 07	20 04	3 02	21 08
196		.0165	17 12	3 40	17 32	3 20	17 57	2 56	18 33	2 22	19 36	1 24
15 SA	12 05 59	+21 30	5 29	18 43	5 09	19 02	4 44	19 28	4 08	20 04	3 04	21 06
197		.0164	18 01	4 27	18 21	4 06	18 47	3 41	19 24	3 04	20 27	2 01
16 SU	12 06 04	+21 21	5 30	18 42	5 10	19 02	4 45	19 27	4 09	20 03	3 06	21 05
198	13 55 FM	.0164	18 48	5 16	19 08	4 55	19 33	4 30	20 07	3 53	21 06	2 50
17 MO	12 06 09	+21 11	5 30	18 42	5 10	19 02	4 45	19 26	4 10	20 02	3 08	21 03
199		.0163	19 33	6 06	19 51	5 47	20 13	5 23	20 44	4 49	21 36	3 51
18 TU	12 06 14	+21 00	5 30	18 42	5 11	19 01	4 46	19 26	4 11	20 00	3 10	21 01
200		.0163	20 15	6 57	20 31	6 40	20 49	6 19	21 15	5 49	21 57	5 00
19 WE	12 06 17	+20 50	5 31	18 42	5 11	19 01	4 47	19 25	4 12	19 59	3 12	20 59
201		.0162	20 55	7 48	21 07	7 34	21 22	7 17	21 42	6 53	22 14	6 13
20 TH	12 06 21	+20 39	5 31	18 41	5 12	19 00	4 48	19 24	4 14	19 58	3 14	20 57
202		.0161	21 34	8 39	21 42	8 29	21 52	8 16	22 06	7 58	22 27	7 30
21 FR	12 06 24	+20 27	5 31	18 41	5 13	19 00	4 49	19 24	4 15	19 57	3 16	20 55
203		.0161	22 11	9 30	22 16	9 24	22 21	9 16	22 28	9 05	22 39	8 48
22 SA	12 06 26	+20 15	5 32	18 41	5 13	18 59	4 50	19 23	4 16	19 56	3 18	20 53
204		.0160	22 49	10 22	22 49	10 19	22 49	10 17	22 49	10 13	22 50	10 07
23 SU	12 06 28	+20 03	5 32	18 41	5 14	18 59	4 50	19 22	4 17	19 55	3 21	20 51
205		.0159	23 27	11 14	23 23	11 16	23 18	11 19	23 11	11 22	23 01	11 28
24 MO	12 06 29	+19 51	5 33	18 40	5 14	18 58	4 51	19 21	4 19	19 53	3 23	20 48
206	11 02 LQ	.0158	none	12 08	none	12 15	23 49	12 23	23 35	12 34	23 14	12 51
25 TU	12 06 29	+19 38	5 33	18 40	5 15	18 58	4 52	19 20	4 20	19 52	3 25	20 46
207		.0157	0 08	13 05	0 00	13 16	none	13 30	none	13 48	23 30	14 18
26 WE	12 06 29	+19 25	5 33	18 39	5 16	18 57	4 53	19 19	4 21	19 51	3 27	20 44
208		.0157	0 53	14 05	0 40	14 20	0 24	14 39	0 03	15 05	23 52	15 47
27 TH	12 06 29	+19 11	5 34	18 39	5 16	18 56	4 54	19 18	4 23	19 49	3 30	20 42
209		.0156	1 42	15 08	1 25	15 26	1 05	15 49	0 37	16 21	none	17 16
28 FR	12 06 27	+18 58	5 34	18 39	5 17	18 56	4 55	19 18	4 24	19 48	3 32	20 39
210		.0155	2 36	16 13	2 17	16 33	1 53	16 58	1 20	17 34	0 24	18 36
29 SA	12 06 26	+18 44	5 34	18 38	5 17	18 55	4 56	19 17	4 25	19 47	3 34	20 37
211		.0153	3 37	17 17	3 16	17 37	2 50	18 03	2 14	18 39	1 11	19 41
30 SU	12 06 23	+18 29	5 35	18 38	5 18	18 54	4 57	19 16	4 27	19 45	3 37	20 35
212		.0152	4 42	18 18	4 21	18 36	3 56	19 00	3 20	19 33	2 19	20 27
31 MO	12 06 20	+18 15	5 35	18 37	5 19	18 54	4 58	19 15	4 28	19 44	3 39	20 32
213	2 25 NM	.0151	5 48	19 14	5 30	19 30	5 08	19 49	4 36	20 16	3 44	20 59

AUGUST 2000

8th MONTH   **31 DAYS**

All times are Coordinated Universal Time (Greenwich Mean Time)

NOTE: Degrees are North Latitude. For each latitude (20°– 60°), numbers on first line indicate Sun.

Numbers on second line indicate Moon.

Moon Phases: FM = Full Moon: LQ = Last (Waning) Quarter: NM = New Moon, FQ = First (Waxing) Quarter

Sun's distance is in Astronomical Units

CAUTION: Must be converted to local time. For instructions see "Calculation of Rise Times."

Day of month, of week, of year	Sun on Meridian / Moon Phase (h m s)	Sun's Declination (° ´) / Distance	20° RISE Sun/Moon (h m)	20° SET Sun/Moon (h m)	30° RISE Sun/Moon (h m)	30° SET Sun/Moon (h m)	40° RISE Sun/Moon (h m)	40° SET Sun/Moon (h m)	50° RISE Sun/Moon (h m)	50° SET Sun/Moon (h m)	60° RISE Sun/Moon (h m)	60° SET Sun/Moon (h m)
1 TU	12 06 17	+18 00	5 36	18 37	5 19	18 53	4 59	19 13	4 30	19 42	3 41	20 30
214		.0150	6 54	20 05	6 39	20 16	6 22	20 31	5 57	20 50	5 17	21 20
2 WE	12 06 12	+17 44	5 36	18 36	5 20	18 52	4 59	19 12	4 31	19 40	3 44	20 27
215		.0149	7 57	20 51	7 47	20 58	7 35	21 07	7 18	21 19	6 52	21 37
3 TH	12 06 08	+17 29	5 36	18 36	5 20	18 52	5 00	19 11	4 32	19 39	3 46	20 25
216		.0147	8 56	21 33	8 51	21 36	8 45	21 39	8 37	21 44	8 23	21 50
4 FR	12 06 02	+17 13	5 37	18 35	5 21	18 51	5 01	19 10	4 34	19 37	3 48	20 22
217		.0146	9 53	22 14	9 53	22 12	9 53	22 10	9 52	22 07	9 51	22 02
5 SA	12 05 56	+16 57	5 37	18 35	5 22	18 50	5 02	19 09	4 35	19 36	3 51	20 19
218		.0144	10 48	22 53	10 52	22 47	10 57	22 39	11 04	22 29	11 15	22 14
6 SU	12 05 50	+16 40	5 37	18 34	5 22	18 49	5 03	19 08	4 37	19 34	3 53	20 17
219		.0143	11 41	23 32	11 50	23 22	12 00	23 09	12 14	22 53	12 37	22 26
7 MO	12 05 42	+16 24	5 38	18 34	5 23	18 48	5 04	19 07	4 38	19 32	3 56	20 14
220	1 02 FQ	.0141	12 34	none	12 46	23 58	13 01	23 41	13 22	23 18	13 56	22 41
8 TU	12 05 34	+16 07	5 38	18 33	5 23	18 47	5 05	19 05	4 40	19 31	3 58	20 12
221		.0140	13 25	0 12	13 41	none	14 00	none	14 28	23 47	15 12	23 00
9 WE	12 05 26	+15 50	5 38	18 32	5 24	18 47	5 06	19 04	4 41	19 29	4 00	20 09
222		.0138	14 17	0 54	14 35	0 37	14 58	0 16	15 30	none	16 24	23 25
10 TH	12 05 17	+15 32	5 39	18 32	5 25	18 46	5 07	19 03	4 43	19 27	4 03	20 06
223		.0136	15 07	1 37	15 27	1 18	15 52	0 55	16 28	0 21	17 29	none
11 FR	12 05 07	+15 14	5 39	18 31	5 25	18 45	5 08	19 02	4 44	19 25	4 05	20 03
224		.0135	15 57	2 23	16 18	2 03	16 44	1 38	17 20	1 02	18 24	0 00
12 SA	12 04 57	+14 57	5 39	18 30	5 26	18 44	5 09	19 00	4 46	19 23	4 08	20 01
225		.0133	16 45	3 12	17 05	2 51	17 30	2 25	18 06	1 49	19 07	0 45
13 SU	12 04 46	+14 38	5 39	18 30	5 26	18 43	5 10	18 59	4 47	19 22	4 10	19 58
226		.0131	17 31	4 01	17 49	3 42	18 13	3 17	18 45	2 42	19 39	1 42
14 MO	12 04 35	+14 20	5 40	18 29	5 27	18 42	5 11	18 58	4 48	19 20	4 12	19 55
227		.0129	18 14	4 53	18 30	4 35	18 50	4 13	19 18	3 41	20 03	2 48
15 TU	12 04 23	+14 01	5 40	18 28	5 27	18 41	5 12	18 56	4 50	19 18	4 15	19 52
228	5 13 FM	.0128	18 55	5 44	19 08	5 29	19 24	5 10	19 46	4 44	20 21	4 02
16 WE	12 04 10	+13 42	5 40	18 28	5 28	18 40	5 13	18 55	4 51	19 16	4 17	19 49
229		.0126	19 34	6 36	19 44	6 24	19 55	6 10	20 11	5 50	20 35	5 18
17 TH	12 03 58	+13 23	5 41	18 27	5 29	18 39	5 14	18 54	4 53	19 14	4 20	19 47
230		.0124	20 12	7 27	20 18	7 19	20 24	7 10	20 33	6 57	20 47	6 36
18 FR	12 03 44	+13 04	5 41	18 26	5 29	18 38	5 15	18 52	4 54	19 12	4 22	19 44
231		.0122	20 50	8 18	20 51	8 15	20 53	8 11	20 55	8 05	20 58	7 55
19 SA	12 03 30	+12 45	5 41	18 26	5 30	18 37	5 16	18 51	4 56	19 10	4 24	19 41
232		.0120	21 28	9 11	21 25	9 11	21 21	9 12	21 16	9 14	21 09	9 16
20 SU	12 03 16	+12 25	5 41	18 25	5 30	18 36	5 16	18 49	4 57	19 08	4 27	19 38
233		.0118	22 07	10 04	22 00	10 09	21 51	10 16	21 39	10 24	21 21	10 38
21 MO	12 03 01	+12 05	5 42	18 24	5 31	18 35	5 17	18 48	4 59	19 06	4 29	19 35
234		.0116	22 49	10 59	22 38	11 08	22 24	11 20	22 05	11 37	21 36	12 02
22 TU	12 02 46	+11 45	5 42	18 23	5 31	18 34	5 18	18 46	5 00	19 04	4 32	19 32
235	18 51 LQ	.0114	23 35	11 56	23 20	12 10	23 01	12 27	22 35	12 51	21 54	13 29
23 WE	12 02 30	+11 25	5 42	18 22	5 32	18 32	5 19	18 45	5 02	19 02	4 34	19 29
236		.0112	none	12 56	none	13 13	23 45	13 35	23 13	14 05	22 21	14 55
24 TH	12 02 14	+11 04	5 42	18 22	5 33	18 31	5 20	18 43	5 03	19 00	4 36	19 26
237		.0110	0 26	13 58	0 08	14 18	none	14 42	none	15 17	23 00	16 17
25 FR	12 01 58	+10 44	5 43	18 21	5 33	18 30	5 21	18 42	5 05	18 58	4 39	19 24
238		.0108	1 22	15 00	1 02	15 21	0 36	15 47	0 01	16 24	23 57	17 27
26 SA	12 01 41	+10 23	5 43	18 20	5 34	18 29	5 22	18 40	5 06	18 56	4 41	19 21
239		.0106	2 23	16 01	2 02	16 21	1 36	16 46	1 00	17 21	none	18 20
27 SU	12 01 23	+10 02	5 43	18 19	5 34	18 28	5 23	18 39	5 08	18 54	4 44	19 18
240		.0104	3 27	16 58	3 08	17 16	2 44	17 37	2 10	18 07	1 12	18 57
28 MO	12 01 06	+9 41	5 43	18 18	5 35	18 27	5 24	18 37	5 09	18 52	4 46	19 15
241		.0101	4 32	17 51	4 16	18 05	3 56	18 22	3 28	18 45	2 41	19 22
29 TU	12 00 48	+9 20	5 44	18 18	5 35	18 26	5 25	18 36	5 11	18 50	4 48	19 12
242	10 19 NM	.0099	5 37	18 40	5 24	18 49	5 09	19 01	4 49	19 16	4 16	19 40
30 WE	12 00 30	+8 58	5 44	18 17	5 36	18 25	5 26	18 34	5 12	18 48	4 51	19 09
243		.0097	6 38	19 24	6 31	19 29	6 22	19 35	6 09	19 43	5 50	19 55
31 TH	12 00 11	+8 37	5 44	18 16	5 37	18 23	5 27	18 33	5 14	18 46	4 53	19 06
244		.0094	7 38	20 06	7 35	20 07	7 32	20 77	7 28	20 07	7 21	20 07

SEPTEMBER 2000

9th MONTH　　　　　　　　　　　　　　　　　　　　　　　　**30 DAYS**

All times are Coordinated Universal Time (Greenwich Mean Time)

NOTE: Degrees are North Latitude. For each latitude (20°– 60°), numbers on first line indicate Sun.
Numbers on second line indicate Moon.

Moon Phases: FM = Full Moon: LQ = Last (Waning) Quarter: NM = New Moon, FQ = First (Waxing) Quarter

Sun's distance is in Astronomical Units

CAUTION: Must be converted to local time. For instructions see "Calculation of Rise Times."

Day of month, of week, of year	Sun on Meridian / Moon Phase (h m s)	Sun's Declination ° ' / Distance	20° RISE Sun/Moon (h m)	20° SET Sun/Moon (h m)	30° RISE Sun/Moon (h m)	30° SET Sun/Moon (h m)	40° RISE Sun/Moon (h m)	40° SET Sun/Moon (h m)	50° RISE Sun/Moon (h m)	50° SET Sun/Moon (h m)	60° RISE Sun/Moon (h m)	60° SET Sun/Moon (h m)
1 FR	11 59 52	+8 15	5 44	18 15	5 37	18 22	5 28	18 31	5 15	18 43	455	19 03
245		.0092	8 35	20 47	8 37	20 43	8 40	20 37	8 43	20 30	8 49	20 19
2 SA	11 59 33	+7 53	5 45	18 14	5 38	18 21	5 29	18 30	5 17	18 41	4 58	19 00
246		.0090	9 30	21 27	9 37	21 18	9 45	21 08	9 57	20 54	10 14	20 32
3 SU	11 59 13	+7 31	5 45	18 13	5 38	18 20	5 30	18 28	5 18	18 39	5 00	18 57
247		.0087	10 24	22 07	10 35	21 55	10 49	21 40	11 07	21 19	11 37	20 46
4 MO	11 58 54	+7 09	5 45	18 12	5 39	18 19	5 31	18 26	5 20	18 37	5 02	18 54
248		.0085	11 17	22 49	11 32	22 33	11 50	22 14	12 15	21 47	12 56	21 03
5 TU	11 58 34	+6 47	5 45	18 12	5 39	18 17	5 32	18 25	5 21	18 35	5 05	18 51
249	16 27 FQ	.0082	12 10	23 33	12 27	23 14	12 49	22 51	13 20	22 19	14 12	21 26
6 WE	11 58 13	+6 24	5 46	18 11	5 40	18 16	5 33	18 23	5 23	18 33	5 07	18 48
250		.0080	13 01	none	13 21	23 58	13 45	23 33	14 20	22 57	15 20	21 57
7 TH	11 57 53	+6 02	5 46	18 10	5 40	18 15	5 34	18 22	5 24	18 31	5 10	18 45
251		.0077	13 52	0 18	14 12	none	14 38	none	15 15	23 42	16 19	22 38
8 FR	11 57 32	+5 40	5 46	18 09	5 41	18 14	5 34	18 20	5 26	18 28	5 12	18 42
252		.0074	14 40	1 06	15 01	0 45	15 27	0 19	16 03	none	17 06	23 31
9 SA	11 57 11	+5 17	5 46	18 08	5 41	18 13	5 35	18 18	5 27	18 26	5 14	18 39
253		.0072	15 27	1 55	15 46	1 35	16 10	1 10	16 44	0 33	17 42	none
10 SU	11 56 50	+4 54	5 46	18 07	5 42	18 11	5 36	18 17	5 29	18 24	5 17	18 36
254		.0069	16 11	2 46	16 28	2 27	16 50	2 04	17 19	1 31	18 08	0 35
11 MO	11 56 29	+4 31	5 47	18 06	5 42	18 10	5 37	18 15	5 30	18 22	5 19	18 33
255		.0066	16 53	3 38	17 07	3 21	17 25	3 01	17 49	2 33	18 28	1 46
12 TU	11 56 08	+4 09	5 47	18 05	5 43	18 09	5 38	18 13	5 32	18 20	5 21	18 30
256		.0064	17 33	4 29	17 44	4 16	17 57	4 00	18 15	3 38	18 43	3 03
13 WE	11 55 46	+3 46	5 47	18 04	5 44	18 08	5 39	18 12	5 33	18 17	5 24	18 26
257	19 37 FM	.0061	18 12	5 21	18 18	5 12	18 27	5 01	18 38	4 46	18 55	4 21
14 TH	11 55 25	+3 23	5 47	18 03	5 44	18 06	5 40	18 10	5 35	18 15	5 26	18 23
258		.0058	18 50	6 13	18 52	6 09	18 55	6 03	19 00	5 54	19 07	5 41
15 FR	11 55 04	+3 00	5 47	18 02	5 45	18 05	5 41	18 08	5 36	18 13	5 28	18 20
259		.0056	19 28	7 06	19 26	7 06	19 24	7 05	19 21	7 04	19 17	7 03
16 SA	11 54 42	+2 37	5 48	18 02	5 45	18 04	5 42	18 07	5 38	18 11	5 31	18 17
260		.0053	20 07	8 00	20 01	8 04	19 53	8 09	19 44	8 15	19 29	8 25
17 SU	11 54 21	+2 13	5 48	18 01	5 46	18 03	5 43	18 05	5 39	18 09	5 33	18 14
261		.0050	20 48	8 55	20 38	9 03	20 25	9 14	20 08	9 28	19 42	9 50
18 MO	11 53 59	+1 50	5 48	18 00	5 46	18 01	5 44	18 03	5 41	18 06	5 35	18 11
262		.0048	21 33	9 52	21 19	10 04	21 01	10 20	20 37	10 42	19 59	11 17
19 TU	11 53 38	+1 27	5 48	17 59	5 47	18 00	5 45	18 02	5 42	18 04	5 38	18 08
263		.0045	22 22	10 51	22 04	11 07	21 42	11 27	21 12	11 56	20 22	12 43
20 WE	11 53 16	+1 04	5 48	17 58	5 47	17 59	5 46	18 00	5 44	18 02	5 40	18 05
264		.0042	23 15	11 51	22 55	12 10	22 30	12 34	21 55	13 08	20 55	14 06
21 TH	11 52 55	+0 40	5 49	17 57	5 48	17 58	5 47	17 58	5 45	18 00	5 42	18 02
265	1 28 LQ	.0040	none	12 52	23 52	13 12	23 26	13 39	22 49	14 15	21 44	15 19
22 FR	11 52 34	+0 17	5 49	17 56	5 48	17 56	5 48	17 57	5 47	17 58	5 45	17 59
266		.0037	0 13	13 51	none	14 12	none	14 38	23 53	15 14	22 51	16 17
23 SA	11 52 13	−0 06	5 49	17 55	5 49	17 55	5 49	17 55	5 48	17 55	5 47	17 56
267		.0034	1 14	14 48	0 54	15 07	0 29	15 30	none	16 03	none	16 57
24 SU	11 51 52	−0 30	5 49	17 54	5 49	17 54	5 50	17 53	5 50	17 53	5 49	17 53
268		.0031	2 17	15 41	1 59	15 57	1 37	16 16	1 06	16 43	0 14	17 25
25 MO	11 51 32	−0 53	5 50	17 53	5 50	17 53	5 51	17 52	5 51	17 51	5 52	17 50
269		.0029	3 20	16 30	3 06	16 42	2 48	16 56	2 24	17 15	1 45	17 45
26 TU	11 51 11	−1 16	5 50	17 52	5 51	17 51	5 52	17 50	5 53	17 49	5 54	17 47
270		.0026	4 22	17 15	4 12	17 22	4 00	17 31	3 44	17 43	3 18	18 00
27 WE	11 50 51	−1 40	5 50	17 51	5 51	17 50	5 53	17 49	5 54	17 47	5 56	17 44
271	19 53 NM	.0023	5 21	17 58	5 17	18 00	5 10	18 03	5 02	18 07	4 50	18 13
28 TH	11 50 31	−2 03	5 50	17 51	5 52	17 49	5 53	17 47	5 56	17 44	5 59	17 41
272		.0020	6 19	18 39	6 19	18 37	6 19	18 34	6 19	18 30	6 19	18 25
29 FR	11 50 11	−2 27	5 50	17 50	5 52	17 48	5 54	17 45	5 57	17 42	6 01	17 38
273		.0017	7 16	19 20	7 21	19 13	7 26	19 05	7 34	18 53	7 47	18 36
30 SA	11 49 52	−2 50	5 51	17 49	5 53	17 46	5 55	17 44	5 59	17 40	6 04	17 35
274		.0014	8 11	20 00	8 21	19 49	8 32	19 36	8 47	19 18	9 12	18 49

OCTOBER 2000

10th MONTH   **31 DAYS**

All times are Coordinated Universal Time (Greenwich Mean Time)

NOTE: Degrees are North Latitude. For each latitude (20°– 60°), numbers on first line indicate Sun.
Numbers on second line indicate Moon.

Moon Phases: FM = Full Moon: LQ = Last (Waning) Quarter: NM = New Moon, FQ = First (Waxing) Quarter

Sun's distance is in Astronomical Units

CAUTION: Must be converted to local time. For instructions see "Calculation of Rise Times."

Day of month, of week, of year	Sun on Meridian / Moon Phase (h m s)	Sun's Declination / Distance (° ′)	20° RISE Sun/Moon (h m)	20° SET Sun/Moon (h m)	30° RISE Sun/Moon (h m)	30° SET Sun/Moon (h m)	40° RISE Sun/Moon (h m)	40° SET Sun/Moon (h m)	50° RISE Sun/Moon (h m)	50° SET Sun/Moon (h m)	60° RISE Sun/Moon (h m)	60° SET Sun/Moon (h m)
1 SU	11 49 32	−3 13	5 51	17 48	5 53	17 45	5 56	17 42	6 00	17 38	6 06	17 32
275		.0012	9 06	20 42	9 19	20 28	9 35	20 10	9 58	19 45	10 35	19 05
2 MO	11 49 13	−3 36	5 51	17 47	5 54	17 44	5 57	17 40	6 02	17 36	6 08	17 29
276		.0009	10 00	21 26	10 16	21 08	10 37	20 46	11 06	20 16	11 54	19 25
3 TU	11 48 55	−4 00	5 51	17 46	5 55	17 43	5 58	17 39	6 03	17 34	6 11	17 26
277		.0006	10 53	22 11	11 12	21 51	11 36	21 26	12 10	20 52	13 08	19 52
4 WE	11 48 36	−4 23	5 52	17 45	5 55	17 42	5 59	17 37	6 05	17 31	6 13	17 23
278		.0003	11 44	22 58	12 05	22 37	12 31	22 11	13 08	21 34	14 12	20 29
5 TH	11 48 18	−4 46	5 52	17 44	5 56	17 40	6 00	17 36	6 06	17 29	6 16	17 20
279	10 59 FQ	.0000	12 34	23 47	12 55	23 27	13 21	23 00	13 59	22 23	15 04	21 18
6 FR	11 48 01	−5 09	5 52	17 44	5 56	17 39	6 01	17 34	6 08	17 27	6 18	17 17
280		.9997	13 21	none	13 42	none	14 07	23 53	14 43	23 18	15 44	22 18
7 SA	11 47 43	−5 32	5 52	17 43	5 57	17 38	6 02	17 32	6 10	17 25	6 20	17 14
281		.9994	14 06	0 38	14 25	0 18	14 48	none	15 20	none	16 13	23 27
8 SU	11 47 26	−5 55	5 53	17 42	5 58	17 37	6 03	17 31	6 11	17 23	6 23	17 11
282		.9991	14 49	1 29	15 05	1 11	15 24	0 50	15 51	0 19	16 34	none
9 MO	11 47 10	−6 18	5 53	17 41	5 58	17 36	6 04	17 29	6 13	17 21	6 25	17 08
283		.9988	15 29	2 20	15 42	2 06	15 57	1 48	16 18	1 23	16 50	0 42
10 TU	11 46 54	−6 40	5 53	17 40	5 59	17 35	6 05	17 28	6 14	17 19	6 28	17 05
284		.9985	16 08	3 12	16 17	3 01	16 27	2 48	16 42	2 30	17 03	2 00
11 WE	11 46 38	−7 03	5 54	17 39	5 59	17 33	6 06	17 26	6 16	17 17	6 30	17 02
285		.9982	16 47	4 04	16 51	3 58	16 56	3 50	17 04	3 38	17 15	3 20
12 TH	11 46 23	−7 26	5 54	17 39	6 00	17 32	6 07	17 25	6 17	17 15	6 33	16 59
286		.9979	17 25	4 57	17 25	4 55	17 25	4 52	17 25	4 48	17 25	4 42
13 FR	11 46 09	−7 48	5 54	17 38	6 01	17 31	6 09	17 23	6 19	17 12	6 35	16 56
287	8 53 FM	.9977	18 04	5 51	18 00	5 54	17 54	5 56	17 47	6 00	17 36	6 06
14 SA	11 45 55	−8 10	5 55	17 37	6 01	17 30	6 10	17 22	6 21	17 10	6 37	16 53
288		.9974	18 46	6 47	18 37	6 54	18 26	7 02	18 11	7 14	17 48	7 32
15 SU	11 45 41	−8 33	5 55	17 36	6 02	17 29	6 11	17 20	6 22	17 08	6 40	16 50
289		.9971	19 30	7 44	19 17	7 56	19 00	8 10	18 38	8 30	18 04	9 01
16 MO	11 45 28	−8 55	5 55	17 36	6 03	17 28	6 12	17 19	6 24	17 06	6 42	16 47
290		.9968	20 18	8 44	20 01	9 00	19 40	9 19	19 11	9 46	18 24	10 30
17 TU	11 45 16	−9 17	5 55	17 35	6 03	17 27	6 13	17 17	6 25	17 04	6 45	16 45
291		.9965	21 11	9 45	20 51	10 04	20 26	10 28	19 52	11 01	18 54	11 57
18 WE	11 45 05	−9 39	5 56	17 34	6 04	17 26	6 14	17 16	6 27	17 02	6 47	16 42
292		.9962	22 07	10 47	21 46	11 08	21 20	11 34	20 43	12 11	19 38	13 15
19 TH	11 44 54	−10 00	5 56	17 33	6 05	17 25	6 15	17 14	6 29	17 00	6 50	16 39
293		.9960	23 08	11 47	22 47	12 08	22 21	12 34	21 44	13 12	20 39	14 17
20 FR	11 44 43	−10 22	5 57	17 33	6 05	17 24	6 16	17 13	6 30	16 58	6 52	16 36
294	7 59 LQ	.9957	none	12 44	23 50	13 04	23 27	13 28	22 53	14 03	21 57	15 01
21 SA	11 44 34	−10 43	5 57	17 32	6 06	17 23	6 17	17 12	6 32	16 56	6 55	16 33
295		.9954	0 09	13 37	none	13 54	none	14 15	none	14 44	23 24	15 32
22 SU	11 44 25	−11 05	5 57	17 31	6 07	17 22	6 18	17 10	6 33	16 55	6 57	16 30
296		.9952	1 11	14 26	0 55	14 39	0 36	14 56	0 09	15 18	none	15 53
23 MO	11 44 16	−11 26	5 58	17 31	6 07	17 21	6 19	17 09	6 35	16 53	7 00	16 28
297		.9949	2 11	15 11	2 00	15 20	1 46	15 31	1 26	15 46	0 55	16 08
24 TU	11 44 09	−11 47	5 58	17 30	6 08	17 20	6 20	17 07	6 37	16 51	7 02	16 25
298		.9946	3 10	15 53	3 03	15 58	2 55	16 03	2 43	16 10	2 25	16 21
25 WE	11 44 02	−12 08	5 58	17 29	6 09	17 19	6 21	17 06	6 38	16 49	7 05	16 22
299		.9943	4 07	16 34	4 05	16 34	4 03	16 33	3 59	16 33	3 54	16 32
26 TH	11 43 56	−12 28	5 59	17 29	6 10	17 18	6 23	17 05	6 40	16 47	7 07	16 19
300		.9941	5 03	17 14	5 06	17 09	5 09	17 03	5 14	16 55	5 21	16 43
27 FR	11 43 50	−12 48	5 59	17 28	6 10	17 17	6 24	17 04	6 42	16 45	7 10	16 17
301	7 58 NM	.9938	5 59	17 54	6 06	17 45	6 15	17 33	6 27	17 18	6 47	16 55
28 SA	11 43 46	−13 09	6 00	17 28	6 11	17 16	6 25	17 02	6 43	16 43	7 13	16 14
302		.9935	6 54	18 35	7 05	18 22	7 20	18 06	7 39	17 44	8 11	17 08
29 SU	11 43 42	−13 29	6 00	17 27	6 12	17 15	6 26	17 01	6 45	16 42	7 15	16 11
303		.9933	7 48	19 18	8 04	19 01	8 23	18 41	8 49	18 12	9 33	17 26
30 MO	11 43 39	−13 48	6 01	17 26	6 12	17 14	6 27	17 00	6 47	16 40	7 18	16 09
304		.9930	8 42	20 03	9 01	19 44	9 23	19 20	9 56	18 46	10 51	17 49
31 TU	11 43 36	−14 08	6 01	17 26	6 13	17 14	6 28	16 59	6 48	16 38	7 20	16 06
305		.9927	9 35	20 50	9 55	20 29	10 21	20 03	10 57	19 26	12 01	18 22

NOVEMBER 2000

11th MONTH   **30 DAYS**

All times are Coordinated Universal Time (Greenwich Mean Time)

NOTE: Degrees are North Latitude. For each latitude (20°– 60°), numbers on first line indicate Sun.

Numbers on second line indicate Moon.

Moon Phases: FM = Full Moon: LQ = Last (Waning) Quarter: NM = New Moon, FQ = First (Waxing) Quarter

Sun's distance is in Astronomical Units

CAUTION: Must be converted to local time. For instructions see "Calculation of Rise Times."

Day of month, of week, of year	Sun on Meridian / Moon Phase (h m s)	Sun's Declination (° ′) / Distance	20° RISE Sun/Moon	20° SET Sun/Moon	30° RISE Sun/Moon	30° SET Sun/Moon	40° RISE Sun/Moon	40° SET Sun/Moon	50° RISE Sun/Moon	50° SET Sun/Moon	60° RISE Sun/Moon	60° SET Sun/Moon
1 WE	11 43 35	−14 27	6 02	17 25	6 14	17 13	6 29	16 57	6 50	16 36	7 23	16 03
306		.9925	10 26	21 39	10 47	21 17	11 14	20 51	11 52	20 13	12 59	19 05
2 TH	11 43 34	−14 46	6 02	17 25	6 15	17 12	6 30	16 56	6 52	16 35	7 25	16 01
307		.9922	11 15	22 29	11 36	22 08	12 02	21 42	12 39	21 06	13 44	20 01
3 FR	11 43 34	−15 05	6 03	17 24	6 16	17 11	6 32	16 55	6 53	16 33	7 28	15 58
308		.9920	12 01	23 20	12 20	23 01	12 45	22 37	13 19	22 04	14 17	21 07
4 SA	11 43 35	−15 24	6 03	17 24	6 16	17 10	6 33	16 54	6 55	16 31	7 31	15 56
309	7 27 FQ	.9917	12 44	none	13 01	23 55	13 23	23 35	13 52	23 07	14 41	22 20
5 SU	11 43 36	−15 42	6 04	17 23	6 17	17 10	6 34	16 53	6 57	16 30	7 33	15 53
310		.9914	13 25	0 11	13 39	none	13 57	none	14 20	none	14 58	23 37
6 MO	11 43 39	−16 00	6 04	17 23	6 18	17 09	6 35	16 52	6 58	16 28	7 36	15 51
311		.9912	14 04	1 02	14 15	0 49	14 27	0 33	14 45	0 12	15 12	none
7 TU	11 43 42	−16 18	6 05	17 23	6 19	17 08	6 36	16 51	7 00	16 27	7 38	15 48
312		.9909	14 42	1 53	14 48	1 44	14 56	1 34	15 07	1 19	15 23	0 55
8 WE	11 43 46	−16 36	6 05	17 22	6 20	17 08	6 37	16 50	7 02	16 25	7 41	15 46
313		.9907	15 20	2 45	15 22	2 41	15 24	2 35	15 28	2 28	15 33	2 16
9 TH	11 43 51	−16 53	6 06	17 22	6 20	17 07	6 38	16 49	7 03	16 24	7 43	15 44
314		.9904	15 58	3 38	15 56	3 38	15 53	3 39	15 49	3 39	15 44	3 39
10 FR	11 43 57	−17 10	6 06	17 22	6 21	17 06	6 40	16 48	7 05	16 22	7 46	15 41
315		.9902	16 39	4 33	16 32	4 38	16 23	4 44	16 12	4 52	15 55	5 04
11 SA	11 44 04	−17 27	6 07	17 21	6 22	17 06	6 41	16 47	7 07	16 21	7 48	15 39
316	21 15 FM	.9900	17 22	5 31	17 11	5 40	16 57	5 52	16 38	6 08	16 08	6 33
12 SU	11 44 11	−17 43	6 07	17 21	6 23	17 05	6 42	16 46	7 08	16 20	7 51	15 37
317		.9897	18 10	6 31	17 54	6 45	17 35	7 02	17 09	7 26	16 26	8 05
13 MO	11 44 20	−17 59	6 08	17 21	6 24	17 05	6 43	16 45	7 10	16 18	7 54	15 34
318		.9895	19 02	7 33	18 43	7 51	18 20	8 13	17 47	8 45	16 52	9 37
14 TU	11 44 29	−18 15	6 08	17 20	6 24	17 04	6 44	16 44	7 12	16 17	7 56	15 32
319		.9893	19 59	8 37	19 38	8 58	19 12	9 23	18 35	9 59	17 31	11 02
15 WE	11 44 39	−18 30	6 09	17 20	6 25	17 04	6 45	16 44	7 13	16 16	7 59	15 30
320		.9891	21 00	9 40	20 39	10 01	20 12	10 28	19 34	11 06	18 27	12 13
16 TH	11 44 50	−18 45	6 10	17 20	6 26	17 03	6 46	16 43	7 15	16 14	8 01	15 28
321		.9889	22 03	10 40	21 43	11 00	21 18	11 26	20 43	12 02	19 42	13 05
17 FR	11 45 01	−19 00	6 10	17 20	6 27	17 03	6 48	16 42	7 16	16 13	8 04	15 26
322		.9886	23 05	11 35	22 48	11 53	22 27	12 16	21 57	12 47	21 08	13 39
18 SA	11 45 14	−19 14	6 11	17 20	6 28	17 02	6 49	16 41	7 18	16 12	8 06	15 24
323	15 24 LQ	.9884	none	12 25	23 53	12 40	23 37	12 58	23 14	13 23	22 38	14 02
19 SU	11 45 27	−19 29	6 11	17 19	6 29	17 02	6 50	16 41	7 20	16 11	8 08	15 22
324		.9882	0 06	13 11	none	13 21	none	13 34	none	13 52	none	14 19
20 MO	11 45 42	−19 42	6 12	17 19	6 29	17 02	6 51	16 40	7 21	16 10	8 11	15 20
325		.9880	1 05	13 53	0 56	13 59	0 46	14 07	0 31	14 16	0 08	14 31
21 TU	11 45 57	−19 56	6 13	17 19	6 30	17 01	6 52	16 39	7 23	16 09	8 13	15 18
326		.9878	2 01	14 33	1 57	14 35	1 53	14 36	1 46	14 39	1 36	14 42
22 WE	11 46 13	−20 09	6 13	17 19	6 31	17 01	6 53	16 39	7 24	16 08	8 16	15 16
327		.9877	2 56	15 12	2 57	15 09	2 58	15 05	3 00	15 00	3 02	14 52
23 TH	11 46 29	−20 21	6 14	17 19	6 32	17 01	6 54	16 38	7 26	16 07	8 18	15 14
328		.9875	3 51	15 51	3 56	15 44	4 03	15 34	4 12	15 22	4 27	15 03
24 FR	11 46 47	−20 34	6 14	17 19	6 33	17 01	6 55	16 38	7 27	16 06	8 20	15 13
329		.9873	4 45	16 31	4 55	16 19	5 07	16 05	5 24	15 46	5 51	15 15
25 SA	11 47 05	−20 46	6 15	17 19	6 34	17 00	6 57	16 37	7 29	16 05	8 23	15 11
330	23 11 NM	.9871	5 39	17 13	5 53	16 57	6 10	16 38	6 34	16 12	7 13	15 30
26 SU	11 47 24	−20 57	6 16	17 19	6 34	17 00	6 58	16 37	7 30	16 04	8 25	15 09
331		.9869	6 33	17 56	6 50	17 38	7 12	17 15	7 42	16 43	8 33	15 50
27 MO	11 47 44	−21 08	6 16	17 19	6 35	17 00	6 59	16 36	7 32	16 04	8 27	15 08
332		.9867	7 26	18 43	7 46	18 22	8 11	17 57	8 46	17 20	9 47	16 18
28 TU	11 48 04	−21 19	6 17	17 19	6 36	17 00	7 00	16 36	7 33	16 03	8 29	15 06
333		.9866	8 18	19 31	8 39	19 10	9 06	18 43	9 44	18 04	10 51	16 57
29 WE	11 48 26	−21 29	6 18	17 19	6 37	17 00	7 01	16 36	7 34	16 02	8 31	15 05
334		.9864	9 08	20 21	9 30	20 00	9 57	19 33	10 35	18 55	11 42	17 48
30 TH	11 48 47	−21 39	6 18	17 19	6 38	17 00	7 02	16 36	7 36	16 02	8 33	15 04
335		.9862	9 56	21 11	10 16	20 52	10 42	20 27	11 18	19 51	12 20	18 50

DECEMBER 2000

12th MONTH   **31 DAYS**

All times are Coordinated Universal Time (Greenwich Mean Time)

NOTE: Degrees are North Latitude. For each latitude (20°– 60°), numbers on first line indicate Sun.

Numbers on second line indicate Moon.

Moon Phases: FM = Full Moon: LQ = Last (Waning) Quarte-r: NM = New Moon, FQ = First (Waxing) Quarter

Sun's distance is in Astronomical Units

CAUTION: Must be converted to local time. For instructions see "Calculation of Rise Times."

Day of month, of week, of year	Sun on Meridian / Moon Phase (h m s) / Distance	Sun's Decli–nation (° ') / Distance	20° RISE Sun/Moon	20° SET Sun/Moon	30° RISE Sun/Moon	30° SET Sun/Moon	40° RISE Sun/Moon	40° SET Sun/Moon	50° RISE Sun/Moon	50° SET Sun/Moon	60° RISE Sun/Moon	60° SET Sun/Moon
1 FR	11 49 10	−21 49	6 19	17 19	6 38	17 00	7 03	16 35	7 37	16 01	8 35	15 02
336		.9860	10 40	22 02	10 58	21 45	11 21	21 23	11 53	20 52	12 47	20 01
2 SA	11 49 33	−21 58	6 19	17 20	6 39	17 00	7 04	16 35	7 38	16 00	8 37	15 01
337		.9859	11 22	22 53	11 37	22 39	11 56	22 21	12 23	21 56	13 06	21 16
3 SU	11 49 57	−22 07	6 20	17 20	6 40	17 00	7 05	16 35	7 40	16 00	8 39	15 00
338		.9857	12 01	23 43	12 13	23 33	12 28	23 19	12 48	23 01	13 21	22 32
4 MO	11 50 21	−22 15	6 21	17 20	6 41	17 00	7 06	16 35	7 41	16 00	8 41	14 59
339	3 55 FQ	.9856	12 38	none	12 47	none	12 57	none	13 11	none	13 32	23 51
5 TU	11 50 46	−22 23	6 21	17 20	6 41	17 00	7 07	16 35	7 42	15 59	8 43	14 58
340		.9854	13 15	0 34	13 19	0 27	13 24	0 19	13 31	0 08	13 42	none
6 WE	11 51 11	−22 30	6 22	17 20	6 42	17 00	7 08	16 35	7 43	15 59	8 45	14 57
341		.9853	13 52	1 25	13 52	1 23	13 52	1 20	13 52	1 17	13 52	1 11
7 TH	11 51 37	−22 37	6 23	17 21	6 43	17 00	7 08	16 35	7 44	15 59	8 47	14 56
342		.9851	14 30	2 18	14 26	2 20	14 20	2 23	14 13	2 27	14 02	2 33
8 FR	11 52 03	−22 43	6 23	17 21	6 44	17 00	7 09	16 35	7 45	15 58	8 48	14 56
343		.9850	15 12	3 13	15 03	3 20	14 51	3 29	14 37	3 41	14 14	3 59
9 SA	11 52 30	−22 49	6 24	17 21	6 44	17 00	7 10	16 35	7 47	15 58	8 50	14 55
344		.9849	15 57	4 12	15 43	4 23	15 27	4 38	15 04	4 58	14 29	5 29
10 SU	11 52 57	−22 55	6 24	17 21	6 45	17 01	7 11	16 35	7 48	15 58	8 51	14 54
345		.9847	16 47	5 13	16 30	5 29	16 08	5 49	15 38	6 17	14 50	7 03
11 MO	11 53 24	−23 00	6 25	17 22	6 46	17 01	7 12	16 35	7 49	15 58	8 53	14 54
346	9 03 FM	.9846	17 43	6 18	17 23	6 37	16 58	7 01	16 22	7 36	15 22	8 34
12 TU	11 53 52	−23 05	6 26	17 22	6 46	17 01	7 13	16 35	7 50	15 58	8 54	14 54
347		.9845	18 45	7 23	18 23	7 45	17 56	8 11	17 18	8 49	16 10	9 56
13 WE	11 54 20	−23 09	6 26	17 22	6 47	17 01	7 13	16 35	7 50	15 58	8 55	14 53
348		.9844	19 49	8 27	19 28	8 48	19 02	9 15	18 25	9 53	17 20	10 59
14 TH	11 54 49	−23 13	6 27	17 23	6 48	17 02	7 14	16 35	7 51	15 58	8 56	14 53
349		.9843	20 54	9 27	20 36	9 46	20 13	10 10	19 41	10 44	18 46	11 41
15 FR	11 55 18	−23 16	6 27	17 23	6 48	17 02	7 15	16 36	7 52	15 58	8 57	14 53
350		.9842	21 58	10 21	21 43	10 37	21 25	10 57	21 00	11 25	20 19	12 09
16 SA	11 55 47	−23 19	6 28	17 24	6 49	17 02	7 15	16 36	7 53	15 59	8 58	14 53
351		.9841	22 59	11 09	22 49	11 22	22 36	11 36	22 19	11 56	21 52	12 28
17 SU	11 56 16	−23 21	6 28	17 24	6 50	17 03	7 16	16 36	7 54	15 59	8 59	14 53
352		.9840	23 57	11 54	23 52	12 01	23 45	12 10	23 36	12 23	23 22	12 42
18 MO	11 56 46	−23 23	6 29	17 25	6 50	17 03	7 17	16 37	7 54	15 59	9 00	14 53
353	0 41 LQ	.9840	none	12 34	none	12 37	none	12 41	none	12 46	none	12 53
19 TU	11 57 15	−23 25	6 30	17 25	6 51	17 04	7 17	16 37	7 55	16 00	9 01	14 54
354		.9839	0 53	13 13	0 52	13 12	0 51	13 10	0 50	13 07	0 49	13 03
20 WE	11 57 45	−23 26	6 30	17 25	6 51	17 04	7 18	16 38	7 56	16 00	9 02	14 54
355		.9838	1 47	13 52	1 51	13 46	1 56	13 38	2 03	13 28	2 13	13 13
21 TH	11 58 15	−23 26	6 31	17 26	6 52	17 05	7 18	16 38	7 56	16 00	9 02	14 54
356		.9838	2 40	14 31	2 49	14 20	2 59	14 08	3 13	13 51	3 36	13 24
22 FR	11 58 45	−23 26	6 31	17 26	6 52	17 05	7 19	16 39	7 57	16 01	9 03	14 55
357		.9837	3 33	15 11	3 46	14 57	4 01	14 40	4 23	14 16	4 58	13 38
23 SA	11 59 15	−23 26	6 32	17 27	6 53	17 06	7 19	16 39	7 57	16 02	9 03	14 56
358		.9836	4 27	15 53	4 43	15 36	5 03	15 14	5 31	14 45	6 18	13 55
24 SU	11 59 45	−23 25	6 32	17 28	6 53	17 06	7 20	16 40	7 57	16 02	9 03	14 56
359		.9836	5 20	16 38	5 39	16 18	6 02	15 54	6 36	15 19	7 34	14 20
25 MO	12 00 15	−23 24	6 32	17 28	6 54	17 07	7 20	16 40	7 58	16 03	9 03	14 57
360	17 22 NM	.9835	6 12	17 25	6 33	17 04	6 59	16 38	7 36	16 00	8 42	14 53
26 TU	12 00 45	−23 22	6 33	17 29	6 54	17 07	7 21	16 41	7 58	16 04	9 04	14 58
361		.9835	7 03	18 15	7 24	17 53	7 51	17 26	8 30	16 48	9 38	15 40
27 WE	12 01 14	−23 19	6 33	17 29	6 54	17 08	7 21	16 42	7 58	16 04	9 04	14 59
362		.9835	7 51	19 05	8 12	18 45	8 39	18 19	9 16	17 42	10 21	16 38
28 TH	12 01 44	−23 17	6 34	17 30	6 55	17 09	7 21	16 42	7 58	16 05	9 03	15 00
363		.9834	8 37	19 56	8 56	19 38	9 20	19 14	9 54	18 42	10 51	17 46
29 FR	12 02 13	−23 13	6 34	17 30	6 55	17 09	7 21	16 43	7 58	16 06	9 03	15 01
364		.9834	9 19	20 47	9 36	20 31	9 57	20 12	10 26	19 45	11 13	19 00
30 SA	12 02 42	−23 10	6 34	17 31	6 55	17 10	7 22	16 44	7 59	16 07	9 03	15 03
365		.9834	9 59	21 37	10 13	21 25	10 29	21 10	10 52	20 49	11 29	20 15
31 SU	12 03 11	−23 05	6 35	17 32	6 56	17 11	7 22	16 45	7 59	16 08	9 03	15 04
366		.9833	10 37	22 27	10 47	22 18	10 59	22 08	11 15	21 54	11 41	21 32

Perpetual Calendar

The number shown for each year indicates which Gregorian calendar to use. For 1583-1802, see "Gregorian Calendar" on page 330. For 1803-20, use numbers for 1983-2000, respectively. For Julian Calendar, see "Julian Calendar" on page 330.

Year	No.	Year	No.	Year	No.	Year	No.	Year	No.	Year	No.	Year	No.	Year	No.	Year	No.	Year	No.
1821	2	1847	6	1873	4	1899	1	1925	5	1951	2	1977	7	2003	4	2029	2	2055	6
1822	3	1848	14	1874	5	1900	2	1926	6	1952	10	1978	1	2004	12	2030	3	2056	14
1823	4	1849	2	1875	6	1901	3	1927	7	1953	5	1979	2	2005	7	2031	4	2057	2
1824	12	1850	3	1876	14	1902	4	1928	8	1954	6	1980	10	2006	1	2032	12	2058	3
1825	7	1851	4	1877	2	1903	5	1929	3	1955	7	1981	5	2007	2	2033	7	2059	4
1826	1	1852	12	1878	3	1904	13	1930	4	1956	8	1982	6	2008	10	2034	1	2060	12
1827	2	1853	7	1879	4	1905	1	1931	5	1957	3	1983	7	2009	5	2035	2	2061	7
1828	10	1854	1	1880	12	1906	2	1932	13	1958	4	1984	8	2010	6	2036	10	2062	1
1829	5	1855	2	1881	7	1907	3	1933	1	1959	5	1985	3	2011	7	2037	5	2063	2
1830	6	1856	10	1882	1	1908	11	1934	2	1960	13	1986	4	2012	8	2038	6	2064	10
1831	7	1857	5	1883	2	1909	6	1935	3	1961	1	1987	5	2013	3	2039	7	2065	5
1832	8	1858	6	1884	10	1910	7	1936	11	1962	2	1988	13	2014	4	2040	8	2066	6
1833	3	1859	7	1885	5	1911	1	1937	6	1963	3	1989	1	2015	5	2041	3	2067	7
1834	4	1860	8	1886	6	1912	9	1938	7	1964	11	1990	2	2016	13	2042	4	2068	8
1835	5	1861	3	1887	7	1913	4	1939	1	1965	6	1991	3	2017	1	2043	5	2069	3
1836	13	1862	4	1888	8	1914	5	1940	9	1966	7	1992	11	2018	2	2044	13	2070	4
1837	1	1863	5	1889	3	1915	6	1941	4	1967	1	1993	6	2019	3	2045	1	2071	5
1838	2	1864	13	1890	4	1916	14	1942	5	1968	9	1994	7	2020	11	2046	2	2072	13
1839	3	1865	1	1891	5	1917	2	1943	6	1969	4	1995	1	2021	6	2047	3	2073	1
1840	11	1866	2	1892	13	1918	3	1944	14	1970	5	1996	9	2022	7	2048	11	2074	2
1841	6	1867	3	1893	1	1919	4	1945	2	1971	6	1997	4	2023	1	2049	6	2075	3
1842	7	1868	11	1894	2	1920	12	1946	3	1972	14	1998	5	2024	9	2050	7	2076	11
1843	1	1869	6	1895	3	1921	7	1947	4	1973	2	1999	6	2025	4	2051	1	2077	6
1844	9	1870	7	1896	11	1922	1	1948	12	1974	3	2000	14	2026	5	2052	9	2078	7
1845	4	1871	1	1897	6	1923	2	1949	7	1975	4	2001	2	2027	6	2053	4	2079	1
1846	5	1872	9	1898	7	1924	10	1950	1	1976	12	2002	3	2028	14	2054	5	2080	9

The page also contains reference monthly calendar grids for the numbered calendars 1, 2, 3, 4, 5, 6 and for the years 1999, 2001, and 2002. Each block shows the twelve months (JANUARY through DECEMBER) with day-of-week columns S M T W T F S.

Calendar configuration 7

JANUARY · FEBRUARY · MARCH · APRIL · MAY · JUNE · JULY · AUGUST · SEPTEMBER · OCTOBER · NOVEMBER · DECEMBER

Calendar configuration 8

JANUARY · FEBRUARY · MARCH · APRIL · MAY · JUNE · JULY · AUGUST · SEPTEMBER · OCTOBER · NOVEMBER · DECEMBER

Calendar configuration 9

JANUARY · FEBRUARY · MARCH · APRIL · MAY · JUNE · JULY · AUGUST · SEPTEMBER · OCTOBER · NOVEMBER · DECEMBER

Calendar configuration 10

JANUARY · FEBRUARY · MARCH · APRIL · MAY · JUNE · JULY · AUGUST · SEPTEMBER · OCTOBER · NOVEMBER · DECEMBER

Calendar configuration 11

JANUARY · FEBRUARY · MARCH · APRIL · MAY · JUNE · JULY · AUGUST · SEPTEMBER · OCTOBER · NOVEMBER · DECEMBER

Calendar configuration 12

JANUARY · FEBRUARY · MARCH · APRIL · MAY · JUNE · JULY · AUGUST · SEPTEMBER · OCTOBER · NOVEMBER · DECEMBER

Calendar configuration 13

JANUARY · FEBRUARY · MARCH · APRIL · MAY · JUNE · JULY · AUGUST · SEPTEMBER · OCTOBER · NOVEMBER · DECEMBER

Calendar configuration 14 / 2000

JANUARY · FEBRUARY · MARCH · APRIL · MAY · JUNE · JULY · AUGUST · SEPTEMBER · OCTOBER · NOVEMBER · DECEMBER

Julian and Gregorian Calendars; Leap Year; Century

Calendars based on the movements of the sun and moon have been used since ancient times, but none has been perfect. The **Julian calendar**, under which Western nations measured time until AD 1582, was authorized by Julius Caesar in 46 BC, the year 709 of Rome. His expert was a Greek, Sosigenes. The Julian calendar, on the assumption that the length of the true year was 365 1/4 days, gave every 4th year 366 days. St. Bede the Venerable, an Anglo-Saxon monk, announced in AD 730 that the 365 1/4-day Julian year was 11 min, 14 sec too long, a cumulative error of about a day every 128 years, but nothing was done about this for more than 800 years.

By 1582 the accumulated error was estimated to amount to 10 days. In that year Pope Gregory XIII decreed that the day following Oct. 4, 1582, should be called Oct. 15, thus dropping 10 days and initiating what became known as the **Gregorian calendar**.

However, with common years 365 days and a 366-day leap year every 4th year, the error in the length of the year would have recurred at the rate of a little more than 3 days every 400 years. Therefore, 3 of every 4 centesimal years (years ending in 00) were made common years, not leap years. Under this plan, 1600 and 2000 are leap years; 1700, 1800, and 1900 are not. **Leap years** are those years divisible by 4, except centesimal years, which are common unless divisible by 400.

The Gregorian calendar was adopted at once by France, Italy, Spain, Portugal, and Luxembourg. Within 2 years most German Catholic states, Belgium, and parts of Switzerland and the Netherlands were brought under the new calendar, and Hungary followed in 1587. The rest of the Netherlands, along with Denmark and the German Protestant states, made the change in 1699-1700. (German Protestants retained the Julian calendar's reckoning of the movable feast of Easter until 1776.)

The British government imposed the Gregorian calendar on all its possessions, including the American colonies, in 1752, decreeing that the day following Sept. 2, 1752, should be called Sept. 14, a loss of 11 days. All dates preceding were marked OS, for Old Style. In addition, New Year's Day was moved to Jan. 1 from Mar. 25 (under the old reckoning, for example, Mar. 24, 1700, had been followed by Mar. 25, 1701). Thus George Washington's birthdate, which was Feb. 11, 1731, OS, became Feb. 22, 1732, NS (New Style). In 1753 Sweden also went Gregorian, although it retained the Julian calendar's rules for Easter until 1844.

In 1793 the French revolutionary government adopted a calendar of 12 months of 30 days with 5 extra days in September of each common year and a 6th every 4th year. Napoleon reinstated the Gregorian calendar in 1806.

The Gregorian system later spread to non-European regions, first in the European colonies and then in independent countries, replacing traditional calendars at least for official purposes. Japan in 1873, Egypt in 1875, China in 1912, and Turkey in 1925 made the change, usually in conjunction with political upheaval. In China, the republican government began reckoning years from its 1911 founding. After 1949, the Communists adopted the Common, or Christian Era, year count, even for the traditional lunar calendar.

In 1918 the Soviet Union decreed that the day after Jan. 31, 1918, OS, would be Feb. 14, 1918, NS. Greece changed over in 1923. For the first time in history, all major cultures now have one calendar. (The Russian Orthodox Church, however, has retained the Julian calendar, as have various Middle Eastern Christian sects.)

To convert from the Julian to the Gregorian calendar, add 10 days to dates Oct. 5, 1582, through Feb. 28, 1700; after that date add 11 days through Feb. 28, 1800; 12 days through Feb. 28, 1900; and 13 days through Feb. 28, 2100.

A **century** consists of 100 consecutive years. The 1st century AD may be said to have run from the years 1 through 100. The 20th century by this reckoning would consist of the years 1901 through 2000 and would technically end Dec. 31, 2000, as would the millennium. The 21st century would thus technically begin Jan. 1, 2001.

Julian Calendar

To find which of the 14 calendars of the Perpetual Calendar applies to any year, starting Jan. 1, under the Julian system, find the century for the desired year in the 3 leftmost columns below. Read across and find the year in the 4 top rows. Then read down. The number in the intersection is the calendar designation for that year.

Year (last 2 figures of desired year)

Century			00	01 29 57 85	02 30 58 86	03 31 59 87	04 32 60 88	05 33 61 89	06 34 62 90	07 35 63 91	08 36 64 92	09 37 65 93	10 38 66 94	11 39 67 95	12 40 68 96	13 41 69 97	14 42 70 98	15 43 71 99	16 44 72	17 45 73	18 46 74	19 47 75	20 48 76	21 49 77	22 50 78	23 51 79	24 52 80	25 53 81	26 54 82	27 55 83	28 56 84
0	700	1400	12	7	1	2	10	5	6	7	8	3	4	5	13	1	2	3	11	6	7	1	9	4	5	6	14	2	3	4	12
100	800	1500	11	6	7	1	9	4	5	6	14	2	3	4	12	7	1	2	10	5	6	7	8	3	4	5	13	1	2	3	11
200	900	1600	10	5	6	7	8	3	4	5	13	1	2	3	11	6	7	1	9	4	5	6	14	2	3	4	12	7	1	2	10
300	1000	1700	9	4	5	6	14	2	3	4	12	7	1	2	10	5	6	7	8	3	4	5	13	1	2	3	11	6	7	1	9
400	1100	1800	8	3	4	5	13	1	2	3	11	6	7	1	9	4	5	6	14	2	3	4	12	7	1	2	10	5	6	7	8
500	1200	1900	14	2	3	4	12	7	1	2	10	5	6	7	8	3	4	5	13	1	2	3	11	6	7	1	9	4	5	6	14
600	1300	2000	13	1	2	3	11	6	7	1	9	4	5	6	14	2	3	4	12	7	1	2	10	5	6	7	8	3	4	5	13

Gregorian Calendar

Choose the desired year from the table below or from the Perpetual Calendar (for years 1803 to 2080). The number after each year designates which calendar to use for that year, as shown in the Perpetual Calendar. (The Gregorian calendar was inaugurated Oct. 15, 1582. From that date to Dec. 31, 1582, use calendar 6.)

1583-1802

1583	7	1603	4	1623	1	1643	5	1663	2	1683	6	1703	2	1723	6	1743	3	1763	7	1783	4
1584	8	1604	12	1624	9	1644	13	1664	10	1684	14	1704	10	1724	14	1744	11	1764	8	1784	12
1585	3	1605	7	1625	4	1645	1	1665	5	1685	2	1705	5	1725	2	1745	6	1765	3	1785	7
1586	4	1606	1	1626	5	1646	2	1666	6	1686	3	1706	6	1726	3	1746	7	1766	4	1786	1
1587	5	1607	2	1627	6	1647	3	1667	7	1687	4	1707	7	1727	4	1747	1	1767	5	1787	2
1588	13	1608	10	1628	14	1648	11	1668	8	1688	12	1708	8	1728	12	1748	9	1768	13	1788	10
1589	1	1609	5	1629	2	1649	6	1669	3	1689	7	1709	3	1729	7	1749	4	1769	1	1789	5
1590	2	1610	6	1630	3	1650	7	1670	4	1690	1	1710	4	1730	1	1750	5	1770	2	1790	6
1591	3	1611	7	1631	4	1651	1	1671	5	1691	2	1711	5	1731	2	1751	6	1771	3	1791	7
1592	11	1612	8	1632	12	1652	9	1672	13	1692	10	1712	13	1732	10	1752	14	1772	11	1792	8
1593	6	1613	3	1633	7	1653	4	1673	1	1693	5	1713	1	1733	5	1753	2	1773	6	1793	3
1594	7	1614	4	1634	1	1654	5	1674	2	1694	6	1714	2	1734	6	1754	3	1774	7	1794	4
1595	1	1615	5	1635	2	1655	6	1675	3	1695	7	1715	3	1735	7	1755	4	1775	1	1795	5
1596	9	1616	13	1636	10	1656	14	1676	11	1696	8	1716	11	1736	8	1756	12	1776	9	1796	13
1597	4	1617	1	1637	5	1657	2	1677	6	1697	3	1717	6	1737	3	1757	7	1777	4	1797	1
1598	5	1618	2	1638	6	1658	3	1678	7	1698	4	1718	7	1738	4	1758	1	1778	5	1798	2
1599	6	1619	3	1639	7	1659	4	1679	1	1699	5	1719	1	1739	5	1759	2	1779	6	1799	3
1600	14	1620	11	1640	8	1660	12	1680	9	1700	2	1720	9	1740	13	1760	10	1780	14	1800	4
1601	2	1621	6	1641	3	1661	7	1681	4	1701	7	1721	4	1741	1	1761	5	1781	2	1801	5
1602	3	1622	7	1642	4	1662	1	1682	5	1702	1	1722	5	1742	2	1762	6	1782	3	1802	6

The Julian Period

How many days have you lived? To determine this, multiply your age by 365, add the number of days since your last birthday, and account for all leap years. Chances are your calculations will go wrong somewhere. Astronomers, however, find it convenient to express dates and time intervals in days rather than in years, months, and days. This is done by placing events within the Julian period.

The Julian period was devised in 1582 by the French classical scholar Joseph Scaliger (1540-1609), and it was named after his father, Julius Caesar Scaliger, not after the Julian calendar as might be supposed.

Scaliger began Julian Day (JD) #1 at noon, Jan. 1, 4713 BC, the most recent time that 3 major chronological cycles began on the same day: (1) the 28-year solar cycle, after which dates in the Julian calendar (e.g., Feb. 11) return to the same days of the week (e.g., Monday); (2) the 19-year lunar cycle, after which the phases of the moon return to the same dates of the year; and (3) the 15-year indiction cycle, used in ancient Rome to regulate taxes. It will take 7,980 years to complete the period, the product of 28, 19, and 15.

Noon of Dec. 31, 1999, marks the beginning of JD 2,451,544; that many days will have passed since the start of the Julian period. The JD at noon of any date in 2000 may be found by adding to this figure the day of the year for that date, which can be obtained from the left half of the "How Far Apart Are Two Dates?" chart.

How Far Apart Are Two Dates?

This table covers a period of 2 years. To use, find the **number** for each date and subtract the smaller from the larger. Example— for days from Feb. 10, 1998, to Dec. 15, 1999, subtract 41 from 714; the result is 673. For leap years, such as 2000, one day must be added; thus Feb. 10, 1999, and Dec. 15, 2000, are 674 days apart.

	First Year													Second Year											
Date	Jan.	Feb.	Mar.	April	May	June	July	Aug.	Sept.	Oct.	Nov.	Dec.	Date	Jan.	Feb.	Mar.	April	May	June	July	Aug.	Sept.	Oct.	Nov.	Dec.
1	1	32	60	91	121	152	182	213	244	274	305	335	1	366	397	425	456	486	517	547	578	609	639	670	700
2	2	33	61	92	122	153	183	214	245	275	306	336	2	367	398	426	457	487	518	548	579	610	640	671	701
3	3	34	62	93	123	154	184	215	246	276	307	337	3	368	399	427	458	488	519	549	580	611	641	672	702
4	4	35	63	94	124	155	185	216	247	277	308	338	4	369	400	428	459	489	520	550	581	612	642	673	703
5	5	36	64	95	125	156	186	217	248	278	309	339	5	370	401	429	460	490	521	551	582	613	643	674	704
6	6	37	65	96	126	157	187	218	249	279	310	340	6	371	402	430	461	491	522	552	583	614	644	675	705
7	7	38	66	97	127	158	188	219	250	280	311	341	7	372	403	431	462	492	523	553	584	615	645	676	706
8	8	39	67	98	128	159	189	220	251	281	312	342	8	373	404	432	463	493	524	554	585	616	646	677	707
9	9	40	68	99	129	160	190	221	252	282	313	343	9	374	405	433	464	494	525	555	586	617	647	678	708
10	10	41	69	100	130	161	191	222	253	283	314	344	10	375	406	434	465	495	526	556	587	618	648	679	709
11	11	42	70	101	131	162	192	223	254	284	315	345	11	376	407	435	466	496	527	557	588	619	649	680	710
12	12	43	71	102	132	163	193	224	255	285	316	346	12	377	408	436	467	497	528	558	589	620	650	681	711
13	13	44	72	103	133	164	194	225	256	286	317	347	13	378	409	437	468	498	529	559	590	621	651	682	712
14	14	45	73	104	134	165	195	226	257	287	318	348	14	379	410	438	469	499	530	560	591	622	652	683	713
15	15	46	74	105	135	166	196	227	258	288	319	349	15	380	411	439	470	500	531	561	592	623	653	684	714
16	16	47	75	106	136	167	197	228	259	289	320	350	16	381	412	440	471	501	532	562	593	624	654	685	715
17	17	48	76	107	137	168	198	229	260	290	321	351	17	382	413	441	472	502	533	563	594	625	655	686	716
18	18	49	77	108	138	169	199	230	261	291	322	352	18	383	414	442	473	503	534	564	595	626	656	687	717
19	19	50	78	109	139	170	200	231	262	292	323	353	19	384	415	443	474	504	535	565	596	627	657	688	718
20	20	51	79	110	140	171	201	232	263	293	324	354	20	385	416	444	475	505	536	566	597	628	658	689	719
21	21	52	80	111	141	172	202	233	264	294	325	355	21	386	417	445	476	506	537	567	598	629	659	690	720
22	22	53	81	112	142	173	203	234	265	295	326	356	22	387	418	446	477	507	538	568	599	630	660	691	721
23	23	54	82	113	143	174	204	235	266	296	327	357	23	388	419	447	478	508	539	569	600	631	661	692	722
24	24	55	83	114	144	175	205	236	267	297	328	358	24	389	420	448	479	509	540	570	601	632	662	693	723
25	25	56	84	115	145	176	206	237	268	298	329	359	25	390	421	449	480	510	541	571	602	633	663	694	724
26	26	57	85	116	146	177	207	238	269	299	330	360	26	391	422	450	481	511	542	572	603	634	664	695	725
27	27	58	86	117	147	178	208	239	270	300	331	361	27	392	423	451	482	512	543	573	604	635	665	696	726
28	28	59	87	118	148	179	209	240	271	301	332	362	28	393	424	452	483	513	544	574	605	636	666	697	727
29	29	—	88	119	149	180	210	241	272	302	333	363	29	394	—	453	484	514	545	575	606	637	667	698	728
30	30	—	89	120	150	181	211	242	273	303	334	364	30	395	—	454	485	515	546	576	607	638	668	699	729
31	31	—	90	—	151	—	212	243	—	304	—	365	31	396	—	455	—	516	—	577	608	—	669	—	730

Chinese Calendar, Asian Festivals

Source: Chinese Information and Culture Center, New York, NY

The Chinese calendar (like the Islamic calendar; see Religious Information section) is a lunar calendar. It is divided into 12 months of 29 or 30 days (compensating for the lunar month's mean duration of 29 days, 12 hr, 44.05 min). This calendar is synchronized with the solar year by the addition of extra months at fixed intervals.

The Chinese calendar runs on a 60-year cycle. The cycles 1876-1935 and 1936-95, with the years grouped under their 12 animal designations, are printed below, along with the first 24 years of the current cycle. It began in 1996 and will last until 2055. The year 2000 (Lunar Year 4698) is found in the 5th column, under Dragon , and is known as a Year of the Dragon. Readers can find the animal name for the year of their birth, marriage, etc., in the same chart. (Note: The first 3-7 weeks of each Western year belong to the previous Chinese year and animal designation.)

Both the Western (Gregorian) and traditional lunar calendars are used publicly in China and in North and South Korea, and 2 New Year's celebrations are held. In Taiwan, in overseas Chinese communities, and in Vietnam, the lunar calendar is used only to set the dates for traditional festivals, with the Gregorian system in general use.

The 4-day Chinese New Year, Hsin Nien, the 3-day Vietnamese New Year festival, Tet, and the 3-to-4-day Korean festival, Suhl, begin at the 2d new moon after the winter solstice. The new moon in the Far East, which is west of the International Date Line, may be one day later than the new moon in the U.S. The festivals may start, therefore, anywhere between Jan. 21 and Feb. 19 of the Gregorian calendar. Feb. 5 marks the start of the new Chinese year in 2000.

Rat	Ox	Tiger	Hare (Rabbit)	Dragon	Snake	Horse	Sheep (Goat)	Monkey	Rooster	Dog	Pig
1876	1877	1878	1879	1880	1881	1882	1883	1884	1885	1886	1887
1888	1889	1890	1891	1892	1893	1894	1895	1896	1897	1898	1899
1900	1901	1902	1903	1904	1905	1906	1907	1908	1909	1910	1911
1912	1913	1914	1915	1916	1917	1918	1919	1920	1921	1922	1923
1924	1925	1926	1927	1928	1929	1930	1931	1932	1933	1934	1935
1936	1937	1938	1939	1940	1941	1942	1943	1944	1945	1946	1947
1948	1949	1950	1951	1952	1953	1954	1955	1956	1957	1958	1959
1960	1961	1962	1963	1964	1965	1966	1967	1968	1969	1970	1971
1972	1973	1974	1975	1976	1977	1978	1979	1980	1981	1982	1983
1984	1985	1986	1987	1988	1989	1990	1991	1992	1993	1994	1995
1996	1997	1998	1999	2000	2001	2002	2003	2004	2005	2006	2007
2008	2009	2010	2011	2012	2013	2014	2015	2016	2017	2018	2019

Standard Time, Daylight Saving Time, and Others

Source: National Imagery and Mapping Agency; U.S. Dept. of Transportation

Standard Time

Standard Time is reckoned from the Prime Meridian of Longitude in Greenwich, England. The world is divided into 24 zones, each 15 deg of arc, or one hour in time apart. The Greenwich meridian (0 deg) extends through the center of the initial zone, and the zones to the east are numbered from 1 to 12, with the prefix "minus" indicating the number of hours to be subtracted to obtain Greenwich Time. Each zone extends 7.5 deg on either side of its central meridian.

Westward zones are similarly numbered, but prefixed "plus," showing the number of hours that must be added to get Greenwich Time. Although these zones apply generally to sea areas, the Standard Time maintained in many countries does not coincide with zone time. A graphical representation of the zones is shown on the Standard Time Zone Chart of the World (WOBZC76) published by the National Imagery and Mapping Agency. This chart is available from the National Ocean Service (NOS), 6501 Lafayette Avenue, Riverdale, MD 20737-1199; telephone: (800) 638-8972.

The U.S. and possessions are divided into 10 Standard Time zones. Each zone is approximately 15 deg of longitude in width. All places in each zone use, instead of their own local time, the time counted from the transit of the "mean sun" across the Standard Time meridian that passes near the middle of that zone. These time zones are designated as Atlantic, Eastern, Central, Mountain, Pacific, Alaska, Hawaii-Aleutian, Samoa, Wake Island, and Guam; the time in these zones is reckoned from the 60th, 75th, 90th, 105th, 120th, 135th, 150th, and 165th meridians west of Greenwich and the 165th and 150th meridians east of Greenwich. The time zone line wanders to conform to local geographical regions. The time in the various zones in the U.S. and U.S. territories west of Greenwich is earlier than Greenwich Time by 4, 5, 6, 7, 8, 9, 10, and 11 hours, respectively. However, Wake Island and Guam cross the International Date Line and are 12 and 10 hours later than Greenwich Time, respectively.

24-Hour Time

Twenty-four-hour time is widely used in scientific work throughout the world. In the U.S. it is also used in operations of the armed forces. In Europe it is frequently used by the transportation networks in preference to the 12-hour AM and PM system. With the 24-hour system the day begins at midnight, and times are designated 0000 through 2359.

International Date Line

The Date Line, approximately coinciding with the 180th meridian, separates the calendar dates. The date must be advanced one day when crossing in a westerly direction and set back one day when crossing in an easterly direction. The Date Line frequently deviates from the 180th meridian because of decisions made by individual nations affected. The line is deflected eastward through the Bering Strait and westward of the Aleutians to prevent separating these areas by date. The line is deflected eastward of the Tonga and New Zealand Islands in the South Pacific for the same reason. More recently it was deflected much farther eastward to include all of Kiribati. The line is established by international custom; there is no international authority prescribing its exact course.

Daylight Saving Time

Daylight Saving Time is achieved by advancing the clock one hour. Daylight Saving Time in the U.S. begins each year at 2 AM on the first Sunday in Apr. and ends at 2 AM on the last Sunday in Oct.

Daylight Saving Time was first observed in the U.S. during World War I, and then again during World War II. In the intervening years, some states and communities observed Daylight Saving Time, using whatever beginning and ending dates they chose. In 1966, Congress passed the Uniform Time Act, which provided that any state or territory that chooses to observe Daylight Saving Time must begin and end on the federal dates. Any state could, by law, exempt itself; a 1972 amendment to the act authorized states split by time zones to observe Daylight Saving Time in one time zone and standard time in the other time zone. Currently, Arizona, Hawaii, the eastern time zone portion of Indiana, Puerto Rico, the U.S. Virgin Islands, and American Samoa do not observe Daylight Saving Time.

Congress and the secretary of transportation both have authority to change time zone boundaries. Since 1966 there have been a number of changes to U.S. time zone boundaries. In addition, efforts to conserve energy have prompted various changes in the times that Daylight Saving Time is observed.

International Usage

Adjusting clock time so as to gain the added daylight on summer evenings is common throughout the world.

Canada, which extends over 6 time zones, generally observes Daylight Saving Time from the first Sunday of Apr. until the last Sunday of Oct. Saskatchewan remains on standard time all year. Communities elsewhere in Canada also may exempt themselves from Daylight Saving Time. Mexico, which occupies 3 time zones, observes Daylight Saving Time during the same period as most of Canada.

Member nations of the European Union (EU) observe a "summer-time period," the EU's version of Daylight Saving Time, from the last Sunday of Mar. until the last Sunday in Oct.

Russia, which extends over 11 time zones, maintains its Standard Time 1 hour fast for its zone designation. Additionally, it proclaims Daylight Saving Time from the last Sunday in Mar. until the 4th Sunday in Oct.

China, which extends across 5 time zones, has decreed that the entire country be placed on Greenwich Time plus 8 hours. Daylight Saving Time is not observed. Japan, which lies within one time zone, also does not modify its legal time during the summer months.

Many countries in the Southern Hemisphere maintain Daylight Saving Time, generally from Oct. to Mar.; however, most countries near the equator do not deviate from Standard Time.

Standard Time Differences—World Cities

The time indicated in the table is fixed by law and is called the legal time or, more generally, Standard Time. Use of Daylight Saving Time varies widely. * Indicates morning of the following day. At 12:00 noon, Eastern Standard Time, the Standard Time (in 24-hour time) in selected cities is as follows:

City	H	M	City	H	M	City	H	M	City	H	M
Addis Ababa....	20	00	Casablanca.....	17	00	Madrid.........	18	00	Sarajevo.......	18	00
Amsterdam.....	18	00	Copenhagen....	18	00	Manila.........	1	00*	Seoul..........	2	00*
Athens........	19	00	Dhaka.........	23	00	Mecca.........	20	00	Shanghai......	1	00*
Auckland.......	5	00*	Dublin.........	17	00	Melbourne......	3	00*	Singapore.....	1	00*
Baghdad.......	20	00	Geneva........	18	00	Montevideo.....	14	00	Stockholm.....	18	00
Bangkok.......	0	00*	Helsinki.......	19	00	Moscow........	20	00	Sydney........	3	00*
Beijing........	1	00*	Ho Chi Minh City.	0	00*	Munich........	18	00	Taipei.........	1	00*
Belfast........	17	00	Hong Kong......	1	00*	Nagasaki.......	2	00*	Tashkent.......	22	00
Berlin.........	18	00	Istanbul.......	19	00	Nairobi........	20	00	Tehran........	20	30
Bogotá........	12	00	Jakarta.........	0	00*	New Delhi......	22	30	Tel Aviv.......	19	00
Bombay (Mumbai)	22	30	Jerusalem......	19	00	Oslo..........	18	00	Tokyo.........	2	00*
Brussels.......	18	00	Johannesburg...	19	00	Paris.........	18	00	Vladivostok.....	3	00*
Bucharest......	19	00	Karachi........	22	00	Prague.........	18	00	Vienna........	18	00
Budapest......	18	00	Kathmandu.....	22	45	Quito.........	12	00	Warsaw........	18	00
Buenos Aires...	14	00	Kiev..........	19	00	Rio de Janeiro...	14	00	Wellington......	5	00*
Cairo..........	19	00	Lagos.........	18	00	Rome.........	18	00	Yangon (Rangoon)	23	30
Calcutta.......	22	30	Lima..........	12	00	St. Petersburg..	20	00	Yokohama......	2	00*
Cape Town.....	19	00	Lisbon........	17	00	Santiago.......	13	00	Zurich........	18	00
Caracas.......	13	00	London........	17	00						

Standard Time Differences—North American Cities

At 12:00 noon, Eastern Standard Time, the Standard Time in selected North American cities is as follows:

City	Time		City	Time		City	Time	
Akron, OH	12 00	Noon	Galveston, TX	11 00	AM	Philadelphia, PA	12 00	Noon
Albuquerque, NM	10 00	AM	Grand Rapids, MI	12 00	Noon	*Phoenix, AZ	10 00	AM
Atlanta, GA.	12 00	Noon	Halifax, NS	1 00	PM	Pierre, SD	11 00	AM
Austin, TX	11 00	AM	Hartford, CT	12 00	Noon	Pittsburgh, PA	12 00	Noon
Baltimore, MD	12 00	Noon	Havana, Cuba	12 00	Noon	Portland, ME	12 00	Noon
Birmingham, AL	11 00	AM	Helena, MT	10 00	AM	Portland, OR	9 00	AM
Bismarck, ND	11 00	AM	*Honolulu, HI	7 00	AM	Providence, RI	12 00	Noon
Boise, ID	10 00	AM	Houston, TX	11 00	AM	Quebec, Que.	12 00	Noon
Boston, MA	12 00	Noon	*Indianapolis, IN	12 00	Noon	*Regina, Sask.	11 00	AM
Buffalo, NY	12 00	Noon	Jacksonville, FL	12 00	Noon	Reno, NV.	9 00	AM
Butte, MT	10 00	AM	Juneau, AK	8 00	AM	Richmond, VA	12 00	Noon
Calgary, Alta.	10 00	AM	Kansas City, MO	11 00	AM	Rochester, NY	12 00	Noon
Charleston, SC.	12 00	Noon	*Kingston, Jamaica	12 00	Noon	Sacramento, CA	9 00	AM
Charleston, WV	12 00	Noon	Knoxville, TN.	12 00	Noon	St. John's, Nfld	1 30	PM
Charlotte, NC	12 00	Noon	Lexington, KY	12 00	Noon	St. Louis, MO	11 00	AM
Charlottetown, PEI	1 00	PM	Lincoln, NE	11 00	AM	St. Paul, MN	11 00	AM
Chattanooga, TN	12 00	Noon	Little Rock, AR	11 00	AM	Salt Lake City, UT	10 00	AM
Cheyenne, WY	10 00	AM	Los Angeles, CA	9 00	AM	San Antonio, TX	11 00	AM
Chicago, IL	11 00	AM	Louisville, KY	12 00	Noon	San Diego, CA	9 00	AM
Cleveland, OH	12 00	Noon	Mexico City, Mexico	11 00	AM	San Francisco, CA	9 00	AM
Colorado Spr., CO	10 00	AM	Memphis, TN	11 00	AM	*San Juan, PR	1 00	PM
Columbus, OH	12 00	Noon	Miami, FL	12 00	Noon	Santa Fe, NM	10 00	AM
Dallas, TX	11 00	AM	Milwaukee, WI.	11 00	AM	Savannah, GA.	12 00	Noon
*Dawson, Yuk.	9 00	AM	Minneapolis, MN	11 00	AM	Seattle, WA	9 00	AM
Dayton, OH	12 00	Noon	Mobile, AL.	11 00	AM	Shreveport, LA	11 00	AM
Denver, CO	10 00	AM	Montreal, Que.	12 00	Noon	Sioux Falls, SD	11 00	AM
Des Moines, IA.	11 00	AM	Nashville, TN	11 00	AM	Spokane, WA	9 00	AM
Detroit, MI	12 00	Noon	Nassau, Bahamas	12 00	Noon	Tampa, FL.	12 00	Noon
Duluth, MN.	11 00	AM	New Haven, CT.	12 00	Noon	Toledo, OH	12 00	Noon
Edmonton, Alta.	10 00	AM	New Orleans, LA.	11 00	AM	Topeka, KS	11 00	AM
El Paso, TX	10 00	AM	New York, NY	12 00	Noon	Toronto, Ont	12 00	Noon
Erie, PA	12 00	Noon	Nome, AK	8 00	AM	*Tucson, AZ	10 00	AM
Evansville, IN	11 00	AM	Norfolk, VA	12 00	Noon	Tulsa, OK	11 00	AM
Fairbanks, AK.	8 00	AM	Oklahoma City, OK	11 00	AM	Vancouver, BC	9 00	AM
Flint, MI	12 00	Noon	Omaha, NE.	11 00	AM	Washington, DC	12 00	Noon
*Fort Wayne, IN	12 00	Noon	Ottawa, Ont.	12 00	Noon	Wichita, KS	11 00	AM
Fort Worth, TX	11 00	AM	*Panama City, Panama	12 00	Noon	Wilmington, DE	12 00	Noon
Frankfort, KY	12 00	Noon	Peoria, IL	11 00	AM	Winnipeg, Man.	11 00	AM

Note: This same table can be used for Daylight Saving Time when it is in effect, but allowance must be made for cities that do not observe it; they are marked with an asterisk (*). Daylight Saving Time is one hour later than Standard Time.

U.S. Legal or Public Holidays, 2000

Technically, the U.S. observes no national holidays; each state has jurisdiction over its holidays, which are designated by legislative enactment or executive proclamation. The president and the U.S. Congress can legally designate holidays only for the District of Columbia and for federal employees. In practice, however, most states observe the federal legal public holidays. Federal legal public holidays are New Year's Day, Martin Luther King Jr.'s Birthday, Washington's Birthday (often called Presidents' Day), Memorial Day, Independence Day, Labor Day, Columbus Day, Veterans Day, Thanksgiving, and Christmas.

Chief Legal or Public Holidays

When a holiday falls on a Saturday or a Sunday, it is usually observed on the preceding Friday or the following Monday. For some holidays, government and business closing practices vary. In most states, the office of the secretary of state can provide details for holiday closings.

The following will be legal or public holidays in most states in 2000:

Jan. 1 (Sat.) — New Year's Day
Jan. 17 (3d Mon. in Jan.) — Martin Luther King Jr.'s Birthday
Feb. 12 (Sat.) — Lincoln's Birthday
Feb. 21 (3d Mon. in Feb.) — Washington's Birthday, or Presidents' Day, or Washington-Lincoln Day

May 29 (last Mon. in May) — Memorial Day, or Decoration Day
July 4 (Tues.) — Independence Day
Sept. 4 (1st Mon. in Sept.) — Labor Day
Nov. 11 (Sat.) — Veterans Day
Nov. 23 (4th Thurs. in Nov.) — Thanksgiving
Dec. 25 (Mon.) — Christmas Day
In some states these also will be holidays in 2000:
Apr. 21 (Fri.) — Good Friday (In some states, observed for half or part of day.)
Oct. 9 (2d Mon. in Oct.) — Columbus Day, or Discoverers' Day, or Pioneers' Day
Nov. 7 (1st Tues. after 1st Mon. in Nov.) — Election Day

Selected International Holidays, 2000

Jan. 31 — Australia Day obsvd., Australia
Feb. 5 — Constitution Day, Mexico
Mar. 4-7 — Carnival, Brazil
Mar. 13 — Commonwealth Day, Canada, Great Britain
Mar. 17 — St. Patrick's Day, Ireland
Mar. 21 — Benito Juarez's Birthday, Mexico
Apr. 8 — Buddha's Birthday, Korea, Japan
Apr. 23 — National Sovereignty Day, Turkey
May 5 — Cinco de Mayo (Battle of Puebla Day), Mexico
May 17 — Constitution Day, Norway
May 22 — Victoria Day, Canada
June 6 — Dragon Boat Festival, China
June 23 — Midsummer Eve, Baltics, Scandinavia
July 1 — Canada Day, Canada

July 14 — Bastille Day, France
Aug. 30 — St. Rose of Lima, Peru
Sept. 4 — Labor Day, Canada
Sept. 16 — Independence Day, Mexico
Sept. 19 — St. Gennaro, Italy
Oct. 3 — German Unification Day, Germany
Oct. 9 — Thanksgiving Day, Canada
Oct. 12 — Día de la Raza, Mexico
Nov. 1-2 — Day of the Dead, Mexico
Nov. 5 — Guy Fawkes Day, Great Britain
Nov. 11 — Remembrance Day, Canada
Dec. 12 — Jamhuri Day, Kenya; Guadalupe Day, Mexico
Dec. 26 — Boxing Day, Australia, Canada, Great Britain, New Zealand

NOTED PERSONALITIES
Widely Known Americans of the Present

Political leaders, journalists, and other widely known living persons. This list excludes many in categories listed elsewhere in Noted Personalities, such as Writers of the Present and Entertainment Personalities, or in the Sports section.

Roger Ailes, b 5/15/40 (Warren, OH), TV exec.
Madeleine K. Albright, b 5/15/37 (Prague, Czech.), sec. of state.
Lamar Alexander, b 7/3/40 (Maryville, TN), former TN gov., presid. candidate.
Stephen E. Ambrose, b 1/10/36 (Decatur, IL), historian.
Walter H. Annenberg, b 3/13/08 (Milwaukee, WI), publisher, philanthropist.
Roone Arledge, b 7/8/31 (Forest Hills, NY), TV exec.
Richard K. Armey, b 7/7/40 (Cando, ND), House majority leader.
Neil Armstrong, b 8/5/30 (Wapakoneta, OH), former astronaut.
Bruce Babbitt, b 6/27/38 (Los Angeles), interior sec.
F. Lee Bailey, b 6/10/33 (Waltham, MA), attorney.
Russell Baker, b 8/14/25 (Loudoun Co., VA), columnist.
Dave Barry, b 7/3/47 (Armonk, NY), humorist.
Marion Barry, b 3/6/36 (Itta Bena, MS), former Wash., DC, mayor.
Gary Bauer, b 1956 (Covington, KY), political activist, pres. contender.
William J. Bennett, b 7/31/43 (Brooklyn, NY), author, former education sec.
Lloyd Bentsen, b 2/11/21 (Mission, TX), former senator, treasury sec., vice-presid. nominee.
Samuel "Sandy" Berger, b 10/28/45 (Sharon, CT), national security adviser.
Joseph R. Biden Jr., b 11/20/42 (Scranton, PA), senator.
James H. Billington, b 6/1/29 (Bryn Mawr, PA), librarian of Congress.
Harry A. Blackmun, b 11/12/08 (Nashville, IL), former Sup. Ct. justice.
Julian Bond, b 1/14/40 (Nashville, TN), civil rights leader.
David Bonior, b 6/6/45 (Detroit), House minority whip.
Daniel Boorstin, b 10/1/14 (Atlanta), historian, former librarian of Congress.
Barbara Boxer, b 11/11/40 (Brooklyn, NY), senator.
Bill Bradley, b 7/28/43 (Crystal City, MO), former senator, basketball player; presid. candidate.
Ed Bradley, b 6/22/41 (Philadelphia), TV journalist.
James Brady, b 9/17/44 (Grand Rapids, MI), former presid. press sec., gun control advocate.
Jimmy Breslin, b 10/17/30 (Jamaica, NY), columnist, author.
Stephen Breyer, b 8/15/38 (San Francisco), Sup. Ct. justice.
David Brinkley, b 7/10/20 (Wilmington, NC), TV journalist.
David Broder, b 9/11/29 (Chicago Heights, IL), journalist.
Tom Brokaw, b 2/6/40 (Webster, SD), TV journalist.
Joyce Brothers, b 9/20/28 (NY City), psychologist.
Edmund G. ("Jerry") Brown Jr., b 4/7/38 (San Francisco), Oakland mayor, former CA gov., pres. candidate.
Willie Brown, b 3/20/34 (Mineola, TX), San Francisco mayor.
Carol M. Browner, b 12/16/55, (Miami, FL), EPA head.
Pat Buchanan, b 11/2/38 (Wash., DC), journalist, presid. candidate.
Art Buchwald, b 10/20/25 (Mt. Vernon, NY), humorist.
William F. Buckley Jr., b 11/24/25 (NY City), columnist, author.
Warren Buffett, b 8/30/30 (Omaha), investor.
Barbara Bush, b 6/8/25 (Rye, NY), former first lady.
George Bush, b 6/12/24 (Milton, MA), former president.
George W. Bush, b 7/6/46 (New Haven, CT), TX gov.; presid. candidate.
Jeb Bush, b 2/11/53 (Houston), FL governor.
Robert Byrd, b 11/20/17 (N. Wilkesboro, NC), senator, former majority leader.
Jimmy Carter, b 10/1/24 (Plains, GA), former president.
Rosalynn Carter, b 8/18/27 (Plains, GA), former first lady.
James Carville Jr., b 10/25/44 (Fort Benning, GA), political consultant.
Steve Case, b 8/21/58 (Honolulu, HI), America Online exec.
Julia Child, b 8/15/12 (Pasadena, CA), TV chef, author.
Noam Chomsky, b 12/7/28 (Philadelphia), linguist; activist.
Connie Chung, b 8/20/46 (Wash., DC), TV journalist.
Liz Claiborne, b 3/31/29 (Brussels, Belg.), fashion designer.
Wesley Clark, b 12/23/44 (Chicago), NATO commander in Europe.
Bill Clinton, b 8/19/46 (Hope, AR), U.S. president.
Chelsea Clinton, b 2/27/80 (Little Rock, AR), daughter of Pres. Clinton and Hillary Rodham Clinton.
Hillary Rodham Clinton, b 10/26/47 (Chicago), first lady.
Johnnie L. Cochran Jr., b 10/2/37 (Shreveport, LA), attorney.
William Cohen, b 8/28/40 (Bangor, ME), defense sec.
Joan Ganz Cooney, b 10/30/29 (Phoenix, AZ), children's TV producer.
Bob Costas, b 3/22/52 (NY City), TV journalist.
Katie Couric, b 1/7/57 (Wash., DC), TV journalist.
Walter Cronkite, b 11/4/16 (St. Joseph, MO), TV journalist.
Andrew Cuomo, b 12/6/57 (NY City), HUD sec.
Mario Cuomo, b 6/15/32 (Queens, NY), former NY gov.

Richard M. Daley, b 4/24/42 (Chicago), Chicago mayor.
William M. Daley, b 8/9/48 (Chicago), commerce sec.
Thomas Daschle, b 12/9/47 (Aberdeen, SD), Senate minority leader.
Gray Davis, b 12/26/42 (NY City), CA governor.
Tom DeLay, b 4/8/47 (Laredo, TX), House majority whip.
Alan Dershowitz, b 9/1/38 (Brooklyn, NY), attorney.
Barry Diller, b 2/2/42 (San Francisco), TV exec.
Christopher Dodd, b 5/27/44 (Willimantic, CT), senator.
Elizabeth Hanford Dole, b 7/29/36 (Salisbury, NC), Red Cross pres.; former transp. sec., labor sec.; presid. candidate
Robert Dole, b 7/22/23 (Russell, KS), former Senate majority leader, presid. nominee.
Pete Domenici, b 5/7/32 (Albuquerque, NM), senator.
Sam Donaldson, b 3/11/34 (El Paso, TX), TV journalist.
Michael S. Dukakis, b 11/3/33 (Boston), former MA gov., presid. nominee.
Roger Ebert, b 6/18/42 (Urbana, IL), film critic.
Marian Wright Edelman, b 6/6/39 (Bennettsville, SC), children's rights advocate.
Michael Eisner, b 3/7/42 (NY City), Disney Co. exec.
John Engler, b 10/12/48 (Mount Pleasant, ME), MI gov.
James Fallows, b 8/2/49 (Philadelphia), journalist.
Jerry Falwell, b 8/11/33 (Lynchburg, VA), TV evangelist, religious educator.
Louis Farrakhan, b 5/11/33 (NY City), Nation of Islam leader.
Dianne Feinstein, b 6/22/33 (San Francisco), senator.
Geraldine Ferraro, b 8/26/35 (Newburgh, NY), former U.S. representative, vice-presid. nominee.
Larry Flynt, b 11/1/42 (Magoffin Co., KY), publisher.
Shelby Foote, b 11/17/16 (Greenville, MS), historian.
Malcolm "Steve" Forbes Jr., b 7/18/47 (Morristown, NJ), publisher, presid. candidate.
Betty Ford, b 4/8/18 (Chicago), former first lady.
Gerald R. Ford, b 7/14/13 (Omaha), former president.
Louis J. Freeh, b 1/6/50 (Jersey City, NJ), FBI director.
Betty Friedan, b 2/4/21 (Peoria, IL), author, feminist.
Milton Friedman, b 7/31/12 (Brooklyn, NY), economist.
John Kenneth Galbraith, b 10/15/08 (Iona Station, Ont.), economist.
Bill Gates, b 10/28/55 (Seattle), Microsoft exec.
Henry Louis Gates Jr., b 9/16/50 (Keyser, WV), scholar.
David Geffen, b 2/21/43 (Brooklyn, NY), entertainment exec.
Richard Gephardt, b 1/31/41 (St. Louis, MO), House minority leader.
Louis Gerstner, b 3/1/42 (Mineola, NY), IBM exec.
Newt Gingrich, b 6/17/43 (Harrisburg, PA), former House Speaker.
Ruth Bader Ginsburg, b 3/15/33 (Bklyn, NY), Sup. Ct. justice.
Rudolph Giuliani, b 5/28/44 (NY City), NY City mayor.
John Glenn, b 7/18/21 (Cambridge, OH), former senator, astronaut.
Dan Glickman, b 11/24/44 (Wichita, KS), agriculture sec.
Doris Kearns Goodwin, b 1/4/43 (Rockville Centre, NY), historian, TV commentator.
Berry Gordy, b 11/28/29 (Detroit), Motown founder.
Al Gore Jr., b 3/31/48 (Wash., DC), U.S. vice president; presid. candidate.
Tipper Gore, b 8/19/48 (Wash., DC), wife of vice president.
Stephen Jay Gould, b 9/10/41 (NY City), biologist, author.
Billy Graham, b 11/7/18 (Charlotte, NC), evangelist.
Katharine Graham, b 6/16/17 (NY City), newspaper publisher.
Phil Gramm, b 7/8/42 (Ft. Benning, GA), senator, former presid. contender.
Jeff Greenfield, b 6/10/43 (NY City), TV journalist.
Alan Greenspan, b 3/6/26 (NY City), Fed chairman.
Andrew Grove, b 9/2/36 (Budapest, Hungary), Intel exec.
Bryant Gumbel, b 9/29/48 (New Orleans), TV journalist.
David Halberstam, b 4/10/34 (NY City), journalist, author.
Paul Harvey, b 9/4/18 (Tulsa, OK), radio journalist.
Orrin Hatch, b 3/22/34 (Homestead Park, PA), senator.
J. Dennis Hastert, b 1/2/42 (Aurora, IL), House Speaker.
Hugh Hefner, b 4/9/26 (Chicago), publisher.
Jesse Helms, b 10/18/21 (Monroe, NC), senator.
Leona Helmsley, b c1920 (NY City), real estate exec.
Heloise, b 4/15/51 (Waco, TX), advice columnist.
Alexis Herman, b 7/16/47 (Mobile, AL), labor sec.
Anita Hill, b 7/10/56 (Morris, OK), legal scholar, complainant against Clarence Thomas.
James P. Hoffa, b 5/19/41, (Detroit), labor leader.
Richard Holbrooke, b 4/24/41 (NY City), U.S. rep. to UN.
H. Wayne Huizenga, b 12/29/39 (Evergreen Park, IL), entrepreneur, sports exec.
Kay Bailey Hutchison, b 7/22/43 (Galveston, TX), senator.
Henry J. Hyde, b 4/18/24 (Chicago), U.S. representative.
Lee Iacocca, b 10/15/24 (Allentown, PA), former auto exec.
Carl Icahn, b 1936 (Queens, NY), financier.

Don Imus, b 7/23/40 (?) (Riverside, CA), radio-TV talk-show host.
Patricia Ireland, b 10/19/45 (Oak Park, IL), feminist leader.
Molly Ivins, b 1944 (Texas), columnist.
Rev. Jesse Jackson, b 10/8/41 (Greenville, SC), civil rights leader, former presid. contender.
Steven Jobs, b 2/24/55 (California), Apple Computer exec.
Lady Bird Johnson, b 12/22/12 (Karnack, TX), former first lady.
Vernon E. Jordan Jr., b 8/15/35 (Atlanta), attorney, presid. adviser, former civil rights leader.
John R. Kasich, b 5/13/52 (McKees Rocks, PA), U.S. representative.
Donna Karan, b 10/2/48 (Forest Hills, NY), fashion designer.
Jeffrey Katzenberg, b 1950 (NY City), entertainment exec.
Jack Kemp, b 7/13/35 (Los Angeles), former vice-presid. nominee, HUD sec., pro football quarterback.
Anthony Kennedy, b 7/23/36 (Sacramento, CA), Sup. Ct. justice.
Caroline Kennedy Schlossberg, b 11/27/57 (Boston), author, daughter of Pres. Kennedy.
Edward M. Kennedy, b 2/22/32 (Brookline, MA), senator.
Jack Kevorkian, b 5/26/28 (Pontiac, MI), physican, assisted-suicide activist.
Coretta Scott King, b 4/27/27 (Marion, AL), civil rights leader, widow of Martin Luther King Jr.
Larry King, b 11/19/33 (Brooklyn, NY), TV journalist.
Michael Kinsley, b 3/9/51 (Detroit), journalist, editor.
Jeane J. Kirkpatrick, b 11/19/26 (Duncan, OK), political scientist, former ambassador to UN.
Henry Kissinger, b 5/27/23 (Fuerth, Germany), former sec. of state, national security adviser, Nobel Peace Prize winner.
Calvin Klein, b 11/19/42 (NY City), fashion designer.
Joe Klein, b 9/7/46 (New York), journalist, author.
Philip H. Knight, b 2/24/38 (Oregon), CEO of Nike.
Edward I. Koch, b 12/12/24 (NY City), former NY City mayor.
C. Everett Koop, b 10/14/16 (Brooklyn, NY), former surgeon general.
Ted Koppel, b 2/8/40 (Lancashire, England), TV journalist.
Brian Lamb, b 10/9/41 (Lafayette, IN), cable TV exec., journalist.
Ann Landers, b 7/4/18 (Sioux City, IA), advice columnist.
Estee Lauder, b 9/1/08 (NY City), founder, cosmetics and fragrance firm.
Matt Lauer, b 1957 (NY City), TV journalist.
Ralph Lauren, b 10/14/39 (Bronx, NY), fashion designer.
Norman Lear, b 7/27/22 (New Haven, CT), TV producer, political activist.
Jim Lehrer, b 5/19/34 (Wichita, KS), TV journalist, author.
Monica Lewinsky, b 7/23/73 (San Francisco), former White House intern, key figure in White House scandal.
Rush Limbaugh, b 1/12/51 (Cape Girardeau, MO), radio talk-show host.
Anne Morrow Lindbergh, b 1906 (Englewood, NJ), author, former aviator, widow of Charles A. Lindbergh.
Joseph Lockhart, b 1959 (Suffern, NY), presid. press sec.
Frank Lorenzo, b 5/19/40 (NY City), airline exec.
Trent Lott, b 10/9/41 (Grenada, MS), Senate majority leader.
Shannon Lucid, b 1/14/43 (Shanghai, China), astronaut.
Richard G. Lugar, b 4/4/32 (Indianapolis), senator.
Connie Mack, b 10/29/40 (Philadelphia), senator.
Mary Matalin, b 8/19/53 (Chicago), political commentator.
John McCain III, b 8/29/36 (Panama Canal Zone), senator; presid. candidate.
Michael McCurry, b 10/27/54 (Charleston, SC), former White House press sec.
George McGovern, b 7/19/22 (Avon, SD), former senator, presid. candidate.
John McLaughlin, b 3/29/27 (Providence, RI), TV journalist.
Robert S. McNamara, b 6/9/16 (San Francisco), former defense sec., World Bank head, author.
Kweisi Mfume, b 10/24/48 (Baltimore), civil rights leader, former U.S. representative.
Kate Millett, b 9/14/34 (St. Paul, MN), author, feminist.
George Mitchell, b 8/20/33, (Waterville, ME), former Senate majority leader, N. Ireland peace negotiator.
Walter Mondale, b 1/5/28 (Ceylon, MN), former senator, presid. nominee.
Marc Morial, b 1/3/58 (New Orleans), New Orleans mayor.
Bill Moyers, b 6/5/34 (Hugo, OK), TV journalist, author.
Daniel P. Moynihan, b 3/16/27 (Tulsa, OK), senator, author.
Rupert Murdoch, b 3/11/31 (Melbourne, Aust.), media exec.
Ralph Nader, b 2/27/34 (Winsted, CT), consumer advocate.
Don Nickles, b 12/6/48 (Ponca City, OK), Senate Majority whip.
Oliver North, b 10/7/43 (San Antonio, TX), radio talk-show host, former National Security Council aide.
Eleanor Holmes Norton, b 6/13/37 (Wash., DC), U.S. House delegate.
Robert Novak, b 2/26/31 (Joliet, IL), journalist.
Sam Nunn, b 9/8/38 (Perry, GA), former senator.
Sandra Day O'Connor, b 3/26/30 (El Paso, TX), Sup. Ct. justice.
Michael Ovitz, b 12/4/46 (Encino, CA), entertainment exec.
Leon F. Panetta, b 6/28/38 (Monterey, CA), former White House chief of staff, U.S. representative.
Rosa Parks, b 2/4/13 (Tuskegee, AL), civil rights activist.

George Pataki, b 6/24/45 (Peekskill, NY), NY gov.
Jane Pauley, b 10/31/50 (Indianapolis), TV journalist.
H. Ross Perot, b 6/27/30 (Texarkana, TX), entrepreneur, former presid. nominee.
George Plimpton, b 3/18/27 (NY City), author, editor.
Norman Podhoretz, b 1/16/30 (NY City), author, editor.
Colin Powell, b 4/5/37 (NY City), former Joint Chiefs of Staff chairman, national security adviser.
Dan Quayle, b 2/4/47 (Indianapolis), former U.S. vice president, senator, presid. candidate.
Dan Rather, b 10/31/31 (Wharton, TX), TV journalist.
Nancy Reagan, b 7/6/23 (NY City), former first lady.
Ronald Reagan, b 2/6/11 (Tampico, IL), former president.
Sumner Redstone, b 5/27/23 (Boston), media exec.
Ralph Reed, b 6/24/61 (Portsmouth, VA), political adviser.
William Rehnquist, b 10/1/24 (Milwaukee), Sup. Ct. chief justice.
Robert B. Reich, b 6/24/46 (Scranton, PA), economist, former labor sec.
Janet Reno, b 7/21/38 (Miami, FL), attorney general.
Ann Richards, b 9/3/33 (Waco, TX), former TX gov.
Bill Richardson, b 11/15/47 (Pasadena, CA), energy sec., former UN ambassador, congressman.
Sally K. Ride, b 5/26/51 (Encino, CA), former astronaut.
Richard Riley, b 1/2/33 (Greenville, SC), education sec.
Richard Riordan, b 1930 (Flushing, NY), Los Angeles mayor.
Cokie Roberts, b 12/27/43 (New Orleans), TV journalist.
Oral Roberts, b 1/24/18 (nr. Ada, OK), TV evangelist, educator.
Pat Robertson, b 3/22/30 (Lexington, VA), religious broadcasting exec., former presid. candidate.
David Rockefeller, b 6/12/15 (NY City), banker.
John D. "Jay" Rockefeller 4th, b 6/18/37 (NY City), senator, former WV gov.
Laurance S. Rockefeller, b 5/26/10 (NY City), philanthropist.
Roy Romer, b 10/31/38 (Garden City, KS), former CO gov.
Andy Rooney, b 1/14/19 (Albany, NY), TV commentator.
Robert Rubin, b 8/29/38 (NY City), treasury sec.
Louis Rukeyser, b 1/30/33 (NY City), TV journalist, financial analyst.
Tim Russert, b 5/7/50 (Buffalo, NY), TV journalist.
William Safire, b 12/17/29 (NY City), columnist.
Vidal Sassoon, b 1/17/28 (London, Eng.), hairstylist, entrepreneur.
Diane Sawyer, b 12/22/45 (Glasgow, KY), TV journalist.
Antonin Scalia, b 3/11/36 (Trenton, NJ), Sup. Ct. justice.
Arthur Schlesinger Jr., b 10/15/17 (Columbus, OH), historian.
Kurt L. Schmoke, b 12/1/49 (Baltimore), Baltimore mayor.
Phyllis Schlafly, b 8/15/24 (St. Louis, MO) political activist.
Patricia Schroeder, b 7/30/40 (Portland, OR), former U.S. representative.
Robert Schuller, b 9/16/26 (Alton, IA), TV evangelist.
Charles Schumer, b. 11/23/50 (Brooklyn, NY), senator.
H. Norman Schwarzkopf, b 8/22/34 (Trenton, NJ), former military leader.
Allan H. ("Bud") Selig, b 7/30/34 (Milwaukee), baseball commissioner.
Donna E. Shalala, b 2/14/41 (Cleveland), sec. of health and human services.
Bernard Shaw, b 1940 (Chicago), TV journalist.
Henry Hugh Shelton, b 1/2/42 (Speed, NC), chairman of Joint Chiefs of Staff.
Maria Shriver, b 11/6/55 (Chicago), TV journalist.
George P. Shultz, b 12/13/20 (NY City), former sec. of state, other cabinet posts.
O. J. Simpson, b 7/9/47 (San Francisco), former football star, murder defendant.
Gene Siskel, b 1/26/46 (Chicago), film critic.
Rodney Slater, b 2/23/55 (Tutwyler, MS), transportation sec.
Liz Smith, b 2/2/23 (Ft. Worth, TX), gossip columnist.
David H. Souter, b 9/17/39 (Melrose, MA), Sup. Ct. justice.
George Soros, b 8/12/30 (Budapest, Hungary), financier, philanthropist.
Arlen Specter, b 2/12/30 (Wichita, KS), senator.
Kenneth Starr, b 7/21/46 (Vernon, TX), independent counsel.
George Steinbrenner, b 7/4/30 (Rocky River, OH), NY Yankees owner.
Gloria Steinem, b 3/25/34 (Toledo, OH), author, feminist.
George Stephanopoulos, b 2/10/61 (Fall River, MA), TV journalist, former presid. adviser.
David J. Stern, b 9/22/42 (NY City), basketball comm.
John Paul Stevens, b 4/20/20 (Chicago), Sup. Ct. justice.
Martha Stewart, b 8/3/41 (Nutley, NJ), homemaking adviser, entrepreneur.
Lawrence H. Summers, b 11/30/54 (New Haven, CT), treas. sec.
John J. Sweeney, b 5/5/34 (NY City), labor leader.
Arthur Ochs Sulzberger Jr., b 9/22/51 (Mt. Kisco, NY), newspaper publisher.
John H. Sununu, b 7/2/39 (Havana, Cuba), political commentator, former White House chief of staff.
Paul Tagliabue, b 11/24/40 (Jersey City, NJ), football comm.

George Tenet, b 1/5/53 (Queens, NY), CIA director.
Clarence Thomas, b 6/23/48 (Savannah, GA), Sup. Ct. justice.
Helen Thomas, b 8/4/20 (Winchester, KY), journalist.
R. David Thomas, b 7/2/32 (Atlantic City, NJ), Wendy's founder.
Fred Thompson, b 8/19/42 (Sheffield, AL), senator.
Hunter S. Thompson, b 7/18/37 (Louisville, KY), journalist.
Tommy G. Thompson, b 11/19/41 (Elroy, WI), WI gov.
J. Strom Thurmond, b 12/5/02 (Edgefield, SC), senator.
Laurence Tisch, b 3/15/23 (NY City), entertainment exec.
Margaret Truman, b 2/17/24 (Independence, MO), author, daughter of Pres. Truman.
Donald Trump, b 1946 (NY City), real estate exec.
Ted Turner, b 11/19/38 (Cincinnati), TV exec, philanthropist.
Peter Ueberroth, b 9/2/37 (Chicago), sports & travel exec.
Jack Valenti, b 9/5/21 (Houston, TX), movie industry exec.
Jesse Ventura, b 7/15/51 (Minneapolis), MN governor, former wrestler.
Abigail Van Buren, b 7/4/18 (Sioux City, IA), advice columnist.
George Voinovich, b 7/13/31, (Cleveland), former OH gov.
Mike Wallace, b 5/9/18 (Brookline, MA), TV journalist.

Barbara Walters, b 9/25/31 (Boston), TV journalist.
J. C. Watts Jr., b 11/18/57 (Eufaula, OK), U.S. representative, Republican Conference chair.
Andrew Weil, b 6/8/42 (Philadelphia), health adviser.
Caspar Weinberger, b 8/18/17 (San Francisco), business exec, former defense sec., other cabinet posts.
Jann Wenner, b 1/7/46 (NY City), publisher.
Cornel West, b 6/23/53 (Tulsa, OK), scholar, critic.
Togo G. West Jr., b 6/21/42 (Winston-Salem, NC), sec. of veterans affairs.
Ruth Westheimer, b. 1928 (Germany), human sexuality expert.
Christine Todd Whitman, b 9/26/46 (New York), NJ gov.
Elie Wiesel, b 9/30/28 (Sighet, Romania), scholar, author, Nobel Peace Prize winner.
L. Douglas Wilder, b 1/17/31 (Richmond, VA), former VA gov.
George Will, b 5/4/41 (Champaign, IL), journalist, author.
Jody Williams, b 10/9/50 (Brattleboro, VT), anti-landmine activist, Nobel Peace Prize winner.
Pete Wilson, b 8/23/33 (Lake Forest, IL), CA gov.
Bob Woodward, b 3/26/43 (Geneva, IL), journalist, author.

Noted African-Americans of the Past
See also other categories.

Ralph David Abernathy, 1926-90, organizer, 1957, pres., 1968, Southern Christian Leadership Conf.
Crispus Attucks, c1723-70, leader of group of colonists that clashed with British soldiers in 1770 Boston Massacre.
Benjamin Banneker, 1731-1806, inventor, astronomer, mathematician, gazetteer.
James P. Beckwourth, 1798-c1867, western fur trader, scout; Beckwourth Pass in N California named for him.
Mary McCleod Bethune, 1875-1955, adviser to FDR and Truman; founder, pres., Bethune-Cookman College.
Henry Blair, 19th cent., pioneer inventor; obtained patents for a corn-planter, 1834, and cotton-planter, 1836.
Edward Bouchet, 1852-1918, first black to earn a PhD at a U.S. university (Yale, 1876); in Phi Beta Kappa.
Tom Bradley, 1917-98, first African-American mayor of L.A.
Sterling A. Brown, 1901-89, poet, literature professor; helped establish African-American literary criticism.
William Wells Brown, 1815-84, novelist, dramatist; first African American to publish a novel.
Ralph Bunche, 1904-71, first black to win the Nobel Peace Prize, 1950; undersecretary of the UN, 1950.
George Washington Carver, 1864-1943, botanist, chemist, and educator; revolutionized the economy of the South.
Charles Waddell Chesnutt, 1858-1932, author known for his short stories, including *The Conjure Woman.*
Eldridge Cleaver, 1935-98, revolutionary social critic; former "minister of information" for Black Panthers; *Soul on Ice.*
James Cleveland, 1931-91, composer, musician, singer; first black gospel artist to appear at Carnegie Hall.
Countee Cullen, 1903-46, poet, prominent in the Harlem Renaissance of the 1920s; *The Black Christ.*
Benjamin O. Davis Sr., 1877-1970, first African-American general, 1940, in U.S. Army.
William L. Dawson, 1886-1970, Illinois congressman, first black chairman of a major U.S. House committee.
Aaron Douglas, 1900-79, "father of black American art."
Frederick Douglass, 1817-95, author, editor, orator, diplomat; edited abolitionist weekly *The North Star.*
St. Clair Drake, 1911-90, black studies pioneer, *Black Metropolis* (1945, with Horace R. Cayton).
Charles Richard Drew, 1904-50, physician, pioneered in development of blood banks.
William Edward Burghardt (W.E.B.) Du Bois, 1868-1963, historian, sociologist; an NAACP founder, 1909.
Paul Laurence Dunbar, 1872-1906, poet, novelist; won fame with *Lyrics of Lowly Life,* 1896.
Jean Baptiste Point du Sable, c1750-1818, pioneer trader and first settler of Chicago, 1779.
Henry O. Flipper, 1856-1940, first African-American to graduate, 1877, from West Point.
Marcus Garvey, 1887-1940, founded Universal Negro Improvement Assn., 1911.
Ewart Guinier, 1911-90, trade unionist; first chairman of Harvard Univ.'s Dept. of African American Studies.
Prince Hall, 1735-1807, activist; founded black Freemasonry; served in American Revolutionary war.
Jupiter Hammon, c1720-1800, poet; first African-American to have his works published, 1761.
Lorraine Hansberry, 1930-65, playwright; won New York Drama Critics Circle Award, 1959; *A Raisin in the Sun.*
William H. Hastie, 1904-76, first black federal judge, appointed 1937; governor of Virgin Islands, 1946-49.
Matthew A. Henson, 1866-1955, member of Peary's 1909 expedition to the North Pole; placed U.S. flag at the pole.
Chester Himes, 1909-84, novelist; *Cotton Comes to Harlem.*
William A. Hinton, 1883-1959, physician, developed tests for syphilis; first black prof., 1949, at Harvard Med. School.

Charles Hamilton Houston, 1895-1950, lawyer, Howard University instructor, champion of minority rights.
Langston Hughes, 1902-67, poet, lyric writer, author; a major influence in 1920s Harlem Renaissance.
Daniel James Jr., 1920-78, first black 4-star general, 1975; commander, North American Air Defense Command.
Henry Johnson, 1897-1929, first American decorated by France in WW1 with the Croix de Guerre.
James Weldon Johnson, 1871-1938, poet, novelist, diplomat; lyricist for *Lift Every Voice and Sing.*
Barbara Jordan, 1936-96, congresswoman, orator, educator.
Ernest Everett Just, 1883-1941, marine biologist; studied egg development; author, *Biology of Cell Surfaces,* 1941.
Rev. Martin Luther King Jr., 1929-68, civil rights leader; led 1956 Montgomery, AL, boycott; founder, pres., Southern Christian Leadership Conference, 1957; Nobel laureate (1964); assassinated.
Lewis H. Latimer, 1848-1928, associate of Edison; supervised installation of first electric street lighting in NYC.
Mickey Leland, 1944-89, U.S. representative from Texas, 1978 until death; chairman of Congressional Black Caucus.
Henry Lewis, 1932-1996, (U.S.) conductor; first black conductor and musical director of major American orchestra.
Malcolm X (Little), 1925-65, Black Muslim, black nationalist leader; promoted black pride; assassinated.
Thurgood Marshall, 1908-93, first black U.S. solicitor general, 1965; first black justice of U.S. Sup. Ct., 1967-91.
Jan Matzeliger, 1852-89, invented lasting machine, patented 1883, which revolutionized the shoe industry.
Benjamin Mays, 1895-1984, educator, civil rights leader; headed Morehouse College, 1940-67.
Ronald McNair, 1950-86, physicist, astronaut; killed in *Challenger* explosion.
Dorie Miller, 1919-43, Navy hero of Pearl Harbor attack.
Willard Motley, 1912-65, novelist; *Knock on Any Door.*
Elijah Muhammad, 1897-1975, founded Black Muslims, 1931.
Pedro Alonzo Niño, navigator of Columbus's Niña, 1492.
Frederick D. Patterson, 1901-88, founder of United Negro College Fund, 1944.
Harold R. Perry, 1916-91, first black American Roman Catholic bishop in the 20th cent.
Adam Clayton Powell Jr., 1908-72, early civil rights leader, congressman, 1945-69.
Joseph H. Rainey, 1832-87, first black elected to U.S. House, 1869, from South Carolina.
A. Philip Randolph, 1889-1979, organized Brotherhood of Sleeping Car Porters, 1925; an organizer of 1941 and 1963 March on Washington movements.
Hiram R. Revels, 1822-1901, first African-American U.S. senator, elected in Mississippi; served 1870-71.
Norbert Rillieux, 1806-94, invented a vacuum pan evaporator, 1846, revolutionizing sugar-refining industry.
Paul Robeson, 1898-1976, actor, singer, civil rights activist; graduated first in class at Rutgers, 1918.
Jackie Robinson, 1919-72, first African-American baseball player to play in the major leagues, 1947, and be inducted into the Baseball Hall of Fame, 1962.
Bayard Rustin, 1910-87, an organizer of the 1963 March on Washington; exec. director, A. Philip Randolph Institute.
Peter Salem, at the Battle of Bunker Hill, June 17, 1775, shot and killed British commander Maj. John Pitcairn.
Carl Stokes, 1927-1996, first black mayor of a major American city (Cleveland), 1967-72.
Willard Townsend, 1895-1957, organized the United Transport Service Employees (redcaps), 1935.
Sojourner Truth, 1797-1883, preacher, abolitionist; worked for black educational opportunity.

Harriet Tubman, 1823-1913, Underground Railroad conductor, nurse and spy for Union Army in the Civil War.

Nat Turner, 1800-31, led most significant of more than 200 slave revolts in U.S., in Southampton, VA; hanged.

Booker T. Washington, 1856-1915, founder, 1881, and first pres. of Tuskegee Institute; *Up From Slavery.*

Harold Washington, 1922-87, first black mayor of Chicago.

Robert C. Weaver, 1907-97, first African-Amerian appointed to cabinet; secretary of HUD.

Phillis Wheatley, c1753-84, poet; 2d American woman and first black woman to be published, 1770.

Walter White, 1893-1955, exec. sec., NAACP, 1931-55.

Roy Wilkins, 1901-81, exec. director, NAACP, 1955-77.

Daniel Hale Williams, 1858-1931, surgeon; performed one of first two open-heart operations, 1893.

Granville T. Woods, 1856-1910, invented third-rail system now used in subways, and automatic air brake.

Carter G. Woodson, 1875-1950, historian; founded Assn. for the Study of Negro Life and History.

Frank Yerby, 1916-91, first best-selling African-American novelist; *The Foxes of Harrow.*

Coleman A. Young, 1918-97, first Afr.-Amer. mayor of Detroit, 1974-93.

Selected Architects and Some of Their Achievements

Max Abramovitz, b 1908, Avery Fisher Hall, NYC; U.S. Steel Bldg. (now USX Towers), Pittsburgh, PA.

Henry Bacon, 1866-1924, Lincoln Memorial, Wash., DC.

Pietro Belluschi, 1899-1994, Juilliard School, Lincoln Center, Pan Am, now MetLife, Bldg. (with Walter Gropius), NYC.

Marcel Breuer, 1902-81, Whitney Museum of American Art (with Hamilton Smith), NYC.

Charles Bulfinch, 1763-1844, State House, Boston; Capitol (part), Wash., DC.

Gordon Bunshaft, 1909-90, Lever House, Park Ave, NYC; Hirshhorn Museum, Wash., DC.

Daniel H. Burnham, 1846-1912, Union Station, Wash. DC; Flatiron Bldg., NYC.

Irwin Chanin, 1892-1988, theaters, skyscrapers, NYC.

Lucio Costa, 1902-98, master plan for city of Brasilia, with Oscar Niemeyer.

Ralph Adams Cram, 1863-1942, Cath. of St. John the Divine, NYC; U.S. Military Acad. (part), West Point, NY.

R. Buckminster Fuller, 1895-1983, U.S. Pavilion (geodesic domes), Expo 67, Montreal.

Frank O. Gehry, b 1929, Hollywood Bowl Shell (phase I), Los Angeles, CA.

Cass Gilbert, 1859-1934, Custom House, Woolworth Bldg., NYC; Supreme Court Bldg., Wash., DC.

Bertram G. Goodhue, 1869-1924, Capitol, Lincoln, NE; St. Thomas's Church, St. Bartholomew's Church, NYC.

Walter Gropius, 1883-1969, Pan Am Bldg. (now MetLife Bldg.) (with Pietro Belluschi), NYC.

Lawrence Halprin, b 1916, Ghirardelli Sq., San Francisco; Nicollet Mall, Minneapolis; FDR Memorial, Wash., DC.

Peter Harrison, 1716-75, Touro Synagogue, Redwood Library, Newport, RI.

Wallace K. Harrison, 1895-1981, Metropolitan Opera House, Lincoln Center, NYC.

Thomas Hastings, 1860-1929, NY Public Library (with John Carrère), Frick Mansion, NYC.

James Hoban, 1762-1831, White House, Wash., DC.

Raymond Hood, 1881-1934, Rockefeller Center (part), Daily News, NYC; Tribune, Chicago, IL.

Richard M. Hunt, 1827-95, Metropolitan Museum (part), NYC; National Observatory, Wash., DC.

William Le Baron Jenney, 1832-1907, Home Insurance (demolished 1931), Chicago, IL.

Philip C. Johnson, b 1906, AT&T headquarters (now 550 Madison Ave.), NYC; Transco Tower, Houston, TX.

Albert Kahn, 1869-1942, General Motors Bldg., Detroit, MI.

Louis Kahn, 1901-74, Salk Laboratory, La Jolla, CA; Yale Art Gallery, New Haven, CT.

Christopher Grant LaFarge, 1862-1938, Roman Catholic Chapel, West Point, NY.

Benjamin H. Latrobe, 1764-1820, Capitol (part), Wash., DC; State Capitol Bldg., Richmond, VA.

Le Corbusier, (Charles-Edouard Jeanneret), 1887-1965, Salvation Army Hostel and Swiss Dormitory, both Paris; master plan for cities of Algiers and Buenos Aires.

William Lescaze, 1896-1969, Philadelphia Savings Fund Society; Borg-Warner Bldg., Chicago.

Maya Lin, b 1959, Vietnam Veterans Memorial, Wash., DC.

Charles Rennie Mackintosh, 1868-1928, Glasgow School of Art; Hill House, Helensburgh.

Bernard R. Maybeck, 1862-1957, Hearst Hall, Univ. of CA, Berkeley; First Church of Christ Scientist, Berkeley, CA.

Charles F. McKim, 1847-1909, Public Library, Boston; Columbia Univ. (part), NYC.

Charles M. McKim, b 1920, KUHT-TV Transmitter Bldg., Lutheran Church of the Redeemer, Houston, TX.

Richard Meier, b 1934, Getty Center Museum, Los Angeles, CA; High Museum of Art, Atlanta, GA.

Ludwig Mies van der Rohe, 1886-1969, Seagram Bldg., (with Philip C. Johnson), NYC; National Gallery, Berlin.

Robert Mills, 1781-1855, Washington Monument, Wash., DC.

Charles Moore, 1925-93, Sea Ranch, near San Francisco; Piazza d'Italia, New Orleans, LA.

Richard J. Neutra, 1892-1970, Mathematics Park, Princeton, NJ; Orange Co. Courthouse, Santa Ana, CA.

Oscar Niemeyer, b 1907, government buildings, Brasilia Palace Hotel, all Brasilia.

Gyo Obata, b 1923, Natl. Air & Space Museum, Smithsonian Inst., Wash., DC; Dallas-Ft. Worth Airport.

Frederick L. Olmsted, 1822-1903, Central Park, NYC; Fairmount Park, Philadelphia, PA.

I(eoh) M(ing) Pei, b 1917, East Wing, Natl. Gallery of Art, Wash., DC; Pyramid, The Louvre, Paris; Rock & Roll Hall of Fame and Museum, Cleveland, OH.

Cesar Pelli, b 1926, World Financial Center, Carnegie Hall Tower, NYC; Petronas Twin Towers, Malaysia.

William Pereira, 1909-85, Cape Canaveral; Transamerica Bldg., San Francisco, CA.

John Russell Pope, 1874-1937, National Gallery, Wash., DC.

John Portman, b 1924, Peachtree Center, Atlanta, GA.

George Browne Post, 1837-1913, NY Stock Exchange; Capitol, Madison, WI.

James Renwick Jr., 1818-95, Grace Church, St. Patrick's Cath., NYC.; Corcoran (now Renwick) Gallery, Wash., DC.

Henry H. Richardson, 1838-86, Trinity Church, Boston, MA.

Kevin Roche, b 1922, Oakland Museum, Oakland, CA; Fine Arts Center, University of Massachusetts, Amherst.

James Gamble Rogers, 1867-1947, Columbia-Presbyterian Medical Center, NYC; Northwestern Univ., Evanston, IL.

John Wellborn Root, 1887-1963, Palmolive Bldg., Chicago; Hotel Statler, Wash., DC.

Paul Rudolph, 1918-97, Jewitt Art Center, Wellesley Colllege, MA; Art & Architecture Bldg., Yale Univ., New Haven, CT.

Eero Saarinen, 1910-61, Gateway to the West Arch, St. Louis, MO; Trans World Flight Center, NYC.

Louis Skidmore, 1897-1962, Atomic Energy Commission town site, Oak Ridge, TN; Terrace Plaza Hotel, Cincinnati, OH.

Clarence S. Stein, 1882-1975, Temple Emanu-El, NYC.

Edward Durell Stone, 1902-78, U.S. Embassy, New Delhi, India; (H. Hartford) Gallery of Modern Art, NYC.

Louis H. Sullivan, 1856-1924, Auditorium Bldg., Chicago, IL.

Richard Upjohn, 1802-78, Trinity Church, NYC.

Max O. Urbahn, 1912-95, Vehicle Assembly Bldg., Cape Canaveral, FL.

Ralph T. Walker, 1889-1973, NY Telephone Bldg. (now NYNEX); IBM Research Lab, Poughkeepsie, NY.

Roland A. Wank, 1898-1970, Cincinnati Union Terminal, OH; head architect (1933-44), Tennessee Valley Authority.

Stanford White, 1853-1906, Washington Arch in Washington Square Park, first Madison Square Garden, NYC.

Frank Lloyd Wright, 1867-1959, Imperial Hotel, Tokyo; Guggenheim Museum, NYC; Marin County Civic Center, San Francisco; Kaufmann "Fallingwater" house, Bear Run, PA.; Taliesen West, Scottsdale, AZ.

William Wurster, 1895-1973, Ghirardelli Sq., San Francisco; Cowell College, UC, Berkeley, CA.

Minoru Yamasaki, 1912-86, World Trade Center, NYC.

Noted Artists, Photographers, and Sculptors of the Past

Artists are painters unless otherwise indicated.

Berenice Abbott, 1898-1991, (U.S.) photographer. Documentary of New York City, *Changing New York* (1939).

Ansel Easton Adams, 1902-84, (U.S.) photographer. Landscapes of the American Southwest.

Washington Allston, 1779-1843, (U.S.) landscapist. *Belshazzar's Feast.*

Albrecht Altdorfer, 1480-1538, (Ger.) landscapist.

Andrea del Sarto, 1486-1530, (It.) frescoes. *Madonna of the Harpies.*

Fra Angelico, c1400-55, (It.) Renaissance muralist. *Madonna of the Linen Drapers' Guild.*

Diane Arbus, 1923-71, (U.S.) photographer. Disturbing images.

Alexsandr Archipenko, 1887-1964, (U.S.) sculptor. *Boxing Match, Medranos.*

Eugène Atget, 1856-1927, (Fr.) photographer. Paris life.

John James Audubon, 1785-1851, (U.S.) *Birds of America.*

Hans Baldung-Grien, 1484-1545, (Ger.) *Todentanz.*

Ernst Barlach, 1870-1938, (Ger.) Expressionist sculptor. *Man Drawing a Sword.*

Frederic-Auguste Bartholdi, 1834-1904, (Fr.) *Liberty Enlightening the World, Lion of Belfort.*

Fra Bartolommeo, 1472-1517, (It.) *Vision of St. Bernard.*
Aubrey Beardsley, 1872-98, (Br.) illustrator. *Salome, Lysistrata, Morte d'Arthur, Volpone.*
Max Beckmann, 1884-1950, (Ger.) Expressionist. *The Descent From the Cross.*
Gentile Bellini, 1426-1507, (It.) Renaissance. *Procession in St. Mark's Square.*
Giovanni Bellini, 1428-1516, (It.) *St. Francis in Ecstasy.*
Jacopo Bellini, 1400-70, (It.) *Crucifixion.*
George Wesley Bellows, 1882-1925, (U.S.) sports artist, portraitist, landscapist. *Stag at Sharkey's, Edith Clavell.*
Thomas Hart Benton, 1889-1975, (U.S.) American regionalist. *Threshing Wheat, Arts of the West.*
Gianlorenzo Bernini, 1598-1680, (It.) Baroque sculpture. *The Assumption.*
Albert Bierstadt, 1830-1902, (U.S.) landscapist. *The Rocky Mountains, Mount Corcoran.*
George Caleb Bingham, 1811-79, (U.S.) *Fur Traders Descending the Missouri.*
William Blake, 1752-1827, (Br.) engraver. *Book of Job, Songs of Innocence, Songs of Experience.*
Rosa Bonheur, 1822-99, (Fr.) *The Horse Fair.*
Pierre Bonnard, 1867-1947, (Fr.) Intimist. *The Breakfast Room, Girl in a Straw Hat.*
Gutzon Borglum, 1871-1941, (U.S.) sculptor. Mt. Rushmore Memorial.
Hieronymus Bosch, 1450-1516, (Flem.) religious allegories. *The Crowning With Thorns.*
Sandro Botticelli, 1444-1510, (It.) Renaissance. *Birth of Venus, Adoration of the Magi, Guiliano de'Medici.*
Margaret Bourke-White, 1906-71, (U.S.) photographer, photojournalist. WW2, USSR, rural South during the Depression.
Mathew Brady, c1823-96, (U.S.) photographer. Official photographer of the Civil War.
Constantin Brancusi, 1876-1957, (Romanian-Fr.) Nonobjective sculptor. *Flying Turtle, The Kiss.*
Georges Braque, 1882-1963, (Fr.) Cubist. *Violin and Palette.*
Pieter Bruegel the Elder, c1525-69, (Flem.) *The Peasant Dance, Hunters in the Snow, Magpie on the Gallows.*
Pieter Bruegel the Younger, 1564-1638, (Flem.) *Village Fair, The Crucifixion.*
Edward Burne-Jones, 1833-98, (Br.) Pre-Raphaelite artist-craftsman. *The Mirror of Venus.*
Alexander Calder, 1898-1976, (U.S.) sculptor. *Lobster Trap and Fish Tail.*
Julia Cameron, 1815-79, (Br.) photographer. Considered one of the most important portraitists of the 19th cent.
Robert Capa (Andrei Friedmann), 1913-54, (Hung.-U.S.) photographer. War photojournalist; invasion of Normandy.
Michelangelo Merisi da Caravaggio, 1573-1610, (It.) Baroque. *The Supper at Emmaus.*
Emily Carr, 1871-1945, (Can.) landscapist. *Blunden Harbour, Big Raven, Rushing Sea of Undergrowth.*
Carlo Carrà, 1881-1966, (It.) Metaphysical school. *Lot's Daughters, The Enchanted Room.*
Mary Cassatt, 1844-1926, (U.S.) Impressionist. *The Cup of Tea, Woman Bathing, The Boating Party.*
George Catlin, 1796-1872, (U.S.) American Indian life. *Gallery of Indians, Buffalo Dance.*
Benvenuto Cellini, 1500-71, (It.) Mannerist sculptor, goldsmith. *Perseus and Medusa.*
Paul Cézanne, 1839-1906, (Fr.) *Card Players, Mont-Sainte-Victoire With Large Pine Trees.*
Marc Chagall, 1887-1985, (Russ.) Jewish life and folklore. *I and the Village, The Praying Jew.*
Jean Simeon Chardin, 1699-1779, (Fr.) still lifes. *The Kiss, The Grace.*
Frederick Church, 1826-1900, (U.S.) Hudson River school. *Niagara, Andes of Ecuador.*
Giovanni Cimabue, 1240-1302, (It.) Byzantine mosaicist. *Madonna Enthroned With St. Francis.*
Claude Lorrain (Claude Gellé), 1600-82, (Fr.) ideal-landscapist. *The Enchanted Castle.*
Thomas Cole, 1801-48, (U.S.) Hudson River school. *The Ox-Bow, In the Catskills.*
John Constable, 1776-1837, (Br.) landscapist. *Salisbury Cathedral From the Bishop's Grounds.*
John Singleton Copley, 1738-1815, (U.S.) portraitist. *Samuel Adams, Watson and the Shark.*
Lovis Corinth, 1858-1925, (Ger.) Expressionist. *Apocalypse.*
Jean-Baptiste-Camille Corot, 1796-1875, (Fr.) landscapist. *Souvenir de Mortefontaine, Pastorale.*
Correggio, 1494-1534, (It.) Renaissance muralist. *Mystic Marriages of St. Catherine.*
Gustave Courbet, 1819-77, (Fr.) Realist. *The Artist's Studio.*
Lucas Cranach the Elder, 1472-1553, (Ger.) Protestant Reformation portraitist. *Luther.*
Imogen Cunningham, 1883-1976, (U.S.) photographer, portraitist. Plant photography.
Nathaniel Currier, 1813-88, and **James M. Ives,** 1824-95, (both U.S.) lithographers. *A Midnight Race on the Mississippi, American Forest Scene—Maple Sugaring.*

John Steuart Curry, 1897-1946, (U.S.) Americana, murals. *Baptism in Kansas.*
Salvador Dalí, 1904-89, (Sp.) Surrealist. *Persistence of Memory, The Crucifixion.*
Honoré Daumier, 1808-79, (Fr.) caricaturist. *The Third-Class Carriage.*
Jacques-Louis David, 1748-1825, (Fr.) Neoclassicist. *The Oath of the Horatii.*
Arthur Davies, 1862-1928, (U.S.) Romantic landscapist. *Unicorns, Leda and the Dioscuri.*
Willem de Kooning, 1904-1997, (Dutch-U.S.) abstract expressionist. *Excavation, Woman I, Door to the River.*
Edgar Degas, 1834-1917, (Fr.) *The Ballet Class.*
Eugène Delacroix, 1798-1863, (Fr.) Romantic. *Massacre at Chios, Liberty Leading the People.*
Paul Delaroche, 1797-1856, (Fr.) historical themes. *Children of Edward IV.*
Luca Della Robbia, 1400-82, (It.) Renaissance terracotta artist. *Cantoria* (singing gallery), Florence cathedral.
Donatello, 1386-1466, (It.) Renaissance sculptor. *David, Gattamelata.*
Jean Dubuffet, 1902-85, (Fr.) painter, sculptor, printmaker. *Group of Four Trees.*
Marcel Duchamp, 1887-1968, (Fr.) Dada artist. *Nude Descending a Staircase, No. 2.*
Raoul Dufy, 1877-1953, (Fr.) Fauvist. *Chateau and Horses.*
Asher Brown Durand, 1796-1886, (U.S.) Hudson River school. *Kindred Spirits.*
Albrecht Dürer, 1471-1528, (Ger.) Renaissance painter, engraver, woodcuts. *St. Jerome in His Study, Melencolia I.*
Anthony van Dyck, 1599-1641, (Flem.) Baroque portraitist. *Portrait of Charles I Hunting.*
Thomas Eakins, 1844-1916, (U.S.) Realist. *The Gross Clinic.*
Alfred Eisenstaedt, 1898-1995, (Ger.-U.S.) photographer, photojournalist. Famous for V-J Day, Aug. 14, 1945, photo of sailor and nurse in Times Square, NYC.
Peter Henry Emerson, 1856-1936, (Br.) photographer. Promoted photography as an independent art form.
Jacob Epstein, 1880-1959, (Br.) religious and allegorical sculptor. *Genesis, Ecce Homo.*
Jan van Eyck, c1390-1441, (Flem.) naturalistic panels. *Adoration of the Lamb.*
Roger Fenton, 1819-68, (Br.) photographer. Crimean War.
Anselm Feuerbach, 1829-80, (Ger.) Romantic Classicist. *Judgment of Paris, Iphigenia.*
John Bernard Flannagan, 1895-1942, (U.S.) animal sculptor. *Triumph of the Egg.*
Jean-Honoré Fragonard, 1732-1806, (Fr.) Rococo. *The Swing.*
Daniel Chester French, 1850-1931, (U.S.) *The Minute Man of Concord;* seated *Lincoln,* Lincoln Memorial, Wash., DC.
Caspar David Friedrich, 1774-1840, (Ger.) Romantic landscapes. *Man and Woman Gazing at the Moon.*
Thomas Gainsborough, 1727-88, (Br.) portraitist. *The Blue Boy, The Watering Place, Orpin the Parish Clerk.*
Alexander Gardner, 1821-82, (U.S.) photographer. Civil War; railroad construction; Great Plains Indians.
Paul Gauguin, 1848-1903, (Fr.) Post-impressionist. *The Tahitians, Spirit of the Dead Watching.*
Lorenzo Ghiberti, 1378-1455, (It.) Renaissance sculptor. Gates of Paradise baptistery doors, Florence.
Alberto Giacometti, 1901-66, (Swiss) attenuated sculptures of solitary figures. *Man Pointing.*
Giorgione, c1477-1510, (It.) Renaissance. *The Tempest.*
Giotto di Bondone, 1267-1337, (It.) Renaissance. *Presentation of Christ in the Temple.*
François Girardon, 1628-1715, (Fr.) Baroque sculptor of classical themes. *Apollo Tended by the Nymphs.*
Vincent van Gogh, 1853-90, (Dutch) *The Starry Night, L'Arlesienne, Bedroom at Arles, Self-Portrait.*
Arshile Gorky, 1905-48, (U.S.) Surrealist. *The Liver Is the Cock's Comb.*
Francisco de Goya y Lucientes, 1746-1828, (Sp.) *The Naked Maja, The Disasters of War* (etchings).
El Greco, 1541-1614, (Sp.) *View of Toledo, Assumption of the Virgin.*
Horatio Greenough, 1805-52, (U.S.) Neo-classical sculptor.
Matthias Grünewald, 1480-1528, (Ger.) mystical religious themes. *The Resurrection.*
Frans Hals, c1580-1666, (Dutch) portraitist. *Laughing Cavalier, Gypsy Girl.*
Austin Hansen, 1910-96, (U.S.) photographer. Harlem, NY, life.
Childe Hassam, 1859-1935, (U.S.) Impressionist. *Southwest Wind, July 14 Rue Daunon.*
Edward Hicks, 1780-1849, (U.S.) folk painter. *The Peaceable Kingdom.*
Lewis Wickes Hine, 1874-1940, (U.S.) photographer. Studies of immigrants, children in industry.
Hans Hofmann, 1880-1966, (U.S.) early abstract Expressionist. *Spring, The Gate.*
William Hogarth, 1697-1764, (Br.) caricaturist. *The Rake's Progress.*
Katsushika Hokusai, 1760-1849, (Jpn.) printmaker. *Crabs.*

Hans Holbein the Elder, 1460-1524, (Ger.) late Gothic. *Presentation of Christ in the Temple.*

Hans Holbein the Younger, 1497-1543, (Ger.) portraitist. *Henry VIII, The French Ambassadors.*

Winslow Homer, 1836-1910, (U.S.) naturalist painter, marine themes. *Marine Coast, High Cliff.*

Edward Hopper, 1882-1967, (U.S.) realistic urban scenes. *Nighthawks, House by the Railroad.*

Jean-Auguste-Dominique Ingres, 1780-1867, (Fr.) Classicist. *Valpincon Bather.*

George Inness, 1825-94, (U.S.) luminous landscapist. *Delaware Water Gap.*

William Henry Jackson, 1843-1942, (U.S.) photographer. American West, building of Union Pacific Railroad.

Donald Judd, 1928-94, (U.S.) sculptor, major Minimalist.

Frida Kahlo, 1907-54, (Mex.) painter; *Self-Portrait With Monkey.*

Vasily Kandinsky, 1866-1944, (Russ.) Abstractionist. *Capricious Forms, Improvisation 38 (second version).*

Paul Klee, 1879-1940, (Swiss) Abstractionist. *Twittering Machine, Pastoral, Death and Fire.*

Gustav Klimt, 1862-1918, (Austrian) cofounder of Vienna Secession Movement, *The Kiss.*

Oscar Kokoschka, 1886-1980, (Austrian) Expressionist. *View of Prague, Harbor of Marseilles.*

Kathe Kollwitz, 1867-1945, (Ger.) printmaker, social justice themes. *The Peasant War.*

Gaston Lachaise, 1882-1935, (U.S.) figurative sculptor. *Standing Woman.*

John La Farge, 1835-1910, (U.S.) muralist. *Red and White Peonies, The Ascension.*

Sir Edwin (Henry) Landseer, 1802-73, (Br.) painter, sculptor. *Shoeing, Rout of Comus.*

Dorothea Lange, 1895-1965, (U.S.), photographer. Depression photographs, migrant farm workers.

Fernand Léger, 1881-1955, (Fr.) machine art. *The Cyclists.*

Leonardo da Vinci, 1452-1519, (It.) *Mona Lisa, Last Supper, The Annunciation.*

Emanuel Leutze, 1816-68, (U.S.) historical themes. *Washington Crossing the Delaware.*

Roy Lichtenstein, 1923-97, (U.S.) pop artist.

Jacques Lipchitz, 1891-1973, (Fr.) Cubist sculptor. *Harpist.*

Filippino Lippi, 1457-1504, (It.) Renaissance.

Fra Filippo Lippi, 1406-69, (It.) Renaissance. *Coronation of the Virgin, Madonna and Child With Angels.*

Morris Louis, 1912-62, (U.S.) abstract Expressionist. *Signa, Stripes, Alpha-Phi.*

Aristide Maillol, 1861-1944, (Fr.) sculptor. *L'Harmonie.*

Édouard Manet, 1832-83, (Fr.) forerunner of Impressionism. *Luncheon on the Grass, Olympia.*

Andrea Mantegna, 1431-1506, (It.) Renaissance frescoes. *Triumph of Caesar.*

Franz Marc, 1880-1916, (Ger.) Expressionist. *Blue Horses.*

John Marin, 1870-1953, (U.S.) Expressionist seascapes. *Maine Island.*

Reginald Marsh, 1898-1954, (U.S.) satirical artist. *Tattoo and Haircut.*

Masaccio, 1401-28, (It.) Renaissance. *The Tribute Money.*

Henri Matisse, 1869-1954, (Fr.) Fauvist. *Woman With the Hat.*

Michelangelo Buonarroti, 1475-1564, (It.) *Pietà, David, Moses, The Last Judgment,* Sistine Chapel ceiling.

Jean-Francois Millet, 1814-75, (Fr.) painter of peasant subjects. *The Gleaners, The Man With a Hoe.*

Joan Miró, 1893-1983, (Sp.) Exuberant colors, playful images. Catalan landscape, *Dutch Interior.*

Amedeo Modigliani, 1884-1920, (It.) *Reclining Nude.*

Piet Mondrian, 1872-1944, (Dutch) Abstractionist. *Composition With Red, Yellow and Blue.*

Claude Monet, 1840-1926, (Fr.) Impressionist. *The Bridge at Argenteuil, Haystacks.*

Henry Moore, 1898-1986, (Br.) sculptor of large-scale, abstract works. *Reclining Figure* (several).

Gustave Moreau, 1826-98, (Fr.) Symbolist. *The Apparition, Dance of Salome.*

James Wilson Morrice, 1865-1924, (Can.) landscapist. *The Ferry, Quebec, Venice, Looking Over the Lagoon.*

William Morris, 1834-1896, (Br.) decorative artist, leader of the Arts and Crafts movement.

Grandma Moses, 1860-1961, (U.S.) folk painter. *Out for the Christmas Trees, Thanksgiving Turkey.*

Edvard Munch, 1863-1944, (Nor.) Expressionist. *The Cry.*

Bartolome Murillo, 1618-82, (Sp.) Baroque religious artist. *Vision of St. Anthony, The Two Trinities.*

Eadweard Muybridge, 1830-1904, (Br.-U.S.) photographer. Studies of motion, *Animal Locomotion.*

Nadar (Gaspar-Félix Tournachon), 1820-1910, (Fr.) photographer, caricaturist, portraitist. Invented photo-essay.

Barnett Newman, 1905-70, (U.S.) abstract Expressionist. *Stations of the Cross.*

Isamu Noguchi, 1904-88, (U.S.) abstract sculptor, designer. *Kouros, BirdC(MU),* sculptural gardens.

Georgia O'Keeffe, 1887-1986, (U.S.) Southwest motifs. *Cow's Skull: Red, White, and Blue, The Shelton With Sunspots.*

José Clemente Orozco, 1883-1949, (Mex.) frescoes. *House of Tears, Pre-Columbian Golden Age.*

Timothy H. O'Sullivan, 1840-82, (U.S.) Civil War photographer.

Charles Willson Peale, 1741-1827, (U.S.) Amer. Revolutionary portraitist. *The Staircase Group,* U.S. presidents.

Rembrandt Peale, 1778-1860, (U.S.) portraitist. Thomas Jefferson.

Pietro Perugino, 1446-1523, (It.) Renaissance. *Delivery of the Keys to St. Peter.*

Pablo Picasso, 1881-1973, (Sp.) painter, sculptor. *Guernica; Dove; Head of a Woman; Head of a Bull, Metamorphosis.*

Piero della Francesca, c1415-92, (It.) Renaissance. *Duke of Urbino, Flagellation of Christ.*

Camille Pissarro, 1830-1903, (Fr.) Impressionist. *Boulevard des Italiens, Morning, Sunlight; Bather in the Woods.*

Jackson Pollock, 1912-56, (U.S.) abstract Expressionist. *Autumn Rhythm.*

Nicolas Poussin, 1594-1665, (Fr.) Baroque pictorial classicism. *St. John on Patmos.*

Maurice B. Prendergast, c1860-1924, (U.S.) Post-impressionist water colorist. *Umbrellas in the Rain.*

Pierre-Paul Prud'hon, 1758-1823, (Fr.) Romanticist. *Crime Pursued by Vengeance and Justice.*

Pierre Cecile Puvis de Chavannes, 1824-98, (Fr.) muralist. *The Poor Fisherman.*

Raphael Sanzio, 1483-1520, (It.) Renaissance. *Disputa, School of Athens, Sistine Madonna.*

Man Ray, 1890-1976, (U.S.) Dada artist. *Observing Time, The Lovers, Marquis de Sade.*

Odilon Redon, 1840-1916, (Fr.) Symbolist painter, lithographer. *In the Dream, Vase of Flowers.*

Rembrandt van Rijn, 1606-69, (Dutch) *The Bridal Couple, The Night Watch.*

Frederic Remington, 1861-1909, (U.S.) painter, sculptor. Portrayer of the American West, *Bronco Buster.*

Pierre-Auguste Renoir, 1841-1919, (Fr.) Impressionist. *The Luncheon of the Boating Party, Dance in the Country.*

Joshua Reynolds, 1723-92, (Br.) portraitist. *Mrs. Siddons As the Tragic Muse.*

Diego Rivera, 1886-1957, (Mex.) frescoes. *The Fecund Earth.*

Henry Peach Robinson, 1830-1901 (Br.) photographer. A leader of "high art" photography.

Norman Rockwell, 1894-1978, (U.S.) painter, illustrator. *Saturday Evening Post* covers.

Auguste Rodin, 1840-1917, (Fr.) sculptor. *The Thinker.*

Mark Rothko, 1903-70, (U.S.) abstract Expressionist. *Light, Earth and Blue.*

Georges Rouault, 1871-1958, (Fr.) Expressionist. *Three Judges.*

Henri Rousseau, 1844-1910, (Fr.) primitive exotic themes. *The Snake Charmer.*

Theodore Rousseau, 1812-67, (Swiss-Fr.) landscapist. *Under the Birches, Evening.*

Peter Paul Rubens, 1577-1640, (Flem.) Baroque. *Mystic Marriage of St. Catherine.*

Jacob van Ruisdael, c1628-82, (Dutch) landscapist. *Jewish Cemetery.*

Charles M. Russell, 1866-1926, (U.S.) Western life.

Salomon van Ruysdael, c1600-70, (Dutch) landscapist. *River with Ferry-Boat.*

Albert Pinkham Ryder, 1847-1917, (U.S.) seascapes and allegories. *Toilers of the Sea.*

Augustus Saint-Gaudens, 1848-1907, (U.S.) memorial statues. *Farragut, Mrs. Henry Adams (Grief).*

Andrea Sansovino, 1460-1529, (It.) Renaissance sculptor. *Baptism of Christ.*

Jacopo Sansovino, 1486-1570, (It.) Renaissance sculptor. *St. John the Baptist.*

John Singer Sargent, 1856-1925, (U.S.) Edwardian society portraitist. *The Wyndham Sisters, Madam X.*

Georges Seurat, 1859-91, (Fr.) Pointillist. *Sunday Afternoon on the Island of La Grande Jatte.*

Gino Severini, 1883-1966, (It.) Futurist and Cubist. *Dynamic Hieroglyph of the Bal Tabarin.*

Ben Shahn, 1898-1969, (U.S.) social and political themes. Sacco and Vanzetti series, *Seurat's Lunch, Handball.*

Charles Sheeler, 1883-1965, (U.S.) abstractionist.

David Alfaro Siqueiros, 1896-1974, (Mex.) political muralist. *March of Humanity.*

David Smith, 1906-65, (U.S.) welded metal sculpture. *Hudson River Landscape, Zig, Cubi* series.

Edward Steichen, 1879-1973, (U.S.) photographer. Credited with transforming photography into an art form.

Alfred Stieglitz, 1864-1946, (U.S.) photographer, editor; helped create acceptance of photography as art.

Paul Strand, 1890-1976, (U.S.) photographer. People, nature, landscapes.

Gilbert Stuart, 1755-1828, (U.S.) portraitist. George Washington, Thomas Jefferson, James Madison.

Thomas Sully, 1783-1872, (U.S.) portraitist. *Col. Thomas Handasyd Perkins, The Passage of the Delaware.*

William Henry Fox Talbot, 1800-77, (Br.) photographer. *Pencil of Nature*, early photographically illustrated book.

George Tames, 1919-94, (U.S.) photographer. Chronicled presidents, political leaders.

Yves Tanguy, 1900-55, (Fr.) Surrealist. *Rose of the Four Winds, Mama, Papa Is Wounded!*

Giovanni Battista Tiepolo, 1696-1770, (It.) Rococo frescoes. *The Crucifixion.*

Jacopo Tintoretto, 1518-94, (It.) Mannerist. *The Last Supper.*

Titian, c1485-1576, (It.) Renaissance. *Venus and the Lute Player, The Bacchanal.*

Jose Rey Toledo, 1916-94, (U.S.) Native American artist. Captured the essence of tribal dances on canvas.

Henri de Toulouse-Lautrec, 1864-1901, (Fr.) *At the Moulin Rouge.*

John Trumbull, 1756-1843, (U.S.) historical themes. *The Declaration of Independence.*

J(oseph) M(allord) W(illiam) Turner, 1775-1851, (Br.) Romantic landscapist. *Snow Storm.*

Paolo Uccello, 1397-1475, (It.) Gothic-Renaissance. *The Rout of San Romano.*

Maurice Utrillo, 1883-1955, (Fr.) Impressionist. *Sacre-Coeur de Montmartre.*

John Vanderlyn, 1775-1852, (U.S.) Neo-classicist. *Ariadne Asleep on the Island of Naxos.*

Diego Velázquez, 1599-1660, (Sp.) Baroque. *Las Meninas, Portrait of Juan de Pareja.*

Jan Vermeer, 1632-75, (Dutch) interior genre subjects. *Young Woman With a Water Jug.*

Paolo Veronese, 1528-88, (It.) devotional themes, vastly peopled canvases. *The Temptation of St. Anthony.*

Andrea del Verrocchio, 1435-88, (It.) Florentine sculptor. *Colleoni.*

Maurice de Vlaminck, 1876-1958, (Fr.) Fauvist landscapist. *Red Trees.*

Andy Warhol, 1928-87, (U.S.) Pop Art. *Campbell's Soup Cans, Marilyn Diptych.*

Antoine Watteau, 1684-1721, (Fr.) Rococo painter of "scenes of gallantry." *The Embarkation for Cythera.*

George Frederic Watts, 1817-1904, (Br.) painter and sculptor of grandiose allegorical themes. *Hope.*

Benjamin West, 1738-1820, (U.S.) realistic historical themes. *Death of General Wolfe.*

Edward Weston, 1886-1958, (U.S.) photographer. Landscapes of American West.

James Abbott McNeill Whistler, 1834-1903, (U.S.) *Arrangement in Grey and Black, No. 1: The Artist's Mother.*

Archibald M. Willard, 1836-1918, (U.S.) *The Spirit of '76.*

Grant Wood, 1891-1942, (U.S.) Midwestern regionalist. *American Gothic, Daughters of Revolution.*

Ossip Zadkine, 1890-1967, (Russ.) School of Paris sculptor. *The Destroyed City, Musicians, Christ.*

Noted Business Leaders, Industrialists, and Philanthropists of the Past

Elizabeth Arden (F. N. Graham), 1884-1966, (U.S.) Canadian-born founder of cosmetics empire.

Philip D. Armour, 1832-1901, (U.S.) industrialist; streamlined meatpacking.

John Jacob Astor, 1763-1848, (U.S.) German-born fur trader, banker, real estate magnate; at death, richest in U.S.

Francis W. Ayer, 1848-1923, (U.S.) ad industry pioneer.

August Belmont, 1816-90, (U.S.) German-born financier.

James B. (Diamond Jim) Brady, 1856-1917, (U.S.) financier, philanthropist, legendary bon vivant.

Adolphus Busch, 1839-1913, (U.S.) German-born businessman; established brewery empire.

Asa Candler, 1851-1929, (U.S.) founded Coca-Cola Co.

Andrew Carnegie, 1835-1919, (U.S.) Scottish-born industrialist; philanthropist; founded Carnegie Steel Co.

Tom Carvel, 1908-89, (Gr.-U.S.) founded ice cream chain.

William Colgate, 1783-1857, (Br.-U.S.) Br.-born businessman, philanthropist; founded soap-making empire.

Jay Cooke, 1821-1905, (U.S.) financier; sold $1 billion in Union bonds during Civil War.

Peter Cooper, 1791-1883, (U.S.) industrialist, inventor, philanthropist; founded Cooper Union (1859).

Ezra Cornell, 1807-74, (U.S.) businessman, philanthropist; headed Western Union, established university.

Erastus Corning, 1794-1872, (U.S.) financier; headed N.Y. Central.

Charles Crocker, 1822-88, (U.S.) railroad builder, financier.

Samuel Cunard, 1787-1865, (Can.) pioneered trans-Atlantic steam navigation.

Marcus Daly, 1841-1900, (U.S.) Irish-born copper magnate.

W. Edwards Deming, 1900-93, (U.S.) quality-control expert who revolutionized Japanese manufacturing.

Walt Disney, 1901-66, (U.S.) pioneer in cinema animation; built entertainment empire.

Herbert H. Dow, 1866-1930, (U.S.) founder of chemical co.

James Duke, 1856-1925, (U.S.) founded American Tobacco, Duke Univ.

Eleuthere I. du Pont, 1771-1834, (Fr.-U.S.) gunpowder manufacturer; founded one of the largest business empires.

Thomas E. Durant, 1820-85, (U.S.) railroad official, financier.

William C. Durant, 1861-1947, (U.S.) industrialist; formed General Motors.

George Eastman, 1854-1932, (U.S.) inventor; manufacturer of photographic equipment.

Marshall Field, 1834-1906, (U.S.) merchant; founded Chicago's largest department store.

Harvey Firestone, 1868-1938, (U.S.) founded tire company.

Avery Fisher, 1906-94, (U.S.) industrialist, philanthropist, founded Fisher electronics.

Henry M. Flagler, 1830-1913, (U.S.) financier; helped form Standard Oil; developed Florida as resort state.

Malcolm Forbes, 1919-90, (U.S.) magazine publisher.

Henry Ford, 1863-1947, (U.S.) auto maker; developed first popular low-priced car.

Henry Ford 2d, 1917-87, (U.S.) headed auto company founded by grandfather.

Henry C. Frick, 1849-1919, (U.S.) steel and coke magnate; had prominent role in development of U.S. Steel.

Jakob Fugger (Jakob the Rich), 1459-1525, (Ger.) headed leading banking, trading house, in 16th-cent. Europe.

Alfred C. Fuller, 1885-1973, (U.S.) Canadian-born businessman; founded brush company.

Elbert H. Gary, 1846-1927, (U.S.) one of the organizers of U.S. Steel; chaired board of directors, 1903-27.

Jean Paul Getty, 1892-1976, (U.S.) founded oil empire.

Amadeo P. Giannini, 1870-1949, (U.S.) founded Bank of America.

Stephen Girard, 1750-1831, (U.S.) French-born financier, philanthropist; richest man in U.S. at his death.

Jay Gould, 1836-92, (U.S.) railroad magnate, financier.

Hetty Green, 1834-1916, (U.S.) financier, the "witch of Wall St."; richest woman in U.S. in her day.

William Gregg, 1800-67, (U.S.) launched textile industry in S.

Meyer Guggenheim, 1828-1905, (U.S.) Swiss-born merchant, philanthropist; built merchandising, mining empires.

Armand Hammer, 1898-1990, (U.S.) headed Occidental Petroleum; promoted U.S.-Soviet ties.

Edward H. Harriman, 1848-1909, (U.S.) railroad financier, administrator; headed Union Pacific.

Henry J. Heinz, 1844-1919, (U.S.) founded food empire.

James J. Hill, 1838-1916, (U.S.) Canadian-born railroad magnate, financier; founded Great Northern Railway.

Conrad N. Hilton, 1888-1979, (U.S.) hotel chain founder.

Howard Hughes, 1905-76, (U.S.) industrialist, aviator, movie maker.

H. L. Hunt, 1889-1974, (U.S.) oil magnate.

Collis P. Huntington, 1821-1900, (U.S.) railroad magnate.

Henry E. Huntington, 1850-1927, (U.S.) railroad builder, philanthropist.

Walter L. Jacobs, 1898-1985, (U.S.) founder of the first rental car agency, which later became Hertz.

Howard Johnson, 1896-1972, (U.S.) founded restaurants.

Henry J. Kaiser, 1882-1967, (U.S.) industrialist; built empire in steel, aluminum.

Minor C. Keith, 1848-1929, (U.S.) railroad magnate; founded United Fruit Co.

Will K. Kellogg, 1860-1951, (U.S.) businessman, philanthropist; founded breakfast food co.

Richard King, 1825-85, (U.S.) cattleman; founded half-million-acre King Ranch in Texas.

William S. Knudsen, 1879-1948, (U.S.) Danish-born auto industry executive.

Samuel H. Kress, 1863-1955, (U.S.) businessman, art collector, philanthropist; founded "dime store" chain.

Ray A. Kroc, 1902-84, (U.S.) founded fast-food chain, McDonald's Corporation.

Alfred Krupp, 1812-87, (Ger.) armaments magnate.

William Levitt, 1907-94, (U.S.) industrialist, "suburb maker".

Thomas Lipton, 1850-1931, (Scot.) merchant, tea empire.

James McGill, 1744-1813, (Scot.-Can.) founded university.

Andrew W. Mellon, 1855-1937, (U.S.) financier, industrialist; benefactor of National Gallery of Art.

Charles E. Merrill, 1885-1956, (U.S.) financier; developed firm of Merrill Lynch.

John Pierpont Morgan, 1837-1913, (U.S.) most powerful figure in finance and industry at the turn of the cent.

Malcolm Muir, 1885-1979, (U.S.) created *Business Week* magazine; headed *Newsweek,* 1937-61.

Samuel Newhouse, 1895-1979, (U.S.) publishing and broadcasting magnate; built communications empire.

Aristotle Onassis, 1906-75, (Gr.) shipping magnate.

William S. Paley, 1901-90, (U.S.) built CBS communic. empire.

George Peabody, 1795-1869, (U.S.) merchant, financier, philanthropist.

James C. Penney, 1875-1971, (U.S.) businessman; developed department store chain.

William C. Procter, 1862-1934, (U.S.) headed soap co.

John D. Rockefeller, 1839-1937, (U.S.) industrialist; established Standard Oil.

John D. Rockefeller Jr., 1874-1960, (U.S.) philanthropist; established foundation; provided land for UN.
Meyer A. Rothschild, 1743-1812, (Ger.) founded international banking house.
Thomas Fortune Ryan, 1851-1928, (U.S.) financier; a founder of American Tobacco.
David Sarnoff, 1891-1971, (U.S.) broadcasting pioneer; established first radio network, NBC.
Richard Sears, 1863-1914, (U.S.) founded mail-order co.
Werner von Siemens, 1816-92, (Ger.) industrialist; inventor.
Alfred P. Sloan, 1875-1966, (U.S.) industrialist, philanthropist; headed General Motors.
A. Leland Stanford, 1824-93, (U.S.) railroad official, philanthropist; founded university.
Nathan Straus, 1848-1931, (U.S.) German-born merchant, philanthropist; headed Macy's.
Levi Strauss, c1829-1902, (U.S.) pants manufacturer.
Clement Studebaker, 1831-1901, (U.S.) wagon, carriage (maker).
Gustavus Swift, 1839-1903, (U.S.) pioneer meatpacker.
Gerard Swope, 1872-1957, (U.S.) industrialist, economist; headed General Electric.
James Walter Thompson, 1847-1928, (U.S.) ad executive.
Alice Tully, 1902-93, (U.S.) philanthropist, arts patron.

Theodore N. Vail, 1845-1920, (U.S.) organized Bell Telephone system; headed AT&T.
Cornelius Vanderbilt, 1794-1877, (U.S.) financier; established steamship, railroad empires.
Henry Villard, 1835-1900, (U.S.) German-born railroad executive, financier.
George Westinghouse, 1846-1914, (US) inventor, manufacturer; organized Westinghouse Electric Co., 1886.
Charles R. Walgreen, 1873-1939, (U.S.) founded drugstore chain.
DeWitt Wallace, 1889-1981, (U.S.) and **Lila Wallace,** 1889-1984, (U.S.) cofounders of *Reader's Digest* magazine.
Sam Walton, 1918-92, (U.S.) founder of Wal-Mart stores.
John Wanamaker, 1838-1922, (U.S.) pioneered department-store merchandising.
Aaron Montgomery Ward, 1843-1913, (U.S.) established first mail-order firm.
Thomas J. Watson, 1874-1956, (U.S.) IBM head, 1914-56.
John Hay Whitney, 1905-82, (U.S.) publisher, sportsman, philanthropist.
Charles E. Wilson, 1890-1961, (U.S.) auto industry exec., public official.
Frank W. Woolworth, 1852-1919, (U.S.) created 5 & 10 chain.
William Wrigley Jr., 1861-1932, (U.S.) founded chewing gum co.

Noted American Cartoonists

Reviewed by Lucy Shelton Caswell, Professor and Curator, Cartoon Research Library, Ohio State University

Scott Adams, b 1957, Dilbert.
Charles Addams, 1912-88, macabre cartoons.
Brad Anderson, b 1924, Marmaduke.
Sergio Aragones, b 1937, *MAD Magazine.*
Peter Arno, 1904-68, *The New Yorker.*
Tex Avery, 1908-80, animator, Bugs Bunny, Porky Pig.
George Baker, 1915-75, The Sad Sack.
Carl Barks, b 1901, Donald Duck comic books.
C. C. Beck, 1910-89, Captain Marvel.
Jim Berry, b 1932, Berry's World.
Herb Block (Herblock), b 1909, political cartoonist.
George Booth, b 1926, *The New Yorker.*
Berkeley Breathed, b 1957, Bloom County.
Dik Browne, 1917-89, Hi & Lois, Hagar the Horrible.
Marjorie Buell, 1904-93, Little Lulu.
Ernie Bushmiller, 1905-82, Nancy.
Milton Caniff, 1907-88, Terry & the Pirates, Steve Canyon.
Al Capp, 1909-79, Li'l Abner.
Roz Chast, b 1954, *The New Yorker.*
Paul Conrad, 1924, political cartoonist.
Roy Crane, 1901-77, Captain Easy, Buz Sawyer.
Robert Crumb, b 1943, underground cartoonist.
Shamus Culhane, 1908-96, animator.
Jay N. Darling (Ding), 1876-1962, political cartoonist.
Jack Davis, b 1926, *MAD Magazine.*
Jim Davis, b 1945, Garfield.
Billy DeBeck, 1890-1942, Barney Google.
Rudolph Dirks, 1877-1968, The Katzenjammer Kids.
Walt Disney, 1901-66, produced animated cartoons, created Mickey Mouse, Donald Duck.
Steve Ditko, b 1927, Spider-Man.
Mort Drucker, b 1929, *MAD Magazine.*
Will Eisner, b 1917, The Spirit.
Jules Feiffer, b 1929, political cartoonist.
Bud Fisher, 1884-1954, Mutt & Jeff.
Ham Fisher, 1900-55, Joe Palooka.
Max Fleischer, 1883-1972, Betty Boop.
Hal Foster, 1892-1982, Tarzan, Prince Valiant.
Fontaine Fox, 1884-1964, Toonerville Folks.
Isadore "Friz" Freleng, 1905-95, animator, Yosemite Sam, Porky Pig, Sylvester and Tweety Bird.
Rube Goldberg, 1883-1970, Boob McNutt.
Chester Gould, 1900-85, Dick Tracy.
Harold Gray, 1894-1968, Little Orphan Annie.
Matt Groening, b 1954, Life in Hell, The Simpsons.
Cathy Guisewite, b 1950, Cathy.
Bill Hanna, b 1910, & **Joe Barbera,** b 1911, animators, Tom & Jerry, Yogi Bear, Flintstones.
Johnny Hart, b 1931, BC, Wizard of Id.
Oliver Harrington, 1912-95, Bootsie.
Alfred Harvey, 1913-94, created Casper the Friendly Ghost.
Jimmy Hatlo, 1898-1963, Little Iodine.
John Held Jr., 1889-1958, Jazz Age.
George Herriman, 1881-1944, Krazy Kat.
Harry Hershfield, 1885-1974, Abie the Agent.
Al Hirschfeld, b 1903, *N.Y. Times* theater caricaturist.
Burne Hogarth, 1911-96, Tarzan.
Helen Hokinson, 1900-49, *The New Yorker.*
Nicole Hollander, b 1939, Sylvia.
Lynn Johnston, b 1947, For Better or For Worse.
Chuck Jones, b 1912, animator, Bugs Bunny, Porky Pig.
Mike Judge, b. 1962, Beavis and Butt-head, King of the Hill.
Bob Kane, b 1916-98, Batman.

Bil Keane, b 1922, The Family Circus.
Walt Kelly, 1913-73, Pogo.
Hank Ketcham, b 1920, Dennis the Menace.
Ted Key, b 1912, Hazel.
Frank King, 1883-1969, Gasoline Alley.
Jack Kirby, 1917-94, Fantastic Four, The Incredible Hulk.
Rollin Kirby, 1875-1952, political cartoonist.
B(ernard) Kliban, 1935-91, cat books.
Edward Koren, b 1935, *The New Yorker.*
Harvey Kurtzman, 1921-93, *MAD Magazine.*
Walter Lantz, 1900-94, Woody Woodpecker.
Gary Larson, b 1950, The Far Side.
Mell Lazarus, b 1929, Momma, Miss Peach.
Stan Lee, b 1922, Marvel Comics.
David Levine, b 1926, *N.Y. Review of Books* caricatures.
Doug Marlette, b 1949, political cartoonist, Kudzu.
Don Martin, b 1931, *MAD Magazine.*
Bill Mauldin, b 1921, political cartoonist.
Jeff MacNelly, b 1947, political cartoonist, Shoe.
Winsor McCay, 1872-1934, Little Nemo.
John T. McCutcheon, 1870-1949, political cartoonist.
George McManus, 1884-1954, Bringing Up Father.
Dale Messick, b 1906, Brenda Starr.
Norman Mingo, 1896-1980, Alfred E. Neuman.
Bob Montana, 1920-75, Archie.
Dick Moores, 1909-86, Gasoline Alley.
Willard Mullin, 1902-78, sports cartoonist; Dodgers "Bum," Mets "Kid."
Russell Myers, b 1938, Broom Hilda.
Thomas Nast, 1840-1902, political cartoonist; Republican elephant.
Pat Oliphant, b 1935, political cartoonist.
Frederick Burr Opper, 1857-1937, Happy Hooligan.
Richard Outcault, 1863-1928, Yellow Kid, Buster Brown.
Mike Peters, b 1943, cartoonist, Mother Goose & Grimm.
George Price, 1901-95, *The New Yorker.*
Antonio Prohias, 1921(?)-98, Spy vs. Spy.
Alex Raymond, 1909-56, Flash Gordon, Jungle Jim.
Forrest (Bud) Sagendorf, 1915-94, Popeye.
Art Sansom, 1920-91, The Born Loser.
Charles Schulz, b 1922, Peanuts.
Elzie C. Segar, 1894-1938, Popeye.
Joe Shuster, 1914-92, & **Jerry Siegel,** 1914-96, Superman.
Sidney Smith, 1887-1935, The Gumps.
Otto Soglow, 1900-75, Little King, Canyon Kiddies.
Art Spiegelman, b 1948, Raw, Maus.
William Steig, b 1907, *The New Yorker.*
Paul Szep, b 1941, political cartoonist.
James Swinnerton, 1875-1974, Little Jimmy.
Paul Terry, 1887-1971, animator of Mighty Mouse.
Bob Thaves, b 1924, Frank and Ernest.
James Thurber, 1894-61, *The New Yorker.*
Garry Trudeau, b 1948, Doonesbury.
Mort Walker, b 1923, Beetle Bailey.
Bill Watterson, b 1958, Calvin and Hobbes.
Russ Westover, 1887-1966, Tillie the Toiler.
Signe Wilkinson, b 1950, political cartoonist.
Frank Willard, 1893-1958, Moon Mullins.
J. R. Williams, 1888-1957, The Willets Family, Out Our Way.
Gahan Wilson, b 1930, *The New Yorker.*
Tom Wilson, b 1931, Ziggy.
Art Young, 1866-1943, political cartoonist.
Chic Young, 1901-73, Blondie.

Noted Economists, Educators, Historians, and Social Scientists of the Past

For Psychologists see Noted Scientists of the Past.

Brooks Adams, 1848-1927, (U.S.) historian, political theoretician; *The Law of Civilization and Decay.*

Henry Adams, 1838-1918, (U.S.) historian; *History of the United States of America, The Education of Henry Adams.*

Francis Bacon, 1561-1626, (Eng.) philosopher, essayist, and statesman; championed observation and induction.

George Bancroft, 1800-91, (U.S.) historian; wrote 10-volume *History of the United States.*

Jack Barbash, 1911-94, (U.S.) labor economist who helped create the AFL-CIO.

Henry Barnard, 1811-1900, (U.S.) public school reformer.

Charles A. Beard, 1874-1948, (U.S.) historian; *The Economic Basis of Politics.*

Bede (the Venerable), c673-735, (Br.) scholar, historian; Ecclesiastical History of the English People.

Ruth Benedict, 1887-1948, (U.S.) anthropologist; studied Indian tribes of the Southwest.

Sir Isaiah Berlin, 1909-97, (Br.) philosopher, historian; *The Age of Enlightenment.*

Louis Blanc, 1811-82, (Fr.) Socialist leader and historian.

Sarah G. Blanding, 1899-1985, (U.S.) head of Vassar College, 1946-64.

Leonard Bloomfield, 1887-1949, (U.S.) linguist; *Language.*

Franz Boas, 1858-1942, (U.S.) German-born anthropologist; studied American Indians.

Van Wyck Brooks, 1886-1963, (U.S.) historian; critic of New England culture, especially literature.

Edmund Burke, 1729-97, (Ir.) British parliamentarian and political philosopher; Reflections on the Revolution in France.

Nicholas Murray Butler, 1862-1947, (U.S.) educator; headed Columbia Univ., 1902-45; Nobel Peace Prize, 1931.

Joseph Campbell, 1904-87, (U.S.) author, editor, teacher; wrote books on mythology, folklore.

Thomas Carlyle, 1795-1881, (Sc.) historian, critic; *Sartor Resartus, Past and Present, The French Revolution.*

Edward Channing, 1856-1931, (U.S.) historian; wrote 6-volume *History of the United States.*

Henry Steele Commager, 1902-98, (U.S.) historian, educator; wrote *The Growth of the American Republic.*

John R. Commons, 1862-1945, (U.S.) economist, labor historian; *Legal Foundations of Capitalism.*

James B. Conant, 1893-1978, (U.S.) educator, diplomat; *The American High School Today.*

Benedetto Croce, 1866-1952, (It.) philosopher, statesman, and historian; *Philosophy of the Spirit.*

Bernard A. De Voto, 1897-1955, (U.S.) historian; wrote trilogy on American West; edited Mark Twain manuscripts.

Melvil Dewey, 1851-1931, (U.S.) devised decimal system of library-book classification.

Emile Durkheim, 1858-1917, (Fr.) a founder of modern sociology; *The Rules of Sociological Method.*

Charles Eliot, 1834-1926, (U.S.) educator, Harvard president.

Friedrich Engels, 1820-95, (Ger.) political writer; with Marx wrote the *Communist Manifesto.*

John Fiske, 1842-1901, (U.S.) historian and lecturer; popularized Darwinian theory of evolution.

Charles Fourier, 1772-1837, (Fr.) utopian socialist.

Giovanni Gentile, 1875-1944, (It.) philosopher, educator; reformed Italian educational system.

Sir James George Frazer, 1854-1941, (Br.) anthropologist; studied myth in religion; *The Golden Bough.*

Henry George, 1839-97, (U.S.) economist, reformer; led single-tax movement.

Edward Gibbon, 1737-94, (Br.) historian; *The History of the Decline and Fall of the Roman Empire.*

Francesco Guicciardini, 1483-1540, (It.) historian; *Storia d'Italia,* principal historical work of the 16th cent.

Thomas Hobbes, 1588-1679, (Eng.) philosopher, political theorist; *Leviathan.*

Richard Hofstadter, 1916-70, (U.S.) historian; *The Age of Reform.*

John Holt, 1924-85, (U.S.) educator and author.

John Maynard Keynes, 1883-1946, (Br.) economist; principal advocate of deficit spending.

Russell Kirk, 1918-94, (U.S.), social philosopher; *The Conservative Mind.*

Alfred L. Kroeber, 1876-1960, (U.S.) cultural anthropologist; studied Indians of North and South America.

James L. Laughlin, 1850-1933, (U.S.) economist; helped establish Federal Reserve System.

Lucien Lévy-Bruhl, 1857-1939, (Fr.) philosopher; studied the psychology of primitive societies; *Primitive Mentality.*

John Locke, 1632-1704, (Eng.) philosopher and political theorist; *Two Treatises of Government.*

Thomas B. Macaulay, 1800-59, (Br.) historian, statesman.

Niccolò Machiavelli, 1469-1527, (It.) writer, statesman. *The Prince.*

Bronislaw Malinowski, 1884-1942, (Pol.) considered the father of social anthropology.

Thomas R. Malthus, 1766-1834, (Br.) economist; famed for *Essay on the Principle of Population.*

Horace Mann, 1796-1859, (U.S.) pioneered modern public school system.

Karl Mannheim, 1893-1947, (Hung.) sociologist, historian; *Ideology and Utopia.*

Karl Marx, 1818-83, (Ger.) political theorist, proponent of Communism; *Communist Manifesto, Das Kapital.*

Giuseppe Mazzini, 1805-72, (It.) political philosopher.

William H. McGuffey, 1800-73, (U.S.) whose *Reader* was a mainstay of 19th-cent. U.S. public education.

George H. Mead, 1863-1931, (U.S.) philosopher, social psychologist.

Margaret Mead, 1901-78, (U.S.) cultural anthropologist; popularized field; *Coming of Age in Samoa.*

Alexander Meiklejohn, 1872-1964, (U.S.) Br.-born educator; championed academic freedom and experimental curricula.

James Mill, 1773-1836, (Sc.) philosopher, historian, economist; a proponent of utilitarianism.

Perry G. Miller, 1905-63, (U.S.) historian; interpreted 17th-cent. New England.

Theodor Mommsen, 1817-1903, (Ger.) historian; *The History of Rome.*

Charles-Louis Montesquieu, 1689-1755, (Fr.) social philosopher; *The Spirit of Laws.*

Maria Montessori, 1870-1952, (It.) educator, physician; started Montessori method of student self-motivation.

Samuel Eliot Morison, 1887-1976, (U.S.) historian; chronicled voyages of early explorers.

Lewis Mumford, 1895-1990, (U.S.) sociologist, critic; *The Culture of Cities.*

Gunnar Myrdal, 1898-1987, (Swed.) economist, social scientist; *Asian Drama: An Inquiry Into the Poverty of Nations.*

Joseph Needham, 1900-95, (Br.) scientific historian; *Science and Civilization in China.*

Allan Nevins, 1890-1971, (U.S.) historian, biographer; *The Ordeal of the Union.*

José Ortega y Gasset, 1883-1955, (Sp.) philosopher; advocated control by elite, *The Revolt of the Masses.*

Robert Owen, 1771-1858, (Br.) political philosopher, reformer; pioneer in cooperative movement.

Thomas (Tom) Paine, 1737-1809, (U.S.) political theorist, writer. *Common Sense.*

Vilfredo Pareto, 1848-1923, (It.) economist, sociologist.

Francis Parkman, 1823-93, (U.S.) historian; *France and England in North America.*

Elizabeth P. Peabody, 1804-94, (U.S.) education pioneer; founded 1st kindergarten in U.S., 1860.

William Prescott, 1796-1859, (U.S.) early American historian; *The Conquest of Peru.*

Pierre Joseph Proudhon, 1809-65, (Fr.) social theorist; father of anarchism; *The Philosophy of Property.*

François Quesnay, 1694-1774, (Fr.) economic theorist.

David Ricardo, 1772-1823, (Br.) economic theorist; advocated free international trade.

Jean-Jacques Rousseau, 1712-78, (Fr.) social philosopher; the father of romantic sensibility; *Confessions.*

Edward Sapir, 1884-1939, (Ger.-U.S.) anthropologist; studied ethnology and linguistics of U.S. Indian groups.

Ferdinand de Saussure, 1857-1913, (Swiss) a founder of modern linguistics.

Hjalmar Schacht, 1877-1970, (Ger.) economist.

Joseph Schumpeter, 1883-1950, (Czech.-U.S.) economist, sociologist.

Elizabeth Seton, 1774-1821, (U.S.) nun; est. parochial school education in U.S.; first native-born American saint.

George Simmel, 1858-1918, (Ger.) sociologist, philosopher; helped establish German sociology.

Adam Smith, 1723-90, (Br.) economist; advocated laissez-faire economy, free trade; *The Wealth of Nations.*

Jared Sparks, 1789-1866, (U.S.) historian, educator, editor; *The Library of American Biography.*

Oswald Spengler, 1880-1936, (Ger.) philosopher and historian; *The Decline of the West.*

William G. Sumner, 1840-1910, (U.S.) social scientist, economist; laissez-faire economy, Social Darwinism.

Hippolyte Taine, 1828-93, (Fr.) historian; basis of naturalistic school; *The Origins of Contemporary France.*

A(lan) J(ohn) P(ercivale) Taylor, 1906-89, (Br.) historian; *The Origins of the Second World War.*

Nikolaas Tinbergen, 1907-88, (Dutch-Br.) ethologist; pioneer in study of animal behavior.

Alexis de Tocqueville, 1805-59, (Fr.) political scientist, historian; *Democracy in America.*

Francis E. Townsend, 1867-1960, (U.S.) led old-age pension movement, 1933.

Arnold Toynbee, 1889-1975, (Br.) historian; *A Study of History,* sweeping analysis of hist. of civilizations.

George Trevelyan, 1838-1928, (Br.) historian, statesman; favored "literary" over "scientific" history; *History of England.*

Barbara Tuchman, 1912-89, (U.S.) author of popular history books, *The Guns of August, The March of Folly.*

Frederick J. Turner, 1861-1932, (U.S.) historian, educator; *The Frontier in American History.*

Thorstein B. Veblen, 1857-1929, (U.S.) economist, social philosopher; *The Theory of the Leisure Class.*

Giovanni Vico, 1668-1744, (It.) historian, philosopher; regarded by many as first modern historian; *New Science.*

Izaak Walton, 1593-1683, (Eng.) wrote biographies; political-philosophical study of fishing, *The Compleat Angler.*

Sidney J., 1859-1947, and **Beatrice,** 1858-1943, **Webb,** (Br.) leading figures in Fabian Society and Labor Party.

Max Weber, 1864-1920, (Ger.) sociologist; *The Protestant Ethic and the Spirit of Capitalism.*

Emma Hart Willard, 1787-1870, (U.S.) pioneered higher education for women.

Noted American Journalists of the Past

Reviewed by Dean Mills, Dean, Missouri School of Journalism

See also Business Leaders, Cartoonists, Writers of the Past.

Franklin P. Adams (F.P.A.), 1881-1960, humorist; wrote column "The Conning Tower."

Joseph W. Alsop, 1910-89, and **Stewart Alsop,** 1914-74, Washington-based political analysts, columnists.

Brooks Atkinson, 1894-1984, theater critic.

James Gordon Bennett, 1795-1872, editor and publisher; founded *N.Y. Herald.*

James Gordon Bennett, 1841-1918, succeeded father, financed expeditions, founded afternoon paper.

Elias Boudinot, d 1839, founding editor of first Native American newspaper in U.S., *Cherokee Phoenix* (1828-34).

Margaret Bourke-White, 1904-71, photojournalist.

Arthur Brisbane, 1864-1936, editor; helped introduce "yellow journalism" with sensational, simply written articles.

Heywood Broun, 1888-1939, author, columnist; founded American Newspaper Guild.

John Campbell, 1653-1728, published *Boston News-Letter,* first continuing newspaper in the American colonies.

Jimmy Cannon, 1909-73, syndicated sports columnist.

John Chancellor, 1927-96, TV journalist; anchored *NBC Nightly News.*

Harry Chandler, 1864-1944, *Los Angeles Times* publisher, 1917-41; made it a dominant force.

Marquis Childs, 1903-90, reporter and columnist for *St. Louis Post-Dispatch* and United Feature syndicate.

Elizabeth Cochrane (Nellie Bly), pioneer woman journalist, investig. reporter, noted for series on trip around the world.

Charles Collingwood, 1917-85, CBS news correspondent, foreign affairs reporter, documentary host.

Howard Cosell, 1920-95, TV and radio sportscaster.

Gardner Cowles, 1861-1946, publisher; founder of Cowles newspaper chain.

Cyrus Curtis, 1850-1933, publisher of *Saturday Evening Post, Ladies Home Journal, Country Gentleman.*

Charles Anderson Dana, 1819-97, editor, publisher; made *N.Y. Sun* famous for its news reporting.

Elmer (Holmes) Davis, 1890-1958, *N.Y. Times* editorial writer; radio commentator.

Richard Harding Davis, 1864-1916, war correspondent, travel writer, fiction writer.

Benjamin Day, 1810-89, published *N.Y. Sun* beginning in 1833, introducing penny press to the U.S.

Frederick Douglass, 1817-95, ex-slave, social reformer, newspaper editor.

Finley Peter Dunne, 1867-1936, humorist, social critic, wrote "Mr. Dooley" columns.

Mary Baker Eddy, 1821-1910, founded Christian Science movement and *Christian Science Monitor.*

Marshall Field III, 1893-1956, retail magnate, *Chicago Sun* founder.

Doris Fleeson, 1901-70, war correspondent, columnist.

James Franklin, 1697-1735, printer, pioneer journalist, publisher of *New England Courant* and *Rhode Island Gazette.*

Fred W. Friendly, 1915-98, radio, TV reporter, announcer, producer, executive, collaborator with Edward R. Murrow.

Margaret Fuller, 1810-50, social reformer, transcendentalist, critic and foreign correspondent for *N.Y. Tribune.*

Frank E. Gannett, 1876-1957, founded newspaper chain.

William Lloyd Garrison, 1805-79, abolitionist; publisher of *The Liberator.*

Elizabeth Meriwether Gilmer (Dorothy Dix), 1861-1951, reporter, pioneer of the advice column genre.

Edwin Lawrence Godkin, 1831-1902, founder of *The Nation,* editor of *N.Y. Evening Post.*

Horace Greeley, 1811-72, editor and politician; founded *N.Y. Tribune.*

Gilbert Hovey Grosvenor, 1875-1966, longtime editor of *National Geographic* magazine.

John Gunther, 1901-70, *Chicago Daily News* foreign correspondent, author.

Sarah Josepha Buell Hale, 1788-1879, writer, first female magazine editor; edited *Ladies' Magazine* (later *Godey's Lady's Book*).

William Randolph Hearst, 1863-1951, founder of Hearst newspaper chain and one of the pioneer yellow journalists.

Gabriel Heatter, 1890-1972, radio commentator.

John Hersey, 1914-98, foreign correspondent for *Time, Life,* and *The New Yorker;* author.

Marguerite Higgins, 1920-66, reporter, war correspondent.

Hedda Hopper, 1885-1966, Hollywood gossip columnist.

Roy Howard, 1883-1964, editor, executive, Scripps-Howard papers and United Press (later United Press International).

Chet (Chester Robert) Huntley, 1911-74, co-anchor of NBC's *Huntley-Brinkley Report.*

Ralph Ingersoll, 1900-85, editor, *Fortune, Time, Life* exec.

H. V. (Hans von) Kaltenborn, 1878-1965, radio commentator, reporter.

Murray Kempton, 1917-97, reporter, columnist for magazines and newspapers, including *N.Y. Post.*

John S. Knight, 1894-1981, editor, publisher; founded Knight newspaper group, which merged into Knight-Ridder.

Joseph Kraft, 1942-86, foreign policy columnist.

Arthur Krock, 1886-1974, *N.Y. Times* political writer, Washington bureau chief.

Charles Kuralt, 1934-97, TV anchor and host of CBS "On the Road" feature stories about life in the U.S.

David Lawrence, 1888-1973, reporter, columnist, publisher; founded *U.S. News & World Report.*

Frank Leslie, 1821-80, engraver and publisher of newspapers and magazines, notably *Leslie's Illustrated Newspaper.*

A(bbott) J(oseph) Liebling, 1904-63, foreign correspondent, critic, principally with *The New Yorker.*

Walter Lippmann, 1889-1974, political analyst, social critic, columnist, author.

Peter Lisagor, 1915-76, Washington bureau chief, *Chicago Daily News;* broadcast commentator.

David Ross Locke, 1833-88, humorist, satirist under pseudonym P.V. Nasby; owned *Toledo (Ohio) Blade.*

Elijah Parish Lovejoy, 1802-37, abolitionist editor in St. Louis and in Alton, IL; killed by proslavery mob.

Clare Booth Luce, 1903-87, war correspondent for *Life;* diplomat, playwright.

Henry R. Luce, 1898-1967, founded *Time, Fortune, Life, Sports Illustrated.*

C(harles) K(enny) McClatchy, 1858-1936 founder of McClatchy newspaper chain.

Samuel McClure, 1857-1949, founder (1893) of *McClure's Magazine,* famous for its investigative reporting.

Anne O'Hare McCormick, 1889-1954, foreign correspondent, first woman on *N.Y. Times* editorial board.

Robert R. McCormick, 1880-1955, editor, publisher, executive of *Chicago Tribune* and *N.Y. Daily News.*

Dwight Macdonald, 1906-1982, reporter, social critic for *The New Yorker, The Nation, Esquire.*

Ralph McGill, 1893-1969, crusading editor and publisher of *Atlanta Constitution.*

O(scar) O(dd) McIntyre, 1884-1938, feature writer, syndicated columnist concentrating on everyday life in New York City.

Don Marquis, 1878-1937, humor columnist for *N.Y. Sun* and *N.Y. Tribune;* wrote "archy and mehitabel" stories.

Robert Maynard, 1937-97, first African-American editor and then owner of major U.S. paper, the *Oakland Tribune.*

Joseph Medill, 1823-99, longtime *editor of Chicago Tribune.*

H(enry) L(ouis) Mencken, 1880-1956, reporter, editor, columnist with *Baltimore Sun* papers; anti-establishment viewpoint.

Edwin Meredith, 1876-1928, founder of magazine company.

Frank A. Munsey, 1854-1925, owner, editor, and publisher of newspapers and magazines, including *Munsey's Magazine.*

Edward R. Murrow, 1908-65, broadcast reporter, executive; reported from Britain in WW2; hosted *See It Now, Person to Person.*

William Rockhill Nelson, 1841-1915, cofounder, editor, and publisher, *Kansas City Star.*

Adolph S. Ochs, 1858-1935, publisher; built *N.Y. Times* into a leading newspaper.

Louella Parsons, 1881-1972, Hollywood gossip columnist.

Alicia Patterson, 1906-63, reporter, editor, and cofounder of *Newsday.*

Drew (Andrew Russell) Pearson, 1879-1969, investigative reporter and columnist.

(James) Westbrook Pegler, 1894-1969, reporter, columnist.

Shirley Povich, 1905-98, sports columnist.

Joseph Pulitzer, 1847-1911, *N.Y. World* publisher; founded Columbia Journalism School, Pulitzer Prizes.

Joseph Pulitzer II, 1885-1955, longtime *St. Louis Post-Dispatch* editor, publisher; built it into major paper.

Ernie (Ernest Taylor) Pyle, 1900-45, reporter, war correspondent; killed in WW2.

Henry Raymond, 1820-69, cofounder, editor, *N.Y. Times*; made it model of objective journalism.

Harry Reasoner, 1923-91, TV reporter, anchor.

John Reed, 1887-1920, reporter, foreign correspondent famous for coverage of Bolshevik Revolution.

Whitelaw Reid, 1837-1912, longtime editor, *N.Y. Tribune.*

James Reston, 1909-95 *N.Y. Times* political reporter, columnist.

Frank Reynolds, 1923-83, TV reporter, anchor.

(Henry) Grantland Rice, 1880-1954, sportswriter.

Jacob Riis, 1849-1914, reporter, photographer; exposed slum conditions in *How the Other Half Lives.*

Max Robinson, 1939-88, TV journalist, first African-American to anchor network news, 1978.

Harold Ross, 1892-1951, founder, editor, The *New Yorker.*

Mike Royko, 1932-97, columnist for *Chicago Sun-Times* and *Chicago Tribune.*

(Alfred) Damon Runyon, 1884-1946, sportswriter, columnist; stories collected in *Guys and Dolls.*

John B. Russwurm, 1799-1851, cofounded (1827) nation's first black newspaper, *Freedom's Journal*, in NYC.

Adela Rogers St. Johns, 1894-1988, reporter, sportswriter for Hearst newspapers.

Harrison Salisbury, 1908-93, reporter, foreign correspondent; a Soviet specialist.

E(dward) W(lyllis) Scripps, 1854-1926, founded first large U.S. newspaper chain, pioneered syndication.

Eric Sevareid, 1912-92, war correspondent, radio newscaster, TV commentator.

William L. Shirer, 1904-93, broadcaster, foreign correspondent; wrote *The Rise and Fall of the Third Reich.*

Red (Walter) Smith, 1905-82, sportswriter.

Edgar P. Snow, 1905-71, correspondent, expert on Chinese Communist movement.

Lawrence Spivak, 1900-94, co-creator, moderator, producer of *Meet the Press.*

(Joseph) Lincoln Steffens, 1866-1936, muckraking journalist.

I(sidor) F(einstein) Stone, 1907-89, one-man editor of *I.F. Stone's Weekly.*

Arthur Hays Sulzberger, 1891-1968, longtime publisher of *N.Y. Times.*

C(yrus) L(eo) Sulzberger, 1912-93, *N.Y. Times* foreign correspondent and columnist.

David Susskind, 1920-87, TV producer, public affairs talk-show host (*Open End*).

John Cameron Swayze, 1906-95, newscaster, anchor of *Camel News Caravan.*

Herbert Bayard Swope, 1882-1958, war correspondent and editor of *N.Y. World.*

Ida Tarbell, 1857-1944, muckraking journalist.

Isaiah Thomas, 1750-1831, printer, publisher, cofounder of revolutionary journal, *Massachusetts Spy.*

Lowell Thomas, 1892-1981, radio newscaster, world traveler.

Dorothy Thompson, 1894-1961, foreign correspondent, columnist, radio commentator.

Ida Bell Wells-Barnett, 1862-1931, African-American reporter, editor, anti-lynching crusader.

William Allen White, 1868-1944, editor, publisher; made *Emporia* (KS) *Gazette* known worldwide.

Walter Winchell, 1897-1972, reporter, columnist, broadcaster of celebrity news.

John Peter Zenger, 1697-1746, printer and journalist; acquitted in precedent-setting libel suit (1735).

Noted Military and Naval Leaders of the Past

Reviewed by Alan C. Aimone, USMA Library

Creighton Abrams, 1914-74, (U.S.) commanded forces in Vietnam, 1968-72.

Alexander the Great, 356-323 B.C., (Maced.) conquered Persia and much of the world known to Europeans.

Harold Alexander, 1891-1969, (Br.) led Allied invasion of Italy, 1943, WW2.

Ethan Allen, 1738-89, (U.S.) headed Green Mountain Boys; captured Ft. Ticonderoga, 1775, Amer. Rev.

Edmund Allenby, 1861-1936, (Br.) in Boer War, WW1; led Egyptian expeditionary force, 1917-18.

Benedict Arnold, 1741-1801, (U.S.) victorious at Saratoga; tried to betray West Point to British, Amer. Rev.

Henry "Hap" Arnold, 1886-1950, (U.S.) commanded Army Air Force in WW2.

John Barry, 1745-1803, (U.S.) won numerous sea battles during Amer. Rev.

Belisarius, c505-565, (Byzant.) won remarkable victories for Byzantine Emperor Justinian I.

Pierre Beauregard, 1818-93, (U.S.) Confed. general, ordered bombardment of Ft. Sumter that began Civil War.

Gebhard von Blücher, 1742-1819, (Ger.) helped defeat Napoleon at Waterloo.

Napoleon Bonaparte, 1769-1821, (Fr.) defeated Russia and Austria at Austerlitz, 1805; invaded Russia, 1812; defeated at Waterloo, 1815.

Edward Braddock, 1695-1755, (Br.) commanded forces in French and Indian War.

Omar N. Bradley, 1893-1981, (U.S.) headed U.S. ground troops in Normandy invasion, 1944, WW2.

John Burgoyne, 1722-92, (Br.) defeated at Saratoga, Amer. Rev.

Julius Caesar, 100-44 BC (Rom.) general and politician; conquered N Gaul; overthrew Roman Republic.

Claire Lee Chennault, 1893-1958, (U.S.) headed Flying Tigers in WW2.

Mark W. Clark, 1896-1984, (U.S.) helped plan N African invasion in WW2; commander of UN forces, Korean War.

Karl von Clausewitz, 1780-1831, (Pruss.) military theorist.

Lucius D. Clay, 1897-1978, (U.S.) led Berlin airlift, 1948-49.

Henry Clinton, 1738-95, (Br.) commander of forces in Amer. Rev., 1778-81.

Cochise, c1815-74, (Nat. Am.) chief of Chiricahua band of Apache Indians in Southwest.

Charles Cornwallis, 1738-1805, (Br.) victorious at Brandywine, 1777; surrendered at Yorktown, Amer. Rev.

Hernan Cortes, 1485-1547, (Sp.) led Spanish conquistadors in the defeat of the Aztec empire, 1519-28.

Crazy Horse, 1849-77, (Nat. Am.) Sioux war chief victorious at battle of Little Bighorn.

George Armstrong Custer, 1839-76, (U.S.) U.S. army officer defeated and killed at battle of Little Bighorn.

Moshe Dayan, 1915-81, (Isr.) directed campaigns in the 1967, 1973 Arab-Israeli wars.

Stephen Decatur, 1779-1820, (U.S.) naval hero of Barbary wars, War of 1812.

Anton Denikin, 1872-1947, (Russ.) led White forces in Russian civil war.

George Dewey, 1837-1917, (U.S.) destroyed Spanish fleet at Manila, 1898, Span.-Amer. War.

Karl Doenitz, 1891-1980, (Ger.) submarine com. in chief and naval commander, WW2.

Hugh C. Dowding, 1883-1970, (Br.) headed RAF, 1936-40, WW2.

Jubal Early, 1816-94, (U.S.) Confed. general, led raid on Washington, 1864, Civil War.

Dwight D. Eisenhower, 1890-1969, (U.S.) commanded Allied forces in Europe, WW2.

David Farragut, 1801-70, (U.S.) Union admiral, captured New Orleans, Mobile Bay, Civil War.

Ferdinand Foch, 1851-1929, (Fr.) headed victorious Allied armies, 1918, WW1.

Nathan Bedford Forrest, 1821-77, (U.S.) Confed. general, led raids against Union supply lines, Civil War.

Frederick the Great, 1712-86, (Pruss.) led Prussia in Seven Years' War.

Horatio Gates, 1728-1806, (U.S.) commanded army at Saratoga, Amer. Rev.

Genghis Khan, 1162-1227, (Mongol) unified Mongol tribes and subjugated much of Asia, 1206-21.

Geronimo, 1829-1909, (Nat. Am.) leader of Chiricahua band of Apache Indians.

Charles G. Gordon, 1833-85, (Br.) led forces in China, Crimean War; killed at Khartoum.

Ulysses S. Grant, 1822-85, (U.S.) headed Union army, Civil War, 1864-65; forced Lee's surrender, 1865.

Nathanael Greene, 1742-86, (U.S.) defeated British in Southern campaign, 1780-81.

Heinz Guderian, 1888-1953, (Ger.) tank theorist, led panzer forces in Poland, France, Russia, WW2.

Che (Ernesto) Guevara, 1928-67, (Arg.) guerrilla leader; prominent in Cuban revolution; killed in Bolivia.

Gustavus Adolphus, 1594-1632, (Swed.) King; military tactician reformer; led forces in Thirty Years' War.

Douglas Haig, 1861-1928, (Br.) led British armies in France, 1915-18, WW1.

William F. Halsey, 1882-1959, (U.S.) defeated Japanese fleet at Leyte Gulf, 1944, WW2.

Hannibal, 247-183 B.C., (Carthag.) invaded Rome, crossing Alps, in Second Punic War, 218-201 B.C.

Sir Arthur Travers Harris, 1895-1984, (Br.) led Britain's WW2 bomber command.

Richard Howe, 1726-99, (Br.) commanded navy in Amer. Rev., 1776-78; June 1 victory against French, 1794.

William Howe, 1729-1814, (Br.) commanded forces in Amer. Rev., 1776-78.

Isaac Hull, 1773-1843, (U.S.) sunk British frigate Guerriere, War of 1812.

Thomas (Stonewall) Jackson, 1824-63, (U.S.) Confed. general, led Shenandoah Valley campaign, Civil War.

Joseph Joffre, 1852-1931, (Fr.) headed Allied armies, won Battle of the Marne, 1914, WW1.

Chief Joseph, c1840-1904, (Nat. Am.) chief of the Nez Percé, led his tribe across 3 states seeking refuge in Canada; surrendered about 30 mi from Canadian border.

John Paul Jones, 1747-92, (U.S.) commanded Bonhomme Richard in victory over Serapis, Amer. Rev., 1779.

Stephen Kearny, 1794-1848, (U.S.) headed Army of the West in Mexican War.

Albert Kesselring, 1885-1960 (Ger.) field marshal who led the defense of Italy in WW2.

Ernest J. King, 1878-1956, (U.S.) key WW2 naval strategist.

Horatio H. Kitchener, 1850-1916, (Br.) led forces in Boer War; victorious at Khartoum; organized army in WW1.

Henry Knox, 1750-1806, (U.S.) general in Amer. Rev.; first sec. of war under U.S. Constitution.

Lavrenti Kornilov, 1870-1918, (Russ.) commander-in-chief, 1917; led counter-revolutionary march on Petrograd.

Thaddeus Kosciusko, 1746-1817, (Pol.) aided Amer. Rev.

Walter Krueger, 1881-1967, (U.S.) led Sixth Army in WW2 in Southwest Pacific.

Mikhail Kutuzov, 1745-1813, (Russ.) fought French at Borodino, Napoleonic Wars, 1812; abandoned Moscow; forced French retreat.

Marquis de Lafayette, 1757-1834, (Fr.) fought in, secured French aid for Amer. Rev.

T(homas) E. Lawrence (of Arabia), 1888-1935, (Br.) organized revolt of Arabs against Turks in WW1.

Henry (Light-Horse Harry) Lee, 1756-1818, (U.S.) cavalry officer in Amer. Rev.

Robert E. Lee, 1807-70, (U.S.) Confed. general defeated at Gettysburg, Civil War; surrendered to Grant, 1865.

Curtis LeMay, 1906-90, (U.S.) Air Force commander in WW2, Korean War, and Vietnam War.

Lyman Lemnitzer, 1899-1988, (U.S.) WW2 hero, later general, chairman of Joint Chiefs of Staff.

James Longstreet, 1821-1904, (U.S.) aided Lee at Gettysburg, Civil War.

Maurice, Count of Nassau, 1567-1625, (Dutch) military innovator; led forces in Thirty Years' War.

Douglas MacArthur, 1880-1964, (U.S.) commanded forces in SW Pacific in WW2; headed occupation forces in Japan, 1945-51; UN commander in Korean War.

Erich von Manstein, 1887-1973, (Ger.) served WW1–2, planned inv. of France (1940), convicted of war crimes.

Carl Gustaf Mannerheim, 1867-1951, (Finn.) army officer and pres. of Finland 1944-46.

Francis Marion, 1733-95, (U.S.) led guerrilla actions in South Carolina during Amer. Rev.

Duke of Marlborough, 1650-1722, (Br.) led forces against Louis XIV in War of the Spanish Succession.

George C. Marshall, 1880-1959, (U.S.) chief of staff in WW2; authored Marshall Plan.

George B. McClellan, 1826-85, (U.S.) Union general, commanded Army of the Potomac, 1861-62, Civil War.

George Meade, 1815-72, (U.S.) commanded Union forces at Gettysburg, Civil War.

Billy Mitchell, 1879-1936, (U.S.) WW1 air-power advocate; court-martialed for insubordination, later vindicated.

Helmuth von Moltke, 1800-91, (Ger.) victorious in Austro-Prussian, Franco-Prussian wars.

Louis de Montcalm, 1712-59, (Fr.) headed troops in Canada, French and Indian War; defeated at Quebec, 1759.

Bernard Law Montgomery, 1887-1976, (Br.) stopped German offensive at Alamein, 1942, WW2; helped plan Normandy.

Daniel Morgan, 1736-1802, (U.S.) victorious at Cowpens, 1781, Amer. Rev.

Louis Mountbatten, 1900-79, (Br.) Supreme Allied Commander of SE Asia, 1943-46, WW2.

Joachim Murat, 1767-1815, (Fr.) led cavalry at Marengo, Austerlitz, and Jena, Napoleonic Wars.

Horatio Nelson, 1758-1805, (Br.) naval commander, destroyed French fleet at Trafalgar.

Michel Ney, 1769-1815, (Fr.) commanded forces in Switz., Aust., Russ., Napoleonic Wars; defeated at Waterloo.

Chester Nimitz, 1885-1966, (U.S.) commander of naval forces in Pacific in WW2.

George S. Patton, 1885-1945, (U.S.) led assault on Sicily, 1943, Third Army invasion of Europe, WW2.

Oliver Perry, 1785-1819, (U.S.) won Battle of Lake Erie in War of 1812.

John Pershing, 1860-1948, (U.S.) commanded Mexican border campaign, 1916, Amer. Expeditionary Force, WW1.

Henri Philippe Pétain, 1856-1951, (Fr.) defended Verdun, 1916; headed Vichy government in WW2.

George E. Pickett, 1825-75, (U.S.) Confed. general famed for "charge" at Gettysburg, Civil War.

Charles Portal, 1893-1971, (Br.) chief of staff, Royal Air Force, 1940-45, led in Battle of Britain.

Hyman Rickover, 1900-86, (U.S.) father of nuclear navy.

Matthew Bunker Ridgway, 1895-1993, (U.S.) commanded Allied ground forces in Korean War.

Erwin Rommel, 1891-1944, (Ger.) headed Afrika Korps, WW2.

Gerd von Rundstedt, 1875-1953, (Ger.) supreme commander in West, 1942-45, WW2.

Aleksandr Samsonov, 1859-1914, (Russ.) led invasion of E Prussia, WW1, defeated at Tannenberg, 1914.

Winfield Scott, 1786-1866, (U.S.) hero of War of 1812; headed forces in Mexican War, took Mexico City.

Philip Sheridan, 1831-88, (U.S.) Union cavalry officer, headed Army of the Shenandoah, 1864-65, Civil War.

William T. Sherman, 1820-91, (U.S.) Union general, sacked Atlanta during "march to the sea," 1864, Civil War.

Carl Spaatz, 1891-1974, (U.S.) directed strategic bombing against Germany, later Japan, in WW2.

Raymond Spruance, 1886-1969, (U.S.) victorious at Midway Island, 1942, WW2.

Joseph W. Stilwell, 1883-1946, (U.S.) headed forces in the China, Burma, India theater in WW2.

J.E.B. Stuart, 1833-64, (U.S.) Confed. cavalry commander, Civil War.

Aleksandr Suvorov, 1729-1800, (Rus.) victorious commander of Allied Russian and Austrian armies against Ottoman Turks in Russo-Turkish War.

George H. Thomas, 1816-70, (U.S.) saved Union army at Chattanooga, 1863; won at Nashville, 1864, Civil War.

Semyon Timoshenko, 1895-1970, (USSR) defended Moscow, Stalingrad, WW2; led winter offensive, 1942-43.

Alfred von Tirpitz, 1849-1930, (Ger.) responsible for submarine blockade in WW1.

Sebastien Le Prestre de Vauban, 1633-1707, (Fr.) innovative military engineer and theorist.

Jonathan M. Wainwright, 1883-1953, (U.S.) forced to surrender on Corregidor, 1942, WW2.

George Washington, 1732-99, (U.S.) led Continental army, 1775-83, Amer. Rev.

Archibald Wavell, 1883-1950, (Br.) commanded forces in N and E Africa, and SE Asia in WW2.

Anthony Wayne, 1745-96, (U.S.) captured Stony Point, 1779, Amer. Rev.

Duke of Wellington, 1769-1852, (Br.) defeated Napoleon at Waterloo, 1815.

James Wolfe, 1727-59, (Br.) captured Quebec from French, 1759, French and Indian War.

Isoroku Yamamoto, 1884-1943, (Jpn.) com. in chief of Japanese fleet and naval planner before and during WW2.

Georgi Zhukov, 1895-1974, (Russ.) defended Moscow, 1941, led assault on Berlin, 1945, WW2.

Noted Philosophers and Religious Figures of the Past

For other Greeks and Romans, see Historical Figures chapter.

Lyman Abbott, 1835-1922, (U.S.) clergyman, reformer; advocate of Christian Socialism.

Pierre Abelard, 1079-1142, (Fr.) philosopher, theologian, teacher; used dialectic method to support Christian beliefs.

Felix Adler, 1851-1933, (U.S.) German-born founder of the Ethical Culture Society.

(St.) Anselm, c1033-1109, (It.) philosopher-theologian, church leader; "ontological argument" for God's existence.

(St.) Thomas Aquinas, 1225-74, (It.) preeminent medieval philosopher-theologian; *Summa Theologica.*

Aristotle, 384-322 BC, (Gr.) pioneering wide-ranging philosopher, logician, ethician, naturalist.

(St.) Augustine, 354-430, (N Africa) philosopher, theologian, bishop; *Confessions, City of God, On the Trinity.*

J. L. Austin, 1911-60, (Br.) ordinary-language philosopher.

Averroes (Ibn Rushd), 1126-98, (Sp.) Islamic philosopher, physician.

Avicenna (Ibn Sina), 980-1037, (Iran.) Islamic philosopher, scientist.

A(lfred) J(ules) Ayer, 1910-89, (Br.) philosopher; logical positivist; *Language, Truth, and Logic.*

Roger Bacon, c1214-94, (Eng.) philosopher and scientist.

Bahaullah (Mirza Husayn Ali), 1817-92, (Pers.) founder of Bahá'í faith.

Karl Barth, 1886-1968, (Swiss) theologian; a leading force in 20th-cent. Protestantism.

Thomas à Becket, 1118-70, (Eng.) archbishop of Canterbury; opposed Henry II; murdered by King's men.

(St.) Benedict, c480-547, (It.) founded the Benedictines.

Jeremy Bentham, 1748-1832, (Br.) philosopher, reformer; enunciated utilitarianism.

Henri Bergson, 1859-1941, (Fr.) philosopher of evolution.

George Berkeley, 1685-1753, (Ir.) idealist philosopher, churchman.

John Biddle, 1615-62, (Eng.) founder of English Unitarianism.

Jakob Boehme, 1575-1624, (Ger.) theosophist and mystic.

Dietrich Bonhoeffer, 1906-1945 (Ger.) Lutheran theologian, pastor; executed as opponent of Nazis.

William Brewster, 1567-1644, (Eng.) headed Pilgrims.

Emil Brunner, 1889-1966, (Swiss) Protestant theologian.

Giordano Bruno, 1548-1600, (It.) philosopher, pantheist.

Martin Buber, 1878-1965, (Ger.) Jewish philosopher, theologian; *I and Thou.*

Buddha (Siddhartha Gautama), c563-c483 BC, (Indian) philosopher; founded Buddhism.

John Calvin, 1509-64, (Fr.) theologian; a key figure in the Protestant Reformation.

Rudolph Carnap, 1891-1970, (U.S.) German-born analytic philosopher; a founder of logical positivism.

William Ellery Channing, 1780-1842, (U.S.) clergyman; early spokesman for Unitarianism.

Auguste Comte, 1798-1857, (Fr.) philosopher; originated positivism.

Confucius, 551-479 BC, (Chin.) founder of Confucianism.

John Cotton, 1584-1652, (Eng.) Puritan theologian.

Thomas Cranmer, 1489-1556, (Eng.) churchman; wrote much of *Book of Common Prayer.*

René Descartes, 1596-1650, (Fr.) philosopher, mathematician; "father of modern philosophy." *Discourse on Method, Meditations on First Philosophy.*

John Dewey, 1859-1952, (U.S.) philosopher, educator; instrumentalist theory of knowledge; helped inaugurate progressive education movement.

Denis Diderot, 1713-84, (Fr.) philosopher, encyclopedist.

John Duns Scotus, c1266-1308, (Sc.) Franciscan philosopher and theologian.

Mary Baker Eddy, 1821-1910, (U.S.) founder of Christian Science; *Science and Health.*

Jonathan Edwards, 1703-58, (U.S.) preacher, theologian.

(Desiderius) Erasmus, c1466-1536, (Dutch) Renaissance humanist; *On the Freedom of the Will.*

Johann Fichte, 1762-1814, (Ger.) idealist philosopher.

Michel Foucault, 1926-84, (Fr.) structuralist philosopher, historian.

George Fox, 1624-91, (Br.) founder of Society of Friends.

(St.) Francis of Assisi, 1182-1226, (It.) founded Franciscans.

al-Ghazali, 1058-1111, Islamic philosopher.

Georg W. F. Hegel, 1770-1831, (Ger.) idealist philosopher; *Phenomenology of Mind.*

Martin Heidegger, 1889-1976, (Ger.) existentialist philosopher; affected many fields; *Being and Time.*

Johann G. Herder, 1744-1803, (Ger.) philosopher, cultural historian; a founder of German Romanticism.

Thomas Hobbes, 1588-1679, (Eng.) philosopher, political theorist; *Leviathan.*

David Hume, 1711-76, (Sc.) leading empiricist philosopher; *Enquiry Concerning Human Understanding.*

Jan Hus, 1369-1415, (Czech.) religious reformer.

Edmund Husserl, 1859-1938, (Ger.) philosopher; founded the phenomenological movement.

Thomas Huxley, 1825-95, (Br.) philosopher, educator.

William Inge, 1860-1954, (Br.) theologian; explored mystic aspects of Christianity.

William James, 1842-1910, (U.S.) philosopher, psychologist; pragmatist; studied religious experience.

Karl Jaspers, 1883-1969, (Ger.) existentialist philosopher.

Joan of Arc, 1412-1431, (Fr.) national heroine and a patron saint of France; key figure in the Hundred Years' War.

Immanuel Kant, 1724-1804, (Ger.) philosopher; founder of modern critical philosophy; *Critique of Pure Reason.*

Thomas à Kempis, c1380-1471, (Ger.) monk, devotional writer; *Imitation of Christ.* attributed to him.

Soren Kierkegaard, 1813-55, (Dan.) religious philosopher; pre-existentialist; *Either/Or, The Sickness Unto Death.*

John Knox, 1505-72, (Sc.) leader of the Protestant Reformation in Scotland.

Lao-Tzu, 604-531 BC, (Chin.) philosopher; considered the founder of the Taoist religion.

Gottfried von Leibniz, 1646-1716, (Ger.) rationalistic philosopher, logician, mathematician.

John Locke, 1632-1704, (Eng.) political theorist, empiricist philosopher; *Essay Concerning Human Understanding.*

(St.) Ignatius Loyola, 1491-1556, (Sp.) founder of the Jesuits; *Spiritual Exercises.*

Martin Luther, 1483-1546, (Ger.) leader of the Protestant Reformation, founded Lutheran church.

Jean-Francois Lyotard, 1924-98, (Fr.) postmodern philosopher, lecturer; *The Post-Modern Condition.*

Maimonides, 1135-1204, (Sp.) major Jewish philosopher.

Gabriel Marcel, 1889-1973, (Fr.) Roman Catholic existentialist philosopher, dramatist, and critic,

Jacques Maritain, 1882-1973, (Fr.) Neo-Thomist philosopher.

Cotton Mather, 1663-1728, (U.S.) defender of orthodox Puritanism; founded Yale, 1701.

Philipp Melanchthon, 1497-1560, (Ger.) theologian, humanist; an important voice in the Reformation.

Maurice Merleau-Ponty, 1908-61, (Fr.) existentialist philosopher; *Phenomenology of Perception.*

Thomas Merton, 1915-68, (U.S.) Trappist monk, spiritual writer; *The Seven Storey Mountain.*

John Stuart Mill, 1806-73, (Br.) philosopher, economist; libertarian political theorist; *Utilitarianism.*

Muhammad, c570-632, (Arab) the prophet of Islam.

Dwight Moody, 1837-99, (U.S.) evangelist.

G(eorge) E(dward) Moore, 1873-1958, (Br.) philosopher; *Principia Ethica,* "A Defense of Common Sense."

Elijah Muhammad, 1897-1975, (U.S.) Leader of the Black Muslim sect.

Heinrich Muhlenberg, 1711-87, (Ger.) organized the Lutheran Church in America.

John H. Newman, 1801-90, (Br.) Roman Catholic convert, cardinal; led Oxford Movement; *Apologia pro Vita Sua.*

Reinhold Niebuhr, 1892-1971, (U.S.) Protestant theologian.

Friedrich Nietzsche, 1844-1900, (Ger.) philosopher; *The Birth of Tragedy, Beyond Good and Evil, Thus Spake Zarathustra.*

Blaise Pascal, 1623-62, (Fr.) philosopher, mathematician; *Pensées.*

(St.) Patrick, c389-c461, (Br.) brought Christianity to Ireland.

(St.) Paul, ?-c67, a key proponent of Christianity; his epistles are first Christian theological writing.

Norman Vincent Peale, 1898-1993, (U.S.) minister, author; *The Power of Positive Thinking.*

C(harles) S. Peirce, 1839-1914, (U.S.) philosopher, logician; originated concept of pragmatism, 1878.

Plato, c428-347 BC, (Gr.) philosopher; wrote classic Socratic dialogues; argued for universal truths and independent reality of ideas or forms; *Republic.*

Plotinus, 205-70, (Rom.) a founder of neo-Platonism; *Enneads.*

Josiah Royce, 1855-1916, (U.S.) idealist philosopher.

Bertrand Russell, 1872-1970, (Br.) philosopher, logician; one of the founders of modern logic; a prolific popular writer.

Charles T. Russell, 1852-1916, (U.S.) founder of Jehovah's Witnesses.

Gilbert Ryle, 1900-76, (Br.) analytic philosopher; *The Concept of Mind.*

George Santayana, 1863-1952, (U.S.) philosopher, writer, critic; *The Sense of Beauty, The Realms of Being.*

Jean-Paul Sartre, 1905-80, (Fr.) philosopher, novelist, playwright. *Nausea, No Exit, Being and Nothingness.*

Friedrich von Schelling, 1775-1854, (Ger.) philosopher of romantic movement.

Friedrich Schleiermacher, 1768-1834, (Ger.) theologian; a founder of modern Protestant theology.

Arthur Schopenhauer, 1788-1860, (Ger.) philosopher; *The World as Will and Idea.*

Albert Schweitzer, 1875-1965, (Ger.) theologian, social philosopher, medical missionary.

Joseph Smith, 1805-44, (U.S.) founded Latter-Day Saints (Mormon) movement, 1830.

Socrates, 469-399 BC, (Gr.) influential philosopher immortalized by Plato.

Herbert Spencer, 1820-1903, (Br.) philosopher of evolution.

Baruch de Spinoza, 1632-77, (Dutch) rationalist philosopher; *Ethics.*

Billy Sunday, 1862-1935, (U.S.) evangelist.

Pierre Teilhard de Chardin, 1881-1955, (Fr.) Jesuit priest, paleontologist, philosopher-theologian; *The Divine Milieu.*

Daisetz Teitaro Suzuki, 1870-1966, (Jpn.) Buddhist scholar.

(St.) Theresa of Lisieux, 1873-97, (Fr.) Carmelite nun revered for everyday sanctity; *The Story of a Soul.*

Emanuel Swedenborg, 1688-1772, (Swed.) philosopher, mystic.

Paul Tillich, 1886-1965, (U.S.) German-born philosopher and theologian; brought depth psychology to Protestantism.

John Wesley, 1703-91, (Br.) theologian, evangelist; founded Methodism.

Alfred North Whitehead, 1861-1947, (Br.) philosopher, mathematician; *Process and Reality.*

William of Occam, c1285-c1349 (Eng.) medieval scholastic philosopher; nominalist.

Roger Williams, c1603-83, (U.S.) clergyman; championed religious freedom and separation of church and state.

Ludwig Wittgenstein, 1889-1951, (Austrian) philosopher; major influence on contemporary language philosophy; *Tractatus Logico-Philosophicus, Philosophical Investigations.*

John Woolman, 1720-72, (U.S.) Quaker social reformer, abolitionist, writer; *The Journal.*

John Wycliffe, 1320-84, (Eng.) theologian, reformer.

(St.) Francis Xavier, 1506-52, (Sp.) Jesuit missionary, "Apostle of the Indies."

Brigham Young, 1801-77, (U.S.) Mormon leader after Smith's assassination; colonized Utah.

Huldrych Zwingli, 1484-1531, (Swiss) theologian; led Swiss Protestant Reformation.

Noted Political Leaders of the Past

(U.S. presidents, vice presidents, Supreme Ct. justices, signers of Decl. of Indep. listed elsewhere.)

Abu Bakr, 573-634, Muslim leader, first caliph, chosen successor to Muhammad.

Dean Acheson, 1893-1971, (U.S.) sec. of state; architect of cold war foreign policy.

Samuel Adams, 1722-1803, (U.S.) patriot, Boston Tea Party firebrand.

Konrad Adenauer, 1876-1967, (Ger.) West German chancellor.

Emilio Aguinaldo, 1869-1964, (Philip.) revolutionary; fought against Spain and the U.S.

Akbar, 1542-1605, greatest Mogul emperor of India.

Salvador Allende Gossens, 1908-1973, (Chilean) Marxist pres. 1970-73; ousted and died in coup.

Herbert H. Asquith, 1852-1928, (Br.) liberal prime min.; instituted major social reform.

Atahualpa, ?-1533, Inca (ruling chief) of Peru.

Kemal Ataturk, 1881-1938, (Turk.) founded modern Turkey.

Clement Attlee, 1883-1967, (Br.) Labour party leader, prime min.; enacted natl. health, nationalized many industries.

Stephen F. Austin, 1793-1836, (U.S.) led Texas colonization.

Mikhail Bakunin, 1814-76, (Russ.) revolutionary; leading exponent of anarchism.

Arthur J. Balfour, 1848-1930, (Br.) foreign sec. under Lloyd George; issued Balfour Declaration backing Zionism.

Bernard M. Baruch, 1870-1965, (U.S.) financier, govt. adviser.

Fulgencio Batista y Zaldívar, 1901-73, (Cub.) Cuban pres. (1940-44, 1952-59) and dictator, overthrown by Castro.

Lord Beaverbrook, 1879-1964, (Br.) financier, statesman, newspaper owner.

Menachem Begin, 1913-92, (Isr.) Israeli prime min., shared 1978 Nobel Peace Prize.

Eduard Benes, 1884-1948, (Czech.) pres. during interwar and post-WW2 eras.

David Ben-Gurion, 1886-1973, (Isr.) first prime min. of Israel, 1948-53, 1955-63.

Thomas Hart Benton, 1782-1858, (U.S.) Missouri senator; championed agrarian interests and westward expansion.

Aneurin Bevan, 1897-1960, (Br.) Labour party leader.

Ernest Bevin, 1881-1951, (Br.) Labour party leader, foreign minister; helped lay foundation for NATO.

Otto von Bismarck, 1815-98, (Ger.) statesman known as the Iron Chancellor; uniter of Germany, 1870.

James G. Blaine, 1830-93, (U.S.) Republican politician, diplomat; influential in Pan-American movement.

Léon Blum, 1872-1950, (Fr.) socialist leader, writer; headed first Popular Front government.

Simón Bolívar, 1783-1830, (Venez.) S. Amer. Revolutionary who liberated much of the continent from Spanish rule.

William E. Borah, 1865-1940, (U.S.) isolationist senator; helped block U.S. membership in League of Nations.

Cesare Borgia, 1476-1507, (It.) soldier, politician; an outstanding figure of the Italian Renaissance.

Willy Brandt, 1913-92, (Ger.) statesman, chancellor of West Germany, 1969-74; promoted East/West peace, *Ostpolitik.*

Leonid Brezhnev, 1906-82, (USSR) Soviet leader, 1964-82.

Aristide Briand, 1862-1932, (Fr.) foreign min.; chief architect of Locarno Pact and anti-war Kellogg-Briand Pact.

William Jennings Bryan, 1860-1925, (U.S.) Democratic, populist leader, orator; 3 times lost race for presidency.

William C. Bullitt, 1891-1967, (U.S.) diplomat; first ambassador to USSR, ambassador to France.

Ralph Bunche, 1904-71, (U.S.) a founder and key diplomat of United Nations for more than 20 years.

John C. Calhoun, 1782-1850, (U.S.) political leader; champion of states' rights and a symbol of the Old South.

Robert Castlereagh, 1769-1822, (Br.) foreign sec.; guided Grand Alliance against Napoleon.

Camillo Benso Cavour, 1810-61, (It.) statesman; largely responsible for uniting Italy under the House of Savoy.

Nicolae Ceausescu, 1918-89, (Roman.) Communist leader, head of state 1967-89.

Austen Chamberlain, 1863-1937, (Br.) statesman; won the Nobel Peace Prize, helped finalize Locarno Treaties, both 1925.

Neville Chamberlain, 1869-1940, (Br.) Conservative prime min. whose appeasement of Hitler led to Munich Pact.

Chiang Kai-shek, 1887-1975, (Chin.) Nationalist Chinese pres. whose government was driven from mainland to Taiwan.

Winston Churchill, 1874-1965, (Br.) prime min., soldier, author; guided Britain through WW2.

Galeazzo Ciano, 1903-44, (It.) fascist foreign minister; helped create Rome-Berlin Axis, executed by Mussolini.

Henry Clay, 1777-1852, (U.S.) "The Great Compromiser," one of the most influential pre-Civil War political leaders.

Georges Clemenceau, 1841-1929, (Fr.) twice prem., Wilson's antagonist at Paris Peace Conference after WW1.

DeWitt Clinton, 1769-1828, (U.S.) political leader; responsible for promoting idea of the Erie Canal.

Robert Clive, 1725-74, (Br.) first administrator of Bengal; laid foundation for British Empire in India.

Jean Baptiste Colbert, 1619-83, (Fr.) statesman; influential under Louis XIV, created the French navy.

Oliver Cromwell, 1599-1658, (Br.) Lord Protector of England, led parliamentary forces during Civil War.

Curzon of Kedleston, 1859-1925, (Br.) viceroy of India, foreign sec.; major force in post-WW1 world.

Édouard Daladier, 1884-1970, (Fr.) Radical Socialist politician, arrested by Vichy, interned by Germans until 1945.

Richard J. Daley, 1902-1976, (U.S.) Chicago mayor.

Georges Danton, 1759-94, (Fr.) leading French Rev. figure.

Jefferson Davis, 1808-89, (U.S.) pres. of the Confederacy.

Charles G. Dawes, 1865-1951, (U.S.) statesman, banker; advanced plan to stabilize post-WW1 German finances.

Alcide De Gasperi, 1881-1954, (It.) prime min.; founder of Christian Democratic party.

Charles De Gaulle, 1890-1970, (Fr.) general, statesman; first pres. of the Fifth Republic.

Deng Xiaoping, 1904-97, (Chin.) "paramount leader" of China; backed economic modernization.

Eamon De Valera, 1882-1975, (Ir.-U.S.) led fight for Irish independence.

Thomas E. Dewey, 1902-71, (U.S.) NY governor; twice loser in try for presidency.

Ngo Dinh Diem, 1901-63, (Viet.) South Vietnamese pres.; assassinated in government takeover.

Everett M. Dirksen, 1896-1969, (U.S.) Senate Republican minority leader, orator.

Benjamin Disraeli, 1804-81, (Br.) prime min.; considered founder of modern Conservative party.

Engelbert Dollfuss, 1892-1934, (Austrian) chancellor; assassinated by Austrian Nazis.

Andrea Doria, 1466-1560, (It.) Genoese admiral, statesman; called "Father of Peace" and "Liberator of Genoa."

Stephen A. Douglas, 1813-61, (U.S.) Democratic leader, orator; opposed Lincoln for the presidency.

Alexander Dubcek, 1921-92, (Czech.) statesman whose attempted liberalization was crushed, 1968.

John Foster Dulles, 1888-1959, (U.S.) sec. of state under Eisenhower, cold war policy-maker.

Friedrich Ebert, 1871-1925, (Ger.) Social Democratic movement leader; 1st pres., Weimar Republic, 1919-25.

Sir Anthony Eden, 1897-1977, (Br.) foreign sec., prime min. during Suez invasion of 1956.

Ludwig Erhard, 1897-1977, (Ger.) economist, West German chancellor; led nation's economic rise after WW2.

Hamilton Fish, 1808-93, (U.S.) sec. of state, successfully mediated disputes with Great Britain, Latin America.

James V. Forrestal, 1892-1949, (U.S.) sec. of navy, first sec. of defense.

Francisco Franco, 1892-1975, (Sp.) leader of rebel forces during Spanish Civil War and dictator of Spain.

Benjamin Franklin, 1706-90, (U.S.) printer, publisher, author, inventor, scientist, diplomat.

Louis de Frontenac, 1620-98, (Fr.) governor of New France (Canada); encouraged explorations, fought Iroquois.

J. William Fulbright, 1905-95, (U.S.) U.S. senator; leading figure in U.S. foreign policy during cold war years.

Hugh Gaitskell, 1906-63, (Br.) Labour party leader; major force in reversing its stand for unilateral disarmament.

Albert Gallatin, 1761-1849, (U.S.) sec. of treasury; instrumental in negotiating end of War of 1812.

Léon Gambetta, 1838-82, (Fr.) statesman, politician; one of the founders of the Third Republic.

Indira Gandhi, 1917-84, (In.) daughter of Jawaharlal Nehru, prime min. of India, 1966-77, 1980-84; assassinated.

Mohandas K. Gandhi, 1869-1948, (In.) political leader, ascetic; led movement against British rule; assassinated.

Giuseppe Garibaldi, 1807-82, (It.) patriot, soldier; a leader in the Risorgimento, Italian unification movement.

William E. Gladstone, 1809-98, (Br.) prime min. 4 times; dominant force of Liberal party from 1868 to 1894.

Paul Joseph Goebbels, 1897-1945, (Ger.) Nazi propagandist, master of mass psychology.

Barry Goldwater, 1909-98 (U.S.) conservative U.S. senator and 1964 Republican presid. nominee.

Klement Gottwald, 1896-1953, (Czech.) Communist leader; ushered Communism into his country.

Alexander Hamilton, 1755-1804, (U.S.) first treasury sec.; champion of strong central government.

Dag Hammarskjold, 1905-61, (Swed.) statesman; UN sec.-general.

Hassan II, King, 1929-99, (Moroc.), ruler since 1962.

John Hay, 1838-1905, (U.S.) sec. of state; primarily associated with Open Door Policy toward China.

Patrick Henry, 1736-99, (U.S.) major revolutionary figure, remarkable orator.

Édouard Herriot, 1872-1957, (Fr.) Radical Socialist leader; twice prem., pres. of National Assembly.

Theodor Herzl, 1860-1904, (Hung.) founded modern Zionism.

Heinrich Himmler, 1900-45, (Ger.) head of Nazi SS and Gestapo.

Paul von Hindenburg, 1847-1934, (Ger.) field marshal, WW1; 2d pres. of Weimar Republic, 1925-34.

Adolf Hitler, 1889-1945, (Ger.) dictator; built Nazism, launched WW2, presided over the Holocaust.

Ho Chi Minh, 1890-1969, (Viet.) N Vietnamese pres., Vietnamese Communist leader.

Harry L. Hopkins, 1890-1946, (U.S.) New Deal administrator; closest adviser to FDR during WW2.

Edward M. House, 1858-1938, (U.S.) diplomat; confidential adviser to Woodrow Wilson.

Samuel Houston, 1793-1863, (U.S.) leader of struggle to win control of Texas from Mexico.

Cordell Hull, 1871-1955, (U.S.) sec. of state, 1933-44; initiated reciprocal trade to lower tariffs, helped organize UN.

Hubert H. Humphrey, 1911-78, (U.S.) Minnesota Democrat; senator; vice pres., pres. candidate.

Hussein, King, 1935-99 (Jordan), peacemaker; ruler of Jordan, 1952-99.

Jinnah, Muhammad Ali, 1876-1948, (Pak.) founder, first governor-general of Pakistan.

Benito Juarez, 1806-72, (Mex.) rallied his country against foreign threats, sought to create democratic, federal republic.

Constantine Karamanlis, 1907-98, (Gr.) Greek prime min. (1955-63, 1974-80); restored democracy; later president.

Frank B. Kellogg, 1856-1937, (U.S.) sec. of state; negotiated Kellogg-Briand Pact to outlaw war.

Robert F. Kennedy, 1925-68, (U.S.) attorney general, senator; assassinated while seeking presidency.

Aleksandr Kerensky, 1881-1970, (Russ.) headed provisional government after Feb. 1917 revolution.

Ayatollah Ruhollah Khomeini, 1900-89, (Iranian) religious-political leader, spearheaded overthrow of shah, 1979.

Nikita Khrushchev, 1894-1971, (USSR) prem., first sec. of Communist party; initiated de-Stalinization.

Kim Il Sung, 1912-94, (Korean) N Korean dictator, 1948-94.

Lajos Kossuth, 1802-94, (Hung.) principal figure in 1848 Hungarian revolution.

Pyotr Kropotkin, 1842-1921, (Russ.) anarchist; championed the peasants but opposed Bolshevism.

Kublai Khan, c1215-94, Mongol emperor; founder of Yüan dynasty in China.

Béla Kun, 1886-c1939, (Hung.) member of 3d Communist Internat.; tried to foment worldwide revolution.

Robert M. LaFollette, 1855-1925, (U.S.) Wisconsin public official; leader of progressive movement.

Fiorello La Guardia, 1882-1947, (U.S.) colorful NY City reform mayor.

Pierre Laval, 1883-1945, (Fr.) politician, Vichy foreign min.; executed for treason.

Andrew Bonar Law, 1858-1923, (Br.) Conservative party politician; led opposition to Irish home rule.

Vladimir Ilyich Lenin (Ulyanov), 1870-1924, (Russ.) revolutionary; founded Bolshevism; Soviet leader 1917-24.

Ferdinand de Lesseps, 1805-94, (Fr.) diplomat, engineer; conceived idea of Suez Canal.

Rene Levesque, 1922-87, (Can.) prem. of Quebec, 1976-85; led unsuccessful separartist campaign.

Maxim Litvinov, 1876-1951, (Pol.-Russ.) revolutionary, commissar of foreign affairs; favored cooperation with West.

Liu Shaoqi, c1898-1974, (Chin.) Communist leader; fell from grace during Cultural Revolution.

David Lloyd George, 1863-1945, (Br.) Liberal party prime min.; laid foundations for modern welfare state.

Henry Cabot Lodge, 1850-1924, (U.S.) Republican senator; led opposition to participation in League of Nations.

Huey P. Long, 1893-1935, (U.S.) Louisiana political demagogue, governor; assassinated.

Rosa Luxemburg, 1871-1919, (Ger.) revolutionary; leader of the German Social Democratic party and Spartacus party.

J. Ramsay MacDonald, 1866-1937, (Br.) first Labour party prime min. of Great Britain.

Harold Macmillan, 1895-1986, (Br.) prime min. of Great Britain, 1957-63.

Joseph R. McCarthy, 1908-57, (U.S.) senator, extremist in searching out alleged Communists and pro-Communists.

Makarios III, 1913-77, (Cypriot) Greek Orthodox archbishop; first pres. of Cyprus.

Mao Zedong, 1893-1976, (Chin.) chief Chinese Marxist theorist, revolutionary, political leader; led Chinese revolution establishing his nation as Communist state.

Jean Paul Marat, 1743-93, (Fr.) revolutionary, politician; identified with radical Jacobins; assassinated.

José Martí, 1853-95, (Cub.) patriot, poet; leader of Cuban struggle for independence.

Jan Masaryk, 1886-1948, (Czech.) foreign min.; died by mysterious alleged suicide following Communist coup.

Thomas G. Masaryk, 1850-1937, (Czech.) statesman, philosopher; first pres. of Czechoslovak Republic.

Jules Mazarin, 1602-61, (Fr.) cardinal, statesman; prime min. under Louis XIII and queen regent Anne of Austria.

Giuseppe Mazzini, 1805-72, (It.), reformer dedicated to Risorgimento movement for renewal of Italy.

Tom Mboya, 1930-69, (Kenyan) political leader; instrumental in securing independence for Kenya.

Cosimo I de' Medici, 1519-74, (It.) Duke of Florence, grand duke of Tuscany.

Lorenzo de' Medici, the Magnificent, 1449-92, (It.) merchant prince; a towering figure in Italian Renaissance.

Catherine de Médicis, 1519-89, (Fr.) queen consort of Henry II, regent of France; influential in Catholic-Huguenot wars.

Golda Meir, 1898-1978, (Isr.) a founder of the state of Israel and prime min., 1969-74.

Klemens W. N. L. Metternich, 1773-1859, (Austrian) statesman; arbiter of post-Napoleonic Europe.

François Mitterrand, 1916-96, (Fr.) pres. of France, 1981-95.

Mobutu Sese Seko, 1930-97, (Zaire) longtime ruler of Zaire (now Congo) (1965-97); exiled after rebellion.

Guy Mollet, 1905-75, (Fr.) socialist politician, resistance leader.

Henry Morgenthau Jr., 1891-1967, (U.S.) sec. of treasury; fundraiser for New Deal and U.S. WW2 activities.

Gouverneur Morris, 1752-1816, (U.S.) statesman, diplomat. financial expert, helped plan decimal coinage.

Benito Mussolini, 1883-1945, (It.) dictator and leader of the Italian fascist state; assassinated.

Imre Nagy, c1896-1958, (Hung.) Communist prem.; assassinated after Soviets crushed 1956 uprising.

Gamal Abdel Nasser, 1918-70, (Egypt.) leader of Arab unification, 2d Egyptian pres.

Jawaharlal Nehru, 1889-1964, (In.) prime min.; guided India through its early years of independence.

Kwame Nkrumah, 1909-72, (Ghan.) 1st prime min., 1957-60, and pres., 1960-66, of Ghana.

Frederick North, 1732-92, (Br.) prime min.; his inept policies led to loss of American colonies.

Daniel O'Connell, 1775-1847, (Ir.) political leader; known as The Liberator.

Omar, c581-644, Muslim leader; 2d caliph, led Islam to become an imperial power.

Thomas P. (Tip) O'Neill Jr., 1912-94, (U.S.) U.S. congressman, Speaker of the House, 1977-86.

Ignace Paderewski, 1860-1941, (Pol.) statesman, pianist; composer, briefly prime min., an ardent patriot.

Viscount Palmerston, 1784-1865, (Br.) Whig-Liberal prime min., foreign min.; embodied British nationalism.

Andreas George Papandreou, 1919-1996, (Gk.) leftist politician, served 2 times as prem. (1981-89, 1993-96).

Georgios Papandreou, 1888-1968, (Gk.) Republican politician; served 3 times as prime min.

Franz von Papen, 1879-1969, (Ger.) politician; major role in overthrow of Weimar Republic and rise of Hitler.

Charles Stewart Parnell, 1846-1891, (Ir.) nationalist leader; "uncrowned king of Ireland."

Lester Pearson, 1897-1972, (Can.) diplomat, Liberal party leader, prime min.

Robert Peel, 1788-1850, (Br.) reformist prime min., founder of Conservative party.

Eva (Evita) Perón, 1919-52 (Arg.) highly influential 2d wife of Juan Perón.

Juan Perón, 1895-1974, (Arg.) dynamic pres. of Argentina (1946-55, 1973-74).

Joseph Pilsudski, 1867-1935, (Pol.) statesman; instrumental in reestablishing Polish state in the 20th cent.

Charles Pinckney, 1757-1824, (U.S.) founding father; his Pinckney plan largely incorporated into Constitution.

Christian Pineau, 1905-95, (Fr.) leader of French Resistance during WW2; French foreign minister, 1956-58.

William Pitt, the Elder, 1708-78, (Br.) statesman; the "Great Commoner," transformed Britain into imperial power.

William Pitt, the Younger, 1759-1806, (Br.) prime min. during French Revolutionary wars.

Georgi Plekhanov, 1857-1918, (Russ.) revolutionary, social philosopher; called "father of Russian Marxism."

Raymond Poincaré, 1860-1934, (Fr.) 9th pres. of the Republic; advocated harsh punishment of Germany after WW1.

Pol Pot, 1925-98, (Camb.) leader of Khmer Rouge; ruled Cambodia, 1975-79; responsible for mass deaths.

Grigori Potemkin, 1739-91, (Russ.) field marshal; favorite of Catherine II.

Yitzhak Rabin, 1922-95, (Isr.) military, political leader; prime min. of Israel, 1974-77, 1992-95; assassinated.

Edmund Randolph, 1753-1813, (U.S.) attorney; prominent in drafting, ratification of constitution.

John Randolph, 1773-1833, (U.S.) Southern planter; strong advocate of states' rights.

Jeannette Rankin, 1880-1973, (U.S.) pacifist; first woman member of U.S. Congress.

Walter Rathenau, 1867-1922, (Ger.) industrialist, statesman.

Sam Rayburn, 1882-1961, (U.S.) Democratic leader; representative for 47 years, House Speaker for 17.

Paul Reynaud, 1878-1966, (Fr.) statesman; prem. in 1940 at the time of France's defeat by Germany.

Syngman Rhee, 1875-1965, (Korean) first pres. of S Korea.

Cecil Rhodes, 1853-1902, (Br.) imperialist, industrial magnate; established Rhodes scholarships in his will.

Cardinal de Richelieu, 1585-1642, (Fr.) statesman, known as "red eminence;" chief minister to Louis XIII.

Maximilien Robespierre, 1758-94, (Fr.) leading figure in French Revolution and Reign of Terror.

Nelson Rockefeller, 1908-79, (U.S.) Republican governor of NY, 1959-73; U.S. vice pres., 1974-77.

George W. Romney, 1907-95, (U.S.) auto exec.; 3-term Republican governor of Michigan.

Eleanor Roosevelt, 1884-1962, (U.S.) influential First Lady, humanitarian, UN diplomat.

Elihu Root, 1845-1937, (U.S.) lawyer, statesman, diplomat; leading Republican supporter of the League of Nations.

Dean Rusk, 1909-95, (U.S.) statesman; sec. of state, 1961-69.

John Russell, 1792-1878, (Br.) Liberal prime min. during the Irish potato famine.

Anwar al-Sadat, 1918-81, (Egypt.) pres., 1970-1981, promoted peace with Israel; Nobel laureate; assassinated.

António de Salazar, 1889-1970, (Port.) longtime dictator.

José de San Martin, 1778-1850, S Amer. revolutionary; protector of Peru.

Eisaku Sato, 1901-75, (Jpn.) prime min.; presided over Japan's post-WW2 emergence as major world power.

Abdul Aziz Ibn Saud, c1880-1953, king of Saudi Arabia, 1932-53.

Philipp Scheidemann, 1865-1939, (Ger.) Social Democratic leader; first chancellor of the German republic.

Robert Schuman, 1886-1963, (Fr.) statesman; founded European Coal and Steel Community.

Carl Schurz, 1829-1906, (U.S.) German-American political leader, journalist, orator, dedicated reformer.

Kurt Schuschnigg, 1897-1977, (Austrian) chancellor; unsuccessful in stopping Austria's annexation by Germany.

William H. Seward, 1801-72, (U.S.) anti-slavery activist; as U.S. sec. of state purchased Alaska.

Carlo Sforza, 1872-1952, (It.) foreign min., anti-fascist.

Sitting Bull, c1831-90, (Nat. Am.) Sioux leader in Battle of Little Bighorn over George A. Custer, 1876.

Alfred E. Smith, 1873-1944, (U.S.) NY Democratic governor; first Roman Catholic to run for presidency.

Margaret Chase Smith, 1897-1995, (U.S.) congresswoman, senator; 1st woman elected to both houses of Congress.

Jan C. Smuts, 1870-1950, (S. African) statesman, philosopher, soldier, prime min.

Paul Henri Spaak, 1899-1972, (Belg.) statesman, socialist leader.

Joseph Stalin, 1879-1953, (USSR) Soviet dictator, 1924-53; instituted forced collectivization, massive purges, and labor camps, causing millions of deaths.

Edwin M. Stanton, 1814-69, (U.S.) sec. of war, 1862-68.

Edward R. Stettinius Jr., 1900-49, (U.S.) industrialist, sec. of state who coordinated aid to WW2 allies.

Adlai E. Stevenson, 1900-65, (U.S.) Democratic leader, diplomat, Illinois governor, presidenial candidate.

Henry L. Stimson, 1867-1950, (U.S.) statesman; served in 5 administrations, foreign policy adviser in 30s and 40s.

Gustav Stresemann, 1878-1929, (Ger.) chancellor, foreign minister; strove to regain friendship for post-WW1 Germany.

Sukarno, 1901-70, (Indon.) dictatorial first pres. of the Indonesian republic.

Sun Yat-sen, 1866-1925, (Chin.) revolutionary; leader of Kuomintang, regarded as the father of modern China.

Robert A. Taft, 1889-1953, (U.S.) conservative Senate leader, called "Mr. Republican."

Charles de Talleyrand, 1754-1838, (Fr.) statesman, diplomat; the major force of the Congress of Vienna of 1814-15.

U Thant, 1909-74 (Bur.) statesman, UN sec.-general.

Norman M. Thomas, 1884-1968, (U.S.) social reformer; 6 times Socialist party presidential candidate.

Josip Broz Tito, 1892-1980, (Yug.) pres. of Yugoslavia from 1953, WW2 guerrilla chief, postwar rival of Stalin.

Palmiro Togliatti, 1893-1964, (It.) major leader of Italian Communist party.

Hideki Tojo, 1885-1948, (Jpn.) statesman, soldier; prime min. during most of WW2.

François Toussaint L'Ouverture, c1744-1803, (Haitian) patriot, martyr; thwarted French colonial aims.

Leon Trotsky, 1879-1940, (Russ.) revolutionary, founded Red Army, expelled from party in conflict with Stalin; assassinated.

Rafael L. Trujillo Molina, 1891-1961, (Dom.) dictator of Dominican Republic, 1930-61; assassinated.

Moise K. Tshombe, 1919-69, (Cong.) pres. of secessionist Katanga, prem. of Congo.

William M. Tweed, 1823-78, (U.S.) politician; absolute leader of Tammany Hall, NYC's Democratic political machine.

Walter Ulbricht, 1893-1973, (Ger.) Communist leader of German Democratic Republic.

Arthur H. Vandenberg, 1884-1951, (U.S.) senator; proponent of bipartisan anti-Communist foreign policy.

Eleutherios Venizelos, 1864-1936, (Gk.) most prominent Greek statesman of early 20th cent.

Hendrik F. Verwoerd, 1901-66, (S. African) prime min.; rigorously applied apartheid policy despite protest.

George Wallace, 1919-98, (U.S.) former segregationist governor of Alabama and presid. candidate.

Robert Walpole, 1676-1745, (Br.) statesman; generally considered Britain's first prime min.

Daniel Webster, 1782-1852, (U.S.) orator, politician; advocate of business interests during Jacksonian agrarianism.

Chaim Weizmann, 1874-1952, (Russ.-Isr.) Zionist leader, scientist; first Israeli pres.

Wendell L. Willkie, 1892-1944, (U.S.) Republican who tried to unseat FDR when he ran for his 3d term.

Harold Wilson, 1916-95, (Br.) Labour party leader; prime min., 1964-70, 1974-76.

Emiliano Zapata, c1879-1919, (Mex.) revolutionary; major influence on modern Mexico.

Todor Zhivkov, 1911-98, (Bulg.) Communist ruler of Bulgaria from 1954 until ousted in a 1989 coup.

Zhou Enlai, 1898-1976, (Chin.) diplomat, prime min.; a leading figure of the Chinese Communist party.

Noted Scientists of the Past

Revised by Peter Barker, Prof. & Chair, Dept. of the Hist. of Science, Univ. of Oklahoma

For pre-modern scientists see also Philosophers and Religious Figures of the Past and Historical Figures chapter.

Albertus Magnus, c1200-1280, (Ger.) theologian, philosopher; helped found medieval study of natural science.

Alhazen (Ibn al-Haytham), c965-ca.1040, mathematician, astronomer; optical theorist.

Andre-Marie Ampère, 1775-1836, (Fr.) mathematician, chemist; founder of electrodynamics.

John V. Atanasoff, 1903-95, (U.S.) physicist; co-invented Atanasoff-Berry Computer (1939-41), regarded in law as the original "automatic electronic digital computer".

Amedeo Avogadro, 1776-1856, (It.) chemist, physicist; proposed that equal volumes of gas contain equal numbers of molecules, permitting determination of molecular weights.

John Bardeen, 1908-91, (U.S.) double Nobel laureate in physics (transistor, 1956; superconductivity, 1972).

A. H. Becquerel, 1852-1908, (Fr.) physicist; discovered radioactivity in uranium (1896).

Alexander Graham Bell, 1847-1922, (U.S.) inventor; first to patent and commercially exploit the telephone (1876).

Daniel Bernoulli, 1700-82, (Swiss) mathematician; developed fluid dynamics and kinetic theory of gases.

Clifford Berry, 1918-1963, (U.S.) collaborated with Atanasoff on the ABC computer (1939-41).

Jöns Jakob Berzelius, 1779-1848, (Swed.) chemist; developed modern chemical symbols and formulas, discovered selenium and thorium.

Henry Bessemer, 1813-98, (Br.) engineer; invented Bessemer steel-making process.

Bruno Bettelheim, 1903-90, (Austrian-U.S.) psychoanalyst specializing in autistic and other disturbed children; *Uses of Enchantment* (1976).

Louis Blériot, 1872-1936, (Fr.) engineer; monoplane pioneer, first Channel flight (1909).

Franz Boas, 1858-1942, (Ger.-U.S.) founded modern anthropology; studied Pacific Coast tribes.

Niels Bohr, 1885-1962, (Dan.) atomic and nuclear physicist; founded quantum mechanics.

Max Born, 1882-1970, (Ger.) atomic and nuclear physicist; helped develop quantum mechanics.

Satyendranath Bose, 1894-1974, (Indian) physicist; forerunner of modern quantum theory for integral-spin particles.

Louis de Broglie, 1892-1987, (Fr.) physicist; proposed quantum wave-particle duality.

Robert Bunsen, 1811-99, (Ger.) chemist; pioneered spectroscopic analysis; discovered rubidium, caesium.

Luther Burbank, 1849-1926, (U.S.) naturalist; developed plant breeding into a modern science.

Vannevar Bush, 1890-1974, (U.S.) electrical engineer; developed differential analyzer, an early analogue computer; headed WWII Office of Scientific Res. and Dev.

Marvin Camras, 1916-95, (U.S.) inventor, electrical engineer; invented magnetic tape recording.

Alexis Carrel, 1873-1944, (Fr.) surgeon, biologist; developed methods of suturing blood vessels and transplanting organs.

Rachel Carson, 1907-64, (U.S.) marine biologist, environmentalist; *Silent Spring* (1962).

George Washington Carver, c1864-1943, (U.S.) agricultural scientist, nutritionist; improved and pioneered new uses for peanuts and sweet potatoes.

James Chadwick, 1891-1974, (Br.) physicist; discovered the neutron (1932); led British Manhattan Project group in U.S. (1943-45).

Daryl Chapin, 1906-95, (U.S.) physicist; with Calvin Fuller and Gerald Pearson of the solar energy cell (1954).

Albert Claude, 1898-1983, (Belg.) a founder of modern cell biology; determined role of mitochondria.

Nicolaus Copernicus, 1473-1543, (Pol.) first modern astronomer to propose sun as center of the planets' motions.

Jacques Yves Cousteau, 1910-1997, (Fr.) oceanographer; coinventor, with E. Gagnan, of the Aqualung (1943).

Seymour Cray, 1925-96, (U.S.) computer industry pioneer; developed supercomputers.

Marie, 1867-1934 (Pol.-Fr.) and **Pierre Curie**, 1859-1906, (Fr.) physical chemists; pioneer investigators of radioactivity, discovered radium and polonium (1898).

Gottlieb Daimler, 1834-1900, (Ger.) engineer, inventor; pioneer automobile manufacturer.

John Dalton, 1766-1844, (Br.) chemist, physicist; formulated atomic theory, made first table of atomic weights.

Charles Darwin, 1809-82, (Br.) naturalist; established theory of organic evolution; *Origin of Species* (1859).

Lee De Forest, 1873-1961, (U.S.) inventor of triode, pioneer in wireless telegraphy, sound pictures, television.

Max Delbruck, 1906-81, (Ger.-U.S.) founded molecular biology.

Rudolf Diesel, 1858-1913, (Ger.) mechanical engineer; patented Diesel engine (1892).

Theodosius Dobzhansky, 1900-75, (Russ.-U.S.) biologist; reconciled genetics and natural selection contributing to "modern synthesis" in evolution.

Christian Doppler, 1803-53, (Austrian) physicist; showed change in wave frequency caused by motion of source, now known as Doppler effect.

J. Presper Eckert Jr., 1919-95, (U.S.) co-inventor, with Mauchly, of the ENIAC computer (1943-45).

Thomas A. Edison, 1847-1931, (U.S.) inventor; held more than 1,000 patents, including incandescent electric lamp.

Paul Ehrlich, 1854-1915, (Ger.) medical researcher in immunology and bacteriology; pioneered antitoxin production.

Albert Einstein, 1879-1955, (Ger.-U.S.) theoretical physicist; founded relativity theory, replacing Newton's theories of space, time, and gravity. Proved E=mc2 (1905).

John F. Enders, 1897-1985, (U.S.) virologist, helped discover vaccines against polio, measles, mumps and chicken pox.

Erik Erikson, 1902-94, (U.S.) psychoanalyst, author; theory of developmental stages of life, *Childhood and Society* (1950).

Leonhard Euler, 1707-83, (Swiss) mathematician, physicist; pioneer of calculus, revived ideas of Fermat.

Gabriel Fahrenheit, 1686-1736, (Ger.) physicist; improved thermometers and introduced Fahrenheit temperature scale.

Michael Faraday, 1791-1867, (Br.) chemist, physicist; discovered electrical induction and invented dynamo (1831).

Philo T. Farnsworth, 1906-71, (U.S.) inventor; built first television system (San Francisco, 1928).

Pierre de Fermat, 1601-65, (Fr.) mathematician; founded modern theory of numbers.

Enrico Fermi, 1901-54, (It.-U.S.) nuclear physicist; demonstrated first controlled chain reaction (Chicago, 1942).

Richard Feynman, 1918-88, (U.S.) theoretical physicist, author; founder of Quantum Electrodynamics (QED).

Alexander Fleming, 1881-1955, (Br.) bacteriologist; discovered penicillin (1928).

Jean B. J. Fourier, 1768-1830, (fr.) introduced method of analysis in math and physics known as Fourier Series.

Sigmund Freud, 1856-1939, (Austrian) psychiatrist; founder of psychoanalysis. *Interpretation of Dreams* (1901).

Erich Fromm, 1900-1980, (U.S.) psychoanalyst. *Man for Himself* (1947).

Galileo Galilei, 1564-1642, (It.) physicist; used telescope to vindicate Copernicus, founded modern science of motion.

Luigi Galvani, 1737-98, (It.) physiologist; studied electricity in living organisms.

Carl Friedrich Gauss, 1777-1855, (Ger.) math. physicist; completed work of Fermat and Euler in number theory.

Joseph Gay-Lussac, 1778-1850, (Fr.) chemist, physicist; investigated behavior of gases, discovered boron.

Josiah W. Gibbs, 1839-1903, (U.S.) theoretical physicist, chemist; founded chemical thermodynamics.

Robert H. Goddard, 1882-1945, (U.S.) physicist; invented liquid fuel rocket (1926).

George W. Goethals, 1858-1928, (U.S.) chief engineer who completed Panama Canal (1907-14).

William C. Gorgas, 1854-1920, (U.S.) physician; pioneer in prevention of yellow fever and malaria.

Ernest Haeckel, 1834-1919, (Ger.) zoologist, evolutionist; early Darwinist, introduced concept of "ecology."

Otto Hahn, 1879-1968, (Ger.) chemist; with Meitner discovered nuclear fission (1938).

Edmund Halley, 1656-1742, (Br.) astronomer; predicted return of 1682 comet ("Halley's Comet") in 1759.

William Harvey, 1578-1657, (Br.) physician, anatomist; discovered circulation of the blood (1628).

Werner Heisenberg, 1901-76, (Ger.) physicist; developed matrix mechanics and uncertainty principle (1927).

Hermann von Helmholtz, 1821-94, (Ger.) physicist, physiologist; formulated principle of conservation of energy.

William Herschel, 1738-1822, (Ger.-Br.) astronomer; discovered Uranus (1781).

Heinrich Hertz, 1857-94, (Ger.) physicist; discovered radio waves and photo-electric effect (1886-7).

David Hilbert, 1862-1943, (Ger.) mathematician; contributed to algebra, calculus and foundational studies (formalism).

Edwin P. Hubble, 1889-1953, (U.S.) astronomer; discovered observational evidence of expanding universe.

Alexander von Humboldt, 1769-1859, (Ger.) naturalist, author; explored S America, created ecology.

Edward Jenner, 1749-1823, (Br.) physician; pioneered vaccination, introduced term "virus."

James Joule, 1818-89, (Br.) physicist; found relation between heat and mechanical energy (conservation of energy).

Carl Jung, 1875-1961, (Swiss) psychiatrist; founder of analytical psychology.

Sister Elizabeth Kenny, 1886-1952, (Austral.) nurse; developed treatment for polio.

Johannes Kepler, 1571-1630, (Ger.) astronomer; discovered laws of planetary motion.

Al-Khawarizmi, early 9th cent., (Arab.), mathematician; regarded as founder of algebra.

Robert Koch, 1843-1910 (Ger.) bacteriologist; isolated bacterial causes of tuberculosis and other diseases.

Georges Köhler, 1946-95, (Ger.) immunologist; with Cesar Milstein he developed monoclonal antibody technique.

Jacques Lacan, 1901-81, (Fr.) controversial influential psychoanalyst.

Joseph Lagrange, 1736-1813, (Fr.) geometer, astronomer; showed that gravity of earth and moon cancels creating stable points in space around them.

Jean B. Lamarck, 1744-1829, (Fr.) naturalist; forerunner of Darwin in evolutionary theory.

Pierre Simon de Laplace, 1749-1827, (Fr.) astronomer, physicist; proposed nebular origin for solar system.

Antoine Lavoisier, 1743-94, (Fr.) a founder of mod. chemistry.

Ernest O. Lawrence, 1901-58, (U.S.) physicist; invented the cyclotron.

Jerome Lejeune, 1927-94, (Fr.) geneticist; discovered chromosomal cause of Down syndrome (1959).

Louis 1903-72, and **Mary Leakey**, 1913-96, (Br.) early hominid paleoanthropologists; discovered remains in Africa.

Anton van Leeuwenhoek, 1632-1723, (Dutch) founder of microscopy.

Kurt Lewin, 1890-1947, (Ger.-U.S.) social psychologist; studied human motivation and group dynamics.

Justus von Liebig, 1803-73, (Ger.) founded quantitative organic chemistry.

Joseph Lister, 1827-1912, (Br.) physician; pioneered antiseptic surgery.

Konrad Lorenz, 1903-89, (Austrian) ethologist; pioneer in study of animal behavior.

Percival Lowell, 1855-1916, (U.S.) astronomer; predicted the existence of Pluto.

Louis, 1864-1948, and **Auguste Lumière**, 1862-1954, (Fr.) invented cinematograph and made first motion picture (1895).

Guglielmo Marconi, 1874-1937, (It.) physicist; developed wireless telegraphy.

John W. Mauchly, 1907-80, (U.S.) co-inventor, with Eckert, of computer ENIAC (1943-45).

James Clerk Maxwell, 1831-79, (Br.) physicist; unified electricity and magnetism; electromagnetic theory of light.

Maria Goeppert Mayer, 1906-72, (Ger.-U.S.) physicist; developed shell model of atomic nuclei.

Barbara McClintock, 1902-92, (U.S.) geneticist; showed that some genetic elements are mobile.

Lise Meitner, 1878-1968, (Austrian) co-discoverer, with Hahn, of nuclear fission (1938).

Gregor J. Mendel, 1822-84, (Austrian) botanist, monk; his experiments became the foundation of modern genetics.

Dmitri Mendeleyev, 1834-1907, (Russ.) chemist; established Periodic Table of the Elements.

Franz Mesmer, 1734-1815, (Ger.) physician; introduced hypnotherapy.

Albert A. Michelson, 1852-1931, (U.S.) physicist; invented interferometer.

Robert A. Millikan, 1868-1953, (U.S.) physicist; measured electronic charge.

Thomas Hunt Morgan, 1866-1945, (U.S.) geneticist, embryologist; established role of chromosomes in heredity.

Isaac Newton, 1642-1727, (Br.) natural philosopher; discovered laws of gravitation, motion; with Leibniz, founded calculus.

Robert N. Noyce, 1927-90, (U.S.) invented microchip.

J. Robert Oppenheimer, 1904-67, (U.S.) physicist; scientific director of Manhattan project.

Wilhelm Ostwald, 1853-1932, (Ger.) chemist, philosopher; main founder of modern physical chemistry.

Louis Pasteur, 1822-95, (Fr.) chemist; showed that germs cause disease and fermentation, originated pasteurization.

Linus C. Pauling, 1901-94, (U.S.) chemist; studied chemical bonds; campaigned for nuclear disarmament.

Jean Piaget, 1896-1980, (Swiss) psychologist; four-stage theory of intellectual development in children.

Max Planck, 1858-1947, (Ger.) physicist; introduced quantum hypothesis (1900).

Walter S. Reed, 1851-1902, (U.S.) army physician; proved mosquitoes transmit yellow fever.

Theodor Reik, 1888-1969, (Austrian-U.S.) psychoanalyst, major Freudian disciple.

Bernhard Riemann, 1826-66, (Ger.) mathematician; developed non-Euclidean geometry used by Einstein.

Wilhelm Roentgen, 1845-1923, (Ger.) physicist; discovered X-rays (1895).

Carl Rogers, 1902-87, (U.S.) psychotherapist, author; originated nondirective therapy.

Ernest Rutherford, 1871-1937, (Br.) physicist; pioneer investigator of radioactivity, identified the atomic nucleus.

Albert B. Sabin, 1906-93, (Russ.-U.S.), developed oral polio live-virus vaccine.

Carl Sagan, 1934-96, (U.S.) astronomer, author.

Jonas Salk, 1914-95, (U.S.) developed first successful polio vaccine, widely used in U.S. after 1955.

Giovanni Schiaparelli, 1835-1910, (It.) astronomer; reported canals on Mars.

Erwin Schrödinger, 1887-1961, (Austrian) physicist; developed wave equation for quantum systems.

Glenn T. Seaborg, 1912-99, (U.S.) chemist, Nobel Prize winner (1951); codiscoverer of plutonium.

Harlow Shapley, 1885-1972, (U.S.) astronomer; mapped galactic clusters and position of Sun in our own galaxy.

B(urrhus) F(rederick) Skinner, 1904-89, (U.S.) psychologist; leading advocate of behaviorism.

Roger W. Sperry, 1913-94, (U.S.) neuorobiologist; established different functions of right and left sides of brain.

Benjamin Spock, 1903-98, (U.S.) pediatrician, child care expert; *Common Sense Book of Baby and Child Care.*

Charles P. Steinmetz, 1865-1923, (Ger.-U.S.) electrical engineer; developed basic ideas on alternating current.

Leo Szilard, 1898-1964, (Hung.-U.S.) physicist; helped on Manhattan project, later opposed nuclear weapons.

Nikola Tesla, 1856-1943, (Serb.-U.S.) invented many electrical devices including a.c. dynamos, transformers and motors.

William Thomson (Lord Kelvin), 1824-1907, (Br.) physicist; aided in success of transatlantic telegraph cable (1865); proposed Kelvin absolute temperature scale.

Alan Turing, 1912-54, (Br.) mathematician; helped develop basis for computers.

Rudolf Virchow, 1821-1902, (Ger.) pathologist; pioneered the modern theory that diseases affect the body through cells.

Alessandro Volta, 1745-1827, (It.) physicist; pioneer in electricity.

Werner von Braun, 1912-77, (Ger.-U.S.) developed rockets for warfare and space exploration.

John Von Neumann, 1903-57, (Hung.-U.S.) mathematician; originated game theory; basic design for modern computers.

Alfred Russell Wallace, 1823-1913, (Br.) naturalist; proposed concept of evolution independently of Darwin.

John B. Watson, 1878-1958, (U.S.) psychologist; a founder of behaviorism.

James E. Watt, 1736-1819, (Br.) mechanical engineer, inventor; invented modern steam engine (1765).

Alfred L. Wegener, 1880-1930, (Ger.) meteorologist, geophysicist; postulated continental drift.

Norbert Wiener, 1894-1964, (U.S.) mathematician; founder of cybernetics.

Sewall Wright, 1889-1988, (U.S.) evolutionary theorist; helped found population genetics.

Wilhelm Wundt, 1832-1920, (Ger.) founder of experimental psychology.

Ferdinand von Zeppelin, 1838-1917, (Ger.) soldier, aeronaut, airship designer.

Noted Social Reformers, Activists, and Humanitarians of the Past

Jane Addams, 1860-1935, (U.S.) cofounder of Hull House; won Nobel Peace Prize, 1931.

Susan B. Anthony, 1820-1906, (U.S.) a leader in temperance, anti-slavery, and woman suffrage movements.

Thomas Barnardo, 1845-1905, (Br.) social reformer; pioneered in care of destitute children.

Clara Barton, 1821-1912, (U.S.) organized Amer. Red Cross.

Henry Ward Beecher, 1813-87, (U.S.) clergyman, abolitionist.

Amelia Bloomer, 1818-94, (U.S.) suffragette, social reformer.

William Booth, 1829-1912, (Br.) founded Salvation Army.

John Brown, 1800-59, (U.S.) abolitionist who led murder of 5 pro-slavery men, was hanged.

Frances Xavier (Mother) Cabrini, 1850-1917, (It.-U.S.) Italian-born nun; founded charitable institutions; first American canonized as a saint, 1946.

Carrie Chapman Catt, 1859-1947, (U.S.) suffragette.

Cesar Chavez, 1927-93, (U.S.) labor leader; helped establish United Farm Workers of America.

Clarence Darrow, 1857-1938, (U.S.) lawyer; defender of "underdog," opponent of capital punishment.

Dorothy Day, 1897-1980, (U.S.) founder of Catholic Worker movement.

Eugene V. Debs, 1855-1926, (U.S.) labor leader; led Pullman strike, 1894; 4-time Socialist presidential candidate.

Dorothea Dix, 1802-87, (U.S.) crusader for mentally ill.

Thomas Dooley, 1927-61, (U.S.) "jungle doctor," noted for efforts to supply medical aid to developing countries.

Marjory Stoneman Douglas, 1890-1998, (U.S.) writer and environmentalist; campaigned to save Florida Everglades.

William Lloyd Garrison, 1805-79, (U.S.) abolitionist.

Emma Goldman, 1869-1940, (Russ.-U.S.) published anarchist *Mother Earth*, birth-control advocate.

Samuel Gompers, 1850-1924, (U.S.) labor leader.

Michael Harrington, 1928-89, (U.S.) exposed poverty in affluent U.S. in *The Other America*, 1963.

Sidney Hillman, 1887-1946, (U.S.) labor leader; helped organize CIO.

Samuel G. Howe, 1801-76, (U.S.) social reformer; changed public attitudes toward the handicapped.

Helen Keller, 1880-1968, (U.S.) crusader for better treatment for the handicapped; deaf and blind herself.

Maggie Kuhn, 1905-95, (U.S.) founded Gray Panthers, 1970.

William Kunstler, 1919-95, (U.S.) civil liberties attorney.

John L. Lewis, 1880-1969, (U.S.) labor leader; headed United Mine Workers, 1920-60.

Karl Menninger, 1893-1990, (U.S.) with brother William founded Menninger Clinic and Menninger Foundation.

Lucretia Mott, 1793-1880, (U.S.) reformer, pioneer feminist.

Philip Murray, 1886-1952, (U.S.) Scottish-born labor leader.

Florence Nightingale, 1820-1910, (Br.) founder of modern nursing.

Emmeline Pankhurst, 1858-1928, (Br.) woman suffragist.

Walter Reuther, 1907-70, (U.S.) labor leader; headed UAW.

Jacob Riis, 1849-1914, (U.S.) crusader for urban reforms.

Margaret Sanger, 1883-1966, (U.S.) social reformer; pioneered the birth-control movement.

Earl of Shaftesbury, 1801-85, (Br.) social reformer.

Elizabeth Cady Stanton, 1815-1902, (U.S.) woman suffrage pioneer.

Lucy Stone, 1818-93, (U.S.) feminist, abolitionist.

Mother Teresa of Calcutta, 1910-97, (Alban.) nun; founded order to care for sick, dying poor; 1979 Nobel Peace Prize.

Philip Vera Cruz, 1905-94, (Filipino-U.S.) helped to found the United Farm Workers Union.

William Wilberforce, 1759-1833, (Br.) social reformer; prominent in struggle to abolish the slave trade.

Frances E. Willard, 1839-98, (U.S.) temperance, women's rights leader.

Mary Wollstonecraft, 1759-97, (Br.) wrote *Vindication of the Rights of Women.*

Noted Writers of the Present

Name (Birthplace)	Birthdate
Chinua Achebe (Ogidi, Nigeria)	11/16/30
Edward Albee (Wash., DC)	3/12/28
Jorge Amado (Bahia, Brazil)	8/1/12
Martin Amis (Oxford, Eng.)	8/25/49
Maya Angelou (St. Louis, MO)	4/4/28
Oscar Arias Sanchez (Heredia, Costa Rica)	9/13/41
John Ashbery (Rochester, NY)	1927
Margaret Atwood (Ottawa, Ont.)	11/18/39
Louis Auchincloss (Lawrence, NY)	9/27/17
John Barth (Cambridge, MD)	5/27/30
Ann Beattie (Wash., DC)	9/7/47
Saul Bellow (Lachine, Que.)	7/10/15
Peter Benchley (NYC)	5/8/40
Thomas Berger (Cincinnati, OH)	7/20/24
Judy Blume (Elizabeth, NJ)	2/12/38
Ray Bradbury (Waukegan, IL)	8/22/20
Barbara Taylor Bradford (Leeds, Eng.)	5/10/33
Gwendolyn Brooks (Topeka, KS)	6/7/17
Hortense Calisher (NYC)	12/20/11
Camilo Jose Cela (Galicia, Spain)	5/11/16
Tom Clancy (Baltimore, MD)	1947
Mary Higgins Clark (NYC)	12/24/31
Beverly Cleary (McMinnville, OR)	4/12/16
Evan S. Connell (Kansas City, MO)	8/17/24
Pat Conroy (Atlanta, GA)	10/26/45
Robin Cook (NYC)	5/4/40
Harry Crews (Alma, GA)	6/6/35
Michael Crichton (Chicago, IL)	10/23/42
Michael Cunningham (Ohio)	1952
Janet Dailey (Storm Lake, IA)	5/21/44
Joan Didion (Sacramento, CA)	12/5/34
E. L. Doctorow (NYC)	1/6/31
Takako Doi (Hyogo, Jap.)	11/30/28
Rita Dove (Akron, OH)	8/28/52
John Gregory Dunne (Hartford, CT)	5/25/32
James Ellroy (Los Angeles)	3/4/48
Louise Erdrich (Little Falls, MN)	7/6/54
Laura Esquivel (Mexico City, Mexico)	1950
Howard Fast (NYC)	11/11/14
Horton Foote (Wharton, TX)	3/14/16
Dario Fo (San Giano, Italy)	3/26/26
Frederick Forsyth (Ashford, Eng.)	1938
Paula Fox (NYC)	4/22/23
Marilyn French (NYC)	11/21/29
Brian Friel (Omagh, Ire.)	11/9/29
Carlos Fuentes (Mexico City, Mex.)	11/11/28
Gabriel Garcia Marquez (Aracata, Colombia)	3/6/28
Frank Gilroy (NYC)	10/13/25
Gail Godwin (Birmingham, AL)	6/18/37
William Goldman (Chicago, IL)	8/12/31
Nadine Gordimer (Springs, S. Africa)	11/20/23
Mary Gordon (Long Island, NY)	12/8/49
Sue Grafton (Louisville, KY)	4/24/40
Günter Grass (Danzig, Ger.)	10/16/27
Shirley Ann Grau (New Orleans, LA)	7/8/29
John Grisham (Jonesboro, AR)	2/8/55
John Guare (NYC)	2/5/38
Arthur Hailey (Luton, Eng.)	4/5/20
Robert Hass (San Francisco, CA)	1941
Vaclav Havel (Prague, Czech.)	10/5/36
Seamus Heaney (N. Ireland)	1939
Joseph Heller (Brooklyn, NY)	5/1/23
Mark Helprin (NYC)	6/28/47
S. E. Hinton (Tulsa, OK)	1948
Ted Hughes (Mytholmroyd, Eng.)	8/17/30
John Irving (Exeter, NH)	3/2/42
John Jakes (Chicago, IL)	3/31/32
P. D. James (Oxford, Eng.)	8/3/20
Erica Jong (NYC)	3/26/42
Garrison Keillor (Anoka, MN)	8/7/42
Thomas Keneally (Sydney, Austral.)	10/7/35
William Kennedy (Albany, NY)	1/16/28
Stephen King (Portland, ME)	9/21/47
Barbara Kingsolver (Annapolis, MD)	4/8/55
Maxine Hong Kingston (Stockton, CA)	10/27/40
Galway Kinnell (Providence, RI)	2/1/27
John Knowles (Fairmont, WV)	9/16/26
Kenneth Koch (Cincinnati, OH)	2/27/25
Dean Koontz (Everett, PA)	7/9/45
Judith Krantz (NYC)	1/9/28
Maxine Kumin (Philadelphia, PA)	6/6/25

Name (Birthplace)	Birthdate
John Le Carré (Poole, Eng.)	10/19/31
Ursula LeGuin (Berkeley, CA)	10/21/29
Madeleine L'Engle (NYC)	11/29/18
Elmore Leonard (New Orleans, LA)	10/11/25
Doris Lessing (Kermanshah, Persia)	10/22/19
Ira Levin (NYC)	8/27/29
Robert Ludlum (NYC)	5/25/27
Alison Lurie (Chicago, IL)	9/3/26
Nagib Mahfouz (Cairo, Egypt)	12/11/11
Norman Mailer (Long Branch, NJ)	1/31/23
David Mamet (Chicago, IL)	11/30/47
Bobbie Ann Mason (nr. Mayfield, KY)	5/1/40
Cormac McCarthy (Providence, RI)	7/20/33
Frank McCourt (Brooklyn, NY)	1930
Colleen McCullough (Wellington, N.S.W.)	6/1/37
Thomas McGuane (Wyandotte, MI)	12/11/39
Larry McMurtry (Wichita Falls, TX)	6/3/36
Arthur Miller (NYC)	10/17/15
Czeslaw Milosz (Seteiniai, Lithuania)	6/30/11
Toni Morrison (Lorain, OH)	2/18/31
Walter Mosley (Los Angeles, CA)	1952
Alice Munro (Wingham, Ont.)	7/10/31
V. S. Naipaul (Port-of-Spain, Trin.)	8/17/32
Joyce Carol Oates (Lockport, NY)	6/16/38
Tim O'Brien (Austin, MN)	10/1/46
Kenzaburo Oe (Ose-mura, Japan)	1935
Cynthia Ozick (NYC)	4/17/28
Grace Paley (NYC)	12/11/22
Marge Piercy (Detroit, MI)	3/31/36
Robert Pinsky (Long Branch, NJ)	10/20/40
Harold Pinter (London, Eng.)	10/10/30
Chaim Potok (NYC)	2/17/29
Reynolds Price (Macon, NC)	2/1/33
E. Annie Proulx (Norwich, CT)	8/22/35
Thomas Pynchon (Glen Cove, NY)	5/8/37
David Rabe (Dubuque, IA)	3/10/40
Ishmael Reed (Chattanooga, TN)	2/22/38
Ruth Rendell (England)	2/17/30
Anne Rice (New Orleans, LA)	10/14/41
Adrienne Rich (Baltimore, MD)	5/16/29
Philip Roth (Newark, NJ)	3/19/33
Salman Rushdie (Bombay, India)	6/19/47
J. D. Salinger (NYC)	1/1/19
Jose Saramago (Azinhaga, Portugal)	1922
Maurice Sendak (NYC)	6/10/28
Sidney Sheldon (Chicago, IL)	2/11/17
Sam Shepard (Ft. Sheridan, IL)	11/5/43
Carol Shields (Oak Park, IL)	6/2/35
Claude Simon (Tananarive, Madagascar)	1913
Neil Simon (NYC)	7/4/27
Jane Smiley (Los Angeles, CA)	9/26/49
Aleksandr Solzhenitsyn (Kislovodsk, Russia)	12/11/18
Susan Sontag (NYC)	1/28/33
Wole Soyinka (Abeokuta, Nigeria)	7/13/34
Mickey Spillane (Brooklyn, NY)	3/9/18
Danielle Steel (NYC)	8/14/47
Richard Stern (NYC)	2/25/28
Mary Stewart (Sunderland, Eng.)	9/17/16
Tom Stoppard (Zlin, Czech.)	7/13/37
William Styron (Newport News, VA)	6/11/25
Wislawa Szymborska (Kornik, Poland)	7/2/23
Amy Tan (Oakland, CA)	2/19/52
Paul Theroux (Medford, MA)	4/10/41
Scott F. Turow (Chicago, IL)	4/12/49
Anne Tyler (Minneapolis, MN)	10/25/41
John Updike (Shillington, PA)	3/18/32
Leon Uris (Baltimore, MD)	8/3/24
Gore Vidal (West Point, NY)	10/3/25
Paula Vogel (Wash. DC)	11/16/51
Kurt Vonnegut Jr. (Indianapolis, IN)	11/11/22
Derek Walcott (Castries, Saint Lucia)	1930
Alice Walker (Eatonton, GA)	2/9/44
Robert James Waller (Rockford, IA)	8/1/39
Joseph Wambaugh (East Pittsburgh, PA)	1/22/37
Wendy Wasserstein (NYC)	10/18/50
Eudora Welty (Jackson, MS)	4/13/09
August Wilson (Pittsburgh, PA)	4/27/45
Lanford Wilson (Lebanon, MO)	4/13/37
Tom Wolfe (Richmond, VA)	3/2/31
Tobias Wolff (Birmingham, AL)	6/19/45
Herman Wouk (NYC)	5/27/15

Poets Laureate

There is no record of the origin of the office of Poet Laureate of England. Henry III (1216-72) reportedly had a Versificator Regis, or King's Poet, paid 100 shillings a year. Other poets said to have filled the role include Geoffrey Chaucer (d 1400), Edmund Spenser (d 1599), Ben Jonson (d 1637), and Sir William d'Avenant (d 1668).

The first official English poet laureate was John Dryden, appointed 1668, for life (as was customary). Then came Thomas Shadwell, in 1689; Nahum Tate, 1692; Nicholas Rowe, 1715; Rev. Laurence Eusden, 1718; Colley Cibber, 1730; William Whitehead,

1757; Rev. Thomas Warton, 1785; Henry James Pye, 1790; Robert Southey, 1813; William Wordsworth, 1843; Alfred, Lord Tennyson, 1850; Alfred Austin, 1896; Robert Bridges, 1913; John Masefield, 1930; C. Day Lewis, 1968; Sir John Betjeman, 1972; Ted Hughes, 1984; Andrew Motion, 1999.

In U.S., appointment is by Librarian of Congress and is not for life: Robert Penn Warren, 1986; Richard Wilbur, 1987; Howard Nemerov, 1988; Mark Strand, 1990; Joseph Brodsky, 1991; Mona Van Duyn, 1992; Rita Dove, 1993; Robert Hass, 1995; Robert Pinsky, 1997.

Noted Writers of the Past

See also Greeks and Romans in Historical Figures chapter; Noted American Journalists.

Alice Adams, 1926-99, (U.S.) novelist, short-story writer. *Superior Woman.*

James Agee, 1909-55, (U.S.) novelist. *A Death in the Family.*

Conrad Aiken, 1889-1973, (U.S.) poet, critic. *Ushant.*

Louisa May Alcott, 1832-88, (U.S.) novelist. *Little Women.*

Sholom Aleichem, 1859-1916, (Russ.) Yiddish writer. *Tevye's Daughter, The Old Country.*

Vicente Aleixandre, 1898-1984, (Sp.) poet. *La destrucción o el amor, Dialogolos del conocimiento.*

Horatio Alger, 1832-1899, (U.S.) "rags-to-riches" books.

Eric Ambler, 1909-98, (Br.) suspense novelist. *A Coffin for Dimitrios.*

Kingsley Amis, 1922-95, (Br.) novelist, critic. *Lucky Jim.*

Hans Christian Andersen, 1805-75, (Dan.) author of fairy tales. *The Ugly Duckling.*

Maxwell Anderson, 1888-1959, (U.S.) playwright. *What Price Glory?, High Tor, Winterset, Key Largo.*

Sherwood Anderson, 1876-1941, (U.S.) short-story writer. "Death in the Woods;" *Winesburg, Ohio.*

Matthew Arnold, 1822-88, (Br.) poet, critic. "Thrysis," "Dover Beach," *Culture and Anarchy.*

Isaac Asimov, 1920-92, (U.S.) versatile writer, espec. of science-fiction. *I Robot.*

W(ystan) H(ugh) Auden, 1907-73, (Br.) poet, playwright, literary critic. "The Age of Anxiety."

Jane Austen, 1775-1817, (Br.) novelist. *Pride and Prejudice, Sense and Sensibility, Emma, Mansfield Park.*

Isaac Babel, 1894-1941, (Russ.) short-story writer, playwright. *Odessa Tales, Red Cavalry.*

James Baldwin, 1924-87, author, playwright. *The Fire Next Time, Blues for Mister Charlie.*

Honoré de Balzac, 1799-1850, (Fr.) novelist. *Le Père Goriot, Cousine Bette, Eugénie Grandet.*

James M. Barrie, 1860-1937, (Br.) playwright, novelist. *Peter Pan, Dear Brutus, What Every Woman Knows.*

Charles Baudelaire, 1821-67, (Fr.) poet. *Les Fleurs du Mal.*

L(yman) Frank Baum, 1856-1919, (U.S.) writer. *Wizard of Oz* series.

Simone de Beauvoir, 1908-86, (Fr.) novelist, essayist. *The Second Sex, Memoirs of a Dutiful Daughter.*

Samuel Beckett, 1906-89, (Ir.) novelist, playwright. *Waiting for Godot, Endgame* (plays); *Murphy, Watt, Molloy* (novels).

Brendan Behan, 1923-64, (Ir.) playwright. *The Quare Fellow, The Hostage, Borstal Boy.*

Robert Benchley, 1889-1945, (U.S.) humorist.

Stephen Vincent Benét, 1898-1943, (U.S.) poet, novelist. *John Brown's Body.*

John Berryman, 1914-72, (U.S.) poet. *Homage to Mistress Bradstreet.*

Ambrose Bierce, 1842-1914, (U.S.) short-story writer, journalist. *In the Midst of Life, The Devil's Dictionary.*

Elizabeth Bishop, 1911-79, (U.S.) poet. *North and South—A Cold Spring.*

William Blake, 1757-1827, (Br.) poet, artist. *Songs of Innocence, Songs of Experience, The Marriage of Heaven and Hell.*

Giovanni Boccaccio, 1313-75, (It.) poet. *Decameron.*

Heinrich Böll, 1917-85, (Ger.) novelist, short-story writer. *Group Portrait With Lady.*

Jorge Luis Borges, 1900-86, (Arg.) short-story writer, poet, essayist. *Labyrinths.*

James Boswell, 1740-95, (Sc.) biographer. *The Life of Samuel Johnson.*

Pierre Boulle, (1913-94), (Fr.) novelist. *The Bridge Over the River Kwai, Planet of the Apes.*

Anne Bradstreet, c1612-72, (U.S.) poet. *The Tenth Muse Lately Sprung Up in America.*

Bertolt Brecht, 1898-1956, (Ger.) dramatist, poet. *The Threepenny Opera, Mother Courage and Her Children.*

Charlotte Brontë, 1816-55, (Br.) novelist. *Jane Eyre.*

Emily Brontë, 1818-48, (Br.) novelist. *Wuthering Heights.*

Elizabeth Barrett Browning, 1806-61, (Br.) poet. *Sonnets From the Portuguese, Aurora Leigh.*

Joseph Brodsky, 1940-96, (Russ.-U.S.) poet. *A Part of Speech, Less Than One, To Urania.*

Robert Browning, 1812-89, (Br.) poet. "My Last Duchess," "Fra Lippo Lippi," *The Ring and The Book.*

Pearl S. Buck, 1892-1973, (U.S.) novelist. *The Good Earth.*

Mikhail Bulgakov, 1891-1940, (Russ.) novelist, playwright. *The Heart of a Dog, The Master and Margarita.*

John Bunyan, 1628-88, (Br.) writer. *Pilgrim's Progress.*

Anthony Burgess, 1917-93, (Br.) author. *A Clockwork Orange.*

Frances Hodgson Burnett, 1849-1924, (Br.-U.S.) novelist. *The Secret Garden.*

Robert Burns, 1759-96, (Sc.) poet. "Flow Gently, Sweet Afton," "My Heart's in the Highlands," "Auld Lang Syne."

Edgar Rice Burroughs, 1875-1950, (U.S.) novelist. *Tarzan of the Apes.*

William S. Burroughs, 1914-97, (U.S.) novelist. *Naked Lunch.*

George Gordon, Lord Byron, 1788-1824, (Br.) poet. *Don Juan, Childe Harold, Manfred, Cain.*

Italo Calvino, 1923-85, (It.) novelist, short-story writer. *If on a Winter's Night a Traveler.*

Albert Camus, 1913-60, (Fr.) writer. *The Stranger, The Fall.*

Karel Capek, 1890-1938, (Czech.) playwright, novelist, essayist. *R.U.R. (Rossum's Universal Robots).*

Truman Capote, 1924-84, (U.S.) author. *Other Voices, Other Rooms, Breakfast at Tiffany's, In Cold Blood.*

Lewis Carroll (Charles Dodgson), 1832-98, (Br.) writer, mathematician. *Alice's Adventures in Wonderland.*

Giacomo Casanova, 1725-98, (It.) adventurer, memoirist.

Willa Cather, 1873-1947, (U.S.) novelist. *O Pioneers!, My Ántonia, Death Comes for the Archbishop.*

Miguel de Cervantes Saavedra, 1547-1616, (Sp.) novelist, dramatist, poet. *Don Quixote.*

Raymond Chandler, 1888-1959, (U.S.) writer of detective fiction. Philip Marlowe series.

Geoffrey Chaucer, c1340-1400, (Br.) poet. *The Canterbury Tales, Troilus and Criseyde.*

John Cheever, 1912-82, (U.S.) novelist, short-story writer. *The Wapshot Scandal,* "The Country Husband."

Anton Chekhov, 1860-1904, (Russ.) short-story writer, dramatist. *Uncle Vanya, The Cherry Orchard, The Three Sisters.*

G(ilbert) K(eith) Chesterton, 1874-1936, (Br.) critic, novelist, relig. apologist. Father Brown series of mysteries.

Kate Chopin, 1851-1904, (U.S.) writer. *The Awakening.*

Agatha Christie, 1890-1976, (Br.) mystery writer; created Miss Marple, Hercule Poirot; *And Then There Were None.*

James Clavell, 1924-94, (Br.-U.S.) novelist. *Shogun, King Rat.*

Jean Cocteau, 1889-1963, (Fr.) writer, visual artist, filmmaker. *The Beauty and the Beast, Les Enfants Terribles.*

Samuel Taylor Coleridge, 1772-1834, (Br.) poet, critic. "Kubla Khan," "The Rime of the Ancient Mariner."

(Sidonie) Colette, 1873-1954, (Fr.) novelist. *Claudine, Gigi.*

Wilkie Collins, 1824-89, (Br.) Novelist. *The Moonstone.*

Joseph Conrad, 1857-1924, (Br.) novelist. *Lord Jim, Heart of Darkness, The Nigger of the Narcissus.*

James Fenimore Cooper, 1789-1851, (U.S.) novelist. *Leatherstocking Tales, The Last of the Mohicans.*

Pierre Corneille, 1606-84, (Fr.) dramatist. *Medeé, Le Cid.*

Hart Crane, 1899-1932, (U.S.) poet. "The Bridge."

Stephen Crane, 1871-1900, (U.S.) novelist, short-story writer. *The Red Badge of Courage,* "The Open Boat."

E. E. Cummings, 1894-1962, (U.S.) poet. *Tulips and Chimneys.*

Roald Dahl, 1916-90, (Br.-U.S.) writer. *Charlie and the Chocolate Factory, James and the Giant Peach.*

Gabriele D'Annunzio, 1863-1938, (It.) poet, novelist, dramatist. *The Child of Pleasure, The Intruder, The Victim.*

Dante Alighieri, 1265-1321, (It.) poet. *The Divine Comedy.*

Robertson Davies, 1913-95, (Can.) novelist, playwright, essayist. Salterton, Deptford, and Cornish trilogies.

Daniel Defoe, 1660-1731, (Br.) writer. *Robinson Crusoe, Moll Flanders, Journal of the Plague Year.*

Peter De Vries, 1910-93, (U.S.) journalist, writer. *The Tunnel of Love, Let Me Count the Ways.*

Charles Dickens, 1812-70, (Br.) novelist. *David Copperfield, Oliver Twist, Great Expectations, A Tale of Two Cities.*

James Dickey, 1923-1997, (U.S.) poet, novelist. *Deliverance.*

Emily Dickinson, 1830-86, (U.S.) lyric poet. "Because I could not stop for Death . . .," "Success is counted sweetest . . ."

Isak Dinesen (Karen Blixen), 1885-1962, (Dan.) author. *Out of Africa, Seven Gothic Tales, Winter's Tales.*

John Donne, 1573-1631, (Br.) poet, divine. *Songs and Sonnets.*

José Donoso, 1924-96, (Chil.) surreal novelist and short-story writer. *The Obscene Bird of Night.*

John Dos Passos, 1896-1970, (U.S.) novelist. *U.S.A.*

Fyodor Dostoyevsky, 1821-81, (Russ.) novelist. *Crime and Punishment, The Brothers Karamazov, The Possessed.*

Arthur Conan Doyle, 1859-1930, (Br.) novelist. Sherlock Holmes mystery stories.

Theodore Dreiser, 1871-1945, (U.S.) novelist. *An American Tragedy, Sister Carrie.*

John Dryden, 1631-1700, (Br.) poet, dramatist, critic. *All for Love, Mac Flecknoe, Absalom and Achitophel.*

Alexandre Dumas, 1802-70, (Fr.) novelist, dramatist. *The Three Musketeers, The Count of Monte Cristo.*

Alexandre Dumas (fils), 1824-95, (Fr.) dramatist, novelist. *La Dame aux Camélias, Le Demi-Monde.*

Lawrence Durrell, 1912-90, (Br.) novelist, poet. *Alexandria Quartet.*

Ilya G. Ehrenburg, 1891-1967, (Russ.) writer. *The Thaw.*

George Eliot (Mary Ann Evans or Marian Evans), 1819-80, (Br.) novelist. *Silas Marner, Middlemarch.*

T(homas) S(tearns) Eliot, 1888-1965, (Br.) poet, critic. *The Waste Land,* "The Love Song of J. Alfred Prufrock."

Stanley Elkin, 1930-95, (U.S.) novelist, short story writer. *George Mills.*

Ralph Ellison, 1914-94, (U.S.), writer. *Invisible Man.*

Ralph Waldo Emerson, 1803-82, (U.S.) poet, essayist. "Brahma," "Nature," "The Over-Soul," "Self-Reliance."

James T. Farrell, 1904-79, (U.S.) novelist. *Studs Lonigan.*

William Faulkner, 1897-1962, (U.S.) novelist. *Sanctuary, Light in August, The Sound and the Fury, Absalom, Absalom!*

Edna Ferber, 1887-1968, (U.S.) novelist, short-story writer, playwright. *So Big, Cimarron, Show Boat.*

Henry Fielding, 1707-54, (Br.) novelist. *Tom Jones.*

F(rancis) Scott Fitzgerald, 1896-1940, (U.S.) short-story writer, novelist. *The Great Gatsby, Tender Is the Night.*

Gustave Flaubert, 1821-80, (Fr.) novelist. *Madame Bovary.*

Ian Fleming, 1908-64, (Br.) novelist; James Bond spy thrillers.

Ford Madox Ford, 1873-1939, (Br.) novelist, critic, poet. *The Good Soldier.*

C(ecil) S(cott) Forester, 1899-1966, (Br.) writer. Horatio Hornblower books.

E(dward) M(organ) Forster, 1879-1970, (Br.) novelist. *A Passage to India, Howards End.*

Anatole France, 1844-1924, (Fr.) writer. *Penguin Island, My Friend's Book, The Crime of Sylvestre Bonnard.*

Robert Frost, 1874-1963, (U.S.) poet. "Birches," "Fire and Ice," "Stopping by Woods on a Snowy Evening."

William Gaddis, 1922-98, (U.S.) novelist. *The Recognitions.*

John Galsworthy, 1867-1933, (Br.) novelist, dramatist. *The Forsyte Saga.*

Erle Stanley Gardner, 1889-1970, (U.S.) mystery writer; created Perry Mason.

Jean Genet, 1911-86, (Fr.) playwright, novelist. *The Maids.*

Kahlil Gibran, 1883-1931, (Lebanese-U.S.) mystical novelist, essayist, poet. *The Prophet.*

André Gide, 1869-1951, (Fr.) writer. *The Immoralist, The Pastoral Symphony, Strait Is the Gate.*

Allen Ginsberg, 1926-1997, (U.S.) Beat poet. "Howl."

Jean Giraudoux, 1882-1944, (Fr.) novelist, dramatist. *Electra, The Madwoman of Chaillot, Ondine, Tiger at the Gate.*

Johann Wolfgang von Goethe, 1749-1832, (Ger.) poet, dramatist, novelist. *Faust, Sorrows of Young Werther.*

Nikolai Gogol, 1809-52, (Russ.) short-story writer, dramatist, novelist. *Dead Souls, The Inspector General.*

William Golding, 1911-93, (Br.) novelist. *Lord of the Flies.*

Oliver Goldsmith, 1728-74, (Br.-Ir.) dramatist, novelist. *The Vicar of Wakefield, She Stoops to Conquer.*

Maxim Gorky, 1868-1936, (Russ.) dramatist, novelist. *The Lower Depths.*

Robert Graves, 1895-1985, (Br.) poet, classical scholar, novelist. *I, Claudius; The White Goddess.*

Thomas Gray, 1716-71, (Br.) poet. "Elegy Written in a Country Churchyard," "The Progress of Poesy."

Julien Green, 1900-98, (U.S.-Fr.) expatriate American, French novelist. *Moira, Each Man in His Darkness.*

Graham Greene, 1904-91, (Br.) novelist. *The Power and the Glory, The Heart of the Matter, The Ministry of Fear.*

Zane Grey, 1872-1939, (U.S.) writer of western stories.

Jakob Grimm, 1785-1863, (Ger.) philologist, folklorist; with brother **Wilhelm,** 1786-1859, collected *Grimm's Fairy Tales.*

Alex Haley, 1921-92, (U.S.) author. *Roots.*

Dashiell Hammett, 1894-1961, (U.S.) detective-story writer; created Sam Spade. *The Maltese Falcon, The Thin Man.*

Knute Hamsun, 1859-1952 (Nor.) novelist. *Hunger.*

Thomas Hardy, 1840-1928, (Br.) novelist, poet. *The Return of the Native, Tess of the D'Urbervilles, Jude the Obscure.*

Joel Chandler Harris, 1848-1908, (U.S.) short-story writer. Uncle Remus series.

Moss Hart, 1904-61, (U.S.) playwright. *Once in a Lifetime, You Can't Take It With You, The Man Who Came to Dinner.*

Bret Harte, 1836-1902, (U.S.) short-story writer, poet. *The Luck of Roaring Camp.*

Jaroslav Hasek, 1883-1923, (Czech.) writer, playwright. *The Good Soldier Schweik.*

John Hawkes, 1925-98, (U.S.) experimental fiction writer. *The Goose on the Grave, Blood Oranges.*

Nathaniel Hawthorne, 1804-64, (U.S.) novelist, short-story writer. *The Scarlet Letter,* "Young Goodman Brown."

Heinrich Heine, 1797-1856, (Ger.) poet. *Book of Songs.*

Lillian Hellman, 1905-84, (U.S.) playwright, author of memoirs. *The Little Foxes, An Unfinished Woman, Pentimento.*

Ernest Hemingway, 1899-1961, (U.S.) novelist, short-story writer. *A Farewell to Arms, For Whom the Bell Tolls.*

O. Henry (W. S. Porter), 1862-1910, (U.S.) short-story writer. "The Gift of the Magi."

George Herbert, 1593-1633, (Br.) poet. "The Altar," "Easter Wings."

Zbigniew Herbert, 1924-98, (Pol.) poet. "Apollo and Marsyas."

Robert Herrick, 1591-1674, (Br.) poet. "To the Virgins to Make Much of Time."

James Herriot (James Alfred Wight), 1916-95, (Br.) novelist, veterinarian. *All Creatures Great and Small.*

John Hersey, 1914-93, (U.S.) novelist, journalist. *Hiroshima, A Bell for Adano.*

Hermann Hesse, 1877-1962, (Ger.) novelist, poet. *Death and the Lover, Steppenwolf, Siddhartha.*

James Hilton, 1900-54, (Br.) novelist. *Lost Horizon.*

Oliver Wendell Holmes, 1809-94, (U.S.) poet, novelist. *The Autocrat of the Breakfast-Table.*

Gerard Manley Hopkins, 1844-89, (Br.) poet. "Pied Beauty."

A(lfred) E. Housman, 1859-1936, (Br.) poet. *A Shropshire Lad.*

William Dean Howells, 1837-1920, (U.S.) novelist, critic. *The Rise of Silas Lapham.*

Langston Hughes, 1902-67, (U.S.) poet, playwright. *The Weary Blues, One-Way Ticket, Shakespeare in Harlem.*

Ted Hughes, 1930-98, (Br.) British poet laureate, 1984-98. *Crow, The Hawk in the Rain.*

Victor Hugo, 1802-85, (Fr.) poet, dramatist, novelist. *Notre Dame de Paris, Les Misérables.*

Zora Neale Hurston, 1903-60, (U.S.) novelist, folklorist. *Their Eyes Were Watching God, Mules and Men.*

Aldous Huxley, 1894-1963, (Br.) writer. *Brave New World.*

Henrik Ibsen, 1828-1906, (Nor.) dramatist, poet. *A Doll's House, Ghosts, The Wild Duck, Hedda Gabler.*

William Inge, 1913-73, (U.S.) playwright. *Picnic; Come Back, Little Sheba; Bus Stop.*

Eugene Ionesco, 1910-94, (Fr.) surrealist dramatist. *The Bald Soprano, The Chairs.*

Washington Irving, 1783-1859, (U.S.) writer. "Rip Van Winkle," "The Legend of Sleepy Hollow."

Christopher Isherwood, 1904-1986, (Br.) novelist, playwright. *The Berlin Stories.*

Shirley Jackson, 1919-65, (U.S.) writer. "The Lottery."

Henry James, 1843-1916, (U.S.) novelist, short-story writer, critic. *The Portrait of a Lady, The Ambassadors, Daisy Miller.*

Robinson Jeffers, 1887-1962, (U.S.) poet, dramatist. *Tamar and Other Poems, Medea.*

Samuel Johnson, 1709-84, (Br.) author, scholar, critic. *Dictionary of the English Language, Vanity of Human Wishes.*

Ben Jonson, 1572-1637, (Br.) dramatist, poet. *Volpone.*

James Joyce, 1882-1941, (Ir.) writer. *Ulysses, Dubliners, A Portrait of the Artist as a Young Man, Finnegans Wake.*

Ernst Junger, 1895-1998, (Ger.) novelist, essayist. *The Peace, On the Marble Cliff.*

Franz Kafka, 1883-1924, (Ger.) novelist, short-story writer. *The Trial, The Castle, The Metamorphosis.*

George S. Kaufman, 1889-1961, (U.S.) playwright. *The Man Who Came to Dinner, You Can't Take It With You, Stage Door.*

Nikos Kazantzakis, 1883-1957, (Gk.) novelist. *Zorba the Greek, A Greek Passion.*

Alfred Kazin, 1915-98 (U.S.) author, critic, teacher. *On Native Grounds.*

John Keats, 1795-1821, (Br.) poet. "Ode on a Grecian Urn," "Ode to a Nightingale," "La Belle Dame Sans Merci."

Jack Kerouac, 1922-1969, (U.S.) author, Beat poet. *On the Road, The Dharma Bums,* "Mexico City Blues."

Joyce Kilmer, 1886-1918, (U.S.) poet. "Trees."

Rudyard Kipling, 1865-1936, (Br.) author, poet. "The White Man's Burden," "Gunga Din," *The Jungle Book.*

Jean de la Fontaine, 1621-95, (Fr.) poet. *Fables choisies.*

Pär Lagerkvist, 1891-1974, (Swed.) poet, dramatist, novelist. *Barabbas, The Sybil.*

Selma Lagerlöf, 1858-1940, (Swed.) novelist. *Jerusalem, The Ring of the Lowenskolds.*

Alphonse de Lamartine, 1790-1869, (Fr.) poet, novelist, statesman. *Méditations poétiques.*

Charles Lamb, 1775-1834, (Br.) essayist. *Specimens of English Dramatic Poets, Essays of Elia.*

Giuseppe di Lampedusa, 1896-1957, (It.) novelist. *The Leopard.*

William Langland, c1332-1400, (Eng.) poet. *Piers Plowman.*

Ring Lardner, 1885-1933, (U.S.) short-story writer, humorist.

Louis L'Amour, 1908-88, (U.S.) western author, screenwriter. *Hondo, The Cherokee Trail.*

D(avid) H(erbert) Lawrence, 1885-1930, (Br.) novelist. *Sons and Lovers, Women in Love, Lady Chatterley's Lover.*

Halldor Laxness, 1902-98, (Icelandic) novelist. *Iceland's Bell.*

Mikhail Lermontov, 1814-41, (Russ.) novelist, poet. "Demon," *Hero of Our Time.*

Alain-René Lesage, 1668-1747, (Fr.) novelist. *Gil Blas de Santillane.*

Gotthold Lessing, 1729-81, (Ger.) dramatist, philosopher, critic. *Miss Sara Sampson, Minna von Barnhelm.*

C(live) S(taples) Lewis, 1898-1963, (Br.) critic, novelist, religious writer. *Allegory of Love; The Lion, the Witch and the Wardrobe; Out of the Silent Planet.*

Sinclair Lewis, 1885-1951, (U.S.) novelist. *Babbitt, Main Street, Arrowsmith, Dodsworth.*

Vachel Lindsay, 1879-1931, (U.S.) poet. *General William Booth Enters Into Heaven, The Congo.*

Hugh Lofting, 1886-1947, (Br.) writer. Dr. Doolittle series.

Jack London, 1876-1916, (U.S.) novelist, journalist. *Call of the Wild, The Sea-Wolf, White Fang.*

Henry Wadsworth Longfellow, 1807-82, (U.S.) poet. *Evangeline, The Song of Hiawatha.*

Amy Lowell, 1874-1925, (U.S.) poet, critic. "Lilacs."

James Russell Lowell, 1819-91, (U.S.) poet, editor. *Poems, The Biglow Papers.*

Robert Lowell, 1917-77, (U.S.) poet. "Lord Weary's Castle."

Archibald MacLeish, 1892-1982, (U.S.) poet. *Conquistador.*

Bernard Malamud, 1914-86, (U.S.) short-story writer, novelist. "The Magic Barrel," *The Assistant, The Fixer.*

Stéphane Mallarmé, 1842-98, (Fr.) poet. *Poésies.*

Sir Thomas Malory, ?-1471, (Br.) writer. *Morte d'Arthur.*

Andre Malraux, 1901-76, (Fr.) novelist. *Man's Fate.*

Osip Mandelstam, 1891-1938, (Russ.) poet. *Stone, Tristia.*

Thomas Mann, 1875-1955, (Ger.) novelist, essayist. *Buddenbrooks, The Magic Mountain,* "Death in Venice."

Katherine Mansfield, 1888-1923, (Br.) short-story writer. "Bliss."

Christopher Marlowe, 1564-93, (Br.) dramatist, poet. *Tamburlaine the Great, Dr. Faustus, The Jew of Malta.*

Andrew Marvell, 1621-78, (Br.) poet. "To His Coy Mistress."

John Masefield, 1878-1967, (Br.) poet. "Sea Fever," "Cargoes," *Salt Water Ballads.*

Edgar Lee Masters, 1869-1950, (U.S.) poet, biographer. *Spoon River Anthology.*

W(illiam) Somerset Maugham, 1874-1965, (Br.) author. *Of Human Bondage, The Moon and Sixpence.*

Guy de Maupassant, 1850-93, (Fr.) novelist, short-story writer. "A Life," "Bel-Ami," "The Necklace."

François Mauriac, 1885-1970, (Fr.) novelist, dramatist. *Viper's Tangle, The Kiss to the Leper.*

Vladimir Mayakovsky, 1893-1930, (Russ.) poet, dramatist. *The Cloud in Trousers.*

Mary McCarthy, 1912-89, (U.S.) critic, novelist, memoirist. *Memories of a Catholic Girlhood.*

Carson McCullers, 1917-67, (U.S.) novelist. *The Heart Is a Lonely Hunter, Member of the Wedding.*

Herman Melville, 1819-91, (U.S.) novelist, poet. *Moby-Dick, Typee, Billy Budd, Omoo.*

George Meredith, 1828-1909, (Br.) novelist, poet. *The Ordeal of Richard Feverel, The Egoist.*

Prosper Mérimée, 1803-70, (Fr.) author. *Carmen.*

James Merrill, 1926-95, (U.S.) poet. *Divine Comedies.*

James Michener, 1907-97, (U.S.) novelist. *Tales of the South Pacific.*

Edna St. Vincent Millay, 1892-1950, (U.S.) poet. *The Harp Weaver and Other Poems.*

Henry Miller, 1891-1980, (U.S.) erotic novelist. *Tropic of Cancer.*

A(lan) A(lexander) Milne, 1882-1956, (Br.) author. *Winnie-the-Pooh.*

John Milton, 1608-74, (Br.) poet, writer. *Paradise Lost, Comus, Lycidas, Areopagitica.*

Mishima Yukio (Hiraoka Kimitake), 1925-70, (Jpn.) writer. *Confessions of a Mask.*

Gabriela Mistral, 1889-1957, (Chil.) poet. *Sonnets of Death.*

Margaret Mitchell, 1900-49, (U.S.) novelist. *Gone With the Wind.*

Jean Baptiste Molière, 1622-73, (Fr.) dramatist. *Le Tartuffe, Le Misanthrope, Le Bourgeois Gentilhomme.*

Ferenc Molnár, 1878-1952, (Hung.) dramatist, novelist. *Liliom, The Guardsman, The Swan.*

Michel de Montaigne, 1533-92, (Fr.) essayist. *Essais.*

Eugenio Montale, 1896-1981, (It.) poet.

(Ir. U.S.) novelist. *The Lonely Passion of Judith Herne.*

Brian Moore, 1921-99, (Ir.-U.S.) novelist. *The Lonely Passion of Judith Hearne.*

Clement C. Moore, 1779-1863, (U.S.) poet, educator. "A Visit From Saint Nicholas."

Marianne Moore, 1887-1972, (U.S.) poet.

Alberto Moravia, 1907-90, (It.) novelist, short-story writer. *The Time of Indifference.*

Sir Thomas More, 1478-1535, (Br.) writer, statesman, saint. *Utopia.*

Wright Morris, 1910-98 (U.S.) novelist. *My Uncle Dudley.*

Murasaki Shikibu, c978-1026, (Jpn.) novelist. *The Tale of Genji.*

Iris Murdoch, 1919-99 (Br.), novelist, philosopher. *The Sea, The Sea.*

Alfred de Musset, 1810-57, (Fr.) poet, dramatist. *La Confession d'un Enfant du Siècle.*

Vladimir Nabokov, 1899-1977, (Russ.-U.S.) novelist. *Lolita, Pale Fire.*

Ogden Nash, 1902-71, (U.S.) poet of light verse.

Pablo Neruda, 1904-73, (Chil.) poet. *Twenty Love Poems and One Song of Despair, Toward the Splendid City.*

Sean O'Casey, 1884-1964, (Ir.) dramatist. *Juno and the Paycock, The Plough and the Stars.*

Frank O'Connor (Michael Donovan), 1903-66, (Ir.) short-story writer. "Guests of a Nation."

Flannery O'Connor, 1925-64, (U.S.) novelist, short-story writer. *Wise Blood,* "A Good Man Is Hard to Find."

Clifford Odets, 1906-63, (U.S.) playwright. *Waiting for Lefty, Awake and Sing, Golden Boy, The Country Girl.*

John O'Hara, 1905-70, (U.S.) novelist, short-story writer. *From the Terrace, Appointment in Samarra, Pal Joey.*

Omar Khayyam, c1028-1122, (Per.) poet. *Rubaiyat.*

Eugene O'Neill, 1888-1953, (U.S.) playwright. *Emperor Jones, Anna Christie, Long Day's Journey Into Night.*

George Orwell, 1903-50, (Br.) novelist, essayist. *Animal Farm, Nineteen Eighty-Four.*

John Osborne, 1929-95, (Br.) dramatist, novelist. *Look Back in Anger, The Entertainer.*

Wilfred Owen, 1893-1918 (Br.) poet. "Dulce et Decorum Est."

Dorothy Parker, 1893-1967, (U.S.) poet, short-story writer. *Enough Rope, Laments for the Living.*

Boris Pasternak, 1890-1960, (Russ.) poet, novelist. *Doctor Zhivago.*

Octavio Paz, 1914-98, (Mex.) poet, essayist. *The Labyrinth of Solitude, They Shall Not Pass!, The Sun Stone.*

Samuel Pepys, 1633-1703, (Br.) public official, diarist.

S(idney) J(oseph) Perelman, 1904-79, (U.S.) humorist. *The Road to Miltown, Under the Spreading Atrophy.*

Charles Perrault, 1628-1703, (Fr.) writer. *Tales From Mother Goose (Sleeping Beauty, Cinderella).*

Petrarch (Francesco Petrarca), 1304-74, (It.) poet. *Africa, Trionfi, Canzoniere.*

Luigi Pirandello, 1867-1936, (It.) novelist, dramatist. *Six Characters in Search of an Author.*

Sylvia Plath, 1932-63, (U.S.) author, poet. *The Bell Jar.*

Edgar Allan Poe, 1809-49, (U.S.) poet, short-story writer, critic. "Annabel Lee," "The Raven," "The Purloined Letter."

Alexander Pope, 1688-1744, (Br.) poet. *The Rape of the Lock, The Dunciad, An Essay on Man.*

Katherine Anne Porter, 1890-1980, (U.S.) novelist, short-story writer. *Ship of Fools.*

Ezra Pound, 1885-1972, (U.S.) poet. *Cantos.*

J(ohn) B. Priestley, 1894-1984, (Br.) novelist, dramatist. *The Good Companions.*

Marcel Proust, 1871-1922, (Fr.) novelist. *Remembrance of Things Past.*

Aleksandr Pushkin, 1799-1837, (Russ.) poet, novelist. *Boris Godunov, Eugene Onegin.*

Mario Puzo, 1920-99, (U.S.) novelist. *The Godfather.*

François Rabelais, 1495-1553, (Fr.) writer. *Gargantua.*

Jean Racine, 1639-99, (Fr.) dramatist. *Andromaque, Phèdre, Bérénice, Britannicus.*

Ayn Rand, 1905-82, (Russ.-U.S.) novelist, moral theorist. *The Fountainhead, Atlas Shrugged.*

Terence Rattigun, 1911-77, (Br.) playwright. *Separate Tables, The Browning Version.*

Erich Maria Remarque, 1898-1970, (Ger.-U.S.) novelist. *All Quiet on the Western Front.*

Samuel Richardson, 1689-1761, (Br.) novelist. *Pamela; or Virtue Rewarded.*

Rainer Maria Rilke, 1875-1926, (Ger.) poet. *Life and Songs, Duino Elegies, Poems From the Book of Hours.*

Arthur Rimbaud, 1854-91, (Fr.) poet. *A Season in Hell.*

Edwin Arlington Robinson, 1869-1935, (U.S.) poet. "Richard Cory," "Miniver Cheevy," *Merlin.*

Theodore Roethke, 1908-63, (U.S.) poet. *Open House, The Waking, The Far Field.*

Romain Rolland, 1866-1944, (Fr.) novelist, biographer. *Jean-Christophe.*

Pierre de Ronsard, 1524-85, (Fr.) poet. *Sonnets pour Hélène, La Franciade.*

Christina Rossetti, 1830-94, (Br.) poet. "When I Am Dead, My Dearest."

Dante Gabriel Rossetti, 1828-82, (Br.) poet, painter. "The Blessed Damozel."

Edmond Rostand, 1868-1918, (Fr.) poet, dramatist. *Cyrano de Bergerac.*

Damon Runyon, 1880-1946, (U.S.) short-story writer, journalist. *Guys and Dolls, Blue Plate Special.*

John Ruskin, 1819-1900, (Br.) critic, social theorist. *Modern Painters, The Seven Lamps of Architecture.*

Antoine de Saint-Exupéry, 1900-44, (Fr.) writer. *Wind, Sand and Stars, The Little Prince.*

Saki, or H(ector) H(ugh) Munro, 1870-1916, (Br.) writer. *The Chronicles of Clovis.*

George Sand (Amandine Lucie Aurore Dupin), 1804-76, (Fr.) novelist. *Indiana, Consuelo.*

Carl Sandburg, 1878-1967, (U.S.) poet. *The People, Yes; Chicago Poems, Smoke and Steel, Harvest Poems.*

William Saroyan, 1908-81, (U.S.) playwright, novelist. *The Time of Your Life, The Human Comedy.*

May Sarton, 1914-95, (Belg.-U.S.) poet, novelist. *Encounter in April, Anger.*

Dorothy L. Sayers, 1893-1957, (Br.) mystery writer; created Lord Peter Wimsey.

Richard Scarry, 1920-94, (U.S.) author of children's books. *Richard Scarry's Best Story Book Ever.*

Friedrich von Schiller, 1759-1805, (Ger.) dramatist, poet, historian. *Don Carlos, Maria Stuart, Wilhelm Tell.*

Sir Walter Scott, 1771-1832, (Sc.) novelist, poet. *Ivanhoe.*

Jaroslav Seifert, 1902-86, (Czech.) poet.

Dr. Seuss (Theodor Seuss Geisel), 1904-91, (U.S.) children's book author and illustrator. *The Cat in the Hat.*

William Shakespeare, 1564-1616, (Br.) dramatist, poet. *Romeo and Juliet, Hamlet, King Lear, Julius Caesar,* sonnets.

George Bernard Shaw, 1856-1950, (Ir.-Br.) playwright, critic. *St. Joan, Pygmalion, Major Barbara, Man and Superman.*

Mary Wollstonecraft Shelley, 1797-1851, (Br.) novelist, feminist. *Frankenstein, The Last Man.*

Percy Bysshe Shelley, 1792-1822, (Br.) poet. *Prometheus Unbound, Adonais,* "Ode to the West Wind," "To a Skylark."

Richard B. Sheridan, 1751-1816, (Br.) dramatist. *The Rivals, School for Scandal.*

Robert Sherwood, 1896-1955, (U.S.) playwright, biographer. *The Petrified Forest, Abe Lincoln in Illinois.*

Mikhail Sholokhov, 1906-84, (Russ.) writer. *The Silent Don.*

Upton Sinclair, 1878-1968, (U.S.) novelist. *The Jungle.*

Isaac Bashevis Singer, 1904-91, (Pol.-U.S.) novelist, short-story writer, in Yiddish. *The Magician of Lublin.*

C(harles) P(ercy) Snow, 1905-80, (Br.) novelist, scientist. *Strangers and Brothers, Corridors of Power.*

Stephen Spender, 1909-95, (Br.) poet, critic, novelist. *Twenty Poems,* "Elegy for Margaret."

Edmund Spenser, 1552-99, (Br.) poet. *The Faerie Queen.*

Christina Stead, 1903-83, (Austral.) novelist, short-story writer. *The Man Who Loved Children.*

Richard Steele, 1672-1729, (Br.) essayist, playwright, began the *Tatler* and *Spectator. The Conscious Lovers.*

Gertrude Stein, 1874-1946, (U.S.) writer. *Three Lives.*

John Steinbeck, 1902-68, (U.S.) novelist. *The Grapes of Wrath, Of Mice and Men, The Winter of Our Discontent.*

Stendhal (Marie Henri Beyle), 1783-1842, (Fr.) novelist. *The Red and the Black, The Charterhouse of Parma.*

Laurence Sterne, 1713-68, (Br.) novelist. *Tristram Shandy.*

Wallace Stevens, 1879-1955, (U.S.) poet. *Harmonium, The Man With the Blue Guitar, Notes Toward a Supreme Fiction.*

Robert Louis Stevenson, 1850-94, (Br.) novelist, poet, essayist. *Treasure Island, A Child's Garden of Verses.*

Bram Stoker, 1845-1910, (Br.) writer. *Dracula.*

Rex Stout, 1886-1975, (U.S.) mystery writer; created Nero Wolfe.

Harriet Beecher Stowe, 1811-96, (U.S.) novelist. *Uncle Tom's Cabin.*

Lytton Strachey, 1880-1932, (Br.) biographer, critic. *Eminent Victorians, Queen Victoria, Elizabeth and Essex.*

August Strindberg, 1849-1912, (Swed.) dramatist, novelist. *The Father, Miss Julie, The Creditors.*

Jonathan Swift, 1667-1745, (Br.) satirist, poet. *Gulliver's Travels,* "A Modest Proposal."

Algernon C. Swinburne, 1837-1909, (Br.) writer. *Atalanta in Calydon.*

John M. Synge, 1871-1909, (Ir.) poet, dramatist. *Riders to the Sea, The Playboy of the Western World.*

Rabindranath Tagore, 1861-1941, (In.) author, poet. *Sadhana, The Realization of Life, Gitanjali.*

Booth Tarkington, 1869-1946, (U.S.) novelist. *Seventeen.*

Peter Taylor, 1917-94, (U.S.) novelist. *A Summons to Memphis.*

Sara Teasdale, 1884-1933, (U.S.) poet. *Helen of Troy and Other Poems, Rivers to the Sea.*

Alfred, Lord Tennyson, 1809-92, (Br.) poet. *Idylls of the King, In Memoriam,* "The Charge of the Light Brigade."

William Makepeace Thackeray, 1811-63, (Br.) novelist. *Vanity Fair, Henry Esmond, Pendennis.*

Dylan Thomas, 1914-53, (Welsh) poet. *Under Milk Wood, A Child's Christmas in Wales.*

Henry David Thoreau, 1817-62, (U.S.) writer, philosopher, naturalist. *Walden,* "Civil Disobedience."

James Thurber, 1894-1961, (U.S.) humorist; "The Secret Life of Walter Mitty," *My Life and Hard Times.*

J(ohn) R(onald) R(euel) Tolkien, 1892-1973, (Br.) writer. *The Hobbit, Lord of the Rings* trilogy.

Leo Tolstoy, 1828-1910, (Russ.) novelist, short-story writer. *War and Peace, Anna Karenina,* "The Death of Ivan Ilyich."

Anthony Trollope, 1815-82, (Br.) novelist. *The Warden, Barchester Towers,* the Palliser novels.

Ivan Turgenev, 1818-83, (Russ.) novelist, short-story writer. *Fathers and Sons, First Love, A Month in the Country.*

Amos Tutuola, 1920-97, (Nigerian) novelist. *The Palm-Wine Drunkard, My Life in the Bush of Ghosts.*

Mark Twain (Samuel Clemens), 1835-1910, (U.S.) novelist, humorist. *The Adventures of Huckleberry Finn.*

Sigrid Undset, 1881-1949, (Nor.) novelist, poet. *Kristin Lavransdatter.*

Paul Valéry, 1871-1945, (Fr.) poet, critic. *La Jeune Parque, The Graveyard by the Sea.*

Jules Verne, 1828-1905, (Fr.) novelist. *Twenty Thousand Leagues Under the Sea.*

François Villon, 1431-63?, (Fr.) poet. *The Lays, The Grand Testament.*

Voltaire (F.M. Arouet), 1694-1778, (Fr.) writer of "philosophical romances"; philosopher, historian; *Candide.*

Robert Penn Warren, 1905-89, (U.S.) novelist, poet, critic. *All the King's Men.*

Evelyn Waugh, 1903-66, (Br.) novelist. *The Loved One, Brideshead Revisited, A Handful of Dust.*

H(erbert) G(eorge) Wells, 1866-1946, (Br.) novelist. *The Time Machine, The Invisible Man, The War of the Worlds.*

Rebecca West, 1893-1983, (Br.) novelist, critic, journalist. *Black Lamb and Grey Falcon.*

Edith Wharton, 1862-1937, (U.S.) novelist. *The Age of Innocence, The House of Mirth, Ethan Frome.*

E(lwyn) B(rooks) White, 1899-1985, (U.S.) essayist, novelist. *Charlotte's Web, Stuart Little.*

Patrick White, 1912-90, (Austral.) novelist. *The Tree of Man.*

T(erence) H(anbury) White, 1906-64, (Br.) author. *The Once and Future King, A Book of Beasts.*

Walt Whitman, 1819-92, (U.S.) poet. *Leaves of Grass.*

John Greenleaf Whittier, 1807-92, (U.S.) poet, journalist. *Snow-Bound.*

Oscar Wilde, 1854-1900, (Ir.) novelist, playwright. *The Picture of Dorian Gray, The Importance of Being Earnest.*

Laura Ingalls Wilder, 1867-1957, (U.S.) novelist. Little House on the Prairie series of children's books.

Thornton Wilder, 1897-1975, (U.S.) playwright. *Our Town, The Skin of Our Teeth, The Matchmaker.*

Tennessee Williams, 1911-83, (U.S.) playwright. *A Streetcar Named Desire, Cat on a Hot Tin Roof, The Glass Menagerie.*

William Carlos Williams, 1883-1963, (U.S.) poet, physician. *Tempers, Al Que Quiere! Paterson,* "This Is Just to Say."

Edmund Wilson, 1895-1972, (U.S.) critic, novelist. *Axel's Castle, To the Finland Station.*

P(elham) G(renville) Wodehouse, 1881-1975, (Br.-U.S.) humorist. The "Jeeves" novels, *Anything Goes.*

Thomas Wolfe, 1900-38, (U.S.) novelist. *Look Homeward, Angel; You Can't Go Home Again.*

Virginia Woolf, 1882-1941, (Br.) novelist, essayist. *Mrs. Dalloway, To the Lighthouse, A Room of One's Own.*

William Wordsworth, 1770-1850, (Br.) poet. "Tintern Abbey," "Ode: Intimations of Immortality," *The Prelude.*

Richard Wright, 1908-60, novelist, short-story writer. *Native Son, Black Boy, Uncle Tom's Children.*

Elinor Wylie, 1885-1928, (U.S.) poet. *Nets to Catch the Wind.*

William Butler Yeats, 1865-1939, (Ir.) poet, playwright. "The Second Coming," *The Wild Swans at Coole.*

Émile Zola, 1840-1902, (Fr.) novelist. *Nana, Thérèse Raquin.*

Noted Figures of the Past in Dance

Source: Reviewed by Gary Parks, Reviews editor, *Dance* magazine

Alvin Ailey, 1931-89, (U.S.) modern dancer, choreographer; melded modern dance and Afro-Caribbean techniques.

Frederick Ashton, 1904-88, (Br.) ballet choreographer; director of Great Britain's Royal Ballet, 1963-70.

Fred Astaire, 1899-1987, (U.S.) dancer, actor; teamed with dancer/actress **Ginger Rogers** (1911-95) in movie musicals.

George Balanchine, 1904-83, (Russ.-U.S.) ballet choreographer, teacher; most influential exponent of the neoclassical style; founded, with Lincoln Kirstein, School of American Ballet and New York City Ballet.

Carlo Blasis, 1803-78, (It.) ballet dancer, choreographer, writer; his teaching methods are standards of classical dance.

August Bournonville, 1805-79, (Dan.) ballet dancer, choreographer, teacher; developed a distinctly Danish style known for its exuberance and lightness.

Gisella Caccialanza, 1914-97, (U.S.) ballerina, charter member of Balanchine's American Ballet.

Enrico Cecchetti, 1850-1928, (It.) ballet dancer, leading dancer of Russia's Imperial Ballet; his technique was basis for Britain's Imperial Soc. of Teachers of Dancing.

Gower Champion, 1921-80, (U.S.) dancer, choreographer, director; with his wife **Marge,** b 1923, (U.S.) choreographed, danced in Broadway musicals and films.

John Cranko, 1927-73, (S. African) choreographer; created narrative ballets based on literary works.

Agnes de Mille, 1909-93, (U.S.) ballerina, choreographer; known for using American themes, she choreographed the ballet *Rodeo* and the musical *Oklahoma*.

Sergei Diaghilev, 1872-1929, (Russ.) impresario; founded Les Ballet Russes; saw ballet as an art unifying dance, drama, music, and decor.

Alexandra Danilova, 1903-97, (Russ.) ballerina; noted teacher at the School of American Ballet.

Isadora Duncan, 1877-1927, (U.S.) expressive dancer who united free movement with serious music; one of the founders of modern dance.

Fanny Elssler, 1810-84, (Austrian) ballerina of the Romantic era; known for dramatic skill, sensual style.

Michel Fokine, 1880-1942, (Russ.) ballet dancer, choreographer, teacher; rejected strict classicism in favor of dramatically expressive style.

Margot Fonteyn, 1919-91, (Br.) prima ballerina, Royal Ballet of Great Britain; famed performance partner of Rudolf Nureyev.

Bob Fosse, 1927-87, (U.S.) jazz dancer, choreographer, director; Broadway musicals and film.

Serge Golovine, 1924-98, (Fr.) ballet dancer with Grand Ballet du Marquis de Cuevas; choreographer.

Martha Graham, 1893-1991, (U.S.) modern dancer, choreographer; created and codified her own dramatic technique.

Martha Hill, 1901-95, (U.S.) educator; leading figure in modern dance; founded American Dance Festival.

Doris Humphrey, 1895-1958, (U.S.) modern dancer, choreographer, writer, teacher; known for her intellect and choreographic range.

Robert Joffrey, 1930-88, (U.S.) ballet dancer, choreographer; cofounded with **Gerald Arpino,** b 1928, (U.S.), the Joffrey Ballet.

Kurt Jooss, 1901-79, (Ger.) choreographer, teacher; created expressionist works using modern and classical techniques.

Tamara Karsavina, 1885-1978, (Russ.) prima ballerina of Russia's Imperial Ballet and Diaghilev's Ballets Russes; partner of Nijinsky.

Nora Kaye, 1920-87, (U.S.) ballerina with Metropolitan Opera Ballet and Ballet Theater (now American Ballet Theatre).

Lincoln Kirstein, 1907-96 (U.S.) brought ballet as an art form to U.S.; founded, with George Balanchine, School of American Ballet and New York City Ballet.

Serge Lifar, 1905-86, (Russ.-Fr.) prem. danseur, choreographer; director of dance at Paris Opera, 1930-45, 1947-58.

José Limón, 1908-72, (Mex.-U.S.) modern dancer, choreographer, teacher; developed technique based on Humphrey.

Catherine Littlefield, 1908-51, (U.S.) ballerina, choreographer, teacher; pioneer of American ballet.

Léonide Massine, 1896-1979, (Russ.-U.S.) ballet dancer, choreographer; his "symphonic ballet" used concert music previously thought unsuitable for dance.

Kenneth MacMillan, 1929-92, (Br.) dancer, choreographer; directed Royal Ballet of Great Britain 1970-77.

Vaslav Nijinsky, 1890-50, (Russ.) prem. danseur, choreographer; leading member of Diaghilev's Ballets Russes; his ballets were revolutionary for their time.

Alwin Nikolais, 1910-93, (U.S.) modern choreographer; created dance theater utilizing mixed media effects.

Jean-George Noverre, 1727-1810, (Fr.) ballet choreographer, teacher, writer; his theories on dramatic ballet still influential; called the "Shakespeare of the dance."

Rudolf Nureyev, 1938-93, (Russ.) prem. danseur, choreographer; leading male dancer of his generation; director of dance at Paris Opera, 1983-89.

Ruth Page, 1903-91, (U.S.) ballerina, choreographer; danced and directed ballet at Chicago Lyric Opera.

Anna Pavlova, 1881-1931, (Russ.) prima ballerina; toured with her own company to world acclaim.

Marius Petipa, 1818-1910, (Fr.) ballet dancer, choreographer; ballet master of the Imperial Ballet; established Russian classicism as leading style of late 19th cent.

Pearl Primus, 1919-95, (Trinidad-U.S.) modern dancer, choreographer, scholar; combined African, Caribbean, and African-American styles.

Jerome Robbins, 1918-98, (U.S.) choreographer, director, dancer; *The King and I, West Side Story, Fiddler on the Roof.*

Bill (Bojangles) Robinson, 1878-1949, (U.S.) tap dancer; called King of Tapology, he won fame on stage and screen rare for an African-American of his era.

Ruth St. Denis, 1877-1968, (U.S.) interpretive dancer, choreographer, teacher; touring widely, she influenced many early modern dancers.

Ted Shawn, 1891-1972, (U.S.) modern dancer, choreographer; teamed with Ruth St. Denis to form Denishawn dance company and school.

Marie Taglioni, 1804-84, (It.) ballerina, teacher; in title role of *La Sylphide* established image of the ethereal ballerina.

Antony Tudor, 1908-87, (Br.) choreographer, teacher; exponent of the "psychological ballet."

Galina Ulanova, 1910-98, (Rus.) revered ballerina with Bolshoi Ballet.

Mary Wigman, 1886-1973, (Ger.) modern dancer, choreographer, teacher; influenced European expressionist dance.

Agrippina Vaganova, 1879-1951, (Russ.) ballet teacher, director; codified Soviet ballet technique that developed virtuosity.

Composers of Classical and Avant Garde Music

Carl Philipp Emanuel Bach, 1714-88, (Ger.) Cantatas, passions, numerous keyboard and instrumental works.

Johann Christian Bach, 1735-82, (Ger.) Concertos, operas, sonatas.

Johann Sebastian Bach, 1685-1750, (Ger.) St. Matthew Passion, The Well-Tempered Clavier.

Samuel Barber, 1910-81, (U.S.) Adagio for Strings, Vanessa.

Béla Bartók, 1881-1945, (Hung.) Concerto for Orchestra, The Miraculous Mandarin.

Amy Beach (Mrs. H. H. A. Beach), 1867-1944, (U.S.) The Year's at the Spring, Fireflies, The Chambered Nautilus.

Ludwig van Beethoven, 1770-1827, (Ger.) Concertos (Emperor), sonatas (Moonlight, Pathetique), 9 symphonies.

Vincenzo Bellini, 1801-35, (It.) I Puritani, La Sonnambula, Norma.

Alban Berg, 1885-1935, (Austrian) Wozzeck, Lulu.

Hector Berlioz, 1803-69, (Fr.) Damnation of Faust, Symphonie Fantastique, Requiem.

Leonard Bernstein, 1918-90, (U.S.) Chichester Psalms, Jeremiah Symphony, Mass.

Georges Bizet, 1838-75, (Fr.) Carmen, Pearl Fishers.

Ernest Bloch, 1880-1959, (Swiss-U.S.) Macbeth (opera), Schelomo, Voice in the Wilderness.

Luigi Boccherini, 1743-1805, (It.) Chamber music and guitar pieces.

Alexander Borodin, 1833-87, (Russ.) Prince Igor, In the Steppes of Central Asia, Polovtzian Dances.

Pierre Boulez, b 1925, (Fr.) LeVisage nuptial, Edats/Multiple, Domaines.

Johannes Brahms, 1833-97, (Ger.) Liebeslieder Waltzes, Acad. Festival Overture, chamber music, 4 symphonies.

Benjamin Britten, 1913-76, (Br.) Peter Grimes, Turn of the Screw, A Ceremony of Carols, War Requiem.

Anton Bruckner, 1824-96, (Austrian) 9 symphonies.

Dietrich Buxtehude, 1637-1707, (Dan.) Organ works, vocal music.

William Byrd, 1543-1623, (Br.) Masses, motets.

John Cage, 1912-92, (U.S.) Winter Music, Fontana Mix.

Emmanuel Chabrier, 1841-94, (Fr.) Le Roi Malgré Lui, Espana.

Gustave Charpentier, 1860-1956, (Fr.) Louise.

Frédéric Chopin, 1810-49, (Pol.) Mazurkas, waltzes, etudes, nocturnes, polonaises, sonatas.

Aaron Copland, 1900-90, (U.S.) Appalachian Spring, Fanfare for the Common Man, Lincoln Portrait.

Claude Debussy, 1862-1918, (Fr.) Pelleas et Melisande, La Mer, Prelude to the Afternoon of a Faun.

Gaetano Donizetti, 1797-1848, (It.) Elixir of Love, Lucia di Lammermoor, Daughter of the Regiment.

Paul Dukas, 1865-1935, (Fr.) Sorcerer's Apprentice.

Antonin Dvorak, 1841-1904, (Czech.) Songs My Mother Taught Me, Symphony in E Minor (From the New World).

Edward Elgar, 1857-1934, (Br.) Enigma Variations, Pomp and Circumstance.

Manuel de Falla, 1876-1946, (Sp.) El Amor Brujo, La Vida Breve, The Three-Cornered Hat.

Gabriel Fauré, 1845-1924, (Fr.) Requiem, Elègie for Cello and Piano.

Cesar Franck, 1822-90, (Belg.) Symphony in D minor, Violin Sonata.

George Gershwin, 1898-1937, (U.S.) Rhapsody in Blue, An American in Paris, Porgy and Bess.

Philip Glass, b 1937, (U.S.) Einstein on the Beach, The Voyage.

Mikhail Glinka, 1804-57, (Russ.) A Life for the Tsar, Ruslan and Ludmilla.

Christoph W. Gluck, 1714-87, (Ger.) Alceste, Iphigènie en Tauride.

Charles Gounod, 1818-93, (Fr.) Faust, Romeo and Juliet.

Edvard Grieg, 1843-1907, (Nor.) Peer Gynt Suite, Concerto in A minor for piano.

George Frideric Handel, 1685-1759, (Ger.-Br.) Messiah, Water Music.
Howard Hanson, 1896-1981, (U.S.) Symphonies No. 1 (Nordic) and No. 2 (Romantic).
Roy Harris, 1898-1979, (U.S.) Symphonies.
(Franz) Joseph Haydn, 1732-1809, (Austrian) Symphonies (Clock, London, Toy), chamber music, oratorios.
Paul Hindemith, 1895-1963, (U.S.) Mathis der Maler.
Gustav Holst, 1874-1934, (Br.) The Planets.
Arthur Honegger, 1892-1955, (Fr.) Judith, Le Roi David, Pacific 231.
Alan Hovhaness, b 1911, (U.S.) Symphonies, Magnificat.
Engelbert Humperdinck, 1854-1921, (Ger.) Hansel and Gretel.
Charles Ives, 1874-1954, (U.S.) Concord Sonata, symphonies.
Aram Khachaturian, 1903-78, (Russ.) Ballets, piano pieces, Sabre Dance.
Zoltán Kodaly, 1882-1967, (Hung.) Háry János, Psalmus Hungaricus.
Fritz Kreisler, 1875-1962, (Austrian) Caprice Viennois, Tambourin Chinois.
Edouard Lalo, 1823-92, (Fr.) Symphonie Espagnole.
Ruggero Leoncavallo, 1857-1919, (It.) Pagliacci.
Franz Liszt, 1811-86, (Hung.) 20 Hungarian rhapsodies, symphonic poems.
Edward MacDowell, 1861-1908, (U.S.) To a Wild Rose.
Gustav Mahler, 1860-1911, (Austrian) Das Lied von der Erde; 9 complete symphonies.
Pietro Mascagni, 1863-1945, (It.) Cavalleria Rusticana.
Jules Massenet, 1842-1912, (Fr.) Manon, Le Cid, Thaïs.
Felix Mendelssohn, 1809-47, (Ger.) A Midsummer Night's Dream, Songs Without Words, violin concerto.
Gian-Carlo Menotti, b 1911, (It.-U.S.) The Medium, The Consul, Amahl and the Night Visitors.
Claudio Monteverdi, 1567-1643, (It.) Opera, masses, madrigals.
Modest Moussorgsky, 1839-81, (Russ.) Boris Godunov, Pictures at an Exhibition.
Wolfgang Amadeus Mozart, 1756-91, (Austrian) Chamber music, concertos, operas (Magic Flute, Marriage of Figaro), 41 symphonies.
Jacques Offenbach, 1819-80, (Fr.) Tales of Hoffmann.
Carl Orff, 1895-1982, (Ger.) Carmina Burana.
Johann Pachelbel, 1653-1706, (Ger.) Canon and Fugue in D major.
Ignacy Paderewski, 1860-1941, (Pol.) Minuet in G.
Niccolò Paganini, 1782-1840, (It.) Caprices for violin solo.
Giovanni Palestrina, c1525-94, (It.) Masses, madrigals.
Krzystof Penderecki, b 1933, (Pol.) Psalmus, Polymorphia, De natura sonoris.
Francis Poulenc, 1899-1963, (Fr.) Dialogues des Carmèlites.
Mel Powell, 1923-98, (U.S.) Duplicates: A Concerto for Two Pianos and Orchestra, Cantilena Concertante.
Sergei Prokofiev, 1891-1953, (Russ.) Classical Symphony, Love for Three Oranges, Peter and the Wolf.

Giacomo Puccini, 1858-1924, (It.) La Boheme, Manon Lescaut, Tosca, Madama Butterfly.
Henry Purcell, 1659-95, (Eng.) Dido and Aeneas.
Sergei Rachmaninoff, 1873-1943, (Russ.) Concertos, preludes (Prelude in C sharp minor), symphonies.
Maurice Ravel, 1875-1937, (Fr.) Bolèro, Daphnis et Chloè, Piano Concerto in D for Left Hand Alone.
Nikolai Rimsky-Korsakov, 1844-1908, (Russ.) Golden Cockerel, Scheherazade, Flight of the Bumblebee.
Gioacchino Rossini, 1792-1868, (It.) Barber of Seville, Othello, William Tell.
Camille Saint-Saëns, 1835-1921, (Fr.) Carnival of Animals (The Swan), Samson and Delilah, Danse Macabre.
Alessandro Scarlatti, 1660-1725, (It.) Cantatas, oratorios, operas.
Domenico Scarlatti, 1685-1757, (It.) Harpsichord works.
Alfred Schnittke, 1934-98, (Sov.-Ger.) Life With an Idiot.
Arnold Schoenberg, 1874-1951, (Austrian) Pelleas and Melisande, Pierrot Lunaire, Verklärte Nacht.
Franz Schubert, 1797-1828, (Austrian) Chamber music (Trout Quintet), lieder, symphonies (Unfinished).
Robert Schumann, 1810-56, (Ger.) Die Frauenliebe und Leben, Träumerei.
Dimitri Shostakovich, 1906-75, (Russ.) Symphonies, Lady Macbeth of the District Mzensk.
Jean Sibelius, 1865-1957, (Finn.) Finlandia.
Bedrich Smetana, 1824-84, (Czech.) The Bartered Bride.
Karlheinz Stockhausen, b 1928, (Ger.) KontraPunkte, Kontakte for Electronic Instruments.
Richard Strauss, 1864-1949, (Ger.) Salome, Elektra, Der Rosenkavalier, Thus Spake Zarathustra.
Igor Stravinsky, 1882-1971, (Russ.) Noah and the Flood, The Rake's Progress, The Rite of Spring.
Toru Takemitsu, 1930-96, (Jpn.) Requiem for Strings, Dorian Horizon.
Peter I. Tchaikovsky, 1840-93, (Russ.) Nutcracker, Swan Lake, The Sleeping Beauty.
Virgil Thomson, 1896-1989, (U.S.) Opera, film music, Four Saints in Three Acts.
Dmitri Tiomkin, 1894-1979, (Russ.-U.S.) film scores, including High Noon.
Sir Michael Tippett, 1905-98, (Br.) A Child of Our Time, The Midsummer Marriage, The Knot Garden.
Ralph Vaughan Williams, 1872-1958, (Eng.) Fantasiz on a Theme by Thomas Tallis, symphonies, vocal music.
Giuseppe Verdi, 1813-1901, (It.) Aida, Rigoletto, Don Carlo, Il Trovatore, La Traviata, Falstaff, Macbeth.
Heitor Villa-Lobos, 1887-1959, (Brazil) Bachianas Brasileiras.
Antonio Vivaldi, 1678-1741, (It.) Concerto grossos (The Four Seasons).
Richard Wagner, 1813-83, (Ger.) Rienzi, Tannhäuser, Lohengrin, Tristan und Isolde.
Carl Maria von Weber, 1786-1826, (Ger.) Der Freischutz.

Composers of Operettas, Musicals, and Popular Music

Richard Adler, b 1921, (U.S.) Pajama Game; Damn Yankees.
Milton Ager, 1893-1979, (U.S.) I Wonder What's Become of Sally; Hard Hearted Hannah; Ain't She Sweet?
Arthur Altman, 1910-94, (U.S.) All or Nothing at All.
Leroy Anderson, 1908-75, (U.S.) Sleigh Ride, Blue Tango, Syncopated Clock.
Paul Anka, b 1941, (Can.) My Way; Tonight Show theme.
Harold Arlen, 1905-86, (U.S.) Stormy Weather; Over the Rainbow; Blues in the Night; That Old Black Magic.
Burt Bacharach, b 1928, (U.S.) Raindrops Keep Fallin' on My Head; Walk on By; What the World Needs Now Is Love.
Ernest Ball, 1878-1927, (U.S.) Mother Machree; When Irish Eyes Are Smiling.
Irving Berlin, 1888-1989, (U.S.) Annie Get Your Gun; Call Me Madam; God Bless America; White Christmas.
Leonard Bernstein, 1918-90, (U.S.) On the Town; Wonderful Town; Candide; West Side Story.
Eubie Blake, 1883-1983, (U.S.) Shuffle Along; I'm Just Wild About Harry.
Jerry Bock, b 1928, (U.S.) Mr. Wonderful; Fiorello; Fiddler on the Roof; The Rothschilds.
Carrie Jacobs Bond, 1862-1946, (U.S.) I Love You Truly.
Nacio Herb Brown, 1896-1964, (U.S.) Singing in the Rain; You Were Meant for Me; All I Do Is Dream of You.
Hoagy Carmichael, 1899-1981, (U.S.) Stardust; Georgia on My Mind; Old Buttermilk Sky.
George M. Cohan, 1878-1942, (U.S.) Give My Regards to Broadway; You're a Grand Old Flag; Over There.
Cy Coleman, b 1929, (U.S.) Sweet Charity; Witchcraft.
John Frederick Coots, 1897-?, (U.S.) Santa Claus Is Coming to Town; You Go to My Head; For All We Know.
Noel Coward, 1899-1973, (Br.) Bitter Sweet; Mad Dogs and Englishmen; Mad About the Boy.
Neil Diamond, b 1941, (U.S.) I'm a Believer; Sweet Caroline.
Walter Donaldson, 1893-1947, (U.S.) My Buddy; Carolina in the Morning; Makin' Whoopee.
Vernon Duke, 1903-69, (U.S.) April in Paris.

Bob Dylan, b 1941, (U.S.) Blowin' in the Wind.
Gus Edwards, 1879-1945, (U.S.) School Days; By the Light of the Silvery Moon; In My Merry Oldsmobile.
Sherman Edwards, 1919-81, (U.S.) See You in September; Wonderful! Wonderful!
Duke Ellington, 1899-1974, (U.S.) Sophisticated Lady; Satin Doll; It Don't Mean a Thing; Solitude.
Sammy Fain, 1902-89, (U.S.) I'll Be Seeing You; Love Is a Many-Splendored Thing.
Fred Fisher, 1875-1942, (U.S.) Peg O' My Heart; Chicago.
Stephen Collins Foster, 1826-64, (U.S.) My Old Kentucky Home; Old Folks at Home.
Rudolf Friml, 1879-1972, (Czech-U.S.) The Firefly; Rose Marie; Vagabond King; Bird of Paradise.
John Gay, 1685-1732, (Br.) The Beggar's Opera.
George Gershwin, 1898-1937, (U.S.) Someone to Watch Over Me; I've Got a Crush on You; Embraceable You.
Morton Gould, 1913-96, (U.S.) Fall River Suite, Holocaust Suite, Spirituals for Orchestra, Stringmusic.
Ferde Grofe, 1892-1972, (U.S.) Grand Canyon Suite.
Marvin Hamlisch, b 1944, (U.S.) The Way We Were; Nobody Does It Better; A Chorus Line.
Ray Henderson, 1896-1970, (U.S.) George White's Scandals; That Old Gang of Mine; Five Foot Two, Eyes of Blue.
Victor Herbert, 1859-1924, (Ir.-U.S.) Mlle. Modiste; Babes in Toyland; The Red Mill; Naughty Marietta; Sweethearts.
Jerry Herman, b 1933, (U.S.) Hello Dolly; Mame.
Brian Holland, b 1941, **Lamont Dozier,** b 1941, **Eddie Holland,** b 1939, (all U.S.) Heat Wave; Stop! In the Name of Love; Baby, I Need Your Loving.
Antonio Carlos Jobim, 1927-94, (Brazil) The Girl From Ipanema; Desafinado; One Note Samba.
Billy (William Martin) Joel, b 1949, (U.S.) Just the Way You Are; Honesty; Piano Man.
Scott Joplin, 1868-1917, (U.S.) Maple Leaf Rag; Treemonisha.
John Kander, b 1927, (U.S.) Cabaret; Chicago; Funny Lady.

Jerome Kern, 1885-1945, (U.S.) *Sally; Sunny; Show Boat.*

Carole King, b 1942, (U.S.) *Will You Love Me Tomorrow?; Natural Woman; One Fine Day; Up on the Roof.*

Burton Lane, 1912-1997, (U.S.) *Finian's Rainbow.*

Franz Lehar, 1870-1948, (Hung.) *Merry Widow.*

Jerry Leiber, & Mike Stoller, both b 1933, (both U.S.) Hound Dog; Searchin'; Yakety Yak; Love Me Tender.

Mitch Leigh, b 1928, (U.S.) *Man of La Mancha.*

John Lennon, 1940-80, & **Paul McCartney,** b 1942, (both Br.) I Want to Hold Your Hand; She Loves You.

Andrew Lloyd Webber, b 1948, (Br.) *Jesus Christ Superstar; Evita; Cats; The Phantom of the Opera.*

Frank Loesser, 1910-69, (U.S.) *Guys and Dolls; Where's Charley?; The Most Happy Fella; How to Succeed....*

Frederick Loewe, 1901-88, (Austrian-U.S.) *Brigadoon; Paint Your Wagon; My Fair Lady; Camelot.*

Henry Mancini, 1924-94, (U.S.) Moon River; Days of Wine and Roses; Pink Panther Theme.

Barry Mann, b 1939, & **Cynthia Weil,** b 1937, (both U.S.) You've Lost That Loving Feeling.

Jimmy McHugh, 1894-1969, (U.S.) Don't Blame Me; I'm in the Mood for Love; I Feel a Song Coming On.

Alan Menken, b 1950, (U.S.) *Little Shop of Horrors.*

Joseph Meyer, 1894-1987, (U.S.) If You Knew Susie; California, Here I Come; Crazy Rhythm.

Chauncey Olcott, 1858-1932, (U.S.) Mother Machree.

Jerome "Doc" Pomus, 1925-91, (U.S.) Save the Last Dance for Me; A Teenager in Love.

Cole Porter, 1893-1964, (U.S.) *Anything Goes; Kiss Me Kate; Can Can; Silk Stockings.*

Smokey Robinson, b 1940, (U.S.) Shop Around; My Guy; My Girl; Get Ready.

Richard Rodgers, 1902-79, (U.S.) *Oklahoma!; Carousel; South Pacific; The King and I; The Sound of Music.*

Sigmund Romberg, 1887-1951, (Hung.) *Maytime; The Student Prince; Desert Song; Blossom Time.*

Harold Rome, 1908-93, (U.S.) *Pins and Needles; Call Me Mister; Wish You Were Here; Fanny; Destry Rides Again.*

Vincent Rose, b 1880-1944, (U.S.) Avalon; Whispering; Blueberry Hill.

Harry Ruby, 1895-1974, (U.S.) Three Little Words; Who's Sorry Now?

Arthur Schwartz, 1900-84, (U.S.) *The Band Wagon;* Dancing in the Dark; By Myself; That's Entertainment.

Neil Sedaka, b 1939, (U.S.) Breaking Up Is Hard to Do.

Paul Simon, b 1942, (U.S.) Sounds of Silence; I Am a Rock; Mrs. Robinson; Bridge Over Troubled Waters.

Stephen Sondheim, b 1930, (U.S.) *A Little Night Music; Company; Sweeney Todd; Sunday in the Park With George.*

John Philip Sousa, 1854-1932, (U.S.) *El Capitan;* Stars and Stripes Forever.

Oskar Straus, 1870-1954, (Austrian) *Chocolate Soldier.*

Johann Strauss, 1825-99, (Austrian) *Gypsy Baron; Die Fledermaus;* waltzes: Blue Danube; Artist's Life.

Charles Strouse, b 1928, (U.S.) *Bye Bye, Birdie; Annie.*

Jule Styne, 1905-94, (Br.-U.S.) *Gentlemen Prefer Blondes; Bells Are Ringing; Gypsy; Funny Girl.*

Arthur S. Sullivan, 1842-1900, (Br.) *H.M.S. Pinafore; Pirates of Penzance; The Mikado.*

Deems Taylor, 1885-1966, (U.S.) *Peter Ibbetson.*

Harry Tobias, 1905-94, (U.S.) *I'll Keep the Lovelight Burning.*

Egbert van Alstyne, 1882-1951, (U.S.) In the Shade of the Old Apple Tree; Memories; Pretty Baby.

Jimmy Van Heusen, 1913-90, (U.S.) Moonlight Becomes You; Swinging on a Star; All the Way; Love and Marriage.

Albert von Tilzer, 1878-1956, (U.S.) I'll Be With You in Apple Blossom Time; Take Me Out to the Ball Game.

Harry von Tilzer, 1872-1946, (U.S.) Only a Bird in a Gilded Cage; On a Sunday Afternoon.

Fats Waller, 1904-43, (U.S.) Honeysuckle Rose; Ain't Misbehavin'.

Harry Warren, 1893-1981, (U.S.) You're My Everything; We're in the Money; I Only Have Eyes for You.

Jimmy Webb, b 1946, (U.S.) Up, Up and Away; By the Time I Get to Phoenix; Didn't We?; Wichita Lineman.

Kurt Weill, 1900-50, (Ger.-U.S.) *Threepenny Opera; Lady in the Dark; Knickerbocker Holiday; One Touch of Venus.*

Percy Wenrich, 1887-1952, (U.S.) When You Wore a Tulip; Moonlight Bay; Put On Your Old Gray Bonnet.

Richard A. Whiting, 1891-1938, (U.S.) Till We Meet Again; Sleepytime Gal; Beyond the Blue Horizon; My Ideal.

John Williams, b 1932, (U.S.) *Jaws; E.T.; Star Wars* series; *Raiders of the Lost Ark* series.

Meredith Willson, 1902-84, (U.S.) *The Music Man.*

Stevie Wonder, b 1950, (U.S.) You Are the Sunshine of My Life; Signed, Sealed, Delivered, I'm Yours.

Vincent Youmans, 1898-1946, (U.S.) *Two Little Girls in Blue; Wildflower; No, No, Nanette; Hit the Deck; Rainbow; Smiles.*

Lyricists

Howard Ashman, 1950-91, (U.S.) Little Shop of Horrors; The Little Mermaid.

Johnny Burke, 1908-84, (U.S.) Misty; Imagination.

Irving Caesar, 1895-1996, (U.S.) Swanee; Tea for Two; Just a Gigolo.

Sammy Cahn, 1913-93, (U.S.) High Hopes; Love and Marriage; The Second Time Around; It's Magic.

Leonard Cohen, b 1934, (Can.) Suzanne; Stranger Song.

Betty Comden, b 1919, (U.S.) and **Adolph Green,** b 1915, (U.S.) The Party's Over; Just in Time; New York, New York.

Hal David, b 1921, (U.S.) What the World Needs Now Is Love.

Buddy De Sylva, 1895-1950, (U.S.) When Day Is Done; Look for the Silver Lining; April Showers.

Howard Dietz, 1896-1983, (U.S.) Dancing in the Dark; You and the Night and the Music; That's Entertainment.

Al Dubin, 1891-1945, (U.S.) Tiptoe Through the Tulips; Anniversary Waltz; Lullaby of Broadway.

Fred Ebb, b 1936, (U.S.) Cabaret; Zorba; Woman of the Year.

Dorothy Fields, 1905-74, (U.S.) On the Sunny Side of the Street; Don't Blame Me; The Way You Look Tonight.

Ira Gershwin, 1896-1983, (U.S.) The Man I Love; Fascinating Rhythm; S'Wonderful; Embraceable You.

William S. Gilbert, 1836-1911, (Br.) The Mikado; H.M.S. Pinafore; Pirates of Penzance.

Gerry Goffin, b 1939, (U.S.) Will You Love Me Tomorrow; Take Good Care of My Baby; Up on the Roof.

Mack Gordon, 1905-59, (Pol.-U.S.) You'll Never Know; The More I See You; Chattanooga Choo-Choo.

Oscar Hammerstein II, 1895-1960, (U.S.) Ol' Man River; Oklahoma; Carousel.

E. Y. (Yip) Harburg, 1898-1981, (U.S.) Brother, Can You Spare a Dime; April in Paris; Over the Rainbow.

Lorenz Hart, 1895-1943, (U.S.) Isn't It Romantic; Blue Moon; Lover; Manhattan; My Funny Valentine.

DuBose Heyward, 1885-1940, (U.S.) Summertime.

Gus Kahn, 1886-1941, (U.S.) Memories; Ain't We Got Fun.

Alan J. Lerner, 1918-86, (U.S.) Brigadoon; My Fair Lady; Camelot; Gigi; On a Clear Day You Can See Forever.

Johnny Mercer, 1909-76, (U.S.) Blues in the Night; Come Rain or Come Shine; Laura; That Old Black Magic.

Bob Merrill, 1921-98, (U.S.) People; (How Much Is That) Doggie in the Window.

Jack Norworth, 1879-1959, (U.S.) Take Me Out to the Ball Game; Shine On Harvest Moon.

Mitchell Parish, 1901-93, (U.S.) Stairway to the Stars; Stardust.

Andy Razaf, 1895-1973, (U.S.) Honeysuckle Rose; Ain't Misbehavin'; S'posin'.

Leo Robin, 1900-84, (U.S.) Thanks for the Memory; Hooray for Love; Diamonds Are a Girl's Best Friend.

Paul Francis Webster, 1907-84, (U.S.) Secret Love; The Shadow of Your Smile; Love Is a Many-Splendored Thing.

Jack Yellen, 1892-1991, (U.S.) Down by the O-Hi-O; Ain't She Sweet; Happy Days Are Here Again.

Notable Opera Singers of the Past

Frances Alda, 1883-1952, (NZ) soprano

Paul Althouse, 1889-1954, (U.S.) tenor

Pasquale Amato, 1878-1942, (It.) baritone

Marian Anderson, 1902-93, (U.S.) contralto

Jussi Björling, 1911-60, (Swed.) tenor

Lucrezia Bori, 1887-1960, (It.) soprano

Maria Callas, 1923-77, (U.S.) soprano

Emma Calvé, 1858-1942, (Fr.) soprano

Enrico Caruso, 1873-1921, (It.) tenor

Feodor Chaliapin, 1873-1938, (Russ.) bass

Boris Christoff, 1914-93, (Bulg.) bass

Richard Crooks, 1900-72, (U.S.) tenor

Giuseppe De Luca, 1876-1950, (It.) baritone

Edouard De Reszke, 1853-1917, (Pol.) bass

Jean De Reszke, 1850-1925, (Pol.) tenor

Emmy Destinn, 1878-1930, (Czech.) soprano

Todd Duncan, 1903-98, (U.S.) baritone

Emma Eames, 1865-1952, (U.S.) soprano

Geraldine Farrar, 1882-1967, (U.S.) soprano

Kirsten Flagstad, 1895-1962, (Nor.) soprano

Olive Fremstad, 1871-1951, (Swed.-U.S.) soprano

Amelita Galli-Curci, 1882-1963, (It.) soprano

Mary Garden, 1874-1967, (Br.) soprano

Beniamino Gigli, 1890-1957, (It.) tenor

Tito Gobbi, 1913-84, (It.) baritone

Frieda Hempel, 1885-1955, (Ger.) soprano

Maria Jeritza, 1887-1982, (Czech.) soprano

Alexander Kipnis, 1891-1978, (Russ.-U.S.) bass

Lilli Lehmann, 1848-1929, (Ger.) soprano

Lotte Lehmann, 1888-1976, (Ger.-U.S.) soprano

Jenny Lind, 1820-87, (Swed.) soprano
John McCormack, 1884-1945, (Ir.) tenor
Blanche Marchesi, 1863-1940, (Fr.) soprano
Nellie Melba, 1861-1931, (Austral.) soprano.
Lauritz Melchior, 1890-1973, (Dan.) tenor
Zinka Milanov, 1906-89, (Yugo.) soprano
Lillian Nordica, 1857-1914, (U.S.) soprano
Adelina Patti, 1843-1919, (It.) soprano
Peter Pears, 1910-86, (Eng.) tenor
Jan Peerce, 1904-84, (U.S.) tenor
Ezio Pinza, 1892-1957, (It.) bass

Lily Pons, 1898-1976, (Fr.) soprano
Rosa Ponselle, 1897-1981, (U.S.) soprano
Hermann Prey, 1929-98, (Ger.) baritone.
Marcella Sembrich, 1858-1935, (Pol.) soprano
Eleanor Steber, 1916-90, (U.S.) soprano
Ferrucio Tagliavini, 1913-95, (It.) tenor
Luisa Tetrazzini, 1871-1940, (It.) soprano
Lawrence Tibbett, 1896-1960, (U.S.) baritone
Richard Tucker, 1913-75, (U.S.) tenor
Pauline Viardot, 1821-1910, (Fr.) mezzo-soprano
Leonard Warren, 1911-60, (U.S.) baritone

Noted Blues and Jazz Artists of the Past

Julian "Cannonball" Adderley, 1928-75, alto sax
Louis "Satchmo" Armstrong, 1900-71, trumpet, singer; "scat" vocals
Mildred Bailey, 1907-51, blues singer
Chet Baker, 1929-88, trumpet
Count Basie, 1904-84, orchestra leader, piano
Sidney Bechet, 1897-1959, early innovator, soprano sax
Bix Beiderbecke, 1903-31, cornet, piano, composer
Tommy Benford, 1906-94, drummer
Bunny Berigan, 1909-42, trumpet, singer
Barney Bigard, 1906-80, clarinet
Ed Blackwell, 1929-92, drummer
Jimmy Blanton, 1921-42, bass
Charles "Buddy" Bolden, 1868-1931, cornet; formed first jazz band.
Big Bill Broonzy, 1893-1958, blues singer, guitar
Clifford Brown, 1930-56, trumpet
Don Byas, 1912-72, tenor sax
Cab Calloway, 1907-94, band leader
Harry Carney, 1910-74, baritone sax
Betty Carter, 1930-98, jazz singer
Sidney Catlett, 1910-51, drums
Doc Cheatham, 1905-97, trumpet
Don Cherry, 1937-95, lyrical jazz trumpet
Charlie Christian, 1919-42, guitar
Kenny Clarke, 1914-85, modern drums
Buck Clayton, 1911-91, trumpet, arranger
James Cleveland, 1931-91, gospel singer
Al Cohn, 1925-88, tenor sax, composer
Cozy Cole, 1909-81, drums
Johnny Coles, 1926-96, trumpet
John Coltrane, 1926-67, tenor sax innovator
Eddie Condon, 1904-73, guitar, band leader; Dixieland
Tadd Dameron, 1917-65, piano, composer
Eddie "Lockjaw" Davis, 1921-86, tenor sax
Miles Davis, 1926-91, trumpet; pioneer of cool jazz
Wild Bill Davison, 1906-89, cornet, early Chicago jazz
Paul Desmond, 1924-77, alto sax
Vic Dickenson, 1906-84, trombone, composer
Willie Dixon, 1915-92, songwriter, blues, "You Shook Me"
Warren "Baby" Dodds, 1898-1959, Dixieland drummer
Johnny Dodds, 1892-1940, clarinet
Jimmy Dorsey, 1904-57, clarinet, alto sax; band leader
Tommy Dorsey, 1905-56, trombone; band leader
Roy Eldridge, 1911-89, trumpet, drums, singer
Duke Ellington, 1899-1974, piano, band leader, composer
Bill Evans, 1929-80, piano
Gil Evans, 1912-88, composer, arranger, piano
Tal Farlow, 1921-98, jazz guitarist
Ella Fitzgerald, 1918-1996, jazz vocalist, "first lady of song"
"Red" Garland, 1923-84, piano
Erroll Garner, 1921-77, piano, composer, "Misty"
Stan Getz, 1927-91, tenor sax
Dizzy Gillespie, 1917-93, trumpet, composer; bop developer
Benny Goodman, 1909-86, clarinet; band, combo leader
Dexter Gordon, 1923-90, tenor sax, bop-derived style
Stéphane Grappelli, 1908-97, violin
Bobby Hackett, 1915-76, trumpet, cornet
W. C. Handy, 1873-1958, composer, "St. Louis Blues"
Coleman Hawkins, 1904-69, tenor sax, "Body and Soul"
Fletcher Henderson, 1898-1952, orchestra leader, arranger
Woody Herman, 1913-87, clarinet, alto sax, band leader
Jay C. Higginbotham, 1906-73, trombone
Earl "Fatha" Hines, 1905-83, piano, songwriter
Al Hirt, 1922-99, trumpet
Johnny Hodges, 1906-70, alto sax
Billie Holiday, 1915-59, blues singer, "Strange Fruit"
Sam "Lightnin" Hopkins, 1912-82, blues singer, guitarist
Howlin' Wolf, 1910-1976, blues singer, harmonica, guitar
Mahalia Jackson, 1911-72, gospel singer
Blind Lemon Jefferson, 1897-1930, blues singer, guitar
Little Willie John, 1937-68, singer, songwriter
Bunk Johnson, 1879-1949, cornet, trumpet
James P. Johnson, 1891-1955, piano, composer
Robert Johnson, 1912-38, blues songwriter, singer, guitarist
Jo Jones, 1911-85, drums
Philly Joe Jones, 1923-85, drums

Thad Jones, 1923-86, trumpet, cornet
Scott Joplin, 1868-1917, ragtime composer
Louis Jordan, 1908-75, singer, alto sax
Stan Kenton, 1912-79, orchestra leader, composer, piano
Albert King, 1923-92, blues guitarist
Gene Krupa, 1909-73, drums, band and combo leader
Scott LaFaro, 1936-61, bass
Huddie Ledbetter (Lead Belly), 1888-1949, blues singer, guitar
Mel Lewis, 1929-90, drummer, orchestra leader
Jimmie Lunceford, 1902-47, band leader, sax
Jimmy McPartland, 1907-91, trumpet
Carmen McRae, 1920-94, jazz singer
Glenn Miller, 1904-44, trombone, dance band leader
Charles Mingus, 1922-79, bass, composer, combo leader
Thelonious Monk, 1920-82, piano, composer, combo leader; bop developer
Wes Montgomery, 1925-68, guitar
"Jelly Roll" Morton, 1885-1941, composer, piano, singer
Bennie Moten, 1894-1935, piano
Gerry Mulligan, 1927-96, baritone sax, songwriter, "cool school"
Turk Murphy, 1915-87, trombone, band leader
Theodore "Fats" Navarro, 1923-50, trumpet
Red Nichols, 1905-65, cornet, combo leader
King Oliver, 1885-1938, cornet, band leader; Louis Armstrong
Sy Oliver, 1910-88, Swing Era arranger, composer, conductor
Kid Ory, 1886-1973, trombone, "Muskrat Ramble"
Charlie "Bird" Parker, 1920-55, alto sax, noted jazz improviser
Joe Pass, 1929-94, guitarist
Art Pepper, 1925-82, alto sax
Oscar Pettiford, 1922-60, a leading bop-era bassist
Bud Powell, 1924-66, piano; modern jazz pioneer
Louis Prima, 1911-78, singer, band leader.
Don Pullen, 1942-95, piano; percussive pianist
Sun Ra, 1915?-93, bandleader, pianist, composer
Gertrude "Ma" Rainey, 1886-1939, blues singer
Don Redman, 1900-64, composer, arranger
Django Reinhardt, 1910-53, guitar; influenced Amer. jazz
Buddy Rich, 1917-87, drums, band leader
Red Rodney, 1928-94, trumpeter
Frank Rosolino, 1926-78, trombone
Jimmy Rowles, 1918-96, jazz composer, accompanist
Jimmy Rushing, 1903-72, blues singer
Pee Wee Russell, 1906-69, clarinet
Zoot Sims, 1925-85, tenor, alto sax, clarinet
Zutty Singleton, 1898-1975, Dixieland drummer
Bessie Smith, 1894-1937, blues singer
Clarence "Pinetop" Smith, 1904-29, piano, singer; pioneer of boogie woogie
Willie "The Lion" Smith, 1897-1973, stride style pianist
Muggsy Spanier, 1906-67, cornet, band leader
Billy Strayhorn, 1915-67, composer, piano
Sonny Stitt, 1924-82, alto, tenor sax
Art Tatum, 1910-56, piano; technical virtuoso
Art Taylor, 1929-95, jazz drummer, bandleader
Jack Teagarden, 1905-64, trombone, singer
Mel Torme, 1925-99, "Velvet Fog", singer
Dave Tough, 1908-48, drums
Lennie Tristano, 1919-78, piano, composer
Joe Turner, 1911-85, blues singer
Sarah Vaughan, 1924-90, singer
Joe Venuti, 1904-78, first great jazz violinist
T-Bone Walker, 1910-75, guitarist; electric blues guitar
Thomas "Fats" Waller, 1904-43, piano, singer, composer
Dinah Washington, 1924-63, singer
Ethel Waters, 1896-1977, jazz and blues singer
Muddy Waters, 1915-83, blues singer, songwriter
Johnny Watson, 1935-96, rhythm and blues guitarist
Chick Webb, 1902-39, band leader, drums
Ben Webster, 1909-73, tenor sax
Junior Wells, 1934-98, blues singer, harmonica
Paul Whiteman, 1890-1967, jazz orchestra leader
Charles "Cootie" Williams, 1908-85, trumpet, band leader
Mary Lou Williams, 1914-81, piano, composer
Teddy Wilson, 1912-86, piano, composer
Kai Winding, 1922-83, trombone, composer
Jimmy Yancey, 1894-1951, piano
Lester "Pres" Young, 1909-59, tenor sax, composer

Noted Country Music Artists of the Past

Roy Acuff, 1903-92, fiddler, singer, songwriter; "Wabash Cannon Ball"

Autry, Gene, 1907-98, first great singing movie cowboy; "Back in the Saddle Again"

Boudleaux Bryant, 1920-87, songwriter, singer; "Hey Joe"

Carter Family (original members, **"Mother" Maybelle** 1909-78; **A.P.,** 1891-1960, **Sara,** 1898-1979) "Wildwood Flower"

Patsy Cline, 1932-63, singer; "Crazy"

Vernon Dalhart, 1883-1948, singer; "The Death of Floyd Collins"

John Denver, 1943-97, singer, songwriter; "Rocky Mountain High"

Jimmy Driftwood, 1907-98, singer, songwriter; "The Battle of New Orleans"

Lester Flatt, 1914-79, singer, guitarist; "Foggy Mountain Breakdown"

Red Foley, 1910-68, singer; "Chattanoogie Shoe Shine Boy""

Tennessee Ernie Ford, 1919-91, singer, TV host; "Sixteen Tons"

Lefty Frizzell, 1928-75, singer, guitarist; "Long Black Veil"

Grandpa Jones, 1913-98, singer, banjo player, comic; Grand Ole Opry and "Hee Haw"

Uncle Dave Macon, 1870-1952, singer, banjo player, comedian

Roger Miller, 1936-92, singer, songwriter; "King of the Road"

Bill Monroe, 1911-96, singer, songwriter, and mandolin player, "father of Bluegrass music"; "Mule Skinner Blues"

Montana, Patsy, 1908-96, yodeling/singing cowgirl; "I Want To Be a Cowboy's Sweetheart"

Minnie Pearl, 1912-96, comedienne, Grand Ole Opry star

Jim Reeves, 1923-64, singer, songwriter; "Four Walls"

Charlie Rich (Silver Fox), 1932-95, singer, songwriter; "The Most Beautiful Girl"

Tex Ritter, 1905-74, singer, songwriter; "Jingle, Jangle, Jingle"

Marty Robbins, 1925-82, singer, songwriter; "A White Sport Coat and a Pink Carnation"

Jimmie Rodgers, 1897-1933, singer, songwriter; "T for Texas"

Rogers, Roy (Leonard Slye), 1911-98, singer, actor; "King of the Cowboys"

Fred Rose, 1898-1954, songwriter, singer, producer, "Blue Eyes Cryin' in the Rain"

Original Sons of the Pioneers, Leonard Slye (Roy Rogers), 1911-98, Bob Nolan, 1908-80, singers, songwriters, "Tumbling Tumbleweeds"; Tim Spencer, 1905-74, singer, songwriter, "Careless Kisses"; Hugh Farr, 1903-80, Karl Farr, 1909-61, Lloyd Perryman, 1917-77, singers

Merle Travis, 1917-83, singer, guitarist, songwriter; "Divorce Me C.O.D."

Ernest Tubb, 1914-84, singer, songwriter and guitarist; "Walking the Floor Over You"

Conway Twitty, 1933-93, singer, songwriter; "Hello Darlin' "

Dottie West, 1932-91, singer, songwriter; "Here Comes My Baby"

Hank Williams Sr., 1923-53, singer, songwriter; "Your Cheatin' Heart"

Bob Wills, 1905-75, Western Swing fiddler, singer, bandleader, songwriter; "New San Antonio Rose"

Tammy Wynette, 1942-98, singer; "Stand By Your Man"

Noted Rock and Roll, Rhythm and Blues, and Rap Artists

Titles in quotation marks are singles; others are albums.

AC/DC: "Back in Black"
Bryan Adams: "Cuts Like a Knife"
Aerosmith: "Sweet Emotion"
*__The Allman Brothers Band (1995):__ "Ramblin' Man"
*__The Animals (1994):__ "House of the Rising Sun"
Paul Anka: "Lonely Boy"
Fiona Apple: "Criminal"
The Association: "Cherish"
Frankie Avalon: "Venus"
Backstreet Boys: "Everybody"
Erykah Badu: "On and On"
*__La Vern Baker (1991):__ "I Cried a Tear"
*__Hank Ballard[1] and the Midnighters (1990):__ "Work With Me, Annie"
*__The Band (1994):__ "The Weight"
*__The Beach Boys (1988):__ "Good Vibrations"
Beastie Boys: "(You Gotta) Fight for Your Right (to Party)"
*__The Beatles (1988):__ *Sgt. Pepper's Lonely Hearts Club Band*
Beck: "Loser"
*__The Bee Gees (1997):__ "Stayin' Alive"
Pat Benatar: "Hit Me With Your Best Shot"
*__Chuck Berry (1986):__ "Johnny B. Goode"
The Big Bopper: "Chantilly Lace"
Bjork: "Human Behavior"
Black Sabbath: "Paranoid"
*__Bobby "Blue" Bland (1992):__ "Turn On Your Love Light"
Mary J. Blige: *My Life*
Blind Faith: "Can't Find My Way Home"
Blondie: "Heart of Glass"
Blood, Sweat, and Tears: "Spinning Wheel"
Gary "U.S." Bonds: "Quarter to Three"
Bon Jovi: "Livin' on a Prayer"
*__Booker T. and the M.G.'s (1992):__ "Green Onions"
Earl Bostic: "Flamingo"
*__David Bowie (1996):__ "Space Oddity"
Boyz II Men: "I'll Make Love to You"
Toni Braxton: "Un-Break My Heart"
*__James Brown (1986):__ "Papa's Got a Brand New Bag"
*__Ruth Brown (1993):__ "Lucky Lips"
Jackson Browne: "Doctor My Eyes"
*__Buffalo Springfield (1997):__ "For What It's Worth"
Jimmy Buffet: "Margaritaville"
Bush: "Glycerine"
*__The Byrds (1991):__ "Turn! Turn! Turn!"
Mariah Carey: "Vision of Love"
The Cars: "Shake It Up"
*__Johnny Cash (1992):__ "I Walk the Line"
*__Ray Charles (1986):__ "Georgia on My Mind"
Cheap Trick: "Surrender"
Chubby Checker: "The Twist"
Chicago: "Saturday in the Park"
Eric Clapton: "Layla"

The Clash: "Rock the Casbah"
*__The Coasters (1987):__ "Yakety Yak"
*__Eddie Cochran (1987):__ "Summertime Blues"
Joe Cocker: "With a Little Help From My Friends"
Phil Collins: "Against All Odds"
*__Sam Cooke (1986):__ "You Send Me"
Coolio: "Gangsta's Paradise"
Alice Cooper: "School's Out"
Elvis Costello: "Alison"
*__Cream (1993):__ "Sunshine of Your Love"
*__Creedence Clearwater Revival (1993):__ "Proud Mary"
*__Crosby, Stills, and Nash (1997):__ "Suite: Judy Blue Eyes"
Sheryl Crow: "All I Want to Do"
The Cure: "Boys Don't Cry"
The Crystals: "Da Doo Ron Ron"
Cypress Hill: "Insane in the Brain"
Danny and the Juniors: "At the Hop"
*__Bobby Darin (1990):__ "Splish Splash"
Spencer Davis Group: "Gimme Some Lovin' "
Deep Purple: "Smoke on the Water"
Def Leppard: "Photograph"
Depeche Mode: "Strange Love"
*__Bo Diddley (1987):__ "Who Do You Love?"
*__Dion[1] and the Belmonts (1989):__ "A Teenager in Love"
Celine Dion: "Because You Loved Me"
Dire Straits: "Money for Nothing"
*__Fats Domino (1986):__ "Blueberry Hill"
Donovan: "Mellow Yellow"
The Doobie Brothers: "What a Fool Believes"
*__The Doors (1993):__ "Light My Fire"
*__The Drifters (1988):__ "Save the Last Dance for Me"
Duran Duran: "Hungry Like the Wolf"
*__Bob Dylan (1988):__ "Like a Rolling Stone"
*__The Eagles (1998):__ "Hotel California"
Earth, Wind, and Fire: "Shining Star"
*__Duane Eddy (1994):__ "Rebel-Rouser"
Emerson, Lake, and Palmer: "Lucky Man"
En Vogue: "Hold On"
The Eurythmics: "Sweet Dreams (Are Made of This)"
*__The Everly Brothers (1986):__ "Wake Up, Little Susie"
The Five Satins: "In the Still of the Night"
*__Fleetwood Mac (1998):__ *Rumours*
*__The Four Seasons (1990):__ "Sherry"
*__The Four Tops (1990):__ "I Can't Help Myself (Sugar Pie, Honey Bunch)"
*__Aretha Franklin (1987):__ "Respect"
Peter Gabriel: "Shock the Monkey"
Marvin Gaye (1987): "I Heard It Through the Grapevine"
Genesis: "No Reply at All"
Grand Funk Railroad: "We're an American Band"
Grand Master Flash and the Furious Five: "The Message"
*__The Grateful Dead (1994):__ "Uncle John's Band"

*Al Green (1995): "Let's Stay Together"
Greenday: "Time of Your Life"
The Guess Who: "American Woman"
Guns N' Roses: "Sweet Child o' Mine"
*Bill Haley[1] and His Comets (1987): "Rock Around the Clock"
Hall and Oates: "Kiss on My List"
Hanson: "MMMBop"
Heart: "Barracuda"
Hootie and the Blowfish: *Cracked Rear* View
Whitney Houston: "I Will Always Love You"
*The Impressions (1991): "For Your Precious Love"
INXS: "Need You Tonight"
*The Isley Brothers (1992): "It's Your Thing"
*The Jackson Five (1997): "ABC"
Janet Jackson: *Rhythm Nation*
Michael Jackson: *Thriller*
*Etta James (1993): "Tell Mama"
Tommy James & The Shondells: "Crimson and Clover"
Jay and the Americans: "This Magic Moment"
*Jefferson Airplane (1996): "White Rabbit"
Jethro Tull: *Aqualung*
Joan Jett: "I Love Rock 'n' Roll"
Jewel: "You Were Meant for Me"
*Billy Joel (1999): "Piano Man"
*Elton John (1994): "Candle in the Wind"
*Little Willie John (1996): "Sleep"
*Janis Joplin (1995): "Me and Bobby McGee"
K.C. and the Sunshine Band: "Get Down Tonight"
*B.B. King (1987): "The Thrill Is Gone"
Carole King: *Tapestry*
*The Kinks (1990): "You Really Got Me"
Kiss: "Rock 'n' Roll All Night"
*Gladys Knight and the Pips (1996): "Midnight Train to Georgia"
*Led Zeppelin (1995): "Stairway to Heaven"
Brenda Lee: "I'm Sorry"
*John Lennon (1994): "Imagine"
*Jerry Lee Lewis (1986): "Whole Lotta Shakin' Going On"
Little Anthony and the Imperials: "Tears on My Pillow"
*Little Richard (1986): "Tutti Frutti"
L. L. Cool J: "Mama Said Knock You Out"
The Lovin' Spoonful: "Summer in the City"
*Frankie Lymon and the Teenagers (1993): "Why Do Fools Fall in Love?"
Lynyrd Skynyrd: "Free Bird"
Madonna: "Material Girl"
*The Mamas and the Papas (1998): "Monday, Monday"
Marilyn Manson: "Beautiful People"
*Bob Marley (1994): *Exodus*
*Martha and the Vandellas (1995): "Dancin' in the Streets"
The Marvelettes: "Please, Mr. Postman"
Dave Matthews Band: "Don't Drink the Water"
*Curtis Mayfield (1999): "Superfly"
*Paul McCartney (1999): "Band on the Run"
Don McLean: "American Pie"
*Clyde McPhatter (1987): "A Lover's Question"
Meat Loaf: "Paradise by the Dashboard Light"
John (Cougar) Mellencamp: "Jack and Diane"
Men at Work: "Who Can It Be Now?"
Metallica: "Enter Sandman"
George Michael: "Faith"
*Joni Mitchell (1997): "Big Yellow Taxi"
The Monkees: "I'm a Believer"
Moody Blues: "Nights in White Satin"
Alanis Morissette: "Ironic"
*Van Morrison (1993): "Brown-Eyed Girl"
*Ricky Nelson (1987): "Hello, Mary Lou"
Nine Inch Nails: "Closer"
Nirvana: *Nevermind*
The Notorious B.I.G.: "Mo Money Mo Problems"
Oasis: "Wonderwall"
*Roy Orbison (1987): "Oh, Pretty Woman"
Ozzy Osbourne: "Crazy Train"
*Parliament/Funkadelic (1997): "One Nation Under a Groove"
Pearl Jam: "Jeremy"
*Carl Perkins (1987): "Blue Suede Shoes"
Peter, Paul, and Mary: "Leaving on a Jet Plane"
Tom Petty and the Heartbreakers: "Refugee"
*Wilson Pickett (1991): "Land of 1,000 Dances"
*Pink Floyd (1996): *The Wall*
*The Platters (1990): "The Great Pretender"
The Police: "Every Breath You Take"
Poco: "Crazy Love"
Iggy Pop: "Lust for Life"
*Elvis Presley (1986): "Love Me Tender"

The Pretenders: "Brass in Pocket"
*Lloyd Price (1998): "Stagger Lee"
Prince (The Artist): "Purple Rain"
Procol Harum: "A Whiter Shade of Pale"
Public Enemy: "Fight the Power"
Puff Daddy and the Family: *No Way Out*
Queen: "Bohemian Rhapsody"
The Ramones: "I Wanna Be Sedated"
*Otis Redding (1989): "(Sittin' on) the Dock of the Bay"
Red Hot Chili Peppers: "Under the Bridge"
*Jimmy Reed (1991): "Ain't That Loving You, Baby?"
Lou Reed: "Walk on the Wild Side"
R.E.M.: "Losing My Religion"
The Righteous Brothers: "You've Lost That Lovin' Feelin' "
Johnny Rivers: "Poor Side of Town"
*Smokey Robinson[1] and the Miracles (1987): "Shop Around"
*The Rolling Stones (1989): "Satisfaction"
The Ronettes: "Be My Baby"
Linda Ronstadt: "You're No Good"
Run-D.M.C.: "Raisin' Hell"
Salt-N-Pepa: "Shoop"
*Sam and Dave (1992): "Soul Man"
*Santana (1998): "Black Magic Woman"
Seal: "Kiss From a Rose"
Neil Sedaka: "Breaking Up Is Hard to Do"
The Sex Pistols: "Anarchy in the U.K."
Tupac Shakur: "How Do U Want It"
*Del Shannon (1999): "Runaway"
*The Shirelles (1996): "Soldier Boy"
Carly Simon: "You're So Vain"
Paul Simon: "50 Ways to Leave Your Lover"
*Simon and Garfunkel (1990): "Bridge Over Troubled Water"
*Sly and the Family Stone (1993): "Everyday People"
Smashing Pumpkins: "Today"
Patti Smith: "Because the Night"
Will Smith: "Gettin' Jiggy With It"
Sonic Youth: "Bull in the Heather"
Soundgarden: "Black Hole Sun"
Spice Girls: "Wannabe"
*Dusty Springfield (1999): "I Only Want to Be With You"
*Bruce Springsteen (1999): "Born to Run"
Squeeze: "Tempted"
*Staple Singers (1999): "I'll Take You There"
Steely Dan: "Rikki Don't Lose That Number"
Steppenwolf: "Born to Be Wild"
*Rod Stewart (1994): "Maggie Mae"
Sting: "If You Love Somebody, Set Them Free"
The Sugar Hill Gang: "Rapper's Delight"
Donna Summer: "Bad Girls"
*The Supremes (1988): "Stop! In the Name of Love"
Talking Heads: "Once in a Lifetime"
James Taylor: "You've Got a Friend"
*The Temptations (1989): "My Girl"
Three Dog Night: "Joy to the World"
TLC: "Waterfalls"
T. Rex: "Bang a Gong (Get It On)"
*Big Joe Turner (1987): "Shake, Rattle & Roll"
*Ike and Tina Turner (1991): "Proud Mary"
*Tina Turner (1991): "What's Love Got to Do With It?"
The Turtles: "Happy Together"
U2: "With or Without You"
Usher: "You Make Me Wanna"
Ritchie Valens: "La Bamba"
Van Halen: "Running With the Devil"
Stevie Ray Vaughan: "Crossfire"
*The Velvet Underground (1996): "Sweet Jane"
*Gene Vincent[1] (1998): "Be-Bop-A-Lula"
Tom Waits: "Downtown Train"
The Wallflowers: "One Headlight"
Dionne Warwick: "I Say a Little Prayer"
*Muddy Waters (1987): "I Can't Be Satisfied"
Mary Wells: "My Guy"
*The Who (1990): *Tommy*
*Jackie Wilson (1987): "That's Why"
*Stevie Wonder (1989): "You Are the Sunshine of My Life"
Wu-Tang Clan: "Protect Ya Neck"
*The Yardbirds (1992): "For Your Love"
Yes: "Roundabout"
*Neil Young (1995): "Down by the River"
*The Young Rascals/The Rascals (1997): "Good Lovin' "
*Frank Zappa[1]/Mothers of Invention (1995): *Sheik Yerbouti*
ZZ Top: "Legs"

* Inducted into Rock and Roll Hall of Fame as performer between 1986 and 1999; year is in parentheses. (1)Only individual performer is in Rock and Roll Hall of Fame.

Entertainers of the Present

living actors, musicians, dancers, singers, producers, directors, radio-TV performers

Name	Birthplace	Birthdate
Abbado, Claudio	Milan, Italy	6/26/33
Abdul, Paula	San Fernando, CA	6/19/62
Abraham, F. Murray	Pittsburgh, PA	10/24/39
Adams, Bryan	Kingston, Ontario	11/5/59
Adams, Don	New York, NY	4/19/26
Adams, Edie	Kingston, PA	4/16/29
Adams, Joey	New York, NY	1/6/11
Adams, Mason	New York, NY	2/26/19
Adjani, Isabelle	Paris, France	6/27/55
Affleck, Ben	Berkeley, CA	8/15/72
Agar, John	Chicago, IL	1/31/21
Agutter, Jenny	London, England	12/20/52
Aiello, Danny	New York, NY	6/20/33
Aimee, Anouk	Paris, France	4/27/34
Albanese, Licia	Bari, Italy	7/22/13
Alberghetti, Anna Maria	Pesaro, Italy	5/15/36
Albert, Eddie	Rock Island, IL	4/22/08
Albert, Marv	New York, NY	6/12/43
Alda, Alan	New York, NY	1/28/36
Alexander, Jane	Boston, MA	10/28/39
Alexander, Jason	Newark, NJ	9/23/59
Allen, Debbie	Houston, TX	1/16/50
Allen, Joan	Rochelle, IL	8/20/56
Allen, Karen	Carrollton, IL	10/5/51
Allen, Steve	New York, NY	12/26/21
Allen, Tim	Denver, CO	6/13/53
Allen, Woody	Brooklyn, NY	12/1/35
Alley, Kirstie	Wichita, KS	1/12/51
Allman, Gregg	Nashville, TN.	12/7/47
Allyson, June	New York, NY	10/7/17
Alonso, Maria Conchita	Cienfuegos, Cuba	6/29/57
Alpert, Herb	Los Angeles, CA	3/31/35
Altman, Robert	Kansas City, MO	2/20/25
Ames, Ed	Boston, MA	7/9/27
Amos, John	Newark, NJ	12/27/42
Amos, Tori	North Carolina	8/22/64
Anderson, Gillian	Chicago, IL	8/9/68
Anderson, Harry	Newport, RI	10/14/49
Anderson, Ian	Dunfermline, Scotland	8/10/47
Anderson, Kevin	Illinois	1/13/60
Anderson, Loni	St. Paul, MN	8/5/46
Anderson, Lynn	Grand Forks, ND	9/26/47
Anderson, Melissa Sue	Berkeley, CA	9/26/62
Anderson, Richard	Long Branch, NJ	8/8/26
Anderson, Richard Dean	Minneapolis, MN	1/23/50
Andersson, Bibi	Stockholm, Sweden	11/11/35
Andress, Ursula	Bern, Switzerland	3/19/36
Andrews, Anthony	London, England	1/12/48
Andrews, Julie	Walton, England	10/1/35
Andrews, Patty	Minneapolis, MN	2/16/20
Aniston, Jennifer	Sherman Oaks, CA	2/11/69
Anka, Paul	Ottawa, Ontario	7/30/41
Ann-Margret	Stockholm, Sweden	4/28/41
Antonioni, Michelangelo	Ferrara, Italy	9/29/12
Apple, Fiona	New York, NY	9/13/77
Applegate, Christina	Los Angeles, CA	11/25/72
Archer, Anne	Los Angeles, CA	8/25/47
Arkin, Adam	Brooklyn, NY	8/19/56
Arkin, Alan	New York, NY	3/26/34
Arnaz, Desi, Jr.	Los Angeles, CA	1/19/53
Arnaz, Lucie	Los Angeles, CA	7/17/51
Arness, James	Minneapolis, MN	5/26/23
Arnold, Eddy	Henderson, TN	5/15/18
Arnold, Tom	Ottumwa, IA	3/6/59
Arquette, Patricia	New York, NY	4/8/68
Arquette, Rosanna	New York, NY	8/10/59
Arroyo, Martina	New York, NY	2/2/37
Arthur, Beatrice	New York, NY	5/13/23
Ashley, Elizabeth	Ocala, FL	8/30/41
Asner, Ed	Kansas City, MO	11/15/29
Assante, Armand	New York, NY	10/4/49
Astin, John	Baltimore, MD	3/30/30
Atkins, Chet	Luttrell, TN	6/20/24
Atkinson, Rowan	Newcastle-Upon-Tyne, Eng.	1/6/55
Attenborough, Richard	Cambridge, England	8/29/23
Auberjonois, Rene	New York, NY	6/1/40
Aumont, Jean-Pierre	Paris, France	1/5/09
Austin, Patti	New York, NY	8/10/48
Autry, Alan	Shreveport, LA	7/31/52
Avalon, Frankie	Philadelphia, PA	9/18/39
Axton, Hoyt	Duncan, OK	3/25/38
Aykroyd, Dan	Ottawa, Ontario	7/1/52
Azaria, Hank	Forest Hills, NY	4/25/64
Aznavour, Charles	Paris, France	5/22/24
Babyface	Indianapolis, IN	4/10/59
Bacall, Lauren	New York, NY	9/16/24
Bacon, Kevin	Philadelphia, PA	7/8/58
Badu, Erykah	Dallas, TX	2/26/71
Baez, Joan	Staten Island, NY	1/9/41
Bain, Conrad	Lethbridge, Alberta	2/4/23
Baio, Scott	Brooklyn, NY	9/22/61
Baker, Anita	Toledo, OH	1/26/58
Baker, Carroll	Johnstown, PA	5/28/31
Baker, Diane	Hollywood, CA	2/25/38
Baker, Joe Don	Groesbeck, TX	2/12/36
Baker, Kathy	Midland, TX.	6/8/50
Bakula, Scott	St. Louis, MO	10/9/55
Baldwin, Alec	Massapequa, NY	4/3/58
Baldwin, Daniel	Massapequa, NY	10/5/60
Baldwin, Stephen	Massapequa, NY	5/12/66
Baldwin, William	Massapequa, NY	2/21/63
Ballard, Kaye	Cleveland, OH	11/20/26
Bancroft, Anne	New York, NY	9/17/31
Banderas, Antonio	Málaga, Spain	8/10/60
Banks, Tyra	Los Angeles, CA	12/4/73
Bannon, Jack	Los Angeles, CA	6/14/40
Baranski, Christine	Buffalo, NY	5/2/52
Barbeau, Adrienne	Sacramento, CA	6/11/45
Bardot, Brigitte	Paris, France	9/28/34
Barker, Bob	Darrington, WA	12/12/23
Barkin, Ellen	New York, NY	4/16/55
Barrie, Barbara	Chicago, IL	5/23/31
Barry, Gene	New York, NY	6/14/19
Barty, Billy	Millsboro, PA	10/25/24
Barrymore, Drew	Los Angeles, CA	2/22/75
Bartoli, Cecilia	Rome, Italy	6/4/66
Baryshnikov, Mikhail	Riga, Latvia	1/28/48
Basinger, Kim	Athens, GA	12/8/53
Bass, Lance	Mississippi	5/4/79
Bassett, Angela	New York, NY	8/16/58
Bassey, Shirley	Cardiff, Wales	1/8/37
Bateman, Jason	Rye, NY	1/14/69
Bateman, Justine	Rye, NY	2/19/66
Bates, Alan	Allestree, England	2/17/34
Bates, Kathy	Memphis, TN.	6/28/48
Battle, Kathleen	Portsmouth, OH	8/13/48
Baxter, Meredith	Los Angeles, CA	6/21/47
Bean, Orson	Burlington, VT	7/22/28
Beatty, Ned	Louisville, KY	7/6/37
Beatty, Warren	Richmond, VA	3/30/37
Beck (Hansen)	Los Angeles, CA	7/8/70
Beck, Jeff	Surrey, England	6/24/44
Beck, John	Chicago, IL	1/28/43
Bedelia, Bonnie	New York, NY	3/25/48
Begley, Ed, Jr.	Los Angeles, CA	9/16/49
Belafonte, Harry	New York, NY	3/1/27
Bel Geddes, Barbara	New York, NY	10/31/22
Bello, Maria	Norristown, PA	4/18/67
Belmondo, Jean-Paul	Neuilly-sur-Seine, France	4/9/33
Belushi, Jim	Chicago, IL	6/15/54
Belzer, Richard	Bridgeport, CT	8/4/44
Benatar, Pat	Brooklyn, NY	1/10/53
Benedict, Dirk	Helena, MT.	3/1/45
Benigni, Roberto	Misericordia, Italy	10/27/52
Bening, Annette	Topeka, KS	5/29/58
Benjamin, Richard	New York, NY	5/22/38
Bennett, Tony	New York, NY	8/3/26
Benson, George	Pittsburgh, PA	3/22/43
Benson, Robby	Dallas, TX	1/21/56
Berenger, Tom	Chicago, IL	5/31/50
Bergen, Candice	Beverly Hills, CA	5/9/46
Bergen, Polly	Knoxville, TN.	7/14/30
Bergman, Ingmar	Uppsala, Sweden	7/14/18
Berle, Milton	New York, NY	7/12/08
Berlinger, Warren	Brooklyn, NY	8/31/37
Berman, Lazar	Leningrad, Russia	2/26/30
Berman, Shelley	Chicago, IL	2/3/26
Bernard, Crystal	Dallas, TX	9/30/64
Bernhard, Sandra	Flint, MI	6/6/55
Bernsen, Corbin	N. Hollywood, CA	9/7/54
Berry, Chuck	St. Louis, MO	10/18/26
Berry, Halle	Cleveland, OH	8/14/68
Berry, Ken	Moline, IL	11/3/33
Bertinelli, Valerie	Wilmington, DE	4/23/60
Bialik, Mayim	San Diego, CA	12/12/75
Bertolucci, Bernardo	Parma, Italy	3/16/41
Bikel, Theodore	Vienna, Austria	5/2/24
Billingsley, Barbara	Los Angeles, CA	12/22/22
Binoche, Juliette	Paris, France	4/9/64
Birney, David	Washington, DC	4/23/39
Bishop, Joey	Bronx, NY	2/3/18
Bisset, Jacqueline	Weybridge, England	9/13/44
Bissett, Josie	Seattle, WA	10/5/69

Name	Birthplace	Birthdate	Name	Birthplace	Birthdate
Björk (Gudmundsdottir)	Rheinberg, Iceland	10/21/66	Buttons, Red	New York, NY	2/5/19
Black, Clint	Katy, TX	2/4/62	Buzzi, Ruth	Westerly, RI	7/24/36
Black, Karen	Park Ridge, IL	7/1/42	Byrne, David	Dumbarton, Scotland	5/14/52
Blades, Ruben	Panama City, Panama	7/16/48	Byrne, Gabriel	Dublin, Ireland	5/12/50
Blair, Janet	Altoona, PA	4/23/21			
Blair, Linda	St. Louis, MO	1/22/59	Caan, James	New York, NY	3/26/39
Blair, Selma	Southfield, MI	6/23/72	Caballe, Montserrat	Barcelona, Spain	4/12/33
Blake, Robert	Nutley, NJ	9/18/33	Caesar, Sid	Yonkers, NY	9/8/22
Blanchett, Cate	Melbourne, Australia	1969	Cage, Nicolas	Long Beach, CA	1/7/64
Bledsoe, Tempestt	Chicago, IL	8/1/73	Cain, Dean	Mt. Clemens, MI	7/31/66
Blethyn, Brenda	Kent, England	2/20/46	Caine, Michael	London, England	3/14/33
Blige, Mary J.	Bronx, NY	1/11/71	Caldwell, Sarah	Maryville, MO	3/6/24
Bloom, Claire	London, England	2/15/31	Caldwell, Zoe	Melbourne, Australia	9/14/33
Blyth, Ann	Mt. Kisco, NY	8/16/28	Cameron, James	Kapuskasiny, Ontario	8/16/54
Bochco, Steven	New York, NY	12/16/43	Cameron, Kirk	Panorama City, CA	10/12/70
Bogdanovich, Peter	Kingston, NY	7/30/39	Camp, Hamilton	London, England	10/30/34
Bogosian, Eric	Boston, MA	4/24/53	Campanella, Joseph	New York, NY	11/21/27
Bologna, Joseph	Brooklyn, NY	12/30/38	Campbell, Bruce	Royal Oak, MI	6/22/58
Bolton, Michael	New Haven, CT	2/26/53	Campbell, Glen	Billstown, AR	4/22/36
Bonet, Lisa	San Francisco, CA	11/16/67	Campbell, Naomi	London, England	5/22/70
Bonham Carter, Helena	London, England	5/23/66	Campbell, Neve	Toronto, Ontario	10/3/73
Bon Jovi, Jon	Sayreville, NJ	3/2/62	Campion, Jane	Wellington, New Zealand	1955
Bono (Vox)	Dublin, Ireland	5/10/60	Cannell, Stephen J.	Los Angeles, CA	2/5/42
Boone, Debby	Hackensack, NJ	9/22/56	Cannon, Dyan	Tacoma, WA	1/4/37
Boone, Pat	Jacksonville, FL	6/1/34	Capshaw, Kate	Ft. Worth, TX	11/3/53
Boreanaz, David	Buffalo, NY	5/16/71	Cardinale, Claudia	Tunis, Tunisia	4/15/39
Borge, Victor	Copenhagen, Denmark	1/3/09	Carey, Drew	Cleveland, OH	5/23/58
Borgnine, Ernest	Hamden, CT	1/24/17	Carey, Mariah	Huntington, NY	3/27/70
Bosson, Barbara	Charleroi, PA	11/1/39	Cariou, Len	Winnipeg, Canada	9/30/39
Bosco, Philip	Jersey City, NJ	9/26/30	Carlin, George	New York, NY	5/12/37
Bosley, Tom	Chicago, IL	10/1/27	Carlisle Hart, Kitty	New Orleans, LA	9/3/15
Bostwick, Barry	San Mateo, CA	2/24/45	Carlyle, Robert	Glasgow, Scotland	4/14/61
Bottoms, Timothy	Santa Barbara, CA	8/30/51	Carmen, Eric	Cleveland, OH	8/11/49
Bowie, David	London, England	1/8/47	Carney, Art	Mt. Vernon, NY	11/4/18
Boxleitner, Bruce	Elgin, IL	5/12/50	Carpenter, John	Carthage, NY	1/16/48
Boy George	London, England	6/14/61	Carpenter, Mary Chapin	Princeton, NJ	2/21/58
Boyle, Peter	Philadelphia, PA	10/18/33	Caron, Leslie	Boulogne, France	7/1/31
Bracco, Lorraine	Brooklyn, NY	10/2/55	Carr, Vikki	El Paso, TX	7/19/41
Bracken, Eddie	New York, NY	2/7/20	Carradine, David	Hollywood, CA	10/8/36
Branagh, Kenneth	Belfast, N. Ireland	12/10/60	Carradine, Keith	San Mateo, CA	8/8/49
Brandauer, Klaus Maria	Steiermark, Austria	6/22/44	Carreras, Jose	Barcelona, Spain	12/5/46
Brando, Marlon	Omaha, NE	4/3/24	Carrere, Tia	Honolulu, HI	1/2/66
Brandy (Norwood)	McComb, MS	2/11/79	Carrey, Jim	Toronto, Ontario	1/17/62
Braschi, Nicoletta	Gesena, Italy	1960	Carroll, Diahann	Bronx, NY	7/17/35
Braugher, Andre	Chicago, Il.	7/1/62	Carroll, Pat	Shreveport, LA	5/5/27
Braxton, Toni	Severn, MD	10/7/66	Carson, Johnny	Corning, IA	10/23/25
Brennan, Eileen	Los Angeles, CA	9/3/35	Carson, Lisa Nicole	Brooklyn, NY	7/12/69
Brenner, David	Philadelphia, PA	2/4/45	Carter, Benny	New York, NY	8/8/07
Brewer, Teresa	Toledo, OH	5/7/31	Carter, Dixie	McLemoresville, TN	5/25/39
Bridges, Beau	Hollywood, CA	12/9/41	Carter, Jack	New York, NY	6/24/23
Bridges, Jeff	Los Angeles, CA	12/4/49	Carter, June	Maces Spring, VA	6/23/29
Brightman, Sarah	Berkhamstead, England	8/14/60	Carter, Lynda	Phoenix, AZ	7/24/51
Brimley, Wilford	Salt Lake City, UT	9/27/34	Carter, Nell	Birmingham, AL	9/13/48
Brinkley, Christie	Malibu, CA	2/2/54	Carter, Ron	Royal Oak Twp, MI	5/4/37
Broderick, Matthew	New York, NY	3/21/62	Carter, Nick	Jamestown, NY	1/28/80
Brolin, James	Los Angeles, CA	7/18/40	Cartwright, Nancy	Ohio	1959
Bronson, Charles	Ehrenfeld, PA	11/3/22	Caruso, David	Forest Hills, NY	1/17/56
Brooks, Albert	Beverly Hills, CA	7/22/47	Carvey, Dana	Missoula, MT.	4/2/55
Brooks, Foster	Louisville, KY	5/11/12	Casadesus, Gaby	Marseilles, France	8/9/01
Brooks, Garth	Tulsa, OK	2/7/62	Cash, Johnny	Kingsland, AR	2/26/32
Brooks, James L	North Bergen, NJ	5/9/40	Cash, Rosanne	Memphis, TN.	5/24/55
Brooks, Mel	New York, NY	6/28/26	Cassidy, David	New York, NY	4/12/50
Brosnan, Pierce	Co. Meath, Ireland	5/16/53	Castellaneta, Dan	Chicago, IL	1958
Brown, Blair	Washington, DC	1948	Cates, Phoebe	New York, NY	7/16/63
Brown, Bobby	Boston, MA	2/5/69	Cathbert, Lacey	Purvis, MS.	9/30/82
Brown, Bryan	Sydney, Australia	6/23/47	Cavett, Dick	Gibbon, NE	11/19/36
Brown, James	Pulaski, TN (?)	6/17/28(?)	Chamberlain, Richard	Beverly Hills, CA	3/31/35
Brown, Les	Reinerton, PA	3/14/12	Chan, Jackie	Hong Kong	4/7/54
Browne, Jackson	Heidelberg, Germany	10/9/48	Channing, Carol	Seattle, WA	1/31/23
Browne, Roscoe Lee	Woodbury, NJ	5/2/25	Channing, Stockard	New York, NY	2/13/44
Brubeck, Dave	Concord, CA	12/6/20	Chaplin, Geraldine	Santa Monica, CA.	7/31/44
Bryson, Peabo	Greenville, SC	4/13/51	Chapman, Tracy	Cleveland, OH	3/30/64
Buckley, Betty	Ft. Worth, TX	7/3/47	Charisse, Cyd	Amarillo, TX	3/8/21
Buffett, Jimmy	Pascagoula, MS	12/25/46	Charles, Ray	Albany, GA	9/23/30
Bujold, Genevieve	Montreal, Quebec	7/1/42	Charo	Murcia, Spain	1/15/51
Bullock, Sandra	Arlington, VA	7/26/64	Chase, Chevy	New York, NY	10/8/43
Bumbry, Grace	St. Louis, MO	1/4/37	Chasez, Joshua (J.C.)	Washington, DC	8/8/76
Burghoff, Gary	Bristol, CT	5/24/40	Cheadle, Don	Kansas City, MO	11/29/64
Burke, Delta	Orlando, FL	7/30/56	Checker, Chubby	Philadelphia, PA	10/3/41
Burnett, Carol	San Antonio, TX	4/26/33	Cher	El Centro, CA	5/20/46
Burns, Edward	Valley Stream, NY	1/29/68	Chiklis, Michael	Lowell, MA	8/30/63
Burrows, Darren E.	Winfield, KS	9/12/66	Chong, Rae Dawn	Vancouver, Canada	2/28/62
Burstyn, Ellen	Detroit, MI	12/7/32	Chong, Thomas	Edmonton, Alberta	5/24/38
Burton, LeVar	Landstuhl, W Germany	2/16/57	Chow Yun-Fat	Hong Kong	5/18/55
Burton, Tim	Burbank, CA	8/25/58	Christensen, Helena	Copenhagen, Denmark	12/25/68
Buscemi, Steve	Brooklyn, NY	12/13/57	Christie, Julie	Assam, India	4/14/40
Busey, Gary	Goose Creek, TX	6/29/44	Christopher, William	Evanston, IL	10/20/32
Busfield, Timothy	Lansing, MI	6/12/57	Church, Thomas Haden	El Paso, TX	6/17/61
Butler, Brett	Montgomery, AL	1/30/58	Clapton, Eric	Surrey, England	3/30/45

Name	Birthplace	Birthdate	Name	Birthplace	Birthdate
Clark, Dick	Mt. Vernon, NY	11/30/29	Dale, Jim	Rothwell, England	8/15/35
Clark, Petula	Ewell, Surrey, England	11/15/32	Dalton, Abby	Las Vegas, NV	8/15/32
Clark, Roy	Meherrin, VA	4/15/33	Dalton, Timothy	Colwyn Bay, Wales	3/21/44
Clay, Andrew Dice	Brooklyn, NY	9/29/58	Daltrey, Roger	London, England	3/1/44
Clayburgh, Jill	New York, NY	4/30/44	Daly, Timothy	Suffern, NY	3/1/58
Cleese, John	Weston-Super-Mare, Eng.	10/27/39	Daly, Tyne	Madison, WI	2/21/47
Cliburn, Van	Shreveport, LA	7/12/34	Damon, Matt	Cambridge, MA	10/8/70
Clooney, George	Lexington, KY	5/6/61	Damone, Vic	Brooklyn, NY	6/12/28
Clooney, Rosemary	Maysville, KY	5/23/28	Danes, Claire	New York, NY	4/12/79
Close, Glenn	Greenwich, CT	3/19/47	D'Angelo, Beverly	Columbus, OH	11/15/54
Coburn, James	Laurel, NE	8/31/28	Dangerfield, Rodney	Babylon, NY	11/22/21
Coca, Imogene	Philadelphia, PA	11/18/08	Daniels, Charlie	Wilmington, NC	10/28/36
Coen, Ethan	St. Louis Park, MN	9/21/57	Daniels, Jeff	Georgia	2/19/55
Coen, Joel	St. Louis Park, MN	11/29/54	Daniels, William	Brooklyn, NY	3/31/27
Cole, Gary	Park Ridge, IL	9/20/57	Danner, Blythe	Philadelphia, PA	2/3/44
Cole, Natalie	Los Angeles, CA	2/6/50	Danson, Ted	San Diego, CA	12/29/47
Cole, Olivia	Memphis, TN	11/26/42	Danza, Tony	New York, NY	4/21/51
Cole, Paula	Manchester, CT	4/5/68	Darby, Kim	Hollywood, CA	7/8/48
Coleman, Dabney	Austin, TX	1/3/32	David, Larry	Brooklyn, NY	1947
Coleman, Gary	Zion, IL	2/8/68	Davidson, John	Pittsburgh, PA	12/13/41
Coleman, Ornette	Fort Worth, TX	3/9/30	Davis, Ann B.	Schenectady, NY	5/5/26
Collins, Joan	London, England	5/23/33	Davis, Clifton	Chicago, IL	10/4/45
Collins, Judy	Seattle, WA	5/1/39	Davis, Geena	Wareham, MA	1/21/57
Collins, Pauline	Exmouth, England	9/3/40	Davis, Judy	Perth, Australia	1955
Collins, Phil	London, England	1/30/51	Davis, Mac	Lubbock, TX	1/21/42
Colvin, Shawn	Vermillion, SD	1/10/56	Davis, Ossie	Cogdell, GA	12/18/17
Combs, Sean "Puffy"	Harlem, NY	11/9/69	Dawber, Pam	Farmington Hills, MI	10/18/51
Comden, Betty	Brooklyn, NY	5/3/19	Dawson, Richard	Hampshire, England	11/20/32
Como, Perry	Canonsburg, PA	5/18/12	Day, Doris	Cincinnati, OH	4/3/24
Connery, Sean	Edinburgh, Scotland	8/25/30	Day, Laraine	Roosevelt, UT	10/13/20
Connick, Harry, Jr.	New Orleans, LA	9/11/67	Day-Lewis, Daniel	London, England	4/29/57
Conniff, Ray	Attleboro, MA	11/6/16	Dean, Jimmy	Plainview, TX	8/10/28
Connors, Mike	Fresno, CA	8/15/25	Dearie, Blossom	E. Durham, NY	4/28/26
Conrad, Robert	Chicago, IL	3/1/35	De Camp, Rosemary	Prescott, AZ	11/14/10
Constantine, Michael	Reading, PA	5/22/27	DeCarlo, Yvonne	Vancouver, BC	9/1/22
Conti, Tom	Paisley, Scotland	11/22/41	Dee, Frances	Los Angeles, CA	11/26/07
Conway, Tim	Willoughby, OH	12/15/33	Dee, Ruby	Cleveland, OH	10/27/23
Cook, Barbara	Atlanta, GA	10/25/27	Dee, Sandra	Bayonne, NJ	4/23/42
Cooke, Alistair	Manchester, England	11/20/08	DeFranco, Buddy	Camden, NJ	2/17/23
Coolidge, Rita	Nashville, TN	5/1/45	DeGeneres, Ellen	Metairie, LA	1/26/58
Coolio	Los Angeles, CA	8/1/63	DeHaven, Gloria	Los Angeles, CA	7/23/25
Cooper, Alice	Detroit, MI	2/4/48	De Havilland, Olivia	Tokyo, Japan	7/1/16
Cooper, Jackie	Los Angeles, CA	9/15/21	Delaney, Kim	Philadelphia, PA	11/29/64
Copperfield, David	Metuchen, NJ	9/16/56	Delany, Dana	New York, NY	3/11/56
Coppola, Francis Ford	Detroit, MI	4/7/39	DeLaurentiis, Dino	Torre Annunziata, Italy	8/8/19
Corbin, Barry	Lamesa, TX	10/16/40	Delon, Alain	Sceaux, France	11/8/35
Cord, Alex	New York, NY	8/3/31	DeLuise, Dom	Brooklyn, NY	8/1/33
Corea, Chick	Chelsea, MA	6/12/41	Demme, Jonathan	Rockville Centre, NY	2/22/44
Corelli, Franco	Ancona, Italy	4/8/23	DeMornay, Rebecca	Santa Rosa, CA	11/29/61
Corey, Jeff	New York, NY	8/10/14	Dench, Judi	York, England	12/9/34
Corley, Pat	Dallas, TX	6/1/30	Deneuve, Catherine	Paris, France	10/22/43
Cosby, Bill	Philadelphia, PA	7/12/37	De Niro, Robert	New York, NY	8/17/43
Costas, Bob	New York, NY	3/22/52	Dennehy, Brian	Bridgeport, CT	7/9/38
Costello, Elvis	London, England	8/25/54	Denver, Bob	New Rochelle, NY	1/9/35
Costner, Kevin	Compton, CA	1/18/55	DePalma, Brian	Newark, NJ	9/11/40
Courtenay, Tom	Hull, England	2/25/37	Depardieu, Gerard	Chateauroux, France	12/27/48
Cox, Courteney	Birmingham, AL	6/15/64	Depp, Johnny	Owensboro, KY	6/9/63
Coyote, Peter	New York, NY	10/10/42	Derek, Bo	Long Beach, CA	11/20/56
Cox, Ronny	Cloudcroft, NM	8/23/38	Dern, Bruce	Chicago, IL	6/4/36
Crain, Jeanne	Barstow, CA	5/25/25	Dern, Laura	Santa Monica, CA	2/1/67
Crawford, Cindy	DeKalb, IL	2/20/66	Devane, William	Albany, NY	9/5/37
Crawford, Michael	Salisbury, England	1/19/42	DeVito, Danny	Neptune, NJ	11/17/44
Crenna, Richard	Los Angeles, CA	11/30/26	DeWitt, Joyce	Wheeling, WV	4/23/49
Crespin, Regine	Marseilles, France	2/23/26	Dey, Susan	Pekin, IL	12/10/52
Cronyn, Hume	London, Ontario	7/18/11	Diamond, Neil	Brooklyn, NY	1/24/41
Crosby, David	Los Angeles, CA	8/14/41	Diaz, Cameron	San Diego, CA	8/30/72
Cross, Ben	London, England	12/16/47	DiCaprio, Leonardo	Los Angeles, CA	11/11/74
Crouse, Lindsay	New York, NY	5/12/48	Dick, Andy	Charleston, SC	12/21/66
Crow, Sheryl	Kennett, MO	2/11/63	Dickinson, Angie	Kulm, ND	9/30/31
Crowe, Cameron	Palm Springs, CA	7/13/57	Diddley, Bo	McComb, MS	12/20/28
Crowe, Russell	New Zealand	4/7/64	Diller, Phyllis	Lima, OH	7/17/17
Crowell, Rodney	Houston, TX	8/17/50	Dillman, Bradford	San Francisco, CA	4/14/30
Cruise, Tom	Syracuse, NY	7/3/62	Dion, Celine	Charlemagne, Quebec	3/30/68
Crystal, Billy	Long Beach, NY	3/14/47	Dillon, Matt	New Rochelle, NY	2/18/64
Culkin, Macaulay	New York, NY	8/26/80	Dobson, Kevin	New York, NY	3/18/44
Cullum, John	Knoxville, TN	3/2/30	Dogg, Snoop	Long Beach, CA	10/20/72
Culp, Robert	Oakland, CA	8/16/30	Doherty, Shannen	Memphis, TN	4/21/71
Cummings, Constance	Seattle, WA	5/15/10	Dolenz, Mickey	Los Angeles, CA	3/8/45
Curry, Tim	Cheshire, England	4/19/46	Domingo, Placido	Madrid, Spain	1/21/41
Curtin, Jane	Cambridge, MA	9/6/47	Domino, Fats	New Orleans, LA	2/26/28
Curtis, Jamie Lee	Los Angeles, CA	11/22/58	Donahue, Phil	Cleveland, OH	12/21/35
Curtis, Keene	Salt Lake City, UT	2/15/23	Donahue, Troy	New York, NY	1/27/36
Curtis, Tony	New York, NY	6/3/25	D'Onofrio, Vincent	Brooklyn, NY	6/30/59
Cusack, Joan	Evanston, IL	10/11/62	Donovan (Leitch)	Glasgow, Scotland	2/10/46
Cusack, John	Evanston, IL	6/28/66	Dorn, Michael	Luling, TX	12/5/52
Cyrus, Billy Ray	Flatwoods, KY	8/25/61	Dorough, Howie	Orlando, FL	8/22/73
			Dotrice, Roy	Guernsey, England	5/26/23
Dafoe, Willem	Appleton, WI	7/22/55	Douglas, Kirk	Amsterdam, NY	12/9/16
Dahl, Arlene	Minneapolis, MN	8/11/28	Douglas, Michael	New Brunswick, NJ	9/25/44

Name	Birthplace	Birthdate
Dow, Tony	Holywood, CA	3/17/45
Down, Lesley-Ann	London, England	3/17/54
Downey, Robert, Jr.	New York, NY	4/4/65
Downey, Roma	Derry, Northern Ireland	5/6/60
Downs, Hugh	Akron, OH	2/14/21
Drescher, Fran	Queens, NY	9/30/57
Drew, Ellen	Kansas City, MO	11/23/15
Dreyfuss, Richard	Brooklyn, NY	10/29/47
Driver, Minnie	London, England	1/31/71
Dryer, Fred	Hawthorne, CA	7/6/46
Duchovny, David	New York, NY	8/7/60
Duffy, Julia	Minneapolis, MN	6/27/51
Duffy, Patrick	Townsend, MT	3/17/49
Dukakis, Olympia	Lowell, MA	6/20/31
Duke, Patty	New York, NY	12/14/46
Dukes, David	San Francisco, CA	6/6/45
Dullea, Keir	Cleveland, OH	5/30/36
Dunaway, Faye	Bascom, FL	1/14/41
Duncan, Sandy	Henderson, TX	2/20/46
Dunham, Katherine	Joliet, IL	6/22/10
Dunne, Griffin	New York, NY	6/8/55
Dunst, Kirsten	New Jersey	4/30/82
Durbin, Deanna	Winnipeg, Manitoba	12/4/21
Durning, Charles	Highland Falls, NY	2/28/23
Dussault, Nancy	Pensacola, FL	6/30/36
Dutton, Charles S.	Baltimore, MD	1/30/51
Duvall, Robert	San Diego, CA	1/5/31
Duvall, Shelley	Houston, TX	7/7/49
Dylan, Bob	Duluth, MN	5/24/41
Dylan, Jakob	New York, NY	12/9/69
Dysart, Richard	Augusta, ME	3/30/29
Easton, Sheena	Bellshill, Scotland	4/27/59
Eastwood, Clint	San Francisco, CA	5/31/30
Ebert, Roger	Urbana, IL	6/18/42
Ebsen, Buddy	Belleville, IL	4/2/08
Eden, Barbara	Tucson, AZ	8/23/34
Edwards, Anthony	Santa Barbara, CA	7/19/63
Edwards, Blake	Tulsa, OK	7/26/22
Edwards, Ralph	Merino, CO	6/13/13
Eichhorn, Lisa	Reading, PA	2/4/52
Eikenberry, Jill	New Haven, CT	1/21/47
Ekberg, Anita	Malmo, Sweden	9/29/31
Ekland, Britt	Stockholm, Sweden	10/6/42
Elam, Jack	Miami, AZ	11/13/16
Electra, Carmen	Cincinnati, OH	4/20/73
Elfman, Jenna	Los Angeles, CA	9/30/71
Elizondo, Hector	New York, NY	12/22/36
Elliott, Bob	Boston, MA	3/26/23
Elliott, Chris	New York, NY	1960
Elliott, Sam	Sacramento, CA	8/9/44
Elvira	Manhattan, KS	9/17/51
Enberg, Dick	Auburn Hills, MI	1/5/35
Englund, Robert	Hollywood, CA	6/6/48
Enya	Gweedore, Ireland	5/17/61
Ephron, Nora	New York, NY	5/19/41
Estefan, Gloria	Havana, Cuba	9/1/57
Estevez, Emilio	New York, NY	5/12/62
Estrada, Erik	New York, NY	3/16/49
Etheridge, Melissa	Leavenworth, KS	5/29/61
Evans, Dale	Uvalde, TX	10/31/12
Evans, Linda	Hartford, CT	11/18/42
Evans, Robert	New York, NY	6/29/30
Everett, Chad	South Bend, IN	6/11/36
Everett, Rupert	Norfolk, England	5/29/59
Everly, Don	Brownie, KY	2/1/37
Everly, Phil	Chicago, IL	1/19/39
Evigan, Greg	South Amboy, NJ	10/14/53
Fabares, Shelley	Santa Monica, CA	1/19/42
Fabian (Forte)	Philadelphia, PA	2/6/43
Fabio	Milan, Italy	3/15/61
Fabray, Nanette	San Diego, CA	10/27/20
Fairbanks, Douglas, Jr.	New York, NY	12/9/09
Fairchild, Morgan	Dallas, TX	2/3/50
Falana, Lola	Philadelphia, PA	9/11/46
Falk, Peter	New York, NY	9/16/27
Farentino, James	Brooklyn, NY	2/24/38
Fargo, Donna	Mt. Airy, NC	11/10/45
Farina, Dennis	Chicago, IL	2/29/44
Farr, Jamie	Toledo, OH	7/1/34
Farrell, Eileen	Willimantic, CT	2/13/20
Farrell, Mike	St. Paul, MN	2/6/39
Farrow, Mia	Los Angeles, CA	2/9/45
Fatone, Joey	New York, NY	1/29/76
Faustino, David	California	3/3/74
Fawcett, Farrah	Corpus Christi, TX	2/2/47
Feinstein, Michael	Columbus, OH	9/7/56
Feldon, Barbara	Pittsburgh, PA	3/12/41
Feliciano, Jose	Lares, Puerto Rico	9/10/45
Feldshuh, Tovah	New York, NY	12/27/53
Fenn, Sherilyn	Detroit, MI	2/1/65
Ferrell, Conchata	Charleston, WV	3/28/43
Ferrer, Mel	Elberon, NJ	8/25/17
Fiedler, John	Platteville, WI	2/3/25
Field, Sally	Pasadena, CA	11/6/46
Fiennes, Joseph	Salisbury, England	5/27/70
Fiennes, Ralph	Suffolk, England	12/22/62
Finney, Albert	Salford, England	5/9/36
Fiorentino, Linda	Philadelphia, PA	3/9/60
Firth, Colin	Grayshott, England	9/10/60
Firth, Peter	Yorkshire, England	10/27/53
Fischer-Dieskau, Dietrich	Berlin, Germany	5/28/25
Fishburne, Laurence	Augusta, GA	7/30/61
Fisher, Carrie	Beverly Hills, CA	10/21/56
Fisher, Eddie	Philadelphia, PA	8/10/28
Fitzgerald, Geraldine	Dublin, Ireland	11/24/13
Flack, Roberta	Black Mountain, NC	2/10/39
Flanagan, Fionnula	Dublin, Ireland	12/10/41
Fleming, Rhonda	Hollywood, CA	8/10/23
Fletcher, Louise	Birmingham, AL	7/22/34
Flockhart, Calista	Freeport, IL	11/11/64
Foch, Nina	Leyden, Netherlands	4/20/24
Fogelberg, Dan	Peoria, IL	8/13/51
Fogerty, John	Berkeley, CA	5/28/45
Foley, Dave	Toronto, Ontario	1/4/63
Fonda, Bridget	Los Angeles, CA	1/27/64
Fonda, Jane	New York, NY	12/21/37
Fonda, Peter	New York, NY	2/23/40
Fontaine, Joan	Tokyo, Japan	10/22/17
Ford, Faith	Alexandria, LA	9/14/64
Ford, Glenn	Quebec, Canada	5/1/16
Ford, Harrison	Chicago, IL	7/13/42
Forman, Milos	Caslav, Czechoslovakia	2/18/32
Forsythe, John	Penns Grove, NJ	1/29/18
Foster, Jodie	New York, NY	11/19/62
Fox, James	London, England	5/19/39
Fox, Matthew	Crowheart, WY	7/14/66
Fox, Michael J.	Edmonton, Alberta	6/9/61
Fox, Vivica A.	Indianapolis, IN	7/30/64
Foxworth, Robert	Houston, TX	11/1/41
Foxworthy, Jeff	Atlanta, GA	9/6/57
Foxx, Jamie	Terrell, TX	12/13/67
Frampton, Peter	Kent, England	4/22/50
Franciosa, Anthony	New York, NY	10/25/28
Francis, Anne	Ossining, NY	9/16/30
Francis, Arlene	Boston, MA	10/20/08
Francis, Connie	Newark, NJ	12/12/38
Franken, Al	New York, NY	5/21/51
Frankenheimer, John	Malba, NY	2/19/30
Franklin, Aretha	Memphis, TN	3/25/42
Franklin, Bonnie	Santa Monica, CA	1/6/44
Franz, Dennis	Maywood, IL	10/28/44
Fraser, Brendan	Indianapolis, IN	12/3/67
Freeman, Al, Jr.	San Antonio, TX	3/21/34
Freeman, Mona	Baltimore, MD	6/9/26
Freeman, Morgan	Memphis, TN	6/1/37
Fricker, Brenda	Dublin, Ireland	2/17/45
Friedkin, William	Chicago, IL	8/29/35
Frost, David	Tenterden, England	4/7/39
Fry, Stephen	London, England	8/24/57
Fuentes, Daisy	Havana, Cuba	11/17/66
Funicello, Annette	Utica, NY	10/22/42
Gabor, Zsa Zsa	Budapest, Hungary	2/6/17
Gabriel, John	Niagara Falls, NY	5/25/31
Gabriel, Peter	London, England	2/13/50
Galway, James	Belfast, Ireland	12/8/39
Garagiola, Joe	St. Louis, MO	2/12/26
Garcia, Andy	Havana, Cuba	4/12/56
Garofalo, Janeane	New Jersey	9/28/64
Garfunkel, Art	New York, NY	11/5/41
Garland, Beverly	Santa Cruz, CA	10/17/26
Garner, James	Norman, OK	4/7/28
Garr, Teri	Lakewood, OH	12/11/45
Garrett, Betty	St. Joseph, MO	5/23/19
Garth, Jennie	Champaign, IL	4/3/72
Gatlin, Larry	Seminole, TX	5/2/48
Gavin, John	Los Angeles, CA	4/8/28
Gayle, Crystal	Paintsville, KY	1/9/51
Gaynor, Mitzi	Chicago, IL	9/4/30
Gazzara, Ben	New York, NY	8/28/30
Geary, Anthony	Coalville, UT	5/29/47
Geary, Cynthia	Jackson, MS	3/21/66
Gedda, Nicolai	Stockholm, Sweden	7/11/25
Gellar, Sarah Michelle	New York, NY	4/14/77
Gere, Richard	Philadelphia, PA	8/31/49
Getty, Estelle	New York, NY	7/25/24

Name	Birthplace	Birthdate
Ghostley, Alice	Eve, MO	8/14/26
Giannini, Giancarlo	Spezia, Italy	8/1/42
Gibb, Barry	Isle of Man, England	9/1/46
Gibb, Maurice	Manchester, England	12/22/49
Gibb, Robin	Manchester, England	12/22/49
Gibbons, Leeza	South Carolina	3/26/57
Gibbs, Marla	Chicago, IL	6/14/31
Gibson, Deborah	New York, NY	8/31/70
Gibson, Henry	Germantown, PA	9/21/35
Gibson, Mel	Peekskill, NY	1/3/56
Gibson, Thomas	Charleston, SC	7/3/62
Gielgud, John	London, England	4/14/04
Gifford, Frank	Santa Monica, CA	8/16/30
Gifford, Kathie Lee	Paris, France	8/16/53
Gilbert, Sara	Santa Monica, CA	1/29/75
Gilbert, Melissa	Los Angeles, CA	5/8/64
Gilberto, Astrud	Salvador, Brazil	3/30/40
Gill, Vince	Norman, OK	4/12/57
Gillette, Anita	Baltimore, MD	8/16/38
Gilley, Mickey	Natchez, MS	3/9/36
Gilpin, Peri	Waco, TX	5/27/63
Ginty, Robert	New York, NY	11/14/48
Givens, Robin	New York, NY	11/27/64
Glaser, Paul Michael	Cambridge, MA	3/25/42
Glenn, Scott	Pittsburgh, PA	1/26/42
Gless, Sharon	Los Angeles, CA	5/31/43
Glover, Crispin	New York, NY	9/20/64
Glover, Danny	San Francisco, CA	7/22/47
Glover, Savion	Newark, NJ	1973
Godard, Jean Luc	Paris, France	12/3/30
Goldberg, Whoopi	New York, NY	11/13/49
Goldblum, Jeff	Pittsburgh, PA	10/22/52
Goldthwait, Bobcat	Syracuse, NY	5/1/62
Goldwyn, Tony	Los Angeles, CA	5/20/60
Gooding, Cuba, Jr.	Bronx, NY	1/2/68
Goodman, John	St. Louis, MO	6/20/52
Gordon-Levitt, Joseph	Los Angeles, CA	2/17/81
Gorme, Eydie	Bronx, NY	8/16/32
Gorshin, Frank	Pittsburgh, PA	4/5/34
Gossett, Louis, Jr.	Brooklyn, NY	5/27/36
Gould, Elliott	Brooklyn, NY	8/29/38
Gould, Harold	Schenectady, NY	12/10/23
Goulet, Robert	Lawrence, MA	11/26/33
Gowdy, Curt	Green River, WY	7/31/19
Graham, Heather	Milwaukee, WI	1/29/70
Graham, Virginia	Chicago, IL	7/4/12
Grammer, Kelsey	St. Thomas, Virgin Isl.	2/20/55
Granger, Farley	San Jose, CA	7/1/25
Grant, Amy	Augusta, GA	12/25/60
Grant, Hugh	London, England	9/9/60
Grant, Lee	New York, NY	10/31/29
Graves, Peter	Minneapolis, MN	3/18/26
Gray, Linda	Santa Monica, CA	9/12/40
Gray, Spaulding	Barrington, RI	6/5/41
Grayson, Kathryn	Winston-Salem, NC	2/9/22
Greco, Jose	Abruzzi, Italy	12/23/18
Green, Adolph	New York, NY	12/2/15
Green, Al	Forrest City, AR	4/13/46
Green, Seth	Philadelphia, PA	2/8/74
Greene, Shecky	Chicago, IL	4/8/26
Greenwood, Bruce	Quebec, Canada	8/12/56
Greer, Jane	Washington, DC	9/9/24
Gregory, Cynthia	Los Angeles, CA	7/8/46
Gregory, Dick	St. Louis, MO	10/12/32
Gregory, James	Bronx, NY	12/23/11
Grey, Jennifer	New York, NY	3/22/60
Grey, Joel	Cleveland, OH	4/11/32
Grier, David Alan	Detroit, MI	6/30/55
Grier, Pam	Winston-Salem, NC	5/26/49
Griffin, Merv	San Mateo, CA	7/6/25
Griffith, Andy	Mount Airy, NC	6/1/26
Griffith, Melanie	New York, NY	8/9/57
Grimes, Tammy	Lynn, MA	1/30/34
Grizzard, George	Roanoke Rapids, NC	4/1/28
Grodin, Charles	Pittsburgh, PA	4/21/35
Grosbard, Ulu	Antwerp, Belgium	1/19/29
Gross, Michael	Chicago, IL	6/21/47
Guest, Christopher	New York, NY	2/5/48
Guillaume, Robert	St. Louis, MO	11/30/37
Guinness, Alec	London, England	4/2/14
Gumbel, Greg	New Orleans, LA	5/3/46
Guthrie, Arlo	New York, NY	7/10/47
Guttenberg, Steve	New York, NY	8/24/58
Guy, Buddy	Lettsworth, LA	7/30/36
Guy, Jasmine	Boston, MA	3/10/64
Hackett, Buddy	Brooklyn, NY	8/31/24
Hackman, Gene	San Bernardino, CA	1/30/30
Hagen, Uta	Gottingen, Germany	6/12/19
Haggard, Merle	Bakersfield, CA	4/6/37
Hagman, Larry	Weatherford, TX	9/21/31
Haid, Charles	San Francisco, CA	6/2/44
Haines, Connie	Savannah, GA	1/20/22
Hale, Barbara	DeKalb, IL	4/18/22
Hall, Arsenio	Cleveland, OH	2/12/55
Hall, Daryl	Pottstown, PA	10/11/48
Hall, Deidre	Milwaukee, WI	10/31/48
Hall, Monty	Winnipeg, Manitoba	8/25/25
Hall, Tom T.	Olive Hill, KY	5/25/36
Halliwell, Geri	Hertfordshire, England	8/6/72
Hamill, Mark	Oakland, CA	9/25/51
Hamilton, George	Memphis, TN	8/12/39
Hamilton, Linda	Salisbury, MD	9/26/56
Hamlin, Harry	Pasadena, CA	10/30/51
Hammer	Oakland, CA	3/29/63
Hampton, Lionel	Louisville, KY	4/20/08
Hancock, Herbie	Chicago, IL	4/12/40
Hanks, Tom	Oakland, CA	7/9/56
Hannah, Daryl	Chicago, IL	12/3/60
Hanson, Curtis	Los Angeles, CA	3/24/45
Hanson, Isaac	Tulsa, OK	11/17/80
Hanson, Taylor	Tulsa, OK	3/14/83
Hanson, Zac	Arlington, VA	10/22/85
Hardison, Kadeem	New York, NY	7/24/66
Harewood, Dorian	Dayton, OH	8/6/51
Harmon, Mark	Burbank, CA	9/2/51
Harper, Jessica	Chicago, IL	10/10/49
Harper, Tess	Mammoth Springs, AR	8/15/50
Harper, Valerie	Suffern, NY	8/22/40
Harrelson, Woody	Midland, TX	7/23/61
Harrington, Pat	New York, NY	8/13/29
Harris, Barbara	Evanston, IL	7/25/35
Harris, Ed	Englewood, NJ	11/28/50
Harris, Emmylou	Birmingham, AL	4/2/47
Harris, Julie	Grosse Pte. Park, MI	12/2/25
Harris, Neil Patrick	Albuquerque, NM	6/15/73
Harris, Richard	Co. Limerick, Ireland	10/1/33
Harris, Rosemary	Ashby, England	9/19/30
Harrison, George	Liverpool, England	2/25/43
Harrison, Gregory	Avalon, CA	5/31/50
Harry, Deborah	Miami, FL	7/1/45
Hart, Mary	Madison, SD	11/8/51
Hart, Melissa Joan	Sayville, NY	4/18/76
Hartley, Hal	Lindenhurst, NY	11/3/59
Hartley, Mariette	New York, NY	6/21/40
Hartman, David	Pawtucket, RI	5/19/35
Hartman, Lisa	Houston, TX	6/1/56
Hasselhoff, David	Baltimore, MD	7/17/52
Hatcher, Teri	Sunnyvale, CA	12/8/64
Hauer, Rutger	Breukelen, Netherlands	1/23/44
Haver, June	Rock Island, IL	6/10/26
Havoc, June	Seattle, WA	11/8/16
Hawke, Ethan	Austin, TX	11/6/70
Hawn, Goldie	Washington, DC	11/21/45
Hayden, Melissa	Toronto, Ontario	4/25/23
Hayek, Salma	Coatzacoalcos, Mexico	9/2/68
Hayes, Isaac	Covington, TN	8/20/42
Hays, Robert	Bethesda, MD	7/24/47
Heard, John	Washington, DC	3/7/45
Hearn, George	Memphis, TN	1935
Heche, Anne	Aurora, OH	5/25/69
Heckart, Eileen	Columbus, OH	3/29/19
Hedren, Tippi	New Ulm, MN	1/19/35
Helfgott, David	Melbourne, Australia	5/19/47
Helmond, Katherine	Galveston, TX	7/5/34
Hemingway, Mariel	Mill Valley, CA	11/21/61
Hemmings, David	Guildford, England	11/18/41
Hemsley, Sherman	Philadelphia, PA	2/1/38
Henderson, Florence	Dale, IN	2/14/34
Henderson, Skitch	Halstad, MN	1/27/18
Henley, Don	Gilmer, TX	7/22/47
Henner, Marilu	Chicago, IL	4/6/52
Henning, Doug	Ft. Garry, Manitoba	5/3/47
Henry, Buck	New York, NY	12/9/30
Hepburn, Katharine	Hartford, CT	5/12/07
Herman, Pee-Wee	Peekskill, NY	8/27/52
Herrmann, Edward	Washington, DC	7/21/43
Hershey, Barbara	Los Angeles, CA	2/5/48
Hesseman, Howard	Lebanon, OR	2/27/40
Heston, Charlton	Evanston, IL	10/4/24
Hewett, Christopher	Sussex, England	4/5/22
Hewitt, Jennifer Love	Waco, TX	2/21/79
Hildegarde	Adell, WI	2/1/06
Hill, Arthur	Melfort, Sask.	8/1/22
Hill, Faith	Jackson, MS	9/21/67
Hill, George Roy	Minneapolis, MN	12/20/22
Hill, Lauryn	South Orange, NJ	5/25/75
Hill, Steven	Seattle, WA	2/24/22

Name	Birthplace	Birthdate
Hiller, Wendy	Stockport, England	8/15/12
Hillerman, John	Denison, TX	12/30/32
Hines, Gregory	New York, NY	2/14/46
Hines, Roy	Boston, MA	3/13/26
Hines, Jerome	Hollywood, CA	11/8/21
Hingle, Pat	Miami, FL	7/19/24
Hirsch, Judd	New York, NY	3/15/35
Ho, Don	Kakaako, Oahu, HI	8/13/30
Hoffman, Dustin	Los Angeles, CA	8/8/37
Hogan, Paul	New South Wales, Australia	10/8/39
Holbrook, Hal	Cleveland, OH	2/17/25
Holder, Geoffrey	Trinidad	8/1/30
Holliday, Polly	Jasper, AL	8/2/37
Holliman, Earl	Delhi, LA	9/11/28
Holly, Lauren	Bristol, PA	10/28/63
Holm, Celeste	New York, NY	4/29/19
Holmes, Katie	Toledo, OH	12/18/78
Hooker, John Lee	Clarksdale, MS	8/22/17
Hooks, Jan	Decatur, GA	4/23/57
Hope, Bob	London, England	5/29/03
Hopkins, Anthony	Port Talbot, South Wales	12/31/37
Hopkins, Bo	Greenville, SC	2/2/42
Hopkins, Telma	Louisville, KY	10/28/48
Hopper, Dennis	Dodge City, KS	5/17/36
Horne, Lena	Brooklyn, NY	6/30/17
Horne, Marilyn	Bradford, PA	1/16/34
Hornsby, Bruce	Williamsburg, VA	11/23/54
Horsley, Lee	Muleshoe, TX	5/15/55
Hoskins, Bob	Suffolk, England	10/26/42
Houston, Whitney	E Orange, NJ	8/9/63
Howard, Ken	El Centro, CA	3/28/44
Howard, Ron	Duncan, OK	3/1/54
Howell, C. Thomas	Los Angeles, CA	12/7/66
Howes, Sally Ann	London, England	7/20/30
Hughes, Barnard	Bedford Hills, NY	7/16/15
Hulce, Tom	Whitewater, WI	12/6/53
Humperdinck, Engelbert	Madras, India	5/3/36
Hunt, Helen	Los Angeles, CA	6/15/63
Hunt, Linda	Morristown, NJ	4/2/45
Hunter, Holly	Conyers, GA	3/20/58
Hunter, Kim	Detroit, MI	11/12/22
Hunter, Tab	New York, NY	7/11/31
Hurley, Elizabeth	Hampshire, England	6/10/65
Hurt, John	Chesterfield, England	1/22/40
Hurt, Mary Beth	Marshalltown, IA	9/26/46
Hurt, William	Washington, DC	3/20/50
Hussey, Ruth	Providence, RI	10/30/14
Huston, Anjelica	Santa Monica, CA	7/8/51
Hutton, Betty	Battle Creek, MI	2/26/21
Hutton, Lauren	Charleston, SC	11/17/44
Hutton, Timothy	Malibu, CA	8/16/60
Hyman, Earle	Rocky Mount, NC	10/11/26
Ian, Janis	New York, NY	4/7/51
Ice Cube	Los Angeles, CA	6/15/69
Ice-T	Newark, NJ	2/16/58
Idle, Eric	Durham, England	3/29/43
Idol, Billy	London, England	11/30/55
Iman	Mogadishu, Somalia	7/25/55
Iglesias, Enrique	Madrid, Spain	5/8/75
Iglesias, Julio	Madrid, Spain	9/23/43
Iman	Mogadishu, Somalia	7/22/55
Imbruglia, Natalie	Australia	2/4/75
Imus, Don	Riverside, CA	7/23/40
Ireland, Kathy	Santa Barbara, CA	3/8/63
Ingram, James	Akron, OH	2/16/56
Irons, Jeremy	Cowes, England	9/19/48
Irving, Amy	Palo Alto, CA	9/10/53
Irving, George S.	Springfield, MA	11/1/22
Ivey, Judith	El Paso, TX	9/4/51
Ivory, James	Berkeley, CA	6/7/28
Jackee	Winston-Salem, NC	8/14/56
Jackson, Anne	Allegheny, PA	9/3/25
Jackson, Glenda	Liverpool, England	5/9/36
Jackson, Janet	Gary, IN	5/16/66
Jackson, Jermaine	Gary, IN	12/11/54
Jackson, Jonathan	Orlando, FL	5/11/82
Jackson, Joshua	Vancouver, Brit. Columbia	6/11/78
Jackson, Kate	Birmingham, AL	10/29/48
Jackson, La Toya	Gary, IN	5/29/56
Jackson, Michael	Gary, IN	8/29/58
Jackson, Milt	Detroit, MI	1/1/22
Jackson, Samuel L.	Chattanooga, TN	12/21/48
Jacobi, Derek	London, England	10/22/38
Jagger, Mick	Dartford, England	7/26/43
James, Etta	Los Angeles, CA	1938
Janis, Conrad	New York, NY	2/11/28
Jardine, Al	Lima, OH	9/3/42
Jarmusch, Jim	Akron, OH	1/22/53
Jarreau, Al	Milwaukee, WI	3/12/40
Jarrette, Keith	Allentown, PA	5/8/45
Jeffreys, Anne	Goldsboro, NC	1/26/23
Jennings, Waylon	Littlefield, TX	6/15/37
Jeter, Michael	Lawrenceburg, TN	8/20/52
Jett, Joan	Philadelphia, PA	9/22/60
Jewel (Kilcher)	Payson, UT	5/3/74
Jewison, Norman	Toronto, Ontario	7/21/26
Jillian, Ann	Cambridge, MA	1/29/50
Joel, Billy	Bronx, NY	5/9/49
John, Elton	Middlesex, England	3/25/47
Johns, Glynis	Durban, S Africa	10/5/23
Johnson, Arte	Benton Harbor, MI	1/20/29
Johnson, Beverly	Buffalo, NY	10/13/52
Johnson, Don	Flatt Creek, MO	12/15/49
Johnson, J. J.	Indianapolis, IN	1/22/24
Johnson, Van	Newport, RI	8/25/16
Johnston, Bruce	Chicago, IL	6/24/44
Johnston, Kristen	Washington, DC	9/20/67
Jolie, Angelina	Los Angeles, CA	6/4/75
Jones, Charlie	Ft. Smith, AR	11/9/30
Jones, Davy	Manchester, England	12/30/45
Jones, Dean	Morgan City, AL	1/25/35
Jones, Elvin	Pontiac, MI	9/9/27
Jones, George	Saratoga, TX	9/12/31
Jones, Grace	Spanishtown, Jamaica	5/19/52
Jones, Jack	Hollywood, CA	1/14/38
Jones, James Earl	Tate Co., MS	1/17/31
Jones, Jennifer	Tulsa, OK	3/2/19
Jones, Quincy	Chicago, IL	3/14/33
Jones, Shirley	Smithton, PA	3/31/34
Jones, Tom	Pontypridd, Wales	6/7/40
Jones, Tommy Lee	San Saba, TX	9/15/46
Jourdan, Louis	Marseilles, France	6/19/19
Jovovich, Milla	Kiev, Ukraine	12/19/75
Judd, Ashley	Los Angeles, CA	4/19/68
Judd, Naomi	Ashland, KY	1/11/46
Judd, Wynonna	Ashland, KY	5/3/64
Jump, Gordon	Dayton, OH	4/1/32
Kahn, Madeline	Boston, MA	9/29/42
Kanaly, Steve	Burbank, CA	3/14/46
Kane, Carol	Cleveland, OH	6/18/52
Kaplan, Gabe	Brooklyn, NY	3/31/45
Karlen, John	New York, NY	5/28/33
Karn, Richard	Seattle, WA	2/17/56
Karras, Alex	Gary, IN	7/15/35
Kasem, Casey	Detroit, MI	4/27/33
Kavner, Julie	Los Angeles, CA	9/7/51
Kazan, Elia	Istanbul, Turkey	9/7/09
Kazan, Lainie	New York, NY	5/15/42
Keach, Stacy	Savannah, GA	6/2/41
Keaton, Diane	Santa Ana, CA	1/5/46
Keaton, Michael	Pittsburgh, PA	9/9/51
Keel, Howard	Gillespie, IL	4/13/17
Keeshan, Bob	Lynbrook, NY	6/27/27
Keitel, Harvey	Brooklyn, NY	5/13/39
Keith, David	Knoxville, TN	5/8/54
Keith, Penelope	Sutton, Surrey, Eng.	4/2/40
Kellerman, Sally	Long Beach, CA	6/2/37
Kennedy, George	New York, NY	2/18/25
Kennedy, Jayne	Washington, DC	11/27/51
Kenny G	Seattle, WA	6/5/56
Kent, Allegra	Los Angeles, CA	8/11/37
Kercheval, Ken	Wolcottville, IN	7/15/35
Kerns, Joanna	San Francisco, CA	2/12/53
Kerr, Deborah	Helensburgh, Scotland	9/30/21
Kessel, Barney	Muskogee, OK	10/17/23
Khan, Chaka	Great Lakes, IL	3/23/53
Kidder, Margot	Yellowknife, N.W.T.	10/17/48
Kidman, Nicole	Honolulu, HI	6/20/67
Kilborn, Craig	Hastings, MN	8/24/62
Kilmer, Val	Los Angeles, CA	12/31/59
Kimbrough, Charles	St. Paul, MN	5/23/36
King, Alan	Brooklyn, NY	12/26/27
King, B. B.	Itta Bena, MS	9/16/25
King, Carole	Brooklyn, NY	2/9/42
King, Larry	Brooklyn, NY	11/19/33
King, Perry	Alliance, OH	4/30/48
Kingsley, Ben	Yorkshire, England	12/31/43
Kinnear, Greg	Logansport, IN	6/17/63
Kinski, Nastassja	Berlin, W. Germany	1/24/60
Kirby, Bruno	New York, NY	4/28/49
Kirby, Durward	Covington, KY	8/24/12
Kirkland, Gelsey	Bethlehem, PA	12/29/53
Kirkpatrick, Chris	Pennsylvania	10/17/71
Kitt, Eartha	North, SC	1/17/27
Klein, Robert	New York, NY	2/8/42

Name	Birthplace	Birthdate
Klemperer, Werner	Cologne, Germany	3/22/19
Kline, Kevin	St. Louis, MO	10/24/47
Klugman, Jack	Philadelphia, PA	4/27/22
Knight, Gladys	Atlanta, GA	5/28/44
Knight, Shirley	Goessel, KS	7/5/36
Knight, Wayne	Cartersville, GA.	8/7/55
Knotts, Don	Morgantown, WV	7/21/24
Konitz, Lee	Chicago, IL	10/13/27
Kopell, Bernie	New York, NY	6/21/33
Korman, Harvey	Chicago, IL	2/15/27
Kotto, Yaphet	New York, NY	11/15/37
Krakowski, Jane	Parsippany, NJ	1969
Kramer, Stanley	New York, NY	9/29/13
Kristofferson, Kris	Brownsville, TX	6/22/36
Kudrow, Lisa	Encino, CA	5/30/63
Kurtz, Swoosie	Omaha, NE	9/6/44
LaBelle, Patti	Philadelphia, PA	5/24/44
Ladd, Cheryl	Huron, SD	7/12/51
Ladd, Diane	Meridian, MS.	11/29/32
Lahti, Christine	Detroit, MI	4/5/50
Laine, Cleo	Middlesex, England	10/28/27
Laine, Frankie	Chicago, IL	3/30/13
Lake, Ricki	New York, NY	9/21/68
Lamarr, Hedy	Vienna, Austria	11/9/13
Lamas, Lorenzo	Santa Monica, CA.	1/20/58
Lambert, Christopher	New York, NY	3/29/57
Landau, Martin	New York, NY	6/20/34
Landis, John	Chicago, IL	8/3/50
Lane, Diane	New York, NY	1/22/63
Lane, Nathan	Jersey City, NJ	2/3/56
lang, k.d.	Consort, Alberta	11/2/61
Lang, Stephen	New York, NY	7/11/52
Lange, Hope	Redding Ridge, CT	11/28/31
Lange, Jessica	Cloquet, MN	4/20/49
Langella, Frank	Bayonne, NJ	1/1/40
Langford, Frances	Lakeland, FL	4/4/13
Lansbury, Angela	London, England.	10/16/25
LaPaglia, Anthony	Adelaide, Australia	1/31/59
Laredo, Ruth	Detroit, MI	11/20/37
Larroquette, John	New Orleans, LA.	11/25/47
LaSalle, Eriq	Hartford, CT	6/23/63
Lauper, Cyndi	New York, NY	6/20/53
Laurie, Piper	Detroit, MI	1/22/32
Lavin, Linda	Portland, ME.	10/15/37
Lawless, Lucy	Mount Albert, New Zealand	3/28/68
Lawrence, Carol	Melrose Park, IL	9/5/34
Lawrence, Joey	Montgomery, PA	4/20/76
Lawrence, Martin	Frankfurt, Germany	4/16/65
Lawrence, Steve	Brooklyn, NY	7/8/35
Lawrence, Vicki	Inglewood, CA.	3/26/49
Leach, Robin	London, England.	8/29/41
Leachman, Cloris	Des Moines, IA	4/4/26
Lear, Norman	New Haven, CT.	7/27/22
Leary, Denis	Boston, MA	8/18/57
Learned, Michael	Washington, DC	4/9/39
LeBlanc, Matt	Newton, MA	7/25/67
LeBon, Simon	Bushey, England	10/27/58
Lee, Ang	Taiwan	10/23/54
Lee, Brenda	Atlanta, GA	12/11/44
Lee, Christopher	London, England.	5/27/22
Lee, Michele	Los Angeles, CA	6/24/42
Lee, Pamela Anderson	Comox, Canada	7/1/67
Lee, Peggy	Jamestown, ND.	5/26/20
Lee, Spike	Atlanta, GA	3/20/57
Leeves, Jane	London, England.	4/18/62
Legrand, Michel	Paris, France.	2/24/32
Leguizamo, John	Bogota, Colombia	7/22/65
Leibman, Ron	New York, NY	10/11/37
Leigh, Janet	Merced, CA.	7/6/27
Leigh, Jennifer Jason	Los Angeles, CA.	2/5/62
Leighton, Laura	Iowa City, IA	3/14/69
Lemmon, Jack	Boston, MA.	2/8/25
Lennox, Annie	Aberdeen, Scotland	12/25/54
Leno, Jay	New Rochelle, NY.	4/28/50
Leonard, Robert Sean	Westwood, NJ.	2/28/69
Leoni, Tea	New York, NY	2/25/66
Leslie, Joan	Detroit, MI	1/26/25
Leto, Jared	Bossier City, LA.	12/26/71
Letterman, David	Indianapolis, IN	4/12/47
Levine, James	Cincinnati, OH.	6/23/43
Levinson, Barry	Baltimore, MD	6/2/32
Lewis, Al	New York, NY	4/30/10
Lewis, Huey	New York, NY	7/5/51
Lewis, Jerry	Newark, NJ	3/16/26
Lewis, Jerry Lee	Ferriday, LA.	9/29/35
Lewis, John	La Grange, IL	5/30/20
Lewis, Juliette	San Fernando Valley, CA	6/21/73
Lewis, Richard	New York, NY	6/29/47
Light, Judith	Trenton, NJ	2/9/50
Lightfoot, Gordon	Orillia, Ontario	11/17/38
Linden, Hal	New York, NY	3/20/31
Linkletter, Art	Saskatchewan, Canada	7/17/12
Linn-Baker, Mark	St. Louis, MO	6/17/53
Liotta, Ray	Newark, NJ	12/18/55
Lithgow, John	Rochester, NY.	10/19/45
Little, Rich	Ottawa, Ontario	11/26/38
Little Richard	Macon, GA	12/5/32
Littrell, Brian	Lexington, KY	2/20/75
L. L. Cool J.	New York, NY	1/14/68
Lloyd, Christopher	Stamford, CT.	10/22/38
Lloyd, Emily	England	9/29/70
Lloyd Webber, Andrew	London, England.	3/22/48
Locke, Sondra	Shelbyville, TN	5/28/47
Lockhart, June	New York, NY	6/25/25
Locklear, Heather	Los Angeles, CA	9/25/61
Loggia, Robert	New York, NY	1/3/30
Loggins, Kenny	Everett, WA.	1/17/47
Lollobrigida, Gina	Subiaco, Italy	7/4/27
Lom, Herbert	Prague, Czechoslovakia	1/9/17
Long, Shelley	Ft. Wayne, IN	8/23/49
Lopez, Jennifer	Bronx, NY	7/24/70
Loren, Sophia	Rome, Italy	9/20/34
Loring, Gloria	New York, NY	12/10/46
Loudon, Dorothy	Boston, MA.	9/17/33
Louis-Dreyfus, Julia	New York, NY	1/13/61
Love, Courtney	San Francisco, CA	7/9/64
Love, Mike	Los Angeles, CA.	3/15/41
Lovett, Lyle	Klein, TX	11/1/57
Lovitz, Jon	Tarzana, CA	7/21/57
Loveless, Patty	Pikeville, KY	1/4/57
Lowe, Rob	Charlottesville, VA.	3/17/64
Lucas, George	Modesto, CA.	5/14/44
Lucci, Susan	Scarsdale, NY.	12/23/48
Luckinbill, Laurence	Ft. Smith, AR.	11/21/34
Ludwig, Christa	Berlin, Germany	3/16/28
Lumet, Sidney	Philadelphia, PA	6/25/24
LuPone, Patti	Northport, NY	4/21/49
Lynch, David	Missoula, MT.	1/20/46
Lynley, Carol	New York, NY	2/13/42
Lynn, Loretta	Butcher Hollow, KY	4/14/35
Ma, Yo Yo	Paris, France	10/7/55
Maazel, Lorin	Paris, France.	3/6/30
MacArthur, James	Los Angeles, CA.	12/8/37
MacCorkindale, Simon	Cambridge, England.	2/12/52
MacDowell, Andie	Gaffney, SC.	4/21/58
MacGraw, Ali	Pound Ridge, NY	4/1/38
MacLachlan, Kyle	Yakima, WA.	2/22/59
MacLaine, Shirley	Richmond, VA.	4/24/34
MacLeod, Gavin	Mt. Kisco, NY	2/28/30
MacNee, Patrick	London, England.	2/6/22
MacNeil, Cornell	Minneapolis, MN	9/24/22
MacNicol, Peter	Dallas, TX	4/10/54
MacPherson, Elle	Sydney, Australia.	3/29/64
Macchio, Ralph	Long Island, NY	11/4/62
Macy, Bill	Revere, MA.	5/18/22
Macy, William H.	Miami, FL	3/13/50
Madden, John	Austin, MN	4/10/36
Madigan, Amy	Chicago, IL	9/11/51
Madonna (Ciccone)	Bay City, MI	8/16/58
Maher, Bill	Rivervale, NJ.	1/20/56
Mahoney, John	Manchester, England	6/20/40
Majors, Lee	Wyandotte, MI	4/23/40
Malden, Karl	Chicago, IL	3/22/13
Malick, Terrence	Ottawa, IL	11/30/43
Malick, Wendie	Buffalo, NY	12/13/50
Malkovich, John	Christopher, IL.	12/9/53
Malone, Dorothy	Chicago, IL	1/30/25
Manchester, Melissa	Bronx, NY	2/15/51
Mandel, Howie	Toronto, Ontario	11/29/55
Mandrell, Barbara	Houston, TX	12/25/48
Mangione, Chuck	Rochester, NY.	11/29/40
Manilow, Barry	New York, NY	6/17/46
Mann, Herbie	New York, NY	4/16/30
Manoff, Dinah	New York, NY	1/25/58
Manson, Marilyn	Canton, OH.	1/5/69
Mantegna, Joe	Chicago, IL	11/13/47
Marceau, Marcel	Strasbourg, France	3/22/23
Marchand, Nancy	Buffalo, NY	6/19/28
Margulies, Julianna	Spring Valley, NY	6/8/66
Marin, Cheech	Los Angeles, CA.	7/13/46
Marinaro, Ed	New York, NY	3/31/50
Markova, Alicia	London, England.	12/1/10
Marriner, Neville	Lincoln, England	4/15/24
Marsalis, Branford	New Orleans, LA.	8/26/60
Marsalis, Wynton	New Orleans, LA.	10/18/61
Marsh, Jean	London, England.	7/1/34

Name	Birthplace	Birthdate
Marshall, Garry	New York, NY	11/13/34
Marshall, Penny	New York, NY	10/15/43
Marshall, Peter	Huntington, WV	3/30/27
Martin, Dick	Detroit, MI	1/30/23
Martin, Kellie	Riverside, CA	10/16/75
Martin, Ricky	San Juan, Puerto Rico	12/24/71
Martin, Steve	Waco, TX	1945
Martin, Tony	San Francisco, CA	12/25/13
Martins, Peter	Copenhagen, Denmark	10/27/46
Mason, Jackie	Sheboygan, WI	6/9/31
Mason, Marsha	St. Louis, MO	4/3/42
Masterson, Mary Stuart	Los Angeles, CA	6/28/66
Mastrantonio, Mary Elizabeth	Lombard, IL	11/17/58
Masur, Kurt	Brieg, Germany	7/18/27
Masur, Richard	New York, NY	11/20/48
Mathers, Jerry	Sioux City, IA	6/2/48
Matheson, Tim	Glendale, CA	12/31/47
Mathis, Johnny	San Francisco, CA	9/30/35
Matlin, Marlee	Morton Grove, IL	8/24/65
Matthau, Walter	New York, NY	10/1/20
Matthews, Dave	Johannesburg, South Africa	1/9/67
May, Elaine	Philadelphia, PA	4/21/32
Mayfield, Curtis	Chicago, IL	6/3/42
Mayo, Virginia	St. Louis, MO	11/30/20
Mazursky, Paul	Brooklyn, NY	4/25/30
McArdle, Andrea	Philadelphia, PA	11/5/63
McBride, Patricia	Teaneck, NJ	8/23/42
McCallum, David	Glasgow, Scotland	9/19/33
McCambridge, Mercedes	Joliet, IL	3/17/18
McCarthy, Andrew	Westfield, NJ	11/29/62
McCarthy, Jenny	Chicago, IL	11/1/72
McCarthy, Kevin	Seattle, WA	2/15/14
McCartney, Paul	Liverpool, England	6/18/42
McCarver, Tim	Memphis, TN	10/16/41
McClanahan, Rue	Healdton, OK	2/21/36
McConaughey, Matthew	Uvalde, Texas	11/4/69
McCoo, Marilyn	Jersey City, NJ	9/30/43
McCormack, Mary	Plainsfield, NJ	4/8/69
McDermott, Dylan	Waterbury, CT	10/26/62
McDonnell, Mary	Wilkes-Barre, PA	1952
McDormand, Frances	Illinois	6/23/57
McDowell, Malcolm	Leeds, England	6/13/43
McEntire, Reba	McAlester, OK	3/28/55
McFerrin, Bobby	New York, NY	3/11/50
McGavin, Darren	Spokane, WA	5/7/22
McGillis, Kelly	Newport Beach, CA	7/9/57
McGoohan, Patrick	New York, NY	3/19/28
McGovern, Elizabeth	Evanston, IL	7/18/61
McGovern, Maureen	Youngstown, OH	7/27/49
McGraw, Tim	Delhi, LA	5/1/67
McGregor, Ewan	Crieff, Scotland	3/31/71
McGuire, Al	New York, NY	9/7/31
McGuire, Dorothy	Omaha, NE	6/14/19
McKean, Michael	New York, NY	10/17/47
McKechnie, Donna	Pontiac, MI	11/16/42
McKellen, Ian	Burnley, England	5/25/39
McLachlan, Sarah	Halifax, Nova Scotia	1/28/68
McLean, A.J.	West Palm Beach, FL	1/9/78
McMahon, Ed	Detroit, MI	3/6/23
McNichol, Kristy	Los Angeles, CA	9/11/62
McPartland, Marian	Stough, England	3/20/20
McRaney, Gerald	Collins, MS	8/19/48
Meadows, Jayne	Wu Chang, China	9/27/20
Meara, Anne	New York, NY	9/20/29
Meat Loaf	Dallas, TX	9/27/47
Mehta, Zubin	Bombay, India	4/29/36
Mellencamp, John	Seymour, IN	10/7/51
Mendes, Sergio	Niteroi, Brazil	2/11/41
Mercer, Marian	Akron, OH	11/26/35
Merchant, Natalie	Jamestown, NY	10/26/63
Merrick, David	St. Louis, MO	11/27/12
Merrill, Dina	New York, NY	12/9/25
Merrill, Robert	Brooklyn, NY	6/4/19
Metcalf, Laurie	Carbondale, IL	6/16/55
Michael, George	Watford, England	6/26/63
Michaels, Al	New York, NY	11/12/44
Michaels, Lorne	Toronto, Canada	11/17/44
Midler, Bette	Honolulu, HI	12/1/45
Midori	Osaka, Japan	10/25/71
Milano, Alyssa	New York, NY	12/19/72
Miles, Sarah	Ingatestone, England	12/31/41
Miles, Vera	near Boise City, OK	8/23/29
Miller, Ann	Houston, TX	4/12/19
Miller, Dennis	Pittsburgh, PA	11/3/53
Miller, Mitch	Rochester, NY	7/4/11
Miller, Penelope Ann	Los Angeles, CA	1/13/64
Mills, Donna	Chicago, IL	12/11/42
Mills, John	Suffolk, England	2/22/08
Milner, Martin	Detroit, MI	12/28/27
Milnes, Sherrill	Downers Grove, IL	1/10/35
Milsap, Ronnie	Robinsville, NC	1/16/44
Minghella, Anthony	Isle of Wight, England	1/6/54
Minnelli, Liza	Los Angeles, CA	3/12/46
Mirren, Helen	London, England	7/2/46
Mitchell, Joni	McLeod, Alberta	11/7/43
Mr. T.	Chicago, IL	5/21/52
Modine, Matthew	Loma Linda, CA	3/22/59
Moffat, Donald	Plymouth, England	12/26/30
Moffo, Anna	Wayne, PA	6/27/27
Molinaro, Al	Kenosha, WI	6/24/19
Moll, Richard	Pasadena, CA	1/13/43
Monica, Arnold	College Park, GA	10/24/80
Montalban, Ricardo	Mexico City, Mexico	11/25/20
Moody, Ron	London, England	1/8/24
Moore, Clayton	Chicago, IL	9/14/14
Moore, Demi	Roswell, NM	11/11/62
Moore, Dudley	London, England	4/19/35
Moore, Julianne	Boston, MA	12/30/60
Moore, Mary Tyler	Brooklyn, NY	12/29/36
Moore, Melba	New York, NY	10/29/45
Moore, Roger	London, England	10/14/27
Moore, Terry	Los Angeles, CA	1/1/29
Moranis, Rick	Toronto, Ontario	4/18/53
Moreau, Jeanne	Paris, France	1/23/28
Moreno, Rita	Humacao, PR	12/11/31
Morgan, Harry	Detroit, MI	4/10/15
Moriarty, Michael	Detroit, MI	4/5/41
Morissette, Alanis	Ottawa, Ontario	6/1/74
Morita, Pat	Isleton, CA	6/28/32
Morris, Garrett	New Orleans, LA	2/1/37
Morris, Howard	New York, NY	9/4/25
Morrison, Van	Belfast, N. Ireland	8/31/45
Morrissey	Manchester, England	5/22/59
Morrow, Rob	New Rochelle, NY	9/21/62
Morse, David	Hamilton, MA	10/11/53
Morse, Robert	Newton, MA	5/18/31
Morton, Joe	New York, NY	10/18/47
Moses, William	Los Angeles, CA	11/17/59
Moss, Kate	London, England	1/16/74
Muldaur, Diana	New York, NY	8/19/38
Mulgrew, Kate	Dubuque, IA	4/29/55
Mull, Martin	Chicago, IL	8/18/43
Mueller-Stahl, Armin	Tilsit, E. Prussia	12/17/20
Mulligan, Richard	New York, NY	11/13/32
Mulroney, Dermot	Alexandria, VA	10/31/63
Munsel, Patrice	Spokane, WA	5/14/25
Murphy, Ben	Jonesboro, AR	3/6/42
Murphy, Eddie	Brooklyn, NY	4/3/61
Murphy, Michael	Los Angeles, CA	5/5/38
Murray, Anne	Springhill, Nova Scotia	6/20/45
Murray, Bill	Evanston, IL	9/21/50
Murray, Don	Hollywood, CA	7/31/29
Musburger, Brent	Portland, OR	5/26/39
Muti, Riccardo	Naples, Italy	7/28/41
Myers, Mike	Toronto, Ontario	5/25/63
Nabors, Jim	Sylacauga, AL	6/12/33
Nash, Graham	Blackpool, England	2/2/42
Naughton, James	Middletown, CT	7/6/46
Neal, Patricia	Packard, KY	1/20/26
Nealon, Kevin	Bridgeport, CT	11/18/53
Neeson, Liam	Ballymena, N. Ireland	6/7/52
Neill, Sam	Ulster, N. Ireland	9/14/47
Nelligan, Kate	London, Ontario	3/16/51
Nelson, Craig T.	Spokane, WA	4/4/46
Nelson, Ed	New Orleans, LA	12/21/28
Nelson, Judd	Portland, ME	11/28/59
Nelson, Tracy	Santa Monica, CA	10/25/63
Nelson, Willie	Abbott, TX	4/30/33
Nero, Peter	New York, NY	5/22/34
Nesmith, Mike	Dallas, TX	12/30/42
Nettleton, Lois	Oak Park, IL	8/16/31
Neuwirth, Bebe	Princeton, NJ	12/31/58
Neville, Aaron	New Orleans, LA	1/24/41
Newhart, Bob	Oak Park, IL	9/5/29
Newman, Paul	Cleveland, OH	1/26/25
Newman, Randy	Los Angeles, CA	11/28/43
Newton, Wayne	Norfolk, VA	4/3/42
Newton-John, Olivia	Cambridge, England	9/26/47
Nicholas, Denise	Detroit, MI	7/12/44
Nicholas, Fayard	Philadelphia, PA	10/20/14
Nicholas, Harold	Philadelphia, PA	3/27/24
Nichols, Mike	Berlin, Germany	11/6/31
Nicholson, Jack	Neptune, NJ	4/28/37
Nicks, Stevie	Phoenix, AZ	5/26/48
Nielsen, Leslie	Regina, Sask.	2/11/26
Nilsson, Birgit	Karup, Sweden	5/17/18

Name	Birthplace	Birthdate
Nimoy, Leonard	Boston, MA	3/26/31
Nolte, Nick	Omaha, NE	2/8/40
Noone, Peter	Manchester, England	11/5/47
Norman, Jessye	Augusta, GA	9/15/45
Norris, Chuck	Ryan, OK	3/10/40
North, Sheree	Los Angeles, CA	1/17/33
Norton, Edward	Columbia, MD	1969
Noth, Christopher	Madison, WI	11/13/57
Novak, Kim	Chicago, IL	2/13/33
Nuyen, France	Marseille, France	7/31/39
Oates, John	New York, NY	4/7/48
O'Brian, Hugh	Rochester, NY	4/19/25
O'Brien, Conan	Brookline, MA	4/18/63
O'Brien, Margaret	San Diego, CA	1/15/37
Ocean, Billy	Fyzabad, Trinidad	1/21/50
O'Connor, Carroll	New York, NY	8/2/24
O'Connor, Donald	Chicago, IL	8/28/25
O'Connor, Sinead	Dublin, Ireland	12/8/66
Odetta	Birmingham, AL	12/31/30
O'Donnell, Chris	Winnetka, IL	6/26/70
O'Donnell, Rosie	Commack, NY	3/21/62
O'Hara, Maureen	Dublin, Ireland	8/17/20
O'Herlihy, Dan	Wexford, Ireland	5/1/19
Oldman, Gary	London, England	3/21/58
Olin, Ken	Chicago, IL	7/30/54
Olin, Lena	Stockholm, Sweden	3/22/55
Olmos, Edward James	E. Los Angeles, CA	2/24/47
Olsen, Ashley	California	6/13/86
Olsen, Mary-Kate	California	6/13/86
Olsen, Merlin	Logan, UT	9/15/40
O'Neal, Ryan	Los Angeles, CA	4/20/41
O'Neal, Tatum	Los Angeles, CA	11/5/63
O'Neill, Ed	Youngstown, OH	4/12/46
Ontkean, Michael	Vancouver, B.C.	1/24/46
Orbach, Jerry	New York, NY	10/20/35
Orlando, Tony	New York, NY	4/3/44
Ormond, Julia	Epsom, England	1/4/65
Osbourne, Ozzy	Birmingham, England	12/3/48
O'Shea, Milo	Dublin, Ireland	6/2/26
Oslin, K.T.	Crosset, AR	1942
Osmond, Donny	Ogden, UT	12/9/57
Osmond, Marie	Ogden, UT	10/13/59
O'Toole, Annette	Houston, TX	4/1/53
O'Toole, Peter	Connemara, Ireland	8/2/32
Owens, Buck	Sherman, TX	8/12/29
Oz, Frank	Herford, England	5/25/44
Ozawa, Seiji	Shenyang, China	9/1/35
Paar, Jack	Canton, OH	5/1/18
Pacino, Al	New York, NY	4/25/40
Packer, Billy	Wellsville, NY	2/25/40
Page, Bettie	Kingsport, TN	4/22/23
Page, Jimmy	Heston, England	1/9/44
Page, Patti	Claremore, OK	11/8/27
Paget, Debra	Denver, CO	8/19/33
Paige, Janis	Tacoma, WA	9/16/22
Palance, Jack	Lattimer, PA	2/18/20
Palin, Michael	Sheffield, England	5/5/43
Palmer, Betsy	East Chicago, IN	11/1/29
Palmer, Geoffrey	London, England	6/4/27
Palmer, Robert	Bately, England	1/19/49
Palminteri, Chazz	Bronx, NY	5/15/51
Paltrow, Gwyneth	Los Angeles, CA	9/28/73
Papas, Irene	Chiliomedion, Greece	3/9/26
Paquin, Anna	Wellington, New Zealand	6/24/82
Parker, Alan	London, England	2/14/44
Parker, Eleanor	Cedarville, OH	6/26/22
Parker, Fess	Ft. Worth, TX	8/16/25
Parker, Jameson	Baltimore, MD	11/18/47
Parker, Jean	Deer Lodge, MT	8/11/15
Parker, Mary-Louise	Fort Jackson, SC	8/2/64
Parker, Sarah Jessica	Nelsonville, OH	3/25/65
Parsons, Estelle	Lynn, MA	11/20/27
Parton, Dolly	Sevierville, TN	1/19/46
Patinkin, Mandy	Chicago, IL	11/30/52
Patric, Jason	Queens, NY	6/17/66
Pavarotti, Luciano	Modena, Italy	10/12/35
Paxton, Bill	Fort Worth, TX	5/17/55
Paycheck, Johnny	Greenfield, OH	5/31/41
Peck, Gregory	La Jolla, CA	4/5/16
Pendergrass, Teddy	Philadelphia, PA	3/26/50
Penn, Arthur	Philadelphia, PA	9/27/22
Penn, Robin Wright	Dallas, TX	4/8/66
Penn, Sean	Burbank, CA	8/17/60
Penny, Joe	London, England	9/14/56
Perez, Rosie	Brooklyn, NY	9/6/64
Perkins, Elizabeth	New York, NY	11/18/60
Perlman, Itzhak	Tel Aviv, Israel	8/31/45

Name	Birthplace	Birthdate
Perlman, Rhea	Brooklyn, NY	3/31/48
Perlman, Ron	New York, NY	4/13/50
Perrine, Valerie	Galveston, TX	9/3/43
Perry, Luke	Fredericktown, OH	10/11/66
Perry, Mathew	Williamstown, MA	8/19/69
Persoff, Nehemiah	Jerusalem, Israel	8/14/20
Pesci, Joe	Newark, NJ	2/9/43
Peters, Bernadette	New York, NY	2/28/48
Peters, Brock	New York, NY	7/2/27
Peters, Jean	Canton, OH	10/15/26
Peters, Roberta	New York, NY	5/4/30
Peterson, Oscar	Montreal, Quebec	8/15/25
Petty, Tom	Gainesville, FL	10/20/53
Pfeiffer, Michelle	Santa Ana, CA	4/29/57
Philbin, Regis	New York, NY	8/25/34
Phillippe, Ryan	New Castle, DE	9/10/75
Phillips, Lou Diamond	Philippines	2/17/62
Phillips, Mackenzie	Alexandria, VA	11/10/59
Phillips, Michelle	Long Beach, CA	6/4/44
Phoenix, Joaquin	Puerto Rico	10/28/74
Pickett, Wilson	Prattville, AL	3/18/41
Pierce, David Hyde	Albany, NY	4/3/59
Pinchot, Bronson	New York, NY	5/20/59
Pinkett Smith, Jada	Baltimore, MD	8/18/71
Pirner, David	Green Bay, WI	4/16/64
Piscopo, Joe	Passaic, NJ	6/17/51
Pitt, Brad	Shawnee, OK	12/18/64
Plant, Robert	W. Bromwich, England	8/20/48
Pleshette, Suzanne	New York, NY	1/31/37
Plowright, Joan	Brigg, England	10/28/29
Plummer, Amanda	New York, NY	3/23/57
Plummer, Christopher	Toronto, Ontario	12/13/27
Poitier, Sidney	Miami, FL	2/20/27
Polanski, Roman	Paris, France	8/18/33
Pollack, Sydney	Lafayette, IN	7/1/34
Ponti, Carlo	Milan, Italy	12/11/13
Pop, Iggy	Ann Arbor, MI	4/21/47
Portman, Natalie	Jerusalem, Israel	6/9/81
Posey, Parker	Baltimore, MD	11/8/64
Post, Markie	Palo Alto, CA	11/4/50
Poston, Tom	Columbus, OH	10/17/27
Potts, Annie	Nashville, TN	10/28/52
Povich, Maury	Washington, DC	1/17/39
Powell, Jane	Portland, OR	4/1/28
Powers, Stefanie	Hollywood, CA	11/2/42
Prentiss, Paula	San Antonio, TX	3/4/39
Presley, Priscilla	New York, NY	5/24/46
Preston, Billy	Houston, TX	9/9/46
Previn, Andre	Berlin, Germany	4/6/29
Price, Leontyne	Laurel, MS	2/10/27
Price, Ray	Perryville, TX	1/12/26
Pride, Charley	Sledge, MS	3/18/39
Priestley, Jason	Vancouver, Brit. Columbia	8/28/69
Prince (The Artist)	Minneapolis, MN	6/7/58
Principal, Victoria	Fukuoka, Japan	1/3/50
Prinze, Freddie, Jr.	Albuquerque, NM	3/8/76
Prosky, Robert	Philadelphia, PA	12/13/30
Pryce, Jonathan	Wales	6/1/47
Pryor, Richard	Peoria, IL	12/1/40
Puente, Tito	New York, NY	4/20/23
Pulliam, Keshia Knight	Newark, NJ	4/9/79
Pullman, Bill	Hornell, NY	12/17/54
Purcell, Sarah	Richmond, IN	10/8/48
Quaid, Dennis	Houston, TX	4/9/54
Quaid, Randy	Houston, TX	10/1/50
Queen Latifah	East Orange, NJ	3/18/70
Quinn, Aidan	Chicago, IL	3/8/59
Quinn, Anthony	Chihuahua, Mexico	4/21/15
Quinn, Martha	Albany, NY	5/11/59
Rachins, Alan	Cambridge, MA	10/10/47
Rae, Charlotte	Milwaukee, WI	4/22/26
Raffi	Cairo, Italy	7/8/48
Rainer, Luise	Vienna, Austria	1/12/10
Raitt, Bonnie	Burbank, CA	11/8/49
Ramey, Samuel	Colby, KS	3/28/42
Ramone, Dee Dee	Berlin, Germany	9/18/52
Ramone, Joey	Forest Hills, NY	5/19/51
Ramone, Johnny	Long Island, NY	10/8/51
Ramone, Tommy	Budapest, Hungary	1/29/52
Rampal, Jean-Pierre	Marseilles, France	1/7/22
Randall, Tony	Tulsa, OK	2/26/20
Randolph, John	New York, NY	6/1/15
Randolph, Joyce	Detroit, MI	10/21/25
Raphael, Sally Jessy	Easton, PA	2/25/43
Rashad, Phylicia	Houston, TX	6/17/48
Ratzenberger, John	Bridgeport, CT	4/6/47
Rawls, Lou	Chicago, IL	12/1/36

Name	Birthplace	Birthdate	Name	Birthplace	Birthdate
Reagan, Ronald	Tampico, IL	2/6/11	Russell, Ken	Southampton, England	7/3/27
Reddy, Helen	Melbourne, Australia	10/25/41	Russell, Keri	Fountain Valley, CA	3/23/76
Redford, Robert	Santa Monica, CA	8/18/37	Russell, Kurt	Springfield, MA	3/17/51
Redgrave, Lynn	London, England	3/8/43	Russell, Mark	Buffalo, NY	8/23/32
Redgrave, Vanessa	London, England	1/30/37	Russell, Leon	Lawton, OK	4/2/41
Reed, Jerry	Atlanta, GA	3/20/37	Russell, Nipsey	Atlanta, GA	10/13/24
Reed, Lou	Long Island, NY	3/2/43	Russell, Theresa	San Diego, CA	3/20/57
Reed, Rex	Ft. Worth, TX	10/2/38	Russo, Rene	Burbank, CA	2/17/54
Reese, Della	Detroit, MI	7/6/31	Rutherford, Ann	Toronto, Ontario	11/2/20
Reeve, Christopher	New York, NY	9/25/52	Ruttan, Susan	Oregon City, OR	9/16/50
Reeves, Keanu	Beirut, Lebanon	9/2/64	Ryan, Meg	Fairfield, CT	11/19/61
Regalbuto, Joe	New York, NY	8/24/49	Ryan, Roz	Detroit, MI	7/7/51
Reid, Tim	Norfolk, VA	12/19/44	Rydell, Bobby	Philadelphia, PA	4/26/42
Reilly, Charles Nelson	New York, NY	1/13/31	Ryder, Winona	Winona, MN	10/29/71
Reiner, Carl	Bronx, NY	3/20/22			
Reiner, Rob	Bronx, NY	3/6/45	Sabato, Antonio, Jr.	Italy	2/29/72
Reinhold, Judge	Wilmington, DE	5/21/56	Sade	Ibadan, Nigeria	1/16/59
Reinking, Ann	Seattle, WA	11/10/50	Sagal, Katie	Los Angeles, CA	1956
Reiser, Paul	New York, NY	3/30/57	Saget, Bob	Philadelphia, PA	5/17/56
Reitman, Ivan	Czechoslovakia	10/27/46	Sahl, Mort	Montreal, Quebec	5/11/27
Resnik, Regina	New York, NY	8/30/24	Saint, Eva Marie	Newark, NJ	7/4/24
Reynolds, Burt	Waycross, GA	2/11/36	St. James, Susan	Los Angeles, CA	8/14/46
Reynolds, Debbie	El Paso, TX	4/1/32	St. John, Jill	Los Angeles, CA	8/19/40
Reznor, Trent	Mercer, PA	5/17/65	Sajak, Pat	Chicago, IL	10/26/47
Rhames, Ving	New York, NY	5/12/61	Saks, Gene	New York, NY	11/8/21
Rhymes, Busta	Brooklyn, NY	5/20/72	Sales, Soupy	Franklinton, NC	1/8/26
Ricci, Christina	Santa Monica, CA	2/12/80	Samms, Emma	London, England	8/28/60
Richards, Keith	Kent, England	12/18/43	Sandler, Adam	Brooklyn, NY	9/9/66
Richards, Michael	Culver City, CA	7/21/49	Sands, Julian	Yorkshire, England	1/15/58
Richardson, Ian	Edinburgh, Scotland	4/7/34	Sanford, Isabel	New York, NY	8/29/17
Richardson, Kevin	Lexington, KY	10/3/72	San Giacomo, Laura	Hoboken, NJ	11/14/62
Richardson, Miranda	Lancashire, England	3/3/58	Sarandon, Susan	New York, NY	10/4/46
Richardson, Natasha	London, England	5/11/63	Sarnoff, Dorothy	New York, NY	5/25/17
Richardson, Patricia	Bethesda, MD	2/23/51	Sartain, Gailard	Tulsa, OK	9/18/46
Richie, Lionel	Tuskegee, AL	6/20/50	Savage, Ben	Chicago, IL	9/13/80
Rickles, Don	New York, NY	5/8/26	Savage, Fred	Highland Park, IL	7/9/76
Rickman, Alan	Hammersmith, England	2/21/46	Saxon, John	Brooklyn, NY	8/5/35
Riegert, Peter	New York, NY	4/11/47	Sayles, John	Schenectady, NY	9/28/50
Rigg, Diana	Doncaster, England	7/20/38	Scaggs, Boz	Dallas, TX	6/8/44
Rimes, LeAnn	Jackson, MS	8/28/82	Scales, Prunella	Surrey, England	1933
Ringwald, Molly	Roseville, CA	2/18/68	Scalia, Jack	Brooklyn, NY	11/10/51
Ritter, John	Burbank, CA	9/17/48	Schallert, William	Los Angeles, CA	7/6/22
Rivera, Chita	Washington, DC	1/23/33	Scheider, Roy	Orange, NJ	11/10/32
Rivera, Geraldo	New York, NY	7/4/43	Schell, Maria	Vienna, Austria	1/15/26
Rivers, Joan	Brooklyn, NY	6/8/37	Schell, Maximilian	Vienna, Austria	12/8/30
Roach, Max	Elizabeth City, NC	1/10/24	Schenkel, Chris	Bippus, IN	8/21/23
Robards, Jason, Jr.	Chicago, IL	7/26/22	Schiffer, Claudia	Rheinbach, Germany	8/25/70
Robbins, Tim	W. Covina, CA	10/16/58	Schneider, John	Mt. Kisco, NY	4/8/54
Roberts, Doris	St. Louis, MO	11/4/29	Schneider, Rob	San Francisco, CA	10/31/64
Roberts, Eric	Biloxi, MS	4/18/56	Schroder, Rick	Staten Island, NY	4/3/70
Roberts, Julia	Smyrna, GA	10/28/67	Schwarzenegger, Arnold	Graz, Austria	7/30/47
Roberts, Pernell	Waycross, GA	5/18/30	Schwarzkopf, Elisabeth	Jarotschin, Poland	12/9/15
Roberts, Tony	New York, NY	10/22/39	Schwimmer, David	Queens, NY	11/12/67
Robertson, Cliff	La Jolla, CA	9/9/25	Sciorra, Annabella	New York, NY	3/24/64
Robertson, Dale	Harrah, OK	7/14/23	Scofield, Paul	Hurst, Pierpont, England	1/21/22
Robinson, Smokey	Detroit, MI	2/19/40	Scolari, Peter	New Rochelle, IL	9/12/54
Roche, Eugene	Boston, MA	9/22/28	Scorsese, Martin	New York, NY	11/17/42
Rock, Chris	South Carolina	2/7/66	Scott, Lizabeth	Scranton, PA	9/29/22
Rodgers, Jimmy	Camas, WA	9/18/33	Scott, Martha	Jamesport, MO	9/22/14
Rodriquez, Johnny	Sabinal, TX	12/10/51	Scott Thomas, Kristin	Cornwall, England	1960
Rogers, Fred	Latrobe, PA	3/20/28	Scotto, Renata	Savona, Italy	2/24/35
Rogers, Kenny	Houston, TX	8/21/38	Scully, Vin	New York, NY	11/29/27
Rogers, Mimi	Coral Gables, FL	1/27/56	Seagal, Steven	Lansing, MI	4/10/51
Rogers, Wayne	Birmingham, AL	4/7/33	Secor, Kyle	Tacoma, WA	5/31/60
Rollins, Sonny	New York, NY	9/7/29	Sedaka, Neil	New York, NY	3/13/39
Roman, Ruth	Lynn, MA	12/22/24	Seeger, Pete	New York, NY	5/3/19
Romano, Ray	New York, NY	12/21/57	Segal, George	Great Neck, NY	2/13/34
Ronstadt, Linda	Tucson, AZ	7/15/46	Seidelman, Susan	Philadelphia, PA	12/11/52
Rooney, Mickey	Brooklyn, NY	9/23/20	Seinfeld, Jerry	New York, NY	4/29/54
Rose, Axl	Lafayette, IN	2/6/62	Sellecca, Connie	New York, NY	5/25/55
Rose Marie	New York, NY	8/15/25	Selleck, Tom	Detroit, MI	1/29/45
Roseanne	Salt Lake City, UT	11/3/52	Severinsen, Doc	Arlington, OR	7/7/27
Ross, Diana	Detroit, MI	3/26/44	Sewell, Rufus	London, England	10/29/67
Ross, Katharine	Hollywood, CA	1/29/42	Seymour, Jane	Middlesex, England	2/15/51
Rossdale, Gavin (Bush)	London, England	10/30/67	Shackelford, Ted	Oklahoma City, OK	6/23/46
Ross, Marion	Albert Lea, MN	10/25/28	Shaffer, Paul	Thunder Bay, Ontario	11/28/49
Rossellini, Isabella	Rome, Italy	6/18/52	Shandling, Garry	Chicago, IL	11/29/49
Rostropovich, Mstislav	Baku, Azerbaijan	3/12/27	Shankar, Ravi	Benares, India	4/7/20
Roth, David Lee	Bloomington, IN	10/10/55	Sharif, Omar	Alexandria, Egypt	4/10/32
Roth, Tim	London, England	5/14/61	Shatner, William	Montreal, Quebec	3/22/31
Rotten, Johnny	England	1/31/56	Shaughnessy, Charles	London, England	2/9/55
Rourke, Mickey	Schenectady, NY	7/16/53	Shaver, Helen	St. Thomas, Ontario	2/24/51
Routledge, Patricia	Birkenhead, England	2/17/29	Shaw, Artie	New York, NY	5/23/10
Rowlands, Gena	Cambria, WI	6/19/30	Shea, John	N. Conway, NH	4/14/49
Rudner, Rita	Coconut Grove, FL	9/17/56	Shearer, Harry	Los Angeles, CA	12/23/43
Ruehl, Mercedes	Queens, NY	2/28/48	Shearer, Moira	Scotland	1/17/26
Rush, Barbara	Denver, CO	1/4/30	Shearing, George	London, England	8/13/19
Rush, Geoffrey	Toowoomba, Australia	1951	Sheedy, Ally	New York, NY	6/12/62
Russell, Jane	Bemidji, MN	6/21/21	Sheen, Charlie	Los Angeles, CA	9/3/65

Name	Birthplace	Birthdate	Name	Birthplace	Birthdate
Sheen, Martin	Dayton, OH	8/3/40	Sternhagen, Frances	Washington, DC	1/13/30
Shelley, Carole	London, England	8/16/39	Stevens, Andrew	Memphis, TN	6/10/55
Shepard, Sam	Ft. Sheridan, IL	11/5/43	Stevens, Cat	London, England	7/21/48
Shepherd, Cybill	Memphis, TN	2/18/49	Stevens, Connie	Brooklyn, NY	8/8/38
Sheridan, Nicollette	Northington, England	11/21/63	Stevens, Rise	New York, NY	6/11/13
Shields, Brooke	New York, NY	5/31/65	Stevens, Stella	Yazoo City, MS	10/1/36
Shire, Talia	New York, NY	4/25/46	Stevenson, Parker	Philadelphia, PA	6/4/52
Short, Bobby	Danville, IL	9/15/24	Stewart, French	Albuquerque, NM	2/20/64
Short, Martin	Hamilton, Ontario	3/26/50	Stewart, Jon	Lawrence, NJ	1963
Show, Grant	Detroit, MI	4/27/62	Stewart, Patrick	Mirfield, England	7/13/40
Shue, Andrew	South Orange, NJ	2/20/67	Stewart, Rod	London, England	1/10/45
Shue, Elisabeth	Wilmington, DE	6/10/63	Stiers, David Ogden	Peoria, IL	10/31/42
Shull, Richard B.	Evanston, IL	2/24/29	Stiller, Ben	New York, NY	11/30/65
Siepi, Cesare	Milan, Italy	2/10/23	Stiller, Jerry	New York, NY	6/8/29
Sikking, James B.	Los Angeles, CA	3/5/34	Stills, Stephen	Dallas, TX	1/3/45
Sills, Beverly	Brooklyn, NY	5/25/29	Sting	Newcastle, England	10/2/51
Silver, Ron	New York, NY	7/2/46	Stipe, Michael	Decatur, GA	1/4/60
Silverman, Jonathan	Los Angeles, CA	8/5/66	Stockwell, Dean	Hollywood, CA	3/5/36
Silverstone, Alicia	San Francisco, CA	10/4/76	Stoltz, Eric	American Samoa	9/30/61
Simmons, Gene	Haifa, Israel	8/25/49	Stone, Dee Wallace	Kansas City, KS	12/14/48
Simmons, Jean	London, England	1/31/29	Stone, Oliver	New York, NY	9/15/46
Simmons, Richard	New Orleans, LA	7/12/48	Stone, Sharon	Meadville, PA	3/10/58
Simon, Carly	New York, NY	6/25/45	Stookey, Paul	Baltimore, MD	12/30/37
Simon, Paul	Newark, NJ	10/13/41	Storch, Larry	New York, NY	1/8/23
Simone, Nina	Tyron, NC	2/21/33	Storm, Gale	Bloomington, TX	4/5/22
Sinatra, Nancy	Jersey City, NJ	6/8/40	Stowe, Madeleine	Los Angeles, CA	8/18/58
Sinbad	Benton Harbor, MI	11/10/56	Straight, Beatrice	Old Westbury, NY	8/2/18
Sinise, Gary	Blue Island, IL	3/7/55	Strait, George	Pearsall, TX	5/18/52
Singleton, John	Los Angeles, CA	1/6/68	Strasser, Robin	New York, NY	5/7/45
Singleton, Penny	Philadelphia, PA	9/15/08	Stratas, Teresa	Toronto, Ontario	5/26/38
Skerritt, Tom	Detroit, MI	8/25/33	Strauss, Peter	New York, NY	2/20/47
Slater, Christian	New York, NY	8/19/69	Streep, Meryl	Summit, NJ	6/22/49
Slater, Helen	Massapequa, NY	12/14/63	Streisand, Barbra	Brooklyn, NY	4/24/42
Slezak, Erika	Hollywood, CA	8/5/46	Stringfield, Sherry	Colorado Springs, CO	6/24/67
Slick, Grace	Chicago, IL	10/30/39	Stritch, Elaine	Detroit, MI	2/2/26
Smirnoff, Yakov	Odessa, Ukraine	1/24/51	Struthers, Sally	Portland, OR	7/28/48
Smith, Allison	New York, NY	12/9/69	Stuart, Gloria	Santa Monica, CA	7/4/10
Smith, Jaclyn	Houston, TX	10/26/47	Stuarti, Enzo	Rome, Italy	3/3/25
Smith, Keely	Norfolk, VA	3/9/35	Sullivan, Susan	New York, NY	11/18/44
Smith, Kevin	Red Bank, NJ	8/2/70	Sumac, Yma	Ichocan, Peru	9/10/27
Smith, Maggie	Ilford, England	12/28/34	Summer, Donna	Boston, MA	12/31/48
Smith, Will	Philadelphia, PA	9/25/68	Sutherland, Donald	St. John, New Brunswick	7/17/34
Smits, Jimmy	New York, NY	7/9/55	Sutherland, Joan	Sydney, Australia	11/7/26
Smothers, Dick	New York, NY	11/20/39	Sutherland, Kiefer	London, England	12/20/66
Smothers, Tom	New York, NY	2/2/37	Swayze, Patrick	Houston, TX	8/18/54
Snipes, Wesley	Orlando, FL	7/31/63	Swit, Loretta	Passaic, NJ	11/4/37
Snow, Hank	Nova Scotia, Canada	5/9/14			
Snyder, Tom	Milwaukee, WI	5/12/36	Takei, George	Los Angeles, CA	4/20/39
Somers, Suzanne	San Bruno, CA	10/16/46	Tallchief, Maria	Fairfax, OK	1/24/25
Sommer, Elke	Berlin, Germany	11/5/41	Tamblyn, Russ	Los Angeles, CA	12/30/34
Sorbo, Kevin	Mound, MN	9/24/58	Tarantino, Quentin	Knoxville, TN	3/27/63
Sorvino, Mira	Tenafly, NJ	9/28/70	Taylor, Billy	Greenville, SC	7/24/21
Sorvino, Paul	Brooklyn, NY	4/13/39	Taylor, Buck	Hollywood, CA	5/13/38
Sothern, Ann	Valley City, ND	1/22/09	Taylor, Elizabeth	London, England	2/27/32
Soul, David	Chicago, IL	8/28/43	Taylor, James	Boston, MA	3/12/48
Spacek, Sissy	Quitman, TX	12/25/49	Taylor, Rip	Washington, DC	1/13/30
Spacey, Kevin	S. Orange, NJ	7/26/59	Taylor, Rod	Sydney, Australia	1/11/29
Spade, David	Birmingham, MI	7/22/65	Taymor, Julie	Newton, MA	12/15/52
Spader, James	Boston, MA	2/7/60	Te Kanawa, Kiri	Gisborne, New Zealand	3/6/44
Spano, Joe	San Francisco, CA	7/7/46	Tebaldi, Renata	Pesaro, Italy	2/1/22
Spears, Britney	Kentwood, LA	12/2/81	Temple Black, Shirley	Santa Monica, CA	4/23/28
Spector, Phil	Bronx, NY	12/25/40	Tennant, Victoria	London, England	9/30/50
Spelling, Aaron	Dallas, TX	4/22/28	Tennille, Toni	Montgomery, AL	5/8/43
Spelling, Tori	Los Angeles, CA	5/16/73	Tesh, John	Garden City, NY	7/9/52
Spielberg, Steven	Cincinnati, OH	12/18/47	Tharp, Twyla	Portland, IN	7/1/41
Spiner, Brent	Houston, TX	2/2/49	Thaxter, Phyllis	Portland, ME	11/20/21
Springer, Jerry	London, England	2/13/44	Thicke, Alan	Kirkland Lake, Ontario	3/1/47
Springfield, Rick	Sydney, Australia	8/23/49	Thiessen, Tiffani-Amber	Long Beach, CA	1/23/74
Springsteen, Bruce	Freehold, NJ	9/23/49	Thomas, Jay	New Orleans, LA	7/12/48
Stack, Robert	Los Angeles, CA	1/13/19	Thomas, Jonathan Taylor	Bethlehem, PA	9/8/81
Stafford, Jo	Coalinga, CA	11/12/18	Thomas, Marlo	Detroit, MI	11/21/43
Stahl, Richard	Detroit, MI	1/4/32	Thomas, Michael Tilson	Hollywood, CA	12/21/44
Stallone, Sylvester	New York, NY	7/6/46	Thomas, Philip Michael	Columbus, OH	5/26/49
Stamos, John	Cypress, CA	8/19/63	Thomas, Richard	New York, NY	6/13/51
Stamp, Terence	Stepney, England	7/22/39	Thompson, Emma	London, England	4/15/59
Stang, Arnold	New York, NY	9/28/25	Thompson, Jack	Sydney, Australia	8/31/40
Stanley, Kim	Tularosa, NM	2/11/25	Thompson, Lea	Rochester, MN	5/31/61
Stanton, Harry Dean	West Irvine, KY	7/14/26	Thompson, Sada	Des Moines, IA	9/27/29
Stapleton, Jean	New York, NY	1/19/23	Thorne-Smith, Courtney	San Francisco, CA	11/8/68
Stapleton, Maureen	Troy, NY	6/21/25	Thornton, Billy Bob	Hot Springs, AR	8/4/55
Starr, Ringo	Liverpool, England	7/7/40	Thurman, Uma	Boston, MA	4/29/70
Steenburgen, Mary	Newport, AR	2/8/53	Tiegs, Cheryl	Minnesota	9/25/47
Stefani, Gwen	Anaheim, CA	10/3/69	Tillis, Mel	Tampa, FL	8/8/32
Steiger, Rod	W. Hampton, NY	4/14/25	Tilly, Jennifer	Los Angeles, CA	9/6/61
Stein, Ben	Washington, DC	11/25/44	Tilly, Meg	Texada, B.C.	2/14/60
Stephens, James	Mt. Kisco, NY	5/18/51	Timberlake, Justin	Memphis, TN	1/31/81
Stern, Daniel	Stamford, CT	5/28/57	Todd, Richard	Dublin, Ireland	6/11/19
Stern, Howard	New York, NY	1/12/54	Tomei, Marisa	New York, NY	12/4/64
Stern, Isaacv	Kreminiecz, Russia	7/21/20	Tomlin, Lily	Detroit, MI	9/1/39

Name	Birthplace	Birthdate	Name	Birthplace	Birthdate
Tomlinson, David	Scotland	5/7/17	Warner, Malcolm-Jamal	Jersey City, NJ	8/18/70
Tork, Peter	Washington, DC	2/13/44	Warren, Lesley Ann	New York, NY	8/16/46
Torn, Rip	Temple, TX	2/6/31	Warrick, Ruth	St. Joseph, MO	6/29/16
Townsend, Robert	Chicago, IL	2/6/57	Warwick, Dionne	East Orange, NJ	12/12/41
Townshend, Peter	Chiswick, England	5/19/45	Washington, Denzel	Mt. Vernon, NY	12/28/54
Travanti, Daniel J.	Kenosha, WI	3/7/40	Waters, John	Baltimore, MD	4/22/46
Travers, Mary	Louisville, KY	11/9/36	Waters, Roger	Great Bookham, England	9/9/44
Travis, Nancy	New York, NY	9/21/61	Waterston, Sam	Cambridge, MA	11/15/40
Travis, Randy	Marshville, NC	5/4/59	Watson, Emily	London, England	1/14/67
Travolta, John	Englewood, NJ	2/18/54	Watts, Andre	Nuremberg, Germany	6/20/46
Trebek, Alex	Sudbury, Ontario	7/22/40	Wayans, Damon	New York, NY	9/4/60
Trevor, Claire	New York, NY	3/8/09	Wayans, Keenen Ivory	New York, NY	6/8/58
Tritt, Travis	Marietta, GA	2/9/63	Waxman, Al	Toronto, Ontario	3/2/35
Tucker, Michael	Baltimore, MD	2/6/44	Weathers, Carl	New Orleans, LA	1/14/48
Tucker, Tanya	Seminole, TX	10/10/58	Weaver, Dennis	Joplin, MO	6/4/24
Tune, Tommy	Wichita Falls, TX	2/28/39	Weaver, Fritz	Pittsburgh, PA	1/19/26
Turlington, Christy	San Francisco, CA	1/2/69	Weaver, Sigourney	New York, NY	10/8/49
Turner, Janine	Lincoln, NE	12/6/62	Weir, Peter	Sydney, Australia	8/8/44
Turner, Kathleen	Springfield, MO	6/19/54	Weitz, Bruce	Norwalk, CT	5/27/43
Turner, Tina	Brownsville, TN	11/26/39	Welch, Raquel	Chicago, IL	9/5/40
Turturro, John	Brooklyn, NY	2/28/57	Weld, Tuesday	New York, NY	8/27/43
Twain, Shania	Windsor, Ontario	8/28/65	Wells, Kitty	Nashville, TN	8/30/19
Twiggy	London, England	9/19/46	Wendt, George	Chicago, IL	10/17/48
Tyler, Liv	Portland, ME	7/1/77	West, Adam	Walla Walla, WA	9/19/29
Tyler, Steven	Boston, MA	3/26/48	Wettig, Patricia	Cincinnati, OH	12/4/51
Tyson, Cicely	New York, NY	12/19/33	Whalley-Kilmer, Joanne	Manchester, England	8/25/64
			Wheaton, Wil	Burbank, CA	7/29/72
Uecker, Bob	Milwaukee, WI	1/26/35	Whitaker, Forest	Longview, TX	7/15/61
Uggams, Leslie	New York, NY	5/25/43	White, Barry	Galveston, TX	9/12/44
Ullman, Tracey	Slough, England	12/30/59	White, Betty	Oak Park, IL	1/17/22
Ullmann, Liv	Tokyo, Japan	12/16/38	White, Jaleel	Los Angeles, CA	11/27/76
Ulrich, Skeet	North Carolina	1/20/70	White, Vanna	N. Myrtle Beach, SC	2/18/57
Underwood, Blair	Tacoma, WA	8/25/64	Whiting, Margaret	Detroit, MI	7/22/24
Urich, Robert	Toronto, Ohio	12/19/47	Whitman, Stuart	San Francisco, CA	2/1/26
Usher (Raymond IV)	Chattanooga, TN	10/14/79	Whitmore, James	White Plains, NY	10/1/21
Ustinov, Peter	London, England	4/16/21	Widmark, Richard	Sunrise, MN	12/26/14
			Wiest, Dianne	Kansas City, MO	3/28/48
Vaccaro, Brenda	Brooklyn, NY	11/18/39	Wilder, Billy	Vienna, Austria	6/22/06
Vale, Jerry	New York, NY	7/8/31	Wilder, Gene	Milwaukee, WI	6/11/35
Valente, Caterina	Paris, France	1/14/31	Williams, Andy	Wall Lake, IA	12/3/30
Valli, Frankie	Newark, NJ	5/3/37	Williams, Barry	Santa Monica, CA	9/30/54
Van Ark, Joan	New York, NY	6/16/43	Williams, Billy Dee	New York, NY	4/6/37
Vance, Courtney B.	Detroit, MI	3/12/60	Williams, Cindy	Van Nuys, CA	8/22/47
Vandross, Luther	New York, NY	4/20/51	Williams, Esther	Los Angeles, CA	8/8/23
Van Damme, Jean-			Williams, Hal	Columbus, OH	12/14/38
Claude	Brussels, Belgium	10/18/60	Williams, Hank, Jr.	Shreveport, LA	5/26/49
Van Der Beek, James	Chesire, CT	3/8/77	Williams, JoBeth	Houston, TX	1949
Van Doren, Mamie	Rowena, SD	2/6/36	Williams, Michelle	Kalcspell, MT	9/9/80
Van Dyke, Dick	West Plains, MO	12/13/25	Williams, Montel	Baltimore, MD	7/3/56
Van Dyke, Jerry	Danville, IL	7/27/31	Williams, Paul	Omaha, NE	9/19/40
Van Halen, Eddie	Nijmegen, Netherlands	1/26/57	Williams, Robin	Chicago, IL	7/21/52
Van Patten, Dick	New York, NY	12/9/28	Williams, Treat	Rowayton, CT	12/1/51
Van Peebles, Mario	Mexico	1/15/57	Williams, Vanessa	New York, NY	3/18/63
Van Sant, Gus	Louisville, KY	7/24/52	Williamson, Kevin	Bern, NC	3/14/65
Vaughn, Robert	New York, NY	11/22/32	Williamson, Nicol	Hamilton, Scotland	9/14/38
Vaughn, Vince	Minneapolis, MN	1970	Willis, Bruce	W. Germany	3/19/55
Vedder, Eddie	Evanston, IL	12/23/65	Wilson, Brian	Hawthorne, CA	6/20/42
Verdon, Gwen	Los Angeles, CA	1/13/25	Wilson, Demond	Valdosta, GA	10/13/46
Vereen, Ben	Miami, FL	10/10/46	Wilson, Elizabeth	Grand Rapids, MI	4/4/25
Verrett, Shirley	New Orleans, LA	5/31/31	Wilson, Nancy	Chillicothe, OH	2/20/37
Vickers, Jon	Prince Albert, Sask.	10/26/26	Windom, William	New York, NY	9/28/23
Vigoda, Abe	New York, NY	2/24/21	Winfield, Paul	Los Angeles, CA	5/22/41
Vincent, Jan-Michael	Denver, CO	7/15/44	Winfrey, Oprah	Kosciusko, MS	1/29/54
Vinson, Helen	Beaumont, TX	9/17/07	Winger, Debra	Cleveland, OH	5/16/55
Vinton, Bobby	Canonsburg, PA	4/16/35	Winkler, Henry	New York, NY	10/30/45
Vitale, Dick	East Rutherford, NJ	6/9/40	Winningham, Mare	Phoenix, AZ	5/6/59
Voight, Jon	Yonkers, NY	12/29/38	Winslet, Kate	Reading, England	10/5/75
Von Stade, Frederica	Somerville, NJ	6/1/45	Winter, Johnny	Beaumont, TX	2/23/44
Von Sydow, Max	Lund, Sweden	4/10/29	Winters, Jonathan	Dayton, OH	11/11/25
			Winters, Shelley	St. Louis, MO	8/18/22
Wagner, Jack	Washington, MO	10/3/59	Winwood, Steve	Birmingham, England	5/12/48
Wagner, Lindsay	Los Angeles, CA	6/22/49	Wiseman, Joseph	Montreal, Quebec	5/15/18
Wagner, Robert	Detroit, MI	2/10/30	Withers, Jane	Atlanta, GA	4/12/26
Wahl, Ken	Chicago, IL	2/14/56	Witherspoon, Reese	Nashville, TN	4/22/76
Wahlberg, Mark	Dorchester, MA	6/5/71	Witt, Alicia	Worcester, MA	8/21/75
Wain, Bea	Bronx, NY	4/30/17	Wolf, Scott	Boston, MA	6/4/68
Waite, Ralph	White Plains, NY	6/22/29	Wonder, Stevie	Saginaw, MI	5/13/50
Waits, Tom	Pomona, CA	12/7/49	Wong, Faye	Beijing, China	8/8/69
Walden, Robert	New York, NY	9/25/43	Woo, John	Guangzhau, China	5/1/46
Walken, Christopher	New York, NY	3/31/43	Wood, Elijah	Cedar Rapids, IA	1/28/81
Wallace, Marcia	Creston, IA	11/1/42	Woodard, Alfre	Tulsa, OK	11/2/53
Wallach, Eli	Brooklyn, NY	12/7/15	Woods, James	Vernal, NJ	4/18/47
Walston, Ray	Laurel, MS	11/2/24	Woodward, Edward	Croyden, England	6/1/30
Walter, Jessica	New York, NY	1/31/44	Woodward, Joanne	Thomasville, GA	2/27/30
Ward, Fred	San Diego, CA	1943	Wopat, Tom	Lodi, WI	9/9/50
Ward, Sela	Meridian, MS	8/11/56	Worth, Irene	Nebraska	6/23/16
Ward, Simon	London, England	10/19/41	Wray, Fay	Alberta, Canada	9/10/07
Warden, Jack	Newark, NJ	9/18/20	Wright, Martha	Seattle, WA	3/23/26
Warfield, Marsha	Chicago, IL	3/5/54	Wright, Max	Detroit, MI	8/2/43

Name	Birthplace	Birthdate	Name	Birthplace	Birthdate
Wright, Steven	New York, NY	12/6/55	Young, Burt	New York, NY	4/30/40
Wright, Teresa	New York, NY	10/27/18	Young, Loretta	Salt Lake City, UT	1/6/13
Wyatt, Jane	Campgaw, NJ	8/10/11	Young, Neil	Toronto, Ontario	11/12/45
Wyle, Noah	Hollywood, CA	6/4/71	Young, Sean	Louisville, KY	11/20/59
Wyman, Bill	London, England	10/24/36			
Wyman, Jane	St. Joseph, MO	1/4/14	Zane, Billy	Chicago, IL	2/24/66
			Zeffirelli, Franco	Florence, Italy	2/12/23
Yankovic, Weird Al	Lynwood, CA	10/23/59	Zellweger, Renee	Katy, TX.	1969
Yanni	Kalamata, Greece	11/4/54	Zemeckis, Robert	Chicago, IL	5/14/51
Yarborough, Glenn	Milwaukee, WI	1/12/30	Zerbe, Anthony	Long Beach, CA	5/20/36
Yarrow, Peter	New York, NY	5/31/38	Zeta-Jones, Catherine	Swansea, Wales	9/25/69
Yearwood, Trisha	Monticello, GA	9/19/64	Zimbalist, Efrem, Jr.	New York, NY	11/30/23
Yoakam, Dwight	Pikesville, KY	10/23/56	Zimbalist, Stephanie	Encino, CA	10/8/56
York, Michael	Fulmer, England	3/27/42	Zimmer, Kim.	Grand Rapids, MI	2/2/55
York, Susannah	London, England	1/9/42	Zukerman, Pinchas	Tel Aviv, Israel	7/16/48
Young, Alan	Northumberland, England	11/19/19	Zuniga, Daphne	San Francisco, CA	10/28/62

Entertainment Personalities of the Past

See also other lists for some deceased entertainers not included here.

Name	Born	Died	Name	Born	Died	Name	Born	Died
Abbott, Bud	1895	1974	Baxter, Anne	1923	1985	Broderick, Helen	1891	1959
Abbott, George	1887	1995	Baxter, Warner	1889	1951	Brown, Joe E.	1892	1973
Acuff, Roy	1903	1992	Beatty, Clyde	1904	1965	Bruce, Lenny	1926	1966
Adams, Maude	1872	1953	Beaumont, Hugh	1909	1982	Bruce, Nigel	1895	1953
Adler, Jacob P	1855	1926	Beavers, Louise	1902	1962	Bruce, Virginia	1910	1982
Adler, Luther	1903	1984	Beery, Noah, Sr.	1884	1946	Brynner, Yul	1915	1985
Adoree, Renee	1898	1933	Beery, Noah, Jr.	1913	1994	Buchanan, Edgar	1903	1979
Aherne, Brian	1902	1986	Beery, Wallace	1889	1949	Buñuel, Luis	1900	1983
Ailey, Alvin	1931	1989	Begley, Ed	1901	1970	Buono, Victor	1938	1982
Akins, Claude	1918	1994	Bellamy, Ralph	1904	1991	Burke, Billie	1885	1970
Albertson, Frank	1909	1964	Belushi, John	1949	1982	Burnette, Smiley	1911	1967
Albertson, Jack	1907	1981	Benaderet, Bea	1906	1968	Burns, George	1896	1996
Alda, Robert	1914	1986	Bendix, William	1906	1964	Burr, Raymond	1917	1993
Allen, Fred	1894	1956	Bennett, Constance	1904	1965	Burton, Richard	1925	1984
Allen, Gracie	1906	1964	Bennett, Joan	1910	1990	Busch, Mae	1897	1946
Allen, Mel	1913	1996	Bennett, Michael	1943	1987	Bushman, Francis X.	1883	1966
Allgood, Sara	1883	1950	Benny, Jack	1894	1974	Butterworth, Charles	1896	1946
Ameche, Don	1908	1993	Benzell, Mimi	1924	1970	Byington, Spring	1893	1971
Ames, Leon	1903	1993	Beradino, John	1917	1996			
Amsterdam, Morey	1909?	1996	Berg, Gertrude	1899	1966	Cabot, Bruce	1904	1972
Anderson, Judith	1897	1992	Bergen, Edgar	1903	1978	Cabot, Sebastian	1918	1977
Anderson, Marian	1902	1993	Bergman, Ingrid	1915	1982	Cagney, James	1899	1986
Andrews, Dana	1909	1992	Berkeley, Busby	1895	1976	Calhern, Louis	1895	1956
Andrews, Laverne	1913	1967	Bernardi, Herschel	1923	1986	Calhoun, Rory	1923	1999
Andrews, Maxine	1918	1995	Bernhardt, Sarah	1844	1923	Callas, Maria	1923	1977
Angeli, Pier	1933	1971	Bernie, Ben	1893	1943	Calloway, Cab	1907	1994
Anita Louise	1915	1970	Bessell, Ted	1939	1996	Cambridge, Godfrey	1933	1976
Arbuckle, Fatty (Roscoe)	1887	1933	Bickford, Charles	1889	1967	Campbell, Mrs. Patrick	1865	1940
Arden, Eve	1908	1990	Big Bopper, The	1930	1959	Candy, John	1950	1994
Arlen, Richard	1900	1976	Bing, Rudolf	1902	1997	Cantin, Has	1911	1993
Arliss, George	1868	1946	Bissell, Whit	1909	1996	Cantor, Eddie	1892	1964
Armetta, Henry	1888	1945	Bixby, Bill	1934	1993	Capra, Frank	1897	1991
Armstrong, Louis	1900	1971	Bjoerling, Jussi	1911	1960	Carey, Harry	1878	1947
Arnaz, Desi	1917	1986	Blackmer, Sidney	1895	1973	Carey, Macdonald	1913	1994
Arnold, Edward	1890	1956	Blake, Amanda	1931	1989	Carpenter, Karen	1950	1983
Arquette, Cliff	1905	1974	Blaine, Vivian	1921	1995	Carradine, John	1906	1988
Arthur, Jean	1900	1991	Blanc, Mel	1908	1989	Carrillo, Leo	1880	1961
Ashcroft, Peggy	1907	1991	Blocker, Dan	1928	1972	Carroll, Leo G.	1892	1972
Astaire, Fred	1899	1987	Blondell, Joan	1909	1979	Carroll, Madeleine	1906	1987
Astor, Mary	1906	1987	Blore, Eric	1888	1959	Carroll, Nancy	1905	1965
Atwill, Lionel	1885	1946	Blue, Ben	1901	1975	Carson, Jack	1910	1963
Auer, Mischa	1905	1967	Blyden, Larry	1925	1975	Caruso, Enrico	1873	1921
Austin, Gene	1900	1972	Bogarde, Dirk	1920	1999	Casals, Pablo	1876	1973
Autry, Gene	1907	1998	Bogart, Humphrey	1899	1957	Cass, Peggy	1924	1999
Ayres, Lew	1908	1996	Boland, Mary	1880	1965	Cassidy, Jack	1927	1976
			Boles, John	1895	1969	Cassavetes, John	1929	1989
Backus, Jim	1913	1989	Bolger, Ray	1904	1987	Castle, Irene	1893	1969
Bailey, Pearl	1918	1990	Bond, Ward	1903	1960	Castle, Vernon	1887	1918
Bainter, Fay	1892	1968	Bondi, Beulah	1892	1981	Caulfield, Joan	1922	1991
Baker, Josephine	1906	1975	Bono, Sonny	1935	1998	Chaliapin, Feodor	1873	1938
Balanchine, George	1904	1983	Boone, Richard	1917	1981	Champion, Gower	1919	1980
Ball, Lucille	1911	1989	Booth, Edwin	1833	1893	Chandler, Jeff	1918	1961
Balsam, Martin	1919	1996	Booth, Junius Brutus	1796	1852	Chaney, Lon	1883	1930
Bancroft, George	1882	1956	Booth, Shirley	1898	1992	Chaney, Lon, Jr.	1905	1973
Bankhead, Tallulah	1903	1968	Bow, Clara	1905	1965	Chapin, Harry	1942	1981
Banks, Leslie	1890	1952	Bowes, Maj. Edward	1874	1946	Chaplin, Charles	1889	1977
Bara, Theda	1890	1955	Bowman, Lee	1914	1979	Chase, Ilka	1905	1978
Barnes, Binnie	1903	1998	Boyd, Stephen	1928	1977	Chatterton, Ruth	1893	1961
Barnum, Phineas T.	1810	1891	Boyd, William	1898	1972	Cherrill, Virginia	1908	1996
Barrymore, Ethel	1879	1959	Boyer, Charles	1899	1978	Chevalier, Maurice	1888	1972
Barrymore, John	1882	1942	Brady, Alice	1893	1939	Clair, René	1898	1981
Barrymore, Lionel	1878	1954	Brand, Neville	1921	1992	Clark, Bobby	1888	1960
Barrymore, Maurice	1848	1905	Brazzi, Rossano	1916	1994	Clark, Dane	1913	1998
Barthelmess, Richard	1897	1963	Brennan, Walter	1894	1974	Clark, Fred	1914	1968
Bartholomew, Freddie	1924	1992	Brent, George	1904	1979	Clift, Montgomery	1920	1966
Bartok, Eva	1926	1998	Brett, Jeremy	1935	1995	Cline, Patsy	1932	1963
Basehart, Richard	1914	1984	Brice, Fanny	1891	1951	Clyde, Andy	1892	1967
Basie, Count	1904	1984	Bridges, Lloyd	1913	1998	Cobain, Kurt	1967	1994

Name	Born	Died
Cobb, Lee J.	1911	1976
Coburn, Charles	1877	1961
Cochran, Steve	1917	1965
Cohan, George M.	1878	1942
Cohen, Myron	1902	1986
Colbert, Claudette	1903	1996
Cole, Nat "King"	1919	1965
Collins, Ray	1890	1965
Colman, Ronald	1891	1958
Columbo, Russ	1908	1934
Connors, Chuck	1921	1992
Conrad, William	1920	1994
Conried, Hans	1917	1982
Conte, Richard	1911	1975
Convy, Bert	1933	1991
Conway, Tom	1904	1967
Coogan, Jackie	1914	1984
Cook, Elisha, Jr.	1904	1995
Cooke, Sam	1935	1964
Cooper, Gary	1901	1961
Cooper, Gladys	1888	1971
Cooper, Melville	1896	1973
Corby, Ellen	1913	1999
Cornell, Katharine	1893	1974
Correll, Charles ("Andy")	1890	1972
Costello, Dolores	1905	1979
Costello, Lou	1906	1959
Cotten, Joseph	1905	1994
Coward, Noel	1899	1973
Cox, Wally	1924	1973
Crabbe, Buster	1908	1983
Crane, Bob	1928	1978
Crawford, Broderick	1911	1986
Crawford, Joan	1904	1977
Crews, Laura Hope	1880	1942
Crisp, Donald	1880	1974
Croce, Jim	1942	1973
Crosby, Bing	1904	1977
Crothers, Scatman	1910	1986
Cugat, Xavier	1900	1990
Cukor, George	1899	1983
Cullen, Bill	1920	1990
Cummings, Robert	1908	1990
Currie, Finlay	1878	1968
Cushing, Peter	1913	1994
Dailey, Dan	1914	1978
Dandridge, Dorothy	1923	1965
Daniell, Henry	1894	1963
Daniels, Bebe	1901	1971
Darin, Bobby	1936	1973
Darnell, Linda	1921	1965
Darwell, Jane	1879	1967
Da Silva, Howard	1909	1986
Davenport, Harry	1866	1949
Davies, Marion	1897	1961
Davis, Bette	1908	1989
Davis, Joan	1907	1961
Davis, Sammy Jr.	1925	1990
Day, Dennis	1917	1988
Dean, James	1931	1955
Defore, Don	1917	1993
Dekker, Albert	1905	1968
Del Rio, Dolores	1908	1983
Demarest, William	1892	1983
DeMille, Agnes	1905	1993
DeMille, Cecil B.	1881	1959
Denison, Michael	1915	1998
Denning, Richard	1914	1998
Dennis, Sandy	1937	1992
Denny, Reginald	1891	1967
Denver, John	1943	1997
Derek, John	1926	1998
DeSica, Vittorio	1901	1974
Devine, Andy	1905	1977
Dewhurst, Colleen	1924	1991
De Wilde, Brandon	1942	1972
De Wolfe, Billy	1907	1974
Diamond, Selma	1920	1985
Dietrich, Marlene	1901	1992
Digges, Dudley	1879	1947
Disney, Walt	1901	1966
Dix, Richard	1894	1949
Donat, Robert	1905	1958
Donlevy, Brian	1901?	1972
Dors, Diana	1931	1984
Douglas, Melvyn	1901	1981
Douglas, Paul	1907	1959
Dove, Billie	1900	1998
Doyle, David	1929	1997
Drake, Alfred	1914	1992
Draper, Ruth	1889	1956
Dresser, Louise	1881	1965
Dressler, Marie	1869	1934
Drew, Mrs. John	1820	1897
Dru, Joanne	1923	1996
Duchin, Eddy	1909	1951
Duff, Howard	1917	1990
Dumbrille, Douglass	1890	1974
Dumont, Margaret	1889	1965
Duncan, Isadora	1878	1927
Dunn, James	1905	1967
Dunne, Irene	1898	1990
Dunnock, Mildred	1904	1991
Durante, Jimmy	1893	1980
Duryea, Dan	1907	1968
Duse, Eleanora	1858	1924
Eagels, Jeanne	1894	1929
Eckstine, Billy	1914	1993
Eddy, Nelson	1901	1967
Edelman, Herb	1933	1996
Edwards, Cliff	1897	1971
Edwards, Gus	1879	1945
Edwards, Vince	1928	1996
Egan, Richard	1923	1987
Ellington, Duke	1899	1974
Elliot, Cass	1941	1974
Elman, Mischa	1891	1967
Errol, Leon	1881	1951
Evans, Edith	1888	1976
Evans, Maurice	1901	1989
Ewell, Tom	1909	1994
Fairbanks, Douglas	1883	1939
Farley, Chris	1964	1997
Farmer, Frances	1914	1970
Farnum, Dustin	1870	1929
Farnum, William	1876	1953
Farrar, Geraldine	1882	1967
Farrell, Charles	1901	1990
Farrell, Glenda	1904	1971
Fassbinder, Rainer Werner	1946	1982
Fay, Frank	1897	1961
Faye, Alice	1912	1998
Fazenda, Louise	1895	1962
Feld, Fritz	1900	1993
Feldman, Marty	1933	1982
Fell, Norman	1924	1998
Fellini, Federico	1920	1993
Fenneman, George	1919	1997
Ferrer, Jose	1912	1992
Fetchit, Stepin	1898	1985
Fiedler, Arthur	1894	1979
Field, Betty	1918	1973
Fields, Gracie	1898	1979
Fields, W.C.	1879	1946
Fields, Totie	1931	1978
Finch, Peter	1916	1977
Fine, Larry	1902	1975
Firkusny, Rudolf	1912	1994
Fiske, Minnie Maddern	1865	1932
Fitzgerald, Barry	1888	1961
Flagstad, Kirsten	1895	1962
Fleming, Eric	1925	1966
Flippen, Jay C.	1900	1971
Flynn, Errol	1909	1959
Flynn, Joe	1925	1974
Foley, Red	1910	1968
Fonda, Henry	1905	1982
Fontaine, Frank	1920	1978
Fontanne, Lynn	1887	1983
Fonteyn, Margot	1919	1991
Ford, John	1895	1973
Ford, Paul	1901	1976
Ford, Tennessee Ernie	1919	1991
Ford, Wallace	1899	1966
Fosse, Bob	1927	1987
Foster, Phil	1914	1985
Foster, Preston	1901	1970
Foxx, Redd	1922	1991
Foy, Eddie	1857	1928
Franchi, Sergio	1933?	1990
Francis, Kay	1903	1968
Franciscus, James	1934	1991
Frann, Mary	1943	1998
Frawley, William	1893	1966
Frederick, Pauline	1885	1938
Friganza, Trixie	1870	1955
Frisco, Joe	1890	1958
Froman, Jane	1907	1980
Fuller, Samuel	1912	1997
Funt, Allen	1914	1999
Furness, Betty	1916	1994
Gabin, Jean	1904	1976
Gable, Clark	1901	1960
Gabor, Eva	1920	1995
Garbo, Greta	1905	1990
Garcia, Jerry	1942	1995
Gardenia, Vincent	1922	1992
Gardner, Ava	1922	1990
Garfield, John	1913	1952
Garland, Judy	1922	1969
Garson, Greer	1904	1996
Gaye, Marvin	1939	1984
Gaynor, Janet	1906	1984
Geer, Will	1902	1978
George, Gladys	1900	1954
Gibb, Andy	1958	1988
Gibson, Hoot	1892	1962
Gilbert, Billy	1894	1971
Gilbert, John	1895	1936
Gilford, Jack	1907	1990
Gillette, William	1855	1937
Gingold, Hermione	1897	1987
Gish, Dorothy	1898	1968
Gish, Lillian	1893	1993
Gleason, Jackie	1916	1987
Gleason, James	1886	1959
Gluck, Alma	1884	1938
Gobel, George	1919	1991
Goddard, Paulette	1905	1990
Godfrey, Arthur	1903	1983
Godunov, Alexander	1949	1995
Goldwyn, Samuel	1882	1974
Gomez, Thomas	1905	1971
Goodman, Benny	1909	1986
Gorcey, Leo	1915	1969
Gordon, Gale	1906	1995
Gordon, Ruth	1896	1985
Gosden, Freeman ("Amos")	1899	1982
Gottschalk, Ferdinand	1869	1944
Gottschalk, Louis	1829	1869
Gould, Glenn	1932	1982
Gould, Morton	1913	1996
Grable, Betty	1916	1973
Graham, Martha	1894	1991
Grahame, Gloria	1925	1981
Granger, Stewart	1913	1993
Grant, Cary	1904	1986
Granville, Bonita	1923	1988
Greene, Lorne	1915	1987
Greenstreet, Sydney	1879	1954
Griffith, David Wark	1874	1948
Griffith, Hugh	1912	1980
Guardino, Harry	1925	1995
Guthrie, Woody	1912	1967
Gwenn, Edmund	1875	1959
Gwynne, Fred	1926	1993
Hale, Alan	1892	1950
Hale, Alan, Jr.	1918	1990
Haley, Bill	1925	1981
Haley, Jack	1899	1979
Hall, Huntz	1919	1999
Hamilton, Margaret	1902	1985
Hammerstein, Oscar	1847	1919
Hardwicke, Cedric	1893	1964
Hardy, Oliver	1892	1957
Harlow, Jean	1911	1937
Harris, Phil	1904	1995
Harrison, Rex	1908	1990
Hart, William S.	1870	1946
Hartman, Phil	1948	1998
Harvey, Laurence	1928	1973
Hawkins, Jack	1910	1973
Hayakawa, Sessue	1890	1973
Hayden, Sterling	1916	1986
Hayes, Gabby	1885	1969
Hayes, Helen	1900	1993
Hayes, Peter Lind	1915	1998
Hayward, Leland	1902	1971
Hayward, Louis	1909	1985
Hayward, Susan	1917	1975
Hayworth, Rita	1918	1987
Head, Edith	1907	1981

Name	Born	Died	Name	Born	Died	Name	Born	Died
Healy, Ted	1896	1937	Kane, Helen	1910	1966	Long, Richard	1927	1974
Heflin, Van	1910	1971	Karloff, Boris	1887	1969	Lopez, Vincent	1895	1975
Heifetz, Jascha	1901	1987	Karns, Roscoe	1893	1970	Lord, Jack	1920?	1998
Held, Anna	1873	1918	Kaufman, Andy	1949	1984	Lorne, Marion	1888	1968
Hemingway, Margaux	1955	1996	Kaye, Danny	1913	1987	Lorre, Peter	1904	1964
Hendrix, Jimi	1942	1970	Kaye, Stubby	1918	1997	Lovejoy, Frank	1912	1962
Henie, Sonja	1912	1969	Kean, Charles	1811	1868	Lowe, Edmund	1890	1971
Henreid, Paul	1908	1992	Kean, Mrs. Charles	1806	1880	Loy, Myrna	1905	1993
Henson, Jim	1936	1990	Kean, Edmund	1787	1833	Lubitsch, Ernst	1892	1947
Hepburn, Audrey	1929	1993	Keaton, Buster	1895	1966	Lugosi, Bela	1882	1956
Hersholt, Jean	1886	1956	Keeler, Ruby	1910	1993	Lukas, Paul	1894	1971
Hickey, William	1928	1997	Keith, Brian	1921	1997	Lundigan, William	1914	1975
Hickson, Joan	1906	1998	Kellaway, Cecil	1894	1973	Lunt, Alfred	1892	1977
Hill, Benny	1925	1992	Kelley, DeForest	1920	1999	Lupino, Ida	1918	1995
Hirt, Al	1922	1999	Kelly, Emmett	1898	1979	Lynde, Paul	1926	1982
Hitchcock, Alfred	1899	1980	Kelly, Gene	1912	1996	Lynn, Diana	1926	1971
Hobson, Valerie	1917	1998	Kelly, Grace	1929	1982			
Hodiak, John	1914	1955	Kelly, Jack	1927	1992	MacDonald, Jeanette	1903	1965
Holden, Fay	1894	1973	Kelly, Nancy	1921	1985	Mack, Ted	1904	1976
Holden, William	1918	1981	Kelly, Patsy	1910	1981	MacLane, Barton	1902	1969
Holliday, Judy	1922	1965	Kelton, Pert	1907	1968	MacMurray, Fred	1908	1991
Holloway, Sterling	1905	1992	Kendall, Kay	1926	1959	MacRae, Gordon	1921	1986
Holly, Buddy	1936	1959	Kennedy, Arthur	1914	1990	Macready, George	1909	1973
Holt, Jack	1888	1951	Kennedy, Edgar	1890	1948	Madison, Guy	1922	1996
Holt, Tim	1918	1973	Kibbee, Guy	1886	1956	Magnani, Anna	1908	1973
Homolka, Oscar	1898	1978	Kilbride, Percy	1888	1964	Mancini, Henry	1924	1994
Hoon, Shannon	1967	1995	Kiley, Richard	1922	1999	Main, Marjorie	1890	1975
Hopkins, Miriam	1902	1972	Knight, Ted	1923	1986	Malle, Louis	1932	1995
Hopper, DeWolf	1858	1935	Kostelanetz, Andre	1901	1980	Mansfield, Jayne	1932	1967
Hopper, William	1915	1970	Kovacs, Ernie	1919	1962	Mantovani, Annunzio	1905	1980
Horowitz, Vladimir	1904	1989	Kruger, Otto	1885	1974	Marais, Jean	1913	1998
Horton, Edward Everett	1886	1970	Kubrick, Stanley	1928	1999	March, Fredric	1897	1975
Houdini, Harry	1874	1926	Kulp, Nancy	1921	1991	March, Hal	1920	1970
Houseman, John	1902	1988	Kurosawa, Akira	1910	1998	Marley, Bob	1945	1981
Howard (Horwitz), Curly	1903	1952				Marshall, Brenda	1915	1992
Howard, Eugene	1881	1965	Ladd, Alan	1913	1964	Marshall, E.G.	1910	1998
Howard, Joe	1867	1961	Lahr, Bert	1895	1967	Marshall, Herbert	1890	1966
Howard, Leslie	1890	1943	Lake, Arthur	1905	1987	Martin, Dean	1917	1995
Howard (Horwitz), Moe	1897	1975	Lake, Veronica	1919	1973	Martin, Mary	1913	1990
Howard (Horwitz), Shemp	1895	1955	Lamas, Fernando	1915	1982	Martin, Ross	1920	1981
			Lamour, Dorothy	1914	1996	Marvin, Lee	1924	1987
Howard, Tom	1885	1955	Lancaster, Burt	1913	1994	Marx, Arthur (Harpo)	1888	1964
Howard, Trevor	1916	1988	Lanchester, Elsa	1902	1986	Marx, Herbert (Zeppo)	1901	1979
Howard, Willie	1885	1949	Lane, Pricilla	1917	1995	Marx, Julius (Groucho)	1890	1977
Hudson, Rock	1925	1985	Landis, Carole	1919	1948	Marx, Leonard (Chico)	1886	1961
Hull, Henry	1890	1977	Landis, Jessie Royce	1904	1972	Marx, Milton (Gummo)	1893	1977
Hull, Josephine	1886	1957	Landon, Michael	1936	1991	Mason, James	1909	1984
Humphrey, Doris	1895	1958	Lang, Fritz	1890	1976	Massey, Daniel	1933	1998
Hunter, Jeffrey	1925	1969	Langdon, Harry	1884	1944	Massey, Raymond	1896	1983
Hunter, Ross	1921	1996	Langtry, Lillie	1853	1929	Mastroianni, Marcello	1924	1996
Husing, Ted	1901	1962	Lanza, Mario	1921	1959	Mature, Victor	1916	1999
Huston, John	1906	1987	LaRue, Lash (Alfred)	1917	1996	Maxwell, Marilyn	1921	1972
Huston, Walter	1884	1950	Lauder, Harry	1870	1950	Mayer, Louis B.	1885	1957
Hutchence, Michael	1960	1997	Laughton, Charles	1899	1962	Maynard, Ken	1895	1973
Hutton, Jim	1934	1979	Laurel, Stan	1890	1965	Mazurki, Mike	1909	1990
Hutton, Robert	1920	1994	Lawford, Peter	1923	1984	McCartney, Linda	1941	1998
Hyde-White, Wilfrid	1903	1991	Lawrence, Gertrude	1898	1952	McClure, Doug	1935	1995
			Lean, David	1908	1991	McCormack, John	1884	1945
Ingram, Rex	1895	1969	Lee, Bernard	1908	1981	McCrea, Joel	1905	1990
Iturbi, Jose	1895	1980	Lee, Bruce	1940	1973	McDaniel, Hattie	1895	1952
Ireland, Jill	1936	1990	Lee, Canada	1907	1952	McDowall, Roddy	1928	1998
Ireland, John	1915	1992	Lee, Gypsy Rose	1914	1970	McFarland, George "Spanky"	1928	1993
Irving, Henry	1838	1905	LeGallienne, Eva	1899	1991	McHugh, Frank	1899	1981
Ives, Burl	1909	1995	Lehmann, Lotte	1888	1976	McIntire, John	1907	1991
			Leigh, Vivien	1913	1967	McLaglen, Victor	1883	1959
Jackson, Joe	1875	1942	Leighton, Margaret	1922	1976	McMahon, Horace	1907	1971
Jackson, Mahalia	1911	1972	Lennon, John	1940	1980	McNeill, Don	1907	1979
Jaeckel, Richard	1926	1997	Lenya, Lotte	1898	1981	McQueen, Butterfly	1911	1995
Jaffe, Sam	1891	1984	Leonard, Eddie	1870	1941	McQueen, Steve	1930	1980
Jagger, Dean	1903	1991	Leonard, Sheldon	1907	1997	Meadows, Audrey	1924	1996
James, Dennis	1917	1997	LeRoy, Mervyn	1900	1987	Medford, Kay	1920	1980
James, Harry	1916	1983	Levant, Oscar	1906	1972	Meek, Donald	1880	1946
Janis, Elsie	1889	1956	Levene, Sam	1905	1980	Meeker, Ralph	1920	1989
Jannings, Emil	1886	1950	Levenson, Sam	1911	1980	Melba, Nellie	1861	1931
Janssen, David	1930	1980	Lewis, Joe E.	1902	1971	Melchior, Lauritz	1890	1973
Jenkins, Allen	1900	1974	Lewis, Shari	1934	1998	Menjou, Adolphe	1890	1963
Jessel, George	1898	1981	Lewis, Ted	1892	1971	Menken, Helen	1902	1966
Johnson, Ben	1918	1996	Liberace	1919	1987	Menuhin, Yehudi	1916	1999
Johnson, Chic	1892	1962	Lind, Jenny	1820	1887	Mercouri, Melina	1925	1994
Jolson, Al	1886	1950	Lindfors, Viveca	1920	1995	Mercury, Freddie	1946	1991
Jones, Brian	1942	1969	Lindley, Audra	1918	1997	Meredith, Burgess	1909	1997
Jones, Buck	1889	1942	Lillie, Beatrice	1894	1989	Merman, Ethel	1908	1984
Jones, Carolyn	1933	1983	Little, Cleavon	1939	1992	Merrill, Gary	1915	1990
Jones, Henry	1912	1999	Lloyd, Harold	1893	1971	Mifune, Toshiro	1920	1997
Jones, Spike	1911	1965	Lloyd, Marie	1870	1922	Milland, Ray	1905	1986
Joplin, Janis	1943	1970	Lockhart, Gene	1891	1957	Miller, Glenn	1904	1944
Jory, Victor	1902	1982	Logan, Ella	1913	1969	Miller, Marilyn	1898	1936
Joslyn, Allyn	1905	1981	Lombard, Carole	1909	1942	Mills, Harry	1913	1982
Julia, Raul	1940	1994	Lombardo, Guy	1902	1977			

Name	Born	Died	Name	Born	Died	Name	Born	Died
Minnevitch, Borrah	1903	1955	Parks, Larry	1914	1975	Rogers, Roy	1911	1998
Mineo, Sal	1939	1976	Pasternack, Josef A.	1881	1940	Rogers, Will	1879	1935
Miranda, Carmen	1913	1955	Pastor, Tony (Vaudevillian)	1837	1908	Roland, Gilbert	1905	1994
Mitchell, Cameron	1918	1994	Pastor, Tony (Bandleader)	1907	1969	Rolle, Esther	1920?	1998
Mitchell, Thomas	1892	1962	Patti, Adelina	1843	1919	Rollins, Howard	1950	1996
Mitchum, Robert	1917	1997	Patti, Carlotta	1840	1889	Romero, Cesar	1907	1994
Mix, Tom	1880	1940	Patrick, Gail	1911	1980	Rooney, Pat	1880	1962
Monica, Corbett	1930	1998	Pavlova, Anna	1885	1931	Rose, Billy	1899	1966
Monroe, Marilyn	1926	1962	Payne, John	1912	1989	Rossellini, Roberto	1906	1977
Monroe, Vaughn	1911	1973	Pearl, Minnie	1912	1996	Rowan, Dan	1922	1987
Montand, Yves	1921	1991	Peerce, Jan	1904	1984	Rubinstein, Artur	1887	1982
Montez, Maria	1917	1951	Pendleton, Nat	1899	1967	Ruggles, Charles	1886	1970
Montgomery, Elizabeth	1933	1995	Penner, Joe	1905	1941	Russell, Gail	1924	1961
Montgomery, Robert	1904	1981	Peppard, George	1928	1994	Russell, Lillian	1861	1922
Moore, Colleen	1900	1988	Perkins, Anthony	1932	1992	Russell, Rosalind	1911	1976
Moore, Grace	1901	1947	Perkins, Carl	1932	1998	Rutherford, Margaret	1892	1972
Moore, Garry	1914	1993	Peters, Susan	1921	1952	Ryan, Irene	1903	1973
Moore, Victor	1876	1962	Phoenix, River	1970	1993	Ryan, Robert	1909	1973
Moorehead, Agnes	1906	1974	Piaf, Edith	1915	1963			
Morgan, Dennis	1910	1994	Pickens, Slim	1919	1983	Sargent, Dick	1933	1994
Morgan, Frank	1890	1949	Pickford, Mary	1893	1979	St. Denis, Ruth	1877	1968
Morgan, Helen	1900	1941	Pidgeon, Walter	1897	1984	Sakall, S.Z.	1884	1955
Morgan, Henry	1915	1994	Pinza, Ezio	1892	1957	Sale (Chic), Charles	1885	1936
Morley, Robert	1908	1992	Pitts, Zasu	1898	1963	Sanders, George	1906	1972
Morris, Chester	1901	1970	Plato, Dana	1964	1999	Savalas, Telly	1924	1994
Morris, Greg	1934	1996	Pleasence, Donald	1919	1995	Schildkraut, Joseph	1895	1964
Morris, Wayne	1914	1959	Pons, Lily	1904	1976	Schipa, Tito	1889	1965
Morrison, Jim	1943	1971	Ponselle, Rosa	1897	1981	Schnabel, Artur	1882	1951
Morrow, Vic	1932	1982	Powell, Dick	1904	1963	Scott, George C.	1927	1999
Mostel, Zero	1915	1977	Powell, Eleanor	1912	1982	Scott, Hazel	1920	1981
Mowbray, Alan	1897	1969	Powell, William	1892	1984	Scott, Randolph	1898	1987
Mulhare, Edward	1923	1997	Power, Tyrone	1913	1958	Scott, Zachary	1914	1965
Mulligan, Gerry	1927	1996	Preminger, Otto	1905	1986	Scott-Siddons, Mrs.	1843	1896
Muni, Paul	1895	1967	Presley, Elvis	1935	1977	Seberg, Jean	1938	1979
Munshin, Jules	1915	1970	Preston, Robert	1918	1987	Seeley, Blossom	1892	1974
Murphy, Audie	1924	1971	Price, Vincent	1911	1993	Segovia, Andres	1893	1987
Murphy, George	1902	1992	Prima, Louis	1911	1978	Selena	1971	1995
Murray, Arthur	1895	1991	Prinze, Freddie	1954	1977	Sellers, Peter	1925	1980
Murray, Kathryn	1906	1999	Prowse, Juliet	1936	1996	Selznick, David O.	1902	1965
Murray, Mae	1885	1965	Pyle, Denver	1920	1997	Sennett, Mack	1884	1960
						Serling, Rod	1924	1975
Nagel, Conrad	1896	1970	Quayle, Anthony	1913	1989	Shakur, Tupac	1971	1996
Naish, J. Carroll	1900	1973	Questel, Mae	1908	1998	Shaw, Robert (actor)	1927	1978
Naldi, Nita	1898	1961	Quintero, José	1924	1999	Shaw, Robert (conductor)	1916	1999
Nance, Jack	1943	1997				Shawn, Ted	1891	1972
Natwick, Mildred	1908	1994	Rabb, Ellis	1930	1998	Shean, Al	1868	1949
Negri, Pola	1897	1987	Rabbit, Eddie	1941	1998	Shearer, Norma	1902	1983
Nelson, Harriet (Hilliard)	1914	1994	Radner, Gilda	1946	1989	Sheridan, Ann	1915	1967
Nelson, Ozzie	1906	1975	Raft, George	1895	1980	Shore, Dinah	1917	1994
Nelson, Rick	1940	1985	Rains, Claude	1890	1967	Shubert, Lee	1875	1953
Nesbit, Evelyn	1885	1967	Ralston, Esther	1902	1994	Siddons, Mrs. Sarah	1755	1831
Newley, Anthony	1931	1999	Rathbone, Basil	1892	1967	Sidney, Sylvia	1910	1999
Nijinsky, Vaslav	1890	1950	Ratoff, Gregory	1897	1960	Signoret, Simone	1921	1985
Nilsson, Anna Q.	1893	1974	Ray, Aldo	1926	1991	Silvers, Phil	1912	1985
Niven, David	1909	1983	Ray, Johnnie	1927	1990	Sim, Alastair	1900	1976
Nolan, Lloyd	1902	1985	Raye, Martha	1916	1994	Sinatra, Frank	1915	1998
Normand, Mabel	1894	1930	Raymond, Gene	1908	1998	Sinclair, Madge	1938	1995
Notorious B.I.G.	1972	1997	Redding, Otis	1941	1967	Siskel, Gene	1946	1999
Novarro, Ramon	1899	1968	Redgrave, Michael	1908	1985	Sitka, Emil	1914	1998
Nureyev, Rudolf	1938	1993	Reed, Donna	1921	1986	Sjostrom, Victor	1879	1960
			Reed, Oliver	1938	1999	Skelton, Red	1913	1997
Oakie, Jack	1903	1978	Reed, Robert	1932	1992	Skinner, Otis	1858	1942
Oakley, Annie	1860	1926	Reeves, George	1914	1959	Smith, Alexis	1921	1992
Oates, Warren	1928	1982	Reinhardt, Max	1873	1943	Smith, Buffalo Bob	1917	1998
Oberon, Merle	1911	1979	Remick, Lee	1935	1991	Smith, C. Aubrey	1863	1948
O'Brien, Edmond	1915	1985	Renaldo, Duncan	1904	1980	Smith, Kate	1907	1986
O'Brien, Pat	1899	1983	Rennie, Michael	1909	1971	Solti, George	1912	1997
O'Connell, Arthur	1908	1981	Renoir, Jean	1894	1979	Sondergaard, Gale	1899	1985
O'Connell, Helen	1921	1993	Rettig, Tommy	1941	1996	Sousa, John Philip	1854	1932
O'Connor, Una	1880	1959	Reynolds, Marjorie	1923	1997	Sparks, Ned	1884	1957
O'Keefe, Dennis	1908	1968	Rich, Charlie	1932	1995	Springfield, Dusty	1939	1999
Oland, Warner	1880	1938	Richardson, Ralph	1902	1983	Stander, Lionel	1908	1994
Olcott, Chauncey	1860	1932	Riddle, Nelson	1921	1985	Stanwyck, Barbara	1907	1990
Oliver, Edna May	1883	1942	Ritchard, Cyril	1898	1977	Stevens, Inger	1934	1970
Olivier, Laurence	1907	1989	Ritter, Tex	1907	1974	Stevens, Mark	1916	1994
Olsen, Ole	1892	1963	Ritter, Thelma	1905	1969	Stevenson, McLean	1929	1996
O'Neill, James	1849	1920	Ritz, Al	1901	1965	Stewart, James	1908	1997
Orbison, Roy	1936	1988	Ritz, Harry	1906	1986	Stickney, Dorothy	1896	1998
Ormandy, Eugene	1899	1985	Ritz, Jimmy	1903	1985	Stokowski, Leopold	1882	1977
O'Sullivan, Maureen	1911	1998	Robbins, Jerome	1918	1998	Stone, Lewis	1879	1953
Ouspenskaya, Maria	1876	1949	Robbins, Marty	1925	1982	Stone, Milburn	1904	1980
Owen, Reginald	1887	1972	Robeson, Paul	1898	1976	Strasberg, Lee	1901	1999
			Robinson, Bill	1878	1949	Strasberg, Susan	1938	1999
Paderewski, Ignace	1860	1941	Robinson, Edward G.	1893	1973	Sturges, Preston	1898	1959
Page, Geraldine	1924	1987	Rochester (E. Anderson)	1905	1977	Sullavan, Margaret	1911	1960
Pallette, Eugene	1889	1954	Roddenberry, Gene	1921	1991	Sullivan, Barry	1912	1994
Palmer, Lilli	1914	1986	Rodgers, Jimmie	1897	1933	Sullivan, Ed	1902	1974
Pangborn, Franklin	1894	1958	Rogers, Buddy	1904	1999	Sullivan, Francis L.	1903	1956
Parks, Bert	1914	1992	Rogers, Ginger	1911	1995	Summerville, Slim	1892	1946

Name	Born	Died
Swanson, Gloria	1899	1983
Swarthout, Gladys	1904	1969
Switzer, Carl "Alfalfa"	1926	1959
Talbot, Lyle	1904	1996
Talmadge, Norma	1893	1957
Tamiroff, Akim	1899	1972
Tandy, Jessica	1909	1994
Tanguay, Eva	1878	1947
Tati, Jacques	1908	1982
Taylor, Deems	1885	1966
Taylor, Dub	1907	1994
Taylor, Estelle	1899	1958
Taylor, Laurette	1887	1946
Taylor, Robert	1911	1969
Terry, Ellen	1847	1928
Thalberg, Irving	1899	1936
Thomas, Danny	1912	1991
Thomas, John Charles	1892	1960
Thorndike, Sybil	1882	1976
Tibbett, Lawrence	1896	1960
Tierney, Gene	1920	1991
Tiny Tim	1932?	1996
Tippett, Sir Michael	1905	1998
Todd, Michael	1909	1958
Tone, Franchot	1903	1968
Torme, Mel	1925	1999
Toscanini, Arturo	1867	1957
Tracy, Lee	1898	1968
Tracy, Spencer	1900	1967
Traubel, Helen	1903	1972
Travers, Henry	1874	1965
Treacher, Arthur	1894	1975
Tree, Herbert Beerbohm	1853	1917
Truex, Ernest	1890	1973
Truffaut, Francois	1932	1984
Tucker, Forrest	1919	1986
Tucker, Richard	1913	1975
Tucker, Sophie	1884	1966
Turner, Lana	1920	1995
Turpin, Ben	1874	1940

Name	Born	Died
Twelvetrees, Helen	1908	1959
Twitty, Conway	1933	1993
Valens, Ritchie	1941	1959
Valentino, Rudolph	1895	1926
Vallee, Rudy	1901	1986
Vance, Vivian	1912	1979
Van Fleet, Jo	1922	1996
Vaughan, Sarah	1924	1990
Veidt, Conrad	1893	1943
Velez, Lupe	1908	1944
Vera-Ellen	1926	1981
Vincent, Gene	1935	1971
Vicious, Sid	1958	1979
Von Stroheim, Erich	1885	1957
Von Zell, Harry	1906	1981
Walker, Junior	1942	1995
Walker, Nancy	1922	1992
Walker, Robert	1918	1951
Wallenda, Karl	1905	1978
Walsh, J. T.	1943	1998
Walsh, Raoul	1887	1980
Walter, Bruno	1876	1962
Ward, Helen	1916	1998
Waring, Fred	1900	1984
Warner, H. B.	1876	1958
Washington, Dinah	1924	1963
Waters, Ethel	1896	1977
Wayne, David	1914	1995
Wayne, John	1907	1979
Webb, Clifton	1891	1966
Webb, Jack	1920	1982
Weems, Ted	1901	1963
Weissmuller, Johnny	1904	1984
Welk, Lawrence	1903	1992
Welles, Orson	1915	1985
Wellman, William	1896	1975
Werner, Oskar	1922	1984
West, Mae	1892	1980
Weston, Jack	1924	1996
Whale, James	1889	1957

Name	Born	Died
Wheeler, Bert	1895	1968
White, Jesse	1919	1997
White, Pearl	1889	1938
Whiteman, Paul	1891	1967
Whitty, May	1865	1948
Wickes, Mary	1910	1995
Wilde, Cornel	1918	1989
Wilding, Michael	1912	1979
Williams, Bert	1877	1922
Williams, Guy	1924	1989
Williams, Hank Sr.	1923	1953
Wills, Bob	1905	1975
Wills, Chill	1903	1978
Wilson, Carl	1946	1998
Wilson, Dennis	1944	1983
Wilson, Flip	1933	1998
Wilson, Marie	1917	1972
Winninger, Charles	1884	1969
Withers, Grant	1904	1959
Wong, Anna May	1907	1961
Wood, Natalie	1938	1981
Wood, Peggy	1892	1978
Woolley, Monty	1888	1963
Wyler, William	1902	1981
Wynette, Tammy	1942	1998
Wynn, Ed	1886	1966
Wynn, Keenan	1916	1986
Yankovic, Frank	1915	1998
York, Dick	1929	1992
Young, Clara Kimball	1890	1960
Young, Gig	1913	1978
Young, Robert	1907	1998
Young, Roland	1887	1953
Youngman, Henny	1906	1998
Zanuck, Darryl F.	1902	1979
Zappa, Frank	1940	1993
Zinneman, Fred	1907	1997
Ziegfeld, Florenz	1869	1932
Zukor, Adolph	1873	1976

Original Names of Selected Entertainers

EDIE ADAMS: Elizabeth Edith Enke
EDDIE ALBERT: Edward Albert Heimberger
ALAN ALDA: Alphonso D'Abruzzo
JASON ALEXANDER: Jay Greenspan
FRED ALLEN: John Sullivan
WOODY ALLEN: Allen Konigsberg
JUNE ALLYSON: Ella Geisman
JULIE ANDREWS: Julia Wells
EVE ARDEN: Eunice Quedens
BEATRICE ARTHUR: Bernice Frankel
JEAN ARTHUR: Gladys Greene
FRED ASTAIRE: Frederick Austerlitz
BABYFACE: Kenneth Edmonds
LAUREN BACALL: Betty Joan Perske
ERYKAH BADU: Erica Wright
ANNE BANCROFT: Anna Maria Italiano
GENE BARRY: Eugene Klass
PAT BENATAR: Patricia Andrejewski
TONY BENNETT: Anthony Benedetto
BUSBY BERKELEY: William Berkeley Enos
IRVING BERLIN: Israel Baline
JACK BENNY: Benjamin Kubelsky
JOEY BISHOP: Joseph Gottlieb
THE BIG BOPPER: Jiles Perry "J.P." Richardson
BONO (VOX): Paul Hewson
VICTOR BORGE: Borge Rosenbaum
DAVID BOWIE: David Robert Jones
BOY GEORGE: George Alan O'Dowd
FANNY BRICE: Fanny Borach
CHARLES BRONSON: Charles Buchinski
ALBERT BROOKS: Albert Einstein
MEL BROOKS: Melvin Kaminsky
GEORGE BURNS: Nathan Birnbaum
ELLEN BURSTYN: Edna Gilhooley
RICHARD BURTON: Richard Jenkins
RED BUTTONS: Aaron Chwatt
NICOLAS CAGE: Nicholas Coppola
MICHAEL CAINE: Maurice Micklewhite
MARIA CALLAS: Maria Kalogeropoulos
DIAHANN CARROLL: Carol Diahann Johnson

JACKIE CHAN: Chan Kwong-Sung
CYD CHARISSE: Tula Finklea
RAY CHARLES: Ray Charles Robinson
CHUBBY CHECKER: Ernest Evans
CHER: Cherilyn Sarkisian
PATSY CLINE: Virginia Patterson Hensley
LEE J. COBB: Leo Jacoby
CLAUDETTE COLBERT: Lily Chauchoin
ALICE COOPER: Vincent Furnier
DAVID COPPERFIELD: David Kotkin
HOWARD COSELL: Howard Cohen
ELVIS COSTELLO: Declan McManus
LOU COSTELLO: Louis Cristillo
PETER COYOTE: Peter Cohon
MICHAEL CRAWFORD: Michael Dumble-Smith
TOM CRUISE: Thomas Mapother IV
TONY CURTIS: Bernard Schwartz
VIC DAMONE: Vito Farinola
RODNEY DANGERFIELD: Jacob Cohen
BOBBY DARIN: Walden Robert Cassotto
DORIS DAY: Doris von Kappelhoff
YVONNE DE CARLO: Peggy Middleton
SANDRA DEE: Alexandra Zuck
JOHN DENVER: Henry John Deutschendorf Jr.
BO DEREK: Mary Cathleen Collins
DANNY DEVITO: Daniel Michaeli
ANGIE DICKINSON: Angeline Brown
BO DIDDLEY: Elias Bates
PHYLLIS DILLER: Phyllis Driver
KIRK DOUGLAS: Issur Danielovitch
MELVYN DOUGLAS: Melvyn Hesselberg
BOB DYLAN: Robert Zimmerman
SHEENA EASTON: Sheena Shirley Orr
BARBARA EDEN: Barbara Huffman
ELVIRA: Cassandra Peterson
RON ELY: Ronald Pierce
ENYA: Eithne Ni Bhraonian
DALE EVANS: Frances Smith
CHAD EVERETT: Raymond Cramton

DOUGLAS FAIRBANKS: Douglas Ullman
MORGAN FAIRCHILD: Patsy McClenny
JAMIE FARR: Jameel Farah
ALICE FAYE: Alice Jeanne Leppert
STEPIN FETCHIT: Lincoln Perry
W.C. FIELDS: William Claude Dukenfield
BARRY FITZGERALD: William Shields
JOAN FONTAINE: Joan de Havilland
JOHN FORD: Sean O'Fearna
JOHN FORSYTHE: John Freund
JODIE FOSTER: Alicia Christian Foster
REDD FOXX: John Sanford
ANTHONY FRANCIOSA: Anthony Papaleo
ARLENE FRANCIS: Arlene Kazanjian
CONNIE FRANCIS: Concetta Franconero
GRETA GARBO: Greta Gustafsson
VINCENT GARDENIA: Vincent Scognamiglio
JOHN GARFIELD: Julius Garfinkle
JUDY GARLAND: Frances Gumm
JAMES GARNER: James Bumgarner
CRYSTAL GAYLE: Brenda Gayle Webb
KATHIE LEE GIFFORD: Kathie Epstein
WHOOPI GOLDBERG: Caryn Johnson
EYDIE GORME: Edith Gormezano
STEWART GRANGER: James Stewart
CARY GRANT: Archibald Leach
LEE GRANT: Lyova Rosenthal
JOEL GREY: Joe Katz
ROBERT GUILLAUME: Robert Williams
BUDDY HACKETT: Leonard Hacker
HAMMER: Stanley Kirk Burrell
JEAN HARLOW: Harlean Carpentier
REX HARRISON: Reginald Carey
LAURENCE HARVEY: Larushka Skikne
HELEN HAYES: Helen Brown
SUSAN HAYWARD: Edythe Marriner
RITA HAYWORTH: Margarita Cansino
PEE-WEE HERMAN: Paul Reubenfeld
WILLIAM HOLDEN: William Beedle
BILLIE HOLIDAY: Eleanora Fagan

JUDY HOLLIDAY: Judith Tuvim
HARRY HOUDINI: Ehrich Weiss
LESLIE HOWARD: Leslie Stainer
ROCK HUDSON: Roy Scherer Jr. (later Fitzgerald)
ENGELBERT HUMPERDINCK: Arnold Dorsey
KIM HUNTER: Janet Cole
BETTY HUTTON: Betty Thornberg
ICE CUBE: O'Shea Jackson
ICE-T: Tracy Morrow
BILLY IDOL: William Broad
DAVID JANSSEN: David Meyer
JAY-Z: Shawn Carter
ANN JILLIAN: Anne Nauseda
ELTON JOHN: Reginald Dwight
DON JOHNSON: Donald Wayne
AL JOLSON: Asa Yoelson
JENNIFER JONES: Phylis Isley
TOM JONES: Thomas Woodward
LOUIS JOURDAN: Louis Gendre
WYNONNA JUDD: Christina Ciminella
BORIS KARLOFF: William Henry Pratt
DANNY KAYE: David Kaminsky
DIANE KEATON: Diane Hall
MICHAEL KEATON: Michael Douglas
CHAKA KHAN: Yvette Stevens
CAROLE KING: Carole Klein
LARRY KING: Larry Zeigler
BEN KINGSLEY: Krishna Banji
NASTASSJA KINSKI: Nastassja Naksyznyski
TED KNIGHT: Tadeus Wladyslaw Konopka
CHERYL LADD: Cheryl Stoppelmoor
VERONICA LAKE: Constance Ockleman
HEDY LAMARR: Hedwig Kiesler
DOROTHY LAMOUR: Mary Leta Dorothy Slaton
MICHAEL LANDON: Eugene Orowitz
MARIO LANZA: Alfredo Cocozza
QUEEN LATIFAH: Dana Owens
STAN LAUREL: Arthur Jefferson
STEVE LAWRENCE: Sidney Leibowitz
BRENDA LEE: Brenda Mae Tarpley
GYPSY ROSE LEE: Rose Louise Hovick
MICHELLE LEE: Michelle Dusiak
PEGGY LEE: Norma Egstrom
JANET LEIGH: Jeanette Morrison
VIVIEN LEIGH: Vivian Hartley
HUEY LEWIS: Hugh Cregg
JERRY LEWIS: Joseph Levitch
HAL LINDEN: Harold Lipshitz
CAROLE LOMBARD: Jane Peters
JACK LORD: John Joseph Ryan
SOPHIA LOREN: Sophia Scicolone
PETER LORRE: Laszio Lowenstein
MYRNA LOY: Myrna Williams
BELA LUGOSI: Bela Ferenc Blasko
MOMS MABLEY: Loretta Mary Aitken
SHIRLEY MACLAINE: Shirley Beaty
ELLE MACPHERSON: Eleanor Gow

LEE MAJORS: Harvey Lee Yeary 2d
KARL MALDEN: Mladen Sekulovich
BARRY MANILOW: Barry Alan Pincus
JAYNE MANSFIELD: Vera Jane Palmer
MARILYN MANSON: Brian Warner
FREDRIC MARCH: Frederick Bickel
PETER MARSHALL: Pierre LaCock
WALTER MATTHAU: Walter Matuschanskayasky
DEAN MARTIN: Dino Crocetti
MEAT LOAF: Marvin Lee Aday
FREDDIE MERCURY: Frederick Bulsara
ETHEL MERMAN: Ethel Zimmerman
GEORGE MICHAEL: Georgios Panayiotou
RAY MILLAND: Reginald Truscott-Jones
ANN MILLER: Lucille Collier
JONI MITCHELL: Roberta Joan Anderson
MARILYN MONROE: Norma Jean Mortenson (later Baker)
YVES MONTAND: Ivo Livi
RON MOODY: Ronald Moodnick
DEMI MOORE: Demetria Guynes
GARRY MOORE: Thomas Garrison Morfit
RITA MORENO: Rosita Alverio
HARRY MORGAN: Harry Bratsburg
MR. T: Lawrence Tero
PAUL MUNI: Muni Weisenfreund
MIKE NICHOLS: Michael Igor Peschowsky
CHUCK NORRIS: Carlos Ray
NOTORIOUS B.I.G.: Christopher Wallace
HUGH O'BRIAN: Hugh Krampke
MAUREEN O'HARA: Maureen Fitzsimons
PATTI PAGE: Clara Ann Fowler
JACK PALANCE: Walter Palanuik
BERT PARKS: Bert Jacobson
MINNIE PEARL: Sarah Ophelia Cannon
BERNADETTE PETERS: Bernadette Lazzaro
EDITH PIAF: Edith Gassion
SLIM PICKENS: Louis Lindley
MARY PICKFORD: Gladys Smith
STEFANIE POWERS: Stefania Federkiewicz
PAULA PRENTISS: Paula Ragusa
ROBERT PRESTON: Robert Preston Meservey
PRINCE (THE ARTIST): Prince Rogers Nelson
DEE DEE RAMONE: Douglas Colvin
JOEY RAMONE: Jeffrey Hyman
JOHNNY RAMONE: John Cummings
TOMMY RAMONE: Tom Erdelyi
TONY RANDALL: Leonard Rosenberg
JOHNNIE RAY: John Alvin
MARTHA RAYE: Margaret O'Reed
DONNA REED: Donna Belle Mullenger

DELLA REESE: Delloreese Patricia Early
BUSTER RHYMES: Trevor Smith Jr.
JOAN RIVERS: Joan Sandra Molinsky
EDWARD G. ROBINSON: Emmanuel Goldenberg
GINGER ROGERS: Virginia McMath
ROY ROGERS: Leonard Franklin Slye
MICKEY ROONEY: Joe Yule Jr.
JOHNNY ROTTEN: John Lydon
LILLIAN RUSSELL: Helen Leonard
MEG RYAN: Margaret Hyra
WINONA RYDER: Winona Horowitz
SOUPY SALES: Milton Hines
SUSAN SARANDON: Susan Tomaling
SEAL: Samuel Sealhenry
RANDOLPH SCOTT: George Randolph Crane
JANE SEYMOUR: Joyce Frankenberg
OMAR SHARIF: Michael Shalhoub
CHARLIE SHEEN: Carlos Irwin Estevez
MARTIN SHEEN: Ramon Estevez
BEVERLY SILLS: Belle Silverman
TALIA SHIRE: Talia Coppola
PHIL SILVERS: Philip Silversmith
SINBAD: David Atkins
"BUFFALO BOB" SMITH: Robert Schmidt
SNAP DOG: Calvin Broadus
ANN SOTHERN: Harriette Lake
ROBERT STACK: Robert Modini
BARBARA STANWYCK: Ruby Stevens
JEAN STAPLETON: Jeanne Murray
RINGO STARR: Richard Starkey
CONNIE STEVENS: Concetta Ingolia
STING: Gordon Sumner
DONNA SUMMER: La Donna Gaines
RIP TAYLOR: Charles Elmer Jr.
ROBERT TAYLOR: Spangler Brugh
DANNY THOMAS: Muzyad Yakhoob, later Amos Jacobs
TINY TIM: Herbert Khaury
RIP TORN: Elmore Rual Torn Jr.
RANDY TRAVIS: Randy Traywick
SOPHIE TUCKER: Sophia Kalish
TINA TURNER: Annie Mae Bullock
TWIGGY: Leslie Hornby
CONWAY TWITTY: Harold Lloyd Jenkins
RUDOLPH VALENTINO: Rudolpho D'Antonguolla
FRANKIE VALLI: Frank Castelluccio
SID VICIOUS: John Simon Ritchie
JOHN WAYNE: Marion Morrison
CLIFTON WEBB: Webb Hollenbeck
RAQUEL WELCH: Raquel Tejada
GENE WILDER: Jerome Silberman
SHELLEY WINTERS: Shirley Schrift
STEVIE WONDER: Stevland Morris
NATALIE WOOD: Natasha Nikolaevna Gurdin
JANE WYMAN: Sarah Jane Fulks
GIG YOUNG: Byron Barr

Selected Royal Families of Europe

	Birthdate		Birthdate		Birthdate		Birthdate
BELGIUM		**LUXEMBOURG**		**NORWAY**		**UNITED KINGDOM**	
King Albert II	6/6/34	Grand Duke Jean	1/5/21	King Harald V	2/21/37	Queen Elizabeth,	
Queen Paola	9/11/37	Grand Duchess		Queen Sonja	7/4/37	Queen Mother	8/4/00
Prince Philippe	4/15/60	Joséphine-Charlotte	10/11/27	Princess Märtha		Queen Elizabeth II	4/21/26
Princess Astrid	6/5/62	Princess Marie-Astrid	2/17/54	Louise	9/22/71	Prince Philip	6/10/21
Prince Laurent	10/19/63	Prince Henri	4/16/55	Crown Prince		Prince Charles	11/14/48
		Princes Jean	5/15/57	Haakon	7/20/73	Prince William	6/21/82
DENMARK		Princess Margaretha	5/15/57			Prince Henry,	
Queen Margrethe II	4/16/40	Prince Guillaume	5/1/63	**SPAIN**		or Harry	9/15/84
Prince Henrik	6/11/34			King Juan Carlos I	1/5/38	Princess Anne	8/15/50
Prince Frederik	5/26/68	**MONACO**		Queen Sofía	11/2/38	Prince Andrew	2/19/60
Prince Joachim	6/7/69	Prince Rainier III	5/31/23	Princess Elena	12/20/63	Princess Beatrice	8/8/88
Princess Alexandra	6/30/64	Prince Albert	3/14/58	Princess Cristina	6/13/65	Princess Eugenie	3/23/90
		Princess Caroline	1/23/57	Crown Prince Felipe	1/30/68	Prince Edward	3/10/64
LIECHTENSTEIN		Princess Stephanie	2/1/65			Princess Margaret	8/21/30
Prince Hans-Adam II	2/14/45			**SWEDEN**			
Princess Marie	4/14/40	**NETHERLANDS**		King Carl XVI			
Crown Prince Alois	6/11/68	Queen Beatrix	1/31/38	Gustav	4/30/46		
Prince Maximilian	5/16/69	Prince Claus	6/9/26	Queen Silvia	12/23/43		
Prince Constantin	3/15/72	Prince Willem-		Crown Princess			
Princess Tatjana	4/10/73	Alexander	4/27/67	Victoria	7/14/77		
		Prince Johan Friso	9/25/68	Prince Carl Philip	5/13/79		
		Prince Constantijn	10/11/69	Princess Madeleine	6/10/82		

UNITED STATES POPULATION

A Profile of America's Diversity—The View From the Census Bureau, 1999

by Kenneth Prewitt

Director, Bureau of the Census, U.S. Department of Commerce

As Americans draw nearer to the end of the century and millennium, the country in which they live is more diverse than ever in its 223-year history. The American population today is a rich mosaic of national origins, spanning a broader age spectrum and exhibiting a more diverse range of living arrangements than ever before, as illustrated by Census Bureau demographic data. Here are a few examples:

Racial and Ethnic Composition

On Aug. 1, 1999, there were an estimated 273 mil people living in the United States; on Census Day, Apr. 1, 1990, 9 years earlier, the nation's population was 249 mil. Of the 1999 population, an estimated 225 mil (82%) were White; 35 mil (13%) were Black or African American; Asians and Pacific Islanders numbered 11 mil (4%); and the American Indian and Alaska Native population was about 2 mil (1%). An estimated 31 mil (12%) were of Hispanic origin—people of Hispanic origin may be of any race. About 196 mil (or 72%) classified themselves as non-Hispanic White.

The Hispanic population increased by more than 9.1 mil people over this 9-year period—more than any other group. Non-Hispanic Whites, with more than 7.9 mil people added, were the runners-up.

Age Structure

On Aug. 1, 1999, 70 mil U.S. residents (26%) were under 18 years old. At the other end of the continuum, 35 mil (13%) were 65 or older. At the farthest tip were 60,712 centenarians (people who are 100 or older). This figure represents a 70% increase from the 1990 estimate of 35,817. Projections indicate that, by the middle of the 21st century, the number of centenarians will exceed 800,000.

Other rapidly growing age groups include people 85 years of age and older, with their population increasing 37% over the 1990-99 period, and the 45-64 group, which increased by 27%, reflecting the aging of the "Baby Boomer" generation.

Because these older age groups are growing so quickly, the median age—with half of all Americans above and half below—reached 35.2 years in 1998, the highest it has ever been. West Virginia's population continued to be the nation's oldest, with a median age of 38.6 years; Utah was the youngest state, with a median age of 26.7 years.

Income Spectrum

The nation's families had a median income of $46,737 as of 1998. While 13% of all families had incomes of $100,000 and over, 10% were below the government's official poverty line of $16,660 for a family of 4.

Although a wide variation in income levels persisted, there has been a rising "income tide" in recent years: in 1998, median family income, even after adjusting for inflation, rose for the 5th consecutive year.

Marriage and Families

About 56% of American adults in 1998 were married and living with their spouse. Another 24% had never married, 7% were widowed, and 10% were divorced.

Of the 103 mil households in the United States, 69% included or constituted a family—that is, 2 or more people related by blood, marriage, or adoption. The remaining households consisted of a person living alone (26%) or 2 or more unrelated people (5%).

About half (49%) of all families included parents and children under 18. All in all, 36% could be considered "traditional" families, that is, consisting of a married

couple with children. Since 1970, these traditional families have declined significantly as a percentage of all families, dropping 14 percentage points. However, their percentage has dropped only 1 point since 1990.

While the number of single mothers (9.8 mil) remained about the same from 1995 to 1998, the number of single fathers rose from 1.7 mil to 2.1 mil. About, 28% of children under 18 years of age lived with just 1 parent in 1998 (around 23% with their mother only, 4% with their father only), while 68% lived with both parents and 4% with other relatives or people not related to them. Nearly 6% of all children under 18 lived in their grandparents' home.

Education

While 17% of Americans ages 25 and older lacked a high school diploma as of 1998, 24% had a bachelor's degree or higher, and 8% had a graduate degree.

Education certainly pays off. In 1997, adults age 18 and over with a bachelor's degree earned an average of $40,478 a year, while those with only a high school diploma made $22,895. Advanced degree-holders earned about $63,229 a year, while those without a high school diploma averaged $16,124.

From 1940, when the Census Bureau began collecting statistics on educational attainment, until the early 1990s, figures showed that among young adults ages 25 to 29, men were more likely than women to have a bachelor's degree or more. Then the situation began to reverse itself so that, by 1998, 29% of women—as compared to 26% of men in this age range—had achieved this level of education.

The Foreign-Born Population

In 1998, the foreign-born population of the United States numbered about 25.2 mil persons, or 9.3% of the total population.

Seven mil persons, or about 28% of the foreign born in the U.S. in 1998, were born in Mexico. Other leading countries of origin included the Philippines (1.2 mil), China (1.0 mil), Cuba (994,000), Vietnam (989,000), and El Salvador (723,000).

Population Growth

Some parts of the nation were growing much faster than others. The fastest growth, as usual, was concentrated in the West, where the population rose 1.6% between 1997 and 1998. Close behind was the South (1.3%). Growing more slowly were the Midwest (0.4%) and the Northeast (0.3%).

Nevada remained the nation's fastest-growing state for the 13th straight year, with its population increasing 4.1% between 1997 and 1998. Nevada's population had climbed by a staggering 45.4% since Apr. 1, 1990, from about 1.2 mil to more than 1.7 mil. Arizona was 2d in population growth during the recent 1-year period, with a 2.5% increase, followed by Georgia and Colorado (2.0% each) and Texas (1.9%). The fastest-growing state in the Midwest—and 17th nationally—was Kansas, at 1.1%; New Hampshire, meanwhile, took the honors in the Northeast with a 1.1% population increase; it was the only state in the region to grow faster than the national average.

Of the nation's 2,426 counties with at least 10,000 people in 1998, 4 of the 10 fastest-growing were in Georgia, 2 each were in Colorado and Texas, and 1 each was in Nevada and Virginia. Forsyth County, GA, near Atlanta, led the country with a population increase of 13.0% from 1997 to 1998. Douglas County, CO, south of

Denver, ranked 2d, growing by 11.2%. Loudoun County, VA, near Washington, DC, was 3d, with an 8.2% rate of growth. For the whole period 1990-98, however, Douglas has been the nation's fastest-growing county, with a population increase of 133.4%.

Among the 10 counties with the largest population gains between 1997 and 1998, there were 4 in California (Los Angeles, Orange, San Diego, and Riverside); 3 in Texas (Harris, Dallas, and Tarrant); and 1 in Arizona (Maricopa), in Nevada (Clark), and in Florida (Broward). Los Angeles County was the biggest gainer from 1997 to 1998, with an increase of 97,027 people. However, Maricopa has added the most people since 1990— 661,974.

The story for the largest U.S. cities was much the same: each of the 10 with the most rapid growth in 1990-98 was in the West or South, with 2 each in Arizona, California, Florida, Nevada, and Texas. Henderson, NV, Chandler, AZ, and Pembroke, FL, topped the list, with 1990-98 population increases of 135%, 78%, and 76%, respectively. The fastest-growing city of this size in either the Northeast or Midwest was Naperville, IL; with a 37% population increase, it barely missed the national top 10.

Crowding

Despite the West's rapid population growth, it still has plenty of wide-open space. In fact, the 3 states with the fewest number of people per square mile were all Western. Alaska topped the list, with only 1 resident per square mile in 1998, followed by Wyoming, with 5, and Montana, with 6. On the other hand, with 1,094 residents per square mile, the most densely populated state was New Jersey. The runners-up were Rhode Island (946) and Massachusetts (784).

International Trade

America's diversity extends to the list of its leading trading partners. These include Canada ($117 bil in combined merchandise imports and exports during the first 4 months of 1999), Japan ($61 bil), Mexico ($59 bil), Germany ($26 bil), China ($26 bil), the United Kingdom ($25 bil), Taiwan ($16 bil), South Korea ($16 bil), France ($15 bil), and Italy ($10 bil).

The Economy

As a possible harbinger of the future, the U.S. economy near the century's end showed a definite shift toward white-collar industries. According to the 1997 Economic Census (conducted every 5 years), manufacturing employment stood at 17 mil people, more than any other sector. At the same time, however, the economy was becoming increasingly service-oriented. Retail trade accounted for more than 14 mil employees, followed closely by health care and social assistance industries, with just under 14 mil employees.

With society becoming increasingly information-driven, there are now more than 3 mil people helping to provide information services, such as publishing, motion pictures, broadcasting, and telecommunications. Another 5 mil work in professional, scientific, and technical services industries that provide the human capital that drives our information economy. And businesses rely on the more than 7 mil people employed in administrative and support services, such as employment agencies, employee leasing services, phone centers, telemarketing bureaus, and travel agencies.

In addition, there are nearly 2 mil people who work in arts, entertainment, and recreation industries, such as amusement parks, golf courses, skiing facilities, and museums.

Preparations for Census 2000 in 1999

Source: Bureau of the Census, U.S. Dept. of Commerce

On Jan. 25, 1999, a little more than 2 months after hearing oral arguments on the issue, the U.S. Supreme Court ruled that the 1976 Census Act prohibits the use of statistical sampling to determine the U.S. population for purposes of congressional apportionment. The High Court's ruling set in motion a period of contention in the Congress, leaving doubts for a time as to whether Census 2000 could even be conducted as scheduled, on Apr. 1, 2000.

The Clinton administration read the Court's ruling as still allowing the use of sampling for other, nonapportionment purposes, such as the redrawing of state legislative districts and the allocation of more than $180 bil a year in federal program funds to state and local communities, as long as the Census Bureau determines that adjusting population totals based on a postcensus survey is technically feasible. The Census Bureau did, in fact, reach such a conclusion.

By mid-1999, the Republican-controlled House had backed away from threats not to fund the departments of State, Justice, and Commerce (the parent agency of the Census Bureau) beyond June 15. The 3 departments were tied together for budget purposes. At the same time, opponents of statistical sampling indicated that any future efforts to modify the Census Bureau's 2000 plans would be channeled through the courts, rather than through the appropriations process.

Following the Supreme Court's action, the Census Bureau asked for and received an additional $44 mil for Fiscal Year 1999 to plan both a traditional census for purposes of apportionment and a postcensus survey of 300,000 housing units (called the Accuracy and Coverage Evaluation [A.C.E.] survey), which will allow the Bureau to correct census raw totals for nonapportionment purposes.

The Census Bureau's request for an additional $1.7 bil for Fiscal Year 2000, on top of the $2.9 bil it had originally sought, initially became swept up in the larger debate over a proposed tax cut and federal spending limits. The bulk of the additional money, the Census Bureau said, would go to pay for 360,000 additional enumerators (for a total work force of 860,000); the enumerators job will be to track down an anticipated 45 mil households that do not mail back their questionnaires. Senate and House conferees were working to resolve this funding issue as Congress prepared to begin its fall 1999 session.

Early in 1999, the Census Bureau released the results of its Census 2000 Dress Rehearsal, held in 3 sites in 1998: the city of Columbia, SC, and 11 surrounding counties; Menominee County, WI; and Sacramento, CA. These showed that the race and ethnic undercounts that have plagued the census in modern times persisted. Hispanics, African Americans, and American Indians again were missed the most frequently in proportion to population. The estimated overall undercount rates were: 6.3% in Sacramento, 3.0% in Menominee, and 9% in South Carolina.

The dress rehearsal was the first chance the Census Bureau had to test a new federal policy allowing respondents to mark more than 1 race category if desired. In Sacramento, 5.4% of the respondents marked more than 1 race category, but in Menominee only 1.2% did so, and in South Carolina, the percentage was only 0.8%.

Meanwhile, precensus activities, consisting mostly of preparing a complete and accurate address list, moved into high gear. Tens of thousands of "listers" fanned out across the country to list or verify 92.5 mil addresses in separate operations called "address listing" and "block canvassing." In late summer, the Census Bureau field-checked challenges to its address list by local or tribal governments that volunteered to check lists for accuracy, in a program called Local Update of Census Addresses.

By July, the first 2 of 4 data-capture centers, equipped with the latest questionnaire scanning devices and other high-tech machinery, were inaugurated in Essex, MD, outside Baltimore and at the Census Bureau's National Processing Center in Jeffersonville, IN. The other 2 were scheduled to open later in the year, at Pomona, CA, and Phoenix, AZ. The 4 new centers were expected to process about 120 mil census questionnaires in all.

Also in 1999, the Census Bureau opened or planned to open 520 local census offices (which will report to 12 existing regional census centers); hired 135,000 temporary workers for precensus activities (mostly address listing), including nearly 5,000 former welfare recipients; signed more than 25,000 regional and 500 national partnership agreements (partner organizations, both public and private, will play a major role in grassroots promotion and recruitment); and launched the Census in Schools project, with mailings of census-related classroom materials to administrators and teachers in more than 40,000 elementary, middle, and high schools. By September, the Bureau had printed more than half of the 426 mil questionnaires to be mailed, hand-delivered, or placed in Be Counted centers around the country in March, 2000.

Other Census 2000 developments in 1999 included an increase in the amount of money earmarked for paid advertising (from $100 mil to $166 mil), space on census questionnaires for answers from 6 persons instead of 5, and independent listing of about 2 mil housing units for the Accuracy and Coverage Evaluation. There was also an increase in the number of language-assistance guides, from 33 to 49 languages (questionnaires themselves will be printed in 5 non-English languages: Spanish, Chinese, Vietnamese, Korean, and Tagalog).

Race and Hispanic Origin for the U.S., 1990 and 1980

Source: Bureau of the Census, U.S. Dept. of Commerce

	1990 CENSUS		1980 CENSUS		% CHANGE
	Number	Percent	Number	Percent	1980-90
ALL PERSONS	248,709,873[1]	100.0	226,545,805	100.0	9.8
White	199,686,070	80.3	188,371,622	83.1	6.0
Black	29,986,060	12.1	26,495,025	11.7	13.2
American Indian, Eskimo, or Aleut	1,959,234	0.8	1,420,400	0.6	37.9
American Indian	1,878,285	0.8	1,364,033	0.6	37.7
Eskimo	57,152	0.0	42,162	0.0	35.6
Aleut	23,797	0.0	14,205	0.0	67.5
Asian-Pacific Islander	7,273,662	2.9	3,500,439[2]	1.5	107.8
Chinese	1,645,472	0.7	806,040	0.4	104.1
Filipino	1,406,770	0.6	774,652	0.3	81.6
Japanese	847,562	0.3	700,974	0.3	20.9
Asian Indian	815,447	0.3	361,531	0.2	125.6
Korean	798,849	0.3	354,593	0.2	125.3
Vietnamese	614,547	0.2	261,729	0.1	134.8
Hawaiian	211,014	0.1	166,814	0.1	26.5
Samoan	62,964	0.0	41,948	0.0	50.1
Guamanian	49,345	0.0	32,158	0.0	53.4
Other Asian-Pacific Islander	821,692	0.3	NA	NA	NA
Other race	9,804,847	3.9	6,758,319	3.0	45.1
PERSONS OF HISPANIC ORIGIN[3]	**22,354,059**	**9.0**	**14,608,673**	**6.4**	**53.0**
Mexican	13,495,938	5.4	8,740,439	3.9	54.4
Puerto Rican	2,727,754	1.1	2,013,945	0.9	35.4
Cuban	1,043,932	0.4	803,226	0.4	30.0
Other Hispanic	5,086,435	2.0	3,051,063	1.3	66.7

NA=Not available. (1) The race data are based on the U.S. population as tabulated in the 1990 census. Figures do not reflect corrections to the 1990 population census; the corrected 1990 U.S. population is 248,765,170. (2) The 1980 count of 3,500,439 Asian-Pacific Islanders, based on 100% tabulations, includes only the 9 Asian-Pacific Islander groups listed separately in the 1980 race item. A figure of 3,726,440, from sample tabulations, is more comparable to the 1990 count since it includes those groups. (3) Persons of Hispanic origin may be of any race.

Estimated Population of American Colonies, 1630-1780

Source: Bureau of the Census, U.S. Dept. of Commerce; in thousands

Colony	1780	1770	1750	1740	1720	1700	1690	1670	1650	1630
TOTAL	2,780.4	2,148.1	1,170.8	905.6	466.2	250.9	210.4	111.9	50.4	4.6
Maine (counties)[1]	49.1	31.3	...	...	...	...	...	...	1.0	0.4
New Hampshire[2]	87.8	62.4	27.5	23.3	9.4	5.0	4.2	1.8	1.3	0.5
Vermont[3]	47.6	10.0	...	...	...	...	...	...	...	...
Plymouth and Massachusetts[1,2,4]	268.6	235.3	188.0	151.6	91.0	55.9	56.9	35.3	15.6	0.9
Rhode Island[2]	52.9	58.2	33.2	25.3	11.7	5.9	4.2	2.2	0.8	...
Connecticut[2]	206.7	183.9	111.3	89.6	58.8	26.0	21.6	12.6	4.1	...
New York[2]	210.5	162.9	76.7	63.7	36.9	19.1	13.9	5.8	4.1	0.4
New Jersey[2]	139.6	117.4	71.4	51.4	29.8	14.0	8.0	1.0	...	...
Pennsylvania[2]	327.3	240.1	119.7	85.6	31.0	18.0	11.4	...	...	...
Delaware[2]	45.4	35.5	28.7	19.9	5.4	2.5	1.5	0.7	0.2	...
Maryland[2]	245.5	202.6	141.1	116.1	66.1	29.6	24.0	13.2	4.5	...
Virginia[2]	538.0	447.0	231.0	180.4	87.8	58.6	53.0	35.3	18.7	2.5
North Carolina[2]	270.1	197.2	73.0	51.8	21.3	10.7	7.6	3.8	...	...
South Carolina[2]	180.0	124.2	64.0	45.0	17.0	5.7	3.9	0.2	...	...
Georgia[2]	56.1	23.4	5.2	2.0	...	...	...	...	...	...
Kentucky[5]	45.0	15.7	...	...	...	...	...	...	...	...
Tennessee[6]	10.0	1.0	...	...	...	...	...	...	...	...

(1) For 1660-1750, Maine counties are included with Massachusetts. Maine was part of Massachusetts until it became a separate state in 1820. (2) One of the original 13 states. (3) Admitted to statehood in 1791. (4) Plymouth became a part of the Province of Massachusetts in 1691. (5) Admitted to statehood in 1792. (6) Admitted to statehood in 1796.

U.S. Population by Official

STATE	1790[1]	1800[1]	1810[1]	1820	1830	1840	1850	1860	1870	1880	1890
AL...	....	1	9	127,901	309,527	590,756	771,623	964,201	996,992	1,262,505	1,513,401
AK..	....	....	....	....	....	....	....	....	....	33,426	32,052
AZ..	....	....	....	....	....	....	....	....	9,658	40,440	88,243
AR..	....	....	1	14,273	30,388	97,574	209,897	435,450	484,471	802,525	1,128,211
CA..	....	....	....	....	....	....	92,597	379,994	560,247	864,694	1,213,398
CO..	....	....	....	....	....	....	....	34,277	39,864	194,327	413,249
CT..	238	251	262	275,248	297,675	309,978	370,792	460,147	537,454	622,700	746,258
DE..	59	64	73	72,749	76,748	78,085	91,532	112,216	125,015	146,608	168,493
DC..	....	8	16	23,336	30,261	33,745	51,687	75,080	131,700	177,624	230,392
FL...	....	....	....	....	34,730	54,477	87,445	140,424	187,748	269,493	391,422
GA..	83	163	252	340,989	516,823	691,392	906,185	1,057,286	1,184,109	1,542,180	1,837,353
HI..	....	....	....	....	....	....	....	....	....	....	....
ID...	....	....	....	....	....	....	....	....	14,999	32,610	88,548
IL...	....	....	12	55,211	157,445	476,183	851,470	1,711,951	2,539,891	3,077,871	3,826,352
IN...	....	6	25	147,178	343,031	685,866	988,416	1,350,428	1,680,637	1,978,301	2,192,404
IA...	....	....	....	....	....	43,112	192,214	674,913	1,194,020	1,624,615	1,912,297
KS..	....	....	....	....	....	....	....	107,206	364,399	996,096	1,428,108
KY..	74	221	407	564,317	687,917	779,828	982,405	1,155,684	1,321,011	1,648,690	1,858,635
LA..	....	....	77	153,407	215,739	352,411	517,762	708,002	726,915	939,946	1,118,588
ME..	97	152	229	298,335	399,455	501,793	583,169	628,279	626,915	648,936	661,086
MD..	320	342	381	407,350	447,040	470,019	583,034	687,049	780,894	934,943	1,042,390
MA..	379	423	472	523,287	610,408	737,699	994,514	1,231,066	1,457,351	1,783,085	2,238,947
MI...	....	....	5	8,896	31,639	212,267	397,654	749,113	1,184,059	1,636,937	2,093,890
MN..	....	....	....	....	....	....	6,077	172,023	439,706	780,773	1,310,283
MS..	....	8	31	75,448	136,621	375,651	606,526	791,305	827,922	1,131,597	1,289,600
MO..	....	....	20	66,586	140,455	383,702	682,044	1,182,012	1,721,295	2,168,380	2,679,185
MT..	....	....	....	....	....	....	....	....	20,595	39,159	142,924
NE..	....	....	....	....	....	....	....	28,841	122,993	452,402	1,062,656
NV..	....	....	....	....	....	....	....	6,857	42,491	62,266	47,355
NH..	142	184	214	244,161	269,328	284,574	317,976	326,073	318,300	346,991	376,530
NJ...	184	211	246	277,575	320,823	373,306	489,555	672,035	906,096	1,131,116	1,444,933
NM..	....	....	....	....	....	....	61,547	93,516	91,874	119,565	160,282
NY..	340	589	959	1,372,812	1,918,608	2,428,921	3,097,394	3,880,735	4,382,759	5,082,871	6,003,174
NC..	394	478	556	638,829	737,987	753,419	869,039	992,622	1,071,361	1,399,750	1,617,949
ND..	....	....	....	....	....	....	....	....	2,405[2]	36,909	190,983
OH..	....	45	231	581,434	937,903	1,519,467	1,980,329	2,339,511	2,665,260	3,198,062	3,672,329
OK..	....	....	....	....	....	....	....	....	....	....	258,657
OR..	....	....	....	....	....	....	12,093	52,465	90,923	174,768	317,704
PA...	434	602	810	1,049,458	1,348,233	1,724,033	2,311,786	2,906,215	3,521,951	4,282,891	5,258,113
RI...	69	69	77	83,059	97,199	108,830	147,545	174,620	217,353	276,531	345,506
SC..	249	346	415	502,741	581,185	594,398	668,507	703,708	705,606	995,577	1,151,149
SD..	....	....	....	....	....	....	....	4,837[2]	11,776[2]	98,268	348,600
TN..	36	106	262	422,823	681,904	829,210	1,002,717	1,109,801	1,258,520	1,542,359	1,767,518
TX..	....	....	....	....	....	....	212,592	604,215	818,579	1,591,749	2,235,527
UT..	....	....	....	....	....	....	11,380	40,273	86,786	143,963	210,779
VT...	85	154	218	235,981	280,652	291,948	314,120	315,098	330,551	332,286	332,422
VA...	692	808	878	938,261	1,044,054	1,025,227	1,119,348	1,219,630	1,225,163	1,512,565	1,655,980
WA..	....	....	....	....	....	....	1,201	11,594	23,955	75,116	357,232
WV..	56	79	105	136,808	176,924	224,537	302,313	376,688	442,014	618,457	762,794
WI...	....	....	....	....	....	30,945	305,391	775,881	1,054,670	1,315,497	1,693,330
WY..	....	....	....	....	....	....	....	....	9,118	20,789	62,555
U.S.	3,929	5,308	7,240	9,638,453	12,866,020[3]	17,068,953[3]	23,191,876	31,443,321	38,558,371	50,189,209	62,979,766

Note: Where possible, population shown is that of the 1990 area of the state. Members of the Armed Forces overseas or other U.S. nationals abroad are not included. Totals have been revised to include corrections of initial tabulated counts. (1) Totals for 1790, 1800, and 1810 are in thousands. (2) 1860 figure is for Dakota Territory; 1870 figures are for parts of Dakota Territory. (3) U.S. total includes persons (5,318 in 1830 and 6,100 in 1840) on public ships in the service of the U.S. not credited to any region, division, or state.

Congressional Apportionment

Source: Bureau of the Census, U.S. Dept. of Commerce

State	1990	1980	State	1990	1980	State	1990	1980	State	1990	1980	State	1990	1980
AL..	7	7	HI..	2	2	MA..	10	11	NM..	3	3	SD..	1	1
AK..	1	1	ID..	2	2	MI..	16	18	NY..	31	34	TN..	9	9
AZ..	6	5	IL..	20	22	MN..	8	8	NC..	12	11	TX..	30	27
AR..	4	4	IN..	10	10	MS..	5	5	ND..	1	1	UT..	3	3
CA..	52	45	IA..	5	6	MO..	9	9	OH..	19	21	VT..	1	1
CO..	6	6	KS..	4	5	MT..	1	2	OK..	6	6	VA..	11	10
CT..	6	6	KY..	6	7	NE..	3	3	OR..	5	5	WA..	9	8
DE..	1	1	LA..	7	8	NV..	2	2	PA..	21	23	WV..	3	4
FL..	23	19	ME..	2	2	NH..	2	2	RI..	2	2	WI..	9	9
GA.	11	10	MD.	8	8	NJ..	13	14	SC..	6	6	WY..	1	1
												TOTALS	435	435

The Constitution, in Article 1, Section 2, provided for a census of the population every 10 years to establish a basis for apportionment of representatives among the states. This apportionment largely determines the number of electoral votes allotted to each state.

The number of representatives of each state in Congress is determined by the state's population, but each state is entitled to one representative regardless of population. A congressional apportionment has been made after each decennial census except that of 1920.

Under provisions of a law that became effective Nov. 15, 1941, representatives are apportioned by the method of equal proportions. In the application of this method, the apportionment is made so that the average population per representative has the least possible variation between one state and any other. The first House of Representatives, in 1789, had 65 members, as provided by the Constitution. As the population grew, the number of representatives was increased, but the total membership has been fixed at 435 since the apportionment based on the 1910 census.

Census, 1790-1990

1900	1910	1920	1930	1940	1950	1960	1970	1980	1990
1,828,697	2,138,093	2,348,174	2,646,248	2,832,961	3,061,743	3,266,740	3,444,354	3,894,025	4,040,389
63,592	64,356	55,036	59,278	72,524	128,643	226,167	302,583	401,851	550,043
122,931	204,354	334,162	435,573	499,261	749,587	1,302,161	1,775,399	2,716,546	3,665,339
1,311,564	1,574,449	1,752,204	1,854,482	1,949,387	1,909,511	1,786,272	1,923,322	2,286,357	2,350,624
1,485,053	2,377,549	3,426,861	5,677,251	6,907,387	10,586,223	15,717,204	19,971,069	23,667,764	29,785,857
539,700	799,024	939,629	1,035,791	1,123,296	1,325,089	1,753,947	2,209,596	2,889,735	3,294,473
908,420	1,114,756	1,380,631	1,606,903	1,709,242	2,007,280	2,535,234	3,032,217	3,107,564	3,287,116
184,735	202,322	223,003	238,380	266,505	318,085	446,292	548,104	594,338	666,168
278,718	331,069	437,571	486,869	663,091	802,178	763,956	756,668	638,432	606,900
528,542	752,619	968,470	1,468,211	1,897,414	2,771,305	4,951,560	6,791,418	9,746,961	12,938,071
2,216,331	2,609,121	2,895,832	2,908,506	3,123,723	3,444,578	3,943,116	4,587,930	5,462,982	6,478,149
154,001	191,874	255,881	368,300	422,770	499,794	632,772	769,913	964,691	1,108,229
161,772	325,594	431,866	445,032	524,873	588,637	667,191	713,015	944,127	1,006,734
4,821,550	5,638,591	6,485,280	7,630,654	7,897,241	8,712,176	10,081,158	11,110,285	11,427,409	11,430,602
2,516,462	2,700,876	2,930,390	3,238,503	3,427,796	3,934,224	4,662,498	5,195,392	5,490,214	5,544,156
2,231,853	2,224,771	2,404,021	2,470,939	2,538,268	2,621,073	2,757,537	2,825,368	2,913,808	2,776,831
1,470,495	1,690,949	1,769,257	1,880,999	1,801,028	1,905,299	2,178,611	2,249,071	2,364,236	2,477,588
2,147,174	2,289,905	2,416,630	2,614,589	2,845,627	2,944,806	3,038,156	3,220,711	3,660,324	3,686,892
1,381,625	1,656,388	1,798,509	2,101,593	2,363,880	2,683,516	3,257,022	3,644,637	4,206,116	4,221,826
694,466	742,371	768,014	797,423	847,226	913,774	969,265	993,722	1,125,043	1,227,928
1,188,044	1,295,346	1,449,661	1,631,526	1,821,244	2,343,001	3,100,689	3,923,897	4,216,933	4,780,753
2,805,346	3,366,416	3,852,356	4,249,614	4,316,721	4,690,514	5,148,578	5,689,170	5,737,093	6,016,425
2,420,982	2,810,173	3,668,412	4,842,325	5,256,106	6,371,766	7,823,194	8,881,826	9,262,044	9,295,287
1,751,394	2,075,708	2,387,125	2,563,953	2,792,300	2,982,483	3,413,864	3,806,103	4,075,970	4,375,665
1,551,270	1,797,114	1,790,618	2,009,821	2,183,796	2,178,914	2,178,141	2,216,994	2,520,770	2,575,475
3,106,665	3,293,335	3,404,055	3,629,367	3,784,664	3,954,653	4,319,813	4,677,623	4,916,766	5,116,901
243,329	376,053	548,889	537,606	559,456	591,024	674,767	694,409	786,690	799,065
1,066,300	1,192,214	1,296,372	1,377,963	1,315,834	1,325,510	1,411,330	1,485,333	1,569,825	1,578,417
42,335	81,875	77,407	91,058	110,247	160,083	285,278	488,738	800,508	1,201,675
411,588	430,572	443,083	465,293	491,524	533,242	606,921	737,681	920,610	1,109,252
1,883,669	2,537,167	3,155,900	4,041,334	4,160,165	4,835,329	6,066,782	7,171,112	7,365,011	7,747,750
195,310	327,301	360,350	423,317	531,818	681,187	951,023	1,017,055	1,303,302	1,515,069
7,268,894	9,113,614	10,385,227	12,588,066	13,479,142	14,830,192	16,782,304	18,241,391	17,558,165	17,990,778
1,893,810	2,206,287	2,559,123	3,170,276	3,571,623	4,061,929	4,556,155	5,084,411	5,880,095	6,632,448
319,146	577,056	646,872	680,845	641,935	619,636	632,446	617,792	652,717	638,800
4,157,545	4,767,121	5,759,394	6,646,697	6,907,612	7,946,627	9,706,397	10,657,423	10,797,603	10,847,115
790,391	1,657,155	2,028,283	2,396,040	2,336,434	2,233,351	2,328,284	2,559,463	3,025,487	3,145,576
413,536	672,765	783,389	953,786	1,089,684	1,521,341	1,768,687	2,091,533	2,633,156	2,842,337
6,302,115	7,665,111	8,720,017	9,631,350	9,900,180	10,498,012	11,319,366	11,800,766	11,864,720	11,882,842
428,556	542,610	604,397	687,497	713,346	791,896	859,488	949,723	947,154	1,003,464
1,340,316	1,515,400	1,683,724	1,738,765	1,899,804	2,117,027	2,382,594	2,590,713	3,120,729	3,486,310
401,570	583,888	636,547	692,849	642,961	652,740	680,514	666,257	690,768	696,004
2,020,616	2,184,789	2,337,885	2,616,556	2,915,841	3,291,718	3,567,089	3,926,018	4,591,023	4,877,203
3,048,710	3,896,542	4,663,228	5,824,715	6,414,824	7,711,194	9,579,677	11,198,655	14,225,513	16,986,335
276,749	373,351	449,396	507,847	550,310	688,862	890,627	1,059,273	1,461,037	1,722,850
343,641	355,956	352,428	359,611	359,231	377,747	389,881	444,732	511,456	562,758
1,854,184	2,061,612	2,309,187	2,421,851	2,677,773	3,318,680	3,966,949	4,651,448	5,346,797	6,189,197
518,103	1,141,990	1,356,621	1,563,396	1,736,191	2,378,963	2,853,214	3,413,244	4,132,353	4,866,669
958,800	1,221,119	1,463,701	1,729,205	1,901,974	2,005,552	1,860,421	1,744,237	1,950,186	1,793,477
2,069,042	2,333,860	2,632,067	2,939,006	3,137,587	3,434,575	3,951,777	4,417,821	4,705,642	4,891,769
92,531	145,965	194,402	225,565	250,742	290,529	330,066	332,416	469,557	453,589
76,212,168	92,228,496	106,021,537	123,202,624	132,164,569	151,325,798	179,323,175	203,302,031	226,542,203	248,765,170

U.S. Center of Population, 1790-1990

Source: Bureau of the Census, U.S. Dept. of Commerce

The U.S. Center of Population is considered here as the center of population gravity, or that point upon which the U.S. would balance if it were a rigid plane without weight and the population distributed thereon, with each individual assumed to have equal weight and to exert an influence on a central point proportional to his or her distance from that point. The 1990 center is 818.6 miles from the 1790 center of population and is 39.5 miles SW of the 1980 center.

YEAR	N Lat °	N Lat ′	N Lat ″	W Long °	W Long ′	W Long ″	APPROXIMATE LOCATION
1790	39	16	30	76	11	12	23 miles east of Baltimore, MD
1800	39	16	6	76	56	30	18 miles west of Baltimore, MD
1810	39	11	30	77	37	12	40 miles northwest by west of Washington, DC (in VA)
1820	39	5	42	78	33	0	16 miles east of Moorefield, WV[1]
1830	38	57	54	79	16	54	19 miles west-southwest of Moorefield, WV[1]
1840	39	2	0	80	18	0	16 miles south of Clarksburg, WV[1]
1850	38	59	0	81	19	0	23 miles southeast of Parkersburg, WV[1]
1860	39	0	24	82	48	48	20 miles south by east of Chillicothe, OH
1870	39	12	0	83	35	42	48 miles east by north of Cincinnati, OH
1880	39	4	8	84	39	40	8 miles west by south of Cincinnati, OH (in KY)
1890	39	11	56	85	32	53	20 miles east of Columbus, IN
1900	39	9	36	85	48	54	6 miles southeast of Columbus, IN
1910	39	10	12	86	32	20	In the city of Bloomington, IN
1920	39	10	21	86	43	15	8 miles south-southeast of Spencer, Owen Co., IN
1930	39	3	45	87	8	6	3 miles northeast of Linton, Greene Co., IN
1940	38	56	54	87	22	35	2 miles southeast by east of Carlisle, Haddon township, Sullivan Co., IN
1950 (inc. Alaska & Hawaii)	38	48	15	88	22	8	3 miles northeast of Louisville, Clay Co., IL
1960	38	35	58	89	12	35	6 1/2 miles northwest of Centralia, Clinton Co., IL
1970	38	27	47	89	42	22	5 miles east southeast of Mascoutah, St. Clair Co., IL
1980	38	8	13	90	34	26	1/4 mile west of De Soto, Jefferson Co., MO
1990	37	52	20	91	12	55	9.7 miles northwest of Steelville, MO

(1) West Virginia was set off from Virginia on Dec. 31, 1862, and was admitted as a state on June 20, 1863.

U.S. Area and Population, 1790-1990

Source: Bureau of the Census, U.S. Dept. of Commerce

| Census date | AREA (sq mi) | | | POPULATION | | | |
	Gross Area	Land Area	Water Area	Number	Per sq mi of land	Increase over preceding census Number	%
1990 (Apr. 1)	3,787,319[1]	3,536,278	251,041[1]	248,765,170	70.3	22,222,967	9.8
1980 (Apr. 1)	3,618,770	3,539,289	79,481	226,542,203	64.0	23,240,172	11.4
1970 (Apr. 1)	3,618,770	3,536,855	81,915	203,302,031	57.5	23,978,856	13.4
1960 (Apr. 1)	3,618,770	3,540,911	77,859	179,323,175	50.6	27,997,377	18.5
1950 (Apr. 1)	3,618,770	3,552,206	66,564	151,325,798	42.6	19,161,229	14.5
1940 (Apr. 1)	3,618,770	3,554,608	64,162	132,164,569	37.2	8,961,945	7.3
1930 (Apr. 1)	3,618,770	3,551,608	67,162	123,202,624	34.7	17,181,087	16.2
1920 (Jan. 1)	3,618,770	3,546,931	71,839	106,021,537	29.9	13,793,041	15.0
1910 (Apr. 15)	3,618,770	3,547,045	71,725	92,228,496	26.0	16,016,328	21.0
1900 (June 1)...............	3,618,770	3,547,314	71,456	76,212,168	21.5	13,232,402	21.0
1890 (June 1)...............	3,612,299	3,540,705	71,594	62,979,766	17.8	12,790,557	25.5
1880 (June 1)...............	3,612,299	3,540,705	71,594	50,189,209	14.2	11,630,838	30.2
1870 (June 1)...............	3,612,299	3,540,705	71,594	38,558,371	10.9	7,115,050	22.6
1860 (June 1)...............	3,021,295	2,969,640	51,655	31,443,321	10.6	8,251,445	35.6
1850 (June 1)...............	2,991,655	2,940,042	51,613	23,191,876	7.9	6,122,423	35.9
1840 (June 1)...............	1,792,552	1,749,462	43,090	17,068,953[2]	9.8	4,203,433	32.7
1830 (June 1)...............	1,792,552	1,749,462	43,090	12,866,020[2]	7.4	3,227,567	33.5
1820 (June 1)...............	1,792,552	1,749,462	43,090	9,638,453	5.5	2,398,572	33.1
1810 (Aug. 6)	1,722,685	1,681,828	40,857	7,239,881	4.3	1,931,398	36.4
1800 (Aug. 4)	891,364	864,746	26,618	5,308,483	6.1	1,379,269	35.1
1790 (Aug. 2)	891,364	864,746	26,618	3,929,214	4.5	—	—

(1) Includes inland, coastal, Great Lakes, and territorial water. Data for prior years cover inland water only. (2) U.S. total includes persons (5,318 in 1830 and 6,100 in 1840) on public ships in the service of the U.S. not credited to any region, division, or state.
NOTE: Percent changes are computed on the basis of change in population since the preceding census date, so the period covered is not always exactly 10 years. Population density figures given for various years represent the area within the boundaries of the U.S. that was under the jurisdiction on the date in question—including, in some cases, considerable areas not organized or settled and not actually covered by the census. In 1870, for example, Alaska was not covered by the census. Population figures shown here may reflect corrections made to the initial tabulated census counts.

Population, by Sex, Race, Residence, and Median Age, 1790-1999

Source: Bureau of the Census, U.S. Dept. of Commerce

(in thousands, except as indicated)

| | SEX | | RACE | | | | RESIDENCE | | MEDIAN AGE (years) | | |
	Male	Female	White	Black Number	Black Percent	Other	Urban	Rural	All races	White	Black
Conterminous U.S.[1]											
1790 (Aug. 2)	NA	NA	3,172	757	19.3	NA	202	3,728	NA	NA	NA
1810 (Aug. 6)	NA	NA	5,862	1,378	19.0	NA	525	6,714	NA	16.0	NA
1820 (Aug. 7)	4,897	4,742	7,867	1,772	18.4	NA	693	8,945	16.7	16.6	17.2
1840 (June 1)	8,689	8,381	14,196	2,874	16.8	NA	1,845	15,224	17.8	17.9	17.6
1860 (June 1)	16,085	15,358	26,923	4,442	14.1	79	6,217	25,227	19.4	19.7	17.5
1870 (June 1)	19,494	19,065	33,589	4,880	12.7	89	9,902	28,656	20.2	20.4	18.5
1880 (June 1)	25,519	24,637	43,403	6,581	13.1	172	14,130	36,026	20.9	21.4	18.0
1890 (June 1)	32,237	30,711	55,101	7,489	11.9	358	22,106	40,841	22.0	22.5	17.8
1900 (June 1)	38,816	37,178	66,809	8,834	11.6	351	30,160	45,835	22.9	23.4	19.4
1920 (Jan. 1)	53,900	51,810	94,821	10,463	9.9	427	54,158	51,553	25.3	25.5	22.3
1930 (Apr. 1)	62,137	60,638	110,287	11,891	9.7	597	68,955	53,820	26.5	26.9	23.5
1940 (Apr. 1)	66,062	65,608	118,215	12,866	9.8	589	74,424	57,246	29.0	29.5	25.3
United States											
1950 (Apr. 1)	74,833	75,864	135,150	15,045	9.9	1,131	96,467	54,230	30.2	30.7	26.2
1960 (Apr. 1)	88,331	90,992	158,832	18,872	10.5	1,620	125,269	54,054	29.5	30.3	23.5
1970 (Apr. 1)[2]	98,912	104,300	177,749	22,580	11.1	2,883	149,647	53,565	28.1	28.9	22.4
1980 (Apr. 1)[3]	110,053	116,493	194,713	26,683	11.8	5,150	167,051	59,495	30.0	30.9	24.9
1985 (July 1, est.)	115,730	122,194	202,031	28,569	12.0	7,324	NA	NA	31.4	32.3	26.6
1990 (Apr. 1)[4]	121,239	127,470	199,686	29,986	12.1	9,233	187,053	61,656	32.9	34.4	28.1
1991 (July 1, est.)	122,984	129,122	210,979	31,107	12.3	10,020	NA	NA	33.1	34.1	28.1
1992 (July 1, est.)	124,506	130,496	212,885	31,670	12.4	10,446	NA	NA	33.4	34.4	28.5
1993 (July 1, est.)	125,938	131,858	214,760	32,168	12.5	10,867	NA	NA	33.7	34.7	28.7
1994 (July 1, est.)	127,216	133,076	216,413	32,653	12.5	11,227	NA	NA	34.0	35.0	29.0
1995 (July 1, est.)	128,569	134,321	218,149	33,095	12.6	11,646	NA	NA	34.3	35.3	29.2
1996 (July 1, est.)	129,746	135,434	219,686	33,514	12.6	11,979	NA	NA	34.6	35.7	29.5
1997 (July 1, est.)	131,018	136,618	221,334	33,947	12.7	12,355	NA	NA	34.9	36.0	29.7
1998 (July 1, est.)	132,263	137,766	222,932	34,370	12.7	12,727	NA	NA	35.3	36.3	29.9
1999 (July 1, est.)	133,352	139,526	224,692	34,903	12.8	13,283	NA	NA	35.5	36.6	30.1

NA=Not available. **NOTE:** Urban and rural definitions may change from census to census. The figures in this table have been adjusted to be consistent with the 1990 urban and rural definitions. (1) Excludes Alaska and Hawaii. (2) The revised 1970 resident population count is 203,302,031, which incorporates changes due to errors found after tabulations were completed. The race and sex data shown here reflect the official 1970 census count; the residence data come from the tabulated count. (3) The race data shown for Apr. 1, 1980, have been modified. (4) The data shown are based on the U.S. population as tabulated in the 1990 census. Figures do not reflect corrections to the 1990 population. The corrected 1990 U.S. population is 248,765,170.

Immigrants Admitted, by Top 30 Metropolitan Areas of Intended Residence, 1997

Source: Immigration and Naturalization Service, U.S. Dept. of Justice
(fiscal year 1997)

Metropolitan Statistical Area	Number	Percentage	Metropolitan Statistical Area	Number	Percentage
New York, NY	107,434	13.5	Fort Lauderdale, FL	10,646	1.3
Los Angeles–Long Beach, CA	62,314	7.8	Detroit, MI	10,019	1.3
Miami, FL	45,707	5.7	Atlanta, GA	9,823	1.2
Chicago, IL	35,386	4.4	Bergen–Passaic, NJ	9,788	1.2
Washington, DC–MD–VA	31,444	3.7	Riverside–San Bernardino, CA	9,518	1.2
Orange County, CA	18,190	2.3	Nassau–Suffolk, NY	9,167	1.1
Houston, TX	17,439	2.2	Sacramento, CA	7,654	1.0
San Jose, CA	17,374	2.2	Jersey City, NJ	7,529	0.9
San Francisco, CA	16,892	2.1	Minneapolis–St. Paul, MN–WI	6,859	0.9
Oakland, CA	15,723	2.0	Portland–Vancouver, OR–WA	6,320	0.8
San Diego, CA	14,758	1.8	Middlesex–Somerset– Hunterdon, NJ	6,081	0.8
Boston–Lawrence–Lowell– Brockton, MA	13,937	1.7	West Palm Beach–Boca Raton, FL	5,858	0.7
Dallas, TX	11,061	1.4	Orlando, FL	5,374	0.7
Philadelphia, PA–NJ	10,858	1.4	Honolulu, HI	5,326	0.7
Newark, NJ	10,801	1.4			
Seattle–Bellevue–Everett, WA	10,692	1.3	**TOTAL immigrants admitted to U.S.**	**798,378**	**100.0**

Immigrants Admitted, by State of Intended Residence, 1997

Source: Immigration and Naturalization Service, U.S. Dept. of Justice
(fiscal year 1997)

STATE	Number of immigrants	STATE	Number of immigrants	STATE	Number of immigrants	STATE	Number of immigrants
AL	1,613	IA	2,766	NJ	41,184	VT	627
AK	1,060	KS	2,828	NM	2,610	VA	19,277
AZ	8,632	KY	1,939	NY	123,716	WA	18,656
AR	1,428	LA	3,319	NC	5,935	WV	418
CA	203,305	ME	817	ND	535	WI	3,175
CO	7,506	MD	19,090	OH	8,189	WY	252
CT	9,528	MA	17,317	OK	3,157	Other:	
DE	1,148	MI	14,727	OR	7,699	Guam	2,083
DC	3,373	MN	8,233	PA	14,553	N Mariana Isls.	103
FL	82,318	MS	1,118	RI	2,543	Puerto Rico	4,884
GA	12,623	MO	4,190	SC	2,446	Virgin Isls.	1,110
HI	6,867	MT	375	SD	490	Armed Service Post	93
ID	1,447	NE	2,270	TN	4,357	Other or unknown	7
IL	38,128	NV	6,541	TX	57,897		
IN	3,892	NH	1,143	UT	2,840	**TOTAL**	**798,378**

U.S. Foreign-Born Population, 1998

Source: Bureau of the Census, U.S. Dept. of Commerce

Percentage of U.S. Population That Is Foreign-Born, 1900-98

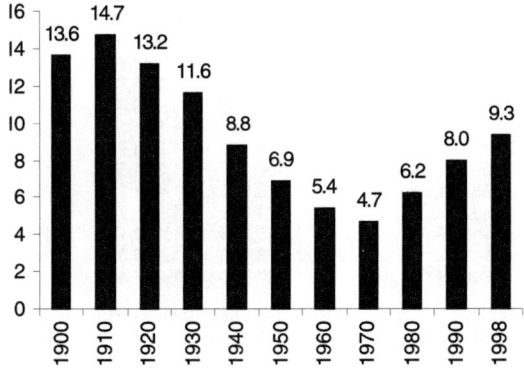

Highest-Ranking Countries of Birth of U.S. Foreign-Born Population, 1998

Country	Number (in thousands)
Mexico	7,119
Philippines	1,207
China (incl. Hong Kong)	1,022
Cuba	994
Vietnam	989
El Salvador	723
India	722
Dominican Republic	635
Great Britain	622
Korea	589
ALL COUNTRIES	**26,281**

Population by State, 1990-98

Source: Bureau of the Census, U.S. Dept. of Commerce

Rank	State	1998 population	1990 population	Percentage change 1990-98	Rank	State	1998 population	1990 population	Percentage change 1990-98
1.	CA	32,666,550	29,785,857	9.7	27.	OK	3,346,713	3,145,576	6.4
2.	TX	19,759,614	16,986,335	16.3	28.	OR	3,281,974	2,842,337	15.5
3.	NY	18,175,301	17,990,778	1.0	29.	CT	3,274,069	3,287,116	−0.4
4.	FL	14,915,980	12,938,071	15.3	30.	IA	2,862,447	2,776,831	3.1
5.	IL	12,045,326	11,430,602	5.4	31.	MS	2,752,092	2,575,475	6.9
6.	PA	12,001,451	11,882,842	1.0	32.	KS	2,629,067	2,477,588	6.1
7.	OH	11,209,493	10,847,115	3.3	33.	AR	2,538,303	2,350,624	8.0
8.	MI	9,817,242	9,295,287	5.6	34.	UT	2,099,758	1,722,850	21.9
9.	NJ	8,115,011	7,747,750	4.7	35.	WV	1,811,156	1,793,477	1.0
10.	GA	7,642,207	6,478,149	18.0	36.	NV	1,746,898	1,201,675	45.4
11.	NC	7,546,493	6,632,448	13.8	37.	NM	1,736,931	1,515,069	14.6
12.	VA	6,791,345	6,189,197	9.7	38.	NE	1,662,719	1,578,417	5.3
13.	MA	6,147,132	6,016,425	2.2	39.	ME	1,244,250	1,227,928	1.3
14.	IN	5,899,195	5,544,156	6.4	40.	ID	1,228,684	1,006,734	22.0
15.	WA	5,689,263	4,866,669	16.9	41.	HI	1,193,001	1,108,229	7.6
16.	MO	5,438,559	5,116,901	6.3	42.	NH	1,185,048	1,109,252	6.8
17.	TN	5,430,621	4,877,203	11.3	43.	RI	988,480	1,003,464	−1.5
18.	WI	5,223,500	4,891,769	6.8	44.	MT	880,453	799,065	10.2
19.	MD	5,134,808	4,780,753	7.4	45.	DE	743,603	666,168	11.6
20.	MN	4,725,419	4,375,665	8.0	46.	SD	738,171	696,004	6.1
21.	AZ	4,668,631	3,665,339	27.4	47.	ND	638,244	638,800	−0.1
22.	LA	4,368,967	4,221,826	3.5	48.	AK	614,010	550,043	11.6
23.	AL	4,351,999	4,040,389	7.7	49.	VT	590,883	562,758	5.0
24.	CO	3,970,971	3,294,473	20.5	50.	DC	523,124	606,900	−13.8
25.	KY	3,936,499	3,686,892	6.8	51.	WY	480,907	453,589	6.0
26.	SC	3,835,962	3,486,310	10.0		U.S.	270,298,524	248,765,170	8.7

Note: Population figures for 1990 include corrections to the original tabulated population.

Density of Population by State, 1920-90

Source: Bureau of the Census, U.S. Dept. of Commerce

(per square mile, land area only)

STATE	1920	1960	1980	1990	STATE	1920	1960	1980	1990	STATE	1920	1960	1980	1990
AL...	45.8	64.2	76.6	79.6	LA...	39.6	72.2	94.5	96.9	OH ..	141.4	236.6	263.3	264.9
AK*..	0.1	0.4	0.7	1.0	ME ..	25.7	31.3	36.3	39.8	OK ..	29.2	33.8	44.1	45.8
AZ...	2.9	11.5	23.9	32.3	MD ..	145.8	313.5	428.7	489.2	OR ..	8.2	18.4	27.4	29.6
AR ..	33.4	34.2	43.9	45.1	MA ..	479.2	657.3	733.3	767.6	PA ..	194.5	251.4	264.3	265.1
CA ..	22.0	100.4	151.4	190.8	MI...	63.8	137.7	162.6	163.6	RI ...	566.4	819.3	897.8	960.3
CO ..	9.1	16.9	27.9	31.8	MN ..	29.5	43.1	51.2	55.0	SC ..	55.2	78.7	103.4	115.8
CT ..	286.4	520.6	637.8	678.4	MS ..	38.6	46.0	53.4	54.9	SD ..	8.3	9.0	9.1	9.2
DE ..	113.5	225.2	307.6	340.8	MO ..	49.5	62.6	71.3	74.3	TN ..	56.1	86.2	111.6	118.3
DC ..	7,292.9	12,523.9	10,132.3	9,882.8	MT ..	3.8	4.6	5.4	5.5	TX...	17.8	36.4	54.3	64.9
FL ..	17.7	91.5	180.0	239.6	NE ..	16.9	18.4	20.5	20.5	UT...	5.5	10.8	17.8	21.0
GA ..	49.3	67.8	94.1	111.9	NV ..	0.7	2.6	7.3	10.9	VT...	38.6	42.0	55.2	60.8
HI*...	39.9	98.5	150.1	172.5	NH ..	49.1	67.2	102.4	123.7	VA...	57.4	99.6	134.7	156.3
ID ...	5.2	8.1	11.5	12.2	NJ...	420.0	805.5	986.2	1,042.0	WA ..	20.3	42.8	62.1	73.1
IL ...	115.7	180.4	205.3	205.6	NM ..	2.9	7.8	10.7	12.5	WV ..	60.9	77.2	80.8	74.5
IN ...	81.3	128.8	152.8	154.6	NY ..	217.9	350.6	370.6	381.0	WI...	47.6	72.6	86.5	90.1
IA ...	43.2	49.2	52.1	49.7	NC ..	52.5	93.2	120.4	136.1	WY ..	2.0	3.4	4.9	4.7
KS ..	21.6	26.6	28.9	30.3	ND ..	9.2	9.1	9.4	9.3	**U.S. Tot.**	**29.9***	**50.6**	**64.0**	**70.3**
KY ..	60.1	76.2	92.3	92.8										

(*) For purposes of comparison, Alaska and Hawaii are included in above tabulation for 1920, even though not states then.

25 Largest Counties, by Population, 1990-98

Source: Bureau of the Census, U.S. Dept of Commerce

COUNTY	1998 population	1990 population	Percentage change, 1990-98	COUNTY	1998 population	1990 population	Percentage change, 1990-98
Los Angeles, CA	9,213,533	8,863,052	4.0	San Bernardino, CA ..	1,635,234	1,418,380	15.3
Cook, IL	5,189,689	5,105,044	1.7	New York, NY	1,550,649	1,487,536	4.2
Harris, TX...........	3,206,063	2,818,101	13.8	Broward, FL........	1,503,407	1,255,531	19.7
Maricopa, AZ	2,784,075	2,122,101	31.2	Riverside, CA........	1,478,838	1,170,413	26.4
San Diego, CA	2,780,592	2,498,016	11.3	Philadelphia, PA	1,436,287	1,585,577	−9.4
Orange, CA	2,721,701	2,410,668	12.9	Middlesex, MA.......	1,424,116	1,398,468	1.8
Kings, NY	2,267,942	2,300,664	−1.4	Alameda, CA........	1,400,322	1,304,346	7.4
Miami-Dade, FL	2,152,437	1,937,194	11.1	Cuyahoga, OH.......	1,380,696	1,412,140	−2.2
Wayne, MI	2,118,129	2,111,687	0.3	Suffolk, NY.........	1,371,269	1,321,768	3.7
Dallas, TX..........	2,050,865	1,852,810	10.7	Tarrant, TX..........	1,355,273	1,170,103	15.8
Queens, NY	1,998,853	1,951,598	2.4	Bexar, TX...........	1,353,052	1,185,394	14.1
King, WA	1,654,876	1,507,305	9.8	Nassau, NY	1,302,220	1,287,444	1.1
Santa Clara, CA.....	1,641,215	1,497,577	9.6				

Note: The following are the **smallest counties** by 1998 population: Yellowstone National Park, MT (55); Kalawao County, HI (74); Loving County, TX (114); King County, TX (363); Arthur County, NE (428); Kenedy County, TX (438); Petroleum County, MT (499); San Juan County, CO (530); McPherson County, NE (563); and Blaine County, NE (578).

Metropolitan Areas, 1990-96

Source: Bureau of the Census, U.S. Dept. of Commerce

(CMSAs and MSAs of more than 600,000 persons listed by 1996 population estimates)

Metropolitan statistical areas (MSAs) are defined for federal statistical use by the Office of Management and Budget (OMB), with technical assistance from the Bureau of the Census. Most individual metropolitan areas with populations over 1 million may, under specified circumstances, be subdivided into component Primary Metropolitan Statistical Areas (PMSAs), in which case the area as a whole is designated a Consolidated Metropolitan Statistical Area (CMSA).

Effective June 30, 1996, the Office of Management and Budget designated 258 MSAs, 76 PMSAs, and 19 CMSAs for the U.S. and Puerto Rico, based on standards published in the Federal Register on March 30, 1990, as applied to 1990 census data.

CMSAs and MSAs	Population 1990	Population 1996	Percentage change 1990-96
New York-Northern New Jersey-Long Island, NY-NJ-CT-PA CMSA	19,549,649	19,938,492	2.0
Los Angeles-Riverside-Orange County, CA CMSA	14,531,529	15,495,155	6.6
Chicago-Gary-Kenosha, IL-IN-WI CMSA	8,239,820	8,599,774	4.4
Washington-Baltimore, DC-MD-VA-WV CMSA	6,726,395	7,164,519	6.5
San Francisco-Oakland-San Jose, CA CMSA	6,249,881	6,605,428	5.7
Philadelphia-Wilmington-Atlantic City, PA-NJ-DE-MD CMSA	5,893,019	5,973,463	1.4
Boston-Worcester-Lawrence, MA-NH-ME-CT CMSA	5,455,403	5,563,475	2.0
Detroit-Ann Arbor-Flint, MI CMSA	5,187,171	5,284,171	1.9
Dallas–Fort Worth, TX CMSA	4,037,282	4,574,561	13.3
Houston–Galveston–Brazoria, TX CMSA	3,731,029	4,253,428	14.0
Atlanta, GA	2,959,500	3,541,230	19.7
Miami-Fort Lauderdale, FL CMSA	3,192,725	3,514,403	10.1
Seattle–Tacoma–Bremerton, WA CMSA	2,970,300	3,320,829	11.8
Cleveland-Akron, OH CMSA	2,859,644	2,913,430	1.9
Minneapolis–St. Paul, MN-WI	2,538,776	2,765,116	8.9
Phoenix-Mesa, AZ	2,238,498	2,746,703	22.7
San Diego, CA	2,498,016	2,655,463	6.3
St. Louis, MO-IL	2,492,348	2,548,238	2.2
Pittsburgh, PA	2,394,811	2,379,411	−0.6
Denver–Boulder–Greeley, CO CMSA	1,980,140	2,277,401	15.0
Tampa–St. Petersburg–Clearwater, FL	2,067,959	2,199,231	6.3
Portland–Salem, OR–WA CMSA	1,793,476	2,078,357	15.9
Cincinnati–Hamilton, OH–KY–IN, CMSA	1,817,569	1,920,931	5.7
Kansas City, MO-KS	1,582,874	1,690,343	6.8
Milwaukee–Racine, WI CMSA	1,607,183	1,642,658	2.2
Sacramento–Yolo, CA CMSA	1,481,220	1,632,133	10.2
Norfolk–Virginia Beach–Newport News, VA–NC	1,444,710	1,540,252	6.6
Indianapolis, IN	1,380,491	1,492,297	8.1
San Antonio, TX	1,324,749	1,490,111	12.5
Columbus, OH	1,345,450	1,447,646	7.6
Orlando, FL	1,224,844	1,417,291	15.7
Charlotte-Gastonia-Rock Hill, NC-SC	1,162,140	1,321,068	13.7
New Orleans, LA	1,285,262	1,312,890	2.1
Salt Lake City-Ogden, UT	1,072,227	1,217,842	13.6
Las Vegas, NV-AZ	852,646	1,201,073	40.9
Buffalo-Niagara Falls, NY	1,189,340	1,175,240	−1.2
Hartford, CT	1,157,585	1,144,574	−1.1
Greensboro–Winston-Salem–High Point, NC	1,050,304	1,141,238	8.7
Providence-Fall River-Warwick, RI-MA	1,134,350	1,124,044	−0.9
Nashville, TN	985,026	1,117,178	13.4
Rochester, NY	1,062,470	1,088,037	2.4
Memphis, TN–AR–MS	1,007,306	1,078,151	7.0
Austin-San Marcos, TX	846,227	1,041,330	23.1
Oklahoma City, OK	958,839	1,026,657	7.1
Raleigh-Durham-Chapel Hill, NC	858,485	1,025,253	19.4
Grand Rapids-Muskegon-Holland, MI	937,891	1,015,099	8.2
Jacksonville, FL	906,727	1,008,633	11.2
West Palm Beach-Boca Raton, FL	863,503	992,840	15.0
Louisville, KY-IN	949,012	991,765	4.5
Dayton-Springfield, OH	951,270	950,661	−0.1
Richmond–Petersburg, VA	865,640	935,174	8.0
Greenville-Spartanburg-Anderson, SC	830,539	896,679	8.0
Birmingham, AL	839,942	894,702	6.5
Albany-Schenectady-Troy, NY	861,623	878,527	2.0
Honolulu, HI	836,231	871,766	4.2
Fresno, CA	755,580	861,753	14.1
Tucson, AZ	666,957	767,873	15.1
Tulsa, OK	708,954	756,493	6.7
Syracuse, NY	742,237	745,691	0.5
El Paso, TX	591,610	684,446	15.7
Omaha, NE–IA	639,580	681,698	6.6
Albuquerque, NM	589,131	670,092	13.7
Knoxville, TN	585,960	649,277	10.8
Scranton–Wilkes-Barre–Hazleton, PA	638,524	628,073	−1.6
Bakersfield, CA	544,981	622,729	14.3
Harrisburg-Lebanon-Carlisle, PA	587,986	614,755	4.6
Allentown-Bethlehem-Easton, PA	595,081	614,304	3.2
Toledo, OH	614,128	611,417	−0.4

Final 1990 census figures showed that the nation in that year had 40 metropolitan areas of at least 1 mil population, including 5 that had reached that size since 1980. The 40 areas had 132.9 mil people, or 53.4% of the U.S. population in 1990. It is estimated that since the 1990 census, the populations of at least 7 additional metropolitan areas (Las Vegas, NV–AZ; Nashville, TN; Austin–San Marcos, TX; Oklahoma City, OK; Raleigh–Durham–Chapel Hill, NC; Grand Rapids–Muskegon–Holland, MI; and Jacksonville, FL) have increased to more than 1 mil. By 1996, 56.1% of the population lived in metropolitan areas that had a total of at least 1 mil inhabitants.

Some 211.8 mil people resided in metropolitan areas in 1996, an increase of more than 13.6 mil (6.9%) since 1990. The population outside metropolitan areas totaled 53.5 mil in 1996, up 3.0 mil (5.8%) from 1990. The metropolitan population in 1996 was 79.8% of the U.S. total, compared with 79.7% in 1990 and 76.2% in 1980.

Population of 100 Largest U.S. Cities, 1850-1998

Source: Bureau of the Census, U.S. Dept. of Commerce (100 most populous cities ranked by July 1, 1998, population estimates)

Rank	City	1998	1990	1980	1970	1950	1900	1850
1.	New York, NY	7,420,166	7,322,564	7,071,639	7,895,563	7,891,957	3,437,202	696,115
2.	Los Angeles, CA	3,597,556	3,485,557	2,968,528	2,811,801	1,970,358	102,479	1,610
3.	Chicago, IL	2,802,079	2,783,726	3,005,072	3,369,357	3,620,962	1,698,575	29,963
4.	Houston, TX	1,786,691	1,654,348	1,595,138	1,233,535	596,163	44,633	2,396
5.	Philadelphia, PA	1,436,287	1,585,577	1,688,210	1,949,996	2,071,605	1,293,697	121,376
6.	San Diego, CA	1,220,666	1,110,623	875,538	697,471	334,387	17,700	...
7.	Phoenix, AZ	1,198,064	988,015	789,704	584,303	106,818	5,544	...
8.	San Antonio, TX	1,114,130	976,514	785,940	654,153	408,442	53,321	3,488
9.	Dallas, TX	1,075,894	1,007,618	904,599	844,401	434,462	42,638	...
10.	Detroit, MI	970,196	1,027,974	1,203,368	1,514,063	1,849,568	285,704	21,019
11.	San Jose, CA	861,284	782,224	629,400	459,913	95,280	21,500	...
12.	San Francisco, CA	745,774	723,959	678,974	715,674	775,357	342,782	34,776
13.	Indianapolis, IN[1]	741,304	731,278	700,807	736,856	427,173	169,164	8,091
14.	Jacksonville, FL[1]	693,630	635,230	540,920	504,265	204,517	28,429	1,045
15.	Columbus, OH	670,234	632,945	565,021	540,025	375,901	125,560	17,882
16.	Baltimore, MD	645,593	736,014	786,741	905,787	949,708	508,957	169,054
17.	El Paso, TX	615,032	515,342	425,259	322,261	130,485	15,906	...
18.	Memphis, TN	603,507	618,652	646,174	623,988	396,000	102,320	8,841
19.	Milwaukee, WI	578,364	628,088	636,297	717,372	637,392	285,315	20,061
20.	Boston, MA	555,447	574,283	562,994	641,071	801,444	560,892	136,881
21.	Austin, TX	552,434	472,020	345,890	253,539	132,459	22,258	629
22.	Seattle, WA	536,978	516,259	493,846	530,831	467,591	80,671	...
23.	Washington, DC	523,124	606,900	638,432	756,668	802,178	278,718	40,001
24.	Nashville, TN[1]	510,274	488,366	455,651	426,029	174,307	80,865	10,165
25.	Charlotte, NC	504,637	419,558	315,474	241,420	134,042	18,091	1,065
26.	Portland, OR	503,891	485,975	368,148	379,967	373,628	90,426	...
27.	Denver, CO	499,055	467,610	492,686	514,678	415,786	133,859	...
28.	Cleveland, OH	495,817	505,616	573,822	750,879	914,808	381,768	17,034
29.	Fort Worth, TX	491,801	447,619	385,164	393,455	278,778	26,688	...
30.	Oklahoma City, OK	472,221	444,724	404,014	368,164	243,504	10,037	...
31.	New Orleans, LA	465,538	496,938	557,927	593,471	570,445	287,104	116,375
32.	Tucson, AZ	460,466	415,444	330,537	262,933	45,454	7,531	...
33.	Kansas City, MO	441,574	434,829	448,028	507,330	456,622	163,752	...
34.	Virginia Beach, VA	432,380	393,089	262,199	172,106	5,390	...	...
35.	Long Beach, CA	430,905	429,321	361,498	358,879	250,767	2,252	...
36.	Albuquerque, NM	419,311	384,915	332,920	244,501	96,815	6,238	...
37.	Las Vegas, NV	404,288	258,877	164,674	125,787	24,624	...	...
38.	Sacramento, CA	404,168	369,365	275,741	257,105	137,572	29,282	6,820
39.	Atlanta, GA	403,819	393,929	425,022	495,039	331,314	89,872	2,572
40.	Fresno, CA	398,133	354,091	217,491	165,655	91,669	12,470	...
41.	Honolulu, HI[2]	395,789	377,059	365,048	324,871	248,034	39,306	...
42.	Tulsa, OK	381,393	367,302	360,919	330,350	182,740	1,390	...
43.	Omaha, NE	371,291	344,463	313,939	346,929	251,117	102,555	...
44.	Miami, FL	368,624	358,648	346,681	334,859	249,276	1,681	...
45.	Oakland, CA	365,874	372,242	339,337	361,561	384,575	66,960	...
46.	Mesa, AZ	360,076	289,199	152,404	63,049	16,790	722	...
47.	Minneapolis, MN	351,731	368,383	370,951	434,400	521,718	202,718	...
48.	Colorado Springs, CO	344,987	280,430	215,105	135,517	45,472	21,085	...
49.	Pittsburgh, PA	340,520	369,879	423,959	520,089	676,806	321,616	46,601
50.	St. Louis, MO	339,316	396,685	452,801	622,236	856,796	575,238	77,860
51.	Cincinnati, OH	336,400	364,114	385,409	453,514	503,998	325,902	115,435
52.	Wichita, KS	329,211	304,017	279,838	276,554	168,279	24,671	...
53.	Toledo, OH	312,174	332,943	354,635	383,062	303,616	131,822	3,829
54.	Arlington, TX	306,497	261,717	160,113	90,229	7,692	1,079	...
55.	Santa Ana, CA	305,965	293,827	204,023	155,710	45,533	4,933	...
56.	Buffalo, NY	300,717	328,175	357,870	462,768	580,132	352,387	42,261
57.	Anaheim, CA	295,153	266,406	219,494	166,408	14,556	1,456	...
58.	Tampa, FL	289,156	280,015	271,577	277,714	124,681	15,839	...
59.	Corpus Christi, TX	281,453	257,428	232,134	204,525	108,287	4,703	...
60.	Newark, NJ	267,823	275,221	329,248	381,930	438,776	246,070	38,894
61.	Riverside, CA	262,140	226,546	170,591	140,089	46,764	7,973	...
62.	Raleigh, NC	259,423	218,859	150,255	122,830	65,679	13,643	4,518
63.	St. Paul, MN	257,284	272,235	270,230	309,866	311,349	163,065	1,112
64.	Louisville, KY	255,045	269,555	298,694	361,706	369,129	204,731	43,194
65.	Anchorage, AK	254,982	226,338	174,431	48,081	11,254	...	...
66.	Birmingham, AL	252,997	265,347	284,413	300,910	326,037	38,415	...
67.	Aurora, CO	250,604	222,103	158,588	74,974	11,421	202	...
68.	Lexington, KY	241,749	225,366	204,165	108,137	55,534	26,369	8,159
69.	Stockton, CA	240,143	210,943	148,283	109,963	70,853	17,506	...
70.	St. Petersburg, FL	236,029	240,318	238,647	216,159	96,738	1,575	...
71.	Jersey City, NJ	232,429	228,517	223,532	260,350	299,017	206,433	6,856
72.	Plano, TX	219,486	127,885	72,331	17,872	2,126	1,304	...
73.	Rochester, NY	216,887	230,356	241,741	295,011	332,488	162,608	36,403
74.	Akron, OH	215,712	223,019	237,177	275,425	274,605	42,728	3,266
75.	Norfolk, VA	215,215	261,250	266,979	307,951	213,513	46,624	14,326
76.	Lincoln, NE	213,088	191,972	171,932	149,518	98,884	40,169	...
77.	Baton Rouge, LA	211,551	219,531	220,394	165,921	125,629	11,269	3,905
78.	Hialeah, FL	211,392	188,008	145,254	102,452	19,676	...	...
79.	Bakersfield, CA	210,284	176,264	105,611	69,515	34,784	4,836	...
80.	Madison, WI	209,306	190,766	170,616	171,809	96,056	19,164	1,525
81.	Fremont, CA	204,298	173,339	131,945	100,869	...	...	...
82.	Mobile, AL	202,181	199,973	200,452	190,026	129,009	38,469	20,515
83.	Chesapeake, VA	199,564	151,982	114,486	89,580	...	...	...
84.	Greensboro, NC	197,910	185,125	155,642	144,076	74,389	10,035	...
85.	Montgomery, AL	197,014	190,350	177,857	133,386	106,525	30,346	8,728
86.	Scottsdale, AZ	195,394	130,099	88,364	67,823	2,032	...	...

Rank	City	1998	1990	1980	1970	1950	1900	1850
87.	Huntington Beach, CA	195,316	181,519	170,505	115,960	5,237	...	...
88.	Richmond, VA	194,173	202,798	219,214	249,332	230,310	85,050	27,570
89.	Glendale, AZ	193,482	147,070	96,988	36,228	8,179	...	...
90.	Garland, TX	193,408	180,635	138,857	81,437	10,571	819	...
91.	Des Moines, IA.	191,293	193,189	191,003	201,404	177,965	62,139	...
92.	Lubbock, TX.	190,974	186,206	174,361	149,101	71,747	...	...
93.	Yonkers, NY	190,153	188,082	195,351	204,297	152,798	47,931	...
94.	Jackson, MS.	188,419	202,062	202,895	153,968	98,271	7,816	1,881
95.	Shreveport, LA	188,319	198,518	206,989	182,064	127,206	16,013	1,728
96.	Augusta, GA[1]	187,689	186,616	47,532	59,864	71,508	39,441	9,448
97.	San Bernardino, CA	186,402	170,036	118,794	104,251	63,058	6,150	...
98.	Fort Wayne, IN	185,716	195,680	172,391	178,269	133,607	45,115	4,282
99.	Grand Rapids, MI	185,437	189,126	181,843	197,649	176,515	87,565	2,686
100.	Glendale, CA	185,086	180,038	139,060	133,000	96,000	...	...

NOTE: The Apr. 1, 1990, census counts include subsequent revisions and take account of subsequent and geographic changes. (1) Indianapolis, IN; Jacksonville, FL; Nashville, TN; and Augusta, GA, are parts of consolidated city-county governments. Populations of other incorporated places in the county have been excluded from the population totals shown here. For years that predate the establishment of a consolidated city-county government, city population is shown. (2) Locations in Hawaii are called "census designated places (CDPs)." Although these areas are not incorporated, they are recognized for census purposes as large urban places. Honolulu CDP is coextensive with Honolulu Judicial District within the city and county of Honolulu.

Projections of Total Population, by Age, 1995-2050
Source: Bureau of the Census, U.S. Dept. of Commerce

Age	1995 Population[1]	1995 Percentage distribution	2000 Population[1]	2000 Percentage distribution	2010 Population[1]	2010 Percentage distribution	2050 Population[1]	2050 Percentage distribution
TOTAL	262,820	100.0	274,634	100.0	297,716	100.0	393,931	100.0
Under 5 years.	19,591	7.5	18,987	6.9	20,012	6.7	27,106	6.9
5-13 years	34,378	13.1	36,043	13.1	35,605	12.0	47,804	12.1
14-17 years	14,773	5.6	15,752	5.7	16,894	5.7	21,207	5.4
18-24 years	24,926	9.5	26,258	9.6	30,138	10.1	36,333	9.2
25-34 years	40,863	15.5	37,233	13.6	38,292	12.9	49,365	12.5
35-44 years	42,514	16.2	44,659	16.3	38,521	12.9	47,393	12.0
45-54 years	31,092	11.8	37,030	13.5	43,564	14.6	43,494	11.0
55-64 years	21,139	8.0	23,961	8.7	35,283	11.9	42,368	10.8
65 years and over. .	33,543	12.8	34,709	12.6	39,408	13.2	78,859	20.0
85 years and over. .	3,634	1.4	4,259	1.6	5,671	1.9	18,223	4.6
100 years and over.	54	0.0	72	0.0	131	0.0	834	0.2

NOTE: All figures shown are for July 1 of the given year, exclude Armed Forces overseas, and are middle series population projections. For the series shown, different assumptions were made regarding fertility rates (lifetime births per woman), life expectancy, and immigration in the coming decades. Assumptions were based on July 1 estimates of U.S. population consistent with the 1990 decennial census, as enumerated. Yearly net immigration was assumed to be 820,000. Percentage distribution may not equal 100, because of overlapping categories shown and rounding. (1) In thousands.

U.S. Population, by Age, Sex, and Household, 1990
Source: Bureau of the Census, U.S. Dept. of Commerce; 1990 Census

Total population	**248,709,873[1]**
AGE	
Under 5 years.	18,354,443
5 to 17 years	45,249,989
18 to 20 years	11,726,868
21 to 24 years	15,010,898
25 to 44 years	80,754,835
45 to 64 years	25,223,086
55 to 59 years	10,531,756
60 to 64 years	10,616,167
65 to 74 years	18,106,558
75 to 84 years	10,055,108
85 years and over.	3,080,165
Median age	32.9
Under 18 years.	63,604,432
Percentage of total population	25.6
65 years and over.	31,241,831
Percentage of total population	12.6

SEX	
Male.	121,239,418
Female.	127,470,455
HOUSEHOLDS BY TYPES	
TOTAL HOUSEHOLDS.	**91,947,410**
Family households (families)	64,517,947
Married-couple families	50,708,322
Percentage of total households.	55.1
Other family, male householder.	3,143,582
Other family, female householder	10,666,043
Nonfamily households	27,429,463
Percentage of total households.	29.8
Householder living alone	22,580,420
Householder 65 years and over	8,824,845
Persons living in households.	242,012,129
Persons per household	2.63
Persons living in group quarters	6,697,744
Institutionalized persons	3,334,018
Other persons in group quarters	3,363,726

(1) Data shown are based on the U.S. population as of Apr. 1, as tabulated in the 1990 census, and do not reflect corrections to the 1990 population. The corrected 1990 U.S. population is 248,765,170.

M I L L E N N I U M   F A C T   B O X

The Aging U.S. Population
Source: Bureau of the Census, U.S. Dept. of Commerce

The U.S. population continues to age. The elderly population (people age 65 and over) increased 10-fold from 1900 to 1990, compared with only a 3-fold increase for those under age 65. The population 85 and older increased 25-fold from 1900 to 1990.

Year[1]	65 AND OVER Number[2]	65 AND OVER Percent	85 AND OVER Number[2]	85 AND OVER Percent	Year[1]	65 AND OVER Number[2]	65 AND OVER Percent	85 AND OVER Number[2]	85 AND OVER Percent
1900 . . .	3,080	4.1	122	0.2	1950 . .	12,269	8.1	577	0.4
1910 . . .	3,949	4.3	167	0.2	1960 . .	16,560	9.2	929	0.5
1920 . . .	4,933	4.7	210	0.2	1970 . .	19,980	9.8	1,409	0.7
1930 . . .	6,634	5.4	272	0.2	1980 . .	25,550	11.3	2,240	1.0
1940 . . .	9,019	6.8	365	0.3	1990 . .	31,079	12.5	3,021	1.2

NOTE: Figures for 1900 to 1950 exclude Alaska and Hawaii. (1) Date of Census. (2) Resident population, in thousands.

U.S. Households, by Type, 1960-98

Source: Bureau of the Census, U.S. Dept. of Commerce

(as of Mar.)

YEAR	Total U.S. households[1]	Married-couple households[1]	Unmarried-couple households[1]	YEAR	Total U.S. households[1]	Married-couple households[1]	Unmarried-couple households[1]
1960	52,799	39,254	439	1989	92,830	52,100	2,764
1970	63,401	44,728	523	1990	93,347	52,317	2,856
1980	80,776	49,112	1,589	1991	94,312	52,147	3,039
1981	82,368	49,294	1,808	1992	95,669	52,457	3,308
1982	83,527	49,630	1,863	1993	96,426	53,090	3,510
1983	83,918	49,908	1,891	1994	97,107	53,171	3,661
1984	85,407	50,090	1,988	1995	98,990	53,858	3,668
1985	86,789	50,350	1,983	1996	99,627	53,567	3,958
1986	88,458	50,933	2,220	1997	101,018	53,604	4,130
1987	89,479	51,537	2,334	1998	102,528	54,317	4,236
1988	91,124	51,675	2,588				

(1) Numbers in thousands.

Living Arrangements of Children, 1970-98

Source: Bureau of the Census, U.S. Dept. of Commerce

(as of Mar.; excludes persons under 18 years of age who maintained households or resided in group quarters)

Race, Hispanic origin, and year	Number (1,000)	BOTH PARENTS	MOTHER ONLY Total	MOTHER ONLY Divorced	MOTHER ONLY Married Spouse absent	MOTHER ONLY Single[1]	MOTHER ONLY Widowed	FATHER ONLY	NEITHER PARENT
White									
1970	58,790	90	8	3	3	Z	2	1	2
1980	52,242	83	14	7	4	1	2	2	2
1990	51,390	79	16	8	4	3	1	3	2
1995	55,315	76	18	8	5	4	1	3	3
1996	55,709	75	18	8	5	5	1	4	3
1997	55,868	75	18	8	4	5	1	4	3
1998	56,118	74	18	8	4	5	1	5	3
Black									
1970	9,422	59	30	5	16	4	4	2	10
1980	9,375	42	44	11	16	13	4	2	12
1990	10,018	38	51	10	12	27	2	4	8
1995	11,301	33	52	11	11	29	2	4	11
1996	11,434	33	53	9	11	31	2	4	9
1997	11,369	35	52	9	11	31	1	5	8
1998	11,407	36	51	9	9	32	1	4	9
Hispanic[2]									
1970	4,006[3]	78	NA	NA	NA	NA	NA	NA	NA
1980	5,459	75	20	6	8	4	2	2	4
1990	7,174	67	27	7	10	8	2	3	3
1995	9,842	63	28	8	9	10	1	4	4
1996	10,251	62	29	7	9	11	1	4	5
1997	10,525	64	27	7	7	12	1	4	5
1998	10,857	64	27	6	8	12	1	4	5

NA = Not available. Z = Less than 0.5%. (1) Never married. (2) Hispanic persons may be of any race. (3) All persons under 18 years old.

Grandchildren Living in the Home of Their Grandparents, 1970-98

Source: Bureau of the Census, U.S. Dept. of Commerce
(numbers in thousands)

YEAR	Total children under 18	Total	WITH PARENT(S) PRESENT Both parents present	WITH PARENT(S) PRESENT Mother only present	WITH PARENT(S) PRESENT Father only present	Without parent(s) present
1998	71,377	3,989	503	1,827	241	1,417
1997	70,983	3,894	554	1,785	247	1,309
1996	70,908	4,060	467	1,943	220	1,431
1995	70,254	3,965	427	1,876	195	1,466
1994	69,508	3,735	436	1,764	175	1,359
1993	66,893	3,368	475	1,647	229	1,017
1992	65,965	3,253	502	1,740	144	867
1991	65,093	3,320	559	1,674	151	937
1990	64,137	3,155	467	1,563	191	935
1980	63,369	2,306	310	922	86	988
1970	69,276	2,214	363	817	78	957

Poverty Level by Family Size, 1996-98

Source: Bureau of the Census, U.S. Dept. of Commerce

	1996	1997	1998		1996	1997	1998
1 person	$7,995	$8,183	$8,316	3 persons	$12,516	$12,802	$13,003
Under 65 years	8,163	8,350	8,480	4 persons	16,036	16,400	16,660
65 years and over	7,525	7,698	7,818	5 persons	18,952	19,380	19,680
2 persons	10,233	10,473	10,634	6 persons	21,389	21,886	22,228
Householder under 65 years	10,564	10,805	10,972	7 persons	24,268	24,802	25,257
Householder 65 years				8 persons	27,091	27,593	28,166
and over	9,491	9,712	9,862	9 persons or more	31,971	32,566	33,339

Poverty Rate

Source: Bureau of the Census, U.S. Dept. of Commerce

The poverty rate is the proportion of the population whose income falls below the government's official poverty level, which is adjusted each year for inflation. The national poverty rate was 12.7% in 1998, a decrease from the 1997 rate of 13.3%. Children remained overrepresented among the poor, with a poverty rate of 18.9%. The elderly were slightly underrepresented, at 10.5%.

Poverty by Family Status, Sex, and Race, 1986-98

Source: Bureau of the Census, U.S. Dept. of Commerce

(numbers in thousands)

	1998 No.	1998 %[1]	1997 No.	1997 %[1]	1996 No.	1996 %[1]	1990 No.	1990 %[1]	1986 No.	1986 %[1]
TOTAL POOR	34,476	12.7	35,574	13.3	36,529	13.7	33,585	13.5	32,370	13.6
In families	25,370	11.2	26,217	11.6	27,376	12.2	25,232	12.0	24,754	12.0
Head of household	7,186	10.0	7,324	10.3	7,708	11.0	7,098	10.7	7,023	10.9
Related children	12,845	18.3	13,422	19.2	13,764	19.8	12,715	19.9	12,257	19.8
Unrelated individuals	8,478	19.9	8,687	20.8	8,452	20.8	7,446	20.7	6,846	21.6
In families, female householder, no husband present	12,907	33.1	13,494	35.1	13,796	35.8	12,578	37.2	11,944	38.3
Head of household	3,831	29.9	3,995	31.6	4,167	32.6	3,768	33.4	3,613	34.6
Related children	7,627	46.1	7,928	49.0	7,990	49.3	7,363	53.4	6,943	54.4
Unrelated female individuals	5,013	22.6	5,240	24.0	5,145	24.2	4,589	24.0	4,311	25.1
All other families	12,463	6.6	12,723	6.8	13,580	7.3	12,654	7.1	12,811	7.3
Head of household	3,355	5.7	3,329	5.7	3,541	6.2	3,330	6.0	3,410	6.3
Related children	5,218	9.7	5,494	10.2	5,774	10.9	5,352	10.7	5,313	10.8
Unrelated male individuals	3,465	17.0	3,447	17.4	3,308	17.0	2,857	16.9	2,536	17.5
TOTAL WHITE POOR	23,454	10.5	24,396	11.0	24,650	11.2	22,326	10.7	22,183	11.0
In families	16,549	8.9	17,258	9.3	17,621	9.6	15,916	9.0	16,393	9.4
Head of household	4,829	8.0	4,990	8.4	5,059	8.6	4,622	8.1	4,811	8.6
Related children	7,935	14.4	8,441	15.4	8,488	15.5	7,696	15.1	7,714	15.3
Female householder, no spouse present	2,123	24.9	2,305	27.7	2,276	27.3	2,010	26.8	2,041	28.2
Unrelated individuals	6,386	18.0	6,593	18.9	6,463	18.9	5,739	18.6	5,198	19.2
TOTAL BLACK POOR	9,091	26.1	9,116	26.5	9,694	28.4	9,837	31.9	8,983	31.1
In families	7,259	24.7	7,386	25.5	7,993	27.6	8,160	31.0	7,410	29.7
Head of household	1,981	23.4	1,985	23.6	2,206	26.1	2,193	29.3	1,987	28.0
Related children	4,073	36.4	4,116	36.8	4,411	39.5	4,412	44.2	4,039	42.7
Female householder, no spouse present	1,557	40.8	1,563	39.8	1,724	43.7	1,648	48.1	1,488	50.1
Unrelated individuals	1,752	32.5	1,645	31.0	1,606	32.2	1,491	35.1	1,431	38.5

(1) Percentage of total U.S. population in each category who fell below poverty level and are enumerated here. For example, of all persons in families in 1998, 11.2%, or 25,370,000, were poor.

Persons Below Poverty Level, 1960-98

Source: Bureau of the Census, U.S. Dept. of Commerce

YEAR	Number below poverty level (in millions) All races[1]	White	Black	Hispanic origin[2]	Percentage below poverty level All races[1]	White	Black	Hispanic origin[2]	Avg. income cutoffs for family of 4 at poverty level[3]
1960	39.9	28.3	NA	NA	22.2	17.8	NA	NA	$3,022
1970	25.4	17.5	7.5	NA	12.6	9.9	33.5	NA	3,968
1980	29.3	19.7	8.6	3.5	13.0	10.2	32.5	25.7	8,414
1990	33.6	22.3	9.8	6.0	13.5	10.7	31.9	28.1	13,359
1991	35.7	23.7	10.2	6.3	14.2	11.3	32.7	28.7	13,924
1992	38.0	25.3	10.8	7.6	14.8	11.9	33.4	29.6	14,335
1993	39.3	26.2	10.9	8.1	15.1	12.2	33.1	30.6	14,763
1994	38.1	25.4	10.2	8.4	14.5	11.7	30.6	30.7	15,141
1995	36.4	24.4	9.9	8.6	13.8	11.2	29.3	30.3	15,569
1996	36.5	24.7	9.7	8.7	13.7	11.2	28.4	29.4	16,036
1997	35.6	24.4	9.1	8.3	13.3	11.0	26.5	27.1	16,400
1998	34.5	23.5	9.1	8.1	12.7	10.5	26.1	25.6	16,660

NA = Not available. **NOTE:** Because of a change in the definition of poverty, data prior to 1980 are not directly comparable to data since 1980. (1) Includes other races not shown separately. (2) Persons of Hispanic origin may be of any race. (3) Figures for 1960-80 represent only nonfarm families.

Persons in Poverty, by State, 1997-98

Source: Bureau of the Census, U.S. Dept. of Commerce

STATE	1998 Percentage	1997 Percentage	STATE	1998 Percentage	1997 Percentage	STATE	1998 Percentage	1997 Percentage	STATE	1998 Percentage	1997 Percentage
AL	15.1	14.8	IL	10.6	11.6	MT	16.1	16.3	RI	12.2	11.9
AK	9.1	8.5	IN	9.1	8.2	NE	11.1	10.0	SC	13.4	13.1
AZ	16.9	18.8	IA	9.3	9.6	NV	10.8	9.6	SD	13.7	14.1
AR	17.2	18.4	KS	9.6	10.4	NH	9.4	7.7	TN	13.9	15.1
CA	16.0	16.8	KY	14.7	16.4	NJ	8.9	9.2	TX	15.9	16.7
CO	8.7	9.4	LA	17.7	18.4	NM	20.8	23.4	UT	8.9	8.3
CT	9.0	10.1	ME	10.2	10.7	NY	16.6	16.6	VT	9.6	10.9
DE	10.0	9.1	MD	7.8	9.3	NC	12.7	11.8	VA	10.8	12.5
DC	22.0	23.0	MA	10.4	11.2	ND	14.4	12.3	WA	9.1	10.5
FL	13.7	14.3	MI	10.6	10.7	OH	11.1	11.8	WV	17.1	17.5
GA	14.0	14.7	MN	10.0	9.7	OK	13.9	15.2	WI	8.5	8.5
HI	12.4	13.0	MS	17.1	18.6	OR	13.3	11.7	WY	12.1	12.7
ID	13.8	13.3	MO	10.8	10.6	PA	11.2	11.4			

Block Grants for Temporary Assistance for Needy Families, Fiscal Year 1998

Source: Admin. for Children and Families, Off. of Planning, Research, and Evaluation, U.S Dept. of Health and Human Services

STATE[1]	Total federal and state TANF expenditures[2]	Average monthly number of families	Average monthly number of recipients	Average monthly number of children[3]	Average monthly payment per Family	Average monthly payment per Recipient
Alabama	$96,356	23,309	56,853	46,230	$344.49	$141.23
Alaska	100,836	10,159	30,997	20,260	827.15	271.09
Arizona	230,106	40,163	112,567	49,532	477.44	170.35
Arkansas	52,193	13,844	34,859	26,758	314.18	124.77
California	5,453,531	722,868	2,126,975	1,498,796	628.69	213.67
Colorado	161,469	21,185	57,294	44,070	635.16	234.86
Connecticut	427,745	48,129	129,286	85,997	740.62	275.71
Delaware	53,770	7,199	18,896	14,706	622.44	237.13
District of Columbia	135,021	21,258	57,015	41,570	529.29	197.35
Florida	553,773	107,951	277,961	212,267	427.49	166.02
Georgia	337,969	75,000	193,725	150,177	375.52	145.38
Hawaii	124,027	17,031	47,401	32,404	606.86	218.05
Idaho	14,591	1,918	4,428	3,035	634.02	274.60
Illinois	984,152	169,735	506,580	382,922	483.18	161.89
Indiana	134,842	40,059	114,406	79,187	280.51	98.22
Iowa	145,542	25,191	68,700	45,742	481.46	176.54
Kansas	125,670	14,136	36,892	26,338	740.83	283.87
Kentucky	177,816	52,642	127,305	90,483	281.49	116.40
Louisiana	103,665	48,228	136,421	126,585	179.12	63.32
Maine	95,375	15,697	41,574	27,317	506.34	191.18
Maryland	298,809	47,388	125,163	89,663	525.46	198.95
Massachusetts	668,318	66,490	175,751	120,081	837.61	316.89
Michigan	966,044	123,693	359,627	250,011	650.84	223.85
Minnesota	312,486	48,248	143,897	98,175	539.72	180.97
Mississippi	93,436	23,722	60,097	47,242	328.23	129.56
Missouri	260,233	60,019	155,376	119,773	361.32	139.57
Montana	44,175	6,356	18,895	13,972	579.22	194.82
Nebraska	61,500	12,960	36,372	27,325	395.44	140.91
Nevada	64,020	10,383	26,839	19,615	513.81	198.78
New Hampshire	64,684	6,857	16,576	10,360	786.13	325.18
New Jersey	477,183	80,489	213,707	152,892	494.05	186.07
New Mexico	120,170	22,053	68,997	43,873	454.11	145.14
New York	3,279,628	336,858	914,627	645,788	811.33	298.81
North Carolina	375,263	77,915	187,156	135,300	401.36	167.09
North Dakota	30,310	3,322	8,892	6,370	760.34	284.06
Ohio	604,330	151,527	397,555	254,111	332.36	126.68
Oklahoma	86,231	24,462	71,439	44,693	293.76	100.59
Oregon	206,778	18,242	47,917	34,047	944.61	359.61
Pennsylvania	817,839	127,715	366,276	271,679	533.64	186.07
Rhode Island	158,488	19,290	54,343	29,872	684.69	243.04
South Carolina	104,389	25,291	66,226	49,659	343.96	131.35
South Dakota	20,400	3,837	10,093	7,756	443.04	168.43
Tennessee	208,246	57,372	148,540	104,400	302.48	116.83
Texas	509,836	145,232	401,200	284,338	292.54	105.90
Utah	87,549	9,396	28,689	21,742	776.47	254.30
Vermont	57,929	7,366	20,306	12,490	655.34	237.74
Virginia	227,043	43,269	104,688	74,013	437.27	180.73
Washington	538,834	77,762	214,701	140,891	577.44	209.14
West Virginia	55,585	19,674	53,796	34,059	235.44	86.11
Wisconsin	211,133	13,965	52,493	33,058	1,259.91	335.18
Wyoming	9,173	1,249	3,064	2,139	611.89	249.52
U.S. TOTALS	**$20,528,491**	**3,148,102**	**8,733,435**	**6,183,763**	**$543.41**	**$195.88**

NOTE: Data are preliminary. Under 1996 legislation, the Aid to Families With Dependent Children (AFDC) program was converted to a state block-grant program (Temporary Assistance to Needy Families [TANF]). Conversion dates varied from state to state, starting from Sept. 30, 1996. As of July 1, 1997, all states were operating under the new program. (1) Data for Guam, Puerto Rico, and the Virgin Islands not available as of Sept. 1999. (2) In thousands. (3) Number of child recipients based on sample data for some states.

Distribution of TANF[1] Adult Recipients by Employment Status, 1997-98

Source: Admin. for Children and Families, Off. of Planning, Research, and Evaluation, U.S Dept. of Health and Human Services

STATE	Adults	% Employed	STATE	Adults	% Employed	STATE	Adults	% Employed
AL	12,987	23.9	LA	36,448	16.5	OK	17,345	25.2
AK	10,873	31.7	ME	13,832	25.1	OR	16,514	6.3
AZ	29,344	38.8	MD	36,532	14.4	PA	120,856	22.2
AR	8,636	13.6	MA	53,317	20.4	RI	18,091	24.5
CA	611,799	30.8	MI	110,175	43.8	SC	17,193	22.7
CO	16,215	17.3	MN	46,410	31.8	SD	2,426	14.7
CT	41,634	49.3	MS	14,054	6.9	TN	39,751	21.5
DE	5,341	22.3	MO	46,885	9.3	TX	116,503	3.9
DC	18,931	2.2	MT	7,544	16.8	UT	11,339	26.4
FL	70,901	28.3	NE	11,131	15.5	VT	7,646	27.3
GA	49,089	12.9	NV	7,795	21.5	VA	38,634	6.9
HI	15,511	24.3	NH	4,919	24.3	WA	74,002	26.2
ID	1,153	18.3	NJ	61,361	11.3	WV	21,670	9.0
IL	153,321	25.5	NM	21,127	12.6	WI	13,703	15.4
IN	31,683	17.5	NY	321,961	14.2	WY	739	16.8
IA	22,789	29.7	NC	51,299	13.4			
KS	10,479	17.8	ND	2,647	15.6	**U.S.**	**2,631,142**	**22.8**
KY	38,732	15.2	OH	112,551	26.0			

(1) TANF = the state block grant program known as Temporary Assistance for Needy Families.

U.S. Places of 5,000 or More Population—With ZIP and Area Codes

Source: U.S. Bureau of the Census, Dept. of Commerce; Bellcore; Lockheed Martin

The following is a list of places of 5,000 or more inhabitants recognized by the Bureau of the Census, U.S. Dept. of Commerce, based on 1998 population estimates. Also given are 1990 census populations. This list includes **places that are incorporated** under the laws of their respective states as cities, boroughs, towns, and villages, as well as boroughs in Alaska and towns in the 6 New England states (Connecticut, Maine, Massachusetts, New Hampshire, Rhode Island, and Vermont), New York, and Wisconsin.

Places that the Census Bureau designates as **"census designated places"** (CDPs) are also included. These communities, marked (c), are statistically compatible with incorporated communities because of their population density. CDP boundaries can change from one census to another. Hawaii is the only state that has no incorporated places recognized by the Census Bureau; all places shown for Hawaii are CDPs. 1998 estimates are not available for most CDPs.

This list also includes, in *italics,* **minor civil divisions (MCDs)** for Connecticut, Maine, Massachusetts, New Hampshire, Rhode Island, and Vermont. MCDs are not incorporated under the laws of the state and not recognized by the Census Bureau as CDPs, but are often the primary political or administrative divisions of a county. These areas may also serve as general-purpose local governments.

An **asterisk** (*) denotes that the ZIP code given is for general delivery; named streets and/or post office boxes within the community may differ. Consult the local postmaster for the correct ZIP code for specific addresses within the community.

Area codes, given in parentheses, refer only to home and business telephone numbers. Some regions have 2 area codes intermixed; these are known as overlays. States where this occurs are noted. When 2 or more area codes are listed for one place, consult local operators for assistance. Area codes based on latest information available as of Sept. 1999. For a listing in numerical order of all area codes in the U.S., Canada, and the Caribbean, see Consumer Information.

For some places listed, no area code and/or ZIP code is available.

Alabama

ZIP	Place		1998	1990
35005	Adamsville	(205)	5,026	5,161
35007	Alabaster	(205)	23,760	14,619
*35950	Albertville	(256)	16,867	14,507
*35010	Alexander City	(256)	16,024	14,917
36420	Andalusia	(334)	8,907	9,269
*36201	Anniston	(256)	25,524	26,638
35016	Arab	(256)	7,467	6,321
*35611	Athens	(256)	19,720	16,901
*36502	Atmore	(334)	8,169	8,046
35954	Attalla	(256)	6,847	6,859
*36830	Auburn	(334)	40,425	5,161
36507	Bay Minette	(334)	8,166	7,168
*35020	Bessemer	(205)	30,841	33,581
*35203	Birmingham	(205)	252,997	265,347
*35957	Boaz	(256)	7,610	6,928
*36426	Brewton	(334)	6,002	5,885
35220	Center Point (c)	(205)	—	22,658
36671	Chickasaw	(334)	6,289	6,651
35044	Childersburg	(256)	5,049	4,579
*35045	Clanton	(205)	8,733	7,669
*35055	Cullman	(256)	14,437	13,367
36322	Daleville	(334)	5,349	5,117
36526	Daphne	(334)	15,687	11,291
*35601	Decatur	(256)	54,694	49,917
36732	Demopolis	(334)	7,569	7,512
*36302	Dothan	(334)	57,069	54,131
*36330	Enterprise	(334)	21,663	20,119
*36027	Eufaula	(334)	13,463	13,220
35064	Fairfield	(205)	11,183	12,200
*36532	Fairhope	(334)	12,734	9,189
*35630	Florence	(256)	39,098	36,426
*36535	Foley	(334)	7,292	4,937
35214	Forestdale (c)	(205)	—	10,395
*35967	Fort Payne	(256)	12,648	11,838
36362	Fort Rucker (c)	(334)	—	7,593
35068	Fultondale	(205)	6,636	6,400
*35901	Gadsden	(256)	42,158	42,523
35071	Gardendale	(205)	9,728	9,251
35905	Glencoe	(256)	5,122	4,687
36037	Greenville	(334)	7,594	7,847
35976	Guntersville	(256)	8,026	7,038
35570	Hamilton	(205)	6,303	6,171
35640	Hartselle	(256)	12,431	11,114
35080	Helena	(205)	9,518	4,303
35259	Homewood	(205)	22,452	22,644
*35244	Hoover	(205)	59,551	39,988
35023	Hueytown	(205)	14,978	15,280
*35801	Huntsville	(256)	175,979	159,880
35210	Irondale	(205)	9,628	9,458
36545	Jackson	(334)	6,223	5,819
36265	Jacksonville	(256)	10,053	10,283
*35501	Jasper	(205)	14,110	13,553
36863	Lanett	(334)	8,623	8,985
35094	Leeds	(205)	10,750	10,009
*35758	Madison	(256)	25,400	14,792
35228	Midfield	(205)	5,150	5,559
36054	Millbrook	(334)	10,824	6,046
*36601	Mobile	(334)	202,181	199,973
*36460	Monroeville	(334)	6,924	6,993
*36104	Montgomery	(334)	197,014	190,350
35004	Moody	(205)	6,470	4,921
35253	Mountain Brook	(205)	18,497	19,810
*35661	Muscle Shoals	(256)	10,966	9,611
*35476	Northport	(205)	20,247	17,297
35121	Oneonta	(205)	5,479	4,844
*36801	Opelika	(334)	24,490	22,122
36467	Opp	(334)	6,933	7,011
36203	Oxford	(256)	11,031	9,537
*36360	Ozark	(334)	12,660	13,030
35124	Pelham	(205)	14,146	9,356
*35125	Pell City	(205)	10,399	7,945
*36867	Phenix City	(334)	27,353	25,311

ZIP	Place		1998	1990
36272	Piedmont	(256)	5,183	5,347
35126	Pinson-Clay-Chalkville (c)	(205)	—	10,987
35127	Pleasant Grove	(205)	9,054	8,458
*36067	Prattville	(334)	25,769	19,816
36610	Prichard	(334)	32,610	34,320
35906	Rainbow City	(256)	8,694	7,667
36274	Roanoke	(334)	6,181	6,362
*35653	Russellville	(256)	8,497	7,812
36201	Saks (c)	(256)	—	11,138
36571	Saraland	(334)	12,976	11,784
36572	Satsuma	(334)	5,973	5,194
*35768	Scottsboro	(256)	14,215	13,786
*36701	Selma	(334)	22,037	23,755
35660	Sheffield	(256)	10,001	10,380
35901	Southside	(256)	6,933	5,580
*35150	Sylacauga	(256)	12,518	12,520
*35160	Talladega	(256)	17,449	18,175
36078	Tallassee	(334)	5,297	5,112
35217	Tarrant	(205)	7,291	8,046
*36582	Theodore (c)	(334)	—	6,509
36619	Tillman's Corner (c)	(334)	—	17,988
*36081	Troy	(334)	13,487	13,051
35173	Trussville	(205)	11,516	8,283
*35401	Tuscaloosa	(205)	83,376	77,866
35674	Tuscumbia	(256)	8,298	8,413
36083	Tuskegee	(334)	10,989	12,257
*36854	Valley	(334)	9,282	9,556
35266	Vestavia Hills	(205)	21,838	19,550
*36092	Wetumpka	(334)	6,228	4,670

Alaska (907)

ZIP	Place	1998	1990
*99501	Anchorage	254,982	226,338
99559	Bethel	6,342	4,674
*99708	College (c)	—	11,249
99702	Eielson AFB (c)	—	5,251
*99701	Fairbanks	33,295	30,843
*99801	Juneau	30,191	26,751
99611	Kenai	7,943	6,327
*99901	Ketchikan	7,543	8,263
*99615	Kodiak	7,720	6,365
99639	Ninilchik (c)	—	10,523
99835	Sitka	8,338	8,588
*99654	Wasilla	5,791	4,028

Arizona

ZIP	Place		1998	1990
*85220	Apache Junction	(480)	21,235	18,092
85323	Avondale	(623)	27,580	17,595
85603	Bisbee	(520)	6,358	6,288
85326	Buckeye	(623)	5,184	4,436
*86442	Bullhead City	(520)	28,152	21,951
86322	Camp Verde	(520)	7,945	6,243
*85222	Casa Grande	(520)	23,003	19,076
*85225	Chandler	(480)	160,329	89,862
86503	Chinle (c)	(520)	—	5,059
86323	Chino Valley	(520)	6,797	4,837
85228	Coolidge	(520)	7,707	6,934
86326	Cottonwood	(520)	7,786	5,918
86326	Cottonwood-Verde Village (c)	(520)	—	7,037
*85607	Douglas	(520)	15,208	13,908
85335	El Mirage	(623)	5,940	5,001
85231	Eloy	(520)	8,117	7,211
*86004	Flagstaff	(520)	56,657	45,857
85232	Florence	(520)	11,911	7,321
85726	Flowing Wells (c)	(520)	—	14,013
.....	Fortuna Foothills (c)	(520)	—	7,737
*85268	Fountain Hills	(480)	19,159	10,030
*85299	Gilbert	(480)	88,840	29,149
*85301	Glendale	(623)	193,482	147,070
*85501	Globe	(520)	6,756	6,062
85338	Goodyear	(623)	15,262	6,258
*85622	Green Valley (c)	(520)	—	13,231

ZIP	Place		1998	1990
85283	Guadalupe	(480)	5,758	5,458
86025	Holbrook	(520)	5,672	4,686
*86401	Kingman	(520)	18,369	13,208
*86403	Lake Havasu City	(520)	40,495	24,363
85653	Marana	(520)	7,197	2,565
*85201	Mesa	(480)	360,076	289,199
*86440	Mohave Valley (c)	(520)	—	6,962
.....	New Kingman-Butler (c)	(520)	—	11,627
*85621	Nogales	(520)	22,042	19,489
85737	Oro Valley	(520)	21,411	9,024
86040	Page	(520)	7,900	6,598
85253	Paradise Valley	(480)	14,544	11,903
*85541	Payson	(520)	11,978	8,377
*85345	Peoria	(623)	87,048	51,080
*85034	Phoenix	(602)	1,198,064	988,015
*86301	Prescott	(520)	34,129	26,592
*86314	Prescott Valley	(520)	18,873	8,904
*85546	Safford	(520)	9,254	7,359
85349	San Luis	(520)	12,149	4,212
*85251	Scottsdale	(480)	195,394	130,099
*86336	Sedona	(520)	9,905	7,720
*85901	Show Low	(520)	6,861	5,020
*85635	Sierra Vista	(520)	38,068	32,983
85635	Sierra Vista Southeast (c)	(520)	—	9,237
85350	Somerton	(520)	6,930	5,293
85713	South Tucson	(520)	5,601	5,171
*85351	Sun City (c)	(623)	—	38,126
*85351	Sun City West (c)	(623)	—	15,997
85248	Sun Lakes (c)	(480)	—	6,578
*85374	Surprise	(623)	14,849	7,122
*85285	Tempe	(480)	167,622	141,993
85353	Tolleson	(623)	5,121	4,436
86045	Tuba City (c)	(520)	460,466	7,323
*85726	Tucson	(520)	—	415,444
*85390	Wickenburg	(520)	5,366	4,515
86047	Winslow	(520)	10,684	9,279
*85364	Yuma	(520)	62,433	56,966

Arkansas

ZIP	Place		1998	1990
71923	Arkadelphia	(870)	10,407	10,407
*72501	Batesville	(870)	9,595	9,595
72012	Beebe	(501)	5,417	5,417
*72714	Bella Vista (c)	(501)	—	9,083
*72015	Benton	(501)	23,121	18,177
72712	Bentonville	(501)	19,691	11,257
*72315	Blytheville	(870)	18,566	22,523
*72022	Bryant	(501)	9,155	5,940
72023	Cabot	(501)	14,445	8,319
*71701	Camden	(870)	13,205	14,701
72830	Clarksville	(501)	7,128	5,833
*72032	Conway	(501)	39,164	26,481
71635	Crossett	(870)	6,097	6,282
71639	Dumas	(870)	5,039	5,520
*71730	El Dorado	(870)	21,848	23,146
*72701	Fayetteville	(501)	53,300	42,247
*72335	Forrest City	(870)	13,064	13,364
*72901	Fort Smith	(501)	75,637	72,798
72936	Greenwood	(501)	6,148	3,984
*72601	Harrison	(870)	11,594	9,936
72543	Heber Springs	(501)	6,543	5,628
72342	Helena	(870)	6,970	7,491
*71801	Hope	(870)	9,775	9,768
*71901	Hot Springs	(501)	37,961	33,095
*71909	Hot Springs Village (c)	(501)	—	6,361
*72076	Jacksonville	(501)	28,840	29,101
*72401	Jonesboro	(870)	52,250	46,535
*72201	Little Rock	(501)	175,303	175,727
*71753	Magnolia	(870)	10,739	11,151
72104	Malvern	(501)	9,738	9,236
*72360	Marianna	(870)	5,263	6,033
72364	Marion	(870)	6,459	4,405
72113	Maumelle	(501)	8,761	6,714
71953	Mena	(501)	6,050	5,475
*71655	Monticello	(870)	8,316	8,119
72110	Morrilton	(501)	6,540	6,551
*72653	Mountain Home	(870)	10,129	9,027
72112	Newport	(870)	6,763	7,459
*72114	North Little Rock	(501)	59,184	61,829
72370	Osceola	(870)	8,188	9,165
*72450	Paragould	(870)	21,971	18,540
*71601	Pine Bluff	(870)	52,968	57,140
72455	Pocahontas	(870)	6,549	6,151
*72756	Rogers	(501)	37,073	24,692
*72801	Russellville	(501)	25,340	21,260
*72143	Searcy	(501)	18,217	15,180
72120	Sherwood	(501)	20,965	18,890
72761	Siloam Springs	(501)	10,734	8,151
*72764	Springdale	(501)	40,287	29,945
72160	Stuttgart	(870)	9,797	10,420
71854	Texarkana	(870)	23,693	22,631
72472	Trumann	(870)	6,519	6,346
*72956	Van Buren	(501)	19,271	14,899
71671	Warren	(870)	5,991	6,455
72390	West Helena	(870)	9,443	10,137
*72301	West Memphis	(870)	26,581	28,259
72396	Wynne	(870)	8,618	8,187

California

Area code (341) overlays area code (510). Area code (424) overlays area code (310). Area code (628) overlays area code (415). Area code (657) overlays area code (714). Area code (669) overlays area code (408). Area code (752) overlays area code (909). See introductory note.

Area code (935) goes into effect on June 10, 2000. Before then use (619).

Area code (951) goes into effect on Feb. 12, 2000. Before then use (909).

ZIP	Place		1998	1990
92301	Adelanto	(760)	15,291	6,815
*91376	Agoura Hills	(818)	20,822	20,396
*94501	Alameda	(510)	78,695	73,979
94507	Alamo (c)	(925)	—	12,277
94706	Albany	(510)	17,205	16,327
*91802	Alhambra	(323)/(626)	84,124	82,087
92656	Aliso Viejo (c)	(949)	—	7,612
90249	Alondra Park (c)	(310)	—	12,215
*91901	Alpine (c) (San Diego)	(935)	—	9,695
*91003	Altadena (c)	(626)	—	42,658
95945	Alta Sierra (c)	(530)	—	5,709
94589	American Canyon	(707)	8,025	7,734
*92803	Anaheim	(909)	295,153	266,406
96007	Anderson	(530)	8,782	8,299
*94509	Antioch	(925)	81,428	62,195
*92307	Apple Valley	(760)	56,440	46,079
*95003	Aptos (c)	(831)	—	9,061
*91006	Arcadia	(626)	50,157	48,284
*95521	Arcata	(707)	16,164	15,211
95825	Arden-Arcade (c)	(916)	—	92,040
*93420	Arroyo Grande	(805)	15,258	14,432
*90701	Artesia	(562)	15,927	15,464
93203	Arvin	(661)	11,167	9,286
94577	Ashland (c)	(510)	—	16,590
*93422	Atascadero	(805)	24,636	23,138
94027	Atherton	(650)	7,744	7,163
95301	Atwater	(209)	24,301	22,282
*95603	Auburn	(530)	12,386	10,653
95201	August (c)	(209)	—	6,376
93204	Avenal	(559)	11,704	9,770
91746	Avocado Heights (c)	(626)	—	14,232
91702	Azusa	(626)	42,624	41,203
*93302	Bakersfield	(661)	210,284	176,264
91706	Baldwin Park	(626)	71,953	69,330
92220	Banning	(951)	26,743	20,572
*92312	Barstow	(760)	23,164	21,472
94565	Bay Point (c)	(925)	—	17,453
93402	Baywood-Los Osos (c)	(805)	—	14,377
95903	Beale AFB (c)	(530)	—	6,912
92223	Beaumont	(951)	10,913	9,685
90201	Bell	(323)	35,204	34,365
*90706	Bellflower	(323)	63,609	61,815
90202	Bell Gardens	(213)/(323)/(562)	44,383	42,315
94002	Belmont	(650)	26,221	24,165
94510	Benicia	(707)	26,851	24,437
95005	Ben Lomond (c)	(831)	—	7,884
*94704	Berkeley	(510)	108,101	102,724
*90210	Beverly Hills	(213)/(310)/(323)	32,400	31,971
92315	Big Bear Lake	(909)	5,843	5,351
94506	Black Hawk (c)	(925)	—	6,199
92316	Bloomington (c)	(909)	—	15,116
*92225	Blythe	(760)	13,566	10,835
93637	Bonadella Ranchos-Madera Ranchos (c)	(559)	—	5,705
*91902	Bonita (c)	(935)	—	12,542
92021	Bostonia (c)	(935)	—	13,670
95006	Boulder Creek (c)	(831)	—	6,725
95416	Boyes Hot Springs (c)	(707)	—	5,973
92227	Brawley	(760)	22,953	18,923
*92822	Brea	(562)/(714)	35,566	32,873
94513	Brentwood	(925)	17,617	7,563
*90622	Buena Park	(714)	73,373	68,784
*91510	Burbank	(818)	97,430	93,649
*94010	Burlingame	(650)	28,097	26,666
*91372	Calabasas	(818)	17,725	16,577
*92231	Calexico	(760)	26,562	18,633
*93504	California City	(661)	9,167	5,955
92320	Calimesa	(909)	8,380	6,654
92233	Calipatria	(760)	7,461	2,701
*93010	Camarillo	(805)	59,348	52,297
93428	Cambria (c)	(805)	—	5,382
95682	Cameron Park (c)	(530)	—	11,897
*95008	Campbell	(408)	38,619	36,088
92055	Camp Pendleton North (c)	(949)	—	10,373
92055	Camp Pendleton South (c)	(949)	—	11,299
92587	Canyon Lake	(951)	12,450	9,991
95010	Capitola	(831)	10,581	10,171
*92008	Carlsbad	(760)	74,732	63,292
*95608	Carmichael (c)	(916)	—	48,702
*93013	Carpinteria	(805)	14,139	13,747
*90745	Carson	(310)	87,647	83,995
92077	Casa de Oro-Mt. Helix (c)	(935)	—	30,727
*94546	Castro Valley (c)	(510)	—	48,619
*92235	Cathedral City	(760)	37,638	30,085
95307	Ceres	(209)	31,929	26,413
90703	Cerritos	(562)	53,883	53,244
91724	Charter Oak (c)	(626)	—	8,858
94541	Cherryland (c)	(510)	—	11,088
92223	Cherry Valley (c)	(909)	—	5,945
*95926	Chico	(530)	46,915	39,970
*91708	Chino	(909)	65,766	59,682
91709	Chino Hills	(909)	45,073	37,868

ZIP	Place		1998	1990
93610	Chowchilla	(559)	10,145	5,930
*91910	Chula Vista	(619)/(935)	160,553	135,160
91702	Citrus (c)	(626)	—	9,481
*95621	Citrus Heights (c)	(916)	—	107,439
91711	Claremont	(909)	33,757	32,610
94517	Clayton	(925)	8,506	7,317
95422	Clearlake	(707)	11,916	11,804
95425	Cloverdale	(707)	5,854	4,924
*93612	Clovis	(559)	63,962	50,323
92236	Coachella	(760)	22,589	16,896
93210	Coalinga	(559)	9,727	8,212
92324	Colton	(909)	44,675	40,213
95932	Colusa	(530)	5,495	4,934
90022	Commerce	(323)/(562)	13,060	12,135
*90221	Compton	(310)	92,269	90,454
*94520	Concord	(925)	117,708	111,308
93212	Corcoran	(559)	17,128	13,360
96021	Corning	(530)	6,170	5,870
*91718	Corona	(951)	112,815	75,943
*92138	Coronado	(935)	25,915	26,540
*94925	Corte Madera	(415)	8,425	8,272
*92628	Costa Mesa	(714)/(949)	102,348	96,357
94931	Cotati	(707)	6,446	5,714
94556	Country Club (c)	(209)	—	9,325
*91722	Covina	(626)	44,492	43,332
95531	Crescent City	(707)	7,886	6,343
92325	Crestline (c)	(909)	—	8,594
90201	Cudahy	(323)	23,458	22,817
*90230	Culver City	(230)/(310)/(323)	39,704	38,793
*95014	Cupertino	(408)	45,095	39,967
90630	Cypress	(714)	47,888	42,655
*94015	Daly City	(415)/(650)	99,231	92,088
92629	Dana Point	(949)	34,453	31,896
*94526	Danville	(925)	40,253	31,306
*95616	Davis	(530)	54,405	46,322
90250	Del Aire (c)	(310)	—	8,040
*93215	Delano	(661)	34,280	22,762
92014	Del Mar	(858)	5,455	4,860
93953	Del Monte Forest (c)	(831)	—	5,069
*92240	Desert Hot Springs	(760)	15,074	11,668
91765	Diamond Bar	(909)	54,470	53,672
93618	Dinuba	(559)	14,912	12,743
94514	Discovery Bay (c)	(925)	—	5,351
95620	Dixon	(707)	14,489	10,417
*90241	Downey	(562)	93,653	91,444
*91009	Duarte	(626)	21,473	20,716
94568	Dublin	(925)	28,001	23,229
93219	Earlimart (c)	(661)	—	5,881
90220	East Compton (c)	(310)	—	7,967
.....	East Foothills (c)		—	14,898
92343	East Hemet (c)	(909)	—	17,611
90638	East La Mirada (c)	(562)	—	9,367
90022	East Los Angeles (c)	(323)/(562)	—	126,379
94303	East Palo Alto	(650)	24,880	23,451
91117	East Pasadena (c)		—	5,910
93257	East Porterville (c)	(559)	—	5,790
.....	East San Gabriel (c)	(626)	—	12,736
93523	Edwards AFB (c)	(661)	—	7,423
*92020	El Cajon	(935)	94,259	88,918
*92244	El Centro	(760)	37,363	31,405
94530	El Cerrito	(510)	24,044	22,869
95762	El Dorado Hills (c)	(916)	—	6,395
*95624	Elk Grove (c)	(916)	—	17,483
*91734	El Monte	(626)	111,653	106,162
*93446	El Paso de Robles	(805)	20,656	18,583
93030	El Rio (c)	(805)	—	6,419
90245	El Segundo	(310)	15,652	15,223
*94802	El Sobrante (c)	(510)	—	9,852
92630	El Toro (c)	(949)	—	62,685
92709	El Toro Station (c)	(949)	—	6,869
*94617	Emeryville	(510)	7,082	5,740
*92024	Encinitas	(760)	59,943	55,406
95320	Escalon	(209)	5,570	4,437
*92025	Escondido	(760)	120,578	108,648
*95501	Eureka	(707)	25,600	27,025
93221	Exeter	(559)	8,476	7,276
*94930	Fairfax	(415)	6,819	6,931
94533	Fairfield	(707)	89,854	78,650
95628	Fair Oaks (c) (Sacramento)	(916)	—	26,867
96052	Fairview (c) (Trinity)	(530)	—	9,045
*92028	Fallbrook (c)	(760)	—	22,095
93223	Farmersville	(559)	7,654	6,235
95018	Felton (c)	(831)	—	5,350
*93015	Fillmore	(805)	13,183	11,992
93622	Firebaugh	(209)	5,853	4,429
90001	Florence-Graham (c)	(323)	—	57,147
95828	Florin (c)	(916)	—	24,330
*95630	Folsom	(916)	45,067	29,802
*92334	Fontana	(909)	109,777	87,535
95841	Foothill Farms (c)	(916)	—	17,135
95437	Fort Bragg	(707)	6,049	6,078
95540	Fortuna	(707)	9,497	8,788
94404	Foster City	(650)	30,441	28,176
*92728	Fountain Valley	(714)	56,679	53,691
95019	Freedom (c)	(831)	—	8,361
*94537	Fremont	(510)	204,298	173,339
*93706	Fresno	(559)	398,133	354,091
*92834	Fullerton	(714)	121,954	114,144
95632	Galt	(209)	16,869	8,889
*90247	Gardena	(310)	53,642	51,481
95205	Garden Acres (c)	(209)	—	8,547
*92842	Garden Grove	(714)	151,264	142,965
92394	George AFB (c)	(760)	—	5,085
*95020	Gilroy	(408)	37,731	31,487
92509	Glen Avon (c)	(951)	—	12,663
*91209	Glendale	(323)/(626)/(818)	185,086	180,038
*91741	Glendora	(626)	49,811	47,832
93561	Golden Hills (c)	(661)	—	5,423
93926	Gonzales	(831)	6,016	4,660
92324	Grand Terrace	(909)	12,225	10,946
*95945	Grass Valley	(530)	9,848	9,048
93308	Greenacres (c)	(661)	—	7,379
93927	Greenfield (Monterey)	(831)	10,040	7,464
93433	Grover Beach	(805)	11,893	11,602
93434	Guadalupe	(805)	5,726	5,479
91745	Hacienda Heights (c)	(626)	—	52,354
94019	Half Moon Bay	(650)	10,622	8,886
*93230	Hanford	(559)	37,151	30,463
90716	Hawaiian Gardens	(323)	13,785	13,639
*90250	Hawthorne	(213)/(310)/(323)	73,413	71,349
*94544	Hayward	(510)	128,872	114,705
95448	Healdsburg	(707)	9,959	9,469
*92546	Hemet	(951)	52,781	43,366
94547	Hercules	(510)	19,797	16,829
90254	Hermosa Beach	(310)	18,777	18,219
*92340	Hesperia	(760)	62,309	50,418
92346	Highland	(909)	42,032	34,439
94010	Hillsborough	(650)	11,514	10,667
*95023	Hollister	(831)	28,403	19,318
92250	Holtville	(760)	5,697	4,820
91720	Home Gardens (c)	(951)	—	7,780
*92647	Huntington Beach	(714)	195,316	181,519
90255	Huntington Park	(323)	58,209	56,129
93234	Huron	(559)	5,649	4,766
92251	Imperial	(760)	7,044	4,113
*91932	Imperial Beach	(935)	28,749	26,512
*92201	Indio	(760)	45,023	36,850
*90301	Inglewood	(213)/(310)/(323)	111,618	109,602
.....	Interlaken (c)	(831)	—	6,404
95640	Ione	(209)	6,620	6,516
*92619	Irvine	(714)/(949)	136,446	110,330
93117	Isla Vista (c)	(805)	—	20,395
94914	Kentfield (c)	(415)	—	6,030
93630	Kerman	(559)	7,042	5,448
93930	King City	(831)	9,145	7,634
93631	Kingsburg	(559)	8,682	7,245
*91011	La Canada Flintridge	(818)	19,907	19,378
*91224	La Crescenta-Montrose (c)	(818)	—	16,968
90045	Ladera Heights (c)	(310)	—	6,316
94549	Lafayette	(925)	27,138	23,366
.....	Laguna (c)		—	9,828
*92652	Laguna Beach	(949)	25,076	23,170
*92654	Laguna Hills	(949)	32,128	22,719
*92607	Laguna Niguel	(949)	53,615	44,723
*90631	La Habra	(562)/(949)	54,294	51,263
90631	La Habra Heights	(562)	6,479	6,226
92352	Lake Arrowhead (c)	(909)	—	6,539
*92531	Lake Elsinore	(951)	28,203	19,733
92630	Lake Forest	(714)	79,923	56,036
92530	Lakeland Village (c)	(951)	—	5,159
93535	Lake Los Angeles (c)	(661)	—	7,977
92040	Lakeside (c)	(935)	—	39,412
*90714	Lakewood	(562)	76,222	73,553
*91941	La Mesa	(935)	55,986	52,911
*90638	La Mirada	(562)/(714)	44,509	40,452
93241	Lamont (c)	(661)	—	11,517
*93539	Lancaster	(661)	118,518	97,300
90623	La Palma	(562)/(714)	16,465	15,392
*91747	La Puente	(626)	38,742	36,955
92253	La Quinta	(760)	20,230	11,215
95401	La Riviera (c)	(916)	—	10,986
95403	Larkfield-Wikiup (c)	(707)	—	6,779
*94939	Larkspur	(415)	11,438	11,068
95330	Lathrop	(209)	8,949	6,841
91750	La Verne	(909)	32,350	30,843
*90260	Lawndale	(310)	28,992	27,331
*91945	Lemon Grove	(619)/(935)	26,039	23,984
93245	Lemoore	(559)	17,489	13,622
90304	Lennox (c)	(310)	—	22,757
95648	Lincoln	(916)	9,409	7,248
95901	Linda (c)	(530)	—	13,033
93247	Lindsay	(559)	8,858	8,338
95062	Live Oak (c) (Santa Cruz)	(831)	—	15,212
95953	Live Oak (Sutter)	(530)	5,139	4,320
*94550	Livermore	(925)	72,284	56,741
95334	Livingston	(209)	10,103	7,317
*95240	Lodi	(209)	56,173	51,874
92354	Loma Linda	(909)	22,466	18,470
90717	Lomita	(213)	19,932	19,442
*93436	Lompoc	(805)	41,169	37,649
*90801	Long Beach	(310)/(562)	430,905	429,321
95650	Loomis	(916)	6,421	5,705
*90720	Los Alamitos	(562)/(949)	12,774	11,788
*94022	Los Altos	(650)	27,927	26,599
94022	Los Altos Hills	(650)	8,179	7,514
*90086	Los Angeles	(213)/(310)/(323)/(818)	3,597,556	3,485,557
93635	Los Banos	(209)	20,109	14,519
*95030	Los Gatos	(408)	29,122	27,357
91709	Los Serranos (c)	(909)	—	7,099

ZIP	Place	1998	1990
94903	Lucas Valley-Marinwood (c) .. (415)	—	5,982
90262	Lynwood .... (213)/(310)/(323)	63,360	61,945
93250	Mc Farland................(661)	7,016	7,005
95521	McKinleyville (c) ....(707)	—	10,749
*93638	Madera ....(559)	36,645	29,283
93637	Madera Acres (c)...........(559)	—	5,245
95954	Magalia (c)................(530)	—	8,987
*90265	Malibu ....(310)	12,556	11,730
*90266	Manhattan Beach ....(310)	33,937	32,063
*95336	Manteca.................(209)	47,424	40,773
92518	March AFB (c)............(951)	—	5,523
93393	Marina ....(831)	47,424	26,512
*90291	Marina Del Rey (c) ....(310)	—	7,431
94553	Martinez ....(925)	34,497	31,800
95901	Marysville ....(530)	11,791	12,324
90270	Maywood................(323)	28,263	27,893
93640	Mendota ....(559)	7,360	6,821
*94025	Menlo Park ....(650)	30,083	28,403
92359	Mentone (c).............(909)	—	5,675
*95340	Merced ....(209)	59,380	56,155
94030	Millbrae ....(650)	21,853	20,414
*94941	Mill Valley.............(415)	12,949	13,029
*95035	Milpitas...............(408)	60,738	50,690
91752	Mira Loma (c) ....(951)	—	15,786
93641	Mira Monte (c)..........(805)	—	7,744
*92690	Mission Viejo ....(949)	95,440	79,464
*95350	Modesto................(209)	182,016	164,746
*91017	Monrovia ....(626)	37,458	35,733
91763	Montclair ....(909)	30,377	28,434
90640	Montebello.............(323)	60,530	59,564
*93940	Monterey ....(831)	31,106	31,954
*91754	Monterey Park.... (323)/(626)/(818)	62,531	60,738
*93021	Moorpark..............(805)	29,991	25,494
*94556	Moraga ....(925)	18,365	15,987
*92552	Moreno Valley..........(951)	144,613	118,779
*95037	Morgan Hill ....(408)	30,730	23,928
93442	Morro Bay ....(805)	10,054	9,664
*94041	Mountain View..........(650)	72,192	67,365
*92564	Murrieta...............(951)	24,097	18,557
92405	Muscoy (c).............(714)	—	7,541
*94558	Napa..................(707)	66,548	61,865
*91950	National City ....(935)	54,994	54,249
92363	Needles................(760)	6,496	5,191
94560	Newark................(510)	43,134	37,861
95360	Newman ....(209)	5,779	4,158
*92658	Newport Beach ....(949)	72,416	66,643
93444	Nipomo (c).............(805)	—	7,109
91760	Norco ....(951)	26,453	23,302
95603	North Auburn (c) ....(530)	—	10,301
94025	North Fair Oaks (c) .....(650)	—	13,912
95660	North Highlands (c).......(916)	—	42,105
*90650	Norwalk ....(562)	97,518	94,279
*94947	Novato.................(415)	48,667	47,585
95361	Oakdale ....(209)	14,947	11,978
*94617	Oakland................(510)	365,874	372,242
94561	Oakley (c) ....(925)	—	18,374
93445	Oceano (c).............(805)	—	6,169
*92056	Oceanside ....(760)	152,367	128,090
93308	Oildale (c) ....(661)	—	26,553
*93023	Ojai ....(805)	7,940	7,613
95961	Olivehurst (c)...........(530)	—	9,738
*91761	Ontario................(909)	147,188	133,179
95060	Opal Cliffs (c).........(831)	—	5,940
*92863	Orange ....(714)	123,820	110,658
93646	Orange Cove ....(559)	7,270	5,604
95662	Orangevale (c)..........(916)	—	26,266
94563	Orinda ....(925)	19,339	16,642
95963	Orland ....(530)	5,681	5,052
93647	Orosi (c)...............(559)	—	5,486
*95965	Oroville ....(530)	12,106	11,885
95965	Oroville East (c) ....(530)	—	8,462
*93030	Oxnard.................(805)	154,622	142,560
94044	Pacifica................(650)	40,677	37,670
93950	Pacific Grove ....(831)	15,860	16,117
95968	Palermo (c) ....(530)	—	5,260
*93590	Palmdale...............(661)	100,157	73,314
*92260	Palm Desert............(760)	29,413	23,252
.....	Palm Desert Country (c)	—	5,626
*92262	Palm Springs...........(760)	43,942	40,144
*94303	Palo Alto ....(650)	59,098	55,900
90274	Palos Verdes Estates ....(310)	13,853	13,512
*95969	Paradise ....(530)	25,781	25,401
90723	Paramount..............(562)	51,131	47,669
95823	Parkway-So. Sacramento (c) .(916)	—	31,903
93648	Parlier ....(559)	10,260	7,938
*91109	Pasadena.....(323)/(626)/(818)	134,587	131,586
	Paso Robles. *See El Paso de Robles*		
95363	Patterson................(209)	9,871	8,626
92509	Pedley (c) ....(951)	—	8,869
*92572	Perris.................(951)	32,632	21,500
*94952	Petaluma...............(707)	50,913	43,166
*90660	Pico Rivera ....(562)	60,683	59,177
94611	Piedmont...............(510)	11,013	10,602
94564	Pinole ....(510)	19,339	17,460
*93449	Pismo Beach............(805)	8,291	7,669
94565	Pittsburg ....(925)	52,796	47,607
*92871	Placentia ....(714)	47,153	41,259
95667	Placerville ....(530)	9,791	8,286
94523	Pleasant Hill ....(925)	33,311	31,583
*94566	Pleasanton ....(925)	64,039	50,570
*91769	Pomona................(909)	135,659	131,700
*93257	Porterville.............(559)	35,602	29,521
*93041	Port Hueneme...........(805)	20,658	20,322
*92064	Poway.................(858)	49,110	43,396
93907	Prunedale (c)...........(831)	—	7,393
*93551	Quartz Hill (c) ....(661)	—	9,626
92065	Ramona (c).............(760)	—	13,040
*95670	Rancho Cordova (c) ......(916)	—	48,731
*91729	Rancho Cucamonga ......(909)	120,047	101,409
92270	Rancho Mirage..........(760)	11,359	9,778
90275	Rancho Palos Verdes ......(310)	42,655	41,667
91941	Rancho San Diego (c).....(935)	—	6,977
92688	Rancho Santa Margarita (c) .(949)	—	11,390
96080	Red Bluff ....(530)	13,199	12,363
*96049	Redding................(530)	77,944	66,176
*92373	Redlands...............(909)	67,309	62,667
*90277	Redondo Beach ....(310)	63,075	60,167
*94063	Redwood City ....(650)	73,438	66,072
93654	Reedley................(559)	18,868	15,791
*92377	Rialto.................(909)	83,933	72,395
*94802	Richmond ....(510)	93,470	86,019
*93556	Ridgecrest.............(760)	30,030	28,295
95003	Rio Del Mar (c) ....(831)	—	8,919
95673	Rio Linda (c) ....(916)	—	9,481
95366	Ripon.................(209)	9,288	7,455
95367	Riverbank ....(209)	14,409	8,591
*92502	Riverside...............(951)	262,140	226,546
*95677	Rocklin ....(916)	30,882	18,806
94572	Rodeo (c) ....(510)	—	7,589
*94928	Rohnert Park ....(707)	40,995	36,326
90274	Rolling Hills Estates ....(310)	7,999	7,789
93560	Rosamond (c) ....(661)	—	7,430
95401	Roseland (c) ....(707)	—	8,779
91770	Rosemead..............(626)	53,186	51,638
95826	Rosemont (c)...........(916)	—	22,851
*95678	Roseville ....(916)	71,609	44,685
90720	Rossmoor (c) ....(714)	—	9,893
91748	Rowland Heights (c) ......(818)	—	42,647
92519	Rubidoux (c)...........(951)	—	24,367
*95814	Sacramento............(916)	404,168	369,365
94574	Saint Helena ....(707)	5,694	4,990
*93907	Salinas ....(831)	121,458	108,777
*94960	San Anselmo ....(415)	11,525	11,735
*92401	San Bernardino.........(909)	186,402	170,036
94066	San Bruno ....(650)	40,710	38,961
*93001	San Buenaventura (Ventura).. (805)	98,366	92,557
94070	San Carlos ....(650)	28,089	26,382
*92674	San Clemente ....(949)	46,495	41,100
*92138	San Diego.......(619)/(858)/(935)	1,220,666	1,110,623
92065	San Diego Country Estates (c) (760)	—	6,874
91773	San Dimas.............(909)	34,221	32,398
*91341	San Fernando ....(818)	23,048	22,580
*94142	San Francisco..........(415)	745,774	723,959
*91778	San Gabriel............(626)	37,964	37,120
93657	Sanger.................(559)	18,135	16,839
*92581	San Jacinto ....(951)	24,330	17,614
*95113	San Jose ....(408)	861,284	782,224
*92690	San Juan Capistrano.......(949)	30,098	26,183
*94577	San Leandro ....(510)	74,387	68,223
94580	San Lorenzo (c).........(510)	—	19,987
*93401	San Luis Obispo ....(805)	42,928	41,958
*92069	San Marcos............(760)	49,584	38,974
*91109	San Marino ....(626)	12,966	12,959
*94402	San Mateo.............(650)	91,282	85,619
94806	San Pablo ....(510)	26,952	25,158
*94915	San Rafael ....(415)	51,057	48,410
94583	San Ramon ....(925)	42,449	35,303
*92711	Santa Ana ....(714)/(949)	305,955	293,827
*93102	Santa Barbara..........(805)	86,645	85,571
*95050	Santa Clara............(408)	100,370	93,613
*91380	Santa Clarita..........(661)	127,001	120,050
*95060	Santa Cruz ....(831)	52,853	49,711
90670	Santa Fe Springs........(562)	15,493	15,520
*93454	Santa Maria............(805)	68,121	61,552
*90401	Santa Monica ....(310)	89,522	86,905
*93060	Santa Paula............(805)	26,725	25,062
*95402	Santa Rosa............(707)	126,891	113,261
*92071	Santee.................(935)	57,740	52,902
*95070	Saratoga ....(408)	29,883	28,061
*94965	Sausalito ....(415)	7,041	7,152
*95066	Scotts Valley ....(831)	10,075	8,667
90740	Seal Beach ....(714)	26,162	25,098
93955	Seaside................(831)	28,258	38,826
*95472	Sebastopol ....(707)	7,609	7,008
93662	Selma ....(559)	17,403	14,757
93263	Shafter................(661)	11,154	9,404
*96019	Shasta Lake ....(916)	9,462	8,821
*91025	Sierra Madre ....(626)	10,960	10,762
90806	Signal Hill ....(562)	8,809	8,371
*93065	Simi Valley............(805)	110,463	100,218
92075	Solana Beach ....(858)	13,696	12,956
93960	Soledad...............(831)	21,159	13,426
95476	Sonoma................(707)	9,065	8,168
95073	Soquel (c) ....(831)	—	9,188
91733	South El Monte ....(626)	21,284	20,850
90280	South Gate ....(323)/(562)	88,384	86,284
*96151	South Lake Tahoe........(530)	23,519	21,586
95965	South Oroville (c) ....(530)	—	7,463
*91030	South Pasadena ......(213)/(323)/(626)/(818)	24,097	23,936

ZIP	Place		1998	1990
*94080	South San Francisco	(650)	58,829	54,312
91770	South San Gabriel (c)	(626)	—	7,700
91744	South San Jose Hills (c)	(626)	—	17,814
90605	South Whittier (c)	(562)	—	49,514
95991	South Yuba (c)	(530)	—	8,816
*91977	Spring Valley (c)	(935)	—	55,331
94309	Stanford (c)	(650)	—	18,097
90680	Stanton	(714)	33,038	30,491
*95208	Stockton	(209)	240,143	210,943
94585	Suisun City	(707)	26,903	22,704
*92586	Sun City (c)	(951)	127,444	117,324
*94086	Sunnyvale	(408)	17,422	12,130
*96130	Susanville	(530)	6,533	5,902
93268	Taft	(661)	127,444	30,491
94941	Tamalpais-Homestead Valley (c)	(415)	—	9,601
*93581	Tehachapi	(661)	6,508	6,182
*92589	Temecula	(951)	44,271	27,177
91780	Temple City	(626)	32,029	31,153
95965	Thermalito (c)	(530)	—	5,646
*91359	Thousand Oaks	(805)	117,199	104,381
94920	Tiburon	(415)	8,139	7,554
*90503	Torrance	(310)	137,533	133,107
*95376	Tracy	(209)	47,643	33,558
*96161	Truckee	(916)	10,213	8,848
*93274	Tulare	(559)	40,935	33,249
*95380	Turlock	(209)	50,266	42,224
*92781	Tustin	(714)/(949)	64,370	50,689
92705	Tustin Foothills (c)	(714)	—	24,358
*92277	Twentynine Palms	(760)	14,271	11,821
92278	Twentynine Palms Base (c)	(760)	—	10,606
95060	Twin Lakes (c)	(831)	—	5,379
95482	Ukiah	(707)	14,694	14,612
94587	Union City	(510)	64,085	53,762
*91785	Upland	(909)	67,826	63,374
*95687	Vacaville	(707)	83,362	71,476
91744	Valinda (c)	(626)	—	18,735
*94590	Vallejo	(707)	111,539	109,199
92343	Valle Vista (c)	(909)	—	8,751
93437	Vandenberg AFB (c)	(805)	—	9,846
93436	Vandenberg Village (c)	(805)	—	5,971
	Ventura. See San Buenaventura			
*92393	Victorville	(760)	68,914	50,103
90043	View Park-Windsor Hills (c)	(310)	—	11,769
92861	Villa Park	(714)	6,745	6,299
.....	Vincent (c)		—	13,713
*93291	Visalia	(559)	89,308	75,659
*92083	Vista	(760)	80,909	71,861
*91788	Walnut	(909)	31,110	29,105
*94596	Walnut Creek	(925)	64,306	60,569
90255	Walnut Park (c)	(213)	—	14,722
93280	Wasco	(661)	20,075	12,412
95386	Waterford	(209)	6,951	4,771
*95076	Watsonville	(831)	33,352	31,099
90044	West Athens (c)	(310)	—	8,859
90502	West Carson (c)	(323)	—	20,143
90247	West Compton (c)	(310)	—	5,451
*91790	West Covina	(626)	99,455	96,226
90069	West Hollywood	(310)/(323)	36,325	36,118
*91359	Westlake Village	(805)	7,892	7,455
*92685	Westminster	(714)	84,042	78,293
90047	Westmont (c)	(323)	—	31,044
91746	West Puente Valley (c)	(626)	—	20,254
*95691	West Sacramento	(916)	29,744	28,898
*90606	West Whittier-Los Nietos (c)	(562)	—	24,164
*90605	Whittier	(562)	79,135	77,671
92595	Wildomar (c)	(951)	—	10,411
90222	Willowbrook (c)	(323)	—	32,772
95988	Willows	(530)	6,194	5,988
95492	Windsor	(707)	13,335	12,002
95694	Winters	(707)	5,212	4,639
95388	Winton (c)	(209)	—	7,559
92502	Woodcrest (c)	(951)	—	7,796
93286	Woodlake	(559)	6,642	5,678
*95695	Woodland	(530)	43,600	40,230
94062	Woodside	(650)	5,553	5,034
*92885	Yorba Linda	(714)	60,156	52,422
96097	Yreka	(530)	6,935	6,948
*95991	Yuba City	(530)	32,994	27,385
92399	Yucaipa	(909)	36,791	32,819
*92286	Yucca Valley	(760)	18,988	16,539

Colorado

Area code (720) overlays area code (303). See introductory note.

ZIP	Place		1998	1990
*80840	Air Force Academy (c)	(719)	—	9,062
81101	Alamosa	(719)	7,677	7,579
80401	Applewood (c)	(303)	—	11,069
*80004	Arvada	(303)	97,610	89,261
*81611	Aspen	(970)	5,188	5,049
*80017	Aurora	(303)	250,604	222,103
80908	Black Forest (c)	(719)	—	8,143
*80302	Boulder	(303)	90,543	85,127
80601	Brighton	(303)	16,841	14,203
*80020	Broomfield	(303)	34,391	24,638
*81212	Canon City	(719)	15,239	12,687
80104	Castle Rock	(303)	14,798	8,710
80120	Castlewood (c)	(303)	—	24,392
80110	Cherry Hills Village	(303)	6,398	5,245

ZIP	Place		1998	1990
81220	Cimarron Hills (c)	(719)	—	11,160
81520	Clifton (c)	(970)	—	12,671
*80903	Colorado Springs	(719)	344,987	280,430
80120	Columbine (c)	(303)	—	23,969
*80022	Commerce City	(303)	17,355	16,466
81321	Cortez	(970)	9,024	7,284
*81625	Craig	(970)	8,734	8,091
*80202	Denver	(303)	499,055	467,610
80022	Derby (c)	(303)	—	6,043
*81301	Durango	(970)	13,854	12,439
*80110	Englewood	(303)	31,593	29,396
80620	Evans	(970)	7,595	5,876
*80439	Evergreen (c)	(303)	—	7,582
80221	Federal Heights	(303)	11,572	9,342
80913	Fort Carson (c)	(719)	—	11,309
*80525	Fort Collins	(970)	108,905	87,491
80621	Fort Lupton	(970)	5,871	5,159
80701	Fort Morgan	(970)	10,049	9,068
80817	Fountain	(719)	13,900	10,754
81504	Fruitvale (c)	(303)	—	5,222
81522	Gateway (c)	(970)	—	7,510
*81601	Glenwood Springs	(970)	7,946	6,561
*80401	Golden	(303)	15,259	13,127
*81501	Grand Junction	(970)	41,265	32,893
*80631	Greeley	(970)	70,434	60,454
*80111	Greenwood Village	(303)	14,449	7,589
80501	Gunbarrel (c)	(303)	—	9,388
*81230	Gunnison	(970)	5,333	4,636
80163	Highlands Ranch (c)	(303)	—	10,181
80127	Ken Caryl (c)	(303)	—	24,391
80026	Lafayette	(303)	20,487	14,708
81050	La Junta	(719)	7,950	7,678
*80226	Lakewood	(303)	136,883	126,475
81052	Lamar	(719)	8,477	8,343
*80126	Littleton	(303)	41,059	33,711
*80501	Longmont	(303)	62,078	51,976
80027	Louisville	(303)	17,871	12,363
*80538	Loveland	(970)	47,116	37,357
80829	Manitou Springs	(719)	5,438	4,535
*81401	Montrose	(970)	11,451	8,854
80233	Northglenn	(303)	29,892	27,195
80649	Orchard Mesa (c)	(303)	—	5,977
*80134	Parker	(303)	11,802	5,450
*81003	Pueblo	(719)	99,406	98,640
81503	Redlands (c)	(970)	—	9,355
81650	Rifle	(970)	15,248	4,858
81201	Salida	(719)	107,301	4,737
80911	Security-Widefield (c)	(719)	—	23,822
80110	Sheridan	(303)	5,595	4,976
80221	Sherrelwood (c)	(303)	—	16,636
80122	Southglenn (c)	(303)	—	43,087
*80477	Steamboat Springs	(970)	6,510	6,695
80751	Sterling	(970)	10,431	10,362
80906	Stratmoor (c)	(719)	—	5,854
80027	Superior	(303)	—	
80229	Thornton	(303)	74,139	55,031
81082	Trinidad	(719)	8,740	8,580
80229	Welby (c)	(303)	—	10,218
80030	Westminster	(303)	95,691	74,619
80221	Westminster East (c)	(303)	—	5,197
*80033	Wheat Ridge	(303)	29,870	29,419
80550	Windsor	(970)	7,853	5,062
*80863	Woodland Park	(719)	6,253	4,610

Connecticut

See introductory note.

ZIP	Place		1998	1990
06401	Ansonia	(203)	17,716	18,403
06001	Avon	(860)	14,093	13,937
06403	Beacon Falls	(203)	5,198	5,083
06037	Berlin	(860)	17,246	16,787
06801	Bethel	(203)	17,874	17,541
06002	Bloomfield	(860)	19,000	19,483
06405	Branford	(203)	27,146	27,603
*06602	Bridgeport	(203)	137,425	141,686
*06010	Bristol	(860)	59,158	60,640
06804	Brookfield	(203)	14,664	14,113
06234	Brooklyn	(860)	6,909	6,681
06013	Burlington	(860)	7,892	7,026
06019	Canton	(860)	8,115	8,268
06040	Central Manchester (c)	(860)	—	30,934
06410	Cheshire	(203)	26,471	25,684
06413	Clinton	(860)	13,111	12,767
06415	Colchester	(860)	12,709	10,980
06340	Conning Towers-Nautilus Park (c)	(860)	—	10,013
06238	Coventry	(860)	11,077	10,063
06416	Cromwell	(860)	12,589	12,286
*06810	Danbury	(203)	65,829	65,585
06820	Darien	(203)	18,085	18,196
06418	Derby	(203)	11,942	12,199
06422	Durham	(860)	6,555	5,732
06423	East Haddam	(860)	7,490	6,676
06424	East Hampton	(860)	11,052	10,428
*06101	East Hartford	(860)	47,369	50,452
06512	East Haven	(203)	26,740	26,144
06333	East Lyme	(860)	15,824	15,340
06612	Easton	(203)	6,745	6,303
06088	East Windsor	(860)	10,026	10,081

ZIP	Place		1998	1990
06029	Ellington	(860)	11,741	11,197
*06082	Enfield	(860)	43,099	45,532
06426	Essex	(860)	6,143	5,904
*06430	Fairfield	(203)	53,740	53,418
*06032	Farmington	(860)	21,161	20,608
06033	Glastonbury Center (c)	(860)	—	7,082
06033	Glastonbury	(860)	28,832	27,901
06035	Granby	(860)	9,592	9,369
*06830	Greenwich	(203)	58,332	58,441
06351	Griswold	(860)	10,509	10,384
*06340	Groton	(860)	9,394	9,837
06340	Groton Town	(860)	41,284	45,144
06437	Guilford	(203)	20,239	19,848
06438	Haddam	(860)	7,210	6,769
*06514	Hamden	(203)	53,011	52,434
*06101	Hartford	(860)	131,523	139,739
06791	Harwinton	(860)	5,405	5,228
06082	Hazardville (c)	(860)	—	5,179
06248	Hebron	(860)	8,043	7,079
06037	Kensington (c)	(860)	—	8,306
06239	Killingly	(860)	16,057	15,889
06419	Killingworth	(860)	5,694	4,814
06249	Lebanon	(860)	6,271	6,041
06339	Ledyard	(860)	14,462	14,913
06759	Litchfield	(860)	8,747	8,365
06443	Madison	(203)	16,197	15,485
*06040	Manchester	(860)	51,657	51,618
06250	Mansfield	(860)	19,061	21,103
06447	Marlborough	(860)	5,754	5,535
*06450	Meriden	(203)	56,667	59,479
06762	Middlebury	(203)	6,069	6,145
06457	Middletown	(860)	43,640	42,762
06460	Milford	(203)	48,254	48,168
06468	Monroe	(203)	18,566	16,896
06353	Montville	(860)	16,618	16,673
06770	Naugatuck	(203)	30,231	30,625
*06050	New Britain	(860)	70,492	75,491
06840	New Canaan	(203)	18,067	17,864
06812	New Fairfield	(203)	13,517	12,911
06057	New Hartford	(860)	6,123	5,769
*06511	New Haven	(203)	123,189	130,474
*06101	Newington	(860)	28,346	29,208
06320	New London	(860)	23,869	28,540
06776	New Milford	(860)	25,512	23,629
06470	Newtown	(203)	23,469	20,779
06471	North Branford	(203)	13,967	12,996
06473	North Haven	(203)	22,148	22,247
*06856	Norwalk	(203)	78,064	78,331
06360	Norwich	(860)	34,931	37,391
06779	Oakville (c)	(860)	—	8,741
06371	Old Lyme	(860)	6,446	6,535
06475	Old Saybrook	(860)	9,739	9,552
06477	Orange	(203)	12,426	12,830
06478	Oxford	(203)	9,279	8,685
06379	Pawcatuck (c)	(860)	—	5,289
06374	Plainfield	(860)	14,593	14,363
06062	Plainville	(860)	16,770	17,392
06782	Plymouth	(860)	12,040	11,822
06480	Portland	(860)	8,772	8,418
06365	Preston	(860)	5,070	5,006
06712	Prospect	(203)	8,137	7,775
06260	Putnam (c)	(860)	—	6,835
06260	Putnam	(860)	8,930	9,031
06896	Redding	(203)	8,068	7,927
06877	Ridgefield Center (c)	(203)	—	6,363
06877	Ridgefield	(203)	21,679	20,919
06067	Rocky Hill	(860)	16,509	16,554
06483	Seymour	(203)	14,193	14,288
06484	Shelton	(203)	37,180	35,418
06082	Sherwood Manor (c)	(860)	—	6,357
06070	Simsbury	(860)	21,782	22,023
06071	Somers	(860)	9,358	9,108
06488	Southbury	(203)	16,370	15,818
06489	Southington	(860)	38,391	38,518
06074	South Windsor	(860)	22,500	22,090
06082	Southwood Acres (c)	(860)	—	8,963
06075	Stafford	(860)	11,497	11,091
*06904	Stamford	(203)	110,056	108,056
06378	Stonington	(860)	16,806	16,919
06268	Storrs (c)	(860)	—	12,198
*06602	Stratford	(203)	49,096	49,389
06078	Suffield	(860)	11,216	11,427
06786	Terryville (c)	(860)	—	5,426
06787	Thomaston	(860)	7,273	6,947
06277	Thompson	(860)	8,975	8,668
06082	Thompsonville (c)	(860)	—	8,458
06084	Tolland	(860)	11,994	11,001
06790	Torrington	(860)	34,529	33,687
06611	Trumbull	(203)	33,291	32,016
06066	Vernon	(860)	29,414	29,841
06492	Wallingford	(203)	40,798	40,822
*06702	Waterbury	(203)	106,412	108,961
06385	Waterford	(860)	17,971	17,930
06795	Watertown	(860)	21,470	20,456
06498	Westbrook	(860)	5,524	5,414
*06101	West Hartford	(860)	56,795	60,110
06516	West Haven	(203)	52,153	54,021
06883	Weston	(203)	8,780	8,648
*06880	Westport	(203)	24,185	24,410
*06101	Wethersfield	(860)	25,179	25,651

ZIP	Place		1998	1990
06226	Willimantic (c)	(860)	—	14,746
06279	Willington	(860)	6,117	5,979
06897	Wilton	(203)	16,329	15,989
06094	Winchester	(860)	11,440	11,524
06280	Windham	(860)	21,598	22,039
06095	Windsor	(860)	27,663	27,817
06096	Windsor Locks	(860)	12,076	12,358
06098	Winsted	(860)	—	8,254
06716	Wolcott	(203)	14,145	13,700
06525	Woodbridge	(203)	8,051	7,924
06798	Woodbury	(203)	8,508	8,131
06281	Woodstock	(860)	6,496	6,008

Delaware (302)

ZIP	Place		1998	1990
19713	Brookside (c)		—	15,307
19703	Claymont (c)		—	9,800
*19901	Dover		30,369	27,630
19809	Edgemoor (c)		—	5,853
19805	Elsmere		5,764	5,935
19963	Milford		6,665	6,032
*19711	Newark		28,000	26,463
19800	Pike Creek (c)		—	10,163
19973	Seaford		6,600	5,689
19977	Smyrna		5,652	5,231
19804	Stanton (c)		—	5,028
19803	Talleyville (c)		—	6,346
*19899	Wilmington		71,678	71,529
19720	Wilmington Manor (c)		—	8,568

District of Columbia (202)

ZIP	Place		1998	1990
*20090	Washington		523,124	606,900

Florida

Area code (407) is overlayed by area code (321), except in De Bary.
Area code (786) overlays area code (305). See introductory note.

ZIP	Place		1998	1990
*32615	Alachua	(904)	5,732	4,667
*32714	Altamonte Springs	(407)	39,278	35,167
.....	Andover (c)		—	6,251
33572	Apollo Beach (c)	(813)	—	6,025
*32712	Apopka	(407)	19,657	13,611
*34266	Arcadia	(863)	6,134	6,488
32233	Atlantic Beach	(904)	12,960	11,636
33823	Auburndale	(863)	9,648	8,846
*33160	Aventura (c)	(305)	—	14,914
*33825	Avon Park	(863)	7,951	8,078
32857	Azalea Park (c)	(407)	—	8,926
*33830	Bartow	(863)	15,025	14,716
.....	Bay Hill (c)		—	5,346
34667	Bayonet Point (c)	(727)	—	21,860
33505	Bayshore Gardens (c)	(941)	—	17,062
33589	Beacon Square (c)	(727)	—	6,265
34233	Bee Ridge (c)	(941)	—	6,406
32073	Bellair-Meadowbrook Terrace (c)	(904)	—	15,606
33430	Belle Glade	(561)	17,224	16,177
*32802	Belle Isle	(407)	6,422	5,272
*34420	Belleview (c)	(352)	—	19,386
*34461	Beverly Hills (c)	(352)	—	6,163
*33509	Bloomingdale (c)	(813)	—	13,912
.....	Boca Del Mar (c)		—	17,754
*33431	Boca Raton	(561)	71,761	61,486
*34135	Bonita Springs (c)	(941)	—	13,600
*33436	Boynton Beach	(561)	53,607	46,284
*34206	Bradenton	(941)	47,049	43,769
*33509	Brandon (c)	(813)	—	57,985
32503	Brent (c)	(850)	—	21,624
33317	Broadview Park (c)	(954)	—	6,109
33313	Broadview-Pompano Park (c)	(954)	—	5,230
*34601	Brooksville	(352)	8,561	7,589
33311	Browardale (c)	(954)	—	6,257
33142	Brownsville (c)	(305)	—	15,607
34743	Buena Ventura Lakes (c)		—	14,148
32404	Callaway	(850)	12,780	12,253
32920	Cape Canaveral	(321)	8,626	8,014
*33909	Cape Coral	(941)	91,180	74,991
33055	Carol City (c)	(305)	—	53,331
33688	Carrollwood (c)	(813)	—	7,195
*33601	Carrollwood Village (c)	(813)	—	15,051
*32707	Casselberry	(407)	24,768	20,736
33401	Century Village (c)	(305)	—	8,363
*33758	Clearwater	(727)	101,474	98,669
*34711	Clermont	(352)	9,287	6,910
33440	Clewiston	(863)	6,458	6,085
*32922	Cocoa	(321)	18,508	17,710
*32931	Cocoa Beach	(321)	12,548	12,123
32922	Cocoa West (c)	(321)	—	6,160
*33097	Coconut Creek	(954)	37,437	27,269
33064	Collier Manor-Cresthaven (c)	(954)	—	7,322
33801	Combee Settlement (c)	(863)	—	5,463
32809	Conway (c)	(407)	—	13,159
33328	Cooper City	(954)	29,207	21,335
*33114	Coral Gables	(305)	40,858	40,091
*33075	Coral Springs	(954)	111,744	78,864
33157	Coral Terrace (c)	(305)	—	23,255
*32536	Crestview	(850)	12,556	9,886

ZIP	Place		1998	1990
33803	Crystal Lake (c)	(863)	—	5,300
33157	Cutler (c)	(305)	—	16,201
33157	Cutler Ridge (c)	(305)	—	21,268
33884	Cypress Gardens (c)	(863)	—	9,188
33919	Cypress Lake (c)	(941)	—	10,491
*33525	Dade City	(352)	6,012	5,633
33004	Dania	(954)	15,162	13,183
33329	Davie	(954)	62,061	47,143
*32114	Daytona Beach	(904)	65,136	61,991
32713	De Bary	(407)	11,026	9,327
*33441	Deerfield Beach	(954)	50,921	46,997
*32433	DeFuniak Springs	(850)	5,239	5,200
*32720	De Land	(904)	18,769	16,622
*33444	Delray Beach	(561)	53,618	47,184
33617	Del Rio (c)	(813)	—	8,248
*32738	Deltona	(407)	58,168	49,429
*32541	Destin	(850)	11,021	8,090
.....	Doctor Phillips (c)	(407)	—	7,963
*34698	Dunedin	(727)	34,990	34,427
33610	East Lake-Orient Park (c)	(813)	—	6,171
33940	East Naples (c)	(941)	—	22,951
*32132	Edgewater	(904)	17,757	15,351
32542	Eglin AFB (c)	(850)	—	8,347
33614	Egypt Lake (c)	(813)	—	14,580
34680	Elfers (c)	(727)	—	12,356
*34295	Englewood (c)	(941)	—	15,025
32534	Ensley (c)	(850)	—	16,362
*32726	Eustis	(352)	15,137	12,856
32804	Fairview Shores (c)	(305)	—	13,192
*32034	Fernandina Beach	(904)	10,408	8,765
32730	Fern Park (c)	(407)	—	8,294
32514	Ferry Pass (c)	(850)	—	26,301
33034	Florida City	(305)	7,059	5,978
32960	Florida Ridge (c)	(561)	—	12,218
32714	Forest City (c)	(407)	—	10,638
.....	Forest Island Park (c)		—	5,988
*33310	Fort Lauderdale	(954)	153,728	149,238
33841	Fort Meade	(863)	5,262	5,151
*33902	Fort Myers	(941)	45,697	44,947
*33931	Fort Myers Beach (c)	(941)	—	9,284
*33922	Fort Myers Shores (c)	(941)	—	5,460
*34981	Fort Pierce	(561)	36,341	36,830
33453	Fort Pierce North (c)	(561)	—	5,833
34982	Fort Pierce South (c)	(561)	—	5,320
*32548	Fort Walton Beach	(850)	21,501	21,407
*32043	Fruit Cove (c)	(904)	—	5,904
34230	Fruitville (c)	(941)	—	9,808
*32602	Gainesville	(352)	92,648	91,482
33801	Gibsonia (c)	(863)	—	5,168
33534	Gibsonton (c)	(813)	—	7,706
32960	Gifford (c)	(561)	—	6,278
33138	Gladeview (c)	(954)	—	15,637
33143	Glenvar Heights (c)	(305)	—	14,823
34116	Golden Gate (c)	(941)	—	14,148
33055	Golden Glades (c)	(305)	—	25,474
32733	Goldenrod (c)	(407)	—	12,362
32560	Gonzalez (c)	(850)	—	7,669
33170	Goulds (c)	(305)	—	7,284
.....	Greater Northdale (c)		—	16,318
33454	Greenacres	(561)	25,811	18,683
32043	Green Cove Springs	(904)	5,363	4,497
*32561	Gulf Breeze	(850)	8,090	5,530
33581	Gulf Gate Estates (c)	(941)	—	11,622
33737	Gulfport	(727)	11,549	11,709
*33844	Haines City	(863)	12,684	11,683
*33009	Hallandale	(305)/(954)	31,260	30,997
.....	Hammocks (c)		—	10,897
.....	Hamptons at Boca Raton (c)		—	11,686
*33010	Hialeah	(305)	211,392	188,008
33016	Hialeah Gardens	(305)	17,076	7,727
.....	Highpoint (c)		—	13,818
*33455	Hobe Sound (c)	(561)	—	11,507
*34689	Holiday (c)	(727)	—	19,360
32125	Holly Hill	(904)	11,529	11,141
*33022	Hollywood	(954)	130,026	121,720
*33030	Homestead	(305)	29,072	26,694
33039	Homestead AFB (c)	(305)	—	5,153
34447	Homosassa Springs (c)	(352)	—	6,271
*34668	Hudson (c)	(727)	—	7,344
*34142	Immokalee (c)	(941)	—	14,120
32937	Indian Harbour Beach	(321)	7,639	6,933
*34450	Inverness	(352)	6,936	5,797
33880	Inwood (c)	(863)	—	6,824
.....	Iona (c)	(941)	—	9,565
33162	Ives Estates (c)	(305)	—	13,531
*32203	Jacksonville	(904)	693,630	635,230
*32250	Jacksonville Beach	(904)	20,643	17,839
33880	Jan Phyl Village (c)	(863)	—	5,308
33568	Jasmine Estates (c)	(727)	—	17,136
*34957	Jensen Beach (c)	(561)	—	9,884
*33458	Jupiter	(561)	30,970	26,753
33183	Kendale Lakes (c)	(305)	—	48,524
33256	Kendall (c)	(305)	—	87,271
.....	Kendall Lakes West (c)	(305)	—	6,038
33149	Key Biscayne (c)	(305)	9,890	8,854
33037	Key Largo (c)	(305)	—	11,336
*33040	Key West	(305)	25,701	24,832
*33573	Kings Point (c)	(305)	—	12,422
*34744	Kissimmee	(407)	38,542	30,337
*32159	Lady Lake	(352)	12,701	8,071
*32055	Lake City	(904)	10,756	9,626
*33804	Lakeland	(863)	74,204	70,576
33801	Lakeland Highlands (c)	(863)	—	9,972
32569	Lake Lorraine (c)	(850)	—	6,779
33054	Lake Lucerne (c)	(305)	—	9,478
33612	Lake Magdalene (c)	(813)	—	15,973
*32746	Lake Mary	(407)	9,449	5,929
33403	Lake Park	(561)	6,897	6,704
.....	Lakes by the Bay (c)		—	5,615
32073	Lakeside (c)	(904)	—	29,137
*33853	Lake Wales	(863)	9,998	9,670
34951	Lakewood Park (c)	(561)	—	7,211
*33461	Lake Worth	(561)	29,116	28,564
34639	Land O'Lakes (c)	(813)	—	7,892
33465	Lantana	(561)	8,708	8,392
*33770	Largo	(727)	66,264	65,910
33313	Lauderdale Lakes	(954)	28,235	27,341
33313	Lauderhill	(954)	50,814	49,015
34272	Laurel (c)	(941)	—	8,245
33714	Lealman (c)	(727)	—	21,748
*34748	Leesburg	(352)	16,911	14,783
*33936	Lehigh Acres (c)	(941)	—	13,611
33033	Leisure City (c)	(305)	—	19,379
33074	Lighthouse Point	(954)	10,706	10,378
33177	Lindgren Acres (c)	(305)	—	22,290
*32060	Live Oak	(904)	7,154	6,332
32860	Lockhart (c)	(407)	—	11,636
34228	Longboat Key	(941)	6,310	5,937
*32750	Longwood	(407)	14,004	13,316
*33549	Lutz (c)	(813)	—	10,552
32444	Lynn Haven	(850)	12,604	9,270
.....	McGregor (c)		—	6,504
*32751	Maitland	(407)	9,013	8,932
33550	Mango (c)	(813)	—	8,700
33050	Marathon (c)	(305)	—	8,857
*33937	Marco (c)	(941)	51,268	9,493
33093	Margate	(954)	6,445	42,985
*32446	Marianna	(850)	69,057	6,292
*32901	Melbourne	(321)	51,268	60,034
32666	Melrose Park (c)	(954)	—	6,477
33561	Memphis (c)	(941)	—	6,760
*32953	Merritt Island (c)	(321)	—	32,886
*33101	Miami	(305)	368,624	92,639
*33152	Miami Beach	(305)	97,053	10,084
33023	Miami Gardens-Utopia-Carver (c)	(954)	—	7,448
33014	Miami Lakes (c)	(305)	—	12,750
33153	Miami Shores (c)	(305)	9,983	10,084
33266	Miami Springs (c)	(305)	13,416	13,268
32976	Micco (c)	(561)	—	8,757
*32068	Middleburg (c)	(904)	—	6,223
*32570	Milton	(850)	7,692	7,216
32754	Mims (c)	(321)	—	9,412
33023	Miramar	(954)	57,215	40,663
*32757	Mount Dora	(352)	9,329	7,294
32526	Myrtle Grove (c)	(850)	—	17,402
*34102	Naples	(941)	19,404	19,505
34102	Naples Park (c)	(941)	—	8,002
33092	Naranja (c)	(305)	—	5,790
32266	Neptune Beach	(904)	6,978	6,816
*34653	New Port Richey	(727)	15,024	14,044
33552	New Port Richey East (c)	(727)	—	9,683
*32168	New Smyrna Beach	(904)	18,167	16,549
*32578	Niceville	(850)	11,973	10,509
33269	Norland (c)	(305)	—	22,109
33308	North Andrews Gardens (c)	(954)	—	9,002
33141	North Bay Village (c)	(305)	5,445	5,383
33918	North Fort Myers (c)	(941)	—	30,027
33068	North Lauderdale	(954)	29,453	26,473
33261	North Miami	(305)	50,772	50,001
33160	North Miami Beach	(305)	35,554	35,361
33940	North Naples (c)	(941)	—	13,422
33408	North Palm Beach	(561)	12,398	11,538
*34287	North Port	(941)	16,307	11,973
34234	North Sarasota (c)	(941)	—	6,702
33307	Oakland Park	(305)	28,476	26,326
33860	Oak Ridge (c)	(407)	—	15,388
*34478	Ocala	(352)	47,035	42,045
32548	Ocean City (c)	(850)	—	5,422
34761	Ocoee	(407)	21,089	12,778
33163	Ojus (c)	(305)	—	15,519
34677	Oldsmar	(813)	10,287	8,361
33265	Olympia Heights (c)	(305)	—	37,792
*33054	Opa-Locka	(305)	15,378	15,283
33054	Opa-Locka North (c)	(305)	—	6,568
*32763	Orange City (c)	(904)	6,165	5,372
*32073	Orange Park (c)	(904)	10,105	9,488
*32802	Orlando	(407)	181,175	164,674
32861	Orlo Vista (c)	(407)	—	5,990
*32174	Ormond Beach	(904)	33,060	29,721
32074	Ormond By-The-Sea (c)	(904)	—	8,157
*32765	Oviedo	(407)	22,162	11,114
32571	Pace (c)	(850)	—	6,277
.....	Page Park-Pine Manor (c)		—	5,116
33476	Pahokee	(561)	7,311	6,822
*32177	Palatka	(904)	10,891	10,447
*32905	Palm Bay	(321)	77,486	62,543
33480	Palm Beach	(561)	9,830	9,814
33408	Palm Beach Gardens	(561)	34,880	24,139
*32135	Palm Coast (c)	(904)	—	14,287
*34221	Palmetto	(941)	10,605	9,268

ZIP	Place		1998	1990
33157	Palmetto Estates (c)	(305)	—	12,293
*34683	Palm Harbor (c)	(727)	—	50,256
*33601	Palm River-Clair Mel (c)	(813)	—	13,691
33460	Palm Springs	(561)	9,998	9,763
33012	Palm Springs North (c)	(305)	—	5,300
32082	Palm Valley (c)	(904)	—	9,960
*32401	Panama City	(850)	39,477	34,396
32417	Panama City Beach	(850)	5,303	4,051
32404	Parker	(850)	5,377	4,598
33060	Parkland	(954)	12,348	3,773
33021	Pembroke Park	(954)	5,070	4,933
33029	Pembroke Pines	(954)	115,361	65,566
*32502	Pensacola	(850)	58,193	59,198
33257	Perrine (c)	(305)	—	15,576
*32347	Perry	(850)	7,328	7,151
32859	Pine Castle (c)	(407)	—	8,276
32858	Pine Hills (c)	(407)	—	35,322
.....	Pine Island Ridge (c)	(954)	—	5,244
*33781	Pinellas Park	(727)	44,179	43,571
33168	Pinewood (c)	(305)	—	15,518
33318	Plantation	(954)	81,424	66,814
*33566	Plant City	(813)	27,093	22,754
*33060	Pompano Beach	(954)	75,982	72,411
33064	Pompano Beach Highlands (c)	(954)		17,915
*33952	Port Charlotte (c)	(941)	—	41,535
32129	Port Orange	(904)	43,020	35,399
32027	Port St. John (o)	(321)	0,900	
*34981	Port St. Lucie	(561)	79,351	55,761
34992	Port Salerno (c)	(561)	—	7,786
*33032	Princeton (c)	(305)	—	7,073
*33950	Punta Gorda	(941)	13,280	10,637
*32351	Quincy	(850)	7,646	7,452
33156	Richmond Heights (c)	(305)	—	8,583
33312	Riverland (c)	(954)	—	5,376
*33569	Riverview (c)	(813)	—	6,478
33419	Riviera Beach	(561)	30,050	27,646
*32955	Rockledge	(321)	19,416	16,023
33411	Royal Palm Beach	(561)	19,170	15,532
33570	Ruskin (c)	(813)	—	6,046
34695	Safety Harbor	(727)	16,146	15,120
*32084	Saint Augustine	(904)	12,573	11,695
*34769	Saint Cloud	(407)	15,193	12,684
*33733	Saint Petersburg	(727)	236,029	240,318
*33736	Saint Petersburg Beach	(727)	9,928	9,200
33912	San Carlos Park (c)	(941)	—	11,785
33432	Sandalfoot Cove (c)	(305)	—	14,214
*32771	Sanford	(407)	36,951	32,387
33957	Sanibel	(941)	5,534	5,468
*34230	Sarasota	(941)	51,035	50,897
33577	Sarasota Springs (c)	(941)	—	16,088
32937	Satellite Beach	(321)	10,128	9,889
33055	Scott Lake (c)	(305)	—	14,588
*32958	Sebastian	(561)	13,942	10,248
*33870	Sebring	(863)	8,684	8,841
*33584	Seffner (c)	(813)	—	5,371
*33770	Seminole	(813)	9,767	9,251
*34242	Siesta Key (c)	(941)	—	7,772
34472	Silver Springs Shores (c)	(352)	—	6,421
32809	Sky Lake (c)	(407)	—	6,202
32703	South Apopka (c)	(407)	—	6,360
33505	South Bradenton (c)	(941)	—	20,398
32121	South Daytona	(904)	13,381	12,488
34277	Southgate (c)	(941)	—	7,324
34233	South Gate Ridge (c)	(941)	—	5,924
33243	South Miami	(305)	10,710	10,404
33157	South Miami Heights (c)	(305)	—	30,030
33707	South Pasadena	(727)	5,562	5,644
32937	South Patrick Shores (c)	(321)	—	10,249
34230	South Sarasota (c)	(941)	—	5,298
33595	South Venice (c)	(941)	—	11,951
32401	Springfield	(904)	9,110	8,719
*34601	Spring Hill (c)	(352)	—	31,117
32091	Starke	(904)	5,556	5,226
*34994	Stuart	(561)	12,385	11,936
*33573	Sun City Center (c)	(813)	—	8,326
33160	Sunny Isles (c)	(305)	—	11,772
33345	Sunrise	(954)	80,338	65,683
33283	Sunset (c)	(305)	—	15,810
33144	Sweetwater	(305)	14,370	13,909
*32301	Tallahassee	(850)	136,628	124,773
33320	Tamarac	(954)	52,929	44,822
33144	Tamiami (c)	(305)	—	33,845
*33601	Tampa	(813)	289,156	280,015
*34689	Tarpon Springs	(727)	19,016	17,874
32778	Tavares	(352)	8,868	7,488
33687	Temple Terrace	(813)	17,713	16,444
33469	Tequesta	(561)	5,064	4,499
*32780	Titusville	(321)	41,533	39,394
32685	Town 'n' Country (c)	(813)	—	60,946
33706	Treasure Island	(727)	7,027	7,266
32867	Union Park (c)	(407)	—	6,890
33620	University West (c)	(813)	—	23,760
32401	Upper Grand Lagoon (c)	(850)	—	7,855
32580	Valparaiso	(850)	6,615	6,316
*34285	Venice	(941)	17,686	17,052
33595	Venice Gardens (c)	(941)	—	7,701
*32960	Vero Beach	(561)	16,387	17,350
32960	Vero Beach South (c)	(561)	—	16,973
.....	Villages of Oriole (c)	(561)	—	5,698
33901	Villas (c)	(941)	—	9,898

ZIP	Place		1998	1990
32507	Warrington (c)	(850)	—	16,040
33314	Washington Park (c)	(954)	—	6,930
32791	Wekiva Springs (c)	(407)	—	23,026
33414	Wellington (c)	(561)	—	20,670
33155	Westchester (c)	(305)	—	29,883
.....	Westgate-Belvedere Homes (c)		—	6,880
33138	West Little River (c)	(305)	—	33,575
32912	West Melbourne	(321)	9,627	8,398
33144	West Miami	(305)	5,738	5,727
*33416	West Palm Beach	(561)	76,308	67,764
.....	West Park (c)		—	10,347
32505	West Pensacola (c)	(850)	—	22,107
33168	Westview (c)	(305)	—	9,668
33165	Westwood Lakes (c)	(305)	—	11,522
.....	Whiskey Creek (c)		—	5,061
33305	Wilton Manors	(954)	12,147	11,804
33803	Winston (c)	(813)	—	9,118
*34787	Winter Garden	(407)	11,871	9,863
*33880	Winter Haven	(863)	25,724	24,725
*32789	Winter Park	(407)	23,377	24,260
*32707	Winter Springs	(407)	28,606	22,151
32547	Wright (c)	(904)	—	18,945
*32097	Yulee (c)	(904)	—	6,915
*33540	Zephyrhills	(813)	9,311	8,220

Georgia

Area code (678) overlays area code (770). See introductory note.

ZIP	Place		1998	1990
*30101	Acworth	(770)	8,923	4,519
31620	Adel	(912)	5,204	5,093
*31706	Albany	(912)	77,545	78,804
*30004	Alpharetta	(770)	24,831	13,002
31709	Americus	(912)	16,887	16,516
*30603	Athens[1]	(706)	89,361	86,522
*30301	Atlanta	(404)	403,819	393,929
30011	Auburn	(770)	5,442	3,139
*30903	Augusta[2]	(706)	187,689	186,616
30168	Austell	(770)	5,032	4,173
*31717	Bainbridge	(912)	10,941	10,803
30032	Belvedere Park (c)	(404)	—	18,089
31723	Blakely	(912)	5,754	5,595
*31520	Brunswick	(912)	15,525	16,433
*30518	Buford	(404)	9,880	8,771
31728	Cairo	(912)	9,312	9,035
*30701	Calhoun	(706)	7,948	7,135
31730	Camilla	(912)	5,303	5,124
30032	Candler-McAfee (c)	(404)	—	29,491
*30114	Canton	(770)	7,880	4,817
*30117	Carrollton	(770)	16,538	16,029
*30120	Cartersville	(770)	12,988	12,037
*30125	Cedartown	(770)	7,733	7,976
30366	Chamblee	(404)	7,693	7,668
30021	Clarkston	(404)	5,859	5,385
30337	College Park	(404)	20,300	20,645
*31908	Columbus	(706)	182,828	178,683
30288	Conley (c)	(404)	—	5,528
*30013	Conyers	(404)	7,551	7,380
*31015	Cordele	(912)	10,763	10,833
.....	Country Club Estates (c)		—	7,500
*30014	Covington	(770)	10,056	9,860
30040	Cumming	(770)	5,132	2,828
*30720	Dalton	(706)	23,127	22,218
31742	Dawson	(912)	5,612	5,295
*30030	Decatur (DeKalb)	(404)	17,414	17,304
31520	Dock Junction (c)	(912)	—	7,094
30362	Doraville	(404)	8,377	7,626
*31533	Douglas	(912)	10,973	10,464
*30134	Douglasville	(404)	16,073	11,635
30333	Druid Hills (c)	(404)	—	12,174
*31021	Dublin	(912)	17,193	16,312
*30096	Duluth	(404)	17,722	9,821
30356	Dunwoody (c)	(404)	—	26,302
30364	East Point	(404)	33,670	34,595
31024	Eatonton	(706)	7,047	6,479
30809	Evans (c)	(706)	—	13,713
30060	Fair Oaks (c)	(404)	—	6,996
30535	Fairview (c)	(706)	—	6,444
*30214	Fayetteville	(404)	9,347	5,827
31750	Fitzgerald	(912)	8,972	8,901
*30297	Forest Park	(404)	16,999	16,958
31905	Fort Benning South (c)	(706)	—	14,617
30905	Fort Gordon (c)	(706)	—	9,140
30742	Fort Oglethorpe	(706)	6,415	5,880
*31313	Fort Stewart (c)	(912)	—	13,774
31030	Fort Valley	(912)	8,171	8,198
30605	Gaines School (c)	(706)	—	11,354
*30501	Gainesville	(770)	19,900	17,885
31418	Garden City	(912)	7,439	7,410
31754	Georgetown (c)	(912)	—	5,554
30316	Gresham Park (c)	(404)	—	9,000
*30223	Griffin	(770)	21,052	21,325
30813	Grovetown	(706)	5,279	3,596
30228	Hampton	(770)	5,007	2,694
30354	Hapeville	(404)	5,258	5,483
*31313	Hinesville	(912)	26,435	21,596
*31546	Jesup	(912)	9,584	8,958
*30144	Kennesaw	(404)	15,655	8,936
31548	Kingsland	(912)	11,584	6,089
30728	La Fayette	(706)	6,777	6,655

ZIP	Place		1998	1990
*30240	La Grange	(706)	25,111	25,574
30741	Lakeview (c)	(706)	—	5,237
*30045	Lawrenceville	(404)	20,008	17,250
*30047	Lilburn	(404)	11,239	9,295
30122	Lithia Springs (c)	(404)	—	11,403
30052	Loganville	(770)	5,120	3,180
30126	Mableton (c)	(404)	—	25,725
*31201	Macon	(912)	114,336	107,365
*30060	Marietta	(404)	51,362	44,129
30917	Martinez (c)	(706)	—	33,731
31061	Milledgeville	(912)	17,917	17,727
*30655	Monroe	(770)	10,444	9,759
*30260	Morrow	(404)	5,133	5,168
*31768	Moultrie	(912)	15,635	14,865
30087	Mountain Park (c)	(404)	—	11,025
31639	Nashville	(912)	5,311	4,782
*30263	Newnan	(770)	14,027	12,497
*30071	Norcross	(404)	6,612	5,947
30319	North Atlanta (c)	(404)	—	27,812
30033	North Decatur (c)	(404)	—	13,936
30033	North Druid Hills (c)	(404)	—	14,170
30032	Panthersville (c)	(404)	—	9,874
30269	Peachtree City	(404)	31,086	19,027
31069	Perry	(912)	9,910	9,452
31322	Pooler	(912)	5,434	4,649
30127	Powder Springs	(404)	10,836	6,862
31643	Quitman	(912)	5,034	5,292
30074	Redan (c)	(404)	—	24,376
31324	Richmond Hill	(912)	6,020	2,934
*30274	Riverdale	(404)	10,202	9,495
*30161	Rome	(706)	30,899	30,425
*30077	Roswell	(404)	57,102	47,986
31558	Saint Marys	(912)	13,823	8,204
31522	Saint Simons Island (c)	(912)	—	12,026
31082	Sandersville	(912)	6,580	6,290
30358	Sandy Springs (c)	(404)	—	67,842
*31402	Savannah	(912)	131,674	137,812
30079	Scottdale (c)	(404)	—	8,636
*30080	Smyrna	(404)	34,855	32,453
*30078	Snellville	(404)	15,418	12,084
30901	South Augusta (c)	(706)	—	55,998
*30458	Statesboro	(912)	21,309	20,770
30281	Stockbridge	(404)	5,251	3,359
*30086	Stone Mountain	(404)	7,053	6,544
30518	Sugar Hill	(404)	7,783	4,519
30747	Summerville	(706)	5,222	5,025
30024	Suwanee	(770)	6,872	2,412
30401	Swainsboro	(912)	7,196	7,361
31791	Sylvester	(912)	6,395	6,023
30286	Thomaston	(706)	9,062	9,127
*31792	Thomasville	(912)	17,565	17,554
30824	Thomson	(706)	6,723	6,862
*31794	Tifton	(912)	14,205	14,215
*30577	Toccoa	(706)	8,780	8,720
*30084	Tucker (c)	(404)	—	25,781
30291	Union City	(404)	10,358	9,347
*31603	Valdosta	(912)	41,816	40,038
*30474	Vidalia	(912)	11,697	11,118
30180	Villa Rica	(770)	6,921	6,542
30339	Vinings (c)	(404)	—	7,417
*31088	Warner Robins	(912)	45,559	43,861
*31501	Waycross	(912)	15,821	16,410
30830	Waynesboro	(706)	5,549	5,669
30901	West Augusta (c)	(706)	—	27,637
31410	Wilmington Island (c)	(912)	—	11,230
30680	Winder	(770)	8,441	7,373
*30188	Woodstock	(770)	5,730	4,361

(1) Athens merged with Clarke County in 1991. The 1998 and 1990 populations are for all of Clarke County except for Winterville and Bogart, which are part of the county but are also separate incorporated places. (2) Augusta merged with Richmond County in 1996. The 1998 and 1990 populations are for all of Richmond County except for Blythe and Hephzibah, which are part of the county but are also separate incorporated places.

Hawaii (808)

ZIP	Place	1998	1990
96701	Aiea (c)	—	8,906
96818	Aliamanu (c)	—	8,835
96706	Ewa Beach (c)	—	14,315
.....	Halawa (c)	—	13,408
96744	Heeia (c)	—	5,010
96853	Hickam Housing (c)	—	6,553
*96720	Hilo (c)	—	37,808
*96820	Honolulu (c)	395,789	377,059
*96732	Kahului (c)	—	16,889
96734	Kailua (c)	—	9,126
96863	Kailua (c)	—	36,818
96744	Kaneohe (c)	—	35,448
.....	Kaneohe Station (c)	—	11,662
96746	Kapaa (c)	—	8,149
96753	Kihei (c)	—	11,107
*96761	Lahaina (c)	—	9,073
96762	Laie (c)	—	5,577
96766	Lihue (c)	—	5,536
96792	Maili (c)	—	6,059
96792	Makaha (c)	—	7,990
96706	Makakilo (c)	—	9,828
96768	Makawao (c)	—	5,405
96789	Mililani Town (c)	—	29,359

ZIP	Place	1998	1990
96792	Nanakuli (c)	—	9,575
96782	Pearl City (c)	—	30,993
96788	Pukalani (c)	—	5,879
96786	Schofield Barracks (c)	—	19,597
.....	Village Park (c)	—	7,407
96786	Wahiawa (c)	—	17,386
96792	Waianae (c)	—	8,758
96793	Wailuku (c)	—	10,688
.....	Waimalu (c)	—	29,967
96796	Waimea (c)	—	5,972
96797	Waipahu (c)	—	31,435
96797	Waipio (c)	—	11,812
96786	Waipio Acres (c)	—	5,304

Idaho (208)

ZIP	Place	1998	1990
83401	Ammon	6,274	5,002
83221	Blackfoot	10,453	9,646
*83707	Boise City	157,452	126,685
83318	Burley	9,376	8,702
*83605	Caldwell	22,340	18,586
83202	Chubbuck	9,054	7,794
*83814	Coeur d'Alene	32,565	24,561
83616	Eagle	7,739	3,327
83617	Emmett	5,421	4,601
83714	Garden City	9,254	6,369
83333	Hailey	5,554	3,575
83835	Hayden	8,745	4,888
*83402	Idaho Falls	48,122	43,973
83338	Jerome	7,453	6,529
83501	Lewiston	30,363	28,082
*83642	Meridian	25,377	9,596
83843	Moscow	19,312	18,398
83647	Mountain Home	10,202	7,913
83648	Mountain Home AFB (c)	—	5,936
*83653	Nampa	41,951	28,365
83661	Payette	6,806	5,672
*83201	Pocatello	53,074	46,117
*83854	Post Falls	15,732	7,349
83440	Rexburg	14,303	14,298
83350	Rupert	5,415	5,455
83864	Sandpoint	7,598	5,561
*83301	Twin Falls	33,296	27,634
83672	Weiser	5,370	4,571

Illinois

Area code (224) overlays area code (847). See introductory note.

ZIP	Place		1998	1990
60101	Addison	(630)	34,074	32,053
60102	Algonquin	(847)	20,093	11,764
60803	Alsip	(708)	19,378	18,227
62002	Alton	(618)	31,457	33,060
60002	Antioch	(847)	7,923	6,105
*60005	Arlington Heights	(847)	76,522	75,463
*60505	Aurora	(630)	124,736	99,672
*60010	Barrington	(847)	9,848	9,538
60103	Bartlett	(630)	34,511	19,395
61607	Bartonville	(309)	6,365	6,555
60510	Batavia	(630)	22,306	17,076
60085	Beach Park	(847)	10,320	9,492
62618	Beardstown	(217)	5,045	5,270
*62220	Belleville	(618)	40,734	42,806
60104	Bellwood	(708)	19,932	20,241
61008	Belvidere	(815)	18,445	16,059
60106	Bensenville	(630)	18,105	17,767
62812	Benton	(618)	7,194	7,216
60402	Berwyn	(708)	43,030	45,426
62010	Bethalto	(618)	9,957	9,507
60108	Bloomingdale	(630)	19,995	16,614
*61701	Bloomington	(309)	58,841	51,889
60406	Blue Island	(708)	20,585	21,203
*60440	Bolingbrook	(630)	54,288	40,843
60538	Boulder Hill (c)	(630)	—	8,894
60914	Bourbonnais	(815)	15,511	13,929
60915	Bradley	(815)	12,604	10,954
60455	Bridgeview	(708)	15,487	14,402
60153	Broadview	(708)	8,269	8,538
60513	Brookfield	(708)	18,155	18,876
60089	Buffalo Grove	(847)	41,857	36,417
60459	Burbank	(708)	27,807	27,600
60521	Burr Ridge	(630)	10,379	8,247
62206	Cahokia	(618)	16,149	17,550
60409	Calumet City	(708)	36,916	37,840
60643	Calumet Park	(708)	8,401	8,418
61520	Canton	(309)	13,820	13,959
*62901	Carbondale	(618)	26,454	27,033
62626	Carlinville	(217)	5,673	5,416
62821	Carmi	(618)	5,627	5,735
*60188	Carol Stream	(630)	36,968	31,759
60110	Carpentersville	(847)	27,271	23,049
60013	Cary	(847)	14,069	10,025
62801	Centralia	(618)	14,229	14,476
62206	Centreville	(618)	7,188	7,489
*61821	Champaign	(217)	64,280	63,502
60410	Channahon	(815)	6,953	4,266
61920	Charleston	(217)	20,437	20,398
62629	Chatham	(217)	7,444	6,074
62233	Chester	(618)	7,639	8,204

ZIP	Place		1998	1990
*60607	Chicago	(312)/(773)	2,802,079	2,783,726
*60411	Chicago Heights	(708)	31,635	32,966
60415	Chicago Ridge	(708)	14,091	13,643
61523	Chillicothe	(309)	5,947	5,959
60804	Cicero	(708)	71,289	67,436
60514	Clarendon Hills	(630)	7,474	6,994
61727	Clinton	(217)	7,314	7,437
62234	Collinsville	(618)	23,308	22,424
62236	Columbia	(618)	6,743	5,524
60478	Country Club Hills	(708)	16,433	15,431
60525	Countryside	(708)	6,083	5,961
60435	Crest Hill	(815)	12,821	10,999
60445	Crestwood	(708)	11,658	10,823
60417	Crete	(708)	8,049	6,773
61610	Creve Coeur	(309)	5,877	5,938
*60014	Crystal Lake	(815)	33,078	24,692
*61832	Danville	(217)	31,761	33,828
60561	Darien	(630)	23,629	20,556
*62525	Decatur	(217)	79,972	83,900
60015	Deerfield	(847)	18,802	17,327
60115	De Kalb	(815)	36,094	35,076
*60018	Des Plaines	(847)	55,272	53,414
61021	Dixon	(815)	15,374	15,134
60419	Dolton	(708)	23,882	23,956
*60515	Downers Grove	(630)	51,716	47,464
62832	Du Quoin	(618)	6,529	6,697
62024	East Alton	(618)	6,700	7,063
61244	East Moline	(309)	20,205	20,147
61611	East Peoria	(309)	22,117	21,378
*62201	East St. Louis	(618)	37,390	40,944
62025	Edwardsville	(618)	16,961	14,582
62401	Effingham	(217)	12,820	11,927
*60120	Elgin	(847)	87,507	77,014
*60009	Elk Grove Village	(847)	34,693	33,429
60126	Elmhurst	(630)	43,505	42,029
60707	Elmwood Park	(708)	22,461	23,206
*60201	Evanston	(847)	71,928	73,233
60805	Evergreen Park	(708)	20,389	20,874
62208	Fairview Heights	(618)	14,795	14,768
60422	Flossmoor	(708)	9,175	8,651
60130	Forest Park	(708)	14,301	14,918
60020	Fox Lake	(847)	8,745	7,539
60423	Frankfort	(815)	10,123	7,180
.....	Frankfort Square (c)	(815)	—	6,227
60131	Franklin Park	(847)	17,941	18,485
61032	Freeport	(815)	25,806	25,840
60030	Gages Lake (c)	(847)	—	8,349
*61401	Galesburg	(309)	32,791	33,530
61254	Geneseo	(309)	6,251	5,990
60134	Geneva	(630)	18,382	12,625
62034	Glen Carbon	(618)	10,012	7,774
60022	Glencoe	(847)	8,407	8,499
60139	Glendale Heights	(630)	30,277	27,915
*60137	Glen Ellyn	(630)	25,956	24,919
60025	Glenview	(847)	39,873	38,436
60425	Glenwood	(708)	9,079	9,289
62035	Godfrey	(618)	17,340	15,675
.....	Goodings Grove (c)	(815)	—	14,054
62040	Granite City	(618)	31,078	32,766
60030	Grayslake	(847)	15,853	7,388
62246	Greenville	(618)	5,456	5,108
60031	Gurnee	(847)	25,016	13,715
60103	Hanover Park	(630)	36,027	32,918
62946	Harrisburg	(618)	9,135	9,318
60033	Harvard	(815)	6,874	5,975
60426	Harvey	(708)	28,756	29,771
60656	Harwood Heights	(708)	7,946	7,680
60047	Hawthorn Woods	(847)	5,817	4,423
60429	Hazel Crest	(708)	13,859	13,334
62948	Herrin	(618)	11,107	10,857
60457	Hickory Hills	(708)	14,113	13,021
62249	Highland	(618)	8,218	7,546
60035	Highland Park	(847)	31,310	30,575
60040	Highwood	(847)	5,130	5,331
60162	Hillside	(708)	7,524	7,672
*60521	Hinsdale	(630)	16,589	16,029
*60195	Hoffman Estates	(847)	48,516	46,363
60430	Homewood	(708)	19,536	19,278
60942	Hoopeston	(217)	5,481	5,871
60067	Inverness	(847)	6,830	6,516
60042	Island Lake	(847)	7,689	4,449
60143	Itasca	(630)	8,422	6,947
*62650	Jacksonville	(217)	18,239	19,327
62052	Jerseyville	(618)	7,508	7,382
*60436	Joliet	(815)	92,285	77,217
60458	Justice	(708)	11,528	11,137
60901	Kankakee	(815)	26,456	27,541
61443	Kewanee	(309)	12,481	12,969
60525	La Grange	(708)	15,002	15,362
60526	La Grange Park	(708)	12,463	12,861
60044	Lake Bluff	(847)	5,613	5,486
60045	Lake Forest	(847)	19,128	17,836
60102	Lake in the Hills	(847)	20,417	5,882
60047	Lake Zurich	(847)	17,181	14,927
60438	Lansing	(708)	28,512	28,131
61301	La Salle	(815)	9,526	9,717
60439	Lemont	(630)	10,544	7,359
*60048	Libertyville	(847)	19,976	19,174
62656	Lincoln	(217)	14,966	15,418
60069	Lincolnshire	(847)	6,139	4,928
60645	Lincolnwood	(847)	11,277	11,365
60046	Lindenhurst	(847)	10,602	8,044
60532	Lisle	(630)	20,820	19,584
62056	Litchfield	(217)	6,632	6,883
60441	Lockport	(815)	13,401	9,401
60148	Lombard	(630)	42,215	39,408
60047	Long Grove	(847)	6,388	4,747
*61130	Loves Park	(815)	18,183	15,457
60411	Lynwood	(708)	7,677	6,535
60534	Lyons (Cook)	(708)	9,700	9,828
*60050	McHenry	(815)	19,451	16,343
61115	Machesney Park	(815)	19,831	19,042
61455	Macomb	(309)	17,778	19,952
60950	Manteno	(815)	5,351	3,709
60152	Marengo	(815)	5,385	4,768
62959	Marion	(618)	15,810	14,597
60426	Markham (Cook)	(708)	12,971	13,136
62258	Mascoutah	(618)	5,603	5,511
60443	Matteson	(708)	12,490	11,378
61938	Mattoon	(217)	18,115	18,441
60153	Maywood	(708)	25,833	27,139
*60160	Melrose Park	(708)	20,400	20,859
61342	Mendota	(815)	7,194	7,017
62960	Metropolis	(618)	6,870	6,734
60445	Midlothian	(708)	14,865	14,372
61264	Milan	(309)	5,907	5,753
60448	Mokena	(708)	12,715	6,128
*61265	Moline	(309)	41,919	43,080
61462	Monmouth	(309)	9,425	9,489
60538	Montgomery	(630)	5,642	4,487
60450	Morris	(815)	11,477	10,274
61550	Morton	(309)	14,742	13,799
60053	Morton Grove	(847)	22,180	22,373
62863	Mount Carmel	(618)	7,882	8,287
60056	Mount Prospect	(847)	53,581	53,168
62864	Mount Vernon	(618)	16,850	17,082
60060	Mundelein	(847)	28,518	21,224
62966	Murphysboro	(618)	8,950	9,176
*60540	Naperville	(630)	117,091	85,806
60451	New Lenox	(815)	14,830	9,698
60714	Niles	(847)	29,502	28,375
61761	Normal	(309)	44,221	40,023
60634	Norridge	(708)	14,311	14,459
60542	North Aurora	(630)	8,899	6,010
*60062	Northbrook	(708)	33,107	32,565
60064	North Chicago	(847)	32,175	34,978
60093	Northfield	(847)	5,422	4,924
60164	Northlake	(708)	11,847	12,505
60546	North Riverside	(708)	6,042	6,180
60521	Oak Brook	(630)	9,428	9,087
60452	Oak Forest	(708)	27,718	26,202
*60303	Oak Lawn	(708)	57,730	56,182
*60303	Oak Park	(708)	50,646	53,648
62269	O'Fallon	(618)	19,414	16,064
62450	Olney	(618)	8,824	8,873
60477	Orland Hills	(708)	6,505	5,510
*60462	Orland Park	(708)	47,856	35,720
60543	Oswego	(630)	10,536	3,949
61350	Ottawa	(815)	18,026	17,574
*60067	Palatine	(847)	45,513	41,554
60463	Palos Heights	(708)	12,164	11,478
60465	Palos Hills	(708)	18,732	17,803
62557	Pana	(217)	5,650	5,796
61944	Paris	(217)	9,024	9,105
60085	Park City	(847)	5,537	4,677
60466	Park Forest	(708)	24,365	24,656
60068	Park Ridge	(847)	37,390	37,075
61554	Pekin	(309)	31,958	32,254
*61601	Peoria	(309)	111,148	113,508
61603	Peoria Heights	(309)	6,555	6,930
61354	Peru	(815)	9,333	9,302
60544	Plainfield	(815)	8,808	4,557
60545	Plano	(630)	5,696	5,104
61764	Pontiac	(815)	11,440	11,428
61356	Princeton	(815)	6,796	7,197
60070	Prospect Heights	(847)	15,398	15,236
*62301	Quincy	(217)	39,918	39,682
61866	Rantoul	(217)	13,945	17,212
60471	Richton Park	(708)	11,720	10,523
60827	Riverdale	(708)	13,220	13,671
60305	River Forest	(708)	11,130	11,669
60171	River Grove	(708)	9,663	9,961
60546	Riverside	(708)	8,318	8,774
60472	Robbins	(708)	7,246	7,498
62454	Robinson	(618)	6,394	6,740
61068	Rochelle	(815)	9,287	8,769
61071	Rock Falls	(815)	9,357	9,669
*61125	Rockford	(815)	143,656	142,815
*61201	Rock Island	(309)	38,714	40,630
60008	Rolling Meadows	(847)	22,844	22,598
60446	Romeoville	(815)	19,015	14,101
60172	Roselle	(630)	23,627	20,803
60073	Round Lake	(847)	5,053	3,550
60073	Round Lake Beach	(847)	23,140	16,406
*60174	Saint Charles	(630)	26,516	22,636
62881	Salem	(618)	7,624	7,470
60548	Sandwich	(815)	6,043	5,607
60411	Sauk Village	(708)	10,973	10,734
*60194	Schaumburg	(847)	74,481	68,586
60176	Schiller Park	(847)	10,941	11,189

ZIP	Place		1998	1990
62225	Scott AFB (c)	(618)	—	7,245
60436	Shorewood	(815)	8,038	6,264
61282	Silvis	(309)	6,987	6,926
*60077	Skokie	(847)	58,628	59,432
60177	South Elgin	(847)	14,910	7,474
60473	South Holland	(708)	21,794	22,105
*62703	Springfield	(217)	117,098	105,412
61362	Spring Valley	(815)	5,044	5,246
60475	Steger	(708)	9,949	9,251
61081	Sterling	(815)	14,623	15,142
60402	Stickney	(708)	5,923	5,678
60107	Streamwood	(630)	34,984	31,197
61364	Streator	(815)	13,726	14,121
60501	Summit	(708)	9,599	9,971
62221	Swansea	(618)	8,927	8,201
60178	Sycamore	(815)	11,227	9,896
62568	Taylorville	(217)	11,236	11,133
60477	Tinley Park	(708)	45,825	37,115
62294	Troy	(618)	7,670	6,194
60466	University Park	(708)	6,435	6,204
*61801	Urbana	(217)	34,872	36,383
62471	Vandalia	(618)	6,583	6,114
60061	Vernon Hills	(847)	18,441	15,319
60181	Villa Park	(630)	22,635	22,279
60555	Warrenville	(630)	13,508	11,389
61571	Washington	(309)	10,611	10,136
62204	Washington Park	(618)	6,864	7,431
62298	Waterloo	(618)	6,331	5,030
60970	Watseka	(815)	5,479	5,424
60084	Wauconda	(847)	8,759	6,294
*60085	Waukegan	(847)	75,999	69,481
60154	Westchester	(708)	17,476	17,301
*60185	West Chicago	(630)	17,865	14,808
60558	Western Springs	(708)	12,435	11,956
62896	West Frankfort	(618)	8,249	8,526
60559	Westmont	(630)	22,654	21,402
61604	West Peoria	(309)	5,483	5,307
*60187	Wheaton	(630)	55,308	51,441
60090	Wheeling	(847)	30,564	29,911
60514	Willowbrook	(630)	9,089	8,651
60091	Wilmette	(847)	26,219	26,694
60481	Wilmington	(815)	5,422	4,743
60190	Winfield	(630)	8,502	7,096
60093	Winnetka	(847)	11,853	12,210
60096	Winthrop Harbor	(847)	7,142	6,240
60097	Wonder Lake (c)	(815)	—	6,664
60191	Wood Dale	(630)	13,402	12,394
60517	Woodridge	(630)	29,382	26,359
62095	Wood River	(618)	11,000	11,490
60098	Woodstock	(815)	17,734	14,368
60482	Worth	(708)	11,152	11,208
60560	Yorkville	(630)	5,953	3,974
60099	Zion	(847)	22,518	19,783

Indiana

ZIP	Place		1998	1990
46001	Alexandria	(765)	5,611	5,709
*46011	Anderson	(765)	58,528	59,518
46703	Angola	(219)	6,577	5,851
46706	Auburn	(219)	11,015	9,386
47006	Batesville	(812)	5,401	4,720
47421	Bedford	(812)	14,619	13,817
46107	Beech Grove	(317)	13,246	13,383
*47408	Bloomington	(812)	65,065	62,735
46714	Bluffton	(219)	9,503	9,104
47601	Boonville	(812)	6,519	6,686
47834	Brazil	(812)	8,058	7,640
46506	Bremen	(219)	5,029	4,725
46112	Brownsburg	(317)	11,779	7,751
*46032	Carmel	(317)	42,074	25,380
46303	Cedar Lake	(219)	9,261	8,885
47111	Charlestown	(812)	6,006	5,889
46304	Chesterton	(219)	10,163	9,118
47129	Clarksville (Clark)	(812)	19,688	19,838
46725	Columbia City	(219)	6,951	5,883
*47201	Columbus	(812)	32,250	33,948
47331	Connersville	(765)	15,266	15,550
47933	Crawfordsville	(765)	14,108	13,584
46307	Crown Point	(219)	19,403	17,728
46229	Cumberland	(317)	5,046	4,557
46122	Danville	(317)	5,154	4,345
46733	Decatur	(219)	9,023	8,642
46514	Dunlap (c)	(219)	—	5,705
46311	Dyer	(219)	13,485	10,923
46312	East Chicago	(219)	30,885	33,892
*46515	Elkhart	(219)	43,673	44,661
46036	Elwood	(765)	8,783	9,494
*47708	Evansville	(812)	122,779	126,272
46038	Fishers	(317)	25,591	7,189
*46802	Fort Wayne	(219)	185,716	195,680
46041	Frankfort	(765)	15,291	14,754
46131	Franklin	(317)	17,259	12,932
46738	Garrett	(219)	5,162	5,349
*46401	Gary	(219)	108,469	116,646
46933	Gas City	(765)	5,586	6,311
*46526	Goshen	(219)	25,262	23,794
46530	Granger (c)	(219)	—	20,241
46135	Greencastle	(765)	9,531	8,984
46140	Greenfield	(317)	13,869	11,657

ZIP	Place		1998	1990
47240	Greensburg	(812)	10,360	9,286
*46142	Greenwood	(317)	33,419	26,507
46319	Griffith	(219)	17,816	17,914
*46320	Hammond	(219)	78,212	84,236
47348	Hartford City	(765)	6,814	6,960
46322	Highland	(219)	23,730	23,696
46342	Hobart	(219)	24,841	24,440
47542	Huntingburg	(812)	5,256	5,236
46750	Huntington	(219)	15,469	16,389
*46206	Indianapolis	(317)	741,304	731,278
*47546	Jasper	(812)	11,174	10,030
*47130	Jeffersonville	(812)	26,018	24,016
46755	Kendallville	(219)	9,349	7,984
*46902	Kokomo	(765)	45,149	44,996
*47901	Lafayette	(765)	44,583	45,933
.....	Lakes of the Four Seasons (c)		—	6,556
46405	Lake Station	(219)	13,903	13,899
*46350	La Porte	(219)	20,226	21,507
46226	Lawrence	(317)	34,561	26,849
46052	Lebanon	(765)	13,840	12,059
47441	Linton	(812)	6,355	5,814
46947	Logansport	(219)	15,831	16,865
46356	Lowell	(219)	7,355	6,430
47250	Madison	(812)	12,510	12,006
*46952	Marion	(765)	28,812	32,607
46151	Martinsville	(765)	12,096	11,677
*46401	Merrillville	(219)	30,571	27,257
*46360	Michigan City	(219)	32,626	33,822
*46544	Mishawaka	(219)	45,310	42,635
47446	Mitchell	(812)	5,105	4,669
47960	Monticello	(219)	5,510	5,237
46158	Mooresville	(317)	8,682	5,779
47620	Mount Vernon	(812)	6,558	7,217
*47302	Muncie	(765)	67,476	71,170
46321	Munster	(219)	20,485	19,949
46550	Nappanee	(219)	5,917	5,474
*47150	New Albany	(812)	38,265	36,322
47362	New Castle	(765)	16,932	17,753
46774	New Haven	(219)	13,809	11,234
*46060	Noblesville	(317)	25,983	17,655
46962	North Manchester	(219)	6,395	6,383
47265	North Vernon	(812)	5,811	5,129
47130	Oak Park (c)	(812)	—	5,630
46970	Peru	(765)	11,324	12,843
46168	Plainfield	(317)	17,739	14,953
46563	Plymouth	(219)	10,140	8,291
46368	Portage	(219)	33,030	29,062
47371	Portland	(219)	5,899	6,483
47670	Princeton	(812)	7,131	8,127
47978	Rensselaer	(219)	5,262	5,045
*47374	Richmond	(765)	37,091	38,705
46975	Rochester	(219)	6,623	5,969
46173	Rushville	(765)	5,481	5,533
46373	Saint John	(219)	8,053	4,921
47167	Salem	(812)	6,385	5,619
46375	Schererville	(219)	24,062	20,155
47170	Scottsburg	(812)	5,755	5,334
47172	Sellersburg	(812)	6,056	5,936
47274	Seymour	(812)	17,026	15,605
46176	Shelbyville	(765)	16,562	15,347
*46624	South Bend	(219)	99,417	105,511
46383	South Haven (c)	(219)	—	6,112
46224	Speedway	(317)	12,213	13,092
47586	Tell City	(812)	8,045	8,088
*47808	Terre Haute	(812)	53,355	57,475
*46383	Valparaiso	(219)	25,931	24,414
47591	Vincennes	(812)	18,875	19,867
46992	Wabash	(219)	11,138	12,127
*46580	Warsaw	(219)	10,797	10,968
47501	Washington	(812)	10,949	10,864
46074	Westfield	(317)	9,988	3,304
*46580	West Lafayette	(765)	27,975	26,144
46391	Westville	(219)	5,701	5,255
47394	Winchester	(765)	5,137	5,095
46077	Zionsville	(317)	7,192	6,207

Iowa

ZIP	Place		1998	1990
50511	Algona	(515)	5,766	6,015
50009	Altoona	(515)	9,567	7,242
*50010	Ames	(515)	48,415	47,198
52205	Anamosa	(319)	5,594	5,100
50021	Ankeny	(515)	25,086	18,482
50022	Atlantic	(712)	7,098	7,432
52722	Bettendorf	(319)	31,737	28,139
*50036	Boone	(515)	12,754	12,392
52601	Burlington	(319)	26,855	27,208
51401	Carroll	(712)	10,331	9,579
50613	Cedar Falls	(319)	34,721	34,298
*52401	Cedar Rapids	(319)	114,563	108,772
52544	Centerville	(515)	5,448	5,936
50616	Charles City	(515)	7,521	7,878
51012	Cherokee	(712)	5,515	6,026
51632	Clarinda	(712)	5,725	5,104
50428	Clear Lake	(515)	8,250	8,183
*52732	Clinton	(319)	27,626	29,201
50325	Clive	(515)	11,125	7,446
52241	Coralville	(319)	12,688	10,347
*51501	Council Bluffs	(712)	56,312	54,315

ZIP	Place		1998	1990
50801	Creston	(515)	7,645	7,911
*52802	Davenport	(319)	96,842	95,333
52101	Decorah	(319)	8,309	8,063
51442	Denison	(712)	6,533	6,604
*50318	Des Moines	(515)	191,293	193,189
*52001	Dubuque	(319)	56,467	57,538
51334	Estherville	(712)	6,375	6,720
52556	Fairfield	(515)	10,322	9,955
50501	Fort Dodge	(515)	24,738	26,057
52627	Fort Madison	(319)	11,332	11,614
51534	Glenwood	(712)	5,244	4,960
50112	Grinnell	(515)	8,744	8,902
*51537	Harlan	(712)	5,128	5,148
52233	Hiawatha	(319)	6,670	5,354
50644	Independence	(319)	5,888	5,972
50125	Indianola	(515)	13,023	11,340
*52240	Iowa City	(319)	60,897	59,735
50126	Iowa Falls	(515)	5,137	5,435
50131	Johnston	(515)	6,906	4,702
52632	Keokuk	(319)	12,179	12,451
50138	Knoxville	(515)	8,164	8,232
51031	Le Mars	(712)	8,922	8,454
52057	Manchester	(319)	5,469	5,137
52060	Maquoketa	(319)	6,049	6,130
52302	Marion	(319)	23,777	20,422
50158	Marshalltown	(515)	25,201	25,178
*50401	Mason City	(515)	28,718	29,040
52641	Mount Pleasant	(319)	8,257	7,959
52761	Muscatine	(319)	22,932	22,881
50201	Nevada	(515)	6,126	6,009
50208	Newton	(515)	15,371	14,799
50211	Norwalk	(515)	6,678	5,726
50662	Oelwein	(319)	6,495	6,691
51041	Orange City	(712)	5,382	4,940
52577	Oskaloosa	(515)	10,673	10,600
52501	Ottumwa	(515)	23,854	24,488
50219	Pella	(515)	9,525	9,270
50220	Perry	(515)	7,301	6,652
*51566	Red Oak	(712)	6,227	6,264
51601	Shenandoah	(712)	5,404	5,572
51250	Sioux Center	(712)	5,848	5,074
*51101	Sioux City	(712)	82,697	80,505
51301	Spencer	(712)	11,170	11,066
50588	Storm Lake	(712)	8,970	8,769
*50318	Urbandale	(515)	27,907	23,775
52349	Vinton	(319)	5,514	5,103
52353	Washington	(319)	7,304	7,074
*50701	Waterloo	(319)	63,703	66,467
50677	Waverly	(319)	8,762	8,539
50595	Webster City	(515)	7,755	7,894
*50265	West Des Moines	(515)	42,333	31,702

Kansas

ZIP	Place		1998	1990
67410	Abilene	(785)	6,519	6,242
67002	Andover	(316)	5,964	4,204
67005	Arkansas City	(316)	12,300	12,762
66002	Atchison	(913)	10,594	10,656
67010	Augusta	(316)	8,839	7,848
66952	Bel Aire	(316)	5,515	3,695
66012	Bonner Springs	(913)	6,724	6,413
66720	Chanute	(316)	9,082	9,488
67337	Coffeyville	(316)	12,031	12,917
67701	Colby	(785)	5,318	5,510
66901	Concordia	(785)	5,594	6,152
67037	Derby	(316)	18,327	14,691
67801	Dodge City	(316)	22,456	21,129
67042	El Dorado	(316)	13,078	11,495
66801	Emporia	(316)	24,462	25,512
66442	Fort Riley North (c)	(785)	—	12,848
66701	Fort Scott	(316)	8,315	8,362
67846	Garden City	(316)	26,039	24,097
66030	Gardner	(913)	6,563	4,277
67530	Great Bend	(316)	14,461	15,427
67601	Hays	(785)	18,866	18,632
67060	Haysville	(316)	8,922	8,364
*67501	Hutchinson	(316)	39,016	39,308
67301	Independence	(316)	9,588	10,030
66749	Iola	(316)	6,171	6,351
66441	Junction City	(785)	16,970	20,642
*66102	Kansas City	(913)	141,297	151,521
66043	Lansing	(913)	8,582	7,120
*66044	Lawrence	(785)	74,244	65,608
66048	Leavenworth	(913)	39,227	38,495
66209	Leawood	(913)	25,886	19,693
66214	Lenexa	(913)	38,826	34,110
*67901	Liberal	(316)	17,486	16,573
67460	McPherson	(316)	13,284	12,422
*66502	Manhattan	(785)	41,318	43,081
66202	Merriam	(913)	12,103	11,819
66203	Mission	(913)	9,478	9,504
67110	Mulvane	(316)	5,134	4,683
67114	Newton	(316)	18,070	16,700
*66061	Olathe	(913)	85,035	63,402
66067	Ottawa	(785)	11,963	10,667
66204	Overland Park	(913)	139,685	111,790
67219	Park City	(316)	5,607	5,081
67357	Parsons	(316)	11,163	11,919
66762	Pittsburg	(316)	18,508	17,789

ZIP	Place		1998	1990
66208	Prairie Village	(913)	23,365	23,186
67124	Pratt	(316)	6,525	6,687
66205	Roeland Park	(913)	7,644	7,706
*67401	Salina	(785)	44,022	42,299
66203	Shawnee	(913)	45,250	37,962
*66601	Topeka	(785)	118,977	119,883
67880	Ulysses	(316)	6,217	5,474
67152	Wellington	(316)	8,470	8,517
*67202	Wichita	(316)	329,211	304,017
67156	Winfield	(316)	11,533	11,931

Kentucky

ZIP	Place		1998	1990
41001	Alexandria	(606)	7,899	5,592
*41101	Ashland	(606)	22,402	23,622
40004	Bardstown	(502)	7,879	6,712
41073	Bellevue	(606)	6,133	6,997
40403	Berea	(606)	10,341	9,129
*42101	Bowling Green	(270)	44,822	41,688
40261	Buechel (c)	(502)	—	7,081
41005	Burlington (c)	(606)	—	6,070
*42718	Campbellsville	(270)	10,776	9,592
*40701	Corbin	(606)	8,000	7,644
*41011	Covington	(606)	40,389	43,646
41031	Cynthiana	(606)	6,350	6,497
*40422	Danville	(606)	16,470	14,454
41074	Dayton	(606)	5,748	6,576
40243	Douglass Hills	(502)	5,793	5,431
41017	Edgewood	(606)	8,521	8,143
*42701	Elizabethtown	(270)	19,905	18,167
41018	Elsmere	(606)	8,043	6,847
41018	Erlanger	(606)	16,900	15,979
40118	Fairdale (c)	(502)	—	6,563
40291	Fern Creek (c.)	(502)	—	16,406
41139	Flatwoods	(606)	7,933	7,799
*41042	Florence	(606)	19,501	18,586
42223	Fort Campbell North (c)	(270)	—	18,861
40121	Fort Knox (c)	(270)	—	21,495
41017	Fort Mitchell	(606)	7,033	7,438
41075	Fort Thomas	(606)	14,929	16,032
41011	Fort Wright	(606)	6,567	6,404
*40601	Frankfort	(502)	26,418	26,535
*42134	Franklin	(270)	8,109	7,607
40324	Georgetown	(502)	14,365	11,414
*42141	Glasgow	(270)	14,062	12,777
40330	Harrodsburg	(606)	7,862	7,335
*41701	Hazard	(606)	5,397	5,416
*42420	Henderson	(270)	26,457	25,945
41076	Highland Heights	(606)	6,325	4,223
40228	Highview (c)	(502)	—	14,814
40229	Hillview	(502)	7,383	6,119
*42240	Hopkinsville	(270)	32,045	29,809
41051	Independence	(606)	13,745	10,444
40269	Jeffersontown	(502)	25,678	23,223
40031	La Grange	(502)	5,405	3,901
40342	Lawrenceburg	(502)	7,949	5,911
40033	Lebanon	(270)	5,754	5,695
*42754	Leitchfield	(270)	5,459	4,965
*40507	Lexington	(606)	241,749	225,366
*40741	London	(606)	7,045	5,757
*40232	Louisville	(502)	255,045	269,555
40252	Lyndon	(502)	8,539	8,037
42431	Madisonville	(270)	19,034	18,693
42066	Mayfield	(270)	10,370	9,935
41056	Maysville	(606)	8,479	8,113
40965	Middlesboro	(606)	10,397	11,328
40253	Middletown	(502)	5,303	5,016
42633	Monticello	(606)	5,720	5,357
40351	Morehead	(606)	8,830	8,357
40353	Mount Sterling	(606)	5,465	5,362
40047	Mount Washington	(502)	6,423	5,256
42071	Murray	(270)	15,905	14,442
40218	Newburg (c)	(502)	—	21,647
*41071	Newport	(606)	16,455	18,871
*40356	Nicholasville	(606)	17,099	13,603
42262	Oak Grove	(502)	5,840	2,863
40259	Okolona (c)	(502)	—	18,902
*42301	Owensboro	(270)	54,041	53,577
*42003	Paducah	(270)	25,883	27,256
*40361	Paris	(606)	8,898	8,730
*41501	Pikeville	(606)	6,243	6,324
40268	Pleasure Ridge Park (c)	(502)	—	25,131
42445	Princeton	(270)	6,929	6,940
*40160	Radcliff	(502)	19,472	19,778
*40475	Richmond	(606)	27,644	21,183
42276	Russellville	(270)	7,869	7,454
40216	Saint Dennis (c)	(502)	—	10,326
*40206	Saint Matthews	(502)	16,583	15,691
*40066	Shelbyville	(502)	6,849	6,155
40165	Shepherdsville	(502)	5,919	4,805
40256	Shively	(502)	16,608	15,535
*42501	Somerset	(606)	12,618	10,735
41015	Taylor Mill	(606)	7,084	5,530
40272	Valley Station (c)	(502)	—	22,840
40383	Versailles	(606)	8,233	7,269
41016	Villa Hills	(606)	7,588	7,370
41101	Westwoods (c)	(606)	—	5,300
*40769	Williamsburg	(606)	6,008	5,493
*40391	Winchester	(606)	15,937	15,799

Louisiana

ZIP	Place		1998	1990
*70510	Abbeville	(337)	11,402	11,769
*71301	Alexandria	(318)	45,800	49,049
70032	Arabi (c)	(504)	—	8,787
70094	Avondale (c)	(504)	—	5,813
*70714	Baker	(225)	13,009	13,087
*71220	Bastrop	(318)	13,516	13,916
*70821	Baton Rouge	(225)	211,551	219,531
70360	Bayou Cane (c)	(504)	—	15,876
70037	Belle Chasse (c)	(504)	—	8,512
*70427	Bogalusa	(504)	13,444	14,280
*71111	Bossier City	(318)	56,637	52,721
70517	Breaux Bridge	(337)	6,913	6,694
70094	Bridge City (c)	(504)	—	8,327
70811	Brownfields (c)	(225)	—	5,229
71291	Brownsville-Bawcomville (c)	(318)	—	7,397
70520	Carencro	(337)	6,015	5,518
*70043	Chalmette (c)	(504)	—	31,860
71291	Claiborne (c)	(318)	—	8,300
*70433	Covington	(504)	8,677	7,691
*70526	Crowley	(337)	13,554	13,983
70345	Cut Off (c)	(504)	—	5,325
*70726	Denham Springs	(225)	9,110	8,381
70634	De Ridder	(337)	11,152	10,475
70047	Destrehan (c)	(504)	—	8,031
70346	Donaldsonville	(225)	8,941	7,949
70072	Estelle (c)	(504)	—	14,091
70535	Eunice	(337)	11,112	11,162
71459	Fort Polk South (c)	(337)	—	10,911
70538	Franklin	(337)	8,593	9,004
70820	Gardere (c)	(225)	—	7,209
*70737	Gonzales	(225)	8,240	7,208
71245	Grambling	(318)	5,271	5,713
*70053	Gretna	(504)	16,569	17,208
*70401	Hammond	(504)	16,617	15,871
70123	Harahan	(504)	9,838	9,927
*70058	Harvey (c)	(504)	—	21,222
*70360	Houma	(504)	29,964	30,495
70544	Jeanerette	(337)	6,374	6,205
70502	Jefferson (c)	(504)	—	14,521
70546	Jennings	(337)	11,314	11,305
*70062	Kenner	(504)	71,641	72,033
70445	Lacombe (c)	(504)	—	6,523
*70501	Lafayette	(337)	113,615	101,865
*70601	Lake Charles	(337)	70,766	70,580
*70068	La Place (c)	(504)	—	24,194
70373	Larose (c)	(504)	—	5,772
*71446	Leesville	(337)	5,533	7,638
*70471	Mandeville	(504)	9,006	7,474
71052	Mansfield	(318)	5,021	5,389
71351	Marksville	(318)	5,427	5,526
*70072	Marrero (c)	(504)	—	36,671
70075	Meraux (c)	(504)	—	8,849
70812	Merrydale (c)	(225)	—	10,395
*70009	Metairie (c)	(504)	—	149,428
*71055	Minden	(318)	13,309	13,661
*71207	Monroe	(318)	53,612	54,909
*70380	Morgan City	(504)	13,671	14,531
70612	Moss Bluff (c)	(337)	—	8,039
*71457	Natchitoches	(318)	16,713	16,609
*70560	New Iberia	(337)	32,664	31,828
*70140	New Orleans	(504)	465,538	496,938
70760	New Roads	(225)	5,274	5,303
71463	Oakdale	(318)	7,769	6,837
70808	Oak Hills Place (c)	(225)	—	5,479
*70570	Opelousas	(337)	18,984	19,091
70392	Patterson	(504)	5,373	5,166
*71360	Pineville	(318)	14,030	15,308
*70764	Plaquemine	(225)	6,312	7,101
70454	Ponchatoula	(504)	5,920	5,499
70767	Port Allen	(225)	6,028	6,277
70601	Prien (c)	(337)	—	6,448
70394	Raceland (c)	(504)	—	5,564
70578	Rayne	(337)	8,459	8,502
71037	Red Chute (c)	(318)	—	5,431
70084	Reserve (c)	(504)	—	8,847
70123	River Ridge (c)	(504)	—	14,800
*71270	Ruston	(318)	19,615	20,071
70582	Saint Martinville	(337)	7,352	7,226
*70087	Saint Rose (c)	(504)	—	6,259
70583	Scott	(337)	5,744	4,912
70817	Shenandoah (c)		—	13,429
*71102	Shreveport	(318)	188,319	198,518
*70458	Slidell	(504)	26,123	24,124
71075	Springhill	(318)	5,514	5,668
*70663	Sulphur	(337)	21,065	20,125
*71282	Tallulah	(318)	8,911	8,526
70056	Terrytown (c)	(504)	—	23,787
*70301	Thibodaux	(504)	14,175	14,125
70053	Timberlane (c)	(504)	—	12,614
70809	Village Saint George (c)	(225)	—	6,242
70586	Ville Platte	(337)	8,722	9,037
70092	Violet (c)	(504)	—	8,574
70094	Waggaman (c)	(504)	—	9,405
*71291	West Monroe	(318)	13,901	14,096
*70094	Westwego	(504)	11,004	11,218
71483	Winnfield	(318)	6,381	6,138
71295	Winnsboro	(318)	5,536	5,755
70791	Zachary	(225)	10,353	9,036

Maine (207)

See introductory note.

ZIP	Place		1998	1990
*04210	Auburn		22,617	24,309
*04330	Augusta		19,978	21,325
*04401	Bangor		30,508	33,181
04530	Bath		9,661	9,799
04915	Belfast		6,639	6,355
03901	Berwick		6,335	5,995
*04005	Biddeford		20,851	20,710
04412	Brewer		8,462	9,021
04011	Brunswick Center (c)		—	14,683
04011	Brunswick		20,778	20,906
04093	Buxton		7,299	6,494
04843	Camden		5,037	5,060
04107	Cape Elizabeth		9,022	8,854
04736	Caribou		7,910	9,415
04021	Cumberland		6,528	5,836
04930	Dexter		5,024	4,419
03903	Eliot		5,792	5,329
04605	Ellsworth		6,283	5,975
04937	Fairfield		6,485	6,718
04105	Falmouth		8,611	7,610
04938	Farmington		7,709	7,436
04032	Freeport		7,541	6,905
04345	Gardiner		6,292	6,746
04038	Gorham		13,296	11,856
04039	Gray		6,645	5,904
04444	Hampden		6,034	5,974
04079	Harpswell		5,022	5,012
04730	Houlton Center (c)		—	5,627
04730	Houlton		5,644	6,613
04239	Jay		5,586	5,080
04043	Kennebunk		9,101	8,004
03904	Kittery Center (c)		—	5,151
03904	Kittery		9,276	9,372
*04240	Lewiston		36,186	39,757
04750	Limestone		7,932	9,922
04457	Lincoln		5,287	5,587
04250	Lisbon		9,235	9,457
04751	Loring AFB (c)		—	7,829
04462	Millinocket Center (c)		—	6,922
04462	Millinocket		6,328	6,956
04963	Oakland		5,273	5,595
04064	Old Orchard Beach Center (c)		—	7,789
04064	Old Orchard Beach		7,756	7,789
04468	Old Town		7,762	8,317
04473	Orono Center (c)		—	9,789
04473	Orono		8,395	10,573
*04101	Portland		62,786	64,157
04769	Presque Isle		8,872	10,550
04841	Rockland		7,865	7,972
04276	Rumford Compact (c)		—	5,419
04276	Rumford		6,760	7,078
04072	Saco		16,068	15,181
04073	Sanford Center (c)		—	10,296
04073	Sanford		20,995	20,463
*04074	Scarborough		14,790	12,518
04976	Skowhegan Center (c)		—	6,990
04976	Skowhegan		9,946	8,725
03908	South Berwick		6,277	5,877
*04101	South Portland		22,810	23,163
04084	Standish		8,519	7,678
04086	Topsham		9,171	8,746
04572	Waldoboro		5,464	4,601
04087	Waterboro		5,464	4,510
*04901	Waterville		16,263	17,173
04090	Wells		8,340	7,778
*04092	Westbrook		16,679	16,121
04062	Windham		14,249	13,020
04901	Winslow Center (c)		—	5,436
04901	Winslow		7,886	7,997
04364	Winthrop		5,952	5,968
04096	Yarmouth		8,134	7,862
03909	York		10,452	9,818

Maryland

Area code (240) overlays area code (301).
Area code (443) overlays area code (410). See introductory note.

ZIP	Place		1998	1990
21005	Aberdeen Proving Ground (c)	(410)	—	5,267
21001	Aberdeen	(410)	13,278	13,087
20783	Adelphi (c)	(301)	—	13,524
20762	Andrews AFB (c)	(410)	—	10,228
*21401	Annapolis	(410)	33,585	33,195
21227	Arbutus (c)	(410)	—	19,750
21012	Arnold (c)	(410)	—	20,261
20916	Aspen Hill (c)	(301)	—	45,494
21220	Ballenger Creek (c)	(410)	—	5,546
*21203	Baltimore	(410)	645,593	736,014
21050	Bel Air North (c)	(410)	—	14,880
21014	Bel Air South (c)	(410)	—	26,421
*21014	Bel Air	(410)	9,385	8,942
*20705	Beltsville (c)	(301)	—	14,476
20814	Bethesda (c)	(301)	—	62,936
20710	Bladensburg	(301)	8,674	8,064
*20715	Bowie	(301)	40,704	37,642
21220	Bowleys Quarters (c)	(410)	—	5,595
21225	Brooklyn Park (c)	(410)	—	10,987
21716	Brunswick	(301)	6,295	5,117

ZIP	Place		1998	1990
20866	Burtonsville (c)	(301)	—	5,853
20818	Cabin John (c)	(301)	—	5,341
20619	California (c)	(410)	—	7,626
20705	Calverton (c)	(301)	—	12,046
21613	Cambridge	(410)	10,734	11,514
20748	Camp Springs (c)	(301)	—	16,392
21401	Cape St. Clair (c)	(410)	—	7,878
21234	Carney (c)	(410)	—	25,578
21228	Catonsville (c)	(410)	—	35,233
20657	Chesapeake Ranch Estates (c)	(301)	—	5,423
20784	Cheverly	(301)	6,458	6,023
*20814	Chevy Chase (c)	(301)	—	8,559
20783	Chillum (c)	(301)	—	31,309
20735	Clinton (c)	(301)	—	19,987
20904	Cloverly (c)	(301)	—	7,904
21030	Cockeysville (c)	(410)	—	18,668
20914	Colesville (c)	(301)	—	18,819
*20740	College Park	(301)	25,855	23,714
*21045	Columbia (c)	(410)/(301)	—	75,883
20743	Coral Hills (c)	(410)	—	11,032
21114	Crofton (c)	(410)	—	12,781
*21502	Cumberland	(301)	21,521	23,712
20872	Damascus (c)	(301)	—	9,817
*20747	District Heights	(301)	7,240	6,711
21222	Dundalk (c)	(410)	—	65,800
20737	East Riverdale (c)	(301)	—	14,187
21601	Easton	(410)	10,713	9,372
21219	Edgemere (c)	(410)	—	9,226
21040	Edgewood (c)	(410)	—	23,903
21784	Eldersburg (c)	(410)	—	9,720
21227	Elkridge (c)	(410)	—	12,953
*21921	Elkton	(410)	10,870	9,073
*21043	Ellicott City (c)	(410)	—	41,396
21221	Essex (c)	(410)	—	40,872
20904	Fairland (c)	(301)	—	19,828
21047	Fallston (c)	(410)	—	5,730
21061	Ferndale (c)	(410)	—	16,355
20747	Forestville (c)	(301)	—	16,731
20755	Fort Meade (c)	(301)	—	12,509
*20744	Fort Washington (c)	(301)	—	24,032
*21701	Frederick	(301)	47,468	40,186
20744	Friendly (c)	(301)	—	9,028
21532	Frostburg	(301)	7,623	8,069
*20877	Gaithersburg	(301)	46,980	39,676
21055	Garrison (c)	(410)	—	5,045
*20874	Germantown (c)	(301)	—	41,145
*21061	Glen Burnie (c)	(410)	—	37,305
20706	Glenarden	(301)	5,429	5,025
20769	Glenn Dale (c)	(301)	—	9,689
20772	Greater Upper Marlboro (c)		—	11,528
21122	Green Haven (c)	(410)	—	14,416
21771	Green Valley (c)	(301)	—	9,424
*20770	Greenbelt	(301)	22,076	20,561
*21740	Hagerstown	(301)	34,105	35,306
21740	Halfway (c)	(301)	—	8,873
21078	Havre de Grace	(410)	10,482	8,952
20903	Hillandale (c)	(301)	—	10,318
20748	Hillcrest Heights (c)	(301)	—	17,136
*20780	Hyattsville	(301)	14,812	13,864
20794	Jessup (c)	(410)	—	6,537
21085	Joppatowne (c)	(410)	—	11,084
20785	Kentland (c)	(301)	—	7,967
20772	Kettering (c)	(301)	—	9,901
20646	La Plata	(301)	6,663	5,841
21122	Lake Shore (c)	(410)	—	13,269
20785	Landover (c)	(301)	—	5,052
20787	Langley Park (c)	(301)	—	17,474
20706	Lanham-Seabrook (c)	(301)	—	16,792
21227	Lansdowne-Baltimore Highlands (c)		—	15,509
20772	Largo (c)	(301)	—	9,475
*20707	Laurel	(301)	18,825	19,086
20653	Lexington Park (c)	(410)	—	9,943
21090	Linthicum (c)	(410)	—	7,547
21207	Lochearn (c)	(410)	—	25,240
21037	Londontowne (c)	(410)	—	6,992
21784	Long Meadow (c)	(410)	—	5,594
*21093	Lutherville-Timonium (c)	(410)	—	16,442
02738	Marion	(508)	5,197	4,496
20748	Marlow Heights (c)	(301)	—	5,885
20772	Marlton (c)	(301)	—	5,523
20707	Maryland City (c)	(301)	—	6,813
21093	Mays Chapel (c)	(410)	—	10,132
21220	Middle River (c)	(410)	—	24,616
21207	Milford Mill (c)	(410)	—	22,547
20717	Mitchellville (c)	(301)	—	12,593
20886	Montgomery Village (c)	(301)	—	32,315
21771	Mount Airy	(301)/(410)	5,380	3,730
20712	Mount Rainier	(301)	8,483	7,954
21402	Naval Academy (c)	(410)	—	5,420
20784	New Carrollton	(301)	12,978	12,002
20815	North Bethesda (c)	(301)	—	29,656
20895	North Kensington (c)	(301)	—	8,607
20707	North Laurel (c)	(301)	—	15,008
20878	North Potomac (c)	(301)	—	18,456
*21842	Ocean City	(410)	5,095	5,146
21113	Odenton (c)	(410)	—	12,833
*20832	Olney (c)	(301)	—	23,019
21206	Overlea (c)	(410)	—	12,137
21117	Owings Mills (c)	(410)	—	9,474
*20750	Oxon Hill-Glassmanor (c)	(301)	—	35,794
20785	Palmer Park (c)	(301)	—	7,019
21234	Parkville (c)	(410)	—	31,617
21401	Parole (c)	(410)	—	10,054
*21122	Pasadena (c)	(410)	—	10,012
21128	Perry Hall (c)	(410)	—	22,723
21282	Pikesville (c)	(410)	—	24,815
*20850	Potomac (c)	(301)	—	45,634
21227	Pumphrey (c)	(410)	—	5,483
21133	Randallstown (c)	(301)	—	26,277
.....	Redland (c)	(301)	—	16,145
21136	Reisterstown (c)	(410)	—	19,314
*20737	Riverdale	(301)	5,164	4,843
21122	Riviera Beach (c)	(410)	—	11,376
*20850	Rockville	(301)	46,788	44,830
20772	Rosaryville (c)	(301)	—	8,976
21237	Rosedale (c)	(410)	—	18,703
.....	Rossmoor (c)		—	6,182
21221	Rossville (c)	(410)	—	9,492
20602	Saint Charles (c)	(301)	—	28,717
*21801	Salisbury	(410)	20,884	20,592
20763	Savage-Guilford (c)	(410)	—	9,669
20743	Seat Pleasant	(301)	5,754	5,359
21144	Severn (c)	(410)	—	24,499
21146	Severna Park (c)	(410)	—	25,879
*20907	Silver Spring (c)	(301)	—	76,046
21061	South Gate (c)	(410)	—	27,564
20895	South Kensington (c)	(301)	—	8,777
20707	South Laurel (c)	(301)	—	18,591
*20752	Suitland-Silver Hills (c)	(301)	—	35,111
*20913	Takoma Park	(301)	18,238	16,724
*20748	Temple Hills (c)	(301)	—	6,865
*21202	Towson (c)	(410)	—	49,445
*20602	Waldorf (c)	(301)	—	15,058
20743	Walker Mill (c)	(301)	—	10,920
21793	Walkersville	(301)	5,235	4,145
*21157	Westminster	(410)	15,776	13,060
20902	Wheaton-Glenmont (c)	(301)	—	53,720
21162	White Marsh (c)	(410)	—	8,183
20903	White Oak (c)	(301)	—	18,671
21207	Woodlawn (c) (Baltimore)	(410)	—	32,907
21284	Woodlawn (c) (Prince George's)	(410)	—	5,329

Massachusetts

See introductory note.

ZIP	Place		1998	1990
02351	*Abington*	(781)	14,876	13,817
01720	*Acton*	(978)	19,206	17,872
02743	*Acushnet*	(508)	10,111	9,554
01220	*Adams*	(413)	8,768	9,445
01220	Adams Center (c)	(413)	—	6,356
01001	*Agawam*	(413)	26,738	27,323
01913	*Amesbury*	(978)	16,076	14,997
01913	Amesbury Center (c)	(978)	—	12,109
*01002	*Amherst*	(413)	35,252	35,228
*01002	Amherst Center (c)	(413)	—	17,824
01810	*Andover*	(978)	—	8,242
01810	Andover	(978)	31,424	29,151
*02205	*Arlington*	(781)	43,431	44,630
01430	*Ashburnham*	(978)	5,577	5,433
01721	*Ashland*	(508)	13,482	12,066
01331	*Athol*	(978)	11,161	11,451
01331	Athol Center (c)	(978)	—	8,732
02703	*Attleboro*	(508)	39,557	38,383
01501	*Auburn*	(508)	15,580	15,005
01432	*Ayer*	(978)	7,515	6,871
02630	Barnstable	(508)	45,187	40,949
01730	*Bedford*	(781)	13,947	12,996
01007	*Belchertown*	(413)	11,946	10,579
02019	*Bellingham*	(508)	15,864	14,877
02478	*Belmont*	(781)	23,907	24,720
02779	*Berkley*	(508)	5,395	4,237
01915	*Beverly*	(978)	39,037	38,195
*01821	*Billerica*	(978)	39,554	37,609
01504	*Blackstone*	(508)	8,386	8,023
*02205	Boston	(617)	555,447	574,283
02532	*Bourne*	(508)	18,007	16,064
01921	*Boxford*	(978)	9,041	6,266
*02205	*Braintree*	(781)	34,906	33,836
02631	*Brewster*	(508)	9,637	8,440
02324	*Bridgewater*	(508)	24,536	21,249
*02303	Brockton	(508)	93,173	92,788
*02205	*Brookline*	(617)	53,911	54,718
01803	*Burlington*	(781)	23,694	23,302
*02139	Cambridge	(617)	93,352	95,802
02021	*Canton*	(781)	20,677	18,530
02330	*Carver*	(508)	11,647	10,590
*02632	Centerville (c)	(508)	—	9,190
01507	*Charlton*	(508)	10,345	9,556
02633	*Chatham*	(508)	7,098	6,579
01824	*Chelmsford*	(978)	33,776	32,383
02150	*Chelsea*	(617)	27,426	28,710
*01020	Chicopee	(413)	54,049	56,632
01510	*Clinton*	(978)	13,053	13,222
01778	*Cochituate* (c)	(508)	—	6,046
02025	*Cohasset*	(781)	7,094	7,075
01742	*Concord*	(978)	17,867	17,076
*01226	*Dalton*	(413)	6,854	7,155

ZIP	Place		1998	1990
01923	Danvers	(978)	25,188	24,174
02714	Dartmouth	(508)	28,503	27,244
*02026	Dedham	(781)	23,721	23,782
02638	Dennis	(508)	14,693	13,864
02715	Dighton	(508)	5,937	5,631
01516	Douglas	(508)	6,634	5,438
02030	Dover.	(508)	5,481	4,915
01826	Dracut	(978)	28,136	25,594
01571	Dudley	(508)	9,802	9,540
*02332	Duxbury	(781)	15,353	13,895
02333	East Bridgewater	(508)	12,584	11,104
02536	East Falmouth (c)	(508)	—	5,577
01028	East Longmeadow	(413)	13,960	13,367
02642	Eastham	(508)	5,033	4,462
01027	Easthampton	(413)	15,627	15,537
02334	Easton	(508)	21,311	19,807
02149	Everett	(617)	34,922	35,701
02719	Fairhaven	(508)	15,937	16,132
*02722	Fall River	(508)	90,654	92,703
*02540	Falmouth	(508)	31,431	27,960
01420	Fitchburg	(978)	40,011	41,194
01433	Fort Devens (c)	(978)	—	8,973
02035	Foxborough	(508)	16,388	14,637
*01701	Framingham	(508)	64,646	64,989
02038	Franklin Center (c).	(508)	—	9,965
02038	Franklin	(508)	28,353	22,095
02702	Freetown	(508)	8,834	8,522
01440	Gardner	(978)	20,261	20,125
01833	Georgetown	(978)	7,384	6,384
*01930	Gloucester	(978)	29,657	28,716
01519	Grafton	(508)	13,742	13,035
01033	Granby	(413)	5,865	5,565
01230	Great Barrington	(413)	7,592	7,725
01301	Greenfield	(413)	18,267	18,666
01301	Greenfield Center (c).	(413)	—	14,016
01450	Groton	(978)	9,205	7,511
01834	Groveland	(978)	5,841	5,214
02338	Halifax	(781)	7,163	6,526
01936	Hamilton	(978)	7,545	7,280
02339	Hanover	(781)	13,278	11,912
02341	Hanson	(781)	9,742	9,028
01451	Harvard	(978)	12,399	12,329
02645	Harwich	(508)	11,765	10,275
*01830	Haverhill.	(978)	55,321	51,418
02043	Hingham	(781)	20,439	19,821
02343	Holbrook	(781)	11,125	11,041
01520	Holden	(508)	15,182	14,628
01746	Holliston	(508)	13,576	12,926
*01040	Holyoke	(413)	40,964	43,704
01747	Hopedale	(508)	5,654	5,666
01748	Hopkinton	(508)	11,351	9,191
01749	Hudson	(978)	17,803	17,233
01749	Hudson Center (c).	(978)	—	14,267
02045	Hull	(781)	10,528	10,466
02601	Hyannis (c)	(508)	—	14,120
01938	Ipswich	(978)	12,656	11,873
02364	Kingston	(781)	10,983	9,045
02347	Lakeville	(508)	8,900	7,785
01523	Lancaster	(978)	6,685	6,661
*01842	Lawrence	(978)	69,420	70,207
01238	Lee	(413)	5,657	5,849
01524	Leicester	(508)	10,442	10,191
01240	Lenox	(413)	5,180	5,069
01453	Leominster.	(978)	40,208	38,145
*02205	Lexington	(781)	29,594	28,974
01773	Lincoln	(781)	7,921	7,666
01460	Littleton	(978)	7,936	7,051
*01028	Longmeadow	(413)	14,710	15,467
*01853	Lowell	(978)	101,075	103,439
01056	Ludlow	(413)	18,957	18,820
01462	Lunenburg	(978)	9,473	9,117
*01901	Lynn.	(781)	81,075	81,245
01940	Lynnfield	(781)	11,359	11,049
02148	Malden.	(781)	52,644	53,884
01944	Manchester-by-the-Sea.	(978)	5,465	5,286
02048	Mansfield	(508)	19,244	16,568
01945	Marblehead	(781)	20,103	19,971
02738	Marion.	(508)	5,197	4,496
01752	Marlborough	(508)	33,278	31,813
02050	Marshfield	(781)	23,538	21,531
02648	Marstons Mills (c)	(508)	—	8,017
02649	Mashpee	(508)	9,343	7,884
02739	Mattapoisett	(508)	6,333	5,850
01754	Maynard	(978)	10,462	10,325
02052	Medfield	(508)	11,726	10,531
*02155	Medford	(781)	55,981	57,407
02053	Medway	(508)	11,738	9,931
02176	Melrose	(781)	27,376	28,150
01860	Merrimac	(978)	5,966	5,166
01844	Methuen	(978)	41,988	39,990
02346	Middleborough	(508)	19,702	17,867
02346	Middleborough Center (c)	(508)	—	6,837
01949	Middleton	(978)	6,040	4,921
01757	Milford	(508)	25,586	25,355
01757	Milford Center (c).	(508)	—	23,339
01527	Millbury	(508)	12,382	12,228
02054	Millis	(508)	8,110	7,613
02186	Milton	(617)	25,662	25,725
01057	Monson	(413)	8,002	7,776
01351	Montague	(413)	8,293	8,316

ZIP	Place		1998	1990
*02584	Nantucket	(508)	7,844	6,012
*01760	Natick	(508)	31,491	30,510
*02205	Needham	(781)	27,924	27,557
*02740	New Bedford	(508)	96,353	99,922
01951	Newbury	(978)	6,168	5,623
01950	Newburyport	(978)	16,808	16,317
*02205	Newton	(617)	80,345	82,585
02056	Norfolk	(508)	10,553	9,259
01247	North Adams	(413)	15,496	16,797
01059	North Amherst (c)	(413)	—	6,239
01845	North Andover	(978)	28,680	29,289
*02760	North Attleborough	(508)	25,065	22,792
01864	North Reading	(978)	25,908	25,038
*01060	Northampton	(413)	13,258	11,929
01532	Northborough	(508)	14,036	13,371
01534	Northbridge	(508)	13,219	12,002
02766	Norton	(508)	16,097	14,265
02061	Norwell	(781)	9,925	9,279
02062	Norwood	(781)	28,824	28,700
01364	Orange	(978)	7,454	7,312
02653	Orleans	(508)	6,362	5,838
01540	Oxford	(508)	13,318	12,588
01540	Oxford Center (c)	(508)	—	5,969
01069	Palmer	(413)	11,858	12,054
*01960	Peabody	(978)	49,204	47,264
02359	Pembroke	(781)	16,621	14,544
01463	Pepperell	(978)	10,964	10,098
01866	Pinehurst (c)	(978)	—	6,614
*01201	Pittsfield.	(413)	45,513	48,622
02762	Plainville	(508)	7,354	6,871
*02360	Plymouth	(508)	49,810	45,608
*02360	Plymouth Center (c)	(508)	—	7,258
*02205	Quincy.	(617)	85,752	84,985
02368	Randolph	(781)	30,567	30,093
02767	Raynham	(508)	10,789	9,867
01867	Reading	(781)	23,371	22,539
02769	Rehoboth	(508)	9,601	8,656
02151	Revere.	(781)	41,663	42,786
02370	Rockland	(781)	17,730	16,123
01966	Rockport	(978)	7,644	7,482
01969	Rowley	(978)	5,343	4,452
01543	Rutland	(508)	5,459	4,936
*01970	Salem	(978)	38,351	38,091
01952	Salisbury	(978)	7,238	6,882
02563	Sandwich	(508)	18,746	15,489
01906	Saugus	(781)	26,576	25,549
02066	Scituate	(781)	17,577	16,786
02771	Seekonk	(508)	13,339	13,046
02067	Sharon	(781)	16,942	15,517
01464	Shirley	(978)	7,674	6,118
01545	Shrewsbury	(508)	27,791	24,146
*02722	Somerset	(508)	17,710	17,655
*02205	Somerville	(617)	74,100	76,210
01002	South Amherst (c)	(413)	—	5,053
01075	South Hadley	(413)	17,097	16,685
02664	South Yarmouth (c)	(508)	—	10,358
01772	Southborough	(508)	7,798	6,628
01550	Southbridge	(508)	17,460	17,816
01550	Southbridge Center (c)	(508)	—	13,631
01077	Southwick	(413)	8,311	7,667
01562	Spencer	(508)	12,432	11,645
01562	Spencer Center (c)	(508)	—	6,306
*01101	Springfield.	(413)	148,144	156,983
01564	Sterling	(978)	7,154	6,481
02180	Stoneham	(781)	22,254	22,203
02072	Stoughton	(781)	27,664	26,777
01775	Stow	(978)	5,842	5,328
01566	Sturbridge	(508)	8,057	7,775
01776	Sudbury	(978)	15,550	14,358
01590	Sutton	(508)	7,597	6,824
01907	Swampscott	(781)	13,868	13,650
02777	Swansea	(508)	15,554	15,411
02780	Taunton	(508)	52,553	49,832
01468	Templeton	(978)	7,116	6,438
01876	Tewksbury	(978)	29,070	27,266
01983	Topsfield	(978)	6,257	5,754
01469	Townsend	(978)	9,169	8,496
01879	Tyngsborough	(978)	10,296	8,642
01568	Upton	(508)	5,524	4,677
01569	Uxbridge	(508)	11,321	10,415
01880	Wakefield	(781)	24,772	24,825
02081	Walpole	(508)	22,640	20,223
*02205	Waltham	(781)	58,540	57,878
01082	Ware	(413)	9,727	9,808
01082	Ware Center (c).	(413)	—	6,533
02571	Wareham	(508)	19,756	19,232
*02205	Watertown.	(781)	32,435	33,284
01778	Wayland	(508)	12,343	11,874
01570	Webster	(508)	16,115	16,196
01570	Webster Center (c)	(508)	—	11,849
*02205	Wellesley	(781)	26,789	26,615
01583	West Boylston	(508)	6,726	6,611
02379	West Bridgewater	(508)	6,742	6,389
01742	West Concord (c)	(978)	—	5,761
*01089	West Springfield	(413)	25,900	27,537
02673	West Yarmouth (c).	(508)	—	5,409
01581	Westborough	(508)	15,428	14,133
*01085	Westfield.	(413)	37,570	38,372
01886	Westford	(978)	19,559	16,392
01473	Westminster	(978)	6,707	6,191

ZIP	Place	1998	1990
02493	Weston .... (781)	10,651	10,200
02790	Westport .... (508)	14,156	13,852
02090	Westwood .... (781)	13,160	12,557
*02205	Weymouth .... (781)	54,903	54,063
01588	Whitinsville (c) .... (508)	—	5,639
02382	Whitman .... (781)	14,229	13,240
01095	Wilbraham .... (413)	12,419	12,635
01267	Williamstown .... (413)	7,948	8,220
01887	Wilmington .... (978)	20,593	17,651
01475	Winchendon .... (978)	9,176	8,805
01890	Winchester .... (781)	20,339	20,267
02152	Winthrop .... (617)	17,179	18,127
*01801	Woburn .... (781)	37,070	35,943
*01613	Worcester .... (508)	166,535	169,759
02093	Wrentham .... (508)	10,259	9,006
02675	Yarmouth .... (508)	22,797	21,174

Michigan

Area code (586) overlays area code (810). See introductory note.

ZIP	Place	1998	1990
49221	Adrian .... (517)	22,086	22,097
49224	Albion .... (517)	9,765	10,066
49401	Allendale (c) .... (616)	—	6,950
48101	Allen Park .... (313)	31,764	31,092
48801	Alma .... (517)	9,230	9,034
49707	Alpena .... (517)	11,581	11,354
*48106	Ann Arbor .... (734)	109,967	109,608
*48321	Auburn Hills .... (248)	19,310	17,076
*49016	Battle Creek .... (616)	53,496	53,516
*48707	Bay City .... (517)	35,485	38,936
48505	Beecher (c) .... (517)	—	14,465
48809	Belding .... (616)	6,129	5,969
*49022	Benton Harbor .... (616)	11,885	12,818
49022	Benton Heights (c) ...... (616)	—	5,465
48072	Berkley .... (248)	16,620	16,960
48025	Beverly Hills .... (248)	10,289	10,610
49307	Big Rapids .... (231)	10,610	12,603
*48012	Birmingham .... (248)	19,991	19,997
48301	Bloomfield (c) .... (810)	—	42,137
48722	Bridgeport (c) .... (517)	—	8,569
*48116	Brighton .... (810)	6,458	5,686
48601	Buena Vista (c) .... (517)	—	8,196
*48501	Burton .... (810)	27,230	27,437
49601	Cadillac .... (231)	10,439	10,104
*48185	Canton (c) .... (734)	—	57,047
48724	Carrollton (c) .... (517)	—	6,521
48015	Center Line .... (810)	8,795	9,026
48813	Charlotte .... (517)	7,841	8,083
49721	Cheboygan .... (231)	5,440	4,997
48017	Clawson .... (248)	13,648	13,874
*48046	Clinton (c) .... (517)	—	85,866
49036	Coldwater .... (517)	9,564	9,607
49321	Comstock Park (c) ...... (616)	—	6,530
49508	Cutlerville (c) .... (616)	—	11,228
48423	Davison .... (810)	5,525	5,693
*48120	Dearborn .... (313)	91,691	89,286
*48127	Dearborn Heights .... (313)	59,805	60,838
*48231	Detroit .... (313)	970,196	1,027,974
49047	Dowagiac .... (616)	5,913	6,418
49506	East Grand Rapids ...... (616)	10,318	10,807
*48826	East Lansing .... (517)	46,509	50,677
48021	Eastpointe .... (810)	34,145	35,283
49001	Eastwood (c) .... (616)	—	6,340
48229	Ecorse .... (313)	11,913	12,180
49829	Escanaba .... (906)	13,280	13,659
49022	Fair Plain (c) .... (616)	—	8,051
*48333	Farmington .... (248)	9,945	10,170
48333	Farmington Hills .... (248)	79,784	74,614
48430	Fenton .... (810)	10,072	8,434
48220	Ferndale .... (248)	24,458	25,084
48134	Flat Rock .... (734)	8,978	7,290
*48501	Flint .... (810)	131,668	140,925
48433	Flushing .... (810)	8,344	8,542
49506	Forest Hills (c) .... (616)	—	16,690
48026	Fraser .... (810)	15,490	13,899
*48135	Garden City .... (734)	32,750	31,846
48439	Grand Blanc .... (810)	8,111	7,760
49417	Grand Haven .... (616)	11,982	11,951
48837	Grand Ledge .... (517)	7,751	7,562
*49501	Grand Rapids .... (616)	185,437	189,126
*49418	Grandville .... (616)	16,483	15,624
48838	Greenville .... (616)	8,452	8,101
48138	Grosse Ile (c) .... (734)	—	9,781
*48231	Grosse Pointe .... (313)	5,770	5,681
48230	Grosse Pointe Farms ..... (313)	10,372	10,092
48230	Grosse Pointe Park ...... (313)	13,037	12,857
48230	Grosse Pointe Woods ..... (313)	18,021	17,715
48212	Hamtramck .... (313)	18,041	18,372
48225	Harper Woods .... (313)	15,094	14,903
48625	Harrison (c) .... (517)	—	24,685
48840	Haslett (c) .... (517)	—	10,230
49058	Hastings .... (616)	6,260	6,549
48030	Hazel Park .... (248)	19,683	20,051
48203	Highland Park .... (313)	19,293	20,121
49242	Hillsdale .... (517)	8,210	8,175
*49423	Holland .... (616)	33,249	30,745
48442	Holly .... (248)	6,385	5,595
48842	Holt (c) .... (517)	—	11,744
49931	Houghton .... (906)	7,215	7,498

ZIP	Place	1998	1990
*48844	Howell .... (517)	9,180	8,147
49426	Hudsonville .... (616)	6,893	6,170
48070	Huntington Woods .... (248)	6,251	6,419
48141	Inkster .... (313)/(734)	31,160	30,772
48846	Ionia .... (616)	10,848	10,349
49801	Iron Mountain .... (906)	8,644	8,525
49938	Ironwood .... (906)	6,300	6,849
49849	Ishpeming .... (906)	6,129	7,200
*49204	Jackson .... (517)	35,183	37,425
*49428	Jenison (c) .... (616)	—	17,882
*49001	Kalamazoo .... (616)	76,241	80,277
49518	Kentwood .... (616)	42,316	37,826
49802	Kingsford .... (906)	5,079	5,480
49843	K.I. Sawyer AFB (c). ..... (906)	—	6,577
48144	Lambertville (c) .... (734)	—	7,860
*48901	Lansing .... (517)	127,825	127,321
48446	Lapeer .... (810)	8,029	7,759
48146	Lincoln Park .... (313)	42,283	41,832
*48150	Livonia .... (734)	101,358	100,850
49431	Ludington .... (231)	8,904	8,507
48071	Madison Heights .... (248)	31,854	32,196
49660	Manistee .... (231)	6,306	6,734
49855	Marquette .... (906)	19,147	21,977
49068	Marshall............... (616)	7,328	6,941
48040	Marysville .... (810)	9,383	8,515
48854	Mason .... (517)	7,233	6,768
48122	Melvindale .... (313)	11,251	11,216
49858	Menominee .... (906)	8,689	9,398
*48640	Midland .... (517)	39,956	38,053
*48381	Milford .... (248)	6,425	5,500
*48161	Monroe .... (734)	21,981	22,902
*48046	Mount Clemens .... (810)	17,723	18,405
*48804	Mount Pleasant .... (517)	23,351	23,299
*49440	Muskegon .... (231)	39,017	39,809
49444	Muskegon Heights .... (231)	12,395	13,176
*48047	New Baltimore .... (810)	6,582	5,798
49120	Niles .... (616)	11,899	12,458
49505	Northview (c) .... (616)	—	13,712
48167	Northville .... (248)	6,447	6,226
49441	Norton Shores .... (231)	22,919	21,755
*48376	Novi .... (248)	44,760	32,998
48237	Oak Park .... (248)	29,595	30,468
*48805	Okemos (c) .... (517)	—	20,216
48867	Owosso .... (517)	15,617	16,322
49770	Petoskey .... (231)	7,663	6,056
48170	Plymouth .... (734)	9,731	9,560
48170	Plymouth Township (c) ... (734)	—	23,646
*48343	Pontiac .... (248)	68,916	71,136
*49081	Portage .... (616)	43,707	41,042
*48061	Port Huron .... (810)	32,256	33,694
*48231	Redford (c) .... (313)	—	54,387
48218	River Rouge .... (313)	10,835	11,314
48192	Riverview .... (734)	14,797	13,894
*48308	Rochester .... (248)	8,207	7,130
48306	Rochester Hills .... (248)	67,413	61,766
48174	Romulus .... (313)/(734)	25,326	22,897
48066	Roseville .... (810)	51,390	51,412
*48068	Royal Oak .... (248)	64,290	65,410
*48605	Saginaw .... (517)	63,464	69,512
48604	Saginaw Township North (c) .. (517)	—	23,018
48603	Saginaw Township South (c) .. (517)	—	13,987
48079	Saint Clair .... (810)	5,581	5,116
*48080	Saint Clair Shores .... (313)	66,056	68,107
48879	Saint Johns .... (517)	7,414	7,392
48085	Saint Joseph .... (616)	9,264	9,214
48176	Saline .... (734)	7,441	6,663
49783	Sault Sainte Marie .... (906)	15,385	14,689
49455	Shelby (c) .... (231)	—	48,655
48609	Shields (c) .... (517)	—	6,634
*48037	Southfield .... (248)	75,104	75,727
48195	Southgate .... (734)	32,375	30,771
49090	South Haven .... (616)	5,296	5,563
48178	South Lyon .... (248)	9,135	6,479
48161	South Monroe (c) .... (734)	—	5,266
49015	Springfield .... (248)	5,638	5,582
*48311	Sterling Heights .... (810)	124,339	117,810
49091	Sturgis .... (616)	10,297	10,130
48180	Taylor. .... (313)/(734)	72,551	70,811
49286	Tecumseh .... (517)	8,350	7,462
48182	Temperance (c) .... (734)	—	6,542
49093	Three Rivers .... (616)	7,229	7,464
*49684	Traverse City .... (231)	15,158	15,155
48183	Trenton .... (734)	21,708	20,586
*48099	Troy .... (248)	79,303	72,884
49504	Walker................ (616)	19,813	17,279
48390	Walled Lake. .... (248)	7,085	6,278
*48090	Warren .... (810)	142,455	144,864
*48329	Waterford (c) .... (248)	—	66,692
48917	Waverly (c) .... (517)	—	15,614
48184	Wayne .... (734)	20,880	19,899
*48325	West Bloomfield (c) ...... (248)	—	54,843
*48185	Westland .... (313)/(734)	86,227	84,724
49019	Westwood (c) .... (616)	—	8,957
48393	Wixom .... (248)	10,981	8,550
48183	Woodhaven .... (734)	13,736	11,631
48753	Wurtsmith AFB (c). ...... (517)	—	5,080
48192	Wyandotte .... (734)	31,884	30,938
49509	Wyoming .... (616)	68,671	63,891
*48197	Ypsilanti............... (734)	22,923	24,846
49464	Zeeland .... (616)	6,066	5,417

Minnesota

ZIP	Place		1998	1990
56007	Albert Lea	(507)	17,593	18,310
56308	Alexandria	(320)	8,351	8,029
55304	Andover	(612)	23,918	15,216
*55303	Anoka	(612)	17,996	17,192
55124	Apple Valley	(612)	45,428	34,598
55112	Arden Hills	(651)	9,915	9,199
55912	Austin	(507)	21,482	21,926
56425	Baxter	(218)	5,414	3,695
*56601	Bemidji	(218)	12,591	11,165
55014	Blaine	(612)/(651)	44,960	38,975
*55420	Bloomington	(612)	86,186	86,335
56401	Brainerd	(218)	13,323	12,353
55429	Brooklyn Center	(612)	27,851	28,887
55443	Brooklyn Park	(612)	63,115	56,381
55313	Buffalo	(612)	9,090	7,302
*55337	Burnsville	(612)/(651)	59,334	51,288
55008	Cambridge	(612)	5,448	5,094
55316	Champlin	(612)	21,116	16,849
55317	Chanhassen	(612)	18,185	11,736
55318	Chaska	(612)	15,348	11,339
55720	Cloquet	(218)	10,868	10,885
55421	Columbia Heights	(612)	18,285	18,910
55433	Coon Rapids	(612)	63,674	52,978
55340	Corcoran	(612)	5,816	5,199
55016	Cottage Grove	(651)	31,250	22,935
56716	Crookston	(218)	7,534	8,119
55428	Crystal	(612)	23,040	23,788
55327	Dayton	(612)	5,097	4,443
*56501	Detroit Lakes	(218)	7,286	7,141
*55806	Duluth	(218)	81,228	85,493
*55121	Eagan	(612)/(651)	60,042	47,409
55005	East Bethel	(612)	10,302	8,050
56721	East Grand Forks	(218)	8,574	8,658
*55344	Eden Prairie	(612)	50,279	39,311
55424	Edina	(612)	45,894	46,075
55330	Elk River	(612)	16,129	11,143
56031	Fairmont	(507)	10,862	11,265
55113	Falcon Heights	(651)	5,258	5,380
55021	Faribault	(507)	18,645	17,085
55024	Farmington	(612)/(651)	10,166	5,940
*56537	Fergus Falls	(218)	13,706	12,362
55025	Forest Lake	(651)	6,634	5,833
55432	Fridley	(612)	27,974	28,335
55427	Golden Valley	(612)	20,349	20,971
*55744	Grand Rapids	(218)	8,166	7,976
*55304	Ham Lake	(612)	11,793	8,924
55033	Hastings	(651)	17,454	15,478
55810	Hermantown	(218)	7,809	6,761
*55746	Hibbing	(218)	17,383	18,046
*55343	Hopkins	(612)	16,279	16,529
55038	Hugo	(651)	5,959	4,417
55350	Hutchinson	(320)	12,521	11,459
56649	International Falls	(218)	7,836	8,325
*55349	Inver Grove Heights	(651)	29,292	22,477
55042	Lake Elmo	(651)	6,825	5,900
55044	Lakeville	(612)	39,166	24,854
55014	Lino Lakes	(651)	14,469	8,807
55355	Litchfield	(320)	6,174	6,041
55117	Little Canada	(651)	9,585	8,971
56345	Little Falls	(320)	7,519	7,371
55115	Mahtomedi	(651)	7,278	5,633
*56001	Mankato	(507)	30,780	31,459
55311	Maple Grove	(612)	46,932	38,736
55109	Maplewood	(651)	34,970	30,954
56258	Marshall	(507)	12,117	12,023
55118	Mendota Heights	(651)	11,529	9,388
*55440	Minneapolis	(612)	351,731	368,383
55345	Minnetonka	(612)	50,952	48,370
56265	Montevideo	(320)	5,290	5,499
*55362	Monticello	(612)	7,154	5,045
*56560	Moorhead	(218)	33,082	32,295
56267	Morris	(320)	5,453	5,613
55364	Mound	(612)	9,717	9,634
55112	Mounds View	(651)	12,874	12,541
55112	New Brighton	(651)	22,845	22,207
54427	New Hope	(612)	21,204	21,853
56073	New Ulm	(507)	13,491	13,132
55056	North Branch	(612)/(651)	7,732	4,267
56001	North Mankato	(507)	16,174	14,684
55109	North Saint Paul	(651)	11,595	10,662
55057	Northfield	(507)	12,570	12,376
.....	Oak Grove	(612)	26,663	18,377
55128	Oakdale	(651)	6,767	5,488
55323	Orono	(612)	7,547	7,285
.....	Otsego	(612)	6,231	5,219
55060	Owatonna	(507)	20,599	19,386
*55446	Plymouth	(612)	61,509	50,889
55372	Prior Lake	(612)	14,864	11,482
55303	Ramsey	(612)	18,226	12,408
55066	Red Wing	(651)	15,843	15,134
55423	Richfield	(612)	34,040	35,710
55422	Robbinsdale	(612)	13,993	14,396
*55901	Rochester	(507)	78,173	70,729
55068	Rosemount	(612)/(651)	13,249	8,622
55113	Roseville	(651)	34,465	33,485
55418	Saint Anthony	(651)	7,804	7,727
*56301	Saint Cloud	(320)	50,745	48,812
55426	Saint Louis Park	(612)	42,387	43,787

ZIP	Place		1998	1990
55071	Saint Paul Park	(651)	257,284	272,235
*55101	Saint Paul	(651)	5,111	4,965
56082	Saint Peter	(507)	9,688	9,481
56377	Sartell	(320)	8,363	5,409
56379	Sauk Rapids	(320)	9,370	7,823
56378	Savage	(612)	17,151	9,906
55379	Shakopee	(612)	16,553	11,739
55126	Shoreview	(651)	26,157	24,587
55331	Shorewood	(612)	7,040	5,913
55075	South Saint Paul	(651)	19,827	20,197
55432	Spring Lake Park	(612)	7,319	6,532
*55082	Stillwater	(651)	15,801	13,882
56701	Thief River Falls	(218)	8,273	8,010
55127	Vadnais Heights	(651)	13,366	11,041
*55792	Virginia	(218)	8,839	9,432
55387	Waconia	(612)	5,246	3,498
56387	Waite Park	(320)	5,897	5,020
56093	Waseca	(507)	8,328	8,385
55118	West Saint Paul	(651)	19,228	19,248
*55110	White Bear Lake	(651)	25,999	24,622
56201	Willmar	(320)	18,805	17,531
55987	Winona	(507)	24,187	25,435
55125	Woodbury	(651)	40,431	20,075
56187	Worthington	(507)	9,977	9,977

Mississippi

ZIP	Place		1998	1990
39730	Aberdeen	(662)	6,915	6,837
38821	Amory	(662)	7,144	7,093
38606	Batesville	(662)	7,416	6,403
*39520	Bay Saint Louis	(228)	9,841	8,063
*39530	Biloxi	(228)	47,316	46,319
38829	Booneville	(662)	8,387	7,955
*39042	Brandon	(601)	14,612	11,089
*39601	Brookhaven	(601)	10,649	10,243
39046	Canton	(601)	12,221	11,723
38614	Clarksdale	(662)	20,461	21,180
*38732	Cleveland	(662)	14,834	15,384
*39056	Clinton	(601)	22,067	21,847
39429	Columbia	(601)	6,935	6,815
*39701	Columbus	(662)	22,297	23,799
*38834	Corinth	(662)	12,204	11,820
39059	Crystal Springs	(601)	5,832	5,643
39532	D'Iberville	(228)	8,211	6,566
39074	Forest	(601)	5,324	5,062
39553	Gautier	(228)	11,139	10,088
*38701	Greenville	(662)	42,042	45,226
*38930	Greenwood	(662)	18,218	18,906
*38901	Grenada	(662)	11,161	10,864
39564	Gulf Hills (c)	(228)	—	5,004
*39501	Gulfport	(228)	64,762	64,045
*39401	Hattiesburg	(601)	48,806	45,325
*38635	Holly Springs	(662)	7,195	7,261
38637	Horn Lake	(662)	13,885	9,069
38751	Indianola	(662)	11,514	11,809
*39205	Jackson	(601)	188,419	202,062
39090	Kosciusko	(662)	6,774	6,986
*39440	Laurel	(601)	18,299	18,827
38756	Leland	(662)	5,970	6,366
39560	Long Beach	(228)	16,776	15,804
39339	Louisville	(662)	7,085	7,165
*39648	McComb	(601)	11,746	11,797
*39110	Madison	(601)	12,618	7,471
*39302	Meridian	(601)	40,255	41,036
*39563	Moss Point	(228)	18,095	17,837
*39120	Natchez	(601)	18,277	19,460
38652	New Albany	(662)	7,238	6,775
*39564	Ocean Springs	(228)	16,519	15,221
38654	Olive Branch	(662)	12,063	3,567
39567	Orange Grove (c)	(228)	—	15,676
38655	Oxford	(662)	12,096	10,026
*39567	Pascagoula	(228)	27,163	25,899
39571	Pass Christian	(228)	6,190	5,557
39288	Pearl	(601)	23,287	19,588
39465	Petal	(601)	8,888	7,883
39350	Philadelphia	(601)	7,725	6,758
39466	Picayune	(601)	12,058	10,633
38863	Pontotoc	(662)	5,219	4,570
39218	Richland	(601)	5,794	4,014
*39157	Ridgeland	(601)	16,545	11,714
38663	Ripley	(662)	5,623	5,371
39533	Saint Martin (c)	(228)	—	6,349
38668	Senatobia	(601)	5,428	4,772
38671	Southaven	(662)	23,434	18,705
*39759	Starkville	(662)	23,434	18,705
*38801	Tupelo	(662)	20,184	18,458
*39180	Vicksburg	(601)	35,589	30,685
39576	Waveland	(228)	27,221	26,886
39367	Waynesboro	(601)	6,986	5,369
.....	West Hattiesburg (c)	(601)	5,400	5,143
39773	West Point	(662)	8,848	8,489
38967	Winona	(662)	5,647	5,965
39194	Yazoo City	(662)	11,941	12,427

Missouri

ZIP	Place		1998	1990
63123	Affton (c)	(314)	—	21,106
63010	Arnold	(636)	21,052	18,828

ZIP	Place		1998	1990
65605	Aurora	(417)	6,856	6,459
*63011	Ballwin	(636)	25,909	27,054
63137	Bellefontaine Neighbors	(314)	10,041	10,918
64012	Belton	(816)	21,778	18,145
63134	Berkeley	(314)	10,432	12,250
63031	Black Jack	(314)	6,311	6,131
*64015	Blue Springs	(816)	44,433	40,103
65613	Bolivar	(417)	8,248	6,845
65233	Boonville	(660)	7,593	7,095
63144	Brentwood	(314)	7,493	8,150
63044	Bridgeton	(314)	16,243	17,732
64429	Cameron	(816)	7,925	6,782
*63701	Cape Girardeau	(573)	35,596	34,475
64834	Carl Junction	(417)	5,405	4,123
64836	Carthage	(417)	11,360	10,747
63830	Caruthersville	(573)	6,986	7,389
*63017	Chesterfield	(636)	46,033	42,325
64601	Chillicothe	(660)	8,305	8,799
63105	Clayton	(314)	13,289	13,926
64735	Clinton	(660)	9,248	8,703
*65201	Columbia	(573)	78,915	69,133
63128	Concord (c)	(314)	—	19,859
63126	Crestwood	(314)	12,274	11,229
63141	Creve Coeur	(314)	11,880	12,289
63020	De Soto	(636)	5,944	5,993
63131	Des Peres	(636)	7,872	8,395
63841	Dexter	(573)	7,605	7,506
63011	Ellisville	(636)	7,824	7,183
63025	Eureka	(636)	5,860	4,683
64024	Excelsior Springs	(816)	11,424	10,373
63640	Farmington	(573)	13,849	11,596
63135	Ferguson	(314)	20,490	22,290
63028	Festus	(636)	8,696	8,105
*63033	Florissant	(314)	47,069	51,038
65473	Fort Leonard Wood (c)	(573)	—	15,863
65251	Fulton	(573)	11,330	10,033
64118	Gladstone	(816)	28,043	26,243
65254	Glasgow Village (c)	(573)	—	5,199
63122	Glendale	(314)	5,465	5,945
64030	Grandview	(816)	23,703	24,973
63401	Hannibal	(573)	17,728	18,004
64701	Harrisonville	(816)	8,795	7,696
63042	Hazelwood	(314)	14,391	15,512
*64050	Independence	(816)	116,832	112,301
63755	Jackson	(573)	11,258	9,256
*65101	Jefferson City	(573)	34,911	35,517
63136	Jennings	(314)	14,694	15,841
*64801	Joplin	(417)	44,612	41,175
*64108	Kansas City	(816)	441,574	434,829
63857	Kennett	(573)	10,621	10,941
63501	Kirksville	(660)	16,979	17,152
63122	Kirkwood	(314)	26,804	28,318
63124	Ladue (St. Louis Co.)	(314)	8,203	8,795
63367	Lake Saint Louis	(636)	9,319	7,536
65536	Lebanon	(417)	11,704	9,983
*64063	Lee's Summit	(816)	66,623	46,418
63125	Lemay (c)	(314)	—	18,005
*64068	Liberty	(816)	25,592	20,459
63552	Macon	(660)	5,428	5,571
63011	Manchester	(636)	7,071	6,506
63143	Maplewood	(314)	9,009	9,962
65340	Marshall	(660)	12,163	12,711
65706	Marshfield	(417)	5,611	4,374
63043	Maryland Heights	(314)	23,470	25,440
64468	Maryville	(816)	10,012	10,663
63129	Mehlville (c)	(314)	—	27,557
65265	Mexico	(573)	11,250	11,290
65270	Moberly	(660)	12,037	12,839
65708	Monett	(417)	7,464	6,529
65711	Mountain Grove	(417)	5,018	4,193
63026	Murphy (c)	(636)	—	9,342
64850	Neosho	(417)	9,531	9,254
64772	Nevada	(417)	8,173	8,597
65714	Nixa	(417)	11,483	4,893
63129	Oakville (c)	(314)	—	31,750
63366	O'Fallon	(636)	35,019	17,427
63132	Olivette	(314)	6,970	7,573
63114	Overland	(314)	16,386	17,987
65721	Ozark	(417)	8,164	4,401
63601	Park Hills	(573)	8,074	7,866
63775	Perryville	(573)	7,480	6,933
64080	Pleasant Hill	(816)	5,045	3,827
*63901	Poplar Bluff	(573)	17,029	16,841
64083	Raymore	(816)	9,752	5,592
64133	Raytown	(816)	28,372	30,601
65738	Republic	(417)	7,008	6,290
64085	Richmond	(816)	5,945	5,738
63117	Richmond Heights	(314)	9,463	10,448
*65401	Rolla	(573)	16,027	14,090
63074	Saint Ann	(314)	13,507	14,449
*63301	Saint Charles	(636)	58,166	50,634
63114	Saint John	(314)	6,814	7,502
*64501	Saint Joseph	(816)	69,622	71,852
*63166	Saint Louis	(314)	339,316	396,685
63376	Saint Peters	(636)	50,297	40,660
63126	Sappington (c)	(314)	—	10,917
*65301	Sedalia	(660)	20,447	19,800
63119	Shrewsbury	(314)	6,125	6,416
63801	Sikeston	(573)	17,792	17,641
63138	Spanish Lake (c)	(314)	—	20,322

ZIP	Place		1998	1990
*65801	Springfield	(417)	142,898	140,494
63080	Sullivan	(573)	6,258	5,661
63127	Sunset Hills	(314)	5,391	4,915
63006	Town and Country	(314)	10,909	10,944
64683	Trenton	(660)	5,774	6,129
63379	Troy	(314)	5,548	3,811
63084	Union	(636)	6,630	6,196
63130	University City	(314)	36,858	40,087
63088	Valley Park	(636)	5,799	4,165
64093	Warrensburg	(660)	17,429	15,244
63090	Washington	(636)	12,282	11,367
64870	Webb City	(417)	8,829	7,538
63119	Webster Groves	(314)	21,332	22,992
63385	Wentzville	(636)	5,599	4,640
65775	West Plains	(417)	11,135	9,214
*63011	Wildwood	(314)	18,123	16,742

Montana (406)

ZIP	Place	1998	1990
59711	Anaconda	9,999	10,356
59714	Belgrade	5,018	3,422
*59101	Billings	91,750	81,125
*59718	Bozeman	29,936	22,660
*59701	Butte	33,994	33,336
*59401	Great Falls	56,395	55,125
59501	Havre	10,015	10,201
*59601	Helena	28,306	24,609
.....	Helena Valley West Central (c)	—	6,327
*59901	Kalispell	16,089	11,917
59044	Laurel	6,027	5,686
59457	Lewistown	6,159	6,097
59047	Livingston	7,348	6,701
59402	Malmstrom AFB (c)	—	5,938
59301	Miles City	8,685	8,461
*59801	Missoula	52,239	42,918
59801	Orchard Homes (c)	—	10,317
59937	Whitefish	5,875	4,368

Nebraska

ZIP	Place		1998	1990
69301	Alliance	(308)	9,602	9,765
68310	Beatrice	(402)	12,376	12,352
*68108	Bellevue	(402)	44,047	39,240
*68008	Blair	(402)	7,566	6,860
69337	Chadron	(308)	5,760	5,588
68108	Chalco (c)	(402)	—	7,337
*68601	Columbus	(402)	20,898	19,480
68333	Crete	(402)	5,136	4,841
*68025	Fremont	(402)	24,429	23,680
69341	Gering	(308)	7,791	7,946
*68802	Grand Island	(308)	41,392	39,487
*68901	Hastings	(402)	21,356	22,837
68949	Holdrege	(308)	5,918	5,671
*68847	Kearney	(308)	27,968	24,396
68128	La Vista	(402)	11,864	9,992
68850	Lexington	(308)	8,976	6,600
*68501	Lincoln	(402)	213,088	191,972
69001	McCook	(308)	7,779	8,112
68410	Nebraska City	(402)	6,839	6,547
*68701	Norfolk	(402)	23,476	21,476
*69101	North Platte	(308)	23,307	22,605
68113	Offutt AFB West (c)	(402)	—	10,883
69153	Ogallala	(308)	5,066	5,095
*68005	Omaha	(402)	371,291	344,463
*68046	Papillion	(402)	20,603	13,892
68048	Plattsmouth	(402)	7,186	6,415
68127	Ralston	(402)	6,219	6,236
*69361	Scottsbluff	(308)	14,294	13,711
68434	Seward	(402)	6,123	5,641
69162	Sidney	(308)	5,993	5,959
68776	South Sioux City	(402)	11,415	9,677
68787	Wayne	(402)	5,373	5,142
68467	York	(402)	7,974	7,940

Nevada

ZIP	Place		1998	1990
*89005	Boulder City	(702)	14,166	12,567
*89701	Carson City	(775)	49,301	40,443
89112	East Las Vegas (c)	(702)	—	11,087
*89801	Elko	(775)	19,204	14,836
.....	Enterprise (c)		—	6,412
*89406	Fallon	(775)	8,587	6,430
89408	Fernley (c)	(775)	—	5,164
89410	Gardnerville Ranchos (c)	(775)	—	7,455
*89015	Henderson	(702)	152,717	64,948
*89450	Incline Village-Crystal Bay (c)	(775)	—	7,119
*89125	Las Vegas	(702)	404,288	258,877
*89024	Mesquite	(702)	10,125	1,871
89191	Nellis AFB (c)	(702)	—	8,377
*89030	North Las Vegas	(702)	94,218	47,849
*89041	Pahrump (c)	(775)	—	7,424
89109	Paradise (c)	(775)	—	124,682
*89501	Reno	(775)	163,334	134,230
*89431	Sparks	(775)	62,432	53,367
89815	Spring Creek (c)	(702)	—	5,866
.....	Spring Valley (c)	(702)	—	51,726
89110	Sunrise Manor (c)	(702)	—	95,362

ZIP	Place		1998	1990
89433	Sun Valley (c)	(775)	—	11,391
89101	Winchester (c)	(702)	—	23,365
*89445	Winnemucca	(775)	9,404	6,473

New Hampshire (603)

See introductory note.

ZIP	Place	1998	1990
03031	Amherst	10,325	9,068
03811	Atkinson	6,579	5,188
03825	Barrington	7,039	6,164
03110	Bedford	16,266	12,563
03220	Belmont	6,267	5,796
03570	Berlin	10,120	11,824
03304	Bow	6,606	5,500
03743	Claremont	13,868	13,902
*03301	Concord	37,444	36,006
03818	Conway	8,823	7,940
03038	Derry Compact (c)	—	20,446
03038	Derry	31,871	29,603
*03820	Dover	25,953	25,042
03824	Durham Compact (c)	—	9,236
03824	Durham	11,223	11,818
03042	Epping	5,827	5,162
03833	Exeter Compact (c)	—	9,556
03833	Exeter	13,770	12,481
03835	Farmington	5,920	5,739
03235	Franklin	8,351	8,304
03246	Gilford	6,199	5,867
03045	Goffstown	16,011	14,621
03841	Hampstead	7,737	6,732
*03842	Hampton Compact (c)	—	7,989
*03842	Hampton	13,080	12,278
03755	Hanover Compact (c)	—	6,538
03755	Hanover	9,876	9,212
03049	Hollis	6,855	5,705
03106	Hooksett	9,792	9,002
03229	Hopkinton	5,018	4,806
03051	Hudson	21,916	19,530
03452	Jaffrey	5,291	5,361
03431	Keene	22,313	22,430
03848	Kingston	6,015	5,591
*03246	Laconia	16,435	15,743
*03766	Lebanon	12,461	12,183
03052	Litchfield	6,906	5,516
03561	Littleton	6,003	5,827
03053	Londonderry Compact (c)	—	10,114
03053	Londonderry	22,441	19,781
*03103	Manchester	102,524	99,332
03253	Meredith	5,169	4,837
03054	Merrimack	24,224	22,156
03055	Milford Compact (c)	—	8,015
03055	Milford	12,775	11,795
*03060	Nashua	82,169	79,662
03857	Newmarket	7,549	7,157
03773	Newport	6,196	6,110
03076	Pelham	11,051	9,408
03275	Pembroke	6,604	6,561
03458	Peterborough	5,594	5,239
03865	Plaistow	7,940	7,316
03264	Plymouth	5,941	5,811
*03801	Portsmouth	25,388	25,925
03077	Raymond	9,714	8,713
03461	Rindge	5,212	4,941
*03867	Rochester	27,869	26,630
03079	Salem	27,830	25,746
03874	Seabrook	7,027	6,503
03878	Somersworth	11,525	11,249
03885	Stratham	5,870	4,955
03275	Suncook (c)	—	5,214
03446	Swanzey	6,678	6,236
03281	Weare	7,139	6,193
03087	Windham	10,181	9,000
03894	Wolfeboro	5,430	4,807

New Jersey

ZIP	Place		1998	1990
08201	Absecon	(609)	7,817	7,298
07401	Allendale	(201)	6,725	5,900
07712	Asbury Park	(732)	17,057	16,799
*08401	Atlantic City	(609)	38,063	37,986
08106	Audubon	(856)	8,848	9,205
07001	Avenel (c)	(732)	—	15,504
08007	Barrington	(856)	7,185	6,792
07002	Bayonne	(201)	61,051	61,444
08722	Beachwood	(732)	10,130	9,324
07109	Belleville (c)	(973)	—	34,213
*08031	Bellmawr	(856)	12,243	12,603
07719	Belmar	(732)	5,904	5,877
07621	Bergenfield	(201)	24,827	24,458
07922	Berkeley Heights Twp. (c)	(908)	—	11,980
08009	Berlin	(856)	6,162	5,672
07924	Bernardsville	(908)	7,085	6,597
08012	Blackwood (c)	(856)	—	5,120
07003	Bloomfield (c)	(973)	—	45,061
07403	Bloomingdale	(973)	8,151	7,530
07603	Bogota	(201)	7,956	7,824
07005	Boonton	(973)	8,545	8,343
08805	Bound Brook	(732)	9,672	9,487

ZIP	Place		1998	1990
*08723	Brick Twp. (c)	(732)	—	66,473
08302	Bridgeton	(856)	18,096	18,942
08807	Bridgewater Twp. (c)	(732)/(908)	—	32,509
08203	Brigantine	(609)	11,599	11,354
08015	Browns Mills (c)	(609)	—	11,429
07828	Budd Lake (c)	(973)	—	7,272
08016	Burlington	(609)	9,486	9,835
07405	Butler	(973)	7,759	7,392
*07006	Caldwell	(973)	7,315	7,542
*08101	Camden	(856)	83,546	87,492
07072	Carlstadt	(201)	5,695	5,510
08069	Carney's Point Twp. (c)	(856)	—	8,443
07008	Carteret	(732)	19,094	19,025
07009	Cedar Grove Twp. (c) (Essex)	(973)	—	12,053
07928	Chatham	(973)	8,071	8,007
*08034	Cherry Hill Twp. (c)	(856)	—	69,319
08077	Cinnaminson Twp. (c)	(856)	—	14,583
07066	Clark Twp. (c)	(732)/(908)	—	14,629
08312	Clayton	(856)	7,018	6,155
08021	Clementon	(856)	5,399	5,601
07010	Cliffside Park	(201)	21,141	20,393
*07015	Clifton	(973)	76,180	71,984
07624	Closter	(201)	8,574	8,094
08108	Collingswood	(856)	14,582	15,289
07067	Colonia (c)	(732)	—	18,238
07016	Cranford Twp. (c)	(908)	—	22,633
07626	Cresskill	(201)	7,882	7,558
08759	Crestwood Village (c)	(732)	—	8,030
*07801	Dover	(973)	15,462	15,115
07628	Dumont	(201)	17,631	17,187
08812	Dunellen	(732)	6,602	6,528
08816	East Brunswick Twp. (c)	(732)	—	43,548
07936	East Hanover Twp. (c)	(973)	—	9,926
*07019	East Orange	(973)	69,598	73,552
07073	East Rutherford	(201)/(973)	8,107	7,902
*07724	Eatontown	(732)	14,077	13,800
07020	Edgewater	(201)	5,764	5,001
08010	Edgewater Park Twp. (c)	(609)	—	8,388
*08818	Edison Twp. (c)	(732)/(908)	—	88,680
*07207	Elizabeth	(908)	110,661	110,002
07407	Elmwood Park	(201)	18,299	17,623
07630	Emerson	(201)	7,174	6,930
07631	Englewood	(201)	25,321	24,850
07632	Englewood Cliffs	(201)	5,898	5,634
08618	Ewing Twp. (c)	(609)	—	34,185
07004	Fairfield (c)	(973)	—	7,615
07704	Fair Haven	(732)	5,467	5,270
07410	Fair Lawn	(201)/(973)	31,091	30,548
07022	Fairview (Bergen)	(201)	11,252	10,733
07023	Fanwood	(908)	7,140	7,115
08518	Florence-Roebling (c)	(609)	—	8,564
07932	Florham Park	(973)	9,060	8,521
08863	Fords (c)	(732)	—	14,392
08640	Fort Dix (c)	(609)	—	10,205
07024	Fort Lee	(201)	33,989	31,997
07416	Franklin	(973)	5,243	4,977
07417	Franklin Lakes	(201)	10,575	9,873
*08873	Franklin Twp. (Somerset) (c)	(732)(908)	—	42,780
07728	Freehold	(732)	10,850	10,742
07026	Garfield	(201)	27,262	26,727
08753	Gilford Park (c)	(732)	—	8,668
08028	Glassboro	(856)	17,588	15,614
08029	Glendora (c)	(856)	—	5,201
07028	Glen Ridge	(973)	6,739	7,076
07452	Glen Rock	(201)	11,149	10,883
08030	Gloucester City	(856)	12,121	12,649
07093	Guttenberg	(201)	8,374	8,268
*07602	Hackensack	(201)	37,813	37,049
07840	Hackettstown	(908)	8,526	8,120
08033	Haddonfield	(856)	11,170	11,633
08035	Haddon Heights	(856)	7,520	7,860
*07508	Haledon	(973)	7,321	6,951
*08609	Hamilton Twp. (Mercer) (c)	(609)	—	86,553
08037	Hammonton	(609)	12,447	12,208
07981	Hanover Twp. (c)	(973)	—	11,538
07029	Harrison	(973)	13,383	13,425
07604	Hasbrouck Heights	(201)	11,704	11,488
*07506	Hawthorne	(973)	18,304	17,084
07730	Hazlet Twp. (c)	(732)	—	21,976
08904	Highland Park (Middlesex)	(732)	13,266	13,279
07642	Hillsdale	(201)	10,142	9,750
07205	Hillside Twp. (c)	(908)/(973)	—	21,044
07030	Hoboken	(201)	33,354	33,397
08753	Holiday City-Berkeley (c)	(732)	—	14,293
.....	Holiday City South (c)	(732)	—	5,452
07843	Hopatcong	(973)	16,241	15,586
08525	Hopewell Twp. (Mercer) (c)	(609)	—	11,590
07111	Irvington (c)	(973)	—	59,774
08830	Iselin (c)	(732)	—	16,141
08527	Jackson Twp. (c)	(732)	—	33,283
08831	Jamesburg	(732)	5,748	5,294
*07303	Jersey City	(201)	232,429	228,517
07734	Keansburg	(732)	11,166	11,069
07032	Kearny	(201)/(973)	35,441	34,874
08824	Kendall Park (c)	(908)	—	7,127
07033	Kenilworth	(908)	7,701	7,574
07735	Keyport	(732)	7,739	7,586
07405	Kinnelon	(973)	9,126	8,470
07871	Lake Mohawk (c)	(973)	—	8,930
08701	Lakewood (c)	(732)	—	26,095

ZIP	Place	1998	1990
08879	Laurence Harbor (c) (732)	—	6,361
08648	Lawrenceville (c) (609)	—	6,446
*08733	Leisure Village West-Pine Lake Park (c) (732)	—	10,139
07605	Leonia (201)	8,555	8,365
07035	Lincoln Park (973)	11,270	10,978
07738	Lincroft (c) (732)	—	6,193
07036	Linden (732)/(908)	37,204	36,701
08021	Lindenwold (856)	18,093	18,734
08221	Linwood (609)	7,109	6,866
07424	Little Falls Twp. (c) (973)	—	11,294
07643	Little Ferry (201)	10,176	9,989
07739	Little Silver (732)	6,207	5,721
07039	Livingston Twp. (c) (973)	—	26,609
07644	Lodi (201)/(973)	22,917	22,355
07740	Long Branch (732)	28,905	28,658
*07946	Long Hill Twp. (c) (973)	—	7,826
07071	Lyndhurst Twp. (c) (201)	—	18,262
08641	McGuire AFB (c) (609)	—	7,580
07940	Madison (973)	15,828	15,850
08859	Madison Park (c) (732)	—	7,490
*07430	Mahwah Twp. (c) (201)	—	17,905
08736	Manasquan (732)	5,528	5,369
08835	Manville (908)	10,899	10,567
08052	Maple Shade Twp. (c) (856)	—	19,211
07040	Maplewood Twp. (c) (973)	—	21,756
08402	Margate City (609)	8,542	8,431
07746	Marlboro Twp. (c) (732)	—	27,974
08053	Marlton (c) (856)	—	10,228
07747	Matawan (732)	9,487	9,239
07607	Maywood (201)	9,694	9,536
08619	Mercerville-Hamilton Sq. (c) (609)	—	26,873
08840	Metuchen (732)	13,038	12,804
08846	Middlesex (732)	13,217	13,055
07748	Middletown Twp. (c) (732)	—	68,183
07432	Midland Park (201)	7,189	7,047
07041	Millburn Twp. (c) (973)	—	18,630
08850	Milltown (Middlesex) (732)	7,022	6,968
08332	Millville (856)	26,359	25,992
08094	Monroe Twp. (Gloucester) (c) (856)	—	26,703
*07042	Montclair (c) (973)	—	37,729
07645	Montvale (201)	7,290	6,946
07045	Montville Twp. (c) (973)	—	15,600
08057	Moorestown-Lenola (c) (856)	—	13,242
07950	Morris Plains (973)	5,298	5,219
*07960	Morristown (973)	16,629	16,189
07092	Mountainside (908)	6,685	6,657
08060	Mount Holly Twp. (c) (609)	—	10,639
08087	Mystic Island (c) (609)	—	7,400
07753	Neptune City (732)	5,060	4,997
*07753	Neptune Twp. (c) (732)	—	28,148
*07102	Newark (973)	267,823	275,221
*08901	New Brunswick (732)	41,768	41,711
07646	New Milford (201)	16,425	15,990
07974	New Providence (908)	11,885	11,439
07860	Newton (973)	7,928	7,521
07031	North Arlington (201)	14,128	13,790
07047	North Bergen Twp. (c) (201)	—	48,414
08902	North Brunswick Twp. (c) (732)	—	31,287
07006	North Caldwell (973)	6,503	6,706
08225	Northfield (609)	7,434	7,305
07508	North Haledon (973)	8,623	7,987
07060	North Plainfield (908)	19,067	18,820
07648	Norwood (201)	5,790	4,858
07110	Nutley (c) (973)	—	27,099
07436	Oakland (201)	12,478	11,997
*08758	Ocean Twp. (Ocean) (c) (609)/(732)	—	5,416
*08050	Ocean Acres (c) (609)	—	5,587
08226	Ocean City (609)	15,760	15,512
07757	Oceanport (732)	6,301	6,146
08857	Old Bridge (c) (732)	—	22,151
08857	Old Bridge Twp. (c) (732)	—	56,493
07675	Old Tappan (201)	5,580	4,254
07649	Oradell (201)	8,182	8,024
*07051	Orange (c) (973)	—	29,925
07650	Palisades Park (201)	15,060	14,536
08065	Palmyra (856)	6,916	7,056
*07652	Paramus (201)	26,103	25,004
07656	Park Ridge (201)	8,594	8,102
07054	Parsippany-Troy Hills Twp. (c) (973)	—	48,478
07055	Passaic (973)	60,817	58,041
*07510	Paterson (973)	148,212	140,891
08066	Paulsboro (856)	6,361	6,577
08110	Pennsauken Twp. (c) (856)	—	34,738
08070	Pennsville Center (c) (856)	—	12,218
07440	Pequannock Twp. (c) (973)	—	12,844
*08861	Perth Amboy (732)	42,481	41,967
08865	Phillipsburg (908)	15,533	15,757
08021	Pine Hill (856)	10,468	9,854
*08854	Piscataway Twp. (c) (732)/(908)	—	47,089
08071	Pitman (856)	9,078	9,365
*07061	Plainfield (908)	46,414	46,577
08232	Pleasantville (609)	16,619	16,027
08742	Point Pleasant (732)	19,349	18,177
08742	Point Pleasant Beach (732)	5,371	5,112
07442	Pompton Lakes (973)	11,180	10,539
*08540	Princeton (609)	11,814	12,016
07508	Prospect Park (973)	5,268	5,053
07065	Rahway (732)	25,336	25,325
08057	Ramblewood (c) (856)	—	6,181
07446	Ramsey (201)	14,480	13,228
07869	Randolph Twp. (c) (973)	—	19,974
08869	Raritan (908)	6,319	5,798
07701	Red Bank (732)	10,858	10,636
07657	Ridgefield (201)	10,183	9,996
07660	Ridgefield Park (201)	12,603	12,454
*07451	Ridgewood (201)/(973)	24,577	24,152
07456	Ringwood (973)	13,504	12,623
07661	River Edge (201)	10,862	10,603
08075	Riverside Twp. (c) (856)	—	7,974
07675	River Vale (c) (201)	—	9,410
07726	Robertsville (c) (732)	—	9,841
07662	Rochelle Park Twp. (c) (201)	—	5,587
07866	Rockaway (973)	6,476	6,243
07068	Roseland (973)	5,289	4,847
07203	Roselle (908)	20,297	20,314
07204	Roselle Park (908)	12,771	12,805
07760	Rumson (732)	6,866	6,701
08078	Runnemede (856)	8,840	9,042
07070	Rutherford (201)	18,116	17,790
07663	Saddle Brook Twp. (c) (201)/(973)	—	13,296
08079	Salem (856)	6,521	6,883
*08872	Sayreville (732)	38,042	34,998
07076	Scotch Plains Twp. (c) (732)/(908)	—	21,150
*07094	Secaucus (201)	13,975	14,061
08753	Silverton (c) (732)	—	9,175
08083	Somerdale (856)	5,443	5,440
*08873	Somerset (c) (732)	—	22,070
08244	Somers Point (609)	11,159	11,216
08876	Somerville (908)	11,777	11,632
08879	South Amboy (732)	7,864	7,851
07079	South Orange Twp. (c) (973)	—	16,390
07080	South Plainfield (732)/(908)	20,903	20,489
08882	South River (732)	14,045	13,692
07871	Sparta Twp. (c) (973)	—	15,157
08884	Spotswood (732)	8,255	7,983
07081	Springfield Twp. (c) (908)/(973)	—	13,420
07762	Spring Lake Heights (732)	5,436	5,341
08084	Stratford (856)	7,376	7,614
07747	Strathmore (c) (732)	—	7,060
07876	Succasunna-Kenvil (c) (201)	—	11,781
*07901	Summit (908)	19,706	19,757
07666	Teaneck Twp. (c) (201)	—	37,825
07670	Tenafly (201)	13,595	13,326
07724	Tinton Falls (732)	15,795	12,361
*08753	Toms River (c) (732)	—	7,524
*07512	Totowa (973)	10,930	10,177
*08650	Trenton (609)	84,494	88,675
08520	Twin Rivers (c) (609)	—	7,715
07083	Union Twp. (Union) (c) (908)	—	50,024
07735	Union Beach (732)	6,467	6,156
07087	Union City (201)	57,621	58,012
07458	Upper Saddle River (201)	7,642	7,198
08406	Ventnor City (609)	10,857	11,005
07044	Verona (c) (973)	—	13,597
08251	Villas (c) (609)	—	8,136
*08360	Vineland (856)	—	54,780
07463	Waldwick (201)	10,097	9,757
07057	Wallington (201)/(973)	11,099	10,828
07465	Wanaque (201)/(973)	10,520	9,711
07882	Washington (908)	6,427	6,474
07675	Washington Twp.(Bergen) (c) (201)	—	9,245
07060	Watchung (908)	5,412	5,110
*07470	Wayne Twp. (c) (973)	—	47,025
07087	Weehawken Twp. (c) (201)	—	12,385
07007	West Caldwell (c) (973)	—	10,422
*07091	Westfield (732)/(908)	29,297	28,870
07728	West Freehold (c) (732)	—	11,166
07764	West Long Branch (732)	7,959	7,690
07480	West Milford Twp. (c) (973)	—	25,430
07093	West New York (201)	38,020	38,125
07052	West Orange (c) (973)	—	39,103
07424	West Paterson (973)	11,704	10,982
07675	Westwood (201)	10,779	10,446
07885	Wharton (973)	5,694	5,405
08610	White Horse (c) (609)	—	9,397
07886	White Meadow Lake (c) (973)	—	8,002
08094	Williamstown (c) (856)	—	10,891
08046	Willingboro Twp. (c) (609)	—	36,291
08095	Winslow Twp. (c) (856)	—	30,087
07095	Woodbridge (c) (732)	—	17,434
07095	Woodbridge Twp. (c) (732)	—	93,092
08096	Woodbury (856)	10,520	10,904
07675	Woodcliff Lake (201)	5,909	5,303
07075	Wood-Ridge (201)/(973)	7,656	7,506
07481	Wyckoff Twp. (c) (201)	—	15,372
08620	Yardville-Groveville (c) (609)	—	9,248
07726	Yorketown (c) (609)	—	6,313

New Mexico (505)

ZIP	Place	1998	1990
*88310	Alamogordo	28,312	27,596
*87101	Albuquerque	419,311	384,915
88021	Anthony (c)	—	5,160
*88210	Artesia	10,973	10,610

ZIP	Place	1998	1990
87410	Aztec	6,134	5,480
87002	Belen	7,936	6,547
87004	Bernalillo	7,570	5,864
87413	Bloomfield	6,260	5,214
87068	Bosque Farms	5,645	3,791
*88220	Carlsbad	26,315	24,952
*88101	Clovis	32,394	30,954
87048	Corrales	6,633	5,453
*88030	Deming	14,517	11,422
*87532	Espanola	8,994	8,389
*87401	Farmington	39,028	33,997
*87301	Gallup	20,120	19,157
87020	Grants	9,294	8,626
*88240	Hobbs	27,156	29,121
88330	Holloman AFB (c)	—	5,891
*88001	Las Cruces	76,102	62,360
87701	Las Vegas	16,487	14,753
87544	Los Alamos (c)	—	11,455
87031	Los Lunas	7,805	6,013
87107	Los Ranchos de Albuquerque	5,019	5,075
88260	Lovington	9,841	9,322
87107	North Valley (c)	—	12,507
87114	Paradise Hills (c)	—	5,513
88130	Portales	10,857	10,690
87740	Raton	7,466	7,372
*87124	Rio Rancho	50,041	32,512
*88201	Roswell	47,624	44,260
*88345	Ruidoso	6,065	4,600
87115	Sandia (c)	—	6,742
*87501	Santa Fe	67,879	56,537
87420	Shiprock (c)	—	7,687
*88061	Silver City	12,064	10,683
87801	Socorro	8,616	8,159
87105	South Valley (c)	—	35,701
88063	Sunland Park	9,591	8,179
87571	Taos	5,389	4,413
87901	Truth or Consequences	6,564	6,221
88401	Tucumcari	5,825	6,827
87544	White Rock (c)	—	6,192
87327	Zuni Pueblo (c)	—	5,857

New York

Area code (347) overlays area code (718).
Area code (646) overlays area code (212). See introductory note.

ZIP	Place		1998	1990
10901	Airmont	(914)	7,657	7,674
*12201	Albany	(518)	94,305	100,031
11507	Albertson (c)	(516)	—	5,166
14411	Albion	(716)	6,742	5,863
14226	Amherst	(716)	110,788	111,711
*11701	Amityville	(516)/(631)	9,166	9,286
12010	Amsterdam	(518)	19,176	20,714
12603	Arlington (c)	(914)	—	11,948
*13021	Auburn	(315)	29,145	31,258
11702	Babylon	(631)	12,001	12,249
11510	Baldwin (c)	(516)	—	22,719
11510	Baldwin Harbor (c)	(516)	—	7,899
13027	Baldwinsville	(315)	6,605	6,591
12020	Ballston Spa	(518)	5,498	5,194
*14020	Batavia	(716)	15,784	16,310
14810	Bath	(607)	5,676	5,801
11706	Bay Shore (c)	(631)	—	21,279
11705	Bayport (c)	(631)	—	7,702
11709	Bayville	(516)	7,280	7,193
11751	Baywood (c)	(631)	—	7,351
12508	Beacon	(914)	13,215	13,243
11710	Bellmore (c)	(516)	—	16,438
11714	Bethpage (c)	(516)	—	15,761
*13902	Binghamton	(607)	46,760	53,008
11716	Bohemia (c)	(631)	—	9,556
11717	Brentwood (c)	(631)	—	45,218
10510	Briarcliff Manor	(914)	7,743	7,070
14610	Brighton (c)	(716)	—	34,455
14420	Brockport	(716)	8,122	8,749
10708	Bronxville	(914)	5,954	6,028
*14240	Buffalo	(716)	300,717	328,175
*14424	Canandaigua	(716)	10,658	10,725
13617	Canton	(315)	6,003	6,379
11514	Carle Place (c)	(516)	—	5,107
11516	Cedarhurst	(516)	5,679	5,716
11934	Center Moriches (c)	(631)	—	5,987
11720	Centereach (c)	(631)	—	26,720
11721	Centerport (Suffolk) (c)	(631)	—	5,333
11722	Central Islip (c)	(516)	—	26,028
14225	Cheektowaga (c)	(716)	—	84,387
10977	Chestnut Ridge	(914)	7,949	7,517
13037	Chittenango	(315)	5,134	4,734
12065	Clifton Park	(518)	33,281	30,117
12047	Cohoes	(518)	16,333	16,825
12205	Colonie	(518)	8,255	8,019
11725	Commack (c)	(631)	—	36,124
10920	Congers (c)	(914)	—	8,003
11726	Copiague (c)	(631)	—	20,769
11727	Coram (c)	(631)	—	30,111
14830	Corning	(607)	11,080	11,938
13045	Cortland	(607)	18,409	19,801
*10520	Croton-on-Hudson	(914)	7,213	7,018
11729	Deer Park (c)	(631)	—	28,840
12054	Delmar (c)	(518)	—	8,360

ZIP	Place		1998	1990
14043	Depew	(716)	16,675	17,673
13214	DeWitt (c)	(315)	—	8,244
11746	Dix Hills (c)	(631)	—	25,849
10522	Dobbs Ferry	(914)	10,070	9,940
14048	Dunkirk	(716)	12,952	13,989
14052	East Aurora	(716)	6,462	6,647
12302	East Glenville (c)	(518)	—	6,518
11576	East Hills	(516)	6,759	6,746
11730	East Islip (c)	(631)	—	14,325
11758	East Massapequa (c)	(516)	—	19,550
11554	East Meadow (c)	(516)	—	36,909
11731	East Northport (c)	(631)	—	20,411
11772	East Patchogue (c)	(631)	—	20,195
14445	East Rochester	(716)	6,532	6,932
11518	East Rockaway	(516)	10,155	10,152
11786	East Shoreham (c)	(631)	—	5,461
10709	Eastchester (c)	(914)	—	18,537
*14901	Elmira	(607)	31,367	33,724
11003	Elmont (c)	(516)	—	28,612
11731	Elwood (c)	(631)	—	10,916
*13760	Endicott	(607)	12,001	13,531
13762	Endwell (c)	(607)	—	12,602
13219	Fairmount (c)	(315)	—	12,266
14450	Fairport	(716)	5,712	5,943
11735	Farmingdale	(516)	8,096	8,022
11738	Farmingville (c)	(631)	—	14,842
*11001	Floral Park	(516)	15,849	15,947
13603	Fort Drum (c)	(315)	—	11,578
11768	Fort Salonga (c)	(631)	—	9,176
11010	Franklin Square (Nassau) (c)	(516)	—	28,205
14063	Fredonia	(716)	10,016	10,436
11520	Freeport	(516)	39,963	39,894
13069	Fulton	(315)	12,195	12,929
*11530	Garden City	(516)	21,616	21,675
11040	Garden City Park (c)	(516)	—	7,437
14624	Gates-North Gates (c)	(716)	—	14,995
14454	Geneseo	(716)	7,413	7,187
14456	Geneva	(315)	13,720	14,143
11542	Glen Cove	(516)	24,935	24,149
12801	Glens Falls North (c)	(518)	—	7,978
12801	Glens Falls	(518)	14,497	15,023
12078	Gloversville	(518)	15,848	16,656
10924	Goshen	(914)	5,164	5,255
13642	Gouverneur	(315)	5,209	4,604
*11021	Great Neck	(516)	8,874	8,745
11020	Great Neck Plaza	(516)	5,973	5,897
14616	Greece (c)	(716)	—	15,632
11740	Greenlawn (c)	(631)	—	13,208
*10583	Greenville (Westchester) (c)	(914)	—	9,528
14075	Hamburg	(716)	9,899	10,442
11946	Hampton Bays (c)	(631)	—	7,893
10528	Harrison	(914)	24,027	23,308
10530	Hartsdale (c)	(914)	—	9,587
10706	Hastings-on-Hudson	(914)	8,003	8,000
*11788	Hauppauge (c)	(631)	—	19,750
10927	Haverstraw	(914)	9,462	9,438
*11551	Hempstead	(516)	46,698	45,982
13350	Herkimer	(315)	7,329	7,945
11557	Hewlett (c)	(516)	—	6,620
*11802	Hicksville (c)	(516)	—	40,174
10977	Hillcrest (c)	(914)	—	6,447
14468	Hilton	(716)	5,550	5,216
11741	Holbrook (c)	(631)	—	25,273
11742	Holtsville (c)	(631)	—	14,972
14843	Hornell	(607)	9,146	9,877
*14845	Horseheads	(607)	6,568	6,802
12839	Hudson Falls	(518)	7,335	7,651
12534	Hudson	(518)	7,841	8,034
11743	Huntington (c)	(631)	—	18,243
11746	Huntington Station (c)	(631)	—	28,247
13357	Ilion	(315)	8,132	8,888
11096	Inwood (c)	(516)	—	7,767
14617	Irondequoit (c)	(716)	—	52,322
10533	Irvington	(914)	6,460	6,348
11751	Islip (c)	(631)	—	18,924
11752	Islip Terrace (c)	(631)	—	5,530
*14850	Ithaca	(607)	28,172	29,541
*14702	Jamestown	(716)	32,166	34,681
10535	Jefferson Valley-Yorktown (c)	(914)	—	14,118
11753	Jericho (Nassau) (c)	(516)	—	13,141
13790	Johnson City	(607)	14,962	16,578
12095	Johnstown	(518)	8,532	9,058
14217	Kenmore	(716)	15,911	17,180
11754	Kings Park (c)	(631)	—	17,773
*12401	Kingston	(914)	21,860	23,095
10950	Kiryas Joel	(914)	9,986	7,437
14218	Lackawanna	(716)	19,220	20,585
10512	Lake Carmel (c)	(914)	—	8,489
11755	Lake Grove	(631)	9,929	9,612
11779	Lake Ronkonkoma (c)	(631)	—	18,997
11552	Lakeview (c)	(516)	—	5,476
14086	Lancaster	(716)	11,087	11,940
10538	Larchmont	(914)	6,128	6,181
12110	Latham (c)	(518)	—	10,131
11559	Lawrence	(516)	6,516	6,513
11756	Levittown (c)	(516)	—	53,286
11757	Lindenhurst	(631)	26,433	26,879
13365	Little Falls	(315)	5,326	5,829
*14094	Lockport	(716)	22,650	24,426
11561	Long Beach	(516)	34,244	33,510

ZIP	Place	1998	1990
12211	Loudonville (c) (518)	—	10,822
11563	Lynbrook (516)	19,341	19,208
10541	Mahopac (c) (914)	—	7,755
12953	Malone (518)	7,109	6,777
11565	Malverne (516)	9,021	9,054
10543	Mamaroneck (914)	17,394	17,325
11030	Manhasset (c) (516)	—	7,718
11050	Manorhaven (516)	5,816	5,672
11949	Manorville (c) (631)	—	6,198
11758	Massapequa (c) (516)	—	22,018
11762	Massapequa Park (516)	18,108	18,044
13662	Massena (315)	11,257	11,716
11950	Mastic (c) (631)	—	13,778
11951	Mastic Beach (c) (631)	—	10,293
13211	Mattydale (c) (315)	—	6,418
12118	Mechanicville (518)	5,151	5,249
11763	Medford (c) (631)	—	21,274
14103	Medina (716)	6,676	6,686
11747	Melville (c) (631)	—	12,586
11566	Merrick (c) (516)	—	23,042
11953	Middle Island (c) (631)	—	7,848
*10940	Middletown (914)	23,953	24,160
11764	Miller Place (c) (631)	—	9,315
11501	Mineola (516)	18,942	19,005
10950	Monroe (914)	7,693	6,672
10952	Monsey (c) (914)	—	13,986
12701	Monticello (914)	6,383	6,597
10970	Mount Ivy (c) (914)	—	6,013
10549	Mount Kisco (914)	9,262	9,108
11766	Mount Sinai (c) (631)	—	8,023
*10551	Mount Vernon (914)	66,824	67,153
12590	Myers Corner (c) (914)	—	5,599
10954	Nanuet (c) (914)	—	14,065
11767	Nesconset (c) (631)	—	10,712
11590	New Cassel (c) (516)	—	10,257
10956	New City (c) (914)	—	33,673
*11040	New Hyde Park (516)	9,770	9,728
12561	New Paltz (914)	5,219	5,470
*10802	New Rochelle (914)	67,225	67,265
*12550	New Windsor Center (c) (914)	—	8,898
*10001	New York (212)/(718)	7,420,166	7,322,564
14513	Newark (315)	9,756	9,849
*12550	Newburgh (914)	26,114	26,454
*14302	Niagara Falls (716)	56,768	61,840
11701	North Amityville (c) (631)	—	13,849
11703	North Babylon (c) (631)	—	18,081
11706	North Bay Shore (c) (631)	—	12,799
11710	North Bellmore (c) (516)	—	19,707
11713	North Bellport (c) (631)	—	8,182
11757	North Lindenhurst (c) (631)	—	10,563
11758	North Massapequa (c) (516)	—	19,365
11566	North Merrick (c) (516)	—	12,113
11040	North New Hyde Park (c) (516)	—	14,359
11772	North Patchogue (c) (631)	—	7,374
11768	Northport (631)	7,418	7,572
13212	North Syracuse (315)	7,055	7,363
14120	North Tonawanda (716)	32,947	34,989
11580	North Valley Stream (c) (516)	—	14,574
11793	North Wantagh (c) (516)	—	12,276
13815	Norwich (607)	7,048	7,613
10960	Nyack (914)	6,614	6,558
11769	Oakdale (c) (631)	—	7,875
11572	Oceanside (c) (516)	—	32,423
13669	Ogdensburg (315)	12,759	13,521
11804	Old Bethpage (c) (516)	—	5,610
14760	Olean (716)	16,170	16,946
13421	Oneida (315)	10,854	10,850
13820	Oneonta (607)	12,965	13,954
12550	Orange Lake (c) (914)	—	5,196
10562	Ossining (914)	23,010	22,582
13126	Oswego (315)	18,054	19,195
11771	Oyster Bay (c) (516)	—	6,687
11772	Patchogue (631)	11,014	11,060
10965	Pearl River (c) (914)	—	15,314
10566	Peekskill (914)	21,111	19,536
10803	Pelham Manor (914)	6,340	6,413
10803	Pelham (914)	5,409	5,443
11714	Plainedge (c) (516)	—	8,739
11803	Plainview (c) (516)	—	26,207
12903	Plattsburgh AFB (c) (518)	—	5,483
*12901	Plattsburgh (518)	18,678	21,255
10570	Pleasantville (914)	6,766	6,592
10573	Port Chester (914)	24,777	24,728
11777	Port Jefferson (631)	7,793	7,455
11776	Port Jefferson Station (c) (631)	—	7,232
12771	Port Jervis (914)	8,778	9,060
11050	Port Washington (c) (516)	—	15,387
13676	Potsdam (315)	9,491	10,251
*12601	Poughkeepsie (914)	27,669	28,844
12144	Rensselaer (518)	7,741	8,255
11961	Ridge (c) (631)	—	11,734
11901	Riverhead (c) (631)	—	8,814
*14692	Rochester (716)	216,887	230,356
*11571	Rockville Centre (516)	24,639	24,727
11778	Rocky Point (c) (631)	—	8,596
12205	Roessleville (c) (518)	—	10,753
*13440	Rome (315)	39,792	44,350
11779	Ronkonkoma (c) (631)	—	20,391
11575	Roosevelt (c) (516)	—	15,030
11577	Roslyn Heights (c) (516)	—	6,405
12303	Rotterdam (c) (518)	—	21,228
10573	Rye Brook (914)	8,535	7,765
10580	Rye (914)	15,326	14,936
11780	Saint James (c) (631)	—	12,703
14779	Salamanca (716)	6,202	6,566
13454	Salisbury (c) (315)	—	12,226
12983	Saranac Lake (518)	5,114	5,377
12866	Saratoga Springs (518)	25,140	25,001
11782	Sayville (c) (631)	—	16,550
10583	Scarsdale (914)	17,802	16,987
*12301	Schenectady (518)	61,698	65,566
10940	Scotchtown (c) (914)	—	8,765
12302	Scotia (518)	7,349	7,359
11579	Sea Cliff (516)	5,013	5,054
11783	Seaford (c) (516)	—	15,597
11507	Searingtown (c) (516)	—	5,020
11784	Selden (c) (631)	—	20,608
13148	Seneca Falls (315)	7,005	7,370
11733	Setauket-East Setauket (c) (516)	—	13,634
11967	Shirley (Suffolk) (c) (631)	—	22,936
10591	Sleepy Hollow[1] (914)	8,180	8,152
11787	Smithtown (c) (631)	—	25,638
13209	Solvay (315)	6,262	6,717
11789	Sound Beach (c) (631)	—	9,102
11735	South Farmingdale (c) (516)	—	15,377
14850	South Hill (c) (607)	—	5,423
11746	South Huntington (c) (631)	—	9,624
14094	South Lockport (c) (716)	—	7,112
11581	South Valley Stream (c) (516)	—	5,328
11971	Southold (c) (631)	—	5,192
14904	Southport (c) (607)	—	7,753
10977	Spring Valley (914)	22,105	21,802
*11790	Stony Brook (c) (631)	—	13,726
10980	Stony Point (c) (Rockland) (914)	—	10,587
10901	Suffern (914)	11,059	11,055
11791	Syosset (c) (516)	—	18,967
*13220	Syracuse (315)	152,215	163,860
10983	Tappan (c) (914)	—	6,867
10591	Tarrytown (914)	11,228	10,739
11776	Terryville (c) (631)	—	10,275
10984	Thiells (c) (914)	—	5,204
10594	Thornwood (c) (914)	—	7,025
*14150	Tonawanda (c) (716)	—	65,284
*14150	Tonawanda (716)	15,901	17,284
*12180	Troy (518)	51,320	54,269
10707	Tuckahoe (914)	6,409	6,302
11553	Uniondale (c) (516)	—	20,328
*13504	Utica (315)	59,334	68,637
10989	Valley Cottage (c) (914)	—	9,007
*11582	Valley Stream (516)	33,891	33,946
11792	Wading River (c) (631)	—	5,317
12586	Walden (914)	6,205	5,836
11793	Wantagh (c) (516)	—	18,567
10990	Warwick (914)	6,834	5,984
10992	Washingtonville (914)	5,695	4,906
*13601	Watertown (315)	27,759	29,429
12189	Watervliet (518)	10,342	11,061
14580	Webster (716)	5,228	5,464
14895	Wellsville (716)	5,065	5,241
*11704	West Babylon (c) (631)	—	42,410
14905	West Elmira (c) (607)	—	5,218
12801	West Glens Falls (c) (518)	—	5,964
10993	West Haverstraw (914)	10,091	9,183
11552	West Hempstead (c) (516)	—	17,689
11743	West Hills (c) (631)	—	5,849
11795	West Islip (c) (631)	—	28,419
*10996	West Point (c) (914)	—	8,024
14224	West Seneca (c) (716)	—	47,866
11590	Westbury (Nassau) (516)	13,189	13,060
12203	Westmere (c) (518)	—	6,750
13219	Westvale (c) (315)	—	5,952
11798	Wheatley Heights (c) (631)	—	5,027
*10602	White Plains (914)	49,944	48,718
14231	Williamsville (716)	5,153	5,583
11596	Williston Park (516)	7,457	7,516
11797	Woodbury (c) (516)	—	8,008
11598	Woodmere (c) (516)	—	15,578
11798	Wyandach (c) (631)	—	8,950
*10702	Yonkers (914)	190,153	188,082
10598	Yorktown Heights (c) (914)	—	7,690

(1) North Tarrytown changed its name to Sleepy Hollow on Dec. 12, 1996.

North Carolina

ZIP	Place	1998	1990
*28001	Albemarle (704)	15,612	14,940
27502	Apex (919)	14,528	4,789
27263	Archdale (336)	7,728	6,975
*27203	Asheboro (336)	17,551	16,362
*28802	Asheville (828)	63,031	63,379
28012	Belmont (704)	8,467	8,434
28711	Black Mountain (828)	7,620	7,156
28607	Boone (828)	13,600	12,949
28712	Brevard (828)	5,644	5,452
*27215	Burlington (336)	40,531	39,498
*28547	Camp Lejeune (c) (910)	—	36,716
27510	Carrboro (919)	14,733	12,134
*27511	Cary (919)	82,071	44,394
*27514	Chapel Hill (919)	42,865	38,711

ZIP	Place		1998	1990
*28204	Charlotte	(704)	504,637	419,558
27520	Clayton	(919)	6,528	4,756
27012	Clemmons	(336)	7,008	5,982
*28328	Clinton	(910)	9,125	8,385
*28025	Concord	(704)	34,617	29,591
28613	Conover	(828)	5,829	5,311
*28334	Dunn	(910)	9,563	9,258
*27701	Durham	(919)	153,513	138,894
*27288	Eden	(336)	14,661	15,238
27932	Edenton	(252)	5,177	5,268
*27909	Elizabeth City	(252)	16,704	16,087
*28302	Fayetteville	(910)	77,295	75,850
28043	Forest City	(828)	7,385	7,475
28307	Fort Bragg (c)	(910)	—	34,744
27526	Fuquay-Varina	(919)	7,478	4,447
27529	Garner	(919)	17,511	14,716
*28052	Gastonia	(704)	56,977	54,725
*27530	Goldsboro	(919)	40,909	40,736
27253	Graham	(336)	11,806	10,368
*27420	Greensboro	(336)	197,910	185,125
*27834	Greenville	(252)	57,005	46,274
28540	Half Moon (c)	(910)	—	6,306
28345	Hamlet	(910)	6,812	6,722
28532	Havelock	(252)	20,274	20,300
27536	Henderson	(252)	14,934	15,655
*28739	Hendersonville	(828)	7,309	7,284
*28603	Hickory	(828)	31,523	28,474
*27260	High Point	(336)	76,117	69,428
27540	Holly Springs	(919)	7,054	1,203
28348	Hope Mills	(910)	9,798	8,272
*28540	Jacksonville	(910)	68,380	78,031
*28081	Kannapolis	(704)	36,975	31,592
*27284	Kernersville	(336)	11,580	11,860
27021	King	(336)	5,764	4,059
28086	Kings Mountain	(704)	9,399	8,768
*28502	Kinston	(252)	24,470	25,295
*28352	Laurinburg	(910)	15,998	16,131
28645	Lenoir	(828)	16,434	16,337
27023	Lewisville	(336)	7,624	6,433
*27292	Lexington	(336)	16,150	16,583
*28092	Lincolnton	(704)	7,550	6,955
28461	Long Beach	(910)	5,729	3,816
*28358	Lumberton	(910)	19,076	18,656
28403	Masonboro (c)	(910)	—	7,010
*28105	Matthews	(704)	17,119	13,756
27302	Mebane	(919)	6,518	4,754
28227	Mint Hill	(704)	17,000	13,637
*28110	Monroe	(704)	23,792	18,623
*28115	Mooresville	(704)	11,621	9,563
28557	Morehead City	(252)	7,160	6,473
*28655	Morganton	(828)	14,892	15,085
27030	Mount Airy	(336)	7,282	7,156
28120	Mount Holly	(704)	7,813	7,710
*28562	New Bern	(252)	21,770	20,728
27604	New Hope (Wake) (c)	(704)	—	5,694
28540	New River Station (c)	(910)	—	9,732
28658	Newton	(828)	12,026	11,134
27565	Oxford	(919)	8,286	7,965
*28374	Pinehurst	(910)	8,908	5,825
28399	Piney Green (c)	(910)	—	8,999
*27611	Raleigh	(919)	259,423	218,859
*27320	Reidsville	(336)	13,840	14,085
27870	Roanoke Rapids	(252)	15,521	15,722
*28379	Rockingham	(910)	8,819	9,399
*27801	Rocky Mount	(252)	56,901	53,078
27573	Roxboro	(336)	7,497	7,332
28601	Saint Stephens (c)	(828)	—	8,734
*28144	Salisbury	(704)	25,100	23,626
*27330	Sanford	(919)	21,784	18,881
28403	Seagate (c)	(910)	—	5,444
*28150	Shelby	(704)	15,568	15,460
.....	Smith Creek (c)	(910)	—	7,461
27577	Smithfield	(919)	10,839	10,180
28052	South Gastonia (c)	(704)	—	5,487
*28387	Southern Pines	(910)	10,412	9,213
28390	Spring Lake	(910)	7,820	7,552
*28677	Statesville	(704)	20,121	20,647
27886	Tarboro	(252)	10,082	11,037
*27360	Thomasville	(336)	18,070	15,915
27370	Trinity (c)	(336)	—	5,469
*27587	Wake Forest	(919)	10,156	5,832
27889	Washington	(252)	9,263	9,160
28786	Waynesville	(828)	7,755	7,282
28472	Whiteville	(910)	5,272	5,340
27892	Williamston	(252)	5,836	5,870
*28402	Wilmington	(910)	68,062	55,530
*27893	Wilson	(252)	40,192	38,400
*27102	Winston-Salem	(336)	164,316	162,292

North Dakota (701)

ZIP	Place	1998	1990
*58501	Bismarck	54,040	49,272
58301	Devils Lake	7,450	7,782
*58601	Dickinson	16,221	16,097
*58102	Fargo	86,718	74,084
58237	Grafton	5,117	4,884
*58201	Grand Forks	47,327	49,417
*58201	Grand Forks AFB (c)	—	9,343
*58401	Jamestown	14,713	15,571
58554	Mandan	15,860	15,177
*58701	Minot	35,286	34,544
*58701	Minot AFB (c)	6,862	7,163
58072	Valley City	9,322	8,751
*58075	Wahpeton	14,091	12,287
58078	West Fargo	12,446	13,136
*58801	Williston	14,713	15,571

Ohio

ZIP	Place		1998	1990
45810	Ada	(419)	5,524	5,428
*44309	Akron	(330)	215,712	223,019
44601	Alliance	(330)	22,448	23,376
44001	Amherst	(440)	11,364	10,332
44805	Ashland	(419)	21,521	20,079
*44004	Ashtabula	(440)	21,472	21,633
45701	Athens	(740)	21,706	21,265
44202	Aurora	(330)	11,530	9,192
44515	Austintown (c)	(330)	—	32,371
44011	Avon	(440)	10,615	7,337
44012	Avon Lake	(440)	17,171	15,066
44203	Barberton	(330)	27,097	27,623
44140	Bay Village	(440)	15,859	17,000
44122	Beachwood	(216)	10,955	10,644
45434	Beavercreek	(937)	40,014	33,626
44146	Bedford	(216)/(440)	13,800	14,822
44146	Bedford Heights	(216)/(440)	11,471	12,131
43906	Bellaire	(740)	5,603	6,028
45305	Bellbrook	(937)	7,033	6,511
43311	Bellefontaine	(937)	13,037	12,126
44811	Bellevue	(419)	8,051	8,157
45714	Belpre	(740)	7,016	6,796
44017	Berea	(440)	18,380	19,051
43209	Bexley	(614)	12,216	13,088
43004	Blacklick Estates (c)	(614)	—	10,080
45242	Blue Ash	(513)	12,374	11,923
44513	Boardman (c)	(330)	—	38,596
43402	Bowling Green	(419)	28,200	28,303
44141	Brecksville	(440)	12,623	11,818
45211	Bridgetown North (c)	(513)	—	11,748
44147	Broadview Heights	(440)	14,187	12,219
44144	Brooklyn	(216)	11,462	11,706
44142	Brook Park	(216)/(440)	22,084	22,865
44212	Brunswick	(330)	32,634	28,218
43506	Bryan	(419)	8,510	8,348
44820	Bucyrus	(419)	13,192	13,496
43725	Cambridge	(740)	11,791	11,748
44405	Campbell	(330)	9,246	10,038
44406	Canfield	(330)	5,459	5,409
*44711	Canton	(330)	79,259	84,161
45005	Carlisle	(937)	5,057	4,872
45822	Celina	(419)	10,615	9,945
*45441	Centerville (Montgomery)	(937)	23,035	21,082
44024	Chardon	(440)	5,010	4,446
45211	Cheviot	(513)	8,863	9,616
45601	Chillicothe	(740)	22,275	21,923
*45202	Cincinnati	(513)	336,400	364,114
43113	Circleville	(740)	12,107	11,666
*44101	Cleveland	(216)	495,817	505,616
44118	Cleveland Heights	(216)	53,533	54,052
43410	Clyde	(419)	5,990	6,087
44408	Columbiana	(330)	6,868	4,961
*43216	Columbus	(614)	670,234	632,945
44030	Conneaut	(440)	12,836	13,241
44410	Cortland	(330)	6,396	5,652
43812	Coshocton	(740)	12,183	12,193
45238	Covedale (c)	(513)	—	6,669
*44222	Cuyahoga Falls	(330)	49,913	48,950
*45401	Dayton	(937)	167,475	182,011
45236	Deer Park	(513)	5,675	6,181
43512	Defiance	(419)	16,458	16,787
43015	Delaware	(740)	18,962	19,966
45833	Delphos	(419)	6,800	7,093
45247	Dent (c)	(513)	—	6,416
44622	Dover (Tuscarawas)	(330)	11,844	11,329
45427	Drexel (c)	(937)	—	5,143
45663	Dry Run (c)	(614)	—	5,389
*43016	Dublin	(614)/(740)	25,506	16,366
44112	East Cleveland	(216)	29,937	33,096
*44094	Eastlake	(440)	20,976	21,161
43920	East Liverpool	(330)	13,151	13,654
44413	East Palestine	(330)	5,255	5,168
45320	Eaton	(937)	7,827	7,396
44004	Edgewood (c)	(440)	—	5,189
*44035	Elyria	(440)	56,278	56,746
45322	Englewood	(937)	11,870	11,402
*44117	Euclid	(216)	50,644	54,875
45324	Fairborn	(937)	31,390	31,300
*45011	Fairfield	(513)	41,765	39,709
44334	Fairlawn	(330)	6,327	5,779
44126	Fairview Park	(440)	16,897	18,028
*45839	Findlay	(419)	37,132	35,703
45224	Finneytown (c)	(513)	—	13,096
45405	Forest Park	(513)	19,442	18,621
45230	Forestville (c)	(513)	—	9,185
45426	Fort McKinley (c)	(937)	—	9,740
44830	Fostoria	(419)	14,379	14,971
45005	Franklin	(513)	11,664	11,026
43420	Fremont	(419)	17,010	17,619

ZIP	Place		1998	1990
43230	Gahanna	(614)	31,579	23,898
44833	Galion	(419)	11,396	11,859
45631	Gallipolis	(740)	5,045	4,831
44125	Garfield Heights	(216)	29,160	31,739
44041	Geneva	(440)	6,883	6,597
44420	Girard	(330)	10,872	11,304
43212	Grandview Heights	(614)	6,559	7,010
44232	Green	(330)	21,975	19,179
45123	Greenfield	(937)	5,586	5,172
45331	Greenville	(937)	13,099	12,863
45253	Groesbeck (c)	(513)	—	6,684
43123	Grove City	(614)	18,938	19,661
*45011	Hamilton	(513)	61,808	61,438
45030	Harrison	(513)	9,566	7,520
43056	Heath	(740)	8,332	7,231
44134	Highland Heights	(440)	7,144	6,249
43026	Hilliard	(614)/(740)	19,934	11,794
45133	Hillsboro	(937)	7,350	6,235
44484	Howland Center (c)	(330)	—	6,732
44425	Hubbard	(330)	8,078	8,248
45424	Huber Heights	(937)	36,997	38,696
43081	Huber Ridge (c)	(614)	—	5,255
44236	Hudson Village	(330)	21,226	17,128
44839	Huron	(419)	7,279	7,067
44131	Independence (Cuyahoga)	(216)/(440)	6,690	6,500
45011	Indian Springs	(513)	12,112	9,648
45638	Ironton	(740)	12,724	12,751
45640	Jackson	(740)	5,999	6,167
*44240	Kent	(330)	26,833	28,835
43326	Kenton	(419)	8,345	8,356
43606	Kenwood (c)	(513)	—	7,469
45429	Kettering	(937)	57,205	60,569
44094	Kirtland	(440)	6,538	5,881
44107	Lakewood	(216)	55,682	59,718
43130	Lancaster	(740)	37,701	34,507
45039	Landen (c)	(513)	—	9,263
45036	Lebanon (Warren)	(513)	13,802	10,461
*45802	Lima	(419)	42,382	45,553
43228	Lincoln Village (c)	(614)	—	9,958
43138	Logan	(740)	7,604	6,725
43140	London	(614)/(740)	8,507	7,807
*44052	Lorain	(440)	68,857	71,245
44641	Louisville	(330)	8,381	8,087
45140	Loveland	(513)	11,780	10,122
44124	Lyndhurst	(216)/(440)	15,109	15,982
44056	Macedonia	(330)	9,191	7,509
.....	Mack South (c)		—	5,767
45243	Madeira	(513)	8,754	9,141
*44901	Mansfield	(419)	49,802	50,627
44137	Maple Heights	(216)	25,302	27,089
45750	Marietta	(740)	14,857	15,026
*43302	Marion	(740)	32,281	34,075
43935	Martins Ferry	(740)	7,447	8,003
43040	Marysville	(937)	13,010	10,362
45040	Mason	(513)	18,850	11,450
*44646	Massillon	(330)	30,894	30,969
43537	Maumee	(419)	14,955	15,561
44124	Mayfield Heights	(440)	18,519	19,847
*44256	Medina	(330)	22,928	19,231
*44060	Mentor	(440)	49,227	47,491
44060	Mentor-on-the-Lake	(216)	8,509	8,271
*45343	Miamisburg	(937)	18,304	17,834
44130	Middleburg Heights	(216)/(440)	14,877	14,702
*45042	Middletown	(513)	48,590	46,758
45150	Milford	(513)	5,978	5,660
45050	Monroe	(513)	6,270	5,380
45242	Montgomery	(513)	9,645	9,733
45439	Moraine	(937)	7,568	5,989
45231	Mount Healthy	(513)	7,028	7,580
43050	Mount Vernon	(740)	14,973	14,550
44262	Munroe Falls	(330)	5,466	5,359
43545	Napoleon	(419)	9,189	8,884
*43055	Newark	(740)	48,245	44,396
45344	New Carlisle	(937)	6,157	6,049
43764	New Lexington	(740)	5,391	5,117
44663	New Philadelphia	(330)	16,615	15,698
44446	Niles	(330)	20,593	21,128
45239	Northbrook (c)	(513)	—	11,471
44720	North Canton	(330)	15,601	14,904
45239	North College Hill	(513)	10,363	11,002
45251	Northgate (c)	(513)	—	7,864
44057	North Madison (c)	(440)	—	8,699
44070	North Olmsted	(440)	33,546	34,204
45502	Northridge (c) (Clark)	(937)	—	5,939
45414	Northridge (c) (Montgomery)	(937)	—	9,448
44039	North Ridgeville	(440)	23,411	21,564
44133	North Royalton	(440)	25,016	23,197
45322	Northview (c)	(937)	—	10,337
43619	Northwood	(419)	5,915	5,506
44203	Norton	(330)	12,007	11,477
44857	Norwalk	(419)	15,694	14,731
45212	Norwood	(513)	21,450	23,674
44146	Oakwood (Cuyahoga)	(440)	8,343	8,957
44074	Oberlin	(440)	7,787	8,191
44138	Olmsted Falls	(440)	7,334	6,741
*45054	Oregon	(419)	19,136	18,334
44667	Orrville	(330)	8,227	7,955
45431	Overlook-Page Manor (c)	(937)	—	13,242
45056	Oxford	(513)	18,789	19,013
44077	Painesville	(440)	15,896	15,769

ZIP	Place		1998	1990
44129	Parma	(216)/(440)	83,347	87,876
44130	Parma Heights	(216)/(440)	20,624	21,448
44124	Pepper Pike	(216)/(440)	6,105	6,185
44646	Perry Heights (c)	(330)	—	9,055
*43551	Perrysburg	(419)	14,411	12,551
43147	Pickerington	(614)/(740)	8,927	5,668
45356	Piqua	(937)	19,810	20,612
44319	Portage Lakes (c)	(330)	—	13,373
43452	Port Clinton	(419)	7,123	7,106
45662	Portsmouth	(740)	22,213	22,676
44266	Ravenna	(330)	11,961	12,069
45215	Reading	(513)	11,488	12,038
43068	Reynoldsburg	(614)/(740)	29,473	25,748
44143	Richmond Heights	(216)/(440)	9,520	9,611
44270	Rittman	(330)	6,568	6,147
44116	Rocky River	(440)	19,506	20,410
43460	Rossford	(419)	6,015	5,861
43950	Saint Clairsville	(740)	5,196	5,136
45885	Saint Marys	(419)	8,373	8,441
44460	Salem	(330)	11,894	12,233
*44870	Sandusky	(419)	28,223	29,764
44870	Sandusky South (c)	(419)	—	6,336
44131	Seven Hills	(216)/(440)	12,276	12,339
44122	Shaker Heights	(216)	28,116	30,955
*45241	Sharonville	(513)	13,870	13,121
44054	Sheffield Lake	(440)	9,915	9,825
44875	Shelby	(419)	9,449	9,610
44878	Shiloh (c)	(419)	—	11,607
45365	Sidney	(937)	19,197	18,710
45236	Silverton	(513)	5,598	5,859
44139	Solon	(440)	20,017	18,548
44121	South Euclid	(216)	22,355	23,866
45066	Springboro	(513)	10,520	6,574
45246	Springdale	(513)	10,081	10,621
*45501	Springfield	(937)	65,568	70,487
*43952	Steubenville	(740)	20,224	22,125
44224	Stow	(330)	31,357	27,998
44241	Streetsboro	(330)	11,996	9,932
44136	Strongsville	(440)	41,304	35,308
44471	Struthers	(330)	12,342	12,284
43560	Sylvania	(419)	17,664	17,489
44278	Tallmadge	(330)	16,054	14,870
45243	The Village of Indian Hill	(513)	5,355	5,383
44883	Tiffin	(419)	18,007	18,604
45371	Tipp City	(937)	6,308	6,483
*43601	Toledo	(419)	312,174	332,943
43964	Toronto	(740)	5,662	6,127
45067	Trenton	(513)	7,712	6,189
45426	Trotwood	(937)	27,964	29,358
45373	Troy	(937)	21,672	19,478
44087	Twinsburg	(330)	15,179	9,606
44683	Uhrichsville	(740)	5,662	5,604
45322	Union	(937)	5,946	5,531
44122	University Heights	(216)	13,409	14,787
43221	Upper Arlington	(614)	31,699	34,128
43351	Upper Sandusky	(419)	5,992	5,906
43078	Urbana	(937)	11,142	11,353
45377	Vandalia	(937)	14,021	13,872
45891	Van Wert	(419)	10,608	10,922
44089	Vermilion	(440)	11,344	11,127
*44281	Wadsworth	(330)	17,567	15,718
45895	Wapakoneta	(419)	9,317	9,214
*44481	Warren	(330)	46,866	50,793
44122	Warrensville Heights	(216)	14,822	15,884
43160	Washington	(740)	13,283	13,080
43567	Wauseon	(419)	6,944	6,322
45692	Wellston	(740)	7,326	6,049
45449	West Carrollton City	(937)	13,709	14,403
*43081	Westerville	(614)	33,437	30,269
44145	Westlake	(440)	29,740	27,018
45694	Wheelersburg (c)	(740)	—	5,113
43213	Whitehall	(614)	19,237	20,572
45239	White Oak (c)	(513)	—	12,430
44092	Wickliffe	(440)	13,953	14,558
44890	Willard	(419)	6,652	6,210
*44094	Willoughby	(440)	21,494	20,510
44094	Willoughby Hills	(440)	8,832	8,427
*44095	Willowick	(440)	14,448	15,269
45177	Wilmington	(937)	11,866	11,199
45459	Woodbourne-Hyde Park (c)	(937)	—	7,837
44691	Wooster	(330)	23,609	22,427
43085	Worthington	(614)	14,103	14,869
45433	Wright-Patterson AFB (c)	(937)	—	8,579
45215	Wyoming	(513)	7,425	8,128
45385	Xenia	(937)	24,994	24,836
*44501	Youngstown	(330)	84,650	95,732
*43701	Zanesville	(740)	26,831	26,778

Oklahoma

ZIP	Place		1998	1990
*74820	Ada	(405)	15,313	15,765
*73521	Altus	(405)	21,552	21,910
73005	Anadarko	(405)	6,782	6,586
*73401	Ardmore	(405)	23,436	23,079
*74003	Bartlesville	(918)	33,672	34,256
73008	Bethany	(405)	20,269	20,075
74008	Bixby	(918)	12,694	9,502
74631	Blackwell	(405)	7,134	7,538
*74012	Broken Arrow	(918)	72,564	58,082

ZIP	Place		1998	1990
*73018	Chickasha	(405)	16,180	14,988
73020	Choctaw	(405)	9,797	8,545
*74017	Claremore	(918)	20,085	13,280
73601	Clinton	(405)	8,707	9,298
74429	Coweta	(918)	6,646	6,159
74023	Cushing	(918)	7,606	7,218
73115	Del City	(405)	23,817	23,928
*73533	Duncan	(405)	21,816	21,732
*74701	Durant	(405)	13,187	12,929
*73034	Edmond	(405)	64,962	52,310
*73644	Elk City	(405)	11,062	10,428
73036	El Reno	(405)	15,786	15,414
*73701	Enid	(405)	45,234	45,309
73503	Fort Sill (c)	(405)	—	12,107
74033	Glenpool	(918)	7,934	6,688
*74344	Grove	(918)	5,438	4,020
73044	Guthrie	(405)	10,281	10,440
73942	Guymon	(405)	8,999	7,803
74437	Henryetta	(918)	6,002	5,872
74743	Hugo	(405)	5,974	5,978
74745	Idabel	(405)	7,278	6,957
74037	Jenks	(918)	9,245	7,484
*73501	Lawton	(405)	81,107	80,561
*74501	McAlester	(918)	17,074	16,739
*74354	Miami	(918)	12,760	13,142
73140	Midwest City	(405)	54,037	52,267
73153	Moore	(405)	45,318	40,318
*74401	Muskogee	(918)	38,386	37,708
73064	Mustang	(405)	12,409	10,434
73065	Newcastle	(405)	5,381	4,214
73068	Nobel	(405)	5,170	4,710
*73069	Norman	(405)	93,019	80,071
*73125	Oklahoma City	(405)	472,221	444,724
74447	Okmulgee	(918)	13,981	13,441
74055	Owasso	(918)	15,032	11,151
73075	Pauls Valley	(405)	5,964	6,150
73077	Perry	(405)	5,136	4,978
*74601	Ponca City	(405)	25,943	26,359
74953	Poteau	(918)	7,755	7,210
74361	Pryor Creek	(918)	9,056	8,327
73080	Purcell	(405)	5,198	4,784
74955	Sallisaw	(918)	7,910	7,122
74063	Sand Springs	(918)	17,255	15,339
*74066	Sapulpa	(918)	19,844	18,074
*74868	Seminole	(405)	6,698	7,071
*74801	Shawnee	(405)	27,008	26,017
74070	Skiatook	(918)	5,331	4,910
*74074	Stillwater	(405)	38,765	36,676
*74464	Tahlequah	(918)	12,336	10,586
74873	Tecumseh	(405)	5,856	5,750
73156	The Village	(405)	10,289	10,353
*74103	Tulsa	(918)	381,393	367,302
74301	Vinita	(918)	5,741	5,804
*74467	Wagoner	(918)	7,281	6,894
73123	Warr Acres	(405)	9,291	9,288
73096	Weatherford	(405)	9,598	10,124
*73801	Woodward	(405)	12,034	12,340
*73099	Yukon	(405)	22,897	20,935

Oregon

Area code (971) overlays area code (503). See introductory note.

ZIP	Place		1998	1990
97321	Albany	(541)	38,832	33,523
*97006	Aloha (c)	(503)	—	34,284
97601	Altamont (c)	(541)	—	18,591
97520	Ashland	(541)	18,095	16,252
97103	Astoria	(503)	9,676	10,069
97814	Baker City	(541)	9,765	9,140
*97005	Beaverton	(503)	62,111	53,307
*97701	Bend	(541)	34,321	23,740
97415	Brookings	(541)	5,106	4,400
97013	Canby	(503)	12,084	8,990
97225	Cedar Hills (c)	(503)	—	9,294
97291	Cedar Mill (c)	(503)	—	9,697
97502	Central Point	(541)	10,583	7,512
97058	City of the Dalles	(541)	11,211	11,021
97420	Coos Bay	(541)	15,229	15,076
97113	Cornelius	(503)	7,560	6,148
*97333	Corvallis	(541)	50,202	44,757
97424	Cottage Grove	(541)	7,635	7,403
97338	Dallas	(503)	12,331	9,422
*97440	Eugene	(541)	128,240	112,733
97024	Fairview	(503)	5,625	2,588
97439	Florence	(541)	6,434	5,171
97116	Forest Grove	(503)	15,200	13,559
97301	Four Corners (c)	(503)	—	12,156
97223	Garden Home-Whitford (c)	(503)	—	6,652
97027	Gladstone	(503)	11,762	10,152
*97526	Grants Pass	(541)	21,366	17,503
97470	Green (c)	(541)	—	5,076
*97030	Gresham	(503)	85,021	68,285
97303	Hayesville (c)	(503)	—	14,318
97230	Hazelwood (c)	(503)	—	11,480
97838	Hermiston	(541)	11,514	10,047
*97123	Hillsboro	(503)	61,111	37,598
97031	Hood River	(541)	5,278	4,632
97351	Independence	(503)	5,970	4,425
97222	Jennings Lodge (c)	(503)	—	6,530
97307	Keizer	(503)	28,967	21,884

ZIP	Place		1998	1990
*97601	Klamath Falls	(541)	18,538	17,737
97850	La Grande	(541)	12,060	11,766
*97034	Lake Oswego	(503)	34,704	30,576
97355	Lebanon	(541)	12,471	10,950
97367	Lincoln City	(541)	6,910	5,903
97128	McMinnville	(503)	24,086	17,894
*97501	Medford	(541)	57,156	47,021
97862	Milton-Freewater	(541)	5,861	5,533
97269	Milwaukie	(503)	19,895	18,670
97361	Monmouth	(503)	7,830	6,288
97132	Newberg	(503)	16,962	13,086
97365	Newport	(541)	9,992	8,437
97459	North Bend	(541)	9,720	9,614
97477	North Springfield (c)	(541)	—	5,451
97268	Oak Grove (c)	(503)	—	12,576
.....	Oak Hills (c)		—	6,450
.....	Oatfield (c)		—	15,348
97914	Ontario	(541)	10,848	9,394
97045	Oregon City	(503)	20,940	14,698
97801	Pendleton	(541)	16,060	15,142
*97208	Portland	(503)	503,891	485,975
97236	Powellhurst-Centennial (c)	(503)	—	28,756
97754	Prineville	(541)	6,098	5,355
97225	Raleigh Hills (c)	(503)	—	6,066
97756	Redmond	(541)	11,728	7,165
*97404	River Road (c)	(541)	—	9,443
.....	Rockcreek (c)		—	8,282
97470	Roseburg	(541)	19,289	18,389
97470	Roseburg North (c)	(541)	—	6,831
97051	Saint Helens	(503)	8,808	7,535
*97309	Salem	(503)	126,702	107,793
97055	Sandy	(503)	5,098	4,154
97401	Santa Clara (c)	(541)	—	12,834
97138	Seaside	(503)	5,668	5,359
97140	Sherwood	(503)	8,875	3,093
97381	Silverton	(503)	6,654	5,635
*97477	Springfield	(541)	50,682	44,664
97383	Stayton	(503)	6,387	5,011
97479	Sutherlin	(541)	5,825	5,020
97386	Sweet Home	(541)	7,444	6,850
97281	Tigard	(503)	36,920	29,435
97060	Troutdale	(503)	13,576	7,852
97062	Tualatin	(503)	19,978	14,664
97225	West Haven-Sylvan (c)	(503)	—	6,009
97068	West Linn	(503)	21,202	16,389
*97225	West Slope (c)	(503)	—	7,959
97503	White City (c)	(541)	—	5,891
97070	Wilsonville	(503)	13,124	7,510
97071	Woodburn	(503)	14,981	13,404

Pennsylvania

Area code (267) overlays area code (215).
Area code (484) overlays area code (610). See introductory note.

ZIP	Place		1998	1990
15001	Aliquippa	(724)	12,448	13,374
*18105	Allentown (Lehigh)	(610)	100,757	105,301
*16603	Altoona	(814)	49,226	51,881
19002	Ambler	(215)	6,465	6,609
15003	Ambridge	(724)	7,553	8,133
18403	Archbald	(570)	6,344	6,291
19003	Ardmore (c)	(610)	—	12,646
15068	Arnold	(724)	5,681	6,113
19407	Audubon (c)	(610)	—	6,328
15202	Avalon	(412)	5,278	5,784
15234	Baldwin	(412)	20,512	21,923
18013	Bangor	(610)	5,101	5,383
15010	Beaver Falls	(724)	10,104	10,687
16823	Bellefonte	(814)	6,112	6,358
15202	Bellevue	(412)	8,329	9,126
18603	Berwick	(570)	10,389	10,976
15102	Bethel Park	(412)	32,869	33,823
*18016	Bethlehem	(610)	69,383	71,427
18447	Blakely	(570)	6,705	7,222
17815	Bloomsburg	(570)	12,495	12,439
19422	Blue Bell (c)	(215)/(610)	—	6,091
19061	Boothwyn (c)	(610)	—	5,069
16701	Bradford	(814)	9,449	9,625
15227	Brentwood	(412)	9,924	10,823
15017	Bridgeville	(412)	5,116	5,445
19007	Bristol	(215)	10,142	10,405
19015	Brookhaven	(610)	8,268	8,570
19008	Broomall (c)	(610)	—	10,930
*16001	Butler	(724)	14,871	15,714
15419	California	(724)	5,229	5,748
*17011	Camp Hill	(717)	7,405	7,831
15317	Canonsburg	(724)	8,639	9,200
18407	Carbondale	(570)	9,631	10,664
17013	Carlisle	(717)	17,720	18,419
15106	Carnegie	(412)	8,499	9,278
15108	Carnot-Moon (c)	(412)	—	10,187
15234	Castle Shannon	(412)	8,478	9,135
18032	Catasauqua	(610)	6,379	6,662
17201	Chambersburg	(717)	17,295	16,647
*19013	Chester	(610)	40,221	41,856
19013	Chester Twp. (c)	(610)	—	5,399
15025	Clairton	(412)	8,763	9,656
16214	Clarion	(814)	6,436	6,457
18411	Clarks Summit	(570)	5,126	5,433
16830	Clearfield	(814)	6,383	6,633

ZIP	Place		1998	1990
19018	Clifton Heights	(610)	6,802	7,111
19320	Coatesville	(610)	10,687	11,038
19426	Collegeville	(610)	5,061	4,227
19023	Collingdale	(610)	8,742	9,175
17109	Colonial Park (c) (Dauphin)	(717)	—	13,777
17512	Columbia	(717)	10,465	10,701
15425	Connellsville	(724)	8,610	9,229
19428	Conshohocken	(610)	8,176	8,064
15108	Coraopolis	(412)	6,204	6,747
16407	Corry	(814)	6,832	7,216
15205	Crafton	(412)	6,550	7,188
19021	Croydon (c)	(215)	—	9,967
19023	Darby	(610)	10,658	11,140
19036	Darby Twp. (c)	(610)	—	10,955
19333	Devon-Berwyn (c)	(610)	—	5,019
18519	Dickson City	(570)	5,785	6,276
15033	Donora	(724)	5,542	5,928
15216	Dormont	(412)	8,848	9,772
19335	Downingtown	(610)	7,767	7,749
18901	Doylestown	(215)	8,380	8,575
19026	Drexel Hill (c)	(610)	—	29,744
15801	Du Bois	(814)	7,941	8,286
18512	Dunmore	(570)	14,253	15,403
15110	Duquesne	(412)	7,684	8,525
19401	East Norriton (c)	(610)	—	13,324
*18042	Easton	(610)	25,361	26,276
18301	East Stroudsburg	(570)	10,056	8,781
17402	East York (c)	(717)	—	8,487
15005	Economy	(724)	9,737	9,305
16412	Edinboro	(814)	6,800	7,736
17022	Elizabethtown	(717)	10,619	9,952
16117	Ellwood City	(724)	8,343	8,894
18049	Emmaus	(610)	11,409	11,157
17025	Enola (c)	(717)	—	5,961
17522	Ephrata	(717)	13,000	12,133
*16501	Erie	(814)	102,640	108,718
18643	Exeter	(570)	6,004	5,691
19030	Fairless Hills (c)	(215)	—	9,026
16121	Farrell	(724)	6,462	6,835
19053	Feasterville-Trevose (c)	(215)	—	6,696
16063	Fernway (c)	(724)	—	9,072
19032	Folcroft	(610)	7,268	7,506
19033	Folsom (c)	(610)	—	8,173
15221	Forest Hills	(412)	6,809	7,335
15238	Fox Chapel	(412)	5,280	5,319
16323	Franklin	(814)	6,845	7,329
15143	Franklin Park	(412)	11,070	10,109
18052	Fullerton (c)	(610)	—	13,127
17325	Gettysburg	(717)	7,376	7,025
15045	Glassport	(412)	5,106	5,582
19036	Glenolden	(610)	7,069	7,260
19038	Glenside (c)	(215)	—	8,704
15601	Greensburg	(724)	15,531	16,318
16125	Greenville	(724)	6,349	6,734
16127	Grove City	(412)	8,056	8,240
15101	Hampton Twp. (c) (Allegheny)	(412)	—	15,568
17331	Hanover	(717)	14,269	14,399
19438	Harleysville (c)	(215)	—	7,405
*17105	Harrisburg	(717)	49,502	52,376
15065	Harrison Twp. (c) (Allegheny)	(412)	—	11,763
19040	Hatboro	(215)	7,274	7,382
18201	Hazleton	(570)	22,542	24,730
18055	Hellertown	(610)	5,549	5,662
16148	Hermitage	(724)	16,274	15,260
17033	Hershey (c)	(717)	—	11,860
16648	Hollidaysburg	(814)	5,337	5,624
16001	Homeacre-Lyndora (c)	(724)	—	7,511
18431	Honesdale	(717)	5,182	4,972
19044	Horsham (c)	(215)	—	15,051
16652	Huntingdon	(814)	7,071	6,843
15701	Indiana	(724)	14,399	15,174
15644	Jeannette	(724)	10,449	11,221
15344	Jefferson	(412)	9,585	9,533
18229	Jim Thorpe	(570)	5,100	5,048
*15907	Johnstown	(814)	25,390	28,124
15108	Kennedy Twp. (c)	(412)	—	7,152
19348	Kennett Square	(610)	5,161	5,218
19406	King of Prussia (c)	(610)	—	18,406
18704	Kingston	(570)	13,146	14,507
19443	Kulpsville (c)	(215)	—	5,183
*17604	Lancaster	(717)	52,951	55,551
19446	Lansdale	(215)	15,936	16,362
19050	Lansdowne	(610)	11,151	11,712
15650	Latrobe	(724)	9,050	9,265
17540	Leacock-Leola-Bareville (c)	(717)	—	5,685
*17042	Lebanon	(717)	23,442	24,800
18235	Lehighton	(610)	5,726	5,914
*19055	Levittown (c)	(215)	—	55,362
17837	Lewisburg	(570)	5,527	5,785
17044	Lewistown (Mifflin)	(717)	8,707	9,341
17112	Linglestown (c)	(717)	—	5,862
19353	Lionville-Marchwood (c)	(610)	—	6,468
17543	Lititz	(717)	8,635	8,280
17745	Lock Haven	(570)	9,012	9,230
17011	Lower Allen (c)	(717)	—	6,329
15068	Lower Burrell	(724)	12,211	12,251
15237	McCandless Twp. (c)	(412)	—	28,781
*15134	McKeesport	(412)	23,089	26,016
15136	McKees Rocks	(412)	7,007	7,691
19002	Maple Glen (c)	(215)	—	5,881

ZIP	Place		1998	1990
16335	Meadville	(814)	14,004	14,318
17055	Mechanicsburg	(717)	9,032	9,452
*19063	Media	(610)	5,723	5,957
17057	Middletown (Dauphin)	(717)	8,915	9,254
18017	Middletown (c) (Northampton)	(610)	—	6,866
17551	Millersville	(717)	7,778	8,099
17847	Milton	(570)	6,327	6,746
15061	Monaca	(724)	6,444	6,739
15062	Monessen	(724)	9,124	9,901
18936	Montgomeryville (c)	(215)	—	9,114
18507	Moosic	(570)	5,377	5,397
19067	Morrisville (Bucks)	(215)	9,461	9,765
17851	Mount Carmel	(570)	6,500	7,196
17552	Mount Joy	(717)	6,535	6,398
15228	Mount Lebanon (c)	(412)	—	34,414
15120	Munhall	(412)	12,019	13,158
15146	Municipality of Monroeville	(412)	27,964	29,169
15668	Municipality of Murrysville	(724)	19,143	17,240
18634	Nanticoke	(570)	11,122	12,267
18064	Nazareth	(610)	5,416	5,713
19086	Nether Providence Twp. (c)	(610)	—	12,730
15066	New Brighton	(724)	6,394	6,854
*16108	New Castle	(724)	26,178	28,334
17070	New Cumberland	(717)	7,287	7,665
15068	New Kensington	(724)	14,805	15,894
*19403	Norristown	(610)	29,763	30,754
18067	Northampton	(610)	8,937	8,717
15104	North Braddock	(412)	6,494	7,036
15137	North Versailles (c)	(412)	—	13,294
16421	Northwest Harborcreek (c)	(814)	—	7,485
19074	Norwood (Delaware)	(610)	6,103	6,162
15139	Oakmont (Allegheny)	(412)	6,673	6,961
15238	O'Hara (c)	(412)	—	9,096
16301	Oil City	(814)	11,185	11,949
18518	Old Forge	(570)	8,576	8,834
19075	Oreland (c)	(215)	—	5,695
18071	Palmerton	(610)	5,231	5,394
17078	Palmyra	(717)	7,230	6,910
19301	Paoli (c)	(610)	—	5,277
16801	Park Forest Village (c)	(814)	—	6,703
17331	Parkville (c)	(717)	—	5,009
15235	Penn Hills (c)	(412)	—	57,632
19096	Penn Wynne (c)	(610)	—	5,807
18944	Perkasie	(215)	8,112	7,878
*19104	Philadelphia	(215)	1,436,287	1,585,577
19460	Phoenixville	(610)	15,207	15,066
*15233	Pittsburgh	(412)	340,520	369,879
*18640	Pittston	(570)	8,662	9,389
15236	Pleasant Hills	(412)	8,277	8,884
15239	Plum	(412)	26,469	25,609
18651	Plymouth	(570)	6,424	7,134
19462	Plymouth Meeting (c)	(610)	—	6,241
*19464	Pottstown	(610)	21,465	21,831
17901	Pottsville	(570)	15,374	16,603
17109	Progress (c)	(717)	—	9,654
19076	Prospect Park	(610)	6,540	6,764
15767	Punxsutawney	(814)	6,736	6,782
18951	Quakertown	(215)	8,920	8,982
19087	Radnor Twp. (c)	(610)	—	27,676
*19612	Reading	(610)	74,762	78,380
17356	Red Lion	(717)	6,033	6,130
18954	Richboro (c)	(215)	—	5,141
19078	Ridley Park	(610)	7,317	7,592
15136	Robinson (Allegheny) (c)	(412)	—	10,830
15237	Ross Twp. (c)	(412)	—	35,102
15857	Saint Marys	(814)	13,842	14,020
19464	Sanatoga (c)	(610)	—	3,723
18840	Sayre	(570)	5,528	5,791
17972	Schuylkill Haven	(570)	5,340	5,610
15106	Scott Twp. (c)	(412)	—	20,413
*18505	Scranton	(570)	74,683	81,805
17870	Selinsgrove	(570)	5,407	5,384
15116	Shaler Twp. (c)	(412)	—	33,694
17872	Shamokin	(570)	8,228	9,184
16146	Sharon	(724)	16,373	17,533
19079	Sharon Hill	(610)	5,540	5,771
17976	Shenandoah	(570)	5,952	6,221
17404	Shiloh (c)	(717)	—	5,315
17257	Shippensburg	(717)	5,467	5,331
15501	Somerset	(814)	6,194	6,454
18964	Souderton	(215)	6,294	5,957
15129	South Park Twp. (c)	(814)	—	14,292
17701	South Williamsport	(570)	6,154	6,496
19064	Springfield (c) (Delaware)	(610)	—	25,326
*16804	State College	(814)	39,550	38,981
15136	Stowe Twp. (c)	(412)	—	9,202
18360	Stroudsburg	(570)	5,928	5,312
16323	Sugarcreek	(814)	5,368	5,532
17801	Sunbury	(570)	10,483	11,591
19081	Swarthmore	(610)	5,976	6,157
15218	Swissvale	(412)	9,701	10,637
18704	Swoyersville	(570)	5,268	5,630
18252	Tamaqua	(570)	7,314	7,943
15084	Tarentum	(724)	5,154	5,674
18517	Taylor	(570)	6,545	6,941
16354	Titusville	(814)	6,367	6,434
19401	Trooper (c)	(610)	—	7,370
15145	Turtle Creek	(412)	6,027	6,556
16686	Tyrone	(814)	5,574	5,743
15401	Uniontown (Fayette)	(724)	11,197	12,034

ZIP	Place		1998	1990
19063	Upper Providence Twp. (c) ...	(610)	—	9,477
15241	Upper Saint Clair (c)	(412)	—	19,023
15690	Vandergrift	(724)	5,443	5,904
19013	Village Green-Green Ridge (c)	(610)	—	9,026
16365	Warren..................	(814)	9,980	11,122
15301	Washington	(724)	14,805	15,864
17268	Waynesboro	(717)	9,891	9,578
17315	Weigelstown (c)..........	(717)	—	8,665
*19380	West Chester............	(610)	17,988	18,041
19380	West Goshen (c)	(610)	—	8,948
*15122	West Mifflin.............	(412)	22,111	23,644
15905	Westmont...............	(814)	5,382	5,789
19401	West Norriton (c).........	(610)	—	15,209
18643	West Pittston	(570)	5,128	5,590
15229	West View	(412)	7,154	7,734
15227	Whitehall (Allegheny)	(412)	13,728	14,451
15131	White Oak	(412)	8,223	8,761
*18703	Wilkes-Barre	(570)	42,828	47,523
15221	Wilkinsburg	(412)	19,128	21,080
15145	Wilkins Twp. (c)	(412)	—	7,487
*17701	Williamsport............	(570)	29,891	31,933
19090	Willow Grove (c)(Montgomery)	(215)	—	16,325
17584	Willow Street (c)	(717)	—	5,817
15025	Wilson	(412)	7,521	7,830
19094	Woodlyn (c).............	(610)	—	10,151
19038	Wyndmoor (c)	(215)	—	5,682
19610	Wyomissing	(610)	7,527	7,332
19050	Yeadon	(610)	11,452	11,980
*17405	York...................	(717)	39,978	42,192

Rhode Island (401)

See introductory note.

ZIP	Place	1998	1990
02806	*Barrington*	15,816	15,849
02809	*Bristol*	21,893	21,625
02830	*Burrillville*	15,938	16,230
02863	Central Falls	16,364	17,637
02813	*Charlestown*	7,165	6,478
02816	*Coventry*	32,397	31,083
*02904	Cranston	74,521	76,060
02864	*Cumberland*	29,445	29,038
02864	Cumberland Hill (c)	—	6,379
02818	*East Greenwich*	12,306	11,865
02914	East Providence	47,882	50,380
02822	*Exeter*	6,124	5,461
02814	*Glocester*	9,300	9,227
02828	Greenville (c)	—	8,303
02833	*Hopkinton*	7,708	6,873
02835	*Jamestown*	5,061	4,999
02919	*Johnston*	26,636	26,542
02881	Kingston (c)	—	6,504
02865	*Lincoln*	19,020	18,045
02842	*Middletown*	19,187	19,460
02882	*Narragansett*	15,888	15,004
02840	Newport	24,279	28,227
02843	Newport East (c)	—	11,080
02852	*North Kingstown*	26,236	23,786
02908	*North Providence*	30,932	32,090
02896	*North Smithfield*	10,635	10,497
02859	Pascoag (c)	—	5,011
*02860	Pawtucket	68,169	72,644
02871	*Portsmouth*	16,771	16,857
*02904	Providence	150,890	160,728
02812	*Richmond*	6,783	5,351
02857	*Scituate*	10,099	9,796
02917	*Smithfield*	18,763	19,163
02879	*South Kingstown*	26,705	24,612
02878	Tiverton (c)	—	7,259
02878	*Tiverton*	14,223	14,312
02864	Valley Falls (c)	—	11,175
*02879	Wakefield-Peacedale (c)	—	7,134
02885	*Warren*	11,405	11,385
*02886	Warwick	84,094	85,427
02891	*Westerly*	23,075	21,605
02891	Westerly Center (c)	—	16,477
02893	*West Warwick*	28,692	29,268
02895	Woonsocket	41,034	43,877

South Carolina

ZIP	Place		1998	1990
29620	Abbeville	(864)	5,281	5,778
*29801	Aiken.................	(803)	22,861	20,386
*29621	Anderson..............	(864)	26,098	26,385
29812	Barnwell..............	(803)	5,298	5,255
.....	Batesburg-Leesville........	(803)	6,207	6,107
*29902	Beaufort...............	(843)	9,956	9,576
29627	Belton	(864)	5,498	4,646
29841	Belvedere (c)..........	(803)	—	6,133
29512	Bennettsville	(843)	8,980	10,095
29611	Berea (c)	(864)	—	13,535
29115	Brookdale (c)..........	(803)	—	5,339
29902	Burton (c)	(843)	—	6,917
29020	Camden...............	(803)	6,254	6,696
29033	Cayce	(803)	11,936	10,824
*29402	Charleston.............	(843)	87,044	88,256
29520	Cheraw	(843)	5,736	5,553
29706	Chester	(803)	6,982	7,158
*29631	Clemson	(864)	12,336	11,145

ZIP	Place		1998	1990
29325	Clinton...............	(864)	9,386	9,603
*29201	Columbia..............	(803)	110,840	110,734
*29526	Conway	(843)	9,993	9,819
*29532	Darlington	(843)	6,992	7,310
29204	Dentsville (c)..........	(803)	—	11,839
29536	Dillon	(843)	6,714	6,829
*29640	Easley	(864)	17,703	15,179
*29501	Florence	(843)	29,511	29,913
29206	Forest Acres...........	(803)	7,002	7,181
*29715	Fort Mill	(803)	5,790	4,930
29644	Fountain Inn	(864)	5,216	4,388
*29341	Gaffney	(864)	13,078	13,149
29605	Gantt (c)	(864)	—	13,891
29576	Garden City (c)	(843)	—	6,305
*29442	Georgetown............	(843)	9,085	9,517
29445	Goose Creek...........	(843)	26,673	24,692
*29602	Greenville.............	(864)	56,436	58,256
*29646	Greenwood	(864)	19,536	20,807
*29650	Greer.................	(864)	12,965	10,322
29406	Hanahan	(843)	13,047	13,176
*29550	Hartsville	(843)	8,277	8,372
*29928	Hilton Head Island......	(843)	30,377	23,694
29621	Homeland Park (c)	(864)	—	6,569
29063	Irmo.................	(803)	10,850	11,284
29456	Ladson (c).............	(843)	—	13,540
29560	Lake City	(843)	6,909	7,153
*29720	Lancaster	(803)	8,676	8,914
29360	Laurens	(864)	9,403	9,694
*29072	Lexington	(803)	7,027	4,046
29571	Marion................	(843)	7,328	7,658
29662	Mauldin	(864)	14,330	11,662
29461	Moncks Corner.........	(843)	5,903	5,599
*29465	Mount Pleasant.........	(843)	41,330	30,108
29574	Mullins...............	(843)	5,436	5,910
*29575	Myrtle Beach	(803)	25,284	24,848
29108	Newberry..............	(803)	9,860	10,543
*29841	North Augusta..........	(803)	16,307	15,684
*29410	North Charleston........	(843)	68,072	70,304
*29582	North Myrtle Beach	(843)	9,542	8,731
29565	Oak Grove (c)	(803)	—	7,173
*29115	Orangeburg............	(803)	12,733	13,772
.....	Parker (c).............		—	11,072
29905	Parris Island (c)........	(843)	—	7,172
29072	Red Bank (c)...........	(803)	—	5,950
29020	Red Hill (c)	(843)	—	6,112
*29730	Rock Hill	(803)	46,218	42,112
29417	Saint Andrews (c)	(843)	—	25,692
29609	Sans Souci (c)..........	(864)	—	7,612
*29678	Seneca	(864)	8,177	7,726
29210	Seven Oaks (c).........	(803)	—	15,722
*29681	Simpsonville	(864)	11,661	11,744
29577	Socastee (c)	(843)	40,954	43,479
*29306	Spartanburg...........	(864)	24,292	22,519
*29483	Summerville	(843)	40,518	40,977
*29150	Sumter................	(803)	40,954	43,479
29687	Taylors (c)	(864)	—	19,619
29379	Union.................	(864)	9,501	9,840
29607	Wade Hampton (c)	(864)	—	20,014
29488	Walterboro............	(843)	5,319	5,595
29611	Welcome (c)	(864)	—	6,560
*29169	West Columbia	(803)	10,941	10,974
29206	Woodfield (c)..........	(803)	—	8,862
29745	York.................	(803)	7,747	6,709

South Dakota (605)

ZIP	Place	1998	1990
*57401	Aberdeen..............	24,865	24,995
57005	Brandon...............	5,042	3,545
57006	Brookings	17,138	16,270
57706	Ellsworth AFB (c)	—	7,017
57350	Huron	11,778	12,448
57042	Madison	6,650	6,257
57301	Mitchell	14,386	13,798
57501	Pierre.................	13,267	12,906
*57701	Rapid City	57,513	54,523
57701	Rapid Valley (c)........	—	5,968
*57101	Sioux Falls............	116,762	100,836
57783	Spearfish..............	8,851	6,966
57785	Sturgis................	5,087	5,537
57069	Vermillion.............	11,967	10,034
57201	Watertown.............	19,909	17,623
57078	Yankton	14,325	12,703

Tennessee

ZIP	Place		1998	1990
37701	Alcoa.................	(865)	7,348	6,400
*37303	Athens................	(423)	13,486	12,054
38184	Bartlett...............	(901)	35,391	27,038
37660	Bloomingdale (c).......	(423)	—	10,953
38008	Bolivar	(901)	5,934	5,969
*37027	Brentwood	(615)	23,331	16,392
37621	Bristol	(423)	23,109	23,421
38012	Brownsville	(901)	10,003	10,017
*37401	Chattanooga...........	(423)	147,790	152,393
37642	Church Hill............	(423)	6,160	5,208
*37040	Clarksville	(615)	97,978	75,542
*37311	Cleveland	(423)	35,454	32,236
*37716	Clinton...............	(865)	9,188	8,960

ZIP	Place		1998	1990
37315	Collegedale	(423)	6,089	5,048
*38017	Collierville	(901)	23,720	14,501
37663	Colonial Heights (c)	(423)	—	6,716
*38401	Columbia	(615)	31,865	28,583
*38501	Cookeville	(615)	25,471	21,744
38019	Covington	(901)	8,162	7,487
*38555	Crossville	(615)	9,778	6,930
37321	Dayton	(423)	6,263	5,671
*37055	Dickson	(615)	11,996	10,487
*38024	Dyersburg	(901)	16,422	16,321
37801	Eagleton Village (c)	(865)	—	5,169
37411	East Brainerd (c)	(423)	—	11,594
37412	East Ridge	(423)	19,885	21,101
*37643	Elizabethton	(423)	13,211	13,087
37650	Erwin	(423)	5,362	5,318
37062	Fairview	(615)	5,733	4,210
37922	Farragut	(865)	16,805	12,802
37334	Fayetteville	(615)	7,373	7,158
*37064	Franklin	(615)	30,925	20,098
37066	Gallatin	(615)	21,608	18,794
*38138	Germantown	(901)	37,587	33,159
*37072	Goodlettsville	(615)	13,325	11,219
*37743	Greeneville	(423)	13,973	13,532
37215	Green Hills (c)	(615)	—	6,763
38040	Halls (c)	(901)	—	6,450
37748	Harriman	(865)	7,005	7,119
37341	Harrison (c)	(423)	—	7,191
38340	Henderson	(901)	5,631	4,760
*37075	Hendersonville	(615)	38,625	32,188
38343	Humboldt	(901)	9,664	9,651
*38301	Jackson	(901)	51,115	49,145
37760	Jefferson City	(865)	8,105	5,875
*37601	Johnson City	(423)	57,079	50,354
*37662	Kingsport	(423)	41,139	40,457
37763	Kingston	(423)	5,004	4,552
*37950	Knoxville	(865)	165,540	169,761
37766	La Follette	(423)	7,628	7,201
37086	La Vergne	(615)	15,072	7,496
38464	Lawrenceburg	(615)	11,107	10,397
*37087	Lebanon	(615)	17,282	15,208
*37771	Lenoir City	(865)	7,081	6,147
37091	Lewisburg	(615)	11,096	9,879
38351	Lexington	(901)	6,849	5,810
37352	Lynchburg	(615)	5,196	4,721
38201	McKenzie	(901)	5,245	5,168
*37110	McMinnville	(615)	12,319	11,194
*37355	Manchester	(615)	8,707	7,709
38237	Martin	(901)	8,915	8,588
*37804	Maryville	(865)	23,308	19,208
*38101	Memphis	(901)	603,507	618,652
37343	Middle Valley (c)	(423)	—	12,255
38358	Milan	(901)	7,532	7,512
*38053	Millington	(901)	18,677	17,866
*37813	Morristown	(423)	23,068	22,513
*37122	Mount Juliet	(615)	8,534	5,389
*37130	Murfreesboro	(615)	58,430	44,922
*37202	Nashville	(615)	510,274	488,366
*37821	Newport	(423)	8,038	7,123
*37830	Oak Ridge	(865)	27,045	27,310
38242	Paris	(901)	9,814	9,332
37148	Portland	(615)	7,318	5,539
37849	Powell (c)	(865)	—	7,534
38478	Pulaski	(615)	8,842	7,916
37415	Red Bank	(423)	11,498	12,320
38063	Ripley	(901)	6,564	6,634
37854	Rockwood	(865)	5,392	5,348
38372	Savannah	(901)	6,675	6,547
*37862	Sevierville	(865)	10,662	7,178
37865	Seymour (c)	(865)	—	7,026
*37160	Shelbyville	(615)	16,149	14,042
37377	Signal Mountain	(423)	6,892	7,034
37167	Smyrna	(615)	22,686	14,720
*37379	Soddy-Daisy	(423)	8,976	8,240
37311	South Cleveland (c)	(423)	—	5,372
38583	Sparta	(931)	5,102	4,681
37172	Springfield	(615)	12,952	11,227
37174	Spring Hill	(931)	5,355	1,464
37874	Sweetwater	(423)	5,450	5,066
37388	Tullahoma	(615)	19,153	16,761
*38261	Union City	(901)	10,147	10,513
37188	White House	(615)	5,965	2,987
37398	Winchester	(615)	6,638	6,305

Texas

Area codes (972) and (469) overlay area code (214).
Area codes (281) and (832) overlay area code (713). See introductory note.

ZIP	Place		1998	1990
*79604	Abilene	(915)	108,257	106,707
75001	Addison	(214)	12,276	8,783
78516	Alamo	(956)	11,078	8,352
78209	Alamo Heights	(210)	6,839	6,502
77039	Aldine (c)	(713)	—	11,133
*78332	Alice	(361)	20,532	19,788
*75002	Allen	(214)	38,941	19,315
*79830	Alpine	(915)	5,803	5,622
*77511	Alvin	(713)	20,797	19,220
*79105	Amarillo	(806)	171,207	157,571
78750	Anderson Mill (c)		—	9,468
79714	Andrews	(915)	10,271	10,678

ZIP	Place		1998	1990
*77515	Angleton	(409)	20,518	17,140
*78336	Aransas Pass	(361)	8,188	7,180
*76004	Arlington	(817)	306,497	261,717
75751	Athens	(903)	11,822	10,982
75551	Atlanta	(214)	5,571	6,118
*78712	Austin	(512)	552,434	472,020
*76020	Azle	(817)	10,345	8,868
77518	Bacliff (c)	(409)	—	5,549
75180	Balch Springs	(214)	18,435	17,406
78602	Bastrop	(512)	5,276	4,044
*77414	Bay City	(409)	18,386	18,170
*77520	Baytown	(713)	68,588	63,843
*77707	Beaumont	(409)	109,841	114,323
*76021	Bedford	(817)	50,148	43,762
*78102	Beeville	(361)	13,736	13,547
*77401	Bellaire	(713)	15,506	13,844
76715	Bellmead	(254)	9,147	8,336
76513	Belton	(254)	15,639	12,463
76126	Benbrook	(817)	21,742	19,564
*79720	Big Spring	(915)	22,382	23,093
*78006	Boerne	(830)	6,170	4,361
75418	Bonham	(903)	7,306	6,688
*79007	Borger	(806)	14,444	15,675
76230	Bowie	(940)	5,412	4,990
76825	Brady	(915)	5,879	5,946
76424	Breckenridge	(254)	5,804	5,665
*77833	Brenham	(409)	13,661	11,952
77611	Bridge City	(409)	8,231	8,010
79316	Brownfield	(806)	8,977	9,560
*78520	Brownsville	(956)	137,883	107,027
*76801	Brownwood	(915)	19,235	18,387
78717	Brushy Creek (c)	(903)	—	5,833
*77801	Bryan	(409)	58,763	55,002
76354	Burkburnett	(940)	10,673	10,145
*76028	Burleson	(817)	20,817	16,113
76520	Cameron	(254)	5,951	5,635
79015	Canyon	(806)	13,346	11,365
78130	Canyon Lake (c)	(830)	—	9,975
78834	Carrizo Springs	(830)	5,779	5,745
*75006	Carrollton	(214)	100,463	82,169
75633	Carthage	(903)	6,813	6,496
*75104	Cedar Hill	(214)	28,248	19,988
*78613	Cedar Park	(512)	18,371	5,161
77530	Channelview (c)	(713)	—	25,564
79201	Childress	(940)	5,221	5,055
*76031	Cleburne	(817)	25,033	22,205
*77327	Cleveland	(713)	7,507	7,124
77015	Cloverleaf (c)	(713)	—	18,230
77531	Clute	(409)	9,835	9,467
76834	Coleman	(915)	5,295	5,410
*77840	College Station	(409)	59,742	52,443
76034	Colleyville	(817)	20,030	12,724
79512	Colorado City	(915)	6,528	4,749
*75428	Commerce	(903)	7,205	6,825
*77301	Conroe	(409)	35,353	27,675
78109	Converse	(210)	11,415	8,887
75019	Coppell	(214)	28,940	16,881
76522	Copperas Cove	(254)	30,946	24,079
76205	Corinth	(940)	8,076	3,944
*78469	Corpus Christi	(361)	281,453	257,428
*75110	Corsicana	(903)	23,184	22,911
75835	Crockett	(409)	7,084	7,024
76036	Crowley	(817)	7,747	6,974
78839	Crystal City	(830)	8,152	8,263
77954	Cuero	(361)	6,491	6,700
79022	Dalhart	(806)	7,053	6,246
*75221	Dallas	(214)	1,075,894	1,007,618
77535	Dayton	(409)	6,214	5,042
76234	Decatur	(214)	5,284	4,245
78016	Devine	(830)	5,130	3,928
77536	Deer Park	(713)	30,575	27,424
*78840	Del Rio	(830)	34,990	30,705
*75020	Denison	(903)	22,170	21,505
*76201	Denton	(940)	76,933	66,270
*75115	De Soto	(214)	35,686	30,544
75941	Diboll	(409)	5,250	4,341
77539	Dickinson	(713)	12,828	11,692
78537	Donna	(956)	15,193	12,652
79029	Dumas	(806)	13,821	12,871
*75138	Duncanville	(214)	36,160	35,008
76135	Eagle Mountain (c)	(817)	—	5,847
*78852	Eagle Pass	(830)	28,713	20,651
*78539	Edinburg	(956)	40,579	31,091
77957	Edna	(361)	6,141	5,436
77437	El Campo	(409)	10,643	10,511
78621	Elgin	(512)	6,158	4,846
*79910	El Paso	(915)	615,032	515,342
78543	Elsa	(956)	6,225	5,242
*75119	Ennis	(214)	15,902	13,869
*76039	Euless	(817)	45,249	38,149
76140	Everman	(817)	5,888	5,672
79838	Fabens (c)	(915)	—	5,599
78355	Falfurrias	(361)	5,966	5,788
75381	Farmers Branch	(214)	26,156	24,250
.....	First Colony (c)		—	18,327
78114	Floresville	(830)	7,023	5,247
*75067	Flower Mound	(214)	44,338	15,527
76119	Forest Hill	(817)	12,040	11,482
75126	Forney	(214)	5,740	4,070
79906	Fort Bliss (c)	(915)	—	13,915

ZIP	Place	1998	1990
76544	Fort Hood (c) (254)	—	35,580
79735	Fort Stockton (915)	8,301	8,524
*76161	Fort Worth (817)	491,801	447,619
78624	Fredericksburg (830)	8,847	6,934
*77541	Freeport (409)	11,594	11,389
*77546	Friendswood (713)	28,897	22,814
*75034	Frisco (214)	26,304	6,138
*76240	Gainesville (940)	14,760	14,256
77547	Galena Park (713)	10,409	10,033
*77550	Galveston (409)	59,567	59,067
*75040	Garland (214)	193,408	180,635
76528	Gatesville (254)	12,003	11,492
*78626	Georgetown (512)	28,790	14,840
75644	Gilmer (903)	5,516	4,824
75647	Gladewater (903)	6,600	6,027
75115	Glenn Heights (214)	5,745	4,564
78629	Gonzales (830)	6,618	6,527
76450	Graham (940)	8,750	8,986
*76048	Granbury (817)	5,626	4,045
*75051	Grand Prairie (214)	113,329	99,606
*76051	Grapevine (817)	40,299	29,407
*75401	Greenville (903)	25,051	23,071
77619	Groves (409)	16,523	16,744
76117	Haltom City (817)	37,061	32,856
76548	Harker Heights (254)	17,347	12,932
*78550	Harlingen (956)	58,210	48,746
*75652	Henderson (903)	11,280	11,139
79045	Hereford (806)	14,667	14,745
76643	Hewitt (254)	11,205	8,983
78557	Hidalgo (956)	6,169	3,292
75205	Highland Park (214)	9,038	8,739
77562	Highlands (c) (713)	—	6,632
75067	Highland Village (214)	12,253	7,027
76645	Hillsboro (254)	7,897	7,072
77563	Hitchcock (409)	6,288	5,868
78861	Hondo (830)	8,446	6,018
*77052	Houston (713)	1,786,691	1,654,348
*77338	Humble (713)	13,341	12,060
*77340	Huntsville (409)	31,706	30,628
*76053	Hurst (817)	37,266	33,574
78362	Ingleside (361)	10,257	5,696
76367	Iowa Park (940)	6,363	6,072
*75015	Irving (214)	178,253	155,037
77029	Jacinto City (713)	9,766	9,343
75766	Jacksonville (214)	13,012	12,765
75951	Jasper (409)	7,838	7,160
77040	Jersey Village (713)	5,707	4,826
78729	Jollyville (c) (512)	—	15,206
*77449	Katy (713)	10,792	8,004
75142	Kaufman (214)	6,529	5,251
*76248	Keller (817)	23,352	13,683
79745	Kermit (915)	6,338	6,875
*78028	Kerrville (830)	21,031	17,384
*75662	Kilgore (903)	11,363	11,066
*76540	Killeen (254)	80,720	63,535
*78363	Kingsville (361)	25,211	25,276
77325	Kingwood (c) (713)	—	37,397
78219	Kirby (210)	8,835	8,326
78236	Lackland AFB (c) (210)	—	9,352
78559	La Feria (956)	5,073	4,360
75065	Lake Dallas (940)	5,532	3,656
77566	Lake Jackson (409)	26,394	22,771
78734	Lakeway (512)	5,532	4,044
77568	La Marque (409)	14,723	14,120
79331	Lamesa (806)	10,050	10,809
76550	Lampasas (512)	8,077	6,382
*75146	Lancaster (214)	24,216	22,117
*77571	La Porte (713)	32,999	27,923
*78041	Laredo (956)	175,783	122,893
*77573	League City (713)	43,633	30,159
*78641	Leander (512)	8,265	3,354
78268	Leon Valley (210)	10,348	9,581
*79336	Levelland (806)	13,596	13,986
*75067	Lewisville (214)	72,466	46,521
77575	Liberty (713)	8,173	7,690
79339	Littlefield (806)	6,244	6,489
78233	Live Oak (210)	10,807	10,023
77351	Livingston (409)	7,699	5,019
78644	Lockhart (512)	11,602	9,205
*75606	Longview (903)	75,576	70,311
*79408	Lubbock (806)	190,974	186,206
*75901	Lufkin (409)	33,253	30,210
78648	Luling (830)	5,279	4,661
77657	Lumberton (409)	8,144	6,640
*78501	McAllen (956)	106,822	84,021
*75070	McKinney (214)	40,404	21,283
76063	Mansfield (817)	23,567	15,615
*78654	Marble Falls (512)	5,656	4,007
76661	Marlin (254)	6,344	6,386
*75670	Marshall (903)	23,548	23,682
78368	Mathis (361)	5,808	5,423
77477	Meadows (713)	6,621	4,606
78570	Mercedes (956)	14,531	12,694
*75149	Mesquite (214)	114,632	101,484
76667	Mexia (254)	6,572	6,933
*79701	Midland (915)	99,621	89,343
76065	Midlothian (214)	7,155	5,040
*76067	Mineral Wells (940)	14,825	14,935
*78572	Mission (956)	40,083	28,653
.....	Mission Bend (c)	—	24,945
*77489	Missouri City (713)	62,371	36,143
79756	Monahans (915)	6,851	8,101
*75455	Mount Pleasant (903)	13,037	12,291
*75961	Nacogdoches (409)	30,755	30,872
77868	Navasota (409)	7,816	6,296
77627	Nederland (409)	16,774	16,192
75570	New Boston (903)	5,265	5,057
*78130	New Braunfels (830)	36,526	27,334
*76161	North Richland Hills (817)	54,622	45,895
*79761	Odessa (915)	91,572	89,699
*77630	Orange (409)	18,524	19,370
*75801	Palestine (903)	18,931	18,042
*79065	Pampa (806)	18,704	19,959
*75460	Paris (903)	25,513	24,799
*77501	Pasadena (713)	133,964	119,604
*77581	Pearland (713)	29,164	18,927
78061	Pearsall (830)	7,338	6,924
78721	Pecan Grove (c)	—	9,502
79772	Pecos (915)	10,757	12,069
79070	Perryton (806)	7,416	7,619
*78660	Pflugerville (512)	9,337	4,444
78577	Pharr (956)	42,318	32,921
*79072	Plainview (806)	22,697	21,698
*75074	Plano (214)	219,486	127,885
78064	Pleasanton (830)	9,335	7,678
*77640	Port Arthur (409)	56,827	58,551
78578	Port Isabel (956)	5,144	4,467
78374	Portland (361)	14,682	12,224
77979	Port Lavaca (361)	11,908	10,886
77651	Port Neches (409)	13,225	12,908
77580	Raymondville (956)	9,596	8,880
76028	Rendon (c) (817)	—	7,658
*75080	Richardson (214)	86,020	74,840
76118	Richland Hills (817)	8,564	7,978
*77469	Richmond (713)	14,307	10,042
78582	Rio Grande City (956)	14,886	10,725
76219	River Oaks (817)	6,822	6,580
76701	Robinson (254)	8,162	7,111
78380	Robstown (361)	13,115	12,849
76567	Rockdale (512)	5,408	5,235
*78382	Rockport (361)	7,191	5,619
*75087	Rockwall (214)	15,668	10,486
78584	Roma (956)	11,216	8,059
77471	Rosenberg (713)	29,081	20,183
*78681	Round Rock (512)	60,686	30,923
*75088	Rowlett (214)	39,030	23,260
75048	Sachse (214)	7,390	5,346
76179	Saginaw (817)	11,238	8,551
*76902	San Angelo (915)	88,233	84,462
*78265	San Antonio (210)	1,114,130	976,514
78586	San Benito (956)	23,317	20,125
78384	San Diego (361)	5,081	4,983
78589	San Juan (956)	18,157	12,561
*78666	San Marcos (512)	39,491	28,738
*77510	Santa Fe (713)	9,833	8,429
78154	Schertz (210)	16,521	10,597
77586	Seabrook (713)	9,183	6,685
75159	Seagoville (214)	9,997	8,969
77474	Sealy (409)	5,461	4,541
*78155	Seguin (830)	21,719	18,692
79360	Seminole (915)	6,750	6,342
*75090	Sherman (903)	34,044	31,584
77656	Silsbee (409)	6,712	6,368
78387	Sinton (361)	6,625	5,549
79364	Slaton (806)	5,953	6,078
*79549	Snyder (915)	11,502	12,195
79910	Socorro (915)	27,085	22,995
77587	South Houston (713)	15,240	14,207
76092	Southlake (817)	16,552	7,082
*77373	Spring (c) (713)	—	33,111
*77477	Stafford (713)	18,870	8,395
76401	Stephenville (254)	15,262	13,502
*77478	Sugar Land (713)	51,725	33,712
*75482	Sulphur Springs (903)	14,616	14,062
79556	Sweetwater (915)	11,862	11,967
76574	Taylor (512)	14,690	11,472
*76501	Temple (254)	49,427	46,150
*75160	Terrell (214)	14,498	12,490
*75501	Texarkana (903)	31,485	32,294
*77590	Texas City (409)	42,488	40,822
75056	The Colony (214)	27,440	22,113
77387	The Woodlands (c) (713)	—	29,205
*77375	Tomball (713)	7,744	6,370
.....	Town West (c)	—	6,166
76262	Trophy Club (817)	5,369	3,922
*75702	Tyler (903)	83,908	75,450
*78148	Universal City (830)	15,354	13,057
76308	University Park (214)	23,018	22,259
*78801	Uvalde (830)	16,214	14,729
*76384	Vernon (940)	10,784	12,001
*77901	Victoria (361)	61,882	55,076
*77662	Vidor (409)	10,956	10,935
*76702	Waco (254)	108,272	103,590
75501	Wake Village (903)	5,316	4,761
76148	Watauga (817)	23,213	20,009
*75165	Waxahachie (214)	22,038	17,984
*76086	Weatherford (817)	18,572	14,804
78728	Wells Branch (c)	—	7,094
*78596	Weslaco (956)	27,630	22,739
79764	West Odessa (c) (915)	—	16,568

ZIP	Place		1998	1990
77005	West University Place	(713)	14,103	12,920
77488	Wharton	(409)	9,237	9,011
75791	Whitehouse	(903)	5,410	4,018
75693	White Oak	(903)	5,897	5,136
76108	White Settlement	(817)	15,956	15,472
*76307	Wichita Falls	(940)	99,236	96,259
78239	Windcrest	(210)	5,662	5,331
76712	Woodway	(254)	9,887	8,695
75098	Wylie	(214)	11,959	8,716
77995	Yoakum	(361)	5,517	5,611
78076	Zapata (c)	(956)	—	7,119

Utah

ZIP	Place		1998	1990
84004	Alpine	(801)	5,418	3,492
84003	American Fork	(801)	19,215	15,722
*84010	Bountiful	(801)	40,427	37,544
84302	Brigham City	(435)	16,960	15,644
84109	Canyon Rim (c)	(801)	—	10,527
*84720	Cedar City	(435)	18,953	13,443
84014	Centerville	(801)	14,811	11,500
*84015	Clearfield	(801)	25,877	21,435
84015	Clinton	(801)	11,514	7,945
84121	Cottonwood Heights (c)	(801)	—	28,766
84121	Cottonwood West (c)	(801)	—	17,476
84020	Draper	(801)	19,147	7,143
84109	East Millcreek (c)	(801)	—	21,184
84025	Farmington	(801)	11,175	9,049
84029	Grantsville	(801)	5,528	4,500
84032	Heber	(801)	5,872	4,782
84003	Highland	(801)	6,315	5,007
84117	Holladay-Cottonwood (c)	(801)	—	14,095
84737	Hurricane	(435)	7,193	3,915
84319	Hyrum	(435)	5,452	4,829
84037	Kaysville	(801)	19,118	13,961
84118	Kearns (c)	(801)	—	28,374
*84041	Layton	(801)	55,112	41,784
84043	Lehi	(801)	15,297	8,475
84042	Lindon	(801)	6,380	3,818
.....	Little Cottonwood Creek Valley (c)	(801)	—	5,042
*84321	Logan	(435)	40,272	32,771
84044	Magna (c)	(801)	—	17,829
84047	Midvale	(801)	11,628	11,886
84109	Millcreek (c)	(801)	—	32,230
84117	Mount Olympus (c)	(801)	—	7,413
84157	Murray	(801)	33,167	31,274
84341	North Logan	(435)	6,051	3,775
84404	North Ogden	(801)	14,811	11,593
84054	North Salt Lake	(801)	8,469	6,464
*84401	Ogden	(801)	66,507	63,943
.....	Oquirrh (c)	(801)	—	7,593
*84057	Orem	(801)	78,937	67,561
*84060	Park City	(801)	6,504	4,468
84651	Payson	(801)	10,951	9,510
84062	Pleasant Grove	(801)	20,491	13,476
84404	Pleasant View	(801)	5,076	3,597
84501	Price	(435)	8,834	8,712
*84601	Provo	(801)	110,419	86,835
84701	Richfield	(435)	6,880	5,593
84403	Riverdale	(801)	7,520	6,419
84065	Riverton	(801)	20,410	11,261
84067	Roy	(801)	31,441	24,560
*84770	Saint George	(435)	46,186	28,572
*84101	Salt Lake City	(801)	174,348	159,928
*84070	Sandy	(801)	99,186	75,240
84335	Smithfield	(435)	7,123	5,566
84095	South Jordan	(801)	26,414	12,215
84403	South Ogden	(801)	14,671	12,105
84165	South Salt Lake	(801)	9,957	10,129
84660	Spanish Fork	(801)	15,555	11,272
84663	Springville	(801)	15,944	13,950
84015	Sunset	(801)	5,060	5,128
84075	Syracuse	(801)	7,540	4,658
84107	Taylorsville	(801)	56,753	51,550
84074	Tooele	(435)	16,748	13,887
84337	Tremonton	(435)	5,116	4,262
84047	Union (c)	(801)	—	13,684
*84078	Vernal	(435)	7,366	6,640
84780	Washington	(435)	6,906	4,198
84403	Washington Terrace	(801)	8,821	8,189
84087	West Bountiful	(801)	5,053	4,477
*84084	West Jordan	(801)	60,804	42,915
84015	West Point	(801)	6,195	4,258
84170	West Valley City	(801)	99,372	86,969
84070	White City (c)	(801)	—	6,506
84087	Woods Cross	(801)	5,887	5,384

Vermont (802)

See introductory note.

ZIP	Place	1998	1990
05641	Barre	9,066	9,482
05641	*Barre*	7,716	7,411
05201	Bennington	16,069	16,451
05201	Bennington (c)	—	9,532
*05301	Brattleboro Center (c)	—	8,612
*05301	*Brattleboro*	11,932	12,241
*05401	*Burlington*	38,453	39,127

ZIP	Place	1998	1990
*05446	*Colchester*	16,275	14,731
05451	*Essex*	18,076	16,498
*05452	*Essex Junction*	8,705	8,396
05047	*Hartford*	9,461	9,404
05849	*Lyndon*	5,561	5,371
*05753	*Middlebury*	8,244	8,034
05468	*Milton*	9,863	8,404
*05602	*Montpelier*	7,734	8,247
05661	*Morristown*	5,303	4,733
05663	*Northfield*	5,823	5,610
05101	*Rockingham*	5,330	5,484
*05701	*Rutland*	17,348	18,230
05478	Saint Albans	7,308	7,339
05478	*Saint Albans*	5,273	4,606
05819	*Saint Johnsbury*	7,362	7,608
05482	*Shelburne*	6,789	5,871
*05401	South Burlington	14,037	12,809
05156	*Springfield*	9,294	9,579
05488	*Swanton*	6,191	5,636
05495	*Williston*	6,868	4,887
05404	*Winooski*	6,619	6,649

Virginia

Area code (571) overlays area code (703). See introductory note.

ZIP	Place		1998	1990
*24210	Abingdon	(540)	7,684	7,003
*22313	Alexandria	(703)	118,300	111,182
22003	Annandale	(540)	—	50,975
22554	Aquia Harbour (c)	(703)	—	6,308
*22210	Arlington (c)	(703)	177,275	170,897
23005	Ashland	(804)	7,394	5,864
*22041	Bailey's Crossroads (c)	(703)	—	19,507
24523	Bedford	(540)	6,317	6,177
22306	Belle Haven (c)	(757)	—	6,427
23234	Bellwood (c)	(804)	—	6,178
23234	Bensley (c)	(804)	—	5,093
*24060	Blacksburg	(540)	33,651	34,590
24605	Bluefield	(540)	5,155	5,363
23235	Bon Air (c)	(804)	—	16,413
*24203	Bristol	(540)	17,486	18,426
24416	Buena Vista	(540)	6,288	6,406
.....	Bull Run (c)	(540)	—	5,525
*22150	Burke (c)	(703)	—	57,734
24018	Cave Spring (c)	(540)	—	24,053
*20120	Centreville (c)	(703)	—	26,585
*20151	Chantilly (c)	(703)	—	29,337
*22906	Charlottesville	(804)	38,223	40,475
*23320	Chesapeake	(757)	199,564	151,982
*23831	Chester (c)	(804)	—	14,986
*24073	Christiansburg	(540)	16,153	15,004
24078	Collinsville (c)	(540)	—	7,280
23834	Colonial Heights	(804)	16,955	16,064
22901	Commonwealth (c)	(804)	—	5,538
.....	Countryside (c)		—	8,349
24426	Covington	(540)	6,857	7,198
22701	Culpeper	(540)	8,964	8,581
22193	Dale City (c)	(540)	—	47,170
*24541	Danville	(804)	50,868	53,056
23228	Dumbarton (c)	(804)	—	8,526
22027	Dunn Loring (c)	(703)	—	6,509
23222	East Highland Park (c)	(804)	—	11,850
23847	Emporia	(804)	5,474	5,479
23803	Ettrick (c)	(804)	—	5,290
*22030	Fairfax	(703)	20,697	19,894
*22046	Falls Church	(703)	10,042	9,522
23901	Farmville	(804)	6,800	6,505
24551	Forest (c)	(804)	—	5,624
22060	Fort Belvoir (c)	(703)	—	8,590
22308	Fort Hunt (c)	(703)	—	12,989
23801	Fort Lee (c)	(804)	—	6,895
22310	Franconia (c)	(703)	—	19,882
23851	Franklin	(757)	8,685	7,864
*22404	Fredericksburg	(540)	21,686	19,027
22630	Front Royal	(540)	13,464	11,880
24333	Galax	(540)	6,864	6,699
*23060	Glen Allen (c)	(804)	—	9,010
23062	Gloucester Point (c)	(804)	—	8,509
22066	Great Falls (c)	(703)	—	6,945
22306	Groveton (c)	(703)	—	19,997
*23670	Hampton	(757)	136,968	133,811
*22801	Harrisonburg	(540)	33,434	30,707
*20170	Herndon	(703)	19,197	16,139
23075	Highland Springs (c)	(804)	—	13,823
24019	Hollins (c)	(540)	—	13,305
23860	Hopewell	(804)	22,529	23,101
22303	Huntington (c)	(703)	—	7,489
22306	Hybla Valley (c)	(703)	—	15,491
22043	Idylwood (c)	(703)	—	14,710
22042	Jefferson (c)	(703)	—	25,782
22041	Lake Barcroft (c)	(703)	—	8,686
22191	Lake Ridge (c)	(540)	—	23,862
23228	Lakeside (c)	(804)	—	12,081
23060	Laurel (c)	(804)	—	13,011
*20175	Leesburg	(703)	27,009	16,202
24450	Lexington	(540)	7,360	6,959
22312	Lincolnia (c)	(703)	—	13,041
*22079	Lorton (c)	(703)	—	15,385
*24506	Lynchburg	(804)	65,473	66,049
*22101	McLean (c)	(703)	—	38,168

ZIP	Place		1998	1990
24572	Madison Heights (c)	(804)	—	11,700
*20110	Manassas	(703)	35,336	27,957
20113	Manassas Park	(703)	8,711	6,734
22030	Mantua (c)	(703)	—	6,804
24354	Marion	(540)	6,519	6,630
*24112	Martinsville	(540)	15,668	16,162
*23111	Mechanicsville (c)	(804)	—	22,027
*22116	Merrifield (c)	(703)	—	8,399
.....	Montclair (c)		—	11,399
23231	Montrose (c)	(804)	—	6,405
22121	Mount Vernon (c)	(703)	—	27,485
22122	Newington (c)	(703)	—	17,965
*23607	Newport News	(757)	178,615	171,439
*23501	Norfolk	(757)	215,215	261,250
22151	North Springfield (c)	(703)	—	8,996
22124	Oakton (c)	(703)	—	24,610
*23804	Petersburg	(804)	34,724	37,027
22043	Pimmit Hills (c)	(703)	—	6,019
23662	Poquoson	(757)	11,455	11,005
*23707	Portsmouth	(757)	98,936	103,910
24301	Pulaski	(540)	9,566	9,985
22134	Quantico Station (c)	(703)	—	7,425
*24141	Radford	(540)	15,734	15,940
*20190	Reston (c)	(703)	—	48,556
*23232	Richmond	(804)	194,173	202,798
22901	Rio (c)	(804)	—	5,133
*24022	Roanoke	(540)	93,749	96,509
24281	Rose Hill (c)	(540)	—	12,675
24153	Salem	(540)	24,679	23,797
22044	Seven Corners (c)	(703)	—	7,280
*23430	Smithfield	(757)	5,297	4,686
24592	South Boston	(804)	6,665	6,997
*22150	Springfield (c)	(703)	—	23,706
*24402	Staunton	(540)	23,346	24,461
*20164	Sterling (c)	(703)	—	20,512
24477	Stuarts Draft (c)	(540)	—	5,087
23162	Sudley (c)	(540)	—	7,321
*23434	Suffolk	(757)	62,703	52,143
22170	Sugarland Run (c)	(703)	—	9,357
24502	Timberlake (c)	(804)	—	10,314
23229	Tuckahoe (c)	(804)	—	42,629
22101	Tysons Corner (c)	(703)	—	13,124
22901	University Heights (c)	(804)	—	6,900
*22180	Vienna (c)	(703)	16,867	14,852
24179	Vinton	(540)	7,175	7,643
*23450	Virginia Beach	(757)	432,380	393,089
*20186	Warrenton	(540)	5,732	4,882
22980	Waynesboro	(540)	18,561	18,549
22110	West Gate (c)	(703)	—	6,565
22152	West Springfield (c)	(703)	—	28,126
*23185	Williamsburg	(757)	11,971	11,409
*22601	Winchester	(540)	22,659	21,947
24592	Wolf Trap (c)	(703)	—	13,133
*22191	Woodbridge (c)	(540)	—	26,401
24382	Wytheville	(540)	8,038	8,036
22110	Yorkshire (c)	(703)	—	5,699

Washington

ZIP	Place		1998	1990
98520	Aberdeen	(360)	16,326	16,565
98036	Alderwood Manor-Bothell North	(425)	—	22,945
98221	Anacortes	(360)	14,880	11,451
98223	Arlington	(360)	6,500	4,037
98335	Artondale (c)	(253)	—	7,141
*98002	Auburn	(253)	37,615	33,650
98604	Battle Ground	(360)	5,414	3,758
*98009	Bellevue	(425)	104,052	95,213
*98225	Bellingham	(360)	61,894	52,179
98390	Bonney Lake	(360)	9,767	7,494
*98011	Bothell	(425)	18,062	12,575
*98337	Bremerton	(360)	39,540	38,142
98036	Brier	(425)	6,667	5,633
98178	Bryn Mawr-Skyway (c)	(206)	—	12,514
98166	Burien	(206)	27,018	27,507
98233	Burlington	(360)	5,904	4,349
98607	Camas	(360)	11,130	6,762
98055	Cascade-Fairwood (c)	(425)	—	30,107
98684	Cascade Park East (c)	(425)	—	6,996
98684	Cascade Park West (c)	(425)	—	6,656
98531	Centralia	(360)	13,176	12,101
98532	Chehalis	(360)	6,655	6,527
99004	Cheney	(509)	8,116	7,723
99403	Clarkston	(509)	7,230	6,753
99324	College Place	(509)	7,190	6,308
99114	Colville	(509)	5,003	4,360
99218	Country Homes (c)	(509)	—	5,126
98042	Covington-Sawyer-Wilderness (c)		—	24,321
98198	Des Moines	(206)	21,425	20,830
99213	Dishman (c)	(509)	—	9,671
.....	East Hill-Meridian (c)		—	42,696
98366	East Port Orchard (c)	(360)	—	5,409
98056	East Renton Highlands (c)	(425)	—	13,218
98802	East Wenatchee	(509)	5,129	3,886
98801	East Wenatchee Bench (c)	(509)	—	12,539
98371	Edgewood	(253)	10,629	8,702
*98020	Edmonds	(425)	33,086	30,743
98387	Elk Plain (c)		—	12,197
98926	Ellensburg	(509)	14,419	12,360
.....	Ellsworth North (c)		—	5,796
98022	Enumclaw	(360)	9,602	7,243
98823	Ephrata	(509)	6,473	5,349
99210	Esperance (c)	(509)	—	11,236
*98201	Everett	(425)	88,625	70,937
98411	Evergreen (c)		—	11,249
99218	Fairwood (c)	(509)	—	5,807
*98002	Federal Way	(253)	74,254	67,535
98248	Ferndale	(360)	7,725	5,398
98466	Fircrest	(253)	5,341	5,270
98597	Five Corners (c)		—	6,776
98433	Fort Lewis (c)	(253)	—	22,224
98930	Grandview	(509)	8,124	7,169
.....	Harbour Pointe (c)		—	9,107
98660	Hazel Dell North (c)	(360)	—	6,924
98665	Hazel Dell South (c)	(360)	—	5,796
98550	Hoquiam	(360)	8,987	8,972
98011	Inglewood-Finn Hill (c)	(425)	—	29,132
*98027	Issaquah	(425)	10,103	7,786
98626	Kelso	(360)	12,246	11,767
98028	Kenmore (c)	(425)	—	8,917
*99336	Kennewick	(509)	50,316	42,148
*98031	Kent	(253)/(425)	45,066	37,960
98033	Kingsgate (c)	(425)	—	14,259
*98033	Kirkland	(425)	45,724	40,059
98509	Lacey	(360)	29,114	19,279
98155	Lake Forest North (c)	(206)	—	8,002
98002	Lakeland North (c)	(253)	—	14,402
98002	Lakeland South (c)	(253)	—	9,027
98036	Lake Serene-North Lynnwood (c)	(425)	—	14,290
98665	Lake Shore (c)	(360)	—	6,268
98258	Lake Stevens	(425)	5,584	3,435
98259	Lakewood	(253)	65,933	55,937
.....	Lea Hill (c)		—	6,876
98632	Longview	(360)	33,800	31,499
98264	Lynden	(360)	8,617	5,709
*98046	Lynnwood	(425)	32,942	28,637
98012	Martha Lake (c)	(425)	—	10,155
*98270	Marysville	(360)	18,702	12,248
98040	Mercer Island	(206)	21,351	20,816
98444	Midland (c)	(253)	—	5,587
98082	Mill Creek	(425)	9,503	7,180
98354	Milton	(253)	5,811	4,995
98661	Minnehaha (c)	(360)	—	9,661
98272	Monroe	(360)	7,483	4,275
98837	Moses Lake	(509)	14,759	11,235
98043	Mountlake Terrace	(425)	20,879	19,320
*98273	Mount Vernon	(360)	22,688	17,647
98275	Mukilteo	(425)	14,620	11,575
98059	Newcastle	(425)	5,355	4,649
98006	Newport Hills (c)	(425)	—	14,736
98166	Normandy Park	(206)	6,807	6,794
98155	North City-Ridgecrest (c)	(206)	—	13,832
.....	North Creek-Canyon Park (c)		—	23,236
98166	North Hill (c)	(206)	—	5,706
98270	North Marysville (c)	(425)	—	18,711
98277	Oak Harbor	(360)	20,599	17,176
*98501	Olympia	(360)	39,188	33,729
99214	Opportunity (c)	(509)	—	22,326
98662	Orchards North (c)	(360)	—	6,479
98662	Orchards South (c)	(360)	—	12,956
99027	Otis Orchards-East Farms (c)	(360)	—	5,811
98047	Pacific	(253)	5,905	4,622
.....	Paine Field-Lake Stickney (c)		—	18,670
98444	Parkland (c)	(253)	—	20,882
98366	Parkwood (c)	(360)	—	6,853
*99301	Pasco	(509)	27,366	20,337
98027	Pine Lake (c)	(425)	—	13,940
*98362	Port Angeles	(360)	18,769	17,710
*98366	Port Orchard	(360)	6,377	4,984
98368	Port Townsend	(360)	8,259	7,001
98370	Poulsbo	(360)	6,219	4,848
98390	Prairie Ridge (c)		—	8,278
*99163	Pullman	(509)	24,950	23,478
*98371	Puyallup	(253)	29,042	23,878
*98052	Redmond	(425)	44,084	35,800
*98058	Renton	(425)	47,463	41,688
99352	Richland	(509)	37,291	32,315
98160	Richmond Beach-Innis Arden (c)	(206)	—	7,242
98113	Richmond Highlands (c)	(206)	—	26,037
98188	Riverton-Boulevard Park (c)	(206)	—	15,337
.....	Sahalee (c)		—	13,951
98686	Salmon Creek (c)	(360)	—	11,989
*98148	Seatac	(206)	22,647	22,760
*98101	Seattle	(206)/(425)	536,978	516,259
98284	Sedro Woolley	(360)	7,744	6,333
98942	Selah	(509)	6,490	5,113
98584	Shelton	(360)	8,125	7,241
98155	Sheridan Beach (c)	(206)	—	6,518
*98133	Shoreline	(206)	52,116	46,979
*98315	Silverdale (c)	(360)	—	7,660
98201	Silver Lake-Fircrest (c)	(360)	—	24,474
*98290	Snohomish	(360)	8,693	6,499
98373	South Hill (c)	(253)	—	12,963
98387	Spanaway (c)	(253)	—	15,001
*99210	Spokane	(509)	184,058	177,165
98388	Steilacoom	(253)	6,121	5,728
*98371	Summit (c)	(253)	—	6,312

ZIP	Place		1998	1990
98390	Sumner	(253)	8,251	7,535
98944	Sunnyside	(509)	12,940	11,238
*98402	Tacoma	(253)	179,814	176,664
98501	Tanglewilde-Thompson Place (c)	(360)	—	6,061
98948	Toppenish	(509)	7,956	7,419
98138	Tukwila	(206)	14,572	14,506
98501	Tumwater	(360)	11,488	9,976
98467	University Place	(253)	32,219	26,724
*98661	Vancouver	(360)	73,526	62,065
98662	Vancouver Mall (c)	(360)	—	6,938
99037	Veradale (c)	(509)	—	7,836
99362	Walla Walla	(509)	28,721	26,482
.....	Waller (c)		—	6,415
98671	Washougal	(360)	6,065	4,764
*98801	Wenatchee	(509)	23,918	21,746
.....	West Lake Sammamish (c)		—	6,087
98258	West Lake Stevens (c)	(425)	—	12,453
99301	West Pasco (c)	(509)	—	7,312
99353	West Richland	(509)	7,049	3,962
99181	West Valley (c)		—	6,594
98166	White Center-Shorewood (c)	(206)	—	20,531
98072	Woodinville	(425)	8,789	7,628
98032	Woodmont Beach (c)	(253)	—	7,493
*98903	Yakima	(509)	64,967	58,427

West Virginia (304)

ZIP	Place	1998	1990
*25801	Beckley	18,187	18,274
24701	Bluefield	12,047	12,756
26330	Bridgeport	7,385	6,837
26201	Buckhannon	5,940	5,909
*25301	Charleston	55,056	57,287
*26301	Clarksburg	17,011	17,970
25301	Cross Lanes (c)	—	10,878
25064	Dunbar	8,372	8,697
26241	Elkins	7,583	7,494
*26554	Fairmont	19,088	20,210
26354	Grafton	5,454	5,524
*25704	Huntington	52,571	54,844
25526	Hurricane	5,371	4,461
26726	Keyser	5,295	5,870
*25401	Martinsburg	15,049	14,073
*26505	Morgantown	26,751	25,879
26041	Moundsville	9,860	10,753
26155	New Martinsville	6,461	6,705
25143	Nitro	6,606	6,851
25901	Oak Hill	6,788	6,812
*26101	Parkersburg	31,715	33,862
.....	Pea Ridge (c)	—	6,535
25550	Point Pleasant	5,038	4,996
24740	Princeton	6,741	7,043
25177	Saint Albans	11,867	12,241
25303	South Charleston	13,148	13,645
25569	Teays Valley (c)	—	8,436
26105	Vienna	11,285	10,862
26062	Weirton	21,206	22,124
26003	Wheeling	32,541	34,882

Wisconsin

ZIP	Place		1998	1990
54301	Allouez	(920)	14,514	14,431
54720	Altoona	(715)	6,580	5,889
54409	Antigo	(715)	8,488	8,284
*59411	Appleton	(920)	65,514	65,695
54806	Ashland	(715)	8,695	8,695
54304	Ashwaubenon	(920)	17,325	16,376
53913	Baraboo	(608)	9,797	9,203
53916	Beaver Dam	(920)	14,603	14,196
54311	Bellevue Town (c)	(920)	—	7,541
*53511	Beloit	(608)	35,157	35,571
54923	Berlin	(920)	5,441	5,371
*53045	Brookfield	(262)	37,747	35,184
53209	Brown Deer	(414)	11,999	12,236
53105	Burlington	(262)	9,984	8,851
53012	Cedarburg	(262)	10,559	10,086
54729	Chippewa Falls	(715)	12,708	12,749
53110	Cudahy	(414)	18,108	18,659
53532	De Forest	(608)	6,865	4,882
53018	Delafield	(262)	6,188	5,347
53115	Delavan	(262)	7,517	6,073
54115	De Pere	(920)	19,479	16,594
*54703	Eau Claire	(715)	59,200	56,806
53121	Elkhorn	(262)	6,616	5,337
53122	Elm Grove	(262)	5,958	6,261
53714	Fitchburg	(608)	19,500	15,648
*54935	Fond du Lac	(920)	39,724	37,755
53538	Fort Atkinson	(920)	10,873	10,213
53217	Fox Point	(414)	6,810	7,238
53132	Franklin	(414)	27,579	21,855
53022	Germantown	(262)	17,269	13,658
53209	Glendale	(414)	13,533	14,088
53024	Grafton	(262)	9,985	9,340
*54303	Green Bay	(920)	97,789	96,466
53129	Greendale	(414)	15,032	15,128
53220	Greenfield	(414)	34,497	33,403
53130	Hales Corners	(414)	7,335	7,623
53027	Hartford	(262)	9,515	8,188
53029	Hartland	(262)	8,024	6,906
54303	Howard	(920)	13,228	9,874
54016	Hudson	(715)	7,992	6,378
*53545	Janesville	(608)	59,149	52,210
53549	Jefferson	(920)	6,664	6,078
54130	Kaukauna	(920)	12,150	11,982
*53140	Kenosha	(262)	87,849	80,426
54136	Kimberly	(920)	5,788	5,406
*54601	La Crosse	(608)	49,075	51,140
53147	Lake Geneva	(262)	6,579	5,979
54140	Little Chute	(920)	10,236	9,207
53558	McFarland	(608)	6,204	5,232
*53714	Madison	(608)	209,306	190,766
*54220	Manitowoc	(920)	33,067	32,521
54143	Marinette	(715)	12,025	11,843
54449	Marshfield	(715)	19,666	19,293
54952	Menasha	(920)	15,412	14,711
*53051	Menomonee Falls	(262)	31,386	26,840
54751	Menomonie	(715)	14,727	13,547
53097	Mequon	(414)	21,938	18,885
54452	Merrill	(715)	10,298	9,860
53562	Middleton	(608)	15,694	13,785
*53201	Milwaukee	(414)	578,364	628,088
53716	Monona	(608)	8,797	8,637
53566	Monroe	(608)	10,762	10,241
53572	Mount Horeb	(608)	5,249	4,182
53149	Mukwonago	(262)	5,896	4,495
53150	Muskego	(414)	21,589	16,813
*54956	Neenah	(920)	23,580	23,219
*53186	New Berlin	(262)	37,230	33,592
54961	New London	(920)	7,175	6,658
54017	New Richmond	(715)	5,964	5,106
53154	Oak Creek	(414)	27,219	19,513
53066	Oconomowoc	(262)	11,385	10,993
54650	Onalaska	(608)	14,751	12,201
53575	Oregon	(608)	6,946	4,519
*54901	Oshkosh	(920)	57,955	55,006
53072	Pewaukee	(262)	7,287	5,287
53818	Platteville	(608)	9,877	9,862
53158	Pleasant Prairie	(262)	14,611	12,037
54467	Plover	(715)	10,369	8,176
53073	Plymouth	(920)	7,544	6,769
53901	Portage	(608)	9,338	8,640
53074	Port Washington	(262)	10,542	9,338
53821	Prairie du Chien	(608)	5,753	5,657
*53401	Racine	(262)	81,095	84,298
53959	Reedsburg	(608)	7,059	5,834
54501	Rhinelander	(715)	7,721	7,382
54868	Rice Lake	(715)	8,259	7,998
54971	Ripon	(920)	7,389	7,241
54022	River Falls	(715)	11,726	10,610
53235	Saint Francis	(414)	9,268	9,245
54166	Shawano	(715)	7,846	7,598
*53081	Sheboygan	(920)	49,377	49,587
53085	Sheboygan Falls	(920)	6,248	5,823
53211	Shorewood	(414)	12,777	14,116
53172	South Milwaukee	(414)	20,466	20,958
54656	Sparta	(608)	8,308	7,788
54481	Stevens Point	(715)	22,196	23,002
53589	Stoughton	(608)	11,542	8,786
54235	Sturgeon Bay	(920)	9,360	9,176
53590	Sun Prairie	(608)	19,763	15,352
54880	Superior	(715)	27,142	27,134
53089	Sussex	(262)	8,389	5,039
54660	Tomah	(608)	8,059	7,572
54241	Two Rivers	(920)	13,029	13,030
53593	Verona	(608)	6,648	5,374
*53094	Watertown	(920)	20,641	19,142
*53186	Waukesha	(262)	61,989	56,894
53597	Waunakee	(608)	8,877	5,897
54981	Waupaca	(715)	5,813	4,946
53963	Waupun	(920)	9,904	8,844
*54403	Wausau	(715)	36,353	37,060
53213	Wauwatosa	(414)	45,850	49,366
53214	West Allis	(414)	59,974	63,221
*53095	West Bend	(262)	28,495	24,470
54476	Weston (c)	(715)	—	9,714
53217	Whitefish Bay	(414)	12,978	14,272
53190	Whitewater	(262)	13,251	12,636
*54494	Wisconsin Rapids	(715)	18,475	18,245

Wyoming (307)

ZIP	Place	1998	1990
*82609	Casper	48,283	46,765
*82009	Cheyenne	53,640	50,008
82414	Cody	8,807	7,897
82633	Douglas	5,655	5,076
*82930	Evanston	11,475	10,904
*82716	Gillette	19,463	17,545
82935	Green River	13,059	12,711
*83002	Jackson	5,817	4,708
82520	Lander	7,378	7,023
*82072	Laramie	25,035	26,687
82435	Powell	5,608	5,292
82301	Rawlins	8,747	9,380
82501	Riverton	10,126	9,202
*82901	Rock Springs	19,408	19,050
82801	Sheridan	14,591	13,904
82240	Torrington	6,024	5,651
82401	Worland	5,989	5,742

Populations and Areas of Counties and States

Source: U.S. Bureau of the Census, Dept. of Commerce; World Almanac research

State population figures are estimates for July 1, 1998. For counties, July 1, 1998, population estimates and Apr. 1, 1990, decennial census figures are given. County areas may not add to total state areas because of rounding.

Alabama

(67 counties, 50,750 sq mi land; pop. 4,351,999)

County	County seat or courthouse	1998 Pop.	1990 Pop.	Land area sq mi
Autauga	Prattville	42,095	34,222	596
Baldwin	Bay Minette	132,828	98,280	1,597
Barbour	Clayton	26,895	25,417	885
Bibb	Centreville	18,926	16,576	622
Blount	Oneonta	46,266	39,248	646
Bullock	Union Springs	11,311	11,042	625
Butler	Greenville	21,695	21,892	777
Calhoun	Anniston	117,018	116,032	609
Chambers	Lafayette	36,713	36,876	597
Cherokee	Centre	21,833	19,543	553
Chilton	Clanton	36,918	32,458	694
Choctaw	Butler	15,917	16,018	914
Clarke	Grove Hill	28,499	27,240	1,239
Clay	Ashland	13,970	13,252	605
Cleburne	Heflin	14,308	12,730	560
Coffee	Elba	42,436	40,240	679
Colbert	Tuscumbia	52,946	51,666	595
Conecuh	Evergreen	13,976	14,054	851
Coosa	Rockford	11,658	11,063	653
Covington	Andalusia	37,402	36,478	1,035
Crenshaw	Luverne	13,636	13,635	610
Cullman	Cullman	74,994	67,613	739
Dale	Ozark	48,872	49,633	561
Dallas	Selma	46,768	48,130	981
De Kalb	Fort Payne	58,454	54,651	778
Elmore	Wetumpka	61,993	49,210	622
Escambia	Brewton	36,740	35,518	948
Etowah	Gadsden	103,975	99,840	535
Fayette	Fayette	18,133	17,962	628
Franklin	Russellville	29,682	27,814	636
Geneva	Geneva	24,944	23,647	576
Greene	Eutaw	9,880	10,153	646
Hale	Greensboro	16,744	15,498	644
Henry	Abbeville	15,836	15,374	562
Houston	Dothan	85,877	81,331	580
Jackson	Scottsboro	51,329	47,796	1,079
Jefferson	Birmingham	659,524	651,520	1,113
Lamar	Vernon	15,731	15,715	605
Lauderdale	Florence	84,325	79,661	670
Lawrence	Moulton	33,447	31,513	693
Lee	Opelika	100,444	87,146	609
Limestone	Athens	62,241	54,135	568
Lowndes	Hayneville	12,984	12,658	718
Macon	Tuskegee	22,951	24,928	611
Madison	Huntsville	278,187	238,912	805
Marengo	Linden	23,378	23,084	977
Marion	Hamilton	30,986	29,830	742
Marshall	Guntersville	80,346	70,832	567
Mobile	Mobile	399,429	378,643	1,233
Monroe	Monroeville	23,965	23,968	1,026
Montgomery	Montgomery	217,693	209,085	790
Morgan	Decatur	109,369	100,043	582
Perry	Marion	12,667	12,759	720
Pickens	Carrollton	21,089	20,699	882
Pike	Troy	28,646	27,595	671
Randolph	Wedowee	19,923	19,881	581
Russell	Phenix City	50,387	46,860	641
Saint Clair	Ashville & Pell City	62,003	49,811	634
Shelby	Columbiana	140,715	99,363	795
Sumter	Livingston	15,766	16,174	905
Talladega	Talladega	76,633	74,109	740
Tallapoosa	Dadeville	40,606	38,826	718
Tuscaloosa	Tuscaloosa	160,768	150,522	1,325
Walker	Jasper	71,027	67,670	795
Washington	Chatom	17,677	16,694	1,081
Wilcox	Camden	13,468	13,568	889
Winston	Double Springs	24,157	22,053	615

Alaska

(27 divisions, 570,374 sq mi land; pop. 614,010)

Census Division	1998 Pop.	1990 Pop.	Land area sq mi
Aleutians East Borough	2,253	2,464	6,985
Aleutians West Census Area	3,879	9,478	4,402
Anchorage Borough	254,982	226,338	1,698
Bethel Census Area	15,967	13,656	41,087
Bristol Bay Borough	1,356	1,410	519
Denali Borough	1,970	1,764	12,719
Dillingham Census Area	4,534	4,012	18,467
Fairbanks North Star Borough	84,217	77,720	7,362
Haines Borough	2,225	2,117	2,357
Juneau Borough	30,191	26,751	2,594
Kenai Peninsula Borough	48,008	40,802	16,079
Ketchikan Gateway Borough	13,443	13,828	1,220
Kodiak Island Borough	14,520	13,309	6,463
Lake and Peninsula Borough	1,699	1,668	23,632
Matanuska-Susitna Borough	56,258	39,683	24,694
Nome Census Area	9,016	8,288	23,013
North Slope Borough	7,152	5,979	87,861
Northwest Arctic Borough	6,758	6,113	35,863
Prince of Wales-Outer Ketchikan Census Area	7,037	6,278	7,325
Sitka Borough	8,338	8,588	2,882
Skagway-Hoonah-Angoon Census Area	3,834	3,680	8,012
Southeast Fairbanks Census Area	5,713	5,913	25,110
Valdez-Cordova Census Area	10,279	9,952	36,945
Wade Hampton Census Area	6,812	5,791	17,124
Wrangell-Petersburg Census Area	6,818	7,042	5,809
Yakutat Borough	799	705	4,865
Yukon-Koyukuk Census Area	5,952	6,714	145,287

Arizona

(15 counties, 113,642 sq mi land; pop. 4,668,631)

County	County seat or courthouse	1998 Pop.	1990 Pop.	Land area sq mi
Apache	Saint Johns	68,782	61,591	11,206
Cochise	Bisbee	112,564	97,624	6,170
Coconino	Flagstaff	114,171	96,591	18,619
Gila	Globe	48,974	40,216	4,768
Graham	Safford	31,696	26,554	4,630
Greenlee	Clifton	9,304	8,008	1,847
La Paz	Parker	14,880	13,844	4,500
Maricopa	Phoenix	2,784,075	2,122,101	9,204
Mohave	Kingman	130,618	93,497	13,312
Navajo	Holbrook	96,997	77,674	9,954
Pima	Tucson	790,755	666,957	9,187
Pinal	Florence	146,929	116,397	5,370
Santa Cruz	Nogales	38,116	29,676	1,238
Yavapai	Prescott	148,511	107,714	8,124
Yuma	Yuma	132,259	106,895	5,514

Arkansas

(75 counties, 52,075 sq mi land; pop. 2,538,303)

County	County seat or courthouse	1998 Pop.	1990 Pop.	Land area sq mi
Arkansas	DeWitt & Stuttgart	20,787	21,653	989
Ashley	Hamburg	24,448	24,319	921
Baxter	Mountain Home	36,402	31,186	554
Benton	Bentonville	134,162	97,499	843
Boone	Harrison	31,872	28,297	591
Bradley	Warren	11,433	11,793	651
Calhoun	Hampton	5,729	5,826	628
Carroll	Berryville & Eureka Springs	22,534	18,654	634
Chicot	Lake Village	14,841	15,713	644
Clark	Arkadelphia	21,933	21,437	866
Clay	Corning & Piggott	17,223	18,107	639
Cleburne	Heber Springs	22,923	19,411	553
Cleveland	Rison	8,452	7,781	598
Columbia	Magnolia	25,060	25,691	766
Conway	Morrilton	19,920	19,151	556
Craighead	Jonesboro & Lake City	77,500	68,956	711
Crawford	Van Buren	50,334	42,493	596
Crittenden	Marion	49,905	49,939	611
Cross	Wynne	19,564	19,225	616
Dallas	Fordyce	9,060	9,614	668
Desha	Arkansas City	15,110	16,798	765
Drew	Monticello	17,575	17,369	828
Faulkner	Conway	78,382	60,006	647
Franklin	Charleston & Ozark	16,932	14,897	610
Fulton	Salem	10,901	10,037	618
Garland	Hot Springs	83,976	73,397	678
Grant	Sheridan	15,897	13,948	632
Greene	Paragould	36,192	31,804	578
Hempstead	Hope	22,113	21,621	729
Hot Spring	Malvern	29,035	26,115	615
Howard	Nashville	13,724	13,569	588
Independence	Batesville	33,054	31,192	764
Izard	Melbourne	13,093	11,364	581
Jackson	Newport	17,783	18,944	634
Jefferson	Pine Bluff	81,556	85,487	885
Johnson	Clarksville	21,403	18,221	662
Lafayette	Lewisville	8,926	9,643	527
Lawrence	Walnut Ridge	17,304	17,455	587
Lee	Marianna	12,406	13,053	602
Lincoln	Star City	14,274	13,690	561
Little River	Ashdown	13,206	13,966	532
Logan	Booneville & Paris	21,173	20,557	710

County	County seat or courthouse	1998 Pop.	1990 Pop.	Land area sq mi
Lonoke	Lonoke	50,156	39,268	766
Madison	Huntsville	13,224	11,618	837
Marion	Yellville	14,918	12,001	598
Miller	Texarkana	39,857	38,467	624
Mississippi	Blytheville & Osceola	50,635	57,525	898
Monroe	Clarendon	10,200	11,333	607
Montgomery	Mount Ida	8,655	7,841	781
Nevada	Prescott	10,034	10,101	620
Newton	Jasper	8,180	7,666	823
Ouachita	Camden	27,921	30,574	733
Perry	Perryville	9,640	7,969	551
Phillips	Helena	27,363	28,830	693
Pike	Murfreesboro	10,592	10,086	603
Poinsett	Harrisburg	24,750	24,664	758
Polk	Mena	19,662	17,347	860
Pope	Russellville	52,059	45,883	812
Prairie	Des Arc & De Valls Bluff	9,410	9,518	646
Pulaski	Little Rock	350,345	349,569	771
Randolph	Pocahontas	17,802	16,558	652
Saint Francis	Forrest City	28,162	28,497	634
Saline	Benton	77,412	64,183	725
Scott	Waldron	10,686	10,205	894
Searcy	Marshall	7,761	7,841	667
Sebastian	Fort Smlth & Greenwood	106,180	99,590	536
Sevier	De Queen	14,623	13,637	564
Sharp	Ash Flat	16,993	14,109	604
Stone	Mountain View	11,154	9,775	607
Union	El Dorado	45,304	46,719	1,039
Van Buren	Clinton	15,550	14,008	712
Washington	Fayetteville	138,454	113,409	950
White	Searcy	64,526	54,676	1,034
Woodruff	Augusta	8,888	9,520	587
Yell	Danville & Dardanelle	19,110	17,759	928

California
(58 counties, 155,973 sq mi land; pop. 32,666,550)

County	County seat or courthouse	1998 Pop.	1990 Pop.	Land area sq mi
Alameda	Oakland	1,400,322	1,304,346	738
Alpine	Markleeville	1,209	1,113	739
Amador	Jackson	33,334	30,039	593
Butte	Oroville	194,597	182,120	1,640
Calaveras	San Andreas	39,830	31,998	1,020
Colusa	Colusa	18,572	16,275	1,151
Contra Costa	Martinez	918,200	803,732	720
Del Norte	Crescent City	27,000	23,460	1,008
El Dorado	Placerville	158,502	125,995	1,712
Fresno	Fresno	755,730	667,490	5,963
Glenn	Willows	26,234	24,798	1,315
Humboldt	Eureka	122,262	119,118	3,573
Imperial	El Centro	144,051	109,303	4,175
Inyo	Independence	18,125	18,281	10,192
Kern	Bakersfield	631,459	544,981	8,142
Kings	Hanford	118,866	101,469	1,390
Lake	Lakeport	55,147	50,631	1,259
Lassen	Susanville	33,285	27,598	4,558
Los Angeles	Los Angeles	9,213,533	8,863,052	4,060
Madera	Madera	114,748	88,090	2,138
Marin	San Rafael	236,770	230,096	520
Mariposa	Mariposa	15,877	14,302	1,451
Mendocino	Ukiah	83,734	80,345	3,509
Merced	Merced	197,730	178,403	1,929
Modoc	Alturas	9,398	9,678	3,944
Mono	Bridgeport	10,288	9,956	3,045
Monterey	Salinas	365,605	355,660	3,322
Napa	Napa	119,288	110,765	754
Nevada	Nevada City	91,334	78,510	958
Orange	Santa Ana	2,721,701	2,410,668	790
Placer	Auburn	229,259	172,796	1,404
Plumas	Quincy	20,370	19,739	2,554
Riverside	Riverside	1,478,838	1,170,413	7,208
Sacramento	Sacramento	1,144,202	1,041,219	966
San Benito	Hollister	48,774	36,697	1,389
San Bernardino	San Bernardino	1,635,234	1,418,380	20,062
San Diego	San Diego	2,780,592	2,498,016	4,205
San Francisco	San Francisco	745,774	723,959	47
San Joaquin	Stockton	550,445	480,628	1,399
San Luis Obispo	San Luis Obispo	234,366	217,162	3,305
San Mateo	Redwood City	700,765	649,623	449
Santa Barbara	Santa Barbara	389,502	369,608	2,739
Santa Clara	San Jose	1,641,215	1,497,577	1,291
Santa Cruz	Santa Cruz	242,994	229,734	446
Shasta	Redding	164,349	147,036	3,786
Sierra	Downieville	3,380	3,318	953
Siskiyou	Yreka	44,044	43,531	6,287
Solano	Fairfield	377,415	339,471	828
Sonoma	Santa Rosa	433,304	388,222	1,576
Stanislaus	Modesto	426,460	370,522	1,495
Sutter	Yuba City	76,976	64,415	603
Tehama	Red Bluff	54,073	49,625	2,951
Trinity	Weaverville	13,117	13,063	3,179
Tulare	Visalia	355,240	311,921	4,824
Tuolumne	Sonora	53,248	48,456	2,236
Ventura	Ventura	731,967	669,016	1,846

County	County seat or courthouse	1998 Pop.	1990 Pop.	Land area sq mi
Yolo	Woodland	153,849	141,210	1,012
Yuba	Marysville	60,067	58,228	631

Colorado
(63 counties, 103,729 sq mi land; pop. 3,970,971)

County	County seat or courthouse	1998 Pop.	1990 Pop.	Land area sq mi
Adams	Brighton	323,853	265,038	1,192
Alamosa	Alamosa	14,448	13,617	723
Arapahoe	Littleton	473,168	391,511	803
Archuleta	Pagosa Springs	9,113	5,345	1,349
Baca	Springfield	4,365	4,556	2,556
Bent	Las Animas	5,497	5,048	1,514
Boulder	Boulder	267,274	225,339	743
Chaffee	Salida	15,075	12,684	1,014
Cheyenne	Cheyenne Wells	2,346	2,397	1,782
Clear Creek	Georgetown	9,001	7,619	396
Conejos	Conejos	7,972	7,453	1,287
Costilla	San Luis	3,641	3,190	1,227
Crowley	Ordway	4,310	3,946	789
Custer	Westcliffe	3,449	1,926	739
Delta	Delta	26,619	20,980	1,142
Denver	Denver	499,055	467,610	153
Dolores	Dove Creek	1,822	1,504	1,067
Douglas	Castle Rock	140,975	60,391	840
Eagle	Eagle	33,538	21,928	1,688
Elbert	Kiowa	18,600	9,646	1,851
El Paso	Colorado Springs	490,378	397,014	2,127
Fremont	Canon City	43,904	32,273	1,533
Garfield	Glenwood Springs	39,301	29,974	2,948
Gilpin	Central City	4,188	3,070	150
Grand	Hot Sulphur Springs	10,050	7,966	1,850
Gunnison	Gunnison	12,456	10,273	3,239
Hinsdale	Lake City	737	467	1,118
Huerfano	Walsenburg	6,813	6,009	1,591
Jackson	Walden	1,535	1,605	1,613
Jefferson	Golden	501,591	438,430	772
Kiowa	Eads	1,633	1,688	1,771
Kit Carson	Burlington	7,313	7,140	2,161
Lake	Leadville	6,391	6,007	377
La Plata	Durango	40,413	32,284	1,692
Larimer	Fort Collins	231,221	186,136	2,601
Las Animas	Trinidad	14,573	13,765	4,773
Lincoln	Hugo	5,729	4,529	2,586
Logan	Sterling	17,890	17,567	1,839
Mesa	Grand Junction	112,891	93,145	3,328
Mineral	Creede	694	558	876
Moffat	Craig	12,535	11,357	4,743
Montezuma	Cortez	22,465	18,672	2,037
Montrose	Montrose	30,764	24,423	2,241
Morgan	Fort Morgan	25,087	21,939	1,286
Otero	La Junta	20,671	20,185	1,263
Ouray	Ouray	3,313	2,295	542
Park	Fairplay	13,399	7,174	2,201
Phillips	Holyoke	4,325	4,189	688
Pitkin	Aspen	13,423	12,661	970
Prowers	Lamar	13,729	13,347	1,641
Pueblo	Pueblo	134,867	123,051	2,389
Rio Blanco	Meeker	6,265	6,051	3,221
Rio Grande	Del Norte	11,453	10,770	913
Routt	Steamboat Springs	17,514	14,088	2,362
Saguache	Saguache	6,076	4,619	3,169
San Juan	Silverton	530	745	388
San Miguel	Tellurideew	5,437	3,653	1,287
Sedgwick	Julesburg	2,547	2,690	548
Summite	Breckenridge	18,749	12,881	608
Teller	Cripple Creek	20,606	12,468	557
Washington	Akron	4,576	4,812	2,521
Weld	Greeley	159,429	131,821	3,993
Yuma	Wray	9,389	8,954	2,366

Connecticut
(8 counties, 4,845 sq mi land; pop. 3,274,069)

County	County seat or courthouse	1998 Pop.	1990 Pop.	Land area sq mi
Fairfield	Bridgeport	838,362	827,645	626
Hartford	Hartford	828,200	851,783	736
Litchfield	Litchfield	181,277	174,092	920
Middlesex	Middletown	150,034	143,196	369
New Haven	New Haven	793,504	804,219	606
New London	Norwich	245,740	254,957	666
Tolland	Rockville	131,831	128,699	410
Windham	Putnam	105,121	102,525	513

Delaware
(3 counties, 1,955 sq mi land; pop. 743,603)

County	County seat or courthouse	1998 Pop.	1990 Pop.	Land area sq mi
Kent	Dover	124,089	110,993	591
New Castle	Wilmington	482,807	441,946	426
Sussex	Georgetown	136,707	113,229	938

District of Columbia

(61 sq mi land; pop. 523,124)

Florida

(67 counties, 53,937 sq mi land; pop. 14,915,980)

County	County seat or courthouse	1998 Pop.	1990 Pop.	Land area sq mi
Alachua	Gainesville	198,662	181,596	874
Baker	Macclenny	21,103	18,486	585
Bay	Panama City	146,999	126,994	764
Bradford	Starke	24,777	22,515	293
Brevard	Titusville	466,093	398,978	1,019
Broward	Fort Lauderdale	1,503,407	1,255,531	1,209
Calhoun	Blountstown	12,420	11,011	567
Charlotte	Punta Gorda	134,899	110,975	694
Citrus	Inverness	114,068	93,513	584
Clay	Green Cove Springs	137,455	105,986	601
Collier	Naples	199,436	152,099	2,026
Columbia	Lake City	52,956	42,613	797
De Soto	Arcadia	24,820	23,865	637
Dixie	Cross City	12,959	10,585	704
Duval	Jacksonville	735,733	672,971	774
Escambia	Pensacola	282,303	262,798	664
Flagler	Bunnell	47,455	28,701	485
Franklin	Apalachicola	10,079	8,967	534
Gadsden	Quincy	44,043	41,116	516
Gilchrist	Trenton	13,791	9,667	349
Glades	Moore Haven	8,492	7,591	774
Gulf	Port Saint Joe	13,476	11,504	565
Hamilton	Jasper	12,651	10,930	515
Hardee	Wauchula	21,046	19,499	637
Hendry	La Belle	29,357	25,773	1,153
Hernando	Brooksville	127,227	101,115	478
Highlands	Sebring	75,206	68,432	1,029
Hillsborough	Tampa	925,277	834,054	1,051
Holmes	Bonifay	18,622	15,778	483
Indian River	Vero Beach	99,155	90,208	503
Jackson	Marianna	45,660	41,375	916
Jefferson	Monticello	12,952	11,296	598
Lafayette	Mayo	6,325	5,578	543
Lake	Tavares	202,207	152,104	953
Lee	Fort Myers	392,895	335,113	804
Leon	Tallahassee	216,978	192,493	667
Levy	Bronson	31,796	25,912	1,118
Liberty	Bristol	6,759	5,569	836
Madison	Madison	17,652	16,569	692
Manatee	Bradenton	239,682	211,707	741
Marion	Ocala	241,513	194,835	1,579
Martin	Stuart	115,940	100,900	556
Miami-Dade	Miami	2,152,437	1,937,194	1,945
Monroe	Key West	81,203	78,024	997
Nassau	Fernandina Beach	55,349	43,941	652
Okaloosa	Crestview	169,289	143,777	936
Okeechobee	Okeechobee	31,158	29,627	774
Orange	Orlando	805,837	677,491	908
Osceola	Kissimmee	145,666	107,728	1,322
Palm Beach	West Palm Beach	1,032,625	863,503	1,974
Pasco	New Port Richey	325,824	281,131	745
Pinellas	Clearwater	878,231	851,659	280
Polk	Bartow	452,584	405,382	1,875
Putnam	Palatka	70,419	65,070	722
Saint Johns	Saint Augustine	116,147	83,829	609
Saint Lucie	Fort Pierce	179,178	150,171	573
Santa Rosa	Milton	117,322	81,608	1,016
Sarasota	Sarasota	303,400	277,776	572
Seminole	Sanford	350,859	287,521	308
Sumter	Bushnell	40,426	31,577	546
Suwannee	Live Oak	32,665	26,780	688
Taylor	Perry	18,849	17,111	1,042
Union	Lake Butler	12,423	10,252	240
Volusia	De Land	423,409	370,737	1,106
Wakulla	Crawfordville	18,652	14,202	607
Walton	De Funiak Springs	37,410	27,759	1,058
Washington	Chipley	20,292	16,919	580

Georgia

(159 counties, 57,919 sq mi land; pop. 7,642,207)

County	County seat or courthouse	1998 Pop.	1990 Pop.	Land area sq mi
Appling	Baxley	16,493	15,744	509
Atkinson	Pearson	7,138	6,213	338
Bacon	Alma	10,375	9,566	285
Baker	Newton	3,673	3,615	343
Baldwin	Milledgeville	41,968	39,530	259
Banks	Homer	12,798	10,308	234
Barrow	Winder	40,344	29,721	162
Bartow	Cartersville	71,929	55,915	460
Ben Hill	Fitzgerald	17,496	16,245	252
Berrien	Nashville	16,353	14,153	453
Bibb	Macon	156,086	150,137	250
Bleckley	Cochran	11,185	10,430	217
Brantley	Nahunta	13,571	11,077	444
Brooks	Quitman	16,000	15,398	494
Bryan	Pembroke	23,482	15,438	442
Bulloch	Statesboro	50,614	43,125	683
Burke	Waynesboro	22,854	20,579	831
Butts	Jackson	17,837	15,326	187
Calhoun	Morgan	5,053	5,013	280
Camden	Woodbine	47,443	30,167	630
Candler	Metter	9,078	7,744	247
Carroll	Carrollton	83,021	71,422	499
Catoosa	Ringgold	50,547	42,464	162
Charlton	Folkston	9,442	8,496	781
Chatham	Savannah	225,543	216,774	440
Chattahoochee	Cusseta	16,679	16,934	249
Chattooga	Summerville	22,813	22,242	314
Cherokee	Canton	134,498	90,204	424
Clarke	Athens	90,630	87,594	121
Clay	Fort Gaines	3,453	3,364	195
Clayton	Jonesboro	208,999	181,436	143
Clinch	Homerville	6,660	6,160	809
Cobb	Marietta	566,203	447,745	340
Coffee	Douglas	34,298	29,592	599
Colquitt	Moultrie	40,156	36,645	552
Columbia	Appling	91,118	66,031	290
Cook	Adel	15,011	13,456	229
Coweta	Newnan	85,028	53,853	443
Crawford	Knoxville	10,667	8,991	325
Crisp	Cordele	20,725	20,011	274
Dade	Trenton	15,058	13,147	174
Dawson	Dawsonville	14,851	9,429	211
Decatur	Bainbridge	27,035	25,517	597
De Kalb	Decatur	593,850	546,171	268
Dodge	Eastman	18,108	17,607	501
Dooly	Vienna	10,388	9,901	393
Dougherty	Albany	95,309	96,321	330
Douglas	Douglasville	89,843	71,120	199
Early	Blakely	12,197	11,854	511
Echols	Statenville	2,401	2,334	404
Effingham	Springfield	36,483	25,687	480
Elbert	Elberton	19,335	18,949	369
Emanuel	Swainsboro	21,023	20,546	686
Evans	Claxton	9,949	8,724	185
Fannin	Blue Ridge	18,622	15,992	386
Fayette	Fayetteville	88,609	62,415	197
Floyd	Rome	85,185	81,251	513
Forsyth	Cumming	86,130	44,083	226
Franklin	Carnesville	19,080	16,650	263
Fulton	Atlanta	739,367	648,779	529
Gilmer	Ellijay	18,672	13,368	427
Glascock	Gibson	2,512	2,357	144
Glynn	Brunswick	67,320	62,496	422
Gordon	Calhoun	41,052	35,067	355
Grady	Cairo	21,501	20,279	458
Greene	Greensboro	13,651	11,793	388
Gwinnett	Lawrenceville	522,095	352,910	433
Habersham	Clarkesville	31,858	27,622	278
Hall	Gainesville	119,210	95,434	394
Hancock	Sparta	9,134	8,908	473
Haralson	Buchanan	24,653	21,966	282
Harris	Hamilton	22,315	17,788	464
Hart	Hartwell	21,833	19,712	232
Heard	Franklin	10,082	8,628	296
Henry	McDonough	104,667	58,741	323
Houston	Perry	105,808	89,208	377
Irwin	Ocilla	8,982	8,649	357
Jackson	Jefferson	37,641	30,005	342
Jasper	Monticello	10,155	8,453	371
Jeff Davis	Hazlehurst	12,751	12,032	333
Jefferson	Louisville	17,767	17,408	528
Jenkins	Millen	8,447	8,247	350
Johnson	Wrightsville	8,316	8,329	304
Jones	Gray	23,020	20,739	394
Lamar	Barnesville	14,706	13,038	185
Lanier	Lakeland	6,986	5,531	187
Laurens	Dublin	43,772	39,988	813
Lee	Leesburg	22,767	16,250	356
Liberty	Hinesville	59,162	52,745	519
Lincoln	Lincolnton	8,276	7,442	211
Long	Ludowici	8,585	6,202	401
Lowndes	Valdosta	85,231	75,981	504
Lumpkin	Dahlonega	18,981	14,573	285
McDuffie	Thomson	21,770	20,119	260
McIntosh	Darien	10,018	8,634	434
Macon	Oglethorpe	13,244	13,114	403
Madison	Danielsville	24,312	21,050	284
Marion	Buena Vista	6,712	5,590	367
Meriwether	Greenville	23,112	22,411	503
Miller	Colquitt	6,409	6,280	283
Mitchell	Camilla	21,176	20,275	512
Monroe	Forsyth	19,645	17,113	396
Montgomery	Mount Vernon	7,741	7,379	245
Morgan	Madison	15,091	12,883	350
Murray	Chatsworth	32,682	26,147	344
Muscogee	Columbus	182,752	179,280	216
Newton	Covington	57,847	41,808	276
Oconee	Watkinsville	23,737	17,618	186
Oglethorpe	Lexington	11,418	9,763	441
Paulding	Dallas	73,534	41,611	314
Peach	Fort Valley	24,462	21,189	151

County	County seat or courthouse	1998 Pop.	1990 Pop.	Land area sq mi
Pickens	Jasper	19,679	14,432	232
Pierce	Blackshear	15,794	13,328	343
Pike	Zebulon	12,645	10,224	218
Polk	Cedartown	36,308	33,815	311
Pulaski	Hawkinsville	8,401	8,108	247
Putnam	Eatonton	17,559	14,137	345
Quitman	Georgetown	2,486	2,210	152
Rabun	Clayton	13,406	11,648	371
Randolph	Cuthbert	7,881	8,023	429
Richmond	Augusta	191,329	189,719	324
Rockdale	Conyers	68,305	54,091	131
Schley	Ellaville	3,945	3,590	168
Screven	Sylvania	14,431	13,842	649
Seminole	Donalsonville	9,788	9,010	238
Spalding	Griffin	57,626	54,457	198
Stephens	Toccoa	25,421	23,436	179
Stewart	Lumpkin	5,468	5,654	459
Sumter	Americus	31,324	30,232	485
Talbot	Talbotton	6,935	6,524	393
Taliaferro	Crawfordville	1,908	1,915	195
Tattnall	Reidsville	18,975	17,722	484
Taylor	Butler	8,306	7,642	378
Telfair	MacRae	11,558	11,000	441
Terrell	Dawson	11,146	10,653	336
Thomas	Thomasville	42,953	38,943	548
Tift	Tifton	36,673	34,998	265
Toombs	Lyons	25,828	24,072	367
Towns	Hiawassee	8,529	6,754	167
Treutlen	Soperton	6,003	5,994	201
Troup	La Grange	58,783	55,532	414
Turner	Ashburn	9,160	8,703	286
Twiggs	Jeffersonville	10,126	9,806	360
Union	Blairsville	16,519	11,993	323
Upson	Thomaston	27,075	26,300	326
Walker	La Fayette	63,082	58,340	446
Walton	Monroe	54,485	38,586	329
Ware	Waycross	35,364	35,471	903
Warren	Warrenton	6,059	6,078	286
Washington	Sandersville	20,033	19,112	681
Wayne	Jesup	25,437	22,356	645
Webster	Preston	2,193	2,263	210
Wheeler	Alamo	4,875	4,903	298
White	Cleveland	17,457	13,006	242
Whitfield	Dalton	82,039	72,462	290
Wilcox	Abbeville	7,365	7,008	380
Wilkes	Washington	10,568	10,597	471
Wilkinson	Irwinton	10,838	10,228	447
Worth	Sylvester	22,485	19,744	570

Hawaii

(5 counties, 6,423 sq mi land; pop. 1,193,001)

County	County seat or courthouse	1998 Pop.	1990 Pop.	Land area sq mi
Hawaii	Hilo	143,135	120,317	4,028
Honolulu	Honolulu	872,478	836,231	600
Kalawao[1]		74	130	13
Kauai	Lihue	56,603	51,177	623
Maui	Wailuku	120,711	100,374	1,159

(1) Administered by state government.

Idaho

(44 counties, 82,751 sq mi land; pop. 1,228,684)

County	County seat or courthouse	1998 Pop.	1990 Pop.	Land area sq mi
Ada	Boise	275,687	205,775	1,055
Adams	Council	3,804	3,254	1,365
Bannock	Pocatello	74,866	66,026	1,113
Bear Lake	Paris	6,539	6,084	971
Benewah	Saint Maries	9,119	7,937	776
Bingham	Blackfoot	41,820	37,583	2,095
Blaine	Hailey	17,200	13,552	2,645
Boise	Idaho City	5,114	3,509	1,903
Bonner	Sandpoint	35,226	26,622	1,738
Bonneville	Idaho Falls	80,672	72,207	1,869
Boundary	Bonners Ferry	9,800	8,332	1,269
Butte	Arco	3,033	2,918	2,233
Camas	Fairfield	846	727	1,075
Canyon	Caldwell	120,266	90,076	590
Caribou	Soda Springs	7,426	6,963	1,766
Cassia	Burley	21,359	19,532	2,567
Clark	Dubois	873	762	1,765
Clearwater	Orofino	9,310	8,505	2,462
Custer	Challis	4,107	4,133	4,926
Elmore	Mountain Home	25,173	21,205	3,078
Franklin	Preston	11,106	9,232	666
Fremont	Saint Anthony	11,897	10,937	1,867
Gem	Emmett	14,816	11,844	563
Gooding	Gooding	13,626	11,633	731
Idaho	Grangeville	15,066	13,768	8,485
Jefferson	Rigby	19,118	16,543	1,095
Jerome	Jerome	17,962	15,138	600
Kootenai	Coeur d'Alene	101,390	69,795	1,245

County	County seat or courthouse	1998 Pop.	1990 Pop.	Land area sq mi
Latah	Moscow	32,051	30,617	1,077
Lemhi	Salmon	8,030	6,899	4,564
Lewis	Nez Perce	4,007	3,516	479
Lincoln	Shoshone	3,792	3,308	1,206
Madison	Rexberg	23,569	23,674	472
Minidoka	Rupert	20,207	19,361	760
Nez Perce	Lewiston	36,852	33,754	849
Oneida	Malad City	4,051	3,492	1,200
Owyhee	Murphy	10,277	8,392	7,678
Payette	Payette	20,519	16,434	408
Power	American Falls	8,309	7,086	1,406
Shoshone	Wallace	13,870	13,931	2,634
Teton	Driggs	5,488	3,439	450
Twin Falls	Twin Falls	62,265	53,580	1,925
Valley	Cascade	8,005	6,109	3,678
Washington	Weiser	10,171	8,550	1,456

Illinois

(102 counties, 55,593 sq mi land; pop. 12,045,326)

County	County seat or courthouse	1998 Pop.	1990 Pop.	Land area sq mi
Adams	Quincy	67,105	66,090	857
Alexander	Cairo	9,745	10,626	236
Bond	Greenville	15,858	14,991	380
Boone	Belvidere	38,734	30,806	281
Brown	Mount Sterling	6,822	5,836	306
Bureau	Princeton	35,530	35,688	869
Calhoun	Hardin	4,981	5,322	254
Carroll	Mount Carroll	16,941	16,805	444
Cass	Virginia	13,266	13,437	376
Champaign	Urbana	167,788	173,025	997
Christian	Taylorville	34,543	34,418	709
Clark	Marshall	16,534	15,921	502
Clay	Louisville	14,485	14,460	469
Clinton	Carlyle	35,591	33,944	474
Coles	Charleston	51,103	51,644	508
Cook	Chicago	5,189,689	5,105,044	946
Crawford	Robinson	20,954	19,464	444
Cumberland	Toledo	11,124	10,670	346
De Kalb	Sycamore	84,169	77,932	634
De Witt	Clinton	16,796	16,516	398
Douglas	Tuscola	19,915	19,464	417
Du Page	Wheaton	880,491	781,689	334
Edgar	Paris	19,652	19,595	624
Edwards	Albion	6,950	7,440	222
Effingham	Effingham	33,504	31,704	479
Fayette	Vandalia	21,972	20,893	717
Ford	Paxton	14,084	14,275	486
Franklin	Benton	40,476	40,319	412
Fulton	Lewistown	38,746	38,080	866
Gallatin	Shawneetown	6,642	6,909	324
Greene	Carrollton	15,549	15,317	543
Grundy	Morris	36,686	32,337	420
Hamilton	McLeansboro	8,611	8,499	435
Hancock	Carthage	21,088	21,373	795
Hardin	Elizabethtown	4,902	5,189	178
Henderson	Oquawka	8,601	8,096	379
Henry	Cambridge	51,580	51,159	823
Iroquois	Watseka	31,243	30,787	1,117
Jackson	Murphysboro	60,410	61,067	588
Jasper	Newton	10,647	10,609	494
Jefferson	Mount Vernon	37,373	37,020	571
Jersey	Jerseyville	21,373	20,539	369
Jo Daviess	Galena	21,468	21,821	601
Johnson	Vienna	13,283	11,347	346
Kane	Geneva	391,241	317,471	521
Kankakee	Kankakee	102,107	96,255	678
Kendall	Yorkville	51,817	39,413	321
Knox	Galesburg	55,526	56,393	716
Lake	Waukegan	605,116	516,418	448
La Salle	Ottawa	110,189	106,913	1,135
Lawrence	Lawrenceville	15,343	15,972	372
Lee	Dixon	36,021	34,392	725
Livingston	Pontiac	39,702	39,301	1,044
Logan	Lincoln	31,289	30,798	618
McDonough	Macomb	33,917	35,244	589
McHenry	Woodstock	240,945	183,241	604
McLean	Bloomington	142,652	129,180	1,184
Macon	Decatur	113,772	117,206	581
Macoupin	Carlinville	48,872	47,679	864
Madison	Edwardsville	259,351	249,238	725
Marion	Salem	41,883	41,561	572
Marshall	Lacon	12,882	12,846	386
Mason	Havana	16,837	16,269	539
Massac	Metropolis	15,584	14,752	239
Menard	Petersburg	12,469	11,164	314
Mercer	Aledo	17,640	17,290	561
Monroe	Waterloo	26,586	22,422	388
Montgomery	Hillsboro	31,390	30,728	704
Morgan	Jacksonville	35,346	36,397	569
Moultrie	Sullivan	14,410	13,930	336
Ogle	Oregon	50,511	45,957	759
Peoria	Peoria	181,609	182,827	620
Perry	Pinckneyville	21,048	21,412	441
Piatt	Monticello	16,400	15,548	440

County	County seat or courthouse	1998 Pop.	1990 Pop.	Land area sq mi
Pike	Pittsfield	17,341	17,577	830
Pope	Golconda	4,808	4,373	371
Pulaski	Mound City	7,203	7,523	201
Putnam	Hennepin	5,826	5,730	160
Randolph	Chester	33,489	34,583	578
Richland	Olney	16,769	16,545	360
Rock Island	Rock Island	147,642	148,723	427
Saint Clair	Belleville	261,941	262,852	664
Saline	Harrisburg	26,149	26,551	383
Sangamon	Springfield	191,378	178,386	868
Schuyler	Rushville	7,632	7,498	437
Scott	Winchester	5,610	5,644	251
Shelby	Shelbyville	22,731	22,261	759
Stark	Toulon	6,290	6,534	288
Stephenson	Freeport	48,951	48,052	564
Tazewell	Pekin	127,958	123,692	649
Union	Jonesboro	17,996	17,619	416
Vermilion	Danville	84,204	88,257	899
Wabash	Mount Carmel	12,630	13,111	224
Warren	Monmouth	18,824	19,181	543
Washington	Nashville	15,367	14,965	563
Wayne	Fairfield	16,989	17,241	714
White	Carmi	15,646	16,522	495
Whiteside	Morrison	59,623	60,186	685
Will	Joliet	459,189	357,313	837
Williamson	Marion	60,819	57,733	424
Winnebago	Rockford	267,642	252,913	514
Woodford	Eureka	35,212	32,653	528

Indiana

(92 counties, 35,870 sq mi land; pop. 5,899,195)

County	County seat or courthouse	1998 Pop.	1990 Pop.	Land area sq mi
Adams	Decatur	33,083	31,095	339
Allen	Fort Wayne	314,218	300,836	657
Bartholomew	Columbus	69,579	63,657	407
Benton	Fowler	9,725	9,441	406
Blackford	Hartford City	13,910	14,067	165
Boone	Lebanon	43,843	38,147	423
Brown	Nashville	15,982	14,080	312
Carroll	Delphi	20,010	18,809	372
Cass	Logansport	38,685	38,413	413
Clark	Jeffersonville	93,805	87,774	375
Clay	Brazil	26,637	24,705	358
Clinton	Frankfort	33,215	30,974	405
Crawford	English	10,582	9,914	306
Daviess	Washington	28,987	27,533	431
Dearborn	Lawrenceburg	47,206	38,835	305
Decatur	Greensburg	25,562	23,645	373
De Kalb	Auburn	39,330	35,324	363
Delaware	Muncie	116,828	119,659	393
Dubois	Jasper	39,682	36,616	430
Elkhart	Goshen	172,310	156,198	464
Fayette	Connersville	25,969	26,015	215
Floyd	New Albany	71,990	64,404	148
Fountain	Covington	18,348	17,808	396
Franklin	Brookville	21,808	19,580	386
Fulton	Rochester	20,620	18,840	369
Gibson	Princeton	32,149	31,913	489
Grant	Marion	72,570	74,169	414
Greene	Bloomfield	33,467	30,410	542
Hamilton	Noblesville	162,597	108,936	398
Hancock	Greenfield	54,524	45,527	306
Harrison	Corydon	34,730	29,890	485
Hendricks	Danville	95,146	75,717	408
Henry	New Castle	48,785	48,139	393
Howard	Kokomo	83,452	80,827	293
Huntington	Huntington	37,259	35,427	383
Jackson	Brownstown	40,992	37,730	509
Jasper	Rensselaer	29,260	24,960	560
Jay	Portland	21,729	21,512	384
Jefferson	Madison	31,466	29,797	361
Jennings	Vernon	27,789	23,661	377
Johnson	Franklin	109,368	88,109	320
Knox	Vincennes	39,388	39,884	516
Kosciusko	Warsaw	71,207	65,294	538
Lagrange	Lagrange	33,484	29,477	380
Lake	Crown Point	478,323	475,594	497
La Porte	La Porte	109,461	107,066	598
Lawrence	Bedford	45,615	42,836	449
Madison	Anderson	131,360	130,669	452
Marion	Indianapolis	813,405	797,159	396
Marshall	Plymouth	45,444	42,182	444
Martin	Shoals	10,531	10,369	336
Miami	Peru	33,543	36,897	376
Monroe	Bloomington	115,130	108,978	394
Montgomery	Crawfordsville	36,337	34,436	505
Morgan	Martinsville	65,500	55,920	407
Newton	Kentland	14,734	13,551	402
Noble	Albion	42,626	37,877	411
Ohio	Rising Sun	5,423	5,315	87
Orange	Paoli	19,606	18,409	400
Owen	Spencer	20,419	17,281	385
Parke	Rockville	16,720	15,410	445
Perry	Cannelton	19,350	19,107	381

County	County seat or courthouse	1998 Pop.	1990 Pop.	Land area sq mi
Pike	Petersburg	12,882	12,509	336
Porter	Valparaiso	145,726	128,932	418
Posey	Mount Vernon	26,512	25,968	409
Pulaski	Winamac	13,257	12,643	434
Putnam	Greencastle	34,468	30,315	480
Randolph	Winchester	27,628	27,148	453
Ripley	Versailles	27,205	24,616	446
Rush	Rushville	18,307	18,129	408
Saint Joseph	South Bend	258,088	247,052	457
Scott	Scottsburg	22,939	20,991	190
Shelby	Shelbyville	43,451	40,307	413
Spencer	Rockport	20,937	19,490	399
Starke	Knox	23,968	22,747	309
Steuben	Angola	31,450	27,446	309
Sullivan	Sullivan	19,270	18,993	447
Switzerland	Vevay	8,893	7,738	221
Tippecanoe	Lafayette	139,005	130,598	500
Tipton	Tipton	16,724	16,119	260
Union	Liberty	7,263	6,976	162
Vanderburgh	Evansville	168,179	165,058	235
Vermillion	Newport	16,908	16,773	257
Vigo	Terre Haute	105,083	106,107	403
Wabash	Wabash	34,537	35,069	413
Warren	Williamsport	8,251	8,176	365
Warrick	Boonville	51,609	44,920	384
Washington	Salem	27,900	23,717	515
Wayne	Richmond	71,313	71,951	404
Wells	Bluffton	26,842	25,948	370
White	Monticello	25,338	23,265	505
Whitley	Columbia City	30,459	27,651	336

Iowa

(99 counties, 55,875 sq mi land; pop. 2,862,447)

County	County seat or courthouse	1998 Pop.	1990 Pop.	Land area sq mi
Adair	Greenfield	8,064	8,409	569
Adams	Corning	4,352	4,866	424
Allamakee	Waukon	13,989	13,855	640
Appanoose	Centerville	13,595	13,743	496
Audubon	Audubon	6,784	7,334	443
Benton	Vinton	25,418	22,429	717
Black Hawk	Waterloo	121,121	123,798	567
Boone	Boone	26,233	25,186	572
Bremer	Waverly	23,411	22,813	438
Buchanan	Independence	21,190	20,844	571
Buena Vista	Storm Lake	19,454	19,965	575
Butler	Allison	15,693	15,731	580
Calhoun	Rockwell City	11,378	11,508	570
Carroll	Carroll	21,706	21,423	569
Cass	Atlantic	14,591	15,128	564
Cedar	Tipton	17,977	17,444	580
Cerro Gordo	Mason City	46,193	46,733	568
Cherokee	Cherokee	13,191	14,098	577
Chickasaw	New Hampton	13,441	13,295	505
Clarke	Osceola	8,362	8,287	431
Clay	Spencer	17,532	17,585	569
Clayton	Elkader	18,722	19,054	779
Clinton	Clinton	49,897	51,040	695
Crawford	Denison	16,446	16,775	714
Dallas	Adel	36,900	29,755	587
Davis	Bloomfield	8,403	8,312	503
Decatur	Leon	8,220	8,338	532
Delaware	Manchester	18,578	18,035	578
Des Moines	Burlington	41,944	42,614	416
Dickinson	Spirit Lake	16,236	14,909	381
Dubuque	Dubuque	87,806	86,403	608
Emmet	Estherville	10,887	11,569	396
Fayette	West Union	21,761	21,843	731
Floyd	Charles City	16,353	17,058	501
Franklin	Hampton	10,863	11,364	583
Fremont	Sidney	7,746	8,226	511
Greene	Jefferson	10,065	10,045	568
Grundy	Grundy Center	12,183	12,029	503
Guthrie	Guthrie Center	11,571	10,935	591
Hamilton	Webster City	16,011	16,071	577
Hancock	Garner	12,044	12,638	571
Hardin	Eldora	18,462	19,094	569
Harrison	Logan	15,364	14,730	697
Henry	Mount Pleasant	19,983	19,226	435
Howard	Cresco	9,689	9,809	473
Humboldt	Dakota City	10,323	10,756	434
Ida	Ida Grove	7,912	8,365	432
Iowa	Marengo	15,550	14,630	587
Jackson	Maquoketa	20,078	19,950	636
Jasper	Newton	35,961	34,795	730
Jefferson	Fairfield	17,113	16,310	435
Johnson	Iowa City	102,724	96,119	615
Jones	Anamosa	20,349	19,444	575
Keokuk	Sigourney	11,499	11,624	579
Kossuth	Algona	17,738	18,591	973
Lee	Fort Madison & Keokuk	38,471	38,687	517
Linn	Cedar Rapids	182,651	168,767	718
Louisa	Wapello	11,938	11,592	402
Lucas	Chariton	9,152	9,070	431

County	County seat or courthouse	1998 Pop.	1990 Pop.	Land area sq mi
Lyon	Rock Rapids	12,012	11,952	588
Madison	Winterset	13,872	12,483	561
Mahaska	Oskaloosa	21,901	21,532	571
Marion	Knoxville	31,357	30,001	554
Marshall	Marshalltown	38,732	38,276	572
Mills	Glenwood	14,477	13,202	437
Mitchell	Osage	11,012	10,928	469
Monona	Onawa	10,110	10,034	693
Monroe	Albia	8,041	8,114	433
Montgomery	Red Oak	11,910	12,076	424
Muscatine	Muscatine	41,126	39,907	439
O'Brien	Primghar	14,910	15,444	573
Osceola	Sibley	6,980	7,267	399
Page	Clarinda	17,269	16,870	535
Palo Alto	Emmetsburg	10,017	10,669	564
Plymouth	Le Mars	24,825	23,388	864
Pocahontas	Pocahontas	8,777	9,525	578
Polk	Des Moines	359,826	327,140	570
Pottawattamie	Council Bluffs	86,174	82,628	954
Poweshiek	Montezuma	18,865	19,033	585
Ringgold	Mount Ayr	5,354	5,420	538
Sac	Sac City	11,931	12,324	576
Scott	Davenport	158,591	150,973	458
Shelby	Harlan	12,978	13,230	591
Sioux	Orange City	31,280	29,903	768
Story	Nevada	75,268	74,252	573
Tama	Toledo	17,739	17,419	721
Taylor	Bedford	7,153	7,114	534
Union	Creston	12,554	12,750	424
Van Buren	Keosauqua	7,886	7,676	485
Wapello	Ottumwa	35,440	35,696	432
Warren	Indianola	40,196	36,033	572
Washington	Washington	20,967	19,612	569
Wayne	Corydon	6,659	7,067	526
Webster	Fort Dodge	38,705	40,342	715
Winnebago	Forest City	11,931	12,122	401
Winneshiek	Decorah	20,934	20,847	690
Woodbury	Sioux City	101,672	98,276	873
Worth	Northwood	7,779	7,991	400
Wright	Clarion	14,003	14,269	581

Kansas

(105 counties, 81,823 sq mi land; pop. 2,629,067)

County	County seat or courthouse	1998 Pop.	1990 Pop.	Land area sq mi
Allen	Iola	14,556	14,638	503
Anderson	Garnett	8,060	7,803	583
Atchison	Atchison	16,908	16,932	432
Barber	Medicine Lodge	5,342	5,874	1,134
Barton	Great Bend	27,641	29,382	894
Bourbon	Fort Scott	15,260	14,966	637
Brown	Hiawatha	11,070	11,128	571
Butler	El Dorado	61,932	50,580	1,428
Chase	Cottonwood Falls	2,950	3,021	776
Chautauqua	Sedan	4,360	4,407	642
Cherokee	Columbus	22,552	21,374	587
Cheyenne	Saint Francis	3,174	3,243	1,020
Clark	Ashland	2,361	2,418	975
Clay	Clay Center	9,148	9,158	644
Cloud	Concordia	10,027	11,023	716
Coffey	Burlington	8,696	8,404	630
Comanche	Coldwater	2,012	2,313	788
Cowley	Winfield	36,319	36,915	1,126
Crawford	Girard	36,360	35,582	593
Decatur	Oberlin	3,456	4,021	894
Dickinson	Abilene	19,742	18,958	848
Doniphan	Troy	7,856	8,134	392
Douglas	Lawrence	93,137	81,798	457
Edwards	Kinsley	3,312	3,787	622
Elk	Howard	3,351	3,327	648
Ellis	Hays	26,309	26,004	900
Ellsworth	Ellsworth	6,285	6,586	716
Finney	Garden City	36,514	33,070	1,300
Ford	Dodge City	29,382	27,463	1,099
Franklin	Ottawa	24,768	21,994	574
Geary	Junction City	25,370	30,453	384
Gove	Gove	3,054	3,231	1,072
Graham	Hill City	3,204	3,543	898
Grant	Ulysses	8,012	7,159	575
Gray	Cimarron	5,595	5,396	869
Greeley	Tribune	1,704	1,774	778
Greenwood	Eureka	8,139	7,847	1,140
Hamilton	Syracuse	2,343	2,388	997
Harper	Anthony	6,430	7,124	802
Harvey	Newton	34,361	31,028	539
Haskell	Sublette	3,976	3,886	577
Hodgeman	Jetmore	2,209	2,177	860
Jackson	Holton	12,130	11,525	657
Jefferson	Oskaloosa	18,243	15,905	536
Jewell	Mankato	3,867	4,251	909
Johnson	Olathe	429,563	355,021	477
Kearny	Lakin	4,177	4,027	870
Kingman	Kingman	8,543	8,292	864
Kiowa	Greensburg	3,470	3,660	722
Labette	Oswego	23,030	23,693	649
Lane	Dighton	2,264	2,375	717
Leavenworth	Leavenworth	71,299	64,371	463
Lincoln	Lincoln	3,338	3,653	719
Linn	Mound City	9,158	8,254	599
Logan	Oakley	2,987	3,081	1,073
Lyon	Emporia	33,920	34,732	851
McPherson	McPherson	28,630	27,268	900
Marion	Marion	13,593	12,888	943
Marshall	Marysville	11,006	11,705	903
Meade	Meade	4,424	4,247	979
Miami	Paola	26,597	23,466	577
Mitchell	Beloit	6,936	7,203	700
Montgomery	Independence	37,089	38,816	645
Morris	Council Grove	6,169	6,198	697
Morton	Elkhart	3,440	3,480	730
Nemaha	Seneca	10,132	10,446	719
Neosho	Erie	16,760	17,035	572
Ness	Ness City	3,607	4,033	1,075
Norton	Norton	5,752	5,947	878
Osage	Lyndon	17,139	15,248	704
Osborne	Osborne	4,712	4,867	893
Ottawa	Minneapolis	5,905	5,634	721
Pawnee	Larned	7,437	7,555	754
Phillips	Phillipsburg	6,080	6,590	886
Pottawatomie	Westmoreland	18,691	16,128	844
Pratt	Pratt	9,700	9,702	735
Rawlins	Atwood	3,125	3,404	1,070
Reno	Hutchinson	63,211	62,389	1,255
Republic	Belleville	6,102	6,482	717
Rice	Lyons	10,360	10,610	727
Riley	Manhattan	63,615	67,139	610
Rooks	Stockton	5,660	6,039	888
Rush	LaCrosse	3,413	3,842	718
Russell	Russell	7,558	7,835	885
Saline	Salina	51,617	49,301	720
Scott	Scott City	5,018	5,289	718
Sedgwick	Wichita	448,050	403,662	1,000
Seward	Liberal	19,984	18,743	640
Shawnee	Topeka	165,348	160,976	550
Sheridan	Hoxie	2,741	3,043	896
Sherman	Goodland	6,511	6,926	1,056
Smith	Smith Center	4,588	5,078	896
Stafford	Saint John	5,000	5,365	792
Stanton	Johnson	2,265	2,333	680
Stevens	Hugoton	5,371	5,048	728
Sumner	Wellington	27,043	25,841	1,182
Thomas	Colby	8,037	8,258	1,075
Trego	WaKeeney	3,283	3,694	888
Wabaunsee	Alma	6,651	6,603	798
Wallace	Sharon Springs	1,802	1,821	914
Washington	Washington	6,490	7,073	899
Wichita	Leoti	2,643	2,758	719
Wilson	Fredonia	10,218	10,289	574
Woodson	Yates Center	3,983	4,116	501
Wyandotte	Kansas City	152,355	162,026	151

Kentucky

(120 counties, 39,732 sq mi land; pop. 3,936,499)

County	County seat or courthouse	1998 Pop.	1990 Pop.	Land area sq mi
Adair	Columbia	16,447	15,360	407
Allen	Scottsville	16,555	14,628	346
Anderson	Lawrenceburg	18,587	14,571	203
Ballard	Wickliffe	8,488	7,902	251
Barren	Glasgow	36,979	34,001	491
Bath	Owingsville	10,570	9,692	279
Bell	Pineville	29,133	31,506	361
Boone	Burlington	79,671	57,589	246
Bourbon	Paris	19,368	19,236	291
Boyd	Catlettsburg	49,543	51,150	160
Boyle	Danville	27,186	25,641	182
Bracken	Brooksville	8,455	7,766	203
Breathitt	Jackson	15,686	15,703	495
Breckinridge	Hardinsburg	17,465	16,312	572
Bullitt	Shepherdsville	59,304	47,567	299
Butler	Morgantown	11,926	11,245	428
Caldwell	Princeton	13,314	13,232	347
Calloway	Murray	33,478	30,735	386
Campbell	Newport	87,381	83,866	152
Carlisle	Bardwell	5,320	5,238	193
Carroll	Carrollton	9,603	9,292	130
Carter	Grayson	26,848	24,340	411
Casey	Liberty	14,773	14,211	446
Christian	Hopkinsville	72,493	68,941	721
Clark	Winchester	31,978	29,496	254
Clay	Manchester	22,799	21,746	471
Clinton	Albany	9,346	9,135	198
Crittenden	Marion	9,574	9,196	362
Cumberland	Burkesville	6,828	6,784	306
Daviess	Owensboro	91,139	87,189	462
Edmonson	Brownsville	11,353	10,357	303
Elliott	Sandy Hook	6,602	6,455	234
Estill	Irvine	15,588	14,614	254
Fayette	Lexington	241,749	225,366	285
Fleming	Flemingsburg	13,441	12,292	351

County	County seat or courthouse	1998 Pop.	1990 Pop.	Land area sq mi
Floyd	Prestonsburg	43,340	43,586	394
Franklin	Frankfort	46,438	44,143	211
Fulton	Hickman	7,542	8,271	209
Gallatin	Warsaw	7,182	5,393	99
Garrard	Lancaster	13,916	11,579	231
Grant	Williamstown	20,347	15,737	260
Graves	Mayfield	35,847	33,550	556
Grayson	Leitchfield	23,763	21,050	504
Green	Greensburg	10,650	10,371	289
Greenup	Greenup	36,874	36,742	346
Hancock	Hawesville	8,941	7,864	189
Hardin	Elizabethtown	91,462	89,240	628
Harlan	Harlan	34,950	36,574	467
Harrison	Cynthiana	17,565	16,248	310
Hart	Munfordville	16,738	14,890	416
Henderson	Henderson	44,457	43,044	440
Henry	New Castle	14,765	12,823	289
Hickman	Clinton	5,247	5,566	245
Hopkins	Madisonville	46,364	46,126	551
Jackson	McKee	12,908	11,955	346
Jefferson	Louisville	672,104	665,123	385
Jessamine	Nicholasville	36,533	30,508	173
Johnson	Paintsville	24,022	23,248	262
Kenton	Covington	146,732	142,031	163
Knott	Hindman	17,989	17,906	352
Knox	Barbourville	31,890	29,676	388
Larue	Hodgenville	13,058	11,679	263
Laurel	London	50,734	43,438	436
Lawrence	Louisa	15,647	13,998	419
Lee	Beattyville	8,021	7,422	210
Leslie	Hyden	13,582	13,642	404
Letcher	Whitesburg	26,185	27,000	339
Lewis	Vanceburg	13,584	13,029	485
Lincoln	Stanford	22,367	20,045	337
Livingston	Smithland	9,432	9,062	316
Logan	Russellville	26,145	24,416	556
Lyon	Eddyville	8,052	6,624	216
McCracken	Paducah	64,460	62,879	251
McCreary	Whitley City	16,659	15,603	428
McLean	Calhoun	9,845	9,628	254
Madison	Richmond	66,502	57,508	441
Magoffin	Salyersville	13,838	13,077	310
Marion	Lebanon	17,018	16,499	347
Marshall	Benton	30,312	27,205	305
Martin	Inez	12,120	12,526	231
Mason	Maysville	17,021	16,666	241
Meade	Brandenburg	28,809	24,170	309
Menifee	Frenchburg	5,736	5,092	204
Mercer	Harrodsburg	20,704	19,148	251
Metcalfe	Edmonton	9,561	8,963	291
Monroe	Tompkinsville	11,201	11,401	331
Montgomery	Mount Sterling	20,932	19,561	199
Morgan	West Liberty	13,559	11,648	381
Muhlenberg	Greenville	32,173	31,318	475
Nelson	Bardstown	35,884	29,710	423
Nicholas	Carlisle	6,998	6,725	197
Ohio	Hartford	22,005	21,105	594
Oldham	La Grange	44,395	33,263	189
Owen	Owenton	10,264	9,035	352
Owsley	Booneville	5,404	5,036	198
Pendleton	Falmouth	13,703	12,036	280
Perry	Hazard	31,049	30,283	342
Pike	Pikeville	72,121	72,584	788
Powell	Stanton	12,945	11,686	180
Pulaski	Somerset	56,294	49,489	662
Robertson	Mount Olivet	2,209	2,124	100
Rockcastle	Mount Vernon	15,951	14,803	318
Rowan	Morehead	22,196	20,353	281
Russell	Jamestown	16,233	14,716	254
Scott	Georgetown	30,685	23,867	285
Shelby	Shelbyville	29,583	24,824	384
Simpson	Franklin	16,401	15,145	236
Spencer	Taylorsville	9,660	6,801	186
Taylor	Campbellsville	22,943	21,146	270
Todd	Elkton	11,222	10,940	376
Trigg	Cadiz	12,399	10,361	443
Trimble	Bedford	7,621	6,090	149
Union	Morganfield	16,577	16,557	345
Warren	Bowling Green	87,323	77,720	545
Washington	Springfield	10,918	10,441	301
Wayne	Monticello	19,107	17,468	459
Webster	Dixon	13,482	13,955	335
Whitley	Williamsburg	35,938	33,326	440
Wolfe	Campton	7,366	6,503	223
Woodford	Versailles	22,830	19,955	191

Louisiana

(64 parishes, 43,566 sq mi land; pop. 4,368,967)

Parish	Parish seat or courthouse	1998 Pop.	1990 Pop.	Land area sq mi
Acadia	Crowley	57,721	55,882	655
Allen	Oberlin	23,888	21,226	765
Ascension	Donaldsonville	71,628	58,214	292
Assumption	Napoleonville	23,015	22,753	339
Avoyelles	Marksville	40,846	39,159	833
Beauregard	De Ridder	31,976	30,083	1,160
Bienville	Arcadia	15,814	16,201	811
Bossier	Benton	93,463	86,088	839
Caddo	Shreveport	242,471	248,253	882
Calcasieu	Lake Charles	180,330	168,134	1,071
Caldwell	Columbia	10,364	9,806	530
Cameron	Cameron	9,063	9,260	1,313
Catahoula	Harrisonburg	11,064	11,065	704
Claiborne	Homer	16,919	17,405	755
Concordia	Vidalia	20,749	20,828	696
De Soto	Mansfield	24,921	25,699	877
East Baton Rouge	Baton Rouge	394,714	380,105	456
East Carroll	Lake Providence	8,905	9,709	422
East Feliciana	Clinton	20,847	19,211	453
Evangeline	Ville Platte	34,097	33,274	664
Franklin	Winnsboro	22,163	22,387	623
Grant	Colfax	18,990	17,526	645
Iberia	New Iberia	73,154	68,297	575
Iberville	Plaquemine	31,173	31,049	619
Jackson	Jonesboro	15,566	15,924	570
Jefferson	Gretna	450,933	448,306	306
Jefferson Davis	Jennings	31,607	30,722	652
Lafayette	Lafayette	186,631	164,762	270
Lafourche	Thibodaux	89,324	85,860	1,085
La Salle	Jena	13,661	13,662	624
Lincoln	Ruston	41,560	41,745	471
Livingston	Livingston	88,104	70,523	648
Madison	Tallulah	12,808	12,463	624
Morehouse	Bastrop	31,477	31,938	794
Natchitoches	Natchitoches	37,018	37,199	1,256
Orleans	New Orleans	465,538	496,938	181
Ouachita	Monroe	146,979	142,191	611
Plaquemines	Pointe a la Hache	26,293	25,575	845
Pointe Coupee	New Roads	23,565	22,540	557
Rapides	Alexandria	126,763	131,556	1,323
Red River	Coushatta	9,599	9,518	389
Richland	Rayville	21,022	20,629	559
Sabine	Many	23,824	22,646	865
Saint Bernard	Chalmette	66,074	66,631	465
Saint Charles	Hahnville	48,278	42,437	284
Saint Helena	Greensburg	9,580	9,874	408
Saint James	Convent	21,132	20,879	246
Saint John the Baptist	Edgard	42,261	39,996	219
Saint Landry	Opelousas	83,816	80,312	929
Saint Martin	Saint Martinville	47,550	44,097	740
Saint Mary	Franklin	57,174	58,086	613
Saint Tammany	Covington	188,936	144,500	854
Tangipahoa	Amite	96,983	85,709	790
Tensas	Saint Joseph	6,631	7,103	603
Terrebonne	Houma	104,534	96,982	1,255
Union	Farmerville	21,989	20,796	878
Vermilion	Abbeville	52,090	50,055	1,174
Vernon	Leesville	51,570	61,961	1,329
Washington	Franklinton	43,059	43,185	670
Webster	Minden	42,707	41,989	596
West Baton Rouge	Port Allen	20,683	19,419	191
West Carroll	Oak Grove	12,213	12,093	359
West Feliciana	Saint Francisville	13,446	12,915	406
Winn	Winnfield	17,714	16,496	951

Maine

(16 counties, 30,865 sq mi land; pop. 1,244,250)

County	County seat or courthouse	1998 Pop.	1990 Pop.	Land area sq mi
Androscoggin	Auburn	101,280	105,259	470
Aroostook	Houlton	76,085	86,936	6,672
Cumberland	Portland	253,582	243,135	836
Franklin	Farmington	28,933	29,008	1,698
Hancock	Ellsworth	49,932	46,948	1,589
Kennebec	Augusta	115,207	115,904	868
Knox	Rockland	37,847	36,310	366
Lincoln	Wiscasset	31,815	30,357	456
Oxford	South Paris	53,673	52,602	2,078
Penobscot	Bangor	142,323	146,601	3,396
Piscataquis	Dover-Foxcroft	18,282	18,653	3,967
Sagadahoc	Bath	35,779	33,535	254
Somerset	Skowhegan	52,380	49,767	3,927
Waldo	Belfast	36,465	33,018	730
Washington	Machias	35,502	35,308	2,569
York	Alfred	175,165	164,587	991

Maryland

(23 counties, 1 ind. city, 9,775 sq mi land; pop. 5,134,808)

County	County seat or courthouse	1998 Pop.	1990 Pop.	Land area sq mi
Allegany	Cumberland	71,333	74,946	425
Anne Arundel	Annapolis	476,060	427,239	416
Baltimore	Towson	721,874	692,134	599
Calvert	Prince Frederick	71,877	51,372	215
Caroline	Denton	29,489	27,035	320

County	County seat or courthouse	1998 Pop.	1990 Pop.	Land area sq mi
Carroll	Westminster	149,697	123,372	449
Cecil	Elkton	82,522	71,347	348
Charles	La Plata	117,963	101,154	461
Dorchester	Cambridge	29,503	30,236	558
Frederick	Frederick	186,777	150,208	663
Garrett	Oakland	29,238	28,138	648
Harford	Bel Air	214,668	182,132	440
Howard	Ellicott City	236,388	187,328	252
Kent	Chestertown	18,925	17,842	279
Montgomery	Rockville	840,879	762,207	495
Prince George's	Upper Marlboro	777,811	723,373	486
Queen Anne's	Centreville	39,672	33,953	372
Saint Mary's	Leonardtown	87,670	75,974	361
Somerset	Princess Anne	24,296	23,440	327
Talbot	Easton	33,065	30,549	269
Washington	Hagerstown	127,352	121,393	458
Wicomico	Salisbury	79,367	74,339	377
Worcester	Snow Hill	42,789	35,028	473
Independent City				
Baltimore		645,593	736,014	81

Massachusetts

(14 counties, 7,838 sq mi land; pop. 6,147,132)

County	County seat or courthouse	1998 Pop.	1990 Pop.	Land area sq mi
Barnstable	Barnstable	208,418	186,605	396
Berkshire	Pittsfield	133,038	139,352	931
Bristol	Taunton	517,543	506,325	556
Dukes	Edgartown	13,888	11,639	104
Essex	Salem	698,806	670,080	498
Franklin	Greenfield	70,597	70,092	702
Hampden	Springfield	439,609	456,310	619
Hampshire	Northampton	149,384	146,568	529
Middlesex	East Cambridge	1,424,116	1,398,468	824
Nantucket	Nantucket	7,844	6,012	48
Norfolk	Dedham	642,705	616,087	400
Plymouth	Plymouth	467,588	435,276	661
Suffolk	Boston	641,715	663,906	59
Worcester	Worcester	731,881	709,705	1,513

Michigan

(83 counties, 56,809 sq mi land; pop. 9,817,242)

County	County seat or courthouse	1998 Pop.	1990 Pop.	Land area sq mi
Alcona	Harrisville	11,108	10,145	675
Alger	Munising	9,887	8,972	918
Allegan	Allegan	101,662	90,509	828
Alpena	Alpena	30,405	30,605	574
Antrim	Bellaire	21,522	18,185	477
Arenac	Standish	16,413	14,906	367
Baraga	L'Anse	8,413	7,954	904
Barry	Hastings	54,535	50,057	556
Bay	Bay City	110,048	111,723	444
Benzie	Beulah	14,678	12,200	321
Berrien	Saint Joseph	160,245	161,378	571
Branch	Coldwater	43,634	41,502	507
Calhoun	Marshall	141,005	135,982	709
Cass	Cassopolis	49,693	49,477	492
Charlevoix	Charlevoix	24,436	21,468	417
Cheboygan	Cheboygan	23,738	21,398	716
Chippewa	Sault Sainte Marie	37,968	34,604	1,561
Clare	Harrison	29,578	24,952	567
Clinton	Saint Johns	63,379	57,893	572
Crawford	Grayling	14,150	12,260	558
Delta	Escanaba	38,947	37,780	1,170
Dickinson	Iron Mountain	27,074	26,831	766
Eaton	Charlotte	101,090	92,879	577
Emmet	Petoskey	28,677	25,040	468
Genesee	Flint	436,084	430,459	640
Gladwin	Gladwin	25,333	21,896	507
Gogebic	Bessemer	17,097	18,052	1,102
Grand Traverse	Traverse City	74,134	64,273	465
Gratiot	Ithaca	40,126	38,982	570
Hillsdale	Hillsdale	46,614	43,431	599
Houghton	Houghton	35,719	35,446	1,012
Huron	Bad Axe	35,303	34,951	837
Ingham	Mason	285,214	281,912	559
Ionia	Ionia	61,700	57,024	573
Iosco	Tawas City	25,111	30,209	549
Iron	Crystal Falls	12,883	13,175	1,167
Isabella	Mount Pleasant	58,026	54,624	574
Jackson	Jackson	156,157	149,756	707
Kalamazoo	Kalamazoo	229,660	223,411	562
Kalkaska	Kalkaska	15,568	13,497	561
Kent	Grand Rapids	545,166	500,631	856
Keweenaw	Eagle River	2,077	1,701	541
Lake	Baldwin	10,475	8,583	568
Lapeer	Lapeer	88,270	74,768	654
Leelanau	Leland	19,142	16,527	349
Lenawee	Adrian	98,412	91,476	751
Livingston	Howell	146,165	115,645	568
Luce	Newberry	6,640	5,763	903
Mackinac	Saint Ignace	11,097	10,674	1,022

County	County seat or courthouse	1998 Pop.	1990 Pop.	Land area sq mi
Macomb	Mount Clemens	787,698	717,400	480
Manistee	Manistee	23,330	21,265	544
Marquette	Marquette	61,565	70,887	1,821
Mason	Ludington	27,950	25,537	495
Mecosta	Big Rapids	40,006	37,308	556
Menominee	Menominee	24,468	24,920	1,044
Midland	Midland	81,842	75,651	521
Missaukee	Lake City	13,892	12,147	567
Monroe	Monroe	143,499	133,600	551
Montcalm	Stanton	60,559	53,059	708
Montmorency	Atlanta	10,011	8,936	548
Muskegon	Muskegon	166,748	158,983	509
Newaygo	White Cloud	45,784	38,206	842
Oakland	Pontiac	1,176,488	1,083,592	873
Oceana	Hart	24,833	22,455	541
Ogemaw	West Branch	21,193	18,681	564
Ontonagon	Ontonagon	7,878	8,854	1,312
Osceola	Reed City	22,106	20,146	566
Oscoda	Mio	8,882	7,842	565
Otsego	Gaylord	22,129	17,957	515
Ottawa	Grand Haven	224,357	187,768	566
Presque Isle	Rogers City	14,424	13,743	660
Roscommon	Roscommon	23,467	19,776	521
Saginaw	Saginaw	210,101	211,946	809
Saint Clair	Port Huron	159,769	145,607	725
Saint Joseph	Centreville	61,226	58,913	504
Sanilac	Sandusky	42,975	39,928	964
Schoolcraft	Manistique	8,805	8,302	1,178
Shiawassee	Corunna	72,569	69,770	539
Tuscola	Caro	58,181	55,498	813
Van Buren	Paw Paw	75,666	70,060	611
Washtenaw	Ann Arbor	303,069	282,937	710
Wayne	Detroit	2,118,129	2,111,687	614
Wexford	Cadillac	29,185	26,360	566

Minnesota

(87 counties, 79,617 sq mi land; pop. 4,725,419)

County	County seat or courthouse	1998 Pop.	1990 Pop.	Land area sq mi
Aitkin	Aitkin	14,152	12,425	1,819
Anoka	Anoka	292,181	243,641	424
Becker	Detroit Lakes	29,381	27,881	1,311
Beltrami	Bemidji	38,729	34,384	2,505
Benton	Foley	34,128	30,185	408
Big Stone	Ortonville	5,654	6,285	497
Blue Earth	Mankato	53,767	54,044	752
Brown	New Ulm	27,037	26,984	611
Carlton	Carlton	30,817	29,259	860
Carver	Chaska	64,674	47,915	357
Cass	Walker	26,465	21,791	2,018
Chippewa	Montevideo	13,053	13,228	583
Chisago	Center City	40,852	30,521	418
Clay	Moorhead	51,599	50,422	1,045
Clearwater	Bagley	8,285	8,309	995
Cook	Grand Marais	4,792	3,868	1,451
Cottonwood	Windom	12,045	12,694	640
Crow Wing	Brainerd	51,681	44,249	997
Dakota	Hastings	342,528	275,189	570
Dodge	Mantorville	17,209	15,731	440
Douglas	Alexandria	31,045	28,674	634
Faribault	Blue Earth	16,244	16,937	714
Fillmore	Preston	20,793	20,777	861
Freeborn	Albert Lea	31,584	33,060	708
Goodhue	Red Wing	43,137	40,690	759
Grant	Elbow Lake	6,178	6,246	547
Hennepin	Minneapolis	1,059,669	1,032,431	557
Houston	Caledonia	19,267	18,497	558
Hubbard	Park Rapids	16,935	14,939	923
Isanti	Cambridge	30,121	25,921	439
Itasca	Grand Rapids	43,857	40,863	2,665
Jackson	Jackson	11,529	11,677	702
Kanabec	Mora	14,173	12,802	525
Kandiyohi	Willmar	41,086	38,761	796
Kittson	Hallock	5,322	5,767	1,097
Koochiching	International Falls	15,538	16,299	3,102
Lac qui Parle	Madison	8,022	8,924	765
Lake	Two Harbors	10,566	10,415	2,099
Lake of the Woods	Baudette	4,563	4,076	1,297
Le Sueur	Le Center	25,320	23,239	449
Lincoln	Ivanhoe	6,459	6,890	537
Lyon	Marshall	24,339	24,789	714
McLeod	Glencoe	34,017	32,030	492
Mahnomen	Mahnomen	5,077	5,044	556
Marshall	Warren	10,313	10,993	1,772
Martin	Fairmont	21,984	22,914	709
Meeker	Litchfield	21,735	20,846	609
Mille Lacs	Milaca	21,044	18,670	575
Morrison	Little Falls	30,543	29,604	1,125
Mower	Austin	37,039	37,385	712
Murray	Slayton	9,517	9,660	705
Nicollet	Saint Peter	29,600	28,076	452
Nobles	Worthington	19,312	20,098	716
Norman	Ada	7,535	7,975	876
Olmsted	Rochester	116,702	106,470	653

County	County seat or courthouse	1998 Pop.	1990 Pop.	Land area sq mi
Otter Tail	Fergus Falls	54,911	50,714	1,980
Pennington	Thief River Falls	13,562	13,306	617
Pine	Pine City	23,916	21,264	1,411
Pipestone	Pipestone	10,092	10,491	466
Polk	Crookston	30,954	32,589	1,971
Pope	Glenwood	10,886	10,745	670
Ramsey	Saint Paul	485,636	485,783	156
Red Lake	Red Lake Falls	4,270	4,525	432
Redwood	Redwood Falls	16,489	17,254	880
Renville	Olivia	16,923	17,673	983
Rice	Faribault	54,106	49,183	498
Rock	Luverne	9,743	9,806	483
Roseau	Roseau	16,120	15,026	1,663
Saint Louis	Duluth	193,431	198,213	6,226
Scott	Shakopee	79,031	57,846	357
Sherburne	Elk River	60,391	41,945	437
Sibley	Gaylord	14,573	14,366	589
Stearns	Saint Cloud	128,094	119,324	1,345
Steele	Owatonna	31,736	30,729	430
Stevens	Morris	10,136	10,634	562
Swift	Benson	10,804	10,724	744
Todd	Long Prairie	24,020	23,363	942
Traverse	Wheaton	4,248	4,463	574
Wabasha	Wabasha	20,943	19,744	525
Wadena	Wadena	13,145	13,154	536
Waseca	Waseca	18,178	18,079	423
Washington	Stillwater	196,486	145,858	392
Watonwan	Saint James	11,470	11,682	435
Wilkin	Breckenridge	7,312	7,516	752
Winona	Winona	48,080	47,828	626
Wright	Buffalo	85,123	68,710	661
Yellow Medicine	Granite Falls	11,416	11,684	758

Mississippi

(82 counties, 46,914 sq mi land; pop. 2,752,092)

County	County seat or courthouse	1998 Pop.	1990 Pop.	Land area sq mi
Adams	Natchez	34,225	35,356	460
Alcorn	Corinth	32,716	31,722	400
Amite	Liberty	13,752	13,328	730
Attala	Kosciusko	18,404	18,481	735
Benton	Ashland	8,140	8,046	407
Bolivar	Cleveland & Rosedale	40,318	41,875	876
Calhoun	Pittsboro	14,822	14,908	587
Carroll	Carrollton & Vaiden	9,995	9,237	628
Chickasaw	Houston & Okolona	18,013	18,085	502
Choctaw	Ackerman	9,385	9,071	419
Claiborne	Port Gibson	11,662	11,370	487
Clarke	Quitman	18,231	17,313	691
Clay	West Point	21,637	21,120	409
Coahoma	Clarksdale	31,089	31,665	554
Copiah	Hazlehurst	28,944	27,592	777
Covington	Collins	17,802	16,527	414
De Soto	Hernando	96,897	67,910	478
Forrest	Hattiesburg	74,364	68,314	467
Franklin	Meadville	8,319	8,377	565
George	Lucedale	19,645	16,673	478
Greene	Leakesville	11,766	10,220	713
Grenada	Grenada	22,427	21,555	422
Hancock	Bay Saint Louis	40,327	31,760	477
Harrison	Gulfport	177,981	165,365	581
Hinds	Jackson & Raymond	247,144	254,441	869
Holmes	Lexington	21,522	21,604	756
Humphreys	Belzoni	11,344	12,134	418
Issaquena	Mayersville	1,629	1,909	413
Itawamba	Fulton	21,072	20,017	532
Jackson	Pascagoula	130,910	115,243	727
Jasper	Bay Springs & Paulding	17,672	17,114	676
Jefferson	Fayette	8,427	8,653	519
Jefferson Davis	Prentiss	13,860	14,051	408
Jones	Ellisville & Laurel	63,461	62,031	694
Kemper	De Kalb	10,575	10,356	766
Lafayette	Oxford	34,555	31,826	631
Lamar	Purvis	36,888	30,424	497
Lauderdale	Meridian	76,143	75,555	704
Lawrence	Monticello	13,053	12,458	431
Leake	Carthage	19,372	18,436	583
Lee	Tupelo	74,637	65,579	450
Leflore	Greenwood	36,951	37,341	592
Lincoln	Brookhaven	31,771	30,278	586
Lowndes	Columbus	61,208	59,308	502
Madison	Canton	72,857	53,794	719
Marion	Columbia	26,386	25,544	542
Marshall	Holly Springs	32,296	30,361	706
Monroe	Aberdeen	38,263	36,582	764
Montgomery	Winona	12,425	12,387	407
Neshoba	Philadelphia	27,653	24,800	570
Newton	Decatur	21,516	20,291	578
Noxubee	Macon	12,366	12,604	695
Oktibbeha	Starkville	39,291	38,375	458
Panola	Batesville & Sardis	33,400	29,996	684
Pearl River	Poplarville	46,862	38,714	812
Perry	New Augusta	11,798	10,865	647
Pike	Magnolia	37,920	36,882	409

County	County seat or courthouse	1998 Pop.	1990 Pop.	Land area sq mi
Pontotoc	Pontotoc	25,397	22,237	497
Prentiss	Booneville	24,295	23,278	415
Quitman	Marks	9,914	10,490	405
Rankin	Brandon	109,613	87,161	775
Scott	Forest	25,001	24,137	609
Sharkey	Rolling Fork	6,650	7,066	428
Simpson	Mendenhall	25,338	23,953	589
Smith	Raleigh	15,296	14,798	636
Stone	Wiggins	13,166	10,750	445
Sunflower	Indianola	34,577	35,129	694
Tallahatchie	Charleston & Sumner	14,893	15,210	644
Tate	Senatobia	23,923	21,432	405
Tippah	Ripley	21,031	19,523	458
Tishomingo	Iuka	18,654	17,683	424
Tunica	Tunica	8,039	8,164	455
Union	New Albany	23,828	22,085	416
Walthall	Tylertown	14,369	14,352	404
Warren	Vicksburg	49,404	47,880	587
Washington	Greenville	65,264	67,935	724
Wayne	Waynesboro	20,368	19,517	810
Webster	Walthall	10,547	10,222	423
Wilkinson	Woodville	9,174	9,678	677
Winston	Louisville	19,387	19,433	607
Yalobusha	Coffeeville & Water Valley	12,366	12,033	467
Yazoo	Yazoo City	25,510	25,506	920

Missouri

(114 cos., 1 ind. city, 68,898 sq mi land; pop. 5,438,559)

County	County seat or courthouse	1998 Pop.	1990 Pop.	Land area sq mi
Adair	Kirksville	24,286	24,577	568
Andrew	Savannah	15,562	14,632	435
Atchison	Rockport	6,999	7,457	545
Audrain	Mexico	23,573	23,599	693
Barry	Cassville	33,120	27,547	779
Barton	Lamar	12,078	11,312	594
Bates	Butler	15,770	15,025	849
Benton	Warsaw	17,040	13,859	706
Bollinger	Marble Hill	11,513	10,619	621
Boone	Columbia	129,098	112,379	685
Buchanan	Saint Joseph	81,776	83,083	410
Butler	Poplar Buff	40,561	38,765	698
Caldwell	Kingston	8,838	8,380	429
Callaway	Fulton	37,437	32,809	839
Camden	Camdenton	33,952	27,495	655
Cape Girardeau	Jackson	66,314	61,633	579
Carroll	Carrollton	10,217	10,748	695
Carter	Van Buren	6,387	5,515	508
Cass	Harrisonville	80,520	63,808	699
Cedar	Stockton	13,215	12,093	476
Chariton	Keytesville	8,621	9,202	756
Christian	Ozark	48,997	32,644	563
Clark	Kahoka	7,467	7,547	507
Clay	Liberty	176,206	153,411	397
Clinton	Plattsburg	19,070	16,595	419
Cole	Jefferson City	69,307	63,579	392
Cooper	Boonville	16,029	14,835	565
Crawford	Steelville	22,165	19,173	743
Dade	Greenfield	7,892	7,449	490
Dallas	Buffalo	15,245	12,646	542
Daviess	Gallatin	7,842	7,865	567
De Kalb	Maysville	11,129	9,967	424
Dent	Salem	14,103	13,702	754
Douglas	Ava	12,422	11,876	815
Dunklin	Kennett	32,700	33,112	546
Franklin	Union	91,763	80,603	922
Gasconade	Hermann	14,890	14,006	520
Gentry	Albany	6,938	6,854	492
Greene	Springfield	226,758	207,949	675
Grundy	Trenton	10,159	10,536	436
Harrison	Bethany	8,506	8,469	725
Henry	Clinton	21,232	20,044	703
Hickory	Hermitage	8,617	7,335	399
Holt	Oregon	5,554	6,034	462
Howard	Fayette	9,741	9,631	466
Howell	West Plains	35,776	31,447	928
Iron	Ironton	10,871	10,726	551
Jackson	Independence	654,986	633,234	605
Jasper	Carthage	99,532	90,465	640
Jefferson	Hillsboro	195,675	171,380	657
Johnson	Warrensburg	47,644	42,514	831
Knox	Edina	4,355	4,482	506
Laclede	Lebanon	31,029	27,158	766
Lafayette	Lexington	32,653	31,107	629
Lawrence	Mount Vernon	33,122	30,236	613
Lewis	Monticello	10,199	10,233	505
Lincoln	Troy	36,556	28,892	631
Linn	Linneus	13,808	13,885	620
Livingston	Chillicothe	14,151	14,592	535
McDonald	Pineville	19,887	16,938	540
Macon	Macon	15,278	15,345	804
Madison	Fredericktown	11,481	11,127	497
Maries	Vienna	8,473	7,976	528

County	County seat or courthouse	1998 Pop.	1990 Pop.	Land area sq mi
Marion	Palmyra	27,771	27,682	438
Mercer	Princeton	4,003	3,723	455
Miller	Tuscumbia	22,422	20,700	592
Mississippi	Charleston	13,395	14,442	413
Moniteau	California	13,263	12,298	417
Monroe	Paris	9,021	9,104	646
Montgomery	Montgomery City	12,074	11,355	539
Morgan	Versailles	18,434	15,574	598
New Madrid	New Madrid	20,370	20,928	678
Newton	Neosho	49,152	44,445	627
Nodaway	Maryville	20,777	21,709	877
Oregon	Alton	10,164	9,470	792
Osage	Linn	12,425	12,018	606
Ozark	Gainesville	9,897	8,598	747
Pemiscot	Caruthersville	21,516	21,921	493
Perry	Perryville	17,410	16,648	475
Pettis	Sedalia	37,069	35,437	685
Phelps	Rolla	38,592	35,248	673
Pike	Bowling Green	16,347	15,969	673
Platte	Platte City	70,068	57,867	420
Polk	Bolivar	25,530	21,826	637
Pulaski	Waynesville	38,507	41,307	547
Putnam	Unionville	4,912	5,079	518
Ralls	New London	8,813	8,476	471
Randolph	Huntsville	24,024	24,370	482
Ray	Richmond	23,708	21,968	570
Reynolds	Centerville	6,624	6,661	811
Ripley	Doniphan	14,072	12,303	630
Saint Charles	Saint Charles	272,353	212,751	561
Saint Clair	Osceola	9,080	8,457	677
Sainte Genevieve	Sainte Genevieve	17,503	16,037	502
Saint Francois	Farmington	55,517	48,904	450
Saint Louis	Clayton	998,696	993,508	508
Saline	Marshall	22,703	23,523	756
Schuyler	Lancaster	4,443	4,236	308
Scotland	Memphis	4,814	4,822	439
Scott	Benton	40,262	39,376	421
Shannon	Eminence	8,252	7,613	1,004
Shelby	Shelbyville	6,802	6,942	501
Stoddard	Bloomfield	29,623	28,895	827
Stone	Galena	26,807	19,078	463
Sullivan	Milan	7,040	6,326	651
Taney	Forsyth	34,504	25,561	632
Texas	Houston	22,357	21,476	1,179
Vernon	Nevada	19,436	19,041	834
Warren	Warrenton	24,600	19,534	432
Washington	Potosi	22,966	20,380	760
Wayne	Greenville	13,059	11,543	761
Webster	Marshfield	29,108	23,753	593
Worth	Grant City	2,295	2,440	267
Wright	Hartville	19,578	16,758	682
Independent City				
Saint Louis		339,316	396,685	62

Montana

(56 counties, 145,556 sq mi land; pop. 880,453)

County	County seat or courthouse	1998 Pop.	1990 Pop.	Land area sq mi
Beaverhead	Dillon	8,867	8,424	5,543
Big Horn	Hardin	12,631	11,337	4,995
Blaine	Chinook	7,148	6,728	4,226
Broadwater	Townsend	4,132	3,318	1,192
Carbon	Red Lodge	9,444	8,080	2,048
Carter	Ekalaka	1,537	1,503	3,340
Cascade	Great Falls	78,983	77,691	2,698
Chouteau	Fort Benton	5,187	5,452	3,973
Custer	Miles City	12,035	11,697	3,783
Daniels	Scobey	2,001	2,266	1,426
Dawson	Glendive	8,849	9,505	2,373
Deer Lodge	Anaconda	9,999	10,356	737
Fallon	Baker	2,941	3,103	1,620
Fergus	Lewistown	12,271	12,083	4,339
Flathead	Kalispell	71,831	59,218	5,099
Gallatin	Bozeman	62,545	50,463	2,507
Garfield	Jordan	1,393	1,589	4,668
Glacier	Cut Bank	12,540	12,121	2,995
Golden Valley	Ryegete	1,041	912	1,175
Granite	Philipsburg	2,667	2,548	1,728
Hill	Havre	17,373	17,654	2,896
Jefferson	Boulder	10,087	7,939	1,657
Judith Basin	Stanford	2,294	2,282	1,870
Lake	Polson	25,648	21,041	1,494
Lewis & Clark	Helena	53,655	47,495	3,461
Liberty	Chester	2,323	2,295	1,430
Lincoln	Libby	18,696	17,481	3,613
McCone	Circle	1,964	2,276	2,643
Madison	Virginia City	6,875	5,989	3,587
Meagher	White Sulphur Springs	1,797	1,819	2,392
Mineral	Superior	3,748	3,315	1,220
Missoula	Missoula	88,989	78,687	2,598
Musselshell	Roundup	4,605	4,106	1,867
Park	Livingston	15,829	14,484	2,656
Petroleum	Winnett	499	519	1,654
Phillips	Malta	4,821	5,163	5,140
Pondera	Conrad	6,402	6,433	1,625
Powder River	Broadus	1,826	2,090	3,297
Powell	Deer Lodge	7,000	6,620	2,326
Prairie	Terry	1,333	1,383	1,737
Ravalli	Hamilton	35,156	25,010	2,394
Richland	Sidney	10,105	10,716	2,084
Roosevelt	Wolf Point	10,987	10,999	2,356
Rosebud	Forsyth	10,050	10,505	5,012
Sanders	Thompson Falls	10,185	8,669	2,762
Sheridan	Plentywood	4,269	4,732	1,677
Silver Bow	Butte	34,560	33,941	718
Stillwater	Columbus	8,069	6,536	1,795
Sweet Grass	Big Timber	3,407	3,154	1,855
Teton	Choteau	6,333	6,271	2,273
Toole	Shelby	4,727	5,046	1,911
Treasure	Hysham	870	874	979
Valley	Glasgow	8,195	8,239	4,921
Wheatland	Harlowton	2,373	2,246	1,423
Wibaux	Wibaux	1,148	1,191	889
Yellowstone	Billings	126,158	113,419	2,635
Yellowstone National Park[1]	NA	55	52	245

NA=Not applicable. (1) The area of Yellowstone National Park in Montana is not included in any county.

Nebraska

(93 counties, 76,878 sq mi land; pop. 1,662,719)

County	County seat or courthouse	1998 Pop.	1990 Pop.	Land area sq mi
Adams	Hastings	29,464	29,625	563
Antelope	Neligh	7,181	7,965	857
Arthur	Arthur	428	462	715
Banner	Harrisburg	878	852	746
Blaine	Brewster	578	675	711
Boone	Albion	6,377	6,667	687
Box Butte	Alliance	12,832	13,130	1,075
Boyd	Butte	2,565	2,835	540
Brown	Ainsworth	3,553	3,657	1,221
Buffalo	Kearney	40,596	37,447	968
Burt	Tekamah	7,998	7,868	493
Butler	David City	8,680	8,601	584
Cass	Plattsmouth	24,486	21,318	559
Cedar	Hartington	9,650	10,131	740
Chase	Imperial	4,248	4,381	895
Cherry	Valentine	6,326	6,307	5,961
Cheyenne	Sidney	9,476	9,494	1,196
Clay	Clay Center	7,147	7,123	573
Colfax	Schuyler	10,716	9,139	413
Cuming	West Point	9,993	10,117	572
Custer	Broken Bow	12,026	12,270	2,576
Dakota	Dakota City	18,792	16,742	264
Dawes	Chadron	8,979	9,021	1,396
Dawson	Lexington	23,183	19,940	1,013
Deuel	Chappell	2,029	2,237	440
Dixon	Ponca	6,300	6,143	476
Dodge	Fremont	35,333	34,500	535
Douglas	Omaha	443,794	416,444	331
Dundy	Benkelman	2,302	2,582	920
Fillmore	Geneva	6,929	7,103	577
Franklin	Franklin	3,730	3,938	576
Frontier	Stockville	3,082	3,101	975
Furnas	Beaver City	5,381	5,553	718
Gage	Beatrice	22,666	22,794	855
Garden	Oshkosh	2,138	2,460	1,705
Garfield	Burwell	2,039	2,141	570
Gosper	Elwood	2,329	1,928	458
Grant	Hyannis	763	769	776
Greeley	Greeley	2,850	3,006	570
Hall	Grand Island	51,851	48,925	546
Hamilton	Aurora	9,471	8,862	544
Harlan	Alma	3,748	3,810	553
Hayes	Hayes Center	1,069	1,222	713
Hitchcock	Trenton	3,442	3,750	710
Holt	O'Neill	12,042	12,599	2,413
Hooker	Mullen	702	793	721
Howard	Saint Paul	6,458	6,057	570
Jefferson	Fairbury	8,378	8,759	573
Johnson	Tecumseh	4,564	4,673	376
Kearney	Minden	6,853	6,629	516
Keith	Ogallala	8,665	8,584	1,061
Keya Paha	Springview	972	1,029	773
Kimball	Kimball	4,082	4,108	952
Knox	Center	9,216	9,564	1,108
Lancaster	Lincoln	235,589	213,641	839
Lincoln	North Platte	33,515	32,508	2,564
Logan	Stapleton	880	878	571
Loup	Taylor	666	683	570
McPherson	Tryon	563	546	859
Madison	Madison	34,585	32,655	573
Merrick	Central City	8,052	8,049	485
Morrill	Bridgeport	5,455	5,423	1,424
Nance	Fullerton	4,099	4,275	441
Nemaha	Auburn	7,697	7,980	409
Nuckolls	Nelson	5,226	5,786	575
Otoe	Nebraska City	14,787	14,252	616

County	County seat or courthouse	1998 Pop.	1990 Pop.	Land area sq mi
Pawnee	Pawnee City	3,131	3,317	432
Perkins	Grant	3,171	3,367	883
Phelps	Holdrege	9,908	9,715	540
Pierce	Pierce	7,914	7,827	574
Platte	Columbus	30,737	29,820	678
Polk	Osceola	5,631	5,668	439
Red Willow	McCook	11,255	11,705	717
Richardson	Falls City	9,420	9,937	554
Rock	Bassett	1,743	2,019	1,009
Saline	Wilber	12,966	12,715	575
Sarpy	Papillion	120,785	102,583	241
Saunders	Wahoo	19,245	18,285	754
Scotts Bluff	Gering	36,109	36,025	739
Seward	Seward	16,299	15,450	575
Sheridan	Rushville	6,454	6,750	2,441
Sherman	Loup City	3,432	3,718	566
Sioux	Harrison	1,486	1,549	2,067
Stanton	Stanton	6,215	6,244	430
Thayer	Hebron	6,277	6,635	575
Thomas	Thedford	797	851	713
Thurston	Pender	7,181	6,936	394
Valley	Ord	4,602	5,169	568
Washington	Blair	18,661	16,607	391
Wayne	Wayne	9,400	9,364	444
Webster	Red Cloud	4,019	4,279	575
Wheeler	Bartlett	952	948	575
York	York	14,512	14,428	576

Nevada

(16 counties, 1 ind. city, 109,806 sq mi land; pop. 1,746,898)

County	County seat or courthouse	1998 Pop.	1990 Pop.	Land area sq mi
Churchill	Fallon	23,293	17,938	4,929
Clark	Las Vegas	1,162,129	741,368	7,911
Douglas	Minden	37,051	27,637	710
Elko	Elko	46,084	33,463	17,182
Esmeralda	Goldfield	1,135	1,344	3,589
Eureka	Eureka	1,994	1,547	4,176
Humboldt	Winnemucca	18,145	12,844	9,648
Lander	Battle Mountain	6,987	6,266	5,494
Lincoln	Pioche	4,220	3,775	10,635
Lyon	Yerington	30,072	20,001	1,994
Mineral	Hawthorne	5,463	6,475	3,757
Nye	Tonopah	28,799	17,781	18,147
Pershing	Lovelock	5,434	4,336	6,009
Storey	Virginia City	3,053	2,526	264
Washoe	Reno	313,660	254,667	6,343
White Pine	Ely	10,078	9,264	8,877
Independent City				
Carson City		49,301	40,443	144

New Hampshire

(10 counties, 8,969 sq mi land; pop. 1,185,048)

County	County seat or courthouse	1998 Pop.	1990 Pop.	Land area sq mi
Belknap	Laconia	52,481	49,216	401
Carroll	Ossipee	39,346	35,410	934
Cheshire	Keene	71,828	70,121	708
Coos	Lancaster	32,875	34,828	1,801
Grafton	Woodsville	78,277	74,929	1,714
Hillsborough	Nashua	363,031	335,838	877
Merrimack	Concord	127,381	120,240	935
Rockingham	Exeter	271,152	245,845	695
Strafford	Dover	108,650	104,233	369
Sullivan	Newport	40,027	38,592	537

New Jersey

(21 counties, 7,419 sq mi land; pop. 8,115,011)

County	County seat or courthouse	1998 Pop.	1990 Pop.	Land area sq mi
Atlantic	Mays Landing	238,047	224,327	561
Bergen	Hackensack	858,529	825,380	234
Burlington	Mount Holly	420,323	395,066	805
Camden	Camden	505,204	502,824	222
Cape May	Cape May Courthouse	98,069	95,089	255
Cumberland	Bridgeton	140,341	138,053	489
Essex	Newark	750,273	777,964	126
Gloucester	Woodbury	247,897	230,082	325
Hudson	Jersey City	557,159	553,099	47
Hunterdon	Flemington	122,428	107,802	430
Mercer	Trenton	331,629	325,824	226
Middlesex	New Brunswick	716,176	671,811	311
Monmouth	Freehold	603,434	553,093	472
Morris	Morristown	459,896	421,361	469
Ocean	Toms River	489,819	433,203	636
Passaic	Paterson	485,737	470,864	185
Salem	Salem	64,912	65,294	338
Somerset	Somerville	282,900	240,245	305
Sussex	Newton	143,030	130,943	521
Union	Elizabeth	500,608	493,819	103
Warren	Belvidere	98,600	91,607	358

New Mexico

(33 counties, 121,364 sq mi land; pop. 1,736,931)

County	County seat or courthouse	1998 Pop.	1990 Pop.	Land area sq mi
Bernalillo	Albuquerque	525,958	480,577	1,166
Catron	Reserve	2,845	2,563	6,928
Chaves	Roswell	62,505	57,849	6,071
Cibola	Grants	26,250	23,794	4,540
Colfax	Raton	13,597	12,925	3,757
Curry	Clovis	45,290	42,207	1,406
DeBaca	Fort Sumner	2,389	2,252	2,325
Dona Ana	Las Cruces	169,165	135,510	3,807
Eddy	Carlsbad	53,543	48,605	4,182
Grant	Silver City	31,612	27,676	3,966
Guadalupe	Santa Rosa	4,050	4,156	3,031
Harding	Mosquero	899	987	2,126
Hidalgo	Lordsburg	6,210	5,958	3,446
Lea	Lovington	56,091	55,765	4,393
Lincoln	Carrizozo	16,400	12,219	4,831
Los Alamos	Los Alamos	18,344	18,115	109
Luna	Deming	24,070	18,110	2,965
McKinley	Gallup	67,558	60,686	5,449
Mora	Mora	4,861	4,264	1,931
Otero	Alamogordo	54,630	51,928	6,627
Quay	Tucumcari	10,024	10,823	2,875
Rio Arriba	Tierra Amarilla	37,787	34,365	5,858
Roosevelt	Portales	18,185	16,702	2,449
Sandoval	Bernalillo	88,049	63,319	3,710
San Juan	Aztec	106,020	91,605	5,514
San Miguel	Las Vegas	28,996	25,743	4,717
Santa Fe	Santa Fe	123,386	98,928	1,909
Sierra	Truth or Consequences	11,025	9,912	4,181
Socorro	Socorro	16,333	14,764	6,647
Taos	Taos	26,815	23,118	2,203
Torrance	Estancia	15,433	10,285	3,345
Union	Clayton	3,985	4,124	3,830
Valencia	Los Lunas	64,626	45,235	1,068

New York

(62 counties, 47,224 sq mi land; pop. 18,175,301)

County	County seat or courthouse	1998 Pop.	1990 Pop.	Land area sq mi
Albany	Albany	292,586	292,793	524
Allegany	Belmont	50,997	50,470	1,030
Bronx[1]	Bronx	1,195,599	1,203,789	42
Broome	Binghamton	196,545	212,160	707
Cattaraugus	Little Valley	85,086	84,234	1,310
Cayuga	Auburn	81,264	82,313	693
Chautauqua	Mayville	138,103	141,895	1,062
Chemung	Elmira	92,021	95,195	408
Chenango	Norwich	51,052	51,768	894
Clinton	Plattsburgh	79,970	85,969	1,039
Columbia	Hudson	63,221	62,982	636
Cortland	Cortland	48,033	48,963	500
Delaware	Delhi	46,086	47,225	1,446
Dutchess	Poughkeepsie	265,317	259,462	802
Erie	Buffalo	934,471	968,584	1,045
Essex	Elizabethtown	37,548	37,152	1,797
Franklin	Malone	48,582	46,540	1,632
Fulton	Johnstown	52,914	54,191	496
Genesee	Batavia	60,654	60,060	494
Greene	Catskill	47,807	44,739	648
Hamilton	Lake Pleasant	5,193	5,279	1,721
Herkimer	Herkimer	64,049	65,809	1,412
Jefferson	Watertown	111,050	110,943	1,272
Kings[1]	Brooklyn	2,267,942	2,300,664	71
Lewis	Lowville	27,494	26,796	1,276
Livingston	Geneseo	66,000	62,372	632
Madison	Wampsville	71,069	69,166	656
Monroe	Rochester	716,072	713,968	659
Montgomery	Fonda	50,755	51,981	405
Nassau	Mineola	1,302,220	1,287,444	287
New York[1]	New York	1,550,649	1,487,536	28
Niagara	Lockport	218,070	220,756	523
Oneida	Utica	230,628	250,836	1,213
Onondaga	Syracuse	458,301	468,973	780
Ontario	Canandaigua	99,662	95,101	644
Orange	Goshen	329,220	307,647	816
Orleans	Albion	44,518	41,846	391
Oswego	Oswego	124,006	121,785	953
Otsego	Cooperstown	60,788	60,517	1,003
Putnam	Carmel	93,358	83,941	232
Queens[1]	Jamaica	1,998,853	1,951,598	109
Rensselaer	Troy	152,689	154,429	654
Richmond[1]	Saint George	407,123	378,977	59
Rockland	New City	281,338	265,475	174
Saint Lawrence	Canton	113,688	111,974	2,686
Saratoga	Ballston Spa	197,606	181,276	812
Schenectady	Schenectady	145,530	149,285	206
Schoharie	Schoharie	32,438	31,859	622
Schuyler	Watkins Glen	19,125	18,662	329
Seneca	Ovid & Waterloo	31,943	33,683	325
Steuben	Bath	97,950	99,088	1,393
Suffolk	Riverhead	1,371,269	1,321,768	911
Sullivan	Monticello	69,111	69,277	970
Tioga	Owego	52,477	52,337	519

County	County seat or courthouse	1998 Pop.	1990 Pop.	Land area sq mi
Tompkins	Ithaca	96,020	94,097	476
Ulster	Kingston	166,351	165,304	1,127
Warren	Lake George	61,261	59,209	870
Washington	Hudson Falls	60,481	59,330	836
Wayne	Lyons	94,977	89,123	604
Westchester	White Plains	897,920	874,866	433
Wyoming	Warsaw	44,049	42,507	593
Yates	Penn Yan	24,202	22,810	338

(1) New York City consists of 5 counties: Bronx, Kings (Brooklyn), New York (Manhattan), Queens, and Richmond (Staten Island).

North Carolina

(100 counties, 48,718 sq mi land; pop. 7,546,493)

County	County seat or courthouse	1998 Pop.	1990 Pop.	Land area sq mi
Alamance	Graham	119,397	108,213	431
Alexander	Taylorsville	31,280	27,544	260
Alleghany	Sparta	9,842	9,590	235
Anson	Wadesboro	24,354	23,474	532
Ashe	Jefferson	24,025	22,209	426
Avery	Newland	15,742	14,867	247
Beaufort	Washington	44,531	42,283	828
Bertie	Windsor	20,459	20,388	699
Bladen	Elizabethtown	30,717	28,663	875
Brunswick	Bolivia	68,416	50,985	855
Buncombe	Asheville	194,873	174,819	656
Burke	Morganton	82,687	75,740	507
Cabarrus	Concord	120,057	98,935	364
Caldwell	Lenoir	76,096	70,709	472
Camden	Camden	6,878	5,904	241
Carteret	Beaufort	60,054	52,553	531
Caswell	Yanceyville	22,215	20,693	426
Catawba	Newton	132,545	118,412	400
Chatham	Pittsboro	45,406	38,759	683
Cherokee	Murphy	22,758	20,170	455
Chowan	Edenton	14,191	13,506	173
Clay	Hayesville	8,575	7,155	215
Cleveland	Shelby	92,753	84,713	464
Columbus	Whiteville	52,634	49,587	937
Craven	New Bern	88,129	81,613	696
Cumberland	Fayetteville	284,629	274,713	653
Currituck	Currituck	17,908	13,736	262
Dare	Manteo	28,952	22,746	382
Davidson	Lexington	141,178	126,677	552
Davie	Mocksville	32,095	27,859	265
Duplin	Kenansville	42,993	39,995	818
Durham	Durham	202,411	181,855	291
Edgecombe	Tarboro	55,199	56,692	505
Forsyth	Winston-Salem	287,701	265,878	410
Franklin	Louisburg	44,743	36,414	492
Gaston	Gastonia	184,247	175,093	357
Gates	Gatesville	10,070	9,305	341
Graham	Robbinsville	7,647	7,196	292
Granville	Oxford	42,908	38,341	531
Greene	Snow Hill	18,308	15,384	265
Guilford	Greensboro	387,722	347,420	650
Halifax	Halifax	56,433	55,516	725
Harnett	Lillington	82,391	67,833	595
Haywood	Waynesville	51,422	46,942	554
Henderson	Hendersonville	80,822	69,285	374
Hertford	Winton	22,289	22,523	354
Hoke	Raeford	30,424	22,856	391
Hyde	Swan Quarter	5,612	5,411	613
Iredell	Statesville	113,247	92,935	574
Jackson	Sylva	30,210	26,846	491
Johnston	Smithfield	106,582	81,306	792
Jones	Trenton	9,456	9,414	473
Lee	Sanford	49,328	41,370	257
Lenoir	Kinston	59,046	57,274	400
Lincoln	Lincolnton	58,093	50,319	299
McDowell	Marion	40,048	35,681	442
Macon	Franklin	28,338	23,499	517
Madison	Marshall	18,756	16,953	449
Martin	Williamston	26,192	25,078	463
Mecklenburg	Charlotte	630,848	511,481	527
Mitchell	Bakersville	14,831	14,433	222
Montgomery	Troy	24,080	23,352	491
Moore	Carthage	71,394	59,000	699
Nash	Nashville	90,968	76,677	540
New Hanover	Wilmington	149,832	120,284	199
Northampton	Jackson	21,184	20,798	536
Onslow	Jacksonville	142,358	149,838	767
Orange	Hillsborough	110,116	93,851	400
Pamlico	Bayboro	12,345	11,368	337
Pasquotank	Elizabeth City	35,474	31,298	227
Pender	Burgaw	39,510	28,855	871
Perquimans	Hertford	11,282	10,447	247
Person	Roxboro	33,647	30,180	392
Pitt	Greenville	126,630	108,480	652
Polk	Columbus	16,835	14,416	238
Randolph	Asheboro	121,289	106,546	788
Richmond	Rockingham	46,221	44,518	474
Robeson	Lumberton	115,589	105,170	949
Rockingham	Wentworth	90,039	86,064	567
Rowan	Salisbury	125,505	110,605	511
Rutherford	Rutherfordton	60,842	56,919	564
Sampson	Clinton	52,438	47,297	946
Scotland	Laurinburg	35,802	33,763	319
Stanly	Albemarle	56,083	51,765	395
Stokes	Danbury	43,292	37,223	452
Surry	Dobson	67,052	61,704	537
Swain	Bryson City	12,300	11,268	528
Transylvania	Brevard	28,481	25,520	378
Tyrrell	Columbia	3,734	3,856	390
Union	Monroe	110,017	84,210	637
Vance	Henderson	42,155	38,892	254
Wake	Raleigh	570,615	426,300	834
Warren	Warrenton	18,297	17,265	429
Washington	Plymouth	13,615	13,997	348
Watauga	Boone	40,965	36,952	313
Wayne	Goldsboro	112,227	104,666	553
Wilkes	Wilkesboro	62,837	59,393	757
Wilson	Wilson	68,188	66,061	371
Yadkin	Yadkinville	34,955	30,488	336
Yancey	Burnsville	16,607	15,419	312

North Dakota

(53 counties, 68,994 sq mi land; pop. 638,244)

County	County seat or courthouse	1998 Pop.	1990 Pop.	Land area sq mi
Adams	Hettinger	2,714	3,174	988
Barnes	Valley City	11,958	12,545	1,492
Benson	Minnewaukan	6,893	7,198	1,389
Billings	Medora	1,058	1,108	1,152
Bottineau	Bottineau	7,226	8,011	1,669
Bowman	Bowman	3,317	3,596	1,162
Burke	Bowbells	2,266	3,002	1,104
Burleigh	Bismarck	66,867	60,131	1,633
Cass	Fargo	116,832	102,874	1,766
Cavalier	Langdon	5,011	6,064	1,489
Dickey	Ellendale	5,644	6,107	1,131
Divide	Crosby	2,366	2,899	1,259
Dunn	Manning	3,560	4,005	2,010
Eddy	New Rockford	2,847	2,951	632
Emmons	Linton	4,311	4,830	1,510
Foster	Carrington	3,802	3,983	635
Golden Valley	Beach	1,876	2,108	1,002
Grand Forks	Grand Forks	66,869	70,683	1,438
Grant	Carson	2,969	3,549	1,660
Griggs	Cooperstown	2,842	3,303	709
Hettinger	Mott	2,924	3,445	1,132
Kidder	Steele	2,877	3,332	1,352
La Moure	La Moure	4,759	5,383	1,147
Logan	Napoleon	2,355	2,847	993
McHenry	Towner	6,076	6,528	1,874
McIntosh	Ashley	3,442	4,021	975
McKenzie	Watford City	5,682	6,383	2,742
McLean	Washburn	9,704	10,457	2,110
Mercer	Stanton	9,418	9,808	1,045
Morton	Mandan	24,575	23,700	1,926
Mountrail	Stanley	6,633	7,021	1,824
Nelson	Lakota	3,716	4,410	982
Oliver	Center	2,215	2,381	724
Pembina	Cavalier	8,485	9,238	1,119
Pierce	Rugby	4,623	5,052	1,018
Ramsey	Devils Lake	12,120	12,681	1,186
Ransom	Lisbon	5,776	5,921	863
Renville	Mohall	2,808	3,160	875
Richland	Wahpeton	18,272	18,148	1,437
Rolette	Rolla	14,219	12,772	903
Sargent	Forman	4,457	4,549	859
Sheridan	McClusky	1,694	2,148	972
Sioux	Fort Yates	4,192	3,761	1,094
Slope	Amidon	865	907	1,218
Stark	Dickinson	22,780	22,832	1,338
Steele	Finley	2,263	2,420	712
Stutsman	Jamestown	20,964	22,241	2,222
Towner	Cando	3,018	3,627	1,025
Traill	Hillsboro	8,544	8,752	862
Walsh	Grafton	13,532	13,840	1,282
Ward	Minot	58,678	57,921	2,013
Wells	Fessenden	5,200	5,864	1,271
Williams	Williston	20,150	21,129	2,071

Ohio

(88 counties, 40,953 sq mi land; pop. 11,209,493)

County	County seat or courthouse	1998 Pop.	1990 Pop.	Land area sq mi
Adams	West Union	28,587	25,371	584
Allen	Lima	107,139	109,755	405
Ashland	Ashland	52,237	47,507	424
Ashtabula	Jefferson	103,300	99,821	703
Athens	Athens	61,490	59,549	507
Auglaize	Wapakoneta	47,103	44,585	401
Belmont	Saint Clairsville	69,175	71,074	537
Brown	Georgetown	40,795	34,966	492
Butler	Hamilton	330,428	291,479	467
Carroll	Carrollton	29,095	26,521	395

County	County seat or courthouse	1998 Pop.	1990 Pop.	Land area sq mi
Champaign	Urbana	38,182	36,019	429
Clark	Springfield	145,341	147,548	400
Clermont	Batavia	175,960	150,167	452
Clinton	Wilmington	39,979	35,417	411
Columbiana	Lisbon	111,521	108,276	533
Coshocton	Coshocton	36,115	35,427	564
Crawford	Bucyrus	47,217	47,870	402
Cuyahoga	Cleveland	1,380,696	1,412,140	458
Darke	Greenville	54,180	53,619	600
Defiance	Defiance	39,824	39,350	411
Delaware	Delaware	92,209	66,929	443
Erie	Sandusky	78,279	76,779	255
Fairfield	Lancaster	123,998	103,472	506
Fayette	Washington Courthouse	28,493	27,466	407
Franklin	Columbus	1,021,194	961,437	540
Fulton	Wauseon	41,895	38,498	407
Gallia	Gallipolis	33,422	30,954	469
Geauga	Chardon	88,788	81,129	404
Greene	Xenia	146,607	136,731	415
Guernsey	Cambridge	40,994	39,024	522
Hamilton	Cincinnati	847,403	866,228	407
Hancock	Findlay	68,922	65,536	531
Hardin	Kenton	31,725	31,111	470
Harrison	Cadiz	16,097	16,085	404
Henry	Napoleon	29,923	29,108	417
Highland	Hillsboro	40,364	35,728	553
Hocking	Logan	29,004	25,533	423
Holmes	Millersburg	37,841	32,849	423
Huron	Norwalk	60,293	56,240	493
Jackson	Jackson	32,563	30,230	420
Jefferson	Steubenville	74,558	80,298	410
Knox	Mount Vernon	53,309	47,473	527
Lake	Painesville	223,779	215,499	228
Lawrence	Ironton	64,427	61,834	455
Licking	Newark	136,896	128,300	687
Logan	Bellefontaine	46,204	42,310	459
Lorain	Elyria	282,149	271,126	493
Lucas	Toledo	448,542	462,361	340
Madison	London	41,576	37,068	465
Mahoning	Youngstown	255,165	264,806	415
Marion	Marion	64,774	64,274	404
Medina	Medina	144,019	122,354	422
Meigs	Pomeroy	24,006	22,987	430
Mercer	Celina	41,198	39,443	463
Miami	Troy	98,147	93,182	407
Monroe	Woodsfield	15,357	15,497	456
Montgomery	Dayton	558,427	573,809	462
Morgan	McConnelsville	14,536	14,194	418
Morrow	Mount Gilead	31,467	27,749	406
Muskingum	Zanesville	84,470	82,068	665
Noble	Caldwell	12,343	11,336	399
Ottawa	Port Clinton	40,983	40,029	255
Paulding	Paulding	20,078	20,488	416
Perry	New Lexington	34,290	31,557	410
Pickaway	Circleville	53,731	48,244	502
Pike	Waverly	27,775	24,249	442
Portage	Ravenna	151,222	142,585	492
Preble	Eaton	43,226	40,113	425
Putnam	Ottawa	35,255	33,819	484
Richland	Mansfield	127,342	126,137	497
Ross	Chillicothe	75,473	69,330	689
Sandusky	Fremont	62,216	61,963	409
Scioto	Portsmouth	80,355	80,327	612
Seneca	Tiffin	60,099	59,733	551
Shelby	Sidney	47,457	44,915	409
Stark	Canton	373,112	367,585	576
Summit	Akron	537,730	514,990	413
Trumbull	Warren	225,066	227,813	616
Tuscarawas	New Philadelphia	88,608	84,090	568
Union	Marysville	39,494	31,969	437
Van Wert	Van Wert	30,200	30,464	410
Vinton	McArthur	12,158	11,098	414
Warren	Lebanon	146,033	113,927	400
Washington	Marietta	63,413	62,254	635
Wayne	Wooster	110,125	101,461	555
Williams	Bryan	38,001	36,956	422
Wood	Bowling Green	119,498	113,269	617
Wyandot	Upper Sandusky	22,826	22,254	406

Oklahoma

(77 counties, 68,679 sq mi land; pop. 3,346,713)

County	County seat or courthouse	1998 Pop.	1990 Pop.	Land area sq mi
Adair	Stillwell	20,349	18,421	576
Alfalfa	Cherokee	6,044	6,416	867
Atoka	Atoka	13,237	12,778	978
Beaver	Beaver	6,056	6,023	1,815
Beckham	Sayre	19,584	18,812	902
Blaine	Watonga	10,513	11,470	929
Bryan	Durant	34,690	32,089	909
Caddo	Anadarko	30,981	29,550	1,278
Canadian	El Reno	85,463	74,409	900
Carter	Ardmore	44,503	42,919	824
Cherokee	Tahlequah	39,138	34,049	751
Choctaw	Hugo	15,077	15,302	774
Cimarron	Boise City	2,959	3,301	1,835
Cleveland	Norman	201,110	174,253	536
Coal	Coalgate	6,009	5,780	518
Comanche	Lawton	113,508	111,486	1,069
Cotton	Walters	6,705	6,651	637
Craig	Vinita	14,450	14,104	761
Creek	Sapulpa	67,142	60,915	956
Custer	Arapaho	25,493	26,897	987
Delaware	Jay	34,154	28,070	741
Dewey	Taloga	4,928	5,551	1,000
Ellis	Arnett	4,291	4,497	1,229
Garfield	Enid	56,859	56,735	1,059
Garvin	Pauls Valley	27,044	26,605	809
Grady	Chickasha	45,934	41,747	1,101
Grant	Medford	5,338	5,689	1,001
Greer	Mangum	6,366	6,559	639
Harmon	Hollis	3,479	3,793	538
Harper	Buffalo	3,596	4,063	1,039
Haskell	Stigler	11,368	10,940	577
Hughes	Holdenville	14,081	13,014	807
Jackson	Altus	28,771	28,764	803
Jefferson	Waurika	6,583	7,010	759
Johnston	Tishomingo	10,346	10,032	645
Kay	Newkirk	46,698	48,056	919
Kingfisher	Kingfisher	13,528	13,212	903
Kiowa	Hobart	10,613	11,347	1,015
Latimer	Wilburton	10,321	10,333	722
Le Flore	Poteau	46,564	43,270	1,586
Lincoln	Chandler	31,361	29,216	959
Logan	Guthrie	30,970	29,011	745
Love	Marietta	8,536	7,788	515
McClain	Purcell	26,224	22,795	570
McCurtain	Idabel	34,783	33,433	1,852
McIntosh	Eufaula	19,050	16,779	620
Major	Fairview	7,829	8,055	957
Marshall	Madill	12,326	10,829	371
Mayes	Pryor	37,638	33,366	656
Murray	Sulphur	12,335	12,042	418
Muskogee	Muskogee	70,004	68,078	814
Noble	Perry	11,425	11,045	732
Nowata	Nowata	9,969	9,992	565
Okfuskee	Okemah	11,402	11,551	625
Oklahoma	Oklahoma City	632,988	599,611	709
Okmulgee	Okmulgee	38,860	36,490	697
Osage	Pawhuska	42,838	41,645	2,251
Ottawa	Miami	30,944	30,561	471
Pawnee	Pawnee	16,438	15,575	570
Payne	Stillwater	65,109	61,507	686
Pittsburg	McAlester	42,798	40,950	1,306
Pontotoc	Ada	34,591	34,119	720
Pottawatomie	Shawnee	62,244	58,760	788
Pushmataha	Antlers	11,584	10,997	1,397
Roger Mills	Cheyenne	3,580	4,147	1,142
Rogers	Claremore	68,128	55,170	675
Seminole	Wewoka	24,770	25,412	633
Sequoyah	Sallisaw	37,531	33,828	674
Stephens	Duncan	43,410	42,299	877
Texas	Guymon	18,640	16,419	2,037
Tillman	Frederick	9,503	10,384	872
Tulsa	Tulsa	543,539	503,341	570
Wagoner	Wagoner	55,259	47,883	563
Washington	Bartlesville	47,519	48,066	417
Washita	Cordell	11,796	11,441	1,004
Woods	Alva	8,366	9,103	1,287
Woodward	Woodward	18,553	18,976	1,242

Oregon

(36 counties, 96,002 sq mi land; pop. 3,281,974)

County	County seat or courthouse	1998 Pop.	1990 Pop.	Land area sq mi
Baker	Baker City	16,448	15,317	3,068
Benton	Corvallis	77,755	70,811	677
Clackamas	Oregon City	334,732	278,850	1,868
Clatsop	Astoria	35,424	33,301	827
Columbia	Saint Helens	44,416	37,557	657
Coos	Coquille	62,162	60,273	1,601
Crook	Prineville	17,236	14,111	2,980
Curry	Gold Beach	21,157	19,327	1,627
Deschutes	Bend	105,640	74,976	3,018
Douglas	Roseburg	101,837	94,649	5,037
Gilliam	Condon	2,023	1,717	1,204
Grant	Canyon City	8,075	7,853	4,529
Harney	Burns	7,198	7,060	10,135
Hood River	Hood River	19,553	16,903	522
Jackson	Medford	173,123	146,387	2,785
Jefferson	Madras	16,627	13,676	1,781
Josephine	Grants Pass	74,377	62,649	1,640
Klamath	Klamath Falls	63,185	57,702	5,945
Lake	Lakeview	7,152	7,186	8,136
Lane	Eugene	314,068	282,912	4,554
Lincoln	Newport	45,368	38,889	980
Linn	Albany	104,464	91,227	2,291
Malheur	Vale	28,542	26,038	9,888

County	County seat or courthouse	1998 Pop.	1990 Pop.	Land area sq mi
Marion	Salem	268,541	228,483	1,185
Morrow	Heppner	9,985	7,625	2,033
Multnomah	Portland	631,082	583,887	435
Polk	Dallas	61,560	49,541	741
Sherman	Moro	1,789	1,918	823
Tillamook	Tillamook	24,356	21,570	1,102
Umatilla	Pendleton	65,495	59,249	3,215
Union	La Grande	24,829	23,598	2,037
Wallowa	Enterprise	7,368	6,911	3,145
Wasco	The Dalles	23,059	21,683	2,381
Washington	Hillsboro	399,697	311,554	724
Wheeler	Fossil	1,566	1,396	1,715
Yamhill	McMinnville	82,085	65,551	716

Pennsylvania

(67 counties, 44,820 sq mi land; pop. 12,001,451)

County	County seat or courthouse	1998 Pop.	1990 Pop.	Land area sq mi
Adams	Gettysburg	86,537	78,274	520
Allegheny	Pittsburgh	1,268,446	1,336,449	730
Armstrong	Kittanning	73,181	73,478	654
Beaver	Beaver	184,406	186,093	435
Bedford	Bedford	49,373	47,919	1,015
Berks	Reading	355,956	336,523	859
Blair	Hollidaysburg	130,615	130,542	526
Bradford	Towanda	62,459	60,967	1,151
Bucks	Doylestown	587,942	541,174	608
Butler	Butler	170,785	152,013	789
Cambria	Ebensburg	156,080	163,062	688
Cameron	Emporium	5,620	5,913	397
Carbon	Jim Thorpe	58,857	56,846	383
Centre	Bellefonte	132,700	124,812	1,108
Chester	West Chester	421,686	376,396	756
Clarion	Clarion	41,841	41,699	603
Clearfield	Clearfield	80,752	78,097	1,147
Clinton	Lock Haven	37,000	37,182	891
Columbia	Bloomsburg	64,120	63,202	486
Crawford	Meadville	89,415	86,170	1,013
Cumberland	Carlisle	208,634	195,257	550
Dauphin	Harrisburg	245,579	237,813	525
Delaware	Media	542,593	547,651	184
Elk	Ridgway	34,540	34,878	829
Erie	Erie	276,401	275,572	802
Fayette	Uniontown	144,847	145,351	790
Forest	Tionesta	5,002	4,802	428
Franklin	Chambersburg	128,002	121,082	772
Fulton	McConnellsburg	14,498	13,837	438
Greene	Waynesburg	40,742	39,550	576
Huntingdon	Huntingdon	44,599	44,164	875
Indiana	Indiana	88,567	89,994	830
Jefferson	Brookville	46,250	46,083	656
Juniata	Mifflintown	22,101	20,625	392
Lackawanna	Scranton	208,455	219,097	459
Lancaster	Lancaster	456,414	422,822	949
Lawrence	New Castle	94,887	96,246	361
Lebanon	Lebanon	117,434	113,744	362
Lehigh	Allentown	299,341	291,130	347
Luzerne	Wilkes-Barre	313,767	328,149	891
Lycoming	Williamsport	117,308	118,710	1,235
McKean	Smethport	46,500	47,131	982
Mercer	Mercer	121,938	121,003	672
Mifflin	Lewistown	46,961	46,197	411
Monroe	Stroudsburg	125,583	95,709	607
Montgomery	Norristown	719,718	678,193	483
Montour	Danville	17,730	17,735	131
Northampton	Easton	258,679	247,105	374
Northumberland	Sunbury	94,017	96,771	460
Perry	New Bloomfield	44,384	41,172	554
Philadelphia	Philadelphia	1,436,287	1,585,577	135
Pike	Milford	40,172	27,966	547
Potter	Coudersport	17,184	16,717	1,081
Schuylkill	Pottsville	148,266	152,585	779
Snyder	Middleburg	38,226	36,680	331
Somerset	Somerset	80,267	78,218	1,075
Sullivan	Laporte	6,107	6,104	450
Susquehanna	Montrose	42,144	40,380	823
Tioga	Wellsboro	41,606	41,126	1,134
Union	Lewisburg	40,897	36,176	317
Venango	Franklin	57,844	59,381	675
Warren	Warren	43,910	45,049	884
Washington	Washington	205,566	204,584	857
Wayne	Honesdale	45,226	39,944	729
Westmoreland	Greensburg	372,103	370,321	1,023
Wyoming	Tunkhannock	29,149	28,076	397
York	York	373,255	339,574	905

Rhode Island

(5 counties, 1,045 sq mi land; pop. 988,480)

County	County seat or courthouse	1998 Pop.	1990 Pop.	Land area sq mi
Bristol	Bristol	49,114	48,859	25
Kent	East Greenwich	161,811	161,135	170
Newport	Newport	82,868	87,194	104
Providence	Providence	574,038	596,270	413
Washington	West Kingston	120,649	110,006	333

South Carolina

(46 counties, 30,111 sq mi land; pop. 3,835,962)

County	County seat or courthouse	1998 Pop.	1990 Pop.	Land area sq mi
Abbeville	Abbeville	24,632	23,862	508
Aiken	Aiken	134,051	120,991	1,073
Allendale	Allendale	11,460	11,722	408
Anderson	Anderson	160,791	145,177	718
Bamberg	Bamberg	16,498	16,902	393
Barnwell	Barnwell	21,766	20,293	549
Beaufort	Beaufort	108,959	86,425	587
Berkeley	Moncks Corner	136,544	128,776	1,100
Calhoun	Saint Matthews	14,051	12,753	380
Charleston	Charleston	316,482	295,041	917
Cherokee	Gaffney	49,170	44,506	393
Chester	Chester	34,401	32,170	581
Chesterfield	Chesterfield	41,080	38,575	799
Clarendon	Manning	30,814	28,450	607
Colleton	Walterboro	37,364	34,377	1,057
Darlington	Darlington	66,366	61,851	562
Dillon	Dillon	29,747	29,114	405
Dorchester	Saint George	88,133	83,060	575
Edgefield	Edgefield	20,003	18,360	502
Fairfield	Winnsboro	22,394	22,295	687
Florence	Florence	124,904	114,344	799
Georgetown	Georgetown	53,727	46,302	815
Greenville	Greenville	353,845	320,167	792
Greenwood	Greenwood	63,623	59,567	456
Hampton	Hampton	19,200	18,191	560
Horry	Conway	174,762	144,053	1,134
Jasper	Ridgeland	16,995	15,487	654
Kershaw	Camden	48,593	43,599	726
Lancaster	Lancaster	58,887	54,516	549
Laurens	Laurens	63,249	58,092	713
Lee	Bishopville	20,399	18,437	410
Lexington	Lexington	205,260	167,611	701
McCormick	McCormick	9,545	8,868	360
Marion	Marion	34,610	33,899	489
Marlboro	Bennettsville	29,589	29,716	480
Newberry	Newberry	34,462	33,172	631
Oconee	Walhalla	64,059	57,494	625
Orangeburg	Orangeburg	87,865	84,803	1,106
Pickens	Pickens	107,087	93,896	497
Richland	Columbia	307,056	286,321	757
Saluda	Saluda	17,025	16,357	451
Spartanburg	Spartanburg	247,458	226,793	811
Sumter	Sumter	107,127	101,276	666
Union	Union	30,495	30,337	514
Williamsburg	Kingstree	37,121	36,815	934
York	York	154,313	131,497	683

South Dakota

(66 counties, 75,896 sq mi land; pop. 738,171)

County	County seat or courthouse	1998 Pop.	1990 Pop.	Land area sq mi
Aurora	Plankinton	2,975	3,135	708
Beadle	Huron	17,183	18,253	1,259
Bennett	Martin	3,389	3,206	1,185
Bon Homme	Tyndall	7,696	7,089	563
Brookings	Brookings	25,989	25,207	795
Brown	Aberdeen	35,433	35,580	1,713
Brule	Chamberlain	5,555	5,485	819
Buffalo	Gannvalley	1,738	1,759	471
Butte	Belle Fourche	9,018	7,914	2,249
Campbell	Mound City	1,917	1,965	736
Charles Mix	Lake Andes	9,337	9,131	1,098
Clark	Clark	4,337	4,403	958
Clay	Vermillion	15,167	13,186	412
Codington	Watertown	25,456	22,698	688
Corson	McIntosh	4,190	4,195	2,473
Custer	Custer	6,930	6,179	1,558
Davison	Mitchell	18,006	17,503	436
Day	Webster	6,400	6,978	1,029
Deuel	Clear Lake	4,512	4,522	624
Dewey	Timber Lake	5,821	5,523	2,303
Douglas	Armour	3,553	3,746	434
Edmunds	Ipswich	4,219	4,356	1,146
Fall River	Hot Springs	7,133	7,353	1,740
Faulk	Faulkton	2,521	2,744	1,000
Grant	Milbank	8,063	8,372	683
Gregory	Burke	4,948	5,359	1,016
Haakon	Philip	2,353	2,624	1,813
Hamlin	Hayti	5,335	4,974	511
Hand	Miller	4,144	4,272	1,437

County	County seat or courthouse	1998 Pop.	1990 Pop.	Land area sq mi
Hanson	Alexandria	2,935	2,994	435
Harding	Buffalo	1,476	1,669	2,671
Hughes	Pierre	15,373	14,817	741
Hutchinson	Olivet	8,045	8,262	813
Hyde	Highmore	1,605	1,696	861
Jackson	Kadoka	2,904	2,811	1,869
Jerauld	Wessington Springs	2,222	2,425	530
Jones	Murdo	1,219	1,324	971
Kingsbury	De Smet	5,712	5,925	838
Lake	Madison	11,139	10,550	563
Lawrence	Deadwood	22,509	20,655	800
Lincoln	Canton	20,411	15,427	578
Lyman	Kennebec	3,768	3,638	1,640
McCook	Salem	5,598	5,688	575
McPherson	Leola	2,738	3,228	1,137
Marshall	Britton	4,563	4,844	839
Meade	Sturgis	21,911	21,878	3,471
Mellette	White River	2,029	2,137	1,307
Miner	Howard	2,799	3,272	570
Minnehaha	Sioux Falls	143,011	123,809	809
Moody	Flandreau	6,505	6,507	520
Pennington	Rapid City	87,702	81,343	2,776
Perkins	Bison	3,505	3,932	2,872
Potter	Gettysburg	2,857	3,190	867
Roberts	Sisseton	9,786	9,914	1,101
Sanborn	Woonsocket	2,740	2,833	569
Shannon	(Attached to Fall River)	12,183	9,902	2,094
Spink	Redfield	7,572	7,981	1,504
Stanley	Fort Pierre	2,929	2,453	1,443
Sully	Onida	1,470	1,589	1,007
Todd	(Attached to Tripp)	9,247	8,352	1,388
Tripp	Winner	6,737	6,924	1,614
Turner	Parker	8,631	8,576	617
Union	Elk Point	12,213	10,189	460
Walworth	Selby	5,582	6,087	708
Yankton	Yankton	21,051	19,252	522
Ziebach	Dupree	2,176	2,220	1,963

Tennessee

(95 counties, 41,219 sq mi land; pop. 5,430,621)

County	County seat or courthouse	1998 Pop.	1990 Pop.	Land area sq mi
Anderson	Clinton	71,116	68,250	338
Bedford	Shelbyville	34,533	30,411	474
Benton	Camden	16,328	14,524	395
Bledsoe	Pikeville	10,795	9,669	406
Blount	Maryville	101,295	85,969	559
Bradley	Cleveland	83,292	73,712	329
Campbell	Jacksboro	38,241	35,079	480
Cannon	Woodbury	12,139	10,467	266
Carroll	Huntingdon	29,115	27,514	599
Carter	Elizabethton	53,323	51,505	341
Cheatham	Ashland City	35,344	27,140	303
Chester	Henderson	14,700	12,819	289
Claiborne	Tazewell	29,529	26,137	434
Clay	Celina	7,255	7,238	236
Cocke	Newport	31,968	29,141	434
Coffee	Manchester	45,767	40,339	429
Crockett	Alamo	13,959	13,378	265
Cumberland	Crossville	44,291	34,736	682
Davidson	Nashville	533,967	510,786	502
Decatur	Decaturville	10,807	10,472	334
De Kalb	Smithville	15,943	14,360	305
Dickson	Charlotte	42,254	35,061	490
Dyer	Dyersburg	36,782	34,854	511
Fayette	Somerville	30,457	25,559	705
Fentress	Jamestown	16,184	14,669	499
Franklin	Winchester	37,465	34,725	553
Gibson	Trenton	48,186	46,315	603
Giles	Pulaski	28,925	25,741	611
Grainger	Rutledge	19,829	17,095	280
Greene	Greeneville	60,502	55,853	622
Grundy	Altamont	14,138	13,362	361
Hamblen	Morristown	54,050	50,480	161
Hamilton	Chattanooga	294,745	285,536	543
Hancock	Sneedville	6,778	6,739	222
Hardeman	Bolivar	24,895	23,377	668
Hardin	Savannah	24,961	22,633	578
Hawkins	Rogersville	49,719	44,565	487
Haywood	Brownsville	19,525	19,437	533
Henderson	Lexington	24,424	21,844	520
Henry	Paris	30,066	27,888	562
Hickman	Centerville	20,553	16,754	613
Houston	Erin	7,853	7,018	200
Humphreys	Waverly	17,059	15,813	532
Jackson	Gainesboro	9,629	9,297	309
Jefferson	Dandridge	43,663	33,016	274
Johnson	Mountain City	16,755	13,766	299
Knox	Knoxville	366,846	335,749	509
Lake	Tiptonville	8,171	7,129	163
Lauderdale	Ripley	24,206	23,491	471
Lawrence	Lawrenceburg	39,358	35,303	617
Lewis	Hohenwald	10,868	9,247	282
Lincoln	Fayetteville	29,761	28,157	570
Loudon	Loudon	39,052	31,255	229
McMinn	Athens	46,283	42,383	430
McNairy	Selmer	24,048	22,422	560
Macon	Lafayette	18,181	15,906	307
Madison	Jackson	85,954	77,982	557
Marion	Jasper	26,851	24,860	500
Marshall	Lewisburg	26,302	21,539	375
Maury	Columbia	69,633	54,812	613
Meigs	Decatur	9,955	8,033	195
Monroe	Madisonville	34,830	30,541	635
Montgomery	Clarksville	127,265	100,498	539
Moore	Lynchburg	5,196	4,721	129
Morgan	Wartburg	18,775	17,300	522
Obion	Union City	32,219	31,717	545
Overton	Livingston	19,557	17,636	433
Perry	Linden	7,508	6,612	415
Pickett	Byrdstown	4,629	4,548	163
Polk	Benton	14,883	13,643	435
Putnam	Cookeville	59,143	51,373	401
Rhea	Dayton	27,836	24,344	316
Roane	Kingston	50,026	47,227	361
Robertson	Springfield	53,077	41,492	477
Rutherford	Murfreesboro	166,035	118,570	619
Scott	Huntsville	20,044	18,358	532
Sequatchie	Dunlap	10,367	8,863	266
Sevier	Sevierville	64,505	51,043	592
Shelby	Memphis	868,825	826,330	755
Smith	Carthage	16,368	14,143	314
Stewart	Dover	11,545	9,479	458
Sullivan	Blountville	150,617	143,596	413
Sumner	Gallatin	124,056	103,281	529
Tipton	Covington	47,343	37,568	459
Trousdale	Hartsville	6,844	5,920	114
Unicoi	Erwin	17,216	16,549	186
Union	Maynardville	16,260	13,694	224
Van Buren	Spencer	5,071	4,846	274
Warren	McMinnville	36,160	32,992	433
Washington	Jonesboro	102,211	92,315	326
Wayne	Waynesboro	16,495	13,955	734
Weakley	Dresden	32,942	31,972	580
White	Sparta	22,708	20,090	377
Williamson	Franklin	117,569	81,021	583
Wilson	Lebanon	83,923	67,675	571

Texas

(254 counties, 261,914 sq mi land; pop. 19,759,614)

County	County seat or courthouse	1998 Pop.	1990 Pop.	Land area sq mi
Anderson	Palestine	52,352	48,024	1,071
Andrews	Andrews	13,976	14,338	1,501
Angelina	Lufkin	77,351	69,884	802
Aransas	Rockport	22,910	17,892	252
Archer	Archer City	8,333	7,973	910
Armstrong	Claude	2,164	2,021	914
Atascosa	Jourdanton	36,471	30,533	1,232
Austin	Bellville	23,439	19,832	653
Bailey	Muleshoe	6,907	7,064	827
Bandera	Bandera	15,754	10,562	792
Bastrop	Bastrop	50,390	38,263	889
Baylor	Seymour	4,152	4,385	871
Bee	Beeville	27,718	25,135	880
Bell	Belton	223,468	191,073	1,059
Bexar	San Antonio	1,353,052	1,185,394	1,247
Blanco	Johnson City	8,400	5,972	711
Borden	Gail	758	799	899
Bosque	Meridian	16,557	15,125	989
Bowie	Boston	83,509	81,665	888
Brazoria	Angleton	230,335	191,707	1,387
Brazos	Bryan	133,407	121,862	586
Brewster	Alpine	8,893	8,653	6,193
Briscoe	Silverton	1,888	1,971	900
Brooks	Falfurrias	8,501	8,204	943
Brown	Brownwood	37,051	34,371	944
Burleson	Caldwell	15,652	13,625	666
Burnet	Burnet	32,195	22,677	995
Caldwell	Lockhart	32,447	26,392	546
Calhoun	Port Lavaca	20,596	19,053	512
Callahan	Baird	12,796	11,859	899
Cameron	Brownsville	326,449	260,120	906
Camp	Pittsburg	10,962	9,904	198
Carson	Panhandle	6,696	6,576	923
Cass	Linden	30,828	29,982	938
Castro	Dimmitt	8,357	9,070	898
Chambers	Anahuac	23,743	20,088	599
Cherokee	Rusk	42,947	41,049	1,052
Childress	Childress	7,532	5,953	710
Clay	Henrietta	10,567	10,024	1,098
Cochran	Morton	3,952	4,377	775
Coke	Robert Lee	3,367	3,424	899
Coleman	Coleman	9,541	9,710	1,273
Collin	McKinney	428,803	264,036	848
Collingsworth	Wellington	3,287	3,573	919
Colorado	Columbus	19,021	18,383	963
Comal	New Braunfels	73,391	51,832	562
Comanche	Comanche	13,568	13,381	938
Concho	Paint Rock	3,119	3,044	992

County	County seat or courthouse	1998 Pop.	1990 Pop.	Land area sq mi
Cooke	Gainesville	32,837	30,777	874
Coryell	Gatesville	77,981	64,226	1,052
Cottle	Paducah	1,922	2,247	901
Crane	Crane	4,510	4,652	786
Crockett	Ozona	4,602	4,078	2,808
Crosby	Crosbyton	7,215	7,304	900
Culberson	Van Horn	3,050	3,407	3,813
Dallam	Dalhart	6,602	5,461	1,505
Dallas	Dallas	2,050,865	1,852,810	880
Dawson	Lamesa	14,700	14,349	902
Deaf Smith	Hereford	19,061	19,153	1,497
Delta	Cooper	4,945	4,857	277
Denton	Denton	384,020	273,525	889
DeWitt	Cuero	19,661	18,840	909
Dickens	Dickens	2,242	2,571	904
Dimmit	Carrizo Springs	10,364	10,433	1,331
Donley	Clarendon	3,822	3,696	930
Duval	San Diego	13,662	12,918	1,793
Eastland	Eastland	17,591	18,488	926
Ector	Odessa	125,729	118,934	901
Edwards	Rocksprings	3,779	2,266	2,120
Ellis	Waxahachie	103,638	85,167	940
El Paso	El Paso	703,127	591,610	1,013
Erath	Stephenville	31,562	27,991	1,086
Falls	Marlin	17,434	17,712	769
Fannin	Bonham	28,129	24,804	892
Fayette	La Grange	21,414	20,095	950
Fisher	Roby	4,241	4,842	901
Floyd	Floydada	8,191	8,497	992
Foard	Crowell	1,699	1,794	707
Fort Bend	Richmond	337,798	225,421	875
Franklin	Mount Vernon	9,676	7,802	286
Freestone	Fairfield	17,675	15,818	885
Frio	Pearsall	15,757	13,472	1,133
Gaines	Seminole	14,992	14,123	1,502
Galveston	Galveston	245,556	217,396	399
Garza	Post	4,608	5,143	896
Gillespie	Fredericksburg	20,045	17,204	1,061
Glasscock	Garden City	1,396	1,447	901
Goliad	Goliad	6,998	5,980	854
Gonzales	Gonzales	17,551	17,205	1,068
Gray	Pampa	23,603	23,967	928
Grayson	Sherman	102,815	95,019	934
Gregg	Longview	113,330	104,948	274
Grimes	Anderson	23,293	18,828	794
Guadalupe	Seguin	80,472	64,873	711
Hale	Plainview	36,676	34,671	1,005
Hall	Memphis	3,644	3,905	903
Hamilton	Hamilton	7,603	7,733	836
Hansford	Spearman	5,347	5,848	920
Hardeman	Quanah	4,591	5,283	695
Hardin	Kountze	48,758	41,320	894
Harris	Houston	3,206,063	2,818,101	1,729
Harrison	Marshall	59,773	57,483	899
Hartley	Channing	5,102	3,634	1,462
Haskell	Haskell	6,158	6,820	903
Hays	San Marcos	88,536	65,614	678
Hemphill	Canadian	3,529	3,720	910
Henderson	Athens	68,757	58,543	874
Hidalgo	Edinburg	522,204	383,545	1,569
Hill	Hillsboro	30,534	27,146	962
Hockley	Levelland	23,788	24,199	908
Hood	Granbury	37,194	28,981	422
Hopkins	Sulphur Springs	30,512	28,833	785
Houston	Crockett	21,901	21,375	1,231
Howard	Big Spring	32,051	32,343	903
Hudspeth	Sierra Blanca	3,250	2,915	4,571
Hunt	Greenville	70,893	64,343	841
Hutchinson	Stinnett	24,077	25,689	887
Irion	Mertzon	1,739	1,629	1,052
Jack	Jacksboro	7,340	6,981	917
Jackson	Edna	13,685	13,039	830
Jasper	Jasper	33,437	31,102	938
Jeff Davis	Fort Davis	2,356	1,946	2,265
Jefferson	Beaumont	241,901	239,389	904
Jim Hogg	Hebbronville	5,007	5,109	1,136
Jim Wells	Alice	40,028	37,679	865
Johnson	Cleburne	118,125	97,165	729
Jones	Anson	18,669	16,490	931
Karnes	Karnes City	12,358	12,455	750
Kaufman	Kaufman	65,736	52,220	786
Kendall	Boerne	21,222	14,589	662
Kenedy	Sarita	438	460	1,457
Kent	Jayton	880	1,010	902
Kerr	Kerrville	43,248	36,304	1,106
Kimble	Junction	4,124	4,122	1,251
King	Guthrie	363	354	912
Kinney	Brackettville	3,482	3,119	1,364
Kleberg	Kingsville	30,163	30,274	871
Knox	Benjamin	4,252	4,837	854
Lamar	Paris	46,045	43,949	917
Lamb	Littlefield	14,760	15,072	1,016
Lampasas	Lampasas	17,775	13,521	712
La Salle	Cotulla	6,034	5,254	1,489
Lavaca	Hallettsville	18,813	18,690	970
Lee	Giddings	14,916	12,854	629
Leon	Centerville	14,489	12,665	1,072
Liberty	Liberty	65,078	52,726	1,160
Limestone	Groesbeck	20,930	20,946	909
Lipscomb	Lipscomb	2,973	3,143	932
Live Oak	George West	10,137	9,556	1,036
Llano	Llano	13,480	11,631	935
Loving	Mentone	114	107	673
Lubbock	Lubbock	229,471	222,636	900
Lynn	Tahoka	6,706	6,758	892
McCulloch	Brady	8,751	8,778	1,069
McLennan	Waco	203,446	189,123	1,042
McMullen	Tilden	790	817	1,113
Madison	Madisonville	11,889	10,931	470
Marion	Jefferson	10,886	9,984	381
Martin	Stanton	5,043	4,956	915
Mason	Mason	3,692	3,423	932
Matagorda	Bay City	37,965	36,928	1,115
Maverick	Eagle Pass	48,131	36,378	1,280
Medina	Hondo	37,685	27,312	1,328
Menard	Menard	2,336	2,252	902
Midland	Midland	119,647	106,611	900
Milam	Cameron	24,286	22,946	1,017
Mills	Goldthwaite	4,723	4,531	748
Mitchell	Colorado City	9,708	8,016	910
Montague	Montague	18,539	17,274	931
Montgomery	Conroe	271,788	182,201	1,044
Moore	Dumas	19,686	17,865	900
Morris	Daingerfield	13,358	13,200	255
Motley	Matador	1,305	1,532	989
Nacogdoches	Nacogdoches	56,220	54,753	947
Navarro	Corsicana	41,738	39,926	1,071
Newton	Newton	14,243	13,569	933
Nolan	Sweetwater	16,501	16,594	912
Nueces	Corpus Christi	316,340	291,145	836
Ochiltree	Perryton	8,827	9,128	918
Oldham	Vega	2,153	2,278	1,501
Orange	Orange	84,905	80,509	356
Palo Pinto	Palo Pinto	25,756	25,055	953
Panola	Carthage	23,070	22,035	801
Parker	Weatherford	81,985	64,785	904
Parmer	Farwell	10,302	9,863	882
Pecos	Fort Stockton	16,003	14,675	4,764
Polk	Livingston	50,309	30,687	1,057
Potter	Amarillo	108,943	97,841	909
Presidio	Marfa	8,636	6,637	3,856
Rains	Emory	8,618	6,715	232
Randall	Canyon	99,664	89,673	915
Reagan	Big Lake	4,203	4,514	1,175
Real	Leakey	2,687	2,412	700
Red River	Clarksville	13,731	14,317	1,050
Reeves	Pecos	14,478	15,852	2,636
Refugio	Refugio	7,907	7,976	770
Roberts	Miami	939	1,025	924
Robertson	Franklin	15,527	15,511	855
Rockwall	Rockwall	37,174	25,604	129
Runnels	Ballinger	11,507	11,294	1,055
Rusk	Henderson	45,877	43,735	924
Sabine	Hemphill	10,551	9,586	490
San Augustine	San Augustine	8,086	7,999	528
San Jacinto	Coldspring	21,768	16,372	571
San Patricio	Sinton	71,393	58,749	692
San Saba	San Saba	5,615	5,401	1,135
Schleicher	Eldorado	2,984	2,990	1,311
Scurry	Snyder	18,073	18,634	903
Shackelford	Albany	3,303	3,316	914
Shelby	Center	22,748	22,034	794
Sherman	Stratford	2,864	2,858	923
Smith	Tyler	168,783	151,309	929
Somervell	Glen Rose	6,421	5,360	187
Starr	Rio Grande City	55,906	40,518	1,223
Stephens	Breckenridge	9,811	9,010	895
Sterling	Sterling City	1,364	1,438	923
Stonewall	Aspermont	1,783	2,013	919
Sutton	Sonora	4,463	4,135	1,454
Swisher	Tulia	8,301	8,133	901
Tarrant	Fort Worth	1,355,273	1,170,103	864
Taylor	Abilene	122,016	119,655	916
Terrell	Sanderson	1,181	1,410	2,358
Terry	Brownfield	12,896	13,218	890
Throckmorton	Throckmorton	1,727	1,880	912
Titus	Mount Pleasant	25,422	24,009	411
Tom Green	San Angelo	102,775	98,458	1,522
Travis	Austin	710,626	576,407	989
Trinity	Groveton	12,613	11,445	693
Tyler	Woodville	20,408	16,646	923
Upshur	Gilmer	35,885	31,370	588
Upton	Rankin	3,749	4,447	1,242
Uvalde	Uvalde	25,565	23,340	1,557
Val Verde	Del Rio	43,831	38,721	3,171
Van Zandt	Canton	44,037	37,944	849
Victoria	Victoria	82,650	74,361	883
Walker	Huntsville	54,972	50,917	788
Waller	Hempstead	27,218	23,389	514
Ward	Monahans	11,801	13,115	836
Washington	Brenham	29,127	26,154	609
Webb	Laredo	188,166	133,239	3,357
Wharton	Wharton	40,133	39,955	1,090
Wheeler	Wheeler	5,293	5,879	914

County	County seat or courthouse	1998 Pop.	1990 Pop.	Land area sq mi
Wichita	Wichita Falls	128,904	122,378	628
Wilbarger	Vernon	13,711	15,121	971
Willacy	Raymondville	19,622	17,705	597
Williamson	Georgetown	223,910	139,551	1,124
Wilson	Floresville	31,423	22,650	807
Winkler	Kermit	7,964	8,626	841
Wise	Decatur	44,135	34,679	905
Wood	Quitman	34,321	29,380	650
Yoakum	Plains	8,010	8,786	800
Young	Graham	17,697	18,126	922
Zapata	Zapata	11,491	9,279	997
Zavala	Crystal City	11,927	12,162	1,299

Utah

(29 counties, 82,168 sq mi land; pop. 2,099,758)

County	County seat or courthouse	1998 Pop.	1990 Pop.	Land area sq mi
Beaver	Beaver	5,896	4,765	2,590
Box Elder	Brigham City	41,949	36,485	5,724
Cache	Logan	86,949	70,183	1,165
Carbon	Price	20,966	20,228	1,479
Daggett	Manila	737	690	698
Davis	Farmington	233,013	187,941	305
Duchesne	Duchesne	14,481	12,645	3,238
Emery	Castle Dale	10,989	10,332	4,452
Garfield	Panguitch	4,272	3,980	5,175
Grand	Moab	8,068	6,620	3,682
Iron	Parowan	28,659	20,789	3,299
Juab	Nephi	7,572	5,817	3,392
Kane	Kanab	6,200	5,169	3,992
Millard	Fillmore	12,249	11,333	6,590
Morgan	Morgan	7,022	5,528	609
Piute	Junction	1,402	1,277	758
Rich	Randolph	1,834	1,725	1,029
Salt Lake	Salt Lake City	850,667	725,956	737
San Juan	Monticello	13,711	12,621	7,821
Sanpete	Manti	21,452	16,259	1,588
Sevier	Richfield	18,452	15,431	1,910
Summit	Coalville	26,746	15,518	1,871
Tooele	Tooele	33,351	26,601	6,946
Uintah	Vernal	25,660	22,211	4,477
Utah	Provo	335,635	263,590	1,998
Wasatch	Heber City	13,267	10,089	1,181
Washington	Saint George	82,115	48,560	2,427
Wayne	Loa	2,379	2,177	2,461
Weber	Ogden	184,065	158,330	576

Vermont

(14 counties, 9,249 sq mi land; pop. 590,883)

County	County seat or courthouse	1998 Pop.	1990 Pop.	Land area sq mi
Addison	Middlebury	35,168	32,953	770
Bennington	Bennington	35,968	35,845	676
Caledonia	Saint Johnsbury	28,529	27,846	651
Chittenden	Burlington	142,642	131,761	539
Essex	Guildhall	6,580	6,405	665
Franklin	Saint Albans	44,017	39,980	637
Grand Isle	North Hero	6,236	5,318	83
Lamoille	Hyde Park	21,597	19,735	461
Orange	Chelsea	27,924	26,149	689
Orleans	Newport	25,296	24,053	697
Rutland	Rutland	62,524	62,142	932
Washington	Montpelier	56,308	54,928	690
Windham	Newfane	42,650	41,588	789
Windsor	Woodstock	55,444	54,055	971

Virginia

(95 counties, 40 ind. cities, 39,598 sq mi land; pop. 6,791,345)

County	County seat or courthouse	1998 Pop.	1990 Pop.	Land area sq mi
Accomack	Accomac	32,245	31,703	455
Albemarle	Charlottesville	78,401	68,172	723
Alleghany	Covington	12,146	12,969	446
Amelia	Amelia Courthouse	10,367	8,787	357
Amherst	Amherst	30,042	28,578	475
Appomattox	Appomattox	13,134	12,298	334
Arlington	Arlington	177,275	170,897	26
Augusta	Staunton	61,775	54,677	972
Bath	Warm Springs	4,891	4,799	532
Bedford	Bedford	55,872	45,552	755
Bland	Bland	6,748	6,514	359
Botetourt	Fincastle	28,561	24,992	543
Brunswick	Lawrenceville	16,716	15,987	566
Buchanan	Grundy	28,929	31,333	504
Buckingham	Buckingham	14,639	12,873	581
Campbell	Rustburg	50,335	47,572	505
Caroline	Bowling Green	22,053	19,217	533
Carroll	Hillsville	27,873	26,565	477
Charles City	Charles City	7,092	6,282	183

County	County seat or courthouse	1998 Pop.	1990 Pop.	Land area sq mi
Charlotte	Charlotte Courthouse	12,259	11,688	475
Chesterfield	Chesterfield	245,915	209,564	426
Clarke	Berryville	12,779	12,101	177
Craig	New Castle	4,882	4,372	330
Culpeper	Culpeper	33,083	27,791	381
Cumberland	Cumberland	7,851	7,825	299
Dickenson	Clintwood	16,894	17,620	333
Dinwiddie	Dinwiddie	24,657	22,319	504
Essex	Tappahannock	9,127	8,689	258
Fairfax	Fairfax	929,239	818,358	396
Fauquier	Warrenton	54,109	48,860	650
Floyd	Floyd	13,091	11,965	382
Fluvanna	Palmyra	18,575	12,429	287
Franklin	Rocky Mount	44,538	39,549	692
Frederick	Winchester	55,229	45,723	415
Giles	Pearisburg	16,242	16,366	358
Gloucester	Gloucester	35,081	30,131	217
Goochland	Goochland	17,823	14,163	285
Grayson	Independence	16,118	16,278	443
Greene	Stanardsville	13,991	10,297	157
Greensville	Emporia	11,281	8,630	296
Halifax	Halifax	36,863	36,030	820
Hanover	Hanover	81,975	63,306	473
Henrico	Henrico	246,052	217,849	238
Henry	Martinsville	55,627	56,942	382
Highland	Monterey	2,499	2,635	416
Isle of Wight	Isle of Wight	29,252	25,053	316
James City	Williamsburg	44,233	34,970	143
King and Queen	King and Queen Courthouse	6,529	6,289	316
King George	King George	17,236	13,527	180
King William	King William	12,768	10,913	275
Lancaster	Lancaster	11,373	10,896	133
Lee	Jonesville	23,815	24,496	437
Loudoun	Leesburg	143,940	86,129	520
Louisa	Louisa	24,675	20,325	498
Lunenburg	Lunenburg	12,043	11,419	432
Madison	Madison	12,697	11,949	322
Mathews	Mathews	9,073	8,348	86
Mecklenburg	Boydton	31,047	29,241	624
Middlesex	Saluda	9,630	8,653	130
Montgomery	Christiansburg	75,878	73,913	388
Nelson	Lovingston	13,917	12,778	472
New Kent	New Kent	13,052	10,445	210
Northampton	Eastville	12,709	13,061	207
Northumberland	Heathsville	11,513	10,524	192
Nottoway	Nottoway	14,999	14,993	315
Orange	Orange	25,408	21,421	342
Page	Luray	22,989	21,690	311
Patrick	Stuart	18,441	17,473	483
Pittsylvania	Chatham	57,384	55,672	971
Powhatan	Powhatan	21,950	15,328	261
Prince Edward	Farmville	19,028	17,320	353
Prince George	Prince George	30,135	27,394	266
Prince William	Manassas	259,827	215,677	338
Pulaski	Pulaski	34,539	34,496	321
Rappahannock	Washington	7,269	6,622	267
Richmond	Warsaw	8,665	7,273	192
Roanoke	Salem	80,839	79,294	251
Rockbridge	Lexington	19,557	18,350	600
Rockingham	Harrisonburg	63,214	57,482	851
Russell	Lebanon	29,049	28,667	475
Scott	Gate City	22,605	23,204	537
Shenandoah	Woodstock	34,663	31,636	512
Smyth	Marion	32,757	32,370	452
Southampton	Courtland	17,450	17,550	600
Spotsylvania	Spotsylvania	83,692	57,403	401
Stafford	Stafford	87,055	61,236	270
Surry	Surry	6,471	6,145	279
Sussex	Sussex	9,925	10,248	491
Tazewell	Tazewell	46,766	45,960	520
Warren	Front Royal	30,126	26,142	214
Washington	Abingdon	49,168	45,887	564
Westmoreland	Montross	16,282	15,480	229
Wise	Wise	38,599	39,573	403
Wythe	Wytheville	26,268	25,471	463
York	Yorktown	58,789	42,434	106
Independent Cities				
Alexandria		118,300	111,182	15
Bedford		6,317	6,177	7
Bristol		17,486	18,426	12
Buena Vista		6,288	6,406	7
Charlottesville		38,223	40,475	10
Chesapeake		199,564	151,982	341
Clifton Forge		4,342	4,679	3
Colonial Heights		16,955	16,064	8
Covington		6,857	7,198	4
Danville		50,868	53,056	43
Emporia		5,474	5,479	7
Fairfax		20,697	19,894	6
Falls Church		10,042	9,522	2
Franklin		8,685	7,864	8
Fredericksburg		21,686	19,027	11
Galax		6,864	6,699	8
Hampton		136,968	133,811	52
Harrisonburg		33,434	30,707	18
Hopewell		22,529	23,101	10

County	County seat or courthouse	1998 Pop.	1990 Pop.	Land area sq mi
Lexington		7,360	6,959	3
Lynchburg		65,473	66,049	49
Manassas		35,336	27,957	10
Manassas Park		8,711	6,734	2
Martinsville		15,668	16,162	11
Newport News		178,615	171,439	68
Norfolk		215,215	261,250	54
Norton		4,155	4,247	7
Petersburg		34,724	37,027	23
Poquoson		11,455	11,005	16
Portsmouth		98,936	103,910	33
Radford		15,734	15,940	10
Richmond		194,173	202,798	60
Roanoke		93,749	96,509	43
Salem		24,679	23,797	15
Staunton		23,346	24,461	20
Suffolk		62,703	52,143	400
Virginia Beach		432,380	393,089	248
Waynesboro		18,561	18,549	14
Williamsburg		11,971	11,409	9
Winchester		22,659	21,947	9

Washington

(39 counties, 66,581 sq mi land; pop. 5,689,263)

County	County seat or courthouse	1998 Pop.	1990 Pop.	Land area sq mi
Adams	Ritzville	15,324	13,603	1,925
Asotin	Asotin	21,264	17,605	636
Benton	Prosser	136,250	112,560	1,703
Chelan	Wenatchee	60,052	52,250	2,922
Clallam	Port Angeles	64,169	56,210	1,745
Clark	Vancouver	326,943	238,053	628
Columbia	Dayton	4,156	4,024	869
Cowlitz	Kelso	91,574	82,119	1,139
Douglas	Waterville	33,631	26,205	1,821
Ferry	Republic	7,170	6,295	2,204
Franklin	Pasco	46,459	37,473	1,242
Garfield	Pomeroy	2,330	2,248	711
Grant	Ephrata	70,545	54,798	2,676
Grays Harbor	Montesano	67,739	64,175	1,917
Island	Coupeville	70,319	60,195	209
Jefferson	Port Townsend	26,232	20,406	1,809
King	Seattle	1,654,876	1,507,305	2,126
Kitsap	Port Orchard	232,623	189,731	396
Kittitas	Ellensburg	31,714	26,725	2,297
Klickitat	Goldendale	19,295	16,616	1,873
Lewis	Chehalis	68,163	59,358	2,408
Lincoln	Davenport	9,734	8,864	2,311
Mason	Shelton	49,867	38,341	961
Okanogan	Okanogan	38,237	33,350	5,268
Pacific	South Bend	20,802	18,882	975
Pend Oreille	Newport	11,526	8,915	1,401
Pierce	Tacoma	676,505	586,203	1,676
San Juan	Friday Harbor	12,493	10,035	175
Skagit	Mount Vernon	99,357	79,545	1,735
Skamania	Stevenson	9,805	8,289	1,657
Snohomish	Everett	587,783	465,628	2,090
Spokane	Spokane	408,669	361,333	1,764
Stevens	Colville	39,464	30,948	2,478
Thurston	Olympia	202,255	161,238	727
Wahkiakum	Cathlamet	3,857	3,327	264
Walla Walla	Walla Walla	53,702	48,439	1,271
Whatcom	Bellingham	156,830	127,780	2,120
Whitman	Colfax	39,487	38,775	2,159
Yakima	Yakima	218,062	188,823	4,296

West Virginia

(55 counties, 24,087 sq mi land; pop. 1,811,156)

County	County seat or courthouse	1998 Pop.	1990 Pop.	Land area sq mi
Barbour	Philippi	16,152	15,699	341
Berkeley	Martinsburg	70,970	59,253	321
Boone	Madison	26,118	25,870	503
Braxton	Sutton	13,185	12,998	514
Brooke	Wellsburg	26,004	26,992	89
Cabell	Huntington	94,273	96,827	282
Calhoun	Grantsville	7,940	7,885	281
Clay	Clay	10,530	9,983	342
Doddridge	West Union	7,554	6,994	321
Fayette	Fayetteville	47,930	47,952	664
Gilmer	Glenville	7,130	7,669	340
Grant	Petersburg	11,098	10,428	477
Greenbrier	Lewisburg	35,383	34,693	1,021
Hampshire	Romney	19,041	16,498	642
Hancock	New Cumberland	33,973	35,233	83
Hardy	Moorefield	11,829	10,977	583
Harrison	Clarksburg	70,891	69,371	416
Jackson	Ripley	27,972	25,938	466
Jefferson	Charles Town	41,368	35,926	210
Kanawha	Charleston	202,011	207,619	903
Lewis	Weston	17,427	17,223	389
Lincoln	Hamlin	22,192	21,382	438
Logan	Logan	41,080	43,032	454
McDowell	Welch	29,916	35,233	535
Marion	Fairmont	56,318	57,249	310
Marshall	Moundsville	35,441	37,356	307
Mason	Point Pleasant	25,869	25,178	432
Mercer	Princeton	63,794	64,980	421
Mineral	Keyser	26,737	26,697	328
Mingo	Williamson	31,926	33,739	423
Monongalia	Morgantown	77,505	75,509	361
Monroe	Union	13,205	12,406	473
Morgan	Berkeley Springs	13,640	12,128	229
Nicholas	Summersville	27,595	26,775	649
Ohio	Wheeling	48,287	50,871	106
Pendleton	Franklin	8,062	8,054	698
Pleasants	St. Marys	7,421	7,546	131
Pocahontas	Marlinton	9,268	9,008	940
Preston	Kingwood	29,811	29,037	648
Putnam	Winfield	51,164	42,835	346
Raleigh	Beckley	79,066	76,819	607
Randolph	Elkins	28,658	27,803	1,040
Ritchie	Harrisville	10,356	10,233	454
Roane	Spencer	15,342	15,120	484
Summers	Hinton	13,146	14,204	361
Taylor	Grafton	15,326	15,144	173
Tucker	Parsons	7,631	7,728	419
Tyler	Middlebourne	9,835	9,796	258
Upshur	Buckhannon	23,526	22,867	355
Wayne	Wayne	41,957	41,636	506
Webster	Webster Springs	10,230	10,729	556
Wetzel	New Martinsville	18,256	19,258	359
Wirt	Elizabeth	5,669	5,192	233
Wood	Parkersburg	86,768	86,915	367
Wyoming	Pineville	27,380	28,990	501

Wisconsin

(72 counties, 54,314 sq mi land; pop. 5,223,500)

County	County seat or courthouse	1998 Pop.	1990 Pop.	Land area sq mi
Adams	Friendship	18,492	15,682	648
Ashland	Ashland	16,474	16,307	1,044
Barron	Barron	43,872	40,750	863
Bayfield	Washburn	15,151	14,008	1,476
Brown	Green Bay	215,373	194,594	529
Buffalo	Alma	14,298	13,584	685
Burnett	Siren	14,646	13,084	822
Calumet	Chilton	38,377	34,291	320
Chippewa	Chippewa Falls	54,574	52,360	1,011
Clark	Neillsville	33,147	31,647	1,216
Columbia	Portage	51,152	45,088	774
Crawford	Prairie du Chien	16,576	15,940	573
Dane	Madison	424,586	367,085	1,202
Dodge	Juneau	83,261	76,559	882
Door	Sturgeon Bay	27,027	25,690	483
Douglas	Superior	43,033	41,758	1,309
Dunn	Menomonie	38,977	35,909	852
Eau Claire	Eau Claire	89,287	85,183	638
Florence	Florence	5,199	4,590	488
Fond du Lac	Fond du Lac	94,690	90,083	723
Forest	Crandon	9,645	8,776	1,014
Grant	Lancaster	49,340	49,266	1,148
Green	Monroe	33,404	30,339	584
Green Lake	Green Lake	19,438	18,651	354
Iowa	Dodgeville	22,422	20,150	763
Iron	Hurley	6,350	6,153	757
Jackson	Black River Falls	17,735	16,588	987
Jefferson	Jefferson	73,550	67,783	557
Juneau	Mauston	23,822	21,650	768
Kenosha	Kenosha	144,339	128,181	273
Kewaunee	Kewaunee	19,806	18,878	343
La Crosse	La Crosse	102,565	97,904	453
Lafayette	Darlington	16,261	16,074	634
Langlade	Antigo	20,466	19,505	873
Lincoln	Merrill	29,727	26,993	883
Manitowoc	Manitowoc	82,412	80,421	592
Marathon	Wausau	123,223	115,400	1,545
Marinette	Marinette	43,033	40,548	1,402
Marquette	Montello	15,101	12,321	456
Menominee	Keshena	4,779	3,890	358
Milwaukee	Milwaukee	911,713	959,275	242
Monroe	Sparta	39,532	36,633	901
Oconto	Oconto	34,014	30,226	998
Oneida	Rhinelander	35,672	31,679	1,125
Outagamie	Appleton	156,269	140,510	640
Ozaukee	Port Washington	81,076	72,831	232
Pepin	Durand	7,118	7,107	232
Pierce	Ellsworth	35,606	32,765	577
Polk	Balsam Lake	38,786	34,773	917
Portage	Stevens Point	64,752	61,405	806
Price	Phillips	15,813	15,600	1,253
Racine	Racine	186,119	175,034	333
Richland	Richland Center	17,891	17,521	586
Rock	Janesville	150,736	139,510	721
Rusk	Ladysmith	15,252	15,079	913
Saint Croix	Hudson	58,936	50,251	722
Sauk	Baraboo	53,369	46,975	838
Sawyer	Hayward	16,110	14,181	1,257

County	County seat or courthouse	1998 Pop.	1990 Pop.	Land area sq mi
Shawano	Shawano	38,756	37,157	893
Sheboygan	Sheboygan	110,170	103,877	514
Taylor	Medford	19,313	18,901	975
Trempealeau	Whitehall	26,469	25,263	734
Vernon	Viroqua	27,343	25,617	795
Vilas	Eagle River	21,277	17,707	873
Walworth	Elkhorn	85,353	75,000	555
Washburn	Shell Lake	15,421	13,772	810
Washington	West Bend	113,906	95,328	431
Waukesha	Waukesha	353,110	304,715	556
Waupaca	Waupaca	50,545	46,104	751
Waushara	Wautoma	21,609	19,385	626
Winnebago	Oshkosh	149,818	140,320	439
Wood	Wisconsin Rapids	76,036	73,605	793

Wyoming

(23 counties, 97,105 sq mi land; pop. 480,907)

County	County seat or courthouse	1998 Pop.	1990 Pop.	Land area sq mi
Albany	Laramie	29,185	30,797	4,274
Big Horn	Basin	11,380	10,525	3,137
Campbell	Gillette	32,465	29,370	4,797
Carbon	Rawlins	15,575	16,659	7,897
Converse	Douglas	12,337	11,128	4,255
Crook	Sundance	5,829	5,294	2,859
Fremont	Lander	36,044	33,662	9,183
Goshen	Torrington	12,886	12,373	2,226
Hot Springs	Thermopolis	4,727	4,809	2,004
Johnson	Buffalo	6,824	6,145	4,166
Laramie	Cheyenne	78,872	73,142	2,686
Lincoln	Kemmerer	13,876	12,625	4,069
Natrona	Casper	63,341	61,226	5,340
Niobrara	Lusk	2,706	2,499	2,626
Park	Cody	25,782	23,178	6,943
Platte	Wheatland	8,626	8,145	2,085
Sheridan	Sheridan	25,165	23,562	2,523
Sublette	Pinedale	5,738	4,843	4,882
Sweetwater	Green River	39,780	38,823	10,426
Teton	Jackson	14,163	11,173	4,008
Unita	Evanston	20,465	18,705	2,082
Washakie	Worland	8,669	8,388	2,240
Weston	Newcastle	6,472	6,518	2,398

Population of Outlying Areas

Source: Bureau of the Census, U.S. Dept. of Commerce; World Almanac research

Population estimates for July 1, 1997, are given for Puerto Rican municipios; all other population counts and all land area figures are from the U.S. census conducted on Apr. 1, 1990. Because only selected areas are shown, the population and land area figures may not equal the total reported. ZIP codes with an asterisk (*) are general delivery ZIP codes. Consult the local postmaster for more specific delivery information. Wake Atoll, Johnston Atoll, and Midway Atoll receive mail through APO and FPO addresses. U.S. outlying areas not listed in this table may not receive U.S. mail delivery.

Commonwealth of Puerto Rico

ZIP code	Municipio	1997 Pop.	Land area sq mi	ZIP code	Municipio	1997 Pop.	Land area sq mi	ZIP code	Municipio	1997 Pop.	Land area sq mi
00601	Adjuntas	20,525	67	00650	Florida	9,052	10	00719	Naranjito	29,457	27
00602	Aguada	38,791	31	00653	Guánica	21,847	37	00720	Orocovis	24,690	64
*00605	Aguadilla	67,010	37	*00785	Guayama	43,216	65	00723	Patillas	21,725	47
00703	Aguas Buenas	30,193	31	00656	Guayanilla	27,830	42	00624	Peñuelas	26,858	45
00705	Aibonito	28,178	31	*00970	Guaynabo	104,901	27	*00732	Ponce	189,900	116
00610	Añasco	27,463	39	00778	Gurabo	33,053	28	00678	Quebradillas	26,109	23
*00613	Arecibo	102,773	126	00659	Hatillo	39,621	42	00677	Rincón	13,760	14
00714	Arroyo	19,905	15	00660	Hormigüeros	16,052	11	00745	Río Grande	50,825	61
00617	Barceloneta	26,644	23	*00791	Humacao	58,918	45	00637	Sabana Grande	24,644	36
00794	Barranquitas	28,587	34	00662	Isabela	41,728	55	00751	Salinas	29,755	69
*00958	Bayamón	233,784	44	00664	Jayuya	16,612	45	00683	San Germán	36,984	55
00623	Cabo Rojo	47,105	70	00795	Juana Díaz	50,848	60	*00902	San Juan	436,334	48
*00726	Caguas	141,871	59	00777	Juncos	42,753	27	00754	San Lorenzo	36,314	53
00627	Camuy	32,135	46	00667	Lajas	26,663	60	00685	San Sebastián	43,989	70
00729	Canóvanas	50,786	33	00669	Lares	32,381	62	00757	Santa Isabel	19,454	34
*00984	Carolina	189,853	45	00670	Las Marías	9,929	46	*00954	Toa Alta	61,113	27
*00963	Cataño	32,414	5	00771	Las Piedras	31,252	34	*00950	Toa Baja	92,947	23
*00737	Cayey	50,600	52	00772	Loíza	28,372	19	*00976	Trujillo Alto	75,658	21
00735	Ceiba	18,571	29	00773	Luquillo	19,025	26	00641	Utuado	35,012	114
00638	Ciales	19,568	67	00674	Manatí	40,190	45	00692	Vega Alta	36,064	28
00739	Cidra	49,440	36	00606	Maricao	6,361	37	*00694	Vega Baja	61,906	46
00769	Coamo	36,856	78	00707	Maunabo	13,605	21	00765	Vieques	9,311	51
00782	Comerío	20,965	28	*00681	Mayagüez	100,001	78	00766	Villalba	23,213	36
00783	Corozal	36,957	43	00676	Moca	37,544	50	00767	Yabucoa	41,848	55
00775	Culebra	1,738	12	00687	Morovis	33,380	39	00698	Yauco	44,176	68
00646	Dorado	32,985	23	00718	Naguabo	25,943	52	**TOTAL**		**3,827,038**	**3,427**
00650	Florida	9,052	10								

Commonwealth of the Northern Mariana Islands

ZIP code	Municipality	1990 Pop.	Land area sq mi	ZIP code	Municipality	1990 Pop.	Land area sq mi	ZIP code	1990 Pop.	Land area sq mi
96950	Northern Islands	36	60	96950	Saipan	38,896	47			
96951	Rota	2,295	33	96952	Tinian	2,118	39	**TOTAL**	**43,345**	**179**

Other U.S. External Territories

ZIP code	Location	1990 Pop.	Land area sq mi	ZIP code	Location	1990 Pop.	Land area sq mi	ZIP code	Location	1990 Pop.	Land area sq mi
	American Samoa			96912	Dededo	31,728	30	96929	Yigo	14,213	35
96799	American Samoa	46,773	77	96917	Inarajan	2,469	6	96914	Yona	5,338	20
	Guam			96923	Mangilao	10,483	6	**TOTAL**		**133,152**	**210**
*96913	Agaña	1,139	1	96916	Merizo	1,742	19				
96919	Agaña Hts.	3,646	1	96927	Mongmong-Toto-Maite	5,845	10		**Virgin Islands**		
96928	Agat	4,960	10	96925	Piti	1,827	6	00820	Saint Croix	50,139	83
96922	Asan	2,070	6	96915	Santa Rita	11,857	2	*00820	Christiansted	2,555	
*96913	Barrigada	8,846	9	96926	Sinajana	2,658	7	*00841	Frederiksted	1,064	
96924	Chalan-Pago-Ordot	4,451	6	96930	Talofofo	2,310	17	*00830	Saint John	3,504	20
				*96913	Tamuning	16,673	1	*00801	Saint Thomas	48,166	31
				96918	Umatac	897	17	00801	Charlotte Amalie	12,331	
								TOTAL		**101,809**	**134**

CITIES OF THE U.S.

Sources: Bureau of the Census: population, with rank in parentheses, estimated as of July 1998, and population growth. Bureau of Labor Statistics: employment (1998). Bureau of Economic Analysis: per capita personal income (1997).

Included here are the 100 most populous cities, based on July 1998 Census Bureau estimates (inc.=incorporated; est.=established). Most data are for the city proper. Some statistics, where noted, apply to the whole MSA (Metropolitan Statistical Area).
Note: Websites are as of Aug. 1999 and subject to change.

Akron, Ohio

Population: 215,712 (74); **Pop. density:** 3,468 per sq. mi; **Pop. growth (1990-98):** −3.3%. **Area:** 62.2 sq. mi. **Employment:** 105,432 employed, 5.6% unemployed. **Per capita income (MSA):** $24,849; % increase, 1996-97: 5.2.
History: settled 1825; inc. as city 1865; located on Ohio-Erie Canal and is a port of entry; since 1870, rubber capital of U.S.
Transportation: 1 airport; major trucking industry; Conrail, Amtrak; metro transit system. **Communications:** 7 radio stations. **Medical facilities:** 4 hosp.; specialized children's treatment center. **Educational facilities:** 4 univ. and colleges; 68 pub. schools. **Further information:** Akron Regional Development Board, Cascade Plaza, Akron, OH 44308.
Websites: http://www.ci.akron.oh.us
http://www.ardb.org

Albuquerque, New Mexico

Population: 419,311 (36); **Pop. density:** 3,172 per sq. mi; **Pop. growth (1990-98):** 8.9%. **Area:** 132.2 sq. mi. **Employment:** 229,074 employed, 4.3% unemployed. **Per capita income (MSA):** $22,987; % increase, 1996-97: 4.8.
History: founded 1706 by the Spanish; inc. 1890.
Transportation: 1 intl. airport; 1 railroad; 1 bus line. **Communications:** 8 TV, 40 radio stations. **Medical facilities:** 6 major hosp. **Educational facilities:** 1 univ., 13 colleges. **Further information:** Albuquerque Convention & Visitors Bureau, PO Box 26866, Albuquerque, NM 87125-6866.
Websites: http://www.abqcvb.org
http://www.cabq.org

Anaheim, California

Population: 295,153 (57); **Pop. density:** 6,663 per sq. mi; **Pop. growth (1990-98):** 10.8%. **Area:** 44.3 sq. mi. **Employment:** 153,108 employed, 3.4% unemployed. **Per capita income (MSA):** $30,115; % increase, 1996-97: 6.8.
History: founded 1857; inc. 1870; now known as home of Disneyland and the Mighty Ducks of Anaheim.
Transportation: access to 3 municipal airports; 4 railroads; Greyhound buses (MSA). **Communications:** 12 TV, 4 radio stations (MSA). **Medical facilities:** 5 hosp.; 4 medical centers (MSA). **Educational facilities:** 13 univ. and colleges; 47 elem., 10 junior high, 11 high schools (MSA). **Further information:** Chamber of Commerce, 100 South Anaheim Blvd., Ste. 300, Anaheim, CA 92805.
Website: http://www.anaheim.net

Anchorage, Alaska

Population: 254,982 (65); **Pop. density:** 150 per sq. mi; **Pop. growth (1990-98):** 12.7%. **Area:** 1,697.6 sq. mi. **Employment:** 135,778 employed, 4.1% unemployed. **Per capita income (MSA):** $29,765; % increase, 1996-97: 4.4.
History: founded 1914 as a construction camp for railroad; HQ of Alaska Defense Command, WWII; severely damaged in earthquake 1964, but now rebuilt and currently population center of Alaska.
Transportation: 1 intl. airport; 1 railroad; transit system. **Communications:** 9 TV, 22 radio stations. **Medical facilities:** 4 hosp. **Educational facilities:** 5 univ., 3 colleges, 89 pub. schools. **Further information:** Chamber of Commerce, 441 W. 5th Ave., Ste. 300, Anchorage, AK 99501-2309.
Websites: http://www.ci.anchorage.ak.us
http://www.anchoragechamber.org

Arlington, Texas

Population: 306,497 (54); **Pop. density:** 3,296 per sq. mi; **Pop. growth (1990-98):** 17.1%. **Area:** 93 sq. mi. **Employment:** 180,361 employed, 2.9% unemployed. **Per capita income (MSA):** $25,150; % increase, 1996-97: 8.1.
History: settled in 1840s between Dallas and Ft. Worth; inc. 1884.
Transportation: Dallas/Ft. Worth airport is 20 min. away; 11 railway lines; intercity transport system in planning stage. **Communications:** 11 TV, 44 radio stations. **Medical facilities:** 2 hosp. **Educational facilities:** 1 univ., 1 junior college; 60 pub. schools. **Further information:** The Arlington Chamber, 316 W. Main St., Arlington, TX 76010.
Websites: http://www.ci.arlington.tx.us
http://www.chamber.arlingtontx.com

Atlanta, Georgia

Population: 403,819 (39); **Pop. density:** 3,064 per sq. mi; **Pop. growth (1990-98):** 2.5%. **Area:** 131.8 sq. mi. **Employment:** 208,056 employed, 5.6% unemployed. **Per capita income (MSA):** $28,253; % increase, 1996-97: 7.7.
History: founded as "Terminus" 1837; renamed Atlanta 1845; inc. 1847; played major role in Civil War; became permanent state capital 1877; birthplace of civil rights movement; host to 1996 Centennial Olympic Games.
Transportation: 1 intl. airport; 3 railroad lines; MARTA bus and rapid rail service. **Communications:** 11 TV, 49 radio stations; 26 cable TV cos. **Medical facilities:** 61 hosp.; VA hosp.; U.S. Centers for Disease Control and Prevention; American Cancer Society. **Educational facilities:** 43 colleges, univ., seminaries, junior colleges, 102 pub. schools. **Further information:** Metro Atlanta Chamber of Commerce, 235 Intl. Blvd. NW, Atlanta, GA 30301.
Website: http://www.metroatlantachamber.com

Aurora, Colorado

Population: 250,604 (67); **Pop. density:** 1,891 per sq. mi; **Pop. growth (1990-98):** 12.8%. **Area:** 140.66 sq. mi. **Employment:** 154,024 employed, 3.0% unemployed. **Per capita income (MSA):** $30,743; % increase, 1996-97: 8.1.
History: located 5 mi east of Denver; early growth stimulated by presence of military bases; fast-growing trade center.
Transportation: adjacent to new Denver Intl. Airport; 1 airport; bus system. **Communications:** 1 TV station. **Medical facilities:** 2 private hosp. **Educational facilities:** 1 univ., 1 community college, 2 technical colleges; 86 pub. schools. **Further information:** Aurora Planning Dept., 1470 S. Havana St., Rm. 608, Aurora, CO 80012.
Websites: http://www.ci.aurora.co.us
http://www.mktplace.net/aurora/chamber

Augusta, Georgia

Population: 187,689 (96); **Pop. density:** 579 per sq. mi. **Pop. growth (1990-98):** 0.6%. **Area:** 324.1 sq. mi. **Employment:** 75,434 employed, 7.1% unemployed. **Per capita income (MSA):** $20,821; % increase 1996-97: 4.3.
History: founded 1736 as colonial trading post; one of the few pre-Civil War manufacturing centers in the South. Augusta National Golf Club, home of Masters Tournament, founded 1933.
Transportation: 1 regional, 1 local airport; Savannah River. **Communications:** 4 TV, 20+ radio stations; 2 major cable TV providers. **Medical facilities:** 9 hosp., including Eisenhower Army Medical Center. **Educational facilities:** 1 univ., 1 medical college, 1 tech. institute; 40 elem., 10 middle, 9 pub. high schools. **Further information:** Augusta Metropolitan Convention and Visitors Bureau, PO Box 1331, Augusta, GA 30903.
Websites: http://www.augustaga.org
http://www.augustagausa.com

Austin, Texas

Population: 552,434 (21); **Pop. density:** 2,536 per sq. mi; **Pop. growth (1990-98):** 17.0%. **Area:** 217.8 sq. mi. **Employment:** 357,455 employed, 2.9% unemployed. **Per capita income (MSA):** $25,420; % increase, 1996-97: 10.6.
History: first permanent settlement 1835; capital of Rep. of Texas 1838; named after Stephen Austin; inc. 1840.
Transportation: 1 intl. airport; 4 railroads. **Communications:** 7 TV, 20 radio stations. **Medical facilities:** 12 hosp. **Educational facilities:** 8 univ. and colleges. **Further information:** Chamber of Commerce, PO Box 1967, Austin, TX 78767.
Websites: http://www.ci.austin.tx.us
http://www.austin-chamber.org

Bakersfield, California

Population: 210,284 (79); **Pop. density:** 2,291 per sq. mi; **Pop. growth (1990-98):** 19.3%. **Area:** 91.8 sq. mi. **Employment:** 88,705 employed, 9.0% unemployed. **Per capita income (MSA):** $18,319; % increase, 1996-97: 4.0.
History: named after Col. Thomas Baker, an early settler; inc. 1898.
Transportation: 1 airport; 3 railroads; Amtrak; Greyhound buses; local bus system. **Communications:** 5 TV, 34 radio stations. **Medical facilities:** 6 major hosp.; 9 convalescent, 1 psychiatric, 3 physical rehab., 5 urgent care facilities; 3 clinics.

Educational facilities: 1 univ., 1 community college, 9 vocational schools, 1 adult school, 1 college of law, 56 elem., 17 junior high, 14 high schools. **Further information:** Greater Bakersfield Chamber of Commerce, 1033 Truxtun Ave., PO Box 1947, Bakersfield, CA 93303.
Website: http://www.bakersfield.org/chamber

Baltimore, Maryland

Population: 645,593 (16); **Pop. density:** 7,990 per sq. mi; **Pop. growth (1990-98):** −12.3%. **Area:** 80.8 sq. mi. **Employment:** 276,623 employed, 9.0% unemployed. **Per capita income (MSA):** $27,770; % increase, 1996-97: 5.5.
History: founded by Maryland legislature 1729; inc. 1797; bombing of Ft. McHenry (1814) inspired Francis Scott Key to write "Star-Spangled Banner"; birthplace of America's railroads 1828; rebuilt after fire 1904; site of National Aquarium 1981.
Transportation: 1 major airport; 3 railroads; bus system; subway system; light rail system; Inner Harbor water taxi system; 2 underwater tunnels. **Communications:** 5 TV, 33 radio stations. **Medical facilities:** 29 hosp.; 2 major medical centers. **Educational facilities:** over 30 univ. and colleges; 183 pub. schools. **Further information:** Greater Baltimore Committee, 111 S. Calvert St., Ste. 1700, Baltimore, MD 21202-6180.
Websites: http://www.ci.baltimore.md.us
http://www.gbc.org

Baton Rouge, Louisiana

Population: 211,551 (77); **Pop. density:** 2,863 per sq. mi; **Pop. growth (1990-98):** −3.6%. **Area:** 73.9 sq. mi. **Employment:** 112,599 employed, 5.0% unemployed. **Per capita income (MSA):** $22,408; % increase, 1996-97: 3.7.
History: claimed by Spain at time of Louisiana Purchase 1803; est. independence by rebellion 1810; inc. as town 1817; became state capital 1849; Union-held most of Civil War.
Transportation: 1 airport, 5 airlines; 1 bus line; 3 railroad trunk lines. **Communications:** 5 TV, 19 radio stations. **Medical facilities:** 5 hosp. **Educational facilities:** 2 univ.; 92 pub., 39 private schools. **Further information:** Chamber of Commerce, PO Box 3217, Baton Rouge, LA 70821.
Websites: http://www.baton-rouge.com/BatonRouge
http://www.brbusiness.org
http://www.brchamber.org

Birmingham, Alabama

Population: 252,997 (66); **Pop. density:** 1,704 per sq. mi; **Pop. growth (1990-98):** −4.7%. **Area:** 148.5 sq. mi. **Employment:** 125,551 employed, 4.4% unemployed. **Per capita income (MSA):** $24,898; % increase, 1996-97: 5.1.
History: settled as a result of discovery of elements needed for steel production; inc. 1871; named after Great Britain's steelmaking center.
Transportation: 1 airport; 4 major rail freight lines, Amtrak; 1 bus line; 75 truck line terminals; 5 air cargo cos.; 7 barge lines; 4 interstate highways. **Communications:** 7 TV, 30 radio stations; 1 educational TV, 1 educational radio station. **Medical facilities:** Univ. of Alabama at Birmingham Medical Center; VA hosp. with organ transplant program; 15 other hosp. **Educational facilities:** 2 univ., 2 colleges, 3 junior colleges. **Further information:** Chamber of Commerce, 2027 First Ave. N, Birmingham, AL 35203.
Websites: http://www.birminghamchamber.com
http://www.ci.bham.al.us

Boston, Massachusetts

Population: 555,447 (20); **Pop. density:** 11,476 per sq. mi; **Pop. growth (1990-98):** −3.3%. **Area:** 48.4 sq. mi. **Employment:** 288,659 employed, 3.4% unemployed. **Per capita income (MSA):** $31,808; % increase, 1996-97: 6.3.
History: settled 1630 by John Winthrop; capital of Mass. Bay Colony; figured strongly in Am. Revolution, earning distinction as the "Cradle of Liberty"; inc. 1822.
Transportation: 1 major airport; 2 railroads; city rail and subway system; 3 underwater tunnels; port. **Communications:** 12 TV, 21 radio stations. **Medical facilities:** 31 hosp.; 8 major medical research centers. **Educational facilities:** 30 univ. and colleges. **Further information:** Greater Boston Convention and Visitors Bureau, 2 Copley Pl., Suite 105, Boston, MA 02116.
Websites: http://www.bostonusa.com
http://www.gbcc.org

Buffalo, New York

Population: 300,717 (56); **Pop. density:** 7,407 per sq. mi; **Pop. growth (1990-98):** −8.4%. **Area:** 40.6 sq. mi. **Employment:** 130,812 employed, 8.5% unemployed. **Per capita income (MSA):** $24,099; % increase, 1996-97: 3.1.

History: founded 1790 by the Dutch; raided twice by British, War of 1812; served as western terminus for Erie Canal, became a center for trade and manufacturing; inc. 1832; last stop on the Underground Railroad; key point for Canada-U.S. political, trade, and social relations.
Transportation: 1 intl. airport; 4 Class I railroads; Amtrak metro rail system; water service to Great Lakes-St. Lawrence Seaway system and Atlantic seaboard. **Communications:** 8 TV, 23 radio stations. **Medical facilities:** 14 hosp., 37 research centers. **Educational facilities:** 12 colleges and univ.; 111 pub. and private schools. **Further information:** Buffalo Niagara Partnership, 300 Main Place Tower, Buffalo, NY 14202-3797.
Websites: http://www.ci.buffalo.ny.us
http://www.thepartnership.org
http://www.rin.buffalo.edu

Charlotte, North Carolina

Population: 504,637 (25); **Pop. density:** 2,895 per sq. mi; **Pop. growth (1990-98):** 20.3%. **Area:** 174.3 sq. mi. **Employment:** 261,870 employed, 2.5% unemployed. **Per capita income (MSA):** $26,480; % increase, 1996-97: 7.5.
History: settled by Scotch-Irish immigrants 1740s; inc. 1768 and named after Queen Charlotte, George III's wife; scene of first major U.S. gold discovery 1799.
Transportation: 1 airport; 2 major railway lines; 2 bus lines; 238 trucking firms. **Communications:** 7 TV, 26 radio stations. **Medical facilities:** 12 hosp., 1 medical center. **Educational facilities:** 4 univ., 8 colleges, 85 elem. schools, 28 middle schools, 14 high schools. **Further information:** Chamber of Commerce, PO Box 32785, Charlotte, NC 28232.
Website: http://www.charlottechamber.org

Chesapeake, Virginia

Population: 199,564 (83); **Pop. density:** 586 per sq. mi; **Pop. growth (1990-98):** 31.3%. **Area:** 340.7 sq. mi. **Employment:** 100,226 employed, 2.7% unemployed. **Per capita income (MSA):** $21,983; % increase, 1996-97: 4.7.
History: region settled in 1620s with first English colonies on banks of Elizabeth River; home to Great Dismal Swamp Canal, first envisioned by George Washington in 1763; Battle of Great Bridge fought here Dec. 1775; inc. as a city 1963.
Transportation: Amtrak, freight rail service; bus service; 1 intl. airport, 2 regional airports; deepwater ports. **Communications:** 1 daily newspaper; 9 TV, 48 radio stations. **Medical facilities:** 1 hosp. **Educational facilities:** 9 colleges and univ.; 42 pub. schools. **Further information:** Hampton Roads, Chesapeake Div. Chamber of Commerce, 400 Volvo Pky., Chesapeake, VA 23320.
Website: http://www.chesapeake.va.us

Chicago, Illinois

Population: 2,802,079 (3); **Pop. density:** 12,333 per sq. mi; **Pop. growth (1990-98):** 0.7%. **Area:** 227.2 sq. mi. **Employment:** 1,230,164 employed, 5.7% unemployed. **Per capita income (MSA):** $30,717; % increase, 1996-97: 5.7.
History: site acquired from Indians 1795; significant white settlement began with opening of Erie Canal 1825; chartered as city 1837; boomed with arrival of railroads from east and canal to Mississippi R.; about one-third of city destroyed by fire 1871; major grain and livestock market.
Transportation: 3 airports; major railroad system, trucking industry. **Communications:** 9 TV, 31 radio stations. **Medical facilities:** over 123 hosp. **Educational facilities:** 95 insts. of higher learning. **Further information:** Chicagoland Chamber of Commerce, 1 IBM Plaza, Ste. 2800, Chicago, IL 60611.
Websites: http://www.ci.chi.il.us
http://www.chicagolandchamber.org

Cincinnati, Ohio

Population: 336,400 (51); **Pop. density:** 4,358 per sq. mi; **Pop. growth (1990-98):** −7.6%. **Area:** 77.2 sq. mi. **Employment:** 165,704 employed, 4.9% unemployed. **Per capita income (MSA):** $26,373; % increase, 1996-97: 6.0.
History: founded 1788 and named after the Society of Cincinnati, an organization of Revolutionary War officers; chartered as village 1802; inc. as city 1819.
Transportation: 1 intl. airport; 3 railroads; 1 bus system. **Communications:** 9 TV, 27 radio stations. **Medical facilities:** 32 hosp.; Children's Hosp. Medical Center; VA hosp. **Educational facilities:** 4 univ.; 11 colleges, 8 technical & 2-year colleges. **Further information:** Chamber of Commerce, 300 Carew Tower, 441 Vine St., Cincinnati, OH 45202.
Websites: http://www.gccc.com
http://www.cincinnatigov.com

Cleveland, Ohio

Population: 495,817 (28); **Pop. density:** 6,439 per sq. mi; **Pop. growth (1990-98):** –1.9%. **Area:** 77 sq. mi. **Employment:** 186,523 employed, 8.5% unemployed. **Per capita income (MSA):** $27,314; % increase, 1996-97: 4.7.

History: surveyed in 1796; given recognition as village 1815, inc. as city 1836; annexed Ohio City 1854.

Transportation: 1 intl. airport; rail service; major port; rapid transit system. **Communications:** 9 TV, 21 radio stations. **Medical facilities:** 14 hosp. **Educational facilities:** 8 univ. and colleges; 127 pub. schools. **Further information:** Greater Cleveland Growth Assn., 200 Tower City Center, 50 Pub. Square, Cleveland, OH 44113-2291.

Websites: http://www.cleveland.oh.us
http://www.clevelandgrowth.com

Colorado Springs, Colorado

Population: 344,987 (48); **Pop. density:** 1,883 per sq. mi; **Pop. growth (1990-98):** 23.0%. **Area:** 183.2 sq. mi. **Employment:** 180,835 employed, 4.5% unemployed. **Per capita income (MSA):** $23,493; % increase, 1996-97: 7.2.

History: city founded in 1871 at the foot of Pike's Peak; inc. 1872.

Transportation: 1 municipal airport; 1 bus line. **Communications:** 9 TV, 28 radio stations. **Medical facilities:** 7 hosp. **Educational facilities:** 11 univ., 12 colleges. **Further information:** Chamber of Commerce, PO Box B, Colorado Springs, CO 80901.

Websites: http://www.coloradosprings-travel.com/ cscvb
http://www.introColoradoSprings
http://www.cscc.org

Columbus, Ohio

Population: 670,234 (15); **Pop. density:** 3,511 per sq. mi; **Pop. growth (1990-98):** 5.9%. **Area:** 190.9 sq. mi. **Employment:** 371,384 employed, 3.0% unemployed. **Per capita income (MSA):** $25,728; % increase, 1996-97: 6.0.

History: first settlement 1797; laid out as new capital 1812 with current name; became city 1834.

Transportation: 6 airports; 3 railroads; 4 intercity bus lines. **Communications:** 8 TV, 25 radio stations. **Medical facilities:** 18 hosp. **Educational facilities:** 11 univ. and colleges; 8 technical/2-year schools; 141 pub. schools (89 elem., 26 middle, 17 high, 9 magnet). **Further information:** Chamber of Commerce, 37 N. High St., Columbus, OH 43215.

Website: http://www.columbus.org

Corpus Christi, Texas

Population: 281,453 (59); **Pop. density:** 2,085 per sq. mi; **Pop. growth (1990-98):** 9.3%. **Area:** 135 sq. mi. **Employment:** 123,869 employed, 6.7% unemployed. **Per capita income (MSA):** $19,781; % increase, 1996-97: 5.6.

History: settled 1839 and inc. 1852.

Transportation: 1 intl. airport; 2 bus lines, metro bus system; 3 freight railroads. **Communications:** 6 TV, 17 radio stations. **Medical facilities:** 14 hosp. including a children's center. **Educational facilities:** 1 univ., 1 college. **Further information:** Greater Corpus Christi Business Alliance, PO Box 640, Corpus Christi, TX 78403.

Website: http://www.cctexas.org

Dallas, Texas

Population: 1,075,894 (9); **Pop. density:** 3,142 per sq. mi; **Pop. growth (1990-98):** 6.8%. **Area:** 342.4 sq. mi. **Employment:** 632,854 employed, 4.2% unemployed. **Per capita income (MSA):** $30,481; % increase, 1996-97: 9.5.

History: first settled 1841; platted 1846; inc. 1871; developed as the financial and commercial center of Southwest; headquarters of regional Federal Reserve Bank; major center for distribution and high-tech manufacturing.

Transportation: 1 intl. airport, 1 regional airport; Amtrak; transit system. **Communications:** 14 TV, 66 radio stations. **Medical facilities:** 16 general hosp.; major medical center. **Educational facilities:** 219 pub. schools, 11 univ. and colleges, 3 community college campuses. **Further information:** Greater Dallas Chamber, Resource Center, 1201 Elm, Ste. 2000, Dallas, TX 75270.

Websites: http://www.dallaschamber.org
http://www.ci.dallas.tx.us

Denver, Colorado

Population: 499,055 (27); **Pop. density:** 3,255 per sq. mi; **Pop. growth (1990-98):** 6.7%. **Area:** 153.3 sq. mi. **Employment:** 272,450 employed, 4.1% unemployed. **Per capita income (MSA):** $30,743; % increase, 1996-97: 8.1.

History: settled 1858 by gold prospectors and miners; inc. 1861; became territorial capital 1867; growth spurred by gold and silver boom; became financial, industrial, cultural center of Rocky Mt. region.

Transportation: 1 intl. airport, 3 corporate reliever airports; 5 rail freight lines, Amtrak; 1 bus line. **Communications:** 14 TV, 29 radio stations. **Medical facilities:** 20 hosp. **Educational facilities:** 15 four-yr. colleges and univ.; 8 two-yr. and community colleges. **Further information:** Denver Metro Chamber of Commerce, 1445 Market St., Denver, CO 80202-1729.

Website: http://www.den-chamber.org

Des Moines, Iowa

Population: 191,293 (91); **Pop. density:** 2,540 per sq. mi; **Pop. growth (1990-98):** –1.0%. **Area:** 75.3 sq. mi. **Employment:** 117,603 employed, 2.7% unemployed. **Per capita income (MSA):** $27,403; % increase, 1996-97: 5.9.

History: Fort Des Moines built 1843; settled and inc. 1851; chartered as city 1857.

Transportation: 1 intl. airport; 4 bus lines; 4 railroads; metro bus system. **Communications:** 5 TV, 17 radio stations. **Medical facilities:** 8 hosp. **Educational facilities:** 2 univ., 5 colleges. **Further information:** Greater Des Moines Chamber of Commerce Federation, 601 Locust St., Ste. 100, Des Moines, IA 50309.

Websites: http://www.dmchamber.com
http://www.ci.des-moines.ia.us

Detroit, Michigan

Population: 970,196 (10); **Pop. density:** 6,995 per sq. mi; **Pop. growth (1990-98):** –5.6%. **Area:** 138.7 sq. mi. **Employment:** 365,606 employed, 7.2% unemployed. **Per capita income (MSA):** $27,619; % increase, 1996-97: 4.4.

History: founded by French 1701; controlled by British 1760; acquired by U.S. 1796; destroyed by fire 1805; inc. as city 1824; capital of state 1837-47; auto manufacturing began 1899.

Transportation: 1 intl. airport; 5 railroads; major intl. port; pub. transit system. **Communications:** 11 TV, 37 radio stations. **Medical facilities:** 28 hosp.; major medical center. **Educational facilities:** 19 univ. and colleges. **Further information:** Detroit Regional Chamber, One Woodward Ave., PO Box 33840, Detroit, MI 48232-0840.

Website: http://www.detroitchamber.com

El Paso, Texas

Population: 615,032 (17); **Pop. density:** 2,506 per sq. mi; **Pop. growth (1990-98):** 19.3%. **Area:** 245.4 sq. mi. **Employment:** 234,432 employed, 9.8% unemployed. **Per capita income (MSA):** $15,216; % increase, 1996-97: 6.2.

History: first settled 1827; inc. 1873; arrival of railroad 1881 boosted city's population and industries.

Transportation: 1 intl. airport; 3 rail providers; 2 interstate highways; 4 intl. ports of entry. **Communications:** 12 TV, 20 radio stations. **Medical facilities:** 6 hosp.; 3 rehabilitation, 11 specialty centers. **Educational facilities:** 1 univ., 3 colleges (1 grad. only). **Further information:** Greater El Paso Chamber of Commerce, 10 Civic Center Plaza, El Paso, TX 79901.

Website: http://www.elpaso.org

Fort Wayne, Indiana

Population: 185,716 (98); **Pop. density:** 2,962 per sq. mi; **Pop. growth (1990-98):** –5.1%. **Area:** 62.7 sq. mi. **Employment:** 95,160 employed, 3.5% unemployed. **Per capita income (MSA):** $24,891; % increase, 1996-97: 5.3.

History: French fort 1680; U.S. fort 1794; settled by 1832; inc. 1840 prior to Wabash-Erie canal completion 1843.

Transportation: 2 airports; 3 railroads; 6 bus lines. **Communications:** 5 TV, 13 radio stations. **Medical facilities:** 3 regional hosp.; VA hosp. **Educational facilities:** 5 univ., 4 colleges, 2 bus. schools; 80 pub. schools. **Further information:** Chamber of Commerce, 826 Ewing Street, Fort Wayne, IN 46802-2182.

Websites: http://www.ft-wayne.in.us
http://www.fwchamber.org

Fort Worth, Texas

Population: 491,801 (29); **Pop. density:** 1,750 per sq. mi; **Pop. growth (1990-98):** 9.9%. **Area:** 281.1 sq. mi. **Employment:** 255,106 employed, 4.3% unemployed. **Per capita income (MSA):** $25,150; % increase, 1996-97: 8.1.

History: established as military post 1849; inc. 1873; oil discovered 1917.

Transportation: 1 intl. airport; 9 major railroads, Amtrak; local bus service; 2 transcontinental, 2 intrastate bus lines. **Communications:** 14 TV, 11 local radio stations. **Medical facilities:** 25 hosp.; 1 children's hosp.; 4 government hosp. **Educational facil-**

ities: 8 univ. and colleges. **Further information:** Chamber of Commerce, 777 Taylor St. #900, Fort Worth, TX 76102. **Website:** http://www.fortworthchamber.com

Fremont, California

Population: 204,298 (81); **Pop. density:** 2,653 per sq. mi; **Pop. growth (1990-98):** 17.9%. **Area:** 77 sq. mi. **Employment:** 103,084 employed, 2.8% unemployed. **Per capita income (MSA):** $31,338; % increase, 1996-97: 6.7.
History: area first settled by Spanish 1769; inc. 1956 with consolidation of 5 communities.
Transportation: intracity bus line; Bay Area Rapid Transit System (southern terminal). **Communications:** 1 radio station. **Medical facilities:** 1 hosp. **Educational facilities:** 1 community college; 43 pub. schools. **Further information:** Chamber of Commerce, 39488 Stevenson Place, Suite 100, Fremont, CA 94539.
Website: http://www.fremontbusiness.com

Fresno, California

Population: 398,133 (40); **Pop. density:** 4,017 per sq. mi; **Pop. growth (1990-98):** 12.4%. **Area:** 99.1 sq. mi. **Employment:** 166,854 employed, 12.8% unemployed. **Per capita income (MSA):** $18,958; % increase, 1996-97: 3.3.
History: founded 1872; inc. as city 1885.
Transportation: 1 municipal airport; Amtrak; 1 bus line; intracity bus system. **Communications:** 13 TV, 23 radio stations. **Medical facilities:** 17 general hosp. **Educational facilities:** 9 colleges; 102 pub. schools. **Further information:** Chamber of Commerce, PO Box 1469, Fresno, CA 93716-1469.
Websites: http://fresno-online.com/cvb
http://www.fresnochamber.com

Garland, Texas

Population: 193,408 (90); **Pop. density:** 3,375 per sq. mi; **Pop. growth (1990-98):** 7.1%. **Area:** 57.3 sq. mi. **Employment:** 119,387 employed, 2.8% unemployed. **Per capita income (MSA):** $30,481; % increase, 1996-97: 9.5.
History: settled 1850s; inc. 1891.
Transportation: 30 min. from Dallas/Ft. Worth Intl. Airport; 2 railroads. **Communications:** 14 local TV (Dallas/Ft. Worth), 25+ radio stations. **Medical facilities:** 2 hosp.; 329 beds. **Educational facilities:** 3 univ., 2 community colleges; 59 pub. schools. **Further information:** Chamber of Commerce, 914 S. Garland Ave., Garland, TX 75040.
Website: http://www.garlandchamber.com

Glendale, Arizona

Population: 193,482 (89); **Pop. density:** 3,424 per sq. mi; **Pop. growth (1990-98):** 31.6%. **Area:** 56.5 sq. mi. **Employment:** 103,285 employed, 2.7% unemployed. **Per capita income (MSA):** $24,137; % increase, 1996-97: 8.2.
History: est. 1892; inc. 1910.
Transportation: 1 local airport, 30 min. from Phoenix Sky Harbor Intl. Airport. **Communications:** 12 TV stations, 40 radio stations. **Medical facilities:** 3 hosp. **Educational facilities:** 12 institutes of higher education, 9 pub. school districts. **Further information:** Chamber of Commerce, PO Box 249, 7105 N. 59th Ave., Glendale, AZ 85311.
Website: http://www.ci.glendale.az.us

Glendale, California

Population: 185,086 (100); **Pop. density:** 6,068 per sq. mi; **Pop. growth (1990-98):** 2.8%. **Area:** 30.5 sq. mi. **Employment:** 89,870 employed, 6.2% unemployed. **Per capita income (MSA):** $25,719; % increase, 1996-97: 4.8.
History: became a town in 1887; inc. 1906.
Transportation: near Los Angeles Intl. airport; commuter trains, Amtrak; bus system. **Communications:** 21 TV, 70 radio stations. **Medical facilities:** 3 hosp; other facilities. **Educational facilities:** 1 community college; 26 pub. schools. **Further information:** City of Glendale Public Information Officer, 613 E. Broadway, Glendale, CA 91206.
Website: http://www.ci.glendale.ca.us

Grand Rapids, Michigan

Population: 185,437 (99); **Pop. density:** 4,186 per sq. mi; **Pop. growth (1990-98):** −2.0%. **Area:** 44.3 sq. mi. **Employment:** 108,030 employed, 4.1% unemployed. **Per capita income (MSA):** $24,960; % increase, 1996-97: 6.1.
History: originally site of Ottawa Indian village; trading post 1826; became lumbering center and incorporated city 1850.
Transportation: 1 intl. airport; 3 rail carriers; Amtrak, Greyhound bus line; transit bus system. **Communications:** 7 TV, 34 radio stations. **Medical facilities:** 11 hosp. **Educational facili-**

ties: 15 colleges; 19 pub. schools, 9 charter schools. **Further information:** Chamber of Commerce, 111 Pearl St. NW, Grand Rapids, MI 49503
Website: http://www.grandrapids.org

Greensboro, North Carolina

Population: 197,910 (84); **Pop. density:** 2,480 per sq. mi; **Pop. growth (1990-98):** 6.9%. **Area:** 79.8 sq. mi. **Employment:** 110,324 employed, 2.8% unemployed. **Per capita income (MSA):** $25,441; % increase, 1996-97: 5.8.
History: settled 1749; site of Revolutionary War conflict 1781 between Nathanael Greene and Cornwallis; inc. 1807.
Transportation: 1 regional airport; 2 railroads; Trailways/Greyhound bus service. **Communications:** all cable TV stations; 11 radio stations. **Medical facilities:** 4 hosp. **Educational facilities:** 2 univ., 3 colleges; 94 pub. schools. **Further information:** Chamber of Commerce, PO Box 3246, Greensboro, NC 27402.
Websites: http://www.ci.greensboro.nc.us
http://www.greensboro.org

Hialeah, Florida

Population: 211,392 (78); **Pop. density:** 9,609 per sq. mi; **Pop. growth (1990-98):** 12.4%. **Area:** 22 sq. mi. **Employment:** 96,577 employed, 6.7% unemployed. **Per capita income (MSA):** $21,688; % increase, 1996-97: 3.4.
History: founded 1917, inc. 1925; industrial and residential city NW of Miami; Hialeah Park Horse Racing Track.
Transportation: 5 mi from Miami Intl. Airport; access to Port of Miami; Amtrak; 2 rail freight lines; Metrorail, Metrobus systems. **Communications:** 5 TV, 7 radio stations. **Medical facilities:** 4 hosp. (30 more in the area). **Educational facilities:** 8 univ. and colleges, 25 pub., 39 private schools. **Further information:** Hialeah-Dade Development, Inc., 501 Palm Ave., Hialeah, FL 33010.
Website: http://www.ci.hialeah.us

Honolulu, Hawaii

Population: 395,789 (41); **Pop. density:** 4,780 per sq. mi; **Pop. growth (1990-98):** 5.0%. **Area:** 82.8 sq. mi. **Employment (MSA):** 406,097 employed, 5.4% unemployed. **Per capita income (MSA):** $27,259; % increase, 1996-97: 2.3.
History: harbor entered by Europeans 1778; declared capital of kingdom by King Kamehameha III 1850; Pearl Harbor naval base attacked by Japanese Dec. 7, 1941.
Transportation: 1 major airport; large, active port for passengers and cargo. **Communications:** 10 TV, 30 radio stations. **Medical facilities:** 13 major medical centers. **Educational facilities:** 4 univ., 5 colleges; 253 pub. schools, 98 private schools. **Further information:** Hawaii Visitors and Convention Bureau, 2270 Kalakaua Avenue, Honolulu, HI 96815.
Websites: http://www.co.honolulu.hi.us
http://www.gohawaii.com

Houston, Texas

Population: 1,786,691 (4); **Pop. density:** 3,309 per sq. mi; **Pop. growth (1990-98):** 8.0%. **Area:** 539.9 sq. mi. **Employment (MSA):** 969,755 employed, 5.0% unemployed. **Per capita income (MSA):** $28,977; % increase, 1996-97: 8.7.
History: founded 1836; inc. 1837; capital of Repub. of Texas 1837-39; developed rapidly after construction of channel to Gulf of Mexico 1914; world center of oil and natural gas technology.
Transportation: 3 commercial airports; 2 mainline railroads; major bus transit system; major intl. port. **Communications:** 15 TV, 54 radio stations. **Medical facilities:** 62 hosp.; major medical center. **Educational facilities:** 24 univ. and colleges. **Further information:** Greater Houston Partnership, 1200 Smith St., Houston, TX 77002-4309.
Websites: http://www.houston.org
http://www.ci.houston.tx.us

Huntington Beach, California

Population: 195,316 (87); **Pop. density:** 7,398 per sq. **Pop. growth (1990-98):** 7.6%. **Area:** 26.4 sq. mi. **Employment:** 116,945 employed, 2.2% unemployed. **Per capita income (MSA):** $30,115; % increase, 1996-97: 6.8.
History: settled in early 1880s; inc. 1909; oil discovered 1920, led to city's development.
Transportation: 1 railroad; 2 bus lines. **Communications:** 2 TV stations. **Medical facilities:** 2 hosp. **Educational facilities:** 1 community college; 34 pub. schools. **Further information:** Chamber of Commerce, Seacliff Office Park, 2100 Main St., #200, Huntington Beach, CA 92648.
Websites: http://www.thebeach.com/cities/hb
http://www.hbchamber.org

Indianapolis, Indiana

Population: 741,304 (13); **Pop. density:** 2,049 per sq. mi;
Pop. growth (1990-98): 1.4%. **Area:** 361.7 sq. mi. **Employment:** 409,378 employed, 2.9% unemployed. **Per capita in-
come (MSA):** $26,662; % increase, 1996-97: 5.7.
History: settled 1820; became capital 1825.
Transportation: 1 intl. airport; 5 railroads; 3 interstate bus
lines. **Communications:** 10 TV, 27 radio stations. **Medical facilities:** 17 hosp.; 1 major medical and research center. **Educational facilities:** 8 univ. and colleges; major pub. library system.
Further information: Chamber of Commerce, 320 N. Meridian
St., Indianapolis, IN 46204.
Websites: http://www.ci.indianapolis.in.us
http://www.indychamber.com

Jackson, Mississippi

Population: 188,419 (94); **Pop. density:** 1,729 per sq. mi;
Pop. growth (1990-98): −6.8%. **Area:** 109 sq. mi. **Employment:** 94,977 employed, 4.5% unemployed. **Per capita income
(MSA):** $21,057; % increase, 1996-97: 4.9.
History: originally known as Le Fleur's Bluff; selected as capital
1822 and named for Andrew Jackson; inc. 1823; scene of secession convention 1861; captured by Sherman 1863.
Transportation: 7 airlines; 1 bus line; 2 railroads; 2 freight
carriers. **Communications:** 7 TV, 28 radio stations. **Medical facilities:** 12 hosp. incl. a VA facility. **Educational facilities:** 2
univ., 4 colleges; 60 pub. schools. **Further information:** Metro
Jackson Chamber of Commerce, PO Box 22548, Jackson, MS
39225-2548.
Website: http://www.metrojackson.com

Jacksonville, Florida

Population: 693,630 (14); **Pop. density:** 914 per sq. mi; **Pop.
growth (1990-98):** 9.2%. **Area:** 758.7 sq. mi. **Employment:**
367,357 employed, 3.2% unemployed. **Per capita income
(MSA):** $24,751; % increase, 1996-97: 6.9.
History: settled 1816 as Cowford; renamed after Andrew
Jackson 1822; inc. 1832; rechartered 1851; scene of conflicts in
Seminole and Civil wars.
Transportation: 1 intl. airport; 3 railroads; 2 interstate bus
lines. **Communications:** 6 TV, 34 radio stations. **Medical facilities:** 11 hosp. **Educational facilities:** 3 univ., 5 colleges; 148
pub. schools, 90 private schools. **Further information:** Chamber of Commerce, 3 Independent Drive, Jacksonville, FL 32202.
Websites: http://www.jacksonvillechamber.org
http://wwwjacksonville.com
http://www.coj.net

Jersey City, New Jersey

Population: 232,429 (71); **Pop. density:** 15,599 per sq. mi;
Pop. growth (1990-98): 1.7%. **Area:** 14.9 sq. mi. **Employment:**
101,543 employed, 9.0% unemployed. **Per capita income
(MSA):** $24,943; % increase, 1996-97: 3.5.
History: site bought from Indians 1630; chartered as town by
British 1668; scene of Revolutionary War conflict 1779; chartered
under present name 1838; important station on Underground
Railroad.
Transportation: Intercity bus and subway system; ferry ser-
vice to Manhattan. **Communications:** see New York, NY. **Med-
ical facilities:** 4 hosp. **Educational facilities:** 3 colleges.
Further information: Hudson County Chamber of Commerce,
253 Washington St., Jersey City, NJ 07302.
Website: http://www.jerseycitynet.com

Kansas City, Missouri

Population: 441,574 (33); **Pop. density:** 1,418 per sq. mi;
Pop. growth (1990-98): 1.6%. **Area:** 311.5 sq. mi. **Employment:** 243,358 employed, 4.7% unemployed. **Per capita in-
come (MSA):** $26,627; % increase, 1996-97: 6.0.
History: settled by 1838 at confluence of the Missouri and
Kansas rivers; inc. 1851.
Transportation: 1 intl. airport; a major rail center; 191 trunk
lines; several barge cos. **Communications:** 7 TV, 29 radio stations. **Medical facilities:** 14 hosp.; VA facility. **Educational facilities:** 9 univ. and colleges. **Further information:** Greater
Kansas City Chamber of Commerce, 911 Main St., Ste. 2600,
Kansas City, MO 64105.
Websites: http://www.kansascity.com
http://www.kcchamber.com
http://www.kcmo.org

Las Vegas, Nevada

Population: 404,288 (37); **Pop. density:** 4,853 per sq. mi;
Pop. growth (1990-98): 56.2%. **Area:** 84 sq. mi. **Employment:**

210,980 employed, 4.1% unemployed. **Per capita income
(MSA):** $25,250; % increase, 1996-97: 8.3.
History: occupied by Mormons 1855-57; bought by railroad
1903; city of Las Vegas inc. 1911; gambling legalized 1931.
Transportation: 1 intl. airport; 2 railroads; bus system. **Communications:** 7 TV, 30 radio stations. **Medical facilities:** 12
hosp. **Educational facilities:** 1 univ., 5 colleges; 219 pub.
schools in area. **Further information:** Chamber of Commerce,
3720 Howard Hughes Parkway, Las Vegas, NV 89109.
Website: http://www.lvchamber.com

Lexington, Kentucky

Population: 241,749 (68); **Pop. density:** 850 per sq. mi; **Pop.
growth (1990-98):** 7.3%. **Area:** 284.5 sq. mi. **Employment:**
138,694 employed, 2.0% unemployed. **Per capita income
(MSA):** $24,838; % increase, 1996-97: 7.4.
History: site was founded and named in 1775 by hunters who
heard of the Revolutionary War battle at Lexington, Mass.; settled 1779; chartered 1782; inc. as a city 1832.
Transportation: 7 comm. airlines; 2 railroads; city buses.
Communications: 5 TV, 21 radio stations. **Medical facilities:** 5
general, 5 specialized hosp. **Educational facilities:** 2 univ., 4
colleges. **Further information:** Greater Lexington Chamber of
Commerce, 330 E. Main St., Lexington, KY 40507.
Website: http://www.lexchamber.com

Lincoln, Nebraska

Population: 213,088 (76); **Pop. density:** 3,366 per sq. mi;
Pop. growth (1990-98): 11.0%. **Area:** 63.3 sq. mi. **Employment:** 126,192 employed, 2.4% unemployed. **Per capita income (MSA):** $24,602; % increase, 1996-97: 5.9.
History: originally called Lancaster; chosen state capital
1867, renamed after Abraham Lincoln; inc. 1869.
Transportation: 1 airport; Greyhound; Amtrak, 2 railroads.
Communications: 2 TV, 13 radio stations. **Medical facilities:** 5
hosp. including VA, rehabilitation facilities. **Educational facilities:** 3 univ., 3 voc.-tech./business colleges; 48 pub., 15 private
schools. **Further information:** Chamber of Commerce, PO Box
83006, Lincoln, NE 68501.
Websites: http://www.lincoln.org
http://www.lcoc.com

Long Beach, California

Population: 430,905 (35); **Pop. density:** 8,618 per sq. mi;
Pop. growth (1990-98): 0.4%. **Area:** 50 sq. mi. **Employment:**
203,358 employed, 6.1% unemployed. **Per capita income
(MSA):** $25,719; % increase, 1996-97: 4.8.
History: settled as early as 1784 by Spanish; by 1884 present
site developed on harbor; inc. 1888; oil discovered 1921.
Transportation: 1 airport; 3 railroads; major intl. port; 4 bus
co. with 40 bus lines, light rail service. **Communications:** 1 radio
station, 1 CATV franchise. **Medical facilities:** 10 hosp. **Educational facilities:** 1 univ., 1 community college (2 campuses); 87
pub. schools in district. **Further information:** Long Beach City
Hall, 333 W. Ocean Blvd., Long Beach, CA 90802
Websites: http://www.ci.long-beach.ca.us
http://www.lbchamber.com

Los Angeles, California

Population: 3,597,556 (2); **Pop. density:** 7,666 per sq. mi;
Pop. growth (1990-98): 3.2%. **Area:** 469.3 sq. mi. **Employment:** 1,723,367 employed, 7.4% unemployed. **Per capita income (MSA):** $25,719 % increase, 1996-97: 4.8.
History: founded by Spanish 1781; captured by U.S. 1846;
inc. 1850; Hollywood a district of L.A.
Transportation: 1 intl. airport; 3 railroads; major freeway system; intracity transit system. **Communications:** 21 TV, 70 radio
stations. **Medical facilities:** 822 hosp. and clinics in metro. area.
Educational facilities: 192 univ. and colleges (incl. junior, community, and other); 1,678 pub. schools; 1,470 private schools.
Further information: Los Angeles Area Chamber of Commerce, 350 S. Bixel St., PO Box 513696, Los Angeles, CA
90051-1696.
Websites: http://www.ci.la.ca.us
http://www.lachamber.org

Louisville, Kentucky

Population: 255,045 (64); **Pop. density:** 4,107 per sq. mi;
Pop. growth (1990-98): −5.4%. **Area:** 62.1 sq. mi. **Employment:** 125,903 employed, 4.3% unemployed. **Per capita income (MSA):** $25,493; % increase, 1996-97: 5.4.
History: settled 1778; named for Louis XVI of France; inc.
1828; base for Union forces in Civil War.
Transportation: 1 municipal airport, 1 private-craft airport; 1
terminal, 4 trunk-line railroads; metro bus line, Greyhound sta-

tion; 5 barge lines. **Communications:** 5 TV, 21 radio stations, 2 educational. **Medical facilities:** 23 hosp. **Educational facilities:** 10 univ. and colleges, 9 business colleges and technical schools. **Further information:** Greater Louisville, Inc. Metro Chamber of Commerce, 600 W. Main St., Louisville, KY 40202.

Website: http://www.greaterlouisville.com

Lubbock, Texas

Population: 190,974 (92); **Pop. density:** 1,835 per sq. mi; **Pop. growth (1990-98):** 2.6%. **Area:** 104.1 sq. mi. **Employment:** 100,739 employed, 3.4% unemployed. **Per capita income (MSA):** $22,032; % increase, 1996-97: 4.7.

History: settled 1879; laid out 1891; inc. 1909 through merger of two towns.

Transportation: 1 intl. airport; 2 railroads, bus line. **Communications:** 5 TV, 18 radio stations. **Medical facilities:** 7 hosp. **Educational facilities:** 3 univ., 1 junior college; 51 pub. schools. **Further information:** Chamber of Commerce, PO Box 561, Lubbock, TX 79408.

Websites: http://www.ci.lubbock.tx.us
http://www.lubbock.org

Madison, Wisconsin

Population: 209,306 (80); **Pop. density:** 3,621 per sq. mi; **Pop. growth (1990-98):** 9.7%. **Area:** 57.8 sq. mi. **Employment:** 129,224 employed, 1.6% unemployed. **Per capita income (MSA):** $27,361; % increase, 1996-97: 5.4.

History: first white settlement 1832; selected as site for state capital, named after James Madison, 1836; chartered 1856.

Transportation: 1 airport, 7 airlines; 1 intracity, 3 intercity bus systems; 3 freight rail lines. **Communications:** 5 TV, 24 radio stations, 3 cable providers. **Medical facilities:** 6 hosp., 92 clinics. **Educational facilities:** 7 colleges and univ., including main branch of Univ. of Wisconsin; 29 elem. schools, 12 middle schools, 5 high schools. **Further information:** Greater Madison Chamber of Commerce, PO Box 71, Madison, WI 53701-0071.

Websites: http://www.ci.madison.wi.us
http://www.greatermadisonchamber.com

Memphis, Tennessee

Population: 603,507 (18); **Pop. density:** 2,142 per sq. mi; **Pop. growth (1990-98):** −2.4%. **Area:** 281.8 sq. mi. **Employment:** 302,490 employed, 4.5% unemployed. **Per capita income (MSA):** $25,905; % increase, 1996-97: 5.5.

History: French, Spanish, and U.S. forts by 1797; settled by 1819; inc. as town 1826, as city 1840; surrendered charter to state 1879 after yellow fever epidemics; rechartered as city 1893.

Transportation: 1 intl. airport; 5 railroads; bus system. **Communications:** 7 TV, 32 radio stations. **Medical facilities:** 19 hosp. **Educational facilities:** 17 univ. and colleges; 209 pub., 76 private schools. **Further information:** Memphis Area Chamber of Commerce, 22 N. Front St., Ste. 200, PO Box 224, Memphis, TN 38101-0224.

Websites: http://www.memphis.acn.net
http://www.memphis.tn.us
http://www.memphischamber.com

Mesa, Arizona

Population: 360,076 (46); **Pop. density:** 3,316 per sq. mi; **Pop. growth (1990-98):** 24.5%. **Area:** 108.6 sq. mi. **Employment:** 190,168 employed, 2.3% unemployed. **Per capita income (MSA):** $24,137; % increase, 1996-97: 8.2.

History: founded by Mormons 1878; inc. 1883; 13 mi. from Phoenix; population boomed fivefold 1960-80.

Transportation: near Sky Harbor Intl. Airport in Phoenix; 2 railroads; bus line. **Medical facilities:** 4 major hosp. **Educational facilities:** 1 univ., 3 colleges; 70 pub. schools. **Further information:** Convention and Visitor's Bureau, 120 N. Center, Mesa, AZ 85201.

Websites: http://www.ci.mesa.az.us
http://www.mesacvb.com

Miami, Florida

Population: 368,624 (44); **Pop. density:** 10,355 per sq. mi; **Pop. growth (1990-98):** 2.8%. **Area:** 35.6 sq. mi. **Employment:** 162,951 employed, 9.3% unemployed. **Per capita income (MSA):** $21,688; % increase, 1996-97: 3.4.

History: site of fort 1836; settlement began 1870; inc. 1896, modern city developed into financial and recreation center; land speculation in 1920s added to city's growth, as did Cuban, Central and South American, and Haitian immigration since 1960.

Transportation: 1 intl. airport; seaport; Amtrak, transit rail system; 2 bus lines; 65 truck lines. **Communications:** 9 commercial, 2 educational TV stations; 41 radio stations. **Medical fa-

cilities:** 36 hosp.; VA hosp. **Educational facilities:** 6 univ. and colleges. **Further information:** Miami-Dade Dept. of Planning, Development, and Regulation, Research Div., 111 NW 1st St., Ste. 1220, Miami, FL 33128.

Websites: http://ci.miami.fl.us
http://www.greatermiami.com
http://www.metro-dade.com

Milwaukee, Wisconsin

Population: 578,364 (19); **Pop. density:** 6,018 per sq. mi; **Pop. growth (1990-98):** −7.9%. **Area:** 96.1 sq. mi. **Employment:** 278,161 employed, 5.2% unemployed. **Per capita income (MSA):** $28,176; % increase, 1996-97: 5.4.

History: Indian trading post by 1674; settlement began 1835; inc. as city 1848; famous beer industry.

Transportation: 1 intl. airport; 3 railroads; major port; 4 bus lines. **Communications:** 12 TV, 37 radio stations. **Medical facilities:** 9 hosp.; major medical center. **Educational facilities:** 10 univ. and colleges, 159 pub. schools. **Further information:** Metropolitan Milwaukee Association of Commerce, 756 N. Milwaukee Street, Milwaukee, WI 53202.

Websites: http://www.ci.mil.wi.us
http://www.milwaukee.org

Minneapolis, Minnesota

Population: 351,731 (47); **Pop. density:** 6,407 per sq. mi; **Pop. growth (1990-98):** −4.5%. **Area:** 54.9 sq. mi. **Employment:** 203,461 employed, 2.6% unemployed. **Per capita income (MSA):** $30,123; % increase, 1996-97: 6.1.

History: site visited by Hennepin 1680; included in area of military reservations 1819; inc. 1867.

Transportation: 1 intl. airport; 5 railroads. **Communications:** 7 TV, 30 radio stations. **Medical facilities:** 7 hosp., incl. leading heart hosp. at Univ. of Minnesota. **Educational facilities:** 10 univ. and colleges; 802 pub., 26 private schools. **Further information:** City of Minneapolis Office of Pub. Affairs, 323M City Hall, 350 S. 5th St., Minneapolis, MN 55415.

Website: http://www.ci.minneapolis.mn.us

Mobile, Alabama

Population: 202,181 (82); **Pop. density:** 1,713 per sq. mi; **Pop. growth (1990-98):** 1.1%. **Area:** 118 sq. mi. **Employment:** 100,623 employed, 4.6% unemployed. **Per capita income (MSA):** $20,119; % increase, 1996-97: 5.4.

History: settled by French 1711; occupied by U.S. 1813; inc. as town 1814, as city 1819; only seaport of Alabama.

Transportation: 4 rail freight lines, Amtrak; 3 airlines; 65 truck lines; leading river system. **Communications:** 7 TV, 21 radio stations. **Medical facilities:** 9 hosp. **Educational facilities:** 3 univ., 3 colleges. **Further information:** Chamber of Commerce, PO Box 2187, Mobile, AL 36652.

Websites: http://www.ci.mobile.al.us
http://www.mobcham.org

Montgomery, Alabama

Population: 197,014 (85); **Pop. density:** 1,459 per sq. mi; **Pop. growth (1990-98):** 3.5%. **Area:** 135 sq. mi. **Employment:** 97,694 employed, 3.4% unemployed. **Per capita income (MSA):** $22,498; % increase, 1996-97: 4.6.

History: inc. as town 1819, as city 1837; became state capital 1846; first capital of Confederacy 1861.

Transportation: 4 airlines; 2 railroads; 2 bus lines; Alabama R. navigable to Gulf of Mexico. **Communications:** 4 TV, 2 CATV, 1 public TV, 16 radio stations. **Medical facilities:** 4 major hosp.; VA and 32 clinics. **Educational facilities:** 6 colleges and univ.; 48 pub., 30 private schools. **Further information:** Montgomery Area Chamber of Commerce, PO Box 79, Montgomery, AL 36101.

Website: http://www.montgomerychamber.com

Nashville, Tennessee

Population: 510,274 (24); **Pop. density:** 1,078 per sq. mi; **Pop. growth (1990-98):** 4.5%. **Area:** 473.3 sq. mi. **Employment:** 299,277 employed, 2.5% unemployed. **Per capita income (MSA):** $27,324; % increase, 1996-97: 7.1.

History: settled 1779; first chartered 1806; became permanent state capital 1843; home of Grand Ole Opry.

Transportation: 1 airport; 1 railroad; bus line; transit system of buses and trolleys. **Communications:** 11 TV, 34 radio stations. **Medical facilities:** 14 hosp.; VA and speech-hearing center. **Educational facilities:** 16 universities and colleges, 129 pub. schools. **Further information:** Chamber of Commerce, 161 4th Ave., Nashville, TN 37219.

Website: http://www.nashvillechamber.com

Newark, New Jersey

Population: 267,823 (60); **Pop. density:** 11,253 per sq. mi; **Pop. growth (1990-98):** –2.7%. **Area:** 23.8 sq. mi. **Employment:** 100,126 employed, 9.6% unemployed. **Per capita income (MSA):** $35,038; % increase, 1996-97: 5.0.

History: settled by Puritans 1666; used as supply base by Washington 1776; inc. as town 1833, as city 1836.

Transportation: 1 intl. airport; 1 intl. seaport, 3 railroads; bus system; subways. **Communications:** 5 TV, 5 radio stations within city limits, 1 daily newspaper, 8 weekly papers. **Medical facilities:** 6 hosp. **Educational facilities:** 5 univ. and colleges; 71 pub. schools, 40 private schools. **Further information:** Regional Business Partnership, 744 Broad St., Newark, NJ 07102-3802.

Websites: http://www.ci.newark.nj.us
http://www.rbp.org
http://www.4newark.com

New Orleans, Louisiana

Population: 465,538 (31); **Pop. density:** 2,578 per sq. mi; **Pop. growth (1990-98):** –6.3%. **Area:** 180.6 sq. mi. **Employment:** 190,413 employed, 6.0% unemployed. **Per capita income (MSA):** $23,148; % increase, 1996-97: 5.0.

History: founded by French 1718; became major seaport on Mississippi R.; acquired by U.S. as part of Louisiana Purchase 1803; inc. as city 1805; Battle of New Orleans was last battle of War of 1812.

Transportation: 2 airports; major railroad center; major intl. port. **Communications:** 9 TV, 27 radio stations. **Medical facilities:** numerous hosp.; major research center. **Educational facilities:** 23 univ. and colleges. **Further information:** New Orleans Metropolitan Convention & Visitors Bureau, Inc., 1520 Sugar Bowl Dr., New Orleans, LA 70112.

Website: http://www.neworleanscvb.com

New York, New York

Population: 7,420,166 (1); **Pop. density:** 24,021 per sq. mi; **Pop. growth (1990-98):** 1.3%. **Area:** 308.9 sq. mi. **Employment:** 3,168,860 employed, 8.0% unemployed. **Per capita income (MSA):** $34,459; % increase, 1996-97: 4.8.

History: trading post established by Henry Hudson 1609; British took control from Dutch 1664 and named city New York; briefly U.S. capital; Washington inaugurated as president 1789; under new charter, 1898, city expanded to include 5 boroughs: The Bronx, Brooklyn, Queens, and Staten Island, as well as Manhattan.

Transportation: 3 intl. airports serve area; 2 rail terminals; major subway network; ferry system; 4 underwater tunnels. **Communications:** 13 TV, 117 radio stations. **Medical facilities:** 81 hosp.; 5 academic medical centers. **Educational facilities:** 92 univ. and colleges; 1,100 pub. schools. **Further information:** Convention and Visitors Bureau, 810 Seventh Ave., New York, NY 10019.

Websites: http://www.ci.nyc.ny.us
http://www.nycvisit.com

Norfolk, Virginia

Population: 215,215 (75); **Pop. density:** 4,000 per sq. mi; **Pop. growth (1990-98):** –17.6%. **Area:** 53.8 sq. mi. **Employment:** 81,164 employed, 5.2% unemployed. **Per capita income (MSA):** $21,983; % increase, 1996-97: 4.7.

History: founded 1682; burned by patriots to prevent capture by British during Revolutionary War; rebuilt and inc. as town 1805, as city 1845; site of world's largest naval base.

Transportation: 1 intl. airport; 2 railroads; Amtrak; bus system. **Communications:** 13 TV, 6 city-access TV, 27 radio stations. **Medical facilities:** 6 hosp. **Educational facilities:** 2 univ., 1 college, 1 medical school; 58 pub. schools. **Further information:** Norfolk Convention and Visitors Bureau, 232 E. Main St., Norfolk, VA 23510.

Website: http://www.norfolk.va.us

Oakland, California

Population: 365,874 (45); **Pop. density:** 6,522 per sq. mi; **Pop. growth (1990-98):** –1.7%. **Area:** 56.1 sq. mi. **Employment:** 174,003 employed, 6.5% unemployed. **Per capita income (MSA):** $31,338; % increase, 1996-97: 6.7.

History: area settled by Spanish 1820; inc. as city under present name 1854.

Transportation: 1 intl. airport; western terminus for 2 railroads; underground, 75-mi underwater subway. **Communications:** 1 TV, 3 radio stations in city. **Medical facilities:** 10 hosp.

in MSA. **Educational facilities:** 8 East Bay colleges and univ.; 81 pub. schools. **Further information:** Oakland Metropolitan Chamber of Commerce, 475 14th St., Oakland, CA 94612-1903.

Websites: http://oakweb.ci.oakland.ca.us
http://www.oaklandchamber.com
http://www.oaklandnet.com

Oklahoma City, Oklahoma

Population: 472,221 (30); **Pop. density:** 776 per sq. mi; **Pop. growth (1990-98):** 6.2%. **Area:** 608.2 sq. mi. **Employment:** 235,924 employed, 4.1% unemployed. **Per capita income (MSA):** $21,659; % increase, 1996-97: 4.5.

History: settled during land rush in Midwest 1889; inc. 1890; became capital 1910; oil discovered 1928.

Transportation: 1 intl. airport; 3 railroads; pub. transit system; 5 major bus lines. **Communications:** 8 TV, 24 radio stations. **Medical facilities:** 20 hosp. **Educational facilities:** 17 univ. and colleges; 83 pub., 37 private schools. **Further information:** Chamber of Commerce, Economic Development Division, 123 Park Ave., Oklahoma City, OK 73102.

Websites: http://www.okcchamber.com
http://www.okccvb.org
http://www.ocbn.org

Omaha, Nebraska

Population: 371,291 (43); **Pop. density:** 3,691 per sq. mi; **Pop. growth (1990-98):** 7.8%. **Area:** 100.6 sq. mi. **Employment:** 197,434 employed, 3.0% unemployed. **Per capita income (MSA):** $26,570; % increase, 1996-97: 6.9.

History: founded 1854; inc. 1857; large food-processing, tele-communications, information-processing center; home of more than 20 insurance companies.

Transportation: 12 major airlines; 4 major railroads; intercity bus line. **Communications:** 8 TV, 22 radio stations. **Medical facilities:** 16 hosp.; institute for cancer research. **Educational facilities:** 5 univ., 4 colleges; 243 pub., 78 private schools. **Further information:** Greater Omaha Chamber of Commerce, 1301 Harney St., Omaha, NE 68102.

Websites: http://www.ci.omaha.ne.us
http://www.accessomaha.com

Philadelphia, Pennsylvania

Population: 1,436,287 (5); **Pop. density:** 10,631 per sq. mi; **Pop. growth (1990-98):** –9.4%. **Area:** 135.1 sq. mi. **Employment:** 600,410 employed, 6.2% unemployed. **Per capita income (MSA):** $29,347; % increase, 1996-97: 4.7.

History: first settled by Swedes 1638; Swedes surrendered to Dutch 1654; settled by English and Scottish Quakers 1678; named Philadelphia 1682; chartered 1701; Continental Congresses convened 1774, 1775; Declaration of Independence signed here 1776; national capital 1790-1800; state capital 1683-1799.

Transportation: 1 major airport; 3 railroads; major freshwater port; subway, el, rail commuter, bus, and streetcar system. **Communications:** 2 major daily newspapers, 11 TV, 45 radio stations. **Medical facilities:** 47 hosp. **Educational facilities:** 25 degree-granting institutions; 10 community college campuses. **Further information:** Greater Philadelphia Chamber of Commerce, Business Information Center, 200 South Broad St., Suite 700, Philadelphia PA 19102.

Websites: http://www.phila.gov
http://www.gpcc.com
http://www.philly.com

Phoenix, Arizona

Population: 1,198,064 (7); **Pop. density:** 2,528 per sq. mi; **Pop. growth (1990-98):** 21.3%. **Area:** 473.9 sq. mi. **Employment:** 674,833 employed, 2.9% unemployed. **Per capita income (MSA):** $24,137; % increase, 1996-97: 8.2.

History: settled 1870; inc. as city 1881; became territorial capital 1889.

Transportation: 1 intl. airport; 5 railroads; transcontinental bus line; pub. transit system. **Communications:** 13 TV, 45 radio stations. **Medical facilities:** 20 hosp., 1 medical research center. **Educational facilities:** 88 institutions of higher learning; 186 pub. schools (143 elem., 19 junior high, 24 high schools). **Further information:** Chamber of Commerce, 201 N. Central Ave., 27th fl., Phoenix, AZ 85073.

Websites: http://www.ci.phoenix.az.us
http://www.phoenixchamber.com

Pittsburgh, Pennsylvania

Population: 340,520 (49); **Pop. density:** 6,124 per sq. mi; **Pop. growth (1990-98):** –7.9%. **Area:** 55.6 sq. mi. **Employment:** 154,163 employed, 4.6% unemployed. **Per capita income (MSA):** $26,243; % increase, 1996-97: 4.1.

History: settled around Ft. Pitt 1758; inc. as city 1816; has one of the largest inland ports; by Civil War, already a center for iron production.

Transportation: 1 intl. airport; 20 railroads; 2 bus lines; trolley/subway system. **Communications:** 6 TV, 26 radio stations. **Medical facilities:** 35 hosp.; VA installation. **Educational facilities:** 3 univ., 6 colleges; 86 pub. schools. **Further information:** Greater Pittsburgh Convention & Visitors Bureau, 4 Gateway Ctr., Pittsburgh, PA 15222.

Website: http://www.visitpittsburgh.com

Plano, TX

Population: 219,486 (72); **Pop. density:** 2,962 per sq. mi; **Pop. growth (1990-98):** 71.6%. **Area:** 74.1 sq. mi. **Employment:** 121,675 employed, 1.8% unemployed. **Per capita income (MSA):** $30,481; % increase, 1996-97: 9.5.

History: settled 1846; inc. as city 1873.

Transportation: 1 bus line. **Communications:** 2 TV, 2 radio stations. **Medical facilities:** 6 hosp. **Educational facilities:** 2 institutes of higher learning, 49 pub. schools. **Further information:** Plano Chamber of Commerce, 1200 East 15th St., PO Drawer 940287, Plano, TX 75094.

Website: http://www.planocc.org

Portland, Oregon

Population: 503,891 (26); **Pop. density:** 4,041 per sq. mi; **Pop. growth (1990-98):** 3.7%. **Area:** 124.7 sq. mi. **Employment:** 263,523 employed, 5.1% unemployed. **Per capita income (MSA):** $27,388; % increase, 1996-97: 7.6.

History: settled by pioneers 1845; developed as trading center, aided by California Gold Rush 1849; city chartered 1851.

Transportation: 1 intl. airport; 2 major rail freight lines, Amtrak; 2 intercity bus lines; 27-mi. frontage freshwater port; mass transit bus and rail system. **Communications:** 7 TV, 37 radio stations. **Medical facilities:** 16 hosp.; VA hosp. **Educational facilities:** 26 univ. and colleges, 1 community college. **Further information:** Portland Metropolitan Chamber of Commerce, 221 N.W. 2d Ave., Portland, OR 9720.

Website: http://www.pdxchamber.org

Raleigh, North Carolina

Population: 259,423 (62); **Pop. density:** 2,945 per sq. mi; **Pop. growth (1990-98):** 18.5%. **Area:** 88.1 sq. mi. **Employment:** 162,587 employed, 1.8% unemployed. **Per capita income (MSA):** $27,711; % increase, 1996-97: 9.1.

History: named after Sir Walter Raleigh; site chosen for capital 1788; laid out 1792; inc. 1795; occupied by Gen. Sherman 1865.

Transportation: 1 intl. airport, 14 airlines, 8 commuter airlines; 3 railroads; 2 bus lines. **Communications:** 8 TV, 31 radio stations. **Medical facilities:** 6 hosp. **Educational facilities:** 6 univ. and colleges; 1 community college; 1067 pub. schools (county). **Further information:** Chamber of Commerce, 800 S. Salisbury St., PO Box 2978, Raleigh, NC 27602.

Websites: http://www.raleigh.acn.net
                http://www.raleighchamber.org

Richmond, Virginia

Population: 194,173 (88); **Pop. density:** 3,231 per sq. mi; **Pop. growth (1990-98):** –4.3%. **Area:** 60.1 sq. mi. **Employment:** 92,590 employed, 3.8% unemployed. **Per capita income (MSA):** $27,797; % increase, 1996-97: 5.9.

History: first settled 1607; became capital of Commonwealth of Virginia, 1779; attacked by British under Benedict Arnold 1781; inc. as city 1782; capital of Confederate States of America, 1861-65.

Transportation: 1 intl. airport; 4 railroads; 3 intracity bus lines; deepwater terminal accessible to oceangoing ships. **Communications:** 6 TV, 26 radio stations. **Medical facilities:** Medical Coll. of Virginia renowned for heart and kidney transplants; 19 other hosp. incl. VA facility. **Educational facilities:** 9 univ. and colleges; 173 pub., 45 private schools. **Further information:** Chamber of Commerce, PO Box 12280, Richmond, VA 23241.

Websites: http://www.ci.richmond.va.us
                http://www.grcc.com

Riverside, California

Population: 262,140 (61); **Pop. density:** 3,374 per sq. mi; **Pop. growth (1990-98):** 15.7%. **Area:** 77.7 sq. mi. **Employ-**ment: 130,461 employed, 6.7% unemployed. **Per capita income (MSA):** $19,604; % increase, 1996-97: 5.2.

History: founded 1870; inc. 1886; known for its citrus industry; home of the parent navel orange.

Transportation: municipal airport, intl. airport nearby; rail freight lines, commuter line; trolley/bus system. **Communications:** 15 TV, 47 radio stations. **Medical facilities:** 3 hosp.; many clinics. **Educational facilities:** 3 univ., 1 community college. **Further information:** Chamber of Commerce, 3985 University Ave., Riverside, CA 92501.

Websites: http://www.ci.riverside.ca.us
                http://www.riverside-chamber.com

Rochester, New York

Population: 216,887 (73); **Pop. density:** 6,058 per sq. mi; **Pop. growth (1990-98):** –5.8%. **Area:** 35.8 sq. mi. **Employment:** 106,442 employed, 6.4% unemployed. **Per capita income (MSA):** $26,170; % increase, 1996-97: 3.5.

History: first permanent settlement 1812; inc. as village 1817, as city 1834; developed as Erie Canal town.

Transportation: 1 intl. airport; Amtrak; 3 bus lines; intracity transit service; Port of Rochester. **Communications:** 6 TV, 18 radio stations. **Medical facilities:** 8 general hosp. **Educational facilities:** 10 colleges, 3 community colleges. **Further information:** Greater Rochester Metro Chamber of Commerce, 55 St. Paul St., Rochester, NY 14604-1391.

Websites: http://www.rochester.lib.ny.us/cityhall
                http://www.rnychamber.com

Sacramento, California

Population: 404,168 (38); **Pop. density:** 4,197 per sq. mi; **Pop. growth (1990-98):** 9.4%. **Area:** 96.3 sq. mi. **Employment:** 179,385 employed, 6.1% unemployed. **Per capita income (MSA):** $25,335; % increase, 1996-97: 6.1.

History: settled 1839; important trading center during California Gold Rush 1840s; became state capital 1854.

Transportation: international, executive, and cargo airports; 2 mainline transcontinental rail carriers; bus and light rail system; Port of Sacramento. **Communications:** 8 TV, 34 radio stations; 3 cable TV cos. **Medical facilities:** 12 major hosp. **Educational facilities:** 7 colleges and univ., 5 private colleges and univ., 5 community colleges, 81 pub. schools. **Further information:** Metro Chamber of Commerce, 917 7th St., Sacramento, CA 95814.

Websites: http://www.ci.sacramento.ca.us
                http://www.sacog.org
                http://www.metrochamber.org

St. Louis, Missouri

Population: 339,316 (50); **Pop. density:** 5,482 per sq. mi; **Pop. growth (1990-98):** –14.5%. **Area:** 61.9 sq. mi. **Employment:** 144,630 employed, 7.8% unemployed. **Per capita income (MSA):** $27,177; % increase, 1996-97: 5.6.

History: founded 1764 as a fur trading post by French; acquired by U.S. 1803; chartered as city 1822; lies on Mississippi R., near confluence with Missouri R.

Transportation: 1 intl. airport; major rail center, 17 trunk-line railroads; major inland port; 14 bus lines; 14 barge lines. **Communications:** 7 TV, 35 radio stations. **Medical facilities:** 65 hosp. **Educational facilities:** 6 univ., 25 colleges and seminaries, 65 elem., 23 middle, 9 high schools. **Further information:** St. Louis Community Development Agency, 1015 Locust St., Ste. 1200, St. Louis, MO 63101.

Websites: http://www.st-louis.mo.us
                http://stlouis.missouri.org

St. Paul, Minnesota

Population: 257,284 (63); **Pop. density:** 4,873 per sq. mi; **Pop. growth (1990-98):** –5.5%. **Area:** 52.8 sq. mi. **Employment:** 138,826 employed, 2.6% unemployed. **Per capita income (MSA):** $30,123; % increase, 1996-97: 6.1.

History: founded in early 1840s as "Pig's Eye Landing"; became capital of the Minnesota territory 1849 and chartered as St. Paul 1854.

Transportation: 1 intl., 1 business airport; 6 major rail lines; 2 interstate bus lines; pub. transit system. **Communications:** 9 TV, 47 radio stations. **Medical facilities:** 6 hosp. **Educational facilities:** 5 univ., 4 colleges; 1 technical, 1 law school, 1 art and design college; 65 public, 39 private schools. **Further information:** St. Paul Area Chamber of Commerce, 332 Minnesota St., Ste. N-205, St. Paul, MN 55101.

Website: http://www.ci.stpaul.mn.us

St. Petersburg, Florida

Population: 236,029 (70); **Pop. density:** 3,987 per sq. mi; **Pop. growth (1990-98):** −1.8%. **Area:** 59.2 sq. mi. **Employment:** 128,741 employed, 3.5% unemployed. **Per capita income (MSA):** $24,879; % increase, 1996-97: 6.6.

History: founded 1888; inc. 1892.

Transportation: 2 airports (1 intl.); Amtrak bus connection; county-wide public bus system; 1 cruise port. **Communications:** 15 TV, 58 radio stations. **Medical facilities:** 4 major hosp.; VA hosp. **Educational facilities:** 1 univ., 1 college, 1 law school, 1 junior college; 126 pub. schools. **Further information:** St. Petersburg Area Chamber of Commerce, PO Box 1371, St. Petersburg, FL 33731.

Website: http://www.stpete.com

San Antonio, Texas

Population: 1,114,130 (8); **Pop. density:** 3,346 per sq. mi; **Pop. growth (1990-98):** 14.1%. **Area:** 333 sq. mi. **Employment:** 502,009 employed, 4.1% unemployed. **Per capita income (MSA):** $22,379; % increase, 1996-97: 6.9.

History: first Spanish garrison 1718; Battle at the Alamo fought here 1836; city subsequently captured by Texans; inc. 1837.

Transportation: 1 intl. airport; 4 railroads; 3 bus lines; pub. transit system. **Communications:** 9 TV, 34 radio stations. **Medical facilities:** 36 hosp.; major medical center. **Educational facilities:** 18 univ. and colleges; 16 pub. school districts. **Further information:** Chamber of Commerce, 602 E. Commerce, PO Box 1628, San Antonio, TX 78296.

Websites: http://www.tristero.com/usa/tx/
http://www.ci.sat.tx.us
http://www.sachamber.org

San Bernardino, California

Population: 186,402 (97); **Pop. density:** 3,383 per sq. mi; **Pop. growth (1990-98):** 7.6%. **Area:** 55.1 sq. mi. **Employment:** 70,952 employed, 8.2% unemployed. **Per capita income (MSA):** $19,604; % increase, 1996-97: 5.2.

History: first explored in 1774 by Viceroy of Mexico; Spanish missionaries settled here 1810; Mormons est. first permanent settlement 1852; inc. 1854.

Transportation: 1 intl. airport; Amtrak; Metrolink; BNSF rail transit system; Omnitrans bus systems. **Communications:** 2 TV, 7 radio stations. **Medical facilities:** 2 hosp. **Educational facilities:** 1 univ., 1 community college; 59 pub. schools. **Further information:** San Bernardino Area Chamber of Commerce, 546 W. 6th St., PO Box 658, San Bernardino, CA 92402.

Website: http://www.ci.san-bernardino.ca.us

San Diego, California

Population: 1,220,666 (6); **Pop. density:** 3,767 per sq. mi; **Pop. growth (1990-98):** 9.9%. **Area:** 324 sq. mi. **Employment:** 583,365 employed, 3.6% unemployed. **Per capita income (MSA):** $24,965; % increase, 1996-97: 6.4.

History: claimed by the Spanish 1542; first mission est. 1769; scene of conflict during Mexican-American War 1846; inc. 1850.

Transportation: 1 major airport; 1 railroad; major freeway system; bus system; trolley system. **Communications:** 8 TV, 22 radio stations, 2 cable providers. **Medical facilities:** 28 hosp. **Educational facilities:** 10 univ., 7 colleges; 176 pub. schools. **Further information:** Greater San Diego Chamber of Commerce, 402 W. Broadway, Ste. 1000, San Diego, CA 92101.

Websites: http://www.sannet.gov
http://www.sdchamber.org

San Francisco, California

Population: 745,774 (12); **Pop. density:** 15,969 per sq. mi; **Pop. growth (1990-98):** 3.0%. **Area:** 46.7 sq. mi. **Employment:** 401,337 employed, 3.7% unemployed. **Per capita income (MSA):** $41,128; % increase, 1996-97: 7.0.

History: nearby Farallon Islands sighted by Spanish 1542; city settled by 1776; claimed by U.S. 1846; became a major city during California Gold Rush 1849; inc. as city 1850; earthquake devastated city 1906.

Transportation: 1 major airport; intracity railway system; 2 railway transit systems; bus and railroad service; ferry system; 1 underwater tunnel. **Communications:** 10 TV; 15 radio stations. **Medical facilities:** 16 medical centers. **Educational facilities:** 16 univ. and colleges, 111 pub. schools, 5 charter schools. **Further information:** Convention & Visitors Bureau, 201 3d St., Ste. 900, San Francisco, CA 94103.

Websites: http://www.ci.sf.ca.us
http://www.sfchamber.com

San Jose, California

Population: 861,284 (11); **Pop. density:** 5,028 per sq. mi; **Pop. growth (1990-98):** 10.1%. **Area:** 171.3 sq. mi. **Employment:** 470,958 employed, 3.8% unemployed. **Per capita income (MSA):** $37,856; % increase, 1996-97: 10.3.

History: founded by the Spanish 1777 between San Francisco and Monterey; state cap. 1849-51; inc. 1850.

Transportation: 1 intl. airport; 2 railroads; bus system. **Communications:** 4 TV, 14 radio stations. **Medical facilities:** 6 hosp. **Educational facilities:** 3 univ. and colleges. **Further information:** Chamber of Commerce, 310 S. First St., San Jose, CA 95113.

Websites: http://www.ipac.net/csj
http://www.sjchamber.com

Santa Ana, California

Population: 305,955 (55); **Pop. density:** 11,290 per sq. mi; **Pop. growth (1990-98):** 4.1%. **Area:** 27.2 sq. mi. **Employment:** 151,883 employed, 5.2% unemployed. **Per capita income (MSA):** $30,115; % increase, 1996-97: 6.8.

History: founded 1869; inc. as city 1886.

Transportation: 1 airport; 5 major freeways including main Los Angeles-San Diego artery; Amtrak. **Communications:** 14 TV, 28 radio stations. **Medical facilities:** 4 hosp. **Educational facilities:** 1 community college. **Further information:** Santa Ana Chamber of Commerce, 1055 N. Main, Suite 904, Santa Ana, CA 92701.

Website: http://www.santaanacc.com

Scottsdale, Arizona

Population: 195,394 (86); **Pop. density:** 1,056 per sq. mi; **Pop. growth (1990-98):** 50.2%. **Area:** 185 sq. mi. **Employment:** 98,614 employed, 1.9% unemployed. **Per capita income (MSA):** $24,137; % increase, 1996-97: 8.2.

History: founded 1888 by Army Chaplain Winfield Scott; inc. June 25, 1951; Frank Lloyd Wright built winter home here (Taliesin West); slogan "West's Most Western Town," by Mayor Malcolm White adopted 1951.

Transportation: 1 intl., 1 local airport; bus system; local transit system; taxi system. **Communications:** 12 TV, 45 radio stations. **Medical facilities:** 33 general hospitals; Mayo Clinic. **Educational facilities:** 5 univ., 9 colleges; 521 pub. schools. **Further information:** Scottsdale Convention and Visitors Bureau, 7343 Scottsdale Rd., Scottsdale, AZ 85251.

Websites: http://www.ci.scottsdale.az.us
http://www.scottsdalecvb.com

Seattle, Washington

Population: 536,978 (22); **Pop. density:** 6,400 per sq. mi; **Pop. growth (1990-98):** 4.0%. **Area:** 83.9 sq. mi. **Employment:** 344,248 employed, 3.6% unemployed. **Per capita income (MSA):** $33,373; % increase, 1996-97: 10.3.

History: settled 1851; inc. 1869; suffered severe fire 1889; played prominent role during Alaska Gold Rush 1897; growth followed opening of Panama Canal 1914; center of aircraft industry WWII.

Transportation: 1 intl. airport; 2 railroads; ferries serve Puget Sound, Alaska, Canada. **Communications:** 7 TV, 39 radio stations. **Medical facilities:** 40 hosp. **Educational facilities:** 7 univ., 6 colleges, 11 community colleges. **Further information:** Greater Seattle Chamber of Commerce, 1301 5th Ave., Ste. 2400, Seattle, WA 98101-2603.

Websites: http://www.ci.seattle.wa.us
http://www.seattlechamber.com

Shreveport, Louisiana

Population: 188,319 (95); **Pop. density:** 1,910 per sq. mi; **Pop. growth (1990-98):** −5.1%. **Area:** 98.6 sq. mi. **Employment:** 90,741 employed, 6.7% unemployed. **Per capita income (MSA):** $21,259; % increase, 1996-97: 3.6.

History: founded 1833 near site of a 160-mi logjam cleared by Capt. Henry Shreve; inc. 1839; oil discovered 1905.

Transportation: 2 airports; 3 bus lines. **Communications:** 6 TV, 20 radio stations. **Medical facilities:** 16 hosp. **Educational facilities:** 4 univ., 1 college; 74 pub. schools. **Further information:** Chamber of Commerce, PO Box 20074, 400 Edwards St., Shreveport, LA 71120.

Website: http://www.shreveportchamber.org

Stockton, California

Population: 240,143 (69); **Pop. density:** 4,565 per sq. mi; **Pop. growth (1990-98):** 13.8%. **Area:** 52.6 sq. mi. **Employment:** 90,694 employed, 12.4% unemployed. **Per capita income (MSA):** $20,092; % increase, 1996-97: 5.9.

History: site purchased 1842; settled 1847; inc. 1850; chief distributing point for agric. products of San Joaquin Valley. **Transportation:** 1 airport; deepwater inland seaport; 4 railroads; 2 bus lines, county bus system. **Communications:** 5 TV stations. **Medical facilities:** 4 hosp.; regional burn, cancer, heart centers. **Educational facilities:** 6 univ. and colleges; 58 pub. schools. **Further information:** Chamber of Commerce, 445 W. Weber Ave., Ste. 220, Stockton, CA 95203.
Websites: http://www.ci.stockton.ca.us
http://www.stocktonchamber.org

Tampa, Florida

Population: 289,156 (58); **Pop. density:** 2,660 per sq. mi; **Pop. growth (1990-98):** 3.3%. **Area:** 108.7 sq. mi. **Employment:** 162,475 employed, 3.5% unemployed. **Per capita income (MSA):** $24,879; % increase, 1996-97: 6.6.
History: U.S. army fort on site 1824; inc. 1855; Ybor City National Historical Landmark district.
Transportation: 1 intl. airport; Port of Tampa; CSX rail, bus system. **Communications:** 12 TV, 30 radio stations. **Medical facilities:** 17 hosp. **Educational facilities:** 3 univ. and colleges; 183 pub. schools. **Further information:** Chamber of Commerce, 401 E. Jackson St., PO Box 420, Tampa, FL 33601.
Website: http://www.tampachamber.com

Toledo, Ohio

Population: 312,174 (53); **Pop. density:** 3,873 per sq. mi; **Pop. growth (1990-98):** −6.2%. **Area:** 80.6 sq. mi. **Employment:** 148,982 employed, 6.4% unemployed. **Per capita income (MSA):** $24,315; % increase, 1996-97: 3.9.
History: site of Ft. Industry 1794; Battles of Ft. Meigs and Ft. Timbers 1812; figured in "Toledo War" 1835-36 between Ohio and Michigan over borders; inc. 1837.
Transportation: 5 major airlines; 4 railroads; 58 motor freight lines; 5 interstate bus lines. **Communications:** 6 TV, 15 radio stations. **Medical facilities:** 7 major hosp. complexes. **Educational facilities:** 7 univ. and colleges. **Further information:** Toledo Area Chamber of Commerce, 300 Madison Ave., Ste. 200, Toledo, OH 43604.
Website: http://www.toledochamber.com

Tucson, Arizona

Population: 460,466 (32); **Pop. density:** 2,946 per sq. mi; **Pop. growth (1990-98):** 10.8%. **Area:** 156.3 sq. mi. **Employment:** 224,115 employed, 3.0% unemployed. **Per capita income (MSA):** $21,068; % increase, 1996-97: 5.0.
History: settled 1775 by Spanish as a presidio; acquired by U.S. in Gadsden Purchase 1853; inc. 1877.
Transportation: 1 intl. airport; 2 railroads; bus system. **Communications:** 9 TV, 27 radio stations. **Medical facilities:** 14 hosp. **Educational facilities:** 3 univ., 1 college; 165 pub. schools. **Further information:** Chamber of Commerce, PO Box 991, Tucson, AZ 85702.
Websites: http://www.ci.tucson.az.us
http://www.tucsonchamber.org

Tulsa, Oklahoma

Population: 381,393 (42); **Pop. density:** 2,078 per sq. mi; **Pop. growth (1990-98):** 3.8%. **Area:** 183.5 sq. mi. **Employment:** 210,640 employed, 3.6% unemployed. **Per capita income (MSA):** $24,206; % increase, 1996-97: 6.9.
History: settled in 1830s by Creek Indians; modern town founded 1882 and inc. 1898; oil discovered early 20th century.
Transportation: 1 intl. airport; 5 rail lines; 5 bus lines; transit bus system. **Communications:** 43 TV, 26 radio stations. **Medical facilities:** 10 hosp. **Educational facilities:** 8 univ. and colleges; 83 pub., 39 private schools. **Further information:**

Metropolitan Tulsa Chamber of Commerce, 616 S. Boston Ave., Ste. 100, Tulsa, OK 74119-1298.
Website: http://www.tulsachamber.com

Virginia Beach, Virginia

Population: 432,380 (34); **Pop. density:** 1,741 per sq. mi; **Pop. growth (1990-98):** 10.0%. **Area:** 248.3 sq. mi. **Employment:** 206,742 employed, 2.8% unemployed. **Per capita income (MSA):** $21,983; % increase, 1996-97: 4.7.
History: area founded by Capt. John Smith 1607; formed by merger with Princess Anne Co. 1963.
Transportation: 1 airport; 2 railroads; 2 bus lines; pub. transit system. **Communications:** 8 TV, 44 radio stations. **Medical facilities:** 2 hosp. **Educational facilities:** 1 univ., 2 colleges; 84 pub. schools. **Further information:** Virginia Beach Dept. of Economic Development, One Columbus Center, Ste. 300, Virginia Beach, VA 23462.
Website: http://www.virginia-beach.va.us

Washington, District of Columbia

Population: 523,124 (23); **Pop. density:** 8,520 per sq. mi; **Pop. growth (1990-98):** −13.8%. **Area:** 61.4 sq. mi. **Employment:** 243,699 employed, 8.8% unemployed. **Per capita income (MSA):** $33,433; % increase, 1996-97: 5.9.
History: U.S. capital; site at Potomac R. chosen by George Washington 1790 on land ceded from VA and MD (portion S of Potomac returned to VA 1846); Congress first met 1800; inc. 1802; sacked by British, War of 1812.
Transportation: 3 intl. airports in area; Amtrak, 6 other passenger & cargo rail lines; Metrobus/Metrorail transit system; bus line. **Communications:** 5 TV, 61 radio stations. **Medical facilities:** 16 hosp. **Educational facilities:** 10 univ. and colleges. **Further information:** DC Chamber of Commerce, 1301 Pennsylvania Ave. NW, Ste. 309, Washington, DC 20004.
Websites: http://www.ci.washington.dc.us
http://www.dcchamber.org

Wichita, Kansas

Population: 329,211 (52); **Pop. density:** 2,860 per sq. mi; **Pop. growth (1990-98):** 8.3%. **Area:** 115.1 sq. mi. **Employment:** 174,479 employed, 3.6% unemployed. **Per capita income (MSA):** $24,434; % increase, 1996-97: 7.0.
History: founded 1864; inc. 1871.
Transportation: 2 airports; 3 major rail freight lines; 2 bus lines. **Communications:** 5 TV, 26 radio stations. **Medical facilities:** 7 hosp., 2 psychiatric rehab. centers. **Educational facilities:** 3 univ., 1 medical school; 96 pub. schools. **Further information:** Chamber of Commerce, 350 W. Douglas Ave., Wichita, KS 67202.
Websites: http://www.wichitakansas.org
http://www.twsu.edu/~cedbrwww

Yonkers, New York

Population: 190,153 (93); **Pop. density:** 10,506 per sq. mi; **Pop. growth (1990-98):** 1.1%. **Area:** 18.1 sq. mi. **Employment:** 87,011 employed, 4.5% unemployed. **Per capita income (MSA):** $34,459; % increase, 1996-97: 4.8.
History: founded 1641 by the Dutch; inc. as town 1855; chartered as city 1872; directly north of NYC.
Transportation: intracity bus system; rail service. **Communications:** see New York, NY. **Medical facilities:** 3 hosp. **Educational facilities:** 1 college; 32 pub. schools. **Further information:** Chamber of Commerce, 20 S. Broadway, 12th fl., Yonkers, NY 10701.
Websites: http://www.ci.yonkers.ny.us
http://www.yonkerschamber.com

M I L L E N N I U M   F A C T   B O X

Top 10 Cities in 1900

In 1900, as in 1998, New York was the biggest U.S. city in population. Chicago, which in 1998 was 3d, ranked 2d in 1900. Los Angeles, which in 1998 ranked 2d, was far down the list in 1900, ranking 36th, with a population of 102,479.

Here are the top ten cities in 1900, with the population of each.

1. New York, N.Y.	3,437,202		6. Baltimore, Md.	508,957
2. Chicago, Ill.	1,698,575		7. Cleveland, Oh	381,768
3. Philadelphia, Pa.	1,293,697		8. Buffalo, N.Y.	352,387
4. St. Louis, Mo.	575,238		9. San Francisco, Cal.	342,782
5. Boston, Mass.	560,892		10. Cincinnati, Oh	325,902

WORLD EXPLORATION AND GEOGRAPHY

Early Explorers of the Western Hemisphere

Reviewed by Susan Skomal, PhD, editor, American Anthropological Assn., and Paul B. Frederic, PhD, prof. of geography, Univ. of Maine.

The first people to discover the New World, or western hemisphere, are believed to have traveled across a "land bridge" from Siberia to Alaska, an isthmus since broken by the Bering Strait. From Alaska, these early Native Americans could then have spread through North, Central, and South America. This theory is supported by archaeological and genetic evidence; a theory that the first Americans came by sea, landing in S America, is no longer seriously considered by archaeologists, partly because of a lack of evidence of such early habitation in Polynesia.

In 1997, archaeologists confirmed evidence of human habitation in the Americas at least 12,500 years ago at a site in Chile known as Monte Verde. This site predates a previously discovered site in Clovis, NM, by over 1,000 years. The findings raise questions concerning the migratory path of these peoples, since a glacier covered most of N America from 20,000 to 13,000 years ago or later. The migration may have taken place in an ice-free corridor or along the west coast, perhaps in vessels along the water. Or perhaps people had spread to S America before the coming of the ice.

At first, these early Americans were hunters, using flint weapons and tools. In Mexico, about 7000-6000 BC, they founded farming cultures and developed crops, such as corn and squash. Eventually they created complex civilizations—the Olmec, Toltec, Aztec, Maya, and, in S America, the Inca. Carbon-14 tests show that humans lived about 8000 BC near what are now Front Royal, VA; Kanawha, WV; and Dutchess Quarry, NY. The Hopewell Culture, based on farming, flourished about 1000 BC; remains of it are seen today in large mounds in Ohio and other states.

Norsemen (Norwegian Vikings sailing out of Iceland and Greenland), led by Leif Ericson, are credited by most scholars with having been the first Europeans to reach America, with at least 5 voyages occurring about AD 1000 to areas they called Helluland, Markland, and Vinland—possibly what are known today as Labrador, Nova Scotia or Newfoundland, and New England. L'Anse aux Meadows, on the N tip of Newfoundland, is the only documented settlement.

Sustained contact between the hemispheres began with the first voyage of Christopher Columbus (born Cristoforo Colombo, c 1451, in or near Genoa, Italy). Columbus made trips to the New World while sailing for the Spanish.

His earliest voyage began when he left Palos, Spain, Aug. 3, 1492, with 88 (est.) men and landed at San Salvador (Watling Islands, Bahamas) on Oct. 12, 1492. His fleet consisted of 3 vessels—the *Niña*, *Pinta*, and *Santa María*. Stops also were made on Cuba and Hispaniola. A 2d expedition left Cadiz, Spain, Sept. 25, 1493, with 17 ships and 1,500 men, and reached the Lesser Antilles Nov. 3. His 3d voyage brought him from Sanlucar, Spain (May 30, 1498, with 6 ships), to the N coast of S America. A few years later a 4th voyage reached the mainland of Central America, after leaving Cadiz, Spain, May 9, 1502. Columbus died in 1506 convinced he had reached Asia by sailing west from Europe.

In N America, John and Sebastian Cabot, Italian explorers sailing for the English, reached Newfoundland and possibly Nova Scotia in 1497. John's 2d voyage (1498), seeking a new trade route to Asia, resulted in the loss of his entire fleet. During this period exploration in the western hemisphere was dominated by Spain and Portugal. In 1497 and 1499 Amerigo Vespucci (whom the Americas are named for), an Italian explorer sailing for the Spanish, passed along the N and E coasts of S America. He was the first to argue that the newly discovered lands were a continent other than Asia. The basic geography of the hemisphere became well understood by the early 1800s, as explorers from many countries helped fill in the map.

Year	Explorer	Nationality (employer, if different)	Area reached or explored
c1000	Leif Ericson	Norse	Newfoundland
1492-1502	Christopher Columbus	Italian (Spanish)	West Indies, S. and C. America
1497	John and Sebastian Cabot	Italian (English)	Atlantic Canada
1497-98	Vasco da Gama	Portuguese	Cape of Good Hope (Africa), India
1497-99	Amerigo Vespucci	Italian (Spanish)	E and N Coast of S. America
1499	Alonso de Ojeda	Spanish	N South American coast, Venezuela
1500, Feb.	Vicente Yañez Pinzon	Spanish	S. American coast, Amazon R.
1500, Apr.	Pedro Álvarez Cabral	Portuguese	Brazil
1500-02	Gaspar Corte-Real	Portuguese	Labrador
1501	Rodrigo de Bastidas	Spanish	Central America
1513	Vasco Nunez de Balboa	Spanish	Panama, Pacific Ocean
1513	Juan Ponce de Leon	Spanish	Florida, Yucatán Peninsula
1515	Juan de Solis	Spanish	Río de la Plata
1519	Alonso de Pineda	Spanish	Mouth of Mississippi R.
1519	Hernando Cortes	Spanish	Mexico
1519-20	Ferdinand Magellan	Portuguese (Spanish)	Straits of Magellan, Tierra del Fuego
1524	Giovanni da Verrazano	Italian (French)	Atlantic coast, inc. New York harbor
1528	Cabeza de Vaca	Spanish	Texas coast and interior
1532	Francisco Pizarro	Spanish	Peru
1534	Jacques Cartier	French	Canada, Gulf of St. Lawrence
1536	Pedro de Mendoza	Spanish	Buenos Aires
1539	Francisco de Ulloa	Spanish	California coast
1539-41	Hernando de Soto	Spanish	Mississippi R., near Memphis
1539	Marcos de Niza	Italian (Spanish)	SW United States
1540	Francisco de Coronado	Spanish	SW United States
1540	Hernando Alarcon	Spanish	Colorado R.
1540	Garcia de Lopez Cardenas	Spanish	Colorado, Grand Canyon
1541	Francisco de Orellana	Spanish	Amazon R.
1542	Juan Rodriguez Cabrillo	Portuguese (Spanish)	W Mexico, San Diego harbor
1565	Pedro Menéndez de Aviles	Spanish	St. Augustine, FL
1576	Sir Martin Frobisher	English	Frobisher's Bay, Canada
1577-80	Sir Francis Drake	English	California coast
1582	Antonio de Espejo	Spanish	Southwest U.S. (New Mexico)
1584	Amadas & Barlow (for Raleigh)	English	Virginia
1585-87	Sir Walter Raleigh's men	English	Roanoke Isl., NC
1595	Sir Walter Raleigh	English	Orinoco R.
1603-09	Samuel de Champlain	French	Canadian interior, Lake Champlain
1607	Capt. John Smith	English	Atlantic coast
1609-10	Henry Hudson	English (Dutch)	Hudson R., Hudson Bay
1634	Jean Nicolet	French	Lake Michigan, Wisconsin
1673	Jacques Marquette, Louis Jolliet	French	Mississippi R., S to Arkansas
1682	Robert Cavelier, sieur de La Salle	French	Mississippi R., S to Gulf of Mexico
1727-29	Vitus Bering	Danish (Russian)	Bering Strait and Alaska
1789	Sir Alexander Mackenzie	Canadian	NW Canada
1804-06	Meriwether Lewis and William Clark	American	Missouri R., Rocky Mts., Columbia R.

Arctic Exploration

Early Explorers

1587 —John Davis (Eng.). Davis Strait to Sanderson's Hope, 72°12′ N.

1596 — Willem Barents and Jacob van Heemskerck (Holland). Discovered Bear Isl., touched NW tip of Spitsbergen, 79°49′ N, rounded Novaya Zemlya, wintered at Ice Haven.

1607 — Henry Hudson (Eng.). North along Greenland's E coast to Cape Hold-with-Hope, 73°30′, then N of Spitsbergen to 80°23′. Returning he explored Hudson's Touches (Jan Mayen).

1616 — William Baffin and Robert Bylot (Eng.). Baffin Bay to Smith Sound.

1728 — Vitus Bering (Russ.). Proved Asia and America are separated, by sailing through strait that now bears his name.

1733-40 — Great Northern Expedition (Russ.). Surveyed Siberian Arctic coast.

1741 — Vitus Bering (Russ.). Sighted Alaska, named Mount St. Elias. His lieutenant, Chirikof, explored coast.

1771 — Samuel Hearne (Hudson's Bay Co.). Overland from Prince of Wales Fort (Churchill) on Hudson Bay to mouth of Coppermine R.

1778 —James Cook (Brit.). Through Bering Strait to Icy Cape, AK, and North Cape, Siberia.

1789 — Alexander Mackenzie (North West Co., Brit.). Montreal to mouth of Mackenzie River.

1806 — William Scoresby (Brit.). N of Spitsbergen to 81°30′.

1820-23 — Ferdinand von Wrangel (Russ.). Surveyed Siberian Arctic coast. His exploration joined James Cook's at North Cape, confirming separation of the continents.

1878-79 —(Nils) Adolf Erik Nordenskjöld (Swed.). The first to navigate the Northeast Passage—an ocean route connecting Europe's North Sea, along the Arctic coast of Asia and through the Bering Sea, to the Pacific Ocean.

1881 — The U.S. steamer *Jeannette*, led by Lt. Cmdr. George W. DeLong, was trapped in ice and crushed, June 1881. DeLong and 11 others died; 12 survived.

1888 — Fridtjof Nansen (Nor.) crossed Greenland's icecap.

1893-96 — Nansen in *Fram* drifted from New Siberian Isls. to Spitsbergen; tried polar dash in 1895, reached Franz Josef Land, 86°14′ N.

1897 — Salomon A. Andrée (Sweden) and 2 others started in balloon from Spitsbergen, July 11, to drift across pole to U.S., and disappeared. Aug. 6, 1930, their bodies were found on White Isl., 82°57′ N, 29°52′ E.

1903-6 —Roald Amundsen (Nor.) first sailed the Northwest Passage—an ocean route linking the Atlantic Ocean to the Pacific via Canada's marine waterways.

North Pole Exploration

Robert E. Peary explored Greenland's coast, 1891-92; tried for North Pole, 1893. In 1900 he reached N limit of Greenland and 83°50′ N; in 1902 he reached 84°06′ N; in 1906 he went from Ellesmere Isl. to 87°06′ N. He sailed in the *Roosevelt*, July 1908, to winter off Cape Sheridan, Grant Land. The dash for the North Pole began Mar. 1 from Cape Columbia, Ellesmere Isl. Peary reportedly reached the pole, 90° N, Apr. 6, 1909; however, subsequent research suggests that he may have miscalculated and fallen short of his goal by c. 30-60 mi. Peary had several supporting groups carrying supplies until the last group turned back at 87°47′ N. Peary, Matthew Henson, and 4 Eskimos proceeded with dog teams and sleds. They were said to have crossed the pole several times, then built an igloo there and remained 36 hours. Started south, Apr. 7 at 4 PM, for Cape Columbia.

1914 — Donald MacMillan (U.S.). Northwest, 200 mi, from Axel Heiberg Isl. to seek Peary's Crocker Land.

1915-17 — Vihjalmur Stefansson (Can.). Discovered Borden, Brock, Meighen, and Lougheed Isls.

1918-20 — Amundsen sailed the Northeast Passage.

1925 — Amundsen and Lincoln Ellsworth (U.S.) reached 87°44′ N in attempt to fly to North Pole from Spitsbergen.

1926 —Richard E. Byrd and Floyd Bennett (U.S.) reputedly flew over North Pole, May 9. (Claim to have reached the Pole is in dispute, however.)

1926 — Amundsen, Ellsworth, and Umberto Nobile (It.) flew from Spitsbergen over North Pole May 12, to Teller, AK, in dirigible *Norge*.

1928 — Nobile crossed North Pole in airship, May 24; crashed, May 25. Amundsen died attempting a rescue.

North Pole Exploration Records

On Aug. 3, 1958, the *Nautilus*, under Comdr. William R. Anderson, became the first ship to cross the North Pole beneath the Arctic ice.

In Aug. 1960, the nuclear-powered U.S. submarine *Seadragon* (Comdr. George P. Steele 2d) made the first E-W underwater transit through the Northwest Passage. Traveling submerged for the most part, it took 6 days to make the 850-mi trek from Baffin Bay to the Beaufort Sea.

On Aug. 16, 1977, the Soviet nuclear icebreaker *Arktika* reached the North Pole, becoming the first surface ship to break through the Arctic ice pack.

On Apr. 30, 1978, Naomi Uemura (Jap.) became the first person to reach the North Pole alone, traveling by dog sled in a 54-day, 600-mi trek over the frozen Arctic.

In Apr. 1982, Sir Ranulph Fiennes and Charles Burton, Brit. explorers, reached the North Pole and became the first to circle the earth from pole to pole. They had reached the South Pole 16 months earlier. The 52,000-mi trek took 3 years, involved 23 people, and cost an estimated $18 mil.

On May 2, 1986, 6 explorers reached the North Pole assisted only by dogs. They became the first to reach the pole without aerial logistics support since at least 1909. The explorers, Amer. Will Steger, Paul Schurke, Anne Bancroft, and Geoff Carroll, and Can. Brent Boddy and Richard Weber, completed the 500-mi journey in 56 days.

On June 15, 1995, Weber and Russ. Mikhail Malakhov became the first pair to make it to the pole and back without any mechanical assistance. The 940-mi trip, made entirely on skis, took 121 days.

Antarctic Exploration

Antarctica has been approached since 1773-75, when Capt. James Cook (Brit.) reached 71° 10′ S. Many sea and landmarks bear names of early explorers. Fabian von Bellingshausen (Russ.) discovered Peter I and Alexander I Isls., 1819-21. Nathaniel Palmer (U.S.) traveled throughout Palmer Peninsula, 60° W, 1820, without realizing that this was a continent. Capt. John Davis (U.S.) made the first known landing on the continent on Feb. 7, 1821. Later, in 1823, James Weddell (Brit.) found Weddell Sea, 74° 15′ S, the southernmost point that had been reached.

First to announce existence of the continent of Antarctica was Charles Wilkes (U.S.), who followed the coast for 1,500 mi, 1840. Adelie Coast, 140° E, was found by Dumont d'Urville (Fr.), 1840. Ross Ice Shelf was found by James Clark Ross (Brit.), 1841-42.

1895 — Leonard Kristensen (Nor.) landed a party on the coast of Victoria Land. They were the first ashore on the main continental mass. C. E. Borchgrevink, a member of that party, returned in 1899 with a Brit. expedition, first to winter on Antarctica.

1902-4 — Robert F. Scott (Brit.) explored Edward VII Peninsula. He reached 82° 17′ S, 146° 33′ E from McMurdo Sound.

1908-9 — Ernest Shackleton (Brit.) introduced the use of Manchurian ponies in Antarctic sledging. He reached 88° 23′ S, discovering a route on to the plateau by way of the Beardmore Glacier and pioneering the way to the pole.

1911 — Roald Amundsen (Nor.) with 4 men and dog teams reached the South Pole, Dec. 14.

1912 — Scott reached the pole from Ross Isl., Jan. 18, with 4 companions. None of Scott's party survived. Their bodies and expedition notes were found, Nov. 12.

1928 — First person to use an airplane over Antarctica was Sir George Hubert Wilkins (Austral.).

1929 — Richard E. Byrd (U.S.) established Little America on Bay of Whales. On 1,600-mi airplane flight begun Nov. 28, he crossed South Pole, Nov. 29, with 3 others.

1934-35 — Byrd led 2d expedition to Little America, explored 450,000 sq mi, wintered alone at weather station, 80°08′ S.

1934-37 — John Rymill led British Graham Land expedition; discovered Palmer Penin. is part of mainland.

1935 — Lincoln Ellsworth (U.S.) flew S along E Coast of Palmer Penin., then crossed continent to Little America, making 4 landings on unprepared terrain in bad weather.

1939-41 — U.S. Antarctic Service Expedition built West Base on Ross Ice Shelf under Paul Siple, and East Base on Palmer Peninsula under Richard Black. U.S. Navy plane flights discovered about 150,000 sq mi of new land.

1940 — Byrd charted most of coast between Ross Sea and Palmer Penin.

1946-47 — U.S. Navy undertook Operation Highjump, commanded by Byrd, included 13 ships and 4,000 men. Airplanes photomapped coastline and penetrated beyond pole.

1946-48 — Ronne Antarctic Research Expedition Comdr., Finn Ronne, USNR, determined the Antarctic to be only one continent with no strait between Weddell Sea and Ross Sea; explored 250,000 sq mi of land by flights to 79° S. Mrs. Ronne and Mrs. H. Darlington were the first women to winter on Antarctica.

1955-57 — U.S. Navy's Operation Deep Freeze led by Adm. Byrd. Supporting U.S. scientific efforts for the International Geophysical Year (IGY), the operation was commanded by Rear Adm. George Dufek. It established 5 coastal stations fronting the Indian, Pacific, and Atlantic oceans and also 3 interior stations; explored more than 1,000,000 sq mi in Wilkes Land.

1957-58 — During the IGY, July 1957 through Dec. 1958, scientists from 12 countries conducted Antarctic research at a network of some 60 stations on Antarctica.

Dr. Vivian E. Fuchs led a 12-person Trans-Antarctic Expedition on the first land crossing of Antarctica. Starting from the Weddell Sea, they reached Scott Station, Mar. 2, 1958, after traveling 2,158 mi in 98 days.

1958 — A group of 5 U.S. scientists led by Edward C. Thiel, seismologist, moving by tractor from Ellsworth Station on Weddell Sea, identified a huge mountain range, 5,000 ft above the ice sheet and 9,000 ft above sea level. The range, originally seen by a Navy plane, was named the Dufek Massif, for Rear Adm. George Dufek.

1959 — Argentina, Australia, Belgium, Chile, France, Japan, New Zealand, Norway, South Africa, U.S.S.R., U.K, and U.S. signed a treaty suspending territorial claims for 30 yrs. and reserving the continent, S of 60° S, for research.

1961-62 — Scientists discovered the Bentley Trench, running from Ross Ice Shelf into Marie Byrd Land, near the end of the Ellsworth Mts., toward the Weddell Sea.

1962 — First nuclear power plant began operation at McMurdo Sound.

1963 — On Feb. 22, a U.S. plane made the longest nonstop flight ever in the South Pole area, covering 3,600 mi in 10 hr. The flight was from McMurdo Station S past the pole to Shackleton Mts., SE to the "Area of Inaccessibility," and back to McMurdo Station.

1964 — A Brit. survey team was landed by helicopter on Cook Island, the first recorded visit since 1775.

1964 — New Zealanders mapped the mountain area from Cape Adare W some 400 mi to Pennell Glacier.

1985 — Igor A. Zotikov, a Russian researcher, discovered sediments in the Ross Ice Shelf that seem to support the continental drift theory. Research by the Ocean Drilling Project off the Queen Maud Land coast indicated that the ice sheets of E Antarctica are 37 million yrs. old.

1989 — Victoria Murden and Shirley Metz became both the first women and the first Americans to reach the South Pole overland when they arrived with 9 others on Jan. 17, 1989. The 51-day trek on skis covered 740 mi.

1991 — 24 nations approved a protocol to the 1959 Antarctica Treaty, Oct. 4. New conservation provisions, including banning oil and other mineral exploration for 50 yrs.

1995 — On Dec. 22, a Norwegian, Borge Ousland, reached the South Pole in the fastest time on skis: 44 days.

1996-97 — Ousland became 1st person to traverse Antarctica alone; reached South Pole Dec. 19, 1996; traveled 1,675 mi in 64 days, ending Jan. 18, 1997.

Volcanoes

Sources: *Volcanoes of the World*, Geoscience Press; Global Volcanism Network, Smithsonian Institution

Roughly 540 volcanoes are known to have erupted during historical times. Nearly 75% of these historically active volcanoes lie along the so-called Ring of Fire, running along the W coast of the Americas from the southern tip of Chile to Alaska, down the E coast of Asia from Kamchatka to Indonesia, and continuing from New Guinea to New Zealand. The Ring of Fire marks the boundary between the mobile tectonic plates underlying the Pacific Ocean and those of the surrounding continents. Other active regions occur along rift zones, where plates pull apart, as in Iceland, or where molten material moves up from the mantle over local "hot spots," as in Hawaii. The vast majority of the earth's volcanism occurs at submarine rift zones. For more information on volcanoes, see the website at http://www.nmnh.si.edu/gvp

Notable Volcanic Eruptions

Approximately 7,000 years ago, Mazama, a 9,900-ft volcano in southern Oregon, erupted violently, ejecting large amounts of ash and pumice and voluminous pyroclastic flows. The ash spread over the entire northwestern U.S. and as far away as Saskatchewan, Can. During the eruption, the top of the mountain collapsed, leaving a caldera 6 mi across and about a half mile deep, which filled with rainwater to form what is now called Crater Lake.

In AD 79, Vesuvio or Vesuvius, a 4,190-ft volcano overlooking Naples Bay, became active after several centuries of apparent quiescence. On Aug. 24 of that year, a heated mud and ash flow swept down the mountain, engulfing the cities of Pompeii, Herculaneum, and Stabiae with debris more than 60 ft deep. About 10% of the population of the 3 towns were killed.

In 1883, an eruption similar to the Mazama eruption occurred on the island of Krakatau. At least 2,000 people died in pyroclastic flows on Aug. 26. The next day, the 2,640-ft peak of the volcano collapsed to 1,000 ft below sea level, sinking most of the island and killing over 3,000. A tsunami (tidal wave) generated by the collapse killed more than 31,000 people in Java and Sumatra, and eventually reached England. Ash from the eruption colored sunsets around the world for 2 years. A similar, even more powerful eruption had taken place 68 years earlier at Mt. Tambora on the Indonesian island of Sumbawa.

Date	Volcano	Deaths (est.)	Date	Volcano	Deaths (est.)
Aug. 24, AD 79	Mt. Vesuvius, Italy	16,000	May 8, 1902	Mt. Pelée, Martinique	28,000
1586	Kelut, Java, Indon.	10,000	Jan. 30, 1911	Mt. Taal, Phil.	1,400
Dec. 15, 1631	Mt. Vesuvius, Italy	4,000	May 19, 1919	Mt. Kelud, Java, Indon.	5,000
Aug. 12, 1772	Mt. Papandayan, Java, Indon.	3,000	Jan. 17-21, 1951	Mt. Lamington, New Guinea	3,000
June 8, 1783	Laki, Iceland	9,350	May 18, 1980	Mt. St. Helens, U.S.	57
May 21, 1792	Mt. Unzen, Japan	14,500	Mar. 28, 1982	El Chichon, Mex.	1,880
Apr. 10-12, 1815	Mt. Tambora, Sumbawa,	92,000[1]	Nov. 13, 1985	Nevado del Ruiz, Colombia	23,000
Aug. 26-28, 1883	Krakatau, Indon.	36,000	Aug. 21, 1986	Lake Nyos, Cameroon	1,700
Apr. 24, 1902	Santa María, Guatemala	1,000[2]	June 15, 1991	Mt. Pinatubo, Luzon, Phil.	800

(1) Of these, 10,000 were directly related to the eruption; an additional 82,000 were the result of starvation and disease brought on by the event. (2) An additional 3,000 deaths due to a malaria outbreak are sometimes attributed to the eruption.

Notable Active Volcanoes

Active volcanoes display a wide range of activity. In this table, years are given for last display of eruptive activity, as of mid-1999; list does not include submarine volcanoes. An eruption may involve explosive ejection of new or old fragmental material, escape of liquid lava, or both. Volcanoes are listed by height, which does not reflect eruptive magnitude.

Name (latest eruption)		Height (ft)
Africa		
Mt. Cameroon (1999)	Cameroon	13,435
Nyiragongo (1994)	Congo	11,400
Nyamuragira (1998)	Congo	10,028
Ol Doinyo Lengai (1996)	Tanzania	9,469
Fogo (1995)	Cape Verde Isls.	9,281
Karthala (1991)	Comoros.	8,000
Piton de la Fournaise (1998)	Réunion Isl., Indian O.	5,981
Lake Nyos (1986)	Cameroon	3,011
Erta-Ale (1998)	Ethiopia	1,650
Antarctica		
Erebus (1998)	Ross Isl	12,450
Deception Island (1970)	S. Shetland Isl.	1,890
Asia-Oceania		
Kliuchevskoi (1999)	Kamchatka, Russia	15,863
Kerinci (1998)	Sumatra, Indon.	12,467
Fuji (1708)	Honshu, Japan	12,388
Tolbachik (1999)	Kamchatka, Russia	12,080
Semeru (1999)	Java, Indon.	12,060
Slamet (1999)	Java, Indon.	11,247
Raung (1997)	Java, Indon.	10,932
Shiveluch (1999)	Kamchatka, Russia	10,771
On-take (1980)	Honshu, Japan	10,049
Mayon (1999)	Luzon, Phil.	9,991
Merapi (1999)	Java, Indon.	9,550
Bezymianny (1999)	Kamchatka, Russia	9,455
Ruapehu (1996)	New Zealand	9,175
Peuet Sague (1999)	Sumatra, Indon.	9,120
Heard (1993)	Indian Ocean	9,006
Baitoushan (1702)	China/Korea.	9,003
Asama (1990)	Honshu, Japan	8,300
Niigata Yake-yama (1989)	Honshu, Japan	8,111
Canlaon (1997)	Negros Isls., Phil.	8,070
Alaid (1996)	Kuril Isl., Russia.	7,674
Ulawun (1993)	Papua New Guinea	7,532
Ngauruhoe (1977)	New Zealand	7,515
Chokai (1974)	Honshu, Japan	7,300
Galunggung (1984)	Java, Indon.	7,113
Azuma (1977)	Honshu, Japan	6,700
Bagana (1998)	Papua New Guinea	6,558
Sangeang Api (1988)	Lesser Sunda Isl., Indon.	6,351
Nasu (1963)	Honshu, Japan	6,210
Tiatia (1981)	Kuril Isl., Russia.	6,013
Soputan (1996)	Sulawesi, Indon.	5,994
Bandai (1888)	Honshu, Japan	5,968
Manam (1999)	Papua New Guinea	5,928
Kuju (1996)	Kyushu, Japan.	5,866
Karangetang-Api Siau (1995)	Sangihe, Indon.	5,853
Kelud (1990)	Java, Indon.	5,679
Adatara (1996)	Honshu, Japan	5,636
Gamalama (1993)	Halmahera, Indon.	5,627
Kirishima (1992)	Kyushu, Japan.	5,577
Gamkonora (1987)	Halmahera, Indon.	5,364
Pinatubo (1995)	Luzon, Phil.	5,249
Aso (1995)	Kyushu, Japan.	5,223
Lokon-Empung (1992)	Sulawesi, Indon.	5,187
Bulusan (1995)	Luzon, Phil.	5,115
Sarychev Peak (1989)	Kuril Isl., Russia.	4,960
Karkar (1979)	Papua New Guinea	4,920
Akan (1998)	Hokkaido, Japan	4,917
Akademia (1996)	Kamchatka, Russia	4,875
Karymsky (1999)	Kamchatka, Russia	4,875
Lopevi (1999)	Vanuatu	4,755
Akita-Yake-yama (1998)	Japan.	4,482
Unzen (1996)	Kyushu, Japan.	4,462
Ambrym (1999)	Vanuatu	4,376
Langila (1999)	Papua New Guinea	4,363
Awu (1992)	Sangihe Isl., Indon.	4,350
Sakura-jima (1999)	Kyushu, Japan.	3,665
Komaga-take (1996)	Hokkaido, Japan	3,740
Dukono (1998)	Halmahera, Indonesia	3,566
Krakatau (1999)	Indonesia.	2,667
Suwanose-jima (1998)	Kyushu, Japan.	2,621
Gaua (1982)	Vanuatu	2,614
Oshima (1990)	Izu Isls., Japan.	2,487
Usu (1982)	Hokkaido, Japan.	2,398
Rabaul (1999)	Papua New Guinea	2,257
Pagan (1993)	N. Mariana Isl.	1,870
Yasur (1998)	Tanna Island, Vanuatu	1,184

Name (latest eruption)		Height (ft)
White Island (1999)	Bay of Plenty, New Zealand	1,053
Taal (1977)	Luzon, Phil.	984
McDonald Island (1999)	Indian Ocn., Australia	610
Central America—Caribbean		
Acatenango (1972)	Guatemala	12,992
Tacana (1986)	Guatemala	12,400
Santa María (1998)	Guatemala	12,375
Fuego (1999)	Guatemala	12,346
Irazú (1965)	Costa Rica	11,260
Turrialba (1866)	Costa Rica	10,958
Póas (1994)	Costa Rica	8,884
Pacaya (1999)	Guatemala	8,373
San Miguel (1986)	El Salvador	6,994
Rincón de la Vieja (1998)	Costa Rica	6,286
San Cristobal (1997)	Nicaragua	5,725
Arenal (1999)	Costa Rica	5,436
Concepción (1986)	Nicaragua	5,282
Soufrière Guadeloupe (1977)	Guadeloupe	4,813
Pelee (1932)	Martinique.	4,583
Momotombo (1905)	Nicaragua	4,127
Soufriere St. Vincent (1979)	St. Vincent	3,865
Soufriere Hills (1999)	Montserrat	3,001
Masaya (1998)	Nicaragua	2,083
South America		
Llullaillaco (1877)	Chile	22,057
Guallatiri (1960)	Chile	19,918
Cotopaxi (1940)	Ecuador	19,347
El Misti (1870?)	Peru	19,101
Tupungatito (1986)	Chile	18,504
Láscar (1995)	Chile	18,346
Ruiz (1991)	Colombia	17,457
Sangay (1998)	Ecuador	17,021
Irruputuncu (1995)	Chile	16,939
Guagua Pichincha (1999)	Ecuador	15,696
Purace (1977)	Colombia	15,601
Galeras (1993)	Colombia	14,029
Llaima (1995)	Chile	10,253
Villarrica (1998)	Chile	9,340
Cerro Hudson (1991)	Chile	8,580
Fernandina (1995)	Galapagos Isls., Ecuador	4,905
Mid-Pacific		
Mauna Loa (1984)	Hawaii, HI	13,680
Kilauea (1999)	Hawaii, HI	4,009
Mid-Atlantic Ridge		
Jan Mayen (1985)	N. Atlantic Ocn., Norway.	7,470
Grímsvötn (1998)	Iceland	5,659
Hekla (1991)	Iceland	4,892
Krafla (1984)	Iceland	2,145
Europe		
Etna (1999)	Italy.	11,053
Vesuvius (1944)	Italy.	4,203
Stromboli (1999)	Italy.	3,038
Santorini (1950)	Greece	1,850
North America		
Pico de Orizaba (1687)	Mexico	18,555
Popocatépetl (1999)	Mexico	17,930
Rainier (1894?)	Washington.	14,410
Wrangell (1907?)	Alaska.	14,163
Shasta (1786)	California	14,162
Colima (1999)	Mexico	12,361
Redoubt (1990)	Alaska.	10,197
Iliamna (1953)	Alaska.	10,016
Shishaldin (1995)	Aleutian Isl., AK	9,373
Pavlof (1997)	Alaska.	8,264
St. Helens (1991)	Washington.	8,363
Veniaminof (1995)	Alaska.	8,225
El Chichon (1982)	Mexico	7,300
Novarupta (Katmai) (1912)	Alaska.	6,715
Makushin (1987)	Aleutian Isl., AK	6,680
Great Sitkin (1974)	Aleutian Isl., AK	5,710
Cleveland (1994)	Aleutian Isl., AK	5,675
Gareloi (1989)	Aleutian Isl., AK	5,161
Atka (1996)	Aleutian Isl., AK	5,029
Korovin (1998)	Aleutian Isl., AK	4,852
Akutan (1992)	Aleutian Isl., AK	4,275
Kiska (1990)	Aleutian Isl., AK	4,275
Augustine (1986)	Alaska.	3,999
Okmok (1997)	Aleutian Isl., AK	3,520
Seguam (1993)	Aleutian Isl., AK	3,458

Mountains

Height of Mount Everest

Mt. Everest was considered 29,002 ft when Edmund Hillary and Tenzing Norgay scaled it in 1953. This triangulation figure had been accepted since 1850. In 1954 the Surveyor General of the Republic of India set the height at 29,028 ft, plus or minus 10 ft because of snow; this figure is used below. The National Geographic Society accepts it, but many mountaineering groups still use 29,002 ft.

In 1987, new calculations based on satellite measurements suggested that the Himalayan peak K-2 rose 29,064 ft above sea level and that Mt. Everest is 800 ft higher. The National Geographic Society kept to the figure of 29,028 ft.

United States, Canada, Mexico

Name	Place	Height (ft)	Name	Place	Height (ft)	Name	Place	Height (ft)
McKinley	AK	20,320	Alverstone	AK-Yukon	14,565	Shavano	CO	14,229
Logan	Yukon	19,850	Browne Tower	AK	14,530	Belford	CO	14,197
Pico de Orizaba	Mexico	18,555	Whitney	CA	14,494	Princeton	CO	14,197
St. Elias	AK-Yukon	18,008	Elbert	CO	14,433	Crestone Needle	CO	14,197
Popocatépetl	Mexico	17,930	Massive	CO	14,421	Yale	CO	14,196
Foraker	AK	17,400	Harvard	CO	14,420	Bross	CO	14,172
Iztaccihuatl	Mexico	17,343	Rainier	WA	14,410	Kit Carson	CO	14,165
Lucania	Yukon	17,147	University Peak	AK	14,410	Wrangell	AK	14,163
King	Yukon	16,971	Williamson	CA	14,375	Shasta	CA	14,162
Steele	Yukon	16,644	La Plata Peak	CO	14,361	El Diente Peak	CO	14,159
Bona	AK	16,550	Blanca Peak	CO	14,345	Point Success	WA	14,158
Blackburn	AK	16,390	Uncompahgre Peak	CO	14,309	Maroon Peak	CO	14,156
Kennedy	AK	16,286	Crestone Peak	CO	14,294	Tabeguache	CO	14,155
Sanford	AK	16,237	Lincoln	CO	14,286	Oxford	CO	14,153
Vancouver	AK-Yukon	15,979	Grays Peak	CO	14,270	Sill	CA	14,153
South Buttress	AK	15,885	Antero	CO	14,269	Sneffels	CO	14,150
Wood	Yukon	15,885	Torreys Peak	CO	14,267	Democrat	CO	14,148
Churchill	AK	15,638	Castle Peak	CO	14,265	Capitol Peak	CO	14,130
Fairweather	AK-BC	15,300	Quandary Peak	CO	14,265	Liberty Cap	WA	14,112
Zinantecatl (Toluca)	Mexico	15,016	Evans	CO	14,264	Pikes Peak	CO	14,110
Hubbard	AK-Yukon	15,015	Longs Peak	CO	14,255	Snowmass	CO	14,092
Bear	AK	14,831	McArthur	Yukon	14,253	Russell	CA	14,088
Walsh	Yukon	14,780	Wilson	CO	14,246	Eolus	CO	14,083
East Buttress	AK	14,730	White Mt. Peak	CA	14,246	Windom	CO	14,082
Matlalcueyetl	Mexico	14,636	North Palisade	CA	14,242	Columbia	CO	14,073
Hunter	AK	14,573	Cameron	CO	14,238	Augusta	AK	14,070

South America

Peak, country	Height (ft)	Peak, country	Height (ft)	Peak, country	Height (ft)
Aconcagua, Argentina	22,834	Coropuna, Peru	21,083	Solo, Argentina	20,492
Ojos del Salado, Arg.-Chile	22,572	Laudo, Argentina	20,997	Polleras, Argentina	20,456
Bonete, Argentina	22,546	Ancohuma, Bolivia	20,958	Pular, Chile	20,423
Tupungato, Argentina-Chile	22,310	Ausangate, Peru	20,945	Chani, Argentina	20,341
Pissis, Argentina	22,241	Toro, Argentina-Chile	20,932	Aucanquilcha, Chile	20,295
Mercedario, Argentina	22,211	Illampu, Bolivia	20,873	Juncal, Argentina-Chile	20,276
Huascaran, Peru	22,205	Tres Cruces, Argentina-Chile	20,853	Negro, Argentina	20,184
Llullaillaco, Argentina-Chile	22,057	Huandoy, Peru	20,852	Quela, Argentina	20,128
El Libertador, Argentina	22,047	Parinacota, Bolivia-Chile	20,768	Condoriri, Bolivia	20,095
Cachi, Argentina	22,047	Tortolas, Argentina-Chile	20,745	Palermo, Argentina	20,079
Incahuasi, Argentina-Chile	21,720	Ampato, Peru	20,702	Solimana, Peru	20,068
Yerupaja, Peru	21,709	El Condor, Argentina	20,669	San Juan, Argentina-Chile	20,049
Galan, Argentina	21,654	Salcantay, Peru	20,574	Sierra Nevada, Arg.-Chile	20,023
El Muerto, Argentina-Chile	21,457	Chimborazo, Ecuador	20,561	Antofalla, Argentina	20,013
Sajama, Bolivia	21,391	Huancarhuas, Peru	20,531	Marmolejo, Argentina-Chile	20,013
Nacimiento, Argentina	21,302	Famatina, Argentina	20,505	Chachani, Peru	19,931
Illimani, Bolivia	21,201	Pumasillo, Peru	20,492		

The highest point in the West Indies is in the Dominican Republic, Pico Duarte (10,417 ft).

Africa

Peak, country/island	Height (ft)	Peak, country/island	Height (ft)	Peak, country/island	Height (ft)
Kilimanjaro, Tanzania	19,340	Meru, Tanzania	14,979	Guna, Ethiopia	13,881
Kenya, Kenya	17,058	Karisimbi, Congo-Rwanda	14,787	Gughe, Ethiopia	13,780
Margherita Pk., Uganda-Congo	16,763	Elgon, Kenya-Uganda	14,178	Toubkal, Morocco	13,661
Ras Dashan, Ethiopia	15,158	Batu, Ethiopia	14,131	Cameroon, Cameroon	13,435

Australia, New Zealand, SE Asian Islands

Peak, country/island	Height (ft)	Peak, country/island	Height (ft)	Peak, country/island	Height (ft)
Jaya, New Guinea	16,500	Wilhelm, New Guinea	14,793	Cook, New Zealand	12,349
Trikora, New Guinea	15,585	Kinabalu, Malaysia	13,455	Semeru, Java, Indon.	12,060
Mandala, New Guinea	15,420	Kerinci, Sumatra, Indon.	12,467	Kosciusko, Australia	7,310

Europe

Peak, country	Height (ft)	Peak, country	Height (ft)	Peak, country	Height (ft)
Alps		Dent D'Herens, Switz.	13,686	Gletscherhorn, Switz.	13,068
Mont Blanc, Fr.-It.	15,771	Breithorn, It., Switz.	13,665	Schalihorn, Switz.	13,040
Monte Rosa (highest peak of group), Switz.	15,203	Bishorn, Switz.	13,645	Scerscen, Switz.	13,028
		Jungfrau, Switz.	13,642	Eiger, Switz.	13,025
Dom, Switz.	14,911	Ecrins, Fr.	13,461	Jagerhorn, Switz.	13,024
Liskamm, It., Switz.	14,852	Monch, Switz.	13,448	Rottalhorn, Switz.	13,022
Weisshorn, Switz.	14,780	Pollux, Switz.	13,422	**Pyrenees**	
Taschhorn, Switz.	14,733	Schreckhorn, Switz.	13,379	Aneto, Sp.	11,168
Matterhorn, It., Switz.	14,690	Ober Gabelhorn, Switz.	13,330	Posets, Sp.	11,073
Dent Blanche, Switz.	14,293	Gran Paradiso, It.	13,323	Perdido, Sp.	11,007
Nadelhorn, Switz.	14,196	Bernina, It., Switz.	13,284	Vignemale, Fr.-Sp.	10,820
Grand Combin, Switz.	14,154	Fiescherhorn, Switz.	13,283	Long, Sp.	10,479
Lenzpitze, Switz.	14,088	Grunhorn, Switz.	13,266	Estats, Sp.	10,304
Finsteraarhorn, Switz.	14,022	Lauteraarhorn, Switz.	13,261	Montcalm, Sp.	10,105
Castor, Switz.	13,865	Durrenhorn, Switz.	13,238	**Caucasus (Europe-Asia)**	
Zinalrothorn, Switz.	13,849	Allalinhorn, Switz.	13,213	Elbrus, Russia	18,510
Hohberghom, Switz.	13,842	Weissmies, Switz.	13,199	Shkhara, Georgia	17,064
Alphubel, Switz.	13,799	Lagginhorn, Switz.	13,156	Dykh Tau, Russia	17,054
Rimpfischhom, Switz.	13,776	Zupo, Switz.	13,120	Kashtan Tau, Russia	16,877
Aletschorn, Switz.	13,763	Fletschhorn, Switz.	13,110	Janqi, Georgia	16,565
Strahlhorn, Switz.	13,747	Adlerhorn, Switz.	13,081	Kazbek, Georgia	16,558

Asia (Mainland)

Peak	Place	Height (ft)	Peak	Place	Height (ft)	Peak	Place	Height (ft)
Everest	Nepal-Tibet	29,028	Kungur	Xinjiang	25,325	Badrinath	India	23,420
K2 (Godwin Austen)	Kashmir	28,250	Tirich Mir	Pakistan	25,230	Nunkun	Kashmir	23,410
Kanchenjunga	India-Nepal	28,208	Makalu II	Nepal-Tibet	25,120	Lenin Peak	Tajikistan	23,405
Lhotse I (Everest)	Nepal-Tibet	27,923	Minya Konka	China	24,900	Pyramid	India-Nepal	23,400
Makalu I	Nepal-Tibet	27,824	Kula Gangri	Bhutan-Tibet	24,784	Api	Nepal	23,399
Lhotse II (Everest)	Nepal-Tibet	27,560	Changtzu (Everest)	Nepal-Tibet	24,780	Pauhunri	India-Tibet	23,385
Dhaulagiri	Nepal	26,810	Muz Tagh Ata	Xinjiang	24,757	Trisul	India	23,360
Manaslu I	Nepal	26,760	Skyang Kangri	Kashmir	24,750	Kangto	India-Tibet	23,260
Cho Oyu	Nepal-Tibet	26,750	Ismail Semani Peak	Tajikistan	24,590	Nyenchhe		
Nanga Parbat	Kashmir	26,660	Jongsang Peak	India-Nepal	24,472	Thanglha	Tibet	23,255
Annapurna I	Nepal	26,504	Jengish Chokusu	Xinjiang-Kyrgyzstan	24,406	Trisuli	India	23,210
Gasherbrum	Kashmir	26,470	Sia Kangri	Kashmir	24,350	Pumori	Nepal-Tibet	23,190
Broad	Kashmir	26,400	Haramosh Peak	Pakistan	24,270	Dunagiri	India	23,184
Gosainthan	Tibet	26,287	Istoro Nal	Pakistan	24,240	Lombo Kangra	Tibet	23,165
Annapurna II	Nepal	26,041	Tent Peak	India-Nepal	24,165	Saipal	Nepal	23,100
Gyachung Kang	Nepal-Tibet	25,910	Chomo Lhari	Bhutan-Tibet	24,040	Macha Pucchare	Nepal	22,958
Disteghil Sar	Kashmir	25,868	Chamlang	Nepal	24,012	Numbar	Nepal	22,817
Himalchuli	Nepal	25,801	Kabru	India-Nepal	24,002	Kanjiroba	Nepal	22,580
Nuptse (Everest)	Nepal-Tibet	25,726	Alung Gangri	Tibet	24,000	Ama Dablam	Nepal	22,350
Masherbrum	Kashmir	25,660	Baltoro Kangri	Kashmir	23,990	Cho Polu	Nepal	22,093
Nanda Devi	India	25,645	Mussu Shan	Xinjiang	23,890	Lingtren	Nepal-Tibet	21,972
Rakaposhi	Kashmir	25,550	Mana	India	23,860	Khumbutse	Nepal-Tibet	21,785
Kamet	India-Tibet	25,447	Baruntse	Nepal	23,688	Hlako Gangri	Tibet	21,266
Namcha Barwa	Tibet	25,445	Nepal Peak	India-Nepal	23,500	Mt. Grosvenor	China	21,190
Gurla Mandhata	Tibet	25,355	Amne Machin	China	23,490	Thagchhab Gangri	Tibet	20,970
Ulugh Muz Tagh	Xinjiang-Tibet	25,340	Gauri Sankar	Nepal-Tibet	23,440	Damavand	Iran	18,606
						Ararat	Turkey	16,804

Antarctica

Peak	Height (ft)	Peak	Height (ft)	Peak	Height (ft)
Vinson Massif	16,864	Miller	13,650	Falla	12,549
Tyree	16,290	Long Gables	13,620	Rucker	12,520
Shinn	15,750	Dickerson	13,517	Goldthwait	12,510
Gardner	15,375	Giovinetto	13,412	Morris	12,500
Epperly	15,100	Wade	13,400	Erebus	12,450
Kirkpatrick	14,855	Fisher	13,386	Campbell	12,434
Elizabeth	14,698	Fridtjof Nansen	13,350	Don Pedro Christophersen	12,355
Markham	14,290	Wexler	13,202	Lysaght	12,326
Bell	14,117	Lister	13,200	Huggins	12,247
Mackellar	14,098	Shear	13,100	Sabine	12,200
Anderson	13,957	Odishaw	13,008	Astor	12,175
Bentley	13,934	Donaldson	12,894	Mohl	12,172
Kaplan	13,878	Ray	12,808	Frankes	12,064
Andrew Jackson	13,750	Sellery	12,779	Jones	12,040
Sidley	13,720	Waterman	12,730	Gjelsvik	12,008
Ostenso	13,710	Anne	12,703	Coman	12,000
Minto	13,668	Press	12,566		

Some Notable U.S. Mountains

Name	Place	Height (ft)	Name	Place	Height (ft)	Name	Place	Height (ft)
Gannett Peak	WY	13,804	Adams	WA	12,277	Clingmans Dome	NC-TN	6,643
Grand Teton	WY	13,766	San Gorgonio	CA	11,502	Washington	NH	6,288
Kings	UT	13,528	Hood	OR	11,239	Rogers	VA	5,729
Cloud	WY	13,175	Lassen	CA	10,457	Marcy	NY	5,344
Wheeler	NM	13,161	Granite	CA	10,321	Katahdin	ME	5,268
Boundary	NV	13,140	Guadalupe	TX	8,749	Spruce Knob	WV	4,861
Granite	MT	12,799	Olympus	WA	7,965	Mansfield	VT	4,393
Borah	ID	12,662	Harney	SD	7,242	Black Mountain	KY	4,145
Humphreys	AZ	12,633	Mitchell	NC	6,684			

Important Islands and Their Areas

Reviewed by Laurel Duda, Marine Biological Laboratory/Woods Hole Oceanographic Inst. Library.

Figure in parentheses shows rank among the world's 10 largest individual islands. Because some islands have not been surveyed accurately, some areas shown are estimates. Figures are for total areas in square miles. Some "islands" listed are island groups. Only the largest islands in a group are listed individually. Only islands 10 sq. miles or larger are listed.

Antarctica

Island	Area
Adelaide	1,400
Alexander	16,700
Berkner	18,500
Roosevelt	2,900

Arctic Ocean

Island	Area
Akimiski, Nunavut	1,159
Amund Ringnes, Nun.	2,029
Axel Heiberg, Nun.	16,671
Baffin, (5)	195,928
Banks, Northwest Territories	27,038
Bathurst, Nun.	6,194
Bolshevik, Russia	4,368
Bolshoy Lyakhovsky, Russia	1,776
Borden, NWT., Nun.	1,079
Bylot, Nun.	4,273
Coats, Nun.	2,123
Cornwallis, Nun.	2,701
Devon, Nun.	21,331
Disko, Greenland	3,312
Ellef Ringnes, Nun.	4,361
Ellesmere, Nun. (10)	75,767
Faddayevskiy, Russia	1,930
Franz Josef Land, Russia	8,000
Iturup (Etorofu), Russia	2,596
King William, Nun.	5,062
Komsomolets, Russia	3,477
Mackenzie King, NWT	1,949
Mansel, Nun.	1,228
Melville, NWT, Nun.	16,274
Milne Land, Greenland	1,400
New Siberian Islands, Russia	14,500
Kotelnyy, Russia	4,504
Novaya Zemlya, Russia (2 isls.)	31,730
Oktyabrskoy, Russia	5,471
Prince Charles, NWT	3,676
Prince of Wales, Nun.	12,872
Prince Patrick, NWT	6,119
Somerset, Nun.	9,570
Southampton, Nun.	15,913
Svalbard (tot. group)	23,957
Nordaustlandet	5,410
Spitsbergen	15,060
Traill, Greenland	1,300
Victoria, NWT, Nun. (9)	83,897
Wrangel, Russia	2,800

Atlantic Ocean

Island	Area
Anticosti, Canada	3,068
Ascension, UK	34
Azores, Portugal (tot. group)	868
Faial	67
San Miguel	291
Bahama Isls., Bahama (tot. group)	5,382
Andros, Bahamas	2,300
Bermuda Islands, UK	20
Bioko Isl., Equatorial Guinea	785
Block Islands, RI, US	21
Canary Islands, Spain (tot. group)	2,807
Fuerteventura	688
Gran Canaria	592
Tenerife	795
Cape Breton, Canada	3,981
Cape Verde Islands	1,557
Caviana, Para, Brazil	1,918
Channel Islands, UK (tot. group)	75
Guernsey	24
Jersey	45
Faroe Islands, Denmark	540
Falkland Islands, UK (tot. group)	4,700
East Falkland	2,550
West Falkland	1,750
Great Britain, UK (8)	84,200
Greenland, Denmark (1)	840,000
Gurupa, Para, Brazil	1,878
Hebrides, Scotland	2,744
Iceland	39,699
Ireland (tot. group)	32,589
Irish Republic	27,137
Northern Ireland	5,452
Isle of Man, UK	227
Isle of Wight, England	147
Long Island, NY, US	1,320
Madeira Islands, Portugal	306

Island	Area
Marajo, Brazil	15,444
Martha's Vineyard, MA, US	89
Mount Desert, ME, US	104
Nantucket, MA, US	45
Newfoundland, Canada	42,031
Orkney Islands, Scotland	390
Prince Edward, Canada	2,185
St. Helena, UK	47
Shetland Islands, Scotland	587
Skye, Scotland	670
South Georgia, UK	1,450
Tierra del Fuego, Chile, Arg.	18,800
Tristan da Cunha, UK	40

Baltic Sea

Island	Area
Aland Islands, Finland	590
Bornholm, Denmark	227
Gotland, Sweden	1,159

Caribbean Sea

Island	Area
Antigua	108
Aruba, Netherlands	75
Barbados	166
Cuba	42,804
Isle of Youth	926
Cayman Islands	100
Curacao, Netherlands	171
Dominica	290
Guadeloupe, France	687
Hispaniola (Haiti and Dominican Rep)	29,389
Jamaica	4,244
Martinique, France	436
Puerto Rico, US	3,339
Tobago	116
Trinidad	1,864
Virgin Islands, UK	59
Virgin Islands, US	134

East Indies

Island	Area
Bali, Indonesia	2,171
Bangka, Indonesia	4,375
Borneo, Indonesia-Malaysia-Brunei (3)	280,100
Bougainville, Papua New Guinea	3,880
Buru, Indonesia	3,670
Celebes, Indonesia	69,000
Flores, Indonesia	5,500
Halmahera, Indonesia	6,865
Java (Jawa), Indonesia	48,900
Madura, Indonesia	2,113
Moluccas, Indonesia	32,307
New Britain, Papua New Guinea	14,093
New Guinea, Indon.-PNG (2)	306,000
New Ireland, PNG	3,707
Seram, Indonesia	6,621
Sumba, Indonesia	4,306
Sumbawa, Indonesia	5,965
Sumatra, Indonesia (6)	165,000
Timor, Indonesia	13,094
Yos Sudarsa, Indonesia	4,500

Indian Ocean

Island	Area
Andaman Isls., India	2,500
Kerguelen	2,247
Madagascar (4)	226,658
Mauritius	720
Pemba, Tanzania	380
Reunion, France	970
Seychelles	176
Sri Lanka	25,332
Zanzibar, Tanzania	640

Mediterranean Sea

Island	Area
Balearic Isls., Spain	1,927
Corfu, Greece	229
Corsica, France	3,369
Crete, Greece	3,189
Cyprus	3,572
Elba, Italy	86
Euboea, Greece	1,411
Malta	95
Rhodes, Greece	540
Sardinia, Italy	9,301
Sicily, Italy	9,926

Pacific Ocean

Island	Area
Admiralty, AK, US	1,709
Aleutian Isls., AK, US (tot. group)	6,912
Adak	275
Amchitka	116
Attu	350
Kanaga	142
Kiska	106
Tanaga	195
Umnak	686
Unalaska	1,051
Unimak	1,571
Baranof, AK, US	1,636
Chichagof, AK, US	2,062
Chiloe, Chile	3,241
Christmas, Kiribati	94
Diomede, Big, Russia	11
Easter Isl., Chile	69
Fiji (tot. group)	7,056
Vanua Levu	2,242
Viti Levu	4,109
Galapagos Isls., Ecuador	3,043
Graham Isl., British Columbia	2,456
Guadalcanal, Solomon Isls.	2,180
Guam, US	210
Hainan, China	13,000
Hawaiian Isls., HI, US (tot. group)	6,428
Hawaii	4,028
Oahu	600
Hong Kong, China	31
Hoste, Chile	1,590
Japan (tot. group)	145,850
Hokkaido	30,144
Honshu (7)	87,805
Kyushu	14,114
Okinawa	459
Shikoku	7,049
Kangaroo, South Australia	1,680
Kodiak, AK, US	3,485
Kupreanof, AK, US	1,084
Marquesas Isls., France	492
Marshall Isls.	70
Melville, Northern Territory, Aus.	2,240
Micronesia	271
New Caledonia, France	6,530
New Zealand (tot. group)	104,454
Chatham Isls.	372
North	44,204
South	58,384
Stewart	674
North Mariana Isls., US.	179
Nunivak, AK, US	1,600
Palau	188
Philippines (tot. group)	115,860
Leyte	2,787
Luzon	40,680
Mindanao	36,775
Mindoro	3,690
Negros	4,907
Palawan	4,554
Panay	4,446
Samar	5,050
Prince of Wales, AK, US	2,770
Revillagigedo, AK, US	1,134
Riesco, Chile	1,973
St. Lawrence, AK, US	1,780
Sakhalin, Russia	29,500
Samoa Isls. (tot. group)	1,177
American Samoa, US	77
Tutuila, US	55
Savail, Samoa	659
Upolu, Samoa	432
Santa Catalina, CA, US.	75
Santa Ines, Chile	1,407
Tahiti, France	402
Taiwan, China (tot. group)	13,969
Jinmen Dao (Quemoy)	56
Tasmania, Australia	26,178
Tonga Isls.	290
Vancouver Isl., Brit. Columbia	12,079
Vanuatu	4,707
Wellington, Chile	2,549

Persian Gulf

Island	Area
Bahrain	217

Areas and Average Depths of Oceans, Seas, and Gulfs

Geographers and mapmakers recognize 4 major bodies of water: the Pacific, the Atlantic, the Indian, and the Arctic oceans. The Atlantic and Pacific oceans are considered divided at the equator into the N and S Atlantic and the N and S Pacific. The Arctic Ocean is the name for waters N of the continental landmasses in the region of the Arctic Circle.

	Area (sq mi)	Avg. depth (ft)		Area (sq mi)	Avg. depth (ft)
Pacific Ocean	64,186,300	12,925	Hudson Bay	281,900	305
Atlantic Ocean	33,420,000	11,730	East China Sea	256,600	620
Indian Ocean	28,350,500	12,598	Andaman Sea	218,100	3,667
Arctic Ocean	5,105,700	3,407	Black Sea	196,100	3,906
South China Sea	1,148,500	4,802	Red Sea	174,900	1,764
Caribbean Sea	971,400	8,448	North Sea	164,900	308
Mediterranean Sea.	969,100	4,926	Baltic Sea	147,500	180
Bering Sea	873,000	4,893	Yellow Sea.	113,500	121
Gulf of Mexico	582,100	5,297	Persian Gulf.	88,800	328
Sea of Okhotsk.	537,500	3,192	Gulf of California	59,100	2,375
Sea of Japan	391,100	5,468			

Principal Ocean Depths

Source: National Imagery and Mapping Agency, U.S. Dept. of Defense

Name of area	Location		Depth		
	(lat.)	(long.)	(meters)	(fathoms)	(ft)
Pacific Ocean					
Mariana Trench	11 deg. 22 min. N	142 deg. 36 min. E	10,924	5,973	35,840
Tonga Trench	23 deg. 16 min. S	174 deg. 44 min. W	10,800	5,906	35,433
Philippine Trench	10 deg. 38 min. N	126 deg. 36 min. E	10,057	5,499	32,995
Kermadec Trench	31 deg. 53 min. N	177 deg. 21 min. W	10,047	5,494	32,963
Bonin Trench	24 deg. 30 min. N	143 deg. 24 min. E	9,994	5,464	32,788
Kuril Trench	44 deg. 15 min. N	150 deg. 34 min. E	9,750	5,331	31,988
Izu Trench.	31 deg. 05 min. N	142 deg. 10 min. E	9,695	5,301	31,808
New Britain Trench	06 deg. 19 min. S	153 deg. 45 min. E	8,940	4,888	29,331
Yap Trench	08 deg. 33 min. N	138 deg. 02 min. E	8,527	4,663	27,976
Japan Trench	36 deg. 08 min. N	142 deg. 43 min. E	8,412	4,600	27,599
Peru-Chile Trench.	23 deg. 18 min. S	71 deg. 14 min. W	8,064	4,409	26,457
Palau Trench	07 deg. 52 min. N	134 deg. 56 min. E	8,054	4,404	26,424
Aleutian Trench	50 deg. 51 min. N	177 deg. 11 min. E	7,679	4,199	25,194
New Hebrides Trench.	20 deg. 36 min. S	168 deg. 37 min. E	7,570	4,139	24,836
North Ryukyu Trench	24 deg. 00 min. N	126 deg. 48 min. E	7,181	3,927	23,560
Mid. America Trench	14 deg. 02 min. N	93 deg. 39 min. W	6,662	3,643	21,857
Atlantic Ocean					
Puerto Rico Trench	19 deg 55 min. N	65 deg. 27 min. W	8,605	4,705	28,232
S Sandwich Trench	55 deg. 42 min. N	25 deg. 56 min. E	8,325	4,552	27,313
Romanche Gap	0 deg. 13 min. S	18 deg. 26 min. W	7,728	4,226	25,354
Cayman Trench	19 deg. 12 min. N	80 deg. 00 min. W	7,535	4,120	24,721
Brazil Basin	09 deg. 10 min. S	23 deg. 02 min. W	6,119	3,346	20,076
Indian Ocean					
Java Trench	10 deg. 19 min. S	109 deg. 58 min. E	7,125	3,896	23,376
Ob' Trench	09 deg. 45 min. S	67 deg. 18 min. E	6,874	3,759	22,553
Diamantina Trench	35 deg. 50 min. S	105 deg. 14 min. E	6,602	3,610	21,660
Vema Trench	09 deg. 08 min. S	67 deg. 15 min. E	6,402	3,501	21,004
Agulhas Basin	45 deg. 20 min. S	26 deg. 50 min. E	6,195	3,387	20,325
Arctic Ocean					
Eurasia Basin.	82 deg. 23 min. N	19 deg. 31 min. E	5,450	2,980	17,881
Mediterranean Sea					
Ionian Basin	36 deg. 32 min. N	21 deg. 06 min. E	5,150	2,816	16,896

Note: Greater depths have been reported in some areas but are not officially confirmed by research vessels.

Latitude, Longitude, and Altitude of World Cities

Source: National Imagery Mapping Agency, U.S. Dept. of Defense

City	Lat.		Long.		Alt. (ft)	City	Lat.		Long.		Alt. (ft)
	°	′	°	′			°	′	°	′	
Athens, Greece	37	59 N	23	44 E	300	Mexico City, Mexico	19	24 N	99	09 W	7,347
Bangkok, Thailand	13	45 N	100	31 E	0	Moscow, Russia	55	45 N	37	35 E	394
Beijing, China.	39	56 N	116	24 E	600	New Delhi, India	28	36 N	77	12 E	770
Berlin, Germany	52	31 N	13	25 E	110	Panama City, Panama	08	58 N	79	32 W	0
Bogotá, Colombia.	04	36 N	74	05 W	8,660	Paris, France	48	52 N	02	20 E	300
Bombay (Mumbai), India .	18	58 N	72	50 E	27	Quito, Ecuador.	00	13 S	78	30 W	9,222
Buenos Aires, Argentina .	34	36 S	58	28 W	0	Rio de Janeiro, Brazil	22	43 S	43	13 W	30
Cairo, Egypt	30	03 N	31	15 E	381	Rome, Italy	41	53 N	12	30 E	95
Jakarta, Indonesia	06	10 S	106	48 E	26	Santiago, Chile	33	27 S	70	40 W	4,921
Jerusalem, Israel	31	46 N	35	14 E	2,500	Seoul, South Korea	37	34 N	127	00 E	34
Johannesburg, So. Afr.. . . .	26	12 S	28	05 E	5,740	Sydney, Australia.	33	53 S	151	12 E	25
Kathmandu, Nepal	27	43 N	85	19 E	4,500	Tehran, Iran	35	40 N	51	26 E	3,937
Kiev, Ukraine	50	26 N	30	31 E	587	Tokyo, Japan	35	42 N	139	46 E	30
London, UK (Greenwich) .	51	30 N	00	00	245	Warsaw, Poland.	52	15 N	21	00 E	360
Manila, Philippines	14	35 N	120	00 E	0	Wellington, New Zealand. .	41	18 S	174	47 E	0

Latitude, Longitude, and Altitude of U.S. and Canadian Cities

Source: U.S. geographic positions, U.S. altitudes provided by Geological Survey, U.S. Dept. of the Interior. Canadian geographic positions and altitudes provided by the Canada Flight Supplement, Natural Resources Canada.

City	Lat. N °	′	″	Long. W °	′	″	Elev. (ft)
Abilene, TX	32	26	55	99	43	58	1,718
Akron, OH	41	4	53	81	31	9	1,050
Albany, NY	42	39	9	73	45	24	20
Albuquerque, NM	35	5	4	106	39	2	4,955
Alert, N.W.T.	82	31	04	62	16	50	100
Allentown, PA	40	36	30	75	29	26	350
Amarillo, TX	35	13	19	101	49	51	3,685
Anchorage, AK	61	13	5	149	54	1	101
Ann Arbor, MI	42	16	15	83	43	35	880
Asheville, NC	35	36	3	82	33	15	2,134
Ashland, KY	38	28	42	82	38	17	558
Atlanta, GA	33	44	56	84	23	17	1,050
Atlantic City, NJ	39	21	51	74	25	24	8
Augusta, GA	33	28	15	81	58	30	414
Augusta, ME	44	18	38	69	46	48	45
Austin, TX	30	16	1	97	44	34	501
Bakersfield, CA	35	22	24	119	1	4	408
Baltimore, MD	39	17	25	76	36	45	100
Bangor, ME	44	48	4	68	46	42	158
Baton Rouge, LA	30	27	2	91	9	16	53
Battle Creek, MI	42	19	16	85	10	47	820
Bay City, MI	43	35	40	83	53	20	595
Beaumont, TX	30	5	9	94	6	6	20
Belleville, Ont.	44	11	32	77	18	34	320
Bellingham, WA	48	45	35	122	29	13	100
Berkeley, CA	37	52	18	122	16	18	150
Billings, MT	45	47	0	108	30	0	3,124
Biloxi, MS	30	23	45	88	53	7	25
Binghamton, NY	42	5	55	75	55	6	865
Birmingham, AL	33	31	14	86	48	9	600
Bismarck, ND	46	48	30	100	47	0	1,700
Bloomington, IL	40	29	3	88	59	37	829
Boise, ID	43	36	49	116	12	9	2,730
Boston, MA	42	21	30	71	3	37	20
Bowling Green, KY	36	59	25	86	26	37	510
Brandon, Man.	49	54	35	99	57	03	1,343
Brantford, Ont.	43	07	53	80	20	33	815
Brattleboro, VT	42	51	3	72	33	30	240
Bridgeport, CT	41	10	1	73	12	19	10
Brockton, MA	42	5	0	71	1	8	112
Buffalo, NY	42	53	11	78	52	43	585
Burlington, Ont.	43	26	33	79	51	03	640
Burlington, VT	44	28	33	73	12	45	113
Butte, MT	46	0	14	112	32	2	5,549
Calgary, Alta.	51	06	50	114	01	13	3,557
Cambridge, MA	42	22	30	71	6	22	30
Canton, OH	40	47	56	81	22	43	1,100
Carson City, NV	39	9	50	119	45	59	4,730
Cedar Rapids, IA	42	0	30	91	38	38	730
Central Islip, NY	40	47	26	73	12	8	88
Champaign, IL	40	6	59	88	14	36	740
Charleston, SC	32	46	35	79	55	52	118
Charleston, WV	38	20	59	81	37	58	606
Charlotte, NC	35	13	37	80	50	36	850
Charlottetown, P.E.I.	46	17	24	63	07	16	160
Chattanooga, TN	35	2	44	85	18	35	685
Cheyenne, WY	41	8	24	104	49	11	6,067
Chicago, IL	41	51	0	87	39	0	596
Churchill, Man.	58	44	14	94	03	26	94
Cincinnati, OH	39	9	43	84	27	25	683
Cleveland, OH	41	29	58	81	41	44	690
Colorado Springs, CO	38	50	2	104	49	15	6,008
Columbia, MO	38	57	6	92	20	2	758
Columbia, SC	34	0	2	81	2	6	314
Columbus, GA	32	27	39	84	59	16	300
Columbus, OH	39	57	40	82	59	56	800
Concord, NH	43	12	29	71	32	17	288
Corpus Christi, TX	27	48	1	97	23	46	35
Dallas, TX	32	47	0	96	48	0	463
Dawson, Yukon	64	02	35	139	07	40	1,214
Dayton, OH	39	45	32	84	11	30	750
Daytona Beach, FL	29	12	38	81	1	23	10
Decatur, IL	39	50	25	88	57	17	670
Denver, CO	39	44	21	104	59	3	5,260
Des Moines, IA	41	36	2	93	36	32	803
Detroit, MI	42	19	53	83	2	45	585
Dodge City, KS	37	45	10	100	1	0	2,550
Dubuque, IA	42	30	2	90	39	52	620
Duluth, MN	46	47	0	92	6	23	610
Durham, NC	35	59	38	78	53	56	394
Eau Claire, WI	44	48	41	91	29	54	850
Edmonton, Alta.	53	34	21	113	31	14	2,200
Elizabeth, NJ	40	39	50	74	12	40	38
El Paso, TX	31	45	31	106	29	11	3,695
Enid, OK	36	23	44	97	52	41	1,246
Erie, PA	42	7	45	80	5	7	650
Eugene, OR	44	3	8	123	5	8	419
Eureka, CA	40	48	8	124	9	45	44
Evansville, IN	37	58	29	87	33	21	388
Fairbanks, AK	64	50	16	147	42	59	440
Fall River, MA	41	42	5	71	9	20	200
Fargo, ND	46	52	38	96	47	22	900
Flagstaff, AZ	35	11	53	111	39	2	6,900
Flint, MI	43	0	45	83	41	15	750
Ft. Smith, AR	35	23	9	94	23	54	446
Ft. Wayne, IN	41	7	50	85	7	44	781
Ft. Worth, TX	32	43	31	97	19	14	670
Fredericton, N.B.	45	52	10	66	31	54	67
Fresno, CA	36	44	52	119	46	17	296
Gadsden, AL	34	0	51	86	0	24	554
Gainesville, FL	29	39	5	82	19	30	183
Gallup, NM	35	31	41	108	44	31	6,508
Galveston, TX	29	18	4	94	47	51	10
Gary, IN	41	35	36	87	20	47	600
Grand Junction, CO	39	3	50	108	33	0	4,597
Grand Rapids, MI.	42	57	48	85	40	5	610
Great Falls, MT	47	30	1	111	18	0	3,334
Green Bay, WI	44	31	9	88	1	11	594
Greensboro, NC	36	4	21	79	47	32	770
Greenville, SC	34	51	9	82	23	39	966
Guelph, Ont.	43	33	0	80	16	0	1,100
Gulfport, MS	30	22	2	89	5	34	25
Halifax, N.S.	44	52	51	63	30	31	477
Hamilton, OH	39	23	58	84	33	41	600
Hamilton, Ont.	43	10	19	79	55	53	780
Harrisburg, PA	40	16	25	76	53	5	320
Hartford, CT	41	45	49	72	41	8	40
Helena, MT	46	35	34	112	2	7	4,090
Hilo, HI	19	43	47	155	5	24	38
Honolulu, HI	21	18	25	157	51	30	18
Houston, TX	29	45	47	95	21	47	40
Huntsville, AL	34	43	49	86	35	10	641
Indianapolis, IN	39	46	6	86	9	29	717
Iowa City, IA	41	39	40	91	31	48	685
Jackson, MI	42	14	45	84	24	5	940
Jackson, MS	32	17	55	90	11	5	294
Jacksonville, FL	30	19	55	81	39	21	12
Jersey City, NJ	40	43	41	74	4	41	83
Johnstown, PA	40	16	42	76	19	0	521
Joplin, MO	37	5	3	94	30	47	990
Juneau, AK	58	18	7	134	25	11	50
Kalamazoo, MI	42	17	30	85	35	14	755
Kansas City, KS	39	6	51	94	37	38	750
Kansas City, MO	39	5	59	94	34	42	740
Kenosha, WI	42	35	5	87	49	16	610
Key West, FL	24	33	19	81	46	58	8
Kingston, Ont.	44	13	31	76	35	49	305
Kitchener, Ont.	43	27	32	80	23	04	1,040
Knoxville, TN	35	57	38	83	55	15	889
Lafayette, IN	40	25	0	86	52	31	567
Lancaster, PA	40	2	16	76	18	21	368
Lansing, MI	42	43	57	84	33	20	830
Laredo, TX	27	30	22	99	30	26	414
Las Vegas, NV	36	10	30	115	8	11	2,000
Lawrence, MA	42	42	25	71	9	49	50
Lethbridge, Alta.	49	37	49	112	47	59	3,047
Lexington, KY	37	59	19	84	28	40	955
Lihue, HI	21	58	52	159	22	16	206
Lima, OH	40	44	33	84	6	19	875
Lincoln, NE	40	48	0	96	40	0	1,150
Little Rock, AR	34	44	47	92	17	22	350
London, Ont.	42	57	31	81	13	33	875
Los Angeles, CA	34	3	8	118	14	34	330

City	Lat. N °	'	"	Long. W °	'	"	Elev. (ft)
Louisville, KY	38	15	15	85	45	34	462
Lowell, MA	42	38	0	71	19	0	102
Lubbock, TX	33	34	40	101	51	17	3,195
Macon, GA	32	50	26	83	37	57	400
Madison, WI	43	4	23	89	24	4	863
Manchester, NH	42	59	44	71	27	19	175
Marshall, TX	32	32	41	94	22	2	410
Medicine Hat, Alta.	50	01	08	110	43	15	2,352
Memphis, TN	35	8	58	90	2	56	254
Meriden, CT	41	32	17	72	48	27	190
Miami, FL	25	46	26	80	11	38	11
Milwaukee, WI	43	2	20	87	54	23	634
Minneapolis, MN	44	58	48	93	15	49	815
Minot, ND	48	13	57	101	17	45	1,555
Mobile, AL	30	41	39	88	2	35	16
Moncton, N.B.	46	06	44	64	40	57	232
Montgomery, AL	32	22	0	86	18	0	250
Montpelier, VT	44	15	36	72	34	33	525
Montréal, Que.	45	41	06	73	55	52	221
Moose Jaw, Sask.	50	19	48	105	33	29	1,892
Muncie, IN	40	11	36	85	23	11	952
Nashville, TN	36	9	57	86	47	4	440
Natchez, MS	31	33	37	91	24	11	230
Newark, NJ	40	44	8	74	10	22	95
New Britain, CT	41	39	40	72	46	48	200
New Haven, CT	41	18	29	72	55	43	40
New Orleans, LA	29	57	16	90	4	30	11
New York, NY	40	42	51	74	0	23	55
Niagara Falls, Ont.	43	07	0	79	04	0	589
Nome, AK	64	30	4	165	24	23	25
Norfolk, VA	36	50	48	76	17	8	10
North Bay, Ont.	46	26	0	79	28	0	1,200
Oakland, CA	37	48	16	122	16	11	42
Ogden, UT	41	13	23	111	58	23	4,299
Oklahoma City, OK	35	28	3	97	30	58	1,195
Omaha, NE	41	15	31	95	56	15	1,040
Orlando, FL	28	32	17	81	22	46	106
Ottawa, Ont.	45	19	09	76	01	20	382
Paducah, KY	37	5	0	88	36	0	345
Pasadena, CA	34	8	52	118	8	37	865
Paterson, NJ	40	55	0	74	10	20	70
Pensacola, FL	30	25	16	87	13	1	32
Peoria, IL	40	41	37	89	35	20	470
Peterborough, Ont.	44	13	48	78	21	48	628
Philadelphia, PA	39	57	8	75	9	51	40
Phoenix, AZ	33	26	54	112	4	24	1,090
Pierre, SD	44	22	6	100	21	2	1,484
Pittsburgh, PA	40	26	26	79	59	46	770
Pittsfield, MA	42	27	0	73	14	45	1,039
Pocatello, ID	42	52	17	112	26	41	4,464
Pt. Arthur, TX	29	53	55	93	55	43	10
Portland, ME	43	39	41	70	15	21	25
Portland, OR	45	31	25	122	40	30	50
Portsmouth, NH	43	4	18	70	45	47	21
Portsmouth, VA	36	50	7	76	17	55	10
Prince Rupert, B.C.	54	17	10	130	26	41	116
Providence, RI	41	49	26	71	24	48	80
Provo, UT	40	14	2	111	39	28	4,549
Pueblo, CO	38	15	16	104	36	31	4,662
Québec City, Que.	46	47	36	71	23	29	244
Racine, WI	42	43	34	87	46	58	630
Raleigh, NC	35	46	19	78	38	20	350
Rapid City, SD	44	4	50	103	13	50	3,247
Reading, PA	40	20	8	75	55	38	266
Regina, Sask.	50	25	55	104	39	57	1,894
Reno, NV	39	31	47	119	48	46	4,498
Richmond, VA	37	33	13	77	27	38	190
Roanoke, VA	37	16	15	79	56	30	940
Rochester, MN	44	1	18	92	28	11	990
Rochester, NY	43	9	17	77	36	57	515
Rockford, IL	42	16	16	89	5	38	715
Sacramento, CA	38	34	54	121	29	36	20
Saginaw, MI	43	25	10	83	57	3	595
St. Catharines, Ont.	43	11	30	79	10	18	321
St. Cloud, MN	45	33	39	94	9	44	1,040
St. John, N.B.	45	18	58	65	53	25	357
St. John's, Nfld.	47	37	07	52	45	07	461
St. Joseph, MO	39	46	7	94	50	47	850
St. Louis, MO	38	37	38	90	11	52	455
St. Paul, MN.	44	56	40	93	5	35	780
St. Petersburg, FL	27	46	14	82	40	44	44
Salem, OR	44	56	35	123	2	2	154
Salina, KS	38	50	25	97	36	40	1,225
Salt Lake City, UT	40	45	39	111	53	25	4,266
San Antonio, TX	29	25	26	98	29	36	650
San Bernardino, CA	34	6	30	117	17	20	1,200
San Diego, CA	32	42	55	117	9	23	40
San Francisco, CA	37	46	30	122	25	6	63
San Jose, CA	37	20	22	121	53	38	87
San Juan, P.R.	18	28	6	66	6	22	8
Santa Barbara, CA	34	25	15	119	41	50	50
Santa Cruz, CA	36	58	27	122	1	47	20
Santa Fe, NM	35	41	13	105	56	14	6,989
Sarasota, FL	27	20	10	82	31	51	27
Saskatoon, Sask.	52	10	15	106	41	59	1,653
Sault Ste. Marie, Ont.	46	29	06	84	30	34	630
Savannah, GA	32	5	0	81	6	0	42
Schenectady, NY	42	48	51	73	56	24	245
Seattle, WA	47	36	23	122	19	51	350
Sheboygan, WI	43	45	3	87	42	52	630
Sherbrooke, Que.	45	26	17	71	41	26	792
Sheridan, WY	44	47	50	106	57	20	3,742
Shreveport, LA	32	31	30	93	45	0	209
Sioux City, IA	42	30	0	96	24	0	1,117
Sioux Falls, SD	43	33	0	96	42	0	1,442
South Bend, IN	41	41	0	86	15	0	725
Spartanburg, SC	34	56	58	81	55	56	816
Spokane, WA	47	39	32	117	25	30	2,000
Springfield, IL	39	48	6	89	38	37	610
Springfield, MA	42	6	5	72	35	25	70
Springfield, MO	37	12	55	93	17	53	1,300
Springfield, OH	39	55	27	83	48	32	1,000
Stamford, CT	41	3	12	73	32	21	35
Steubenville, OH	40	22	11	80	38	3	1,060
Stockton, CA	37	57	28	121	17	23	15
Sudbury, Ont.	46	37	30	80	47	56	1,140
Superior, WI	46	43	15	92	6	14	642
Sydney, N.S.	46	09	41	60	02	52	203
Syracuse, NY	43	2	53	76	8	52	400
Tacoma, WA	47	15	11	122	26	35	380
Tallahassee, FL	30	26	17	84	16	51	188
Tampa, FL	27	56	50	82	27	31	48
Terre Haute, IN	39	28	0	87	24	50	501
Texarkana, TX	33	25	30	94	2	51	324
Thunder Bay, Ont.	48	22	19	89	19	26	653
Timmins, Ont	48	34	11	81	22	36	967
Toledo, OH	41	39	50	83	33	19	615
Topeka, KS	39	2	54	95	40	40	1,000
Toronto, Ont.	43	37	39	79	23	46	251
Trenton, NJ	40	13	1	74	44	36	54
Trois-Rivières, Que.	46	21	10	72	40	46	198
Troy, NY	42	43	42	73	41	32	35
Tucson, AZ	32	13	18	110	55	33	2,390
Tulsa, OK	36	9	14	95	59	33	804
Urbana, IL	40	6	38	88	12	26	725
Utica, NY	43	6	3	75	13	59	415
Vancouver, B.C.	49	11	42	123	10	55	14
Victoria, B.C.	48	38	49	123	25	33	63
Waco, TX	31	32	57	97	8	47	405
Walla Walla, WA	46	3	53	118	20	31	1,000
Washington, DC	38	53	42	77	2	12	25
Waterloo, IA	42	29	34	92	20	34	850
West Palm Beach, FL	26	42	54	80	3	13	21
Wheeling, WV	40	3	50	80	43	16	672
Whitehorse, Yukon	60	42	36	135	04	06	2,305
White Plains, NY	41	2	2	73	45	48	220
Wichita, KS	37	41	32	97	20	14	1,305
Wilkes-Barre, PA	41	14	45	75	52	54	550
Wilmington, DE	39	44	45	75	32	49	100
Wilmington, NC	34	13	32	77	56	42	50
Windsor, Ont.	42	16	29	82	57	30	622
Winnipeg, Man.	49	54	39	97	14	36	783
Winston-Salem, NC	36	5	59	80	14	40	912
Worcester, MA	42	15	45	71	48	10	480
Yakima, WA	46	36	8	120	30	17	1,066
Yellowknife, N.W.T.	62	27	46	114	26	25	675
Youngstown, OH	41	5	59	80	38	59	861
Yuma, AZ	32	43	31	114	37	25	160
Zanesville, OH	39	56	25	82	0	48	710

Principal World Rivers

Reviewed by Laurel Duda, Marine Biological Laboratory, Woods Hole Oceanogr. Inst. Library. For N American rivers, see separate table.

Africa

River	Outflow	Length (mi)
Chari	Lake Chad	500
Congo	Atlantic Ocean	2,900
Gambia	Atlantic Ocean	700
Kasai	Congo River	1,000
Limpopo	Indian Ocean	1,100
Lualaba	Congo River	1,100
Niger	Gulf of Guinea	2,590
Nile	Mediterranean	4,160
Okavango	Okavango Delta	1,000
Orange	Atlantic Ocean	1,300
Senegal	Atlantic Ocean	1,020
Ubangi	Congo River	660
Zambezi	Indian Ocean	1,700

Asia

River	Outflow	Length (mi)
Amu Darya	Aral Sea	1,550
Amur	Tatar Strait	1,780
Angara	Yenisey River	1,151
Brahmaputra	Bay of Bengal	1,800
Chang	East China Sea	3,964
Euphrates	Shatt al-Arab	1,700
Ganges	Bay of Bengal	1,560
Godavari	Bay of Bengal	900
Hsi (see Xi)		
Huang	Yellow Sea	3,395
Indus	Arabian Sea	1,800
Irrawaddy	Andaman Sea	1,337
Jordan	Dead Sea	200
Kolyma	Arctic Ocean	1,323
Krishna	Bay of Bengal	800
Kura	Caspian Sea	848
Lena	Laptev Sea	2,734
Mekong	South China Sea	2,700
Narbada (see Narmada)		
Narmada	Arabian Sea	800
Ob	Gulf of Ob	2,268
Ob-Irtysh	Gulf of Ob	3,362
Salween	Gulf of Martaban	1,500
Songhua	Amur River	1,150
Sungari	Amur River	1,197
Sutlej	Indus River	900
Syr	Aral Sea	1,370
Tarim	Lop Nor Basin	1,261
Tigris	Shatt al-Arab	1,180
Xi	South China Sea	1,200
Yamuna	Ganges River	855
Yangtze (see Chang)		
Yellow (see Huang)		
Yenisey	Kara Sea	2,543

Australia

River	Outflow	Length (mi)
Murray-Darling	Indian Ocean	2,310
Murrumbidgee	Murray River	981

Europe

River	Outflow	Length (mi)
Bug, Northern	Wisla	481
Bug, Southern	Dnieper River	532
Danube	Black Sea	1,776
Don	Sea of Azov	1,224
Dnieper	Black Sea	1,420
Dniester	Black Sea	877
Drava	Danube River	447
Dvina, North	White Sea	824
Dvina, West	Gulf of Riga	634
Ebro	Mediterranean	565
Elbe	North Sea	724
Garonne	Bay of Biscay	357
Kama	Volga River	1,122
Loire	Bay of Biscay	634
Mame	Seine River	326
Meuse	North Sea	580
Oder	Baltic Sea	567
Oka	Volga River	932
Pechora	Barents Sea	1,124
Po	Adriatic Sea	405
Rhine	North Sea	820
Rhone	Gulf of Lions	505
Seine	English Channel	496
Shannon	Atlantic Ocean	230
Tagus	Atlantic Ocean	626
Thames	North Sea	210
Tiber	Tyrrhenian Sea	252
Tisza	Danube River	600
Ural	Caspian Sea	1,575
Volga	Caspian Sea	2,290
Weser	North Sea	454
Wisla	Gulf of Gdansk	675

South America

River	Outflow	Length (mi)
Amazon	Atlantic Ocean	4,000
Araguaia	Tocantins River	1,100
Iça (see Putumayo)		
Iguaça	Parana River	808
Japura	Amazon River	1,750
Madeira	Amazon River	2,013
Magdalena	Caribbean Sea	956
Negro	Amazon River	1,400
Orinoco	Atlantic Ocean	1,600
Paraguay	Parana River	1,584
Parana	Rio de la Plata	2,485
Pilcomayo	Paraguay River	1,000
Purus	Amazon River	2,100
Putumayo	Amazon River	1,000
Rio de la Plata	Atlantic Ocean	150
Rio Roosevelt	Aripuana	400
Sao Francisco	Atlantic Ocean	1,988
Tocantins	Para River	1,677
Ucayali	Marañón River	910
Uruguay	Rio de la Plata	1,000
Xingu	Amazon River	1,300

Major Rivers in North America

Reviewed by Laurel Duda, Marine Biological Laboratory, Woods Hole Oceanographic Inst. Library

River	Source or upper limit of length	Outflow	Length (mi)
Alabama	Gilmer County, GA	Mobile River	729
Albany	Lake St. Joseph, Ontario	James Bay	610
Allegheny	Potter County, PA	Ohio River	325
Altamaha-Ocrnulgee	Junction of Yellow and South Rivers, Newton County, GA	Atlantic Ocean	392
Apalachicola-Chattahoochee	Towns County, GA	Gulf of Mexico	524
Arkansas	Lake County, CO	Mississippi River	1,459
Assiniboine	Eastern Saskatchewan	Red River	450
Attawapiskat	Attawapiskat, Ontario	James Bay	465
Back (NWT)	Contwoyto Lake	Chantrey Inlet, Arctic Ocean	605
Big Black (MS)	Webster County, MS	Mississippi River	330
Brazos	Junction of Salt and Double Mountain Forks, Stonewall County, TX	Gulf of Mexico	950
Canadian	Las Animas County, CO	Arkansas River	906
Cedar (IA)	Dodge County, MN	Iowa River	329
Cheyenne	Junction of Antelope Creek and Dry Fork, Converse County, WY	Missouri River	290
Churchill, Man.	Methy Lake, Saskatchewan	Hudson Bay	1,000
Cimarron	Colfax County, NM	Arkansas River	600
Colorado (AZ)	Rocky Mountain Natl. Park, CO (90 mi in Mexico)	Gulf of California	1,450
Colorado (TX)	West Texas	Matagorda Bay	862
Columbia	Columbia Lake, British Columbia	Pacific Ocean, bet. OR and WA	1,243
Columbia, Upper	Columbia Lake, British Columbia	To mouth of Snake River	890
Connecticut	Third Connecticut Lake, NH	Long Island Sound, CT	407
Coppermine (NWT)	Lac de Gras	Coronation Gulf, Arctic Ocean	525
Cumberland	Letcher County, KY	Ohio River	720
Delaware	Schoharie County, NY	Liston Point, Delaware Bay	390
Fraser	Near Mount Robson (on Continental Divide)	Strait of Georgia	850
Gila	Catron County, NM	Colorado River	649
Green (UT-WY)	Junction of Wells and Trail Creeks, SubletteCounty, WY	Colorado River	730
Hamilton (Lab.)	Lake Ashuanipi	Atlantic Ocean	532
Hudson	Henderson Lake, Essex County, NY	Upper NY Bay	306
Illinois	St. Joseph County, IN	Mississippi River	420
James (ND-SD)	Wells County, ND	Missouri River	710
James (VA)	Junction of Jackson and Cowpasture Rivers, Botetourt County, VA	Hampton Roads	340
Kanawha-New	Junction of North and South Forks of New River, NC	Ohio River	352
Kentucky	Junction of North and Middle Forks, Lee County, KY	Ohio River	259
Klamath	Lake Ewauna, Klamath Falls, OR	Pacific Ocean	250
Kootenay	Kootenay Lake, British Columbia	Columbia River	485
Koyukuk	Endicott Mountains, AK	Yukon River	470
Kuskokwim	Alaska Range	Kuskokwim Bay	724

River	Source or upper limit of length	Outflow	Length (mi)
Liard	Southern Yukon, AK	Mackenzie River	693
Little Missouri	Crook County, WY	Missouri River	560
Mackenzie	Great Slave Lake, N.W.T.	Arctic Ocean	1,060
Milk	Junction of North and South Forks, Alberta	Missouri River	625
Minnesota	Big Stone Lake, MN	Mississippi River	332
Mississippi	Lake Itasca, MN	Gulf of Mexico	2,340
Mississippi-Missouri-Red Rock	Source of Red Rock, Beaverhead Co., MT	Gulf of Mexico	3,710
Missouri	Junction of Jefferson, Madison, and Gallatin Rivers, Madison County, MT	Mississippi River	2,315
Missouri-Red Rock	Source of Red Rock, Beaverhead Co., MT	Mississippi River	2,540
Mobile-Alabama-Coosa	Gilmer County, GA	Mobile Bay	774
Nelson (Man.)	Lake Winnipeg	Hudson Bay	410
Neosho	Morris County, KS	Arkansas River, OK	460
Niobrara	Niobrara County, WY	Missouri River, NE	431
North Canadian	Union County, NM	Canadian River, OK	800
North Platte	Junction of Grizzly and Little Grizzly Creeks, Jackson County, CO	Platte River, NE	618
Ohio	Junction of Allegheny and Monongahela Rivers, Pittsburgh, PA	Mississippi River	981
Ohio-Allegheny	Potter County, PA	Mississippi River	1,310
Osage	East-central Kansas	Missouri River	500
Ottawa	Lake Capimitchigama	St. Lawrence River	790
Ouachita	Polk County, AR	Black River	605
Peace	Stikine Mountains, B.C.	Slave River	1,210
Pearl	Neshoba County, MS	Gulf of Mexico	411
Pecos	Mora County, NM	Rio Grande	926
Pee Dee-Yadkin	Watauga County, NC	Winyah Bay	435
Pend Oreille-Clark Fork	Near Butte, MT	Columbia River	531
Platte	Junction of North and South Platte Rivers, NE	Missouri River	310
Porcupine	Ogilvie Mountains, AK	Yukon River, AK	569
Potomac	Garrett County, MD	Chesapeake Bay	383
Powder	Junction of South and Middle Forks, WY	Yellowstone River	375
Red (OK-TX-LA)	Curry County, NM	Mississippi River	1,290
Red River of the North	Junction of Otter Tail and Bois de Sioux Rivers, Wilkin County, MN	Lake Winnipeg	545
Republican	Junction of North Fork and Arikaree River, NE	Kansas River	445
Rio Grande	San Juan County, CO	Gulf of Mexico	1,900
Roanoke	Junction of N and S Forks, Montgomery Co., VA	Albemarle Sound	380
Rock (IL-WI)	Dodge County, WI	Mississippi River	300
Sabine	Junction of S and Caddo Forks, Hunt County, TX	Sabine Lake	380
Sacramento	Siskiyou County, CA	Suisun Bay	377
St. Francis	Iron County, MO	Mississippi River	425
St. John	Northwestern Maine	Bay of Fundy	418
St. Lawrence	Lake Ontario	Gulf of St. Lawrence, Atlantic Ocean	800
Saguenay	Lake St. John, Quebec	St. Lawrence River	434
Salmon (ID)	Custer County, ID	Snake River	420
San Joaquin	Junction of S and Middle Forks, Madera Co., CA	Suisun Bay	350
San Juan	Silver Lake, Archuleta County, CO	Colorado River	360
Santee-Wateree-Catawba	McDowell County, NC	Atlantic Ocean	538
Saskatchewan, North	Rocky Mountains	Saskatchewan R.	800
Saskatchewan, South	Rocky Mountains	Saskatchewan R.	865
Savannah	Junction of Seneca and Tugaloo Rivers, Anderson County, SC	Atlantic Ocean, GA-SC	314
Savern (Ont.)	Sandy Lake	Hudson Bay	610
Smoky Hill	Cheyenne County, CO	Kansas River, KS	540
Snake	Teton County, WY	Columbia River, WA	1,038
South Platte	Junction of S and Middle Forks, Park County, CO	Platte River	424
Susitna	Alaska Range	Cook Inlet	313
Susquehanna	Huyden Creek, Otsego County, NY	Chesapeake Bay	447
Tallahatchie	Tippah County, MS	Yazoo River	301
Tanana	Wrangell Mountains, AK	Yukon River	659
Tennessee	Junction of French Broad and Holston Rivers	Ohio River	652
Tennessee-French Broad	Courthouse Creek, Transylvania County, NC	Ohio River	886
Tombigbee	Prentiss County, MS	Mobile River	525
Trinity	North of Dallas, TX	Galveston Bay	360
Wabash	Darke County, OH	Ohio River	512
Washita	Hemphill County, TX	Red River, OK	500
White (AR-MO)	Madison County, AR	Mississippi River	722
Willamette	Douglas County, OR	Columbia River	309
Wind-Bighorn	Junction of Wind and Little Wind Rivers, Fremont Co., WY (Source of Wind R. is Togwotee Pass, Teton Co., WY)	Yellowstone River	338
Wisconsin	Lac Vieux Desert, Vilas County, WI	Mississippi River	430
Yellowstone	Park County, WY	Missouri River	682
Yukon	McNeil R., Yukon Territory	Bering Sea	1,979

Highest and Lowest Continental Altitudes

Source: National Geographic Society

Continent	Highest point	Elev. (ft)	Lowest point	ft below sea level
Asia	Mount Everest, Nepal-Tibet	29,028	Dead Sea, Israel-Jordan	1,312
South America	Mount Aconcagua, Argentina	22,834	Valdes Peninsula, Argentina	131
North America	Mount McKinley, AK	20,320	Death Valley, California	282
Africa	Kilimanjaro, Tanzania	19,340	Lake Assal, Djibouti	512
Europe	Mount Elbrus, Russia	18,510	Caspian Sea, Russia, Azerbaijan	92
Antarctica	Vinson Massif	16,864	Bentley Subglacial Trench	8,327[1]
Australia	Mount Kosciusko, New South Wales	7,310	Lake Eyre, South Australia	52

(1) Estimated level of the continental floor. Lower points that have yet to be discovered may exist further beneath the ice.

Major Natural Lakes of the World

Source: Geological Survey, U.S. Dept. of the Interior

A lake is generally defined as a body of water surrounded by land. By this definition some bodies of water that are called seas, such as the Caspian Sea and the Aral Sea, are really lakes. In the following table, the word *lake* is omitted when it is part of the name.

Name	Continent	Area (sq mi)	Length (mi)	Maximum depth (ft)	Elevation (ft)
Caspian Sea	Asia-Europe	143,244	760	3,363	−92
Superior	North America	31,700	350	1,330	600
Victoria	Africa	26,828	250	270	3,720
Aral Sea	Asia	24,904[1]	280	220	174
Huron	North America	23,000	206	750	579
Michigan	North America	22,300	307	923	579
Tanganyika	Africa	12,700	420	4,823	2,534
Baykal	Asia	12,162	395	5,315	1,493
Great Bear	North America	12,096	192	1,463	512
Nyasa (Malawi)	Africa	11,150	360	2,280	1,550
Great Slave	North America	11,031	298	2,015	513
Erie	North America	9,910	241	210	570
Winnipeg	North America	9,417	266	60	713
Ontario	North America	7,340	193	802	245
Balkhash	Asia	7,115	376	85	1,115
Ladoga	Europe	6,835	124	738	13
Chad	Africa	6,300	175	24	787
Maracaibo	South America	5,217	133	115	sea level
Onega	Europe	3,710	145	328	108
Eyre	Australia	3,600[2]	90	4	−52
Volta	Africa	3,276	250	...	...
Titicaca	South America	3,200	122	922	12,500
Nicaragua	Central America	3,100	102	230	102
Athabasca	North America	3,064	208	407	700
Reindeer	North America	2,568	143	720	1,106
Turkana (Rudolf)	Africa	2,473	154	240	1,230
Issyk Kul	Asia	2,355	115	2,303	5,279
Torrens	Australia	2,230	130	...	92
Vanern	Europe	2,156	91	328	144
Nettilling	North America	2,140	67	...	95
Winnipegosis	North America	2,075	141	38	830
Albert	Africa	2,075	100	168	2,030
Kariba	Africa	2,050	175	390	1,590
Nipigon	North America	1,872	72	540	1,050
Gairdner	Australia	1,840	90	...	112
Urmia	Asia	1,815	90	49	4,180
Manitoba	North America	1,799	140	12	813

(1) Probably less because of the diversion of feeder rivers. (2) Approximate figure, subject to great seasonal variation.

The Great Lakes

Source: National Ocean Service, U.S. Dept. of Commerce

The Great Lakes form the world's largest body of fresh water, and with their connecting waterways are the largest inland water transportation unit. Draining the great North Central basin of the U.S., they enable shipping to reach the Atlantic via their outlet, the St. Lawrence R., and to reach the Gulf of Mexico via the Illinois Waterway, from Lake Michigan to the Mississippi R. A 3d outlet connects with the Hudson R. and then the Atlantic via the New York State Barge Canal System. Traffic on the Illinois Waterway and the N.Y. State Barge Canal System is limited to recreational boating and small shipping vessels.

Only one of the lakes, Lake Michigan, is wholly in the U.S.; the others are shared with Canada. Ships move from the shores of Lake Superior to Whitefish Bay at the E end of the lake, then through the Soo (Sault Ste. Marie) locks, through the St. Mary's R. and into Lake Huron. To reach Gary and the Port of Indiana and South Chicago, IL, ships move W from Lake Huron to Lake Michigan through the Straits of Mackinac. Lake Superior is 601 ft above low water datum at Rimouski, Quebec, on the International Great Lakes Datum (1985). From Duluth, MN, to the E end of Lake Ontario is 1,156 mi.

	Superior	Michigan	Huron	Erie	Ontario
Length in mi	350	307	206	241	193
Breadth in mi	160	118	183	57	53
Deepest soundings in ft	1,333	923	750	210	802
Volume of water in cu mi	2,935	1,180	850	116	393
Area (sq mi) water surface—U.S.	20,600	22,300	9,100	4,980	3,560
Canada	11,100	.....	13,900	4,930	3,990
Area (sq mi) entire drainage basin—U.S.	16,900	45,600	16,200	18,000	15,200
Canada	32,400	.....	35,500	4,720	12,100
TOTAL AREA (sq mi) U.S. and Canada	**81,000**	**67,900**	**74,700**	**32,630**	**34,850**
Low water datum above mean water level at Rimouski, Quebec, avg. level in ft (1985)	601.10	577.50	577.50	569.20	243.30
Latitude, N	46° 25 min	41° 37 min	43° 00 min	41° 23 min	43° 11 min
	49° 00 min	46° 06 min	46° 17 min	42° 52 min	44° 15 min
Longitude, W	84° 22 min	84° 45 min	79° 43 min	78° 51 min	76° 03 min
	92° 06 min	88° 02 min	84° 45 min	83° 29 min	79° 53 min
National boundary line in mi	282.8	None	260.8	251.5	174.6
United States shoreline (mainland only) mi	863	1,400	580	431	300

Famous Waterfalls

Source: National Geographic Society

The earth has thousands of waterfalls, some of considerable magnitude. Their relative importance is determined not only by height but also by volume of flow, steadiness of flow, crest width, whether the water drops sheerly or over a sloping surface, and whether it descends in one leap or in a succession of leaps. A series of low falls flowing over a considerable distance is known as a **cascade**.

Estimated mean annual flow, in cubic feet per second, of major waterfalls are as follows: Niagara, 212,200; Paulo Afonso, 100,000; Urubupunga, 97,000; Iguazu, 61,000; Patos-Maribondo, 53,000; Victoria, 35,400; and Kaieteur, 23,400.

Elevation = total drop in feet in one or more leaps. #=falls of more than one leap; *= falls that diminish greatly seasonally; **= falls that reduce to a trickle or are dry for part of each year. If the river names are not shown, they are same as the falls. R. = river; (C) = cascade type.

Name and location	Elevation (ft)
Africa	
Angola	
Ruacana, Cuene R.	406
Ethiopia	
Fincha	508
Lesotho	
Maletsunyane*	630
Zimbabwe-Zambia	
Victoria, Zambezi R.*	343
South Africa	
Augrabies, Orange R.*	480
Tugela#.	2,014
Tanzania-Zambia	
Kalambo*	726
Asia	
India	
Cauvery*	330
Jog (Gersoppa),Sharavathi R.*	830
Japan	
Kegon, Daiya R.*	330
Australia	
New South Wales	
Wentworth	614
Wollomombi	1,100
Queensland	
Tully	885
Wallaman, Stony Cr.#.	1,137
New Zealand	
Helena	890
Sutherland, Arthur R.#	1,904
Europe	
Austria	
Gastein#.	492
Gavarnie*	1,385
Great Britain	
Scotland	
Glomach.	370
Wales	
Rhaiadr.	240
Italy	
Frua, Toce R. (C)	470
Norway	
Mardalsfossen (Northern)	1,535
Mardalsfossen (Southern)#	2,149
Skjeggedal, Nybuai R.#**	1,378
Skykje**	984
Vetti, Morka-Koldedola R.	900
Sweden	
Handol#.	427
Switzerland	
Giessbach (C)	984
Reichenbach#	656
Simmen#	459
Staubbach	984
Trummelbach#	1,312
North America	
Canada	
Alberta	
Panther, Nigel Cr.	600
British Columbia	
Della#	1,443
Takakkaw, Daly Glacier#	1,200
Quebec	
Montmorency	274
Canada—United States	
Niagara: American.	182
Horseshoe.	173
United States	
California	
Feather, Fall R.*	640
Yosemite National Park	
Bridalveil*	620
Illilouette*	370
Nevada, Merced R.*	594
Ribbon**	1,612
Silver Strand, Meadow Br.**	1,170
Vernal, Merced R. *	317
Yosemite#**	2,425
Colorado	
Seven, South Cheyenne Cr.#	300
Hawaii	
Akaka, Kolekole Str.	442
Idaho	
Shoshone, Snake R.**	212
Kentucky	
Cumberland	68
Maryland	
Great, Potomac R. (C) *	71
Minnesota	
Minnehaha**	53
New Jersey	
Passaic.	70
New York	
Taughannock*	215
Oregon	
Multnomah#	620
Tennessee	
Fall Creek	256
Washington	
Mt. Rainier Natl. Park	
Sluiskin, Paradise R.	300
Snoqualmie**	268
Wisconsin	
Big Manitou, Black R. (C)*	165
Wyoming	
Yellowstone Natl. Pk. Tower	132
Yellowstone (upper)*	109
Yellowstone (lower)*	308
Mexico	
El Salo	218
South America	
Argentina-Brazil	
Iguazu	230
Brazil	
Glass	1,325
Patos-Maribondo, Grande R.	115
Paulo Afonso, Sao Francisco R. .	275
Urubupunga, Parana R.	39
Colombia	
Catarata de Candelas,Cusiana R.	984
Tequendama, Bogota R.*	427
Ecuador	
Agoyan, Pastaza R.*	200
Guyana	
Kaieteur, Potaro R.	741
Great, Kamarang R.	1,600
Marina, Ipobe R.#.	500
Venezuela	
Angel#*	3,212
Cuquenan	2,000

Notable Deserts of the World

Arabian (Eastern), 70,000 sq mi in Egypt between the Nile R. and Red Sea, extending southward into Sudan

Atacama, 600-mi-long area rich in nitrate and copper deposits in N Chile

Chihuahuan, 140,000 sq mi in TX, NM, AZ, and Mexico

Dasht-e Kauir, approx. 300 mi long by approx. 100 mi wide in N central Iran

Dasht-e Lut, 20,000 sq mi in E Iran

Death Valley, 3,300 sq mi in CA and NV

Gibson, 120,000 sq mi in the interior of W Australia

Gobi, 500,000 sq mi in Mongolia and China

Great Sandy, 150,000 sq mi in W Australia

Great Victoria, 150,000 sq mi in SW Australia

Kalahari, 225,000 sq mi in S Africa

Kara Kum, 120,000 sq mi in Turkmenistan

Kyzyl Kum, 100,000 sq mi in Kazakhstan and Uzbekistan

Libyan, 450,000 sq mi in the Sahara, extending from Libya through SW Egypt into Sudan

Mojave, 15,000 sq mi in southern CA

Namib, long narrow area (varies from 30-100 mi wide) extending 800 mi along SW coast of Africa

Nubian, 100,000 sq mi in the Sahara in NE Sudan

Patagonia, 300,000 sq mi in S Argentina

Painted Desert, section of high plateau in northern AZ extending 150 mi

Rub al-Khali (Empty Quarter), 250,000 sq mi in the S Arabian Peninsula

Sahara, 3,500,000 sq mi in N Africa, extending westward to the Atlantic. Largest desert in the world

Sonoran, 70,000 sq mi in southwestern AZ and southeastern CA extending into NW Mexico

Syrian, 100,000-sq-mi arid wasteland extending over much of N Saudi Arabia, E Jordan, S Syria, and W Iraq

Taklimakan, 140,000 sq mi in Xinjiang Prov., China

Thar (Great Indian), 100,000-sq-mi arid area extending 400 mi along India-Pakistan border

PRESIDENTIAL ELECTIONS

Popular and Electoral Vote, 1992 and 1996

Source: Voter News Service; Federal Election Commission; totals are official.

	1996 Electoral Vote			1996 Democrat	Republican	Reform	1992 Electoral Vote			1992 Democrat	Republican	Independent
State	Clinton	Dole	Perot	Clinton	Dole	Perot	Clinton	Bush	Perot	Clinton	Bush	Perot
AL	0	9	0	662,165	769,044	92,149	0	9	0	690,080	804,283	183,109
AK	0	3	0	80,380	122,746	26,333	0	3	0	78,294	102,000	73,481
AZ	8	0	0	653,288	622,073	112,072	0	8	0	543,050	572,086	353,741
AR	6	0	0	475,171	325,416	69,884	6	0	0	505,823	337,324	99,132
CA	54	0	0	5,119,835	3,828,380	697,847	54	0	0	5,121,325	3,630,574	2,296,006
CO	0	8	0	671,152	691,848	99,629	8	0	0	629,681	562,850	366,010
CT	8	0	0	735,740	483,109	139,523	8	0	0	682,318	578,313	348,771
DE	3	0	0	140,355	99,062	28,719	3	0	0	126,054	102,313	59,213
DC	3	0	0	158,220	17,339	3,611	3	0	0	192,619	20,698	9,681
FL	25	0	0	2,545,968	2,243,324	483,776	0	25	0	2,071,651	2,171,781	1,052,481
GA	0	13	0	1,053,849	1,080,843	146,337	13	0	0	1,008,966	995,252	309,657
HI	4	0	0	205,012	113,943	27,358	4	0	0	179,310	136,822	53,003
ID	0	4	0	165,443	256,595	62,518	0	4	0	137,013	202,645	130,395
IL	22	0	0	2,341,744	1,587,021	346,408	22	0	0	2,453,350	1,734,096	840,515
IN	0	12	0	887,424	1,006,693	224,299	0	12	0	848,420	989,375	455,934
IA	7	0	0	620,258	492,644	105,159	7	0	0	586,353	504,891	253,468
KS	0	6	0	387,659	583,245	92,639	0	6	0	390,434	449,951	312,358
KY	8	0	0	636,614	623,283	120,396	8	0	0	665,104	617,178	203,944
LA	9	0	0	927,837	712,586	123,293	9	0	0	815,971	733,386	211,478
ME	4	0	0	312,788	186,378	85,970	4	0	0	263,420	206,504	206,820
MD	10	0	0	966,207	681,530	115,812	10	0	0	988,571	707,094	281,414
MA	12	0	0	1,571,509	718,058	227,206	12	0	0	1,318,639	805,039	630,731
MI	18	0	0	1,989,653	1,481,212	336,670	18	0	0	1,871,182	1,554,940	824,813
MN	10	0	0	1,120,438	766,476	257,704	10	0	0	1,020,997	747,841	562,506
MS	0	7	0	394,022	439,838	52,222	0	7	0	400,258	487,793	85,626
MO	11	0	0	1,025,935	890,016	217,188	11	0	0	1,053,873	811,159	518,741
MT	0	3	0	167,922	179,652	55,229	3	0	0	154,507	144,207	107,225
NE	0	5	0	236,761	363,467	71,278	0	5	0	216,864	343,678	174,104
NV	4	0	0	203,974	199,244	43,986	4	0	0	189,148	175,828	132,580
NH	4	0	0	246,166	196,486	48,387	4	0	0	209,040	202,484	121,337
NJ	15	0	0	1,652,361	1,103,099	262,134	15	0	0	1,436,206	1,356,865	521,829
NM	5	0	0	273,495	232,751	32,257	5	0	0	261,617	212,824	91,895
NY	33	0	0	3,756,177	1,933,492	503,458	33	0	0	3,444,450	2,346,649	1,090,721
NC	0	14	0	1,107,849	1,225,938	168,059	0	14	0	1,114,042	1,134,661	357,864
ND	0	3	0	106,905	125,050	32,515	0	3	0	99,168	136,244	71,084
OH	21	0	0	2,148,222	1,859,883	483,207	21	0	0	1,984,942	1,894,310	1,036,426
OK	0	8	0	488,105	582,315	130,788	0	8	0	473,066	592,929	319,878
OR	7	0	0	649,641	538,152	121,221	7	0	0	621,314	475,757	354,091
PA	23	0	0	2,215,819	1,801,169	430,984	23	0	0	2,239,164	1,791,841	902,667
RI	4	0	0	233,050	104,683	43,723	4	0	0	213,299	131,601	105,045
SC	0	8	0	506,283	573,458	64,386	0	8	0	479,514	577,507	138,872
SD	0	3	0	139,333	150,543	31,250	0	3	0	124,888	136,718	73,295
TN	11	0	0	909,146	863,530	105,918	11	0	0	933,521	841,300	199,968
TX	0	32	0	2,459,683	2,736,167	378,537	0	32	0	2,281,815	2,496,071	1,354,781
UT	0	5	0	221,633	361,911	66,461	0	5	0	183,429	322,632	203,400
VT	3	0	0	137,894	80,352	31,024	3	0	0	133,590	88,122	65,985
VA	0	13	0	1,091,060	1,138,350	159,861	0	13	0	1,038,650	1,150,517	348,639
WA	11	0	0	1,123,323	840,712	201,003	11	0	0	993,037	731,234	541,780
WV	5	0	0	327,812	233,946	71,639	5	0	0	331,001	241,974	108,829
WI	11	0	0	1,071,971	845,029	227,339	11	0	0	1,041,066	930,855	544,479
WY	0	3	0	77,934	105,388	25,928	0	3	0	68,160	79,347	51,263
Total	379	159	0	47,401,185	39,197,469	8,085,294	370	168	0	44,908,254	39,102,343	19,741,065

PRESIDENTIAL ELECTION RETURNS BY COUNTIES

All results are official. Results for New England states are for selected cities or towns because county results are not available. Totals are always statewide. D-Democrat; R-Republican; RF-Reform; I-Independent. (In 1996, Ross Perot was listed on the ballot in some states as "Independent.")

Source: Voter News Service; Federal Election Commission; Alaska Division of Elections

Alabama

County	1996 Clinton (D)	Dole (R)	Perot (RF)	1992 Clinton (D)	Bush (R)	Perot (I)
Autauga	5,015	9,509	813	4,819	8,715	1,916
Baldwin	12,776	29,487	4,520	12,195	26,270	7,656
Barbour	4,787	3,627	515	4,836	4,475	1,020
Bibb	2,775	3,037	455	2,900	3,124	686
Blount	5,061	9,056	985	5,433	8,882	1,949
Bullock	3,078	1,154	111	3,259	1,253	266
Butler	3,828	3,352	538	4,021	3,494	867
Calhoun	15,725	18,088	2,613	16,453	20,623	4,717
Chambers	5,515	4,707	812	5,938	5,682	1,427
Cherokee	4,399	3,048	899	4,222	2,745	846
Chilton	5,354	7,910	929	4,946	8,126	1,363
Choctaw	4,074	2,623	413	3,941	3,069	489
Clarke	4,831	4,785	478	5,023	5,495	872
Clay	2,306	2,694	538	2,073	2,859	652
Cleburne	1,737	2,063	385	2,144	2,425	630
Coffee	5,168	7,805	1,042	5,776	7,591	2,021
Colbert	10,226	8,305	1,696	12,206	8,073	2,098
Conecuh	2,903	2,093	445	3,155	2,463	552
Coosa	2,121	1,721	262	2,330	1,973	476
Covington	4,543	6,035	1,098	5,004	6,840	1,880
Crenshaw	2,172	1,939	317	2,404	2,339	485
Cullman	9,544	14,308	2,440	10,451	14,411	4,113
Dale	4,732	8,288	1,216	5,098	8,123	2,423
Dallas	10,507	6,612	477	11,053	7,394	1,110
DeKalb	6,544	9,823	1,609	8,245	10,519	2,741
Elmore	6,530	12,937	1,368	6,223	11,356	2,765
Escambia	4,651	5,214	867	4,809	5,955	1,616
Etowah	17,976	16,835	2,529	20,558	17,467	4,277
Fayette	3,381	3,191	590	3,830	3,604	1,012
Franklin	5,028	4,449	966	5,953	4,794	1,075
Geneva	3,174	4,725	857	3,622	4,843	1,323
Greene	3,526	796	55	3,865	805	194
Hale	3,372	1,893	190	3,481	2,001	486
Henry	3,019	3,082	515	2,804	2,970	667
Houston	8,791	17,476	1,653	8,857	17,360	3,492
Jackson	8,204	5,650	1,573	10,628	5,711	2,462
Jefferson	120,208	130,980	7,997	125,889	149,832	22,191
Lamar	2,843	2,955	597	2,849	3,262	763
Lauderdale	13,619	14,058	2,574	15,936	13,728	4,009
Lawrence	5,254	3,893	964	6,364	3,576	1,624
Lee	12,919	17,985	1,949	13,770	16,885	4,572
Limestone	8,045	10,862	1,659	8,087	9,862	3,584
Lowndes	3,970	1,369	72	3,500	1,328	284
Macon	7,018	987	150	7,253	1,134	283
Madison	42,259	50,390	7,437	38,974	51,444	16,989
Marengo	4,899	4,013	337	5,632	4,470	919
Marion	5,049	4,742	979	6,167	5,692	1,389
Marshall	8,722	12,323	2,150	10,421	12,249	3,795
Mobile	54,749	66,775	7,555	54,962	72,935	15,105
Monroe	3,815	4,382	486	3,872	4,919	759
Montgomery	38,382	37,784	2,036	37,342	40,742	7,647
Morgan	14,616	21,765	3,348	15,091	21,073	7,683
Perry	4,053	1,703	119	3,712	1,829	213
Pickens	4,018	3,322	403	3,783	3,634	690
Pike	4,514	5,281	503	4,688	5,423	1,024
Randolph	3,023	3,304	603	3,318	3,813	919
Russell	7,834	5,025	792	8,647	5,587	1,360
St. Clair	6,187	12,762	1,417	6,517	12,447	2,614
Shelby	11,280	37,090	2,035	10,317	32,736	5,022
Sumter	4,706	1,561	172	4,810	1,807	388
Talladega	10,385	10,931	1,335	10,695	12,661	2,629
Tallapoosa	6,071	7,627	1,038	5,703	8,140	1,562
Tuscaloosa	23,067	27,939	3,048	23,495	27,454	7,011
Walker	12,929	9,837	2,012	14,831	11,301	3,344
Washington	3,935	2,900	819	4,046	3,270	829
Wilcox	3,303	1,454	71	3,439	1,671	174
Winston	3,120	4,728	723	3,415	5,550	1,110
Totals	**662,165**	**769,044**	**92,149**	**690,080**	**804,283**	**183,109**

Alabama Vote Since 1948

1948, Thurmond, States' Rights, 171,443; Dewey, Rep., 40,930; Wallace, Prog., 1,522; Watson, Proh., 1,085.

1952, Eisenhower, Rep., 149,231; Stevenson, Dem., 275,075; Hamblen, Proh., 1,814.

1956, Stevenson, Dem., 290,844; Eisenhower, Rep., 195,694; Independent electors, 20,323.

1960, Kennedy, Dem., 324,050; Nixon, Rep., 237,981; Faubus, States' Rights, 4,367; Decker, Proh., 2,106; King, Afro-Americans, 1,485; scattering, 236.

1964, Dem. (electors unpledged), 209,848; Goldwater, Rep., 479,085; scattering, 105.

1968, Nixon, Rep., 146,923; Humphrey, Dem., 196,579; Wallace, 3d Party, 691,425; Munn, Proh., 4,022.

1972, Nixon, Rep., 728,701; McGovern, Dem., 219,108 plus 37,815 Natl. Demo. Party of Alabama; Schmitz, Conservative, 11,918; Munn., Proh., 8,551.

1976, Carter, Dem., 659,170; Ford, Rep., 504,070; Maddox, Amer. Ind., 9,198; Bubar, Proh., 6,669; Hall, Com., 1,954; MacBride, Libertarian, 1,481.

1980, Reagan, Rep., 654,192; Carter, Dem., 636,730; Anderson, Independent, 16,481; Rarick, Amer. Ind., 15,010; Clark, Libertarian, 13,318; Bubar, Statesman, 1,743; Hall, Com., 1,629; DeBerry, Soc. Workers, 1,303; McReynolds, Socialist, 1,006; Commoner, Citizens, 517.

1984, Reagan, Rep., 872,849; Mondale, Dem., 551,899; Bergland, Libertarian, 9,504.

1988, Bush, Rep., 815,576; Dukakis, Dem., 549,506; Paul, Lib., 8,460; Fulani, Ind., 3,311.

1992, Bush, Rep., 804,283; Clinton, Dem., 690,080; Perot, Ind., 183,109; Marrou, Libertarian, 5,737; Fulani, New Alliance, 2,161.

1996, Dole, Rep., 769,044; Clinton, Dem., 662,165; Perot, Ind. (Ref.), 92,149; Browne, Libertarian, 5,290; Phillips, Ind., 2,365; Hagelin, Natural Law, 1,697; Harris, Ind., 516.

Alaska

Election District[1]	1996 Clinton (D)	Dole (R)	Perot (RF)	1992 Clinton (D)	Bush (R)	Perot (I)
No. 1	1,480	4,209	696	2,055	2,495	2,120
No. 2	2,563	3,247	912	2,565	2,916	2,137
No. 3	3,724	2,671	654	4,064	2,447	1,424
No. 4	3,037	3,336	694	2,688	2,894	1,561
No. 5	2,148	2,564	826	2,095	1,844	1,684
No. 6	1,576	2,707	557	1,546	2,345	1,748
No. 7	2,177	3,517	907	2,088	2,173	2,244
No. 8	1,643	3,624	826	1,509	2,499	2,325
No. 9	1,334	3,459	727	1,540	2,349	2,368
No. 10	2,203	4,184	642	1,947	3,548	1,899
No. 11	1,946	3,073	603	2,009	2,730	2,081
No. 12	1,825	3,568	543	1,831	2,999	2,039
No. 13	2,780	3,270	608	3,001	2,963	1,907
No. 14	1,471	3,005	458	1,423	3,013	1,599
No. 15	2,178	1,974	552	2,389	1,842	1,591
No. 16	1,629	1,328	414	1,814	1,375	1,320
No. 17	1,868	3,284	633	1,749	2,623	1,958
No. 18	2,708	4,245	694	2,483	3,629	2,134
No. 19	2,014	3,159	636	1,931	2,539	1,840
No. 20	2,144	3,025	545	2,383	2,914	1,823
No. 21	2,228	2,553	557	2,386	2,437	1,693
No. 22	2,511	3,887	624	2,253	3,164	1,713
No. 23	1,071	2,127	388	1,139	2,127	1,217
No. 24	1,914	3,653	548	1,876	3,441	1,930
No. 25	1,629	4,099	691	1,513	3,197	2,122
No. 26	1,519	3,913	883	1,439	2,675	2,419
No. 27	1,887	4,384	1,122	1,625	2,757	2,401
No. 28	1,645	4,202	1,333	1,522	2,459	2,825
No. 29	3,023	3,012	658	3,216	2,205	2,026
No. 30	1,794	2,785	601	1,860	2,434	1,912
No. 31	1,903	2,721	684	1,969	2,223	1,992
No. 32	1,275	2,736	675	1,150	2,339	1,724
No. 33	1,852	4,089	759	1,712	3,100	2,278
No. 34	1,388	3,677	734	1,455	3,408	2,201
No. 35	1,447	3,016	875	1,572	2,525	2,139
No. 36	2,321	1,992	453	1,748	2,081	1,322
No. 37	2,134	1,835	456	1,822	1,689	925
No. 38	2,436	1,716	393	1,897	2,011	850
No. 39	2,692	1,618	404	1,797	1,777	860
No. 40	1,260	1,280	368	1,211	1,786	1,122
Totals	**80,377**	**122,744**	**26,333**	**78,294**	**102,000**	**73,481**

(1) 1992 and 1996 results are not comparable because of a 1994 reapportionment of districts.

Alaska Vote Since 1960

1960, Kennedy, Dem., 29,809; Nixon, Rep., 30,953.

1964, Johnson, Dem., 44,329; Goldwater, Rep., 22,930.

1968, Nixon, Rep., 37,600; Humphrey, Dem., 35,411; Wallace, 3d Party, 10,024.

1972, Nixon, Rep., 55,349; McGovern, Dem., 32,967; Schmitz, Amer., 6,903.

1976, Carter, Dem., 44,058; Ford, Rep., 71,555; MacBride, Libertarian, 6,785.

1980, Reagan, Rep., 86,112; Carter, Dem., 41,842; Clark, Libertarian, 18,479; Anderson, Ind., 11,155; write-in, 857.

1984, Reagan, Rep., 138,377; Mondale, Dem., 62,007; Bergland, Libertarian, 6,378.

1988, Bush, Rep., 119,251; Dukakis, Dem., 72,584; Paul, Lib., 5,484; Fulani, New Alliance, 1,024.

1992, Bush, Rep., 102,000; Clinton, Dem., 78,294; Perot, Ind., 73,481; Gritz, Populist/America First, 1,379; Marrou, Libertarian, 1,378.

1996, Dole, Rep., 122,746; Clinton, Dem., 80,380; Perot, Ref., 26,333; Nader, Green, 7,597; Browne, Libertarian, 2,276; Phillips, Taxpayers, 925; Hagelin, Natural Law, 729.

Arizona

	1996			1992		
County	Clinton (D)	Dole (R)	Perot (RF)	Clinton (D)	Bush (R)	Perot (I)
Apache	12,394	4,761	1,296	11,218	4,588	1,979
Cochise	13,782	14,365	3,346	12,701	12,202	7,857
Coconino	20,475	13,638	3,666	18,888	13,769	9,363
Gila	8,577	6,407	2,211	7,571	5,781	4,694
Graham	3,938	4,222	1,034	3,391	4,169	1,860
Greenlee	1,755	1,159	426	1,695	1,451	794
La Paz	1,964	1,902	597	1,808	1,599	1,488
Maricopa	363,991	386,015	58,479	285,457	360,049	221,475
Mohave	16,629	17,997	6,369	13,255	13,684	12,706
Navajo	12,912	9,262	2,461	10,882	7,994	4,787
Pima	137,983	104,121	18,809	128,569	97,036	53,925
Pinal	19,579	13,034	3,972	15,468	11,669	9,231
Santa Cruz	5,241	2,256	600	3,512	3,024	1,447
Yavapai	21,801	29,921	6,649	18,268	23,419	16,409
Yuma	12,267	13,013	2,157	10,367	11,652	5,726
Totals	653,288	622,073	112,072	543,050	572,086	353,741

Arizona Vote Since 1948

1948, Truman, Dem., 95,251; Dewey, Rep., 77,597; Wallace, Prog., 3,310; Watson, Proh., 786; Teichert, Soc. Labor, 121.

1952, Eisenhower, Rep., 152,042; Stevenson, Dem., 108,528.

1956, Eisenhower, Rep., 176,990; Stevenson, Dem., 112,880; Andrews, Ind. 303.

1960, Kennedy, Dem., 176,781; Nixon, Rep., 221,241; Hass, Soc. Labor, 469.

1964, Johnson, Dem., 237,753; Goldwater, Rep., 242,535; Hass, Soc. Labor, 482.

1968, Nixon, Rep., 266,721; Humphrey, Dem., 170,514; Wallace, 3d Party, 46,573; McCarthy, New Party, 2,751; Halstead, Soc. Workers, 85; Cleaver, Peace and Freedom, 217; Blomen, Soc. Labor, 75.

1972, Nixon, Rep., 402,812; McGovern, Dem., 198,540; Schmitz, Amer., 21,208; Soc. Workers, 30,945. Because of ballot peculiarities in 3 counties (particularly Pima), thousands of voters cast ballots for the Soc. Workers Party *and* one of the major candidates. Court ordered both votes counted as official.

1976, Carter, Dem., 295,602; Ford, Rep., 418,642; McCarthy, Ind., 19,229; MacBride, Libertarian, 7,647; Camejo, Soc. Workers, 928; Anderson, Amer., 564; Maddox, Amer. Ind., 85.

1980, Reagan, Rep., 529,688; Carter, Dem., 246,843; Anderson, Ind., 76,952; Clark, Libertarian, 18,784; De Berry, Soc. Workers, 1,100; Commoner, Citizens, 551; Hall, Com., 25; Griswold, Workers World, 2.

1984, Reagan, Rep., 681,416; Mondale, Dem., 333,854; Bergland, Libertarian, 10,585.

1988, Bush, Rep., 702,541; Dukakis, Dem., 454,029; Paul, Lib., 13,351; Fulani, New Alliance, 1,662.

1992, Bush, Rep., 572,086; Clinton, Dem., 543,050; Perot, Ind., 353,741; Gritz, Populist/America First, 8,141; Marrou, Libertarian, 6,759; Hagelin, Natural Law, 2,267.

1996, Clinton, Dem., 653,288; Dole, Rep., 622,073; Perot, Ref., 112,072; Browne, Libertarian, 14,358.

Arkansas

	1996			1992		
County	Clinton (D)	Dole (R)	Perot (RF)	Clinton (D)	Bush (R)	Perot (I)
Arkansas	4,220	1,910	463	4,709	2,594	639
Ashley	5,011	2,428	704	5,876	2,686	931
Baxter	6,703	6,877	1,572	6,991	5,640	2,938
Benton	17,205	23,748	4,147	15,774	21,126	6,128
Boone	5,745	6,093	1,132	6,128	6,094	2,079
Bradley	2,566	1,146	221	2,954	1,482	391
Calhoun	1,306	727	237	1,389	1,047	257
Carroll	3,689	3,957	986	3,769	3,535	1,500
Chicot	3,090	1,056	233	3,504	1,242	347
Clark	5,281	2,112	567	5,767	2,403	714
Clay	3,848	1,512	464	4,848	1,647	568
Cleburne	4,475	3,807	1,021	5,090	3,580	1,263
Cleveland	1,741	990	268	1,893	1,127	337
Columbia	4,730	3,376	678	4,747	3,702	1,090
Conway	4,055	2,307	746	4,898	2,719	803
Craighead	13,284	9,210	1,778	13,931	9,104	2,274
Crawford	6,749	7,182	1,683	6,656	6,882	2,442
Crittenden	8,415	4,673	554	9,683	5,910	848
Cross	3,631	2,000	466	4,058	2,303	602
Dallas	2,118	1,041	236	2,107	1,458	345
Desha	3,230	978	247	3,815	1,279	392
Drew	3,570	1,657	395	3,748	1,938	596
Faulkner	12,032	10,178	1,528	13,000	9,491	2,437
Franklin	3,269	2,246	626	3,217	2,495	987
Fulton	2,361	1,351	455	2,827	1,258	631
Garland	19,211	13,662	2,769	18,811	12,886	3,475
Grant	2,948	1,925	557	3,190	2,272	702
Greene	6,622	3,757	1,014	7,541	3,510	1,213
Hempstead	4,983	2,021	501	5,476	2,387	1,022
Hot Spring	6,002	2,864	1,123	6,308	3,036	1,209
Howard	2,741	1,478	369	2,764	1,728	466
Independence	6,240	4,021	1,126	7,083	4,232	1,444
Izard	2,818	1,678	541	3,419	1,532	606
Jackson	4,304	1,525	611	4,944	1,864	673
Jefferson	19,701	6,330	1,284	21,819	7,525	2,067
Johnson	3,585	2,367	757	3,951	2,563	1,013
Lafayette	2,466	971	374	2,273	1,188	504
Lawrence	3,652	1,823	609	4,146	2,124	636
Lee	3,267	1,013	257	3,436	1,293	308
Lincoln	2,517	907	221	2,805	1,142	390
Little River	3,183	1,409	480	3,327	1,483	890
Logan	3,832	2,966	1,048	3,995	3,408	1,220
Lonoke	8,049	6,414	1,369	7,963	6,253	1,554
Madison	2,504	2,303	461	2,415	2,238	598
Marion	2,735	2,312	764	2,757	2,023	1,327
Miller	6,469	4,874	1,043	7,050	5,273	2,249
Mississippi	8,301	3,919	1,016	10,046	4,697	981
Monroe	2,247	973	202	2,578	1,324	355
Montgomery	1,830	1,137	427	1,904	1,205	576
Nevada	2,279	976	345	2,242	1,217	455
Newton	1,631	1,927	498	1,765	1,730	608
Ouachita	6,635	3,136	733	7,411	3,711	1,238
Perry	1,873	1,143	395	1,906	1,162	412
Phillips	5,715	2,205	461	6,456	2,695	634
Pike	2,362	1,401	441	2,168	1,577	472
Poinsett	4,686	2,034	647	5,341	2,425	761
Polk	2,824	2,852	876	3,162	2,757	1,225
Pope	8,433	8,243	1,891	7,704	8,056	1,989
Prairie	2,211	1,025	305	2,366	1,154	434
Pulaski	75,084	44,780	6,014	79,482	47,789	8,751
Randolph	3,213	1,789	561	3,921	1,766	578
St. Francis	5,562	2,523	506	6,548	3,289	766
Saline	14,027	11,695	2,612	12,671	10,105	2,751
Scott	2,259	1,426	513	2,228	1,695	610
Searcy	1,669	1,786	381	1,679	1,772	503
Sebastian	15,514	16,482	2,899	16,570	16,817	6,023
Sevier	2,553	1,379	446	2,558	1,592	643
Sharp	3,573	2,635	687	3,761	2,486	921
Stone	2,227	1,526	579	2,622	1,672	697
Union	8,373	6,053	1,073	8,786	7,305	1,919
Van Buren	3,521	2,345	830	3,819	2,612	888
Washington	20,419	19,476	3,133	22,029	20,292	5,304
White	10,204	8,659	1,828	10,494	8,538	2,366
Woodruff	2,044	598	186	2,589	676	227
Yell	3,749	2,111	714	4,165	2,506	940
Totals	475,171	325,416	69,884	505,823	337,324	99,132

Arkansas Vote Since 1948

1948, Truman, Dem., 149,659; Dewey, Rep., 50,959; Thurmond, States' Rights, 40,068; Thomas, Soc., 1,037; Wallace, Prog., 751; Watson, Proh., 1.

1952, Eisenhower, Rep., 177,155; Stevenson, Dem., 226,300; Hamblen, Proh., 886; MacArthur, Christian Nationalist, 458; Hass, Soc. Labor, 1.

1956, Stevenson, Dem., 213,277; Eisenhower, Rep., 186,287; Andrews, Ind., 7,008.

1960, Kennedy, Dem., 215,049; Nixon, Rep., 184,508; Natl. States' Rights, 28,952.

1964, Johnson, Dem., 314,197; Goldwater, Rep., 243,264; Kasper, Natl. States' Rights, 2,965.

1968, Nixon, Rep., 189,062; Humphrey, Dem., 184,901; Wallace, 3d Party, 235,627.

1972, Nixon, Rep., 445,751; McGovern, Dem., 198,899; Schmitz, Amer., 3,016.

1976, Carter, Dem., 498,604; Ford, Rep., 267,903; McCarthy, Ind., 639; Anderson, Amer., 389.

1980, Reagan, Rep., 403,164; Carter, Dem., 398,041; Anderson, Ind., 22,468; Clark, Libertarian, 8,970; Commoner, Citizens, 2,345; Bubar, Statesman, 1,350; Hall, Com., 1,244.

1984, Reagan, Rep., 534,774; Mondale, Dem., 338,646; Bergland, Libertarian, 2,220.

1988, Bush, Rep., 466,578; Dukakis, Dem., 349,237; Duke, Chr. Pop., 5,146; Paul, Lib., 3,297.

1992, Clinton, Dem., 505,823; Bush, Rep., 337,324; Perot, Ind., 99,132; Phillips, U.S. Taxpayers, 1,437; Marrou, Libertarian, 1,261; Fulani, New Alliance, 1,022.

1996, Clinton, Dem., 475,171; Dole, Rep., 325,416; Perot, Ref., 69,884; Nader, Ind., 3,649; Browne, Ind., 3,076; Phillips, Ind., 2,065; Forbes, Ind., 932; Collins, Ind., 823; Masters, Ind., 749; Hagelin, Ind., 729; Moorehead, Ind., 747; Hollis, Ind., 538; Dodge, Ind., 483.

California

	1996			1992		
County	Clinton (D)	Dole (R)	Perot (RF)	Clinton (D)	Bush (R)	Perot (I)
Alameda	303,903	106,581	24,270	334,224	109,292	81,643

County	1996 Clinton (D)	Dole (R)	Perot (RF)	1992 Clinton (D)	Bush (R)	Perot (I)
Alpine	258	264	63	215	222	186
Amador . . .	5,868	6,870	1,267	5,286	5,477	4,553
Butte.	30,651	38,961	6,393	32,489	31,608	20,231
Calaveras. .	6,646	8,279	1,612	5,989	6,006	4,848
Colusa	2,054	3,047	404	1,798	2,589	1,206
Contra Costa	196,512	123,954	20,416	194,960	112,965	72,518
Del Norte . .	3,652	3,670	1,225	3,639	3,083	2,575
El Dorado . .	22,957	32,759	5,077	21,012	25,906	17,503
Fresno	94,448	98,813	10,962	92,418	89,137	36,299
Glenn	2,841	5,041	788	2,666	3,812	2,278
Humboldt. .	24,628	19,803	5,811	28,854	18,299	12,340
Imperial . . .	14,591	9,705	1,778	11,109	9,759	4,247
Inyo.	2,601	3,924	811	2,695	3,689	1,999
Kern	62,658	92,151	13,452	60,510	80,762	36,891
Kings	11,254	12,368	1,745	9,982	10,673	4,899
Lake	10,432	7,458	2,539	10,548	6,678	5,797
Lassen	3,318	5,194	1,080	3,388	3,836	3,004
Los Angeles	1,430,629	746,544	157,752	1,446,529	799,607	488,624
Madera	11,254	16,510	2,192	10,863	13,066	6,156
Marin	67,406	32,714	6,559	76,158	30,479	22,986
Mariposa . .	2,920	3,976	729	3,023	2,982	2,211
Mendocino.	14,952	9,765	3,685	18,344	7,958	9,753
Merced	21,786	20,847	3,427	20,133	17,981	10,914
Modoc	1,368	2,285	528	1,489	1,803	1,269
Mono	1,580	1,882	447	1,489	1,570	1,248
Monterey . .	57,700	39,794	7,240	54,861	36,461	24,472
Napa.	24,588	17,439	4,254	24,215	15,662	13,150
Nevada	15,369	21,784	3,330	15,433	17,343	11,072
Orange	327,485	446,717	66,195	306,930	426,613	232,394
Placer.	34,981	49,808	6,542	30,783	38,298	21,741
Plumas	3,540	4,905	919	3,742	3,599	2,551
Riverside . .	168,579	178,611	35,481	166,241	159,457	102,233
Sacramento	203,019	166,049	23,856	197,540	160,366	91,412
San Benito.	7,030	5,384	1,044	5,354	4,112	3,182
San Bernardino	183,372	180,135	39,330	183,634	176,563	109,183
San Diego .	389,964	402,876	63,037	367,397	352,125	259,249
San Francisco .	209,777	45,479	9,659	233,263	57,352	29,018
San Joaquin	67,253	65,131	9,692	63,655	58,355	31,205
San Luis Obispo . . .	40,395	46,733	8,204	40,136	36,384	27,314
San Mateo .	152,304	73,508	15,047	149,232	75,080	50,465
Santa Barbara . .	70,650	63,915	9,457	69,215	57,375	35,105
Santa Clara	297,639	168,291	34,908	296,265	170,870	128,895
Santa Cruz.	58,250	27,766	6,555	66,183	24,916	21,615
Shasta	20,848	34,736	5,875	21,605	28,190	17,990
Sierra	573	877	170	653	691	519
Siskiyou . . .	7,022	8,653	1,879	8,254	6,660	5,567
Solano	64,644	40,742	8,682	64,320	38,883	27,851
Sonoma . . .	100,738	53,555	13,862	104,334	47,619	43,859
Stanislaus .	53,738	52,403	8,360	52,415	47,275	27,651
Sutter	8,504	14,264	1,533	7,883	12,956	4,881
Tehama . . .	7,290	10,292	2,325	7,508	7,419	5,884
Trinity	2,203	2,530	856	1,967	1,886	2,092
Tulare	32,669	46,272	5,106	31,188	40,482	16,430
Tuolumne. .	8,950	10,386	1,925	9,216	8,525	6,294
Ventura . . .	110,772	109,202	23,054	99,011	94,911	71,844
Yolo.	33,033	18,807	3,150	33,297	17,574	11,073
Yuba	5,789	7,971	1,480	5,785	7,333	3,637
Totals	5,119,835	3,828,380	697,847	5,121,325	3,630,574	2,296,006

California Vote Since 1948

1948, Truman, Dem., 1,913,134; Dewey, Rep., 1,895,269; Wallace, Prog., 190,381; Watson, Proh., 16,926; Thomas, Soc., 3,459; Thurmond, States' Rights, 1,228; Teichert, Soc. Labor, 195; Dobbs, Soc. Workers, 133.

1952, Eisenhower, Rep., 2,897,310; Stevenson, Dem., 2,197,548; Hallinan, Prog., 24,106; Hamblen, Proh., 15,653; MacArthur, (Tenny Ticket) 3,326; (Kellems Ticket) 178; Hass, Soc. Labor, 273; Hoopes, Soc., 206; scattered, 3,249.

1956, Eisenhower, Rep., 3,027,668; Stevenson, Dem., 2,420,136; Holtwick, Proh., 11,119; Andrews, Constitution, 6,087; Hass, Soc. Labor, 300; Hoopes, Soc., 123; Dobbs, Soc. Workers, 96; Smith, Christian Natl., 8.

1960, Kennedy, Dem., 3,224,099; Nixon, Rep., 3,259,722; Decker, Proh., 21,706; Hass, Soc. Labor, 1,051.

1964, Johnson, Dem., 4,171,877; Goldwater, Rep., 2,879,108; Hass, Soc. Labor, 489; DeBerry, Soc. Workers, 378; Munn, Proh., 305; Hensley, Universal, 19.

1968, Nixon, Rep., 3,467,664; Humphrey, Dem., 3,244,318; Wallace, 3d Party, 487,270; Peace and Freedom, 27,707; McCarthy, Alternative, 20,721; Gregory, write-in, 3,230; Mitchell, Com., 260; Munn, Proh., 59; Blomen, Soc. Labor, 341; Soeters, Defense, 17.

1972, Nixon, Rep., 4,602,096; McGovern, Dem., 3,475,847; Schmitz, Amer., 232,554; Spock, Peace and Freedom, 55,167; Hall, Com., 373; Hospers, Libertarian, 980; Munn, Proh., 53; Fisher, Soc. Labor, 197; Jenness, Soc. Workers, 574; Green, Universal, 21.

1976, Carter, Dem., 3,742,284; Ford, Rep., 3,882,244; MacBride, Libertarian, 56,388; Maddox, Amer. Ind., 51,098; Wright, People's, 41,731; Camejo, Soc. Workers, 17,259; Hall, Com., 12,766; write-in, McCarthy, 58,412; other write-in, 4,935.

1980, Reagan, Rep. 4,524,858; Carter, Dem., 3,083,661; Anderson, Ind., 739,833; Clark, Libertarian, 148,434; Commoner, Ind., 61,063; Smith, Peace and Freedom, 18,116; Rarick, Amer. Ind., 9,856.

1984, Reagan, Rep. 5,305,410; Mondale, Dem., 3,815,947; Bergland, Libertarian, 48,400.

1988, Bush, Rep., 5,054,917; Dukakis, Dem., 4,702,233; Paul, Lib., 70,105; Fulani, Ind., 31,181.

1992, Clinton, Dem., 5,121,325; Bush, Rep., 3,630,575; Perot, Ind., 2,296,006; Marrou, Libertarian, 48,139; Daniels, Ind., 18,597; Phillips, U.S. Taxpayers, 12,711.

1996, Clinton, Dem., 5,119,835; Dole, Rep., 3,828,380; Perot, Ref., 697,847; Nader, Green, 237,016; Browne, Libertarian, 73,600; Feinland, Peace & Freedom, 25,332; Phillips, Amer. Ind., 21,202; Hagelin, Natural Law, 15,403.

Colorado

County	1996 Clinton (D)	Dole (R)	Perot (RF)	1992 Clinton (D)	Bush (R)	Perot (I)
Adams	48,314	36,666	7,206	45,357	30,856	26,379
Alamosa.	2,330	2,038	437	1,928	1,572	1,089
Arapahoe.	68,306	82,778	8,476	66,607	72,221	44,363
Archuleta	997	1,963	360	819	1,242	741
Baca.	659	1,321	203	726	1,240	647
Bent	1,046	917	209	985	759	506
Boulder	63,316	41,922	6,840	64,567	33,553	27,762
Chaffee	2,768	3,052	538	2,284	2,419	1,549
Cheyenne	328	739	91	301	615	292
Clear Creek. . . .	1,863	1,746	365	1,744	1,356	1,308
Conejos	1,726	1,149	245	1,705	1,160	578
Costilla	1,168	333	112	1,180	366	199
Crowley	559	680	114	570	602	276
Custer	412	920	164	343	651	368
Delta.	3,584	6,047	1,060	3,424	4,359	2,627
Denver	120,312	58,529	8,777	121,961	55,418	37,298
Dolores	276	417	95	242	315	285
Douglas	16,232	32,120	2,662	9,991	18,592	11,329
Eagle	5,094	4,637	1,193	3,870	3,100	3,821
Elbert	1,894	4,125	507	1,237	2,205	1,567
El Paso	55,822	102,403	11,175	45,827	86,044	34,346
Fremont	5,344	7,437	1,438	5,356	5,961	3,709
Garfield	5,722	6,281	1,562	5,082	4,404	4,408
Gilpin	799	682	184	726	462	545
Grand.	2,012	2,264	473	1,678	1,763	1,454
Gunnison	2,812	2,230	570	2,389	1,662	1,671
Hinsdale.	185	289	56	151	188	136
Huerfano	1,483	996	210	1,224	685	385
Jackson	222	486	107	216	422	326
Jefferson	89,494	101,517	12,967	80,834	82,705	58,404
Kiowa	246	549	74	290	472	267
Kit Carson	1,073	2,068	235	925	1,801	919
Lake	1,338	728	274	1,426	605	863
La Plata	6,509	8,057	1,403	5,913	5,522	4,083
Larimer	40,965	45,935	6,823	38,232	35,995	24,879
Las Animas	3,611	1,905	427	3,847	1,739	953
Lincoln	729	1,272	164	640	1,079	581
Logan.	2,765	4,032	609	2,718	3,420	2,184
Mesa	17,114	24,761	3,707	15,162	18,169	10,474
Mineral.	192	179	69	171	159	117
Moffat.	1,635	2,466	649	1,386	1,809	1,875
Montezuma	2,578	4,175	827	2,270	3,124	2,205
Montrose	4,019	6,730	1,187	3,713	4,847	3,093
Morgan	3,347	4,557	687	2,985	3,724	2,175
Otero	3,386	3,356	581	3,485	3,120	1,590
Ouray	569	984	167	461	653	466
Park	1,844	2,661	534	1,307	1,530	1,396
Philips	706	1,284	156	692	1,075	525
Pitkin	3,949	1,969	535	3,820	1,686	1,907
Prowers	1,745	2,504	342	1,770	2,371	1,184
Pueblo	28,791	17,402	3,374	30,261	16,120	9,841
Rio Blanco	731	1,697	243	778	1,231	794
Rio Grande	1,720	2,129	379	1,541	1,927	1,043
Routt	3,660	3,019	859	3,188	2,358	2,564
Saguache	969	712	160	1,011	675	471
San Juan	133	153	50	147	118	183
San Miguel	1,535	773	231	1,380	628	634
Sedgwick	519	715	101	397	447	295
Summit	3,970	3,261	823	3,344	2,256	2,715
Teller	2,312	4,458	707	1,873	3,050	1,927
Washington	649	1,566	190	660	1,266	671
Weld.	21,325	26,518	4,347	19,295	20,958	13,571
Yuma	1,439	2,589	319	1,269	2,019	1,197
Totals.	671,152	691,848	99,629	629,681	562,850	366,010

Colorado Vote Since 1948

1948, Truman, Dem., 267,288; Dewey, Rep., 239,714; Wallace, Prog., 6,115; Thomas, Soc., 1,678; Dobbs, Soc. Workers, 228; Teichert, Soc. Labor, 214.

1952, Eisenhower, Rep., 379,782; Stevenson, Dem., 245,504; MacArthur, Constitution, 2,181; Hallinan, Prog., 1,919; Hoopes, Soc., 365; Hass, Soc. Labor, 352.

1956, Eisenhower, Rep., 394,479; Stevenson, Dem., 263,997; Hass, Soc. Lab., 3,308; Andrews, Ind., 759; Hoopes, Soc., 531.

1960, Kennedy, Dem., 330,629; Nixon, Rep., 402,242; Hass, Soc. Labor, 2,803; Dobbs, Soc. Workers, 572.

1964, Johnson, Dem., 476,024; Goldwater, Rep., 296,767; Hass, Soc. Labor, 302; DeBerry, Soc. Workers, 2,537; Munn, Proh., 1,356.

1968, Nixon, Rep., 409,345; Humphrey, Dem., 335,174; Wallace, 3d Party, 60,813; Blomen, Soc. Labor, 3,016; Gregory, New-party, 1,393; Munn, Proh., 275; Halstead, Soc. Workers, 235.

1972, Nixon, Rep., 597,189; McGovern, Dem., 329,980; Fisher, Soc. Labor, 4,361; Hospers, Libertarian, 1,111; Hall, Com., 432; Jenness, Soc. Workers, 555; Munn, Proh., 467; Schmitz, Amer., 17,269; Spock, Peoples, 2,403.

1976, Carter, Dem., 460,353; Ford, Rep., 584,367; McCarthy, Ind., 26,107; MacBride, Libertarian, 5,330; Bubar, Proh., 2,882.

1980, Reagan, Rep., 652,264; Carter, Dem., 367,973; Anderson, Ind., 130,633; Clark, Libertarian, 25,744; Commoner, Citizens, 5,614; Bubar, Statesman, 1,180; Pulley, Socialist, 520; Hall, Com., 487.

1984, Reagan, Rep., 821,817; Mondale, Dem., 454,975; Bergland, Libertarian, 11,257.

1988, Bush, Rep., 728,177; Dukakis, Dem., 621,453; Paul, Lib., 15,482; Dodge, Proh., 4,604.

1992, Clinton, Dem., 629,681; Bush, Rep., 562,850; Perot, Ind., 366,010; Marrou, Libertarian, 8,669; Fulani, New Alliance, 1,608.

1996, Dole, Rep., 691,848; Clinton, Dem., 671,152; Perot, Ref., 99,629; Nader, Green, 25,070; Browne, Libertarian, 12,392; Collins, Ind., 2,809; Phillips, Amer. Constitution, 2,813; Hagelin, Natural Law, 2,547; Hollis, Soc., 669; Moorehead, Workers World, 599; Templin, Amer., 557; Dodge, Proh., 375; Harris, Soc. Workers, 244.

Connecticut

City	1996			1992		
	Clinton	Dole	Perot	Clinton	Bush	Perot
	(D)	(R)	(RF)	(D)	(R)	(I)
Bridgeport	22,883	6,785	2,367	22,321	13,149	6,263
Bristol........	13,616	6,560	3,049	11,872	8,407	7,890
Danbury......	12,102	7,965	2,158	9,909	10,310	5,517
Fairfield......	12,639	12,314	2,092	12,099	13,968	5,941
Greenwich	11,622	14,308	1,437	11,893	15,885	4,584
Hartford	22,929	3,082	1,010	26,971	6,180	3,390
New Britain. ...	14,322	4,911	1,717	14,159	7,040	4,983
New Haven ...	26,161	4,822	1,555	29,774	8,931	4,130
Norwalk	17,354	10,800	2,237	16,488	14,743	6,046
Stamford......	25,005	14,696	2,595	23,185	19,809	6,763
Waterbury	18,901	12,075	3,169	16,366	16,155	9,188
West Hartford	19,037	10,781	1,890	19,623	12,266	5,017
Other	519,169	374,010	114,247	467,658	431,470	279,059
Totals........	**735,740**	**483,109**	**139,523**	**682,318**	**578,313**	**348,771**

Connecticut Vote Since 1948

1948, Truman, Dem., 423,297; Dewey, Rep., 437,754; Wallace, Prog., 13,713; Thomas, Soc., 6,964; Teichert, Soc. Labor, 1,184; Dobbs, Soc. Workers, 606.

1952, Eisenhower, Rep., 611,012; Stevenson, Dem., 481,649; Hoopes, Soc., 2,244; Hallinan, Peoples, 1,466; Hass, Soc. Labor, 535; write-in, 5.

1956, Eisenhower, Rep., 711,837; Stevenson, Dem., 405,079; scattered, 205.

1960, Kennedy, Dem., 657,055; Nixon, Rep., 565,813.

1964, Johnson, Dem., 826,269; Goldwater, Rep., 390,996; scattered, 1,313.

1968, Nixon, Rep., 556,721; Humphrey, Dem., 621,561; Wallace, 3d Party, 76,650; scattered, 1,300.

1972, Nixon, Rep., 810,763; McGovern, Dem., 555,498; Schmitz, Amer., 17,239; scattered, 777.

1976, Carter, Dem., 647,895; Ford, Rep., 719,261; Maddox, George Wallace Party, 7,101; LaRouche, U.S. Labor, 1,789.

1980, Reagan, Rep., 677,210; Carter, Dem., 541,732; Anderson, Ind., 171,807; Clark, Libertarian, 8,570; Commoner, Citizens, 6,130; scattered, 836.

1984, Reagan, Rep., 890,877; Mondale, Dem., 569,597.

1988, Bush, Rep., 750,241; Dukakis, Dem., 676,584; Paul, Lib., 14,071; Fulani, New Alliance, 2,491.

1992, Clinton, Dem., 682,318; Bush, Rep., 578,313; Perot, Ind., 348,771; Marrou, Libertarian, 5,391; Fulani, New Alliance, 1,363.

1996, Clinton, Dem., 735,740; Dole, Rep., 483,109; Perot, Ref., 139,523; Nader, Green, 24,321; Browne, Libertarian, 5,788; Phillips, Concerned Citizens, 2,425; Hagelin, Natural Law, 1,703.

Delaware

County	1996			1992		
	Clinton	Dole	Perot	Clinton	Bush	Perot
	(D)	(R)	(RF)	(D)	(R)	(I)
Kent	18,327	15,932	4,705	15,364	15,562	8,916
New Castle ...	98,837	60,943	17,748	91,516	66,311	37,581
Sussex	23,191	22,187	6,266	19,174	20,440	12,716
Totals........	**140,355**	**99,062**	**28,719**	**126,054**	**102,313**	**59,213**

Delaware Vote Since 1948

1948, Truman, Dem., 67,813; Dewey, Rep., 69,688; Wallace, Prog., 1,050; Watson, Proh., 343; Thomas, Soc., 250; Teichert, Soc. Labor, 29.

1952, Eisenhower, Rep., 90,059; Stevenson, Dem., 83,315; Hass, Soc. Labor, 242; Hamblen, Proh., 234; Hallinan, Prog., 155; Hoopes, Soc., 20.

1956, Eisenhower, Rep., 98,057; Stevenson, Dem., 79,421; Oltwick, Proh., 400; Hass, Soc. Labor, 110.

1960, Kennedy, Dem., 99,590; Nixon, Rep., 96,373; Faubus, States' Rights, 354; Decker, Proh., 284; Hass, Soc. Labor, 82.

1964, Johnson, Dem., 122,704; Goldwater, Rep., 78,078; Hass, Soc. Labor, 113; Munn, Proh., 425.

1968, Nixon, Rep., 96,714; Humphrey, Dem., 89,194; Wallace, 3d Party, 28,459.

1972, Nixon, Rep., 140,357; McGovern, Dem., 92,283; Schmitz, Amer., 2,638; Munn, Proh., 238.

1976, Carter, Dem., 122,596; Ford, Rep., 109,831; McCarthy, non-partisan, 2,437; Anderson, Amer., 645; LaRouche, U.S. Labor, 136; Bubar, Proh., 103; Levin, Soc. Labor, 86.

1980, Reagan, Rep., 111,252; Carter, Dem., 105,754; Anderson, Ind., 16,288; Clark, Libertarian, 1,974; Greaves, Amer., 400.

1984, Reagan, Rep., 152,190; Mondale, Dem., 101,656; Bergland, Libertarian, 268.

1988, Bush, Rep., 139,639; Dukakis, Dem., 108,647; Paul, Lib., 1,162; Fulani, New Alliance, 443.

1992, Clinton, Dem., 126,054; Bush, Rep., 102,313; Perot, Ind., 59,213; Fulani, New Alliance, 1,105.

1996, Clinton, Dem., 140,355; Dole, Rep., 99,062; Perot, Ind. (Ref.), 28,719; Browne, Libertarian, 2,052; Phillips, Taxpayers, 348; Hagelin, Natural Law, 274.

District of Columbia

	1996			1992		
	Clinton	Dole	Perot	Clinton	Bush	Perot
	(D)	(R)	(RF)	(D)	(R)	(I)
Totals.........	158,220	17,339	3,611	192,619	20,698	9,681

District of Columbia Vote Since 1964

1964, Johnson, Dem., 169,796; Goldwater, Rep., 28,801.

1968, Nixon, Rep., 31,012; Humphrey, Dem., 139, 566.

1972, Nixon, Rep., 35,226; McGovern, Dem., 127,627; Reed, Soc. Workers, 316; Hall, Com., 252.

1976, Carter, Dem., 137,818; Ford, Rep., 27,873; Camejo, Soc. Workers, 545; MacBride, Libertarian, 274; Hall, Com., 219; LaRouche, U.S. Labor, 157.

1980, Reagan, Rep., 23,313; Carter, Dem., 130,231; Anderson, Ind., 16,131; Commoner, Citizens, 1,826; Clark, Libertarian, 1,104; Hall, Com., 369; DeBerry, Soc. Labor, 173; Griswold, Workers World, 52; write-ins, 690.

1984, Mondale, Dem., 180,408; Reagan, Rep., 29,009; Bergland, Libertarian, 279.

1988, Bush, Rep., 27,590; Dukakis, Dem., 159,407; Fulani, New Alliance, 2,901; Paul, Lib., 554.

1992, Clinton, Dem., 192,619; Bush, Rep., 20,698; Perot, Ind., 9,681; Fulani, New Alliance, 1,459; Daniels, Ind., 1,186.

1996, Clinton, Dem., 158,220; Dole, Rep., 17,339; Perot, Ref., 3,611; Nader, Green, 4,780; Browne, Libertarian, 588; Hagelin, Natural Law, 283; Harris, Soc. Workers, 257.

Florida

County	1996			1992		
	Clinton	Dole	Perot	Clinton	Bush	Perot
	(D)	(R)	(RF)	(D)	(R)	(I)
Alachua	40,144	25,303	8,072	37,876	22,806	15,293
Baker.......	2,273	3,684	667	1,974	3,417	1,315
Bay.........	17,020	28,290	5,922	12,830	22,820	9,702
Bradford....	3,356	4,038	819	3,040	3,671	1,572
Brevard	80,416	87,980	25,249	61,070	84,545	49,491
Broward.....	320,736	142,834	38,964	276,309	164,782	90,923
Calhoun.....	1,794	1,717	630	1,665	1,721	1,176
Charlotte	27,121	27,836	7,783	22,904	24,302	14,711
Citrus	22,042	20,114	7,244	15,935	16,402	12,310
Clay	13,246	30,332	3,281	10,597	26,313	8,414
Collier	23,182	42,590	6,320	18,794	38,447	14,514
Columbia....	6,691	7,588	1,970	5,526	6,489	2,906
Dade[1].......	317,378	209,634	24,722	254,444	235,149	53,957
De Soto	3,219	3,272	965	2,646	3,070	1,687
Dixie........	1,731	1,398	652	1,855	1,401	1,094
Duval	112,258	126,867	13,844	92,010	123,480	33,335
Escambia....	37,768	60,839	8,587	32,018	52,775	19,868
Flagler	9,583	8,232	2,185	6,692	6,241	3,387
Franklin	2,095	1,563	878	1,534	1,660	1,143
Gadsden	9,405	3,813	938	8,478	3,975	1,871
Gilchrist	1,985	1,939	841	1,511	1,395	1,090
Glades	1,530	1,361	521	1,305	1,185	878
Gulf.........	2,480	2,424	1,054	1,938	2,650	1,245

County	1996 Clinton (D)	Dole (R)	Perot (RF)	1992 Clinton (D)	Bush (R)	Perot (I)
Hamilton....	1,734	1,518	406	1,622	1,402	695
Hardee.....	2,417	2,926	851	2,017	2,898	1,498
Hendry.....	3,882	3,855	1,135	2,690	3,279	2,032
Hernando...	28,520	22,039	7,272	19,171	17,896	11,845
Highlands...	14,244	15,608	3,739	11,234	14,497	6,592
Hillsborough	144,223	136,621	25,154	115,261	130,611	63,037
Holmes.....	2,310	3,248	1,208	1,877	3,196	1,426
Indian River	16,373	22,709	4,635	12,359	19,137	12,375
Jackson.....	6,665	7,187	1,602	5,481	6,720	2,447
Jefferson....	2,543	1,851	393	2,270	1,506	894
Lafayette....	829	1,166	316	866	1,037	612
Lake.......	29,750	35,089	8,813	23,199	30,818	15,606
Lee........	65,692	80,882	18,389	53,656	73,423	38,446
Leon.......	50,058	33,914	6,672	47,770	31,964	17,207
Levy.......	4,938	4,299	1,774	4,330	3,796	2,784
Liberty.....	868	913	376	820	1,126	617
Madison....	2,791	2,195	578	2,644	2,006	1,174
Manatee....	41,835	44,059	10,360	33,826	42,708	23,282
Marion.....	37,033	41,397	11,340	30,823	35,438	20,524
Martin......	20,851	28,516	5,005	14,778	24,768	13,433
Monroe.....	15,219	12,021	4,817	10,435	9,891	8,306
Nassau.....	7,276	12,134	1,657	5,497	9,364	3,251
Okaloosa....	16,434	40,631	5,432	12,003	32,755	16,649
Okeechobee	4,824	3,415	1,666	3,418	3,298	2,645
Orange.....	105,513	106,026	18,191	82,656	108,738	44,827
Osceola....	21,870	18,335	6,091	15,009	19,139	11,021
Palm Beach	230,621	133,762	30,739	187,840	140,317	76,223
Pasco......	66,472	48,346	18,011	53,125	47,721	34,650
Pinellas....	184,728	152,125	36,990	160,217	158,733	101,150
Polk.......	66,735	67,943	14,991	51,442	65,952	28,198
Putnam.....	12,008	9,781	3,272	10,707	8,909	5,975
St. Johns...	16,713	27,311	4,205	12,284	20,173	7,397
St. Lucie....	36,168	28,892	8,482	23,873	24,397	19,813
Santa Rosa	10,923	26,244	4,957	6,526	17,229	8,735
Sarasota....	63,648	69,198	14,939	54,536	66,831	34,281
Seminole ...	45,051	59,778	9,357	35,649	57,085	24,477
Sumter.....	7,014	5,960	2,375	5,027	4,366	2,901
Suwannee ..	4,479	5,742	1,874	3,985	4,571	2,790
Taylor......	3,583	3,188	1,140	2,568	2,693	1,929
Union......	1,388	1,636	425	1,247	1,543	770
Volusia	78,905	63,067	17,319	65,213	59,155	30,813
Wakulla....	3,054	2,931	1,091	2,319	2,586	1,790
Walton	5,341	7,706	2,342	3,886	5,719	3,886
Washington	2,992	3,522	1,287	2,544	3,694	1,596
Totals......	2,545,968	2,243,324	483,776	2,071,651	2,171,781	1,052,481

(1) In 1997, Dade County changed its name to Miami-Dade County.

Florida Vote Since 1948

1948, Truman, Dem., 281,988; Dewey, Rep., 194,280; Thurmond, States' Rights, 89,755; Wallace, Prog., 11,620.

1952, Eisenhower, Rep., 544,036; Stevenson, Dem., 444,950; scattered, 351.

1956, Eisenhower, Rep., 643,849; Stevenson, Dem., 480,371.

1960, Kennedy, Dem., 748,700; Nixon, Rep., 795,476.

1964, Johnson, Dem., 948,540; Goldwater, Rep., 905,941.

1968, Nixon, Rep., 886,804; Humphrey, Dem., 676,794; Wallace, 3d Party, 624,207.

1972, Nixon, Rep., 1,857,759; McGovern, Dem., 718,117; scattered, 7,407.

1976, Carter, Dem., 1,636,000; Ford, Rep., 1,469,531; McCarthy, Ind., 23,643; Anderson, Amer., 21,325.

1980, Reagan, Rep., 2,046,951; Carter, Dem., 1,419,475; Anderson, Ind., 189,692; Clark, Libertarian, 30,524; write-ins, 285.

1984, Reagan, Rep., 2,728,775; Mondale, Dem., 1,448,344.

1988, Bush, Rep., 2,616,597; Dukakis, Dem., 1,655,851; Paul, Lib., 19,796; Fulani, New Alliance, 6,655.

1992, Bush, Rep., 2,171,781; Clinton, Dem., 2,071,651; Perot, Ind., 1,052,481; Marrou, Libertarian, 15,068.

1996, Clinton, Dem., 2,545,968; Dole, Rep., 2,243,324; Perot, Ref., 483,776; Browne, Libertarian, 23,312.

Georgia

County	1996 Clinton (D)	Dole (R)	Perot (RF)	1992 Clinton (D)	Bush (R)	Perot (I)
Appling....	2,070	2,572	446	2,455	2,514	1,047
Atkinson....	823	784	215	1,056	779	342
Bacon......	1,360	1,580	402	1,423	1,301	604
Baker......	955	408	105	864	391	210
Baldwin....	5,740	4,570	849	5,813	4,262	1,679
Banks.....	1,536	1,925	595	1,530	1,551	583
Barrow.....	3,928	5,342	942	3,991	4,328	1,633
Bartow.....	6,853	9,250	1,770	6,675	7,742	2,500
Ben Hill....	2,198	1,516	358	2,348	1,476	619
Berrien.....	2,066	1,950	525	2,103	1,637	796
Bibb	26,727	20,778	2,268	28,070	19,847	6,021
Bleckley	1,365	1,632	300	1,710	1,570	662
Brantley....	1,494	1,738	386	1,883	1,541	840
Brooks	1,977	1,738	314	1,895	1,779	630
Bryan	2,152	3,577	513	2,031	2,789	1,095
Bulloch	5,396	6,646	939	4,903	5,690	2,020

County	1996 Clinton (D)	Dole (R)	Perot (RF)	1992 Clinton (D)	Bush (R)	Perot (I)
Burke	3,915	2,590	389	3,647	2,390	807
Butts........	2,271	2,027	416	2,448	1,768	619
Calhoun.....	1,217	541	106	1,301	464	248
Camden.....	3,644	4,222	572	2,952	3,517	1,077
Candler	1,097	1,131	264	1,192	1,014	541
Carroll	8,438	11,157	2,002	8,404	10,750	3,358
Catoosa....	5,185	8,237	1,257	4,817	7,599	2,290
Charlton.....	1,368	1,374	280	1,127	1,333	427
Chatham	35,781	31,987	3,028	31,533	31,925	8,269
Chattahoo-chee	565	398	115	604	413	177
Chattooga ...	3,003	2,513	796	2,976	2,439	965
Cherokee....	10,802	24,527	2,872	8,113	16,054	4,950
Clarke......	15,206	10,504	1,201	15,403	10,459	2,987
Clay.......	787	293	62	778	264	155
Clayton......	30,687	20,625	3,494	25,890	23,965	7,942
Clinch......	973	789	182	759	790	286
Cobb	73,750	114,188	10,438	63,960	103,734	28,747
Coffee......	3,407	3,934	711	3,275	3,778	1,256
Colquitt.....	4,135	4,847	977	3,891	4,680	1,682
Columbia....	8,601	21,291	1,709	7,115	16,657	4,379
Cook.......	1,780	1,354	267	1,731	1,318	537
Coweta.....	7,794	13,058	1,949	7,093	9,814	3,587
Crawford ...	1,534	1,290	270	1,648	974	549
Crisp.......	2,504	2,321	445	2,610	2,253	823
Dade.......	1,737	2,295	618	1,782	2,191	823
Dawson.....	1,434	2,343	473	1,399	1,696	790
Decatur.....	3,245	3,035	497	3,198	3,142	1,068
DeKalb	137,903	60,255	6,742	124,559	70,282	19,741
Dodge	2,696	2,478	587	3,002	2,287	978
Dooly	1,951	990	207	1,993	1,034	350
Dougherty ...	15,600	11,144	1,072	15,236	12,455	3,178
Douglas.....	9,631	14,495	2,109	8,869	13,349	4,362
Early.......	1,648	1,374	246	1,970	1,457	652
Echols......	308	335	97	312	361	238
Effingham ...	3,031	5,022	769	2,690	3,814	1,443
Elbert......	2,900	2,393	552	3,025	2,372	757
Emanuel	2,947	2,451	450	2,951	2,662	755
Evans.......	1,117	1,206	204	1,230	1,244	480
Fannin......	2,741	3,373	782	2,902	3,255	1,028
Fayette	9,875	21,005	2,016	8,430	17,576	5,598
Floyd	10,464	12,426	2,345	11,614	12,378	3,779
Forsyth.....	5,957	15,013	1,889	4,936	8,652	3,453
Franklin.....	2,338	2,364	665	2,505	2,391	1,014
Fulton	143,306	89,809	7,720	147,459	85,451	23,578
Gilmer	2,464	3,121	725	2,311	2,661	879
Glascock	348	532	128	316	516	180
Glynn	8,058	12,305	1,137	8,581	11,242	3,053
Gordon......	4,239	5,232	1,284	4,103	5,265	1,818
Grady.......	2,862	2,674	633	2,520	2,370	1,126
Greene......	2,115	1,702	173	2,259	1,307	483
Gwinnett	53,819	96,610	10,236	44,253	81,822	23,926
Habersham ..	3,170	4,730	1,149	3,098	4,569	1,444
Hall.........	10,362	19,280	2,321	11,214	16,108	5,043
Hancock.....	2,135	438	71	2,461	506	189
Haralson	2,850	3,260	808	3,281	3,142	1,167
Harris.......	2,779	3,829	489	2,679	3,316	954
Hart........	3,486	2,884	767	3,614	2,607	1,376
Heard.......	1,248	1,170	406	1,456	1,190	617
Henry.......	9,498	16,968	2,320	7,817	12,634	3,769
Houston.....	12,760	17,050	2,730	12,270	14,119	6,263
Irwin........	1,225	1,085	224	1,366	973	465
Jackson.....	3,746	4,782	899	3,792	3,976	1,381
Jasper......	1,553	1,423	243	1,485	1,153	373
Jeff Davis ...	1,576	1,796	428	2,031	1,947	958
Jefferson....	3,404	2,077	298	3,220	2,077	685
Jenkins......	1,336	955	166	1,401	929	394
Johnson.....	1,194	815	242	1,473	1,314	502
Jones.......	3,195	3,272	497	3,338	2,770	1,159
Lamar......	2,125	1,988	409	2,065	1,707	600
Lanier......	818	519	160	811	600	298
Laurens.....	5,792	6,118	818	6,184	6,146	1,602
Lee........	2,005	3,983	506	1,811	3,061	1,024
Liberty	4,462	3,042	580	3,853	2,832	1,176
Lincoln.....	1,334	1,391	208	1,327	1,149	479
Long........	936	791	236	874	719	355
Lowndes	9,470	10,578	1,518	9,019	10,276	2,864
Lumpkin	1,949	2,576	588	2,010	1,972	1,035
McDuffie	2,725	3,254	395	2,640	2,955	860
McIntosh	1,927	1,219	293	1,925	1,027	550
Macon	2,618	1,006	159	2,491	944	363
Madison.....	2,571	3,992	868	2,393	3,351	1,129
Marion......	977	678	159	1,145	711	198
Meriwether...	3,492	2,259	480	4,002	2,364	942
Miller.......	909	847	235	934	826	455
Mitchell......	3,165	2,033	372	3,052	1,917	818
Monroe.....	2,768	3,054	488	2,774	2,423	949
Montgomery	1,233	1,163	284	1,185	1,009	416
Morgan	2,111	2,118	364	2,057	1,797	596
Murray	2,861	3,289	938	2,764	3,256	1,186
Muscogee ..	24,867	19,360	1,891	25,476	21,386	4,327
Newton......	6,759	7,274	1,258	5,811	5,804	1,998
Oconee.....	2,992	5,116	615	2,745	4,125	1,182
Oglethorpe...	1,570	1,826	369	1,491	1,590	620
Paulding.....	5,699	10,152	1,603	5,212	7,180	2,654
Peach.......	3,582	2,676	471	3,677	2,327	947
Pickens	2,693	3,041	783	2,359	2,332	1,037

County	1996 Clinton (D)	Dole (R)	Perot (RF)	1992 Clinton (D)	Bush (R)	Perot (I)
Pierce......	1,420	2,319	333	1,852	1,899	708
Pike........	1,474	2,054	357	1,651	1,822	623
Polk........	4,298	4,130	1,076	4,872	4,158	1,598
Pulaski.....	1,554	1,196	268	1,756	1,075	614
Putnam.....	2,340	2,306	474	2,149	1,756	775
Quitman....	514	224	59	523	284	113
Rabun......	1,943	2,213	585	1,878	1,902	825
Randolph...	1,438	816	126	1,756	887	315
Richmond...	30,738	23,670	2,310	28,910	24,227	6,290
Rockdale ...	7,656	13,006	1,750	7,003	11,945	3,664
Schley.....	576	470	123	601	511	180
Screven	2,087	1,862	263	1,940	1,705	709
Seminole ...	1,265	1,003	250	1,193	850	468
Spalding....	6,017	7,376	1,059	6,392	7,262	2,044
Stephens ...	3,072	3,890	979	2,976	4,047	1,448
Stewart....	1,537	525	152	1,540	1,186	175
Sumter	4,239	3,358	451	4,489	3,616	1,046
Talbot	1,579	652	111	1,768	671	238
Taliaferro...	615	235	36	755	269	80
Tattnall	2,369	2,518	541	2,360	2,566	996
Taylor	1,450	1,002	195	1,508	1,078	281
Telfair	1,856	1,143	322	2,238	1,324	613
Terrell	1,509	1,111	129	1,942	1,143	384
Thomas	5,183	5,649	667	4,841	5,500	1,591
Tift........	4,198	5,613	728	3,930	4,485	1,139
Toombs	2,763	3,646	602	2,648	3,609	1,210
Towns......	1,664	2,030	459	1,487	1,674	537
Treutlen	912	723	122	1,116	898	318
Troup	5,940	8,716	1,090	6,412	8,118	2,488
Turner......	1,272	924	246	1,669	936	370
Twiggs	1,927	958	210	2,097	853	432
Union	2,175	2,685	622	2,304	2,533	804
Upson......	3,491	3,783	731	3,740	4,053	1,186
Walker	6,743	8,817	1,969	6,217	8,489	2,748
Walton	5,618	7,934	1,323	4,821	5,619	1,923
Ware.......	4,171	4,746	636	4,573	4,573	1,263
Warren.....	1,230	735	83	1,239	751	180
Washington	4,057	2,348	488	3,508	2,384	820
Wayne	2,734	3,709	665	3,052	3,381	1,107
Webster	529	235	59	600	208	103
Wheeler	751	460	141	880	601	214
White	1,864	2,959	556	1,756	2,477	981
Whitfield	7,720	12,368	1,637	7,335	12,003	2,866
Wilcox......	1,067	882	171	1,365	916	433
Wilkes......	1,971	1,417	184	1,955	1,535	464
Wilkinson ...	2,278	1,332	287	2,286	1,232	520
Worth	2,300	2,752	521	2,578	2,344	905
Totals......	1,053,849	1,080,843	146,337	1,008,966	995,252	309,657

Georgia Vote Since 1948

1948, Truman, Dem., 254,646; Dewey, Rep., 76,691; Thurmond, States' Rights, 85,055; Wallace, Prog., 1,636; Watson, Proh., 732.

1952, Eisenhower, Rep., 198,979; Stevenson, Dem., 456,823; Liberty Party, 1.

1956, Stevenson, Dem., 444,388; Eisenhower, Rep., 222,778; Andrews, Ind., write-in, 1,754.

1960, Kennedy, Dem., 458,638; Nixon, Rep., 274,472; write-in, 239.

1964, Johnson, Dem., 522,557; Goldwater, Rep., 616,600.

1968, Nixon, Rep., 380,111; Humphrey, Dem., 334,440; Wallace, 3d Party, 535,550; write-in, 162.

1972, Nixon, Rep., 881,496; McGovern, Dem., 289,529; scattered, 2,935; Schmitz, Amer., 812.

1976, Carter, Dem., 979,409; Ford, Rep., 483,743; write-in, 4,306.

1980, Reagan, Rep., 654,168; Carter, Dem., 890,955; Anderson, Ind., 36,055; Clark, Libertarian, 15,627.

1984, Reagan, Rep., 1,068,722; Mondale, Dem., 706,628.

1988, Bush, Rep., 1,081,331; Dukakis, Dem., 714,792; Paul, Lib., 8,435; Fulani, New Alliance, 5,099.

1992, Clinton, Dem., 1,008,966; Bush, Rep., 995,252; Perot, Ind., 309,657; Marrou, Libertarian, 7,110.

1996, Dole, Rep., 1,080,843; Clinton, Dem., 1,053,849; Perot, Ref., 146,337; Browne, Libertarian, 17,870.

Hawaii

County	1996 Clinton (D)	Dole (R)	Perot (RF)	1992 Clinton (D)	Bush (R)	Perot (I)
Hawaii.......	27,262	13,516	5,137	25,725	15,460	8,889
Honolulu......	143,793	85,779	17,389	123,908	103,937	35,728
Kauai	13,357	5,325	1,568	10,715	6,274	1,756
Maui	20,600	9,323	3,264	18,962	11,151	6,630
Totals........	205,012	113,943	27,358	179,310	136,822	53,003

Hawaii Vote Since 1960

1960, Kennedy, Dem., 92,410; Nixon, Rep., 92,295.

1964, Johnson, Dem., 163,249; Goldwater, Rep., 44,022.

1968, Nixon, Rep., 91,425; Humphrey, Dem., 141,324; Wallace, 3d Party, 3,469.

1972, Nixon, Rep., 168,865; McGovern, Dem., 101,409.

1976, Carter, Dem., 147,375; Ford, Rep., 140,003; MacBride, Libertarian, 3,923.

1980, Reagan, Rep., 130,112; Carter, Dem., 135,879; Anderson, Ind., 32,021; Clark, Libertarian, 3,269; Commoner, Citizens, 1,548; Hall, Com., 458.

1984, Reagan, Rep., 184,934; Mondale, Dem., 147,098; Bergland, Libertarian, 2,167.

1988, Bush, Rep., 158,625; Dukakis, Dem., 192,364; Paul, Lib., 1,999; Fulani, New Alliance, 1,003.

1992, Clinton, Dem., 179,310; Bush, Rep., 136,822; Perot, Ind., 53,003; Gritz, Populist/America First, 1,452; Marrou, Libertarian, 1,119.

1996, Clinton, Dem., 205,012; Dole, Rep., 113,943; Perot, Ref., 27,358; Nader, Green, 10,386; Browne, Libertarian, 2,493; Hagelin, Natural Law, 570; Phillips, Taxpayers, 358.

Idaho

County	1996 Clinton (D)	Dole (R)	Perot (RF)	1992 Clinton (D)	Bush (R)	Perot (I)
Ada..........	43,040	61,811	11,171	31,941	49,000	28,192
Adams	537	1,053	311	457	754	695
Bannock.......	12,806	14,058	4,158	11,091	12,016	8,116
Bear Lake	805	1,583	396	562	1,419	684
Benewah	1,488	1,667	701	1,270	1,223	1,165
Bingham	4,304	8,391	2,021	3,565	7,333	4,144
Blaine........	3,840	3,003	1,193	2,865	2,243	2,831
Boise	879	1,576	440	623	912	754
Bonner.......	5,294	6,207	2,669	4,995	3,937	4,645
Bonneville	9,013	19,977	3,921	7,014	16,557	10,241
Boundary.....	1,194	1,937	626	1,095	1,479	1,136
Butte.........	507	741	233	433	602	392
Camas	156	283	95	134	202	145
Canyon	11,800	23,988	3,956	9,095	19,220	8,974
Caribou	841	1,740	501	562	1,350	1,088
Cassia	1,596	4,663	976	1,351	4,052	1,785
Clark.........	117	266	45	95	195	119
Clearwater	1,507	1,658	650	1,433	1,152	1,098
Custer	635	1,249	400	564	829	729
Elmore........	2,324	3,668	845	1,858	3,087	1,867
Franklin	807	2,435	589	524	2,115	890
Fremont	1,114	3,042	630	903	2,333	1,349
Gem..........	1,968	3,362	833	1,609	2,455	1,555
Gooding......	1,503	2,637	980	1,530	2,178	1,591
Idaho........	1,979	3,871	1,083	1,974	2,709	1,900
Jefferson	1,427	4,925	994	978	3,471	2,164
Jerome	1,679	3,358	1,014	1,739	2,972	1,768
Kootenai......	13,627	18,740	6,083	11,553	13,065	11,261
Latah	7,741	6,311	1,828	7,233	5,353	3,602
Lemhi	1,015	2,334	461	996	1,540	1,175
Lewis	674	861	316	674	593	491
Lincoln	478	744	319	514	656	441
Madison	1,216	5,706	744	741	4,591	1,920
Minidoka	1,977	4,008	977	1,815	3,304	1,875
Nez Perce	7,491	6,675	2,385	7,069	5,431	4,363
Oneida	429	993	285	351	713	590
Owyhee.......	895	2,033	354	686	1,469	862
Payette.......	2,119	3,901	906	1,656	2,895	2,055
Power........	1,070	1,501	344	837	1,352	697
Shoshone	2,981	1,588	1,283	3,182	1,441	1,878
Teton	866	1,251	326	472	762	608
Twin Falls.....	6,826	12,393	3,383	6,593	10,335	6,043
Valley........	1,564	2,089	568	1,259	1,548	1,313
Washington	1,314	2,318	525	1,122	1,802	1,204
Totals.........	164,443	256,595	62,518	137,013	202,645	130,395

Idaho Vote Since 1948

1948, Truman, Dem., 107,370; Dewey, Rep., 101,514; Wallace, Prog., 4,972; Watson, Proh., 628; Thomas, Soc., 332.

1952, Eisenhower, Rep., 180,707; Stevenson, Dem., 95,081; Hallinan, Prog., 443; write-in, 23.

1956, Eisenhower, Rep., 166,979; Stevenson, Dem., 105,868; Andrews, Ind., 126; write-in, 16.

1960, Kennedy, Dem., 138,853; Nixon, Rep., 161,597.

1964, Johnson, Dem., 148,920; Goldwater, Rep., 143,557.

1968, Nixon, Rep., 165,369; Humphrey, Dem., 89,273; Wallace, 3d Party, 36,541.

1972, Nixon, Rep., 199,384; McGovern, Dem., 80,826; Schmitz, Amer., 28,869; Spock, Peoples, 903.

1976, Carter, Dem., 126,549; Ford, Rep., 204,151; Maddox, Amer., 5,935; MacBride, Libertarian, 3,558; LaRouche, U.S. Labor, 739.

1980, Reagan, Rep., 290,699; Carter, Dem., 110,192; Anderson, Ind., 27,058; Clark, Libertarian, 8,425; Rarick, Amer., 1,057.

1984, Reagan, Rep., 297,523; Mondale, Dem., 108,510; Bergland, Libertarian, 2,823.

1988, Bush, Rep., 253,881; Dukakis, Dem., 147,272; Paul, Lib., 5,313; Fulani, Ind., 2,502.

1992, Clinton, Dem., 137,013; Bush, Rep., 202,645; Perot, Ind., 130,395; Gritz, Populist/America First, 10,281; Marrou, Libertarian, 1,167.

1996, Dole, Rep., 256,595; Clinton, Dem., 165,443; Perot, Ref., 62,518; Browne, Libertarian, 3,325; Phillips, Taxpayers, 2,230; Hagelin, Natural Law, 1,600.

Illinois

	1996			1992		
	Clinton	Dole	Perot	Clinton	Bush	Perot
County	(D)	(R)	(RF)	(D)	(R)	(I)
Adams	11,336	13,836	3,069	11,748	13,529	6,157
Alexander..	2,753	1,212	321	2,566	1,301	474
Bond......	3,213	3,018	685	3,428	2,715	1,373
Boone.....	5,345	6,181	1,377	5,114	5,589	2,880
Brown.....	997	1,053	237	1,146	1,029	504
Bureau	7,651	6,528	1,798	7,551	6,836	3,465
Calhoun ..	1,676	941	363	1,519	745	532
Carroll....	2,926	3,029	792	2,854	3,297	1,502
Cass......	2,834	2,214	589	3,200	2,162	1,072
Champaign	32,454	28,232	4,806	35,003	27,096	13,571
Christian..	7,431	5,563	1,727	9,042	5,087	3,401
Clark......	2,995	3,409	781	3,338	3,175	1,450
Clay	2,750	2,703	719	2,962	2,471	1,193
Clinton	6,104	6,065	1,580	6,686	5,771	3,315
Coles	8,950	8,038	2,137	9,402	8,098	4,707
Cook......	1,153,289	461,557	96,633	1,249,533	605,300	281,999
Crawford...	3,627	3,965	1,057	3,964	3,606	2,062
Cumberland	1,776	2,002	657	2,111	1,860	1,209
DeKalb	12,715	12,380	3,009	13,744	12,655	7,680
DeWitt	2,878	2,978	694	3,009	3,164	1,543
Douglas ...	2,955	3,272	740	3,341	3,309	1,600
DuPage ...	129,709	164,630	27,419	114,564	178,271	76,839
Edgar	3,552	3,746	935	4,014	3,790	1,930
Edwards...	1,089	1,613	384	1,299	1,601	634
Effingham..	4,825	7,696	1,555	5,221	6,329	3,354
Fayette	3,887	3,881	964	4,833	3,508	1,730
Ford	2,065	3,077	590	2,175	3,046	1,222
Franklin....	9,814	5,354	2,096	12,744	5,504	3,180
Fulton.....	8,857	5,155	1,610	9,725	5,062	2,874
Gallatin....	2,113	856	527	2,371	990	568
Greene.....	2,734	2,245	903	3,164	2,391	1,461
Grundy....	6,759	6,177	1,860	6,122	6,346	3,724
Hamilton..	2,242	1,677	560	2,582	1,521	862
Hancock...	4,001	3,961	1,148	4,213	3,714	2,091
Hardin....	1,323	790	485	1,665	985	515
Henderson.	1,953	1,233	408	2,013	1,310	715
Henry.....	11,201	8,393	2,194	11,077	8,989	4,231
Iroquois ...	4,559	6,564	1,522	4,440	6,948	3,073
Jackson ...	12,214	7,422	2,082	13,373	6,899	3,995
Jasper.....	2,038	2,234	641	2,284	1,996	1,160
Jefferson ..	7,263	5,937	1,647	8,665	5,497	3,403
Jersey.....	4,275	3,211	1,186	4,749	2,933	2,363
Jo Daviess.	4,171	3,915	1,131	4,044	4,249	2,102
Johnson....	2,009	2,241	640	2,299	2,124	944
Kane......	47,902	54,375	11,270	44,568	55,684	27,179
Kankakee..	16,820	14,595	3,574	17,229	15,411	7,264
Kendall....	6,499	8,958	2,055	5,423	8,521	4,394
Knox......	12,487	7,822	2,096	12,524	8,331	4,351
Lake	93,315	93,149	16,640	81,693	99,000	42,384
LaSalle....	21,643	15,299	5,259	23,276	16,078	10,434
Lawrence ..	2,871	2,568	916	3,270	2,681	1,498
Lee	5,895	6,677	1,520	5,530	6,652	3,191
Livingston..	5,641	7,653	1,409	6,007	8,004	3,029
Logan.....	4,618	6,518	1,141	5,169	6,567	2,420
McDonough	5,632	5,049	1,217	5,814	5,297	2,770
McHenry ..	31,240	41,136	10,082	24,783	41,356	21,817
McLean ...	22,708	26,428	3,816	23,090	25,726	10,282
Macon	24,256	18,161	4,540	27,449	18,684	9,236
Macoupin..	11,107	7,235	2,532	12,050	6,518	5,018
Madison ...	53,568	35,758	10,121	58,484	32,167	23,110
Marion	7,792	5,999	1,825	9,669	5,764	3,407
Marshall ...	2,640	2,453	586	2,819	2,491	1,169
Mason	3,385	2,430	600	3,969	2,473	1,245
Massac....	2,841	2,507	675	3,347	2,754	892
Menard....	2,204	3,106	534	2,264	2,834	1,179
Mercer	4,278	2,688	889	3,990	2,983	1,535
Monroe....	4,798	5,350	1,276	4,894	4,807	2,813
Montgomery	6,338	4,770	1,436	7,424	4,407	2,956
Morgan....	6,150	6,352	1,633	6,351	6,566	3,317
Moultrie ...	2,629	2,199	596	3,056	2,065	1,322
Ogle	6,765	9,558	1,876	6,512	9,008	4,455
Peoria.....	37,383	30,990	5,220	38,099	30,718	12,195
Perry......	5,347	3,237	1,262	6,009	3,105	1,955
Piatt	3,274	3,265	818	3,520	3,076	1,822
Pike.......	3,604	3,225	1,039	4,016	3,342	1,643
Pope......	915	850	277	1,063	951	391
Pulaski ...	1,524	1,036	235	1,987	1,169	379
Putnam....	1,425	987	322	1,574	969	752
Randolph ..	7,419	5,422	1,698	8,529	4,899	3,092
Richland...	2,679	3,137	927	3,286	3,053	1,689
Rock Island	34,822	20,626	5,135	37,412	23,212	10,416
St. Clair...	53,405	33,066	7,027	57,625	31,951	17,592
Saline.....	6,156	3,693	1,752	7,258	3,667	2,302
Sangamon.	38,902	42,174	6,446	40,052	39,641	16,861
Schuyler ..	1,636	1,597	483	1,650	1,512	815
Scott......	1,012	1,112	396	1,057	1,132	588
Shelby	4,249	4,215	1,262	5,101	3,631	2,401
Stark......	1,262	1,278	312	1,336	1,384	625
Stephenson	7,145	8,871	1,940	7,899	9,005	4,677
Tazewell...	24,139	24,395	4,814	26,428	23,469	9,927
Union	4,252	3,147	832	4,681	3,003	1,373

	1996			1992		
	Clinton	Dole	Perot	Clinton	Bush	Perot
County	(D)	(R)	(RF)	(D)	(R)	(I)
Vermilion ...	15,525	12,015	3,577	18,383	11,703	8,162
Wabash ...	2,177	2,381	683	2,436	2,485	1,302
Warren.....	3,500	2,974	742	3,661	3,325	1,436
Washington.	2,744	3,339	790	2,986	3,003	1,542
Wayne	3,054	4,029	999	3,332	3,809	1,702
White	3,553	2,878	888	4,308	3,057	1,428
Whiteside..	11,913	8,859	2,436	12,329	10,146	4,589
Will	69,354	62,506	15,485	59,633	58,337	32,788
Williamson..	12,510	9,734	2,877	14,361	9,462	4,779
Winnebago.	46,264	44,479	8,192	48,298	42,221	21,227
Woodford...	5,270	8,527	1,170	5,490	8,032	2,733
Totals......	2,341,744	1,587,021	346,408	2,453,350	1,734,096	840,515

Illinois Vote Since 1948

1948, Truman, Dem., 1,994,715; Dewey, Rep., 1,961,103; Watson, Proh., 11,959; Thomas, Soc., 11,522; Teichert, Soc. Labor, 3,118.

1952, Eisenhower, Rep., 2,457,327; Stevenson, Dem., 2,013,920; Hass, Soc. Labor, 9,363; write-in, 448.

1956, Eisenhower, Rep., 2,623,327; Stevenson, Dem., 1,775,682; Hass, Soc. Labor, 8,342; write-in, 56.

1960, Kennedy, Dem., 2,377,846; Nixon, Rep., 2,368,988; Hass, Soc. Labor, 10,560; write-in, 15.

1964, Johnson, Dem., 2,796,833; Goldwater, Rep., 1,905,946; write-in, 62.

1968, Nixon, Rep., 2,174,774; Humphrey, Dem., 2,039,814; Wallace, 3d Party, 390,958; Blomen, Soc. Labor, 13,878; write-in, 325.

1972, Nixon, Rep. 2,788,179; McGovern, Dem., 1,913,472; Fisher, Soc. Labor, 12,344; Schmitz, Amer., 2,471; Hall, Com., 4,541; others, 2,229.

1976, Carter, Dem., 2,271,295; Ford, Rep., 2,364,269; McCarthy, Ind., 55,939; Hall, Com., 9,250; MacBride, Libertarian, 8,057; Camejo, Soc. Workers, 3,615; Levin, Soc. Labor, 2,422; LaRouche, U.S. Labor, 2,018; write-in, 1,968.

1980, Reagan, Rep., 2,358,049; Carter, Dem., 1,981,413; Anderson, Ind., 346,754; Clark, Libertarian, 38,939; Commoner, Citizens, 10,692; Hall, Com., 9,711; Griswold, Workers World, 2,257; DeBerry, Soc. Workers, 1,302; write-ins, 604.

1984, Reagan, Rep., 2,707,103; Mondale, Dem., 2,086,499; Bergland, Libertarian, 10,086.

1988, Bush, Rep., 2,310,939; Dukakis, Dem., 2,215,940; Paul, Lib., 14,944; Fulani, Solid., 10,276.

1992, Clinton, Dem., 2,453,350; Bush, Rep., 1,734,096; Perot, Ind., 840,515; Marrou, Libertarian, 9,218; Fulani, New Alliance, 5,267; Gritz, Populist/America First, 3,577; Hagelin, Natural Law, 2,751; Warren, Soc. Workers, 1,361.

1996, Clinton, Dem., 2,341,744; Dole, Rep., 1,587,021; Perot, Ref., 346,408; Browne, Libertarian, 22,548; Phillips, Taxpayers, 7,606; Hagelin, Natural Law, 4,606.

Indiana

	1996			1992		
	Clinton	Dole	Perot	Clinton	Bush	Perot
County	(D)	(R)	(RF)	(D)	(R)	(I)
Adams........	4,247	6,960	1,346	3,708	6,078	2,865
Allen........	41,450	59,255	8,808	39,629	55,003	25,809
Bartholomew...	9,301	13,188	2,815	8,284	13,146	5,882
Benton........	1,311	1,947	609	1,221	2,030	1,056
Blackford	2,335	2,070	681	2,088	2,347	1,319
Boone	4,625	11,338	1,498	3,982	9,485	3,826
Brown........	2,413	2,988	802	2,029	2,633	1,635
Carroll	2,747	4,062	1,171	2,561	3,800	2,173
Cass.........	5,419	8,020	2,029	4,757	7,421	3,944
Clark........	17,799	14,396	3,578	17,460	13,333	5,653
Clay	3,605	4,858	1,406	3,306	4,696	2,134
Clinton	3,949	6,156	1,355	3,490	6,141	2,535
Crawford	2,324	1,759	700	2,260	1,903	819
Daviess	3,230	5,531	994	3,201	5,591	1,695
Dearborn	6,269	8,318	1,731	5,116	6,974	3,384
Decatur	3,190	4,782	1,389	2,774	5,195	2,299
Dekalb	4,840	6,851	1,534	4,652	6,682	3,554
Delaware	20,385	18,126	6,042	19,556	20,473	10,453
Dubois	6,499	6,840	1,777	5,878	6,785	3,195
Elkhart	16,598	28,770	5,133	14,660	27,920	9,450
Fayette	3,822	4,091	1,137	3,969	4,376	2,299
Floyd	13,814	12,473	2,609	13,166	11,932	4,421
Fountain	2,327	3,984	1,033	2,829	3,391	2,162
Franklin	2,808	4,167	943	2,456	3,831	1,858
Fulton........	2,956	3,934	1,143	2,552	3,982	1,963
Gibson	6,488	5,392	1,585	6,909	5,172	2,680
Grant	9,818	13,443	3,008	9,211	13,806	5,597
Greene	5,277	5,746	1,690	5,431	5,410	2,610
Hamilton	14,153	42,792	4,234	10,215	34,622	10,365
Hancock	6,123	12,907	2,258	4,752	11,072	4,752
Harrison	5,900	6,073	1,839	5,768	5,403	2,469
Hendricks......	9,392	22,293	3,405	7,071	18,373	7,519
Henry........	7,667	8,537	2,381	6,794	8,720	4,416
Howard	11,999	16,771	4,172	10,288	15,306	8,575
Huntington.....	4,287	8,275	1,400	3,855	9,093	2,967

County	1996 Clinton (D)	Dole (R)	Perot (RF)	1992 Clinton (D)	Bush (R)	Perot (I)
Jackson	5,150	5,883	1,590	5,663	7,246	3,148
Jasper	3,554	5,173	1,271	3,033	4,809	2,019
Jay	3,356	3,584	1,022	3,208	3,609	1,994
Jefferson	5,441	4,827	1,438	5,510	4,937	2,565
Jennings	4,223	4,461	1,629	3,471	4,392	2,370
Johnson	11,278	23,733	3,975	8,712	20,353	8,246
Knox	7,003	6,395	2,022	6,718	6,683	3,719
Kosciusko	6,166	15,084	2,531	5,307	14,179	5,115
LaGrange	2,704	4,033	949	2,093	3,584	1,736
Lake	100,198	47,873	15,051	102,778	53,867	28,635
LaPorte	19,879	14,106	5,133	17,717	14,962	9,641
Lawrence	5,703	8,107	2,063	5,557	7,712	3,452
Madison	23,772	23,151	6,447	22,276	23,479	13,100
Marion	124,448	133,329	21,358	122,234	141,369	57,878
Marshall	5,486	8,158	1,698	4,912	8,048	3,522
Martin	1,848	2,281	485	2,018	2,523	883
Miami	4,260	6,719	1,657	3,967	6,416	3,428
Monroe	18,531	16,744	3,179	19,712	16,661	6,943
Montgomery	3,825	7,705	1,766	3,371	7,602	3,511
Morgan	5,812	12,872	2,755	4,690	10,939	5,375
Newton	1,897	2,075	801	1,757	2,295	1,274
Noble	5,101	6,782	1,521	4,411	5,883	3,328
Ohio	1,083	1,098	281	970	1,009	527
Orange	3,016	3,355	938	2,948	3,738	1,296
Owen	2,244	3,056	874	2,207	2,753	1,563
Parke	2,453	3,151	981	2,429	2,953	1,696
Perry	4,427	2,554	913	4,829	2,973	1,560
Pike	2,780	2,174	884	2,960	2,156	1,238
Porter	24,044	22,931	7,169	21,022	22,644	13,096
Posey	4,965	4,638	1,304	4,632	4,435	2,357
Pulaski	2,010	2,693	634	1,950	2,712	1,214
Putnam	3,962	5,958	1,619	3,487	5,341	3,174
Randolph	4,087	4,708	1,557	3,870	4,937	2,939
Ripley	4,097	5,303	1,216	3,480	5,033	2,406
Rush	2,578	3,827	973	2,168	3,873	1,948
St. Joseph	45,704	38,281	8,379	46,203	38,934	18,828
Scott	3,798	2,620	760	4,085	2,649	1,092
Shelby	5,374	7,778	1,874	4,560	8,075	3,521
Spencer	4,058	3,770	739	4,301	3,789	1,464
Starke	3,854	3,108	1,096	3,695	3,100	1,885
Steuben	4,124	5,513	1,390	3,630	4,868	2,896
Sullivan	4,076	3,207	1,178	4,211	3,052	1,857
Switzerland	1,496	1,266	403	1,535	1,211	636
Tippecanoe	17,232	22,556	5,394	17,343	23,050	9,684
Tipton	2,478	3,980	861	2,125	3,906	1,816
Union	1,019	1,334	364	898	1,394	664
Vanderburgh	30,934	28,509	6,132	33,799	30,271	12,513
Vermillion	3,251	2,334	1,029	3,652	2,360	1,794
Vigo	17,974	15,751	4,508	18,050	15,834	8,141
Wabash	4,577	6,990	1,294	4,518	7,062	3,424
Warren	1,394	1,678	560	1,367	1,601	1,020
Warrick	9,285	9,221	2,471	8,612	8,087	3,862
Washington	3,819	4,066	1,264	4,092	4,043	1,846
Wayne	10,905	12,188	2,525	9,960	12,221	5,095
Wells	3,752	6,322	1,157	3,282	5,799	2,890
White	3,396	4,642	1,610	2,988	4,622	2,582
Whitley	4,176	5,965	1,392	3,569	5,217	3,195
Totals	887,424	1,006,693	224,299	848,420	989,375	455,934

Indiana Vote Since 1948

1948, Truman, Dem., 807,833; Dewey, Rep., 821,079; Watson, Proh., 14,711; Wallace, Prog., 9,649; Thomas, Soc., 2,179; Teichert, Soc. Labor, 763.

1952, Eisenhower, Rep., 1,136,259; Stevenson, Dem., 801,530; Hamblen, Proh., 15,335; Hallinan, Prog., 1,222; Hass, Soc. Labor, 979.

1956, Eisenhower, Rep., 1,182,811; Stevenson, Dem., 783,908; Holtwick, Proh., 6,554; Hass, Soc. Labor, 1,334.

1960, Kennedy, Dem., 952,358; Nixon, Rep., 1,175,120; Decker, Proh., 6,746; Hass, Soc. Labor, 1,136.

1964, Johnson, Dem., 1,170,848; Goldwater, Rep., 911,118; Munn, Proh., 8,266; Hass, Soc. Labor, 1,374.

1968, Nixon, Rep., 1,067,885; Humphrey, Dem., 806,659; Wallace, 3d Party, 243,108; Munn, Proh., 4,616; Halstead, Soc. Workers, 1,293; Gregory, write-in, 36.

1972, Nixon, Rep., 1,405,154; McGovern, Dem., 708,568; Reed, Soc. Workers, 5,575; Fisher, Soc. Labor, 1,688; Spock, Peace and Freedom, 4,544.

1976, Carter, Dem., 1,014,714; Ford, Rep., 1,185,958; Anderson, Amer., 14,048; Camejo, Soc. Workers, 5,695; LaRouche, U.S. Labor, 1,947.

1980, Reagan, Rep., 1,255,656; Carter, Dem., 844,197; Anderson, Ind., 111,639; Clark, Libertarian, 19,627; Commoner, Citizens, 4,852; Greaves, Amer., 4,750; Hall, Com., 702; DeBerry, Soc., 610.

1984, Reagan, Rep., 1,377,230; Mondale, Dem., 841,481; Bergland, Libertarian, 6,741.

1988, Bush, Rep., 1,297,763; Dukakis, Dem., 860,643; Fulani, New Alliance, 10,215.

1992, Bush, Rep., 989,375; Clinton, Dem., 848,420; Perot, Ind., 455,934; Marrou, Libertarian, 7,936; Fulani, New Alliance, 2,583.

1996, Dole, Rep., 1,006,693; Clinton, Dem., 887,424; Perot, Ref., 224,299; Browne, Libertarian, 15,632.

Iowa

County	1996 Clinton (D)	Dole (R)	Perot (RF)	1992 Clinton (D)	Bush (R)	Perot (I)
Adair	1,802	1,655	458	1,655	1,713	814
Adams	1,070	920	320	1,034	863	679
Allamakee	2,551	2,457	680	2,362	2,627	1,543
Appanoose	2,747	2,233	554	2,810	2,346	1,161
Audubon	1,827	1,314	314	1,589	1,373	887
Benton	5,546	3,835	846	4,467	3,469	2,454
Black Hawk	29,651	19,322	3,623	29,584	21,398	10,182
Boone	6,446	4,293	987	5,913	4,148	2,070
Bremer	5,023	4,213	862	4,774	4,482	2,338
Buchanan	4,997	3,043	836	4,166	3,313	2,126
Buena Vista	3,420	3,636	831	3,374	3,863	1,955
Butler	3,061	3,036	489	2,548	3,209	1,333
Calhoun	2,193	2,077	462	2,140	2,169	946
Carroll	4,333	3,392	998	3,800	3,439	2,192
Cass	2,616	3,384	809	2,231	3,176	1,608
Cedar	3,856	2,966	756	3,296	2,965	1,945
Cerro Gordo	11,943	7,427	1,689	11,415	8,250	4,498
Cherokee	2,853	2,629	834	2,590	2,768	1,503
Chickasaw	3,355	2,191	759	2,913	2,129	1,566
Clarke	2,053	1,401	440	1,921	1,417	899
Clay	3,659	3,129	802	3,346	3,011	1,964
Clayton	4,284	2,944	912	3,742	3,044	2,309
Clinton	11,481	7,624	2,300	11,683	8,746	4,414
Crawford	3,140	2,686	847	3,004	2,693	1,905
Dallas	8,017	6,647	1,198	6,554	5,587	2,665
Davis	1,894	1,445	382	1,962	1,344	718
Decatur	1,846	1,287	452	1,866	1,316	786
Delaware	3,704	3,065	679	3,093	3,195	2,144
Des Moines	10,761	5,778	1,792	11,309	6,378	3,386
Dickinson	3,562	3,129	901	3,106	3,196	1,974
Dubuque	20,839	13,391	3,304	20,539	14,007	8,208
Emmet	2,270	1,641	470	2,239	1,749	1,010
Fayette	4,832	3,848	890	4,412	3,879	2,493
Floyd	3,769	2,379	689	3,688	2,404	1,611
Franklin	2,232	2,054	417	2,049	2,137	1,045
Fremont	1,481	1,576	480	1,422	1,459	1,003
Greene	2,519	1,861	396	2,422	1,952	956
Grundy	2,322	2,928	401	1,895	3,160	1,069
Guthrie	2,552	2,034	515	2,234	1,962	1,216
Hamilton	3,455	3,109	661	3,262	3,031	1,348
Hancock	2,399	2,353	529	2,175	2,428	1,170
Hardin	4,053	3,505	713	3,792	3,590	1,547
Harrison	2,576	3,070	820	2,349	2,763	1,691
Henry	3,798	3,478	914	3,544	3,435	1,522
Howard	2,303	1,528	555	2,099	1,516	1,193
Humboldt	2,080	2,236	590	1,765	2,299	1,093
Ida	1,589	1,684	436	1,449	1,714	1,061
Iowa	3,354	3,042	575	2,560	2,656	1,709
Jackson	4,609	2,827	936	4,421	2,673	2,096
Jasper	8,776	6,414	1,263	8,120	6,866	2,972
Jefferson	2,597	2,541	571	2,562	2,541	1,241
Johnson	27,888	13,402	2,313	28,656	14,041	8,625
Jones	4,668	3,083	765	3,508	3,071	2,306
Keokuk	2,545	2,080	432	2,329	1,981	1,238
Kossuth	4,031	3,477	932	3,660	3,464	1,906
Lee	8,831	4,932	1,734	9,366	4,777	2,920
Linn	45,497	30,958	5,607	38,567	30,215	19,643
Louisa	2,081	1,565	590	2,091	1,691	1,044
Lucas	2,168	1,586	433	2,072	1,734	848
Lyon	1,489	3,396	422	1,331	3,272	1,068
Madison	3,070	2,550	654	2,525	2,421	1,168
Mahaska	3,737	4,473	656	3,714	4,953	1,508
Marion	5,978	6,100	871	5,531	6,062	1,896
Marshall	8,669	7,017	1,455	8,303	6,784	3,100
Mills	2,068	2,958	683	1,798	2,699	1,638
Mitchell	2,596	1,877	563	2,177	1,933	1,199
Monona	1,952	1,674	580	1,939	1,660	1,231
Monroe	1,884	1,272	329	1,829	1,323	612
Montgomery	1,912	2,583	663	1,599	2,404	1,341
Muscatine	7,674	5,858	1,705	7,089	6,087	3,583
O'Brien	2,236	3,877	578	2,122	3,869	1,557
Osceola	1,010	1,736	274	990	1,756	813
Page	2,220	4,032	753	1,951	3,670	1,669
Palo Alto	2,371	1,817	477	2,374	1,789	1,186
Plymouth	3,745	5,117	997	3,171	5,196	2,039
Pocahontas	1,981	1,707	478	1,919	1,743	942
Polk	83,877	60,884	9,516	78,585	63,708	24,155
Pottawattamie	13,276	15,648	3,534	13,228	15,671	8,035
Poweshiek	4,183	3,221	681	4,056	3,245	1,680
Ringgold	1,439	967	310	1,341	967	551
Sac	2,170	2,209	579	1,896	2,138	1,157
Scott	32,694	26,751	4,991	33,765	28,844	11,423
Shelby	2,176	3,056	652	2,094	2,809	1,614
Sioux	2,392	10,864	718	2,226	10,637	1,771
Story	17,234	12,468	2,091	17,118	12,702	6,275
Tama	3,994	2,986	713	3,573	2,948	1,748
Taylor	1,458	1,419	379	1,430	1,200	910
Union	2,787	2,156	660	2,565	2,224	1,280

County	1996 Clinton (D)	Dole (R)	Perot (RF)	1992 Clinton (D)	Bush (R)	Perot (I)
Van Buren	1,536	1,460	347	1,464	1,418	811
Wapello	8,437	4,828	1,376	8,670	4,852	2,513
Warren	9,120	6,905	1,267	8,612	7,242	3,217
Washington ...	3,828	3,600	636	3,384	3,576	1,994
Wayne	1,650	1,295	310	1,632	1,299	642
Webster	8,380	6,275	1,580	8,562	6,992	3,272
Winnebago ...	2,679	2,211	590	2,322	2,407	1,329
Winneshiek ...	4,122	3,532	973	3,791	3,331	2,416
Woodbury	17,224	16,368	3,436	17,398	18,148	7,182
Worth	2,293	1,284	403	2,009	1,382	1,044
Wright.	2,912	2,473	536	2,776	2,708	1,151
Totals........	**620,258**	**492,644**	**105,159**	**586,353**	**504,891**	**253,468**

Iowa Vote Since 1948

1948, Truman, Dem., 522,380; Dewey, Rep., 494,018; Wallace, Prog., 12,125; Teichert, Soc. Labor, 4,274; Watson, Proh., 3,382; Thomas, Soc., 1,829; Dobbs, Soc. Workers, 26.

1952, Eisenhower, Rep., 808,906; Stevenson, Dem., 451,513; Hallinan, Prog., 5,085; Hamblen, Proh., 2,882; Hoopes, Soc., 219; Hass, Soc. Labor, 139; scattering, 29.

1956, Eisenhower, Rep., 729,187; Stevenson, Dem., 501,858; Andrews (A.C.P. of Iowa), 3,202; Hoopes, Soc., 192; Hass, Soc. Labor, 125.

1960, Kennedy, Dem., 550,565; Nixon, Rep., 722,381; Hass, Soc. Labor, 230; write-in, 634.

1964, Johnson, Dem., 733,030; Goldwater, Rep., 449,148; Hass, Soc. Labor, 182; DeBerry, Soc. Workers, 159; Munn, Proh., 1,902.

1968, Nixon, Rep., 619,106; Humphrey, Dem., 476,699; Wallace, 3d Party, 66,422; Munn, Proh., 362; Halstead, Soc. Workers, 3,377; Cleaver, Peace and Freedom, 1,332; Blomen, Soc. Labor, 241.

1972, Nixon, Rep., 706,207; McGovern, Dem., 496,206; Schmitz, Amer., 22,056; Jenness, Soc. Workers, 488; Fisher, Soc. Labor, 195; Hall, Com., 272; Green, Universal, 199; scattered, 321.

1976, Carter, Dem., 619,931; Ford, Rep., 632,863; McCarthy, Ind., 20,051; Anderson, Amer., 3,040; MacBride, Libertarian, 1,452.

1980, Reagan, Rep., 676,026; Carter, Dem., 508,672; Anderson, Ind., 115,633; Clark, Libertarian, 13,123; Commoner, Citizens, 2,273; McReynolds, Socialist, 534; Hall, Com., 298; DeBerry, Soc. Workers, 244; Greaves, Amer., 189; Bubar, Statesman, 150; scattering, 519.

1984, Reagan, Rep., 703,088; Mondale, Dem., 605,620; Bergland, Libertarian, 1,844.

1988, Bush, Rep., 545,355; Dukakis, Dem., 670,557; LaRouche, Ind., 3,526; Paul, Lib., 2,494.

1992, Clinton, Dem., 586,353; Bush, Rep., 504,891; Perot, Ind., 253,468; Hagelin, Natural Law, 3,079; Gritz, Populist/America First, 1,177; Marrou, Libertarian, 1,076.

1996, Clinton, Dem., 620,258; Dole, Rep., 492,644; Perot, Ref., 105,159; Nader, Green, 6,550; Hagelin, Natural Law, 3,349; Browne, Libertarian, 2,315; Phillips, Taxpayers, 2,229; Harris, Soc. Workers, 331.

Kansas

County	1996 Clinton (D)	Dole (R)	Perot (RF)	1992 Clinton (D)	Bush (R)	Perot (I)
Allen	2,299	2,797	793	2,312	2,351	1,746
Anderson	1,367	1,636	449	1,178	1,218	1,282
Atchison	2,926	2,828	727	2,959	2,521	2,020
Barber	730	1,696	279	759	1,225	893
Barton	3,121	7,855	1,004	3,846	5,113	4,574
Bourbon	2,491	3,318	760	2,509	2,876	1,763
Brown........	1,529	2,688	497	1,476	2,203	1,603
Butler	7,294	13,979	2,274	7,029	9,166	7,355
Chase........	496	778	259	470	610	600
Chautauqua..	568	1,142	222	598	853	607
Cherokee	3,771	4,138	1,072	4,083	3,589	2,067
Cheyenne.....	422	1,211	174	407	863	477
Clark........	334	855	109	293	676	341
Clay	963	2,793	389	947	2,198	1,434
Cloud........	1,615	2,743	609	1,720	2,131	1,578
Coffey........	1,118	2,369	572	1,021	1,824	1,443
Comanche....	298	691	133	325	636	324
Cowley.......	5,588	7,872	1,904	5,405	5,422	4,911
Crawford.....	7,504	6,447	1,785	7,366	5,468	3,706
Decatur......	417	1,255	156	576	940	565
Dickinson	2,423	5,174	888	2,518	3,851	2,833
Doniphan	1,050	1,962	0	1,177	1,579	1,200
Douglas	18,116	16,116	2,630	19,439	12,944	9,630
Edwards......	539	1,088	180	567	769	584
Elk..........	488	933	206	485	748	503
Ellis.........	4,142	6,809	894	4,544	3,985	3,887
Ellsworth......	899	2,078	245	1,010	1,197	1,020

County	1996 Clinton (D)	Dole (R)	Perot (RF)	1992 Clinton (D)	Bush (R)	Perot (I)
Finney	2,420	6,188	805	2,612	5,278	3,011
Ford	2,628	5,681	914	2,635	4,342	3,341
Franklin	3,552	5,007	1,184	2,968	3,699	3,184
Geary.........	2,444	3,686	618	2,559	2,928	2,057
Gove	351	1,123	141	379	792	532
Graham	432	1,031	152	554	752	603
Grant	633	1,772	250	619	1,561	835
Gray.........	404	1,457	164	443	1,039	686
Greeley	161	567	47	191	504	175
Greenwood	1,108	1,932	552	1,262	1,411	1,167
Hamilton	342	811	84	386	716	271
Harper	836	1,941	355	845	1,371	1,151
Harvey	4,918	8,382	1,023	5,047	6,259	3,653
Haskell........	304	1,143	96	336	1,023	462
Hodgeman.....	251	808	99	258	625	343
Jackson	1,983	2,682	735	1,639	1,970	1,927
Jefferson	2,757	3,781	1,030	2,538	2,569	2,642
Jewell........	417	1,374	188	546	1,050	698
Johnson	68,129	110,368	10,425	59,573	85,418	49,136
Kearny	335	1,041	106	384	943	376
Kingman	1,006	2,659	409	1,100	1,680	1,370
Kiowa........	331	1,264	170	355	1,057	475
Labette.......	3,931	4,283	1,091	4,196	3,368	2,577
Lane.........	271	865	86	265	674	356
Leavenworth ...	9,098	10,778	2,419	8,077	7,738	7,306
Lincoln	528	1,372	212	612	893	657
Linn	1,590	2,077	535	1,353	1,413	1,358
Logan........	296	1,155	112	355	905	446
Lyon........	4,884	6,612	1,584	4,811	5,090	4,717
McPherson	3,536	8,142	1,115	3,645	5,745	3,561
Marion	1,673	4,173	492	1,627	3,142	1,557
Marshall......	1,932	2,811	713	2,022	2,030	1,786
Meade	426	1,443	173	430	1,135	592
Miami	4,237	5,256	1,339	3,835	3,528	3,701
Mitchell.......	833	2,435	246	938	1,601	1,098
Montgomery ...	5,269	7,428	1,528	5,453	6,848	3,570
Morris........	965	1,553	451	957	1,071	1,071
Morton.......	376	1,073	124	398	915	350
Nemaha	1,648	3,014	676	1,580	2,220	1,804
Neosho	2,527	3,409	907	2,799	2,926	2,136
Ness..........	428	1,336	186	565	967	678
Norton	640	1,814	265	779	1,469	815
Osage	2,502	3,487	1,101	2,297	2,561	2,532
Osborne.......	608	1,582	191	779	1,003	819
Ottawa	752	1,846	261	764	1,284	762
Pawnee	932	1,927	275	1,118	1,357	1,097
Phillips	758	2,005	242	843	1,579	955
Pottawatomie...	1,997	4,504	1,035	2,099	3,106	2,759
Pratt	1,367	2,591	408	1,466	1,779	1,528
Rawlins	335	1,393	146	393	1,023	517
Reno	9,108	14,275	2,661	9,257	11,377	7,636
Republic......	688	2,283	268	939	1,767	1,084
Rice	1,434	2,842	482	1,555	2,158	1,543
Riley	6,746	11,113	1,478	7,933	8,394	5,387
Rooks........	650	1,864	251	771	1,249	1,063
Rush..........	547	1,239	185	689	756	665
Russell........	705	3,347	164	1,178	1,434	1,395
Saline.........	7,728	12,475	2,192	7,890	8,565	7,108
Scott........	458	1,750	160	480	1,426	621
Sedgwick	59,643	93,397	11,875	62,670	75,577	47,238
Seward........	1,309	3,812	396	1,488	3,477	1,818
Shawnee	32,803	34,845	7,304	31,972	29,344	20,653
Sheridan	264	1,053	95	347	739	546
Sherman	736	2,110	220	810	1,630	828
Smith	638	1,628	213	789	1,236	816
Stafford	651	1,604	276	777	1,064	910
Stanton	189	628	60	224	556	214
Stevens	405	1,548	213	390	1,408	674
Sumner	3,638	5,952	1,260	3,564	4,087	3,887
Thomas	866	2,725	295	932	1,849	1,129
Trego	548	1,205	209	608	727	574
Wabaunsee....	966	1,884	479	851	1,254	1,258
Wallace	160	738	65	164	679	219
Washington	804	2,397	326	893	1,740	1,054
Wichita	239	796	80	241	681	303
Wilson	1,297	2,458	562	1,331	1,925	1,365
Woodson	598	953	290	590	662	604
Wyandotte	31,252	14,011	3,931	34,397	12,872	13,620
Totals.........	**387,659**	**583,245**	**92,639**	**390,434**	**449,951**	**312,358**

Kansas Vote Since 1948

1948, Truman, Dem., 351,902; Dewey, Rep., 423,039; Watson, Proh., 6,468; Wallace, Prog., 4,603; Thomas, Soc., 2,807.

1952, Eisenhower, Rep., 616,302; Stevenson, Dem., 273,296; Hamblen, Proh., 6,038; Hoopes, Soc., 530.

1956, Eisenhower, Rep., 566,878; Stevenson, Dem., 296,317; Holtwick, Proh., 3,048.

1960, Kennedy, Dem., 363,213; Nixon, Rep., 561,474; Decker, Proh., 4,138.

1964, Johnson, Dem., 464,028; Goldwater, Rep., 386,579; Munn, Proh., 5,393; Hass, Soc. Labor, 1,901.

1968, Nixon, Rep., 478,674; Humphrey, Dem., 302,996; Wallace, 3d Party, 88,921; Munn, Proh., 2,192.

1972, Nixon, Rep., 619,812; McGovern, Dem., 270,287; Schmitz, Conservative, 21,808; Munn, Proh., 4,188.

1976, Carter, Dem., 430,421; Ford, Rep., 502,752; McCarthy, Ind., 13,185; Anderson, Amer., 4,724; MacBride, Libertarian, 3,242; Maddox, Conservative, 2,118; Bubar, Proh., 1,403.

1980, Reagan, Rep., 566,812; Carter, Dem., 326,150; Anderson, Ind., 68,231; Clark, Libertarian, 14,470; Shelton, Amer., 1,555; Hall, Com., 967; Bubar, Statesman, 821; Rarick, Conservative, 789.

1984, Reagan, Rep., 674,646; Mondale, Dem., 332,471; Bergland, Libertarian, 3,585.

1988, Bush, Rep., 554,049; Dukakis, Dem., 422,636; Paul, Ind., 12,553; Fulani, Ind., 3,806.

1992, Clinton, Dem., 390,434; Bush, Rep., 449,951; Perot, Ind., 312,358; Marrou, Libertarian, 4,314.

1996, Dole, Rep., 583,245; Clinton, Dem., 387,659; Perot, Ref., 92,639; Browne, Libertarian, 4,557; Phillips, Ind., 3,519; Hagelin, Ind., 1,655.

Kentucky

	1996			1992		
County	Clinton (D)	Dole (R)	Perot (RF)	Clinton (D)	Bush (R)	Perot (I)
Adair	1,821	3,876	790	2,044	3,740	617
Allen	1,781	3,032	393	2,040	2,747	606
Anderson	2,898	2,972	751	2,491	2,731	1,219
Ballard	2,255	1,064	411	2,268	1,108	500
Barren	5,044	5,700	1,065	5,688	5,467	1,778
Bath	1,886	1,229	428	2,229	1,259	694
Bell	5,058	3,917	940	5,745	4,501	1,193
Boone	8,379	15,085	1,900	6,514	12,306	4,676
Bourbon	3,030	2,592	603	2,895	2,707	1,290
Boyd	9,668	7,054	2,070	10,496	7,387	3,195
Boyle	3,877	4,157	709	3,894	4,019	1,335
Bracken	1,055	1,371	271	1,259	1,162	500
Breathitt	3,106	1,058	397	3,496	1,303	515
Breckinridge	2,956	3,151	670	3,113	2,941	945
Bullitt	7,651	8,697	1,973	7,830	7,745	3,333
Butler	1,260	2,531	348	1,468	2,729	596
Caldwell	2,434	2,067	637	3,000	1,966	670
Calloway	5,281	4,989	1,223	6,181	4,654	1,853
Campbell	11,957	16,640	2,312	10,673	16,382	5,659
Carlisle	1,355	816	245	1,383	844	309
Carroll	1,689	1,170	351	2,119	1,046	566
Carter	3,728	3,240	781	4,224	3,305	989
Casey	1,106	3,187	525	1,409	3,317	542
Christian	6,843	8,285	1,064	6,709	7,737	1,789
Clark	4,987	4,739	1,095	4,892	4,625	1,955
Clay	2,135	3,716	478	2,012	4,747	648
Clinton	1,072	2,521	350	1,241	2,830	348
Crittenden	1,480	1,509	400	1,740	1,576	495
Cumberland	753	1,654	227	917	1,866	268
Daviess	15,366	15,844	3,344	16,592	14,936	5,112
Edmonson	1,595	2,619	298	1,653	2,486	438
Elliott	1,298	421	284	1,796	444	273
Estill	1,724	2,220	479	1,837	2,453	736
Fayette	43,632	42,950	5,345	38,306	41,908	14,215
Fleming	1,913	2,313	522	2,257	2,045	815
Floyd	9,655	3,139	1,518	13,351	3,540	1,723
Franklin	11,251	7,132	1,873	9,896	7,591	3,340
Fulton	1,614	863	223	1,813	1,073	306
Gallatin	1,189	838	299	1,171	699	445
Garrard	1,486	2,540	337	1,730	2,359	697
Grant	2,541	2,697	661	2,097	2,128	1,149
Graves	6,991	5,130	1,596	8,001	5,311	1,943
Grayson	2,716	4,249	677	2,909	4,533	993
Green	1,285	2,763	475	1,760	2,709	500
Greenup	6,883	5,370	1,627	7,214	4,975	2,188
Hancock	1,547	1,356	418	1,714	1,261	551
Hardin	11,031	12,642	2,815	9,417	12,299	4,026
Harlan	5,874	3,337	884	6,796	3,970	1,391
Harrison	2,934	2,433	801	2,795	2,148	1,225
Hart	2,527	2,701	501	2,852	2,401	579
Henderson	8,051	5,092	1,556	8,270	5,125	2,678
Henry	2,324	2,110	564	2,838	1,640	720
Hickman	1,220	695	247	1,296	861	294
Hopkins	7,239	6,363	1,512	8,881	6,032	2,565
Jackson	960	3,045	299	776	3,398	341
Jefferson	144,207	114,860	19,413	152,728	116,566	39,822
Jessamine	4,428	6,686	1,040	3,764	6,474	2,059
Johnson	3,348	3,262	1,010	3,669	3,614	1,118
Kenton	19,407	28,579	3,680	16,344	27,261	9,336
Knott	4,842	1,201	517	5,500	1,243	560
Knox	3,736	4,502	811	3,787	5,011	972
Larue	2,040	2,140	469	2,190	2,154	582
Laurel	4,306	9,454	1,211	4,560	8,583	1,859
Lawrence	2,195	1,812	481	2,400	2,084	557
Lee	1,023	1,302	181	1,170	1,617	356
Leslie	1,466	2,296	304	1,591	2,879	450
Letcher	4,160	2,222	782	5,817	3,011	1,206
Lewis	1,415	2,365	561	1,713	2,493	673
Lincoln	2,550	3,006	526	2,532	2,624	762
Livingston	2,228	1,258	449	2,386	1,339	578

	1996			1992		
County	Clinton (D)	Dole (R)	Perot (RF)	Clinton (D)	Bush (R)	Perot (I)
Logan	4,181	3,888	704	4,064	3,710	1,043
Lyon	1,641	999	284	1,583	820	293
McCracken	12,670	10,221	2,268	13,341	10,657	3,077
McCreary	1,710	2,527	488	1,934	3,588	624
McLean	1,834	1,368	385	2,223	1,355	529
Madison	8,142	9,212	1,613	8,005	8,719	3,038
Magoffin	2,249	1,434	337	3,261	1,992	440
Marion	2,922	2,013	757	3,403	2,091	805
Marshall	6,054	4,579	1,391	6,576	4,368	1,773
Martin	1,807	1,612	401	1,715	1,961	393
Mason	2,444	2,588	484	2,657	2,432	916
Meade	3,653	2,855	912	3,387	2,641	1,298
Menifee	979	608	179	1,311	557	254
Mercer	3,179	3,264	738	3,010	3,211	1,298
Metcalfe	1,349	1,651	355	1,703	1,683	409
Monroe	1,114	3,300	415	1,515	3,776	480
Montgomery	3,372	2,681	705	3,686	2,590	1,308
Morgan	1,843	1,439	380	2,655	1,239	498
Muhlenberg	6,564	3,569	1,218	7,901	3,551	1,624
Nelson	5,392	4,645	1,067	5,437	4,495	1,638
Nicholas	1,092	950	265	1,341	894	513
Ohio	3,487	3,475	1,076	4,022	3,385	1,423
Oldham	6,202	10,477	1,521	5,457	8,263	2,855
Owen	1,603	1,709	454	1,830	1,108	613
Owsley	647	920	153	678	1,437	209
Pendleton	1,926	2,177	462	1,740	1,810	1,086
Perry	6,015	3,382	894	6,619	4,128	1,308
Pike	14,126	7,160	2,148	17,358	8,212	2,444
Powell	2,156	1,526	523	2,323	1,809	874
Pulaski	5,340	11,945	1,420	5,465	11,423	2,449
Robertson	360	368	117	439	329	170
Rockcastle	1,160	3,106	338	1,144	3,287	446
Rowan	3,215	2,309	724	3,558	2,469	1,212
Russell	1,582	4,017	837	1,950	4,641	673
Scott	4,258	4,349	977	3,639	3,810	1,800
Shelby	4,629	5,307	780	4,398	4,550	1,451
Simpson	2,749	2,186	401	2,834	2,280	708
Spencer	1,404	1,614	341	1,383	1,305	466
Taylor	2,897	4,573	829	3,518	4,319	1,044
Todd	1,744	1,912	424	1,858	1,691	612
Trigg	2,087	1,975	394	2,438	1,820	573
Trimble	1,245	999	308	1,413	789	413
Union	2,913	1,554	598	3,325	1,605	794
Warren	11,642	15,784	1,835	11,529	14,748	3,533
Washington	1,639	2,116	383	2,008	2,098	542
Wayne	2,422	3,122	481	2,516	3,412	560
Webster	2,852	1,568	660	3,380	1,408	854
Whitley	4,174	5,402	1,027	4,600	5,998	1,533
Wolfe	1,297	772	202	1,674	697	297
Woodford	3,910	4,270	746	3,161	3,992	1,535
Totals	636,614	623,283	120,396	665,104	617,178	203,944

Kentucky Vote Since 1948

1948, Truman, Dem., 466,756; Dewey, Rep., 341,210; Thurmond, States' Rights, 10,411; Wallace, Prog., 1,567; Thomas, Soc., 1,284; Watson, Proh., 1,245; Teichert, Soc. Labor, 185.

1952, Eisenhower, Rep., 495,029; Stevenson, Dem., 495,729; Hamblen, Proh., 1,161; Hass, Soc. Labor, 893; Hallinan, Proh., 336.

1956, Eisenhower, Rep., 572,192; Stevenson, Dem., 476,453; Byrd, States' Rights, 2,657; Holtwick, Proh., 2,145; Hass, Soc. Labor, 358.

1960, Kennedy, Dem., 521,855; Nixon, Rep., 602,607.

1964, Johnson, Dem., 669,659; Goldwater, Rep., 372,977; Kasper, Natl. States Rights, 3,469.

1968, Nixon, Rep., 462,411; Humphrey, Dem., 397,547; Wallace, 3d Party, 193,098; Halstead, Soc. Workers, 2,843.

1972, Nixon, Rep., 676,446; McGovern, Dem., 371,159; Schmitz, Amer., 17,627; Jenness, Soc. Workers, 685; Hall, Com., 464; Spock, Peoples, 1,118.

1976, Carter, Dem., 615,717; Ford, Rep., 531,852; Anderson, Amer., 8,308; McCarthy, Ind., 6,837; Maddox, Amer. Ind., 2,328; MacBride, Libertarian, 814.

1980, Reagan, Rep., 635,274; Carter, Dem., 616,417; Anderson, Ind., 31,127; Clark, Libertarian, 5,531; McCormack, Respect For Life, 4,233; Commoner, Citizens, 1,304; Pulley, Socialist, 393; Hall, Com., 348.

1984, Reagan, Rep., 815,345; Mondale, Dem., 536,756.

1988, Bush, Rep., 734,281; Dukakis, Dem., 580,368; Duke, Pop., 4,494; Paul, Lib., 2,118.

1992, Clinton, Dem., 665,104; Bush, Rep., 617,178; Perot, Ind., 203,944; Marrou, Libertarian, 4,513.

1996, Clinton, Dem., 636,614; Dole, Rep., 623,283; Perot, Ref., 120,396; Browne, Libertarian, 4,009; Phillips, Taxpayers, 2,204; Hagelin, Natural Law, 1,493.

Louisiana

| | 1996 | | | 1992 | | |
Parish	Clinton (D)	Dole (R)	Perot (RF)	Clinton (D)	Bush (R)	Perot (I)
Acadia	12,300	9,246	2,234	12,276	9,017	3,145
Allen	4,930	2,589	1,187	5,626	3,069	1,245
Ascension	15,263	10,885	3,027	13,036	10,275	4,295
Assumption ...	6,416	2,698	904	5,639	2,928	1,358
Avoyelles	9,689	4,433	1,937	8,696	4,851	2,139
Beauregard ...	4,925	5,526	1,834	5,037	5,119	2,103
Bienville	4,335	2,402	457	3,899	2,412	832
Bossier	15,504	16,852	2,660	11,313	15,628	4,863
Caddo	55,543	38,445	4,821	47,733	42,665	11,830
Calcasieu	38,238	26,494	8,281	33,570	24,847	10,980
Caldwell	2,117	1,842	514	2,061	1,752	653
Cameron	2,103	1,365	594	1,985	1,329	995
Catahoula....	2,692	1,770	615	2,570	1,976	773
Claiborne	3,609	2,500	530	3,263	2,599	926
Concordia.....	4,565	3,134	855	4,283	3,223	1,317
DeSoto.......	6,221	3,526	646	5,671	3,643	1,358
E. Baton Rouge	83,493	77,811	7,990	68,622	81,072	16,102
East Carroll ...	2,149	1,008	186	1,835	1,142	283
East Feliciana .	4,714	2,949	660	4,093	2,813	932
Evangeline	7,847	5,278	1,447	8,564	5,147	2,124
Franklin.......	4,076	3,961	814	4,127	3,889	1,311
Grant	2,980	3,117	1,055	3,122	3,214	1,174
Iberia	15,087	12,014	2,448	13,040	11,905	4,337
Iberville......	9,553	4,031	1,076	8,218	5,211	1,543
Jackson	3,368	3,030	571	3,370	3,072	882
Jefferson	80,407	92,820	9,667	64,302	100,493	21,278
Jefferson Davis	6,897	4,311	1,543	7,022	4,513	2,221
Lafayette.....	32,504	36,419	4,631	28,583	32,406	9,124
Lafourche.....	18,810	12,105	2,984	16,182	12,744	5,077
LaSalle.......	2,543	2,925	947	2,389	3,068	993
Lincoln	7,903	6,973	761	7,205	7,220	1,751
Livingston.....	13,276	16,159	4,150	11,499	14,808	4,971
Madison	3,085	1,591	315	2,773	1,702	469
Morehouse....	6,160	5,193	963	6,013	5,364	1,727
Natchitoches ..	8,296	5,471	1,053	6,974	5,694	1,606
Orleans	144,720	39,576	3,805	133,261	52,019	10,889
Ouachita......	24,525	28,559	3,586	20,835	27,600	6,612
Plaquemines ..	5,348	4,493	856	4,467	5,018	1,729
Pointe Coupee.	6,835	3,545	845	6,512	3,563	1,157
Rapides	23,004	21,548	4,670	20,873	22,783	6,599
Red River.....	2,641	1,344	268	2,360	1,649	566
Richland......	4,143	3,765	645	3,706	3,808	1,054
Sabine	4,263	3,543	1,043	4,173	3,564	1,219
St. Bernard....	14,312	13,549	2,664	12,305	16,131	4,308
St. Charles....	10,612	9,316	1,307	8,810	9,158	2,593
St. Helena	3,692	1,455	417	3,416	1,515	589
St. James.....	7,247	2,832	608	6,609	3,339	993
St. John the Baptist	9,937	6,025	966	8,977	6,730	1,922
St. Landry.....	20,636	12,273	2,311	20,383	11,882	4,266
St. Martin	12,492	6,296	1,607	11,252	5,909	2,573
St. Mary	12,402	8,018	1,850	10,648	8,792	3,257
St. Tammany ..	24,281	44,761	4,741	19,735	37,839	9,005
Tangipahoa ...	18,617	15,517	3,144	15,194	14,128	4,612
Tensas	1,882	1,000	176	1,666	1,153	353
Terrebonne....	18,550	13,944	3,359	13,325	14,662	5,505
Union	4,260	4,418	696	4,005	4,434	1,209
Vermilion	12,609	7,653	1,954	12,324	7,062	3,127
Vernon	6,195	5,449	2,068	6,005	5,912	2,313
Washington ...	9,603	6,642	1,643	9,095	7,227	2,303
Webster	9,688	6,153	1,324	8,380	6,640	2,629
W. Baton Rouge	5,697	3,254	799	5,131	3,522	1,249
West Carroll ..	1,853	2,366	461	2,068	2,082	771
W. Feliciana ...	2,416	1,616	388	2,328	1,501	516
Winn.........	3,779	2,803	735	3,537	2,932	843
Totals........	927,837	712,586	123,293	815,971	733,386	211,478

Louisiana Vote Since 1948

1948, Thurmond, States' Rights, 204,290; Truman, Dem., 136,344; Dewey, Rep., 72,657; Wallace, Prog., 3,035.

1952, Eisenhower, Rep., 306,925; Stevenson, Dem., 345,027.

1956, Eisenhower, Rep., 329,047; Stevenson, Dem., 243,977; Andrews, States' Rights, 44,520.

1960, Kennedy, Dem., 407,339; Nixon, Rep., 230,890; States' Rights (unpledged), 169,572.

1964, Johnson, Dem., 387,068; Goldwater, Rep., 509,225.

1968, Nixon, Rep., 257,535; Humphrey, Dem., 309,615; Wallace, 3d Party, 530,300.

1972, Nixon, Rep., 686,852; McGovern, Dem., 298,142; Schmitz, Amer., 52,099; Jenness, Soc. Workers, 14,398.

1976, Carter, Dem., 661,365; Ford, Rep., 587,446; Maddox, Amer., 10,058; Hall, Com., 7,417; McCarthy, Ind., 6,588; MacBride, Libertarian, 3,325.

1980, Reagan, Rep., 792,853; Carter, Dem., 708,453; Anderson, Ind., 26,345; Rarick, Amer. Ind., 10,333; Clark, Libertarian, 8,240; Commoner, Citizens, 1,584; DeBerry, Soc. Work., 783.

1984, Reagan, Rep., 1,037,299; Mondale, Dem., 651,586; Bergland, Libertarian, 1,876.

1988, Bush, Rep., 883,702; Dukakis, Dem., 717,460; Duke, Pop., 18,612; Paul, Lib., 4,115.

1992, Clinton, Dem., 815,971; Bush, Rep., 733,386; Perot, Ind., 211,478; Gritz, Populist/America First, 18,545; Marrou, Libertarian, 3,155; Daniels, Ind., 1,663; Phillips, U.S. Taxpayers, 1,552; Fulani, New Alliance, 1,434; LaRouche, Ind., 1,136.

1996, Clinton, Dem., 927,837; Dole, Rep., 712,586; Perot, Ref., 123,293; Browne, Libertarian, 7,499; Nader, Liberty, Ecology, Community, 4,719; Phillips, Taxpayers, 3,366; Hagelin, Natural Law, 2,981; Moorehead, Workers World, 1,678.

Maine

| | 1996 | | | 1992 | | |
City	Clinton (D)	Dole (R)	Perot (RF)	Clinton (D)	Bush (R)	Perot (I)
Auburn........	5,750	3,060	1,484	5,025	3,653	3,964
Augusta.......	5,307	2,353	1,100	4,657	3,003	3,002
Bangor........	7,609	4,476	1,399	6,826	5,185	4,689
Biddeford.....	5,653	1,768	1,019	4,945	2,533	2,717
Brunswick	5,258	2,850	841	4,686	3,058	2,282
Gorham	2,990	2,269	710	2,516	2,422	2,015
Lewiston......	10,275	3,182	2,113	9,265	4,372	6,180
Orono........	2,748	1,106	369	2,813	1,336	1,502
Portland	19,755	7,178	2,255	19,510	8,660	6,910
Presque Isle ...	2,015	1,491	594	1,750	1,709	1,318
Saco..........	4,506	2,140	834	4,000	2,769	2,303
Sanford	4,368	2,239	1,524	3,854	3,030	3,215
Scarborough ...	3,906	3,214	805	2,941	3,235	2,033
S. Portland....	6,777	3,241	906	5,933	3,999	2,734
Waterville......	4,219	1,478	750	3,868	1,832	2,257
Westbrook.....	4,373	2,186	864	3,665	2,904	2,512
Windham......	3,251	2,396	898	2,444	2,603	2,250
York	2,970	3,225	649	2,445	2,740	1,648
Other	211,058	137,226	66,856	172,277	147,461	153,289
Totals.........	312,788	186,378	85,970	263,420	206,504	206,820

Maine Vote Since 1948

1948, Truman, Dem., 111,916; Dewey, Rep., 150,234; Wallace, Prog., 1,884; Thomas, Soc., 547; Teichert, Soc. Labor, 206.

1952, Eisenhower, Rep., 232,353; Stevenson, Dem., 118,806; Hallinan, Prog., 332; Hass, Soc. Labor, 156; Hoopes, Soc., 138; scattered, 1.

1956, Eisenhower, Rep., 249,238; Stevenson, Dem., 102,468.

1960, Kennedy, Dem., 181,159; Nixon, Rep., 240,608.

1964, Johnson, Dem., 262,264; Goldwater, Rep., 118,701.

1968, Nixon, Rep., 169,254; Humphrey, Dem., 217,312; Wallace, 3d Party, 6,370.

1972, Nixon, Rep., 256,458; McGovern, Dem., 160,584; scattered, 229.

1976, Carter, Dem., 232,279; Ford, Rep., 236,320; McCarthy, Ind., 10,874; Bubar, Proh., 3,495.

1980, Reagan, Rep., 238,522; Carter, Dem., 220,974; Anderson, Ind., 53,327; Clark, Libertarian, 5,119; Commoner, Citizens, 4,394; Hall, Com., 591; write-ins, 84.

1984, Reagan, Rep., 336,500; Mondale, Dem., 214,515.

1988, Bush, Rep., 307,131; Dukakis, Dem., 243,569; Paul, Lib., 2,700; Fulani, New Alliance, 1,405.

1992, Clinton, Dem., 263,420; Perot, Ind., 206,820; Bush, Rep., 206,504; Marrou, Libertarian, 1,681.

1996, Clinton, Dem., 312,788; Dole, Rep., 186,378; Perot, Ref., 85,970; Nader, Green, 15,279; Browne, Libertarian, 2,996; Phillips, Taxpayers, 1,517; Hagelin, Natural Law, 825.

Maryland

| | 1996 | | | 1992 | | |
County	Clinton (D)	Dole (R)	Perot (RF)	Clinton (D)	Bush (R)	Perot (I)
Allegany........	11,025	12,136	2,652	11,501	13,862	5,081
Anne Arundel ..	72,147	83,574	14,287	68,629	81,467	35,191
Baltimore......	132,599	114,449	20,393	143,498	126,728	51,757
Calvert........	10,008	11,509	1,932	8,619	10,026	4,499
Caroline.......	3,251	3,874	947	2,822	3,856	1,729
Carroll	17,122	30,316	4,873	15,447	28,405	10,965
Cecil	10,144	10,885	3,124	10,232	10,784	6,115
Charles	15,890	17,432	2,333	14,498	17,293	6,501
Dorchester.....	4,613	4,337	1,008	3,933	4,934	2,010
Frederick	25,081	34,494	4,989	21,848	31,290	11,373
Garrett	3,121	5,400	1,200	2,856	5,714	1,987
Harford........	29,779	39,686	7,939	27,164	36,593	17,002
Howard........	47,569	40,849	6,011	44,763	38,594	16,182
Kent	3,207	3,055	676	3,093	3,094	1,411
Montgomery ...	198,807	117,730	14,450	199,757	119,705	41,971
Prince George's	176,612	52,697	9,153	168,691	62,955	23,355
Queen Anne's ..	5,054	7,147	1,312	4,668	6,829	2,958
St. Mary's.....	9,988	11,835	1,827	8,931	11,485	4,550
Somerset......	3,557	2,919	613	3,210	3,450	1,230
Talbot.........	6,697	6,914	944	4,642	6,771	2,233
Washington	16,481	21,434	3,934	16,495	21,977	7,537
Wicomico......	12,303	12,687	2,160	11,481	13,560	5,140

County	1996 Clinton (D)	Dole (R)	Perot (RF)	1992 Clinton (D)	Bush (R)	Perot (I)
Worcester.....	7,587	7,621	1,612	6,040	7,237	3,256
City						
Baltimore	145,441	28,467	7,473	185,753	40,725	17,381
Totals.........	966,207	681,530	115,812	988,571	707,094	281,414

Maryland Vote Since 1948

1948, Truman, Dem., 286,521; Dewey, Rep., 294,814; Wallace, Prog., 9,983; Thomas, Soc., 2,941; Thurmond, States' Rights, 2,476; Wright, write-in, 2,294.

1952, Eisenhower, Rep., 499,424; Stevenson, Dem., 395,337; Hallinan, Prog., 7,313.

1956, Eisenhower, Rep., 559,738; Stevenson, Dem., 372,613.

1960, Kennedy, Dem., 565,800; Nixon, Rep., 489,538.

1964, Johnson, Dem., 730,912; Goldwater, Rep., 385,495; write-in, 50.

1968, Nixon, Rep., 517,995; Humphrey, Dem., 538,310; Wallace, 3d Party, 178,734.

1972, Nixon, Rep., 829,305; McGovern, Dem., 505,781; Schmitz, Amer., 18,726.

1976, Carter, Dem., 759,612; Ford, Rep., 672,661.

1980, Reagan, Rep., 680,606; Carter, Dem., 726,161; Anderson, Ind., 119,537; Clark, Libertarian, 14,192.

1984, Reagan, Rep., 879,918; Mondale, Dem., 787,935; Bergland, Libertarian, 5,721.

1988, Bush, Rep., 876,167; Dukakis, Dem., 826,304; Paul, Lib., 6,748; Fulani, New Alliance, 5,115.

1992, Clinton, Dem., 988,571; Bush, Rep., 707,094; Perot, Ind., 281,414; Marrou, Libertarian, 4,715; Fulani, New Alliance, 2,786.

1996, Clinton, Dem., 966,207; Dole, Rep., 681,530; Perot, Ref., 115,812; Browne, Libertarian, 8,765; Phillips, Taxpayers, 3,402; Hagelin, Natural Law, 2,517.

Massachusetts

City	1996 Clinton (D)	Dole (R)	Perot (RF)	1992 Clinton (D)	Bush (R)	Perot (I)
Boston	125,529	33,366	8,428	114,260	41,868	25,189
Brockton...	16,361	6,972	2,738	13,209	8,863	7,579
Brookline ..	18,812	4,579	799	19,848	4,892	2,629
Cambridge .	29,913	4,976	1,415	30,737	5,847	4,106
Chicopee ..	14,203	5,188	2,495	11,433	6,138	6,452
Fall River .	22,796	4,287	2,612	18,652	5,456	6,922
Framingham	16,836	6,669	1,700	15,165	8,114	6,089
Lawrence..	8,615	2,804	1,096	7,698	5,079	3,245
Lowell.....	16,912	5,896	2,911	14,492	8,467	8,893
Lynn......	18,370	5,634	2,726	15,275	7,350	7,665
Medford ...	16,639	5,844	1,741	14,690	7,690	5,480
New Bedford...	23,620	4,151	2,547	20,880	5,255	6,965
Newton....	30,005	8,499	1,674	29,136	9,623	5,685
Quincy	23,182	9,824	3,066	18,891	12,306	9,068
Somerville .	20,206	3,983	1,455	19,792	5,883	4,416
Springfield .	31,266	9,110	3,407	27,302	12,200	10,361
Waltham...	13,607	5,830	1,663	11,333	7,365	5,092
Weymouth .	13,536	6,904	2,181	10,762	7,849	6,552
Worcester..	35,607	12,879	3,925	32,326	17,228	10,488
Other	1,075,494	570,663	178,627	872,758	617,566	487,855
Totals.....	1,571,509	718,058	227,206	1,318,639	805,039	630,731

Massachusetts Vote Since 1948

1948, Truman, Dem., 1,151,788; Dewey, Rep., 909,370; Wallace, Prog., 38,157; Teichert, Soc. Labor, 5,535; Watson, Proh., 1,663.

1952, Eisenhower, Rep., 1,292,325; Stevenson, Dem., 1,083,525; Hallinan, Prog., 4,636; Hass, Soc. Labor, 1,957; Hamblen, Proh., 886; scattered, 69; blanks, 41,150.

1956, Eisenhower, Rep., 1,393,197; Stevenson, Dem., 948,190; Hass, Soc. Labor, 5,573; Holtwick, Proh., 1,205; others, 341.

1960, Kennedy, Dem., 1,487,174; Nixon, Rep., 976,750; Hass, Soc. Labor, 3,892; Decker, Proh., 1,633; others, 31; blank and void, 26,024.

1964, Johnson, Dem., 1,786,422; Goldwater, Rep., 549,727; Hass, Soc. Labor, 4,755; Munn, Proh., 3,735; scattered, 159; blank, 48,104.

1968, Nixon, Rep., 766,844; Humphrey, Dem., 1,469,218; Wallace, 3d Party, 87,088; Blomen, Soc. Labor, 6,180; Munn, Proh., 2,369; scattered, 53; blanks, 25,394.

1972, Nixon, Rep., 1,112,078; McGovern, Dem., 1,332,540; Jenness, Soc. Workers, 10,600; Fisher, Soc. Labor, 129; Schmitz, Amer., 2,877; Spock, Peoples, 101; Hall, Com., 46; Hospers, Libertarian, 43; scattered, 8.

1976, Carter, Dem., 1,429,475; Ford, Rep., 1,030,276; McCarthy, Ind., 65,637; Camejo, Soc. Workers, 8,138; Anderson, Amer., 7,555; La Rouche, U.S. Labor, 4,922; MacBride, Libertarian, 135.

1980, Reagan, Rep., 1,057,631; Carter, Dem., 1,053,802; Anderson, Ind., 382,539; Clark, Libertarian, 22,038; DeBerry, Soc. Workers, 3,735; Commoner, Citizens, 2,056; McReynolds, Soc., 62; Bubar, Statesman, 34; Griswold, Workers World, 19; scattered, 2,382.

1984, Reagan, Rep., 1,310,936; Mondale, Dem., 1,239,606.

1988, Bush, Rep., 1,194,635; Dukakis, Dem., 1,401,415; Paul, Lib., 24,251; Fulani, New Alliance, 9,561.

1992, Clinton, Dem., 1,318,639; Bush, Rep., 805,039; Perot, Ind., 630,731; Marrou, Libertarian, 9,021; Fulani, New Alliance, 3,172; Phillips, U.S. Taxpayers, 2,218; Hagelin, Natural Law, 1,812; LaRouche, Ind., 1,027.

1996, Clinton, Dem., 1,571,509; Dole, Rep., 718,058; Perot, Ref., 227,206; Browne, Libertarian, 20,424; Hagelin, Natural Law, 5,183; Moorehead, Workers World, 3,276.

Michigan

County	1996 Clinton (D)	Dole (R)	Perot (RF)	1992 Clinton (D)	Bush (R)	Perot (I)
Alcona	2,619	2,227	669	2,383	2,247	1,117
Alger......	2,229	1,429	537	2,144	1,471	941
Allegan	14,361	20,859	3,269	12,823	19,077	8,742
Alpena	7,114	4,525	1,730	6,894	4,878	3,236
Antrim	4,226	4,630	1,129	3,431	3,984	2,528
Arenac	3,472	2,247	844	3,244	2,330	1,608
Baraga	1,601	1,209	460	1,695	1,160	754
Barry	9,467	11,139	2,282	8,652	9,489	6,303
Bay.......	27,835	16,038	5,410	26,492	16,383	11,258
Benzie	3,081	2,856	763	2,715	2,438	1,657
Berrien	24,614	28,254	5,958	25,840	29,252	14,056
Branch....	6,567	6,321	1,779	5,850	5,976	4,683
Calhoun ..	26,287	20,953	4,765	25,542	19,791	13,058
Cass......	8,207	7,373	2,241	8,047	7,391	4,756
Charlevoix .	4,689	4,864	1,303	4,063	4,017	3,360
Cheboygan	5,018	4,244	1,462	4,459	3,864	2,495
Chippewa..	6,532	5,137	1,453	5,434	5,462	2,706
Clare	6,311	3,742	1,531	5,346	3,916	2,812
Clinton	11,945	13,694	2,698	10,116	12,216	7,877
Crawford ..	2,666	2,157	840	2,252	2,193	1,442
Delta......	8,561	5,925	1,543	8,387	6,027	3,485
Dickinson .	5,614	4,408	1,478	5,689	4,273	3,022
Eaton.....	19,781	20,092	4,378	16,752	18,669	12,208
Emmet	4,892	6,002	1,512	4,245	5,312	3,576
Genesee ..	106,065	49,332	17,671	105,156	47,834	46,259
Gladwin ..	5,494	3,670	1,466	4,457	3,616	2,649
Gogebic ..	4,436	2,769	917	4,792	2,838	1,543
Grand Traverse ..	12,987	16,355	3,527	11,148	13,629	9,495
Gratiot	6,793	6,214	1,762	5,678	6,280	3,866
Hillsdale ..	5,955	7,947	2,262	5,244	7,579	4,968
Houghton..	5,957	5,941	1,584	6,558	5,575	2,945
Huron.....	6,827	6,126	1,811	6,023	6,491	4,064
Ingham....	63,584	43,096	8,640	61,596	43,926	27,683
Ionia......	9,261	9,574	2,354	8,370	9,135	6,211
Iosco	6,240	4,410	1,710	5,369	4,912	3,131
Iron.......	3,232	2,014	755	3,648	1,971	1,344
Isabella....	9,635	7,460	2,069	8,784	7,706	5,434
Jackson ...	24,633	24,987	5,968	23,686	25,424	15,194
Kalamazoo .	45,644	40,703	5,867	43,568	38,035	21,666
Kalkaska ..	2,666	2,455	922	2,297	2,173	1,915
Kent	85,912	121,335	14,120	82,305	115,285	43,707
Keweenaw.	572	491	169	582	378	212
Lake	2,606	1,213	552	2,351	1,194	981
Lapeer	14,308	13,369	4,793	11,982	12,326	10,541
Leelanau ..	4,019	5,155	924	3,445	3,993	2,685
Lenawee ..	16,924	14,168	4,167	15,399	14,297	9,517
Livingston .	22,517	30,598	6,337	17,851	27,539	15,971
Luce	1,107	964	366	972	958	660
Mackinac .	2,700	2,281	742	2,293	2,278	1,379
Macomb..	151,430	120,616	29,859	130,732	147,795	67,954
Manistee ..	5,383	3,807	1,230	5,193	3,491	2,923
Marquette .	15,168	8,805	2,492	16,038	9,665	5,768
Mason	5,597	5,066	1,525	4,829	5,102	3,096
Mecosta ..	6,370	5,289	1,373	6,097	6,047	3,612
Menominee .	4,880	4,038	1,205	4,559	3,995	2,487
Midland ...	15,177	16,547	3,964	13,382	16,149	8,945
Missaukee .	2,256	3,012	719	1,893	2,829	1,306
Monroe ...	26,072	19,678	6,315	24,957	20,250	13,551
Montcalm .	10,053	8,679	2,530	8,730	8,420	5,504
Montmorency	2,120	1,760	682	1,903	1,794	1,077
Muskegon .	35,328	21,873	5,794	32,515	23,769	15,268
Newaygo ..	7,614	7,868	2,047	6,455	7,333	4,056
Oakland...	241,884	219,855	36,709	214,733	242,160	94,911
Oceana ...	4,419	3,947	1,286	3,846	3,944	2,713
Ogemaw ..	4,725	2,904	1,369	4,016	2,936	2,122
Ontonagon .	2,080	1,523	604	2,451	1,463	805
Osceola ...	4,085	3,855	1,068	3,529	3,606	2,199
Oscoda ...	1,652	1,545	503	1,471	1,583	755
Otsego	3,351	3,638	1,280	3,129	3,393	2,635
Ottawa	27,024	61,436	6,275	22,180	56,862	16,855
Presque Isle	3,449	2,463	932	3,308	2,398	1,612
Roscommon .	6,092	4,135	1,539	5,243	4,170	2,551
Saginaw ...	47,579	31,577	8,081	43,819	32,103	20,523
St. Clair ..	28,881	22,495	8,134	23,385	24,508	18,523
St. Joseph .	8,529	9,764	2,319	7,817	9,836	6,209

County	1996 Clinton (D)	Dole (R)	Perot (RF)	1992 Clinton (D)	Bush (R)	Perot (I)
Sanilac....	7,092	7,821	2,265	5,868	7,891	4,894
Schoolcraft.	2,187	1,200	460	2,139	1,253	721
Shiawassee	14,662	11,714	3,703	12,629	10,930	8,632
Tuscola....	10,314	9,154	3,013	9,138	8,636	6,765
Van Buren.	13,355	11,347	2,946	12,466	10,357	7,255
Washtenaw	73,106	40,097	8,020	73,325	41,386	21,889
Wayne....	504,466	175,886	43,554	508,464	227,002	102,074
Wexford.	5,510	4,866	1,386	4,894	4,696	2,923
Totals.....	1,989,653	1,481,212	336,670	1,871,182	1,554,940	824,813

Michigan Vote Since 1948

1948, Truman, Dem., 1,003,448; Dewey, Rep., 1,038,595; Wallace, Prog., 46,515; Watson, Proh., 13,052; Thomas, Soc., 6,063; Teichert, Soc. Labor, 1,263; Dobbs, Soc. Workers, 672.

1952, Eisenhower, Rep., 1,551,529; Stevenson, Dem., 1,230,657; Hamblen, Proh., 10,331; Hallinan, Prog., 3,922; Hass, Soc. Labor, 1,495; Dobbs, Soc. Workers, 655; scattered, 3.

1956, Eisenhower, Rep., 1,713,647; Stevenson, Dem., 1,359,898; Holtwick, Proh., 6,923.

1960, Kennedy, Dem., 1,687,269; Nixon, Rep., 1,620,428; Dobbs, Soc. Workers, 4,347; Decker, Proh., 2,029; Daly, Tax Cut, 1,767; Hass, Soc. Labor, 1,718; Ind. Amer., 539.

1964, Johnson, Dem., 2,136,615; Goldwater, Rep., 1,060,152; DeBerry, Soc. Workers, 3,817; Hass, Soc. Labor, 1,704; Proh. (no candidate listed), 699; scattering, 145.

1968, Nixon, Rep., 1,370,665; Humphrey, Dem., 1,593,082; Wallace, 3d Party, 331,968; Halstead, Soc. Workers, 4,099; Blomen, Soc. Labor, 1,762; Cleaver, New Politics, 4,585; Munn, Proh., 60; scattering, 29.

1972, Nixon, Rep., 1,961,721; McGovern, Dem., 1,459,435; Schmitz, Amer., 63,321; Fisher, Soc. Labor, 2,437; Jenness, Soc. Workers, 1,603; Hall, Com., 1,210.

1976, Carter, Dem., 1,696,714; Ford, Rep., 1,893,742; McCarthy, Ind., 47,905; MacBride, Libertarian, 5,406; Wright, People's, 3,504; Camejo, Soc. Workers, 1,804; LaRouche, U.S. Labor, 1,366; Levin, Soc. Labor, 1,148; scattering, 2,160.

1980, Reagan, Rep., 1,915,225; Carter, Dem., 1,661,532; Anderson, Ind., 275,223; Clark, Libertarian, 41,597; Commoner, Citizens, 11,930; Hall, Com., 3,262; Griswold, Workers World, 30; Greaves, Amer., 21; Bubar, Statesman, 9.

1984, Reagan, Rep., 2,251,571; Mondale, Dem., 1,529,638; Bergland, Libertarian, 10,055.

1988, Bush, Rep., 1,965,486; Dukakis, Dem., 1,675,783; Paul, Lib., 18,336; Fulani, Ind., 2,513.

1992, Clinton, Dem., 1,871,182; Bush, Rep., 1,554,940; Perot, Ind., 824,813; Marrou, Libertarian, 10,175; Phillips, U.S. Taxpayers, 8,263; Hagelin, Natural Law, 2,954.

1996, Clinton, Dem., 1,989,653; Dole, Rep., 1,481,212; Perot, Ref., 336,670; Browne, Libertarian, 27,670; Hagelin, Natural Law, 4,254; Moorehead, Workers World, 3,153; White, Soc. Equality, 1,554.

Minnesota

County	1996 Clinton (D)	Dole (R)	Perot (RF)	1992 Clinton (D)	Bush (R)	Perot (I)
Aitkin.......	3,810	2,327	1,155	3,400	2,151	1,951
Anoka......	63,756	41,745	16,448	54,621	39,458	35,140
Becker.....	5,911	5,461	1,813	4,958	5,430	3,238
Beltrami	8,006	5,806	1,635	7,210	5,204	3,473
Benton	6,006	4,835	2,133	5,156	5,053	4,048
Big Stone ...	1,619	990	368	1,610	1,052	740
Blue Earth ..	12,420	9,082	3,324	11,531	8,813	7,299
Brown......	4,864	5,580	1,786	4,278	5,390	3,845
Carlton	8,052	4,034	1,591	7,736	3,922	3,005
Carver	11,554	12,380	3,781	8,349	10,201	7,942
Cass	5,437	4,791	1,620	4,901	4,276	2,939
Chippewa...	3,178	2,119	782	2,929	2,143	1,505
Chisago	8,611	5,984	2,812	7,077	4,813	5,098
Clay	10,476	8,764	1,733	9,845	9,666	3,835
Clearwater ..	1,578	1,423	471	1,587	1,315	841
Cook	1,169	1,010	246	1,005	878	704
Cottonwood .	2,737	2,633	741	2,382	2,481	1,749
Crow Wing ..	11,156	10,095	3,423	8,896	9,112	6,367
Dakota	77,297	57,244	17,095	63,660	52,312	40,244
Dodge.....	3,233	2,888	1,223	2,620	3,049	2,231
Douglas	6,450	6,747	2,093	5,252	6,356	4,138
Faribault ...	3,817	3,272	1,103	3,339	3,439	2,322
Fillmore	4,732	3,466	1,575	3,977	3,583	3,011
Freeborn ...	8,458	5,166	2,226	7,759	5,089	4,878
Goodhue ...	9,931	7,293	2,806	7,916	7,321	5,790
Grant	1,806	1,284	434	1,521	1,261	885
Hennepin ..	285,126	173,887	47,663	278,648	179,581	123,659
Houston	4,153	3,674	1,439	3,744	3,853	2,697
Hubbard ...	3,802	3,593	1,141	3,362	3,227	1,949
Isanti.......	6,041	4,450	2,242	5,386	3,988	3,898
Itasca	10,706	6,506	2,889	9,621	5,952	5,147

County	1996 Clinton (D)	Dole (R)	Perot (RF)	1992 Clinton (D)	Bush (R)	Perot (I)
Jackson	2,727	2,153	908	2,481	1,824	1,918
Kanabec	2,927	1,924	996	2,532	1,876	1,836
Kandiyohi....	9,009	7,119	2,229	7,914	6,784	4,869
Kittson	1,394	1,055	270	1,307	1,098	558
Koochiching..	3,472	2,080	1,098	3,474	1,954	1,993
LacQuiParle..	2,420	1,447	561	2,342	1,435	1,163
Lake	3,388	1,684	752	3,415	1,465	1,437
Lake of the Woods	888	814	287	794	762	629
Le Sueur ...	5,457	3,902	1,699	4,662	3,858	3,363
Lincoln	1,641	1,199	504	1,555	1,084	967
Lyon	5,062	4,932	1,351	4,481	4,591	3,180
McLeod	6,027	5,474	2,402	4,919	5,422	4,933
Mahnomen ..	1,026	877	270	1,035	854	483
Marshall	2,333	2,068	710	2,309	2,136	1,306
Martin......	4,718	4,303	1,405	4,019	4,438	3,089
Meeker.....	4,531	3,428	1,571	3,861	3,497	3,120
Mille Lacs ..	4,336	2,948	1,467	3,648	2,814	2,615
Morrison.....	5,728	5,054	2,310	5,588	5,038	3,710
Mower	10,413	4,994	2,464	9,935	5,147	5,001
Murray	2,173	1,907	753	1,993	1,609	1,588
Nicollet	6,772	5,057	1,737	6,055	5,091	3,799
Nobles	4,106	3,769	1,132	3,756	3,548	2,586
Norman	1,875	1,392	425	1,784	1,541	776
Olmsted	22,857	22,860	5,640	19,039	23,404	13,806
Otter Tail....	10,519	11,808	3,191	9,176	11,074	6,274
Pennington ..	2,814	2,129	910	2,578	2,155	1,598
Pine	5,432	3,080	1,597	4,929	2,841	2,952
Pipestone....	1,999	2,096	599	1,773	1,953	1,429
Polk	6,369	5,563	1,502	5,850	5,817	3,176
Pope	2,803	1,992	665	2,619	1,886	1,390
Ramsey	133,878	66,954	20,351	130,932	68,206	50,757
Red Lake ...	1,053	695	334	1,020	691	472
Redwood....	2,997	3,700	1,053	2,740	3,408	2,710
Renville	3,956	2,887	1,311	3,414	2,852	2,598
Rice	12,821	7,016	2,872	10,908	7,015	6,057
Rock........	2,142	2,169	554	2,006	2,065	1,244
Roseau	2,759	2,988	1,081	2,346	2,785	2,099
St. Louis....	60,736	25,553	11,308	61,813	24,579	21,714
Scott	14,657	12,734	4,886	11,225	10,936	9,881
Sherburne ..	10,551	8,699	3,665	7,843	7,339	6,534
Sibley	2,769	2,590	1,226	2,421	2,315	2,407
Stearns	24,238	21,474	8,150	21,451	22,502	14,834
Steele.......	6,974	5,617	2,197	5,152	5,964	4,542
Stevens	2,741	2,141	467	2,466	2,229	1,086
Swift	3,054	1,541	690	2,980	1,603	1,359
Todd	4,520	4,078	1,958	4,059	3,990	2,976
Traverse	1,135	775	295	1,053	841	582
Wabasha	4,523	3,452	1,474	3,736	3,397	3,012
Wadena.....	2,480	2,696	801	2,340	2,492	1,535
Waseca	3,819	3,171	1,385	3,146	3,118	2,621
Washington ..	45,119	31,219	10,106	35,820	26,568	22,585
Watonwan ...	2,534	1,997	711	2,100	1,871	1,574
Wilkin	1,319	1,508	358	1,122	1,626	748
Winona	10,272	7,955	2,907	9,707	8,585	5,993
Wright	15,542	13,224	5,550	12,465	11,650	10,829
Yellow Medicine	2,741	2,006	818	2,593	1,909	1,645
Totals.......	1,120,438	766,476	257,704	1,020,997	747,841	562,506

Minnesota Vote Since 1948

1948, Truman, Dem., 692,966; Dewey, Rep., 483,617; Wallace, Prog., 27,866; Thomas, Soc., 4,646; Teichert, Soc. Labor, 2,525; Dobbs, Soc. Workers, 606.

1952, Eisenhower, Rep., 763,211; Stevenson, Dem., 608,458; Hallinan, Prog., 2,666; Hass, Soc. Labor, 2,383; Hamblen, Proh., 2,147; Dobbs, Soc. Workers, 618.

1956, Eisenhower, Rep., 719,302; Stevenson, Dem., 617,525; Hass, Soc. Labor (Ind. Gov.), 2,080; Dobbs, Soc. Workers, 1,098.

1960, Kennedy, Dem., 779,933; Nixon, Rep., 757,915; Dobbs, Soc. Workers, 3,077; Industrial Gov., 962.

1964, Johnson, Dem., 991,117; Goldwater, Rep., 559,624; DeBerry, Soc. Workers, 1,177; Hass, Industrial Gov., 2,544.

1968, Nixon, Rep., 658,643; Humphrey, Dem., 857,738; Wallace, 3d Party, 68,931; scattered, 2,443; Halstead, Soc. Workers, 808; Blomen, Ind. Gov't., 285; Mitchell, Com., 415; Cleaver, Peace, 935; McCarthy, write-in, 585; scattered, 170.

1972, Nixon, Rep., 898,269; McGovern, Dem., 802,346; Schmitz, Amer., 31,407; Spock, Peoples, 2,805; Fisher, Soc. Labor, 4,261; Jenness, Soc. Workers, 940; Hall, Com., 662; scattered, 962.

1976, Carter, Dem., 1,070,440; Ford, Rep., 819,395; McCarthy, Ind., 35,490; Anderson, Amer., 13,592; Camejo, Soc. Workers, 4,149; MacBride, Libertarian, 3,529; Hall, Com., 1,092.

1980, Reagan, Rep., 873,268; Carter, Dem., 954,173; Anderson, Ind., 174,997; Clark, Libertarian, 31,593; Commoner, Citizens, 8,406; Hall, Com., 1,117; DeBerry, Soc. Workers, 711; Griswold, Workers World, 698; McReynolds, Soc., 536; write-ins, 281.

1984, Reagan, Rep., 1,032,603; Mondale, Dem., 1,036,364; Bergland, Libertarian, 2,996.

1988, Bush, Rep., 962,337; Dukakis, Dem., 1,109,471; McCarthy, Minn. Prog., 5,403; Paul, Lib., 5,109.

1992, Clinton, Dem., 1,020,997; Bush, Rep., 747,841; Perot, Ind., 562,506; Marrou, Libertarian, 3,373; Gritz, Populist/America First, 3,363; Hagelin, Natural Law, 1,406.

1996, Clinton, Dem., 1,120,438; Dole, Rep., 766,476; Perot, Ref., 257,704; Nader, Green, 24,908; Browne, Libertarian, 8,271; Peron, Grass Roots, 4,898; Phillips, Taxpayers, 3,416; Hagelin, Natural Law, 1,808; Birrenbach, Ind. Grass Roots, 787; Harris, Soc. Workers, 684; White, Soc. Equality, 347.

Mississippi

County	1996 Clinton (D)	Dole (R)	Perot (RF)	1992 Clinton (D)	Bush (R)	Perot (I)
Adams	8,218	5,378	779	8,255	5,831	1,753
Alcorn	4,964	4,960	929	6,373	6,249	1,349
Amite	2,824	2,521	351	2,608	2,561	498
Attala	3,092	3,130	383	3,015	3,520	529
Benton	1,944	993	209	2,402	1,253	293
Bolivar	8,670	4,027	320	8,801	4,752	593
Calhoun	2,178	2,470	351	2,462	3,191	607
Carroll.	2,041	2,629	245	1,182	1,695	200
Chickasaw	2,971	2,535	401	3,220	3,150	629
Choctaw	1,247	1,715	247	1,435	2,026	298
Claiborne	3,739	784	103	3,302	935	161
Clarke.	2,337	3,470	366	2,259	4,207	450
Clay	4,267	2,948	337	4,620	3,297	626
Coahoma	5,776	3,441	256	6,409	4,120	518
Copiah	4,415	4,138	375	4,397	4,600	409
Covington.	2,628	3,219	417	2,775	3,525	654
DeSoto	10,282	18,135	2,399	8,833	16,104	2,569
Forrest	7,965	11,278	1,094	8,333	12,432	1,909
Franklin.	1,381	1,586	329	1,587	1,942	393
George	1,888	3,311	710	2,650	4,141	1,335
Greene	1,347	1,947	322	1,664	2,406	559
Grenada	4,402	4,527	470	4,203	4,721	609
Hancock	4,303	5,820	1,143	4,651	6,422	2,302
Harrison	18,775	25,486	3,726	15,268	25,049	6,855
Hinds	45,410	35,653	2,929	43,434	45,031	5,341
Holmes	4,720	1,536	140	4,092	1,694	203
Humphreys . . .	2,305	1,382	110	2,696	1,721	258
Issaquena . . .	546	269	42	550	298	79
Itawamba	2,987	3,490	732	3,635	4,142	918
Jackson	13,598	24,918	2,947	13,017	25,321	6,484
Jasper.	3,170	2,615	353	3,059	2,789	568
Jefferson	2,531	489	89	2,796	562	156
Jefferson Davis	2,663	1,890	264	2,991	2,228	382
Jones	7,360	13,020	1,362	8,035	13,824	2,523
Kemper.	2,048	1,439	188	2,243	1,830	278
Lafayette.	4,646	4,753	580	5,224	5,251	861
Lamar.	3,169	8,609	925	3,208	8,259	1,543
Lauderdale. . . .	8,668	15,055	1,036	8,489	17,098	1,659
Lawrence	2,481	2,392	471	2,582	2,689	765
Leake	2,902	3,017	406	3,333	3,943	497
Lee	8,438	11,815	1,361	7,710	12,231	2,041
Leflore	6,853	4,456	240	6,374	5,298	611
Lincoln	4,294	5,960	778	4,744	7,040	1,281
Lowndes.	6,220	9,169	750	6,552	10,509	1,716
Madison	9,354	14,467	759	9,386	12,810	1,478
Marion	4,334	5,023	585	4,654	5,776	1,162
Marshall	7,521	3,272	482	7,913	3,847	689
Monroe	5,184	5,206	889	4,933	5,994	1,255
Montgomery . . .	1,970	1,943	197	2,076	2,324	370
Neshoba.	2,646	4,545	560	3,090	6,135	794
Newton.	2,163	4,223	464	2,146	5,128	494
Noxubee.	2,801	1,287	119	3,188	1,623	203
Oktibbeha. . . .	5,923	6,142	395	5,726	6,381	984
Panola	5,408	3,701	513	6,066	4,644	729
Pearl River	4,892	8,212	1,190	4,683	7,726	2,352
Perry.	1,413	2,178	450	1,490	2,538	462
Pike.	6,302	5,403	683	6,279	6,005	1,380
Pontotoc.	2,597	4,289	774	2,965	4,595	777
Prentiss	3,053	3,473	574	3,385	4,317	781
Quitman.	2,186	1,121	126	2,422	1,451	210
Rankin	8,614	24,585	2,093	8,155	24,537	3,454
Scott	3,163	4,018	466	3,349	5,268	691
Sharkey	1,566	906	70	1,526	1,008	145
Simpson.	2,851	4,455	525	3,213	5,358	726
Smith	1,858	3,371	522	1,968	4,106	680
Stone	1,551	2,288	417	1,447	2,295	447
Sunflower.	4,960	2,926	290	5,050	3,726	600
Tallahatchie . . .	2,990	1,676	251	2,902	2,213	380
Tate.	3,195	3,694	406	3,519	4,196	634
Tippah	2,992	3,249	661	3,475	4,444	802
Tishomingo . . .	2,709	2,766	609	3,910	3,393	751
Tunica.	1,263	557	55	1,451	693	96
Union	3,316	4,375	788	3,714	5,173	816
Walthall.	2,240	2,239	444	2,476	2,728	711
Warren.	8,774	9,261	1,259	8,175	10,209	2,146
Washington . . .	10,053	6,762	437	10,588	7,598	795
Wayne	2,652	3,219	595	3,064	3,874	824
Webster	1,379	2,254	255	1,746	2,791	444
Wilkinson	2,807	1,016	226	3,210	1,399	307
Winston	3,488	3,498	434	3,953	4,311	688
Yalobusha	2,437	1,711	332	2,617	2,179	438
Yazoo	4,754	4,152	362	4,880	5,113	669
Totals.	**394,022**	**439,838**	**52,222**	**400,258**	**487,793**	**85,626**

Mississippi Vote Since 1948

1948, Thurmond, States' Rights, 167,538; Truman, Dem., 19,384; Dewey, Rep., 5,043; Wallace, Prog., 225.

1952, Eisenhower, Ind. vote pledged to Rep. candidate, 112,966; Stevenson, Dem., 172,566.

1956, Eisenhower, Rep., 56,372; Stevenson, Dem., 144,498; Black and Tan Grand Old Party, 4,313; total, 60,685; Byrd, Ind., 42,966.

1960, Kennedy, Dem., 108,362; Democratic unpledged electors, 116,248; Nixon, Rep., 73,561. Mississippi's victorious slate of 8 unpledged Democratic electors cast their votes for Sen. Harry F. Byrd (D, VA).

1964, Johnson, Dem., 52,618; Goldwater, Rep., 356,528.

1968, Nixon, Rep., 88,516; Humphrey, Dem., 150,644; Wallace, 3d Party, 415,349.

1972, Nixon, Rep., 505,125; McGovern, Dem., 126,782; Schmitz, Amer., 11,598; Jenness, Soc. Workers, 2,458.

1976, Carter, Dem., 381,309; Ford, Rep., 366,846; Anderson, Amer., 6,678; McCarthy, Ind., 4,074; Maddox, Ind., 4,049; Camejo, Soc. Workers, 2,805; MacBride, Libertarian, 2,609.

1980, Reagan, Rep., 441,089; Carter, Dem., 429,281; Anderson, Ind., 12,036; Clark, Libertarian, 5,465; Griswold, Workers World, 2,402; Pulley, Soc. Workers, 2,347.

1984, Reagan, Rep., 582,377; Mondale, Dem., 352,192; Bergland, Libertarian, 2,336.

1988, Bush, Rep., 557,890; Dukakis, Dem., 363,921; Duke, Ind., 4,232; Paul, Lib., 3,329.

1992, Bush, Rep., 487,793; Clinton, Dem., 400,258; Perot, Ind., 85,626; Fulani, New Alliance, 2,625; Marrou, Libertarian, 2,154; Phillips, U.S. Taxpayers, 1,652; Hagelin, Natural Law, 1,140.

1996, Dole, Rep., 439,838; Clinton, Dem., 394,022; Perot, Ind. (Ref.), 52,222; Browne, Libertarian, 2,809; Phillips, Taxpayers, 2,314; Hagelin, Natural Law, 1,447; Collins, Ind., 1,205.

Missouri

County	1996 Clinton (D)	Dole (R)	Perot (RF)	1992 Clinton (D)	Bush (R)	Perot (I)
Adair.	4,441	4,656	1,170	4,232	4,141	2,224
Andrew. . . .	2,807	3,281	964	2,675	2,652	2,151
Atchison. . .	1,266	1,327	367	1,208	1,140	840
Audrain. . . .	4,690	3,955	1,046	4,731	3,798	2,099
Barry	4,352	5,855	1,494	4,791	5,565	2,381
Barton	1,625	2,812	563	1,433	2,775	971
Bates	3,224	2,904	949	2,993	2,499	2,225
Benton	2,996	2,895	764	3,195	2,511	1,551
Bollinger. . .	2,044	2,420	506	2,150	2,289	909
Boone	24,984	22,047	4,083	26,176	19,405	12,040
Buchanan .	15,848	12,610	4,248	16,570	11,275	9,404
Butler	5,780	6,996	1,414	6,602	6,450	2,189
Caldwell. . .	1,487	1,464	468	1,456	1,295	1,283
Callaway . .	5,880	5,567	1,530	5,799	4,880	3,266
Camden. . .	5,566	7,190	1,809	5,140	5,554	3,891
Cape Girardeau.	9,957	15,557	1,861	9,605	13,464	5,199
Carroll.	2,080	1,839	580	2,100	1,774	1,495
Carter.	1,172	1,180	301	1,169	1,101	405
Cass.	11,743	13,495	3,474	10,246	10,349	9,216
Cedar.	2,027	2,484	658	2,064	2,085	1,173
Chariton . . .	2,072	1,508	423	2,141	1,378	1,067
Christian. . .	6,627	9,477	2,301	6,242	7,422	3,422
Clark.	1,749	1,081	458	1,815	1,039	725
Clay	32,603	28,935	7,048	30,565	23,798	20,951
Clinton	3,445	2,780	848	3,400	2,391	2,423
Cole	10,857	16,140	2,121	10,201	15,270	5,770
Cooper. . . .	2,753	2,900	891	2,709	2,867	1,735
Crawford . .	3,349	2,990	1,223	3,515	2,831	2,002
Dade	1,243	1,822	447	1,332	1,577	834
Dallas	2,277	2,554	787	2,533	2,116	1,392
Daviess . . .	1,534	1,321	466	1,477	1,107	1,143
DeKalb . . .	1,679	1,627	492	1,630	1,318	1,207
Dent.	2,234	2,542	693	2,689	2,125	1,049
Douglas . . .	1,744	2,601	775	2,126	2,569	1,081
Dunklin. . . .	5,428	3,766	934	6,277	4,024	1,166
Franklin . . .	13,908	13,715	5,517	13,431	11,477	11,043
Gasconade .	2,104	2,997	820	1,952	2,690	1,672
Gentry	1,493	1,361	416	1,519	1,272	921
Greene. . . .	39,300	48,193	8,569	41,137	46,457	17,770
Grundy. . . .	2,073	1,883	631	1,968	1,749	1,372
Harrison. . .	1,628	1,737	484	1,590	1,563	1,059
Henry	4,579	3,260	1,231	4,232	2,681	2,807
Hickory. . . .	1,858	1,491	531	1,929	1,259	864
Holt.	1,144	1,323	314	1,050	1,202	781
Howard . . .	2,014	1,545	568	2,085	1,253	1,090
Howell	5,261	5,991	2,066	5,492	5,360	2,650
Iron.	2,221	1,328	568	2,507	1,276	841
Jackson . . .	140,317	85,534	21,047	145,999	78,611	66,142
Jasper	11,462	18,361	3,545	11,727	17,592	6,440
Jefferson . .	32,073	23,877	8,893	32,569	20,637	20,057
Johnson. . .	6,220	6,276	1,911	5,546	5,032	4,578
Knox.	891	862	254	1,010	724	523

County	1996 Clinton (D)	Dole (R)	Perot (RF)	1992 Clinton (D)	Bush (R)	Perot (I)
Laclede....	4,047	5,887	1,459	4,179	5,176	2,852
Lafayette..	6,118	5,489	1,516	5,213	4,651	3,561
Lawrence..	4,465	6,099	1,613	4,666	5,608	2,570
Lewis.....	2,050	1,453	644	2,196	1,461	892
Lincoln....	5,644	4,897	1,881	5,453	3,718	3,572
Linn......	2,967	2,097	781	2,916	1,967	1,524
Livingston..	2,913	2,384	777	2,505	2,370	1,976
McDonald..	1,980	3,008	923	2,281	3,010	1,551
Macon....	2,937	2,634	848	3,194	2,256	1,697
Madison...	2,351	1,595	625	2,501	1,673	899
Maries	1,540	1,560	516	1,732	1,356	915
Marion	4,924	4,653	1,082	5,156	4,762	1,841
Mercer	700	660	208	843	626	378
Miller......	3,110	4,387	1,185	2,905	4,175	2,391
Mississippi.	3,235	1,595	380	3,226	1,675	776
Moniteau ..	2,129	2,603	693	2,018	2,566	1,499
Monroe....	1,938	1,333	532	2,060	1,153	969
Montgomery	2,277	2,124	772	2,063	1,974	1,266
Morgan....	3,006	3,059	1,006	2,906	2,819	2,028
New Madrid	4,451	2,417	663	4,883	2,431	962
Newton....	5,840	10,067	1,995	5,987	8,804	3,567
Nodaway ..	3,966	3,362	1,043	3,723	3,147	2,484
Oregon....	1,795	1,502	475	2,258	1,402	564
Osage.....	2,045	2,890	608	1,860	2,784	1,423
Ozark.....	1,445	1,882	595	1,581	1,772	906
Pemiscot ..	3,371	1,820	458	3,924	2,161	670
Perry......	2,517	3,427	777	2,525	3,205	1,498
Pettis.....	6,057	7,336	1,716	5,314	6,823	4,278
Phelps	6,405	6,990	1,703	6,852	6,040	3,774
Pike.......	3,495	2,209	916	3,609	2,255	1,464
Platte	12,705	13,332	3,035	10,920	9,380	9,062
Polk......	3,307	4,521	1,169	3,316	3,465	1,879
Pulaski	3,783	4,089	1,141	4,113	3,793	2,057
Putnam....	857	1,091	276	838	1,143	522
Ralls......	1,998	1,513	520	2,158	1,349	880
Randolph..	4,502	3,274	1,130	4,951	3,025	2,212
Ray.......	4,714	2,884	1,113	4,457	2,563	2,567
Reynolds ..	1,631	903	386	2,014	776	532
Ripley.....	2,081	1,988	530	2,300	1,814	739
St. Charles.	41,369	47,705	11,591	37,263	38,673	30,351
St. Clair....	1,974	1,815	650	1,965	1,555	1,083
St. Francois	9,034	6,200	2,266	9,367	5,889	3,635
St. Louis...	225,524	196,096	34,850	235,760	188,285	109,099
Ste. Genevieve	3,597	2,078	942	3,795	1,780	1,547
Saline.....	4,765	2,931	1,090	4,643	2,688	2,815
Schuyler...	857	777	287	936	742	487
Scotland...	990	773	326	1,070	798	617
Scott......	7,011	6,641	1,483	7,452	6,265	2,763
Shannon...	1,882	1,339	524	2,135	1,224	579
Shelby	1,410	1,213	413	1,435	1,169	786
Stoddard ..	4,883	5,020	1,185	5,720	4,608	1,977
Stone	3,497	5,223	1,353	3,256	4,035	1,884
Sullivan....	1,402	1,275	340	1,510	1,326	596
Taney	4,623	6,844	1,580	4,682	6,081	2,395
Texas	3,897	4,065	1,335	4,597	3,470	1,900
Vernon	3,363	3,123	1,135	3,546	2,851	1,890
Warren	3,443	3,768	1,254	3,213	2,953	2,471
Washington	4,315	2,259	1,169	4,211	2,157	1,618
Wayne	2,754	2,172	674	3,073	2,101	837
Webster...	3,855	4,958	1,214	4,149	4,361	2,108
Worth	572	540	150	599	483	328
Wright.....	2,280	3,754	890	2,814	3,427	1,425
City St. Louis..	91,233	22,121	7,276	102,356	25,441	18,864
Totals.....	1,025,935	890,016	217,188	1,053,873	811,159	518,741

Missouri Vote Since 1948

1948, Truman, Dem., 917,315; Dewey, Rep., 655,039; Wallace, Prog., 3,998; Thomas, Soc., 2,222.

1952, Eisenhower, Rep., 959,429; Stevenson, Dem., 929,830; Hallinan, Prog., 987; Hamblen, Proh., 885; MacArthur, Christian Nationalist, 302; America First, 233; Hoopes, Soc., 227; Hass, Soc. Labor, 169.

1956, Stevenson, Dem., 918,273; Eisenhower, Rep., 914,299.

1960, Kennedy, Dem., 972,201; Nixon, Rep., 962,221.

1964, Johnson, Dem., 1,164,344; Goldwater, Rep., 653,535.

1968, Nixon, Rep., 811,932; Humphrey, Dem., 791,444; Wallace, 3d Party, 206,126.

1972, Nixon, Rep., 1,154,058; McGovern, Dem., 698,531.

1976, Carter, Dem., 999,163; Ford, Rep., 928,808; McCarthy, Ind., 24,329.

1980, Reagan, Rep., 1,074,181; Carter, Dem., 931,182; Anderson, Ind., 77,920; Clark, Libertarian, 14,422; DeBerry, Soc. Workers, 1,515; Commoner, Citizens, 573; write-ins, 31.

1984, Reagan, Rep., 1,274,188; Mondale, Dem., 848,583.

1988, Bush, Rep., 1,084,953; Dukakis, Dem., 1,001,619; Fulani, New Alliance, 6,656; Paul, write-in, 434.

1992, Clinton, Dem., 1,053,873; Bush, Rep., 811,159; Perot, Ind., 518,741; Marrou, Libertarian, 7,497.

1996, Clinton, Dem., 1,025,935; Dole, Rep., 890,016; Perot, Ref., 217,188; Phillips, Taxpayers, 11,521; Browne, Libertarian, 10,522; Hagelin, Natural Law, 2,287.

Montana

County	1996 Clinton (D)	Dole (R)	Perot (RF)	1992 Clinton (D)	Bush (R)	Perot (I)
Beaverhead....	1,164	2,414	412	1,098	1,746	1,202
Big Horn.......	2,453	1,336	424	2,154	1,377	840
Blaine........	1,316	1,127	435	1,355	971	699
Broadwater	603	1,029	318	491	830	505
Carbon........	1,854	2,147	713	1,549	1,562	1,482
Carter........	150	522	89	154	497	220
Cascade	15,707	14,291	4,749	14,719	12,494	9,151
Chouteau	1,039	1,660	434	959	1,380	870
Custer	2,115	2,467	695	1,968	2,105	1,505
Daniels........	510	558	240	457	496	402
Dawson	1,903	1,890	842	1,785	1,679	1,370
Deer Lodge ...	3,331	883	772	3,174	832	1,207
Fallon	452	778	276	446	731	427
Fergus	1,866	3,671	605	1,615	2,736	1,934
Flathead......	10,452	16,542	4,786	9,746	11,699	9,109
Gallatin.......	10,972	14,559	3,146	9,535	11,109	7,711
Garfield	107	562	69	125	403	281
Glacier	2,292	1,270	491	2,076	1,222	997
Golden Valley	128	284	73	142	192	157
Granite........	429	733	228	358	556	386
Hill	3,517	2,601	950	3,618	2,408	2,017
Jefferson	1,775	2,248	729	1,415	1,541	1,172
Judith Basin...	452	753	126	409	610	415
Lake	4,195	4,723	1,804	3,938	3,596	2,878
Lewis & Clark ..	11,535	11,665	3,140	11,117	9,351	5,560
Liberty	379	634	144	321	512	363
Lincoln	2,705	3,552	1,425	2,765	2,799	2,637
McCone.......	390	615	244	424	528	395
Madison	955	1,984	516	779	1,415	1,043
Meagher	281	505	142	260	422	310
Mineral	658	549	383	664	403	543
Missoula	21,874	16,034	5,586	20,347	12,898	9,735
Musselshell	652	1,121	291	648	876	691
Park	2,564	3,837	959	2,258	2,846	2,182
Petroleum	62	186	36	61	135	95
Phillips........	705	1,392	401	634	1,026	949
Pondera.......	1,123	1,438	383	1,046	1,252	855
Powder River...	236	663	137	258	547	340
Powell........	952	1,274	531	989	1,058	872
Prairie........	259	417	99	260	412	179
Ravalli	5,200	8,138	2,731	4,644	5,392	4,573
Richland.......	1,614	2,021	906	1,440	1,760	1,525
Roosevelt.....	2,118	1,209	645	1,827	1,212	1,089
Rosebud	1,681	1,413	547	1,669	1,130	1,099
Sanders	1,573	2,043	990	1,689	1,361	1,378
Sheridan	1,187	832	408	1,077	795	782
Silver Bow ...	11,199	3,909	2,447	9,960	3,491	4,570
Stillwater	1,282	1,871	618	1,178	1,390	1,056
Sweet Grass ...	469	1,109	186	395	880	507
Teton	1,188	1,701	416	1,043	1,364	969
Toole	874	1,203	386	854	943	903
Treasure.......	171	237	87	157	206	178
Valley	1,674	1,838	645	1,715	1,497	1,320
Wheatland....	391	563	127	384	478	284
Wibaux........	197	284	128	195	234	173
Yellowstone ...	22,992	26,367	6,139	20,163	22,822	13,133
Totals.........	167,922	179,652	55,229	154,507	144,207	107,225

Montana Vote Since 1948

1948, Truman, Dem., 119,071; Dewey, Rep., 96,770; Wallace, Prog., 7,313; Thomas, Soc., 695; Watson, Proh., 429.

1952, Eisenhower, Rep., 157,394; Stevenson, Dem., 106,213; Hallinan, Prog., 723; Hamblen, Proh., 548; Hoopes, Soc., 159.

1956, Eisenhower, Rep., 154,933; Stevenson, Dem., 116,238.

1960, Kennedy, Dem., 134,891; Nixon, Rep., 141,841; Decker, Proh., 456; Dobbs, Soc. Workers, 391.

1964, Johnson, Dem., 164,246; Goldwater, Rep., 113,032; Kasper, Natl. States' Rights, 519; Munn, Proh., 499; DeBerry, Soc. Workers, 332.

1968, Nixon, Rep., 138,835; Humphrey, Dem., 114,117; Wallace, 3d Party, 20,015; Halstead, Soc. Workers, 457; Munn, Proh., 510; Caton, New Reform, 470.

1972, Nixon, Rep., 183,976; McGovern, Dem., 120,197; Schmitz, Amer., 13,430.

1976, Carter, Dem., 149,259; Ford, Rep., 173,703; Anderson, Amer., 5,772.

1980, Reagan, Rep., 206,814; Carter, Dem., 118,032; Anderson, Ind., 29,281; Clark, Libertarian, 9,825.

1984, Reagan, Rep., 232,450; Mondale, Dem., 146,742; Bergland, Libertarian, 5,185.

1988, Bush, Rep., 190,412; Dukakis, Dem., 168,936; Paul, Lib., 5,047; Fulani, New Alliance, 1,279.

1992, Clinton, Dem., 154,507; Bush, Rep., 144,207; Perot, Ind., 107,225; Gritz, Populist/America First, 3,658.

1996, Dole, Rep., 179,652; Clinton, Dem., 167,922; Perot, Ref., 55,229; Browne, Libertarian, 2,526; Hagelin, Natural Law, 1,754.

Nebraska

County	1996 Clinton (D)	Dole (R)	Perot (RF)	1992 Clinton (D)	Bush (R)	Perot (I)
Adams	3,935	6,924	1,513	3,445	6,346	3,273
Antelope	884	2,005	457	650	1,979	1,134
Arthur	25	187	46	18	148	97
Banner	62	309	30	68	284	128
Blaine	53	284	39	64	256	130
Boone	806	1,695	424	604	1,588	956
Box Butte	1,782	2,458	695	1,935	2,198	1,508
Boyd	372	778	181	353	744	468
Brown	359	1,105	289	311	999	525
Buffalo	4,277	10,004	1,484	3,742	9,708	4,083
Burt	1,237	1,707	497	1,224	1,667	1,009
Butler	1,099	2,042	512	1,087	1,881	1,157
Cass	3,477	4,878	1,239	2,949	4,314	2,657
Cedar	1,218	2,171	739	1,007	1,981	1,507
Chase	365	1,277	197	398	1,000	674
Cherry	551	1,905	332	563	1,707	730
Cheyenne	1,059	2,571	287	967	2,197	1,061
Clay	880	1,982	425	802	1,818	952
Colfax	1,065	1,954	492	1,011	1,915	1,197
Cuming	1,033	2,520	503	835	2,711	1,192
Custer	1,293	3,453	615	1,126	3,180	1,492
Dakota	2,632	2,592	721	2,322	2,771	1,307
Dawes	1,108	1,991	442	987	1,961	1,103
Dawson	2,180	4,794	1,044	1,739	4,710	2,305
Deuel	245	629	111	232	558	327
Dixon	931	1,478	414	830	1,484	726
Dodge	5,181	7,484	1,894	4,665	7,269	4,432
Douglas	70,708	92,334	14,863	67,003	93,421	38,641
Dundy	224	752	112	244	664	332
Fillmore	1,058	1,696	321	988	1,495	993
Franklin	483	1,013	215	477	967	527
Frontier	310	901	169	302	785	479
Furnas	663	1,475	207	624	1,365	804
Gage	4,008	4,413	1,346	3,309	3,995	2,726
Garden	279	851	155	212	697	385
Garfield	249	625	111	221	595	270
Gosper	275	609	150	254	492	297
Grant	84	258	55	75	247	124
Greeley	472	642	155	435	587	395
Hall	6,708	10,183	2,403	5,519	9,264	5,822
Hamilton	1,172	2,623	457	992	2,379	1,213
Harlan	520	1,120	203	488	991	623
Hayes	87	439	39	85	362	207
Hitchcock	409	977	173	359	824	540
Holt	1,107	3,436	677	835	3,131	1,714
Hooker	115	308	83	70	283	102
Howard	853	1,294	417	778	1,138	940
Jefferson	1,520	1,979	495	1,506	1,783	1,177
Johnson	770	1,009	309	822	885	642
Kearney	782	1,953	296	644	1,751	844
Keith	830	2,504	460	731	2,019	1,130
Keya Paha	94	385	47	105	368	158
Kimball	527	1,011	212	408	931	440
Knox	1,266	2,123	531	968	2,112	1,166
Lancaster	43,339	44,812	8,595	41,207	41,400	21,783
Lincoln	5,165	7,482	2,043	5,142	7,025	3,384
Logan	79	294	72	80	271	98
Loup	74	229	28	58	233	96
McPherson	50	233	33	49	217	62
Madison	3,047	7,965	1,554	2,352	7,851	3,486
Merrick	997	2,084	449	864	1,854	1,072
Morrill	620	1,296	262	577	1,184	752
Nance	585	892	238	559	851	569
Nemaha	1,232	1,888	485	1,110	1,696	1,020
Nuckolls	757	1,383	306	834	1,277	825
Otoe	2,279	3,290	877	2,038	2,960	1,800
Pawnee	580	766	207	566	670	565
Perkins	352	1,018	163	300	842	522
Phelps	1,071	3,015	465	829	2,748	1,298
Pierce	697	1,923	446	611	1,853	1,084
Platte	3,010	7,948	1,353	2,409	7,712	3,656
Polk	750	1,504	268	661	1,435	812
Red Willow	1,365	3,112	499	1,164	2,488	1,660
Richardson	1,517	2,089	633	1,513	2,050	1,356
Rock	180	564	135	162	588	233
Saline	2,523	1,945	689	2,425	1,740	1,576
Sarpy	12,806	23,023	3,722	10,720	20,482	9,270
Saunders	2,777	4,514	1,223	2,509	4,037	2,567
Scotts Bluff	4,547	7,641	1,251	4,173	7,213	3,514
Seward	2,432	3,479	745	2,118	3,044	1,722
Sheridan	573	1,834	289	535	1,698	751
Sherman	567	822	266	568	736	582
Sioux	138	551	75	148	445	206
Stanton	577	1,457	386	496	1,274	786
Thayer	933	1,698	334	923	1,387	1,077
Thomas	64	303	62	69	283	115

County	1996 Clinton (D)	Dole (R)	Perot (RF)	1992 Clinton (D)	Bush (R)	Perot (I)
Thurston	962	835	293	865	898	487
Valley	758	1,346	274	716	1,173	693
Washington	2,248	4,391	971	2,108	4,035	2,148
Wayne	1,048	2,150	440	921	2,122	1,047
Webster	621	1,094	236	624	972	657
Wheeler	106	241	69	88	246	127
York	1,653	4,266	559	1,385	3,783	1,825
Totals	236,761	363,467	71,278	216,864	343,678	174,104

Nebraska Vote Since 1948

1948, Truman, Dem., 224,165; Dewey, Rep., 264,774.

1952, Eisenhower, Rep., 421,603; Stevenson, Dem., 188,057.

1956, Eisenhower, Rep., 378,108; Stevenson, Dem., 199,029.

1960, Kennedy, Dem., 232,542; Nixon, Rep., 380,553.

1964, Johnson, Dem., 307,307; Goldwater, Rep., 276,847.

1968, Nixon, Rep., 321,163; Humphrey, Dem., 170,784; Wallace, 3d Party, 44,904.

1972, Nixon, Rep., 406,298; McGovern, Dem., 169,991; scattered, 817.

1976, Carter, Dem., 233,287; Ford, Rep., 359,219; McCarthy, Ind., 9,383; Maddox, Amer. Ind., 3,378; MacBride, Libertarian, 1,476.

1980, Reagan, Rep., 419,214; Carter, Dem., 166,424; Anderson, Ind., 44,854; Clark, Libertarian, 9,041.

1984, Reagan, Rep., 459,135; Mondale, Dem., 187,475; Bergland, Libertarian, 2,075.

1988, Bush, Rep., 397,956; Dukakis, Dem., 259,235; Paul, Lib., 2,534; Fulani, New Alliance, 1,740.

1992, Bush, Rep., 343,678; Clinton, Dem., 216,864; Perot, Ind., 174,104; Marrou, Libertarian, 1,340.

1996, Dole, Rep., 363,467; Clinton, Dem., 236,761; Perot, Ref., 71,278; Browne, Libertarian, 2,792; Phillips, Ind., 1,928; Hagelin, Natural Law, 1,189.

Nevada

County	1996 Clinton (D)	Dole (R)	Perot (RF)	1992 Clinton (D)	Bush (R)	Perot (I)
Churchill	2,282	4,369	821	1,770	3,789	1,964
Clark	127,963	103,431	23,177	124,586	97,403	75,364
Douglas	5,109	8,828	1,486	3,928	6,182	4,814
Elko	3,149	6,512	1,539	2,782	5,208	3,628
Esmeralda	140	277	91	118	221	220
Eureka	158	412	90	129	330	214
Humboldt	1,467	2,334	603	810	1,505	1,149
Lander	660	1,107	361	423	885	652
Lincoln	499	936	255	511	890	394
Lyon	3,419	4,753	1,104	2,777	3,509	2,716
Mineral	1,068	814	361	909	918	746
Nye	3,300	3,979	1,544	2,561	2,743	2,501
Pershing	565	743	203	467	643	429
Storey	614	705	244	488	458	550
Washoe	44,915	49,477	9,970	39,500	42,636	30,974
White Pine	1,397	1,399	546	1,354	1,206	1,070
City						
Carson City	7,269	9,168	1,591	6,035	7,302	5,195
Totals	203,974	199,244	43,986	189,148	175,828	132,580

Nevada Vote Since 1948

1948, Truman, Dem., 31,291; Dewey, Rep., 29,357; Wallace, Prog., 1,469.

1952, Eisenhower, Rep., 50,502; Stevenson, Dem., 31,688.

1956, Eisenhower, Rep., 56,049; Stevenson, Dem., 40,640.

1960, Kennedy, Dem., 54,880; Nixon, Rep., 52,387.

1964, Johnson, Dem., 79,339; Goldwater, Rep., 56,094.

1968, Nixon, Rep., 73,188; Humphrey, Dem., 60,598; Wallace, 3d Party, 20,432.

1972, Nixon, Rep., 115,750; McGovern, Dem., 66,016.

1976, Carter, Dem., 92,479; Ford, Rep., 101,273; MacBride, Libertarian, 1,519; Maddox, Amer. Ind., 1,497; scattered, 5,108.

1980, Reagan, Rep., 155,017; Carter, Dem., 66,666; Anderson, Ind., 17,651; Clark, Libertarian, 4,358.

1984, Reagan, Rep., 188,770; Mondale, Dem., 91,655; Bergland, Libertarian, 2,292.

1988, Bush, Rep., 206,040; Dukakis, Dem., 132,738; Paul, Lib., 3,520; Fulani, New Alliance, 835.

1992, Clinton, Dem., 189,148; Bush, Rep., 175,828; Perot, Ind., 132,580; Gritz, Populist/America First, 2,892; Marrou, Libertarian, 1,835.

1996, Clinton, Dem., 203,974; Dole, Rep., 199,244; Perot, Ref., 43,986; "None of These Candidates," 5,608; Nader, Green, 4,730; Browne, Libertarian, 4,460; Phillips, Ind. Amer., 1,732; Hagelin, Natural Law, 545.

New Hampshire

	1996			1992		
	Clinton	Dole	Perot	Clinton	Bush	Perot
City	(D)	(R)	(RF)	(D)	(R)	(I)
Concord....	9,719	5,082	1,164	8,325	5,651	2,843
Derry......	4,814	4,503	1,083	3,962	4,750	3,363
Dover......	6,332	3,752	930	5,449	4,197	2,246
Hudson.....	3,841	3,167	976	3,053	3,315	2,774
Keene......	5,401	2,910	621	5,210	3,257	1,736
Laconia....	2,865	2,842	508	2,390	3,033	1,496
Londonderry.	3,666	4,076	838	2,915	3,960	2,532
Manchester..	20,185	14,704	3,053	16,627	16,298	7,441
Merrimack..	4,934	4,499	949	3,764	4,410	2,787
Nashua....	16,584	11,479	2,858	14,777	12,514	8,306
Portsmouth.	6,343	3,014	661	6,132	3,563	2,088
Rochester...	5,489	3,650	1,108	4,588	4,272	2,541
Salem......	5,164	4,257	1,241	4,184	4,864	3,382
Other......	150,829	128,551	32,397	127,664	128,400	77,802
Totals........	246,166	196,486	48,387	209,040	202,484	121,337

New Hampshire Vote Since 1948

1948, Truman, Dem., 107,995; Dewey, Rep., 121,299; Wallace, Prog., 1,970; Thomas, Soc., 86; Teichert, Soc. Labor, 83; Thurmond, States' Rights, 7.

1952, Eisenhower, Rep., 166,287; Stevenson, Dem., 106,663.

1956, Eisenhower, Rep., 176,519; Stevenson, Dem., 90,364; Andrews, Const., 111.

1960, Kennedy, Dem., 137,772; Nixon, Rep., 157,989.

1964, Johnson, Dem., 182,065; Goldwater, Rep., 104,029.

1968, Nixon, Rep., 154,903; Humphrey, Dem., 130,589; Wallace, 3d Party, 11,173; New Party, 421; Halstead. Soc. Workers, 104.

1972, Nixon, Rep., 213,724; McGovern, Dem., 116,435; Schmitz, Amer., 3,386; Jenness, Soc. Workers, 368; scattered, 142.

1976, Carter, Dem., 147,645; Ford, Rep., 185,935; McCarthy, Ind., 4,095; MacBride, Libertarian, 936; Reagan, write-in, 388; La Rouche, U.S. Labor, 186; Camejo, Soc. Workers, 161; Levin, Soc. Labor, 66; scattered, 215.

1980, Reagan, Rep., 221,705; Carter, Dem., 108,864; Anderson, Ind., 49,693; Clark, Libertarian, 2,067; Commoner, Citizens, 1,325; Hall, Com., 129; Griswold, Workers World, 76; DeBerry, Soc. Workers, 72; scattered, 68.

1984, Reagan, Rep., 267,051; Mondale, Dem., 120,377; Bergland, Libertarian, 735.

1988, Bush, Rep., 281,537; Dukakis, Dem., 163,696; Paul, Lib., 4,502; Fulani, New Alliance, 790.

1992, Clinton, Dem., 209,040; Bush, Rep., 202,484; Perot, Ind., 121,337; Marrou, Libertarian, 3,548.

1996, Clinton, Dem., 246,166; Dole, Rep., 196,486; Perot, Ref., 48,387; Browne, Libertarian, 4,214; Phillips, Taxpayers, 1,344.

New Jersey

	1996			1992		
	Clinton	Dole	Perot	Clinton	Bush	Perot
County	(D)	(R)	(RF)	(D)	(R)	(I)
Atlantic....	44,434	29,538	8,261	39,633	34,279	15,890
Bergen....	191,085	141,164	25,512	171,104	178,223	52,082
Burlington.	85,086	57,337	18,407	72,845	63,709	35,322
Camden...	114,962	52,791	17,433	104,915	67,205	37,144
Cape May.	19,849	19,357	4,978	17,324	21,502	9,798
Cumberland	25,444	14,744	5,348	22,220	19,253	9,901
Essex.....	175,387	65,172	9,513	158,130	89,146	26,961
Gloucester.	51,928	32,138	14,361	42,425	37,335	24,132
Hudson...	116,121	38,288	8,965	99,799	66,505	14,569
Hunterdon.	18,446	26,379	5,686	15,423	25,130	12,736
Mercer....	77,641	40,559	10,536	71,383	50,473	22,503
Middlesex.	145,201	82,433	24,643	128,824	108,701	45,055
Monmouth.	120,414	99,975	22,754	101,750	117,715	45,445
Morris.....	81,092	95,830	15,299	67,593	108,431	32,447
Ocean.....	94,243	82,830	22,864	75,431	95,984	41,668
Passaic...	85,879	53,584	10,944	70,030	71,147	21,494
Salem.....	12,044	9,294	4,124	10,062	10,363	7,274
Somerset..	50,673	51,868	8,377	42,867	56,044	21,014
Sussex....	19,525	26,746	6,705	14,775	29,510	12,537
Union.....	108,102	65,912	12,432	96,671	87,742	23,991
Warren....	14,805	17,160	4,992	13,002	18,468	9,866
Totals....	1,652,361	1,103,099	262,134	1,436,206	1,356,865	521,829

New Jersey Vote Since 1948

1948, Truman, Dem., 895,455; Dewey, Rep., 981,124; Wallace, Prog., 42,683; Watson, Proh., 10,593; Thomas, Soc., 10,521; Dobbs, Soc. Workers, 5,825; Teichert, Soc. Labor, 3,354.

1952, Eisenhower, Rep., 1,373,613; Stevenson, Dem., 1,015,902; Hoopes, Soc., 8,593; Hass, Soc. Labor, 5,815; Hallinan, Prog., 5,589; Krajewski, Poor Man's, 4,203; Dobbs, Soc. Workers, 3,850; Hamblen, Proh., 989.

1956, Eisenhower, Rep., 1,606,942; Stevenson Dem., 850,337; Holtwick, Proh., 9,147; Hass, Soc. Labor, 6,736; Andrews,

Cons., 5,317; Dobbs, Soc. Workers, 4,004; Krajewski, Amer. Third Party, 1,829.

1960, Kennedy, Dem., 1,385,415; Nixon, Rep., 1,363,324; Dobbs, Soc. Workers, 11,402; Lee, Cons., 8,708; Hass, Soc. Labor, 4,262.

1964, Johnson, Dem., 1,867,671; Goldwater, Rep., 963,843; DeBerry, Soc. Workers, 8,181; Hass, Soc. Labor, 7,075.

1968, Nixon, Rep., 1,325,467; Humphrey, Dem., 1,264,206; Wallace, 3d Party, 262,187; Halstead, Soc. Workers, 8,667; Gregory, Peace and Freedom, 8,084; Blomen, Soc. Labor, 6,784.

1972, Nixon, Rep., 1,845,502; McGovern, Dem., 1,102,211; Schmitz, Amer., 34,378; Spock, Peoples, 5,355; Fisher, Soc. Labor, 4,544; Jenness, Soc. Workers, 2,233; Mahalchik, Amer. First, 1,743; Hall, Com., 1,263.

1976, Carter, Dem., 1,444,653; Ford, Rep., 1,509,688; McCarthy, Ind., 32,717; MacBride, Libertarian, 9,449; Maddox, Amer., 7,716; Levin, Soc. Labor, 3,686; Hall, Com., 1,662; LaRouche, U.S. Labor, 1,650; Camejo, Soc. Workers, 1,184; Wright, People's, 1,044; Bubar, Proh., 554; Zeidler, Soc., 469.

1980, Reagan, Rep., 1,546,557; Carter, Dem., 1,147,364; Anderson, Ind., 234,632; Clark, Libertarian, 20,652; Commoner, Citizens, 8,203; McCormack, Right to Life, 3,927; Lynen, Middle Class, 3,694; Hall, Com., 2,555; Pulley, Soc. Workers, 2,198; McReynolds, Soc., 1,973; Gahres, Down With Lawyers, 1,718; Griswold, Workers World, 1,288; Wendelken, Ind., 923.

1984, Reagan, Rep., 1,933,630; Mondale, Dem., 1,261,323; Bergland, Libertarian, 6,416.

1988, Bush, Rep., 1,740,604; Dukakis, Dem., 1,317,541; Lewin, Peace and Freedom, 9,953; Paul, Lib., 8,421.

1992, Clinton, Dem., 1,436,206; Bush, Rep., 1,356,865; Perot, Ind., 521,829; Marrou, Libertarian, 6,822; Fulani, New Alliance, 3,513; Phillips, U.S. Taxpayers, 2,670; LaRouche, Ind., 2,095; Warren, Soc. Workers, 2,011; Daniels, Ind., 1,996; Gritz, Populist/America First, 1,867; Hagelin, Natural Law, 1,353.

1996, Clinton, Dem., 1,652,361; Dole, Rep., 1,103,099; Perot, Ref., 262,134; Nader, Green, 32,465; Browne, Libertarian, 14,763; Hagelin, Natural Law, 3,887; Phillips, Taxpayers, 3,440; Harris, Soc. Workers, 1,837; Moorehead, Workers World, 1,337; White, Soc. Equality, 537.

New Mexico

	1996			1992		
	Clinton	Dole	Perot	Clinton	Bush	Perot
County	(D)	(R)	(RF)	(D)	(R)	(I)
Bernalillo....	88,140	78,832	8,708	90,863	77,304	31,241
Catron......	423	923	114	465	771	289
Chaves.....	7,014	9,991	1,271	6,360	8,872	3,590
Cibola......	4,030	2,245	488	3,334	2,051	847
Colfax......	2,659	1,975	411	2,607	1,730	871
Curry.......	4,116	7,378	842	3,699	6,831	2,056
De Baca....	509	489	86	451	526	204
Dona Ana...	22,766	17,541	2,269	19,894	16,308	7,682
Eddy.......	8,959	8,534	1,297	7,409	7,313	3,430
Grant.......	5,860	3,993	778	5,603	2,917	1,685
Guadalupe...	1,208	436	79	1,225	691	173
Harding.....	264	321	28	268	312	98
Hidalgo.....	943	789	209	995	871	442
Lea.........	5,393	7,661	1,465	5,047	7,921	3,233
Lincoln......	2,209	3,396	666	1,730	2,669	1,431
Los Alamos..	3,983	4,999	560	3,897	4,320	2,339
Luna.......	3,001	2,616	598	2,637	2,166	1,445
McKinley....	10,124	4,470	650	9,405	4,720	1,304
Mora.......	1,646	561	131	1,555	668	188
Otero.......	5,938	9,065	1,096	5,377	7,481	3,257
Quay.......	1,830	1,943	377	1,758	1,759	755
Rio Arriba...	7,965	2,551	469	7,832	2,680	984
Roosevelt....	2,097	3,245	467	2,172	3,215	1,085
Sandoval...	13,081	11,015	1,482	10,951	8,491	3,954
San Juan...	12,070	17,478	2,355	11,302	13,415	5,351
San Miguel..	6,995	1,938	405	6,186	2,183	965
Santa Fe....	26,349	10,857	1,846	27,189	9,684	5,656
Sierra.......	2,154	2,140	431	1,771	1,562	1,055
Socorro.....	3,374	2,315	455	2,908	2,186	918
Taos........	6,635	2,126	545	7,051	2,260	1,300
Torrance....	2,072	2,154	332	1,662	1,667	810
Union.......	519	995	125	519	975	355
Valencia.....	9,169	7,779	1,222	7,495	6,305	2,902
Totals.......	273,495	232,751	32,257	261,617	212,824	91,895

New Mexico Vote Since 1948

1948, Truman, Dem., 105,464; Dewey, Rep., 80,303; Wallace, Prog., 1,037; Watson, Proh., 127; Thomas, Soc., 83; Teichert, Soc. Labor, 49.

1952, Eisenhower, Rep., 132,170; Stevenson, Dem., 105,661; Hamblen, Proh., 297; Hallinan, Ind. Prog., 225; MacArthur, Christian National, 220; Hass, Soc. Labor, 35.

1956, Eisenhower, Rep., 146,788; Stevenson, Dem., 106,098; Holtwick, Proh., 607; Andrews, Ind., 364; Hass, Soc. Labor, 69.

1960, Kennedy, Dem., 156,027; Nixon, Rep., 153,733; Decker, Proh., 777; Hass, Soc. Labor, 570.

1964, Johnson, Dem., 194,017; Goldwater, Rep., 131,838; Hass, Soc. Labor, 1,217; Munn, Proh., 543.

1968, Nixon, Rep., 169,692; Humphrey, Dem., 130,081; Wallace, 3d Party, 25,737; Chavez, 1,519; Halstead, Soc. Workers, 252.

1972, Nixon, Rep., 235,606; McGovern, Dem., 141,084; Schmitz, Amer., 8,767; Jenness, Soc. Workers, 474.

1976, Carter, Dem., 201,148; Ford, Rep., 211,419; Camejo, Soc. Workers, 2,462; MacBride, Libertarian, 1,110; Zeidler, Soc., 240; Bubar, Proh., 211.

1980, Reagan, Rep., 250,779; Carter, Dem., 167,826; Anderson, Ind., 29,459; Clark, Libertarian, 4,365; Commoner, Citizens, 2,202; Bubar, Statesman, 1,281; Pulley, Soc. Workers, 325.

1984, Reagan, Rep., 307,101; Mondale, Dem., 201,769; Bergland, Libertarian, 4,459.

1988, Bush, Rep., 270,341; Dukakis, Dem., 244,497; Paul, Lib., 3,268; Fulani, New Alliance, 2,237.

1992, Clinton, Dem., 261,617; Bush, Rep., 212,824; Perot, Ind., 91,895; Marrou, Libertarian, 1,615.

1996, Clinton, Dem., 273,495; Dole, Rep., 232,751; Perot, Ref., 32,257; Nader, Green, 13,218; Browne, Libertarian, 2,996; Phillips, Taxpayers, 713; Hagelin, Natural Law, 644.

New York

County	1996 Clinton (D)	Dole (R)	Perot (RF)	1992 Clinton (D)	Bush (R)	Perot (I)
Albany	85,993	39,785	11,957	80,641	49,452	24,064
Allegany...	6,621	8,107	2,730	4,848	8,976	4,703
Bronx.....	248,276	30,435	7,186	225,038	63,310	15,115
Broome ...	44,407	31,327	9,114	43,444	34,653	21,280
Cattaraugus	13,029	12,971	5,151	10,150	13,944	10,662
Cayuga ...	15,879	11,093	4,420	13,088	12,065	10,279
Chautauqua	26,831	21,261	7,484	22,645	21,222	18,455
Chemung..	16,977	14,287	3,967	15,099	16,088	7,493
Chenango .	8,797	7,319	2,822	8,017	8,114	5,356
Clinton	15,386	9,759	3,488	12,881	13,455	5,389
Columbia .	12,910	10,324	3,466	11,368	11,568	5,829
Cortland...	9,130	7,606	2,398	7,815	7,782	5,098
Delaware ..	8,724	7,684	2,601	7,152	8,829	4,404
Dutchess ..	47,339	41,929	12,294	41,655	46,709	26,320
Erie	224,554	132,343	45,679	196,233	129,444	123,358
Essex.....	7,893	6,379	2,363	6,717	8,278	3,784
Franklin ...	8,494	5,072	2,499	7,654	6,635	3,857
Fulton.....	9,779	7,881	3,214	8,400	9,137	5,120
Genesee ..	10,074	10,821	2,996	8,071	11,663	6,192
Greene....	8,251	8,712	2,790	6,924	9,390	4,689
Hamilton ..	1,228	1,841	492	963	2,038	793
Herkimer ..	11,910	10,085	4,235	10,880	12,052	6,866
Jefferson ..	16,783	12,362	4,561	13,380	14,227	9,461
Kings	432,232	81,406	15,031	411,183	133,344	33,014
Lewis	4,402	3,965	1,669	3,676	4,101	3,164
Livingston .	10,868	10,981	2,889	8,648	12,112	5,775
Madison...	11,832	11,324	3,379	10,099	11,293	7,391
Monroe ...	164,858	115,694	23,936	141,502	134,021	63,229
Montgomery	10,485	7,172	3,253	9,509	8,802	5,020
Nassau ...	303,587	196,820	36,122	282,593	246,881	77,097
New York ..	394,131	67,839	11,144	416,142	84,501	27,689
Niagara ...	44,203	31,438	12,564	35,649	30,401	30,126
Oneida	44,399	37,996	11,296	40,966	43,806	22,717
Onondaga .	100,190	73,771	17,602	90,645	77,642	45,175
Ontario....	19,156	17,237	4,391	16,064	18,995	9,571
Orange....	54,995	45,956	11,778	45,946	53,493	22,499
Orleans ...	6,233	6,865	1,986	4,927	7,468	4,275
Oswego ...	20,440	17,159	7,499	16,990	18,530	14,853
Otsego....	11,470	8,774	3,217	10,471	10,141	5,841
Putnam ...	16,173	17,452	4,032	14,048	18,934	8,011
Queens ...	372,925	107,650	22,288	349,520	157,561	46,014
Rensselaer	34,273	23,482	8,405	29,793	28,937	15,198
Richmond ..	64,684	52,207	8,968	56,901	70,707	19,678
Rockland ..	63,127	40,395	6,798	56,759	49,608	15,026
St. Lawrence.	21,798	10,827	5,309	18,197	13,901	9,758
Saratoga ..	39,832	34,337	10,141	33,011	36,917	19,091
Schenectady	35,404	22,106	7,865	32,335	26,258	14,838
Schoharie .	5,902	5,353	1,796	4,997	5,678	3,327
Schuyler...	3,303	3,134	1,037	2,859	3,226	2,051
Seneca	6,825	5,004	1,889	5,810	5,432	3,660
Steuben ...	14,481	17,710	5,496	12,043	19,761	9,378
Suffolk	261,828	182,510	52,209	220,811	229,467	112,973
Sullivan ...	15,052	9,321	3,453	13,717	11,396	6,336
Tioga	8,769	9,416	2,721	7,791	9,287	5,867
Tompkins ..	20,772	11,532	2,623	23,197	11,520	6,704
Ulster	35,852	26,212	9,246	32,886	29,223	17,952
Warren....	11,603	11,152	3,623	9,820	12,260	6,401
Washington	9,572	8,954	3,648	8,429	10,305	6,143
Wayne	15,145	15,837	4,619	11,866	18,019	9,188
Westchester	196,310	123,719	18,028	184,300	151,990	39,933
Wyoming ..	5,735	7,477	2,411	4,045	7,324	4,837
Yates	4,066	3,925	1,190	3,242	4,366	2,354
Totals	**3,756,177**	**1,933,492**	**503,458**	**3,444,450**	**2,346,649**	**1,090,721**

New York Vote Since 1948

1948, Truman, Dem., 2,557,642; Liberal, 222,562; total, 2,780,204; Dewey, Rep., 2,841,163; Wallace, Amer. Lab., 509,559; Thomas, Soc., 40,879; Teichert, Ind. Gov't., 2,729; Dobbs, Soc. Workers, 2,675.

1952, Eisenhower, Rep., 3,952,815; Stevenson, Dem., 2,687,890; Liberal, 416,711; total, 3,104,601; Hallinan, Amer. Lab., 64,211; Hoopes, Soc., 2,664; Dobbs, Soc. Workers, 2,212; Hass, Ind. Gov't., 1,560; scattering, 178; blank and void, 87,813.

1956, Eisenhower, Rep., 4,340,340; Stevenson, Dem., 2,458,212; Liberal, 292,557; total, 2,750,769; write-in votes for Andrews, 1,027; Werdel, 492; Hass, 150; Hoopes, 82; others, 476.

1960, Kennedy, Dem., 3,423,909; Liberal, 406,176; total, 3,830,085; Nixon, Rep., 3,446,419; Dobbs, Soc. Workers, 14,319; scattering, 256; blank and void, 88,896.

1964, Johnson, Dem., 4,913,156; Goldwater, Rep., 2,243,559; Hass, Soc. Labor, 6,085; DeBerry, Soc. Workers, 3,215; scattering, 188; blank and void, 151,383.

1968, Nixon, Rep., 3,007,932; Humphrey, Dem., 3,378,470; Wallace, 3d Party, 358,864; Blomen, Soc. Labor, 8,432; Halstead, Soc. Workers, 11,851; Gregory, Freedom and Peace, 24,517; blank, void, and scattering, 171,624.

1972, Nixon, Rep., 3,824,642; Cons., 368,136; McGovern, Dem., 2,767,956; Liberal, 183,128; Reed, Soc. Workers, 7,797; Fisher, Soc. Labor, 4,530; Hall, Com., 5,641; blank, void, or scattered, 161,641.

1976, Carter, Dem., 3,389,558; Ford, Rep., 3,100,791; MacBride, Libertarian, 12,197; Hall, Com., 10,270; Camejo, Soc. Workers, 6,996; LaRouche, U.S. Labor, 5,413; blank, void, or scattered, 143,037.

1980, Reagan, Rep., 2,893,831; Carter, Dem., 2,728,372; Anderson, Ind., 467,801; Clark, Libertarian, 52,648; McCormack, Right To Life, 24,159; Commoner, Citizens, 23,186; Hall, Com., 7,414; DeBerry, Soc. Workers, 2,068; Griswold, Workers World, 1,416; scattering, 1,064.

1984, Reagan, Rep., 3,664,763; Mondale, Dem., 3,119,609; Bergland, Libertarian, 11,949.

1988, Bush, Rep., 3,081,871; Dukakis, Dem., 3,347,882; Marra, Right to Life, 20,497; Fulani, New Alliance, 15,845.

1992, Clinton, Dem., 3,444,450; Bush, Rep., 2,346,649; Perot, Ind., 1,090,721; Warren, Soc. Workers, 15,472; Marrou, Libertarian, 13,451; Fulani, New Alliance, 11,318; Hagelin, Natural Law, 4,420.

1996, Clinton, Dem., 3,756,177; Dole, Rep., 1,933,492; Perot, Ind. (Ref.), 503,458; Nader, Green, 75,956; Phillips, Right to Life, 23,580; Browne, Libertarian, 12,220; Hagelin, Natural Law, 5,011; Harris, Soc. Workers, 2,762; Moorehead, Workers World, 3,473.

North Carolina

County	1996 Clinton (D)	Dole (R)	Perot (RF)	1992 Clinton (D)	Bush (R)	Perot (I)
Alamance..	15,814	22,461	3,395	15,521	20,637	6,444
Alexander .	3,955	6,748	1,004	4,849	6,764	2,002
Alleghany..	1,801	1,936	458	2,271	1,853	600
Anson	4,890	2,193	512	5,269	2,334	921
Ashe	3,825	5,203	865	4,624	5,200	1,220
Avery.....	1,586	3,870	655	1,755	3,895	1,123
Beaufort...	6,172	8,154	834	6,445	7,337	2,174
Bertie.....	4,202	1,745	316	4,382	1,756	600
Bladen	4,952	3,335	655	5,700	3,214	1,248
Brunswick .	10,041	10,065	1,815	10,117	8,833	3,349
Buncombe..	31,658	30,518	6,254	32,955	30,892	11,481
Burke.....	11,678	13,853	2,654	12,565	13,397	4,124
Cabarrus ..	14,447	23,035	3,626	13,513	21,281	6,251
Caldwell...	8,050	12,653	2,099	9,033	12,543	3,965
Camden...	1,186	1,074	293	1,153	1,039	479
Carteret...	7,566	11,721	1,467	8,028	10,334	3,401
Caswell ...	4,312	3,310	510	4,725	2,793	827
Catawba ..	15,601	26,898	3,629	16,334	25,466	7,523
Chatham ..	9,353	7,731	1,113	9,520	6,568	2,425
Cherokee..	3,129	3,883	785	3,686	4,021	1,040
Chowan....	2,239	1,659	359	2,136	1,661	700
Clay	1,462	1,769	387	1,600	1,890	465
Cleveland. .	12,728	13,474	1,931	13,037	13,650	3,784
Columbus ..	9,019	6,017	1,170	11,469	5,462	1,963
Craven....	10,317	13,264	1,528	9,998	11,575	3,679
Cumberland	32,739	29,804	3,776	30,291	27,139	6,792
Currituck ..	2,277	2,569	770	1,935	2,188	1,163
Dare.....	4,522	4,977	1,258	3,925	4,357	2,388
Davidson ..	13,593	24,797	3,698	16,462	24,869	8,324
Davie.....	3,525	8,141	915	3,675	6,796	1,903
Duplin	6,179	5,432	766	6,816	5,286	1,636
Durham ...	49,186	27,825	3,122	47,331	27,581	7,504
Edgecombe.	10,568	6,010	660	11,174	6,275	2,175

County	1996 Clinton (D)	Dole (R)	Perot (RF)	1992 Clinton (D)	Bush (R)	Perot (I)
Forsyth....	46,543	59,160	5,747	49,006	52,787	14,262
Franklin ...	6,448	5,648	891	6,517	4,669	2,062
Gaston....	19,458	33,149	3,921	19,121	34,714	7,490
Gates.....	2,155	1,072	307	2,206	1,158	466
Graham ...	1,210	1,801	270	1,551	1,919	403
Granville .	6,747	5,498	432	6,178	4,538	1,321
Greene....	2,224	2,689	280	2,768	2,180	780
Guilford ..	69,208	67,727	9,739	66,319	60,140	19,601
Halifax	9,551	5,700	816	9,960	5,769	2,047
Harnett....	8,767	11,596	1,287	8,473	9,751	2,684
Haywood ..	9,350	7,995	2,594	10,385	7,292	3,303
Henderson.	10,626	19,182	2,679	10,747	17,010	5,260
Hertford ...	4,856	1,823	356	4,609	2,208	846
Hoke	3,510	1,914	481	3,730	1,711	887
Hyde	1,109	782	143	1,206	740	340
Iredell.....	13,102	21,163	2,970	13,263	19,411	6,204
Jackson ...	5,211	4,244	970	5,753	4,275	1,516
Johnston ..	11,175	18,704	2,163	11,284	15,418	4,939
Jones.....	1,829	1,682	197	1,962	1,438	444
Lee.......	6,290	7,321	980	5,852	6,658	2,125
Lenoir.....	8,635	9,433	822	8,793	8,932	2,107
Lincoln....	7,721	11,439	1,619	8,150	11,018	3,142
McDowell..	4,553	6,407	1,275	5,309	6,090	1,881
Macon	4,209	5,267	1,121	4,624	4,797	1,829
Madison...	3,333	3,110	538	3,980	3,121	857
Martin	4,500	3,590	445	4,069	2,958	981
Mecklenburg	103,429	97,719	10,473	97,065	99,496	31,283
Mitchell ...	1,496	3,874	549	1,727	4,405	877
Montgomery	3,856	3,379	587	4,422	3,543	1,185
Moore	9,847	14,760	1,761	9,649	12,448	4,448
Nash	11,142	15,309	1,751	10,809	14,446	4,544
New Hanover..	22,839	27,889	3,615	20,291	24,338	7,401
Northampton	5,207	1,881	402	5,195	1,845	916
Onslow....	8,685	13,396	1,857	8,045	11,842	4,387
Orange....	28,674	15,053	1,534	28,595	13,009	5,535
Pamlico ...	2,204	2,270	297	2,229	1,929	809
Pasquotank	4,233	2,999	565	4,709	3,419	1,434
Pender....	5,409	5,538	945	5,825	4,857	1,725
Perquimans	2,069	1,561	369	1,818	1,429	624
Person....	4,540	4,883	591	4,323	4,460	1,431
Pitt	17,555	18,227	2,037	17,959	16,609	5,262
Polk	2,704	3,516	493	2,939	3,448	1,134
Randolph .	10,783	23,030	3,593	11,274	20,697	6,870
Richmond .	7,564	3,973	1,230	9,163	4,356	2,015
Robeson ..	17,361	8,146	2,105	19,378	7,777	3,277
Rockingham	12,096	14,255	2,528	13,880	12,678	4,671
Rowan	13,461	22,754	2,902	14,308	21,297	7,053
Rutherford .	7,162	9,792	1,585	7,855	9,748	2,695
Sampson ..	8,150	8,241	825	8,698	8,007	1,852
Scotland...	4,870	2,858	548	5,175	2,980	1,196
Stanly.....	7,131	11,446	1,690	7,735	11,030	2,855
Stokes	4,769	9,471	1,025	6,463	7,979	2,183
Surry	7,303	11,117	1,538	9,392	10,866	3,164
Swain	1,869	1,444	401	2,117	1,640	568
Transylvania	4,842	6,734	1,183	5,120	5,984	2,006
Tyrrell.....	908	488	112	928	553	189
Union	11,525	18,802	2,477	10,789	16,542	4,601
Vance.....	6,385	4,651	575	6,598	4,747	1,444
Wake	103,574	108,780	11,811	88,979	86,798	31,140
Warren	4,141	1,861	319	4,656	1,767	693
Washington	2,790	1,562	171	2,902	1,780	563
Watauga ..	7,349	8,146	1,415	8,262	7,899	3,007
Wayne	11,580	16,588	1,178	10,307	14,247	2,798
Wilkes	6,793	12,395	1,967	7,991	12,547	3,307
Wilson	9,779	10,518	1,100	10,105	10,176	2,630
Yadkin	2,927	8,439	913	3,913	7,311	1,725
Yancey	3,956	3,973	720	4,285	3,994	917
Totals.....	1,107,849	1,225,938	168,059	1,114,042	1,134,661	357,864

North Carolina Vote Since 1948

1948, Truman, Dem., 459,070; Dewey, Rep., 258,572; Thurmond, States' Rights, 69,652; Wallace, Prog., 3,915.

1952, Eisenhower, Rep., 558,107; Stevenson, Dem., 652,803.

1956, Eisenhower, Rep., 575,062; Stevenson, Dem., 590,530.

1960, Kennedy, Dem., 713,136; Nixon, Rep., 655,420.

1964, Johnson, Dem., 800,139; Goldwater, Rep., 624,844.

1968, Nixon, Rep., 627,192; Humphrey, Dem., 464,113; Wallace, 3d Party, 496,188.

1972, Nixon, Rep., 1,054,889; McGovern, Dem., 438,705; Schmitz, Amer., 25,018.

1976, Carter, Dem., 927,365; Ford, Rep., 741,960; Anderson, Amer., 5,607; MacBride, Libertarian, 2,219; LaRouche, U.S. Labor, 755.

1980, Reagan, Rep., 915,018; Carter, Dem., 875,635; Anderson, Ind., 52,800; Clark, Libertarian, 9,677; Commoner, Citizens, 2,287; DeBerry, Soc. Workers, 416.

1984, Reagan, Rep., 1,346,481; Mondale, Dem., 824,287; Bergland, Libertarian, 3,794.

1988, Bush, Rep., 1,237,258; Dukakis, Dem., 890,167; Fulani, New Alliance, 5,682; Paul, write-in, 1,263.

1992, Clinton, Dem., 1,114,042; Bush, Rep., 1,134,661; Perot, Ind., 357,864; Marrou, Libertarian, 5,171.

1996, Dole, Rep., 1,225,938; Clinton, Dem., 1,107,849; Perot, Ref., 168,059; Browne, Libertarian, 8,740; Hagelin, Natural Law, 2,771.

North Dakota

County	1996 Clinton (D)	Dole (R)	Perot (RF)	1992 Clinton (D)	Bush (R)	Perot (I)
Adams......	366	575	200	469	647	499
Barnes......	2,317	2,449	666	2,124	2,728	1,568
Benson......	1,059	850	252	1,126	874	610
Billings......	116	281	107	123	279	270
Bottineau....	1,280	1,682	536	1,266	1,787	1,036
Bowman	489	710	261	506	712	678
Burke......	416	483	176	458	551	506
Burleigh	10,679	15,464	3,535	8,940	16,484	6,780
Cass	21,693	24,238	4,116	18,077	25,312	9,513
Cavalier.....	941	1,188	326	866	1,527	723
Dickey	953	1,418	276	918	1,514	616
Divide	637	488	209	634	515	456
Dunn	587	830	304	667	784	637
Eddy	553	517	201	575	591	432
Emmons	544	1,148	441	595	1,047	774
Foster	664	801	265	565	803	556
Golden Valley	235	520	163	255	503	352
Grand Forks .	11,376	11,606	2,663	10,930	13,705	6,349
Grant	300	760	295	415	900	629
Griggs	670	731	162	647	773	330
Hettinger	418	765	238	465	854	500
Kidder	434	691	242	468	739	489
La Moure....	880	1,220	276	797	1,270	679
Logan	360	705	254	383	703	390
McHenry	1,096	1,187	453	1,173	1,321	886
McIntosh ...	470	1,005	295	450	1,134	454
McKenzie....	928	1,338	428	787	1,324	969
McLean	1,759	1,988	618	1,808	2,124	1,330
Mercer	1,300	1,953	764	1,323	2,274	1,378
Morton......	3,745	4,699	1,566	3,594	5,042	2,787
Mountrail ...	1,277	965	360	1,393	1,017	861
Nelson......	827	745	206	841	864	486
Oliver.......	333	499	183	306	503	407
Pembina	1,191	1,678	400	1,186	1,917	991
Pierce	671	1,017	270	761	1,099	554
Ramsey	2,123	2,077	549	2,008	2,516	1,507
Ransom	1,199	920	303	1,166	1,102	625
Renville	562	576	210	580	655	429
Richland	2,890	3,345	782	2,688	3,873	1,698
Rolette.....	2,299	823	448	2,002	895	660
Sargent	1,003	814	241	961	816	463
Sheridan ...	252	566	121	276	589	304
Sioux	393	207	82	463	264	244
Slope	123	260	60	145	226	162
Stark	3,095	4,086	1,456	3,003	4,491	3,123
Steele	620	486	115	598	503	267
Stutsman....	3,589	3,784	1,141	3,313	4,039	2,580
Towner	649	542	187	748	600	402
Traill........	1,822	1,820	380	1,638	2,019	875
Walsh	2,082	2,222	599	1,936	2,544	1,384
Ward	8,660	10,546	2,587	7,856	12,056	5,856
Wells	962	1,192	373	888	1,171	850
Williams.....	3,018	3,590	1,174	3,008	3,664	3,180
Totals	106,905	125,050	32,515	99,168	136,244	71,084

North Dakota Vote Since 1948

1948, Truman, Dem., 95,812; Dewey, Rep., 115,139; Wallace, Prog., 8,391; Thomas, Soc., 1,000; Thurmond, States' Rights, 374.

1952, Eisenhower, Rep., 191,712; Stevenson, Dem., 76,694; MacArthur, Christian Nationalist, 1,075; Hallinan, Prog., 344; Hamblen, Proh., 302.

1956, Eisenhower, Rep., 156,766; Stevenson, Dem., 96,742; Andrews, Amer., 483.

1960, Kennedy, Dem., 123,963; Nixon, Rep., 154,310; Dobbs, Soc. Workers, 158.

1964, Johnson, Dem., 149,784; Goldwater, Rep., 108,207; DeBerry, Soc. Workers, 224; Munn, Proh., 174.

1968, Nixon, Rep., 138,669; Humphrey, Dem., 94,769; Wallace, 3d Party, 14,244; Halstead, Soc. Workers, 128; Munn, Prohibition, 38; Troxell, Ind., 34.

1972, Nixon, Rep., 174,109; McGovern, Dem., 100,384; Jenness, Soc. Workers, 288; Hall, Com., 87; Schmitz, Amer., 5,646.

1976, Carter, Dem., 136,078; Ford, Rep., 153,470; Anderson, Amer., 3,698; McCarthy, Ind., 2,952; Maddox, Amer. Ind., 269; MacBride, Libertarian, 256; scattering, 371.

1980, Reagan, Rep., 193,695; Carter, Dem., 79,189; Anderson, Ind., 23,640; Clark, Libertarian, 3,743; Commoner, Libertarian, 429; McLain, Natl. People's League, 296; Greaves, Amer., 235; Hall, Com., 93; DeBerry, Soc. Workers, 89; McReynolds, Soc., 82; Bubar, Statesman, 54.

1984, Reagan, Rep., 200,336; Mondale, Dem., 104,429; Bergland, Libertarian, 703.

1988, Bush, Rep., 166,559; Dukakis, Dem., 127,739; Paul, Lib., 1,315; LaRouche, Natl. Econ. Recovery, 905.

1992, Clinton, Dem., 99,168; Bush, Rep., 136,244; Perot, Ind., 71,084.

1996, Dole, Rep., 125,050; Clinton, Dem., 106,905; Perot, Ref., 32,515; Browne, Libertarian, 847; Phillips, Ind., 745; Hagelin, Natural Law, 349.

	1996			1992		
	Clinton	Dole	Perot	Clinton	Bush	Perot
County	(D)	(R)	(RF)	(D)	(R)	(I)
Trumbull...	55,604	24,811	13,563	54,591	25,831	26,791
Tuscarawas.	15,244	13,388	5,682	14,787	13,179	8,785
Union.....	4,989	8,290	1,596	3,465	7,818	3,433
Van Wert..	4,453	6,999	1,487	3,822	7,227	3,102
Vinton....	2,350	1,673	728	2,308	1,975	1,050
Warren....	17,089	33,210	4,689	13,542	27,998	11,115
Washington	10,945	11,965	2,832	10,380	12,204	5,415
Wayne.....	14,850	19,628	5,771	13,953	18,350	9,482
Williams...	5,524	7,747	2,121	4,862	7,614	4,902
Wood.....	23,183	20,518	5,065	20,754	20,579	11,682
Wyandot ..	3,677	4,473	1,347	3,031	4,411	2,929
Totals.....	2,148,222	1,859,883	483,207	1,984,942	1,894,310	1,036,426

Ohio Vote Since 1948

1948, Truman, Dem., 1,452,791; Dewey, Rep., 1,445,684; Wallace, Prog., 37,596.

1952, Eisenhower, Rep., 2,100,391; Stevenson, Dem., 1,600,367.

1956, Eisenhower, Rep., 2,262,610; Stevenson, Dem., 1,439,655.

1960, Kennedy, Dem., 1,944,248; Nixon, Rep., 2,217,611.

1964, Johnson, Dem., 2,498,331; Goldwater, Rep., 1,470,865.

1968, Nixon, Rep., 1,791,014; Humphrey, Dem., 1,700,586; Wallace, 3d Party, 467,495; Gregory, 372; Munn, Proh., 19; Blomen, Soc. Labor, 120; Halstead, Soc. Workers, 69; Mitchell, Com., 23.

1972, Nixon, Rep., 2,441,827; McGovern, Dem., 1,558,889; Fisher, Soc. Labor, 7,107; Hall, Com., 6,437; Schmitz, Amer., 80,067; Wallace, Ind., 460.

1976, Carter, Dem., 2,011,621; Ford, Rep., 2,000,505; McCarthy, Ind., 58,258; Maddox, Amer. Ind., 15,529; MacBride, Libertarian, 8,961; Hall, Com., 7,817; Camejo, Soc. Workers, 4,717; LaRouche, U.S. Labor, 4,335; scattered, 130.

1980, Reagan, Rep., 2,206,545; Carter, Dem., 1,752,414; Anderson, Ind., 254,472; Clark, Libertarian, 49,033; Commoner, Citizens, 8,564; Hall, Com., 4,729; Congress, Ind., 4,029; Griswold, Workers World, 3,790; Bubar, Statesman, 27.

1984, Reagan, Rep., 2,678,559; Mondale, Dem., 1,825,440; Bergland, Libertarian, 5,886.

1988, Bush, Rep., 2,416,549; Dukakis, Dem., 1,939,629; Fulani, Ind., 12,017; Paul, Ind., 11,926.

1992, Clinton, Dem., 1,984,942; Bush, Rep., 1,894,310; Perot, Ind., 1,036,426; Marrou, Libertarian, 7,252; Fulani, New Alliance, 6,413; Gritz, Populist/America First, 4,699; Hagelin, Natural Law, 3,437; LaRouche, Ind., 2,446.

1996, Clinton, Dem., 2,148,222; Dole, Rep., 1,859,883; Perot, Ref., 483,207; Browne, Ind., 12,851; Moorehead, Ind., 10,813; Hagelin, Natural Law, 9,120; Phillips, Ind., 7,361.

Ohio

	1996			1992		
	Clinton	Dole	Perot	Clinton	Bush	Perot
County	(D)	(R)	(RF)	(D)	(R)	(I)
Adams	4,317	4,763	1,223	3,998	4,722	1,993
Allen......	15,529	24,325	3,799	13,777	25,322	8,131
Ashland ...	6,573	10,402	2,630	5,985	9,864	4,950
Ashtabula..	19,341	13,287	5,700	18,843	13,254	10,765
Athens....	13,418	7,154	2,777	13,423	7,184	5,074
Auglaize...	6,652	10,169	2,641	4,960	10,455	4,840
Belmont ...	17,705	8,213	4,452	18,527	8,614	6,142
Brown	6,318	6,970	1,941	5,540	5,912	3,676
Butler.....	43,690	67,023	10,540	39,682	63,375	27,527
Carroll	4,792	4,449	2,445	4,731	4,224	3,434
Champaign	5,990	6,568	2,219	5,201	7,004	3,992
Clark	27,890	22,297	7,083	26,692	24,011	12,571
Clermont ..	21,329	36,457	5,795	17,558	32,065	14,279
Clinton	5,303	7,504	1,588	4,638	7,290	3,402
Columbiana	20,716	15,386	7,127	19,765	15,016	12,611
Coshocton .	6,005	6,018	2,183	6,212	5,705	4,081
Crawford ..	7,449	8,730	3,072	6,351	8,618	5,764
Cuyahoga .	341,357	163,770	50,691	337,548	187,186	112,352
Darke.....	8,871	10,798	3,168	7,016	11,098	6,217
Defiance ..	6,343	7,469	1,929	5,735	7,195	4,187
Delaware ..	13,463	24,123	3,471	9,263	18,225	9,244
Erie	16,730	12,204	4,225	14,531	12,459	8,720
Fairfield ...	18,821	26,850	4,660	14,249	24,125	12,246
Fayette....	3,665	4,831	1,047	2,976	4,916	2,162
Franklin ...	192,795	178,412	25,400	176,656	186,324	79,049
Fulton.....	6,662	8,703	2,412	5,576	8,358	4,798
Gallia	5,386	5,135	1,839	5,350	5,776	2,549
Geauga ...	14,143	19,662	4,848	11,466	18,200	10,577
Greene....	25,082	30,677	5,246	20,139	27,651	11,459
Guernsey..	6,731	5,970	2,251	6,428	5,749	4,103
Hamilton ..	160,458	186,493	21,335	148,409	192,447	60,145
Hancock...	9,334	17,252	2,904	7,944	16,821	7,002
Hardin	4,930	5,506	1,365	4,364	5,851	2,867
Harrison...	3,721	2,310	1,302	3,830	2,289	1,679
Henry.....	4,762	6,385	1,550	3,933	6,196	3,178
Highland ..	5,837	7,102	1,629	4,866	7,020	3,315
Hocking ...	4,646	4,017	1,564	3,935	3,761	2,831
Holmes ...	2,531	5,213	1,276	1,969	5,079	1,945
Huron.....	8,858	8,750	3,338	7,930	9,480	6,751
Jackson ...	5,538	4,922	1,529	5,016	5,422	2,389
Jefferson ..	19,402	10,212	4,748	20,978	10,764	6,910
Knox	7,562	10,159	2,138	7,259	9,044	5,282
Lake......	43,186	40,974	12,507	37,682	40,766	26,878
Lawrence ..	11,595	8,832	3,232	12,325	10,044	4,536
Licking ...	22,624	28,276	6,516	18,898	26,918	13,806
Logan.....	6,397	8,325	2,264	4,889	9,364	4,472
Lorain.....	55,744	34,937	14,889	50,962	36,803	30,425
Lucas.....	104,911	58,120	17,282	99,989	63,297	38,108
Madison ...	5,072	6,871	1,368	3,998	6,865	3,170
Mahoning..	72,716	31,397	13,213	64,731	31,191	29,417
Marion	10,482	11,112	2,897	9,444	11,675	6,471
Medina....	23,727	26,120	8,700	18,995	24,090	17,290
Meigs.....	4,275	3,622	1,453	4,226	3,916	2,098
Mercer....	6,300	8,832	2,361	4,883	8,683	4,913
Miami.....	15,540	19,509	4,599	12,547	19,741	10,544
Monroe ...	3,914	1,856	1,128	4,235	1,823	1,505
Montgomery	115,416	95,391	18,298	108,017	104,751	47,854
Morgan ...	2,385	2,566	922	2,402	2,719	1,551
Morrow....	4,627	5,655	1,745	3,907	5,208	3,623
Muskingum	13,813	13,861	4,880	11,670	14,168	8,731
Noble	2,366	2,183	899	2,201	2,223	1,429
Ottawa ...	9,321	6,991	2,438	8,128	6,782	4,832
Paulding...	3,449	3,760	1,292	3,293	3,652	2,510
Perry	5,819	4,606	1,854	4,972	4,712	3,810
Pickaway ..	7,042	8,666	1,702	5,765	8,690	4,319
Pike	5,542	3,759	1,402	5,057	4,094	2,192
Portage ...	29,441	18,939	9,178	26,325	18,447	17,065
Preble	6,611	8,139	2,235	5,557	8,023	4,460
Putnam ...	4,972	9,294	1,767	3,962	9,338	3,648
Richland...	20,832	23,697	6,613	19,606	23,532	13,370
Ross	12,649	10,286	2,648	10,452	10,825	5,616
Sandusky..	11,547	10,033	3,617	9,878	10,772	6,682
Scioto.....	15,041	11,679	4,418	14,715	11,931	6,860
Seneca	10,044	9,713	3,498	9,280	9,763	6,967
Shelby	6,729	8,773	2,686	5,262	8,854	5,835
Stark	73,437	60,212	23,004	70,064	61,863	42,413
Summit ...	112,050	73,555	27,723	107,881	77,530	55,151

Oklahoma

	1996			1992		
	Clinton	Dole	Perot	Clinton	Bush	Perot
County	(D)	(R)	(RF)	(D)	(R)	(I)
Adair	2,792	2,956	751	2,645	2,994	914
Alfalfa.......	796	1,504	348	741	1,567	722
Atoka	2,281	1,542	532	2,336	1,561	1,255
Beaver......	515	1,893	199	580	1,699	565
Beckham	2,797	2,912	817	2,947	2,913	1,929
Blaine	1,832	2,127	563	1,564	2,209	1,258
Bryan.......	5,962	3,943	1,396	6,259	3,452	3,713
Caddo	4,844	3,422	1,358	4,861	3,664	2,911
Canadian....	8,977	18,139	3,297	7,215	16,756	8,985
Carter	6,979	6,769	1,997	7,171	5,947	5,188
Cherokee....	6,817	5,046	1,777	6,794	4,977	3,297
Choctaw	3,198	1,580	589	3,413	1,641	1,298
Cimarron ...	361	986	102	395	965	254
Cleveland....	26,038	36,457	6,785	24,404	35,561	20,352
Coal........	1,205	734	323	1,448	714	618
Comanche...	12,841	14,461	2,819	12,237	15,704	7,463
Cotton	1,258	1,042	381	1,314	910	853
Craig	2,649	2,058	758	2,780	2,106	1,316
Creek	9,674	9,861	2,837	9,118	10,055	5,984
Custer	4,027	4,723	1,101	3,540	5,362	2,741
Delaware	5,094	5,230	1,573	4,842	4,840	2,689
Dewey	816	1,179	292	845	1,244	684
Ellis	619	1,090	279	594	1,072	632
Garfield	7,504	11,712	2,523	6,720	13,095	5,559
Garvin	4,639	3,745	1,345	4,811	3,983	3,014
Grady.......	6,256	7,228	2,048	6,177	6,997	4,528
Grant	867	1,382	384	864	1,311	871
Greer	1,240	905	361	1,162	964	640
Harmon	729	448	143	783	496	326
Harper......	511	1,036	219	486	1,038	501
Haskell......	2,762	1,442	590	3,069	1,461	995
Hughes	2,748	1,510	730	2,850	1,522	1,158
Jackson	3,245	4,422	892	3,273	3,893	2,227
Jefferson	1,430	865	337	1,580	671	758
Johnston	1,998	1,229	532	2,096	1,191	1,040
Kay	6,882	9,741	2,785	6,643	9,115	6,984

County	1996 Clinton (D)	Dole (R)	Perot (RF)	1992 Clinton (D)	Bush (R)	Perot (I)
Kingfisher....	1,626	3,423	621	1,379	3,479	1,534
Kiowa......	1,973	1,638	510	2,143	1,635	1,114
Latimer......	2,222	1,189	578	2,606	1,212	1,049
Le Flore....	6,831	5,689	1,721	7,843	5,850	3,021
Lincoln.....	4,332	5,243	1,500	3,904	5,315	3,160
Logan......	4,854	5,949	1,410	4,453	6,071	3,239
Love......	1,675	1,224	385	1,708	922	1,033
McClain....	3,753	4,363	1,289	3,378	4,377	2,996
McCurtain ...	4,350	3,892	1,483	5,082	3,519	2,852
McIntosh	4,219	2,400	1,044	4,184	2,225	1,469
Major.......	900	2,188	410	731	2,154	857
Marshall.....	2,624	1,605	663	2,519	1,478	1,486
Mayes......	6,377	5,268	1,617	6,432	5,445	3,235
Murray.....	2,620	1,712	723	2,594	1,536	1,447
Muskogee....	12,963	8,974	3,163	13,619	8,782	5,454
Noble......	1,756	2,318	694	1,333	2,474	1,449
Nowata....	1,788	1,457	586	1,912	1,531	1,063
Okfuskee....	2,074	1,380	536	2,141	1,580	889
Oklahoma ...	80,438	120,429	18,411	76,271	126,788	56,139
Okmulgee....	7,555	4,246	1,487	7,767	4,586	3,013
Osage.....	7,342	5,827	1,938	6,894	5,891	4,477
Ottawa.....	5,844	4,127	1,496	6,304	4,141	2,721
Pawnee.....	2,663	2,560	756	2,612	2,675	1,656
Payne......	9,985	11,686	2,472	9,886	13,032	7,852
Pittsburg....	8,475	5,966	2,217	8,523	5,659	4,594
Pontotoc.....	6,470	5,366	1,712	6,350	5,206	3,916
Pottawatamia.	9,141	9,802	2,724	8,616	10,350	6,520
Pushmataha.	2,270	1,458	588	2,553	1,319	1,000
Roger Mills...	733	959	233	767	890	505
Rogers.....	9,544	12,883	3,022	8,257	12,455	7,101
Seminole....	4,225	2,935	1,041	4,624	3,253	2,330
Sequoyah....	5,665	4,733	1,673	6,092	4,925	2,486
Stephens....	7,248	8,144	2,312	7,644	7,085	5,692
Texas......	1,408	4,139	518	1,487	4,059	1,417
Tillman......	1,827	1,368	471	1,749	1,377	1,039
Tulsa......	76,924	111,243	18,201	71,165	117,465	49,760
Wagoner	7,749	9,392	2,357	7,041	9,053	5,381
Washington ..	6,732	11,605	2,255	6,593	11,342	5,664
Washita....	1,913	1,994	748	1,929	1,912	1,468
Woods.....	1,431	2,151	497	1,361	2,225	1,167
Woodward...	2,403	4,093	963	2,063	4,006	2,411
Totals	488,105	582,315	130,788	473,066	592,929	319,878

Oklahoma Vote Since 1948

1948, Truman, Dem., 452,782; Dewey, Rep., 268,817.

1952, Eisenhower, Rep., 518,045; Stevenson, Dem., 430,939.

1956, Eisenhower, Rep., 473,769; Stevenson, Dem., 385,581.

1960, Kennedy, Dem., 370,111; Nixon, Rep., 533,039.

1964, Johnson, Dem., 519,834; Goldwater, Rep., 412,665.

1968, Nixon, Rep., 449,697; Humphrey, Dem., 301,658; Wallace, 3d Party, 191,731.

1972, Nixon, Rep., 759,025; McGovern, Dem., 247,147; Schmitz, Amer., 23,728.

1976, Carter, Dem., 532,442; Ford, Rep., 545,708; McCarthy, Ind., 14,101.

1980, Reagan, Rep., 695,570; Carter, Dem., 402,026; Anderson, Ind., 38,284; Clark, Libertarian, 13,828.

1984, Reagan, Rep., 861,530; Mondale, Dem., 385,080; Bergland, Libertarian, 9,066.

1988, Bush, Rep., 678,367; Dukakis, Dem., 483,423; Paul, Lib., 6,261; Fulani, New Alliance, 2,985.

1992, Clinton, Dem., 473,066; Bush, Rep., 592,929; Perot, Ind., 319,878; Marrou, Libertarian, 4,486.

1996, Dole, Rep., 582,315; Clinton, Dem., 488,105; Perot, Ref., 130,788; Browne, Libertarian, 5,505.

Oregon

County	1996 Clinton (D)	Dole (R)	Perot (RF)	1992 Clinton (D)	Bush (R)	Perot (I)
Baker.......	2,547	3,975	900	2,395	2,862	2,191
Benton......	17,211	12,450	2,445	17,966	11,550	8,103
Clackamas...	67,709	59,443	12,304	60,310	53,724	39,776
Clatsop	7,732	5,334	1,582	7,700	4,683	4,316
Columbia....	9,275	6,205	2,330	8,298	5,227	5,670
Coos	12,171	10,886	3,460	12,072	9,284	7,989
Crook......	2,607	3,250	948	2,508	2,703	2,024
Curry......	4,202	4,790	1,560	3,841	3,809	3,310
Deschutes ...	17,151	21,135	5,306	15,693	15,655	12,293
Douglas....	15,250	21,855	4,465	14,137	19,011	12,377
Gilliam.....	485	398	143	374	377	283
Grant......	1,180	2,110	432	1,135	1,496	1,302
Harney.....	980	1,948	506	973	1,350	1,024
Hood River...	3,654	2,794	721	3,106	2,453	2,235
Jackson	29,230	33,771	7,470	29,146	28,704	18,633
Jefferson	2,555	2,634	813	2,161	1,962	1,741
Josephine ...	11,113	16,048	3,546	11,007	13,003	8,426
Klamath	7,207	12,116	2,538	7,918	11,864	6,636
Lake........	962	2,239	385	1,019	1,791	980
Lane........	69,461	48,253	11,498	74,083	41,789	34,906
Lincoln......	10,552	6,717	2,269	9,603	5,716	6,127
Linn......	17,041	18,331	4,773	15,399	16,461	13,256
Malheur.....	2,827	6,045	844	2,539	5,374	2,654
Marion.....	48,637	46,415	8,802	41,137	42,145	26,156
Morrow.....	1,426	1,381	455	1,174	1,187	1,089
Multnomah...	159,878	71,094	17,536	165,081	72,326	58,236
Polk.......	10,942	11,478	2,093	9,551	10,082	5,818
Sherman	444	476	126	362	424	326
Tillamook....	5,775	3,884	1,263	5,040	3,359	2,997
Umatilla	8,774	9,703	2,500	6,787	7,095	5,581
Union......	4,379	5,414	1,241	3,990	4,223	3,305
Wallowa	1,321	2,379	483	1,203	1,630	1,209
Wasco......	4,967	3,662	1,004	4,663	3,242	3,008
Washington ..	76,619	65,221	11,446	67,528	57,146	41,575
Wheeler.....	299	481	121	267	357	227
Yamhill......	13,078	13,900	2,913	11,148	11,693	8,312
Totals	649,641	538,152	121,221	621,314	475,757	354,091

Oregon Vote Since 1948

1948, Truman, Dem., 243,147; Dewey, Rep., 260,904; Wallace, Prog., 14,978; Thomas, Soc., 5,051.

1952, Eisenhower, Rep., 420,815; Stevenson, Dem., 270,579; Hallinan, Ind., 3,665.

1956, Eisenhower, Rep., 406,393; Stevenson, Dem., 329,204.

1960, Kennedy, Dem., 367,402; Nixon, Rep., 408,060.

1964, Johnson, Dem., 501,017; Goldwater, Rep., 282,779; write-in, 2,509.

1968, Nixon, Rep., 408,433; Humphrey, Dem., 358,866; Wallace, 3d Party, 49,683; write-in, McCarthy, 1,496; N. Rockefeller, 69; others, 1,075.

1972, Nixon, Rep., 486,686; McGovern, Dem., 392,760; Schmitz, Amer., 46,211; write-in, 2,289.

1976, Carter, Dem., 490,407; Ford, Rep., 492,120; McCarthy, Ind., 40,207; write-in, 7,142.

1980, Reagan, Rep., 571,044; Carter, Dem., 456,890; Anderson, Ind., 112,389; Clark, Libertarian, 25,838; Commoner, Citizens, 13,642; scattered, 1,713.

1984, Reagan, Rep., 658,700; Mondale, Dem., 536,479.

1988, Bush, Rep., 560,126; Dukakis, Dem., 616,206; Paul, Lib., 14,811; Fulani, Ind., 6,487.

1992, Clinton, Dem., 621,314; Bush, Rep., 475,757; Perot, Ind., 354,091; Marrou, Libertarian, 4,277; Fulani, New Alliance, 3,030.

1996, Clinton, Dem., 649,641; Dole, Rep., 538,152; Perot, Ref., 121,221; Nader, Pacific, 49,415; Browne, Libertarian, 8,903; Phillips, Taxpayers, 3,379; Hagelin, Natural Law, 2,798; Hollis, Soc., 1,922.

Pennsylvania

County	1996 Clinton (D)	Dole (R)	Perot (RF)	1992 Clinton (D)	Bush (R)	Perot (I)
Adams....	10,774	15,338	3,186	9,576	13,552	6,313
Allegheny..	284,480	204,067	42,309	324,004	183,035	103,470
Armstrong .	11,130	11,052	3,452	12,995	9,122	6,166
Beaver....	39,578	26,048	8,276	44,877	21,361	15,954
Bedford....	5,954	10,064	2,041	5,840	9,216	3,731
Berks.....	49,887	56,289	13,788	46,031	52,939	31,663
Blair......	15,036	21,282	4,014	14,857	21,447	8,284
Bradford...	7,736	10,393	2,712	6,903	10,221	5,452
Bucks.....	103,313	94,899	24,544	97,902	94,584	53,931
Butler.....	21,990	32,038	6,145	22,303	23,656	15,013
Cambria...	30,391	20,341	7,837	34,334	20,770	11,070
Cameron...	822	1,113	283	824	1,173	676
Carbon....	9,457	7,193	2,992	9,072	7,243	5,222
Centre	21,145	20,935	4,173	21,177	20,478	9,356
Chester	64,783	77,029	14,067	59,643	74,002	34,536
Clarion....	5,954	6,916	2,064	5,584	6,477	3,619
Clearfield ..	11,991	12,987	3,758	12,247	11,553	6,989
Clinton....	5,658	4,293	1,424	5,397	4,471	2,654
Columbia..	8,379	8,234	3,654	8,261	9,742	5,683
Crawford ..	12,943	14,659	3,519	12,813	14,112	7,392
Cumberland	28,749	43,943	5,669	26,635	43,447	14,344
Dauphin...	40,936	44,417	6,967	36,990	45,479	16,063
Delaware..	115,946	92,628	21,883	111,210	108,587	43,728
Elk.......	5,749	4,889	2,293	5,018	4,908	3,885
Erie......	57,508	39,884	10,386	56,381	39,283	21,510
Fayette....	26,359	14,019	5,722	30,577	12,820	10,162
Forest	964	902	325	890	801	448
Franklin ...	14,980	25,392	4,127	13,440	23,387	6,941
Fulton	1,620	2,665	554	1,588	2,558	869
Greene	7,620	4,002	2,052	8,438	3,482	3,186
Huntingdon .	5,285	7,324	1,813	5,153	7,249	3,273
Indiana ...	13,868	12,874	3,674	15,194	10,966	7,089
Jefferson ..	5,846	8,156	2,322	5,998	7,271	4,403
Juniata....	2,896	4,128	911	2,601	3,980	1,819
Lackawanna	46,377	26,930	8,189	46,054	33,443	15,667
Lancaster ..	49,120	92,875	11,601	44,255	88,447	26,807
Lawrence..	18,993	13,088	4,002	20,830	12,359	7,950

County	1996 Clinton (D)	Dole (R)	Perot (RF)	1992 Clinton (D)	Bush (R)	Perot (I)
Lebanon...	14,187	21,885	4,235	12,350	21,512	9,005
Lehigh	48,568	45,103	10,947	46,711	42,631	24,853
Luzerne ...	60,174	43,577	12,424	56,623	49,285	21,007
Lycoming ..	13,516	21,535	3,855	13,315	20,536	9,170
McKean ...	5,509	6,838	2,350	5,331	6,965	4,019
Mercer	23,003	17,213	5,108	23,264	16,081	10,277
Mifflin	5,327	6,888	1,392	4,946	6,300	3,382
Monroe ...	16,547	17,326	4,650	13,468	14,557	9,257
Montgomery	143,664	121,047	24,392	136,572	125,704	53,738
Montour ..	2,183	2,785	784	2,150	3,096	1,373
Northampton	43,959	35,726	9,848	42,203	34,429	20,234
North- umberland	13,418	13,551	5,173	12,814	15,057	7,782
Perry	4,611	8,156	1,609	4,086	7,871	3,334
Philadelphia	412,988	85,345	29,329	434,904	133,328	65,455
Pike	5,509	6,697	1,873	4,382	6,084	3,019
Potter	2,146	3,714	925	1,892	3,452	1,687
Schuylkill ..	24,860	22,920	8,471	23,679	25,780	13,398
Snyder....	3,405	6,742	1,451	2,952	6,934	2,686
Somerset ..	12,719	14,735	3,968	12,493	13,858	6,333
Sullivan ...	1,071	1,352	418	1,030	1,340	731
Susque- hanna ..	5,912	7,354	2,266	5,368	7,356	3,946
Tioga	4,961	7,382	1,993	4,868	7,823	3,804
Union	3,658	6,570	1,431	3,623	6,362	2,255
Venango ..	8,205	8,398	2,777	8,230	8,545	4,695
Warren ...	7,291	7,056	2,504	6,972	6,585	4,795
Washington	40,952	27,777	8,661	46,143	21,977	16,083
Wayne	5,928	8,077	2,126	4,817	8,184	3,727
Westmore- land ..	63,686	62,058	16,230	69,817	47,315	37,036
Wyoming ..	4,049	4,888	1,414	3,158	5,143	2,525
York	49,596	65,188	11,652	46,113	60,130	27,743
Totals	**2,215,819**	**1,801,169**	**430,984**	**2,239,164**	**1,791,841**	**902,667**

Pennsylvania Vote Since 1948

1948, Truman, Dem., 1,752,426; Dewey, Rep., 1,902,197; Wallace, Prog., 55,161; Thomas, Soc., 11,325; Watson, Proh., 10,338; Dobbs, Militant Workers, 2,133; Teichert, Ind. Gov., 1,461.

1952, Eisenhower, Rep., 2,415,789; Stevenson, Dem., 2,146,269; Hamblen, Proh., 8,771; Hallinan, Prog., 4,200; Hoopes, Soc., 2,684; Dobbs, Militant Workers, 1,502; Hass, Ind. Gov., 1,347; scattered, 155.

1956, Eisenhower, Rep., 2,585,252; Stevenson, Dem., 1,981,769; Hass, Soc. Labor, 7,447; Dobbs, Militant Workers, 2,035.

1960, Kennedy, Dem., 2,556,282; Nixon, Rep., 2,439,956; Hass, Soc. Labor, 7,185; Dobbs, Soc. Workers, 2,678; scattering, 440.

1964, Johnson, Dem., 3,130,954; Goldwater, Rep., 1,673,657; DeBerry, Soc. Workers, 10,456; Hass, Soc. Labor, 5,092; scattering, 2,531.

1968, Nixon, Rep., 2,090,017; Humphrey, Dem., 2,259,405; Wallace, 3d Party, 378,582; Blomen, Soc. Labor, 4,977; Halstead, Soc. Workers, 4,862; Gregory, Peace and Freedom, 7,821; others, 2,264.

1972, Nixon, Rep., 2,714,521; McGovern, Dem., 1,796,951; Schmitz, Amer., 70,593; Jenness, Soc. Workers, 4,639; Hall, Com., 2,686; others, 2,715.

1976, Carter, Dem., 2,328,677; Ford, Rep., 2,205,604; McCarthy, Ind., 50,584; Maddox, Constitution, 25,344; Camejo, Soc. Workers, 3,009; LaRouche, U.S. Labor, 2,744; Hall, Com., 1,891; others, 2,934.

1980, Reagan, Rep., 2,261,872; Carter, Dem., 1,937,540; Anderson, Ind., 292,921; Clark, Libertarian, 33,263; DeBerry, Soc. Workers, 20,291; Commoner, Consumer, 10,430; Hall, Com., 5,184.

1984, Reagan, Rep., 2,584,323; Mondale, Dem., 2,228,131; Bergland, Libertarian, 6,982.

1988, Bush, Rep., 2,300,087; Dukakis, Dem., 2,194,944; McCarthy, Consumer, 19,158; Paul, Lib., 12,051.

1992, Clinton, Dem., 2,239,164; Bush, Rep., 1,791,841; Perot, Ind., 902,667; Marrou, Libertarian, 21,477; Fulani, New Alliance, 4,661.

1996, Clinton, Dem., 2,215,819; Dole, Rep., 1,801,169; Perot, Ref., 430,984; Browne, Libertarian, 28,000; Phillips, Constitutional, 19,552; Hagelin, Natural Law, 5,783.

Rhode Island

City	1996 Clinton (D)	Dole (R)	Perot (RF)	1992 Clinton (D)	Bush (R)	Perot (I)
Cranston	20,901	9,098	3,457	18,589	12,450	8,331
East Providence .	12,846	4,199	1,971	11,701	5,843	4,661
Pawtucket ...	14,719	3,877	2,508	14,177	6,322	6,244
Providence...	29,450	7,068	2,733	32,536	11,519	7,816
Warwick.....	23,152	10,414	4,541	20,504	13,348	10,526
Other.......	131,982	70,027	28,513	115,792	82,119	67,467
Totals	**233,050**	**104,683**	**43,723**	**213,299**	**131,601**	**105,045**

Rhode Island Vote Since 1948

1948, Truman, Dem., 188,736; Dewey, Rep., 135,787; Wallace, Prog., 2,619; Thomas, Soc., 429; Teichert, Soc. Labor, 131.

1952, Eisenhower, Rep., 210,935; Stevenson, Dem., 203,293; Hallinan, Prog., 187; Hass, Soc. Labor, 83.

1956, Eisenhower, Rep., 225,819; Stevenson, Dem., 161,790.

1960, Kennedy, Dem., 258,032; Nixon, Rep., 147,502.

1964, Johnson, Dem., 315,463; Goldwater, Rep., 74,615.

1968, Nixon, Rep., 122,359; Humphrey, Dem., 246,518; Wallace, 3d Party, 15,678; Halstead, Soc. Workers, 383.

1972, Nixon, Rep., 220,383; McGovern, Dem., 194,645; Jenness, Soc. Workers, 729.

1976, Carter, Dem., 227,636; Ford, Rep., 181,249; MacBride, Libertarian, 715; Camejo, Soc. Workers, 462; Hall, Com., 334; Levin, Soc. Labor, 188.

1980, Reagan, Rep., 154,793; Carter, Dem., 198,342; Anderson, Ind., 59,819; Clark, Libertarian, 2,458; Hall, Com., 218; McReynolds, Soc., 170; DeBerry, Soc. Workers, 90; Griswold, Workers World, 77.

1984, Reagan, Rep., 212,080; Mondale, Dem., 197,106; Bergland, Libertarian, 277.

1988, Bush, Rep., 177,761; Dukakis, Dem., 225,123; Paul, Lib., 825; Fulani, New Alliance, 280.

1992, Clinton, Dem., 213,299; Bush, Rep., 131,601; Perot, Ind., 105,045; Fulani, New Alliance, 1,878.

1996, Clinton, Dem., 233,050; Dole, Rep., 104,683; Perot, Ref., 43,723; Nader, Green, 6,040; Browne, Libertarian, 1,109; Phillips, Taxpayers, 1,021; Hagelin, Natural Law, 435; Moorehead, Workers World, 186.

South Carolina

County	1996 Clinton (D)	Dole (R)	Perot (RF)	1992 Clinton (D)	Bush (R)	Perot (I)
Abbeville	3,493	3,054	537	3,968	3,317	1,036
Aiken	14,314	26,539	1,984	14,802	25,731	6,056
Allendale	2,222	941	87	2,159	1,049	212
Anderson	17,460	24,137	3,896	16,072	24,793	6,966
Bamberg	3,380	1,715	192	3,426	1,906	360
Barnwell	3,620	3,808	310	3,344	4,026	752
Beaufort....	15,764	17,575	1,838	11,466	14,735	4,966
Berkeley	13,358	17,691	1,922	12,533	18,048	4,632
Calhoun.....	2,716	2,520	316	2,770	2,418	564
Charleston...	43,571	48,675	3,514	40,095	47,403	10,354
Cherokee....	5,821	6,689	1,064	5,453	6,887	2,186
Chester	5,108	3,157	758	5,458	3,451	1,350
Chesterfield ..	5,734	4,028	768	5,691	4,183	1,315
Clarendon ...	5,930	3,841	395	6,033	4,147	744
Colleton.....	5,329	4,462	550	5,455	4,545	1,245
Darlington ...	8,943	8,220	898	9,090	8,912	1,863
Dillon.......	3,992	2,774	275	4,953	3,575	831
Dorchester...	9,931	15,283	1,591	9,160	15,004	3,648
Edgefield	3,576	3,640	244	3,433	3,339	596
Fairfield	4,719	2,414	284	4,867	2,518	652
Florence	15,004	18,490	1,563	15,569	19,802	3,499
Georgetown..	8,298	7,023	950	7,494	6,870	1,840
Greenville ...	41,605	71,210	6,761	34,651	65,066	13,699
Greenwood ..	8,193	8,865	985	7,621	9,079	2,101
Hampton	4,828	2,111	344	4,332	2,402	564
Horry	23,722	26,159	4,446	18,896	23,489	8,472
Jasper	4,053	2,024	348	3,453	1,725	549
Kershaw.....	6,764	8,513	996	6,585	8,499	2,150
Lancaster....	8,752	7,544	1,598	8,307	7,757	2,563
Laurens	7,055	8,057	1,341	6,638	8,347	2,157
Lee	3,588	1,973	320	4,454	2,730	611
Lexington....	18,907	39,658	3,703	18,312	41,759	8,652
McCormick ..	1,858	1,104	148	1,846	899	295
Marion	6,359	3,595	356	5,843	3,647	822
Marlboro	5,348	2,148	494	5,111	2,526	895
Newberry ...	4,804	5,670	682	4,896	5,980	1,393
Oconee	7,398	10,503	1,961	6,617	10,379	3,405
Orangeburg..	18,610	10,494	1,112	18,440	11,328	2,383
Pickens	8,369	17,151	2,211	8,275	17,008	4,128
Richland	52,222	39,092	3,158	53,648	43,744	7,918
Saluda......	2,486	2,825	371	2,393	2,968	833
Spartanburg..	26,814	35,972	3,885	25,488	37,707	8,900
Sumter......	12,198	12,080	933	11,852	12,576	2,062
Union.......	5,407	3,855	749	4,644	4,647	1,371
Williamsburg .	6,987	3,957	375	8,077	5,289	864
York........	16,873	22,222	3,173	15,844	21,297	6,418
Totals	**506,283**	**573,458**	**64,386**	**479,514**	**577,507**	**138,872**

South Carolina Vote Since 1948

1948, Thurmond, States' Rights, 102,607; Truman, Dem., 34,423; Dewey, Rep., 5,386; Wallace, Prog., 154; Thomas, Soc., 1.

1952, Eisenhower ran on two tickets. Under state law vote cast for two Eisenhower slates of electors could not be combined. Eisenhower, Ind., 158,289; Rep., 9,793; total, 168,082; Stevenson, Dem., 173,004; Hamblen, Proh., 1.

1956, Eisenhower, Rep., 75,700; Stevenson, Dem., 136,372; Byrd, Ind., 88,509; Andrews, Ind., 2.

1960, Kennedy, Dem., 198,129; Nixon, Rep., 188,558; write-in, 1.

1964, Johnson, Dem., 215,700; Goldwater, Rep., 309,048; write-ins: Nixon, 1, Wallace, 5; Powell, 1; Thurmond, 1.

1968, Nixon, Rep., 254,062; Humphrey, Dem., 197,486; Wallace, 3d Party, 215,430.

1972, Nixon, Rep., 477,044; McGovern, Dem., 184,559; United Citizens, 2,265; Schmitz, Amer., 10,075; write-in, 17.

1976, Carter, Dem., 450,807; Ford, Rep., 346,149; Anderson, Amer., 2,996; Maddox, Amer. Ind., 1,950; write-in, 681.

1980, Reagan, Rep., 439,277; Carter, Dem., 428,220; Anderson, Ind., 13,868; Clark, Libertarian, 4,807; Rarick, Amer. Ind., 2,086.

1984, Reagan, Rep., 615,539; Mondale, Dem., 344,459; Bergland, Libertarian, 4,359.

1988, Bush, Rep., 606,443; Dukakis, Dem., 370,554; Paul, Lib., 4,935; Fulani, United Citizens, 4,077.

1992, Clinton, Dem., 479,514; Bush, Rep., 577,507; Perot, Ind., 138,872; Marrou, Libertarian, 2,719; Phillips, U.S. Taxpayers, 2,680; Fulani, New Alliance, 1,235.

1996, Dole, Rep., 573,458; Clinton, Dem., 506,283; Perot, Ref./Patriot, 64,386; Browne, Libertarian, 4,271; Phillips, Taxpayers, 2,043; Hagelin, Natural Law, 1,248.

South Dakota

| | 1996 | | | 1992 | | |
County	Clinton (D)	Dole (R)	Perot (RF)	Clinton (D)	Bush (R)	Perot (I)
Aurora	664	709	199	680	594	435
Beadle	3,984	3,670	842	3,925	3,363	1,819
Bennett	507	539	93	413	556	221
Bon Homme	1,569	1,428	391	1,294	1,212	836
Brookings	5,105	5,112	979	4,645	4,698	2,614
Brown	7,913	6,801	1,622	7,521	6,665	3,812
Brule	1,091	981	281	1,060	908	687
Buffalo	465	134	35	282	137	72
Butte	1,132	1,947	541	973	1,674	1,039
Campbell	202	623	140	222	574	252
Chas. Mix.	1,913	1,711	390	1,639	1,570	886
Clark	956	998	272	799	803	761
Clay	2,980	2,008	505	2,826	1,869	1,303
Codington	4,722	4,995	1,239	3,701	3,943	3,262
Corson	539	533	216	444	483	321
Custer	1,122	1,740	418	1,078	1,422	845
Davison	3,364	3,371	737	3,285	3,111	1,706
Day	1,840	1,282	395	1,578	1,161	973
Deuel	1,090	955	275	880	778	761
Dewey	1,114	657	195	766	642	340
Douglas	524	1,210	161	481	1,175	403
Edmunds	973	1,055	263	894	944	415
Fall River	1,357	1,636	417	1,416	1,533	792
Faulk	493	726	165	488	658	281
Grant	1,805	1,782	471	1,484	1,595	1,018
Gregory	923	1,208	286	879	1,027	688
Haakon	284	887	110	209	860	245
Hamlin	1,101	1,352	285	826	1,133	774
Hand	803	1,187	250	785	1,130	624
Hanson	541	801	170	566	522	341
Harding	151	537	90	139	515	225
Hughes	2,788	4,469	531	2,578	4,325	1,160
Hutchinson	1,285	2,177	409	1,211	2,002	920
Hyde	309	493	95	301	440	211
Jackson/ Washabaugh	423	646	88	351	627	184
Jerauld	656	530	151	600	518	346
Jones	184	463	75	166	454	154
Kingsbury	1,357	1,297	320	1,267	1,113	744
Lake	2,526	1,966	593	2,388	1,890	1,299
Lawrence	3,568	4,430	1,308	3,157	3,770	2,673
Lincoln	3,643	4,201	682	2,943	3,365	1,593
Lyman	646	726	130	486	669	311
McCook	1,166	1,292	245	1,167	1,177	617
McPherson	463	1,080	182	478	945	322
Marshall	1,185	861	189	1,056	810	427
Meade	2,960	4,984	1,133	2,694	4,724	2,611
Mellette	302	417	67	277	417	140
Miner	739	571	170	698	543	332
Minnehaha	29,790	27,432	4,425	27,016	25,081	11,496
Moody	1,443	1,024	284	1,473	898	715
Pennington	12,784	19,293	3,149	11,106	18,052	8,358
Perkins	460	983	225	566	872	541
Potter	534	979	181	493	901	375
Roberts	2,186	1,646	474	1,716	1,437	954
Sanborn	647	630	151	632	595	376
Shannon	1,926	253	87	1,267	225	137
Spink	1,636	1,651	360	1,732	1,527	839

| | 1996 | | | 1992 | | |
County	Clinton (D)	Dole (R)	Perot (RF)	Clinton (D)	Bush (R)	Perot (I)
Stanley	454	795	121	427	719	240
Sully	321	592	106	273	565	167
Todd	1,380	482	108	915	456	246
Tripp	1,088	1,680	337	1,046	1,459	848
Turner	1,682	1,970	385	1,507	1,906	867
Union	2,378	2,234	555	2,210	1,784	1,085
Walworth	939	1,461	366	829	1,439	628
Yankton	3,775	3,885	1,073	3,404	3,430	2,511
Ziebach	483	375	62	280	328	117
Totals	139,333	150,543	31,250	124,888	136,718	73,295

South Dakota Vote Since 1948

1948, Truman, Dem., 117,653; Dewey, Rep., 129,651; Wallace, Prog., 2,801.

1952, Eisenhower, Rep., 203,857; Stevenson, Dem., 90,426.

1956, Eisenhower, Rep., 171,569; Stevenson, Dem., 122,288.

1960, Kennedy, Dem., 128,070; Nixon, Rep., 178,417.

1964, Johnson, Dem., 163,010; Goldwater, Rep., 130,108.

1968, Nixon, Rep., 149,841; Humphrey, Dem., 118,023; Wallace, 3d Party, 13,400.

1972, Nixon, Rep., 166,476; McGovern, Dem., 139,945; Jenness, Soc. Workers, 994.

1976, Carter, Dem., 147,068; Ford, Rep., 151,505; MacBride, Libertarian, 1,619; Hall, Com., 318; Camejo, Soc. Workers, 168.

1980, Reagan, Rep., 198,343; Carter, Dem., 103,855; Anderson, Ind., 21,431; Clark, Libertarian, 3,824; Pulley, Soc. Workers, 250.

1984, Reagan, Rep., 200,267; Mondale, Dem., 116,113.

1988, Bush, Rep., 165,415; Dukakis, Dem., 145,560; Paul, Lib., 1,060; Fulani, New Alliance, 730.

1992, Clinton, Dem., 124,888; Bush, Rep., 136,718; Perot, Ind., 73,295.

1996, Dole, Rep., 150,543; Clinton, Dem., 139,333; Perot, Ref., 31,250; Browne, Libertarian, 1,472; Phillips, Taxpayers, 912; Hagelin, Natural Law, 316.

Tennessee

| | 1996 | | | 1992 | | |
County	Clinton (D)	Dole (R)	Perot (RF)	Clinton (D)	Bush (R)	Perot (I)
Anderson	13,457	11,943	1,817	13,482	11,838	3,149
Bedford	5,735	4,634	823	5,978	3,836	1,541
Benton	4,341	2,395	663	3,896	1,625	559
Bledsoe	1,621	1,626	251	1,884	1,776	352
Blount	14,687	19,310	2,556	14,655	18,415	4,468
Bradley	9,095	15,478	1,856	9,889	16,528	3,212
Campbell	6,122	4,393	785	6,756	4,897	1,240
Cannon	2,318	1,468	361	2,593	1,229	495
Carroll	4,912	4,206	697	5,741	4,842	1,139
Carter	6,218	10,540	1,383	6,502	10,712	1,898
Cheatham	4,883	4,283	705	4,817	3,496	1,433
Chester	1,922	2,746	203	2,317	2,834	439
Claiborne	3,861	4,023	727	4,509	4,065	860
Clay	1,559	1,108	316	1,922	1,072	223
Cocke	3,326	4,481	798	3,495	5,298	1,124
Coffee	7,951	7,038	1,205	8,534	6,047	2,420
Crockett	2,256	1,872	201	2,657	2,180	507
Cumberland	6,676	8,096	1,399	6,393	7,116	2,200
Davidson	110,805	78,453	9,018	106,355	76,567	20,184
Decatur	2,262	1,712	229	2,633	1,667	351
De Kalb	3,213	1,696	342	4,382	1,714	608
Dickson	7,458	5,283	996	7,863	4,450	1,730
Dyer	5,602	5,059	676	5,845	5,668	1,241
Fayette	4,655	4,406	416	4,211	3,713	657
Fentress	2,332	2,307	386	2,730	2,391	606
Franklin	6,929	5,296	1,057	7,773	4,507	1,837
Gibson	8,851	6,614	891	9,555	7,161	1,536
Giles	4,948	3,269	733	5,601	2,827	1,309
Grainger	2,162	2,875	382	2,242	2,772	513
Greene	6,885	9,779	1,604	7,857	9,912	2,930
Grundy	2,596	1,094	326	2,997	1,004	366
Hamblen	7,006	9,797	1,106	7,114	8,898	1,760
Hamilton	48,008	55,205	6,699	46,770	53,476	14,400
Hancock	760	1,259	116	1,000	1,274	151
Hardeman	4,859	2,961	346	4,832	3,122	594
Hardin	3,508	3,980	594	3,922	3,875	734
Hawkins	6,367	8,164	1,282	6,623	7,758	1,847
Haywood	3,565	2,293	154	3,511	2,518	331
Henderson	2,841	4,002	408	3,502	4,719	785
Henry	6,153	4,272	992	6,797	3,661	1,588
Hickman	3,917	2,002	460	4,093	1,820	795
Houston	1,868	742	182	2,012	648	280
Humphreys	3,675	1,892	423	3,875	1,641	609
Jackson	2,889	944	289	3,208	708	332
Jefferson	4,688	6,446	882	4,740	6,184	1,385
Johnson	1,698	3,137	491	1,781	3,170	574
Knox	61,158	70,761	6,402	59,702	66,607	15,669
Lake	1,273	589	110	1,449	680	151
Lauderdale	4,349	2,481	308	4,452	2,928	561
Lawrence	6,188	6,115	973	6,816	5,608	1,403
Lewis	1,971	1,298	316	2,491	1,218	434

County	1996 Clinton (D)	Dole (R)	Perot (RF)	1992 Clinton (D)	Bush (R)	Perot (I)
Lincoln	4,361	4,551	761	5,063	3,814	1,371
Loudon	5,552	7,097	889	5,414	6,444	1,602
McMinn	5,987	7,655	1,033	6,682	7,453	1,812
McNairy	4,050	3,960	519	4,691	4,093	774
Macon	2,240	2,481	421	2,961	2,299	443
Madison	13,577	14,908	968	13,629	14,869	2,634
Marion	5,194	3,166	768	5,589	3,262	1,186
Marshall	4,447	2,781	603	4,491	2,516	1,050
Maury	10,367	8,737	1,366	9,997	7,440	2,821
Meigs	1,476	1,228	245	1,673	1,355	453
Monroe	4,872	5,257	713	5,384	6,025	936
Montgomery	16,498	15,133	1,781	14,507	13,011	3,753
Moore	935	846	177	1,151	661	327
Morgan	2,767	2,070	446	3,190	2,306	658
Obion	6,226	4,310	932	6,497	4,812	1,494
Overton	3,800	1,756	431	4,489	1,657	468
Perry	1,444	747	178	1,889	708	317
Pickett	901	1,046	116	1,144	1,094	121
Polk	2,450	1,910	377	2,583	1,584	419
Putnam	10,047	9,093	1,487	10,858	7,998	2,473
Rhea	3,969	4,476	694	4,289	4,860	1,163
Roane	9,744	9,044	1,438	9,812	8,719	2,396
Robertson	8,465	6,685	993	8,498	5,271	1,978
Rutherford	22,815	24,565	3,787	21,084	18,877	7,005
Scott	2,506	2,646	431	2,730	3,011	643
Sequatchie	1,598	1,391	288	1,754	1,381	405
Sevier	7,136	11,847	1,650	6,719	11,714	2,760
Shelby	179,663	136,315	8,307	191,322	153,310	20,223
Smith	3,812	1,857	346	5,061	1,482	486
Stewart	2,962	1,306	386	2,779	1,046	487
Sullivan	20,571	29,296	3,555	20,935	28,801	6,730
Sumner	19,205	20,863	2,783	19,387	17,401	5,177
Tipton	6,596	7,585	799	5,652	6,757	1,279
Trousdale	1,615	683	190	1,846	565	243
Unicoi	2,131	3,122	447	2,375	3,344	709
Union	2,421	2,253	385	2,478	2,274	580
Van Buren	1,010	504	128	1,329	555	191
Warren	6,389	4,226	917	7,189	3,704	1,415
Washington	13,259	18,960	2,237	13,071	18,206	4,002
Wayne	1,574	2,715	323	1,868	2,955	424
Weakley	5,657	4,622	873	5,691	4,800	1,355
White	3,592	2,498	505	4,102	2,118	821
Williamson	15,231	27,699	2,071	13,053	22,015	5,026
Wilson	13,655	13,817	1,841	13,861	12,061	3,848
Totals	**909,146**	**863,530**	**105,918**	**933,521**	**841,300**	**199,968**

Tennessee Vote Since 1948

1948, Truman, Dem., 270,402; Dewey, Rep., 202,914; Thurmond, States' Rights, 73,815; Wallace, Prog., 1,864; Thomas, Soc., 1,288.

1952, Eisenhower, Rep., 446,147; Stevenson, Dem., 443,710; Hamblen, Proh., 1,432; Hallinan, Prog., 885; MacArthur, Christian Nationalist, 379.

1956, Eisenhower, Rep., 462,288; Stevenson, Dem., 456,507; Andrews, Ind., 19,820; Holtwick, Proh., 789.

1960, Kennedy, Dem., 481,453; Nixon, Rep., 556,577; Faubus, States' Rights, 11,304; Decker, Proh., 2,458.

1964, Johnson, Dem., 635,047; Goldwater, Rep., 508,965; write-in, 34.

1968, Nixon, Rep., 472,592; Humphrey, Dem., 351,233; Wallace, 3d Party, 424,792.

1972, Nixon, Rep., 813,147; McGovern, Dem., 357,293; Schmitz, Amer., 30,373; write-in, 369.

1976, Carter, Dem., 825,879; Ford, Rep., 633,969; Anderson, Amer., 5,769; McCarthy, Ind., 5,004; Maddox, Amer. Ind., 2,303; MacBride, Libertarian, 1,375; Hall, Com., 547; LaRouche, U.S. Labor, 512; Bubar, Proh., 442; Miller, Ind., 316; write-in, 230.

1980, Reagan, Rep., 787,761; Carter, Dem., 783,051; Anderson, Ind., 35,991; Clark, Libertarian, 7,116; Commoner, Citizens, 1,112; Bubar, Statesman, 521; McReynolds, Soc., 519; Hall, Com., 503; DeBerry, Soc. Workers, 490; Griswold, Workers World, 400; write-ins, 152.

1984, Reagan, Rep., 990,212; Mondale, Dem., 711,714; Bergland, Libertarian, 3,072.

1988, Bush, Rep., 947,233; Dukakis, Dem., 679,794; Paul, Ind., 2,041; Duke, Ind., 1,807.

1992, Clinton, Dem., 933,521; Bush, Rep., 841,300; Perot, Ind., 199,968; Marrou, Libertarian, 1,847.

1996, Clinton, Dem., 909,146; Dole, Rep., 863,530; Perot, Ind. (Ref.), 105,918; Nader, Ind., 6,427; Browne, Ind., 5,020; Phillips, Ind., 1,818; Collins, Ind., 688; Hagelin, Ind., 636; Michael, Ind., 408; Dodge, Ind., 324.

Texas

County	1996 Clinton (D)	Dole (R)	Perot (RF)	1992 Clinton (D)	Bush (R)	Perot (I)
Anderson	5,693	6,458	1,170	5,322	5,598	3,519
Andrews	1,181	2,360	431	1,081	2,266	875
Angelina	11,346	11,789	2,160	10,318	9,722	6,204
Aransas	2,964	3,769	655	2,246	2,826	1,676
Archer	1,235	1,974	437	1,284	1,560	1,106
Armstrong	272	582	75	278	561	187
Atascosa	4,259	4,102	813	3,766	3,806	2,035
Austin	2,719	4,669	577	2,278	4,015	1,585
Bailey	706	1,246	109	677	1,308	376
Bandera	1,383	3,700	520	1,059	2,674	1,537
Bastrop	6,773	6,323	1,342	6,252	4,980	3,240
Baylor	955	860	262	990	611	529
Bee	4,561	3,611	539	4,083	3,633	1,367
Bell	22,638	30,348	3,666	18,684	24,936	11,026
Bexar	180,308	161,619	17,822	172,513	168,816	72,110
Blanco	1,028	1,919	330	891	1,370	830
Borden	93	194	45	106	184	87
Bosque	2,427	2,840	739	2,173	2,300	1,999
Bowie	13,657	12,750	2,760	11,825	11,776	6,659
Brazoria	22,959	36,392	5,869	21,861	30,384	18,954
Brazos	13,968	22,082	2,215	14,819	23,943	10,372
Brewster	1,643	1,438	299	1,383	1,127	712
Briscoe	408	416	65	430	360	164
Brooks	2,945	413	108	2,856	585	318
Brown	4,138	6,524	1,081	4,264	5,313	3,034
Burleson	2,419	2,174	347	2,511	2,013	1,179
Burnet	4,123	5,744	1,108	3,638	4,272	2,865
Caldwell	3,961	3,239	545	3,794	2,749	1,776
Calhoun	2,753	2,832	507	2,550	2,640	1,579
Callahan	1,666	2,480	534	1,694	2,134	1,452
Cameron	34,891	18,434	2,760	29,435	20,123	9,286
Camp	1,912	1,488	252	1,938	1,219	821
Carson	742	1,742	227	825	1,647	578
Cass	5,691	4,066	1,038	5,476	3,999	2,168
Castro	1,107	1,231	144	1,113	1,307	485
Chambers	2,876	4,101	818	2,832	3,398	2,122
Cherokee	5,185	6,483	971	5,003	5,847	3,273
Childress	719	1,072	165	881	1,033	421
Clay	1,690	1,997	465	1,919	1,586	1,397
Cochran	541	667	127	454	750	255
Coke	595	790	157	580	640	393
Coleman	1,488	1,793	349	1,579	1,462	1,095
Collin	37,854	83,750	10,443	24,508	60,514	43,287
Collingsworth	581	729	118	635	697	265
Colorado	2,795	3,381	574	2,442	3,286	1,421
Comal	7,132	16,763	1,903	6,312	12,651	5,841
Comanche	2,138	2,123	511	2,296	1,666	1,281
Concho	434	488	107	489	414	329
Cooke	3,782	7,320	1,150	3,105	5,299	4,658
Coryell	5,300	7,143	1,443	4,157	6,144	3,974
Cottle	404	331	77	542	245	235
Crane	616	984	201	514	918	412
Crockett	684	714	147	653	623	368
Crosby	1,122	968	189	1,010	1,006	313
Culberson	804	329	99	424	251	171
Dallam	483	970	170	434	922	325
Dallas	255,766	260,058	36,759	231,412	256,007	170,571
Dawson	1,612	2,319	232	1,639	2,691	518
Deaf Smith	1,655	3,051	310	1,642	3,137	772
Delta	849	744	146	864	599	551
Denton	36,138	65,313	9,294	27,891	48,492	39,653
DeWitt	2,074	3,577	483	2,127	3,238	1,346
Dickens	509	421	117	536	373	250
Dimmit	2,242	604	128	3,172	844	361
Donley	495	988	97	578	893	260
Duval	3,958	543	136	4,006	698	326
Eastland	2,594	3,272	705	2,738	2,830	1,698
Ector	12,017	17,746	2,511	11,130	18,161	6,668
Edwards	437	511	60	254	460	171
Ellis	10,832	16,046	2,750	9,537	13,564	10,303
El Paso	83,964	43,255	6,300	07,715	47,224	19,738
Erath	3,664	4,750	1,134	3,531	3,835	3,046
Falls	3,256	2,260	479	2,761	1,826	1,185
Fannin	4,276	3,495	980	4,164	2,510	2,919
Fayette	3,119	4,195	708	2,923	3,789	2,088
Fisher	1,142	537	170	1,242	539	442
Floyd	986	1,530	126	947	1,676	385
Foard	355	166	52	435	207	152
Fort Bend	38,163	49,945	4,363	29,992	41,039	16,853
Franklin	1,484	1,575	386	1,338	1,058	942
Freestone	2,630	2,888	568	2,445	2,316	1,596
Frio	2,593	1,225	253	2,377	1,275	654
Gaines	1,012	1,812	353	1,095	2,138	696
Galveston	38,458	35,251	5,897	38,623	31,303	20,103
Garza	703	946	103	558	982	345
Gillespie	1,655	5,867	542	1,600	4,712	2,018
Glasscock	70	382	30	100	379	93
Goliad	1,135	1,335	148	1,069	1,236	521
Gonzales	2,110	2,687	354	2,006	2,502	1,018
Gray	2,114	6,102	568	2,426	6,105	1,810
Grayson	14,338	17,169	3,745	12,547	12,322	13,327
Gregg	13,659	21,611	2,079	12,797	20,542	8,437
Grimes	2,584	2,564	538	2,594	2,402	1,213
Guadalupe	8,079	14,254	1,811	6,567	10,818	5,618
Hale	3,204	5,905	605	2,761	6,098	1,357
Hall	750	626	94	819	631	263
Hamilton	1,200	1,493	323	1,100	1,232	921
Hansford	343	1,493	105	345	1,660	398

County	1996 Clinton (D)	1996 Dole (R)	1996 Perot (RF)	1992 Clinton (D)	1992 Bush (R)	1992 Perot (I)
Hardeman .	750	610	168	954	614	362
Hardin	7,179	8,529	2,112	6,753	5,885	4,129
Harris	386,726	421,462	42,364	360,171	406,778	172,922
Harrison...	10,307	9,835	1,427	9,538	8,733	4,371
Hartley ...	463	1,242	101	406	1,081	308
Haskell....	1,374	966	225	1,438	852	562
Hays	11,580	12,865	1,990	10,842	10,008	6,252
Hemphill...	344	986	104	479	989	232
Henderson.	10,085	10,345	2,274	9,105	8,368	6,746
Hidalgo ...	56,335	24,437	3,536	51,205	26,976	9,757
Hill	3,988	4,401	1,052	3,929	3,669	2,752
Hockley ...	2,170	4,230	519	2,301	4,261	1,291
Hood	5,459	7,575	1,445	4,359	5,313	4,457
Hopkins ...	4,522	4,341	1,034	4,085	3,398	3,147
Houston ...	3,383	3,443	585	3,250	3,067	1,690
Howard ...	3,732	5,007	1,037	3,735	5,129	1,984
Hudspeth..	427	367	92	364	325	178
Hunt......	8,801	10,746	2,225	7,452	9,739	7,387
Hutchinson.	2,553	6,350	864	2,833	6,034	1,993
Irion	213	386	86	256	283	290
Jack	1,019	1,162	301	1,254	1,041	1,045
Jackson ...	1,785	2,533	309	1,722	2,451	976
Jasper	5,039	4,523	1,041	5,658	3,870	2,539
Jeff Davis..	370	482	99	321	360	187
Jefferson ..	45,854	32,821	5,314	48,405	29,622	17,242
Jim Hogg ..	1,437	307	64	1,520	478	107
Jim Wells..	7,116	2,989	430	7,812	3,311	1,413
Johnson...	12,817	16,246	3,250	12,030	13,473	11,573
Jones	2,422	2,351	614	2,400	2,088	1,436
Karnes....	2,154	1,869	291	1,897	1,990	802
Kaufman ..	7,383	8,697	1,831	6,498	6,578	5,913
Kendall....	2,092	5,940	620	1,374	4,162	1,773
Kenedy....	133	71	4	87	69	18
Kent	260	187	67	271	175	163
Kerr	4,192	11,173	1,236	3,707	8,787	3,790
Kimble	521	898	131	467	790	354
King	46	97	29	54	79	56
Kinney	503	650	97	598	634	299
Kleberg ...	5,136	3,391	431	5,109	3,897	1,470
Knox	785	599	149	854	521	438
Lamar	6,075	6,393	1,198	6,328	5,778	4,093
Lamb	1,683	2,593	283	1,737	2,998	709
Lampasas .	1,819	3,008	509	1,508	2,233	1,432
LaSalle....	1,522	570	85	1,522	586	211
Lavaca....	2,575	3,697	551	2,700	3,362	1,696
Lee.......	2,008	2,354	421	1,847	2,108	1,088
Leon......	2,217	2,839	499	2,042	2,212	1,251
Liberty	6,877	7,784	2,011	7,036	6,959	4,311
Limestone .	3,236	2,691	693	3,188	2,358	1,505
Lipscomb..	357	869	115	338	839	270
Live Oak ..	1,372	1,929	292	1,345	1,805	806
Llano	2,633	4,290	762	2,409	3,056	1,799
Loving	14	48	15	20	31	45
Lubbock ..	22,786	47,304	3,996	22,240	48,847	11,618
Lynn......	903	1,151	136	902	1,233	291
McCulloch .	1,231	1,465	296	1,393	1,108	986
McLennan .	27,050	30,666	5,131	25,903	28,473	15,505
McMullen..	117	274	35	78	274	89
Madison...	1,470	1,576	293	1,553	1,544	778
Marion	2,028	1,260	353	2,156	1,245	882
Martin	643	973	140	641	986	356
Mason	618	949	151	570	776	364
Matagorda .	5,374	5,876	1,190	4,759	5,328	3,045
Maverick ..	5,307	1,050	202	4,540	2,002	771
Medina....	3,880	5,710	715	3,650	4,912	2,167
Menard ...	490	443	102	553	354	367
Midland ...	9,513	25,382	2,079	9,160	24,143	7,880
Milam.....	3,869	3,019	657	3,542	2,414	1,495
Mills	748	1,044	230	753	702	530
Mitchell ...	1,213	949	232	1,353	1,128	604
Montague..	2,718	3,029	842	2,885	2,304	2,330
Montgomery	20,722	51,011	6,065	18,551	39,976	19,203
Moore	1,358	3,353	359	1,361	3,147	976
Morris.....	2,973	1,449	402	3,028	1,400	1,138
Motley	164	380	56	256	446	117
Nacog-doches..	7,641	10,361	1,352	6,937	9,864	4,803
Navarro ...	6,078	5,236	1,140	6,006	4,897	3,800
Newton ...	2,554	1,409	474	3,249	1,212	1,032
Nolan	2,582	2,166	613	2,490	1,993	1,455
Nueces ...	50,009	37,470	5,103	46,317	36,781	17,374
Ochiltree ..	467	2,448	167	557	2,419	576
Oldham ...	213	583	77	225	583	177
Orange....	13,741	12,560	2,836	15,305	9,793	7,321
Palo Pinto .	3,938	3,666	1,011	3,392	2,852	3,010
Panola	4,168	4,008	777	3,950	3,473	1,906
Parker	9,447	14,580	2,703	7,934	10,321	9,148
Parmer....	676	2,042	160	637	1,829	564
Pecos.....	1,816	1,730	369	1,778	1,836	895
Polk	6,360	6,473	1,347	5,942	5,390	2,884
Potter.....	9,273	14,995	1,799	9,527	13,510	4,655
Presidio ...	1,205	383	111	1,189	400	290
Rains.....	1,265	1,123	335	1,108	975	890
Randall ...	9,177	28,266	1,985	9,119	24,971	6,340
Reagan ...	407	645	101	337	651	259
Real......	414	845	178	463	787	386
Red River..	2,339	1,783	433	2,686	1,735	1,228

County	1996 Clinton (D)	1996 Dole (R)	1996 Perot (RF)	1992 Clinton (D)	1992 Bush (R)	1992 Perot (I)
Reeves ...	2,279	1,007	245	2,569	1,244	734
Refugio ...	1,635	1,376	222	1,531	1,469	716
Roberts ...	122	421	40	126	391	99
Robertson .	2,912	1,944	315	2,927	1,707	963
Rockwall ..	3,289	8,319	1,121	2,397	6,427	4,393
Runnels ...	1,417	1,941	396	1,401	1,653	1,279
Rusk	5,988	8,423	1,072	5,391	7,560	3,575
Sabine	1,913	1,660	334	2,288	1,490	894
San Augustine	1,924	1,296	324	1,737	1,243	667
San Jacinto	2,771	2,878	810	2,846	2,494	1,653
San Patricio	8,132	7,678	1,085	8,202	7,456	3,178
San Saba .	726	991	194	716	723	660
Schleicher .	505	587	111	420	452	355
Scurry	2,099	2,929	813	1,609	2,670	1,826
Shackelford	502	792	169	484	623	422
Shelby	3,720	3,482	815	3,986	3,217	1,487
Sherman ..	243	809	89	261	851	256
Smith.....	18,265	32,171	2,933	17,514	27,753	13,569
Somervell .	993	1,099	273	782	872	903
Starr.....	6,312	756	157	7,668	1,209	345
Stephens..	1,218	1,714	336	1,115	1,573	1,062
Sterling ...	186	394	86	127	322	182
Stonewall..	487	323	105	561	242	322
Sutton	508	688	102	524	687	387
Swisher ...	1,224	1,159	195	1,413	989	541
Tarrant....	170,431	208,312	28,715	156,230	183,387	129,998
Taylor.....	13,213	23,682	2,912	12,382	22,614	10,331
Terrell.....	278	185	47	325	176	128
Terry	1,272	2,013	269	1,461	2,309	619
Throckmorton...	285	360	90	401	389	228
Titus......	3,725	3,438	744	3,625	3,024	2,146
Tom Green.	11,782	18,112	2,757	11,437	14,989	10,244
Travis.....	128,970	98,454	14,008	130,546	88,105	56,158
Trinity	2,774	2,058	460	2,784	1,988	1,133
Tyler......	3,340	2,804	645	3,465	2,357	1,529
Upshur....	5,032	5,174	1,086	4,776	4,511	2,896
Upton.....	424	685	88	489	908	313
Uvalde....	3,397	3,494	403	3,482	3,635	1,387
Val Verde..	5,623	4,357	548	4,748	4,102	2,093
Van Zandt..	5,752	7,453	1,756	5,310	5,810	5,239
Victoria ...	8,238	14,457	1,197	7,604	13,086	5,136
Walker	6,088	7,177	1,186	5,619	6,662	3,619
Waller	4,535	3,559	499	4,270	3,065	1,692
Ward	1,644	1,620	446	1,695	1,769	948
Washington	3,460	6,319	601	3,283	5,817	1,738
Webb.....	18,997	4,712	936	14,509	7,789	2,517
Wharton...	5,176	6,163	871	4,643	5,503	2,624
Wheeler...	750	1,355	174	938	1,458	367
Wichita ...	15,775	20,495	3,371	17,021	17,956	11,478
Wilbarger..	1,730	2,037	465	1,924	1,959	1,453
Willacy....	3,789	1,332	241	3,359	1,490	652
Williamson.	24,175	36,836	4,931	19,437	26,208	15,415
Wilson	3,713	4,530	760	3,711	3,766	2,105
Winkler ...	872	1,009	218	942	1,173	582
Wise	5,056	6,330	1,516	4,478	4,555	4,485
Wood.....	4,711	6,228	1,184	4,084	4,708	3,494
Yoakum ...	738	1,485	218	595	1,486	484
Young	2,394	3,647	639	2,464	2,894	2,302
Zapata....	1,786	521	131	2,052	866	326
Zavala....	2,629	463	91	3,058	571	237
Totals	**2,459,683**	**2,736,167**	**378,537**	**2,281,815**	**2,496,071**	**1,354,781**

Texas Vote Since 1948

1948, Truman, Dem., 750,700; Dewey, Rep., 282,240; Thurmond, States' Rights, 106,909; Wallace, Prog., 3,764; Watson, Proh., 2,758; Thomas, Soc., 874.

1952, Eisenhower, Rep., 1,102,878; Stevenson, Dem., 969,228; Hamblen, Proh., 1,983; MacArthur, Christian Nationalist, 833; MacArthur, Constitution, 730; Hallinan, Prog., 294.

1956, Eisenhower, Rep., 1,080,619; Stevenson, Dem., 859,958; Andrews, Ind., 14,591.

1960, Kennedy, Dem., 1,167,932; Nixon, Rep., 1,121,699; Sullivan, Constitution, 18,169; Decker, Proh., 3,870; write-in, 15.

1964, Johnson, Dem., 1,663,185; Goldwater, Rep., 958,566; Lightburn, Constitution, 5,060.

1968, Nixon, Rep., 1,227,844; Humphrey, Dem., 1,266,804; Wallace, 3d Party, 584,269; write-in, 489.

1972, Nixon, Rep., 2,298,896; McGovern, Dem., 1,154,289; Schmitz, Amer., 6,039; Jenness, Soc. Workers, 8,664; others, 3,393.

1976, Carter, Dem., 2,082,319; Ford, Rep., 1,953,300; McCarthy, Ind., 20,118; Anderson, Amer., 11,442; Camejo, Soc. Workers, 1,723; write-in, 2,982.

1980, Reagan, Rep., 2,510,705; Carter, Dem., 1,881,147; Anderson, Ind., 111,613; Clark, Libertarian, 37,643; write-in, 528.

1984, Reagan, Rep., 3,433,428; Mondale, Dem., 1,949,276.

1988, Bush, Rep., 3,036,829; Dukakis, Dem., 2,352,748; Paul, Lib., 30,355; Fulani, New Alliance, 7,208.

1992, Clinton, Dem., 2,281,815; Bush, Rep., 2,496,071; Perot, Ind., 1,354,781; Marrou, Libertarian, 19,699.

1996, Dole, Rep., 2,736,167; Clinton, Dem., 2,459,683; Perot, Ind. (Ref.), 378,537; Browne, Libertarian, 20,256; Phillips, Taxpayers, 7,472; Hagelin, Natural Law, 4,422.

Utah

| | 1996 | | | 1992 | | |
County	Clinton (D)	Dole (R)	Perot (RF)	Clinton (D)	Bush (R)	Perot (I)
Beaver	687	1,164	217	668	1,040	330
Box Elder	3,170	8,373	1,578	2,186	7,712	4,507
Cache	6,595	16,832	2,399	4,973	15,971	8,032
Carbon	4,172	2,343	952	4,480	2,038	2,002
Daggett	131	237	55	122	172	117
Davis	19,301	42,768	7,495	14,924	39,087	24,105
Duchesne	892	2,648	566	772	1,983	1,229
Emery	1,371	2,033	663	1,349	1,643	1,138
Garfield	283	1,330	222	309	1,235	355
Grand	1,199	1,384	432	1,160	1,100	991
Iron	1,887	6,550	716	1,537	5,616	1,693
Juab	928	1,290	353	823	1,237	616
Kane	304	1,682	290	295	1,241	534
Millard	945	2,681	505	742	2,496	1,064
Morgan	859	1,659	337	520	1,339	851
Piute	176	475	59	169	429	146
Rich	179	523	88	154	525	187
Salt Lake	117,951	127,951	27,620	100,082	117,247	91,968
San Juan	1,675	2,139	271	1,639	2,004	576
Sanpete	1,568	3,631	801	1,302	2,995	1,742
Sevier	1,327	4,031	670	1,039	3,160	1,671
Summit	4,177	3,867	971	3,013	3,133	3,060
Tooele	3,992	3,881	1,244	3,270	3,676	3,011
Uintah	1,714	4,743	899	1,374	3,505	2,250
Utah	18,291	69,653	8,106	14,090	61,398	24,558
Wasatch	1,374	2,222	558	1,042	1,822	1,234
Washington	4,816	17,637	2,069	3,364	11,310	4,623
Wayne	265	741	121	236	706	251
Weber	21,404	27,443	6,204	17,795	26,812	20,559
Totals	221,633	361,911	66,461	183,429	322,632	203,400

Utah Vote Since 1948

1948, Truman, Dem., 149,151; Dewey, Rep., 124,402; Wallace, Prog., 2,679; Dobbs, Soc. Workers, 73.

1952, Eisenhower, Rep., 194,190; Stevenson, Dem., 135,364.

1956, Eisenhower, Rep., 215,631; Stevenson, Dem., 118,364.

1960, Kennedy, Dem., 169,248; Nixon, Rep., 205,361; Dobbs, Soc. Workers, 100.

1964, Johnson, Dem., 219,628; Goldwater, Rep., 181,785.

1968, Nixon, Rep., 238,728; Humphrey, Dem., 156,665; Wallace, 3d Party, 26,906; Halstead, Soc. Workers, 89; Peace and Freedom, 180.

1972, Nixon, Rep., 323,643; McGovern, Dem., 126,284; Schmitz, Amer., 28,549.

1976, Carter, Dem., 182,110; Ford, Rep., 337,908; Anderson, Amer., 13,304; McCarthy, Ind., 3,907; MacBride, Libertarian, 2,438; Maddox, Amer. Ind., 1,162; Camejo, Soc. Workers, 268; Hall, Com., 121.

1980, Reagan, Rep., 439,687; Carter, Dem., 124,266; Anderson, Ind., 30,284; Clark, Libertarian, 7,226; Commoner, Citizens, 1,009; Greaves, Amer., 965; Rarick, Amer. Ind., 522; Hall, Com., 139; DeBerry, Soc. Workers, 124.

1984, Reagan, Rep., 469,105; Mondale, Dem., 155,369; Bergland, Libertarian, 2,447.

1988, Bush, Rep., 428,442; Dukakis, Dem., 207,352; Paul, Lib., 7,473; Dennis, Amer., 2,158.

1992, Clinton, Dem., 183,429; Bush, Rep., 322,632; Perot, Ind., 203,400; Gritz, Populist/America First, 28,602; Marrou, Libertarian, 1,900; Hagelin, Natural Law, 1,319; LaRouche, Ind., 1,089.

1996, Dole, Rep., 361,911; Clinton, Dem., 221,633; Perot, Ref., 66,461; Nader, Green, 4,615; Browne, Libertarian, 4,129; Phillips, Taxpayers, 2,601; Templin, Ind. Amer., 1,290; Crane, Ind., 1,101; Hagelin, Natural Law, 1,085; Moorehead, Workers World, 298; Harris, Soc. Workers, 235; Dodge, Proh., 111.

Vermont

| | 1996 | | | 1992 | | |
City	Clinton (D)	Dole (R)	Perot (RF)	Clinton (D)	Bush (R)	Perot (I)
Barre City	1,890	1,107	376	1,807	1,508	1,035
Bennington	3,454	1,654	960	3,646	2,151	1,536
Brattleboro	3,016	1,195	395	3,519	1,447	847
Burlington	11,600	3,762	1,309	12,508	4,462	3,241
Colchester	3,314	2,035	769	2,966	1,997	1,739
Essex	4,063	2,944	796	3,825	2,960	2,302
Hartford	2,106	1,290	400	2,034	1,564	793
Montpelier	2,458	1,118	269	2,490	1,407	657
Rutland City	3,817	2,320	741	3,888	2,915	1,722
S. Burlington	3,929	2,274	548	3,730	2,131	1,359
Springfield	2,267	1,189	565	2,179	1,468	1,091
Other	95,980	59,464	23,900	90,998	64,112	49,663
Totals	137,894	80,352	31,024	133,590	88,122	65,985

Vermont Vote Since 1948

1948, Truman, Dem., 45,557; Dewey, Rep., 75,926; Wallace, Prog., 1,279; Thomas, Soc., 585.

1952, Eisenhower, Rep., 109,717; Stevenson, Dem., 43,355; Hallinan, Prog., 282; Hoopes, Soc., 185.

1956, Eisenhower, Rep., 110,390; Stevenson, Dem., 42,549; scattered, 39.

1960, Kennedy, Dem., 69,186; Nixon, Rep., 98,131.

1964, Johnson, Dem., 107,674; Goldwater, Rep., 54,868.

1968, Nixon, Rep., 85,142; Humphrey, Dem., 70,255; Wallace, 3d Party, 5,104; Halstead, Soc. Workers, 295; Gregory, New Party, 579.

1972, Nixon, Rep., 117,149; McGovern, Dem., 68,174; Spock, Liberty Union, 1,010; Jenness, Soc. Workers, 296; scattered, 318.

1976, Carter, Dem., 77,798; Carter, Ind. Vermonter, 991; Ford, Rep., 100,387; McCarthy, Ind., 4,001; Camejo, Soc. Workers, 430; LaRouche, U.S. Labor, 196; scattered, 99.

1980, Reagan, Rep., 94,598; Carter, Dem., 81,891; Anderson, Ind., 31,760; Commoner, Citizens, 2,316; Clark, Libertarian, 1,900; McReynolds, Liberty Union, 136; Hall, Com., 118; DeBerry, Soc. Workers, 75; scattering, 413.

1984, Reagan, Rep., 135,865; Mondale, Dem., 95,730; Bergland, Libertarian, 1,002.

1988, Bush, Rep., 124,331; Dukakis, Dem., 115,775; Paul, Lib., 1,000; LaRouche, Ind., 275.

1992, Clinton, Dem., 133,590; Bush, Rep., 88,122; Perot, Ind., 65,985.

1996, Clinton, Dem., 137,894; Dole, Rep., 80,352; Perot, Ref., 31,024; Nader, Green, 5,585; Browne, Libertarian, 1,183; Hagelin, Natural Law, 498; Peron, Grass Roots, 480; Phillips, Taxpayers, 382; Hollis, Liberty Union, 292; Harris, Soc. Workers, 199.

Virginia

| | 1996 | | | 1992 | | |
County	Clinton (D)	Dole (R)	Perot (RF)	Clinton (D)	Bush (R)	Perot (I)
Accomack	5,220	5,013	1,218	4,950	5,666	2,304
Albemarle	14,089	15,243	1,533	13,886	13,894	3,855
Alleghany	2,398	2,015	607	2,396	2,294	926
Amelia	1,625	2,119	323	1,534	2,062	574
Amherst	4,864	5,094	835	4,101	5,482	1,268
Appomattox	2,239	2,625	510	1,919	2,830	801
Arlington	45,573	26,106	2,782	47,756	26,376	7,992
Augusta	5,965	13,458	1,916	5,190	12,896	3,397
Bath	922	847	247	855	1,075	354
Bedford	7,786	11,955	1,976	6,792	10,496	3,251
Bland	939	1,167	385	1,001	1,368	408
Botetourt	4,576	6,404	1,138	4,349	5,904	1,819
Brunswick	3,442	2,059	340	3,687	2,480	479
Buchanan	6,551	2,785	858	7,405	3,297	815
Buckingham	2,374	1,974	392	2,193	2,368	459
Campbell	6,788	10,273	1,505	5,999	10,931	2,553
Caroline	3,897	2,816	521	3,770	2,947	965
Carroll	3,611	5,088	1,158	3,790	5,664	1,388
Charles City	1,842	729	178	2,010	729	251
Charlotte	2,007	2,103	431	2,098	2,293	640
Chesterfield	30,220	56,650	6,004	28,028	56,626	16,898
Clarke	1,906	2,201	379	1,811	1,994	802
Craig	895	979	262	965	1,008	304
Culpeper	3,907	5,688	787	3,444	5,226	1,640
Cumberland	1,303	1,544	275	1,284	1,643	372
Dickenson	3,913	2,229	660	4,839	2,574	660
Dinwiddie	3,871	3,503	666	3,624	3,648	1,198
Essex	1,668	1,627	188	1,583	1,897	382
Fairfax	170,150	176,033	16,134	160,186	170,488	53,012
Fauquier	6,759	11,063	1,287	6,600	10,497	3,464
Floyd	1,909	2,374	545	2,026	2,575	672
Fluvanna	2,676	3,442	457	2,134	2,811	871
Franklin	7,300	7,382	2,015	6,590	6,724	2,232
Frederick	5,976	10,608	1,599	4,942	9,425	2,981
Giles	3,196	2,566	841	3,346	3,023	1,142
Gloucester	4,710	6,447	1,266	4,058	6,461	2,640
Goochland	2,784	4,119	424	2,589	3,834	994
Grayson	2,661	3,004	675	2,615	3,378	860
Greene	1,440	2,351	346	1,353	2,265	627
Greensville	2,381	1,176	263	2,237	1,335	360
Halifax	5,599	6,490	876	4,752	5,199	1,140
Hanover	9,880	22,086	2,447	8,021	20,336	5,674
Henrico	41,121	54,430	5,920	36,807	56,910	14,720
Henry	9,061	9,110	2,370	9,296	9,005	3,212
Highland	446	631	134	494	686	212
Isle of Wight	4,952	5,416	893	4,380	5,370	1,536
James City	7,247	10,120	1,116	6,536	8,781	2,675
King and Queen	1,393	1,073	213	1,811	2,570	918
King George	1,875	2,597	341	1,363	1,206	323
King William	1,765	2,346	339	1,822	2,591	758
Lancaster	1,844	2,709	324	1,812	2,841	739
Lee	4,444	3,225	822	5,215	3,504	1,002
Loudoun	19,942	25,715	3,082	14,462	19,290	7,391
Louisa	3,761	3,768	693	3,399	3,461	1,381
Lunenburg	1,995	2,063	299	2,082	2,227	505

County	1996 Clinton (D)	Dole (R)	Perot (RF)	1992 Clinton (D)	Bush (R)	Perot (I)
Madison...	1,734	2,296	360	1,700	2,341	653
Mathews..	1,602	2,206	403	1,402	2,179	884
Mecklenburg	4,408	4,933	789	4,273	5,401	1,128
Middlesex .	1,704	2,141	350	1,597	2,224	768
Montgomery	10,867	10,517	2,594	10,658	10,606	3,449
Nelson	2,782	1,988	411	2,586	2,159	748
New Kent..	1,859	2,852	520	1,738	2,708	1,017
Northampton	2,569	1,763	522	2,568	2,088	844
Northumberland ..	1,957	2,605	375	1,862	2,667	729
Nottoway ..	2,327	2,416	346	2,411	2,610	606
Orange....	3,590	4,435	750	3,348	4,092	1,425
Page	2,868	3,876	640	3,010	4,203	1,163
Patrick	2,301	3,547	719	2,465	3,521	1,026
Pittsylvania	7,681	12,117	1,469	7,675	11,467	2,296
Powhatan..	2,254	4,679	626	1,950	3,832	1,232
Prince Edward ..	2,678	2,530	403	2,775	2,858	635
Prince George ..	3,498	5,216	698	3,087	4,799	1,459
Prince William...	33,462	39,292	4,881	26,486	35,432	13,190
Pulaski....	5,333	5,387	1,399	5,633	6,148	2,066
Rappahannock.....	1,405	1,505	213	1,273	1,410	487
Richmond..	1,101	1,424	201	1,034	1,609	366
Roanoke ..	15,387	20,700	2,934	14,704	20,667	5,477
Rockbridge.	3,116	3,274	760	2,908	3,228	1,254
Rockingham	5,867	14,035	1,318	5,407	13,016	2,839
Russell....	5,437	3,706	862	6,480	3,891	958
Scott	3,449	4,086	798	3,979	4,515	957
Shenandoah	4,224	7,440	1,353	3,956	7,746	2,063
Smyth	4,990	4,966	1,407	4,924	6,128	1,618
Southampton	3,454	2,275	564	3,199	2,844	754
Spotsylvania	10,342	13,786	1,860	8,133	11,829	3,918
Stafford ...	9,902	14,098	1,856	7,718	12,528	4,481
Surry	1,753	944	181	1,823	1,046	364
Sussex	2,089	1,378	256	2,193	1,527	446
Tazewell...	7,500	6,131	1,554	8,586	6,375	1,872
Warren....	3,814	4,657	904	3,554	4,319	1,650
Washington	6,939	9,098	1,654	7,269	9,150	2,288
Wesmoreland	2,949	2,333	427	2,758	2,554	818
Wise.....	6,712	4,660	1,478	7,681	5,144	1,835
Wythe	3,275	4,274	955	3,616	5,121	1,557
York	7,731	11,396	1,469	6,218	10,197	3,426
Cities						
Alexandria .	27,968	15,554	1,472	30,784	16,700	4,934
Bedford ...	1,065	990	212	963	1,091	313
Bristol.....	2,586	2,983	429	2,948	3,616	851
Buena Vista	1,090	713	216	1,023	849	291
Charlottesville	7,916	4,091	565	8,685	4,705	1,397
Chesapeake	28,713	29,251	4,456	23,495	28,909	9,237
Clifton Forge	974	486	147	958	632	251
Colonial Heights .	1,782	4,632	518	1,721	5,298	1,312
Covington .	1,394	763	255	1,442	995	402
Danville ...	8,168	9,254	762	8,134	9,584	1,679
Emporia ...	1,103	835	98	1,048	1,094	157
Fairfax	3,909	4,319	422	3,884	4,333	1,439
Falls Church	2,375	1,644	202	2,864	1,912	599
Franklin ...	1,962	1,200	201	1,696	1,347	272
Fredericksburg ...	3,215	2,579	300	3,266	2,819	738
Galax	1,033	910	221	957	1,087	276
Hampton ..	24,493	16,596	2,783	23,395	19,219	6,581
Harrisonburg.....	3,346	4,945	434	3,414	4,935	1,162
Hopewell ..	2,868	3,493	550	2,863	3,818	1,227
Lexington ..	1,059	850	112	1,128	894	228
Lynchburg .	10,281	11,441	1,155	9,587	12,518	2,545
Manassas .	4,378	5,799	670	3,647	5,453	1,971
Manassas Park.....	748	916	151	567	792	356
Martinsville	2,941	2,446	387	3,073	2,690	748
Newport News	27,678	23,072	3,090	25,743	26,779	8,217
Norfolk	37,655	18,693	3,435	37,602	22,362	8,732
Norton	802	416	138	871	472	182
Petersburg.	8,105	2,261	423	8,671	3,125	834
Poquoson..	1,409	3,422	400	1,086	3,354	960
Portsmouth.	22,150	10,686	2,238	20,416	12,575	4,360
Radford ...	2,113	1,742	381	2,183	1,996	582
Richmond .	42,273	20,993	2,762	47,642	24,341	6,992
Roanoke ..	17,282	12,283	2,169	17,724	13,443	3,753
Salem	4,282	4,936	796	4,028	5,143	1,430
S. Boston[1].	—	—	—	1,051	1,435	252
Staunton ..	3,162	4,526	605	2,851	4,989	1,146
Suffolk	10,827	8,572	1,266	9,196	8,697	2,150
Virginia Beach ...	52,142	63,741	9,328	44,294	68,936	24,087
Waynesboro	2,398	3,466	462	2,302	3,758	961
Williamsburg.	1,820	1,560	162	1,856	1,349	445
Winchester.	3,027	3,681	434	2,768	3,833	1,048
Totals	1,091,060	1,138,350	159,861	1,038,650	1,150,517	348,639

(1) South Boston merged with Halifax County in July 1996.

Virginia Vote Since 1948

1948, Truman, Dem., 200,786; Dewey, Rep., 172,070; Thurmond, States' Rights, 43,393; Wallace, Prog., 2,047; Thomas, Soc., 726; Teichert, Soc. Labor, 234.

1952, Eisenhower, Rep., 349,037; Stevenson, Dem., 268,677; Hass, Soc. Labor, 1,160; Hoopes, Soc. Dem., 504; Hallinan, Prog., 311.

1956, Eisenhower, Rep., 386,459; Stevenson, Dem., 267,760; Andrews, States' Rights, 42,964; Hoopes, Soc. Dem., 444; Hass, Soc. Labor, 351.

1960, Kennedy, Dem., 362,327; Nixon, Rep., 404,521; Coiner, Cons., 4,204; Hass, Soc. Labor, 397.

1964, Johnson, Dem., 558,038; Goldwater, Rep., 481,334; Hass, Soc. Labor, 2,895.

1968, Nixon, Rep., 590,319; Humphrey, Dem., 442,387; Wallace, 3d Party, *320,272; Blomen, Soc. Labor, 4,671; Munn, Proh., 601; Gregory, Peace and Freedom, 1,680.
   *10,561 votes for Wallace were omitted in the count.

1972, Nixon, Rep., 988,493; McGovern, Dem., 438,887; Schmitz, Amer., 19,721; Fisher, Soc. Labor, 9,918.

1976, Carter, Dem., 813,896; Ford, Rep., 836,554; Camejo, Soc. Workers, 17,802; Anderson, Amer., 16,686; LaRouche, U.S. Labor, 7,508; MacBride, Libertarian, 4,648.

1980, Reagan, Rep., 989,609; Carter, Dem., 752,174; Anderson, Ind., 95,418; Commoner, Citizens, 14,024; Clark, Libertarian, 12,821; DeBerry, Soc. Workers, 1,986.

1984, Reagan, Rep., 1,337,078; Mondale, Dem., 796,250.

1988, Bush, Rep., 1,309,162; Dukakis, Dem., 859,799; Fulani, Ind., 14,312; Paul, Lib., 8,336.

1992, Clinton, Dem., 1,038,650; Bush, Rep., 1,150,517; Perot, Ind., 348,639; LaRouche, Ind., 11,937; Marrou, Libertarian, 5,730; Fulani, New Alliance, 3,192.

1996, Dole, Rep., 1,138,350; Clinton, Dem., 1,091,060; Perot, Ref., 159,861; Phillips, Taxpayers, 13,687; Browne, Libertarian, 9,174; Hagelin, Natural Law, 4,510.

Washington

County	1996 Clinton (D)	Dole (R)	Perot (RF)	1992 Clinton (D)	Bush (R)	Perot (I)
Adams	1,740	2,356	448	1,449	2,087	1,010
Asotin	3,349	2,860	936	3,239	2,425	1,849
Benton	20,783	26,664	5,311	16,459	22,883	12,878
Chelan	8,595	12,363	2,332	7,860	10,716	4,606
Clallam ...	12,585	12,432	3,187	10,820	9,765	7,775
Clark	52,254	46,794	9,663	42,648	36,906	26,163
Columbia ..	743	948	228	668	761	466
Cowlitz....	18,054	11,221	3,441	15,052	10,000	9,246
Douglas ...	3,913	5,682	1,132	3,731	4,920	2,315
Ferry	1,197	1,091	408	963	773	762
Franklin ...	4,961	5,946	992	3,743	4,486	2,597
Garfield ...	497	623	117	473	620	222
Grant	8,065	10,895	2,496	7,278	9,503	4,898
Grays Harbor...	14,082	7,635	3,757	12,599	6,904	7,460
Island.....	12,157	12,387	2,787	9,555	9,526	7,889
Jefferson ..	7,145	4,607	1,385	6,148	3,467	3,168
King......	417,846	232,811	51,309	391,050	212,986	167,216
Kitsap	44,167	35,304	8,769	34,442	29,340	23,873
Kittitas ...	5,707	5,224	1,214	5,432	4,078	2,778
Klickitat ...	3,214	2,662	875	2,758	2,085	1,938
Lewis	10,331	13,238	3,373	7,810	12,316	6,684
Lincoln	1,806	2,587	518	1,653	2,152	1,098
Mason	10,088	7,149	2,816	8,076	5,776	5,577
Okanogan .	4,810	5,890	1,797	5,015	4,265	3,541
Pacific	5,095	2,598	1,131	4,587	2,243	2,351
Pend Oreille	2,126	2,012	709	1,798	1,528	1,340
Pierce	120,893	89,295	22,051	102,243	77,410	59,523
San Juan ..	3,663	2,523	508	3,353	1,901	1,776
Skagit	18,295	16,937	4,818	15,936	13,388	10,973
Skamania ..	1,724	1,387	450	1,474	1,102	1,050
Snohomish	109,624	81,885	22,731	88,643	69,137	65,838
Spokane ...	71,727	66,628	16,532	69,526	59,984	38,251
Stevens ...	5,591	7,524	2,158	4,960	5,706	3,769
Thurston ..	45,522	29,835	7,622	38,293	25,643	19,551
Wahkiakum	924	619	215	696	488	584
Walla Walla	8,038	9,085	1,894	7,325	7,894	4,507
Whatcom .	29,074	27,153	4,854	26,619	23,801	12,455
Whitman ..	7,262	6,734	1,315	7,637	6,428	3,220
Yakima ...	25,676	27,668	4,724	21,026	25,841	10,583
Totals	1,123,323	840,712	201,003	993,037	731,234	541,780

Washington Vote Since 1948

1948, Truman, Dem., 476,165; Dewey, Rep., 386,315; Wallace, Prog., 31,692; Watson, Proh., 6,117; Thomas, Soc., 3,534; Teichert, Soc. Labor, 1,133; Dobbs, Soc. Workers, 103.

1952, Eisenhower, Rep., 599,107; Stevenson, Dem., 492,845; MacArthur, Christian Nationalist, 7,290; Hallinan, Prog., 2,460; Hass, Soc. Labor, 633; Hoopes, Soc., 254; Dobbs, Soc. Workers, 119.

1956, Eisenhower, Rep., 620,430; Stevenson, Dem., 523,002; Hass, Soc. Labor, 7,457.

1960, Kennedy, Dem., 599,298; Nixon, Rep., 629,273; Hass, Soc. Labor, 10,895; Curtis, Constitution, 1,401; Dobbs, Soc. Workers, 705.

1964, Johnson, Dem., 779,699; Goldwater, Rep., 470,366; Hass, Soc. Labor, 7,772; DeBerry, Freedom Soc., 537.

1968, Nixon, Rep., 588,510; Humphrey, Dem., 616,037; Wallace, 3d Party, 96,990; Blomen, Soc. Labor, 488; Cleaver, Peace and Freedom, 1,609; Halstead, Soc. Workers, 270; Mitchell, Free Ballot, 377.

1972, Nixon, Rep., 837,135; McGovern, Dem., 568,334; Schmitz, Amer., 58,906; Spock, Ind., 2,644; Fisher, Soc. Labor, 1,102; Jenness, Soc. Workers, 623; Hall, Com., 566; Hospers, Libertarian, 1,537.

1976, Carter, Dem., 717,323; Ford, Rep., 777,732; McCarthy, Ind., 36,986; Maddox, Amer. Ind., 8,585; Anderson, Amer., 5,046; MacBride, Libertarian, 5,042; Wright, People's, 1,124; Camejo, Soc. Workers, 905; LaRouche, U.S. Labor, 903; Hall, Com., 817; Levin, Soc. Labor, 713; Zeidler, Soc., 358.

1980, Reagan, Rep., 865,244; Carter, Dem., 650,193; Anderson, Ind., 185,073; Clark, Libertarian, 29,213; Commoner, Citizens, 9,403; DeBerry, Soc. Workers, 1,137; McReynolds, Soc., 956; Hall, Com., 834; Griswold, Workers World, 341.

1984, Reagan, Rep., 1,051,670; Mondale, Dem., 798,352; Bergland, Libertarian, 8,844.

1988, Bush, Rep., 903,835; Dukakis, Dem., 933,516; Paul, Lib., 17,240; LaRouche, Ind., 4,412.

1992, Clinton, Dem., 993,037; Bush, Rep., 731,234; Perot, Ind., 541,780; Marrou, Libertarian, 7,533; Gritz, Populist/America First, 4,854; Hagelin, Natural Law, 2,456; Phillips, U.S. Taxpayers, 2,354; Fulani, New Alliance, 1,776; Daniels, Ind., 1,171.

1996, Clinton, Dem., 1,123,323; Dole, Rep., 840,712; Perot, Ref., 201,003; Nader, Ind., 60,322; Browne, Libertarian, 12,522; Hagelin, Natural Law, 6,076; Phillips, Taxpayers, 4,578; Collins, Ind., 2,374; Moorehead, Workers World, 2,189; Harris, Soc. Workers, 738.

West Virginia

County	1996 Clinton (D)	Dole (R)	Perot (RF)	1992 Clinton (D)	Bush (R)	Perot (I)
Barbour	3,076	2,155	784	3,467	2,322	1,153
Berkeley	8,321	9,859	2,291	7,159	9,134	3,645
Boone	6,048	1,917	927	6,576	2,021	1,037
Braxton	3,001	1,441	527	3,396	1,535	823
Brooke	5,338	2,741	1,375	5,693	2,582	2,103
Cabell	16,277	13,179	2,968	15,111	13,203	5,311
Calhoun	1,402	1,000	307	1,627	1,095	537
Clay	2,074	1,137	355	1,928	1,255	462
Doddridge	865	1,335	382	968	1,500	515
Fayette	9,471	3,669	1,552	9,574	3,991	2,002
Gilmer	1,390	933	316	1,576	1,085	484
Grant	1,206	2,599	481	1,011	2,762	519
Greenbrier	6,286	4,434	1,418	5,784	4,442	1,898
Hampshire	2,335	2,814	605	2,365	2,767	1,022
Hancock	7,521	4,268	2,158	7,830	3,897	3,267
Hardy	1,911	1,895	438	1,917	2,144	602
Harrison	14,746	8,857	3,135	15,480	9,687	5,131
Jackson	4,882	4,235	1,295	5,102	4,192	1,908
Jefferson	6,361	5,287	1,307	5,363	4,656	2,114
Kanawha	40,357	29,311	6,412	38,315	31,358	11,778
Lewis	2,868	2,285	974	2,931	2,413	1,197
Lincoln	4,994	2,530	696	4,502	2,637	787
Logan	10,840	2,627	1,532	11,095	3,336	1,835
McDowell	5,989	1,550	655	7,019	1,941	803
Marion	12,994	6,160	2,881	14,042	6,380	4,736
Marshall	7,045	4,460	2,202	7,298	4,463	3,402
Mason	5,284	3,581	1,533	5,331	3,808	2,045
Mercer	8,721	7,768	2,141	9,511	7,888	2,817
Mineral	3,487	4,380	1,170	3,992	4,837	1,884
Mingo	7,584	2,229	1,020	7,342	2,584	915
Monongalia	13,406	10,189	3,040	14,142	9,831	4,576
Monroe	2,382	2,131	559	2,418	2,311	685
Morgan	1,929	2,599	513	1,854	2,585	886
Nicholas	4,769	2,649	1,071	5,042	2,959	1,495
Ohio	8,781	7,267	2,065	9,522	7,421	3,632
Pendleton	1,591	1,431	276	1,626	1,589	362
Pleasants	1,478	1,265	416	1,387	1,248	731
Pocahontas	1,796	1,242	426	1,741	1,401	627
Preston	4,237	4,257	1,760	3,933	4,429	2,109
Putnam	8,029	8,803	1,901	6,817	7,653	2,910
Raleigh	12,547	8,628	2,355	13,171	8,700	3,247
Randolph	5,469	3,348	1,184	5,097	3,496	1,582
Ritchie	1,385	1,906	522	1,474	2,184	745
Roane	2,572	2,069	622	2,607	2,207	1,009
Summers	2,397	1,505	438	2,650	1,652	565
Taylor	2,692	1,977	844	2,843	2,022	1,242
Tucker	1,649	1,217	424	1,805	1,261	550
Tyler	1,459	734	563	1,587	1,593	1,013
Upshur	3,052	3,325	1,031	3,161	3,505	1,558
Wayne	8,300	5,492	1,633	8,392	5,729	2,199
Webster	2,292	654	369	2,320	811	436
Wetzel	3,209	2,037	1,004	3,753	2,271	1,550
Wirt	906	928	280	1,043	939	394
Wood	13,261	15,502	3,694	13,529	15,441	6,998
Wyoming	5,550	2,155	812	5,782	2,821	996
Totals	327,812	233,946	71,639	331,001	241,974	108,829

West Virginia Vote Since 1948

1948, Truman, Dem., 429,188; Dewey, Rep., 316,251; Wallace, Prog., 3,311.

1952, Eisenhower, Rep., 419,970; Stevenson, Dem., 453,578.

1956, Eisenhower, Rep., 449,297; Stevenson, Dem., 381,534.

1960, Kennedy, Dem., 441,786; Nixon, Rep., 395,995.

1964, Johnson, Dem., 538,087; Goldwater, Rep., 253,953.

1968, Nixon, Rep., 307,555; Humphrey, Dem., 374,091; Wallace, 3d Party, 72,560.

1972, Nixon, Rep., 484,964; McGovern, Dem., 277,435.

1976, Carter, Dem., 435,864; Ford, Rep., 314,726.

1980, Reagan, Rep., 334,206; Carter, Dem., 367,462; Anderson, Ind., 31,691; Clark, Libertarian, 4,356.

1984, Reagan, Rep., 405,483; Mondale, Dem., 328,125.

1988, Bush, Rep., 310,065; Dukakis, Dem., 341,016; Fulani, New Alliance, 2,230.

1992, Clinton, Dem., 331,001; Bush, Rep., 241,974; Perot, Ind., 108,829; Marrou, Libertarian, 1,873.

1996, Clinton, Dem., 327,812; Dole, Rep., 233,946; Perot, Ref., 71,639; Browne, Libertarian, 3,062.

Wisconsin

County	1996 Clinton (D)	Dole (R)	Perot (RF)	1992 Clinton (D)	Bush (R)	Perot (I)
Adams	4,119	2,450	1,122	3,539	2,465	2,003
Ashland	3,808	1,863	861	4,213	2,372	1,746
Barron	8,025	6,158	2,692	8,063	6,572	5,479
Bayfield	3,895	2,250	899	3,873	2,393	1,786
Brown	42,823	38,563	8,036	37,513	42,352	22,395
Buffalo	2,681	1,800	972	2,996	2,029	1,889
Burnet	3,625	2,452	962	3,172	2,340	1,855
Calumet	6,940	7,049	2,112	5,701	7,541	5,055
Chippewa	9,647	7,520	3,567	10,487	8,215	6,408
Clark	5,540	4,622	2,486	5,540	4,977	4,284
Columbia	10,336	8,377	2,377	9,348	9,099	5,439
Crawford	3,658	2,249	1,060	3,540	2,390	1,797
Dane	109,347	59,487	12,436	114,724	61,957	31,874
Dodge	12,625	12,890	3,322	11,438	14,971	9,136
Door	5,590	4,948	1,475	4,735	5,468	3,506
Douglas	10,976	5,167	2,001	12,319	5,679	4,150
Dunn	7,536	4,917	2,555	7,965	5,283	4,809
Eau Claire	20,298	13,900	5,160	21,221	15,915	9,783
Florence	869	927	316	978	942	719
Fond du Lac	15,542	16,488	4,204	13,757	19,785	10,660
Forest	2,092	1,166	678	1,904	1,393	1,062
Grant	9,203	7,021	2,648	8,914	7,678	6,405
Green	6,136	4,697	1,534	5,467	4,887	3,735
Green Lake	3,152	3,565	1,025	2,772	3,897	2,827
Iowa	4,690	2,866	1,071	4,467	3,288	2,341
Iron	1,725	1,260	469	1,762	1,273	835
Jackson	3,705	2,262	1,163	3,681	2,644	2,040
Jefferson	13,188	12,681	3,177	11,593	13,072	7,960
Juneau	4,331	3,226	1,393	4,177	4,051	2,670
Kenosha	27,964	18,296	6,507	27,341	19,854	14,232
Kewaunee	4,311	3,431	1,161	4,050	3,570	2,700
La Crosse	23,647	16,482	4,844	22,838	18,891	10,224
La Fayette	3,261	2,172	944	3,143	2,582	2,079
Langlade	4,074	3,206	1,249	3,630	3,890	2,444
Lincoln	6,166	4,076	1,800	5,297	4,321	3,605
Manitowoc	16,750	13,239	3,941	15,903	14,008	11,179
Marathon	24,012	19,874	6,749	21,482	20,948	14,600
Marinette	8,413	7,231	2,367	7,626	7,984	5,412
Marquette	2,859	2,208	915	2,533	2,322	1,818
Menominee	992	307	107	691	244	221
Milwaukee	216,620	119,407	26,027	235,521	151,314	76,039
Monroe	6,924	5,299	2,081	6,427	6,118	4,183
Oconto	6,723	5,389	1,655	5,898	5,720	4,405
Oneida	7,619	6,339	2,604	7,160	6,725	4,782
Outagamie	28,815	27,758	7,235	23,735	30,370	18,479
Ozaukee	13,269	22,078	2,774	11,879	22,805	8,002
Pepin	1,585	1,007	456	1,673	1,098	781
Pierce	7,970	4,599	2,074	7,824	4,844	4,492
Polk	8,334	5,387	2,369	7,746	5,446	4,753
Portage	15,901	9,631	3,410	15,553	10,914	7,083
Price	3,523	2,545	1,218	3,575	2,654	2,286

County	1996 Clinton (D)	Dole (R)	Perot (RF)	1992 Clinton (D)	Bush (R)	Perot (I)
Racine	38,567	30,107	7,611	34,875	32,310	20,227
Richland. . .	3,502	2,642	901	3,458	3,144	1,899
Rock	32,450	20,096	6,800	31,154	21,942	15,700
Rusk	2,941	2,219	1,331	3,376	2,430	2,085
St. Croix. . .	11,384	8,253	3,180	10,281	8,114	7,125
Sauk	9,889	7,448	2,448	9,128	8,886	5,280
Sawyer. . . .	2,773	2,603	962	2,796	2,658	1,861
Shawano . .	6,850	6,396	2,071	6,062	7,253	4,540
Sheboygan	22,022	20,067	4,157	20,568	22,526	11,295
Taylor.	3,253	3,108	1,457	3,305	3,415	2,590
Trem-pealeau . .	5,848	3,035	1,688	6,218	3,577	3,160
Vernon	5,572	3,796	1,523	5,673	4,072	2,890
Vilas.	4,226	4,496	1,548	3,764	4,616	2,827
Walworth . .	13,283	15,099	3,729	11,825	15,727	9,029
Washburn .	3,231	2,703	920	3,080	2,586	1,978
Washington	17,154	25,829	4,786	13,339	22,739	13,045
Waukesha .	57,354	91,729	13,109	50,270	91,461	36,622
Waupaca . .	7,800	8,679	2,464	6,666	10,252	6,088
Waushara .	3,824	3,573	1,264	3,402	4,045	2,829
Winnebago	29,564	27,880	6,531	27,234	33,709	16,140
Wood	14,650	12,666	4,599	13,208	13,843	8,822
Totals	1,071,971	845,029	227,339	1,041,066	930,855	544,479

Wisconsin Vote Since 1948

1948, Truman, Dem., 647,310; Dewey, Rep., 590,959; Wallace, Prog., 25,282; Thomas, Soc., 12,547; Teichert, Soc. Labor, 399; Dobbs, Soc. Workers, 303.

1952, Eisenhower, Rep., 979,744; Stevenson, Dem., 622,175; Hallinan, Ind., 2,174; Dobbs, Ind., 1,350; Hoopes, Ind., 1,157; Hass, Ind., 770.

1956, Eisenhower, Rep., 954,844; Stevenson, Dem., 586,768; Andrews, Ind., 6,918; Hoopes, Soc., 754; Hass, Soc. Labor, 710; Dobbs, Soc. Workers, 564.

1960, Kennedy, Dem., 830,805; Nixon, Rep., 895,175; Dobbs, Soc. Workers, 1,792; Hass, Soc. Labor, 1,310.

1964, Johnson, Dem., 1,050,424; Goldwater, Rep., 638,495; DeBerry, Soc. Workers, 1,692; Hass, Soc. Labor, 1,204.

1968, Nixon, Rep., 809,997; Humphrey, Dem., 748,804; Wallace, 3d Party, 127,835; Blomen, Soc. Labor, 1,338; Halstead, Soc. Workers, 1,222; scattered, 2,342.

1972 Nixon, Rep., 989,430; McGovern, Dem., 810,174; Schmitz, Amer., 47,525; Spock, Ind., 2,701; Fisher, Soc. Labor, 998; Hall, Com., 663; Reed, Ind., 506; scattered, 893.

1976, Carter, Dem., 1,040,232; Ford, Rep., 1,004,987; McCarthy, Ind., 34,943; Maddox, Amer. Ind., 8,552; Zeidler, Soc., 4,298; MacBride, Libertarian, 3,814; Camejo, Soc. Workers, 1,691; Wright, People's, 943; Hall, Com., 749; LaRouche, U.S. Lab., 738; Levin, Soc. Labor, 389; scattered, 2,839.

1980, Reagan, Rep., 1,088,845; Carter, Dem., 981,584; Anderson, Ind., 160,657; Clark, Libertarian, 29,135; Commoner, Citizens, 7,767; Rarick, Constitution, 1,519; McReynolds, Soc., 808; Hall, Com., 772; Griswold, Workers World, 414; DeBerry, Soc. Workers, 383; scattering, 1,337.

1984, Reagan, Rep., 1,198,584; Mondale, Dem., 995,740; Bergland, Libertarian, 4,883.

1988, Bush, Rep., 1,047,499; Dukakis, Dem., 1,126,794; Paul, Lib., 5,157; Duke, Pop., 3,056.

1992, Clinton, Dem., 1,041,066; Bush, Rep., 930,855; Perot, Ind., 544,479; Marrou, Libertarian, 2,877; Gritz, Populist/America

First, 2,311; Daniels, Ind., 1,883; Phillips, U.S. Taxpayers, 1,772; Hagelin, Natural Law, 1,070.

1996, Clinton, Dem., 1,071,971; Dole, Rep., 845,029; Perot, Ref., 227,339; Nader, Green, 28,723; Phillips, Taxpayers, 8,811; Browne, Libertarian, 7,929; Hagelin, Natural Law, 1,379; Moorehead, Workers World, 1,333; Hollis, Soc., 848; Harris, Soc. Workers, 483.

Wyoming

County	1996 Clinton (D)	Dole (R)	Perot (RF)	1992 Clinton (D)	Bush (R)	Perot (I)
Albany	6,399	5,967	1,333	5,713	4,176	2,862
Big Horn	1,438	2,821	545	1,216	2,216	1,236
Campbell. . . .	3,468	6,382	1,954	2,709	5,315	3,133
Carbon.	2,690	2,930	855	2,737	2,320	1,579
Converse. . . .	1,520	2,702	639	1,307	2,159	1,260
Crook.	651	1,698	394	568	1,377	718
Fremont.	5,445	7,554	1,840	4,765	5,387	3,594
Goshen	1,923	2,989	547	1,754	2,395	1,144
Hot Springs . .	779	1,348	287	740	978	652
Johnson.	815	2,071	378	656	1,614	844
Laramie	13,676	16,924	2,958	12,177	12,890	6,607
Lincoln.	1,803	3,764	906	1,430	2,595	1,495
Natrona	11,240	13,182	3,524	9,817	9,717	7,647
Niobrara.	325	757	209	298	635	355
Park.	3,240	7,430	1,318	2,771	5,218	3,145
Platte.	1,631	2,155	579	1,398	1,668	956
Sheridan	4,594	5,892	1,414	4,139	4,303	3,035
Sublette	677	1,829	401	536	1,168	828
Sweetwater . .	7,088	5,591	2,792	6,417	4,476	3,879
Teton	4,042	3,918	839	3,120	2,854	2,340
Uinta	2,414	3,471	1,242	2,047	2,701	2,041
Washakie. . . .	1,205	2,250	470	1,118	1,720	1,084
Weston	871	1,763	504	727	1,465	829
Totals	77,934	105,388	25,928	68,160	79,347	51,263

Wyoming Vote Since 1948

1948, Truman, Dem., 52,354; Dewey, Rep., 47,947; Wallace, Prog., 931; Thomas, Soc., 137; Teichert, Soc. Labor, 56.

1952, Eisenhower, Rep., 81,047; Stevenson, Dem., 47,934; Hamblen, Proh., 194; Hoopes, Soc., 40; Haas, Soc. Labor, 36.

1956, Eisenhower, Rep., 74,573; Stevenson, Dem., 49,554.

1960, Kennedy, Dem., 63,331; Nixon, Rep., 77,451.

1964, Johnson, Dem., 80,718; Goldwater, Rep., 61,998.

1968, Nixon, Rep., 70,927; Humphrey, Dem., 45,173; Wallace, 3d Party, 11,105.

1972, Nixon, Rep., 100,464; McGovern, Dem., 44,358; Schmitz, Amer., 748.

1976, Carter, Dem., 62,239; Ford, Rep., 92,717; McCarthy, Ind., 624; Reagan, Ind., 307; Anderson, Amer., 290; MacBride, Libertarian, 89; Brown, Ind., 47; Maddox, Amer. Ind., 30.

1980, Reagan, Rep., 110,700; Carter, Dem., 49,427; Anderson, Ind., 12,072; Clark, Libertarian, 4,514.

1984, Reagan, Rep., 133,241; Mondale, Dem., 53,370; Bergland, Libertarian, 2,357.

1988, Bush, Rep., 106,867; Dukakis, Dem., 67,113; Paul, Lib., 2,026; Fulani, New Alliance, 545.

1992, Clinton, Dem., 68,160; Bush, Rep., 79,347; Perot, Ind., 51,263.

1996, Dole, Rep., 105,388; Clinton, Dem., 77,934; Perot, Ind. (Ref.), 25,928; Browne, Libertarian, 1,739; Hagelin, Natural Law, 582.

1996 Official Presidential General Election Results

Source: Voter News Service; *Congressional Quarterly*

Candidate (Party)	Popular Vote	Percent of Popular Vote	Candidate (Party)	Popular Vote	Percent of Popular Vote
Bill Clinton (Democrat)	47,401,185	49.25	Jerome White (Socialist Equality) . . .	2,438	.00
Bob Dole (Republican)	39,197,469	40.73	Diane Beall Templin (American)	1,847	.00
Ross Perot (Reform).	8,085,294	8.40	Earl Dodge (Prohibition)	1,293	.00
Ralph Nader (Green)	651,771	.68	A. Peter Crane (Independent).	1,101	.00
Harry Browne (Libertarian).	485,120	.50	Ralph Forbes (Independent).	932	.00
Howard Phillips (U.S. Taxpayers) . . .	182,924	.19	John Birrenbach (Independent Grass Roots).	787	.00
John Hagelin (Natural Law)	112,978	.12	Isabell Masters (Independent)	749	.00
Monica Moorehead (Workers World)	29,082	.03	Steve Michael (Independent)	408	.00
Marsha Feinland (Peace & Freedom)	25,332	.03	Write-in	24,475	.03
James Harris (Socialist Workers) . . .	8,286	.01	None of These Candidates (Nevada)	5,608	.01
Charles Collins (Independent)	7,899	.01	**TOTAL**	**96,236,625**	**100**
Dennis Peron (Grass Roots)	5,378	.01			
Mary Cal Hollis (Socialist)	4,269	.00			

Note: Party designations may vary from one state to another.

Voter Turnout in Presidential Elections, 1932-96

Source: Federal Election Commission; Commission for Study of American Electorate; *Congressional Quarterly*

Candidates	Voter Participation (% of voting-age population)	Candidates	Voter Participation (% of voting-age population)
1932 Roosevelt-Hoover	52.4	1968 Humphrey-Nixon	60.9
1936 Roosevelt-Landon	56.0	1972 McGovern-Nixon	55.2[1]
1940 Roosevelt-Willkie	58.9	1976 Carter-Ford	53.5
1944 Roosevelt-Dewey	56.0	1980 Carter-Reagan	54.0
1948 Truman-Dewey	51.1	1984 Mondale-Reagan	53.1
1952 Stevenson-Eisenhower	61.6	1988 Dukakis-Bush	50.2
1956 Stevenson-Eisenhower	59.3	1992 Clinton-Bush-Perot	55.9
1960 Kennedy-Nixon	62.8	1996 Clinton-Dole-Perot	49.0
1964 Johnson-Goldwater	61.9		

(1) The sharp drop in 1972 followed the expansion of eligibility with the enfranchisement of 18- to 20-year-olds.

Electoral Votes for President

(based on 1990 Census)

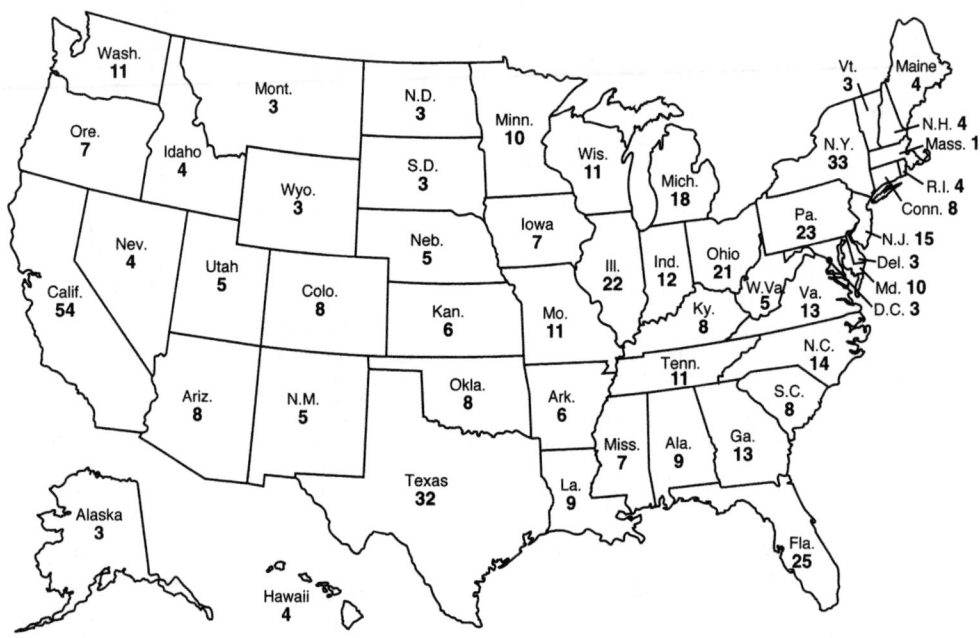

The Electoral College

The president and the vice president are the only elective federal officials not chosen by direct vote of the people. They are elected by the members of the Electoral College, an institution provided for in the U.S. Constitution.

On presidential election day, the first Tuesday after the first Monday in Nov. of every 4th year, each state chooses as many electors as it has senators and representatives in Congress. In 1964, for the first time, as provided by the 23d Amendment to the Constitution, the District of Columbia voted for 3 electors. Thus, with 100 senators and 435 representatives, there are 538 members of the Electoral College, with a majority of 270 electoral votes needed to elect the president and vice president.

Although political parties were not part of the original plan created by the Founding Fathers, today political parties customarily nominate their lists of electors at their respective state conventions. Some states print names of the candidates for president and vice president at the top of the Nov. ballot; others list only the electors' names. In either case, the electors of the party receiving the highest vote are elected.

The electors meet on the first Monday after the 2d Wednesday in Dec. in their respective state capitals or in some other place prescribed by state legislatures. By long-established custom, they vote for their party nominees, although this is not required by law.

The Constitution requires electors to cast a ballot for at least one person who is not an inhabitant of that elector's home state. This ensures that presidential and vice presidential candidates from the same party will not be from the same state. Also, an elector cannot be a member of Congress or hold federal office.

Certified and sealed lists of the votes of the electors in each state are sent to the president of the U.S. Senate, who then opens them in the presence of the members of the Senate and House of Representatives in a joint session held in early Jan., and the electoral votes of all the states are then officially counted. (The count was made on Jan. 9 in 1997.)

If no candidate for president has a majority, the House of Representatives chooses a president from the top 3 candidates, with all representatives from each state combining to cast one vote for that state. The House decided the outcome of the 1800 and 1824 presidential elections. If no candidate for vice president has a majority, the Senate chooses from the top 2, with the senators voting as individuals. The Senate chose the vice president following the 1836 election.

Under the electoral college system, a candidate who fails to be the top vote getter in the popular vote still may win a majority of electoral votes. This happened in the elections of 1876 and 1888.

In the 1996 election, Bill Clinton won 379 electoral votes, and Bob Dole won 159.

Third-Party and Independent Presidential Candidates

Although many "third party" candidates or independents have pursued the presidency, only 9 of these have polled more than a million votes. In most elections since 1860, fewer than one vote in 20 has been cast for a third-party candidate. In only 5 presidential elections since then have all non-major-party candidates combined polled more than 10% of the vote. The major vote getters in those elections were James B. Weaver (People's Party), 1892; former President Theodore Roosevelt (Progressive Party), 1912; Robert M. La Follette (Progressive Party), 1924; George C. Wallace (American Independent Party), 1968; and H. Ross Perot, as an independent in 1992.

Roosevelt outpolled the Republican candidate, William Howard Taft, in 1912, capturing 28% of the popular vote

and 88 electoral votes. In 1948, Strom Thurmond was able to capture 39 electoral votes (from 5 Southern states); however, all third parties received only 5.75% of the popular vote in the election. Twenty years later, George Wallace's popularity in the same region allowed him to get 46 electoral votes and 13.5% of the popular vote. In 1992 Perot was able to capture 19% of the popular vote; however, he did not win a single state. In 1996, Perot (as the candidate of his newly formed Reform Party) won 8% of the vote; all third-party candidates combined won 10%.

Despite the difficulty in winning the presidency, independent and third-party candidates sometimes succeed in winning other offices and often bring the attention of all presidential candidates combined to particular issues.

Notable Third Party and Independent Campaigns by Year

Party	Presidential nominee	Year	Issues	Strength in ...
Anti-Masonic	William Wirt	1832	Against secret societies and oaths	PA, VT
Liberty	James G. Birney	1844	Anti-slavery	North
Free Soil	Martin Van Buren	1848	Anti-slavery	NY, OH
American (Know-Nothing)	Millard Fillmore	1856	Anti-immigrant	Northeast, South
Greenback	Peter Cooper	1876	For "cheap money," labor rights	National
Greenback	James B. Weaver	1880	For "cheap money," labor rights	National
Prohibition	John P. St. John	1884	Anti-liquor	National
People's (Populists)	James B. Weaver	1892	For "cheap money," end of national banks	South, West
Socialist	Eugene V. Debs	1900-12; 1920	For public ownership	National
Progressive (Bull Moose)	Theodore Roosevelt	1912	Against high tariffs	Midwest, West
Progressive	Robert M. La Follette	1924	Farmer and labor rights	Midwest, West
Socialist	Norman Thomas	1928-48	Liberal reforms	National
Union	William Lemke	1936	Anti-New Deal	National
States' Rights (Dixiecrats)	Strom Thurmond	1948	For states' rights	South
Progressive	Henry A. Wallace	1948	Anti-cold war	NY, CA
American Independent	George C. Wallace	1968	For states' rights	South
American	John G. Schmitz	1972	For "law and order"	Far West, OH, LA
None (Independent)	John B. Anderson	1980	A 3d choice	National
None (Independent)	H. Ross Perot	1992	Federal budget deficit	National
Reform	H. Ross Perot	1996	Deficit; campaign finance	National

Major-Party Nominees for President and Vice President

Asterisk (*) denotes winning ticket

	Democratic		Republican	
Year	President	Vice President	President	Vice President
1856	James Buchanan*	John Breckinridge	John Frémont	William Dayton
1860	Stephen A. Douglas (1)	Herschel V. Johnson	Abraham Lincoln*	Hannibal Hamlin
1864	George McClellan	G.H. Pendleton	Abraham Lincoln*	Andrew Johnson
1868	Horatio Seymour	Francis Blair	Ulysses S. Grant*	Schuyler Colfax
1872	Horace Greeley	B. Gratz Brown	Ulysses S. Grant*	Henry Wilson
1876	Samuel J. Tilden	Thomas Hendricks	Rutherford B. Hayes*	William Wheeler
1880	Winfield Hancock	William English	James A. Garfield*	Chester A. Arthur
1884	Grover Cleveland*	Thomas Hendricks	James Blaine	John Logan
1888	Grover Cleveland	A.G. Thurman	Benjamin Harrison*	Levi Morton
1892	Grover Cleveland*	Adlai Stevenson	Benjamin Harrison	Whitelaw Reid
1896	William J. Bryan	Arthur Sewall	William McKinley*	Garret Hobart
1900	William J. Bryan	Adlai Stevenson	William McKinley*	Theodore Roosevelt
1904	Alton Parker	Henry Davis	Theodore Roosevelt*	Charles Fairbanks
1908	William J. Bryan	John Kern	William H. Taft*	James Sherman
1912	Woodrow Wilson*	Thomas Marshall	William H. Taft	James Sherman(2)
1916	Woodrow Wilson*	Thomas Marshall	Charles Hughes	Charles Fairbanks
1920	James M. Cox	Franklin D. Roosevelt	Warren G. Harding*	Calvin Coolidge
1924	John W. Davis	Charles W. Bryan	Calvin Coolidge*	Charles G. Dawes
1928	Alfred E. Smith	Joseph T. Robinson	Herbert Hoover*	Charles Curtis
1932	Franklin D. Roosevelt*	John N. Garner	Herbert Hoover	Charles Curtis
1936	Franklin D. Roosevelt*	John N. Garner	Alfred M. Landon	Frank Knox
1940	Franklin D. Roosevelt*	Henry A. Wallace	Wendell L. Willkie	Charles McNary
1944	Franklin D. Roosevelt*	Harry S. Truman	Thomas E. Dewey	John W. Bricker
1948	Harry S. Truman*	Alben W. Barkley	Thomas E. Dewey	Earl Warren
1952	Adlai E. Stevenson	John J. Sparkman	Dwight D. Eisenhower*	Richard M. Nixon
1956	Adlai E. Stevenson	Estes Kefauver	Dwight D. Eisenhower*	Richard M. Nixon
1960	John F. Kennedy*	Lyndon B. Johnson	Richard M. Nixon	Henry Cabot Lodge
1964	Lyndon B. Johnson*	Hubert H. Humphrey	Barry M. Goldwater	William E. Miller
1968	Hubert H. Humphrey	Edmund S. Muskie	Richard M. Nixon*	Spiro T. Agnew
1972	George S. McGovern	R. Sargent Shriver Jr.	Richard M. Nixon*	Spiro T. Agnew
1976	Jimmy Carter*	Walter F. Mondale	Gerald R. Ford	Bob Dole
1980	Jimmy Carter	Walter F. Mondale	Ronald Reagan*	George Bush
1984	Walter F. Mondale	Geraldine Ferraro	Ronald Reagan*	George Bush
1988	Michael S. Dukakis	Lloyd Bentsen	George Bush*	Dan Quayle
1992	Bill Clinton*	Al Gore	George Bush	Dan Quayle
1996	Bill Clinton*	Al Gore	Bob Dole	Jack Kemp

(1) Douglas and Johnson were nominated at the Baltimore convention. An earlier convention in Charleston, SC, failed to reach a consensus and resulted in a split in the party. The Southern faction of the Democrats nominated John Breckinridge for president and Joseph Lane for vice president. (2) Died Oct. 30; replaced on ballot by Nicholas Butler.

Popular and Electoral Vote for President

(D) Democrat; (DR) Democratic Republican; (F) Federalist; (LR) Liberal Republican; (NR) National Republican; (P) People's; (PR) Progressive; (R) Republican; (RF) Reform; (SR) States' Rights; (W) Whig; Asterisk (*)—See notes.

Year	President elected	Popular	Elec.	Major losing candidate(s)	Popular	Elec.
1789	George Washington (F)	Unknown	69	No opposition	—	—
1792	George Washington (F)	Unknown	132	No opposition	—	—
1796	John Adams (F)	Unknown	71	Thomas Jefferson (DR)	Unknown	68
1800*	Thomas Jefferson (DR)	Unknown	73	Aaron Burr (DR)	Unknown	73
1804	Thomas Jefferson (DR)	Unknown	162	Charles Pinckney (F)	Unknown	14
1808	James Madison (DR)	Unknown	122	Charles Pinckney (F)	Unknown	47
1812	James Madison (DR)	Unknown	128	DeWitt Clinton (F)	Unknown	89
1816	James Monroe (DR)	Unknown	183	Rufus King (F)	Unknown	34
1820	James Monroe (DR)	Unknown	231	John Quincy Adams (DR)	Unknown	1
1824*	John Quincy Adams (DR)	105,321	84	Andrew Jackson (DR)	155,872	99
				Henry Clay (DR)	46,587	37
				William H. Crawford (DR)	44,282	41
1828	Andrew Jackson (D)	647,231	178	John Quincy Adams (NR)	509,097	83
1832	Andrew Jackson (D)	687,502	219	Henry Clay (NR)	530,189	49
1836	Martin Van Buren (D)	762,678	170	William H. Harrison (W)	548,007	73
1840	William H. Harrison (W)	1,275,017	234	Martin Van Buren (D)	1,128,702	60
1844	James K. Polk (D)	1,337,243	170	Henry Clay (W)	1,299,068	105
1848	Zachary Taylor (W)	1,360,101	163	Lewis Cass (D)	1,220,544	127
				Martin Van Buren (Free Soil)	291,501	—
1852	Franklin Pierce (D)	1,601,474	254	Winfield Scott (W)	1,386,578	42
1856	James Buchanan (D)	1,927,995	174	John C. Fremont (R)	1,391,555	114
				Millard Fillmore (American)	873,053	8
1860	Abraham Lincoln (R)	1,866,352	180	Stephen A. Douglas (D)	1,375,157	12
				John C. Breckinridge (D)	845,763	72
				John Bell (Const. Union)	589,581	39
1864	Abraham Lincoln (R)	2,216,067	212	George McClellan (D)	1,808,725	21
1868	Ulysses S. Grant (R)	3,015,071	214	Horatio Seymour (D)	2,709,615	80
1872*	Ulysses S. Grant (R)	3,597,070	286	Horace Greeley (D-LR)*	2,834,079	—
1876*	Rutherford B. Hayes (R)	4,033,950	185	Samuel J. Tilden (D)	4,284,757	184
1880	James A. Garfield (R)	4,449,053	214	Winfield S. Hancock (D)	4,442,030	155
1884	Grover Cleveland (D)	4,911,017	219	James G. Blaine (R)	4,848,334	182
1888*	Benjamin Harrison (R)	5,444,337	233	Grover Cleveland (D)	5,540,050	168
1892	Grover Cleveland (D)	5,554,414	277	Benjamin Harrison (R)	5,190,802	145
				James Weaver (P)	1,027,329	22
1896	William McKinley (R)	7,035,638	271	William J. Bryan (D-P)	6,467,946	176
1900	William McKinley (R)	7,219,530	292	William J. Bryan (D)	6,358,071	155
1904	Theodore Roosevelt (R)	7,628,834	336	Alton B. Parker (D)	5,084,491	140
1908	William H. Taft (R)	7,679,006	321	William J. Bryan (D)	6,409,106	162
1912	Woodrow Wilson (D)	6,286,214	435	Theodore Roosevelt (PR)	4,216,020	88
				William H. Taft (R)	3,483,922	8
1916	Woodrow Wilson (D)	9,129,606	277	Charles E. Hughes (R)	8,538,221	254
1920	Warren G. Harding (R)	16,152,200	404	James M. Cox (D)	9,147,353	127
1924	Calvin Coolidge (R)	15,725,016	382	John W. Davis (D)	8,385,586	136
				Robert M. La Follette (PR)	4,822,856	13
1928	Herbert Hoover (R)	21,392,190	444	Alfred E. Smith (D)	15,016,443	87
1932	Franklin D. Roosevelt (D)	22,821,857	472	Herbert Hoover (R)	15,761,841	59
1936	Franklin D. Roosevelt (D)	27,751,597	523	Alfred Landon (R)	16,679,583	8
1940	Franklin D. Roosevelt (D)	27,243,466	449	Wendell Willkie (R)	22,304,755	82
1944	Franklin D. Roosevelt (D)	25,602,505	432	Thomas E. Dewey (R)	22,006,278	99
1948	Harry S. Truman (D)	24,105,812	303	Thomas E. Dewey (R)	21,970,065	189
				Strom Thurmond (SR)	1,169,021	39
				Henry A. Wallace (PR)	1,157,172	—
1952	Dwight D. Eisenhower (R)	33,936,252	442	Adlai E. Stevenson (D)	27,314,992	89
1956*	Dwight D. Eisenhower (R)	35,585,316	457	Adlai E. Stevenson (D)	26,031,322	73
1960*	John F. Kennedy (D)	34,227,096	303	Richard M. Nixon (R)	34,108,546	219
1964	Lyndon B. Johnson (D)	43,126,506	486	Barry M. Goldwater (R)	27,176,799	52
1968	Richard M. Nixon (R)	31,785,480	301	Hubert H. Humphrey (D)	31,275,166	191
				George C. Wallace (3d party)	9,906,473	46
1972*	Richard M. Nixon (R)	47,165,234	520	George S. McGovern (D)	29,170,774	17
1976*	Jimmy Carter (D)	40,828,929	297	Gerald R. Ford (R)	39,148,940	240
1980	Ronald Reagan (R)	43,899,248	489	Jimmy Carter (D)	35,481,435	49
				John B. Anderson (independent)	5,719,437	—
1984	Ronald Reagan (R)	54,281,858	525	Walter F. Mondale (D)	37,457,215	13
1988*	George Bush (R)	48,881,221	426	Michael S. Dukakis (D)	41,805,422	111
1992	Bill Clinton (D)	44,908,254	370	George Bush (R)	39,102,343	168
				H. Ross Perot (independent)	19,741,065	—
1996	Bill Clinton (D)	47,401,185	379	Bob Dole (R)	39,197,469	159
				H. Ross Perot (RF)	8,085,294	—

1800—Elected by House of Representatives because of tied electoral vote. **1824**—Elected by House of Representatives because no candidate had polled a majority. By 1824, the Democratic Republicans had become a loose coalition of competing political groups. By 1828, the supporters of Jackson were known as Democrats, and the John Q. Adams and Henry Clay supporters as National Republicans. **1872**—Greeley died Nov. 29, 1872. His electoral votes were split among 4 individuals. **1876**—FL, LA, OR, and SC election returns were disputed. Congress in joint session (Mar. 2, 1877) declared Hayes and Wheeler elected president and vice president. **1888**—Cleveland had more popular votes than Harrison, but since Harrison won 233 electoral votes against 168 for Cleveland, Harrison won the presidency. **1956**—Democrats elected 74 electors, but one from Alabama refused to vote for Stevenson. **1960**—Sen. Harry F. Byrd (D, VA) received 15 electoral votes. **1972**—John Hospers of California received one vote from an elector of Virginia. **1976**—Ronald Reagan of CA received one vote from an elector of Washington. **1988**—Sen. Lloyd Bentsen (D, TX) received 1 vote from an elector of West Virginia.

Presidents of the U.S.

No.	Name	Politics	Born	in	Inaug.	at age	Died	at age
1.	George Washington	Fed.	1732, Feb. 22	VA	1789	57	1799, Dec. 14	67
2.	John Adams	Fed.	1735, Oct. 30	MA	1797	61	1826, July 4	90
3.	Thomas Jefferson	Dem.-Rep.	1743, Apr. 13	VA	1801	57	1826, July 4	83
4.	James Madison	Dem.-Rep.	1751, Mar. 16	VA	1809	57	1836, June 28	85
5.	James Monroe	Dem.-Rep.	1758, Apr. 28	VA	1817	58	1831, July 4	73
6.	John Quincy Adams	Dem.-Rep.	1767, July 11	MA	1825	57	1848, Feb. 23	80
7.	Andrew Jackson	Dem.	1767, Mar. 15	SC	1829	61	1845, June 8	78
8.	Martin Van Buren	Dem.	1782, Dec. 5	NY	1837	54	1862, July 24	79
9.	William Henry Harrison	Whig	1773, Feb. 9	VA	1841	68	1841, Apr. 4	68
10.	John Tyler	Whig	1790, Mar. 29	VA	1841	51	1862, Jan. 18	71
11.	James Knox Polk	Dem.	1795, Nov. 2	NC	1845	49	1849, June 15	53
12.	Zachary Taylor	Whig	1784, Nov. 24	VA	1849	64	1850, July 9	65
13.	Millard Fillmore	Whig	1800, Jan. 7	NY	1850	50	1874, Mar. 8	74
14.	Franklin Pierce	Dem.	1804, Nov. 23	NH	1853	48	1869, Oct. 8	64
15.	James Buchanan	Dem.	1791, Apr. 23	PA	1857	65	1868, June 1	77
16.	Abraham Lincoln	Rep.	1809, Feb. 12	KY	1861	52	1865, Apr. 15	56
17.	Andrew Johnson	(1)	1808, Dec. 29	NC	1865	56	1875, July 31	66
18.	Ulysses Simpson Grant	Rep.	1822, Apr. 27	OH	1869	46	1885, July 23	63
19.	Rutherford Birchard Hayes	Rep.	1822, Oct. 4	OH	1877	54	1893, Jan. 17	70
20.	James Abram Garfield	Rep.	1831, Nov. 19	OH	1881	49	1881, Sept. 19	49
21.	Chester Alan Arthur	Rep.	1830, Oct. 5	VT	1881	50	1886, Nov. 18	56
22.	Grover Cleveland	Dem.	1837, Mar. 18	NJ	1885	47	1908, June 24	71
23.	Benjamin Harrison	Rep.	1833, Aug. 20	OH	1889	55	1901, Mar. 13	67
24.	Grover Cleveland	Dem.	1837, Mar. 18	NJ	1893	55	1908, June 24	71
25.	William McKinley	Rep.	1843, Jan. 29	OH	1897	54	1901, Sept. 14	58
26.	Theodore Roosevelt	Rep.	1858, Oct. 27	NY	1901	42	1919, Jan. 6	60
27.	William Howard Taft	Rep.	1857, Sept. 15	OH	1909	51	1930, Mar. 8	72
28.	Woodrow Wilson	Dem.	1856, Dec. 28	VA	1913	56	1924, Feb. 3	67
29.	Warren Gamaliel Harding	Rep.	1865, Nov. 2	OH	1921	55	1923, Aug. 2	57
30.	Calvin Coolidge	Rep.	1872, July 4	VT	1923	51	1933, Jan. 5	60
31.	Herbert Clark Hoover	Rep.	1874, Aug. 10	IA	1929	54	1964, Oct. 20	90
32.	Franklin Delano Roosevelt	Dem.	1882, Jan. 30	NY	1933	51	1945, Apr. 12	63
33.	Harry S. Truman	Dem.	1884, May 8	MO	1945	60	1972, Dec. 26	88
34.	Dwight David Eisenhower	Rep.	1890, Oct. 14	TX	1953	62	1969, Mar. 28	78
35.	John Fitzgerald Kennedy	Dem.	1917, May 29	MA	1961	43	1963, Nov. 22	46
36.	Lyndon Baines Johnson	Dem.	1908, Aug. 27	TX	1963	55	1973, Jan. 22	64
37.	Richard Milhous Nixon (2)	Rep.	1913, Jan. 9	CA	1969	56	1994, Apr. 22	81
38.	Gerald Rudolph Ford	Rep.	1913, July 14	NE	1974	61		
39.	Jimmy Carter	Dem.	1924, Oct. 1	GA	1977	52		
40.	Ronald Reagan	Rep.	1911, Feb. 6	IL	1981	69		
41.	George Bush	Rep.	1924, June 12	MA	1989	64		
42.	Bill Clinton	Dem.	1946, Aug. 19	AR	1993	46		

(1) Andrew Johnson was a Democrat, nominated vice president by Republicans, and elected with Lincoln on National Union ticket.
(2) Resigned Aug. 9, 1974.

U.S. Presidents, Vice Presidents, Congresses

	President	Service		Vice President	Congress
1.	George Washington	Apr. 30, 1789—Mar. 3, 1797	1.	John Adams	1, 2, 3, 4
2.	John Adams	Mar. 4, 1797—Mar. 3, 1801	2.	Thomas Jefferson	5, 6
3.	Thomas Jefferson	Mar. 4, 1801—Mar. 3, 1805	3.	Aaron Burr	7, 8
	"	Mar. 4, 1805—Mar. 3, 1809	4.	George Clinton	9, 10
4.	James Madison	Mar. 4, 1809—Mar. 3, 1813		" (1)	11, 12
	"	Mar. 4, 1813—Mar. 3, 1817	5.	Elbridge Gerry (2)	13, 14
5.	James Monroe	Mar. 4, 1817—Mar. 3, 1825	6.	Daniel D. Tompkins	15, 16, 17, 18
6.	John Quincy Adams	Mar. 4, 1825—Mar. 3, 1829	7.	John C. Calhoun	19, 20
7.	Andrew Jackson	Mar. 4, 1829—Mar. 3, 1833		" (3)	21, 22
	"	Mar. 4, 1833—Mar. 3, 1837	8.	Martin Van Buren	23, 24
8.	Martin Van Buren	Mar. 4, 1837—Mar. 3, 1841	9.	Richard M. Johnson	25, 26
9.	William Henry Harrison(4)	Mar. 4, 1841—Apr. 4, 1841	10.	John Tyler	27
10.	John Tyler	Apr. 6, 1841—Mar. 3, 1845			27, 28
11.	James K. Polk	Mar. 4, 1845—Mar. 3, 1849	11.	George M. Dallas	29, 30
12.	Zachary Taylor(4)	Mar. 5, 1849—July 9, 1850	12.	Millard Fillmore	31
13.	Millard Fillmore	July 10, 1850—Mar. 3, 1853			31, 32
14.	Franklin Pierce	Mar. 4, 1853—Mar. 3, 1857	13.	William R. King (5)	33, 34
15.	James Buchanan	Mar. 4, 1857—Mar. 3, 1861	14.	John C. Breckinridge	35, 36
16.	Abraham Lincoln	Mar. 4, 1861—Mar. 3, 1865	15.	Hannibal Hamlin	37, 38
	" (4)	Mar. 4, 1865—Apr. 15, 1865	16.	Andrew Johnson	39
17.	Andrew Johnson	Apr. 15, 1865—Mar. 3, 1869			39, 40
18.	Ulysses S. Grant	Mar. 4, 1869—Mar. 3, 1873	17.	Schuyler Colfax	41, 42
	"	Mar. 4, 1873—Mar. 3, 1877	18.	Henry Wilson (6)	43, 44
19.	Rutherford B. Hayes	Mar. 4, 1877—Mar. 3, 1881	19.	William A. Wheeler	45, 46
20.	James A. Garfield(4)	Mar. 4, 1881—Sept. 19, 1881	20.	Chester A. Arthur	47
21.	Chester A. Arthur	Sept. 20, 1881—Mar. 3, 1885			47, 48
22.	Grover Cleveland(7)	Mar. 4, 1885—Mar. 3, 1889	21.	Thomas A. Hendricks(8)	49, 50
23.	Benjamin Harrison	Mar. 4, 1889—Mar. 3, 1893	22.	Levi P. Morton	51, 52
24.	Grover Cleveland(7)	Mar. 4, 1893—Mar. 3, 1897	23.	Adlai E. Stevenson	53, 54
25.	William McKinley	Mar. 4, 1897—Mar. 3, 1901	24.	Garret A. Hobart(9)	55, 56
	" (4)	Mar. 4, 1901—Sept. 14, 1901	25.	Theodore Roosevelt	57
26.	Theodore Roosevelt	Sept. 14, 1901—Mar. 3, 1905			57, 58
	"	Mar. 4, 1905—Mar. 3, 1909	26.	Charles W. Fairbanks	59, 60
27.	William H. Taft	Mar. 4, 1909—Mar. 3, 1913	27.	James S. Sherman (10)	61, 62

	President	Service		Vice President	Congress
28.	Woodrow Wilson	Mar. 4, 1913—Mar. 3, 1921	28.	Thomas R. Marshall	63, 64, 65, 66
29.	Warren G. Harding (4)	Mar. 4, 1921—Aug. 2, 1923	29.	Calvin Coolidge	67
30.	Calvin Coolidge	Aug. 3, 1923—Mar. 3, 1925			68
	"	Mar. 4, 1925—Mar. 3, 1929	30.	Charles G. Dawes	69, 70
31.	Herbert C. Hoover	Mar. 4, 1929—Mar. 3, 1933	31.	Charles Curtis	71, 72
32.	Franklin D. Roosevelt (11)	Mar. 4, 1933—Jan. 20, 1941	32.	John N. Garner	73, 74, 75, 76
	" (4)	Jan. 20, 1941—Jan. 20, 1945	33.	Henry A. Wallace	77, 78
		Jan. 20, 1945—Apr. 12, 1945	34.	Harry S. Truman	79
33.	Harry S. Truman	Apr. 12, 1945—Jan. 20, 1949			79, 80
	"	Jan. 20, 1949—Jan. 20, 1953	35.	Alben W. Barkley	81, 82
34.	Dwight D. Eisenhower	Jan. 20, 1953—Jan. 20, 1961	36.	Richard M. Nixon	83, 84, 85, 86
35.	John F. Kennedy(4)	Jan. 20, 1961—Nov. 22, 1963	37.	Lyndon B. Johnson	87, 88
36.	Lyndon B. Johnson	Nov. 22, 1963—Jan. 20, 1965			88
	"	Jan. 20, 1965—Jan. 20, 1969	38.	Hubert H. Humphrey	89, 90
37.	Richard M. Nixon	Jan. 20, 1969—Jan. 20, 1973	39.	Spiro T. Agnew (12)	91, 92, 93
	" (13)	Jan. 20, 1973—Aug. 9, 1974	40.	Gerald R. Ford (14)	93
38.	Gerald R. Ford (15)	Aug. 9, 1974—Jan. 20, 1977	41.	Nelson A. Rockefeller (16)	93, 94
39.	Jimmy (James Earl) Carter	Jan. 20, 1977—Jan. 20, 1981	42.	Walter F. Mondale	95, 96
40.	Ronald Reagan	Jan. 20, 1981—Jan. 20, 1989	43.	George Bush	97, 98, 99, 100
41.	George Bush	Jan. 20, 1989—Jan. 20, 1993	44.	Dan Quayle	101, 102
42.	Bill Clinton	Jan. 20, 1993—	45.	Al Gore	103, 104, 105, 106

(1) Died Apr. 20, 1812. (2) Died Nov. 23, 1814. (3) Resigned Dec. 28, 1832, to become U.S. senator. (4) Died in office. (5) Died Apr. 18, 1853. (6) Died Nov. 22, 1875. (7) Terms not consecutive. (8) Died Nov. 25, 1885. (9) Died Nov. 21, 1899. (10) Died Oct. 30, 1912. (11) First president to be inaugurated under 20th Amendment, Jan. 20, 1937. (12) Resigned Oct. 10, 1973. (13) Resigned Aug. 9, 1974. (14) First nonelected vice president, chosen under 25th Amendment procedure. (15) First president never elected president or vice president. (16) Second nonelected vice president, chosen under 25th Amendment.

Vice Presidents of the U.S.

The numerals given vice presidents do not coincide with those given presidents, because some presidents had none and some had more than one.

	Name	Birthplace	Year	Home	Inaug.	Politics	Place of death	Year	Age
1.	John Adams	Quincy, MA	1735	MA	1789	Fed.	Quincy, MA	1826	90
2.	Thomas Jefferson	Shadwell, VA	1743	VA	1797	Dem.-Rep.	Monticello, VA	1826	83
3.	Aaron Burr	Newark, NJ	1756	NY	1801	Dem.-Rep.	Staten Island, NY	1836	80
4.	George Clinton	Ulster Co., NY	1739	NY	1805	Dem.-Rep.	Washington, DC	1812	73
5.	Elbridge Gerry	Marblehead, MA	1744	MA	1813	Dem.-Rep.	Washington, DC	1814	70
6.	Daniel D. Tompkins	Scarsdale, NY	1774	NY	1817	Dem.-Rep.	Staten Island, NY	1825	51
7.	John C. Calhoun (1)	Abbeville, SC	1782	SC	1825	Dem.-Rep.	Washington, DC	1850	68
8.	Martin Van Buren	Kinderhook, NY	1782	NY	1833	Dem.	Kinderhook, NY	1862	79
9.	Richard M. Johnson (2)	Louisville, KY	1780	KY	1837	Dem.	Frankfort, KY	1850	70
10.	John Tyler	Greenway, VA	1790	VA	1841	Whig	Richmond, VA	1862	71
11.	George M. Dallas	Philadelphia, PA	1792	PA	1845	Dem.	Philadelphia, PA	1864	72
12.	Millard Fillmore	Summerhill, NY	1800	NY	1849	Whig	Buffalo, NY	1874	74
13.	William R. King	Sampson Co., NC	1786	AL	1853	Dem.	Dallas Co., AL	1853	67
14.	John C. Breckinridge	Lexington, KY	1821	KY	1857	Dem.	Lexington, KY	1875	54
15.	Hannibal Hamlin	Paris, ME	1809	ME	1861	Rep.	Bangor, ME	1891	81
16.	Andrew Johnson	Raleigh, NC	1808	TN	1865	(3)	Carter Co., TN	1875	66
17.	Schuyler Colfax	New York, NY	1823	IN	1869	Rep.	Mankato, MN	1885	62
18.	Henry Wilson	Farmington, NH	1812	MA	1873	Rep.	Washington, DC	1875	63
19.	William A. Wheeler	Malone, NY	1819	NY	1877	Rep.	Malone, NY	1887	68
20.	Chester A. Arthur	Fairfield, VT	1830	NY	1881	Rep.	New York,NY	1886	57
21.	Thomas A. Hendricks	Muskingum Co., OH	1819	IN	1885	Dem.	Indianapolis, IN	1885	66
22.	Levi P. Morton	Shoreham, VT	1824	NY	1889	Rep.	Rhinebeck, NY	1920	96
23.	Adlai E. Stevenson (4)	Christian Co., KY	1835	IL	1893	Dem.	Chicago, IL	1914	78
24.	Garret A. Hobart	Long Branch, NJ	1844	NJ	1897	Rep.	Paterson, NJ	1899	55
25.	Theodore Roosevelt	New York, NY	1858	NY	1901	Rep.	Oyster Bay, NY	1919	60
26.	Charles W. Fairbanks	Unionville Centre, OH	1852	IN	1905	Rep.	Indianapolis, IN	1918	66
27.	James S. Sherman	Utica, NY	1855	NY	1909	Rep.	Utica, NY	1912	57
28.	Thomas R. Marshall	N. Manchester, IN	1854	IN	1913	Dem.	Washington, DC	1925	71
29.	Calvin Coolidge	Plymouth, VT	1872	MA	1921	Rep.	Northampton, MA	1933	60
30.	Charles G. Dawes	Marietta, OH	1865	IL	1925	Rep.	Evanston, IL	1951	85
31.	Charles Curtis	Topeka, KS	1860	KS	1929	Rep.	Washington, DC	1936	76
32.	John Nance Garner	Red River Co., TX	1868	TX	1933	Dem.	Uvalde, TX	1967	98
33.	Henry Agard Wallace	Adair County, IA	1888	IA	1941	Dem.	Danbury, CT	1965	77
34.	Harry S. Truman	Lamar, MO	1884	MO	1945	Dem.	Kansas City, MO	1972	88
35.	Alben W. Barkley	Graves County, KY	1877	KY	1949	Dem.	Lexington, VA	1956	78
36.	Richard M. Nixon	Yorba Linda, CA	1913	CA	1953	Rep.	New York, NY	1994	81
37.	Lyndon B. Johnson	Johnson City, TX	1908	TX	1961	Dem.	San Antonio, TX	1973	64
38.	Hubert H. Humphrey	Wallace, SD	1911	MN	1965	Dem.	Waverly, MN	1978	66
39.	Spiro T. Agnew(5)	Baltimore, MD	1918	MD	1969	Rep.	Berlin, MD	1996	77
40.	Gerald R. Ford(6)	Omaha, NE	1913	MI	1973	Rep.			
41.	Nelson A. Rockefeller(7)	Bar Harbor, ME	1908	NY	1974	Rep.	New York, NY	1979	70
42.	Walter F. Mondale	Ceylon, MN	1928	MN	1977	Dem.			
43.	George Bush	Milton, MA	1924	TX	1981	Rep.			
44.	Dan Quayle	Indianapolis, IN	1947	IN	1989	Rep.			
45.	Al Gore	Washington, DC	1948	TN	1993	Dem.			

(1) John C. Calhoun resigned Dec. 28, 1832, having been elected to the Senate to fill a vacancy. (2) Richard M. Johnson was the only vice president to be chosen by the Senate because of a tied vote in the Electoral College. (3) Andrew Johnson was a Democrat, nominated vice president by Republicans, and elected with Lincoln on the National Union Ticket. (4) Adlai E. Stevenson, 23d vice president, was grandfather of Democratic candidate for president in 1952 and 1956. (5) Resigned Oct. 10, 1973. (6) First nonelected vice president, chosen under 25th Amendment procedure. (7) Second nonelected vice president, chosen under 25th Amendment procedure.

AFGHANISTAN

ALBANIA

ALGERIA

ANDORRA

ANGOLA

ANTIGUA AND BARBUDA

ARGENTINA

ARMENIA

AUSTRALIA

AUSTRIA

AZERBAIJAN

THE BAHAMAS

BAHRAIN

BANGLADESH

BARBADOS

BELARUS

BELGIUM

BELIZE

BENIN

BHUTAN

BOLIVIA

BOSNIA AND HERZEGOVINA

BOTSWANA

BRAZIL

BRUNEI

BULGARIA

BURKINA FASO

BURUNDI

CAMBODIA

CAMEROON

CANADA

CAPE VERDE

CENTRAL AFRICAN REPUBLIC

CHAD

CHILE

CHINA

COLOMBIA

COMOROS

CONGO, DEM. REP. OF THE

CONGO REPUBLIC

COSTA RICA

CÔTE D'IVOIRE

CROATIA

CUBA

CYPRUS

CZECH REPUBLIC

DENMARK

DJIBOUTI

DOMINICA

DOMINICAN REPUBLIC

ECUADOR

EGYPT

EL SALVADOR

EQUATORIAL GUINEA

ERITREA

ESTONIA	ETHIOPIA	FIJI	FINLAND	FRANCE
GABON	THE GAMBIA	GEORGIA	GERMANY	GHANA
GREECE	GRENADA	GUATEMALA	GUINEA	GUINEA-BISSAU
GUYANA	HAITI	HONDURAS	HUNGARY	ICELAND
INDIA	INDONESIA	IRAN	IRAQ	IRELAND
ISRAEL	ITALY	JAMAICA	JAPAN	JORDAN
KAZAKHSTAN	KENYA	KIRIBATI	NORTH KOREA	SOUTH KOREA
KUWAIT	KYRGYZSTAN	LAOS	LATVIA	LEBANON
LESOTHO	LIBERIA	LIBYA	LIECHTENSTEIN	LITHUANIA
LUXEMBOURG	MACEDONIA	MADAGASCAR	MALAWI	MALAYSIA
MALDIVES	MALI	MALTA	MARSHALL ISLANDS	MAURITANIA

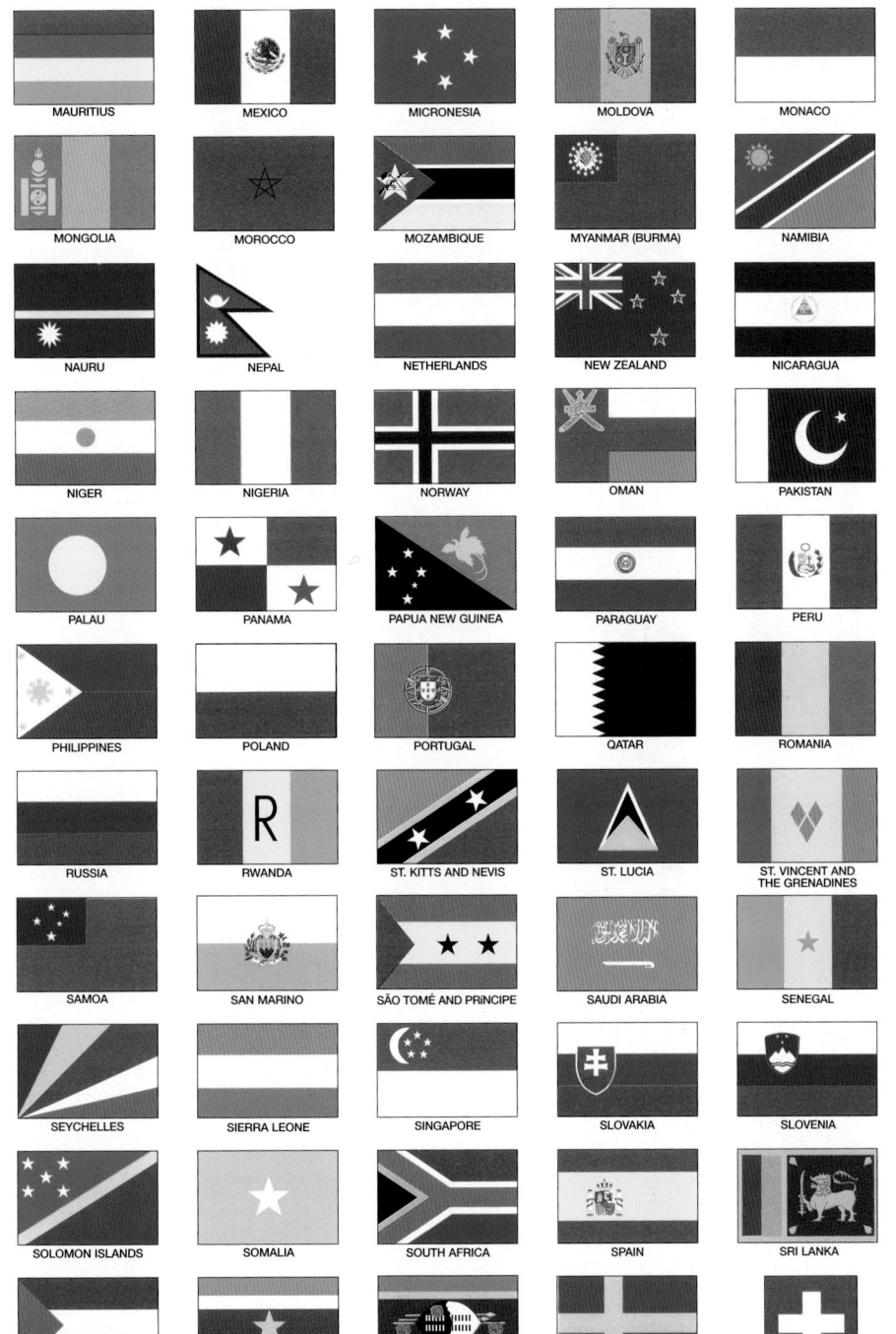

MAURITIUS   MEXICO   MICRONESIA   MOLDOVA   MONACO

MONGOLIA   MOROCCO   MOZAMBIQUE   MYANMAR (BURMA)   NAMIBIA

NAURU   NEPAL   NETHERLANDS   NEW ZEALAND   NICARAGUA

NIGER   NIGERIA   NORWAY   OMAN   PAKISTAN

PALAU   PANAMA   PAPUA NEW GUINEA   PARAGUAY   PERU

PHILIPPINES   POLAND   PORTUGAL   QATAR   ROMANIA

RUSSIA   RWANDA   ST. KITTS AND NEVIS   ST. LUCIA   ST. VINCENT AND THE GRENADINES

SAMOA   SAN MARINO   SÃO TOMÉ AND PRINCIPE   SAUDI ARABIA   SENEGAL

SEYCHELLES   SIERRA LEONE   SINGAPORE   SLOVAKIA   SLOVENIA

SOLOMON ISLANDS   SOMALIA   SOUTH AFRICA   SPAIN   SRI LANKA

SUDAN   SURINAME   SWAZILAND   SWEDEN   SWITZERLAND

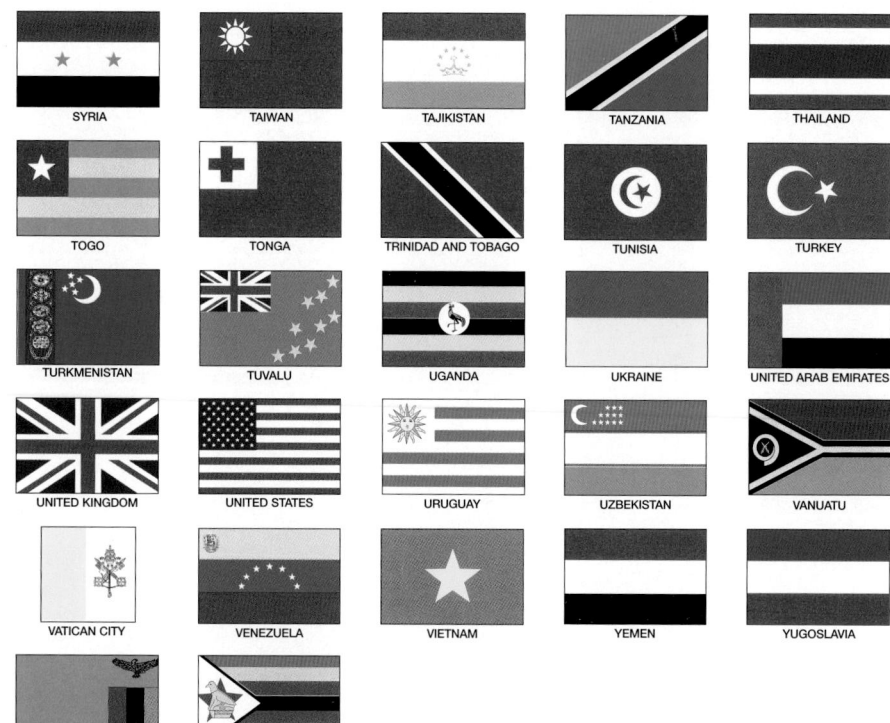

SYRIA	TAIWAN	TAJIKISTAN	TANZANIA	THAILAND
TOGO	TONGA	TRINIDAD AND TOBAGO	TUNISIA	TURKEY
TURKMENISTAN	TUVALU	UGANDA	UKRAINE	UNITED ARAB EMIRATES
UNITED KINGDOM	UNITED STATES	URUGUAY	UZBEKISTAN	VANUATU
VATICAN CITY	VENEZUELA	VIETNAM	YEMEN	YUGOSLAVIA
ZAMBIA	ZIMBABWE			

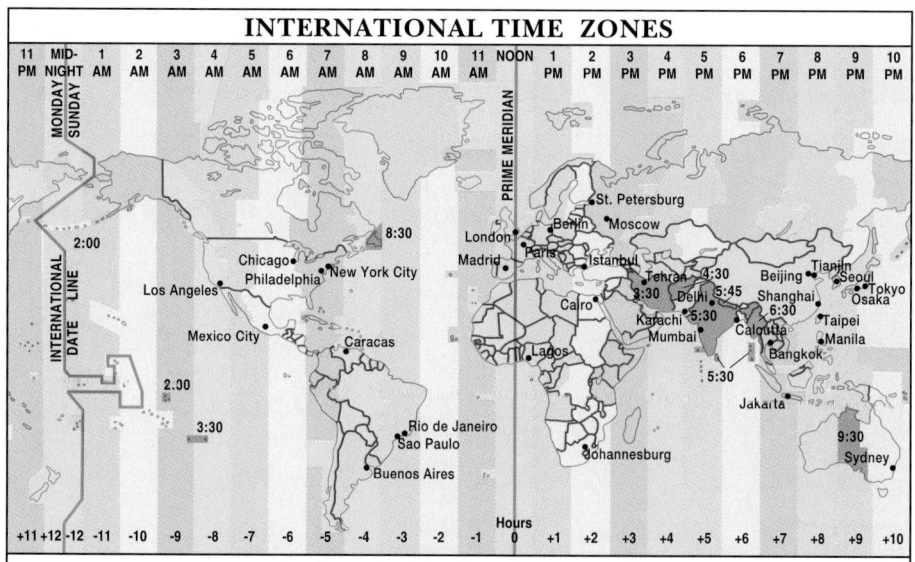

INTERNATIONAL TIME ZONES

The world is divided into 24 time zones, each 15° longitude wide. The longitudinal meridian passing through Greenwich, England, is the starting point, and is called the *prime meridian*. The 12th zone is divided by the 180th meridian (International Date Line). When the line is crossed going west, the date is advanced one day; when crossed going east, the date becomes a day earlier.

© MapQuest.com, Inc.

508

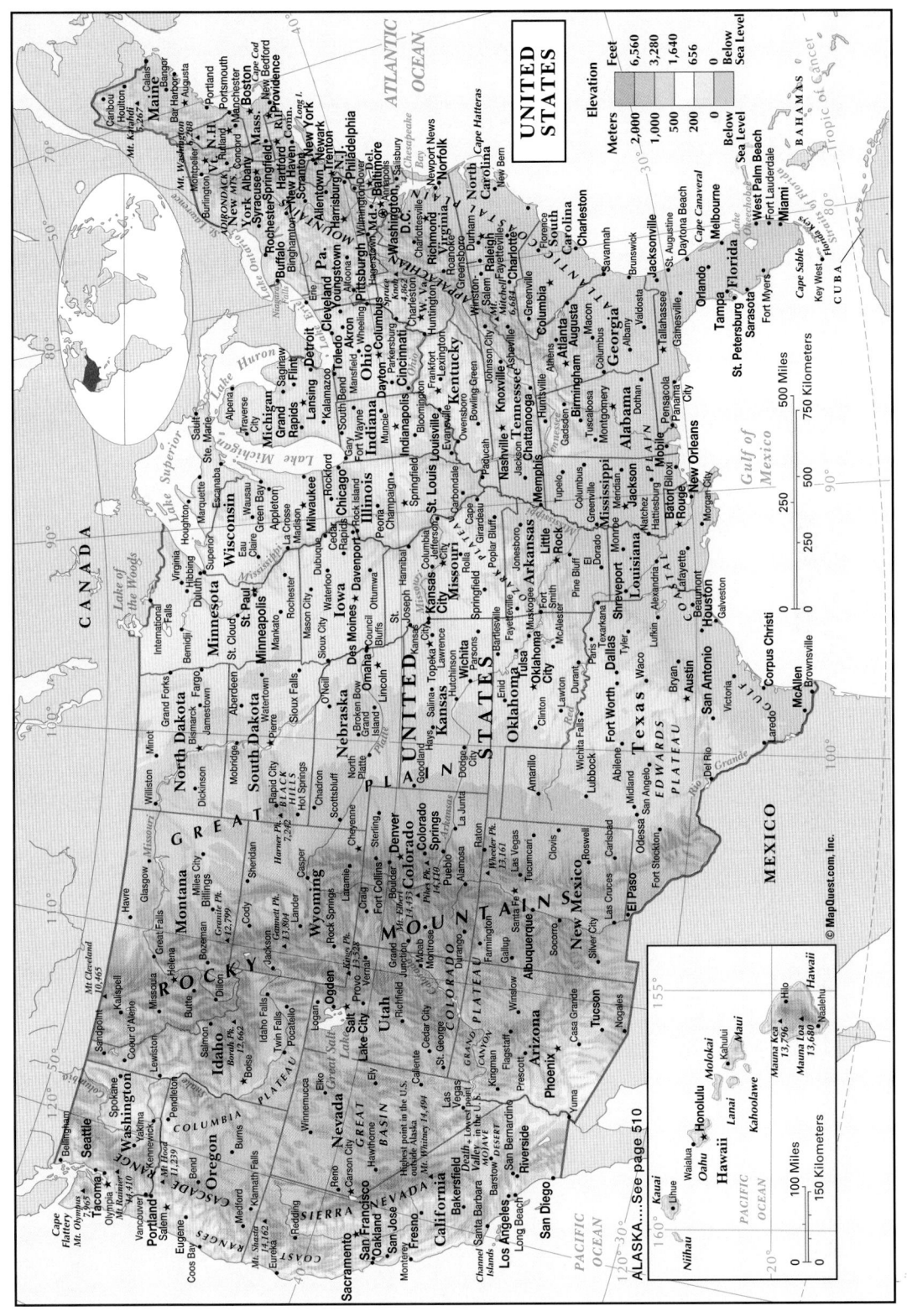

UNITED STATES

Elevation

Meters	Feet
2,000	6,560
1,000	3,280
500	1,640
200	656
Below Sea Level	0 Below Sea Level

© MapQuest.com, Inc.

ALASKA....See page 510

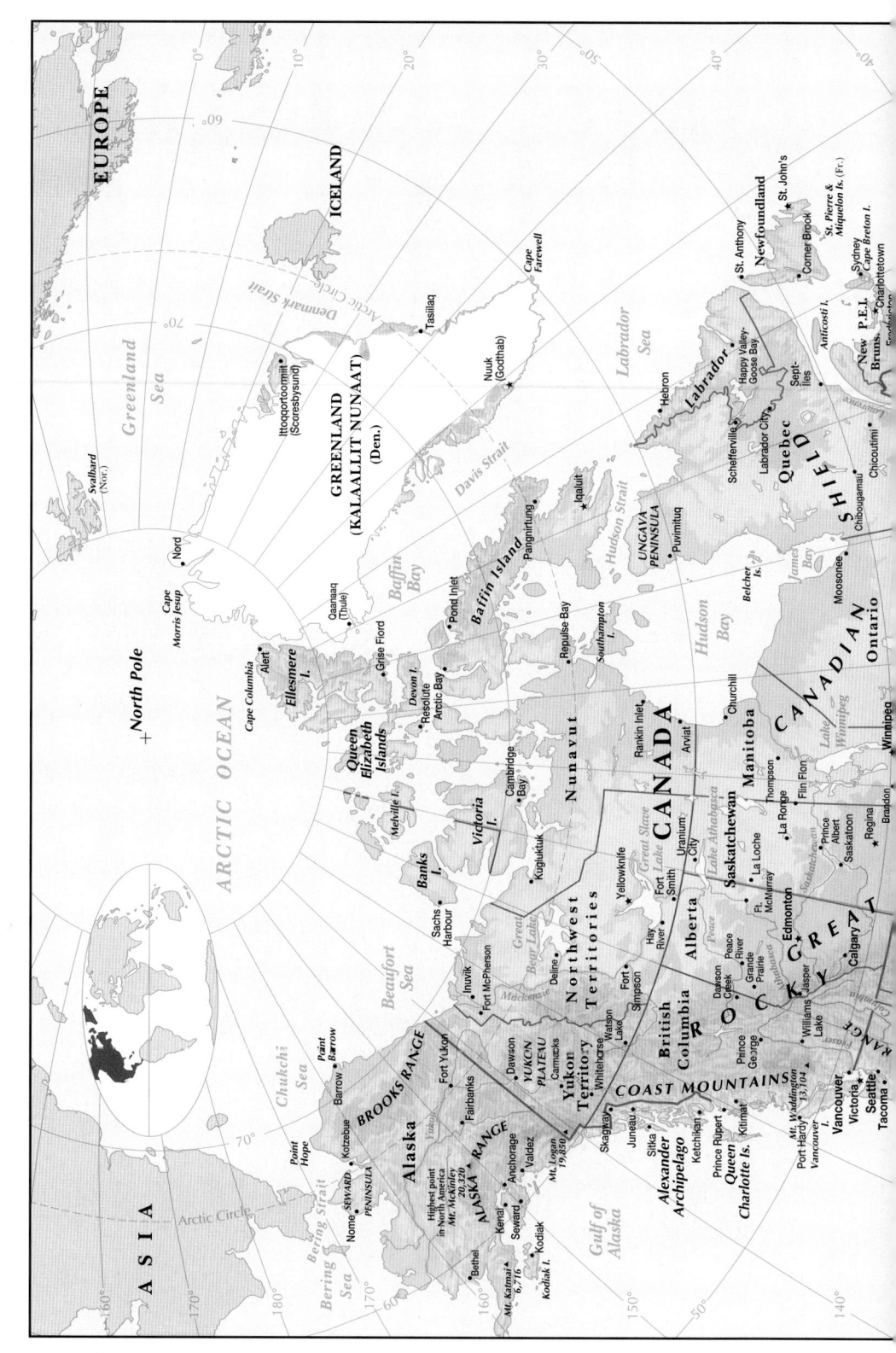

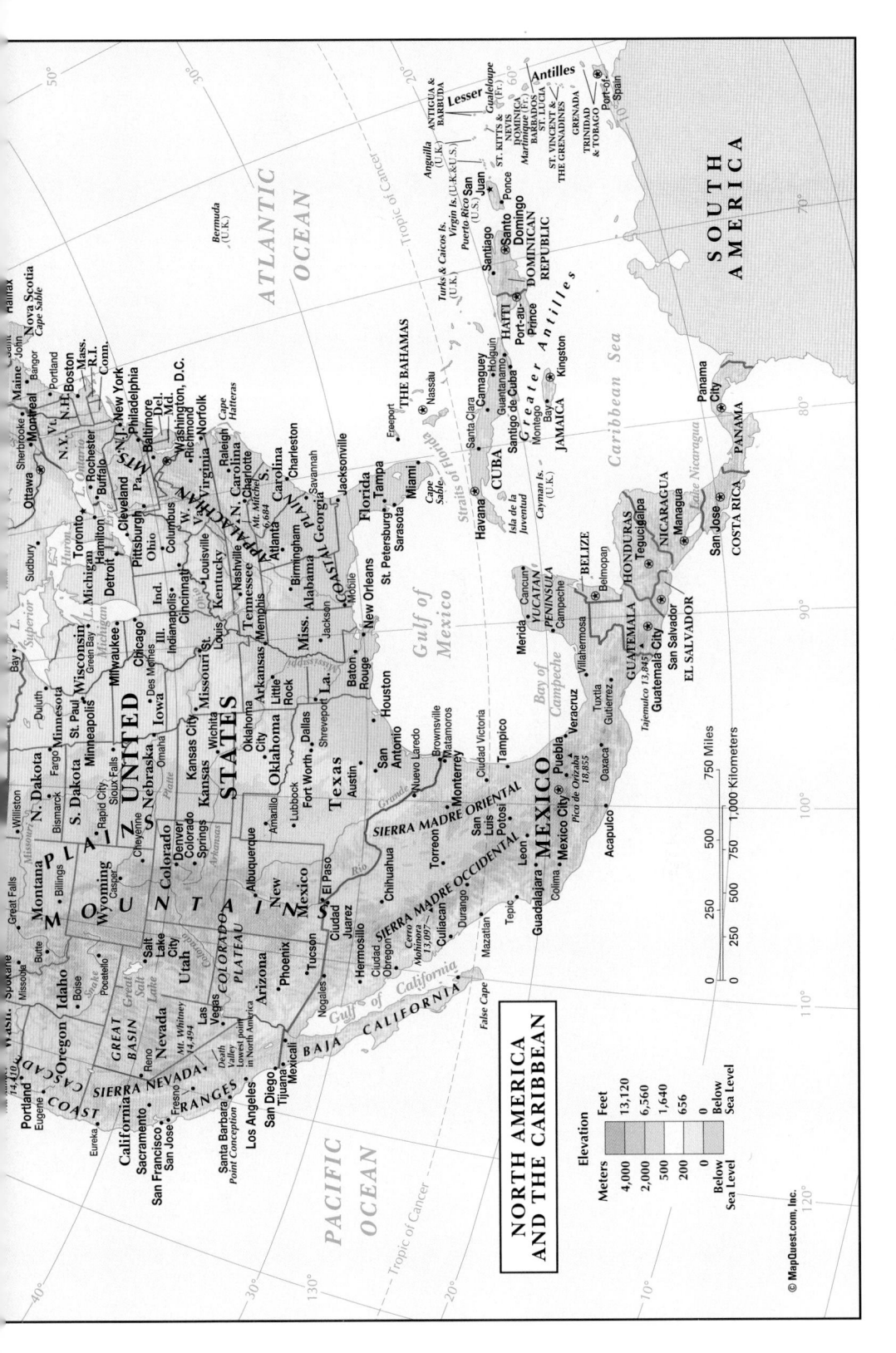

NORTH AMERICA
AND THE CARIBBEAN

Elevation

Meters	Feet
4,000	13,120
2,000	6,560
500	1,640
200	656
0	0
Below Sea Level	Below Sea Level

© MapQuest.com, Inc.

ATLANTIC OCEAN

PACIFIC OCEAN

SOUTH AMERICA

Caribbean Sea

Gulf of Mexico

UNITED STATES

MEXICO

Tropic of Cancer

0   250   500   750 Miles
0   250   500   750   1,000 Kilometers

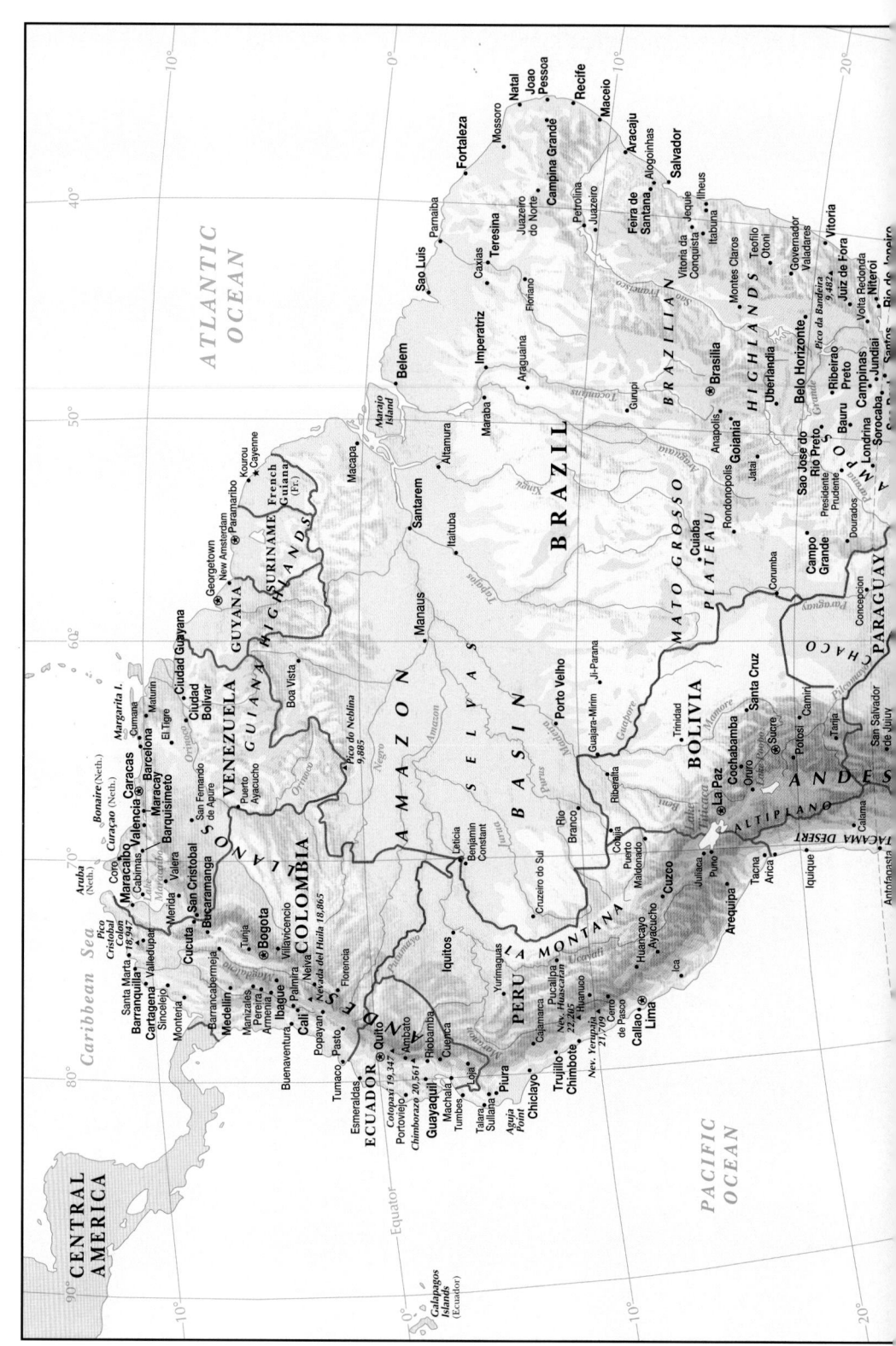

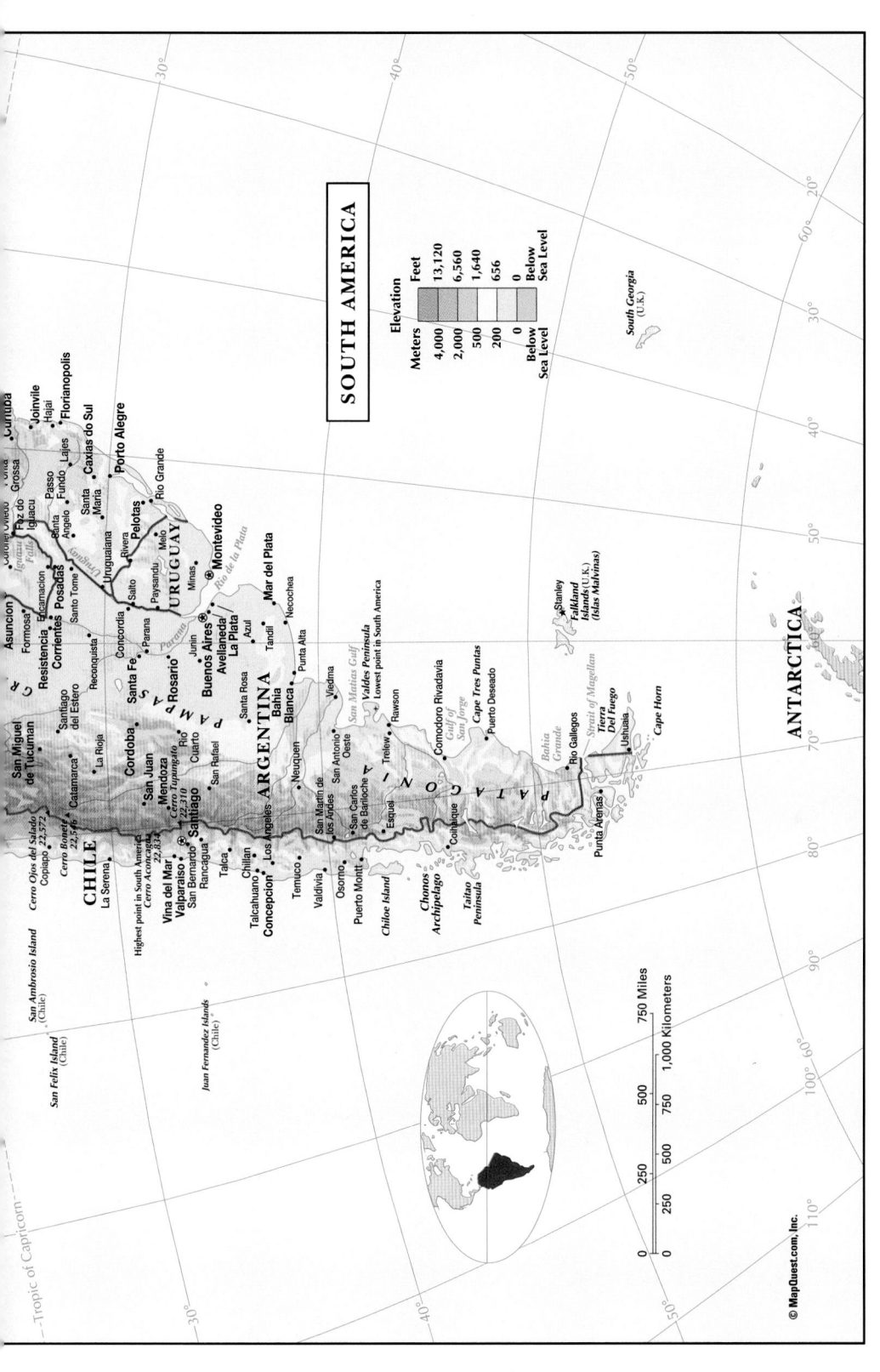

SOUTH AMERICA

EUROPE

Elevation

Meters		Feet
4,000		13,120
2,000		6,560
500		1,640
200		656
0		0
Below Sea Level		Below Sea Level

GREENLAND
(KALAALLIT NUNAAT)
(Denmark)

Arctic Circle

Isafjordhur

Akureyn
Keflavik
ICELAND
Reykjavik

Seydhisfjordhur

Narvik

Bodo

Norwegian Sea

Namsos

Torshavn *Faroe Islands* (Den.)

Trondheim
Molde
Alesund

Ostersund
Sundsvall

Shetland Islands (U.K.)

NORWAY   SWEDEN

Bergen

Haugesund   Oslo   Borlange
Stavanger   Drammen   Uppsala
Skien   Karlstad   Orebro

Orkney Islands
Hebrides
Thurso
Inverness

Kristiansand   Stockholm
Norrkoping
Jonkoping   Olar

Scotland
Aberdeen
Dundee

ATLANTIC OCEAN

Alborg   Vaxjo
Jutland   Arhus   Halmstad   Helsingborg
Esbjerg   DENMARK   Odense   Malmo
Copenhagen

Londonderry
Northern Ireland
Belfast

Glasgow
Ayr   Edinburgh

North Sea

Baltic
Bornholm (Den.)

Londonderry
Galway

IRELAND   Dublin
Limerick

UNITED   Newcastle
KINGDOM

Kiel   Rostock
Lubeck   Szczecin   Gdansk
Hamburg   NORTHERN

Liverpool   Leeds
Manchester   Kingston upon Hull
Cork   Waterford   Sheffield

Cardiff   Coventry   Norwich
Wales Swansea
Bristol   England

Bremen   Bydgoszcz
Groningen   Hannover   Berlin   Poznar
NETHERLANDS   Bielefeld   Madgeburg   POLANI
Amsterdam   Munster   GERMANY   Leipzig   Wroclaw
The Hague   Essen   Kassel   Erfurt   Dresden   Liberec   Walbrzyc
Rotterdam   Cologne   Chemnitz   Prague   Ostrava

Plymouth   London
Portsmouth   Dover

Land's End

English Channel

Channel Is. (U.K.)
Le Havre
Brest

BELGIUM Liege   Bonn   Pizen   Brno
Antwerp   Wiesbaden   CZECH. REP.
Brussels   Frankfurt   Regensburg   Bratislava
Lille   LUXEMBOURG   Mannheim   Nurnberg   Vienna
Luxembourg   Saarbrucken   Augsburg   Munich   Linz   HU
Caen   Rouen   Nancy   Stuttgart   Salzburg   AUSTRIA   Gyor
Rennes   Le Mans   Paris   Strasbourg   Basel   Innsbruck   Graz
Orleans   Dijon   Bern   LIECHTENSTEIN   Klagenfurt   Pecs
Nantes   Tours   FRANCE   Zurich   Udine   SLOVENIA
Limoges   Geneva SWITZERLAND   Bergamo   Trieste   Ljubljana   Zagreb
A Coruna   Clermont-Ferrand   Lyon   Mt. Blanc   Matterhorn   Verona   Rijeka   CROATIA
Vigo   Bordeaux   Saint-Etienne   15,771   Milan   Venice   DINAR   BOS.
Gijon   Santander   Grenoble   Torino   HERZ
Leon   Bilbao   Donostia-   Avignon   Nice   Genoa   Florence   Sarajevo
Braga   Vitoria-Gasteiz   San Sebastian   Toulouse   Marseille   MONACO   Pisa   SAN MARINO
Porto   Pamplona   Montpellier   Toulon   Perugia   Ancona   Split
Coimbra   Valladolid   PYRENEES   Corsica (Fr.)   Elba   Adriatic   Dubrovnik
IBERIAN   Pico de   Nice   Ajaccio   VATICAN CITY
Salamanca   Aneto   ANDORRA   Rome
PORTUGAL   11,168   Barcelona   ITALY   Foggia
Lisbon   Madrid   Zaragoza   Tarragona   Corsica (Fr.)   Vesuvius   Bari
Setubal   Toledo   Castellon de la Plana   Sassari   Naples   4,202   Taranto
Badajoz   Valencia   Majorca   Sardinia (It.)   Salerno
SPAIN   Palma de   Minorca
PENINSULA   Mallorca   Cagliari
Cape   Cordoba   Alicante   Balearic Is. (Sp.)   Tyrrhenian Sea   Ionia
St. Vincent   Seville   Murcia   Cartagena   Sea
Cadiz   Malaga   Granada   Almeria   Palermo   Messina
Strait of   Gibraltar   Etna   Reggio di
Gibraltar   (U.K.)   11,053   Calabria
Mediterranean   Sicily (It.)   Catania

AFRICA

0	250	500 Miles
0	250	500	750 Kilometers

MALTA Valletta   Sea

Barents Sea

North Cape
Hammerfest
Vardo
Tromso
Ivalo
Kiruna
LAPLAND
Murmansk
Apatity
KOLA PENINSULA
Novaya Zemlya
Naryan-Mar
Pechora
Ukhta
URAL
ASIA

Luleau
Skelleftea
Oulu
Rovaniemi
Belomorsk
Arkhangelsk
White Sea
Syktyvkar
Berezniki
RUSSIA
MOUNTAINS

FINLAND
Vaasa
Kuopio
Jyvaskyla
Lahti
Tampere
Pori
Turku
Kotka
Helsinki
Aland Is. (Fin.)
Gulf of Finland
Tallinn
ESTONIA
Tartu
Pskov
Lake Ladoga
Petrozavodsk
Lake Onega
Kotlas
Dvina
Kirov
Izhevsk
Perm
Ufa
Kama
Sterlitzmak

St. Petersburg
Cherepovets
Vologda
Rybinsk
Yaroslavl
Ivanovo
Kostroma
Yoshkar Ola
Nizhniy Novgorod
Kazan
Cheboksary
Naberezhnye Chelny

Velikiy Novgorod
Tver
Vladimir
Ulyanovsk
Tolyatti
Orsk
Orenburg
Ural

Riga
LATVIA
Liepaja
Klaipeda
LITHUANIA
Kaunas Vilnius
(RUSSIA)
Kaliningrad
Daugavpils
Vitsyebsk
Moscow
Kaluga
Ryazan
Saransk
Tula
Penza
Samara

EUROPEAN PLAIN
Bialystok
Warsaw
Hrodna
Babruysk
Mahilyow
Minsk
BELARUS
Homyel
Smolensk
Bryansk
Lipetsk
Tambov
Saratov

Radom
Lodz
Kielce
Lublin
Brest
Pinsk
Chernihiv
Sumy
Kursk
Belgorod
Voronezh

Katowice
Krakow
Lviv
Zhytomyr
Kiev (Kyiv)
Cherkasy
Kharkiv
Poltava
Luhansk
Volgograd
KAZAKHSTAN

CARPATHIAN MOUNTAINS
Chernivtsi
Vinnytsia
UKRAINE
Dnipropetrovsk
Donetsk
Horlivka
Don
Astrakhan

SLOVAKIA
Kosice
Bystrica
Miskolc
Oradea
MOLDOVA
Iasi
Chisinau
Kryvyi Rih
Zaporizhzhia
Mariupol
Rostov-na-Donu

Budapest
Debrecen
HARY
Kecskemet
Szeged
Cluj-Napoca
Mykolaiv
Odesa
Sea of Azov
Stavropol
Caspian

ROMANIA
Timisoara
Galati
Brasov
Ploiesti
CRIMEA PENINSULA
Krasnodar
Mt. Elbrus 18,510
Highest point in Europe
Nalchik
Vladikavkaz
Grozny
Makhachkala
Sea

Novi Sad
Belgrade
Serbia
Bucharest
Craiova
Ruse
Constanta
Sevastopol
Simferopol
CAUCASUS MTS.

YUGOSLAVIA
Nis
Danube
BULGARIA
Pleven
Varna
Black Sea

Podgorica
BALKAN
Shkoder
Skopje
Sofia
Stara Zagora
Burgas
Plovdiv

F.Y.R. MAC.
Durres
Tirana
ALBANIA
Vlore
PENINSULA
Kavala
Thessaloniki
Istanbul
TURKEY

Olympus 9,570
Larisa
Ioannina
Volos
Aegean Sea
Dardanelles

GREECE
Patras
Athens
Corinth
Peloponnesus
Kalamai
Sparta
Cyclades
Rhodes (Gr.)

Khania
Crete (Gr.)
Iraklion
Sea of Crete

ASIA

© MapQuest.com, Inc.

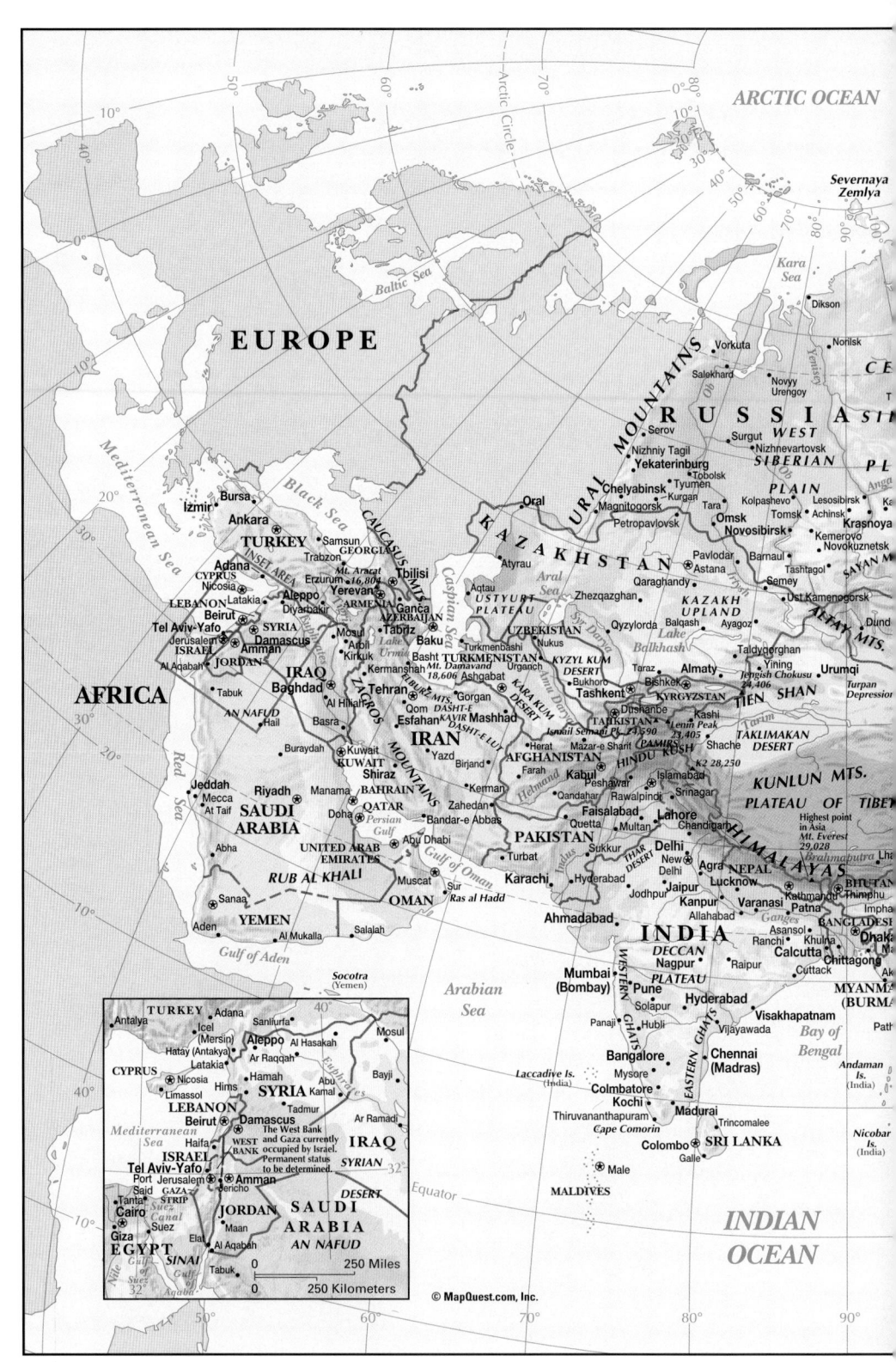

ARCTIC OCEAN

Severnaya
Zemlya

Kara
Sea

Dikson

Norilsk

Vorkuta

Baltic Sea

EUROPE

Salekhard

Novyy
Urengoy

CE

Yenisey

Serov

URAL MOUNTAINS

Nizhniy Tagil

Yekaterinburg

Surgut

Nizhnevartovsk

RUSSIA

WEST
SIBERIAN

SI

Chelyabinsk

Tyumen

Tobolsk

PLAIN

Anga

Izmir

Bursa

Ankara

Black Sea

TURKEY

Samsun

Trabzon

Erzurum

GEORGIA

Mt. Ararat
16,804

Tbilisi

Magnitogorsk

Kurgan

Oral

KAZAKHSTAN

Tara

Omsk

Novosibirsk

Kolpashevo

Tomsk

Achinsk

Krasnoya

Kemerovo
Novokuznetsk

Adana

CYPRUS

Nicosia

Aleppo

Diyarbakir

Latakia

Yerevan

ARMENIA

Ganca

Caspian Sea

Atyrau

Aral
Sea

Petropavlovsk

Pavlodar

Astana

Barnaul

Semey

Tashtagol

SAYAN

Ust Kamenogorsk

ALTAY MTS.

Dund

LEBANON

Beirut

Tel Aviv-Yafo

SYRIA

Baku

AZERBAIJAN

Aqtau

USTYURT
PLATEAU

Aqtobe

Qaraghandy

Balqash

KAZAKH
UPLAND

Ayagoz

Taldyqorghan

Jerusalem

ISRAEL

Damascus

Amman

JORDAN

Al Aqabah

Mosul

Kirkuk

Urmia

Basht

Tabriz

Lake
Urmia

Mashhad

TURKMENISTAN

Nukus

Urganch

UZBEKISTAN

Syr Darya

Qyzylorda

Taraz

Bishkek

Lake
Balkhash

Yining

Almaty

Jenggish Chokusu
24,406

Urumqi

Turpan
Depression

AFRICA

Tabuk

Hail

AN NAFUD

Kermanshah

Mt. Damavand
18,606

Ashgabat

Gorgan

Tehran

Qom

DASHT-E
KAVIR

KARAKUM
DESERT

Amu Darya

KYZYL KUM
DESERT

Bukhoro

Tashkent

Dushanbe

TAJIKISTAN

Ismail Semani Pk. 24,590

KYRGYZSTAN

TIEN SHAN

Kashi

Lenin Peak
23,405

PAMIRS

Shache

TAKLIMAKAN
DESERT

ZAGROS MTS.

Baghdad

Al Hillah

Basra

Esfahan

Yazd

DASHT-E LUT

IRAN

Birjand

Herat

Mazar-e Sharif

HINDU KUSH

K2 28,250

KUNLUN MTS.

Buraydah

Kuwait

KUWAIT

Shiraz

Kerman

Farah

Kabul

Peshawar

Islamabad

Qandahar

Rawalpindi

Srinagar

PLATEAU OF TIBET

Jeddah

Mecca

At Taif

Riyadh

BAHRAIN

Manama

QATAR

Doha

Persian
Gulf

Zahedan

Bandar-e Abbas

Quetta

Multan

AFGHANISTAN

Faisalabad

Lahore

Chandigarh

Highest point
in Asia
Mt. Everest
29,028

HIMALAYAS

Brahmaputra

Lha

Abha

SAUDI
ARABIA

UNITED ARAB
EMIRATES

Abu Dhabi

Gulf of Oman

PAKISTAN

THAR
DESERT

Sukkur

Delhi

New
Delhi

Agra

NEPAL

Kathmandu

Lucknow

Thimphu

BHUTAN

Sanaa

RUB AL KHALI

Muscat

Sur

Ras al Hadd

OMAN

Turbat

Karachi

Hyderabad

Jodhpur

Jaipur

Kanpur

Varanasi

Allahabad

Ganges

Patna

BANGLADESH

Dhaka

Aden

YEMEN

Al Mukalla

Salalah

Gulf of Aden

Ahmadabad

INDIA

DECCAN
PLATEAU

Nagpur

Raipur

Ranchi

Asansol

Khulna

Calcutta

Chittagong

Cuttack

MYANMAR
(BURMA)

Ak

Socotra
(Yemen)

Arabian
Sea

Mumbai
(Bombay)

Pune

Solapur

Hyderabad

WESTERN GHATS

Panaji

Hubli

Vijayawada

Visakhapatnam

Bay of
Bengal

Path

Andaman
Is.
(India)

Laccadive Is.
(India)

Bangalore

Mysore

Coimbatore

Kochi

Chennai
(Madras)

EASTERN GHATS

Thiruvananthapuram

Cape Comorin

Madurai

Trincomalee

Colombo

Galle

SRI LANKA

Nicobar
Is.
(India)

MALDIVES

Male

Equator

INDIAN
OCEAN

Inset map

TURKEY

Antalya

Icel
(Mersin)

Adana

Sanliurfa

Hatay (Antakya)

Aleppo

Al Hasakah

Mosul

CYPRUS

Nicosia

Latakia

Limassol

Ar Raqqah

Hamah

Euphrates

Hims

SYRIA

Abu
Kamal

Bayji

LEBANON

Beirut

Damascus

Tadmur

Ar Ramadi

Mediterranean
Sea

Haifa

The West Bank
and Gaza currently
occupied by Israel.
Permanent status
to be determined.

IRAQ

SYRIAN

ISRAEL

Tel Aviv-Yafo

WEST
BANK

Jericho

Port
Said

Jerusalem

GAZA
STRIP

Amman

DESERT

EGYPT

Cairo

Tanta

Suez
Canal

Giza

Suez

SINAI

JORDAN

Maan

Al Aqabah

SAUDI
ARABIA

AN NAFUD

Nile

Gulf of
Suez

Elat

Tabuk

Gulf of
Aqaba

250 Miles
0

250 Kilometers
0

© MapQuest.com, Inc.

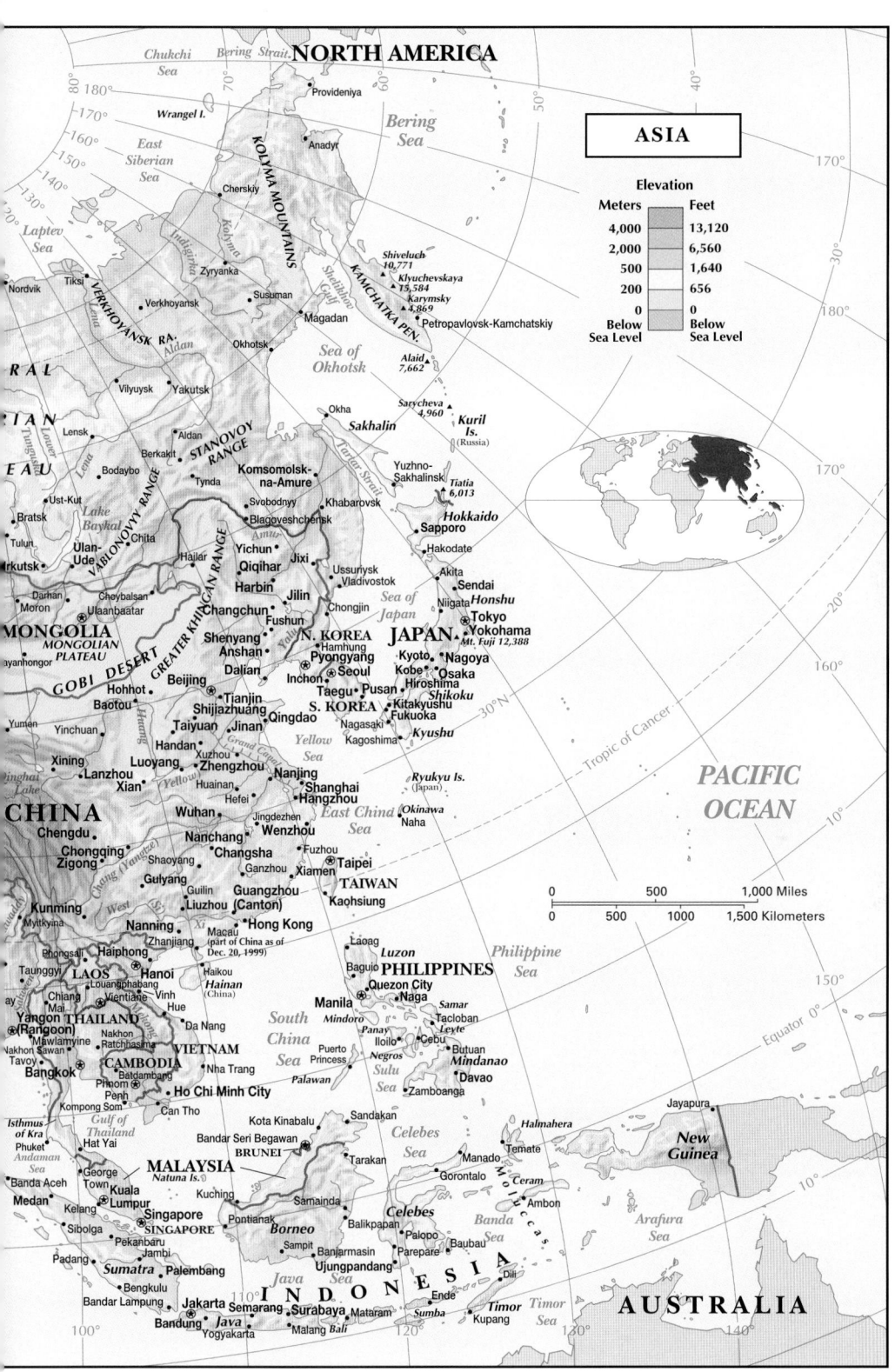

NORTH AMERICA

Chukchi
Sea

Bering Strait

Provideniya

Wrangel I.

East
Siberian
Sea

Anadyr

Bering
Sea

Nordvik

Tiksi

Laptev
Sea

KOLYMA MOUNTAINS

Cherskiy

Kolyma

Indigirka

Zyryanka

VERKHOYANSK RA.

Lena

Verkhoyansk

Susuman

Shiveluch
10,771

Klyuchevskaya
15,584

Karymsky
4,869

Magadan

KAMCHATKA PEN.

Petropavlovsk-Kamchatskiy

Okhotsk

Sea of
Okhotsk

Aldan

Vilyuysk

Yakutsk

Alaid
7,662

ASIA

Elevation

Meters		Feet
4,000		13,120
2,000		6,560
500		1,640
200		656
0		0
Below Sea Level		Below Sea Level

RAL

IAN

EAU

Lensk

Tungusr

Lena

Ust-Kut

Bratsk

Tulun

rkutsk

Lake
Baikal

Bodaybo

Berkakit

Aldan

STANOVOY RANGE

Tynda

Komsomolsk-
na-Amure

Svobodnyy

Okha

Sarycheva
4,960

Sakhalin

Kuril
Is.
(Russia)

Blagoveshchensk

Khabarovsk

Tiatia
6,013

Ulan-
Ude

Darhan
Moron

Chita

Choybalsan

Ulaanbaatar

YABLONOVYY RANGE

Hailar

Yichun

Qiqihar

GREATER KHINGAN RANGE

Harbin

Jixi

Jilin

Ussuriysk

Vladivostok

Chongjin

Hamhung

Yuzhno-
Sakhalinsk

Hokkaido

Sapporo

Hakodate

Akita

Sendai

Niigata

Honshu

Tokyo

MONGOLIA

MONGOLIAN
PLATEAU

ayanhongor

Changchun

Fushun

Shenyang

Anshan

Dalian

N. KOREA

Pyongyang

Seoul

GOBI DESERT

Hohhot

Baotou

Yumen

Yinchuan

Beijing

Tianjin

Shijiazhuang

Jinan

Qingdao

Taiyuan

Inchon

Taegu

Pusan

S. KOREA

JAPAN

Kyoto

Kobe

Nagasaki

Hiroshima

Kagoshima

Kitakyushu

Fukuoka

Shikoku

Kyushu

Osaka

Nagoya

Yokohama

Mt. Fuji 12,388

Sea of
Japan

Xining

Lanzhou

Xian

Handan

Xuzhou

Luoyang

Zhengzhou

Huainan

Nanjing

Hefei

Shanghai

Hangzhou

Yellow
Sea

Ryukyu Is.
(Japan)

Jinghai
Lake

Yellow
Hwang

Grand Canal

CHINA

Chengdu

Chongqing

Zigong

Wuhan

Nanchang

Changsha

Shaoyang

Jingdezhen

Wenzhou

Fuzhou

Ganzhou

East China
Sea

Okinawa

Naha

Kunming

Myitkyina

Guiyang

Guilin

Liuzhou

Guangzhou
(Canton)

Taipei

Xiamen

TAIWAN

Kaohsiung

Chang (Yangtze)

West

Xi

Zhanjiang

Macau
(part of China as of
Dec. 20, 1999)

Hong Kong

PACIFIC
OCEAN

Nanning

Haiphong

Laoag

Baguio

Luzon

PHILIPPINES

Philippine
Sea

Phongsali

LAOS

Louangphabang

Hanoi

Hainan
(China)

Haikou

Quezon City

Manila

Naga

Samar

Chiang
Mai

Vientiane

Vinh

Hue

THAILAND

Yangon
(Rangoon)

Nakhon
Sawan

Nakhon
Ratchasima

VIETNAM

Da Nang

South
China
Sea

Mindoro

Panay

Iloilo

Tacloban

Leyte

Cebu

Butuan

Mawlamyine

Tavoy

Bangkok

CAMBODIA

Battambang

Phnom
Penh

Ho Chi Minh City

Puerto
Princess

Palawan

Negros

Mindanao

Davao

Zamboanga

Sulu
Sea

Tropic of Cancer

Kompong Som

Can Tho

Kota Kinabalu

Sandakan

Halmahera

Jayapura

Isthmus
of Kra

Phuket

Andaman
Sea

Banda Aceh

Medan

Gulf of
Thailand

Hat Yai

George
Town

MALAYSIA

Natuna Is.

Bandar Seri Begawan

BRUNEI

Tarakan

Celebes
Sea

Temate

Manado

Gorontalo

New
Guinea

Kelang

Kuala
Lumpur

Kuching

Samainda

Ceram

Ambon

Arafura
Sea

Sibolga

Singapore

SINGAPORE

Pontianak

Borneo

Balikpapan

Celebes

Palopo

Baubau

Molucas

Banda
Sea

Padang

Pekanbaru

Jambi

Sampit

Banjarmasin

Ujungpandang

Parepare

Sumatra

Palembang

Java
Sea

Timor
Sea

AUSTRALIA

Bengkulu

Bandar Lampung

Jakarta

Semarang

Surabaya

Mataram

Sumba

Ende

Dili

Kupang

Timor

INDONESIA

Bandung

Java

Yogyakarta

Malang

Bali

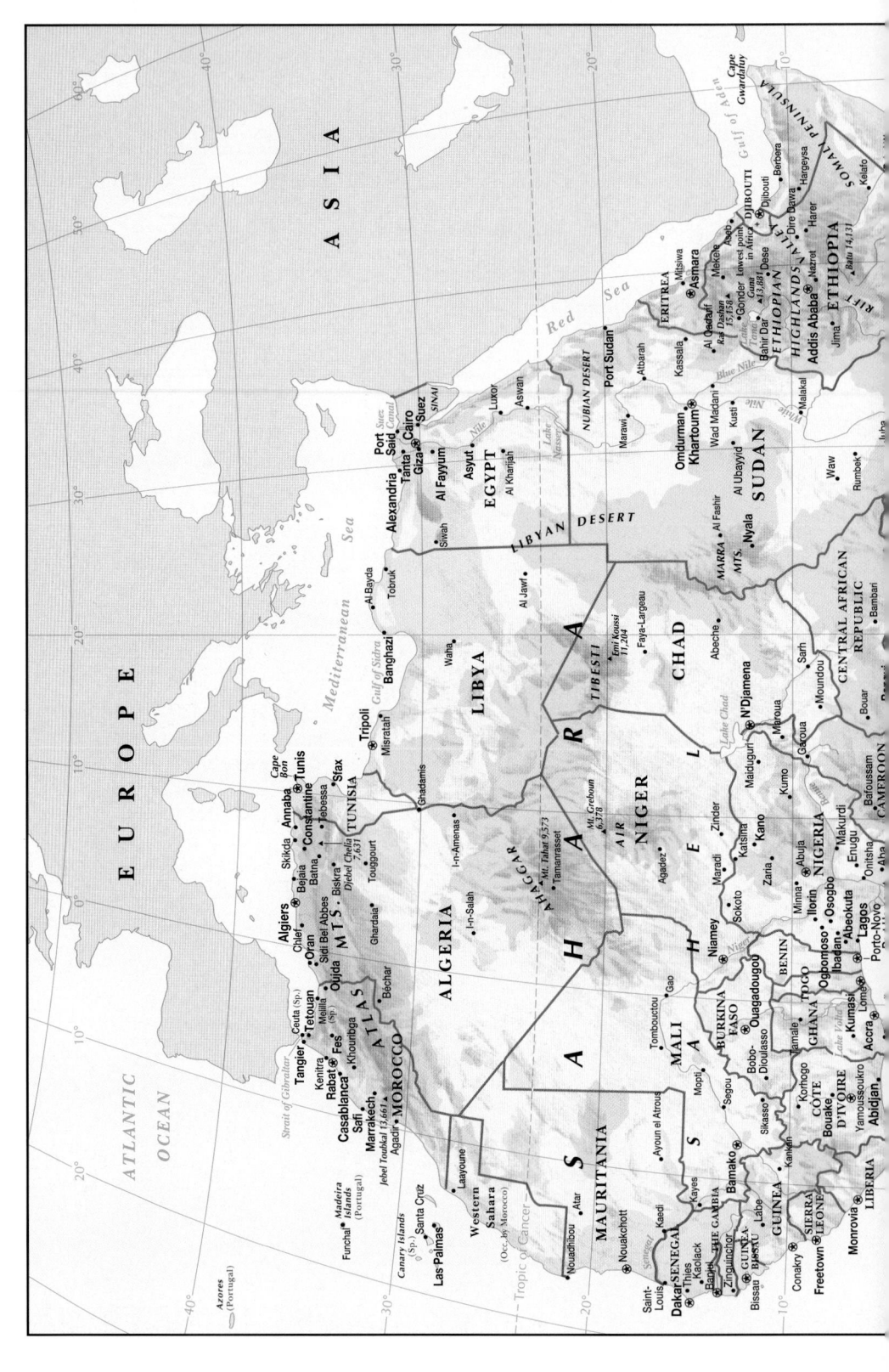

ATLANTIC
OCEAN

EUROPE

ASIA

Mediterranean Sea

Red Sea

Gulf of Aden
Cape Gwardafuy

SOMALI PENINSULA

Azores
(Portugal)

Madeira
Islands
(Portugal)

Funchal

Canary Islands
(Sp.)
Santa Cruz
Las Palmas

Strait of Gibraltar
Ceuta (Sp.)
Tangier
Tetouan
Melilla (Sp.)
Rabat Kenitra
Casablanca Fes Oujda
Safi Khouribga
Marrakech
Agadir MOROCCO
Jebel Toubkal 13,661

Gulf of Tunis
Cape Bon
Tunis
Annaba
Skikda Constantine
Algiers Bejaia Tebessa
Chlef Batna Sfax
Oran Biskra TUNISIA
Sidi Bel Abbes Djebel Chelia 7,631
Béchar Touggourt
Ghardaia Toggourt
In-Salah

A T L A S   M T S.

Layoune

Western
Sahara
(Occ. by Morocco)

Tropic of Cancer

MAURITANIA
Nouadhibou Atar
Nouakchott
Kaedi

Saint-
Louis
Dakar SENEGAL
Thies Kaolack
THE GAMBIA
Banjul
Zinguinchor GUINEA-BISSAU
Bissau
Conakry
Freetown SIERRA LEONE
Monrovia LIBERIA

ALGERIA

Ghadamis

In-Amenas

A H A G G A R
Mt. Tahat 9,573
Tamanrasset

S A H A R A

Ayoun el Atrous
Ayoun el Atrous

MALI

Kayes
Bamako
Bafing
Labe
Kankan
GUINEA

Bobo-
Dioulasso

Segou
Sikasso
Korhogo
COTE
D'IVOIRE
Bouake
Yamoussoukro
Abidjan

Tombouctou
Mopti

Gao

BURKINA
FASO
Ouagadougou

GHANA
Lake Volta
Kumasi
Accra

Niamey
NIGER
Agadez

AIR
Mt. Grebour 6,378

TIBESTI
Emi Koussi 11,204
Faya-Largeau

LIBYA

Tripoli
Misratah
Banghazi
Waha

Al Bayda
Tobruk

LIBYAN DESERT

EGYPT

Siwah
Al Kharijah

Alexandria
Tanta Cairo
Giza Suez
Al Fayyum
Asyut
Luxor
Aswan

Port Said
Port Said
SINAI
Suez

NUBIAN DESERT

Port Sudan

Marawi
Atbarah
Kassala
Al Qadarif
Wad Madani

SUDAN

Omdurman
Khartoum
Al Ubayyid
Al Fashir
Nyala
Waw
Rumbek
Malakal
Juba

MARRA
MTS.

Abeche

CHAD

N'Djamena
Lake Chad
Mundou
Sarh

CENTRAL AFRICAN
REPUBLIC
Bouar
Bambari

Maiduguri
Kumo
Katsina
Kano
Zinder Zaria
Maradi Sokoto
NIGERIA
Minna Abuja
Ilorin Enugu
Ogbomoso Osogbo Makurdi
Ibadan
Abeokuta
Lagos
Porto-Novo
TOGO
BENIN
Lome
CAMEROON
Batoussam
Gwa
Foumban

ERITREA
Asmara
Mitsiwa
Mekele
Al Qadaref

Bahir Dar
Gonder
Ras Dashan 15,158
Gunu 13,804 lowest point in Africa
Blue Nile
Lake Tana

ETHIOPIAN HIGHLANDS
Addis Ababa ETHIOPIA
Jima
Nazret
Dese
Aba 14,131

DJIBOUTI
Djibouti
Dire Dawa
Harer
Hargeysa
Berbera
Kelafo

RIFT VALLEY

White Nile
Kusti

N I G E R

S A H E L

Niger
Gulf of Guinea

518

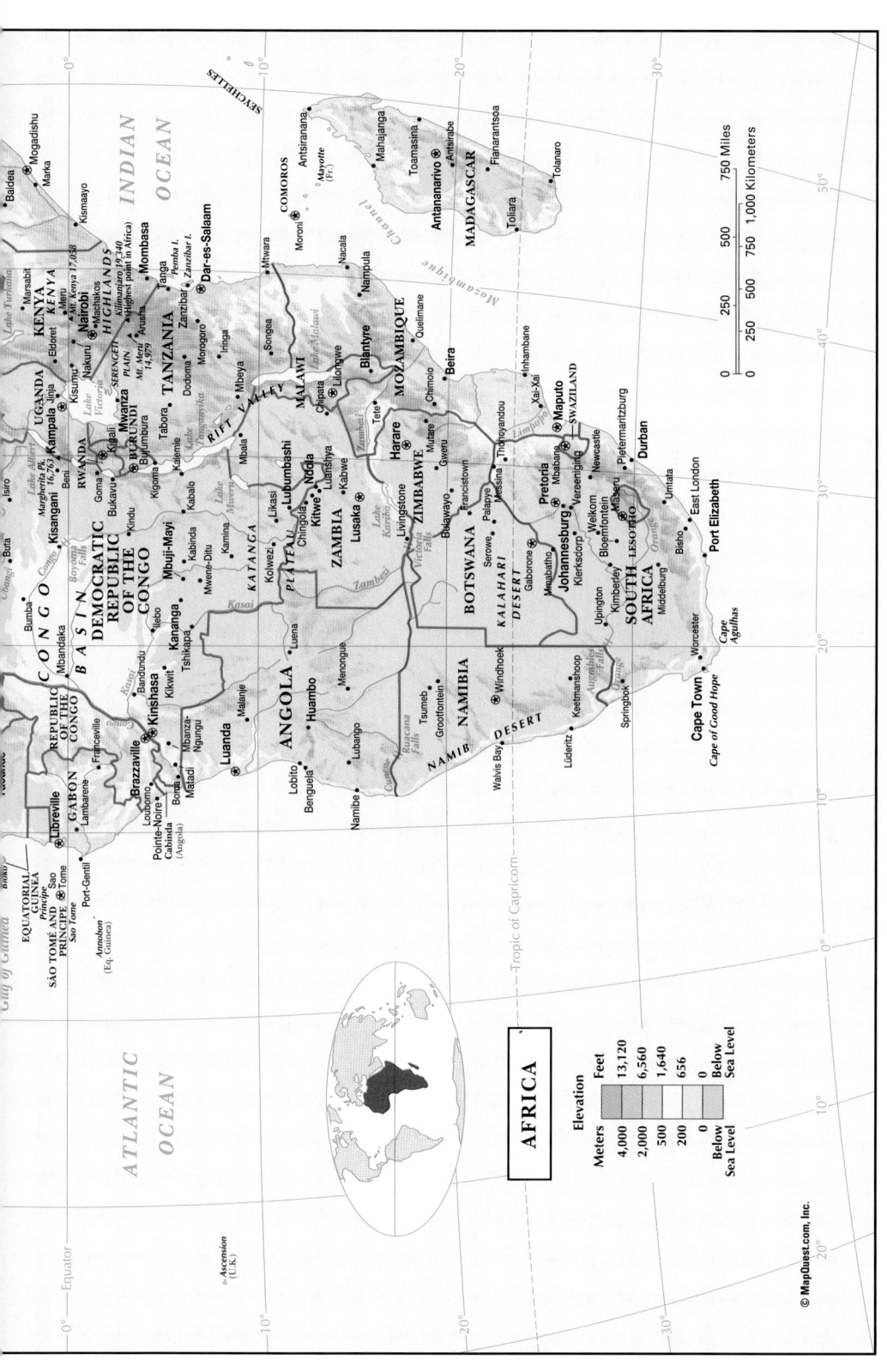

AFRICA

Elevation

Meters	Feet
4,000	13,120
2,000	6,560
500	1,640
200	656
0	0
Below Sea Level	Below Sea Level

© MapQuest.com, Inc.

INDIAN OCEAN

ATLANTIC OCEAN

Gulf of Guinea

MADAGASCAR

KENYA

UGANDA

TANZANIA

RWANDA

BURUNDI

DEMOCRATIC REPUBLIC OF THE CONGO

REPUBLIC OF THE CONGO

GABON

EQUATORIAL GUINEA

SÃO TOMÉ AND PRÍNCIPE

ANGOLA

ZAMBIA

MALAWI

MOZAMBIQUE

ZIMBABWE

NAMIBIA

BOTSWANA

SOUTH AFRICA

LESOTHO

SWAZILAND

COMOROS

SEYCHELLES

CONGO BASIN

KATANGA

KALAHARI DESERT

NAMIB DESERT

SERENGETI PLAIN

RIFT VALLEY

PLATEAU

Mozambique Channel

Tropic of Capricorn

Equator

Mogadishu
Marka
Baidoa
Kismaayo
Marsabit
Nairobi
Nakuru
Eldoret
Kisumu
Kampala
Jinja
Kigali
Bujumbura
Mwanza
Mombasa
Tanga
Zanzibar I.
Pemba I.
Dar-es-Salaam
Dodoma
Morogoro
Iringa
Mbeya
Tabora
Kigoma
Kalemie
Antsiranana
Mahajanga
Toamasina
Antananarivo
Antsirabe
Fianarantsoa
Toliara
Tolanaro
Moroni
Mayotte (Fr.)
Nacala
Nampula
Quelimane
Mtwara
Songea
Lilongwe
Blantyre
Chipata
Beira
Chimoio
Tete
Harare
Mutare
Gweru
Bulawayo
Francistown
Xai-Xai
Inhambane
Maputo
Mbabane
Manzini
Pretoria
Johannesburg
Vereeniging
Klerksdorp
Welkom
Maseru
Bloemfontein
Kimberley
Pietermaritzburg
Newcastle
Durban
East London
Umtata
Bisho
Port Elizabeth
Cape Town
Cape of Good Hope
Cape Agulhas
Worcester
Springbok
Lüderitz
Keetmanshoop
Upington
Middelburg
Gaborone
Serowe
Mahalapye
Palapye
Messina
Thohoyandou
Windhoek
Walvis Bay
Tsumeb
Grootfontein
Namibe
Lobito
Benguela
Lubango
Menongue
Malanje
Luanda
Kinshasa
Brazzaville
Libreville
Port-Gentil
Lambaréné
Franceville
Loubomo
Pointe-Noire
Cabinda (Angola)
Matadi
Boma
Mbanza-Ngungu
Bandundu
Kikwit
Tshikapa
Kananga
Mbuji-Mayi
Ilebo
Kindu
Bukavu
Goma
Beni
Isiro
Bumba
Mbandaka
Butar
Lusaka
Livingstone
Kabwe
Lubumbashi
Ndola
Kitwe
Chingola
Kolwezi
Likasi
Kabinda
Mwene-Ditu
Kamina
Kabalo
Kolwezi
Lilongwe
Lubango

São Tomé
Annobon (Eq. Guinea)
Ascension (U.K.)

Lake Turkana
Lake Victoria
Lake Albert
Lake Tanganyika
Lake Mweru
Lake Kariba
Lake Malawi

Mt. Kenya 17,058
Mt. Meru 14,979
Kilimanjaro 19,340
*Highest point in Africa
Margherita Pk. 16,763

KENYA HIGHLANDS

Victoria Falls
Ruacana Falls
Augrabies Falls

Congo
Kasai
Zambezi
Limpopo
Orange

750 Miles
1,000 Kilometers

0   250   500   750
0   250   500

519

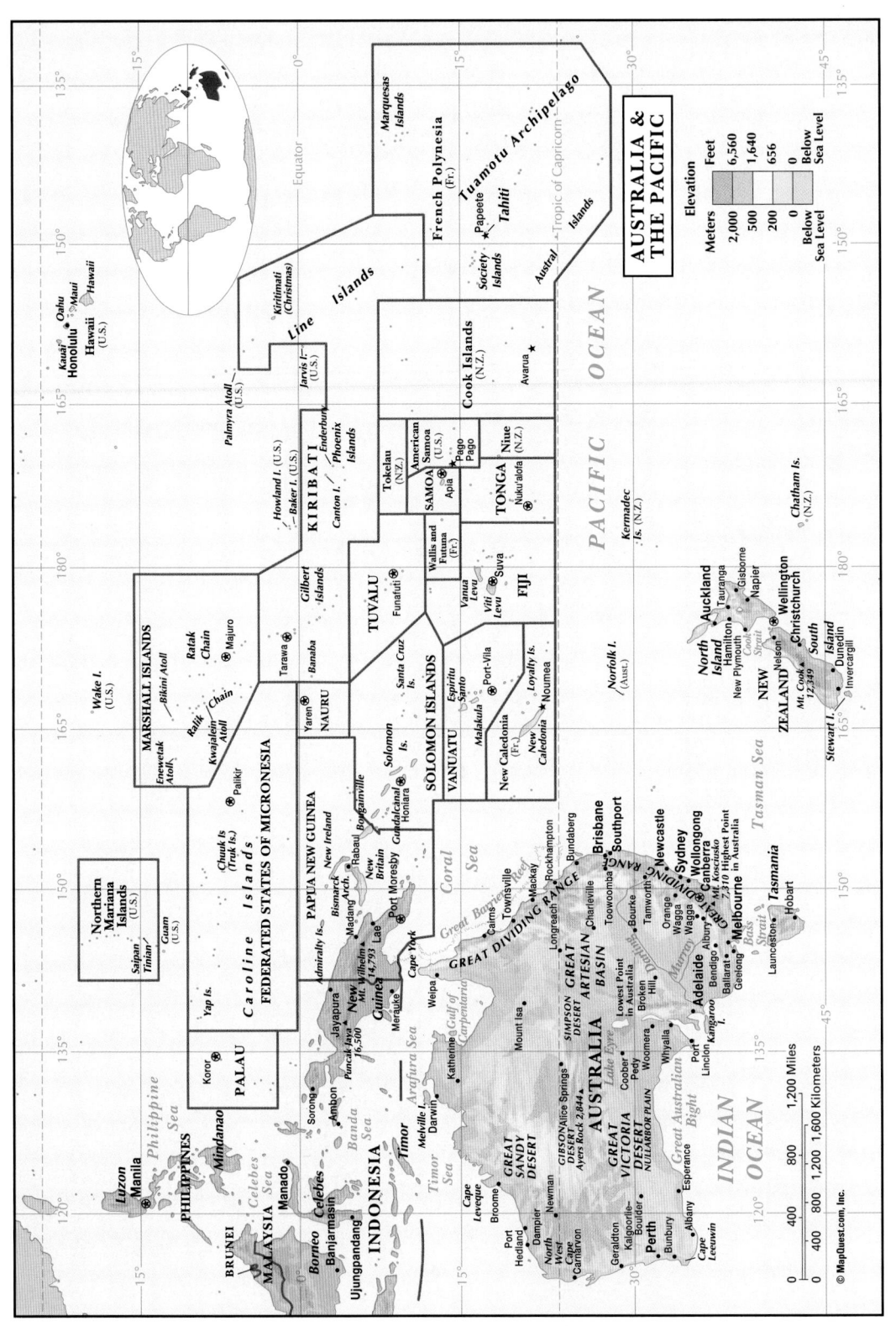

AUSTRALIA &
THE PACIFIC

Elevation

Meters	Feet
2,000	6,560
500	1,640
200	656
0	0
Below Sea Level	Below Sea Level

PACIFIC OCEAN

INDIAN OCEAN

AUSTRALIA

NEW ZEALAND

PAPUA NEW GUINEA

FEDERATED STATES OF MICRONESIA

MARSHALL ISLANDS

KIRIBATI

TUVALU

FIJI

VANUATU

SOLOMON ISLANDS

NAURU

PALAU

TONGA

SAMOA

TUAMOTU Archipelago

French Polynesia (Fr.)

Cook Islands (N.Z.)

Northern Mariana Islands (U.S.)

INDONESIA

PHILIPPINES

MALAYSIA

BRUNEI

© MapQuest.com, Inc.

520

UNITED STATES HISTORY

1492
Christopher Columbus and crew sighted land **Oct. 12** in the present-day Bahamas.

1497
John Cabot explored northeast coast to Delaware.

1513
Juan Ponce de León explored Florida coast.

1524
Giovanni da Verrazano led French expedition along coast from Carolina north to Nova Scotia; entered New York harbor.

1539
Hernando de Soto landed in Florida **May 28;** crossed Mississippi River, **1541.**

1540
Francisco Vásquez de Coronado explored Southwest north of Rio Grande. Hernando de Alarcón reached Colorado River; Don Garcia Lopez de Cardenas reached Grand Canyon. Others explored California coast.

1565
St. Augustine, FL, founded **Sept. 8** by Pedro Menéndez. Razed by Francis Drake **1586.**

1579
Francis Drake entered San Francisco Bay and claimed region for Britain.

1607
Capt. John Smith and 105 cavaliers in 3 ships landed on Virginia coast, started first permanent English settlement in New World at **Jamestown** in **May.**

1609
Henry Hudson, English explorer of Northwest Passage, employed by Dutch, sailed into New York harbor in **Sept.,** and up Hudson to Albany. **Samuel de Champlain** explored Lake Champlain, just to the north.

Spaniards settled **Santa Fe, NM.**

1619
House of Burgesses, first representative assembly in New World, elected **July 30** at Jamestown, VA.

First black laborers—indentured servants—in English N. American colonies, landed by Dutch at Jamestown in **Aug.** Chattel slavery legally recognized, **1650.**

1620
Plymouth Pilgrims, Puritan separatists, left Plymouth, England, **Sept. 16** on *Mayflower.* They reached Cape Cod **Nov. 19,** explored coast; 103 passengers landed **Dec. 26** at Plymouth. **Mayflower Compact** was agreement to form a government and abide by its laws. Half of colony died during harsh winter.

1624
Dutch colonies started in Albany and in New York area, where **New Netherland** was established in **May.**

1626
Peter Minuit bought Manhattan for Dutch from Man-a-hat-a Indians during summer for goods valued at $24; named island **New Amsterdam.**

1630
Settlement of **Boston** established by Massachusetts colonists led by John Winthrop.

1634
Maryland, founded as a Catholic colony, under a charter granted to Lord Baltimore. Religious toleration granted **1649.**

1636
Roger Williams founded Providence, RI, **June,** as a democratically ruled colony with separation of church and state. Charter granted, **1644.**

Harvard College founded **Oct. 28,** now oldest in U.S.; grammar school, compulsory education established at Boston.

1640
First book was printed in America, the so-called Bay Psalm Book.

1647
Liberal constitution drafted in Rhode Island.

1660
British Parliament passed First **Navigation Act Dec. 1,** regulating colonial commerce to suit English needs.

1664
British troops Sept. 8 seized **New Netherland** from Dutch. Charles II granted New Netherland and city of New Amsterdam to brother, Duke of York; both renamed **New York.** Dutch recaptured colony **1673,** but ceded it to Britain **Nov. 10, 1674.**

1673
Jacques **Marquette** and Louis **Jolliet** reached the upper **Mississippi** and traveled down it.

1676
Nathaniel Bacon led planters against autocratic British Gov. Sir William Berkeley, burned Jamestown, VA, **Sept. 19.** Rebellion collapsed when Bacon died; 23 followers executed.

Bloody **Indian war** in New England ended **Aug. 12.** King Philip, Wampanoag chief, and Narragansett Indians killed.

1682
Robert Cavelier, Sieur de La Salle, claimed lower Mississippi River country for France, called it Louisiana **Apr. 9.** Had French outposts built in Illinois and Texas, **1684.** Killed during mutiny **Mar. 19, 1687.**

William Penn arrived in **Pennsylvania.**

1683
William Penn signed treaty with Delaware Indians and made payment for Pennsylvania lands.

1692
Witchcraft delusion at Salem, MA; 20 alleged witches executed by special court.

1696
Capt. William Kidd settled in America, was hired by British to fight pirates and take booty, but himself became one. Arrested and sent to England; hanged **1701.**

1699
French settlements made in Mississippi, Louisiana.

1704
Indians attacked Deerfield, MA, Feb. 28-29; killed 40, carried off 100.

Boston News Letter, **first regular newspaper,** started by John Campbell, postmaster. (An earlier paper, *Publick Occurences,* was suppressed after one issue **1690.)**

1709
British-Colonial troops captured French fort, Port Royal, Nova Scotia, in **Queen Anne's War 1701-13.** France yielded Nova Scotia by treaty **1713.**

1712
Slaves revolted in New York **Apr. 6.** Six committed suicide; 21 were executed. Second rising, **1741;** 13 slaves hanged, 13 burned, 71 deported.

1716
First theater in colonies opened in Williamsburg, VA.

1726
Poor people **rioted** in Philadelphia.

Great Awakening religious revival began.

1732
Benjamin Franklin published the first *Poor Richard's Almanac;* published annually to **1757.**

Last of the 13 colonies, **Georgia,** chartered.

1735
Editor **John Peter Zenger acquitted** in New York of libeling British governor by criticizing his conduct in office.

1740-41
Capt. Vitus Bering reached Alaska.

1744
King George's War pitted British and colonials vs. French. Colonials captured Louisburg, Cape Breton Is., **June 17, 1745.** Returned to France **1748** by Treaty of Aix-la-Chapelle.

1752
Benjamin Franklin, flying kite in thunderstorm, proved lightning is electricity **June 15;** invented lightning rod.

1754
French and Indian War began when French occupied Ft. Duquesne (Pittsburgh). British moved Acadian French from Nova Scotia to Louisiana **Oct. 8, 1755.** British captured Québec **Sept. 18, 1759,** in battles in which French Gen. Joseph de Montcalm and British Gen. James Wolfe were killed. Peace pact signed **Feb. 10, 1763.** French lost Canada and Midwest.

1764
Sugar Act placed duties on lumber, foodstuffs, molasses, and rum in colonies, to pay French and Indian War debts.

1765
Stamp Act, enacted by Parliament **Mar. 22,** required revenue stamps to help fund royal troops. Nine colonies, at **Stamp Act Congress** in New York **Oct. 7-25,** adopted Declaration of Rights. Stamp Act **repealed Mar. 17, 1766.**

1767
Townshend Acts levied taxes on glass, painter's lead, paper, and tea. In **1770** all duties except those on tea were repealed.
1770
British troops fired **Mar. 5** into Boston mob, killed 5 including **Crispus Attucks,** a black man, reportedly leader of group; later called **Boston Massacre.**
1773
East India Co. tea ships turned back at Boston, New York, and Philadelphia in **May.** Cargo ship burned at Annapolis **Oct. 14;** cargo thrown overboard at **Boston Tea Party Dec. 16,** to protest the tea tax.
1774
"Intolerable Acts" of Parliament curtailed Massachusetts self-rule; barred use of Boston harbor till tea was paid for.
First Continental Congress held in Philadelphia **Sept. 5-Oct. 26;** called for civil disobedience against British.
Rhode Island abolished slavery.
1775
Patrick Henry addressed Virginia convention, **Mar. 23,** said "Give me liberty or give me death."
Paul Revere and William Dawes on night of **Apr. 18** rode to alert Patriots that British were on their way to Concord to destroy arms. At Lexington, MA, **Apr. 19,** Minutemen lost 8. On return from Concord, British took 273 casualties.
Col. Ethan Allen (joined by Col. Benedict Arnold) captured **Ft. Ticonderoga, NY, May 10;** also Crown Point. Colonials headed for **Bunker Hill,** fortified Breed's Hill, Charlestown, MA. Repulsed British under Gen. William Howe twice before retreating **June 17;** called Battle of Bunker Hill.
Continental Congress **June 15** named **George Washington** commander in chief.
1776
France and Spain each agreed **May 2** to provide arms.
In Continental Congress **June 7,** Richard Henry Lee (VA) moved "that these united colonies are and of right ought to be free and independent states." Resolution adopted July 2. **Declaration of Independence** approved **July 4.**
Col. William Moultrie's batteries at **Charleston, SC,** repulsed British sea attack **June 28.** Washington, with 10,000 men, lost **Battle of Long Island Aug. 27;** evacuated New York.
Nathan Hale executed as spy by British **Sept. 22.**
Brig. Gen. Arnold's **Lake Champlain** fleet was defeated at Valcour **Oct. 11,** but British returned to Canada. Howe failed to destroy Washington's army at **White Plains Oct. 28.** Hessians captured Ft. Washington, Manhattan, and 3,000 men **Nov. 16;** captured Ft. Lee, NJ, **Nov. 18.**
Washington, in Pennsylvania, recrossed **Delaware River Dec. 25-26,** defeated Hessians at Trenton, NJ, **Dec. 26.**
1777
Washington defeated Lord Cornwallis at **Princeton Jan. 3.** Continental Congress adopted Stars and Stripes.
Maj. Gen. John Burgoyne, force of 8,000 from Canada, captured **Ft. Ticonderoga July 6.** Americans beat back Burgoyne at Bemis Heights **Oct. 7,** cut off British escape route. Burgoyne surrendered 5,000 men at **Saratoga, NY, Oct. 17.**
Articles of Confederation adopted by Continental Congress **Nov. 15;** ratified **Mar. 1,** by last state, Maryland.
France recognized independence of 13 colonies **Dec. 17.**
1778
France signed treaty of aid with U.S. **Feb. 6.** Sent fleet; British evacuated Philadelphia in consequence **June 18.**
1779
John Paul Jones on the *Bonhomme Richard* defeated *Serapis* in British North Sea waters **Sept. 23.**
1780
Charleston, SC, fell to the British **May 12,** but a British force was defeated near **Kings Mountain, NC, Oct. 7** by militiamen.
Benedict Arnold found to be a traitor **Sept. 23.** Arnold escaped, made brigadier general in British army.
1781
Bank of North America incorporated **May 26.**
Cornwallis, sapped by Patriot victories, retired to **Yorktown, VA.** Adm. Francois Joseph de Grasse landed 3,000 French and stopped British fleet in Hampton Roads. Washington and Jean Baptiste de Rochambeau joined forces, arrived near Williamsburg **Sept. 26.** Siege of Cornwallis began **Oct. 6; Cornwallis surrendered Oct. 19.**

1782
New **British** cabinet agreed **in March** to **recognize U.S.** independence. Preliminary agreement signed in Paris **Nov. 30.**
1783
Massachusetts Supreme Court declared **slavery** illegal in that state.
Britain, U.S. signed Paris **peace treaty Sept. 3** recognizing American independence (Congress ratified it **Jan. 14, 1784**).
Washington ordered army disbanded Nov. 3, bade farewell to his officers at Fraunces Tavern, New York City, **Dec. 4.**
Noah Webster published *American Spelling Book.*
1784
Thomas Jefferson's proposal to **ban slavery** in new territory after 1802 was narrowly defeated **Mar. 1.**
First successful daily newspaper, *Pennsylvania Packet & General Advertiser*, published **Sept. 21.**
1786
Delegates from 5 states at **Annapolis, MD, Sept. 11-14** asked Congress to call convention in Philadelphia to write practical constitution for the 13 states.
1787
Shays's Rebellion of debt-ridden farmers in Massachusetts failed **Jan. 25.**
Northwest Ordinance adopted **July 13** by Continental Congress for Northwest Territory, N of Ohio River, W of New York; made rules for statehood. Guaranteed freedom of religion, support for schools, no slavery.
Constitutional convention opened at Philadelphia **May 25** with Washington presiding. Constitution accepted by delegates **Sept. 17;** ratification by 9th state, New Hampshire, **June 21, 1788,** meant adoption; declared in effect **Mar. 4, 1789.**
1789
George Washington chosen president by all electors voting (73 eligible, 69 voting, 4 absent); John Adams, vice president, got 34 votes. First Congress met at Federal Hall, New York City, **Mar. 4.** Washington inaugurated there **Apr. 30.** Supreme Court created by Federal Judiciary Act **Sept. 24.** Congress submitted Bill of Rights to states **Sept. 25.**
1790
Congress, **Mar. 1,** authorized decennial **U.S. census; Naturalization Act** (2-year residency) passed **Mar. 26.**
Congress met in Philadelphia, new temporary capital, **Dec. 6.**
1791
Bill of Rights went into effect **Dec. 15.**
1792
Coinage Act established **U.S. Mint** in Philadelphia **Apr. 2.**
Gen. **"Mad" Anthony Wayne** made commander in Ohio-Indiana area, trained "American Legion," established string of forts. Routed Indians at Fallen Timbers on Maumee River **Aug. 20, 1794,** checked British at Fort Miami, OH.
White House cornerstone laid **Oct. 13.**
1793
Eli Whitney invented **cotton gin,** reviving Southern slavery.
1794
Whiskey Rebellion, W Pennsylvania farmers protesting liquor tax of **1791,** was suppressed by federal militia **Sept. 1794.**
1795
U.S. bought peace from **Algerian pirates** by paying $1 mil ransom for 115 seamen **Sept. 5,** followed by annual tributes.
Gen. Wayne signed peace with Indians at Fort Greenville.
University of North Carolina became first operating state university.
1796
Washington's Farewell Address as president delivered **Sept. 19.** Gave strong warnings against permanent alliances with foreign powers, big public debt, large military establishment, and devices of "small, artful, enterprising minority."
1797
U.S. **frigate** *United States* launched at Philadelphia **July 10;** *Constellation* at Baltimore **Sept. 7;** *Constitution* (Old Ironsides) at Boston **Sept. 20.**
1798
Alien & Sedition Acts passed by Federalists **June-July;** intended to silence political opposition.
War with France threatened over French raids on U.S. shipping and rejection of U.S. diplomats. Navy (45 ships) and 365 privateers captured 84 French ships. USS *Constellation* took French warship *Insurgente* **1799.** Napoleon stopped French raids after becoming First Consul.

1800
Federal government moved to **Washington, DC.**

1801
John Marshall named Supreme Court chief justice, **Jan. 20.**
Tripoli declared war June 10 against U.S., which refused added tribute to commerce-raiding Arab corsairs. Land and naval campaigns forced Tripoli to negotiate **peace June 4, 1805.**

1803
Supreme Court, in **Marbury v Madison** case, for the first time overturned a U.S. law **Feb. 24.**
Napoleon sold all of **Louisiana,** stretching to Canadian border, to U.S., for $11,250,000 in bonds, plus $3,750,000 indemnities to American citizens with claims against France. U.S. took title **Dec. 20.** Purchase doubled U.S. area.

1804
Lewis and Clark expedition ordered by Pres. Thomas Jefferson to explore what is now northwest U.S. Started from St. Louis **May 14;** ended **Sept. 23, 1806.**
Vice Pres. **Aaron Burr shot Alexander Hamilton** in a duel **July 11** in Weehawken, NJ; Hamilton died the next day.

1807
Robert Fulton made first practical steamboat trip; left New York City **Aug. 17,** reached Albany, 150 mi, in 32 hr.
Embargo Act banned all trade with foreign countries, forbidding ships to set sail for foreign ports **Dec. 22.**

1808
Slave importation outlawed. Some 250,000 slaves were illegally imported **1808-60.**

1811
William Henry Harrison, governor of Indiana, defeated Indians under the Prophet, in battle of **Tippecanoe Nov. 7.**
Cumberland Road begun at Cumberland, MD; became important route to West.

1812
War of 1812 had 3 main causes: Britain seized U.S. ships trading with France; Britain seized 4,000 naturalized U.S. sailors by **1810;** Britain armed Indians who raided western border. U.S. stopped trade with Europe **1807** and **1809.** Trade with Britain only was stopped **1810.**
Unaware that Britain had raised the blockade against France 2 days before, **Congress declared war June 18.**
USS *Essex* captured *Alert* **Aug. 13;** USS *Constitution* destroyed *Guerriere* **Aug. 19;** USS *Wasp* took *Frolic* **Oct. 18;** USS *United States* defeated *Macedonian* off Azores **Oct. 25;** *Constitution* beat *Java* **Dec. 29.** British took Detroit **Aug. 16.**

1813
Oliver H. Perry defeated British fleet at Battle of Lake Erie, **Sept. 10.** U.S. won Battle of the Thames, Ontario, **Oct. 5,** but failed in Canadian invasion attempts. York (Toronto) and Buffalo were burned.

1814
British landed in Maryland in Aug., defeated U.S. force **Aug. 24, burned Capitol and White House.** Maryland militia stopped British advance **Sept. 12.** Bombardment of Ft. McHenry, Baltimore, for 25 hours, **Sept. 13-14,** by British fleet failed; Francis Scott Key wrote words to **"Star Spangled Banner."**
U.S. won naval Battle of **Lake Champlain Sept. 11.** Peace treaty signed at Ghent **Dec. 24.**

1815
Some 5,300 British, unaware of peace treaty, attacked U.S. entrenchments near **New Orleans, Jan. 8.** British had more than 2,000 casualties; Americans lost 71.
U.S. flotilla finally ended piracy by **Algiers, Tunis, Tripoli** by **Aug. 6.**

1816
Second **Bank of the U.S.** chartered.

1817
Rush-Bagot treaty signed **Apr. 28-29;** limited U.S., British armaments on the Great Lakes.
William Cullen Bryant's poem "Thanatopsis" published.

1819
Spain ceded **Florida** to U.S. **Feb. 22.**
American steamship *Savannah* made first part-steam-powered, part-sail-powered crossing of Atlantic, Savannah, GA, to Liverpool, England, 29 days.

1820
First organized **immigration of blacks to Africa** from U.S. began with 86 free blacks sailing **Feb.** to Sierra Leone.

Henry Clay's **Missouri Compromise** bill passed by Congress **Mar. 3.** Slavery was allowed in Missouri, but not elsewhere west of the Mississippi River north of 36° 30´ latitude (the southern line of Missouri). Repealed **1854.**

1821
Emma Willard founded Troy Female Seminary, first U.S. women's college.

1823
Monroe Doctrine, opposing European intervention in the Americas, enunciated by Pres. James Monroe **Dec. 2.**

1824
Pawtucket, RI, **weavers strike,** first such action by women.

1825
After a deadlocked election, John Quincy Adams was elected president by the U.S. House, **Feb. 9.**
Erie Canal opened; first boat left Buffalo **Oct. 26,** reached New York City **Nov. 4.** Canal cost $7 mil but opened Great Lakes area, made New York City chief Atlantic port.
John Stevens, of Hoboken, NJ, built and operated first experimental **steam locomotive** in U.S.

1826
Thomas Jefferson and John Adams both died **July 4.**

1828
South Carolina **Dec. 19** declared the right of state **nullification of federal laws,** opposing the "Tariff of Abominations."
Noah Webster published his *American Dictionary of the English Language.*
Baltimore & Ohio, 1st U.S. passenger railroad, was begun **July 4.**

1829
Andrew Jackson inaugurated as president, **Mar. 4.**

1830
Mormon church organized by Joseph Smith in Fayette, NY, **Apr. 6.**

1831
William Lloyd Garrison began abolitionist newspaper *The Liberator,* **Jan. 1.**
Nat Turner, black slave in Virginia, led local slave rebellion, starting **Aug. 21;** 57 whites killed. Troops called in, 100 slaves killed, Turner captured, tried, and hanged **Nov. 11.**

1832
Black Hawk War (IL-WI) **Apr.-Sept.** pushed Sauk and Fox Indians west across Mississippi.
South Carolina convention passed **Ordinance of Nullification Nov. 24** against permanent tariff, threatening to withdraw from Union. Congress **Feb. 1833** passed compromise tariff act, whereupon South Carolina repealed its act.

1833
Oberlin College became first in U.S. to adopt coeducation.

1835
Seminole Indians in Florida under Osceola began attacks **Nov. 1,** protesting forced removal. The unpopular 8-year war ended **Aug. 14, 1842;** Indians were sent to Oklahoma.
Texas proclaimed right to secede from Mexico; Sam Houston put in command of Texas army, **Nov. 2-4.**
Gold discovered on **Cherokee land** in Georgia. Indians forced to cede lands **Dec. 20** and to cross Mississippi.
Halley's Comet passed by the Earth.

1836
Texans besieged in Alamo in San Antonio by Mexicans under Santa Anna **Feb. 23-Mar. 6;** entire garrison killed. Texas independence declared, **Mar. 2.** At San Jacinto **Apr. 21,** Sam Houston and Texans defeated Mexicans.
Marcus Whitman, H. H. Spaulding, and wives reached Fort Walla Walla on Columbia River, OR. **First white women to cross plains.**

1838
Cherokee Indians made **"Trail of Tears,"** removed from Georgia to Oklahoma starting **Oct.**

1841
First emigrant **wagon train for California,** 47 persons, left Independence, MO, **May 1,** reached California **Nov. 4.**
Brook Farm commune set up by New England Transcendentalist intellectuals. Lasted to **1846.**

1842
Webster-Ashburton Treaty signed **Aug. 9,** fixing the U.S.-Canada border in Maine and Minnesota.
First use of **anesthetic** (sulfuric ether gas).
Settlement of Oregon began via **Oregon Trail.**

1843
More than 1,000 settlers left Independence, MO, for Oregon **May 22,** arrived **Oct.**

1844
First message over first **telegraph line** sent **May 24** by inventor Samuel F.B. Morse from Washington to Baltimore: "What hath God wrought!"

1845
Texas Congress **voted for annexation** by U.S. **July 4.** U.S. Congress admitted Texas to Union **Dec. 29.**

Edgar Allan Poe's poem "The Raven" published.

1846
Mexican War began after Pres. James K. Polk ordered Gen. Zachary Taylor to seize disputed Texan land settled by Mexicans. After border clash, U.S. declared war **May 13;** Mexico **May 23.**

Bear flag of Republic of California raised by American settlers at Sonoma **June 14.**

About 12,000 U.S. troops took Vera Cruz **Mar. 27, 1847,** and Mexico City **Sept. 14, 1847.** By treaty, signed **Feb. 2, 1848,** war was ended, and Mexico ceded claims to Texas, California, and other territory.

Treaty with Britain **June 15** set **boundary in Oregon** territory at 49th parallel (extension of existing line). Expansionists had used slogan "54× 40′ or fight."

Mormons, after violent clashes with settlers over polygamy, left Nauvoo, IL, for West under Brigham Young; settled **July 1847** at **Salt Lake City, UT.**

Elias Howe invented **sewing machine.**

1847
First **adhesive U.S. postage stamps** on sale **July 1;** Benjamin Franklin 5¢, Washington 10¢.

Ralph Waldo Emerson published first book of poems; **Henry Wadsworth Longfellow** published *Evangeline.*

1848
Gold discovered Jan. 24 in California; 80,000 prospectors emigrated in **1849.**

Lucretia Mott and Elizabeth Cady Stanton led **Seneca Falls, NY, Women's Rights Convention July 19-20.**

1850
Sen. Henry Clay's **Compromise of 1850** admitted California as 31st state **Sept. 9,** with slavery forbidden; made Utah and New Mexico territories; made Fugitive Slave Law more harsh; ended District of Columbia slave trade.

Nathaniel Hawthorne's *The Scarlet Letter* published.

1851
Herman Melville's *Moby-Dick* published.

1852
Uncle Tom's Cabin, by **Harriet Beecher Stowe,** published.

1853
Comm. Matthew C. Perry, U.S.N., received by Japan, **July 14; negotiated treaty to open Japan** to U.S. ships.

1854
Republican Party formed at Ripon, WI, **Feb. 28.** Opposed Kansas-Nebraska Act (became law **May 30**), which left issue of slavery to vote of settlers.

Henry David Thoreau published *Walden.*

Treaty ratified with Mexico **Apr. 25,** providing for purchase of a strip of land **(Gadsden Purchase).**

1855
Walt Whitman published *Leaves of Grass.*

First railroad train crossed Mississippi on the river's first bridge, Rock Island, IL, Davenport, IA, **Apr. 21.**

1856
Republican Party's first nominee for president, **John C. Fremont,** defeated. Abraham Lincoln made 50 speeches for him.

Lawrence, KS, sacked **May 21** by proslavery group; abolitionist **John Brown** led antislavery men against Missourians at **Osawatomie, KS, Aug. 30.**

1857
Dred Scott decision by Supreme Court **Mar. 6** held that slaves did not become free in a free state, Congress could not bar slavery from a territory, and blacks could not be citizens.

1858
First **Atlantic cable** was completed, by Cyrus W. Field **Aug. 5.**

Lincoln-Douglas debates in Illinois **Aug. 21-Oct. 15.**

1859
First commercially productive **oil well,** drilled near Titusville, PA, by Edwin L. Drake **Aug. 27.**

Abolitionist **John Brown,** with 21 men, seized U.S. Armory at **Harpers Ferry Oct. 16.** U.S. Marines captured raiders, killing several. Brown was hanged for treason **Dec. 2.**

1860
Approximately 20,000 **New England shoe workers** went on strike **Feb. 22** and won higher wages.

Abraham Lincoln, Republican, elected president **Nov. 6** in 4-way race.

First **Pony Express** between Sacramento, CA, and St. Joseph, MO, started **Apr. 3;** service ended **Oct. 24, 1861,** when first transcontinental telegraph line was completed.

1861
Seven southern states set up **Confederate States of America Feb. 8,** with Jefferson Davis as president, captured federal arsenals and forts. **Civil War** began as Confederates fired on **Ft. Sumter** in Charleston, SC, **Apr. 12,** capturing it **Apr. 14.**

Pres. **Lincoln called for 75,000 volunteers Apr. 15.** By **May,** 11 states had seceded. Lincoln blockaded Southern ports **Apr. 19,** cutting off vital exports, aid.

Confederates repelled Union forces at first **Battle of Bull Run July 21.**

First **transcontinental telegraph** was put in operation.

1862
Homestead Act approved **May 20;** it granted free family farms to settlers.

Land Grant Act approved **July 7,** providing for public land sale to benefit agricultural education; eventually led to establishment of state university systems.

Union forces were victorious in Western campaigns, took **New Orleans May 1.** Battles in East were inconclusive.

1863
Pres. Lincoln issued **Emancipation Proclamation Jan. 1,** freeing "all slaves in areas still in rebellion."

Entire **Mississippi River** was in Union hands by **July 4.** Union forces won a major victory at **Gettysburg, PA, July 1-3.** Lincoln read his **Gettysburg Address Nov. 19.**

In **draft riots** in New York City about 1,000 were killed or wounded; some blacks were hanged by mobs **July 13-16.**

1864
Gen. William Tecumseh **Sherman marched through Georgia,** taking Atlanta **Sept. 1,** Savannah **Dec. 22.**

Sand Creek massacre of Cheyenne and Arapaho Indians **Nov. 29.** Cavalry attacked Indians awaiting surrender terms.

1865
Gen. **Robert E. Lee surrendered** 27,800 Confederate troops to Gen. Ulysses S. Grant at Appomattox Court House, VA, **Apr. 9.** J. E. Johnston surrendered 31,200 to Sherman at Durham Station, NC, **Apr. 18.** Last rebel troops surrendered **May 26.**

Pres. **Lincoln was shot Apr. 14** by John Wilkes Booth in Ford's Theater, Washington, DC; died the following morning. Vice Pres. **Andrew Johnson** was sworn in as president. Booth was hunted down; fatally wounded, perhaps by his own hand, **Apr. 26.** Four co-conspirators were hanged **July 7.**

13th Amendment, abolishing slavery, ratified **Dec. 6.**

1866
Ku Klux Klan formed secretly in South to terrorize blacks who voted. Disbanded **1869-71.** A 2d Klan organized **1915.**

Congress took control of Southern Reconstruction, backed freedmen's rights.

1867
Alaska sold to U.S. by Russia for $7.2 mil **Mar. 30** through efforts of Sec. of State William H. Seward.

Horatio Alger published first book, *Ragged Dick.*

The **Grange** was organized **Dec. 4,** to protect farmer interests.

1868
The World Almanac, a publication of the *New York World,* appeared for the first time.

Pres. **Johnson** tried to remove Edwin M. Stanton, secretary of war; was impeached by House **Feb. 24** for violation of Tenure of Office Act; acquitted by Senate Mar.-May.

1869
Financial **"Black Friday"** in New York **Sept. 24;** caused by attempt to "corner" gold.

Transcontinental railroad completed; golden spike driven at Promontory, UT, **May 10,** marking the junction of Central Pacific and Union Pacific.

Knights of Labor formed in Philadelphia. By **1886,** this labor union had 700,000 members nationally.

Woman suffrage law passed in Wyoming Territory **Dec. 10.**

1871
Great fire destroyed **Chicago Oct. 8-11.**

1872
Amnesty Act restored civil rights to citizens of the South **May 22** except for 500 Confederate leaders.

Congress founded first national park—**Yellowstone.**

1873
First U.S. **postal card** issued **May 1.**

Banks failed, panic began in **Sept.** Depression lasted 5 years.

"Boss" William Tweed of New York City convicted **Nov. 19** of stealing public funds. He died in jail in **1878.**

New York's Bellevue Hospital started **first nursing school.**

1875
Congress passed **Civil Rights Act Mar. 1,** giving equal rights to blacks in public accommodations and jury duty. Act invalidated in **1883** by Supreme Court.

First **Kentucky Derby** held **May 17.**

1876
Samuel J. Tilden, Democrat, received majority of popular votes for president over **Rutherford B. Hayes,** Republican, but 22 electoral votes were in dispute; issue left to Congress. Hayes given presidency in **Feb. 1877** after Republicans agreed to end Reconstruction of South.

Col. **George A. Custer** and 264 soldiers of the 7th Cavalry killed **June 25** in "last stand," Battle of the Little Big Horn, MT, in Sioux Indian War.

1877
Molly Maguires, Irish terrorist society in Scranton, PA, mining areas, was broken up by the hanging, **June 21,** of 11 leaders for murders of mine officials and police.

Pres. Rutherford B. Hayes sent troops in violent national **railroad strike.**

1878
First commercial **telephone** exchange opened, New Haven, CT, **Jan. 28.**

Thomas A. Edison founded **Edison Electric Light Co.** on **Oct. 15.**

1879
F. W. Woolworth opened his first five-and-ten store, in Utica, NY, **Feb. 22.**

Henry George published *Progress & Poverty,* advocating single tax on land.

1881
Pres. **James A. Garfield shot** in Washington, DC, **July 2;** died **Sept. 19.**

Booker T. Washington founded Tuskegee Institute for blacks.

Helen Hunt Jackson published *A Century of Dishonor,* about mistreatment of Indians.

1883
Pendleton Act passed **Jan. 16,** reformed civil service.

Brooklyn Bridge opened **May 24.**

1884
Mark Twain's masterpiece, *The Adventures of Huckleberry Finn,* appeared.

1886
Haymarket riot and bombing, **May 4,** followed bitter labor battles for 8-hour day in Chicago; 7 police and 4 workers died. Eight anarchists found guilty **Aug. 20,** 4 hanged **Nov. 11.**

Geronimo, Apache Indian, finally surrendered **Sept. 4.**

Statue of Liberty dedicated **Oct. 28.**

American Federation of Labor (AFL) formed **Dec. 8** by 25 craft unions.

1888
Great blizzard struck eastern U.S. **Mar. 11-14,** causing about 400 deaths.

1889
U.S. opened Oklahoma to white settlement **Apr. 22;** within 24 hours **claims for 2 mil acres** were staked by 50,000 settlers.

Johnstown, PA, flood May 31; 2,200 lives lost.

1890
Battle of **Wounded Knee, SD, Dec. 29,** the last major conflict between Indians and U.S. troops. About 200 Indian men, women, and children and 29 soldiers were killed.

Sherman Antitrust Act passed **July 2,** began federal effort to curb monopolies.

Jacob Riis published *How the Other Half Lives,* about city slums.

Poems of **Emily Dickinson** published posthumously.

1891
Forest Reserve Act Mar. 3 let president close public forest land to settlement for establishment of national parks.

1892
Ellis Island, in New York Bay, opened **Jan. 1** to receive immigrants.

Homestead, PA, strike at Carnegie steel mills; 7 guards and 11 strikers and spectators shot to death **July 6;** setback for unions.

1893
Financial panic began, led to 4-year depression.

1894
Thomas A. Edison's kinetoscope (motion pictures) (invented **1887**) given first public showing **Apr. 14.**

The **Pullman strike** began **May 11** at a railroad car plant in Chicago.

Jacob S. Coxey led army of unemployed from the Midwest, reaching Washington, DC, **Apr. 30.** Coxey arrested **May 1** for trespassing on Capitol grounds; his army disbanded.

1896
William Jennings Bryan delivered "Cross of Gold" speech **July 8;** won Democratic Party nomination.

Supreme Court, in **Plessy v. Ferguson,** approved racial segregation under the "separate but equal" doctrine.

1898
U.S. **battleship** *Maine* blown up **Feb. 15** at Havana; 260 killed.

U.S. **blockaded Cuba Apr. 22** in aid of independence forces. U.S. declared war on Spain, **Apr. 24,** destroyed Spanish fleet in Philippines **May 1,** took Guam **June 20.**

Puerto Rico taken by U.S. **July 25-Aug. 12.** Spain agreed **Dec. 10** to cede Philippines, Puerto Rico, and Guam, and approved independence for Cuba.

Annexation of **Hawaii** signed by Pres. William McKinley, **July 7.**

1899
Filipino insurgents, unable to get recognition of independence from U.S., started guerrilla war **Feb. 4.** Their leader, Emilio Aguinaldo, captured **May 23, 1901.** Philippine Insurrection ended **1902.**

U.S. declared **Open Door Policy** to make China an open international market and to preserve its integrity as a nation.

John Dewey published *The School and Society,* advocating "progressive education."

1900
Carry Nation, Kansas antisaloon agitator, began raiding with hatchet.

U.S. helped suppress **"Boxers"** in Beijing.

International Ladies' Garment Workers Union was founded in New York City **June 3.**

1901
Texas had first significant **oil strike, Jan. 10.**

Pres. **McKinley was shot Sept. 6** in Buffalo, NY, by an anarchist, Leon Czolgosz; died **Sept. 14.**

1903
Treaty between U.S. and Colombia to have U.S. dig **Panama Canal** signed **Jan. 22,** rejected by Colombia. Panama declared independence from Colombia with U.S. support **Nov. 3;** recognized by Pres. Theodore Roosevelt **Nov. 6.** U.S., Panama signed canal treaty **Nov. 18.**

Wisconsin set first **direct primary** voting system **May 23.**

First successful flight in heavier-than-air mechanically propelled airplane by **Orville Wright Dec. 17** near Kitty Hawk, NC, 120 ft in 12 secs. Fourth flight same day by **Wilbur Wright,** 852 ft in 59 secs. Improved plane patented, **1906.**

Great Train Robbery, pioneering film, produced.

1904
Ida Tarbell published muckraking *History of Standard Oil.*

1905
First **Rotary Club** founded in Chicago.

1906
San Francisco earthquake and fire **Apr. 18-19** left 503 dead, $350 mil damages.
Pure Food and Drug Act and Meat Inspection Act both passed **June 30.**

1907
Financial panic and depression started **Mar. 13.**
First round-world cruise of U.S. **"Great White Fleet"**; 16 battleships, 12,000 men.

1908
Henry Ford introduced **Model T** car, priced at $850, **Oct. 1.**

1909
Adm. Robert E. Peary claimed to have reached **North Pole Apr. 6** on 6th attempt, accompanied by Matthew Henson, a black man, and 4 Eskimos; may have fallen short.
National Conference on the Negro convened **May 30,** leading to founding of National Association for the Advancement of Colored People.

1910
Boy Scouts of America founded **Feb. 8.**

1911
Supreme Court dissolved **Standard Oil Co. May 15.**
Building holding New York City's **Triangle Shirtwaist Co.** factory caught fire **Mar. 25;** 146 died.
First **transcontinental airplane flight** (with numerous stops) by C. P. Rodgers, New York to Pasadena, CA, **Sept. 17-Nov. 5;** time in air 82 hr, 4 min.

1912
American Girl Guides founded **Mar. 12;** name changed in **1913** to **Girl Scouts.**
U.S. sent Marines **Aug. 14** to **Nicaragua,** which was in default of loans to U.S. and Europe.

1913
NY Armory Show brought modern art to U.S. **Feb. 17.**
U.S. blockaded Mexico in support of revolutionaries.
Charles Beard published his *Economic Interpretation of the Constitution.*
Federal Reserve System was authorized **Dec. 23,** in a major reform of U.S. banking and finance.

1914
Ford Motor Co. raised basic wage rates from $2.40 for 9-hr day to $5 for 8-hr day **Jan. 5.**
When U.S. sailors were arrested at Tampico, Mexico, **Apr. 9,** Atlantic fleet was sent to **Veracruz,** occupied city.
Pres. Woodrow Wilson proclaimed **U.S. neutrality** in the European war **Aug. 4.**
Panama Canal was officially opened **Aug. 15.**
The **Clayton Antitrust Act** was passed **Oct. 15,** strengthening federal antimonopoly powers.

1915
First transcontinental **telephone call,** New York to San Francisco, completed **Jan. 25,** by Alexander Graham Bell and Thomas A. Watson.
British ship *Lusitania* sunk **May 7** by German submarine; 128 American passengers lost (Germany had warned passengers in advance). As a result of U.S. campaign, Germany issued apology and promise of payments **Oct. 5.** Pres. Wilson asked for a military fund increase **Dec. 7.**
U.S. troops landed in **Haiti July 28.** Haiti became a virtual U.S. protectorate under **Sept. 16** treaty.

1916
Gen. John J. **Pershing entered Mexico** to pursue Francisco (Pancho) Villa, who had raided U.S. border areas. Forces withdrawn **Feb. 5, 1917.**
Rural Credits Act passed **July 17,** followed by Warehouse Act **Aug. 11;** both provided financial aid to farmers.
Bomb exploded during **San Francisco** Preparedness Day parade **July 22,** killed 10. Thomas J. Mooney, labor organizer, and Warren K. Billings, shoe worker, were convicted **1917;** both later pardoned.
U.S. bought **Virgin Islands** from Denmark **Aug. 4.**
Jeannette Rankin (R, MT) elected as **first-ever female** member of U.S. **House.**
U.S. established military government in the **Dominican Republic Nov. 29.**
Trade and loans to **European allies** soared during the year.
Carl Sandburg published *Chicago Poems.*

1917
Germany, suffering from British blockade, declared almost unrestricted **submarine warfare Jan. 31.** U.S. cut diplomatic ties with Germany **Feb. 3,** and formally declared war **Apr. 6.**
Conscription law was passed **May 18.** First U.S. troops arrived in Europe **June 26.**
18th **(Prohibition)** Amendment to the Constitution was submitted to the states by Congress **Dec. 18.** On **Jan. 16, 1919,** the 36th state (Nevada) ratified it.

1918
Pres. Wilson set out his **14 Points** as basis for peace **Jan. 8.**
More than 1 mil **American troops** were in Europe by **July.** Allied counteroffensive launched at Château-Thierry **July 18.** War ended with signing of armistice **Nov. 11.**
Influenza epidemic killed an estimated 20 mil worldwide, 548,000 in U.S.

1919
First **transatlantic flight,** by U.S. Navy seaplane, left Rockaway, NY, **May 8,** stopped at Newfoundland, Azores, Lisbon **May 27.**
Boston police strike Sept. 9; National Guard breaks strike.
Sherwood Anderson published *Winesburg, Ohio.*
About 250 **alien radicals** were deported **Dec. 22.**

1920
In national **Red Scare,** some 2,700 Communists, anarchists, and other radicals were arrested **Jan.-May.**
Senate refused **Mar. 19** to ratify the **League of Nations Covenant.**
Radicals Nicola **Sacco** and Bartolomeo **Vanzetti** accused of killing 2 men in Massachusetts payroll holdup **Apr. 15.** Found guilty **1921.** A 6-year campaign for their release failed, and both were executed **Aug. 23, 1927.** Controversial verdict repudiated **1977,** by proclamation of Massachusetts Gov. Michael Dukakis.
First regular licensed **radio broadcasting** begun **Aug. 20.**
19th Amendment ratified **Aug. 18,** giving women right to vote.
League of Women Voters founded.
Wall St., New York City, **bomb** explosion killed 30, injured 100, did $2 mil damage **Sept. 16.**
Sinclair Lewis's *Main Street,* **F. Scott Fitzgerald's** *This Side of Paradise* published.

1921
Congress sharply curbed **immigration,** set national quota system **May 19.**
Joint congressional resolution declaring **peace with Germany, Austria,** and **Hungary** signed **July 2** by Pres. Warren G. Harding; treaties were signed in **Aug.**
Limitation of Armaments Conference met in Washington, DC, **Nov. 12-Feb. 6, 1922.** Major powers agreed to curtail naval construction, outlaw poison gas, restrict submarine attacks on merchant vessels, respect integrity of China.
Ku Klux Klan began revival with violence against Catholics in North, South, and Midwest.

1922
Violence during **coal-mine strike** at Herrin, IL, **June 22-23** cost 36 lives, including those of 21 nonunion miners.
Reader's Digest founded.

1923
First **sound-on-film motion picture,** *Phonofilm,* shown at Rivoli Theater, New York City, beginning in **April.**

1924
Law approved by Congress **June 15** making all **Indians citizens.**
Nellie Tayloe Ross elected governor of Wyoming **Nov. 9** as nation's first woman governor. **Miriam (Ma) Ferguson** elected governor of Texas **Nov. 9;** installed **Jan. 20, 1925.**
George Gershwin wrote *Rhapsody in Blue.*

1925
John T. Scopes found guilty of having taught **evolution** in Dayton, TN, high school, fined $100 and costs **July 24.**

1926
Dr. Robert H. Goddard demonstrated practicality of **rockets Mar. 16** at Auburn, MA, with first liquid-fuel rocket; rocket traveled 184 ft in 2.5 sec.
Congress established **Army Air Corps July 2.**
Air Commerce Act passed **Nov. 2,** providing federal aid for airlines and airports.
Ernest Hemingway's *The Sun Also Rises* published.

1927

About 1,000 **marines landed in China Mar. 5** to protect property in civil war.

Capt. **Charles A. Lindbergh** left Roosevelt Field, NY, **May 20** alone in plane *Spirit of St. Louis* on first New York-Paris nonstop flight. Reached Le Bourget airfield **May 21,** 3,610 mi in 331/2 hours.

The Jazz Singer, with **Al Jolson,** demonstrated part-talking pictures in New York City **Oct. 6.**

Show Boat opened in New York **Dec. 27.**

O. E. Rolvaag published *Giants in the Earth.*

1928

Herbert Hoover elected president, defeating New York Gov. **Alfred E. Smith,** a Catholic.

Amelia Earhart became first woman to fly the Atlantic, **June 17.**

1929

"St. Valentine's Day massacre" in Chicago **Feb. 14;** gangsters killed 7 rivals.

Farm price stability aided by **Agricultural Marketing Act,** passed **June 15.**

Albert B. Fall, former secretary of the interior, was convicted of accepting bribe of $100,000 in the leasing of the **Elk Hills (Teapot Dome)** naval oil reserve; sentenced **Nov. 1** to a year in prison and fined $100,000.

Stock market crash Oct. 29 marked end of past prosperity as stock prices plummeted. Stock losses for 1929-31 estimated at $50 bil; worst American depression began.

Thomas Wolfe published *Look Homeward, Angel.* **William Faulkner** published *The Sound and the Fury.*

1930

London **Naval Reduction Treaty** signed by U.S., Britain, Italy, France, and Japan **Apr. 22;** in effect **Jan. 1, 1931;** expired **Dec. 31, 1936.**

Hawley-Smoot Tariff signed; rate hikes slash world trade.

1931

Empire State Building opened in New York City **May 1.**

Al Capone was convicted of tax evasion **Oct. 17.**

Pearl Buck published *The Good Earth.*

1932

Reconstruction Finance Corp. established **Jan. 22** to stimulate banking and business. Unemployment at 12 mil.

19-month-old **Charles Lindbergh Jr. was kidnapped Mar. 1;** found dead **May 12.** Bruno Hauptmann found guilty in trial **Jan.-Feb. 1935;** executed **Apr. 3, 1936.**

Bonus March on Washington, DC, launched **May 29** by World War I veterans demanding Congress pay their bonus in full.

Franklin D. Roosevelt elected president for the first time.

1933

Pres. Roosevelt named **Frances Perkins** U.S. secretary of labor; first woman in U.S. cabinet.

All **banks in the U.S. were ordered closed** by Pres. Roosevelt **Mar. 6.**

In a "100 days" special session, **Mar. 9-June 16,** Congress passed **New Deal** social and economic measures, including measures to regulate banks, distribute funds to the jobless, create jobs, raise agricultural prices, and set wage and production standards for industry.

Tennessee Valley Authority created by act of Congress, **May 18.**

Gold standard dropped by U.S.; announced by Pres. Roosevelt **Apr. 19,** ratified by Congress **June 5.**

Prohibition ended in the U.S. as 36th state ratified 21st Amendment **Dec. 5.**

U.S. foreswore armed intervention in **western hemisphere** nations **Dec. 26.**

1934

U.S. troops pulled out of **Haiti Aug. 6.**

1935

Works Progress Administration **(WPA)** instituted **May 6.** Rural Electrification Administration created **May 11.** National Industrial Recovery Act struck down by Supreme Court **May 27.**

Comedian **Will Rogers** and aviator **Wiley Post killed Aug. 15** in Alaska plane crash.

Social Security Act passed by Congress **Aug. 14.**

Huey Long, senator from Louisiana and national political leader, **assassinated Sept. 8.**

Porgy and Bess opened **Oct. 10** in New York.

Committee for Industrial Organization (CIO; later Congress of Industrial Organizations) formed to expand industrial unionism **Nov. 9.**

1936

Boulder Dam completed.

Margaret Mitchell published *Gone With the Wind.*

1937

Joe Louis knocked out James J. Braddock, became world heavyweight champ **June 22.**

Amelia Earhart, aviator, and copilot Fred Noonan lost **July 2** near Howland Island, in the Pacific.

Pres. Roosevelt asked for 6 additional Supreme Court justices; **"packing" plan** defeated.

Auto, steel labor unions won first big contracts.

1938

Naval Expansion Act passed **May 17.**

National minimum wage enacted **June 25.**

Orson Welles radio dramatization of **Martian invasion,** *War of the Worlds,* caused nationwide scare **Oct. 30.**

1939

Pres. Roosevelt asked for **defense budget hike Jan. 5, 12.**

New York World's Fair opened **Apr. 30,** closed **Oct. 31;** reopened **May 11, 1940,** and finally closed **Oct. 21.**

Albert Einstein alerted Pres. Roosevelt to **A-bomb** opportunity in **Aug. 2** letter.

U.S. declared its neutrality in European war **Sept. 5.**

Roosevelt proclaimed a limited **national emergency Sept. 8,** an unlimited emergency **May 27, 1941.** Both ended by Pres. Harry Truman **Apr. 28, 1952.**

John Steinbeck published *Grapes of Wrath.*

1940

Gone With the Wind and *The Wizard of Oz* appeared on screen.

U.S. okayed sale of **surplus war materiel** to Britain **June 3;** announced transfer of 50 overaged destroyers **Sept. 3.**

First **peacetime draft** approved **Sept. 14.**

Richard Wright published *Native Son.*

1941

Four Freedoms termed essential by Pres. Roosevelt in speech to Congress **Jan. 6:** freedom of speech and religion, freedom from want and fear.

Lend-Lease Act signed **Mar. 11** provided $7 bil in military credits for Britain. Lend-Lease for USSR approved in **Nov.**

U.S. occupied **Iceland July 7.**

The **Atlantic Charter,** 8-point declaration of principles, issued by Roosevelt and British Prime Min. Winston Churchill **Aug. 14.**

Japan attacked **Pearl Harbor,** Hawaii, 7:55 am Hawaiian time, **Dec. 7;** 19 ships sunk or damaged, 2,300 dead. U.S. declared war on Japan **Dec. 8,** on Germany and Italy **Dec. 11.**

1942

Japanese troops took Bataan peninsula **Apr. 8,** Corregidor **May 6.**

Federal government forcibly moved 110,000 **Japanese-Americans** from West Coast to detention camps. Exclusion lasted 3 years.

Battle of **Midway June 4-7** was Japan's first major defeat.

Marines landed on **Guadalcanal Aug. 7;** last Japanese not expelled until **Feb. 9, 1943.**

U.S., Britain invaded North Africa **Nov. 8.**

First **nuclear chain reaction** (fission of uranium isotope U-235) produced at University of Chicago, under physicists Arthur Compton, Enrico Fermi, others **Dec. 2.**

1943

Oklahoma! opened **Mar. 31** on Broadway.

All war contractors barred from **racial discrimination, May 27.**

Pres. Roosevelt signed **June 10** pay-as-you-go income tax bill. Starting **July 1** wage and salary earners were subject to a **paycheck withholding** tax.

Race riot in Detroit June 21; 34 dead, 700 injured. Riot in Harlem section of New York City; 6 killed.

U.S., Britain invaded **Sicily July 9,** Italian **mainland Sept. 3.** Marines advanced on **Gilbert Island in Nov.**

1944

U.S., Allied forces invaded Europe at **Normandy June 6** in greatest amphibious landing in history.

GI Bill of Rights signed **June 22,** providing veterans' benefits.

U.S. forces landed on **Leyte,** Philippines, **Oct. 20.**

1945

Yalta Conference met in the Crimea, USSR, **Feb. 4-11.** Roosevelt, Churchill, and Soviet leader Joseph Stalin agreed that their 3 countries, plus France, would occupy Germany and that the Soviet Union would enter war against Japan.

Marines landed on **Iwo Jima Feb. 19,** won control of Iwo Jima **Mar. 16** after heavy casualties. U.S. forces invaded **Okinawa Apr. 1,** captured Okinawa **June 21.**

Pres. Roosevelt, 63, died in Warm Springs, GA, **Apr. 12;** Vice Pres. **Harry S. Truman** became president.

Germany surrendered May 7; May 8 proclaimed V-E Day.

First **atomic bomb,** produced at Los Alamos, NM, exploded at Alamogordo, NM, **July 16.** Bomb dropped on **Hiroshima Aug. 6,** with about 75,000 people killed; bomb dropped on **Nagasaki Aug. 9,** killing about 40,000. Japan agreed to surrender, **Aug. 14;** formally surrendered **Sept. 2.**

At **Potsdam Conference, July 17-Aug. 2,** leaders of U.S., USSR, and Britain agreed on disarmament of Germany, occupation zones, war crimes trials.

U.S. forces entered **Korea** south of 38th parallel to displace Japanese **Sept. 8.**

Gen. Douglas MacArthur took over supervision of Japan **Sept. 9.**

1946

Strike by 400,000 **mine workers** began **Apr. 1;** other industries followed.

Philippines given independence by U.S. **July 4.**

1947

Pres. Truman asked Congress to aid Greece and Turkey to combat Communist terrorism (**Truman Doctrine), Mar. 12.** Approved **May 15.**

UN Security Council voted unanimously **Apr. 2** to place under **U.S. trusteeship** the Pacific islands formerly mandated to Japan.

Jackie Robinson joined the Brooklyn Dodgers **Apr. 11,** breaking the color barrier in major league baseball.

Taft-Hartley Labor Act curbing strikes was vetoed by Truman **June 20;** Congress overrode the veto.

Proposals known as the **Marshall Plan,** under which the U.S. would extend aid to European countries, were made by Sec. of State George C. Marshall **June 5.** Congress authorized some $12 bil in next 4 years.

1948

USSR halted all surface traffic into W. Berlin, **June 23;** in response, U.S. and British troops launched an airlift. Soviet blockade halted **May 12, 1949;** airlift ended **Sept. 30.**

Organization of American States founded **Apr. 30.**

Alger Hiss indicted **Dec. 15** for perjury, after denying he had passed secret documents to Whittaker Chambers for transmission to a Communist spy ring. Convicted **Jan. 21, 1950.**

Pres. Truman reelected Nov. 2, defeating Gov. Thomas E. Dewey in a historic upset.

Kinsey Report on sexuality in the human male published.

1949

U.S. troops withdrawn from **Korea June 29.**

NATO established **Aug. 24** by U.S., Canada, and 10 Western European nations, agreeing that an armed attack against one or more would be considered an attack against all.

Mrs. I. Toguri D'Aquino (**Tokyo Rose** of Japanese wartime broadcasts) was sentenced **Oct. 7** to 10 years in prison for treason. Paroled **1956,** pardoned **1977.**

Eleven leaders of **U.S. Communist Party** convicted **Oct. 14** of advocating violent overthrow of U.S. government; sentenced to prison. Supreme Court upheld convictions **1951.**

1950

Masked bandits robbed **Brink's, Inc.,** Boston express office, **Jan. 17** of $2.8 mil, of which $1.2 mil was in cash. Case solved **1956;** 8 sentenced to life.

Pres. Truman authorized production of the **H-bomb Jan. 31.**

North Korea forces invaded **South Korea June 25.** UN asked for troops to restore peace.

Truman ordered Air Force and Navy to Korea **June 27.** Truman approved ground forces, air strikes against North Korea **June 30.**

U.S. sent 35 military advisers to **South Vietnam June 27,** and agreed to provide military and economic aid to anti-Communist government.

Army seized all railroads Aug. 27 on Truman's order to prevent a general strike; returned to owners in **1952.**

U.S. forces landed at Inchon Sept. 15; UN force took Pyongyang **Oct. 20,** reached China border **Nov. 20;** China sent troops across border **Nov. 26.**

Two members of **Puerto Rican nationalist** movement tried to kill Pres. Truman **Nov. 1.**

U.S. **Dec. 8** banned shipments to **Communist China** and to Asiatic ports trading with it.

1951

Sen. Estes Kefauver led Senate investigation into organized crime.

Julius Rosenberg, his wife, Ethel, and Morton Sobell found guilty **Mar. 29** of conspiracy to commit wartime espionage. Rosenbergs executed **June 19, 1953.** Sobell sentenced to 30 years; released **1969.**

Gen. Douglas MacArthur removed from Korea command **Apr. 11** by Pres. Truman, for having made unauthorized policy statements.

Korea cease-fire talks began in July; lasted 2 years. **Fighting ended July 27, 1953.**

Tariff concessions by the U.S. to the Soviet Union, Communist China, and all Communist-dominated lands were suspended **Aug. 1.**

The **U.S., Australia,** and **New Zealand** signed a mutual security pact **Sept. 1.**

Transcontinental television inaugurated **Sept. 4** with Pres. Truman's address at the Japanese Peace Treaty Conference in San Francisco.

Japanese peace treaty signed in San Francisco **Sept. 8** by U.S., Japan, and 47 other nations.

J. D. Salinger published *Catcher in the Rye.*

1952

U.S. **seizure of nation's steel mills** was ordered by Pres. Truman **Apr. 8** to avert a strike. Ruled illegal by Supreme Court **June 2.**

Peace contract between West Germany, U.S., Great Britain, and France was signed **May 26.**

The last racial and ethnic barriers to naturalization removed, **June 26-27,** with passage of **Immigration and Naturalization Act of 1952.**

First **hydrogen device** explosion **Nov. 1** at Eniwetok Atoll in Pacific.

1953

Pres. Dwight D. Eisenhower announced **May 8** that U.S. had given France $60 mil for **Indochina War.** More aid was announced in **Sept.**

Korean War armistice signed **July 27.**

1954

Nautilus, first atomic-powered submarine, was launched at Groton, CT, **Jan. 21.**

Five members of Congress were wounded in the House **Mar. 1** by 4 **Puerto Rican independence supporters** who fired at random from a spectators' gallery.

Sen. Joseph McCarthy (R, WI) led televised hearings **Apr. 22-June 17** into alleged Communist influence in the Army.

Racial segregation in public schools unanimously ruled unconstitutional by Supreme Court **May 17,** in *Brown* v. *Board of Education of Topeka.*

Southeast Asia Treaty Organization (**SEATO**) formed by defense pact signed in Manila **Sept. 8** by U.S., Britain, France, Australia, New Zealand, Philippines, Pakistan, and Thailand.

Condemnation of **Sen. McCarthy** voted by Senate, 67-22, **Dec. 2** for contempt of Senate subcommittee, abuse of its members, insults to Senate during Army investigation hearings.

1955

U.S. agreed **Feb. 12** to help train **South Vietnamese** army.

Supreme Court ordered **"all deliberate speed"** in integration of public schools **May 31.**

A **summit meeting** of leaders of U.S., Britain, France, and USSR took place **July 18-23** in Geneva, Switzerland.

Rosa Parks refused **Dec. 1** to give her seat to a white man on a **bus in Montgomery, AL.** Bus segregation ordinance declared unconstitutional by a federal court following boycott and NAACP protest.

America's 2 largest labor organizations merged **Dec. 5,** creating the AFL-CIO.

1956

Massive resistance to Supreme Court desegregation rulings was called for **Mar. 12** by 101 Southern congressmen.

Federal-Aid **Highway Act** signed **June 29,** inaugurating interstate highway system.

First transatlantic **telephone cable** activated **Sept. 25.**

1957

Congress approved first **civil rights bill** for blacks since Reconstruction **Apr. 29,** to protect voting rights.

National Guardsmen, called out by Arkansas Gov. Orval Faubus **Sept. 4,** barred 9 black students from entering all-white high school in **Little Rock.** Faubus complied **Sept. 21** with federal court order to remove Guardsmen, but the blacks were ordered to withdraw by local authorities. Pres. Eisenhower sent federal troops **Sept. 24** to enforce court order.

Jack Kerouac published *On the Road.*

1958

First U.S. **earth satellite** to go into orbit, **Explorer I,** launched by Army **Jan. 31** at Cape Canaveral, FL; discovered Van Allen radiation belt.

U.S. Marines sent to **Lebanon** to protect elected government from threatened overthrow **July-Oct.**

First domestic **jet airline** passenger service in U.S. opened by National Airlines **Dec. 10** between New York and Miami.

1959

Alaska admitted as 49th state **Jan. 3; Hawaii** admitted as 50th **Aug. 21.**

St. Lawrence Seaway opened **Apr. 25.**

Soviet **Premier Nikita Khrushchev** paid unprecedented visit to U.S. **Sept. 15-27;** made transcontinental tour.

1960

Sit-ins began **Feb. 1** when 4 black college students in Greensboro, NC, refused to move from a Woolworth lunch counter when denied service. By **Sept. 1961** more than 70,000 students, whites and blacks, had participated in sit-ins.

Congress approved a strong **voting rights act Apr. 21.**

A U.S. **U-2 reconnaissance plane** was shot down in the Soviet Union **May 1;** pilot Gary Powers captured. The incident led to cancellation of an imminent Paris summit conference.

Vice Pres. Richard Nixon and Sen. John F. Kennedy faced each other, **Sept. 26,** in the first in a series of televised **campaign debates. Kennedy defeated Nixon** to win the presidency, **Nov. 8.**

U.S. announced **Dec. 15** it backed rightist group in **Laos,** which took power the next day.

1961

U.S. severed diplomatic and consular relations with **Cuba Jan. 3,** after disputes over nationalizations of U.S. firms, U.S. military presence at Guantanamo base.

Invasion of Cuba's **"Bay of Pigs" Apr. 17** by Cuban exiles trained, armed, and directed by U.S., attempted to overthrow the regime of Premier Fidel Castro, unsuccessfully.

Peace Corps created by executive order, **Mar. 1.**

Commander Alan B. Shepard Jr. was rocketed from Cape Canaveral, FL, 116.5 mi above the earth in a Mercury capsule **May 5,** in first U.S.-crewed suborbital space flight.

"Freedom Rides" from Washington, DC, across deep South were launched **May 20** to **protest segregation** in interstate transportation.

1962

Lt. Col. John H. Glenn Jr. became first American in orbit **Feb. 20** when he circled the earth 3 times in the Mercury capsule *Friendship 7.*

Pres. John F. Kennedy said **Feb. 14** U.S. military advisers in Vietnam would fire if fired upon.

Supreme Court **Mar. 26** backed **"one-man one-vote"** apportionment of seats in state legislatures.

James Meredith became first black student at University of Mississippi **Oct. 1** after 3,000 troops put down riots.

A Soviet **offensive missile buildup in Cuba** was revealed **Oct. 22** by Pres. Kennedy, who ordered a naval and air quarantine on shipment of offensive military equipment to the island. He and Soviet Premier Khrushchev agreed **Oct. 28** on a formula to end the crisis. Kennedy announced **Nov. 2** that Soviet missile bases in Cuba were being dismantled.

Rachel Carson's *Silent Spring* launched environmentalist movement.

1963

Supreme Court ruled **Mar. 18** that all **criminal defendants** must have counsel and that illegally acquired evidence was inadmissible in state as well as federal courts.

University of Alabama **desegregated** after Gov. **George Wallace** stepped aside when confronted by federally deployed National Guard troops, **June 11.**

Civil rights leader **Medgar Evers** was assassinated **June 12.**

Supreme Court ruled, 8-1, **June 17** that laws requiring **recitation of the Lord's Prayer** or Bible verses in public schools were unconstitutional.

A limited **nuclear test-ban treaty** was agreed upon **July 25** by the U.S., the Soviet Union, and Britain.

March on Washington by 200,000 persons **Aug. 28** in support of **black demands** for equal rights. Highlight was "I have a dream" speech by **Dr. Martin Luther King Jr.**

Baptist church in Birmingham, AL, bombed **Sept. 15** in racial violence; 4 black girls killed.

South Vietnam Pres. **Ngo Dinh Diem assassinated Nov. 2;** U.S. had earlier withdrawn support.

Pres. Kennedy shot and fatally wounded Nov. 22 as he rode in a motorcade through downtown Dallas, TX. Vice Pres. **Lyndon B. Johnson sworn in** as president. **Lee Harvey Oswald arrested** and charged with the murder; he was shot and fatally wounded **Nov. 24.** Jack Ruby, a nightclub owner, was convicted of Oswald's murder; he died in **1967,** while awaiting retrial following reversal of his conviction.

Betty Friedan's *Feminine Mystique* ignited the women's movement.

1964

Panama suspended relations with U.S. **Jan. 9** after riots. U.S. offered **Dec. 18** to negotiate a new canal treaty.

Supreme Court ordered **Feb. 17** that **congressional districts** have equal populations.

U.S. reported **May 27** it was sending military planes to **Laos.**

Omnibus **civil rights bill** cleared by Congress **July 2,** signed same day by Pres. Johnson, banning discrimination in voting, jobs, public accommodations.

Three **civil rights workers** were reported missing in Mississippi **June 22;** found buried **Aug. 4.** Twenty-one white men were arrested. On **Oct. 20, 1967,** an all-white federal jury convicted 7 of conspiracy in the slayings.

Bill establishing **Medicare,** government health insurance program for persons over 65, signed **July 30.**

U.S. Congress **Aug. 7** the passed **Tonkin Gulf Resolution,** authorizing presidential action in Vietnam, after N Vietnamese boats reportedly attacked 2 U.S. destroyers **Aug. 2.**

Congress approved **War on Poverty** bill **Aug. 11,** providing for a domestic Peace Corps **(VISTA),** a **Job Corps,** and anti-poverty funding.

The **Warren Commission** released **Sept. 27** a report concluding that Lee Harvey Oswald was solely responsible for the Kennedy assassination.

Pres. Johnson was elected to a full term, **Nov. 3,** defeating Republican **Sen. Barry Goldwater** (AZ) in a landslide.

1965

Pres. Johnson in **Feb.** ordered continuous **bombing of North Vietnam** below 20th parallel.

Malcolm X assassinated **Feb. 21** at New York City rally.

Some 14,000 U.S. troops sent to **Dominican Republic** during civil war **Apr. 28.** All troops withdrawn by next year.

March from Selma to Montgomery, AL, begun Mar. 21 by Rev. Martin Luther King Jr. to demand federal protection of **blacks' voting rights.** New **Voting Rights Act** signed **Aug. 6.**

Los Angeles riot by blacks living in **Watts** area resulted in 34 deaths and $200 mil in property damage **Aug. 11-16.**

National **immigration** quota system abolished **Oct. 3.**

Electric power failure blacked out most of northeastern U.S., parts of 2 Canadian provinces the night of **Nov. 9-10.**

U.S. forces in **S. Vietnam** reached 184,300 by year-end.

1966

U.S. forces began firing into **Cambodia May 1.**

Bombing of Hanoi area of N Vietnam by U.S. planes began **June 29.** By **Dec. 31,** 385,300 U.S. troops were stationed in S Vietnam, plus 60,000 offshore and 33,000 in Thailand.

Medicare began **July 1.**

Edward Brooke (R, MA) elected **Nov. 8** as first black U.S. senator in 85 years.

1967

Black U.S. Rep. **Adam Clayton Powell** (D, NY) was denied **Mar. 1** his seat because of charges he misused government funds. Reelected in **1968**, he was seated, but fined $25,000 and stripped of his seniority.

Pres. Johnson and Soviet Premier Aleksei Kosygin met **June 23 and 25** at **Glassboro State College** in NJ; agreed not to let any crisis push them into war.

The **25th Amendment**, providing for **presidential succession**, was ratified **Feb. 10.**

USS *Liberty,* an intelligence ship, was torpedoed by Israel in the Mediterranean, apparently by accident **June 8;** 34 killed.

Riots by blacks in **Newark, NJ, July 12-17** killed 26, injured 1,500; more than 1,000 arrested. In **Detroit, MI, July 23-30,** more than 40 died; 2,000 injured, 5,000 left homeless by rioting, looting, burning in city's black ghetto.

Thurgood Marshall was sworn in **Oct. 2** as first black U.S. Supreme Court Justice. **Carl B. Stokes** (D, Cleveland) and **Richard G. Hatcher** (D, Gary, IN) were elected first black mayors of major U.S. cities **Nov. 7.**

1968

USS *Pueblo* and 83-man crew seized in Sea of Japan **Jan. 23** by North Koreans; 82 men released **Dec. 22.**

"Tet offensive": Communist troops attacked Saigon, 30 province capitals **Jan. 30,** suffered heavy casualties.

Pres. Johnson **curbed bombing** of North Vietnam **Mar. 31.** Peace talks began in Paris **May 10.** All bombing of North halted **Oct. 31.**

Martin Luther King Jr., 39, assassinated Apr. 4 in Memphis, TN. **James Earl Ray,** an escaped convict, pleaded guilty to the slaying, was sentenced to 99 years.

Sen. Robert F. Kennedy (D, NY), 42, **shot June 5** in Hotel Ambassador, Los Angeles, after celebrating presidential primary victories. Died **June 6.** Sirhan Bishara Sirhan, convicted of murder, **1969;** death sentence commuted to life in prison, **1972.**

Vice Pres. **Hubert Humphrey nominated** for president by Democrats **at national convention in Chicago,** marked by clash between police and **antiwar protesters, Aug. 26-29.**

The Republican nominee, **Richard Nixon, won the presidency,** defeating Hubert Humphrey in a close race **Nov. 5.**

Rep. Shirley Chisholm (D, NY) became the first black woman elected to Congress.

1969

Expanded 4-party **Vietnam peace talks** began **Jan. 18.** U.S. force peaked at 543,400 in April. Withdrawal started **July 8.** Pres. Nixon set Vietnamization policy **Nov. 3.**

U.S. astronaut **Neil Armstrong,** commander of the *Apollo 11* mission, became the first person to **set foot on the moon, July 20;** followed by astronaut **Edwin Aldrin;** astronaut **Michael Collins** remained aboard command module.

Woodstock music festival near Bethel, NY, drew 300,000-500,000 people, **Aug. 15-17.**

Anti-Vietnam War **demonstrations peaked** in U.S.; some 250,000 marched in Washington, DC, **Nov. 15.**

Massacre of hundreds of civilians at **Mylai, South Vietnam,** in 1968 incident reported **Nov. 16.**

1970

United Mine Workers official **Joseph A. Yablonski,** his wife, and their daughter found shot to death **Jan. 5;** UMW chief W. A. (Tony) Boyle later convicted of the killing.

A federal jury **Feb. 18** found the **"Chicago 7"** antiwar activists innocent of conspiring to incite riots during the 1968 **Democratic National Convention.** However, 5 were convicted of crossing state lines with intent to incite riots.

Millions of Americans participated in antipollution demonstrations **Apr. 22** to mark the **first Earth Day.**

U.S. and South Vietnamese forces crossed **Cambodian** borders **Apr. 30** to get at enemy bases. Four students were killed **May 4** at **Kent State** University in Ohio by National Guardsmen during a protest against the war.

Two **women generals,** the first in U.S. history, were named by Pres. Nixon **May 15.**

A **postal reform** measure was signed **Aug. 12,** creating an independent U.S. Postal Service.

1971

Charles Manson and 3 of his cult followers were found guilty **Jan. 25** of first-degree murder in **1969** slaying of actress Sharon Tate and 6 others.

The 26th Amendment, lowering the **voting age to 18** in all elections, was ratified **June 30.**

A court-martial jury **Mar. 29** convicted **Lt. William L. Calley Jr.** of premeditated murder of 22 South Vietnamese at Mylai on **Mar. 16, 1968.** He was sentenced to life imprisonment **Mar. 31.** Sentence was reduced to 20 years **Aug. 20.**

Publication of classified **Pentagon papers** on U.S. involvement in Vietnam was begun **June 13** by the *New York Times.* In a 6-3 vote, U.S. Supreme Court **June 30** upheld the right of the *Times* and the *Washington Post* to publish the documents.

U.S. bombers struck massively in North Vietnam for 5 days starting **Dec. 26** in retaliation for alleged violations of agreements reached prior to the 1968 bombing halt.

1972

Pres. Nixon arrived in **Beijing Feb. 21** for an 8-day visit to China, which he called a "journey for peace."

By a vote of 84 to 8, the Senate, **Mar. 22,** approved banning **discrimination** on the basis of sex, and sent the measure to the states for ratification.

North Vietnamese forces launched the biggest attacks in 4 years across the demilitarized zone **Mar. 30.** The U.S. responded **Apr. 15** by resumption of bombing of Hanoi and Haiphong after a 4-year lull.

Pres. Nixon announced **May 8** the mining of **North Vietnam ports.** Last U.S. combat troops left **Aug. 11.**

Gov. George C. Wallace (AL), campaigning for the presidency at a Laurel, MD, shopping center **May 15, was shot** and seriously wounded. Arthur H. Bremer **convicted Aug. 4,** sentenced to 63 years for shooting Wallace and 3 bystanders.

In **first visit of a U.S. president to Moscow,** Pres. Nixon arrived **May 22** for a week of summit talks with Kremlin leaders that culminated in a landmark **strategic arms pact.**

Five men were arrested **June 17** for breaking into the offices of the Democratic National Committee in the **Watergate** office complex in Washington, DC.

Pres. **Nixon reelected Nov. 7** in a landslide, defeating Democratic Sen. George McGovern (SD).

The **Dow Jones** industrial average closed above 1,000 for the first time, **Nov. 14.**

Full-scale **bombing of North Vietnam** resumed after Paris peace negotiations reached an impasse **Dec. 18.**

1973

Five of 7 defendants in **Watergate** break-in trial pleaded guilty **Jan. 11 and 15;** the other 2 were convicted **Jan. 30.**

In *Roe* v. *Wade,* Supreme Court ruled, 7-2, **Jan. 22,** that states may not ban **abortions** during **first 3 months of pregnancy** and may regulate, but may not ban, abortions during 2d trimester.

Four-party **Vietnam peace pacts** were signed in Paris **Jan. 27,** and North Vietnam released some 590 U.S. prisoners by **Apr. 1.** Last U.S. troops left **Mar. 29.**

End of the military draft announced **Jan. 27.**

Top **Nixon aides** H. R. Haldeman, John D. Ehrlichman, and John Dean and Attorney Gen. Richard Kleindienst **resigned Apr. 30,** amid charges of White House efforts to obstruct justice in the Watergate case.

John Dean, former Nixon counsel, told Senate hearings **June 25** that Nixon, his staff and campaign aides, and the Justice Department had conspired to cover up Watergate facts.

The U.S. officially ceased bombing in **Cambodia** at midnight **Aug. 14** in accord with a June congressional action.

Vice Pres. Spiro T. Agnew Oct. 10 resigned and pleaded no contest to a charge of tax evasion on payments made to him by contractors when he was governor of Maryland. **Gerald R. Ford Oct. 12** became **first appointed vice president** under the 25th Amendment; sworn in **Dec. 6.**

A total ban on **oil exports** to the U.S. was imposed by Arab oil-producing nations **Oct. 19-21** after the outbreak of an Arab-Israeli war. The ban was lifted **Mar. 18, 1974.**

Attorney Gen. Elliot Richardson resigned, and his deputy William D. Ruckelshaus and **Watergate Special Prosecutor Archibald Cox** were **fired** by Pres. Nixon **Oct. 20,** when Cox threatened to secure a judicial ruling that Nixon was violating a court order to give tapes to Judge John Sirica. **Leon Jaworski** named **Nov. 1** by the Nixon administration to succeed Cox.

Congress overrode **Nov. 7** Pres. Nixon's veto of the **war powers** bill, which curbed president's power to commit armed forces to hostilities abroad without congressional approval.

1974

Impeachment hearings opened **May 9** against Pres. Nixon by the House Judiciary Committee.

John D. Ehrlichman and 3 **White House "plumbers"** found guilty **July 12** of conspiring to violate the civil rights of Pentagon Papers leaker Daniel Ellsberg's psychiatrist by breaking into his office.

U.S. Supreme Court ruled, 8-0, **July 24** that Nixon had to turn over **64 tapes** of White House conversations.

House Judiciary Committee, in televised hearings **July 24-30**, recommended 3 **articles of impeachment** against Pres. Nixon. The first, voted 27-11 **July 27,** charged conspiracy to obstruct justice in the Watergate cover-up. The 2d, voted 28-10 **July 29,** charged abuses of power. The 3d, voted 21-17 **July 30,** charged defiance of committee subpoenas. The House voted **Aug. 20,** 412-3, to accept the committee report, which included the impeachment articles.

Pres. Nixon announced his resignation, Aug. 8, and resigned Aug. 9; his support in Congress had begun to collapse **Aug. 5,** after release of tapes implicating him in Watergate cover-up. **Vice Pres. Gerald R. Ford was sworn in Aug. 9** as 38th U.S. president.

A **pardon** to ex-Pres. Nixon for any federal crimes he committed while president issued by Pres. Ford **Sept. 8.**

1975

Found guilty of Watergate cover-up charges Jan. 1 were ex-Atty. Gen. John Mitchell, ex-presidential advisers H. R. Haldeman and John Ehrlichman.

U.S. launched **evacuation of American and some South Vietnamese from Saigon Apr. 29** as Communist forces completed takeover of South Vietnam; **South Vietnamese** government officially **surrendered Apr. 30.**

U.S. merchant ship *Mayaguez* and its crew of 39 were seized by Cambodian forces in Gulf of Siam **May 12.** In rescue operation, U.S. Marines attacked Tang Island, planes bombed air base; Cambodia surrendered ship and crew.

Congress voted $405 mil for **South Vietnam refugees May 16;** 140,000 were flown to the U.S.

Illegal CIA operations, including records on 300,000 persons and groups, and infiltration of agents into black, antiwar, and political movements, described by panel headed by Vice Pres. **Nelson Rockefeller June 10.**

Publishing heiress **Patricia (Patty) Hearst,** kidnapped **Feb. 5, 1974,** by "Symbionese Liberation Army" militants, was captured, in San Francisco **Sept. 18** with others. She was convicted **Mar. 20, 1976,** of bank robbery.

1976

U.S. celebrated **200th anniversary of independence July 4,** with festivals, parades, and New York City's Operation Sail, a gathering of tall ships from around the world.

"Legionnaire's disease" killed 29 persons who attended an American Legion convention **July 21-24** in Philadelphia.

Viking II set down on **Mars'** Utopia Plains **Sept. 3,** following the successful landing by *Viking I* **July 20.**

1977

Pres. Jimmy Carter **Jan. 21** pardoned most Vietnam War **draft evaders.**

Convicted murderer **Gary Gilmore executed** by a Utah firing squad **Jan. 17,** in the first exercise of capital punishment in the U.S. since **1967.**

Pres. Carter signed an act **Aug. 4** creating a new cabinet-level **Energy Department.**

1978

U.S. Senate voted **Apr. 18** to turn over **Panama Canal** to Panama Dec. 31, 1999; **Mar. 16** vote had given approval to a treaty guaranteeing the area's neutrality after the year 2000.

Californians, **June 6,** approved **Proposition 13,** a state constitutional amendment slashing property taxes.

U.S. Supreme Court, **June 28,** ruled against **racial quotas** in *Bakke* v. *University of California.*

1979

Partial meltdown released radioactive material **Mar. 28,** at nuclear reactor on **Three Mile Island** near Middletown, PA.

Federal government announced, **Nov. 1,** a $1.5 bil loan-guarantee plan to aid the ailing **Chrysler Corp.**

1980

Some 90 people, including 63 Americans, **taken hostage, Nov. 4,** at **American embassy in Tehran,** Iran, by militant followers of Ayatollah Khomeini. He demanded return of former Shah Muhammad Reza Pahlavi, who was undergoing medical treatment in New York City.

Pres. Carter announced, **Jan. 4, economic sanctions against the USSR,** in retaliation for Soviet invasion of Afghanistan. At Carter's request, **U.S. Olympic Committee** voted, **Apr. 12,** against U.S. participation in Moscow Summer Olympics.

Eight Americans killed and 5 wounded, **Apr. 24, in ill-fated** attempt to **rescue hostages** held by Iranian militants.

Mt. St. Helens, in Washington state, **erupted, May 18.** The blast, with others **May 25** and **June 12,** left 57 dead.

In a sweeping victory, **Nov. 4, Ronald Reagan** (R) was elected 40th president, defeating incumbent Pres. Carter. Republicans gained control of the Senate.

Former Beatle **John Lennon** was shot and killed, **Dec. 8,** in New York City.

1981

Minutes after **Reagan's inauguration Jan. 20,** the **52 Americans** held **hostage in Iran** for 444 days were freed.

Pres. Reagan was **shot and seriously wounded, Mar. 30,** in Washington, DC; also seriously wounded were a Secret Service agent, a policeman, and Press Sec. **James Brady. John W. Hinckley Jr.** arrested, found not guilty by reason of insanity in **1982,** and committed to mental institution.

World's first reusable spacecraft, the **space shuttle** *Columbia,* was sent into space, **Apr. 12.**

Congress, **July 29,** passed Pres. Reagan's **tax-cut legislation,** expected to save taxpayers $750 bil over 5 years.

Federal air traffic controllers, Aug. 3, began an illegal **nationwide strike.** Most defied a back-to-work order and were dismissed by Pres. Reagan **Aug. 5.**

In a 99-0 vote, the Senate confirmed, **Sept. 21,** appointment of **Sandra Day O'Connor** as an **associate justice of U.S. Supreme Court.** She was the first woman appointed to that body.

1982

The 13-year-old lawsuit against **AT&T** by the **Justice Dept.** was settled **Jan. 8.** AT&T agreed to give up the 22 Bell System companies and was allowed to expand.

The Equal Rights Amendment was defeated after a 10-year struggle for ratification.

In Dec., **unemployment** hit 10.8%, highest since 1940.

A retired dentist, **Dr. Barney B. Clark,** 61, became first recipient of a **permanent artificial heart, Dec. 2.**

1983

On **Apr. 20, Pres. Reagan** signed a compromise bipartisan bill designed to save **Social Security** from bankruptcy.

Sally Ride became the first American **woman** to travel in **space, June 18,** when the **space shuttle** *Challenger* was launched from Cape Canaveral, FL.

On **Sept. 1, a South Korean passenger jet** infringing on Soviet air space was **shot down;** 269 people were killed.

On **Oct. 23,** 241 **U.S. Marines and sailors** were killed in Lebanon when a TNT-laden suicide bomb blew up Marine headquarters at **Beirut** International Airport.

U.S. troops, with a small force from 6 **Caribbean** nations, invaded **Grenada Oct. 25.** After a few days, Grenadian militia and Cuban "construction workers" were overcome, U.S. citizens evacuated, and the **Marxist regime deposed.**

1984

The space shuttle *Challenger* was launched on its 4th trip into space, **Feb. 3.** On **Feb. 7,** Navy Capt. Bruce McCandless, followed by Army Lt. Colonel Robert Stewart, **became first humans to fly free of a spacecraft.**

On **May 7,** American **Vietnam war** veterans reached an out-of-court **settlement with 7 chemical companies** in a class-action suit over the herbicide **Agent Orange.**

Former Vice Pres. **Walter Mondale** won the **Democratic presidential nomination, June 6; he chose Rep. Geraldine Ferraro** (D, NY), as candidate for **vice president.**

Pres. **Reagan** was reelected **Nov. 6** in the greatest Republican **landslide** in history, carrying 49 states.

1985

"**Live Aid**," a rock concert broadcast around the world **July 13,** raised $70 mil for starving peoples of Africa.

On **June 14 a TWA jet was seized** by terrorists after takeoff from Athens; 153 passengers and crew held hostage for 17 days; 1 U.S. serviceman killed.

On **Oct. 7, 4 Palestinian hijackers seized** Italian cruise ship **Achille Lauro** in the Mediterranean and held it hostage for 2 days; one American, Leon Klinghoffer, was killed.

1986

On **Jan. 20,** for the first time, the U.S. officially observed **Martin Luther King Jr. Day.**

Moments after liftoff, **Jan. 28,** the space shuttle *Challenger* exploded, killing 6 astronauts and **Christa McAuliffe,** a New Hampshire teacher, on board.

Congress, overriding Pres. Reagan's veto in **Sept.,** imposed **economic sanctions on South Africa.**

U.S. Senate confirmed, **Sept. 17,** Reagan's nomination of **William Rehnquist** as chief justice and **Antonin Scalia** as associate justice of the Supreme Court.

Congress passed, in late **Sept.,** a major **tax reform law.**

In **congressional races, Nov. 4, Democrats** won a 55-45 Senate majority and enlarged their House majority.

Press reports in early **Nov.** broke first news of the **Iran-contra scandal,** involving secret U.S. sale of arms to Iran.

Ivan Boesky, accused of insider trading, agreed, **Nov. 14,** to plead guilty to an unspecified criminal count he was barred for life from trading securities.

1987

Pres. Reagan produced the nation's first **trillion-dollar budget, Jan. 5.**

The stock market continued to rise, with the **Dow Jones closing** at 2002.25, **Jan. 8,** its **first finish above 2000.**

An **Iraqi warplane missile killed 37 sailors** on the frigate USS *Stark* in the Persian Gulf, **May 17.** Iraq called it an accident.

Public hearings by Senate and House committees investigating the **Iran-contra affair** were held **May-Aug.** Lt. Col. Oliver North said he had believed all his activities were authorized by his superiors. Pres. Reagan, **Aug. 12,** denied knowing of a diversion of funds to the contras.

Wall Street crashed, Oct. 19, with the Dow Jones plummeting a record 508 points.

Pres. **Reagan** and Soviet leader **Mikhail Gorbachev, Dec. 8,** signed a **pact to dismantle** all 1,752 **U.S.** and 859 **Soviet missiles** with a 300- to 3,400-mi. range.

1988

Nearly **1.4 mil illegal aliens** met **May 4** deadline for applying for amnesty under a new federal policy.

A missile, fired from **U.S. Navy warship** *Vincennes,* in the Persian Gulf, mistakenly struck and **destroyed** a commercial **Iranian airliner, July 3,** killing all 290.

George Bush, vice president under Reagan, was **elected** 41st U.S. **president, Nov. 8.** Bush decisively defeated the Democratic nominee, Gov. **Michael Dukakis** (MA).

Drexel Burnham Lambert agreed, **Dec. 21, to plead guilty to 6 violations of federal law,** including insider trading, and **pay penalties of $650 mil,** the largest such settlement ever.

1989

One of the **largest oil spills in U.S. history** occurred after the *Exxon Valdez* struck Bligh Reef in Alaska's Prince William Sound, **Mar. 24.**

Former National Security Council staff member **Oliver North** was convicted, **May 4,** on charges related to **Iran-contra** scandal. Conviction thrown out on appeal in **1991.**

A measure to **rescue the savings and loan industry** was signed into law, **Aug. 9,** by Pres. Bush.

Army Gen. **Colin Powell** was nominated **Aug. 10** by Pres. Bush, as **chairman of the Joint Chiefs of Staff;** he became the first black to hold the post.

Just before a World Series game, **Oct. 17,** an **earthquake struck the San Francisco Bay area,** causing 62 deaths.

L. Douglas Wilder (D) elected governor of Virginia, the **first U.S. black governor** since Reconstruction.

U.S. troops invaded Panama, Dec. 20, overthrowing the government of **Manuel Noriega.** Noriega, wanted by U.S. authorities on drug charges, surrendered **Jan. 3, 1990.**

1990

Pres. Bush signed **Americans With Disabilities Act** on **July 26,** barring discrimination against handicapped.

Justice William Brennan announced, **July 20,** his resignation from the U.S. Supreme Court; his replacement, **Judge David Souter,** was confirmed **Sept. 27.**

Operation Desert Shield forces left for **Saudi Arabia, Aug. 7,** to defend that country following the **invasion** of its neighbor **Kuwait by Iraq,** Aug. 2.

Pres. Bush signed, **Nov. 5,** a bill to **reduce budget deficits** $500 bil over 5 years, by spending curbs and tax hikes.

MILLENNIUM FACT BOX

Population Shifts by Region

Source: U.S. Bureau of the Census

Shown below is the percentage of the total U.S. population living in each region of the country, according to the results of the decennial censuses. Initially, almost all Americans lived in the Northeast and the South. Before long, however, the North Central region began to attract large numbers of settlers and was growing in importance, and by the mid-19th century the West had begun its steady growth. By 1990 it had a larger share of the population than the Northeast.

	Northeast	South	North Central	West		Northeast	South	North Central	West
1790	50.09%	49.91%	—	—	1900	27.62%	32.18%	34.55%	5.65%
1800	49.66%	49.39%	0.96%	—	1910	28.05%	31.87%	32.41%	7.68%
1810	48.16%	47.80%	4.03%	—	1920	27.98%	31.24%	32.09%	8.69%
1820	45.24%	45.85%	8.91%	—	1930	27.94%	30.73%	31.33%	10.00%
1830	43.07%	44.36%	12.51%	—	1940	27.22%	31.53%	30.37%	10.88%
1840	39.61%	40.72%	19.64%	—	1950	26.09%	31.19%	29.38%	13.34%
1850	37.20%	38.73%	23.30%	0.77%	1960	24.91%	30.66%	28.79%	15.64%
1860	33.69%	35.41%	28.93%	1.97%	1970	24.12%	30.89%	27.93%	17.12%
1870	31.90%	31.87%	33.67%	2.57%	1980	21.67%	33.28%	26.00%	19.07%
1880	28.90%	32.91%	34.60%	3.59%	1990	20.43%	34.35%	23.99%	21.22%
1890	27.64%	31.80%	35.58%	4.98%					

1991

The **U.S. and its allies defeated Iraq** in the **Persian Gulf War** and liberated Kuwait, which Iraq had overrun in Aug. **1990.** On **Jan. 17,** the allies launched a devastating **attack on Iraq from the air.** In a **ground war** starting **Feb. 24,** which lasted just 100 hours, the U.S.-led attackers killed or captured many thousands of Iraqi soldiers and sent the rest into retreat before Pres. Bush ordered a cease-fire **Feb. 27.**

U.S. **House bank** ordered closed **Oct. 3** after revelations House members had written 8,331 bad checks.

The **Senate approved, Oct. 15, nomination of Clarence Thomas** to the Supreme Court, despite allegations of sexual harassment against him by **Anita Hill,** a former aide. He became the 2d African-American to serve on the Court, replacing retiring Justice **Thurgood Marshall,** the 1st black to do so.

Charles Keating convicted of securities fraud **Dec. 4.**

1992

Riots swept South-Central Los Angeles Apr. 29, after **jury acquitted 4 white policemen** on all but one count in videotaped 1991 beating of black motorist **Rodney King.** Death toll in the L.A. violence was put at 52.

Bill Clinton (D) was **elected** 42d president, **Nov. 3,** defeating **Pres. Bush** (R) and independent **Ross Perot.**

A UN-sanctioned military force, led by U.S. troops, arrived in **Somalia Dec. 9.**

1993

A bomb exploded in a parking garage beneath the **World Trade Center** in New York City, **Feb. 26,** killing 6 people. Two Islamic militants were convicted in the bombing, **Nov. 12, 1997.** Four men were found guilty, **Mar. 4, 1994.**

Janet Reno became the first woman U.S. attorney general **Mar. 12.**

Four federal agents were killed, Feb. 28, during an unsuccessful raid on the **Branch Davidian compound near Waco, TX.** A 51-day siege of the compound by federal agents ended **Apr. 19,** when the compound **burned down,** leaving more than 70 cult members dead.

Eleven **cult** members were acquitted **Feb. 26, 1994** of charges in the deaths of the federal agents.

A federal jury, **Apr. 17,** found **2 Los Angeles police officers guilty** and 2 not guilty of violating the civil rights of motorist **Rodney King** in 1991 beating incident.

"The Great Flood of 1993" inundated 8 mil acres in 9 Midwestern states in summer, leaving 50 dead.

Pres. Clinton, **July 19,** announced a "don't ask, don't tell, don't pursue" policy for **homosexuals** in the U.S. military.

Vincent Foster, deputy White House counsel, found shot to death in a N Virginia park, an apparent suicide.

Judge Ruth Bader Ginsburg was sworn in, **Aug. 10,** as 107th justice of the Supreme Court.

Pres. Clinton, **Aug. 10,** signed a compromise bill designed to **cut federal budget deficits** $496 bil over 5 years, through spending cuts and new taxes.

The **"Brady Bill,"** a major gun-control measure, was signed into law by Pres. Clinton **Nov. 30.**

1994

North American Free Trade Agreement took effect **Jan. 1.**

A predawn **earthquake struck the Los Angeles area, Jan. 17,** claiming 61 lives and causing heavy damage.

Attorney Gen. Janet Reno **Jan. 20** appointed Robert Fiske independent counsel to probe **Whitewater affair;** under a court ruling he was **replaced Aug. 5 by Kenneth Starr.** Congressional committees, **late July,** began Whitewater hearings.

Byron De La Beckwith convicted **Feb. 5** of the 1963 murder of **civil rights leader Medgar Evers.**

Longtime CIA officer **Aldrich Ames** and his wife were **charged Feb. 21 with spying.** Under a plea bargain, he received life in prison, while she was sentenced to 63 months.

Major league **baseball players went on strike,** following **Aug. 11** games; strike ended **Apr. 25, 1995.**

Senate Majority Leader George Mitchell (D, ME), **Sept. 26,** dropped efforts to pass Clinton's **health-care reform** package.

Republicans won control of Congress in **Nov. 8** elections.

1995

When the 104th Congress opened, **Jan. 4, Sen. Bob Dole** (R, KS) became **Senate majority leader** and **Rep. Newt Gingrich** (R, GA) was elected **House Speaker.** A bill to end Congress's exemption from federal labor laws, first in a series of measures in Republicans' **"Contract With America,"** cleared Congress **Jan. 17;** signed into law **Jan. 23.**

Clinton invoked emergency powers, **Jan. 31,** to extend a **$20 bil loan to** help Mexico avert financial collapse.

The last UN peacekeeping troops withdrew from **Somalia Feb. 28-Mar. 3,** with the aid of U.S. Marines. In **Haiti,** peacekeeping responsibilities were transferred from U.S. to UN forces **Mar. 31,** with the U.S. providing 2,400 soldiers.

A truck **bomb** exploded outside **a federal office building in Oklahoma City Apr. 19, killing 168** people in all.

The U.S. space shuttle *Atlantis* made the first in a series of planned **dockings with** the Russian space station *Mir,* **June 29-July 4.**

A U.S. **F-16 fighter jet** piloted by Air Force Capt. **Scott O'Grady was shot down** over Bosnia and Herzegovina **June 2;** O'Grady was **rescued** by U.S. Marines 6 days later.

At least **800 people died** in the Midwest and Northeast from a **heat wave July 12-17.**

The U.S. announced on **July 11** that it was reestablishing **diplomatic relations with Vietnam.**

Former football star **O. J. Simpson** was found **not guilty Oct. 3** of the **June 1994** murders of his former wife, Nicole Brown Simpson, and her friend Ronald Goldman.

Hundreds of thousands of African-American men participated in **"Million Man March"** and rally in Washington, DC, **Oct. 16,** organized by Rev. Louis Farrakhan.

Billy Dale, discharged head of the White House travel office, was **acquitted of embezzlement Nov. 16.**

The federal **55-mile-per-hour speed limit** was **repealed** by a measure signed **Nov. 28.**

After talks outside Dayton, OH, **warring parties in Bosnia and Herzegovina reached agreement Nov. 21** to end their conflict; treaty was signed **Dec. 14,** after which first of some 20,000 **U.S. peacekeeping troops** arrived in Bosnia.

Five Americans were among 7 **killed, Nov. 13,** when **2 bombs exploded** at a military post **in Riyadh, Saudi Arabia.**

A budget impasse between Congress and Pres. Clinton led to a partial government shutdowns beginning **Nov. 14.** Operations resumed Nov. 20 under continuing resolutions.

The **Dow Jones** industrial average, passed 5,000 **Nov. 21.**

1996

Long-sought records released by White House **Jan. 5** showed **Hillary Rodham Clinton** did 60 hours of work for an S&L linked to **Whitewater** scandal. Responding to a subpoena, she testified **Jan. 26** before a grand jury.

Senate, **Jan. 26,** approved, 87–4, the Second Strategic Arms Reduction Treaty.

On **Feb. 24 Cuban jets shot down 2 unarmed planes** owned by a Cuban exile organization; all 4 persons on the planes were presumed killed.

John Salvi found guilty, **Mar. 18,** in the **1994 murder** of receptionists **at 2 abortion clinics** in Brookline, MA.

Congress, in late **Mar.,** approved a **"line item veto"** bill allowing the president to veto parts of a spending bill while approving the rest. It was later struck down by the Supreme Court, **June 25, 1998.**

U.S. Commerce Sec. **Ron Brown** was killed **Apr. 3** in a plane crash in Croatia.

On **Apr. 3, Theodore Kaczynski** was arrested in Montana; later charged with being the notorious **Unabomber** who had killed 3 people in a series of bombings.

On **Apr. 10,** Pres. Clinton vetoed a bill that would have banned so-called **partial-birth abortions.**

James and Susan McDougal were convicted **May 28** of fraud and conspiracy. Arkansas Gov. **Jim Guy Tucker** was convicted of similar charges by the same jury.

The antitax **Freemen** surrendered to federal authorities **June 13** after an 81-day standoff at a ranch near Jordan, MT. Four of the group's leaders were convicted, **July 8, 1998,** of conspiring to defraud four banks.

Republicans **June 12** chose Sen. **Trent Lott** (MS) as new majority leader to replace Sen. **Robert Dole,** who resigned, **June 11,** to focus on his presidential campaign.

The Republican majority and Democratic minority on the **Senate Whitewater Committee** issued separate final reports, **June 18,** based on their investigations.

A **bomb** exploded at a military complex near Dhahran, **Saudi Arabia, June 25,** killing 19 American servicemen.

TWA Flight 800, bound from New York to Paris, **crashed** into the Atlantic shortly after takeoff **July 17,** killing 230. The crash was later attributed to probable mechanical failure.

On **July 27 a bomb exploded** in an Atlanta park filled with people attending the **Olympics;** one person was directly killed.

A wide-ranging **welfare reform bill** which provided for welfare through block grants to states and ended federal guarantee of subsidies to poor people with children, was signed into law **Aug. 22.**

Shannon Lucid, Sept. 26, completed a space voyage of 188 days, a record for women and for U.S. astronauts.

Pres. Clinton was reelected to 2d term, **Nov. 5,** carrying 31 states and District of Columbia.

1997

Bombs were detonated at **2 abortion clinics** in Tulsa, OK, **Jan. 1,** in Atlanta on **Jan. 16,** and again at the first site in Tulsa on **Jan. 19.** Six people were injured.

Newt Gingrich (R, GA) was reelected Speaker of the U.S. House, **Jan. 7;** he was fined and reprimanded by colleagues for alleged misuse of tax-exempt donations.

Madeleine Albright was sworn in as secretary of state **Jan. 23,** becoming the first woman to head State Dept.

Ennis Cosby, 27, son of comedian **Bill Cosby,** was shot to death **Jan. 16** in Los Angeles, as he changed a flat tire. Mikhail Markhasev was convicted of the murder, **July 7, 1998.**

Harold Nicholson, a former CIA official, pleaded guilty, **Mar. 3,** to spying for Russia.

Thirty-nine members of the **Heaven's Gate religious cult** were found dead in a large house in Rancho Santa Fe, CA, **Mar. 26,** in an apparent mass suicide.

James McDougal, former partner with then-Gov. Bill Clinton in the Whitewater Development Corp., was sentenced **Apr. 14** to 3 years in prison for seeking to enrich himself with fraudulent loans. He died in prison, **Mar. 8, 1998.**

The rising Red River, **Apr. 19,** drove residents of **Grand Forks,** ND, and East Grand Forks, MN, from their homes, many of which were destroyed.

Tiger Woods, a 21-year-old African-American golfer, won the Masters Tournament, **Apr. 13,** achieving many firsts.

White House and congressional negotiators reached an agreement, **May 2,** intended to result in a balanced federal budget by 2002.

Garry Kasparov, the world chess champion, was defeated by a computer, IBM's Deep Blue, in a 6-game match that concluded **May 11** in New York City.

Timothy McVeigh was convicted of conspiracy and murder, **June 2,** in 1995 Oklahoma City bombing.

NASA's *Pathfinder* landed on Mars, **July 4.**

Fashion designer **Gianni Versace** was shot to death in Miami Beach, **July 15,** apparently by serial murderer Andrew Phillip Cunanan, who later committed suicide.

Autumn Jackson, 22, who claimed to be the out-of-wedlock daughter of comedian **Bill Cosby,** was convicted **July 25** of seeking to extort $40 mil from him.

Hundreds of thousands of Christian men from the **Promise Keepers** gathered on the Mall in Washington, DC, **Oct. 4,** to reaffirm faith in God and family values.

On **Oct. 27,** the **Dow Jones** fell 554.26 points, the largest 1-day point decline yet. On **Oct. 28,** the Dow rebounded, surging 337.17 points, the largest-yet single-day point advance.

The **Oct. 30** murder conviction of **Louise Woodward,** a British au pair, was overturned, **Nov. 10,** and her sentence for involuntary manslaughter, was reduced to time served.

Islamic militants **Ramzi Ahmed Yousef** and **Eyad Ismoil Yousef** were convicted, **Nov. 12,** in the 1993 bombing of the World Trade Center in New York City.

On **Nov. 19, Bobbi McCaughey,** 29, in Des Moines, IA, delivered the first set of live septuplets (4 boys, 3 girls) to survive more than a month.

Terry Nichols was convicted **Dec. 23** on some charges related to the 1995 **Oklahoma City bombing**.

1998

It was reported **Jan. 21** that Kenneth Starr, the independent counsel investigating the Whitewater scandal, had evidence of a sexual relationship between Pres. Clinton and onetime White House intern Monica Lewinsky. Clinton publicly denied an affair with Lewinsky.

Theodore Kaczynski, the so-called Unabomber, pleaded guilty **Jan. 22** in connection with California and New Jersey bombings that killed 3 people and injured 2.

A security guard was killed, **Jan. 29,** when a bomb exploded outside an abortion clinic in Birmingham, AL.

The state of Texas, **Feb. 3,** executed its first female convict in 135 years–**Karla Faye Tucker**—who had confessed to, and repented of, 2 pickax murders.

On **Mar. 23,** the movie *Titanic* won 11 Academy Awards, including best picture of 1997, equaling the total won by *Ben-Hur* in 1959.

Mitchell Johnson, 13, and **Andrew Golden, 11,** were arrested, **Mar. 24,** for allegedly killing 4 schoolgirls and a teacher outside a **Jonesboro, AR,** middle school. They were later committed to a juvenile detention center.

A federal judge, **Apr. 1,** dismissed the sexual harassment suit brought against **Pres. Clinton** by **Paula Corbin Jones.**

Viagra, a new prescription drug to treat male impotence, went on sale **Apr. 10.**

On **Apr. 25,** First Lady **Hillary Rodham Clinton** provided videotaped testimony at the White House for the Little Rock, AR, grand jury in the Whitewater case.

In a sealed ruling **May 4,** a U.S. district judge held Pres. Clinton could not invoke executive or attorney-client privilege to prevent aides from testifying.

On **May 7, Daimler-Benz AG,** producer of the Mercedes-Benz luxury automobile, and the **Chrysler Corp.** announced merger plans.

The TV show *Seinfeld* aired its last episode **May 14.**

Kipland Kinkel, 15, was arrested in **Springfield, OR, May 21,** and charged with the shotgun murder of his parents and 2 students at his high school.

In a tentative settlement reported **July 8, Dow Corning** agreed to pay $3.2 billion to 170,000 women who claimed they had become ill from silicone breast implants.

Capitol Police Officer **Jacob Chestnut** and Special Agent **John Gibson** were killed by a gunman, **July 24,** at the Capitol in Washington, DC.

Monica Lewinsky, July 28, agreed to testify before a Whitewater grand jury in return for immunity. On **Aug. 6,** Lewinsky, on the stand, admitted having had a sexual relationship with Clinton, but said she had never been asked to lie. In testimony provided to a grand jury, and in an address to the nation on **Aug. 17, Pres. Clinton** acknowledged having had an inappropriate relationship with **Monica Lewinsky.** On **Sept. 9,** independent counsel **Kenneth Starr** sent to the House what he called "substantial and credible information that may constitute grounds" for impeaching **Pres. Clinton.**

The Senate, **Sept. 18,** sustained Clinton's veto of a ban on partial-birth abortion.

Mark McGwire, Sept. 28, hit his 62nd **home run** of the season, breaking **Roger Maris's** season record.

On **Sept. 30,** Pres. Clinton announced a budget surplus of $70 billion for fiscal year 1998, the first since 1969.

The House Judiciary Committee, **Oct. 5,** voted 21-16 along party lines, to recommend to the full House that the Clinton impeachment investigation proceed. The House concurred **Oct. 8,** voting 258-176; 31 Democrats voted yes.

Matthew Shepard, an openly gay student at the University of Wyoming, died **Oct. 12** from injuries received in an assault.

Dr. Barnett Slepian, an obstetrician who performed abortions, was shot to death in his home near Buffalo, NY, **Oct. 23** by a sniper.

John Glenn, the first U.S. astronaut to orbit Earth, returned to space **Oct. 29-Nov. 7,** aboard the shuttle *Discovery.*

On **Oct. 30** the **Food and Drug Administration** approved the use of **tamoxifen** to help prevent breast cancer in healthy women at high risk of developing it.

In **Nov. 3** offyear elections Democrats gained 5 House seats. **Speaker Newt Gingrich** (R, GA) announced **Nov. 6** he would not seek reelection to his post.

Pres. Clinton, Nov. 13, settled a suit by agreeing to pay $850,000 to **Paula Corbin Jones.** She alleged that he had made an unwanted sexual advance to her in 1991.

The **FBI** reported **Nov. 22** the U.S. murder rate in 1997 was at its lowest since 1967—6.8 per 100,000 residents.

The country's 4 largest tobacco companies, in a settlement, **Nov. 23,** with 46 states, the District of Columbia, and 4 territories, agreed to pay $206 billion over 25 years to cover public health costs related to smoking.

Attorney Gen. **Janet Reno** announced **Nov. 24** she would not appoint an independent counsel to investigate **Vice Pres. Al Gore** in connection with phone calls he had made to raise money for the 1996 presidential campaign.

Mike Espy, former secretary of agriculture, was found not guilty of corruption charges by a federal jury in Washington, DC, **Dec. 2.**

James P. Hoffa became president of the International Brotherhood of Teamsters **Dec. 5.**

On **Dec. 14,** Democratic Party donor **Johnny Chung** was sentenced to probation and community service for illegal campaign contributions.

The U.S. House of Representatives gave its approval, **Dec. 19,** to 2 articles of **impeachment** charging **Pres. Clinton** with perjury and obstruction of justice.

The Mayflower Compact

The threat of James I to "harry them out of the land" sent a little band of religious dissenters from England to Holland in 1608. They were known as Separatists because they wished to cut all ties with the established church. In 1620, some of them, known now as the Pilgrims, joined with a larger group in England to set sail on the *Mayflower* for the New World. A joint stock company financed their venture.

In November, they sighted Cape Cod and decided to land an exploring party at Plymouth Harbor. A rebellious group picked up at Southampton and London troubled the Pilgrim leaders, however, and to control their actions 41 Pilgrims drew up the Mayflower Compact and signed it before going ashore. The voluntary agreement to govern themselves was America's first written constitution. It reads as follows:

In the name of God, Amen. We, whose names are underwritten, the Loyal Subjects of our dread Sovereign Lord, King *James,* by the Grace of God, of *Great Britain, France and Ireland,* King, *Defender of the Faith,* etc.

Having undertaken for the Glory of God, and Advancement of the Christian Faith, and the Honour of our King and Country, a voyage to plant the first colony in the northern Parts of Virginia; do by these Presents, solemnly and mutually in the Presence of God and one of another, covenant and combine ourselves together into a civil Body Politick, for our better Ordering and Preservation, and Furtherance of the Ends aforesaid; And by Virtue hereof to enact, constitute, and frame, such just and equal Laws, Ordinances, Acts, Constitutions and Offices, from time to time, as shall be thought most meet and convenient for the General good of the Colony; unto which we promise all due Submission and Obedience.

In Witness whereof we have hereunto subscribed our names at *Cape Cod* the eleventh of *November,* in the Reign of our Sovereign Lord, King *James* of *England, France* and *Ireland,* the eighteenth, and of *Scotland* the fifty-fourth. *Anno Domini, 1620.*

The Continental Congress: Meetings, Presidents

Meeting places	Dates of meetings	Congress presidents	Date elected
Philadelphia, PA	Sept. 5 to Oct. 26, 1774	Peyton Randolph, VA [1]	Sept. 5, 1774
"	"	Henry Middleton, SC	Oct. 22, 1774
Philadelphia, PA	May 10, 1775 to Dec. 12, 1776	Peyton Randolph, VA	May 10, 1775
"	"	John Hancock, MA	May 24, 1775
Baltimore, MD	Dec. 20, 1776 to Mar. 4, 1777	"	
Philadelphia, PA	Mar. 5 to Sept. 18, 1777		
Lancaster, PA	Sept. 27, 1777 (one day)		
York, PA	Sept. 30, 1777 to June 27, 1778	Henry Laurens, SC	Nov. 1, 1777[4]
Philadelphia, PA	July 2, 1778 to June 21, 1783	John Jay, NY	Dec. 10, 1778
"	"	Samuel Huntington, CT	Sept. 28, 1779
"	"	Thomas McKean, DE	July 10, 1781
"	"	John Hanson, MD [2]	Nov. 5, 1781
"	"	Elias Boudinot, NJ	Nov. 4, 1782
Princeton, NJ	June 30 to Nov. 4, 1783	Thomas Mifflin, PA	Nov. 3, 1783
Annapolis, MD	Nov. 26, 1783 to June 3, 1784	"	
Trenton, NJ	Nov. 1 to Dec. 24, 1784	Richard Henry Lee, VA	Nov. 30, 1784
New York City, NY	Jan. 11 to Nov. 4, 1785		
"	Nov. 7, 1785 to Nov. 3, 1786	John Hancock, MA [3]	Nov. 23, 1785
"	"	Nathaniel Gorham, MA	June 6, 1786
"	Nov. 6, 1786 to Oct. 30, 1787	Arthur St. Clair, PA	Feb. 2, 1787
"	Nov. 5, 1787 to Oct. 21, 1788	Cyrus Griffin, VA	Jan. 22, 1788
"	Nov. 3, 1788 to Mar. 2, 1789		

(1) Resigned Oct. 22, 1774. (2) Titled "President of the United States in Congress Assembled," John Hanson is considered by some the first U.S. president because he was the first to serve under the Articles of Confederation. He was, however, little more than presiding officer of the Congress, which retained full executive power. He could be considered the head of government, but not head of state. (3) Resigned May 29, 1786, without serving, because of illness. (4) Articles of Confederation agreed upon, Nov. 15, 1777; last ratification from Maryland, Mar. 1, 1781.

Patrick Henry's Speech to the Virginia Convention

The following is an excerpt from Patrick Henry's speech to the Virginia Convention on Mar. 23, 1775:

Gentlemen may cry, peace, peace—but there is no peace. The war is actually begun! The next gale that sweeps from the north will bring to our ears the clash of resounding arms! Our brethren are already in the field! Why stand we here idle? What is it that gentlemen wish? What would they have? Is life so dear, or peace so sweet, as to be purchased at the price of chains and slavery? Forbid it, Almighty God! I know not what course others may take; but as for me, give me liberty, or give me death!

How the Declaration of Independence Was Adopted

On June 7, 1776, Richard Henry Lee, who had issued the first call for a congress of the colonies, introduced in the Continental Congress at Philadelphia a resolution declaring "that these United Colonies are, and of right ought to be, free and independent states, that they are absolved from all allegiance to the British Crown, and that all political connection between them and the state of Great Britain is, and ought to be, totally dissolved."

The resolution, seconded by John Adams on behalf of the Massachusetts delegation, came up again on June 10 when a committee of 5, headed by Thomas Jefferson, was appointed to express the purpose of the resolution in a declaration of independence. The others on the committee were John Adams, Benjamin Franklin, Robert R. Livingston, and Roger Sherman.

Drafting the Declaration was assigned to Jefferson, who worked on a portable desk of his own construction in a room at Market and 7th Sts. The committee reported the result on June 28, 1776. The members of the Congress suggested a number of changes, which Jefferson called "deplorable." They didn't approve Jefferson's arraignment of the British people and King George III for encouraging and fostering the slave trade, which Jefferson called "an execrable commerce." They made 86 changes, eliminating 480 words and leaving 1,337. In the final form, capitalization was erratic. Jefferson had written that men were endowed with "inalienable" rights; in the final copy it came out as "unalienable" and has been thus ever since.

The Lee-Adams resolution of independence was adopted by 12 yeas on July 2—the actual date of the act of independence. The Declaration, which explains the act, was adopted July 4, in the evening.

After the Declaration was adopted, July 4, 1776, it was turned over to John Dunlap, printer, to be printed on broadsides. The original copy was lost and one of his broadsides was attached to a page in the journal of the Congress. It was read aloud July 8 in Philadelphia, PA, Easton, PA, and Trenton, NJ. On July 9 at 6 pm it was read by order of Gen. George Washington to the troops assembled on the Common in New York City (City Hall Park).

The Continental Congress of July 19, 1776, adopted the following resolution:

"Resolved, That the Declaration passed on the 4th, be fairly engrossed on parchment with the title and stile of 'The Unanimous Declaration of the thirteen United States of America' and that the same, when engrossed, be signed by every member of Congress."

Not all delegates who signed the engrossed Declaration were present on July 4. Robert Morris (PA), William Williams (CT), and Samuel Chase (MD) signed on Aug. 2; Oliver Wolcott (CT), George Wythe (VA), Richard Henry Lee (VA), and Elbridge Gerry (MA) signed in August and September; Matthew Thornton (NH) joined the Congress Nov. 4 and signed later. Thomas McKean (DE) rejoined Washington's army before signing and said later that he signed in 1781.

Charles Carroll of Carrollton was appointed a delegate by Maryland on July 4, 1776, presented his credentials July 18, and signed the engrossed Declaration on Aug. 2. Born Sept. 19, 1737, he was 95 years old and the last surviving signer when he died on Nov. 14, 1832.

Two Pennsylvania delegates who did not support the Declaration on July 4 were replaced.

The 4 New York delegates did not have authority from their state to vote on July 4. On July 9, the New York state convention authorized its delegates to approve the Declaration, and the Congress was so notified on July 15, 1776. The 4 signed the Declaration on Aug. 2.

The original engrossed Declaration is preserved in the National Archives Building in Washington.

Declaration of Independence

The Declaration of Independence was adopted by the Continental Congress in Philadelphia on July 4, 1776. John Hancock was president of the Congress, and Charles Thomson was secretary. A copy of the Declaration, engrossed on parchment, was signed by members of Congress on and after Aug. 2, 1776. On Jan. 18, 1777, Congress ordered that "an authenticated copy, with the names of the members of Congress subscribing the same, be sent to each of the United States, and that they be desired to have the same put upon record." Authenticated copies were printed in broadside form in Baltimore, where the Continental Congress was then in session. The following text is that of the original printed by John Dunlap at Philadelphia for the Continental Congress. The original is on display at the National Archives in Washington, DC.

IN CONGRESS, July 4, 1776.

A DECLARATION

By the REPRESENTATIVES of the

UNITED STATES OF AMERICA,

In GENERAL CONGRESS assembled

When in the Course of human Events, it becomes necessary for one People to dissolve the Political Bands which have connected them with another, and to assume among the Powers of the Earth, the separate and equal Station to which the Laws of Nature and of Nature's God entitle them, a decent Respect to the Opinions of Mankind requires that they should declare the causes which impel them to the Separation.

We hold these Truths to be self-evident, that all Men are created equal, that they are endowed by their Creator with certain unalienable Rights, that among these are Life, Liberty, and the Pursuit of Happiness—That to secure these Rights, Governments are instituted among Men, deriving their just Powers from the Consent of the Governed, that whenever any Form of Government becomes destructive of these Ends, it is the Right of the People to alter or to abolish it, and to institute new Government, laying its Foundation on such Principles, and organizing its Powers in such Form, as to them shall seem most likely to effect their Safety and Happiness. Prudence, indeed, will dictate that Governments long established should not be changed for light and transient Causes; and accordingly all Experience hath shewn, that Mankind are more disposed to suffer, while Evils are sufferable, than to right themselves by abolishing the Forms to which they are accustomed. But when a long Train of Abuses and Usurpations, pursuing invariably the same Object, evinces a Design to reduce them under absolute Despotism, it is their Right, it is their Duty, to throw off such Government, and to provide new Guards for their future Security. Such has been the patient Sufferance of these Colonies; and such is now the Necessity which constrains them to alter their former Systems of Government. The History of the present King of Great-Britain is a History of repeated Injuries and Usurpations, all having in direct Object the Establishment of an absolute Tyranny over these States. To prove this, let Facts be submitted to a candid World.

He has refused his Assent to Laws, the most wholesome and necessary for the public Good.

He has forbidden his Governors to pass Laws of immediate and pressing Importance, unless suspended in their Operation till his Assent should be obtained; and when so suspended, he has utterly neglected to attend to them.

He has refused to pass other Laws for the Accommodation of large Districts of People, unless those People would relinquish the Right of Representation in the Legislature, a Right inestimable to them, and formidable to Tyrants only.

He has called together Legislative Bodies at Places unusual, uncomfortable, and distant from the Depository of their Public Records, for the sole Purpose of fatiguing them into Compliance with his Measures.

He has dissolved Representative Houses repeatedly, for opposing with manly Firmness his Invasions on the Rights of the People.

He has refused for a long Time, after such Dissolutions, to cause others to be elected; whereby the Legislative Powers, incapable of Annihilation, have returned to the People at large for their exercise; the State remaining in the mean time exposed to all the Dangers of Invasion from without, and Convulsions within.

He has endeavoured to prevent the Population of these States; for that Purpose obstructing the Laws for Naturalization of Foreigners; refusing to pass others to encourage their Migrations hither, and raising the Conditions of new Appropriations of Lands.

He has obstructed the Administration of Justice, by refusing his Assent to Laws for establishing Judiciary Powers.

He has made Judges dependent on his Will alone, for the Tenure of their Offices, and the Amount and payment of their Salaries.

He has erected a Multitude of new Offices, and sent hither Swarms of Officers to harrass our People, and eat out their Substance.

He has kept among us, in Times of Peace, Standing Armies, without the consent of our Legislatures.

He has affected to render the Military independent of, and superior to the Civil Power.

He has combined with others to subject us to a Jurisdiction foreign to our Constitution, and unacknowledged by our Laws; giving his Assent to their Acts of pretended Legislation:

For quartering large Bodies of Armed Troops among us:

For protecting them, by a mock Trial, from Punishment for any Murders which they should commit on the Inhabitants of these States:

For cutting off our Trade with all Parts of the World:

For imposing Taxes on us without our Consent:

For depriving us, in many Cases, of the Benefits of Trial by Jury:

For transporting us beyond Seas to be tried for pretended Offences:

For abolishing the free System of English Laws in a neighbouring Province, establishing therein an arbitrary Government, and enlarging its Boundaries, so as to render it at once an Example and fit Instrument for introducing the same absolute Rule into these Colonies:

For taking away our Charters, abolishing our most valuable Laws, and altering fundamentally the Forms of our Governments:

For suspending our own Legislatures, and declaring themselves invested with Power to legislate for us in all Cases whatsoever.

He has abdicated Government here, by declaring us out of his Protection and waging War against us.

He has plundered our Seas, ravaged our Coasts, burnt our towns, and destroyed the Lives of our People.

He is, at this Time, transporting large Armies of foreign Mercenaries to complete the works of Death, Desolation, and Tyranny, already begun with circumstances of Cruelty and Perfidy, scarcely paralleled in the most barbarous Ages, and totally unworthy the Head of a civilized Nation.

He has constrained our fellow Citizens taken Captive on the high Seas to bear Arms against their Country, to become the Executioners of their Friends and Brethren, or to fall themselves by their Hands.

He has excited domestic Insurrections amongst us, and has endeavoured to bring on the Inhabitants of our Frontiers, the merciless Indian Savages, whose known Rule of Warfare, is an undistinguished Destruction, of all Ages, Sexes and Conditions.

In every stage of these Oppressions we have Petitioned for Redress in the most humble Terms: Our repeated Petitions have been answered only by repeated Injury. A Prince, whose Character is thus marked by every act which may define a Tyrant, is unfit to be the Ruler of a free People.

Nor have we been wanting in Attentions to our British Brethren. We have warned them from Time to Time of Attempts by their Legislature to extend an unwarrantable Jurisdiction over us. We have reminded them of the Circumstances of our Emigration and Settlement here. We have appealed to their native Justice and Magnanimity, and we have conjured them by the Ties of our common Kindred to disavow these Usurpations, which, would inevitably interrupt our Connections and Correspondence. They too have been deaf to the Voice of Justice and of Consanguinity. We must, therefore, acquiesce in the Necessity, which denounces our Separation, and hold them, as we hold the rest of Mankind, Enemies in War, in Peace, Friends.

We, therefore, the Representatives of the UNITED STATES OF AMERICA, in General Congress, Assembled, appealing to the Supreme Judge of the World for the Rectitude of our Intentions, do, in the Name, and by Authority of the good People of these Colonies, solemnly Publish and Declare, That these United Colonies are, and of Right ought to be, Free and Independent States; that they are absolved from all Allegiance to the British Crown, and that all political Connection between them and the State of Great-Britain, is and ought to be totally dissolved; and that as Free and Independent States, they have full Power to levy War, conclude Peace, contract Alliances, establish Commerce, and to do all other Acts and Things which Independent States may of right do. And for the support of this declaration, with a firm Reliance on the Protection of Divine Providence, we mutually pledge to each other our lives, our Fortunes, and our sacred Honor.

JOHN HANCOCK, President

Attest.

CHARLES THOMSON, Secretary.

Signers of the Declaration of Independence

Delegate (state)	Occupation	Birthplace	Born	Died
Adams, John (MA)	Lawyer	Braintree (Quincy), MA	Oct. 30, 1735	July 4, 1826
Adams, Samuel (MA)	Political leader	Boston, MA	Sept. 27, 1722	Oct. 2, 1803
Bartlett, Josiah (NH)	Physician, judge	Amesbury, MA	Nov. 21, 1729	May 19, 1795
Braxton, Carter (VA)	Farmer	Newington Plantation, VA	Sept. 10, 1736	Oct. 10, 1797
Carroll, Chas. of Carrollton (MD)	Lawyer	Annapolis, MD	Sept. 19, 1737	Nov. 14, 1832
Chase, Samuel (MD)	Judge	Princess Anne, MD	Apr. 17, 1741	June 19, 1811
Clark, Abraham (NJ)	Surveyor	Roselle, NJ	Feb. 15, 1726	Sept. 15, 1794
Clymer, George (PA)	Merchant	Philadelphia, PA	Mar. 16, 1739	Jan. 23, 1813
Ellery, William (RI)	Lawyer	Newport, RI	Dec. 22, 1727	Feb. 15, 1820
Floyd, William (NY)	Soldier	Brookhaven, NY	Dec. 17, 1734	Aug. 4, 1821
Franklin, Benjamin (PA)	Printer, publisher	Boston, MA	Jan. 17, 1706	Apr. 17, 1790
Gerry, Elbridge (MA)	Merchant	Marblehead, MA	July 17, 1744	Nov. 23, 1814
Gwinnett, Button (GA)	Merchant	Down Hatherly, England	c. 1735	May 19, 1777
Hall, Lyman (GA)	Physician	Wallingford, CT	Apr. 12, 1724	Oct. 19, 1790
Hancock, John (MA)	Merchant	Braintree (Quincy), MA	Jan. 12, 1737	Oct. 8, 1793
Harrison, Benjamin (VA)	Farmer	Berkeley, VA	Apr. 5, 1726	Apr. 24, 1791

Delegate (state)	Occupation	Birthplace	Born	Died
Hart, John (NJ)	Farmer	Stonington, CT	c. 1711	May 11, 1779
Hewes, Joseph (NC)	Merchant	Princeton, NJ	Jan. 23, 1730	Nov. 10, 1779
Heyward, Thos. Jr. (SC)	Lawyer, farmer	St. Luke's Parish, SC	July 28, 1746	Mar. 6, 1809
Hooper, William (NC)	Lawyer	Boston, MA	June 28, 1742	Oct. 14, 1790
Hopkins, Stephen (RI)	Judge, educator	Providence, RI	Mar. 7, 1707	July 13, 1785
Hopkinson, Francis (NJ)	Judge, author	Philadelphia, PA	Sept. 21, 1737	May 9, 1791
Huntington, Samuel (CT)	Judge	Windham County, CT	July 3, 1731	Jan. 5, 1796
Jefferson, Thomas (VA)	Lawyer	Shadwell, VA	Apr. 13, 1743	July 4, 1826
Lee, Francis Lightfoot (VA)	Farmer	Westmoreland County, VA	Oct. 14, 1734	Jan. 11, 1797
Lee, Richard Henry (VA)	Farmer	Westmoreland County, VA	Jan. 20, 1732	June 19, 1794
Lewis, Francis (NY)	Merchant	Llandaff, Wales	Mar., 1713	Dec. 31, 1802
Livingston, Philip (NY)	Merchant	Albany, NY	Jan. 15, 1716	June 12, 1778
Lynch, Thomas Jr. (SC)	Farmer	Winyah, SC	Aug. 5, 1749	(at sea) 1779
McKean, Thomas (DE)	Lawyer	New London, PA	Mar. 19, 1734	June 24, 1817
Middleton, Arthur (SC)	Farmer	Charleston, SC	June 26, 1742	Jan. 1, 1787
Morris, Lewis (NY)	Farmer	Morrisania (Bronx County), NY.	Apr. 8, 1726	Jan. 22, 1798
Morris, Robert (PA)	Merchant	Liverpool, England	Jan. 20, 1734	May 9, 1806
Morton, John (PA)	Judge	Ridley, PA	1724	Apr., 1777
Nelson, Thos. Jr. (VA)	Farmer	Yorktown, VA	Dec. 26, 1738	Jan. 4, 1789
Paca, William (MD)	Judge	Abingdon, MD	Oct. 31, 1740	Oct. 23, 1799
Paine, Robert Treat (MA)	Judge	Boston, MA	Mar. 11, 1731	May 12, 1814
Penn, John (NC)	Lawyer	Near Port Royal, VA	May 17, 1741	Sept. 14, 1788
Read, George (DE)	Judge	Near North East, MD	Sept. 18, 1733	Sept. 21, 1798
Rodney, Caesar (DE)	Judge	Dover, DE	Oct. 7, 1728	June 29, 1784
Ross, George (PA)	Judge	New Castle, DE	May 10, 1730	July 14, 1779
Rush, Benjamin (PA)	Physician	Byberry, PA (Philadelphia)	Dec. 24, 1745	Apr. 19, 1813
Rutledge, Edward (SC)	Lawyer	Charleston, SC	Nov. 23, 1749	Jan. 23, 1800
Sherman, Roger (CT)	Lawyer	Newton, MA	Apr. 19, 1721	July 23, 1793
Smith, James (PA)	Lawyer	Dublin, Ireland	c. 1719	July 11, 1806
Stockton, Richard (NJ)	Lawyer	Near Princeton, NJ	Oct. 1, 1730	Feb. 28, 1781
Stone, Thomas (MD)	Lawyer	Charles County, MD	1743	Oct. 5, 1787
Taylor, George (PA)	Ironmaster	Ireland	1716	Feb. 23, 1781
Thornton, Matthew (NH)	Physician	Ireland	1714	June 24, 1803
Walton, George (GA)	Judge	Prince Edward County, VA	1741	Feb. 2, 1804
Whipple, William (NH)	Merchant, judge	Kittery, ME.	Jan. 14, 1730	Nov. 28, 1785
Williams, William (CT)	Merchant	Lebanon, CT	Apr. 23, 1731	Aug. 2, 1811
Wilson, James (PA)	Judge	Carskerdo, Scotland	Sept. 14, 1742	Aug. 28, 1798
Witherspoon, John (NJ)	Clergyman, educator	Gifford, Scotland	Feb. 5, 1723	Nov. 15, 1794
Wolcott, Oliver (CT)	Judge	Windsor, CT.	Dec. 1, 1726	Dec. 1, 1797
Wythe, George (VA)	Lawyer	Elizabeth City Co. (Hampton), VA	1726	June 8, 1806

Origin of the Constitution

The War of Independence was conducted by delegates from the original 13 states, called the Congress of the United States of America and known as the Continental Congress. In 1777 the Congress submitted to the legislatures of the states the Articles of Confederation and Perpetual Union, which were ratified by New Hampshire, Massachusetts, Rhode Island, Connecticut, New York, New Jersey, Pennsylvania, Delaware, Virginia, North Carolina, South Carolina, and Georgia and finally, in 1781, by Maryland.

The first article read: "The stile of this confederacy shall be the United States of America." This did not signify a sovereign nation, because the states delegated only those powers they could not handle individually, such as to wage war, make treaties, and contract debts for general expenses (e.g. paying the army). Taxes for payment of such debts were levied by the individual states. The president signed himself "President of the United States in Congress assembled," but here the United States were considered in the plural, a cooperating group.

When the war was won, it became evident that a stronger federal union was needed. The Congress left the initiative to the legislatures. Virginia in Jan. 1786 appointed commissioners to meet with representatives of other states; delegates from Virginia, Delaware, New York, New Jersey, and Pennsylvania met at Annapolis. Alexander Hamilton prepared their call asking delegates from all states to meet in Philadelphia in May 1787 "to render the Constitution of the federal government adequate to the exigencies of the union." Congress endorsed the plan on Feb. 21, 1787. Delegates were appointed by all states except Rhode Island.

The convention met on May 14, 1787. George Washington was chosen president (presiding officer). The states certified 65 delegates, but 10 did not attend. The work was done by 55, not all of whom were present at all sessions. Of the 55 attending delegates, 16 failed to sign, and 39 actually signed Sept. 17, 1787, some with reservations. Some historians have said 74 delegates (9 more than the 65 actually certified) were named and 19 failed to attend. These 9 additional persons refused the appointment, were never delegates, and were never counted as absentees. Washington sent the Constitution to Congress, and that body, Sept. 28, 1787, ordered it sent to the legislatures, "in order to be submitted to a convention of delegates chosen in each state by the people thereof."

The Constitution was ratified by votes of state conventions as follows: Delaware, Dec. 7, 1787, unanimous; Pennsylvania, Dec. 12, 1787, 43 to 23; New Jersey, Dec. 18, 1787, unanimous; Georgia, Jan. 2, 1788, unanimous; Connecticut, Jan. 9, 1788, 128 to 40; Massachusetts, Feb. 6, 1788, 187 to 168; Maryland, Apr. 28, 1788, 63 to 11; South Carolina, May 23, 1788, 149 to 73; New Hampshire, June 21, 1788, 57 to 46; Virginia, June 25, 1788, 89 to 79; New York, July 26, 1788, 30 to 27. Nine states were needed to establish the operation of the Constitution "between the states so ratifying the same," and New Hampshire was the 9th state. The government did not declare the Constitution in effect until the first Wednesday in Mar. 1789, which was Mar. 4. After that, North Carolina ratified it on Nov. 21, 1789, 194 to 77; and Rhode Island, May 29, 1790, 34 to 32. Vermont in convention ratified it on Jan. 10, 1791, and by act of Congress approved on Feb. 18, 1791, was admitted into the Union as the 14th state, Mar. 4, 1791.

Constitution of the United States
The Original 7 Articles

The text of the Constitution given here (with the exception of Amendment XXVII) is taken from the pocket-size edition of the Constitution published by the U.S. Government Printing Office as a result of a U.S. House and Senate resolution to print the Constitution in its original form as amended through July 5, 1971. **Text preceding** each article, section, or amendment is a brief summary, added by *The World Almanac. Text in brackets* indicates that an item has been superseded or amended, or provides background information.

PREAMBLE

We, the People of the United States, in Order to form a more perfect Union, establish Justice, insure domestic Tranquility, provide for the common defence, promote the general Welfare, and secure the Blessings of Liberty to ourselves and our Posterity, do ordain and establish this Constitution for the United States of America.

ARTICLE I.

Section 1—Legislative powers; in whom vested:
All legislative Powers herein granted shall be vested in a Congress of the United States, which shall consist of a Senate and House of Representatives.

Section 2—House of Representatives, how and by whom chosen. Qualifications of a Representative. Representatives and direct taxes, how apportioned. Enumeration. Vacancies to be filled. Power of choosing officers, and of impeachment.
The House of Representatives shall be composed of Members chosen every second Year by the People of the several States, and the Electors in each State shall have the Qualifications requisite for Electors of the most numerous Branch of the State Legislature.

No person shall be a Representative who shall not have attained to the Age of twenty-five Years, and been seven Years a Citizen of the United States, and who shall not, when elected, be an Inhabitant of that State in which he shall be chosen.

[Representatives and direct taxes shall be apportioned among the several States which may be included within this Union, according to their respective Numbers, which shall be determined by adding to the whole Number of free Persons, including those bound to Service for a Term of Years, and excluding Indians not taxed, three-fifths of all other persons.] [The previous sentence was superseded by Amendment XIV, section 2.] The actual Enumeration shall be made within three Years after the first Meeting of the Congress of the United States, and within every subsequent Term of ten Years, in such Manner as they shall by Law direct. The Number of Representatives shall not exceed one for every thirty Thousand, but each State shall have at Least one Representative; and until such enumeration shall be made, the State of New Hampshire shall be entitled to chuse three, Massachusetts eight, Rhode-Island and Providence Plantations one, Connecticut five, New-York six, New Jersey four, Pennsylvania eight, Delaware one, Maryland six, Virginia ten, North Carolina five, South Carolina five, and Georgia three.

When vacancies happen in the Representation from any State, the Executive Authority thereof shall issue Writs of Election to fill such Vacancies.

The House of Representatives shall chuse their Speaker and other Officers; and shall have the sole Power of Impeachment.

Section 3—Senators, how and by whom chosen. How classified. Qualifications of a Senator. President of the Senate, his right to vote. President pro tem., and other officers of the Senate, how chosen. Power to try impeachments. When President is tried, Chief Justice to preside. Sentence.
The Senate of the United States shall be composed of two Senators from each State, *[chosen by the Legislature thereof] [the preceding five words were superseded by Amendment XVII, section 1]* for six Years; and each Senator shall have one Vote.

Immediately after they shall be assembled in Consequence of the first Election, they shall be divided as equally as may be into three Classes. The Seats of the Senators of the first Class shall be vacated at the Expiration of the second Year, of the second Class at the Expiration of the fourth Year, and of the third Class at the Expiration of the Sixth year, so that one-third may be chosen every second Year; *[and if Vacancies happen by Resignation, or otherwise, during the Recess of the Legislature of any State, the Executive thereof may make temporary Appointments until the next Meeting of the Legislature, which shall then fill such Vacancies.] [The words in parentheses were superseded by Amendment XVII, section 2.]*

No person shall be a Senator who shall not have attained to the Age of thirty Years, and been nine Years a Citizen of the United States, and who shall not, when elected, be an Inhabitant of that State for which he shall be chosen.

The Vice President of the United States shall be President of the Senate, but shall have no Vote, unless they be equally divided.

The Senate shall chuse their other Officers, and also a President pro tempore, in the absence of the Vice President, or when he shall exercise the Office of President of the United States.

The Senate shall have the sole Power to try all Impeachments. When sitting for that Purpose, they shall be on Oath or Affirmation. When the President of the United States is tried, the Chief Justice shall preside: And no Person shall be convicted without the Concurrence of two thirds of the Members present.

Judgment in Cases of Impeachment shall not extend further than to removal from Office, and disqualification to hold and enjoy any Office of honor, Trust or Profit under the United States: but the Party convicted shall nevertheless be liable and subject to Indictment, Trial, Judgment and Punishment, according to Law.

Section 4—Times, etc., of holding elections, how prescribed. One session each year.
The Times, Places and Manner of holding Elections for Senators and Representatives, shall be prescribed in each State by the Legislature thereof; but the Congress may at any time by Law make or alter such Regulations, except as to the Place of Chusing Senators.

The Congress shall assemble at least once in every Year, and such Meeting shall *[be on the first Monday in December,] [The words in parentheses were superseded by Amendment XX, section 2.]* unless they shall by Law appoint a different Day.

Section 5—Membership, quorum, adjournments, rules. Power to punish or expel. Journal. Time of adjournments, how limited, etc.
Each House shall be the Judge of the Elections, Returns and Qualifications of its own Members, and a Majority of each shall constitute a Quorum to do Business; but a smaller number may adjourn from day to day, and may be authorized to compel the Attendance of absent Members, in such manner, and under such Penalties as each House may provide.

Each House may determine the Rules of its Proceedings, punish its members for disorderly Behavior, and, with the Concurrence of two thirds, expel a Member.

Each House shall keep a Journal of its Proceedings, and from time to time publish the same, excepting such Parts as may in their Judgment require Secrecy; and the Yeas and Nays of the Members of either House on any question shall, at the Desire of one fifth of those Present, be entered on the Journal.

Neither House, during the Session of Congress, shall, without the Consent of the other, adjourn for more than three days, nor to any other Place than that in which the two Houses shall be sitting.

Section 6—Compensation, privileges, disqualifications in certain cases.

The Senators and Representatives shall receive a Compensation for their Services, to be ascertained by Law, and paid out of the Treasury of the United States. They shall in all Cases, except Treason, Felony and Breach of the Peace, be privileged from Arrest during their Attendance at the Session of their respective Houses, and in going to and returning from the same; and for any Speech or Debate in either House, they shall not be questioned in any other Place.

No Senator or Representative shall, during the Time for which he was elected, be appointed to any civil Office under the Authority of the United States, which shall have been created, or the Emoluments whereof shall have been encreased during such time; and no Person holding any Office under the United States, shall be a Member of either House during his Continuance in Office.

Section 7—House to originate all revenue bills. Veto. Bill may be passed by two-thirds of each House, notwithstanding, etc. Bill, not returned in ten days, to become a law. Provisions as to orders, concurrent resolutions, etc.

All bills for raising Revenue shall originate in the House of Representatives; but the Senate may propose or concur with Amendments as on other Bills.

Every Bill which shall have passed the House of Representatives and the Senate, shall, before it become a Law, be presented to the President of the United States; If he approve he shall sign it, but if not he shall return it, with his Objections to that House in which it shall have originated, who shall enter the Objections at large on their Journal, and proceed to reconsider it. If after such Reconsideration two thirds of that House shall agree to pass the Bill, it shall be sent, together with the Objections, to the other House, by which it shall likewise be reconsidered, and if approved by two thirds of that House, it shall become a Law. But in all such Cases the Votes of both Houses shall be determined by Yeas and Nays, and the Names of the Persons voting for and against the Bill shall be entered on the Journal of each House respectively. If any Bill shall not be returned by the President within ten Days (Sundays excepted) after it shall have been presented to him, the Same shall be a Law, in like Manner as if he had signed it, unless the Congress by their Adjournment prevent its Return, in which Case it shall not be a Law.

Every order, Resolution, or Vote to which the Concurrence of the Senate and House of Representatives may be necessary (except on a question of Adjournment) shall be presented to the President of the United States; and before the Same shall take Effect, shall be approved by him, or being disapproved by him, shall be repassed by two thirds of the Senate and House of Representatives, according to the Rules and Limitations prescribed in the Case of a Bill.

Section 8—Powers of Congress.

The Congress shall have Power To lay and collect Taxes, Duties, Imposts and Excises, to pay the Debts and provide for the common Defence and general Welfare of the United States; but all Duties, Imposts and Excises shall be uniform throughout the United States;

To borrow money on the credit of the United States;

To regulate Commerce with foreign Nations, and among the several States, and with the Indian Tribes;

To establish an uniform Rule of Naturalization, and uniform Laws on the subject of Bankruptcies throughout the United States;

To coin Money, regulate the Value thereof, and of foreign Coin, and fix the Standard of Weights and Measures;

To provide for the Punishment of counterfeiting the Securities and current Coin of the United States;

To establish Post Offices and post Roads;

To promote the Progress of Science and useful Arts, by securing for limited Times to Authors and Inventors the exclusive Right to their respective Writings and Discoveries;

To constitute Tribunals inferior to the supreme Court;

To define and punish Piracies and Felonies committed on the high Seas, and Offenses against the Law of Nations;

To declare War, grant Letters of Marque and Reprisal, and make Rules concerning Captures on Land and Water;

To raise and support Armies, but no Appropriation of Money to that Use shall be for a longer Term than two Years;

To provide and maintain a Navy;

To make Rules for the Government and Regulation of the land and naval Forces;

To provide for calling forth the Militia to execute the Laws of the Union, suppress Insurrections and repel Invasions;

To provide for organizing, arming, and disciplining the Militia, and for governing such Part of them as may be employed in the Service of the United States, reserving to the States respectively, the Appointment of the Officers, and the Authority of training the Militia according to the discipline prescribed by Congress;

To exercise exclusive Legislation in all Cases whatsoever, over such District (not exceeding ten Miles square) as may, by Cession of particular States, and the acceptance of Congress, become the Seat of the Government of the United States, and to exercise like Authority over all Places purchased by the Consent of the Legislature of the State in which the Same shall be, for the Erection of Forts, Magazines, Arsenals, dock-Yards, and other needful Buildings;—And

To make all Laws which shall be necessary and proper for carrying into Execution the foregoing Powers, and all other Powers vested by this Constitution in the Government of the United States, or in any Department or Officer thereof.

Section 9—Provision as to migration or importation of certain persons. Habeas corpus, bills of attainder, etc. Taxes, how apportioned. No export duty. No commercial preference. Money, how drawn from Treasury, etc. No titular nobility. Officers not to receive presents, etc.

The Migration or Importation of such Persons as any of the States now existing shall think proper to admit, shall not be prohibited by the Congress prior to the Year one thousand eight hundred and eight, but a tax or duty may be imposed on such Importation, not exceeding ten dollars for each Person.

The privilege of the Writ of Habeas Corpus shall not be suspended, unless when in Cases of Rebellion or Invasion the public Safety may require it.

No Bill of Attainder or ex post facto Law shall be passed.

No capitation, or other direct, Tax shall be laid, unless in Proportion to the Census or Enumeration herein before directed to be taken. *[Modified by Amendment XVI.]*

No Tax or Duty shall be laid on Articles exported from any State.

No Preference shall be given by any Regulation of Commerce or Revenue to the Ports of one State over those of another: nor shall Vessels bound to, or from, one State, be obliged to enter, clear, or pay Duties in another.

No Money shall be drawn from the Treasury, but in Consequence of Appropriations made by Law; and a regular Statement and Account of the Receipts and Expenditures of all public Money shall be published from time to time.

No Title of Nobility shall be granted by the United States: and no Person holding any Office of Profit or Trust under them, shall, without the Consent of the Congress, accept of any present, Emolument, Office, or Title, of any kind whatever, from any King, Prince, or foreign State.

Section 10—States prohibited from the exercise of certain powers.

No State shall enter into any Treaty, Alliance, or Confederation; grant Letters of Marque and Reprisal; coin Money; emit Bills of Credit; make any Thing but gold and silver Coin a Tender in Payment of Debts; pass any Bill of Attainder, ex post facto Law, or Law impairing the Obligation of Contracts, or grant any Title of Nobility.

No State shall, without the Consent of the Congress, lay any Imposts or Duties on Imports or Exports, except what may be absolutely necessary for executing its inspection Laws: and the net Produce of all Duties and Imposts, laid by any State on Imports or Exports, shall be for the Use of the Treasury of the United States; and all such Laws shall be subject to the Revision and Control of the Congress.

No State shall, without the Consent of Congress, lay any duty of Tonnage, keep Troops, or Ships of War in time of Peace, enter into any Agreement or Compact with another State, or with a foreign Power, or engage in War, unless actually invaded, or in such imminent Danger as will not admit of delay.

ARTICLE II.

Section 1—President: his term of office. Electors of President; number and how appointed. Electors to vote on same day. Qualification of President. On whom his duties devolve in case of his removal, death, etc. President's compensation. His oath of office.

The executive Power shall be vested in a President of the United States of America. He shall hold his Office during the Term of four Years, and, together with the Vice President, chosen for the same Term, be elected, as follows.

Each State shall appoint, in such Manner as the Legislature thereof may direct, a Number of Electors, equal to the whole Number of Senators and Representatives to which the State may be entitled in the Congress: but no Senator or Representative, or Person holding an Office of Trust or Profit under the United States, shall be appointed an Elector.

[The Electors shall meet in their respective States, and vote by Ballot for two persons, of whom one at least shall not be an Inhabitant of the same State with themselves. And they shall make a List of all the Persons voted for, and of the Number of Votes for each; which List they shall sign and certify, and transmit sealed to the Seat of the Government of the United States, directed to the President of the Senate. The President of the Senate shall, in the Presence of the Senate and House of Representatives, open all the Certificates, and the Votes shall then be counted. The Person having the greatest Number of Votes shall be the President, if such Number be a Majority of the whole Number of Electors appointed; and if there be more than one who have such Majority, and have an equal Number of Votes, then the House of Representatives shall immediately chuse by Ballot one of them for President; and if no Person have a Majority, then from the five highest on the List the said House shall in like Manner chuse the President. But in chusing the President, the Votes shall be taken by States, the Representation from each State having one Vote; a quorum for this Purpose shall consist of a Member or Members from two thirds of the States, and a Majority of all the States shall be necessary to a Choice. In every Case, after the Choice of the President, the Person having the greatest Number of Votes of the Electors shall be the Vice President. But if there should remain two or more who have equal Votes, the Senate shall chuse from them by Ballot the Vice-President.]

[This clause was superseded by Amendment XII.]

The Congress may detemine the Time of chusing the Electors, and the Day on which they shall give their Votes; which Day shall be the same throughout the United States.

No person except a natural born Citizen, or a Citizen of the United States, at the time of the Adoption of this Constitution, shall be eligible to the Office of President; neither shall any Person be eligible to that Office who shall not have attained to the Age of thirty-five Years, and been fourteen Years a Resident within the United States.

[For qualification of the Vice President, see Amendment XII.]

In Case of the Removal of the President from Office, or of his Death, Resignation, or Inability to discharge the Powers and Duties of the said Office, the same shall devolve on the Vice President, and the Congress may by Law, provide for the Case of Removal, Death, Resignation or Inability, both of the President and Vice President, declaring what Officer shall then act as President, and such Officer shall act accordingly, until the Disability be removed, or a President shall be elected.

[This clause has been modified by Amendments XX and XXV.]

The President shall, at stated Times, receive for his Services, a Compensation, which shall neither be encreased nor diminished during the Period for which he shall have been elected, and he shall not receive within that Period any other Emolument from the United States, or any of them.

Before he enter on the Execution of his Office, he shall take the following Oath or Affirmation:–"I do solemnly swear (or affirm) that I will faithfully execute the Office of President of the United States, and will to the best of my Ability, preserve, protect and defend the Constitution of the United States."

Section 2—President to be Commander-in-Chief. He may require opinions of cabinet officers, etc., may pardon. Treaty-making power. Nomination of certain officers. When President may fill vacancies.

The President shall be Commander in Chief of the Army and Navy of the United States, and of the Militia of the several States, when called into the actual Service of the United States; he may require the Opinion in writing, of the principal Officer in each of the executive Departments, upon any subject relating to the Duties of their respective Offices, and he shall have Power to Grant Reprieves and Pardons for Offenses against the United States, except in Cases of Impeachment.

He shall have Power, by and with the Advice and Consent of the Senate, to make Treaties, provided two-thirds of the Senators present concur; and he shall nominate, and by and with the Advice and Consent of the Senate, shall appoint Ambassadors, other public Ministers and Consuls, Judges of the supreme Court, and all other Officers of the United States, whose Appointments are not herein otherwise provided for, and which shall be established by Law: but the Congress may by Law vest the Appointment of such inferior Officers, as they think proper, in the President alone, in the Courts of Law, or in the Heads of Departments.

The President shall have Power to fill up all Vacancies that may happen during the Recess of the Senate, by granting Commissions which shall expire at the End of their next Session.

Section 3—President shall communicate to Congress. He may convene and adjourn Congress, in case of disagreement, etc. Shall receive ambassadors, execute laws, and commission officers.

He shall from time to time give to the Congress Information of the State of the Union, and recommend to their Consideration such Measures as he shall judge necessary and expedient; he may, on extraordinary Occasions, convene both Houses, or either of them, and in Case of Disagreement between them, with Respect to the Time of Adjournment, he may adjourn them to such Time as he shall think proper; he shall receive Ambassadors and other public Ministers; he shall take Care that the Laws be faithfully executed, and shall Commission all the Officers of the United States.

Section 4—All civil offices forfeited for certain crimes.

The President, Vice President and all civil Officers of the United States, shall be removed from Office on Impeachment for, and Conviction of, Treason, Bribery, or other high Crimes and Misdemeanors.

ARTICLE III.

Section 1—Judicial powers, Tenure. Compensation.

The judicial Power of the United States, shall be vested in one supreme Court, and in such inferior Courts as the Congress may from time to time ordain and establish. The Judges, both of the supreme and inferior Courts, shall hold their Offices during good Behaviour, and shall, at stated Times, receive for their Services, a Compensation, which shall not be diminished during their Continuance in Office.

Section 2—Judicial power; to what cases it extends. Original jurisdiction of Supreme Court; appellate jurisdiction. Trial by jury, etc. Trial, where.

The judicial Power shall extend to all Cases, in Law and Equity, arising under this Constitution, the Laws of the United States, and Treaties made, or which shall be made, under their Authority;–to all Cases affecting Ambassadors, other public Ministers and Consuls;–to all Cases of admiralty and maritime Jurisdiction;–to Controversies to which the United States shall be a Party;–to Controversies between two or more States;–between a State and Citizens of another State;–between Citizens of different States;–between Citizens of the same State claiming Lands under Grants of different States, and between a State, or the Citizens thereof, and foreign States, Citizens or Subjects.

[This section is modified by Amendment XI.]

In all Cases affecting Ambassadors, other public Ministers and Consuls, and those in which a State shall be Party, the supreme Court shall have original Jurisdiction. In all the other Cases before mentioned, the supreme Court shall have appellate Jurisdiction, both as to Law and Fact, with such Exceptions, and under such Regulations as the Congress shall make.

The trial of all Crimes, except in Cases of Impeachment, shall be by Jury; and such Trial shall be held in the State where the said Crimes shall have been committed; but when not committed within any State, the Trial shall be at such Place or Places as the Congress may by Law have directed.

Section 3—Treason Defined, Proof of, Punishment of.

Treason against the United States, shall consist only in levying War against them, or in adhering to their Enemies, giving them Aid and Comfort. No Person shall be convicted of Treason unless on the Testimony of two Witnesses to the same overt Act, or on Confession in open Court.

The Congress shall have Power to declare the Punishment of Treason, but no Attainder of Treason shall work Corruption of Blood, or Forfeiture except during the Life of the Person attainted.

ARTICLE IV.

Section 1—Each State to give credit to the public acts, etc., of every other State.

Full Faith and Credit shall be given in each State to the public Acts, Records, and judicial Proceedings of every other State. And the Congress may by general Laws prescribe the Manner in which such Acts, Records and Proceedings shall be proved, and the Effect thereof.

Section 2—Privileges of citizens of each State. Fugitives from justice to be delivered up. Persons held to service having escaped, to be delivered up.

The Citizens of each State shall be entitled to all Privileges and Immunities of Citizens in the several States.

A Person charged in any State with Treason, Felony, or other Crime, who shall flee from Justice, and be found in another State, shall on demand of the executive Authority of the State from which he fled, be delivered up, to be removed to the State having Jurisdiction of the Crime.

[No Person held to Service or Labour in one State, under the Laws thereof, escaping into another, shall, in Consequence of any Law or Regulation therein, be discharged from such Service or Labour, but shall be delivered up on Claim of the Party to whom such Service or Labour may be due.] [This clause was superseded by Amendment XIII.]

Section 3—Admission of new States. Power of Congress over territory and other property.

New States may be admitted by the Congress into this Union; but no new State shall be formed or erected within the Jurisdiction of any other State; nor any State be formed by the Junction of two or more States, or parts of States, without the Consent of the Legislatures of the States concerned as well as of the Congress.

The Congress shall have Power to dispose of and make all needful Rules and Regulations respecting the Territory or other Property belonging to the United States; and nothing in this Constitution shall be so construed as to Prejudice any Claims of the United States, or of any particular State.

Section 4—Republican form of government guaranteed. Each state to be protected.

The United States shall guarantee to every State in this Union a Republican Form of Government, and shall protect each of them against Invasion; and on Application of the Legislature, or of the Executive (when the Legislature cannot be convened) against domestic Violence.

ARTICLE V.

Constitution: how amended; proviso.

The Congress, whenever two-thirds of both Houses shall deem it necessary, shall propose Amendments to this Constitution, or, on the Application of the Legislatures of two-thirds of the several States, shall call a Convention for proposing Amendments, which, in either Case, shall be valid to all Intents and Purposes, as part of this Constitution, when ratified by the Legislatures of three-fourths of the several States, or by Conventions in three-fourths thereof, as the one or the other Mode of Ratification may be proposed by the Congress: Provided that no Amendment which may be made prior to the Year One thousand eight hundred and eight shall in any Manner affect the first and fourth Clauses in the Ninth Section of the first Article; and that no State, without its Consent, shall be deprived of its equal Suffrage in the Senate.

ARTICLE VI.

Certain debts, etc., declared valid. Supremacy of Constitution, treaties, and laws of the United States. Oath to support Constitution, by whom taken. No religious test.

All Debts contracted and Engagements entered into, before the Adoption of this Constitution, shall be as valid against the United States under this Constitution, as under the Confederation.

This Constitution, and the Laws of the United States which shall be made in Pursuance thereof; and all Treaties made, or which shall be made, under the Authority of the United States, shall be the supreme Law of the Land; and the Judges in every State shall be bound thereby, any Thing in the Constitution or Laws of any State to the Contrary notwithstanding.

The Senators and Representatives before mentioned, and the Members of the several State Legislatures, and all executive and judicial Officers, both of the United States and of the several States, shall be bound by Oath or Affirmation, to support this Constitution; but no religious Test shall ever be required as a Qualification to any Office or public Trust under the United States.

ARTICLE VII.

What ratification shall establish Constitution.

The Ratification of the Conventions of nine States shall be sufficient for the Establishment of this Constitution between the States so ratifying the Same.

Done in Convention by the Unanimous Consent of the States present the Seventeenth Day of September in the Year of our Lord one thousand seven hundred and Eighty seven and of the Independence of the United States of America the Twelfth.

In Witness whereof We have hereunto subscribed our Names.

Go WASHINGTON, Presidt and deputy from Virginia
New Hampshire—John Langdon, Nicholas Gilman
Massachusetts—Nathaniel Gorham, Rufus King
Connecticut—Wm Saml Johnson, Roger Sherman
New York—Alexander Hamilton
New Jersey—Wil: Livingston, David Brearley, Wm Paterson, Jona: Dayton
Pennsylvania—B Franklin, Thomas Mifflin, Robt. Morris, Geo. Clymer, Thos. FitzSimons, Jared Ingersoll, James Wilson, Gouv Morris
Delaware—Geo: Read, Gunning Bedford jun, John Dickinson, Richard Bassett, Jaco: Broom
Maryland—James McHenry, Dan: of St Thos Jenifer, Danl Carrol
Virginia—John Blair, James Madison Jr.
North Carolina—Wm Blount, Richd. Dobbs Spaight, Hu Williamson
South Carolina—J. Rutledge, Charles Cotesworth Pinckney, Charles Pinckney, Pierce Butler
Georgia—William Few, Abr Baldwin
Attest: William Jackson, Secretary.

Ten Original Amendments: The Bill of Rights

In force Dec. 15, 1791

[The First Congress, at its first session in the City of New York, Sept. 25, 1789, submitted to the states 12 amendments to clarify certain individual and state rights not named in the Constitution. They are generally called the Bill of Rights.

Influential in framing these amendments was the Declaration of Rights of Virginia, written by George Mason (1725-1792) in 1776. Mason, a Virginia delegate to the Constitutional Convention, did not sign the Constitution and opposed its ratification on the ground that it did not sufficiently oppose slavery or safeguard individual rights.

In the preamble to the resolution offering the proposed amendments, Congress said: "The conventions of a number of the States having at the time of their adopting the Constitution, expressed a desire, in order to prevent misconstruction or abuse of its powers, that further declaratory and restrictive clauses should be added, and as extending the ground of public confidence in the government will best insure the beneficent ends of its institution, be it resolved," etc.

Ten of these amendments, now commonly known as one to 10 inclusive, but originally 3 to 12 inclusive, were ratified by the states as follows: New Jersey, Nov. 20, 1789; Maryland, Dec. 19, 1789; North Carolina, Dec. 22, 1789; South Carolina, Jan. 19, 1790; New Hampshire, Jan. 25, 1790; Delaware, Jan. 28, 1790; New York, Feb. 27, 1790; Pennsylvania, Mar. 10, 1790; Rhode Island, June 7, 1790; Vermont, Nov. 3, 1791; Virginia, Dec. 15, 1791; Massachusetts, Mar. 2, 1939; Georgia, Mar. 18, 1939; Connecticut, Apr. 19, 1939. These original 10 ratified amendments follow as Amendments I to X inclusive.

Of the two original proposed amendments that were not ratified promptly by the necessary number of states, the first related to apportionment of Representatives; the second, relating to compensation of members of Congress, was ratified in 1992 and became Amendment 27.]

AMENDMENT I.

Religious establishment prohibited. Freedom of speech, of press, right to assemble and to petition.

Congress shall make no law respecting an establishment of religion, or prohibiting the free exercise thereof; or abridging the freedom of speech, or of the press; or the right of the people peaceably to assemble, and to petition the Government for a redress of grievances.

AMENDMENT II.

Right to keep and bear arms.

A well regulated Militia, being necessary to the security of a free State, the right of the people to keep and bear Arms, shall not be infringed.

AMENDMENT III.

Conditions for quarters for soldiers.

No Soldier shall, in time of peace be quartered in any house, without the consent of the Owner, nor in time of war, but in a manner to be prescribed by law.

AMENDMENT IV.

Protection from unreasonable search and seizure.

The right of the people to be secure in their persons, houses, papers, and effects, against unreasonable searches and seizures, shall not be violated, and no Warrants shall issue, but upon probable cause, supported by Oath or affirmation, and particularly describing the place to be searched, and the persons or things to be seized.

AMENDMENT V.

Provisions concerning prosecution and due process of law. Double jeopardy restriction. Private property not to be taken without compensation.

No person shall be held to answer for a capital, or otherwise infamous crime, unless on a presentment or indictment of a Grand Jury, except in cases arising in the land or naval forces, or in the Militia, when in actual service in time of War or public danger; nor shall any person be subject for the same offence to be twice put in jeopardy of life or limb; nor shall be compelled in any criminal case to be a witness against himself, nor be deprived of life, liberty, or property, without due process of law; nor shall private property be taken for public use, without just compensation.

AMENDMENT VI.

Right to speedy trial, witnesses, etc.

In all criminal prosecutions, the accused shall enjoy the right to a speedy and public trial, by an impartial jury of the State and district wherein the crime shall have been committed, which district shall have been previously ascertained by law, and to be informed of the nature and cause of the accusation; to be confronted with the witnesses against him; to have compulsory process for obtaining witnesses in his favor, and to have the Assistance of Counsel for his defence.

AMENDMENT VII.

Right of trial by jury.

In suits at common law, where the value in controversy shall exceed twenty dollars, the right of trial by jury shall be preserved, and no fact tried by a jury, shall be otherwise reexamined in any Court of the United States, than according to the rules of the common law.

AMENDMENT VIII.

Excessive bail or fines; cruel and unusual punishment.

Excessive bail shall not be required, nor excessive fines imposed, nor cruel and unusual punishments inflicted.

AMENDMENT IX.

Rule of construction of Constitution.

The enumeration in the Constitution, of certain rights, shall not be construed to deny or disparage others retained by the people.

AMENDMENT X.

Rights of States under Constitution.

The powers not delegated to the United States by the Constitution, nor prohibited by it to the States, are reserved to the States respectively, or to the people.

Amendments Since the Bill of Rights

AMENDMENT XI.

Judicial powers construed.

The Judicial power of the United States shall not be construed to extend to any suit in law or equity, commenced or prosecuted against one of the United States by Citizens of another State, or by Citizens or Subjects of any Foreign State.

[This amendment was proposed to the Legislatures of the several States by the Third Congress on March. 4, 1794, and was declared to have been ratified in a message from the President to Congress, dated Jan. 8, 1798.

[It was on Jan. 5, 1798, that Secretary of State Pickering received from 12 of the States authenticated ratifications, and informed President John Adams of that fact.

[As a result of later research in the Department of State, it is now established that Amendment XI became part of the Constitution on Feb. 7, 1795, for on that date it had been ratified by 12 States as follows:

[1. New York, Mar. 27, 1794. 2. Rhode Island, Mar. 31, 1794. 3. Connecticut, May 8, 1794. 4. New Hampshire, June 16, 1794. 5. Massachusetts, June 26, 1794. 6. Vermont, between Oct. 9, 1794, and Nov. 9, 1794. 7. Virginia, Nov. 18, 1794. 8. Georgia, Nov. 29, 1794. 9. Kentucky, Dec. 7, 1794. 10. Maryland, Dec. 26, 1794. 11. Delaware, Jan. 23, 1795. 12. North Carolina, Feb. 7, 1795.

[On June 1, 1796, more than a year after Amendment XI had become a part of the Constitution—but before anyone was officially aware of this—Tennessee had been admitted as a State; but not until Oct. 16, 1797, was a certified copy of the resolution of Congress proposing the amendment sent to the Governor of Tennessee, John Sevier, by Secretary of State Pickering, whose office was then at Trenton, New Jersey, because of the epidemic of yellow fever at Philadelphia; it seems, however, that the Legislature of Tennessee took no action on Amendment XI, owing doubtless to the fact that public announcement of its adoption was made soon thereafter.

[Besides the necessary 12 States, one other, South Carolina, ratified Amendment XI, but this action was not taken until Dec. 4, 1797; the two remaining States, New Jersey and Pennsylvania, failed to ratify.]

AMENDMENT XII.
Manner of choosing President and Vice-President.

[Proposed by Congress Dec. 9, 1803; ratified June 15, 1804.]

The Electors shall meet in their respective states and vote by ballot for President and Vice-President, one of whom, at least, shall not be an inhabitant of the same state with themselves; they shall name in their ballots the person voted for as President, and in distinct ballots the person voted for as Vice-President, and they shall make distinct lists of all persons voted for as President, and of all persons voted for as Vice-President, and of the number of votes for each, which lists they shall sign and certify, and transmit sealed to the seat of the government of the United States, directed to the President of the Senate;–The President of the Senate shall, in presence of the Senate and House of Representatives, open all the certificates and the votes shall then be counted;—The person having the greatest number of votes for President, shall be the President, if such number be a majority of the whole number of Electors appointed; and if no person have such majority, then from the persons having the highest numbers not exceeding three on the list of those voted for as President, the House of Representatives shall choose immediately, by ballot, the President. But in choosing the President, the votes shall be taken by states, the representation from each state having one vote; a quorum for this purpose shall consist of a member or members from two-thirds of the states, and a majority of all the states shall be necessary to a choice. *[And if the House of Representatives shall not choose a President whenever the right of choice shall devolve upon them, before the fourth day of March next following, then the Vice-President shall act as President, as in the case of the death or other constitutional disability of the President.]* *[The words in parentheses were superseded by Amendment XX, section 3.]* The person having the greatest number of votes as Vice-President, shall be the Vice-President, if such number be a majority of the whole number of Electors appointed, and if no person have a majority, then from the two highest numbers on the list, the Senate shall choose the Vice-President; a quorum for the purpose shall consist of two-thirds of the whole number of Senators, and a majority of the whole number shall be necessary to a choice. But no person constitutionally ineligible to the office of President shall be eligible to that of Vice-President of the United States.

THE RECONSTRUCTION AMENDMENTS

[Amendments XIII, XIV, and XV are commonly known as the Reconstruction Amendments, inasmuch as they followed the Civil War, and were drafted by Republicans who were bent on imposing their own policy of reconstruction on the South. Post-bellum legislatures there—Mississippi, South Carolina, Georgia, for example—had set up laws which, it was charged, were contrived to perpetuate Negro slavery under other names.]

AMENDMENT XIII.
Slavery abolished.

[Proposed by Congress Jan. 31, 1865; ratified Dec. 6, 1865. The amendment, when first proposed by a resolution

in Congress, was passed by the Senate, 38 to 6, on Apr. 8, 1864, but was defeated in the House, 95 to 66 on June 15, 1864. On reconsideration by the House, on Jan. 31, 1865, the resolution passed, 119 to 56. It was approved by President Lincoln on Feb. 1, 1865, although the Supreme Court had decided in 1798 that the President has nothing to do with the proposing of amendments to the Constitution, or their adoption.]

1. Neither slavery nor involuntary servitude, except as a punishment for crime whereof the party shall have been duly convicted, shall exist within the United States, or any place subject to their jurisdiction.

2. Congress shall have power to enforce this article by appropriate legislation.

AMENDMENT XIV.
Citizenship rights not to be abridged.

[The following amendment was proposed to the Legislatures of the several states by the 39th Congress, June 13, 1866, ratified July 9, 1868, and declared to have been ratified in a proclamation by the Secretary of State, July 28, 1868.

[The 14th amendment was adopted only by virtue of ratification subsequent to earlier rejections. Newly constituted legislatures in both North Carolina and South Carolina (respectively July 4 and 9, 1868), ratified the proposed amendment, although earlier legislatures had rejected the proposal. The Secretary of State issued a proclamation, which, though doubtful as to the effect of attempted withdrawals by Ohio and New Jersey, entertained no doubt as to the validity of the ratification by North and South Carolina. The following day (July 21, 1868), Congress passed a resolution which declared the 14th Amendment to be a part of the Constitution and directed the Secretary of State so to promulgate it. The Secretary waited, however, until the newly constituted Legislature of Georgia had ratified the amendment, subsequent to an earlier rejection, before the promulgation of the ratification of the new amendment.]

1. All persons born or naturalized in the United States, and subject to the jurisdiction thereof, are citizens of the United States and of the State wherein they reside. No State shall make or enforce any law which shall abridge the privileges or immunities of citizens of the United States; nor shall any State deprive any person of life, liberty, or property, without due process of law; nor deny to any person within its jurisdiction the equal protection of the laws.

2. Representatives shall be apportioned among the several States according to their respective numbers, counting the whole number of persons in each State, excluding Indians not taxed. But when the right to vote at any election for the choice of electors for President and Vice-President of the United States, Representatives in Congress, the Executive and Judicial officers of a State, or the members of the Legislature thereof, is denied to any of the male inhabitants of such State, being twenty-one years of age, and citizens of the United States, or in any way abridged, except for participation in rebellion, or other crime, the basis of representation therein shall be reduced in the proportion which the number of such male citizens shall bear to the whole number of male citizens twenty-one years of age in such State.

3. No person shall be a Senator or Representative in Congress, or elector of President and Vice-President, or hold any office, civil or military, under the United States, or under any State, who, having previously taken an oath, as a member of Congress, or as an officer of the United States, or as a member of any State legislature, or as an executive or judicial officer of any State, to support the Constitution of the United States, shall have engaged in insurrection or rebellion against the same, or given aid or comfort to the enemies thereof. But Congress may by a vote of two-thirds of each House, remove such disability.

4. The validity of the public debt of the United States, authorized by law, including debts incurred for payment of pensions and bounties for services in suppressing insurrection or rebellion, shall not be questioned. But neither the United States nor any State shall assume or pay any debt or obligation incurred in aid of insurrection or rebellion against

the United States, or any claim for the loss or emancipation of any slave; but all such debts, obligations and claims shall be held illegal and void.

The Congress shall have power to enforce, by appropriate legislation, the provisions of this article.

AMENDMENT XV.
Race no bar to voting rights.

[The following amendment was proposed to the legislatures of the several States by the 40th Congress, Feb. 26, 1869, and ratified Feb. 8, 1870.]

1. The right of citizens of the United States to vote shall not be denied or abridged by the United States or by any State on account of race, color, or previous condition of servitude–

2. The Congress shall have power to enforce this article by appropriate legislation.

AMENDMENT XVI.
Income taxes authorized.

[Proposed by Congress July 12, 1909; ratified Feb. 3, 1913.]

The Congress shall have power to lay and collect taxes on incomes, from whatever source derived, without apportionment among the several States, and without regard to any census or enumeration.

AMENDMENT XVII.
United States Senators to be elected by direct popular vote.

[Proposed by Congress May 13, 1912; ratified Apr. 8, 1913.]

The Senate of the United States shall be composed of two Senators from each State, elected by the people thereof, for six years; and each Senator shall have one vote. The electors in each State shall have the qualifications requisite for electors of the most numerous branch of the State legislatures.

When vacancies happen in the representation of any State in the Senate, the executive authority of such State shall issue writs of election to fill such vacancies: *Provided,* That the legislature of any State may empower the executive thereof to make temporary appointments until the people fill the vacancies by election as the legislature may direct.

This amendment shall not be so construed as to affect the election or term of any Senator chosen before it becomes valid as part of the Constitution.

AMENDMENT XVIII.
Liquor prohibition amendment.

[Proposed by Congress Dec. 18, 1917; ratified Jan. 16, 1919. Repealed by Amendment XXI, effective Dec. 5, 1933.]

1. After one year from the ratification of this article the manufacture, sale, or transportation of intoxicating liquors within, the importation thereof into, or the exportation thereof from the United States and all territory subject to the jurisdiction thereof for beverage purposes is hereby prohibited.

2. The Congress and the several States shall have concurrent power to enforce this article by appropriate legislation.

3. This article shall be inoperative unless it shall have been ratified as an amendment to the Constitution by the legislatures of the several States as provided in the Constitution, within seven years from the date of the submission hereof to the States by the Congress.

[The total vote in the Senates of the various States was 1,310 for, 237 against—84.6% dry. In the lower houses of the States the vote was 3,782 for, 1,035 against—78.5% dry.

[The amendment ultimately was adopted by all the States except Connecticut and Rhode Island.]

AMENDMENT XIX.
Giving nationwide suffrage to women.

[Proposed by Congress June 4, 1919; ratified Aug. 18, 1920.]

The right of citizens of the United States to vote shall not be denied or abridged by the United States or by any State on account of sex.

Congress shall have power to enforce this Article by appropriate legislation.

AMENDMENT XX.
Terms of President and Vice President to begin on Jan. 20; those of Senators, Representatives, Jan. 3.

[Proposed by Congress Mar. 2, 1932; ratified Jan. 23, 1933.]

1. The terms of the President and Vice President shall end at noon on the 20th day of January, and the terms of Senators and Representatives at noon on the 3d day of January, of the years in which such terms would have ended if this article had not been ratified; and the terms of their successors shall then begin.

2. The Congress shall assemble at least once in every year, and such meeting shall begin at noon on the 3d day of January, unless they shall by law appoint a different day.

3. If, at the time fixed for the beginning of the term of the President, the President elect shall have died, the Vice President elect shall become President. If a President shall not have been chosen before the time fixed for the beginning of his term, or if the President elect shall have failed to qualify, then the Vice President elect shall act as President until a President shall have qualified; and the Congress may by law provide for the case wherein neither a President elect nor a Vice President elect shall have qualified, declaring who shall then act as President, or the manner in which one who is to act shall be selected, and such person shall act accordingly until a President or Vice President shall have qualified.

4. The Congress may by law provide for the case of the death of any of the persons from whom the House of Representatives may choose a President whenever the right of choice shall have devolved upon them, and for the case of the death of any of the persons from whom the Senate may choose a Vice President whenever the right of choice shall have devolved upon them.

5. Sections 1 and 2 shall take effect on the 15th day of October following the ratification of this article (Oct. 1933).

6. This article shall be inoperative unless it shall have been ratified as an amendment to the Constitution by the legislatures of three-fourths of the several States within seven years from the date of its submission.

AMENDMENT XXI.
Repeal of Amendment XVIII.

[Proposed by Congress Feb. 20, 1933; ratified Dec. 5, 1933.]

1. The eighteenth article of amendment to the Constitution of the United States is hereby repealed.

2. The transportation or importation into any State, Territory, or possession of the United States for delivery or use therein of intoxicating liquors, in violation of the laws thereof, is hereby prohibited.

3. This article shall be inoperative unless it shall have been ratified as an amendment to the Constitution by conventions in the several States, as provided in the Constitution, within seven years from the date of the submission hereof to the States by the Congress.

AMENDMENT XXII.
Limiting Presidential terms of office.

[Proposed by Congress Mar. 24, 1947; ratified Feb. 27, 1951.]

1. No person shall be elected to the office of the President more than twice, and no person who has held the office of President, or acted as President, for more than two years of a term to which some other person was elected President shall be elected to the office of the President more than once. But this Article shall not apply to any person holding the office of President when this Article was proposed by the Congress, and shall not prevent any person who may be holding the office of President, or acting as President, during the term within which this Article becomes operative from holding the office of President or acting as President during the remainder of such term.

2. This article shall be inoperative unless it shall have been ratified as an amendment to the Constitution by the legislatures of three-fourths of the several States within seven years from the date of its submission to the States by the Congress.

AMENDMENT XXIII.
Presidential vote for District of Columbia.

[Proposed by Congress June 16, 1960; ratified Mar. 29, 1961.]
1. The District constituting the seat of Government of the United States shall appoint in such manner as the Congress may direct:

A number of electors of President and Vice President equal to the whole number of Senators and Representatives in Congress to which the District would be entitled if it were a State, but in no event more than the least populous State; they shall be in addition to those appointed by the States, but they shall be considered, for the purposes of the election of President and Vice President, to be electors appointed by a State; and they shall meet in the District and perform such duties as provided by the twelfth article of amendment.

2. The Congress shall have power to enforce this article by appropriate legislation.

AMENDMENT XXIV.
Barring poll tax in federal elections.

[Proposed by Congress Aug. 27, 1962; ratified Jan. 23, 1964.]
1. The right of citizens of the United States to vote in any primary or other election for President or Vice President, for electors for President or Vice President, or for Senator or Representative in Congress, shall not be denied or abridged by the United States or any State by reason of failure to pay any poll tax or other tax.

2. The Congress shall have power to enforce this article by appropriate legislation.

AMENDMENT XXV.
Presidential disability and succession.

[Proposed by Congress July 6, 1965; ratified Feb. 10, 1967.]
1. In case of the removal of the President from office or of his death or resignation, the Vice President shall become President.

2. Whenever there is a vacancy in the office of the Vice President, the President shall nominate a Vice President who shall take office upon confirmation by a majority vote of both houses of Congress.

3. Whenever the President transmits to the President pro tempore of the Senate and the Speaker of the House of Representatives his written declaration that he is unable to dis-charge the powers and duties of his office, and until he transmits to them a written declaration to the contrary, such powers and duties shall be discharged by the Vice President as Acting President.

4. Whenever the Vice President and a majority of either the principal officers of the executive departments or of such other body as Congress may by law provide, transmit to the President pro tempore of the Senate and the Speaker of the House of Representatives their written declaration that the President is unable to discharge the powers and duties of his office, the Vice President shall immediately assume the powers and duties of the office as Acting President.

Thereafter, when the President transmits to the President pro tempore of the Senate and the Speaker of the House of Representatives his written declaration that no inability exists, he shall resume the powers and duties of his office unless the Vice President and a majority of either the prin-cipal officers of the executive department or of such other body as Congress may by law provide, transmit within four days to the President pro tempore of the Senate and the Speaker of the House of Representatives their written dec-laration that the President is unable to discharge the powers and duties of his office. Thereupon Congress shall decide the issue, assembling within forty-eight hours for that pur-pose if not in session. If the Congress, within twenty-one days after receipt of the latter written declaration, or, if Congress is not in session, within twenty-one days after Congress is required to assemble, determines by two-thirds vote of both Houses that the President is unable to dis-charge the powers and duties of his office, the Vice Presi-dent shall continue to discharge the same as Acting President; otherwise, the President shall resume the powers and duties of his office.

AMENDMENT XXVI.
Lowering voting age to 18 years.

[Proposed by Congress Mar. 23, 1971; ratified June 30, 1971.]
1. The right of citizens of the United States, who are eighteen years of age or older, to vote shall not be denied or abridged by the United States or by any State on account of age.

2. The Congress shall have the power to enforce this arti-cle by appropriate legislation.

AMENDMENT XXVII.
Congressional pay.

[Proposed by Congress Sept. 25, 1789; ratified May 7, 1992.]
No law, varying the compensation for the services of the Senators and Representatives, shall take effect, until an elec-tion of Representatives shall have intervened.

How a Bill Becomes a Law

A senator or representative introduces a bill by sending it to the clerk of the House or the Senate, who assigns it a number and title. This procedure is termed the first reading. The clerk then refers the bill to the appropriate Senate or House committee.

If the committee opposes the bill, it will table, or kill, it. Otherwise, the committee holds hearings to listen to opin-ions and facts offered by members and other interested peo-ple. The committee then debates the bill and possibly offers amendments. A vote is taken, and if favorable, the bill is sent back to the clerk of the House or Senate.

The clerk reads the bill to the house—the second reading. Members may then debate the bill and suggest amendments.

After debate and possibly amendment, the bill is given a third reading, simply of the title, and put to a voice or roll-call vote.

If passed, the bill goes to the other house, where it may be defeated or passed, with or without amendments. If defeated, the bill dies. If passed with amendments, a confer-ence committee made up of members of both houses works out the differences and arrives at a compromise.

After passage of the final version by both houses, the bill is sent to the president. If the president signs it, the bill becomes a law. The president may, however, veto the bill by refusing to sign it and sending it back to the house where it originated, with reasons for the veto.

The president's objections are then read and debated, and a roll-call vote is taken. If the bill receives less than a two-thirds majority, it is defeated. If it receives at least two-thirds, it is sent to the other house. If that house also passes it by at least a two-thirds majority, the veto is overridden, and the bill becomes a law.

If the president neither signs nor vetoes the bill within 10 days—not including Sundays—it automatically becomes a law even without the president's signature. However, if Congress has adjourned within those 10 days, the bill is automatically killed; this indirect rejection is termed a pocket veto.

Note: Under "line-item veto" legislation effective Jan. 1, 1997, the president was authorized, under certain circum-stances, to veto a bill in part, but the legislation was found unconstitutional by the Supreme Court, June 25, 1998.

Confederate States and Secession

The American Civil War (1861-65) grew out of sectional disputes over the continued existence of slavery in the South and the contention of Southern legislators that the states retained many rights, including the right to secede.

The war was not fought by state against state but by one federal regime against another, the Confederate government in Richmond assuming control over the economic, political, and military life of the South, under protest from Georgia and South Carolina.

South Carolina voted an ordinance of secession from the Union, repealing its 1788 ratification of the U.S. Constitution on Dec. 20, 1860, to take effect on Dec. 24. Other states seceded in 1861. Their votes in conventions were: Mississippi, Jan. 9, 84-15; Florida, Jan. 10, 62-7; Alabama, Jan. 11, 61-39; Georgia, Jan. 19, 208-89; Louisiana, Jan. 26, 113-17; Texas, Feb. 1, 166-7, ratified by popular vote on Feb. 23 (for 34,794, against 11,325); Virginia, Apr. 17, 88-55, ratified by popular vote on May 23 (for 128,884; against 32,134); Arkansas, May 6, 69-1; Tennessee, May 7, ratified by popu-

lar vote on June 8 (for 104,019, against 47,238); North Carolina, May 21.

Missouri Unionists stopped secession in conventions Feb. 28 and Mar. 9. The legislature condemned secession Mar. 7. Under the protection of Confederate troops, secessionist members of the legislature adopted a resolution of secession at Neosho, Oct. 31. The Confederate Congress seated the secessionists' representatives.

Kentucky did not secede, and its government remained Unionist. In a part of the state occupied by Confederate troops, Kentuckians approved secession, and the Confederate Congress admitted their representatives.

The Maryland legislature voted against secession Apr. 27, 53-13. Delaware did not secede. Western Virginia held conventions at Wheeling, named a pro-Union governor on June 11, 1861, and was admitted to the Union as West Virginia on June 20, 1863. Its constitution provided for gradual abolition of slavery.

Confederate Government

Forty-two delegates from South Carolina, Georgia, Alabama, Mississippi, Louisiana, and Florida met in convention at Montgomery, AL, on Feb. 4, 1861. They adopted a provisional constitution of the Confederate States of America and elected Jefferson Davis (MS) as provisional president and Alexander H. Stephens (GA) as provisional vice president.

A permanent constitution was adopted Mar. 11. It abolished the African slave trade, but it did not bar interstate

commerce in slaves. On July 20 the Congress moved to Richmond, VA. Davis was elected president in October and was inaugurated on Feb. 22, 1862.

The Congress adopted a flag, consisting of a red field with a white stripe, and a blue jack with a circle of white stars. Later the more popular flag was the red field with blue diagonal crossbars that held 13 white stars, for the 11 states in the Confederacy plus Kentucky and Missouri.

Lincoln's Address at Gettysburg, 1863

Fourscore and seven years ago our fathers brought forth on this continent a new nation, conceived in liberty and dedicated to the proposition that all men are created equal.

Now we are engaged in a great civil war, testing whether that nation or any nation so conceived and so dedicated can long endure. We are met on a great battle field of that war. We have come to dedicate a portion of that field, as a final resting-place for those who here gave their lives that that nation might live. It is altogether fitting and proper that we should do this.

But, in a larger sense, we can not dedicate—we can not consecrate—we can not hallow—this ground. The brave men, living and dead, who struggled here, have consecrated

it, far above our poor power to add or detract. The world will little note, nor long remember, what we say here, but it can never forget what they did here. It is for us the living, rather, to be dedicated here to the unfinished work which they who fought here have thus far so nobly advanced. It is rather for us to be here dedicated to the great task remaining before us—that from these honored dead we take increased devotion to that cause for which they gave the last full measure of devotion—that we here highly resolve that these dead shall not have died in vain—that this nation, under God, shall have a new birth of freedom—and that government of the people, by the people, for the people, shall not perish from the earth.

Selected Landmark Decisions of the U.S. Supreme Court

1803: Marbury v. Madison. The Court ruled that Congress exceeded its power in the Judiciary Act of 1789; the Court thus established its power to review acts of Congress and declare invalid those it found in conflict with the Constitution.

1819: McCulloch v. Maryland. The Court ruled that Congress had the authority to charter a national bank, under the Constitution's granting of the power to enact all laws "necessary and proper" to responsibilities of government.

1819: Trustees of Dartmouth College v. Woodward. The Court ruled that a state could not arbitrarily alter the terms of a college's contract. (The Court later used a similar principle to limit the states' ability to interfere with business contracts.)

1857: Dred Scott v. Sanford. The Court declared unconstitutional the already-repealed Missouri Compromise of 1820 because it deprived a person of his or her property—a slave—without due process of law. The Court also ruled that slaves were not citizens of any state nor of the U.S. (The latter part of the decision was overturned by ratification of the 14th Amendment in 1868.)

1896: Plessy v. Ferguson. The Court ruled that a state law requiring federal railroad trains to provide separate but equal facilities for black and white passengers neither infringed upon federal authority to regulate interstate commerce nor violated the 13th and 14th Amendments. (The "separate but equal" doctrine remained effective until the 1954 **Brown v. Board of Education** decision.)

1904: Northern Securities Co. v. U.S. The Court ruled that a holding company formed solely to eliminate competition between two railroad lines was a combination in restraint of trade, violating the federal antitrust act.

1908: Muller v. Oregon. The Court upheld a state law limiting the working hours of women. (Louis D. Brandeis, counsel for the state, cited evidence from social workers, physicians, and factory inspectors that the number of hours women worked affected their health and morals.)

1911: Standard Oil Co. of New Jersey et al. v. U.S. The Court ruled that the Standard Oil Trust must be dissolved because of its unreasonable restraint of trade.

1919: Schenck v. U.S. The Court sustained the Espionage Act of 1917, maintaining that freedom of speech and press could be constrained if "the words used . . . create a clear and present danger. . ."

1925: Gitlow v. New York. The Court ruled that the First Amendment prohibition against government abridgment of the freedom of speech applied to the states as well as to the federal government. The decision was the first of a number of rulings holding that the 14th Amendment extended the guarantees of the Bill of Rights to state action.

1935: Schechter Poultry Corp. v. U.S. The Court ruled that Congress exceeded its authority to delegate legislative powers and to regulate interstate commerce when it enacted the National Industrial Recovery Act, which afforded the U.S. president too much discretionary power.

1951: Dennis et al. v. U.S. The Court upheld convictions under the Smith Act of 1940 for invoking Communist theory that advocated the forcible overthrow of the government. (In the **1957 Yates v. U.S.** decision, the Court moderated this ruling by allowing such advocacy in the abstract, if not connected to action to achieve the goal.)

1954: Brown v. Board of Education of Topeka. The Court ruled that separate public schools for black and white students were inherently unequal, so that state-sanctioned segregation in public schools violated the equal protection guarantee of the 14th Amendment. And in **Bolling v. Sharpe** the Court ruled that the congressionally mandated segregated public school system in the District of Columbia violated the 5th Amendment's due process guarantee of personal liberty. (The Brown ruling also led to abolition of state-sponsored segregation in other public facilities.)

1957: Roth v. U.S., Alberts v. California. The Court ruled obscene material was not protected by First Amendment guarantees of freedom of speech and press, defining obscene as "utterly without redeeming social value" and appealing to "prurient interests" in the view of the average person. This definition was modified in later decisions, and the "average person" standard was replaced by the "local community" standard in **Miller v. California (1973).**

1961: Mapp v. Ohio. The Court ruled that evidence obtained in violation of the 4th Amendment guarantee against unreasonable search and seizure must be excluded from use at state as well as federal trials.

1962: Engel v. Vitale. The Court held that public school officials could not require pupils to recite a state-composed prayer, even if it was nondenominational and voluntary, because this would be an unconstitutional attempt to establish religion.

1962: Baker v. Carr. The Court held that the constitutional challenges to the unequal distribution of voters among legislative districts could be resolved by federal courts.

1963: Gideon v. Wainwright. The Court ruled that state and federal defendants who are charged with serious crimes must have access to an attorney, at state expense if necessary.

1964: New York Times Co. v. Sullivan. The Court ruled that the First Amendment protected the press from libel suits for defamatory reports about public officials unless an injured party could prove that a defamatory report was made out of malice or "reckless disregard" for the truth.

1965: Griswold v. Conn. The Court ruled that a state unconstitutionally interfered with personal privacy in the marriage relationship when it prohibited anyone, including married couples, from using contraceptives.

1966: Miranda v. Arizona. The Court ruled that, under the guarantee of due process, suspects in custody, before being questioned, must be informed that they have the right to remain silent, that anything they say may be used against them, and that they have the right to counsel.

1973: Roe v. Wade, Doe v. Bolton. The Court ruled that the fetus was not a "person" with constitutional rights and that a right to privacy inherent in the 14th Amendment's due process guarantee of personal liberty protected a woman's decision to have an abortion. During the first trimester of pregnancy, the Court maintained, the decision should be left entirely to a woman and her physician. Some regulation of abortion procedures was allowed in the 2d trimester, and some restriction of abortion in the 3d.

1974: U.S. v. Nixon. The Court ruled that neither the separation of powers nor the need to preserve the confidentiality of presidential communications could alone justify an absolute executive privilege of immunity from judicial demands for evidence to be used in a criminal trial.

1976: Gregg v. Georgia, Profitt v. Fla., Jurek v. Texas. The Court held that death, as a punishment for persons convicted of first degree murder, was not in and of itself cruel and unusual punishment in violation of the 8th Amendment. But the Court ruled that the sentencing judge and jury must consider the individual character of the offender and the circumstances of the particular crime.

1978: Regents of Univ. of Calif. v. Bakke. The Court ruled that a special admissions program for a state medical school, under which a set number of places were reserved for minorities, violated the 1964 Civil Rights Act, which forbids excluding anyone, because of race, from a federally funded program. However, the Court ruled that race could be considered as one of a complex of factors.

1986: Bowers v. Hardwick. The Court refused to extend any constitutional right of privacy to homosexual activity, upholding a Georgia law that in effect made such activity a crime. (Although the Georgia law made no distinction between heterosexual or homosexual sodomy, enforcement had been confined to homosexuals; the statute was invalidated by the state supreme court in 1998.) In **Romer v. Evans (1996),** the Court struck down a Colorado constitutional provision that barred legislation protecting homosexuals from discrimination.

1990: Cruzan v. Missouri. The Court ruled that a person had the right to refuse life-sustaining medical treatment. However, the Court also ruled that, before treatment could be withheld from a comatose patient, a state could require "clear and convincing evidence" that the patient would not have wanted to live. And in 2 **1997** rulings, **Washington v. Glucksberg** and **Vacco v. Quill,** the Court ruled that states could ban doctor-assisted suicide.

1995: Adarand Constructors v. Peña. The Court held that federal programs that classify people by race, unless "narrowly tailored" to accomplish a "compelling governmental interest," may deny individuals the right to equal protection. Such federal programs, the Court maintained, must adhere to the same strict standards required of state-run affirmative action programs.

1995: U.S. Term Limits Inc. v. Thorton. The Court ruled that neither states nor Congress could limit terms of members of Congress, since the Constitution reserves to the people the right to choose federal lawmakers.

1997: Clinton v. Jones. Rejecting an appeal by Pres. Clinton in a sexual harassment suit, the Court ruled that a sitting president did not have temporary immunity from a lawsuit for actions outside the realm of official duties.

1997: City of Boerne v. Flores. The Court overturned a 1993 law that banned enforcement of laws that "substantially burden" religious practice unless there is a "compelling need" to do so. The Court held that the act was an unwarranted intrusion by Congress on states' prerogatives and an infringement on the judiciary's role.

1997: Reno v. ACLU. Citing the right to free expression, the Court overturned a provision making it a crime to display or distribute "indecent" or "patently offensive" material on the Internet. In **1998,** however, the Court ruled in **NEA v. Finley** that "general standards of decency" may be used as a criterion in federal arts funding.

1998: Clinton v. City of New York. The Court struck down the Line-Item Veto Act (1996), holding that it unconstitutionally gave the president "the unilateral power to change the text of duly enacted statutes."

1998: Faragher v. City of Boca Raton, Burlington Industries, Inc. v. Ellerth. The Court issued new guidelines for workplace sexual harassment suits, holding employers responsible for misconduct by supervisory employees. And in **Oncale v. Sundowner Offshore Services,** the Court ruled that the law against sexual harassment applies regardless of whether harasser and victim are the same sex.

1999: Dept. of Commerce v. U.S. House. Upholding a challenge to plans for the 2000 census, the Court required an actual head count for apportioning the U.S. House of Representatives, but allowed the use of statistical sampling methods for other purposes, such as the allocation of federal funds.

1999: Alden v Maine, Florida Prepaid v. College Savings Bank, College Savings Bank v. Florida. In a series of rulings, the Court applied the principle of "sovereign immunity" to shield states in large part from being sued under federal law.

Presidential Oath of Office

The Constitution (Article II) directs that the president-elect shall take the following oath or affirmation to be inaugurated as president: "I do solemnly swear [affirm] that I will faithfully execute the office of President of the United States, and will, to the best of my ability, preserve, protect,

and defend the Constitution of the United States." (Custom decrees the addition of the words "So help me God" at the end of the oath when taken by the president-elect, with the left hand on the Bible for the duration of the oath, and the right hand slightly raised.)

Law on Succession to the Presidency

If by reason of death, resignation, removal from office, inability, or failure to qualify there is neither a president nor vice president to discharge the powers and duties of the office of president, then the speaker of the House of Representatives shall upon his resignation as speaker and as representative, act as president. The same rule shall apply in the case of the death, resignation, removal from office, or inability of an individual acting as president.

If at the time when a speaker is to begin the discharge of the powers and duties of the office of president there is no speaker, or the speaker fails to qualify as acting president, then the president pro tempore of the Senate, upon his resignation as president pro tempore and as senator, shall act as president.

An individual acting as president shall continue to act until the expiration of the then current presidential term, except that (1) if his discharge of the powers and duties of the office is founded in whole or in part in the failure of both

the president-elect and the vice president-elect to qualify, then he shall act only until a president or vice president qualifies, and (2) if his discharge of the powers and duties of the office is founded in whole or in part on the inability of the president or vice president, then he shall act only until the removal of the disability of one of such individuals.

If, by reason of death, resignation, removal from office, or failure to qualify, there is no president pro tempore to act as president, then the officer of the United States who is highest on the following list, and who is not under any disability to discharge the powers and duties of president shall act as president; the secretaries of state, treasury, defense, attorney general; secretaries of interior, agriculture, commerce, labor, health and human services, housing and urban development, transportation, energy, education, veterans affairs.

(Legislation approved July 18, 1947; amended Sept. 9, 1965, Oct. 15, 1966, Aug. 4, 1977, and Sept. 27, 1979. See also Constitutional Amendment XXV.)

Origin of the United States National Motto

In God We Trust, designated as the U.S. National Motto by Congress in 1956, originated during the Civil War as an inscription for U. S. coins, although it was used by Francis Scott Key in a slightly different form when he wrote "The Star-Spangled Banner" in 1814. On Nov. 13, 1861, when Union morale had been shaken by battlefield defeats, the Rev. M. R. Watkinson, of Ridleyville, PA, wrote to Secy. of the Treasury Salmon P. Chase. "From my heart I have felt our national shame in

disowning God as not the least of our present national disasters," the minister wrote, suggesting "recognition of the Almighty God in some form on our coins." Secy. Chase ordered designs prepared with the inscription *In God We Trust* and backed coinage legislation that authorized use of this slogan. It first appeared on some U.S. coins in 1864, and disappeared and reappeared on various coins until 1955, when Congress ordered it placed on all paper money and all coins.

The Great Seal of the U.S.

On July 4, 1776, the Continental Congress appointed a committee consisting of Benjamin Franklin, John Adams, and Thomas Jefferson "to bring in a device for a seal of the United States of America." The designs submitted by this and a subsequent committee were considered unacceptable. After many delays, a third committee, appointed early in 1782, presented a design prepared by William Barton. Charles Thomson, the sec-

retary of Congress, suggested certain changes, and Congress finally approved the design on June 20, 1782. The obverse side of the seal shows an American bald eagle. In its mouth is a ribbon bearing the motto *e pluribus unum* (one out of many). In the eagle's talons are the arrows of war and an olive branch of peace. The reverse side shows an unfinished pyramid with an eye (the eye of Providence) above it.

The American's Creed

William Tyler Page, Clerk of the U.S. House of Representatives, wrote "The American's Creed" in 1917. It was accepted by the House on behalf of the American people on April 3, 1918.

"I believe in the United States of America as a government of the people, by the people, for the people; whose just powers are derived from the consent of the governed; a democracy in a republic; a sovereign Nation of many sovereign States; a perfect union, one and inseparable; established upon those principles of freedom, equality, justice, and

humanity for which American patriots sacrificed their lives and fortunes.

"I therefore believe it is my duty to my country to love it, to support its Constitution, to obey its laws, to respect its flag, and to defend it against all enemies."

The Flag of the U.S.—The Stars and Stripes

The 50-star flag of the United States was raised for the first time officially at 12:01 AM on July 4, 1960, at Fort McHenry National Monument in Baltimore, MD. The 50th star had been added for Hawaii; a year earlier the 49th, for Alaska. Before that, no star had been added since 1912, when New Mexico and Arizona were admitted to the Union.

The true history of the Stars and Stripes has become so cluttered by myth and tradition that the facts are difficult, and in some cases impossible, to establish. For example, it is not certain who designed the Stars and Stripes, who made the first such flag, or even whether it ever flew in any sea fight or land battle of the American Revolution.

All agree, however, that the Stars and Stripes originated as the result of a resolution offered by the Marine Committee

of the Second Continental Congress at Philadelphia and adopted on June 14, 1777. It read:

Resolved: that the flag of the United States be thirteen stripes, alternate red and white; that the union be thirteen stars, white in a blue field, representing a new constellation.

Congress gave no hint as to the designer of the flag, no instructions as to the arrangement of the stars, and no information on its appropriate uses. Historians have been unable to find the original flag law.

The resolution establishing the flag was not even published until Sept. 2, 1777. Despite repeated requests, Washington did not get the flags until 1783, after the American Revolution was over. And there is no certainty that they were the Stars and Stripes.

Early Flags

Many historians consider the first flag of the U.S. to have been the Grand Union (sometimes called Great Union) flag, although the Continental Congress never officially adopted it. This flag was a modification of the British Meteor flag, which had the red cross of St. George and the white cross of St. Andrew combined in the blue canton. For the Grand Union flag, 6 horizontal stripes were imposed on the red field, dividing it into 13 alternating red and white stripes. On Jan. 1, 1776, when the Continental Army came into formal existence, this flag was unfurled on Prospect Hill, Somerville, MA. Washington wrote that "we hoisted the Union Flag in compliment to the United Colonies."

One of several flags about which controversy has raged for years is at Easton, PA. Containing the devices of the national flag in reversed order, this flag has been in the public library at Easton for more than 150 years. Some contend that this flag was actually the first Stars and Stripes, first displayed on July 8, 1776. This flag has 13 red and white stripes in the canton, 13 white stars centered in a blue field.

A flag was hastily improvised from garments by the defenders of Fort Schuyler at Rome, NY, Aug. 3-22, 1777. Historians believe it was the Grand Union Flag.

The Sons of Liberty had a flag of 9 red and white stripes, to signify 9 colonies, when they met in New York in 1765 to oppose the Stamp Tax. By 1775, the flag had grown to 13 red and white stripes, with a rattlesnake on it.

At Concord, Apr. 19, 1775, the minutemen from Bedford, MA, are said to have carried a flag having a silver arm with sword on a red field. At Cambridge, MA, the Sons of Liberty used a plain red flag with a green pine tree on it.

In June 1775, Washington went from Philadelphia to Boston to take command of the army, escorted to New York by the Philadelphia Light Horse Troop. It carried a yellow flag that had an elaborate coat of arms—the shield charged with 13 knots, the motto "For These We Strive"—and a canton of 13 blue and silver stripes.

In Feb. 1776, Col. Christopher Gadsden, a member of the Continental Congress, gave the South Carolina Provincial Congress a flag "such as is to be used by the commander-in-chief of the American Navy." It had a yellow field, with a rattlesnake about to strike and the words "Don't Tread on Me."

At the Battle of Bennington, Aug. 16, 1777, patriots used a flag of 7 white and 6 red stripes with a blue canton extending down 9 stripes and showing an arch of 11 white stars over the figure 76 and a star in each of the upper corners. The stars are 7-pointed. This flag is preserved in the Historical Museum at Bennington, VT.

At the Battle of Cowpens, Jan. 17, 1781, the 3d Maryland Regiment is said to have carried a flag of 13 red and white stripes, with a blue canton containing 12 stars in a circle around one star.

Who Designed the Flag? No one knows for certain. Francis Hopkinson, designer of a naval flag, declared he also had designed the flag and in 1781 asked Congress to reimburse him for his services. Congress did not do so. Dumas Malone of Columbia University wrote: "This talented man . . . designed the American flag."

Who Called the Flag "Old Glory"? The flag is said to have been named Old Glory by William Driver, a sea captain of Salem, MA. One legend has it that when he raised the flag on his brig, the *Charles Doggett*, in 1824, he said: "I name thee Old Glory." But his daughter, who presented the flag to the Smithsonian Institution, said he named it at his 21st birthday celebration on Mar. 17, 1824, when his mother presented the homemade flag to him.

The Betsy Ross Legend. The widely publicized legend that Mrs. Betsy Ross made the first Stars and Stripes in June 1776, at the request of a committee composed of George Washington, Robert Morris, and George Ross, an uncle, was first made public in 1870, by a grandson of Mrs. Ross. Historians have been unable to find a historical record of such a meeting or committee.

Adding New Stars

The flag of 1777 was used until 1795. Then, on the admission of Vermont and Kentucky to the Union, Congress passed and Pres. Washington signed an act that after May 1, 1795, the flag should have 15 stripes, alternating red and white, and 15 white stars on a blue field.

When new states were admitted, it became evident that the flag would become burdened with stripes. Congress thereupon ordered that after July 4, 1818, the flag should have 13 stripes, symbolizing the 13 original states; that the union have 20 stars, and that whenever a new state was admitted a new star should be added on the July 4 following admission. No law designates the permanent arrangement of the stars. However, since 1912, when a new state has been admitted, the new design has been announced by executive order. No star is specifically identified with any state.

Code of Etiquette for Display and Use of the U.S. Flag

Reviewed by National Flag Foundation

Although the Stars and Stripes originated in 1777, it was not until 146 years later that there was a serious attempt to establish a uniform code of etiquette for the U.S. flag. On Feb. 15, 1923, the War Department issued a circular on the rules of flag usage. These rules were adopted almost in their entirety June 14, 1923, by a conference of 68 patriotic organizations in Washington, D.C. Finally, on June 22, 1942, a joint resolution of Congress, amended by Public Law 94-344, July 7, 1976, codified "existing rules and customs pertaining to the display and use of the flag . . ."

When to Display the Flag—The flag should be displayed on all days, especially on legal holidays and other special occasions, on official buildings when in use, in or near polling places on election days, and in or near schools when in session. Citizens may fly the flag at any time. It is customary to display it only from sunrise to sunset on buildings and on stationary flagstaffs in the open. It may be displayed at night, however, on special occasions, preferably lighted. The flag now flies over the White House both day and night. It flies over the Senate wing of the Capitol when the Senate is in session and over the House wing when that body is in session. It flies day and night over the east and west fronts of the Capitol, without floodlights at night but receiving illumination from the Capitol Dome. It flies 24 hours a day at several other places, including the Fort McHenry National Monument in Baltimore, where it inspired Francis Scott Key to write "The Star Spangled Banner." The flag also flies 24 hours a day, properly illuminated, at U.S. Customs ports of entry.

Flying the Flag at Half-Staff—Flying the flag at half-staff, that is, halfway up the staff, is a signal of mourning. The flag should be hoisted to the top of the staff for an instant before being lowered to half-staff. It should be hoisted to the peak again before being lowered for the day or night.

As provided by presidential proclamation, the flag should fly at half-staff for 30 days from the day of death of a president or former president; for 10 days from the day of death of a vice president, chief justice or retired chief justice of the U.S., or speaker of the House of Representatives; from day of death until burial of an associate justice of the Supreme Court, cabinet member, former vice president, Senate president pro tempore, or majority or minority Senate or House leader; for a U.S. senator, representative, territorial delegate, or the resident commissioner of Puerto Rico, on day of death and the following day within the metropolitan area of the District of Columbia and from day of death until burial within the decedent's state, congressional district, territory or commonwealth; and for the death of the governor of a state, territory, or possession of the U.S., from day of death until burial.

On Memorial Day, the flag should fly at half-staff until noon and then be raised to the peak. The flag should also fly at half-staff on Korean War Veterans Armistice Day (July 27), National Pearl Harbor Remembrance Day (Dec. 7), and Peace Officers Memorial Day (May 15).

How to Fly the Flag—The flag should be hoisted briskly and lowered ceremoniously and should never be allowed to touch the ground or the floor. When the flag is hung over a sidewalk from a rope extending from a building to a pole, the union should be away from the building. When the flag is hung over the center of a street the union should be to the north in an east-west street and to the east in a north-south street. No other flag may be flown above or, if on the same level, to the right of the U.S. flag, except that at the United Nations Headquarters the UN flag may be placed above flags of all member nations and other national flags may be flown with equal prominence or honor with the flag of the U.S. At services by Navy chaplains at sea, the church pennant may be flown above the flag.

When 2 flags are placed against a wall with crossed staffs, the U.S. flag should be at right—its own right, and its staff should be in front of the staff of the other flag; when a number of flags are grouped and displayed from staffs, it should be at the center and highest point of the group.

Church and Platform Use—In an auditorium, the flag may be displayed flat, above and behind the speaker. When displayed from a staff in a church or in a public auditorium, the flag should hold the position of superior prominence, in advance of the audience, and in the position of honor at the speaker's right as she or he faces the audience. Any other flag so displayed should be placed on the left of the speaker or to the right of the audience.

When the flag is displayed horizontally or vertically against a wall, the stars should be uppermost and at the observer's left.

When used to cover a casket, the flag should be placed so that the union is at the head and over the left shoulder. It should not be lowered into the grave nor touch the ground.

How to Dispose of Worn Flags—When the flag is in such condition that it is no longer a fitting emblem for display, it should be destroyed in a dignified way, preferably by burning.

When to Salute the Flag—All persons present should face the flag, stand at attention, and salute on the following occasions: (1) when the flag is passing in a parade or in a review, (2) during the ceremony of hoisting or lowering, (3) when the national anthem is played, and (4) during the Pledge of Allegiance. Those present in uniform should render the military salute. Those not in uniform should place the right hand over the heart. A man wearing a hat should remove it with his right hand and hold it to his left shoulder during the salute.

Prohibited Uses of the Flag—The flag should not be dipped to any person or thing. (An exception—customarily, ships salute by dipping their colors.) It should never be displayed with the union down save as a distress signal. It should never be carried flat or horizontally, but always aloft and free.

It should not be displayed on a float, an automobile, or a boat except from a staff. It should never be used as a covering for a ceiling, nor have placed on it any word, design, or drawing. It should never be used as a receptacle for carrying anything. It should not be used to cover a statue or a monument.

The flag should never be used for advertising purposes, nor be embroidered on such articles as cushions or handkerchiefs, printed or otherwise impressed on boxes or anything that is designed for temporary use and discard; or used as a costume or athletic uniform. Advertising signs should not be fastened to its staff or halyard.

The flag should never be used as drapery of any sort, never festooned, drawn back, nor up, in folds, but always allowed to fall free. Bunting of blue, white, and red, always arranged with the blue above and the white in the middle, should be used for covering a speaker's desk, draping the front of a platform, and for decoration in general.

An act of Congress approved on Feb. 8, 1917, provided certain penalties for the desecration, mutilation, or improper use of the flag within the District of Columbia. A 1968 federal law provided penalties of as much as a year's imprisonment or a $1,000 fine or both for publicly burning or otherwise desecrating any U.S. flag. In addition, many states have laws against flag desecration. In 1989, the Supreme Court ruled that no laws could prohibit political protesters from burning the flag. The decision had the effect of declaring unconstitutional the flag desecration laws of 48 states, as well as a similar federal statute, in cases of peaceful political expression.

The Supreme Court, in June 1990, declared that a new federal law making it a crime to burn or deface the American flag violated the free-speech guarantee of the First Amendment. The 5-4 decision led to renewed calls in Congress for a constitutional amendment to make it possible to prosecute flag burners.

Pledge of Allegiance to the Flag

I pledge allegiance to the flag of the United States of America and to the republic for which it stands, one nation under God, indivisible, with liberty and justice for all.

This, the current official version of the Pledge of Allegiance, has developed from the original pledge, which was first published in the Sept. 8, 1892, issue of *Youth's Companion*, a weekly magazine then published in Boston. The original pledge contained the phrase "my flag," which was changed more than 30 years later to "flag of the United States of America." A 1954 act of Congress added the words "under God."

The authorship of the pledge had been in dispute for many years. The *Youth's Companion* stated in 1917 that the original draft was written by James B. Upham, an executive of the magazine who died in 1910. A leaflet circulated by the magazine later named Upham as the originator of the draft "afterwards condensed and perfected by him and his associates of the Companion force."

Francis Bellamy, a former member of *Youth's Companion* editorial staff, publicly claimed authorship of the pledge in 1923. In 1939, the United States Flag Association, acting on the advice of a committee named to study the controversy, upheld the claim of Bellamy, who had died 8 years earlier. In 1957 the Library of Congress issued a report attributing the authorship to Bellamy.

The History of the National Anthem

"The Star-Spangled Banner" was ordered played by the military and naval services by Pres. Woodrow Wilson in 1916. It was designated the national anthem by Act of Congress, Mar. 3, 1931. The words were written by Francis Scott Key, of Georgetown, MD, during the bombardment of Fort McHenry, Baltimore, Sept. 13-14, 1814. Key was a lawyer, a graduate of St. John's College, Annapolis, and a volunteer in a light artillery company. When a friend, Dr. Beanes, a Maryland physician, was taken aboard Admiral Cockburn's British squadron for interfering with ground troops, Key and J. S. Skinner, carrying a note from Pres. Madison, went to the fleet under a flag of truce on a cartel ship to ask Beanes's release. Cockburn consented, but as the fleet was about to sail up the Patapsco to bombard Fort McHenry, he detained them, first on HMS *Surprise* and then on a supply ship.

Key witnessed the bombardment from his own vessel. It began at 7 AM, Sept. 13, 1814, and lasted, with intermissions, for 25 hr. The British fired more than 1,500 shells, each weighing as much as 220 lb. They were unable to approach closely because the U.S. had sunk 22 vessels. Only 4 Americans were killed and 24 wounded. A British bomb-ship was disabled.

During the event, Key wrote a stanza on the back of an envelope. Next day at Indian Queen Inn, Baltimore, he wrote out the poem and gave it to his brother-in-law, Judge J. H. Nicholson. Nicholson suggested use of the tune, "Anacreon in Heaven" (attributed to a British composer named John Stafford Smith), and had the poem printed on broadsides, of which 2 survive. On Sept. 20 it appeared in the *Baltimore American*. Later Key made 3 copies; one is in the Library of Congress, and one in the Pennsylvania Historical Society. The copy Key wrote on Sept. 14 remained in the Nicholson family for 93 years. In 1907 it was sold to Henry Walters of Baltimore. In 1934 it was bought at auction by the Walters Art Gallery, Baltimore, for $26,400. In 1953 it was sold to the Maryland Historical Society for the same price.

The flag that Key saw during the bombardment is preserved in the Smithsonian Institution, Washington, DC. It is 30 by 42 ft and has 15 alternating red and white stripes and 15 stars, for the original 13 states plus Kentucky and Vermont. It was made by Mary Young Pickersgill. The Baltimore Flag House, a museum, occupies her premises, which were restored in 1953.

The Star-Spangled Banner

I

Oh, say can you see by the dawn's early light
What so proudly we hailed at the twilight's last gleaming?
Whose broad stripes and bright stars thru the perilous fight,
O'er the ramparts we watched were so gallantly streaming?
And the rocket's red glare, the bombs bursting in air,
Gave proof through the night that our flag was still there.
Oh, say does that star-spangled banner yet wave
O'er the land of the free and the home of the brave?

II

On the shore, dimly seen through the mists of the deep,
Where the foe's haughty host in dread silence reposes,
What is that which the breeze, o'er the towering steep,
As it fitfully blows, half conceals, half discloses?
Now it catches the gleam of the morning's first beam,
In full glory reflected now shines in the stream:
'Tis the star-spangled banner! Oh long may it wave
O'er the land of the free and the home of the brave!

III

And where is that band who so vauntingly swore
That the havoc of war and the battle's confusion,
A home and a country should leave us no more!
Their blood has washed out their foul footsteps' pollution.
No refuge could save the hireling and slave
From the terror of flight, or the gloom of the grave:
And the star-spangled banner in triumph doth wave
O'er the land of the free and the home of the brave!

IV

Oh! thus be it ever, when freemen shall stand
Between their loved home and the war's desolation!
Blest with victory and peace, may the heav'n rescued land
Praise the Power that hath made and preserved us a nation.
Then conquer we must, when our cause it is just,
And this be our motto: "In God is our trust."
And the star-spangled banner in triumph shall wave
O'er the land of the free and the home of the brave!

America (My Country 'Tis of Thee)

First sung in public on July 4, 1831, at a service in the Park Street Church, Boston, the words were written by Rev. Samuel Francis Smith, a Baptist clergyman, who set them to a melody he found in a German songbook, unaware that it was the tune for the British anthem, "God Save the King/Queen."

My country, 'tis of thee,
Sweet land of liberty, Of thee I sing.
Land where my fathers died!
Land of the Pilgrims' pride!
From ev'ry mountainside,
Let freedom ring!

My native country, thee,
Land of the noble free,
Thy name I love.
I love thy rocks and rills,
Thy woods and templed hills;
My heart with rapture thrills
Like that above.

Let music swell the breeze,
And ring from all the trees
Sweet freedom's song.
Let mortal tongues awake;
Let all that breathe partake;
Let rocks their silence break,
The sound prolong.

Our fathers' God, to Thee,
Author of liberty,
To Thee we sing.
Long may our land be bright
With freedom's holy light;
Protect us by Thy might,
Great God, our King!

America, the Beautiful

Words composed by Katharine Lee Bates, a Massachusetts educator and author, in 1893, inspired by the view she experienced atop Pikes Peak. The final form was established in 1911, and it is set to the music of Samuel A. Ward's "Materna."

O beautiful for spacious skies,
For amber waves of grain,
For purple mountain majesties
Above the fruited plain.
America! America! God shed His grace on thee,
And crown thy good with brotherhood
From sea to shining sea.

O beautiful for pilgrim feet
Whose stern impassion'd stress
A thorough-fare for freedom beat
Across the wilderness.
America! America!
God mend thine ev'ry flaw,
Confirm thy soul in self control,
Thy liberty in law.

O beautiful for heroes prov'd
In liberating strife,
Who more than self their country lov'd
And mercy more than life.
America! America!
May God thy gold refine
ill all success be nobleness,
And ev'ry gain divine.

O beautiful for patriot dream
That sees beyond the years,
Thine alabaster cities gleam,
Undimmed by human tears.
America! America!
God shed His grace on thee,
And crown thy good with brotherhood
From sea to shining sea.

The Liberty Bell: Its History and Significance

The Liberty Bell, in Independence National Historical Park, Philadelphia, is an object of great reverence to Americans because of its association with the historic events of the American Revolution.

The original Province bell was ordered by Assembly Speaker and Chairman of the State House Superintendents Isaac Norris and was ordered from Thomas Lester, Whitechapel Foundry, London. It reached Philadelphia at the end of August 1752. It bore an inscription from Leviticus 25:10: "PROCLAIM LIBERTY THROUGHOUT ALL THE LAND UNTO ALL THE INHABITANTS THEREOF."

The bell was cracked by a stroke of its clapper in Sept. 1752 while it hung on a truss in the State House yard for testing. Pass & Stow, Philadelphia founders, recast the bell, adding $1\frac{1}{2}$ ounces of copper to a pound of the original "Whitechapel" metal to reduce its high tone and brittleness. It was found that the bell contained too much copper, injuring its tone, so Pass & Stow recast it again, this time successfully.

In June 1753 the bell was hung in the old wooden steeple of the State House. In use while the Continental Congress was in session in the State House, it rang out in defiance of British tax and trade restrictions, and it proclaimed the Boston Tea Party and the first public reading of the Declaration of Independence.

On Sept. 18, 1777, when the British Army was about to occupy Philadelphia, the Liberty Bell was moved in a baggage train of the American Army to Allentown, PA, where it was hidden in the Zion Reformed Church until June 27, 1778. The bell was moved back to Philadelphia after the British left the city.

In July 1781 the wooden steeple became insecure and had to be taken down. The bell was lowered into the brick section of the tower, where it remained until 1828. Between 1828 and 1844 the old State House bell continued to ring during special occasions. It rang for the last time on Feb. 23, 1846. In 1852 it was placed on exhibition in the Declaration Chamber of Independence Hall.

In 1876, when many thousands of Americans visited Philadelphia for the Centennial Exposition, the bell was placed in its old wooden support in the tower hallway. In 1877 it was hung from the ceiling of the tower by a chain of 13 links. It was returned again to the Declaration Chamber and in 1896 taken back to the tower hall, where it occupied a glass case. In 1915 the case was removed so that the public might touch it. On Jan. 1, 1976, just after midnight to mark the opening of the Bicentennial Year, the bell was moved to a new glass and steel pavilion behind Independence Hall for easier viewing.

The measurements of the bell are: circumference around the lip, 12 ft $1/2$ in; circumference around the crown, 6 ft $11 1/4$ in; lip to the crown, 3 ft; height over the crown, 2 ft 3 in; thickness at lip, 3 in; thickness at crown, $1 1/4$ in; weight, 2,080 lb; length of clapper, 3 ft 2 in.

The specific source of the crack in the bell is unknown.

Statue of Liberty National Monument

Since 1886, the Statue of Liberty Enlightening the World has stood as a symbol of freedom in New York harbor. It also commemorates French-American friendship, for it was given by the people of France and designed by French sculptor Frederic Auguste Bartholdi (1834-1904).

Edouard de Laboulaye, French historian, suggested the French present a monument to the U.S., the latter to provide pedestal and site. Bartholdi visualized a colossal statue at the entrance of New York harbor, welcoming the peoples of the world with the torch of liberty.

On Washington's Birthday, Feb. 22, 1877, Congress approved the use of a site on Bedloe's Island suggested by Bartholdi. This island of 12 acres had been owned in the 17th century by a Walloon named Isaac Bedloe. It was called Bedloe's until Aug. 3, 1956, when Pres. Eisenhower approved a resolution of Congress changing the name to Liberty Island.

The statue was finished on May 21, 1884, and formally presented to the U.S. minister to France, Levi Parsons Morton, July 4, 1884, by Ferdinand de Lesseps, head of the Franco-American Union, promoter of the Panama Canal, and builder of the Suez Canal.

On Aug. 5, 1884, the Americans laid the cornerstone for the pedestal. This was to be built on the foundations of Fort Wood, which had been erected by the government in 1811. The American committee had raised $125,000, but this was found to be inadequate. Joseph Pulitzer, owner of the *New York World*, appealed on Mar. 16, 1885, for general donations. By Aug. 11, 1885, he had raised $100,000.

The statue arrived dismantled, in 214 packing cases, from Rouen, France, in June 1885. The last rivet of the statue was driven on Oct. 28, 1886, when Pres. Grover Cleveland dedicated the monument.

The statue weighs 450,000 lb, or 225 tons. The copper sheeting weighs 200,000 lb. There are 167 steps from the land level to the top of the pedestal, 168 steps inside the statue to the head, and 54 rungs on the ladder leading to the arm that holds the torch.

A $2.5 million building housing the American Museum of Immigration was opened by Pres. Richard Nixon on Sept. 26, 1972, at the base of the statue. It houses a permanent exhibition of photos, posters, and artifacts tracing the history of American immigration. The Statue of Liberty National Monument is administered by the National Park Service.

Two years of restoration work was completed before the statue's centennial celebration on July 4, 1986. Among other repairs, the multimillion dollar project included replacing the 1,600 wrought iron bands that hold the statue's copper skin to its frame, replacing its torch, and installing an elevator.

A 4-day extravaganza of concerts, tall ships, ethnic festivals, and fireworks, July 3-6, 1986, celebrated the 100th anniversary. The festivities included Chief Justice Warren E. Burger's swearing-in of 5,000 new citizens on Ellis Island, while 20,000 others across the country were simultaneously sworn in through a satellite telecast.

The ceremonies were followed by others on Oct. 28, 1986, to mark the statue's exact 100th birthday.

Dimensions of the Statue	Ft.	In.
Height from base to torch (45.3 meters)	151	1
Foundation of pedestal to torch (91.5 meters)	305	1
Heel to top of head	111	1
Length of hand	16	5
Index finger	8	0
Size of finger nail, 13x10 in.		
Head from chin to cranium	17	3
Head thickness from ear to ear	10	0
Length of nose	4	6
Right arm, length	42	0
Right arm, greatest thickness	12	0
Thickness of waist	35	0
Width of mouth	3	0
Tablet, length	23	7
Tablet, width	13	7
Tablet, thickness	2	0

Emma Lazarus's Famous Poem

Engraved on pedestal below the statute.

The New Colossus

Not like the brazen giant of Greek fame,
With conquering limbs astride from land to land;
Here at our sea-washed, sunset gates shall stand
A mighty woman with a torch, whose flame
Is the imprisoned lightning, and her name
Mother of Exiles. From her beacon-hand
Glows world-wide welcome; her mild eyes command
The air-bridged harbor that twin cities frame.
"Keep ancient lands, your storied pomp!" cries she
With silent lips. "Give me your tired, your poor,
Your huddled masses yearning to breathe free,
The wretched refuse of your teeming shore.
Send these, the homeless, tempest-tost to me,
I lift my lamp beside the golden door!"

Ellis Island

Ellis Island was the gateway to America for more than 12 million immigrants between 1892 and 1924. In the late 18th century, Samuel Ellis, a New York City merchant, purchased the island and gave it his name. From Ellis, it passed to New York State, and the U.S. government bought it in 1808. On Jan. 1, 1892 the government opened the first federal immigration center in the U.S. on the island. The $27 1/2$-acre site eventually supported more than 35 buildings, including the Main Building with its Great Hall, in which as many as 5,000 people a day were processed.

Closed as an immigration station in 1954, Ellis Island was proclaimed part of the Statue of Liberty National Monument in 1965 by Pres. Lyndon B. Johnson. After a 6-year $170 million restoration project funded by The Ellis Island Fdn. Inc., Ellis Island was reopened as a museum in 1990. Artifacts, historic photographs and documents, oral histories, and ethnic music depicting 400 years of American immigration are housed in the museum. The museum also includes The American Immigrant Wall of Honor (http://www.wallofhonor.com). With its new Millennium section to open in April 2000, it will hold more than 600,000 names. In 1998, the Supreme Court ruled that nearly 90% of the island (the 24.2 acres which are landfill) lies in New Jersey, while the original 3.3 acres are in New York.

BIOGRAPHIES OF U.S. PRESIDENTS

George Washington (1789-97)

George Washington, first president, Federalist, was born on Feb. 22, 1732, in Wakefield on Pope's Creek, Westmoreland Co., VA, the son of Augustine and Mary Ball Washington. He spent his early childhood on a farm near Fredericksburg. His father died when George was 11. He studied mathematics and surveying, and when he was 16, he went to live with his elder half brother, Lawrence, who built and named Mount Vernon. George surveyed the lands of Thomas Fairfax in the Shenandoah Valley, keeping a diary. He accompanied Lawrence to Barbados, West Indies, where he contracted smallpox and was deeply scarred. Lawrence died in 1752, and George inherited his property. He valued land, and when he died, he owned 70,000 acres in Virginia and 40,000 acres in what is now West Virginia.

Washington's military service began in 1753, when Lt. Gov. Robert Dinwiddie of Virginia sent him on missions deep into Ohio country. He clashed with the French and had to surrender Fort Necessity on July 3, 1754. He was an aide to the British general Edward Braddock and was at his side when the army was ambushed and defeated (July 9, 1755) on a march to Fort Duquesne. He helped take Fort Duquesne from the French in 1758.

After Washington's marriage to Martha Dandridge Custis, a widow, in 1759, he managed his family estate at Mount Vernon. Although not at first for independence, he opposed the repressive measures of the British crown and took charge of the Virginia troops before war broke out. He was made commander of the newly created Continental Army by the Continental Congress on June 15, 1775.

The American victory was due largely to Washington's leadership. He was resourceful, a stern disciplinarian, and the one strong, dependable force for unity. Washington favored a federal government. He became chairman of the Constitutional Convention of 1787 and helped get the Constitution ratified. Unanimously elected president by the Electoral College, he was inaugurated Apr. 30, 1789, on the balcony of New York's Federal Hall.

He was reelected in 1792. Washington made an effort to avoid partisan politics as president.

Refusing to consider a 3d term, he retired to Mount Vernon in March 1797. He suffered acute laryngitis after a ride in snow and rain around his estate, was bled profusely, and died Dec. 14, 1799.

John Adams (1797-1801)

John Adams, 2d president, Federalist, was born on Oct. 30, 1735, in Braintree (now Quincy), MA, the son of John and Susanna Boylston Adams. He was a great-grandson of Henry Adams, who came from England in 1636. He graduated from Harvard in 1755 and then taught school and studied law. He married Abigail Smith in 1764. In 1765 he argued against taxation without representation before the royal governor. In 1770 he successfully defended in court the British soldiers who fired on civilians in the Boston Massacre. He was a delegate to the Continental Congress and a signer of the Declaration of Independence. In 1778, Congress sent Adams and John Jay to join Benjamin Franklin as diplomatic representatives in Europe. Because he ran second to Washington in Electoral College balloting in February 1789, Adams became the nation's first vice president; he was reelected in 1792.

In 1796 Adams was chosen president by the electors. His administration was marked by rivalry with Alexander Hamilton and a crisis in U.S.-French relations. He was extraordinarily unpopular for securing passage of the Alien and Sedition Acts in 1798. His foreign policy contributed significantly to the election of Thomas Jefferson in 1800.

Adams lived for a quarter century after he left office, during which time he wrote extensively. He died July 4, 1826, on the same day as Thomas Jefferson (the 50th anniversary of the Declaration of Independence).

Thomas Jefferson (1801-9)

Thomas Jefferson, 3d president, Democratic-Republican, was born on Apr. 13, 1743, in Shadwell in Goochland (now Albemarle) Co., VA, the son of Peter and Jane Randolph Jefferson. Peter died when Thomas was 14, leaving him 2,750 acres and his slaves. Jefferson attended (1760-62) the College of William and Mary, read Greek and Latin classics, and played the violin. In 1769 he was elected to the Virginia House of Burgesses. In 1770 he began building his home, Monticello, and in 1772 he married Martha Wayles Skelton, a wealthy widow. Jefferson helped establish the Virginia Committee of Correspondence. As a member of the Second Continental Congress he drafted the Declaration of Independence in late June 1776. He also was a member of the Virginia House of Delegates (1776-79) and was first elected governor of Virginia in 1779, succeeding Patrick Henry. He was reelected governor in 1780 but resigned in June 1781 after British troops invaded Virginia. During his term he wrote the statute on religious freedom. After his wife's death in 1782, Jefferson again became a delegate to the Congress, and in 1784 he drafted the report that was the basis for the Ordinances of 1784, 1785, and 1787. He was minister to France from 1785 to 1789, when George Washington appointed him secretary of state.

Jefferson's strong faith in the consent of the governed conflicted with the emphasis on executive control, favored by Alexander Hamilton, secretary of the Treasury, and Jefferson resigned on Dec. 31, 1793. In the 1796 election Jefferson was the Democratic-Republican candidate for president; John Adams won the election, and Jefferson became vice president. In 1800, Jefferson and Aaron Burr received equal Electoral College votes. The House of Representatives elected Jefferson president. Major events of his first term were the Louisiana Purchase (1803) and the Lewis and Clark Expedition. An important development during his second term was passage of the Embargo Act, barring U.S. ships from setting sail to foreign ports. Jefferson established the University of Virginia and designed its buildings. He died July 4, 1826, on the same day as John Adams (the 50th anniversary of the Declaration of Independence).

James Madison (1809-17)

James Madison, 4th president, Democratic-Republican, was born on Mar. 16, 1751, in Port Conway, King George Co., VA, the son of James and Eleanor Rose Conway Madison. Madison graduated from Princeton in 1771. He served in the Virginia Constitutional Convention (1776), and, in 1780, became a delegate to the Second Continental Congress. He was chief recorder at the Constitutional Convention in 1787 and supported ratification in the *Federalist Papers*, written with Alexander Hamilton and John Jay. In 1789, Madison was elected to the House of Representatives, where he helped frame the Bill of Rights and fought against passage of the Alien and Sedition Acts. In the 1790s, he helped found the Democratic-Republican Party, which ultimately became the Democratic Party. He became Jefferson's secretary of state in 1801.

Madison was elected president in 1808. His first term was marked by tensions with Great Britain, and his conduct of foreign policy was criticized by the Federalists and by his own party. Nevertheless, he was reelected in 1812, the year war was declared on Great Britain. The war that many considered a second American revolution ended with a treaty that settled none of the issues. Madison's most important action after the war was demilitarizing the U.S.-Canadian border.

In 1817, Madison retired to his estate, Montpelier, where he served as an elder statesman, "the last of the fathers." He edited his famous papers on the Constitutional Convention and helped found the University of Virginia, of which he became rector in 1826. He died June 28, 1836.

James Monroe (1817-25)

James Monroe, 5th president, Democratic-Republican, was born on Apr. 28, 1758, in Westmoreland Co., VA, the son of Spence and Eliza Jones Monroe. He entered the College of William and Mary in 1774 but left to serve in the 3d Virginia Regiment during the American Revolution. After the war, he studied law with Thomas Jefferson. In 1782 he was elected to the Virginia House of Delegates, and he served (1783-86) as a delegate to the Confederation Congress. He opposed ratification of the Constitution because it lacked a bill of rights. Monroe was elected to the U.S. Senate in 1790. In 1794 President George Washington appointed Monroe minister to France. He served twice as governor of Virginia (1799-1802, 1811). President Jefferson also sent him to France as minister (1803), and from 1803 to 1807 he served as minister to Great Britain.

In 1816 Monroe was elected president; he was reelected in 1820 with all but one Electoral College vote. His administration became known as the Era of Good Feeling. He obtained Florida from Spain, settled boundary disputes with Britain over Canada, and eliminated border forts. He supported the antislavery position that led to the Missouri Compromise. His most significant contribution was the Monroe Doctrine, which opposed European intervention in the Western Hemisphere and became a cornerstone of U.S. foreign policy.

Although Monroe retired to Oak Hill, VA, financial problems forced him to sell his property and move to New York City. He died there on July 4, 1831.

John Quincy Adams (1825-29)

John Quincy Adams, 6th president, independent Federalist, later Democratic-Republican, was born on July 11, 1767, in Braintree (now Quincy), MA, the son of John and Abigail Adams. His father was the 2d president. He studied abroad and at Harvard University from which he graduated in 1787. In 1803, he was elected to the U.S. Senate. President Monroe chose him as his secretary of state in 1817. In this capacity he negotiated the cession of the Floridas from Spain, supported exclusion of slavery in the Missouri Compromise, and helped formulate the Monroe Doctrine.

In 1824 Adams was elected president by the House of Representatives after he failed to win an Electoral College majority. His expansion of executive powers was strongly opposed, and in the 1828 election he lost to Andrew Jackson. In 1831 he entered the House of Representatives and served 17 years with distinction. He opposed slavery, the annexation of Texas, and the Mexican War. He helped establish the Smithsonian Institution. He suffered a stroke in the House and died in the Speaker's Room on Feb. 23, 1848.

Andrew Jackson (1829-37)

Andrew Jackson, 7th president, Democratic-Republican, later a Democrat, was born on Mar. 15, 1767, in the Waxhaw district, on the border of North Carolina and South Carolina, the son of Andrew and Elizabeth Hutchinson Jackson. At the age of 13, he joined the militia to fight in the American Revolution and was captured. Orphaned at the age of 14, Jackson was brought up by a well-to-do uncle. By age 20, he was practicing law, and he later served as prosecuting attorney in Nashville, TN. In 1796 he helped draft the constitution of Tennessee, and for a year he occupied its one seat in the House of Representatives. The next year he served in the U.S. Senate.

In the War of 1812, Jackson crushed (1814) the Creek Indians at Horseshoe Bend, AL, and, with an army consisting chiefly of backwoodsmen, defeated (1815) General Edward Pakenham's British troops at the Battle of New Orleans. In 1818 he briefly invaded Spanish Florida to quell Seminoles and outlaws who harassed frontier settlements. In 1824 he ran for president against John Quincy Adams. Although he won the most popular and electoral votes, he did not have a majority. The House of Representatives decided the election and chose Adams. In the 1828 election, however, Jackson defeated Adams, carrying the West and the South.

As president, Jackson introduced what became known as the spoils system—rewarding party members with government posts. Perhaps his most controversial act, however, was depositing federal funds in so-called pet banks, those directed by Democratic bankers, rather than in the Bank of the United States. "Let the people rule" was his slogan. In 1832, Jackson killed the congressional caucus for nominating presidential candidates and substituted the national convention. When South Carolina refused to collect imports under his protective tariff, he ordered army and naval forces to Charleston. After leaving office in 1837, he retired to the Hermitage, outside Nashville, where he died on June 8, 1845.

Martin Van Buren (1837-41)

Martin Van Buren, 8th president, Democrat, was born on Dec. 5, 1782, in Kinderhook, NY, the son of Abraham and Maria Hoes Van Buren. After attending local schools, he studied law and became a lawyer at the age of 20. A consummate politician, Van Buren began his career in the New York state senate and then served as state attorney general from 1816 to 1819. He was elected to the U.S. Senate in 1821. He helped swing eastern support to Andrew Jackson in the 1828 election and then served as Jackson's secretary of state from 1829 to 1831. In 1832 he was elected vice president. Known as the Little Magician, Van Buren was extremely influential in Jackson's administration. In the election of 1836 he defeated William

Henry Harrison for president and took office as the financial panic of 1837 initiated a nationwide depression. Although he instituted the independent treasury system, his refusal to spend land revenues led to his defeat by William Henry Harrison in the election of 1840. In 1844 he lost the Democratic nomination to James Knox Polk. In 1848 he again ran for president on the Free Soil ticket but lost. He died in Kinderhook on July 24, 1862.

William Henry Harrison (1841)

William Henry Harrison, 9th president, Whig, who served only 31 days, was born on Feb. 9, 1773, in Berkeley, Charles City Co., VA, the son of Benjamin Harrison, a signer of the Declaration of Independence, and of Elizabeth Bassett Harrison. He attended Hampden-Sydney College. Harrison served as secretary of the Northwest Territory in 1798 and was its delegate to the House of Representatives in 1799. He was the first governor of the Indiana Territory and served as superintendent of Indian affairs. With 900 men he put down a Shawnee uprising at Tippecanoe, IN, on Nov. 7, 1811. A generation later, in 1840, he waged a rousing presidential campaign, using the slogan "Tippecanoe and Tyler too." The Tyler of the slogan was his running mate, John Tyler. Although born to one of the wealthiest, most prestigious, and most influential families in Virginia, Harrison was elected president with a "log cabin and hard cider" slogan. He caught pneumonia during the inauguration and died Apr. 4, 1841.

John Tyler (1841-45)

John Tyler, 10th president, independent Whig, was born on Mar. 29, 1790, in Greenway, Charles City Co., VA, the son of John and Mary Armistead Tyler. His father was governor of Virginia (1808-11). Tyler graduated from the College of William and Mary in 1807 and in 1811 was elected to the Virginia legislature. In 1816 he was chosen for the U.S. House of Representatives. He served in the Virginia legislature again from 1823 to 1825, when he was elected governor of Virginia. After a stint in the U.S. Senate (1827-36), he was elected vice president (1840). When William Henry Harrison died only a month after taking office, Tyler succeeded him. Because he was the first person to occupy the presidency without having been elected to that office, he was referred to as "His Accidency." Tyler gained passage of the Preemption Act of 1841, which gave squatters on government land the right to buy 160 acres at the minimum auction price. His last act as president was to sign the resolution annexing Texas. Tyler accepted renomination in 1844 from some Democrats but withdrew in favor of the official party candidate, James K. Polk. He died in Richmond, VA, on Jan. 18, 1862.

James Knox Polk (1845-49)

James Knox Polk, 11th president, Democrat, was born on Nov. 2, 1795, in Mecklenburg Co., NC, the son of Samuel and Jane Knox Polk. He graduated from the University of North Carolina in 1818 and served in the Tennessee state legislature from 1823 to 1825. He served in the U.S. House of Representatives from 1825 to 1839, the last 4 years as Speaker. He was governor of Tennessee from 1839 to 1841. In 1844, after the Democratic National Convention became deadlocked, it nominated Polk, who thus became the nation's first "dark horse" candidate for president. He was nominated primarily because he was known to favor annexation of Texas. As president, Polk reestablished the independent treasury system originated by Van Buren. He was so intent on acquiring California from Mexico that he sent troops under Zachary Taylor to the Mexican border and, when Mexicans attacked, declared that a state of war existed. The Mexican War ended with the annexation of California and much of the Southwest as part of America's "manifest destiny." Polk compromised on the Oregon boundary ("54-40 or fight!") by accepting the 49th parallel and yielding Vancouver Island to the British. A few weeks after leaving office, Polk died in Nashville, TN, on June 15, 1849.

Zachary Taylor (1849-50)

Zachary Taylor, 12th president, Whig, who served only 16 months, was born on Nov. 24, 1784, in Orange Co., VA, the son of Richard and Sarah Strother Taylor. He grew up on his father's plantation near Louisville, KY, where he was educated by private tutors. In 1808 Taylor joined the regular army and was commissioned first lieutenant. He fought in the War of

1812, the Black Hawk War (1832), and the second Seminole War (beginning in 1837). He was called "Old Rough and Ready." In 1846 President Polk sent him with an army to the Rio Grande. When the Mexicans attacked him, Polk declared war. Outnumbered 4-1, Taylor defeated (1847) Santa Anna at Buena Vista. A national hero, he received the Whig nomination in 1848 and was elected president, even though he had never bothered to vote. He resumed the spoils system and, though a slaveholder, worked to admit California as a free state. He fell ill and died in office on July 9, 1850.

Millard Fillmore (1850-53)

Millard Fillmore, 13th president, Whig, was born on Jan. 7, 1800, in Cayuga Co., NY, the son of Nathaniel and Phoebe Millard Fillmore. Although he had little schooling, he became a law clerk at the age of 22 and a year later was admitted to the bar. He was elected to the New York state assembly in 1828 and served until 1831. From 1833 until 1835 and again from 1837 to 1843, he represented his district in the U.S. House of Representatives. He opposed the entrance of Texas as a slave territory and voted for a protective tariff. In 1844 he was defeated for governor of New York. In 1848 he was elected vice president, and he succeeded as president after Taylor's death. Fillmore favored the Compromise of 1850 and signed the Fugitive Slave Law. His policies pleased neither expansionists nor slaveholders, and he was not renominated in 1852. In 1856 he was nominated by the American (Know-Nothing) Party, but despite the support of the Whigs, he was defeated by James Buchanan. He died in Buffalo, NY, on Mar. 8, 1874.

Franklin Pierce (1853-57)

Franklin Pierce, 14th president, Democrat, was born on Nov. 23, 1804, in Hillsboro, NH, the son of Benjamin Pierce, an American Revolutionary War general and governor of New Hampshire, and Anna Kendrick. He graduated from Bowdoin College in 1824 and was admitted to the bar in 1827. He was elected to the New Hampshire state legislature in 1829 and was chosen Speaker in 1831. He went to the U.S. House of Representatives in 1833 and was elected a U.S. senator in 1837. He enlisted in the Mexican War and became brigadier general under Gen. Winfield Scott. In 1852 Pierce was nominated as the Democratic presidential candidate on the 49th ballot. He decisively defeated Gen. Scott, his Whig opponent, in the election. Although against slavery, Pierce was influenced by pro-slavery Southerners. He supported the controversial Kansas-Nebraska Act, which left the question of slavery in the new territories of Kansas and Nebraska to popular vote. Pierce signed a reciprocity treaty with Canada and approved the Gadsden Purchase, a border area on a proposed railroad route, from Mexico. Denied renomination by the Democrats, he spent most of his remaining years in Concord, NH, where he died on Oct. 8, 1869.

James Buchanan (1857-61)

James Buchanan, 15th president, Federalist, later Democrat, was born on Apr. 23, 1791, near Mercersburg, PA, the son of James and Elizabeth Speer Buchanan. He graduated from Dickinson College in 1809 and was admitted to the bar in 1812. He fought in the War of 1812 as a volunteer. He was twice elected to the Pennsylvania general assembly, and in 1821 he entered the U.S. House of Representatives. After briefly serving (1832-33) as minister to Russia, he was elected U.S. senator from Pennsylvania. As Polk's secretary of state (1845-49), he ended the Oregon dispute with Britain and supported the Mexican War and annexation of Texas. As minister to Great Britain, he signed the Ostend Manifesto (1854), declaring a U.S. right to take Cuba by force should efforts to purchase it fail. Nominated by Democrats, Buchanan was elected president in 1856. On slavery he favored popular sovereignty and choice by state constitutions but did not consistently uphold this position. He denied the right of states to secede but opposed coercion and attempted to keep peace by not provoking secessionists. Buchanan left office having failed to deal decisively with the situation. He died at Wheatland, his estate, near Lancaster, PA, on June 1, 1868.

Abraham Lincoln (1861-65)

Abraham Lincoln, 16th president, Republican, was born on Feb. 12, 1809, in a log cabin on a farm then in Hardin Co., KY, now in Larue, the son of Thomas and Nancy Hanks Lincoln. The Lincolns moved to Spencer Co., IN, near Gentryville, when Abe was 7. After Abe's mother died, his father married (1819) Mrs. Sarah Bush Johnston. In 1830 the family moved to Macon Co., IL.

Defeated in 1832 in a race for the state legislature, Lincoln was elected on the Whig ticket 2 years later and served in the lower house from 1834 to 1842. In 1837 Lincoln was admitted to the bar and became partner in a Springfield, IL, law office. He soon won recognition as an effective and resourceful attorney. In 1846, he was elected to the House of Representatives, where he attracted attention during a single term for his opposition to the Mexican War and his position on slavery. In 1856 he campaigned for the newly founded Republican Party, and in 1858 he became its senatorial candidate against Stephen A. Douglas. Although he lost the election, Lincoln gained national recognition from his debates with Douglas.

In 1860, Lincoln was nominated for president by the Republican Party on a platform of restricting slavery. He ran against Douglas, a northern Democrat; John C. Breckinridge, a Southern proslavery Democrat; and John Bell, of the Constitutional Union Party. As a result of Lincoln's winning the election, South Carolina seceded from the Union on Dec. 20, 1860, followed in 1861 by 10 other Southern states.

The Civil War erupted when Fort Sumter, which Lincoln decided to resupply, was attacked by Confederate forces on Apr. 12, 1861. Lincoln called successfully for recruits from the North. On Sept. 22, 1862, 5 days after the Battle of Antietam, Lincoln announced that slaves in territory then in rebellion would be free Jan. 1, 1863, the date of the Emancipation Proclamation. His speeches, including his Gettysburg and Inaugural addresses, are remembered for their eloquence.

Lincoln was reelected, in 1864, over Gen. George B. McClellan, Democrat. Lee surrendered on Apr. 9, 1865. On Apr. 14, Lincoln was shot by actor John Wilkes Booth in Ford's Theater, in Washington, DC. He died the next day.

Andrew Johnson (1865-69)

Andrew Johnson, 17th president, Democrat, was born on Dec. 29, 1808, in Raleigh, NC, the son of Jacob and Mary McDonough Johnson. He was apprenticed to a tailor as a youth, but ran away after two years and eventually settled in Greenville, TN. He became popular with the townspeople and in 1829 was elected councilman and later mayor. In 1835 he was sent to the state general assembly. In 1843 he was elected to the U.S. House of Representatives, where he served for 10 years. Johnson was governor of Tennessee from 1853 to 1857, when he was elected to the U.S. Senate. He supported John C. Breckinridge against Lincoln in the 1860 election. Although Johnson had held slaves, he opposed secession and tried to prevent Tennessee from seceding. In Mar. 1862, Lincoln appointed him military governor of occupied Tennessee.

In 1864, in order to balance Lincoln's ticket with a Southern Democrat, the Republicans nominated Johnson for vice president. He was elected vice president with Lincoln and then succeeded to the presidency upon Lincoln's death. Soon afterward, in a controversy with Congress over the president's power over the South, he proclaimed an amnesty to all Confederates, except certain leaders, if they would ratify the 13th Amendment abolishing slavery. States doing so added anti-Negro provisions that enraged Congress, which restored military control over the South. When Johnson removed Edwin M. Stanton, secretary of war, without notifying the Senate, the House, in Feb. 1868, impeached him. Ostensibly charging him with thereby having violated the Tenure of Office Act, the House was actually responding to his opposition to harsh congressional Reconstruction, expressed in repeated vetoes. He was tried by the Senate, and in May, in two separate votes on different counts, was acquitted, both times by only one vote. Johnson was denied renomination but remained politically active. He was re-elected to the Senate in 1874. Johnson died July 31, 1875, at Carter Station, TN.

Ulysses Simpson Grant (1869-77)

Ulysses S. Grant, 18th president, Republican, was born on Apr. 27, 1822, in Point Pleasant, OH, the son of Jesse R. and Hannah Simpson Grant. The next year the family moved to Georgetown, OH. Grant was named Hiram Ulysses, but on entering West Point in 1839, his name was put down as Ulysses Simpson, and he adopted it. He graduated in 1843. During the Mexican War, Grant served under both Gen. Zachary Taylor and Gen. Winfield Scott. In 1854, he resigned his commission because of loneliness and drinking problems, and in the following years he engaged in generally unsuccessful farming and business ventures. With the start of the Civil War, he was named colonel and then brigadier general of the Illinois Volunteers. He took Forts Henry and Donelson and fought at Shiloh. His brilliant campaign against Vicksburg and his victory at Chattanooga made him so prominent that Lincoln placed him in command of all Union armies. Grant accepted Lee's surrender at Appomattox Court House on Apr. 9, 1865. President Johnson appointed Grant secretary of war when he suspended Stanton, but Grant was not confirmed. He was nominated for president by the Republicans in 1868 and elected over Horatio Seymour, Democrat. The 15th Amendment, amnesty bill, and the peaceful settlement of disputes with Great Britain were events of his administration. The Liberal Republicans and Democrats opposed him with Horace Greeley in the 1872 election, but Grant was reelected. His second administration was marked by many scandals, including widespread corruption in the Treasury Department and the Indian Service. An attempt by the Stalwarts (Old Guard Republicans) to nominate him in 1880 failed. In 1884 the collapse of Grant & Ward, an investment firm in which he was a partner, left him penniless. He wrote his personal memoirs while ill with cancer and completed them shortly before his death at Mt. McGregor, NY, on July 23, 1885.

Rutherford Birchard Hayes (1877-81)

Rutherford B. Hayes, 19th president, Republican, was born on Oct. 4, 1822, in Delaware, OH, the son of Rutherford and Sophia Birchard Hayes. He was reared by his uncle, Sardis Birchard. Hayes graduated from Kenyon College in 1842 and from Harvard Law School in 1845. He practiced law in Lower Sandusky (now Fremont), OH, and was city solicitor of Cincinnati from 1858 to 1861. During the Civil War, he was major of the 23d Ohio Volunteers. He was wounded several times, and by the end of the war he had risen to the rank of brevet major general. While serving (1865-67) in the U.S. House of Representatives, Hayes supported Reconstruction and Johnson's impeachment. He was twice elected governor of Ohio (1867, 1869). After losing a race for the U.S. House in 1872, he was reelected governor of Ohio in 1875. In 1876 he was nominated for president and believed he had lost the election to Samuel J. Tilden, Democrat. But a few Southern states submitted 2 sets of electoral votes, and the result was in dispute. An electoral commission, appointed by Congress and consisting of 8 Republicans and 7 Democrats, awarded all disputed votes to Hayes, allowing him to become president by one electoral vote. Hayes, keeping a promise to southerners, withdrew troops from areas still occupied in the South, ending the era of Reconstruction. He proposed civil service reforms, alienating those favoring the spoils system, and advocated repeal of the Tenure of Office Act restricting presidential power to dismiss officials. He supported sound money and specie payments. Hayes died in Fremont, OH, on Jan. 17, 1893.

James Abram Garfield (1881)

James A. Garfield, 20th president, Republican, was born on Nov. 19, 1831, in Orange, Cuyahoga Co., OH, the son of Abram and Eliza Ballou Garfield. His father died in 1833, and he was reared in poverty by his mother. He worked as a canal bargeman, a farmer, and a carpenter and managed to secure a college education. He taught at Hiram College and later became principal. In 1859 he was elected to the Ohio legislature. Antislavery and antisecession, he volunteered for military service in the Civil War, becoming colonel of the 42d Ohio Infantry and brigadier in 1862. He fought at Shiloh, was chief of staff for Gen. William Starke Rosecrans, and was made major general for gallantry at Chickamauga. He entered Congress as a radical Republican in 1863, calling for execution or exile of Confederate leaders, but he moderated his views after the Civil War. On the electoral commission in 1877 he voted for Hayes against Tilden on strict party lines. He was a senator-elect in 1880 when he became the Republican nominee for president. He was chosen as a compromise over Gen. Grant, James G. Blaine, and John Sherman, and won election despite some bitterness among Grant's supporters. On July 2, 1881, Garfield was shot and seriously wounded by a mentally disturbed office-seeker, Charles J. Guiteau, while entering a railroad station in Washington, DC. He died on Sept. 19, 1881, in Elberon, NJ.

Chester Alan Arthur (1881-85)

Chester A. Arthur, 21st president, Republican, was born on Oct. 5, 1829, in Fairfield, VT, the son of William and Malvina Stone Arthur. He graduated from Union College in 1848, taught school in Vermont, then studied law and opened a practice in New York City. In 1853 he argued in a fugitive slave case that slaves transported through New York state were thereby freed. In 1871, he was appointed to the lucrative post of collector of the Port of New York. President Hayes, an opponent of the spoils system, forced Arthur to resign in 1878. This made the New York machine strong enemies of Hayes. Arthur and the Stalwarts (Old Guard Republicans) tried to nominate Grant for a 3d term in 1880. When Garfield was nominated, Arthur was nominated for vice president in the interests of harmony. Upon Garfield's assassination, Arthur became president. Despite his past connections, he signed civil service reform legislation. Arthur tried to dissuade Congress from enacting the high protective tariff of 1883. He was defeated for renomination in 1884 by James G. Blaine. He died in New York City on Nov. 18, 1886.

Grover Cleveland (1885-89; 1893-97)

(According to a ruling of the State Dept., Grover Cleveland should be counted as both the 22d and the 24th president, because his 2 terms were not consecutive.)

Grover Cleveland, Democrat, was born Stephen Grover Cleveland on Mar. 18, 1837, in Caldwell, NJ, the son of Richard F. and Ann Neal Cleveland. When he was a small boy, his family moved to New York. Prevented by his father's death from attending college, he studied by himself and was admitted to the bar in Buffalo, NY, in 1859. In succession he became assistant district attorney (1863), sheriff (1871), mayor (1881), and governor of New York (1882). He was an independent, honest administrator who hated corruption. He was nominated for president over Tammany Hall opposition in 1884 and defeated Republican James G. Blaine. As president, he enlarged the civil service and vetoed many pension raids on the Treasury. In the 1888 election he was defeated by Benjamin Harrison, although his popular vote was larger. Reelected over Harrison in 1892, he faced a money crisis brought about by a lowered gold reserve, circulation of paper, and exorbitant silver purchases under the Sherman Silver Purchase Act. He obtained a repeal of the Sherman Act, but was unable to secure effective tariff reform. A severe economic depression and labor troubles racked his administration, but he refused to interfere in business matters and rejected Jacob Coxey's demand for unemployment relief. In 1894, he broke the Pullman strike. In 1896, the Democrats repudiated his administration and chose silverite William Jennings Bryan as their candidate. Cleveland died in Princeton, NJ, on June 24, 1908.

Benjamin Harrison (1889-93)

Benjamin Harrison, 23d president, Republican, was born on Aug. 20, 1833, in North Bend, OH, the son of John Scott and Elizabeth Irwin Harrison. His great-grandfather, Benjamin Harrison, was a signer of the Declaration of Independence; his grandfather, William Henry Harrison, was 9th president; his father was a member of Congress. He attended school on his father's farm and graduated from Miami University in Oxford, OH, in 1852. He was admitted to the bar in 1854 and practiced in Indianapolis. During the Civil War, he rose to the rank of brevet brigadier general and fought at Kennesaw Mountain, at Peachtree Creek, at Nashville, and in the Atlanta campaign. He lost the 1876 gubernatorial election in Indiana but succeeded in becoming a U.S. senator in 1881. In 1888 he defeated Cleveland for president despite receiving fewer popular votes. As president, he expanded the pension list and signed the McKinley high tariff bill, the Sherman Antitrust Act, and the Sherman Silver Purchase Act. During his administration, 6 states were admitted to the Union. He was defeated for reelection in 1892. He died in Indianapolis on Mar. 13, 1901.

William McKinley (1897-1901)

William McKinley, 25th president, Republican, was born on Jan. 29, 1843, in Niles, OH, the son of William and Nancy Allison McKinley. McKinley briefly attended Allegheny College. When the Civil War broke out in 1861, he enlisted and served for the duration. He rose to captain and in 1865 was made brevet major. After studying law in Albany, NY, he opened (1867) a law office in Canton, OH. He served twice in the U.S. House of Representatives (1877-83; 1885-91) and led the fight there for the McKinley Tariff, which was passed in 1890. However, he was not reelected to the House as a result. He served two terms (1892-96) as governor of Ohio. In 1896 he was elected president as a proponent of a protective tariff and sound money (gold standard), over William Jennings Bryan, the Democrat and a proponent of free silver. McKinley was reluctant to intervene in Cuba, but the loss of the battleship *Maine* at Havana crystallized opinion. He demanded Spain's withdrawal from Cuba; Spain made some concessions, but Congress announced a state of war as of Apr. 21, 1898. He was reelected in the 1900 campaign, defeating Bryan's anti-imperialist arguments with the promise of a "full dinner pail." McKinley was respected for his conciliatory nature and for his conservative stance on business issues. On Sept. 6, 1901, while welcoming citizens at the Pan-American Exposition, in Buffalo, NY, he was shot by Leon Czolgosz, an anarchist. He died Sept. 14.

Theodore Roosevelt (1901-9)

Theodore Roosevelt, 26th president, Republican, was born on Oct. 27, 1858, in New York City, the son of Theodore and Martha Bulloch Roosevelt. He was a 5th cousin of Franklin D. Roosevelt and an uncle of Eleanor Roosevelt. Roosevelt graduated from Harvard University in 1880. He attended Columbia Law School briefly but abandoned the study of law to enter politics. He was elected to the New York state assembly in 1881 and served until 1884. He spent the next 2 years ranching and hunting in the Dakota Territory. Back in politics in 1886, he ran unsuccessfully for mayor of New York City. He was Civil Service commissioner in Washington, DC, from 1889 to 1895. From 1895 to 1897, he served as New York City's police commissioner. He was assistant secretary of the navy under McKinley. The Spanish-American War made Roosevelt a nationally known figure. He organized the 1st U.S. Volunteer Cavalry (Rough Riders) and, as lieutenant colonel, led the charge up Kettle Hill in San Juan. Elected New York governor in 1898, he fought the spoils system and achieved taxation of corporation franchises.

Nominated for vice president in 1900, he became the nation's youngest president when McKinley was assassinated. He was reelected in 1904. As president he fought corruption of politics by big business, dissolved the Northern Securities Co. and others for violating antitrust laws, intervened in the 1902 coal strike on behalf of the public, obtained the Elkins Law (1903) forbidding rebates to favored corporations, and helped pass the Hepburn Railway Rate Act of 1906 (extending Jurisdiction of the Interstate Commerce Commission). He helped obtain passage of the Pure Food and Drug Act (1906), and employers' liability laws. Roosevelt vigorously organized conservation efforts. He mediated (1905) the peace between Japan and Russia, for which he won the Nobel Peace Prize. He abetted the 1903 revolution in Panama that led to U.S. acquisition of territory for the Panama Canal.

In 1908 Roosevelt obtained the nomination of William H. Taft, who was elected. Feeling that Taft had abandoned his policies, Roosevelt unsuccessfully sought the nomination in 1912. He bolted the party and ran on the Progressive "Bull Moose" ticket against Taft and Woodrow Wilson, splitting the Republicans and ensuring Wilson's election. He was shot during the campaign but recovered. In 1916, after unsuccessfully seeking the presidential nomination for himself, Roosevelt supported the Republican candidate, Charles E. Hughes. A strong friend of Britain, he fought for American intervention in World War I. He wrote some 40 books on many topics; his book *The Winning of the West* is perhaps best known. He died Jan. 6, 1919, at Sagamore Hill, Oyster Bay, NY.

William Howard Taft (1909-13)

William Howard Taft, 27th president, Republican, and 10th chief justice of the U.S., was born on Sept. 15, 1857, in Cincinnati, OH, the son of Alphonso and Louisa Maria Torrey Taft. His father was secretary of war and attorney general in Grant's cabinet and minister to Austria and Russia under Arthur. Taft graduated from Yale in 1878 and from Cincinnati Law School in 1880. After working as a law reporter for Cincinnati newspapers, he served as assistant prosecuting attorney (1881-82), assistant county solicitor (1885), judge, superior court (1887), U.S. solicitor-general (1890), and federal circuit judge (1892). In 1900 he became head of the U.S. Philippines Commission and was the first civil governor of the Philippines (1901-4). In 1904 he served as secretary of war, and in 1906 he was sent to Cuba to help avert a threatened revolution. He was groomed for

the presidency by Theodore Roosevelt and elected over William Jennings Bryan in 1908. Taft vigorously continued Roosevelt's trust-busting, instituted the Department of Labor, and drafted the amendments calling for direct election of senators and the income tax. His tariff and conservation policies angered progressives. Although renominated in 1912, he was opposed by Roosevelt, who ran on the Progressive Party ticket; the result was Democrat Woodrow Wilson's election. Taft, with some reservations, supported the League of Nations. After leaving office, he was professor of constitutional law at Yale (1913-21) and chief justice of the U.S. (1921-30). Taft was the only person in U.S. history to have been both president and chief justice. Illness forced him to resign from the Court in Feb. 1930, and he died in Washington, DC, on Mar. 8, 1930.

Woodrow Wilson (1913-21)

Thomas Woodrow Wilson, 28th president, Democrat, was born on Dec. 28, 1856, in Staunton, VA, the son of Joseph Ruggles and Janet (Jessie) Woodrow Wilson. He grew up in Georgia and South Carolina. He attended Davidson College in North Carolina before graduating from Princeton University in 1879. He studied law at the University of Virginia and then studied political science at Johns Hopkins University, where he received his PhD in 1886. He taught at Bryn Mawr (1885-88) and then at Wesleyan (1888-90) before joining the faculty at Princeton. He was president of Princeton from 1902 until 1910, when he was elected governor of New Jersey. In 1912 he was nominated for president with the aid of William Jennings Bryan, who sought to block James "Champ" Clark and Tammany Hall. Wilson won the election because the Republican vote for Taft was split by the Progressives.

As president, Wilson protected American interests in revolutionary Mexico and fought for American rights on the high seas. He oversaw the creation of the Federal Reserve system, cut the tariff, and developed a reputation as a reformer. His sharp warnings to Germany led to the resignation of his secretary of state, Bryan, a pacifist. In 1916 he was reelected by a slim margin with the slogan, "He kept us out of war," although his attempts to mediate in the war failed. After several American ships had been sunk by the Germans, he secured a declaration of war against Germany on Apr. 6, 1917.

Wilson outlined his peace program on Jan. 8, 1918, in the Fourteen Points, a state paper that had worldwide influence. He enunciated a doctrine of self-determination for the settlement of territorial disputes. The Germans accepted his terms and an armistice on Nov. 11, 1918.

Wilson went to Paris to help negotiate the peace treaty, the crux of which he considered the League of Nations. The Senate demanded reservations that would not make the U.S. subordinate to the votes of other nations in case of war. Wilson refused to consider any reservations and toured the country to get support. He suffered a stroke in Oct. 1919. An invalid for months, he clung to his executive powers while his wife and doctors effectively functioned as president.

Wilson was awarded the 1919 Nobel Peace Prize, but the treaty embodying the League of Nations was ultimately rejected by the Senate in 1920. He left the White House in Mar. 1921. He died in Washington, DC, on Feb. 3, 1924.

Warren Gamaliel Harding (1921-23)

Warren Gamaliel Harding, 29th president, Republican, was born on Nov. 2, 1865, near Corsica (now Blooming Grove), OH, the son of George Tyron and Phoebe Elizabeth Dickerson Harding. He attended Ohio Central College, studied law, and became editor and publisher of a county newspaper. He entered the political arena as state senator (1901-4) and then served as lieutenant governor (1904-6). In 1910 he ran unsuccessfully for governor of Ohio; then in 1914 he was elected to the U.S. Senate. In the Senate he voted for antistrike legislation, woman suffrage, and the Volstead Prohibition Enforcement Act over President Wilson's veto. He opposed the League of Nations. In 1920 he was nominated for president and defeated James M. Cox in the election. The Republicans capitalized on war weariness and fear that Wilson's League of Nations would curtail U.S. sovereignty. Harding stressed a return to "normalcy" and worked for tariff revision and the repeal of excess profits law and high income taxes. His secretary of interior, Albert B. Fall, became involved in the Teapot Dome scandal. As rumors began to circulate about the corruption in his administration, Harding became ill while returning from a trip to Alaska, and he died in San Francisco on Aug. 2, 1923.

Calvin Coolidge (1923-29)

John Calvin Coolidge, 30th president, Republican, was born on July 4, 1872, in Plymouth, VT, the son of John Calvin and Victoria J. Moor Coolidge. Coolidge graduated from Amherst College in 1895. He entered Republican state politics and served as mayor of Northampton, MA, as state senator, as lieutenant governor, and, in 1919, as governor. In Sept. 1919, Coolidge attained national prominence by calling out the state guard in the Boston police strike. He declared: "There is no right to strike against the public safety by anybody, anywhere, anytime." This brought his name before the Republican convention of 1920, where he was nominated for vice president. He succeeded to the presidency on Harding's death. As president, he opposed the League of Nations and the soldiers' bonus bill, which was passed over his veto. In 1924 he was elected by a huge majority. He substantially reduced the national debt. He twice vetoed the McNary-Haugen farm bill, which would have provided relief to financially hard-pressed farmers. With Republicans eager to renominate him, Coolidge simply announced, Aug. 2, 1927: "I do not choose to run for president in 1928." He died in Northampton, MA, on Jan. 5, 1933.

Herbert Clark Hoover (1929-33)

Herbert Hoover, 31st president, Republican, was born on Aug. 10, 1874, in West Branch, IA, the son of Jesse Clark and Hulda Randall Minthorn Hoover. Hoover grew up in Indian Territory (now Oklahoma) and Oregon and graduated from Stanford University with a degree in engineering in 1895. He worked briefly with the U.S. Geological Survey and then managed mines in Australia, Asia, Europe, and Africa. While chief engineer of imperial mines in China, he directed food relief for victims of the Boxer Rebellion. He gained a reputation not only as an engineer but as a humanitarian as he directed the American Relief Committee, London (1914-15) and the U.S. Commission for Relief in Belgium (1915-19). He was U.S. Food Administrator (1917-19), American Relief Administrator (1918-23), and in charge of Russian Relief (1918-23). He served as secretary of commerce under both Harding and Coolidge. Some historians believe that he was the most effective secretary of commerce ever to hold that office.

In 1928 Hoover was elected president over Alfred E. Smith. In 1929 the stock market crashed, and the economy collapsed. During the depression, Hoover inaugurated some government assistance programs, but he was opposed to administration of aid through a federal bureaucracy. As the effects of the depression continued, he was defeated in the 1932 election by Franklin D. Roosevelt. President Truman named him coordinator of the European Food Program (1946) and chairman of the Commission on Organization of the Executive Branch (1947-49; 1953-55). Hoover died in New York City on Oct. 20, 1964.

Franklin Delano Roosevelt (1933-45)

Franklin D. Roosevelt, 32d president, Democrat, was born on Jan. 30, 1882, near Hyde Park, NY, the son of James and Sara Delano Roosevelt. He graduated from Harvard University in 1904. He attended Columbia University Law School without taking a degree and was admitted to the New York state bar in 1907. His political career began when he was elected to the New York state senate in 1910. In 1913 President Wilson appointed him assistant secretary of the navy, a post he held during World War I.

In 1920 Roosevelt ran for vice president with James Cox and was defeated. From 1921 to 1928 he worked in his New York law office and was also vice president of Fidelity & Deposit Co. of Maryland. In Aug. 1921, he was stricken with poliomyelitis, which left his legs paralyzed. As a result of therapy he was able to stand, or walk a few steps, with the aid of leg braces.

Roosevelt served 2 terms as governor of New York (1929-33). In 1932, W. G. McAdoo, pledged to John N. Garner, threw his votes to Roosevelt, who was nominated for president. The depression and the promise to repeal Prohibition ensured his election. He asked for emergency powers, proclaimed the New Deal, and put into effect a vast number of administrative changes. Foremost was the use of public funds for relief and public works, resulting in deficit financing. He greatly expanded the federal government's regulation of business and by an excess profits tax and progressive income taxes produced a redistribution of earnings on an unprecedented scale. The Wagner Act gave labor many advantages in organizing and collective bargaining. He promoted legislation establishing the Social Security system. He was the last president inaugurated on Mar. 4 (1933) and the first inaugurated on Jan. 20 (1937).

Roosevelt was the first president to use radio for "fireside chats." When the Supreme Court nullified some New Deal laws, he sought power to "pack" the Court with additional justices, but Congress refused to give him the authority. He was the first president to break the "no 3d term" tradition (1940) and was elected to a 4th term in 1944, despite failing health. Roosevelt was openly hostile to fascist governments before World War II and launched a lend-lease program on behalf of the Allies. With British Prime Min. Winston Churchill he wrote a declaration of principles to be followed after Nazi defeat (the Atlantic Charter of Aug. 14, 1941) and urged the Four Freedoms (freedom of speech, of worship, from want, from fear) Jan. 6, 1941. When Japan attacked Pearl Harbor on Dec. 7, 1941, the U.S. entered the war. Roosevelt conferred with allied heads of state at Casablanca (Jan. 1943), Quebec (Aug. 1943), Tehran (Nov.-Dec. 1943), Cairo (Nov. and Dec. 1943), and Yalta (Feb. 1945). He did not, however, see the end of the war. He died of a cerebral hemorrhage in Warm Springs, GA, on Apr. 12, 1945.

Harry S. Truman (1945-53)

Harry S. Truman, 33d president, Democrat, was born on May 8, 1884, in Lamar, MO, the son of John Anderson and Martha Ellen Young Truman. A family disagreement on whether his middle name should be Shippe or Solomon, after names of 2 grandfathers, resulted in his using only the middle initial S. After graduating from high school in Independence, MO, he worked (1901) for the *Kansas City Star,* as a railroad timekeeper, and as a clerk in Kansas City banks until about 1905. He ran his family's farm from 1906 to 1917. He served in France during World War I. After the war he opened a haberdashery shop, was a judge on the Jackson Co. Court (1922-24), and attended Kansas City School of Law (1923-25).

Truman was elected to the U.S. Senate in 1934 and reelected in 1940. In 1944, with Roosevelt's backing, he was nominated for vice president and elected. On Roosevelt's death in 1945, Truman became president. In 1948, in a famous upset victory, he defeated Republican Thomas E. Dewey to win election to a new term.

Truman authorized the first uses of the atomic bomb (Hiroshima and Nagasaki, Aug. 6 and 9, 1945), bringing World War II to a rapid end. He was responsible for what came to be called the Truman Doctrine (to aid nations such as Greece and Turkey, threatened by Communist takeover), and his strong commitment to NATO and to the Marshall Plan helped bring them about. In 1948-49, he broke a Soviet blockade of West Berlin with a massive airlift. When Communist North Korea invaded South Korea (June 1950), he won UN approval for a "police action" and sent in forces under Gen. Douglas MacArthur. When MacArthur opposed his policy of limited objectives, Truman removed him.

Truman was responsible for a higher minimum-wage, increased Social Security, and aid-for-housing laws. He died in Kansas City, MO, on Dec. 26, 1972.

Dwight David Eisenhower (1953-61)

Dwight D. Eisenhower, 34th president, Republican, was born on Oct. 14, 1890, in Denison, TX, the son of David Jacob and Ida Elizabeth Stover Eisenhower. He grew up on a small farm in Abilene, KS, and graduated from West Point in 1915. He was on the staff of Gen. Douglas MacArthur in the Philippines from 1935 to 1939. In 1942, he was made commander of Allied forces landing in North Africa; the next year he was made full general. He became supreme Allied commander in Europe that same year and as such led the Normandy invasion (June 6, 1944). He was given the rank of general of the army on Dec. 20, 1944, which was made permanent in 1946. On May 7, 1945, Eisenhower received the surrender of Germany at Rheims. He returned to the U.S. to serve as chief of staff (1945-48). His war memoir, *Crusade in Europe* (1948), was a best-seller.

In 1948 he became president of Columbia University; in 1950 he became Commander of NATO forces.

Eisenhower resigned from the army and was nominated for president by the Republicans in 1952. He defeated Adlai E. Stevenson in the 1952 election and again in 1956. Eisenhower called himself a moderate, favored the "free market system" vs. government price and wage controls, kept government out of labor disputes, reorganized the defense establishment, and promoted missile programs. He continued foreign aid, sped the end of the Korean War, endorsed Taiwan and SE Asia defense treaties, backed the UN in condemning the Anglo-French raid on Egypt, and advocated the "open skies" policy of mutual inspection with the USSR. He sent U.S. troops into Little Rock, AR, in Sept. 1957, during the segregation crisis.

Eisenhower died on Mar. 28, 1969, in Washington, DC.

John Fitzgerald Kennedy (1961-63)

John F. Kennedy, 35th president, Democrat, was born on May 29, 1917, in Brookline, MA, the son of Joseph P. and Rose Fitzgerald Kennedy. He graduated from Harvard University in 1940. While serving in the navy (1941-45), he commanded a PT boat in the Solomons and won the Navy and Marine Corps Medal. In 1956, while recovering from spinal surgery, he wrote *Profiles in Courage,* which won a Pulitzer Prize in 1957. He served in the House of Representatives from 1947 to 1953 and was elected to the Senate in 1952 and again in 1958. In 1960, Kennedy won the Democratic nomination for president and narrowly defeated Republican Vice Pres. Richard M. Nixon. Kennedy was the youngest president ever elected and the first Roman Catholic.

In Apr. 1961, the new Kennedy administration suffered a severe setback when an invasion force of anti-Castro Cubans, trained and directed by the U.S. Central Intelligence Agency, failed to establish a beachhead at the Bay of Pigs in Cuba. By the same token, one of Kennedy's most important acts as president was his successful demand on Oct. 22, 1962, that the Soviet Union dismantle its missile bases in Cuba. Kennedy also defied Soviet attempts to force the Allies out of Berlin. He started the Peace Corps, and he backed civil rights and expanded medical care for the aged. Space exploration was greatly developed during his administration.

On Nov. 22, 1963, Kennedy was assassinated while riding in a motorcade in Dallas, TX.

Lyndon Baines Johnson (1963-69)

Lyndon B. Johnson, 36th president, Democrat, was born on Aug. 27, 1908, near Stonewall, TX, the son of Sam Ealy and Rebekah Baines Johnson. He graduated from Southwest Texas State Teachers College in 1930 and attended Georgetown University Law School. He taught public speaking in Houston (1930-31) and then served as secretary to Rep. R. M. Kleberg (1931-35). In 1937 Johnson won an election to fill the vacancy caused by the death of a U.S. representative and in 1938 was elected to the full term, after which he returned for 4 terms. During 1941 and 1942 he also served in the Navy in the Pacific, earning a Silver Star for bravery. He was elected U.S. senator in 1948 and reelected in 1954. He became Democratic leader of the Senate in 1953. Johnson had strong support for the Democratic presidential nomination at the 1960 convention, where the nominee, John F. Kennedy, asked him to run for vice president. His campaigning helped overcome religious bias against Kennedy in the South.

Johnson became president when Kennedy was assassinated. He was elected to a full term in 1964. Johnson's domestic program was of considerable importance. He won passage of major civil rights, anti-poverty, aid to education, and healthcare (Medicare, Medicaid) legislation—the "Great Society" program. However, his escalation of the war in Vietnam came to overshadow the achievements of his administration. In the face of increasing division in the nation and in his own party over his handling of the war, Johnson declined to seek another term.

Johnson died on Jan. 22, 1973, in San Antonio, TX.

Richard Milhous Nixon (1969-74)

Richard M. Nixon, 37th president, Republican, was born on Jan. 9, 1913, in Yorba Linda, CA, the son of Francis Anthony and Hannah Milhous Nixon. He graduated from Whittier College in 1934 and from Duke University Law School in 1937. After practicing law in Whittier and serving briefly in the Office of Price Administration in 1942, he entered the Navy and served in the South Pacific. Nixon was elected to the House of Representatives in 1946 and 1948. He achieved prominence as the House Un-American Activities Committee member who forced the showdown leading to the Alger Hiss perjury conviction. In 1950 he was elected to the Senate.

Nixon was elected vice president in the Eisenhower landslides of 1952 and 1956. He won the Republican nomination for president in 1960 but was narrowly defeated by John F. Kennedy. He ran unsuccessfully for governor of California in 1962. In 1968 he again won the GOP presidential nomination, then defeated Hubert Humphrey for the presidency.

Nixon appointed 4 Supreme Court justices, including the chief justice, moving the court to the right, and as a "new federalist" sought to shift responsibility to state and local governments. He dramatically altered relations with China, which he visited in 1972—the first president to do so. With foreign affairs adviser Henry Kissinger he pursued détente with the Soviet Union. He began a gradual withdrawal from Vietnam, but U.S. troops remained there through his first term. He ordered an incursion into Cambodia (1970) and the bombing of Hanoi and mining of Haiphong Harbor (1972). Reelected by a large majority in Nov. 1972, he secured a Vietnam cease-fire.

Nixon's 2d term was cut short by scandal, after disclosures relating to a June 1972 burglary of Democratic Party headquarters in the Watergate office complex. After it emerged that most of Nixon's office conversations and calls had been taped, the courts and Congress sought the tapes for criminal proceedings against former White House aides and for a House inquiry into possible impeachment. Nixon claimed executive privilege to keep the tapes secret, but the Supreme Court ruled against him. In late July the House Judiciary Committee recommended adoption of 3 impeachment articles charging him with obstruction of justice, abuse of power, and contempt of Congress. On Aug. 5, he released transcripts of conversations that linked him to cover-up activities. He resigned on Aug. 9, becoming the first president ever to do so. In later years, Nixon emerged as an elder statesman.

Nixon died Apr. 22, 1994, in New York City.

Gerald Rudolph Ford (1974-77)

Gerald R. Ford, 38th president, Republican, was born on July 14, 1913, in Omaha, NE, the son of Leslie and Dorothy Gardner King, and was named Leslie Jr. When he was 2, his parents were divorced, and his mother moved with the boy to Grand Rapids, MI. There she met and married Gerald R. Ford, who formally adopted him and gave him his own name. Ford graduated from the University of Michigan in 1935 and from Yale Law School in 1941. He began practicing law in Grand Rapids, but in 1942 joined the navy and served in the Pacific, leaving the service in 1946 as a lieutenant commander. He entered the House of Representatives in 1949 and spent 25 years in the House, 8 of them as Republican leader.

On Oct. 12, 1973, after Vice President Spiro T. Agnew resigned, Ford was nominated by President Nixon to replace him. It was the first use of the procedures set out in the 25th Amendment. When Nixon resigned, Aug. 9, 1974, Ford became president; he was the only president who was never elected either to the presidency or to the vice presidency. On Sept. 8, in a controversial move, he pardoned Nixon for any federal crimes he might have committed as president. Ford vetoed 48 bills in his first 21 months in office, mostly in the interest of fighting high inflation; he was less successful in curbing high unemployment. In foreign policy, Ford continued to pursue détente. He was narrowly defeated in the 1976 election by Democrat Jimmy Carter.

Jimmy (James Earl) Carter (1977-81)

Jimmy (James Earl) Carter, 39th president, Democrat, was the first president from the Deep South since before the Civil War. He was born on Oct. 1, 1924, in Plains, GA, the son of James and Lillian Gordy Carter.

Carter graduated from the U.S. Naval Academy in 1946 and in 1952 entered the navy's nuclear submarine program as an aide to Capt. (later Adm.) Hyman Rickover. He studied nuclear physics at Union College. Carter's father died in 1953, and he left the navy to take over the family businesses. He served in the Georgia state senate (1963-67) and as governor of Georgia (1971-75). In 1976, Carter won the Democratic nomination and defeated President Gerald R. Ford.

On his first full day in office, Carter pardoned all Vietnam draft evaders. He played a major role in the peace negotiations between Israel and Egypt. However, Carter was widely criticized for the poor state of the economy and was viewed by many as weak in his handling of foreign policy. In Nov. 1979, Iranian student militants attacked the U.S. embassy in Tehran and held members of the embassy staff hostage. Efforts to obtain release of the hostages was a major preoccupation during the rest of his term. He reacted to the Soviet invasion of Afghanistan by imposing a grain embargo and boycotting the Moscow Olympic Games.

Carter was defeated by Ronald Reagan in the 1980 election. Carter administration efforts finally resulted in the release of the hostages on Inauguration Day, 1981, just after Reagan officially became president. After leaving office, Carter was hailed for his humanitarian efforts and took a prominent role in mediating international disputes.

Ronald Wilson Reagan (1981-89)

Ronald Wilson Reagan, 40th president, Republican, was born on Feb. 6, 1911, in Tampico, IL, the son of John Edward and Nellie Wilson Reagan. Reagan graduated from Eureka College in 1932, after which he worked as a sports announcer in Des Moines, IA. He began a successful career as an actor in 1937, starring in numerous movies, and later in television, until the 1960s. He served as president of the Screen Actors Guild from 1947 to 1952 and in 1959-60. Reagan was elected governor of California in 1966 and reelected in 1970.

In 1980, Reagan gained the Republican presidential nomination and won a landslide victory over Jimmy Carter. He was easily reelected in 1984. Reagan successfully forged a bipartisan coalition in Congress, which led to enactment of his program of large-scale tax cuts, cutbacks in many government programs, and a major defense buildup. He signed a Social Security reform bill designed to provide for the long-term solvency of the system. In 1986, he signed into law a major tax-reform bill. He was shot and wounded in an assassination attempt in 1981.

In 1982, the U.S. joined France and Italy in maintaining a peacekeeping force in Beirut, Lebanon, and the next year Reagan sent a task force to invade the island of Grenada after 2 Marxist coups there. Reagan's opposition to international terrorism led to the U.S. bombing of Libyan military installations in 1986. He strongly supported El Salvador, the Nicaraguan contras, and other anti-communist governments and forces throughout the world. He also held 4 summit meetings with Soviet leader Mikhail Gorbachev. At the 1987 meeting in Washington, DC, a historic treaty eliminating short- and medium-range missiles from Europe was signed.

Reagan faced a crisis in 1986-87, when it was revealed that the U.S. had sold weapons to Iran in exchange for release of U.S. hostages being held in Lebanon and that subsequently some of the money was diverted to the Nicaraguan contras (Congress had barred aid to the contras). The scandal led to the resignation of leading White House aides. As Reagan left office in Jan. 1989, the nation was experiencing its 6th consecutive year of economic prosperity. Reagan, however, was unable to control the high budget deficits that plagued him throughout his administration.

In 1994, in a letter to the American people, Reagan revealed that he was suffering from Alzheimer's disease.

George Herbert Walker Bush (1989-93)

George Herbert Walker Bush, 41st president, Republican, was born on June 12, 1924, in Milton, MA, the son of Prescott and Dorothy Walker Bush. He served as a U.S. Navy pilot in World War II. After graduating from Yale University in 1948, he settled in Texas, where, in 1953, he helped found an oil company. After losing a bid for a U.S. Senate seat in Texas in 1964, he was elected to the House of Representatives in 1966 and 1968. He lost a 2d U.S. Senate race in 1970. Subsequently he served as U.S. ambassador to the United Nations (1971-73), headed the U.S. Liaison Office in Beijing (1974-75), and was director of central intelligence (1976-77).

Following an unsuccessful bid for the 1980 Republican presidential nomination, Bush was chosen by Ronald Reagan as his vice presidential running mate. He served as U.S. vice president from 1981 to 1989.

In 1988, Bush gained the Republican presidential nomination and defeated Democrat Michael Dukakis in the November elections. Bush took office faced with the ongoing U.S. budget and trade deficits as well as the rescue of insolvent U.S. savings and loan institutions. He faced a severe budget deficit annually, struggled with military cutbacks in light of reduced cold war tensions, and vetoed abortion-rights legislation. In 1990 he agreed to a budget deficit-reduction plan that included tax hikes.

Bush supported Soviet reforms and Eastern Europe democratization. He was criticized by some for keeping U.S. policy tied closely to Mikhail Gorbachev as the Soviet leader lost power and for underreaction to China's violent repression of pro-democracy demonstrators in 1989. In Dec. 1989, Bush sent troops to Panama; they overthrew the government and captured strongman Gen. Manuel Noriega.

Bush reacted to Iraq's Aug. 1990 invasion of Kuwait by sending U.S. forces to the Persian Gulf area and assembling a UN-backed coalition, including NATO and Arab League members. After a month-long air war, in Feb. 1991, Allied forces retook Kuwait in a 4-day ground assault. The quick victory, with light casualties, gave Bush one of the highest presidential approval ratings in history. His popularity plummeted by the end of 1991, however, as the economy struggled through a prolonged recession. He was defeated by his Democratic opponent, Bill Clinton, in the 1992 election.

Bill (William Jefferson) Clinton (1993-)

Bill Clinton, 42d president, Democrat, was born on Aug. 19, 1946, in Hope, AR, son of William Blythe and Virginia Cassidy Blythe, and was named William Jefferson Blythe IV. Blythe died in an automobile accident before his son was born. His widow married Roger Clinton, and at the age of 16, William Jefferson Blythe IV changed his name to Bill Clinton. Clinton graduated from Georgetown University in 1968, attended Oxford University as a Rhodes scholar, and earned a degree from Yale Law School in 1973.

Clinton worked on George McGovern's 1972 presidential campaign. He taught at the University of Arkansas from 1973 to 1976, when he was elected state attorney general. In 1978, he was elected governor, becoming the nation's youngest. Defeated for reelection in 1980, he was returned to office in 1982, 1984, 1986, and 1990. He married Hillary Rodham in 1975.

Despite attacks on his character, Clinton won most of the 1992 presidential primaries, moving his party toward the center as he tried to broaden his appeal; as the party's presidential nominee he defeated Pres. George Bush in November. In 1993, Clinton won passage of a measure to reduce the federal budget deficit and won congressional approval of the North American Free Trade Agreement. His plan for major health-care reform legislation died in Congress.

After 1994 midterm elections, Clinton faced Republican majorities in both houses of Congress. He followed a centrist course at home, sent troops to Bosnia to help implement a peace settlement, and cultivated relations with Russia and China.

Though accused of improprieties in his involvement in an Arkansas real estate venture (Whitewater) and in other matters, Clinton easily won reelection in 1996. In 1997 he reached agreement with Congress on legislation to balance the federal budget by 2002. In 1998, Clinton became only the 2d U.S. president ever to be impeached by the House of Representatives. Charged with perjury and obstruction of justice in connection with an attempted cover-up of a sexual relationship with a former White House intern, he was acquitted by the Senate in 1999. Despite the scandal he retained wide popularity, aided by a strong economy.

Foreign policy issues were prominent in Clinton's last years in office. Relations with China were strained by a spy scandal and human rights issues. In 1999, the U.S., under Clinton, joined other NATO nations in an aerial bombing campaign that induced Serbian Pres. Slobodan Milosevic to withdraw troops from the Kosovo region, where they had been terrorizing and driving out large numbers of ethnic Albanians.

Wives and Children of the Presidents

Name (Born–died; married)	State	Sons/ daughters
Martha Dandridge Custis Washington (1731-1802; 1759)	VA	None
Abigail Smith Adams (1744-1818; 1764)	MA	3/2
Martha Wayles Skelton Jefferson (1748-82; 1772)	VA	1/5
Dorothea "Dolley" Payne Todd Madison (1768-1849; 1794)	NC	None
Elizabeth Kortright Monroe (1768-1830; 1786)	NY	.../2 (A)
Louisa Catherine Johnson Adams (1775-1852; 1797)	MD(B)	3/1
Rachel Donelson Robards Jackson (1767-1828; 1791)	VA	None
Hannah Hoes Van Buren (1783-1819; 1807)	NY	4/...
Anna Tuthill Symmes Harrison (1775-1864; 1795)	NJ	6/4
Letitia Christian Tyler (1790-1842; 1813)	VA	3/4 (A)
Julia Gardiner Tyler (1820-89; 1844)	NY	5/2
Sarah Childress Polk (1803-91; 1824)	TN	None
Margaret Mackall Smith Taylor (1788-1852; 1810)	MD	1/5
Abigail Powers Fillmore (1798-1853; 1826)	NY	1/1
Caroline Carmichael McIntosh Fillmore (1813-81; 1858)	NJ	None
Jane Means Appleton Pierce (1806-63; 1834)	NH	3/...
Mary Todd Lincoln (1818-82; 1842)	KY	4/...
Eliza McCardle Johnson (1810-76; 1827)	TN	3/2
Julia Boggs Dent Grant (1826-1902; 1848)	MO	3/1
Lucy Ware Webb Hayes (1831-89; 1852)	OH	7/1
Lucretia Rudolph Garfield (1832-1918; 1858)	OH	4/1
Ellen Lewis Herndon Arthur (1837-80; 1859)	VA	2/1
Frances Folsom Cleveland (1864-1947; 1886)	NY	2/3
Caroline Lavinia Scott Harrison (1832-92; 1853)	OH	1/1
Mary Scott Lord Dimmick Harrison (1858-1948; 1896)	PA	.../1
Ida Saxton McKinley (1847-1907; 1871)	OH	.../2
Alice Hathaway Lee Roosevelt (1861-84; 1880)	MA	.../1
Edith Kermit Carow Roosevelt (1861-1948; 1886)	CT	4/1
Helen Herron Taft (1861-1943; 1886)	OH	2/1
Ellen Louise Axson Wilson (1860-1914; 1885)	GA	.../3
Edith Bolling Galt Wilson (1872-1961; 1915)	VA	None
Florence Kling De Wolfe Harding (1860-1924; 1891)	OH	None
Grace Anna Goodhue Coolidge (1879-1957; 1905)	VT	2/...
Lou Henry Hoover (1875-1944; 1899)	IA	2/...
Anna Eleanor Roosevelt Roosevelt (1884-1962; 1905)	NY	4/1 (A)
Elizabeth Virginia "Bess" Wallace Truman (1885-1982; 1919)	MO	.../1
Mamie Geneva Doud Eisenhower (1896-1979; 1916)	IA	1/... (A)
Jacqueline Lee Bouvier Kennedy (1929-94; 1953)	NY	1/1 (A)
Claudia "Lady Bird" Alta Taylor Johnson (1912; 1934)	TX	.../2
Thelma Catherine Patricia Ryan Nixon (1912-1993; 1940)	NV	.../2
Elizabeth Bloomer Warren Ford (1918; 1948)	IL	3/1
Rosalynn Smith Carter (1927; 1946)	GA	3/1
Anne Frances "Nancy" Robbins Davis Reagan (1921; 1952)	NY	1/1 (C)
Barbara Pierce Bush (1925; 1945)	NY	4/2
Hillary Rodham Clinton (1947; 1975)	IL	.../1

James Buchanan, 15th president, was unmarried. (A) plus one infant, deceased. (B) Born in London, father a MD citizen. (C) Pres. Reagan married and divorced Jane Wyman; they had a daughter who died in infancy, a son and daughter who lived past infancy.

First Lady Hillary Rodham Clinton

Hillary Rodham Clinton was born in Chicago, Oct. 26, 1947, to Hugh and Dorothy Rodham. She graduated from Wellesley College and Yale Law School. She married Bill Clinton in 1975, and a daughter, Chelsea, was born in 1980. From 1977 to 1992, she was a partner in the Rose Law Firm in Little Rock, AR. In this capacity she did some work for an S&L linked to the Whitewater scandal. In 1988 and 1991, she was voted one of the "100 Most Influential Lawyers in America" by the *National Law Journal*.

In 1993-94, as first lady, she played a leading role in an unsuccessful effort to reform the U.S. health-care system. In 1995 her book *It Takes a Village*, about the needs of children, was published; her recording of the text won a Grammy in 1997. She was a critic of investigations by independent counsel Kenneth Starr aimed at the president and others, and she expressed loyalty to her husband after he admitted an improper extramarital relationship in Aug. 1998. She expressed interest in moving to the U.S. Senate after the White House, and in 1999 took steps toward seeking a U.S. Senate seat in New York.

Burial Places of the Presidents

President	Burial Place	President	Burial Place	President	Burial Place
Washington	Mt. Vernon, VA	Fillmore	Buffalo, NY	T. Roosevelt	Oyster Bay, NY
J. Adams	Quincy, MA	Pierce	Concord, NH	Taft	Arlington Natl. Cemetery
Jefferson	Charlottesville, VA	Buchanan	Lancaster, PA	Wilson	Wash. Natl. Cathedral
Madison	Montpelier Station, VA	Lincoln	Springfield, IL	Harding	Marion, OH
Monroe	Richmond, VA	A. Johnson	Greeneville, TN	Coolidge	Plymouth, VT
J. Q. Adams	Quincy, MA	Grant	New York, NY	Hoover	West Branch, IA
Jackson	Nashville, TN	Hayes	Fremont, OH	F. Roosevelt	Hyde Park, NY
Van Buren	Kinderhook, NY	Garfield	Cleveland, OH	Truman	Independence, MO
W. H. Harrison	North Bend, OH	Arthur	Albany, NY	Eisenhower	Abilene, KS
Tyler	Richmond, VA	Cleveland	Princeton, NJ	Kennedy	Arlington Natl. Cemetery
Polk	Nashville, TN	B. Harrison	Indianapolis, IN	L. B. Johnson	Johnson City, TX
Taylor	Louisville, KY	McKinley	Canton, OH	Nixon	Yorba Linda, CA

Presidential Facts

- **Oldest president**: Ronald Reagan, who was 77 when he left office
- **Youngest president**: Theodore Roosevelt, who was 42 when sworn in after McKinley's death
- **Only president to serve more than 2 terms**: Franklin D. Roosevelt
- **Only president to serve 2 terms that were not back to back**: Grover Cleveland, both the 22d and the 24th president
- **President who served the shortest term**: William Henry Harrison, who died of pneumonia 31 days after being inaugurated
- **Only president to also serve as chief justice of the U.S.:** William Howard Taft
- **Only president to resign:** Richard Nixon, after a House committee recommended impeachment for Watergate scandal

- **State where the greatest number of presidents were born:** Virginia (8)
- **First president to live in the White House:** John Adams
- **Only president who was never married**: James Buchanan. His niece acted as White House hostess.
- **Only president to serve without having been elected vice president or president in a national election**: Gerald Ford
- **Presidents who died on July 4**: John Adams, Thomas Jefferson, and James Monroe
- **Presidents who died in office**: Eight presidents have died in office. Four of them were assassinated: Abraham Lincoln, James Garfield, William McKinley, and John F. Kennedy. The other four were William Henry Harrison, Zachary Taylor, Warren G. Harding, and Franklin Delano Roosevelt.

Presidential Libraries

The libraries listed here, except for that of Richard Nixon (which is private), are coordinated by the National Archives and Records Administration (Website: http://www.nara.gov/nara/president/overview.html). NARA also has custody of the Nixon presidential historical materials. Materials for presidents before Herbert Hoover are held by private institutions.

Herbert Hoover Library
211 Parkside Dr., PO Box 488
West Branch, IA 52358-0488
PHONE: 319-643-5301
FAX: 319-643-5825
E-MAIL: library@hoover.nara.gov

Franklin D. Roosevelt Library
511 Albany Post Rd.
Hyde Park, NY 12538-1999
PHONE: 914-229-8114
FAX: 914-229-0872
E-MAIL: library@roosevelt.nara.gov

Harry S. Truman Library
500 West U.S. Hwy. 24
Independence, MO 64050-1798
PHONE: 816-833-1400
FAX: 816-833-4368
E-MAIL: library@truman.nara.gov

Dwight D. Eisenhower Library
200 S.E. 4th St.
Abilene, KS 67410-2900
PHONE: 785-263-4751
FAX: 785-263-4218
E-MAIL: library@eisenhower.nara.gov

John Fitzgerald Kennedy Library
Columbia Pt.
Boston, MA 02125-3398
PHONE: 617-929-4500
FAX: 617-929-4538
E-MAIL: library@kennedy.nara.gov

Lyndon Baines Johnson Library
2313 Red River St.
Austin, TX 78705-5702
PHONE: 512-916-5137
FAX: 512-478-9104
E-MAIL: library@johnson.nara.gov

Richard Nixon Library & Birthplace
18001 Yorba Linda Blvd.
Yorba Linda, CA 92886-3949
PHONE: 714-993-3393
FAX: 714-528-0544
WEBSITE: http://www.nixonfoundation.org
E-MAIL: stedman@chapman.edu

Gerald R. Ford Library
1000 Beal Ave.
Ann Arbor, MI 48109-2114
PHONE: 734-741-2218
FAX: 734-741-2341
E-MAIL: library@fordlib.nara.gov

Jimmy Carter Library
441 Freedom Pkwy.
Atlanta, GA 30307-1406
PHONE: 404-331-3942
FAX: 404-730-2215
E-MAIL: library@carter.nara.gov

Ronald Reagan Library
40 Presidential Dr.
Simi Valley, CA 93065-0666
PHONE: 805-522-8444
FAX: 805-522-9621
E-MAIL: library@reagan.nara.gov

George Bush Library
1000 George Bush Dr., West
College Station, TX 77482-0410
PHONE: 409-260-9552
FAX: 409-260-9557
E-MAIL: library@bush.nara.gov

Impeachment in U.S. History

The U.S. Constitution provides for impeachment and, upon conviction, removal from office of federal officials on grounds of "Treason, Bribery, or other high Crimes and Misdemeanors" (Article II, Sect. 4). Impeachment is the bringing of charges by the House of Representatives. It is followed by a Senate trial; a two-thirds vote in the Senate is needed for conviction and removal from office, which does not preclude criminal indictment and trial (Article I, Sect. 2, Para. 5; Sect. 3, Para. 6-7).

In 1868, Pres. Andrew Johnson became the first president to be impeached by the U.S. House; he was tried but not convicted by the Senate. In 1974, articles of impeachment against Pres. Richard Nixon, in connection with the Watergate scandal, were voted by the House Judiciary Committee. However, Nixon resigned Aug. 9, before the full House could vote on impeaching him. In 1998, Pres. Bill Clinton was impeached by the U.S. House in connection with covering up a relationship with a former White House intern; he was tried in the Senate in early 1999 and acquitted. A list of all impeached federal officials follows:

Name	Position Held	Senate Trial Began	Action Taken	Date
William Blount	Senator, TN	Dec. 17, 1798	Charges dismissed	Jan. 14, 1799
John Pickering	District Court Judge, NH	Mar. 3, 1803	Removed from office	Mar. 12, 1804
Samuel Chase	Supreme Court Assoc. Justice	Nov. 30, 1804	Acquitted	Mar. 1, 1805
James H. Peck	District Court Judge, MO	Apr. 26, 1830	Acquitted	Jan. 31, 1831
West H. Humphreys	District Court Judge, TN	May 7, 1862	Removed from office	June 26, 1862
Andrew Johnson	President	Feb. 25, 1868	Acquitted	May 26, 1868
William W. Belknap	Secretary of War	Mar. 3, 1876	Acquitted	Aug. 1, 1876
Charles Swayne	District Court Judge, FL	Dec. 14, 1904	Acquitted	Feb. 27, 1905
Robert W. Archbald	Commerce Court, Assoc. Judge	July 13, 1912	Removed from office	Jan. 13, 1913
George W. English	District Court Judge, IL	Nov. 4, 1926*	Charges dismissed	Nov. 4, 1926
Harold Louderback	District Court Judge, CA	May 15, 1933	Acquitted	May 24, 1933
Halsted L. Ritter	District Court Judge, FL	Apr. 6, 1936	Removed from office	Apr. 17, 1936
Harry E. Claiborne	District Court Judge, NV	Oct. 7, 1986	Removed from office	Oct. 9, 1986
Alcee L. Hastings	District Court Judge, FL	Oct. 18, 1989	Removed from office	Oct. 20, 1989
Walter L. Nixon	District Court Judge, MS	Nov. 1, 1989	Removed from office	Nov. 3, 1989
William J. Clinton	President	Jan. 7, 1999	Acquitted	Feb. 12, 1999

*Date of resignation, after which the impeachment charges were dismissed.

UNITED STATES FACTS

Superlative U.S. Statistics[1]

Source: U.S. Geological Survey, Dept. of the Interior; U.S. Bureau of the Census, Dept. of Commerce; World Almanac research

Area for 50 states and Washington, DC . . .	Total	3,717,796 sq mi
	Land, 3,536,278 sq mi; Water, 181,518 sq mi	
Largest state	Alaska	615,230 sq mi
Smallest state	Rhode Island	1,231 sq mi
Largest county (excluding Alaska)	San Bernardino County, CA	20,106 sq mi
Smallest county	Kalawao, HI	52 sq mi
Largest incorporated city	Sitka, AK	2,881 sq mi
Northernmost city	Barrow, AK	71°17′ N
Northernmost point	Point Barrow, AK	71°23′ N
Southernmost city	Hilo, HI	19°44′ N
Southernmost settlement	Naalehu, HI	19°03′ N
Southernmost point	Ka Lae (South Cape), Island of Hawaii . . . . .18°55′ N (155°41′ W)	
Easternmost city	Eastport, ME	66°59′05″ W
Easternmost settlement[2]	Amchitka Isl., AK	179°15′ E
Easternmost point[2]	Pochnoi Point, on Semisopochnoi Isl., AK	179°46′ E
Westernmost city	Atka, AK	174°12′ W
Westernmost settlement	Adak Station, AK.	176°39′ W
Westernmost point	Amatignak Isl., AK.	179°06′ W
Highest settlement	Climax, CO	11,360 ft
Lowest settlement	Calipatria, CA	−184 ft
Highest point on Atlantic coast	Cadillac Mountain, Mount Desert Isl., ME.	1,530 ft
Oldest national park	Yellowstone National Park (1872), WY, MT, ID	2,219,791 acres
Largest national park	Wrangell-St. Elias, AK.	8,323,618 acres
Highest waterfall	Yosemite Falls—Total in 3 sections	2,425 ft
	Upper Yosemite Fall	1,430 ft
	Cascades in middle section	675 ft
	Lower Yosemite Fall	320 ft
Longest river system	Mississippi-Missouri-Red Rock	3,710 mi
Highest mountain	Mount McKinley, AK	20,320 ft
Lowest point	Death Valley, CA	−282 ft
Deepest lake	Crater Lake, OR	1,932 ft
Rainiest spot	Mount Waialeale, HI . . . . Annual avg rainfall 460 in	
Largest gorge	Grand Canyon, Colorado River, AZ . . . . . . 277 mi long, 600 ft to 18 mi wide, 1 mi deep	
Deepest gorge	Hells Canyon, Snake River, OR-ID	7,900 ft
Strongest surface wind	Mount Washington, NH, recorded 1934	231 mph
Largest dam	New Cornelia Tailings, Ten Mile Wash, AZ[3] . . . . 274,026,000 cu yds material used	
Tallest building	Sears Tower, Chicago, IL	1,450 ft
Largest building	Boeing 747 Manufacturing Plant, Everett, WA . . . . . . .205,600,000 cu ft; covers 47 acres	
Tallest structure	TV tower, Blanchard, ND.	2,063 ft
Longest bridge span	Verrazano-Narrows, NY	4,260 ft
Highest bridge	Royal Gorge, CO. . . . . 1,053 ft above water	
Deepest well	Gas well, Washita County, OK	31,441 ft

The 48 Contiguous States

Area for 48 states and Washington, DC . . .	Total	3,096,107 sq mi[4]
	Land, 2,959,481 sq mi; Water, 136,626 sq mi	
Largest state	Texas	276,277 sq mi
Northernmost city	Bellingham, WA.	48°46′ N
Northernmost settlement	Angle Inlet, MN	49°21′ N
Northernmost point	Northwest Angle, MN	49°23′ N
Southernmost city	Key West, FL.	24°33′ N
Southernmost mainland city	Florida City, FL	25°27′ N
Southernmost point	Key West, FL.	24°33′ N
Easternmost settlement	Lubec, ME.	66°58′49″ W
Easternmost point	West Quoddy Head, ME.	66°57′ W
Westernmost town	La Push, WA	124°38′ W
Westernmost point	Cape Alava, WA	124°44′ W
Highest mountain	Mount Whitney, CA	14,494 ft

(1) All areas are total area, including water, unless otherwise noted. (2) Alaska's Aleutian Islands extend into the eastern hemisphere and thus technically contain the easternmost point and settlement in the U.S. (3) The New Cornelia Tailings Dam is a privately owned industrial dam composed of tailings, remnants of a mining process. (4) Does not add, because of rounding.

Geodetic Datum of North America

In July 1986, the National Oceanic and Atmospheric Administration's National Geodetic Survey (NGS), in cooperation with Canada and Mexico, completed readjustment and redefinition of the system of latitudes and longitudes. The resulting North American Datum of 1983 (NAD 83) replaces the North American Datum of 1927, as well as local reference systems for Hawaii and for Puerto Rico and the Virgin Islands. The change was prompted by Hawaii's increased need for accurate coordinate information. To facilitate use of satellite surveying and navigation systems, such as the Global Positioning System (GPS), the new datum was redefined using the Geodetic Reference System 1980 as the reference ellipsoid because this model more closely approximates the true size and shape of the earth. In addition, the origin of the coordinate system is referenced to the mass center of the earth to coincide with the orbital orientation of the GPS satellites. Positional changes resulting from the datum redefinition can reach 330 ft in the continental U.S., Canada, and Mexico. Changes that exceed 660 ft can be expected in Alaska, Puerto Rico, and the Virgin Islands. Hawaii's coordinates changed about 1,300 ft.

Additional Statistical Information About the U.S.

The annual *Statistical Abstract of the United States,* published by U.S. Dept. of Commerce, contains additional social, political, and economic data about the U.S. For information on this and other printed publications, write to: Superintendent of Documents, Government Printing Office, PO Box 371954, Pittsburgh, PA 15250-7954, or call (202) 512-1800. For information on electronic products, write to: U.S. Dept. of Commerce, Bureau of the Census, PO Box 277943, Atlanta, GA 30384-7943, or call (301) 457-4100. Parts of *The Statistical Abstract* can be viewed on the Internet at http://www.census.gov/statab/www

Highest and Lowest Altitudes in U.S. States and Territories

Source: U.S. Geological Survey, Dept. of the Interior

(Minus sign means below sea level.)

	HIGHEST POINT			LOWEST POINT		
	Name	County	Elev. (ft)	Name	County	Elev. (ft)
Alabama	Cheaha Mountain	Cleburne	2,405	Gulf of Mexico		Sea level
Alaska	Mount McKinley	Denali	20,320	Pacific Ocean		Sea level
Arizona	Humphreys Peak	Coconino	12,633	Colorado R.	Yuma	70
Arkansas	Magazine Mountain	Logan	2,753	Ouachita R.	Ashley-Union	55
California	Mount Whitney	Inyo-Tulare	14,494	Death Valley	Inyo	−282
Colorado	Mount Elbert	Lake	14,433	Arkansas R	Prowers	3,350
Connecticut	Mount Frissell	Litchfield	2,380	Long Island Sound		Sea level
Delaware	On Ebright Road	New Castle	448	Atlantic Ocean		Sea level
Dist. of Columbia	Tenleytown	N W part	410	Potomac R.		1
Florida	Sec. 30, T6N, R20W[1]	Walton	345	Atlantic Ocean		Sea level
Georgia	Brasstown Bald	Towns-Union	4,784	Atlantic Ocean		Sea level
Guam	Mount Lamlam	Agat District	1,332	Pacific Ocean		Sea level
Hawaii	Mauna Kea	Hawaii	13,796	Pacific Ocean		Sea level
Idaho	Borah Peak	Custer	12,662	Snake R.	Nez Perce	710
Illinois	Charles Mound	Jo Daviess	1,235	Mississippi R	Alexander	279
Indiana	Franklin Township	Wayne	1,257	Ohio R	Posey	320
Iowa	Sec. 29, T100N, R41W[1]	Osceola	1,670	Mississippi R	Lee	480
Kansas	Mount Sunflower	Wallace	4,039	Verdigris R	Montgomery	679
Kentucky	Black Mountain	Harlan	4,139	Mississippi R	Fulton	257
Louisiana	Driskill Mountain	Bienville	535	New Orleans	Orleans	−8
Maine	Mount Katahdin	Piscataquis	5,267	Atlantic Ocean		Sea level
Maryland	Backbone Mountain	Garrett	3,360	Atlantic Ocean		Sea level
Massachusetts	Mount Greylock	Berkshire	3,487	Atlantic Ocean		Sea level
Michigan	Mount Arvon	Baraga	1,979	Lake Erie	Monroe	571
Minnesota	Eagle Mountain	Cook	2,301	Lake Superior		600
Mississippi	Woodall Mountain	Tishomingo	806	Gulf of Mexico		Sea level
Missouri	Taum Sauk Mt.	Iron	1,772	St. Francis R	Dunklin	230
Montana	Granite Peak	Park	12,799	Kootenai R	Lincoln	1,800
Nebraska	Johnson Township	Kimball	5,424	Missouri R	Richardson	840
Nevada	Boundary Peak	Esmeralda	13,140	Colorado R.	Clark	479
New Hampshire	Mt. Washington	Coos	6,288	Atlantic Ocean		Sea level
New Jersey	High Point	Sussex	1,803	Atlantic Ocean		Sea level
New Mexico	Wheeler Peak	Taos	13,161	Red Bluff Res.	Eddy	2,842
New York	Mount Marcy	Essex	5,344	Atlantic Ocean		Sea level
North Carolina	Mount Mitchell	Yancey	6,684	Atlantic Ocean		Sea level
North Dakota	White Butte	Slope	3,506	Red R.	Pembina	750
Ohio	Campbell Hill	Logan	1,549	Ohio R	Hamilton	455
Oklahoma	Black Mesa	Cimarron	4,973	Little R	McCurtain	289
Oregon	Mount Hood	Clackamas-Hood R.	11,239	Pacific Ocean		Sea level
Pennsylvania	Mt. Davis	Somerset	3,213	Delaware R	Delaware	Sea level
Puerto Rico	Cerro de Punta	Ponce District	4,390	Atlantic Ocean		Sea level
Rhode Island	Jerimoth Hill	Providence	812	Atlantic Ocean		Sea level
Samoa	Lata Mountain	Tau Island	3,160	Pacific Ocean		Sea level
South Carolina	Sassafras Mountain	Pickens	3,560	Atlantic Ocean		Sea level
South Dakota	Harney Peak	Pennington	7,242	Big Stone Lake	Roberts	966
Tennessee	Clingmans Dome	Sevier	6,643	Mississippi R	Shelby	178
Texas	Guadalupe Peak	Culberson	8,749	Gulf of Mexico		Sea level
Utah	Kings Peak	Duchesne	13,528	Beaverdam Wash.	Washington	2,000
Vermont	Mount Mansfield	Lamoille	4,393	Lake Champlain		95
Virginia	Mount Rogers	Grayson-Smyth	5,729	Atlantic Ocean		Sea level
Virgin Islands	Crown Mountain	St. Thomas Island	1,556	Atlantic Ocean		Sea level
Washington	Mount Rainier West	Pierce	14,410	Pacific Ocean		Sea level
West Virginia	Spruce Knob	Pendleton	4,861	Potomac R	Jefferson	240
Wisconsin	Timms Hill	Price	1,951	Lake Michigan		579
Wyoming	Gannett Peak	Fremont	13,804	Belle Fourche R	Crook	3,099

(1) Sec.=section; T=township; R=range; N=north; W=west.

U.S. Coastline by States

Source: National Oceanic and Atmospheric Administration, U.S. Dept. of Commerce

(in statute miles)

	Coastline[1]	Shoreline[2]		Coastline[1]	Shoreline[2]
ATLANTIC COAST	**2,069**	**28,673**	**GULF COAST**	**1,631**	**17,141**
Connecticut	0	618	Alabama	53	607
Delaware	28	381	Florida	770	5,095
Florida	580	3,331	Louisiana	397	7,721
Georgia	100	2,344	Mississippi	44	359
Maine	228	3,478	Texas	367	3,359
Maryland	31	3,190			
Massachusetts	192	1,519	**PACIFIC COAST**	**7,623**	**40,298**
New Hampshire	13	131	Alaska	5,580	31,383
New Jersey	130	1,792	California	840	3,427
New York	127	1,850	Hawaii	750	1,052
North Carolina	301	3,375	Oregon	296	1,410
Pennsylvania	0	89	Washington	157	3,026
Rhode Island	40	384			
South Carolina	187	2,876	**ARCTIC COAST, ALASKA**	**1,060**	**2,521**
Virginia	112	3,315			
			UNITED STATES	**12,383**	**88,633**

(1) Figures are lengths of general outline of seacoast. Measurements were made with a unit measure of 30 minutes of latitude on charts as near the scale of 1:1,200,000 as possible. Coastline of sounds and bays is included to a point where they narrow to width of unit measure, and includes the distance across at such point. (2) Figures obtained in 1939-40 with a recording instrument on the largest-scale charts and maps then available. Shoreline of outer coast, offshore islands, sounds, bays, rivers, and creeks is included to the head of tidewater or to a point where tidal waters narrow to a width of 100 ft.

States: Settled, Capitals, Entry Into Union, Area, Rank

The 13 colonies that seceded from Great Britain and fought the War of Independence (American Revolution) became the 13 original states. They were (in the order in which they ratified the Constitution): Delaware, Pennsylvania, New Jersey, Georgia, Connecticut, Massachusetts, Maryland, South Carolina, New Hampshire, Virginia, New York, North Carolina, and Rhode Island.

State	Set-tled[1]	Capital	Entered Union Date	Order	Extent in miles Long (approx.	Wide mean)	Area in sq mi Land	Inland Water	Total	Rank in area[2]
AL...	1702	Montgomery	Dec. 14, 1819	22	330	190	50,750	1,486	52,237	30
AK ..	1784	Juneau	Jan. 3, 1959	49	1,480[3]	810	570,374	44,856	615,230	1
AZ...	1776	Phoenix	Feb. 14, 1912	48	400	310	113,642	364	114,006	6
AR ..	1686	Little Rock	June 15, 1836	25	260	240	52,075	1,107	53,182	28
CA ..	1769	Sacramento	Sept. 9, 1850	31	770	250	155,973	2,895	158,869	3
CO ..	1858	Denver	Aug. 1, 1876	38	380	280	103,729	371	104,100	8
CT ..	1634	Hartford	Jan. 9, 1788	5	110	70	4,845	698	5,544	48
DE ..	1638	Dover	Dec. 7, 1787	1	100	30	1,955	442	2,396	49
DC ..	NA	Washington	NA	NA	...	...	61	7	68	51
FL...	1565	Tallahassee	Mar. 3, 1845	27	500	160	53,937	5,991	59,928	23
GA ..	1733	Atlanta	Jan. 2, 1788	4	300	230	57,919	1,058	58,977	24
HI ...	1820	Honolulu	Aug. 21, 1959	50	...	...	6,423	36	6,459	47
ID ...	1842	Boise	July 3, 1890	43	570	300	82,751	823	83,574	14
IL ...	1720	Springfield	Dec. 3, 1818	21	390	210	55,593	2,325	57,918	25
IN ...	1733	Indianapolis	Dec. 11, 1816	19	270	140	35,870	550	36,420	38
IA ...	1788	Des Moines	Dec. 28, 1846	29	310	200	55,875	401	56,276	26
KS ..	1727	Topeka	Jan. 29, 1861	34	400	210	81,823	459	82,282	15
KY ..	1774	Frankfort	June 1, 1792	15	380	140	39,732	679	40,411	37
LA...	1699	Baton Rouge	Apr. 30, 1812	18	380	130	43,566	6,085	49,651	31
ME ..	1624	Augusta	Mar. 15, 1820	23	320	190	30,865	2,876	33,741	39
MD ..	1634	Annapolis	Apr. 28, 1788	7	250	90	9,775	2,522	12,297	42
MA ..	1620	Boston	Feb. 6, 1788	6	190	50	7,838	1,403	9,241	45
MI...	1668	Lansing	Jan. 26, 1837	26	490	240	56,809	39,895	96,705	11
MN ..	1805	St. Paul	May 11, 1858	32	400	250	79,617	7,326	86,943	12
MS ..	1699	Jackson	Dec. 10, 1817	20	340	170	46,914	1,372	48,286	32
MO ..	1735	Jefferson City	Aug. 10, 1821	24	300	240	68,898	811	69,709	21
MT ..	1809	Helena	Nov. 8, 1889	41	630	280	145,556	1,490	147,046	4
NE ..	1823	Lincoln	Mar. 1, 1867	37	430	210	76,878	481	77,358	16
NV ..	1849	Carson City	Oct. 31, 1864	36	490	320	109,806	761	110,567	7
NH ..	1623	Concord	June 21, 1788	9	190	70	8,969	314	9,283	44
NJ...	1660	Trenton	Dec. 18, 1787	3	150	70	7,419	796	8,215	46
NM ..	1610	Santa Fe	Jan. 6, 1912	47	370	343	121,364	234	121,598	5
NY ..	1614	Albany	July 26, 1788	11	330	283	47,224	6,766	53,989	27
NC ..	1660	Raleigh	Nov. 21, 1789	12	500	150	48,718	3,954	52,672	29
ND ..	1812	Bismarck	Nov. 2, 1889	39	340	211	68,994	1,710	70,704	18
OH ..	1788	Columbus	Mar. 1, 1803	17	220	220	40,953	3,875	44,828	34
OK ..	1889	Oklahoma City	Nov. 16, 1907	46	400	220	68,679	1,224	69,903	20
OR ..	1811	Salem	Feb. 14, 1859	33	360	261	96,002	1,129	97,132	10
PA...	1682	Harrisburg	Dec. 12, 1787	2	283	160	44,820	1,239	46,058	33
RI ...	1636	Providence	May 29, 1790	13	40	30	1,045	186	1,231	50
SC ..	1670	Columbia	May 23, 1788	8	260	200	30,111	1,078	31,189	40
SD ..	1859	Pierre	Nov. 2, 1889	40	380	210	75,896	1,225	77,121	17
TN ..	1769	Nashville	June 1, 1796	16	440	120	41,219	926	42,146	36
TX...	1682	Austin	Dec. 29, 1845	28	790	660	261,914	5,363	267,277	2
UT ..	1847	Salt Lake City	Jan. 4, 1896	45	350	270	82,168	2,736	84,904	13
VT...	1724	Montpelier	Mar. 4, 1791	14	160	80	9,249	366	9,615	43
VA...	1607	Richmond	June 25, 1788	10	430	200	39,598	2,729	42,326	35
WA ..	1811	Olympia	Nov. 11, 1889	42	360	240	66,581	4,055	70,637	19
WV ..	1727	Charleston	June 20, 1863	35	240	130	24,087	145	24,231	41
WI...	1766	Madison	May 29, 1848	30	310	260	54,314	11,186	65,499	22
WY ..	1834	Cheyenne	July 10, 1890	44	360	280	97,105	714	97,818	9

Note: Land and water areas may not add to totals because of rounding. NA=Not applicable. (1) First permanent European settlement. (2) Rank is based on total area, including inland and coastal waters. (3) Aleutian Islands and Alexander Archipelago are not considered in these measurements.

The Continental Divide of the U.S.

The Continental Divide of the U.S., also known as the Great Divide, is located at the watershed created by the mountain ranges, or tablelands, of the Rocky Mountains. This watershed separates the waters that drain easterly into the Atlantic Ocean and its marginal seas, such as the Gulf of Mexico, from those waters that drain westerly into the Pacific Ocean. The majority of easterly flowing water in the U.S. drains into the Gulf of Mexico before reaching the Atlantic Ocean. The majority of westerly flowing water, before reaching the Pacific Ocean, drains either through the Columbia River or through the Colorado River, which flows into the Gulf of California before reaching the Pacific Ocean.

The location and route of the Continental Divide across the U.S. can briefly be described as follows:

Beginning at point of crossing the U.S.-Mexican boundary, near long. 108°45′ W, the Divide, in a northerly direction,

crosses New Mexico along the western edge of the Rio Grande drainage basin, entering Colorado near long. 106°41′ W.

From there by a very irregular route north across Colorado along the W summits of the Rio Grande and of the Arkansas, the South Platte, and the North Platte river basins, and across Rocky Mountain National Park, entering Wyoming near long. 106°52′ W.

From there in a northwesterly direction, forming the W rims of the North Platte, the Big Horn, and the Yellowstone river basins, crossing the SW portion of Yellowstone National Park.

From there in a westerly and then a northerly direction forming the common boundary of Idaho and Montana, to a point on said boundary near long. 114°00′ W.

From there northeasterly and northwesterly through Montana and the Glacier National Park, entering Canada near long. 114°04′ W.

Chronological List of Territories, With State Admissions to Union

Source: National Archives and Records Service

Name of territory	Date of Organic Act creating territory	Organic Act effective	Admission as state	Yrs. terr.
Northwest Territory[1]	July 13, 1787	No fixed date	Mar. 1, 1803[2]	16
Territory southwest of River Ohio	May 26, 1790	No fixed date	June 1, 1796[3]	6
Mississippi	Apr. 7, 1798	When president acted	Dec. 10, 1817	19
Indiana	May 7, 1800	July 4, 1800	Dec. 11, 1816	16
Orleans	Mar. 26, 1804	Oct. 1, 1804	Apr. 30, 1812[4]	7
Michigan	Jan. 11, 1805	June 30, 1805	Jan. 26, 1837	31
Louisiana-Missouri[5]	Mar. 3, 1805	July 4, 1805	Aug. 10, 1821	16
Illinois	Feb. 3, 1809	Mar. 1, 1809	Dec. 3, 1818	9
Alabama	Mar. 3, 1817	When MS became a state	Dec. 14, 1819	2
Arkansas	Mar. 2, 1819	July 4, 1819	June 15, 1836	17
Florida	Mar. 30, 1822	No fixed date	Mar. 3, 1845	23
Wisconsin	Apr. 20, 1836	July 3, 1836	May 29, 1848	12
Iowa	June 12, 1838	July 3, 1838	Dec. 28, 1846	8
Oregon	Aug. 14, 1848	Date of act	Feb. 14, 1859	10
Minnesota	Mar. 3, 1849	Date of act	May 11, 1858	9
New Mexico	Sept. 9, 1850	On president's proclamation	Jan. 6, 1912	61
Utah	Sept. 9, 1850	Date of act	Jan. 4, 1896	46
Washington	Mar. 2, 1853	Date of act	Nov. 11, 1889	36
Nebraska	May 30, 1854	Date of act	Mar. 1, 1867	12
Kansas	May 30, 1854	Date of act	Jan. 29, 1861	6
Colorado	Feb. 28, 1861	Date of act	Aug. 1, 1876	15
Nevada	Mar. 2, 1861	Date of act	Oct. 31, 1864	3
Dakota	Mar. 2, 1861	Date of act	Nov. 2, 1889	28
Arizona	Feb. 24, 1863	Date of act	Feb. 14, 1912	49
Idaho	Mar. 3, 1863	Date of act	July 3, 1890	27
Montana	May 26, 1864	Date of act	Nov. 8, 1889	25
Wyoming	July 25, 1868	When officers were qualified	July 10, 1890	22
Alaska[6]	May 17, 1884	No fixed date	Jan. 3, 1959	75
Oklahoma	May 2, 1890	Date of act	Nov. 16, 1907	17
Hawaii	Apr. 30, 1900	June 14, 1900	Aug. 21, 1959	59

(1) Included what is now Ohio, Indiana, Illinois, Michigan, Wisconsin, eastern Minnesota. (2) Whole territory admitted as the state of Ohio. (3) Admitted as the state of Tennessee. (4) Admitted as the state of Louisiana. (5) The organic act for Missouri Territory of June 4, 1812, became effective Dec. 7, 1812. (6) Although the May 17, 1884, act actually constituted Alaska as a district, it was often referred to as a territory, and unofficially administered as such. The Territory of Alaska was legally and formally organized by an act of Aug. 24, 1912.

Geographic Centers, U.S. and Each State

Source: U.S. Geological Survey, Dept. of the Interior

There is no generally accepted definition of geographic center and no uniform method for determining it. Following the U.S. Geological Survey, the geographic center of an area is defined here as the center of gravity of the surface, or that point on which the surface would balance if it were a plane of uniform thickness. All locations in the following list are approximate.

No marked or monumented point has been established by any government agency as the geographic center of the 50 states, the conterminous U.S. (48 states), or the North American continent. However, a group of citizens erected a monument in Lebanon, KS, marking it as geographic center of the conterminous U.S., and a cairn in Rugby, ND, designates that location as the center of the North American continent.

United States, including Alaska and Hawaii—W of Castle Rock, Butte County, South Dakota; lat. 44°58′ N, long. 103°46′ W

Conterminous U.S. (48 states)—Near Lebanon, Smith Co., Kansas, lat. 39°50′ N, long. 98°35′ W

North American continent—6 mi W of Balta, Pierce County, North Dakota; lat. 48°10′ N, long. 100°10′ W

STATE—COUNTY, LOCALITY OF CENTER

Alabama—Chilton, 12 mi SW of Clanton
Alaska—lat. 63°50′ N, long. 152°W; approx. 60 mi NW of Mt. McKinley
Arizona—Yavapai, 55 mi E-SE of Prescott
Arkansas—Pulaski, 12 mi NW of Little Rock
California—Madera, 38 mi E of Madera
Colorado—Park, 30 mi NW of Pikes Peak
Connecticut—Hartford, at East Berlin
Delaware—Kent, 11 mi S of Dover
District of Columbia—Near 4th and L Sts. NW
Florida—Hernando, 12 mi N-NW of Brooksville
Georgia—Twiggs, 18 mi SE of Macon
Hawaii—Hawaii, lat. 20°15′ N, long. 156°20′ W, off Maui Is.
Idaho—Custer, SW of Challis
Illinois—Logan, 28 mi NE of Springfield
Indiana—Boone, 14 mi N-NW of Indianapolis
Iowa—Story, 5 mi NE of Ames
Kansas—Barton, 15 mi NE of Great Bend
Kentucky—Marion, 3 mi N-NW of Lebanon
Louisiana—Avoyelles, 3 mi SE of Marksville
Maine—Piscataquis, 18 mi N of Dover
Maryland—Prince George's, 4.5 mi NW of Davidsonville
Massachusetts—Worcester, N part of city

Michigan—Wexford, 5 mi N-NW of Cadillac
Minnesota—Crow Wing, 10 mi SW of Brainerd
Mississippi—Leake, 9 mi W-NW of Carthage
Missouri—Miller, 20 mi SW of Jefferson City
Montana—Fergus, 11 mi W of Lewistown
Nebraska—Custer, 10 mi NW of Broken Bow
Nevada—Lander, 26 mi SE of Austin
New Hampshire—Belknap, 3 mi E of Ashland
New Jersey—Mercer, 5 mi SE of Trenton
New Mexico—Torrance, 12 mi S-SW of Willard
New York—Madison, 12 mi S of Oneida and 26 mi SW of Utica
North Carolina—Chatham, 10 mi NW of Sanford
North Dakota—Sheridan, 5 mi SW of McClusky
Ohio—Delaware, 25 mi N-NE of Columbus
Oklahoma—Oklahoma, 8 mi N of Oklahoma City
Oregon—Crook, 25 mi S-SE of Prineville
Pennsylvania—Centre, 2.5 mi SW of Bellefonte
Rhode Island—Kent, 1 mi S-SW of Crompton
South Carolina—Richland, 13 mi SE of Columbia
South Dakota—Hughes, 8 mi NE of Pierre
Tennessee—Rutherford, 5 mi NE of Murfreesboro
Texas—McCulloch, 15 mi NE of Brady
Utah—Sanpete, 3 mi N of Manti
Vermont—Washington, 3 mi E of Roxbury
Virginia—Buckingham, 5 mi SW of Buckingham
Washington—Chelan, 10 mi W-SW of Wenatchee
West Virginia—Braxton, 4 mi E of Sutton
Wisconsin—Wood, 9 mi SE of Marshfield
Wyoming—Fremont, 58 mi E-NE of Lander

International Boundary Lines of the U.S.

The length of the N boundary of the conterminous U.S.—the U.S.-Canadian border, excluding Alaska—is 3,987 mi according to the U.S. Geological Survey, Dept. of the Interior. The length of the Alaskan-Canadian border is 1,538 mi. The length of the U.S.-Mexican border, from the Gulf of Mexico to the Pacific Ocean, is approximately 1,933 mi (1963 boundary agreement).

Origins of the Names of U.S. States

Source: State officials, Smithsonian Institution, and Topographic Division, U.S. Geological Survey, Dept. of the Interior

Alabama—Indian for tribal town, later a tribe (Alabamas or Alibamons) of the Creek confederacy.

Alaska—Russian version of Aleutian (Eskimo) word, *alakshak*, for "peninsula," "great lands," or "land that is not an island."

Arizona—Spanish version of Pima Indian word for "little spring place," or Aztec *arizuma*, meaning "silver-bearing."

Arkansas—Algonquin name for the Quapaw Indians, meaning "south wind."

California—Bestowed by the Spanish conquistadors (possibly by Cortez). It was the name of an imaginary island, an earthly paradise, in *Las Serges de Esplandian*, a Spanish romance written by Montalvo in 1510. *Baja California* (Lower California, in Mexico) was first visited by Spanish in 1533. The present U.S. state was called *Alta* (Upper) *California*.

Colorado—From Spanish for "red," first applied to Colorado River.

Connecticut—From Mohican and other Algonquin words meaning "long river place."

Delaware—Named for Lord De La Warr, early governor of Virginia; first applied to river, then to Indian tribe (Lenni-Lenape), and the state.

District of Columbia—For Christopher Columbus, 1791.

Florida—Named by Ponce de Leon *Pascua Florida*, "Flowery Easter," on Easter Sunday, 1513.

Georgia—For King George II of England, by James Oglethorpe, colonial administrator, 1732.

Hawaii—Possibly derived from native word for homeland, *Hawaiki* or *Owhyhee*.

Idaho—Said to be a coined name with an invented meaning: "gem of the mountains"; originally suggested for the Pikes Peak mining territory (Colorado), then applied to the new mining territory of the Pacific Northwest. Another theory suggests *Idaho* may be a Kiowa Apache term for the Comanche.

Illinois—French for *Illini* or "land of *Illini*," Algonquin word meaning "men" or "warriors."

Indiana—Means "land of the Indians."

Iowa—Indian word variously translated as "here I rest" or "beautiful land." Named for the Iowa R., which was named for the Iowa Indians.

Kansas—Sioux word for "south wind people."

Kentucky—Indian word that is variously translated as "dark and bloody ground," "meadowland," and "land of tomorrow."

Louisiana—Part of territory called Louisiana by Sieur de La Salle for French King Louis XIV.

Maine—From Maine, ancient French province. Also: descriptive, referring to the mainland as distinct from the many coastal islands.

Maryland—For Queen Henrietta Maria, wife of Charles I of England.

Massachusetts—From Indian tribe named after "large hill place" identified by Capt. John Smith as being near Milton, MA.

Michigan—From Chippewa words, *mici gama*, meaning "great water," after the lake of the same name.

Minnesota—From Dakota Sioux word meaning "cloudy water" or "sky-tinted water" of the Minnesota River.

Mississippi—Probably Chippewa; *mici zibi*, "great river" or "gathering-in of all the waters." Also: Algonquin word, *messipi*.

Missouri—An Algonquin Indian term meaning "river of the big canoes."

Montana—Latin or Spanish for "mountainous."

Nebraska—From Omaha or Otos Indian word meaning "broad water" or "flat river," describing the Platte River.

Nevada—Spanish, meaning "snow-clad."

New Hampshire—Named, 1629, by Capt. John Mason of Plymouth Council for his home county in England.

New Jersey—The Duke of York, 1664, gave a patent to John Berkeley and Sir George Carteret to be called Nova Caesaria, or New Jersey, after England's Isle of Jersey.

New Mexico—Spaniards in Mexico applied term to land north and west of Rio Grande in the 16th century.

New York—For Duke of York and Albany, who received patent to New Netherland from his brother Charles II and sent an expedition to capture it, 1664.

North Carolina—In 1619 Charles I gave a large patent to Sir Robert Heath to be called Province of Carolana, from *Carolus*, Latin name for Charles. A new patent was granted by Charles II to Earl of Clarendon and others. Divided into North and South Carolina, 1710.

North Dakota—*Dakota* is Sioux for "friend" or "ally."

Ohio—Iroquois word for "fine or good river."

Oklahoma—Choctaw word meaning "red man," proposed by Rev. Allen Wright, Choctaw-speaking Indian.

Oregon—Origin unknown. One theory holds that the name may have been derived from that of the Wisconsin River, shown on a 1715 French map as "Ouaricon-sint."

Pennsylvania—William Penn, the Quaker who was made full proprietor of this area by King Charles II in 1681, suggested "Sylvania," or "woodland," for his tract. The king's government owed Penn's father, Admiral William Penn, 16,000 pounds, and the land was granted as partial settlement. Charles II added the "Penn" to Sylvania, against the desires of the modest proprietor, in honor of the admiral.

Puerto Rico—Spanish for "rich port."

Rhode Island—Exact origin is unknown. One theory notes that Giovanni de Verrazano recorded an island about the size of Rhodes in the Mediterranean in 1524, but others believe the state was named *Roode Eylandt* by Adriaen Block, Dutch explorer, because of its red clay.

South Carolina—See North Carolina.

South Dakota—See North Dakota.

Tennessee—*Tanasi* was the name of Cherokee villages on the Little Tennessee River. From 1784 to 1788 this was the State of Franklin, or Frankland.

Texas—Variant of word used by Caddo and other Indians meaning "friends" or "allies," and applied to them by the Spanish in eastern Texas. Also written *Texias, Tejas, Teysas*.

Utah—From a Navajo word meaning "upper," or "higher up," as applied to a Shoshone tribe called Ute. Spanish form is *Yutta*. The English is *Uta* or *Utah*. Proposed name *Deseret*, "land of honeybees," from Book of Mormon, was rejected by Congress.

Vermont—From French words *vert* (green) and *mont* (mountain). The Green Mountains were said to have been named by Samuel de Champlain. When the state was formed, 1777, Dr. Thomas Young suggested combining *vert* and *mont* into Vermont.

Virginia—Named by Sir Walter Raleigh, who fitted out the expedition of 1584, in honor of Queen Elizabeth, the Virgin Queen of England.

Washington—Named after George Washington. When the bill creating the Territory of Columbia was introduced in the 32d Congress, the name was changed to Washington because of the existence of the District of Columbia.

West Virginia—So named when western counties of Virginia refused to secede from the U.S. in 1863.

Wisconsin—An Indian name, spelled *Ouisconsin* and *Mesconsin* by early chroniclers. Believed to mean "grassy place" in Chippewa. Congress made it *Wisconsin*.

Wyoming—From the Algonquin words for "large prairie place," "at the big plains," or "on the great plain."

Territorial Sea of the U.S.

According to a Dec. 27, 1988, proclamation by Pres. Ronald Reagan: "The territorial sea of the United States henceforth extends to 12 nautical miles from the baselines of the United States determined in accordance with international law. In accordance with international law, as reflected in the applicable provisions of the 1982 United Nations Convention on the Law of the Sea, within the territorial sea of the United States, the ships of all countries enjoy the right of innocent passage and the ships and aircraft of all countries enjoy the right of transit passage through international straits."

Accession of Territory by the U.S.

Source: U.S. Dept. of the Interior; Bureau of the Census, U.S. Dept. of Commerce

	Acquisition date	Land area (sq mi)[1]		Acquisition date	Land area (sq mi)[1]		Acquisition date	Land area (sq mi)[1]
TOTAL U.S.[2]	NA	**3,540,305**	Texas	1845	388,687	Puerto Rico[5]	1899	3,427
50 states and			Oregon Territory.....	1846	286,541	Guam[6]	1899	210
Washington, DC...	NA	3,536,278	Mexican Cession	1848	529,189	American Samoa[7] ...	1900	77
Territory in 1790[3]	NA	895,415	Gadsden Purchase...	1853	29,670	U.S. Virgin Islands ...	1917	134
Louisiana Purchase[4]	1803	909,380	Alaska.............	1867	570,374	N Mariana Islands ...	1986	179
Purchase of Florida ..	1819	58,666	Hawaii.............	1898	6,423	All other[8]	NA	15

NA=not applicable. (1) Area figures from the Bureau of the Census, Apr. 1, 1990. As a result of independent rounding, the sum of these figures does not equal the total. (2) Includes outlying areas. (3) Includes that part of a drainage basin of Red River of the North, S of 49th parallel, sometimes considered part of Louisiana Purchase. (4) Also acquired areas W of the Mississippi River amounting to 22,834 sq mi, but relinquished to Spain 97,150 sq mi, or a net loss of 15,650 sq mi. (5) Ceded by Spain in 1898, ratified in 1899, and became the Commonwealth of Puerto Rico by Act of Congress on July 25, 1952. (6) Acquired 1898; ratified 1899. (7) Acquired 1899; ratified 1900. (8) Consisting of the following islands, with gross areas as indicated in sq mi: Midway (2), Wake (3), Palmyra Atoll (combined area, 5), Navassa (2), Baker, Howland, and Jarvis (combined area, 3), Johnston Atoll (combined area, 1), and Kingman Reef (less than 0.5).

Federally Owned Land, by State, 1997

Source: Office of Governmentwide Policy, General Services Administration; as of Sept. 30, 1997

State	Federal acreage[1]	Total acreage of state[2]	Percentage of federally owned acreage[1]	State	Federal acreage[1]	Total acreage of state[2]	Percentage of federally owned acreage[1]
AL....	1,080,003.8	32,678,400	3.305	MT.....	25,485,346.3	93,271,040	27.324
AK	171,787,844.1	365,481,600	47.003	NE	514,976.6	49,031,680	1.050
AZ....	31,336,525.9	72,688,000	43.111	NV	56,081,559.1	70,264,320	79.815
AR	2,739,635.8	33,599,360	8.154	NH.....	734,448.1	5,768,960	12.731
CA	44,757,473.9	100,206,720	44.665	NJ	101,564.1	4,813,440	2.110
CO	24,128,981.8	56,485,760	36.292	NM.....	26,218,230.1	77,766,400	33.712
CT	6,909.8	3,135,360	0.220	NY	196,621.4	30,680,960	0.641
DE	1,916.0	1,265,920	0.151	NC.....	2,023,456.1	31,402,880	6.460
DC	9,155.2	39,040	23.451	ND	1,412,803.4	44,452,480	3.178
FL.....	2,644,547.2	34,721,280	7.617	OH.....	279,601.5	26,222,080	1.066
GA	1,458,131.9	37,295,360	3.910	OK.....	677,617.5	44,087,680	1.537
HI	350,151.4	4,105,600	8.529	OR.....	31,809,283.0	61,598,720	51.640
ID	32,991,904.9	52,933,120	62.328	PA	622,551.0	28,804,480	2.161
IL	404,913.1	35,795,200	1.131	RI.....	3,100.1	677,120	0.458
IN	394,186.1	23,158,400	1.702	SC.....	934,567.5	19,374,080	4.824
IA	29,742.6	35,860,480	0.083	SD	2,577,446.3	48,881,920	5.273
KS	349,708.0	52,510,720	0.666	TN	1,576,100.7	26,727,680	5.897
KY	1,082,523.7	25,512,320	4.243	TX	2,008,137.6	168,217,600	1.194
LA....	744,782.4	28,867,840	2.580	UT	33,898,254.6	52,696,960	64.327
ME	145,279.4	19,847,680	0.732	VT	376,551.1	5,936,640	6.343
MD	157,419.5	6,319,360	2.491	VA	2,279,369.0	25,496,320	8.940
MA ...	52,108.6	5,034,880	1.035	WA.....	11,938,958.4	42,693,760	27.964
MI	3,979,627.2	36,492,160	10.905	WV.....	1,077,285.4	15,410,560	6.991
MN ...	4,068,850.2	51,205,760	7.946	WI	1,733,205.5	35,011,200	4.950
MS	1,276,358.8	30,222,720	4.223	WY.....	30,878,165.7	62,343,040	49.530
MO	1,658,262.1	44,248,320	3.748	**TOTAL**..	**563,031,190.2**	**2,271,343,360**	**24.791**

Note: Totals do not include inland water. (1) Excludes trust properties. (2) Bureau of the Census, U.S. Dept. of Commerce figures.

Special Recreation Areas Administered by the U.S. Forest Service, 1998

Source: U.S. Forest Service, Dept. of Agriculture

Area name	Location	Estab.	Acres	Area name	Location	Estab.	Acres
Admiralty Island	AK	1980	978,881	Mount Rogers	VA......	1966	114,520
Allegheny	PA	1984	23,063	Mount St. Helens..........	WA	1989	112,593
Arapaho	CO.......	1978	30,690	Newberry	OR	1990	54,822
Beech Creek	OK.......	1988	7,500	North Cascades	WA.....	1984	87,600
Cascade Head	OR.......	1974	6,630	Opal Creek	OR	1996	13,000
Columbia River Gorge	OR-WA ...	1986	63,150	Oregon Dunes	OR	1972	27,212
Coosa Bald................	GA.......	1991	7,100	Pine Ridge................	NE	1986	6,600
Ed Jenkins	GA.......	1991	23,166	Rattlesnake	MT	1980	59,119
Flaming Gorge	WY-UT....	1968	189,825	Sawtooth	ID	1972	729,322
Grand Island..............	MI	1990	12,961	Smith River	CA	1990	305,169
Hells Canyon	ID-OR	1975	536,648	Spring Mt.	NV	1993	316,000
Indian Nations	OK.......	1988	40,051	Spruce Knob-Seneca			
Jemez....................	NM.......	1993	57,000	Rocks	WV	1965	57,237
Misty Fiords	AK	1980	2,293,428	Whiskeytown-Shasta-			
Mono Basin	CA	1984	115,600	Trinity..................	CA	1965	176,367
Mount Baker..............	WA.......	1984	8,473	White Rocks	VT.......	1984	36,400
Mount Pleasant	VA	1994	7,580	Winding Stair Mt..........	OK	1988	25,890

National Parks, Other Areas Administered by National Park Service

Dates when sites were authorized for initial protection by Congress or by presidential proclamation are given in parentheses. If different, the date the area was given its current designation, or was transferred to the National Park Service, follows. Gross area in acres, as of Dec. 31, 1998, follows date(s). More than 83 mil acres of federal land are now administered by the National Park Service.

NATIONAL PARKS

Acadia, ME (1916/1929) 47,738. Includes Mount Desert Isl., half of Isle au Haut, Schoodic Peninsula on mainland. Highest elevation on Eastern seaboard.

American Samoa, AS (1988) 9,000. Features a paleotropical rain forest and a coral reef. No federal facilities.

Arches, UT (1929/1971) 73,379. Contains giant red sandstone arches and other products of erosion.

Badlands, SD (1929/1978) 242,756. Prairie with bison, bighorn, and antelope. Contains animal fossils from 26 to 37 mil years ago.

Big Bend, TX (1935) 801,163. Rio Grande, Chisos Mts.

Biscayne, FL (1968/1980) 172,924. Aquatic park encompassing chain of islands south of Miami.

Bryce Canyon, UT (1923/1928) 35,835. Spectacularly colorful and unusual display of erosion effects.

Canyonlands, UT (1964) 337,598. At junction of Colorado and Green rivers; extensive evidence of prehistoric Indians.

Capitol Reef, UT (1937/1971) 241,904. A 70-mi uplift of sandstone cliffs dissected by high-walled gorges.

Carlsbad Caverns, NM (1923/1930) 46,766. Largest known caverns; not yet fully explored.

Channel Islands, CA (1938/1980) 249,354. Sea lion breeding place, nesting sea birds, unique plants.

Crater Lake, OR (1902) 183,224. Extraordinary blue lake in the crater of Mt. Mazama, a volcano that erupted about 7,700 years ago; deepest U.S. lake.

Death Valley, CA-NV (1933/1994) 3,367,628. Large desert area. Includes the lowest point in the Western Hemisphere; also includes Scottys Castle.

Denali, AK (1917/1980) 4,740,907. Name changed from Mt. McKinley NP. Contains highest mountain in U.S.; wildlife.

Dry Tortugas, FL (1935/1992) 64,700. Formerly Ft. Jefferson National Monument.

Everglades, FL (1934) 1,508,607. Largest remaining subtropical wilderness in continental U.S.

Gates of the Arctic, AK (1978/1984) 7,523,813. Vast wilderness in north central region. Limited federal facilities.

Glacier, MT (1910) 1,013,572. Superb Rocky Mt. scenery, numerous glaciers and glacial lakes. Part of Waterton-Glacier Intl. Peace Park established by U.S. and Canada in 1932.

Glacier Bay, AK (1925/1986) 3,224,794. Great tidewater glaciers that move down mountainsides and break up into the sea; much wildlife.

Grand Canyon, AZ (1893/1919) 1,217,403. Most spec-tacular part of Colorado River's greatest canyon.

Grand Teton, WY (1929) 309,993. Most impressive part of the Teton Mts., winter feeding ground of largest American elk herd.

Great Basin, NV (1922/1986) 77,180. Includes Wheeler Pk., Lexington Arch, and Lehman Caves.

Great Smoky Mountains, NC-TN (1926/1934) 521,621. Largest Eastern mountain range, magnificent forests.

Guadalupe Mountains, TX (1966) 86,416. Extensive Permian limestone fossil reef; tremendous earth fault.

Haleakala, HI (1916/1960) 28,350. Dormant volcano on Maui with large colorful craters.

Hawaii Volcanoes, HI (1916/1961) 209,695. Contains Kilauea and Mauna Loa, active volcanoes.

Hot Springs, AR (1832/1921) 5,549. Bathhouses are furnished with thermal waters from the park's 47 hot springs; these waters are used for bathing and drinking.

Isle Royale, MI (1931) 571,790. Largest island in Lake Superior, noted for its wilderness area and wildlife.

Joshua Tree, CA (1936/1994) 1,022,976. Desert region includes Joshua trees, other plant and animal life.

Katmai, AK (1918/1980) 3,674,530. "Valley of Ten Thousand Smokes," scene of 1912 volcanic eruption.

Kenai Fjords, AK (1978/1980) 669,983. Abundant marine mammals, birdlife; the Harding Icefield, one of the 4 major icecaps in U.S.

Kings Canyon, CA (1890/1940) 461,901. Mountain wilderness, dominated by Kings River Canyons and High Sierra; contains giant sequoias.

Kobuk Valley, AK (1978/1980) 1,750,698. Contains geological and recreational sites. Limited federal facilities.

Lake Clark, AK (1978/1980) 2,619,859. Across Cook Inlet from Anchorage. A scenic wilderness rich in fish and wildlife. Limited federal facilities.

Lassen Volcanic, CA (1907/1916) 106,372. Contains Lassen Peak, recently active volcano, and other volcanic phenomena.

Mammoth Cave, KY (1926/1941) 52,830. 144 mi of surveyed underground passages, beautiful natural formations, river 300 ft below surface.

Mesa Verde, CO (1906) 52,122. Most notable and best preserved prehistoric cliff dwellings in the U.S.

Mount Rainier, WA (1899) 235,613. Greatest single-peak glacial system in the U.S.

North Cascades, WA (1968) 504,781. Spectacular mountainous region with many glaciers, lakes.

Olympic, WA (1909/1938) 922,651. Mountain wilderness containing finest remnant of Pacific Northwest rain forest, active glaciers, Pacific shoreline, rare elk.

Petrified Forest, AZ (1906/1962) 93,533. Extensive petrified wood and Indian artifacts. Contains part of Painted Desert.

Redwood, CA (1968) 112,430. 40 mi of Pacific coastline, groves of ancient redwoods and world's tallest trees.

Rocky Mountain, CO (1915) 265,723. On the Continental Divide; includes peaks over 14,000 ft.

Saguaro, AZ (1933/1994) 91,444. Part of the Sonoran Desert; includes the giant saguaro cacti, unique to the region.

Sequoia, CA (1890) 402,510. Groves of giant sequoias, highest mountain in conterminous U.S.—Mt. Whitney (14,494 ft). World's largest tree.

Shenandoah, VA (1926) 198,182. Portion of the Blue Ridge Mts.; overlooks Shenandoah Valley; Skyline Drive.

Theodore Roosevelt, ND (1947/1978) 70,447. Contains part of T.R.'s ranch and scenic badlands.

Virgin Islands, VI (1956) 14,689. Authorized to cover 75% of St. John Isl. and Hassel Isl.; lush growth, lovely beaches, Carib Indian petroglyphs, evidence of colonial Danes.

Voyageurs, MN (1971) 218,200. Abundant lakes, forests, wildlife, canoeing, boating.

Wind Cave, SD (1903) 28,295. Limestone caverns in Black Hills. Extensive wildlife includes a herd of bison.

Wrangell-St. Elias, AK (1978/1980) 8,323,618. Largest area in park system, most peaks over 16,000 ft, abundant wildlife; day's drive east of Anchorage. Limited federal facilities.

Yellowstone, ID-MT-WY (1872) 2,219,791. World's first national park. World's greatest geyser area has about 10,000 geysers and hot springs; spectacular falls and impressive canyons of the Yellowstone River; grizzly bear, moose, and bison.

Yosemite, CA (1890) 761,266. Yosemite Valley, the nation's highest waterfall, grove of sequoias, and mountains.

Zion, UT (1909/1919) 146,592. Unusual shapes and landscapes have resulted from erosion and faulting; evidence of past volcanic activity; Zion Canyon, with sheer walls ranging up to 2,640 ft, is readily accessible.

NATIONAL HISTORICAL PARKS

Appomattox Court House, VA (1930/1954) 1,775. Where Lee surrendered to Grant.

Boston, MA (1974) 41. Includes Faneuil Hall, Old North Church, Bunker Hill, Paul Revere House.

Cane River Creole (and heritage area), LA (1994) 207. Preserves the Creole culture as it developed along the Cane R.

Chaco Culture, NM (1907/1980) 33,974. Ruins of pueblos built by prehistoric Indians.

Chesapeake and Ohio Canal, MD-DC-WV (1938/1971) 19,553. 184-mi historic canal; DC to Cumberland, MD.

Colonial, VA (1930/1936) 9,349. Includes most of Jamestown Isl., site of first successful English colony; Yorktown, site of Cornwallis's surrender to George Washington; and the Colonial Parkway.

Cumberland Gap, KY-TN-VA (1940) 20,454. Mountain pass of the Wilderness Road, which carried the first great migration of pioneers into America's interior.

Dayton Aviation Heritage, OH (1992) 86. Commemorates the area's aviation heritage.

George Rogers Clark, Vincennes, IN (1966) 26. Commemorates American defeat of British in West during Revolution.

Harpers Ferry, MD-VA-WV (1944/1963) 2,287. At the confluence of the Shenandoah and Potomac rivers, the site of John Brown's 1859 raid on the Army arsenal.

Hopewell Culture, OH (1923/1992) 1,245. Formerly Mound City Group National Monument.

Independence, PA (1948) 45. Contains several properties associated with the American Revolution and the founding of the U.S. Includes Independence Hall.

Jean Laffite (and preserve), LA (1907/1978) 20,020. Includes Chalmette, site of 1815 Battle of New Orleans; French Quarter.

Kalaupapa, HI (1980) 10,779. Molokai's former leper colony site and other historic areas.

Kaloko-Honokohau, HI (1978) 1,161. Preserves the native culture of Hawaii. No federal facilities.

Keweenaw, MI (1992) 1,870. Site of first significant copper mine in U.S. Federal facilities are under development.

Klondike Gold Rush, AK-WA (1976) 13,191. Alaskan Trails in 1898 Gold Rush. Museum in Seattle.

Lowell, MA (1978) 141. Textile mills, canal, 19th-cent. structures; park shows planned city of Industrial Revolution.

Lyndon B. Johnson, TX (1969/1980) 1,570. President's birthplace, boyhood home, ranch.

Marsh-Billings, VT (1992) 643. Boyhood home of pioneer conservationist George Perkins Marsh. No federal facilities.

Minute Man, MA (1959) 967. Where the colonial Minute Men battled the British, Apr. 19, 1775. Also contains Nathaniel Hawthorne's home.

Morristown, NJ (1933) 1,698. Sites of important military encampments during the American Revolution; Washington's headquarters, 1777, 1779-80.

Natchez, MS (1988) 108. Mansions, townhouses, and villas related to history of Natchez.

New Bedford Whaling, MA (1996) 20. Preserves structures and relics associated with the city's 19th-cent. whaling industry.

New Orleans Jazz, LA (1994) 4. Preserves, educates, and interprets jazz as it has evolved in New Orleans.

Nez Perce, ID (1965) 2,123. Illustrates the history and culture of the Nez Perce Indian country (38 separate sites).

Pecos, NM (1965/1990) 6,671. Ruins of ancient Pueblo of Pecos, archaeological sites, and 2 associated Spanish colonial missions from the 17th and 18th centuries.

Pu'uhonua o Honaunau, HI (1955/1978) 182. Until 1819, a sanctuary for Hawaiians vanquished in battle and for those guilty of crimes or breaking taboos.

Salt River Bay (and ecological preserve), St. Croix, VI (1992) 946. The only site known where, 500 years ago, members of a Columbus party landed on what is now territory of the U.S.

San Antonio Missions, TX (1978) 819. Four of finest Spanish missions in U.S., 18th-cent. irrigation system.

San Francisco Maritime, CA (1988) 28. Artifacts, photographs, and historic vessels related to the development of the Pacific Coast.

San Juan Island, WA (1966) 1,752. Commemorates peaceful relations between the U.S., Canada, and Great Britain since the 1872 boundary disputes.

Saratoga, NY (1938) 3,392. Scene of a major 1777 battle that became a turning point in the American Revolution.

Sitka, AK (1910/1972) 107. Scene of last major resistance of the Tlingit Indians to the Russians, 1804.

Tumacacori, AZ (1908/1990) 46. Historic Spanish Catholic mission building stands near the site first visited by Jesuit Father Kino in 1691.

Valley Forge, PA (1976) 3,466. Continental Army campsite in 1777-78 winter.

War in the Pacific, GU (1978) 1,992. Seven distinct units illustrating the Pacific theater of WWII. Limited federal facilities.

Women's Rights, NY (1980) 7. Seneca Falls site where Lucretia Mott, Elizabeth Cady Stanton began rights movement in 1848.

NATIONAL BATTLEFIELDS

Antietam, MD (1890/1978) 3,234. Battle here ended first Confederate invasion of North, Sept. 17, 1862.

Big Hole, MT (1910/1963) 656. Site of major battle with Nez Perce Indians.

Cowpens, SC (1929/1972) 842. American Revolution battlefield.

Fort Donelson, TN-KY (1928/1985) 552. Site of first major Union victory.

Fort Necessity, PA (1931/1961) 903. Site of first battle of French and Indian War.

Monocacy, MD (1934/1976) 1,647. Civil War battle in defense of Washington, DC, fought here, July 9, 1864.

Moores Creek, NC (1926/1980) 88. 1776 battle between Patriots and Loyalists commemorated here.

Petersburg, VA (1926/1962) 2,659. Scene of 10-month Union campaigns, 1864-65.

Stones River, TN (1927/1960) 712. Scene of battle that began federal offensive to trisect the Confederacy.

Tupelo, MS (1929/1961) 1. Site of crucial battle over Sherman's supply line, 1865.

Wilson's Creek, MO (1960/1970) 1,750. Scene of Civil War battle for control of Missouri.

NATIONAL BATTLEFIELD PARKS

Kennesaw Mountain, GA (1917/1935) 2,884. Site of two major battles of Atlanta campaign in Civil War.

Manassas, VA (1940) 5,212. Scene of two battles in Civil War, 1861 and 1862.

Richmond, VA (1936) 1,078. Site of battles defending Confederate capital.

NATIONAL BATTLEFIELD SITE

Brices Cross Roads, MS (1929) 1. Civil War battlefield.

NATIONAL MILITARY PARKS

Chickamauga and Chattanooga, GA-TN (1890) 8,119. Site of major Confederate victory, 1863.

Fredericksburg and Spotsylvania County, VA (1927/1933) 7,917. Sites of several major Civil War battles and campaigns.

Gettysburg, PA (1895/1933) 5,989. Site of decisive Confederate defeat in North and of Gettysburg Address.

Guilford Courthouse, NC (1917/1933) 220. American Revolution battle site.

Horseshoe Bend, AL (1956) 2,040. On Tallapoosa River, where Gen. Andrew Jackson's forces broke the power of the Upper Creek Indian Confederacy.

Kings Mountain, SC (1931/1933) 3,945. Site of American Revolution battle.

Pea Ridge, AR (1956) 4,300. Scene of Civil War battle.

Shiloh, TN (1894/1933) 3,973. Major Civil War battlesite; includes some well-preserved Indian burial mounds.

Vicksburg, MS (1899/1933) 1,736. Union victory gave North control of the Mississippi and split the Confederate forces.

NATIONAL MEMORIALS

Arkansas Post, AR (1960) 389. First permanent French settlement in the lower Mississippi River valley.

Arlington House, the Robert E. Lee Memorial, VA (1925/1972) 28. Lee's home overlooking the Potomac.

Chamizal, El Paso, TX (1966/1974) 55. Commemorates 1963 settlement of 99-year border dispute with Mexico.

Coronado, AZ (1941/1952) 4,750. Commemorates first European exploration of the Southwest.

DeSoto, FL (1948) 27. Commemorates 16th-cent. Spanish explorations.

Federal Hall, NY (1939/1955) 0.45. First seat of U.S. government under the Constitution.

Fort Caroline, FL (1950) 138. On St. Johns River, overlooks site of a French Huguenot colony.

Fort Clatsop, OR (1958) 125. Lewis and Clark encampment, 1805-6.

Franklin Delano Roosevelt, DC (1982) 8. Statues of Pres. Roosevelt and Eleanor Roosevelt, as well as waterfalls and gardens.

General Grant, NY (1958) 0.76. Tomb of Grant and wife.

Hamilton Grange, NY (1962) 0.11. Home of Alexander Hamilton.

Jefferson National Expansion Memorial, St. Louis, MO (1935) 193. Commemorates westward expansion.

Johnstown Flood, PA (1964) 164. Commemorates tragic flood of 1889.

Korean War Veterans, DC (1986) 2. Dedicated in 1995; honors those who served in the Korean War.

Lincoln Boyhood, IN (1962) 200. Lincoln grew up here.

Lincoln Memorial, DC (1911/1933) 107. Marble statue of the 16th U.S. president.

Lyndon B. Johnson Grove on the Potomac, DC (1973) 17. Overlooks the Potomac R.; vista of the Capital.

Mount Rushmore, SD (1925) 1,278. World-famous sculpture of 4 presidents.

Oklahoma City National Memorial, OK (1997) 6. Commemorates site of April 19, 1995, bombing which killed 168.

Perry's Victory and International Peace Memorial, Put-in-Bay, OH (1936/1972) 25. The world's most massive Doric column, constructed 1912-15, promotes pursuit of peace through arbitration and disarmament.

Roger Williams, Providence, RI (1965) 5. Memorial to founder of Rhode Island.

Thaddeus Kosciuszko, PA (1972) 0.02. Memorial to Polish hero of American Revolution.

Theodore Roosevelt Island, DC (1932/1933) 89. Statue of Roosevelt in wooded island sanctuary.

Thomas Jefferson Memorial, DC (1934) 18. Statue of Jefferson in an inscribed circular, colonnaded structure.

USS *Arizona*, HI (1980). 11. Memorializes American losses at Pearl Harbor.

Vietnam Veterans, DC (1980) 2. Black granite wall inscribed with names of those missing or killed in action in the Vietnam War.

Washington Monument, DC (1848/1933) 106. Obelisk honoring the first U.S. president.

Wright Brothers, NC (1927/1953) 428. Site of first powered flight.

NATIONAL HISTORICAL SITES

Abraham Lincoln Birthplace, Hodgenville, KY (1916/1959) 337. Early 17th-cent. cabin.

Adams, Quincy, MA (1946/1952) 14. Home of Pres. John Adams, John Quincy Adams, and celebrated descendants.

Allegheny Portage Railroad, PA (1964) 1,249. Linked the Pennsylvania Canal system and the West.

Andersonville, Andersonville, GA (1970) 495. Noted Civil War prisoner-of-war camp.

Andrew Johnson, Greeneville, TN (1935/1963) 17. Two homes and the tailor shop of the 17th U.S. president.

Bent's Old Fort, CO (1960) 799. Reconstruction of S Plains outpost.

Boston African American, MA (1980) 0.18. Pre-Civil War black history structures.

Brown v. Board of Education, KS (1992) 2. Commemorates the landmark 1954 U.S. Supreme Court decision.

Carl Sandburg Home, Flat Rock, NC (1968) 264. Poet's home.

Charles Pinckney, SC (1988) 28. Statesman's farm.
Christiansted, St. Croix, VI (1952/1961) 27. Commemorates Danish colony.
Clara Barton, MD (1974) 9. Home of founder of American Red Cross.
Edgar Allan Poe, PA (1978/1980) 0.52. Writer's home.
Edison, West Orange, NJ (1955/1962) 21. Inventor's home and laboratory.
Eisenhower, Gettysburg, PA (1967) 690. Home of 34th president.
Eleanor Roosevelt, Hyde Park, NY (1977) 181. The former first lady's personal retreat.
Eugene O'Neill, Danville, CA (1976) 13. Playwright's home.
Ford's Theatre, DC (1866/1970) 0.29. Includes theater, now restored, where Lincoln was assassinated, house where he died, and Lincoln Museum.
Fort Bowie, AZ (1964) 1,000. Focal point of operations against Geronimo and the Apaches.
Fort Davis, TX (1961) 474. Key frontier outpost in West Texas.
Fort Laramie, WY (1938/1960) 833. Military post on Oregon Trail.
Fort Larned, KS (1964/1966) 718. Military post on Santa Fe Trail.
Fort Point, San Francisco, CA (1970) 29. West Coast fortification.
Fort Raleigh, NC (1941) 513. First attempted English settlement in North America.
Fort Scott, KS (1965/1978) 17. Commemorates U.S. frontier of 1840s and '50s.
Fort Smith, AR-OK (1961) 75. Active post during 1817-90.
Fort Union Trading Post, MT-ND (1966) 444. Principal fur-trading post on upper Missouri, 1829-67.
Fort Vancouver, WA (1948/1961) 209. Headquarters for Hudson's Bay Company in 1825. Early political seat.
Frederick Douglass, DC (1962/1988) 9. Home of famous black abolitionist, writer, and orator.
Frederick Law Olmsted, MA (1979) 7. Home of famous city planner.
Friendship Hill, PA (1978) 675. Home of Albert Gallatin, Jefferson's and Madison's secretary of treasury.
Golden Spike, UT (1957) 2,735. Commemorates completion of first transcontinental railroad in 1869.
Grant-Kohrs Ranch, MT (1972) 1,618. Ranch house and part of 19th-cent. ranch.
Hampton, MD (1948) 62. 18th-cent. Georgian mansion.
Harry S. Truman, MO (1983) 7. Home of Pres. Truman after 1919.
Herbert Hoover, West Branch, IA (1965) 187. Birthplace and boyhood home of 31st president.
Home of Franklin D. Roosevelt, Hyde Park, NY (1944) 290. FDR's birthplace, home, and "summer White House."
Hopewell Furnace, PA (1938/1985) 848. 19th-cent. iron-making village.
Hubbell Trading Post, AZ (1965) 160. Still active today.
James A. Garfield, Mentor, OH (1980) 8. Home of 20th president.
Jimmy Carter, GA (1987) 71. Birthplace and home of 39th president.
John Fitzgerald Kennedy, Brookline, MA (1967) 0.09. Birthplace and childhood home of 35th president.
John Muir, Martinez, CA (1964) 345. Home of early conservationist and writer.
Knife River Indian Villages, ND (1974) 1,758. Remnants of villages last occupied by Hidatsa and Mandan Indians.
Lincoln Home, Springfield, IL (1971) 12. Lincoln's residence at the time he was elected 16th president, 1860.
Little Rock Central High School, AR (1998) 18. Commemorates 1957 desegregation during which federal troops had to be called in to protect 9 black students.
Longfellow, Cambridge, MA (1972) 2. Longfellow's home, 1837-82, and Washington's headquarters during Boston siege, 1775-76.
Maggie L. Walker, VA (1978) 1. Richmond home of black leader and bank president, daughter of an ex-slave.
Manzanar, Lone Pine, CA (1992) 814. Commemorates Manzanar War Relocation Ctr., a Japanese-American internment camp during WWII. No federal facilities.
Martin Luther King Jr., Atlanta, GA (1980) 34. Birthplace, grave, and church of the civil rights leader. Limited federal facilities.
Martin Van Buren, NY (1974) 40. Lindenwald, home of 8th president, near Kinderhook.
Mary McLeod Bethune Council House, DC (1982/1991) 0.07. Commemorates Bethune's leadership in the black women's movement.
Nicodemus, KS (1996) 161. Only remaining western town established by African-Americans during Reconstruction.
Ninety Six, SC (1976) 989. Colonial trading village.
Palo Alto Battlefield, TX (1978) 3,357. Scene of first battle of the Mexican War.
Pennsylvania Avenue, DC (1965) Acreage undetermined. Also includes area adjacent to the road between Capitol and

White House, encompassing Ford's Theatre and a number of other federal structures.
Puukohola Heiau, HI (1972) 86. Ruins of temple built by King Kamehameha.
Sagamore Hill, Oyster Bay, NY (1962) 83. Home of Pres. Theodore Roosevelt from 1885 until his death in 1919.
Saint-Gaudens, Cornish, NH (1964) 148. Home, studio, and gardens of American sculptor Augustus Saint-Gaudens.
Saint Paul's Church, NY, NY (1943) 6. Site associated with John Peter Zenger's "freedom of press" trial.
Salem Maritime, MA (1938) 9. Only port never seized from the patriots by the British. Major fishing and whaling port.
San Juan, PR (1949) 75. 16th-cent. Span. fortifications.
Saugus Iron Works, MA (1974) 9. Reconstructed 17th-cent. colonial ironworks.
Springfield Armory, MA (1974) 55. Small-arms manufacturing center for nearly 200 years.
Steamtown, PA (1986) 62. Railyard, roadhouse, repair shops of former Delaware, Lackawanna & Western Railroad.
Theodore Roosevelt Birthplace, New York, NY (1962) 0.11. Reconstructed brownstone.
Theodore Roosevelt Inaugural, Buffalo, NY (1966) 1. Wilcox House where he took oath of office, 1901.
Thomas Stone, MD (1978) 328. Home of signer of Declaration of Independence, built in 1771.
Tuskegee Airmen, AL (1998) 90. Airfield where pilots of all-black air corps unit of WWII received flight training.
Tuskegee Institute, AL (1974) 58. College founded by Booker T. Washington in 1881 for blacks.
Ulysses S. Grant, St. Louis Co., MO (1989) 10. Home of Grant during pre-Civil War years.
Vanderbilt Mansion, Hyde Park, NY (1940) 212. Mansion of 19th-cent. financier.
Washita Battlefield, OK (1996) 315. Scene of Nov. 27, 1868, battle between Plains tribes and the U.S. army.
Weir Farm, Wilton, CT (1990) 61. Home and studio of American impressionist painter J. Alden Weir.
Whitman Mission, WA (1936/1963) 98. Site where Dr. and Mrs. Marcus Whitman ministered to the Indians until slain by them in 1847.
William Howard Taft, Cincinnati, OH (1969) 3. Birthplace and early home of the 27th president.

NATIONAL MONUMENTS

Name	State	Year[1]	Acreage
Agate Fossil Beds	NE	1965	3,055
Alibates Flint Quarries	TX	1965	1,371
Aniakchak[2]	AK	1978	137,176
Aztec Ruins	NM	1923	318
Bandelier	NM	1916	33,677
Black Canyon of the Gunnison	CO	1933	20,766
Booker T. Washington	VA	1956	224
Buck Island Reef	VI	1961	880
Cabrillo	CA	1913	137
Canyon de Chelly	AZ	1931	83,840
Cape Krusenstern[3]	AK	1978	649,182
Capulin Volcano	NM	1916	793
Casa Grande Ruins	AZ	1889	473
Castillo de San Marcos	FL	1924	21
Castle Clinton	NY	1946	1
Cedar Breaks	UT	1933	6,155
Chiricahua	AZ	1924	11,985
Colorado	CO	1911	20,534
Congaree Swamp	SC	1976	21,867
Craters of the Moon	ID	1924	53,440
Devils Postpile	CA	1911	798
Devils Tower	WY	1906	1,347
Dinosaur	CO-UT	1915	210,844
Effigy Mounds	IA	1949	1,481
El Malpais	NM	1987	114,277
El Morro	NM	1906	1,279
Florissant Fossil Beds	CO	1969	5,998
Fort Frederica	GA	1936	241
Fort Matanzas	FL	1924	228
Fort McHenry National Monument and Historic Shrine	MD	1925	43
Fort Pulaski	GA	1924	5,623
Fort Stanwix	NY	1935	16
Fort Sumter	SC	1948	195
Fort Union	NM	1954	721
Fossil Butte	WY	1972	8,198
George Washington Birthplace	VA	1930	550
George Washington Carver	MO	1943	210
Gila Cliff Dwellings	NM	1907	533
Grand Portage	MN	1951	710
Great Sand Dunes	CO	1932	38,662
Hagerman Fossil Beds[3]	ID	1988	4,351
Hohokam Pima[4]	AZ	1972	1,690
Homestead Natl. Monument of America	NE	1936	195
Hovenweep	CO-UT	1923	785

Name	State	Year[1]	Acreage
Jewel Cave	SD	1908	1,274
John Day Fossil Beds	OR	1974	14,057
Lava Beds	CA	1925	46,560
Little Big Horn Battlefield	MT	1879	765
Montezuma Castle	AZ	1906	858
Muir Woods	CA	1908	554
Natural Bridges	UT	1908	7,636
Navajo	AZ	1909	360
Ocmulgee	GA	1934	702
Oregon Caves	OR	1909	488
Organ Pipe Cactus	AZ	1937	330,689
Petroglyph	NM	1990	7,240
Pinnacles	CA	1908	16,265
Pipe Spring	AZ	1923	40
Pipestone	MN	1937	282
Poverty Point[2]	LA	1988	911
Rainbow Bridge[3]	UT	1910	160
Russell Cave	AL	1961	310
Salinas Pueblo Missions	NM	1909	1,071
Scotts Bluff	NE	1919	3,003
Statue of Liberty	NJ-NY	1924	58
Sunset Crater Volcano	AZ	1930	3,040
Timpanogos Cave	UT	1922	250
Tonto	AZ	1907	1,120
Tuzigoot	AZ	1939	801
Walnut Canyon	AZ	1915	3,579
White Sands	NM	1933	143,733
Wupatki	AZ	1924	35,422
Yucca House[4]	CO	1919	34

NATIONAL PRESERVES

Aniakchak	AK	1978	465,603
Bering Land Bridge[3]	AK	1978	2,698,919
Big Cypress	FL	1974	720,573
Big Thicket	TX	1974	97,191
Denali	AK	1917	1,334,200
Gates of the Arctic	AK	1978	948,629
Glacier Bay	AK	1925	58,406
Katmai	AK	1918	418,699
Lake Clark	AK	1978	1,410,642
Little River Canyon[2]	AL	1992	13,633
Mojave	CA	1994	1,553,816
Noatak[3]	AK	1978	6,569,904
Tallgrass Prairie	KS	1996	10,894
Timucuan Ecological & Historic Preserve[3]	FL	1988	46,000
Wrangell-St. Elias	AK	1978	4,852,773
Yukon-Charley Rivers[3]	AK	1978	2,526,512

NATIONAL SEASHORES

Assateague Island	MD-VA	1965	39,723
Canaveral	FL	1975	57,662
Cape Cod	MA	1961	43,614
Cape Hatteras	NC	1937	30,319
Cape Lookout	NC	1966	28,243
Cumberland Island	GA	1972	36,416
Fire Island	NY	1964	19,579
Gulf Islands	FL-MS	1971	137,458
Padre Island	TX	1962	130,434
Point Reyes	CA	1962	71,060

NATIONAL PARKWAYS

Blue Ridge	NC-VA	1933	88,734
George Washington Memorial	VA-MD-DC	1930	7,248
John D. Rockefeller Jr. Mem.	WY	1972	23,777
Natchez Trace	MS-AL-TN	1938	51,747

NATIONAL LAKESHORES

Apostle Islands	WI	1970	69,372

Indiana Dunes	IN	1966	15,136
Pictured Rocks	MI	1966	73,236
Sleeping Bear Dunes	MI	1970	73,236

NATIONAL RESERVES

City of Rocks[3]	ID	1988	14,107
Ebey's Landing[3]	WA	1978	19,000

NATIONAL RIVERS

Big South Fork Natl. R and Recreation Area	KY-TN	1976	125,242
Buffalo	AR	1972	94,328
Mississippi Natl. R and Recreation Area	MN	1988	53,775
New River Gorge	WV	1978	70,902
Niobrara	NE-SD	1991	NA
Ozark	MO	1964	80,790

NATIONAL WILD AND SCENIC RIVERS

Alagnak	AK	1980	30,745
Bluestone[2]	WV	1978	4,310
Delaware	NY-NJ-PA	1978	1,973
Great Egg Harbor	NJ	1992	NA
Missouri	NE-SD	1991	NA
Obed	TN	1976	5,173
Rio Grande[2]	TX	1978	9,600
Saint Croix	MN-WI	1968	67,452
Upper Delaware	NY-PA	1978	75,000

NATIONAL RECREATION AREAS

Amistad	TX	1965	58,500
Bighorn Canyon	MT-WY	1966	120,296
Boston Harbor Islands	MA	1996	1,482
Chattahoochee R.	GA	1978	9,206
Chickasaw	OK	1902	9,889
Curecanti	CO	1965	41,972
Cuyahoga Valley	OH	1974	32,527
Delaware Water Gap	NJ-PA	1965	67,210
Gateway	NJ-NY	1972	26,612
Gauley R.[3]	WV	1988	11,342
Glen Canyon	AZ-UT	1958	1,254,306
Golden Gate	CA	1972	73,690
Lake Chelan	WA	1968	61,958
Lake Mead	AZ-NV	1936	1,495,666
Lake Meredith	TX	1965	44,978
Lake Roosevelt[5]	WA	1946	100,390
Ross Lake	WA	1968	117,575
Santa Monica Mts.[3]	CA	1978	153,726
Whiskeytown	CA	1965	42,503

NATIONAL SCENIC TRAILS

Appalachian	ME to GA	1968	213,548
Natchez Trace	MS-TN	1983	10,995
Potomac Heritage	MD-DC-VA-PA	1983	NA

PARKS (no other classification)

Catoctin Mountain	MD	1954	5,770
Constitution Gardens	DC	1974	52
Fort Washington	MD	1930	341
Greenbelt	MD	1950	1,176
National Capital	DC	1933	6,544
National Mall	DC	1933	146
Piscataway	MD	1961	4,372
Prince William Forest	VA	1948	18,572
Rock Creek	DC	1890	1,754
White House	DC	1933	18
Wolf Trap Farm Park for the Performing Arts	VA	1966	130

INTERNATIONAL HISTORIC SITE

Saint Croix Island[3]	ME	1949	45

NA=Not available. (1) Year first designated. (2) No federal facilities. (3) Limited federal facilities. (4) Not open to the public. (5) Formerly Coulee Dam National Recreation Area.

20 Most-Visited Sites in the National Park System, 1998

Source: National Park Service, Dept. of the Interior

Attendance at all areas administered by the National Park Service in 1998 totaled 286,762,265 recreation visits.

Site (location)	Recreation visits	Site (location)	Recreation visits
Blue Ridge Parkway (NC, VA)	19,026,498	Vietnam Veterans Memorial (DC)	4,687,299
Golden Gate National Recreation Area (CA)	14,046,590	Castle Clinton National Monument (NY)	4,390,268
Great Smoky Mountains National Park (TN, NC)	9,989,395	Lincoln Memorial (DC)	4,368,912
Lake Mead National Recreation Area (AZ, NV)	8,788,055	Gulf Islands National Seashore (FL, MS)	4,293,301
Gateway National Recreation Area (NY, NJ)	7,124,022	Franklin Delano Roosevelt National Memorial (DC)	4,258,807
George Washington Memorial National Parkway (VA, MD, DC)	6,584,802	Grand Canyon National Park (AZ)	4,239,682
Natchez Trace National Parkway (MS, AL, TN)	5,810,094	Yosemite National Park (CA)	3,657,132
Statue of Liberty National Monument (NY, NJ)	5,200,633	Olympic National Park (WA)	3,577,007
Delaware Water Gap National Recreation Area (PA, NJ)	5,019,175	San Francisco Maritime National Historical Park (CA)	3,535,081
Cape Cod National Seashore (MA)	4,804,185	Cuyahoga Valley National Recreation Area (OH)	3,467,107

Federal Indian Reservations and Trust Lands[1]

Source: Tiller Research, Inc., Albuquerque, NM

State	No. of reser.	Tribally owned acreage[2]	Individually owned acreage[2]	No. of persons[3]	Major tribes and/or nations
Alabama	1	230	0	16,506	Poarch Creek
Alaska	1[4]	86,773	1,265,432	85,698	Aleut, Eskimo, Athabascan,[5] Haida, Tlingit, Tsimpshian
Arizona	23	19,775,959	311,579	203,527	Navajo, Apache, Papago, Hopi, Yavapai, Pima
California	96	520,049	66,769	242,164	Hoopa, Paiute, Yurok, Karok, Cherokee
Colorado	2	764,120	2,805	27,776	Ute
Connecticut	1	1,638	0	6,654	Mashantucket Pequot
Florida	4	153,874	0	36,335	Seminole, Miccosukee, Cherokee
Idaho	4	609,622	327,301	13,780	Shoshone, Bannock, Nez Perce
Iowa	1	3,550	0	7,349	Sac and Fox
Kansas	4	7,219	23,763	21,965	Potawatomi, Kickapoo, Iowa
Louisiana	3	415	0	18,541	Chitimacha, Coushatta, Tunica-Biloxi
Maine	3	191,511	0	5,998	Passamaquoddy, Penobscot, Maliseet
Massachusetts	1	157	0	12,241	Wampanoag
Michigan	8	14,411	9,276	55,638	Chippewa, Potawatomi, Ottawa, Cherokee
Minnesota	14	779,138	50,338	49,909	Chippewa, Sioux
Mississippi	1	20,486	0	8,525	Choctaw
Montana	7	2,663,385	2,911,450	47,679	Blackfoot, Crow, Sioux, Assiniboine, Cheyenne
Nebraska	3	23,792	43,208	12,410	Omaha, Winnebago, Santee Sioux
Nevada	19	1,147,088	78,529	19,637	Paiute, Shoshone, Washoe
New Mexico	25	7,252,326	630,293	134,355	Apache, Navajo, Pueblo
New York	8	118,199	0	62,651	Seneca, Mohawk, Onondaga, Oneida
North Carolina	1	56,509	0	80,155	Cherokee, Lumbee
North Dakota	3	214,006	627,289	25,917	Sioux, Chippewa, Mandan, Arikara, Hidatsa
Oklahoma	36[6]	96,839	1,000,165	252,420	Cherokee, Creek, Choctaw, Chickasaw, Osage, Cheyenne, Arapahoe, Kiowa, Comanche
Oregon	7	660,367	135,053	38,496	Warm Springs, Wasco, Paiute, Umatilla, Siletz
Rhode Island	1	1,800	0	4,071	Narragansett
South Carolina	1	639	0	8,246	Catawba
South Dakota	9	2,399,531	2,121,188	50,575	Sioux
Texas	3	4,726	0	65,877	Alabama-Coushatta, Tiwa, Kickapoo
Utah	4	2,286,448	32,838	24,283	Ute, Goshute, Southern Paiute, Navajo
Washington	27	2,250,731	467,785	81,483	Yakama, Lummi, Quinault
Wisconsin	11	338,097	80,345	39,387	Chippewa, Oneida, Winnebago
Wyoming	1	1,958,095	101,537	9,479	Shoshone, Arapahoe

(1) In Oct. 1993, the Bureau of Indian Affairs of the U.S. Dept. of the Interior published in the *Federal Register* (vol. 58, no. 202, pp. 54364-69) a comprehensive listing of 552 "Indian Entities Recognized and Eligible to Receive Services From the United States Bureau of Indian Affairs" (328 in the conterminous 48 states, 224 in Alaska). The term *Indian entities* includes Indian tribes, bands, villages, groups, and pueblos; also included are Eskimo and Aleut villages and tribes. All such entities have a government-to-government relationship with the U.S. Some reservation boundaries transcend state boundaries (e.g., Navajo, which is in Arizona, New Mexico, and Utah). For the purpose of "Number of Reservations," such reservations are counted in the state where their population is predominant and/or tribal headquarters are located. (2) Information provided by the Bureau of Indian Affairs; data current as of 1990. Acreages refer only to lands that are either owned by the tribes and individual members or are held in trust by the U.S. government. Many of these parcels are located off reservations. Not all lands within reservation boundaries are necessarily trust lands. Many are privately owned by tribes, tribal members, or non-Indians; others are the property of various governmental agencies. (3) Total Native American (Indian, Eskimo, or Aleut) population in each state with reservation/trust lands, including those persons living outside the Bureau of Indian Affairs service area. Populations as of 1990. (4) The only federally recognized reservation in Alaska is the Annette Island Reserve. In all other cases, the U.S. government's relationship to Native Americans in Alaska is set out by the Alaska Native Claims Settlement Act of 1971. The act provided for the establishment of regional and village corporations to conduct business for profit and nonprofit purposes; these corporations are also landowners. There are 12 regional corporations, each with organized village corporations, plus one regional corporation for Alaska Natives outside the state. (5) Aleuts and Eskimos are racially and linguistically related. Athabascans are related to the Navajo and Apache Indians. (6) There are 36 tribal entities in Oklahoma, each of which owns land in the state. Because of the way in which the state was formed out of the Oklahoma and Indian territories, the reservation status of land in the state is frequently disputed in both civil and criminal proceedings.

Largest American Indian Tribes

Source: Bureau of the Census, U.S. Dept. of Commerce, as of 1990 census

Tribe	Number	Percent	Tribe	Number	Percent
ALL AMERICAN INDIANS	1,937,391	100.0	Chickasaw	21,522	1.1
Cherokee	369,035	19.0	Tohono O'Odham	16,876	0.9
Navajo	225,298	11.6	Potawatomi	16,719	0.9
Sioux	107,321[1]	5.5	Seminole	15,564	0.8
Chippewa	105,988	5.5	Pima	15,074	0.8
Choctaw	86,231	4.5	Tlingit	14,417	0.7
Pueblo	55,330	2.9	Alaskan Athabaskans	14,198	0.7
Apache	53,330	2.8	Cheyenne	11,809	0.6
Iroquois[2]	52,557	2.7	Comanche	11,437	0.6
Lumbee	50,888	2.6	Paiute	11,369	0.6
Creek	45,872	2.4	Osage	10,430	0.5
Blackfoot	37,992	2.0	Puget Sound Salish	10,384	0.5
Canadian and Latin American	27,179	1.4	Yaqui	9,838	0.5

(1) Any entry from NC with the spelling "Siouan" in the 1990 census was miscoded to count as Sioux. (2) Reporting and/or processing problems in the 1990 census have affected accuracy of the data for this tribe.

WORLD HISTORY

Prehistory: Our Ancestors Emerge

Revised by Susan Skomal, Ph.D., Editor, Anthropology Newsletter, *American Anthropological Association*

Homo sapiens. The precise origins of *Homo sapiens,* the species to which all humans belong, are subject to broad speculation based on a small, but increasing, number of fossils, on genetic and anatomical studies, and on interpretation of the geological record. Most scientists at least agree that humans evolved from apelike primate ancestors in a process that began millions of years ago.

Current theories trace the first hominid (humanlike primate) to Africa, where at least 2 lines of hominids appeared 5 to 7 million years before the present (BP). In one line was *Australopithecus,* a social animal that lived from perhaps 5 million to 3 million years BP, then apparently died out. In the other, human line was *Homo habilis,* a large-brained specimen that walked upright and had a dextrous hand. *Homo habilis* appeared some 2.5 million years BP, lived in semipermanent camps, had a food-gathering economy, and probably produced stone tools.

Homo erectus, the nearest ancestor to humans, appeared in Africa perhaps 2 million years BP and began spreading into Asia and Europe soon after. It had a fairly large brain and a skeletal structure similar to that of modern humans. *Homo erectus* hunted, learned to control fire, and may have had some primitive language skills. Brain development to *Homo sapiens,* then to the subspecies *Homo sapiens sapiens,* occurred between 500,000 and 50,000 years BP in Africa. All modern humans are members of the subspecies *Homo sapiens sapiens.*

Humans have roamed widely over the globe throughout their development. There is increasing evidence that migration from Asia to Australia via the Timor Straits took place as early as 100,000 BP. Evidence of hominids in Siberia dates as early as 300,000 BP. First confirmed evidence for the crossing from Asia to the Americas, by land bridge, dates to the end of the last Ice Age, at 12,500 BP.

Earliest cultures. A variety of cultural modes—in toolmaking, diet, shelter, and possibly social arrangements and spiritual expression—arose as humans adapted to different geographic and climatic zones and the database of knowledge grew. Sites from all over the world show seasonal migration patterns and efficient exploitation of a wide range of plant and animal foods.

Archaeologists recognize 5 basic toolmaking traditions as arising and often coexisting from more than 2.5 million years ago to the near past: (1) the *chopper tradition*—also known as the Oldowan—found in Africa, producing crude chopping tools and simple flake tools; (2) the *biface* or hand-ax tradition, found in Africa, W and S Europe, and S Asia, producing pointed hand axes chipped on both faces for cutting; (3) the *flake tradition,* found in Africa and Europe, producing small cutting and flaking tools; (4) the *blade tradition,* a more efficient technology characteristic of the Upper Paleolithic, found across Eurasia to Siberia and N Africa, producing many usable blades from a single stone; and (5) the *microlith tradition,* found throughout the inhabited world, producing specialized small tools for use as projectile points, in carving softer materials, and in making more complex tools.

Sketchy evidence remains for the stages in increasing control over the environment. Fire was used for heating and cooking by 465,000 BP in W France. Fire-hardened wooden spears, weighted and set with small stone blades, were fashioned by big-game hunters 400,000 years ago in Germany. Scraping tools found at certain sites (200,000-30,000 BP in Europe, N Africa, the Middle East, and Cent. Asia) suggest the treatment of skins for clothing. By the time Australia was settled, human ancestors had learned to navigate in boats over open water. The earliest bone tools found to date were developed 80,000 years ago in the Congo basin by fishermen, who created sophisticated fishing tackle to catch giant catfish.

Early human ancestors included artists and musicians. About 60,000 years ago the earliest immigrants to Australia carved and painted abstract designs on rocks. Painting and decoration flourished, along with stone and ivory sculpture, from 30,000 BP in Europe; more than 200 caves, mainly in S France and N Spain, show remarkable examples of naturalistic wall painting. Other examples have been found in Africa. Proto-religious rites are suggested by these works, and by evidence of ritual burial. A variety of musical instruments, including bone flutes with precisely bored holes, have been found in Paleolithic (early Stone Age) sites going back as far as 40,000-80,000 years BP.

Neolithic advances. Some time after 10,000 BC, among widely separated communities, a series of dramatic technological and social changes occurred, marking the Neolithic, or New Stone, Age. As the world climate became drier and warmer, humans learned to cultivate plants. This in turn encouraged growth of permanent settlements. Animals were domesticated. Manufacture of pottery and cloth began. These techniques permitted a dramatic increase in world population and social complexity, and accelerated humankind's ability to manipulate the environment.

Sites in N, Cent., and S America, SE Europe, and the Middle East show roughly contemporaneous (10,000-8000 BC) evidence of one or more Neolithic traits. Dates near 6000-3000 BC have been given for E and S Asian, W European, and sub-Saharan African Neolithic remains. The variety of crops—field grains, rice, maize, and roots—and varying mix of other characteristics suggest that this adaptation occurred independently in all these regions.

History Begins: 4000-1000 BC

Near Eastern cradle. If history began with writing, the first chapter opened in Mesopotamia, the Tigris-Euphrates river valley. The Sumerians used clay tablets with pictographs to keep records after 4000 BC. A **cuneiform** (wedge-shaped) script evolved by 3000 BC as a full syllabic alphabet. Neighboring peoples adapted the script to their own language.

Sumerian life centered, from 4000 BC, on large cities (Eridu, Ur, Uruk, Nippur, Kish, and Lagash) organized around temples and priestly bureaucracies, with surrounding plains watered by vast irrigation works and worked with traction plows. Sailboats, wheeled vehicles, potter's wheels, and kilns were used. Copper was smelted and tempered from c 4000 BC; bronze was produced not long after. Ores, as well as precious stones and metals, were obtained through long-distance ship and caravan trade. Iron was used from c 2000 BC. Improved ironworking, developed partly by the Hittites, became widespread by 1200 BC.

Sumerian political primacy passed among cities and their kingly dynasties. Semitic-speaking peoples, with cultures derived from the Sumerian, founded a succession of dynasties that ruled in Mesopotamia and neighboring areas for most of 1,800 years; among them were the **Akkadians** (first under Sargon I, c 2350 BC), the Amorites (whose laws, codified by **Hammurabi,** c 1792-1750 BC, have biblical parallels), and the Assyrians, with interludes of rule by the Hittites, Kassites, and Mitanni.

Mesopotamian learning, maintained by scribes and preserved in vast libraries, was practically oriented. Advances in mathematics related mostly to construction, commerce, and administration. Lists of astronomical phenomena, plants, animals, and stones were maintained; medical texts listed ailments and herbal cures. The Sumerians worshiped anthropomorphic gods representing natural forces, such as Anu, god of heaven, and Enlil (Ea), god of water. Sacrifices were made at **ziggurats**—huge stepped temples.

Paleontology: The History of Life

All dates are approximate, and are subject to change based on new fossil finds or new dating techniques,
but the sequence of events is generally accepted. Dates are in years before the present.

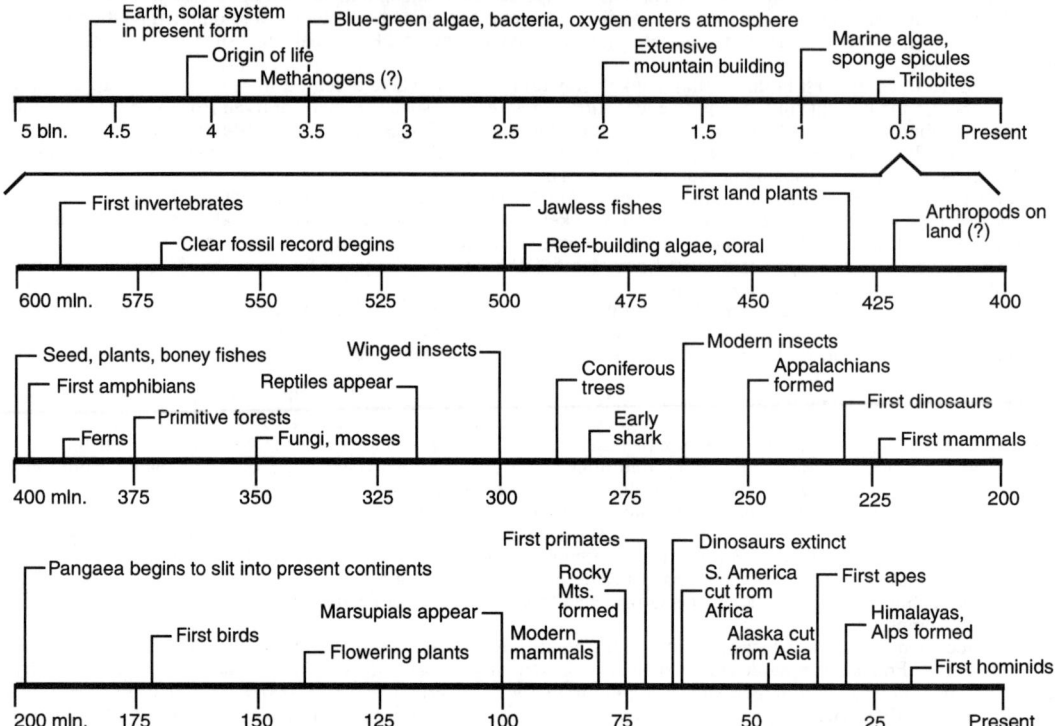

The Syria-Palestine area, site of some of the earliest urban remains (Jericho, 7000 BC), and of the recently uncovered **Ebla** civilization (fl 2500 BC), experienced Egyptian cultural and political influence along with Mesopotamian. The **Phoenician** coast was an active commercial center. A phonetic alphabet was invented here before 1600 BC. It became the ancestor of many other alphabets.

Egypt. Agricultural villages along the Nile River were united by around 3300 BC into 2 kingdoms, Upper and Lower Egypt, which were unified (c 3100 BC) under the pharaoh Menes. A bureaucracy supervised construction of canals and monuments (**pyramids** starting 2700 BC). Control over Nubia to the S was asserted from 2600 BC. Brilliant Old Kingdom Period achievements in architecture, sculpture, and painting, which reached their height during the 3d and 4th Dynasties, set the standards for subsequent Egyptian civilization. **Hieroglyphic writing** appeared by 3200 BC, recording a sophisticated literature that included religious writings, philosophies, history, and science. An ordered hierarchy of gods, including totemistic animal elements, was served by a powerful priesthood in Memphis. The pharaoh was identified with the falcon god Horus. Other trends included belief in an afterlife and short-lived quasi-monotheistic reforms introduced by the pharaoh **Akhenaton** (c 1379-1362 BC).

After a period of dominance by Semitic Hyksos from Asia (c 1700-1550 BC), the New Kingdom established an empire in Syria. Egypt became increasingly embroiled in Asiatic wars and diplomacy. Conquered by Persia in 525 BC, it eventually faded away as an independent culture.

India. An urban civilization with a so-far-undeciphered writing system stretched across the Indus Valley and along the Arabian Sea c 3000-1500 BC. Major sites are Harappa and **Mohenjo-Daro** in Pakistan, well-planned geometric cities with underground sewers and vast granaries. The entire region may have been ruled as a single state. Bronze was used, and arts and crafts were well developed. Religious life apparently took the form of fertility cults. Indus civilization was probably in decline when it was destroyed by **Aryan invaders** from the NW, speaking an Indo-European language from which most languages of Pakistan, N India, and Bangladesh descend. Led by a warrior aristocracy whose legendary deeds are in the **Rig Veda**, the Aryans spread E and S, bringing their sky gods, priestly (Brahman) ritual, and the beginnings of the caste system; local customs and beliefs were assimilated by the conquerors.

Europe. On Crete, the Bronze Age **Minoan civilization** emerged c 2500 BC. A prosperous economy and richly decorative art was supported by seaborne commerce. Mycenae and other cities in mainland Greece and Asia Minor (e.g., **Troy**) preserved elements of the culture until c 1200 BC. Cretan Linear A script (c 2000-1700 BC) remains undeciphered; Linear B script (c 1300-1200 BC) records an early Greek dialect. Unclear is the possible connection between Mycenaean monumental stonework and the megalithic monuments of W Europe, Iberia, and Malta (c 4000-1500 BC).

China. Proto-Chinese neolithic cultures had long covered N and SE China when the first large political state was organized in the N by the **Shang dynasty** (c 1523 BC). Shang kings called themselves Sons of Heaven, and they presided over a cult of human and animal sacrifice to ancestors and nature gods. The Chou dynasty, starting c 1027 BC, expanded the area of the Son of Heaven's dominion, but feudal states exercised most temporal power. A writing system with 2,000 characters was already in use under the Shang, with **pictographs** later supplemented by phonetic characters. Many of its principles and symbols, despite changes in spoken Chinese, were preserved in later writing systems. Technical advances allowed urban specialists to create fine ceramic and jade products, and bronze casting after 1500 BC was the most advanced in the world. Bronze artifacts have recently been discovered in N Thailand dating from 3600 BC, hundreds of years before similar Middle Eastern finds.

Americas. Olmecs settled (1500 BC) on the Gulf coast of Mexico and soon developed the first civilization in the western hemisphere. Temple cities and huge stone sculpture date from 1200 BC. A rudimentary calendar and writing system existed. Olmec religion, centering on a jaguar god, and Olmec art forms influenced all later Meso-American cultures.

Classical Era of Old World Civilizations: 1000 BC-400 BC

Greece. After a period of decline during the Dorian Greek invasions (1200-1000 BC), Greece and the Aegean area developed a unique civilization. Drawing upon Mycenaean traditions, Mesopotamian learning (weights and measures, lunisolar calendar, astronomy, musical scales), the Phoenician alphabet (modified for Greek), and Egyptian art, the revived **Greek city-states** saw a rich elaboration of intellectual life. Homer's epic poems, the *Iliad* and the *Odyssey,* were probably composed around the 8th cent. BC. Long-range commerce was aided by metal coinage (introduced by the Lydians in Asia Minor before 700 BC); colonies were founded around the Mediterranean (Cumae in Italy in 760 BC; Massalia in France c 600 BC) and Black Sea shores.

Philosophy, starting with Ionian speculation on the nature of matter (Thales, c 634-546 BC), continued by other "Pre-Socratics" (e.g., Heraclitus, c 535-415 BC; Parmenides, born c 515 BC), reached a high point in Athens in the rationalist idealism of **Plato** (c 428-347 BC), a disciple of **Socrates** (c 469-399 BC; executed for alleged impiety), and in **Aristotle** (384-322 BC), a pioneer in many fields, from natural sciences to logic, ethics, and metaphysics. The **arts** were highly valued. Architecture culminated in the **Parthenon** (438 BC) by Phidias (fl 490-430 BC). Poetry (Sappho, c 610-580 BC; Pindar, c 518-438 BC) and **drama** (Aeschylus, 525-456 BC; Sophocles, c 496-406 BC; Euripides, c 484-406 BC) thrived. Male beauty and strength, a chief artistic theme, were enhanced at the gymnasium and celebrated at the national games at Olympia. Ruled by local tyrants or **oligarchies,** the Greeks were not politically united, but managed to resist inclusion in the Persian Empire—Persian king Darius was defeated at Marathon (490 BC), his son Xerxes at Salamis (480 BC), and the Persian army at Plataea (479 BC). Local warfare was common; the **Peloponnesian Wars** (431-404 BC) ended in Sparta's victory over Athens. Greek political power waned, but Greek cultural forms spread throughout the ancient world.

Hebrews. Nomadic Hebrew tribes entered Canaan before 1200 BC, settling among other Semitic peoples speaking the same language. They brought from the desert a **monotheistic** faith said to have been revealed to Abraham in Canaan c 1800 BC and Moses at Mt. Sinai c 1250 BC, after the Hebrews' escape from bondage in Egypt. David (r 1000-961 BC) and Solomon (r 961-922 BC) united them in a kingdom that briefly dominated the area. **Phoenicians** to the N founded Mediterranean colonies (Carthage, c 814 BC) and sailed into the Atlantic.

A temple in Jerusalem became the national religious center, with sacrifices performed by a hereditary priesthood. Polytheistic influences, especially of the fertility cult of Baal, were opposed by **prophets** (Elijah, Amos, Isaiah).

Divided into **two kingdoms** after Solomon, the Hebrews were unable to resist the revived Assyrian empire, which conquered Israel, the N kingdom, in 722 BC. Judah, the S kingdom, was conquered in 586 BC by the Babylonians under Nebuchadnezzar II. With the fixing of most of the biblical canon by the mid-4th cent. BC and the emergence of rabbis, Judaism successfully survived the loss of Hebrew autonomy. A Jewish kingdom was revived under the Hasmoneans (168-42 BC).

China. During the **Eastern Chou** dynasty (770-256 BC), Chinese culture spread E to the sea and S to the Yangtze R. Large feudal states on the periphery of the empire contended for preeminence, but continued to recognize the Son of Heaven (king), who retained a purely ritual role enriched with courtly music and dance. In the Age of Warring States (403-221 BC), when the first sections of the **Great Wall** were built, the Ch'in state in the W gained supremacy and finally united all of China.

Iron tools entered China c 500 BC, and casting techniques were advanced, aiding agriculture. Peasants owned their land and owed civil and military service to nobles. China's cities grew in number and size; barter remained the chief trade medium.

Intellectual ferment among noble scribes and officials produced the Classical Age of Chinese literature and philosophy. **Confucius** (551-479 BC) urged a restoration of a supposedly harmonious social order of the past through proper conduct in accordance with one's station and through filial and ceremonial piety. The *Analects* attributed to him are revered throughout E Asia. **Mencius** (d 289 BC) added the view that the Mandate of Heaven can be removed from an unjust dynasty. The Legalists sought to curb the supposed natural wickedness of people through new institutions and harsh laws; they aided the Ch'in rise to power. The Naturalists emphasized the balance of opposites—yin, yang—in the world. **Taoists** sought mystical knowledge through meditation and disengagement.

India. The political and cultural center of India shifted from the Indus to the Ganges River Valley. Buddhism, Jainism, and mystical revisions of orthodox Vedism all developed c 500-300 BC. The *Upanishads,* last part of the *Veda,* urged escape from the physical world. Vedism remained the preserve of the Brahman caste. In contrast, **Buddhism,** founded by Siddarta Gautama (c 563-c 483 BC)—Buddha ("Enlightened One")—appealed to merchants in the urban centers and took hold at first (and most lastingly) on the geographic fringes of Indian civilization. The classic Indian epics were composed in this era: the **Ramayana** perhaps c 300 BC, the **Mahabharata** over a period starting 400 BC.

N India was divided into a large number of monarchies and aristocratic republics, probably derived from tribal groupings, when the Magadha kingdom was formed in Bihar c 542 BC. It soon became the dominant power. The **Maurya dynasty,** founded by Chandragupta c 321 BC, expanded the kingdom, uniting most of N India in a centralized bureaucratic empire. The third Mauryan king, **Asoka** (reigned c 274-236 BC), conquered most of the subcontinent. He converted to Buddhism and inscribed its tenets on pillars throughout India. He downplayed the caste system and tried to end expensive sacrificial rites.

Major Gods & Goddesses of the Classical World

Greek	Roman	Relations	Sphere or Position
Aphrodite	Venus	Daughter of Zeus & Dione	Love
Apollo	——	Son of Zeus & Leto	Healing, poetry, light
Ares	Mars	Son of Zeus & Hera	War
Artemis	Diana	Daughter of Zeus & Leto	Hunting, chastity
Athena	Minerva	Daughter of Zeus & Metis	Wisdom, crafts, war
Cronus	Saturn	Father of Zeus	Titans' ruler
Demeter	Ceres	Sister of Zeus	Agriculture, fertility
Dionysus	Bacchus	Son of Zeus & Semele	Wine, fertility, ecstasy
Eros	Cupid	Son of Ares & Aphrodite	Love
Hades	Pluto	Brother of Zeus	The underworld, death
Hephaestus	Vulcan	Son of Zeus & Hera	Fire
Hera	Juno	Wife & sister of Zeus	Earth
Hermes	Mercury	Son of Zeus & Maia	Travel, commerce, gods' messenger
Hestia	Vesta	Sister of Zeus	The hearth
Pan	——	Son of Hermes & a wood nymph	Forests, flocks, shepherds
Persephone	Proserpina	Daughter of Zeus & Demeter	Grain
Poseidon	Neptune	Brother of Zeus	The sea
Rhea	Ops	Mother of Zeus	The earth
Uranus	Uranus	Father of Titans (elder gods)	The heavens
Zeus	Jupiter	Son of Cronus & Rhea	Ruler of the gods

Before its final decline in India, Buddhism developed into a popular worship of heavenly Bodhisattvas ("enlightened beings"), and it produced a refined architecture (the Great Stupa [shrine] at Sanchi, AD 100) and sculpture (Gandhara reliefs, AD 1-400).

Persia. Aryan peoples (Persians, Medes) dominated the area of present Iran by the beginning of the 1st millennium BC. The prophet **Zoroaster** (born c 628 BC) introduced a dualistic religion in which the forces of good (Ahura Mazda, "Lord of Wisdom") and evil (Ahriam) battle for dominance; individuals are judged by their actions and earn damnation or salvation. Zoroaster's hymns (*Gathas*) are included in the *Avesta*, the Zoroastrian scriptures. A version of this faith became the established religion of the Persian Empire and probably influenced later monotheistic religions.

Africa. Nubia, periodically occupied by Egypt since about 2600 BC, ruled Egypt c 750-661 BC and survived as an independent Egyptianized kingdom (**Kush;** capital Meroe) for 1,000 years. The Iron Age Nok culture flourished c 500 BC- AD 200 on the Benue Plateau of **Nigeria.**

Americas. The Chavin culture controlled N Peru from 900 BC to 200 BC. Its ceremonial centers, featuring the jaguar god, survived long after. Chavin architecture, ceramics, and textiles influenced other Peruvian cultures. **Mayan civilization** began to develop in Central America as early as 1500 BC.

The Seven Wonders of the Ancient World

These ancient works of art and architecture were considered awe-inspiring in splendor and/or size by the Greek and Roman world of the Alexandrian epoch. Later classical writers disagreed as to which works made up the list of Wonders, but the following were usually included:

The Pyramids of Egypt: The only surviving ancient Wonder, these monumental structures of masonry, located at Giza on the W bank of the Nile R above Cairo, were built from c 2700 to 2500 BC as royal tombs. Three—Khufu (Cheops), Khafra (Chephren), and Menkaura (Mycerimus)—were often grouped as the first Wonder of the World. The largest, the Great Pyramid of Khufu, is a solid mass of limestone blocks covering 13 acres. It is estimated to contain 2.3 million blocks of stone, the stones themselves averaging $2^1/_2$ tons and some weighing 30 tons. Its construction reputedly took 100,000 laborers 20 years.

The Hanging Gardens of Babylon: These gardens were laid out on a brick terrace 400 ft square and 75 ft above the ground. To irrigate the plants, screws were turned to lift water from the Euphrates R. The gardens were probably built by King Nebuchadnezzar II about 600 BC. The Walls of Babylon, long, thick, and made of colorfully glazed brick, were also considered by some among the Seven Wonders.

The Pharos (Lighthouse) of Alexandria: This structure was designed about 270 BC, during the reign of Ptolemy II, by the Greek architect Sostratos. Estimates of its height range from 200 to 600 ft.

The Colossus of Rhodes: A bronze statue of the sun god Helios, the Colossus was worked on for 12 years in the third cent. BC by the sculptor Chares. It was probably 120 ft high. A symbol of the city of Rhodes at its height, the statue stood on a promontory overlooking the harbor.

The Temple of Artemis (Diana) at Ephesus: This largest and most complex temple of ancient times was built about 550 BC and was made of marble except for its tile-covered wooden roof. It was begun in honor of a non-Hellenic goddess who later became identified with the Greek goddess of the same name. Ephesus was one of the greatest of the Ionian cities.

The Mausoleum at Halicarnassus: The source of our word *mausoleum*, this marble tomb was built in what is now SE Turkey by Artemisia for her husband Mausolus, king of Caria in Asia Minor, who died in 353 BC. About 135 ft high, the tomb was adorned with the works of 4 sculptors.

The Statue of Zeus (Jupiter) at Olympia: This statue of the king of the gods showed him seated on a throne. His flesh was made of ivory, his robe and ornaments of gold. Reputedly 40 ft high, the statue was made by Phidias and was placed in the great temple of Zeus in the sacred grove of Olympia about 457 BC.

Great Empires Unite the Civilized World: 400 BC-AD 400

Persia and Alexander the Great. Cyrus, ruler of a small kingdom in Persia from 559 BC, united the Persians and Medes within 10 years and conquered Asia Minor and Babylonia in another 10. His son Cambyses, followed by **Darius** (r 522-486 BC), added vast lands to the E and N as far as the Indus Valley and Central Asia, as well as Egypt and Thrace. The whole empire was ruled by an international bureaucracy and army, with Persians holding the chief positions. The resources and styles of all the subject civilizations were exploited to create a rich syncretic art.

The kingdom of Macedon, which under Philip II dominated the Greek world and Egypt, was passed on to his son **Alexander** in 336 BC. Within 13 years, Alexander had conquered all the Persian dominions. Imbued by his tutor Aristotle with Greek ideals, Alexander encouraged Greek colonization, and Greek-style cities were founded. After his death in 323 BC, wars of succession divided the empire into 3 parts—**Macedon,** Egypt (ruled by the **Ptolemies**), and the **Seleucid** Empire.

In the ensuing 300 years (the **Hellenistic Era**), a cosmopolitan Greek-oriented culture permeated the ancient world from W Europe to the borders of India, absorbing native elites everywhere.

Hellenistic philosophy stressed the private individual's search for happiness. The Cynics followed Diogenes (c 372-287 BC), who stressed self-sufficiency and restriction of desires and expressed contempt for luxury and social convention. Zeno (c 335-c 263 BC) and the **Stoics** exalted reason, identified it with virtue, and counseled an ascetic disregard for misfortune. The **Epicureans** tried to build lives of moderate pleasure without political or emotional involvement. Hellenistic arts imitated life realistically, especially in sculpture and literature (comedies of Menander, 342-292 BC).

The sciences thrived, especially at Alexandria, where the Ptolemies financed a great library and museum. Fields of study included mathematics (**Euclid's** geometry, c 300 BC); astronomy (heliocentric theory of Aristarchus, 310-230 BC; Julian calendar, 45 BC; Ptolemy's *Almagest*, c AD 150); geography (world map of Eratosthenes, 276-194 BC); hydraulics (**Archimedes,** 287-212 BC); medicine (Galen, AD 130-200); and chemistry. Inventors refined uses for siphons, valves, gears, springs, screws, levers, cams, and pulleys.

A restored Persian empire under the **Parthians** (N Iranian tribesmen) controlled the eastern Hellenistic world from 250 BC to AD 229. The Parthians and the succeeding Sassanian dynasty (c AD 224- 651) fought with Rome periodically. The **Sassanians** revived Zoroastrianism as a state religion and patronized a nationalistic artistic and scholarly renaissance.

Rome. The city of Rome was founded, according to legend, by Romulus in 753 BC. Through military expansion and colonization, and by granting citizenship to conquered tribes, the city annexed all of Italy S of the Po in the 100-year period before 268 BC. The Latin and other Italic tribes were annexed first, followed by the **Etruscans** (founders of a great civilization, N of Rome) and the Greek colonies in the S. With a large standing army and reserve forces of several hundred thousand, Rome was able to defeat **Carthage** in the 3 **Punic Wars** (264-241, 218-201, 149-146 BC), despite the invasion of Italy (218 BC) by **Hannibal,** thus gaining Sicily and territory in Spain and N Africa.

New provinces were added in the E, as Rome exploited local disputes to conquer Greece and Asia Minor in the 2d cent. BC, and Egypt in the 1st (after the defeat and suicide of **Antony and Cleopatra,** 30 BC). All the Mediterranean civilized world up to the disputed Parthian border was now Roman and remained so for 500 years. Less civilized regions were added to the Empire: Gaul (conquered by **Julius Caesar,** 58-51 BC), Britain (AD 43), and Dacia NE of the Danube (AD 107).

The original aristocratic republican government, with democratic features added in the 5th and 4th cent. BC, deteriorated under the pressures of empire and class conflict (**Gracchus** brothers, social reformers, murdered in 133 BC and 121 BC; slave revolts in 135 BC and 73 BC). After a series of civil wars (Marius vs. Sulla 88-82 BC, Caesar vs. **Pompey** 49-45 BC, triumvirate vs. Caesar's assassins 44-43 BC, Antony vs. Octavian 32-30 BC), the empire came under the rule of a deified monarch (first emperor, **Augustus**, 27 BC-AD 14). Provincials (nearly all granted citizenship by Caracalla, AD 212) came to dominate the army and civil service. Traditional Roman law, systematized and interpreted by independent jurists, and local self-rule in provincial cities were supplanted by a vast tax-collecting bureaucracy in the 3d and 4th cent. The legal rights of women, children, and slaves were strengthened.

Roman innovations in **civil engineering** included water mills, windmills, and rotary mills and use of cement that hardened under water. Monumental architecture (baths, theaters, temples) relied on the arch and the dome. The network of roads (some still standing) stretched 53,000 mi, passing through mountain tunnels as long as 3.5 mi. Aqueducts brought water to cities; underground sewers removed waste.

Roman art and literature were to a large extent derivative of Greek models. Innovations were made in sculpture (naturalistic busts, equestrian statues), decorative wall painting (as at Pompeii), satire (Juvenal, AD 60-127), history (Tacitus, AD 56-120), prose romance (Petronius, d AD 66). Gladiatorial contests dominated public amusements, which were supported by the state.

India. The **Gupta** monarchs reunited N India c AD 320. Their peaceful and prosperous reign saw a revival of Hindu religious thought and Brahman power. The old Vedic traditions were combined with devotion to many indigenous deities (who were seen as manifestations of Vedic gods). **Caste lines** were reinforced, and Buddhism gradually disappeared. The art (often erotic), architecture, and literature of the period, patronized by the Gupta court, are considered among India's finest achievements (Kalidasa, poet and dramatist, fl. c AD 400). Mathematical innovations included use of the zero and decimal numbers. Invasions by White Huns from the NW destroyed the empire c 550. Rich cultures also developed in S India in this era. Emotional Tamil religious poetry aided the Hindu revival. The Pallava kingdom controlled much of S India c 350-880 and helped spread Indian civilization to SE Asia.

China. The Ch'in ruler Shih Huang Ti (r 221-210 BC), known as the First Emperor, centralized political authority in China, standardized the written language, laws, weights, measures, and coinage, and conducted a census, but tried to destroy most philosophical texts. The **Han dynasty** (202 BC-AD 220) instituted the Mandarin bureaucracy, which lasted for 2,000 years. Local officials were selected by examination in the Confucian classics and trained at the imperial university and at provincial schools. The invention of **paper** facilitated this bureaucratic system. Agriculture was promoted, but the peasants bore most of the tax burden. Irrigation was improved, water clocks and sundials were used, astronomy and mathematics thrived, and landscape painting was perfected.

With the expansion S and W (to nearly the present borders of today's China), trade was opened with India, SE Asia, and the Middle East, over sea and caravan routes. Indian missionaries brought Mahayana Buddhism to China by the 1st cent. AD and spawned a variety of sects. Taoism was revived and merged with popular superstitions. Taoist and Buddhist monasteries and convents multiplied in the turbulent centuries after the collapse of the Han dynasty.

Monotheism Spreads: AD 1-750

Roman Empire. Polytheism was practiced in the Roman Empire, and religions indigenous to particular Middle Eastern nations became international. Roman citizens worshiped **Isis** of Egypt, **Mithras** of Persia, **Demeter** of Greece, and the great mother **Cybele** of Phrygia. Their cults centered on mysteries (secret ceremonies) and the promise of an afterlife, symbolized by the death and rebirth of the god. The Jews of the empire preserved their monotheistic religion—Judaism, the world's oldest (c 1300 BC) continuous religion. Its teachings are contained in the Bible (the Old Testament). First-cent. Judaism embraced several sects, including the **Sadducees**, mostly drawn from the Temple priesthood, who were culturally Hellenized; the **Pharisees**, who upheld the full range of traditional customs and practices as of equal weight to literal scriptural law and elaborated synagogue worship; and the **Essenes**, an ascetic, millennarian sect. Messianic fervor led to repeated, unsuccessful rebellions against Rome (66-70, 135). As a result, the Temple in Jerusalem was destroyed and the population decimated; this event marked the beginning of the Diaspora (living in exile). To preserve the faith, a program of codification of law was begun at the academy of Yavneh. The work continued for some 500 years in Palestine and in Babylonia, ending in the final redaction (c 600) of the **Talmud**, a huge collection of legal and moral debates, rulings, liturgy, biblical exegesis, and legendary materials.

Christianity, which emerged as a distinct sect by the 2d half of the 1st cent., is based on the teachings of **Jesus**, whom believers considered the Savior (Messiah or Christ) and son of God. Missionary activities of the Apostles and such early leaders as **Paul of Tarsus** spread the faith. Intermittent persecution, as in Rome under Nero in AD 64, on grounds of suspected disloyalty, failed to disrupt the Christian communities. Each congregation, generally urban and of plebeian character, was tightly organized under a leader (bishop), elders (presbyters or priests), and assistants (deacons). The four **Gospels** (accounts of the life and teachings of Jesus) and the Acts of the Apostles were written down in the late 1st and early 2d cent. and circulated along with letters of Paul and other Christian leaders. An authoritative canon of these writings was not fixed until the 4th cent.

A school for priests was established at Alexandria in the 2d cent. Its teachers (**Origen** c 182-251) helped define doctrine and promote the faith in Greek-style philosophical works. Neoplatonism was given Christian coloration in the writings of Church Fathers such as **Augustine** (354-430). Christian hermits began to associate in monasteries, first in Egypt (St. Pachomius c 290-345), then in other eastern lands, then in the W (**St. Benedict's rule**, 529). Devotion to saints, especially Mary, mother of Jesus, spread. Under **Constantine** (r 306-37), Christianity became in effect the established religion of the Empire. Pagan temples were expropriated, state funds were used to build churches and support the hierarchy, and laws were adjusted in accordance with Christian ideas. Pagan worship was banned by the end of the 4th cent., and severe restrictions were placed on Judaism.

The newly established church was rocked by doctrinal disputes, often exacerbated by regional rivalries. Chief heresies (as defined by church councils, backed by imperial authority) were **Arianism**, which denied the divinity of Jesus; the **Monophysite** position denying the human nature of Christ; **Donatism**, which regarded as invalid any sacraments administered by sinful clergy; and **Pelagianism,** which denied the necessity of unmerited divine aid (grace) for salvation.

Islam. The earliest Arab civilization emerged by the end of the 2d millennium BC in the watered highlands of Yemen. Seaborne and caravan trade in frankincense and myrrh connected the area with the Nile and Fertile Crescent. The Minaean, Sabean (Sheba), and Himyarite states successively held sway. By Muhammad's time (7th cent. AD), the region was a province of Sassanian Persia. In the N, the Nabataean kingdom at Petra and the kingdom of Palmyra were Aramaicized, Romanized, and finally absorbed, as neighboring Judea had been, into the Roman Empire. Nomads shared the central region with a few trading towns and oases. Wars between tribes and raids on communities were common and were celebrated in a poetic tradition that by the 6th cent. helped establish a classic literary Arabic.

About 610, **Muhammad**, a 40-year-old Arab of Mecca, emerged as a prophet to his people. He proclaimed a revelation from the one true God, calling on contemporaries to abandon idolatry and restore the faith of Abraham. He introduced his religion as "Islam," meaning "submission" to the one God, Allah, as a continuation of the biblical faith of Abraham, Moses, and Jesus, all respected as prophets in this system. His teachings, recorded in the **Koran** (al-Qur'an in Arabic), in many ways were inclusive of Abrahamic monotheistic ideas known to the Jews and Christians in Arabia. A key aspect of the Abrahamic connection was insistence on justice in society, which led to severe opposition among the aristocrats in Mecca. As conditions worsened for Muhammad and his followers, he decided in 622 to make a *hijra* (emigration) to Medina, 200 mi to the N. This event marks the beginning of the Muslim lunar calendar. Hostilities between Mecca and Medina increased, and in 629 Muhammad conquered Mecca. By the time of his death in 632, nearly all the Arabian peninsula accepted his political and religious leadership.

After his death the majority of Muslims recognized the leadership of the **caliph** ("successor") Abu Bakr (632-34), followed by Umar (634-44), Uthman (644-56), and Ali (656-60). A minority, the **Shiites**, insisted instead on the leadership of Ali, Muhammad's cousin and son-in-law. By 644, **Muslim rule** over Arabia was confirmed. Muslim armies had threatened the Byzantine and Persian empires, which were weakened by wars and disaffection among subject peoples (including Coptic and Syriac Christians opposed to the Byzantine Orthodox establishment). Syria, Palestine, Egypt, Iraq, and Persia fell to Muslim armies. The new administration assimilated existing systems in the region; hence the conquered peoples participated in running of the empire. The Koran recognized the so-called Peoples of the Book, i.e., Christians, Jews, and Zoroastrians, as tolerated monotheists, and Muslim policy was relatively tolerant to minorities living as "protected" peoples. An expanded tax system, based on conquests of the Persian and Byzantine empires, provided revenue to organize campaigns against neighboring non-Muslim regions.

Disputes over succession, and pious opposition to injustices in society, led to a number of oppositional movements, which also led to the factionalization of Muslim community. The **Shiites** supported leadership candidates descended from Muhammad, believing them to be carriers of some kind of divine authority. The **Kharijites** supported an egalitarian system derived from the Koran, opposing and even engaging in battle against those who did not agree with them.

Under the **Umayyads** (661-750) and **Abbasids** (750-1256), territorial expansion led Muslim armies across N Africa and into Spain (711). Muslim armies in the W were stopped at Tours (France) in 732 by the Frankish ruler **Charles Martel**. Asia Minor, the Indus Valley, and Transoxiana were conquered in the E. The conversion of conquered peoples to Islam was gradual. In many places the official Arabic language supplanted the local tongues. But in the eastern regions the Arab rulers and their armies adopted Persian cultures and language as part of their Muslim identity.

New Peoples Enter World History: 400-900

Barbarian invasions. Germanic tribes infiltrated S and E from their Baltic homeland during the 1st millennium BC, reaching S Germany by 100 BC and the Black Sea by AD 214. Organized into large federated tribes under elected kings, most resisted Roman domination and raided the empire in time of civil war (Goths took Dacia in 214, raided Thrace in 251-69). Germanic troops and commanders dominated the Roman armies by the end of the 4th cent. **Huns**, invaders from Asia, entered Europe in 372, driving more Germans into the W empire. Emperor Valens allowed Visigoths to cross the Danube in 376. Huns under Attila (d 453) raided Gaul, Italy, and the Balkans. The W empire, weakened by overtaxation and social stagnation, was overrun in the 5th cent. Gaul was effectively lost in 406-7, Spain in 409, Britain in 410, Africa in 429-39. Rome was sacked in 410 by Visigoths under Alaric and in 455 by Vandals. The last western emperor, Romulus Augustulus, was deposed in 476 by the Germanic chief Odovacar.

Celts. Celtic cultures, which in pre-Roman times covered most of W Europe, were confined almost entirely to the British Isles after the Germanic invasions. **St. Patrick** completed (c 457-92) the conversion of Ireland. A strong monastic tradition took hold. Irish monastic missionaries in Scotland, England, and the continent (Columba c 521-97; Columban c 543-615) helped restore Christianity after the Germanic invasions. Monasteries became centers of classic and Christian learning and presided over the recording of a Christianized Celtic mythology, elaborated by secular writers and bards. An intricate decorative art style developed, especially in book illumination (Lindisfarne Gospels, c 700; Book of Kells, 8th cent.).

Successor states. The Visigothic kingdom in Spain (from 419) and much of France (to 507) saw continuation of Roman administration, language, and law (Breviary of Alaric, 506) until its destruction by the Muslims (711). The Vandal kingdom in Africa (from 429) was conquered by the Byzantines in 533. Italy was ruled successively by an Ostrogothic kingdom under Byzantine suzerainty (489-554), direct Byzantine government, and German Lombards (568-774). The Lombards divided the peninsula with the Byzantines and papacy under the dynamic reformer **Pope Gregory the Great** (590-604) and successors.

King Clovis (r 481-511) united the Franks on both sides of the Rhine and, after his conversion to Christianity, defeated the Arian heretics, Burgundians (after 500), and Visigoths (507) with the support of native clergy and the papacy. Under the **Merovingian** kings, a feudal system emerged: Power was fragmented among hierarchies of military landowners. Social stratification, which in late Roman times had acquired legal, hereditary sanction, was reinforced. The Carolingians (747-987) expanded the kingdom and restored central power. **Charlemagne** (r 768-814) conquered nearly all the Germanic lands, including Lombard Italy, and was crowned Emperor by Pope Leo III in Rome in 800. A centuries-long decline in commerce and arts was reversed under Charlemagne's patronage. He welcomed Jews to his kingdom, which became a center of Jewish learning (Rashi, 1040-1105). He sponsored the Carolingian Renaissance of learning under the Anglo-Latin scholar Alcuin (c 732-804), who reformed church liturgy.

Byzantine Empire. Under **Diocletian** (r 284-305) the empire had been divided into 2 parts to facilitate administration and defense. **Constantine** founded (330) **Constantinople** (at old Byzantium) as a fully Christian city. Commerce and taxation financed a sumptuous, orientalized court, a class of hereditary bureaucratic families, and magnificent urban construction (Hagia Sophia, 532-37). The city's fortifications and naval innovations repelled assaults by Goths, Huns, Slavs, Bulgars, Avars, Arabs, and Scandinavians. Greek replaced Latin as the official language by c 700. Byzantine art, a solemn, sacral, and stylized variation of late classical styles (mosaics at the Church of San Vitale, Ravenna, Italy 526-48), was a starting point for medieval art in E and W Europe.

Justinian (r 527-65) reconquered parts of Spain, N Africa, and Italy, codified Roman law (Codex Justinianus [529] was medieval Europe's chief legal text), closed the Platonic Academy at Athens, and ordered all pagans to convert. Lombards in Italy and Arabs in Africa retook most of his conquests. The Isaurian dynasty from Anatolia (from 717) and the Macedonian dynasty (867-1054) restored military and commercial power. The Iconoclast controversy (726-843) over the permissibility of images helped alienate the Eastern Church from the papacy.

Abbasid Empire. Baghdad (est. 762), became seat of the **Abbasid dynasty** (est 750), while Ummayads continued to rule in Spain. A brilliant cosmopolitan civilization emerged, inaugurating a Muslim-Arab golden age. Arabic was the lingua franca of the empire; intellectual sources from Persian, Sanskrit, Greek, and Syriac were rendered into Arabic. Christians and Jews equally participated in this translation movement, which also involved interaction between Jewish legal thought and Islamic law, as much as between Christian theology and Muslim scholasticism. Persian-style court life, with art and music, flourished at the court of **Harun al-Rashid** (786-809), celebrated in the masterpiece known to English readers as *The Ara-*

bian Nights. The sciences, medicine, and mathematics were pursued at Baghdad, Cordova, and Cairo (est. 969). The culmination of this intellectual synthesis in Islamic civilization came with the scientific and philosophical works of **Avicenna** (Ibn Sina, 980-1037), **Averroes** (Ibn Rushd, 1126-98), and **Maimonides** (1135-1204), a Jew who wrote in Arabic. This intellectual tradition was translated into Latin and opened a new period in Christian thought.

The decentralization of the Abbasid empire, from 874, led to establishment of various Muslim dynasties under different ethnic groups. Persians, Berbers, and Turks ruled different regions, retaining connection with the Abbasid caliph at the religious level. The Abbasid period also saw various religious movements against the orthodox position held by governing authorities. This situation in religion led to establishment of different legal, theological, and mystical schools of thought. The most influential mass movement was **Sufism**, which aimed at the reaching out of the average individual in quest of a spiritual path. Al-Ghazali (1058-1111) is credited with reconciling personal Sufism with orthodox Sunni tradition.

Africa. Immigrants from Saba in S Arabia helped set up the **Axum** kingdom in Ethiopia in the 1st cent. (their language, Ge'ez, is preserved by the Ethiopian Church). In the 3d cent., when the kingdom became Christianized, it defeated Kushite Meroe and expanded its influence into Yemen. Axum was the center of a vast ivory trade and controlled the Red Sea coast until c 1100. Arab conquest in Egypt cut Axum's political and economic ties with Byzantium.

The Iron Age entered W Africa by the end of the 1st millennium BC. **Ghana**, the first known sub-Saharan state, ruled in the upper Senegal-Niger region c 400-1240, controlling the trade of gold from mines in the S to trans-Sahara caravan routes to the N. The **Bantu** peoples, probably of W African origin, began to spread E and S perhaps 2,000 years ago, displacing the Pygmies and Bushmen of central and S Africa during a 1,500-year period.

Japan. The advanced Neolithic Yayoi period, when irrigation, rice farming, and iron and bronze casting techniques were introduced from China or Korea, persisted to c AD 400. The myriad Japanese states were then united by the **Yamato** clan, under an emperor who acted as chief priest of the animistic Shinto cult. Japanese political and military intervention by the 6th cent. in Korea, then under strong Chinese influence, quickened a Chinese cultural invasion of Japan, bringing Buddhism, the Chinese language (which long remained a literary and governmental medium), Chinese ideographs, and Buddhist styles in painting, sculpture, literature, and architecture (7th cent., Horyu-ji temple at Nara). The Taika Reforms (646) tried unsuccessfully to centralize Japan according to Chinese bureaucratic and Buddhist philosophical values. A nativist reaction against the Buddhist **Nara period** (710-94) ushered in the **Heian period** (794-1185) centered at the new capital, Kyoto. Japanese elegance and simplicity modified Chinese styles in architecture, scroll painting, and literature; the writing system was also simplified. The courtly novel *Tale of Genji* (1010-20) testifies to the enhanced role of women.

Southeast Asia. The historic peoples of SE Asia began arriving some 2,500 years ago from China and Tibet, displacing scattered aborigines. Their agriculture relied on rice and yams. Indian cultural influences were strongest; literacy and Hindu and Buddhist ideas followed the S India-China trade route. From the S tip of Indochina, the kingdom of **Funan** (1st-7th cent.) traded as far W as Persia. It was absorbed by Chenla, itself conquered by the **Khmer Empire** (600-1300). The Khmers, under Hindu god-kings (Suryavarman II, 1113-c 1150), built the monumental Angkor Wat temple center for the royal phallic cult. The **Nam-Viet** kingdom in Annam, dominated by China and Chinese culture for 1,000 years, emerged in the 10th cent., growing at the expense of the Khmers, who also lost ground in the NW to the new, highly organized **Thai** kingdom. On Sumatra, the **Srivijaya** Empire controlled vital sea lanes (7th to 10th cent.). A Buddhist dynasty, the Sailendras, ruled central **Java** (8th-9th cent.), building at Borobudur one of the largest stupas in the world.

China. The Sui dynasty (581-618) ushered in a period of commercial, artistic, and scientific achievement in China, continuing under the **Tang** dynasty (618-906). Inventions like the magnetic compass, gunpowder, the abacus, and printing were introduced or perfected. Medical innovations included cataract surgery. The state, from its cosmopolitan capital, Chang-an, supervised foreign trade, which exchanged Chinese silks, porcelains, and art for spices, ivory, etc., over Central Asian caravan routes and sea routes reaching Africa. A golden age of poetry bequeathed valuable works to later generations (Tu Fu, 712-70; Li Po, 701-62). Landscape painting flourished. Commercial and industrial expansion continued under the **Northern Sung** dynasty (960-1126), facilitated by paper money and credit notes. But commerce never achieved respectability; government monopolies expropriated successful merchants. The population, long stable at 50 million, doubled in 200 years with the introduction of early-ripening rice and the double harvest. In art, native Chinese styles were revived.

Americas. From 300 to 600 a Native American empire stretched from the Valley of Mexico to Guatemala, centering on the huge city **Teotihuacán** (founded 100 BC). To the S, in Guatemala, a high **Mayan** civilization developed (150-900) around hundreds of rural ceremonial centers. The Mayans improved on Olmec writing and the calendar and pursued astronomy and mathematics (using the idea of zero). In South America, a widespread pre-Inca culture grew from **Tiahuanacu**, Bolivia, near Lake Titicaca (Gateway of the Sun, c 700).

Christian Europe Regroups and Expands: 900-1300

Scandinavians. Pagan Danish and Norse (Viking) adventurers, traders, and pirates raided the coasts of the British Isles (Dublin, est. c 831), France, and even the Mediterranean for over 200 years beginning in the late 8th cent. Inland settlement in the W was limited to Great Britain (King Canute, 994-1035) and Normandy, settled (911) under Rollo, as a fief of France. Vikings also reached Iceland (874), Greenland (c 986), and North America (**Leif Eriksson** and others, c 1000). Norse traders (**Varangians**) developed Russian river commerce from the 8th to the 11th cent. and helped set up a state at Kiev in the late 9th cent. Conversion to Christianity occurred in the 10th cent., reaching Sweden 100 years later. In the 11th cent. Norman bands conquered S Italy and Sicily, and Duke **William of Normandy** conquered (1066) England, bringing feudalism and the French language, essential elements in later English civilization.

Central and East Europe. Slavs began to expand from about AD 150 in all directions in Europe, and by the 7th cent. they reached as far S as the Adriatic and Aegean seas. In the Balkan Peninsula they dislocated Romanized local populations or assimilated newcomers (Bulgarians, a Turkic people). The first Slavic states were Moravia (628) in Central Europe and the Bulgarian state (680) in the Balkans. Missions of St. Methodius and Cyril (whose Greek-based cyrillic alphabet is still used by some S and E Slavs) converted (863) Moravia.

The Eastern Slavs, part-civilized under the overlordship of the Turkish-Jewish **Khazar** trading empire (7th-10th cent.), gravitated toward Constantinople by the 9th cent. The **Kievan state** adopted (989) Eastern Christianity under Prince Vladimir. King Boleslav I (992-1025) began **Poland's** long history of eastern conquest. The Magyars (**Hungarians**), in present-day Hungary since 896, accepted (1001) Latin Christianity.

Germany. The German kingdom that emerged after the breakup of Charlemagne's W Empire remained a confederation of largely autonomous states. Otto I, a Saxon who was king from 936, established the **Holy Roman Empire**—a union of Germany and N Italy—in alliance with Pope John XII, who crowned (962) him emperor; he defeated (955) the Magyars. Imperial power was greatest under the **Hohenstaufens** (1138-1254), despite the growing opposition of the papacy, which ruled central Italy, and the Lombard League cities. Frederick II (1194-1250) improved administration and patronized the arts; after his death, German influence was removed from Italy.

Christian Spain. From its N mountain redoubts, Christian rule slowly migrated S through the 11th cent., when Muslim unity collapsed. After the capture (1085) of **Toledo**, the kingdoms of Portugal, Castile, and Aragon undertook repeated crusades of reconquest, finally completed in 1492. Elements of Islamic civilization persisted in recaptured areas, influencing all Western Europe.

Crusades. Pope Urban II called (1095) for a crusade to restore Asia Minor to Byzantium and to regain the Holy Land from the Turks. Some 10 crusades (to 1291) succeeded only in founding 4 temporary Frankish states in the Levant. The 4th crusade sacked (1204) Constantinople. In Rhineland (1096), England (1290), and France (1306), Jews were massacred or expelled, and wars were launched against Christian heretics (**Albigensian** crusade in France, 1229). Trade in eastern luxuries expanded, led by the Venetian naval empire.

Economy. The agricultural base of European life benefited from improvements in **plow design** (c 1000) and by draining of lowlands and clearing of forests, leading to a rural population increase. Towns grew in N Italy, Flanders, and N Germany (Hanseatic League). Improvements in **loom design** permitted factory textile production. **Guilds** dominated urban trades from the 12th cent. Banking (centered in Italy, 12th-15th cent.) facilitated long-distance trade.

The Church. The split between the Eastern and Western churches was formalized in 1054. Western and Central Europe was divided into 500 bishoprics under one united hierarchy, but conflicts between secular and church authorities were frequent (German **Investiture Controversy**, 1075-1122). Clerical power was first strengthened through the international monastic reform begun at Cluny in 910. Popular religious enthusiasm often expressed itself in heretical movements (Waldensians from 1173), but was channeled by the **Dominican** (1215) and **Franciscan** (1223) friars into the religious mainstream.

Arts. **Romanesque** architecture (11th-12th cent.) expanded on late Roman models, using the rounded arch and massed stone to support enlarged basilicas. Painting and sculpture followed Byzantine models. The literature of **chivalry** was exemplified by the epic (*Chanson de Roland*, c 1100) and by courtly love poems of the troubadours of Provence and minnesingers of Germany. **Gothic** architecture emerged in France (choir of St. Denis, c 1040) and spread as French cultural influence predominated in Europe. Rib vaulting and pointed arches were used to combine soaring heights with delicacy, and they freed walls for display of stained glass. Exteriors were covered with painted relief sculpture and embellished with elaborate architectural detail.

Learning. Law, medicine, and philosophy were advanced at independent **universities** (Bologna, late 11th cent.), originally corporations of students and masters. Twelfth-cent. translations of Greek classics, especially Aristotle, encouraged an analytic approach. Scholastic philosophy, from Anselm (1033-1109) to **Aquinas** (1225-74), attempted to understand revelation through reason.

Apogee of Central Asian Power; Islam Grows: 1250-1500

Turks. Turkic peoples, of Central Asian ancestry, were a military threat to the Byzantine and Persian Empires from the 6th cent. After several waves of invasions, during which most of the Turks adopted Islam, the **Seljuk Turks** took (1055) Baghdad. They ruled Persia, Iraq and, after 1071, Asia Minor, where massive numbers of Turks settled. The empire was divided in the 12th cent. into smaller states ruled by Seljuks, Kurds (**Saladin**, c 1137-93), and Mamluks (a military caste of former Turk, Kurd, and Circassian slaves), which governed Egypt and the Middle East until the Ottoman era (c 1290-1922).

Osman I (r c 1290-1326) and succeeding sultans united Anatolian Turkish warriors in a militaristic state that waged holy war against Byzantium and Balkan Christians. Most of the Balkans had been subdued, and Anatolia united, when Constantinople fell (1453). By the mid-16th cent., Hungary, the Middle East, and N Africa had been conquered. The Turkish advance was stopped at Vienna (1529) and at the naval battle of Lepanto (1571) by Spain, Venice, and the papacy.

The Ottoman state was governed in accordance with orthodox Muslim law. Greek, Armenian, and Jewish communities were segregated and were ruled by religious leaders responsible for taxation; they dominated trade. State offices and most army ranks were filled by slaves through a system of child conscription among Christians.

India. Mahmud of Ghazni (971-1030) led repeated Turkish raids into N India. Turkish power was consolidated in 1206 with the start of the **Sultanate at Delhi**. Centralization of state power under the early Delhi sultans went far beyond traditional Indian practice. Muslim rule of most of the subcontinent lasted until the British conquest some 600 years later.

Mongols. **Genghis Khan** (c 1167-1227) first united the feuding Mongol tribes, and built their armies into an effective offensive force around a core of highly mobile cavalry. He and his immediate successors created the largest land empire in history; by 1279 it stretched from the E coast of Asia to the Danube, from the Siberian steppes to the Arabian Sea. East-West trade and contacts were facilitated (Marco Polo, c 1254-1324). The W Mongols were Islamized by 1295; successor states soon lost their Mongol character by assimilation. They were briefly reunited under the Turk Tamerlane (1336-1405).

Kublai Khan ruled China from his new capital Beijing (est. c 1264). Naval campaigns against Japan (1274, 1281) and Java (1293) were defeated, the latter by the Hindu-Buddhist maritime kingdom of Majapahit. The **Yuan** dynasty used Mongols and other foreigners (including Europeans) in official posts and tolerated the return of Nestorian Christianity (suppressed 841-45) and the spread of Islam in the S and W. A native reaction expelled the Mongols in 1367-68.

Russia. The Kievan state in Russia, weakened by the decline of Byzantium and the rise of the Catholic Polish-Lithuanian state, was overrun (1238-40) by the Mongols. Only the northern trading republic of Novgorod remained independent. The grand dukes of Moscow emerged as leaders of a coalition of princes that eventually (by 1481) defeated the Mongols. After the fall of Constantinople in 1453, the **Tsars** (Caesars) at Moscow (from Ivan III, r 1462-1505) set up an independent Russian Orthodox Church. Commerce failed to revive. The isolated Russian state remained agrarian, with the peasant class falling into serfdom.

Persia. A revival of Persian literature, making use of the Arab alphabet and literary forms, began in the 10th cent. (epic of Firdausi, 935-1020). An art revival, influenced by Chinese styles introduced after the Mongols came to power in Iran, began in the 13th cent. Persian cultural and political forms, and often the Persian language, were used for centuries by Turkish and Mongol elites from the Balkans to India. Persian mystics from Rumi (1207-73) to Jami (1414-92) promoted **Sufism** in their poetry.

Africa. Two militant Islamic Berber dynasties emerged from the Sahara to carve out empires from the Sahel to central Spain—the **Almoravids** (c 1050-1140) and the fanatical **Almohads** (c 1125-1269). The Ghanaian empire was replaced in the upper Niger by Mali (c 1230-1340), whose Muslim rulers imported Egyptians to help make **Timbuktu** a center of commerce (in gold, leather, and slaves) and learning. The Songhay empire (to 1590) replaced Mali. To the S, forest kingdoms produced refined artworks (Ife terra cotta, **Benin** bronzes). Other Muslim states in Nigeria (Hausas) and Chad originated in the 11th cent. and continued in some form until the 19th-cent. European conquest. Less-developed Bantu kingdoms existed across central Africa.

Some 40 Muslim Arab-Persian trading colonies and city-states were established all along the E African coast from the 10th cent. (Kilwa, Mogadishu). The interchange with Bantu peoples produced the **Swahili** language and culture. Gold, palm oil, and slaves were brought from the interior, stimulating the growth of the Monamatapa kingdom of the Zambezi (15th cent.). The Christian Ethiopian empire (from 13th cent.) continued the traditions of Axum.

Southeast Asia. Islam was introduced into Malaya and the Indonesian islands by Arab, Persian, and Indian traders. Coastal Muslim cities and states (starting before 1300) soon dominated the interior. Chief among these was the **Malacca** state (c 1400-1511), on the Malay peninsula.

Arts and Statecraft Thrive in Europe: 1350-1600

Italian Renaissance and Humanism. Distinctive Italian achievements in the arts in the late Middle Ages (**Dante,** 1265-1321; Giotto, 1276-1337) led to the vigorous new styles of the Renaissance (14th-16th cent.). Patronized by the rulers of the quarreling petty states of Italy (**Medicis** in Florence and the papacy, c 1400-1737), the plastic arts perfected realistic techniques, including **perspective** (Masaccio, 1401-28, **Leonardo,** 1452-1519). Classical motifs were used in architecture, and increased talent and expense were put into secular buildings. The Florentine dialect was refined as a national literary language (**Petrarch,** 1304-74). Greek refugees from the E strengthened the respect of humanist scholars for the classic sources. Soon an international movement aided by the spread of **printing** (Gutenberg, c 1397(?)-1468), **humanism** was optimistic about the power of human reason (Erasmus of Rotterdam, 1466-1536, **More's** *Utopia*, 1516) and valued individual effort in the arts and in politics (**Machiavelli,** 1469-1527).

France. The French monarchy, strengthened in its repeated struggles with powerful nobles (Burgundy, Flanders, Aquitaine) by alliances with the growing commercial towns, consolidated bureaucratic control under Philip IV (r 1285-1314) and extended French influence into Germany and Italy (popes at Avignon, France, 1309-1417). The **Hundred Years War** (1337-1453) ended English dynastic claims in France (battles of Crécy, 1346, and Poitiers, 1356; Joan of Arc executed, 1431). A French Renaissance, dating from royal invasions (1494, 1499) of Italy, was encouraged at the court of Francis I (r 1515-47), who centralized taxation and law. French vernacular literature consciously asserted its independence (La Pléiade, 1549).

England. The evolution of England's unique political institutions began with the **Magna Carta** (1215), by which King John guaranteed the privileges of nobles and church against the monarchy and assured jury trial. After the **Wars of the Roses** (1455-85), the **Tudor dynasty** reasserted royal prerogatives (Henry VIII, r 1509-47), but the trend toward independent departments and ministerial government also continued. English trade (wool exports from c 1340) was protected by the nation's growing maritime power (**Spanish Armada** destroyed, 1588).

English replaced French and Latin in the late 14th cent. in law and literature (**Chaucer,** c 1340-1400) and English translation of the Bible began (Wycliffe, 1380s). **Elizabeth I** (r 1558-1603) presided over a confident flowering of poetry (Spenser, 1552-99), drama (**Shakespeare,** 1564-1616), and music.

German Empire. From among a welter of minor feudal states, church lands, and independent cities, the **Habsburgs** assembled a far-flung territorial domain, based in Austria from 1276. The family held the title Holy Roman Emperor from 1438 to the Empire's dissolution in 1806, but failed to centralize its domains, leaving Germany disunited for centuries. Resistance to Turkish expansion brought Hungary under Austrian control from the 16th cent. The Netherlands, Luxembourg, and Burgundy were added in 1477, curbing French expansion.

The Flemish painting tradition of naturalism, technical proficiency, and bourgeois subject matter began in the 15th cent. (**Jan Van Eyck,** c 1390-1441), the earliest northern manifestation of the Renaissance. Albrecht **Dürer** (1471-1528) typified the merging of late Gothic and Italian trends in 16th-cent. German art. Imposing civic architecture flourished in the prosperous commercial cities.

Spain. Despite the unification of Castile and Aragon in 1479, the 2 countries retained separate governments, and the nobility, especially in Aragon and Catalonia, retained many privileges. Spanish lands in Italy (Naples, Sicily) and the Netherlands entangled the country in European wars through the mid-17th cent., while explorers, traders, and conquerors built up a Spanish empire in the Americas and the Philippines. From the late 15th cent., a **golden age** of literature and art produced works of social satire (plays of Lope de Vega, 1562-1635; **Cervantes,** 1547-1616), as well as spiritual intensity (**El Greco,** 1541-1614; **Velazquez,** 1599-1660).

Black Death. The bubonic plague reached Europe from the E in 1348, killing as much as half the population by 1350. Labor scarcity forced a rise in wages and brought greater freedom to the peasantry, making possible **peasant uprisings** (Jacquerie in France, 1358; Wat Tyler's rebellion in England, 1381). In the *ciompi* revolt (1378), Florentine wage earners demanded a say in economic and political power.

Explorations. Organized European maritime exploration began, seeking to evade the Venice-Ottoman monopoly of E trade and to promote Christianity. Beginning in 1418, expeditions from Portugal explored the W coast of Africa, until Vasco da Gama rounded the Cape of Good Hope in 1497 and reached India. A Portuguese trading empire was consolidated by the seizure of Goa (1510) and Malacca (1551). Japan was reached in 1542. The voyages of Christopher **Columbus** (1492-1504) uncovered a world new to Europeans, which Spain hastened to subdue. Navigation schools in Spain and Portugal, the development of large sailing ships (carracks), and the invention (c 1475) of the rifle aided European penetration.

Mughals and Safavids. E of the Ottoman Empire, 2 Muslim dynasties ruled unchallenged in the 16th and 17th cent. The Mughal dynasty of India, founded by Persianized Turkish invaders from the NW under Babur, dates from their 1526 conquest of the Delhi Sultanate. The dynasty ruled most of India for more than 200 years, surviving nominally until 1857. **Akbar** (r 1556-1605) consolidated administration at his glorious court, where the Urdu language (Persian-influenced Hindi) developed. Trade relations with Europe increased. Under Shah Jahan (1629-58), a secularized art fusing Hindu and Muslim element flourished in miniature painting and in architecture (**Taj Mahal**). **Sikhism** (founded c 1519) combined elements of both faiths. Suppression of Hindus and Shi'ite Muslims in S India in the late 17th cent. weakened the empire.

Fanatical devotion to the Shi'ite sect characterized the Safavids (1502-1736) of Persia and led to hostilities with the Sunni Ottomans for more than a century. The prosperity and the strength of the empire are evidenced by the mosques at its capital city, **Isfahan.** The Safavids enhanced Iranian national consciousness.

China. The **Ming** emperors (1368-1644), the last native dynasty in China, wielded unprecedented personal power, while the Confucian bureaucracy began to suffer from inertia. European trade (Portuguese monopoly through **Macao** from 1557) was strictly controlled. Jesuit scholars and scientists (Matteo Ricci, 1552-1610) introduced some Western science; their writings familiarized the West with China. Chinese technological inventiveness declined from this era, but the arts thrived, especially painting and ceramics.

Japan. After the decline of the first hereditary shogunate (chief generalship) at **Kamakura** (1185-1333), fragmentation of power accelerated, as did the consequent social mobility. Under Kamakura and the Ashikaga shogunate (1338-1573), the daimyos (lords) and samurai (warriors) grew more powerful and promoted a martial ideology. Japanese pirates and traders plied the China coast. Popular Buddhist movements included the nationalist Nichiren sect (from c 1250) and **Zen** (brought from China, 1191), which stressed meditation and a disciplined esthetic (tea ceremony, gardening, martial arts, No drama).

Reformed Europe Expands Overseas: 1500-1700

Reformation begun. Theological debate and protests against real and perceived clerical corruption existed in the medieval Christian world, expressed by such dissenters as John **Wycliffe** (c 1320-84) and his followers, the Lollards, in England, and **Huss** (burned as a heretic, 1415) in Bohemia.

Martin **Luther** (1483-1546) preached that faith alone leads to salvation, without the mediation of clergy or good works. He attacked the authority of the pope, rejected priestly celibacy, and recommended individual study of the Bible (which he translated c 1525). His 95 Theses (1517) led to his excommunication (1521). John **Calvin** (1509-64) said that God's elect were predestined for salvation and that good conduct and success were signs of election. Calvin in Geneva and John Knox (1505-72) in Scotland established theocratic states.

Henry VIII asserted English national authority and secular power by breaking away (1534) from the Catholic Church. Monastic property was confiscated, and some Protestant doctrines given official sanction.

Religious wars. A century and a half of religious wars began with a S German peasant uprising (1524), repressed with Luther's support. Radical sects—democratic, pacifist, millennarian—arose (Anabaptists ruled Münster in 1534-35) and were suppressed violently. Civil war in France from 1562 between **Huguenots** (Protestant nobles and merchants) and Catholics ended with the 1598 **Edict of Nantes**, tolerating Protestants (revoked 1685). Habsburg attempts to restore Catholicism in Germany were resisted in 25 years of fighting; the 1555 Peace of Augsburg guarantee of religious independence to local princes and cities was confirmed only after the **Thirty Years War** (1618-48), when much of Germany was devastated by local and foreign armies (Sweden, France).

A Catholic Reformation, or **Counter Reformation**, met the Protestant challenge, clearly defining an official theology at the Council of Trent (1545-63). The **Jesuit** order (Society of Jesus), founded in 1534 by Ignatius Loyola (1491-1556), helped reconvert large areas of Poland, Hungary, and S Germany and sent missionaries to the New World, India, and China, while the Inquisition helped suppress heresy in Catholic countries. A revival of piety appeared in the devotional literature (Teresa of Avila, 1515-82) and grandiose Baroque art (Bernini, 1598-1680) of Roman Catholic countries.

Scientific Revolution. The late nominalist thinkers (Ockham, c 1300-49) of Paris and Oxford challenged Aristotelian orthodoxy, allowing for a freer scientific approach. At the same time, metaphysical values, such as the Neoplatonic faith in an orderly, mathematical cosmos, still motivated and directed inquiry. Nicolaus **Copernicus** (1473-1543) promoted the heliocentric theory, which was confirmed when Johannes Kepler (1571-1630) discovered the mathematical laws describing the orbits of the planets. The traditional Christian-Aristotelian belief that heavens and earth were fundamentally different collapsed when **Galileo** (1564-1642) discovered moving sunspots, irregular moon topography, and moons around Jupiter. He and Sir Isaac **Newton** (1642-1727) developed a mechanics that unified cosmic and earthly phenomena. Newton and Gottfried von Leibniz (1646-1716) invented calculus, and René Descartes (1596-1650) invented analytic geometry.

An explosion of **observational science** included the discovery of blood circulation (Harvey, 1578-1657) and microscopic life (Leeuwenhoek, 1632-1723) and advances in anatomy (Vesalius, 1514-64, dissected corpses) and chemistry (Boyle, 1627-91). Scientific research institutes were founded: Florence (1657), London (**Royal Society**, 1660), Paris (1666). Inventions proliferated (Savery's steam engine, 1696).

Arts. Mannerist trends of the High Renaissance (**Michelangelo**, 1475-1564) exploited virtuosity, grace, novelty, and exotic subjects and poses. The notion of artistic genius was promoted, in contrast to the anonymous medieval artisan. Private connoisseurs entered the art market. These trends were elaborated in the 17th cent. **Baroque** era on a grander scale. Dynamic movement in painting and sculpture was emphasized by sharp lighting effects, use of rich materials (colored marble, gilt), and realistic details. Curved facades, broken lines, rich, deep-cut detail, and ceiling decoration characterized Baroque architecture, especially in Germany. Monarchs, princes, and prelates, usually Catholic, used Baroque art to enhance and embellish their authority, as in royal portraits (Velazquez, 1599-1660; Van Dyck, 1599-1641).

National styles emerged. In France, a taste for rectilinear order and serenity (Poussin, 1594-1665), linked to the new rational philosophy, was expressed in classical forms. The influence of **classical values** in French literature (tragedies of **Racine**, 1639-99) gave rise to the "battle of the Ancients and Moderns." New forms included the essay (**Montaigne**, 1533-92) and novel (*Princesse de Cleves*, La Fayette, 1678).

Dutch painting of the 17th cent. was unique in its wide social distribution. The Flemish tradition of undemonstrative realism reached its peak in **Rembrandt** (1606-69) and Jan Vermeer (1632-75).

Economy. European economic expansion was stimulated by the new trade with the East, by New World gold and silver, and by a doubling of population (50 million in 1450, 100 million in 1600). New business and financial techniques were developed and refined, such as joint-stock companies, insurance, and letters of credit and exchange. The Bank of Amsterdam (1609) and the Bank of England (1694) broke the old monopoly of private banking families. The rise of a business mentality was typified by the spread of clock towers in cities in the 14th cent. By the mid-15th cent., portable clocks were available; the first watch was invented in 1502.

By 1650, most governments had adopted the **mercantile system**, in which they sought to amass metallic wealth by protecting their merchants' foreign and colonial trade monopolies. The rise in prices and the new coin-based economy undermined the craft guild and feudal manorial systems. Expanding industries (clothweaving, mining) benefited from technical advances. Coal replaced disappearing wood as the chief fuel; it was used to fuel new 16th-cent. blast furnaces making cast iron.

New World. The **Aztecs** united much of the Meso-American culture area in a militarist empire by 1519, from their capital, Tenochtitlán (pop. 300,000), which was the center of a cult requiring ritual human sacrifice. Most of the civilized areas of South America were ruled by the centralized Inca Empire (1476-1534), stretching 2,000 mi from Ecuador to NW Argentina. Lavish and sophisticated traditions in pottery, weaving, sculpture, and architecture were maintained in both regions.

These empires, beset by revolts, fell in 2 short campaigns to gold-seeking Spanish forces based in the Antilles and Panama. Hernan **Cortes** took Mexico (1519-21); Francisco **Pizarro**, Peru (1532-35). From these centers, land and sea expeditions claimed most of North and South America for Spain. The Indian high cultures did not survive the impact of Christian missionaries and the new upper class of whites and mestizos. In turn, New World silver and such Indian products as potatoes, tobacco, corn, peanuts, chocolate, and rubber exercised a major economic influence on Europe. Although the Spanish administration intermittently concerned itself with the welfare of Indians, the population remained impoverished at most levels. European diseases reduced the native population.

Brazil, which the Portuguese reached in 1500 and settled after 1530, and the Caribbean colonies of several European nations developed a plantation economy where sugarcane, tobacco, cotton, coffee, rice, indigo, and lumber were grown by slaves. From the early 16th to late 19th cent., 10 million Africans were transported to **slavery** in the New World.

Netherlands. The urban, Calvinist N provinces of the Netherlands rebelled (1568) against Habsburg Spain and founded an oligarchic mercantile republic. Their strategic control of the Baltic grain market enabled them to exploit Mediterranean food shortages. Religious refugees—French and Belgian Protestants, Iberian Jews—added to the cosmopolitan commercial talent pool. After Spain absorbed Portugal in 1580, the Dutch seized Portuguese possessions and created a vast, though short-lived commercial empire in Brazil, the Antilles, Africa, India, Ceylon, Malacca, Indonesia, and Taiwan and challenged or supplanted Portuguese traders in China and Japan. Revolution in 1640 restored Portuguese independence.

England. Anglicanism became firmly established under **Elizabeth I** after a brief Catholic interlude under "Bloody Mary" (1553-58). But religious and political conflicts led to a rebellion (1642) by Parliament. Roundheads (Puritans) defeated Cavaliers (Royalists); Charles I was beheaded (1649). The new Commonwealth was ruled as a military dictatorship by Oliver **Cromwell**, who also brutally crushed (1649-51) an Irish rebellion. Conflicts within the Puritan camp (democratic Levelers defeated, 1649) aided the Stuart restoration (1660), but Parliament was strengthened and the peaceful **"Glorious Revolution"** (1688) advanced political and religious liberties (writings of **Locke**, 1632-1704). British privateers (Drake, 1540-96) challenged Spanish control of the New World and penetrated Asian trade routes (Madras taken, 1639). North American colonies (Jamestown, 1607; Plymouth, 1620) provided an outlet for religious dissenters from Europe.

France. Emerging from the religious civil wars in 1628, France regained military and commercial great power status (under the ministries of **Richelieu**, Mazarin, and Colbert). Under **Louis XIV** (reigned 1643-1715), royal absolutism triumphed over nobles and local *parlements* (defeat of Fronde, 1648-53). Permanent colonies were founded in Canada (1608), the Caribbean (1626), and India (1674).

Sweden. Sweden seceded from the Scandinavian Union in 1523. The thinly populated agrarian state (with copper, iron, and timber exports) was united by the Vasa kings, whose conquests by the mid-17th cent. made Sweden the dominant Baltic power. The empire collapsed in the Great Northern War (1700-21).

Poland. After the union with Lithuania in 1447, Poland ruled vast territories from the Baltic to the Black Sea, resisting German and Turkish incursions. Catholic nobles failed to gain the loyalty of their Orthodox Christian subjects in the E; commerce and trades were practiced by German and Jewish immigrants. The bloody 1648-49 Cossack uprising began the kingdom's dismemberment.

China. A new dynasty, the **Manchus,** invaded from the NE, seized power in 1644, and expanded Chinese control to its greatest extent in Central and SE Asia. Trade and diplomatic contact with Europe grew, carefully controlled by China. New crops (sweet potato, maize, peanut) allowed an economic and population growth (pop. 300 million, in 1800). Traditional arts and literature were pursued with increased sophistication (*Dream of the Red Chamber*, novel, mid-18th cent.).

Japan. Tokugawa Ieyasu, shogun from 1603, finally unified and pacified feudal Japan. Hereditary daimyos and samurai monopolized government office and the professions. An urban merchant class grew, literacy spread, and a cultural renaissance occurred (**haiku,** a verse innovation of the poet Basho, 1644-94). Fear of European domination led to persecution of Christian converts from 1597 and to stringent isolation from outside contact from 1640.

Philosophy, Industry, and Revolution: 1700-1800

Science and Reason. Greater faith in human reason and empirical observation as a source of truth and a means to improve the physical and social environment, espoused since the Renaissance (Francis Bacon, 1561-1626), was bolstered by scientific discoveries in spite of theological opposition (Galileo's forced retraction, 1633). René **Descartes** (1596-1650) used a rationalistic approach modeled on geometry to discover "self-evident" truths as a foundation of knowledge. Sir Isaac **Newton** emphasized induction from experimental observation. Baruch de **Spinoza** (1632-77), who called for political and intellectual freedom, developed a systematic rationalistic philosophy in his classic work *Ethics*.

French philosophers assumed leadership of the **Enlightenment** in the 18th cent. Montesquieu (1689-1755) used British history to support his notions of limited government. **Voltaire's** (1694-1778) diaries and novels of exotic travel illustrated the intellectual trends toward secular ethics and relativism. Jean-Jacques **Rousseau's** (1712-1778) radical concepts of the **social contract** and of the inherent goodness of the common man gave impetus to antimonarchical republicanism. The *Encyclopedia* (1751-72, edited by Diderot and d'Alembert), designed as a monument to reason, was largely devoted to practical technology.

In England, ideals of political and religious liberty were connected with empiricist philosophy and science in the followers of Locke. But British empiricism, especially as developed by the skeptical David **Hume** (1711-76), radically reduced the role of reason in philosophy, as did the evolutionary approach to law and politics of Edmund Burke (1729-97) and the utilitarian ethics of Jeremy Bentham (1748-1832). Adam Smith (1723-90) and other **physiocrats** called for a rationalization of economic activity by removing artificial barriers to a supposedly natural free exchange of goods.

German writers participated in the new philosophical trends popularized by Christian von Wolff (1679-1754). Immanuel **Kant's** (1724-1804) transcendental idealism, unifying an empirical epistemology with a priori moral and logical concepts, directed German thought away from skepticism. Italian contributions included work on electricity (Galvani, 1737-98; Volta, 1745-1827), the pioneer historiography of Vico (1668-1744), and writings on penal reform (Beccaria, 1738-94). Benjamin Franklin (1706-90) was celebrated in Europe for his varied achievements.

The growth of the **press** (*Spectator*, 1711-12) and the wide distribution of realistic but sentimental **novels** attested to the increase of a large bourgeois public.

Arts. Rococo art, characterized by extravagant decorative effects, asymmetries copied from organic models, and artificial pastoral subjects, was favored by the continental aristocracy for most of the cent. (Watteau, 1684-1721) and had musical analogies in the ornamentalized polyphony of late Baroque. The **Neoclassical** art after 1750, associated with the new scientific archaeology, was more streamlined and was infused with the supposed moral and geometric rectitude of the Roman Republic (David, 1748-1825). In England, **town planning** on a grand scale began.

Industrial Revolution in England. Agricultural improvements, such as the sowing drill (1701) and livestock breeding, were implemented on the large fields provided by enclosure of common lands by private owners. Profits from agriculture and from colonial and foreign trade (1800 volume, £54 million) were channeled through hundreds of banks and the **Stock Exchange** (est 1773) into new industrial processes.

The Newcomen steam pump (1712) aided coal mining. Coal fueled the new efficient steam engines patented by James Watt in 1769, and coke-smelting produced cheap, sturdy iron for machinery by the 1730s. The **flying shuttle** (1733) and **spinning jenny** (c 1764) were used in the large new cotton textile factories, where women and children were much of the work force. Goods were transported cheaply over **canals** (2,000 mi; built 1760-1800).

American Revolution. The British colonies in North America attracted a mass immigration of religious dissenters and poor people throughout the 17th and 18th cent., coming from the British Isles, Germany, the Netherlands, and other countries. The population reached 3 million non-natives by the 1770s. The small native population was greatly reduced by European diseases and by wars with and between the various colonies. British attempts to control colonial trade and to tax the colonists to pay for the costs of colonial administration and defense clashed with traditions of local self-government and eventually provoked the colonies to rebellion.

Central and East Europe. The monarchs of the three states that dominated E Europe—Austria, Prussia, and Russia—accepted the advice and legitimation of philosophes in creating more modern, centralized institutions in their kingdoms, enlarged by the division (1772-95) of Poland.

Under **Frederick II** (r 1740-86) Prussia, with its efficient modern army, doubled in size. State monopolies and tariff protection fostered industry, and some legal reforms were introduced. Austria's heterogeneous realms were unified under **Maria Theresa** (r 1740-80) and **Joseph II** (r 1780-90). Reforms in education, law, and religion were enacted, and the Austrian serfs were freed (1781). With its defeat in the Seven Years' War in 1763, Austria failed to regain Silesia, which had been seized by Prussia, but it was compensated by expansion to the E and S (Hungary, Slavonia, 1699; Galicia, 1772).

Russia, whose borders continued to expand in all directions, adopted some Western bureaucratic and economic policies under **Peter I** (r 1682-1725) and **Catherine II** (r 1762-96). Trade and cultural contacts with the West multiplied from the new Baltic Sea capital, **St. Petersburg** (est 1703).

French Revolution. The growing French middle class lacked political power and resented aristocratic tax privileges, especially in light of the successful American Revolution. Peasants lacked adequate land and were burdened with feudal obligations to nobles. War with Britain led to the loss of French Canada and drained the treasury, finally forcing the king to call the **Estates-General** in 1789 (first time since 1614), in an atmosphere of food riots (poor crop in 1788).

Aristocratic resistance to absolutism was soon overshadowed by the reformist Third Estate (middle class), which proclaimed itself the **National Constituent Assembly** June 17 and took the "Tennis Court oath" on June 20 to secure a constitution. The storming of the **Bastille** on July 14, 1789, by Parisian artisans was followed by looting and seizure of aristocratic property throughout France. Assembly reforms included abolition of class and regional privileges, a Declaration of Rights, suffrage by taxpayers (75% of males), and the **Civil Constitution of the Clergy** providing for election and loyalty oaths for priests. A republic was declared Sept. 22, 1792, in spite of royalist pressure from Austria and Prussia, which had declared war in April (joined by Britain the next year). Louis XVI was beheaded Jan. 21, 1793, and Queen Marie Antoinette was beheaded Oct. 16, 1793.

Royalist uprisings in La Vendée and military reverses led to institution of a **reign of terror** in which tens of thousands of opponents of the Revolution and criminals were executed. Radical reforms in the **Convention** period (Sept. 1793-Oct. 1795) included the abolition of colonial slavery, economic measures to aid the poor, support of public education, and a short-lived de-Christianization.

Division among radicals (execution of Hebert, Danton, and Robespierre, 1794) aided the ascendancy of a moderate **Directory**, which consolidated military victories. **Napoleon Bonaparte** (1769-1821), a popular young general, exploited political divisions and participated in a coup Nov. 9, 1799, making himself first consul (dictator).

India. Sikh and Hindu rebels (Rajputs, Marathas) and Afghans destroyed the power of the Mughals during the 18th cent. After France's defeat (1763) in the Seven Years' War, Britain was the primary European trade power in India. Its control of inland **Bengal and Bihar** was recognized (1765) by the Mughal shah, who granted the **British East India Co.** (under Clive, 1725-74) the right to collect land revenue there. Despite objections from Parliament (1784 India Act), the company's involvement in local wars and politics led to repeated acquisitions of new territory. The company exported Indian textiles, sugar, and indigo.

Change Gathers Steam: 1800-40

French ideals and empire spread. Inspired by the ideals of the French Revolution, and supported by the expanding French armies, new republican regimes arose near France: the **Batavian** Republic in the Netherlands (1795-1806), the **Helvetic** Republic in Switzerland (1798-1803), the **Cisalpine** Republic in N Italy (1797-1805), the **Ligurian** Republic in Genoa (1797-1805), and the **Parthenopean** Republic in S Italy (1799). A Roman Republic existed briefly in 1798 after Pope Pius VI was arrested by French troops. In Italy and Germany, new nationalist sentiments were stimulated both in imitation of and in reaction to developments in France (anti-French and anti-Jacobin peasant uprisings in Italy, 1796-99).

From 1804, when Napoleon declared himself emperor, to 1812, a succession of military victories (Austerlitz, 1805; Jena, 1806) extended his control over most of Europe, through puppet states (**Confederation of the Rhine** united W German states for the first time and **Grand Duchy of Warsaw** revived Polish national hopes), expansion of the empire, and alliances.

Among the lasting reforms initiated under Napoleon's absolutist reign were: establishment of the Bank of France, centralization of tax collection, codification of law along Roman models (Code Napoléon), and reform and extension of secondary and university education. In an 1801 concordat, the papacy recognized the effective autonomy of the French Catholic Church. Napoleon's continental successes were offset by British victory under Adm. Horatio Nelson in the **Battle of Trafalgar** (1805). Some 400,000 French soldiers were killed in the Napoleonic Wars, along with 600,000 foreign troops.

Last gasp of old regime. The disastrous 1812 invasion of Russia exposed Napoleon's overextension. After Napoleon's 1814 exile at Elba, his armies were defeated (1815) at **Waterloo**, by British and Prussian troops.

At the **Congress of Vienna**, the monarchs and princes of Europe redrew their boundaries, to the advantage of Prussia (in Saxony and the Ruhr), Austria (in Illyria and Venetia), and Russia (in Poland and Finland). British conquest of Dutch and French colonies (S Africa, Ceylon, Mauritius) was recognized, and France, under the restored Bourbons, retained its expanded 1792 borders. The settlement brought 50 years of international peace to Europe.

But the Congress was unable to check the advance of liberal ideals and of nationalism among the smaller European nations. The 1825 **Decembrist uprising** by liberal officers in Russia was easily suppressed. But an independence movement in **Greece**, stirred by commercial prosperity and a cultural revival, succeeded in expelling Ottoman rule by 1831, with the aid of Britain, France, and Russia.

A constitutional monarchy was secured in France by the **1830 Revolution**; Louis Philippe became king. The revolutionary contagion spread to **Belgium**, which gained its independence (1830) from the Dutch monarchy, to **Poland**, whose rebellion was defeated (1830-31) by Russia, and to Germany.

Romanticism. A new style in intellectual and artistic life began to replace Neoclassicism and Rococo after the mid-18th cent. By the early 19th cent., this style, Romanticism, had prevailed in the European world.

Rousseau had begun the reaction against rationalism; in education (*Émile*, 1762) he stressed subjective spontaneity over regularized instruction. German writers (Lessing, 1729-81; Herder, 1744-1803) favorably compared the German folk song to classical forms and began a cult of Shakespeare, whose passion and "natural" wisdom was a model for the romantic *Sturm und Drang* (Storm and Stress) movement. **Goethe's** *Sorrows of Young Werther* (1774) set the model for the tragic, passionate genius.

A new interest in **Gothic architecture** in England after 1760 (Walpole, 1717-97) spread through Europe, associated with an aesthetic Christian and mystic revival (**Blake**, 1757-1827). Celtic, Norse, and German mythology and folk tales were revived or imitated (Macpherson's Ossian translation, 1762; Grimm's Fairy Tales, 1812-22). The medieval revival (Scott's

Ivanhoe, 1819) led to a new interest in history, stressing national differences and organic growth (**Carlyle**, 1795-1881; Michelet, 1798-1874), corresponding to theories of natural evolution (Lamarck's *Philosophie Zoologique*, 1809; Lyell's *Geology*, 1830-33). A reaction against classicism characterized the English **romantic poets** (beginning with **Wordsworth,** 1770-1850). Revolution and war fed an emphasis on freedom and conflict, expressed by both poets (**Byron**, 1788-1824; **Hugo**, 1802-85) and philosophers (**Hegel**, 1770-1831).

Wild gardens replaced the formal French variety, and painters favored rural, stormy, and mountainous landscapes (**Turner**, 1775-1851; **Constable**, 1776-1837). Clothing became freer, with wigs, hoops, and ruffles discarded. Originality and genius were expected in the life as well as the work of inspired artists (Murger's *Scenes from Bohemian Life*, 1847-49). Exotic locales and themes (as in Gothic horror stories) were used in art and literature (Delacroix, 1798-1863; **Poe**, 1809-49).

Music exhibited the new dramatic style and a breakdown of classical forms (**Beethoven,** 1770-1827). The use of folk melodies and modes aided the growth of distinct national traditions (Glinka in Russia, 1804-57).

Latin America. Francois **Toussaint L'Ouverture** led a successful slave revolt in Haiti, which subsequently became the first Latin American state to achieve independence (1804). The mainland Spanish colonies won their independence (1810-24), under such leaders as Simon **Bolivar** (1783-1830). Brazil became an independent empire (1822) under the Portuguese prince regent. A new class of military officers divided power with large landholders and the church.

United States. Heavy immigration and exploitation of ample natural resources fueled rapid economic growth. The spread of the franchise, public education, and antislavery sentiment were signs of a widespread democratic ethic.

China. Failure to keep pace with Western arms technology exposed China to greater European influence and hampered efforts to bar imports of opium, which had damaged Chinese society and drained wealth overseas. In the **Opium War** (1839-42), Britain forced China to expand trade opportunities and to cede Hong Kong.

Triumph of Progress: 1840-80

Idea of Progress. As a result of the cumulative scientific, economic, and political changes of the preceding eras, the idea took hold among literate people in the West that continuing growth and improvement was the usual state of human and natural life.

Darwin's statement of the **theory of evolution** and survival of the fittest (*Origin of Species*, 1859), defended by intellectuals and scientists against theological objections, was taken as confirmation that progress was the natural direction of life. The controversy helped define popular ideas of the dedicated scientist and ever-expanding human knowledge of and control over the world (Foucault's demonstration of earth's rotation, 1851; **Pasteur's** germ theory, 1861).

Liberals following Ricardo (1772-1823) in their faith that unrestrained competition would bring continuous economic expansion sought to adjust political life to the new social realities and believed that unregulated competition of ideas would yield truth (**Mill**, 1806-73). In England, successive reform bills (1832, 1867, 1884) gave representation to the new industrial towns and extended the franchise to the middle and lower classes and to Catholics, Dissenters, and Jews. On both sides of the Atlantic, reformists tried to improve conditions for the mentally ill (**Dix**, 1802-87), women (Anthony, 1820-1906), and prisoners. Slavery was barred in the British Empire (1833), the U.S. (1865), and Brazil (1888).

Socialist theories based on ideas of human perfectibility or progress were widely disseminated. Utopian socialists such as Saint-Simon (1760-1825) envisaged an orderly, just society directed by a technocratic elite. A model factory town, New Lanark, Scotland, was set up by utopian Robert Owen (1771-1858), and communal experiments were tried in the U.S. (most notably, Brook Farm, Mass., 1841-47). Bakunin's (1814-76) anarchism represented the opposite utopian extreme of total freedom. Karl **Marx** (1818-83) posited the inevitable triumph of socialism in industrial countries through a dialectical process of class conflict.

Spread of industry. The technical processes and managerial innovations of the English industrial revolution spread to Europe (especially Germany) and the U.S., causing an explosion of industrial production, demand for raw materials, and competition for markets. Inventors, both trained and self-educated, provided the means for larger-scale production (Bessemer steel, 1856; sewing machine, 1846). Many inventions were shown at the 1851 London Great Exhibition at the **Crystal Palace,** the theme of which was universal prosperity.

Local specialization and long-distance trade were aided by a revolution in transportation and communication. Railroads were first introduced in the 1820s in England and the U.S. More than 150,000 mi of track had been laid worldwide by 1880, with another 100,000 mi laid in the next decade. Steamships were improved (*Savannah* crossed Atlantic, 1819). The **telegraph**, perfected by 1844 (Morse), connected the Old and New Worlds by cable in 1866 and quickened the pace of international commerce and politics. The first commercial **telephone** exchange went into operation in the U.S. in 1878.

The new class of industrial workers, uprooted from their rural homes, lacked job security and suffered from dangerous overcrowded conditions at work and at home. Many responded by organizing **trade unions** (legalized in England, 1824; France, 1884). The U.S. Knights of Labor had 700,000 members by 1886. The First International (1864-76) tried to unite workers internationally around a Marxist program. The quasi-Socialist Paris Commune uprising (1871) was violently suppressed. Factory Acts to reduce child labor and regulate conditions were passed (1833-50 in England). Social security measures were introduced by the Bismarck regime (1883-89) in Germany.

Revolutions of 1848. Among the causes of the continent-wide revolutions were an international collapse of credit and resulting unemployment, bad harvests in 1845-47, and a cholera epidemic. The new urban proletariat and expanding bourgeoisie demanded a greater political role. Republics were proclaimed in France, Rome, and Venice. Nationalist feelings reached fever pitch in the Habsburg empire, as Hungary declared independence under Kossuth, as a Slav Congress demanded equality, and as Piedmont tried to drive Austria from Lombardy. A national liberal assembly at Frankfurt called for German unification.

But riots fueled bourgeois fears of socialism (**Marx and Engels**, *Communist Manifesto*, 1848), and peasants remained conservative. The old establishment—the Papacy, the Habsburgs with the help of the Czarist Russian army —was able to rout the revolutionaries by 1849. The French Republic succumbed to a renewed monarchy by 1852 (Emperor Napoleon III).

Great nations unified. Using the "blood and iron" tactics of Bismarck from 1862, Prussia controlled N Germany by 1867 (war with Denmark, 1864; Austria, 1866). After defeating France in 1870 (annexation of Alsace-Lorraine), it won the allegiance of S German states. A new **German Empire** was proclaimed (1871). **Italy**, inspired by Giuseppe Mazzini (1805-72) and Giuseppe Garibaldi (1807-82), was unified by the reformed Piedmont kingdom through uprisings, plebiscites, and war.

The **U.S.**, its area expanded after the 1846-48 Mexican War, defeated (1861-65) a secession attempt by slave states. The Canadian provinces were united in an autonomous **Dominion of Canada** (1867). Control in **India** was removed from the East India Co. and centralized under British administration after the 1857-58 Sepoy rebellion, laying the groundwork for the modern Indian State. Queen Victoria was named Empress of India (1876).

Europe dominates Asia. The Ottoman Empire began to collapse in the face of Balkan nationalisms and European imperial incursions in N Africa (**Suez Canal**, 1869). The Turks had lost control of most of both regions by 1882. Russia completed its expansion S by 1884 (despite the temporary setback of the **Crimean War** with Turkey, Britain, and France, 1853-56), taking Turkestan, all the Caucasus, and Chinese areas in the E and sponsoring Balkan Slavs against the Turks. A succession of reformist and reactionary regimes presided over a slow modernization (serfs freed, 1861). Persian independence suffered as Russia and British India competed for influence.

China was forced to sign a series of unequal treaties with European powers and Japan. Overpopulation and an inefficient dynasty brought misery and caused rebellions (Taiping, Muslims) leaving tens of millions dead. **Japan** was forced by the U.S. (Commodore Perry's visits, 1853-54) and Europe to end its isolation. The Meiji restoration (1868) gave power to a Westernizing oligarchy. Intensified empire-building gave Burma to Britain (1824-85) and Indochina to France (1862-95). Christian missionary activity followed imperial and trade expansion in Asia.

Respectability. The fine arts were expected to reflect and encourage the good morals and manners among the Victorians. Prudery, exaggerated delicacy, and familial piety were heralded by **Bowdler's** expurgated edition (1818) of Shakespeare. Government-supported mass education sought to inculcate a work ethic as a means to escape poverty (**Horatio Alger**, 1832-99).

The official **Beaux Arts** school in Paris set an international style of imposing public buildings (Paris Opera, 1861-74; Vienna Opera, 1861-69) and uplifting statues (Bartholdi's Statue of Liberty, 1884). Realist painting, influenced by photography (Daguerre, 1837), appealed to a new mass audience with social or historical narrative (Wilkie, 1785-1841; Poynter, 1836-1919) or with serious religious, moral, or social messages (pre-Raphaelites, Millet's *Angelus*, 1858) often drawn from ordinary life. The **Impressionists** (Monet, 1840-1926; Pissarro, 1830-1903; Renoir, 1841-1919) rejected the formalism, sentimentality, and precise techniques of academic art in favor of a spontaneous, undetailed rendering of the world through careful representation of the effect of natural light on objects.

Realistic **novelists** presented the full panorama of social classes and personalities, but retained sentimentality and moral judgment (**Dickens**, 1812-70; **Eliot**, 1819-80; **Tolstoy**, 1828-1910; **Balzac**, 1799-1850).

Veneer of Stability: 1880-1900

Imperialism triumphant. The vast **African** interior, visited by European explorers (Barth, 1821-65; Livingstone, 1813-73), was conquered by the European powers in rapid, competitive thrusts from their coastal bases after 1880, mostly for domestic political and international strategic reasons. W African Muslim kingdoms (Fulani), Arab slave traders (Zanzibar), and Bantu military confederations (Zulu) were alike subdued. Only Christian Ethiopia (defeat of Italy, 1896) and Liberia resisted successfully. France (W Africa) and Britain ("Cape to Cairo," **Boer War**, 1899-1902) were the major beneficiaries. The ideology of "the white man's burden" (Kipling, *Barrack Room Ballads*, 1892) or of a "civilizing mission" (France) justified the conquests.

W European foreign capital investment soared to nearly $40 billion by 1914, but most was in E Europe (France, Germany), the Americas (Britain), and the Europeans' colonies. The foundation of the modern interdependent world economy was laid, with cartels dominating raw material trade.

An industrious world. Industrial and technological proficiency characterized the 2 new great powers—Germany and the U.S. Coal and iron deposits enabled Germany to reach 2d or 3d place status in iron, steel, and shipbuilding by the 1900s. German electrical and chemical industries were world leaders. The U.S. post-Civil War boom (interrupted by "panics"—1884, 1893, 1896) was shaped by massive immigration from S and E Europe from 1880, government subsidy of railroads, and huge private monopolies (Standard Oil, 1870; U.S. Steel, 1901). The **Spanish-American War**, 1898 (Philippine Insurrection, 1899-1902), and the **Open Door policy** in China (1899) made the U.S. a world power.

England led in **urbanization** (72% by 1890), with **London** the world capital of finance, insurance, and shipping. Sewer systems (Paris, 1850s), electric subways (London, 1890), parks, and bargain department stores helped improve living standards for most of the urban population of the industrial world.

Westernization of Asia. Asian reaction to European economic, military, and religious incursions took the form of imitation of Western techniques and adoption of Western ideas of progress and freedom. The Chinese "self-strengthening" movement of the 1860s and 1870s included rail, port, and arsenal improvements and metal and textile mills. Reformers such as **K'ang Yu-wei** (1858-1927) won liberalizing reforms in 1898, right after the European and Japanese "scramble for concessions."

A universal education system in Japan and importation of foreign industrial, scientific, and military experts aided Japan's unprecedented rapid modernization after 1868, under the authoritarian Meiji regime. Japan's victory in the **Sino-Japanese War** (1894-95) put Formosa and Korea in its power.

In India, the British alliance with the remaining princely states masked reform sentiment among the Westernized urban elite; higher education had been conducted largely in English for 50 years. The **Indian National Congress**, founded in 1885, demanded a larger government role for Indians.

MILLENNIUM FACT BOX

The Year 1900 on Record

The World Almanac 1901 contained a one-page section with a Record of Events in 1900. Among the world events listed were the following:

Jan. 20. John Ruskin died.

April 3. Queen Victoria started on her visit to Ireland.

April 14. The Paris International Exposition was formally opened by President Loubet.

April 26. Hull and a part of Ottawa, Canada, were destroyed by fire; 12,000 persons homeless, and $15,000,000 property loss. Seven lives lost.

June 15. The Prince de Joinville, last surviving son of King Louis Philippe, of France, died.

June 20. Baron von Ketteler, German Minister to China, was murdered by a mob in Peking.

July 5. Democratic National Convention at Kansas City, Mo., nominated Bryan and Stevenson.

July 9. General Porfirio Diaz was re-elected President of Mexico.

July 13. The Earl of Hopetown was appointed Governor of the new Commonwealth of Australia.

July 30. King Humbert of Italy was assassinated by Angelo Bresci at Monza, Italy.

Aug. 14. Rain in the famine regions of India.

Sept. 1. Lord Roberts proclaimed the Transvaal British territory.

***Fin-de-siècle* sophistication. Naturalist** writers pushed realism to its extreme limits, adopting a quasi-scientific attitude and writing about formerly taboo subjects such as sex, crime, extreme poverty, and corruption (Flaubert, 1821-80; Zola, 1840-1902; Hardy, 1840-1928). Unseen or repressed psychological motivations were explored in the clinical and theoretical works of Sigmund **Freud** (1856-1939) and in works of fiction (**Dostoyevsky,** 1821-81; James, 1843-1916; Schnitzler, 1862-1931; others).

A contempt for bourgeois life or a desire to shock a complacent audience was shared by the French **symbolist** poets (Verlaine, 1844-96; Rimbaud, 1854-91), by neopagan English writers (Swinburne, 1837-1909), by continental dramatists (**Ibsen,** 1828-1906), and by satirists (**Wilde,** 1854-1900). The German philosopher Friedrich **Nietzsche** (1844-1900) was influential in his elitism and pessimism.

Postimpressionist art neglected long-cherished conventions of representation (Cézanne, 1839-1906) and showed a willingness to learn from primitive and non-European art (Gauguin, 1848-1903; Japanese prints).

Racism. Gobineau (1816-82) gave a pseudobiological foundation to modern racist theories, which spread in the latter 19th cent., along with **Social Darwinism,** the belief that societies are and should be organized as a struggle for survival of the fittest. The medieval period was interpreted as an era of natural Germanic rule (Chamberlain, 1855-1927), and notions of superiority were associated with German national aspirations (Treitschke, 1834-96). **Anti-Semitism,** with a new racist rationale, became a significant political force in Germany (Anti-Semitic Petition, 1880), Austria (Lueger, 1844-1910), and France (Dreyfus case, 1894-1906).

Last Respite: 1900-9

Alliances. While the peace of Europe (and its dependencies) continued to hold (1907 **Hague Conference** extended the rules of war and international arbitration procedures), imperial rivalries, protectionist trade practices (in Germany and France), and the escalating arms race (British *Dreadnought* battleship launched; Germany widens Kiel canal, 1906) exacerbated minor disputes (German-French Moroccan "crises," 1905, 1911).

Security was sought through alliances: **Triple Alliance** (Germany, Austria-Hungary, Italy; renewed in 1902 and 1907); Anglo-Japanese Alliance (1902), Franco-Russian Alliance (1899), **Entente Cordiale** (Britain, France, 1904), Anglo-Russian Treaty (1907), German-Ottoman friendship.

Ottomans decline. The inefficient, corrupt Ottoman government was unable to resist further loss of territory. Nearly all European lands were lost in 1912 to Serbia, Greece, Montenegro, and Bulgaria. Italy took Libya and the Dodecanese islands the same year, and Britain took Kuwait (1899) and the Sinai (1906). The **Young Turk** revolution in 1908 forced the sultan to restore a constitution, and it introduced some social reform, industrialization, and secularization.

British Empire. British trade and cultural influence remained dominant in the empire, but constitutional reforms presaged its eventual dissolution: The colonies of **Australia** were united in 1901 under a self-governing commonwealth. **New Zealand** acquired dominion status in 1907. The old Boer republics joined Cape Colony and Natal in the self-governing **Union of South Africa** in 1910.

The 1909 Indian Councils Act enhanced the role of elected province legislatures in **India.** The Muslim League (founded 1906) sought separate communal representation.

East Asia. Japan exploited its growing industrial power to expand its empire. Victory in the 1904-5 war against Russia (naval battle of Tsushima, 1905) assured Japan's domination of **Korea** (annexed 1910) and Manchuria (Port Arthur taken, 1905).

In China, central authority began to crumble (empress died, 1908). Reforms (Confucian exam system ended 1905, modernization of the army, building of railroads) were inadequate, and secret societies of reformers and nationalists, inspired by the Westernized **Sun Yat-sen** (1866-1925) fomented periodic uprisings in the S.

Siam, whose independence had been guaranteed by Britain and France in 1896, was split into spheres of influence by those countries in 1907.

Russia. The population of the Russian Empire approached 150 million in 1900. Reforms in education, in law, and in local institutions (*zemstvos*) and an industrial boom starting in the 1880s (oil, railroads) created the beginnings of a modern state, despite the autocratic tsarist regime. Liberals (1903 Union of Liberation), Socialists (Social Democrats founded 1898, Bolsheviks split off 1903), and populists (Social Revolutionaries founded 1901) were periodically repressed, and national minorities were persecuted (anti-Jewish pogroms, 1903, 1905-6).

An industrial crisis after 1900 and harvest failures aggravated poverty among urban workers, and the 1904-5 defeat by Japan (which checked Russia's Asian expansion) sparked **the Revolution of 1905-6.** A **Duma** (parliament) was created, and an agricultural reform (under Stolypin, prime minister 1906-11) created a large class of land-owning peasants (kulaks).

The world shrinks. Developments in transportation and communication and mass population movements helped create an awareness of an interdependent world. Early **automobiles** (Daimler, Benz, 1885) were experimental or were designed as luxuries. Assembly-line mass production (Ford Motor Co., 1903) made the invention practicable, and by 1910 nearly 500,000 motor vehicles were registered in the U.S. alone. **Heavier-than-air flights** began in 1903 in the U.S. (Wright brothers), preceded by glider, balloon, and model plane advances in several countries. Trade was advanced by improvements in **ship design** (gyrocompass, 1910), speed (*Lusitania* crossed Atlantic in 5 days, 1907), and reach (Panama Canal begun, 1904).

The first transatlantic **radio** telegraphic transmission occurred in 1901, 6 years after Marconi discovered radio. Radio transmission of human speech had been made in 1900. Telegraphic transmission of photos was achieved in 1904, lending immediacy to news reports. **Phonographs,** popularized by Caruso's recordings (starting 1902), made for quick international spread of musical styles (ragtime). **Motion pictures,** perfected in the 1890s (Dickson, Lumière brothers), became a popular and artistic medium after 1900; newsreels appeared in 1909.

Emigration from crowded European centers soared in the decade: 9 million migrated to the U.S., and millions more went to Siberia, Canada, Argentina, Australia, South Africa, and Algeria. Some 70 million Europeans emigrated in the cent. before 1914. Several million Chinese, Indians, and Japanese migrated to SE Asia, where their urban skills often enabled them to take a predominant economic role.

Social reform. The social and economic problems of the poor were kept in the public eye by realist fiction writers (Dreiser's *Sister Carrie*, 1900; Gorky's *Lower Depths*, 1902; Sinclair's *The Jungle*, 1906), journalists (U.S. **muckrakers**—Steffens, Tarbell), and artists (Ashcan school). Frequent labor strikes and occasional assassinations by anarchists or radicals (Empress Elizabeth of Austria, 1898; King Umberto I of Italy, 1900; U.S. Pres. McKinley, 1901; Russian Interior Minister Plehve, 1904; Portugal's King Carlos, 1908) added to social tension and fear of revolution.

But democratic reformism prevailed. In Germany, Bernstein's (1850-1932) **revisionist Marxism,** downgrading revolution, was accepted by the powerful Social Democrats and trade unions. The British Fabian Society (the Webbs, Shaw) and the Labour Party (founded 1906) worked for reforms such as Social Security and union rights (1906), while woman suffragists grew more militant. U.S. **progressives** fought big business (Pure Food and Drug Act, 1906). In France, the 10-hour work day (1904) and separation of church and state (1905) were reform victories, as was universal suffrage in Austria (1907).

Arts. An unprecedented period of experimentation, centered in France, produced several new **painting** styles: Fauvism exploited bold color areas (Matisse, *Woman With Hat*, 1905); expressionism reflected powerful inner emotions (the Brücke group, 1905); cubism combined several views of an object on one flat surface (Picasso's *Demoiselles*, 1906-7); futurism tried to depict speed and motion (Italian Futurist Manifesto, 1910). **Architects** explored new uses of steel structures, with facades either neoclassical (Adler and Sullivan in U.S.); curvilinear Art Nouveau (Gaudi's Casa Mila, 1905-10); or functionally streamlined (Wright's Robie House, 1909).

Music and dance shared the experimental spirit. Ruth St. Denis (1877-1968) and Isadora Duncan (1878-1927) pioneered modern dance, while Sergei Diaghilev in Paris revitalized classic ballet from 1909. Composers explored atonal music (Debussy, 1862-1918) and dissonance (Schoenberg, 1874-1951) or revolutionized classical forms (Stravinsky, 1882-1971), often showing jazz or folk music influences.

War and Revolution: 1910-19

War threatens. Germany under Wilhelm II sought a political and imperial role consonant with its industrial strength, challenging Britain's world supremacy and threatening France, which was still resenting the loss (1871) of Alsace-Lorraine. Austria wanted to curb an expanded Serbia (after 1912) and the threat it posed to its own Slav lands. Russia feared Austrian and German political and economic aims in the Balkans and Turkey. An accelerated arms race resulted: The German standing army rose to more than 2 million men by 1914. Russia and France had more than a million each, and Austria and the British Empire nearly a million each. Dozens of enormous battleships were built by the powers after 1906.

The **assassination of Austrian Archduke Franz Ferdinand** by a Serbian, June 28, 1914, was the pretext for war. The system of alliances made the conflict Europe-wide; Germany's invasion of Belgium to outflank France forced Britain to enter the war. Patriotic fervor was nearly unanimous among all classes in most countries.

World War I. German forces were stopped in France in one month. The rival armies dug **trench networks**. Artillery and improved machine guns prevented either side from any lasting advance despite repeated assaults (600,000 dead at **Verdun**, Feb.-July 1916). Poison gas, used by Germany in 1915, proved ineffective. The entrance of more than 1 million U.S. troops tipped the balance after mid-1917, forcing Germany to sue for peace the next year. The formal armistice was signed on Nov. 11, 1918.

In the E, the Russian armies were thrown back (battle of **Tannenberg**, Aug. 20, 1914), and the war grew unpopular in Russia. An allied attempt to relieve Russia through Turkey failed (**Gallipoli**, 1915). The **Russian Revolution** (1917) abolished the monarchy. The new Bolshevik regime signed the capitulatory Brest-Litovsk peace in March 1918. Italy entered the war on the allied side in May 1915 but was pushed back by Oct. 1917. A renewed offensive with Allied aid in Oct.-Nov. 1918 forced Austria to surrender.

The British Navy successfully blockaded Germany, which responded with submarine U-boat attacks; **unrestricted submarine warfare** against neutrals after Jan. 1917 helped bring the U.S. into the war. Other battlefields included Palestine and Mesopotamia, both of which Britain wrested from the Turks in 1917, and the African and Pacific colonies of Germany, most of which fell to Britain, France, Australia, Japan, and South Africa.

From 1916, the civilian populations and economies of both sides were mobilized to an unprecedented degree. Hardships intensified among fighting nations in 1917 (French mutiny crushed in May). More than 10 million soldiers died in the war.

Settlement. At the **Paris Peace Conference** (Jan.-June 1919), concluded by the **Treaty of Versailles**, and in subsequent negotiations and local wars (Russian-Polish War, 1920), the map of Europe was redrawn with a nod to U.S. Pres. Wilson's principle of self-determination. Austria and Hungary were separated, and much of their land was given to Yugoslavia (formerly Serbia), Romania, Italy, and the newly independent Poland and Czechoslovakia. Germany lost territory in the W, N, and E, while Finland and the Baltic states were detached from Russia. Turkey lost nearly all its Arab lands to British-sponsored Arab states or to direct French and British rule. Belgium's sovereignty was recognized.

A huge **reparations** burden and partial demilitarization were imposed on Germany. Pres. Wilson obtained approval for a League of Nations, but the U.S. Senate refused to allow the U.S. to join.

Russian revolution. Military defeats and high casualties caused a contagious lack of confidence in Tsar Nicholas, who was forced to abdicate Mar. 1917. A liberal provisional government failed to end the war, and massive desertions, riots, and fighting between factions followed. A moderate socialist government under Aleksandr Kerensky was overthrown (Nov. 1917) in a violent coup by the **Bolsheviks** in Petrograd under **Lenin**, who later disbanded the elected Constituent Assembly.

The Bolsheviks brutally suppressed all opposition and ended the war with Germany in Mar. 1918. **Civil war** broke out in the summer between the Red Army, including the Bolsheviks and their supporters, and monarchists, anarchists, nationalities (Ukrainians, Georgians, Poles), and others. Small U.S., British, French, and Japanese units also opposed the Bolsheviks (1918-19; Japan in Vladivostok to 1922). The civil war, anarchy, and pogroms devastated the country until the 1920 Red Army victory. The wartime total monopoly of political, economic, and police power by the Communist Party leadership was retained.

Other European revolutions. An unpopular monarchy in **Portugal** was overthrown in 1910. The new republic took severe anticlerical measures in 1911.

After a century of Home Rule agitation, during which **Ireland** was devastated by famine (1 million dead, 1846-47) and emigration, republican militants staged an unsuccessful uprising in Dublin during Easter 1916. The execution of the leaders and mass arrests by the British won popular support for the rebels. The Irish Free State, comprising all but the 6 N counties, achieved dominion status in 1922.

In the aftermath of the world war, radical revolutions were attempted in Germany (**Spartacist** uprising, Jan. 1919), **Hungary** (Kun regime, 1919), and elsewhere. All were suppressed or failed for lack of support.

Chinese revolution. The Manchu Dynasty was overthrown and a republic proclaimed in Oct. 1911. First Pres. Sun Yat-sen resigned in favor of strongman Yuan Shih-k'ai. Sun organized the parliamentarian **Kuomintang** party.

Students launched protests on May 4, 1919, against League of Nations concessions in China to Japan. Nationalist, liberal, and socialist ideas and political groups spread. The **Communist Party** was founded in 1921. A Communist regime took power in Mongolia with Soviet support in 1921.

India restive. Indian objections to British rule erupted in nationalist riots as well as in the nonviolent tactics of Mahatma **Gandhi** (1869-1948). Nearly 400 unarmed demonstrators were shot at **Amritsar** in Apr. 1919. Britain approved limited self-rule that year.

Mexican revolution. Under the long Diaz dictatorship (1877-1911) the economy advanced, but Indian and mestizo lands were confiscated, and concessions to foreigners (mostly U.S.) damaged the middle class. A **revolution in 1910** led to civil wars and U.S. intervention (1914, 1916-17). Land reform and a more democratic constitution (1917) were achieved.

The Seven Natural Wonders of the Modern World

This list names features of significance widely noted by world travelers during recent centuries.

Mt. Everest: The highest peak in the world, Mt. Everest is in S central Asia, in the Himalaya range, on the frontier of Nepal and Tibet. Controversy surrounds its actual elevation. A 1954 Indian government survey placed it at 29,028 ft above sea level; however, more recent surveys cast some doubt on this figure. The summit was first scaled in 1953.

Victoria Falls: This 343-ft waterfall is on the Zambezi R in S central Africa on the border between Zimbabwe and Zambia. The river here is about 1 mi wide. A railroad bridge, completed in 1905, spans the gorge below the falls.

The Grand Canyon: This exceptionally deep (more than 1 mi) and extremely beautiful steep-walled chasm in NW Arizona is about 217 mi long and up to 18 mi wide. Excavated by the Colorado R, it is of relatively recent origin; apparently, erosion began a little more than a million years ago. The canyon contains towering buttes, mesas, and valleys within its main gorge.

Paricutin: This volcano is one of the world's youngest. It was discovered in 1943 west of Mexico City.

The Harbor at Rio de Janeiro, Brazil (as seen from the sea): One of the world's most beautiful natural harbors, the harbor at Rio is surrounded by low mountain ranges whose spurs extend almost to the waterside, and thus divide the city.

The Northern Lights: Also known as aurora borealis, the Northern Lights consists of rapidly shifting patches and dancing columns of light of various hues. The aurora assumes an endless variety of forms, including the arch, the band, filaments and streamers at right angles to the arch or band, the corona, clouds, the glow, and curtains, fans, flames, or streamers of various shapes.

The Great Barrier Reef: This chain of coral reefs is in the Coral Sea, off the E coast of Queensland, Australia. It is the largest known deposit of coral and extends in a NW direction more than 1200 mi. The reef serves as a barrier to disturbances in the Coral Sea, thus affording a sheltered passage for ships.

The Aftermath of War: 1920-29

U.S. Easy credit, technological ingenuity, and war-related industrial decline in Europe caused a long economic boom, in which ownership of the new products—**autos, phones, radios**—became democratized. Prosperity, an increase in women workers, woman suffrage (1920), and drastic change in fashion (flappers, mannish bob for women, clean-shaven men) created a wide perception of social change, despite prohibition of alcoholic beverages (1919-33). Union membership and strikes increased. Fear of radicals led to Palmer raids (1919-20) and the Sacco/Vanzetti case (1921-27).

Europe sorts itself out. Germany's liberal **Weimar constitution** (1919) could not guarantee a stable government in the face of rightist violence (Rathenau assassinated, 1922) and Communist refusal to cooperate with Socialists. Reparations and Allied occupation of the Rhineland caused staggering inflation that destroyed middle-class savings, but economic expansion resumed after mid-decade, aided by U.S. loans. A sophisticated, **innovative culture** developed in architecture and design (Bauhaus, 1919-28), film (Lang, *M*, 1931), painting (Grosz), music (Weill, *Threepenny Opera*, 1928), theater (Brecht, *A Man's a Man*, 1926), criticism (Benjamin), philosophy (Jung), and fashion. This culture was considered decadent and socially disruptive by rightists.

England elected its first Labour governments (Jan. 1924, June 1929). A 10-day general strike in support of coal miners failed in May 1926. In **Italy**, strikes, political chaos, and violence by small Fascist bands culminated in the Oct. 1922 Fascist March on Rome, which established Mussolini's dictatorship. Strikes were outlawed (1926), and Italian influence was pressed in the Balkans (Albania a protectorate, 1926). A conservative dictatorship was also established in **Portugal** in a 1926 military coup.

Czechoslovakia, the only stable democracy to emerge from the war in Central or East Europe, faced opposition from Germans (in the Sudetenland), Ruthenians, and some Slovaks. As the industrial heartland of the old Habsburg empire, it remained fairly prosperous. With French backing, it formed the Little Entente with Yugoslavia (1920) and Romania (1921) to block Austrian or Hungarian irredentism. Hungary remained dominated by the landholding classes and expansionist feeling. Croats and Slovenes in **Yugoslavia** demanded a federal state until King Alexander I proclaimed (1929) a royal dictatorship. Poland faced nationality problems as well (Germans, Ukrainians, Jews); Pilsudski ruled as dictator from 1926. The Baltic states were threatened by traditionally dominant ethnic Germans and by Soviet-supported Communists.

An economic collapse and famine in **Russia** (1921-22) claimed 5 million lives. The New Economic Policy (1921) allowed land ownership by peasants and some private commerce and industry. Stalin was absolute ruler within 4 years of Lenin's death (1924). He inaugurated a brutal collectivization program (1929-32) and used foreign Communist parties for Soviet state advantage.

Internationalism. Revulsion against World War I led to pacifist agitation, to the Kellogg-Briand Pact renouncing aggressive war (1928), and to **naval disarmament** pacts (Washington, 1922; London, 1930). But the League of Nations was able to arbitrate only minor disputes (Greece-Bulgaria, 1925).

Middle East. Mustafa Kemal (**Ataturk**) led **Turkish** nationalists in resisting Italian, French, and Greek military advances (1919-23). The sultanate was abolished (1922), and elaborate reforms were passed, including secularization of law and adoption of the Latin alphabet. Ethnic conflict led to persecution of **Armenians** (more than 1 million dead in 1915, 1 million expelled), Greeks (forced Greek-Turk population exchange, 1923), and Kurds (1925 uprising).

With evacuation of the Turks from **Arab** lands, the puritanical Wahabi dynasty of E Arabia conquered (1919-25) what is now Saudi Arabia. British, French, and Arab dynastic and nationalist maneuvering resulted in the creation of 2 more Arab monarchies in 1921—Iraq and Transjordan (both under British control)—and 2 French mandates—Syria and Lebanon. Jewish immigration into British-mandated **Palestine**, inspired by the Zionist movement, was resisted by Arabs, at times violently (1921, 1929 massacres).

Reza Khan ruled **Persia** after his 1921 coup (shah from 1925), centralized control, and created the trappings of a modern secular state.

China. The Kuomintang under **Chiang Kai-shek** (1887-1975) subdued the warlords by 1928. The Communists were brutally suppressed after their alliance with the Kuomintang was broken in 1927. Relative peace thereafter allowed for industrial and financial improvements, with some Russian, British, and U.S. cooperation.

Arts. Nearly all bounds of subject matter, style, and attitude were broken in the arts of the period. **Abstract** art first took inspiration from natural forms or narrative themes (Kandinsky from 1911) and then worked free of any representational aims (Malevich's suprematism, 1915-19; Mondrian's geometric style from 1917). The **Dada** movement (from 1916) mocked artistic pretension with absurd collages and constructions (Arp, Tzara, from 1916). Paradox, illusion, and psychological taboos were exploited by **surrealists** by the latter 1920s (Dali, Magritte). Architectural schools celebrated industrial values, whether vigorous abstract constructivism (Tatlin, *Monument to 3rd International*, 1919) or the machined, streamlined **Bauhaus** style, which was extended to many design fields (Helvetica typeface).

Prose writers explored revolutionary narrative modes related to dreams (Kafka's *Trial*, 1925), internal monologue (Joyce's **Ulysses**, 1922), and word play (Stein's *Making of Americans*, 1925). Poets and novelists wrote of modern alienation (Eliot's *Waste Land*, 1922) and aimlessness (Lost Generation).

Sciences. Scientific specialization prevailed by the 20th cent. Advances in knowledge and technological aptitude increased with the geometric rise in the number of practitioners. Physicists challenged common-sense views of causality, observation, and a mechanistic universe, putting science further beyond popular grasp (**Einstein's** general theory of relativity, 1915; Bohr's quantum mechanics, 1913; Heisenberg's uncertainty principle, 1927).

Rise of Totalitarians: 1930-39

Depression. A worldwide financial panic and economic depression began with the Oct. 1929 U.S. stock market crash and the May 1931 failure of the Austrian Credit-Anstalt. A credit crunch caused international bankruptcies and **unemployment**: 12 million jobless by 1932 in the U.S., 5.6 million in Germany, 2.7 million in England. Governments responded with **tariff restrictions** (Smoot-Hawley Act, 1930; Ottawa Imperial Conference, 1932), which dried up world trade. Government public works programs were vitiated by deflationary budget balancing.

Germany. Years of agitation by violent extremists were brought to a head by the Depression. Nazi leader Adolf **Hitler** was named chancellor in Jan. 1933 and given dictatorial power by the Reichstag in March. Opposition parties were disbanded, strikes banned, and all aspects of economic, cultural, and religious life were brought under central government and Nazi party control and manipulated by sophisticated propaganda. Severe persecution of Jews began (**Nuremberg Laws**, Sept. 1935). Many Jews, political opponents, and others were sent to concentration camps (Dachau, 1933), where thousands died or were killed. Public works, renewed conscription (1935), arms production, and a 4-year plan (1936) all but ended unemployment.

Hitler's expansionism started with reincorporation of the Saar (1935), occupation of the **Rhineland** (Mar. 1936), and annexation of Austria (Mar. 1938). At **Munich** (Sept. 1938) an indecisive Britain and France sanctioned German dismemberment of Czechoslovakia.

Russia. Urbanization and education advanced. Rapid industrialization was achieved through successive **5-year plans** starting in 1928, using severe labor discipline and mass forced labor. Industry was financed by a decline in living standards and exploitation of agriculture, which was almost totally collectivized by the early 1930s (*kolkhoz*, collective farm; *sovkhoz*, state farm, often in newly worked lands). Successive **purges** increased the role of professionals and management at the expense of workers. Millions perished in a series of manufactured disasters: extermination (1929-34) of kulaks (peasant landowners), severe famine (1932-33), party purges and show trials (Great Purge, 1936-38), suppression of nationalities, and poor conditions in labor camps.

Spain. An industrial revolution during World War I created an urban proletariat, which was attracted to socialism and anarchism; Catalan nationalists challenged central authority. The 5 years after King Alfonso left Spain in Apr. 1931 were dominated by tension between intermittent leftist and anticlerical governments and clericals, monarchists, and other rightists. Anarchist and Communist rebellions were crushed, but a July 1936 extreme right rebellion led by Gen. Francisco **Franco** and aided by Nazi Germany and Fascist Italy succeeded, after a 3-year **civil war** (more than 1 million dead in battles and atrocities). The war polarized international public opinion.

Italy. Despite propaganda for the ideal of the Corporate State, few domestic reforms were attempted. An entente with Hungary and Austria (Mar. 1934), a pact with Germany and Japan (Nov. 1937), and intervention by 50,000-75,000 troops in Spain (1936-39) sealed Italy's identification with the fascist bloc (anti-Semitic laws after Mar. 1938). Ethiopia was conquered (1935-36), and Albania annexed (Jan. 1939) in conscious imitation of ancient Rome.

East Europe. Repressive regimes fought for power against an active opposition (liberals, socialists, Communists, peasants, Nazis). Minority groups and Jews were restricted within national boundaries that did not coincide with ethnic population patterns. In the destruction of **Czechoslovakia**, Hungary occupied S Slovakia (Nov. 1938) and Ruthenia (Mar. 1939), and a pro-Nazi regime took power in the rest of Slovakia. Other boundary disputes (e.g., Poland-Lithuania, Yugoslavia-Bulgaria, Romania-Hungary) doomed attempts to build joint fronts against Germany or Russia. Economic depression was severe.

East Asia. After a period of liberalism in **Japan**, nativist militarists dominated the government with peasant support. Manchuria was seized (Sept. 1931-Feb. 1932), and a puppet state was set up (Manchukuo). Adjacent Jehol (Inner Mongolia) was occupied in 1933. China proper was invaded in July 1937; large areas were conquered by Oct. 1938. Hundreds of thousands of rapes, murders, and other atrocities were attributed to the Japanese.

In **China** Communist forces left Kuomintang-besieged strongholds in the S in a Long March (1934-35) to the N. The Kuomintang-Communist civil war was suspended in Jan. 1937 in the face of threatening Japan.

The democracies. The Roosevelt Administration, in office Mar. 1933, embarked on an extensive program of **New Deal** social reform and economic stimulation, including protection for labor unions (heavy industries organized), Social Security, public works, wage-and-hour laws, and assistance to farmers. Isolationist sentiment (1937 Neutrality Act) prevented U.S. intervention in Europe, but military expenditures were increased in 1939.

French political instability and polarization prevented resolution of economic and international security questions. The **Popular Front** government under Leon Blum (June 1936-Apr. 1938) passed social reforms (40-hour week) and raised arms spending. National coalition governments, which ruled Britain from Aug. 1931, brought some economic recovery but failed to define a consistent international policy until Chamberlain's government (from May 1937), which practiced deliberate **appeasement** of Germany and Italy.

India. Twenty years of agitation for autonomy and then for independence (Gandhi's **salt march**, 1930) achieved some constitutional reform (extended provincial powers, 1935) despite Muslim-Hindu strife. Social issues assumed prominence with peasant uprisings (1921), strikes (1928), Gandhi's efforts for untouchables (1932 "fast unto death"), and social and agrarian reform by the provinces after 1937.

Arts. The streamlined, geometric design motifs of Art Deco (from 1925) prevailed through the 1930s. **Abstract art** flourished (Moore sculptures from 1931) alongside a new **realism** related to social and political concerns (Socialist Realism, the official Soviet style from 1934; Mexican muralist Rivera, 1886-1957; and Orozco, 1883-1949), which were also expressed in fiction and poetry (Steinbeck's *Grapes of Wrath*, 1939; Sandburg's *The People, Yes*, 1936). Modern architecture (International Style, 1932) was unchallenged in its use of artificial materials (concrete, glass), lack of decoration, and monumentality (Rockefeller Center, 1929-40). U.S.-made films captured a worldwide audience with their larger-than-life fantasies *(Gone With the Wind, The Wizard of Oz,* both 1939).

War, Hot and Cold: 1940-49

War in Europe. The Nazi-Soviet nonaggression pact (Aug. 1939) freed Germany to attack Poland (Sept.). Britain and France, which had guaranteed Polish independence, declared war on Germany. Russia seized E Poland (Sept.), attacked Finland (Nov.), and took the Baltic states (July 1940). Mobile German forces staged *blitzkrieg* attacks during Apr.-June 1940, conquering neutral Denmark, Norway, and the Low Countries and defeating France; 350,000 British and French troops were evacuated at **Dunkirk** (May). The **Battle of Britain** (June-Dec. 1940) denied Germany air superiority. German-Italian campaigns won the Balkans by Apr. 1941. Three million Axis troops **invaded Russia** in June 1941, marching through Ukraine to the Caucasus, and through White Russia and the Baltic republics to Moscow and Leningrad.

Russian winter counterthrusts (1941-42 and 1942-43) stopped the German advance (**Stalingrad,** Sept. 1942-Feb. 1943). With British and U.S. Lend-Lease aid and sustaining great casualties, the Russians drove the Axis from all E Europe and the Balkans in the next 2 years. Invasions of N Africa (Nov. 1942), Italy (Sept. 1943), and **Normandy** (launched on D-Day, June 6, 1944) brought U.S., British, Free French, and allied troops to Germany by spring 1945. Germany surrendered May 7, 1945.

War in Asia-Pacific. Japan occupied Indochina in Sept. 1940, dominated Thailand in Dec. 1941, and attacked Hawaii (**Pearl Harbor**), the Philippines, Hong Kong, and Malaya on Dec. 7, 1941 (precipitating U.S. entrance into the war). Indonesia was attacked in Jan. 1942, and Burma was conquered in Mar. 1942. The Battle of **Midway** (June 1942) turned back the Japanese advance. "Island-hopping" battles (**Guadalcanal,** Aug. 1942-Jan. 1943; **Leyte Gulf,** Oct. 1944; **Iwo Jima,** Feb.-Mar. 1945; **Okinawa,** Apr. 1945) and massive bombing raids on Japan from June 1944 wore out Japanese defenses. U.S. atom bombs, dropped Aug. 6 and 9 on **Hiroshima** and Nagasaki, forced Japan to agree, on Aug. 14, to surrender; formal surrender was on Sept. 2, 1945.

Atrocities. The war brought 20th-cent. cruelty to its peak. The Nazi regime systematically killed an estimated 5-6 million Jews, including some 3 million who died in death camps (e.g., **Auschwitz**). Gypsies, political opponents, sick and retarded people, and others deemed undesirable were also murdered by the Nazis, as were vast numbers of Slavs, especially leaders.

Civilian deaths. German bombs killed 70,000 British civilians. More than 100,000 Chinese civilians were killed by Japanese forces in the capture and occupation of Nanking. Severe retaliation by the Soviet army, E European partisans, Free French, and others took a heavy toll. U.S. and British bombing of Germany killed hundreds of thousands, as did U.S. bombing of Japan (80,000-200,000 at Hiroshima alone). Some 45 million people lost their lives in the war.

Settlement. The **United Nations** charter was signed in San Francisco on June 26, 1945, by 50 nations. The International Tribunal at **Nuremberg** convicted 22 German leaders for war crimes in Sept. 1946; 23 Japanese leaders were convicted in Nov. 1948. Postwar border changes included large gains in territory for the USSR, losses for Germany, a shift to the W in Polish borders, and minor losses for Italy. Communist regimes, supported by Soviet troops, took power in most of E Europe, including Soviet-occupied Germany (GDR proclaimed Oct. 1949). Japan lost all overseas lands.

Recovery. Basic political and social changes were imposed on Japan and W Germany by the western allies (Japan constitution adopted, Nov. 1946; W German basic law, May 1949). U.S. **Marshall Plan** aid ($12 billion, 1947-51) spurred W European economic recovery after a period of severe inflation and strikes in Europe and the U.S. The British Labour Party introduced a national health service and nationalized basic industries in 1946.

Cold War. Western fears of further Soviet advances (Cominform formed in Oct. 1947; Czechoslovakia coup, Feb. 1948; Berlin blockade, Apr. 1948-Sept. 1949) led to the formation of **NATO**. Civil War in Greece and Soviet pressure on Turkey led to U.S. aid under the **Truman Doctrine** (Mar. 1947). Other anti-Communist security pacts were the Organization of American States (Apr. 1948) and the SE Asia Treaty Organization (Sept. 1954). A new wave of **Soviet purges** and repression intensified in the last years of Stalin's rule, extending to E Europe (Slansky trial in Czechoslovakia, 1951). Only Yugoslavia resisted Soviet control (expelled by Cominform, June 1948; U.S. aid, June 1949).

China, Korea. Communist forces emerged from World War II strengthened by the Soviet takeover of industrial Manchuria. In 4 years of fighting, the Kuomintang was driven from the mainland; the People's Republic was proclaimed Oct. 1, 1949. Korea was divided by USSR and U.S. occupation forces. Separate republics were proclaimed in the 2 zones in Aug.-Sept. 1948.

India. India and Pakistan became independent dominions on Aug. 15, 1947. Millions of Hindu and Muslim refugees were created by the partition; riots (1946-47) took hundreds of thousands of lives; Mahatma **Gandhi** was assassinated in Jan. 1948. Burma became completely independent in Jan. 1948; Ceylon took dominion status in Feb.

Middle East. The UN approved partition of Palestine into Jewish and Arab states. **Israel** was proclaimed a state, May 14, 1948. Arabs rejected partition, but failed to defeat Israel in war (May 1948-July 1949). Immigration from Europe and the Middle East swelled Israel's Jewish population. British and French forces left Lebanon and Syria in 1946. Transjordan occupied most of Arab Palestine.

Southeast Asia. Communists and others fought against restoration of French rule in Indochina from 1946; a non-Communist government was recognized by France in Mar. 1949, but fighting continued. Both Indonesia and the Philippines became independent; the former in 1949 after 4 years of war with Netherlands, the latter in 1946. Philippine economic and military ties with the U.S. remained strong; a Communist-led peasant rising was checked in 1948.

Arts. New York became the center of the world art market; **abstract expressionism** was the chief mode (Pollock from 1943, de Kooning from 1947). Literature and philosophy explored **existentialism** (Camus's *The Stranger*, 1942; Sartre's *Being and Nothingness*, 1943). Non-Western attempts to revive or create regional styles (Senghor's Négritude, Mishima's novels) only confirmed the emergence of a universal culture. Radio and phonograph records spread American popular music (swing, bebop) around the world.

The American Decade: 1950-59

Polite decolonization. The peaceful decline of European political and military power in Asia and Africa accelerated in the 1950s. Nearly all of **N Africa** was freed by 1956, but France fought a bitter war to retain Algeria, with its large European minority, until 1962. **Ghana**, independent in 1957, led a parade of new black African nations (more than 2 dozen by 1962), which altered the political character of the UN. Ethnic disputes often exploded in the new nations after decolonization (UN troops in Cyprus, 1964; **Nigerian civil war**, 1967-70). Leaders of the new states, mostly sharing socialist ideologies, tried to create an Afro-Asian bloc (Bandung Conference, 1955), but Western economic influence and U.S. political ties remained strong (Baghdad Pact, 1955).

Trade. World trade volume soared, in an atmosphere of monetary stability assured by international accords (**Bretton Woods,** 1944). In Europe, economic integration advanced (**European Economic Community,** 1957; European Free Trade Association, 1960). Comecon (1949) coordinated the economies of Soviet-bloc countries.

U.S. Economic growth produced an abundance of consumer goods (9.3 million motor vehicles sold, 1955). Suburban housing tracts changed life patterns for middle and working classes (Levittown, 1946-51). Pres. Dwight **Eisenhower's** landslide election victories (1952, 1956) reflected consensus politics. Senate condemnation of Senator Joseph **McCarthy** (Dec. 1954) curbed the political abuse of anti-Communism. A system of alliances and military bases bolstered U.S. influence on all continents. Trade and payments surpluses were balanced by overseas investments and foreign aid ($50 billion, 1950-59).

USSR. In the "thaw" after Stalin's death in 1953, relations with the West improved (evacuation of Vienna, Geneva summit conference, both 1955). Repression of scientific and cultural life eased, and many prisoners were freed or rehabilitated culminating in **de-Stalinization** (1956). **Nikita Khrushchev's** leadership aimed at consumer sector growth, but farm production lagged, despite the virgin lands program (from 1954). Soviet crushing of the 1956 Hungarian revolution, the 1960 U-2 spy plane episode, and other incidents renewed East-West tension and domestic curbs.

East Europe. Resentment of Russian domination and Stalinist repression combined with nationalist, economic, and religious factors to produce periodic violence. E Berlin workers rioted (1953), Polish workers rioted in Poznan (June 1956), and a broad-based **revolution** broke out in **Hungary** (Oct. 1956). All were suppressed by Soviet force or threats (at least 7,000 dead in Hungary). But Poland was allowed to restore private ownership of farms, and a degree of personal and economic freedom returned to Hungary. Yugoslavia experimented with worker self-management and a market economy.

Korea. The 1945 division of Korea along the 38th parallel left industry in the N, which was organized into a militant regime and armed by the USSR. The S was politically disunited. More than 60,000 N Korean troops invaded the S on June 25, 1950. The U.S., backed by the UN Security Council, sent troops. UN troops reached the Chinese border in Nov. Some 200,000 Chinese troops crossed the Yalu R. and drove back UN forces. By spring 1951 battle lines had become stabilized near the original 38th parallel border, but heavy fighting continued. Finally, an armistice was signed on July 27, 1953. U.S. troops remained in the S, and U.S. economic and military aid continued. The war stimulated rapid economic recovery in Japan.

China. Starting in 1952, industry, agriculture, and social institutions were forcibly collectivized. In a massive purge, as many as several million people were executed as Kuomintang supporters or as class and political enemies. The **Great Leap Forward** (1958-60) unsuccessfully tried to force the pace of development by substituting labor for investment.

Indochina. Ho Chi Minh's forces, aided by the USSR and the new Chinese Communist government, fought French and pro-French Vietnamese forces to a standstill and captured the strategic **Dienbienphu** camp in May 1954. The Geneva Agreements divided Vietnam in half pending elections (never held) and recognized Laos and Cambodia as independent. The U.S. aided the anti-Communist Republic of Vietnam in the S.

Middle East. Arab revolutions placed leftist, militantly nationalist regimes in power in Egypt (1952) and Iraq (1958). But Arab unity attempts failed (United Arab Republic joined Egypt, Syria, Yemen, 1958-61). Arab refusal to recognize Israel (Arab League economic blockade began Sept. 1951) led to a permanent state of war, with repeated incidents (Gaza, 1955). Israel occupied Sinai, and Britain and France took (Oct. 1956) the Suez Canal, but were replaced by the UN Emergency Force. The Mossadegh government in Iran nationalized (May 1951) the British-owned oil industry in May, but was overthrown (Aug. 1953) in a U.S.-aided coup.

Latin America. Argentinian dictator Juan **Perón**, in office 1946, enforced land reform, some nationalization, welfare state measures, and curbs on the Roman Catholic Church, and crushed opposition. A Sept. 1955 coup deposed Perón. The 1952 revolution in Bolivia brought land reform, nationalization of tin mines, and improvement in the status of Indians, who nevertheless remained poor. The Batista regime in Cuba was overthrown (Jan. 1959) by Fidel **Castro**, who imposed a Communist dictatorship, aligned Cuba with the USSR, but improved education and health care. A U.S.-backed anti-Castro invasion (**Bay of Pigs**, Apr. 1961) was crushed. Self-government advanced in the British Caribbean.

Technology. Large outlays on research and development in the U.S. and the USSR focused on military applications (H-bomb in U.S., 1952; USSR, 1953; Britain, 1957; intercontinental missiles, late 1950s). Soviet launching of the **Sputnik** satellite (Oct. 4, 1957) spurred increases in U.S. science education funds (National Defense Education Act).

Literature and film. Alienation from social and literary conventions reached an extreme in the theater of the absurd (Beckett's *Waiting for Godot*, 1952), the "new novel" (Robbe-Grillet's *Voyeur*, 1955), and avant-garde film (Antonioni's *L'Avventura*, 1960). U.S. beatniks (Kerouac's *On the Road*, 1957) and others rejected the supposed conformism of Americans (Riesman's *The Lonely Crowd*, 1950).

Rising Expectations: 1960-69

Economic boom. The longest sustained economic boom on record spanned almost the entire decade in the capitalist world; the closely watched GNP figure doubled (1960-70) in the U.S., fueled by Vietnam War–related budget deficits. The **General Agreement on Tariffs and Trade** (1967) stimulated W European prosperity, which spread to peripheral areas (Spain, Italy, E Germany). Japan became a top economic power. Foreign investment aided the industrialization of Brazil. There were limited Soviet economic reform attempts.

Reform and radicalization. Pres. John F. **Kennedy**, inaugurated 1961, emphasized youthful idealism and vigor; his assassination Nov. 22, 1963, was a national trauma. A series of political and social reform movements took root in the U.S., later spreading to other countries. Blacks demonstrated nonviolently and with partial success against segregation and poverty (1963 March on Washington; 1964 **Civil Rights Act**), but some urban ghettos erupted in extensive riots (Watts, 1965; Detroit, 1967; Martin Luther King assassination, Apr. 4, 1968). New concern for the poor (Harrington's *Other America*, 1963) helped lead to Pres. Lyndon Johnson's **"Great Society"** programs (Medicare, Water Quality Act, Higher Education Act, all 1965). Concern with the **environment** surged (Carson's *Silent Spring*, 1962). **Feminism** revived as a cultural and political movement (Friedan's *Feminine Mystique*, 1963; National Organization for Women founded 1966), and a movement for homosexual rights emerged (Stonewall riot in NYC, 1969). Pope John XXIII called the **Second Vatican Council** (1962-65), which liberalized Roman Catholic liturgy and some other aspects of Catholicism.

Opposition to U.S. involvement in Vietnam, especially among university students (**Moratorium** protest, Nov. 1969), turned violent (Weatherman Chicago riots, Oct. 1969). **New Left** and Marxist theories became popular, and membership in radical groups (Students for a Democratic Society, Black Panthers) increased. Maoist groups, especially in Europe, called for total transformation of society. In France, students sparked a nationwide strike affecting 10 million workers in May-June 1968, but an electoral reaction barred revolutionary change.

Arts and styles. The boundary between fine and popular arts was blurred to some extent by Pop Art (Warhol) and rock musicals (*Hair*, 1968). Informality and exaggeration prevailed in fashion (beards, miniskirts). A nonpolitical "counterculture" developed, rejecting traditional bourgeois life goals and personal habits, and use of marijuana and hallucinogens spread (**Woodstock** festival, Aug. 1969). Indian influence was felt in religion (Ram Dass) and fashion, and The **Beatles,** who brought unprecedented sophistication to rock music, became for many a symbol of the decade.

Science. Achievements in space (**humans on the moon,** July 1969) and electronics (lasers, integrated circuits) encouraged a faith in scientific solutions to problems in agriculture ("green revolution"), medicine (heart transplants, 1967), and other areas. Harmful technology, it was believed, could be controlled (1963 nuclear weapon test ban treaty, 1968 nonproliferation treaty).

China. Mao's revolutionary militancy caused disputes with the USSR under "revisionist" Khrushchev, starting in 1960. The 2 powers exchanged fire in 1969 border disputes. China used force to capture (1962) areas disputed with India. The **"Great Proletarian Cultural Revolution"** tried to impose a utopian egalitarian program in China and spread revolution abroad; political struggle, often violent, convulsed China in 1965-68.

Indochina. Communist-led guerrillas aided by N Vietnam fought from 1960 against the S Vietnam government of Ngo Dinh Diem (killed 1963). The U.S. military role increased after the 1964 **Tonkin Gulf** incident. U.S. forces peaked at 543,400 in Apr. 1969. Massive numbers of N Vietnamese troops also fought. Laotian and Cambodian neutrality were threatened by Communist insurgencies, with N Vietnamese aid, and U.S. intrigues.

Third World. A bloc of authoritarian leftist regimes among the newly independent nations emerged in political opposition to the U.S.-led Western alliance and came to dominate the conference of nonaligned nations (Belgrade, 1961; Cairo, 1964; Lusaka, 1970). Soviet political ties and military bases were established in Cuba, Egypt, Algeria, Guinea, and other countries whose leaders were regarded as revolutionary heroes by opposition groups in pro-Western or colonial countries. Some leaders were ousted in coups by pro-Western groups—Zaire's Patrice Lumumba (killed 1961), Ghana's Kwame Nkrumah (exiled 1966), and Indonesia's Sukarno (effectively ousted in 1965 after a Communist coup failed).

Middle East. Arab-Israeli tension erupted into a brief war June 1967. Israel emerged from the war as a major regional power. Military shipments before and after the war brought much of the Arab world into the Soviet political sphere. Most Arab states broke U.S. diplomatic ties, while Communist countries cut their ties to Israel. Intra-Arab disputes continued: Egypt and Saudi Arabia supported rival factions in a bloody Yemen civil war 1962-70; Lebanese troops fought Palestinian commandos 1969.

East Europe. To stop the large-scale exodus of citizens, E German authorities built (Aug. 1961) a **fortified wall across Berlin.** Soviet sway in the Balkans was weakened by Albania's support of China (USSR broke ties in Dec. 1961) and Romania's assertion (1964) of industrial and foreign policy autonomy. Liberalization (spring 1968) in Czechoslovakia was crushed with massive force by troops of 5 Warsaw Pact countries. W German treaties (1970) with the USSR and Poland facilitated the transfer of German technology and confirmed postwar boundaries.

Disillusionment: 1970-79

U.S.: Caution and neoconservatism. A relatively sluggish economy, energy and resource shortages, and environmental problems contributed to a **"limits of growth"** philosophy. Suspicion of science and technology killed or delayed major projects (supersonic transport dropped, 1971; Seabrook nuclear power plant protests, 1977-78) and was fed by the Three Mile Island nuclear reactor accident (Mar. 1979).

There were signs of growing mistrust of big government and weakened support for new social policies. School busing and racial quotas were opposed (Bakke decision, June 1978); the proposed Equal Rights Amendment for women languished; civil rights legislation aimed at protecting homosexuals was opposed (Dade County referendum, June 1977).

Completion of Communist forces' takeover of **South Vietnam** (evacuation of U.S. civilians, Apr. 1975), revelations of Central Intelligence Agency misdeeds (Rockefeller Commission report, June 1975), and **Watergate** scandals (Nixon resigned in Aug. 1974) reduced faith in U.S. moral and material capacity to influence world affairs. Revelations of Soviet crimes (Solzhenitsyn's *Gulag Archipelago,* 1974) and Soviet intervention in Africa helped foster a revival of anti-Communist sentiment.

Economy sluggish. The 1960s boom faltered in the 1970s; a severe recession in the U.S. and Europe (1974-75) followed a huge oil price hike (Dec. 1973). Monetary instability (U.S. cut ties to gold in Aug. 1971), the decline of the dollar, and protectionist moves by industrial countries (1977-78) threatened trade. Business investment and spending for research declined. Severe inflation plagued many countries (25% in Britain, 1975; 18% in U.S., 1979).

China picks up pieces. After the 1976 deaths of Mao Zedong and Zhou Enlai, struggle for the leadership succession was won by pragmatists. A nationwide purge of orthodox Maoists was carried out, and the **Gang of Four,** led by Mao's widow, Chiang Ching, arrested. The new leaders freed more than 100,000 political prisoners and reduced public adulation of Mao. Political and trade ties were expanded with Japan, Europe, and the U.S. in the late 1970s, as relations worsened with the USSR, Cuba, and Vietnam (4-week invasion by China, 1979). Ideological guidelines in industry, science, education, and the armed forces, which the ruling faction said had caused chaos and decline, were reversed (bonuses to workers, Dec. 1977; exams for college entrance, Oct. 1977). Severe restrictions on cultural expression were eased.

Some Famous Dates of the Second Millennium

Below are a few dates marking key events in the history of the world since AD1000.

1066	William, Duke of Normandy, conquered England.	1815	Napoleon was defeated at Waterloo.
1095	Pope Urban II called for the First Crusade.	1821	Simon Bolivar freed Venezuela from Spanish rule, in a campaign that led to widespread independence in South America.
1211	Genghis Khan invaded China, as he built the largest empire in history.		
1215	England's King John accepted the Magna Carta, limiting royal power.	1854	Japan opened its trade to the West after Commodore Matthew Perry arrived with gunships in Tokyo Bay.
1325	The Aztecs founded their capital city of Tenochtitlan.	1869	The Suez Canal opened.
1348	The Black Death (bubonic plague) reached Europe from the East.	1914	The assassination of Austrian Archduke Franz Ferdinand precipitated World War I.
1453	Constantinople fell to the Ottoman Turks.	1917	The Bolsheviks took power in Russia in a violent coup.
1455	Johann Gutenberg printed 200 Bibles, launching a technological revolution.	1933	Adolf Hitler assumed power in Germany.
1492	Christopher Columbus reached the New World.	1945	The U.S. dropped atom bombs on Hiroshima and Nagasaki, precipitating the surrender of Japan and end of World War II.
1517	Martin Luther made public his Ninety-five Theses, starting the Protestant Reformation.		
1769	James Watt patented the steam engine, initiating the Industrial Revolution.	1949	The People's Republic of China was established, after the defeat of Nationalist forces.
1776	The American colonies declared independence.	1969	Neil Armstrong became the first human to walk on the Moon.
1789	The French Revolution was inaugurated with the storming of the Bastille.		
1796	Edward Jenner discovered a vaccine for smallpox, laying the foundation for modern immunology.	1989	The Berlin Wall was opened, heralding the end of the Cold War and the coming collapse of the Soviet Union.

Europe. European unity moves (EEC-EFTA trade accord, 1972) faltered as economic problems appeared (Britain floated pound, 1972; France floated franc, 1974). Germany and Switzerland curbed guest workers from southern Europe. Greece and Turkey quarreled over Cyprus and Aegean oil rights.

All non-Communist Europe was under democratic rule after free elections were held (June 1976) in **Spain** 7 months after the death of Franco. The conservative, colonialist regime in **Portugal** was overthrown in Apr. 1974. In **Greece** the 7-year-old military dictatorship yielded power in 1974. Northern Europe, though ruled mostly by Socialists (**Swedish** Socialists unseated in 1976 after 44 years in power), turned more conservative. The **British** Labour government imposed (1975) wage curbs and suspended nationalization schemes. Terrorism in **Germany** (1972 Munich Olympics killings) led to laws curbing some civil liberties. **French** "new philosophers" rejected leftist ideologies, and the shaky Socialist-Communist coalition lost a 1978 election bid.

Religion and politics. The improvement in **Muslim** countries' political fortunes by the 1950s (with the exception of Central Asia under Soviet and Chinese rule) and the growth of Arab oil wealth were followed by a resurgence of traditional religious fervor. Libyan dictator Muammar al-Qaddafi mixed Islamic laws with socialism and called for Muslim return to Spain and Sicily. The illegal Muslim Brotherhood in **Egypt** was accused of violence, while extreme groups bombed (1977) theaters to protest secular values.

In **Turkey**, the National Salvation Party was the first Islamic group to share (1974) power since secularization in the 1920s. In **Iran, Ayatollah Ruhollah Khomeini**, led a revolution that deposed the secular shah (Jan. 1979) and created an Islamic republic there. Religiously motivated Muslims took part in an insurrection in Saudi Arabia that briefly seized (1979) the Grand Mosque in Mecca. Muslim puritan opposition to **Pakistan** Pres. Zulfikar Ali-Bhutto helped lead to his overthrow in July 1977. Muslim solidarity, however, could not prevent Pakistan's eastern province (**Bangladesh**) from declaring (Dec. 1971) independence after a bloody civil war.

Muslim and Hindu resentment of coerced sterilization in **India** helped defeat the Gandhi government, which was replaced (Mar. 1977) by a coalition including religious Hindu parties. Muslims in the S **Philippines**, aided by Libya, rebelled against central rule from 1973.

The Buddhist Soka Gakkai movement launched (1964) the Komeito party in **Japan,** which became a major opposition party in 1972 and 1976 elections.

Evangelical Protestant groups grew in the U.S. A revival of interest in Orthodox Christianity occurred among **Russian** intellectuals (Solzhenitsyn). The secularist **Israeli** Labor party, after decades of rule, was ousted in 1977 by conservatives led by Menachem Begin; religious militants founded settlements on the disputed West Bank, part of biblically promised Israel. U.S. Reform Judaism revived many previously discarded traditional practices.

Old-fashioned religious wars raged intermittently in **Northern Ireland** (Catholic vs. Protestant, 1969-   ) and **Lebanon** (Christian vs. Muslim, 1975-   ), while religious militancy complicated the Israel-Arab dispute (1973 Israel-Arab war). Despite a 1979 **peace treaty between Egypt and Israel,** increased militancy on the West Bank impeded further progress.

Latin America. Repressive conservative regimes strengthened their hold on most of the continent, with a violent coup against the elected (Sept. 1973) Allende government in **Chile,** a 1976 military coup in **Argentina,** and coups against reformist regimes in **Bolivia** (1971, 1979) and **Peru** (1976). In Central America increasing liberal and leftist militancy led to the ouster (1979) of the Somoza regime of **Nicaragua** and to civil conflict in **El Salvador.**

Indochina. Communist victories in Vietnam, Cambodia, and Laos by May 1975 led to new turmoil. The **Pol Pot regime** ordered millions of city-dwellers to resettle in rural areas, in a program of forced labor, combined with terrorism, that cost more than 1 million lives (1975-79) and caused hundreds of thousands of ethnic Chinese and others to flee Vietnam ("boat people," 1979). The Vietnamese invasion of Cambodia swelled the refugee population and contributed to widespread starvation in that devastated country.

Russian expansion. Soviet influence, checked in some countries (troops ousted by Egypt, 1972), was projected farther afield, often with the use of Cuban troops (Angola, 1975-89; Ethiopia, 1977-88) and aided by a growing navy, a merchant fleet, and international banking ability. **Détente** with the West—1972 Berlin pact, 1972 strategic arms pact (**SALT**)—gave way to a more antagonistic relationship in the late 1970s, exacerbated by the Soviet invasion (1979) of **Afghanistan.**

Africa. The last remaining European colonies were granted independence (**Spanish Sahara,** 1976; **Djibouti,** 1977) and, after 10 years of civil war and many negotiation sessions, a black government took over (1979) in Zimbabwe (Rhodesia); white domination remained in **South Africa**. Great power involvement in local wars (Russia in **Angola, Ethiopia**; France in **Chad, Zaire, Mauritania**) and the use of tens of thousands of Cuban troops were denounced by some African leaders. Ethnic or tribal clashes made Africa a locus of sustained warfare during the late 1970s.

Arts. Traditional modes of painting, architecture, and music received increased popular and critical attention in the 1970s. These more conservative styles coexisted with modernist works in an atmosphere of increased variety and tolerance.

Revitalization of Capitalism, Demand for Democracy: 1980-89

USSR, Eastern Europe. A troublesome 1980-85 for the USSR was followed by 5 years of astonishing change: the surrender of the Communist monopoly, remaking of the Soviet state, and the beginning of the disintegration of the Soviet empire. After the deaths of Leonid **Brezhnev** (1982) and 2 successors (Andropov in 1984 and Chernenko in 1985), the harsh treatment of dissent and restriction of emigration, and the Soviet invasion (Dec. 1979) of Afghanistan, Gen. Sec. Mikhail **Gorbachev** (in office 1985-1991) promoted *glasnost* and *perestroika*—economic, political, and social reform. Supported by the Communist Party (July 1988), he signed (Dec. 1987) the INF disarmament treaty, and he pledged (1988) to cut the military budget. Military withdrawal from Afghanistan was completed in Feb. 1989, democratization was not hindered in Poland and Hungary, and the Soviet people chose (Mar. 1989) part of the new Congress from competing candidates. By decade's end the **Cold War** appeared to be fading away.

In **Poland, Solidarity**, the labor union founded (1980) by Lech **Walesa**, was outlawed in 1982 and then legalized in 1988, after years of unrest. Poland's first free election since the Communist takeover brought Solidarity victory (June 1989); Tadeusz Mazowiecki, a Walesa adviser, became (Aug. 1989) prime minister in a government with the Communists. In the fall of 1989 the failure of Marxist economies in **Hungary, East Germany, Czechoslovakia, Bulgaria,** and **Romania** brought the collapse of the Communist monopoly and a demand for democracy. In a historic step, the **Berlin Wall** was opened in Nov. 1989.

U.S. "The Reagan Years" (1981-88) brought the **longest economic boom** yet in U.S. history via budget and tax cuts, deregulation, "junk bond" financing, leveraged buyouts, and mergers and takeovers. However, there was a stock market crash (Oct. 1987), and federal budget deficits and the trade deficit increased. Foreign policy showed a **strong anti-Communist stance**, via increased defense spending, aid to anti-Communists in Central America, invasion of Cuba-threatened Grenada, and championing of the MX missile system and "Star Wars" missile defense program. Four Reagan-Gorbachev summits (1985-88) climaxed in the INF treaty (1987), as the Cold War began to wind down. The Iran-contra affair (North's TV testimony, July 1987) was a major political scandal. Homelessness and drug abuse (especially "crack" cocaine) were growing social problems. In 1988, Vice Pres. George Bush was elected to succeed Ronald Reagan as president.

Middle East. The Middle East remained militarily unstable, with sharp divisions along economic, political, racial, and religious lines. In **Iran**, the Islamic revolution of 1979 created a strong anti-U.S. stance (hostage crisis, Nov. 1979-Jan. 1981). In Sept. 1980, **Iraq** repudiated its border agreement with Iran and began major hostilities that led to an 8-year war in which millions were killed.

Libya's support for international terrorism induced the U.S. to close (May 1981) its diplomatic mission there and embargo (Mar. 1982) Libyan oil. The U.S. accused Libyan leader Muammar al-Qaddafi of aiding (Dec. 1985) terrorists in Rome and of Vienna airport attacks, and retaliated by bombing Libya (Apr. 1986).

Israel affirmed (July 1980) all Jerusalem as its capital, destroyed (1981) an Iraqi atomic reactor, and invaded (1982) Lebanon, forcing the PLO to agree to withdraw. A **Palestinian uprising**, including women and children hurling rocks and bottles at troops, began (Dec. 1987) in Israeli-occupied Gaza and spread to the West Bank; troops responded with force, killing 300 by the end of 1988, with 6,000 more in detention camps.

Israeli withdrawal from **Lebanon** began in Feb. 1985 and ended in June 1985, as Lebanon continued torn by military and political conflict. Artillery duels (Mar.-Apr. 1989) between Christian East Beirut and Muslim West Beirut left 200 dead and 700 wounded. At decade's end, violence still dominated.

Latin America. In **Nicaragua**, the leftist Sandinista National Liberation Front, in power after the 1979 civil war, faced problems as a result of Nicaragua's military aid to leftist guerrillas in El Salvador and U.S. backing of antigovernment contras. The U.S. CIA admitted (1984) having directed the mining of Nicaraguan ports, and the U.S. sent humanitarian (1985) and military (1986) aid. Profits from secret arms sales to Iran were found (1987) diverted to contras. Cease-fire talks between the Sandinista government and contras came in 1988, and elections were held in Feb. 1990.

In **El Salvador**, a military coup (Oct. 1979) failed to halt extreme right-wing violence and left-wing terrorism. Archbishop Oscar Romero was assassinated in Mar. 1980; from Jan. to June some 4,000 civilians reportedly were killed in the civil unrest. In 1984, newly elected Pres. José Napoleon Duarte worked to stem human rights abuses, but violence continued.

In **Chile**, Gen. Augusto Pinochet yielded the presidency after a democratic election (Dec. 1989), but remained as head of the army. He had ruled the country since 1973, imposing harsh measures against leftists and dissidents; at the same time he introduced economic programs that restored prosperity to Chile.

Africa. 1980-85 marked a rapid decline in the economies of virtually all African countries, a result of accelerating desertification, the world economic recession, heavy indebtedness to overseas creditors, rapid population growth, and political instability. Some 60 million Africans faced prolonged hunger in 1981; much of Africa had one of the worst droughts ever in 1983, and by year's end **150 million faced near-famine**. "Live Aid," a marathon rock concert, was presented in July 1985, and the U.S. and Western nations sent aid in Sept. 1985. Economic hardship fueled political unrest and coups. Wars in Ethiopia and Sudan and military strife in several other nations continued. AIDS took a heavy toll.

South Africa. Antiapartheid sentiment gathered force; demonstrations and violent police response grew. South African white voters approved (Nov. 1983) the first constitution to give Coloureds and Asians a voice, while still excluding blacks (70% of the population). The U.S. imposed economic sanctions in Aug. 1985, and 11 Western nations followed in September. P. W. Botha, 1980s president, was succeeded by F. W. **de Klerk**, in Sept. 1989, who promised "evolutionary" change via negotiation with the black population.

China. During the 1980s the Communist government and paramount leader **Deng Xiaoping** pursued **far-reaching changes**, expanding commercial and technical ties to the industrialized world and increasing the role of market forces in stimulating urban development. Apr. 1989 brought new demands for political reforms; student demonstrators camped out in Tiananmen Sq., Beijing, in a massive peaceful protest. Some 100,000 students and workers marched, and at least 20 other cities saw protests. In response, martial law was imposed; army troops crushed the demonstration in and around Tiananmen Square on June 3-4, with death toll estimates at 500-7,000, up to 10,000 dissidents arrested, 31 people tried and executed. The conciliatory Communist Party chief was ousted; the Politburo adopted (July) reforms against official corruption.

Japan. Japan's relations with other nations, especially the U.S., were dominated by **trade imbalances favoring Japan**. In 1985 the U.S. trade deficit with Japan was $49.7 billion, one-third of the total U.S. trade deficit. After Japan was found (Apr. 1986) to sell semiconductors and computer memory chips below cost, the U.S. was assured a "fair share" of the market, but charged (Mar. 1987) Japan with failing to live up to the agreement.

European Community. With the addition of Greece, Portugal, and Spain, the EC became a **common market of more than 300 million people**, the West's largest trading entity. Margaret **Thatcher** became the first British prime minister in the 20th century to win a 3d consecutive term (1987). France elected (1981) its first socialist president, François **Mitterrand**, who was reelected in 1988. Italy elected (1983) its first socialist premier, Bettino **Craxi**.

International terrorism. With the 1979 overthrow of the shah of Iran, terrorism became a prominent political tactic. It increased through the 1980s, but with fewer high-profile attacks after 1985. In 1979-81, Iranian militants held 52 Americans hostage in Iran for 444 days; in 1983 a TNT-laden suicide terrorist blew up U.S. Marine headquarters in Beirut, killing 241 Americans, and a truck bomb blew up a French paratroop barracks, killing 58. The *Achille Lauro* cruise ship was hijacked in 1986, and an American passenger killed; the U.S. subsequently intercepted the Egyptian plane flying the terrorists to safety. Incidents rose to 700 in 1985, and to 1,000 in 1988. **Assassinated leaders** included Egypt's Pres. Anwar al-**Sadat** (1981), India's Prime Min. Indira **Gandhi** (1984), and Lebanese Premier Rashid **Karami** (1987).

Post–Cold War World: 1990-98

Soviet Empire breakup. The world community witnessed the extraordinary spectacle of a superpower's disintegration when the **Soviet Union** broke apart into 15 independent states. The 1980s had already seen internal reforms and a decline of Communist power both within the Soviet Union and in Eastern Europe. The Soviet breakup began in earnest with the declarations of independence adopted by the Baltic republics of **Lithuania, Latvia,** and **Estonia** during an abortive coup against reformist leader Mikhail **Gorbachev** (Aug. 1991). The other republics soon took the same step. In Dec. 1991, **Russia, Ukraine,** and **Belarus** declared the Soviet Union dead; Gorbachev resigned, and the Soviet Parliament went out of existence. Most of the former republics formed a loose confederation called the **Commonwealth of Independent States**. The Warsaw Pact and the Council for Mutual Economic Assistance (Comecon) were disbanded. **Russia** remained the predominant country after the breakup, but its people soon suffered severe economic hardship as the nation, under Pres. Boris **Yeltsin**, moved to revamp the economy and to adopt a free market system. In Oct. 1993, **anti-Yeltsin forces** occupied the Parliament building and were ousted by the army; about 140 people died in the fighting.

The Muslim republic of **Chechnya** declared independence from the rest of Russia, but this was met with an invasion by Russian troops (Dec. 1994). Vicious fighting continued for almost 21 months. A cease-fire finally took hold (1996), and after the Russian withdrawal, Chechnya held elections, making a former guerrilla leader president (1997).

Europe. Yugoslavia broke apart, and hostilities ensued among the republics along ethnic and religious lines. **Croatia, Slovenia,** and **Macedonia** declared independence (1991), followed by **Bosnia-Herzegovina** (1992). **Serbia** and **Montenegro** remained as the republic of Yugoslavia. Bitter fighting followed, especially in Bosnia, where Serbs reportedly engaged in "ethnic cleansing" of the Muslim population; a peace plan (Dayton accord), brokered by the United States, was signed by **Bosnia, Serbia,** and **Croatia** (Dec. 1995), with NATO troops responsible for policing its implementation. In **Yugoslavia** during 1997 and 1998, President Slobodan Milosevic pursued a violent crackdown in the Kosovo region against Kosovo Liberation Army separatists and other ethnic Albanians.

The two **Germanys** were reunited after 45 years (Oct. 1990). The union was greeted with jubilation, but stresses became apparent when free market principles were applied to the aging East German industries, resulting in many plant closings and rising unemployment. German chancellor Helmut Kohl, a Christian Democrat, lost power after 16 years, in Sept. 1998 elections; Gerhard Schroeder, a Social Democrat, took over. Czechoslovakia broke apart peacefully (Jan. 1993), becoming the **Czech Republic** and **Slovakia.** In **Poland,** Lech **Walesa** was elected president (Dec. 1991) but was defeated in his bid for a 2d term (Nov. 1995).

NATO approved the Partnership for Peace Program (Jan. 1994) coordinating the defense of **Eastern** and **Central European** countries; Russia joined the program later that year. NATO signed a pact with **Russia** (1997) providing for NATO expansion into the former Soviet-bloc countries; a similar treaty was set up with **Ukraine.** (The **Czech Republic, Hungary,** and **Poland** became members in Jan. 1999.) Efforts toward European unity continued with adoption of a single market (Jan. 1993) and conversion of the European Community to the **European Union** as the Maestricht Treaty took effect (Nov. 1993). Agreement was reached for 11 EU members to participate in Economic and Monetary Union, adopting a common currency **(euro)** for some purposes in Jan. 1999, with the euro to go in common circulation in 2002.

An intraparty revolt forced Margaret **Thatcher** out as prime minister of **Great Britain,** to be succeeded by John Major (Nov. 1990); 7 years later, Major suffered an overwhelming defeat at the hands of the new Labour Party leader, Tony Blair (May 1997). The divorce of Prince Charles and Princess Diana, followed by the death of Diana in a car accident (Aug. 1997), made headlines around the world. Talks on **peace** in **Northern Ireland** that included participation of Sinn Fein, political arm of the IRA, led to a ground-breaking peace plan, approved in an all-Ireland vote (May 1998). In **Scotland** voters overwhelmingly approved establishment of a regional legislature (1997), and in **Wales** voters narrowly approved establishment of a local assembly (1997). In a historic innovation, the Church of England ordained 32 women as priests (Mar. 1994).

Middle East. In Aug. 1990, **Iraq's Saddam Hussein** ordered his troops to invade **Kuwait.** The UN approved military action in response (Nov. 1990), and U.S. Pres. George **Bush** put together an international military force. U.S. and allied planes bombed Iraq (Jan. 1991) and launched a land attack, crushing the invasion (Feb. 1991). After Iraq formally accepted a cease-fire (Apr. 1991), U.S. troops withdrew, but "no-fly" zones were set up over northern Iraq to protect the Kurds and over southern Iraq to protect Shiite Muslims. The UN imposed sanctions on Iraq for failure to abide by the cease-fire. Iraq's reported failure to cooperate with UN arms inspectors seeking to eliminate "weapons of mass destruction" led to repeated air strikes by the U.S. and Britain.

Last Western hostages were freed in **Lebanon,** June 1992. **Israel** and the **Palestine Liberation Organization** signed a peace accord (Sept. 1993) providing for Palestinian self-government in the West Bank and Gaza Strip. Prime Minister Yitzhak **Rabin** and Foreign Minister Shimon Peres of Israel and Yasir **Arafat** of the PLO received the Nobel Peace Prize for their efforts (1994). Six Arab nations relaxed their boycott against Israel (1994), and Israel and **Jordan** signed a peace treaty (Oct. 1994). Rabin was assassinated (Nov. 1995) by an Israeli opponent of the peace process. After new elections (May 1996), Benjamin Netanyahu became prime minister and adopted a harder line on implementation of the peace accord. Arafat was elected to the presidency of the Palestinian Authority (Jan. 1996). A long-delayed interim agreement (the Wye Memorandum) on Israel military withdrawal from part of the West Bank was reached Oct. 1998. The Netanyahu government fell in Dec. 1998 (a Labour government took power after May 1999 elections).

Asia. Hong Kong was returned to **China** (July 1997) after being a British colony for 156 years. China, which emerged in the decade as a major developing economic power, had agreed to follow a policy of "one country, two systems" in Hong Kong. **Jiang Zemin,** general secretary of the Chinese Communist Party, assumed the additional post of president of China (Mar. 1993) and was the apparent key leader after the death of paramount leader **Deng Xiaoping** (Feb. 1997). China released from prison—and exiled—some well-known dissidents but continued to be criticized for detentions and other alleged human rights abuses, including persecutions of Christians and forced abortions.

After years of prosperity, **Thailand, Indonesia,** and **South Korea** in 1997 began to suffer economic reverses that had a worldwide ripple effect. These countries received billion-dollar IMF bailout packages. In **Indonesia,** protests over mismanagement led to the resignation of Pres. **Suharto** (May 1998) after 32 years of nearly autocratic rule. In **South Korea,** former dissident Kim Dae Jung was elected president (Dec. 1997). Two previous presidents, Roh Tae Woo and Chun Doo Hwan, were convicted of crimes committed in office but were given amnesty by the new president.

In **Japan** members of a religious cult, released the nerve gas sarin on 5 Tokyo subway cars, killing 12 people and injuring more than 5,500 (Mar. 1995). Tamil rebels continued their armed conflict in **Sri Lanka.** In **Afghanistan** the Taliban, an extreme Islamic fundamentalist group, gained control of Kabul (Sept. 1996) and, eventually, most of the country.

In **North Korea,** longtime dictator **Kim Il Sung** died (July 1994), to be succeeded by his son. In the same year the country signed an agreement with the United States setting a timetable for North Korea to eliminate its nuclear program. The country also suffered a severe drought, and many were estimated to have died of starvation. Uneasy relations between **India** and **Pakistan** reached a new level when both nations conducted nuclear tests in 1998.

Africa. South Africa was transformed as the white-dominated government abandoned apartheid and the country made the transition to a nonracial democratic government. Pres. F. W. **de Klerk** released Nelson **Mandela** from prison (Feb. 1990), after he had been held by the government for 27 years, and lifted a ban on the African National Congress. The white government repealed its apartheid laws (1990, 1991). Mandela was elected president (Apr. 1994), and a new constitution became law (Dec. 1996).

The decades-long rule of Mobutu Sese Seko in **Zaire** came to an end (May 1997) at the hands of rebel forces led by Laurent Kabila; an ailing Mobutu fled the country and soon after died. Kabila changed the country name back to **Democratic Republic of the Congo**; conditions remained unstable. After the presidents of **Burundi** and **Rwanda** were killed in an airplane crash (Apr. 1994), violence erupted in Rwanda between Hutu and Tutsi factions; tens of thousands were slain. The conflict spread to refugee camps in neighboring Zaire and Burundi. Factional fighting also erupted in **Somalia** after Pres. Muhammad Siad Barre was ousted (Jan. 1991). The UN sent a U.S.-led peacekeeping force, but it was unsuccessful in restoring order. Some soldiers of the peacekeeping force were killed, including 23 Pakistanis (June 1993) and 18 U.S. Rangers (Oct. 1993). The UN ended its mission (Mar. 1995) with no durable formal government in place. **Liberia** endured factional fighting that lasted almost 5 years and claimed 150,000 lives; a cease-fire was concluded in Aug. 1995. The World Health Organization reported (1995) that Africa accounted for 70% of AIDS cases worldwide.

World Population Growth: AD 1-1999

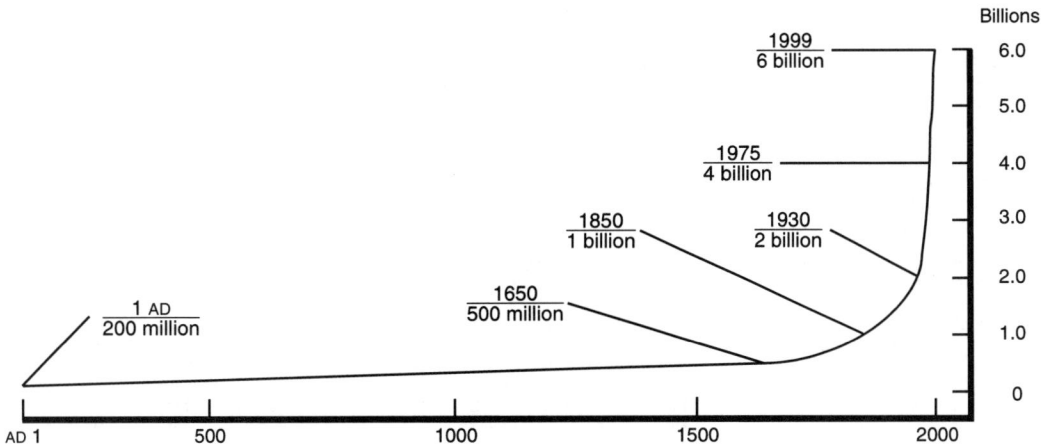

A 16-year civil war came to an end in **Angola** (May 1991) when the government signed a peace accord with the rebel UNITA faction. The country remained unstable, however, even after the inauguration of a national unity government (Apr. 1997). **Namibia** officially became independent in Mar. 1990. Claimed by South Africa since 1919 and placed under UN authority in 1971, it had long been a focus of colonial rivalries. In **Algeria,** the army cancelled a 2d round of parliamentary elections (Jan. 1992) after the Islamic party won a first round. Islamic fundamentalists then began a terrorist campaign that, along with killings by pro-government squads, eventually claimed thousands of lives.

North America. The **North American Free Trade Agreement** (NAFTA), liberalizing trade between the United States, Canada, and Mexico, went into effect Jan. 1, 1994. In **Canada**, the Progressive Conservative Party suffered a crushing defeat in general elections (Oct. 1993), and liberal Jean Chrétien became prime minister.

In the **United States**, in the 1992 presidential election, Democrat Bill **Clinton** defeated Pres. Bush, but in 1994 congressional elections Republicans gained control of both houses of Congress. Congress passed legislation under which federal protection for welfare recipients was ended and funds turned over to the states for their own programs. Clinton reached agreement with Congress on measures to eliminate the federal budget deficit. Clinton won reelection in 1996, but the new administration was plagued by continuing scandals, probed by special prosecutors.

The U.S. Army and Navy were torn by sexual scandals involving abuse of women personnel. The **CIA** suffered embarrassment with the discovery of espionage by agents (Aldrich Ames, Harold Nicholson).

In **Mexico,** Ernesto Zedillo of the ruling PRI party was elected president (July 1994) after the party's first candidate was assassinated. The country soon faced a crisis affecting the value of the peso, but recovered with the help of a bailout package from the U.S. A peasant revolt spearheaded by the **Zapatista National Liberation Army** erupted in the state of Chiapas (Jan. 1994). The initial outbreak was suppressed, but the movement remained in existence.

Central America. In **Haiti**, Jean-Bertrand Aristide was elected president (Dec. 1990) but was ousted in a military coup after 9 months in office. The UN approved a U.S.-led invasion to restore the elected leader; shortly before troops arrived, a delegation headed by former U.S. Pres. Jimmy Carter arranged (Sept. 1994) for the junta to step aside for Aristide. In **Nicaragua**, Violetta Chamarro defeated Daniel Ortega in the presidential election (Feb. 1990), thus ousting the Sandinistas. In **Panama**, U.S. troops invaded and overthrew the government of Manuel Noriega (Dec. 1989), who was wanted on drug charges; Noriega was captured Jan. 1990. In **El Salvador** (1991) and **Guatemala** (1996) the governments signed agreements with rebel factions to end long-running civil conflicts.

South America. Alberto Fujimori was elected president of **Peru** in June 1990 and, despite his suppression of the constitution (1992), was reelected in 1995. Peru succeeded in capturing (Sept. 1992) the leader of the Shining Path guerrilla movement. Leftist guerrillas took hostages at an ambassador's residence in Lima (Dec. 1996); one hostage was killed during a government assault rescuing the rest (Apr. 1997). In **Brazil**, Fernando Henrique Cardoso was elected president (Oct. 1994) and reelected in 1998 amid a growing economic slump; the IMF announced a $42 billion aid package (Nov. 1998). A UN Conference on Environment and Development, or **Earth Summit**, was held (June 1992) in **Rio de Janeiro,** with delegates from 178 nations.

Crime and terrorism. A court in Rostov-on-Don, **Russia**, convicted and sentenced to death (Oct. 1992) Andrei Chikatilo, a 56-year-old former schoolteacher reputed to be the worst serial killer in modern times. He was estimated to have killed at least 52 young women and children.

Terrorism, often linked to Mideastern sources and with the U.S. as object, continued. A terrorist bomb exploded in a garage beneath New York City's World Trade Center, killing 6 people (Feb. 1993). Bombings of a U.S. military training center (Nov. 1995) and a barracks holding U.S. airmen (June 1996), both in **Saudi Arabia,** killed 7 and 19, respectively. Bombs exploded outside U.S. embassies in Kenya and Tanzania, Aug. 1998, killing over 220 people; the U.S. retaliated with missiles fired at alleged terrorist-linked sites in Afghanistan and Sudan. In an act not associated with Mideastern terrorists, the Alfred P. Murrah Federal Building in Oklahoma City, OK, was destroyed by a bomb that killed 168 people (Apr. 1995).

Science. The powerful **Hubble Space Telescope** was launched in Apr. 1990; flaws in its mirrors and solar panels were repaired by space-walking astronauts (Dec. 1993). The U.S. space shuttle *Atlantis* docked with the orbiting Russian space station *Mir* (June 1995) for the first time, in the first of several joint missions in a spirit of post-Cold-War cooperation. In Nov. 1998 the first component for a new **International Space Station** was launched into space from Kazakhstan.

Scottish scientist Ian Wilmot announced (Feb. 1997) the **cloning** of a sheep, nicknamed Dolly—the first mammal successfully cloned from a cell from an adult animal.

HISTORICAL FIGURES
Ancient Greeks and Latins
Greeks

Aeschines, orator, 389-314 BC
Aeschylus, dramatist, 525-456 BC
Aesop, fableist, c620-c560 BC
Alcibiades, politician, 450-404 BC
Anacreon, poet, c582-c485 BC
Anaxagoras, philosopher, c500-428 BC
Anaximander, philosopher, 611-546 BC
Anaximenes, philosopher, c570-500 BC
Antiphon, speechwriter, c480-411 BC
Apollonius, mathematician, c265-170 BC
Archimedes, math. 287-212 BC
Aristophanes, dramatist, c448-380 BC
Aristotle, philosopher, 384-322 BC
Athenaeus, scholar, fl. c200
Callicrates, architect, fl. 5th cent. BC
Callimachus, poet, c305-240 BC
Cratinus, comic dramatist, 520-421 BC
Democritus, philosopher, c460-370 BC
Demosthenes, orator, 384-322 BC
Diodorus, historian, fl. 20 BC

Diogenes, philosopher, 372-c287 BC
Dionysius, historian, d. c7 BC
Empedocles, philosopher, c490-430 BC
Epicharmus, dramatist, c530-440 BC
Epictetus, philosopher, c55-c135
Epicurus, philosopher, 341-270 BC
Eratosthenes, scientist, 276-194 BC
Euclid, mathematician, fl. c300 BC
Euripides, dramatist, c484-406 BC
Galen, physician, 130-200
Heraclitus, philosopher, c540-c475 BC
Herodotus, historian, c484-420 BC
Hesiod, poet, 8th cent.
Hippocrates, physician, c460-377 BC
Homer, poet, fl. c700 BC(?)
Isocrates, orator, 436-338 BC
Menander, dramatist, 342-292 BC
Parmenides, philosopher, b c515 BC
Pericles, statesman, c495-429 BC
Phidias, sculptor, c500-435 BC

Pindar, poet, c518-c438 BC
Plato, philosopher, c428-347 BC
Plutarch, biographer, c46-120
Polybius, historian, c200-c118 BC
Praxiteles, sculptor, 400-330 BC
Pythagoras, phil., math., c580-c500 BC
Sappho, poet, c610-c580 BC
Simonides, poet, 556-c468 BC
Socrates, philosopher, 469-399 BC
Solon, statesman, 640-560 BC
Sophocles, dramatist, c496-406 BC
Strabo, geographer, c63 BC-ad24
Thales, philosopher, c634-546 BC
Themistocles, politician, c524-c460 BC
Theocritus, poet, c310-250 BC
Theophrastus, phil., c372-c287 BC
Thucydides, historian, fl. 5th cent. BC
Timon, philosopher, c320-c230 BC
Xenophon, historian, c434-c355 BC
Zeno, philosopher, c495-c430 BC

Latins

Ammianus, historian, c330-395
Apuleius, satirist, c124-c170
Boethius, scholar, c480-524
Caesar, Julius, leader, 100-44 BC
Catilina, politician, c108-62 BC
Cato (Elder), statesman, 234-149 BC
Catullus, poet, c84-54 BC
Cicero, orator, 106-43 BC
Claudian, poet, c370-c404
Ennius, poet, 239-170 BC
Gellius, author, c130-c165
Horace, poet, 65-8 BC

Juvenal, satirist, 60-127
Livy, historian, 59 BC-ad17
Lucan, poet, 39-65
Lucilius, poet, c180-c102 BC
Lucretius, poet, c99-c55 BC
Martial, epigrammatist, c38-c103
Nepos, historian, c100-c25 BC
Ovid, poet, 43 BC-AD17
Persius, satirist, 34-62
Plautus, dramatist, c254-c184 BC
Pliny the Elder, scholar, 23-79
Pliny the Younger, author, 62-113

Quintilian, rhetorician, c35-c97
Sallust, historian, 86-34 BC
Seneca, philosopher, 4 BC-ad65
Silius, poet, c25-101
Statius, poet, c45-c96
Suetonius, biographer, c69-c122
Tacitus, historian, 56-120
Terence, dramatist, 185-c159 BC
Tibullus, poet, c55-c19 BC
Virgil, poet, 70-19 BC
Vitruvius, architect, fl. 1st cent. BC

Rulers of England and Great Britain

Name		Began	Died	Age	Rgd
ENGLAND					
Saxons and Danes					
Egbert	King of Wessex, won allegiance of all English	829	839	—	10
Ethelwulf	Son, King of Wessex, Sussex, Kent, Essex	839	858	—	19
Ethelbald	Son of Ethelwulf, displaced father in Wessex	858	860	—	2
Ethelbert	2d son of Ethelwulf, united Kent and Wessex	860	866	—	6
Ethelred I	3d son, King of Wessex, fought Danes	866	871	—	5
Alfred	The Great, 4th son, defeated Danes	871	899	52	28
Edward	The Elder, Alfred's son, united English, claimed Scotland	899	924	55	25
Athelstan	The Glorious, Edward's son, King of Mercia, Wessex	924	940	45	16
Edmund	3d son of Edward, King of Wessex, Mercia	940	946	25	6
Edred	4th son of Edward	946	955	32	9
Edwy	The Fair, eldest son of Edmund, King of Wessex	955	959	18	3
Edgar	The Peaceful, 2d son of Edmund, ruled all English	959	975	32	17
Edward	The Martyr, eldest son of Edgar, murdered by stepmother	975	978	17	4
Ethelred II	The Unready, 2d son of Edgar, married Emma of Normandy	978	1016	48	37
Edmund II	Ironside, son of Ethelred II, King of London	1016	1016	27	0
Canute	The Dane, gave Wessex to Edmund, married Emma	1016	1035	40	19
Harold I	Harefoot, natural son of Canute	1035	1040	—	5
Hardecanute	Son of Canute by Emma, Danish King	1040	1042	24	2
Edward	The Confessor, son of Ethelred II (canonized 1161)	1042	1066	62	24
Harold II	Edward's brother-in-law, last Saxon King	1066	1066	44	0
House of Normandy					
William I	The Conqueror, defeated Harold at Hastings	1066	1087	60	21
William II	Rufus, 3d son of William I, killed by arrow	1087	1100	43	13
Henry I	Beauclerc, youngest son of William I	1100	1135	67	35
House of Blois					
Stephen	Son of Adela, daughter of William I, and Count of Blois	1135	1154	50	19
House of Plantagenet					
Henry II	Son of Geoffrey Plantagenet (Angevin) by Matilda, daughter of Henry I	1154	1189	56	35
Richard I	Coeur de Lion, son of Henry II, crusader	1189	1199	42	10
John	Lackland, son of Henry II, signed Magna Carta, 1215	1199	1216	50	17
Henry III	Son of John, acceded at 9, under regency until 1227	1216	1272	65	56
Edward I	Son of Henry III	1272	1307	68	35
Edward II	Son of Edward I, deposed by Parliament, 1327	1307	1327	43	20
Edward III	Of Windsor, son of Edward II	1327	1377	65	50
Richard II	Grandson of Edward III, minor until 1389, deposed 1399	1377	1400	33	22
House of Lancaster					
Henry IV	Son of John of Gaunt, Duke of Lancaster, son of Edward III	1399	1413	47	13
Henry V	Son of Henry IV, victor of Agincourt	1413	1422	34	9
Henry VI	Son of Henry V, deposed 1461, died in Tower	1422	1471	49	39

Name		Began	Died	Age	Rgd
House of York					
Edward IV	Great-great-grandson of Edward III, son of Duke of York	1461	1483	40	22
Edward V	Son of Edward IV, murdered in Tower of London	1483	1483	13	0
Richard III	Brother of Edward IV, fell at Bosworth Field	1483	1485	32	2
House of Tudor					
Henry VII	Son of Edmund Tudor, Earl of Richmond, whose father had married the widow of Henry V; descended from Edward III through his mother, Margaret Beaufort via John of Gaunt. By marriage with daughter of Edward IV he united Lancaster and York	1485	1509	53	24
Henry VIII	Son of Henry VII, by Elizabeth, daughter of Edward IV.	1509	1547	56	38
Edward VI	Son of Henry VIII, by Jane Seymour, his 3d queen. Ruled under regents. Was forced to name Lady Jane Grey his successor. Council of State proclaimed her queen July 10, 1553. Mary Tudor won Council, was proclaimed queen July 19, 1553. Mary had Lady Jane Grey beheaded for treason, Feb., 1554	1547	1553	16	6
Mary I	Daughter of Henry VIII, by Catherine of Aragon	1553	1558	43	5
Elizabeth I	Daughter of Henry VIII, by Anne Boleyn	1558	1603	69	44
GREAT BRITAIN **House of Stuart**					
James I	James VI of Scotland, son of Mary, Queen of Scots. *First to call himself King of Great Britain. This became official with the Act of Union, 1707*	1603	1625	59	22
Charles I	Only surviving son of James I; beheaded Jan. 30, 1649	1625	1649	48	24
Commonwealth, 1649–1660 Council of State, 1649; Protectorate, 1653					
The Cromwells	Oliver Cromwell, Lord Protector	1653	1658	59	—
	Richard Cromwell, son, Lord Protector, resigned May 25, 1659	1658	1712	86	—
House of Stuart (Restored)					
Charles II	Eldest son of Charles I, died without issue	1660	1685	55	25
James II	2d son of Charles I. Deposed 1688. Interregnum Dec. 11, 1688, to Feb. 13, 1689	1685	1701	68	3
William III	Son of William, Prince of Orange, by Mary, daughter of Charles I	1689	1702	51	13
and Mary II	Eldest daughter of James II and wife of William III		1694	33	6
Anne	2d daughter of James II	1702	1714	49	12
House of Hanover					
George I	Son of Elector of Hanover, by Sophia, granddaughter of James I	1714	1727	67	13
George II	Only son of George I, married Caroline of Brandenburg	1727	1760	77	33
George III	Grandson of George II, married Charlotte of Mecklenburg	1760	1820	81	59
George IV	Eldest son of George III, Prince Regent, from Feb. 1811	1820	1830	67	10
William IV	3d son of George III, married Adelaide of Saxe-Meiningen	1830	1837	71	7
Victoria	Daughter of Edward, 4th son of George III; married (1840) Prince Albert of Saxe-Coburg and Gotha, who became Prince Consort	1837	1901	81	63
House of Saxe-Coburg and Gotha					
Edward VII	Eldest son of Victoria, married Alexandra, Princess of Denmark	1901	1910	68	9
House of Windsor *Name Adopted July 17, 1917*					
George V	2d son of Edward VII, married Princess Mary of Teck	1910	1936	70	25
Edward VIII	Eldest son of George V; acceded Jan. 20, 1936, abdicated Dec. 11	1936	1972	77	1
George VI	2d son of George V; married Lady Elizabeth Bowes-Lyon	1936	1952	56	15
Elizabeth II	Elder daughter of George VI, acceded Feb. 6, 1952	1952	—	—	—

Rulers of Scotland

Kenneth I MacAlpin was the first Scot to rule both Scots and Picts, AD 846.

Duncan I was the first general ruler, 1034. Macbeth seized the kingdom 1040, was slain by Duncan's son, Malcolm III MacDuncan (Canmore), 1057.

Malcolm married Margaret, Saxon princess who had fled from the Normans. Queen Margaret introduced English language and English monastic customs. She was canonized, 1250. Her son Edgar, 1097, moved the court to Edinburgh. His brothers Alexander I and David I succeeded. Malcolm IV, the Maiden, 1153, grandson of David I, was followed by his brother, William the Lion, 1165, whose son was Alexander II, 1214. The latter's son, Alexander III, 1249, defeated the Norse and regained the Hebrides. When he died, 1286, his granddaughter, Margaret, child of Eric of Norway and grandniece of Edward I of England, known as the Maid of Norway, was chosen ruler, but died 1290, aged 8.

John Baliol, 1292-1296. (Interregnum, 10 years.)

Robert Bruce (The Bruce), 1306-1329, victor at Bannockburn, 1314.

David II, only son of Robert Bruce, ruled 1329-1371.

Robert II, 1371-1390, grandson of Robert Bruce, son of Walter, the Steward of Scotland, was called The Steward, first of the so-called Stuart line.

Robert III, son of Robert II, 1390-1406.

James I, son of Robert III, 1406-1437.

James II, son of James I, 1437-1460.

James III, eldest son of James II, 1460-1488.

James IV, eldest son of James III, 1488-1513.

James V, eldest son of James IV, 1513-1542.

Mary, daughter of James V, born 1542, became queen when one week old; was crowned 1543. Married, 1558, Francis, son of Henry II of France, who became king 1559, died 1560. Mary ruled Scots 1561 until abdication, 1567. She also married Henry Stewart, Lord Darnley (1565), and James, Earl of Bothwell (1567). Imprisoned by Elizabeth I, Mary was beheaded 1587.

James VI, 1566-1625, son of Mary and Lord Darnley, became King of England on death of Elizabeth in 1603. Although the thrones were thus united, the legislative union of Scotland and England was not effected until the Act of Union, May 1, 1707.

Prime Ministers of Great Britain

Designations in parentheses describe each government;
W=Whig; T=Tory; Cl=Coalition; P=Peelite; L=Liberal; C=Conservative; La=Labour.

Sir Robert Walpole (W) 1721-1742	Earl Grey (W) 1830-1834	Herbert H. Asquith (Li) 1908-1915
Earl of Wilmington (W) 1742-1743	Viscount Melbourne (W) 1834	Herbert H. Asquith (Cl) 1915-1916
Henry Pelham (W) 1743-1754	Sir Robert Peel (T) 1834-1835	David Lloyd George (Cl) 1916-1922
Duke of Newcastle (W) 1754-1756	Viscount Melbourne (W) 1835-1841	Andrew Bonar Law (C) 1922-1923
Duke of Devonshire (W) 1756-1757	Sir Robert Peel (T) 1841-1846	Stanley Baldwin (C) 1923-1924
Duke of Newcastle (W) 1757-1762	Lord (later Earl) John Russell (W)1846-1852	James Ramsay MacDonald (La) . 1924
Earl of Bute (T). 1762-1763	Earl of Derby (T) 1852	Stanley Baldwin (C) 1924-1929
George Grenville (W). 1763-1765	Earl of Aberdeen (P). 1852-1855	James Ramsay MacDonald (La) . 1929-1931
Marquess of Rockingham (W) . . 1765-1766	Viscount Palmerston (Li) 1855-1858	James Ramsay MacDonald (Cl) . 1931-1935
William Pitt the Elder (Earl of	Earl of Derby (C) 1858-1859	Stanley Baldwin (Cl) 1935-1937
Chatham) (W). 1766-1768	Viscount Palmerston (Li) 1859-1865	Neville Chamberlain (Cl) 1937-1940
Duke of Grafton (W) 1768-1770	Earl Russell (Li) 1865-1866	Winston Churchill (Cl) 1940-1945
Frederick North (Lord North) (T)1770-1782	Earl of Derby (C) 1866-1868	Winston Churchill (C). 1945
Marquess of Rockingham (W) . . 1782	Benjamin Disraeli (C). 1868	Clement Attlee (La). 1945-1951
Earl of Shelburne (W). 1782-1783	William E. Gladstone (Li). 1868-1874	Sir Winston Churchill (C) 1951-1955
Duke of Portland (Cl) 1783	Benjamin Disraeli (C). 1874-1880	Sir Anthony Eden (C) 1955-1957
William Pitt the Younger (T) . . . 1783-1801	William E. Gladstone (Li). 1880-1885	Harold Macmillan (C) 1957-1963
Henry Addington (T) 1801-1804	Marquess of Salisbury (C) 1885-1886	Sir Alec Douglas-Home (C) 1963-1964
William Pitt the Younger (T) . . . 1804-1806	William E. Gladstone (Li). 1886	Harold Wilson (La) 1964-1970
William Wyndham Grenville,	Viscount Palmerston (Li) 1859-1865	Neville Chamberlain (Cl) 1937-1940
Baron Grenville (W). 1806-1807	Marquess of Salisbury (C) 1886-1892	Edward Heath (C) 1970-1974
Duke of Portland (T). 1807-1809	William E. Gladstone (Li). 1892-1894	Harold Wilson (La) 1974-1976
Spencer Perceval (T). 1809-1812	Earl of Rosebery (Li) 1894-1895	James Callaghan (La) 1976-1979
Earl of Liverpool (T) 1812-1827	Marquess of Salisbury (C) 1895-1902	Margaret Thatcher (C). 1979-1990
George Canning (T) 1827	Arthur J. Balfour (C). 1902-1905	John Major (C). 1990-1997
Viscount Goderich (T) 1827-1828	Sir Henry Campbell Bannerman (Li)1905-1908	Tony Blair (La) 1997-
Duke of Wellington (T). 1828-1830		

Historical Periods of Japan

Yamato	c. 300-592	Conquest of Yamato plain c. AD 300.	**Muromachi**	1392-1573	Unification of Southern and Northern Courts, 1392.
Asuka	592-710	Accession of Empress Suiko, 592.	**Sengoku**	1467-1600	Beginning of the Onin war, 1467.
Nara	710-794	Completion of Heijo (Nara),710; the capital moves to Nagaoka, 784.	**Momoyama**	1573-1603	Oda Nobunaga enters Kyoto, 1568; Nobunaga deposes last Ashikaga shogun, 1573; Tokugawa Ieyasu victor at Sekigahara, 1600.
Heian	794-1185	Completion of Heian (Kyoto), 794.			
Fujiwara	858-1160	Fujiwara-no-Yoshifusa becomes regent, 858.	**Edo**	1603-1867	Ieyasu becomes shogun, 1603.
Taira	1160-1185	Taira-no-Kiyomori assumes control, 1160; Minamoto-no-Yoritomo victor over Taira, 1185.	**Meiji**	1868-1912	Enthronement of Emperor Mutsuhito (Meiji), 1867; Meiji Restoration and Charter Oath, 1868.
Kamakura	1192-1333	Yoritomo becomes shogun, 1192.	**Taisho**	1912-1926	Accession of Emperor Yoshihito, 1912.
Namboku	1334-1392	Restoration of Emperor Godaigo, 1334; Southern Court established by Godaigo at Yoshino, 1336.	**Showa**	1926-1989	Accession of Emperor Hirohito, 1926.
Ashikaga	1338-1573	Ashikaga Takauji becomes shogun, 1338.	**Heisei**	1989-	Accession of Emperor Akihito, 1989.

Rulers of France: Kings, Queens, Presidents

Caesar to Charlemagne

Julius Caesar subdued the Gauls, native tribes of Gaul (France), 58 to 51 BC. The Romans ruled 500 years. The Franks, a Teutonic tribe, reached the Somme from the East c. AD 250. By the 5th century the Merovingian Franks ousted the Romans. In 451, with the help of Visigoths, Burgundians and others, they defeated Attila and the Huns at Chalons-sur-Marne.

Childeric I became leader of the Merovingians 458. His son Clovis I (Chlodwig, Ludwig, Louis), crowned 481, founded the dynasty. After defeating the Alemanni (Germans) 496, he was baptized a Christian and made Paris his capital. His line ruled until Childeric III was deposed, 751.

The West Merovingians were called Neustrians, the eastern Austrasians. Pepin of Herstal (687-714), major domus,

or head of the palace, of Austrasia, took over Neustria as dux (leader) of the Franks. Pepin's son, Charles, called Martel (the Hammer), defeated the Saracens at Tours-Poitiers, 732; was succeeded by his son, Pepin the Short, 741, who deposed Childeric III and ruled as king until 768.

His son, Charlemagne, or Charles the Great (742-814), became king of the Franks, 768, with his brother Carloman, who died 771. Charlemagne ruled France, Germany, parts of Italy, Spain, and Austria, and enforced Christianity. Crowned Emperor of the Romans by Pope Leo III in St. Peter's, Rome, Dec. 25, 800. Succeeded by son, Louis I the Pious, 814. At death, 840, Louis left empire to sons, Lothair (Roman emperor); Pepin I (king of Aquitaine); Louis II (of Germany); Charles the Bald (France). They quarreled and, by the peace of Verdun, 843, divided the empire.

The date preceding each entry is year of accession.

The Carolingians

843 Charles I (the Bald); Roman Emperor, 875
877 Louis II (the Stammerer), son
879 Louis III (died 882) and Carloman, brothers
885 Charles II (the Fat); Roman Emperor, 881
888 Eudes (Odo), elected by nobles
898 Charles III (the Simple), son of Louis II, defeated by
922 Robert, brother of Eudes, killed in war
923 Rudolph (Raoul), Duke of Burgundy
936 Louis IV, son of Charles III
954 Lothair, son, aged 13, defeated by Capet
986 Louis V (the Sluggard), left no heirs

The Capets

987 Hugh Capet, son of Hugh the Great
996 Robert II (the Wise), his son
1031 Henry I, his son
1060 Philip I (the Fair), son
1108 Louis VI (the Fat), son
1137 Louis VII (the Younger), son
1180 Philip II (Augustus), son, crowned at Reims
1223 Louis VIII (the Lion), son
1226 Louis IX, son, crusader; Louis IX (1214-1270) reigned 44 years, arbitrated disputes with English King Henry III; led crusades, 1248 (captured in Egypt 1250) and 1270, when he died of plague in Tunis. Canonized 1297 as St. Louis.
1270 Philip III (the Hardy), son
1285 Philip IV (the Fair), son, king at 17
1314 Louis X (the Headstrong), son. His posthumous son, John I, lived only 7 days
1316 Philip V (the Tall), brother of Louis X
1322 Charles IV (the Fair), brother of Louis X

House of Valois

1328 Philip VI (of Valois), grandson of Philip III
1350 John II (the Good), his son, retired to England
1364 Charles V (the Wise), son
1380 Charles VI (the Beloved), son
1422 Charles VII (the Victorious), son. In 1429 Joan of Arc (Jeanne d'Arc) promised Charles to oust the English, who occupied northern France. Joan won at Orleans and Patay and had Charles crowned at Reims, July 17, 1429. Joan was captured May 24, 1430, and executed May 30, 1431, at Rouen for heresy. Charles ordered her rehabilitation, effected 1455.
1461 Louis XI (the Cruel), son, civil reformer
1483 Charles VIII (the Affable), son
1498 Louis XII, great-grandson of Charles V
1515 Francis I, of Angouleme, nephew, son-in-law. Francis I (1494-1547) reigned 32 years, fought 4 big wars, was patron of the arts, aided Cellini, del Sarto, Leonardo da Vinci, Rabelais, embellished Fontainebleau.
1547 Henry II, son, killed at a joust in a tournament. He was the husband of Catherine de Medicis (1519-1589) and the lover of Diane de Poitiers (1499-1566). Catherine was born in Florence, daughter of Lorenzo de Medici. By her marriage to Henry II she became the mother of Francis II, Charles IX, Henry III and Queen Margaret (Reine Margot), wife of Henry IV. She persuaded Charles IX to order the massacre of Huguenots on the Feast of St. Bartholomew, Aug. 24, 1572, the day her daughter was married to Henry of Navarre.
1559 Francis II, son. In 1548, Mary, Queen of Scots since infancy, was betrothed when 6 to Francis, aged 4. They were married 1558. Francis died 1560, aged 16; Mary ruled Scotland, abdicated 1567.
1560 Charles IX, brother
1574 Henry III, brother, assassinated

House of Bourbon

1589 Henry IV, of Navarre, assassinated. Henry IV made ene-mies when he gave tolerance to Protestants by Edict of Nantes, 1598. He was grandson of Queen Margaret of Navarre, literary patron. He married Margaret of Valois, daughter of Henry II and Catherine de Medicis; was divorced; in 1600 married Marie de Medicis, who became Regent of France, 1610-1617, for her son, Louis XIII, but was exiled by Richelieu, 1631.

1610 Louis XIII (the Just), son. Louis XIII (1601-1643) married Anne of Austria. His ministers were Cardinals Richelieu and Mazarin.
1643 Louis XIV (The Grand Monarch), son. Louis XIV was king 72 years. He exhausted a prosperous country in wars for thrones and territory. By revoking the Edict of Nantes (1685) he caused the emigration of the Huguenots. He said: "I am the state."
1715 Louis XV, great-grandson. Louis XV married a Polish princess; lost Canada to the English. His favorites, Mme. Pompadour and Mme. Du Barry, influenced policies. Noted for saying "After me, the deluge."
1774 Louis XVI, grandson; married Marie Antoinette, daughter of Empress Maria Therese of Austria. King and queen beheaded by Revolution, 1793. Their son, called Louis XVII, died in prison, never ruled.

First Republic

1792 National Convention of the French Revolution
1795 Directory, under Barras and others
1799 Consulate, Napoleon Bonaparte, first consul. Elected consul for life, 1802.

First Empire

1804 Napoleon I (Napoleon Bonaparte), emperor. Josephine (de Beauharnais), empress, 1804-1809; Marie Louise, empress, 1810-1814. Her son, Francois (1811-1832), titular King of Rome, later Duke de Reichstadt and "Napoleon II," never ruled. Napoleon abdicated 1814, died 1821.

Bourbons Restored

1814 Louis XVIII, king; brother of Louis XVI
1824 Charles X, brother; reactionary; deposed by the July Revolution, 1830

House of Orleans

1830 Louis-Philippe, the "citizen king"

Second Republic

1848 Louis Napoleon Bonaparte, president, nephew of Napoleon I.

Second Empire

1852 Napoleon III (Louis Napoleon Bonaparte), emperor; Eugenie (de Montijo), empress. Lost Franco-Prussian war, deposed 1870. Son, Prince Imperial (1856-1879), died in Zulu War. Eugenie died 1920.

Third Republic—Presidents

1871 Thiers, Louis Adolphe (1797-1877)
1873 MacMahon, Marshal Patrice M. de (1808-1893)
1879 Grevy, Paul J. (1807-1891)
1887 Sadi-Carnot, M. (1837-1894), assassinated
1894 Casimir-Perier, Jean P. P. (1847-1907)
1895 Faure, François Felix (1841-1899)
1899 Loubet, Emile (1838-1929)
1906 Fallieres, C. Armand (1841-1931)
1913 Poincare, Raymond (1860-1934)
1920 Deschanel, Paul (1856-1922)
1920 Millerand, Alexandre (1859-1943)
1924 Doumergue, Gaston (1863-1937)
1931 Doumer, Paul (1857-1932), assassinated
1932 Lebrun, Albert (1871-1950), resigned 1940
1940 Vichy govt. under German armistice: Henri Philippe Petain (1856-1951), Chief of State, 1940-1944. Provisional govt. after liberation: Charles de Gaulle (1890-1970), Oct. 1944-Jan. 21, 1946; Felix Gouin (1884-1977), Jan. 23, 1946; Georges Bidault (1899-1983), June 24, 1946.

Fourth Republic—Presidents

1947 Auriol, Vincent (1884-1966)
1954 Coty, Rene (1882-1962)

Fifth Republic—Presidents

1959 De Gaulle, Charles Andre J. M. (1890-1970)
1969 Pompidou, Georges (1911-1974)
1974 Giscard d'Estaing, Valery (1926-   )
1981 Mitterrand, François (1916-1996)
1995 Chirac, Jacques (1932-   )

Rulers of Middle Europe; Rise and Fall of Dynasties; Rulers of Germany

Carolingian Dynasty

Charles the Great, or Charlemagne, ruled France, Italy, and Middle Europe; established Ostmark (later Austria); crowned Roman emperor by pope in Rome, AD 800; died 814.

Louis I (Ludwig) the Pious, son; crowned by Charlemagne 814; died 840.

Louis I, the German, son; succeeded to East Francia (Germany) 843-876.

Charles the Fat, son; inherited East Francia and West Francia (France) 876, reunited empire, crowned emperor by pope 881, deposed 887.

Arnulf, nephew, 887-899. Partition of empire.

Louis the Child, 899-911, last direct descendant of Charlemagne.

Conrad I, duke of Franconia, first elected German king, 911-918, founded House of Franconia.

Saxon Dynasty; First Reich

Henry I, the Fowler, duke of Saxony, 919-936.

Otto I, the Great, 936-973, son; crowned Holy Roman Emperor by pope, 962.

Otto II, 973-983, son; failed to oust Greeks and Arabs from Sicily.

Otto III, 983-1002, son; crowned emperor at 16.

Henry II, the Saint, duke of Bavaria, 1002-1024, great-grandson of Otto the Great.

House of Franconia

Conrad II, 1024-1039, elected king of Germany.

Henry III, the Black, 1039-1056, son; deposed 3 popes; annexed Burgundy.

Henry IV, 1056-1106, son; regency by his mother, Agnes of Poitou. Banned by Pope Gregory VII, he did penance at Canossa.

Henry V, 1106-1125, son; last of Salic House.

Lothair, duke of Saxony, 1125-1137. Crowned emperor in Rome, 1134.

House of Hohenstaufen

Conrad III, duke of Swabia, 1138-1152. In 2d Crusade.

Frederick I, Barbarossa, 1152-1190; Conrad's nephew.

Henry VI, 1190-1196, took lower Italy from Normans. Son became king of Sicily.

Philip of Swabia, 1197-1208, brother.

Otto IV, of House of Welf, 1198-1215; deposed.

Frederick II, 1215-1250, son of Henry VI; king of Sicily; crowned king of Jerusalem in 5th Crusade.

Conrad IV, 1250-1254, son; lost lower Italy to Charles of Anjou.

Conradin, 1252-1268, son, king of Jerusalem and Sicily, beheaded. Last Hohenstaufen.

Interregnum, 1254-1273, Rise of the Electors.

Transition

Rudolph I of Hapsburg, 1273-1291, defeated King Ottocar II of Bohemia. Bequeathed duchy of Austria to eldest son, Albert.

Adolph of Nassau, 1292-1298, killed in war with Albert of Austria.

Albert I, king of Germany, 1298-1308, son of Rudolph.

Henry VII, of Luxemburg, 1308-1313, crowned emperor in Rome. Seized Bohemia, 1310.

Louis IV of Bavaria (Wittelsbach), 1314-1347. Also elected was Frederick of Austria, 1314-1330 (Hapsburg). Abolition of papal sanction for election of Holy Roman Emperor.

Charles IV, of Luxemburg, 1347-1378, grandson of Henry VII, German emperor and king of Bohemia, Lombardy, Burgundy; took Mark of Brandenburg.

Wenceslaus, 1378-1400, deposed.

Rupert, Duke of Palatine, 1400-1410.

Sigismund, 1411-1437.

Hungary

Stephen I, house of Arpad, 997-1038. Crowned king 1000; converted Magyars; canonized 1083. After several centuries of feuds Charles Robert of Anjou became Charles I, 1308-1342.

Louis I, the Great, son, 1342-1382; joint ruler of Poland with Casimir III, 1370. Defeated Turks.

Mary, daughter, 1382-1395, ruled with husband. Sigismund of Luxemburg, 1387-1437, also king of Bohemia. As bro. of Wenceslaus he succeeded Rupert as Holy Roman Emperor, 1410.

Albert, 1438-1439, son-in-law of Sigismund; also Roman emperor as Albert II (see under Hapsburg).

Ulaszlo I of Poland, 1440-1444.

Ladislaus V, posthumous son of Albert II, 1444-1457. John Hunyadi (Hunyadi Janos), governor (1446-1452), fought Turks, Czechs; died 1456.

Matthias I (Corvinus), son of Hunyadi, 1458-1490. Shared rule of Bohemia, captured Vienna, 1485, annexed Austria, Styria, Carinthia.

Ulaszlo II (king of Bohemia), 1490-1516.

Louis II, son, aged 10, 1516-1526. Wars with Suleiman, Turk. In 1527 Hungary split between Ferdinand I, Archduke of Austria, bro.-in-law of Louis II, and John Zapolya of Transylvania. After Turkish invasion, 1547, Hungary split between Ferdinand, Prince John Sigismund (Transylvania), and the Turks.

House of Hapsburg

Albert V of Austria, Hapsburg, crowned king of Hungary, Jan. 1438, Roman emperor, March 1438, as Albert II; died 1439.

Frederick III, cousin, 1440-1493. Fought Turks.

Maximilian I, son, 1493-1519. Assumed title of Holy Roman Emperor (German), 1493.

Charles V, grandson, 1519-1556. King of Spain with mother co-regent; crowned Roman emperor at Aix, 1520. Confronted Luther at Worms; attempted church reform and religious conciliation; abdicated 1556.

Ferdinand I, king of Bohemia, 1526, of Hungary, 1527; disputed German king, 1531. Crowned Roman emperor on abdication of brother Charles V, 1556.

Maximilian II, son, 1564-1576.

Rudolph II, son, 1576-1612.

Matthias, brother, 1612-1619, king of Bohemia and Hungary.

Ferdinand II of Styria, king of Bohemia, 1617, of Hungary, 1618, Roman emperor, 1619. Bohemian Protestants deposed him, elected Frederick V of Palatine, starting Thirty Years War.

Ferdinand III, son, king of Hungary, 1625, Bohemia, 1627, Roman emperor, 1637. Peace of Westphalia, 1648, ended war.

Leopold I, 1658-1705; Joseph I, 1705-1711; Charles VI, 1711-1740.

Maria Theresa, daughter, 1740-1780, Archduchess of Austria, queen of Hungary; ousted pretender, Charles VII, crowned 1742; in 1745 obtained election of her husband Francis I as Roman emperor and co-regent (d. 1765). Fought Seven Years' War with Frederick II of Prussia. Mother of Marie Antoinette.

Joseph II, son, 1765-1790, Roman emperor, reformer; powers restricted by Empress Maria Theresa until her death, 1780. First partition of Poland. Leopold II, 1790-1792.

Francis II, son, 1792-1835. Fought Napoleon. Proclaimed first hereditary emperor of Austria, 1804. Forced to abdicate as Roman emperor, 1806; last use of title. Ferdinand I, son, 1835-1848, abdicated during revolution.

Austro-Hungarian Monarchy

Francis Joseph I, nephew, 1848-1916, emperor of Austria, king of Hungary. Dual monarchy of Austria-Hungary formed, 1867. After assassination of heir, Archduke Francis Ferdinand, June 28, 1914, Austrian diplomacy precipitated World War I.

Charles I, grand-nephew, 1916-1918, last emperor of Austria and king of Hungary. Abdicated Nov. 11-13, 1918, died 1922.

Rulers of Prussia

Nucleus of Prussia was the Mark of Brandenburg. First margrave Albert the Bear (Albrecht), 1134-1170. First Hohenzollern margrave was Frederick, burgrave of Nuremberg, 1417-1440.

Frederick William, 1640-1688, the Great Elector. Son, Frederick III, 1688-1713, crowned King Frederick of Prussia, 1701.

Frederick William I, son, 1713-1740.

Frederick II, the Great, son, 1740-1786, annexed Silesia, part of Austria.

Frederick William II, nephew, 1786-1797.

Frederick William III, son, 1797-1840. Napoleonic wars.

Frederick William IV, son, 1840-1861. Uprising of 1848 and first parliament and constitution.

Second and Third Reich

William I, 1861-1888, brother. Annexation of Schleswig and Hanover; Franco-Prussian war, 1870-1871, proclamation of German Reich, Jan. 18, 1871, at Versailles; William, German emperor (Deutscher Kaiser), Bismarck, chancellor.

Frederick III, son, 1888.

William II, son, 1888-1918. Led Germany in World War I, abdicated as German emperor and king of Prussia, Nov. 9, 1918. Died in exile in Netherlands, June 4, 1941. Minor rulers of Bavaria, Saxony, Wurttemberg also abdicated.

Germany proclaimed republic at Weimar, July 1, 1919. Presidents included: Frederick Ebert, 1919-1925; Paul von Hindenburg-Beneckendorff, 1925, reelected 1932, d. Aug. 2, 1934. Adolf Hitler, chancellor, chosen successor as Leader-Chancellor (Fuehrer-Reichskanzler) of Third Reich. Annexed Austria, Mar. 1938. Precipitated World War II, 1939-1945. Suicide Apr. 30, 1945.

Germany After 1945

Following World War II, Germany was split between democratic West and Soviet-dominated East. West German chancellors: Konrad Adenauer, 1949-1963; Ludwig Erhard, 1963-1966; Kurt Georg Kiesinger, 1966-1969; Willy Brandt, 1969-1974; Helmut Schmidt, 1974-1982; Helmut Kohl, 1982-1990. East German Communist party leaders: Walter Ulbricht, 1946-1971; Erich Honecker, 1971-1989; Egon Krenz, 1989-1990.

Germany reunited Oct. 3, 1990. Post-reunification chancellors: Helmut Kohl, 1990-1998; Gerhard Schröder, 1998- .

Rulers of Poland

House of Piasts

Miesko I, 962?-992; Poland Christianized 966. Expansion under 3 Boleslavs: I, 992-1025, son, crowned king 1024; II, 1058-1079, great-grandson, exiled after killing bishop Stanislav who became chief patron saint of Poland; III, 1106-1138, nephew, divided Poland among 4 sons, eldest suzerain.

1138-1306, feudal division. 1226 founding in Prussia of military order Teutonic Knights. 1226 invasion by Tartars/Mongols.

Vladislav I, 1306-1333, reunited most Polish territories, crowned king 1320. Casimir III the Great, 1333-1370, son, developed economic, cultural life, foreign policy.

House of Anjou

Louis I, 1370-1382, nephew/was also Louis I of Hungary.

Jadwiga, 1384-1399, daughter, married 1386 Jagiello, Grand Duke of Lithuania.

House of Jagiellonians

Vladislav II, 1386-1434, Christianized Lithuania, founded personal union between Poland & Lithuania. Defeated 1410 Teutonic Knights at Grunwald.

Vladislav III, 1434-1444, son, simultaneously king of Hungary. Fought Turks, killed 1444 in battle of Varna.

Casimir IV, 1446-1492, brother, competed with Hapsburgs, put son Vladislav on throne of Bohemia, later also of Hungary (Ulaszlo II).

Sigismund I, 1506-1548, son, patronized science and arts, his and son's reign "Golden Age."

Sigismund II, 1548-1572, son, established 1569 real union of Poland and Lithuania (lasted until 1795).

Elective Kings

Polish nobles in 1572 proclaimed Poland a republic headed by king to be elected by whole nobility.

Stephen Batory, 1576-1586, duke of Transylvania, married Ann, sister of Sigismund II August. Fought Russians.

Sigismund III Vasa, 1587-1632, nephew of Sigismund II. 1592-1598 also king of Sweden. His generals fought Russians, Turks.

Vladislav II Vasa, 1632-1648, son. Fought Russians.

John II Casimir Vasa, 1648-1668, brother. Fought Cossacks, Swedes, Russians, Turks, Tatars (the "Deluge"). Abdicated 1668.

John III Sobieski, 1674-1696. Won Vienna from besieging Turks, 1683.

Stanislav II, 1764-1795, last king. Encouraged reforms; 1791 1st modern Constitution in Europe. 1772, 1793, 1795 Poland partitioned among Russia, Prussia, Austria. Unsuccessful insurrection against foreign invasion 1794 under Kosciuszko, American-Polish general.

1795-1918: Poland Under Foreign Rule

1807-1815 Grand Duchy of Warsaw created by Napoleon I, Frederick August of Saxony grand duke.

1815 Congress of Vienna proclaimed part of Poland "Kingdom" in personal union with Russia.

Polish uprisings: 1830 against Russia; 1846, 1848 against Austria; 1863 against Russia—all repressed.

1918-1939: Second Republic

1918-1922 Head of State Jozef Pilsudski. Presidents: Gabriel Narutowicz 1922, assassinated; Stanislav Wojciechowski 1922-1926, had to abdicate after Pilsudski's coup d'état; Ignacy Moscicki, 1926-1939, ruled (with Pilsudski until his death, 1935) as virtual dictator.

1939-1945: Poland Under Foreign Occupation

Nazi aggression Sept. 1939. Polish government-in-exile, first in France, then in England. Vladislav Raczkiewicz president; Gen. Vladislav Sikorski, then Stanislav Mikolajczyk, prime ministers. Soviet-sponsored Polish Committee of National Liberation proclaimed at Lublin July 1944, transformed into government Jan. 1, 1945.

Poland After 1945

In the late 1940s, Poland came increasingly under Soviet control. Communist party ruled in Poland until Aug. 1989, when democratic Solidarity party gained control of government. Solidarity leader Lech Walesa was elected president, Nov. 1990; succeeded by former Communist Aleksander Kwasniewski, Nov. 1995.

Rulers of Denmark, Sweden, Norway

Denmark

Earliest rulers invaded Britain; King Canute, who ruled in London 1016-1035, was most famous. The Valdemars furnished kings until the 15th century. In 1282 the Danes won the first national assembly, Danehof, from King Erik V.

Most redoubtable medieval character was Margaret, daughter of Valdemar IV, born 1353, married at 10 to King Haakon VI of Norway. In 1376 she had her first infant son Olaf made king of Denmark. After his death, 1387, she was regent of Denmark and Norway. In 1388 Sweden accepted her as sovereign. In 1389 she made her grand-nephew, Duke Erik of Pomerania, titular king of Denmark, Sweden, and Norway, with herself as regent. In 1397 she effected the Union of Kalmar of the three kingdoms and had Erik VII crowned. In 1439 the three kingdoms deposed him and elected, 1440, Christopher of Bavaria king (Christopher III). On his death, 1448, the union broke up.

Succeeding rulers were unable to enforce their claims as rulers of Sweden until 1520, when Christian II conquered Sweden. He was thrown out 1522, and in 1523 Gustavus Vasa united Sweden. Denmark continued to dominate Norway until the Napoleonic wars, when Frederick VI, 1808-1839, joined the Napoleonic cause after Britain had destroyed the Danish fleet, 1807. In 1814 he was forced to cede Norway to Sweden and Helgoland to Britain, receiving Lauenburg. Successors Christian VIII, 1839; Frederick VII, 1848; Christian IX, 1863; Frederick VIII, 1906; Christian X, 1912; Frederick IX, 1947; Margrethe II, 1972.

Sweden

Early kings ruled at Uppsala, but did not dominate the country. Sverker, c1130-c1156, united the Swedes and Goths. In 1435 Sweden obtained the Riksdag, or parliament. After the Union of Kalmar, 1397, the Danes either ruled or harried the country until Christian II of Denmark conquered it anew, 1520. This led to a rising under Gustavus Vasa, who ruled Sweden 1523-1560, and established an independent kingdom. Charles IX, 1599-1611, crowned 1604, conquered Moscow. Gustavus II Adolphus, 1611-1632, was called the Lion of the North. Later rulers: Christina, 1632; Charles X Gustavus, 1654; Charles XI, 1660; Charles XII (invader of Russia and Poland, defeated at Poltava, June 28, 1709), 1697; Ulrika Eleanora, sister, elected queen 1718; Frederick I (of Hesse), her husband, 1720; Adolphus Frederick, 1751; Gustavus III, 1771; Gustavus IV Adolphus, 1792; Charles XIII, 1809. (Union with Norway began 1814.) Charles XIV John, 1818 (he was Jean Bernadotte, Napoleon's Prince of Ponte Corvo, elected 1810 to succeed Charles XIII); he founded the present dynasty: Oscar I, 1844; Charles XV, 1859; Oscar II, 1872; Gustavus V, 1907; Gustav VI Adolf, 1950; Carl XVI Gustaf, 1973.

Norway

Overcoming many rivals, Harald Haarfager, 872-930, conquered Norway, Orkneys, and Shetlands; Olaf I, great-grandson, 995-1000, brought Christianity into Norway, Iceland, and Greenland. In 1035 Magnus the Good also became king of Denmark. Haakon V, 1299-1319, had married his daughter to Erik of Sweden. Their son, Magnus, became ruler of Norway and Sweden at 6. His son, Haakon VI, married Margaret of Denmark; their son Olaf IV became king of Norway and Denmark, followed by Margaret's regency and the Union of Kalmar, 1397.

In 1450 Norway became subservient to Denmark. Christian IV, 1588-1648, founded Christiania, now Oslo. After Napoleonic wars, when Denmark ceded Norway to Sweden, a strong nationalist movement forced recognition of Norway as an independent kingdom united with Sweden under the Swedish kings, 1814-1905. In 1905 the union was dissolved and Prince Charles of Denmark became Haakon VII. He died Sept. 21, 1957; succeeded by son, Olav V. Olav V died Jan. 17, 1991; succeeded by son, Harald V.

Rulers of the Netherlands and Belgium

The Netherlands (Holland)

William Frederick, Prince of Orange, led a revolt against French rule, 1813; crowned king, 1815. Belgium seceded Oct. 4, 1830, after a revolt. The secession was ratified by the two kingdoms by treaty, Apr. 19, 1839.

Succession: William II, son, 1840; William III, son, 1849; Wilhelmina, daughter of William III and his 2d wife Princess Emma of Waldeck, 1890; Wilhelmina abdicated, Sept. 4, 1948, in favor of daughter, Juliana. Juliana abdicated, Apr. 30, 1980, in favor of daughter, Beatrix.

Belgium

A national congress elected Prince Leopold of Saxe-Coburg as king; he took the throne July 21, 1831, as Leopold I.

Succession: Leopold II, son, 1865; Albert I, nephew of Leopold II, 1909; Leopold III, son of Albert, 1934; Prince Charles, Regent 1944; Leopold returned 1950, yielded powers to son Baudouin, Prince Royal, Aug. 6, 1950, abdicated July 16, 1951. Baudouin I took throne July 17, 1951, died July 31, 1993; succeeded by brother, Albert II.

Roman Rulers

From Romulus to the end of the Empire in the West. Rulers in the East sat in Constantinople and, for a brief period, in Nicaea, until the capture of Constantinople by the Turks in 1453, when Byzantium was succeeded by the Ottoman Empire.

BC The Kingdom	81 Domitianus	314 Maximinus II, Constantinus I, Licinius
753 Romulus (Quirinus)	96 Nerva	314 Constantinus I and Licinius
716 Numa Pompilius	81 Domitianus	324 Constantinus I (the Great)
673 Tullus Hostilius	96 Nerva	337 Constantinus II, Constans I,
640 Ancus Marcius	98 Trajanus	Constantius II
616 L. Tarquinius Priscus	117 Hadrianus	340 Constantius II and Constans I
578 Servius Tullius	138 Antoninus Pius	350 Constantius II
534 L. Tarquinius Superbus	161 Marcus Aurelius and Lucius Verus	361 Julianus II (the Apostate)
The Republic	169 Marcus Aurelius (alone)	363 Jovianus
509 Consulate established	180 Commodus	**West (Rome) and**
509 Quaestorship instituted	193 Pertinax; Julianus I	**East (Constantinople)**
498 Dictatorship introduced	193 Septimius Severus	364 Valentinianus I (West) and
494 Plebeian Tribunate created	211 Caracalla and Geta	Valens (East)
494 Plebeian Aedileship created	212 Caracalla (alone)	367 Valentinianus I with Gratianus (West)
444 Consular Tribunate organized	217 Macrinus	and Valens (East)
435 Censorship instituted	218 Elagabalus (Heliogabalus)	375 Gratianus with Valentinianus
366 Praetorship established	222 Alexander Severus	II (West) and Valens (East)
366 Curule Aedileship created	235 Maximinus I (the Thracian)	378 Gratianus with Valentinianus II (West),
362 Military Tribunate elected	238 Gordianus I and Gordianus II;	Theodosius I (East)
326 Proconsulate introduced	Pupienus and Balbinus	383 Valentinianus II (West) and
311 Naval Duumvirate elected	238 Gordianus III	Theodosius I (East)
217 Dictatorship of Fabius Maximus	244 Philippus (the Arabian)	394 Theodosius I (the Great)
133 Tribunate of Tiberius Gracchus	249 Decius	395 Honorius (West) and Arcadius (East)
123 Tribunate of Gaius Gracchus	251 Gallus and Volusianus	408 Honorius (West) and
82 Dictatorship of Sulla	253 Aemilianus	Theodosius II (East)
60 First Triumvirate formed (Caesar,	253 Valerianus and Gallienus	423 Valentinianus III (West) and
Pompeius, Crassus)	258 Gallienus (alone)	Theodosius II (East)
46 Dictatorship of Caesar	268 Claudius Gothicus	450 Valentinianus III (West)
43 Second Triumvirate formed	270 Quintillus	and Marcianus (East)
(Octavianus, Antonius, Lepidus)	270 Aurelianus	455 Maximus (West), Avitus
The Empire	275 Tacitus	(West); Marcianus (East)
27 Augustus (Gaius Julius Caesar	276 Florianus	456 Avitus (West), Marcianus (East)
Octavianus)	276 Probus	457 Majorianus (West), Leo I (East)
AD	282 Carus	461 Severus II (West), Leo I (East)
14 Tiberius I	283 Carinus and Numerianus	467 Anthemius (West), Leo I (East)
37 Gaius Caesar (Caligula)	286 Diocletianus and Maximianus	472 Olybrius (West), Leo I (East)
41 Claudius I	305 Galerius and Constantius I	473 Glycerius (West), Leo I (East)
54 Nero	306 Galerius, Maximinus II, Severus I	474 Julius Nepos (West), Leo II (East)
68 Galba	307 Galerius, Maximinus II,	475 Romulus Augustulus (West) and
69 Galba; Otho, Vitellius	Constantinus I, Licinius, Maxentius	Zeno (East)
69 Vespasianus	311 Maximinus II, Constantinus I,	476 End of Empire in West; Odovacar, King,
79 Titus	Licinius, Maxentius	drops title of Emperor; murdered by
		King Theodoric of Ostrogoths, 493

Rulers of Modern Italy

After the fall of Napoleon in 1814, the Congress of Vienna, 1815, restored Italy as a political patchwork, comprising the Kingdom of Naples and Sicily, the Papal States, and smaller units. Piedmont and Genoa were awarded to Sardinia, ruled by King Victor Emmanuel I of Savoy.

United Italy emerged under the leadership of Camillo, Count di Cavour (1810-1861), Sardinian prime minister. Agitation was led by Giuseppe Mazzini (1805-1872) and Giuseppe Garibaldi (1807-1882), soldier; Victor Emmanuel I abdicated 1821. After a brief regency for a brother, Charles Albert was king 1831-1849, abdicating when defeated by the Austrians at Novara. Succeeded by Victor Emmanuel II, 1849-1861.

In 1859 France forced Austria to cede Lombardy to Sardinia, which gave rights to Savoy and Nice to France. In 1860 Garibaldi led 1,000 volunteers in a spectacular cam-

paign, took Sicily and expelled the King of Naples. In 1860 the House of Savoy annexed Tuscany, Parma, Modena, Romagna, the Two Sicilys, the Marches, and Umbria. Victor Emmanuel assumed the title of King of Italy at Turin Mar. 17, 1861.

In 1866, Victor Emmanuel allied with Prussia in the Austro-Prussian War, and with Prussia's victory received Venetia. On Sept. 20, 1870, his troops under Gen. Raffaele entered Rome and took over the Papal States, ending temporal power of the Roman Catholic Church.

Succession: Umberto I, 1878, assassinated 1900; Victor Emmanuel III, 1900, abdicated 1946, died 1947; Humbert II, 1946, ruled a month. In 1921 Benito Mussolini (1883-1945) formed the Fascist party; he became prime minister Oct. 31, 1922. He entered World War II as an ally of Hitler. He was deposed July 25, 1943.

At a plebiscite June 2, 1946, Italy voted for a republic; Premier Alcide de Gasperi became chief of state June 13, 1946. On June 28, 1946, the Constituent Assembly elected Enrico de Nicola, Liberal, provisional president. Successive presidents: Luigi Einaudi, elected May 11, 1948; Giovanni Gronchi, Apr. 29, 1955; Antonio Segni, May 6, 1962; Giuseppe Saragat, Dec. 28, 1964; Giovanni Leone, Dec. 29, 1971; Alessandro Pertini, July 9, 1978; Francesco Cossiga, July 9, 1985; Oscar Luigi Scalfaro, May 28, 1992.

Rulers of Spain

From 8th to 11th centuries Spain was dominated by the Moors (Arabs and Berbers). The Christian reconquest established small kingdoms (Asturias, Aragon, Castile, Catalonia, Leon, Navarre, and Valencia). In 1474 Isabella, b. 1451, became Queen of Castile & Leon. Her husband, Ferdinand, b. 1452, inherited Aragon 1479, with Catalonia, Valencia, and the Balearic Islands, became Ferdinand V of Castile. By Isabella's request Pope Sixtus IV established the Inquisition, 1478. Last Moorish kingdom, Granada, fell 1492. Columbus opened New World of colonies, 1492. Isabella died 1504, succeeded by her daughter, Juana "the Mad," but Ferdinand ruled until his death 1516.

Charles I, b. 1500, son of Juana, grandson of Ferdinand and Isabella, and of Maximilian I of Hapsburg; succeeded later as Holy Roman Emperor, Charles V, 1520; abdicated 1556. Philip II, son, 1556-1598, inherited only Spanish throne; conquered Portugal, fought Turks, sent Armada vs. England. Married to Mary I of England, 1554-1558. Succession: Philip III, 1598-1621; Philip IV, 1621-1665; Charles II, 1665-1700, left Spain to Philip of Anjou, grandson of Louis XIV, who as Philip V, 1700-1746, founded Bourbon dynasty; Ferdinand VI, 1746-1759; Charles III, 1759-1788; Charles IV, 1788-1808, abdicated.

Napoleon now dominated politics and made his brother Joseph King of Spain 1808, but the Spanish ousted him in 1813. Ferdinand VII, 1808, 1814-1833, lost American colonies; succeeded by daughter Isabella II, aged 3, with wife Maria Christina of Naples regent until 1843. Isabella deposed by revolution 1868. Elected king by the Cortes, Amadeo of Savoy, 1870; abdicated 1873. First republic, 1873-74. Alphonso XII, son of Isabella, 1875-85. His posthumous son was Alphonso XIII, with his mother, Queen Maria Christina regent; Spanish-American war, Spain lost Cuba, gave up Puerto Rico, Philippines, Sulu Is., Marianas. Alphonso took throne 1902, aged 16, married British Princess Victoria Eugenia of Battenberg. Dictatorship of Primo de Rivera, 1923-30, precipitated revolution of 1931. Alphonso agreed to leave without formal abdication. Monarchy abolished; the second republic established, with socialist backing. Niceto Alcala Zamora was president until 1936, when Manuel Azaña was chosen.

In July 1936, the army in Morocco revolted against the government and General Francisco Franco led the troops into Spain. The revolution succeeded by Feb. 1939, when Azaña resigned. Franco became chief of state, with provisions that if he was incapacitated, the Regency Council by two-thirds vote could propose a king to the Cortes, which needed to have a two-thirds majority to elect him.

Alphonso XIII died in Rome Feb. 28, 1941, aged 54. His property and citizenship had been restored.

A law restoring the monarchy was approved in a 1947 referendum. Prince Juan Carlos, b. 1938, grandson of Alphonso XIII, was designated by Franco and the Cortes (Parliament) in 1969 as future king and chief of state. Franco died in office, Nov. 20, 1975; Juan Carlos proclaimed king, Nov. 22.

Leaders in the South American Wars of Liberation

Simon Bolivar (1783-1830), Jose Francisco de San Martin (1778-1850), and Francisco Antonio Gabriel Miranda (1750-1816) are among the heroes of the early 19th century struggles of South American nations to free themselves from Spain. All three, and their contemporaries, operated in periods of factional strife, during which soldiers and civilians suffered.

Miranda, a Venezuelan, who had served with the French in the American Revolution and commanded parts of the French Revolutionary armies in the Netherlands, attempted to start a revolt in Venezuela in 1806 and failed. In 1810, with British and American backing, he returned and was briefly a dictator, until the British withdrew their support. In 1812 he was overcome by the royalists in Venezuela and taken prisoner, dying in a Spanish prison in 1816.

San Martin was born in Argentina and during 1789-1811 served in campaigns of the Spanish armies in Europe and Africa. He first joined the independence movement in Argentina in 1812 and in 1817 invaded Chile with 4,000 men over the mountain passes. Here he and Gen. Bernardo O'Higgins (1778-1842) defeated the Spaniards at Chacabuco, 1817, and O'Higgins was named Liberator and became first director of Chile, 1817-23. In 1821 San Martin occupied Lima and Callao, Peru, and became protector of Peru.

Bolivar, the greatest leader of South American liberation from Spain, was born in Venezuela, the son of an aristocratic family. He first served under Miranda in 1812 and in 1813 captured Caracas, where he was named Liberator. Forced out next year by civil strife, he led a campaign that captured Bogota in 1814. In 1817 he was again in control of Venezuela and was named dictator. He organized Nueva Granada with the help of General Francisco de Paula Santander (1792-1840). By joining Nueva Granada, Venezuela, and the area that is now Panama and Ecuador, the republic of Colombia was formed, with Bolivar president. After numerous setbacks he decisively defeated the Spaniards in the second battle of Carabobo, Venezuela, June 24, 1821.

In May, 1822, Gen. Antonio Jose de Sucre, Bolivar's lieutenant, took Quito. Bolivar went to Guayaquil to confer with San Martin, who resigned as protector of Peru and withdrew from politics. With a new army of Colombians and Peruvians Bolivar defeated the Spaniards in a battle at Junín in 1824 and cleared Peru.

De Sucre organized Charcas (Upper Peru) as Republica Bolivar (now Bolivia) and acted as president in place of Bolivar, who wrote its constitution. De Sucre defeated the Spanish faction of Peru at Ayacucho, Dec. 19, 1824.

Continued civil strife finally caused the Colombian federation to break apart. Santander turned against Bolivar, but the latter defeated him and banished him. In 1828 Bolivar gave up the presidency he had held precariously for 14 years. He became ill from tuberculosis and died Dec. 17, 1830. He is buried in the national pantheon in Caracas.

Rulers of Russia; Leaders of the USSR and Russian Federation

First ruler to consolidate Slavic tribes was Rurik, leader of the Russians who established himself at Novgorod, AD 862. He and his immediate successors had Scandinavian affiliations. They moved to Kiev after 972 and ruled as Dukes of Kiev. In 988 Vladimir was converted and adopted the Byzantine Greek Orthodox service, later modified by Slav influences. Important as organizer and lawgiver was Yaroslav, 1019-1054, whose daughters married kings of Norway, Hungary, and France. His grandson, Vladimir II (Monomakh), 1113-1125, was progenitor of several rulers, but in 1169 Andrew Bogolubski overthrew Kiev and began the line known as Grand Dukes of Vladimir.

Of the Grand Dukes of Vladimir, Alexander Nevsky, 1246-1263, had a son, Daniel, first to be called Duke of

Muscovy (Moscow), who ruled 1294-1303. His successors became Grand Dukes of Muscovy. After Dmitri III Donskoi defeated the Tatars in 1380, they also became Grand Dukes of all Russia. Independence of the Tatars and considerable territorial expansion were achieved under Ivan III, 1462-1505.

Tsars of Muscovy—Ivan III was referred to in church ritual as Tsar. He married Sofia, niece of the last Byzantine emperor. His successor, Basil III, died in 1533 when Basil's son Ivan was only 3. He became Ivan IV, "the Terrible"; crowned 1547 as Tsar of all the Russias, ruled until 1584. Under the weak rule of his son, Feodor I, 1584-1598, Boris Godunov had control. The dynasty died, and after years of tribal strife and intervention by Polish and Swedish armies, the Russians united under 17-year-old Michael Romanov, distantly related to the first wife of Ivan IV. He ruled 1613-1645 and established the Romanov line. Fourth ruler after Michael was Peter I.

Tsars, or Emperors, of Russia (Romanovs)—Peter I, 1682-1725, known as Peter the Great, took title of Emperor in 1721. His successors and dates of accession were: Catherine, his widow, 1725; Peter II, his grandson, 1727; Anne, Duchess of Courland, 1730, daughter of Peter the Great's brother, Tsar Ivan V; Ivan VI, 1740, great-grandson of Ivan V, child, kept in prison and murdered 1764; Elizabeth, daughter of Peter I, 1741; Peter III, grandson of Peter I, 1761, deposed 1762 for his consort, Catherine II, former princess of Anhalt Zerbst (Germany) who is known as Catherine the Great; Paul I, her son, 1796, killed 1801; Alexander I, son of Paul, 1801, defeated Napoleon; Nicholas I, his brother, 1825; Alexander II, son of Nicholas, 1855, assassinated 1881 by terrorists; Alexander III, son, 1881. Nicholas II, son, 1894-1917, last Tsar of Russia, was forced to abdicate by the Revolution that followed losses to Germany in WWI. The Tsar, the Empress, the Tsarevich (Crown Prince), and the Tsar's 4 daughters were murdered by the Bolsheviks in Yekaterinburg, July 16, 1918.

Provisional Government—Prince Georgi Lvov and Alexander Kerensky, premiers, 1917.

Union of Soviet Socialist Republics

Bolshevik Revolution, Nov. 7, 1917, removed Kerensky from power; council of People's Commissars formed, Lenin (Vladimir Ilyich Ulyanov) became premier. Lenin died Jan. 21, 1924. Aleksei Rykov (executed 1938) and V. M. Molotov held the office, but actual ruler was Joseph Stalin (Joseph Vissarionovich Djugashvili), general secretary of the Central Committee of the Communist Party. Stalin became president of the Council of Ministers (premier) May 7, 1941, died Mar. 5, 1953. Succeeded by Georgi M. Malenkov, as head of the Council and premier, and Nikita S. Khrushchev, first secretary of the Central Committee. Malenkov resigned Feb. 8, 1955, became deputy premier, was dropped July 3, 1957. Marshal Nikolai A. Bulganin became premier Feb. 8, 1955; was demoted and Khrushchev became premier Mar. 27, 1958.

Khrushchev was ousted Oct. 14-15, 1964, replaced by Leonid I. Brezhnev as first secretary of the party and by Aleksei N. Kosygin as premier. On June 16, 1977, Brezhnev also took office as president. He died Nov. 10, 1982; 2 days later the Central Committee elected former KGB head Yuri V. Andropov president. Andropov died Feb. 9, 1984; on Feb. 13, Konstantin U. Chernenko chosen by Central Committee as its general secretary. Chernenko died Mar. 10, 1985; on Mar. 11, he was succeeded as general secretary by Mikhail Gorbachev, who replaced Andrei Gromyko as president on Oct. 1, 1988. Gorbachev resigned Dec. 25, 1991, and the Soviet Union officially disbanded the next day. A loose Commonwealth of Independent States, made up of most of the 15 former Soviet constituent republics, was created.

Post-Soviet Russia

After adopting a degree of sovereignty, the Russian Republic had held elections in June 1991. Boris Yeltsin was sworn in, July 10, 1991, as Russia's first elected president. With the Dec. 1991 dissolution of the Soviet Union, Russia (officially Russian Federation) became a founding member of the Commonwealth of Independent States.

Governments of China

(Until 221 BC and frequently thereafter, China was not a unified state. Where dynastic dates overlap, the rulers or events referred to appeared in different areas of China.)

Hsia	c1994BC -	c1523BC	Tang (a golden age of Chinese culture; capital: Xian)	618 -	906
Shang	c1523 -	c1028	Five Dynasties (Yellow River basin)	902 -	960
Western Chou	c1027 -	770	Ten Kingdoms (southern China)	907 -	979
Eastern Chou	770 -	256	Liao (Khitan Mongols; capital at site of Beijing)	947 -	1125
Warring States	403 -	222			
Ch'in (first unified empire)	221 -	206	Sung	960 -	1279
Han	202BC - AD	220	Northern Sung (reunified central and southern China)	960 -	1126
Western Han (expanded Chinese state beyond the Yellow and Yangtze River valleys)	202BC - AD	9	Western Hsai (non-Chinese rulers in northwest)	990 -	1227
Hsin (Wang Mang, usurper)	AD 9 -	23	Chin (Tatars; drove Sung out of central China)	1115 -	1234
Eastern Han (expanded Chinese state into Indochina and Turkestan)	25 -	220	Yuan (Mongols; Kublai Khan est. capital at site of Beijing, c. 1264)	1271 -	1368
Three Kingdoms (Wei, Shu, Wu)	220 -	265	Ming (China reunified under Chinese rule; capital: Nanjing, then Beijing in 1420)	1368 -	1644
Chin (western)	265 -	317	Ch'ing (Manchus, descendents of Tatars)	1644 -	1911
(eastern)	317 -	420	Republic (disunity; provincial rulers, warlords)	1912 -	1949
Northern Dynasties (followed several short-lived governments by Turks, Mongols, etc.)	386 -	581	People's Republic of China	1949 -	—
Southern Dynasties (capital: Nanjing)	420 -	589			
Sui (reunified China)	581 -	618			

Leaders of China Since 1949

Mao Zedong	Chairman, Central People's Administrative Council, Communist Party (CPC), 1949-1976
Zhou Enlai	Premier, foreign minister, 1949-1976
Deng Xiaoping	Vice Premier, 1952-1966, 1973-1976, 1977-1980; "paramount leader," 1978-1997
Liu Shaoqi	President, 1959-1969
Hua Guofeng	Premier, 1976-1980; CPC Chairman, 1976-1981
Zhao Ziyang	Premier, 1980-1988; CPC General Secretary, 1987-1989
Hu Yaobang	CPC Chairman, 1981-1982; CPC General Secretary, 1982-1987
Li Xiannian	President, 1983-1988
Yang Shangkun	President, 1988-1993
Li Peng	Premier, 1988-98
Jiang Zemin	CPC General Secretary, 1989-; President, 1993-
Zhu Rongi	Premier, 1998-

Invention	Date	Inventor	Nationality
Adding machine	1642	Pascal	French
Adding machine	1885	Burroughs	U.S.
Aerosol spray	1926	Rotheim	Norwegian
Airbag	1974	General Motors	U.S.
Air brake	1868	Westinghouse	U.S.
Air conditioning	1902	Carrier	U.S.
Air pump	1654	Guericke	German
Airplane, automatic pilot	1912	Sperry	U.S.
Airplane, experimental	1896	Langley	U.S.
Airplane, hydro	1911	Curtiss	U.S.
Airplane jet engine	1939	Ohain	German
Airplane with motor	1903	Wright Bros.	U.S.
Airship	1852	Giffard	French
Airship, rigid dirigible	1900	Zeppelin	German
Arc welder	1919	Thomson	U.S.
Aspartame	1965	Schlatter	U.S.
Autogyro	1920	de la Cierva	Spanish
Automobile, differential gear	1885	Benz	German
Automobile, electric	1892	Morrison	U.S.
Automobile, exp'mtl	1864	Marcus	Austrian
Automobile, gasoline	1889	Daimler	German
Automobile, gasoline	1892	Duryea	U.S.
Automobile magneto	1897	Bosch	German
Automobile muffler	1904	Pope	U.S.
Automobile self-starter	1911	Kettering	U.S.
Babbitt metal	1839	Babbitt	U.S.
Bakelite	1907	Baekeland	Belgium, U.S.
Balloon	1783	Montgolfier	French
Barometer	1643	Torricelli	Italian
Bicycle, modern	1885	Starley	English
Bifocal lens	1780	Franklin	U.S.
Block signals, railway	1867	Hall	U.S.
Bomb, depth	1916	Tait	U.S.
Bottle machine	1895	Owens	U.S.
Braille printing	1829	Braille	French
Bubble gum	1928	Diemer	U.S.
Burner, gas	1855	Bunsen	German
Calculating machine	1833	Babbage	English
Calculator, electronic pocket	1972	Merryman, Van Tassel	U.S.
Camera, Kodak	1888	Eastman, Walker	U.S
Camera, Polaroid Land	1948	Land	U.S.
Car coupler	1873	Janney	U.S.
Carburetor, gasoline	1893	Maybach	German
Card time recorder	1894	Cooper	U.S.
Carding machine	1797	Whittemore	U.S.
Carpet sweeper	1876	Bissell	U.S.
Cash register	1879	Ritty	U.S.
Cassette, audio	1963	Philips Co.	Dutch
Cassette, videotape	1969	Sony	Japanese
Cathode-ray tube	1897	Braun	German
CAT, or CT, scan	1973	Hounsfield	English
Cellophane	1908	Brandenberger	Swiss
Celluloid	1870	Hyatt	U.S.
Cement, Portland	1824	Aspdin	English
Chronometer	1761	Harrison	English
Circuit breaker	1925	Hilliard	U.S.
Circuit, integrated	1959	Kilby, Noyce, Texas Instr.	U.S.
Clock, pendulum	1657	Huygens	Dutch
Coaxial cable system	1929	Affel, Espensched.	U.S.
Coke oven	1893	Hoffman	Austrian
Compressed air rock drill	1871	Ingersoll	U.S.
Comptometer	1887	Felt	U.S.
Computer, automatic sequence	1944	Aiken, et al.	U.S.
Computer, electronic	1942	Atanasoff, Berry	U.S.
Computer, laptop	1987	Sinclair	English
Computer, mini	1960	Digital Corp	U.S.
Condenser microphone (telephone)	1916	Wente	U.S.
Contact lens, corneal	1948	Tuohy	U.S.
Contraceptive, oral	1954	Pincus, Rock	U.S.
Corn, hybrid	1917	Jones	U.S.
Correction fluid	1951	Nesmith	U.S.
Cotton gin	1793	Whitney	U.S.
Cream separator	1878	DeLaval	Swedish
Cultivator, disc	1878	Mallon	U.S.
Cystoscope	1878	Nitze	German
Diesel engine	1895	Diesel	German
Disc, compact	1972	RCA	U.S.
Disc player, compact	1979	Sony, Philips Co.	Japan, Dutch
Disk, floppy	1970	IBM	U.S.
Disk, video	1972	Philips Co.	Dutch
Dynamite	1866	Nobel	Swedish
Dynamo, continuous current	1871	Gramme	Belgian
Dynamo, hydrogen cooled	1915	Schuler	U.S.
Electric battery	1800	Volta	Italian
Electric fan	1882	Wheeler	U.S.
Electrocardiograph	1903	Einthoven	Dutch
Electroencephalograph	1929	Berger	German
Electromagnet	1824	Sturgeon	English
Electron spectrometer	1944	Deutsch, Elliott, Evans	U.S.
Electron tube multigrid	1913	Langmuir	U.S.
Electroplating	1805	Brugnatelli	Italian
Electrostatic generator	1929	Van de Graaff	U.S.
Elevator brake	1852	Otis	U.S.
Elevator, push button	1922	Larson	U.S.
Engine, automatic transmission	1910	Fottinger	German
Engine, coal-gas 4-cycle	1876	Otto	German
Engine, compression ignition	1883	Daimler	German
Engine, electric ignition	1883	Benz	German
Engine, gas, compound	1926	Eickemeyer	U.S.
Engine, gasoline	1872	Brayton, Geo.	U.S.
Engine, gasoline	1889	Daimler	German
Engine, jet	1930	Whittle	English
Engine, steam, piston	1705	Newcomen	English
Engine, steam, piston	1769	Watt	Scottish
Engraving, half-tone	1852	Talbot	U.S.
Fiberglass	1938	Owens-Corning	U.S.
Fiber optics	1955	Kapany	English
Filament, tungsten	1913	Coolidge	U.S.
Flanged rail	1831	Stevens	U.S.
Flatiron, electric	1882	Seely	U.S.
Food, frozen	1923	Birdseye	U.S.
Freon	1930	Midgley, et al.	U.S.
Furnace (for steel)	1858	Siemens	German
Galvanometer	1820	Sweigger	German
Gas discharge tube	1922	Hull	U.S.
Gas lighting	1792	Murdoch	Scottish
Gas mantle	1885	Welsbach	Austrian
Gasoline (lead ethyl)	1922	Midgley	U.S.
Gasoline, cracked	1913	Burton	U.S.
Gasoline, high octane	1930	Ipatieff	Russian
Geiger counter	1913	Geiger	German
Glass, laminated safety	1909	Benedictus	French
Glider	1853	Cayley	English
Gun, breechloader	1811	Thornton	U.S.
Gun, Browning	1897	Browning	U.S.
Gun, magazine	1875	Hotchkiss	U.S.
Gun, silencer	1908	Maxim, H.P.	U.S.
Guncotton	1847	Schoenbein	German
Gyrocompass	1911	Sperry	U.S.
Gyroscope	1852	Foucault	French
Harvester-thresher	1818	Lane	U.S.
Heart, artificial	1982	Jarvik	U.S.
Helicopter	1939	Sikorsky	U.S.
Hydrometer	1768	Baume	French
Iron lung	1928	Drinker, Slaw	U.S.
Kaleidoscope	1817	Brewster	Scottish
Kinetoscope	1889	Edison	U.S.
Lacquer, nitrocellulose	1921	Flaherty	U.S.
Lamp, arc	1847	Staite	English
Lamp, fluorescent	1938	General Electric, Westinghouse	U.S.
Lamp, incandescent	1879	Edison	U.S.
Lamp, incand., frosted	1924	Pipkin	U.S.
Lamp, incand., gas	1913	Langmuir	U.S.
Lamp, klieg	1911	Kliegl, A. & J.	U.S.
Lamp, mercury vapor	1912	Hewitt	U.S.
Lamp, miner's safety	1816	Davy	English
Lamp, neon	1909	Claude	French
Lathe, turret	1845	Fitch	U.S.
Launderette	1934	Cantrell	U.S.
Lens, achromatic	1758	Dollond	English
Lens, fused bifocal	1908	Borsch	U.S.
Leyden jar (condenser)	1745	von Kleist	German
Lightning rod	1752	Franklin	U.S.
Linoleum	1860	Walton	English
Linotype	1884	Mergenthaler	U.S.
Lock, cylinder	1851	Yale	U.S.
Locomotive, electric	1851	Vail	U.S.
Locomotive, exp'mtl	1802	Trevithick	English
Locomotive, exp'mtl	1812	Fenton, et al.	English
Locomotive, exp'mtl	1813	Hedley	English
Locomotive, exp'mtl	1814	Stephenson	English
Locomotive, practical	1829	Stephenson	English
Locomotive, 1st U.S.	1830	Cooper, P.	U.S.
Loom, power	1785	Cartwright	English
Loudspeaker, dynamic	1924	Rice, Kellogg	U.S.
Machine gun	1862	Gatling	U.S.
Machine gun, improved	1872	Hotchkiss	U.S.
Machine gun (Maxim)	1883	Maxim, H.S.	U.S., Eng.
Magnet, electro	1828	Henry	U.S.
Mantle, gas	1885	Welsbach	Austrian
Mason jar	1858	Mason, J.	U.S.
Match, friction	1827	Walker, J.	English
Mercerized textiles	1843	Mercer, J.	English
Meter, induction	1888	Shallenberg	U.S.
Metronome	1816	Malezel	German
Microcomputer	1973	Truong, et al.	French
Micrometer	1636	Gascoigne	English
Microphone	1877	Berliner	U.S.
Microprocessor	1971	Intel Corp.	U.S.
Microscope, compound	1590	Janssen	Dutch
Microscope, electronic	1931	Knoll, Ruska	German
Microscope, field ion	1951	Mueller	German
Microwave oven	1947	Spencer	U.S.
Monitor, warship	1861	Ericsson	U.S.
Monotype	1887	Lanston	U.S.
Motor, AC	1892	Tesla	U.S.
Motor, DC	1837	Davenport	U.S.
Motor, induction	1887	Tesla	U.S.
Motorcycle	1885	Daimler	German

Invention	Date	Inventor	Nationality
Movie machine	1894	Jenkins	U.S.
Movie, panoramic	1952	Waller	U.S.
Movie, talking	1927	Warner Bros.	U.S.
Mower, lawn	1831	Budding, Ferrabee	English
Mowing machine	1822	Bailey	U.S.
Neoprene	1930	Carothers	U.S.
Nylon	1937	Du Pont lab.	U.S.
Nylon synthetic	1930	Carothers	U.S.
Oil cracking furnace	1891	Gavrilov	Russian
Oil filled power cable	1921	Emanueli	Italian
Oleomargarine	1869	Mege-Mouries	French
Ophthalmoscope	1851	Helmholtz	German
Pacemaker	1952	Zoll	U.S.
Paper	105	Ts'ai	Chinese
Paper clip	1900	Waaler	Norwegian
Paper machine	1809	Dickinson	U.S.
Parachute	1785	Blanchard	French
Pen, ballpoint	1888	Loud	U.S.
Pen, fountain	1884	Waterman	U.S.
Pen, steel	1780	Harrison	English
Pendulum	1583	Galileo	Italian
Percussion cap	1807	Forsythe	Scottish
Phonograph	1877	Edison	U.S.
Photo, color	1892	Ives	U.S.
Photo film, celluloid	1893	Reichenbach	U.S.
Photo film, transparent	1884	Eastman, Goodwin	U.S.
Photoelectric cell	1895	Elster	German
Photocopier	1938	Carlson	U.S.
Photographic paper	1835	Talbot	English
Photography	1816	Niepce	French
Photography	1835	Talbot	English
Photography	1835	Daguerre	French
Photophone	1880	Bell	U.S.-Scot.
Phototelegraphy	1925	Bell Labs.	U.S.
Piano	1709	Cristofori	Italian
Piano, player	1863	Fourneaux	French
Pin, safety	1849	Hunt	U.S.
Pistol (revolver)	1836	Colt	U.S.
Plow, cast iron	1785	Ransome	English
Plow, disc.	1896	Hardy	U.S.
Pneumatic hammer	1890	King	U.S.
Post-it note	1974	Fry	U.S.
Powder, smokeless	1884	Vieille	French
Printing press, rotary	1845	Hoe	U.S.
Printing press, web	1865	Bullock	U.S.
Propeller, screw	1804	Stevens	U.S.
Propeller, screw	1837	Ericsson	Swedish
Pulsars	1967	Bell	English
Punch card accounting	1889	Hollerith	U.S.
Quasars	1963	Schmidt	U.S.
Radar	1940	Watson-Watt	Scottish
Radio, magnetic detector	1902	Marconi	Italian
Radio, signals	1895	Marconi	Italian
Radio amplifier	1906	De Forest	U.S.
Radio beacon	1928	Donovan	U.S.
Radio crystal oscillator	1918	Nicolson	U.S.
Radio receiver, cascade tuning	1913	Alexanderson	U.S.
Radio receiver, heterodyne	1913	Fessenden	U.S.
Radio transmitter triode modulation	1914	Alexanderson	U.S.
Radio tube diode	1905	Fleming	English
Radio tube oscillator	1915	De Forest	U.S.
Radio tube triode	1906	De Forest	U.S.
Radio FM, 2-path	1933	Armstrong	U.S.
Rayon (acetate)	1895	Cross	English
Rayon (cuprammonium)	1890	Despeissis	French
Rayon (nitrocellulose)	1884	Chardonnet	French
Razor, electric	1917	Schick	U.S.
Razor, safety	1895	Gillette	U.S.
Reaper	1834	McCormick	U.S.
Record, cylinder	1887	Bell, Tainter	U.S.
Record, disc.	1887	Berliner	U.S.
Record, long playing	1947	Goldmark	U.S.
Record, wax cylinder	1888	Edison	U.S.
Refrigerator car	1868	David	U.S.
Resin, synthetic	1931	Hill	English
Richter scale	1935	Richter	U.S.
Rifle, repeating	1860	Spencer	U.S.
Rocket engine	1926	Goddard	U.S.
Rubber, vulcanized	1839	Goodyear	U.S.
Saccharin	1879	Remsen, Fahlberg	U.S.
Saw, band	1808	Newberry	English
Saw, circular	1777	Miller	English
Scotch tape	1930	Drew	U.S.
Seat belt	1959	Volvo	Swedish
Sewing machine	1846	Howe	U.S.
Shoe-lasting machine	1883	Matzeliger	U.S.
Shoe-sewing machine	1860	McKay	U.S.
Shrapnel shell	1784	Shrapnel	English
Shuttle, flying	1733	Kay	English
Sleeping-car	1865	Pullman	U.S.
Slide rule	1620	Oughtred	English
Soap, hardwater	1928	Bertsch	German
Spectroscope	1859	Kirchoff, Bunsen	German
Spectroscope (mass)	1918	Dempster	U.S.
Spinning jenny	c.1764	Hargreaves	English
Spinning mule	1779	Crompton	English
Steamboat, exp'mtl	1778	Jouffroy	French
Steamboat, exp'mtl	1785	Fitch	U.S.
Steamboat, exp'mtl	1787	Rumsey	U.S.
Steamboat, exp'mtl	1788	Miller	Scottish
Steamboat, exp'mtl	1803	Fulton	U.S.
Steamboat, exp'mtl	1804	Stevens	U.S.
Steamboat, practical	1802	Symington	Scottish
Steamboat, practical	1807	Fulton	U.S.
Steam car	1770	Cugnot	French
Steam turbine	1884	Parsons	English
Steel (converter)	1856	Bessemer	English
Steel alloy	1891	Harvey	U.S.
Steel alloy, high-speed	1901	Taylor, White	U.S.
Steel, electric	1900	Heroult	French
Steel, manganese	1884	Hadfield	English
Steel, stainless	1916	Brearley	English
Stereoscope	1838	Wheatstone	English
Stethoscope	1819	Laennec	French
Stethoscope, binaural	1840	Cammann	U.S.
Stock ticker	1870	Edison	U.S.
Storage battery, rechargeable	1859	Plante	French
Stove, electric	1896	Hadaway	U.S.
Submarine	1891	Holland	U.S.
Submarine, even keel	1894	Lake	U.S.
Submarine, torpedo	1776	Bushnell	U.S.
Superconductivity	1957	Bardeen, Cooper, Schreiffer	U.S.
Synthesizer	1964	Moog	U.S.
Tank, military	1914	Swinton	English
Tape recorder, magnetic	1899	Poulsen	Danish
Teflon	1938	Du Pont	U.S.
Telegraph, magnetic	1837	Morse	U.S.
Telegraph, quadruplex	1864	Edison	U.S.
Telegraph, railroad	1887	Woods	U.S.
Telegraph, wireless high frequency	1895	Marconi	Italian
Telephone	1876	Bell	U.S.-Scot.
Telephone, automatic	1891	Strowger	U.S.
Telephone, cellular	1947	Bell Labs	U.S.
Telephone, radio	1900	Poulsen, Fessenden	Danish
Telephone, radio	1906	De Forest	U.S.
Telephone, radio, l. d.	1915	AT&T	U.S.
Telephone, recording	1898	Poulsen	Danish
Telephone, wireless	1899	Collins	U.S.
Telephone amplifier	1912	De Forest	U.S.
Telescope	1608	Lippershey	Neth.
Telescope	1609	Galileo	Italian
Telescope, astronomical	1611	Kepler	German
Teletype	1928	Morkrum, Kleinschmidt	U.S.
Television, color	1928	Baird	Scottish
Television, electronic	1927	Farnsworth	U.S.
Television, iconoscope	1923	Zworykin	U.S.
Television, mech. scanner	1923	Baird	Scottish
Thermometer	1593	Galileo	Italian
Thermometer	1730	Reaumur	French
Thermometer, mercury	1714	Fahrenheit	German
Time, self-regulator	1918	Bryce	U.S.
Time recorder	1890	Bundy	U.S.
Tire, double-tube	1845	Thomson	Scottish
Tire, pneumatic	1888	Dunlop	Scottish
Toaster, automatic	1918	Strite	U.S.
Toilet, flush	1589	Harington	English
Tool, pneumatic	1865	Law	English
Torpedo, marine	1804	Fulton	U.S.
Tractor, crawler	1904	Holt	U.S.
Transformer, AC	1885	Stanley	U.S.
Transistor	1948	Shockley, Brattain, Bardeen	U.S.
Trolley car, electric	1884-87	Van DePoele, Sprague	U.S.
Tungsten, ductile	1912	Coolidge	U.S.
Tupperware	1945	Tupper	U.S.
Turbine, gas	1849	Bourdin	French
Turbine, hydraulic	1849	Francis	U.S.
Turbine, steam	1884	Parsons	English
Type, movable	1447	Gutenberg	German
Typewriter	1867	Sholes, Soule, Glidden	U.S.
Vacuum cleaner, electric	1907	Spangler	U.S.
Vacuum evaporating pan	1846	Rillieux	U.S.
Velcro	1948	de Mestral	Swiss
Video game ("Pong")	1972	Buschnel	U.S.
Video home system (VHS)	1975	Matsushita, JVC	Japan
Washer, electric	1901	Fisher	U.S.
Welding, atomic hydrogen	1924	Langmuir, Palmer	U.S.
Welding, electric	1877	Thomson	U.S.
Windshield wiper	1903	Anderson	U.S.
Wind tunnel	1912	Eiffel	French
Wire, barbed	1874	Glidden	U.S.
Wire, barbed	1875	Haish	U.S.
Wrench, double-acting	1913	Owen	U.S.
X-ray tube	1913	Coolidge	U.S.
Zeppelin	1900	Zeppelin	German

MILLENNIUM FACT BOX

Science and Technology in 1900

In 1900, a number of important advances took place in science and technology; many of these are still significant in our lives.

The escalator was 1st manufactured
The 1st book of postage stamps was issued
Eastman Kodak introduced the Brownie camera
Sigmund Freud published *The Interpretation of Dreams*
Karl Landsteiner developed the modern classification of 4 blood types
The paper clip was patented
Max Planck formulated his quantum theory

Count Ferdinand von Zeppelin's airship made its 1st flight, carrying 5 people
Friedrich Ernst Dorn discovered the element radon
A wall-mounted telephone with separate ear and mouth-piece was introduced
The yellow fever virus was shown to be transmitted by the *Aëdes aegypti* mosquito
X rays were 1st used in the treatment of cancer

Discoveries and Innovations: Chemistry, Physics, Biology, Medicine

	Date	Discoverer	Nationality
Acetylene gas	1862	Berthelot	French
ACTH	1927	Evans, Long	U.S.
Adrenalin	1901	Takamine	Japan
Aluminum, electrolytic process	1886	Hall	U.S.
Aluminum, isolated	1825	Oersted	Danish
Anesthesia, ether	1842	Long	U.S.
Anesthesia, local	1885	Koller	Austrian
Anesthesia, spinal	1898	Bier	German
Aniline dye	1856	Perkin	English
Anti-rabies	1885	Pasteur	French
Antiseptic surgery	1867	Lister	English
Antitoxin, diphtheria	1891	Von Behring	German
Argyrol	1897	Bayer	German
Arsphenamine	1910	Ehrlich	German
Aspirin	1853	Gerhardt	French
Atabrine	1932	Mietzsch, et al.	German
Atomic numbers	1913	Moseley	English
Atomic theory	1803	Dalton	English
Atomic time clock	1948	Lyons	U.S.
Atomic time clock, cesium beam	1948	Essen	English
Atom-smashing theory	1919	Rutherford	English
Bacitracin	1943	Johnson, Meleneyl	U.S.
Bacteria, description	1676	Leeuwenhoek	Dutch
Barbital	1903	Fischer	German
Bleaching powder	1798	Tennant	English
Blood, circulation	1628	Harvey	English
Blood plasma storage (blood banks)	1940	Drew	U.S.
Bordeaux mixture	1885	Millardet	French
Bromine from the sea	1826	Balard	French
Calcium carbide	1888	Wilson	U.S.
Calculus	1670	Newton	English
Camphor synthetic	1896	Haller	French
Canning (food)	1804	Appert	French
Carbomycin	1952	Tanner	U.S.
Carbon oxides	1925	Fisher	German
Chemotherapy	1909	Ehrlich	German
Chloramphenicol	1947	Burkholder	U.S.
Chlorine	1774	Scheele	Swedish
Chloroform	1831	Guthrie, S.	U.S.
Chlortetracycline	1948	Duggen	U.S.
Classification of plants and animals	1735	Linnaeus	Swedish
Cloning, mammal	1996	Wilmut, et al.	Scottish
Cocaine	1860	Niermann	German
Combustion explained	1777	Lavoisier	French
Conditioned reflex	1914	Pavlov	Russian
Cortisone	1936	Kendall	U.S.
Cortisone, synthesis	1946	Sarett	U.S.
Cosmic rays	1910	Gockel	Swiss
Cyanamide	1905	Frank, Caro	German
Cyclotron	1930	Lawrence	U.S.
DDT (not applied as insecticide until 1939)	1874	Zeidler	German
Deuterium	1932	Urey, Brickwedde, Murphy	U.S.
DNA (structure)	1951	Crick	English
		Watson	U.S.
		Wilkins	English
Electric resistance, law of	1827	Ohm	German
Electric waves	1888	Hertz	German
Electrolysis	1852	Faraday	English
Electromagnetism	1819	Oersted	Danish
Electron	1897	Thomson, J.	English
Electron diffraction	1936	Thomson, G.	English
		Davisson	U.S.
Electroshock treatment	1938	Cerletti, Bini	Italian
Erythromycin	1952	McGuire	U.S.

	Date	Discoverer	Nationality
Evolution, natural selection	1858	Darwin	English
Falling bodies, law of	1590	Galileo	Italian
Gases, law of combining volumes	1808	Gay-Lussac	French
Geometry, analytic	1619	Descartes	French
Gold, cyanide process for extraction	1887	MacArthur, Forest	British
Gravitation, law	1687	Newton	English
Holograph	1948	Gabor	British
Human heart transplant	1967	Barnard	S. African
Human immunodeficiency virus identified	1984	Mortagnier	French
		Gallo	U.S.
Indigo, synthesis of	1880	Baeyer	German
Induction, electric	1830	Henry	U.S.
Insulin	1922	Banting, Best, Macleod	Canadian, Scottish
Intelligence testing	1905	Binet, Simon	French
In vitro fertilization	1978	Steptoe, Edwards	English
Isoniazid	1952	Hoffmann-LaRoche	U.S.
		Domagk	German
Isotopes, theory	1912	Soddy	English
Laser (light amplification by stimulated emission of radiation)	1957	Gould	U.S.
Light, velocity	1675	Roemer	Danish
Light, wave theory	1690	Huygens	Dutch
Lithography	1796	Senefelder	Bohemian
Lobotomy	1935	Egas Moniz	Portuguese
Logarithms	1614	Napier	Scottish
LSD-25	1943	Hoffman	Swiss
Mendelian laws	1866	Mendel	Austrian
Mercator projection (map)	1568	Mercator (Kremer)	Flemish
Methanol	1661	Boyle	Irish
Milk condensation	1853	Borden	U.S.
Molecular hypothesis	1811	Avogadro	Italian
Motion, laws of	1687	Newton	English
Neomycin	1949	Waksman, Lechevalier	U.S.
Neutron	1932	Chadwick	English
Nitric acid	1648	Glauber	German
Nitric oxide	1772	Priestley	English
Nitroglycerin	1846	Sobrero	Italian
Oil cracking process	1891	Dewar	U.S.
Oxygen	1774	Priestley	English
Oxytetracycline	1950	Finlay, et al.	U.S.
Ozone	1840	Schonbein	German
Paper, sulfite process	1867	Tilghman	U.S.
Paper, wood pulp, sulfate process	1884	Dahl	German
Penicillin	1928	Fleming	Scottish
practical use	1941	Florey, Chain	English
Periodic law and table of elements	1869	Mendeleyev	Russian
Physostigmine synthesis	1935	Julian	U.S.
Pill, birth-control	1954	Pincus, Rock	U.S.
Planetary motion, laws	1609	Kepler	German
Plutonium fission	1940	Kennedy, Wahl, Seaborg, Segre	U.S.
Polymyxin	1947	Ainsworth	English
Positron	1932	Anderson	U.S.
Proton	1919	Rutherford	N. Zealand
Psychoanalysis	1900	Freud	Austrian
Quantum theory	1900	Planck	German
Quasars	1963	Matthews, Sandage	U.S.
Quinine synthetic	1946	Woodward, Doering	U.S.
Radioactivity	1896	Becquerel	French
Radiocarbon dating	1947	Libby	U.S.
Radium	1898	Curie, Pierre	French
		Curie, Marie	Pol.-Fr.
Relativity theory	1905	Einstein	German

	Date	Discoverer	Nationality		Date	Discoverer	Nationality
Reserpine	1949	Jal Vaikl	Indian	Uranium fission, atomic reactor	1942	Fermi, Szilard	U.S.
Schick test	1913	Schick	U.S.	Vaccine, measles	1963	Enders	U.S.
Silicon	1823	Berzelius	Swedish	Vaccine, meningitis		Gordon, et al.,	
Smallpox eradication	1979	World Health Org.	UN	(first conjugate)	1987	Connaught Lab.	U.S.
Streptomycin	1944	Waksman, et al	U.S.	Vaccine, polio	1954	Salk	U.S.
Sulfanilamide	1935	Bovet, Trefouel	French	Vaccine, polio, oral	1960	Sabin	U.S.
Sulfanilamide theory	1908	Gelmo	German	Vaccine, rabies	1885	Pasteur	French
Sulfapyridine	1938	Ewins, Phelps	English	Vaccine, smallpox	1796	Jenner	English
Sulfathiazole	1939	Fosbinder, Walter	U.S.	Vaccine, typhus	1909	Nicolle	French
Sulfuric acid	1831	Phillips	English	Vaccine, varicella	1974	Takahashi	Japan
Sulfuric acid, lead	1746	Roebuck	English	Van Allen belts, radiation	1958	Van Allen	U.S.
Syphilis test	1906	Wassermann	German	Vitamin A	1913	McCollum, Davis	U.S.
Thiacetazone	1950	Belmisch, Mietzsch, Domagk	German	Vitamin B	1916	McCollum	U.S.
				Vitamin C	1928	Szent-Gyorgyi, King	U.S.
Tuberculin	1890	Koch	German	Vitamin D	1922	McCollum	U.S.
Uranium fission theory	1939	Hahn, Meitner, Strassmann,	German	Vitamin K	1935	Dam, Doisy	U.S.
		Bohr	Danish	Xerography	1938	Carlson	U.S.
		Fermi	Italian	X ray	1895	Roentgen	German
		Einstein, Pegram, Wheeler	U.S.				

Top 20 Corporations Receiving U.S. Patents in 1998

Source: *Technology Assessment and Forecast Report*, U.S. Patent and Trademark Office, U.S. Department of Commerce

Rank	Company	Number of patents	Rank	Company	Number of patents
1.	International Business Machines Corp.	2,657	11.	Mitsubishi Denki K. K.	1,080
2.	Canon K. K.	1,928	12.	Matsushita Electric Industrial Co., Ltd.	1,034
3.	NEC Corp.	1,627	13.	Lucent Technologies Inc.	928
4.	Motorola, Inc.	1,406	14.	Hewlett-Packard Company	805
5.	Sony Corp.	1,316	15.	Xerox Corp.	769
6.	Samsung Electronics Co., Ltd.	1,304	16.	General Electric Company	729
7.	Fujitsu Ltd.	1,189	17.	U.S. Philips Corp.	725
8.	Toshiba Corp.	1,170	18.	Intel Corp.	701
9.	Eastman Kodak Company	1,124	19.	Siemens Aktiengesellschaft	626
10.	Hitachi, Ltd.	1,094	20.	Texas Instruments, Inc.	611

Breaking the Sound Barrier; Speed of Sound

The prefix **Mach** is used to describe supersonic speed. It was named for Ernst Mach (1838-1916), a Czech-born Austrian physicist, who contributed to the study of sound. When a plane moves at the speed of sound, it is Mach 1. When the plane is moving at twice the speed of sound, it is Mach 2. When it is moving below the speed of sound, the speed can be designated accordingly—for example, Mach 0.90. Mach may be defined as the ratio of the velocity of a rocket or a jet to the velocity of sound in the medium being considered.

When a plane passes the sound barrier—flying faster than sound travels—listeners in the area hear thunderclaps, but the pilot of the plane does not hear them.

Sound is produced by vibrations of an object and is transmitted by alternate increase and decrease in pressures that radiate outward through a material media of molecules—somewhat like waves spreading out on a pond after a rock has been tossed into it.

The **frequency of sound** is determined by the number of times the vibrating waves undulate per second and is measured in cycles per second. The slower the cycle of waves, the lower the frequency. As frequencies increase, the sound is higher in pitch.

Sound is audible to human beings only if the frequency falls within a certain range. The human ear is usually not sensitive to frequencies of fewer than 20 vibrations per second or greater than about 20,000 vibrations per second—although this range varies among individuals. Any sound at a pitch higher than the human ear can hear is termed ultrasonic.

Intensity, or loudness, is the strength of the pressure of these radiating waves and is measured in decibels. The human ear responds to intensity in a range from zero to 120 decibels. Any sound with a pressure of more than 120 decibels is painful to the human ear.

The **speed of sound** is generally defined as 1,088 feet per second at sea level at 32° F. It varies in other temperatures and in different media. Sound travels faster in water than in air, and even faster in iron and steel. It takes about 5 seconds to travel a mile in air, and 1 second to move a mile under water, and $1/3$ second to move a mile in iron. Sound travels through ice-cold vapor at approximately 4,708 feet per second; for other media, speeds are: ice-cold water, 4,938; granite, 12,960; hardwood, 12,620; brick, 11,960; glass, 16,410 to 19,690; silver, 8,658; gold, 5,717.

Light; Colors of the Spectrum

Light, a form of electromagnetic radiation similar to radiant heat, radio waves, and X rays, is emitted from a source in straight lines and spreads out over a larger and larger area as it travels; the light per unit area diminishes as the square of the distance.

The English mathematician and physicist Sir Isaac Newton (1642-1727) described light as an **emission of particles**; the Dutch astronomer, mathematician, and physicist Christiaan Huygens (1629-95) developed the theory that light travels by a **wave motion**. It is now believed that these 2 theories are essentially complemen-

tary, and the development of quantum theory has led to results where light acts like a series of particles in some experiments and like a wave in others.

The **speed of light** was first measured in a laboratory experiment by the French physicist Armand Hippolyte Louis Fizeau (1819-96). Today the speed of light is known very precisely as 299,792.458 km per sec (or 186,282.396 mi per sec) in a vacuum. The velocity of light in air varies slightly with color, averaging about 3% less than in a vacuum; the speed in water is about 25% less, and in glass, 33% less.

Color sensations are produced through the excitation of the retina of the eye by light vibrating at different frequencies. The different colors of the spectrum may be produced by viewing a light beam that is refracted by passage through a prism, which breaks the light into its wavelengths.

Customarily, the **primary colors** of the spectrum are taken to be the 6 monochromatic colors that occupy relatively large areas of the spectrum: red, orange, yellow, green, blue, and violet. However, Newton named a 7th color, indigo, situated between blue and violet on the spectrum. Aubert estimated (1865) the solar spectrum to contain approximately 1,000 distinguishable hues; of the hues, according to Rood (1881), 2 million tints and shades can be distinguished. Luckiesh stated (1915) that 55 distinctly different hues have been seen in a single spectrum.

Many physicists recognize only 3 primary colors: red, yellow, and blue (Mayer, 1775); red, green, and violet (Thomas Young, 1801); or red, green, and blue (Clerk Maxwell, 1860).

The color sensation of **black** is due to complete lack of stimulation of the retina, that of **white** to complete stimulation. The **infrared and ultraviolet rays**, below the red (long) end of the spectrum and above the violet (short) end respectively, are invisible to the naked eye. Heat is the principal effect of the infrared rays, and chemical action that of the ultraviolet rays.

Weight or Mass of Water

Weight, at 20° C

1	cubic inch	0.0360 pound		1	U.S. gallon	8.33 pounds
12	cubic inches	0.433 pound		13.45	U.S. gallons	112.0 pounds
1	cubic foot	62.4 pounds		269.0	U.S. gallons	2240.0 pounds
1	cubic foot	7.48052 U.S. gal			**Mass, at 4° C (Maximum Density)**	
1.8	cubic feet	112.0 pounds		1	cubic centimeter	1 gram
35.96	cubic feet	2240.0 pounds		1	liter	1 kilogram
				1	cubic meter	1 metric ton

Density of Gases and Vapors

at 0° C and 760 mmHg; kilograms per cubic meter

Gas	Mass	Gas	Mass	Gas	Mass
Acetylene	1.171	Ethylene	1.260	Methyl fluoride	1.545
Air	1.293	Fluorine	1.696	Mono methylamine	1.38
Ammonia	0.759	Helium	0.178	Neon	0.900
Argon	1.784	Hydrogen	0.090	Nitric oxide	1.341
Arsine	3.48	Hydrogen bromide	3.50	Nitrogen	1.250
Butane-iso	2.60	Hydrogen chloride	1.639	Nitrosyl chloride	2.99
Butane-n	2.519	Hydrogen iodide	5.724	Nitrous oxide	1.997
Carbon dioxide	1.977	Hydrogen selenide	3.66	Oxygen	1.429
Carbon monoxide	1.250	Hydrogen sulfide	1.539	Phosphine	1.48
Carbon oxysulfide	2.72	Krypton	3.745	Propane	2.020
Chlorine	3.214	Methane	0.717	Silicon tetrafluoride	4.67
Chlorine monoxide	3.89	Methyl chloride	2.25	Sulfur dioxide	2.927
Ethane	1.356	Methyl ether	2.091	Xenon	5.897

Chemical Elements, Atomic Weights, Discoverers

Source: Darleane C. Hoffman, Ph.D., Lawrence Berkeley National Laboratory and Department of Chemistry, University of California, Berkeley

Atomic weights, based on the exact number 12 as the assigned atomic mass of the principal isotope of carbon, carbon 12, are provided through the courtesy of the International Union of Pure and Applied Chemistry (IUPAC) and Butterworth Scientific Publications. For the radioactive elements, with the exception of uranium and thorium, the mass number of either the isotope of longest half-life (*) or the better known isotope (**) is given.

Chemical element	Symbol	Atomic number	Atomic weight	Year discov.	Discoverer
Actinium	Ac	89	227.03	1899	Debierne
Aluminum	Al	13	26.9815	1825	Oersted
Americium	Am	95	243*	1944	Seaborg, et al.
Antimony	Sb	51	121.75	1450	Valentine
Argon	Ar	18	39.948	1894	Rayleigh, Ramsay
Arsenic	As	33	74.9216	13th c.	Albertus Magnus
Astatine	At	85	210*	1940	Corson, et al.
Barium	Ba	56	137.33	1808	Davy
Berkelium	Bk	97	247*	1949	Thompson, Ghiorso, Seaborg
Beryllium	Be	4	9.0122	1798	Vauquelin
Bismuth	Bi	83	208.980	15th c.	Valentine
Bohrium	Bh	107	264*	1981	Münzenberg, et al.
Boron	B	5	10.811a	1808	Gay-Lussac, Thenard
Bromine	Br	35	79.904b	1826	Balard
Cadmium	Cd	48	112.41	1817	Stromeyer
Calcium	Ca	20	40.08	1808	Davy
Californium	Cf	98	251*	1950	Thompson, et al.
Carbon	C	6	12.01115a	BC	unknown
Cerium	Ce	58	140.12	1803	Klaproth
Cesium	Cs	55	132.905	1860	Bunsen, Kirchhoff
Chlorine	Cl	17	35.453b	1774	Scheele
Chromium	Cr	24	51.996b	1797	Vauquelin
Cobalt	Co	27	58.9332	1735	Brandt
Copper	Cu	29	63.546b	BC	unknown
Curium	Cm	96	247*	1944	Seaborg, James, Ghiorso
Dysprosium	Dy	66	162.50*	1886	Boisbaudran
Einsteinium	Es	99	252*	1952	Ghiorso, et al.
Erbium	Er	68	167.26	1843	Mosander
Europium	Eu	63	151.96	1901	Demarcay
Fermium	Fm	100	257*	1953	Ghiorso, et al.
Fluorine	F	9	18.9984	1771	Scheele
Francium	Fr	87	223*	1939	Perey
Gadolinium	Gd	64	157.25	1886	Marignac
Gallium	Ga	31	69.72	1875	Boisbaudran

Chemical element	Symbol	Atomic number	Atomic weight	Year discov.	Discoverer
Germanium	Ge	32	72.59	1886	Winkler
Gold	Au	79	196.967	BC	unknown
Hafnium	Hf	72	178.49	1923	Coster, Hevesy
Hahnium[1]	Ha	105	262*	1970	Ghiorso, et al.
Hassium	Hs	108	269*	1984	Münzenberg, et al.
Helium	He	2	4.0026	1868	Janssen, Lockyer
Holmium	Ho	67	164.930	1878	Soret, Delafontaine
Hydrogen	H	1	1.00797a	1766	Cavendish
Indium	In	49	114.82	1863	Reich, Richter
Iodine	I	53	126.9044	1811	Courtois
Iridium	Ir	77	192.22	1804	Tennant
Iron	Fe	26	55.847b	BC	unknown
Krypton	Kr	36	83.80	1898	Ramsay, Travers
Lanthanum	La	57	138.91	1839	Mosander
Lawrencium	Lr	103	262*	1961	Ghiorso, et al.
Lead	Pb	82	207.19	BC	unknown
Lithium	Li	3	6.939	1817	Arfvedson
Lutetium	Lu	71	174.97	1907	Welsbach, Urbain
Magnesium	Mg	12	24.312	1829	Bussy
Manganese	Mn	25	54.9380	1774	Gahn
Meitnerium	Mt	109	268*	1982	Münzenberg, et al.
Mendelevium	Md	101	258*	1955	Ghiorso, et al.
Mercury	Hg	80	200.59	BC	unknown
Molybdenum	Mo	42	95.94	1782	Hjelm
Neodymium	Nd	60	144.24	1885	Welsbach
Neon	Ne	10	20.183	1898	Ramsay, Travers
Neptunium	Np	93	237.05*	1940	McMillan, Abelson
Nickel	Ni	28	58.70	1751	Cronstedt
Niobium[2]	Nb	41	92.906	1801	Hatchett
Nitrogen	N	7	14.0067	1772	Rutherford
Nobelium	No	102	259*	1958	Ghiorso, et al.
Osmium	Os	76	190.2	1804	Tennant
Oxygen	O	8	15.9994a	1774	Priestley, Scheele
Palladium	Pd	46	106.4	1803	Wollaston
Phosphorus	P	15	30.9738	1669	Brand
Platinum	Pt	78	195.09	1735	Ulloa
Plutonium	Pu	94	244*	1941	Seaborg, et al.
Polonium	Po	84	210**	1898	P. and M. Curie
Potassium	K	19	39.102	1807	Davy
Praseodymium	Pr	59	140.907	1885	Welsbach
Promethium	Pm	61	147**	1945	Glendenin, Marinsky, Coryell
Protactinium	Pa	91	231.04*	1917	Hahn, Meitner
Radium	Ra	88	226.03*	1898	P. and M. Curie, Bemont
Radon	Rn	86	222*	1900	Dorn
Rhenium	Re	75	186.21	1925	Noddack, Tacke, Berg
Rhodium	Rh	45	102.905	1803	Wollaston
Rubidium	Rb	37	85.47	1861	Bunsen, Kirchhoff
Ruthenium	Ru	44	101.07	1845	Klaus
Rutherfordium	Rf	104	261*	1969	Ghiorso, et al.
Samarium	Sm	62	150.35	1879	Boisbaudran
Scandium	Sc	21	44.956	1879	Nilson
Seaborgium	Sg	106	266*	1974	Ghiorso, et al.
Selenium	Se	34	78.96	1817	Berzelius
Silicon	Si	14	28.086a	1823	Berzelius
Silver	Ag	47	107.868b	BC	unknown
Sodium	Na	11	22.9898	1807	Davy
Strontium	Sr	38	87.62	1790	Crawford
Sulfur	S	16	32.064a	BC	unknown
Tantalum	Ta	73	180.948	1802	Ekeberg
Technetium	Tc	43	99**	1937	Perrier, Segre
Tellurium	Te	52	127.60	1782	Von Reichenstein
Terbium	Tb	65	158.9324	1843	Mosander
Thallium	Tl	81	204.37	1861	Crookes
Thorium	Th	90	232.038	1828	Berzelius
Thulium	Tm	69	168.934	1879	Cleve
Tin	Sn	50	118.69	BC	unknown
Titanium	Ti	22	47.90	1791	Gregor
Tungsten (Wolfram)	W	74	183.85	1783	d'Elhujar
Uranium	U	92	238.03	1789	Klaproth
Vanadium	V	23	50.942	1830	Sefstrom
Xenon	Xe	54	131.30	1898	Ramsay, Travers
Ytterbium	Yb	70	173.04	1878	Marignac
Yttrium	Y	39	88.905	1794	Gadolin
Zinc	Zn	30	65.37	BC	unknown
Zirconium	Zr	40	91.22	1789	Klaproth

Note: 109 elements are listed here. In addition, elements 110-112 were discovered at the Gesellschaft für Schwerionenforschung (GSI) at Darmstadt, Germany, by a team led by Dr. Sigurd Hofmann; these elements have not yet been named. Elements 110 and 111, discovered in 1994-95, have atomic weights 273 and 272, respectively; element 112, discovered in 1996, has atomic weight 277. In May 1999, scientists at Lawrence Berkeley National Laboratory reported evidence for 2 new superheavy elements: 118 with mass number 293 and 116 with mass number 289. They also reported element 114 with mass number 285. A Dubna (Russia)/Lawrence Livermore National Laboratory group reported evidence in Jan. 1999 for element 114 with mass number 289. As of Sept. 1999, neither report had been confirmed by another group. (1) The name Dubnium (Db) has been approved by IUPAC, but the name Hahnium is commonly used in the U.S. (2) Formerly Columbium. (a) Atomic weights so designated are known to be variable because of natural variations in isotopic composition. The observed ranges are: hydrogen ±0.0001; boron ±0.003; carbon ±0.005; oxygen ±0.0001; silicon ±0.001; sulfur ±0.003. (b) Atomic weights so designated are believed to have the following experimental uncertainties: chlorine ±0.001; chromium ±0.001; iron ±0.003; copper ±0.001; bromine ±0.001; silver ±0.001.

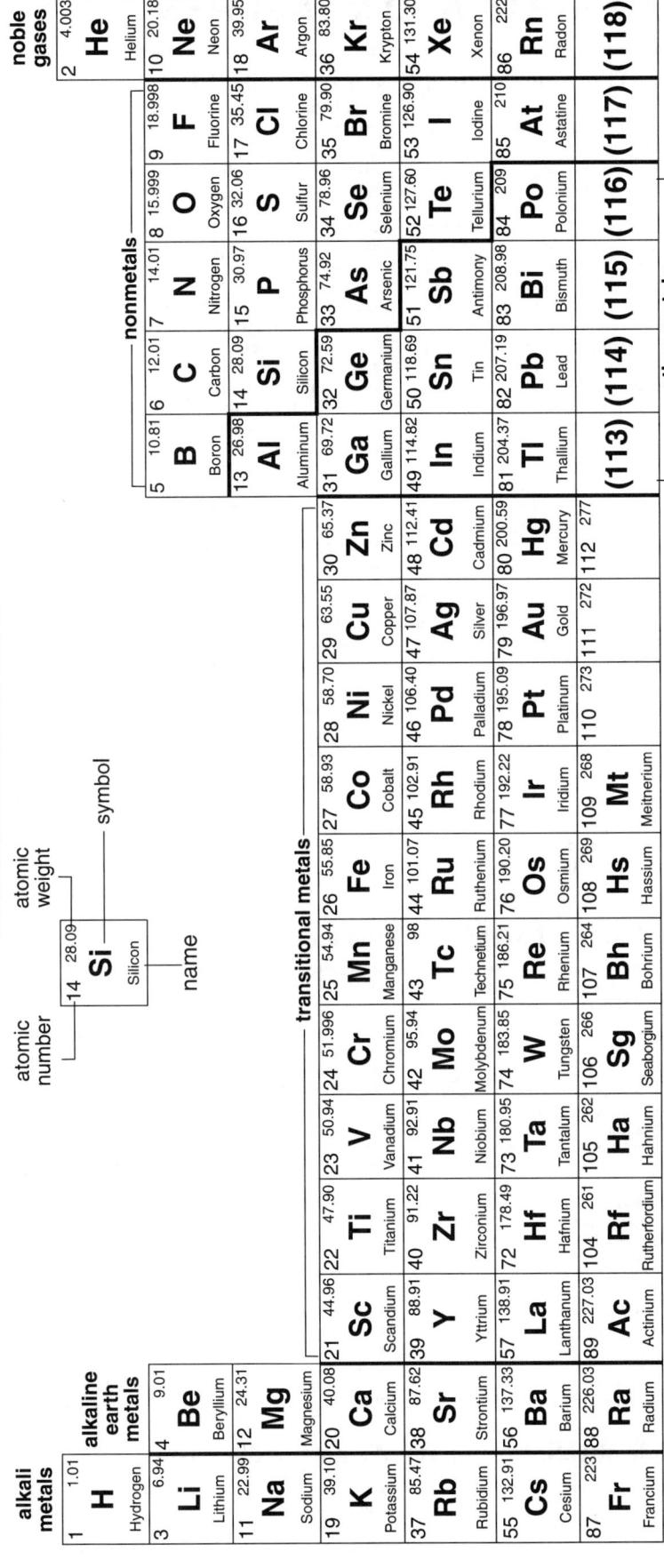

Periodic Table of the Elements

Source: © 1996 Lawrence Berkeley National Laboratory
Parentheses indicate undiscovered elements.

WEIGHTS AND MEASURES

Source: National Institute of Standards and Technology, U.S. Dept. of Commerce

The International System of Units (SI)

Two systems of weights and measures coexist in the U.S. today: the U.S. Customary System and the International System of Units (SI, after the initials of Système International). SI, commonly identified with the metric system, is actually a more complete, coherent version of it. Throughout U.S. history, the Customary System (inherited from, but now different from, the British Imperial System) has been generally used; federal and state legislation has given it, through implication, standing as the primary weights and measures system. The metric system, however, is the only system that Congress has ever specifically sanctioned. An 1866 law reads:

It shall be lawful throughout the United States of America to employ the weights and measures of the metric system; and no contract or dealing, or pleading in any court, shall be deemed invalid or liable to objection because the weights or measures expressed or referred to therein are weights or measures of the metric system.

Since that time, use of the metric system in the U.S. has slowly and steadily increased, particularly in the scientific community, in the pharmaceutical industry, and in the manufacturing sector—the last motivated by the practice in international commerce, in which the metric system is now predominantly used.

On Feb. 10, 1964, the National Bureau of Standards (now known as the National Institute of Standards and Technology) issued the following statement:

Henceforth it shall be the policy of the National Bureau of Standards to use the units of the International System (SI), as adopted by the 11th General Conference on Weights and Measures (October 1960), except when the use of these units would obviously impair communication or reduce the usefulness of a report.

On Dec. 23, 1975, Pres. Gerald R. Ford signed the Metric Conversion Act of 1975. It defines the metric system as being the International System of Units as interpreted in the U.S. by the secretary of commerce. The Trade Act of 1988 and other legislation declare the metric system the preferred system of weights and measures for U.S. trade and commerce, call for the federal government to adopt metric specifications, and mandate the Commerce Dept. to oversee the program. However, the metric system has still not become the system of choice for most Americans' daily use.

The following 7 units serve as the base units for the International System: **length**—meter; **mass**—kilogram; **time**—second; **electric current**—ampere; **thermodynamic temperature**—kelvin; **amount of substance**—mole; and **luminous intensity**—candela.

Prefixes

The following prefixes, in combination with the basic unit names, provide the multiples and submultiples in the International System. For example, the unit name *meter*, with the prefix *kilo* added, produces *kilometer*, meaning "1,000 meters."

Prefix	Symbol	Multiples	Equivalent	Prefix	Symbol	Submultiples	Equivalent
yotta	Y	10^{24}	septillionfold	deci	d	10^{-1}	tenth part
zetta	Z	10^{21}	sextillionfold	centi	c	10^{-2}	hundredth part
exa	E	10^{18}	quintillionfold	milli	m	10^{-3}	thousandth part
peta	P	10^{15}	quadrillionfold	micro	μ	10^{-6}	millionth part
tera	T	10^{12}	trillionfold	nano	n	10^{-9}	billionth part
giga	G	10^{9}	billionfold	pico	p	10^{-12}	trillionth part
mega	M	10^{6}	millionfold	femto	f	10^{-15}	quadrillionth part
kilo	k	10^{3}	thousandfold	atto	a	10^{-18}	quintillionth part
hecto	h	10^{2}	hundredfold	zepto	z	10^{-21}	sextillionth part
deka	da	10	tenfold	yocto	y	10^{-24}	septillionth part

Tables of Metric Weights and Measures

(**Note:** The SI generally uses the term *mass* instead of *weight*. Mass is a measure of an object's inertial property, or the amount of matter it contains. Weight is a measure of the force exerted on an object by gravity or the force needed to support it. Also, the SI does not make a distinction between "dry volume" and "liquid volume.")

Length

10 millimeters (mm) . . . .= 1 centimeter (cm)
10 centimeters. . . . .= 1 decimeter (dm) = 100 millimeters
10 decimeters . . . .= 1 meter (m) = 1,000 millimeters
10 meters. . . . .= 1 dekameter (dam)
10 dekameters . . . .= 1 hectometer (hm) = 100 meters
10 hectometers . . . .= 1 kilometer (km) = 1,000 meters

Area

100 square millimeters (mm^2) = 1 square centimeter (cm^2)
10,000 square centimeters . . .= 1 square meter (m^2) = 1,000,000 square millimeters
100 square meters . . . .= 1 are (a)
100 ares . . . .= 1 hectare (ha) = 10,000 square meters
100 hectares . . . .= 1 square kilometer (km^2) = 1,000,000 square meters

Volume

10 milliliters (mL) . . . .= 1 centiliter (cL)
10 centiliters . . . .= 1 deciliter (dL) = 100 milliliters
10 deciliters . . . .= 1 liter (L) = 1,000 milliliters
10 liters. . . . .= 1 dekaliter (daL)
10 dekaliters . . . .= 1 hectoliter (hL) = 100 liters
10 hectoliters . . . .= 1 kiloliter (kL) = 1,000 liters

Volume (Cubic Measure)

1,000 cubic millimeters (mm^3) = 1 cubic centimeter (cm^3)
1,000 cubic centimeters . . . . . = 1 cubic decimeter (dm^3) = 1,000,000 cubic millimeters
1,000 cubic decimeters. . . . . . = 1 cubic meter (m^3) = 1 stere = 1,000,000 cubic centimeters = 1,000,000,000 cubic millimeters

Weight (Mass)

10 milligrams (mg) . . . . = 1 centigram (cg)
10 centigrams. . . . . = 1 decigram (dg) = 100 milligrams
10 decigrams . . . . = 1 gram (g) = 1,000 milligrams
10 grams . . . . = 1 dekagram (dag)
10 dekagrams . . . . = 1 hectogram (hg) = 100 grams
10 hectograms . . . . = 1 kilogram (kg) = 1,000 grams
1,000 kilograms . . . . = 1 metric ton (t)

Table of U.S. Customary Weights and Measures

Length

12 inches (in) . . . .= 1 foot (ft)
3 feet. . . . .= 1 yard (yd)
5$\frac{1}{2}$ yards . . . .= 1 rod (rd), pole, or perch (16 $\frac{1}{2}$ feet)

40 rods . . . . = 1 furlong (fur) = 220 yards = 660 feet
8 furlongs . . . . = 1 statute mile (mi) = 1,760 yards = 5,280 feet
3 miles . . . . = 1 league = 5,280 yards = 15,840 feet
6076.11549 feet . . . . = 1 international nautical mile

Volume (Liquid Measure)

When necessary to distinguish the liquid pint or quart from the dry pint or quart, the word *liquid* or the abbreviation *liq* is used in combination with the name or abbreviation of the liquid unit.

4 gills (gi) . . . . = 1 pint (pt) = 28.875 cubic inches
2 pints . . . . = 1 quart (qt) = 57.75 cubic inches
4 quarts . . . . = 1 gallon (gal) = 231 cubic inches
= 8 pints = 32 gills

Volume (Dry Measure)

When necessary to distinguish the dry pint or quart from the liquid pint or quart, the word *dry* is used in combination with the name or abbreviation of the dry unit.

2 pints (pt) . . . . = 1 quart (qt) = 67.2006 cubic inches
8 quarts . . . . = 1 peck (pk) = 537.605 cubic inches
= 16 pints
4 pecks . . . . = 1 bushel (bu) = 2,150.42 cubic inches
= 32 quarts

Area

Squares and cubes of units are sometimes abbreviated by using superscripts. For example, ft^2 means square foot, and ft^3 means cubic foot.

144 square inches . . . . = 1 square foot (ft^2)
9 square feet . . . . = 1 square yard (yd^2) = 1,296 square inches
30 1/4 square yards . . = 1 square rod (rd^2) = 272 $^1/_4$ square feet
160 square rods . . . . . = 1 acre = 4,840 square yards
= 43,560 square feet
640 acres . . . . = 1 square mile (mi^2)
1 mile square . . . . = 1 section (of land)
6 miles square . . . . . . = 1 township = 36 sections = 36 square miles

Cubic Measure

1 cubic foot (ft^3) . . . . = 1,728 cubic inches (in^3)
27 cubic feet . . . . = 1 cubic yard (yd^3)

Gunter's, or Surveyor's, Chain Measure

7.92 inches (in) . . . . . . = 1 link
100 links . . . . = 1 chain (ch) = 4 rods = 66 feet
80 chains . . . . = 1 statute mile (mi) = 320 rods = 5,280 feet

Avoirdupois Weight

When necessary to distinguish the avoirdupois ounce or pound from the troy ounce or pound, the word *avoirdupois* or the abbreviation *avdp* is used in combination with the name or abbreviation of the avoirdupois unit. The *grain* is the same in avoirdupois and troy weight.

27 11/32 grains . . . . = 1 dram (dr)
16 drams . . . . = 1 ounce (oz) = 437 1/2 grains
16 ounces . . . . = 1 pound (lb) = 256 drams
= 7,000 grains
100 pounds . . . . = 1 hundredweight (cwt)*
20 hundredweights . . . . = 1 ton = 2,000 pounds*

In *gross* or *long* measure, the following values are recognized.
112 pounds . . . . = 1 gross or long hundredweight*
20 gross or long
hundredweights . . . . = 1 gross or long ton = 2,240 pounds*

*When the terms *hundredweight* and *ton* are used unmodified, they are commonly understood to mean the 100-pound hundredweight and the 2,000-pound ton, respectively; these units may be designated *net* or *short* when necessary to distinguish them from the corresponding units in gross or long measure.

Troy Weight

24 grains . . . . = 1 pennyweight (dwt)
20 pennyweights . . . . = 1 ounce troy (oz t) = 480 grains
12 ounces troy . . . . = 1 pound troy (lb t)
= 240 pennyweights = 5,760 grains

Tables of Equivalents

In this table it is necessary to distinguish between the *international* and the *survey* foot. The international foot, defined in 1959 as exactly equal to 0.3048 meter, is shorter than the old survey foot by exactly 2 parts in 1 million. The survey foot is still used in data expressed in feet in geodetic surveys within the U.S. In this table the survey foot is indicated with capital letters.

When the name of a unit is enclosed in brackets, e.g., [1 hand], either (1) the unit is not in general current use in the U.S. or (2) the unit is believed to be based on custom and usage rather than on formal definition.

Equivalents involving decimals are, in most instances, rounded to the 3d decimal place; exact equivalents are so designated.

Lengths

1 angstrom (Å) . . . . = 0.1 nanometer (exactly)
= 0.000 1 micrometer (exactly)
= 0.000 000 1 millimeter (exactly)
= 0.000 000 004 inch
1 cable's length . . . . = 120 fathoms (exactly)
= 720 FEET (exactly)
= 219 meters
1 centimeter (cm) . . . . = 0.3937 inch
1 chain (ch) (Gunter's or
surveyor's) . . . . = 66 FEET (exactly)
= 20.1168 meters
= 100 feet
1 chain (engineer's) . . . . = 30.48 meters (exactly)
1 decimeter (dm) . . . . = 3.937 inches
1 degree (geographical) . . . . = 364,566.929 feet
= 69.047 miles (avg.)
= 111.123 kilometers (avg.)
-of latitude . . . . = 68.708 miles at equator
= 69.403 miles at poles
-of latitude . . . . = 69.171 miles at equator
1 dekameter (dam) . . . . = 32.808 feet
1 fathom . . . . = 6 FEET (exactly)
= 1.8288 meters
1 foot (ft) . . . . = 0.3048 meters (exactly)
= 10 chains (surveyors) (exactly)
1 furlong (fur) . . . . = 660 FEET (exactly)
= 1/8 statute mile (exactly)
= 201.168 meters
[1 hand] (height measure for
horses from ground to top
of shoulders) . . . . = 4 inches
1 inch (in) . . . . = 2.54 centimeters (exactly)
1 kilometer (km) . . . . = 0.621371 mile = 3,280.8 feet
1 league (land) . . . . = 3 statute miles (exactly)
= 4.828 kilometers

1 link (Gunter's or surveyor's) . . . = 7.92 inches (exactly)
= 0.201 meter
1 link (engineer's) . . . . = 1 foot = 0.305 meter
1 meter (m) . . . . = 39.37 inches = 1.09361 yards
1 micrometer (μm) . . . . = 0.001 millimeter (exactly)
[the Greek letter mu] . . . . = 0.00003937 inch
1 mil . . . . = 0.001 inch (exactly)
= 0.0254 millimeter (exactly)
1 mile (mi) (statute or land) . . . . = 5,280 FEET (exactly)
= 1.609344 kilometers (exactly)
1 international nautical mile (nmi) = 1.852 kilometers (exactly)
= 1.150779 statute miles
= 6,076.11549 feet
1 millimeter (mm) . . . . = 0.03937 inch
1 nanometer (nm) . . . . = 0.001 micrometer (exactly)
= 0.00000003937 inch
1 pica (typography) . . . . = 12 points
1 point (typography) . . . . = 0.013 837 inch (exactly)
= 0.351 millimeter
1 rod (rd), pole, or perch . . . . . . = 16$^1/_2$ FEET (exactly)
= 5.029 meters
1 yard (yd) . . . . = 0.9144 meter (exactly)

Areas or Surfaces

1 acre . . . . = 43,560 square FEET (exactly)
= 4,840 square yards
= 0.405 hectare
1 are (a) . . . . = 119.599 square yards
= 0.025 acre
1 bolt (cloth measure):
length . . . . = 100 yards (on modern looms)
width . . . . = 45 or 60 inches
1 hectare (ha) . . . . = 2.471 acres
[1 square (building)] . . . . = 100 square feet
1 square centimeter (cm^2) . . . = 0.155 square inch
1 square decimeter (dm^2) . . . . = 15.500 square inches
1 square foot (ft^2) . . . . = 929.030 square centimeters

1 square inch (in²)..........= 6.4516 square centimeters
    (exactly)
1 square kilometer (km²) ....= 247.104 acres
    = 0.386102 square mile
1 square meter (m²) ....= 1.196 square yards
    = 10.764 square feet
1 square mile (mi²)........= 258.999 hectares
1 square millimeter (mm²) ...= 0.002 square inch
1 square rod (rd²), sq. pole, or
    sq. perch...........= 25.293 square meters
1 square yard (yd²)........= 0.836127 square meter

Capacities or Volumes

1 barrel (bbl), liquid   = 31 to 42 gallons*

*There are a variety of "barrels" established by law or usage. For example: federal taxes on fermented liquors are based on a barrel of 31 gallons; many state laws fix the "barrel for liquids" as 31¹/₂ gallons; one state fixes a 36-gallon barrel for cistern measurement; federal law recognizes a 40-gallon barrel for "proof spirits"; by custom, 42 gallons constitute a barrel of crude oil or petroleum products for statistical purposes, and this equivalent is recognized "for liquids" by 4 states.

1 barrel (bbl), standard for fruits,
    vegetables, and other dry com-
    modities except dry cranberries ..= 7,056 cubic inches
    =1 barrel (bbl), standard for
    fruits,
1 barrel (bbl), standard, cranberry ..= 86 ⁴⁵/₆₄ dry quarts
    = 2.709 bushels, struck measure
    = 5,826 cubic inches
1 board foot (lumber measure) ....= a foot-square board 1 inch thick
1 bushel (bu) (U.S.)   = 2,150.42 cubic inches (exactly)
    (struck measure) ....= 35.239 liters
[1 bushel, heaped (U.S.)] ....= 2,747.715 cubic inches
    = 1.278 bushels, struck measure*
*Frequently recognized as 1¹/₄ bushels, struck measure.
[1 bushel (bu) (British)   = 1.032 U.S. bushels, struck
    Imperial) (struck measure)] .....   measure
    = 2,219.36 cubic inches
    = 2,219.36 cubic inches
1 cord (cd) firewood ....= 128 cubic feet (exactly)
1 cubic centimeter (cm³)..........= 0.061 cubic inch
1 cubic decimeter (dm³) ....= 61.024 cubic inches
1 cubic inch (in³) ....= 0.554 fluid ounce
    =4.433 fluid drams
    =16.387 cubic centimeters
1 cubic foot (ft³) ....= 7.481 gallons
    = 28.317 cubic decimeters
1 cubic meter (m³) ....= 1.308 cubic yards
1 cubic yard (yd³) ....= 0.765 cubic meter
1 cup, measuring ....= 8 fluid ounces (exactly)
    = ¹/₂ liquid pint (exactly)
[1 dram, fluid (fl dr) (British)] .....= 0.961 U.S. fluid dram
    = 0.217 cubic inch
    = 3.552 milliliters
1 dekaliter (daL) ....= 2.642 gallons
    = 1.135 pecks
1 gallon (gal) (U.S.) ....= 231 cubic inches (exactly)
    = 3.785 liters
    = 0.833 British gallon
    = 128 U.S. fluid ounces (exactly)
[1 gallon (gal) British Imperial] ....= 277.42 cubic inches
    = 1.201 U.S. gallons
    = 4.546 liters
    = 160 British fluid ounces
    (exactly)
1 gill (gi) ....= 7.219 cubic inches = 4 fluid
    ounces (exactly)
    = 0.118 liter
1 hectoliter (hL) ....= 26.418 gallons
    = 2.838 bushels
1 liter (L) (1 cubic decimeter exactly) = 1.057 liquid quarts
    = 0.908 dry quart
    = 61.024 cubic inches
1 milliliter (mL) (1 cu cm exactly) ..= 0.271 fluid dram
    = 16.231 minims
    = 0.061 cubic inch

1 ounce, liquid (U.S.) ....= 1.805 cubic inches
    = 29.574 milliliters
    = 1.041 British fluid ounces
[1 ounce, fluid (fl oz) (British)] ....= 0.961 U.S. fluid ounce
    = 1.734 cubic inches
    = 28.412 milliliters
1 peck (pk) ....= 8.810 liters
1 pint (pt), dry ....= 33.600 cubic inches
    = 0.551 liter
1 pint (pt), liquid ....= 28.875 cubic inches (exactly)
    = 0.473 liter
1 quart (qt), dry (U.S.) ....= 67.201 cubic inches
    = 1.101 liters
    = 0.969 British quart
1 quart (qt), liquid (U.S.) ....= 57.75 cubic in (exactly)
    = 0.946 liter
    = 0.833 British quart
[1 quart (qt) (British)] ....= 69.354 cubic inches
    = 1.032 U.S. dry quarts
    = 1.201 U.S. liquid quarts
1 tablespoon ....= 3 teaspoons*(exactly)
    = 4 fluid drams
    = ¹/₂ fluid ounce (exactly)
1 teaspoon ....= ¹/₃ tablespoon*(exactly)
    = 1¹/₃ fluid drams*

*The equivalent "1 teaspoon=1¹/₃ fluid drams" has been found to correspond more closely with the actual capacities of teaspoons in use than the equivalent "1 teaspoon=1 fluid dram" which is given by many dictionaries.

Weights or Masses

1 assay ton** (AT)   = 29.167 grams

**Used in assaying. The assay ton bears the same relation to the milligram that a ton of 2,000 pounds avoirdupois bears to the ounce troy; hence, the weight in milligrams of precious metal obtained from one assay ton of ore gives directly the number of troy ounces to the net ton.

1 bale (cotton measure) ....= 500 pounds in U.S.
    = 750 pounds in Egypt
1 carat (c) ....= 200 milligrams (exactly)
    = 3.086 grains
1 dram avoirdupois (dr avdp) ...= 27¹¹/₃₂ (= 27.344) grains
    = 1.772 grams
1 gamma (g) ....= 1 microgram (exactly),
    see below
1 grain ....= 64.7989 milligrams
1 gram ....= 15.432 grains
    = 0.035 ounce, avoirdupois
1 hundredweight, gross or long***
    (gross cwt) ....= 112 pounds (exactly)
    = 50.802 kilograms
1 hundredweight, net or
    short (cwt or net cwt) ......= 100 pounds (exactly)
    = 45.359 kilograms
1 kilogram (kg) ....= 2.20462 pounds
1 microgram (μg) ....= 0.000001 gram (exactly)
1 milligram (mg) ....= 0.015 grain
1 ounce, avoirdupois (oz avdp) ..= 437.5 grains (exactly)
    = 0.911 troy ounce
    = 28.3495 grams
1 ounce, troy (oz t) ....= 480 grains (exactly)
    = 1.097 avoirdupois ounces
    = 31.103 grams
1 pennyweight (dwt) ....= 1.555 grams
1 pound, avoirdupois (lb avdp) ..= 7,000 grains (exactly)
    = 1.215 troy pounds
    = 453.59237 grams (exactly)
1 pound, troy (lb t) ....= 5,760 grains (exactly)
    = 0.823 pound, avoirdupois
    = 373.242 grams
1 ton, gross or long*** (gross ton) = 2,240 pounds (exactly)
    = 1.12 net tons (exactly)
    = 1.016 metric tons

***The gross or long ton and hundredweight are used commercially in the U.S. to only a limited extent, usually in restricted industrial fields. These units are the same as the British ton and hundredweight.

1 ton, metric (t) ....= 2,204.623 pounds
    = 0.984 gross ton
    = 1.102 net tons
1 ton, net or short
    (sh ton) ....= 2,000 pounds (exactly)
    = 0.893 gross ton
    = 0.907 metric ton

Tables of Interrelation of Units of Measurement

Units of length and area of the international and survey measures are included in the following tables.

1 international foot = 0.999 998 survey foot (exactly)
1 survey foot = 1200/3937 meter (exactly)
1 international foot = 2 × 0.0254 meter (exactly)

*Indicates exact values.

Units of Length

Units	Inches	Links	Feet	Yards	Rods	Chains	Miles	Cm	Meters
1 inch=	1*	0.126 263	0.083 333	0.027 778	0.005 051	0.001 263	0.000 016	2.54*	0.025 4*
1 link=	7.92*	1*	0.66*	0.22*	0.04*	0.01	0.000 125	20.117	0.201 168
1 foot=	12*	1.515 152	1*	0.333 333	0.060 606	0.015 152	0.000 189	30.48*	0.304 8*
1 yard=	36*	4.545 45	3*	1*	0.181 818	0.045 455	0.000 568	91.44*	0.914 4*
1 rod=	198*	25*	16.5*	5.5*	1*	0.25*	0.003 125*	502.92	5.029 2
1 chain=	792*	100*	66*	22*	4*	1*	0.012 5*	2011.68	20.116 8
1 mile=	63 360*	8000*	5280*	1760*	320*	80*	1*	160 934.4*	1609.344*
1 cm=	0.3937	0.049 710	0.032 808	0.010 936	0.001 988	0.000 497	0.000 006	1*	0.01*
1 meter=	39.37	4.970 960	3.280 840	1.093 613	0.198 838	0.049 710	0.000 621	100*	1*

Units of Area

Units	Sq. inches	Sq. links	Sq. feet	Sq. yards	Sq. rods	Sq. chains
1 sq. inch=	1*	0.015 942 3	0.006 944	0.000 771 605	0.000 025 5	0.000 001 594
1 sq. link=	62.726 4	1*	0.435 6*	0.0484*	0.0016*	0.000 1
1 sq. foot=	144*	2.295 684	1*	0.111 111 1	0.003 673 09	0.000 229 568
1 sq. yard=	1296*	20.661 16	9*	1*	0.033 057 85	0.002 066 12
1 sq. rod=	39 204*	625*	272.25*	30.25*	1*	0.062 5
1 sq. chain=	627 264*	10 000*	4356*	484*	16*	1*
1 acre=	6 272 640*	100 000*	43 560*	4840*	160*	10*
1 sq. mile=	4 014 489 600*	64 000 000*	27 878 400*	3 097 600*	102 400*	6400*
1 sq. cm=	0.155 000 3	0.002 471 05	0.001 076	0.000 119 599	0.000 003 954	0.000 000 247
1 sq. meter=	1550.003	24.710 44	10.763 91	1.195 990	0.039 536 70	0.002 471 044
1 hectare=	15 500 031	247 104	107 639.1	11 959.90	395.367 0	24.710 44*

Units	Acres	Sq. miles	Sq. cm	Sq. meters	Hectares
1 sq. inch=	0.000 000 159 423	0.000 000 000 249 10	6.451 6*	0.000 645 16*	0.000 000 065
1 sq. link=	0.000 01*	0.000 000 015 625*	404.685 642 24	0.040 468 56	0.000 004 047
1 sq. foot=	0.000 022 956 84	0.000 000 035 870 06	929.034 1	0.092 903 41	0.000 009 290
1 sq. yard=	0.000 206 611 6	0.000 000 322 830 6	8361.273 6*	0.836 127 36	0.000 083 613
1 sq. rod=	0.006 25*	0.000 009 765 625*	252 929.5	25.292 95	0.002 529 295
1 sq. chain=	0.1*	0.000 156 25*		404.687 3	0.040 468 73
1 acre=	1*	0.001 562 5*		4046.873	0.404 687 3
1 sq. mile=	640*	1*		2 589 988.11	258.998 811 034
1 sq. cm=	0.000 000 024 711	0.000 000 000 038 610	1*	0.000 1*	0.000 000 01*
1 sq. meter=	0.000 247 104 4	0.000 000 386 102 2	10 000*	1*	0.0001*
1 hectare=	2.471 044	0.003 861 006	100 000 000*	10 000*	1*

Units of Weight or Mass Not Greater Than Pounds and Kilograms

Units	Grains	Pennyweights	Avdp drams	Avdp ounces
1 grain=	1*	0.041 666 67	0.036 571 43	0.002 285 71
1 pennyweight=	24*	1*	0.877 714 3	0.054 857 14
1 dram avdp=	27.343 75*	1.139 323	1*	0.062 5*
1 ounce avdp=	437.5*	18.229 17	16*	1*
1 ounce troy=	480*	20*	17.554 29	1.097 143
1 pound troy=	5760*	240*	210.651 4	13.165 71
1 pound avdp=	7000*	291.666 7	256*	16*
1 milligram=	0.015 432	0.000 643 015	0.000 564 383	0.000 035 274
1 gram=	15.432 36	0.643 014 9	0.564 383 4	0.035 273 96
1 kilogram=	15 432.36	643.014 9	564.383 4	35.273 96

Units	Troy ounces	Troy pounds	Avdp pounds	Milligrams	Grams	Kilograms
1 grain=	0.002 083 33	0.000 173 611	0.000 142 857	64.798 91*	0.064 798 91*	0.000 064 799
1 pennywt.=	0.05*	0.004 166 667	0.003 428 571	1555.173 84*	1.555 173 84*	0.001 555 174
1 dram avdp=	0.056 966 15	0.004 747 179	0.003 906 25	1771.845 195	1.771 845 195	0.001 771 845
1 oz avdp=	0.911 458 3	0.075 954 86	0.062 5	28 349.523 125*	28.349 523 125*	0.028 349 52
1 oz troy=	1*	0.083 333 333	0.068 571 43	31 103.476 8*	31.103 476 8*	0.031 103 48
1 lb troy=	12*	1*	0.822 857 1	373 241.721 6*	373.241 721 6*	0.373 241 722
1 lb avdp=	14.583 33	1.215 278	1*	453 592.37*	453.592 37*	0.453 592 37*
1 milligram=	0.000 032 151	0.000 002 679	0.000 002 205	1*	0.001*	0.000 001*
1 gram=	0.032 150 75	0.002 679 229	0.002 204 623	1000*	1*	0.001*
1 kilogram=	32.150 75	2.679 229	2.204 623	1 000 000*	1000*	1*

Units of Weight or Mass Not Less Than Avoirdupois Ounces

Units	Avdp oz	Avdp lb	Short cwt	Short tons	Long tons	Kilograms	Metric tons
1 oz avdp=	1*	0.0625*	0.000 625*	0.000 031 25*	0.000 027 902	0.028 349 523	0.000 028 350
1 lb avdp=	16*	1*	0.01*	0.000 5*	0.000 446 429	0.453 592 37*	0.000 453 592
1 sh cwt=	1600*	100*	1*	0.05*	0.044 642 86	45.359 237*	0.045 359 237*
1 sh ton=	32 000*	2000*	20*	1*	0.892 857 1	907.184 74*	0.907 184 74*
1 long ton=	35 840*	2240*	22.4*	1.12*	1*	1 016.046 908 8*	1.016 046 909
1 kg=	35.273 96	2.204 623	0.022 046 23	0.001 102 311	0.000 984 207	1*	0.001*
1 metric ton=	35 273.96	2204.623	22.046 23	1.102 311	0.984 206 5	1000*	1*

Units of Volume

Units	Cubic inches	Cubic feet	Cubic yards	Cubic cm	Cubic dm	Cubic meters
1 cubic inch=	1*	0.000 578 704	0.000 021 433	16.387 064*	0.016 387	0.000 016 387
1 cubic foot=	1728*	1*	0.037 037 04	28 316.846 592*	28.316 847	0.028 316 847
1 cubic yard=	46 656*	27*	1*	764 554.857 984*	764.554 858	0.764 554 858
1 cubic cm=	0.061 023 74	0.000 035 315	0.000 001 308	1*	0.001*	0.000 001*
1 cubic dm=	61.023 74	0.035 314 67	0.001 307 951	1000*	1*	0.001*
1 cubic meter=	61 023.74	35.314 67	1.307 951	1 000 000*	1000*	1*

Units of Capacity (Liquid Measure)

Units	Minims	Fluid drams	Fluid ounces	Gills	Liquid pint
1 minim=	1*	0.016 666 7	0.002 083 33	0.000 520 833	0.000 130 208
1 fluid dram=	60*	1*	0.125*	0.031 25*	0.007 812 5*
1 fluid ounce=	480*	8*	1*	0.25*	0.062 5*
1 gill=	1920*	32*	4*	1*	0.25*
1 liquid pint=	7680*	128*	16*	4*	1*
1 liquid quart=	15 360*	256*	32*	8*	2*
1 gallon=	61 440*	1024*	128*	32*	8*
1 cubic inch=	265.974	4.432 900	0.554 112 6	0.138 528 1	0.034 632 03
1 cubic foot=	459 603.1	7660.052	957.506 5	239.376 6	59.844 16
1 liter=	16 230.73	270.512 18	33.814 02	8.453 506	2.113 376

Units	Liquid quarts	Gallons	Cubic inches	Cubic feet	Liters
1 minim=	0.000 065 104 17	0.000 016 276 04	0.003 759 766	0.000 002 175 790	0.000 061 611 52
1 flu. dram=	0.003 906 25*	0.000 976 562 5*	0.225 585 9	0.000 130 547 4	0.003 696 691
1 fluid oz=	0.031 25*	0.007 812 5*	1.804 687 5*	0.001 044 379	0.029 573 53
1 gill=	0.125*	0.031 25*	7.218 75*	0.004 177 517	0.118 294 118
1 liquid pt=	0.5*	0.125*	28.875*	0.016 710 07	0.473 176 473
1 liquid qt=	1*	0.25*	57.75*	0.033 420 14	0.946 352 946
1 gallon=	4*	1*	231*	0.133 680 6	3.785 411 784
1 cubic inch=	0.017 316 02	0.004 329 004	1*	0.000 578 703 7	0.016 387 064
1 cubic foot=	29.922 08	7.480 519	1728*	1*	28.316 846 592*
1 liter=	1.056 688	0.264 172 05	61.023 74	0.035 314 67	1*

Units of Capacity (Dry Measure)

Units	Dry pints	Dry quarts	Pecks	Bushels	Cubic in.	Liters
1 dry pint=	1*	0.5*	0.062 5*	0.015 625*	33.600 312 5*	0.550 610 47
1 dry quart=	2*	1*	0.125*	0.031 25*	67.200 625*	1.101 220 9
1 peck=	16*	8*	1*	0.25*	537.605*	8.809 767 5
1 bushel=	64*	32*	4*	1*	2150.42*	35.239 07
1 cubic inch=	0.029 761 6	0.014 880 8	0.001 860 10	0.000 465 025	1*	0.016 387 06
1 liter=	1.816 166	0.908 083	0.113 510 37	0.028 377 59	61.023 74	1*

Miscellaneous Measures

Caliber—the diameter of a gun bore. In the U.S., caliber is traditionally expressed in hundredths of inches, e.g., .22 or .30. In Britain, caliber is often expressed in thousandths of inches, e.g., .270 or .465. Now it is commonly expressed in millimeters, e.g., the 5.56 mm M16 rifle. Heavier weapons' caliber has long been expressed in millimeters, e.g., the 81 mm mortar, the 105 mm howitzer (light), the 155 mm howitzer (medium or heavy). Naval guns' caliber refers to the barrel length as a multiple of the bore diameter. A 5-inch, 50-caliber naval gun has a 5-inch bore and a barrel length of 250 inches.

Karat or carat—a measure of fineness for gold equal to $1/24$ part of pure gold in an alloy. Thus 24-karat gold is pure; 18-karat gold is $1/4$ alloy. (A *carat* is also a unit of weight for precious stones, equal to 200 milligrams.)

Decibel (dB)—a measure of the relative loudness or intensity of sound. A 20-decibel sound is 10 times louder than a 10-decibel sound; 30 decibels is 100 times louder; 40 decibels is 1,000 times louder, etc. One decibel is the smallest difference between sounds detectable by the human ear. A 120-decibel sound is painful.

10 decibels	– a light whisper
20	– quiet conversation
30	– normal conversation
40	– light traffic
50	– typewriter, loud conversation
60	– noisy office
70	– normal traffic, quiet train
80	– rock music, subway
90	– heavy traffic, thunder
100	– jet plane at takeoff

Em—a printer's measure designating the square width of any given type size. Thus, an em of 10-point type is 10 points. An en is half an em.

Gauge—a measure of shotgun bore diameter. Gauge numbers originally referred to the number of lead balls just fitting the gun barrel diameter required to make a pound. Thus, a 16-gauge shotgun's bore was smaller than a 12-gauge shotgun's. Today, an international agreement assigns millimeter measures to each gauge, e.g.:

Gauge	Bore diameter (in mm)
6	23.34
10	19.67
12	18.52
14	17.60
16	16.81
20	15.90

Horsepower—the power needed to lift 550 pounds 1 foot in 1 second or to lift 33,000 pounds 1 foot in 1 minute. Equivalent to 746 watts or 2,546.0756 Btu/h.

Knot—a measure of the speed of ships. A knot equals 1 nautical mile per hour.

Quire—25 sheets of paper

Ream—500 sheets of paper

Electrical Units

The **watt** is the unit of power (electrical, mechanical, thermal, etc.). Electrical power is given by the product of the voltage and the current.

Energy is sold by the **joule,** but in common practice the billing of electrical energy is expressed in terms of the **kilowatt-hour,** which is 3,600,000 joules or 3.6 megajoules.

The **horsepower** is a nonmetric unit sometimes used in mechanics. It is equal to 746 watts.

The **ohm** is the unit of electrical resistance and represents the physical property of a conductor that offers a resistance to the flow of electricity, permitting just 1 ampere to flow at 1 volt of pressure.

Ancient Measures

Biblical

Cubit	=	21.8 inches
Omer	=	0.45 peck
	=	3.964 liters
Ephah	=	10 omers
Shekel	=	0.497 ounce
	=	14.1 grams

Greek

Cubit	=	18.3 inches
Stadion	=	607.2 or 622 feet
Obolos	=	715.38 milligrams
Drachma	=	4.2923 grams
Mina	=	0.9463 pound
Talent	=	60 mina

Roman

Cubit	=	17.5 inches
Stadium	=	202 yards
As, libra, pondus	=	325.971 grams
	=	0.71864 pound

Spirits Measures

Pony	= 0.5 jigger	Quart	= 32 shots	For champagne only:	
Shot.	= 0.666 jigger		= 1.25 fifths	Rehoboam	= 3 magnums
	= 1.0 ounce	Magnum	= 2 quarts	Methuselah	= 4 magnums
Jigger	= 1.5 shots		= 2.49797 bottles	Salmanazar	= 6 magnums
Pint	= 16 shots		(wine)	Balthazar	= 8 magnums
	= 0.625 fifth	For champagne and brandy only:		Nebuchadnezzar . .	= 10 magnums
Fifth.	= 25.6 shots	Jeroboam	= 6.4 pints		
	= 1.6 pints		= 1.6 magnum	Wine bottle (standard)	= 0.800633 quart
	= 0.8 quart		= 0.8 gallon		= 0.7576778 liter
	= 0.75706 liter				

Temperature Conversion Table

The numbers in the **center column** refer to the temperatures in either degrees Celsius or degrees Fahrenheit that are to be converted. If converting from degrees Fahrenheit to Celsius, refer to the column on the left; if converting from degrees Celsius to Fahrenheit, consult the column on the right.

For temperatures not shown: To convert Fahrenheit to Celsius by formula, subtract 32 degrees and divide by 1.8; to convert Celsius to Fahrenheit, multiply by 1.8 and add 32 degrees.

Note: Although the term *centigrade* is still frequently used, the International Committee on Weights and Measures and the National Institute of Standards and Technology have recommended since 1948 that this scale be called Celsius.

Celsius	Fahrenheit		Celsius		Fahrenheit	Celsius		Fahrenheit
−273.2	−459.7	. . . .	−17.8	0	32	35.0	95	203
−184	−300	. . . .	−12.2	10	50	36.7	98	208.4
−169	−273	− 459.4	− 6.67	20	68	37.8	100	212
−157	−250	− 418	− 1.11	30	86	43	110	230
−129	−200	− 328	4.44	40	104	49	120	248
−101	−150	− 238	10.0	50	122	54	130	266
− 73.3	−100	− 148	15.6	60	140	60	140	284
− 45.6	− 50	− 58	21.1	70	158	66	150	302
− 40.0	− 40	− 40	23.9	75	167	93	200	392
− 34.4	− 30	− 22	26.7	80	176	121	250	482
− 28.9	− 20	− 4	29.4	85	185	149	300	572
− 23.3	− 10	14	32.2	90	194			

Boiling and Freezing Points

Water boils at 212° F (0° C) at sea level. For every 550 feet above sea level, boiling point of water is lower by about 1° F. Methyl alcohol boils at 148° F. Average human oral temperature, 98.6° F. Water freezes at 32° F (100° C).

Compound Interest

Compounded Annually

Principal	Period	4%	5%	6%	7%	8%	9%	10%	12%	14%	16%
$100	1 day	0.011	0.014	0.016	0.019	0.022	0.025	0.027	0.033	0.038	0.044
	1 week. . .	0.077	0.096	0.115	0.134	0.153	0.173	0.192	0.230	0.268	0.307
	6 mos. . . .	2.00	2.50	3.00	3.50	4.00	4.50	5.00	6.00	7.00	8.00
	1 year . . .	4.00	5.00	6.00	7.00	8.00	9.00	10.00	12.00	14.00	16.00
	2 years . .	8.16	10.25	12.36	14.49	16.64	18.81	21.00	25.44	29.96	34.56
	3 years . .	12.49	15.76	19.10	22.50	25.97	29.50	33.10	40.49	48.15	56.09
	4 years . .	16.99	21.55	26.25	31.08	36.05	41.16	46.41	57.35	68.90	81.06
	5 years . .	21.67	27.63	33.82	40.26	46.93	53.86	61.05	76.23	92.54	110.03
	6 years . .	26.53	34.01	41.85	50.07	58.69	67.71	77.16	97.38	119.50	143.64
	7 years . .	31.59	40.71	50.36	60.58	71.38	82.80	94.87	121.07	150.23	182.62
	8 years . .	36.86	47.75	59.38	71.82	85.09	99.26	114.36	147.60	185.26	227.84
	9 years . .	42.33	55.13	68.95	83.85	99.90	117.19	135.79	177.31	225.19	280.30
	10 years . .	48.02	62.89	79.08	96.72	115.89	136.74	159.37	210.58	270.72	341.14
	12 years . .	60.10	79.59	101.22	125.22	151.82	181.27	213.84	289.60	381.79	493.60
	15 years . .	80.09	107.89	139.66	175.90	217.22	264.25	317.72	447.36	613.79	826.55
	20 years . .	119.11	165.33	220.71	286.97	366.10	460.44	572.75	864.63	1,274.35	1,846.08

Common Fractions Reduced to Decimals

8ths	16ths	32ds	64ths		8ths	16ths	32ds	64ths		8ths	16ths	32ds	64ths	
			1	= 0.015625				23	= 0.359375				45	= 0.703125
		1	2	= 0.03125	3	6	12	24	= 0.375			23	46	= 0.71875
			3	= 0.046875				25	= 0.390625				47	= 0.734375
	1	2	4	= 0.0625			13	26	= 0.40625	6	12	24	48	= 0.75
			5	= 0.078125				27	= 0.421875				49	= 0.765625
		3	6	= 0.09375		7	14	28	= 0.4375			25	50	= 0.78125
			7	= 0.109375				29	= 0.453125				51	= 0.796875
1	2	4	8	= 0.125			15	30	= 0.46875		13	26	52	= 0.8125
			9	= 0.140625				31	= 0.484375				53	= 0.828125
		5	10	= 0.15625	4	8	16	32	= 0.5			27	54	= 0.84375
			11	= 0.171875				33	= 0.515625				55	= 0.859375
	3	6	12	= 0.1875			17	34	= 0.53125	7	14	28	56	= 0.875
			13	= 0.203125				35	= 0.546875				57	= 0.890625
		7	14	= 0.21875			18	36	= 0.5625			29	58	= 0.90625
			15	= 0.234375				37	= 0.578125				59	= 0.921875
2	4	8	16	= 0.25			19	38	= 0.59375		15	30	60	= 0.9375
			17	= 0.265625				39	= 0.609375				61	= 0.953125
		9	18	= 0.28125	5	10	20	40	= 0.625				62	= 0.96875
			19	= 0.296875				41	= 0.640625			31		
	5	10	20	= 0.3125			21	42	= 0.65625				63	= 0.984375
			21	= 0.328125				43	= 0.671875					
		11	22	= 0.34375		11	22	44	= 0.6875	8	16	32	64	= 1.0

Measures of Force and Pressure

Dyne = force necessary to accelerate a 1-gram mass 1 centimeter per second squared = 0.000072 poundal

Poundal = force necessary to accelerate a 1-pound mass 1 foot per second squared = 13,825.5 dynes = 0.138255 newtons

Newton = force needed to accelerate a 1-kilogram mass 1 meter per second squared

Pascal (pressure) = 1 newton per square meter = 0.020885 pound per square foot

Atmosphere (air pressure at sea level) = 2,116.102 pounds per square foot = 14.6952 pounds per square inch = 1.0332 kilograms per square centimeter = 101,323 newtons per square meter

Mathematical Formulas

Note: The value of π (the Greek letter pi) is approximately 3.14159265 (equal to the ratio of the circumference of a circle to the diameter). The equivalence is typically rounded further to 3.1416 or 3.14.

To find the CIRCUMFERENCE of a:

Circle — Multiply the diameter by pi.

To find the AREA of a:

Circle — Multiply the square of the radius (equal to ½ the diameter) by pi.

Rectangle — Multiply the length of the base by the height.

Sphere (surface) — Multiply the square of the radius by pi and multiply by 4.

Square — Square the length of one side.

Trapezoid — Add the 2 parallel sides, multiply by the height, and divide by 2.

Triangle — Multiply the base by the height, divide by 2.

To find the VOLUME of a:

Cone — Multiply the square of the radius of the base by pi, multiply by the height, and divide by 3.

Cube — Cube the length of one edge.

Cylinder — Multiply the square of the radius of the base by pi and multiply by the height.

Pyramid — Multiply the area of the base by the height and divide by 3.

Rectangular Prism — Multiply the length by the width by the height.

Sphere — Multiply the cube of the radius by pi, multiply by 4, and divide by 3.

Playing Cards and Dice Chances

5-Card Poker Hands

Hand	Number possible	Odds against
Royal flush	4	649,739 to 1
Other straight flush	36	72,192 to 1
Four of a kind	624	4,164 to 1
Full house	3,744	693 to 1
Flush	5,108	508 to 1
Straight	10,200	254 to 1
Three of a kind	54,912	46 to 1
Two pairs	123,552	20 to 1
One pair	1,098,240	4 to 3 (1.37 to 1)
Nothing	1,302,540	1 to 1
TOTAL	**2,598,960**	

Note: Although there are only 13 4-of-a-kind combinations, the above numbers take into account the total possibilities when a 5th card is figured in to make a 5-card hand.

Dice
(probabilities of consecutive winning plays)

No. consecutive wins	By 7,11, or point	No. consecutive wins	By 7, 11, or point
1	244 in 495	6	1 in 70
2	6 in 25	7	1 in 141
3	3 in 25	8	1 in 287
4	1 in 17	9	1 in 582
5	1 in 34		

Dice
(probabilities on 2 dice)

Total	Odds against (single toss)	Total	Odds against (single toss)
2	35 to 1	8	31 to 5
3	17 to 1	9	8 to 1
4	11 to 1	10	11 to 1
5	8 to 1	11	17 to 1
6	31 to 5	12	35 to 1
7	5 to 1		

Pinochle Auction
(odds against finding in "widow" of 3 cards)

Open places	Odds	Open places	Odds
1	5 to 1 against	4	1½ to 1 for
2	2 to 1 against	5	2 to 1 for
3	Even	6	3 to 1 for

Bridge

The odds—against suit distribution in a hand of 4-4-3-2 are about 4 to 1, against 5-4-2-2 about 8 to 1, against 6-4-2-1 about 20 to 1, against 7-4-1-1 about 254 to 1, against 8-4-1-0 about 2,211 to 1, and against 13-0-0-0 about 158,753,389,899 to 1.

Large Numbers

U.S.	Number of zeros	British[1], French, German	U.S.	Number of zeros	British[1], French, German
million.	6	million	tredecillion	42	septillion
billion.	9	milliard	quattuordecillion	45	1,000 septillion
trillion	12	billion	quindecillion	48	octillion
quadrillion.	15	1,000 billion	sexdecillion	51	1,000 octillion
quintillion	18	trillion	septendecillion	54	nonillion
sextillion	21	1,000 trillion	octodecillion.	57	1,000 nonillion
septillion	24	quadrillion	novemdecillion.	60	decillion
octillion.	27	1,000 quadrillion	vigintillion	63	1,000 decillion
nonillion	30	quintillion	googol	100	googol
decillion	33	1,000 quintillion	centillion	303	—
undecillion	36	sextillion	—	600	centillion
duodecillion	39	1,000 sextillion	googolplex	googol	googolplex

(1) In recent years, it has become more common in Britain to use American terminology for large numbers.

Roman Numerals

I	—	1	VI	—	6	XI	—	11	L	—	50	CD	—	400
II	—	2	VII	—	7	XIX	—	19	LX	—	60	D	—	500
III	—	3	VIII	—	8	XX	—	20	XC	—	90	CM	—	900
IV	—	4	IX	—	9	XXX	—	30	C	—	100	M	—	1,000
V	—	5	X	—	10	XL	—	40	CC	—	200			

Note: The numerals V, X, L, C, D, or M shown with a horizontal line on top denote 1,000 times the original value.

THE INTERNET AND COMPUTERS
What Is the Internet?

The **Internet** is a vast computer network of computer networks. In 1994, 3 million people (most of them in the U.S.) made use of it. By mid-1999, close to 200 million people around the world were using the Internet, and it is estimated that by 2005, 1 billion people may be connected.

Some other facts about the Internet and computing:

- Recent research estimates that there are now more than 800 million pages on the World Wide Web.
- Experts estimate that traffic on the Internet doubles every 9 to 12 months.
- By December 1996, about 627,000 Internet domain names had been registered. By September 1999, over 10 million had been registered.
- Today, 8 hair-thin optical fibers can transmit the equivalent of 90,000 encyclopedia volumes in a second.

The Internet is not owned or funded by any one institution, organization, or government. It has no CEO and is not a commercial service. Its development is guided by the Internet Society (ISOC), composed of volunteers. The ISOC appoints the Internet Architecture Board (IAB), which works out issues of standards, network resources, etc. Another volunteer group, the Internet Engineering Task Force (IETF), handles day-to-day issues of Internet operation.

Practically speaking, the Internet is composed of **people**, **hardware**, and **software**. With the proper equipment on both ends, you can sit at your computer and communicate with someone anyplace in the world. You can also use the Internet to access vast amounts of information, including text, graphics, sound, and video. From your computer you can send e-mail, listen to music, "chat" with people on another continent, do banking, buy stocks, books, flowers or cars, work with others on an electronic whiteboard, and, with the appropriate equipment, video-conference.

How Did the Internet Originate?

The Internet grew out of a series of developments in the academic, governmental, and information technology communities. Listed below are some of the major milestones:

- In 1969, ARPANET, an experimental 4-computer network, was established by the Advanced Research Projects Agency (ARPA) of the U.S. Defense Dept. so that research scientists could communicate.
- By 1971, ARPANET linked about 2 dozen computers ("hosts") at 15 sites, including MIT and Harvard. By 1981, there were over 200 hosts.
- During the 1980s, more and more computers using different operating systems were connected. In 1983, the military portion of ARPANET was moved onto the MILNET, and ARPANET was disbanded in 1990.
- In the late 1980s, the National Science Foundation's NSFNET began its own network and allowed everyone to access it. It was, however, mainly the domain of "techies," computer-science graduates, and professors.
- Legislation in the early 1990s expanded NSFNET, renamed it NREN (National Research and Education Network), and encouraged development of commercial transmission and network services. The mass commercialization of today's Internet is largely a result of such legislation.
- 1991 saw the release of the first **browser**, or software for accessing what would become known as the **World Wide Web**. In 1993, the National Center for Supercomputing Applications released versions of Mosaic (the first graphical Web browser) for Microsoft Windows, for Unix systems running the X Window System, and for the Apple Macintosh.
- In 1994, Netscape Communications released the Netscape Navigator browser, and in 1995, Microsoft released Internet Explorer. Soon these browsers were in head-to-head competition.
- In 1998, the U.S. Justice Dept. and attorneys general from several states filed suit against Microsoft, claiming that the inclusion of Internet Explorer in Windows 98 violated antitrust guidelines. Testimony ended in June 1999; a ruling was expected by year's end.

How Can You Get On the Internet?

First, you need the equipment. Basic Internet access is possible with any computer that has a **modem** connected to a phone line. These days nearly all computers being sold are capable of surfing the World Wide Web. If you have an older, less powerful machine, not able to cope with the most up-to-date version of the Microsoft or Netscape browsers, you can still get on the Web by using an earlier version, or by obtaining one of the other browsers available. You can run the inexpensive and widely praised Opera, for example, on a machine with an 80386 processor and 8 megabytes of RAM memory. Opera, however, does not provide all the bells and whistles that Microsoft and Netscape offer, and as of late 1999 was not yet available for the Macintosh.

Modem speed is critical to your Internet travels. The higher the modem's baud rate, the faster Web pages appear on the screen. For example, a 3.5-minute video clip will download in 46 minutes if you are using a 28.8-kbps modem. With a 128-kbps ISDN line, the same video clip will download in 10 minutes, and with a 10-mbps cable modem, it will take only 8 seconds.

An **Internet service provider** is a company that provides access to the Internet; some also provide content and e-mail. The best-known ISPs are the commercial online services such as America Online, CompuServe (now owned by AOL), and MSN (The Microsoft Network), but many national companies (such as MCI, AT&T, Earthlink, Prodigy, and Mindspring) and local and regional companies also provide Internet access. ISPs generally charge a monthly subscription rate. Some may charge additionally for connect time beyond a specified maximum.

Internet Resources

E-mail. Electronic mail is probably the most widely used resource on the Internet. An e-mail address consists of a **username**, a **service**, and a **domain**. In `Walmanac@aol.com` (The World Almanac's e-mail address), Walmanac is the username, aol the service (in this case, America Online), and com the domain (in this case, a company).

Domains. Domains are identified in the Domain Name System. For years the registration of the most popular Internet addresses (with such domains as com, net, and org) was administered by Network Solutions Inc., under contract to the federal government. A new nonprofit corporation, the Internet Corporation for Assigned Names and Numbers (ICANN), was set up in 1998 to oversee the system, with administration of the registration process to be opened up to more companies. ICANN made an initial selection of additional companies in April 1999. The first to start operations was register.com, which recorded its millionth domain name by September.

Here are the most familiar top-level domains as of mid-1999 (subject to future change and additions):

Domain	What It Is
.com	generally a commercial organization, business, or company
.edu	a 4-year higher-educational institution
.gov	a nonmilitary government entity
.int	an international organization
.mil	a military organization
.net	suggested for a network administration
.org	suggested for a nonprofit organization

Outside the U.S., the final part of a domain name represents the country where the site is located—for example, jp in Japan, uk in the United Kingdom, and ru in Russia.

FAQs. Frequently Asked Questions documents contain the answers to common Internet questions. Reading some of these documents may help Internet newcomers.

FTP. File Transfer Protocol is a method of transferring files on the Internet. Using FTP, you log on to a remote site, view the available files, and copy them to your computer. The address for a site accessible by FTP site begins with ftp.

Newsgroups. Newsgroups, a classic institution of the Internet, are found on the part of the Internet called Usenet. In a newsgroup, messages concerning a particular topic are posted in a public forum. You can simply read the postings, or you can post something yourself.

World Wide Web. The World Wide Web was developed in the early 1990s at the European Center for Nuclear Research as an environment in which scientists in Geneva, Switzerland, could share information. It has evolved into a medium with text, graphics, audio, animation, and video. A **website** address begins with http://. The Web is a graphical environment that can be navigated through **hyperlinks.** From one site you click on hyperlinks to go to related sites.

How the Internet Works

The Internet involves 3 basic elements: server, client, and network. A **server** is a computer program that makes data available to other programs on the same or other computers—it "serves" them. A **client** is a computer that requests data from a server. A **network** is an interconnected system in which multiple computers can communicate, via copper wire, coaxial cable, fiber-optic cable, satellite transmission, etc. The software by which you access Internet resources is the **browser.** When you go to a site on the World Wide Web, you access the site's files.

Here are the steps in opening and accessing a file:
- In the browser, specify the address, or **URL,** of the website.
- The browser sends your request to the Internet service provider's server.
- That server sends the request to the server at the specified URL.
- The file is sent to the ISP's server, which sends the file back to the browser, which displays the file.

How the Internet Is Being Used

The number of Americans accessing the Internet to read news is growing rapidly. In 1996, only 6% went online for news; in 1998, that number increased to 20%. Nearly 90% of Web users access the Internet for news, which is available from around the world and is usually free. Worldwide, some 5,000 newspapers publish online. All but 3 of the top 50 U.S. magazines had a Web presence by January 1998. In addition, more than 800 U.S. television stations have websites.

Communication via e-mail or online chat and the storing and distribution of information are not the only uses to which the Internet is being put. **E-commerce**—the conducting of transactions over the Internet—is skyrocketing. Transactions between companies constitute a significant part of the Internet's growing impact in commerce, since more and more firms are turning to the Web to procure goods and services. The online procurement industry is reportedly doubling in size every year.

People are also **shopping** more online. A survey found that in the first half of 1999, 53% of Web users in the U.S. bought something online, compared to 26% in 1998. You can also sell things over the Internet; hence the surging popularity of online auctions. Sometimes you can actually receive goods online—e.g., computer programs, newspapers, and music recordings. Music of near-CD quality can be obtained (either for free or for a fee) in the popular format known as **MP3,** which compresses the electronic audio file somewhat in order to reduce download time. MP3 files are still large, however, so you will want to have as fast a modem as possible. Check out the site http://www.mp3.com for more information.

More people are booking air flights, cruises, hotel rooms, and rental cars online than ever. Projections by the Travel Industry Association of America suggest that the number of people buying **travel** services online in the U.S. will grow from 41.0 million in 1998 to 48.1 million in 1999 and 71.9 million in 2002, with total online travel revenue rising from $1.9 billion in 1998 to $3.2 billion in 1999, to $8.9 billion in 2002. Airline ticket sales accounted for 90% of all online travel revenues in 1996; by 2002 the proportion is expected to fall to 75%.

If your system has a microphone, speakers, and the right software, you can also use the Internet as a transmission channel for making voice phone calls. Worldwide Internet telephony usage is projected to surge from 310 million minutes in 1998 to almost 3 billion by 2000 and 135 billion by 2004.

Online banking by consumers is still in its infancy, but it is a growing segment of the Internet. Banks that offer online banking allow customers to check their balances, transfer funds, and pay bills while seated at their home computer. Again, for good reason. An online transaction costs the bank much less than a face-to-face interaction with a bank's teller. By the year 2004, 3 to 5 times as many households are expected to be banking online as in 1998.

Another branch of the financial services industry is the shining star of commerce over the Internet. Experts expect the **online brokerage** market to reach total assets of $3 trillion by 2003, up from $415 billion in 1998. By 2003 the number of households trading online is projected to exceed 20.3 million, up from 4.3 million 5 years before.

Safety and Security on the Internet

Common sense dictates some basic security rules.
- Do not give out your phone number, address, or other personal information, unless needed for a transaction at a site you trust.
- Be extremely careful about giving out credit card numbers.
- If you feel someone is being threatening or dangerous, inform your Internet service provider, which can issue a warning or can even withdraw online privileges.

Viruses and Worms. There is always a risk of acquiring a computer **virus.** The consequences of being infected with these malign bits of computer code can vary. Some viruses may merely display a whimsical message on your screen. Some may wreak havoc in your system. Your system can pick up a virus from a program downloaded from the Internet or elsewhere via modem (or received on a floppy disk); in some circumstances a virus can be communicated via e-mail, as was the case with the Melissa virus in early 1999.

You should have antivirus software installed on your computer, keep it up to date, and try to keep abreast of reports of new viruses that may require special attention. Be careful about opening e-mail from unknown correspondents, and if you have programs with a macro capability (macros are bits of auxiliary coding that are meant to play a helpful role but can be taken advantage of by some viruses, such as Melissa), make sure the programs' macro virus protection (if any) is turned on. Keep macros disabled if you do not know what you might want to use them for.

Worms are another type of mischievous cyber creature. Strictly speaking, while viruses propagate by infecting other programs, worms propagate without the help of a carrier program. A **Trojan horse** is a malicious computer code concealed within harmless code or data, but at some point capable of taking control and causing damage.

Filtering. The two major browsers, Netscape Navigator and Internet Explorer, and some search engines contain features that let you filter the content that can be viewed on your computer.

Netparents.org is an association devoted to providing information about such resources. Check out its website at http://www.netparents.org

Where to Start on the Web

Many people have a favorite site that they go to first when logging on to the World Wide Web. (Most browsers permit you to pick your own start site.) A convenient choice for such a site is a **portal,** a gateway site offering a variety of services, such as free e-mail (sometimes free voice mail as well), chat, instant messaging, news services, stock updates, weather reports, real estate listings, yellow pages, people finders, TV and movie listings, shopping, and even tools to create and post your own Web page. Many portals permit you to customize the opening screen. Another common feature is a personal calendar to help you schedule activities.

Leading portals include:

AltaVista	.http://www.altavista.com
AOL	.http://www.aol.com
Excite	.http://www.excite.com
Go/Infoseek . .	.http://www.go.com
HotBot	.http://www.hotbot.com
LookSmart . .	.http://www.looksmart.com
Lycos	.http://www.lycos.com
MSN	.http://www.msn.com
Netscape Netcenter	.http://www.netscape.com
Snap	.http://www.snap.com
Yahoo!	.http://www.yahoo.com

A cornerstone of every portal is a **search engine**, which you can use to locate information on sites throughout a large part of the Internet. (No search engine covers the entire Web completely.) Some portals offer a list of search engines to choose from. Search engines typically allow you to find occurrences of a particular key word (or words). Every search engine uses different methods for finding, indexing, and retrieving information. Some store only the title and URL of sites; others index every word of a site's content. Some give extra weight to words in titles or other key positions. The bigger search engines work with the help of a program called a "spider," "crawler," or "bot." This visits sites across the Web and extracts information that can be used to create the search engine's index.

In addition to a search engine, which requires you to submit key words, some portals also offer a subject guide—a menu-like "directory," generally compiled by humans. You drill down through the hierarchical directory structure to find a subcategory with websites of interest. Yahoo! is a popular example. LookSmart also offers an impressive directory (which Excite makes use of). Another example is the NBC-owned Snap.

Besides the search engines found at the portals listed above, useful search engines and directories include:

Direct Hit (http://www.directhit.com), which calls itself a "Popularity Engine." It monitors the activity of previous Internet searchers on various search engines and thus claims to be able to identify the most relevant and popular sites that meet your request.

FASTSearch (http://www.alltheweb.com) says it searches more of the Web than any other engine—it was the first search engine to cover at least 200 million pages.

Google (http://www.google.com) is also available in the Netscape Netcenter portal. It relies largely on link popularity in ranking the sites it retrieves.

Northern Light (http://www.northernlight.com) organizes retrieved documents according to topic.

Open Directory (http://dmoz.org) aims to cope with the vast size of the Web and produce the most comprehensive directory by using volunteer editors. Its information is used by such services as Netscape, Lycos, and HotBot.

WebCrawler (http://www.webcrawler.com) is a smaller search engine owned by Excite.

You might try a **meta-search engine**, which submits your request to several different search engines at the same time. However, meta-search engines typically do not exhaust each of the individual search engines' databases, and they may be unable to transmit complicated search requests. Among the better-known meta-search engines are **Dogpile** (http://www.dogpile.com), which also offers a directory and portal services; **Inference Find** (http://www.infind.com); and **Metafind** (http://www.metafind.com).

If you would like to find out more information about search engines, go to the site called **Search Engine Watch,** at http://www.searchenginewatch.com

Internet Lingo

The following abbreviations are sometimes used in Internet documents and in e-mail.

BCNU Be seeing you	**GG** Got to go	**LOL** Laughing out loud
BTW By the way	**GOK** God only knows	**OTOH** On the other hand
F2F Face to face; a personal meeting	**HHOK** Ha, ha—only kidding	**PLS** Please
FCOL For crying out loud	**IMHO** In my humble opinion	**ROTFL** Rolling on the floor laughing
FWIW For what it's worth	**IMO** In my opinion	**TAFN** That's all for now
FYI For your information	**J/K** Just kidding	**TTFN** Ta-ta for now

Emoticons, or **smileys**, are a series of typed characters that, when turned sideways, resemble a face and express an emotion. Here are some smileys often encountered on the Internet.

:-)	Smile	:-D	Laugh	:-(	Unhappy	:-b..	Drooling
;-)	Wink	:-*	Kiss	:-o	Shouting	{*}	A hug and a kiss

Internet Directory to Selected Sites

The Websites listed are but a sampling of what is available. For others, see the Where to Get Help directory in the Health chapter, the Business Directory in the Consumer Information chapter, the Sports Directory in the Sports section, and chapters on Travel and Tourism, Associations and Societies, U.S. Cities, U.S. States, the U.S. Government, and Nations of the World. Sites or products are not endorsed by *The World Almanac*.

You must type an address exactly as written, including capital and lowercase letters, and any nonalphanumeric characters. When there is no hyphen or other punctuation at the end of a line in a Website address, no spacing or punctuation should be added. You may be unable to connect to a site because (1) You have mistyped the address, (2) the site is busy, or (3) it has moved or no longer exists.

Online Service Providers
America Online
http://www.aol.com
AT&T WorldNet Service
http://www.att.net
CompuServe
http://www.compuserve.com
EarthLink
http://www.earthlink.net
Erols
http://www.erols.com
Microsoft Network
http://www.msn.com
MindSpring
http://www.mindspring.net
Prodigy
http://www.prodigy.com
WebTV Networks
http://www.webtv.com
Internet Service Providers
http://thelist.internet.com

Security and Screening Information
The National Fraud Information Center
http://www.fraud.org
The Secure Electronic Transaction Standard (general information about electronic commerce)
http://www.visa.com/cgi-bin/vee/nt/ecomm/main.html?2+0

Directories
Bigfoot (e-mail addresses and white page listings)
http://www.bigfoot.com
Four11, the Internet White Pages
http://www.four11.com
InfoSpace, the Ultimate Directory
http://www.infospace.com

People Search
http://www.yahoo.com/search/people
Switchboard, the People and Business Directory
http://www.switchboard.com
WhoWhere?
http://www.whowhere.com

Bookstores
Amazon.com Inc.
http://www.amazon.com
Barnes and Noble
http://www.barnesandnoble.com
Books.Com
http://www.books.com/scripts/news.exe
Borders.Com
http://www.borders.com
The Complete Guide to Online Bookstores
http://www.bookarea.com

Economic Data

Bureau of Economic Analysis
http://www.bea.doc.gov
Bureau of Labor Statistics
http://www.bls.gov
Economics Statistics Briefing Room
http://www.whitehouse.gov/
  fsbr/prices.html
Economy at a Glance
http://stats.bls.gov/
  eag.table.html
Government Information Sharing Project
http://govinfo.kerr.orst.edu
Office of Management and Budget
http://www.access.gpo.
  gov/omb/omb003.html
Statistical Abstract of the United States (a sampling)
http://www.census.gov:80/
  stat_abstract
STAT-USA/Internet (a subscription-based government service)
http://www.stat-usa.gov/
  stat-usa.html

Your Money

Internal Revenue Service
http://www.irs.gov
Wall Street Journal
http://www.wsj.com
American Stock Exchange
http://www.amex.com
E*TRADE
http://www.etrade.com
MarketWatch
http://www.marketwatch.com
NASDAQ
http://www.nasdaq.com
New York Stock Exchange
http://www.nyse.com
Personal Finance Calculators
http://www.financenter.com/
  index.html
Priceline (for offering a price for goods or services)
http://www.priceline.com
Debt Calculator
http://www.uclending.com
Mortgage Calculator
http://www.weichert.com/
  mortgage
Retirement Calculator
http://www.worldi.com/
  index.htm

Job Search Sites

CareerBuilder
http://www.careerbuilder.com
CareerMosaic
http://www.careermosaic.com
Monster.com
http://www.monster.com

Auctions

eBay
http://www.ebay.com
Onsale
http://www.onsale.com
uBid Online Auction
http://www.ubid.com
Yahoo! Auctions
http://auctions.yahoo.com

Electronic Greeting Cards

Blue Mountain Arts
http://www1.bluemountain.com
Egreetings Network
http://www.egreetings.com
Electronic Postcards
http://www.electronic
  postcards.net
Micro-Images Multimedia Greeting Cards
http://www.microimg.com/
  postcards
1001 Postcards
http://www.postcards.org
123 Greetings
http://www.123greetings.com

Audio/Video

Broadcast.com
http://www.broadcast.com
LiveUpdate
http://www.liveupdate.com
MP3.com
http://www.mp3.com
Real Networks
http://www.real.com

Health

CenterWatch Clinical Trials Listing Service
http://www.centerwatch.com
Drkoop.com
http://DrKoop.com
Drugstore.com
http://www.drugstore.com
Healthfinder
http://www.healthfinder.gov
Mayo Clinic Health Oasis
http://www.mayohealth.org
Medscape
http://www.medscape.com
The Merck Manual
http://www.merck.com/pubs
National Food Safety Database
http://www.foodsafety.org/
  index.htm
National Institutes of Health (Health Information)
http://www.nih.gov/health
PlanetRx
http://www.planetrx.com
U.S. National Library of Medicine
http://www.nlm.nih.gov

News

The Associated Press
http://www.ap.org
BBC Online
http://www.bbc.co.uk/home/
  today
Cable News Network
http://www.cnn.com
The New York Times on the Web
http://www.nytimes.com
Reuters
http://www.reuters.com

Weather

National Weather Service Home Page
http://www.nws.noaa.gov
National Center for Environmental Prediction
http://www.ncep.noaa.gov
Storm Prediction Center
http://www.nssl.noaa.gov/
  ~spc
Tropical Prediction Center
http://www.nhc.noaa.gov
Weather Calculator
http://nwselp.epcc.edu/elp/
  wxcalc.html

What's New

Internet Scout Project (latest resources for researchers)
http://scout.cs.wisc.edu/
  index.html
Nerd World: Media (what's new in computer world)
http://www.nerdworld.com/
  whatsnew.html
Netscape's What's New
http://home.netscape.com/
  netcenter/new.html
What's New Too!
http://newtoo.manifest.com
Yahoo! What's New (listing of every new site each day; sometimes thousands)
http://www.yahoo.com/new

Chat Sites

America Online
http://www.aol.com/community/
  chat/allchats.html
Excite
http://www.excite.com/
  communities
The Globe
http://www.theglobe.com
IVILLAGE: The Women's Network
http://www.ivillage.com
Lycos
http://chat.lycos.com
Star Media (in Spanish and Portuguese)
http://www.starmedia.com
Yahoo
http://www.yahoo.com

Sites for Kids

Children's Television Workshop
http://www.ctw.org
ePlay
http://www.ePlay.com
Goosebumps
http://place.scholastic.com/
  goosebumps/index.htm
Judy Blume's Home Base
http://www.judyblume.com/
  home.html
The Newbery Medal
http://www.ala.org/alsc/
  newbery.html
Peace Corps Kids World
http://www.peacecorps.gov/
  kids
Rock and Roll Hall of Fame and Museum
http://www.rockhall.com
Seussville
http://www.randomhouse.com/
  seussville
SuperSite for Kids
http://www.bonus.com
Weekly Reader
http://www.weeklyreader.com
White House for Kids
http://www2.whitehouse.gov/
  WH/kids/html/home.html
Yahooligans (guide to homework help sites)
http://www.yahooligans.com

Sports

Little League Baseball
http://www.littleleague.org
Major League Baseball
http://www.majorleague
  baseball.com
Major League Soccer
http://www.mlsnet.com
National Basketball Association
http://www.nba.com
National Football League
http://www.nfl.com
National Hockey League
http://www.nhl.com
Special Olympics
http://www.specialolympics.org
Women's National Basketball Association
http://www.wnba.com

Genealogy

FamilySearch
http://www.familysearch.org
Genealogy.com
http://www.genealogy.com
National Archives and Records Administration
http://www.nara.gov
RootsWeb
http://www.rootsweb.com
USGenWeb Project
http://www.usgenweb.org

Resources for Families

Babies Online
http://babiesonline.com
BabyCenter
http://babycenter.com
Family.Com
http://family.disney.com _
Family Internet
http://www.familyinternet.com
KidsHealth.org
http://www.kidshealth.org

Kidshop Online
http://www.kidshoponline.com
KidSource Online
http://www.kidsource.com
ParenthoodWeb
http://www.parenthoodweb.com
Parent Soup
http://www.parentsoup.com
ParentsPlace.com
http://parentsplace.com
ParentTime
http://www.parentime.com
Screen It! Entertainment Reviews for Parents
http://www.screenit.com
Zero to Three
http://www.zerotothree.org

Reference

BookWire—The First Place to Look for Book Information
http://www.bookwire.com
CIA Publications and Handbooks
http://odci.gov/cia/publications/pubs.html

Funk & Wagnalls Encyclopedia
http://www.funkandwagnalls.com
Internet Search Tools, The Library of Congress
http://lcweb.loc.gov/global/search.html
Libweb—Library Servers via WWW
http://sunsite.berkeley.edu/Libweb
Liszt, the Mailing List Directory
http://www.liszt.com
Miriam-Webster Network Editions
http://www.m-w.com/book/neteds/neteds.htm
On-line Dictionaries, A Web of
http://www.facstaff.bucknell.edu/rbeard/diction.html
Refdesk
http://www.refdesk.com
Roget's Thesaurus
http://www.thesaurus.com

Most-Visited Websites, Aug. 1999

Source: Media Matrix, Inc.

Rank	Website	Unique visitors (000)	Rank	Website	Unique visitors (000)
1.	http://www.yahoo.com	33,398	11.	http://www.passport.com	13,677
2.	http://www.aol.com	29,333	12.	http://www.angelfire.com	12,156
3.	http://www.msn.com	28,079	13.	http://www.amazon.com	11,618
4.	http://www.geocities.com	21,389	14.	http://www.tripod.com	11,261
5.	http://www.netscape.com	19,809	15.	http://www.altavista.com	9,185
6.	http://www.go.com	18,496	16.	http://www1.bluemountain.com	9,168
7.	http://www.microsoft.com	18,031	17.	http://www.real.com	8,891
8.	http://www.lycos.com	14,857	18.	http://www.ebay.com	8,790
9.	http://www.excite.com	14,149	19.	http://www.xoom.com	8,662
10.	http://www.hotmail.com	13,868	20.	http://www.about.com	8,569

Percent of U.S. Households Using the Internet, by Selected Characteristics, 1998

Source: National Telecommunications and Information Administration, U.S. Dept. of Commerce

	U.S.	Rural	Urban	Central city		U.S.	Rural	Urban	Central city
TOTAL	26.2	22.2	27.5	24.5	**Educational attainment**				
Race					Elementary	3.1	1.8	3.7	3.4
White, not Hispanic	29.8	23.7	32.4	32.3	Some high school	6.3	6.1	6.4	5.2
Black, not Hispanic	11.2	7.1	11.7	10.2	High school diploma or GED	16.3	15.5	16.6	13.7
Hispanic	12.6	9.8	12.9	10.2	Some college	30.2	29.6	30.4	26.4
Age of householder					Bachelor's degree or more	48.9	47.0	49.4	47.7
Under 25 years	20.5	13.3	22.0	22.8	**Region**				
25-34 years	30.1	24.2	31.6	28.8	Northeast	26.7	29.7	25.9	18.7
35-44 years	34.1	30.2	35.4	31.3	Midwest	25.4	21.5	26.9	24.0
45-54 years	35.0	30.8	36.5	30.7	South	23.5	19.0	25.6	22.6
55+ years	14.6	12.4	15.4	13.8	West	31.3	26.2	32.0	31.8

Percent of U.S. Households With a Computer, by Selected Characteristics, 1994, 1998

Source: National Telecommunications and Information Administration, U.S. Dept. of Commerce

	1994	1998	'94-'98 increase		1994	1998	'94-'98 increase
U.S. TOTAL	24.1	42.1	18.0	$10,000-$14,999	8.2	15.9	7.7
Race				$15,000-$19,999	11.7	21.2	9.5
White, not Hispanic	27.1	46.6	19.5	$20,000-$24,999	15.2	25.7	10.5
Black, not Hispanic	10.3	23.2	12.9	$25,000-$34,999	19.8	35.8	16.0
Hispanic	12.3	25.5	13.2	$35,000-$49,999	33.0	50.2	17.2
Age of householder				$50,000-$74,999	46.0	66.3	20.3
Under 25 years	18.1	32.3	14.2	$75,000+	60.9	79.9	19.0
25-34 years	25.1	46.0	20.9	**Household type**			
35-44 years	34.1	54.9	20.8	Married couple with children under 18	46.0	61.8	15.8
45-54 years	33.6	54.7	21.1	Male householder with children under 18	25.8	35.0	9.2
55+ years	12.7	25.8	13.1	Female householder with children under 18	19.3	31.7	12.4
Educational attainment				Family households without children	26.6	43.2	16.6
Elementary	2.6	7.9	5.3	Nonfamily households	15.0	27.5	12.5
Some high school	6.0	15.7	9.7	**Region**			
High school diploma or GED	14.8	31.2	16.4	Northeast	22.9	41.3	18.4
Some college	28.9	49.3	20.4	Midwest	24.1	42.9	18.8
Bachelor's degree or more	48.4	68.7	20.3	South	20.9	38.0	17.1
Annual income				West	30.6	48.9	18.3
Under $5,000	8.4	15.9	7.5				
$5,000-$9,999	6.1	12.3	6.2				

Top-Selling Software, 1999
Source: PC Data, Reston, VA
(based on average U.S. sales, Jan.-June 1999)

All Software
1. TurboTax, Intuit
2. TurboTax Deluxe, Intuit
3. Norton Antivirus 5.0, Symantec
4. Microsoft Windows 98 Upgrade, Microsoft
5. Taxcut 1998 Deluxe Filing Edition, Block Financial
6. TurboTax 30 State, Intuit
7. Sim City 3000, Electronic Arts
8. Quicken, Intuit
9. Print Shop 6.0 Deluxe, Mattel Media
10. Quicken Deluxe, Intuit
11. VirusScan 4.0 Classic, Network Associates
12. Baldur's Gate, Interplay
13. QuickBooks Pro 99, Intuit
14. Norton System Works 2.0, Symantec
15. QuickBooks 99, Intuit
16. Cabela's Big Game Hunter 2, Activision
17. TurboTax State CA, Intuit
18. Blue's ABC Time Activities, Humongous (GT Interactive)
19. Microsoft Encarta Encyclopedia, Microsoft
20. American Greetings Creatacard Plus 3, Mattel Media

Games
1. Sim City 3000, Electronic Arts
2. Baldur's Gate, Interplay
3. Cabela's Big Game Hunter 2, Activision
4. Starcraft Expansion: Brood Wars, Havas Interactive
5. Half-Life, Havas Interactive
6. Monopoly Game, Hasbro Interactive
7. Deer Hunter II 3D, GT Interactive
8. Frogger, Hasbro Interactive
9. Deer Avenger, Havas Interactive
10. Sid Meier's Alpha Centauri, Electronic Arts

Reference Software
1. Microsoft Encarta Encyclopedia, Microsoft
2. World Book Encyclopedia Standard 99, IBM
3. Compton's Interactive Encyclopedia 99 Deluxe, Mattel Media
4. World Book Encyclopedia Family Reference Suite 99, IBM
5. Microsoft Encarta Reference Suite, Microsoft

Home Education Software
1. Blue's ABC Time Activities, Humongous (GT Interactive)
2. Rugrats Movie Activity Challenge, Mattel Media
3. Jumpstart First Grade, Havas Interactive
4. Kid Pix Studio Deluxe, Mattel Media
5. Jumpstart Preschool, Havas Interactive
6. Learn To Speak Spanish 7.0, Mattel Media
7. Jumpstart Second Grade, Havas Interactive
8. Rugrats Adventure Game, Mattel Media
9. Blu's Birthday Adventure, Humongous (GT Interactive)
10. Jumpstart Kindergarten II, Havas Interactive

Personal Productivity Software
1. TurboTax, Intuit
2. TurboTax Deluxe, Intuit
3. Taxcut 1998 Deluxe Filing Edition, Block Financial
4. TurboTax 30 State, Intuit
5. Quicken, Intuit
6. Print Shop 6.0 Deluxe, Mattel Media
7. Quicken Deluxe, Intuit
8. TurboTax State CA, Intuit
9. American Greetings Creatacard Plus 3, Mattel Media
10. Print Shop 6.0, Mattel Media

Business Software
1. Norton Antivirus 5.0, Symantec
2. Microsoft Windows 98 Upgrade, Microsoft
3. VirusScan 4.0 Classic, Network Associates
4. QuickBooks Pro 99, Intuit
5. Norton System Works 2.0, Symantec
6. QuickBooks 99, Intuit
7. Norton System Works 1.0, Symantec
8. McAfee Office, Network Associates
9. pcAnywhere 32 Host/Remote, Symantec
10. Microsoft Plus 98, Microsoft
11. Microsoft Office 97 Upgrade, Microsoft
12. Complete Red Hat Linux 5.2 Deluxe, Macmillan
13. Winfax Pro 9.0, Symantec
14. Norton Utilities 4.0, Symantec
15. Microsoft Windows 98, Microsoft

U.S. Cellular Telephone Subscribership, Dec. 1985-Dec. 1998
Source: The CTIA Semi-Annual Wireless Survey. Used with permission of CTIA.

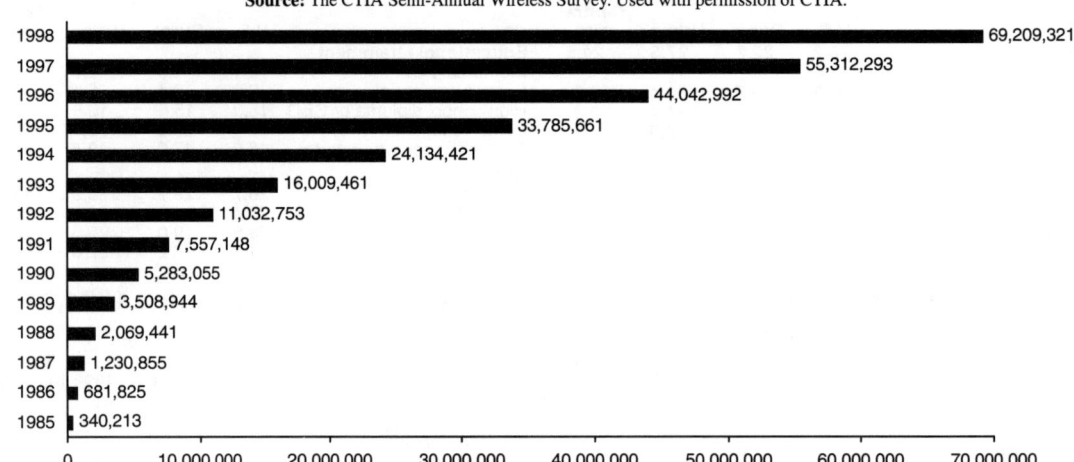

Year	Subscribers
1998	69,209,321
1997	55,312,293
1996	44,042,992
1995	33,785,661
1994	24,134,421
1993	16,009,461
1992	11,032,753
1991	7,557,148
1990	5,283,055
1989	3,508,944
1988	2,069,441
1987	1,230,855
1986	681,825
1985	340,213

Worldwide Use of Cellular Telephones, 1998
Source: Gartner Group's Dataquest

Country/Region	Number of subscribers (millions)	% of population	Country/Region	Number of subscribers (millions)	% of population
Finland	2.9	57	Singapore	1.0	32
Norway	2.1	48	Portugal	3.1	31
Sweden	4.1	46	Australia	5.9	31
Hong Kong	2.9	43	Japan	39.0	31
Israel	2.1	37	South Korea	14.0	30
Italy	20.5	36	Austria	2.3	29
Denmark	1.9	35	U.S.	69.8[1]	26

(1) Estimates from other sources may vary.

Glossary of Computer and Internet Terms

application A program designed to assist in the performance of a specific task, such as word processing, accounting, or inventory management.

artificial intelligence (AI) The branch of computer science concerned with enabling computers to simulate such aspects of human intelligence as speech recognition, deduction, inference, creative response, and the ability to learn from experience.

ASCII Pronounced "askee." An acronym for American Standard Code for Information Interchange, a coding scheme using 7 or 8 bits that assigns numeric values to up to 256 characters, including letters, numerals, punctuation marks, control characters, and other symbols.

backup (noun); back up (verb) As a noun, a duplicate copy of a program, a disk, or data. As a verb, to make a duplicate copy of a program, a disk, or data.

bandwidth Data transfer capacity of a digital communications system.

baud rate Speed at which a modem can transmit data.

BBS An abbreviation for bulletin board system, a computer system equipped with one or more modems or other means of network access that serves as an information and message-passing center for remote users.

bit Short for binary digit; the smallest unit of information handled by a computer. One bit expresses a 1 or a 0 in a binary numeral, or a true or a false logical condition, and is represented physically by an element such as a high or low voltage at one point in a circuit or a small spot on a disk magnetized one way or the other.

boot The process of starting or resetting a computer.

browser *See* **Web browser.**

bug An error in coding or logic that causes a program to malfunction or to produce incorrect results. Also, a recurring physical problem that prevents a system or set of components from working together properly.

bulletin board system *See* **BBS.**

byte A unit of data, today almost always consisting of 8 bits. A byte can represent a single character, such as a letter, a digit, or a punctuation mark.

CD-ROM Acronym for compact disc read-only memory, a form of storage characterized by high capacity (roughly 650 megabytes) and the use of laser optics rather than magnetic means for reading data.

central processing unit (CPU) The computational and control unit of a computer; the device that interprets and executes instructions.

certificate authority An issuer of digital certificates, the cyberspace equivalent of ID cards.

chat room The informal term for a data communication channel that links computers and permits users to "converse", often about a particular subject that interests them, by sending text messages to one another in real time.

chip *See* **integrated circuit.**

client On a local area network, a computer that accesses shared network resources provided by another computer (called a server). *See also* **server.**

computer Any machine that does three things: accepts structured input, processes it according to prescribed rules, and produces the results as output.

CPU *See* **central processing unit.**

crash The failure of either a program or a disk drive. A program crash results in the loss of all unsaved data and can leave the operating system unstable enough to require restarting the computer.

cursor A special on-screen indicator, such as a blinking underline or rectangle, that marks the place of which keystrokes will appear when typed.

cyberspace The universe of environments, such as the Internet, in which persons interact by means of connected computers.

cyberspeak Terminology and language (often jargon, slang, and acronyms) relating to the Internet—computer-connected—environment, that is, cyberspace. Most words prefixed by *cyber-* have the same meaning as their "real-world" counterparts, but specifically indicate their use in the online culture of the Internet and the World Wide Web. Examples: cybercafé, cybercash.

database A file composed of records, each of which contains fields, together with a set of operations for searching, sorting, recombining, and other functions.

data compression A means of reducing the space or bandwidth needed to store or transmit a block of data.

debug To detect, locate, and correct logical or syntactical errors in a program or malfunctions in hardware.

defragger A software utility for reuniting parts of a file that have become fragmented through rewriting and updating.

desktop publishing The use of a computer and specialized software to combine text and graphics to create a document that can be printed on either a laser printer or a typesetting machine.

dial-up access Connection to a data communications network through the public switched telecommunication network.

Digerati Cyberspace populace that can be roughly compared to *literati*. Digerati are renowned as or claiming to be knowledgeable about topics and issues related to the digital revolution; more specifically, they are "in the know" about the Internet and online activities.

digital certificate 1. An assurance that software downloaded from the Internet comes from a reputable source. 2. A user identity card or "driver's license" for cyberspace. Issued by a certificate authority.

directory service A service on a network that returns mail addresses of other users or enables a user to locate hosts and services.

disk A round, flat piece of flexible plastic (floppy disk) or inflexible metal (hard disk) coated with a magnetic material that can be electrically influenced to hold information recorded in digital (binary) format.

disk drive An electromechanical device that reads from and writes to disks.

disk operating system Abbreviated DOS. A generic term describing any operating system that is loaded from disk devices when the system is started or rebooted.

distance learning Broadly, any educational or learning process or system in which the teacher/instructor is separated geographically or in time from his or her students; or in which students are separated from other students or educational resources.

DOS *See* **disk operating system.**

download In communications, to transfer a copy of a file from a remote computer to the requesting computer by means of a modem or network. *See also* **upload.**

dynamic HTML A technology designed to add richness, interactivity, and graphical interest to Web pages by providing those pages with the ability to change and update themselves in response to user actions, without the need for repeated downloads from a server.

encryption The process of encoding data to prevent unauthorized access, especially during transmission. The U.S. National Bureau of Standards created a complex encryption standard, DES (Data Encryption Standard), that provides almost unlimited ways to encrypt documents.

FAQ An abbreviation for Frequently Asked Questions, a document listing common questions and answers on a particular subject. FAQs are often posted on Internet newsgroups where new participants ask the same questions that regular readers have answered many times.

field A location in a record in which a particular type of data is stored.

file A complete, named collection of information, such as a program, a set of data used by a program, or a user-created document.

firewall A security system intended to protect an organization's network against external threats, such as hackers, from another network. *See also* **proxy server.**

flame An abusive or personally insulting e-mail message or newsgroup posting.

format In general, the structure or appearance of a unit of data. As a verb, to change the appearance of selected text or the contents of a selected cell in a spreadsheet.

forum A medium provided by an online service or BBS for users to carry on written discussions of a topic by posting messages and replying to them.

FTP An abbreviation for File Transfer Protocol, the protocol used for copying files to and from remote computer systems on a network using TCP/IP such as the Internet.

gigabyte Abbreviated GB; 1024 megabytes. *See* **megabyte.**

graphical user interface Abbreviated GUI (pronounced "gooey"). A type of environment that represents programs, files, and options by means of icons, menus, and dialog boxes on the screen. The user can select and activate these options by pointing and clicking with a mouse or, often, with the keyboard. *See also* **icon.**

hacker A computerphile—a person who is engrossed in computer technology and programming or who likes to examine the code of operating systems and other programs to see how they work. Also, a person who uses computer expertise for illicit ends, such as for gaining access to computer systems without permission and tampering with programs and data.

hard copy Printed output on paper, film, or other permanent medium.

hit Retrieval of a document, such as a home page, from a website.

home page A document intended to serve as a starting point in a hypertext system, especially the World Wide Web. Also, an entry page for a set of Web pages and other files in a website.

host The main computer in a system of computers or terminals connected by communications links.

HTML An abbreviation for HyperText Markup Language, the markup language used for documents on the World Wide Web.

HTTP An abbreviation for HyperText Transfer Protocol, the client/server protocol used to access information on the Web.

hyperlink A connection between an element in a hypertext document, such as a word, phrase, symbol, or image, and a different element in the document, another hypertext document, a file, or a script. The user activates the link by clicking on the linked element, which is usually highlighted in some way.

hypermedia The integration of any combination of text, graphics, sound, and video into a primarily associative system of information storage and retrieval in which users jump from subject to related subject.

hypertext Text linked together in a complex, nonsequential web of associations in which the user can browse through related topics.

icon A small image displayed on the screen to represent an object that can be manipulated by the user.

import To bring information from one system or program into another.

integrated circuit Also called a chip. A device consisting of a number of connected circuit elements, such as transistors and resistors, fabricated on a single chip of silicon crystal or other semiconductor material.

interactive Characterized by conversational exchange of input and output, as when a user enters a question or command the system immediately responds.

Internet The worldwide collection of networks and gateways that use the TCP/IP suite of protocols to communicate with each other. At the heart of the Internet is a backbone of high-speed data communication lines between major nodes or host computers, consisting of thousands of commercial, government, educational, and other computer systems, that route data and messages.

intranet A TCP/IP network designed for information processing within a company or organization. It usually employs Web pages for information dissemination and Internet applications, such as Web browsers.

IP address Short for Internet Protocol address, a 32-bit (4-byte) binary number that uniquely identifies a host (computer) connected to the Internet to other Internet hosts, for communication through the transfer of packets.

Java A programming language, developed by Sun Microsystems, Inc., that can be run on any platform.

kilobyte Abbreviated K, KB, or Kbyte; 1,024 bytes.

LAN Rhymes with "can." Acronym for local area network, a group of computers and other devices dispersed over a limited area and connected by a link that enables any device to interact with any other on the network.

laptop A small, portable computer that runs on either batteries or AC power, designed for use during travel. Laptops have flat screens and small keyboards. Some weigh as little as 5 pounds.

legacy system A computer, software program, network, or other computer equipment that remains in use after a business or organization installs new systems.

link *See* **hyperlink.**

logon The process of identifying oneself to a computer after connecting to it over a communications line. Also called *login.*

lurk To receive and read articles or messages in a newsgroup or other online conference without contributing anything to the ongoing conversation.

mailing list A list of names and e-mail addresses that are grouped under a single name. When a user places the name of the mailing list in a mail client's To: field, the client automatically sends the same message to the machine where the mailing list resides, and that machine sends the message to all the addresses on the list.

mainframe computer A high-level computer designed for the most intensive computational tasks.

markup language A set of codes in a text file that instruct a printer or video display how to format, index, and link the contents of the file. Examples of markup languages are HyperText Markup Language (HTML), which is used in Web pages, and Standard Generalized Markup Language (SGML), which is used for typesetting and desktop publishing purposes and in electronic documents.

megabyte Abbreviated MB. Usually 1,048,576 bytes (2^{20}); sometimes interpreted as 1 million bytes.

meltdown The complete collapse of a computer network caused by a high level of traffic.

memory Circuitry that allows information to be stored and retrieved. In common usage it refers to the fast semiconductor storage (RAM) directly connected to the processor. *See also* **RAM.**

menu A list of options from which a program user can make a selection in order to perform a desired action, such as choosing a command or applying a format.

microcomputer A computer built around a single-chip microprocessor.

microprocessor A central processing unit (CPU) on a single chip. *See also* **integrated circuit.**

minicomputer A mid-level computer built to perform complex computations while dealing efficiently with input and output from users connected via terminals.

modem A communications device that enables a computer to transmit information over a standard telephone line.

monitor The device on which images generated by the computer's video adapter are displayed.

motherboard The main circuit board containing the primary components of a computer system.

mouse A common pointing device. It has a flat-bottomed casing designed to be gripped by one hand.

multitasking A mode of operation offered by an operating system in which a computer works on more than one task at a time.

Net Short for Internet.

netiquette Short for network etiquette.

netizen A person who participates in online communication through the Internet and other networks, especially conference and chat services.

network A group of computers and associated devices that are connected by communications facilities.

newbie An inexperienced user on the Internet.

newsgroup A forum on the Internet for threaded discussions on a specified range of subjects. A newsgroup consists of articles and follow-up posts. *See* **post, thread.**

online Activated and ready for operating; capable of communicating with or being controlled by a computer.

operating system The software that controls the allocation and usage of hardware resources such as memory, CPU time, disk space, and peripheral devices.

optical scanner An input device that uses light-sensing equipment to scan paper or another medium, translating the pattern of light and dark or color into a digital signal that can be manipulated by either optical character recognition software or graphics software.

packet A unit of information transmitted as a whole from one device to another on a network.

password A unique string of characters that a user types in as an identification code.

PC Abbreviation for personal computer, a microcomputer that conforms to the standard developed by IBM for personal computers, which uses an Intel microprocessor (or one that is compatible).

peripheral A device, such as a disk drive, printer, modem, or joystick, that is connected to a computer and is controlled by the computer's microprocessor.

pixel Short for picture element; also called *pel*. One spot in a rectilinear grid of thousands of such spots that are individually "painted" to form an image produced on the screen by a computer or on paper by a printer.

portal A website that serves as a gateway to the Internet. A portal is a collection of links, content, and services designed to guide users to information they are likely to find interesting—news, weather, entertainment, commerce sites, chat rooms, and so on.

post To submit an article in a newsgroup or other online conference. *See* **thread.**

program A sequence of instructions that can be executed by a computer.

protocol A set of rules or standards designed to enable computers to communicate with one another and to exchange information with as little error as possible.

proxy server A firewall component that manages Internet traffic to and from a local area network and can provide other features, e.g., document caching and access control.

RAM Pronounced "ram." An acronym for random access memory. Semiconductor-based memory that can be read and written by the CPU or other hardware devices.

routing table In data communications, a table of information that provides network hardware (bridges and routers) with the directions needed to forward packets of data to locations on other networks.

search engine On the Internet, a program that searches for keywords in files and documents.

server On a local area network (LAN), a computer running software that controls access to the network and its resources, such as printers and disk drives. On the Internet or other network, a computer or program that responds to commands from a client. *See* **client, LAN.**

sleep mode A power management mode that shuts down all unnecessary computer operations to save energy; also known as suspend mode.

snail mail A phrase popular on the Internet for referring to mail services provided by the United States Postal Service and similar agencies in other countries.

software Computer programs; instructions that make hardware work.

spam An unsolicited e-mail message sent to many recipients at one time, or a news article posted simultaneously to many newsgroups. Electronic junk mail.

spreadsheet program An application commonly used for budgets, forecasting, and other finance-related tasks that organizes data values using cells, where the relationships between cells are defined by formulas.

supercomputer A large, extremely fast, and expensive computer used for complex or sophisticated calculations.

system administrator The person responsible for administering use of a multiuser computer system, communications system, or both.

TCP/IP An abbreviation for Transmission Control Protocol/Internet Protocol, a protocol developed by the Department of Defense for communications between computers. It has become the de facto standard for data transmission over networks, including the Internet.

technophobe A person who is afraid of or dislikes technological advances, especially computers.

telecommute To work in one location (often, at home) and communicate with a main office at a different location through a personal computer.

teleconferencing The use of audio, video, or computer equipment linked through a communications system to enable geographically separated individuals to participate in a meeting or discussion.

teleworker A businessperson who substitutes information technologies for work-related travel. Teleworkers include home-based and small business workers who use computer and communications technologies to interact with customers and/or colleagues.

thread In electronic mail and Internet newsgroups, a series of messages and replies related to a specific topic.

upload In communications, the process of transferring a copy of a file from a local computer to a remote computer by means of a modem or network.

URL An abbreviation for Uniform Resource Locator, an address for a resource on the Internet.

Usenet A worldwide network of Unix systems that has a decentralized administration and is used as a bulletin board system by special-interest discussion groups.

user interface The portion of a program with which a user interacts.

user-friendly Easy to learn and easy to use.

virus An intrusive program that infects computer files by inserting in those files copies of itself.

voice recognition The capability of a computer to understand the spoken word for the purpose of receiving commands and data input from the speaker.

Web *See* **World Wide Web.**

Web browser A client application that enables a user to view HTML documents, follow the hyperlinks among them, transfer files, and execute some programs.

webcasting Popular term for broadcasting information via the World Wide Web, using push and pull technologies to move selected information from a server to a client.

webmaster The person or persons responsible for creating and maintaining a site on the World Wide Web.

website A group of related HTML documents and associated files, scripts, and databases that is served up by an HTTP server on the World Wide Web.

WebTV® Trademark name for technology from Microsoft and WebTV Networks that provide consumers with the ability to access the Internet on a television by means of a set-top box equipped with a modem.

wide area network (WAN) A communications network that connects geographically separated areas.

window In applications and graphical interfaces, a portion of the screen that can contain its own document or message.

word processor A program for manipulating text-based documents; the electronic equivalent of paper, pen, typewriter, eraser, and, most likely, dictionary and thesaurus.

workstation A combination of input, output, and computing hardware used for work by an individual.

World Wide Web (WWW) The total set of interlinked hypertext documents residing on Web, or HTTP, servers all around the world.

WYSIWYG Pronounced "wizzywig." An acronym for "What you see is what you get." A display method that shows documents and graphics characters on the screen as they will appear when printed.

Zip drive A disk drive developed by Iomega that uses 3.5-inch removable disks (Zip disks) capable of storing 100 megabytes of data apiece. See also **disk drive.**

BUILDINGS, BRIDGES, AND TUNNELS
Tallest Buildings in the World

Source: Council on Tall Buildings and Urban Habitat, Lehigh Univ.; Jeff Herzer and Marshall Gerometta, WTB/World's Tallest Buildings, http://www.worldstallest.com

List includes some structures that do not have stories and are not technically considered buildings. Asterisk (*) denotes building still under construction. Double asterisk (**) includes spire. Year is date of completion. Height is in feet.

Name, Year, City, Country	Height	Stories
Petronas Tower I, 1998, Kuala Lumpur, Malaysia	1,483**	88
Petronas Tower II, 1998, Kuala Lumpur, Malaysia	1,483**	88
Sears Tower, 1974, Chicago, United States	1,450	110
Jin Mao Bldg., 1998, Shanghai, China	1,381**	88
One World Trade Center, 1972, New York, U.S.	1,368	110
Two World Trade Center, 1973, New York, U.S.	1,362	110
CITIC Plaza, 1996, Guangzhou, China	1,283**	80
Shun Hing Square, 1996, Shenzhen, China	1,260	81
Empire State Building, 1931, New York, U.S.	1,250	102
Central Plaza, 1992, Hong Kong, China	1,227**	78
Bank of China, 1989, Hong Kong, China	1,209**	70
The Centre, 1998, Hong Kong, China	1,148**	79
T&C Tower, 1998, Kaohsiung, Taiwan	1,140	85
Amoco Bldg., 1973, Chicago, U.S.	1,136	80
John Hancock Center, 1969, Chicago, U.S.	1,127	100
*Burj al Arab Hotel, 1999, Dubai, U.A.E.	1,053**	60
Baiyoke Tower II, 1998, Bangkok, Thailand	1,050	90
Chrysler Bldg., 1930, New York, U.S.	1,046**	77
NationsBank Tower, 1993, Atlanta, U.S.	1,023**	55
*Emirates Towers One, 1999, Dubai, U.A.E.	1,020	55
Library Tower, 1990, Los Angeles, U.S.	1,018	75
Telekom Malaysia Headquarters, 1999, Kuala Lumpur, Malaysia	1,017	55
AT&T Corporate Center, 1989, Chicago, U.S.	1,007**	61
Chase Tower, 1982, Houston, United States	1,002	75
*Emirates Towers Two, 1999, Dubai, U.A.E.	1,001	52
Two Prudential Plaza, 1990, Chicago, U.S.	995**	52
Ryugyong Hotel, 1995, Pyongyang, North Korea	984	105
Commerzbank Tower, 1997, Frankfurt, Germany	981**	63
Wells Fargo Plaza, 1983, Houston, U.S.	972	71
Landmark Tower, 1993, Yokohama, Japan	971	70
311 S. Wacker, 1990, Chicago, United States	961	65
First Canadian Place, 1975, Toronto, Canada	951	72
American International Bldg., 1932, New York, U.S.	951	67
Cheung Kong Centre, 1999, Hong Kong, China	951	70
Key Tower, 1991, Cleveland, U.S.	950**	57
One Liberty Place, 1987, Philadelphia, U.S.	945**	61
Columbia Seafirst Center, 1985, Seattle, U.S.	943	76
The Trump Bldg., 1930, New York, U.S.	927	71
NationsBank Plaza, 1985, Dallas, U.S.	921	72
United Overseas Bank Plaza, 1992, Singapore	919	66
Republic Plaza, 1995, Singapore	919	66
Overseas Union Bank Centre, 1986, Singapore	919	66
Citicorp Center, 1977, New York, U.S.	915	59
Scotia Plaza, 1988, Toronto, Canada	902	68
Williams Tower, 1983, Houston, U.S.	901	64
Renaissance Tower, 1974, Dallas, U.S.	886**	56
900 N. Michigan Ave., 1989, Chicago, U.S.	871	66
NationsBank Corporate Center, 1992, Charlotte, U.S.	871	60
SunTrust Plaza, 1992, Atlanta, U.S.	871	60
BCE Place–Canada Trust Tower, 1990, Toronto	863	51

* under construction; to be completed in 2000. **Includes spires.

Other Notable Tall Buildings in North American Cities

Source: Council on Tall Buildings and Urban Habitat, Lehigh Univ.; Jeff Herzer and Marshall Gerometta, WTB/World's Tallest Buildings, http://www.worldstallest.com

Lists include some structures that do not have stories and are not technically considered "buildings." Height is generally measured from sidewalk to roof, including penthouse and tower if enclosed as integral part of structure; stories generally counted from street level. Asterisk (*) denotes building still under construction. Year in parentheses is date of completion. **Note: A # next to city name indicates city has one or more buildings that are not listed** because they already appear in the table of Tallest Buildings in the World. NA = not available or not applicable.

Atlanta, GA#

Building	Ht. (ft.)	Stories
One Atlantic Center (1988)	820	50
191 Peachtree (1991)	770	50
Westin Peachtree Plaza (1976)	723	73
Georgia Pacific Tower (1981)	697	51
Promenade II/A.T.& T. (1989)	691	40
Bellsouth (1980)	677	47
CLG Grand/Occidental Hotel (1992)	609	53
State of Georgia Tower (1967)	556	44
Marriot Marquis (1985)	554	52
Equitable Bldg (1967)	453	34
101 Marietta Tower (1976)	446	36
National Bank of Ga. (1961)	439	32
A.T.& T. Long Line Bldg. (1975)	433	NA
Bell South Enterprises (1990)	428	28
Atlanta Plaza I (1986)	425	32
Park Place (1986)	420	40
Club Tower Apts. (1989)	410	38
Peachtree Summit No. 1 (1975)	406	31
Coca Cola Headquarters Bldg. (1979)	403	26
Tower Place. (1974)	401	29

Baltimore, MD

Building	Ht. (ft.)	Stories
U. S. F. & G. Co. (1973)	529	40
Maryland National Bank (1929)	509	34
William Donald Schaefer Tower	493	29
Commerce Place (1992)	454	31
World Trade Center (1977)	405	32
100 E. Pratt St. (1992)	400	28

Boston, MA

Building	Ht. (ft.)	Stories
John Hancock Tower. (1976)	788	60
Prudential Tower (1964)	750	52
Federal Reserve Bldg. (1983)	604	32
Boston Company Bldg. (1970)	601	41
One International Place (1987)	600	46
First National Bank of Boston (1971)	591	37
One Financial Center (1984)	590	46
One Post Office Square (1981)	525	40
1 Federal St. (1975)	520	38
Exchange Place (1984)	510	39
Sixty State St. (1977)	509	38
1 Beacon St. (1972)	507	36
New England Merchant Bank (1968)	500	40
U.S. Custom House (1915)	496	32
John Hancock Bldg. (1973)	495	26
State St. Bank (1966)	477	34
99 High St. (1971)	455	32
100 Summer St. (1975)	450	33
Two International Place (1993)	433	35
McCormack Bldg.	401	22
Harbor Towers I, 85 E. India (1971)	400	40
Keystone Custodian Funds (1971)	400	32

Calgary, Alberta

Building	Ht. (ft.)	Stories
Petro Canada Tower (1984)	689	52
Bankers Hall East Tower (1989)	645	50
*Bankers Hall West Tower (2000)	645	50
Canterra Tower (1988)	580	46
First Canadian Centre (1983)	530	43
Calgary Eatons Centre	530	40
Western Canadian Place-N. Tower (1983)	507	41
Scotia Square (1975)	504	38
Nova Bldg., 801 7th Ave. SW (1982)	500	37
Petro-Canada Tower, E. Tower (1983)	469	33
Two Bow Valley Square (1974)	468	39
Home Oil Tower (1976)	463	34
Canada Trust (1991)	462	40
5th & 5th Bldg. (1980)	460	35
Shell Tower (1977)	460	34
Oxford Square South (1976)	449	33
Four Bow Valley Square (1982)	441	37
Esso Plaza I	435	34
Esso Plaza II	435	34
Cascade 300	432	31
T.D. Square.	421	33
Western Canadian Place-S. Tower (1983)	420	32
Family Life Bldg.	410	33
Pan Canadian Bldg., 150 9th Ave. SW (1982)	410	28
Norcen Tower (1976)	408	33
Alberta Stock Exchange	407	33

Charlotte, NC#

Building	Ht. (ft.)	Stories
One First Union Center (1988)	588	42
NCNB Plaza, 101 S. Tryon (1974)	503	40
Interstate Tower, 121 W. Trade St. (1990)	462	32
Interstate/Johnson Lane Bldg. (1997)	447	30
First Union Plaza (1971)	433	32
Wachovia Center, 400 S. Tryon (1974)	420	32

Chicago, IL#

Building	Ht. (ft.)	Stories
Water Tower Place (1976)	859	74
*Park Tower (2000)	853	67
First National Bank Plaza (1969)	850	60
3 First National Plaza (1981)	753	57
Chicago Title & Trust Center (1992)	742	51
Olympia Centre (1986)	725	63
IBM Bldg., 330 N. Wabash Ave. (1973)	695	52
Paine Webber Bldg. (1990)	680	50
One Magnificent Mile (1983)	673	58
R.R. Donnelley Center (1992)	668	50
Daley Center (1965)	662	31
Lake Point Tower (1968)	645	70
Leo Burnett (1989)	635	50
NBC Tower (1989)	627	34
Chicago Place (1991)	608	49
Board of Trade (incl. statue) (1930)	605	44
Prudential Bldg. (1955)	601	41
Heller International Tower (1992)	600	45
CNA Plaza (1972)	600	45
1000 Lake Shore Plaza Apts. (1964)	590	55
Marina City Apts. 1 (1964)	588	61
Marina City Apts. 2 (1964)	588	61
Citicorp Center (1985)	588	41
Mid Continental Plaza (1972)	582	50
North Pier Apartment Tower (1990)	581	61
Madison Plaza (1982)	580	45
Stone Container Bldg. (1983)	575	41
190 S. Lasalle St. (1986)	573	42
Onterie Center (1985)	570	57
Chicago Temple (1923)	568	21
919 N. Michigan Ave. (incl. beacon) (1929)	565	37
Huron Plaza Apts. (1983)	560	56
1 N. Franklin St. (1991)	560	38
Morton International Tower (1990)	560	36
Pittsfield (1927)	557	38
Civic Opera Bldg. (1929)	555	45
Newberry Plaza (1974)	553	53
Boulevard Towers South (1985)	553	44
30 N. LaSalle St. (1975)	553	43
Harbor Poin (1975)	550	54
One S. Wacker Dr. (1983)	550	40
LaSalle National Bank (1934)	535	44
Park Place Tower (1971)	531	56
One LaSalle St. (1930)	530	49
111 E. Chestnut St. (1972)	529	56
Chemical Plaza (1986)	526	37
River Plaza, Rush & Hubbard (1977)	524	56
The Parkshore (1991)	523	61
35 E. Wacker Dr. (1926)	523	40
United Insurance (1962)	522	41
North Harbor Tower (1991)	520	55
Chicago Merchantile Exchange, 10 S. Wacker Dr. (1987)	520	40
Chicago Merchantile Exchange, 30 S. Wacker Dr. (1983)	520	40
Kloczvunski Federal Bldg. (1976)	520	40
Lincoln Tower (1928)	519	42
Harris Bank II (1974)	518	35
One Financial Place (1985)	515	40
Intercontinental Hotel (1929)	513	42
Quaker Tower. (1987)	510	35
200 S. Wacker Dr. (1981)	505	38
Carbide & Carbon (1929)	503	37
1 Superior Place (1999)	502	52
Savings of America Tower (1991)	501	41
Columbus Plaza (1980)	500	49
Xerox Centre (1980)	500	40
Ontario Place	495	51
Presidential Towers, 555 W. Madison St. (1985)	495	49
Presidential Towers, 575 W. Madison St. (1985)	495	49
Presidential Towers, 605 W. Madison St. (1985)	495	49
Presidential Towers, 625 W. Madison St. (1985)	495	49
LaSalle-Wacker (1934)	491	41
American National Bank (1929)	479	40
Park Tower Condos (1974)	476	54
Bankers, 105 W. Adams St. (1927)	476	41
Brunswick Bldg. (1965)	475	37
Britannica Center (1924)	475	37
333 Wacker Dr. (1983)	475	36
American Furniture Mart (1926)	474	30
City Place (1990)	470	40
188 Randolph Tower (1925)	465	45
*The Bristol (2000)	465	42
Tribune Tower (1925)	462	36
The New York (1986)	461	50
Chicago Marriot (1978)	460	45
Olympic Towers (1984)	460	45
Swissotel (1989)	457	43
Equitable Life (1964)	457	35
Roanoke (1915)	452	37
Gateway Center III (1972)	450	35

Cincinnati, OH

Building	Ht. (ft.)	Stories
Carew Tower (1930)	574	48
Central Trust Tower (1913)	495	28
312 Walnut St. (1990)	468	36
Fifth Third Center (1970)	460	31
Atrium Two, 221 E. 4th St. (1984)	428	30
Dubois Tower (1969)	423	32
Chemed Center (1990)	410	32
Cincinnati Commerce Center (1984)	402	29

Cleveland, OH#

Building	Ht. (ft.)	Stories
Terminal Tower (1930)	708	52
BP America (1985)	658	46
Erieview Plaza Tower (1964)	529	40
One Cleveland Cente (1983)	450	31
Bank One Center (1991)	446	38
*Federal Courthouse (2001)	430	24
Justice Center, 1250 Ontario (1976)	420	26
Federal Bldg. (1967)	419	32
National City Center, 1900 E. 9th St. (1980)	410	35

Columbus, OH

Building	Ht. (ft.)	Stories
State Office Tower(1973)	624	41
Leveque-Lincoln Tower (1930)	555	47
Ohio Bureau of Welfare & Commerce (1990)	530	33
Huntington Center	512	37
Verne-Riffe State Office Tower (1988)	503	33
One Nationwide Plaza (1976)	485	40
Franklin County Courthouse	464	27
One Riverside Plaza	456	31
Borden Bldg. (1974)	438	34
Three Nationwide Plaza (1989)	408	29

Dallas, TX#

Building	Ht. (ft.)	Stories
Bank One Center (1987)	787	60
Texas Commerce Plaza (1987)	738	55
Fountain Place (1986)	720	62
Trammel Crow Tower (1984)	686	50
First City Center (1983)	655	50
Thanksgiving Tower (1982)	645	50
Energy Plaza(1983)	629	49
Elm Place (1965)	625	52
Republic Center Tower II (1964)	598	50
One Bell Plaza (1984)	58	37
One Lincoln Plaza (1984)	579	45
Cityplace Center East (1989)	560	42
Reunion Tower (1976)	560	
Adam's Mark Hotel Center (1959)	550	42
Maxus Energy (198)	550	34
21 Bryan St.(1973)	512	40
Olympia & York Tower (1982)	483	36
San Jacinto Tower (1982)	456	33
Republic Center Tower I (1954)	452	36
Renaissance Hotel (1983)	451	29
Adam's Mark Hotel North Tower (198)	448	31
One Dallas Centre (1979)	448	30
One Main Place (1968)	445	34
16 Pacific Bldg. (1964)	434	31
Merchantile National Bank Bldg. (1937)	430	31
Magnolia Bldg. (1923)	430	27
Fidelity Union Tower (1959)	400	33
Mart Hotel	400	29

Dayton, OH

Building	Ht. (ft.)	Stories
Kettering Tower (1970)	405	30
Dayton Arcade Center (1989)	400	20

Denver, CO

Building	Ht. (ft.)	Stories
Republic Plaza (1984)	741	56
US West Tower (1982)	709	54
One Norwest Center (1983)	698	52
1999 Broadway (1985)	544	43
MCI Tower (1981)	522	42
Qwest Tower (1978)	507	40
Amoco Bldg. (1980)	448	36

Building	Ht. (ft.)	Stories
17th St. Plaza (1982)	438	32
First Interstate Tower North (1974)	434	32
Brooks Towers, 1020 15th St. (1968)	420	42
One Denver Place (1981)	416	34
Tabor Center (1984)	408	32
Manville Plaza (1989)	404	29

Detroit, MI

Building	Ht. (ft.)	Stories
Westin Hotel, Renaissance Center I (1977)	739	73
One Detroit Center (1990)	620	45
Penobscot Bldg. (1928)	557	47
Renaissance Center II (1976)	534	39
Renaissance Center III (1976)	534	39
Renaissance Center IV (1976)	534	39
Renaissance Center V (1976)	534	39
Guardian (1928)	485	40
Book Tower (1925)	472	35
Madden Bldg. (1988)	470	29
Prudential (1975)	448	32
Cadillac Tower (1928)	437	40
David Stott Bldg. (1928)	436	38
ANR Bldg. (1962)	430	30
Fisher Bldg. (1928)	420	28
Town Center (1975)	405	32

Dunwoody, GA

Building	Ht. (ft.)	Stories
Concourse Tower #5 (1988)	570	32
Concourse Tower #6 (1991)	553	32
Ravinia #3 (1991)	444	34

Edmonton, Alberta

Building	Ht. (ft.)	Stories
Manulife Place, 10170-101 St. (1983)	479	39
Royal Trust Tower (1973)	476	30
AGT Tower, 10020-100 St. (1971)	441	34
Canada Trust Tower (1982)	440	31
Commerce Place (1990)	409	30

Fort Worth, TX

Building	Ht. (ft.)	Stories
Burnett Plaza (1983)	558	40
Center Tower II (1984)	546	38
UPR Plaza (1982)	525	40
Chase Texas Tower (1982)	475	33
Bank One Tower (1974)	454	36

Hartford, CT

Building	Ht. (ft.)	Stories
City Place (1983)	535	38
Travelers Ins. (1919)	527	34
Goodwin Square (1990)	522	30
Hartford Plaza (1967)	420	22

Honolulu, HI

Building	Ht. (ft.)	Stories
First Hawaiian Bank Bldg. (1996)	435	30
Waterfront Towers (1990)	400	46
Naroru Tower (1991)	400	45
Imperial Plaza (1992)	400	40

Houston, TX#

Building	Ht. (ft.)	Stories
NationsBank Center (1983)	780	56
Texaco Heritage Plaza (1987)	762	53
Southwest Bank of Texas (1980)	748	55
Houston Industries Plaza (1974)	741	53
1600 Smith St. (1984)	732	55
Chevron Tower (1982)	725	52
One Shell Plaza (1970)	714	50
Enron Bldg. (1983)	691	50
Capital National Bank Plaza (1980)	685	50
One Houston Center (1978)	678	47
First City Tower (1984)	662	47
San Felipe Plaza (1984)	625	45
Exxon Bldg. (1962)	606	44
America Tower, (1983)	590	42
Two Houston Center (1974)	579	40
Marathon Oil Tower (1983)	562	41
1415 Louisiana Tower (1983)	550	44
Kellogg Tower (1973)	550	40
Pennzoil Bldg. 1 (1975)	523	36
Pennzoil Bldg. 2, 700 Milam St. (1975)	523	36
Two Allen Center, 1200 Smith St. (1978)	521	36
1201 Louisiana Bldg. (1971)	518	35
The Huntington (1982)	503	34
Tenneco Bldg. (1962)	502	33
Conoco Tower (1973)	465	32
One Allen Center (1974)	452	34
Summit Tower West (1979)	441	31
Coastal Tower (1978)	441	31
Four Leafs Towers I (1982)	439	40
Four Leafs Towers I (1982)	439	40

Building	Ht. (ft.)	Stories
Phoenix Tower (1984)	434	34
Chevron Bldg. (1929)	428	37
The Spires (1984)	426	41
Central Tower (1983)	420	30
First National City Bank (1960)	410	32
Houston Lighting & Power (1968)	410	27
Neils Esperson Bldg., 802 Travis St. (1927)	409	31
Hyatt Regency (1972)	401	34

Indianapolis, IN

Building	Ht. (ft.)	Stories
Bank One Tower (1990)	820	48
American United Life Ins. (1981)	533	37
NBD Bank Tower (1969)	504	37
Market Tower (1988)	415	32
300 N. Meridian Bldg. (1988)	408	28

Jacksonville, FL

Building	Ht. (ft.)	Stories
Barnett Tower (1990)	617	43
Ind. Life & Accident Ins. Co (1975)	535	37
Southern Bell (1983)	435	27
Gulf Life (1967)	432	28

Kansas City, MO

Building	Ht. (ft.)	Stories
One Kansas City Place (1988)	626	42
A.T.& T. Town Pavillion (1986)	590	38
Hyatt Regency (1980)	504	40
Power & Light Bldg. (1931)	476	32
City Hall, (1937)	443	29
Fidelity Bank and Trust Bldg. (1931)	426	35
1201 Walnut (1991)	425	30
Federal Office Bldg. (1962)	413	35
Commerce Tower (1965)	407	32
City Center Square (1977)	402	30

Las Vegas, NV

Building	Ht. (ft.)	Stories
Stratosphere Tower (1996)	1149	NA
New York, New York Hotel and Casino (1997)	525	48
Bellagio Hotel and Casino (1998)	508	36
Venetian Hotel and Casino (1999)	480	35
Mandalay Bay Hotel and Casino (1999)	460	43
Rio Masquerade Tower (1996)	450	40
Paris Hotel and Casino (1999)	440	34
Fitzgeralds Hotel	400	33

Little Rock, AR

Building	Ht. (ft.)	Stories
TCBY Towers (1986)	546	40
First Commercial Bank (1975)	454	30

Los Angeles, CA#

Building	Ht. (ft.)	Stories
First Interstate Bank (1974)	858	62
Two California Plaza (1992)	750	52
So. Cal. Gas Center (1991)	749	52
333 South Hope Bldg. (1975)	743	55
777 Tower (1990)	725	53
Wells Fargo Tower (1983)	723	54
Sanwa Bank Plaza (1989)	717	52
Atlantic Richfield Tower (1971)	699	52
Bank of America Tower (1971)	699	52
444 S. Flower St. (1979)	625	48
A.T.& T. Bldg. (1969)	620	42
One California Plaza (1985)	578	42
Century Plaza Tower 1(1973)	571	44
Century Plaza Tower 2 (1973)	571	44
IBM Tower (1983)	560	44
Citycorp Plaza (1988)	534	41
(1999) Ave. of the Stars (1989)	533	38
Manulife Tower (1990)	517	37
Union Bank Square (1968)	516	40
MCA-Getty	506	36
WTC Bldg. (1987)	496	36
Fox Plaza (1987)	492	36
ARCO Center (1989)	462	33
Equitable Life (1969)	454	34
City Hall (1927)	454	28
Transamerica Center (1965)	452	32
Mutual Life Bldg. (1970)	435	31
Warner Center Plaza III	415	25
Macy's Plaza (1973)	414	33

Louisville, KY

Building	Ht. (ft.)	Stories
Aegon Center (1992)	549	35
National City Tower (1972)	512	40
Citizens Plaza (1971)	420	30
Humana Center (1985)	417	28

Mexico City, Mexico

Building	Ht. (ft.)	Stories
*Torre Mayor	738	55

Building	Ht. (ft.)	Stories
Petrolaos Mexicanos (1984)	702	52
Torre Latino Americana (1952)	597	45
Hotel de Mexico (1972)	573	48
Los Arcos Bosques I (1997)	525	35
Nonoalco Tlatelolco Tower (1962)	417	25

Miami, FL

Building	Ht. (ft.)	Stories
First Union Financial Center (1983)	738	55
NationsBank Tower (1987)	625	47
Metro-Dade Adminstration Bldg. (1985)	510	30
Santa Maria (1997)	508	54
Florida National Tower (1986)	484	35
Portofino Tower (1997)	464	44
One Biscayne Corp. (1972)	456	40
Barnett Tower (1986)	450	33
Courthouse Center (1986)	405	30

Milwaukee, WI

Building	Ht. (ft.)	Stories
Firstar Center (1971)	625	44
Faison Bldg. (1989)	549	37
Milwaukee Center (1987)	426	29
411 Bldg. (1983)	408	30

Minneapolis, MN

Building	Ht. (ft.)	Stories
IDS Center (1973)	775	57
First Bank Place (1992)	774	53
Norwest Center (1988)	773	57
Piper Jaffray Tower (1984)	627	42
Multifoods Tower (1983)	608	51
Dain Bosworth Plaza (1994)	539	40
Pillsbury Center (1981)	530	40
150 South Fifth (1987)	498	36
Anderson Consulting Center (1987)	496	32
*Target Plaza South (2001)	491	33
Plaza VII, 45 S. 7th (1987)	475	36
*US Bank-Piper Jaffray Center (2000)	469	30
AT&T Tower (1991)	464	34
Foshay Tower (1929)	447	32
NW Bell Telephone (1931)	416	26
Hennepin Co. Government Center (1973)	403	24

Montreal, Quebec

Building	Ht. (ft.)	Stories
1000 Rue de la Gauchetiere (1991)	669	51
Marathon (IBM) (1989)	640	47
Place Victoria (1963)	624	47
Place Villa Marie (1962)	616	42
Canadian Imperial Bank (1962)	604	43
Tour McGill College (1992)	519	38
Le Complexe Desjardins Sud (1975)	498	40
Les Cooperants (1987)	479	34
Holiday Inn (1977)	450	38
Place Montreal Trust (1988)	449	32
C.I.L. House (1962)	429	32
Le Complexe Desjardins Est (1975)	428	32
La Tour Laurier	425	36
Port Royal Apts. (1964)	424	33
Chateau Champlain Hotel (1967)	420	38
Tour Terminal (1966)	400	30

Nashville, TN

Building	Ht. (ft.)	Stories
South Central Bell Bldg. (1994)	617	33
3rd National Financial Center (1985)	490	30
National Life & Accident (1970)	452	31
Nashville Life & Casualty	409	30
City Center (1987)	402	27

Newark, NJ

Building	Ht. (ft.)	Stories
Midatlantic National Bank	465	36
Raymond-Commerce	448	36
Park Plaza Bldg.	400	26

New Orleans, LA

Building	Ht. (ft.)	Stories
One Shell Square (1972)	697	51
Place St. Charles (1985)	645	53
Plaza Tower (1969)	531	45
Energy Centre (1984)	530	39
LL&E Tower (1987)	481	36
Sheraton Hotel (1985)	478	47
Marriott Hotel (1972)	450	42
Texaco Bldg. (1983)	442	33
Canal Place One (1979)	439	32
1010 Common (1971)	438	31
World Trade Center (1965)	407	33

New York, NY#

Building	Ht. (ft.)	Stories
Condé Nast Bldg. (1999)	866	48
*Trump World Tower (2001)	861	72
G. E. Bldg. (1933)	850	70
Cityspire (1989)	814	72
Chase Manhattan Plaza (1960)	813	60
MetLife Bldg. (1963)	808	59
Woolworth (1913)	792	57
1 Worldwide Plaza (1989)	778	47
Carnegie Tower (1991)	757	60
Equitable Center West (1985)	752	51
One Penn Plaza (1972)	750	57
1251 Ave. of Americas (1971)	750	54
J.P. Morgan Headquarters(1989)	745	50
1 Liberty Plaza (1973)	743	54
Citibank (1931)	741	57
World Financial Center Tower C (1988)	739	53
One Astor Plaza (1969)	730	54
Metropolitan Tower (1985)	716	68
Chase World Headquarters (1960)	707	52
General Motors (1968)	705	50
Metropolitan Life Tower (1909)	700	50
500 5th Ave. (1931)	697	60
Americas Tower (1992)	692	48
Solow Bldg. (1974)	689	50
Marine Midland Bank (1966)	688	52
Chemical Bank NY Trust. (1963)	687	50
55 Water St. (1972)	687	53
1585 Broadway (1989)	685	42
Four Seasons Hotel (1993)	682	52
Trump International Hotel & Tower (1970)	679	52
Bertelsmann Building (1990)	676	42
McGraw Hill (1972)	674	51
Lincoln (1930)	673	53
Citicorp, Queens (1990)	673	50
Paramount Plaza (1970)	670	48
Trump Tower, 725 5th Ave. (1982)	664	58
Irving Trust (1932)	654	50
599 Lexington Ave. (1986)	653	51
Museum Tower Apts. (1985)	650	58
712 5th Ave. (1990)	650	53
American Brands (1967)	648	47
Sony Bldg. (1983)	648	37
World Financial Center Tower B (1986)	645	51
345 Park Ave. (1968)	634	44
1 New York Plaza (1968)	630	50
Grace Plaza (1974)	630	50
Home Insurance Co. (1966)	630	44
N.Y. Telephone (1970)	630	40
Central Park Place (1988)	628	56
1 Dag Hammarskjold Plaza (1972)	628	49
888 7th Ave. (1971)	628	45
Chanin Bldg. (1929)	625	56
Burlington House (1969)	625	50
Waldorf Astoria (1931)	625	47
Trump Palace (1991)	623	55
Olympic Tower (1976)	620	51
10 E. 40th St. (1929)	620	48
101 Park Ave. (1982)	618	50
RCA Victor Bldg. (1930)	616	50
750 7th Ave. (1989)	615	35
New York Life (1928)	615	33
Tower 49 (1985)	614	44
Penney Bldg. (1964)	609	46
IBM, 590 Madison Ave. (1983)	603	41
3 Lincoln Center (1993)	595	60
Celanese Bldg. (1973)	592	45
Rihga Royal Hotel, 151 W. 54th St. (1990)	590	54
U. S. Court House (1927)	590	37
The Millennium Hilton Hotel (1992)	588	58
Time & Life (1959)	587	48
Jacob K. Javits Federal Bldg. (1967)	587	41
MillenniumTower (1995)	580	54
Stevens Tower (1971)	580	42
Cooper Bregstein (1969)	580	40
Municipal Bldg. (1914)	580	34
520 Madison Ave. (1981)	577	43
1 Madison Square Plaza (1973)	576	42
Park Ave. Plaza (1981)	575	44
World Financial Center Tower A (1986)	575	42
One Financial Square (1987)	575	37
Marriot Marquis Times Square (1985)	574	50
Westavco Bldg. (1967)	574	42
1166 Ave. of Am. (1974)	572	44
Socony Mobil (1956)	572	42
780 3rd Ave. (1983)	570	49
7 World Trade Center (1985)	570	47
Sperry Rand (1963)	570	43
600 3rd Ave. (1971)	570	42
450 Lexington Ave. (1991)	568	38
Paramount Tower (1998)	567	51
1 Bankers Trust Plaza (1974)	565	40
Helmsley Bldg. (1928)	565	35

Building	Ht. (ft.)	Stories
New York Palace Hote (1980)	563	51
30 Broad St. (1932)	562	48
Park Ave. Tower (1986)	561	36
Sherry-Netherland (1927)	560	40
Swiss Bank Tower (1990)	560	36
Continental Can (1962)	557	39
3 Park Ave. (1975)	556	42
Continental Corp. (1983)	555	41
Sperry & Hutchinson (1964)	555	41
*Madison Belvedere (1999)	554	48
Interchem (1970)	552	45
919 3rd Ave. (1970)	550	47
Bell Atlantic. (1974)	550	29
Burroughs Bldg. (1963)	550	44
2 Grand Central Tower (1982)	550	44
Bankers Trust (1963)	547	41
Transportation Bldg. (1928)	546	44
Equitable (1915)	545	36
Galleria (1975)	544	56
17 State St. (1988)	542	41
Blue Cross Blue Shield (1973)	542	38
New York Telephone (1976)	540	42
Paine Webber Bldg. (1959)	540	42
Ritz Tower (1925)	540	41
Bankers Trust (1912)	540	39
Lefcourt Colonial Bldg. (1920)	537	45
1700 Broadway (1969)	533	42
Downtown Athletic Club (1930)	530	45
Nelson Towers (1931)	525	46
Hotel Pierre (1928)	525	44
810 7th Ave. (1970)	525	42
767 3rd Ave. (1980)	525	39
House of Seagram (1958)	525	38
Random House (1969)	522	40
Du Mont Bldg. (1931)	520	42
North American Plywood (1972)	520	41
26 Broadway (1922)	520	31
Newsweek Bldg. (1931)	518	42
964 Third Ave. (1969)	518	39
Sterling Drug Bldg. (1964)	515	41
Citibank (1961)	515	41
Navarre (1930)	513	44
Bank Of New York (1927)	513	31
140 Broadway Reality Corp. (1950)	512	41
Williamsburg Savings Bank (1929)	512	42
International, Rockefeller Center (1935)	512	41
ITT-American (1967)	512	40
1155 Ave. of Am. (1984)	511	40
The Sheffield Apts. (1978)	505	50
United Nations Secretariat Bldg. (1950)	505	39
1 UN Plaza (1976)	505	39
2 UN Plaza (1983)	505	39
The Corinthian (1987)	504	55
2 New York Plaza (1970)	504	40
22 E. 40th St. (1931)	503	43
60 Broad St. (1962)	503	39
Lefcourt National Bldg. (1928)	503	37
Sheraton Centre (1962)	501	51
World Apparel Center (1968)	501	42
Crowne Plaza Hotel (1989)	500	46

Oklahoma City, OK

Building	Ht. (ft.)	Stories
Liberty Tower (1971)	500	36
First National Bank (1974)	493	33
City Place (1985)	440	32
First Oklahoma Tower (1982)	425	31

Omaha, NE

Building	Ht. (ft.)	Stories
*One First National Center (2002)	638	40
Woodmen Tower (1969)	469	30
Enron Bldg. (1960)	400	18

Orlando, FL

Building	Ht. (ft.)	Stories
Sun Bank Center Tower (1988)	441	31
Orange County Courthouse (1997)	416	24
Barnett Bank Center (1988)	409	28

Philadelphia, PA#

Building	Ht. (ft.)	Stories
Two Liberty Place (1989)	848	58
Mellon Bank Center (1990)	792	54
Bell-Atlantic Tower (1991)	739	53
Blue Cross Tower (1990)	700	50
City Hall (incl. statue)(1901)	585	9
Commerce Square #1 (1990)	572	40
Commerce Square #2(1992)	572	40
1818 Market St. (1974)	500	40
Philadelphia Saving Fund Society (1932)	492	39
Provident Mutual Life (1983)	491	40
Centre Square (1973)	490	38

Building	Ht. (ft.)	Stories
5 Penn Center (1970)	488	36
Industrial Valley Bldg. (1969)	482	32
Philadelphia National Bank (1930)	475	25
Two Mellon Plaza (1930)	450	30
Two Logan Square (1988)	435	34
2000 Market St. (1973)	435	29
One Reading Center	417	32
Centre Square (1973)	416	32
Two Girard Plaza (1930)	412	30
Fidelity Bank (1927)	405	30
Lewis Tower (1929)	400	33
One Logan Square (1982)	400	32

Phoenix, AZ

Building	Ht. (ft.)	Stories
Bank One Center (1972)	483	40
Bank of America Tower (1976)	407	31

Pittsburgh, PA

Building	Ht. (ft.)	Stories
USX Tower (1970)	841	64
One Mellon Bank Center (1983)	725	54
One PPG Place (1984)	635	40
Fifth Ave. Place (1987)	616	32
One Oxford Centre (1982)	615	46
Gulf Tower (1932)	582	44
Univ. of Pittsburgh Cath. of Learning (1930)	535	42
Mellon Bank (1951)	520	41
1 Oliver Plaza (1968)	511	40
Grant Bldg. (1928)	485	40
Koppers (1929)	475	34
2 Oliver Plaza	469	32
Equibank (1975)	445	34
CNG Tower (1987)	430	32
Pittsburgh National Bank (1972)	424	30
Alcoa Bldg. (1953)	410	30

Portland, OR

Building	Ht. (ft.)	Stories
U.S. West Tower (1973)	546	40
U. S. Bancorp Tower (1983)	536	42
Koin Tower Plaza	509	35

Providence, RI

Building	Ht. (ft.)	Stories
Fleet Bank Bldg. (1927)	428	26
BankBoston Tower (1973)	410	28

Richmond, VA

Building	Ht. (ft.)	Stories
James Monroe Bldg. (1981)	450	29
United Virginia Bank (1984)	400	24

St. Louis, MO

Building	Ht. (ft.)	Stories
Gateway Arch (1965)	630	NA
Metropolitan Square Tower (1988)	593	42
One Bell Center, 900 Pine St. (1984)	588	44
*Thos. F. Eagleton Federal Courthouse (1999)	557	27
Mercantile Center Tower (1975)	485	35
NationsBank Plaza (1982)	420	31
Laclede Gas Bldg. (1970)	400	31

St. Paul, MN

Building	Ht. (ft.)	Stories
Minnesota World Trade Center (1987)	471	36
Galtier Plaza Jackson Tower (1986)	440	46
First National Bank	417	32

San Antonio, TX

Building	Ht. (ft.)	Stories
Tower of the Americas (1968)	622	
Marriott Rivercenter (1988)	546	38
Weston Centre (1988)	444	32
Tower Life, 310 S. St. Mary's (1929)	404	30

San Diego, CA

Building	Ht. (ft.)	Stories
One American Plaza (1991)	500	34
Symphony Tower (1989)	499	34
Hyatt Regency (1992)	497	40
Emerald-Shapley Center (1991)	450	30
One Harbor Drive (1992)	424	41

San Francisco, CA

Building	Ht. (ft.)	Stories
Transamerica Pyramid (1972)	853	48
Bank of America (1969)	779	52
First Interstate Center (1986)	695	48
101 California St. (1986)	600	48
5 Fremont Center (1983)	600	43
Embarcadero Center No. 4 (1984)	570	45
Embarcadero Center No. 1 (1970)	569	45
One Market Plaza (1976)	565	43
Wells Fargo (1966)	561	43
Standard Oil (1975)	551	39

Building	Ht. (ft.)	Stories
One Sansome-Citicorp (1984)	550	39
Shaklee Bldg. (1979)	537	38
Crocker National Bank (1967)	532	38
Aetna Life (1969)	529	38
First & Market Bldg. (1973)	529	38
Metropolitan Life (1973)	524	38
333 Bush St. (1986)	495	36
Hilton Hotel (1971)	493	46
Pacific Gas & Electric (1970)	492	34
Union Bank (1972)	490	37
Pacific Insurance (1972)	476	34
Bechtel Bldg., Beale St. (1967)	475	33
333 Market Bldg. (1979)	474	33
Hartford Bldg. (1965)	465	33
Mutual Benefit Life (1969)	438	32
Marriott Hotel (1989)	436	38
Russ Bldg. (1928)	435	31
Pacific Bell Headquarters (1925)	435	26
Pacific Gateway (1983)	416	30
Embarcadero Center No. 2 (1974)	412	31
Embarcadero Center No. 3 (1976)	412	31
595 Market Bldg. (1977)	410	31
123 Mission Bldg.	406	28
Embarcadero Center West (1988)	405	33
101 Montgomery St. (1983)	405	29

Seattle, WA#

Building	Ht. (ft.)	Stories
Two Union Square (1989)	743	56
Washington Mutual Tower (1988)	730	55
A.T.& T. Gateway Tower (1990)	722	62
1001 4th Place (1969)	609	50
Space Needle (1962)	605	NA
Pacific First Center (1989)	580	44
First Interstate Center (1983)	574	48
Seafirst 5th Ave. Plaza. (1981)	543	42
Security Pacific Bank Tower (1977)	514	42
Smith Tower (1914)	500	42
Key Tower (1986)	493	40
Federal Office Bldg. (1973)	487	37
Pacific Northwest Bell (1976)	466	33
One Union Square (1981)	456	38
1111 3rd Ave. (1980)	454	35
Washington Plaza, 2nd Tower	448	44
Westin Bldg. (1981)	409	34

Tampa, FL

Building	Ht. (ft.)	Stories
100 N. Tampa (1992)	579	42
NationsBank Plaza (1986)	577	36
Tampa City Center (1981)	537	39
Landmark Center (1992)	525	36
Park Tower (1973)	458	36
NCNB Plaza (1988)	454	33

Toledo, OH

Building	Ht. (ft.)	Stories
Owens-Illinois Corp. (1962)	411	32
Owens-Corning Tower (1970)	400	30

Toronto, Ontario#

Building	Ht. (ft.)	Stories
CN Tower (1976)	1815	NA
BCE Place, Canada Trust Tower (1990)	863	51

Building	Ht. (ft.)	Stories
Commerce Court West (1973)	784	57
TD Centre - Toronto Dominion Bank Tower (1967)	740	56
BCE Place, Bay-Wellington Tower (1991)	705	43
TD Centre - Royal Trust Tower (1969)	600	43
Royal Bank Plaza (1976)	567	41
Manulife Centre (1975)	545	51
TD Centre - Aetna Tower (1985)	515	37
Eaton Centre (1990)	494	34
Workmans Compensation Bldg. (1995)	487	33
Two Bloor West (1974)	486	34
Exchange Tower (1981)	480	36
CIBC-Commerce Court North (1930)	477	34
Simpson Tower (1968)	473	33
Cadillac-Fairview (1982)	465	36
Palace Point (1991)	455	46
Palace Pier (1978)	453	46
Continental Bank (1980)	450	35
Sheraton Centre (1972)	443	43
Hudson's Bay Centre (1974)	442	35
Two Bloor East (1974)	442	34
Royal York Hotel (1929)	439	26
Ernst & Younge Tower (1990)	438	31
Old Toronto Exchange Bldg. (1990)	436	31
Leaside Towers (2 bldgs.) (1970)	423	44
Commercial Union Tower (1974)	420	32
Metro Hall (1991)	420	27
Maple Leaf Mills	419	30
Plaza 2 Hotel	415	41
Sun Life Bldg., 150 King St. (1981)	410	27
Young-Eglington Centre-Triathlon Tower (1975)	408	30
The Fairbanks	405	42

Tulsa, OK

Building	Ht. (ft.)	Stories
Bank of Oklahoma Tower (1975)	667	52
Cityplex Tower	648	60
1st National Bank (1973)	516	41
Mid-Continent Tower	513	36
4th National Bank (1966)	412	33
320 South Boston	400	24

Vancouver, British Columbia

Building	Ht. (ft.)	Stories
Royal Centre Tower (1973)	466	36
Scotia Bank Tower Vancouver Centre (1977)	462	36
Canada Trust Tower	454	35
Scotia Bank Tower (1974)	451	36
*Wall Centre Garden Hotel (2000)	450	46
Bentall IV, 1055 Dunsmir (1981)	450	35
Park Place, 666 Burrard (1984)	450	35
T-D Bank Tower, 700 W. Georgia (1971)	440	30
200 Granville Square (1973)	438	28
Harbour Centre, 555 W. Hastings (1977)	428	32

Winnipeg, Manitoba

Building	Ht. (ft.)	Stories
Toronto Dominion Centre (1990)	420	33
Richardson Bldg., 1 Lombard Place (1969)	406	34

Winston-Salem, NC

Building	Ht. (ft.)	Stories
Wachovia Bldg. (1995)	460	28
Wachovia Bldg. (1965)	410	27

MILLENNIUM FACT BOX

The Ten Most Significant Buildings of the Century

By Richard A. Meier

Richard A. Meier is an internationally renowned architect who has received numerous national Honor Awards from the American Institute of Architects, as well as the Pritzker Prize, considered the field's highest honor.

As we anticipate the advent of a new millennium and century, The World Almanac asked Meier what he considered to be the ten most architecturally significant buildings of the 20th century. Here is his list, from newest to oldest.

THE GETTY CENTER, Los Angeles, CA (1984-98)—
Richard Meier & Partners
NEUE STAATSGALERIE, Stuttgart, Germany (1977-84)—
James Stirling
SALK INSTITUTE, La Jolla, CA (1965)—
Louis I. Kahn
DOMINICAN MONASTERY OF LA TOURETTE, Eveux,
France (1960)—Le Corbusier
EAMES HOUSE, Pacific Palisades, CA (1949)—
Charles Eames

FALLINGWATER, Bear Run, PA (1936)—
Frank Lloyd Wright
VILLA SAVOYE, Poissy, France (1928-31)—
Le Corbusier
LOVELL HEALTH HOUSE, Los Angeles, CA (1927-29)—
Richard Neutra
FRIEDRICHSTRASSE OFFICE BUILDING (1919)[1]; **GLASS
SKYSCRAPER** (1921)[1]—Mies van der Rohe
DOMINO HOUSE SKELETON (1914)—Le Corbusier[2]

(1) Dates are for the designs. They were never built.
(2) A drawing that illustrates his principles for the separation of the structure and the outside wall of the building so that the structure is independent of the exterior wall.

Some Other Notable Tall Buildings in North American Cities

Building	City	Ht. (ft.)	Stories	Building	City	Ht. (ft.)	Stories
Erastus Corning II Tower(1973)	Albany, NY	589	44	Industrial Trust Bldg. (1927)	Providence, RI	428	26
Vehicle Assembly Bldg	Cape Canaveral, FL	552	40	First Natl. Bank	Mobile, AL	420	33
*Dataflux Tower (2000)	Monterrey, Mexico	549	NA	L.D.S. Church Office Bldg.	Salt Lake City, UT	420	30
Marine Midland Ctr. (1970)	Buffalo, NY	529	40	Century 21	Hamilton, Ont.	418	43
State Capitol (1932)	Baton Rouge, LA	460	34	Oakbrook Terrace Tower	Oakbrook, IL.	418	31
South Trust Tower (1986)	Birmingham, AL	454	34	Complexe G (1972)	Quebec City, Que.	415	33
Xerox Tower (1967)	Rochester, NY	443	30	1 Shoreline Plaza, S. Tower	Corpus Christi, TX	411	28
One Summit Square (1981)	Fort Wayne, IN	442	27	Silver Legacy Hotel & Casino	Reno, NV	410	38
State Capitol	Lincoln, NE	432	22	Financial Ctr. (1986)	Lexington, KY	410	30
2 Hanover Sq. (1991)	Raleigh, NC	431	29	Ordway Bldg. (1985)	Oakland, CA	404	28
Union Planters Bank (1965)	Memphis, TN	430	37	Wells Fargo Center (1991)	Sacramento, CA	423	32
Taj Mahal	Atlantic City, NJ	429	51	United American Bank (1977)	Knoxville, TN	400	30

Notable Bridges in America

Source: Federal Highway Administration, Bridge Division, U.S. Dept. of Transportation; World Almanac research

Asterisk (*) designates railroad bridge. Double asterisk (**) designates bridge under construction.
Span of a bridge is the distance between its supports.

Suspension

Year	Bridge	Location	Main span (ft.)
1964	Verrazano-Narrows	New York, NY	4,260
1937	Golden Gate	San Fran. Bay, CA	4,200
1957	Mackinac Straits	Sts. of Mackinac, MI	3,800
1931	Geo. Washington	Hudson R., NY–NJ	3,500
1950	Tacoma Narrows	Tacoma, WA	2,800
1936	San. Fran.-Oakland Bay[1]	San Fran. Bay, CA	2,310
1939	Bronx-Whitestone	East R., NY	2,300
1970	Pierre Laporte	Quebec, Canada	2,190
1960	Seaway Skyway	Ogdensburg, NY	2,150
1968	Del. Memorial	Wilmington, DE	2,150
1957	Walt Whitman	Philadelphia, PA	2,000
1929	Ambassador	Detroit, MI–Can.	1,850
1917	Quebec	Quebec, Canada	1,800
1961	Throgs Neck	Long Is. Sound, NY	1,800
1926	Benjamin Franklin	Philadelphia, PA	1,750
1924	Bear Mt.	Hudson R., NY	1,632
1903	Williamsburg	East R., NY	1,600
1952	Wm. Preston La. Mem.[2]	Sandy Point, MD	1,600
1952	Chesapeake Bay	Sandy Point, MD	1,600
1969	Newport	Narragansett Bay, RI	1,600
1883	Brooklyn	East R., NY	1,595
1939	Lion's Gate	Burrard Inlet, BC	1,550
1930	Mid-Hudson	Poughkeepsie, NY	1,500
1963	Vincent Thomas	L. A. Harbor, CA	1,500
1909	Manhattan	East R., NY	1,470
1955	MacDonald Bridge	Halifax, Nova Scotia	1,447
1970	A. Murray Mackay	Halifax, Nova Scotia	1,400
1936	Triborough	East R., NY	1,380
1931	St. Johns	Portland, OR	1,207
1929	Mount Hope	RI	1,200
1960	Ogdensburg	St. Lawrence R., NY	1,150
1965	Bidwell Bar Bridge	Oroville, CA	1,108
1964	Middle Fork Feather	CA	1,105
1939	Deer Isle	ME	1,080
1931	Simon Kenton Memorial	Ohio R., Maysville, KY	1,060
1936	Ile d'Orleans	St. Lawrence R., Quebec	1,059
1867	John A. Roebling	Ohio R., KY	1,057
1971	Dent	Clearwater Co., ID	1,050
1900	Miampimi	Mexico	1,030
1849	Wheeling	Ohio R., WV	1,010

Cantilever

Year	Bridge	Location	Main span (ft.)
1917	Québec Bridge	St. Lawrence R., Quebec	1,800
1988	Greater New Orleans Bridge	Mississippi R., New Orleans, LA	1,575
1995	Gramercy Bridge	Mississippi R., Gramercy, LA	1,460
1936	Transbay	San Fran. Bay, CA	1,400
1968	Baton Rouge Bridge	Mississippi R., Baton Rouge, LA	1,235
1955	Tappan Zee	Hudson R., NY	1,212
1930	Lewis and Clark	Longview, WA–OR	1,200
1976	Patapsco River	Baltimore , MD	1,200
1909	Queensboro	East R., NY	1,182
1927	Carquinez Strait	CA	1,100
1958	Parallel Span	CA	1,100
1930	Jacques Cartier	Montreal, Quebec	1,097
1968	Isaiah D. Hart	Jacksonville, FL	1,088
1956	Richmond[3]	San Fran. Bay, CA	1,070
1929	Grace Memorial	Charleston, SC	1,050
1980	Newburgh-Beacon	Hudson R., NY	1,000

Year	Bridge	Location	Main span (ft.)
1949	Martin Luther King	St. Louis, MO	963
1975	Caruthersville	Mississippi R., MO–TN	920
1969	Silver Memorial	Pt. Pleasant, WV–OH	900
1977	Saint Marys	Saint Marys, WV–OH	900
1981	Ravenswood	WV	900
1987	Carl Perkins	Ohio R., KY	900
1988	Mississippi R.	Natchez, MS	875
1938	Blue Water	Pt. Huron, MI	871
1972	Mississippi R.	Vicksburg, MS	870
1972	N. Fork American R.	Auburn, CA	862
1940	*Baton Rouge	Mississippi R., LA	848
1899	*Cornwall	St. Lawrence R.	843
1940	Rte. 82	Mississippi R., AR	840
1961	Mississippi R.	Greenville, MS	840
1961	Rte. 49	Mississippi R., AR	840
1963	Brent Spence	KY–OH	830
1940	Mississippi R.	Vicksburg, MS	825
1963	Mississippi R.	Donaldsonville, LA	825
1929	Clark Memorial	Ohio R., KY	820
1961	Campbellton-Cross Pt.	New Brunswick, Can.	815
1932	Washington Mem.	Seattle, WA	800
1935	Rip Van Winkle	Catskill, NY	800
1938	Cairo	Ohio R., IL–KY	800

Simple Truss

Year	Bridge	Location	Main span (ft.)
1976	Chester	Chester, WV	745
1929	Irvin S. Cobb	Ohio R.,IL–KY	716
1922	*Tanana R.	Nenana, AK	700
1967	I-77, Ohio R.	Williamstown, WV	650
1917	MacArthur[4]	St. Louis, IL–MO	647
1992	St. Charles	Missouri R, MO	625
1933	Atchafalaya	Morgan City, LA	608
1924	*Castleton	Hudson R., NY	598
1937	Delaware R.	Easton, PA	550
1930	Swindell Bridge	Pittsburgh, PA	545
1952	Allegheny R. Tpk.	Pittsburgh, PA	534
1930	*Martinez	Martinez, CA	528
1951	Rankin	Pittsburgh, PA	525
1914	Old Brownsville	Brownsville, PA	520
1906	Donora-Webster	Donora-Webster, PA	515
1909	Hulton	Pittsburgh, PA	505
1967	Tanana R.	AK	500

Steel Truss

Year	Bridge	Location	Main span (ft.)
1988	Glade Creek	Raleigh Co., WV	784
1973	Atchafalaya R.	Krotz Springs, LA	780
1972	Piscataqua R.	NH–ME	756
1972	Atchafalaya R.	Simmesport, LA	720
1957	SR-3, Rappahannock R.	Middlesex Co., VA	648
1978	Atchafalaya R.	Morgan City, LA	607
1959	Summit	Summit, DE	600
1969	Reedy Point	Delaware City, DE	600
1938	US-22	Delaware R., NJ	540
1955	Interstate (I-5)	Columbia R., OR–WA	531
1910	McKinley, St. Louis[4]	Mississippi R., MO	517
1972	Mississippi R.	Muscatine, IA	512
1896	Newport	Ohio R., KY	511
1989	US 190, Atchafalaya R.	Krotz Springs, LA	506
1931	Lucy Jefferson Lewis	Cumberland R., KY	500
1958	Lake Oahe	Gettysburg, SD	500
1958	Lake Oahe	Mobridge, SD	500
1970	Lake Koocanusa	Lincoln Co., MT	500

Continuous Truss

Year	Bridge	Location	Main span (ft.)
1966	Columbia R. (Astoria)	OR–WA	1,232
1977	Francis Scott Key	Baltimore, MD	1,200
1981	Ravenswood/Ohio R.	Ravenswood, WV	902
1995	**Central	Ohio R., KY–OH	850
1943	Dubuque	Mississippi R., IA	845
1966	Charles Braga	Fall River, MA	840
1956	Earl C. Clements[5]	Ohio R., IL–KY	825
1929	U.S. 31	Ohio R., IN–KY	820
1953	John E. Mathews	Jacksonville, FL	810
1950	Maurice J. Tobin	Boston, MA	801
1940	Gov. Nice Memorial	Potomac River, MD	800
1957	Kingston-Rhinecliff	Hudson R., NY	800
1992	Mark Clark Expy. I-526	Cooper R., Charleston, SC	800
1986	Rochester-Monaca	Rochester-Monaca, PA	780
1940	U.S. 231	Ohio R., IN	750
1974	Carroll L. Cropper	Ohio R., IN–KY	750
1981	Sewickley	Sewickley, PA	750
1984	13th St. Bridge, Ohio R.	Ashland, KY	740
1959	Monaca-E. Rochester	Monaca-E. Rochester, PA	730
1976	Betsy Ross	Philadelphia, PA	729
1929	U.S. 421	Ohio R., IN–KY	727
1967	Matthew E. Welsh[6]	Mauckport, IN	725
1962	U.S. 41	Ohio R., IN–KY	720
1994	6th St.	Huntington, WV	720
1970	Vanport	Vanport, PA	715
1962	Champlain	Montreal, Que.	707
1962	John F. Kennedy[7]	Ohio R., IN–KY	701
1973	Girard Point	Philadelphia, PA	700
1954	PA Tpk., Delaware R.	Philadelphia, PA	682
1938	Port Arthur-Orange	TX	680
1949	George Platt	Philadelphia, PA	680
1926	Cape Girardeau	Mississippi R., MO	677
1929	'Cincinnati	Ohio R., OH	675
1946	Chester	Mississippi R, IL	670
1970	Gulfgate	Port Arthur, TX	664
1994	Williamstown-Marietta	Williamstown, WV	650
1955	Jefferson City	Missouri R., MO	640
1930	Quincy	Mississippi R., IL	628
1959	US 181, over harbor	Corpus Christi, TX	620
1961	Shippingport	Shippingport, PA	620
1935	Bourne-Sagamore	Cape Cod Canal, MA	616
1965	Clarion R. (I-80)	Clarion, PA	612
1975	Donora-Monessen	Donora-Monessen, PA	608
1957	Blatnik	Duluth, MN	600
1965	Rio Grande Gorge	Taos, NM	600
1991	Hoffstadt Creek	Mt. St. Helens, WA	600

Continuous Box and Plate Girder

Year	Bridge	Location	Main span (ft.)
1967	San Mateo-Hayward #2	San Fran. Bay, CA	750
1976	Intracoastal Canal	Forked Is., LA	750
1977	Intracoastal Canal	Gibbstown, LA	750
1982	Houston Ship Chan	Houston, TX	750
1969	San Diego-Coronado[8]	San Diego Bay, CA	660
1987	Umatilla, Columbia R.	OR–WA	660
1994	Acosta	Jacksonville, FL	630
1981	Douglas	Juneau, AK	620
1976	Wax L. Outlet	Calumet, LA	618
1963	Poplar St.	St. Louis, MO	600
1981	Glenn Jackson (I-205)	Columbia R., OR–WA	600
1976	Stanislaus River	Sonora, CA	580
1982	Illinois R.	Pekin, IL	550
1982	I-440	Arkansas R., AR	540
1980	US-64, Tennessee R.	Savannah, TN	525
1965	McDonald-Cartier	Ottawa, Ont.	520
1988	Mon City	Monongahela, PA	520

Continuous Plate

Year	Bridge	Location	Main span (ft.)
1973	Ship Channel (I-610)	Houston, TX	630
1971	W. Atchafalaya	Henderson, LA	573
1992	State Route 76	Paris, TN	525
1981	Illinois 23	Illinois R., IL	510
1968	Trinity R.	Dallas, TX	480
1978	San Joaquin R.	Antioch, CA	460
1977	Thomas Johnson Mem.	Solomons, MD	451
1967	Mississippi R.	La Crosse, WI	450
1975	I-129	Missouri R., IA–NE	450
1979	Lewis	St. Louis, MO	450
1992	Cuba Landing Bridge	Tennessee R., TN	450

Cable-Stayed

Year	Bridge	Location	Main span (ft.)
1986	Annacis (Alex Fraser)	Vancouver, BC	1,526
1993	Quetzalapa Bridge	Quetzalapa, Mexico	1,391
1988	Dames Point	Jacksonville, FL	1,300
1995	Houston Ship Channel	Baytown, TX	1,250

Year	Bridge	Location	Main span (ft.)
1983	Hale Boggs Memorial	Luling, LA	1,222
1987	Sunshine Skyway	Tampa Bay, FL	1,200
1988	Tampico/Panuco R.	Mexico	1,181
1988	ALRT Fraser River Bridge	Vancouver, BC	1,115
1990	Talmadge Mem.	Savannah, GA	1,100
1993	Mezcala	Mex. City/Acapulco Hwy.	1,024
1978	Pasco-Kennewick	Columbia R., WA	981
1984	Coatzacoalcos R.	Mexico	919
1985	E. Huntington	E. Huntington, WV	900
1987	Bayview Bridge	Quincy, IL	900
1970	Burton Bridge	New Brunswick, Canada	850
1990	Weirton-Steubenville	WV–OH	820
1969	Papineau-Leblanc	Montreal, Que.	790
1991	Cochrane	Mobile, AL	780
1994	Clark Bridge	Alton, IL	756
1995	Chesapeake & Delaware Canal Bridge	Dover-Wilmington, DE	750

I-Beam Girder

Year	Bridge	Location	Main span (ft.)
1980	Interstate 20	Shreveport, LA	438
1988	Route 18	Weston's Mill Pond, NJ	276

Steel Arch

Year	Bridge	Location	Main span (ft.)
1977	New River Gorge	Fayetteville, WV	1,700
1931	Bayonne (Kill Van Kull)	Bayonne, NJ	1,652
1973	Fremont	Portland, OR	1,255
1964	Port Mann	Vancouver, BC	1,200
1967	Trois-Rivieres	St. Lawrence R., Que.	1,100
1967	Lavioleete	Three Rivers, Canada	1,100
1992	Roosevelt Lake	Roosevelt Lake, AZ	1,080
1959	Glen Canyon	Page, AZ	1,028
1962	Lewiston-Queenston	Niagara R., Ont.	1,000
1976	Perrine	Twin Falls, ID	993
1941	Rainbow Bridge	Niagara Falls, NY	984
1917	'Hell Gate	East R., N.Y.	977
1977	Moundsville	Ohio R., WV	912
1992	I-255, Miss. R.	St. Louis, MO	909
1972	I-40, Miss. R.[9]	AR–TN	900

Concrete Arch

Year	Bridge	Location	Main span (ft.)
1993	Natchez Trace Pkwy.	Franklin, TN	582
1993	Lake Street Bridge	St. Paul, MN	556
1971	Selah Creek (twin)	Selah, WA	549
1968	Cowlitz R.	Mossyrock, WA	520
1931	Westinghouse	Pittsburgh, PA.	460
1923	Cappelen	Minneapolis, MN	435
1930	Jack's Run	Pittsburgh, PA.	400

Segmental Concrete

Year	Bridge	Location	Main span (ft.)
1997	Confederation Bridge	Prince Edward Isl., NB	820
1982	Jesse H, Jones Memorial	Houston, TX	750
1978	Shubenacadie River	S. Maitland, Nova Scotia	790
1992	Narragansett Bay Crossing	Jamestown, RI	674
1986	WB I-82 (Columbia R.)	Umatilla, OR.	660
1976	Stanislaus River	Parrets Ferry. CA	640
1992	Jamestown-Verranzzano	Jamestown, RI	636
1981	Gastineau Channel Br.	Juneau, AK	620
1991	Veterans Memorial Centennial Bridge	Coeur d'Alene, ID	520
1974	Pine Valley Creek	Pine Valley, CA.	450
1988	Zilwaukee Bridge (twin)	Zilwaukee, MI	392
1985	Red River Bridge	Boyce, LA.	370

Twin Concrete Trestle[10]

Year	Bridge	Location	Main span (ft.)
1979	I-55/I-10	Manchac, LA	181,157
1969	L. Pontchartrain Cswy.	Mandeville, LA	126,720
1972	Atchafalaya Flwy.	Baton Rouge, LA.	93,984
1963	L. Pontchartrain	Slidell, LA	28,547
1983	*Interstate 310	Kenner, LA	25,925

Concrete Slab Dam[10]

Year	Bridge	Location	Main span (ft.)
1927	Conowingo Dam	MD	4,611
1952	SR-4, Roanoke R.	Mecklenburg Co., VA	2,785
1936	Hoover Dam	Lake Mead, NV	1,324

Drawbridges
Vertical Lift

Year	Bridge	Location	Main span (ft.)
1959	'Arthur Kill	NY–NJ	558
1965	Pennsylvania Railroad	Kirkwood-Mt. Pleasant, DE	548
1935	'Cape Cod Canal	Cape Cod, MA	544
1961	'Delair	Delaware R., NJ	542

Year	Bridge	Location	Main span (ft.)
1931	Burlington-Bristol	Delaware R., NJ–PA	540
1937	Marine Parkway	Jamaica Bay, NY	540
1908	*Willamette R.	Portland, OR	521
1968	Second Narrows	Vancouver, B.C.	493
1912	*A-S-B Fratt.	Kansas City, MO	428
1945	*Harry S Truman	Kansas City, MO	427
1955	Roosevelt Island	East R., NY	418
1980	US-17, James R.	Isle of Wight, Co., VA	415
1932	*M-K-T R.R.	Missouri R., MO	414
1969	Cape Fear Mem.	Wilmington, NC	408

Bascule

Year	Bridge	Location	Main span (ft.)
1940	Lorain	Black R., OH	333
1917	SR-8, Tennessee R.	Chattanooga, TN	306
1956	Duwamish R.	Seattle, WA	300
1955	Chehalis R.	Aberdeen, WA	288
1968	Elizabeth R.	Chesapeake, VA	280
1913	Broadway	Portland, OR	278
1954	Fuller Warren	Jacksonville, FL	267

Swing Bridges

Year	Bridge	Location	Main span (ft.)
1927	Fort Madison[4]	Mississippi R., IA	545
1991	SW. Spokane St.	Seattle, WA	480
1930	Rigolets Pass	New Orleans, LA	400
1950	Douglass Memorial	Washington, DC	386
1945	Lord Delaware	Mattaponi R., VA	252

Swing Span

Year	Bridge	Location	Main span (ft.)
1952	US-17	York R., VA	500
1897	*Duluth	St. Louis Bay, MN	486
1899	*C.M.&N.R.R.	Chicago, IL	474
1913	Rt. 82, Conn-R.	E. Haddam, CT	465
1914	*Coos Bay	OR	458

Floating Pontoon

Year	Bridge	Location	Main span (ft.)
1963	Evergreen Pt.	Seattle, WA	7,578
1961	Hood Canal	Pt. Gamble, WA	6,521
1993	Lacey V. Murrow[11]	Seattle, WA	6,620
1989	Third Lake Washington.	Seattle, WA	5,811

(1) Swing span bridge with 2 spans of 2,310 ft. each. (2) A second bridge in parallel was completed in 1973. (3) The Richmond Bridge has twin spans 1,070 ft. each. (4) Railroad and vehicular bridge. (5) Two spans each 825 ft. (6) Two spans each 707 ft. (7) Two spans each 700 ft. (8) Two spans each 660 ft. (9) Two spans each 900 ft. (10) Length listed is total length of bridge. (11) Replaces the original Lacey V. Murrow bridge, which opened in 1940 and sank in 1990.

Oldest U.S. Bridges in Continuous Use

Built in 1697, the stone-arch Frankford Ave. Bridge crosses Pennypack Creek in Philadelphia, PA. A 3-span bridge with a total length of 75 ft., it was constructed as part of the King's Road, which eventually connected Philadelphia to New York.

The oldest covered bridge, completed in 1827, is the double-span, 278-ft. Haverhill Bath Bridge, which spans the Ammonoosuc River, between the towns of Bath and Haverhill, NH.

Some Notable International Bridges

Span of bridge is the distance between its supports. Asterisk (*) designates under construction as of Sept. 1999.

Suspension

Year	Bridge	Location	Main span (ft.)
1998	Akashi Kaikyo	Japan	6,570
1998	Storebælt (East Bridge)	Denmark	5,328
1981	Humber	England	4,626
1999	Jiangyin Yangtze	China	4,544
1997	Tsing Ma[1]	China	4,518
1997	Hoga Kusten	Sweden	3,970
1988	Minami Bisan-Seto	Japan	3,609
1988	Bosphorus II	Turkey	3,576
1973	Bosphorus I	Turkey	3,524
1999	Kurushima III.	Japan	3,379
1999	Kurushima II	Japan	3,346
1966	Tagus River[2]	Portugal	3,323
1964	Forth Road	Scotland	3,300
1988	Kita Bisan-Seto	Japan	3,248
1966	Severn	England	3,241
1988	Shimotsui Strait.	Japan	3,084

Cantilever

Year	Bridge	Location	Main span (ft.)
1890	Forth[3] (rail)	Scotland	1,710
1974	Nanko	Japan	1,673

Steel Arch

Year	Bridge	Location	Main span (ft.)
1932	Sydney Harbour	Australia	1,650
1967	Zdakov	Czech Republic	1,244
1962	Thatcher	Panama Canal Zone	1,128
1961	Runcorn-Widnes	England	1,082

	Bridge	Location	Main span (ft.)
1935	Birchenough	Zimbabwe	1,080

Concrete Arch

Year	Bridge	Location	Main span (ft.)
1980	Krk I	Croatia	1,280
1964	Gladesville	Australia	1,000
1964	Amizade	Brazil	951
1963	Arrabida	Portugal	886
1943	Sando	Sweden	866

Steel Plate and Box Girder

Year	Bridge	Location	Main span (ft.)
1974	President Costa e Silva	Brazil	984
1956	Sava I	Yugoslavia	856
1966	Zoobrüke	Germany	850

Cable-Stayed

Year	Bridge	Location	Main span (ft.)
1999	Tatara	Japan	2,920
1995	Pont de Normandie	France	2,808
1996	Quingzhou Minjang	China	1,985
1993	Yangpu	China	1,975
1997	Xupu	China	1,936
1998	Meiko Chuo	Japan	1,936
1991	Skarnsundet	Norway	1,739
1995	Tsurumi Tsubasa	Japan	1,673
2000*	Oresund	Denmark/Sweden	1,614
1991	Ikuchi	Japan	1,608
1994	Higashi Kobe	Japan	1,591
1997	Ting Kau	China	1,558

(1) Double-decked road and rail bridge. (2) Railroad and highway bridge. (3) Two spans of 1,710 ft. each.

Underwater Vehicular Tunnels in North America

(more than 5,000 ft. in length; year in parentheses is year of completion)

Name	Location	Waterway	Feet
Brooklyn-Battery (1950) (twin)	New York, NY	East River	9,117
Holland Tunnel (1927) (twin)	New York, NY	Hudson River	8,557
Ted Williams Tunnel (1995)	Boston, MA	Boston Harbor	8,448
Lincoln Tunnel (1937, 1945, 1957) (3 tubes)	New York, NY	Hudson River	8,216
Thimble Shoal Channel (1964)	Northampton Co., VA	Chesapeake Bay	8,187
Chesapeake Channel (1964)	Northampton Co., VA	Chesapeake Bay	7,941
Fort McHenry Tunnel (1985) (twin)	Baltimore, MD	Baltimore Harbor	7,920
Hampton Roads (1957) (twin)	Hampton, VA	Hampton Roads	7,479
Baltimore Harbor Tunnel (1957) (twin)	Baltimore, MD	Patapsco River	7,392
Queens Midtown (1940) (twin)	New York, NY	East River	6,414
Sumner Tunnel (1934)	Boston, MA	Boston Harbor	5,653
Louis-Hippolyte Lafontaine Tunnel	Montreal, Que.	St. Lawrence River	5,280
Detroit-Windsor (1930)	Detroit, MI.	Detroit River	5,160
Callahan Tunnel (1961)	Boston, MA	Boston Harbor	5,070

Land Vehicular Tunnels in the U.S.
Source: Federal Highway Administration
(more than 3,000 ft. in length)

Name	Location	Feet	Name	Location	Feet
E. Johnson Memorial	I-70, CO	8,959	Lehigh (twin)	PA Turnpike	4,379
Eisenhower Memorial	I-70, CO	8,941	Wawona	Yosemite Natl. Park, CA	4,233
Allegheny (twin)	PA Turnpike	6,072	Big Walker Mt.	Bland Co., VA	4,229
Liberty Tubes	Pittsburgh, PA.	5,920	Squirrel Hill	Pittsburgh, PA	4,225
Zion Natl. Park	Rte. 9, UT	5,766	Hanging Lake (twin)	Glenwood Canyon, CO.	4,000
East River Mt.	Mercer Co., VA	5,654	Caldecott (3 tubes)	Oakland, CA	3,616
East River Mt. (twin)	VA–WV	5,412	Fort Pitt (twin)	Pittsburgh, PA	3,560
Tuscarora (twin)	PA Turnpike	5,400	Mount Baker Ridge	Seattle, WA	3,456
Tetsuo Harano (twin)	H-3, HI	5,165	Dingess Tunnel	Mingo Co., WV	3,400
Kittatinny (twin)	PA Turnpike	4,660	Mall Tunnel	Dist. of Columbia.	3,400
Cumberland Gap (twin)	KY–TN	4,600	Cody No. 1	U.S. 14, 16, 20, WY	3,202
Blue Mountain (twin)	PA Turnpike	4,435			

World's Longest Railway Tunnels
Source: Railway Directory & Year Book

Tunnel	Date	Miles	Operating railway	Country
Seikan	1985	33.50	Japanese Railway	Japan
English Channel Tunnel	1994	31.04	Eurotunnel	United Kingdom-France
Dai-shimizu	1979	14.00	Japanese Railway	Japan
Simplon No. 1 and 2	1906, 1922	12.00	Swiss Fed. & Italian St.	Switzerland-Italy
Kanmon	1975	12.00	Japanese Railway	Japan
Apennine	1934	11.00	Italian State	Italy
Rokko	1972	10.00	Japanese Railway	Japan
Mt. MacDonald	1989	9.10	Canadian Pacific	Canada
Gotthard	1882	9.00	Swiss Federal	Switzerland
Lotschberg	1913	9.00	Bern-Lotschberg-Simplon	Switzerland
Hokuriku	1962	9.00	Japanese Railway	Japan
Mont Cenis (Frejus)	1871	8.00	Italian State	France-Italy
Shin-Shimizu	1961	8.00	Japanese Railway	Japan
Aki	1975	8.00	Japanese Railway	Japan
Cascade	1929	8.00	Burlington Northern	United States
Flathead	1970	8.00	Burlington Northern	United States

World's Largest-Capacity Hydro Plants
Source: U.S. Committee on Large Dams of the Intl. Commission on Large Dams, 1999

Rank order	Name	Country	Rated capacity now (MW)	Rated capacity planned (MW)	Rank order	Name	Country	Rated capacity now (MW)	Rated capacity planned (MW)
1.	Turukhansk (Lower Tungu-ska)*	Russia	—	20,000	10.	La Grande 2	Canada	5,328	5,328
2.	Three Gorges Dam*	China	—	18,200	11.	Churchill Falls	Canada	5,225	5,225
3.	Itaipu	Brazil/ Paraguay	7,400	13,320	12.	Xingo	Brazil	3,012	5,020
					13.	Tarbela	Pakistan	1,750	4,678
4.	Grand Coulee	U.S.	6,495	10,830	14.	Bratsk	Russia	4,500	4,500
5.	Guri (Raúl Leoni)	Venezuela	10,300	10,300	14.	Ust-Ilim	Russia	3,675	4,500
6.	Tucuruí	Brazil	2,640	7,260	16.	Cabora Bassa	Mozambique.	2,425	4,150
7.	Sayano-Shushensk*	Russia	—	6,400	17.	Boguchany*	Russia	—	4,000
8.	Corpus Posadas	Argentina/ Paraguay	4,700	6,000	18.	Rogun*	Tajikistan	3,600	3,600
					18.	Oak Creek	U.S.	3,600	3,600
9.	Krasnoyarsk	Russia	6,000	6,000	20.	Paulo Afonso I	Brazil	1,524	3,409

*Planned or under construction.

Major Dams of the World
Source: U.S. Committee on Large Dams of the Intl. Commission on Large Dams, 1999

World's Highest Dams

Rank order	Name	Country	Height above lowest formation(m)	Rank order	Name	Country	Height above lowest formation(m)
1.	Rogun*	Tajikistan	335	11.	Mica	Canada	243
2.	Nurek	Tajikistan	300	12.	Sayano-Shushensk	Russia	242
3.	Grand Dixence	Switzerland	285	13.	Ertan*	China	240
4.	Inguri	Georgia	272	14.	La Esmeralda	Colombia	237
5.	Vajont	Italy	262	15.	Kishau*	India	236
6.	Manuel M. Torres	Mexico	261	16.	El Cajón	Honduras	234
7.	Tehri*	India	261	17.	Chirkei	Russia	233
8.	Alvaro Obregon	Mexico	260	18.	Oroville	U.S.	230
9.	Mauvoisin	Switzerland	250	19.	Bhakra	India	226
10.	Alberto Lleraso*	Colombia	243	20.	Hoover	U.S.	221

* Under construction.

World's Largest-Volume Embankment Dams

Rank order	Name	Country	Volume cubic meters × 1000	Rank order	Name	Country	Volume cubic meters × 1000
1.	Tarbela	Pakistan	148,500	5.	Yacireta*	Argentina	81,000
2.	Fort Peck	U.S.	96,050	6.	Rogun*	Tajikistan	75,500
3.	Tucurui	Brazil	85,200	7.	Oahe	U.S.	70,339
4.	Ataturk*	Turkey	85,000	8.	Guri	Venezuela	70,000

Rank order	Name	Country	Volume cubic meters × 1000	Rank order	Name	Country	Volume cubic meters × 1000
9.	Parambikulam	India	69,165	15.	San Luis	U.S.	59,559
10.	High Island West	China	67,000	16.	Nurek	Tajikistan	58,000
11.	Gardiner	Canada	65,000	17.	Tanda	Pakistan	57,250
12.	Afsluitdijk	Netherlands	63,400	18.	Garrison	U.S.	50,843
13.	Mangla	Pakistan	63,379	19.	Cochiti	U.S.	50,228
14.	Oroville	U.S.	59,635	20.	Oosterschelde	Netherlands	50,000

*Under construction.

World's Largest-Capacity Reservoirs

Source: U.S. Committee on Large Dams of the Intl. Commission on Large Dams, 1999

Rank order	Name	Country	Capacity cubic meters × 100,000	Rank order	Name	Country	Capacity cubic meters × 100,000
1.	Owen Falls	Uganda	204,800	11.	Zeya	Russia	68,400
2.	Kariba	Zimbabwe/ Zambia	180,600	12.	La Grande 2	Canada	61,715
3.	Bratsk	Russia	169,000	13.	La Grande 3	Canada	60,020
4.	High Aswan	Egypt	162,000	14.	Ust-Ilim	Russia	59,300
5.	Akosombo	Ghana	147,960	15.	Boguchany*	Russia	58,200
6.	Daniel Johnson	Canada	141,851	16.	Kuibyshev	Russia	58,000
7.	Xinfeng	China	138,960	17.	Serra de Mesa	Brazil	54,400
8.	Guri	Venezuela	135,000	18.	Caniapiaoau Barrago KA 3	Canada	53,790
9.	W A C Bennett	Canada	74,300	19.	Cahora Bassa	Mozambique	52,000
10.	Krasnoyarsk	Russia	73,300	20.	Bukhtarma	Kazakhstan	49,800

*Under construction.

Major U.S. Dams and Reservoirs

Source: Committee on Register of Dams, Corps of Engineers, U.S. Army, Oct. 1998

Highest U.S. Dams

Rank Order	Dam name	River	State	Type	Height Feet	Height Meters	Year completed
1.	Oroville	Feather	California	E	754	230	1968
2.	Hoover	Colorado	Nevada	A	725	221	1936
3.	Dworshak	N. Fork Clearwater	Idaho	G	718	219	1973
4.	Glen Canyon	Colorado	Arizona	A	708	216	1966
5.	New Bullards Bar	North Yuba	California	A	636	194	1970
6.	New Melones	Stanislaus	California	R	626	191	1979
7.	Swift	Lewis	Washington	E	610	186	1958
8.	Mossyrock	Cowlitz	Washington	A	607	185	1968
9.	Shasta	Sacramento	California	G	600	183	1945
10.	Don Pedro	Tuolumne	California	E	567	173	1971

E= Embankment, Earthfill; R= Embankment, Rockfill; G= Gravity; A= Arch.

Largest U.S. Embankment Dams

Rank Order	Dam name	River	State	Type	Volume Cubic yards × 1000	Volume Cubic meters × 1000	Year completed
1.	Fort Peck	Missouri	Montana	E	125,624	96,050	1937
2.	Oahe	Missouri	South Dakota	E	91,996	70,339	1958
3.	Oroville	Feather	California	E	77,997	59,635	1968
4.	San Luis	San Luis Creek	California	E	77,897	59,559	1967
5.	Garrison	Missouri	North Dakota	E	66,498	50,843	1953
6.	Cochiti	Rio Grande	New Mexico	E	65,693	50,228	1975
7.	Fort Randall	Missouri	South Dakota	E	49,962	38,200	1952
8.	Castaic	Castaic Creek	California	E	43,998	33,640	1973
9.	Ludington P/S	Lake Michigan	Michigan	E	37,699	28,824	1973
10.	Kingsley	N. Platte	Nebraska	E	31,999	24,466	1941

E= Embankment, Earthfill.

Largest U.S. Reservoirs

Rank Order	Dam name, location	Reservoir name	Location	Reservoir capacity Acre-Feet	Reservoir capacity Cubic meters × 1000	Year completed
1.	Hoover, NV	Lake Mead	AZ/NV	28,253,000	34,850,000	1936
2.	Glen Canyon, AZ	Lake Powell	AZ/UT	26,997,000	33,300,000	1966
3.	Garrison, ND	Lake Sakakawea	ND	22,635,000	27,920,000	1953
4.	Oahe, SD	Lake Oahe	ND/SD	22,238,000	27,430,000	1958
5.	Fort Peck, MT	Fort Peck Lake	MT	17,933,000	22,120,000	1937
6.	Grand Coulee, WA	F. D. Roosevelt Lake	WA	9,558,000	11,790,000	1942
7.	Libby, MT	Lake Koocanusa	MT/B.C.	5,813,000	7,170,000	1973
8.	Fort Randall, SD	Lake Francis Case	SD	4,621,000	5,700,000	1952
9.	Shasta, CA	Lake Shasta	CA	4,548,000	5,610,000	1945
10.	Toledo Bend, LA	Toledo Bend Lake	LA/TX	4,475,000	5,520,000	1968

1 acre-foot = 1 acre of water, 1 foot deep

STATES AND OTHER AREAS OF THE U.S.

Sources: Population: Commerce Dept., Bureau of the Census (July 1998 est., including armed forces stationed in the state). Area: Bureau of the Census, Geography Division; forested land: Agriculture Dept., Forest Service. Lumber production: Bureau of the Census, Industry Division; mineral production: Dept. of Interior, Office of Mineral Information; commercial fishing: Commerce Dept., Natl. Marine Fisheries Service; value of construction: McGraw-Hill Information Systems Co., F.W. Dodge Division. Personal per capita income: Commerce Dept., Bureau of Economic Analysis; sales tax: CCH Inc.; unemployment: Labor Dept., Bureau of Labor Statistics. Tourism: Tourism Industries/ITA, Tourism Works for America Report. Lottery figures (not all states have a lottery): *La Fleur's Lottery World.* Finance: Federal Deposit Insurance Corp. Federal employees: Labor Dept., Office of Personnel Management. Energy: Energy Dept., Energy Information Administration. Other information from sources in individual states.

Note: Population density is for land area only. Categories under racial or employment distrib. are not necessarily all-inclusive and may not add to 100%. Hispanic population may be any race. Nonfuel mineral values for some states exclude small amounts to avoid disclosing proprietary data. Famous Persons lists may include some nonnatives associated with the state as well as persons born there. Website addresses listed may not be official state sites and are not endorsed by *The World Almanac;* all are subject to change.

Alabama

Heart of Dixie, Camellia State

People. Population (1998): 4,351,999; rank: 23; **net change** (1990-98): 7.7%. **Pop. density** (1998): 85.8 per sq mi. **Racial distribution** (1990): 73.6% white; 25.3% black. **Hispanic pop.:** 0.6%.

Geography. Total area: 52,237 sq mi; rank: 30. **Land area:** 50,750 sq mi; rank: 28. **Acres forested:** 21,974,000. **Location:** East South Central state extending N-S from Tenn. to the Gulf of Mexico; E of the Mississippi River. **Climate:** long, hot summers; mild winters; generally abundant rainfall. **Topography:** coastal plains, including Prairie Black Belt, give way to hills, broken terrain; highest elevation, 2,407 ft. **Capital:** Montgomery.

Economy. Chief industries: pulp & paper, chemicals, electronics, apparel, textiles, primary metals, lumber and wood products, food processing, fabricated metals, automotive tires, oil and gas exploration. **Chief manuf. goods:** electronics, cast iron & plastic pipe, fabricated steel products, ships, paper products, chemicals, steel, mobile homes, fabrics, poultry processing, soft drinks, furniture, tires. **Chief crops:** cotton, greenhouse & nursery, peanuts, sweet potatoes, potatoes and other vegetables. **Livestock** (Jan. 1999): 1.5 mil cattle/calves; 6,500 sheep/lambs; (Dec. 1998) 190,000 hogs/pigs; (Dec. 1998) 16.6 mil chickens (excl. broilers); (Dec. 1997) 906.2 mil broilers. **Timber/lumber** (1998): pine, hardwoods; 2.5 bil bd. ft. **Nonfuel minerals** (est. 1998): $947 mil; mostly portland cement, crushed stone, lime, sand & gravel, masonry cement. **Commercial fishing** (1998): $47 mil. **Chief port:** Mobile. **Value of construction** (1997): $4.8 bil. **Gross state product** (1997): $103.1 bil. **Employment distrib.** (May 1999): 23.4% trade; 23.9% serv; 19.2% mfg.; 18.1% govt. **Per cap. pers. income** (1998): $21,442. **Sales tax** (1999): 4%. **Unemployment** (1998): 4.2%. **Tourism expends.** (1996): $4.4 bil.

Finance. FDIC-insured commercial banks (1998): 160. **Deposits:** $101.8 bil. **FDIC-insured savings institutions** (1998): 12. **Assets:** $2.2 bil.

Federal govt. civ. employees (Mar. 1998): 38,265. **Avg. salary:** $45,386. **Notable fed. facilities:** George C. Marshall NASA Space Center; Gunter Annex & Maxwell AFB; Ft. Rucker; Ft. McClellan; Natl. Fertilizer Develop. Center; Navy Station & U.S. Corps of Engineers; Redstone Arsenal.

Energy. Electricity production (1998, kWh, by source): Coal: 71.5 bil; Petroleum: 260 mil; Gas: 2.4 bil; Hydroelectric: 10.6 bil; Nuclear: 28.7 bil.

State data. Motto: We dare defend our rights. **Flower:** Camellia. **Bird:** Yellowhammer. **Tree:** Southern Longleaf pine. **Song:** Alabama. **Entered union** Dec. 14, 1819; rank, 22d. **State fair:** Regional and county fairs held in Sept. and Oct.; no state fair.

History. Alabama was inhabited by the Creek, Cherokee, Chickasaw, Alabama, and Choctaw peoples when the Europeans arrived. The first Europeans were Spanish explorers in the early 1500s. The French made the first permanent settlement on Mobile Bay, 1702. France later gave up the entire region to England under the Treaty of Paris, 1763. Spanish forces took control of the Mobile Bay area, 1780, and it remained Spanish until U.S. troops seized the area, 1813. Most of present-day Alabama was held by the Creeks until Gen. Andrew Jackson broke their power, 1814, and they were removed to Oklahoma Territory. The state seceded, 1861, and the Confederate states were organized Feb. 4, at Montgomery, the first capital; it was readmitted, 1868.

Tourist attractions. First White House of the Confederacy, Civil Rights Memorial, Alabama Shakespeare Festival, all Montgomery; Ivy Green, Helen Keller's birthplace, Tuscumbia; Civil Rights Museum, statue of Vulcan, Birmingham; Carver Museum, Tuskegee; W. C. Handy Home & Museum, Florence; Alabama Space and Rocket Center, Huntsville; Moundville State Monument, Moundville; Pike Pioneer Museum, Troy; USS *Alabama* Memorial Park, Mobile; Russell Cave Natl. Monument, near Bridgeport: a detailed record of occupancy by humans from about 10,000 BC to AD 1650.

Famous Alabamians. Hank Aaron, Tallulah Bankhead, Hugo L. Black, Paul "Bear" Bryant, George Washington Carver, Nat King Cole, William C. Handy, Bo Jackson, Helen Keller, Coretta Scott King, Harper Lee, Joe Louis, Willie Mays, John Hunt Morgan, Jesse Owens, George Wallace, Booker T. Washington, Hank Williams.

Tourist information. Bureau of Tourism and Travel, 401 Adams Avenue, Suite 126, Montgomery, AL 36104. **Toll-free travel information.** 1-800-ALABAMA out of state.

Website. http://alaweb.asc.edu

Tourism website. http://www.touralabama.org

Alaska

The Last Frontier
(unofficial)

People. Population (1998): 614,010; rank: 48; **net change** (1990-98): 11.6%. **Pop. density** (1998): 1.1 per sq mi. **Racial distribution** (1990): 75.5% white; 4.1% black; 15.6% Amer. Indian, Eskimo or Aleut; 3.6% Asian or Pacific Is. **Hispanic pop.:** 3.2%.

Geography. Total area: 615,230 sq mi; rank: 1. **Land area:** 570,374 sq mi; rank: 1. **Acres forested:** 129,131,000. **Location:** NW corner of North America, bordered on E by Canada. **Climate:** SE, SW, and central regions, moist and mild; far north extremely dry. Extended summer days, winter nights, throughout. **Topography:** includes Pacific and Arctic mountain systems, central plateau, and Arctic slope. Mt. McKinley, 20,320 ft, is the highest point in North America. **Capital:** Juneau.

Economy. Chief industries: petroleum, tourism, fishing, mining, forestry, transportation, aerospace. **Chief manuf. goods:** fish products, lumber & pulp, furs. **Agriculture: Chief crops:** greenhouse products, barley, oats, hay, potatoes, lettuce, aquaculture. **Livestock** (Jan. 1999): 11,500 cattle/calves; 1,600 sheep/lambs; (Dec. 1998) 2,100 hogs/pigs. **Timber/lumber:** spruce, yellow cedar, hemlock. **Nonfuel minerals** (est. 1998): $911 mil; mostly zinc, lead, gold, sand & gravel, silver. **Commercial fishing** (1998): $951.4 mil. **Chief ports:** Anchorage, Dutch Harbor, Kodiak, Seward, Skagway, Juneau, Sitka, Valdez, Wrangell. **Internat. airports at:** Anchorage, Fairbanks, Juneau. **Value of construction** (1997): $1.0 bil. **Gross state product** (1997): $24.4 bil. **Employment distrib.** (May 1999): 27.2% govt.; 25.1% serv.; 20.8% trade; 5.1% mfg. **Per cap. pers. income** (1998): $25,675. **Sales tax:** none. **Unemployment** (1998): 5.8%. **Tourism expends.** (1996): $1.4 bil.

Finance. FDIC-insured commercial banks (1998): 6. **Deposits:** $3.8 bil. **FDIC-insured savings institutions** (1998): 2. **Assets:** $259 mil.

Federal govt. Fed. civ. employees (Mar. 1998): 11,573. **Avg. salary:** $43,535.

Energy. Electricity production (1998, kWh, by source): Coal: 171 mil; Petroleum: 757 mil; Gas: 2.5 bil; Hydroelectric: 1.1 bil.

State data. Motto: North to the future. **Flower:** Forget-Me-Not. **Bird:** Willow ptarmigan. **Tree:** Sitka spruce. **Song:** Alaska's Flag. **Entered union** Jan. 3, 1959; rank, 49th. **State fair** at Palmer; late Aug.-early Sept.

History. Early inhabitants were the Tlingit-Haida people and tribes of the Athabascan family. The Aleut and Inuit (Eskimo), who arrived about 4,000 years ago from Siberia, lived in the coastal areas. Vitus Bering, a Danish explorer working for Russia, was the first European to land in Alaska, 1741. The first permanent Russian settlement was established on Kodiak Island, 1784. In 1799, the Russian-American Co. controlled the region, and the first chief manager, Aleksandr Baranov, set up headquarters at Archangel, near present-day Sitka. Sec. of State William H. Seward bought Alaska from Russia for $7.2 mil in 1867, a bargain some called "Seward's Folly." In 1896, gold was discovered in the Klondike region, and the famed gold rush began. Alaska became a territory in 1912.

Tourist attractions. Inside Passage; Portage Glacier; Mendenhall Glacier; Ketchikan Totems; Glacier Bay Natl. Park and Preserve; Denali Natl. Park, one of N. America's great wildlife sanctuaries, surrounding Mt. McKinley, N. America's highest peak; Mt. Roberts Tramway, Juneau; Pribilof Islands fur seal rookeries; restored St. Michael's Russian Orthodox Cathedral, Sitka; Katmai Natl. Park & Preserve.

Famous Alaskans. Tom Bodett, Susan Butcher, Ernest Gruening, Gov. Tony Knowles, Sydney Laurence, Libby Riddles, Jefferson "Soapy" Smith.

Tourist information. Alaska Division of Tourism, PO Box 110801, Juneau, AK 99811-0801; 1-907-465-2010.

Website. http://www.state.ak.us

Tourism website. http://www.commerce.state.ak.us/tourism

Arizona

Grand Canyon State

People. Population (1998): 4,668,631; rank: 21; **net change** (1990-98): 27.4%. **Pop. density** (1998): 41.1 per sq mi. **Racial distribution:** (1990): 80.8% white; 3.0% black; 5.6% American Indian. **Hispanic pop.:** 18.8%.

Geography. Total area: 114,006 sq mi; rank: 6. **Land area:** 113,642 sq mi; rank: 6. **Acres forested:** 19,596,000. **Location:** in the southwestern U.S. **Climate:** clear and dry in the southern regions and northern plateau; high central areas have heavy winter snows. **Topography:** Colorado plateau in the N, containing the Grand Canyon; Mexican Highlands running diagonally NW to SE; Sonoran Desert in the SW. **Capital:** Phoenix.

Economy. Chief industries: manufacturing, construction, tourism, mining, agriculture. **Chief manuf. goods:** electronics, printing & publishing, foods, prim. & fabric. metals, aircraft and missiles, apparel. **Chief crops:** cotton, lettuce, cauliflower, broccoli, sorghum, barley, corn, wheat, citrus fruits. **Livestock** (Jan. 1999): 810,000 cattle/calves; 140,000 sheep/lambs; (Dec. 1998) 115,000 hogs/pigs. **Timber/lumber** (1998): pine, fir, spruce; 98 mil bd. ft. **Nonfuel minerals** (est. 1998): $2.82 bil; mostly copper, sand & gravel, cement, molybdenum, lime. **Internat. airports at:** Phoenix, Tucson, Yuma. **Value of construction** (1997): $10.0 bil. **Gross state product** (1997): $121.2 bil. **Employment distrib.** (May 1999): 30.2% services; 24.1% trade; 16.0% govt.; 10.2% mfg. **Per cap. pers. income** (1998): $23,060. **Sales tax** (1999): 5%. **Unemployment** (1998): 4.1%. **Tourism expends.** (1996): $8.4 bil. **Lottery** (1998): total sales: $250.7 mil; net income: $76.5 mil.

Finance. FDIC-insured commercial banks (1998): 43. **Deposits:** $25.1 bil. **FDIC-insured savings institutions** (1998): 3. **Assets:** $688 mil.

Federal govt. Fed. civ. employees (Mar. 1998): 27,445. **Avg. salary:** $40,130. **Notable fed. facilities:** Luke, Davis-Monthan AF bases; Ft. Huachuca Army Base; Yuma Proving Grounds.

Energy. Electricity production (1998, kWh, by source): Coal: 36.2 bil; Petroleum: 61 mil; Gas: 3.5 bil; Hydroelectric: 11.2 bil; Nuclear: 30.3 bil.

State data. Motto: Ditat Deus (God enriches). **Flower:** Blossom of the Saguaro cactus. **Bird:** Cactus wren. **Tree:** Paloverde. **Song:** Arizona. **Entered union** Feb. 14, 1912; rank, 48th. **State fair** at Phoenix; late Oct.-early Nov.

History. Anasazi, Mogollon, and Hohokam civilizations inhabited the area c 300 BC-AD 1300, later Pueblo peoples; Navajo and Apache came c 15th cent. Marcos de Niza, a Franciscan, and Estevanico, a former black slave, explored, 1539; Spanish explorer Francisco Vásquez de Coronado visited, 1540. Eusebio Francisco Kino, a Jesuit missionary, taught Indians 1692-1711, and left missions. Tubac, a Spanish fort, became the first European settlement, 1752. Spain ceded Arizona to Mexico, 1821. The U.S. took over, 1848, af-

ter the Mexican War. The area below the Gila River was obtained from Mexico in the Gadsden Purchase, 1853. Arizona became a territory, 1863. Apache wars ended with Geronimo's surrender, 1886.

Tourist attractions. The Grand Canyon of the Colorado; Painted Desert; Petrified Forest Natl. Park; Canyon de Chelly; Meteor Crater; London Bridge, Lake Havasu City; Biosphere 2, Oracle; Navajo Natl. Monument; Sedona.

Famous Arizonans. Bruce Babbitt, Cochise, Geronimo, Barry Goldwater, Zane Grey, Carl Hayden, George W. P. Hunt, Helen Jacobs, Percival Lowell, William H. Pickering, John J. Rhodes, Morris Udall, Stewart Udall, Frank Lloyd Wright.

Tourist information. Arizona Office of Tourism, Ste. 4015, 2702 N. 3rd St., Phoenix, AZ 85004.

Website. http://www.state.az.us

Tourism website. http://www.arizonaguide.com

Arkansas

The Natural State, The Razorback State

People. Population (1998): 2,538,303; rank: 33; **net change** (1990-98): 8.0%. **Pop. density** (1998): 48.7 per sq mi. **Racial distribution:** (1990): 82.7% white; 15.9% black. **Hispanic pop.:** 0.8%.

Geography. Total area: 53,182 sq mi; rank: 28. **Land area:** 52,075 sq mi; rank: 27. **Acres forested:** 17,864,000. **Location:** in the west south-central U.S. **Climate:** long, hot summers, mild winters; generally abundant rainfall. **Topography:** eastern delta and prairie, southern lowland forests, and the northwestern highlands, which include the Ozark Plateaus. **Capital:** Little Rock.

Economy. Chief industries: manufacturing, agriculture, tourism, forestry. **Chief manuf. goods:** food products, chemicals, lumber, paper, plastics, electric motors, furniture, auto components, airplane parts, apparel, machinery, steel. **Chief crops:** rice, soybeans, cotton, tomatoes, grapes, apples, commercial vegetables, peaches, wheat. **Livestock** (Jan. 1999): 1.8 mil cattle/calves; (Dec. 1997) 750,000 hogs/pigs; (Dec. 1998) 23.5 mil chickens (excl. broilers); (Dec. 1997) 1.2 bil broilers. **Timber/lumber** (1998): oak, hickory, gum, cypress, pine; 2.4 bil bd. ft. **Nonfuel minerals** (est. 1998): $598 mil; mostly bromine, crushed stone, portland cement, sand & gravel. **Chief ports:** Little Rock, Pine Bluff, Osceola, Helena, Fort Smith, Van Buren, Camden, Dardanelle, North Little Rock, West Memphis, Crossett, McGehee, Morrilton. **Value of construction** (1997): $3.0 bil. **Gross state product** (1997): $58.4 bil. **Employment distrib.** (May 1999): 23.7% serv.; 22.9% trade; 22.3% mfg.; 16.4% govt. **Per cap. pers. income** (1998): $20,346. **Sales tax** (1999): 4.625%. **Unemployment** (1998): 5.5%. **Tourism expends.** (1996): $3.3 bil.

Finance. FDIC-insured commercial banks (1998): 202. **Deposits:** $21.5 bil. **FDIC-insured savings institutions** (1998): 12. **Assets:** $3.3 bil.

Federal govt. Fed. civ. employees (Mar. 1998): 10,916. **Avg. salary:** $38,906. **Notable fed. facilities:** Nat'l. Center for Toxicological Research, Jefferson; Pine Bluff Arsenal, Little Rock AFB.

Energy. Electricity production (1998, kWh, by source): Coal: 23.1 bil; Petroleum: 144 mil; Gas: 3.7 bil; Hydroelectric: 3.1 bil; Nuclear: 13.1 bil.

State data. Motto: Regnat Populus (The people rule). **Flower:** Apple blossom. **Bird:** Mockingbird. **Tree:** Pine. **Song:** Arkansas. **Entered union** June 15, 1836; rank, 25th. **State fair** at Little Rock; late Sept.-early Oct.

History. Quapaw, Caddo, Osage, Cherokee, and Choctaw peoples lived in the area at the time of European contact. The first European explorers were de Soto, 1541; Marquette and Jolliet, 1673; and La Salle, 1682. The first settlement was by the French under Henri de Tonty, 1686, at Arkansas Post. In 1762, the area was ceded by France to Spain, then given back again, 1800, and was part of the Louisiana Purchase, 1803. It was made a territory, 1819. Arkansas seceded in 1861, only after the Civil War began; more than 10,000 Arkansans fought on the Union side.

Tourist attractions. Hot Springs Natl. Park (water ranging from 95°F-147°F); Eureka Springs; Ozark Folk Center, Blanchard Caverns, both near Mountain View; Crater of Diamonds (only U.S. diamond mine) near Murfreesboro; Toltec Mounds Archeological State Park, Little Rock; Buffalo Natl.

River; Mid-America Museum, Hot Springs; Pea Ridge National Military Park, Pead Ridge; Tanyard Springs, Morrilton; Wiederkehr Wine Village, Wiederkehr Village.

Famous Arkansans. Daisy Bates, Dee Brown, Paul "Bear" Bryant, Glen Campbell, Johnny Cash, Hattie Caraway, Bill Clinton, "Dizzy" Dean, Orval Faubus, James W. Fulbright, John H. Johnson, John Grisham, Douglas MacArthur, John L. McClellan, James S. McDonnell, Scottie Pippen, Dick Powell, Winthrop Rockefeller, Mary Steenburgen, Edward Durell Stone, Archibald Yell.

Tourist Information. Arkansas Dept. of Parks & Tourism, One Capitol Mall, Little Rock, AR 72201

Toll-free travel information. 1-800-NATURAL.

Website. http://www.state.ar.us

Tourism website. http://www.arkansas.com

California
Golden State

People. Population (1998): 32,666,550; rank: 1; **net change** (1990-98): 9.7%. **Pop. density** (1998): 209.4 per sq mi. **Racial distribution** (1990): 69.0% white; 7.4% black; 9.6% Asian. **Hispanic pop.:** 25.8%.

Geography. Total area: 158,869 sq mi; rank: 3. **Land area:** 155,973 sq mi; rank: 3. **Acres forested:** 37,263,000. **Location:** on western coast of the U.S. **Climate:** moderate temperatures and rainfall along the coast; extremes in the interior. **Topography:** long mountainous coastline; central valley; Sierra Nevada on the east; desert basins of the southern interior; rugged mountains of the north. **Capital:** Sacramento.

Economy. Chief industries: agriculture, tourism, apparel, electronics, telecommunications, entertainment. **Chief manuf. goods:** electronic and electrical equip., computers, industrial machinery, transportation equip. and instruments, food. **Chief farm products:** milk and cream, grapes, cotton, flowers, oranges, rice, nursery products, hay, tomatoes, lettuce, strawberries, almonds, asparagus. **Livestock** (Jan. 1999): 5.0 mil cattle/calves; 810,000 sheep/lambs; (Dec. 1998) 210,000 hogs/pigs; (Dec. 1998) 30.0 mil chickens (excl. broilers); (Dec. 1997) 237.3 mil broilers. **Timber/lumber** (1998): fir, pine, redwood, oak; 3.1 bil bd. ft. **Nonfuel minerals** (est. 1998): $2.97 bil; mostly portland cement, sand & gravel, boron, crushed stone, gold. **Commercial fishing** (1998): $110.7 mil. **Chief ports:** Long Beach, Los Angeles, San Diego, Oakland, San Francisco, Sacramento, Stockton. **Internat. airports at:** Fresno, Los Angeles, Sacramento, San Francisco, San Jose, San Diego. **Value of construction** (1997): $36.7 bil. **Gross state product** (1997): $1.03 tril. **Employment distrib.** (May 1999): 31.6% serv.; 22.7% trade; 14.0% mfg.; 15.9% govt. **Per cap. pers. income** (1998): $27,503. **Sales tax** (1999): 6%. **Unemployment** (1998): 5.9%. **Tourism expends.** (1996): $62.6 bil. **Lottery** (1998): total sales: $2.3 bil; net income: $834.8 mil.

Finance. FDIC-insured commercial banks (1998): 336. **Deposits:** $398.9 bil. **FDIC-insured savings institutions** (1998): 49. **Assets:** $287.4 bil.

Federal govt. Fed. civ. employees (Mar. 1998): 152,868. **Avg. salary:** $45,463. **Notable fed. facilities:** Vandenberg, Beale, Travis, McClellan AF bases; San Francisco Mint.

Energy. Electricity production (1998, kWh, by source): Petroleum: 121 mil; Gas: 26.4 bil; Hydroelectric: 48.7 bil; Nuclear: 34.6 bil.

State data. Motto: Eureka (I have found it). **Flower:** Golden poppy. **Bird:** California valley quail. **Tree:** California redwood. **Song:** I Love You, California. **Entered union** Sept. 9, 1850; rank, 31st. **State fair** at Sacramento; late Aug.-early Sept.

History. Early inhabitants included more than 100 different Native American tribes with multiple dialects. The first European explorers were Cabrillo, 1542, and Drake, 1579. The first settlement was the Spanish Alta California mission at San Diego, 1769, first in a string founded by Franciscan Father Junípero Serra. U.S. traders and settlers arrived in the 19th cent. and staged the Bear Flag revolt, 1846, in protest against Mexican rule; later that year U.S. forces occupied California. At the end of the Mexican War, Mexico ceded the territory to the U.S., 1848; that same year gold was discovered, and the famed gold rush began.

Tourist attractions. The *Queen Mary*, Long Beach; Palomar Mountain; Disneyland, Anaheim; Getty Center, Los Angeles; Tournament of Roses and Rose Bowl, Pasadena; Universal Studios, Hollywood; Long Beach Aquarium of the Pacific; Golden State Museum, Sacramento; San Diego Zoo;

Yosemite Valley; Lassen and Sequoia-Kings Canyon natl. parks; Lake Tahoe; Mojave and Colorado deserts; San Francisco Bay; Napa Valley; Monterey Peninsula; oldest living things on earth believed to be a stand of Bristlecone pines in the Inyo National Forest, est. 4,700 years old; world's tallest tree, 365-ft "National Geographic Society" coast redwood, in Humboldt Redwoods State Park.

Famous Californians. Edmund G. (Pat) Brown, Jerry Brown, Luther Burbank, Ted Danson, Leonardo DiCaprio, Joe DiMaggio, John C. Fremont, Robert Frost, Tom Hanks, Helen Hunt, Bret Harte, William Randolph Hearst, Jack Kemp, Monica Lewinsky, Jack London, Mark McGwire, Aimee Semple McPherson, Marilyn Monroe, John Muir, Richard M. Nixon, George S. Patton Jr., Ronald Reagan, Sally K. Ride, William Saroyan, Father Junípero Serra, Leland Stanford, John Steinbeck, Shirley Temple, Earl Warren, Tiger Woods.

California Division of Tourism. P.O. Box 1499, Sacramento, CA 95812-1499.

Toll-free travel information. 1-800-862-2543.

Website. http://www.state.ca.us/s

Tourism website. http://gocalif.ca.gov

Colorado
Centennial State

People. Population (1998): 3,970,971; rank: 24; **net change** (1990-98): 20.5%. **Pop. density** (1998): 38.3 per sq mi. **Racial distribution** (1990): 88.2% white; 4.0% black. **Hispanic pop.:** 12.9%.

Geography. Total area: 104,100 sq mi; rank: 8. **Land area:** 103,729 sq mi; rank: 8. **Acres forested:** 21,338,000. **Location:** in W central U.S. **Climate:** low relative humidity, abundant sunshine, wide daily, seasonal temp. ranges; alpine conditions in the high mountains. **Topography:** eastern dry high plains; hilly to mountainous central plateau; western Rocky Mountains of high ranges, with broad valleys, deep, narrow canyons. **Capital:** Denver.

Economy. Chief industries: manufacturing, construction, government, tourism, agriculture, aerospace, electronics equipment. **Chief manuf. goods:** computer equip. & instruments, foods, machinery, aerospace products. **Chief crops:** corn, wheat, hay, sugar beets, barley, potatoes, apples, peaches, pears, dry edible beans, sorghum, onions, oats, sunflowers, vegetables. **Livestock** (Jan. 1999): 3.1 mil cattle/calves; 440,000 sheep/lambs; (Dec. 1998) 870,000 hogs/pigs; (Dec. 1998) 4.6 mil chickens (excl. broilers). **Timber/lumber** (1998): oak, ponderosa pine, Douglas fir; 114 mil bd. ft. **Nonfuel minerals** (est. 1998): $604 mil; mostly sand & gravel, portland cement, molybdenum, crushed stone, gold. **Internat. airport at:** Denver. **Value of construction** (1997): $9.2 bil. **Gross state product** (1997): $126.0 bil. **Employment distrib.** (May 1999): 29.9% serv.; 24.3% trade; 15.6% govt.; 9.9% mfg. **Per cap. pers. income** (1998): $28,657. **Sales tax** (1999): 3%. **Unemployment** (1998): 3.8%. **Tourism expends.** (1996): $8.3 bil. **Lottery** (1998): total sales: $374.7 mil; net income: $97.7 mil.

Finance. FDIC-insured commercial banks (1998): 195. **Deposits:** $31.3 bil. **FDIC-insured savings institutions** (1998): 11. **Assets:** $982 bil.

Federal govt. Fed. civ. employees (Mar. 1998): 33,588. **Avg. salary:** $46,119. **Notable fed. facilities:** U.S. Air Force Academy; U.S. Mint; Ft. Carson; Natl. Renewable Energy Labs; U.S. Rail Transportation Test Center; N. American Aerospace Defense Command; Consolidated Space Operations Ctr.; Denver Federal Center; Natl. Center for Atmospheric Research; Natl. Instit. for Standards in Technology; Natl. Oceanic and Atmospheric Administration.

Energy. Electricity production (1998, kWh, by source): Coal: 33.1 bil; Petroleum: 37 mil; Gas: 964 mil; Hydroelectric: 1.4 bil.

State data. Motto: Nil Sine Numine (Nothing Without Providence). **Flower:** Rocky Mountain columbine. **Bird:** Lark bunting. **Tree:** Colorado blue spruce. **Song:** Where the Columbines Grow. **Entered union** Aug. 1, 1876; rank 38th. **State fair** at Pueblo; mid-Aug. - early Sept.

History. Early civilization centered around the Mesa Verde c 2,000 years ago, later, Ute, Pueblo, Cheyenne, and Arapaho peoples lived in the area. The region was claimed by Spain, but passed to France. The U.S. acquired eastern Colorado in the Louisiana Purchase, 1803. Lt. Zebulon M. Pike explored the area, 1806, discovering the peak that bears his name. After the Mexican War, 1846-48, U.S. immigrants settled in the east, former Mexicans in the south. Gold was discovered in 1858, causing a population boom. Displaced Native Americans protested, resulting in the so-called Sand Creek Massacre, 1864, where more than 200 Cheyenne and Arapaho were killed. All Native Americans were later removed to Oklahoma Territory.

Tourist attractions. Rocky Mountain Natl. Park; Aspen Ski Resort; Garden of the Gods, Colorado Springs; Great Sand Dunes, Dinosaur, Black Canyon of the Gunnison, and Colorado natl. monuments; Pikes Peak and Mt. Evans highways; Mesa Verde Natl. Park (ancient Anasazi Indian cliff dwellings); Grand Mesa Natl. Forest; mining towns of Central City, Silverton, Cripple Creek; Burlington's Old Town; Bent's Fort, outside La Junta; Georgetown Loop Historic Mining Railroad Park, Cumbres & Toltec Scenic Railroad; limited stakes gaming in Central City, Blackhawk, Cripple Creek, Ignacio, and Towaoe.

Famous Coloradans. Tim Allen, Frederick Bonfils, Henry Brown, Molly Brown, William N. Byers, M. Scott Carpenter, Jack Dempsey, Mamie Eisenhower, Douglas Fairbanks, Barney Ford, Scott Hamilton, Chief Ourey, "Baby Doe" Tabor, Lowell Thomas, Byron R. White, Paul Whiteman.

State Chamber of Commerce. 1776 Lincoln, Ste. 1200, Denver, CO 80203. Phone: 303-831-7411

Tourist information. Colorado Travel and Tourism Authority, P. O. Box 3524, Englewood, CO 80155.

Toll-free travel information. 1-800-COLORADO.

Website. http://www.state.co.us

Tourism website. http://www.colorado.com

Connecticut

Constitution State, Nutmeg State

People. Population (1998): 3,274,069; rank: 29; **net change** (1990-98): -0.4%. **Pop. density** (1998): 675.8 per sq mi. **Racial distribution** (1990): 87.0% white; 8.3% black. **Hispanic pop.:** 6.5%.

Geography. Total area: 5,544 sq mi; rank: 48. **Land area:** 4,845 sq mi; rank: 48. **Acres forested:** 1,819,000. **Location:** New England state in NE corner of the U.S. **Climate:** moderate; winters avg. slightly below freezing; warm, humid summers. **Topography:** western upland, the Berkshires, in NW, highest elevations; narrow central lowland N-S; hilly eastern upland drained by rivers. **Capital:** Hartford.

Economy. Chief industries: manufacturing, retail trade, government, services, finances, insurance, real estate. **Chief manuf. goods:** aircraft engines and parts, submarines, helicopters, machinery and computer equipment, electronics and electrical equipment, medical instruments, pharmaceuticals. **Chief crops:** nursery stock, Christmas trees, mushrooms, vegetables, sweet corn, tobacco, apples. **Livestock** (Jan. 1999): 64,000 cattle/calves; 5,000 sheep/lambs; (Dec. 1998) 5,000 hogs/pigs; (Dec. 1998) 4.0 mil chickens (excl. broilers). **Timber/lumber** (1998): oak, birch, beech, maple; 50 mil bd. ft. **Nonfuel minerals** (est. 1998): $105 mil; mostly crushed stone, sand & gravel, dimension stone, clays, gemstones. **Commercial fishing** (1998): $34.4 mil. **Chief ports:** New Haven, Bridgeport, New London. **Internat. airport at:** Windsor Locks. **Value of construction** (1997): $3.8 bil. **Gross state product** (1997): $134.6 bil. **Employment distrib.** (May 1999): 31.4% serv.; 21.5% trade; 16.6% mfg.; 13.8% govt. **Per cap. pers. income** (1998): $37,598. **Sales tax** (1999): 6%. **Unemployment** (1998): 3.4%. **Tourism expends.** (1996): $4.3 bil. **Lottery** (1998): total sales: $805.6 mil; net income: $258.1 mil.

Finance. FDIC-insured commercial banks (1998): 28. **Deposits:** $5.1 bil. **FDIC-insured savings institutions** (1998): 49. **Assets:** $40.6 bil.

Federal govt. Fed. civ. employees (Mar. 1998): 7,488. **Avg. salary:** $46,431. **Notable fed. facilities:** U.S. Coast Guard Academy; U.S. Navy Submarine Base.

Energy. Electricity production (1998, kWh, by source): Coal: 1.5 bil; Petroleum: 8.6 bil; Gas: 977 mil; Hydroelectric: 385 mil; Nuclear: 3.2 mil.

State data. Motto: Qui Transtulit Sustinet (He who transplanted still sustains). **Flower:** Mountain laurel. **Bird:** American robin. **Tree:** White oak. **Song:** Yankee Doodle. **Fifth** of the 13 original states to ratify the Constitution, Jan. 9, 1788. **State Fair:** largest fair at Durham, late Sept.; no state fair.

History. At the time of European contact, inhabitants of the area were Algonquian peoples, including the Mohegan and Pequot. Dutch explorer Adriaen Block was the first European visitor, 1614. By 1634, settlers from Plymouth Bay had started colonies along the Connecticut River; in 1637 they defeated the Pequots. The Colony of Connecticut was chartered by England, 1662, adding New Haven, 1665. In the American Revolution, Connecticut Patriots fought in most major campaigns, while Connecticut privateers captured British merchant ships.

Tourist attractions. Mark Twain House, Hartford; Yale University's Art Gallery, Peabody Museum, both in New Haven; Mystic Seaport; Mystic Marine Life Aquarium; P. T. Barnum Museum, Bridgeport; Gillette Castle, Hadlyme; U.S.S. *Nautilus* Memorial, Groton (1st nuclear-powered submarine); Mashantucket Pequot Museum & Research Center, Foxwoods Resort & Casino, both; Ledyard, Mohegan Sun, Uncasville; Lake Compounce, Bristol.

Famous "Nutmeggers." Ethan Allen, Phineas T. Barnum, Samuel Colt, Jonathan Edwards, Nathan Hale, Katharine Hepburn, Isaac Hull, Robert Mitchum, J. Pierpont Morgan, Israel Putnam, Wallace Stevens, Harriet Beecher Stowe, Mark Twain, Noah Webster, Eli Whitney.

Tourist information. Dept. of Economic and Community Development, 505 Hudson St., Hartford, CT 06106.

Toll-free travel information. 1-800-CTBOUND

Website. http://www.state.ct.us

Tourism website. http://www.state.ct.us/tourism

Delaware

First State, Diamond State

People. Population (1998): 743,603; rank: 45; **net change** (1990-98): 11.6%. **Pop. density** (1998): 380.4 per sq mi. **Racial distribution** (1990): 80.3% white; 16.9% black. **Hispanic pop.:** 2.4%.

Geography. Total area: 2,396 sq mi; rank: 49. **Land area:** 1,955 sq mi; rank: 49. **Acres forested:** 398,000. **Location:** occupies the Delmarva Peninsula on the Atlantic coastal plain. **Climate:** moderate. **Topography:** Piedmont plateau to the N, sloping to a near sea-level plain. **Capital:** Dover.

Economy. Chief industries: chemicals, agriculture, finance, poultry, shellfish, tourism, auto assembly, food processing, transportation equipment. **Chief manuf. goods:** nylon, apparel, luggage, foods, autos, processed meats and vegetables, railroad & aircraft equipment. **Chief crops:** soybeans, potatoes, corn, mushrooms, lima beans, green peas, barley, cucumbers, wheat, corn, grain sorghum, greenhouse & nursery. **Livestock** (Jan. 1999): 29,000 cattle/calves; (Dec. 1998) 31,000 hogs/pigs; (Dec. 1998) 495,000 chickens (excl. broilers); (Dec. 1997) 256.9 mil broilers. **Timber/lumber** (1998): hardwoods and softwoods (except for southern yellow pine); 14 mil bd. ft. **Nonfuel minerals** (est. 1998): $11.2 mil; mostly magnesium compounds, sand & gravel, gemstones. **Commercial fishing** (1998): $5.9 mil. **Chief ports:** Wilmington. **Internat. airport at:** Philadelphia/Wilmington. **Value of construction** (1997): $935 mil. **Gross state product** (1997): $31.6 bil. **Employment distrib.** (May 1999): 28.0% serv; 21.8% trade; 14.5% mfg.; 13.4% govt. **Per cap. pers. income** (1998): $29,814. **Sales tax:** none. **Unemployment** (1998): 3.8%. **Tourism expends.** (1996): $1.0 bil. **Lottery** (1998): total sales: $447.2 mil; net income: $164.5 mil.

Finance. FDIC-insured commercial banks (1998): 34. **Deposits:** $54.2 bil. **FDIC-insured savings institutions** (1998): 5. **Assets:** $4.7 bil.

Federal govt. Fed. civ. employees (Mar. 1998): 2,529. **Avg. salary:** $41,053. **Notable fed. facilities:** Dover Air Force Base, Federal Wildlife Refuge, Bombay Hook.

Energy. Electricity production (1998, kWh, by source): Coal: 3.8 bil; Petroleum: 1.2 bil; Gas: 1.3 bil.

State data. Motto: Liberty and independence. **Flower:** Peach blossom. **Bird:** Blue hen chicken. **Tree:** American holly. **Song:** Our Delaware. **First** of original 13 states to ratify the Constitution, Dec. 7, 1787. **State fair** at Harrington; end of July.

History. The Lenni Lenape (Delaware) people lived in the region at the time of European contact. Henry Hudson located the Delaware R., 1609, and in 1610, English explorer Samuel Argall entered Delaware Bay, naming the area after Virginia's governor, Lord De La Warr. The Dutch first settled near present Lewes, 1631, but the colony was destroyed by Indians. Swedes settled at Fort Christina (now Wilmington), 1638. Dutch settled anew, 1651, near New Castle and seized the Swedish settlement, 1655, only to lose all Delaware and

New Netherland to the British, 1664. After 1682, Delaware became part of Pennsylvania, and in 1704 it was granted its own assembly. In 1776, it adopted a constitution as the state of Delaware. Although it remained in the Union during the Civil War, Delaware retained slavery until abolished by the 13th Amendment in 1865.

Tourist attractions. Ft. Christina Monument, site of founding of New Sweden, Holy Trinity (Old Swedes) Church, erected 1698, the oldest Protestant church in the U.S. still in use, Wilmington; Hagley Museum, Winterthur Museum and Gardens, both near Wilmington; historic district, New Castle; John Dickinson "Penman of the Revolution" home, Dover; Rehoboth Beach, "nation's summer capital," Rehoboth; Dover Downs Intl. Speedway.

Famous Delawareans. Thomas F. Bayard, Henry Seidel Canby, E. I. du Pont, John P. Marquand, Howard Pyle, Caesar Rodney.

Chamber of Commerce. 1200 N. Orange St., Ste. 200, Wilmington, DE 19899-0671.

Toll-free travel information. 1-800-441-8846.

Website. http://www.state.de.us

Tourism website. http://www.state.de.us/tourism/intro.htm

Florida
Sunshine State

People. Population (1998): 14,915,980; rank: 4; **net change** (1990-98): 15.3%. **Pop. density** (1998): 276.5 per sq mi. **Racial distribution** (1990): 83.1% white; 13.6% black. **Hispanic pop.:** 12.2%.

Geography. Total area: 59,928 sq mi; rank: 23. **Land area:** 53,937 sq mi; rank: 26. **Acres forested:** 16,549,000. **Location:** peninsula jutting southward 500 mi between the Atlantic and the Gulf of Mexico. **Climate:** subtropical N of Bradenton-Lake Okeechobee-Vero Beach line; tropical S of line. **Topography:** land is flat or rolling; highest point is 345 ft in the NW. **Capital:** Tallahassee.

Economy. Chief industries: tourism, agriculture, manufacturing, construction, services, international trade. **Chief manuf. goods:** electric & electronic equipment, transportation equipment, food, printing & publishing, chemicals, instruments, industrial machinery. **Chief crops:** citrus fruits, vegetables, melons, greenhouse and nursery products, potatoes, sugarcane, strawberries. **Livestock** (Jan. 1999): 1.8 mil cattle/calves; (Dec. 1998) 55,000 hogs/pigs; (Dec. 1998) 12.8 mil chickens (excl. broilers); (Dec. 1997) 132.4 mil broilers. **Timber/lumber** (1998): pine, cypress, cedar; 782 mil bd. ft. **Nonfuel minerals** (est. 1998): $1.96 bil; mostly phosphate rock, crushed stone, portland cement, sand & gravel, titanium. **Commercial fishing** (1998): $188.6 mil. **Chief ports:** Pensacola, Tampa, Manatee, Miami, Port Everglades, Jacksonville, St. Petersburg, Canaveral. **Internat. airports at:** Ft. Lauderdale/Hollywood, Daytona Beach, Ft. Myers, Key West, Jacksonville, Miami, Orlando, St. Petersburg/Clearwater, Panama City, Tampa, Sarasota/Bradenton, West Palm Beach. **Value of construction** (1997): $25.2 bil. **Gross state product** (1997): $380.6 bil. **Employment distrib.** (May 1999): 36.7% services, 25.0% trade, 14.1% govt., 7.2% mfg. **Per cap. pers. income** (1998): $25,852. **Sales tax** (1999): 6%. **Unemployment** (1998): 4.3% **Tourism expends.** (1996): $48.1 bil. **Lottery** (1998): total sales: $2.1 bil; net income: $918.1 mil.

Finance. FDIC-insured commercial banks (1998): 250. **Deposits:** $62.5 bil. **FDIC-insured savings institutions** (1998): 42. **Assets:** $17.9 bil.

Federal govt. Fed. civ. employees (Mar. 1998): 61,314. **Avg. salary:** $43,668. **Notable fed. facilities:** John F. Kennedy Space Center, NASA-Kennedy Space Center's Spaceport USA; Eglin Air Force Base; Pensacola Naval Training Center; MacDill Air Force Base, Tampa.

Energy. Electricity production (1998, kWh, by source): Coal: 65.5 bil; Petroleum: 41.0 bil; Gas: 31.7 bil; Hydroelectric: 199 mil; Nuclear: 31.1 bil.

State data. Motto: In God we trust. **Flower:** Orange blossom. **Bird:** Mockingbird. **Tree:** Sabal palmetto palm. **Song:** Old Folks at Home. **Entered union** Mar. 3, 1845; rank, 27th. **State fair** at Tampa; early Feb.

History. The original inhabitants of Florida included the Timucua, Apalachee, and Calusa peoples. Later the Seminole migrated from Georgia to Florida, becoming dominant there in the early 18th cent. The first European to see Flori-

da was Ponce de León, 1513. France established a colony, Fort Caroline, on the St. John River, 1564. Spain settled St. Augustine, 1565, and Spanish troops massacred most of the French. Britain's Sir Francis Drake burned St. Augustine, 1586. In 1763, Spain ceded Florida to Great Britain, which held the area briefly, 1763-83, before returning it to Spain. After Andrew Jackson led a U.S. invasion, 1818, Spain ceded Florida to the U.S., 1819. The Seminole War, 1835-42, resulted in removal of most Native Americans to Oklahoma Territory. Florida seceded from the Union, 1861, and was readmitted in 1868.

Tourist attractions. Miami Beach; St. Augustine, oldest permanent European settlement in U.S.; Castillo de San Marcos, St. Augustine; Walt Disney World's Magic Kingdom, EPCOT Center, and Disney-MGM Studios, Animal Kingdom all near Orlando; Sea World, Universal Studios, near Orlando; Spaceport USA, Kennedy Space Center; Everglades Natl. Park; Ringling Museum of Art, Ringling Museum of the Circus, both in Sarasota; Cypress Gardens, Winter Haven; Busch Gardens, Tampa; U.S. Astronaut Hall of Fame, Mariana Caverns; Church St. Station, Orlando; Silver Springs, Ocala.

Famous Floridians. Marjory Stoneman Douglas, Henry M. Flagler, James Weldon Johnson, MacKinlay Kantor, Chief Osceola, Claude Pepper, Henry B. Plant, A. Philip Randolph, Marjorie Kinnan Rawlings, Joseph W. Stilwell, Charles P. Summerall.

Tourist information. Visit Florida, P.O. Box 1100, Tallahassee, FL 32302-1100, 1-850-488-5607.

Toll free number. 1-888-735-2872 (1-888-7FLA-USA)

Website. http://www.state.fl.us

Tourism website. http://www.flausa.com

Georgia
Empire State of the South, Peach State

People. Population (1998): 7,642,207; rank: 10; **net change** (1990-98): 18.0%. **Pop. density** (1998): 131.9 per sq mi. **Racial distribution** (1990): 71.0% white; 27.0% black. **Hispanic pop.:** 1.7%.

Geography. Total area: 58,977 sq mi; rank: 24. **Land area:** 57,919 sq mi; rank: 21. **Acres forested:** 24,137,000. **Location:** South Atlantic state. **Climate:** maritime tropical air masses dominate in summer; polar air masses in winter; E central area drier. **Topography:** most southerly of the Blue Ridge Mts. cover NE and N central; central Piedmont extends to the fall line of rivers; coastal plain levels to the coast flatlands. **Capital:** Atlanta.

Economy. Chief industries: services, manufacturing, retail trade. **Chief manuf. goods:** textiles, apparel, food, and kindred products, pulp & paper products. **Chief crops:** peanuts, cotton, corn, tobacco, hay, soybeans. **Livestock** (Jan. 1999): 1.3 mil cattle/calves; (Dec. 1998) 430,000 hogs/pigs; (Dec. 1998) 31.1 mil chickens (excl. broilers); (Dec. 1997) 1.2 bil broilers. **Timber/lumber** (1998): pine, hardwood; 3.2 bil bd. ft. **Nonfuel minerals** (est. 1998): $2.14 bil; mostly clays (kaolin), crushed stone, portland cement, clays (fuller's earth), sand & gravel. **Commercial fishing** (1998): $23.7 mil. **Chief ports:** Savannah, Brunswick. **Internat. airports at:** Atlanta. **Value of construction** (1997): $13.6 bil. **Gross state product** (1997): $229.4 bil. **Employment distrib.** (May 1999): 27.2% serv.; 24.7% trade; 14.5% mfg.; 15.4% govt. **Per cap. pers. income** (1998): $25,020. **Sales tax** (1999): 4%. **Unemployment** (1998): 4.2%. **Tourism expends.** (1996): $12.1 bil. **Lottery** (1998): total sales: $1.7 bil; net income: $555.1 mil.

Finance. FDIC-insured commercial banks (1998): 349. **Deposits:** $47.3 bil. **FDIC-insured savings institutions** (1998): 28. **Assets:** $6.7 bil.

Federal govt. Fed. civ. employees (Mar. 1998): 62,017. **Avg. salary:** $42,141. **Notable fed. facilities:** Dobbins AFB; Fts. Benning, Gordon, Fort Gillen; Fort Stewart; King's Bay Naval Base; Moody Air Force Base; Navy Supply Corps School; McPherson; Fed. Law Enforcement Training Ctr., Glynco, Warner Robins AFB; Centers for Disease Control.

Energy. Electricity production (1998, kWh, by source): Coal: 69.9 bil; Petroleum: 671 mil; Gas: 1.8 bil; Hydroelectric: 5.0 bil; Nuclear: 31.4 bil.

State data. Motto: Wisdom, justice and moderation. **Flower:** Cherokee rose. **Bird:** Brown thrasher. **Tree:** Live oak. **Song:** Georgia On My Mind. **Fourth** of the 13 original states to ratify the Constitution, Jan. 2, 1788. **State fair** at Perry, 5th Friday after Labor Day.

History. Creek and Cherokee peoples were early inhabitants of the region. The earliest known European settlement was the Spanish mission of Santa Catalina, 1566, on Saint Catherines Island. Gen. James Oglethorpe established a colony at Savannah, 1733, for the poor and religiously persecuted. Oglethorpe defeated a Spanish army from Florida at Bloody Marsh, 1742. In the American Revolution, Georgians seized the Savannah armory, 1775, and sent the munitions to the Continental Army. They fought seesaw campaigns with Cornwallis's British troops, twice liberating Augusta and forcing final evacuation by the British from Savannah, 1782. The Cherokee were removed to Oklahoma Territory, 1832-38, and thousands died on the long march, known as the Trail of Tears. Georgia seceded from the Union, 1861, and was invaded by Union forces, 1864, under Gen. William T. Sherman, who took Atlanta, Sept. 2, and proceeded on his famous "march to the sea," ending in Dec., in Savannah. Georgia was readmitted, 1870.

Tourist attractions. State Capitol, Stone Mt. Park, Six Flags Over Georgia, Kennesaw Mt. Natl. Battlefield Park, Martin Luther King Jr. Natl. Historic Site, Underground Atlanta, Jimmy Carter Library & Museum, all Atlanta; Chickamauga and Chattanooga Natl. Military Park, near Dalton; Chattahoochee Natl. Forest; alpine village of Helen; Dahlonega, site of America's first gold rush; Brasstown Bald Mt.; Lake Lanier; Franklin D. Roosevelt's Little White House, Warm Springs; Callaway Gardens, Pine Mt.; Andersonville Natl. Historic Site; Okefenokee Swamp, near Waycross; Jekyll Island; St. Simons Island; Cumberland Island Natl. Seashore; historic riverfront district, Savannah.

Famous Georgians. Griffin Bell, James Bowie, James Brown, Erskine Caldwell, Jimmy Carter, Ray Charles, Lucius D. Clay, Ty Cobb, James Dickey, John C. Fremont, Newt Gingrich, Joel Chandler Harris, "Doc" Holliday, Martin Luther King Jr., Gladys Knight, Sidney Lanier, Little Richard, Juliette Gordon Low, Margaret Mitchell, Flannery O'Connor, Otis Redding, Jackie Robinson, Clarence Thomas, Alice Walker, Joseph Wheeler, Joanne Woodward, Andrew Young.

Chamber of Commerce. 235 International Blvd., Atlanta, GA 30303; (404) 880-9000.

Toll-free travel information. 1-800-VISITGA.

Website. http://www.state.ga.us

Tourism website. http://www.georgia.org

Hawai'i
Aloha State

People. Population (1998): 1,193,001; rank: 41; **net change** (1990-98): 7.6%. **Pop. density** (1998): 185.7 per sq mi. **Racial distribution** (1990): 33.4% white; 2.5% black; 61.8% Asian or Pacific Is. **Hispanic pop.:** 7.3%.

Geography. Total area: 6,459 sq mi; rank: 47. **Land area:** 6,423 sq mi; rank: 47. **Acres forested:** 1,748,000. **Location:** Hawaiian Islands lie in the North Pacific, 2,397 mi SW from San Francisco. **Climate:** subtropical, with wide variations in rainfall; Waialeale, on Kaua'i, wettest spot in U.S. (annual rainfall 460 in.) **Topography:** islands are tops of a chain of submerged volcanic mountains; active volcanoes: Mauna Loa, Kilauea. **Capital:** Honolulu.

Economy. Chief industries: tourism, defense, sugar, pineapples. **Chief manuf. goods:** processed sugar, canned pineapple, clothing, foods, printing & publishing. **Chief crops:** sugar, pineapples, macadamia nuts, fruits, coffee, vegetables, floriculture. **Livestock** (Jan. 1999): 173,000 cattle/calves; (Dec. 1998) 29,000 hogs/pigs; (Dec. 1998) 747,000 chickens (excl. broilers); (Dec. 1997) 1.0 mil broilers. **Nonfuel minerals** (est. 1998): $85.5 mil; mostly crushed stone, portland cement, masonry cement, gemstones. **Commercial fishing** (1998): $62.1 mil. **Chief ports:** Honolulu, Nawiliwili, Barbers Point, Kahului, Hilo. **Internat. airport at:** Honolulu. **Value of construction** (1997): $1.7 bil. **Gross state product** (1997): $38.0 bil. **Employment distrib.** (May 1999): 33.3% serv.; 24.7% trade; 20.9% govt.; 3.0% mfg. **Per cap. pers. income** (1998): $26,137. **Sales tax** (1999): 4%. **Unemployment** (1998): 6.2%. **Tourism expenditures** (1996): $14.0 bil.

Finance. FDIC-insured commercial banks (1998): 12. **Deposits:** $16.6 bil. **FDIC-insured savings institutions** (1998): 3. **Assets:** $6.8 bil.

Federal govt. Fed. civ. employees (Mar. 1998): 19,833. **Avg. salary:** $41,426. **Notable fed. facilities:** Pearl Harbor Naval Shipyard; Hickam AFB; Schofield Barracks; Ft. Shafter; Marine Corps Base-Kaneohe Bay; Barbers Point NAS; Wheeler AFB; Prince Kuhio Federal Building.

Energy. Electricity production (1998, kWh, by source): Petroleum: 6.3 bil; Hydroelectric: 14 mil.

State data. Motto: The life of the land is perpetuated in righteousness. **Flower:** Yellow hibiscus. **Bird:** Hawaiian goose. **Tree:** Kukui (Candlenut). **Song:** Hawai'i Pono'i. **Entered union** Aug. 21, 1959; rank, 50th. **State fair:** at O'ahu, late June.

History. Polynesians from islands 2,000 mi to the south settled the Hawaiian Islands, probably between AD 300 and AD 600. The first European visitor was British captain James Cook, 1778. Between 1790 and 1810, the islands were united politically under the leadership of a native king, Kamehameha I, whose five successors—all bearing the name Kamehameha—ruled the kingdom from his death, 1819, until the end of the dynasty, 1872. Missionaries arrived, 1820, bringing Western culture. King Kamehameha III and his chiefs created the first constitution and a legislature that set up a public school system. Sugar production began, 1835, and it became the dominant industry. In 1893, Queen Liliuokalani was deposed, and a republic was instituted, 1894, headed by Sanford B. Dole. Annexation by the U.S. came in 1898. The Japanese attack on Pearl Harbor, Dec. 7, 1941, brought the U.S. into World War II.

Tourist attractions. Hawaii Volcanoes, Haleakala natl. parks; Natl. Memorial Cemetery of the Pacific, Waikiki Beach, Diamond Head, Honolulu; U.S.S. *Arizona* Memorial, Pearl Harbor; Hanauma Bay; Polynesian Cultural Center, Laie; Nu'uanu Pali; Waimea Canyon; Wailoa and Wailuku River state parks.

Famous Islanders. Bernice Pauahi Bishop, Tia Carrera, Father Damien de Veuster, Don Ho, Duke Kahanamoku, King Kamehameha, Brook Mahealani Lee, Daniel K. Inouye, Jason Scott Lee, Queen Liliuokalani, Bette Midler, Ellison Onizuka.

Chamber of Commerce of Hawaii. 1132 Bishop St., Suite 200, Honolulu, HI 96813; phone: (808) 545-4300.

Toll free travel information. 1-800-464-2924.

Website. http://www.hawaii.gov

Tourism website. http://www.gohawaii.com

Idaho
Gem State

People. Population (1998): 1,228,684; rank: 40; **net change** (1990-98): 22.0%. **Pop. density** (1998): 14.8 per sq mi. **Racial distribution** (1990): 94.4% white; 0.3% black. **Hispanic pop.:** 5.3%.

Geography. Total area: 83,574 sq mi; rank: 14. **Land area:** 82,751 sq mi; rank: 11. **Acres forested:** 21,621,000. **Location:** northwestern Mountain state bordering on British Columbia. **Climate:** tempered by Pacific westerly winds; drier, colder, continental climate in SE; altitude an important factor. **Topography:** Snake R. plains in the S; central region of mountains, canyons, gorges (Hells Canyon, 7,900 ft, deepest in N. America); subalpine northern region. **Capital:** Boise.

Economy. Chief industries: manufacturing, agriculture, tourism, lumber, mining, electronics. **Chief manuf. goods:** electronic components, computer equipment, processed foods, lumber and wood products, chemical products, primary metals, fabricated metal products, machinery. **Chief crops:** potatoes, peas, dry beans, sugar beets, alfalfa seed, lentils, wheat, hops, barley, plums and prunes, mint, onions, corn, cherries, apples, hay. **Livestock** (Jan. 1999): 1.9 mil cattle/calves; 265,000 sheep/lambs; (Dec. 1998) 27,000 hogs/pigs; (Dec. 1998) 1.3 mil chickens (excl. broilers). **Timber/lumber** (1998): yellow, white pine; Douglas fir; white spruce; 1.9 bil bd. ft. **Nonfuel minerals** (est. 1998): $444 mil; mostly phosphate rock, gold, sand & gravel, molybdenum, silver. **Chief port:** Lewiston. **Value of construction** (1997) $1.8 bil. **Gross state product** (1997): $29.1 bil. **Employment distrib.** (May 1999): 25.1% trade; 24.7% serv., 19.8% govt.; 14.2% mfg. **Per cap. pers. income** (1998): $21,081. **Sales tax** (1999): 5%. **Unemployment** (1998): 5.0%. **Tourism expenditures** (1996): $1.9 bil. **Lottery** (1998): total sales: $89.6 mil; net income: $20.6 mil.

Finance. FDIC-insured commercial banks (1998): 17. **Deposits:** $1.6 bil. **FDIC-insured savings institutions** (1998): 3. **Assets:** $533 mil.

Federal govt. Fed. civ. employees (Mar. 1998): 7,336. **Avg. salary:** $42,059. **Notable fed. facilities:** Idaho Natl. Engineering Lab; Mt. Home Air Force Base.

Energy. Electricity production (1998, kWh, by source): Hydroelectric: 12.0 bil.

State data. Motto: Esto Perpetua (It is perpetual). **Flower:** Syringa. **Bird:** Mountain bluebird. **Tree:** White pine. **Song:** Here We Have Idaho. **Entered union** July 3, 1890; rank, 43d. **State fair** at Boise, late Aug.; at Blackfoot, early Sept.

History. Early inhabitants were Shoshone, Northern Paiute, Bannock, and Nez Percé peoples. White exploration of the region began with Lewis and Clark, 1805-6. Next came fur traders, setting up posts, 1809-34, and missionaries, 1830s-50s. Mormons made their first permanent settlement at Franklin, 1860. Idaho's gold rush began the same year and brought thousands of permanent settlers. Most remarkable of the Indian wars was the 1,700-mi trek, 1877, of Chief Joseph and the Nez Percé, pursued by U.S. troops through 3 states and caught just short of the Canadian border. The Idaho territory was organized, 1863. Idaho adopted a progressive constitution and became a state, 1890.

Tourist attractions. Hells Canyon, deepest gorge in N. America; World Center for Birds of Prey; Craters of the Moon; Sun Valley, in Sawtooth Mts.; Crystal Falls Cave; Shoshone Falls; Lava Hot Springs; Lake Pend Oreille; Lake Coeur d'Alene; Sawtooth Natl. Recreation Area; River of No Return Wilderness Area; Redfish Lake.

Famous Idahoans. William E. Borah, Frank Church, Fred T. Dubois, Ezra Pound, Chief Joseph, Sacagawea.

Tourist information. Department of Commerce, 700 W. State St., Boise, ID 83720.

Toll-free travel information. 1-800-VISIT-ID.

Website. http://www.state.id.us

Tourism website. http://www.visitid.org

Illinois
Prairie State

People. Population (1998): 12,045,326; rank: 5; **net change** (1990-98): 5.4%. **Pop. density** (1998): 216.7 per sq mi. **Racial distribution** (1990): 78.3% white; 14.8% black **Hispanic pop.:** 7.9%.

Geography. Total area: 57,918 sq mi; rank: 25. **Land area:** 55,593 sq mi; rank: 24. **Acres forested:** 4,266,000. **Location:** East North Central state; western, southern, and eastern boundaries formed by Mississippi, Ohio, and Wabash rivers, respectively. **Climate:** temperate; typically cold, snowy winters, hot summers. **Topography:** prairie and fertile plains throughout; open hills in the southern region. **Capital:** Springfield.

Economy. Chief industries: services, manufacturing, travel, wholesale and retail trade, finance, insurance, real estate, construction, health care, agriculture. **Chief manuf. goods:** machinery, electric and electronic equipment, prim. & fabric. metals, chemical products, printing & publishing, food and kindred products. **Chief crops:** corn, soybeans, wheat, sorghum, hay. **Livestock** (Jan. 1999): 1.5 mil cattle/calves; 74,000 sheep/lambs; (Dec. 1998) 4.8 mil hogs/pigs; (Dec. 1998) 3.5 mil chickens (excl. broilers). **Timber/lumber** (1998): oak, hickory, maple, cottonwood; 86 mil bd. ft. **Nonfuel minerals** (est. 1998): $862 mil; mostly crushed stone, portland cement, sand & gravel, lime. **Commercial fishing** (1998): $98,000. **Chief ports:** Chicago. **Internat. airport at:** Chicago. **Value of construction** (1997): $12.5 bil. **Gross state product** (1997): $393.6 bil. **Employment distrib.** (May 1999): 30.5% serv.; 22.7% trade; 16.2% mfg.; 13.5% govt. **Per cap. pers. income** (1998): $28,873. **Sales tax** (1999): 6.25%. **Unemployment** (1998): 4.5%. **Tourism expends.** (1996): $18.4 bil. **Lottery** (1998): total sales: $1.6 bil; net income: $536.4 mil.

Finance. FDIC-insured commercial banks (1998): 745. **Deposits:** $207.6 bil. **FDIC-insured savings institutions** (1998): 124. **Assets:** $49.8 bil.

Federal govt. Fed. civ. employees (Mar. 1998): 42,709. **Avg. salary:** $46,735. **Notable fed. facilities:** Fermi Natl. Accelerator Lab; Argonne Natl. Lab; Rock Island Arsenal; Great Lakes, Naval Training Station, Scott AFB.

Energy. Electricity production (1998, kWh, by source): Coal: 70.3 bil; Petroleum: 838 mil; Gas: 4.5 bil; Hydroelectric: 51 mil; Nuclear: 55.6 bil.

State data. Motto: State sovereignty—national union. **Flower:** Native violet. **Bird:** Cardinal. **Tree:** White oak. **Song:** Illinois. **Entered union** Dec. 3, 1818; rank, 21st. **State fair** at Springfield, mid-Aug.; DuQuoin, late Aug.

History. Seminomadic Algonquian peoples, including the Peoria, Illinois, Kaskaskia, and Tamaroa, lived in the region at the time of European contact. Fur traders were the first Europeans in Illinois, followed shortly by Jolliet and Marquette, 1673, and La Salle, 1680, who built a fort near present-day Peoria. The first settlements were French, at Cahokia, near present-day St. Louis, 1699, and Kaskaskia, 1703. France ceded the area to Britain, 1763, and in 1778, American Gen. George Rogers Clark took Kaskaskia from the British without a shot. Defeat of Native American tribes in Black Hawk War, 1832, and growth of railroads brought change to the area. In 1787, it became part of the Northwest Territory. Post-Civil War Illinois became a center for the labor movement as bitter strikes, such as the Haymarket Square riot, occurred in 1885-86.

Tourist attractions. Chicago museums and parks; Lincoln shrines at Springfield, New Salem, Sangamon County; Cahokia Mounds, Collinsville; Starved Rock State Park; Crab Orchard Wildlife Refuge; Mormon settlement at Nauvoo; Fts. Kaskaskia, Chartres, Massac (parks); Shawnee Natl. Forest, Southern Illinois; Illinois State Museum, Springfield; Dickson Mounds Museum, between Havana and Lewistown.

Famous Illinoisans. Jane Addams, Saul Bellow, Jack Benny, Ray Bradbury, Gwendolyn Brooks, William Jennings Bryan, St. Frances Xavier Cabrini, Hillary Rodham Clinton, Clarence Darrow, John Deere, Stephen A. Douglas, James T. Farrell, George W. Ferris, Marshall Field, Betty Friedan, Benny Goodman, Ulysses S. Grant, Ernest Hemingway, Wild Bill Hickok, Henry J. Hyde, Abraham Lincoln, Vachel Lindsay, Edgar Lee Masters, Oscar Mayer, Cyrus McCormick, Ronald Reagan, Carl Sandburg, Adlai Stevenson, Frank Lloyd Wright, Philip Wrigley.

Tourist information. Illinois Dept. of Commerce and Community Affairs, 620 E. Adams St., Springfield, IL 62701.

Toll-free travel information: 1-800-2-CONNECT.

Website. http://www.state.il.us

Tourism website. http://www.enjoyillinois.com

Indiana
Hoosier State

People. Population (1998): 5,899,195; rank: 14; **net change** (1990-98): 6.4%. **Pop. density** (1998): 164.5 per sq mi. **Racial distribution** (1990): 90.6% white; 7.8% black. **Hispanic pop.:** 1.8%.

Geography. Total area: 36,420 sq mi; rank: 38. **Land area:** 35,870 sq mi; rank: 38. **Acres forested:** 4,439,000. **Location:** East North Central state; Lake Michigan on N border. **Climate:** 4 distinct seasons with a temperate climate. **Topography:** hilly southern region; fertile rolling plains of central region; flat, heavily glaciated north; dunes along Lake Michigan shore. **Capital:** Indianapolis.

Economy: Chief industries: manufacturing, services, agriculture, government, wholesale and retail trade, transportation and public utilities. **Chief manuf. goods:** primary metals, transportation equipment, motor vehicles & equip., industrial machinery & equipment, electronic & electric equipment. **Chief crops:** corn, soybeans, wheat, nursery and greenhouse products, vegetables, popcorn, fruit, hay, tobacco, mint. **Livestock** (Jan. 1999): 1.0 mil cattle/calves; 57,000 sheep/lambs; (Dec. 1998) 4.0 mil hogs/pigs; (Dec. 1998) 29.4 mil chickens (excl. broilers). **Timber/lumber** (1998): oak, tulip, beech, sycamore; 380 mil bd. ft. **Nonfuel minerals** (est. 1998): $698 mil; mostly crushed stone, portland cement, sand & gravel, lime, masonry cement. **Commercial fishing** (1997): $327,000 **Chief ports:** Burns Harbor, Portage; Southwind Maritime, Mt. Vernon; Clark Maritime, Jeffersonville. **Internat. airports at:** Indianapolis, Ft. Wayne. **Value of construction** (1997): $9.2 bil. **Gross state product** (1997): $161.7 bil. **Employment distrib.** (May 1999): 24.6% serv; 23.5% trade; 23.2% mfg.; 13.7% govt. **Per cap. pers. income** (1998): $24,219. **Sales tax** (1999): 5%. **Unemployment** (1998): 3.1%. **Tourism expends.** (1996): $5.2 bil. **Lottery** (1998): total sales: $648.2 mil; net income: $195.5 mil.

Finance. FDIC-insured commercial banks (1998): 169. **Deposits:** $54.5 bil. **FDIC-insured savings institutions** (1998): 69. **Assets:** $15.6 bil.

Federal govt. Fed. civ. employees (Mar. 1998): 19,774. **Avg. salary:** $40,970. **Notable fed. facilities:** Naval Air Warfare Center; Ft. Benjamin Harrison; Del. Grissom AFB; Naval Surface Warfare Center.

Energy. Electricity production (1998, kWh, by source): Coal: 110.7 bil; Petroleum: 822 mil; Gas: 775 mil; Hydroelectric: 479 mil.

State data. Motto: Crossroads of America. **Flower:** Peony. **Bird:** Cardinal. **Tree:** Tulip poplar. **Song:** On the Banks of the Wabash, Far Away. **Entered union** Dec. 11, 1816; rank, 19th. **State fair** at Indianapolis; mid-Aug.

History. When the Europeans arrived, Miami, Potawatomi, Kickapoo, Piankashaw, Wea, and Shawnee peoples inhabited the area. A French trading post was built, 1731-32, at Vincennes. La Salle visited the present South Bend area, 1679 and 1681. The first French fort was built near present-day Lafayette, 1717. France ceded the area to Britain, 1763. During the American Revolution, American Gen. George Rogers Clark captured Vincennes, 1778, and defeated British forces, 1779. At war's end, Britain ceded the area to the U.S. Miami Indians defeated U.S. troops twice, 1790, but were beaten, 1794, at Fallen Timbers by Gen. Anthony Wayne. At Tippecanoe, 1811, Gen. William H. Harrison defeated Tecumseh's Indian confederation. The Delaware, Potawatomi, and Miami were moved farther west, 1820-1850.

Tourist attractions. Lincoln Log Cabin Historic Site, near Charleston; George Rogers Clark Park, Vincennes; Wyandotte Cave; Tippecanoe Battlefield Memorial Park; Benjamin Harrison home; Indianapolis 500 raceway and museum, all Indianapolis; Indiana Dunes, near Chesterton; National College Football Hall of Fame, South Bend; Hoosier Nat'l. Forest, south-central Indiana.

Famous "Hoosiers." Larry Bird, Ambrose Burnside, Hoagy Carmichael, Jim Davis, James Dean, Eugene V. Debs, Theodore Dreiser, Paul Dresser, Gil Hodges, David Letterman, Jane Pauley, Cole Porter, Dan Quayle, Gene Stratton Porter, Ernie Pyle, James Whitcomb Riley, Oscar Robertson, Red Skelton, Booth Tarkington, Kurt Vonnegut, Lew Wallace, Wendell L. Wilkie, Wilbur Wright.

Chamber of Commerce. One North Capital, Suite 200, Indianapolis, IN 46204.

Toll-free travel information. 1-800-289-6646.

Website. http://www.ai.org

Tourism website. http://www.state.in.us/tourism

Finance. FDIC-insured commercial banks (1998): 443. **Deposits:** $38.0 bil. **FDIC-insured savings institutions** (1998): 23. **Assets:** $3.2 bil.

Federal govt. Fed. civ. employees (Mar. 1998): 6,897. **Avg. salary:** $40,829.

Energy. Electricity production (1998, kWh, by source): Coal: 31.9 bil; Petroleum: 110 mil; Gas: 412 mil; Hydroelectric: 893 mil; Nuclear: 3.8 bil.

State data. Motto: Our liberties we prize, and our rights we will maintain. **Flower:** Wild rose. **Bird:** Eastern goldfinch. **Tree:** Oak. **Rock:** Geode. **Entered union** Dec. 28, 1846; rank, 29th. **State fair** at Des Moines; mid-Aug.

History. Early inhabitants were Mound Builders who dwelt on Iowa's fertile plains. Later, Woodland tribes including the Iowa and Yankton Sioux lived in the area. The first Europeans, Marquette and Jolliet, gave France its claim to the area, 1673. In 1762, France ceded the region to Spain, but Napoleon took it back, 1800. It became part of the U.S. through the Louisiana Purchase, 1803. Native American Sauk and Fox tribes moved into the area from states farther east but relinquished their land in defeat, after the 1832 uprising led by the Sauk chieftain Black Hawk. By mid-19th cent. they were forced to move on to Kansas. Iowa became a territory in 1838, and entered as a free state, 1846, strongly supporting the Union.

Tourist attractions. Herbert Hoover birthplace and library, West Branch; Effigy Mounds Natl. Monument, prehistoric Indian burial site, Marquette; Amana Colonies; Grant Wood's paintings and memorabilia, Davenport Municipal Art Gallery; Living History Farms, Des Moines; Adventureland, Altoona; Boone & Scenic Valley Railroad, Boone; Greyhound Parks, in Dubuque and Council Bluffs; Prairie Meadows horse racing, Altoona; riverboat cruises and casino gambling, Mississippi and Missouri Rivers; Iowa Great Lakes, Okoboji.

Famous Iowans. Tom Arnold, Johnny Carson, Marquis Childs, Buffalo Bill Cody, Mamie Dowd Eisenhower, George Gallup, Susan Glaspell, James Norman Hall, Harry Hansen, Herbert Hoover, Ann Landers, Glenn Miller, Lillian Russell, Billy Sunday, James A. Van Allen, Carl Van Vechten, Henry Wallace, John Wayne, Meredith Willson, Grant Wood.

Tourist information. Division of Tourism, Iowa Dept. of Economic Development, 200 E. Grand Ave., Des Moines, IA 50309.

Toll-free travel information. 1-800-345-IOWA.

Website. http://www.state.ia.us

Tourism website. http://www.state.ia.us/tourism/index.html

Iowa

Hawkeye State

People. Population (1998): 2,862,447; rank: 30; **net change** (1990-98): 3.1%. **Pop. density** (1998): 51.2 per sq mi. **Racial distribution** (1990): 96.6% white; 1.7% black **Hispanic pop.:** 1.2%.

Geography. Total area: 56,276 sq mi; rank: 26. **Land area:** 55,875 sq mi; rank: 23. **Acres forested:** 2,050,000. **Location:** West North Central state bordered by Mississippi R. on the E and Missouri R. on the W. **Climate:** humid, continental. **Topography:** Watershed from NW to SE; soil especially rich and land level in the N central counties. **Capital:** Des Moines.

Economy. Chief industries: agriculture, communications, construction, finance, insurance, trade, services, manufacturing. **Chief manuf. goods:** processed food products, tires, farm machinery, electronic products, appliances, household furniture, chemicals, fertilizers, auto accessories. **Chief crops:** silage and grain corn, soybeans, oats, hay. **Livestock** (Jan. 1999): 3.7 mil cattle/calves; 260,000 sheep/lambs; (Dec. 1998) 15.3 mil hogs/pigs; (Dec. 1998) 31.3 mil chickens (excl. broilers); (Dec. 1996) 17.2 mil broilers. **Timber/lumber** (1998): red cedar; 76 mil bd. ft. **Nonfuel minerals** (est. 1998): $524 mil; mostly crushed stone, portland cement, sand & gravel, gypsum, lime. **Internat. airport at:** Des Moines. **Value of construction** (1997): $3.2 bil. **Gross state product** (1997): $80.4 bil. **Employment distrib.** (May 1999): 26.4% serv.; 24.5% trade; 18.0% mfg.; 16.1% govt. **Per cap. pers. income** (1998): $23,925. **Sales tax** (1999): 5%. **Unemployment** (1998): 2.8%. **Tourism expends.** (1996): $3.6 bil. **Lottery** (1998): total sales: $173.9 mil; net income: $42.7 mil.

Kansas

Sunflower State

People. Population (1998): 2,629,067; rank: 32; **net change** (1990-98): 6.1%. **Pop. density** (1998): 32.1 per sq mi. **Racial distribution** (1990): 90.1% white; 5.8% black. **Hispanic pop.:** 3.8%.

Geography. Total area: 82,282 sq mi; rank: 15. **Land area:** 81,823 sq mi; rank: 13. **Acres forested:** 1,359,000. **Location:** West North Central state, with Missouri R. on E. **Climate:** temperate but continental, with great extremes between summer and winter. **Topography:** hilly Osage Plains in the E; central region level prairie and hills; high plains in the W. **Capital:** Topeka.

Economy. Chief industries: manufacturing, finance, insurance, real estate, services. **Chief manuf. goods:** transportation equipment, machinery & computer equipment, food and kindred products, printing & publishing. **Chief crops:** wheat, sorghum, corn, hay, soybeans, sunflowers. **Livestock** (Jan. 1999): 6.6 mil cattle/calves; 100,000 sheep/lambs; (Dec. 1998) 1.6 mil hogs/pigs; (Dec. 1998) 1.9 mil chickens (excl. broilers). **Timber/lumber:** (1998) oak, walnut; 12 mil bd. ft. **Nonfuel minerals** (est. 1998): $535 mil; mostly portland cement, salt, crushed stone, helium, sand & gravel. **Chief ports:** Kansas City. **Internat. airports at:** Wichita. **Value of construction** (1997): $3.8 bil. **Gross state product** (1997): $71.8 bil. **Employment distrib.** (May 1999): 25.6% serv.; 24.4% trade; 18.3% govt.; 15.9% mfg. **Per cap. pers. income** (1998): $24,981. **Sales tax** (1999): 4.9%. **Unemployment** (1998): 3.8%. **Tourism expends.** (1996): $3.1 bil. **Lottery** (1998): total sales: $192.0 mil; net income: $59.8 mil.

Finance. FDIC-insured commercial banks (1998): 393. **Deposits:** $28.9 bil. **FDIC-insured savings institutions** (1998): 18. **Assets:** $8.3 bil.

Federal govt. Fed. civ. employees (Mar. 1998): 14,806. Avg. salary: $41,351. Notable fed. facilities: Fts. Riley, Leavenworth; Leavenworth Federal Penitentiary; Colmery-O'Neal Veterans Hospital.

Energy. Electricity production (1998, kWh, by source): Coal: 28.0 bil; Petroleum: 122 mil; Gas: 2.9 bil; Nuclear: 10.4 bil.

State data. Motto: Ad Astra per Aspera (To the stars through difficulties). Flower: Native sunflower. Bird: Western meadowlark. Tree: Cottonwood. Song: Home on the Range. Entered union Jan. 29, 1861; rank, 34th. State fair at Hutchinson; begins Friday after Labor Day.

History. When Coronado first explored the area, Wichita, Pawnee, Kansa, and Osage peoples lived there. These Native Americans—hunters who also farmed—were joined on the Plains by the nomadic Cheyenne, Arapaho, Comanche, and Kiowa about 1800. French explorers established trading between 1682 and 1739, and the U.S. took over most of the area in the Louisiana Purchase, 1803. After 1830, thousands of eastern Native Americans were removed to Kansas. Kansas became a territory, 1854. Violent incidents between pro- and antislavery settlers caused the territory to be known as "Bleeding Kansas." It eventually entered the Union as a free state, 1861. Railroad construction after the war made Abilene and Dodge City terminals of large cattle drives from Texas.

Tourist attractions. Eisenhower Center, Abilene; Agricultural Hall of Fame and Natl. Center, Bonner Springs; Dodge City-Boot Hill & Frontier Town; Old Cowtown Museum, Wichita; Ft. Scott and Ft. Larned, restored 1800s cavalry forts; Kansas Cosmosphere and Space Center, Hutchinson; Woodlands Racetrack, Kansas City; U.S. Cavalry Museum, Ft. Riley; NCAA Visitors Center, Shawnee; Heartland Park Raceway, Topeka.

Famous Kansans. Ed Asner, Roscoe "Fatty" Arbuckle, Thomas Hart Benton, John Brown, George Washington Carver, Wilt Chamberlain, Walter P. Chrysler, Glenn Cunningham, John Stuart Curry, Robert Dole, Amelia Earhart, Wyatt Earp, Dwight D. Eisenhower, Ron Evans, Wild Bill Hickok, Cyrus Holliday, Dennis Hopper, William Inge, Walter Johnson, Nancy Landon Kassebaum, Buster Keaton, Emmett Kelly, Alf Landon, Edgar Lee Masters, Hattie McDaniel, Oscar Micheaux, Carry Nation, Georgia Neese-Gray, Charlie Parker, Gordon Parks, Jim Ryun, Barry Sanders, Vivian Vance, William Allen White, Jess Willard.

Tourist information. Kansas Dept. of Commerce & Housing, Travel and Tourism Div., 700 SW Harrison, Suite 1300, Topeka, KS 66601; 1-913-296-2009.

Toll-free travel information. 1-800-2KANSAS.

Website. http://www.ink.org

Tourism website. http://kansascommerce.com

Kentucky
Bluegrass State

People. Population (1998): 3,936,499; rank: 25; net change (1990-98): 6.8%. Pop. density (1998): 99.1 per sq mi. Racial distribution (1990): 92.0% white; 7.1% black Hispanic pop.: 0.6%.

Geography. Total area: 40,411 sq mi; rank: 37. Land area: 39,732 sq mi; rank: 36. Acres forested: 12,714,000. Location: East South Central state, bordered on N by Illinois, Indiana, Ohio; on E by West Virginia and Virginia; on S by Tennessee; on W by Missouri. Climate: moderate, with plentiful rainfall. Topography: mountainous in E; rounded hills of the Knobs in the N; Bluegrass, heart of state; wooded rocky hillsides of the Pennyroyal; Western Coal Field; the fertile Purchase in the SW. Capital: Frankfort.

Economy. Chief industries: manufacturing, services, finance, insurance and real estate, retail trade, public utilities. Chief manuf. goods: transportation & industrial machinery, apparel, printing & publishing, food products, electric & electronic equipment. Chief crops: tobacco, corn, soybeans. Livestock (Jan. 1999): 2.4 mil cattle/calves; 21,000 sheep/lambs; (Dec. 1998) 520,000 hogs/pigs; (Dec. 1998) 5.3 mil chickens (excl. broilers); (Dec. 1997) 110.6 mil broilers. Timber/lumber (1998): hardwoods, pines; 777 mil bd. ft. Nonfuel minerals (est. 1998): $489 mil; mostly crushed stone, lime, portland cement, sand & gravel, clays. Chief ports: Paducah, Louisville, Covington, Owensboro, Ashland, Henderson County, Lyon County, Hickman-Fulton County. Internat. airports at: Covington and Louisville. Value of construction (1997): $4.8 bil. Gross state product (1997): $100.0 bil. Employment distrib. (May 1999): 25.5% serv.; 23.9% trade; 17.8% mfg.; 16.8% govt. Per cap. pers. income (1998): $21,506. Sales tax (1999): 6%. Unemployment (1998): 4.6%. Tourism expends. (1996): $4.4 bil. Lottery (1998): total sales: $585 mil; net income: $180.6 mil.

Finance. FDIC-insured commercial banks (1998): 261. Deposits: $39.9 bil. FDIC-insured savings institutions (1998): 40. Assets: $3.5 bil.

Federal govt. Fed. civ. employees (Mar. 1998): 20,259. Avg. salary: $37,208. Notable fed. facilities: U.S. Gold Bullion Depository, Fort Knox; Federal Correctional Institution, Lexington.

Energy. Electricity production (1998, kWh, by source): Coal: 82.4 bil; Petroleum: 127 mil; Gas: 496 mil; Hydroelectric: 3.1 bil.

State data. Motto: United we stand, divided we fall. Flower: Goldenrod. Bird: Cardinal. Tree: Tulip Poplar. Song: My Old Kentucky Home. Entered union June 1, 1792; rank, 15th. State fair at Louisville, late Aug.

History. The area was predominantly hunting grounds for Shawnee, Wyandot, Delaware, and Cherokee peoples. Explored by Americans Thomas Walker and Christopher Gist, 1750-51, Kentucky was the first area west of the Alleghenies settled by American pioneers. The first permanent settlement was Harrodsburg, 1774. Daniel Boone blazed the Wilderness Trail through the Cumberland Gap and founded Ft. Boonesborough, 1775. Conflicts with Native Americans, spurred by the British, were unceasing until, during the American Revolution, Gen. George Rogers Clark captured British forts in Indiana and Illinois, 1778. In 1792, Virginia dropped its claims to the region, and it became the 15th state. Although officially a Union state, Kentuckians had divided loyalties during the Civil War and were forced to choose sides; its slaves were freed only after the adoption of the 13th Amendment to the U.S. Constitution, 1865.

Tourist attractions. Kentucky Derby; Louisville; Land Between the Lakes Natl. Recreation Area, Kentucky Lake and Lake Barkley; Mammoth Cave Natl. Park; Echo River, 360 ft below ground; Lake Cumberland; Lincoln's birthplace, Hodgenville; My Old Kentucky Home State Park, Bardstown; Cumberland Gap Natl. Historical Park, Middlesboro; Kentucky Horse Park, Lexington; Shaker Village, Pleasant Hill.

Famous Kentuckians. Muhammad Ali, John James Audubon, Alben W. Barkley, Daniel Boone, Louis D. Brandeis, John C. Breckinridge, Kit Carson, Albert B. "Happy" Chandler, Henry Clay, Jefferson Davis, D. W. Griffith, "Casey" Jones, Abraham Lincoln, Mary Todd Lincoln, Thomas Hunt Morgan, Carry Nation, Col. Harland Sanders, Diane Sawyer, Jesse Stuart, Adlai Stevenson, Zachary Taylor, Robert Penn Warren, Whitney Young Jr.

Tourist Information. Kentucky Dept. of Travel, 500 Mero St., #2200, Frankfort, KY 40601.

Toll-free travel information. 1-800-225-TRIP.

Website. http://www.state.ky.us

Tourism website. http://www.kentuckytourism.com

Louisiana
Pelican State

People. Population (1998): 4,368,967; rank: 22; net change (1990-98): 3.5%. Pop. density (1998): 100.3 per sq mi. Racial distribution (1990): 67.3% white; 30.8% black. Hispanic pop.: 2.2%.

Geography. Total area: 49,651 sq mi; rank: 31. Land area: 43,566 sq mi; rank: 33. Acres forested: 13,864,000. Location: West South Central state on the Gulf Coast. Climate: subtropical, affected by continental weather patterns. Topography: lowlands of marshes and Mississippi R. flood plain; Red R. Valley lowlands; upland hills in the Florida Parishes; average elevation, 100 ft. Capital: Baton Rouge.

Economy. Chief industries: wholesale and retail trade, tourism, manufacturing, construction, transportation, communication, public utilities, finance, insurance, real estate, mining. Chief manuf. goods: chemical products, foods, transportation equipment, electronic equipment, petroleum products, lumber, wood, and paper. Chief crops: soybeans, sugarcane, rice, corn, cotton, sweet potatoes, pecans, sorghum, aquaculture. Livestock (Jan. 1999): 900,000 cattle/calves; 6,500 sheep/lambs; (Dec. 1998) 30,000 hogs/pigs; (Dec. 1998) 2.7 mil chickens (excl. broilers). Timber/lumber (1998): pines, hardwoods, oak; 1.3 bil bd. ft. Nonfuel minerals (est. 1998): $379 mil; mostly salt, sulfur, sand & gravel, crushed stone. **Commercial fishing** (1998): $291.9 mil. **Chief ports:** New Orleans, Baton Rouge, Lake Charles, Port of S. Louisiana (La Place), Shreveport, Plaquemine, St. Bernard, Alexandria. **Internat. airports at:** New Orleans, Alexandria. **Value of construction** (1997): $4.7 bil. **Gross state product** (1997): $124.3 bil. **Employment distrib.** (May 1999): 27.2% serv.; 23.4% trade; 19.1% govt.; 9.9% mfg. **Per cap. pers. income** (1998): $21,346. **Sales tax** (1999): 4%. **Unemployment** (1998): 5.7%. **Tourism expends.** (1996): $7.0 bil. **Lottery** (1998): total sales: $292.9 mil; net income: $107.2 mil.

Finance. FDIC-insured commercial banks (1998): 150. **Deposits:** $39.5 bil. **FDIC-insured savings institutions** (1998): 33. **Assets:** $4.1 bil.

Federal govt. Fed. civ. employees (Mar. 1998): 19,923. **Avg. salary:** $40,662. **Notable federal facilities:** Strategic Petroleum Reserve, Michoud Assembly Plant, Southeast U.S. Agricultural Research Ctr., U.S. Army Corps of Engineers, all New Orleans; Ft. Polk military bases, Barksdale; U.S. Public Service Hospital, Carville; Naval Air Station, Chalmette; V.A. Hospital, Pineville.

Energy. Electricity production (1998, kWh, by source): Coal: 20.8 bil; Petroleum: 600 mil; Gas: 28.3 bil; Nuclear: 16.4 bil.

State data. Motto: Union, justice, and confidence. **Flower:** Magnolia. **Bird:** Eastern brown pelican. **Tree:** Cypress. **Song:** Give Me Louisiana. **Entered union** Apr. 30, 1812; rank, 18th. **State fair** at Shreveport; Oct.

History. Caddo, Tunica, Choctaw, Chitimacha, and Chawash peoples lived in the region at the time of European contact. Europeans Cabeza de Vaca and Panfilo de Narvaez first visited, 1530. The region was claimed for France by La Salle, 1682. The first permanent settlement was by the French at Biloxi, now in Mississippi, 1699. France ceded the region to Spain, 1762, took it back, 1800, and sold it to the U.S., 1803, in the Louisiana Purchase. During the American Revolution, Spanish Louisiana aided the Americans. Admitted as a state in 1812, Louisiana was the scene of the Battle of New Orleans, 1815.

Louisiana Creoles are descendants of early French and/or Spanish settlers. About 4,000 Acadians, French settlers in Nova Scotia, Canada, were forcibly transported by the British to Louisiana in 1755 (an event commemorated in Longfellow's "Evangeline") and settled near Bayou Teche; their descendants became known as Cajuns. Another group, the Islenos, were descendants of Canary Islanders brought to Louisiana by a Spanish governor in 1770. Traces of Spanish and French survive in local dialects.

Tourist attractions. Mardi Gras, French Quarter, Superdome, Dixieland jazz, Aquarium of the Americas, Audubon Zoo & Gardens, all New Orleans; Battle of New Orleans site; Longfellow-Evangeline Memorial Park, St. Martinville; Kent House Museum, Alexandria; Hodges Gardens, Natchitoches, USS Kidd Memorial, Baton Rouge.

Famous Louisianans. Louis Armstrong, Pierre Beauregard, Judah P. Benjamin, Braxton Bragg, Kate Chopin, Lillian Hellman, Grace King, Bob Livingston, Huey Long, Winton Marsalis, Leonidas K. Polk, Anne Rice, Henry Miller Shreve, Edward D. White Jr.

Tourist information. Louisiana Office of Tourism, PO Box 94291, Baton Rouge, LA 70804-9291.
Toll-free travel information. 1-800-677-4082.
Website. http://www.state.la.us
Tourism website. http://www.louisianatravel.com

Maine
Pine Tree State

People. Population (1998): 1,244,250; rank: 39; **net change** (1990-98): 1.3%. **Pop. density** (1998): 40.3 per sq mi. **Racial distribution** (1990): 98.4% white; 0.4% black **Hispanic pop.:** 0.6%.

Geography. Total area: 33,741 sq mi; rank: 39. **Land area:** 30,865 sq mi; rank: 39. **Acres forested:** 17,533,000. **Location:** New England state at northeastern tip of U.S. **Climate:** Southern interior and coastal, influenced by air masses from the S and W; northern clime harsher, avg. over 100 in. snow in winter. **Topography:** Appalachian Mts. extend through state; western borders have rugged terrain; long sand beaches on southern coast; northern coast mainly rocky promontories, peninsulas, fjords. **Capital:** Augusta.

Economy. Chief industries: manufacturing, agriculture, fishing, services, trade, government, finance, insurance, real estate, construction. **Chief manuf. goods:** paper & wood products, transportation equipment. **Chief crops:** potatoes, aquaculture products. **Livestock** (Jan. 1999): 100,000 cattle/calves; 11,000 sheep/lambs; (Dec. 1998) 6,000 hogs/pigs; (Dec. 1998) 6.8 mil chickens (excl. broilers). **Timber/lumber** (1998): pine, spruce, fir; 1.2 bil bd. ft. **Nonfuel minerals** (est. 1998): $76.2 mil; mostly sand & gravel, portland cement, crushed stone, peat, masonry cement. **Commercial fishing** (1998): $216.4 mil. **Chief ports:** Searsport, Portland, Eastport. **Internat. airports at:** Portland, Bangor. **Value of construction** (1997): $1.1 bil. **Gross state product** (1997): $30.2 bil. **Employment distrib.** (May 1999): 30.0% serv.; 25.1% trade; 16.1% govt.; 14.6% mfg. **Per cap. pers. income (1998):** $22,952. **Sales tax** (1999): 5.5%. **Unemployment** (1998): 4.4%. **Tourism expends.** (1996): $1.9 bil. **Lottery** (1998): total sales: $148.9 mil; net income: $41.8 mil.

Finance. FDIC-insured commercial banks (1998): 17. **Deposits:** $4.0 bil. **FDIC-insured savings institutions** (1998): 28. **Assets:** $10.3 bil.

Federal govt. Fed. civ. employees (Mar. 1998): 7,470. **Avg. salary:** $41,421. **Notable fed. facilities:** Kittery Naval Shipyard; Brunswick Naval Air Station.

Energy. Electricity production (1998, kWh, by source): Petroleum: 1.7 bil; Hydroelectric: 1.8 bil.

State data. Motto: Dirigo (I direct). **Flower:** White pine cone and tassel. **Bird:** Chickadee. **Tree:** Eastern white pine. **Song:** State of Maine Song. **Entered union** Mar. 15, 1820; rank, 23d. **State fair:** at Bangor, late July; at Skowhegan, mid-Aug.

History. When the Europeans arrived, Maine was inhabited by Algonquian peoples including the Abnaki, Penobscot, and Passamaquoddy. Maine's rocky coast was believed to have been explored by the Cabots, 1498-99. French settlers arrived, 1604, at the St. Croix River, English, c 1607, on the Kennebec; both settlements failed. Maine was made part of Massachusetts, 1691. In the American Revolution, a Maine regiment fought at Bunker Hill. A British fleet destroyed Falmouth (now Portland), 1775, but the British ship *Margaretta* was captured near Machiasport. In 1820, Maine broke off and became a separate state.

Tourist attractions. Acadia Natl. Park, Bar Harbor, on Mt. Desert Island; Old Orchard Beach; Portland's Old Port; Kennebunkport; Common Ground Country Fair; Portland Headlight; Baxter State Pk.; Freeport/L. L. Bean.

Famous "Down Easters." James G. Blaine, Cyrus H. K. Curtis, Hannibal Hamlin, Sarah Jewett, Stephen King, Henry Wadsworth Longfellow, Sir Hiram and Hudson Maxim, Edna St. Vincent Millay, George Mitchell, Edmund Muskie, Edwin Arlington Robinson, Kate Douglas Wiggin, Ben Ames Williams.

Chamber of Commerce and Industry. Maine Chamber & Business Alliance, 7 Community Dr., Augusta, ME 04330.
Toll-free travel information. 1-888-624-6345 (from within the United States and Canada).
Website. http://www.state.me.us
Tourism website. http://www.visitmaine.com

Maryland
Old Line State, Free State

People. Population (1998): 5,134,808; rank: 19; **net change** (1990-98): 7.4%. **Pop. density** (1998): 525.3 per sq mi. **Racial distribution** (1990): 71.0% white; 24.9% black; 2.9% Asian. **Hispanic pop.:** 2.6%.

Geography. Total area: 12,297 sq mi; rank: 42. **Land area:** 9,775 sq mi; rank: 42. **Acres forested:** 2,700,000. **Location:** South Atlantic state stretching from the Ocean to the Allegheny Mts. **Climate:** continental in the west; humid subtropical in the east. **Topography:** Eastern Shore of coastal plain and Maryland Main of coastal plain, piedmont plateau, and the Blue Ridge, separated by the Chesapeake Bay. **Capital:** Annapolis.

Economy. Chief industries: manufacturing, biotechnology and information technology, services, tourism. **Chief manuf. goods:** electric and electronic equipment; food and kindred products, chemicals and allied products, printed materials. **Chief crops:** greenhouse and nursery products, soybeans, corn. **Livestock** (Jan. 1999): 250,000 cattle/calves; 23,000 sheep/lambs; (Dec. 1998) 65,000 hogs/pigs; (Dec. 1998) 4.6 mil chickens (excl. broilers); (Dec. 1997) 295.3 mil broilers. **Timber/lumber:** (1998) hardwoods; 262 mil bd. ft. **Nonfuel minerals** (est. 1998): $358 mil; mostly crushed stone, portland cement, sand & gravel, masonry cement, dimension stone. **Commercial fishing** (1998): $67.2 mil. **Chief port:** Baltimore. **Internat. airport at:** Baltimore-Washington Intl. **Value of construction** (1997): $5.9 bil. **Gross state product** (1997): $153.8 bil. **Employment distrib.** (May 1999): 34.1% serv.; 23.0% trade; 18.6% govt.; 7.5% mfg. **Per cap. pers. income** (1998): $29,943. **Sales tax** (1999): 5%. **Unemployment** (1998): 4.6%. **Tourism expends.** (1996): $6.4 bil. **Lottery** (1998): total sales: $1.1 bil; net income: $399.9 mil.

Finance. FDIC-insured commercial banks (1998): 80. **Deposits:** $33.5 bil. **FDIC-insured savings institutions** (1998): 66. **Assets:** $8.0 bil.

Federal govt. Fed. civ. employees (Mar. 1998): 101,099. **Avg. salary:** $52,860. **Notable fed. facilities:** U.S. Naval Academy; Natl. Agriculture Research Center; Ft. George G. Meade, Aberdeen Proving Ground; Goddard Space Flight Center; Natl. Institutes of Health; Natl. Institute of Standards & Technology; Food & Drug Administration; Bureau of the Census.

Energy. Electricity production (1998, kWh, by source): Coal: 29.1 bil; Petroleum: 3.3 bil; Gas: 1.1 bil; Hydroelectric: 1.7 bil; Nuclear: 13.3 bil.

State data. Motto: Fatti Maschii, Parole Femine (Manly deeds, womanly words). **Flower:** Black-eyed Susan. **Bird:** Baltimore oriole. **Tree:** White oak. **Song:** Maryland, My Maryland. **Seventh** of the original 13 states to ratify Constitution, Apr. 28, 1788. **State fair** at Timonium; late Aug.-early Sept.

History. Europeans encountered Algonquian-speaking Nanticoke and Piscataway and Iroquois-speaking Susquehannock when they first visited the area. Italian explorer Verrazano visited the Chesapeake region in the early 16th cent. English Capt. John Smith explored and mapped the area, 1608. William Claiborne set up a trading post on Kent Island in Chesapeake Bay, 1631. King Charles I granted land to Cecilius Calvert, Lord Baltimore, 1632; Calvert's brother Leonard, with about 200 settlers, founded St. Marys, 1634. The bravery of Maryland troops in the American Revolution, as at the Battle of Long Island, won the state its nickname "The Old Line State." In the War of 1812, when a British fleet tried to take Ft. McHenry, Marylander Francis Scott Key wrote "The Star Spangled Banner," 1814. Although a slave-holding state, Maryland remained with the Union during the Civil War and was the site of the battle of Antietam, 1862, which halted Gen. Robert E. Lee's march north.

Tourist attractions. The Preakness at Pimlico track, Baltimore; The Maryland Million at Laurel Race Course; Ocean City; restored Ft. McHenry, near which Francis Scott Key wrote "The Star-Spangled Banner"; Edgar Allan Poe house, Ravens Football at Memorial Stadium, Camden Yards, Natl. Aquarium, Harborplace, all Baltimore; Antietam Battlefield, near Hagerstown; South Mountain Battlefield; U.S. Naval Academy, Annapolis; Maryland State House, Annapolis, 1772, the oldest still in legislative use in the U.S.

Famous Marylanders. John Astin, Benjamin Banneker, Tom Clancy, Jonathan Demme, Francis Scott Key, H. L. Mencken, Charles Willson Peale, William Pinkney, Edgar Allan Poe, Babe Ruth, Upton Sinclair, Roger B. Taney, John Waters.

Maryland Dept. of Business & Economic Development. 217 E. Redwood St., Baltimore, MD 21202; (410) 767-6870. **Toll-free travel information.** 1-800-543-1036. **Website.** http://www.state.md.us **Tourism website.** http//www.mdisfun.org

Massachusetts

Bay State, Old Colony

People. Population (1998): 6,147,132; rank: 13; **net change** (1990-98): 2.2%. **Pop. density** (1998): 784.3 per sq mi. **Racial distribution** (1990): 89.8% white; 5.0% black; 2.4% Asian. **Hispanic pop.:** 4.8%.

Geography. Total area: 9,241 sq mi; rank: 45. **Land area:** 7,838 sq mi; rank: 45. **Acres forested:** 3,203,000. **Location:** New England state along Atlantic seaboard. **Climate:** temperate, with colder and drier clime in western region. **Topography:** jagged indented coast from Rhode Island around Cape Cod; flat land yields to stony upland pastures near cen-

tral region and gentle hilly country in west; except in west, land is rocky, sandy, and not fertile. **Capital:** Boston.

Economy. Chief industries: services, trade, manufacturing. **Chief manuf. goods:** electric and electronic equipment, instruments, industrial machinery and equipment, printing and publishing, fabricated metal products. **Chief crops:** cranberries, greenhouse, nursery, vegetables. **Livestock** (Jan. 1999): 57,000 cattle/calves; 8,500 sheep/lambs; (Dec. 1998) 19,500 hogs/pigs; (Dec. 1998) 494,000 chickens (excl. broilers). **Timber/lumber** white pine, oak, other hard woods. **Nonfuel minerals** (est. 1998): $192 mil; mostly crushed stone, sand & gravel, dimension stone, lime, clays. **Commercial fishing** (1998): $204.4 mil. **Chief ports:** Boston, Fall River, New Bedford, Salem, Gloucester, Plymouth. **Internat. airport at:** Boston. **Value of construction** (1997): $10.3 bil. **Gross state product** (1997): $221.0 bil. **Employment distrib.** (May 1999): 36.0% serv.; 22.8% trade; 13.5% mfg.; 12.9% govt. **Per cap. pers. income** (1998): $32,797. **Sales tax** (1999): 5%. **Unemployment** (1998): 3.3%. **Tourism expends.** (1996): $10.2 bil. **Lottery** (1998): total sales: $3.2 bil; net income: $774.9 mil.

Finance. FDIC-insured commercial banks (1998): 44. **Deposits:** $94.5 bil. **FDIC-insured savings institutions** (1998): 190. **Assets:** $61.4 bil.

Federal govt. Fed. civ. employees (Mar. 1998): 25,367. **Avg. salary:** $46,082. **Notable fed. facilities:** Thomas P. O'Neill Jr. Federal Bldg., J.W. McCormack Bldg., John Fitzgerald Kennedy Federal Bldg., Q.M. Laboratory, Natick.

Energy. Electricity production (1998, kWh, by source): Coal: 8.2 bil; Petroleum: 10.0 bil; Gas: 1.8 bil; Hydroelectric: 331 mil; Nuclear: 5.7 bil.

State data. Motto: Ense Petit Placidam Sub Libertate Quietem (By the sword we seek peace, but peace only under liberty). **Flower:** Mayflower. **Bird:** Chickadee. **Tree:** American elm. **Song:** All Hail to Massachusetts. **Sixth** of the original 13 states to ratify Constitution, Feb. 6, 1788. **State Fair** at Topsfield, early Oct.

History. Early inhabitants were the Algonquian, Nauset, Wampanoag, Massachuset, Pennacook, Nipmuc, and Pocumtuc peoples. Pilgrims settled in Plymouth, 1620, giving thanks for their survival with the first Thanksgiving Day, 1621. About 20,000 new settlers arrived, 1630-40. Native American relations with the colonists deteriorated leading to King Philip's War, 1675-76, which the colonists won, ending Native American resistance. Demonstrations against British restrictions set off the Boston Massacre, 1770, and the Boston Tea Party, 1773. The first bloodshed of American Revolution was at Lexington, 1775.

Tourist attractions. Provincetown artists' colony; Cape Cod; Plymouth Rock, Plymouth Plantation, Mayflower II, all Plymouth; Freedom Trail, Museum of Fine Arts, Children's Museum, Museum of Science, New England Aquarium, JFK Library, Boston Ballet, Boston Pops, Boston Symphony Orchestra, all Boston; Tanglewood, Jacob's Pillow Dance Festival, Hancock Shaker Village Berkshire Railway Museum, all in the Berkshires; Salem; Old Sturbridge Village; Deerfield Historic District; Walden Pond; Naismith Memorial Basketball Hall of Fame, Springfield.

Famous "Bay Staters." John Adams, John Quincy Adams, Samuel Adams, Louisa May Alcott, Horatio Alger, Susan B. Anthony, Crispus Attucks, Clara Barton, Alexander Graham Bell, Stephen Breyer, George Bush, John Cheever, E. E. Cummings, Emily Dickinson, Charles Eliot, Ralph Waldo Emerson, William Lloyd Garrison, Edward Everett Hale, John Hancock, Nathaniel Hawthorne, Oliver Wendell Holmes, Winslow Homer, Elias Howe, John F. Kennedy, James Russell Lowell, Cotton Mather, Samuel F. B. Morse, Edgar Allan Poe, Paul Revere, Dr. Seuss, Henry David Thoreau, James McNeil Whistler, John Greenleaf Whittier.

Tourist information. Massachusetts Office of Travel & Tourism, 100 Cambridge St., 13th Floor, Boston, MA 02202. **Toll-free travel information.** 1-800-227-6277. **Website.** http://www.state.ma.us **Tourism website.** http://www.mass-vacation.com

Michigan

Great Lakes State, Wolverine State

People. Population (1998): 9,817,242; rank: 8; **net change** (1990-98): 5.6%. **Pop. density** (1998): 172.8 per sq mi. **Racial distribution** (1990): 83.4% white; 13.9% black. **Hispanic pop.:** 2.2%.

Geography. Total area: 96,705 sq mi; rank: 11. **Land area:** 56,809 sq mi; rank: 22. **Acres forested:** 18,253,000. **Location:** East North Central state bordering on 4 of the 5 Great Lakes, divided into an Upper and Lower Peninsula by the Straits of Mackinac, which link lakes Michigan and Huron. **Climate:** well-defined seasons tempered by the Great Lakes. **Topography:** low rolling hills give way to northern tableland of hilly belts in Lower Peninsula; Upper Peninsula is level in the east, with swampy areas; western region is higher and more rugged. **Capital:** Lansing.

Economy. Chief industries: manufacturing, services, tourism, agriculture, forestry/lumber. **Chief manuf. goods:** automobiles, transportation equipment, machinery, fabricated metals, food products, plastics, office furniture. **Chief crops:** corn, wheat, soybeans, dry beans, hay, potatoes, sweet corn, apples, cherries, sugar beets, blueberries, cucumbers, Niagra grapes. **Livestock** (Jan. 1999): 1.1 mil cattle/calves; 62,000 sheep/lambs; (Dec. 1998) 1.1 mil hogs/pigs; (Dec. 1998) 6.7 mil chickens (excl. broilers); (Dec. 1997) 640,000 broilers. **Timber/lumber** (1998): maple, oak, aspen; 683 mil bd. ft. **Nonfuel minerals** (est. 1998): $1.66 bil; mostly portland cement, iron ore, sand & gravel, magnesium compounds, crushed stone. **Commercial fishing** (1998): $8.9 mil. **Chief ports:** Detroit, Saginaw River, Escanaba, Muskegon, Sault Ste. Marie, Port Huron, Marine City. **Internat. airports at:** Detroit, Grand Rapids, Flint, Kalamazoo, Lansing, Saginaw. **Value of construction** (1997): $10.7 bil. **Gross state product** (1997): $272.6 bil. **Employment distrib.** (May 1999): 27.7% serv.; 23.8% trade; 21.1% mfg.; 14.5% govt. **Per cap. pers. income** (1998): $25,857. **Sales tax** (1999): 6%. **Unemployment** (1998): 3.9%. **Tourism expends.** (1996): $9.6 bil. **Lottery** (1998): total sales: $1.6 bil; net income: $709.2 mil.

Finance. FDIC-insured commercial banks (1998): 165. **Deposits:** $87.0 bil. **FDIC-insured savings institutions** (1998): 24. **Assets:** $30.8 bil.

Federal govt. Fed. civ. employees (Mar. 1998): 21,459. **Avg. salary:** $46,463. **Notable fed. facilities:** Isle Royal, Sleeping Bear Dunes national parks.

Energy. Electricity production (1998, kWh, by source): Coal: 69.1 bil; Petroleum: 1.0 bil; Gas: 2.2 bil; Hydroelectric: 352 mil; Nuclear: 12.5 bil.

State data. Motto: Si Quaeris Peninsulam Amoenam, Circumspice (If you seek a pleasant peninsula, look about you). **Flower:** Apple blossom. **Bird:** Robin. **Tree:** White pine. **Song:** Michigan, My Michigan. **Entered union** Jan. 26, 1837; rank, 26th. **State fair** at Detroit, late Aug.–early Sept.; Upper Peninsula (Escanaba), mid-Aug.

History. Early inhabitants were the Ojibwa, Ottawa, Miami, Potawatomi, and Huron. French fur traders and missionaries visited the region, 1616, set up a mission at Sault Ste. Marie, 1641, and a settlement there, 1668. French settlements were taken over, 1763, by the British, who crushed a Native American uprising led by Ottawa chieftain Pontiac that same year. Treaty of Paris ceded territory to U.S., 1783, but British remained until 1796. The British seized Ft. Mackinac and Detroit, 1812. After Oliver H. Perry's Lake Erie victory and William H. Harrison's victory near the Thames River, 1813, the British retreated to Canada. The opening of the Erie Canal, 1825, and new land laws and Native American cessions led the way for a flood of settlers.

Tourist attractions. Henry Ford Museum, Greenfield Village, both in Dearborn; Michigan Space Center, Jackson; Tahquamenon (Hiawatha) Falls; DeZwaan windmill and Tulip Festival, Holland; "Soo Locks," St. Mary's Falls Ship Canal, Sault Ste. Marie, Kalamazoo Aviation History Museum; Mackinac Island; Kellogg's Cereal City USA, Battle Creek; Museum of African-American History, Motown Historical Museum, both Detroit.

Famous People. Ralph Bunche, Paul de Kruif, Thomas A. Edison, Edna Ferber, Gerald R. Ford, Henry Ford, Aretha Franklin, Edgar Guest, Lee Iacocca, Robert Ingersoll, Magic Johnson, Will Kellogg, Ring Lardner, Elmore Leonard, Charles Lindbergh, Joe Louis, Madonna, Jack Paar, Pontiac, Diana Ross, Glenn Seaborg, Tom Selleck, John Smoltz, Lily Tomlin, Stewart Edward White, Malcolm X.

State Chamber of Commerce. 600 S. Walnut, Lansing, MI 48933. Phone: 517-371-2100

Toll-free travel information. 1-888-784-7328.

Website. http://www.migov.state.mi.us

Tourism website. http://www.michigan.org

Minnesota
North Star State, Gopher State

People. Population (1998): 4,725,419; rank: 20; **net change** (1990-98): 8.0%. **Pop. density** (1998): 59.4 per sq mi. **Racial distribution** (1990): 94.4% white; 2.2% black; 1.8% Asian. **Hispanic pop.:** 1.2%.

Geography. Total area: 86,943 sq mi; rank: 12. **Land area:** 79,617 sq mi; rank: 14. **Acres forested:** 16,718,000. **Location:** West North Central state bounded on the E by Wisconsin and Lake Superior, on the N by Canada, on the W by the Dakotas, and on the S by Iowa. **Climate:** northern part of state lies in the moist Great Lakes storm belt; the western border lies at the edge of the semi-arid Great Plains. **Topography:** central hill and lake region covering approx. half the state; to the NE, rocky ridges and deep lakes; to the NW, flat plain; to the S, rolling plains and deep river valleys. **Capital:** St. Paul.

Economy. Chief industries: agribusiness, forest products, mining, manufacturing, tourism. **Chief manuf. goods:** food, chemical and paper products, industrial machinery, electric and electronic equipment, computers, printing & publishing, scientific and medical instruments, fabricated metal products, forest products. **Chief crops:** corn, soybeans, wheat, sugar beets, hay, barley, potatoes, sunflowers. **Livestock** (Jan. 1999): 2.5 mil cattle/calves; 175,000 sheep/lambs; (Dec. 1998) 5.7 mil hogs/pigs; (Dec. 1998) 15.6 mil chickens (excl. broilers); (Dec. 1997) 46.3 mil broilers. **Timber/lumber** (1998): needle-leaves and hardwoods; 306 mil bd. ft. **Nonfuel minerals** (est. 1998): $1.56 bil; mostly iron ore, sand & gravel, crushed stone, dimension stone. **Commercial fishing** (1998): $224,000. **Chief ports:** Duluth, St. Paul, Minneapolis. **Internat. airport at:** Minneapolis-St. Paul. **Value of construction** (1997): $6.2 bil. **Gross state product** (1997): $149.4 bil. **Employment distrib.** (May 1999): 28.5% serv.; 24.0% trade; 17.0% mfg.; 14.8% govt. **Per cap. pers. income** (1998): $27,510. **Sales tax** (1999): 6.5%. **Unemployment** (1998): 2.5%. **Tourism expends.** (1996): $5.8 bil. **Lottery** (1998): total sales: $372.9 mil; net income: $87.3 mil.

Finance. FDIC-insured commercial banks (1998): 514. **Deposits:** $103.7 bil. **FDIC-insured savings institutions** (1998): 22. **Assets:** $2.5 bil.

Federal govt. Fed. civ. employees (Mar. 1998): 12,945. **Avg. salary:** $45,052.

Energy. Electricity production (1998, kWh, by source): Coal: 29.9 bil; Petroleum: 650 mil; Gas: 652 mil; Hydroelectric: 695 mil; Nuclear: 11.6 bil.

State data. Motto: L'Etoile du Nord (The star of the north). **Flower:** Pink and white lady's-slipper. **Bird:** Common loon. **Tree:** Red pine. **Song:** Hail! Minnesota. **Entered union** May 11, 1858; rank, 32d. **State fair** at St. Paul/Minneapolis; late Aug.-early Sept.

History. Dakota Sioux were early inhabitants of the area, and in the 16th cent., the Ojibwa began moving in from the east. French fur traders Médard Chouart and Pierre Esprit Radisson entered the region in the mid-17th cent. In 1679, French explorer Daniel Greysolon, sieur Duluth, claimed the entire region in the name of France. Britain took the area east of the Mississippi, 1763. The U.S. took over that portion after the American Revolution and in 1803, gained the western area in the Louisiana Purchase. The U.S. built Ft. St. Anthony (now Ft. Snelling), 1819, and in 1837, bought Native American lands, spurring an influx of settlers from the east. In 1849, the Territory of Minnesota was created. Sioux Indians staged a bloody uprising, the Battle of Woods Lake, 1862, and were driven from the state.

Tourist attractions. Minneapolis Institute of Arts, Walker Art Center, Minneapolis Sculpture Garden, Minnehaha Falls (inspiration for Longfellow's Hiawatha), Guthrie Theater, Minneapolis; Ordway Theater, St. Paul; Voyageurs Natl. Park; Mayo Clinic, Rochester; St. Paul Winter Carnival; North Shore (of Lake Superior).

Famous Minnesotans. Warren Burger, William O. Douglas, Bob Dylan, F. Scott Fitzgerald, Judy Garland, Cass Gilbert, Hubert Humphrey, Garrison Keillor, Sister Elizabeth Kenny, Sinclair Lewis, Paul Manship, Roger Maris, E. G. Marshall, William and Charles Mayo, Eugene McCarthy, Walter F. Mondale, Charles Schulz, Harold Stassen, Thorstein Veblen.

Chamber of Commerce. 30 East 7th St., Suite 1700, St. Paul, MN 55101-4901.

Toll-free travel information. 1-800-657-3700.

Website. http://www.state.mn.us

Tourism website. http://www.dted.state.mn.us/explore/explore.html

Mississippi
Magnolia State

People. Population (1998): 2,752,092; rank: 31; **net change** (1990-98): 6.9%. **Pop. density** (1998): 58.7 per sq mi. **Racial distribution** (1990): 63.5% white; 35.6% black. **Hispanic pop.:** 0.6% .

Geography. Total area: 48,286 sq mi; rank: 32. **Land area:** 46,914 sq mi; rank: 31. **Acres forested:** 17,000,000. **Location:** East South Central state bordered on the W by the Mississippi R. and on the S by the Gulf of Mexico. **Climate:** semi-tropical, with abundant rainfall, long growing season, and extreme temperatures unusual. **Topography:** low, fertile delta between the Yazoo and Mississippi rivers; loess bluffs stretching around delta border; sandy gulf coastal terraces followed by piney woods and prairie; rugged, high sandy hills in extreme NE followed by Black Prairie Belt, Pontotoc Ridge, and flatwoods into the north central highlands. **Capital:** Jackson.

Economy. Chief industries: warehousing & distribution, services, manufacturing, government, wholesale and retail trade. **Chief manuf. goods:** chemicals & plastics, food & kindred products, furniture, lumber & wood products, electrical machinery, transportation equipment. **Chief crops:** cotton, rice, soybeans. **Livestock** (Jan. 1999): 1.2 mil cattle/calves; (Dec. 1998) 275,000 hogs/pigs; (Dec. 1998) 11.4 mil chickens (excl. broilers); (Dec. 1997) 720.3 mil broilers. **Timber/lumber** (1998): pine, oak, hardwoods; 2.8 bil bd. ft. **Nonfuel minerals** (est. 1998): $190 mil; mostly sand & gravel, portland cement, clays, crushed stone. **Commercial fishing** (1998): $48.4 mil. **Chief ports:** Pascagoula, Vicksburg, Gulfport, Natchez, Greenville. **Value of construction** (1997): $2.6 bil. **Gross state product** (1997): $58.3 bil. **Employment distrib.** (May 1999): 23.2% serv.; 21.8% trade; 21.1% mfg.; 19.9% govt. **Per cap. pers. income** (1998): $18,958. **Sales tax** (1999): 7%. **Unemployment** (1998): 5.4%. **Tourism expends.** (1996): $3.5 bil.

Finance. FDIC-insured commercial banks (1998): 96. **Deposits:** $21.6 bil. **FDIC-insured savings institutions** (1998): 12. **Assets:** $1.4 bil.

Federal govt. Fed. civ. employees (Mar. 1998): 16,930. **Avg. salary:** $40,776. **Notable fed. facilities:** Columbus, Keesler AF bases; Meridian Naval Air Station, John C. Stennis Space Center; U.S. Army Corps of Engineers Waterway Experiment Station.

Energy. Electricity production (1998, kWh, by source): Coal: 11.7 bil; Petroleum: 5.4 bil; Gas: 5.6 bil; Nuclear: 9.2 bil.

State data. Motto: Virtute et Armis (By valor and arms). **Flower:** Magnolia. **Bird:** Mockingbird. **Tree:** Magnolia. **Song:** Go, Mississippi! **Entered union** Dec. 10, 1817; rank, 20th. **State fair** at Jackson; early Oct.

History. Early inhabitants of the region were Choctaw, Chickasaw, and Natchez peoples. Hernando de Soto explored the area, 1540, and sighted the Mississippi River, 1541. Robert La Salle traced the river from Illinois to its mouth and claimed the entire valley for France, 1682. The first settlement was the French Ft. Maurepas, near Ocean Springs, 1699. The area was ceded to Britain, 1763; American settlers followed. During the American Revolution, Spain seized part of the area, remaining even after the U.S. acquired title at the end of the conflict; Spain finally moved out, 1798. The Territory of Mississippi was formed, 1798. Mississippi seceded, 1861. Union forces captured Corinth and Vicksburg and destroyed Jackson and much of Meridian. Mississippi was readmitted to the Union in 1870.

Tourist attractions. Vicksburg Natl. Military Park and Cemetery, other Civil War sites; Hattiesburg; Natchez Trace; Indian mounds; Antebellum homes; pilgrimages in Natchez and some 25 other cities; Smith Robertson Museum, Mynelle Gardens, both Jackson; Mardi Gras and Shrimp Festival, both in Biloxi; Gulf Islands Natl. Seashore; Casinos on the Mississippi River; the Mississippi Coast.

Famous Mississippians. Dana Andrews, Margaret Walker Alexander, Jimmy Buffett, Hodding Carter III, Bo Diddley, William Faulkner, Brett Favre, Shelby Foote, Morgan Freeman, John Grisham, Fannie Lou Hamer, Jim Henson, Robert Johnson, James Earl Jones, B. B. King, L. Q. C. Lamar, Trent Lott, Gerald McRaney, Willie Morris, Walter Payton, Elvis Presley, Leontyne Price, Charley Pride, LeAnn Rimes, Muddy Waters, Eudora Welty, Tennessee Williams, Oprah Winfrey, Johnny Winter, Richard Wright, Tammy Wynette.

Tourist Information. Dept. of Economic & Community Development. PO Box 849, Jackson, MS 39205-0849. **Toll-free travel information.** 1-800-WARMEST. **Website.** http://www.state.ms.us **Tourism website.** http://www.decd.state.ms.us/tourism.htm

Missouri
Show Me State

People. Population (1998): 5,438,559; rank: 16; **net change** (1990-98): 6.3%. **Pop. density** (1998): 78.9 per sq mi. **Racial distribution** (1990): 87.7% white; 10.7% black. **Hispanic pop.:** 1.2%.

Geography. Total area: 69,709 sq mi; rank: 21. **Land area:** 68,898 sq mi; rank: 18. **Acres forested:** 14,007,000. **Location:** West North Central state near the geographic center of the conterminous U.S.; bordered on the E by the Mississippi R., on the NW by the Missouri R. **Climate:** continental, susceptible to cold Canadian air, moist, warm gulf air, and drier SW air. **Topography:** rolling hills, open, fertile plains, and well-watered prairie N of the Missouri R.; south of the river land is rough and hilly with deep, narrow valleys; alluvial plain in the SE; low elevation in the west. **Capital:** Jefferson City.

Economy. Chief industries: agriculture, manufacturing, aerospace, tourism. **Chief manuf. goods:** transportation equipment, food and related products, electrical and electronic equipment, chemicals. **Chief crops:** soybeans, corn, wheat, hay. **Livestock** (Jan. 1999): 4.4 mil cattle/calves; 85,000 sheep/lambs; (Dec. 1998) 3.3 mil hogs/pigs; (Dec. 1998) 8.8 mil chickens (excl. broilers); (Dec. 1997) 250.0 mil broilers. **Timber/lumber** (1998): oak, hickory; 578 mil bd. ft. **Nonfuel minerals** (est. 1998): $1.36 bil; mostly crushed stone, lead, portland cement, lime, zinc. **Chief ports:** St. Louis, Kansas City. **Internat. airports at:** St. Louis, Kansas City. **Value of construction** (1997): $5.9 bil. **Gross state product** (1997): $152.1 bil. **Employment distrib.** (May 1999): 28.1% serv.; 23.6% trade; 15.6% govt.; 15.3% mfg. **Per cap. pers. income** (1998): $24,427. **Sales tax** (1999): 4.225%. **Unemployment** (1998): 4.2%. **Tourism expends.** (1996): $7.9 bil. **Lottery** (1998): total sales: $494.3 mil; net income: $149.5 mil.

Finance. FDIC-insured commercial banks (1998): 382. **Deposits:** $62.0 bil. **FDIC-insured savings institutions** (1998): 42. **Assets:** $6.6 bil.

Federal govt. Fed. civ. employees (Mar. 1998): 32,096. **Avg. salary:** $40,415. **Notable fed. facilities:** Federal Reserve banks; Ft. Leonard Wood; Jefferson Barracks; Whiteman AFB.

Energy. Electricity production (1998 kWh, by source): Coal: 62.5 bil; Petroleum: 310 mil; Gas: 1.2 bil; Hydroelectric: 2.3 bil; Nuclear: 8.5 bil.

State data. Motto: Salus Populi Suprema Lex Esto (The welfare of the people shall be the supreme law). **Flower:** Hawthorn. **Bird:** Bluebird. **Tree:** Dogwood. **Song:** Missouri Waltz. **Entered union** Aug. 10, 1821; rank, 24th. **State fair** at Sedalia; 3d week in Aug.

History. Early inhabitants of the region were Algonquian Sauk, Fox, and Illinois and Siouan Osage, Missouri, Iowa, and Kansa peoples. Hernando de Soto visited 1541. French hunters and lead miners made the first settlement c 1735, at Ste. Genevieve. The territory was ceded to Spain by the French, 1763, then returned to France, 1800. The U.S. acquired Missouri as part of the Louisiana Purchase, 1803. The influx of white settlers drove Native American tribes to the Kansas and Oklahoma territories; most were gone by 1836. The fur trade and the Santa Fe Trail provided prosperity; St. Louis became the gateway for pioneers heading West. Missouri entered the Union as a slave state, 1821. Though it remained with the Union, pro- and anti-slavery forces battled there during the Civil War.

Tourist attractions. Silver Dollar City, Branson; Mark Twain Area, Hannibal; Pony Express Museum, St. Joseph; Harry S. Truman Library, Independence; Gateway Arch, St. Louis; Worlds of Fun, Kansas City; Lake of the Ozarks; Churchill Mem., Fulton; State Capitol, Jefferson City.

Famous Missourians. Maya Angelou, Robert Altman, Burt Bacharach, Josephine Baker, Scot Bakula, Thomas Hart Benton, Tom Berenger, Chuck Berry, George Caleb Bingham, Daniel Boone, Omar Bradley, Kate Capshaw, Dale Carnegie, George Washington Carver, Bob Costas, Walter Cronkite, Walt Disney, T. S. Eliot, Richard Gephardt, John

Goodman, Betty Grable, Edwin Hubble, Jesse James, Marianne Moore, Reinhold Niebuhr, J. C. Penney, John J. Pershing, Brad Pitt, Joseph Pulitzer, Ginger Rogers, Bess Truman, Harry S. Truman, Kathleen Turner, Tina Turner, Mark Twain, Dick Van Dyke, Tennessee Williams, Lanford Wilson, Shelley Winters, Jane Wyman.

Chamber of Commerce. 428 E. Capitol, Jefferson City, MO 65101.

Toll-free travel information. 1-888-925-3875, ext. 124.

Website. http://www.ecodev.state.mo.us

Tourism website. http://www.missouritourism.org

Montana
Treasure State

People. Population (1998): 880,453; rank: 44; **net change** (1990-98): 10.2%. **Pop. density** (1998): 6.0 per sq mi. **Racial distribution** (1990): 92.7% white; 0.3% black; 6.0% Amer. Indian. **Hispanic pop.:** 1.5%.

Geography. Total area: 147,046 sq mi; rank: 4. **Land area:** 145,556 sq mi; rank: 4. **Acres forested:** 22,512,000. **Location:** Mountain state bounded on the E by the Dakotas, on the S by Wyoming, on the SSW by Idaho, and on the N by Canada. **Climate:** colder, continental climate with low humidity. **Topography:** Rocky Mts. in western third of the state; eastern two-thirds gently rolling northern Great Plains. **Capital:** Helena.

Economy. Chief industries: agriculture, timber, mining, tourism, oil and gas. **Chief manuf. goods:** food products, wood & paper products, primary metals, printing & publishing, petroleum and coal products. **Chief crops:** wheat, barley, sugar beets, hay, oats. **Livestock** (Jan. 1999): 2.6 mil cattle/calves; 380,000 sheep/lambs; (Dec. 1998) 190,000 hogs/pigs; (Dec. 1998) 380,000 chickens (excl. broilers). **Timber/lumber** (1998): Douglas fir, pines, larch; 1.3 bil bd. ft. **Nonfuel minerals** (est. 1998): $500 mil; mostly copper, gold, portland cement, palladium metal, molybdenum. **Internat. airports at:** Great Falls, Billings, Kalispell, Missoula. **Value of construction** (1997): $827 mil. **Gross state product** (1997): $19.2 bil. **Employment distrib.** (May 1999): 29.1% serv.; 26.8% trade; 20.6% govt.; 6.5% mfg. **Per cap. pers. income** (1998): $20,172. **Sales tax:** none. **Unemployment** (1998): 5.6%. **Tourism expends.** (1996): $1.7 bil. **Lottery** (1998): total sales: $29.8 mil; net income: $6.7 mil.

Finance. FDIC-insured commercial banks (1998): 89. **Deposits:** $8.2 bil. **FDIC-insured savings institutions** (1998): 5. **Assets:** $1.5 bil.

Federal govt. Fed. civ. employees (Mar. 1998): 7,689. **Avg. salary:** $40,916. **Notable fed. facilities:** Malmstrom AFB; Ft. Peck, Hungry Horse, Libby, Yellowtail dams; numerous missile silos.

Energy. Electricity production (1998, kWh, by source): Coal: 16.5 bil; Petroleum: 14 mil; Gas: 41 mil; Hydroelectric: 11.1 bil.

State data. Motto: Oro y Plata (Gold and silver). **Flower:** Bitterroot. **Bird:** Western meadowlark. **Tree:** Ponderosa pine. **Song:** Montana. **Entered union** Nov. 8, 1889; rank, 41st. **State fair** at Great Falls; late July-early Aug.

History. Cheyenne, Blackfoot, Crow, Assiniboin, Salish (Flatheads), Kootenai, and Kalispel peoples were early inhabitants of the area. French explorers visited the region, 1742. The U.S. acquired the area partly through the Louisiana Purchase, 1803, partly through explorations of Lewis and Clark, 1805-6. Fur traders and missionaries established posts early 19th cent. Gold was discovered, 1863, and the Montana territory was established, 1864. Indian uprisings reached their peak with the Battle of Little Bighorn, 1876. Chief Joseph and the Nez Percé tribe surrendered here, 1877, after long trek across the state. Mining activity and the coming of the Northern Pacific Railway, 1883, brought population growth. Copper wealth from the Butte pits resulted in the turn of the century "War of Copper Kings" as factions fought for control of "the richest hill on earth."

Tourist attractions. Glacier Natl. Park; Yellowstone Natl. Park; Museum of the Rockies, Bozeman; Museum of the Plains Indian, Blackfeet Reservation, near Browning; Little Bighorn Battlefield Natl. Monument and Custer Natl. Cemetery; Flathead Lake; Helena; Lewis and Clark Caverns State Park, near Whitehall; Lewis and Clark Interpretive Center, Great Falls.

Famous Montanans. Gary Cooper, Marcus Daly, Chet Huntley, Will James, Myrna Loy, Mike Mansfield, Brent Mus-

burger, Jeannette Rankin, Charles M. Russell, Lester Thurow.

Chamber of Commerce. 2030 11th Ave., PO Box 1730, Helena, MT 59624.

Toll-free travel information. 1-800-VISITMT.

Website. http://www.mt.gov

Tourism website. http://travel.mt.gov

Nebraska
Cornhusker State

People. Population (1998): 1,662,719; rank: 38; **net change** (1990-98): 5.3%. **Pop. density** (1998): 21.6 per sq mi. **Racial distribution** (1990): 93.8% white; 3.6% black. **Hispanic pop.:** 2.3%.

Geography. Total area: 77,358 sq mi; rank: 16. **Land area:** 76,878 sq mi; rank: 15. **Acres forested:** 722,000. **Location:** West North Central state with the Missouri R. for a NE and E border. **Climate:** continental semi-arid. **Topography:** till plains of the central lowland in the eastern third rising to the Great Plains and hill country of the north central and NW. **Capital:** Lincoln.

Economy. Chief industries: agriculture, manufacturing. **Chief manuf. goods:** processed foods, industrial machinery, printed materials, electric and electronic equipment, primary and fabricated metal products, transportation equipment. **Chief crops:** corn, sorghum, soybeans, hay, wheat, dry beans, oats, potatoes, sugar beets. **Livestock** (Jan. 1999): 6.7 mil cattle/calves; 105,000 sheep/lambs; (Dec. 1998) 3.4 mil hogs/pigs; (Dec. 1998) 13.2 mil chickens (excl. broilers); (Dec. 1997) 1.6 mil broilers. **Timber/lumber** (1998): oak, hickory, and elm; 25 mil bd. ft. **Nonfuel minerals** (est. 1999): $174 mil; mostly portland cement, sand & gravel, crushed stone, masonry cement, clays. **Chief ports:** Omaha, Sioux City, Brownville, Blair, Plattsmouth, Nebraska City. **Value of construction** (1997): $2.0 bil. **Gross state product** (1997): $48.9 bil. **Employment distrib.** (May 1999): 27.1% serv.; 24.3% trade; 17.1% govt.; 13.3% mfg. **Per cap. pers. income** (1998): $24,754. **Sales tax** (1999): 5%. **Unemployment** (1998): 2.7%. **Tourism expends.** (1996): $2.3 bil. **Lottery** (1998): total sales: $73.9 mil; net income: $19.8 mil.

Finance. FDIC-insured commercial banks (1998): 315. **Deposits:** $23.0 bil. **FDIC-insured savings institutions** (1998): 13. **Assets:** $14.3 bil.

Federal govt. Fed. civ. employees (Mar. 1998): 7,783. **Avg. salary:** $41,910. **Notable fed. facilities:** Offutt AFB.

Energy. Electricity production (1998, kWh, by source): Coal: 18.3 bil; Petroleum: 42 mil; Gas: 400 mil; Hydroelectric: 1.7 mil; Nuclear: 8.3 bil.

State data. Motto: Equality before the law. **Flower:** Goldenrod. **Bird:** Western meadowlark. **Tree:** Cottonwood. **Song:** Beautiful Nebraska. **Entered union** Mar. 1, 1867; rank, 37th. **State fair** at Lincoln; Aug.- Sept.

History. When the Europeans first arrived, Pawnee, Ponca, Omaha, and Oto peoples lived in the region. Spanish and French explorers and fur traders visited the area prior to its acquisition in the Louisiana Purchase, 1803. Lewis and Clark passed through, 1804-6. The first permanent settlement was Bellevue, near Omaha, 1823. The region was gradually settled, despite the 1834 Indian Intercourse Act, which declared Nebraska Indian country and excluded white settlement. Conflicts with settlers eventually forced Native Americans to move to reservations. Many Civil War veterans settled under free land terms of the 1862 Homestead Act; as agriculture grew, struggles followed between homesteaders and ranchers.

Tourist attractions. State Museum (Elephant Hall), State Capitol, both Lincoln; Stuhr Museum of the Prairie Pioneer, Grand Island; Museum of the Fur Trade, Chadron; Henry Doorly Zoo, Joslyn Art Museum, both Omaha; Ashfall Fossil Beds, Strategic Air Command Museum, Ashland; Boys Town, west of Omaha; Arbor Lodge State Park, Nebraska City; Buffalo Bill Ranch State Hist. Park, North Platte; Pioneer Village, Minden; Oregon Trail landmarks; Scotts Bluff Natl. Monument; Chimney Rock Historic Site; Ft. Robinson; Hastings Museum, Hastings.

Famous Nebraskans. Fred Astaire, Marlon Brando, Charles W. Bryan, William Jennings Bryan, Warren Buffett, Johnny Carson, Willa Cather, Dick Cavett, William F. "Buffalo Bill" Cody, Loren Eiseley, Rev. Edward J. Flanagan, Henry Fonda, Gerald R. Ford, Bob Gibson, Rollin Kirby, Harold Lloyd, Malcolm X, J. Sterling Morton, John Neihardt, Nick

Nolte, George Norris, John J. Pershing, Roscoe Pound, Chief Red Cloud, Mari Sandoz, Robert Taylor, Darryl F. Zanuck.

Chamber of Commerce and Industry. 1320 Lincoln Mall, Ste. 201, Lincoln, NE 68508; 402-474-4422

Toll-free travel information. 1-800-228-4307.

Website. http://www.state.ne.us

Tourism website. http://www.visitnebraska.org

Nevada

Sagebrush State, Battle Born State, Silver State

People. Population (1998): 1,746,898; rank: 36; **net change** (1990-98): 45.4%. **Pop. density** (1998): 15.9 per sq mi. **Racial distribution** (1990): 84.3% white; 6.6% black; 3.2% Asian. **Hispanic pop.:** 10.4%.

Geography. Total area: 110,567 sq mi; rank: 7. **Land area:** 109,806 sq mi; rank: 7. **Acres forested:** 8,938,000. **Location:** Mountain state bordered on N by Oregon and Idaho, on E by Utah and Arizona, on SE by Arizona, and on SW and W by California. **Climate:** semi-arid and arid. **Topography:** rugged N-S mountain ranges; highest elevation, Boundary Peak, 13,140 ft; southern area is within the Mojave Desert; lowest elevation, Colorado River at southern tip of state, 479 ft. **Capital:** Carson City.

Economy. Chief industries: gaming, tourism, mining, manufacturing, government, retailing, warehousing, trucking. **Chief manuf. goods:** food products, plastics, chemicals, aerospace products, lawn and garden irrigation equipment, seismic and machinery-monitoring devices. **Chief crops:** hay, alfalfa seed, potatoes, onions, garlic, barley, wheat. **Livestock** (Jan. 1999): 510,000 cattle/calves; 85,000 sheep/lambs; (Dec. 1998) 7,500 hogs/pigs. **Timber/lumber:** piñon, juniper, other pines. **Nonfuel minerals** (est. 1998): $3.1 bil; mostly gold, copper, silver, sand & gravel, diatomite. **Internat. airports at:** Las Vegas, Reno. **Value of construction** (1997): $6.7 bil. **Gross state product** (1997): $57.4 bil. **Employment distrib.** (May 1999): 42.6% serv.; 20.5% trade; 12.3% govt.; 4.4% mfg. **Per cap. pers. income** (1998): $27,200. **Sales tax** (1999): 6.5%. **Unemployment** (1998): 4.3%. **Tourism expends.** (1996): $17.6 bil.

Finance. FDIC-insured commercial banks (1998): 26. **Deposits:** $8.8 bil.

Federal govt. Fed. civ. employees (Mar. 1998): 7,049. **Avg. salary:** $44,594. **Notable fed. facilities:** Nevada Test Site; Hawthorne Army Ammunition Plant; Nellis Air Force Base and Gunnery Range; Fallon Naval Air Station; Palomino Valley Wild Horse and Burro Placement Center.

Energy. Electricity production (1998, kWh, by source): Coal: 17.2 bil; Petroleum: 50 mil; Gas: 6.2 bil; Hydroelectric: 3.2 bil.

State data. Motto: All for our country. **Flower:** Sagebrush. **Bird:** Mountain bluebird. **Trees:** Single-leaf piñon and bristlecone pine. **Song:** Home Means Nevada. **Entered union** Oct. 31, 1864; rank, 36th. **State fair** at Reno; late Aug.

History. Shoshone, Paiute, Bannock, and Washoe peoples lived in the area at the time of European contact. Nevada was first explored by Spaniards, 1776. Hudson's Bay Co. trappers explored the north and central region, 1825; trader Jedediah Smith crossed the state, 1826-27. The area was acquired by the U.S., 1848, at the end of the Mexican War. The first settlement, Mormon Station, now Genoa, was established, 1849. Discovery of the Comstock Lode, rich in gold and silver, 1859, spurred a population boom. In the early 20th cent., Nevada adopted progressive measures such as the initiative, referendum, recall, and woman suffrage.

Tourist attractions. Legalized gambling at: Lake Tahoe, Reno, Las Vegas, Laughlin, Elko County, and elsewhere. Hoover Dam; Lake Mead; Great Basin Natl. Park; Valley of Fire State Park; Virginia City; Red Rock Canyon Natl. Conservation Area; Liberace Museum, the Las Vegas Strip, Guinness World of Records Museum, Lost City Museum, Overton, Lamoille Canyon, Pyramid Lake, all Las Vegas. Skiing near Lake Tahoe.

Famous Nevadans. Walter Van Tilburg Clark, George Ferris, Sarah Winnemucca Hopkins, Paul Laxalt, Dat So La Lee, John William Mackay, Anne Martin, Pat McCarran, Key Pittman, William Morris Stewart.

Tourist information. Commission on Tourism, 5151 S. Carson St., Carson City, NV 89701.

Toll-free travel information. 1-800-638-2328.

Website. http://www.state.nv.us

Tourism website. http://www.travelnevada.com

New Hampshire

Granite State

People. Population (1998): 1,185,048; rank: 42; **net change** (1990-98): 6.8%. **Pop. density** (1998): 132.1 per sq mi. **Racial distribution** (1990): 98.0% white; 0.6% black. **Hispanic pop.:** 1.0%.

Geography. Total area: 9,283 sq mi; rank: 44. **Land area:** 8,969 sq mi; rank: 44. **Acres forested:** 4,981,000. **Location:** New England state bounded on S by Massachusetts, on W by Vermont, on N and NW by Canada, on E by Maine and the Atlantic Ocean. **Climate:** highly varied, due to its nearness to high mountains and ocean. **Topography:** low, rolling coast followed by countless hills and mountains rising out of a central plateau. **Capital:** Concord.

Economy. Chief industries: tourism, manufacturing, agriculture, trade, mining. **Chief manuf. goods:** machinery, electrical and electronic products, plastics, fabricated metal products. **Chief crops:** dairy products, nursery & greenhouse products, hay, vegetables, fruit, maple syrup & sugar products. **Livestock** (Jan. 1999): 48,000 cattle/calves; 7,000 sheep/lambs; (Dec. 1998) 4,000 hogs/pigs; (Dec. 1998) 215,000 chickens (excl. broilers). **Timber/lumber** (1998): white pine, hemlock, oak, birch; 314 mil bd. ft. **Nonfuel minerals** (est. 1998): $53.1 mil; mostly sand & gravel, crushed and dimension stone, gemstones. **Commercial fishing** (1998): $11.2 mil. **Chief ports:** Portsmouth, Hampton, Rye. **Value of construction** (1997): $1.3 bil. **Gross state product** (1997): $38.1 bil. **Employment distrib.** (May 1999): 29.9% serv.; 25.9% trade; 17.7% mfg.; 13.5% govt. **Per cap. pers. income** (1998): $29,022. **Sales tax:** none. **Unemployment** (1998): 2.9%. **Tourism expends.** (1996): $1.8 bil. **Lottery** (1998): total sales: $183.8 mil; net income: $57.5 mil.

Finance. FDIC-insured commercial banks (1998): 19. **Deposits:** $11.1 bil. **FDIC-insured savings institutions** (1998): 20. **Assets:** $8.1 bil.

Federal govt. Fed. civ. employees (Mar. 1998): 3,307. **Avg. salary:** $49,529.

Energy. Electricity production (1998, kWh, by source): Coal: 3.5 bil; Petroleum: 1.4 bil; Gas: 10 mil; Hydroelectric: 975 mil; Nuclear: 8.4 bil.

State data. Motto: Live free or die. **Flower:** Purple lilac. **Bird:** Purple finch. **Tree:** White birch. **Song:** Old New Hampshire. **Ninth** of the original 13 states to ratify the Constitution, June 21, 1788. **State Fair:** Many agricultural fairs statewide, July through Sept.; no State fair

History. Algonquian-speaking peoples, including the Pennacook, lived in the region when the Europeans arrived. The first explorers to visit the area were England's Martin Pring, 1603, and France's Champlain, 1605. The first settlement was Odiorne's Point (now port of Rye), 1623. Native American conflicts were ended, 1759, by Robert Rogers' Rangers. Before the American Revolution, New Hampshire residents seized a British fort at Portsmouth, 1774, and drove the royal governor out, 1775. New Hampshire became the first colony to adopt its own constitution, 1776. Three regiments served in the Continental Army, and scores of privateers raided British shipping.

Tourist attractions. Mt. Washington, highest peak in Northeast; Lake Winnipesaukee; White Mt. National Forest; Crawford, Franconia—famous for the Old Man of the Mountain, described by Hawthorne as the Great Stone Face, Pinkham notches, all White Mt. region; the Flume, a spectacular gorge; the aerial tramway, Cannon Mt.; Strawbery Banke, Portsmouth; Shaker Village, Canterbury; Saint-Gaudens, natl. historic site, Cornish; Mt. Monadnock.

Famous New Hampshirites. Salmon P. Chase, Ralph Adams Cram, Mary Baker Eddy, Daniel Chester French, Robert Frost, Horace Greeley, Sarah Buell Hale, Franklin Pierce, Augustus Saint-Gaudens, David H. Souter, Daniel Webster.

Tourist information. Division of Travel & Tourism Development, PO Box 1856, Concord, NH 03302-1856.

Toll-free travel information. 1-800-386-4664.

Website. http://www.state.nh.us

Tourism website. http://www.visitnh.gov

New Jersey
Garden State

People. Population (1998): 8,115,011; rank: 9; **net change** (1990-98): 4.7%. **Pop. density** (1998): 1,093.8 per sq mi. **Racial distribution** (1990): 79.3% white; 13.4% black; 3.5% Asian. **Hispanic pop.:** 9.6%.

Geography. Total area: 8,215 sq mi; rank: 46. **Land area:** 7,419 sq mi; rank: 46. **Acres forested:** 2,007,000. **Location:** Middle Atlantic state bounded on N and E by New York and Atlantic Ocean, on S and W by Delaware and Pennsylvania. **Climate:** moderate, with marked difference bet. NW and SE extremities. **Topography:** Appalachian Valley in the NW also has highest elevation, High Pt., 1,801 ft; Appalachian Highlands, flat-topped NE-SW mountain ranges; Piedmont Plateau, low plains broken by high ridges (Palisades) rising 400-500 ft; Coastal Plain, covering three-fifths of state in SE, rises from sea level to gentle slopes. **Capital:** Trenton.

Economy. Chief industries: pharmaceuticals/drugs, telecommunications, biotechnology, printing & publishing. **Chief manuf. goods:** chemicals, electronic equipment, food. **Chief crops:** nursery/greenhouse, tomatoes, blueberries, peaches, peppers, cranberries, soybeans. **Livestock** (Jan. 1999): 53,000 cattle/calves; 12,000 sheep/lambs; (Dec. 1998) 24,000 hogs/pigs; (Dec. 1998) 2.1 mil chickens (excl. broilers). **Timber/lumber:** (1998) pine, cedar, mixed hardwoods; 8 mil bd. ft. **Nonfuel minerals** (est. 1998): $301 mil; mostly crushed stone, sand & gravel, greensand marl, peat. **Commercial fishing** (1998): $91 mil. **Chief ports:** Newark, Elizabeth, Hoboken, Camden. **Internat. airport at:** Newark. **Value of construction** (1997): $8.3 bil. **Gross state product** (1997): $294.1 bil. **Employment distrib.** (May 1999): 32.7% serv.; 23.4% trade; 14.8% govt.; 12.1% mfg. **Per cap. pers. income** (1998): $33,937. **Sales tax** (1999): 6%. **Unemployment** (1998): 4.6%. **Tourism expends.** (1996): $13.2 bil. **Lottery** (1998): total sales: $1.6 bil; net income: $643.9 mil.

Finance. FDIC-insured commercial banks (1998): 72. **Deposits:** $74.2 bil. **FDIC-insured savings institutions** (1998): 78. **Assets:** $43.0 bil.

Federal govt. Fed. civ. employees (Mar. 1998): 27,819. **Avg. salary:** $49,429. **Notable fed. facilities:** McGuire AFB; Fort Dix; Fort Monmouth; Picatinny Arsenal; Lakehurst Naval Air Engineering Center.

Energy. Electricity production (1998, kWh, by source): Coal: 5.6 bil; Petroleum: 485 mil; Gas: 2.9 bil; Hydroelectric: −146 mil; Nuclear: 27.1 bil.

State data. Motto: Liberty and prosperity. **Flower:** Purple violet. **Bird:** Eastern goldfinch. **Tree:** Red oak. **Third** of the original 13 states to ratify the Constitution, Dec. 18, 1787. **State fair** at Cherry Hill; late July-early Aug.

History. The Lenni Lenape (Delaware) peoples lived in the region and had mostly peaceful relations with European colonists, who arrived after the explorers Verrazano, 1524, and Hudson, 1609. The first permanent European settlement was Dutch, at Bergen (now Jersey City), 1660. When the British took New Netherland, 1664, the area between the Delaware and Hudson Rivers was given to Lord John Berkeley and Sir George Carteret. During the American Revolution, New Jersey was the scene of nearly 100 battles, large and small, including Trenton, 1776; Princeton, 1777; Monmouth, 1778.

Tourist attractions. 127 mi of beaches; Miss America Pageant, Atlantic City; Grover Cleveland birthplace, Caldwell; Cape May Historic District; Edison Natl. Historic Site, W. Orange; Six Flags Great Adventure, Jackson; Liberty State Park, Jersey City; Meadowlands Sports Complex, E. Rutherford; Pine Barrens wilderness area; Princeton University; numerous Revolutionary War historical sites; State Aquarium, Camden.

Famous New Jerseyans. Count Basie, Judy Blume, Bill Bradley, Jon Bon Jovi, Aaron Burr, Grover Cleveland, James Fenimore Cooper, Stephen Crane, Thomas Edison, Albert Einstein, Allen Ginsberg, Alexander Hamilton, Whitney Houston, Buster Keaton, Joyce Kilmer, George McClellan, Thomas Paine, Dorothy Parker, Molly Pitcher, Paul Robeson, Philip Roth, Wally Schirra, H. Norman Schwarzkopf, Frank Sinatra, Bruce Springsteen, Martha Stewart, Meryl Streep, Walt Whitman, William Carlos Williams, Woodrow Wilson.

Chamber of Commerce. 50 W. State St., Trenton, NJ 08608.

Toll-free travel information. 1-800-JERSEY7.

Website. http://www.state.nj.us

Tourism website. http://www.state.nj.us/travel

New Mexico
Land of Enchantment

People. Population (1998): 1,736,931; rank: 37; **net change** (1990-98): 14.6%. **Pop. density** (1998): 14.3 per sq mi. **Racial distribution** (1990): 75.6% white; 2.0% black; 8.9% Amer. Indian. **Hispanic pop.:** 38.2%.

Geography. Total area: 121,598 sq mi; rank: 5. **Land area:** 121,364 sq mi rank: 5. **Acres forested:** 15,296,000. **Location:** southwestern state bounded by Colorado on the N, Oklahoma, Texas, and Mexico on the E and S, and Arizona on the W. **Climate:** dry, with temperatures rising or falling 5° F with every 1,000 ft elevation. **Topography:** eastern third, Great Plains; central third, Rocky Mts. (85% of the state is over 4,000-ft elevation); western third, high plateau. **Capital:** Santa Fe.

Economy. Chief industries: government, services, trade. **Chief manuf. goods:** foods, machinery, apparel, lumber, printing, transportation equipment, electronics, semiconductors. **Chief crops:** hay, onions, chiles, greenhouse nursery, pecans, cotton. **Livestock** (Jan. 1999): 1.6 mil cattle/calves; 275,000 sheep/lambs; (Dec. 1998) 6,000 hogs/pigs; (Dec. 1998) 1.4 mil chickens (excl. broilers). **Timber/lumber** (1998): ponderosa pine, Douglas fir; 103 mil bd. ft. **Nonfuel minerals** (est. 1998): $860 mil; mostly copper, potash, sand & gravel, portland cement, perlite. **Internat. airports at:** Albuquerque. **Value of construction** (1997): $1.9 bil. **Gross state product** (1997): $45.2 bil. **Employment distrib.** (May 1999): 28.8% serv.; 24.6% govt.; 23.7% trade; 6.0% mfg. **Per cap. pers. income** (1998): $19,936. **Sales tax** (1999): 5%. **Unemployment** (1998): 6.2%. **Tourism expends.** (1996): $3.3 bil. **Lottery** (1998): total sales: $84.9 mil; net income: $20.1 mil.

Finance. FDIC-insured commercial banks (1998): 58. **Deposits:** $11.4 bil. **FDIC-insured savings institutions** (1998): 10. **Assets:** $2.4 bil.

Federal govt. Fed. civ. employees (Mar. 1998): 21,569. **Avg. salary:** $41,935. **Notable fed. facilities:** Kirtland, Cannon, Holloman AF bases; Los Alamos Scientific Laboratory; White Sands Missile Range; Natl. Solar Observatory; Natl. Radio Astronomy Observatory, Sandia National Laboratories.

Energy. Electricity production (1998, kWh, by source): Coal: 27.5 bil; Petroleum: 23 mil; Gas: 3.6 bil; Hydroelectric: 236 mil.

State data. Motto: Crescit Eundo (It grows as it goes). **Flower:** Yucca. **Bird:** Roadrunner. **Tree:** Piñon. **Song:** O, Fair New Mexico; Asi Es Nuevo Mexico. **Entered union** Jan. 6, 1912; rank: 47th. **State fair** at Albuquerque; mid-Sept.

History. Early inhabitants were peoples of the Mogollon and Anasazi civilizations, followed by the Pueblo peoples, Anasazi descendants. The nomadic Navajo and Apache tribes arrived c 15th cent. Franciscan Marcos de Niza and a former black slave, Estevanico, explored the area, 1539, seeking gold. First settlements were at San Juan Pueblo, 1598, and Santa Fe, 1610. Settlers alternately traded and fought with the Apache, Comanche, and Navajo. Trade on the Santa Fe Trail to Missouri started, 1821. The Mexican War was declared in May 1846; Gen. Stephen Kearny took Santa Fe without firing a shot, Aug. 18, 1846, declaring New Mexico part of the U.S. All Hispanic New Mexicans and Pueblo became U.S. citizens by terms of the 1848 treaty ending the war, but Congress denied the area statehood and created the territory of New Mexico, 1850. Pancho Villa raided Columbus, 1916, and U.S. troops were sent to the area. The world's first atomic bomb was exploded near Alamogordo, south of Santa Fe, 1945.

Tourist attractions. Carlsbad Caverns Natl. Park, with the largest natural underground chamber in the world; Santa Fe, oldest capital in U.S.; White Sands Natl. Monument, the largest gypsum deposit in the world; Chaco Culture National Historical Park; Acoma Pueblo, the "sky city," built atop a 357-ft mesa; Taos; Taos Art Colony; Taos Ski Valley; Ute Lake State Park; Shiprock.

Famous New Mexicans. Ben Abruzzo, Maxie Anderson, Billy (the Kid) Bonney, Kit Carson, Bob Foster, Peter Hurd, Archbishop Jean Baptiste Lamy, Nancy Lopez, Bill Mauldin, Georgia O'Keeffe, Kim Stanley, Al Unser, Bobby Unser, Lew Wallace.

Tourist information. New Mexico Dept. of Tourism, PO Box 20002, Santa Fe, NM 87503.

Toll-free travel information. 1-800-545-2040, ext. 751

Website. http://www.state.nm.us

Tourism website. http://www.newmexico.org

New York
Empire State

People. Population (1998): 18,175,301; rank: 3; **net change** (1990-98): 1.0%. **Pop. density** (1998): 384.9 per sq mi. **Racial distribution** (1990): 74.4% white; 15.9% black; 3.9% Asian. **Hispanic pop.:** 12.3%.

Geography. Total area: 53,989 sq mi; rank: 27. **Land area:** 47,224 sq mi; rank: 30. **Acres forested:** 18,713,000. **Location:** Middle Atlantic state, bordered by the New England states, Atlantic Ocean, New Jersey and Pennsylvania, Lakes Ontario and Erie, and Canada. **Climate:** variable; the SE region moderated by the ocean. **Topography:** highest and most rugged mountains in the NE Adirondack upland; St. Lawrence-Champlain lowlands extend from Lake Ontario NE along the Canadian border; Hudson-Mohawk lowland follows the flows of the rivers N and W, 10-30 mi wide; Atlantic coastal plain in the SE; Appalachian Highlands, covering half the state westward from the Hudson Valley, include the Catskill Mts., Finger Lakes; plateau of Erie-Ontario lowlands. **Capital:** Albany.

Economy. Chief industries: manufacturing, finance, communications, tourism, transportation, services. **Principal manufactured goods:** books & periodicals, clothing & apparel, pharmaceuticals, machinery, instruments, toys & sporting goods, electronic equipment, automotive & aircraft components. **Chief crops:** apples, grapes, strawberries, cherries, pears, onions, potatoes, cabbage, sweet corn, green beans, cauliflower, field corn, hay, wheat, oats, dry beans. **Products:** milk, cheese, maple syrup, wine. **Livestock** (Jan. 1999): 1.5 mil cattle/calves; 55,000 sheep/lambs; (Dec. 1998) 60,000 hogs/pigs; (Dec. 1998) 4.7 mil chickens (excl. broilers); (Dec. 1997) 1.4 mil broilers. **Timber/lumber** (1998): birch, sugar and red maple, basswood, hemlock, pine, oak, ash; 567 mil bd. ft. **Nonfuel minerals** (est. 1998): $939 mil; mostly crushed stone, portland cement, salt, sand & gravel, zinc. **Commercial fishing** (1998): $84.3 mil. **Chief ports:** New York, Buffalo, Albany. **Internat. airports at:** New York, Buffalo, Syracuse, Massena, Ogdensburg, Watertown, Niagara Falls, Newburgh. **Value of construction** (1997): $15.8 bil. **Gross state product** (1997): $651.7 bil. **Employment distrib.** (May 1999): 34.5% serv.; 20.1% trade; 17.1% govt.; 10.8% mfg. **Per cap. pers. income** (1998): $31,734. **Sales tax** (1999): 4%. **Unemployment** (1998): 5.6%. **Tourism expends.** (1996): $31.3 bil. **Lottery** (1998): total sales: $3.9 bil; net income: $1.5 bil.

Finance. FDIC-insured commercial banks (1998): 153. **Deposits:** $677.5 bil. **FDIC-insured savings institutions** (1998): 91. **Assets:** $120.1 bil.

Federal govt. Fed. civ. employees (Mar. 1998): 58,255. **Avg. salary:** $44,449. **Notable fed. facilities:** West Point Military Academy; Merchant Marine Academy; Ft. Drum; Rome Labs.; Watervliet Arsenal.

Energy. Electricity production (1998, kWh, by source): Coal: 23.5 bil; Petroleum: 15.5 bil; Gas: 19.9 bil; Hydroelectric: 26.6 bil; Nuclear: 31.3 bil.

State data. Motto: Excelsior (Ever upward). **Flower:** Rose. **Bird:** Bluebird. **Tree:** Sugar maple. **Song:** I Love New York. **Eleventh** of the original 13 states to ratify the Constitution, July 26, 1788. **State fair** at Syracuse; late Aug.-early Sept.

History. Algonquians including the Mahican, Wappinger, and Lenni Lenape inhabited the region, as did the Iroquoian Mohawk, Oneida, Onondaga, Cayuga, and Seneca tribes, who established the League of the Five Nations. In 1609, Henry Hudson visited the river named for him, and Champlain explored the lake named for him. The first permanent settlement was Dutch, near present-day Albany, 1624. New Amsterdam was settled, 1626, at the S tip of Manhattan Island. A British fleet seized New Netherland, 1664. Ninety-two of the 300 or more engagements of the American Revolution were fought in New York, including the Battle of Bemis Heights-Saratoga, 1777, a turning point of the war. Completion of Erie Canal, 1825, established the state as a gateway to the West. The first woman's rights convention was held in Seneca Falls, 1848.

Tourist attractions. New York City; Adirondack and Catskill Mts.; Finger Lakes; Great Lakes; Thousand Islands; Niagara Falls; Saratoga Springs; Philipsburg Manor, Sunnyside (Washington Irving's home), the Dutch Church of Sleepy Hollow, all in Tarrytown area; Corning Glass Center and Steuben factory, Corning; Fenimore House, Natl. Baseball Hall of Fame and Museum, both in Cooperstown; Ft. Ticonderoga overlooking Lakes George and Champlain; Empire State Plaza, Albany; Lake Placid; Franklin D. Roosevelt Natl. Historic Site, including the Roosevelt Library, Hyde Park; Long Island beaches; Theodore Roosevelt estate, Sagamore Hill, Oyster Bay; Turning Stone Casino.

Famous New Yorkers. Woody Allen, Susan B. Anthony, James Baldwin, Lucille Ball, Humphrey Bogart, Benjamin Cardozo, De Witt Clinton, Peter Cooper, Aaron Copland, George Eastman, Millard Fillmore, Lou Gehrig, George and Ira Gershwin, Ruth Bader Ginsburg, Rudolph Giuliani, Jackie Gleason, Julia Ward Howe, Charles Evans Hughes, Washington Irving, Henry and William James, John Jay, Michael Jordan, Edward Koch, Fiorello La Guardia, Herman Melville, J. Pierpont Morgan Jr., Joyce Carol Oates, Eugene O'Neill, Colin Powell, Nancy Reagan, John D. Rockefeller, Nelson Rockefeller, Eleanor Roosevelt, Franklin D. Roosevelt, Theodore Roosevelt, J. D. Salinger, Jerry Seinfeld, Paul Simon, Alfred E. Smith, Elizabeth Cady Stanton, Barbra Streisand, Donald Trump, William (Boss) Tweed, Martin Van Buren, Gore Vidal, Edith Wharton, Walt Whitman.

Tourist information. Empire State Development, Travel Information Center, 1 Commerce Plaza, Albany, NY 12245.

Toll-free travel information. 1-800-CALLNYS from U.S. states and territories and Canada; 1-518-474-4116 from other areas.

Website. http://www.empire.state.ny.us

Tourism website. http://www.iloveny.state.ny.us

North Carolina
Tar Heel State, Old North State

People. Population (1998): 7,546,493; rank: 11; **net change** (1990-98): 13.8%. **Pop. density** (1998): 154.9 per sq mi. **Racial distribution** (1990): 75.6% white; 22.0% black; 1.2% Amer. Indian. **Hispanic pop.:** 1.2%.

Geography. Total area: 52,672 sq mi; rank: 29. **Land area:** 48,718 sq mi; rank: 29. **Acres forested:** 19,278,000. **Location:** South Atlantic state bounded by Virginia, South Carolina, Georgia, Tennessee, and the Atlantic Ocean. **Climate:** sub-tropical in SE, medium-continental in mountain region; tempered by the Gulf Stream and the mountains in W. **Topography:** coastal plain and tidewater, two-fifths of state, extending to the fall line of the rivers; piedmont plateau, another two-fifths, of gentle to rugged hills; southern Appalachian Mts. contains the Blue Ridge and Great Smoky Mts. **Capital:** Raleigh.

Economy. Chief industries: manufacturing, agriculture, tourism. **Chief manuf. goods:** food products, textiles, industrial machinery and equipment, electrical and electronic equipment, furniture, tobacco products, apparel. **Chief crops:** tobacco, cotton, soybeans, corn, food grains, wheat, peanuts, sweet potatoes. **Livestock** (Jan. 1999): 980,000 cattle/calves; 14,500 sheep/lambs; (Dec. 1998) 9.7 mil hogs/pigs; (Dec. 1998) 17.2 mil chickens (excl. broilers); (Dec. 1997) 665.0 mil broilers. **Timber/lumber** (1998): yellow pine, oak, hickory, poplar, maple; 2.4 bil bd. ft. **Nonfuel minerals** (est. 1998: $785 mil; mostly crushed stone, phosphate rock, sand & gravel, clays. **Commercial fishing** (1998): $104.8 mil. **Internat. airports at:** Charlotte/Douglas, Raleigh/Durham. **Chief ports:** Morehead City, Wilmington. **Value of construction** (1997): $14.0 bil. **Gross state product** (1997): $218.9 bil. **Employment distrib.** (May 1999): 25.2% serv.; 22.9% trade; 21.2% mfg.; 15.6% govt. **Per cap. pers. income** (1998): $24,036. **Sales tax** (1999): 4%. **Unemployment** (1998): 3.5%. **Tourism expends.** (1996): $10.2 bil.

Finance. FDIC-insured commercial banks (1998): 67. **Deposits:** $414.5 bil. **FDIC-insured savings institutions** (1998): 56. **Assets:** $8.2 bil.

Federal govt. Fed. civ. employees (Mar. 1998): 30,895. **Avg. salary:** $40,025. **Notable fed. facilities:** Ft. Bragg; Camp LeJeune Marine Base; U.S. EPA Research and Development Labs, Cherry Point Marine Corps Air Station; Natl. Humanities Center; Natl. Inst. of Environmental Health Science; Natl. Center for Health Statistics Lab, Research Triangle Park.

Energy. Electricity production (1998, kWh, by source): Coal: 69.0 bil; Petroleum: 286 mil; Gas: 936 mil; Hydroelectric: 4.1 bil; Nuclear: 38.8 bil.

State data. Motto: Esse Quam Videri (To be rather than to seem). **Flower:** Dogwood. **Bird:** Cardinal. **Tree:** Pine. **Song:** The Old North State. **Twelfth** of the original 13 states to ratify the Constitution, Nov. 21, 1789. **State fair** at Raleigh; mid-Oct.

History. Algonquian, Siouan, and Iroquoian peoples lived in the region at the time of European contact. The first English colony in America was the first of 2 established by Sir Walter Raleigh on Roanoke Island, 1585 and 1587. The first group returned to England; the second, the "Lost Colony," disappeared without a trace. Permanent settlers came from Virginia, c 1660. Roused by British repression, the colonists drove out the royal governor, 1775. The province's congress was the first to vote for independence; ten regiments were furnished to the Continental Army. Cornwallis's forces were defeated at Kings Mountain, 1780, and forced out after Guilford Courthouse, 1781. The state seceded in 1861, and provided more troops to the Confederacy than any other state; readmitted in 1868.

Tourist attractions. Cape Hatteras and Cape Lookout natl. seashores; Great Smoky Mts.; Guilford Courthouse and Moore's Creek parks; 66 American Revolution battle sites; Bennett Place, near Durham, where Gen. Joseph Johnston surrendered the last Confederate army to Gen. William Sherman; Ft. Raleigh, Roanoke Island, where Virginia Dare, first child of English parents in the New World, was born Aug. 18, 1587; Wright Brothers Natl. Memorial, Kitty Hawk; Battleship *North Carolina*, Wilmington; NC Zoo, Asheboro; NC Symphony, NC Museum, Raleigh; Carl Sandburg Home, Hendersonville, Biltmore House & Gardens, Asheville.

Famous North Carolinians. David Brinkley, Shirley Caesar, John Coltrane, Elizabeth Dole, Ava Gardner, Richard J. Gatling, Billy Graham, Andy Griffith, O. Henry, Andrew Jackson, Andrew Johnson, Michael Jordan, Wm. Rufus King, Charles Kuralt, Dolley Madison, Theolonius Monk, Edward R. Murrow, Arnold Palmer, Richard Petty, James K. Polk, Carl Sandburg, Enos Slaughter, Dean Smith, James Taylor, Thomas Wolfe, Orville and Wilbur Wright.

Tourist information. North Carolina Division of Tourism, Film & Sports Development, 301 N. Wilmington St., Raleigh, NC 27601.

Toll-free travel information. 1-800-VISITNC.

Website. http://www.state.nc.us

Tourism website. http://www.visitnc.com

North Dakota
Peace Garden State

People. Population (1998): 638,244; rank: 47; **net change** (1990-98): -0.1%. **Pop. density** (1998): 9.3 per sq mi. **Racial distribution** (1990): 94.6% white; 0.6% black; 4.1% Amer. Indian. **Hispanic pop.:** 0.7%.

Geography. Total area: 70,704 sq mi; rank: 18. **Land area:** 68,994 sq mi; rank: 17. **Acres forested:** 462,000. **Location:** West North Central state, situated exactly in the middle of North America, bounded on the N by Canada, on the E by Minnesota, on the S by South Dakota, on the W by Montana. **Climate:** continental, with a wide range of temperature and moderate rainfall. **Topography:** Central Lowland in the E comprises the flat Red River Valley and the Rolling Drift Prairie; Missouri Plateau of the Great Plains on the W. **Capital:** Bismarck.

Economy. Chief industries: agriculture, mining, tourism, manufacturing, telecommunications, energy, food processing. **Chief manuf. goods:** farm equipment, processed foods, fabricated metal, high-tech. electronics. **Chief crops:** spring wheat, durum, barley, flaxseed, oats, potatoes, dry edible beans, honey, soybeans, sugar beets, sunflowers, hay. **Livestock** (Jan. 1999): 1.9 mil cattle/calves; 134,000 sheep/lambs; (Dec. 1998) 205,000 hogs/pigs; (Dec. 1998) 270,000 chickens (excl. broilers). **Timber/lumber:** (1997) oak, ash, cottonwood, aspen; 1 mil bd. ft. **Nonfuel minerals** (est. 1998): $34.7 mil; mostly sand & gravel, lime, clays, gemstones. **Internat. airports at:** Fargo, Grand Forks, Bismarck, Minot, Pembina, Dunseith. **Value of construction** (1997): $788 mil. **Gross state product** (1997): $15.8 bil. **Employment distrib.** (May 1999): 28.4% serv.; 25.4% trade; 22.1% govt.; 7.5% mfg. **Per cap. pers. income** (1998): $21,675. **Sales tax** (1999): 5%. **Unemployment** (1998): 3.2%. **Tourism expends.** (1996): $1.1 bil.

Finance. FDIC-insured commercial banks (1998): 114. **Deposits:** $8.1 bil. **FDIC-insured savings institutions** (1998): 3. **Assets:** $917 mil.

Federal govt. Fed. civ. employees (Mar. 1998): 4,888. **Avg. salary:** $39,069. **Notable fed. facilities:** Strategic Air Command Base; Northern Prairie Wildlife Research Center;

Garrison Dam; Theodore Roosevelt Natl. Park; Grand Forks Energy Research Center; Ft. Union Natl. Historic Site.

Energy. Electricity production (1998, kWh, by source): Coal: 28.2 bil; Petroleum: 47 mil; Hydroelectric: 2.3 bil.

State data. Motto: Liberty and union, now and forever, one and inseparable. **Flower:** Wild prairie rose. **Bird:** Western meadowlark. **Tree:** American elm. **Song:** North Dakota Hymn. **Entered union** Nov. 2, 1889; rank, 39th. **State fair** at Minot; July.

History. At the time of European contact, the Ojibwa, Yanktonai and Teton Sioux, Mandan, Arikara, and Hidatsa peoples lived in the region. Pierre de Varennes was the first French fur trader in the area, 1738, followed later by the English. The U.S. acquired half the territory in the Louisiana Purchase, 1803. Lewis and Clark built Ft. Mandan, near present-day Stanton, 1804-5, and wintered there. In 1818, American ownership of the other half was confirmed by agreement with Britain. The first permanent settlement was at Pembina, 1812. Missouri River steamboats reached the area, 1832, the first railroad, 1873, bringing many homesteaders. The "bonanza farm" craze of the 1870s-80s attracted many settlers. The state was first to hold a national Presidential primary, 1912.

Tourist attractions. North Dakota Heritage Center, Bismarck; Bonanzaville, Fargo; Ft. Union Trading Post Natl. Historic Site; Lake Sakakawea; Intl. Peace Garden; Theodore Roosevelt Natl. Park, including Elkhorn Ranch, Badlands; Ft. Abraham Lincoln State Park and Museum, near Mandan; Dakota Dinosaur Museum, Dickinson; Knife River Indian Villages-National Historic Site.

Famous North Dakotans. Maxwell Anderson, Angie Dickinson, John Bernard Flannagan, Phil Jackson, Louis L'Amour, Peggy Lee, Eric Sevareid, Vilhjalmur Stefansson, Lawrence Welk.

Greater North Dakota Association (Chamber of Commerce). PO Box 2639, 2000 Schafer St., Bismarck, ND 58501.

Toll-free travel information. 1-800-HELLO-ND

Website. http://www.state.nd.us

Tourism website. http//www.glness.com/tourism

Ohio
Buckeye State

People. Population (1997): 11,209,493; rank: 7; **net change** (1990-98): 3.3%. **Pop. density** (1998): 273.7 per sq mi. **Racial distribution** (1990): 87.8% white; 10.6% black. **Hispanic pop.:** 1.3%.

Geography. Total area: 44,828 sq mi; rank: 34. **Land area:** 40,953 sq mi; rank: 35. **Acres forested:** 7,863,000. **Location:** East North Central state bounded on the N by Michigan and Lake Erie; on the E and S by Pennsylvania, West Virginia, and Kentucky; on the W by Indiana. **Climate:** temperate but variable; weather subject to much precipitation. **Topography:** generally rolling plain; Allegheny plateau in E; Lake Erie plains extend southward; central plains in the W. **Capital:** Columbus.

Economy. Chief industries: manufacturing, trade, services. **Chief manuf. goods:** transportation equipment, machinery, primary and fabricated metal products. **Chief crops:** corn, hay, winter wheat, oats, soybeans. **Livestock** (Jan. 1999): 1.2 mil cattle/calves; 125,000 sheep/lambs; (Dec. 1998) 1.7 mil hogs/pigs; (Dec. 1998) 37.4 mil chickens (excl. broilers); (Dec. 1997) 45.8 mil broilers. **Timber/lumber** (1998): oak, ash, maple, walnut, beech; 382 mil bd. ft. **Nonfuel minerals** (est. 1998): $1.15 bil; mostly crushed stone, sand & gravel, salt, lime, portland cement. **Commercial fishing** (1998): $2.6 mil. **Chief ports:** Toledo, Conneaut, Cleveland, Ashtabula. **Internat. airports at:** Cleveland, Cincinnati, Columbus, Dayton. **Value of construction** (1997): $14.7 bil. **Gross state product** (1997): $320.5 bil. **Employment distrib.** (May 1999): 27.8% serv.; 24.2% trade; 19.7% mfg.; 13.9% govt. **Per cap. pers. income** (1998): $25,134. **Sales tax** (1999): 5%. **Unemployment** (1998): 4.3%. **Tourism expends.** (1996): $11.1 bil. **Lottery** (1998): total sales: $2.2 bil; net income: $783.0 mil.

Finance. FDIC-insured commercial banks (1998): 220. **Deposits:** $170.5 bil. **FDIC-insured savings institutions** (1998): 140. **Assets:** $59.7 bil.

Federal govt. Fed. civ. employees (Mar. 1998): 43,836. **Avg. salary:** $46,801. **Notable fed. facilities:** Wright Patterson AFB; Defense Construction Supply Center; Lewis Research Ctr.; Portsmouth Gaseous Diffusion Plant.

Energy. Electricity production (1998, kWh, by source): Coal: 128.7 bil; Petroleum: 351 mil; Gas: 519 mil; Hydroelectric: 406 mil; Nuclear: 16.5 mil.

State data. Motto: With God, all things are possible. **Flower:** Scarlet carnation. **Bird:** Cardinal. **Tree:** Buckeye. **Song:** Beautiful Ohio. **Entered union** Mar. 1, 1803; rank, 17th. **State fair** at Columbus; Aug.

History. Wyandot, Delaware, Miami, and Shawnee peoples sparsely occupied the area when the first Europeans arrived. La Salle visited the region, 1669, and France claimed the area, 1682. Around 1730, traders from Pennsylvania and Virginia entered the area; the French and their Native American allies sought to drive them out. France ceded its claim, 1763, to Britain. During the American Revolution, George Rogers Clark seized British posts and held the region, until Britain gave up its claim, 1783, in the Treaty of Paris. The region became U.S. territory after the American Revolution. First organized settlement was at Marietta, 1788. Indian warfare ended with Anthony Wayne's victory at Fallen Timbers, 1794. In the War of 1812, Oliver Hazard Perry's victory on Lake Erie and William Henry Harrison's invasion of Canada, 1813, ended British incursions.

Tourist attractions. Mound City Group Natl. Monuments, a group of 24 prehistoric Indian burial mounds; Neil Armstrong Air and Space Museum, Wapakoneta; Air Force Museum, Dayton; Pro Football Hall of Fame, Canton; King's Island amusement park, Mason; Lake Erie Islands, Cedar Point amusement park, both Sandusky; birthplaces, homes of, and memorials to U.S. Pres.s W. H. Harrison, Grant, Garfield, Hayes, McKinley, Harding, Taft, Benjamin Harrison; Amish Region, Tuscarawas/Holmes counties; German Village, Columbus; Sea World, Aurora; Jack Nicklaus Sports Center, Mason; Bob Evans Farm, Rio Grande; Rock and Roll Hall of Fame and Museum, Cleveland.

Famous Ohioans. Sherwood Anderson, Neil Armstrong, George Bellows, Ambrose Bierce, Erma Bombeck, Hart Crane, George Coster, Clarence Darrow, Paul Laurence Dunbar, Thomas Edison, Clark Gable, John Glenn, Bob Hope, William Dean Howells, Toni Morrison, Jack Nicklaus, Jesse Owens, Pontiac, Eddie Rickenbacker, John D. Rockefeller Sr. and Jr., Roy Rogers, Pete Rose, Arthur Schlesinger Jr., Gen. William Sherman, Steven Spielberg, Gloria Steinem, Harriet Beecher Stowe, Charles Taft, Robert A. Taft, William H. Taft, Tecumseh, James Thurber, Orville and Wilbur Wright.

Chamber of Commerce. PO Box 15159. 230 E. Town St., Columbus, OH 43215-0159.

Toll-free travel information. 1-800-BUCKEYE.

Website. http://www.state.oh.us

Tourism website. http://www.ohiotourism.com

Oklahoma
Sooner State

People. Population (1998): 3,346,713; rank: 27; **net change** (1990-98): 6.4%. **Pop. density** (1998): 48.7 per sq mi. **Racial distribution** (1990): 82.1% white; 7.4% black; 8.0% Amer. Indian. **Hispanic pop.:** 2.7%.

Geography. Total area: 69,903 sq mi; rank: 20. **Land area:** 68,679 sq mi; rank: 19. **Acres forested:** 7,539,000. **Location:** West South Central state bounded on the N by Colorado and Kansas; on the E by Missouri and Arkansas; on the S and W by Texas and New Mexico. **Climate:** temperate; southern humid belt merging with colder northern continental; humid eastern and dry western zones. **Topography:** high plains predominate in the W, hills and small mountains in the E; the east central region is dominated by the Arkansas R. Basin, and the Red R. Plains, in the S. **Capital:** Oklahoma City.

Economy. Chief industries: manufacturing, mineral and energy exploration and production, agriculture, services. **Chief manuf. goods:** nonelectrical machinery, transportation equipment, food products, fabricated metal products. **Chief crops:** wheat, cotton, hay, peanuts, grain sorghum, soybeans, corn, pecans. **Livestock** (Jan. 1999): 5.2 mil cattle/calves; 55,000 sheep/lambs; (Dec. 1998) 1.9 mil hogs/pigs; (Dec. 1998) 5.4 mil chickens (excl. broilers); (Dec. 1997) 197.4 mil broilers. **Timber/lumber:** pine, oak, hickory. **Nonfuel minerals** (est. 1998): $408 mil; mostly crushed stone, portland cement, sand & gravel, iodine. **Chief ports:** Ca-

toosa, Muskogee. **Internat. airports at:** Oklahoma City, Tulsa. **Value of construction** (1997): $3.1 bil. **Gross state product** (1997): $76.6 bil. **Employment distrib.** (May 1999): 28.6% serv.; 23.0% trade; 19.1% govt.; 12.6% mfg. **Per cap. pers. income** (1998): $21,072. **Sales tax** (1999): 4.5%. **Unemployment** (1998): 4.5%. **Tourism expends.** (1996): $3.3 bil.

Finance. FDIC-insured commercial banks (1998): 309. **Deposits:** $28.7 bil. **FDIC-insured savings institutions** (1998): 12. **Assets:** $7.9 bil.

Federal govt. Fed. civ. employees (Mar. 1998): 31,036. **Avg. salary:** $40,899. **Notable fed. facilities:** Federal Aviation Agency and Tinker AFB, Oklahoma City; Ft. Sill, Lawton; Altus AFB; Vance AFB.

Energy. Electricity production (1998, kWh, by source): Coal: 31.0 bil; Petroleum: 8 mil; Gas: 17.0 bil; Hydroelectric: 3.4 bil.

State data. Motto: Labor Omnia Vincit (Labor conquers all things). **Flower:** Mistletoe. **Bird:** Scissor-tailed flycatcher. **Tree:** Redbud. **Song:** Oklahoma! **Entered union** Nov. 16, 1907; rank, 46th. **State fair** at Oklahoma City; last 2 full weeks of Sept.

History. The region was sparsely inhabited by Native American tribes when Coronado, the first European, arrived in 1541; in the 16th and 17th cent., French traders visited. Part of the Louisiana Purchase, 1803, Oklahoma was established as Indian Territory (but not given territorial government). It became home to the "Five Civilized Tribes"—Cherokee, Choctaw, Chickasaw, Creek, and Seminole—after the forced removal of Indians from the eastern U.S., 1828-46. The land was also used by Comanche, Osage, and other Plains Indians. As white settlers pressed west, land was opened for homesteading by runs and lottery, the first run on Apr. 22, 1889. The most famous run was to the Cherokee Outlet, 1893.

Tourist attractions. Cherokee Heritage Center, Tahlequah; White Water Bay and Frontier City theme pks., both Oklahoma City; Will Rogers Memorial, Claremore; Natl. Cowboy Hall of Fame and Remington Park Race Track, both Oklahoma City; Ft. Gibson Stockade, near Muskogee; Ouachita Natl. Forest; Tulsa's art deco district; Wichita Mts. Wildlife Refuge, Lawton; Woolaroc Museum & Wildlife Preserve, Bartlesville; Sequoyah's Home Site, near Sallisaw; Philbrook Museum of Art and Gilcrease Museum, both Tulsa.

Famous Oklahomans. Troy Aikman, Carl Albert, Gene Autry, Johnny Bench, Garth Brooks, William "Hopalong Cassidy" Boyd, Lon Chaney, Walter Cronkite, L. Gordon Cooper, Jerome "Dizzy" Dean, Ralph Ellison, John Hope Franklin, James Garner, Geronimo, Woody Guthrie, Paul Harvey, Ron Howard, Gen. Patrick J. Hurley, Jeane Kirkpatrick, Louis L'Amour, Shannon Lucid, Mickey Mantle, Reba McEntire, Wiley Post, Tony Randall, Oral Roberts, Will Rogers, Maria Tallchief, Jim Thorpe, J.C. Watts.

Chamber of Commerce. Chamber of Commerce, 330 NE 10th, Oklahoma City, OK 73104.

Tourism Dept. PO Box 60789, Oklahoma City, OK 73146-0789.

Toll-free travel information. 1-800-652-6552.

Website. http://www.state.ok.us

Tourism website. http://www.otrd.state.ok.us

Oregon
Beaver State

People. Population (1998): 3,281,974; rank: 28; **net change** (1990-98): 15.5%. **Pop. density** (1998): 34.2 per sq mi. **Racial distribution** (1990): 92.8% white; 1.6% black. **Hispanic pop.:** 4.0%.

Geography. Total area: 97,132 sq mi; rank: 10. **Land area:** 96,002 sq mi; rank: 10. **Acres forested:** 27,997,000. **Location:** Pacific state, bounded on N by Washington; on E by Idaho; on S by Nevada and California; on W by the Pacific. **Climate:** coastal mild and humid climate; continental dryness and extreme temperatures in the interior. **Topography:** Coast Range of rugged mountains; fertile Willamette R. Valley to E and S; Cascade Mt. Range of volcanic peaks E of the valley; plateau E of Cascades, remaining two-thirds of state. **Capital:** Salem.

Economy. Chief industries: manufacturing, services, trade, finance, insurance, real estate, government, construction. **Chief manuf. goods:** electronics & semiconductors, lumber & wood products metals, transportation equipment, processed food, paper. **Chief crops:** greenhouse, hay, wheat, grass seed, potatoes, onions, Christmas trees, pears, mint. **Livestock** (Jan. 1999): 1.5 mil cattle/calves; 215,000 sheep/lambs; (Dec. 1998) 30,000 hogs/pigs; (Dec. 1998) 3.5 mil chickens (excl. broilers); (Dec. 1997) 21.7 mil broilers. **Timber/lumber** (1998): Douglas fir, hemlock, ponderosa pine; 5.6 bil bd. ft. **Nonfuel minerals** (est. 1998): $272 mil; mostly sand & gravel, crushed stone, portland cement, diatomite, lime. **Commercial fishing** (1998): $50.3 mil. **Chief ports:** Portland, Astoria, Coos Bay. **Internat. airports at:** Portland, Klamath Falls. **Value of construction** (1997): $6.0 bil. **Gross state product** (1997): $98.4 bil. **Employment distrib.** (May 1999): 27.3% serv.; 24.5% trade; 16.4% govt.; 15.2% mfg. **Per cap. pers. income** (1998): $24,766. **Sales tax:** none. **Unemployment** (1998): 5.6%. **Tourism expends.** (1996): $4.9 bil. **Lottery** (1998): total sales: $717.9 mil; net income: $297.7 mil.

Finance. FDIC-insured commercial banks (1998): 42. **Deposits:** $5.3 bil. **FDIC-insured savings institutions** (1998): 7. **Assets:** $27.8 bil.

Federal govt. Fed. civ. employees (Mar. 1998): 17,586. **Avg. salary:** $43,221. **Notable fed. facilities:** Bonneville Power Administration.

Energy. Electricity production (1998, kWh, by source): Coal: 3.3 bil; Petroleum: 33 mil; Gas: 3.5 bil; Hydroelectric: 39.5 bil.

State data. Motto: She flies with her own wings. **Flower:** Oregon grape. **Bird:** Western meadowlark. **Tree:** Douglas fir. **Song:** Oregon, My Oregon. **Entered union** Feb. 14, 1859; rank, 33d. **State fair** at Salem; 12 days ending with Labor Day.

History. More than 100 Native American tribes inhabited the area at the time of European contact, including the Chinook, Yakima, Cayuse, Modoc, and Nez Percé. Capt. Robert Gray sighted and sailed into the Columbia River, 1792; Lewis and Clark, traveling overland, wintered at its mouth, 1805-6; John Jacob Astor established a trading post in the Columbia River region, 1811. Settlers arrived in the Williamette Valley, 1834. In 1843, the first large wave of settlers arrived via the Oregon Trail. Early in the 20th cent., the "Oregon System"— political reforms that included the initiative, referendum, recall, direct primary, and woman suffrage—was adopted.

Tourist attractions. John Day Fossil Beds Natl. Monument; Columbia River Gorge; Timberline Lodge, Mt. Hood Natl. Forest; Crater Lake Natl. Park; Oregon Dunes Natl. Recreation Area; Ft. Clatsop Natl. Memorial; Oregon Caves Natl. Monument; Oregon Museum of Science and Industry, Portland; Shakespearean Festival, Ashland; High Desert Museum, Bend; Multnomah Falls; Diamond Lake.

Famous Oregonians. Ernest Bloch, Raymond Carver, Ernest Haycox, Chief Joseph, Phil Knight, Edwin Markham, Tom McCall, Dr. John McLoughlin, Joaquin Miller, Bob Packwood, Linus Pauling, John Reed, Alberto Salazar, Mary Decker Slaney, William Simon U'Ren.

Tourist information. Economic Development Department, 775 Summer St. NE, Salem, OR 97310.

Toll-free travel information. 1-800-547-7842.

Website. http://www.state.or.us

Tourism website. http://www.traveloregon.com

Pennsylvania

Keystone State

People. Population (1998): 12,001,451; rank: 6; **net change** (1990-98): 1.0%. **Pop. density** (1998): 267.8 per sq mi. **Racial distribution** (1990): 88.5% white; 9.2% black. **Hispanic pop.:** 2.0%.

Geography. Total area: 46,058 sq mi; rank: 33. **Land area:** 44,820 sq mi; rank: 32. **Acres forested:** 16,969,000. **Location:** Middle Atlantic state, bordered on the E by the Delaware R.; on the S by the Mason-Dixon Line; on the W by West Virginia and Ohio; on the N/NE by Lake Erie and New York. **Climate:** continental with wide fluctuations in seasonal temperatures. **Topography:** Allegheny Mts. run SW to NE, with Piedmont and Coast Plain in the SE triangle; Allegheny Front a diagonal spine across the state's center; N and W rugged plateau falls to Lake Erie Lowland. **Capital:** Harrisburg.

Economy. Chief industries: agribusiness, advanced manufacturing, health care, travel & tourism, depository institutions, biotechnology, printing & publishing, research & con-sulting, trucking & warehousing, transportation by air, engineering & management, legal services. **Chief manuf. goods:** fabricated metal products; industrial machinery & equipment, transportation equipment, rubber & plastics, electronic equipment, chemicals & pharmaceuticals, lumber & wood products, stone, clay, & glass products. **Chief crops:** corn, hay, mushrooms, apples, potatoes, winter wheat, oats, vegetables, tobacco, grapes, peaches. **Livestock** (Jan. 1999): 1.7 mil cattle/calves; 83,000 sheep/lambs; (Dec. 1998) 1.1 mil hogs/pigs; (Dec. 1998) 28.5 mil chickens (excl. broilers); (Dec. 1997) 135.2 mil broilers. **Timber/lumber** (1998): pine, oak, maple; 1.1 bil bd. ft. **Nonfuel minerals** (est. 1998): $1.28 bil; mostly crushed stone, portland cement, lime, sand & gravel, masonry cement. **Commercial fishing** (1998): $105,000. **Chief ports:** Philadelphia, Pittsburgh, Erie. **Internat. airports at:** Allentown, Erie, Harrisburg, Philadelphia, Pittsburgh, Wilkes-Barre/Scranton. **Value of construction** (1997): $10.1 bil. **Gross state product** (1997): $340.0 bil. **Employment distrib.** (May 1999): 32.4% serv.; 22.5% trade; 16.8% mfg.; 12.8% govt. **Per cap. pers. income** (1998): $26,792. **Sales tax** (1999): 6%. **Unemployment** (1998): 4.6%. **Tourism expends.** (1996): $13.1 bil. **Lottery** (1998): total sales: $1.7 bil; net income: $714.6 mil.

Finance. FDIC-insured commercial banks (1998): 197. **Deposits:** $137.0 bil. **FDIC-insured savings institutions** (1998): 116. **Assets:** $60.8 bil.

Federal govt. Fed. civ. employees (Mar. 1998): 62,585. **Avg. salary:** $41,564. **Notable fed. facilities:** Carlisle Barracks; Army War College; Naval Inventory Control Point, Phila. and Mechanicsbrg; Defense Personnel Supply Center, Phila.; Defense Distribution Center, New Cumberland; Tobyhanna Army Depot; Letterkenny Army Depot; NAS Willow Grove; 911th Air Wing, Pittsburgh; Naval Surface Warfare Center, Phila.; Charles E. Kelly Support Facility.

Energy. Electricity production (1998, kWh, by source): Coal: 106.5 bil; Petroleum: 4.1 bil; Gas: 572 mil; Hydroelectric: 1.6 bil; Nuclear: 61.1 bil.

State data. Motto: Virtue, liberty and independence. **Flower:** Mountain laurel. **Bird:** Ruffed grouse. **Tree:** Hemlock. **Second** of the original 13 states to ratify the Constitution, Dec. 12, 1787. **State fair** at Harrisburg; 2d week in Jan. at State Farm Show Building.

History. At the time of European contact, Lenni Lenape (Delaware), Shawnee and Iroquoian Susquehannocks, Erie, and Seneca occupied the region. Swedish explorers established the first permanent settlement, 1643, on Tinicum Island. In 1655, the Dutch seized the settlement but lost it to the British, 1664. The region was given by Charles II to William Penn, 1681. Philadelphia ("brotherly love") was the capital of the colonies during most of the American Revolution, and of the U.S., 1790-1800. Philadelphia was taken by the British, 1777; Washington's troops encamped at Valley Forge in the bitter winter of 1777-78. The Declaration of Independence, 1776, and the Constitution, 1787, were signed in Philadelphia. The Civil War battle of Gettysburg, July 1-3, 1863, marked a turning point, favoring Union forces.

Tourist attractions. Independence Natl. Historic Park, Franklin Institute Science Museum, Philadelphia Museum of Art, all in Philadelphia; Valley Forge Natl. Historic Park; Gettysburg Natl. Military Park; Pennsylvania Dutch Country; Hershey; Duquesne Incline, Carnegie Institute, Heinz Hall, all in Pittsburgh; Pocono Mts.; Pennsylvania's Grand Canyon, Tioga County; Allegheny Natl. Forest; Laurel Highlands; Presque Isle State Park; Fallingwater, Ligonier; Johnstown; SteamTown U.S.A., Scranton; State Flagship Niagara, Erie; Oil Heritage Region, Northwest PA.

Famous Pennsylvanians. Marian Anderson, Maxwell Anderson, George Blanda, James Buchanan, Andrew Carnegie, Rachel Carson, Perry Como, Thomas Eakins, Stephen Foster, Benjamin Franklin, Robert Fulton, Martha Graham, Milton Hershey, Gene Kelly, Grace Kelly (Princess Grace of Monaco), George C. Marshall, Dan Marino, John J. McCloy, Margaret Mead, Andrew W. Mellon, Joe Montana, Stan Musial, Joe Namath, Arnold Palmer, Robert E. Peary, John O'Hara, Mary Roberts Rinehart, Betsy Ross, Will Smith, Jimmy Stewart, Jim Thorpe, Johnny Unitas, John Updike, Honus Wagner, Andy Warhol, Benjamin West.

Chamber of Business and Industry. 417 Walnut St., Harrisburg, PA 17120; 717-255-3252.

Toll-free travel information. 1-800-VISITPA.

Website. http://www.state.pa.us

Tourism website. http://www.state.pa.us/visit

Rhode Island

Little Rhody, Ocean State

People. Population (1998): 988,480; rank: 43; **net change** (1990-98): -1.5%. **Pop. density** (1998): 945.9 per sq mi. **Racial distribution** (1990): 91.4% white; 3.9% black. **Hispanic pop.:** 4.6%.

Geography. Total area: 1,231 sq mi; rank: 50. **Land area:** 1,045 sq mi; rank: 50. **Acres forested:** 401,000. **Location:** New England state. **Climate:** invigorating and changeable. **Topography:** eastern lowlands of Narragansett Basin; western uplands of flat and rolling hills. **Capital:** Providence.

Economy. Chief industries: services, manufacturing. **Chief manuf. goods:** costume jewelry, toys, machinery, textiles, electronics. **Chief crops:** nursery products, turf & vegetable production. **Livestock** (Jan. 1999): 6,000 cattle/calves; (Dec. 1997) 3,000 hogs/pigs; (Dec. 1998) 78,000 chickens (excl. broilers). **Timber/lumber:** (1998) oak; 11 mil bd. ft. **Nonfuel minerals** (est. 1998): $27.8 mil; mostly sand & gravel, crushed stone, gemstones. **Commercial fishing** (1998): $71.1 mil. **Chief ports:** Providence, Quonset Point, Newport. **Value of construction** (1997): $773 mil. **Gross state product** (1997): $27.9 bil. **Employment distrib.** (May 1999): 34.3% services; 21.8% trade; 16.8% mfg.; 13.8% govt. **Per cap. pers. income** (1998): $26,797. **Sales tax** (1999): 7%. **Unemployment** (1998): 4.9%. **Tourism expends.** (1996): $922 mil. **Lottery** (1998): total sales: $634.1 mil; net income: $115.0 mil.

Finance. FDIC-insured commercial banks (1998): 7. **Deposits:** $56.6 bil. **FDIC-insured savings institutions** (1998): 6. **Assets:** $1.7 bil.

Federal govt. Fed. civ. employees (Mar. 1998): 5,996. **Avg. salary:** $47,700. **Notable fed. facilities:** Naval War College; Naval Underwater Warfare Center; Natl. Marine Fisheries Laboratory; EPA Environmental Research Laboratory.

Energy. Electricity production (1998, kWh, by source): Petroleum: 9 mil; Gas: 2.1 bil.

State data. Motto: Hope. **Flower:** Violet. **Bird:** Rhode Island red. **Tree:** Red maple. **Song:** Rhode Island. **Thirteenth** of original 13 states to ratify the Constitution, May 29, 1790. **State fair** at Richmond; mid-Aug.

History. When the Europeans arrived Narragansett, Niantic, Nipmuc, and Wampanoag peoples lived in the region. Verrazano visited the area, 1524. The first permanent settlement was founded at Providence, 1636, by Roger Williams, who was exiled from the Massachusetts Bay Colony; Anne Hutchinson, also exiled, settled Portsmouth, 1638. Quaker and Jewish immigrants seeking freedom of worship began arriving, 1650s-60s. The colonists broke the power of the Narragansett in the Great Swamp Fight, 1675, the decisive battle in King Philip's War. British trade restrictions angered colonists, and they burned the British customs vessel *Gaspee*, 1772. The colony became the first to formally renounce all allegiance to King George III, May 4, 1776. Initially opposed to joining the Union, Rhode Island was the last of the 13 colonies to ratify the Constitution, 1790.

Tourist attractions. Newport mansions; yachting races including Newport to Bermuda; Block Island; Touro Synagogue, oldest in U.S., Newport; first Baptist Church in America, Providence; Slater Mill Historic Site, Pawtucket; Gilbert Stuart birthplace, Saunderstown.

Famous Rhode Islanders. Ambrose Burnside, George M. Cohan, Nelson Eddy, Jabez Gorham, Nathanael Greene, Christopher and Oliver La Farge, Matthew C. and Oliver Hazard Perry, Gilbert Stuart.

Tourist Information. Rhode Island Economic Development Corporation, One W. Exchange St., Providence, RI 02903.

Toll-free travel information. 1-800-556-2484.

Website. http://www.state.ri.us

Tourism website. http://visitrhodeisland.com

South Carolina

Palmetto State

People. Population (1998): 3,835,962; rank: 26; **net change** (1990-98): 10.0%. **Pop. density** (1998): 127.4 per sq mi. **Racial distribution** (1990): 69.0% white; 29.8% black. **Hispanic pop.:** 0.9%.

Geography. Total area: 31,189 sq mi; rank: 40. **Land area:** 30,111 sq mi; rank: 40. **Acres forested:** 12,257,000. **Lo-**

cation: South Atlantic state, bordered by North Carolina on the N; Georgia on the SW and W; the Atlantic Ocean on the E, SE, and S. **Climate:** humid subtropical. **Topography:** Blue Ridge province in NW has highest peaks; piedmont lies between the mountains and the fall line; coastal plain covers two-thirds of the state. **Capital:** Columbia.

Economy. Chief industries: tourism, agriculture, manufacturing. **Chief manuf. goods:** textiles, chemicals and allied products, machinery and fabricated metal products, apparel and related products. **Chief crops:** tobacco, cotton, soybeans, corn, wheat, peaches, tomatoes. **Livestock** (Jan. 1999): 480,000 cattle/calves; (Dec. 1998) 270,000 hogs/pigs; (Dec. 1998) 5.6 mil chickens (excl. broilers); (Dec. 1997) 182.8 mil broilers. **Timber/lumber** (1998): pine, oak; 1.5 bil bd. ft. **Nonfuel minerals** (est. 1998): $589 mil; mostly portland cement, crushed stone, gold, sand & gravel, masonry cement. **Commercial fishing** (1998): $28.3 mil. **Chief ports:** Charleston, Georgetown, Beaufort/ Port Royal. **Internat. airport at:** Charleston. **Value of construction** (1997): $6.0 bil. **Gross state product** (1997): $93.3 bil. **Employment distrib.** (May 1999): 24.2% trade; 23.8% serv.; 19.8% mfg.; 17.1% govt. **Per cap. pers. income** (1998): $21,309. **Sales tax** (1999): 5%. **Unemployment** (1998): 3.8%. **Tourism expends.** (1996): $6.2 bil.

Finance. FDIC-insured commercial banks (1998): 77. **Deposits:** $15.6 bil. **FDIC-insured savings institutions** (1998): 30. **Assets:** $8.0 bil.

Federal govt. Fed. civ. employees (Mar. 1998): 15,920. **Avg. salary:** $40,364. **Notable fed. facilities:** Polaris Submarine Base; Barnwell Nuclear Power Plant; Ft. Jackson; Parris Island; Savannah River Plant.

Energy. Electricity production (1998, kWh, by source): Coal: 32.4 bil.; Petroleum: 331 mil; Gas: 415 mil; Hydroelectric: 2.5 bil; Nuclear: 48.8 bil.

State data. Motto: Dum Spiro Spero (While I breathe, I hope). **Flower:** Yellow jessamine. **Bird:** Carolina wren. **Tree:** Palmetto. **Song:** Carolina. **Eighth** of the original 13 states to ratify the Constitution, May 23, 1788. **State fair** at Columbia; mid-Oct.

History. At the time of European settlement, Cherokee, Catawba, and Muskogean peoples lived in the area. The first English colonists settled near the Ashley River, 1670, and moved to the site of Charleston, 1680. The colonists seized the government, 1775, and the royal governor fled. The British took Charleston, 1780, but were defeated at Kings Mountain that same year, and at Cowpens and Eutaw Springs, 1781. In the 1830s, South Carolinians, angered by federal protective tariffs, adopted the Nullification Doctrine, holding that a state can void an act of Congress. The state was the first to secede from the Union, 1860, and Confederate troops fired on and forced the surrender of U.S. troops at Ft. Sumter, in Charleston Harbor, launching the Civil War. South Carolina was readmitted, 1868.

Tourist attractions. Historic Charleston; Ft. Sumter Natl. Monument, in Charleston Harbor; Charleston Museum, est. 1773, oldest museum in U.S.; Middleton Place, Magnolia Plantation, Cypress Gardens, Drayton Hall, all near Charleston; other gardens at Brookgreen, Edisto, Glencairn; Myrtle Beach; Hilton Head Island; Revolutionary War battle sites; Andrew Jackson State Park & Museum; South Carolina State Museum, Columbia; Riverbanks Zoo, Columbia.

Famous South Carolinians. Charles Bolden, James F. Byrnes, John C. Calhoun, DuBose Heyward, Ernest F. Hollings, Andrew Jackson, Jesse Jackson, James Longstreet, Francis Marion, Ronald McNair, Charles Pinckney, John Rutledge, Thomas Sumter, Strom Thurmond, John B. Watson.

Tourist information. S. Carolina Dept. of Parks, Recreation, & Tourism; 803-734-0122.

Toll-free travel information. 1-800-346-3634.

Website. http://www.state.sc.us

Tourism website. http://www.sccsi.com/sc

South Dakota

Coyote State, Mount Rushmore State

People. Population (1998): 738,171; rank: 46; **net change** (1990-98): 6.1%. **Pop. density** (1998): 9.7 per sq mi. **Racial distribution** (1990): 91.6% white; 0.5% black; 7.3% Amer. Indian. **Hispanic pop.:** 0.8%.

Geography. Total area: 77,121 sq mi; rank: 17. **Land area:** 75,896 sq mi; rank: 16. **Acres forested:** 1,690,000. **Location:** West North Central state bounded on the N by North

Dakota; on the E by Minnesota and Iowa; on the S by Nebraska; on the W by Wyoming and Montana. **Climate:** characterized by extremes of temperature, persistent winds, low precipitation and humidity. **Topography:** Prairie Plains in the E; rolling hills of the Great Plains in the W; the Black Hills, rising 3,500 ft, in the SW corner. **Capital:** Pierre.

Economy. Chief industries: agriculture, services, manufacturing. **Chief manuf. goods:** food and kindred products, machinery, electric and electronic equipment. **Chief crops:** corn, soybeans, oats, wheat, sunflowers, sorghum. **Livestock** (Jan. 1999): 3.9 mil cattle/calves; 420,000 sheep/lambs; (Dec. 1998) 1.4 mil hogs/pigs; (Dec. 1998) 2.6 mil chickens (excl. broilers). **Timber/lumber** ponderosa pine. **Nonfuel minerals** (est. 1998): $269 mil; mostly gold, portland cement, sand & gravel, lime. **Value of construction** (1997): $742 mil. **Gross state product** (1997): $20.2 bil. **Employment distrib.** (May 1999): 26.3% serv.; 24.8% trade; 19.4% govt.; 13.5% mfg. **Per cap. pers. income** (1998): $22,114. **Sales tax** (1999): 4%. **Unemployment** (1998): 2.9%. **Tourism expends.** (1996) $1.0 mil. **Lottery** (1998): total sales: $555.3 mil; net income: $97.6 mil.

Finance. FDIC-insured commercial banks (1998): 104. **Deposits:** $11.7 bil. **FDIC-insured savings institutions** (1998): 4. **Assets:** $893 mil.

Federal govt. Fed. civ. employees (Mar. 1998): 6,624. **Avg. salary:** $38,160. **Notable fed. facilities:** Ellsworth AFB, Corp of Engineers, Nat'l Park Service.

Energy. Electricity production (1998, kWh, by source): Coal: 3.1 bil; Petroleum: 27 mil; Gas: 211 mil; Hydroelectric: 5.8 bil.

State data. Motto: Under God, the people rule. **Flower:** Pasqueflower. **Bird:** Chinese ring-necked pheasant. **Tree:** Black Hills spruce. **Song:** Hail, South Dakota. **Entered union** Nov. 2, 1889; rank, 40th. **State fair** at Huron; late Aug.-early Sept.

History. At the time of first European contact, Mandan, Hidatsa, Arikara and Sioux lived in the area. The French Verendrye brothers explored the region, 1742-43. The U.S. acquired the area, 1803, in the Louisiana Purchase. Lewis and Clark passed through the area, 1804-6. In 1817 a trading post was opened at Fort Pierre, which later became the site of the first European settlement in South Dakota. Gold was discovered, 1874, in the Black Hills on the great Sioux reservation; the "Great Dakota Boom" began in 1879. Conflicts with Native Americans led to the Great Sioux Agreement, 1889, which established reservations and opened up more land for white settlement. The massacre of Native American families at Wounded Knee, 1890, ended Sioux resistance.

Tourist attractions. Black Hills; Mt. Rushmore; Needles Highway; Harney Peak, tallest E. of Rockies; Deadwood, 1876 Gold Rush town; Custer State Park; Jewel Cave Natl. Monument; Badlands Natl. Park "moonscape"; "Great Lakes of S. Dakota"; Ft. Sisseton; Great Plains Zoo & Museum, Sioux Falls; Corn Palace, Mitchell; Wind Cave Natl. Park; Crazy Horse Memorial, mountain carving in progress.

Famous South Dakotans. Sparky Anderson, Tom Brokaw, Crazy Horse, Thomas Daschle, Myron Floren, Mary Hart, Cheryl Ladd, Dr. Ernest O. Lawrence, George McGovern, Billy Mills, Allen Neuharth, Pat O'Brien, Sitting Bull.

Tourist information. Department of Tourism, Capitol Lake Plaza, 711 E. Wells Ave., c/o 500 E. Capitol Ave., Pierre, SD 57501-5070.

Toll-free travel information. 1-800-SDAKOTA.

Website. http://www.state.sd.us

Tourism website. http://www.state.sd.us/tourism

Tennessee

Volunteer State

People. Population (1998): 5,430,621; rank: 17; **net change** (1990-98): 11.3%. **Pop. density** (1998): 131.8 per sq mi. **Racial distribution** (1990): 83.0% white; 16.0% black. **Hispanic pop.:** 0.7%.

Geography. Total area: 42,146 sq mi; rank: 36. **Land area:** 41,219 sq mi; rank: 34. **Acres forested:** 13,612,000. **Location:** East South Central state bounded on the N by Kentucky and Virginia; on the E by North Carolina; on the S by Georgia, Alabama, and Mississippi; on the W by Arkansas

and Missouri. **Climate:** humid continental to the N; humid subtropical to the S. **Topography:** rugged country in the E; the Great Smoky Mts. of the Unakas; low ridges of the Appalachian Valley; the flat Cumberland Plateau; slightly rolling terrain and knobs of the Interior Low Plateau, the largest region; Eastern Gulf Coastal Plain to the W, laced with streams; Mississippi Alluvial Plain, a narrow strip of swamp and flood plain in the extreme W. **Capital:** Nashville.

Economy. Chief industries: manufacturing, trade, services, tourism, finance, insurance, real estate. **Chief manuf. goods:** chemicals, food, transportation equipment, industrial machinery & equipment, fabricated metal products, rubber/plastic products, paper & allied products, printing & publishing. **Chief crops:** tobacco, cotton, lint, soybeans, grain, corn. **Livestock** (Jan. 1999): 2.2 mil cattle/calves; 13,000 sheep/lambs; (Dec. 1998) 300,000 hogs/pigs; (Dec. 1998) 2.3 mil chickens (excl. broilers); (Dec. 1997) 138.6 mil broilers. **Timber/lumber** (1998): red oak, white oak, yellow poplar, hickory; 816 mil bd. ft. **Nonfuel minerals** (est. 1998): $709 mil; mostly crushed stone, zinc, portland cement, sand & gravel, clays. **Chief ports:** Memphis, Nashville, Chattanooga, Knoxville. **Internat. airports at:** Memphis, Nashville. **Value of construction** (1997): $8.2 bil. **Gross state product** (1997): $147.0 bil. **Employment distrib.** (May 1999): 26.9% serv.; 23.9% trade; 19.0% mfg.; 14.4% govt. **Per cap. pers. income** (1998): $23,559. **Sales tax** (1999): 6%. **Unemployment** (1998): 4.2%. **Tourism expends.** (1996): $8.4 bil.

Finance. FDIC-insured commercial banks (1998): 204. **Deposit:** $76.5 bil. **FDIC-insured savings institutions** (1998): 25. **Assets:** $5.6 bil.

Federal govt. Fed. civ. employees (Mar. 1998): 32,648. **Avg. salary:** $43,435. **Notable fed. facilities:** Tennessee Valley Authority; Oak Ridge Nat'l. Laboratories; Arnold Engineering Development Center; Ft. Campbell Army Base; Millington Naval Station.

Energy. Electricity production (1998, kWh, by source): Coal: 55.1 bil; Petroleum: 699 mil; Gas: 551 mil; Hydroelectric: 9.4 bil; Nuclear: 28.4 bil.

State data. Motto: Agriculture and commerce. **Flower:** Iris. **Bird:** Mockingbird. **Tree:** Tulip poplar. **Song:** The Tennessee Waltz. **Entered union** June 1, 1796; rank, 16th. **State fair** at Nashville; mid-Sept.

History. When the first European explorers arrived, Creek and Yuchi peoples lived in the area; the Cherokee moved into the region in the early 18th cent. Spanish explorers first visited the area, 1541. English traders crossed the Great Smokies from the east while France's Marquette and Jolliet sailed down the Mississippi on the west, 1673. The first permanent settlement was by Virginians on the Watauga River, 1769. During the American Revolution, the colonists helped win the Battle of Kings Mountain (NC), 1780, and joined other eastern campaigns. The state seceded from the Union, 1861, and saw many Civil War engagements, but 30,000 soldiers fought for the Union. Tennessee was readmitted in 1866, the only former Confederate state not to have a postwar military government.

Tourist attractions. Reelfoot Lake; Lookout Mountain, Chattanooga; Fall Creek Falls; Great Smoky Mountains Natl. Park; Lost Sea, Sweetwater; Cherokee Natl. Forest; Cumberland Gap Natl. Park; Andrew Jackson's home, the Hermitage, near Nashville; homes of Pres.s Polk and Andrew Johnson; American Museum of Science and Energy, Oak Ridge; Parthenon, Grand Old Opry, Opryland USA, all Nashville; Dollywood theme park, Pigeon Forge; Tennessee Aquarium, Chattanooga; Graceland, home of Elvis Presley, Memphis; Alex Haley Home and Museum, Henning; Casey Jones Home and Museum, Jackson.

Famous Tennesseans. Roy Acuff, Davy Crockett, David Farragut, Ernie Ford, Aretha Franklin, Morgan Freeman, Al Gore Jr., Alex Haley, William C. Handy, Sam Houston, Cordell Hull, Andrew Jackson, Andrew Johnson, Casey Jones, Estes Kefauver, Grace Moore, Dolly Parton, Minnie Pearl, James Polk, Elvis Presley, Dinah Shore, Bessie Smith, Alvin York.

Tourist information. Dept. of Tourist Development, 5th Floor, Rachel Jackson Bldg., 320 6th Ave. N., Nashville, TN 37202.

Toll-free travel information. 1-800-TENN200.

Website. http://www.state.tn.us

Tourism website. http://www.state.tn.us/tourdev

Texas
Lone Star State

People. Population (1998): 19,759,614; rank: 2; **net change** (1990-98): 16.3%. **Pop. density** (1998): 75.4 per sq mi. **Racial distribution** (1990): 75.2% white; 11.9% black. **Hispanic pop.:** 25.5%.

Geography. Total area: 267,277 sq mi; rank: 2. **Land area:** 261,914 sq mi; rank: 2. **Acres forested:** 19,193,000. **Location:** Southwestern state, bounded on the SE by the Gulf of Mexico; on the SW by Mexico, separated by the Rio Grande; surrounding states are Louisiana, Arkansas, Oklahoma, New Mexico. **Climate:** extremely varied; driest region is the Trans-Pecos; wettest is the NE. **Topography:** Gulf Coast Plain in the S and SE; North Central Plains slope upward with some hills; the Great Plains extend over the Panhandle, are broken by low mountains; the Trans-Pecos is the southern extension of the Rockies. **Capital:** Austin.

Economy. Chief industries: manufacturing, trade, oil and gas extraction, services. **Chief manuf. goods:** industrial machinery and equipment, foods, electrical and electronic products, chemicals and allied products, apparel. **Chief crops:** cotton, grains (wheat), sorghum grain, vegetables, citrus and other fruits, greenhouse/nursery, pecans, peanuts. **Chief farm products:** milk, eggs **Livestock** (Jan. 1999): 14.0 mil cattle/calves; 1.4 mil sheep/lambs; (Dec. 1998) 640,000 hogs/pigs; (Dec. 1998) 24.4 mil chickens (excl. broilers); (Dec. 1997) 455.1 mil broilers. **Timber/lumber** (1998): pine, cypress; 1.4 bil bd. ft. **Nonfuel minerals** (est. 1998): $1.92 bil; mostly portland cement, crushed stone, sand & gravel, magnesium metal, salt. **Commercial fishing** (1998): $183.3 mil. **Chief ports:** Houston, Galveston, Brownsville, Beaumont, Port Arthur, Corpus Christi. **Major Internat. airports at:** Houston, Dallas/Ft. Worth, San Antonio. **Value of construction** (1997): $27.2 bil. **Gross state product** (1997): $601.6 bil. **Employment distrib.** (May 1999): 28.6% serv.; 23.5% trade; 16.9% govt.; 12.0% mfg. **Per cap. pers. income** (1998): $24,957. **Sales tax** (1999): 6.25%. **Unemployment** (1998): 4.8%. **Tourism expends.** (1996): $27.6 bil. **Lottery** (1998): total sales: $3.1 bil; net income: $1.2 bil.

Finance. FDIC-insured commercial banks (1998): 799. **Deposits:** $149.2 bil. **FDIC-insured savings institutions** (1998): 52. **Assets:** $52.7 bil.

Federal govt. Fed. civ. employees (Mar. 1998): 108,098. **Avg. salary:** $41,749. **Notable fed. facilities:** Fort Hood, Kelly AFB, and Ft. Sam Houston.

Energy. Electricity production (1998, kWh, by source): Coal: 132.6 bil; Petroleum: 137 mil; Gas: 120.2 bil; Hydroelectric: 1.4 bil; Nuclear: 38.7 bil.

State data. Motto: Friendship. **Flower:** Bluebonnet. **Bird:** Mockingbird. **Tree:** Pecan. **Song:** Texas, Our Texas. **Entered union** Dec. 29, 1845; rank: 28th. **State fair** at Dallas; mid-Oct.

History. At the time of European contact, Native American tribes in the region were numerous and diverse in culture. Coahuiltecan, Karankawa, Caddo, Jumano, and Tonkawa peoples lived in the area, and during the 19th cent., the Apache, Comanche, Cherokee, and Wichita arrived. Spanish explorer Pineda sailed along the Texas coast, 1519; Cabeza de Vaca and Coronado visited the interior, 1541. Spaniards made the first settlement at Ysleta, near El Paso, 1682. Americans moved into the land early in the 19th cent. Mexico, of which Texas was a part, won independence from Spain, 1821; Santa Anna became dictator in 1835; Texans rebelled. Santa Anna wiped out defenders of the Alamo, 1836; Sam Houston's Texans defeated Santa Anna at San Jacinto, and independence was proclaimed that same year. The Republic of Texas, with Sam Houston as its first president, functioned as a nation until 1845, when it was admitted to the Union.

Tourist attractions. Padre Island Natl. Seashore; Big Bend, Guadalupe Mts. natl. parks; The Alamo; Ft. Davis; Six Flags Amusement Park; Sea World and Fiesta Texas, both in San Antonio; San Antonio Missions Natl. Historical Park; Cowgirl Hall of Fame, Fort Worth; Lyndon B. Johnson Natl. Historical Park, marking his birthplace, boyhood home, and ranch, near Johnson City; Lyndon B. Johnson Library and Museum, Austin; Texas State Aquarium, Corpus Christi; Kimball Art Museum, Fort Worth; George Bush Library, College Station.

Famous Texans. Stephen F. Austin, Lloyd Bentsen, James Bowie, Carol Burnett, George Bush, George W. Bush, Joan Crawford, J. Frank Dobie, Dwight D. Eisenhower, Farrah Fawcett, Sam Houston, Howard Hughes, Lyndon B. Johnson, Tommy Lee Jones, Janis Joplin, Barbara Jordan, Mary Martin, Chester Nimitz, Sandra Day O'Connor, H. Ross Perot, Katherine Ann Porter, Dan Rather, Sam Rayburn, Ann Richards, Sissy Spacek, Kenneth Starr, George Strait.

Chamber of Commerce. 900 Congress, Suite 501, Austin, TX 78701.

Toll-free travel information. 1-800-8888TEX.

Website. http://www.state.tx.us

Tourism website. http://www.traveltex.com

Utah
Beehive State

People. Population (1998): 2,099,758; rank: 34; **net change** (1990-98): 21.9%. **Pop. density** (1998): 25.6 per sq mi. **Racial distribution** (1990): 93.8% white; 0.7% black. **Hispanic pop.:** 4.9%.

Geography. Total area: 84,904 sq mi; rank: 13. **Land area:** 82,168 sq mi; rank: 12. **Acres forested:** 16,234,000. **Location:** Middle Rocky Mountain state; its southeastern corner touches Colorado, New Mexico, and Arizona, and is the only spot in the U.S. where 4 states join. **Climate:** arid; ranging from warm desert in SW to alpine in NE. **Topography:** high Colorado plateau is cut by brilliantly colored canyons of the SE; broad, flat, desert-like Great Basin of the W; the Great Salt Lake and Bonneville Salt Flats to the NW; Middle Rockies in the NE run E-W; valleys and plateaus of the Wasatch Front. **Capital:** Salt Lake City.

Economy. Chief industries: services, trade, manufacturing, government, transportation, utilities. **Chief manuf. goods:** medical instruments, electronic components, food products, fabricated metals, transportation equipment, steel and copper. **Chief crops:** hay, corn, wheat, barley, apples, potatoes, cherries, onions, peaches, pears. **Livestock** (Jan. 1999): 890,000 cattle/calves; 400,000 sheep/lambs; (Dec. 1998) 380,000 hogs/pigs; (Dec. 1998) 2.2 mil chickens (excl. broilers). **Timber/lumber:** (1998) aspen, spruce, pine; 27 mil bd. ft. **Nonfuel minerals** (est. 1998): $1.3 bil; mostly copper, gold, molybdenum, magnesium metal, sand & gravel. **Internat. airports at:** Salt Lake City. **Value of construction** (1997): $5.3 bil. **Gross state product** (1997): $55.4 bil. **Employment distrib.** (May 1999): 27.7% serv.; 23.6% trade; 17.0% govt.; 12.8% mfg. **Per cap. pers. income** (1998): $21,019. **Sales tax** (1999): 4.75%. **Unemployment** (1998): 3.8%. **Tourism expends.** (1996): $3.6 bil.

Finance. FDIC-insured commercial banks (1998): 50. **Deposits:** $22.9 bil. **FDIC-insured savings institutions** (1998): 3. **Assets:** $1.3 bil.

Federal govt. Fed. civ. employees (Mar. 1998): 23,044. **Avg. salary:** $39,109. **Notable fed. facilities:** Hill AFB; Tooele Army Depot; IRS Western Service Center.

Energy. Electricity production (1998, kWh, by source): Coal: 33.2 bil; Petroleum: 31 mil; Gas: 463 mil; Hydroelectric: 1.3 bil.

State data. Motto: Industry. **Flower:** Sego lily. **Bird:** Seagull. **Tree:** Blue spruce. **Song:** Utah, We Love Thee. **Entered union** Jan. 4, 1896; rank: 45th. **State fair** at Salt Lake City; Sept.

History. Ute, Gosiute, Southern Paiute, and Navajo peoples lived in the region at the time of European contact. Spanish Franciscans visited the area, 1776; American fur traders followed. Permanent settlement began with the arrival of the Mormons, 1847; they made the arid land bloom and created a prosperous economy. The State of Deseret was organized in 1849, and asked admission to the Union. In 1850, Congress established the region as the territory of Utah, and Brigham Young was appointed governor. The Union and Pacific Railroads met near Promontory, May 10, 1869, creating the first transcontinental railroad. Statehood was not achieved until 1896, after a long period of controversy over the Mormon Church's doctrine of polygamy, which it discontinued in 1890.

Tourist attractions. Temple Square, Mormon Church headquarters, Salt Lake City; Great Salt Lake; Zion National Park, Canyonlands, Bryce Canyon, Arches, and Capitol Reef natl. parks; Dinosaur, Rainbow Bridge, Timpanogos Cave, and Natural Bridges natl. monuments; Lake Powell; Flaming Gorge Natl. Recreation Area.

Famous Utahans. Maude Adams, Ezra Taft Benson, John Moses Browning, Mariner Eccles, Philo Farnsworth, James Fletcher, David M. Kennedy, J. Willard Marriott, Merlin Olsen,

Osmond family, Ivy Baker Priest, George Romney, Brigham Young, Loretta Young.

Tourist information. Utah Travel Council, Council Hall, Salt Lake City, UT 84114; 801-538-1030.

Toll-free travel information. 1-800-200-1160

Website. http://www.state.ut.us

Tourism website. http://www.utah.com

Vermont

Green Mountain State

People. Population (1998): 590,883; rank: 49; **net change** (1990-98): 5.0%. **Pop. density** (1998): 63.9 per sq mi. **Racial distribution** (1990): 98.6% white; 0.3% black; 0.6% Asian. **Hispanic pop.:** 0.7%.

Geography. Total area: 9,615 sq mi; rank: 43. **Land area:** 9,249 sq mi; rank: 43. **Acres forested:** 4,538,000. **Location:** northern New England state. **Climate:** temperate, with considerable temperature extremes; heavy snowfall in mountains. **Topography:** Green Mts. N-S backbone 20-36 mi wide; avg. altitude 1,000 ft. **Capital:** Montpelier.

Economy. Chief industries: manufacturing, tourism, agriculture, trade, finance, insurance, real estate, government. **Chief manuf. goods:** machine tools, furniture, scales, books, computer components, speciality foods. **Chief crops:** dairy products, apples, maple syrup, greenhouse/nursery, vegetables and small fruits. **Livestock** (Jan. 1999): 310,000 cattle/calves; 18,000 sheep/lambs; (Dec. 1998) 3,500 hogs/pigs; (Dec. 1998) 206,000 chickens (excl. broilers). **Timber/lumber** (1998): pine, spruce, fir, hemlock; 255 mil bd. ft. **Nonfuel minerals** (est. 1998): $96 mil; mostly dimension stone, crushed stone, sand & gravel, talc & pyrophyllite, gemstones. **Internat. airport at:** Burlington. **Value of construction** (1997): $622 mil. **Gross state product** (1997): $15.2 bil. **Employment distrib.** (May 1999): 30.7% serv.; 22.3% trade; 16.8% mfg.; 16.2% govt. **Per cap. pers. income** (1998): $24,175. **Sales tax** (1999): 5%. **Unemployment** (1998): 3.4%. **Tourism expends.** (1996): $1.3 bil. **Lottery** (1998): total sales: $74.1 mil; net income: $22.2 mil.

Finance. FDIC-insured commercial banks (1998): 21. **Deposits:** $6.4 bil. **FDIC-insured savings institutions** (1998): 5. **Assets:** $938 mil.

Federal govt. Fed. civ. employees (Mar. 1998): 2,648. **Avg. salary:** $41,498.

Energy. Electricity production (1998, kWh, by source): Petroleum: 41 mil; Gas: 1 mil; Hydroelectric: 848 mil; Nuclear: 3.4 bil.

State data. Motto: Freedom and unity. **Flower:** Red clover. **Bird:** Hermit thrush. **Tree:** Sugar maple. **Song:** Hail, Vermont. **Entered union** Mar. 4, 1791; rank, 14th. **State fair** at Rutland; early Sept.

History. Before the arrival of the Europeans, Abnaki and Mahican peoples lived in the region. Champlain explored the lake that bears his name, 1609. The first American settlement was Ft. Dummer, 1724, near Brattleboro. During the American Revolution, Ethan Allen and the Green Mountain Boys captured Ft. Ticonderoga (NY), 1775; John Stark defeated part of Burgoyne's forces near Bennington, 1777. In the War of 1812, Thomas MacDonough defeated a British fleet on Lake Champlain off Plattsburgh (NY), 1814.

Tourist attractions. Shelburne Museum; Rock of Ages Quarry, Graniteville; Vermont Marble Exhibit, Proctor; Bennington Battle Monument; Pres. Calvin Coolidge homestead, Plymouth; Maple Grove Maple Museum, St. Johnsbury; Ben & Jerry's Factory, Waterbury.

Famous Vermonters. Ethan Allen, Chester A. Arthur, Calvin Coolidge, George Dewey, John Dewey, Stephen A. Douglas, Dorothy Canfield Fisher, James Fisk.

Chamber of Commerce. PO Box 37, Montpelier, VT 05601.

Tourist information. Vermont Dept. of Tourism and Marketing, 6 Baldwin St., Drawer 33, Montpelier, VT 05633-1301.

Toll-free travel information. 1-800-VERMONT

Website. http://www.state.vt.us

Tourism website. http://www.travel-vermont.com

Virginia

Old Dominion

People. Population (1998): 6,791,345; rank: 12; **net change** (1990-98): 9.7%. **Pop. density** (1998): 171.5 per sq mi. **Racial distribution** (1990): 77.4% white; 18.8% black; 2.6% Asian. **Hispanic pop.:** 2.6%.

Geography. Total area: 42,326 sq mi; rank: 35. **Land area:** 39,598 sq mi; rank: 37. **Acres forested:** 15,858,000. **Location:** South Atlantic state bounded by the Atlantic Ocean on the E and surrounded by North Carolina, Tennessee, Kentucky, West Virginia, and Maryland. **Climate:** mild and equable. **Topography:** mountain and valley region in the W, including the Blue Ridge Mts.; rolling piedmont plateau; tidewater, or coastal plain, including the eastern shore. **Capital:** Richmond.

Economy. Chief industries: services, trade, government, manufacturing, tourism, agriculture. **Chief manuf. goods:** food processing, transportation equipment, printing, textiles, electronic & electrical equipment, industrial machinery & equipment, lumber & wood products, chemicals, rubber & plastics, furniture. **Chief crops:** tobacco, grain corn, soybeans, winter wheat, peanuts, lint & seed cotton. **Livestock** (Jan. 1999): 1.7 mil cattle/calves; 57,000 sheep/lambs; (Dec. 1998) 390,000 hogs/pigs; (Dec. 1998) 5.1 mil chickens (excl. broilers); (Dec. 1997) 259.4 mil broilers. **Timber/lumber** (1998): pine and hardwoods; 1.4 bil bd. ft. **Nonfuel minerals** (est. 1998): $679 mil; mostly crushed stone, sand & gravel, portland cement, lime, kyanite. **Commercial fishing** (1998): $112.7 mil. **Chief ports:** Hampton Roads, Richmond, Alexandria. **Internat. airports at:** Norfolk, Dulles, Richmond, Newport News. **Value of construction** (1997): $10.1 bil. **Gross state product** (1997): $211.3 bil. **Employment distrib.** (May 1999): 31.4% serv.; 22.1% trade; 18.1% govt.; 11.6% mfg. **Per cap. pers. income** (1998): $27,385. **Sales tax** (1999): 3.5%. **Unemployment** (1998): 2.9%. **Tourism expends.** (1996): $11.0 bil. **Lottery** (1998): total sales: $914.3 mil; net income: $318.9 mil.

Finance. FDIC-insured commercial banks (1998): 152. **Deposits:** $52.0 bil. **FDIC-insured savings institutions** (1998): 20. **Assets:** $15.5 bil.

Federal govt. Fed. civ. employees (Mar. 1998): 117,273. **Avg. salary:** $49,872. **Notable fed. facilities:** Pentagon; Norfolk Naval Station, Norfolk Naval Air Station; Naval Shipyard; Marine Corps Base; Langley AFB; NASA at Langley.

Energy. Electricity production (1998, kWh, by source): Coal: 31.5 bil; Petroleum: 2.7 bil; Gas: 2.2 bil; Hydroelectric: 256 mil; Nuclear: 27.2 bil.

State data. Motto: Sic Semper Tyrannis (Thus always to tyrants). **Flower:** Dogwood. **Bird:** Cardinal. **Tree:** Dogwood. **Song Emeritus:** Carry Me Back to Old Virginia. **Tenth** of the original 13 states to ratify the Constitution, June 25, 1788. **State fair** at Richmond; late Sept.-early Oct.

History. Living in the area at the time of European contact were the Cherokee and Susquehanna and the Algonquians of the Powhatan Confederacy. English settlers founded Jamestown, 1607. Virginians took over much of the government from royal governor Dunmore, 1775, forcing him to flee. Virginians under George Rogers Clark freed the Ohio-Indiana-Illinois area of British forces. Benedict Arnold burned Richmond and Petersburg for the British, 1781. That same year, Britain's Cornwallis was trapped at Yorktown and surrendered, ending the American Revolution. Virginia seceded from the Union, 1861, and Richmond became the capital of the Confederacy. Hampton Roads, off the Virginia coast, was the site of the famous naval battle of the USS *Monitor* and CSS *Virginia* (Merrimac), 1862. Virginia was readmitted, 1870.

Tourist attractions. Colonial Williamsburg; Busch Gardens, Williamsburg; Wolf Trap Farm, near Falls Church; Arlington Natl. Cemetery; Mt. Vernon, home of George Washington; Jamestown Festival Park; Yorktown; Jefferson's Monticello, Charlottesville; Robert E. Lee's birthplace, Stratford Hall, and grave, Lexington; Appomattox; Shenandoah Natl. Park; Blue Ridge Parkway; Virginia Beach; Paramount's King's Dominion, near Richmond.

Famous Virginians. Richard E. Byrd, James B. Cabell, Henry Clay, Jerry Falwell, William Henry Harrison, Patrick Henry, Thomas Jefferson, Joseph E. Johnston, Robert E. Lee, Meriwether Lewis and William Clark, James Madison, John Marshall, George Mason, James Monroe, Pocahontas, Edgar Allan Poe, John Randolph, Walter Reed, John Smith, William Styron, Zachary Taylor, John Tyler, Maggie Walker, Booker T. Washington, George Washington, Woodrow Wilson.

Chamber of Commerce. 9 South Fifth St., Richmond, VA 23219.

Toll-free travel information. 1-800-VISITVA.

Website. http://www.state.va.us

Tourism website. http://www.virginia.org

Washington
Evergreen State

People. Population (1998): 5,689,263; rank: 15; **net change** (1990-98): 16.9%. **Pop. density** (1998): 85.4 per sq mi. **Racial distribution** (1990): 88.5% white; 3.1% black; 4.3% Asian. **Hispanic pop.:** 4.4%.

Geography. Total area: 70,637 sq mi; rank: 19. **Land area:** 66,581 sq mi; rank: 20. **Acres forested:** 20,483,000. **Location:** Pacific state bordered by Canada on the N; Idaho on the E; Oregon on the S; and the Pacific Ocean on the W. **Climate:** mild, dominated by the Pacific Ocean and protected by the Cascades. **Topography:** Olympic Mts. on NW peninsula; open land along coast to Columbia R.; flat terrain of Puget Sound Lowland; Cascade Mts. region's high peaks to the E; Columbia Basin in central portion; highlands to the NE; mountains to the SE. **Capital:** Olympia.

Economy. Chief industries: advanced technology, aerospace, biotechnology, intl. trade, forestry, tourism, recycling, agriculture & food processing. **Chief manuf. goods:** computer software, aircraft, pulp & paper, lumber and plywood, aluminum, processed fruits and vegetables, machinery, electronics. **Chief crops:** apples, potatoes, hay, farm forest products. **Livestock** (Jan. 1999): 1.2 mil cattle/calves; 50,000 sheep/lambs; (Dec. 1998) 37,000 hogs/pigs; (Dec. 1998) 7.0 mil chickens (excl. broilers); (Dec. 1997) 38.8 mil broilers. **Timber/lumber** (1998): Douglas fir, hemlock, cedar, pine; 4.5 bil bd. ft. **Nonfuel minerals** (est. 1998): $538 mil; mostly sand & gravel, magnesium metal, crushed stone, portland cement, gold. **Commercial fishing** (1998): $123.2 mil. **Chief ports:** Seattle, Tacoma, Vancouver, Kelso-Longview. **Internat. airports at:** Seattle/Tacoma, Spokane, Boeing Field. **Value of construction** (1997): $8.5 bil. **Gross state product** (1997): $172.3 bil. **Employment distrib.** (May 1999): 27.6% serv.; 24.3% trade; 17.8% govt.; 13.9% mfg. **Per cap. pers. income** (1998): $27,961. **Sales tax** (1999): 6.5%. **Unemployment** (1998): 4.8%. **Tourism expends.** (1996): $7.5 bil. **Lottery** (1998): total sales: $454.7 mil; net income: $144.7 mil.

Finance. FDIC-insured commercial banks (1998): 78. **Deposits:** $10.6 bil. **FDIC-insured savings institutions** (1998): 21. **Assets:** $50.6 bil.

Federal govt. Fed. civ. employees (Mar. 1998): 42,817. **Avg. salary:** $44,430. **Notable fed. facilities:** Bonneville Power Admin.; Ft. Lewis; McChord AFB; Hanford Nuclear Reservation; Bremerton Naval Shipyards.

Energy. Electricity production (1998, kWh, by source): Coal: 9.3 bil; Petroleum: 39 mil; Gas: 1.1 bil; Hydroelectric: 79.4 bil; Nuclear: 6.9 bil.

State data. Motto: Alki (By and by). **Flower:** Western rhododendron. **Bird:** Willow goldfinch. **Tree:** Western hemlock. **Song:** Washington, My Home. **Entered union** Nov. 11, 1889; rank, 42d. **State fairs:** 5 area fairs, in Aug. and Sept.; no state fair.

History. At the time of European contact, many Native American tribes lived in the area, including the Nez Percé, Spokan, Yakima, Cayuse, Okanogan, Walla Walla, and Colville peoples, who lived in the interior region, and the Nooksak, Chinook, Nisqually, Clallam, Makah, Quinault, and Puyallup peoples, who inhabited the coastal area. Spain's Bruno Hezeta sailed the coast, 1775. In 1792, British naval officer George Vancouver mapped Puget Sound area, and that same year, American Capt. Robert Gray sailed up the Columbia River. Canadian fur traders set up Spokane House, 1810. Americans under John Jacob Astor established a post at Ft. Okanogan, 1811, and missionary Marcus Whitman settled near Walla Walla, 1836. Final agreement on the border of Washington and Canada was made with Britain, 1846, and Washington became part of the Oregon Territory, 1848. Gold was discovered, 1855.

Tourist attractions. Seattle Waterfront, Seattle Center and Space Needle, Museum of Flight, all Seattle; Mt. Rainier, Olympic, and North Cascades natl. parks; Mt. St. Helens; Puget Sound; San Juan Islands; Grand Coulee Dam; Columbia R. Gorge Natl. Scenic Area; Spokane's Riverfront Park.

Famous Washingtonians. Bing Crosby, William O. Douglas, Bill Gates, Henry M. Jackson, Gary Larson, Mary McCarthy, Robert Motherwell, Edward R. Murrow, Theodore Roethke, Marcus Whitman, Minoru Yamasaki.

Tourist information. WA State Tourism Division, PO Box 42500, Olympia, WA 98504-2500.

Toll-free travel information. 1-800-544-1800. ext. 101
Website. http://access.wa.gov
Tourism website. http://www.tourism.wa.gov

West Virginia
Mountain State

People. Population (1998): 1,811,156; rank: 35; **net change** (1990-98): 1.0%. **Pop. density** (1998): 75.2 per sq mi. **Racial distribution** (1990): 96.2% white; 3.1% black. **Hispanic pop.:** 0.5%.

Geography. Total area: 24,231 sq mi; rank: 41. **Land area:** 24,087 sq mi; rank: 41. **Acres forested:** 12,128,000. **Location:** South Atlantic state bounded on the N by Ohio, Pennsylvania, Maryland; on the S and W by Virginia, Kentucky, Ohio; on the E by Maryland and Virginia. **Climate:** humid continental climate except for marine modification in the lower panhandle. **Topography:** ranging from hilly to mountainous; Allegheny Plateau in the W, covers two-thirds of the state; mountains here are the highest in the state, over 4,000 ft. **Capital:** Charleston.

Economy. Chief industries: manufacturing, services, mining, tourism. **Chief manuf. goods:** machinery, plastic & hardwood prods., fabricated metals, chemicals, aluminum, automotive parts, steel. **Chief crops:** apples, peaches, hay, tobacco, corn, wheat, oats. **Chief farm products:** dairy products, eggs. **Livestock** (Jan. 1999): 440,000 cattle/calves; 40,000 sheep/lambs; (Dec. 1998) 16,000 hogs/pigs; (Dec. 1998) 1.9 mil chickens (excl. broilers); (Dec. 1997) 90.8 mil broilers. **Timber/lumber** (1998): oak, yellow poplar, hickory, walnut, cherry; 737 mil bd. ft. **Nonfuel minerals** (est. 1998): $204 mil; mostly crushed stone, portland cement, sand & gravel, lime, salt. **Chief port:** Huntington. **Value of construction** (1997): $1.2 bil. **Gross state product** (1997): $38.2 bil. **Employment distrib.** (May 1999): 29.1% serv.; 23.0% trade; 19.4% govt.; 11.4% mfg. **Per cap. pers. income** (1998): $19,362. **Sales tax** (1999): 6%. **Unemployment** (1998): 6.6%. **Tourism expends.** (1996): $1.6 bil. **Lottery** (1998): total sales: $305.1 mil; net income: $91.3 mil.

Finance. FDIC-insured commercial banks (1998): 89. **Deposits:** $18.8 bil. **FDIC-insured savings institutions** (1998): 7. **Assets:** $775 bil.

Federal govt. Fed. civ. employees (Mar. 1998): 11,211. **Avg. salary:** $41,694. **Notable fed. facilities:** National Radio Astronomy Observatory; Bureau of Public Debt Bldg.; Harpers Ferry Natl. Park; Correctional Institution for Women; FBI Identification Center.

Energy. Electricity production (1998, kWh, by source): Coal: 89.0 bil; Petroleum: 194 mil; Gas: 42 mil; Hydroelectric: 361 mil.

State data. Motto: Montani Semper Liberi (Mountaineers are always free). **Flower:** Big rhododendron. **Bird:** Cardinal. **Tree:** Sugar maple. **Songs:** The West Virginia Hills; This Is My West Virginia; West Virginia, My Home, Sweet Home. **Entered union** June 20, 1863; rank, 35th. **State fair** at Lewisburg (Fairlea); late Aug.

History. Sparsely inhabited at the time of European contact, the area was primarily Native American hunting grounds. British explorers Thomas Batts and Robert Fallam reached the New River, 1671. Early American explorers included George Washington, 1753, and Daniel Boone. In the fall of 1774, frontiersmen defeated an allied Indian uprising at Point Pleasant. The area was part of Virginia and often objected to rule by the eastern part of the state. When Virginia seceded in 1861, the Wheeling Convention repudiated the act and created a new state, Kanawha, later renamed West Virginia. It was admitted to the Union 1863.

Tourist attractions. Harpers Ferry Natl. Historic Park; Science and Cultural Center, Charleston; White Sulphur (in Greenbrier) and Berkeley Springs mineral water spas; New River Gorge, Fayetteville; Winter Place, Exhibition Coal Mine, both Beckley; Monongahela Natl. Forest; Fenton Glass, Williamstown; Viking Glass, New Martinsville; Blenko Glass, Milton; Sternwheel Regatta, Charleston; Mountain State Forest Festival; Snowshoe Ski Resort, Slaty Fork; Canaan State Park, Davis; Mountain State Arts & Crafts Fair, Ripley; Ogle Bay, Wheeling; White water rafting, several locations.

Famous West Virginians. Newton D. Baker, Pearl Buck, John W. Davis, Thomas "Stonewall" Jackson, Don Knotts, Dwight Whitney Morrow, Michael Owens, Walter Reuther, Cyrus Vance, Charles "Chuck" Yeager.

Tourist information. Dept. of Commerce, West Virginia Division of Tourism, State Capitol, Charleston WV 25305.

Toll-free travel information. 1-800-CALLWVA.
Website. http://www.state.wv.us
Tourism website. http://wvweb.com/www/travel_recreation

Wisconsin
Badger State

People. Population (1998): 5,223,500; rank: 18; **net change** (1990-98): 6.8%. **Pop. density** (1998): 96.2 per sq mi. **Racial distribution** (1990): 92.2% white; 5.0% black. **Hispanic pop.:** 1.9% .

Geography. Total area: 65,499 sq mi; rank: 22. **Land area:** 54,314 sq mi; rank: 25. **Acres forested:** 15,513,000. **Location:** East North Central state, bounded on the N by Lake Superior and Upper Michigan; on the E by Lake Michigan; on the S by Illinois; on the W by the St. Croix and Mississippi rivers. **Climate:** long, cold winters and short, warm summers tempered by the Great Lakes. **Topography:** narrow Lake Superior Lowland plain met by Northern Highland, which slopes gently to the sandy crescent Central Plain; Western Upland in the SW; 3 broad parallel limestone ridges running N-S are separated by wide and shallow lowlands in the SE. **Capital:** Madison.

Economy. Chief industries: services, manufacturing, trade, government, agriculture, tourism. **Chief manuf. goods:** food products, motor vehicles & equip., paper products, medical instruments and supplies, printing, plastics. **Chief crops:** corn, hay, soybeans, potatoes, cranberries, sweet corn, peas, oats, snap beans. **Chief products:** milk, butter, cheese, canned and frozen vegetables. **Livestock** (Jan. 1999): 3.4 mil cattle/calves; 83,000 sheep/lambs; (Dec. 1998) 690,000 hogs/pigs; (Dec. 1998) 4.7 mil chickens (excl. broilers); (Dec. 1997) 32.9 mil broilers. **Timber/lumber** (1998): maple, birch, oak, evergreens; 698 mil bd. ft. **Nonfuel minerals** (est. 1998): $261 mil; mostly crushed stone, sand & gravel, copper, lime. **Commercial fishing** (1998): $4.4 mil. **Chief ports:** Superior, Ashland, Milwaukee, Green Bay, Kewaunee, Pt. Washington, Manitowoc, Sheboygan, Marinette, Kenosha. **Internat. airports at:** Milwaukee. **Value of construction** (1997): $6.1 bil. **Gross state product** (1997): $147.3 bil. **Employment distrib.** (May 1999): 26.2% serv.; 22.3% mfg.; 22.3% trade; 14.4% govt. **Per cap. pers. income** (1998): $25,079. **Sales tax** (1999): 5%. **Unemployment** (1998): 3.4%. **Tourism expends.** (1996): $5.4 bil. **Lottery** (1998): total sales: $418.6 mil; net income: $171.5 mil.

Finance. FDIC-insured commercial banks (1998): 344. **Deposits:** $62.2 bil. **FDIC-insured savings institutions** (1998): 45. **Assets:** $16.6 bil.

Federal govt. Fed. civ. employees (Mar. 1998): 11,269. **Avg. salary:** $40,962. **Notable fed. facilities:** Ft. McCoy.

Energy. Electricity production (1998, kWh, by source): Coal: 39.8 bil; Petroleum: 200 mil; Gas: 1.2 bil; Hydroelectric: 1.5 bil; Nuclear: 9.4 bil.

State data. Motto: Forward. **Flower:** Wood violet. **Bird:** Robin. **Tree:** Sugar maple. **Song:** On, Wisconsin! **Entered union** May 29, 1848; rank, 30th. **State fair** at State Fair Park, West Allis; July-Aug.

History. At the time of European contact, Ojibwa, Menominee, Winnebago, Kickapoo, Sauk, Fox, and Potawatomi peoples inhabited the region. Jean Nicolet was the first European to see the Wisconsin area, arriving in Green Bay, 1634; French missionaries and fur traders followed. The British took over, 1763. The U.S. won the land after the American Revolution, but the British were not ousted until after the War of 1812. Lead miners came next, then farmers. In 1816, the U.S. government built a fort at Prairie du Chien on Wisconsin's border with Iowa. Native Americans in the area rebelled against the seizure of their tribal lands in the Black Hawk War of 1832, but treaties from 1829 to 1848, transferred all land titles in Wisconsin to the U.S. government. Railroads were started in 1851, serving growing wheat harvests and iron mines. Some 96,000 soldiers served the Union cause during the Civil War.

Tourist attractions. Old Wade House and Carriage Museum, Greenbush; Villa Louis, Prairie du Chien; Circus World Museum, Baraboo; Wisconsin Dells; Old World Wisconsin, Eagle; Door County peninsula; Chequamegon and Nicolet national forests; Lake Winnebago; House on the Rock, Dodgeville; Monona Terrace, Madison.

Famous Wisconsinites. Carrie Chapman Catt, Edna Ferber, King Camp Gillette, Harry Houdini, Robert La Follette, Alfred Lunt, Pat O'Brien, Georgia O'Keeffe, William H. Rehnquist, John Ringling, Donald K. "Deke" Slayton, Spencer Tracy, Thorstein Veblen, Orson Welles, Laura Ingalls Wilder, Thornton Wilder, Frank Lloyd Wright.

Tourist information. Wisconsin Dept. of Tourism, 201 W. Washington Ave., PO Box 7976, Madison, WI 53707-7976.

Toll-free travel information. 1-800-432-8747.

Website. http://www.state.wi.us

Tourism website. http://tourism.state.wi.us

Wyoming
Equality State, Cowboy State

People. Population (1998): 480,907; rank: 50; **net change** (1990-98): 6.0%. **Pop. density** (1998): 5.0 per sq mi. **Racial distribution** (1990): 94.2% white; 0.8% black; 2.1% Amer. Indian. **Hispanic pop.:** 5.7%.

Geography. Total area: 97,818 sq mi; rank: 9. **Land area:** 97,105 sq mi; rank: 9. **Acres forested:** 9,966,000. **Location:** Mountain state lying in the high western plateaus of the Great Plains. **Climate:** semi-desert conditions throughout; true desert in the Big Horn and Great Divide basins. **Topography:** the eastern Great Plains rise to the foothills of the Rocky Mts.; the Continental Divide crosses the state from the NW to the SE. **Capital:** Cheyenne.

Economy. Chief industries: mineral extraction, oil, natural gas, tourism and recreation, agriculture. **Chief manuf. goods:** refined petroleum, wood, stone, clay products, foods, electronic devices, sporting apparel, and aircraft. **Chief crops:** wheat, beans, barley, oats, sugar beets, hay. **Livestock** (Jan. 1999): 1.6 mil cattle/calves; 660,000 sheep/lambs; (Dec. 1998) 140,000 hogs/pigs; (Dec. 1998) 17,000 chickens (excl. broilers). **Timber/lumber** (1998): ponderosa & lodgepole pine, Douglas fir, Engelmann spruce; 232 mil bd. ft. **Nonfuel minerals** (est. 1998): $1.06 bil; mostly soda ash, clays, helium, portland cement, crushed stone. **Internat. airport at:** Casper. **Value of construction** (1997): $655 mil. **Gross state product** (1997): $17.6 bil. **Employment distrib.** (May 1999): 25.3% govt.; 23.1% trade; 22.1% serv.; 4.9% mfg. **Per cap. pers. income** (1998): $23,167. **Sales tax** (1999): 4%. **Unemployment** (1998): 4.8%. **Tourism expends.** (1996): $1.5 bil.

Finance. FDIC-insured commercial banks (1998): 52. **Deposits:** $9.1 bil. **FDIC-insured savings institutions** (1998): 4. **Assets:** $364 mil.

Federal govt. Fed. civ. employees (Mar. 1998): 4,380. **Avg. salary:** $40,087. **Notable fed. facilities:** Warren AFB.

Energy. Electricity production (1998, kWh, by source): Coal: 43.3 bil; Petroleum: 43 mil; Gas: 27 mil; Hydroelectric: 1.3 bil.

State data. Motto: Equal Rights. **Flower:** Indian Paintbrush. **Bird:** Western Meadowlark. **Tree:** Plains Cottonwood. **Song:** Wyoming. **Entered union** July 10, 1890; rank, 44th. **State fair** at Douglas; late Aug.

History. Shoshone, Crow, Cheyenne, Oglala Sioux, and Arapaho peoples lived in the area at the time of European contact. France's François and Louis La Verendrye were the first Europeans to see the region, 1743. John Colter, an American, was first to traverse Yellowstone area, 1807-8. Trappers and fur traders followed in the 1820s. Forts Laramie and Bridger became important stops on the pioneer trails to the West Coast. Population grew after the Union Pacific crossed the state, 1868. Women won the vote, for the first time in the U.S., from the Territorial Legislature, 1869. Disputes between large land owners and small ranchers culminated in the Johnson County Cattle War, 1892; federal troops were called in to restore order.

Tourist attractions. Yellowstone Natl. Park, the first U.S. national park, est. 1872; Grand Teton Natl. Park; Natl. Elk Refuge; Devils Tower Natl. Monument; Fort Laramie Natl. Historic Site and nearby pioneer trail ruts; Buffalo Bill Historical Center, Cody; Cheyenne Frontier Days, Cheyenne.

Famous Wyomingites. James Bridger, William F. "Buffalo Bill" Cody, Esther Hobart Morris, Nellie Tayloe Ross.

Tourist information. Division of Tourism & State Marketing, I-25 at College Dr., Cheyenne, WY 82002.

Toll-free travel information. 1-800-CALLWYO.

Website. http://www.state.wy.us

Tourism website. http://www.state.wy.us/state/tourism/tourism.html

District of Columbia

People. Population (1998): 523,124; **net change** (1990-98): -13.8%. **Pop. density** (1998): 8,575.8 per sq mi.

Geography. Total area: 68 sq mi; rank: 51. **Land area:** 61 sq mi; rank: 51. **Location:** at the confluence of the Potomac and Anacostia rivers, flanked by Maryland on the N, E, and SE and by Virginia on the SW. **Climate:** hot humid summers, mild winters. **Topography:** low hills rise toward the N away from the Potomac R. and slope to the S; highest elevation, 410 ft, lowest Potomac R., 1 ft.

Economy. Chief industries: government, service, tourism. **Value of construction** (1997): $673 mil. **Gross state product** (1997): $52.4 bil. **Employment distrib.** (May 1999): 44.8% serv.; 36.6% govt.; 7.9% trade; 2.0% mfg. **Per cap. pers. income** (1998): $37,278. **Sales tax** (1999): 5.75%. **Unemployment** (1998): 8.8%. **Tourism expenditures** (1996): $5.1 bil. **Lottery** (1998) total sales: $226.4 mil; net income: $81.4 mil.

Finance. FDIC-insured commercial banks & trust companies (1998): 7. **Deposits:** $987 mil. **FDIC-insured savings institutions** (1998): 1. **Assets:** $267 mil.

Federal govt. No. of federal employees (Mar. 1998): 142,734. **Avg. salary:** $58,622.

Energy. Electricity production (1998, kWh, by source): Petroleum: 244 mil.

District data. Motto: Justitia omnibus (Justice for all). **Flower:** American beauty rose. **Tree:** Scarlet oak. **Bird:** Wood thrush.

History. The District of Columbia, coextensive with the city of Washington, is the seat of the U.S. federal government. It lies on the west central edge of Maryland on the Potomac River, opposite Virginia. Its area was originally 100 sq mi taken from the sovereignty of Maryland and Virginia. Virginia's portion south of the Potomac was given back to that state in 1846.

The 23d Amendment (1961) granted residents the right to vote for president and vice president for the first time since 1800 and gave them 3 members in the Electoral College. The first such votes were cast in Nov. 1964.

Congress, which has legislative authority over the District under the Constitution, established in 1874 a government of 3 commissioners appointed by the president. The Reorganization Plan of 1967 substituted a single appointive commissioner (also called mayor), assistant, and 9-member City Council. Funds were still appropriated by Congress; residents had no vote in local government, except to elect school board members. In Sept. 1970, Congress approved legislation giving the District one delegate to the House of Representatives, who can vote in committee but not on the floor. The first delegate was elected 1971.

In May 1974 voters approved a congressionally drafted charter giving them the right to elect their own mayor and a 13-member city council; the first took office Jan. 2, 1975. The district won the right to levy taxes; Congress retained power to veto council actions and approve the city budget.

Proposals for a "federal town" for the deliberations of the Continental Congress were made in 1783, 4 years before the adoption of the Constitution. Rivalry between Northern and Southern delegates over the site appeared in the First Congress, 1789. John Adams, presiding officer of the Senate, cast the deciding vote of that body for Germantown, PA. In 1790 Congress compromised by making Philadelphia the temporary capital for 10 years. The Virginia members of the House wanted a permanent capital on the eastern bank of the Potomac, while the Southerners opposed having the nation assume the war debts of the 13 original states as provided under the Assumption Bill, fathered by Alexander Hamilton. Hamilton and Jefferson arranged a compromise: the Virginia men voted for the Assumption Bill, and the Northerners conceded the capital to the Potomac. Pres. Washington chose the site in Oct. 1790 and persuaded landowners to sell their holdings to the government. The capital was named Washington.

Washington appointed Pierre Charles L'Enfant, a Frenchman, to plan the capital on an area not more than 10 mi square. The L'Enfant plan, for streets 100 to 110 ft. wide and one avenue 400 ft. wide and a mile long, seemed grandiose and foolhardy, but Washington endorsed it. When L'Enfant ordered a wealthy landowner to remove his new manor house because it obstructed a vista, and demolished it when the owner refused, Washington stepped in and dismissed the architect. Andrew Ellicott, who was working on surveying the area, finished the official map and design of the city. Ellicott was assisted by Benjamin Banneker, a distinguished black architect and astronomer.

On Sept. 18, 1793, Pres. Washington laid the cornerstone of the north wing of the Capitol. On June 3, 1800, Pres. John Adams moved to Washington, and on June 10, Philadelphia ceased to be the temporary capital. The City of Washington was incorporated in 1802; the District of Columbia was created as a municipal corporation in 1874, embracing Washington, Georgetown, and Washington County.

Tourist attractions: See Washington, DC, Capital of the U.S.

Tourist information. Washington, DC Convention and Visitors Association, 1212 New York Ave. NW, #600, Washington, DC 20005; phone: 202-789-7000.

Website. http://dcpages.ari.net

Tourism website. http://www.washington.org

OUTLYING U.S. AREAS

American Samoa

People. Population (1997 est.): 61,819. **Population growth rate** (1997 est.): 3.72%. **Major ethnic group:** Samoan (Polynesian), Caucasian, Tongan. **Languages:** Samoan, English.

Land area: 77 sq. mi. **Total area:** 90 sq. mi. **Capital:** Pago Pago, Island of Tutuila. **Motto:** Samoa Muamua le Atua (In Samoa, God Is First). **Song:** Amerika Samoa. **Flower:** Paogo (Ula-fala). **Plant:** Ava.

Public education. Student-teacher ratio (1995): 20.0.

Boasting spectacular scenery and delightful South Seas climate, American Samoa is the most southerly of all lands under U.S. sovereignty. It is an unincorporated territory consisting of 7 small islands of the Samoan group: **Tutuila, Aunu'u, Manu'a Group (Ta'u, Olosega, Ofu), Rose,** and **Swains Island.** The islands are 2,300 mi SW of Honolulu.

Economy. Chief industries: tuna processing, trade, services, tourism. **Chief crops:** vegetables, nuts, melons and other fruits. **Livestock** (1990): 179 cattle; 7,580 hogs/pigs; 27,401 chickens.

Finance. FDIC-insured commercial banks (1998): 1. **Deposits:** $48 mil.

A tripartite agreement between Great Britain, Germany, and the U.S. in 1899 gave the U.S. sovereignty over the eastern islands of the Samoan group; these islands became American Samoa. Local chiefs ceded Tutuila and Aunu'u to the U.S. in 1900, and the Manu'a group and Rose in 1904; Swains Island was annexed in 1925. Samoa (Western), comprising the larger islands of the Samoan group, was a New Zealand mandate and UN Trusteeship until it became independent Jan. 1, 1962 (now called Samoa).

Tutuila and Aunu'u have an area of 53 sq mi. Ta'u has an area of 17 sq mi, and the islets of Ofu and Olosega, 5 sq mi with a population of a few thousand. Swains Island has nearly 2 sq mi and a population of about 100.

About 70% of the land is bush and mountains. Chief exports are fish products. Taro, breadfruit, yams, coconuts, pineapples, oranges, and bananas are also produced.

From 1900 to 1951, American Samoa was under the jurisdiction of the U.S. Navy. Since 1951, it has been under the Interior Dept. On Jan. 3, 1978, the first popularly elected Samoan governor and lieutenant governor were inaugurated. Previously, the governor was appointed by the Secretary of the Interior. American Samoa has a bicameral legislature and elects a delegate to the House of Representatives, with no vote except in committees.

The American Samoans are of Polynesian origin. They are nationals of the U.S.; approximately 20,000 live in Hawaii, 65,000 in California and Washington.

Website. http://www.samoanet.com

Guam

Where America's Day Begins

People. Population (1997 est.): 160,595. **Population growth rate** (1997): 2.5%. **Pop. density** (1990): 631.6 per sq mi. **Major ethnic groups** Chamorro, Filipino, Caucasian, Chinese, Japanese, Korean. (Native Guamanians, ethnically Chamorros, are basically of Indonesian stock, with a mixture of Spanish and Filipino; in addition to the official language, they speak the native Chamorro. **Languages:** English, Chamorro, Japanese. **Migration** (1990): About 52% of population were born elsewhere; of these, 48% in Asia, 40% in U.S.

Geography. Total area: 217 sq mi. **Land area:** 210 sq. mi. **Location:** largest and southernmost of the Mariana Islands in the West Pacific, 3,700 mi W of Hawaii. **Climate:** tropical, with temperatures from 70° to 90° F; avg. annual rainfall, about 70 in. **Topography:** coralline limestone plateau in the N; southern chain of low volcanic mountains sloping gently to the W, more steeply to coastal cliffs on the E; general elevation, 500 ft; highest point, Mt. Lamlam, 1,334 ft. **Capital:** Hagatna.

Economy. Chief industries: tourism, U.S. military, construction, banking, printing & publishing. **Chief manuf. goods:** textiles, foods. **Chief crops:** cabbages, eggplants, cucumber, long beans, tomatoes, bananas, coconuts, watermelon, yams, cantaloupe, papayas, maize, sweet potatoes. **Livestock** (1992): 388 cattle; 2,038 hogs/pigs; 12,206 chickens. **Chief port:** Apra Harbor. **Internat. airport at:** Hagatna. **Value of construction** (1994): $614.3 mil. **Employment distrib.** (1994): 64.9% govt.; 20.2% trade; 19.5% serv. **Per capita income** (1992): $10,834. **Unemployment** (1994): 6.7%. **Tourism expends.** (1995): $4.9 bil.

Finance. FDIC-insured commercial banks (1998): 2. **Deposits:** $678 mil. **FDIC-insured savings institutions** (1998): 2. **Assets:** $283 mil.

Federal govt. Federal employees (1990): 7,200. **Notable fed. facilities:** Anderson AFB; naval, air, and port bases.

Public education. Student-teacher ratio (1995): 18.3.

Misc. data. Flower: Puti Tai Nobio (Bougainvillea). **Bird:** Toto (Fruit dove). **Tree:** Ifit (Intsiabijuga). **Song:** Stand Ye Guamanians.

History. Guam was probably settled by voyagers from the Indonesian-Philippine archipelago by 3d cent. BC. Pottery, rice cultivation, and megalithic technology show strong East Asian cultural influence. Centralized, village clan-based communities engaged in agriculture and offshore fishing. The estimated population by the early 16th cent. was 50,000-75,000. Magellan arrived in the Marianas Mar. 6, 1521. They were colonized in 1668 by Spanish missionaries, who named them the Mariana Islands in honor of Maria Anna, queen of Spain. When Spain ceded Guam to the U.S., it sold the other Marianas to Germany. Japan obtained a League of Nations mandate over the German islands in 1919; in Dec. 1941 it seized Guam, which was retaken by the U.S. in July-August 1944.

Guam is a self-governing organized unincorporated U.S. territory. The Organic Act of 1950 provided for a governor, elected to a 4-year term, and a 21-member unicameral legislature, elected biennially by the residents, who are American citizens. In 1970, the first governor was elected. In 1972, a U.S. law gave Guam one delegate to the U.S. House of Representatives who has a voice but no vote, except in committees.

Guam's quest to change its status to a U.S. Commonwealth began in the late 1970s. The Guam Commission on Self-Determination, created in 1984, developed a draft Commonwealth Act. In 1993, legislation proposing a change of status was submitted to the U.S. Congress. In 1994, the U.S. Congress passed legislation transferring 3,200 acres of land on Guam from federal to local control.

Tourist attractions. Tropical climate, oceanic marine environment; annual mid-Aug. Merizo Water Festival; Tarzan Falls; beaches; water sports; duty-free port shopping.

Website. http://www.gov.gu
Tourism website. http://www.chamarro.com

Commonwealth of the Northern Mariana Islands

People. Population (1997 est.): 53,552. **Major ethnic Groups:** Chamorro, Carolinians and other Micronesians, Caucasian, Japanese, Chinese, Korean. **Languages:** English, Chamorro, Carolinian.

Total area: 189 sq. mi. **Land area:** 179 sq. mi. Located in the perpetually warm climes between Guam and the Tropic of Cancer, the 14 islands of the Northern Marianas form a 300-mil-long archipelago. The indigenous population in 1990 was concentrated on the 3 largest of the 6 inhabited islands: **Saipan,** the seat of government and commerce (38,896), **Rota** (2,295), and **Tinian** (2,118).

Economy. Chief industries: trade, services, and tourism. **Chief manuf. goods:** apparel, stone, clay and glass products. **Chief crops:** melons, vegetables, horticulture, fruits and nuts. **Livestock** (1990) 4,513 cattle; 1,260 hogs/pigs; 9,580. **Employment distrib.** (1992): 53% trade; 33% serv.; 8% const.; 6% manuf.

Education. Pupil-teacher ratio (1995): 20.9.

The people of the Northern Marianas are predominantly of Chamorro cultural extraction, although Carolinians and immigrants from other areas of E. Asia and Micronesia have also settled in the islands. English is among the several languages commonly spoken. Pursuant to the Covenant of 1976, which established the Northern Marianas as a commonwealth in political union with the U.S., most of the indigenous population and many domiciliaries of these islands achieved U.S. citizenship on Nov. 3, 1986, when the U.S. terminated its administration of the UN trusteeship as it affected the Northern Marianas. From July 18, 1947, the U.S. had administered the Northern Marianas under a trusteeship agreement with the UN Security Council.

The Northern Mariana Islands has been self-governing since 1978, when a constitution drafted and adopted by the people became effective and a popularly elected bicameral legislature (2-year term), with offices of governor (4-year term) and lieut. governor, was inaugurated.

Commonwealth of Puerto Rico
(Estado Libre Asociado de Puerto Rico)

People. Population (1997 est): 3,828,506 (about 2.7 mil more Puerto Ricans reside in the mainland U.S.); **net change** (1990-96): 7.4% **Pop. density** (1990): 1,035 per sq mi. **Urban** (1990): 66.8%. **Ethnic distribution** (1990): 99.9% Hispanic. **Languages:** Spanish and English are joint official languages.

Geography. Total area: 3,508 sq. mi. **Land area:** 3,339 sq mi. **Location:** island lying between the Atlantic to the N and the Caribbean to the S; it is easternmost of the West Indies group called the Greater Antilles, of which Cuba, Hispaniola, and Jamaica are the larger islands. **Climate:** mild, with a mean temperature of 77° F. **Topography:** mountainous throughout three-fourths of its rectangular area, surrounded by a broken coastal plain; highest peak, Cerro de Punto, 4,390 ft. **Capital:** San Juan.

Economy. Chief industries: manufacturing, service. **Chief manuf. goods:** pharmaceuticals, apparel, electronics & other electric equipment, industrial machinery. **Gross domestic product:** (1995) $28.4 bil. **Chief crops:** coffee, plantains, pineapples, tomatoes, sugarcane, bananas, mangos, ornamental plants. **Livestock** (1996): 370,655 cattle; 182,247 hogs; 12.6 mil poultry. **Nonfuel minerals** (1996): $31.1 mil, mostly portland cement, crushed stone. **Commercial fishing** (1996): $15.7 mil. **Chief ports/river shipping:** San Juan, Ponce, Mayagüez. **Major airports at:** San Juan, Ponce, Mayagüez, Aguadilla. **Value of construction** (1996): $4.1 bil. **Employment distrib.** (1996): 32.6% public admin.; 23.1% serv.; 19.9% trade; 16.3% mfg. **Per capita income** (1996): $7,882. **Unemployment** (1998): 13.3%. **Tourism expends.** (1995): $1.9 mil.

Finance. FDIC-insured commercial banks (1998): 12. **Deposits:** $25.6 bil. **FDIC-insured savings institutions** (1998): 1. **Assets:** $30 mil.

Federal govt. Fed. civ. employees (1997): 13,874. **Notable fed. facilities:** U.S. Naval Station at Roosevelt Roads; P.R. National Guard Training Area at Camp Santiago, and at Ft. Allen, Juana Diaz; Sabana SECA Communications Center (U.S. Navy); U.S. Army Station at Ft. Buchanan.

Energy. Electricity production (1996): 15.9 bil kWh.

Public education. Student-teacher ratio (1995): 16.0. **Min. teachers' salary** (1997): $1,500 monthly.

Misc. data. Motto: Joannes Est Nomen Eius (John is his name). **Flower:** Maga. **Bird:** Reinita. **Tree:** Ceiba. **National anthem:** La Borinqueña.

History. Puerto Rico (or Borinquen, after the original Arawak Indian name, Boriquen) was visited by Columbus on his second voyage, Nov. 19, 1493. In 1508, the Spanish arrived.

Sugarcane was introduced, 1515, and slaves were imported 3 years later. Gold mining petered out, 1570. Spaniards fought off a series of British and Dutch attacks; slavery was abolished, 1873. Under the treaty of Paris, Puerto Rico was ceded to the U.S. after the Spanish-American War, 1898. In 1952 the people voted in favor of Commonwealth status.

The Commonwealth of Puerto Rico is a self-governing part of the U.S. with a primarily Hispanic culture. The island's citizens have virtually the same control over their internal affairs as do the 50 states of the U.S. However, they do not vote in national general elections, only in national primaries.

Puerto Rico is represented in the U.S. House of Representatives by a delegate who has a voice but no vote, except in committees.

No federal income tax is collected from residents on income earned from local sources in Puerto Rico. Nevertheless, as part of the U.S. legal system, Puerto Rico is subject to the provisions of the U.S. Constitution; most federal laws apply as they do in the 50 states.

Puerto Rico's famous "Operation Bootstrap," begun in the late 1940s, succeeded in changing the island from "The Poorhouse of the Caribbean" to an area with the highest per capita income in Latin America. This program encouraged manufacturing and development of the tourist trade by selective tax exemption, low-interest loans, and other incentives. Despite the marked success of Puerto Rico's development efforts over an extended period of time, per capita income in Puerto Rico is low in comparison to that of the U.S.

Tourist attractions. Ponce Museum of Art; Forts El Morro and San Cristobal; Old Walled City of San Juan; Arecibo Observatory; Cordillera Central and state parks; El Yunque Rain Forest; San Juan Cathedral; Porta Coeli Chapel and Museum of Religious Art, Interamerican Univ., San Germán; Condado Convention Center; Casa Blanca, Ponce de León family home, Puerto Rican Family Museum of 16th and 17th centuries, and Fine Arts Center all in San Juan.

Cultural facilities and events. Festival Casals classical music concerts, mid-June; Puerto Rico Symphony Orchestra at Music Conservatory; Botanical Garden and Museum of Anthropology, Art, and History at the University of Puerto Rico; Institute of Puerto Rican Culture, at the Dominican Convent; and many popular festivals.

Famous Puerto Ricans. Julia de Burgos, Marta Casals Istomin, Pablo Casals, José Celso Barbosa, Orlando Cepeda, Roberto Clemente, José de Diego, José Feliciano, Doña Felisa Rincón de Gautier, Luis A. Ferré, José Ferrer, Commodore Diégo E. Hernández, Miguel Hernández Agosto, Rafael Hernández (El Jibarito), Rafael Hernández Colón, Raúl Julía, René Marqués, Concha Meléndez, Rita Moreno, Luis Muñoz Marín, Luis Palés Matos, Adm. Horacio Rivero.

Chamber of Commerce. 100 Tetuán, PO Box S-3789, San Juan, PR 00902.

Website. http://fortaleza.govpr.org

Tourism website. http://www.discoverpuertorico.com

Virgin Islands

St. John, St. Croix, St. Thomas

People. Population (1997 est.): 97,240. **Population growth rate** (1994 est.): –0.52%. **Major ethnic groups:** West Indian, French, Hispanic. **Languages:** English (official), Spanish, Creole.

Geography. Total area: 171 sq mi. **Land area:** 134 sq mi. **Location:** 3 larger and 50 smaller islands and cays in the S and W of the V.I. group (British V.I. colony to the N and E), which is situated 70 mi E of Puerto Rico, located W of the Anegada Passage, a major channel connecting the Atlantic Ocean and the Caribbean Sea. **Climate:** subtropical; the sun tempered by gentle trade winds; humidity is low; average temperature, 78° F. **Topography:** St. Thomas is mainly a ridge of hills running E and W, and has little tillable land; St. Croix rises abruptly in the N but slopes to the S to flatlands and lagoons; St. John has steep, lofty hills and valleys with little level tillable land. **Capital:** Charlotte Amalie, St. Thomas.

Economy. Chief industries: tourism, rum, alumina, petroleum refining, watches, textiles, electronics, printing & publishing. **Chief manuf. goods:** rum, textiles, pharmaceuticals, perfumes, stone, glass & clay products. **Chief crops:** vegetables, horticulture, fruits and nuts. **Livestock** (1992): 7,132 cattle; 1,311 hogs/pigs; 9,087 chickens. **Minerals:** sand, gravel. **Chief ports:** Cruz Bay, St. John; Frederiksted and Christiansted, St. Croix; Charlotte Amalie, St. Thomas. **Internat. airports on:** St. Thomas, St. Croix. **Value of construction** (1992): $168.9 mil. **Employment distrib.** (1992): 50% trade; 43% serv. **Per capita income** (1989): $11,052. **Unemployment** (1992): 2.8%. **Tourism expends.** (1995): $792 mil.

Finance. FDIC-insured savings institutions (1998): 2. **Deposits:** $81 mil. **FDIC-insured savings institutions** (1998): 1. **Assets:** $55 mil.

Energy. Electricity production (1992): 565 mil kWh.

Public education. Student-teacher ratio (1995): 14.0.

Misc. data. Flower: Yellow elder or yellow trumpet, local designation Ginger Thomas. **Bird:** Yellow breast. **Song:** Virgin Islands March.

History. The islands were visited by Columbus in 1493. Spanish forces, 1555, defeated the Caribes and claimed the territory; by 1596 the native population was annihilated. First permanent settlement in the U.S. territory, 1672, by the Danes; U.S. purchased the islands, 1917, for defense purposes.

The Virgin Islands has a republican form of government, headed by a governor and lieut. governor elected, since 1970, by popular vote for 4-year terms. There is a 15-member unicameral legislature, elected by popular vote for a 2-year term. Residents of the V.I. have been U.S. citizens since 1927. Since 1973 they have elected a delegate to the U.S. House of Representatives, who has a voice but no vote, except in committees.

Tourist attractions. Magens Bay, St. Thomas; duty-free shopping; Virgin Islands Natl. Park, beaches, Indian relics, and evidence of colonial Danes.

Tourist information. Dept. of Economic Development & Agriculture: St. Thomas, PO Box 6400, St. Thomas, VI 00801; St. Croix, PO Box 4535, Christiansted, St. Croix 00820.

Website. http://www.usvi.net

Other Islands

Navassa lies between Jamaica and Haiti, 100 mi south of Guantanamo Bay, Cuba, in the Caribbean; it covers about 2 sq mi, is reserved by the U.S. for a lighthouse, and is uninhabited. It is administered by the U.S. Coast Guard.

Wake Atoll, and its neighboring atolls, **Wilkes** and **Peale,** lie in the Pacific Ocean on the direct route from Hawaii to Hong Kong, about 2,300 mi W of Honolulu and 1,290 mi E of Guam. The group is 4.5 mi long, 1.5 mi wide, and totals less than 3 sq mi. in land area. The U.S. flag was hoisted over Wake Atoll, July 4, 1898; formal possession taken Jan. 17, 1899. Wake was administered by the U.S. Air Force, 1972-94. The population consists of about 200 persons.

Midway Atoll, acquired in 1867, consist of 2 atolls, **Sand** and **Eastern,** in N Pacific 1,150 mi. NW of Honolulu, with an area of about 2 sq mi, administered by the U.S. Navy. There is no indigenous population; total pop. is about 450. **Johnston Atoll,** 717 mi WSW of Honolulu, area 1 sq mi, is operated by the Defense Nuclear Agency, and the Fish and Wildlife Service, U.S. Dept. of the Interior; its population is about 1,200. **Kingman Reef,** 920 mi S of Hawaii, is under Navy control. **Howland, Jarvis,** and **Baker Islands,** 1,400-1,650 mi SW of Honolulu, uninhabited since World War II, are under the Interior Dept. **Palmyra** is an atoll about 1,000 mi S of Hawaii, 5 sq mi. Privately owned, it is under the Interior Dept.

WASHINGTON, DC, CAPITAL OF THE U.S.

Most attractions are free. All times are subject to change. For more details call the Washington, DC, Convention and Visitors Association at 202-789-7000, or check out the website at: http://www.washington.org

Bureau of Engraving and Printing

The **Bureau of Engraving and Printing** of the U.S. Treasury Dept. is the headquarters for the making of U.S. paper money. Free 35-minute self-guided tours (tickets required) Mon.-Fri., 9 AM-2 PM year-round; extended hours, June-Aug., 5 PM-6:40 PM. Closed federal holidays. 14th and C Sts. SW. Phone: 202-874-3019.

Website. http://www.moneyfactory.com

Capitol

The **United States Capitol** was originally designed by Dr. William Thornton, an amateur architect, who submitted a plan in 1793 that won him $500 and a city lot.

The south, or House, wing was completed in 1807 under the direction of Benjamin H. Latrobe.

The present Senate and House wings and the iron dome were designed and constructed by Thomas U. Walter, 4th architect of the Capitol, between 1851 and 1863.

The present cast iron dome at its greatest exterior measures 135 ft 5 in., and it is topped by the bronze Statue of Freedom that stands 19^1/$_2$ ft and weighs 14,985 lb. On its base are the words *E Pluribus Unum* (Out of Many, One).

The Capitol is open from 9 AM to 8 PM, March-Aug., and 9 AM to 4:30 PM, Sept.-Feb., daily. It is closed Jan. 1, Thanksgiving Day, and Dec. 25. Tours through the Capitol, including the House and Senate galleries, are conducted Mon.-Sat.

To observe debate in the House or Senate while Congress is in session, individuals living in the U.S. may obtain tickets to the visitor's galleries from their U.S. representative or senator. Visitors from other countries may obtain passes at the Capitol. Between Constitution & Independence Ave., at Pennsylvania Ave. Phone: 202-225-6827.

Website. http://www.aoc.gov

Federal Bureau of Investigation

The **Federal Bureau of Investigation** offers guided one-hour tours of its headquarters, beginning with a videotape presentation. Visitors learn about the history of the FBI and see such things as the weapons confiscated from famous gangsters, photos of the most-wanted fugitives, the DNA laboratory, goods forfeited and seized in narcotics operations, and a sharpshooting demonstration.

Tours are conducted Mon.-Fri., 8:45 AM-4:15 PM, except Jan. 1, Dec. 25, and other federal holidays. Tickets may be obtained at the FBI on day of tour or through a U.S. representative or senator. J. Edgar Hoover Bldg., Pennsylvania Ave., between 9th and 10th Sts. NW. Phone: 202-324-3447.

Website. http://www.fbi.gov

Folger Shakespeare Library

The **Folger Shakespeare Library,** on Capitol Hill, is a research institution holding rare books and manuscripts of the Renaissance period and the largest collection of Shakespearean materials in the world, including 79 copies of the First Folio. The library's museum and performing arts programs are presented in the Elizabethan Theatre, which resembles an innyard theater of Shakespeare's day.

Exhibit may be visited Mon.-Sat., 10 AM-4 PM., 201 E. Capitol St., SE , Phone: 202-544-7077.

Website. http://www.folger.edu

Holocaust Memorial Museum

The **U.S. Holocaust Memorial Museum** opened on Apr. 21, 1993. The museum documents, through permanent and temporary displays, interactive videos, and special lectures, the events of the Holocaust beginning in 1933 and continuing World War II. The permanent exhibition is not recommended for children under the age of 11.

The museum is open daily, 10 AM-5:30 PM, except Yom Kippur and Dec. 25, and extended hours (8 AM-10 PM) Apr. 3-Sept. 2. A limited number of free tickets are available on day of visit; advance tickets may be ordered for a small fee. 100 Raoul Wallenberg Pl. SW. Phone: 202-488-0400.

Website. http://www.ushmm.org/index.html

Jefferson Memorial

Dedicated in 1943, the **Thomas Jefferson Memorial** stands on the south shore of the Tidal Basin in West Potomac Park. It is a circular stone structure, with Vermont marble on the exterior and Georgia white marble inside, and combines architectural elements of the dome of the Pantheon in Rome and the rotunda designed by Jefferson for the University of Virginia.

The memorial, on the south edge of the Tidal Basin, is open daily, 8 AM-midnight. An elevator and curb ramps for the handicapped are in service. Phone: 202-426-6841.

Website. http://www.nps.gov/thje/index2.htm

John F. Kennedy Center

The **John F. Kennedy Center for the Performing Arts,** designated by Congress as the National Cultural Center and the official memorial in Washington, DC, to Pres. John F. Kennedy, opened Sept. 8, 1971. Designed by Edward Durell Stone, the center includes an opera house, a concert hall, several theaters, 2 restaurants, and a library.

Free tours are available daily, 10 AM-1 PM. 2700 F St. NW. Phone: 202-416-8340, or 1-800-444-1324.

Website. http://www.kennedy-center.org

Korean War Veterans Memorial

Dedicated on July 27, 1995, the **Korean War Veterans Memorial** honors all Americans who served in the Korean War. Situated at the west end of the Mall, across the reflecting pool from the Vietnam Memorial, the triangular-shaped stone and steel memorial features a multiservice formation of 19 troops clad in ponchos with the wind at their back, ready for combat. A granite wall, with images of the men and women who served, juts into a pool of water, the Pool of Remembrance, and is inscribed with the words *Freedom Is Not Free.*

The $18 mil memorial, which was funded by private donations, is open 8 AM-midnight. Independence Ave. at Lincoln Memorial. Phone: 202-619-7222.

Website. http://www.nps.gov/kwvm/index2.htm

Library of Congress

Established by and for Congress in 1800, the **Library of Congress** has extended its services over the years to other government agencies and other libraries, to scholars, and to the general public, and it now serves as the national library. It contains more than 80 million items in 470 languages.

The library's exhibit halls are open to the public Mon.-Fri., 8:30 AM-9:30 PM; Sat., 8:30 AM-6 PM. The library is closed Jan. 1 and Dec. 25. 101 Independence Ave., SE. Phone: 202-707-8000.

Website. http://www.loc.gov

Lincoln Memorial

Designed by Henry Bacon, the **Lincoln Memorial** in West Potomac Park, on the axis of the Capitol and the Washington Monument, consists of a large marble hall enclosing a heroic statue of Abraham Lincoln in meditation sitting on a large armchair. The memorial was dedicated on May 30, 1922. The statue was designed by Daniel Chester French and sculpted by French and the Piccirilli brothers. Murals and ornamentation on the bronze ceiling beams are by Jules Guerin. The text of the Gettysburg Address is in the south chamber; that of Lincoln's Second Inaugural speech is in the north chamber. Each is engraved on a stone tablet.

The memorial is open 24 hr daily. An elevator for the handicapped is in service. W. Potomac Park at 23rd St. NW. Phone: 202-619-7222.

Website. http://www.nps.gov/linc/index2.htm

National Archives and Records

Original copies of the Declaration of Independence, the Constitution, and the Bill of Rights are on permanent display in the **National Archives** Exhibition Hall. The National Archives also holds other valuable U.S. government records and historic maps, photographs, and manuscripts.

Central Research and Microfilm Research Rooms are also available to the public for genealogical research.

The Exhibition Hall is open daily, 10 AM-5:30 PM; closed Dec. 25. 7th & Pennsylvania Ave. NW. Phone: 202-501-5000.

Website. http://www.nara.gov

National Gallery of Art

The **National Gallery of Art**, situated on the north side of the Mall facing Constitution Avenue, was established by Congress, Mar. 24, 1937, and opened Mar. 17, 1941. The original West building was designed by John Russell Pope. The East building, opened in 1978, was designed by I. M. Pei. The National Gallery is separate from, but maintains a relationship with, the Smithsonian Institution.

Open daily, 10 AM-5 PM; Sunday, 11 AM-6 PM. Closed Jan. 1 and Dec. 25. 4th & Constitution Ave NW. Phone: 202-737-4215.

Website. http://www.nga.gov

Franklin Delano Roosevelt Memorial

Opened May 2, 1997, by Pres. Bill Clinton, the **FDR Memorial** features 9 bronze sculptural ensembles depicting FDR, Eleanor Roosevelt (the first First Lady to be honored in a national memorial), and events from the Great Depression and World War II. This 7.5-acre memorial is located near the Tidal Basin in a park-like setting and includes waterfalls, quiet pools, and reddish Dakota granite upon which some of Pres. Roosevelt's well-known words are carved. The monument is wheelchair accessible.

Grounds, staffed daily, 8 AM-midnight, except Dec. 25. 1850 W. Basin Dr. SW. Phone: (202) 619-7222.

Website. http://www.nps.gov/fdrm/index2.htm

Smithsonian Institution

The **Smithsonian Institution,** established in 1846, is the world's largest museum complex and consists of 14 museums and the National Zoo. It holds some 100 mil. artifacts and specimens in its trust. Nine museums are on the National Mall between the Washington Monument and the Capitol; 5 other museums and the zoo are elsewhere in Washington (the Cooper-Hewitt Museum and the National Museum of the American Indian, also administered by the Smithsonian, are in New York City). The **Smithsonian Information Center,** is located in "the Castle" on the Mall. Also on the Mall are the **National Museum of American History,** the **National Museum of Natural History,** the **National Air and Space Museum,** the **Hirshhorn Museum and Sculpture Garden,** the **Arthur M. Sackler Gallery,** the **National Museum of African Art,** the **Freer Gallery of Art,** and the **Arts and Industries Building.** Near the Sackler Gallery is the **Enid A. Haupt Garden.** Located nearby are the **National Postal Museum,** the **National Museum of American Art,** the **National Portrait Gallery,** and the **Renwick Gallery.** Farther away, at 1901 Fort Place SE, is the **Anacostia Museum.**

Most museums are open daily, except Dec. 25, 10 AM-5:30 PM. Phone: 202-357-2700.

Website. http://www.si.edu

Vietnam Veterans Memorial

Originally dedicated on Nov. 13, 1982, the **Vietnam Veterans Memorial** is a recognition of the men and women who served in the armed forces in the Vietnam War. On a V-shaped black-granite wall, designed by Maya Ying Lin, are inscribed the names of the more than 58,000 Americans who lost their lives or remain missing.

Since 1982, 2 additions have been made to the Memorial. The 1st, dedicated on Nov. 11, 1984, is the Frederick Hart sculpture *Three Servicemen*. On Nov. 11, 1993, the Vietnam Women's Memorial was dedicated, honoring the more than 11,500 women who served in Vietnam. The bronze sculpture, portraying 3 women helping a wounded male soldier, was designed by Glenna Goodacre.

The memorial is open 24 hr daily. Constitution Ave. & Bacon Dr. NW. Phone: 202-634-1568.

Website. http://www.thevirtualwall.org

Washington Monument

The **Washington Monument,** dedicated in 1885, is a tapering shaft, or obelisk, of white marble, 555 ft, 5^1/8 inches in height and 55 ft, 1^1/2 in. square at base. Eight small windows, 2 on each side, are located at the 500-ft level, where points of interest are indicated.

Open daily (except Dec. 25), 9 AM-4:30 PM; 8 AM-midnight, Apr.-Labor Day. Free timed passes are available; passes are available in advance for a small fee. 15th & Constitution Ave. NW. Phone: 202-426-6841.

Website. http://www.nps.gov/wamo/monument/40links.htm

White House

The **White House,** the President's residence, stands on 18 acres on the south side of Pennsylvania Ave., between the Treasury and the old Executive Office Building. The walls are of sandstone, quarried at Aquia Creek, VA. The exterior walls were painted, causing the building to be termed the "White House." On Aug. 24, 1814, during Madison's administration, the house was burned by the British. James Hoban rebuilt it by Oct. 1817.

The White House is normally open for free self-guided tours Tues.-Sat., 10 AM-noon (passes, necessary mid-March-mid-Sept., are available at White House Visitor's Center, 8 AM-noon, located at 1450 Pennsylvania Ave., NW). Only the public rooms on the ground floor and state floor may be visited. Free reserved tickets for guided congressional tours can be obtained 8 to 10 weeks in advance from your local U.S. representative or senator. 1600 Pennsylvania Ave. Phone: 202-456-7041.

Website. http://www.whitehouse.gov

Attractions Near Washington, DC

Arlington National Cemetery

Arlington National Cemetery, on the former Custis estate in Arlington, VA, is the site of the **Tomb of the Unknowns** and is the final resting place of Pres. John Fitzgerald Kennedy, who was buried there on Nov. 25, 1963. His wife, Jacqueline Bouvier Kennedy Onassis, was buried at the same site on May 23, 1994. An eternal flame burns over the grave site. In an adjacent area is the grave of Pres. Kennedy's brother Sen. Robert F. Kennedy (NY), interred on June 8, 1968. Many other famous Americans are also buried at Arlington, as well as more than 200,000 American soldiers from every major war.

North of the National Cemetery, approximately 350 yd, stands the **U.S. Marine Corps War Memorial**, also known as Iwo Jima. The memorial is a bronze statue of the raising of the U.S. flag on Mt. Suribachi, Feb. 23, 1945, during World War II, executed by Felix de Weldon from the photograph by Joe Rosenthal.

On the southern side of the Memorial Bridge, near the cemetery entrance, a memorial honoring the women in the military was dedicated, Oct. 18, 1997. The **Women in Military Service for America Memorial** is a half-circle granite monument, 30 ft. high and 226 ft. in diameter, with the Great Seal of the United States in the center.

Open daily, 8 AM-5 PM (8 AM-7 PM., Apr.-Sept.), Arlington, VA. Phone: 703-607-8052.

Mount Vernon

Mount Vernon, George Washington's estate, is on the south bank of the Potomac R., 16 mi below Washington, DC, in northern Virginia. The present house is an enlargement of one apparently built on the site by Augustine Washington, who lived there 1735-38. His son Lawrence came there in 1743, and renamed the plantation Mount Vernon in honor of Admiral Vernon, under whom he had served in the West Indies. Lawrence Washington died in 1752 and was succeeded as proprietor by his half-brother, George Washington. The estate has been restored to its 18th-century appearance and includes many original furnishings. Washington and his wife, Martha, are buried on the grounds.

Open 365 days, 8 AM-5 PM, Apr.-Aug., 9 AM-5 PM, Sept., Oct., Mar.; 9 AM-4 PM, Nov.-Feb. Phone: 703-780-2000, or 1-800-429-1520. Admission: adults $8, seniors (62+) $7.50, children (6-11) $4, age 5 and under free.

Website. http://www.mountvernon.org

The Pentagon

The **Pentagon,** headquarters of the Department of Defense, is one of the world's largest office buildings. Situated in Arlington, VA, it houses more than 23,000 employees in offices that occupy 3,707,745 sq ft.

Free tours (about every 2 hrs) are available Mon.-Fri. (excluding federal holidays), starting at 9 AM; last tour begins at 3:20 PM. Arlington, VA (I-395 South to Boundary Channel Drive exit). Phone: 703-695-1776.

Website. http://www.defenselink.mil/pubs/pentagon

AWARDS — MEDALS — PRIZES
The Alfred B. Nobel Prize Winners

Alfred B. Nobel (1833-96), inventor of dynamite, bequeathed $9 mil, the interest to be distributed yearly to those judged to have had most benefited humankind in physics, chemistry, medicine-physiology, literature, and promotion of peace. These prizes were first awarded in 1901. The first Nobel Memorial Prize in Economic Science was awarded in 1969, funded by the central bank of Sweden. If the year is omitted, no award was given. In 1999, each prize was worth more than $1 mil. For 1999 Nobel Prize winners, see separate section, The 1999 Nobel Prizes.

Physics

1998 Robert B. Laughlin, Horst L. Störmer, Daniel C. Tsui, all U.S.
1997 Steven Chu, William D. Phillips, both U.S.; Claude Cohen-Tannoudji, Fr.
1996 David M. Lee, Douglas D. Osheroff, Robert C. Richardson, all U.S.
1995 Martin Perl, Frederick Reines, both U.S.
1994 Bertram N. Brockhouse, Can.; Clifford G. Shull, U.S.
1993 Joseph H. Taylor, Russell A. Hulse, both U.S.
1992 Georges Charpak, Pol.-Fr.
1991 Pierre-Giles de Gennes, Fr.
1990 Richard E. Taylor, Can.; Jerome I. Friedman, Henry W. Kendall, both U.S.
1989 Norman F. Ramsey, U.S.; Hans G. Dehmelt, Ger.-U.S.; Wolfgang Paul, Ger.
1988 Leon M. Lederman, Melvin Schwartz, Jack Steinberger, all U.S.
1987 K. Alex Müller, Swiss; J. Georg Bednorz, Ger.
1986 Ernest Ruska, Ger.; Gerd Binnig, Ger.; Heinrich Rohrer, Swiss
1985 Klaus von Klitzing, Ger.
1984 Carlo Rubbia, It.; Simon van der Meer, Dutch
1983 Subrahmanyan Chandrasekhar, William A. Fowler, both U.S.
1982 Kenneth G. Wilson, U.S.
1981 Nicolaas Bloembergen, Arthur Schaalow, both U.S.; Kai M. Siegbahn, Swed.
1980 James W. Cronin, Val L. Fitch, both U.S.
1979 Steven Weinberg, Sheldon L. Glashow, both U.S.; Abdus Salam, Pakistani
1978 Pyotr Kapitsa, USSR; Arno Penzias, Robert Wilson, both U.S.
1977 John H. Van Vleck, Philip W. Anderson, both U.S.; Nevill F. Mott, Br.
1976 Burton Richter, Samuel C.C. Ting, both U.S.
1975 James Rainwater, U.S.; Ben Mottelson, U.S.-Dan.; Aage Bohr, Dan.

1974 Martin Ryle, Antony Hewish, both Br.
1973 Ivar Giaever, U.S.; Leo Esaki, Jpn.; Brian D. Josephson, Br.
1972 John Bardeen, Leon N. Cooper, John R. Schrieffer, all U.S.
1971 Dennis Gabor, Br.
1970 Louis Neel, Fr.; Hannes Alfven, Swed.
1969 Murray Gell-Mann, U.S.
1968 Luis W. Alvarez, U.S.
1967 Hans A. Bethe, U.S.
1966 Alfred Kastler, Fr.
1965 Richard P. Feynman, Julian S. Schwinger, both U.S.; Shinichiro Tomonaga, Jpn.
1964 Nikolai G. Basov, Aleksander M. Prochorov, both USSR; Charles H. Townes, U.S.
1963 Maria Goeppert-Mayer, Eugene P. Wigner, both U.S.; J. Hans D. Jensen, Ger.
1962 Lev. D. Landau, USSR
1961 Robert Hofstadter, U.S.; Rudolf L. Mossbauer, Ger.
1960 Donald A. Glaser, U.S.
1959 Owen Chamberlain, Emilio G. Segre, both U.S.
1958 Pavel Cherenkov, Ilya Frank, Igor Y. Tamm, all USSR
1957 Tsung-dao Lee, Chen Ning Yang, both U.S.
1956 John Bardeen, Walter H. Brattain, William Shockley, all U.S.
1955 Polykarp Kusch, Willis E. Lamb, both U.S.
1954 Max Born, Br.; Walter Bothe, Ger.
1953 Frits Zernike, Dutch
1952 Felix Bloch, Edward M. Purcell, both U.S.
1951 Sir John D. Cockroft, Br.; Ernest T. S. Walton, Ir.
1950 Cecil F. Powell, Br.
1949 Hideki Yukawa, Jpn.
1948 Patrick M. S. Blackett, Br.
1947 Sir Edward V. Appleton, Br.
1946 Percy W. Bridgman, U.S.
1945 Wolfgang Pauli, U.S.
1944 Isidor Isaac Rabi, U.S.

1943 Otto Stern, U.S.
1939 Ernest O. Lawrence, U.S.
1938 Enrico Fermi, It.-U.S.
1937 Clinton J. Davisson, U.S.; Sir George P. Thomson, Br.
1936 Carl D. Anderson, U.S.; Victor F. Hess, Aus.
1935 Sir James Chadwick, Br.
1933 Paul A. M. Dirac, Br.; Erwin Schrodinger, Austria
1932 Werner Heisenberg, Ger.
1930 Sir Chandrasekhara V. Raman, Indian
1929 Prince Louis-Victor de Broglie, Fr.
1928 Owen W. Richardson, Br.
1927 Arthur H. Compton, U.S.; Charles T. R. Wilson, Br.
1926 Jean B. Perrin, Fr.
1925 James Franck, Gustav Hertz, both Ger.
1924 Karl M. G. Siegbahn, Swed.
1923 Robert A. Millikan, U.S.
1922 Niels Bohr, Dan.
1921 Albert Einstein, Ger.-U.S.
1920 Charles E. Guillaume, Fr.
1919 Johannes Stark, Ger.
1918 Max K. E. L. Planck, Ger.
1917 Charles G. Barkla, Br.
1915 Sir William H. Bragg, Sir William L. Bragg, both Br.
1914 Max von Laue, Ger.
1913 Heike Kamerlingh-Onnes, Dutch
1912 Nils G. Dalen, Swed.
1911 Wilhelm Wien, Ger.
1910 Johannes D. van der Waals, Dutch
1909 Carl F. Braun, Ger.; Guglielmo Marconi, It.
1908 Gabriel Lippmann, Fr.
1907 Albert A. Michelson, U.S.
1906 Sir Joseph J. Thomson, Br.
1905 Philipp E. A. von Lenard, Ger.
1904 John W. Strutt, Lord Rayleigh, Br.
1903 Antoine Henri Becquerel, Pierre Curie, both Fr.; Marie Curie, Pol.-Fr.
1902 Hendrik A. Lorentz, Pieter Zeeman, both Dutch
1901 Wilhelm C. Roentgen, Ger.

Chemistry

1998 Walter Kohn, U.S.; John A. Pople, Br.
1997 Paul D. Boyer, U.S., & John E. Walker, Br.; Jens C. Skou, Dan.
1996 Harold W. Kroto, Br.; Robert F. Curl Jr., Richard E. Smalley, both U.S.
1995 Paul Crutzen, Dutch; Mario Molina, Mex.-U.S.; Sherwood Rowland, U.S.
1994 George A. Olah, U.S.
1993 Kary B. Mullis, U.S.; Michael Smith, Br.-Can.
1992 Rudolph A. Marcus, Can.-U.S.
1991 Richard R. Ernst, Swiss
1990 Elias James Corey, U.S.
1989 Thomas R. Cech, Sidney Altman, both U.S.
1988 Johann Deisenhofer, Robert Huber, Hartmut Michel, all Ger.
1987 Donald J. Cram, Charles J. Pedersen, both U.S.; Jean-Marie Lehn, Fr.
1986 Dudley Herschbach, Yuan T. Lee, both U.S.; John C. Polanyi, Can.
1985 Herbert A. Hauptman, Jerome Karle, both U.S.
1984 Bruce Merrifield, U.S.
1983 Henry Taube, Can.
1982 Aaron Klug, S. Afr.
1981 Kenichi Fukui, Jpn.; Roald Hoffmann, U.S.
1980 Paul Berg, Walter Gilbert, both U.S.; Frederick Sanger, Br.

1979 Herbert C. Brown, U.S.; George Wittig, Ger.
1978 Peter Mitchell, Br.
1977 Ilya Prigogine, Belg.
1976 William N. Lipscomb, U.S.
1975 John Cornforth, Austral.-Br.; Vladimir Prelog, Yugo.-Swiss
1974 Paul J. Flory, U.S.
1973 Ernst Otto Fischer, Ger.; Geoffrey Wilkinson, Br.
1972 Christian B. Anfinsen, Stanford Moore, William H. Stein, all U.S.
1971 Gerhard Herzberg, Canadian
1970 Luis F. Leloir, Arg.
1969 Derek H. R. Barton, Br.; Odd Hassel, Nor.
1968 Lars Onsager, U.S.
1967 Manfred Eigen, Ger.; Ronald G. W. Norrish, George Porter, both Br.
1966 Robert S. Mulliken, U.S.
1965 Robert B. Woodward, U.S.
1964 Dorothy C. Hodgkin, Br.
1963 Giulio Natta, It.; Karl Ziegler, Ger.
1962 John C. Kendrew, Max F. Perutz, both Br.
1961 Melvin Calvin, U.S.
1960 Willard F. Libby, U.S.
1959 Jaroslav Heyrovsky, Czech.
1958 Frederick Sanger, Br.
1957 Sir Alexander R. Todd, Br.
1956 Sir Cyril N. Hinshelwood, Br.; Nikolai N. Semenov, USSR

1955 Vincent du Vigneaud, U.S.
1954 Linus C. Pauling, U.S.
1953 Hermann Staudinger, Ger.
1952 Archer J. P. Martin, Richard L. M. Synge, both Br.
1951 Edwin M. McMillan, Glenn T. Seaborg, both U.S.
1950 Kurt Alder, Otto P. H. Diels, both Ger.
1949 William F. Giauque, U.S.
1948 Arne W. K. Tiselius, Swed.
1947 Sir Robert Robinson, Br.
1946 James B. Sumner, John H. Northrop, Wendell M. Stanley, all U.S.
1945 Artturi I. Virtanen, Fin.
1944 Otto Hahn, Ger.
1943 Georg de Hevesy, Hung.
1939 Adolf F. J. Butenandt, Ger.; Leopold Ruzicka, Swiss
1938 Richard Kuhn, Ger.
1937 Walter N. Haworth, Br.; Paul Karrer, Swiss
1936 Peter J. W. Debye, Dutch
1935 Frederic & Irene Joliot-Curie, both Fr.
1934 Harold C. Urey, U.S.
1932 Irving Langmuir, U.S.
1931 Friedrich Bergius, Karl Bosch, both Ger.
1930 Hans Fischer, Ger.
1929 Sir Arthur Harden, Br.; Hans von Euler-Chelpin, Swed.

1928 Adolf O. R. Windaus, Ger.	1918 Fritz Haber, Ger.	1908 Ernest Rutherford, Br.
1927 Heinrich O. Wieland, Ger.	1915 Richard M. Willstatter, Ger.	1907 Eduard Buchner, Ger.
1926 Theodor Svedberg, Swed.	1914 Theodore W. Richards, U.S.	1906 Henri Moissan, Fr.
1925 Richard A. Zsigmondy, Ger.	1913 Alfred Werner, Swiss	1905 Adolf von Baeyer, Ger.
1923 Fritz Pregl, Austrian	1912 Victor Grignard, Paul Sabatier, both Fr.	1904 Sir William Ramsay, Br.
1922 Francis W. Aston, Br.	1911 Marie Curie, Pol.-Fr.	1903 Svante A. Arrhenius, Swed.
1921 Frederick Soddy, Br.	1910 Otto Wallach, Ger.	1902 Emil Fischer, Ger.
1920 Walther H. Nernst, Ger.	1909 Wilhelm Ostwald, Ger.	1901 Jacobus H. van't Hoff, Dutch

Physiology or Medicine

1998 Robert F. Furchgott, Louis J. Ignarro, Ferid Murad, all U.S.	1973 Karl von Frisch, Ger.; Konrad Lorenz, Ger.-Aus.; Nikolaas Tinbergen, Br.	1946 Hermann J. Muller, U.S.
1997 Stanley B. Prusiner, U.S.	1972 Gerald M. Edelman, U.S.; Rodney R. Porter, Br.	1945 Ernst B. Chain, Sir Alexander Fleming, Sir Howard W. Florey, all Br.
1996 Peter C. Doherty, Austral.; Rolf M. Zinkernagel, Swiss	1971 Earl W. Sutherland Jr., U.S.	1944 Joseph Erlanger, Herbert S. Gasser, both U.S.
1995 Edward B. Lewis, Eric F. Wieschaus, both U.S.; Christiane Nuesslein-Volhard, Ger.	1970 Julius Axelrod, U.S.; Sir Bernard Katz, Br.; Ulf von Euler, Swed.	1943 Henrik C. P. Dam, Dan.; Edward A. Doisy, U.S.
1994 Alfred G. Gilman, Martin Rodbell, both U.S.	1969 Max Delbrück, Alfred D. Hershey, Salvador Luria, all U.S.	1939 Gerhard Domagk, Ger.
1993 Phillip A. Sharp, U.S.; Richard J. Roberts, Br.	1968 Robert W. Holley, H. Gobind Khorana, Marshall W. Nirenberg, all U.S.	1938 Corneille J. F. Heymans, Belg.
1992 Edmond H. Fisher, Edwin G. Krebs, both U.S.	1967 Ragnar Granit, Swed.; Haldan Keffer Hartline, George Wald, both U.S.	1937 Albert Szent-Gyorgyi, Hung.-U.S.
		1936 Sir Henry H. Dale, Br.; Otto Loewi, U.S.
		1935 Hans Spemann, Ger.
1991 Edwin Neher, Bert Sakmann, both Ger.	1966 Charles B. Huggins, Francis Peyton Rous, both U.S.	1934 George R. Minot, William P. Murphy, G. H. Whipple, all U.S.
1990 Joseph E. Murray, E. Donnall Thomas, both U.S.	1965 François Jacob, Andre Lwoff, Jacques Monod, all Fr.	1933 Thomas H. Morgan, U.S.
1989 J. Michael Bishop, Harold E. Varmus, both U.S.	1964 Konrad E. Bloch, U.S.; Feodor Lynen, Ger.	1932 Edgar D. Adrian, Sir Charles S. Sherrington, both Br.
1988 Gertrude B. Elion, George H. Hitchings, both U.S.; Sir James Black, Br.	1963 Sir John C. Eccles, Austral.; Alan L. Hodgkin, Andrew F. Huxley, both Br.	1931 Otto H. Warburg, Ger.
		1930 Karl Landsteiner, U.S.
1987 Susumu Tonegawa, Jpn.	1962 Francis H. C. Crick, Maurice H. F. Wilkins, both Br.; James D. Watson, U.S.	1929 Christiaan Eijkman, Dutch; Sir Frederick G. Hopkins, Br.
1986 Rita Levi-Montalcini, It.-U.S., Stanley Cohen, U.S.	1961 Georg von Bekesy, U.S.	1928 Charles J. H. Nicolle, Fr.
		1927 Julius Wagner-Jauregg, Austrian
1985 Michael S. Brown, Joseph L. Goldstein, both U.S.	1960 Sir F. MacFarlane Burnet, Austral.; Peter B. Medawar, Br.	1926 Johannes A. G. Fibiger, Dan.
1984 Cesar Milstein, Br.-Arg.; Georges J. F. Koehler, Ger.; Niels K. Jerne, Br.-Dan.	1959 Arthur Kornberg, Severo Ochoa, both U.S.	1924 Willem Einthoven, Dutch
		1923 Frederick G. Banting, Can.; John J. R. Macleod, Scot.
1983 Barbara McClintock, U.S.	1958 George W. Beadle, Edward L. Tatum, Joshua Lederberg, all U.S.	
1982 Sune Bergstrom, Bengt Samuelsson, both Swed.; John R. Vane, Br.	1957 Daniel Bovet, It.	1922 Archibald V. Hill, Br.; Otto F. Meyerhof, Ger.
1981 Roger W. Sperry, David H. Hubel, Tosten N. Wiesel, all U.S.	1956 Andre F. Cournand, Dickinson W. Richards Jr., both U.S.; Werner Forssmann, Ger.	1920 Schack A. S. Krogh, Dan.
		1919 Jules Bordet, Belg.
		1914 Robert Barany, Aus.
1980 Baruj Benacerraf, George Snell, both U.S.; Jean Dausset, Fr.	1955 Alex H. T. Theorell, Swed.	1913 Charles R. Richet, Fr.
	1954 John F. Enders, Frederick C. Robbins, Thomas H. Weller, all U.S.	1912 Alexis Carrel, Fr.
1979 Allan M. Cormack, U.S.; Godfrey N. Hounsfield, Br.		1911 Allvar Gullstrand, Swed.
	1953 Hans A. Krebs, Br.; Fritz A. Lipmann, U.S.	1910 Albrecht Kossel, Ger.
1978 Daniel Nathans, Hamilton O. Smith, both U.S.; Werner Arber, Swiss	1952 Selman A. Waksman, U.S.	1909 Emil T. Kocher, Swiss
	1951 Max Theiler, U.S.	1908 Paul Ehrlich, Ger.; Elie Metchnikoff, Fr.
1977 Rosalyn S. Yalow, Roger C.L. Guillemin, Andrew V. Schally, all U.S.	1950 Philip S. Hench, Edward C. Kendall, both U.S.; Tadeus Reichstein, Swiss	1907 Charles L. A. Laveran, Fr.
		1906 Camillo Golgi, It.; Santiago Ramon y Cajal, Span.
1976 Baruch S. Blumberg, Daniel Carleton Gajdusek, both U.S.	1949 Walter R. Hess, Swiss; Antonio Moniz, Port.	
		1905 Robert Koch, Ger.
1975 David Baltimore, Howard Temin, both U.S.; Renato Dulbecco, It.-U.S.	1948 Paul H. Müller, Swiss	1904 Ivan P. Pavlov, Russ.
	1947 Carl F. Cori, Gerty T. Cori, both U.S.; Bernardo A. Houssay, Arg.	1903 Niels R. Finsen, Dan.
1974 Albert Claude, Lux.-U.S.; George Emil Palade, Rom.-U.S.; Christian Rene de Duve, Belg.		1902 Sir Ronald Ross, Br.
		1901 Emil A. von Behring, Ger.

Literature

1998 José Saramago, Por.	1968 Yasunari Kawabata, Jpn.	1932 John Galsworthy, Br.
1997 Dario Fo, It.	1967 Miguel Angel Asturias, Guat.	1931 Erik A. Karlfeldt, Swed.
1996 Wislawa Szymborska, Pol.	1966 Samuel Joseph Agnon, Isr.; Nelly Sachs, Swed.	1930 Sinclair Lewis, U.S.
1995 Seamus Heaney, Ir.		1929 Thomas Mann, Ger.
1994 Kenzaburo Oe, Jpn.	1965 Mikhail Sholokhov, USSR	1928 Sigrid Undset, Nor.
1993 Toni Morrison, U.S.	1964 Jean Paul Sartre, Fr. (declined)	1927 Henri Bergson, Fr.
1992 Derek Walcott, W. Ind.	1963 Giorgos Seferis, Gk.	1926 Grazia Deledda, It.
1991 Nadine Gordimer, S. Afr.	1962 John Steinbeck, U.S.	1925 George Bernard Shaw, Ir.-Br.
1990 Octavio Paz, Mex.	1961 Ivo Andric, Yugo.	1924 Wladyslaw S. Reymont, Pol.
1989 Camilo José Cela, Span.	1960 Saint-John Perse, Fr.	1923 William Butler Yeats, Ir.
1988 Naguib Mahfouz, Egy.	1959 Salvatore Quasimodo, It.	1922 Jacinto Benavente, Span.
1987 Joseph Brodsky, USSR-U.S.	1958 Boris L. Pasternak, USSR (declined)	1921 Anatole France, Fr.
1986 Wole Soyinka, Nig.	1957 Albert Camus, Fr.	1920 Knut Hamsun, Nor.
1985 Claude Simon, Fr.	1956 Juan Ramon Jimenez, Span.	1919 Carl F. G. Spitteler, Swiss
1984 Jaroslav Siefert, Czech.	1955 Halldor K. Laxness, Ice.	1917 Karl A. Gjellerup, Henrik Pontoppidan, both Dan.
1983 William Golding, Br.	1954 Ernest Hemingway, U.S.	
1982 Gabriel Garcia Marquez, Colombian-Mex.	1953 Sir Winston Churchill, Br.	1916 Verner von Heidenstam, Swed.
	1952 Francois Mauriac, Fr.	1915 Romain Rolland, Fr.
1981 Elias Canetti, Bulg.-Br.	1951 Par F. Lagerkvist, Swed.	1913 Rabindranath Tagore, Indian
1980 Czeslaw Milosz, Pol.-U.S.	1950 Bertrand Russell, Br.	1912 Gerhart Hauptmann, Ger.
1979 Odysseus Elytis, Gk.	1949 William Faulkner, U.S.	1911 Maurice Maeterlinck, Belg.
1978 Isaac Bashevis Singer, U.S.	1948 T.S. Eliot, Br.	1910 Paul J. L. Heyse, Ger.
1977 Vicente Aleixandre, Span.	1947 Andre Gide, Fr.	1909 Selma Lagerlof, Swed.
1976 Saul Bellow, U.S.	1946 Hermann Hesse, Ger.-Swiss	1908 Rudolf C. Eucken, Ger.
1975 Eugenio Montale, It.	1945 Gabriela Mistral, Chil.	1907 Rudyard Kipling, Br.
1974 Eyvind Johnson, Harry Edmund Martinson, both Swed.	1944 Johannes V. Jensen, Dan.	1906 Giosue Carducci, It.
	1939 Frans E. Sillanpaa, Fin.	1905 Henryk Sienkiewicz, Pol.
1973 Patrick White, Austral.	1938 Pearl S. Buck, U.S.	1904 Frederic Mistral, Fr.; Jose Echegaray, Span.
1972 Heinrich Böll, Ger.	1937 Roger Martin du Gard, Fr.	
1971 Pablo Neruda, Chil.	1936 Eugene O'Neill, U.S.	1903 Bjornsterne Bjornson, Nor.
1970 Aleksandr I. Solzhenitsyn, USSR	1934 Luigi Pirandello, It.	1902 Theodor Mommsen, Ger.
1969 Samuel Beckett, Ir.	1933 Ivan A. Bunin, USSR	1901 Rene F. A. Sully Prudhomme, Fr.

Peace

1998 John Hume, David Trimble, both N. Ir.	1973 Henry Kissinger, U.S.; Le Duc Tho, N.	1931 Jane Addams, Nicholas Murray
1997 Jody Williams, U.S.; International Campaign to Ban Landmines	Viet. (Tho declined)	Butler, both U.S.
1996 Bishop Carlos Ximenes Belo, José Ramos-Horta, both Timorese	1971 Willy Brandt, Ger.	1930 Nathan Soderblom, Swed.
	1970 Norman E. Borlaug, U.S.	1929 Frank B. Kellogg, U.S.
	1969 Intl. Labor Organization	1927 Ferdinand E. Buisson, Fr.; Ludwig
1995 Joseph Rotblat, Pol.-Br.; Pugwash Conference	1968 Rene Cassin, Fr.	Quidde, Ger.
	1965 UN Children's Fund (UNICEF)	1926 Aristide Briand, Fr.; Gustav
1994 Yasir Arafat, Pal.; Shimon Peres, Yitzhak Rabin, both Isr.	1964 Martin Luther King Jr., U.S.	Stresemann, Ger.
	1963 International Red Cross,	1925 Sir J. Austen Chamberlain, Br.;
1993 Frederik W. de Klerk, Nelson Mandela, both S. Afr.	League of Red Cross Societies	Charles G. Dawes, U.S.
	1962 Linus C. Pauling, U.S.	1922 Fridtjof Nansen, Nor.
1992 Rigoberta Menchú, Guat.	1961 Dag Hammarskjold, Swed.	1921 Karl H. Branting, Swed.;
1991 Aung San Suu Kyi, Myanmarese	1960 Albert J. Luthuli, S. Afr.	Christian L. Lange, Nor.
1990 Mikhail S. Gorbachev, USSR	1959 Philip J. Noel-Baker, Br.	1920 Leon V.A. Bourgeois, Fr.
1989 Dalai Lama, Tibet	1958 Georges Pire, Belg.	1919 Woodrow Wilson, U.S.
1988 UN Peacekeeping Forces	1957 Lester B. Pearson, Can.	1917 International Red Cross
1987 Oscar Arias Sanchez, Costa Rican	1954 Office of UN High Com. for Refugees	1913 Henri La Fontaine, Belg.
1986 Elie Wiesel, Rom.-U.S.	1953 George C. Marshall, U.S.	1912 Elihu Root, U.S.
1985 Intl. Physicians for the Prevention of Nuclear War, U.S.	1952 Albert Schweitzer, Fr.	1911 Tobias M.C. Asser, Dutch; Alfred H.
	1951 Leon Jouhaux, Fr.	Fried, Austrian
1984 Bishop Desmond Tutu, S. Afr.	1950 Ralph J. Bunche, U.S.	1910 Permanent Intl. Peace Bureau
1983 Lech Walesa, Pol.	1949 Lord John Boyd Orr of Brechin	1909 Auguste M. F. Beernaert, Belg.; Paul
1982 Alva Myrdal, Swed.; Alfonso Garcia Robles, Mex.	Mearns, Br.	H. B. B. d'Estournelles de Constant, Fr.
	1947 Friends Service Council, Br.; Amer.	1908 Klas P. Arnoldson, Swed.;
1981 Office of UN High Com. for Refugees	Friends Service Committee, U.S.	Fredrik Bajer, Dan.
1980 Adolfo Perez Esquivel, Arg.	1946 Emily G. Balch, John R. Mott, both U.S.	1907 Ernesto T. Moneta, It.; Louis Renault, Fr.
1979 Mother Teresa of Calcutta, Alb.-Ind.	1945 Cordell Hull, U.S.	1906 Theodore Roosevelt, U.S.
1978 Anwar Sadat, Egy.; Menachem Begin, Isr.	1944 International Red Cross	1905 Baroness Bertha von Suttner,
	1938 Nansen International Office	Austrian
1977 Amnesty International	for Refugees	1904 Institute of International Law
1976 Mairead Corrigan, Betty Williams, both N. Ir.	1937 Viscount Cecil of Chelwood, Br.	1903 Sir William R. Cremer, Br.
	1936 Carlos de Saavedra Lamas, Arg.	1902 Elie Ducommun, Charles A. Gobat,
1975 Andrei Sakharov, USSR	1935 Carl von Ossietzky, Ger.	both Swiss
1974 Eisaku Sato, Jpn.; Sean MacBride, Ir.	1934 Arthur Henderson, Br.	1901 Jean H. Dunant, Swiss; Frederic
	1933 Sir Norman Angell, Br.	Passy, Fr.

Nobel Memorial Prize in Economic Science

1998 Amartya Sen, Indian	1989 Trygve Haavelmo, Nor.	1977 Bertil Ohlin, Swed.; James E. Meade,
1997 Robert C. Merton, U.S.; Myron S. Scholes, Can.-U.S.	1988 Maurice Allais, Fr.	Br.
	1987 Robert M. Solow, U.S.	1976 Milton Friedman, U.S.
1996 James A. Mirrlees, Br.; William Vickrey, Can.-U.S.	1986 James M. Buchanan, U.S.	1975 Tjalling Koopmans, Dutch-U.S.; Leonid Kantorovich, USSR
1995 Robert E. Lucas Jr., U.S.	1985 Franco Modigliani, It.-U.S.	1974 Gunnar Myrdal, Swed.; Friedrich A.
1994 John C. Harsanyi, John F. Nash, both U.S.; Reinhard Selten, Ger.	1984 Richard Stone, Br.	von Hayek, Austrian
	1983 Gerard Debreu, Fr.-U.S.	1973 Wassily Leontief, U.S.
1993 Robert W. Fogel, Douglass C. North, both U.S.	1982 George J. Stigler, U.S.	1972 Kenneth J. Arrow, U.S.; John R. Hicks, Br.
	1981 James Tobin, U.S.	
1992 Gary S. Becker, U.S.	1980 Lawrence R. Klein, U.S.	1971 Simon Kuznets, U.S.
1991 Ronald H. Coase, Br.-U.S.	1979 Theodore W. Schultz, U.S.; Sir Arthur	1970 Paul A. Samuelson, U.S.
1990 Harry M. Markowitz, William F. Sharpe, Merton H. Miller, all U.S.	Lewis, Br.	1969 Ragnar Frisch, Nor.; Jan Tinbergen, Dutch
	1978 Herbert A. Simon, U.S.	

Pulitzer Prizes in Journalism, Letters, and Music

The Pulitzer Prizes were endowed by Joseph Pulitzer (1847-1911), publisher of the *New York World*, in a bequest to Columbia Univ. and have been awarded annually, in years shown, for work the year before. Prizes are now $5,000 in each category, except Meritorious Public Service, for which a medal is given. If a year is omitted, no award was given that year.

Journalism

Meritorious Public Service

1918—New York Times. Also special award to Minna Lewinson and Henry Beetle Hough	**1952**—St. Louis Post-Dispatch
1919—Milwaukee Journal	**1953**—Whiteville (NC) News Reporter; Tabor City (NC) Tribune
1921—Boston Post	**1954**—Newsday (Long Island, NY)
1922—New York World	**1955**—Columbus (GA) Ledger and Sunday Ledger-Enquirer
1923—Memphis (TN) Commercial Appeal	**1956**—Watsonville (CA) Register-Pajaronian
1924—New York World	**1957**—Chicago Daily News
1926—Enquirer-Sun, Columbus, GA	**1958**—Arkansas Gazette, Little Rock
1927—Canton (OH) Daily News	**1959**—Utica (NY) Observer-Dispatch and Utica Daily Press
1928—Indianapolis (IN) Times	**1960**—Los Angeles Times
1929—New York Evening World	**1961**—Amarillo (TX) Globe-Times
1931—Atlanta (GA) Constitution	**1962**—Panama City (FL) News-Herald
1932—Indianapolis (IN) News	**1963**—Chicago Daily News
1933—New York World-Telegram	**1964**—St.Petersburg (FL) Times
1934—Medford (OR) Mail-Tribune	**1965**—Hutchinson (KS) News
1935—Sacramento (CA) Bee	**1966**—Boston Globe
1936—Cedar Rapids (IA) Gazette	**1967**—Louisville (KY) Courier-Journal; Milwaukee Journal
1937—St.Louis Post-Dispatch	**1968**—Riverside (CA) Press-Enterprise
1938—Bismarck (ND) Tribune	**1969**—Los Angeles Times
1939—Miami (FL) Daily News	**1970**—Newsday (Long Island, NY)
1940—Waterbury (CT) Republican and American	**1971**—Winston-Salem (NC) Journal & Sentinel
1941—St.Louis Post-Dispatch	**1972**—New York Times
1942—Los Angeles Times	**1973**—Washington Post
1943—Omaha World Herald	**1974**—Newsday (Long Island, NY)
1944—New York Times	**1975**—Boston Globe
1945—Detroit Free Press	**1976**—Anchorage (AK) Daily News
1946—Scranton (PA) Times	**1977**—Lufkin (TX) News
1947—Baltimore Sun	**1978**—Philadelphia Inquirer
1948—St. Louis Post-Dispatch	**1979**—Point Reyes (CA) Light
1949—Nebraska State Journal	**1980**—Gannett News Service
1950—Chicago Daily News; St. Louis Post-Dispatch	**1981**—Charlotte (NC) Observer
1951—Miami (FL) Herald and Brooklyn Eagle	**1982**—Detroit News
	1983—Jackson (MS) Clarion-Ledger
	1984—Los Angeles Times

1985—Ft. Worth (TX) Star-Telegram
1986—Denver Post
1987—Pittsburgh Press
1988—Charlotte (NC) Observer
1989—Anchorage (AK) Daily News
1990—Philadelphia Inquirer; Washington (NC) Daily News
1991—Des Moines Register
1992—Sacramento (CA) Bee
1993—Miami (FL) Herald
1994—Akron (OH) Beacon Journal
1995—Virgin Islands Daily News, St. Thomas
1996—News & Observer, Raleigh (NC)
1997—New Orleans Times-Picayune
1998—Grand Forks (ND) Herald
1999—Washington Post

Reporting

This category originally embraced all fields. Later, separate categories were made for national and international reporting.
1917—Herbert Bayard Swope, New York World
1918—Harold A. Littledale, New York Evening Post
1920—John J. Leary Jr., New York World
1921—Louis Seibold, New York World
1922—Kirke L. Simpson, Associated Press (AP)
1923—Alva Johnston, New York Times
1924—Magner White, San Diego Sun
1925—James W. Mulroy, Alvin H. Goldstein, Chicago Daily News
1926—William Burke Miller, Louisville (KY) Courier—Journal
1927—John T. Rogers, St. Louis Post Dispatch
1929—Paul Y. Anderson, St. Louis PostDispatch
1930—Russell D. Owens, New York Times. Also $500 to W.O. Dapping, Auburn (NY) Citizen
1931—A.B. MacDonald, Kansas City Star
1932—W.C. Richards, D.D. Martin, J.S. Pooler, F.D. Webb, J.N.W. Sloan, Detroit Free Press
1933—Francis A. Jamieson, AP
1934—Royce Brier, San Francisco Chronicle
1935—William H. Taylor, New York Herald Tribune
1936—Lauren D. Lyman, New York Times
1937—John J. O'Neill, NY Herald Tribune; William L. Laurence, NY Times; Howard W. Blakeslee, AP; Gobind Behari Lal, Universal Service; and David Dietz, Scripps-Howard Newspapers
1938—Raymond Sprigle, Pittsburgh Post-Gazette
1939—Thomas L. Stokes, Scripps-Howard Newspaper Alliance
1940—S. Burton Heath, New York World-Telegram
1941—Westbrook Pegler, New York World-Telegram
1942—Stanton Delaplane, San Francisco Chronicle
1943—George Weller, Chicago Daily News
1944—Paul Schoenstein, New York Journal-American
1945—Jack S. McDowell, San Francisco Call-Bulletin
1946—William L. Laurence, New York Times
1947—Frederick Woltman, New York World-Telegram
1948—George E. Goodwin, Atlanta Journal
1949—Malcolm Johnson, New York Sun
1950—Meyer Berger, New York Times
1951—Edward S. Montgomery, San Francisco Examiner
1952—George de Carvalho, San Francisco Chronicle

(1) General or Spot; (2) Special or Investigative
1953—(1) Providence (RI) Journal and Evening Bulletin; (2) Edward J. Mowery, New York World-Telegram & Sun
1954—(1) Vicksburg (MS) Sunday Post-Herald; (2) Alvin Scott McCoy, Kansas City Star
1955—(1) Mrs. Caro Brown, Alice (TX) Daily Echo; (2) Roland K. Towery, Cuero (TX) Record
1956—(1) Lee Hills, Detroit Free Press; (2) Arthur Daley, New York Times
1957—(1) Salt Lake Tribune; (2) Wallace Turner and William Lambert, Portland Oregonian
1958—(1) Fargo, (ND) Forum; (2) George Beveridge, Washington (DC) Evening Star
1959—(1) Mary Lou Werner, Washington (DC) Evening Star; (2) John Harold Brislin, Scranton (PA) Tribune, and The Scrantonian
1960—(1) Jack Nelson, Atlanta Constitution; (2) Miriam Ottenberg, Washington (DC) Evening Star
1961—(1) Sanche de Gramont, New York Herald Tribune; (2) Edgar May, Buffalo (NY) Evening News
1962—(1) Robert D. Mullins, Deseret News, Salt Lake City; (2) George Bliss, Chicago Tribune
1963—(1) Sylvan Fox, William Longgood, Anthony Shannon, New York World-Telegram & Sun; (2) Oscar Griffin Jr., Pecos (TX) Independent and Enterprise
1964—(1) Norman C. Miller, Wall Street Journal; (2) James V. Magee, Albert V. Gaudiosi, Frederick A. Meyer, Philadelphia Bulletin
1965—(1) Melvin H. Ruder, Hungry Horse News, Columbia Falls, (MT); (2) Gene Goltz, Houston Post
1966—(1) Los Angeles Times staff; (2) John A. Frasca, Tampa (FL) Tribune
1967—(1) Robert V. Cox, Chambersburg (PA) Public Opinion; (2) Gene Miller, Miami (FL) Herald

1968—(1) Detroit Free Press staff; (2) J. Anthony Lukas, New York Times
1969—(1) John Fetterman, Louisville Courier-Journal and Times; (2) Albert L. Delugach, St. Louis Globe Democrat, and Denny Walsh, Life
1970—(1) Thomas Fitzpatrick, Chicago Sun-Times; (2) Harold Eugene Martin, Montgomery Advertiser & Alabama Journal
1971—(1) Akron (OH) Beacon Journal staff; (2) William Hugh Jones, Chicago Tribune
1972—(1) Richard Cooper, John Machacek, Rochester (NY) Times-Union; (2) Timothy Leland, Gerard M. O'Neill, Stephen A. Kurkjian, Anne De Santis, Boston Globe
1973—(1) Chicago Tribune; (2) Sun Newspapers of Omaha
1974—(1) Hugh F. Hough, Arthur M. Petacque, Chicago Sun-Times; (2) William Sherman, New York Daily News
1975—(1) Xenia (OH) Daily Gazette; (2) Indianapolis Star
1976—(1) Gene Miller, Miami (FL) Herald; (2) Chicago Tribune
1977—(1) Margo Huston, Milwaukee Journal; (2) Acel Moore, Wendell Rawls Jr., Philadelphia Inquirer
1978—(1) Richard Whitt, Louisville (KY) Courier-Journal; (2) Anthony R. Dolan, Stamford (CT) Advocate
1979—(1) San Diego (CA) Evening Tribune; (2) Gilbert M. Gaul, Elliot G. Jaspin, Pottsville (PA) Republican
1980—(1) Philadelphia Inquirer; (2) Stephen A. Kurkjian, Alexander B. Hawes Jr., Nils Bruzelius, Joan Vennochi, Robert M. Porterfield, Boston Globe
1981—(1) Longview (WA) Daily News staff; (2) Clark Hallas, Robert B. Lowe, Arizona Daily Star
1982—(1) Kansas City Star, Kansas City Times; (2) Paul Henderson, Seattle Times
1983—(1) Fort Wayne (IN) News-Sentinel; (2) Loretta Tofani, Washington Post
1984—(1) New York Newsday; (2) Boston Globe
1985—(1) Thomas Turcol, Virginian-Pilot and Ledger-Star, Norfolk, VA; (2) William K. Marimow, Philadelphia Inquirer; Lucy Morgan, Jack Reed, St. Petersburg (FL) Times
1986—(1) Edna Buchanan, Miami (FL) Herald; (2) Jeffrey A. Marx, Michael M. York, Lexington (KY) Herald-Leader
1987—(1) Akron (OH) Beacon Journal; (2) Daniel R. Biddle, H.G. Bissinger, Fredric N. Tulsky, Philadelphia Inquirer; John Woestendiek, Philadelphia Inquirer
1988—(1) Alabama Journal; Lawrence (MA) Eagle-Tribune; (2) Walt Bogdanich, Wall Street Journal
1989—(1) Louisville (KY) Courier-Journal; (2) Bill Dedman, Atlanta Journal and Constitution
1990—(1) San Jose (CA) Mercury News; (2) Lon Kilzer, Chris Ison, Minneapolis-St. Paul Star Tribune
1991—(1) Miami (FL) Herald; (2) Joseph T. Hallinan, Susan M. Headden, Indianapolis Star
1992—(1) New York Newsday; (2) Lorraine Adams, Dan Malone, Dallas Morning News
1993—(1) Los Angeles Times; Jeff Brazil, Steve Berry, Orlando (FL) Sentinel
1994—(1) New York Times staff; (2) Providence (RI) Journal-Bulletin staff
1995—(1) Los Angeles Times staff; (2) Brian Donovan, Stephanie Saul, New York Newsday
1996—(1) New York Times, Robert D. McFadden; (2) Orange County (CA) Register staff
1997—(1) Long Island (NY) Newsday, staff; (2) Eric Nalder, Deborah Nelson, Alex Tizon, Seattle Times
1998—(1) Los Angeles Times staff; (2) Gary Cohn, Will Englund, Baltimore Sun
1999—(1) The Hartford Courant staff; (2) The Miami Herald staff

Criticism (1) or Commentary (2)

1970—(1) Ada Louise Huxtable, New York Times; (2) Marquis W. Childs, St. Louis Post-Dispatch
1971—(1) Harold C. Schonberg, New York Times; (2) William A. Caldwell, The Record, Hackensack, NJ
1972—(1) Frank Peters Jr., St. Louis Post-Dispatch; (2) Mike Royko, Chicago Daily News
1973—(1) Ronald Powers, Chicago Sun-Times; (2) David S. Broder, Washington Post
1974—(1) Emily Genauer, New York Newsday; (2) Edwin A. Roberts Jr., National Observer
1975—(1) Roger Ebert, Chicago Sun Times; (2) Mary McGrory, Washington Star
1976—(1) Alan M. Kriegsman, Washington Post; (2) Walter W. (Red) Smith, New York Times
1977—(1) William McPherson, Washington Post; (2) George F. Will, Washington Post Writers Group
1978—(1) Walter Kerr, New York Times; (2) William Safire, New York Times
1979—(1) Paul Gapp, Chicago Tribune; (2) Russell Baker, New York Times
1980—(1) William A. Henry III, Boston Globe; (2) Ellen Goodman, Boston Globe
1981—(1) Jonathan Yardley, Washington Star; (2) Dave Anderson, New York Times

1982—(1) Martin Bernheimer, Los Angeles Times; (2) Art Buchwald, Los Angeles Times Syndicate
1983—(1) Manuela Hoelterhoff, Wall Street Journal; (2) Claude Sitton, Raleigh (NC) News & Observer
1984—(1) Paul Goldberger, New York Times; (2) Vermont Royster, Wall Street Journal
1985—(1) Howard Rosenberg, Los Angeles Times; (2) Murray Kempton, New York Newsday
1986—(1) Donal J. Henahan, New York Times; (2) Jimmy Breslin, New York Daily News
1987—(1) Richard Eder, Los Angeles Times; (2) Charles Krauthammer, Washington Post
1988—(1) Tom Shales, Washington Post; (2) Dave Barry, Miami (FL) Herald
1989—(1) Michael Skube, Raleigh, NC, News & Observer; (2) Clarence Page, Chicago Tribune
1990—(1) Allan Temko, San Francisco Chronicle; (2) Jim Murray, Los Angeles Times
1991—(1) David Shaw, Los Angeles Times; (2) Jim Hoagland, Washington Post
1992—(1) No award; (2) Anna Quindlen, New York Times
1993—(1) Michael Dirda, Washington Post; (2) Liz Balmaseda, Miami (FL) Herald
1994—(1) Lloyd Schwartz, Boston Phoenix; (2) William Raspberry, Washington Post
1995—(1) Margo Jefferson, New York Times; (2) Jim Dwyer, New York Newsday
1996—(1) Robert Campbell, Boston Globe; (2) E.R. Shipp, New York Daily News
1997—(1) Tim Page, Washington Post; (2) Eileen McNamara, Boston Globe
1998—(1) Michiko Kakutani, New York Times; (2) Mike McAlary, New York Daily News
1999—(1) Blair Kamin, Chicago Tribune; (2) Maureen Dowd, New York Times

National Reporting
1942—Louis Stark, New York Times
1944—Dewey L. Fleming, Baltimore Sun
1945—James B. Reston, New York Times
1946—Edward A. Harris, St. Louis Post-Dispatch
1947—Edward T. Folliard, Washington Post
1948—Bert Andrews, New York Herald Tribune; Nat S. Finney, Minneapolis Tribune
1949—Charles P. Trussell, New York Times
1950—Edwin O. Guthman, Seattle Times
1952—Anthony Leviero, New York Times
1953—Don Whitehead, AP
1954—Richard Wilson, Des Moines Register
1955—Anthony Lewis, Washington Daily News
1956—Charles L. Bartlett, Chattanooga (TN) Times
1957—James Reston, New York Times
1958—Relman Morin, AP; Clark Mollenhoff, Des Moines Register & Tribune
1959—Howard Van Smith, Miami (FL) News
1960—Vance Trimble, Scripps-Howard, Washington, DC
1961—Edward R. Cony, Wall Street Journal
1962—Nathan G. Caldwell, Gene S. Graham, Nashville Tennessean
1963—Anthony Lewis, New York Times
1964—Merriman Smith, UPI
1965—Louis M. Kohlmeier, Wall Street Journal
1966—Haynes Johnson, Washington (DC) Evening Star
1967—Monroe Karmin, Stanley Penn, Wall Street Journal
1968—Howard James, Christian Science Monitor; Nathan K. Kotz, Des Moines Register
1969—Robert Cahn, Christian Science Monitor
1970—William J. Eaton, Chicago Daily News
1971—Lucinda Franks, Thomas Powers, UPI
1972—Jack Anderson, United Feature Syndicate
1973—Robert Boyd, Clark Hoyt, Knight Newspapers
1974—James R. Polk, Washington (DC) Star-News; Jack White, Providence (RI) Journal-Bulletin
1975—Donald L. Barlett, James B. Steele, Philadelphia Inquirer
1976—James Risser, Des Moines Register
1977—Walter Mears, AP
1978—Gaylord D. Shaw, Los Angeles Times
1979—James Risser, Des Moines Register
1980—Charles Stafford, Bette Swenson Orsini, St. Petersburg (FL) Times
1981—John M. Crewdson, New York Times
1982—Rick Atkinson, Kansas City Times
1983—Boston Globe
1984—John Noble Wilford, New York Times
1985—Thomas J. Knudson, Des Moines Register
1986—Craig Flournoy, George Rodrigue, Dallas Morning News; Arthur Howe, Philadelphia Inquirer
1987—Miami (FL) Herald; New York Times
1988—Tim Weiner, Philadelphia Inquirer
1989—Donald L. Barlett, James B. Steele, Philadelphia Inquirer
1990—Ross Anderson, Bill Dietrich, Mary Ann Gwinn, Eric Nalder, Seattle Times

1991—Marjie Lundstrom, Rochelle Sharpe, Gannett News Service
1992—Jeff Taylor, Mike McGraw, Kansas City Star
1993—David Maraniss, Washington Post
1994—Eileen Welsome, Albuquerque Tribune
1995—Tony Horwitz, Wall Street Journal
1996—Alix M. Freedman, Wall Street Journal
1997—Wall Street Journal staff
1998—Russell Carollo, Jeff Nesmith, Dayton (OH) Daily News
1999—New York Times staff

International Reporting
1942—Laurence Edmund Allen, AP
1943—Ira Wolfert, North American Newspaper Alliance
1944—Daniel DeLuce, AP
1945—Mark S. Watson, Baltimore Sun
1946—Homer W. Bigart, New York Herald Tribune
1947—Eddy Gilmore, AP
1948—Paul W. Ward, Baltimore Sun
1949—Price Day, Baltimore Sun
1950—Edmund Stevens, Christian Science Monitor
1951—Keyes Beech, Fred Sparks, Chicago Daily News; Homer Bigart, Marguerite Higgins, New York Herald Tribune; Relman Morin, Don Whitehead, AP
1952—John M. Hightower, AP
1953—Austin C. Wehrwein, Milwaukee Journal
1954—Jim G. Lucas, Scripps-Howard Newspapers
1955—Harrison Salisbury, New York Times
1956—William Randolph Hearst Jr., Frank Conniff, Hearst Newspapers; Kingsbury Smith, INS
1957—Russell Jones, UPI
1958—New York Times
1959—Joseph Martin, Philip Santora, New York Daily News
1960—A.M. Rosenthal, New York Times
1961—Lynn Heinzerling, AP
1962—Walter Lippmann, New York Herald Tribune Syndicate
1963—Hal Hendrix, Miami (FL) News
1964—Malcolm W. Browne, AP; David Halberstam, New York Times
1965—J.A. Livingston, Philadelphia Bulletin
1966—Peter Arnett, AP
1967—R. John Hughes, Christian Science Monitor
1968—Alfred Friendly, Washington Post
1969—William Tuohy, Los Angeles Times
1970—Seymour M. Hersh, Dispatch News Service
1971—Jimmie Lee Hoagland, Washington Post
1972—Peter R. Kann, Wall Street Journal
1973—Max Frankel, New York Times
1974—Hedrick Smith, New York Times
1975—William Mullen and Ovie Carter, Chicago Tribune
1976—Sydney H. Schanberg, New York Times
1978—Henry Kamm, New York Times
1979—Richard Ben Cramer, Philadelphia Inquirer
1980—Joel Brinkley, Jay Mather, Louisville (KY) Courier-Journal
1981—Shirley Christian, Miami (FL) Herald
1982—John Darnton, New York Times
1983—Thomas L. Friedman, New York Times; Loren Jenkins, Washington Post
1984—Karen Elliot House, Wall Street Journal
1985—Josh Friedman, Dennis Bell, Ozler Muhammad, New York Newsday
1986—Lewis M. Simons, Pete Carey, Katherine Ellison, San Jose (CA) Mercury News
1987—Michael Parks, Los Angeles Times
1988—Thomas L. Friedman, New York Times
1989—Glenn Frankel, Wash. Post; Bill Keller, NY Times
1990—Nicholas D. Kirstof, Sheryl WuDunn, NY Times
1991—Caryle Murphy, Washington Post; Serge Schmemann, New York Times
1992—Patrick J. Sloyan, New York Newsday
1993—John F. Burns, NY Times; Roy Gutman, NY Newsday
1994—Dallas Morning News team
1995—Mark Fritz, AP
1996—David Rohde, Christian Science Monitor
1997—John F. Burns, New York Times
1998—New York Times staff
1999—Wall Street Journal staff

Washington or Foreign Correspondence
Category was merged with others in 1948.
1929—Paul Scott Mowrer, Chicago Daily News
1930—Leland Stowe, New York Herald Tribune
1931—H.R. Knickerbocker, Philadelphia Public Ledger and New York Evening Post
1932—Walter Duranty, New York Times; Charles G. Ross, St. Louis Post-Dispatch
1933—Edgar Ansel Mowrer, Chicago Daily News
1934—Frederick T. Birchall, New York Times
1935—Arthur Krock, New York Times
1936—Wilfred C. Barber, Chicago Tribune
1937—Anne O'Hare McCormick, New York Times
1938—Arthur Krock, New York Times
1939—Louis P. Lochner, AP
1940—Otto D. Tolischus, New York Times
1941—Bronze plaque to commemorate work of American correspondents on war fronts

1942—Carlos P. Romulo, Philippines Herald
1943—Hanson W. Baldwin, New York Times
1944—Ernest Taylor Pyle, Scripps-Howard Newspaper Alliance
1945—Harold V. (Hal) Boyle, AP
1946—Arnaldo Cortesi, New York Times
1947—Brooks Atkinson, New York Times

Editorial Writing

1917—New York Tribune
1918—Louisville (KY) Courier-Journal
1920—Harvey E. Newbranch, Omaha Evening World-Herald
1922—Frank M. O'Brien, New York Herald
1923—William Allen White, Emporia (KS) Gazette
1924—Frank Buxton, Boston Herald, Special Prize; Frank I. Cobb, New York World
1925—Robert Lathan, Charleston (SC) News and Courier
1926—Edward M. Kingsbury, New York Times
1927—F. Lauriston Bullard, Boston Herald
1928—Grover C. Hall, Montgomery (AL) Advertiser
1929—Louis Isaac Jaffe, Norfolk Virginian-Pilot
1931—Chas. Ryckman, Fremont (NE) Tribune
1933—Kansas City Star
1934—E. P. Chase, Atlantic (IA) News Telegraph
1936—Felix Morley, Washington Post; George B. Parker, Scripps-Howard Newspapers
1937—John W. Owens, Baltimore Sun
1938—W.W. Waymack, Des Moines Register & Tribune
1939—Ronald G. Callvert, Portland Oregonian
1940—Bart Howard, St. Louis Post-Dispatch
1941—Reuben Maury, New York Daily News
1942—Geoffrey Parsons, New York Herald Tribune
1943—Forrest W. Seymour, Des Moines Register & Tribune
1944—Henry J. Haskell, Kansas City Star
1945—George W. Potter, Providence (RI) Journal-Bulletin
1946—Hodding Carter, Greenville (MS) Delta Democrat-Times
1947—William H. Grimes, Wall Street Journal
1948—Virginius Dabney, Richmond (VA) Times-Dispatch
1949—John H. Crider, Boston Herald; Herbert Elliston, Washington Post
1950—Carl M. Saunders, Jackson (MI) Citizen-Patriot
1951—William H. Fitzpatrick, New Orleans States
1952—Louis LaCoss, St. Louis Globe Democrat
1953—Vermont C. Royster, Wall Street Journal
1954—Don Murray, Boston Herald
1955—Royce Howes, Detroit Free Press
1956—Lauren K. Soth, Des Moines Register & Tribune
1957—Buford Boone, Tuscaloosa (AL) News
1958—Harry S. Ashmore, Arkansas Gazette
1959—Ralph McGill, Atlanta Constitution
1960—Lenoir Chambers, Norfolk Virginian-Pilot
1961—William J. Dorvillier, San Juan (Puerto Rico) Star
1962—Thomas M. Storke, Santa Barbara (CA) News-Press
1963—Ira B. Harkey Jr., Pascagoula (MS) Chronicle
1964—Hazel Brannon Smith, Lexington (MS) Advertiser
1965—John R. Harrison, Gainesville (FL) Sun
1966—Robert Lasch, St. Louis Post-Dispatch
1967—Eugene C. Patterson, Atlanta Constitution
1968—John S. Knight, Knight Newspapers
1969—Paul Greenberg, Pine Bluff (AR) Commercial
1970—Philip L. Geyelin, Washington Post
1971—Horance G. Davis Jr., Gainesville (FL) Sun
1972—John Strohmeyer, Bethlehem (PA) Globe-Times
1973—Roger B. Linscott, Berkshire Eagle, Pittsfield, MA
1974—F. Gilman Spencer, Trenton (NJ) Trentonian
1975—John D. Maurice, Charleston (WV) Daily Mail
1976—Philip Kerby, Los Angeles Times
1977—Warren L. Lerude, Foster Church, Norman F. Cardoza, Reno Evening Gazette and Nevada State Journal
1978—Meg Greenfield, Washington Post
1979—Edwin M. Yoder, Washington Star
1980—Robert L. Bartley, Wall Street Journal
1982—Jack Rosenthal, New York Times
1983—Editorial board, Miami Herald
1984—Albert Scardino, Georgia Gazette
1985—Richard Aregood, Philadelphia Daily News
1986—Jack Fuller, Chicago Tribune
1987—Jonathan Freedman, Tribune (San Diego)
1988—Jane Healy, Orlando (FL) Sentinel
1989—Lois Wille, Chicago Tribune
1990—Thomas J. Hylton, Pottstown (PA) Mercury
1991—Ron Casey, Harold Jackson, Joey Kennedy, Birmingham (AL) News
1992—Maria Henson, Lexington (KY) Herald-Leader
1994—R. Bruce Dold, Chicago Tribune
1995—Jeffrey Good, St. Petersburg (FL) Times
1996—Robert B. Semple Jr., New York Times
1997—Michael Gartner, Ames (IA) Daily Tribune
1998—Bernard L. Stein, Riverdale (NY) Press
1999—Editorial Board of the Daily News, New York, NY

Editorial Cartooning

1922—Rollin Kirby, New York World
1924—Jay N. Darling, Des Moines Register
1925—Rollin Kirby, New York World
1926—D. R. Fitzpatrick, St. Louis Post-Dispatch
1927—Nelson Harding, Brooklyn Eagle
1928—Nelson Harding, Brooklyn Eagle
1929—Rollin Kirby, New York World
1930—Charles Macauley, Brooklyn Eagle
1931—Edmund Duffy, Baltimore Sun
1932—John T. McCutcheon, Chicago Tribune
1933—H. M. Talburt, Washington Daily News
1934—Edmund Duffy, Baltimore Sun
1935—Ross A. Lewis, Milwaukee Journal
1937—C. D. Batchelor, New York Daily News
1938—Vaughn Shoemaker, Chicago Daily News
1939—Charles G. Werner, Daily Oklahoman
1940—Edmund Duffy, Baltimore Sun
1941—Jacob Burck, Chicago Times
1942—Herbert L. Block, Newspaper Enterprise Assn.
1943—Jay N. Darling, Des Moines Register
1944—Clifford K. Berryman, Washington Star
1945—Bill Mauldin, United Feature Syndicate
1946—Bruce Alexander Russell, Los Angeles Times
1947—Vaughn Shoemaker, Chicago Daily News
1948—Reuben L. (Rube) Goldberg, New York Sun
1949—Lute Pease, Newark (NJ) Evening News
1950—James T. Berryman, Washington Star
1951—Reginald W. Manning, Arizona Republic
1952—Fred L. Packer, New York Mirror
1953—Edward D. Kuekes, Cleveland Plain Dealer
1954—Herbert L. Block, Washington Post & Times-Herald
1955—Daniel R. Fitzpatrick, St. Louis Post-Dispatch
1956—Robert York, Louisville (KY) Times
1957—Tom Little, Nashville Tennessean
1958—Bruce M. Shanks, Buffalo (NY) Evening News
1959—Bill Mauldin, St. Louis Post-Dispatch
1961—Carey Orr, Chicago Tribune
1962—Edmund S. Valtman, Hartford (CT) Times
1963—Frank Miller, Des Moines Register
1964—Paul Conrad, Denver Post
1966—Don Wright, Miami (FL) News
1967—Patrick B. Oliphant, Denver Post
1968—Eugene Gray Payne, Charlotte (NC) Observer
1969—John Fischetti, Chicago Daily News
1970—Thomas F. Darcy, New York Newsday
1971—Paul Conrad, Los Angeles Times
1972—Jeffrey K. MacNelly, Richmond (VA) News-Leader
1974—Paul Szep, Boston Globe
1975—Garry Trudeau, Universal Press Syndicate
1976—Tony Auth, Philadelphia Inquirer
1977—Paul Szep, Boston Globe
1978—Jeffrey K. MacNelly, Richmond (VA) News Leader
1979—Herbert L. Block, Washington Post
1980—Don Wright, Miami (FL) News
1981—Mike Peters, Dayton (OH) Daily News
1982—Ben Sargent, Austin (TX) American-Statesman
1983—Richard Locher, Chicago Tribune
1984—Paul Conrad, Los Angeles Times
1985—Jeffrey K. MacNelly, Chicago Tribune
1986—Jules Feiffer, Village Voice (NY)
1987—Berke Breathed, Washington Post
1988—Doug Marlette, Atlanta Constitution, Charlotte (NC) Observer
1989—Jack Higgins, Chicago Sun-Times
1990—Tom Toles, Buffalo (NY) News
1991—Jim Borgman, Cincinnati Enquirer
1992—Signe Wilkinson, Philadelphia Daily News
1993—Stephen R. Benson, Arizona Republic
1994—Michael P. Ramirez, Commercial Appeal, Memphis, TN
1995—Mike Luckovich, Atlanta Constitution
1996—Jim Morin, Miami (FL) Herald
1997—Walt Handelsman, New Orleans Times-Picayune
1998—Stephen P. Breen, Asbury Park Press, Neptune, NJ
1999—David Horsey, Seattle Post-Intelligencer

Spot News Photography

1942—Milton Brooks, Detroit News
1943—Frank Noel, AP
1944—Frank Filan, AP; Earl L. Bunker, Omaha World-Herald
1945—Joe Rosenthal, AP
1947—Arnold Hardy, amateur, Atlanta, GA
1948—Frank Cushing, Boston Traveler
1949—Nathaniel Fein, New York Herald Tribune
1950—Bill Crouch, Oakland (CA) Tribune
1951—Max Desfor, AP
1952—John Robinson, Don Ultang, Des Moines Register & Tribune
1953—William M. Gallagher, Flint (MI) Journal
1954—Mrs. Walter M. Schau, amateur
1955—John L. Gaunt Jr., Los Angeles Times
1956—New York Daily News
1957—Harry A. Trask, Boston Traveler
1958—William C. Beall, Washington Daily News
1959—William Seaman, Minneapolis Star
1960—Andrew Lopez, UPI
1961—Yasushi Nagao, Mainichi Newspapers, Tokyo
1962—Paul Vathis, AP
1963—Hector Rondon, La Republica, Caracas, Venezuela
1964—Robert H. Jackson, Dallas Times-Herald

1965—Horst Faas, AP
1966—Kyoichi Sawada, UPI
1967—Jack R. Thornell, AP
1968—Rocco Morabito, Jacksonville (FL) Journal
1969—Edward Adams, AP
1970—Steve Starr, AP
1971—John Paul Filo, Valley Daily News & Daily Dispatch of Tarentum & New Kensington (PA)
1972—Horst Faas, Michel Laurent, AP
1973—Huynh Cong Ut, AP
1974—Anthony K. Roberts, AP
1975—Gerald H. Gay, Seattle Times
1976—Stanley Forman, Boston Herald American
1977—Neal Ulevich, AP; Stanley Forman, Boston Herald American
1978—John H. Blair, UPI
1979—Thomas J. Kelly III, Pottstown (PA) Mercury
1980—UPI
1981—Larry C. Price, Ft. Worth (TX) Star-Telegram
1982—Ron Edmonds, AP
1983—Bill Foley, AP
1984—Stan Grossfeld, Boston Globe
1985—The Register, Santa Ana, CA
1986—Carol Guzy, Michel duCille, Miami (FL) Herald
1987—Kim Komenich, San Francisco Examiner
1988—Scott Shaw, Odessa (TX) American
1989—Ron Olshwanger, St. Louis Post-Dispatch
1990—Oakland (CA) Tribune photo staff
1991—Greg Marinovich, AP
1992—Associated Press staff
1993—Ken Geiger, William Snyder, Dallas Morning News
1994—Paul Watson, Toronto Star
1995—Carol Guzy, Washington Post
1996—Charles Porter IV, AP
1997—Annie Wells, Santa Rosa (CA) Press Democrat
1998—Martha Rial, Pittsburgh Post-Gazette
1999—Associated Press photo staff

Feature Photography
1968—Toshio Sakai, UPI
1969—Moneta Sleet Jr., Ebony
1970—Dallas Kinney, Palm Beach (FL) Post
1971—Jack Dykinga, Chicago Sun-Times
1972—Dave Kennerly, UPI
1973—Brian Lanker, Topeka (KS) Capitol-Journal
1974—Slava Veder, AP
1975—Matthew Lewis, Washington Post
1976—Louisville (KY) Courier-Journal and Louisville Times
1977—Robin Hood, Chattanooga (TN) News-Free Press
1978—J. Ross Baughman, AP
1979—Staff photographers, Boston Herald American
1980—Erwin H. Hagler, Dallas Times-Herald
1981—Taro M. Yamasaki, Detroit Free Press
1982—John H. White, Chicago Sun-Times
1983—James B. Dickman, Dallas Times-Herald
1984—Anthony Suad, Denver Post
1985—Stan Grossfeld, Boston Globe; Larry C. Price, Phila. Inquirer
1986—Tom Gralish, Philadelphia Inquirer
1987—David Peterson, Des Moines Register
1988—Michel duCille, Miami (FL) Herald
1989—Manny Crisostomo, Detroit Free Press
1990—David C. Turnley, Detroit Free Press
1991—William Snyder, Dallas Morning News
1992—John Kaplan, Block Newspapers (Toledo, OH)
1993—AP staff
1994—Kevin Carter, New York Times
1995—AP staff
1996—Stephanie Welsh, Newhouse News Service
1997—Alexander Zemlianichenko, AP
1998—Clarence Williams, Los Angeles Times
1999—Associated Press photo staff

Special Citation
1930—William O. Dapping, Auburn (NY) Citizen
1938—Edmonton (Alberta) Journal, bronze plaque
1941—New York Times

Fiction
1918—Ernest Poole, *His Family*
1919—Booth Tarkington, *The Magnificent Ambersons*
1921—Edith Wharton, *The Age of Innocence*
1922—Booth Tarkington, *Alice Adams*
1923—Willa Cather, *One of Ours*
1924—Margaret Wilson, *The Able McLaughlins*
1925—Edna Ferber, *So Big*
1926—Sinclair Lewis, *Arrowsmith* (refused prize)
1927—Louis Bromfield, *Early Autumn*
1928—Thornton Wilder, *Bridge of San Luis Rey*
1929—Julia M. Peterkin, *Scarlet Sister Mary*
1930—Oliver LaFarge, *Laughing Boy*
1931—Margaret Ayer Barnes, *Years of Grace*
1932—Pearl S. Buck, *The Good Earth*
1933—T. S. Stribling, *The Store*
1934—Caroline Miller, *Lamb in His Bosom*
1935—Josephine W. Johnson, *Now in November*

1944—Byron Price and Mrs. William Allen White
1945—American Press cartographers, for war maps
1947—(Pulitzer centennial year) Columbia Univ. and the Graduate School of Journalism; St. Louis Post-Dispatch
1948—Dr. Frank Diehl Fackenthal
1951—C(yrus) L. Sulzberger, New York Times
1952—Max Kase, New York Journal-American; Kansas City Star
1953—New York Times, Lester Markel
1958—Walter Lippmann, New York Herald Tribune
1964—Gannett Newspapers, "The Road to Integration"
1976—Prof. John Hohenberg, Admin. of Pulitzer Prizes
1978—Richard Lee Strout, Christian Science Monitor and New Republic
1987—Joseph Pulitzer Jr.
1996—Herb Caen, San Francisco Chronicle

Feature Writing
1979—Jon D. Franklin, Baltimore Evening Sun
1980—Madeleine Blais, Miami (FL) Herald Tropic Magazine;
1981—Teresa Carpenter, Village Voice, New York City
1982—Saul Pett, AP
1984—Peter M. Rinearson, Seattle Times
1985—Alice Steinbach, Baltimore Sun
1986—John Camp, St. Paul Pioneer Press & Dispatch
1987—Steve Twomey, Philadelphia Inquirer
1988—Jacqui Banaszynski, St. Paul Pioneer Press Dispatch
1989—David Zucchino, Philadelphia Inquirer
1990—Dave Curtin, Colorado Springs Gazette Telegraph
1991—Sheryl James, St. Petersburg (FL) Times
1992—Howell Raines, New York Times
1993—George Lardner Jr., Washington Post
1994—Isabel Wilkerson, New York Times
1995—Ron Suskind, Wall Street Journal
1996—Rick Bragg, New York Times
1997—Lisa Pollak, Baltimore Sun
1998—Thomas French, St. Petersburg (FL) Times
1999—Angelo B. Henderson, Wall Street Journal

Explanatory Reporting
1985—Jon Franklin, Baltimore Evening Sun
1986—New York Times staff
1987—Jeff Lyon, Peter Gorner, Chicago Tribune
1988—Daniel Hertzberg, James B. Stewart, Wall Street Journal
1989—David Hanners, William Snyder, Karen Blessen, Dallas Morning News
1990—David A. Vise, Steve Coll, Washington Post
1991—Susan C. Faludi, Wall Street Journal
1992—Robert S. Capers, Eric Lipton, Hartford (CT) Courant
1993—Mike Toner, Atlanta Journal-Constitution
1994—Ronald Kotulak, Chicago Tribune
1995—Leon Dash, Lucian Perkins, Washington Post
1996—Laurie Garrett, New York Newsday
1997—Michael Vitez, Ron Cortes, April Saul, Phila. Inquirer
1998—Paul Salopek, Chicago Tribune
1999—Richard Read, Oregonian (Portland)

Specialized Reporting (1985-90)
1985—Randall Savage, Jackie Crosby, Macon (GA) Tel. & News
1986—Andrew Schneider & Mary Pat Flaherty, Pittsburgh Press
1987—Alex S. Jones, New York Times
1988—Dean Baquet, William Gaines, Ann Marie Lipinski, Chicago Tribune
1989—Edward Humes, Orange County (CA) Register
1990—Tamar Stieber, Albuquerque Journal

Beat Reporting
1991—Natalie Angier, New York Times
1992—Deborah Blum, Sacramento (CA) Bee
1993—Paul Ingrassia, Joseph B. White, Wall Street Journal
1994—Eric Freedman, Jim Mitzelfeld, Detroit News
1995—David Shribman, Boston Globe
1996—Bob Keeler, New York Newsday
1997—Byron Acohido, Seattle Times
1998—Linda Greenhouse, New York Times
1999—Chuck Philips and Michael A. Hiltzik, Los Angeles Times

Letters

1936—Harold L. Davis, *Honey in the Horn*
1937—Margaret Mitchell, *Gone With the Wind*
1938—John P. Marquand, *The Late George Apley*
1939—Marjorie Kinnan Rawlings, *The Yearling*
1940—John Steinbeck, *The Grapes of Wrath*
1942—Ellen Glasgow, *In This Our Life*
1943—Upton Sinclair, *Dragon's Teeth*
1944—Martin Flavin, *Journey in the Dark*
1945—John Hersey, *A Bell for Adano*
1947—Robert Penn Warren, *All the King's Men*
1948—James A. Michener, *Tales of the South Pacific*
1949—James Gould Cozzens, *Guard of Honor*
1950—A. B. Guthrie Jr., *The Way West*
1951—Conrad Richter, *The Town*
1952—Herman Wouk, *The Caine Mutiny*
1953—Ernest Hemingway, *The Old Man and the Sea*
1955—William Faulkner, *A Fable*
1956—MacKinlay Kantor, *Andersonville*

1958—James Agee, *A Death in the Family*
1959—Robert Lewis Taylor, *The Travels of Jaimie McPheeters*
1960—Allen Drury, *Advise and Consent*
1961—Harper Lee, *To Kill a Mockingbird*
1962—Edwin O'Connor, *The Edge of Sadness*
1963—William Faulkner, *The Reivers*
1965—Shirley Ann Grau, *The Keepers of the House*
1966—Katherine Anne Porter, *Collected Stories*
1967—Bernard Malamud, *The Fixer*
1968—William Styron, *The Confessions of Nat Turner*
1969—N. Scott Momaday, *House Made of Dawn*
1970—Jean Stafford, *Collected Stories*
1972—Wallace Stegner, *Angle of Repose*
1973—Eudora Welty, *The Optimist's Daughter*
1975—Michael Shaara, *The Killer Angels*
1976—Saul Bellow, *Humboldt's Gift*
1978—James Alan McPherson, *Elbow Room*
1979—John Cheever, *The Stories of John Cheever*
1980—Norman Mailer, *The Executioner's Song*
1981—John Kennedy Toole, *A Confederacy of Dunces*
1982—John Updike, *Rabbit Is Rich*
1983—Alice Walker, *The Color Purple*
1984—William Kennedy, *Ironweed*
1985—Alison Lurie, *Foreign Affairs*
1986—Larry McMurtry, *Lonesome Dove*
1987—Peter Taylor, *A Summons to Memphis*
1988—Toni Morrison, *Beloved*
1989—Anne Tyler, *Breathing Lessons*
1990—Oscar Hijuelos, *The Mambo Kings Play Songs of Love*
1991—John Updike, *Rabbit at Rest*
1992—Jane Smiley, *A Thousand Acres*
1993—Robert Olen Butler, *A Good Scent From a Strange Mountain*
1994—E. Annie Proulx, *The Shipping News*
1995—Carol Shields, *The Stone Diaries*
1996—Richard Ford, *Independence Day*
1997—Steven Millhauser, *Martin Dressler: The Tale of an American Dreamer*
1998—Philip Roth, *American Pastoral*
1999—Michael Cunningham, *The Hours*

Drama

1918—Jesse Lynch Williams, *Why Marry?*
1920—Eugene O'Neill, *Beyond the Horizon*
1921—Zona Gale, *Miss Lulu Bett*
1922—Eugene O'Neill, *Anna Christie*
1923—Owen Davis, *Icebound*
1924—Hatcher Hughes, *Hell-Bent for Heaven*
1925—Sidney Howard, *They Knew What They Wanted*
1926—George Kelly, *Craig's Wife*
1927—Paul Green, *In Abraham's Bosom*
1928—Eugene O'Neill, *Strange Interlude*
1929—Elmer Rice, *Street Scene*
1930—Marc Connelly, *The Green Pastures*
1931—Susan Glaspell, *Alison's House*
1932—George S. Kaufman, Morrie Ryskind, and Ira Gershwin, *Of Thee I Sing*
1933—Maxwell Anderson, *Both Your Houses*
1934—Sidney Kingsley, *Men in White*
1935—Zoe Akins, *The Old Maid*
1936—Robert E. Sherwood, *Idiot's Delight*
1937—George S. Kaufman and Moss Hart, *You Can't Take It With You*
1938—Thornton Wilder, *Our Town*
1939—Robert E. Sherwood, *Abe Lincoln in Illinois*
1940—William Saroyan, *The Time of Your Life*
1941—Robert E. Sherwood, *There Shall Be No Night*
1943—Thornton Wilder, *The Skin of Our Teeth*
1945—Mary Chase, *Harvey*
1946—Russel Crouse and Howard Lindsay, *State of the Union*
1948—Tennessee Williams, *A Streetcar Named Desire*
1949—Arthur Miller, *Death of a Salesman*
1950—Richard Rodgers, Oscar Hammerstein 2d and Joshua Logan, *South Pacific*
1952—Joseph Kramm, *The Shrike*
1953—William Inge, *Picnic*
1954—John Patrick, *Teahouse of the August Moon*
1955—Tennessee Williams, *Cat on a Hot Tin Roof*
1956—Frances Goodrich and Albert Hackett, *The Diary of Anne Frank*
1957—Eugene O'Neill, *Long Day's Journey Into Night*
1958—Ketti Frings, *Look Homeward, Angel*
1959—Archibald MacLeish, *J. B.*
1960—George Abbott, Jerome Weidman, Sheldon Harnick, and Jerry Bock, *Fiorello!*
1961—Tad Mosel, *All the Way Home*
1962—Frank Loesser and Abe Burrows, *How to Succeed in Business Without Really Trying*
1965—Frank D. Gilroy, *The Subject Was Roses*
1967—Edward Albee, *A Delicate Balance*
1969—Howard Sackler, *The Great White Hope*
1970—Charles Gordone, *No Place to Be Somebody*

1971—Paul Zindel, *The Effect of Gamma Rays on Man-in-the-Moon Marigolds*
1973—Jason Miller, *That Championship Season*
1975—Edward Albee, *Seascape*
1976—Michael Bennett, James Kirkwood, Nicholas Dante, Marvin Hamlisch, and Edward Kleban, *A Chorus Line*
1977—Michael Cristofer, *The Shadow Box*
1978—Donald L. Coburn, *The Gin Game*
1979—Sam Shepard, *Buried Child*
1980—Lanford Wilson, *Talley's Folly*
1981—Beth Henley, *Crimes of the Heart*
1982—Charles Fuller, *A Soldier's Play*
1983—Marsha Norman, *'night, Mother*
1984—David Mamet, *Glengarry Glen Ross*
1985—Stephen Sondheim and James Lapine, *Sunday in the Park With George*
1987—August Wilson, *Fences*
1988—Alfred Uhry, *Driving Miss Daisy*
1989—Wendy Wasserstein, *The Heidi Chronicles*
1990—August Wilson, *The Piano Lesson*
1991—Neil Simon, *Lost in Yonkers*
1992—Robert Schenkkan, *The Kentucky Cycle*
1993—Tony Kushner, *Angels in America: Millennium Approaches*
1994—Edward Albee, *Three Tall Women*
1995—Horton Foote, *The Young Man From Atlanta*
1996—Jonathan Larson, *Rent*
1998—Paula Vogel, *How I Learned to Drive*
1999—Margaret Edson, *Wit*

History (U.S.)

1917—J. J. Jusserand, *With Americans of Past and Present Days*
1918—James Ford Rhodes, *History of the Civil War*
1920—Justin H. Smith, *The War With Mexico*
1921—William Sowden Sims, *The Victory at Sea*
1922—James Truslow Adams, *The Founding of New England*
1923—Charles Warren, *The Supreme Court in United States History*
1924—Charles Howard McIlwain, *The American Revolution: A Constitutional Interpretation*
1925—Frederick L. Paxton, *A History of the American Frontier*
1926—Edward Channing, *A History of the U.S.*
1927—Samuel Flagg Bemis, *Pinckney's Treaty*
1928—V. L Parrington, *Main Currents in American Thought*
1929—Fred A. Shannon, *The Organization and Administration of the Union Army, 1861-65*
1930—Claude H. Van Tyne, *The War of Independence*
1931—Bernadotte E. Schmitt, *The Coming of the War, 1914*
1932—Gen. John J. Pershing, *My Experiences in the World War*
1933—Frederick J. Turner, *The Significance of Sections in American History*
1934—Herbert Agar, *The People's Choice*
1935—Charles McLean Andrews, *The Colonial Period of American History*
1936—Andrew C. McLaughlin, *The Constitutional History of the United States*
1937—Van Wyck Brooks, *The Flowering of New England*
1938—Paul Herman Buck, *The Road to Reunion, 1865-1900*
1939—Frank Luther Mott, *A History of American Magazines*
1940—Carl Sandburg, *Abraham Lincoln: The War Years*
1941—Marcus Lee Hansen, *The Atlantic Migration, 1607-1860*
1942—Margaret Leech, *Reveille in Washington*
1943—Esther Forbes, *Paul Revere and the World He Lived In*
1944—Merle Curti, *The Growth of American Thought*
1945—Stephen Bonsal, *Unfinished Business*
1946—Arthur M. Schlesinger Jr., *The Age of Jackson*
1947—James Phinney Baxter 3d, *Scientists Against Time*
1948—Bernard De Voto, *Across the Wide Missouri*
1949—Roy F. Nichols, *The Disruption of American Democracy*
1950—O. W. Larkin, *Art and Life in America*
1951—R. Carlyle Buley, *The Old Northwest: Pioneer Period 1815-1840*
1952—Oscar Handlin, *The Uprooted*
1953—George Dangerfield, *The Era of Good Feelings*
1954—Bruce Catton, *A Stillness at Appomattox*
1955—Paul Horgan, *Great River: The Rio Grande in North American History*
1956—Richard Hofstadter, *The Age of Reform*
1957—George F. Kennan, *Russia Leaves the War*
1958—Bray Hammond, *Banks and Politics in America—From the Revolution to the Civil War*
1959—Leonard D. White and Jean Schneider, *The Republican Era; 1869-1901*
1960—Margaret Leech, *In the Days of McKinley*
1961—Herbert Feis, *Between War and Peace: The Potsdam Conference*
1962—Lawrence H. Gibson, *The Triumphant Empire: Thunderclouds Gather in the West*
1963—Constance McLaughlin Green, *Washington: Village and Capital, 1800-1878*
1964—Sumner Chilton Powell, *Puritan Village: The Formation of a New England Town*
1965—Irwin Unger, *The Greenback Era*
1966—Perry Miller, *Life of the Mind in America*

1967—William H. Goetzmann, *Exploration and Empire: The Explorer and Scientist in the Winning of the American West*
1968—Bernard Bailyn, *The Ideological Origins of the American Revolution*
1969—Leonard W. Levy, *Origin of the Fifth Amendment*
1970—Dean Acheson, *Present at the Creation: My Years in the State Department*
1971—James McGregor Burns, *Roosevelt: The Soldier of Freedom*
1972—Carl N. Degler, *Neither Black nor White*
1973—Michael Kammen, *People of Paradox: An Inquiry Concerning the Origins of American Civilization*
1974—Daniel J. Boorstin, *The Americans: The Democratic Experience*
1975—Dumas Malone, *Jefferson and His Time*
1976—Paul Horgan, *Lamy of Santa Fe*
1977—David M. Potter, *The Impending Crisis*
1978—Alfred D. Chandler Jr., *The Visible Hand: The Managerial Revolution in American Business*
1979—Don E. Fehrenbacher, *The Dred Scott Case: Its Significance in American Law and Politics*
1980—Leon F. Litwack, *Been in the Storm So Long*
1981—Lawrence A. Cremin, *American Education: The National Experience, 1783-1876*
1982—C. Vann Woodward, ed., *Mary Chesnut's Civil War*
1983—Rhys L. Issac, *The Transformation of Virginia, 1740-1790*
1985—Thomas K. McCraw, *Prophets of Regulation*
1986—Walter A. McDougall, *The Heavens and the Earth*
1987—Bernard Bailyn, *Voyagers to the West*
1988—Robert V. Bruce, *The Launching of Modern American Science, 1846-1876*
1989—Taylor Branch, *Parting the Waters: America in the King Years, 1954-63*; and James M. McPherson, *Battle Cry of Freedom: The Civil War Era*
1990—Stanley Karnow, *In Our Image: America's Empire in the Philippines*
1991—Laurel Thatcher Ulrich, *A Midwife's Tale: The Life of Martha Ballard*, based on her diary, 1785-1812
1992—Mark E. Neely Jr., *The Fate of Liberty: Abraham Lincoln and Civil Liberties*
1993—Gordon S. Wood, *The Radicalism of the American Revolution*
1995—Doris Kearns Goodwin, *No Ordinary Time: Franklin and Eleanor Roosevelt: The Home Front in World War II*
1996—Alan Taylor, *William Cooper's Town: Power and Persuasion on the Frontier of the Early American Republic*
1997—Jack N. Rakove, *Original Meanings: Politics and Ideas in the Making of the Constitution*
1998—Edward J. Larson, *Summer for the Gods: The Scopes Trial and America's Continuing Debate Over Science and Religion*
1999—Edwin G. Burrows and Mike Wallace, *Gotham: A History of New York City to 1898*

Biography or Autobiography

1917—Laura E. Richards and Maude Howe Elliott, assisted by Florence Howe Hall, *Julia Ward Howe*
1918—William Cabell Bruce, *Benjamin Franklin, Self-Revealed*
1919—Henry Adams, *The Education of Henry Adams*
1920—Albert J. Beveridge, *The Life of John Marshall*
1921—Edward Bok, *The Americanization of Edward Bok*
1922—Hamlin Garland, *A Daughter of the Middle Border*
1923—Burton J. Hendrick, *The Life and Letters of Walter H. Page*
1924—Michael Pupin, *From Immigrant to Inventor*
1925—M. A. DeWolfe Howe, *Barrett Wendell and His Letters*
1926—Harvey Cushing, *Life of Sir William Osler*
1927—Emory Holloway, *Whitman: An Interpretation in Narrative*
1928—Charles Edward Russell, *The American Orchestra and Theodore Thomas*
1929—Burton J. Hendrick, *The Training of an American: The Earlier Life and Letters of Walter H. Page*
1930—Marquis James, *The Raven* (Sam Houston)
1931—Henry James, *Charles W. Eliot*
1932—Henry F. Pringle, *Theodore Roosevelt*
1933—Allan Nevins, *Grover Cleveland*
1934—Tyler Dennett, *John Hay*
1935—Douglas Southall Freeman, *R. E. Lee*
1936—Ralph Barton Perry, *The Thought and Character of William James*
1937—Allan Nevins, *Hamilton Fish: The Inner History of the Grant Administration*
1938—Divided between Odell Shepard, *Pedlar's Progress* (Bronson Alcott) and Marquis James, *Andrew Jackson*
1939—Carl Van Doren, *Benjamin Franklin*
1940—Ray Stannard Baker, *Woodrow Wilson, Life and Letters*
1941—Ola Elizabeth Winslow, *Jonathan Edwards*
1942—Forrest Wilson, *Crusader in Crinoline* (Harriet Beecher Stowe)
1943—Samuel Eliot Morison, *Admiral of the Ocean Sea* (Christopher Columbus)
1944—Carleton Mabee, *The American Leonardo: The Life of Samuel F. B. Morse*
1945—Russell Blaine Nye, *George Bancroft: Brahmin Rebel.*
1946—Linny Marsh Wolfe, *Son of the Wilderness* (John Muir)
1947—William Allen White, *Autobiography of William Allen White*

1948—Margaret Clapp, *Forgotten First Citizen: John Bigelow*
1949—Robert E. Sherwood, *Roosevelt and Hopkins*
1950—Samuel Flagg Bemis, *John Quincy Adams and the Foundations of American Foreign Policy*
1951—Margaret Louise Coit, *John C. Calhoun: American Portrait*
1952—Merlo J. Pusey, *Charles Evans Hughes*
1953—David J. Mays, *Edmund Pendleton, 1721-1803*
1954—Charles A. Lindbergh, *The Spirit of St. Louis*
1955—William S. White, *The Taft Story*
1956—Talbot F. Hamlin, *Benjamin Henry Latrobe*
1957—John F. Kennedy, *Profiles in Courage*
1958—Douglas Southall Freeman (Vols. I-VI) and John Alexander Carroll and Mary Wells Ashworth (Vol. VII), *George Washington*
1959—Arthur Walworth, *Woodrow Wilson: American Prophet*
1960—Samuel Eliot Morison, *John Paul Jones*
1961—David Donald, *Charles Sumner and the Coming of the Civil War*
1963—Leon Edel, *Henry James: Vols. 2-3*
1964—Walter Jackson Bate, *John Keats*
1965—Ernest Samuels, *Henry Adams*
1966—Arthur M. Schlesinger Jr., *A Thousand Days*
1967—Justin Kaplan, *Mr. Clemens and Mark Twain*
1968—George F. Kennan, *Memoirs (1925-1950)*
1969—B. L. Reid, *The Man From New York: John Quinn and His Friends*
1970—T. Harry Williams, *Huey Long*
1971—Lawrence Thompson, *Robert Frost: The Years of Triumph, 1915-1938*
1972—Joseph P. Lash, *Eleanor and Franklin*
1973—W. A. Swanberg, *Luce and His Empire*
1974—Louis Sheaffer, *O'Neill, Son and Artist*
1975—Robert A. Caro, *The Power Broker: Robert Moses and the Fall of New York*
1976—R.W.B. Lewis, *Edith Wharton: A Biography*
1977—John E. Mack, *A Prince of Our Disorder: The Life of T. E. Lawrence*
1978—Walter Jackson Bate, *Samuel Johnson*
1979—Leonard Baker, *Days of Sorrow and Pain: Leo Baeck and the Berlin Jews*
1980—Edmund Morris, *The Rise of Theodore Roosevelt*
1981—Robert K. Massie, *Peter the Great: His Life and World*
1982—William S. McFeely, *Grant: A Biography*
1983—Russell Baker, *Growing Up*
1984—Louis R. Harlan, *Booker T. Washington*
1985—Kenneth Silverman, *The Life and Times of Cotton Mather*
1986—Elizabeth Frank, *Louise Bogan: A Portrait*
1987—David J. Garrow, *Bearing the Cross: Martin Luther King Jr. and the Southern Christian Leadership Conference*
1988—David Herbert Donald, *Look Homeward: A Life of Thomas Wolfe*
1989—Richard Ellmann, *Oscar Wilde*
1990—Sebastian de Grazia, *Machiavelli in Hell*
1991—Steven Naifeh and Gregory White Smith, *Jackson Pollock: An American Saga*
1992—Lewis B. Puller Jr., *Fortunate Son: The Healing of a Vietnam Vet*
1993—David McCullough, *Truman*
1994—David Levering Lewis, *W.E.B. DuBois: Biography of a Race, 1868-1919*
1995—Joan D. Hedrick, *Harriet Beecher Stowe: A Life*
1996—Jack Miles, *God: A Biography*
1997—Frank McCourt, *Angela's Ashes: A Memoir*
1998—Katharine Graham, *Personal History*
1999—A. Scott Berg, *Lindbergh*

American Poetry

Before 1922, awards were funded by the Poetry Society: **1918**—*Love Songs*, by Sara Teasdale; **1919**—*Old Road to Paradise*, by Margaret Widdemer; *Corn Huskers*, by Carl Sandburg.
1922—Edwin Arlington Robinson, *Collected Poems*
1923—Edna St. Vincent Millay, *The Ballad of the Harp-Weaver; A Few Figs From Thistles; other works*
1924—Robert Frost, *New Hampshire: A Poem With Notes and Grace Notes*
1925—Edwin Arlington Robinson, *The Man Who Died Twice*
1926—Amy Lowell, *What's O'Clock*
1927—Leonora Speyer, *Fiddler's Farewell*
1928—Edwin Arlington Robinson, *Tristram*
1929—Stephen Vincent Benet, *John Brown's Body*
1930—Conrad Aiken, *Selected Poems*
1931—Robert Frost, *Collected Poems*
1932—George Dillon, *The Flowering Stone*
1933—Archibald MacLeish, *Conquistador*
1934—Robert Hillyer, *Collected Verse*
1935—Audrey Wurdemann, *Bright Ambush*
1936—Robert P. Tristram Coffin, *Strange Holiness*
1937—Robert Frost, *A Further Range*
1938—Marya Zaturenska, *Cold Morning Sky*
1939—John Gould Fletcher, *Selected Poems*
1940—Mark Van Doren, *Collected Poems*
1941—Leonard Bacon, *Sunderland Capture*
1942—William Rose Benet, *The Dust Which Is God*
1943—Robert Frost, *A Witness Tree*

1944—Stephen Vincent Benet, *Western Star*
1945—Karl Shapiro, *V-Letter and Other Poems*
1947—Robert Lowell, *Lord Weary's Castle*
1948—W. H. Auden, *The Age of Anxiety*
1949—Peter Viereck, *Terror and Decorum*
1950—Gwendolyn Brooks, *Annie Allen*
1951—Carl Sandburg, *Complete Poems*
1952—Marianne Moore, *Collected Poems*
1953—Archibald MacLeish, *Collected Poems*
1954—Theodore Roethke, *The Waking*
1955—Wallace Stevens, *Collected Poems*
1956—Elizabeth Bishop, *Poems, North and South*
1957—Richard Wilbur, *Things of This World*
1958—Robert Penn Warren, *Promises: Poems 1954-1956*
1959—Stanley Kunitz, *Selected Poems 1928-1958*
1960—W. D. Snodgrass, *Heart's Needle*
1961—Phyllis McGinley, *Times Three: Selected Verse From Three Decades*
1962—Alan Dugan, *Poems*
1963—William Carlos Williams, *Pictures From Breughel*
1964—Louis Simpson, *At the End of the Open Road*
1965—John Berryman, *77 Dream Songs*
1966—Richard Eberhart, *Selected Poems*
1967—Anne Sexton, *Live or Die*
1968—Anthony Hecht, *The Hard Hours*
1969—George Oppen, *Of Being Numerous*
1970—Richard Howard, *Untitled Subjects*
1971—William S. Merwin, *The Carrier of Ladders*
1972—James Wright, *Collected Poems*
1973—Maxine Winokur Kumin, *Up Country*
1974—Robert Lowell, *The Dolphin*
1975—Gary Snyder, *Turtle Island*
1976—John Ashbery, *Self-Portrait in a Convex Mirror*
1977—James Merrill, *Divine Comedies*
1978—Howard Nemerov, *Collected Poems*
1979—Robert Penn Warren, *Now and Then: Poems 1976-1978*
1980—Donald Justice, *Selected Poems*
1981—James Schuyler, *The Morning of the Poem*
1982—Sylvia Plath, *The Collected Poems*
1983—Galway Kinnell, *Selected Poems*
1984—Mary Oliver, *American Primitive*
1985—Carolyn Kizer, *Yin*
1986—Henry Taylor, *The Flying Change*
1987—Rita Dove, *Thomas and Beulah*
1988—William Meredith, *Partial Accounts: New and Selected Poems*
1989—Richard Wilbur, *New and Collected Poems*
1990—Charles Simic, *The World Doesn't End*
1991—Mona Van Duyn, *Near Changes*
1992—James Tate, *Selected Poems*
1993—Louise Glück, *The Wild Iris*
1994—Yusef Komunyakaa, *Neon Vernacular*
1995—Philip Levine, *The Simple Truth*
1996—Jorie Graham, *The Dream of the Unified Field*
1997—Lisel Mueller, *Alive Together: New and Selected Poems*
1998—Charles Wright, *Black Zodiac*
1999—Mark Strand, *Blizzard of One*

General Nonfiction

1962—Theodore H. White, *The Making of the President 1960*
1963—Barbara W. Tuchman, *The Guns of August*
1964—Richard Hofstadter, *Anti-Intellectualism in American Life*
1965—Howard Mumford Jones, *O Strange New World*

1966—Edwin Way Teale, *Wandering Through Winter*
1967—David Brion Davis, *The Problem of Slavery in Western Culture*
1968—Will and Ariel Durant, *Rousseau and Revolution*
1969—Norman Mailer, *The Armies of the Night;* Rene Jules Dubos, *So Human an Animal: How We Are Shaped by Surroundings and Events*
1970—Eric H. Erikson, *Gandhi's Truth*
1971—John Toland, *The Rising Sun*
1972—Barbara W. Tuchman, *Stilwell and the American Experience in China, 1911-1945*
1973—Frances FitzGerald, *Fire in the Lake: The Vietnamese and the Americans in Vietnam;* Robert Coles, *Children of Crisis,* Volumes II & III
1974—Ernest Becker, *The Denial of Death*
1975—Annie Dillard, *Pilgrim at Tinker Creek*
1976—Robert N. Butler, *Why Survive? Being Old in America*
1977—William W. Warner, *Beautiful Swimmers*
1978—Carl Sagan, *The Dragons of Eden*
1979—Edward O. Wilson, *On Human Nature*
1980—Douglas R. Hofstadter, *Gödel, Escher, Bach: An Eternal Golden Braid*
1981—Carl E. Schorske, *Fin-de-Siecle Vienna: Politics and Culture*
1982—Tracy Kidder, *The Soul of a New Machine*
1983—Susan Sheehan, *Is There No Place on Earth for Me?*
1984—Paul Starr, *Social Transformation of American Medicine*
1985—Studs Terkel, *The Good War*
1986—Joseph Lelyveld, *Move Your Shadow;* J. Anthony Lukas, *Common Ground*
1987—David K. Shipler, *Arab and Jew*
1988—Richard Rhodes, *The Making of the Atomic Bomb*
1989—Neil Sheehan, *A Bright Shining Lie: John Paul Vann and America in Vietnam*
1990—Dale Maharidge and Michael Williamson, *And Their Children After Them*
1991—Bert Holldobler and Edward O. Wilson, *The Ants*
1992—Daniel Yergin, *The Prize: The Epic Quest for Oil*
1993—Garry Wills, *Lincoln at Gettysburg*
1994—David Remnick, *Lenin's Tomb: The Last Days of the Soviet Empire*
1995—Jonathan Weiner, *The Beak of the Finch: A Story of Evolution in Our Time*
1996—Tina Rosenberg, *The Haunted Land: Facing Europe's Ghosts After Communism*
1997—Richard Kluger, *Ashes to Ashes: America's Hundred-Year Cigarette War, the Public Health, and the Unabashed Triumph of Philip Morris*
1998—Jared Diamond, *Guns, Germs, and Steel: The Fates of Human Societies*
1999—John McPhee, *Annals of the Former World*

Special Citation

1944—Richard Rodgers and Oscar Hammerstein II, for *Oklahoma!*
1957—Kenneth Roberts, for his historical novels
1960—*The Armada,* by Garrett Mattingly
1961—*American Heritage Picture History of the Civil War*
1973—*George Washington, Vols. I-IV,* by James Thomas Flexner
1977—Alex Haley, for *Roots*
1978—E.B. White
1984—Theodore Seuss Geisel (Dr. Seuss)
1992—Art Spiegelman, for *Maus*

Music

1943—William Schuman, *Secular Cantata No. 2, A Free Song*
1944—Howard Hanson, *Symphony No. 4, Op. 34*
1945—Aaron Copland, *Appalachian Spring*
1946—Leo Sowerby, *The Canticle of the Sun*
1947—Charles E. Ives, *Symphony No. 3*
1948—Walter Piston, *Symphony No. 3*
1949—Virgil Thomson, *Louisiana Story*
1950—Gian-Carlo Menotti, *The Consul*
1951—Douglas Moore, *Giants in the Earth*
1952—Gail Kubik, *Symphony Concertante*
1954—Quincy Porter, *Concerto for Two Pianos and Orchestra*
1955—Gian-Carlo Menotti, *The Saint of Bleecker Street*
1956—Ernest Toch, *Symphony No. 3*
1957—Norman Dello Joio, *Meditations on Ecclesiastes*
1958—Samuel Barber, *Vanessa*
1959—John La Montaine, *Concerto for Piano and Orchestra*
1960—Elliott Carter, *Second String Quartet*
1961—Walter Piston, *Symphony No. 7*
1962—Robert Ward, *The Crucible*
1963—Samuel Barber, *Piano Concerto No. 1*
1966—Leslie Bassett, *Variations for Orchestra*
1967—Leon Kirchner, *Quartet No. 3*
1968—George Crumb, *Echoes of Time and The River*
1969—Karel Husa, *String Quartet No. 3*
1970—Charles W. Wuorinen, *Time's Encomium*
1971—Mario Davidovsky, *Synchronisms No. 6*
1972—Jacob Druckman, *Windows*
1973—Elliott Carter, *String Quartet No. 3*

1974—Donald Martino, *Notturno*
1975—Dominick Argento, *From the Diary of Virginia Woolf*
1976—Ned Rorem, *Air Music*
1977—Richard Wernick, *Visions of Terror and Wonder*
1978—Michael Colgrass, *Deja Vu for Percussion and Orchestra*
1979—Joseph Schwantner, *Aftertones of Infinity*
1980—David Del Tredici, *In Memory of a Summer Day*
1982—Roger Sessions, *Concerto for Orchestra*
1983—Ellen T. Zwilich, *Three Movements for Orchestra*
1984—Bernard Rands, *Canti del Sole*
1985—Stephen Albert, *Symphony, RiverRun*
1986—George Perle, *Wind Quintet IV*
1987—John Harbison, *The Flight Into Egypt*
1988—William Bolcom, *12 New Etudes for Piano*
1989—Roger Reynolds, *Whispers Out of Time*
1990—Mel Powell, *Duplicates: A Concerto for Two Pianos and Orchestra*
1991—Shulamit Ran, *Symphony*
1992—Wayne Peterson, *The Face of the Night, The Heart of the Dark*
1993—Christopher Rouse, *Trombone Concerto*
1994—Gunther Schuller, *Of Reminiscences and Reflections*
1995—Morton Gould, *Stringmusic*
1996—George Walker, *Lilacs*
1997—Wynton Marsalis, *Blood on the Fields*
1998—Aaron Jay Kernis, *String Quartet No. 2*
1999—Melinda Wagner, *Concerto for Flute, Strings and Percussion*   *(continued)*

Special Citation in Music

1974—Roger Sessions
1976—Scott Joplin
1982—Milton Babbitt

1985—William Schuman
1998—George Gershwin
1999—Edward Kennedy "Duke" Ellington

Miscellaneous Book Awards

Year in parentheses is year awarded

Academy of American Poets Awards (1998), Lenore Marshall Prize, $10,000: Mark Jarman; Raiziss/de Palchi Translation Award (fellowship) $20,000: Geoffrey Brock, *Poesie del disamore*; Tanning Prize, $100,000: A. R. Ammons; Academy Fellowship, $20,000 stipend: Charles Simic. (1999) James Laughlin Award, $5,000: Tory Dent, *HIV, Mon Amour*; Landon Translation Award, $1,000: W. D. Snodgrass, *Selected Translations*; Walt Whitman Award, $5,000: Judy Jordan, *Carolina Ghost Woods*

American Academy of Arts and Letters (1999), Gold Medal for Belles Lettres and Criticism: Harold Bloom; Arthur Rense Poetry Prize, $20,000: James McMichael. E. M. Forster Award, $15,000: Nick Hornby; Academy Awards in Literature ($7,500 each): fiction: Susanna Moore, Richard Price, Lee Smith; translation and fiction: Edmund Keeley; poetry: Ron Padgett, Sherod Santos, C.K. Williams; nonfiction: Jon Krakauer. Award of Merit Medal for Drama, medal and $5,000: Romulus Linney; Richard and Hinda Rosenthal Foundation Award, $5,000: Sigrid Nunez, *Mitz: The Marmoset of Bloomsbury;* Sue Kaufman Prize for First Fiction, $2,500: Michael Byers, *The Coast of Good Intentions;* Harold D. Vursell Memorial Award, $5,000: Dava Sobel, *Longitude;* Morton Dauwen Zabel Award for Fiction, $5,000: Kathryn Davis; Michael Braude Award for light verse, $5,000: Thomas M. Disch; Addison Metcalf Award, $5,000: Reginald McKnight; Witter Bynner Poetry Award, $2,500: Brigit Pegeen Kelly; Rome Fellowship in Literature, one-year residence at the American Academy in Rome (1999-2000): Tom Andrews

Booker Prize (1998) by Booker PLC for best novel written in English by a UK, Commonwealth, or South African author: Ian McEwan, *Amsterdam*

Christopher Awards (1999), by The Christophers, for expression of highest values of human spirit, bronze medallion each: David Halberstam, *The Children;* Caroline Alexander with photos by Frank Hurley, *The Endurance: Shackleton's Legendary Antarctic Expedition;* Bernard Haring, *Free and Faithful: My Life in the Catholic Church;* Craig Kielburger with Kevin Major, *Free the Children: A Young Man's Personal Crusade Against Child Labor;* Thomas Cahill, *The Gifts of the Jews: How a Tribe of Desert Nomads Changed the Way Everyone Feels;* Peter Ackroyd, *The Life of Thomas More;* Christopher DeVinck, *Love's*

Harvest: Family, Faith, Friends; Patrick Ahern, *Maurice & Therese: The Story of a Love;* John Lewis with Michael D'Orso, *Walking With the Wind: A Memoir of the Movement*

Golden Kite Awards (1998), by Society of Children's Book Writers and Illustrators. Fiction: Joan Bauer, *Rules of the Road;* nonfiction: Russell Freedman, *Martha Graham: A Dancer's Life;* picture-illustration: Uri Shulevitz, *Snow;* picture book text: Kristine O'Connell George, *Old Elm Speaks: Tree Poems*

Hugo Awards (1999), by the World Science Fiction Convention: Novel: *To Say Nothing of the Dog,* Connie Willis; novella: *Oceanic,* Greg Egan; novelette: *Taklamakan,* Bruce Sterling; short story: *The Very Pulse of the Machine,* Michael Swanwick.

Coretta Scott King Award (1999), by the American Library Assn. for African American authors and illustrators of outstanding books for children and young adults: Author: Angela Johnson, *Heaven;* Illustrator: Michele Wood, *i see the rhythm.*

Lincoln Prize (1999), by Lincoln and Soldiers Institute at Gettysburg College, for lifetime contribution to Civil War studies, $35,000 and a bronze bust of Lincoln: Douglas L. Wilson, *Honor's Voice: The Transformation of Abraham Lincoln.*

National Book Awards (1998), by National Book Foundation, $10,000 each. Fiction: Alice McDermott, *Charming Billy;* nonfiction: Edward Ball, *Slaves in the Family;* poetry: Gerald Stern, *This Time: New and Selected Poems;* young people's literature: Louis Sachar, *Holes;* Medal for Distinguished Contribution to American Letters: John Updike

National Book Critics Circle Awards (1999), Fiction: Alice Munro, *The Love of a Good Woman;* nonfiction: Philip Gourevitch, *We Wish to Inform You That Tomorrow We Will Be Killed With Our Families: Stories From Rwanda;* criticism: Gary Giddins, *Visions of Jazz: The First Century;* biography and autobiography: Sylvia Nasar, *A Beautiful Mind;* poetry: Marie Ponsot, *The Bird Catcher;* Nona Balakian Citation for Excellence in Reviewing: Albert Mobilio

PEN/Faulkner Award (1999), for fiction, $15,000: Michael Cunningham, *The Hours*

Edgar Awards (1999), by the Mystery Writers of America: Grand Master award: P.D. James; best novel: *Mr. White's Confession,* Robert Clark; best short story: "Poachers," Tom Franklin

Newbery Medal Books

The Newbery Medal is awarded annually by the Association for Library Service to Children, a division of the American Library Association, to the author of the most distinguished contribution to American literature for children.

Year Given	Book, Author
1922	*The Story of Mankind,* Hendrik Willem van Loon
1923	*The Voyages of Dr. Dolittle,* Hugh Lofting
1924	*The Dark Frigate,* Charles Boardman Hawes
1925	*Tales From Silver Lands,* Charles Joseph Finger
1926	*Shen of the Sea,* Arthur Bowie Chrisman
1927	*Smoky, the Cowhorse,* Will James
1928	*Gay-Neck,* Dhan Gopal Mukerji
1929	*The Trumpeter of Krakow,* Eric P. Kelly
1930	*Hitty, Her First Hundred Years,* Rachel Field
1931	*The Cat Who Went to Heaven,* Elizabeth Coatsworth
1932	*Waterless Mountain,* Laura Adams Armer
1933	*Young Fu of the Upper Yangtze,* Elizabeth Foreman Lewis
1934	*Invincible Louisa,* Cornelia Lynde Meigs
1935	*Dobry,* Monica Shannon
1936	*Caddie Woodlawn,* Carol Ryrie Brink
1937	*Roller Skates,* Ruth Sawyer
1938	*The White Stag,* Kate Seredy
1939	*Thimble Summer,* Elizabeth Enright
1940	*Daniel Boone,* James Daugherty
1941	*Call It Courage,* Armstrong Sperry
1942	*The Matchlock Gun,* Walter D. Edmonds
1943	*Adam of the Road,* Elizabeth Janet Gray
1944	*Johnny Tremain,* Esther Forbes
1945	*Rabbit Hill,* Robert Lawson
1946	*Strawberry Girl,* Lois Lenski
1947	*Miss Hickory,* Carolyn S. Bailey
1948	*Twenty-One Balloons,* William Pène Du Bois
1949	*King of the Wind,* Marguerite Henry
1950	*The Door in the Wall,* Marguerite de Angeli
1951	*Amos Fortune, Free Man,* Elizabeth Yates
1952	*Ginger Pye,* Eleanor Estes
1953	*Secret of the Andes,* Ann Nolan Clark
1954	*. . . And Now Miguel,* Joseph Krumgold
1955	*The Wheel on the School,* Meindert DeJong
1956	*Carry On, Mr. Bowditch,* Jean Lee Latham

Year Given	Book, Author
1957	*Miracles on Maple Hill,* Virginia Sorensen
1958	*Rifles for Watie,* Harold Keith
1959	*The Witch of Blackbird Pond,* Elizabeth George Speare
1960	*Onion John,* Joseph Krumgold
1961	*Island of the Blue Dolphins,* Scott O'Dell
1962	*The Bronze Bow,* Elizabeth George Speare
1963	*A Wrinkle in Time,* Madeleine L'Engle
1964	*It's Like This, Cat,* Emily Cheney Neville
1965	*Shadow of a Bull,* Maja Wojciechowska
1966	*I, Juan de Pareja,* Elizabeth Borton de Trevino
1967	*Up a Road Slowly,* Irene Hunt
1968	*From the Mixed-Up Files of Mrs. Basil E. Frankweiler,* E. L. Konigsburg
1969	*The High King,* Lloyd Alexander
1970	*Sounder,* William H. Armstrong
1971	*The Summer of the Swans,* Betsy Byars
1972	*Mrs. Frisby and the Rats of NIMH,* Robert C. O'Brien
1973	*Julie of the Wolves,* Jean George
1974	*The Slave Dancer,* Paula Fox
1975	*M. C. Higgins the Great,* Virginia Hamilton
1976	*Grey King,* Susan Cooper
1977	*Roll of Thunder, Hear My Cry,* Mildred D. Taylor
1978	*Bridge to Terabithia,* Katherine Paterson
1979	*The Westing Game,* Ellen Raskin
1980	*A Gathering of Days,* Joan Blos
1981	*Jacob Have I Loved,* Katherine Paterson
1982	*A Visit to William Blake's Inn: Poems for Innocent and Experienced Travelers,* Nancy Willard
1983	*Dicey's Song,* Cynthia Voigt
1984	*Dear Mr. Henshaw,* Beverly Cleary
1985	*The Hero and the Crown,* Robin McKinley
1986	*Sarah, Plain and Tall,* Patricia MacLachlan
1987	*The Whipping Boy,* Sid Fleischman
1988	*Lincoln: A Photobiography,* Russell Freedman
1989	*Joyful Noise: Poems for Two Voices,* Paul Fleischman

Year Given	Book, Author	Year Given	Book, Author
1990	*Number the Stars*, Lois Lowry	1995	*Walk Two Moons*, Sharon Creech
1991	*Maniac Magee*, Jerry Spinelli	1996	*The Midwife's Apprentice*, Karen Cushman
1992	*Shiloh*, Phyllis Reynolds Naylor	1997	*The View From Saturday*, E. L. Konigsburg
1993	*Missing May*, Cynthia Rylant	1998	*Out of the Dust*, Karen Hesse
1994	*The Giver*, Lois Lowry	1999	*Holes*, Louis Sachar

Caldecott Medal Books

The Caldecott Medal is awarded annually by the Association for Library Service to Children, a division of the American Library Association, to the illustrator of the most distinguished American picture book for children.

Year Given	Book, Illustrator	Year Given	Book, Illustrator
1938	*Animals of the Bible*, Dorothy P. Lathrop	1970	*Sylvester and the Magic Pebble*, William Steig
1939	*Mei Li*, Thomas Handforth	1971	*A Story A Story*, Gail E. Haley
1940	*Abraham Lincoln*, Ingri & Edgar Parin d'Aulaire	1972	*One Fine Day*, Nonny Hogrogian
1941	*They Were Strong and Good*, Robert Lawson	1973	*The Funny Little Woman*, Blair Lent
1942	*Make Way for Ducklings*, Robert McCloskey	1974	*Duffy and the Devil*, Margot Zemach
1943	*The Little House*, Virginia Lee Burton	1975	*Arrow to the Sun*, Gerald McDermott
1944	*Many Moons*, Louis Slobodkin	1976	*Why Mosquitoes Buzz in People's Ears*, Leo & Diane Dillon
1945	*Prayer for a Child*, Elizabeth Orton Jones	1977	*Ashanti to Zulu: African Traditions*, Leo & Diane Dillon
1946	*The Rooster Crows*, Maude & Miska Petersham	1978	*Noah's Ark*, Peter Spier
1947	*The Little Island*, Leonard Weisgard	1979	*The Girl Who Loved Wild Horses*, Paul Goble
1948	*White Snow, Bright Snow*, Roger Duvoisin	1980	*Ox-Cart Man*, Barbara Cooney
1949	*The Big Snow*, Berta & Elmer Hader	1981	*Fables*, Arnold Lobel
1950	*Song of the Swallows*, Leo Politi	1982	*Jumanji*, Chris Van Allsburg
1951	*The Egg Tree*, Karherine Milhous	1983	*Shadow*, Marcia Brown
1952	*Finders Keepers*, Nicolas, pseud. (Nicholas Mordvinoff)	1984	*The Glorious Flight: Across the Channel with Louis Bleriot*, Alice and Martin Provensen
1953	*The Biggest Bear*, Lynd Ward	1985	*Saint George and the Dragon*, Trina Schart Hyman
1954	*Madeline's Rescue*, Ludwig Bemelmans	1986	*The Polar Express*, Chris Van Allsburg
1955	*Cinderella, or the Little Glass Slipper*, Marcia Brown	1987	*Hey, Al*, Richard Egielski
1956	*Frog Went A-Courtin'*, Feodor Rojankovsky	1988	*Owl Moon*, John Schoenherr
1957	*A Tree Is Nice*, Marc Simont	1989	*Song and Dance Man*, Stephen Grammell
1958	*Time of Wonder*, Robert McCloskey	1990	*Lon Po Po: A Red-Riding Hood Story From China*, Ed Young
1959	*Chanticleer and the Fox,* Barbara Cooney	1991	*Black and White*, David Macaulay
1960	*Nine Days to Christmas*, Marie Hall Ets	1992	*Tuesday*, David Wiesner
1961	*Baboushka and the Three Kings*, Nicolas Sidjakov	1993	*Mirette on the High Wire*, Emily Arnold McCully
1962	*Once a Mouse*, Marcia Brown	1994	*Grandfather's Journey*, Allen Say
1963	*The Snowy Day*, Ezra Jack Keats	1995	*Smoky Night*, David Diaz
1964	*Where the Wild Things Are*, Maurice Sendak	1996	*Officer Buckle and Gloria*, Peggy Rathmann
1965	*May I Bring a Friend?*, Beni Montressor	1997	*Golem*, David Wisniewski
1966	*Always Room for One More*, Nonny Hogrogian	1998	*Rapunzel*, Paul O. Zelinsky
1967	*Sam, Bang, and Moonshine*, Evaline Ness	1999	*Snowflake Bentley*, Mary Azarian
1968	*Drummer Hoff*, Ed Emberley		
1969	*The Fool of the World and the Flying Ship*, Uri Shulevitz		

Journalism

Year in parentheses is year awarded

National Journalism Awards (1999), by Scripps Howard Foundation, $2,500 each. Editorial writing: David V. Hawpe, *Courier-Journal* (Louisville, KY); human interest writing: Gary Pomerantz, *Atlanta Journal-Constitution*; environmental reporting (over 100,000 circ.): *Seattle Times*; environmental reporting (under 100,000 circ.): *Yakima (WA) Herald-Republic*; public service reporting (over 100,000 circ.): *Philadelphia Inquirer*; public service reporting (under 100,000 circ.): *Charleston (WV) Gazette*; commentary: R. Bruce Dold, *Chicago Tribune*; photojournalism: Patrick Davison, *Denver Rocky Mountain News*; college cartoonist: Audra Ann Furuichi, Univ. of Hawaii at Manoa; distinguished service to literacy: (dual) *Baltimore Sun* and Betty Williford, Elberton, GA; distinguished service to First Amendment: (7 Indiana newspapers), *Times of Northwest Indiana*, *Star Press* (Muncie), *South Bend Tribune*, *Journal-Gazette* (Fort Wayne), *Indianapolis Star-News*, *Evansville Courier & Press*, *Tribune Star* (Terre Haute); business/economics reporting: Richard Read, *Oregonian* (Portland); Electronic journalism. Small market radio: WAMC, Northeast Public Radio, Albany, NY; large market radio: WTN Radio, Nashville, TN; small market TV/cable: WANE-TV, Fort Wayne, IN; large market TV/cable: NewsChannel 8, Washington, DC

National Magazine Awards (1999), by American Society of Magazine Editors and Columbia Univ. Graduate School of Journalism. Gen. excel., circ. over 1 mil: *Vanity Fair*, 400,000 to 1 mil: *Conde Nast Traveler*, 100,000-400,000: *Fast Company*; under 100,000: *I.D.* Single topic issue: *The Oxford American*; spec. interests: *PC Computing*; feature writing: *The American Scholar*; fiction: *Harper's;* design: *ESPN The Magazine;* photography: *Martha Stewart Living;* reporting: *Newsweek;* personal service: *Good Housekeeping;* public interest: *Time;* essays & criticism: *The Atlantic Monthly;* gen. excel., new media: *Cigar Aficionado*

Overseas Press Club Awards (1999), for journalism abroad. Hal Boyle Award (newspaper or wire service reporting): Ken Guggenheim, Niko Price, Associated Press, "Hurricane Mitch"; Bob Considine Award (newspaper or wire-service interpretation): Barton Gellman, *Washington Post*, "Shell Games: Washington, Baghdad and the Hunt for Iraq's Forbidden Weapons"; Robert Capa Gold Medal (photographic reporting requiring exceptional courage and enterprise) James Nachtwey, Magnum for Time, "Indonesia: Descent into Madness"; Olivier Rebbot Award (photography in mags. and books): Ettore Malanca, *Sipa/New York Times Magazine*, "Romania's Lost Boys"; John Faber Award (photography in newspapers and wire services): Eric Mencher, *Philadelphia Inquirer*, "Rwanda: Aftermath of Genocide"; Lowell Thomas Award (radio news or interpretation): Sandy Tolan, *Homelands Productions for Fresh Air*, "The Lemon Tree"; David Kaplan Award (TV spot news reporting): The CNN Team, "Strikes Against Iraq and the Impeachment Hearings"; Edward R. Murrow Award (TV interpretation or documentary): William Cran, Stephanie Tepper, David Fanning, Michael Sullivan, *Frontline/WGBH Boston and InVision Productions*, "Ambush in Mogadishu"; Ed Cunningham Memorial Award (mag. reporting): Mark Danner, *New York Review of Books*, "Yugoslav Wars"; Thomas Nast Award (cartooning): Kevin Kallaugher, *Baltimore Sun*; Morton Frank Award (business reporting): Michael Shari, Joyce Barnathan, Pete Engardio, Dean Foust, Jonathan Moore, Sheri Prasso, Christopher Power, *Business Week*, "Indonesia in Turmoil"; Malcolm Forbes Award (business reporting in newspapers or wire service): Richard Read, *Oregonian*, "The French Fry Connection"; Carl Spielvogel Award (business reporting in broadcast media): Brenda Breslauer, Brian Ross, *ABC News*, "Nazi Stolen Art"; Cornelius Ryan Award (nonfiction book): Philip Gourevitch, *We Wish to Inform You That Tomorrow We Will Be Killed With Our Families: Stories From Rwanda;* Madeline Dane Ross Award (foreign reporting concerned with human condition): Kevin Sullivan, Mary Jordan, Keith Richburg, *Washington Post*, "Shattered Lives: The Asian Middle Class"; Eric and Amy Burger Award (reporting in the broadcast media on human rights): Cynthia McFadden, Beth Osisek, *ABC News: Primetime Live*, "Russian Girls for Sale"; Joe and Laurie Dine Award (reporting in a print medium dealing with human rights): Mark O'Keefe, *Oregonian*, "Christians Under Siege"; Whitman Bassow Award (reporting on internat. environmental issues): Anne Garrels, Loren Jenkins, *National Public Radio*, "Water Series"; Robert Spiers Benjamin Award (reporting on Latin America, any medium): Edwin Garcia, Michelle Levander, Ricardo Sandoval, *San Jose Mercury News*, "Lost in Transit"

George Foster Peabody Awards (1999) by the Univ. of Georgia. Coverage of Africa, National Public Radio; *Sisterhood of Hope,* WHAS Radio, Louisville, KY; *I Must Keep Fightin':* *The Art of Paul Robeson,* National Public Radio; Performance Today, National Public Radio; *The Reckoning,* presented on *Public Eye With Bryant Gumbel,* CBS News, New York; Christiane Amanpour, for international reporting on Cable News Network and CBS News: *60 Minutes; The Olympic Bribery Scandal,* KTVX-TV, Salt Lake City, UT; *FRONTLINE: Washington's Other Scandal,* WGBH/FRONTLINE, Washington Media Associates, and Public Affairs Television; *About Race,* KRON-TV, San Francisco; *The Human Body* and The Learning Channel; *Africans in America: America's Journey Through Slavery,* WGBH-TV, Boston; Travis, ITVS and City People Productions; Frank Lloyd Wright, Florentine Films and WETA-TV, Washington, DC; *When Good Men Do Nothing,* BBC, London and WGBH-TV, Boston; *American Masters: Alexander Calder,* Thirteen/WNET, New York, Florentine Films/Sherman Pictures; *Cold War,* Jeremy Isaacs Productions and CNN Productions, Atlanta; *The American Experience: Riding the Rails,* The American Experience, The American History Project/Out of the Blue Productions Inc., and WGBH Educational Foundation; *Dateline NBC: Checks and Balances,* NBC News, New York; *The American Experience: America 1900,* The American Experience, David Grubin Productions Inc., and WGBH Educational Foundation; *Christopher,* WANE-TV, Fort Wayne, IN; *The Bear,* TVC and Channel 4, London; HBO Sports Documentaries, Home Box Office, New York; *Dr. Katz: Professional Therapist,* Comedy Central, Tom Snyder Productions Inc., and Popular Arts Entertainment, in association with HBO Downtown Productions; *Mobil Masterpiece Theatre: King Lear,* A Chestermead Production for the BBC, London and WGBH-TV, Boston; *Shot Through the Heart,* Home Box Office, New York; *The Baby Dance,* Showtime Networks Inc., Egg Pictures, and Pacific Motion Pictures; *The Practice,* ABC, David E. Kelley Productions; *NYPD Blue: Raging Bulls,* ABC, Steven Bochco Productions; *Ally McBeal,* FOX, David E. Kelley Productions; *The Larry Sanders Show: Flip,* Home Box Office and Brillstein-Grey Entertainment; Linda Ellerbee, Host of "Nick News"; Jac Venza; Robert Halmi Sr.

George Polk Awards (1999), by Long Island Univ., for excellence in journalism. National reporting: Donald L. Bartlett and James B. Steele, *Time* magazine; foreign reporting: Tracy Wilkinson, *Los Angeles Times;* legal reporting: Joe Stephens, *Kansas City Star;* career award: Russell Baker; international reporting: Alix M. Freedman, *Wall Street Journal;* network television reporting: Brian Ross and Rhonda Schwartz, ABC News; book award: *We Wish to Inform You That Tomorrow We Will Be Killed With Our Families,* by Philip Gourevitch; local reporting: Clifford Levy, *New York Times;* commentary: Juan Gonzalez, *New York Daily News;* radio reporting: Amy Goodman and Jeremy Scahill, Pacifica Radio; medical reporting: Robert Whitaker and Dolores Kong, *Boston Globe;* environmental reporting: Gardiner Harris and R.G. Dunlop, *Courier-Journal* (Louisville); economic reporting: Kevin Sullivan, Mary Jordan, and Keith Richburg, *Washington Post*

Reuben Award, by National Cartoonists Society. Best cartoonist of 1998: Will Eisner

The Spingarn Medal

The Spingarn Medal has been awarded annually since 1915 (except in 1938) by the National Assoc. for the Advancement of Colored People for the highest achievement by a black American in the previous year.

1915 Ernest E. Just	1936 John Hope	1958 Daisy Bates and the Little Rock Nine	1979 Rosa L. Parks
1916 Charles Young	1937 Walter White	1959 Edward Kennedy (Duke) Ellington	1980 Dr. Rayford W. Logan
1917 Harry T. Burleigh	1939 Marian Anderson	1960 Langston Hughes	1981 Coleman Young
1918 William S. Braithwaite	1940 Louis T. Wright	1961 Kenneth B. Clark	1982 Dr. Benjamin E. Mays
1919 Archibald H. Grimké	1941 Richard Wright	1962 Robert C. Weaver	1983 Lena Horne
1920 W. E. B. Du Bois	1942 A. Philip Randolph	1963 Medgar W. Evers	1984 Thomas Bradley
1921 Charles S. Gilpin	1943 William H. Hastie	1964 Roy Wilkins	1985 Bill Cosby
1922 Mary B. Talbert	1944 Charles Drew	1965 Leontyne Price	1986 Dr. Benjamin L. Hooks
1923 George W.Carver	1945 Paul Robeson	1966 John H. Johnson	1987 Percy E. Sutton
1924 Roland Hayes	1946 Thurgood Marshall	1967 Edward W. Brooke	1988 Frederick D. Patterson
1925 James W. Johnson	1947 Dr. Percy L. Julian	1968 Sammy Davis Jr.	1989 Jesse Jackson
1926 Carter G. Woodson	1948 Channing H. Tobias	1969 Clarence M. Mitchell Jr.	1990 L. Douglas Wilder
1927 Anthony Overton	1949 Ralph J. Bunche	1970 Jacob Lawrence	1991 Gen. Colin L. Powell
1928 Charles W. Chesnutt	1950 Charles H. Houston	1971 Leon H. Sullivan	1992 Barbara Jordan
1929 Mordecai W. Johnson	1951 Mabel K. Staupers	1972 Gordon Parks	1993 Dorothy I. Height
1930 Henry A. Hunt	1952 Harry T. Moore	1973 Wilson C. Riles	1994 Maya Angelou
1931 Richard B. Harrison	1953 Paul R. Williams	1974 Damon Keith	1995 John Hope Franklin
1932 Robert R. Moton	1954 Theodore K. Lawless	1975 Henry (Hank) Aaron	1996 A. Leon Higginbotham
1933 Max Yergan	1955 Carl Murphy	1976 Alvin Ailey	1997 Carl T. Rowan
1934 William T. B. Williams	1956 Jack R. Robinson	1977 Alex Haley	1998 Myrlie Evers-Williams
1935 Mary McLeod Bethune	1957 Martin Luther King Jr.	1978 Andrew Young	1999 Earl G. Graves Sr.

Miscellaneous Awards

Year in parentheses is year awarded

American Academy of Arts and Letters (1999), Gold Medal for Painting: Robert Rauschenberg; Award for Distinguished Service to the Arts: Harvey Lichtenstein; Medal for Spoken Language: Mario Cuomo; Arnold W. Brunner Memorial Prize in Architecture, $5,000: Fumihiko Maki; Academy Awards, $7,500 each, in Architecture: Eric Owen Moss; in Art: George Condo, Frank Moore, Thomas Nozkowski, Altoon Sulton, Joe Zucker; in Music: Nathan Currier, Michael Gordon, Erica Muhl, Julia Wolfe; Jimmy Ernst Award in Art, $5,000: Dorothea Rockburne; Walter Hinrichsen Award: Edmund Campion; Charles Ives Fellowship in Music, $12,500: Steven Burke; Charles Ives Scholarships in Music, $7,500 each: Roshanne Etezady, Paul Yeon Lee, David Mallamud, Carter Pann, Jason Roth, Robert Zimmerman; Wladimir and Rhoda Lakond Award in Music, $5,000: David Stock; Goddard Lieberson Fellowships in Music, $12,500 each: Tamar Diesendruck, David Sampson; Willard L. Metcalf Award in Art, $5,000: Desirée Alvarez; Richard and Hinda Rosenthal Foundation Award in Art, $5,000: Karin Davie.

National Humanities Medal (formerly Charles Frankel Prize), by National Endowment for the Humanities, for those who have increased public awareness of the humanities, $5,000 each (1998): Stephen E. Ambrose, E.L. Doctorow, Diana L. Eck, Nancye Brown Gaj, Henry Louis Gates Jr., Vartan Gregorian, Ramón Eduardo Ruiz, Arthur M. Schlesinger Jr., Garry Wills

Intel Science Talent Search (formerly given by Westinghouse) (1999), 1st place, $50,000 schol.: Natalia Toro, Boulder, CO; 2d place, $40,000 schol.: David Moore, Silver Spring, MD; 3d place, $30,000 schol.: Keith Winstein, Aurora, IL

National Inventor of the Year Awards (1999), by Intellectual Property Owners: William C. Atkinson; Robert P. Cloutier; Michael L. Wash; Arthur A. Whitfield, Eastman Kodak Company.

John F. Kennedy Center for the Performing Arts Awards (1999), Victor Borge, Sean Connery, Judith Jamison, Jason Robards, Stevie Wonder

1999 Library of the Year Award, by Gale Research, Inc., and Library Journal: The Valley Library at Oregon State University, OR

Presidential Medal of Freedom (1999), by the White House: Lloyd M. Bentsen, Edgar M. Bronfman, Jimmy Carter, Rosalyn Carter, Evelyn Dubrow, Sister M. Isolina Ferre, Gerald R. Ford, Oliver White Hill, Max Kampelman, Helmut Kohl, Edgar Wayburn.

National Medal of the Arts (1998), by the National Endowment for the Arts and the White House: Jacques d'Amboise, Antoine "Fats" Domino, Ramblin' Jack Elliott, Frank Gehry, Barbara Handman, Agnes Martin, Gregory Peck, Roberta Peters, Philip Roth, Sara Lee Corporation, Steppenwolf Theatre Company, Gwen Verdon

Pritzker Architecture Prize (1999), by the Hyatt Foundation, $100,000: Sir Norman Foster, Great Britain

1999 Teacher of the Year, by Council of Chief State School Officers and Scholastic, Inc.: Andy Baumgartner, Augusta, GA

Templeton Prize for Progress in Religion (1999), by Templeton Foundation, about $1.2 million: Ian Barbour

Miss America Winners, 1921-2000

1921	Margaret Gorman, Washington, DC	1966	Deborah Irene Bryant, Overland Park, Kansas
1922-23	Mary Campbell, Columbus, Ohio	1967	Jane Anne Jayroe, Laverne, Oklahoma
1924	Ruth Malcolmson, Philadelphia, Pennsylvania	1968	Debra Dene Barnes, Moran, Kansas
1925	Fay Lamphier, Oakland, California	1969	Judith Anne Ford, Belvidere, Illinois
1926	Norma Smallwood, Tulsa, Oklahoma	1970	Pamela Anne Eldred, Birmingham, Michigan
1927	Lois Delander, Joliet, Illinois	1971	Phyllis Ann George, Denton, Texas
1933	Marion Bergeron, West Haven, Connecticut	1972	Laurie Lea Schaefer, Columbus, Ohio
1935	Henrietta Leaver, Pittsburgh, Pennsylvania	1973	Terry Anne Meeuwsen, DePere, Wisconsin
1936	Rose Coyle, Philadelphia, Pennsylvania	1974	Rebecca Ann King, Denver, Colorado
1937	Bette Cooper, Bertrand Island, New Jersey	1975	Shirley Cothran, Fort Worth, Texas
1938	Marilyn Meseke, Marion, Ohio	1976	Tawney Elaine Godin, Yonkers, New York
1939	Patricia Donnelly, Detroit, Michigan	1977	Dorothy Kathleen Benham, Edina, Minnesota
1940	Frances Marie Burke, Philadelphia, Pennsylvania	1978	Susan Perkins, Columbus, Ohio
1941	Rosemary LaPlanche, Los Angeles, California	1979	Kylene Barker, Galax, Virginia
1942	Jo-Caroll Dennison, Tyler, Texas	1980	Cheryl Prewitt, Ackerman, Mississippi
1943	Jean Bartel, Los Angeles, California	1981	Susan Powell, Elk City, Oklahoma
1944	Venus Ramey, Washington, D.C.	1982	Elizabeth Ward, Russellville, Arkansas
1945	Bess Myerson, New York City, New York	1983	Debra Maffett, Anaheim, California
1946	Marilyn Buferd, Los Angeles, California	1984	Vanessa Williams*, Milwood, New York
1947	Barbara Walker, Memphis, Tennessee		Suzette Charles, Mays Landing, New Jersey
1948	BeBe Shopp, Hopkins, Minnesota	1985	Sharlene Wells, Salt Lake City, Utah
1949	Jacque Mercer, Litchfield, Arizona	1986	Susan Akin, Meridian, Mississippi
1951	Yolande Betbeze, Mobile, Alabama	1987	Kellye Cash, Memphis, Tennessee
1952	Coleen Kay Hutchins, Salt Lake City, Utah	1988	Kaye Lani Rae Rafko, Monroe, Michigan
1953	Neva Jane Langley, Macon, Georgia	1989	Gretchen Carlson, Anoka, Minnesota
1954	Evelyn Margaret Ay, Ephrata, Pennsylvania	1990	Debbye Turner, Columbia, Missouri
1955	Lee Meriwether, San Francisco, California	1991	Marjorie Vincent, Oak Park, Illinois
1956	Sharon Ritchie, Denver, Colorado	1992	Carolyn Suzanne Sapp, Honolulu, Hawaii
1957	Marian McKnight, Manning, South Carolina	1993	Leanza Cornett, Jacksonville, Florida
1958	Marilyn Van Derbur, Denver, Colorado	1994	Kimberly Aiken, Columbia, South Carolina
1959	Mary Ann Mobley, Brandon, Mississippi	1995	Heather Whitestone, Birmingham, Alabama
1960	Lynda Lee Mead, Natchez, Mississippi	1996	Shawntel Smith, Muldrow, Oklahoma
1961	Nancy Fleming, Montague, Michigan	1997	Tara Dawn Holland, Overland Park, Kansas
1962	Maria Fletcher, Asheville, North Carolina	1998	Kate Shindle, Evanston, Illinois
1963	Jacquelyn Mayer, Sandusky, Ohio	1999	Nicole Johnson, Roanoke, Virginia
1964	Donna Axum, El Dorado, Arkansas	2000	Heather Renee French, Maysville, Kentucky
1965	Vonda Kay Van Dyke, Phoenix, Arizona		

* Resigned July 23, 1984.

Entertainment Awards
1998-99 Emmy Awards
Selected Prime-Time Emmy Awards

Drama series: *The Practice*, ABC
Comedy series: *Ally McBeal*, Fox
Miniseries: *Horatio Hornblower*, A&E
Variety, music or comedy series: *Late Show with David Letterman*, CBS
Variety, music or comedy special: *1998 Tony Awards*, CBS
Made-for-television movie: *A Lesson Before Dying*, HBO
Lead actor, drama series: Dennis Franz, *NYPD Blue*, ABC
Lead actress, drama series: Edie Falco, *The Sopranos*, HBO
Lead actor, comedy series: John Lithgow, *3rd Rock From the Sun*, NBC
Lead actress, comedy series: Helen Hunt, *Mad About You*, NBC
Lead actor, miniseries/special: Stanley Tucci, *Winchell*, HBO
Lead actress, miniseries/special: Helen Mirren, *The Passion of Ayn Rand*, Showtime

Supporting actor, drama series: Michael Badalucco, *The Practice*, ABC
Supporting actress, drama series: Holland Taylor, *The Practice*, ABC
Supporting actor, comedy series: David Hyde Pierce, *Frasier*, NBC
Supporting actress, comedy series: Kristen Johnston, *3rd Rock From the Sun*, NBC
Supporting actor, miniseries/special: Peter O'Toole, *Joan of Arc*, CBS
Supporting actress, miniseries/special: Anne Bancroft, *Deep in My Heart*, CBS
Individual performance, variety/music program: John Leguizamo, *John Leguizamo's Freak*, HBO

1998-99 Emmy Awards
Selected Daytime Emmy Awards

Drama series: *General Hospital*, ABC
Actress: Susan Lucci, *All My Children*, ABC
Actor: Anthony Geary, *General Hospital*, ABC
Supporting actress: Sharon Case, *The Young and the Restless*, CBS
Supporting actor: Stuart Damon, *General Hospital*, ABC
Younger actress: Heather Tom, *The Young and the Restless*, CBS
Younger actor: Jonathan Jackson, *General Hospital*, ABC
Directing team: *The Young and the Restless*, CBS
Writing team: *General Hospital*, ABC

Game/audience participation show: *Win Ben Stein's Money*, COM
Game show host: Ben Stein & Jimmy Kimmel, Co-hosts
Preschool children's series: *Sesame Street*, PBS
Children's special: *The Island on Bird Street*, SHO
Children's animated program: *Arthur*, PBS
Performer in a children's series: Fred Rogers, *Mister Rogers' Neighborhood*, PBS
Talk show: *The Rosie O'Donnell Show*, SYN
Talk show host: Rosie O'Donnell

Tony (Antoinette Perry) Awards (for Broadway Theater)
Tony Awards Given in 1999

Play: *Side Man*
Musical: *Fosse*
Book of a musical: *Parade* by Alfred Uhry
Actor, play: Brian Dennehy, *Death of a Salesman*
Actress, play: Judi Dench, *Amy's View*
Actor, musical: Martin Short, *Little Me*
Actress, musical: Bernadette Peters, *Annie Get Your Gun*
Musical score: Jason Robert Brown, *Parade*
Director, play: Robert Falls, *Death of a Salesman*
Director, musical: Matthew Bourne, *Swan Lake*
Play revival: *Death of a Salesman*
Musical revival: *Annie Get Your Gun*
Featured actor, play: Frank Wood, *Side Man*

Featured actress, play: Elizabeth Franz, *Death of a Salesman*
Featured actor, musical: Roger Bart, *You're a Good Man, Charlie Brown*
Featured actress, musical: Kristin Chenoweth, *You're a Good Man Charlie Brown*
Choreography: Matthew Bourne, *Swan Lake*
Costume design: Lez Brotherston, *Swan Lake*
Scenic design: Richard Hoover, *Not About Nightingales*
Lighting design: Andrew Bridge, *Fosse*
Orchestrations: Ralph Burns & Douglas Besterman, *Fosse*
Special Awards: Uta Hagen, Arthur Miller, Isabelle Stevenson, production of *Fool Moon*
Regional Theater: Crossroads Theater, New Brunswick, NJ

1999 Golden Globe Awards
(Awarded for work in 1998)

Film

Drama: *Saving Private Ryan*
Musical/comedy: *Shakespeare in Love*
Actress, drama: Cate Blanchett, *Elizabeth*
Actor, drama: Jim Carrey, *The Truman Show*
Actress, musical/comedy: Gwyneth Paltrow, *Shakespeare in Love*
Actor, musical/comedy: Michael Caine, *Little Voice*
Supp. actress, drama: Lynn Redgrave, *Gods and Monsters*
Supp. actor, drama: Ed Harris, *The Truman Show*
Director: Steven Spielberg, *Saving Private Ryan*
Screenplay: Marc Norman and Tom Stoppard, *Shakespeare in Love*
Foreign-language film: *Central Station* (Brazil)
Original score: Burkhard Dallwitz with additional music by Philip Glass, *The Truman Show*
Original song: "The Prayer," *Quest for Camelot: The Magic Sword*

Cecil B. De Mille award for lifetime achievement: Jack Nicholson

Television

Series, drama: *The Practice*, Fox
Actress, drama: Keri Russell, *Felicity*
Actor, drama: Dylan McDermott, *The Practice*
Series, musical/comedy: *Ally McBeal*, Fox
Actress, musical/comedy: Jenna Elfman, *Dharma and Greg*
Actor, musical/comedy: Michael J. Fox, *Spin City*
Miniseries, movie made for TV: *From the Earth to the Moon*, HBO
Actress, miniseries, movie made for TV: Angelina Jolie, *Gia*
Actor, miniseries, movie made for TV: Stanley Tucci, *Winchell*
Supporting actress, miniseries, movie made for TV: (tie) Camryn Manheim, *The Practice* and Faye Dunaway, *Gia*
Supporting actor, miniseries, movie made for TV: (tie) Don Cheadle, *The Rat Pack* and Gregory Peck, *Moby Dick*

Academy Awards (Oscars) for 1927-1998

1927-28
Picture: *Wings*
Actor: Emil Jannings, *The Way of All Flesh*
Actress: Janet Gaynor, *Seventh Heaven*
Director: Frank Borzage, *Seventh Heaven;* Lewis Milestone, *Two Arabian Knights*

1928-29
Picture: *Broadway Melody*
Actor: Warner Baxter, *In Old Arizona*
Actress: Mary Pickford, *Coquette*
Director: Frank Lloyd, *The Divine Lady*

1929-30
Picture: *All Quiet on the Western Front*
Actor: George Arliss, *Disraeli*
Actress: Norma Shearer, *The Divorcee*
Director: Lewis Milestone, *All Quiet on the Western Front*

1930-31
Picture: *Cimarron*
Actor: Lionel Barrymore, *Free Soul*
Actress: Marie Dressler, *Min and Bill*
Director: Norman Taurog, *Skippy*

1931-32
Picture: *Grand Hotel*
Actor: Fredric March, *Dr. Jekyll and Mr. Hyde;* Wallace Beery, *The Champ* (tie)
Actress: Helen Hayes, *The Sin of Madelon Claudet*
Director: Frank Borzage, *Bad Girl*
Special: Walt Disney, *Mickey Mouse*

1932-33
Picture: *Cavalcade*
Actor: Charles Laughton, *The Private Life of Henry VIII*
Actress: Katharine Hepburn, *Morning Glory*
Director: Frank Lloyd, *Cavalcade*

1934
Picture: *It Happened One Night*
Actor: Clark Gable, *It Happened One Night*
Actress: Claudette Colbert, *It Happened One Night*
Director: Frank Capra, *It Happened One Night*

1935
Picture: *Mutiny on the Bounty*
Actor: Victor McLaglen, *The Informer*
Actress: Bette Davis, *Dangerous*
Director: John Ford, *The Informer*

1936
Picture: *The Great Ziegfeld*
Actor: Paul Muni, *Story of Louis Pasteur*
Actress: Luise Rainer, *The Great Ziegfeld*
Sup. Actor: Walter Brennan, *Come and Get It*
Sup. Actress: Gale Sondergaard, *Anthony Adverse*
Director: Frank Capra, *Mr. Deeds Goes to Town*

1937
Picture: *Life of Emile Zola*
Actor: Spencer Tracy, *Captains Courageous*
Actress: Luise Rainer, *The Good Earth*
Sup. Actor: Joseph Schildkraut, *Life of Emile Zola*
Sup. Actress: Alice Brady, *In Old Chicago*
Director: Leo McCarey, *The Awful Truth*

1938
Picture: *You Can't Take It With You*
Actor: Spencer Tracy, *Boys Town*
Actress: Bette Davis, *Jezebel*
Sup. Actor: Walter Brennan, *Kentucky*
Sup. Actress: Fay Bainter, *Jezebel*
Director: Frank Capra, *You Can't Take It With You*

1939
Picture: *Gone With the Wind*
Actor: Robert Donat, *Goodbye, Mr. Chips*
Actress: Vivien Leigh, *Gone With the Wind*
Sup. Actor: Thomas Mitchell, *Stage Coach*
Sup. Actress: Hattie McDaniel, *Gone With the Wind*
Director: Victor Fleming, *Gone With the Wind*

1940
Picture: *Rebecca*
Actor: James Stewart, *The Philadelphia Story*
Actress: Ginger Rogers, *Kitty Foyle*
Sup. Actor: Walter Brennan, *The Westerner*
Sup. Actress: Jane Darwell, *The Grapes of Wrath*
Director: John Ford, *The Grapes of Wrath*

1941
Picture: *How Green Was My Valley*
Actor: Gary Cooper, *Sergeant York*
Actress: Joan Fontaine, *Suspicion*
Sup. Actor: Donald Crisp, *How Green Was My Valley*
Sup. Actress: Mary Astor, *The Great Lie*
Director: John Ford, *How Green Was My Valley*

1942
Picture: *Mrs. Miniver*
Actor: James Cagney, *Yankee Doodle Dandy*
Actress: Greer Garson, *Mrs. Miniver*
Sup. Actor: Van Heflin, *Johnny Eager*
Sup. Actress: Teresa Wright, *Mrs. Miniver*
Director: William Wyler, *Mrs. Miniver*

1943
Picture: *Casablanca*
Actor: Paul Lukas, *Watch on the Rhine*
Actress: Jennifer Jones, *The Song of Bernadette*
Sup. Actor: Charles Coburn, *The More the Merrier*
Sup. Actress: Katina Paxinou, *For Whom the Bell Tolls*
Director: Michael Curtiz, *Casablanca*

1944
Picture: *Going My Way*
Actor: Bing Crosby, *Going My Way*
Actress: Ingrid Bergman, *Gaslight*
Sup. Actor: Barry Fitzgerald, *Going My Way*
Sup. Actress: Ethel Barrymore, *None But the Lonely Heart*
Director: Leo McCarey, *Going My Way*

1945
Picture: *The Lost Weekend*
Actor: Ray Milland, *The Lost Weekend*

Actress: Joan Crawford, *Mildred Pierce*
Sup. Actor: James Dunn, *A Tree Grows in Brooklyn*
Sup. Actress: Anne Revere, *National Velvet*
Director: Billy Wilder, *The Lost Weekend*

1946
Picture: *The Best Years of Our Lives*
Actor: Fredric March, *The Best Years of Our Lives*
Actress: Olivia de Havilland, *To Each His Own*
Sup. Actor: Harold Russell, *The Best Years of Our Lives*
Sup. Actress: Anne Baxter, *The Razor's Edge*
Director: William Wyler, *The Best Years of Our Lives*

1947
Picture: *Gentleman's Agreement*
Actor: Ronald Colman, *A Double Life*
Actress: Loretta Young, *The Farmer's Daughter*
Sup. Actor: Edmund Gwenn, *Miracle on 34th Street*
Sup. Actress: Celeste Holm, *Gentleman's Agreement*
Director: Elia Kazan, *Gentleman's Agreement*

1948
Picture: *Hamlet*
Actor: Laurence Olivier, *Hamlet*
Actress: Jane Wyman, *Johnny Belinda*
Sup. Actor: Walter Huston, *Treasure of Sierra Madre*
Sup. Actress: Claire Trevor, *Key Largo*
Director: John Huston, *Treasure of Sierra Madre*

1949
Picture: *All the King's Men*
Actor: Broderick Crawford, *All the King's Men*
Actress: Olivia de Havilland, *The Heiress*
Sup. Actor: Dean Jagger, *Twelve O'Clock High*
Sup. Actress: Mercedes McCambridge, *All the King's Men*
Director: Joseph L. Mankiewicz, *Letter to Three Wives*

1950
Picture: *All About Eve*
Actor: Jose Ferrer, *Cyrano de Bergerac*
Actress: Judy Holliday, *Born Yesterday*
Sup. Actor: George Sanders, *All About Eve*
Sup. Actress: Josephine Hull, *Harvey*
Director: Joseph L. Mankiewicz, *All About Eve*

1951
Picture: *An American in Paris*
Actor: Humphrey Bogart, *The African Queen*
Actress: Vivien Leigh, *A Streetcar Named Desire*
Sup. Actor: Karl Malden, *A Streetcar Named Desire*
Sup. Actress: Kim Hunter, *A Streetcar Named Desire*
Director: George Stevens, *A Place in the Sun*

1952
Picture: *The Greatest Show on Earth*
Actor: Gary Cooper, *High Noon*
Actress: Shirley Booth, *Come Back, Little Sheba*
Sup. Actor: Anthony Quinn, *Viva Zapata!*
Sup. Actress: Gloria Grahame, *The Bad and the Beautiful*
Director: John Ford, *The Quiet Man*

1953
Picture: *From Here to Eternity*
Actor: William Holden, *Stalag 17*
Actress: Audrey Hepburn, *Roman Holiday*
Sup. Actor: Frank Sinatra, *From Here to Eternity*
Sup. Actress: Donna Reed, *From Here to Eternity*
Director: Fred Zinnemann, *From Here to Eternity*

1954
Picture: *On the Waterfront*
Actor: Marlon Brando, *On the Waterfront*
Actress: Grace Kelly, *The Country Girl*
Sup. Actor: Edmond O'Brien, *The Barefoot Contessa*
Sup. Actress: Eva Marie Saint, *On the Waterfront*
Director: Elia Kazan, *On the Waterfront*

1955
Picture: *Marty*
Actor: Ernest Borgnine, *Marty*
Actress: Anna Magnani, *The Rose Tattoo*
Sup. Actor: Jack Lemmon, *Mister Roberts*
Sup. Actress: Jo Van Fleet, *East of Eden*
Director: Delbert Mann, *Marty*

1956
Picture: *Around the World in 80 Days*
Actor: Yul Brynner, *The King and I*
Actress: Ingrid Bergman, *Anastasia*
Sup. Actor: Anthony Quinn, *Lust for Life*
Sup. Actress: Dorothy Malone, *Written on the Wind*
Director: George Stevens, *Giant*

1957
Picture: *The Bridge on the River Kwai*
Actor: Alec Guinness, *The Bridge on the River Kwai*
Actress: Joanne Woodward, *The Three Faces of Eve*
Sup. Actor: Red Buttons, *Sayonara*
Sup. Actress: Miyoshi Umeki, *Sayonara*
Director: David Lean, *The Bridge on the River Kwai*

1958
Picture: *Gigi*
Actor: David Niven, *Separate Tables*
Actress: Susan Hayward, *I Want to Live*
Sup. Actor: Burl Ives, *The Big Country*
Sup. Actress: Wendy Hiller, *Separate Tables*
Director: Vincente Minnelli, *Gigi*

1959
Picture: *Ben-Hur*
Actor: Charlton Heston, *Ben-Hur*
Actress: Simone Signoret, *Room at the Top*
Sup. Actor: Hugh Griffith, *Ben-Hur*
Sup. Actress: Shelley Winters, *Diary of Anne Frank*
Director: William Wyler, *Ben-Hur*

1960
Picture: *The Apartment*
Actor: Burt Lancaster, *Elmer Gantry*
Actress: Elizabeth Taylor, *Butterfield 8*
Sup. Actor: Peter Ustinov, *Spartacus*
Sup. Actress: Shirley Jones, *Elmer Gantry*
Director: Billy Wilder, *The Apartment*

1961
Picture: *West Side Story*
Actor: Maximilian Schell, *Judgment at Nuremberg*
Actress: Sophia Loren, *Two Women*
Sup. Actor: George Chakiris, *West Side Story*
Sup. Actress: Rita Moreno, *West Side Story*
Director: Jerome Robbins, Robert Wise, *West Side Story*

1962
Picture: *Lawrence of Arabia*
Actor: Gregory Peck, *To Kill a Mockingbird*
Actress: Anne Bancroft, *The Miracle Worker*
Sup. Actor: Ed Begley, *Sweet Bird of Youth*
Sup. Actress: Patty Duke, *The Miracle Worker*
Director: David Lean, *Lawrence of Arabia*

1963
Picture: *Tom Jones*
Actor: Sidney Poitier, *Lilies of the Field*
Actress: Patricia Neal, *Hud*
Sup. Actor: Melvyn Douglas, *Hud*
Sup. Actress: Margaret Rutherford, *The V.I.P.s*
Director: Tony Richardson, *Tom Jones*

1964
Picture: *My Fair Lady*
Actor: Rex Harrison, *My Fair Lady*
Actress: Julie Andrews, *Mary Poppins*
Sup. Actor: Peter Ustinov, *Topkapi*
Sup. Actress: Lila Kedrova, *Zorba the Greek*
Director: George Cukor, *My Fair Lady*

1965
Picture: *The Sound of Music*
Actor: Lee Marvin, *Cat Ballou*
Actress: Julie Christie, *Darling*
Sup. Actor: Martin Balsam, *A Thousand Clowns*
Sup. Actress: Shelley Winters, *A Patch of Blue*
Director: Robert Wise, *The Sound of Music*

1966
Picture: *A Man for All Seasons*
Actor: Paul Scofield, *A Man for All Seasons*
Actress: Elizabeth Taylor, *Who's Afraid of Virginia Woolf?*
Sup. Actor: Walter Matthau, *The Fortune Cookie*
Sup. Actress: Sandy Dennis, *Who's Afraid of Virginia Woolf?*
Director: Fred Zinnemann, *A Man for All Seasons*

1967
Picture: *In the Heat of the Night*
Actor: Rod Steiger, *In the Heat of the Night*
Actress: Katharine Hepburn, *Guess Who's Coming to Dinner*
Sup. Actor: George Kennedy, *Cool Hand Luke*
Sup. Actress: Estelle Parsons, *Bonnie and Clyde*
Director: Mike Nichols, *The Graduate*

1968
Picture: *Oliver!*
Actor: Cliff Robertson, *Charly*
Actress: Katharine Hepburn, *The Lion in Winter;*
Barbra Streisand, *Funny Girl* (tie)
Sup. Actor: Jack Albertson, *The Subject Was Roses*
Sup. Actress: Ruth Gordon, *Rosemary's Baby*
Director: Sir Carol Reed, *Oliver!*

1969
Picture: *Midnight Cowboy*
Actor: John Wayne, *True Grit*
Actress: Maggie Smith, *The Prime of Miss Jean Brodie*
Sup. Actor: Gig Young, *They Shoot Horses, Don't They?*
Sup. Actress: Goldie Hawn, *Cactus Flower*
Director: John Schlesinger, *Midnight Cowboy*

1970
Picture: *Patton*
Actor: George C. Scott, *Patton* (refused)
Actress: Glenda Jackson, *Women in Love*
Sup. Actor: John Mills, *Ryan's Daughter*
Sup. Actress: Helen Hayes, *Airport*
Director: Franklin Schaffner, *Patton*

1971
Picture: *The French Connection*
Actor: Gene Hackman, *The French Connection*
Actress: Jane Fonda, *Klute*
Sup. Actor: Ben Johnson, *The Last Picture Show*
Sup. Actress: Cloris Leachman, *The Last Picture Show*
Director: William Friedkin, *The French Connection*

1972
Picture: *The Godfather*
Actor: Marlon Brando, *The Godfather* (refused)
Actress: Liza Minnelli, *Cabaret*
Sup. Actor: Joel Grey, *Cabaret*
Sup. Actress: Eileen Heckart, *Butterflies Are Free*
Director: Bob Fosse, *Cabaret*

1973
Picture: *The Sting*
Actor: Jack Lemmon, *Save the Tiger*
Actress: Glenda Jackson, *A Touch of Class*
Sup. Actor: John Houseman, *The Paper Chase*
Sup. Actress: Tatum O'Neal, *Paper Moon*
Director: George Roy Hill, *The Sting*

1974
Picture: *The Godfather, Part II*
Actor: Art Carney, *Harry and Tonto*
Actress: Ellen Burstyn, *Alice Doesn't Live Here Anymore*
Sup. Actor: Robert DeNiro, *The Godfather, Part II*
Sup. Actress: Ingrid Bergman, *Murder on the Orient Express*
Director: Francis Ford Coppola, *The Godfather, Part II*

1975
Picture: *One Flew Over the Cuckoo's Nest*
Actor: Jack Nicholson, *One Flew Over the Cuckoo's Nest*
Actress: Louise Fletcher, *One Flew Over the Cuckoo's Nest*
Sup. Actor: George Burns, *The Sunshine Boys*
Sup. Actress: Lee Grant, *Shampoo*
Director: Milos Forman, *One Flew Over the Cuckoo's Nest*

1976
Picture: *Rocky*
Actor: Peter Finch, *Network*
Actress: Faye Dunaway, *Network*
Sup. Actor: Jason Robards, *All the President's Men*
Sup. Actress: Beatrice Straight, *Network*
Director: John G. Avildsen, *Rocky*

1977
Picture: *Annie Hall*
Actor: Richard Dreyfuss, *The Goodbye Girl*
Actress: Diane Keaton, *Annie Hall*
Sup. Actor: Jason Robards, *Julia*
Sup. Actress: Vanessa Redgrave, *Julia*
Director: Woody Allen, *Annie Hall*

1978
Picture: *The Deer Hunter*
Actor: Jon Voight, *Coming Home*
Actress: Jane Fonda, *Coming Home*
Sup. Actor: Christopher Walken, *The Deer Hunter*
Sup. Actress: Maggie Smith, *California Suite*
Director: Michael Cimino, *The Deer Hunter*

1979
Picture: *Kramer vs. Kramer*
Actor: Dustin Hoffman, *Kramer vs. Kramer*
Actress: Sally Field, *Norma Rae*
Sup. Actor: Melvyn Douglas, *Being There*
Sup. Actress: Meryl Streep, *Kramer vs. Kramer*
Director: Robert Benton, *Kramer vs. Kramer*

1980
Picture: *Ordinary People*
Actor: Robert DeNiro, *Raging Bull*
Actress: Sissy Spacek, *Coal Miner's Daughter*
Sup. Actor: Timothy Hutton, *Ordinary People*
Sup. Actress: Mary Steenburgen, *Melvin & Howard*
Director: Robert Redford, *Ordinary People*

1981
Picture: *Chariots of Fire*
Actor: Henry Fonda, *On Golden Pond*
Actress: Katharine Hepburn, *On Golden Pond*
Sup. Actor: John Gielgud, *Arthur*
Sup. Actress: Maureen Stapleton, *Reds*
Director: Warren Beatty, *Reds*

1982
Picture: *Gandhi*
Actor: Ben Kingsley, *Gandhi*
Actress: Meryl Streep, *Sophie's Choice*
Sup. Actor: Louis Gossett Jr., *An Officer and a Gentleman*
Sup. Actress: Jessica Lange, *Tootsie*
Director: Richard Attenborough, *Gandhi*

1983
Picture: *Terms of Endearment*
Actor: Robert Duvall, *Tender Mercies*
Actress: Shirley MacLaine, *Terms of Endearment*
Sup. Actor: Jack Nicholson, *Terms of Endearment*
Sup. Actress: Linda Hunt, *The Year of Living Dangerously*
Director: James L. Brooks, *Terms of Endearment*

1984
Picture: *Amadeus*
Actor: F. Murray Abraham, *Amadeus*
Actress: Sally Field, *Places in the Heart*
Sup. Actor: Haing S. Ngor, *The Killing Fields*
Sup. Actress: Peggy Ashcroft, *A Passage to India*
Director: Milos Forman, *Amadeus*

1985
Picture: *Out of Africa*
Actor: William Hurt, *Kiss of the Spider Woman*
Actress: Geraldine Page, *The Trip to Bountiful*
Sup. Actor: Don Ameche, *Cocoon*
Sup. Actress: Anjelica Huston, *Prizzi's Honor*
Director: Sydney Pollack, *Out of Africa*

1986
Picture: *Platoon*
Actor: Paul Newman, *The Color of Money*
Actress: Marlee Matlin, *Children of a Lesser God*
Sup. Actor: Michael Caine, *Hannah and Her Sisters*
Sup. Actress: Dianne Wiest, *Hannah and Her Sisters*
Director: Oliver Stone, *Platoon*

1987
Picture: *The Last Emperor*
Actor: Michael Douglas, *Wall Street*
Actress: Cher, *Moonstruck*
Sup. Actor: Sean Connery, *The Untouchables*
Sup. Actress: Olympia Dukakis, *Moonstruck*
Director: Bernardo Bertolucci, *The Last Emperor*

1988
Picture: *Rain Man*
Actor: Dustin Hoffman, *Rain Man*

Actress: Jodie Foster, *The Accused*
Sup. Actor: Kevin Kline, *A Fish Called Wanda*
Sup. Actress: Geena Davis, *The Accidental Tourist*
Director: Barry Levinson, *Rain Man*

1989
Picture: *Driving Miss Daisy*
Actor: Daniel Day-Lewis, *My Left Foot*
Actress: Jessica Tandy, *Driving Miss Daisy*
Sup. Actor: Denzel Washington, *Glory*
Sup. Actress: Brenda Fricker, *My Left Foot*
Director: Oliver Stone, *Born on the Fourth of July*

1990
Picture: *Dances With Wolves*
Actor: Jeremy Irons, *Reversal of Fortune*
Actress: Kathy Bates, *Misery*
Sup. Actor: Joe Pesci, *Goodfellas*
Sup. Actress: Whoopi Goldberg, *Ghost*
Director: Kevin Costner, *Dances With Wolves*

1991
Picture: *The Silence of the Lambs*
Actor: Anthony Hopkins, *The Silence of the Lambs*
Actress: Jodie Foster, *The Silence of the Lambs*
Sup. Actor: Jack Palance, *City Slickers*
Sup. Actress: Mercedes Ruehl, *The Fisher King*
Director: Jonathan Demme, *The Silence of the Lambs*

1992
Picture: *Unforgiven*
Actor: Al Pacino, *Scent of a Woman*
Actress: Emma Thompson, *Howards End*
Sup. Actor: Gene Hackman, *Unforgiven*
Sup. Actress: Marisa Tomei, *My Cousin Vinny*
Director: Clint Eastwood, *Unforgiven*

1993
Picture: *Schindler's List*
Actor: Tom Hanks, *Philadelphia*
Actress: Holly Hunter, *The Piano*
Sup. Actor: Tommy Lee Jones, *The Fugitive*
Sup. Actress: Anna Paquin, *The Piano*
Director: Steven Spielberg, *Schindler's List*

1994
Picture: *Forrest Gump*
Actor: Tom Hanks, *Forrest Gump*
Actress: Jessica Lange, *Blue Sky*
Sup. Actor: Martin Landau, *Ed Wood*
Sup. Actress: Dianne Wiest, *Bullets Over Broadway*
Director: Robert Zemeckis, *Forrest Gump*

1995
Picture: *Braveheart*
Actor: Nicolas Cage, *Leaving Las Vegas*
Actress: Susan Sarandon, *Dead Man Walking*
Sup. Actor: Kevin Spacey, *The Usual Suspects*
Sup. Actress: Mira Sorvino, *Mighty Aphrodite*
Director: Mel Gibson, *Braveheart*

1996
Picture: *The English Patient*

Actor: Geoffrey Rush, *Shine*
Actress: Frances McDormand, *Fargo*
Sup. Actor: Cuba Gooding Jr., *Jerry Maguire*
Sup. Actress: Juliette Binoche, *The English Patient*
Director: Anthony Minghella, *The English Patient*

1997
Picture: *Titanic*
Actor: Jack Nicholson, *As Good As It Gets*
Actress: Helen Hunt, *As Good As It Gets*
Sup. Actor: Robin Williams, *Good Will Hunting*
Sup. Actress: Kim Basinger, *L.A. Confidential*
Director: James Cameron, *Titanic*

1998
Picture: *Shakespeare in Love*
Actor: Roberto Benigni, *Life Is Beautiful*
Actress: Gwyneth Paltrow, *Shakespeare in Love*
Sup. Actor: James Coburn, *Affliction*
Sup. Actress: Judi Dench, *Shakespeare in Love*
Director: Steven Spielberg, *Saving Private Ryan*
Foreign Film: *Life Is Beautiful*, Italy
Original Screenplay: Marc Norman and Tom Stoppard, *Shakespeare in Love*
Adapted Screenplay: Bill Condon, *Gods and Monsters*
Cinematography: Janusz Kaminski, *Saving Private Ryan*
Art Direction: Martin Childs (art director) and Jill Quertier (set decorator), *Shakespeare in Love*
Film Editing: Michael Kahn, *Saving Private Ryan*
Original Song: "When You Believe," *The Prince of Egypt*, Stephen Schwartz
Original musical or comedy score: Stephen Warbeck, *Shakespeare in Love*
Original dramatic score: Nicola Piovani, *Life Is Beautiful*
Costume Design: Sandy Powell, *Shakespeare in Love*
Makeup: Jenny Shircore, *Elizabeth*
Sound: Gary Rydstrom, Gary Summers, Andy Nelson, and Ronald Judkins, *Saving Private Ryan*
Documentary Feature: James Moll and Ken Lipper, *The Last Days*
Documentary Short Subject: Keiko Ibi, *The Personals; Improvisations on Romance in the Golden Years*
Short Film, Live: Kim Magnusson and Anders Thomas Jensen, *Election Night* (Valgaften)
Short Film, Animated: Chris Wedge, *Bunny*
Visual Effects: Joel Hynek, Nicholas Brooks, Stuart Robertson, and Kevin Mack, *What Dreams May Come*
Sound Effects Editing: Gary Rydstrom and Richard Hymns, *Saving Private Ryan*
Thalberg Award: Norman Jewison
Honorary Oscar: Elia Kazan

Other Film Awards
Year in parentheses is year awarded

Cannes Film Festival Awards (1999), Feature Films. Palme d'Or (Golden Palm): *Rosetta*, Luc and Jean-Pierre Dardenne; Grand Prix (Grand Jury Prize): *L'humanite* (Humanity), Bruno Dumont; best actress: (co-winner) Severine Caneele, *L'humanite* and Emilie Dequenne, *Rosetta*; best actor: Emmanuel Schotte, *L'humanite*; best director: Pedro Almodovar, *Todo Sobre Mi Madre* (All About My Mother); best screenplay: Alexandre Sokourov, *Moloch*. Special Jury Prize (daring and originality): *A Carta*, Manoel de Oliveira. Camera d'Or (Golden Camera—first-time director): de Murali Nair, *Marana Simhasanam*; Grand Prix Technique de la Commission Superieure Technique (technical prize in art direction): Chen Kaige, *The Emperor and the Assassin*. Short Films. Palme d'Or: *When the Day Breaks*, Wendy Tilby and Amanda Forbis; Grand Prix: (co-winner) *Stop*, Rodolphe Marconi and *So-Poong*, Song Ilgon.

Directors Guild of America Awards (1999), Feature film: Steven Spielberg, *Saving Private Ryan*; documentary: Jerry Blumenthal, Peter Gilbert, Gordon Quinn, *Vietnam: Long Time Coming*

Sundance Film Festival Awards (1999), Grand Jury Prize: (drama) *Three Seasons*; (docu.) *American Movie*. Directing Award: (drama) Eric Mendelsohn, *Judy Berlin*; (docu.) Barbara Sonneborn, *Regret to Inform*. Waldo Salt Screenwriting Award: Audrey Wells, *Guinevere* and Frank Whaley, *Joe the King* (split). Freedom of Expression Award: *The Black Press: Soldiers Without Swords*. Audience Award: (drama) *Three Seasons*; (docu.) *Genghis Blues*; (world) *Run Lola Run* and *Train of Life* (split). Cinematography Award: (drama) Lisa Rinzler, *Three Seasons*; (docu.) Emiko Omori, *Rabbit in the Moon* and *Regret to Inform*. Filmmakers Trophy: (drama) *Tumbleweeds*; (docu.) *Sing Faster: The Strangehands Ringcycle*. Special Jury Award: *On the Ropes*; (for comedic perf.) Steve Zahn, *Happy, Texas*; (for distinctive vision in filmmaking) *Treasure Island*. Short Filmmaking: (Jury Prize) *More*; (Special Jury Award) *Fishbelly White*. Latin Amer. Cinema Award: (Jury Prize) *Santitos*; (Special Jury Award) *Life Is to Whistle*.

Grammy Awards

Source: National Academy of Recording Arts & Sciences

Selected Grammy Awards for 1998

Record (single): "My Heart Will Go On," Celine Dion
Album: *The Miseducation of Lauryn Hill,* Lauryn Hill
Song: "My Heart Will Go On," James Horner and Will Jennings, songwriters (Celine Dion)
New artist: Lauryn Hill
Female pop vocal perf.: Celine Dion, "My Heart Will Go On"
Male pop vocal perf.: Eric Clapton, "My Father's Eyes"
Pop album: *Ray of Light,* Madonna
Group or duo pop perf. with vocal: The Brian Setzer Orchestra, "Jump Jive An' Wail"
Traditional pop vocal perf.: Live at Carnegie Hall-The 50th Anniversary Concert, Patti Page
Female rock vocal perf.: Alanis Morissette, "Uninvited"
Male rock vocal perf.: Lenny Kravitz, "Fly Away"
Rock perf., duo or group with vocal: Aerosmith, "Pink"
Rock song: "Uninvited," Alanis Morissette
Rock album: *The Globe Sessions,* Sheryl Crow
Female R & B vocal perf.: Lauryn Hill, "Doo Wop (That Thing)"
Male R & B vocal perf.: Stevie Wonder, "St. Louis Blues"
R & B perf., duo or group with vocal: Brandy & Monica, "The Boy Is Mine"
R & B song: "Doo Wop (That Thing)," Lauryn Hill
R & B album: *The Miseducation of Lauryn Hill,* Lauryn Hill
Rap solo perf.: Will Smith, "Gettin' Jiggy Wit It,"
Rap perf., duo or group: Beastie Boys, "Intergalactic"

Rap album: *Vol. 2...Hard Knock Life,* Jay-Z
Jazz vocal perf.: Shirley Horn, "I Remember Miles"
Contemporary jazz perf.: "Imaginary Day," Pat Metheny Group
Contemporary blues album: *Slow Down,* Keb' Mo'
Traditional blues album: *Any Place I'm Going,* Otis Rush
Female country vocal perf.: Shania Twain, "You're Still the One"
Male country vocal perf.: Vince Gill, "If You Ever Have Forever In Mind"
Country perf., duo or group with vocal: Dixie Chicks, "There's Your Trouble"
Country song: "You're Still the One," Robert John "Mutt" Lange and Shania Twain
Country album: *Wide Open Spaces,* Dixie Chicks
Contemporary folk album: *Car Wheels on a Gravel Road,* Lucinda Williams
Traditional folk album: *Long Journey Home,* The Chieftains with Various Artists
Reggae album: *Friends,* Sly and Robbie
Producer, non-classical, Rob Cavallo; classical, Steven Epstein
Opera recording: *Bartok: Bluebeard's Castle;* Pierre Boulez, conductor
Classical vocal perf.: Renee Fleming (soprano), *The Beautiful Voice*
Classical album: *Barber: Prayers of Kierkegaard/Vaughn Williams: Dona Nobis Pacem/Bartok: Cantana Profana*; Robert Shaw, conductor

Grammy Awards for 1958-97

Record (single)	Year	Album
Domenico Modugno, "Nel Blu Dipinto Di Blu (Volare)"	1958	Henry Mancini, *The Music From Peter Gunn*
Bobby Darin, "Mack the Knife"	1959	Frank Sinatra, *Come Dance With Me*
Percy Faith, "Theme From a Summer Place"	1960	Bob Newhart, *Button Down Mind*
Henry Mancini, "Moon River"	1961	Judy Garland, *Judy at Carnegie Hall*
Tony Bennett, "I Left My Heart in San Francisco"	1962	Vaughn Meader, *The First Family*
Henry Mancini, "The Days of Wine and Roses"	1963	Barbra Streisand, *The Barbra Streisand Album*
Stan Getz, Astrud Gilberto, "The Girl From Ipanema"	1964	Stan Getz, Astrud Gilberto, *Getz/Gilberto*
Herb Alpert, "A Taste of Honey"	1965	Frank Sinatra, *September of My Years*
Frank Sinatra, "Strangers in the Night"	1966	Frank Sinatra, *A Man and His Music*
5th Dimension, "Up, Up and Away"	1967	The Beatles, *Sgt. Pepper's Lonely Hearts Club Band*
Simon & Garfunkel, "Mrs. Robinson"	1968	Glen Campbell, *By the Time I Get to Phoenix*
5th Dimension, "Aquarius/Let the Sunshine In"	1969	Blood Sweat and Tears, *Blood, Sweat and Tears*
Simon & Garfunkel, "Bridge Over Troubled Water"	1970	Simon & Garfunkel, *Bridge Over Troubled Water*
Carole King, "It's Too Late"	1971	Carole King, *Tapestry*
Roberta Flack, "The First Time Ever I Saw Your Face"	1972	George Harrison and friends, *The Concert for Bangla Desh*
Roberta Flack, "Killing Me Softly With His Song"	1973	Stevie Wonder, *Innervisions*
Olivia Newton-John, "I Honestly Love You"	1974	Stevie Wonder, *Fulfillingness' First Finale*
Captain & Tennille, "Love Will Keep Us Together"	1975	Paul Simon, *Still Crazy After All These Years*
George Benson, "This Masquerade"	1976	Stevie Wonder, *Songs in the Key of Life*
Eagles, "Hotel California"	1977	Fleetwood Mac, *Rumours*
Billy Joel, "Just the Way You Are"	1978	Bee Gees, *Saturday Night Fever*
The Doobie Brothers, "What a Fool Believes"	1979	Billy Joel, *52nd Street*
Christopher Cross, "Sailing"	1980	Christopher Cross, *Christopher Cross*
Kim Carnes, "Bette Davis Eyes"	1981	John Lennon, Yoko Ono, *Double Fantasy*
Toto, "Rosanna"	1982	Toto, *Toto IV*
Michael Jackson, "Beat It"	1983	Michael Jackson, *Thriller*
Tina Turner, "What's Love Got to Do With It"	1984	Lionel Richie, *Can't Slow Down*
USA for Africa, "We Are the World"	1985	Phil Collins, *No Jacket Required*
Steve Winwood, "Higher Love"	1986	Paul Simon, *Graceland*
Paul Simon, "Graceland"	1987	U2, *The Joshua Tree*
Bobby McFerrin, "Don't Worry, Be Happy"	1988	George Michael, *Faith*
Bette Midler, "Wind Beneath My Wings"	1989	Bonnie Raitt, *Nick of Time*
Phil Collins, "Another Day in Paradise"	1990	Quincy Jones, *Back on the Block*
Natalie Cole, with Nat "King" Cole, "Unforgettable"	1991	Natalie Cole, with Nat "King" Cole, *Unforgettable*
Eric Clapton, "Tears in Heaven"	1992	Eric Clapton, *Unplugged*
Whitney Houston, "I Will Always Love You"	1993	Whitney Houston, *The Bodyguard*
Sheryl Crow, "All I Wanna Do"	1994	Tony Bennett, *MTV Unplugged*
Seal, "Kiss From a Rose"	1995	Alanis Morissette, *Jagged Little Pill*
Eric Clapton, "Change the World"	1996	Celine Dion, *Falling Into You*
Shawn Colvin, "Sunny Came Home"	1997	Bob Dylan, *Time Out of Mind*

1999 MTV Video Music Awards

Video of the Year: Lauryn Hill, "Doo Wop (That Thing)"
Best Male Video: Will Smith, "Miami"
Best Female Video: Lauryn Hill, "Doo Wop (That Thing)"
Best Group Video: TLC, "No Scrubs"
Best Rap Video: Jay-Z featuring Ja & Amil-lion, "Can I Get A..."
Best Dance Video: Ricky Martin, "Livin' La Vida Loca"
Best Pop Video: Ricky Martin, "Livin' La Vida Loca"
Best Rock Video: Korn, "Freak on a Leash"
Best Hip Hop Video: Beastie Boys, "Intergalactic"
Best New Artist: Eminem, "My Name Is"

Breakthrough Video: Fatboy Slim, "Praise You"
Best R&B Video: Lauryn Hill, "Doo Wop (That Thing)"
Best Video From a Film: Madonna, "Beautiful Stranger," from *Austin Powers: The Spy Who Shagged Me*
Best Direction: Fatboy Slim, "Praise You"
Best Choreography: Fatboy Slim, "Praise You"
Best Special Effects: Garbage, "Special"
Best Art Direction: Lauryn Hill, "Doo Wop (That Thing)"
Best Editing: Korn, "Freak on a Leash"
Best Cinematography: Marilyn Manson, "The D"
Viewers' Choice: Backstreet Boys, "I Want It That Way"

RELIGIOUS INFORMATION

Membership of Religious Groups in the U.S.

Source: *1999 Yearbook of American & Canadian Churches*, copyright National Council of the Churches of Christ in the USA; World Almanac research

These membership figures generally are based on reports made by officials of each group, and not on any religious census. Figures from other sources may vary. Many groups keep careful records; others only estimate. Not all groups report annually. Christian church membership figures reported in this table are inclusive and refer to *all* "members," not simply full communicants or confirmed members. Definitions of "member," however, vary from one denomination to another.

The number of houses of worship appears in parentheses. * Indicates that the group declines to make membership figures public. Groups reporting fewer than 5,000 members are not included; where membership numbers are not available, only those groups with 50 or more houses of worship are listed.

Religious Group	Members
Adventist churches:	
Advent Christian Ch. (311)	26,819
Ch. of God Gen. Conf. (Oregon, IL; Morrow, GA) (90)	5,040
Seventh-day Adventists (4,348)	825,654
American Catholic Church (100)	**25,000**
American Rescue Workers (15)	**10,000**
Apostolic Christian Church of America (89)	12,538
Bahá'í Faith (7,000 centers)	**133,000[1]**
Baptist churches:	
American Baptist Assn. (1,705)	250,000
American Baptist Chs. in the U.S.A. (5,830)	1,503,267
Baptist Bible Fellowship Intl. (4,500)	1,200,000
Baptist General Conference (879)	134,795
Baptist Missionary Assn. of America (1,342)	234,334
Conservative Baptist Assn. of America (1,084)	200,000
Free Will Baptists, Natl. Assn. of (2,320)	210,305
General Assn. of General Baptists (790)	72,326
General Assn. of Regular Baptist Chs. (1,440)	115,950
Natl. Baptist Convention, U.S.A., Inc. (33,000)	8,200,000
Natl. Missionary Baptist Convention of America	2,500,000
North American Baptist Conference (268)	43,850
Progressive National Baptist Convention (2,000)	2,500,000
Separate Baptists in Christ (100)	8,000
Southern Baptist Convention (40,887)	15,891,514
Brethren in Christ (199)	**18,424**
Brethren (German Baptists):	
Brethren Ch. (Ashland, OH) (117)	13,856
Church of the Brethren (1,095)	141,400
Grace Brethren Chs., Fellowship of (260)	30,371
Old German Baptist Brethren (57)	5,832
Buddhist Churches of America (60)	**15,750[1]**
Christian Brethren (Plymouth Brethren) (1,150)	**100,000**
Christian Church (Disciples of Christ) (3,818)	**879,436**
Christian Ch. of N.A., Gen. Council (350)	**31,558**
Christian Congregation, Inc. (1,438)	**115,881**
Christian and Missionary Alliance (1,964)	**328,078**
Christian Union, Churches of Christ in (240)	**10,400**
Church Christ in Christian Union (226)	**9,858**
Church of Christ (Holiness) U.S.A. (166)	**10,243**
Church of Christ Scientist (2,200)	*
Church of the United Brethren in Christ (280)	**23,585**
Churches of Christ (14,470)	**1,800,000**
Churches of God:	
Chs. of God, General Conference (342)	31,557
Ch. of God (Anderson, IN) (2,347)	229,302
Ch. of God (Seventh Day), Denver, CO (175)	10,000
Ch. of God by Faith Inc. (145)	8,235
Ch. of God, Mountain Assembly (118)	6,140
Church of the Nazarene (5,118)	**619,576**
Community Churches, Intl. Council of (517)	**250,000**
Congreg. Christian Chs., Nat'l Assoc. of (435)	**68,510**
Conservative Congregational Christian Conference (227)	**38,956**
Eastern Orthodox churches:	
American Carpatho-Russian Orthodox Greek Catholic Ch. (79)	12,998
Antiochian Orthodox Christian Diocese of N.A. (16)	50,000
Apostolic Catholic Assyrian Ch. of the East, N.A. Diocese (22)	120,000
Armenian Apostolic Ch. of America (28)	200,000
Diocese of America, Armenian Church (72)	414,000
Coptic Orthodox Ch. (85)	180,000
Greek Orthodox Diocese of America (523)	1,954,500
Malankara Mar Thoma Syñan Ch. Diocese of N.A. and Europe (65)	30,000
Orthodox Ch. in America (600)	2,000,000
Romanian Orthodox Episcopate of N. America (37)	65,000
Russian Orthodox Church Outside of Russia (177)	*
Syrian Orthodox Ch. of Antioch (21)	32,500

Religious Group	Members
Episcopal Church (7,379)	**2,339,113**
Evangelical Church (134)	**12,430**
Evangelical Congregational Church (148)	**22,957**
Evangelical Covenant Church (622)	**93,414**
Evangelical Free Church of America (1,224)	**242,619**
Friends:	
Evangelical Friends Intl.-N.A. Region (92)	8,666
Friends General Conference (620)	32,000
Friends United Meeting (620)	32,000
Religious Society of Friends (Conservative) (1,200)	104,000
Full Gospel Fellowship of Churches and Ministers Intl. (650)	**195,000**
General Church of the New Jerusalem (34)	**5,424**
Grace Gospel Fellowship (128)	**60,000**
Hindu	**1,285,000[1]**
Independent Fundamental Churches of America (670)	**69,857**
Islam	**5,500,000[1]**
Jehovah's Witnesses (10,883)	**974,719**
Jewish organizations:	
Union of American Hebrew Congregations (Reform) (889)	1,500,000
Union of Orthodox Jewish Congregations of America (800)	1,075,000
United Synagogues of Conservative Judaism, The (755)	1,500,000
Latter-day Saints:	
Ch. of Jesus Christ of Latter-day Saints (Mormon) (10,811)	4,923,100
Reorganized Ch. of Jesus Christ of Latter-day Saints (1,237)	248,523
Liberal Catholic Ch.—Province of the U.S.A. (16)	**6,500**
Lutheran churches:	
Apostolic Lutheran Ch. of America (60)	*
Ch. of the Lutheran Brethren of America (117)	13,530
Ch. of the Lutheran Confession (74)	8,768
Evangelical Lutheran Ch. in America (10,889)	5,185,055
Evangelical Lutheran Synod (136)	22,089
Free Lutheran Congregations, Assn. of (243)	32,659
Latvian Evangelical Lutheran Church in America (74)	16,900
Lutheran Ch.—Missouri Synod (6,215)	2,603,036
Lutheran Chs., American Assn. of (90)	18,704
Wisconsin Evangelical Lutheran Synod (1,240)	411,295
Mennonite churches:	
Beachy Amish Mennonite Chs. (114)	7,853
Church of God in Christ (Mennonite) (102)	11,551
Hutterian Brethren (428)	42,800
Mennonite Brethren Chs., Gen. Conf. (368)	82,130
Mennonite Church (1,004)	90,959
Mennonite Ch., General Conference of (264)	34,731
Old Order Amish Ch. (898)	80,820
Methodist churches:	
African Methodist Episcopal Ch. (8,000)	3,500,000
African Methodist Episcopal Zion Ch. (3,098)	1,252,369
Evangelical Methodist Ch. (123)	8,615
Free Methodist Ch. of North America (1,029)	72,834
Primitive Methodist Ch. in the U.S.A. (78)	6,588
Southern Methodist Ch. (122)	7,992
United Methodist Ch. (36,170)	8,496,047
The Wesleyan Church (1,578)	119,107
Metropolitan Community Churches, Universal Fellowship of (285)	**46,000**
Missionary Church (333)	**47,550**
Moravian churches:	
Moravian Ch. in America, North Province (260)	30,371

Religious Group	Members	Religious Group	Members
Natl. Organization of the New Apostolic Ch. of North America (554)	**41,863**	**Presbyterian churches:**	
		Associate Reformed Presbyterian Ch. (General Synod) (238)	40,060
Pentecostal churches:		Cumberland Presbyterian Ch. (771)	88,068
Apostolic Faith Mission Ch. of God (18)	10,450	Cumberland Presbyterian Ch. in America (152)	15,142
Apostolic Overcoming Holy Catholic Church of God Inc.(146)	12,871	Evangelical Presbyterian Ch. (186)	57,502
Assemblies of God (11,920)	2,494,574	Genl. Assembly of the Korean Presbyterian Church in America (203)	26,988
Bible Church of Christ (6)	6,850	Orthodox Presbyterian Ch. (198)	21,765
Bible Fellowship Church (58)	7,132	Presbyterian Ch. in America (1,340)	279,549
Church of God (Cleveland, TN) (6,060)	753,230	Presbyterian Ch. (U.S.A.) (11,295)	3,610,753
Church of God in Christ (15,300)	5,499,875	Reformed Presbyterian Ch. of N. America (86)	6,105
Church of God of Prophecy (1,908)	76,531	**Reformed churches:**	
Elim Fellowship (90)	*	Christian Reformed Ch. in N. America (723)	196,464
Intl. Ch. of the Foursquare Gospel (1,832)	231,522	Hungarian Reformed Ch. in America (27)	9,780
Intl. Pentecostal Church of Christ (70)	5,311	Netherlands Reformed Congregations (23)	8,753
Intl. Pentecostal Holiness Church (1,681)	170,382	Protestant Reformed Churches in America (27)	6,494
Open Bible Standard Chs. (374)	*	Reformed Ch. in America (949)	305,476
Pentecostal Assemblies of the World Inc. (1,600)	1,000,000	United Church of Christ (6,061)	1,438,181
Pentecostal Church of God (1,230)	111,900	**Reformed Episcopal Church (102)**	**6,084**
Pentecostal Free Will Baptist Ch. (157)	16,000	**Roman Catholic Church (22,728)**	**61,207,914**
United Pentecostal Ch. Intl. (3,790)	*	**Salvation Army (1,264)**	**453,150**
Polish National Catholic Church (143)	**50,000**	**Unitarian Universalist Assn. of N. A. (1043)**	**214,000**

(1) Estimate; figures from other sources may vary.

Headquarters of Selected Religious Groups in the U.S.

Source: *1999 Yearbook of American & Canadian Churches,* copyright National Council of the Churches of Christ in the USA; World Almanac research

Year organized in parentheses

African Methodist Episcopal Church, (1787), 1134 11th St. NW, Washington, DC 20001; Senior Bishop, Bishop John H. Adams

African Methodist Episcopal Zion Church (1796), PO Box 32843, Charlotte, NC 28232; Bishop Nathaniel Jarrett Jr.

American Baptist Churches in the U.S.A. (1907), PO Box 851, Valley Forge, PA 19482; http://www.abc-usa.org; Pres., James B. Johnson

American Hebrew Congregations, Union of, 838 5th Ave., New York, NY 10021; http://www.uahcweb.org; Rabbi Eric Yoffie

American Rescue Workers (1890), 643 Elmira St., Williamsport, PA 17701; http://www.arwus.com; Commander-in-Chief & Pres., Gen. Claude S. Astin Jr., Rev.

Antiochian Orthodox Christian Archdiocese of North America (1895), 358 Mountain Rd., Englewood, NJ 07631; http://www.archdiocese@antiochian.org; Primate, Metropolitan Philip Saliba

Armenian Apostolic Church of America (1887), **Eastern Prelacy:** 138 E. 39th St., New York, NY 10016; http://www.armprelacy.org; Prelate, Bishop Oshagan Choloyan; **Western Prelacy:** 4401 Russel Ave., Los Angeles, CA 90027; Prelate, Bishop Moushegh Mardirossian

Assemblies of God (1914), 1445 Boonville Ave., Springfield, MO 65802; http://www.agifellowship.org; Gen. Supt., Thomas E. Trask

Bahá'í Faith, National Spiritual Assembly of the Bahá'í's of the U.S., 536 Sheridan Rd., Wilmette, IL 60091; http://www.us.bahai.org; Secy. Gen., Dr. Robert Henderson

Baptist Bible Fellowship Intl. (1950), Baptist Bible Fellowship Missions Bldg., 720 E. Kearney St., Springfield, MO 65803; Pres., Sam Davison

Baptist Convention, Southern (1845), 901 Commerce St., Ste. 750, Nashville, TN 37203; http://www.sbcnet.org; Pres., Paige Patterson

Baptist Convention, U.S.A., Inc., National 1700 Baptist World Center Dr., Nashville, TN 37207; Pres. (acting), Rev. S.C. Cureton

Baptist Convention of America, Inc., National (1880), 777 S. R.L. Thornton Freeway, Ste. 205, Dallas, TX 75203; http://www.greatertempleofgod.com; Pres., Dr. E. Edward Jones

Baptist Convention of America, Natl. Missionary (1988), 1404 E. Firestone, Los Angeles, CA 90001; Pres., Dr. W. T. Snead Sr.

Baptist General Conference (1852), 2002 S. Arlington Heights Rd., Arlington Heights, IL 60005; http://www.bgc.bethel.edu; Pres., Dr. Robert Ricker

Brethren in Christ Church (1778), PO Box 290, Grantham, PA 17027; Moderator, Dr. Warren L. Hoffman

Buddhist Churches of America (1899), 1710 Octavia St., San Francisco, CA 94109; Presiding Bishop, Hakubun Watanabe

Christian Church (Disciples of Christ) (1832), 130 E. Washington St., PO Box 1986, Indianapolis, IN 46206; http://www.disciples.org; Gen. Minister and Pres., Richard L. Hamm

Christian Churches and Churches of Christ, 4210 Bridgetown Rd., Box 11326, Cincinnati, OH 45211; http://www.nacc-online.org

Christian Congregation, Inc., The (1887), 804 W. Hemlock St., LaFollette, TN 37766; Gen. Supt., Rev. Ora W. Eads, D.D.

Christian Methodist Episcopal Church (1870), 4466 Elvis Presley Blvd., Memphis, TN 38116; Executive Secretary, Dr. W. Clyde Williams

Christian and Missionary Alliance (1897), PO Box 35000, Colorado Springs, CO 80935; http://www.cmalliance.org; Pres., Rev. Peter N. Nanfelt, D.D.

Christian Reformed Church in North America (1857), 2850 Kalamazoo Ave. SE, Grand Rapids, MI 49560; http://www.crcna.org; Gen. Secy., Dr. David H. Engelhard

Church of the Brethren (1708), 1451 Dundee Ave., Elgin, IL 60120; Moderator, Lowell A. Flory

Church of Christ (1830), PO Box 472, Independence, MO 64051; Council of Apostles Sec., Apostle Smith N. Brickhouse

Church of God (Anderson, IN) (1881), Box 2420, Anderson, IN 46018; http://www.chog.org; Gen. Secy., Edward L. Foggs

Church of God (Cleveland, TN) (1886), PO Box 2430, Cleveland, TN 37320; Gen. Overseer, Paul L. Walker

Church of God in Christ (1907), Mason Temple, 939 Mason St., Memphis, TN 38126; Presiding Bishop, Bishop Chandler D. Owens

Church of Jesus Christ (Bickertonites) (1862), 6th & Lincoln Sts., Monongahela, PA 15063; Pres., Dominic Thomas

Church of Jesus Christ of Latter-day Saints (Mormon), The (1830), 47 E. South Temple St., Salt Lake City, UT 84150; http://www.lds.org; Pres., Gordon B. Hinckley

Church of the Nazarene (1907), 6401 The Paseo, Kansas City, MO 64131; Gen. Secy., Jack Stone

Community Churches, Internat. Council of (1950), 21116 Washington Pkwy., Frankfort, IL 60423; Pres., Rev. Dr. Gregory Smith

Conservative Judaism, United Synagogues of, 155 5th Ave., New York, NY 10010; http://www.uscj.org; Pres., Rabbi Jerome Epstein

Coptic Orthodox Church, 427 West Side Ave., Jersey City, NJ 07304

Cumberland Presbyterian Church (1810), 1978 Union Ave., Memphis, TN 38104; http://www.cumberland.org; Moderator, Rev. Masaharu Asayama

Episcopal Church (1789), 815 Second Ave., New York, NY 10017; http://www.ecusa.anglican.org; Presiding Bishop and Primate, Most Rev. Frank Tracy Griswold III

Evangelical Free Church of America (1884), 901 E. 78th St., Minneapolis, MN 55420; Acting Pres., Rev. William Hamel

Evangelical Lutheran Church in America (1987), 8765 W. Higgins Rd., Chicago, IL 60631; http://www.elca.org; Bishop, Rev. Dr. H. George Anderson

Fellowship of Grace Brethren Churches (1882), PO Box 386, Winona Lake, IN 46590; http://www.grace-brethren.org; Moderator, Dr. Galen Wiley

First Church of Christ, Scientist, The (1879), 175 Huntington Ave., Boston, MA 02115; http://www.tfccs.com; Pres., Thomas J. Black

Free Methodist Church of North America (1860), World Ministries Center, 770 N. High School Rd., Indianapolis, IN 46214

Friends General Conference (1900), 1216 Arch St. 2B, Philadelphia, PA 19107; Gen. Secy., Bruce Birchard

Greek Orthodox Archdiocese of America (1922), 8-10 E. 79th St., New York, NY 10021; http://www.goarch.org; Primate of Greek Orthodox Church in America, Archbishop Demetrios

International Church of the Foursquare Gospel (1927), 1910 W. Sunset Blvd., Ste. 200, PO Box 26902, Los Angeles, CA 90026; http://www.foursquare.org; Pres., Dr. Paul C. Risser

Islamic Society of N.A., 25351 Five Mile Rd., Redford Township, MI 48239; Secy., Nihad Hamed

Jehovah's Witnesses, 25 Columbia Heights, Brooklyn, NY 11201; Pres., Milton G. Henschel

Lutheran Church—Missouri Synod (1847), 1333 S. Kirkwood Rd., St. Louis, MO 63122; http://www.lcms.org; Pres., Dr. A. L. Barry

Mennonite Brethren Churches, General Conference of (1860), 4812 E. Butler Ave., Fresno CA 93727; Moderator, Ed Boschman

Mennonite Church (1893), 421 S. Second St., Ste. 600, Elkhart, IN 46516; http://www.mennonites.org; Moderator, Dwight McFadden, Jr.

Mennonite Church, The General Conference (1860), 722 Main, P.O. Box 347, Newton, KS 67114; http://www2.southwind.net/~gcmc; Moderator, Darrell Fast

Moravian Church in America (1735), **Northern Prov.:** 1021 Center St., PO Box 1245, Bethlehem, PA 18016; http://www.moravian.org; Pres., Rev. R. Burke Johnson; **Southern Prov.:** 459 S. Church St., Winston-Salem, NC 27101; Pres., Rev. Dr. Robert E. Sawyer; **Alaska Prov.:** PO Box 545, Bethel, AK 99559; Pres., Rev. Frank Chingliak

National Baptist Convention, Inc. Progressive (1961), 601 50th St., NE, Washington, DC 20019; http://www.pribc.org; Pres., Dr. Bennett W. Smith Sr.

Orthodox Church in America (1794), PO Box 675, Syosset, NY 11791; http://www.oca.org; Primate, Most Blessed Theodosius

Orthodox Jewish Congregations in America, Union of 11 Broadway., New York, NY 10004; http://www.ou.org; Pres., Mandell I. Ganchrow, M.D.

Pentecostal Assemblies of the World, Inc., 3939 Meadows Dr., Indianapolis, IN 46205; Presiding Bishop, Paul A. Bowers

Presbyterian Church in America (1973), 1852 Century Pl., Atlanta, GA 30345; http://www.pcanet.org; Moderator, Rev. Kennedy Smartt

Presbyterian Church (U.S.A.), (1983), 100 Witherspoon St., Louisville, KY 40202; http://www.pcusa.org; Moderator, Douglas W. Oldenburg

Reformed Church in America (1628), 475 Riverside Dr., New York, NY 10115; http://www.rca.org; Pres., Frederick Kruithg

Reorganized Church of Jesus Christ of Latter-day Saints (1830), PO Box 1059, Independence, MO 64051; Pres. W. Grant McMurray

Roman Catholic Church (1634), National Conference of Catholic Bishops, 3211 Fourth St., Washington, DC 20017; Pres., Bishop Anthony M. Pilla

Romanian Orthodox Episcopate of America (1929), PO Box 309, Grass Lake, MI 49240; http://www.roea.org; Ruling Bishop, His Grace Bishop Nathaniel Popp

Salvation Army (1865), 615 Slaters Lane, Alexandria, VA 22313; National Comdr., Commissioner Robert A. Watson

Seventh-Day Adventist Ch. (1863), 12501 Old Columbia Pike, Silver Spring, MD 20904; Pres., Robert S. Folkenberg

Swedenborgian Church (1792), 48 Sargent St., Newton, MA 02158; http://www.swedenborg.org; Pres., Rev. Ronald P. Brugier

Unitarian Universalist Association of North America (1793), 25 Beacon St., Boston, MA 02108; http://www.uua.org; Pres., John Buehrens

United Church of Christ (1957), 700 Prospect Ave., Cleveland, OH 44115; http://www.apk.net/ucc; Pres., Rev. Paul H. Sherry

United Methodist Church (1968), 1204 Freedom Rd., Cranberry Twp., PA 16066; Pres. Council of Bishops, Bishop George W. Bashore

United Pentecostal Church Intl. (1925), 8855 Dunn Rd., Hazelwood, MO 63042; http://www.upcimain@aol.com; Gen. Superintendent, Rev. Nathaniel A. Urshan

Volunteers of America (1896), 110 S. Union St., Alexandria, VA 22314; Chairperson, Jean Galloway

Wesleyan Church (1968), PO Box 50434, Indianapolis, IN 46250; http://www.wesleyan.org; Gen. Supts., Dr. Earle L. Wilson, Dr. Lee M. Haines, Dr. Thomas E. Armiger

Membership of Religious Groups in Canada

Source: *1999 Yearbook of American and Canadian Churches*; World Almanac research

Figures are generally based on reports of all-inclusive number of "members" by officials of each group. Some groups keep careful records; others only estimate. Not all groups report annually. The number of houses of worship appears in parentheses. *Indicates the group declines to make membership figures public. Groups reporting fewer than 5,000 members are not included. Where membership numbers are not available, only groups with 50 or more houses of worship are listed.

Religious Group	Members
Anglican Church of Canada (2,957)	739,699
Antiochian Orthodox Christian Archdiocese of North America (215)	350,000
Apostolic Church of Pentecost of Canada, Inc. (153)	22,300
Armenian Holy Apostolic Church (Canadian Diocese) (10)	75,000
Associated Gospel Churches (126)	9,284
Bahá'í Faith (1,480)	28,500
Baptist Conference, North American (123)	17,577
Baptist Convention of Ontario and Quebec (386)	57,800
Baptist Ministries, Canadian (1,133)	129,055
Baptist Union of Western Canada (161)	20,006
Christian and Missionary Alliance in Canada (376)	87,197
Christian Brethren (also known as Plymouth Brethren) (600)	50,000
Christian Reformed Church in North America (246)	81,023
Church of God (Cleveland, TN) (115)	8,908
Church of Jesus Christ of Latter-day Saints in Canada (391)	130,000
Church of the Nazarene Canada (165)	11,963
Churches of Christ in Canada (140)	8,000
Estonian Evangelical Lutheran Church (11)	5,089
Evangelical Baptist Churches in Canada, Fellowship of (506)	*
Evangelical Free Church of Canada (133)	22,528
Evangelical Lutheran Church in Canada (650)	198,751
Evangelical Mennonite Conference of Canada (53)	6,508
Evangelical Missionary Church of Canada (145)	12,217
Free Methodist Church in Canada (129)	5,360

Religious Group	Members
Greek Orthodox Diocese of Toronto (Canada) (76)	350,000
Hindu[1]	90,000
Independent Assemblies of God Intl. (Canada) (214)	*
Islam[1]	120,000
Jehovah's Witnesses (1,388)	113,763
Jewish congregations[1] (270+)	250,000
Lutheran Church–Canada (329)	79,844
Mennonite Brethren Churches, Canadian Conference of (208)	31,733
Mennonite Church (Canada) (117)	8,172
Mennonites in Canada, Conference of (223)	35,995
Old Order Amish Church (930)	*
Orthodox Church in America (Canada Section) (606)	1,000,000
Pentecostal Assemblies of Canada (1,100)	218,782
Pentecostal Assemblies of Newfoundland (140)	29,361
Presbyterian Church in Canada (1,012)	211,812
Reformed Church in Canada (41)	6,490
Reformed Churches, Canadian and American (48)	14,727
Reorganized Church of Jesus Christ of Latterday Saints (75)	11,264
Roman Catholic Church in Canada (5,716)	12,498,605
Salvation Army in Canada (370)	95,763
Seventh-Day Adventist Church in Canada (336)	46,962
Southern Baptists, Canadian Convention of (130)	8,228
United Baptist Convention of the Atlantic Provinces (550)	63,787
United Church of Canada (3,820)	1,649,754
United Pentecostal Church in Canada (199)	*
Wesleyan Church of Canada (82)	5,374

(1) Estimates; figures from other sources may vary.

Headquarters of Selected Religious Groups in Canada

Source: *1999 Yearbook of American & Canadian Churches*, copyright National Council of the Churches of Christ in the USA; *World Almanac* research

(Year organized in parentheses)

Anglican Church of Canada (1700), Church House, 600 Jarvis St., Toronto, ON M4Y 2J6; Primate, Most Rev. Michael G. Peers

Bahá'í National Centre of Canada, 7200 Leslie St., Thornhill, ON L3T 6L8; Gen'l.-Secy., Reginald Newkirk

Baptist Ministries, Can., 7185 Millcreek Dr., Mississauga, ON L5N 5R4; http://www.cbmin.org; Pres., Dr. Carmine Moir

Christian and Missionary Alliance in Canada (1887), B, Willowdale, ON M2K 2R6; http://www.cmacan.org; Pres., Dr. Arnold Cook

Church of Jesus Christ of Latter-day Saints (Mormon), The (1830), 50 E. North Temple St., Salt Lake City, UT 84150

Church of the Nazarene in Canada (1902), 20 Regan Rd. Unit 9, Brampton, ON L7A 1C3; http://web.1-888.com.nazarene/national; Natl. Dir., Dr. William E. Stewart

Evangelical Baptist Churches in Canada, Fellowship of (1953), 679 Southgate Dr., Guelph, ON N1G 4S2; Pres., Rev. Terry D. Cuthbert

Evang. Lutheran Church in Canada (1985), 302-393 Portage Ave., Winnipeg, MB R3B 3H6; Bishop, Rev. Telmor G. Sartison

Evang. Missionary Church of Canada (1993), #550 1212 31st Ave. NE, Calgary, AB T2E 7S8; Pres., Rev. Mark Bolender

Greek Orthodox Metropolis of Toronto, 86 Overlea Blvd., Toronto, ON M4H 1C6; http://www.gocanada.org; His Eminence Metropolitan Archbishop Sotirios

Jehovah's Witnesses (1879), Canadian office: Box 4100, Halton Hills, ON L7G 4Y4; Pres., Milton Henschel

Jewish Congress, Canadian (1919), 1590 Ave. Docteur Penfield, Montreal, Que. H3G 1C5; http://www.cjc.ca; Nat. Exec. Dir., Jack Silverstone (Nonreligious umbrella organization of Jewish groups)

Lutheran Church—Canada (1959), 3074 Portage Ave., Winnipeg, MB R3K OY2; Pres., Ralph Mayan

Mennonite Church (1898), 421 S. Second St., Ste. 600, Elkhart, IN 46516; Mod., Dwight McFadden Jr.

Muslim Communities in Canada, Council of, 1250 Ramsey View Ct., Ste. 504, Sudbury, ON P3E 2E7; Director, Mir Iqbal Ali

North American Shi'a Muslim Communities Organization (NASIMCO), 300 John St., PO Box 87629, Dawnhill, ON L3T 7R3; Pres. Ghulamabbas Sajan

Pentecostal Assemblies of Canada (1919), 6745 Century Ave., Mississauga, ON L5N 6P7; http://www.paoc.org; Gen. Supt., Rev. William D. Morrow

Presbyterian Church (1925), 50 Wynford Dr., North York, ON M3C 1J7; http://www.presbycan.cal; Prinicipal Clerk: Rev. Stephen Kendall

Roman Catholic Church (1618), Canadian Conference of Catholic Bishops, 90 Parent Ave., Ottawa, ON K1N 7B1; Pres., Most Rev. M. le Cardinal Jean-Claude Turcotte

Salvation Army (1909), 2 Overlea Blvd., Toronto, ON M4H 1P4; http://www.sallynet.org; Territorial Cmdr., Commissioner Norman Have

Seventh-Day Adventist Church (1901), 1148 King St., E. Oshawa, ON L1H 1H8; Pres., Orville Parchment

Ukrainan Orthodox Church (1918), Office of the Consistory, 9 St. John's Ave., Winnipeg, MB R2W 1G8; http://www.uocc.ca; Primate, Most Rev. Metropolitan Wasyly Fedak

United Brethren Church (1767) 302 Lake St., Huntington, IN 46750; Pres., Rev. Brian Magnus

United Church of Canada (1925), The United Church House, 3250 Bloor St. W., Ste. 300, Etobicoke, ON M8X 2Y4; http://www.uccan.org; Mod., William F. Phipps

Wesleyan Church (1968), The Wesleyan Church Intl. Center, PO Box 50434, Indianapolis, IN 46250; Dist. Supt., Rev. Donald E. Hodgins

Adherents of All Religions by Six Continental Areas, Mid-1998

Source: *1999 Encyclopædia Britannica Book of the Year*

	Africa	Asia	Europe	Latin America	Northern America	Oceania	World
Baha'is	2,263,000	3,260,000	126,000	825,000	753,000	105,000	6,764,000
Buddhists	138,000	348,806,000	1,517,000	622,000	2,445,000	266,000	353,794,000
Chinese folk religionists	33,000	377,795,000	250,000	184,000	839,000	61,000	379,162,000
Christians	356,277,000	283,734,000	558,729,000	462,965,000	256,882,000	24,451,000	1,943,038,000
Roman Catholics	114,316,000	106,399,000	286,124,000	442,808,000	69,536,000	7,318,000	1,026,501,000
Protestants	87,190,000	43,998,000	85,924,000	45,295,000	95,063,000	6,503,000	316,445,000
Orthodox	33,660,000	15,232,000	158,775,000	549,000	4,852,000	675,000	213,743,000
Anglicans	20,551,000	856,000	25,632,000	853,000	3,260,000	5,190,000	63,748,000
Confucianists	0	6,207,000	11,000	0	0	23,000	6,241,000
Ethnic religionists	97,200,000	148,189,000	1,262,000	1,231,000	424,000	259,000	248,565,000
Hindus	2,411,000	755,500,000	1,382,000	785,000	1,266,000	345,000	761,689,000
Jains	65,000	3,850,000	0	0	7,000	0	3,922,000
Jews	230,000	4,139,000	2,530,000	1,121,000	5,996,000	95,000	14,111,000
Mandeans	0	38,000	0	0	0	0	38,000
Muslims	315,000,000	812,000,000	31,401,000	1,624,000	4,349,000	248,000	1,164,622,000
New-Religionists	27,000	98,548,000	155,000	604,000	759,000	51,000	100,144,000
Shintoists	0	2,727,000	0	7,000	55,000	0	2,789,000
Sikhs	53,000	21,531,000	236,000	0	498,000	14,000	22,332,000
Spiritists	3,000	2,000	129,000	11,498,000	148,000	7,000	11,785,000
Zoroastrians	1,000	269,000	0	0	3,000	0	274,000
Other religionists	68,000	11,000	233,000	95,000	585,000	9,000	1,001,000
Nonreligious	4,863,000	600,822,000	108,000,000	15,300,000	27,500,000	3,170,000	759,655,000
Atheists	420,000	121,451,000	23,444,000	2,673,000	1,569,000	356,000	149,913,000

Adherents. As defined and enumerated in *World Christian Encyclopedia* (1982), projected to mid-1998, adjusted for recent data.

Continents. These follow current UN demographic practice, which divides the world into the 6 major areas shown above. "Asia" here includes the former USSR Central Asian republics. "Europe" includes all of Russia and extends eastward to Vladivostok, the Sea of Japan, and the Bering Strait.

Buddhists. 56% Mahayana, 38% Theravada (Hinayana), 6% Tantrayana (Lamaism).

Chinese folk religionists. Followers of traditional Chinese religion (local deities, ancestor veneration, Confucian ethics, Taoism, universism, divination, some Buddhist elements).

Christians. Total Christians include those affiliated with churches not shown, plus other persons professing in censuses or polls to be Christians but not affiliated with any church.

Confucians. Non-Chinese followers of Confucius and Confucianism, mostly Koreans in Korea.

Hindus. 70% Vaishnavites, 25% Shaivites, 2% neo-Hindus and reform Hindus.

Jews. Adherents of Judaism.

Muslims. 83% Sunni Muslims, 16% Shia Muslims (Shi'ites), 1% other schools.

New-Religionists. Followers of Asian 20th-cent. New Religions, New Religious movements, radical new crisis religions, and non-Christian syncretistic mass religions, all founded since 1800 and most since 1945.

Other religionists. Including 70 minor world religions and a large number of spiritist religions, New Age religions, quasi-religions, pseudo religions, parareligions, religious or mystic systems, and religious and semireligious brotherhoods of numerous varieties.

Nonreligious. Persons professing no religion, nonbelievers, agnostics, freethinkers, dereligionized secularists indifferent to all religion.

Atheists. Persons professing atheism, skepticism, disbelief, or irreligion, including antireligious (opposed to all religion).

Episcopal Church Liturgical Colors and Calendar

Source: Church Publishing Incorporated, New York

The liturgical colors in the Episcopal Church are as follows: **White**—from Christmas Day through the First Sunday after Epiphany; Maundy Thursday (as an alternative to crimson at the Eucharist); from the Vigil of Easter to the Day of Pentecost (Whitsunday); Trinity Sunday; Feasts of the Lord (except Holy Cross Day); the Confession of St. Peter; the Conversion of St. Paul; St. Joseph; St. Mary Magdalene; St. Mary the Virgin; St. Michael and All Angels; All Saints' Day; St. John the Evangelist; memorials of other saints who were not martyred; Independence Day and Thanksgiving Day; weddings and funerals. **Red**—the Day of Pentecost; Holy Cross Day; feasts of apostles and evangelists (except those listed above); feasts and memorials of martyrs (including Holy Innocents' Day). **Violet**—Advent and Lent. **Crimson** (dark red)—Holy Week. **Green**—the seasons after Epiphany and after Pentecost. **Black**—optional alternative for funerals. Alternative colors used in some churches: **Blue**—Advent; **Lenten White**—Ash Wednesday to Palm Sunday.

In the Episcopal Church the days of fasting are Ash Wednesday and Good Friday. Other days of special devotion (penitence) are the 40 days of Lent and all Fridays of the year, except those in Christmas and Easter seasons and any Feasts of the Lord that occur on a Friday or during Lent. Ember Days (optional) are days of prayer for the church's ministry. They fall on the Wednesday, Friday, and Saturday after the first Sunday in Lent, the Day of Pentecost, Holy Cross Day, and the Third Sunday of Advent. Rogation Days (also optional), the 3 days before Ascension Day, are days of prayer for God's blessing on the crops, on commerce and industry, and for conservation of the earth's resources.

Days, etc.	1999	2000	2001	2002	2003
Golden Number	5	6	7	8	9
Sunday Letter	C	B & A	G	F	E
Sundays after Epiphany	6	9	8	5	8
Ash Wednesday	Feb. 17	Mar. 8	Feb. 28	Feb. 13	Mar. 5
First Sunday in Lent	Feb. 21	Mar. 12	Mar. 4	Feb. 17	Mar. 9
Passion/Palm Sunday	Mar. 28	Apr. 16	Apr. 8	Mar. 24	Apr. 13
Good Friday	Apr. 2	Apr. 21	Apr. 13	Mar. 29	Apr. 18
Easter Day	Apr. 4	Apr. 23	Apr. 15	Mar. 31	Apr. 20
Ascension Day	May 13	June 1	May 24	May 9	May 29
The Day of Pentecost	May 23	June 11	June 3	May 19	June 8
Trinity Sunday	May 30	June 18	June 10	May 26	June 15
Numbered Proper of 2 Pentecost	#5	#7	#6	#4	#7
First Sunday of Advent	Nov. 28	Dec. 3	Dec. 2	Dec. 1	Nov. 30

Greek Orthodox Movable Ecclesiastical Dates, 1999-2003

This 5-year chart has the dates of feast days and fasting days, which are determined annually on the basis of the date of Holy Pascha (Easter). This ecclesiastical cycle begins with the first day of the Triodion and ends with the Sunday of All Saints, a total of 18 weeks.

	1999	2000	2001	2002	2003
Triodion begins	Jan. 31	Feb. 20	Feb. 4	Feb. 24	Feb. 16
Sat. of Souls	Feb. 13	Mar. 4	Feb. 17	Mar. 9	Mar. 1
Meat Fare	Feb. 14	Mar. 5	Feb. 18	Mar. 10	Mar. 2
2d Sat. of Souls	Feb. 20	Mar 11	Feb. 24	Mar. 16	Mar. 8
Lent Begins	Feb. 22	Mar. 13	Feb. 26	Mar. 18	Mar. 10
St. Theodore—3d Sat. of Souls	Feb. 27	Mar. 18	Mar. 3	Mar. 23	Mar. 15
Sunday of Orthodoxy	Feb. 28	Mar. 19	Mar. 4	Mar. 24	Mar. 16
Sat. of Lazarus	Apr. 3	Apr. 22	Apr. 7	Apr. 27	Apr. 19
Palm Sunday	Apr. 4	Apr. 23	Apr. 8	Apr. 28	Apr. 20
Holy (Good) Friday	Apr. 9	Apr. 28	Apr. 13	May 3	Apr. 25
Western Easter	Apr. 4	Apr. 23	Apr. 15	Mar. 31	Apr. 20
Orthodox Easter	Apr. 11	Apr. 30	Apr. 15	May 5	Apr. 27
Ascension	May 20	June 8	May 24	June 13	June 5
Sat. of Souls	May 29	June 17	June 2	June 22	June 14
Pentecost	May 30	June 18	June 3	June 23	June 15
All Saints	June 6	June 25	June 10	June 30	June 22

Important Islamic Dates, 1999-2004 (1420-24)

Source: Imad-ad-Dean, Inc., Bethesda, MD 20814

The Islamic calendar is a strict lunar calendar reckoned from the year of the Hijra (Muhammad's flight from Mecca to Medina). Each year consists of 12 lunar months of 29 or 30 days beginning and ending with each new moon's visible crescent. Common years have 354 days; leap years have 355 days. Some Muslim countries employ a conventionalized calendar with the leap day added to the last month, Dhûl Hijah, but for religious purposes the leap date is taken into account by tracking each new moon sighting. The dates given below are based on the convention that the first new moon must be seen before the following dawn on the East Coast of the Americas. Actual (local) Western Hemisphere sightings may occur a day later, but never a day earlier, than these dates reflect.

	1999-2000 (1420)	2000-01 (1421)	2001-02 (1422)	2002-03 (1423)	2003-04 (1424)
New Year's Day (Muharram1)	Apr. 17,1999	Apr. 6, 2000	Mar. 26, 2001	Mar. 15, 2002	Mar. 4, 2003
Ashura (Muharram 10)	Apr. 26, 1999	Apr. 15, 2000	Apr. 4, 2001	Mar. 24, 2002	Mar. 13, 2003
Mawlid (Rabi'l 12)	June 26, 1999	June 14, 2000	June 4, 2001	May 24, 2002	May 13, 2003
Ramadan 1	Dec. 9, 1999	Nov. 27, 2000	Nov. 16, 2001	Nov. 6, 2002	Oct. 26, 2003
Id al-Fitr (Shawwal)	Jan. 8, 2000	Dec. 27, 2000	Dec. 16, 2001	Dec. 5, 2002	Nov. 25, 2003
Id al-Adha (Dhûl-Hijjah 10)	Mar. 16, 2000	Mar. 5, 2001	Feb. 22, 2002	Feb. 11, 2003	Feb. 1, 2004

Jewish Holy Days, Festivals, and Fasts, 1999-2003

	1999 (5759-60)		2000 (5760-61)		2001 5761-62)		2002 (5762-63)		2003 (5763-64)	
Tu B'Shvat	Feb. 1	Mon.	Jan. 22	Sat.	Feb. 8	Thu.	Jan. 28	Mon.	Jan. 18	Sat.
Ta'anis Esther (Fast of Esther)	Mar. 1	Mon.	Mar. 20	Mon.	Mar. 8	Thu.	Feb. 25	Mon.	Mar. 17	Mon.
Purim	Mar. 2	Tue.	Mar. 21	Tue.	Mar. 9	Fri.	Feb. 26	Tue.	Mar. 18	Tue.
Pesach (Passover)	Apr. 1	Thu.	Apr. 20	Thu.	Apr. 8	Sun.	Mar. 28	Thu.	Apr. 17	Thu.
	Apr. 8	Thu.	Apr. 27	Thu.	Apr. 15	Sun.	Apr. 4	Sat.	Apr. 24	Thu.
Lag B'Omer	May 4	Tue.	May 23	Tue.	May 11	Fri.	Apr. 30	Tue.	May. 20	Tue.
Shavuot (Pentecost)	May 21	Fri.	June 9	Fri.	May 28	Mon.	May 17	Fri.	June 6	Fri.
	May 22	Sat.	June 10	Sat.	May 29	Tue.	May 18	Sat.	June 7	Sat.
Fast of the 17th Day of Tammuz	July 1	Thu.	July 20	Thu.	July 8	Sun.	June 27	Thu.	July 17	Thu.
Fast of the 9th Day of Av	July 22	Thu.	Aug. 10	Thu.	July 29	Sun.	July 18	Thu.	Aug. 7	Thu.
Rosh Hashanah (Jewish New Year)	Sept. 11	Sat.	Sept. 30	Sat.	Sept. 18	Tue.	Sept. 7	Sat.	Sept. 27	Sat.
	Sept. 12	Sun.	Oct. 1	Sun.	Sept. 19	Wed.	Sept. 8	Sun.	Sept. 28	Sun.
Fast of Gedalya	Sept. 13	Mon.	Oct. 2	Mon.	Sept. 20	Thu.	Sept. 9	Mon.	Sept. 29	Mon.
Yom Kippur (Day of Atonement)	Sept. 20	Mon.	Oct. 9	Mon.	Sept. 27	Thu.	Sept. 16	Mon.	Oct. 6	Mon.
Sukkot	Sept. 25	Sat.	Oct. 14	Sat.	Oct. 2	Tue.	Sept. 21	Sat.	Oct. 11	Sat.
	Oct. 1	Fri.	Oct. 20	Fri.	Oct. 8	Mon.	Sept. 27	Fri.	Oct. 17	Fri.
Shmini Atzeret	Oct. 2	Sat.	Oct. 21	Sat.	Oct. 9	Tue.	Sept. 28	Sat.	Oct. 18	Sat.
	Oct. 3	Sun.	Oct. 22	Sun.	Oct. 10	Wed.	Sept. 29	Sun.	Oct. 19	Sun.
Hanukkah	Dec. 4	Sat.	Dec. 22	Fri.	Dec. 10	Mon.	Nov. 30	Sat.	Dec. 20	Sat.
	Dec. 11	Sat.	Dec. 29	Fri.	Dec. 17	Mon.	Dec. 7	Sat.	Dec. 27	Sat.
Fast of the 10th of Tevet	Dec. 19	Sun.	Jan. 5, 2001	Fri.	Dec. 25	Tue.	Jan. 15, 2003	Sun.	Jan. 4, 2004	Sun.

The months of the Jewish year are: 1) Tishri; 2) Cheshvan (also Marcheshvan); 3) Kislev; 4) Tevet (also Tebeth); 5) Shebat (also Shebhat); 6) Adar; 6a) Adar Sheni (II) added in leap years; 7) Nisan; 8) Iyar; 9) Sivan; 10) Tammuz; 11) Av (also Abh); 12) Elul. All Jewish holy days, etc., begin at sunset on the previous day. *Date changed to avoid Sabbath.

Ash Wednesday and Easter Sunday (Western churches), 1901-2100

Year	Ash Wed.	Easter Sunday	Year	Ash Wed.	Easter Sunday	Year	Ash Wed.	Easter Sunday	Year	Ash Wed.	Easter Sunday
1901	Feb. 20	Apr. 7	1952	Feb. 27	Apr. 13	2003	Mar. 5	Apr. 20	2052	Mar. 6	Apr. 21
1902	Feb. 12	Mar. 30	1953	Feb. 18	Apr. 5	2004	Feb. 25	Apr. 11	2053	Feb. 19	Apr. 6
1903	Feb. 25	Apr. 12	1954	Mar. 3	Apr. 18	2005	Feb. 9	Mar. 27	2054	Feb. 11	Mar. 29
1904	Feb. 17	Apr. 3	1955	Feb. 23	Apr. 10	2006	Mar. 1	Apr. 16	2055	Mar. 3	Apr. 18
1905	Mar. 8	Apr. 23	1956	Feb. 15	Apr. 1	2007	Feb. 21	Apr. 8	2056	Feb. 16	Apr. 2
1906	Feb. 28	Apr. 15	1957	Mar. 6	Apr. 21	2008	Feb. 6	Mar. 23	2057	Mar. 7	Apr. 22
1907	Feb. 13	Mar. 31	1958	Feb. 19	Apr. 6	2009	Feb. 25	Apr. 12	2058	Feb. 27	Apr. 14
1908	Mar. 4	Apr. 19	1959	Feb. 11	Mar. 29	2010	Feb. 17	Apr. 4	2059	Feb. 12	Mar. 30
1909	Feb. 24	Apr. 11	1960	Mar. 2	Apr. 17	2011	Mar. 9	Apr. 24	2060	Mar. 3	Apr. 18
1910	Feb. 9	Mar. 27	1961	Feb. 15	Apr. 2	2012	Feb. 22	Apr. 8	2061	Feb. 23	Apr. 10
1911	Mar. 1	Apr. 16	1962	Mar. 7	Apr. 22	2013	Feb. 13	Mar. 31	2062	Feb. 8	Mar. 26
1912	Feb. 21	Apr. 7	1963	Feb. 27	Apr. 14	2014	Mar. 5	Apr. 20	2063	Feb. 28	Apr. 15
1913	Feb. 5	Mar. 23	1964	Feb. 12	Mar. 29	2015	Feb. 18	Apr. 5	2064	Feb. 20	Apr. 6
1914	Feb. 25	Apr. 12	1965	Mar. 3	Apr. 18	2016	Feb. 10	Mar. 27	2065	Feb. 11	Mar. 29
1915	Feb. 17	Apr. 4	1966	Feb. 23	Apr. 10	2017	Mar. 1	Apr. 16	2066	Feb. 24	Apr. 11
1916	Mar. 8	Apr. 23	1967	Feb. 8	Mar. 26	2018	Feb. 14	Apr. 1	2067	Feb. 16	Apr. 3
1917	Feb. 21	Apr. 8	1968	Feb. 28	Apr. 14	2019	Mar. 6	Apr. 21	2068	Mar. 7	Apr. 22
1918	Feb. 13	Mar. 31	1969	Feb. 19	Apr. 6	2020	Feb. 26	Apr. 12	2069	Feb. 27	Apr. 14
1919	Mar. 5	Apr. 20	1970	Feb. 11	Mar. 29	2021	Feb. 17	Apr. 4	2070	Feb. 12	Mar. 30
1920	Feb. 18	Apr. 4	1971	Feb. 24	Apr. 11	2022	Mar. 2	Apr. 17	2071	Mar. 4	Apr. 19
1921	Feb. 9	Mar. 27	1972	Feb. 16	Apr. 2	2023	Feb. 22	Apr. 9	2072	Feb. 24	Apr. 10
1922	Mar. 1	Apr. 16	1973	Mar. 7	Apr. 22	2024	Feb. 14	Mar. 31	2073	Feb. 8	Mar. 26
1923	Feb. 14	Apr. 1	1974	Feb. 27	Apr. 14	2025	Mar. 5	Apr. 20	2074	Feb. 28	Apr. 15
1924	Mar. 5	Apr. 20	1975	Feb. 12	Mar. 30	2026	Feb. 18	Apr. 5	2075	Feb. 20	Apr. 7
1925	Feb. 25	Apr. 12	1976	Mar. 3	Apr. 18	2027	Feb. 10	Mar. 28	2076	Mar. 4	Apr. 19
1926	Feb. 17	Apr. 4	1977	Feb. 23	Apr. 10	2028	Mar. 1	Apr. 16	2077	Feb. 24	Apr. 11
1927	Mar. 2	Apr. 17	1978	Feb. 8	Mar. 26	2028	Mar. 1	Apr. 16	2078	Feb. 16	Apr. 3
1928	Feb. 22	Apr. 8	1979	Feb. 28	Apr. 15	2029	Feb. 14	Apr. 1	2079	Mar. 8	Apr. 23
1929	Feb. 13	Mar. 31	1980	Feb. 20	Apr. 6	2030	Mar. 6	Apr. 21	2080	Feb. 21	Apr. 7
1930	Mar. 5	Apr. 20	1981	Mar. 4	Apr. 19	2031	Feb. 26	Apr. 13	2081	Feb. 12	Mar. 30
1931	Feb. 18	Apr. 5	1982	Feb. 24	Apr. 11	2032	Feb. 11	Mar. 28	2082	Mar. 4	Apr. 19
1932	Feb. 10	Mar. 27	1983	Feb. 16	Apr. 3	2033	Mar. 2	Apr. 17	2083	Feb. 17	Apr. 4
1933	Mar. 1	Apr. 16	1984	Mar. 7	Apr. 22	2034	Feb. 22	Apr. 9	2084	Feb. 9	Mar. 26
1934	Feb. 14	Apr. 1	1985	Feb. 20	Apr. 7	2035	Feb. 7	Mar. 25	2085	Feb. 28	Apr. 15
1935	Mar. 6	Apr. 21	1986	Feb. 12	Mar. 30	2036	Feb. 27	Apr. 13	2086	Feb. 13	Mar. 31
1936	Feb. 26	Apr. 12	1987	Mar. 4	Apr. 19	2037	Feb. 18	Apr. 5	2087	Mar. 5	Apr. 20
1937	Feb. 10	Mar. 28	1988	Feb. 17	Apr. 3	2038	Mar. 10	Apr. 25	2088	Feb. 25	Apr. 11
1938	Mar. 2	Apr. 17	1989	Feb. 8	Mar. 26	2039	Feb. 23	Apr. 10	2089	Feb. 16	Apr. 3
1939	Feb. 22	Apr. 9	1990	Feb. 28	Apr. 15	2040	Feb. 15	Apr. 1	2090	Mar. 1	Apr. 16
1940	Feb. 7	Mar. 24	1991	Feb. 13	Mar. 31	2041	Mar. 6	Apr. 21	2091	Feb. 21	Apr. 8
1941	Feb. 26	Apr. 13	1992	Mar. 4	Apr. 19	2042	Feb. 19	Apr. 6	2092	Feb. 13	Mar. 30
1942	Feb. 18	Apr. 5	1993	Feb. 24	Apr. 11	2043	Feb. 11	Mar. 29	2093	Feb. 25	Apr. 12
1943	Mar. 10	Apr. 25	1994	Feb. 16	Apr. 3	2044	Mar. 2	Apr. 17	2094	Feb. 17	Apr. 4
1944	Feb. 23	Apr. 9	1995	Mar. 1	Apr. 16	2045	Feb. 22	Apr. 9	2095	Mar. 9	Apr. 24
1945	Feb. 14	Apr. 1	1996	Feb. 21	Apr. 7	2046	Feb. 7	Mar. 25	2096	Feb. 29	Apr. 15
1946	Mar. 6	Apr. 21	1997	Feb. 12	Mar. 30	2047	Feb. 27	Apr. 14	2097	Feb. 13	Mar. 31
1947	Feb. 19	Apr. 6	1998	Feb. 25	Apr. 12	2048	Feb. 19	Apr. 5	2098	Mar. 5	Apr. 20
1948	Feb. 11	Mar. 28	1999	Feb. 17	Apr. 4	2049	Mar. 3	Apr. 18	2099	Feb. 25	Apr. 12
1949	Mar. 2	Apr. 17	2000	Mar. 8	Apr. 23	2050	Feb. 23	Apr. 10	2100	Feb. 10	Mar. 28
1950	Feb. 22	Apr. 9	2001	Feb. 28	Apr. 15	2051	Feb. 15	Apr. 2			
1951	Feb. 7	Mar. 25	2002	Feb. 13	Mar. 31						

The Ten Commandments

According to Judeo-Christian tradition, as related in the Bible, the Ten Commandments were revealed by God to Moses and form the basic moral component of God's covenant with Israel. The Ten Commandments appear in 2 places in the Old Testament—Exodus 20:1-17 and Deuteronomy 5:6-21; the phrasing is similar but not identical.

Following is abridged text of the Ten Commandments in Exodus 20:1-17:

I. I am the Lord your God, who brought you out of the land of Egypt, out of the house of bondage. You shall have no other gods before me.
II. You shall not make for yourself a graven image. You shall not bow down to them or serve them.
III. You shall not take the name of the Lord your God in vain.
IV. Remember the sabbath day, to keep it holy.

V. Honor your father and your mother.
VI. You shall not kill.
VII. You shall not commit adultery.
VIII. You shall not steal.
IX. You shall not bear false witness against your neighbor.
X. You shall not covet.

Most Protestant, Anglican, and Orthodox Christians follow Jewish tradition, which considers the introduction ("I am the Lord . . .") the first commandment and makes the prohibition against idolatry the second. Roman Catholic and Lutheran traditions follow a division used by St. Augustine, which combines I and II and splits the last commandment into 2 that separately prohibit coveting of a neighbor's wife and a neighbor's goods. This arrangement alters the numbering of the other commandments by one.

Books of the Bible

Old Testament—Standard Protestant List

Genesis	II Chronicles	Daniel
Exodus	Ezra	Hosea
Leviticus	Nehemiah	Joel
Numbers	Esther	Amos
Deuteronomy	Job	Obadiah
Joshua	Psalms	Jonah
Judges	Proverbs	Micah
Ruth	Ecclesiastes	Nahum
I Samuel	Song of Solomon	Habakkuk
II Samuel	Isaiah	Zephaniah
I Kings	Jeremiah	Haggai
II Kings	Lamentations	Zechariah
I Chronicles	Ezekiel	Malachi

New Testament List

Matthew	Ephesians	Hebrews
Mark	Phillippians	James
Luke	Colossians	I Peter
John	I Thessalonians	II Peter
Acts	II Thessalonians	I John
Romans	I Timothy	II John
I Corinthians	II Timothy	III John
II Corinthians	Titus	Jude
Galatians	Philemon	Revelation

The standard Protestant Old Testament consists of the same 39 books as in the Bible of Judaism, but the latter is organized differently. The Old Testament used by Roman Catholics has 7 additional "deuterocanonical" books, plus some additional parts of books. The 7 are: **Tobit, Judith, Wisdom, Sirach (Ecclesiasticus), Baruch, I Maccabees,** and **II Maccabees.** Both Catholic and Protestant versions of the New Testament have 27 books, with the same names.

Roman Catholic Hierarchy
Source: U.S. Catholic Conference; as of mid-1999

Supreme Pontiff

At the head of the Roman Catholic Church is the supreme pontiff, Pope John Paul II, Karol Wojtyla, born at Wadowice (Kraków), Poland, May 18, 1920; ordained priest Nov. 1, 1946; appointed bishop July 4, 1958; promoted to archbishop of Kraków Jan. 13, 1964; proclaimed cardinal June 26, 1967; elected pope as successor of Pope John Paul I Oct. 16, 1978; installed as pope Oct. 22, 1978.

College of Cardinals

Members of the Sacred College of Cardinals are chosen by the pope to be his chief assistants and advisers in the administration of the church. Among their duties is the election of the pope when the Holy See becomes vacant.

In its present form, the College of Cardinals dates from the 12th century. The first cardinals, from about the 6th century, were deacons and priests of the leading churches of Rome and were bishops of neighboring dioceses. The title of cardinal was limited to members of the college in 1567. The number of cardinals was set at 70 in 1586 by Pope Sixtus V. From 1959 Pope John XXIII began to increase the number; however, the number of cardinals eligible to participate in papal elections was limited to 120. There were lay cardinals until 1918, when the Code of Canon Law specified that all cardinals must be priests. Pope John XXIII in 1962 established that all cardinals must be bishops. The first age limits were set in 1971 by Pope Paul VI, who decreed that at age 80 cardinals must retire from curial departments and offices and from participation in papal elections.

North American Cardinals

Name	Office	Born	Named Cardinal
Aloysius M. Ambrozic	Archbishop of Toronto	1930	1998
Luis Apone Martinez	Archbishop of San Juan	1922	1973
William W. Baum	Major Penitentiary of Apostolic Penitentiary, the Vatican	1926	1976
Anthony J. Bevilacqua	Archbishop of Philadelphia	1923	1991
G. Emmett Carter[1]	Archbishop emeritus of Toronto	1912	1979
Ernesto Corripio Ahumada[1]	Archbishop emeritus of Mexico	1919	1979
Edouard Gagnon	Pres. of Pontifical Commission of Intl. Eucharistic Congresses	1918	1985
Francis E. George	Archbishop of Chicago	1937	1998
James A. Hickey	Archbishop of Washington, DC	1920	1988
William Henry Keeler	Archbishop of Baltimore	1931	1994
Bernard F. Law	Archbishop of Boston	1931	1985
Roger Mahony	Archbishop of Los Angeles	1936	1991
Adam Joseph Maida	Archbishop of Detroit	1930	1994
John J. O'Connor	Archbishop of New York	1920	1985
Norberto Rivera Carrera	Archbishop of Mexico City	1942	1998
Juan Sandoval Iniquez	Archbishop of Guadalajara	1933	1994
James F. Stafford	President of the Pontifical Council for the Laity	1932	1998
Adolfo Antonio Suarez Rivera	Archbishop of Monterrey	1927	1994
Edmund C. Szoka	Pres. of Prefecture of Economic Affairs of Holy See, the Vatican	1927	1988
Jean-Claude Turcotte	Archbishop of Montreal	1936	1994
Louis-Albert Vachon[1]	Archbishop emeritus of Quebec	1912	1985

(1) Ineligible to take part in papal elections, as of Sept. 1999.

Chronological List of Popes

Source: Annuario Pontificio. Table lists year of accession of each pope.

The Roman Catholic Church named the Apostle Peter as founder of the church in Rome and the first pope. He arrived there c 42, was martyred there c 67, and was ultimately canonized as a saint. **The pope's temporal title is:** Sovereign of the State of Vatican City. **The pope's spiritual titles are:** Bishop of Rome, Vicar of Jesus Christ, Successor of St. Peter, Prince of the Apostles, Supreme Pontiff of the Universal Church, Patriarch of the West, Primate of Italy, Archbishop and Metropolitan of the Roman Province.

The names of antipopes are followed by an *. Antipopes were illegitimate claimants of or pretenders to the papal throne.

Year	Pope	Year	Pope	Year	Pope	Year	Pope
	St. Peter	615	St. Deusdedit or Adeodatus	974	Benedict VII	1316	John XXII
67	St. Linus	619	Boniface V	983	John XIV	1328	Nicholas V*
76	St. Anacletus or Cletus	625	Honorius I	985	John XV	1334	Benedict XII
88	St. Clement I	640	Severinus	996	Gregory V	1342	Clement VI
97	St. Evaristus	640	John IV	997	John XVI*	1352	Innocent VI
105	St. Alexander I	642	Theodore I	1004	John XVIII	1362	Bl. Urban V
115	St. Sixtus I	649	St. Martin I, Martyr	1009	Sergius IV	1370	Gregory XI
125	St. Telesphorus	654	St. Eugene I	1012	Benedict VIII	1378	Urban VI
136	St. Hyginus	657	St. Vitalian	1012	Gregory*	1378	Clement VII*
140	St. Pius I	672	Adeodatus II	1024	John XIX	1389	Boniface IX
155	St. Anicetus	676	Donus	1032	Benedict IX	1394	Benedict XIII*
166	St. Soter	678	St. Agatho	1045	Sylvester III	1404	Innocent VII
175	St. Eleutherius	682	St. Leo II	1045	Benedict IX	1406	Gregory XII
189	St. Victor I	684	St. Benedict II	1045	Gregory VI	1409	Alexander V*
199	St. Zephyrinus	685	John V	1046	Clement II	1410	John XXII*
217	St. Callistus I	686	Conon	1047	Benedict IX	1417	Martin V
217	St. Hippolytus*	687	Theodore*	1048	Damasus II	1431	Eugene IV
222	St. Urban I	687	Paschal*	1049	St. Leo IX	1439	Felix V*
230	St. Pontian	687	St. Sergius I	1055	Victor II	1447	Nicholas V
235	St. Anterus	701	John VI	1057	Stephen IX (X)	1455	Callistus III
236	St. Fabian	705	John VII	1058	Benedict X*	1458	Pius II
251	St. Cornelius	708	Sisinnius	1059	Nicholas II	1464	Paul II
251	Novatian*	708	Constantine	1061	Alexander II	1471	Sixtus IV
253	St. Lucius I	715	St. Gregory II	1061	Honorius II*	1484	Innocent VIII
254	St. Stephen I	731	St. Gregory III	1073	St. Gregory VII	1492	Alexander VI
257	St. Sixtus II	741	St. Zachary	1080	Clement III*	1503	Pius III
259	St. Dionysius	752	Stephen II (III)	1086	Bl. Victor III	1503	Julius II
269	St. Felix I	757	St. Paul I	1088	Bl. Urban II	1513	Leo X
275	St. Eutychian	767	Constantine*	1099	Paschal II	1522	Adrian VI
283	St. Caius	768	Philip*	1100	Theodoric*	1523	Clement VII
296	St. Marcellinus	768	Stephen III (IV)	1102	Albert*	1534	Paul III
308	St. Marcellus I	772	Adrian I	1105	Sylvester IV*	1550	Julius III
309	St. Eusebius	795	St. Leo III	1118	Gelasius II	1555	Marcellus II
311	St. Melchiades	816	Stephen IV (V)	1118	Gregory VIII*	1555	Paul IV
314	St. Sylvester I	817	St. Paschal I	1119	Callistus II	1559	Pius IV
336	St. Marcus	824	Eugene II	1124	Honorius II	1566	St. Pius V
337	St. Julius I	827	Valentine	1124	Celestine II*	1572	Gregory XIII
352	Liberius	827	Gregory IV	1130	Innocent II	1585	Sixtus V
355	Felix II*	844	John*	1130	Anacletus II*	1590	Urban VII
366	St. Damasus I	844	Sergius II	1138	Victor IV*	1590	Gregory XIV
366	Ursinus*	847	St. Leo IV	1143	Celestine II	1591	Innocent IX
384	St. Siricius	855	Benedict III	1144	Lucius II	1592	Clement VIII
399	St. Anastasius I	855	Anastasius*	1145	Bl. Eugene III	1605	Leo XI
401	St. Innocent I	858	St. Nicholas I	1153	Anastasius IV	1605	Paul V
417	St. Zosimus	867	Adrian II	1154	Adrian IV	1621	Gregory XV
418	St. Boniface I	872	John VIII	1159	Alexander III	1623	Urban VIII
418	Eulabus*	882	Marinus I	1159	Victor IV*	1644	Innocent X
422	St. Celestine I	884	St. Adrian III	1164	Paschal III*	1655	Alexander VII
432	St. Sixtus III	885	Stephen V (VI)	1168	Callistus III*	1667	Clement IX
440	St. Leo I	891	Formosus	1179	Innocent III*	1670	Clement X
461	St. Hilary	896	Boniface VI	1181	Lucius III	1676	Bl. Innocent XI
468	St. Simplicius	896	Stephen VI (VII)	1185	Urban III	1689	Alexander VIII
483	St. Felix III (II)	897	Romanus	1187	Clement III	1691	Innocent XII
492	St. Gelasius I	897	Theodore II	1187	Gregory VIII	1700	Clement XI
496	Anastasius II	898	John IX	1191	Celestine III	1721	Innocent XIII
498	St. Symmachus	900	Benedict IV	1198	Innocent III	1724	Benedict XIII
498	Lawrence* (501-505)	903	Leo V	1216	Honorius III	1730	Clement XII
514	St. Hormisdas	903	Christopher*	1227	Gregory IX	1740	Benedict XIV
523	St. John I, Martyr	904	Sergius III	1241	Celestine IV	1758	Clement XIII
526	St. Felix IV (III)	911	Anastasius III	1243	Innocent IV	1769	Clement XIV
530	Boniface II	913	Landus	1254	Alexander IV	1775	Pius VI
530	Dioscorus*	914	John X	1261	Urban IV	1800	Pius VII
533	John II	928	Leo VI	1265	Clement IV	1823	Leo XII
535	St. Agapitus I	928	Stephen VII (VIII)	1271	Bl. Gregory X	1829	Pius VIII
536	St. Silverius, Martyr	931	John XI	1276	Bl. Innocent V	1831	Gregory XVI
537	Vigilius	936	Leo VII	1276	Adrian V	1846	Pius IX
556	Pelagius I	939	Stephen VIII (IX)	1276	John XXI	1878	Leo XIII
561	John III	942	Marinus II	1277	Nicholas III	1903	St. Pius X
575	Benedict I	946	Agapitus II	1281	Martin IV	1914	Benedict XV
579	Pelagius II	955	John XII	1285	Honorius IV	1922	Pius XI
590	St. Gregory I	963	Leo VIII	1288	Nicholas IV	1939	Pius XII
604	Sabinian	964	Benedict V	1294	St. Celestine V	1958	John XXIII
607	Boniface III	965	John XIII	1294	Boniface VIII	1963	Paul VI
608	St. Boniface IV	973	Benedict VI	1303	Bl. Benedict XI	1978	John Paul I
		974	Boniface VII*	1305	Clement V	1978	John Paul II

Major Christian Denominations

Brackets indicate some features that tend to

Denom-ination	Origins	Organization	Authority	Special rites
Baptists	In radical Reformation, objections to infant baptism, demands for church and state separation; John Smyth, English Separatist, in 1609; Roger Williams, 1638, Providence, RI.	Congregational; each local church is autonomous.	Scripture; some Baptists, particularly in the South, interpret the Bible literally.	[Baptism, usually early teen years and after, by total immersion;] Lord's Supper.
Church of Christ (Disciples)	Among evangelical Presbyterians in KY (1804) and PA (1809), in distress over Protestant factionalism and decline of fervor; organized in 1832.	Congregational.	["Where the Scriptures speak, we speak; where the Scriptures are silent, we are silent."]	Adult baptism; Lord's Supper (weekly).
Episco-palians	Henry VIII separated English Catholic Church from Rome, 1534, for political reasons; Protestant Episcopal Church in U.S. founded in 1789.	[Diocesan bishops, in apostolic succession, are elected by parish representatives; the national Church is headed by General Convention and Presiding Bishop; part of the Anglican Communion.]	Scripture as interpreted by tradition, especially 39 Articles (1563); not dogmatic; tri-annual convention of bishops, priests, and laypeople.	Infant baptism, Eucharist, and other sacraments; sacrament taken to be symbolic, but as having real spiritual effect.
Jehovah's Witnesses	Founded in 1870 in PA by Charles Taze Russell; incorporated as Watch Tower Bible and Tract Society of PA, 1884; name Jehovah's Witnesses adopted in 1931.	A governing body located in NY coordinates worldwide activities; each congregation cared for by a body of elders; each Witness considered a minister.	The Bible.	Baptism by immersion; annual Lord's Meal ceremony.
Latter-day Saints (Mormons)	In a vision of the Father and the Son reported by Joseph Smith (1820s) in NY. Smith also reported receiving new scripture on golden tablets: The Book of Mormon.	Theocratic; 1st Presidency (church president, 2 counselors), 12 Apostles preside over international church. Local congregations headed by lay priesthood leaders.	Revelation to living prophet (church president). The Bible, Book of Mormon, and other revelations to Smith and his successors.	Baptism, at age 8; laying on of hands (which confers the gift of the Holy Ghost); Lord's Supper; temple rites: baptism for the dead, marriage for eternity,
Lutherans	Begun by Martin Luther in Wittenberg, Germany in 1517; objection to Catholic doctrine of salvation and sale of indulgences; break complete, 1519.	Varies from congregational to episcopal; in U.S., a combination of regional synods and congregational polities is most common.	Scripture alone. The Book of Concord (1580), which includes the three Ecumenical Creeds, is subscribed to as a correct exposition of Scripture.	Infant baptism; Lord's Supper; Christ's true body and blood present "in, with, and under the bread and wine."
Methodists	Rev. John Wesley began movement in 1738, within Church of England; first U.S. denomination, Baltimore (1784).	Conference and superintendent system; [in United Methodist Church, general superintendents are bishops—not a priestly order, only an office—who are elected for life.]	Scripture as interpreted by tradition, reason, and experience.	Baptism of infants or adults; Lord's Supper commanded; other rites include marriage, ordination, solemnization of personal commitments.
Orthodox	Developed in original Christian proselytizing; broke with Rome in 1054, after centuries of doctrinal disputes and diverging traditions	Synods of bishops in autonomous, usually national, churches elect a patriarch, archbishop, or metropolitan; these men, as a group, are the heads of the church.	Scripture, tradition, and the first 7 church councils up to Nicaea II in 787; bishops in council have authority in doctrine and policy.	Seven sacraments: infant baptism and anointing, Eucharist, ordination, penance, and marriage.
Pente-costal	In Topeka, KS (1901) and Los Angeles (1906), in reaction to perceived loss of evangelical fervor among Methodists and others.	Originally a movement, not a formal organization, Pentecostalism now has a variety of organized forms and continues also as a movement.	Scripture; individual charismatic leaders, the teachings of the Holy Spirit.	[Spirit baptism, especially as shown in "speaking in tongues"; healing and sometimes exorcism;] adult baptism; Lord's Supper.
Presby-terians	In 16th-cent. Calvinist Reformation; differed with Lutherans over sacraments, church government; John Knox founded Scotch Presbyterian church about 1560.	[Highly structured representational system of ministers and laypersons (presbyters) in local, regional, and national bodies (synods)].	Scripture.	Infant baptism; Lord's Supper; bread and wine symbolize Christ's spiritual presence.
Roman Catholics	Traditionally, founded by Jesus who named St. Peter the 1st vicar; developed in early Christian proselytizing, especially after the conversion of imperial Rome in the 4th cent.	[Hierarchy with supreme power vested in pope elected by cardinals;] councils of bishops advise on matters of doctrine and policy.	[The pope, when speaking for the whole church in matters of faith and morals; and tradition (which is partly recorded in Scripture and expressed in church councils).]	Mass; 7 sacraments: baptism, reconciliation, Eucharist, confirmation, marriage, ordination, and anointing of the sick (unction).
United Church of Christ	[By ecumenical union, in 1957, of Congregationalists and Evangelical & Reformed, representing both Calvinist] and [Lutheran traditions.]	Congregational; a General Synod, representative of all congregations, sets general policy.	Scripture.	Infant baptism; Lord's Supper.

How Do They Differ?

distinguish a denomination sharply from others.

Practice	Ethics	Doctrine	Other	Denomi-nation
Worship style varies from staid to evangelistic; extensive missionary activity.	Usually opposed to alcohol and tobacco; some tendency toward a perfectionist ethical standard.	*[No creed; true church is of believers only, who areall equal].*	Believing no authority can stand between the believer and God, the Baptists are strong supporters of church and state separation.	**Baptists**
Tries to avoid any rite not considered part of the 1st-century church; some congregations may reject instrumental music.	Some tendency toward perfectionism; increasing interest in social action programs.	Simple New Testament faith; avoids any elaboration not firmly based on Scripture.	Highly tolerant in doctrinal and religious matters; strongly supportive of scholarly education.	**Church of Christ (Disciples)**
Formal, based on "Book of Common Prayer," updated 1979; services range from austerely simple to highly liturgical.	Tolerant, sometimes permissive; some social action programs.	Scripture; the "historic creeds," which include the Apostles, Nicene, and Athanasian, and the "Book of Common Prayer;" ranges from Anglo-Catholic to low church, with Calvinist influences.	Strongly ecumenical, holding talks with many branches of Christendom.	**Episco-palians**
Meetings are held in Kingdom Halls and members' homes for study and worship; extensive door-to-door visitations.	High moral code; stress on marital fidelity and family values; avoidance of tobacco and blood transfusions.	*[God, by his first creation, Christ, will soon destroy all wickedness; 144,000 faithful ones will rule in heaven with Christ over others on a paradise earth.]*	Total allegiance proclaimed only to God's kingdom or heavenly government by Christ; main periodical, *The Watchtower,* is printed in 115 languages.	**Jehovah's Winesses**
Simple service with prayers, hymns, sermon; private temple ceremonies may be more elaborate.	Temperance; strict moral code; tithing; a strong work ethic with communal self-reliance; strong missionary activity; family emphasis.	Jesus Christ is the Son of God, the Eternal Father. Jesus' atonement saves all humans; those who are obedient to God's laws may become joint-heirs with Christ in God's kingdom.	Mormons believe theirs is the true church of Jesus Christ, restored by God through Joseph Smith. Official name: The Church of Jesus Christ of Latter-day Saints.	**Latter-day Saints (Mormons)**
Relatively simple, formal liturgy with emphasis on the sermon.	Generally conservative in personal and social ethics; doctrine of "2 kingdoms" (worldly and holy) supports conservatism in secular affairs.	Salvation by grace alone through faith; Lutheranism has made major contributions to Protestant theology.	Though still somewhat divided along ethnic lines (German, Swedish, etc.), main divisions are between fundamentalists and liberals.	**Lutherans**
Worship style varies widely by denomination, local church, geography.	Originally pietist and perfectionist; always strong social activist elements.	No distinctive theological development; 25 Articles abridged from Church of England's 39, not binding.	In 1968, The United Methodist Church was formed by the union of The Methodist Church and The Evangelical United Brethren Church.	**Methodists**
[Elaborate liturgy, usually in the vernacular, though extremely traditional; the liturgy is the essence of Orthodoxy; veneration of icons.]	Tolerant; little stress on social action; divorce, remarriage permitted in some cases; bishops are celibate; priests need not be.	Emphasis on Christ's resurrection, rather than crucifixion; the Holy Spirit proceeds from God the Father only.	Orthodox Church in America originally under Patriarch of Moscow, was granted autonomy in 1970; Greek Orthodox do not recognize this autonomy.	**Orthodox**
Loosely structured service with rousing hymns and sermons, culminating in spirit baptism.	Usually, emphasis on perfectionism, with varying degrees of tolerance.	Simple traditional beliefs, usually Protestant, with emphasis on the immediate presence of God in the Holy Spirit.	Once confined to lower-class "holy rollers," Pentecostalism now appears in mainline churches and has established middle-class congregations.	**Pente-costal**
A simple, sober service in which the sermon is central.	Traditionally, a tendency toward strictness, with firm church- and self-discipline; otherwise tolerant.	Emphasizes the sovereignty and justice of God; no longer dogmatic.	Although traces of belief in predestination (that God has foreordained salvation for the"elect") remain, this idea is no longer a central element in Presbyterianism.	**Presby-terians**
Relatively elaborate ritual centered on the Mass; also rosary recitation, novenas, etc.	Traditionally strict, but increasingly tolerant in practice; divorce and remarriage not accepted, but annulments sometimes granted; celibate clergy, except in Eastern rite.	Highly elaborated; salvation by merit gained through grace; dogmatic; special veneration of Mary, the mother of Jesus.	Relatively rapid change followed Vatican Council II; Mass now in vernacular; more stress on social action, tolerance, ecumenism.	**Roman Catholics**
Usually simple services with emphasis on the sermon.	Tolerant; some social action emphasis.	Standard Protestant; "Statement of Faith" (1959) is not binding.	The 2 main churches in the 1957 union represented earlier unions with small groups of almost every Protestant denomination.	**United Church of Christ**

Major Non-Christian World Religions

Source: Reviewed by Anthony Padovano, PhD, STD, prof. of literature & relig. studies, Ramapo College, NJ, adj. prof. of theol., Fordham U., NYC; Islam reviewed by Abdulaziz Sachedina, PhD, prof. of Islamic studies, Univ. of Virginia

Buddhism

Founded: About 525 BC, reportedly near Benares, India.

Founder: Gautama Siddhartha (c 563-483 BC), the Buddha, who achieved enlightenment through intense meditation.

Sacred Texts: The *Tripitaka*, a collection of the Buddha's teachings, rules of monastic life, and philosophical commentaries on the teachings; also a vast body of Buddhist teachings and commentaries, many of which are called *sutras.*

Organization: The basic institution is the *sangha*, or monastic order, through which the traditions are passed to from generation to generation. Monastic life tends to be democratic and anti-authoritarian. Large lay organizations have developed in some sects.

Practice: Varies widely according to the sect, and ranges from austere meditation to magical chanting and elaborate temple rites. Many practices, such as exorcism of devils, reflect pre-Buddhist beliefs.

Divisions: A variety of sects grouped into 3 primary branches: Theravada (sole survivor of the ancient Hinayana schools), which emphasizes the importance of pure thought and deed; Mahayana (includes Zen and Soka-gakkai), which ranges from philosophical schools to belief in the saving grace of higher beings or ritual practices and to practical meditative disciplines; and Tantrism, a combination of belief in ritual magic and sophisticated philosophy.

Location: Throughout Asia, from Sri Lanka to Japan. Zen and Soka-gakkai have several thousand adherents in the U.S.

Beliefs: Life is misery and decay, and there is no ultimate reality in it or behind it. The cycle of endless birth and rebirth continues because of desire and attachment to the unreal "self." Right meditation and deeds will end the cycle and achieve Nirvana, the Void, nothingness.

Hinduism

Founded: About 500 BC by Aryan invaders of India where their Vedic religion intermixed with the practices and beliefs of the natives.

Sacred texts: The *Veda,* including the *Upanishads,* a collection of rituals and mythological and philosophical commentaries; a vast number of epic stories about gods, heroes, and saints, including the *Bhagavadgita,* a part of the *Mahabharata,* and the *Ramayana;* and a great variety of other literature.

Organization: None, strictly speaking. Generally, rituals should be performed or assisted by Brahmins, the priestly caste, but in practice, simpler rituals can be performed by anyone. Brahmins are the final judges of ritual purity, the vital element in Hindu life. Temples and religious organizations are usually presided over by Brahmins.

Practice: A variety of private rituals, primarily passage rites (e.g., initiation, marriage, death, etc.) and daily devotions, and a similar variety of public rites in temples. Of the public rites, the *puja,* a ceremonial dinner for a god, is the most common.

Divisions: There is no concept of orthodoxy in Hinduism, which presents a variety of sects, most of them devoted to the worship of one of the many gods. The 3 major living traditions are those devoted to the gods Vishnu and Shiva and to the goddess Shakti; each is divided into further subsects. Numerous folk beliefs and practices, often in amalgamation with the above groups, exist side by side with sophisticated philosophical schools and exotic cults.

Location: Mainly India, Nepal, Malaysia, Guyana, Suriname, and Sri Lanka.

Beliefs: There is only one divine principle; the many gods are only aspects of that unity. Life in all its forms is an aspect of the divine, but it appears as a separation from the divine, a meaningless cycle of birth and rebirth (*samsara*) determined by the purity or impurity of past deeds (*karma*). To improve one's *karma* or escape *samsara* by pure acts, thought, and/or devotion is the aim of every Hindu.

Islam

Founded: About AD 622 in Mecca, Arabian Peninsula.

Founder: Muhammad (c 570-632), the Prophet.

Sacred texts: The *Koran* (al-Qur'an), the Word of God; *Sunna,* collections of *adth,* describing what Muhammad said or did.

Organization: Since the founder was both a prophet and a statesman, Muslim leadership has combined the civil and moral function of a state. Within the larger community, there are cultural and national groups, held together by a common religious law, the *Shari'a,* enforced uniformly in matters of religion only. In social transactions the community has often departed from traditional formulations. Although Islam is basically egalitarian and suspicious of authoritarianism, Muslim culture tends to be dominated by the conservative spirit of its religious establishment, the *ulema.*

Practice: Besides the general moral guidance that determines everyday life, there are "Five Pillars of Islam": profession of faith (oneness of God and prophethood of Muhammad); prayer 5 times a day; alms (*zakat*) from one's savings and estate; dawn-to-dusk fasting in the month of Ramadan; and once in a lifetime, pilgrimage to Mecca, if possible.

Divisions: There are 2 major groups: the majority known as Sunni and the minority Shiites. Shiites believe in Twelve Imams (perfect teachers) after the Prophet, of whom the last Imam has lived an invisible existence since 874, continuing to guide his community. Sunni Muslims believe in God's overpowering will over their affairs and tend to be predestinarian; Shiites believe in free will and give a substantial role to human reason in daily life. Sufism (mystical dimension of Islam) is prevalent among both Sunni and Shiites. Sufis emphasize personal relation to God and obedience informed by love of God.

Location: W Africa to Philippines, across band including E Africa, Central Asia and W China, India, Malaysia, Indonesia. Islam has several million adherents in North America.

Beliefs: Strictly monotheistic. God is creator of the universe, omnipotent, omniscient, just, forgiving, and merciful. The human is God's highest creation, but weak and egocentric, prone to forget the goal of life, constantly tempted by the Satan, an evil being. God revealed the Koran to Muhammad to guide humanity to truth and justice. Those who repent and sincerely "submit" (literal meaning of "islam") to God attain salvation. The forgiven enter the Paradise, and the wicked burn in Hell.

Judaism

Founded: About 1300 BC.

Founder: Abraham is regarded as the founding patriarch, but the Torah of Moses is the basic source of the teachings.

Sacred Texts: The 5 books of Moses constitute the written Torah. Special sanctity is also assigned other writings of the Hebrew Bible—the teachings of oral Torah are recorded in the Talmud, in the Midrash, and in various commentaries.

Organization: Originally theocratic, Judaism has evolved a congregational polity. The basic institution is the local synagogue, operated by the congregation and led by a rabbi of their choice. Chief rabbis in France and Great Britain have authority only over those who accept it; in Israel, the 2 chief rabbis have civil authority in family law.

Practice: Among traditional practicioners, almost all areas of life are governed by strict religious discipline. Sabbath and holidays are marked by special observances, and attendance at public worship is considered especially important then. Chief annual observances are Passover, celebrating liberation of the Israelites from Egypt and marked by the Seder meal in homes, and the 10 days from Rosh Hashana (New Year) to Yom Kippur (Day of Atonement), a period of fasting and penitence.

Divisions: Judaism is an unbroken spectrum from ultraconservative to ultraliberal, largely reflecting different points of view regarding the binding character of the prohibitions and duties—particularly the dietary and Sabbath observations—traditionally prescribed for the daily life of the Jew.

Location: Almost worldwide, with concentrations in Israel and the U.S.

Beliefs: Strictly monotheistic. God is the creator and absolute ruler of the universe. Men and women are free to choose to rebel against God's rule. God established a particular relationship with the Hebrew people: by obeying a divine law God gave them, they would be a special witness to God's mercy and justice. Judaism stresses ethical behavior (and, among the traditional, careful ritual obedience) as true worship of God.

LANGUAGE

New Words in English

The following words and definitions were provided by Merriam-Webster Inc., publishers of *Merriam-Webster's Collegiate Dictionary, Tenth Edition*. The words or meanings are among those that the Merriam-Webster editors decided had achieved enough currency in English to be added in the 1998 or 1999 copyright revision of the dictionary.

bafflegab: gobbledygook

benchmarking: the study of a competitor's product or business practices in order to improve the performance of one's own company

bioregion: a region whose limits are naturally defined by topographic and biological features (as mountains and ecosystems)

bioremediation: the treatment of pollutants or waste by the use of microorganisms (as bacteria) that break down the undesirable substances

bloviate: to speak or write verbosely or windily

brownfield: a tract of land that has been developed for industrial purposes, polluted, and then abandoned

cellophane noodle: a translucent noodle made from mung beans

cherry-pick: to select the best or most desirable

comfort food: food prepared in a traditional style having a usually nostalgic or sentimental appeal

distance learning: learning that takes place via electronic media linking instructors and students who are not together in a classroom

echinacea: the dried rhizome, roots, or other parts of any of three composite herbs used in folk medicine and some patent medicines especially for a supposed beneficial effect on the immune system

edge city: a suburb that has developed its own political, economic, and commercial base independent of the central city

euro: the basic monetary unit put into use by certain countries of the European Union beginning in 1999

face time: 1: the amount of time one spends appearing on television; 2: time spent in a face-to-face meeting with someone; 3: time spent at one's place of employment beyond normal work hours

farfalle: butterfly-shaped pasta

feng shui: a Chinese practice in which a structure or site is chosen and configured so as to harmonize with the spiritual forces that inhabit it; also: auspicious orientation, placement, or arrangement as determined by feng shui

fusion cuisine: food prepared using techniques and ingredients of various ethnic or regional cuisines

Paralympics: a series of international contests for athletes with disabilities that are associated with and held following the summer and winter Olympic Games

portobello: a cultivated mushroom belonging to a large dark meaty variety of the button mushroom

rightsize: to reduce (as a workforce) to an optimal size

screen saver: a computer program that usually displays various images on the screen of a computer that is on but not in use, to prevent damage to the screen's phosphors

spam: unsolicited usually commercial E-mail sent to a large number of addresses

three-peat: a third consecutive championship

velociraptor: any of a genus of theropod dinosaurs of the late Cretaceous having a long head with a flat snout and a large sickle-shaped claw on the second toe of each foot

waitstaff: the staff of servers at a restaurant

wetware: the human brain or a human being considered especially with respect to human logical and computational capabilities

Eponyms
(words named for people)

Bloody Mary—a vodka and tomato juice drink; after the nickname of Mary I, Queen of England (1553-58), notorious for persecution of Protestants

bloomers—full, loose trousers that are gathered at the knee; after Amelia Bloomer, an American social reformer who advocated (1851) such clothing

bobbies—in Great Britain, police officers; after Sir Robert Peel, the statesman who organized the London police force, 1850

bowdlerize—to delete written matter considered indelicate; after Thomas Bowdler, English editor of an expurgated Shakespeare (1825)

boycott—to avoid trade or dealings with, as a protest; after Charles C. Boycott, an English land agent in County Mayo, Ireland, ostracized in 1880 for refusing to reduce rents

Braille—a system of writing for the blind; after Louis Braille, the French teacher of the blind who invented it (1853)

Casanova—a man who is a promiscuous and unscrupulous lover; after Giovanni Giacomo Casanova (1725-98), an Italian adventurer

chauvinist—excessively patriotic; after Nicolas Chauvin, a character in a 19th-cent. play who is devoted to Napoleon

derby—a stiff felt hat with a dome-shaped crown and rather narrow rolled brim; after Edward Stanley, 12th earl of Derby, who in 1780 founded the Derby horse race, to which these hats are worn

diesel—a type of internal combustion engine or a vehicle driven by it; after Rudolf Diesel (1858-1913), who built the first successful diesel engine

gerrymander—to draw an election district in such a way as to favor a political party; after Elbridge Gerry, who created (1812) just such an election district (shaped like a salamander) during his governorship of Massachusetts

guillotine—a machine for beheading; after Joseph Guillotin, a French physician who proposed its use in 1789 as more humane than hanging

leotard—a close-fitting garment for the torso, worn by dancers, acrobats, and the like; after Julius Leotard, a 19th-cent. French aerial gymnast

sandwich—2 or more slices of bread with a filling in between; after John Montagu, 4th earl of Sandwich (1718-92), who supposedly ate food in this form so that he would not have to leave the gaming table

silhouette—an outline image; from Étienne de Silhouette (1709-67), a close-fisted French finance minister

National Spelling Bee

The Scripps Howard National Spelling Bee, conducted by Scripps Howard Newspapers and other leading newspapers since 1939, was instituted by the Louisville (KY) *Courier-Journal* in 1925. Children under 16 years old and not beyond 8th grade are eligible to compete for cash prizes at the finals, held annually in Washington, DC. The 1999 winners were: 1st place, Nupur Lala, Tampa, FL; 2d place, David Lewandowski, Schererville, IN; 3d place (tie), April DeGideo, Ambler, PA, and George Abraham Thampy, Maryland Heights, MO.

Here are the last words given, and spelled correctly, in each of the years 1980-99 at the national spelling bee.

1980	elucubrate	1985	milieu	1990	fibranne	1995	xanthosis
1981	sarcophagus	1986	odontalgia	1991	antipyretic	1996	vivisepulture
1982	psoriasis	1987	staphylococci	1992	lyceum	1997	euonym
1983	purim	1988	elegiacal	1993	kamikaze	1998	chiaroscurist
1984	luge	1989	spoliator	1994	antediluvian	1999	logorrhea

Foreign Words and Phrases

(L=Latin; F=French; Y=Yiddish; G=Greek; I=Italian; S=Spanish)

ad hoc (L; ad HOK): for the end or purpose at hand

ad infinitum (L; ad in-fi-NITE-um): without end; forever

ad nauseam (L; ad NAWZ-ee-um): to a sickening degree

apropos (F; ap-ruh-POH): relevant

bête noire (F; BET NWAHR): a thing or person viewed with particular dislike or fear

bon appétit (F; BOH nap-uh-teet): have a good meal!

bona fide (L; BOH nuh-fid): (needs diacritical over "I"): genuine; in good faith

carte blanche (F; kahrt BLANNSH): full discretionary power

cause célèbre (F; kawz suh-LEB-ruh): a notorious incident

c'est la vie (F; say lah VEE): that's life

chutzpah (Y; KHOOT-spuh): nerve bordering on arrogance

coup de grâce (F; kooh duh GRAHS): the final blow

coup d'état (F; kooh day TAH): overthrow of an existing government by a small group

crème de la crème (F; KREM duh luh KREM): the best of the best

cum laude/magna cum laude/summa cum laude (L; KUHM loud-ay; MAGN-ya ...; SOO-ma ...): with praise or honor/with great praise or honor/with the highest praise or honor

de facto (L; di FAK-toh): in fact, though not by right

déjà vu (F; DAY-zhah VOOH): the sensation that something happening has happened before

de jure (L; dee JOOR-ee, day YOOR-ay): in accordance with right or law; officially

de rigueur (F; duh ree-GUR): necessary according to convention or etiquette

détente (F; day-TAHNT): an easing of strained relations

éminence grise (F; ay-meh-NAHNN-suh GREEZ): one who wields power behind the scenes

enfant terrible (F; ahnn-FAHNN te-REE-bluh): one whose unconventional behavior causes embarrassment

en masse (F; ahn MAHS): in a large body

ergo (L; ER-goh): therefore

esprit de corps (F; es-PREE duh KAWR): group spirit; feeling of camaraderie

ex post facto (L; eks pohst FAK-toh): retroactive(ly)

fait accompli (F; fayt uh-kom-PLEE): an accomplished fact

faux pas (F; fowe PAH): a social blunder

hoi polloi (G; hoy puh-LOY): the masses

in loco parentis (L; in LOH-koh puh-REN-tis): in place of a parent

in memoriam (L; in muh-MAWR-ee-uhm): in memory of

in situ (L; in SEYE-tyooh): in the original place or position

in toto (L; in TOH-toh): totally

je ne sais quoi (F; zhuh nuh say KWAH): I don't know what; the little something that eludes description

joie de vivre (F; zhwah duh VEEV-ruh): zest for life

mea culpa (L; MAY-uh CUL-puh): through my fault

modus operandi (L; MOH-duhs op-uh-RAN-dee): method of operation

noblesse oblige (F; noh-BLES oh-BLEEZH): the obligation of nobility to help the less fortunate

non compos mentis (L; non KOM-puhs MEN-tis): not of sound mind

nouveau riche (F; nooh-voh REESH): a person newly rich; perhaps one who spends money conspicuously

persona non grata (L; per-SOH-nah non GRAH-tah): unwelcome person

postmortem (L; pohst-MORE-tuhm): after death; autopsy; analysis after an event

prima donna (I; pree-muh DAH-nuh): a principal female opera singer; temperamental person

pro tempore (L; proh TEM-puh-ree): for the time being

que sera sera (S; keh sair-AH sair-AH): what will be will be

quid pro quo (L; kwid proh KWOH): something given or received for something else

raison d'être (F; RAY-zohnn DET-ruh): reason for being

savoir faire (F; sav-wahr-FAIR): dexterity in social affairs

schlemiel (Y; shleh-MEEL): an unlucky, bungling person

semper fidelis (L; SEM-puhr fee-DAY-lis): always faithful

status quo (L; STAY-tus QWOH): the existing order of things

terra firma (L; TER-uh FUR-muh): solid ground

tour de force (F; TOOR duh FAWRS): feat accomplished through great skill

verbatim (L; ver-BAY-tuhm): word for word

vis-à-vis (F; vee-ZUH-VEE): compared with; with regard to; with respect to

Some Common Abbreviations and Acronyms

Acronyms are pronounceable words formed from first letters (or syllables) of other words. Some abbreviations below (e.g., AIDS, NATO) are thus acronyms. Some acronyms are words coined as abbreviations and written in lower case (e.g., "radar," "yuppie"); these are among abbreviations that may be more familiar than the terms they stand for. Acronyms do not have periods; usage for other abbreviations varies, but periods have become less common. Capitalization usage may vary from what is shown here. Italicized words preceding parnthetical definitions below are Latin unless otherwise noted. See also other chapters, including Internet and Computers; Weights and Measures.

AA=Alcoholics Anonymous

AAA=American Automobile Association

AARP=American Association of Retired Persons

ABA=American Bar Association

AC=alternating current

AD=*anno Domini* (in the year of the Lord)

AFL-CIO=American Federation of Labor and Congress of Industrial Organizations

AIDS=acquired immune deficiency syndrome

AM=*ante meridiem* (before noon)

AMA=American Medical Association

anon=anonymous

APO=army post office

ASAP=as soon as possible

ASCAP=American Society of Composers, Authors, and Publishers

ASPCA=American Society for Prevention of Cruelty to Animals

ATM=automated teller machine

Ave.=Avenue

AWOL=absent without leave

BA=Bachelor of Arts

bbl=barrel(s)

BC=before Christ

BCE=before Common Era

bpd=barrels per day

BS=Bachelor of Science

Btu=British thermal unit(s)

bu=bushel(s)

C= Celsius, centigrade

c=*circa* (about), copyright

Capt.=Captain

CE=Common Era

CEO=chief executive officer

CFO=chief financial officer

CIA=Central Intelligence Agency

cm=centimeter(s)

COD=cash (or collect) on delivery

Col.=Colonel

COLA=cost of living allowance

CPA=certified public accountant

Cpl.=Corporal

CPR=cardiopulmonary resuscitation

DA=district attorney

DAR=Daughters of the American Revolution

DC=direct current

DD=Doctor of Divinity

DDS=Doctor of Dental Science (or Surgery)

DNA=deoxyribonucleic acid

DNR=do not resuscitate

DOA=dead on arrival

DWI=driving while intoxicated

ed.=edited, edition, editor

e.g.=*exempli gratia* (for example)

EKG=electrocardiogram

EPA=Environmental Protection Agency

ESP=extrasensory perception

et al.=*et alii* (and others)

etc.=*et cetera* (and so forth)

EU=European Union

F=Fahrenheit

FBI=Federal Bureau of Investigation

FICA=Federal Insurance Contributions Act (Social Security)

FOB=free on board

ft=foot, feet

FY=fiscal year

FYI=for your information

gal=gallon(s)

GB=gigabyte(s)

GDP=gross domestic product
Gen.=General
GIGO=garbage in, garbage out
GNP=gross national product
GOP=Grand Old Party (Republican Party)
Hon.=the Honorable
HOV=high-occupancy vehicle
hr=hour(s)
ht=height
HVAC=heating, ventilating, and air-conditioning
i.e.=*id est* (that is)
IMF=International Monetary Fund
in.=inch(es)
IQ=intelligence quotient
IRA=individual retirement account, Irish Republican Army
IRS=Internal Revenue Service
ISBN=International Standard Book Number
JD=*Juris Doctor* (doctor of laws)
JP=Justice of the Peace
K=Kelvin
k=karat
KB=kilobyte(s)
kg=kilogram(s)
km=kilometer(s)
kw=kilowatt(s)
kwh=kilowatt-hour(s)
l=liter(s)
lb=*libra* (pound or pounds)
Lieut. or Lt.=Lieutenant
LLB=*Legum Baccalaureus* (bachelor of laws)
m=meter(s)
MA=Master of Arts
Maj.=Major
MB=megabyte(s)
MD=*Medicinae Doctor* (doctor of medicine)
MFN=most favored nation

mi=mile(s)
MIA=missing in action
min=minute(s)
ml=milliliter(s)
mm=millimeter(s)
mph=miles per hour
MS=Master of Science
MSG=monosodium glutamate
Msgr.=Monsignor
MVP=most valuable player
NAACP=National Association for the Advancement of Colored People
NASA=National Aeronautics and Space Administration
NAFTA=North American Free Trade Agreement
NATO=North Atlantic Treaty Organization
NB=*nota bene* (note carefully)
NCAA=National Collegiate Athletic Association
no=*numero* (number)
NOW=National Organization for Women
op=*opus* (work)
OPEC=Organization of Petroleum Exporting Countries
oz=ounce(s)
p, pp=page(s)
PAC=political action committee
PC=personal computer
PhD=*Philosophiae Doctor* (doctor of philosophy)
PIN=Personal Identification Number
PM=*post meridiem* (afternoon)
PO=post office
POW=prisoner of war
PS=*post scriptum* (postscript)
pt=part(s), pint(s), point(s)
Pvt.=Private
qt=quart(s)
q.v.=*quod vide* (which see)

radar=radio detecting and ranging
REM=rapid eye movement
Rev.=Reverend
RFD=rural free delivery
RIP=*requiescat in pace* (May he/she rest in peace)
RN=registered nurse
RNA=ribonucleic acid
ROTC=Reserve Officers' Training Corps
rpm=revolutions per minute
RR=railroad
RSVP=*répondez s'il vous plaît* (Fr.) (Please reply)
SASE=self-addressed stamped envelope
sec=second(s)
Sgt.=Sergeant
SIDS=sudden infant death syndrome
S.J.=Society of Jesus (Jesuits)
SRO=standing room only
St.=Saint, Street
TGIF=Thank God It's Friday
UFO=unidentified flying object
UHF=ultrahigh frequency
UNESCO=United Nations Educational, Social, and Cultural Organization
UNICEF=United Nations (International) Children's (Emergency) Fund
UPC=Universal Product Code
USS=United States ship
v (or vs)=*versus* (against)
VCR=videocassette recorder
VHF=very high frequency
VISTA=Volunteers in Service to America
W=watt(s)
Wasp=white Anglo-Saxon Protestant
WHO=World Health Organization
yd=yard(s)
yuppie=young urban professional
ZIP=zone improvement plan (U.S. Postal Service)

Names of the Days

ENGLISH	RUSSIAN	HEBREW	FRENCH	ITALIAN	SPANISH	GERMAN	JAPANESE
Sunday	Voskresenye	Yom rishon	Dimanche	Domenica	Domingo	Sonntag	Nichiyo\bi
Monday	Ponedelnik	Yom sheni	Lundi	Lunedì	Lunes	Montag	Getsuyo\bi
Tuesday	Vtornik	Yom shlishi	Mardi	Martedì	Martes	Dienstag	Kayo\bi
Wednesday	Sreda	Yom ravii	Mercredi	Mercoledì	Miércoles	Mittwoch	Suiyo\bi
Thursday	Chetverg	Yom hamishi	Jeudi	Giovedì	Jueves	Donnerstag	Mokuyo\bi
Friday	Pyatnitsa	Yom shishi	Vendredi	Venerdì	Viernes	Freitag	Kin-yo\bi
Saturday	Subbota	Shabbat	Samedi	Sabato	Sábado	Samstag	Doyo\bi

Names for Animal Young

The young of many animals have come to be called by special names. Many of these are listed below.

bunny: rabbit
calf: cattle, elephant, antelope, rhino, hippo, whale, others
cheeper: grouse, partridge, quail
chick, chicken: fowl
cockerel: rooster
codling, sprag: codfish
colt: horse (male)
cub: lion, bear, shark, fox, others
cygnet: swan

duckling: duck
eaglet: eagle
elver: eel
eyas: hawk, others
fawn: deer
filly: horse (female)
fingerling: fish generally
flapper: wild fowl
fledgling: birds generally
foal: horse, zebra, others
fry: fish generally
gosling: goose
heifer: cow

joey: kangaroo, others
kid: goat
kit: fox, beaver, rabbit, cat
kitten, kitty, catling: cats, other small mammals
lamb, lambkin, cosset, hog: sheep
leveret: hare
nestling: birds generally
owlet: owl
parr, smolt, grilse: salmon
piglet, shoat, farrow, suckling: pig

polliwog, tadpole: frog
poult: turkey
pullet: hen
pup: dog, seal, sea lion, fox
puss, pussy: cat
spike, blinker, tinker: mackerel
squab: pigeon
squeaker: pigeon, others
whelp: dog, tiger, beasts of prey
yearling: cattle, sheep, horse, others

Top 10 First Names of Americans by Decade of Birth

Source: Compiled by Dr. Cleveland Kent Evans, Bellevue University, Bellevue, NE; based on Social Security Administration records

BOYS:

1880-1889	John, William, George, Charles, James, Joseph, Frank, Henry, Harry, Thomas
1890-1899	John, William, James, George, Joseph, Charles, Frank, Harry, Robert, Henry
1900-1909	John, William, James, George, Joseph, Charles, Robert, Frank, Edward, Henry
1910-1919	John, William, James, Robert, Joseph, Charles, George, Edward, Frank, Walter
1920-1929	John, Robert, James, William, Charles, George, Joseph, Richard, Edward, Donald
1930-1939	Robert, James, John, William, Richard, Charles, Donald, George, Thomas, Joseph
1940-1949	James, Robert, John, William, Richard, David, Charles, Thomas, Michael, Ronald
1950-1959	Michael, James, Robert, John, David, William, Steven, Richard, Thomas, Mark
1960-1969	Michael, John, David, James, Robert, Mark, Steven, William, Jeffrey, Richard
1970-1979	Michael, Christopher, Jason, David, James, John, Brian, Robert, Steven, William
1980-1989	Michael, Christopher, Matthew, Joshua, David, Daniel, James, John, Robert, Brian
1990-1998	Michael, Christopher, Matthew, Joshua, Nicholas, Jacob, Andrew, Daniel, Brandon, Tyler

GIRLS:

1880-1889 . . . . Mary, Anna, Elizabeth, Margaret, Emma, Minnie, Bertha, Rose, Alice, Florence
1890-1899 . . . . Mary, Anna, Margaret, Elizabeth, Helen, Florence, Ruth, Rose, Ethel, Marie
1900-1909 . . . . Mary, Helen, Margaret, Anna, Ruth, Catherine, Elizabeth, Dorothy, Marie, Mildred
1910-1919 . . . . Mary, Helen, Dorothy, Margaret, Ruth, Catherine, Mildred, Anna, Elizabeth, Frances
1920-1929 . . . . Mary, Dorothy, Betty, Helen, Margaret, Ruth, Virginia, Catherine, Doris, Frances
1930-1939 . . . . Mary, Betty, Barbara, Shirley, Patricia, Dorothy, Joan, Margaret, Carol, Nancy
1940-1949 . . . . Mary, Linda, Barbara, Patricia, Carol, Sandra, Nancy, Sharon, Judith, Susan
1950-1959 . . . . Deborah, Mary, Linda, Patricia, Susan, Barbara, Karen, Nancy, Donna, Catherine
1960-1969 . . . . Lisa, Deborah, Mary, Karen, Michelle, Susan, Kimberly, Lori, Teresa, Linda
1970-1979 . . . . Jennifer, Michelle, Amy, Melissa, Kimberly, Lisa, Angela, Heather, Kelly, Sarah
1980-1989 . . . . Jessica, Jennifer, Ashley, Sarah, Amanda, Stephanie, Nicole, Melissa, Katherine, Megan
1990-1998 . . . . Ashley, Jessica, Sarah, Brittany, Emily, Kaitlyn, Samantha, Megan, Brianna, Amanda

Origins of Popular American Given Names

Source: Dr. Cleveland Kent Evans, Bellevue University, Bellevue, NE

Boys

Andrew: Gr. *andreios*, "man, manly"

Austin: Eng. form of Lat. *Augustinus*, "magnificent"

Brandon: Eng. place name, "gorse-covered hill"

Brian: Irish, perhaps Celtic *Brigonos*, "high, noble"

Charles: Ger. *ceorl*, "free man"

Christopher: Gr. *Khristophoros*, "bearing Christ [in one's heart]"

Daniel: Heb. "God is my judge"

David: Heb. *Dodavehu*, perhaps "darling"

Donald: Scots Gaelic *Domhnall*, "world rule"

Edward: Old Eng. *Eadweard*, "wealth-guard"

Frank: Ger. "Frenchman"

George: Gr. *georgos*, "soil tiller, farmer"

Harry: Middle Eng. form of Henry

Henry: Ger. *Haimric*, "home-power"

Jacob: Heb. *Yaakov*, "God protects" or "supplanter"

James: Late Lat. *Iacomus*, form of Jacob

Jason: Gr. *Iason*, "healer"

Jeffrey: Norman Fr. , from Ger. *Gaufrid*, "land-peace," or *Gisfrid*, "pledge-peace"

John: Heb. *Yohanan*, "God is gracious"

Joseph: Heb. *Yosef*, "[God] shall add"

Joshua: Heb. *Yoshua*, "God saves"

Mark: Lat. *Marcus*, perhaps "of Mars, the war god"

Matthew: Heb. *Mattathia*, "gift of God"

Michael: Heb. "Who could ever be like God?"

Nicholas: Gr. *Nikolaos*, "victory-people"

Richard: Ger. "power-hardy"

Robert: Ger. *Hrodberht*, "fame-bright"

Ronald: Scots form of Old Norse *Rögnvaldr*, "advice-ruler"

Steven: Gr. *stephanos*, "crown, garland"

Thomas: Aramaic "twin"

Tyler: Old Eng. *tigeler*, "tile layer"

Walter: Ger. *Waldheri*, "rule-army"

William: Ger. *Wilhelm*, "will-helmet"

Zachary: Eng. form of Heb. *Zechariah*, "God has remembered"

Girls

Alice: Old Fr. form of Ger. *Adalheidis*, "noble kind"

Amanda: 17th-cent. invention from Lat., "lovable"

Amy: Old Fr. *Amee*, "beloved"

Angela: Gr. *angelos*, "messenger [of God]"

Anna: Lat. and Gr. form of Hannah

Ashley: Eng. place name, "ash grove"

Barbara: Gr. *barbarus*, "foreign"

Bertha: Ger. *behrt*, "bright"

Betty: 18th-cent. pet form of Elizabeth

Brianna: modern fem. form of Brian

Brittany: place name, Fr. province settled by Britons

Caitlin: Irish form of Katherine

Carol: form of Charles

Deborah: Heb. "bee"

Donna: Ital. "lady"

Doris: Gr. "woman of the Dorian tribe," name of a sea nymph

Dorothy: Gr. *Dorothea*, "gift of God"

Elizabeth: Heb. *Elisheba*, perhaps "God is my oath" or "God is good fortune"

Emily: Roman *Aemilia*, possibly from Lat. *aemulus*, "rival"

Emma: Ger. *ermen*, "whole, entire"

Ethel: Old Eng. *aethel*, "noble"

Florence: Lat. *florens*, "flourishing"

Frances: fem. form of Francis, "a Frenchman"

Haley: Eng. place name, "hay clearing"

Hannah: Heb. "He has favored me"

Heather: Middle Eng. *hathir*, "heather"

Helen: Gr. *Helene*, possibly "sunbeam"

Jennifer: Cornish form of Welsh *Gwenhwyfar*, "fair-smooth"

Jessica: Shakespearean invention, probably fem. form of Jesse, Heb. "God exists"

Joan: Middle Eng. fem. form of John

Judith: Hebrew "Jewish woman"

Kaitlyn: modern American spelling of Caitlin

Karen: Danish form of Katherine

Katherine: from *Aikaterine*, Egyptian name later modified to resemble Gr. *katharos*, "pure"

Kayla: modern invention; or Yiddish form of Kelila, Heb. "crown of laurel"

Kelly: Irish Gaelic *Ceallagh,* perhaps "churchgoer" or "bright-headed"

Kimberly: Eng. place name, "Cyneburgh's clearing"

Linda: Sp. "pretty" or Ger. "tender"

Lisa: pet form of Elizabeth

Lori: pet form of either Lorraine (French "land of Lothar's people") or Laura (Latin "laurel")

Madison: Middle Eng. surname, "son of Madeline or Maud"

Margaret: Gr. *margaron*, "pearl"

Maria: Lat. form of Mary

Marie: Fr. form of Mary

Mary: Eng. form of Heb. *Maryam*, perhaps "seeress" or "wished-for child"

Megan: Welsh form of Margaret

Melissa: Gr. "bee"

Michelle: Fr. fem. form of Michael

Mildred: Old Eng. *Mildthryth*, "mild-strength"

Minnie: Pet form of Wilhelminia, fem. form of William

Nancy: medieval Eng. pet form of Agnes, Gr. *hagnos*, "holy"; later also used as pet form for Ann

Nicole: Fr. fem. form of Nicholas

Patricia: Lat. *Patricius*, "belonging to the noble class"

Rose: Ger. *hros*, "horse," or Lat. *rosa,* "rose"

Ruth: Heb., perhaps "companion"

Samantha: colonial American invention, probably combining Sam from Samuel [Heb. "name of God"] with -antha from Gr. *anthos,* "flower"

Sandra: short form of Alessandra, Ital. fem. of Alexander, Gr. "defend-man"

Sarah: Heb., "princess"

Sharon: Biblical place name, Hebrew "plain"

Shirley: Eng. place name, "bright clearing" or "shire meadow"

Stephanie: Fr. fem. form of Steven

Susan: Eng. form of Heb. *Shoshana*, "lily"

Taylor: Anglo-Norman *taillour*, "tailor"

Teresa: Spanish, perhaps "woman from Therasia"

Virginia: Lat., "virgin-like"

Pen Names

Shalom Aleichem (Solomon J. Rabinowitz)
Woody Allen (Allen Stewart Konigsberg)
Currer, Ellis, and Acton Bell (Charlotte, Emily, and Anne Brontë)
John le Carré (David John Moore Cornwell)
Lewis Carroll (Charles Lutwidge Dodgson)
Colette (Sidonie Gabrielle Colette)
Isak Dinesen (Karen Blixen)
Elia (Charles Lamb)

George Eliot (Mary Ann or Marian Evans)
Maksim Gorky (Aleksey Maksimovich Peshkov)
O. Henry (William Sydney Porter)
James Herriot (James Alfred Wight)
P. D. James (Phyllis Dorothy James White)
[John] Ross Macdonald (Kenneth Millar)
André Maurois (Émile Herzog)
Molière (Jean Baptiste Poquelin)
Frank O'Connor (Michael Donovan)
George Orwell (Eric Arthur Blair)

Ellery Queen (Frederic Dannay and Manfred B. Lee)
Mary Renault (Mary Challans)
Françoise Sagan (Françoise Quoirez)
Saki (Hector Hugh Munro)
George Sand (Amandine Lucie Aurore Dupin)
Dr. Seuss (Theodor Seuss Geisel)
Stendhal (Marie Henri Beyle)
Mark Twain (Samuel Clemens)
Voltaire (François Marie Arouet)
Tom Wolfe (Thomas Kennerly Jr.)

Forms of Address

GOVERNMENT	Address	Salutation
President of the U.S.	The President, The White House, Washington, DC 20500; also, The President and Mrs. ____ or The President and Mr.	Dear Sir or Madam; Mr. President or Madam President; Dear Mr. President or Dear Madam President
U.S. Vice President	The Vice President, The White House, Washington, DC 20500; also, The Vice President and Mrs. ____ or The Vice President and Mr. ____	Dear Sir or Madam; Mr. Vice President or Madam Vice President; Dear Mr. Vice President or Dear Madam Vice President
Chief Justice	The Hon. Firstname Surname, Chief Justice of the U.S., The Supreme Court, Washington, DC 20543	Dear Sir or Madam; Dear Mr. or Madam Chief Justice
Associate Justice	The Hon. Justice Firstname Surname, The Supreme Court, Washington, DC 20543	Dear Sir or Madam; Dear Justice Surname
Judge	The Hon. Firstname Surname, Associate Judge, U.S. District Court	Dear Judge Surname
Attorney General	The Hon. Firstname Surname, Attorney General, Dept. of Justice, Constitution Ave. & 10th St. NW, Washington, DC 20530	Dear Sir or Madam; Dear Mr. or Ms. Attorney General
Cabinet Officer	The Hon. Firstname Surname, Secretary of ____	Dear Mr. or Madam Secretary; or Dear Mr. or Ms. Surname
Senator	The Hon. or Sen. Firstname Surname, U.S. Senate, Washington, DC 20510	Dear Mr. or Madam Senator, or Dear Mr. or Ms. Surname
Representative	The Hon. or Rep. Firstname Surname, House of Representatives, Washington, DC 20515	Dear Mr. or Madam Surname
Speaker of the House	The Hon. Speaker of the House of Representatives, House of Representatives, Washington, DC 20515	Dear Mr. or Madam Speaker
Ambassador, U.S.	The Hon. Firstname Surname, American Ambassador[1]	Sir or Madam; Dear Mr. or Madam Ambassador
Ambassador, Foreign	His or Her Excellency[2] Firstname Surname, Ambassador of _____	Excellency[2] ; Dear Mr. or Madam Ambassador
Governor	The Hon. Firstname Surname, Governor of State; or in some states, His or Her Excellency, the Governor of State	Sir or Madam; Dear Governor Surname
Mayor	The Hon. Firstname Surname, Mayor of City	Sir or Madam; Dear Mayor Surname
MILITARY PERSONNEL		
All Titles	Full or abbreviated rank + full name + comma + abbreviation for branch of service. Example: Adm. John Smith, USN	Dear Rank Surname
RELIGIOUS		
Clergy, Protestant	The Reverend Firstname Surname[3]	Dear Ms. or Mr. Surname
Pope	His Holiness Pope Name or His Holiness the Pope	Your Holiness or Most Holy Father
Priest	The Reverend Firstname Surname or The Reverend Father Surname	Reverend Father, Dear Father Surname, or Dear Father
Rabbi	Rabbi Firstname Surname	Dear Rabbi Surname
ROYALTY AND NOBILITY		
King/Queen	His or Her Majesty, King or Queen of Country	Sir or Madam, or May it please Your Majesty

(1) If in Canada or Latin America, The Ambassador of the United States of America. (2) An American ambassador is not properly addressed as His or Her Excellency. (3) A member of the Protestant clergy who has a doctorate may be so addressed; for example, The Reverend Firstname Surname, DD, and Dear Dr. Surname.

Commonly Misspelled English Words

accidentally	committee	environment	incidentally	miniature	privilege
accommodate	conscientious	existence	independent	misspelled	receive
acknowledgment	conscious	fascinating	indispensable	mysterious	receipt
acquainted	convenience	February	inoculate	necessary	rhythm
all right	deceive	finally	irresistible	noticeable	ridiculous
already	defendant	fluorine	judgment	occasionally	separate
amateur	describe	foreign	laboratory	occurrence	seize
appearance	description	forty	license	opportunity	similar
appropriate	desirable	government	lightning	optimistic	sincerely
bureau	despair	grammar	liquefy	parallel	supersede
business	desperate	harass	maintenance	performance	transferred
character	eliminate	humorous	marriage	permanent	weird
commitment	embarrass	hurrying	millennium	perseverance	Wednesday

Commonly Confused English Words

adverse: unfavorable
averse: opposed

affect: to influence
effect: to bring about

allusion: an indirect reference
illusion: an unreal impression

appraise: to set a value on
apprise: to inform

capital: the seat of government
capitol: building where a legislature meets

complement: to make complete; something that completes
compliment: to praise; praise

denote: to mean
connote: to suggest beyond the explicit meaning

discreet: prudent
discrete: separate, distinct

disinterested: impartial
uninterested: without interest

elicit: to draw or bring out
illicit: illegal

emigrate: to leave for another place of residence
immigrate: to come to another place of residence

grisly: inspiring horror or great fear
grizzly: sprinkled or streaked with gray

historic: important in history
historical: relating to history

imminent: ready to take place
eminent: standing out

imply: to suggest but not explicitly; to entail
infer: to assume or understand information not relayed explicitly

include: used when the items following are part of a whole
comprise: used when the items following are all of a whole

incredible: unbelievable
incredulous: skeptical

ingenious: clever
ingenuous: innocent

oral: spoken, as opposed to written
verbal: relating to language

prostrate: stretched out face down
prostate: relating to prostate gland

The Principal Languages of the World

Source: Prof. Sidney Culbert, 351525, University of Washington, Seattle, WA 98195; data as of mid-1999

Languages Spoken by the Most People

	Speakers (millions)			Speakers (millions)			Speakers (millions)	
	Native[1]	Total		Native[1]	Total		Native[1]	Total
Mandarin	885	1,075	**Arabic**	211	256	**Japanese**	125	126
Hindi	375	496	**Bengali**	210	215	**German**	100	128
Spanish	358	425	**Portuguese**	178	194	**French**	77	129
English	347	514	**Russian**	165	275	**Malay-Indonesian**	58	176

(1) A native speaker is one for whom the language is his or her first language.

Languages Spoken by at Least 1 Million People

Total number (in millions) of speakers (native plus nonnative) of languages spoken by at least 1 million people. A native speaker is one for whom the language is his or her first language. Locations in parentheses are principal areas where the language is spoken.

NOTE: Languages are distinguished here according to criteria commonly accepted by linguists and may sometimes be more narrowly defined than others may suppose. For example, Neapolitan, Piedmontese, Sardinian, Sicilian, and Venetian, all spoken in Italy, are here regarded as distinct languages, so their speakers are not counted as Italian speakers. In some cases nonlinguistic criteria supervene, as in the case of Arabic here, where speakers of many variants are, according to custom, included under one broad term, "Arabic."

Language	Speakers
Achinese (N Sumatra, Indonesia)	3
Afghan (see Pashtu)	
Afrikaans (S. Africa)	6
Akan (or Twi-Fanti) (Ghana)	8
Albanian (Albania; Kosovo, Yugoslavia)	5
Amharic (Ethiopia)	21
Arabic (see also above)	256
Armenian (Armenia)	6
Assamese[1] (India; Bangladesh)	10
Aymara (Bolivia; Peru)	2
Azeri (Azerbaijan)	11
Balinese (Bali, Indonesia)	4
Baluchi (Baluchistan, in SW Pakistan and SE Iran)	5
Bashkir (Bashkortostan, Russia)	1
Batak Toba (Indonesia)	3
Baule (Côte d'Ivoire)	2
Beja (Spoken Arabic dialect group)	5
Bemba (Zambia)	2
Bengali[1] (see also above)	215
Berber[2]	
Beti (Cameroon; Gabon; Eq. Guinea)	2
Bhili (India)	6
Bikol (SE Luzon, Philippines)	4
Brahui (Pakistan)	2
Bugis (Indonesia; Malaysia)	4
Bulgarian (Bulgaria)	9
Burmese (Myanmar)	32
Buyi (S Guizhou, S China)	2
Byelorussian (Belarus)	10
Cantonese (China)	71
Catalan (NE Spain; Balearic Is.; S France; Andorra)	10
Cebuano (Bohol Sea, Philippines)	13
Chagga (Kilimanjaro area, Tanzania)	1
Chiga (Uganda)	1
Chinese[3]	
Chuvash (Chuvash, Russia)	2
Czech (Czech Republic)	12
Danish (Denmark)	5
Dimli (E Cent. Turkey)	1
Dogri (Jammu-Kashmir, CE India)	1
Dong (S Cent. China)	2
Dutch-Flemish (Netherlands; Belgium; NE France)	21
Dyerma (SW Niger)	2
Edo (Bendel, S Nigeria)	1
Efik (incl. Ibibio) (SE Nigeria)	6
English (see also above)	514
Esperanto	2
Estonian (Estonia)	1
Ewe (SE Ghana; S Togo)	3
Fang-Bulu (dialects of Beti)	
Farsi (see Persian)	
Finnish (Finland; Sweden)	6
Fon (S Cent. Benin; S Togo)	1
French (see also above)	129
Fula (or Peulh) (Cameroon; Nigeria)	13
Fulakunda (Senegal; Gambia; Guinea-Bissau)	3
Futa Jalon (Guinea; Sierra Leone)	3
Galician (Galicia, NW Spain)	3
Galla (see Oromo)	
Ganda (or Luganda) (S Uganda)	3
Georgian (Georgia)	4
German (see also above)	128
Gilaki (Gilan, NW Iran)	3
Gogo (Riff Valley, Tanzania)	1
Gondi (Cent. India)	3
Greek (Greece)	12
Guarani (Paraguay)	5
Gujarati[1] (W Cent. India; S Pakistan)	44
Gusii (Kisii District, Nyanza, Kenya)	2
Gypsy (see Romany)	
Hadiyya (Arusi, Ethiopia)	1
Haitian-Creole French	8
Hakka (or Kejia) (SE China)	34
Hausa (N Nigeria; Niger; Cameroon)	40
Haya (Kagera, NW Tanzania)	1
Hebrew (Israel)	5
Hindi[1,4] (see also above)	496
Hmong (S China; SE Asia)	6
Ho (Bihar and Orissa States, India)	1
Hungarian (or Magyar) (Hungary)	14
Iban (Indonesia; Malaysia)	1
Ibibio (see Efik)	
Igbo (or Ibo) (lower Niger, Nigeria)	18
Ijaw (Niger River delta, Nigeria)	2
Ilocano (NW Luzon, Philippines)	8
Indonesian (see Malay-Indonesian)	
IsiXhosa (SW Cape Prov., South Africa)	8
IsiZulu (N. Natal, South Africa; Lesotho)	9
Italian (Italy)	62
Japanese (see also above)	126
Javanese (Java, Indonesia)	64
Kabyle (W Kabylia, N Algeria)	3
Kamba (E Kenya)	3
Kannada (S India)	47
Kanuri (Nigeria; Niger; Chad; Cameroon)	4
Karen (see Sgaw)	
Karo-Dairi (N Sumatra, Indonesia)	1
Kashmiri[1] (N India; NE Pakistan)	4
Kazak (Kazakhstan)	8
Khalka (see Mongolian)	
Khmer (Cambodia; Vietnam; Thai.)	7
Khmer, Northern (Thailand)	1
Kikuyu (or Gekoyo) (WC Kenya)	6
Kituba (Bas-Congo, Bandundu, Congo)	5
Kongo (W Congo[5]; S Congo Rep.; NW Angola)	3
Konkani (Maharashtra and SW India)	2
Korean (Korea; China; Japan)	78
Kurdish (Iran; Iraq; Turkey)	11
Kurukh (or Oraon) (Cent. and E India)	2
Kyrgyz (Kyrgyzstan)	3
Lampung (Sumatra, Indonesia)	2
Lao[6] (Laos)	4
Latvian (Latvia)	2
Lingala (incl. Bangala) (Congo[5])	8
Lithuanian (Lithuania)	4
Luba-Lulua (or Chiluba) (Congo[5])	7
Luba-Shaba (Shaba, Congo[5])	2
Luhya (W Kenya)	3
Luo (Kenya; Nyanza, Tanzania)	4
Luri (SW Iran; Iraq)	4
Lwena (E Angola; W Zambia)	1
Macedonian (Macedonia)	2
Madurese (Madura, Indonesia)	10
Magindanaon (S Philippines)	1
Makassar (S Sulawesi, Indonesia)	2
Makua (S Tanzania; N Mozambique)	2
Malagasy (Madagascar)	
Malay-Indonesian (see also above)	176
Malay, Pattani (SE Thailand)	2
Malayalam[1] (Kerala, S India)	36
Malinke-Bambara-Dyula (W Africa)	9
Mandarin (China; Taiwan; see also above)	1,075
Marathi[1] (Maharashtra, India)	72
Mazandarani (N Iran)	3
Mbundu (Benguela, Angola)	4
Mbundu (Luanda, Angola)	3
Meithei (NE India; Bangladesh)	1
Mende (Sierra Leone)	2

Language		Language		Language	
Meru (Eastern Province, Cent. Tanzania)	1	Russian (see also above)	275	Tamil[1] (Tamil Nadu, India; Sri Lanka)	75
Mien (China; Viet.; Laos; Thailand)	1	Samar-Leyte (Cent. E Philippines)	3	Tartar (Tartarstan, Russia)	8
Min (SE China; Taiwan; Malaysia)	51	Sango (Cent. African Republic)	5	Tausug (Philippines; Malaysia)	1
Minangkabau (W Sumatra, Indon.)	6	Santali (E India; Nepal)	6	Telugu[1] (Andhra Pradesh,	
Moldavian (incl. with Romanian)		Sasak (Lombok, Alas Strait, Indon.)	2	SE India)	75
Mongolian (Mongolia; NE China)	6	Sepedi	4	Temne (central Sierra Leone)	2
Moré (Cent. Burkina Faso)	5	Serbo-Croatian (Croatia; Serbia; other		Thai[6] (Thailand)	53
Nepali (Nepal; NE India; Bhutan)	16	former Yugoslav republics and		Tho (N Vietnam; S China)	2
Ngulu (Mozambique; Malawi)	3	autonomous regions)	21	Thonga (Mozambique; So. Africa)	3
Nkole (Western Prov., Uganda)	2	Sesotho (S. Africa)	4	Tibetan (SW China; N India; Nepal)	4
Norwegian (Norway)	5	Setswana (S. Africa)	2	Tigrinya (S Eritrea; Tigre, Ethiopia)	5
Nung (NE of Hanoi, Vietnam; China)	1	Sgaw (SW Myanmar)	2	Tiv (SE Nigeria; Cameroon)	2
Nupe (Kwara, Niger States, Nigeria)	1	Shan (E Myanmar)	3	Tong (see Dong)	
Nyamwezi-Sukuma (NW Tanzania)	6	Shilha (W Algeria; S Morocco)	2	Tonga (SW Zambia; NW	
Nyanja (Malawi; Zambia;		Shona (Zimbabwe)	7	Zimbabwe)	2
Zimbabwe)	10	Sidamo (Sidamo, S Ethiopia)	2	Tulu (S India)	2
Oriya[1] (Cent. and E India)	31	Sindhi[1] (SE Pakistan; W India)	19	Tumbuka (N Malawi; NE Zambia)	2
Oromo (W Ethiopia; N Kenya)	10	Sinhalese (Sri Lanka)	13	Turkish (Turkey)	62
Pampangan (NW of Manila, Philip.)	2	SiSwati (S. Africa)	1	Turkmen (Turkmenistan;	
Panay-Hiligaynon (Philippines)	7	Slovak (Slovakia)	6	Afghanistan)	5
Pangasinan (Lingayen G., Philip.)	2	Slovene (Slovenia)	2	Twi-Fante (see Akan)	
Pashtu (Pakistan; Afghanistan; Iran)	19	Soga (Busoga, Uganda)	1	Uighur (Xinjiang, NW China)	8
Pedi (see Sotho, Northern)		Somali (Som.; Eth.; Ken.; Djibouti)	5	Ukrainian (Ukraine; Russia;	
Persian (Iran; Afghanistan)	36	Songye (Kasai Or., NW Shaba,		Poland)	49
Polish (Poland)	45	Congo[5])	1	Urdu (Pakistan; India)	106
Portuguese (see also above)	194	Soninke (Mali; countries to W S E)	1	Uzbek (Uzbekistan)	18
Provençal (S France)	2	Sotho, Northern (So. Africa)	4	Vietnamese (Vietnam)	69
Punjabi[1] (Punjab, Pakistan; India)	96	Sotho, Southern (So. Afr.; Lesotho)	4	Wolaytta (SE Ethiopia)	2
Pushto (see Pashtu)		Spanish (see also above)	425	Wolof (Senegal)	7
Quechua A (Peru; Boliv.; Ecuad.; Arg.)	8	Sundanese (Sunda Strait,		Wu (Shanghai region, China)	71
Rejang (SW Sumatra, Indonesia)	1	Indonesia)	26	Yao (see Mien)	
Riff (N Morocco; Algerian coast)	2	Swahili (Kenya; Tanzania; Congo[5];		Yao (Malawi; Tanzania;	
Romanian (Romania; Moldova)	26	Uganda)	50	Mozamb.)	2
Romany[7]	7	Swedish (Sweden; Finland)	9	Yi (S and SW China)	7
Ruanda (Rwanda; Uganda;		Sylhetti (Bangladesh)	5	Yiddish[8]	3
Congo[5])	6	Tagalog (Philippines)	58	Yoruba (SW Nigeria; Zou, Benin)	23
Rundi (Burundi)	6	Tajiki (Tajikistan; Uzbek.; Kyrgyz.)	5	Zande (NE Congo[5]; SW Sudan)	1
		Tamazight (N Morocco; W Algeria)	3	Zhuang (S China)	4

(1) One of the 15 languages under the constitution of India. (2) See Kabyle, Kabyle, Tamazight, Shilha, Riff. (3) See Mandarin, Cantonese, Wu, Min, and Hakka. The "common speech" (Putonghua) or the "national language" (Guoyu) is a standardized form of Mandarin as spoken in the area of Beijing. (4) Hindi and Urdu are essentially the same language, Hindustani. As the official language of Pakistan, it is written in a modified Arabic script and called Urdu. As the official language of India, it is written in the Devanagari script and called Hindi. (5) Congo refers to the Democratic Republic of the Congo (formerly Zaire), as distinct from the smaller Congo Republic. (6) The distinction between some Thai dialects and Lao are political rather than linguistic. (7) Mainly in Cent., E, and SE Europe and Turkey; some in the U.S. (8) Yiddish is usually considered a variant of German, but it has its own standard grammar and dictionaries, has a highly developed literature, and is written in Hebrew characters.

American Manual Alphabet

In the American Manual Alphabet, each letter of the alphabet is represented by a position of the fingers. This system was originally developed in France by Abbe Charles Michel De l'Epee in the late 1700s. It was brought to the U.S. by Laurent Clerce (1785-1869), a Frenchman who taught deaf or hearing-impaired people.

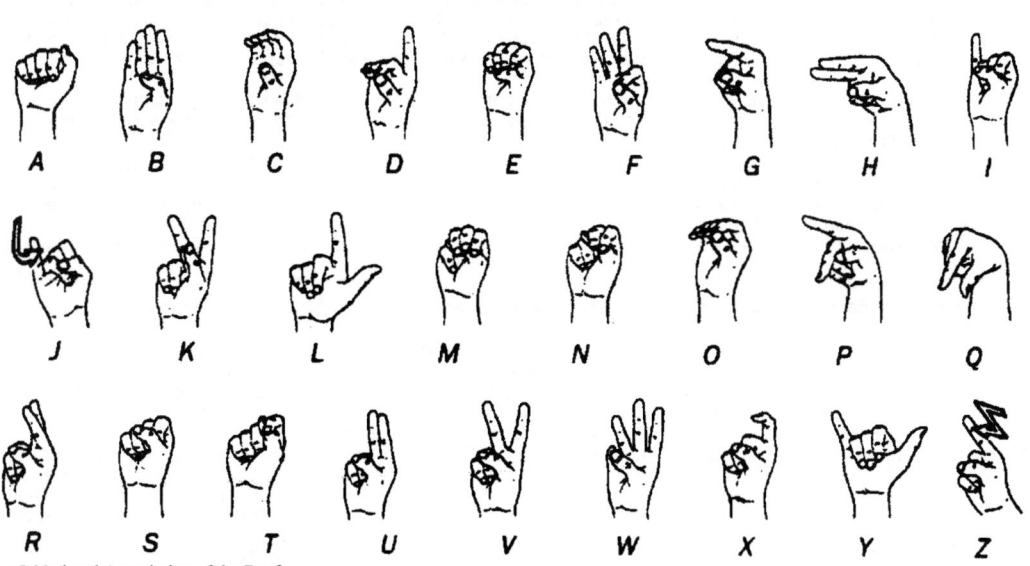

© National Association of the Deaf

TRADE AND TRANSPORTATION
U.S. Trade With Selected Countries and Major Areas, 1998

Source: Office of Trade and Econ. Analysis, U.S. Dept. of Commerce

(in millions of dollars; trade bal. ranked by size of deficit)

COUNTRY/AREA	U.S. trade balance with	Rank	Exports to	Rank	Imports from	Rank
Japan	$-64,014.1	1	$57,831.0	3	$121,845.0	2
China	-56,927.3	2	14,241.3	12	71,168.7	4
Germany	-23,184.6	3	26,657.4	5	49,842.0	5
Canada	-16,652.7	4	156,603.4	1	173,256.1	1
Mexico	-15,856.5	5	78,772.5	2	94,629.0	3
Taiwan	-14,960.3	6	18,164.5	7	33,124.8	7
Italy	-11,968.3	7	8,990.8	17	20,959.1	10
Malaysia	-10,043.0	8	8,957.0	18	19,000.0	11
Thailand	-8,197.8	9	5,238.6	26	13,436.4	13
Korea, South	-7,456.3	10	16,485.5	9	23,941.8	9
Indonesia	-7,041.7	11	2,298.9	38	9,340.6	17
France	-6,287.2	12	17,728.7	8	24,015.9	8
Philippines	-5,210.7	13	6,736.6	21	11,947.3	14
India	-4,672.8	14	3,564.4	32	8,237.2	23
Sweden	-4,025.9	15	3,822.1	30	7,848.0	24
Nigeria	-3,377.2	16	816.8	56	4,194.0	32
Ireland	-2,754.0	17	5,646.8	24	8,400.9	22
Venezuela	-2,665.6	18	6,515.8	22	9,181.4	18
Singapore	-2,662.0	19	15,693.6	10	18,355.7	12
Russia	-2,194.6	20	3,552.6	33	5,747.2	27
Angola	-1,886.2	21	354.7	73	2,240.9	42
Israel	-1,657.1	22	6,983.3	20	8,640.4	20
Sri Lanka	-1,576.0	23	190.4	91	1,766.5	47
Bangladesh	-1,527.5	24	318.4	79	1,845.9	46
Switzerland	-1,442.9	25	7,247.4	19	8,690.3	19
North America	-32,509.1	NA	235,375.9	NA	267,885.1	NA
Organization for Economic Cooperation & Development in Europe (OECD)	-29,268.8	NA	161,757.3	NA	191,026.1	NA
Western Europe	-29,400.2	NA	162,570.7	NA	191,971.0	NA
Euro Area	-27,907.7	NA	102,924.7	NA	130,832.4	NA
European Union (EU)	-27,345.8	NA	149,034.5	NA	176,380.3	NA
European Free Trade Association (EFTA)	-2,871.8	NA	9,200.6	NA	12,072.4	NA
Eastern Europe	-3,465.1	NA	7,437.7	NA	10,902.7	NA
Former Soviet Republics	-2,159.1	NA	4,929.5	NA	7,088.6	NA
Pacific Rim Countries	-160,376.2	NA	167,366.7	NA	327,742.9	NA
Assn. of Southeast Asian Nations (ASEAN)	-33,243.6	NA	39,047.5	NA	72,291.1	NA
Asia—Newly Industrialized Countries	-22,691.5	NA	63,269.0	NA	85,960.5	NA
Asia—South	-7,143.8	NA	4,780.0	NA	13,698.3	NA
Asia—Near East	4,891.1	NA	23,657.0	NA	18,765.9	NA
Latin American Free Trade Assn. (LAFTA)	-6,841.6	NA	120,650.8	NA	127,492.4	NA
20 Latin American Republics	-6,467.8	NA	135,301.5	NA	141,769.3	NA
Central American Common Market	-849.7	NA	8,401.9	NA	9,251.6	NA
South/Central America	13,129.0	NA	63,395.1	NA	50,266.1	NA
North Atlantic Treaty Organization (NATO)	-36,350.0	NA	297,564.0	NA	333,914.0	NA
Organization of Petroleum Exporting Countries (OPEC)	-8,771.1	NA	25,153.6	NA	33,924.7	NA
WORLD TOTAL	$-229,758.4	NA	$682,137.7	NA	$911,896.1	NA

NA – Not applicable. **Note:** Details may not equal totals because of rounding or incomplete enumeration.

Definitions of areas as used in the table:

North America—Canada, Mexico.

OECD—Austria, Belgium, Denmark, Finland, France, Germany, Greece, Iceland, Ireland, Italy, Liechtenstein, Luxembourg, Monaco, Netherlands, Norway, Portugal, San Marino, Spain, Svalbard/Jan Mayen Island, Sweden, Switzerland, Turkey, United Kingdom.

Western Europe—Andorra, Austria, Belgium, Bosnia and Herzegovina, Croatia, Cyprus, Denmark, Faroe Islands, Finland, France, Germany, Gibraltar, Greece, Iceland, Ireland, Italy, Liechtenstein, Luxembourg, Macedonia, Malta and Gozo, Monaco, Netherlands, Norway, Portugal, San Marino, Slovenia, Spain, Svalbard/Jan Mayen Island, Sweden, Switzerland, Turkey, United Kingdom, Vatican City, Yugoslavia.

Euro area—Austria, Belgium, Finland, France, Germany, Ireland, Italy, Luxemboug, Netherlands, Portugal, Spain.

European Union—Belgium, Denmark, France, Germany, Greece, Ireland, Italy, Luxembourg, Netherlands, Portugal, Spain, United Kingdom.

European Free Trade Association —Austria, Finland, Iceland, Liechtenstein, Norway, Sweden, Switzerland.

Eastern Europe—Albania, Armenia, Azerbaijan, Belarus, Bulgaria, Czech Republic, Estonia, Georgia, Hungary, Kazakhstan, Kyrgyzstan, Latvia, Lithuania, Moldova, Poland, Romania, Russia, Slovakia, Tajikistan, Turkmenistan, Ukraine, Uzbekistan.

Former Soviet Republics—Armenia, Azerbaijan, Belarus, Estonia, Georgia, Kazakhstan, Kyrgyzstan, Latvia, Lithuania, Moldova, Russia, Tajikistan, Turkmenistan, Ukraine, Uzbekistan.

Pacific Rim Countries/Territories—Australia, Brunei, China, Hong Kong, Indonesia, Japan, South Korea, Macao, Malaysia, New Zealand, Papua New Guinea, Philippines, Singapore, Taiwan.

ASEAN—Brunei, Indonesia, Malaysia, Philippines, Singapore, Thailand.

Asia—Newly Industrialized Countries—Hong Kong (special administrative region of China), Korea, Singapore, Taiwan.

Asia—South—Afghanistan, Bangladesh, India, Nepal, Pakistan, Sri Lanka.

Asia—Near East—Bahrain, Iran, Iraq, Israel, Jordan, Kuwait, Lebanon, Oman, Qatar, Saudi Arabia, Syria, U.A.E., Yemen.

LAFTA—Argentina, Bolivia, Brazil, Chile, Colombia, Ecuador, Mexico, Paraguay, Peru, Uruguay, Venezuela.

20 Latin American Republics—Argentina, Bolivia, Brazil, Chile, Colombia, Costa Rica, Cuba, Dominican Republic, Ecuador, El Salvador, Guatemala, Haiti, Honduras, Mexico, Nicaragua, Panama, Paraguay, Peru, Uruguay, Venezuela.

Central American Common Market—Costa Rica, El Salvador, Guatemala, Honduras, Nicaragua.

South/Central America—Anguilla, Antigua and Barbuda, Argentina, Aruba, Bahamas, Barbados, Belize, Bermuda, Bolivia, Brazil, British Virgin Islands, Cayman Islands, Chile, Colombia, Costa Rica, Cuba, Dominica, Dominican Republic, Ecuador, El Salvador, Falkland Islands, French Guiana, Grenada, Guadeloupe, Guatemala, Guyana, Haiti, Honduras, Jamaica, Martinique, Montserrat, Netherland Antilles, Nicaragua, Panama, Paraguay, Peru, St. Kitts and Nevis, St. Lucia, St. Vincent and the Grenadines, Suriname, Trinidad and Tobago, Turks and Caicos Islands, Uruguay, Venezuela.

NATO—Belgium, Canada, Denmark, France, Germany, Greece, Iceland, Ireland, Italy, Liechtenstein, Luxembourg, Monaco, Netherlands, Norway, Portugal, San Marino, Spain, Svalbard/Jan Mayan Island, Sweden, Switzerland, Turkey, United Kingdom.

OPEC—Algeria, Gabon, Indonesia, Iran, Iraq, Kuwait, Libya, Nigeria, Qatar, Saudi Arabia, United Arab Emirates, Venezuela.

U.S. Exports and Imports by Principal Commodity Groupings, 1998

Source: Office of Trade and Economic Analysis, U.S. Dept. of Commerce

(millions of dollars)

	Exports	Imports		Exports	Imports
TOTAL	**$682,138**	**$911,896**	Lighting, plumbing	$1,405	$3,393
Agricultural commodities	**50,654**	**35,748**	Metal manufactures	10,699	13,506
Animal feeds	4,054	607	Metalworking machinery	5,270	7,926
Cereal flour	1,274	1,461	Nickel	365	893
Coffee	10	3,069	Optical goods	1,907	2,727
Corn	4,618	142	Paper and paperboard	9,930	12,793
Cotton, raw and linters	2,572	19	Photographic equipment	3,481	5,661
Hides and skins	1,130	110	Plastic articles[1]	5,554	6,137
Meat and preparations	6,427	2,848	Platinum	388	3,051
Oils/fats, vegetable	1,852	1,326	Pottery	103	1,719
Rice	1,207	182	Power generating machinery	28,743	28,160
Soybeans	4,885	54	Printed materials	4,671	3,075
Sugar	3	715	Records/magnetic media	6,057	4,383
Tobacco, unmanufactured	1,459	780	Rubber articles[1]	1,301	1,645
Vegetables and fruit	7,324	8,365	Rubber tires and tubes	2,554	4,095
Wheat	3,690	283	Scientific instruments	24,169	15,500
Manufactured goods	**552,778**	**790,754**	Ships, boats	1,716	1,132
ADP equipment; office machinery	40,735	76,755	Silver and bullion	629	662
Airplane parts	15,035	5,912	Spacecraft	1,102	91
Airplanes	35,242	7,052	Specialized industrial machinery	27,305	22,986
Aluminum	3,606	5,970	Televisions, VCRs, etc.	23,417	42,449
Artwork/antiques	1,140	3,976	Textile yarn, fabric	8,976	12,896
Basketware, etc.	2,583	3,834	Toys/games/sporting goods	3,342	18,695
Chemicals - cosmetics	4,763	2,895	Travel goods	302	3,946
Chemicals - dyeing	3,471	2,472	Vehicles	53,536	119,726
Chemicals - fertilizers	3,231	1,570	Watches/clocks/parts	312	3,210
Chemicals - inorganic	4,709	5,126	Wood manufactures	1,690	5,647
Chemicals - medicinal	9,341	10,908	**Mineral fuels**	**10,251**	**57,323**
Chemicals - other[1]	10,838	4,833	Coal	3,176	724
Chemicals - organic	14,920	18,327	Crude oil	895	37,252
Chemicals - plastics	16,608	8,565	Liquefied propane/butane	203	933
Clothing	8,497	53,742	Mineral fuels, other	2,619	1,132
Copper	1,276	3,046	Natural gas	243	5,273
Electrical machinery	65,575	79,365	Petroleum preparations	2,857	10,947
Footwear	720	13,881	**Selected commodities:**		
Furniture and bedding	4,411	13,339	Alcoholic bev., distilled	385	2,296
Gem diamonds	124	8,497	Cigarettes	4,165	102
General industrial machinery	30,121	28,811	Cork, wood, lumber	4,091	7,616
Glass	2,030	1,774	Crude fertilizers	1,600	1,300
Glassware	709	1,609	Fish and preparations	2,172	8,104
Gold, nonmonetary	5,464	3,587	Metal ores; scrap	3,581	4,099
Iron and steel mill products	5,481	17,159	Pulp and waste paper	3,443	2,442
Jewelry	769	5,270			

Note: Not all products are listed in each commodity group. (1) Those not specified elsewhere.

Trends in U.S. Foreign Trade, 1790-1998

Source: Office of Trade and Economic Analysis, U.S. Dept. of Commerce

In 1790, U.S. exports and imports combined came to $43 million and there was a $3 million trade deficit. By 1998, U.S. exports and imports combined amounted to nearly $2 trillion, and the trade deficit, which had generally been climbing in recent years (after a century of trade surpluses), reached around $230 billion.

(in millions of dollars)

Year	Exports	Imports	Trade Balance	Year	Exports	Imports	Trade Balance
1790	$20	$23	$-3	1905	$1,519	$1,118	$401
1795	48	70	-22	1910	1,745	1,557	188
1800	71	91	-20	1915	2,769	1,674	1,094
1805	96	121	-25	1920	8,228	5,278	2,950
1810	67	85	-19	1925	4,910	4,227	683
1815	53	113	-60	1930	3,843	3,061	782
1820	70	74	-5	1935	2,283	2,047	235
1825	91	90	1	1940	4,021	2,625	1,396
1830	72	63	9	1945	9,806	4,159	5,646
1835	115	137	-22	1950	9,997	8,954	1,043
1840	124	98	25	1955	14,298	11,566	2,732
1845	106	113	-7	1960	19,659	15,073	4,586
1850	144	174	-29	1965	26,742	21,520	5,222
1855	219	258	-39	1970	42,681	40,356	2,325
1860	334	354	-20	1975	107,652	98,503	9,149
1865	166	239	-73	1980	220,626	244,871	-24,245
1870	393	436	-43	1985	213,133	345,276	-132,143
1875	513	533	-20	1990	394,030	495,042	-101,012
1880	836	668	168	1995	584,742	743,445	-158,703
1885	742	578	165	1996	625,075	795,289	-170,214
1890	858	789	69	1997	689,182	870,671	-181,489
1895	808	732	76	1998	682,138	911,896	-229,758
1900	1,394	850	545				

(1) Because of rounding, not all totals add.

The North American Free Trade Agreement (NAFTA)

NAFTA, a comprehensive plan for free trade between the U.S., Canada, and Mexico, took effect on Jan. 1, 1994. Major provisions are as follows:

Agriculture—Tariffs on all farm products are to be eliminated over 15 years. Domestic price-support systems may continue provided they do not distort trade.

Automobiles—After 8 years, at least 62.5% of an automobile's value must have been produced in North America for it to qualify for duty-free status. Tariffs are to be phased out over 10 years.

Banking—U.S. and Canadian banks may acquire Mexican commercial banks accounting for as much as 8% of the industry's capital. All limits on ownership end in 2004.

Disputes—Special judges have jurisdiction to resolve disagreements within strict timetables.

Energy—Mexico continues to bar foreign ownership of its oil fields but, starting in 2004, U.S. and Canadian companies can bid on contracts offered by Mexican oil and electricity monopolies.

Environment—The agreement cannot be used to overrule national and state environmental, health, or safety laws.

Immigration—All 3 countries must ease restrictions on the movement of business executives and professionals.

Jobs—Barriers to limit Mexican migration to U.S. remain.

Patent and copyright protection—Mexico strengthened its laws providing protection to intellectual property.

Tariffs—Tariffs on 10,000 customs goods are to be eliminated over 15 years. One-half of U.S. exports to Mexico are to be considered duty-free within 5 years.

Textiles—A "rule of origin" provision requires most garments to be made from yarn and fabric produced in North America. Most tariffs are being phased out over 5 years.

Trucking—Trucks are to have free access on crossborder routes and throughout the 3 countries by 1999.

U.S. Trade With Canada and Mexico, 1992-98

Source: Office of Trade and Economic Analysis, U.S. Dept. of Commerce

(U.S. exports to, imports from, Canada and Mexico in millions of dollars)

	MEXICO				CANADA		
Year	Exports	Imports	Trade Balance[1]	Year	Exports	Imports	Trade Balance[1]
1992	$40,592	$35,211	$5,381	1992	$90,594	$98,630	$−8,0361
1993	41,581	39,917	1,664	1993	100,444	111,216	−10,772
1994[2]	50,844	49,494	1,350	1994[2]	114,439	128,406	−13,968
1995	46,292	61,685	−15,393	1995	127,226	145,349	−18,123
1996	56,792	74,297	−17,506	1996	134,210	155,893	−21,682
1997	71,388	85,938	−14,549	1997	151,767	167,234	−15,467
1998	78,773	94,629	−15,857	1998	156,603	173,256	−16,653

(1) Totals may not add due to rounding. (2) NAFTA provisions began to take effect Jan. 1, 1994.

50 Busiest U.S. Ports, 1997

Source: Corps of Engineers, Dept. of the Army, U.S. Dept. of Defense

(ports ranked by tonnage handled; all figures in tons)

Rank	Port	Total	Domestic	Foreign	Imports	Exports
1.	South Louisiana, LA, Port of	183,628,353	106,846,289	76,782,064	26,414,598	50,367,466
2.	Houston, TX	165,456,278	62,609,724	102,846,554	72,640,589	30,205,965
3.	New York, NY & NJ	135,266,431	78,552,576	56,713,855	48,122,399	8,591,456
4.	New Orleans, LA	89,441,772	37,003,350	52,438,422	23,924,557	28,513,865
5.	Corpus Christi, T	86,843,760	24,625,068	62,218,692	54,215,016	8,003,676
6.	Baton Rouge, LA	84,023,102	45,616,108	38,406,994	28,575,663	9,831,331
7.	Valdez, AK	73,647,151	70,107,042	3,540,109	36,761	3,503,348
8.	Plaquemines, LA, Port of	63,607,222	46,959,050	16,648,172	5,495,185	11,152,987
9.	Long Beach, CA	57,255,301	18,898,756	38,356,545	21,175,723	17,180,822
10.	Texas City, TX	56,645,675	19,214,997	37,430,678	35,061,311	2,369,367
11.	Tampa, FL	55,333,607	36,729,922	18,603,685	7,255,497	11,348,188
12.	Pittsburgh, PA	51,662,378	51,662,378	0	0	0
13.	Lake Charles, LA	51,278,579	21,568,542	29,710,037	25,190,443	4,519,594
14.	Mobile, AL	49,120,007	24,275,907	24,844,100	13,169,050	11,675,050
15.	Beaumont, TX	48,665,380	15,038,639	33,626,741	28,234,027	5,392,714
16.	Norfolk Harbor, VA	46,322,012	10,904,961	35,417,051	5,933,603	29,483,448
17.	Philadelphia, PA	44,967,869	15,010,358	29,957,511	29,490,353	467,158
18.	Duluth-Superior, MN and WI	41,928,885	31,154,138	10,774,747	847,587	9,927,160
19.	Los Angeles, CA	41,774,252	13,194,710	28,579,542	16,201,305	12,378,237
20.	Baltimore, MD	40,028,849	14,806,775	25,222,074	14,971,414	10,250,660
21.	Port Arthur, TX	37,318,229	7,589,290	29,728,939	26,949,644	2,782,295
22.	St. Louis, MO and IL	31,287,584	31,287,584	0	0	0
23.	Pascagoula, MS	31,270,055	10,020,911	21,249,144	18,081,418	3,167,726
24.	Portland, OR	29,560,776	13,022,044	16,538,732	3,370,151	13,168,581
25.	Seattle, WA	26,564,230	7,913,684	18,650,546	7,848,268	10,802,278
26.	Freeport, TX	26,280,731	5,140,665	21,140,066	19,172,508	1,967,558
27.	Huntington, WV	25,175,459	25,175,459	0	0	0
28.	Chicago, IL	24,867,996	21,005,028	3,862,958	3,271,231	591,737
29.	Paulsboro, NJ	24,391,944	10,695,695	1,369,249	13,516,240	180,009
30.	Richmond, CA	21,705,683	16,484,842	5,220,841	3,446,500	1,774,341
31.	Marcus Hook, PA	21,520,644	9,952,592	11,567,652	11,394,793	172,859
32.	Boston, MA	20,892,983	9,903,527	10,989,456	10,317,904	671,552
33.	Newport News, VA	20,755,282	6,109,423	14,645,859	1,378,593	13,267,266
34.	Tacoma, WA	20,683,326	7,603,646	13,079,680	4,202,695	8,876,985
35.	Port Everglades, FL	19,924,784	11,959,222	7,965,562	6,126,585	1,838,977
36.	Jacksonville, FL	18,186,104	9,295,617	8,890,487	7,398,244	1,492,243
37.	Detroit, MI	18,135,326	12,021,050	6,114,276	5,614,244	500,032
38.	Cleveland, OH	18,113,321	14,773,821	3,339,500	3,165,069	174,431
39.	Memphis, TN	18,015,173	18,015,173	0	0	0
40.	Savannah, GA	17,929,269	3,227,330	14,701,939	7,386,671	7,315,268
41.	Charleston, SC	17,874,161	4,750,915	13,123,246	5,939,086	7,184,160
42.	Indiana Harbor, IN	16,523,799	16,007,025	516,774	457,414	59,360
43.	Portland, ME	16,333,742	1,684,857	14,648,885	14,575,783	73,102
44.	Lorain, OH	15,954,569	15,413,589	540,980	540,980	0
45.	Toledo, OH	14,421,587	7,509,037	6,912,550	1,839,616	5,072,934
46.	San Juan, PR	14,067,151	9,127,545	4,939,606	4,316,317	623,289
47.	Anacortes, WA	13,903,514	12,184,288	1,719,226	672,316	1,046,910
48.	Two Harbors, MN	13,507,844	13,437,728	70,116	0	70,116
49.	Cincinnati, OH	12,878,606	12,878,606	0	0	0
50.	Honolulu, HI	12,703,903	9,666,427	3,037,476	2,890,906	146,570

Major Merchant Fleets of the World

Source: Maritime Administration, U.S. Dept of Commerce

Fleets of oceangoing steam and motor ships totaling 1,000 gross tons or more as of Jan. 1999. Excludes ships operating exclusively on the Great Lakes and inland waterways and special types such as channel ships, icebreakers, cable ships, and merchant ships owned by any military force. Gross tonnage is a volume measurement; each cargo gross ton represents 100 cubic ft of enclosed space. Deadweight (Dwt) tonnage is carrying capacity of a ship in long tons (2,240 lb). Only some major types of vessels are shown separately. Tonnage figures may not add, because of rounding.

(tonnage in thousands)

	ALL VESSELS[1]			Tanker			Dry Bulk			Containership		
	No. of ships	Gross tons	Dwt tons	No. of ships	Gross tons	Dwt tons	No. of ships	Gross tons	Dwt tons	No. of ships	Gross tons	Dwt tons
ALL COUNTRIES	27,825	489,300	755,435	6,781	182,147	317,337	5,726	158,096	277,242	2,382	53,569	61,237
United States	470	12,246	16,741	154	5,189	9,289	15	350	604	91	3,097	3,096
Privately owned . . .	281	9,544	13,213	126	4,698	8,403	15	350	604	87	3,026	3,026
Government owned	189	2,702	3,528	28	491	886	—	—	—	4	71	70
Panama	4,485	95,844	145,769	985	27,396	47,516	1,302	40,474	70,980	475	11,910	13,303
Liberia	1,644	60,462	97,232	698	32,820	57,659	453	16,860	29,932	181	4,421	5,087
Greece	737	24,595	44,072	262	13,398	25,516	307	8,976	16,093	47	1,297	1,406
Bahamas	1,042	26,301	40,944	249	12,952	23,670	158	4,982	8,691	50	989	1,044
Malta	1,312	23,712	39,311	352	10,294	18,758	368	8,605	14,711	45	686	774
Cyprus	1,431	23,068	36,059	179	4,308	7,386	478	11,210	19,423	122	2,335	2,780
Singapore	879	19,884	31,436	384	9,502	16,690	134	4,616	8,512	161	3,100	3,622
Norway (NIS)[2]	655	19,420	30,373	289	11,134	19,642	105	3,930	6,994	5	96	119
China, People's Republic of	1,476	14,845	22,342	248	2,071	3,251	338	6,659	11,115	95	1,357	1,652
Japan	698	13,745	19,694	279	7,109	10,507	172	3,946	7,218	29	890	873
Philippines	534	7,826	12,278	68	182	280	209	5,651	9,736	11	138	178
Saint Vincent & the Grenadines.	784	7,616	11,554	96	1,152	1,968	142	3,146	5,462	28	187	221
Marshall Islands	112	6,342	10,901	43	3,785	7,045	41	1,555	2,809	20	871	937
India	289	6,293	10,586	97	2,936	5,110	124	2,826	4,759	6	83	111
Hong Kong, China . . .	191	6,067	10,190	9	346	642	109	4,190	7,785	40	914	1,024
Turkey.	518	5,767	9,556	73	583	1,039	167	3,942	6,844	13	108	142
Germany	514	7,400	9,243	19	178	272	—	—	—	293	6,280	7,978
China, Republic of (Taiwan)	184	5,305	8,342	17	901	1,554	54	2,346	4,344	75	1,919	2,237
Russia	1,448	6,732	8,244	266	1,549	2,235	113	1,155	1,700	24	268	299
Korea (South)	427	5,098	7,931	106	483	829	105	2,933	5,305	45	808	944
Bermuda.	98	4,776	7,846	33	2,726	4,758	20	1,172	2,248	19	595	589
Italy.	352	5,589	7,741	193	2,242	3,515	33	1,525	2,867	14	377	393
Malaysia	370	5,014	7,338	112	2,187	3,024	58	1,447	2,545	48	667	817
Brazil	168	4,135	6,894	76	1,877	3,171	45	1,789	3,179	6	134	166
Isle of Man	149	4,101	6,756	72	2,496	4,409	22	810	1,494	21	398	474
Denmark (DIS)[3]	309	4,878	6,654	66	1,289	2,182	13	521	967	61	2,478	2,834
Iran	121	3,207	5,637	24	1,624	3,141	45	994	1,676	3	10	12
French Antarctic Territory	73	2,589	4,413	35	1,621	3,088	4	351	641	11	409	462
Kuwait.	46	2,360	3,837	28	1,939	3,341	1	17	27	5	182	191
Netherlands	411	3,387	3,814	59	441	685	4	87	123	39	1,249	1,336
Indonesia	491	2,310	3,497	123	817	1,297	24	358	577	12	86	111
Antigua & Barbuda. . .	414	2,414	3,145	10	28	42	20	294	473	93	1,060	1,346
Thailand	285	1,848	2,970	89	367	661	38	501	836	12	119	160
Norway.	124	1,853	2,952	39	1,553	2,749	6	15	19	—	—	—
Romania.	190	1,950	2,904	8	197	337	38	846	1,378	2	15	17
Belize	414	1,719	2,564	64	350	623	24	247	422	6	51	54
United Kingdom	140	2,459	2,506	55	621	1,051	4	44	69	25	998	1,088
Australia	52	1,775	2,418	15	710	740	27	901	1,499	4	114	128
Egypt	113	1,216	1,935	16	212	368	22	613	1,054	1	14	18
France	52	1,235	1,906	25	881	1,570	1	1	2	4	99	114
Poland	63	1,182	1,889	2	13	19	51	1,091	1,808	—	—	—
Vanuatu	83	1,356	1,809	9	112	160	30	734	1,253	—	—	—
Sweden	167	2,124	1,809	64	534	873	6	25	32	—	—	—
Cayman Islands	73	1,057	1,612	18	215	356	12	455	813	5	59	68
Bulgaria	98	1,058	1,549	11	151	267	35	546	855	5	56	67
Ukraine.	264	1,331	1,495	22	63	95	10	207	342	3	32	28
Saudi Arabia.	60	1,152	1,456	24	344	593	1	12	20	7	222	217
Portugal	123	86	1,357	30	422	703	14	187	333	4	21	26
Croatia	63	792	1,232	5	9	12	21	516	889	5	82	98
Mexico	46	791	1,164	38	656	1,004	—	—	—	4	124	147

(1) Includes combination passenger and cargo ships and other type of vessels not listed separately. (2) Norwegian international ship registry. (3) Danish international ship registry.

World Trade Organization (WTO)

Following World War II, the major economic powers of the world negotiated a set of rules for reducing and limiting trade barriers and for settling trade disputes. These rules were called the General Agreement on Tariffs and Trade (GATT). Headquarters to oversee the administration of the GATT were established in Geneva, Switzerland. Periodically, rounds of multilateral trade negotiations under the GATT were carried out. The 8th round, begun in 1986 in Punta del Este, Uruguay, and dubbed the Uruguay Round, concluded on Dec. 15, 1993, when 117 countries completed a new trade-liberalization agreement. The name for the GATT was changed to the World Trade Organization (WTO), which officially came into being Jan. 1, 1995.

New Passenger Cars Imported Into the U.S., by Country of Origin,[1] 1968-98

Source: Bureau of the Census, Foreign Trade Division

	Japan	Germany[2]	Italy	United Kingdom	Sweden	France	South Korea	Mexico	Canada	Total[3]
1968.....	169,849	707,972	33,843	96,787	52,515	39,551	NA	NA	500,881	1,620,452
1969.....	260,005	642,157	41,569	104,050	41,008	24,457	NA	NA	691,146	1,846,717
1970.....	381,338	674,945	42,523	76,257	57,844	37,114	NA	NA	692,783	2,013,420
1971.....	703,672	770,807	51,469	106,710	61,925	23,316	NA	0	802,281	2,587,484
1972.....	697,788	676,967	64,614	72,038	64,541	14,713	NA	9	842,300	2,485,901
1973.....	624,805	677,465	56,102	64,140	58,626	8,219	NA	4,469	871,557	2,437,345
1974.....	791,791	619,757	107,071	72,512	60,817	21,331	NA	3,914	817,559	2,572,557
1975.....	695,573	370,012	102,344	67,106	51,993	15,647	NA	0	733,766	2,074,653
1976.....	1,128,936	349,804	82,500	77,190	37,466	21,916	NA	0	825,590	2,536,749
1977.....	1,341,530	423,492	55,437	56,889	39,370	19,215	NA	NA	849,814	2,790,144
1978.....	1,563,047	416,231	69,689	54,478	56,140	28,502	NA	6	833,061	3,024,982
1979.....	1,617,328	495,565	72,456	46,911	65,907	27,887	NA	4	677,008	3,005,523
1980.....	1,991,502	338,711	46,899	32,517	61,496	47,386	NA	1	594,770	3,116,448
1981.....	1,911,525	234,052	21,635	12,728	68,042	42,477	NA	1	563,943	2,856,286
1982.....	1,801,185	259,385	9,402	13,023	89,231	50,032	NA	27	702,495	2,926,407
1983.....	1,871,192	239,807	5,442	17,261	114,726	40,823	NA	2	835,665	3,133,836
1984.....	1,948,714	335,032	8,582	19,833	114,854	37,788	NA	NA	1,073,425	3,559,427
1985.....	2,527,467	473,110	8,689	24,474	142,640	42,882	NA	13,647	1,144,805	4,397,679
1986.....	2,618,711	451,699	11,829	27,506	148,700	10,869	169,309	41,983	1,162,226	4,691,297
1987.....	2,417,509	377,542	8,648	50,059	138,565	26,707	399,856	126,266	926,927	4,589,010
1988.....	2,123,051	264,249	6,053	31,636	108,006	15,990	455,741	148,065	1,191,357	4,450,213
1989.....	2,051,525	216,881	9,319	29,378	101,571	4,885	270,609	133,049	1,151,122	4,042,728
1990.....	1,867,794	245,286	11,045	27,271	93,084	1,976	201,475	215,986	1,220,221	3,944,602
1991.....	1,762,347	171,097	2,886	14,862	62,905	1,727	186,740	249,498	1,109,248	3,612,665
1992.....	1,598,919	205,248	1,791	10,997	76,832	65	130,110	266,111	1,119,223	3,447,200
1993.....	1,501,953	180,383	1,178	20,029	58,742	23	122,943	299,634	1,371,856	3,604,361
1994.....	1,488,159	178,774	1,010	28,217	63,867	58	213,962	360,367	1,525,746	3,909,079
1995.....	1,114,360	204,932	1,031	42,450	82,593	14	131,718	462,800	1,552,691	3,624,428
1996.....	1,012,785	234,381	1,125	43,890	86,593	5	140,572	550,620	1,589,980	3,698,604
1997.....	1,387,419	300,013	1,530	43,325	79,725	18	222,539	543,494	1,727,542	4,372,227
1998.....	1,454,581	372,274	1,444	49,080	84,408	11	211,638	583,455	1,825,260	4,655,384

(1) Excludes passenger cars assembled in U.S. foreign trade zones. (2) Figures prior to 1991 are for West Germany. (3) Includes countries not shown separately.

Passenger Car Production, U.S. Plants

Source: Ward's Communications

Series	1997	1998	Series	1997	1998
BMW Z3	62,943	54,802	**Total Cadillac**	**169,914**	**176,032**
Total BMW	**62,943**	**54,802**	Cavalier	332,189	233,806
Cirrus	38,057	47,246	Corvette	24,673	32,046
Sebring	36,519	36,753	Malibu	233,349	252,479
Total Chrysler	**74,576**	**83,999**	Prizm	60,838	45,284
Avenger	31,056	23,459	**Total Chevrolet**	**651,049**	**563,615**
Neon (Dodge)	124,831	124,729	Achieva	58,017	6
Stratus	114,437	115,674	Alero	0	56,429
Viper	1,790	1,216	Aurora	27,714	21,751
Total Dodge	**272,114**	**265,078**	Cutlass	40,100	50,562
Talon	11,069	295	Intrigue	54,177	95,255
Total Eagle	**11,069**	**295**	Olds 88	74,834	68,140
Breeze	74,483	64,270	Supreme	21,518	0
Neon (Plymouth)	86,656	78,372	**Total Oldsmobile**	**276,360**	**292,143**
Prowler	463	2,124	Bonneville	75,650	57,516
Total Plymouth	**161,602**	**144,766**	Grand Am	236,519	184,335
Total Chrysler Corp.	**519,361**	**494,138**	Grand Prix	163,335	136,948
Contour	136,329	129,945	Sunfire	122,797	96,851
Escort	238,981	198,679	**Total Pontiac**	**598,301**	**475,650**
Mustang	119,196	149,129	Saturn	271,471	243,976
Probe	9,564	0	Saturn Ev1	404	125
Taurus	371,861	400,652	**Total Saturn**	**271,875**	**244,101**
Thunderbird	47,073	0	**Total General Motors**	**2,255,081**	**1,965,362**
Total Ford	**923,004**	**878,405**	Acura CL	31,241	30,480
Continental	34,421	36,328	Acura TL	0	31,324
Mark	17,467	6,103	**Total Acura**	**31,241**	**61,804**
Town Car	94,594	110,718	Accord	415,588	424,660
Total Lincoln	**146,482**	**153,149**	Civic	201,439	208,239
Cougar	24,019	73,093	**Total Honda**	**617,027**	**632,899**
Mystique	43,501	39,818	Mazda 626	89,248	94,175
Sable	123,873	111,676	Mazda MX6	1,582	0
Tracer	38,473	27,760	**Total Mazda**	**90,830**	**94,175**
Total Mercury	**229,866**	**252,347**	Eclipse	67,622	50,715
Total Ford Motor Co.	**1,299,352**	**1,283,901**	Galant	42,820	45,917
Legacy	102,180	104,229	**Total Mitsubishi**	**110,442**	**96,632**
Total Subaru	**102,180**	**104,229**	Altima	163,934	162,273
Total Fuji	**102,180**	**104,229**	Nissan 200SX	27,812	6,102
LeSabre	150,275	142,155	Sentra	87,764	54,358
Park Ave	72,260	65,343	**Total Nissan**	**279,510**	**222,733**
Riviera	13,603	6,317	Avalon	79,722	83,718
Skylark	51,444	6	Camry	325,251	297,012
Total Buick	**287,582**	**213,821**	Cavalier (Toyota)	11,940	4,788
ElDorado	21,783	14,397	Corolla	149,041	158,180
Fleetwood Deville	115,264	105,206	**Total Toyota**	**565,954**	**543,698**
Seville	32,867	56,429	**GRAND TOTAL**	**5,933,921**	**5,554,373**

Cars Registered in the U.S., 1900-97[1]

Source: U.S. Dept. of Transportation, Federal Highway Administration

(includes automobiles for public and private use)

Year	Cars Registered	Year	Cars Registered	Year	Cars Registered
1900	8,000	1945	25,796,985	1990	133,700,497
1905	77,400	1950	40,339,077	1991	128,299,601
1910	458,377	1955	52,144,739	1992	126,581,148
1915	2,332,426	1960	61,671,390	1993	127,327,189
1920	8,131,522	1965	75,257,588	1994	127,883,469
1925	17,481,001	1970	89,243,557	1995	128,386,775
1930	23,034,753	1975	106,705,934	1996	129,728,311
1935	22,567,827	1980	121,600,843	1997	129,748,704
1940	27,465,826	1985	127,885,193		

(1) There were no publicly owned vehicles before 1925; statistics also exclude military vehicles for all years. Alaska and Hawaii data included since 1960.

Domestic and Imported Retail Car Sales in the U.S., 1980-98

Source: Ward's Communications

Calendar year	Domestic	IMPORTS				Total U.S. sales	Import %		U.S.-sponsored imports
		From Japan	From Germany	From other countries	Total imports		Total	Japan	
1980..	6,581,307	1,905,968	305,219	186,700	2,397,887	8,979,194	26.7	21.2	223,310
1981..	6,208,760	1,858,896	282,881	185,502	2,327,279	8,536,039	27.3	21.8	174,665
1982..	5,758,586	1,801,969	247,080	174,508	2,223,557	7,982,143	27.9	22.6	139,767
1983..	6,795,295	1,915,621	279,748	191,403	2,386,772	9,182,067	26.0	20.9	136,798
1984..	7,951,523	1,906,206	344,416	188,220	2,438,842	10,390,365	23.5	18.3	116,965
1985..	8,204,542	2,217,837	423,983	195,925	2,837,745	11,042,287	25.7	20.1	206,252
1986..	8,214,897	2,382,614	443,721	418,286	3,244,621	11,459,518	28.3	20.8	314,358
1987..	7,080,858	2,190,405	347,881	657,465	3,195,751	10,276,609	31.1	21.3	348,154
1988..	7,526,038	2,022,602	280,099	700,991	3,003,692	10,529,730	28.5	19.2	393,412
1989..	7,072,902	1,897,143	248,561	553,660	2,699,364	9,772,266	27.6	19.4	340,425
1990..	6,896,888	1,719,384	265,116	418,823	2,403,323	9,300,211	25.8	18.5	296,778
1991..	6,136,757	1,500,309	192,776	344,814	2,037,899	8,174,656	24.9	18.4	280,673
1992..	6,276,557	1,451,766	200,851	283,938	1,936,555	8,213,112	23.6	17.7	228,927
1993..	6,741,667	1,328,445	186,177	261,570	1,776,192	8,517,859	20.9	15.6	185,284
1994..	7,255,303	1,239,450	192,241	303,489	1,735,214	8,990,517	19.3	13.8	95,399
1995..	7,128,712	981,462	207,555	317,269	1,506,257	8,634,964	17.4	11.4	99,657
1996..	7,253,582	726,940	237,984	308,247	1,273,171	8,526,753	14.9	8.5	72,166
1997..	6,916,769	726,104	297,028	332,173	1,355,305	8,272,074	16.4	8.8	87,107
1998..	6,761,381	691,162	367,283	319,653	1,378,098	8,139,479	16.9	8.5	83,509

Sport Utility Vehicle Sales in the U.S., 1988-98

Source: Ward's Communications

In 1988, 960,852 sport utility vehicles (SUVs) were sold in the United States, accounting for 6.3% of all sales of light vehicles (cars, SUVs, minivans, vans, pickup trucks, and trucks under 14,000 lbs.). By 1998, sales of SUVs in the U.S. increased to 2,796,310, accounting for 18.0% of total light vehicle sales.

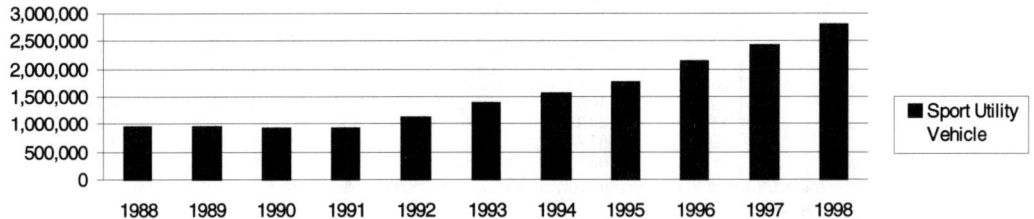

U.S. Light-Vehicle Fuel Efficiency, 1975-99

Source: Environmental Protection Agency, Office of Mobile Sources

Since 1975, both light-duty trucks (SUVs, minivans, vans, and light trucks) and cars have generally become more fuel-efficient, but their fuel efficiency has declined in recent years. In addition, light-duty trucks, which are less fuel-efficient than cars, have come to occupy an increasing proportion of the total light vehicle market, rising from only 19% in 1975 to an estimated 46% by 1999. This increase has been a major factor in the recent decline in the fuel efficiency of the average light vehicle sold.

YEAR	Cars (MPG*)	Light-duty Trucks (MPG*)	All Light Vehicles (MPG*)	YEAR	Cars (MPG*)	Light-duty Trucks (MPG*)	All Light Vehicles (MPG*)
1975.....	15.8	13.7	15.3	1995	28.3	20.5	24.7
1980.....	23.5	18.6	22.5	1996	28.3	20.8	24.8
1985.....	27.0	20.6	25.0	1997	28.4	20.7	24.5
1990.....	27.8	20.7	25.2	1998	28.6	20.6	24.4
1994.....	28.0	20.8	24.6	1999	28.1	20.3	23.8

* MPG value represents city and highway fuel efficiency combined in a 55%/45% ratio.

U.S. Car Sales by Vehicle Size and Type, 1983, 1993, and 1998
Source: Ward's Communications

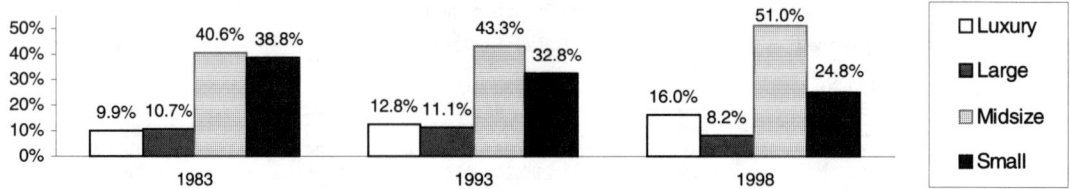

Top-Selling Passenger Cars in the U.S. by Calendar Year, 1994-98
(Domestic and Import)
Source: Ward's Communications

1998

1. Toyota Camry 429,575	8. Saturn 231,786
2. Honda Accord 401,071	9. Chevrolet Malibu 223,703
3. Ford Taurus 371,074	10. Pontiac Grand Am 180,428
4. Honda Civic 334,562	11. Chevrolet Lumina 177,631
5. Ford Escort 291,936	12. Ford Mustang 144,732
6. Chevrolet Cavalier 256,099	13. Nissan Altima 144,451
7. Toyota Corolla 250,501	14. Ford Contour 139,838

15. Buick LeSabre 136,551
16. Buick Century 126,220
17. Pontiac Grand Prix 122,915
18. Dodge Neon 117,964
19. Mercury Grand Marquis 114,162
20. Nissan Maxima 113,843

1997

1. Toyota Camry 397,156	8. Chevrolet Lumina 228,451
2. Honda Accord 384,609	9. Toyota Corolla 218,461
3. Ford Taurus 357,162	10. Pontiac Grand Am 204,078
4. Honda Civic 315,546	11. Chevrolet Malibu 164,654
5. Chevrolet Cavalier 302,161	12. Ford Contour 151,060
6. Ford Escort 283,898	13. Buick LeSabre 150,744
7. Saturn 250,810	14. Nissan Altima 144,483

15. Pontiac Grand Prix 142,018
16. Nissan Maxima 123,215
17. Nissan Seneywtra 122,468
18. Dodge Neon 121,854
19. Dodge Intrepid 118,537
20. Ford Mustang 116,610

1996

1. Ford Taurus 401,049
2. Honda Accord 382,298
3. Toyota Camry 359,433
4. Honda Civic 286,350
5. Ford Escort 284,644
6. Saturn 278,574
7. Chevrolet Cavalier 277,222
8. Chevrolet Lumina 237,973
9. Pontiac Grand Am 222,477
10. Toyota Corolla 209,048

1995

1. Ford Taurus 366,266
2. Honda Accord 341,384
3. Toyota Camry 328,595
4. Honda Civic 289,435
5. Saturn 285,674
6. Ford Escort 285,570
7. Dodge/Plymouth Neon 240,189
8. Pontiac Grand Am 234,226
9. Chevrolet Lumina 214,595
10. Toyota Corolla 213,636

1994

1. Ford Taurus 397,031
2. Honda Accord 367,615
3. Ford Escort 336,967
4. Toyota Camry 321,979
5. Saturn 286,003
6. Honda Civic 267,023
7. Pontiac Grand Am 262,310
8. Chevrolet Corsica/Beretta . . . 222,129
9. Toyota Corolla 210,926
10. Chevrolet Cavalier 187,263

World Motor Vehicle Production, 1950-98
Source: American Automobile Manufacturers Assn.; for 1998: Automotive News Data Center and Marketing Systems GmbH
(in thousands)

Year	United States	Canada	Europe	Japan	Other	World total	U.S. % of world total
1998	12,042	2,568	17,305	10,283	13,161	55,359	21.8
1997	12,119	2,571	17,773	10,975	10,024	53,463	22.7
1996	11,799	2,397	17,550	10,346	9,241	51,332	23.0
1995	11,985	2,408	17,045	10,196	8,349	49,983	24.0
1994	12,263	2,321	16,195	10,554	8,167	49,500	24.8
1993	10,898	2,246	15,208	11,228	7,205	46,785	23.3
1992	9,729	1,961	17,628	12,499	6,269	48,088	20.2
1991	8,811	1,888	17,804	13,245	5,180	46,928	18.8
1990	9,783	1,928	18,866	13,487	4,496	48,554	20.1
1985	11,653	1,933	16,113	12,271	2,939	44,909	25.9
1980	8,010	1,324	15,496	11,043	2,692	38,565	20.8
1970	8,284	1,160	13,049	5,289	1,637	29,419	28.2
1960	7,905	398	6,837	482	866	16,488	47.9
1950	8,006	388	1,991	32	160	10,577	75.7

Note: As far as can be determined, production refers to vehicles locally manufactured. Data for 1998 may not be fully comparable with earlier years because derived from different source.

Motor Vehicle Production by Selected Countries, 1998
Source: Automotive News Data Center and Marketing Systems GmbH

Country	Passenger cars	Trucks	Total	Country	Passenger cars	Trucks	Total
Argentina	392,594	104,851	497,445	Mexico	952,909	474,681	1,427,590
Australia	302,900	26,713	329,613	Netherlands	248,000	40,432	288,432
Austria	83,000	11,890	94,890	Poland	615,300	73,480	688,780
Belgium	1,000,847	98,388	1,099,235	Portugal	187,000	90,285	277,285
Brazil	1,244,463	328,665	1,573,128	Romania	129,000	20,500	149,500
Canada	1,489,294	1,079,197	2,568,491	Russia	957,636	117,036	1,074,672
China	465,139	1,042,572	1,507,711	South Africa	205,100	106,076	311,176
Czech Republic	365,250	69,687	434,937	Spain	2,155,603	596,183	2,751,786
France	2,542,399	367,218	2,909,617	Sweden	361,000	119,354	480,354
Germany	5,348,115	373,114	5,721,229	Taiwan	268,069	107,843	375,912
India	465,245	240,257	705,502	Turkey	276,759	104,565	381,324
Italy	1,385,000	272,321	1,657,321	United Kingdom	1,748,305	238,484	1,986,789
Japan	7,994,491	2,288,921	10,283,412	United States	5,554,312	6,487,853	12,042,165
Korea, South	1,625,125	369,029	1,994,154				
Malaysia	181,331	42,279	223,610	**WORLD TOTAL**	**39,440,103**	**15,918,810**	**55,358,913**

Licensed Drivers, by Age

Source: Federal Highway Administration, U.S. Dept. of Transportation

	1996			1997			1987	Percent change total drivers
Age	Male	Female	Total	Male	Female	Total	Total	1987-97
under 16	15,343	13,651	28,994	17,147	15,426	32,573	88,000[1]	−62.98
16	800,288	750,087	1,550,375	849,960	801,863	1,651,823	1,675,000	−1.38
17	1,198,089	1,114,889	2,312,978	1,243,768	1,167,949	2,411,717	2,478,000	−2.67
18	1,342,413	1,211,750	2,554,163	1,405,100	1,297,378	2,702,477	2,792,000	−3.20
19	1,454,111	1,333,378	2,787,489	1,468,377	1,359,977	2,828,354	3,000,000	−5.72
(19 and under)	4,810,244	4,423,755	9,234,000	4,984,352	4,642,592	9,626,944	10,034,000	−4.05
20	1,497,405	1,386,884	2,884,289	1,541,870	1,450,818	2,992,689	3,109,000	−3.74
21	1,502,939	1,413,063	2,916,002	1,509,678	1,430,610	2,940,287	3,314,000	−11.27
22	1,558,185	1,463,833	3,022,019	1,557,276	1,479,656	3,036,932	3,531,000	−13.99
23	1,604,854	1,513,211	3,118,065	1,600,699	1,520,800	3,121,499	3,819,000	−18.26
24	1,702,213	1,616,730	3,318,942	1,644,341	1,564,545	3,208,886	3,885,000	−17.40
(20-24)	7,865,595	7,393,721	15,259,317	7,853,864	7,446,429	15,300,293	17,657,000	−13.34
25-29	9,356,328	8,945,634	18,301,962	9,291,690	8,904,770	18,196,460	20,577,000	−11.56
30-34	10,120,589	9,870,916	19,991,504	9,955,143	9,694,083	19,649,226	20,029,000	−1.89
35-39	10,520,825	10,438,935	20,959,760	10,580,986	10,485,065	21,066,051	18,018,000	16.91
40-44	9,775,750	9,751,892	19,527,642	10,025,905	10,020,601	20,046,507	14,804,000	35.41
45-49	8,754,215	8,710,255	17,464,470	8,854,435	8,851,735	17,706,170	11,622,000	52.35
50-54	6,839,745	6,763,244	13,602,989	7,310,833	7,237,772	14,548,605	9,952,000	46.18
55-59	5,341,398	5,258,010	10,599,409	5,617,685	5,563,392	11,181,077	9,700,000	15.26
60-64	4,565,061	4,485,658	9,050,719	4,631,544	4,578,222	9,209,766	9,302,000	−0.99
65-69	4,234,471	4,230,660	8,465,131	4,214,778	4,236,057	8,450,835	7,931,000	6.55
70-74	3,604,445	3,749,142	7,353,587	3,654,142	3,830,740	7,484,883	NA	NA
75-79	2,562,955	2,716,477	5,279,432	2,650,038	2,835,297	5,485,335	NA	NA
80-84	1,400,321	1,515,635	2,915,955	1,485,906	1,608,477	3,094,383	NA	NA
85 and over	766,714	766,750	1,533,464	793,805	868,866	1,662,670	NA	NA
TOTAL	**90,518,656**	**89,020,684**	**179,539,340**	**91,905,105**	**90,804,099**	**182,709,204**	**161,816,000**	**12.91**

(1) Comparisons between "licensed" drivers under age 16 in 1987 and in 1997 are not entirely valid because of a change in definition in 1990, which interpreted "licensed" drivers more strictly than before. NA = not available.

Highway Speed Limits, by State

Source: National Motorists Association

Under the National Highway System Designation Act, signed Nov. 28, 1995, by Pres. Bill Clinton, states were allowed to set their own highway speed limits, as of Dec. 8, 1995. Under federal legislation enacted in 1974 during the energy crisis, states had been, in effect, restricted to a National Maximum Speed Limit (NMSL) of 55 miles per hour (raised in 1987 to 65 mph on rural interstates).

New maximum speed limits by state are given in the table below; all speeds are given in miles per hour. Most data current as of Sept. 1, 1999.

For more information visit the National Motorists Association's website at http://www.motorists.org

STATE	CARS Interstate	CARS Other Primary	TRUCKS Interstate	TRUCKS Other Primary	STATE	CARS Interstate	CARS Other Primary	TRUCKS Interstate	TRUCKS Other Primary
AL....	70	65	70	65	MT ...	75	70/65	65[1]	60[1]/55
AK ...	65	55	65	55	NE ...	75	65	75	65
AZ....	75	65	75	65	NV ...	75	70	75	70
AR ...	70	55	65	55	NH ...	65	65	65	65
CA ...	70	65	55	55	NJ ...	65	65	65	65
CO ...	75	65	75	65	NM ...	75	65	75	65
CT ...	65	65	65	65	NY ...	65	65	65[2]	65
DE ...	65	50	65	50	NC ...	70	65	70	65
FL....	70	65	70	65	ND ...	70	65/55	70	65/55
GA ...	70	65	70	65	OH ...	65	65	55	55
HI	55	55	55	55	OK ...	75	70	75	70
ID	75	65	65	65	OR ...	65	55	55	55
IL	65	65	55	55	PA ...	65	55	65	55
IN	65	55	60	55	RI....	65	55	65	55
IA	65	65	65	65	SC ...	65	55	65	55
KS ...	70	65	70	65	SD ...	75	65	75	65
KY ...	65	55	65	55	TN ...	70	65	70	65
LA....	70	65	70	65	TX ...	70/65	70/65	60/55	60/55
ME ...	65	55	65	55	UT ...	75	65	75	65
MD ...	65	55	65	55	VT ...	65	50	65	50
MA ...	65	65	65	65	VA ...	65	55	65	55
MI....	70	70	55	55	WA...	70	55	60	55
MN ...	70	65	70	65	WV...	70	65	70	65
MS ...	70	70	70	70	WI ...	65	65	65	65
MO ...	70	70	70	70	WY ...	75	65	75	65

(1) 55 mph for triple-trailer trucks. (2) 55 mph for double-trailer trucks on the Thruway.
NOTE: Where two speeds are given, the first is for daytime and the second for nighttime. "Daytime" means from one-half hour before sunrise to one-half hour after sunset; "nighttime" means at any other hour.

The Most Popular Colors, by Type of Vehicle, 1998 Model Year
Source: Ward's Communications

Luxury cars		Full size/ intermediate cars		Compact/sports cars		Light trucks and vans	
COLOR	%	COLOR	%	COLOR	%	COLOR	%
Light brown	17.7	Med./dark Green	16.4	Med./dark Green	15.9	White	22.5
White Metallic	12.3	White	15.6	Black	15.0	Med./dark Green	15.5
Black	12.3	Light Brown	14.1	White	14.7	Black	11.5
White	11.3	Silver	11.0	Silver	10.4	Medium Red	7.2
Med./Dark Green	10.0	Black	8.9	Bright Red	9.5	Bright Red	7.1
Silver	9.2	Medium Red	6.5	Light Brown	7.0	Silver	6.2
Medium Red	7.5	Med./Dark Blue	6.0	Medium Red	6.4	Light Brown	6.1
Med./Dark Gray	5.3	Dark Red	4.9	Med./dark Blue	5.3	Med./Dark Blue	4.7
Med./Dark Blue	4.8	Light Blue	3.8	Teal/Aqua	4.0	Dark Red	4.5
Gold	4.8	White Metallic	3.2	Purple	3.4	Med./Dark Brown	3.2

Selected Motor Vehicle Statistics
Source: Federal Highway Administration; U.S. Dept. of Transportation; Insurance Institute for Highway Safety; 1997 figures where not specified.

STATE	Driver's age (Jan 1, 1998) Regular[1]	Driver's age (Jan 1, 1998) Learner's Permit	State gas tax cents/ gal. (July 1, 1999)	Safety belt use law[10]	Licensed drivers per 1,000 resident pop.	Regist. motor vehicles per 1,000 pop.	Licensed drivers per motor vehicle	Gals. of fuel used per vehicle	Miles per gal.	Annual miles driven per vehicle	Vehicle miles per licensed driver
Alabama.....	16	15	18	P	784	850	0.94	807	18.05	14,568	15,783
Alaska	16	14	8	S	732	890	0.84	567	14.26	8,088	9,831
Arizona......	16	15y, 7m	18	S	685	690	1.00	868	15.94	13,837	13,941
Arkansas	16	14	19.5	S	745	648	1.17	1,165	14.77	17,226	14,981
California	17[2]	15	18	P	632	773	0.83	633	18.08	11,450	14,010
Colorado.....	17	15	22	S	729	905	0.81	599	17.89	10,715	13,308
Connecticut ..	16y, 4m[2]	16	32	P	694	813	0.87	592	18.14	10,736	12,577
Delaware	16y, 10m[2]	15y, 10m	23	S	732	839	0.89	680	19.17	13,047	14,946
Dist. of Col....	16	16	20	P	673	442	1.60	808	17.62	14,234	9,338
Florida......	18	15	13.1	S	802	742	1.11	719	17.14	12,324	11,406
Georgia	18	15	7.5	P	676	834	0.82	867	16.72	14,494	18,430
Hawaii.......	15	15	16	P	623	584	1.09	594	19.31	11,470	10,756
Idaho	16[3]	15	25	S	697	893	0.80	747	15.95	11,914	15,263
Illinois.......	17[2]	15	19	S	647	710	0.92	731	16.09	11,764	12,912
Indiana	18	15	15	P	669	912	0.74	734	17.49	12,836	17,489
Iowa	17[2]	14	20	P	685	1,000	0.70	664	14.78	9,814	14,329
Kansas......	16	14	20	S	703	829	0.86	757	16.28	12,325	14,534
Kentucky.....	16y, 6m	16	16.4	S	659	712	0.94	986	16.33	16,097	17,386
Louisiana	17	15	20	P	615	784	0.80	717	15.88	11,387	14,504
Maine	16[2]	15	19	S	725	852	0.87	757	16.53	12,512	14,703
Maryland	17y, 7m[4]	15y, 9m	23.5	P	657	743	0.89	701	17.56	12,311	13,927
Massachusetts	18	16	21	S	718	829	0.88	570	17.46	9,953	11,487
Michigan.....	17	14y, 9m	19	P[9]	691	821	0.86	682	16.77	11,445	13,591
Minnesota....	17[2]	15	20	S	606	838	0.73	691	17.82	12,311	17,029
Mississippi ...	16	15	18.4	S	631	818	0.78	847	16.66	14,110	18,298
Missouri	16	15y, 6m	17	S	693	805	0.87	866	16.72	14,476	16,820
Montana	15	14y, 6m	27	S	754	1,115	0.69	643	14.91	9,587	14,178
Nebraska	17	15	22.8	S	712	910	0.80	742	15.27	11,332	14,486
Nevada......	16[2]	15y, 6m	24.75	S	707	683	1.06	917	15.52	14,236	13,750
New Hampshire .	18	15y, 6m	19.5	No	753	961	0.79	643	15.45	9,936	12,685
New Jersey...	17y, 6m	16	10.5	S	692	722	0.98	754	14.44	10,884	11,354
New Mexico ...	16y, 6m	15	18.5	P	690	876	0.81	805	17.99	14,485	18,368
New York	17[2]	16[6]	28.9	P	581	600	0.99	628	17.69	11,108	11,470
North Carolina	16y, 6m	15	21.2	P	727	779	0.95	798	17.74	14,155	15,167
North Dakota .	16	14	21	S	706	1,085	0.66	694	14.77	10,247	15,753
Ohio	17[2]	15y, 6m	22	S	732	904	0.82	615	16.68	10,257	12,665
Oklahoma....	16	15y, 6m	17	P	687	869	0.81	798	17.99	14,356	18,168
Oregon......	17[7]	15[7]	24	P	702	891	0.80	618	18.06	11,162	14,174
Pennsylvania .	17[2]	16	25.9	S	692	734	0.95	684	16.24	11,107	11,783
Rhode Island .	17y, 7m[2]	16	29	S	689	719	0.97	603	16.52	9,964	10,397
South Carolina	16y, 3m[3]	15	16	S	695	758	0.93	892	16.26	14,501	15,818
South Dakota .	16	14	22	S	710	974	0.75	759	14.56	11,050	15,144
Tennessee ...	16	15	20	S	732	845	0.88	775	17.22	13,347	15,405
Texas	16[2]	15	20	P	660	665	1.03	895	17.18	15,375	15,483
Utah	17[4]	15y, 9m	24.5	S	659	743	0.90	789	16.94	13,366	15,065
Vermont	16[2]	15	20	S	807	842	0.98	794	16.43	13,044	13,601
Virginia......	16[2]	15	17.5	P	728	848	0.87	740	16.64	12,317	14,348
Washington ..	16[2]	15	23	S	715	838	0.86	640	16.96	10,855	12,730
West Virginia .	16	15	25.35	S	708	746	0.99	794	17.03	13,523	14,258
Wisconsin....	16[2]	15y, 6m	25.4	S	710	819	0.88	692	18.57	12,853	14,814
Wyoming	16	15	14	S	735	1,154	0.66	1,012	13.53	13,690	21,476
AVERAGE ...					**683**	**816**	**0.90**	**747**	**16.67**	**12,396**	**14,394**

NOTE: Many states are moving toward graduated licensing systems that phase in full driving privileges. During the learner's phase, driving generally is not permitted unless there is an adult supervisor. In an intermediate phase, young licensees not yet having unrestricted licenses may be allowed to drive unsupervised under certain conditions but not others. (1) Unrestricted operation of private passenger car. (2) Applicants under age 18 must have completed an approved driver education course. (3) Applicants under age 17 must have completed an approved driver education course. (4) Initial applicants of any age must have completed an approved driver education course. (5) Applicants under age 16 must have completed an approved driver education course. (6) Driving in New York City is prohibited; driving in Nassau and Suffolk Counties is limited. (7) Effective 3/1/00. (8) Effective 12/22/99 (9) Effective 4/1/00. (10) As of Sept. 1, 1999. P = officer may stop vehicle for a violation (primary); S = an officer may issue seat belt citation only when vehicle is stopped for another moving violation (secondary).

Road Mileage Between Selected U.S. Cities

	Atlanta	Boston	Chicago	Cincin-nati	Cleve-land	Dallas	Denver	Des Moines	Detroit	Houston
Atlanta, Ga.	...	1,037	674	440	672	795	1,398	870	699	789
Boston, Mass.	1,037	...	963	840	628	1,748	1,949	1,280	695	1,804
Chicago, Ill.	674	963	...	287	335	917	996	327	266	1,067
Cincinnati, Oh.	440	840	287	...	244	920	1,164	571	259	1,029
Cleveland, Oh.	672	628	335	244	...	1,159	1,321	652	170	1,273
Dallas Tex.	795	1,748	917	920	1,159	...	781	684	1,143	243
Denver, Col.	1,398	1,949	996	1,164	1,321	781	...	669	1,253	1,019
Detroit, Mich.	699	695	266	259	170	1,143	1,253	584	...	1,265
Houston, Tex.	789	1,804	1,067	1,029	1,273	243	1,019	905	1,265	...
Indianapolis, Ind. . .	493	906	181	106	294	865	1,058	465	278	987
Kansas City, Mo.	798	1,391	499	591	779	489	600	195	743	710
Los Angeles, Cal. . . .	2,182	2,979	2,054	2,179	2,367	1,387	1,059	1,727	2,311	1,538
Memphis, Tenn. . . .	371	1,296	530	468	712	452	1,040	599	713	561
Milwaukee, Wis.	761	1,050	87	374	422	991	1,029	361	353	1,142
Minneapolis, Minn. . .	1,068	1,368	405	692	740	936	841	252	671	1,157
New Orleans, La. . .	479	1,507	912	786	1,030	496	1,273	978	1,045	356
New York, N.Y.	841	206	802	647	473	1,552	1,771	1,119	637	1,608
Omaha, Neb.	986	1,412	459	693	784	644	537	132	716	865
Philadelphia, Pa. . . .	741	296	738	567	413	1,452	1,691	1,051	573	1,508
Pittsburgh, Pa.	687	561	452	287	129	1,204	1,411	763	287	1,313
Portland Ore.	2,601	3,046	2,083	2,333	2,418	2,009	1,238	1,786	2,349	2,205
St. Louis, Mo.	541	1,141	289	340	529	630	857	333	513	779
San Francisco	2,496	3,095	2,142	2,362	2,467	1,753	1,235	1,815	2,399	1,912
Seattle, Wash.	2,618	2,976	2,013	2,300	2,348	2,078	1,307	1,749	2,279	2,274
Tulsa, Okla.	772	1,537	683	736	925	257	681	443	909	478
Washington, DC. . . .	608	429	671	481	346	1,319	1,616	984	506	1,375

	India-napolis	Kansas City	Los Angeles	Louis-ville	Memphis	Mil-waukee	Minne-apolis	New Orleans	New York	Omaha
Atlanta, Ga.	493	798	2,182	382	371	761	1,068	479	841	986
Boston, Mass.	906	1,391	2,979	941	1,296	1,050	1,368	1,507	206	1,412
Chicago, Ill.	181	499	2,054	292	530	87	405	912	802	459
Cincinnati, Oh.	106	591	2,179	101	468	374	692	786	647	693
Cleveland Oh.	294	779	2,367	345	712	422	740	1,030	473	784
Dallas, Tex.	865	489	1,387	819	452	991	936	496	1,552	644
Denver, Col.	1,058	600	1,059	1,120	1,040	1,029	841	1,273	1,771	537
Detroit, Mich.	278	743	2,311	360	713	353	671	1,045	637	716
Houston, Tex.	987	710	1,538	928	561	1,142	1,157	356	1,608	865
Indianapolis, Ind. . .	...	485	2,073	111	435	268	586	796	713	587
Kansas City, Mo. . . .	485	...	1,589	520	451	537	447	806	1,198	201
Los Angeles, Cal. . . .	2,073	1,589	...	2,108	1,817	2,087	1,889	1,883	2,786	1,595
Memphis, Tenn. . . .	435	451	1,817	367	...	612	826	390	1,100	652
Milwaukee, Wis. . . .	268	537	2,087	379	612	...	332	994	889	493
Minneapolis, Minn. . .	586	447	1,889	697	826	332	...	1,214	1,207	357
New Orleans, La. . .	796	806	1,883	685	390	994	1,214	...	1,311	1,007
New York, N.Y.	713	1,198	2,786	748	1,100	889	1,207	1,311	...	1,251
Omaha, Neb.	587	201	1,595	687	652	493	357	1,007	1,251	...
Philadelphia, Pa. . . .	633	1,118	2,706	668	1,000	825	1,143	1,211	100	1,183
Pittsburgh, Pa.	353	838	2,426	388	752	539	857	1,070	368	895
Portland, Ore.	2,272	1,809	959	2,320	2,259	2,010	1,678	2,505	2,885	1,654
St. Louis, Mo.	235	257	1,845	263	285	363	552	673	948	449
San Francisco	2,293	1,835	379	2,349	2,125	2,175	1,940	2,249	2,934	1,683
Seattle, Wash.	2,194	1,839	1,131	2,305	2,290	1,940	1,608	2,574	2,815	1,638
Tulsa, Okla.	631	248	1,452	659	401	757	695	647	1,344	387
Washington, DC. . . .	558	1,043	2,631	582	867	758	1,076	1,078	233	1,116

	Phila-delphia	Pitts-burgh	Portland	St. Louis	Salt Lake City	San Francisco	Seattle	Toledo	Tulsa	Wash., DC
Atlanta, Ga.	741	687	2,601	541	1,878	2,496	2,618	640	772	608
Boston, Mass.	296	561	3,046	1,141	2,343	3,095	2,976	739	1,537	429
Chicago, Ill.	738	452	2,083	289	1,390	2,142	2,013	232	683	671
Cincinnati, Oh.	567	287	2,333	340	1,610	2,362	2,300	200	736	481
Cleveland Oh.	413	129	2,418	529	1,715	2,467	2,348	111	925	346
Dallas, Tex.	1,452	1,204	2,009	630	1,242	1,753	2,078	1,084	257	1,319
Denver, Col.	1,691	1,411	1,238	857	504	1,235	1,307	1,218	681	1,616
Detroit, Mich.	576	287	2,349	513	1,647	2,399	2,279	59	909	506
Houston, Tex.	1,508	1,313	2,205	779	1,438	1,912	2,274	1,206	478	1,375
Indianapolis, Ind. . . .	633	353	2,272	235	1,504	2,293	2,194	219	631	558
Kansas City, Mo. . . .	1,118	838	1,809	257	1,086	1,835	1,839	687	248	1,043
Los Angeles, Cal. . . .	2,706	2,426	959	1,845	715	379	1,131	2,276	1,452	2,631
Memphis, Tenn.	1,000	752	2,259	285	1,535	2,125	2,290	654	401	867
Milwaukee, Wis.	825	539	2,010	363	1,423	2,175	1,940	319	757	758
Minneapolis, Minn. . .	1,143	857	1,678	552	1,186	1,940	1,608	637	695	1,076
New Orleans, La. . . .	1,211	1,070	2,505	673	1,738	2,249	2,574	986	647	1,078
New York, N.Y.	100	368	2,885	948	2,182	2,934	2,815	578	1,344	233
Omaha, Neb.	1,183	895	1,654	449	931	1,683	1,638	681	387	1,116
Philadelphia, Pa. . . .	...	288	2,821	868	2,114	2,866	2,751	514	1,264	133
Pittsburgh, Pa.	288	...	2,535	588	1,826	2,578	2,465	228	984	221
Portland, Ore.	2,821	2,535	...	2,060	767	636	172	2,315	1,913	2,754
St. Louis, Mo.	868	588	2,060	...	1,337	2,089	2,081	454	396	793
San Francisco	2,866	2,578	636	2,089	752	...	808	2,364	1,760	2,799
Seattle, Wash.	2,751	2,465	172	2,081	836	808	...	2,245	1,982	2,684
Tulsa, Okla.	1,264	984	1,913	396	1,172	1,760	1,982	850	...	1,189
Washington, DC. . . .	133	221	2,754	793	2,047	2,799	2,684	447	1,189	...

Air Distances Between Selected World Cities in Statute Miles

Point-to-point measurements are usually from City Hall.

	Bangkok	Beijing	Berlin	Cairo	Cape Town	Caracas	Chicago	Hong Kong	Honolulu	Lima
Bangkok	...	2,046	5,352	4,523	6,300	10,555	8,570	1,077	6,609	12,244
Beijing	2,046	...	4,584	4,698	8,044	8,950	6,604	1,217	5,077	10,349
Berlin	5,352	4,584	...	1,797	5,961	5,238	4,414	5,443	7,320	6,896
Cairo	4,523	4,698	1,797	...	4,480	6,342	6,141	5,066	8,848	7,726
Cape Town	6,300	8,044	5,961	4,480	...	6,366	8,491	7,376	11,535	6,072
Caracas	10,555	8,950	5,238	6,342	6,366	...	2,495	10,165	6,021	1,707
Chicago	8,570	6,604	4,414	6,141	8,491	2,495	...	7,797	4,256	3,775
Hong Kong	1,077	1,217	5,443	5,066	7,376	10,165	7,797	...	5,556	11,418
Honolulu	6,609	5,077	7,320	8,848	11,535	6,021	4,256	5,556	...	5,947
London	5,944	5,074	583	2,185	5,989	4,655	3,958	5,990	7,240	6,316
Los Angeles	7,637	6,250	5,782	7,520	9,969	3,632	1,745	7,240	2,557	4,171
Madrid	6,337	5,745	1,165	2,087	5,308	4,346	4,189	6,558	7,872	5,907
Melbourne	4,568	5,643	9,918	8,675	6,425	9,717	9,673	4,595	5,505	8,059
Mexico City	9,793	7,753	6,056	7,700	8,519	2,234	1,690	8,788	3,789	2,639
Montreal	8,338	6,519	3,740	5,427	7,922	2,438	745	7,736	4,918	3,970
Moscow	4,389	3,607	1,006	1,803	6,279	6,177	4,987	4,437	7,047	7,862
New York	8,669	6,844	3,979	5,619	7,803	2,120	714	8,060	4,969	3,639
Paris	5,877	5,120	548	1,998	5,786	4,732	4,143	5,990	7,449	6,370
Rio de Janeiro	9,994	10,768	6,209	6,143	3,781	2,804	5,282	11,009	8,288	2,342
Rome	5,494	5,063	737	1,326	5,231	5,195	4,824	5,774	8,040	6,750
San Francisco	7,931	5,918	5,672	7,466	10,248	3,902	1,859	6,905	2,398	4,518
Singapore	883	2,771	6,164	5,137	6,008	11,402	9,372	1,605	6,726	11,689
Stockholm	5,089	4,133	528	2,096	6,423	5,471	4,331	5,063	6,875	7,166
Tokyo	2,865	1,307	5,557	5,958	9,154	8,808	6,314	1,791	3,859	9,631
Warsaw	5,033	4,325	322	1,619	5,935	5,559	4,679	5,147	7,366	7,215
Washington, DC	8,807	6,942	4,181	5,822	7,895	2,047	596	8,155	4,838	3,509

	London	Los Angeles	Madrid	Melbourne	Mexico City	Montreal	Moscow	New Delhi	New York	Paris
Bangkok	5,944	7,637	6,337	4,568	9,793	8,338	4,389	1,813	8,669	5,877
Beijing	5,074	6,250	5,745	5,643	7,753	6,519	3,607	2,353	6,844	5,120
Berlin	583	5,782	1,165	9,918	6,056	3,740	1,006	3,598	3,979	548
Cairo	2,185	7,520	2,087	8,675	7,700	5,427	1,803	2,758	5,619	1,998
Cape Town	5,989	9,969	5,308	6,425	8,519	7,922	6,279	5,769	7,803	5,786
Caracas	4,655	3,632	4,346	9,717	2,234	2,438	6,177	8,833	2,120	4,732
Chicago	3,958	1,745	4,189	9,673	1,690	745	4,987	7,486	714	4,143
Hong Kong	5,990	7,240	6,558	4,595	8,788	7,736	4,437	2,339	8,060	5,990
Honolulu	7,240	2,557	7,872	5,505	3,789	4,918	7,047	7,412	4,969	7,449
London	...	5,439	785	10,500	5,558	3,254	1,564	4,181	3,469	214
Los Angeles	5,439	...	5,848	7,931	1,542	2,427	6,068	7,011	2,451	5,601
Madrid	785	5,848	...	10,758	5,643	3,448	2,147	4,530	3,593	655
Melbourne	10,500	7,931	10,758	...	8,426	10,395	8,950	6,329	10,359	10,430
Mexico City	5,558	1,542	5,643	8,426	...	2,317	6,676	9,120	2,090	5,725
Montreal	3,254	2,427	3,448	10,395	2,317	...	4,401	7,012	331	3,432
Moscow	1,564	6,068	2,147	8,950	6,676	4,401	...	2,698	4,683	1,554
New York	3,469	2,451	3,593	10,359	2,090	331	4,683	7,318	...	3,636
Paris	214	5,601	655	10,430	5,725	3,432	1,554	4,102	3,636	...
Rio de Janeiro	5,750	6,330	5,045	8,226	4,764	5,078	7,170	8,753	4,801	5,684
Rome	895	6,326	851	9,929	6,377	4,104	1,483	3,684	4,293	690
San Francisco	5,367	347	5,803	7,856	1,887	2,543	5,885	7,691	2,572	5,577
Singapore	6,747	8,767	7,080	3,759	10,327	9,203	5,228	2,571	9,534	6,673
Stockholm	942	5,454	1,653	9,630	6,012	3,714	716	3,414	3,986	1,003
Tokyo	5,959	5,470	6,706	5,062	7,035	6,471	4,660	3,638	6,757	6,053
Warsaw	905	5,922	1,427	9,598	6,337	4,022	721	3,277	4,270	852
Washington, DC	3,674	2,300	3,792	10,180	1,885	489	4,876	7,500	205	3,840

	Rio de Janeiro	Rome	San Francisco	Singapore	Stockholm	Tehran	Tokyo	Vienna	Warsaw	Wash., DC
Bangkok	9,994	5,494	7,931	883	5,089	3,391	2,865	5,252	5,033	8,807
Beijing	10,768	5,063	5,918	2,771	4,133	3,490	1,307	4,648	4,325	6,942
Berlin	6,209	737	5,672	6,164	528	2,185	5,557	326	322	4,181
Cairo	6,143	1,326	7,466	5,137	2,096	1,234	5,958	1,481	1,619	5,822
Cape Town	3,781	5,231	10,248	6,008	6,423	5,241	9,154	5,656	5,935	7,895
Caracas	2,804	5,195	3,902	11,402	5,471	7,320	8,808	5,372	5,559	2,047
Chicago	5,282	4,824	1,859	9,372	4,331	6,502	6,314	4,698	4,679	596
Hong Kong	11,009	5,774	6,905	1,605	5,063	3,843	1,791	5,431	5,147	8,155
Honolulu	8,288	8,040	2,398	6,726	6,875	8,070	3,859	7,632	7,366	4,838
London	5,750	895	5,367	6,747	942	2,743	5,959	771	905	3,674
Los Angeles	6,330	6,326	347	8,767	5,454	7,682	5,470	6,108	5,922	2,300
Madrid	5,045	851	5,803	7,080	1,653	2,978	6,706	1,128	1,427	3,792
Melbourne	8,226	9,929	7,856	3,759	9,630	7,826	5,062	9,790	9,598	10,180
Mexico City	4,764	6,377	1,887	10,327	6,012	8,184	7,035	6,320	6,337	1,885
Montreal	5,078	4,104	2,543	9,203	3,714	5,880	6,471	4,009	4,022	489
Moscow	7,170	1,483	5,885	5,228	716	1,532	4,660	1,043	721	4,876
New York	4,801	4,293	2,572	9,534	3,986	6,141	6,757	4,234	4,270	205
Paris	5,684	690	5,577	6,673	1,003	2,625	6,053	645	852	3,840
Rio de Janeiro	...	5,707	6,613	9,785	6,683	7,374	11,532	6,127	6,455	4,779
Rome	5,707	...	6,259	6,229	1,245	2,127	6,142	477	820	4,497
San Francisco	6,613	6,259	...	8,448	5,399	7,362	5,150	5,994	5,854	2,441
Singapore	9,785	6,229	8,448	...	5,936	4,103	3,300	6,035	5,843	9,662
Stockholm	6,683	1,245	5,399	5,936	...	2,173	5,053	780	494	4,183
Tokyo	11,532	6,142	5,150	3,300	5,053	4,775	...	5,689	5,347	6,791
Warsaw	6,455	820	5,854	5,843	494	1,879	5,689	347	...	4,472
Washington, DC	4,779	4,497	2,441	9,662	4,183	6,341	6,791	4,438	4,472	...

TRAVEL AND TOURISM

World Tourism Receipts, 1989-98

Source: World Tourism Organization
(in billions; figures rounded)

Global spending on travel and tourism has more than doubled over the decade as the standard of living for many people in the world has risen appreciably and more countries have become accessible to tourists.

1989 . . . $221	1991 . . . $278	1993 . . . $324	1995 . . . $405	1997 . . . $436
1990 . . . 269	1992 . . . 315	1994 . . . 354	1996 . . . 436	1998 . . . 445

World's Top 10 Tourist Destinations, 1998

Source: World Tourism Organization
(number of arrivals in millions; excluding same-day visitors)

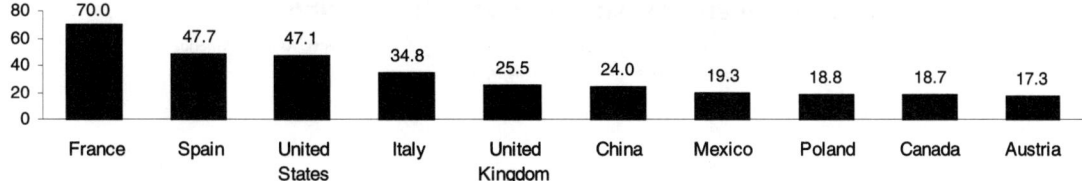

France	Spain	United States	Italy	United Kingdom	China	Mexico	Poland	Canada	Austria
70.0	47.7	47.1	34.8	25.5	24.0	19.3	18.8	18.7	17.3

World's Top 20 Tourism Earners, 1998

Source: World Tourism Organization
International tourism receipts (excluding transportation)
(in millions of dollars)

Rank 1990	1998	Country	Receipts 1998	Rank 1990	1998	Country	Receipts 1998	Rank 1990	1998	Country	Receipts 1998
1	1	United States	$74,240	7	8	Austria	$12,164	11	15	Hong Kong, China	$7,114
3	2	Italy	30,427	9	9	Canada	9,133	23[1]	16	Russian Federation	7,107
2	3	France	29,700	15	10	Australia	8,575	12	17	Singapore	6,501
4	4	Spain	29,585	65	11	Poland	8,400	13	18	Thailand	6,392
5	5	United Kingdom	21,295	21	12	Turkey	8,300	14	19	Netherlands	5,749
6	6	Germany	16,840	8	13	Switzerland	8,208	18	20	South Korea	5,700
25	7	China	12,500	10	14	Mexico	7,850				

(1) Former USSR.

Average Number of Vacation Days, Selected Countries, 1998

Source: World Tourism Organization

Country	Days	Country	Days	Country	Days
Italy	42	Brazil	34	Korea	25
France	37	United Kingdom	28	Japan	25
Germany	35	Canada	26	United States	13

International Travel to the U.S., 1986-98

Source: Tourism Industries, International Trade Administration, Dept. of Commerce
(Visitors each year, in millions; some figures revised, may differ from other sources; 1998 total is preliminary)

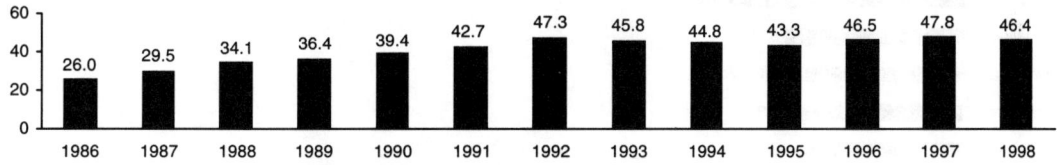

1986	1987	1988	1989	1990	1991	1992	1993	1994	1995	1996	1997	1998
26.0	29.5	34.1	36.4	39.4	42.7	47.3	45.8	44.8	43.3	46.5	47.8	46.4

International Visitors to the U.S., 1996[1]

Source: Tourism Industries, International Trade Administration, Dept. of Commerce

Country of origin	Visitors (thousands)	Expenditures (millions)[2]	Expenditures per visitor	Country of origin	Visitors (thousands)	Expenditures (millions)[2]	Expenditures per visitor
Canada	15,301	$6,763.0	$442	Brazil	891	NA	—
Mexico	8,530	3,001.0	352	South Korea	796	NA	—
Japan	5,047	13,163.0	2,608	Italy	552	$1,440.0	$2,609
United Kingdom	3,105	7,306.0	2,353	Australia	461	1,819.0	3,946
Germany	1,973	4,573.0	2,318				
France	990	2,255.0	2,278	**All Countries**	**46,324**	**$69,908.0[3]**	**$1,509**

(1) Excludes cruise travel. (2) Excludes international passenger fare payments. (3) Does not include international traveler spending on U.S. carriers for transactions made outside the U.S. NA=not available.

Traveler Spending in the U.S., 1987-96

Source: Tourism Industries, International Trade Administration, Dept. of Commerce
(in billions)

	Domestic Travelers	International Travelers		Domestic Travelers	International Travelers		Domestic Travelers	International Travelers
1987	$235	$31	1991	$296	$64	1994	$339	$78
1988	258	38	1992	308	71	1995	360	80
1989	273	47	1993	322	75	1996	383	90[1]
1990	291	58						

(1) Includes international traveler spending on U.S. carriers for transactions made outside the U.S.

U.S. Pleasure Travel Overview, 1998

Source: TIA Research Dept., "Tourism Works for America Report," by Travel Industry Assn. of America

Pleasure travel volume in the U.S. in 1998 amounted to 897.6 million person-trips, up 4% from 1997 (862.4 million person-trips) and up a total of 45% since 1988. It accounted for 71% of all U.S. resident travel in 1998. About 50% of all pleasure travelers visited friends and relatives as the primary purpose of their trip. About 35% traveled for entertainment purposes, and 15% traveled for outdoor recreation purposes, similar to the trend in 1997. As in previous years, a large majority of pleasure travelers continued to travel by auto—car/truck/RV/rental car (83%). A much smaller proportion (13% in 1998) traveled by airplane. About 41% of pleasure travelers stayed in a hotel or motel, with an average length of stay of 3.6 nights per trip, up slightly from 1997. About 40% of pleasure travelers stayed with friends or relatives, which is similar to 1997. The average length of stay per trip with friends or relatives was down slightly from 1997, 3.6 nights vs. 3.8 nights.

U.S. Resident Pleasure Travel Volume, 1986-98

Source: TIA Research Dept., "Tourism Works for America Report," by Travel Industry Assn. of America

(in millions of person-trips of 100 mi or more, one-way)

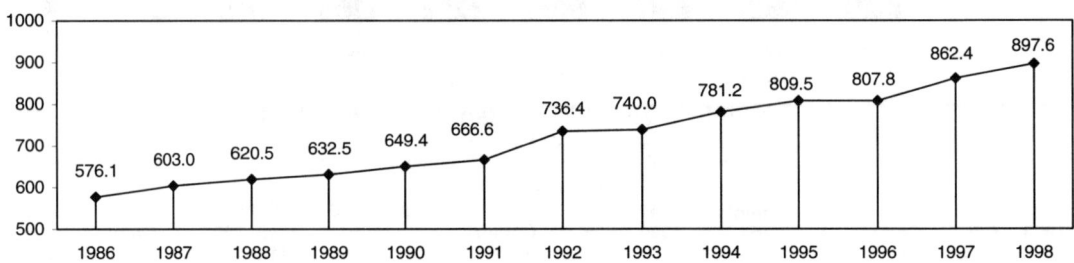

Top U.S. States by Total Traveler Spending, 1997[1]

Source: "Tourism Works for America Report," by Travel Industry Assn. of America; includes spending (in billions of dollars) in states by both domestic and international travelers

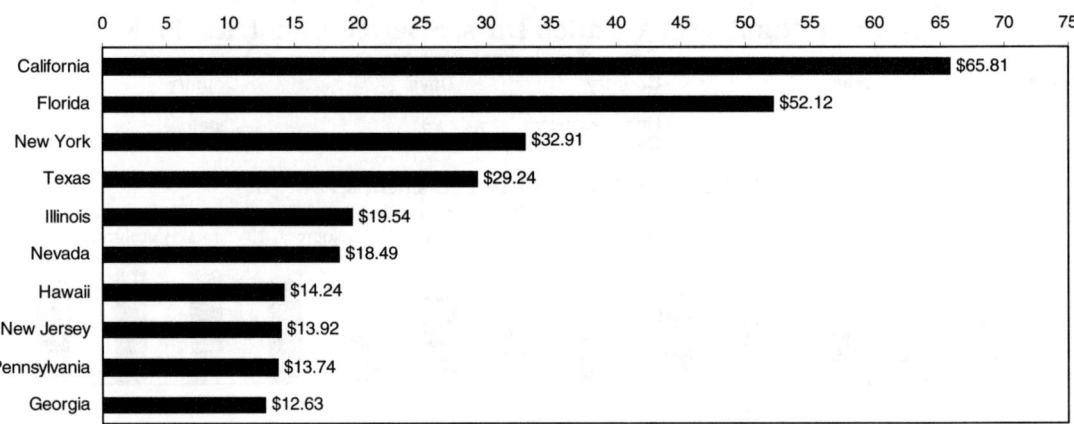

(1) Figures are preliminary.

Travel Websites

The following websites are representative of the travel and tourism industry. Inclusion here does not represent endorsement by *The World Almanac*. Websites listed under "Maps" enable the user to plot a route to a destination.

AIRLINES

American Airlines
http://www.americanair.com
America West Airlines
http://www.americawest.com
Continental Airlines
http://www.flycontinental.com
Delta Air Lines
http://www.delta-air.com
Northwest Airlines
http://www.nwa.com
Southwest Airlines
http://www.iflyswa.com
Trans World Airlines
http://www.twa.com

United Airlines
http://www.ual.com
USAirways
http://www.usair.com

HOTELS/RESORTS

Best Western Int'l.
http://www.bestwestern.com
Choice Hotels Int'l.,
Clarion Hotels & Resorts,
Comfort Inns,
Econo Lodges,
Quality Inns,
Rodeway Inns,
Sleep Inns
http://www.hotelchoice.com

Days Inn of America
http://www.daysinn.com
Doubletree Hotels
http://www.doubletreehotels.com
Embassy Suites
http://www.embassy-suites.com
Four Seasons Hotels
http://www.fshr.com
Hilton Hotels
http://www.hilton.com
Holiday Inn Worldwide
http://www.holiday-inn.com
Hyatt Hotels and Resorts
http://www.hyatt.com
Inter-Continental Hotels
http://www.interconti.com
Loews Hotels
http://www.loewshotels.com
Marriott Int'l.
http://www.marriott.com
Radisson Hotels Int'l.
http://www.radisson.com
Sheraton Hotels & Resorts
http://www.sheraton.com
Westin Hotels & Resorts
http://www.westin.com
Wyndham Hotels & Resorts
http://www.wyndham.com

MAPS
http://www.freetrip.com
http://www.mapquest.com
http://www.mapsonus.com

CRUISE LINES
Carnival Cruise Lines
http://www.carnival.com
Celebrity Cruises
http://www.celebrity-cruises.com
Costa Cruise Lines
http://www.costacruises.com
Cunard Line
http://www.cunardline.com
Holland America Line
http://www.hollandamerica.com
Norwegian Cruise Line

http://www.ncl.com
Princess Cruises
http://www.princesscruises.com
Renaissance Cruises
http://www.renaissancecruises.com
Royal Caribbean Int'l.
http://www.rccl.com
Windjammer Barefoot Cruises
http://www.windjammer.com

TRAINS
Amtrak
http://www.amtrak.com
BC Rail (Canada)
http://www.bcrail.com
Rail Europe
http://www.raileurope.com

CAR RENTALS
Alamo Rent A Car
http://www.goalamo.com
Avis Rent-A-Car
http://www.avis.com
Budget Rent A Car
http://www.budgetrentacar.com
Dollar Rent A Car
http://www.dollarcar.com
Enterprise Rent-A-Car
http://www.enterprise.com
Hertz
http://www.hertz.com
National Car Rental
http://www.nationalcar.com
Rent-A-Wreck
http://www.rent-a-wreck.com
Thrifty Rent-A-Car
http://www.thrifty.com

BUSES
Gray Line Worldwide
http://www.grayline.com
Greyhound Lines
http://www.greyhound.com
Peter Pan Bus Lines
http://www.peterpan-bus.com

Customs Exemptions for Travelers
Source: U.S. Dept. of the Treasury, U.S. Customs Service

U.S. residents returning after a stay abroad of at least 48 hours are usually granted customs exemptions of $400 each (this and all exemptions figured according to fair retail value). The duty-free articles must accompany the traveler at the time of return, be for personal or household use, have been acquired as an incident of the trip, and be properly declared to Customs. No more than 1 liter of alcoholic beverages or more than 100 cigars and 200 cigarettes (1 carton) may be included in the $400 exemption. The exemption for alcoholic beverages holds only if the returning resident is at least 21 years old at the time of arrival. Cuban cigars may be included only if purchased in Cuba.

If a U.S. resident arrives directly or indirectly from a U.S. island possession—American Samoa, Guam, or U.S. Virgin Islands—a customs exemption of $1,200 is allowed. Up to 1,000 cigarettes may be included, but only 200 of them may have been purchased elsewhere. If a U.S. resident returns from any one of the following places, the exemption is $600: Antigua and Barbuda, Aruba, Bahamas, Barbados, Belize, British Virgin Islands, Costa Rica, Dominica, Dominican Republic, El Salvador, Grenada, Guatemala, Guyana, Haiti, Honduras, Jamaica, Montserrat, Netherlands Antilles, Nicaragua, Panama, St. Kitts and Nevis, St. Lucia, St. Vincent and the Grenadines, Trinidad and Tobago.

The $400, $600, or $1,200 exemption may be granted only if the exemption has not been used in whole or part within the preceding 30-day period and only if the stay abroad was for at least 48 hours. The 48-hr absence requirement does not apply to travelers returning from Mexico or U.S. Virgin Islands. Travelers who cannot claim the $400, $600, or $1,200 exemption because of the 30-day or 48-hr provisions may bring in free of duty and tax articles acquired abroad for personal or household use up to a value of $25.

There are also allowances for goods when shipped. Goods shipped for personal use may be imported free of duty and tax if the total value is no more than $200. This exemption does not apply to perfume containing alcohol if it is valued at more than $5 retail, to alcoholic beverages, or to cigars and cigarettes. The $200 mail exemption does not apply to merchandise subject to absolute or tariff-rate quotas unless the item is for personal use. Tailor-made suits ordered from Hong Kong, however, are subject to quota/visa requirements even if imported for personal use.

Bona fide gifts of not more than $100 in value, when shipped, can be received in the U.S. free of duty and tax, provided that the same person does not receive more than $100 in gift shipments in one day. The limit is increased to $200 for bona fide gift items shipped from U.S. Virgin Islands, American Samoa, or Guam. (Shipping of alcoholic beverages, including wine and beer, by mail is prohibited by U.S. postal laws.) These gifts are not declared by the traveler upon return to the U.S.

The U.S. Customs Service booklet "Know Before You Go" answers frequently asked customs questions and is available free by writing U.S. Customs Services, KBYG, PO Box 7407, Washington, DC 20044. Online information can be obtained at the U.S. Customs website—http://www.custom.ustreas.gov

Passports, Health Regulations, and Travel Warnings for Foreign Travel

Source: Bureau of Consular Affairs, U.S. Dept. of State

Passports are issued by the U.S. Department of State to citizens and nationals of the U.S. for the purpose of documenting them for foreign travel and identifying them as U.S. citizens. For U.S. citizens traveling on business or as tourists, especially in Europe, a U.S. passport is often sufficient to gain admission for a limited stay. For many countries, however, a **visa** must also be obtained before entering. It is the responsibility of the traveler to check in advance and obtain any visas where required, from the appropriate embassy or nearest consulate of each country.

Each country has its own specific guidelines concerning length and purpose of visit, etc. Some may require visitors to display proof that they (1) have sufficient funds to stay for the intended time period and (2) have onward/return tickets.

Some countries, including **Canada, Mexico,** and some **Caribbean** islands, do not require a passport or a visa for limited stays. Such countries do require proof of U.S. citizenship, and may have other requirements that must be met. For further information, check with the embassy or nearest consulate of the country you plan to visit.

How to Obtain a Passport

Those who have never been issued a passport in their own name must apply in person before (1) a passport agent; (2) a clerk of any federal court or state court of record or a clerk or judge of a probate court accepting applications; (3) a postal clerk at a post office that is authorized to accept passport applications; or (4) a U.S. diplomatic or consular officer abroad.

A DSP-11 is the correct form to use for those who must apply in person. All persons are required to obtain individual passports in their own name. However, a parent or legal guardian must execute the application for children under 13.

Persons who possess their most recent passport issued within the last 12 years and after their 18th birthday, may be eligible to apply for a new passport by mail. The form DSP-82, *Application for Passport by Mail*, must be filled out and mailed to the address shown on the form, together with the previous passport, 2 recent identical photographs (see below), and a fee of $55. The DSP-82 may not be used if the most recent passport has been altered or mutilated.

Proof of citizenship—A full validity passport previously issued to the applicant or one in which he or she was included will be accepted as proof of U.S. citizenship. If the applicant has no prior passport and was born in the U.S., a certified copy of the birth certificate generally must be presented. It must generally show the given name and surname, the date and place of birth, and that the birth record was filed shortly after birth. A delayed birth certificate (filed more than 1 year after date of birth) is acceptable if it shows that acceptable secondary evidence was used for creating this record.

If a birth certificate is not obtainable, a notice from a state registrar must be submitted stating that no birth record exists. It must be accompanied by the best obtainable secondary evidence, such as a baptismal certificate or hospital birth record.

A naturalized citizen with no previous passport must present a Certificate of Naturalization. A person born abroad claiming U.S. citizenship through either a native-born or a naturalized citizen parent must normally submit a Certificate of Citizenship issued by the Immigration and Naturalization Service or a Consular Report of Birth or Certification of Birth Abroad issued by the Dept. of State. If such a document has not been obtained, evidence of citizenship of the parent(s) through whom citizenship is claimed and evidence that would establish the parent/child relationship must be submitted. Additionally, if citizenship is derived through birth to citizen parent(s), the applicant must submit parents' marriage certificate plus an affidavit from parent(s) showing periods and places of residence or presence in the U.S. and abroad, and specifying periods spent abroad in the employment of the U.S. government, including the armed forces, or with certain international organizations. If citizenship is derived through naturalization of parents, evidence of admission to the U.S. for permanent residence also is required.

It is important to apply for a passport as far in advance as possible. Passport offices are busiest between March and September. It can take several weeks to receive a passport.

Photographs—Passport applicants must submit 2 identical photographs that are recent (normally not more than 6 months old) and that are a good likeness of and satisfactorily identify the applicant. Photographs should be 2 × 2 in. in size. The image size, from bottom of chin to top of head (including hair), should not be less than 1 inch or more than 1-3/8 in. Photographs should be portrait-type prints. They must be clear, front view, full face, with a plain white or off-white background. Photos that depict the applicant as relaxed and smiling are encouraged.

Identity—Applicants must establish their identity to the satisfaction of the authorities. Generally acceptable documents of identity include a previous U.S. passport, a Certificate of Naturalization, a Certificate of Citizenship, a valid driver's license, or a government identification card. Applicants may not use a Social Security card, learner's or temporary driver's license, credit card, or expired ID card. Extremely old documents cannot be used by themselves.

Applicants unable to establish identity must present some documentation in their own name and be accompanied by a person who has known them at least 2 years and is a U.S. citizen or legal U.S. permanent resident alien. That person must sign an affidavit before the individual who executes the application, and must establish his or her own identity.

Fees—For persons under 16 years of age, the basic passport fee is $40. These passports are valid for 5 years from the date of issue. The basic fee is $60 for passports issued to persons 16 and older. These passports are valid for 10 years from date of issuance. To receive a passport within 10 days or less, a $35 expedite fee is required. An additional fee of $10 is charged for the execution of the application. There is no execution fee when using DSP-82, *Application for Passport by Mail*. Applicants eligible to use this form pay only a $55 passport fee.

Passport loss—The loss or theft of a valid passport should be reported immediately in writing to Passport Services, 1111 19th St., NW, Dept. of State, Washington, DC 20524-1705, telephone: (202) 647-0518, or to the nearest passport agency or nearest U.S. embassy or consulate when abroad.

General Information—Visit Passport Services on the internet at http://www.travel.state.gov

Health Regulations

Under the regulations adopted by the World Health Organization, a country may require International Certificates of Vaccination against yellow fever. A cholera immunization may be required for travelers from infected areas. Check with health care providers or your records to see that other immunizations (e.g., for tetanus and polio) are up-to-date.

Prophylactic medication for malaria and certain other preventive measures are advisable for travel to some countries. No immunizations are needed to return to the U.S. An increasing number of countries have regulations regarding AIDS testing, particularly for longtime visitors. Detailed information is included in *Health Information for International Travel*, available from the U.S. Government Printing Office, Washington, DC 20402, for $14. Information may also be obtained from your local health department or physician, or by calling the Centers for Disease Control and Prevention at (404) 332-4559.

General information—The booklets *Passports—Applying for the Easy Way* and *Foreign Entry Requirements* are available for 50¢ each from the Consumer Information Center, Pueblo, CO 81009. For online information, as well as HIV Testing Requirements, go to the Consular Affairs website—http://www.travel.state.gov

Travel Warnings

Travel Warnings are issued when the State Dept. decides, based on relevant information, to recommend that Americans avoid travel to a certain country; these are subject to change. For the latest information, 24 hours a day, dial 202/647-5225 from a touch-tone telephone. As of Oct. 1, 1999, travel warnings were in effect for: Afghanistan, Albania, Algeria, Angola, Bosnia and Herzegovina, Burundi, Central African Republic, Colombia, Congo, Congo Republic, Eritrea, Ethiopia, Guinea-Bissau, Iran, Iraq, Lebanon, Liberia, Libya, Nigeria, Pakistan, Sierra Leone, Somalia, Sudan, Tajikistan, Yemen, and Yugoslavia.

HEALTH
Basic First Aid

Knowing what to do for an injured victim until a doctor or other trained person gets to the accident scene can save a life, especially in cases of stoppage of breathing, severe bleeding, and shock.

People with special medical problems, such as diabetes, cardiovascular disease, epilepsy, or allergy, are urged to wear some sort of emblem identifying the problem, as a safeguard against receiving medication that might be harmful or even fatal. Emblems may be obtained from Medic Alert Foundation, 2323 Colorado Ave., Turlock, CA 95382; 800-344-3226.

It is important to get medical assistance as soon as possible.

Animal bite — Wash wound with soap under running water and apply antibiotic ointment and dressing. When possible, the animal should be caught alive for rabies testing.

Asphyxiation — Start rescue breathing immediately after getting patient to fresh air.

Bleeding — Elevate the wound above the heart if possible. Press hard on wound with sterile compress until bleeding stops. Send for doctor if bleeding is severe.

Burn — If mild, with skin unbroken and no blisters, flush with cool water until pain subsides. Apply a loose sterile dry dressing if necessary. If severe, send for doctor. Apply sterile compresses and keep patient comfortably warm until doctor's arrival. Do not try to clean burn or break blisters.

Chemical in eye — With patient lying down, pour cupfuls of water immediately into corner of eye, letting it run to other side to remove chemicals thoroughly. Cover with sterile compress. Get medical attention immediately. Continue to flush until medical help arrives.

Choking — See **Abdominal Thrust.**

Convulsions — Place person on back on bed or rug. Loosen clothing. Turn head to side. Do not place a blunt object between the patient's teeth. If convulsions do not stop, get medical attention immediately.

Cut (minor) — Apply mild antiseptic and sterile compress after washing with soap under warm running water.

Fainting — If victim feels faint, lower head to knees. Lay patient down on back with head turned to side if he or she becomes unconscious. Elevate the legs 8 to 10 inches. Loosen clothing and open windows. Keep patient lying quietly for at least 15 minutes after he or she regains consciousness. Call doctor if faint lasts for more than a few minutes.

Foreign body in eye — Touch object with moistened corner of handkerchief if it can be seen. If it cannot be seen or does not come out after a few attempts, take patient to doctor. Do not rub the eye.

Frostbite — Handle frostbitten area gently. Do not rub. Soak affected area in water no warmer than 105°F. Do not allow frostbitten area to touch the container. Soak until frostbitten part looks red and feels warm. Loosely bandage. If fingers or toes are frostbitten, put gauze between them.

Heat Stroke and Heat Exhaustion — Remove the patient from the heat. Loosen any tight clothing and apply cool, wet cloths to the skin. Give the victim cool water, to drink slowly. Call an ambulance if the victim refuses water, vomits, or experiences changes in consciousness.

Hypothermia — Move victim to a warm place. Remove wet clothing and dry victim, if necessary. Warm patient gradually by wrapping the person in warm blankets or clothing. Apply heat pads or other heat sources if available, but not directly to the body. Give the victim warm liquids. Call an ambulance if breathing is slowed or stopped or if the pulse is slow or irregular.

Loss of Limb — If a limb is severed, it is important to properly protect the limb so that it can possibly be reattached. After the patient is cared for, the limb should be wrapped in a sterile gauze or clean material and placed in a clean plastic bag, garbage can, or other suitable container. Pack ice around the limb on the OUTSIDE of the bag to keep the limb cold. Call ahead to the hospital to alert staff there of the situation.

Poisoning — Call ambulance and Poison Control Center and follow their directions. Use antidote listed on label if container is found. Do not give the victim any food or drink or induce vomiting, unless specified by the Poison Control Center.

Shock (injury-related) — Keep the victim lying down on back; if uncertain as to his or her injuries, keep the patient flat on the back. Otherwise elevate feet and legs 12 inches. Maintain normal body temperature; if the weather is cold or damp, place blankets or extra clothing over and under the victim; if weather is hot, provide shade.

Snakebite —Wash the injury. Keep the area still and at a lower level than the heart. Keep the victim quiet. Use a snakebite kit if available.

Sprains and fractures — Apply ice to reduce swelling and pain. Do not try to straighten or move broken limbs. Apply a splint to immobilize the injured area if the victim must be transported.

Sting from insect — If possible, remove stinger. Wash the area with soap and water; cover it to keep it clean. Apply a cold pack to reduce pain and swelling. Call physician immediately if body swells or patient collapses.

Unconsciousness — Send for doctor and place person on his or her back. Start rescue breathing if victim stops breathing. Never give food or liquids.

Abdominal Thrust (Heimlich Maneuver)

The American Red Cross and the American Heart Association both agree that the recommended first aid for choking victims is the abdominal thrust, also known as the Heimlich maneuver, after its creator, Dr. Henry Heimlich. Slaps on the back are no longer advised and may even prove detrimental to a choking victim.

- Get behind the victim and wrap your arms around him or her about 1-2 inches above the navel.
- Make a fist with one hand and place it, with the thumb knuckle pressing inward at the abdomen.
- Grasp the fist with the other hand and give upward thrusts until object is removed or help arrives.

Rescue Breathing

Stressing that your breath can save a life, the American Red Cross gives the following directions for rescue breathing if the victim is not breathing:

- Determine consciousness by tapping the victim on the shoulder and asking loudly, "Are you okay?"
- Tilt the victim's head back so that the chin is pointing upward. Do not press on the soft tissue under the chin, as this might obstruct the airway. If you suspect that an accident victim might have neck or back injuries, open the airway by placing the tips of your index and middle fingers on the corners of the person's jaw to lift it forward without tilting the head.
- Place your cheek and ear close to the victim's mouth and nose. Look at the chest to see if it rises and falls. Listen and feel for air to be exhaled for about 5 seconds.
- If there is no breathing, pinch the victim's nostrils shut with the thumb and index finger of your hand that is pressing on the victim's forehead. Another way to prevent leakage of air when the lungs are inflated is to press your cheek against the victim's nose.
- Blow air into the mouth by taking a deep breath and then sealing your mouth tightly around the victim's mouth. Initially, give 2, slow (approx. 1.5 seconds each), full breaths.
- Watch the patient's chest to see if it rises.
- Stop when the chest is expanded. Raise your mouth; turn your head to the side and listen for exhalation.
- Watch the chest to see if it falls. Check pulse. If there is a pulse, continue rescue breathing. If there is no pulse, start CPR.
- Repeat giving 1 breath every 5 seconds until the victim starts breathing.

Note: Infants (up to 1 year) and children (1 to 8 years) should be treated as described above, except for the following:

- Do not tilt the head as far back as an adult's head.
- Both the mouth and nose of an infant should be sealed by the mouth.
- Give breaths to a child once every 3 seconds.
- Blow into the infant's mouth and nose once every 3 seconds with less pressure and volume than for a child.

Nutritive Value of Food (Calories, Proteins, etc.)

Source: *Home and Garden Bulletin No. 72;* U.S. Dept. of Agriculture

FOOD	Measure	Grams	Food Energy (calories)	Protein (grams)	Fat (grams)	Saturated fats (grams)	Carbohydrate (grams)	Calcium (milligrams)	Iron (milligrams)	Sodium (milligrams)	Vitamin A (I.U.)	Ascorbic Acid (milligrams)
DAIRY PRODUCTS												
Cheese, cheddar, cut pieces	1 oz.	28	115	7	9	6.0	T	204	0.2	176	300	0
Cheese, cottage, small curd	1 cup	210	215	26	9	6.0	6	126	0.3	850	340	T
Cheese, cream	1 oz.	28	100	2	10	6.2	1	23	0.3	84	400	0
Cheese, Swiss	1 oz.	28	95	7	7	4.5	1	219	0.2	388	230	0
Half-and-half	1 tbsp.	15	20	T	2	1.1	1	16	T	6	70	T
Cream, sour	1 tbsp.	12	25	T	3	1.6	1	14	T	6	90	T
Milk, whole	1 cup	244	150	8	8	5.1	11	291	0.1	120	310	2
Milk, nonfat (skim)	1 cup	245	85	8	T	0.3	12	302	0.1	126	500	2
Milkshake, chocolate	10 oz.	283	355	9	8	4.8	60	374	0.9	314	240	0
Ice cream, hardened	1 cup	133	270	5	14	8.9	32	176	0.1	116	540	1
Sherbet	1 cup	193	270	2	4	2.4	59	103	0.3	88	190	4
Yogurt, fruit-flavored	8 oz.	227	230	10	2	1.6	43	345	0.2	133	100	1
EGGS												
Fried in margarine	1	46	90	6	7	1.9	1	25	0.7	162	390	0
Hard-cooked	1	50	75	6	5	1.6	1	25	0.6	62	280	0
Scrambled (milk added) in margarine	1	61	100	7	7	2.2	1	44	0.7	171	420	T
FATS & OILS												
Butter, salted	1 tbsp.	14	100	T	11	7.1	T	3	T	116	430	0
Margarine, salted	1 tbsp.	14	100	T	11	2.2	T	4	T	132	460	0
Olive oil	1 tbsp.	14	125	0	14	1.9	0	0	0	0	0	0
Salad dressing, blue cheese	1 tbsp.	15	75	1	8	1.5	1	12	T	164	30	T
Salad dressing, French, regular	1 tbsp.	16	85	T	9	1.4	1	2	T	188	T	T
Salad dressing, French, low calorie	1 tbsp.	16	25	T	2	0.2	2	6	T	306	T	T
Salad dressing, Italian	1 tbsp.	15	80	T	9	1.3	1	1	T	162	30	T
Mayonnaise	1 tbsp.	14	100	T	11	1.7	T	3	0.1	80	40	0
FISH, MEAT, POULTRY												
Clams, raw, meat only	3 oz.	85	65	11	1	0.3	2	59	2.6	102	90	9
Crabmeat, canned	1 cup	135	135	23	3	0.5	1	61	1.1	1,350	50	0
Fish sticks, frozen, reheated	1 fish stick	28	70	6	3	0.8	4	11	0.3	53	20	0
Salmon canned (pink), solids and liquid	3 oz.	85	120	17	5	0.9	0	167	0.7	443	60	0
Sardines, Atlantic, canned in oil, drained solids	3 oz.	85	175	20	9	2.1	0	371	2.6	425	190	0
Shrimp, French fried	3 oz.	85	200	16	10	2.5	11	61	2.0	384	90	0
Trout, broiled, with butter and lemon juice	3 oz.	85	175	21	9	4.1	T	26	1.0	122	230	1
Tuna, canned in oil	3 oz.	85	165	24	7	1.4	0	7	1.6	303	70	0
Bacon, broiled or fried crisp	3 slices	19	110	6	9	3.3	T	2	0.3	303	0	6
Ground beef, broiled, regular	3 oz.	85	245	20	18	6.9	0	9	2.1	70	T	0
Roast beef, relatively lean (lean only)	2.6 oz.	75	135	22	5	1.9	0	3	1.5	46	T	0
Beef steak, lean and fat	3 oz.	85	240	23	15	6.4	0	9	2.6	53	T	0
Beef & vegetable stew	1 cup	245	220	16	11	4.4	15	29	2.9	292	5,690	17
Lamb, chop, broiled loin, lean and fat	2.8 oz.	80	235	22	16	7.3	0	16	1.4	62	T	0
Liver, beef, fried	3 oz.	85	185	23	7	2.5	7	9	5.3	90	30,690	23
Ham, light cure, roasted, lean and fat	3 oz.	85	205	18	14	5.1	0	6	0.7	1,009	0	0
Pork, chop, broiled, lean and fat	3.1 oz.	87	275	24	19	7.0	0	3	0.7	61	10	T
Bologna	2 slices	57	180	7	16	6.1	2	7	0.9	581	0	12
Frankfurter, pork, cooked	1	45	145	5	13	4.8	1	5	0.5	504	0	12
Sausage, pork link, cooked	1 link	13	50	3	4	1.4	T	4	0.2	168	0	T
Veal, cutlet, braised or broiled	3 oz.	85	185	23	9	4.1	0	9	0.8	56	T	0
Chicken, drumstick, fried, bones removed	2.5 oz.	72	195	16	11	3.0	6	12	1.0	194	60	0
Chicken, roasted, half breast, without skin	3 oz.	86	140	27	3	0.9	0	13	0.9	64	20	0
Turkey, roasted, chopped light and dark meat	1 cup	140	240	41	7	2.3	0	35	2.5	98	0	0
Frankfurter, chicken, cooked	1	45	115	6	9	2.5	3	43	0.9	616	60	0
FRUITS & FRUIT PRODUCTS												
Apple, raw, 2-3/4 in. diam.	1	138	80	T	T	0.1	21	10	0.2	T	70	8
Apple juice	1 cup	248	115	T	T	T	29	17	0.9	7	T	2
Apricots, raw	3	106	50	1	T	T	12	15	0.6	1	2,770	11
Banana, raw	1	114	105	1	1	0.2	27	7	0.4	1	90	10
Cherries, sweet, raw	10	68	50	1	1	0.1	11	10	0.3	T	150	5
Cranberry juice cocktail, sweetened	1 cup	253	145	T	T	T	38	8	0.4	10	10	108
Fruit cocktail, canned, in heavy syrup	1 cup	255	185	1	T	T	48	15	0.7	15	520	5
Grapefruit, raw, medium, white	1/2	120	40	1	T	T	10	14	0.1	T	10	41
Grapes, Thompson seedless	10	50	35	T	T	0.1	9	6	0.1	1	40	5
Lemonade, frozen, unsweetened	6 oz.	244	55	1	1	0.1	16	20	0.3	2	30	77
Cantaloupe, 5-in. diam.	1/2	267	95	2	1	0.1	22	29	0.6	24	8,610	113
Orange, 2-5/8 in. diam.	1	131	60	1	T	T	15	52	0.1	T	270	70
Orange juice, frozen, diluted	1 cup	249	110	2	T	T	27	22	0.2	2	190	97
Peach, raw, 2-1/2 in. diam.	1	87	35	1	T	T	10	4	0.1	T	470	6
Raisins, seedless	1 cup	145	435	5	1	0.2	115	71	3.0	17	10	5
Strawberries, whole	1 cup	149	45	1	1	T	10	21	0.6	1	40	84
Tomatoes, raw	1	123	25	1	T	T	5	9	0.6	10	1,390	22
Watermelon, 4 by 8 in. wedge	1 piece	482	155	3	2	0.3	35	39	0.8	10	1,760	46
GRAIN PRODUCTS												
Bagel, plain	1	68	200	7	2	0.3	38	29	1.8	245	0	0
Biscuit, 2 in. diam., from home recipe	1	28	100	2	5	1.2	13	47	0.7	195	10	T
Bread, pita, enriched, white, 6-1/2 in. diam	1 pita	60	165	6	1	0.1	12	15	0.7	124	0	0
Bread, white, enriched	1 slice	25	65	2	1	0.3	12	32	0.7	129	T	T
Bread, whole-wheat	1 slice	28	70	3	1	0.4	13	20	1.0	180	T	T
Oatmeal or rolled oats, without added salt	1 cup	234	145	6	2	0.4	25	19	1.6	2	40	0
Bran flakes (40% bran), added sugar, salt, iron, vitamins	1 oz.	28	90	4	1	0.1	22	14	8.1	264	1,250	0
Corn flakes, added sugar, salt, iron, vitamins	1 oz.	28	110	2	T	T	24	1	1.8	351	1,250	15
Rice, puffed, added iron, thiamine, niacin	1 oz.	28	110	2	T	T	25	4	1.8	340	1,250	15
Wheat, shredded, plain, 1 biscuit or 2/3 cup	1 oz.	28	100	3	1	0.1	23	11	1.2	3	0	0
Bulgur, uncooked	1 cup	170	600	19	3	1.2	129	49	9.5	7	0	0
Cake, angel food, 1/12 of cake	1	53	125	3	T	T	29	44	0.2	269	0	0
Cupcake, 2-1/2 in. diam., with chocolate icing	1	35	120	2	4	1.8	20	21	0.7	92	50	T

FOOD	Measure	Grams	Food Energy (calories)	Protein (grams)	Fat (grams)	Saturated fats (grams)	Carbohydrate (grams)	Calcium (milligrams)	Iron (milligrams)	Sodium (milligrams)	Vitamin A (I.U.)	Ascorbic Acid (milligrams)
Plain sheet cake with white, uncooked frosting, 1/9 of cake.	1	121	445	4	14	4.6	77	61	1.2	275	240	T
Fruitcake, dark, 1/32 of loaf.	1	43	165	2	7	1.5	25	41	1.2	67	50	16
Cake, pound, 1/17 of loaf.	1	29	110	2	5	3.0	15	8	0.5	108	160	0
Cheesecake, 1/12 of 9-in. diam. cake.	1	92	280	5	18	9.9	26	52	0.4	204	230	5
Brownies, with nuts, from commercial recipe	1	25	100	1	4	1.6	16	13	0.6	59	70	T
Cookies, chocolate chip, from home recipe.	4	40	185	2	11	3.9	26	13	1.0	82	20	0
Crackers, graham, 2-1/2 in. squares	2	14	60	1	1	0.4	11	6	0.4	86	0	0
Crackers, saltines	4	12	50	1	1	0.5	9	3	0.5	165	0	0
Danish pastry, round piece	1	57	220	4	12	3.6	26	60	1.1	218	60	T
Doughnut, cake type	1	50	210	3	12	2.8	24	22	1.0	192	20	T
Macaroni, firm stage (hot)	1 cup	130	190	7	1	0.1	39	14	2.1	1	0	0
Muffin, bran, commercial mix.	1	45	140	3	4	1.3	24	27	1.7	385	100	0
Muffin, corn, from home recipe	1	45	145	3	5	1.5	21	66	0.9	169	80	T
Noodles, enriched, cooked	1 cup	160	200	7	2	0.5	37	16	2.6	3	110	0
Pie, apple, 1/6 of pie	1	158	405	3	18	4.6	60	13	1.6	476	50	2
Pie, cherry, 1/6 of pie.	1	158	410	4	18	4.7	61	22	1.6	480	700	0
Pie, lemon meringue, 1/6 of pie	1	140	355	5	14	4.3	53	20	1.4	395	240	4
Pie, pecan, 1/6 of pie.	1	138	575	7	32	4.7	71	65	4.6	305	220	0
Popcorn, air-popped, plain	1 cup	8	30	1	T	T	6	1	0.2	T	10	0
Pretzels, stick.	10	3	10	T	T	T	2	1	0.1	48	0	0
Rolls, enriched, brown & serve	1	28	85	2	2	0.5	14	33	0.8	155	T	T
Rolls, frankfurter & hamburger	1	40	115	3	2	0.5	20	54	1.2	241	T	T
Tortillas, corn.	1	30	65	2	1	0.1	13	42	0.6	1	80	0
LEGUMES, NUTS, SEEDS												
Beans, Black	1 cup	171	225	15	1	0.1	41	47	2.9	1	T	0
Beans, Great Northern, cooked.	1 cup	180	210	14	1	0.1	38	90	4.9	13	0	0
Peanuts, roasted in oil, salted	1 cup	145	840	39	71	9.9	27	125	2.8	626	0	0
Peanut butter	1 tbsp.	16	95	5	8	1.4	3	5	0.3	75	0	0
Refried beans, canned	1 cup	290	295	18	3	0.4	51	141	5.1	1,228	0	17
Tofu.	1 piece	120	85	9	5	0.7	3	108	2.3	8	0	0
Sunflower seeds, hulled.	1 oz.	28	160	6	14	1.5	5	33	1.9	1	10	T
MIXED FOODS												
Chop suey with beef and pork, home recipe	1 cup	250	300	26	17	4.3	13	60	4.8	1,053	600	33
Enchilada.	1	230	235	20	16	7.7	24	97	3.3	1,332	2,720	T
Pizza, cheese, 1/8 of 15 in.-diam. pie	1	120	290	15	9	4.1	39	220	1.6	699	750	2
Spaghetti with meatballs & tomato sauce	1 cup	248	330	19	12	3.9	39	124	3.7	1,009	1,590	22
SUGARS & SWEETS												
Candy, caramels	1 oz.	28	115	1	3	2.2	22	42	0.4	64	T	T
Candy, milk chocolate	1 oz.	28	145	2	9	5.4	16	50	0.4	23	30	T
Fudge, chocolate	1 oz.	28	115	1	3	2.1	21	22	0.3	54	T	T
Gelatin dessert, from prepared powder.	1/2 cup	120	70	2	0	0.0	17	2	T	55	0	0
Candy, hard.	1 oz.	28	110	0	0	0.0	28	T	0.1	7	0	0
Honey	1 tbsp.	21	65	T	0	0.0	17	1	0.1	1	0	T
Jams & Preserves	1 tbsp.	20	55	T	T	0.0	14	4	0.2	2	T	T
Popsicle, 3 fl. oz.	1	95	70	0	0	0.0	18	0	T	11	0	0
Sugar, white, granulated	1 tbsp.	12	45	0	0	0.0	12	T	T	T	0	0
VEGETABLES												
Asparagus, spears, cooked from raw	4 spears	60	15	2	T	T	3	14	0.4	2	500	16
Beans, green, from frozen, cuts.	1 cup	135	35	2	T	T	8	61	1.1	18	710	11
Broccoli, cooked from raw	1 spear	180	50	5	1	0.1	10	82	2.1	20	2,540	113
Cabbage, raw, coarsely shredded or sliced.	1 cup	70	15	1	T	T	4	33	0.4	13	90	33
Carrots, raw, 7-1/2 by 1-1/8 in.	1	72	30	1	T	T	7	19	0.4	25	20,250	7
Cauliflower, cooked, drained, from raw	1 cup	125	30	2	T	T	6	34	0.5	8	20	69
Celery, raw.	1 stalk	40	5	T	T	T	1	14	0.2	35	50	3
Collards, cooked from raw.	1 cup	190	25	2	T	0.1	5	148	0.8	36	4,220	19
Corn, sweet, yellow, cooked from raw	1 ear	77	85	3	1	0.2	19	2	0.5	13	170	5
Eggplant, cooked, steamed	1 cup	96	25	1	T	T	6	6	0.3	3	60	1
Lettuce, iceberg, chopped.	1 cup	55	5	1	T	T	1	10	0.3	5	180	2
Lettuce, looseleaf (such as romaine).	1 cup	56	10	1	T	T	2	38	0.8	5	1,060	10
Mushrooms, raw.	1 cup	70	20	1	T	T	3	4	0.9	3	0	2
Onions, raw, chopped	1 cup	160	55	2	T	0.1	12	40	0.6	3	0	13
Peas, green, frozen, cooked	1 cup	160	125	8	T	0.1	23	38	2.5	139	1,070	16
Potatoes, baked, peeled	1	156	145	3	T	T	34	8	0.5	8	0	20
Potatoes, frozen, French fried (oven-heated)	10	50	110	2	4	2.1	17	5	0.7	16	0	5
Potatoes, mashed, milk added	1 cup	210	160	4	1	0.7	37	55	0.6	636	40	14
Potato chips.	10	20	105	1	7	1.8	10	5	0.2	94	0	8
Potato salad.	1 cup	250	360	7	21	3.6	28	48	1.6	1,323	520	25
Spinach, drained, cooked from raw.	1 cup	180	40	5	T	0.1	7	245	6.4	126	14,740	18
Sweet potatoes, baked in skin, peeled	1	114	115	2	T	T	28	32	0.5	11	24,880	28
Vegetable juice cocktail, canned	1 cup	242	45	2	T	T	11	27	1.0	883	2,830	67
MISCELLANEOUS												
Beer, regular.	12 fl. oz.	360	150	1	0	0.0	13	14	0.1	18	0	0
Gin, rum, vodka, whisky, 86 proof	1-1/2 fl. oz.	42	105	0	0	0.0	T	T	T	T	0	0
Wine, table, white.	3-1/2 fl. oz.	102	80	T	0	0.0	3	9	0.3	5	(1)	0
Cola-type beverage.	12 fl. oz.	369	160	0	0	0.0	41	11	0.2	18	0	0
Ginger ale.	12 fl. oz.	366	125	0	0	0.0	32	11	0.1	29	0	0
Coffee, brewed	6 fl. oz.	180	T	T	T	T	T	4	T	2	0	0
Tea, brewed.	8 fl. oz.	240	T	T	T	T	T	0	T	1	0	0
Catsup.	1 tbsp.	15	15	T	T	T	4	3	0.1	156	210	2
Mustard, prepared, yellow	1 tsp.	5	5	T	T	T	T	4	0.1	63	0	T
Olives, canned, green	4 medium	13	15	T	2	0.2	T	8	0.2	312	40	0
Pickles, dill, whole	1	65	5	T	T	T	1	17	0.7	928	70	4
Relish, finely chopped, sweet	1 tbsp.	15	20	T	T	T	5	3	0.1	107	20	1
Soup, tomato, prepared with milk	1 cup	248	160	6	6	2.9	22	159	1.8	932	850	68
Soup, chicken noodle, prepared with water.	1 cup	241	75	4	2	0.7	9	17	0.8	1,106	710	T
Soup, green pea, prepared with water.	1 cup	250	165	9	3	1.4	27	28	2.0	988	200	2
Soup, vegetarian, prepared with water.	1 cup	241	70	2	2	0.3	12	22	1.1	822	3,010	1

T — Indicates trace. (1) — Value not determined. **NOTE:** Values shown here for these foods may be from several different manufacturers and, therefore, may differ somewhat from the values provided by one source.

Food and Nutrition

The U.S. Dept. of Health and Human Services and the Dept. of Agriculture reissued guidelines in 1996 that offer dietary and exercise advice for children age 2 and over, as well as for adults. Recommended were: (1) no more than 30 percent of calories from fat, or about 65 grams of fat in a 2,000-calorie daily diet; with no more than 10% of calories, or 20 grams of fat, from saturated fats; (2) maximum alcohol consumption of about 1 drink a day for women, 2 for men; (3) daily consumption of vegetables, 3-5 servings; fruits, 2-4; pastas, cereals, or breads, 6-11; milk, 2-3; meat, poultry, fish, beans, and eggs, 2-3. (For vegetables, 1 serving equals about 1 cup raw leafy greens or one-half cup other kinds; fruit, 1 medium apple, banana, or orange, or $3/4$ cup of fruit juice; grains, 1 slice of bread, $1/2$ cup of pasta, or 1 oz. cereal; milk, 1 cup or 1.5 oz. of cheese; meat and poultry, 2-3 oz. cooked lean beef or chicken without skin; cooked dry beans, $1/2$ cup.)

PROTEIN

Proteins, composed of amino acids, are essential to good nutrition. They build, maintain, and repair the body. Best sources: eggs, milk, fish, meat, poultry, soybeans, nuts. High-quality proteins such as eggs, meat, or fish supply all 8 amino acids needed in the diet. Plant foods can be combined to meet protein needs as well: whole grain breads and cereals, rice, oats, soybeans, other beans, split peas, and nuts.

FATS

Fats provide energy by furnishing calories to the body, and they also carry vitamins A, D, E, and K. They are the most concentrated source of energy in the diet. Best sources of polyunsaturated and monounsaturated fats: margarine, vegetable/plant oils, nuts. Meats, cheeses, butter, cream, egg yolks, lard are concentrated sources of saturated fats.

CARBOHYDRATES

Carbohydrates provide energy for body function and activity by supplying immediate calories. The carbohydrate group includes sugars, starches, fiber, and starchy vegetables. Best sources: grains, legumes, potatoes, vegetables, fruits.

FIBER

The portion of plant foods that our bodies cannot digest is known as fiber. There are 2 basic types: *insoluble* ("roughage") and *soluble*. Insoluble fibers help move food materials through the digestive tract; soluble fibers tend to slow them down. Both types absorb water, thus prevent and treat constipation by softening and increasing the bulk of the undigested food components passing through the digestive tract. Soluble fibers have also been reported to be helpful in reducing blood cholesterol levels. Best sources: beans, bran, fruits, whole grains, vegetables.

WATER

Water dissolves and transports other nutrients throughout the body, aiding the processes of digestion, absorption, circulation, and excretion. It helps regulate body temperature.

VITAMINS

Vitamin A—promotes good eyesight and helps keep the skin and mucous membranes resistant to infection. Best sources: liver, sweet potatoes, carrots, kale, cantaloupe, turnip greens, collard greens, broccoli, fortified milk.

Vitamin B1 (thiamine)—prevents beriberi. Essential to carbohydrate metabolism and health of nervous system. Best sources: pork, enriched cereals, grains, soybeans, nuts.

Vitamin B2 (riboflavin)—protects the skin, mouth, eyes, eyelids, and mucous membranes. Essential to protein and energy metabolism. Best sources: milk, meat, poultry, cheese, broccoli, spinach.

Vitamin B6 (pyridoxine)—important in the regulation of the central nervous system and in protein metabolism. Best sources: whole grains, meats, fish, poultry, nuts, brewers' yeast.

Vitamin B12 (cobalamin)—needed to form red blood cells. Best sources: meat, fish, poultry, eggs, dairy products.

Niacin—maintains health of skin, tongue, digestive system. Best sources: poultry, peanuts, fish, enriched flour and bread.

Folic acid (folacin)—required for normal blood cell formation, growth, and reproduction and for important chemical reactions in body cells. Best sources: yeast, orange juice, green leafy vegetables, wheat germ, asparagus, broccoli, nuts.

Other B vitamins—biotin, pantothenic acid.

Vitamin C (ascorbic acid)—maintains collagen, a protein necessary for the formation of skin, ligaments, and bones. It helps heal wounds and mend fractures and aids in resisting some types of viral and bacterial infections. Best sources: citrus fruits and juices, cantaloupe, broccoli, brussels sprouts, potatoes and sweet potatoes, tomatoes, cabbage.

Vitamin D—important for bone development. Best sources: sunlight, fortified milk and milk products, fish-liver oils, egg yolks.

Vitamin E (tocopherol)—helps protect red blood cells. Best sources: vegetable oils, wheat germ, whole grains, eggs, peanuts, margarine, green leafy vegetables.

Vitamin K—necessary for formation of prothrombin, which helps blood to clot. Also made by intestinal bacteria. Best dietary sources: green leafy vegetables, tomatoes.

MINERALS

Calcium—works with phosphorus in building and maintaining bones and teeth. Best sources: milk and milk products, cheese, blackstrap molasses, some types of tofu.

Phosphorus—performs more functions than any other mineral, and plays a part in nearly every chemical reaction in the body. Best sources: cheese, milk, meats, poultry, fish, tofu.

Iron—Necessary for the formation of myoglobin, which is a reservoir of oxygen for muscle tissue, and hemoglobin, which transports oxygen in the blood. Best sources: lean meats, beans, green leafy vegetables, shellfish, enriched breads and cereals, whole grains.

Other minerals—chromium, cobalt, copper, fluorine, iodine, magnesium, manganese, molybdenum, potassium, selenium, sodium, sulfur, and zinc.

Understanding Food Label Claims

Source: Food Labeling Education Information Center, Beltville, MD

The federal Nutrition Labeling and Education Act of 1990 provides that manufacturers can make certain claims on processed food labels only if they meet the definitions specified here:

SUGAR

Sugar free: less than 0.5 g per serving

No added sugar; Without added sugar; No sugar added:
- No sugars added during processing or packing, including ingredients that contain sugars (for example, fruit juices, applesauce, or dried fruit).
- Processing does not increase the sugar content above the amount naturally present in the ingredients. (A functionally insignificant increase in sugars is acceptable from processes used for purposes other than increasing sugar content.)
- The food for which it substitutes normally contains added sugars.

Reduced sugar: at least 25% less sugar than reference food

CALORIES

Low calorie: 40 calories or less per serving; if the serving is 30 g or less or 2 tablespoons or less, 40 calories or less per 50 g of food

Calorie free: under 5 calories per serving

Reduced or Fewer calories: at least 25% fewer calories than reference food

FAT

Fat free: less than 0.5 g of fat per serving

Saturated fat free: less than 0.5 g of saturated fat per serving, and the level of trans fatty acids does not exceed 1% of total fat

Low fat: 3 g or less per serving and, if the serving is 30 g or less or 2 tbs or less, per 50 g of the food

Low saturated fat: 1 g or less per serving and not more than 15% of calories from saturated fatty acids

Reduced or Less fat: at least 25% less per serving than reference food

CHOLESTEROL

Cholesterol free: less than 2 mg of cholesterol and 2 g or less of saturated fat per serving

Low cholesterol: 20 mg or less and 2 g or less of saturated fat per serving and, if the serving is 30 g or less or 2 tbs or less, per 50 g of the food

Reduced or Less cholesterol: at least 25% less than reference food

SODIUM

Sodium free: less than 5 mg per serving

Low sodium: 140 mg or less per serving and, if the serving is 30 g or less or 2 tbs or less, per 50 g of the food

Very low sodium: 35 mg or less per serving and, if the serving is 30 g or less or 2 tbs or less, per 50 g of the food

Reduced or Less sodium: at least 25% less per serving than reference food

FIBER

High fiber: 5 g or more per serving. (Also, must meet low-fat definition, or must state level of total fat.)

Good source of fiber: 2.5 g to 4.9 g per serving

More or Added fiber: at least 2.5 g more per serving than reference food

Dietary Requirements

In April 1998, the Institute of Medicine of the Food and Nutrition Board, National Academy of Sciences, released a report on Dietary Reference Intakes (DRIs), which updated and expanded dietary requirements previously set for thiamin, riboflavin, niacin, B6, folate, B12, pantothenic acid, biotin, and choline. A year earlier, the Institute updated the requirements for calcium, phosphorus, magnesium, vitamin D, and fluoride. All these new values are based on the latest knowledge relevant to optimizing health at all stages of life, not simply protecting against nutritional deficiencies. Reports on other nutrients are under development. In the meantime, the previously established Recommended Dietary Allowances (RDAs) for these nutrients apply.

The new DRIs include 4 categories for daily consumption: **RDA**—the intake that meets the nutrient requirements of almost all (97-98%) healthy individuals in a specified group; **Estimated Average Requirement (EAR)**—the intake that meets the estimated nutrient need of half the individuals in a specified group; **Adequate Intake (AI)**—the intake specified when sufficient broad scientific evidence is not available to calculate an EAR (for healthy breast-fed infants, the AI is the mean intake; the AI for other life stage groups is believed to cover their needs, but lack of data or uncertainty in the data prevent clear specification of this coverage); and **Tolerable Upper Intake Level (UL)**—the maximum intake that is unlikely to pose risks of adverse health effects in almost all healthy individuals in a specified group. RDAs and AIs may both be used as goals for individual intake. The UL is not recommended as a goal.

Recommended Levels for B Vitamins and Choline

Source: Food and Nutrition Board, National Academy of Sciences—Institute of Medicine, 1998

	Thiamin (mg/d)	Ribo-flavin (mg/d)	Niacin (mg/d)[1]	B6 (mg/d)	Folate (µg/d)[2]	B12 (µg/d)	Panto-thenic Acid (mg/d)	Biotin (µg/d)	Choline[3] (µg/d)
Infants									
0-5 mos............	0.2*	0.3*	2*	0.1*	65*	0.4*	1.7*	5*	125*
6-11 mos...........	0.3*	0.4*	3*	0.3*	80*	0.5*	1.8*	6*	150*
Children									
1-3 yrs.............	0.5	0.5	6	0.5	150	0.9	2*	8*	200*
4-8 yrs.............	0.6	0.6	8	0.6	200	1.2	3*	12*	250*
Males									
9-13 yrs............	0.9	0.9	12	1.0	300	1.8	4*	20*	375*
14-18 yrs...........	1.2	1.3	16	1.3	400	2.4	5*	25*	550*
19-30 yrs...........	1.2	1.3	16	1.3	400	2.4	5*	30*	550*
31-50 yrs...........	1.2	1.3	16	1.3	400	2.4	5*	30*	550*
51-70 yrs...........	1.2	1.3	16	1.7	400	2.4[4]	5*	30*	550*
over 70 yrs	1.2	1.3	16	1.7	400	2.4[4]	5*	30*	550*
Females									
9-13 yrs............	0.9	0.9	12	1.0	300	1.8	4*	20*	375*
14-18 yrs...........	1.0	1.0	14	1.2	400[5]	2.4	5*	25*	400*
19-30 yrs...........	1.1	1.1	14	1.3	400[5]	2.4	5*	30*	425*
31-50 yrs...........	1.1	1.1	14	1.3	400[5]	2.4	5*	30*	425*
51-70 yrs...........	1.1	1.1	14	1.5	400[5]	2.4[4]	5*	30*	425*
over 70 yrs	1.1	1.1	14	1.5	400	2.4[4]	5*	30*	425*
Pregnant (all ages)	1.4	1.4	18	1.9	600[6]	2.6	6*	30*	450*
Lactating (all ages)	1.5	1.6	17	2.0	500	2.8	7*	35*	550*

mg/d=milligrams/day. µg=microgram (s). µg/d=micrograms/day. **NOTE:** Adequate Intakes are followed by an asterisk. (1) As niacin equivalents. 1 mg of niacin = 60 mg of tryptophan. (2) As dietary folate equivalents (DFE). 1 DFE = 1 µg food folate = 0.6 µg of folic acid (from fortified food or supplement) consumed with food = 0.5 µg of synthetic (supplemental) folic acid taken on an empty stomach. (3) Although AIs have been set for choline, there is little evidence to assess whether a dietary supply of choline is needed at all stages of the life cycle, and it may be that the body can produce the required amount at some of these stages. (4) Since 10-30% of older people may malabsorb food-bound B12, it is advisable for those over 50 yrs to meet this RDA mainly by taking foods fortified with B12 or a B12-containing supplement. (5) In view of evidence linking folate intake with neural tube defects in the fetus, it is recommended that all women capable of becoming pregnant consume 400 µg of synthetic folic acid from fortified foods and/or supplements in addition to taking in food folate from a varied diet. (6) It is assumed that women will continue taking 400 µg of folic acid until their pregnancy is confirmed and they enter prenatal care, which ordinarily occurs after the critical time for formation of the neural tube.

Recommended Levels for Calcium, Phosphorus, Magnesium, Vitamin D, and Fluoride

Source: Food and Nutrition Board, National Academy of Sciences—Institute of Medicine, 1997

	Calcium		Phosphorus				Magnesium						Vitamin D		Fluoride		
							EAR[1]		RDA[1]		AI[1]			AI[1]		AI[1]		
	AI[1]	UL[2]	EAR[1]	RDA[1]	AI[1]	UL[2]	m	f	m	f	m	f	UL[1,3]	AI[4,5]	UL[4]	m	f	UL[1]
Infants																		
0-6 mos.......	210	ND	—	—	100	ND	—	—	—	—	30	30	ND	5	25	0.01	0.01	0.7
6-12 mos......	270	ND	—	—	275	ND	—	—	—	—	75	75	ND	5	25	0.50	0.50	0.9
Children																		
1-3 yrs........	500	2.5	380	460	—	3.0	65	65	80	80	—	—	65	5	50	0.70	0.70	1.3
4-8 yrs........	800	2.5	405	500	—	3.0	110	110	130	130	—	—	110	5	50	1.10	1.10	2.2
9-13 yrs......	1,300	2.5	1,055	1,250	—	4.0	200	200	240	240	—	—	350	5	50	2.00	2.00	10.0
14-18 yrs.....	1,300	2.5	1,055	1,250	—	4.0	340	300	410	360	—	—	350	5	50	3.20	2.90	10.0
Adults																		
19-30 yrs......	1,000	2.5	580	700	—	3.4	330	255	400	310	—	—	350	5	50	3.80	3.10	10.0
31-50 yrs.....	1,000	2.5	580	700	—	3.4	350	265	420	320	—	—	350	5	50	3.80	3.10	10.0
51-70 yrs.....	1,200	2.5	580	700	—	3.4	350	265	420	320	—	—	350	10	50	3.80	3.10	10.0
over 70 yrs	1,200	2.5	580	700	—	3.0	350	265	420	320	—	—	350	15	50	3.80	3.10	10.0
Pregnant																		
18 yrs or less ..	1,300	2.5	1,055	1,250	—	3.5	—	335	—	400	—	—	350	5	50	—	2.90	10.0
19-30 yrs......	1,000	2.5	580	700	—	3.5	—	290	—	350	—	—	350	—	50	—	3.10	10.0
31-50 yrs.....	1,000	2.5	580	700	—	3.5	—	300	—	360	—	—	350	—	50	—	3.10	10.0
Lactating																		
18 yrs or less ..	1,300	2.5	1,055	1,250	—	4.0	—	300	—	360	—	—	350	5	50	—	2.90	10.0
19-30 yrs.....	1,000	2.5	580	700	—	4.0	—	255	—	310	—	—	350	—	50	—	3.10	10.0
31-50 yrs.....	1,000	2.5	580	700	—	4.0	—	265	—	320	—	—	350	—	50	—	3.10	10.0

m =male. f =female. ND =Not determinable, because of a lack of data on adverse effects in this age group and a concern over body's lack of ability to handle excess amounts. Source of intake in this case should be from food only. (1) mg/day. (2) g/day. (3) The UL for magnesium represents intake from a pharmacological agent only and does not include intake from food and water. (4) microgram/day. (5) In the absence of adequate exposure to sunlight.

Recommended Dietary Allowances (RDAs)

Source: Food and Nutrition Board, National Academy of Sciences—Institute of Medicine, 1989

	Weight (lbs)	Protein (g)	Fat soluble vitamins			Vitamin C[3] (µg)	Minerals			
			Vitamin A[1]	Vitamin E[2]	Vitamin K (µg)		Iron (mg)	Zinc (mg)	Iodine (µg)	Selenium (µg)
Infants . . . to 6 mos.	13	13	375	3	25	0.3	6	5	40	10
6 mos. to 1 yr. .	20	14	375	4	35	0.5	10	5	50	15
Children . . 1-3.	29	16	400	6	50	0.7	10	10	70	20
4-6.	44	24	500	7	75	1.0	10	10	90	20
7-10.	62	28	700	7	100	1.4	10	10	120	30
Males . . . . 11-14.	99	45	1,000	10	150	2.0	12	15	150	40
15-18.	145	59	1,000	10	200	2.0	12	15	150	50
19-24.	160	58	1,000	10	200	2.0	10	15	150	70
25-50.	174	63	1,000	10	200	2.0	10	15	150	70
51+	170	63	1,000	10	200	2.0	10	15	150	70
Females . . 11-14.	101	46	800	8	150	2.0	15	12	150	45
15-18.	120	44	800	8	180	2.0	15	12	150	50
19-24.	128	46	800	8	180	2.0	15	12	150	55
25-50.	138	50	800	8	180	2.0	15	12	150	55
51+	143	50	800	8	180	2.0	10	12	150	55
Pregnant.	—	60	800	10	400	2.2	30	15	175	65
Lactating. . 1st 6 mos.	—	65	1,300	12	280	2.6	15	19	200	75
2d 6 mos	—	62	1,200	11	260	2.6	15	16	200	75

g=grams. µg=micrograms. (1) Retinol equivalents. (2) Milligrams alpha-tocopherol equivalents. (3) Vitamin C is water-soluble.

Weight Guidelines for Adults

Source: *Clinical Guidelines on the Identification, Evaluation, and Treatment of Overweight and Obesity in Adults,*
National Heart, Lung, and Blood Institute, National Institutes of Health, 1998

Guidelines on identification, evaluation, and treatment of overweight and obesity in adults were released in June 1998 by the National Heart, Lung, and Blood Institute (NHLBI), in cooperation with the National Institute of Diabetes and Digestive and Kidney Diseases. The guidelines, based on research into risk factors in heart disease, stroke, and other conditions, define degrees of overweight and obesity in terms of **body mass index (BMI)**, which is based on weight and height and is strongly correlated with total body fat content. A BMI of 25-29 is said to indicate **overweight**; a BMI of 30 or above is said to indicate **obesity**. Weight reduction is advised for persons with a BMI of 25 or higher, about 55% of the adult population. (Previous guidelines have been less stringent.) Factors such as large waist circumference, high blood pressure or cholesterol, and a family history of obesity-related disease may increase risk. The table given here shows the BMI for certain heights and weights. For weight reduction tips, while they last, write to the NHLBI Information Center, PO Box 30105, Bethesda, MD 20824-0105. See also the NHLBI website: http://www.nhlbi.nih.gov/index.htm

Weight (lbs)

Height	HEALTHY						OVERWEIGHT					OBESE							
4'10". . .	91	96	100	105	110	115	119	124	129	134	138	143	148	153	158	162	167	172	177	181
4'11". . .	94	99	104	109	114	119	124	128	133	138	143	148	153	158	163	168	173	178	183	188
5'0". . . .	97	102	107	112	118	123	128	133	138	143	148	153	158	163	168	174	179	184	189	194
5'1". . . .	100	106	111	116	122	127	132	137	143	148	153	158	164	169	174	180	185	190	195	201
5'2". . . .	104	109	115	120	126	131	136	142	147	153	158	164	169	175	180	186	191	196	202	207
5'3". . . .	107	113	118	124	130	135	141	146	152	158	163	169	175	180	186	191	197	203	208	214
5'4". . . .	110	116	122	128	134	140	145	151	157	163	169	174	180	186	192	197	204	209	215	221
5'5". . . .	114	120	126	132	138	144	150	156	162	168	174	180	186	192	198	204	210	216	222	228
5'6". . . .	118	124	130	136	142	148	155	161	167	173	179	186	192	198	204	210	216	223	229	235
5'7". . . .	121	127	134	140	146	153	159	166	172	178	185	191	198	204	211	217	223	230	236	242
5'8". . . .	125	131	138	144	151	158	164	171	177	184	190	197	203	210	216	223	230	236	243	249
5'9". . . .	128	135	142	149	155	162	169	176	182	189	195	203	209	216	223	230	236	243	250	257
5'10". . .	132	139	146	153	160	167	174	181	188	195	202	209	216	222	229	236	243	250	257	264
5'11". . .	136	143	150	157	165	172	179	186	193	200	208	215	222	229	236	243	250	257	265	272
6'0". . . .	140	147	154	162	169	177	184	191	199	206	213	221	228	235	242	250	258	265	272	279
6'1". . . .	144	151	159	166	174	182	189	197	204	212	219	227	235	242	250	257	265	272	280	288
6'2". . . .	148	155	163	171	179	186	194	202	210	218	225	233	241	249	256	264	272	280	287	295
6'3". . . .	152	160	168	176	184	192	200	208	216	224	232	240	248	256	264	272	279	287	295	303
6'4". . . .	156	164	172	180	189	197	205	213	221	230	238	246	254	263	271	279	287	295	304	312
BMI[1]. . .	19	20	21	22	23	24	25	26	27	28	29	30	31	32	33	34	35	36	37	38

(1) The BMI numbers apply to both men and women. Some very muscular people may have a high BMI without health risks.

MILLENNIUM FACT BOX

Food for Thought

Source: Economic Research Service, U.S. Dept. of Agriculture

During the 20th century, Americans began to drink less whole milk and more low-fat or skim milk, and eat less butter and more margarine. Poultry consumption has increased substantially—but consumption of red meat has also risen. The chart below shows U.S. per capita consumption of selected foods in 4 years that span the century.

	Whole milk[1]	Low-fat & skim milk[1]	Butter[2]	Margarine[2]	Red meat[2]	Poultry[2]	Fish & shellfish[2]
1909	26.85	7.30	17.9	1.2	101.7	11.2	11.0
1939	29.24	4.79	17.4	2.3	86.9	11.9	10.8
1969	26.60	5.46	5.6	10.7	129.5	32.9	11.2
1998	8.32	15.38	4.2	8.6	115.6	68.4	14.8

(1) Gallons. (2) Pounds.

U.S. Recommended Childhood Immunization Schedule

Source: Advisory Committee on Immunization Practices (ACIP), Amer. Acad. of Pediatrics, and Amer. Acad. of Family Physicians (AAFP), 1999

Vaccines are listed under the routinely recommended ages[1]. Bars indicate the range of recommended ages. Catch-up immunization should be done when feasible. Bars with double rules indicate vaccines to be given if previously recommended doses were missed or given earlier than the recommended minimum age.

AGE → VACCINE ↓	Birth	1 mo	2 mos	4 mos	6 mos	12 mos	15 mos	18 mos	4-6 yrs	11-12 yrs	14-16 yrs
Hepatitis B[2]	Hep B	Hep B			Hep B					Hep B	
Diphtheria, Tetanus, Pertussis (whooping cough)[3]			DTaP	DTaP	DTaP		DTaP[3]		DTaP	Td	
H. influenzae type b[4]			Hib	Hib	Hib	Hib					
Polio[5]			IPV	IPV	Polio[5]				Polio		
Rotavirus[6]			Rv[6]	Rv[6]	Rv[6]						
Measles, Mumps, Rubella (German measles)[7]						MMR			MMR[7]	MMR[7]	
Varicella (chickenpox)[8]						Var				Var[8]	

(1) This schedule indicates the recommended ages for administration of childhood vaccines. Combination vaccines may be used whenever administration of all components of the vaccine is called for.

(2) **Infants born to mothers who do not have hepatitis B** should receive the 2d dose of hepatitis B vaccine at least 1 month after the 1st. The 3d dose should be given at least 4 months after the 1st and at least 2 months after the 2d, but not before 6 months of age.
Infants born to mothers who have hepatitis B should receive hepatitis B vaccine and 0.5 mL (milliliters) of hepatitis B immune globulin (HBIG) within 12 hours of birth. The 2d dose is recommended at 1-2 months of age, the 3d at 6 months.
Infants born to mothers whose hepatitis B status is unknown should receive hepatitis B vaccine within 12 hours of birth. Maternal blood should be drawn at the time of delivery to determine if the mother has hepatitis B; if so, the infant should receive HBIG as soon as possible (no later than 1 week of age).
All children and adolescents (through 18 years of age) not immunized against hepatitis B may begin the series at any time. Special efforts should be made to immunize children who were born in or whose parents were born in areas of the world in which prevalence of hepatitis B infection is moderate or high.

(3) **DTaP** (diphtheria and tetanus toxoids and acellular pertussis vaccine) is the preferred vaccine for all doses in the immunization series, including completion of the series in children who have received 1 or more doses of whole-cell DTP vaccine. Whole-cell DTP is an acceptable alternative to DTaP. The 4th dose of the vaccine may be given as early as 12 months of age, provided 6 months have elapsed since the 3d dose. Td (tetanus and diphtheria toxoids) is recommended at 11-12 years of age if at least 5 years have elapsed since the last dose of DTP, DTaP, or DT. Boosters are recommended every 10 years.

(4) *Haemophilus influenzae* **type b** is a bacterium that can cause such serious infectious diseases as meningitis and pneumonia. Three *H. influenzae* type b (Hib) conjugate vaccines are licensed for infant use. If PRP-OMP (PedvaxHIB and COMVAX [Merck]) is given at 2 and 4 months of age, a dose at 6 months is not required. Because clinical studies in infants have demonstrated that using some combination products may cause a lower immune response to the Hib vaccine component, DTaP/Hib combination products should not be used for primary immunization in infants at 2, 4, or 6 months of age, unless FDA-approved for these ages.

(5) Two **poliovirus vaccines** are currently licensed in the U.S.: inactivated poliovirus vaccine (IPV) and oral poliovirus vaccine (OPV). The ACIP, AAP, and AAFP now recommend that the 1st 2 doses of poliovirus vaccine should be IPV. The ACIP continues to recommend a sequential schedule of 2 doses of IPV given at ages 2 and 4 months, followed by 2 doses of OPV at 12-18 months and 4-6 years. Use of IPV for all doses is also acceptable and is recommended for people with impaired or weakened immune systems and for their household contacts. OPV is no longer recommended for the 1st 2 doses of the schedule and is acceptable only for special circumstances such as children of parents who do not accept the recommended number of injections, late initiation of immunization which would require an unacceptable number of injections, and imminent travel to areas in which polio is prevalent.

(6) **Rotavirus** (Rv) is a virus that can infect the digestive system and cause gastroenteritis and diarrhea. New information from the CDC indicates there may be an increased risk of intussusception during the 1st few weeks after receiving the RV vaccine. Health care providers temporarily should suspend administration of the RV vaccine to unimmunized and partially immunized children, pending collection and evaluation of additional information. When it is determined that the RV vaccine should be given, the 1st dose of Rv vaccine should not be given before 6 weeks of age, and the minimum interval between doses is 3 weeks. The Rv vaccine series should not be initiated at 7 months of age or older, and all doses should be completed by the 1st birthday.

(7) The 2d dose of the **measles, mumps, rubella vaccine** (MMR) is recommended routinely at 4-6 years of age but may be given at any time, provided that at least 4 weeks have elapsed since the 1st dose and both doses are given beginning at or after 12 months of age. Those who have not previously received the 2d dose should complete the schedule by the age of 11 or 12.

(8) Varicella vaccine (Var) is recommended any time after the 1st birthday for susceptible children (those who lack a reliable history of **chickenpox** and who have not been immunized). Susceptible people 13 years of age or older should receive 2 doses, given at least 4 weeks apart.

Allergies and Asthma

Source: Asthma and Allergy Foundation of America, 1233 20th St., NW, Suite 402, Washington, DC 20036; phone: (800) 7-ASTHMA

One out of every five Americans suffers from **allergies**. People with allergies have extra-sensitive immune systems that react to normally harmless substances. Allergens that sometimes produce this reaction include plant pollens, dust mites, or animal dander; plants such as poison ivy; certain drugs, such as penicillin; and certain foods such as eggs, milk, nuts, or seafood.

The tendency to develop allergies is usually inherited, and allergies usually begin to appear in childhood, but they can show up at any age. Common allergies for infants include food allergies and eczema (patches of dry skin). Older children and adults may often develop allergic rhinitis (hay fever), a reaction to an inhaled allergen; common symptoms include nasal congestion, runny nose, and sneezing.

It is best to avoid contact with the allergen, if feasible. In some cases, medications such as antihistamines are used to decrease the reaction, and there are treatments aimed at gradually desensitizing the patient to the allergen. Other effective allergy treatments include decongestants, eye drops, and ointments.

Some people with allergies also have **asthma**, and allergens are a common asthma trigger. Asthma is a disease of chronic inflammation, affecting the passages that carry air into and out of the lungs. It is most often seen in children but can develop at any age.

People with asthma have inflamed, supersensitive airways that tighten and become filled with mucus during an asthma episode. Wheezing, difficulty in breathing, tightening of the chest, and coughing are common symptoms. Asthma can progress through stages to become life-threatening if not controlled. Emergency symptoms of asthma include a bluish cast to the face and lips, severe anxiety, increased pulse rate, and sweating.

Besides common allergens, tobacco smoke, cold air, and pollution can trigger an asthma attack, as can viral infections or physical exercise that taxes the breathing. Of course, an accurate diagnosis by a physician is important. Although there is no cure for asthma or allergies, they can be controlled with medications and lifestyle changes.

Cancer Prevention

Source: American Cancer Society, 1599 Clifton Road NE, Atlanta, GA 30329-4251; phone: (800) 227-2345

PRIMARY PREVENTION: Modifiable determinants of cancer risk.

Smoking
Lung cancer mortality rates are about 23 times higher for current male smokers, and 13 times higher for current female smokers, than for those who have never smoked. Smoking accounts for about 30% of all cancer deaths in the U.S. Tobacco use is responsible for nearly 1 in 5 deaths in the U.S. Smoking is associated with cancer of the lung, mouth, pharynx, larynx, esophagus, pancreas, uterine cervix, kidney, and bladder.

Nutrition and Diet
Risk for colon, rectum, breast (among postmenopausal women), kidney, prostate, and endometrial cancers increases in obese people. A diet high in fat may be a factor in the development of certain cancers, particularly cancer of the colon and rectum, prostate, and endometrium. High-fiber foods may help reduce risk of colon cancer. Eating 5 or more servings of fruits and vegetables each day, and eating other foods from plant sources (especially grains and beans), may reduce risk for many cancers. Physical activity can help protect against some cancers.

Sunlight
Many of the one million skin cancers that are expected to be diagnosed in 1999 could have been prevented by protection from the sun's rays. Epidemiological evidence shows that sun exposure is a major factor in the development of melanoma and that the incidence rates are increasing around the world.

Alcohol
Heavy drinking, especially when accompanied by cigarette smoking or smokeless tobacco use, increases risk of cancers of the mouth, larynx, pharynx, esophagus, and liver. Studies have also noted an association between alcohol consumption and an increased risk of breast cancer.

Smokeless Tobacco
Use of chewing tobacco or snuff increases risk of cancers of the mouth and pharynx. The excess risk of cancer of the cheek and gum may reach nearly 50-fold among long-term snuff users.

Estrogen
Estrogen treatment to control menopausal symptoms can increase risk of endometrial cancer. However, including progesterone in estrogen replacement therapy helps to minimize this risk. Use of estrogen by menopausal women needs careful discussion by the woman and her physician, while research continues.

Radiation
Excessive exposure to ionizing radiation can increase cancer risk. Medical and dental X rays are adjusted to deliver the lowest dose possible without sacrificing image quality. Excessive radon exposure in the home may increase lung cancer risk, especially in cigarette smokers. If levels are found to be too high, remedial actions should be taken.

Environmental Hazards
Exposure to various chemicals (including benzene, asbestos, vinyl chloride, arsenic, and aflatoxin) increases risk of various cancers. Risk of lung cancer from asbestos is greatly increased when combined with smoking. Pesticides, low-frequency radiation, toxic wastes, and proximity to nuclear power plants have not been proven to cause cancer.

Cancer-Detection Guidelines

Source: American Cancer Society, 1599 Clifton Road NE, Atlanta, GA 30329-4251; phone: (800) 227-2345

SECONDARY PREVENTION: Steps to diagnose a cancer or precursor as early as possible after it has developed.

A cancer-related checkup is recommended every 3 years for people aged 20-40 and every year for people 40 years of age and older. This exam should include health counseling and, depending on a person's age, might include examinations for cancers of the thyroid, oral cavity, skin, lymph nodes, testes, and ovaries, as well as for some nonmalignant diseases. Special tests for certain cancer sites are recommended as outlined below:

Breast Cancer
- Breast self-exam monthly, beginning at age 20.
- Breast clinical physical examination for women aged 20-39, every 3 years; 40 and over, every year.
- Mammography for women aged 40 and over, every year.

Cervical Cancer
Annual Pap test and pelvic exam for women who are or have been sexually active or have reached age 18. After 3 or more consecutive satisfactory normal annual exams, the Pap test may be performed less frequently at the discretion of the physician.

Colorectal Cancer
Beginning at age 50, both men and women should follow this testing schedule:
- Yearly fecal occult blood test, plus flexible sigmoidoscopy and digital rectal examination every 5 years, or
- Colonoscopy and digital rectal examination every 10 years, or
- Double-contrast barium enema and digital rectal examination every 5-10 years.

Endometrial Cancer
Sampling of asymptomatic women at high risk of developing endometrial cancer should begin at menopause and may be indicated at various intervals thereafter, depending on the degree of risk and other factors determined by the physician.

Oral Cancer
Regular checkups by dentists and primary care physicians will show any abnormalities.

Prostate Cancer
Both Prostate-Specific Antigen (PSA) and Digital Rectal Examination (DRE) should be offered annually, beginning at age 50, to men who have at least a 10-year life expectancy, and should be offered to younger men who are at high risk. Information should be provided to patients regarding potential risks and benefits of intervention. Men who choose to undergo screening should begin at age 50. However, men in high-risk groups, such as those with a strong familial predisposition (e.g., 2 or more affected first-degree relatives) or African Americans may begin at a younger age (e.g., 45 years).

Skin Cancer
Adults should practice skin self-exam regularly. Suspicious lesions should be evaluated promptly by a physician.

Breast Cancer

Source: American Cancer Society, Inc., 1599 Clifton Road NE, Atlanta, GA 30329-4251; phone: (800) 227-2345

It is estimated that, in 1999, about 175,000 women and 1,300 men in the United States will be diagnosed with breast cancer, and about 43,300 women and 400 men will die from it. Breast cancer is the second largest cause of cancer death for women in the U.S. (lung cancer ranks first), but mortality rates have been declining, especially among younger women, probably because of earlier detection and improved treatment.

Breast cancer is often manifested first in an abnormality that appears on a **mammogram**. Physical signs and symptoms that show up later, and may be detectable by a woman or her physician, include a breast lump, and, less commonly, thickening, swelling, distortion, or tenderness; skin irritation or dimpling; and nipple pain, scaliness, or retraction. Breast pain is more commonly associated with benign (noncancerous) conditions.

Studies show that **early detection** increases survival and treatment options. The American Cancer Society (ACS) recommends that women 40 and older should have an annual mammogram, have an annual clinical breast exam by a health care professional, and perform monthly breast self-examinations. The ACS recommends that women ages 20-39 should have a clinical breast exam every 3 years and should also perform monthly breast self-examinations. Although most breast lumps that are detected are noncancerous, any suspicious lump needs to be biopsied.

The **risk** for breast cancer increases as a woman ages. The risk is also higher for women with a personal or family history of the disease; early start of menstruation; late onset of menopause; recent use of oral contraceptives or postmenopausal estrogens; no children or no live birth until age 30 or older;

and relatively high education and socioeconomic status. Other risk factors include alcohol consumption and obesity.

Treatment for breast cancer may involve lumpectomy (local removal of a tumor), mastectomy (surgical removal of the breast), radiation therapy, chemotherapy, or hormone therapy. For early-stage breast cancer, long-term survival rates following lumpectomy plus radiation therapy are similar to survival rates after modified radical mastectomy.

Numerous **drugs** that may prevent breast cancer or improve its treatment are being studied. One is **tamoxifen**, a synthetic hormone that blocks the action of estrogen in the breast. Already used for treating breast cancer, it has been shown to reduce the likelihood of developing the disease in women considered at higher than average risk, including women age 60 and older. Unfortunately, tamoxifen also has dangerous side effects, such as increased risk of uterine cancer and blood clots in the lungs. Research is also being done on another drug, **Raloxifene**, which is approved for preventing osteoporosis in postmenopausal women. It is now being directly compared to tamoxifen in a large clinical study to evaluate its effect on breast cancer risk.

A new technique is being developed to determine whether cancer has spread to the lymph nodes. In this procedure, called **sentinel node biopsy**, a radioactive tracer and/or a blue dye are injected into the region of a tumor. The injected material first enters the "sentinel node" (the first lymph node to receive lymph from the tumor). The physician can detect the radioactivity and/or blue color and remove the node to study it. If this sentinel node contains cancer, more lymph nodes are removed. If it is cancer-free, additional lymph node surgery is avoided.

Trends in Daily Use of Cigarettes, for U.S. 8th, 10th, and 12th Graders

Source: *Monitoring the Future*, Univ. of Michigan Inst. for Social Research and National Inst. on Drug Abuse

(percent who smoked daily in last 30 days)

	8th grade					'97-'98 change	10th grade					'97-'98 change	12th grade					'97-'98 change
	1994	1995	1996	1997	1998		1994	1995	1996	1997	1998		1994	1995	1996	1997	1998	
TOTAL......	8.8	9.3	10.4	9.0	8.8	−0.2	14.6	16.3	18.3	18.0	15.8	−2.2	19.4	21.6	22.2	24.6	22.4	−2.2
Sex																		
Male	9.5	9.2	10.5	9.0	8.1	−0.9	15.2	16.3	18.1	17.2	14.7	−2.5	20.4	21.7	22.2	24.8	22.7	−2.1
Female	8.0	9.2	10.1	8.7	9.0	+0.3	13.7	16.1	18.6	18.5	16.8	−1.7	18.1	20.8	21.8	23.6	21.5	−2.1
College plans																		
None or under																		
4 yrs......	22.6	22.5	26.0	25.4	25.2	−0.2	28.9	32.7	34.3	35.4	31.7	−3.7	29.8	33.7	33.2	35.6	34.6	−1.0
Complete																		
4 yrs......	6.8	7.5	8.0	6.9	6.6	−0.3	11.5	13.3	15.5	15.0	12.9	−2.1	15.7	17.4	18.9	20.6	18.4	+2.2
Region																		
Northeast ..	8.6	9.2	11.0	8.8	6.1	−2.7	14.1	15.8	18.8	18.0	18.7	+0.7	21.3	22.5	27.0	29.4	23.4	−6.0
North central	9.4	11.0	12.4	10.3	11.2	+0.9	16.9	17.6	20.6	19.5	17.3	−2.2	23.8	25.7	26.1	28.0	27.8	−0.2
South	9.4	9.4	10.4	9.5	10.2	+0.7	15.5	19.3	20.5	20.5	17.1	−3.4	19.3	21.7	20.5	22.6	21.8	−0.8
West......	7.4	7.0	7.5	6.8	5.8	−1.0	9.7	9.4	10.7	11.1	8.8	+2.3	12.4	14.5	13.8	17.5	15.5	−2.0
Race/Ethnicity [1]																		
White	9.7	10.5	11.7	11.4	10.4	−1.0	16.5	17.6	20.0	21.4	20.3	+1.1	22.9	23.9	25.4	27.8	28.3	+0.5
Black	2.6	2.8	3.2	3.7	3.8	+0.1	3.8	4.7	5.1	5.6	5.8	+0.2	4.9	6.1	7.0	7.2	7.4	+0.2
Hispanic ...	9.0	9.2	8.0	8.1	8.4	+0.3	8.1	9.9	11.6	10.8	9.4	−1.4	10.6	11.6	12.9	14.0	13.0	+0.4

(1) For each of these groups, data for the specified year and previous year have been combined to increase sample size and thus provide a more reliable estimate.

Some Benefits of Quitting Smoking

Source: American Cancer Society, Inc., 1599 Clifton Road NE, Atlanta, GA 30329-4251; phone: (800) 227-2345

Within 20 Minutes
- Blood pressure drops to a level close to that before the last cigarette
- Temperature of hands and feet increases to normal

Within 8 Hours
- Carbon monoxide level in the blood drops to normal

Within 24 Hours
- Chance of heart attack decreases

Within 2 Weeks to 3 Months
- Circulation improves
- Lung function increases up to 30%

Within 1 to 9 Months
- Coughing, sinus congestion, fatigue, and shortness of breath decrease

- Cilia regain normal function in the lungs, increasing the ability to handle mucus, clean the lungs, reduce infection

Within 1 Year
- Excess risk of coronary heart disease is half that of a smoker's

Within 5 Years
- Stroke risk is reduced to that of a nonsmoker 5-15 years after quitting

Within 10 Years
- Lung cancer death rate about half that of a continuing smoker's
- Risk of cancer of the mouth, throat, esophagus, bladder, kidney, and pancreas decreases

Within 15 Years
- Risk of coronary heart disease is that of a nonsmoker's

Diabetes

Source: American Diabetes Association, 1660 Duke St., Alexandria, VA 22314; phone: (800) 342-2383

Diabetes is a chronic disease in which the body does not produce or properly use **insulin**, a hormone needed to convert sugar, starches, and other foods into energy necessary for daily life. Both genetics and environment appear to play roles in the onset of diabetes. This disease, which has no cure, is the 6th-leading cause of death by disease in the U.S. According to death certificate data, diabetes contributed to more than 187,000 deaths in 1995.

In 1997, the American Diabetes Association issued **new guidelines for diagnosing diabetes**. The recommendations include: lowering the acceptable level of blood sugar from 140 mg of glucose/deciliter of blood to 126 mg/deciliter, possibly identifying 2 million more people with the disease; testing all adults 45 years and older, and then every 3 years if normal; and testing at a younger age, or more frequently, in high-risk individuals. The American Diabetes Association believes that detection at an earlier stage will help prevent or delay complications of diabetes.

There are 2 major types of diabetes:
- **Type 1 (formerly known as insulin dependent).** The body produces very little or no insulin; disease most often begins in childhood or early adulthood. People with type 1 diabetes must take daily insulin injections to stay alive.
- **Type 2 (formerly known as non-insulin dependent).** The body does not produce enough or cannot properly use insulin. It is the most common form of the disease (90-95% of cases in people over age 20) and often begins later in life.

Warning Signs of Diabetes

Type 1 Diabetes (usually occurs suddenly):
- frequent urination
- unusual thirst
- extreme hunger
- unusual weight loss
- extreme fatigue
- irritability

Type 2 Diabetes (occurs less suddenly):
- any type 1 symptoms
- frequent infections
- blurred vision
- cuts/bruises slow to heal
- tingling/numbness in hands or feet
- recurring skin, gum, or bladder infections

Complications of Diabetes

More than one-third of all individuals with diabetes do not know that they have the disease until one of its life-threatening complications occurs. Potential complications include:

Blindness. Diabetes is the leading cause of blindness in people ages 20-74. Each year, from 12,000 to 24,000 people lose their sight because of diabetes.

Kidney disease. 10% to 21% of all people with diabetes develop kidney disease. In 1995, more than 27,900 people initiated treatment for end-stage renal disease (kidney failure) because of diabetes.

Amputations. Diabetes is the most frequent cause of nontraumatic lower limb amputations. The risk of a leg amputation is 15 to 40 times greater for a person with diabetes than for the average American. Each year, an estimated 56,000 people lose a foot or leg to complications brought on by diabetes.

Heart disease and stroke. People with diabetes are 2 to 4 times more likely to have heart disease (more than 77,000 deaths due to heart disease annually). And they are 2 to 4 times more likely to suffer a stroke.

Health-care and related costs for the treatment of the disease, added to the cost of lost productivity, total nearly $100 billion annually in the U.S.

Alzheimer's Disease

Source: Alzheimer's Association, 919 N Michigan Ave., Suite 1100, Chicago, IL 60611-1676; phone: (800) 272-3900

Alzheimer's disease is a progressive, degenerative disease of the brain in which brain cells die and are not replaced. It results in impaired memory, thinking, and behavior, and is the most common form of dementing illness. The debilitating nature of the disease renders patients susceptible to infections (such as pneumonia and urinary tract infections) as they become emaciated, incontinent, immobile, or enter a persistent vegetative state.

Alzheimer's disease afflicts an estimated 4 million Americans, striking men and women of all races. Although most people diagnosed with Alzheimer's are older than age 60, the disease can occur even in people in their 40s and 50s. An estimated 10% of those 65 years of age or older, and almost half of those over age 85, have the disease. It is estimated that the cost of diagnosis, treatment, and long-term care for patients with the disease amounts to $100 billion per year in the United States.

The **rate of progression** of Alzheimer's disease from the onset of symptoms until death ranges from 3 to 20 years; the average is 8 years. Eventually, patients become totally incapable of caring for themselves.

Diagnosis is complicated by the lack of a single, simple test to identify the disease. Through a series of diagnostic tests by a qualified physician, possible causes of symptoms, such as depression, drug interactions, nutrient imbalances, or other forms of dementia, such as those associated with stroke, Huntington's disease, Parkinson's disease, Pick's disease, and infections (AIDS, meningitis, syphilis) are ruled out, yielding a diagnosis of Alzheimer's disease that is 80-90% accurate. A definitive diagnosis is possible only with a brain biopsy or an autopsy.

No **treatment** has proven successful in reversing the course of the disease, and providing care for patients with Alzheimer's disease is physically and psychologically demanding. Nearly 70% of those afflicted with the disease live at home and are cared for by family and friends. In the last stages of the disease, it is often necessary for those afflicted to be cared for in a nursing home. Nearly half of all nursing home patients in the United States suffer from Alzheimer's disease.

People with Alzheimer's disease need a safe, stable environment and a regular daily schedule. Physical exercise and social activity are important, as is proper nutrition. A bracelet identifying the person's name and condition may be helpful in case the person wanders away.

The causes of the disease are unknown.

Warning Signs of Alzheimer's Disease

- Recent memory loss that affects job performance
- Inability to learn new information
- Difficulty with everyday tasks such as cooking or dressing oneself
- Inability to remember simple words
- Use of inappropriate words when communicating
- Disorientation of time and place
- Poor or decreased judgment
- Problems with abstract thinking
- Putting objects in inappropriate places
- Rapid changes in mood or behavior
- Increased irritability, anxiety, depression, confusion, and restlessness
- Prolonged loss of initiative

Heart and Blood Vessel Disease

Source: American Heart Association, 7272 Greenville Ave., Dallas, TX 75231-4596; phone: (800) 242-8721

Warning Signs

Of Heart Attack
- Uncomfortable pressure, fullness, squeezing, or pain in the center of the chest lasting 2 minutes or longer
- Pain may radiate to the shoulder, arm, neck, or jaw
- Sweating may accompany pain or discomfort
- Nausea and vomiting also may occur
- Shortness of breath, dizziness, or fainting may accompany other signs

The American Heart Association advises immediate action at the onset of these symptoms. The association points out that more than half of heart attack victims die within 1 hour of the onset of symptoms and before they have reached the hospital.

Of Stroke
- Sudden temporary weakness or numbness of face or limbs on one side of the body
- Temporary loss of speech, or trouble speaking or understanding speech
- Temporary dim or lost vision, especially in one eye
- Unexplained dizziness, unsteadiness, or sudden falls

Some Major Risk Factors

Blood pressure—High blood pressure increases the risk of stroke, heart attack, kidney failure, and congestive heart failure.

Cholesterol—A blood cholesterol level over 240 mg/dl (milligrams of cholesterol per deciliter of blood) approximately doubles the risk of coronary heart disease; about 20% of the U.S. adult population (39.9 mil) have a cholesterol level over 240 mg/dl. Levels between 200 and 240 mg/dl are in a zone of moderate and increasing risk.

Cigarettes—Cigarette smokers have more than twice the risk of heart attack and 2-4 times the risk of sudden cardiac death as nonsmokers. Young smokers have a higher risk for early death from stroke.

Obesity—Using a body mass index (BMI) of 25 and higher for overweight and 30 and higher for obesity, 105.7 mil Americans age 20 and over are overweight and 43.1 mil are obese.

Understanding Blood Pressure

High blood pressure, or hypertension, affects people of all races, sexes, ethnic origins, and ages. Various causes can trigger this often symptomless disease. Since hypertension can increase one's risk for stroke, heart attack, kidney failure, and congestive heart failure, it is recommended that individuals have a blood pressure reading at least once every 2 years (more often if advised by a physician).

A blood pressure reading is really two measurements in one, with one written over the other, such as 122/78. The **upper number (systolic pressure)** represents the amount of pressure in the blood vessels when the heart contracts (beats) and pushes blood through the circulatory system. The **lower number (diastolic pressure)** represents the pressure in the blood vessels between beats, when the heart is resting. According to National Institutes of Health guidelines, normal blood pressure is below 130/85 and "high normal" is between 130/85 and 139/89.

High blood pressure is divided into 3 stages, based upon severity:
- **Stage 1** is from 140/90 through 159/99
- **Stage 2** is from 160/100 through 179/109
- **Stage 3** is 180/110 or greater

The diagnosis of hypertension can be based on either the systolic or the diastolic reading.

High blood pressure usually cannot be cured, but it can be controlled in a variety of ways, including lifestyle modifications and medication. Treatment always should be at the direction and under the supervision of a physician.

Examples of Moderate[1] Amounts of Exercise

Source: *Physical Activity and Health: A Report of the Surgeon General*, U.S. Dept. of Health and Human Services, 1996

ACTIVITY	DURATION[2] (min)	ACTIVITY	DURATION[2] (min)
Washing and waxing a car	45-60	Raking leaves	30
Washing windows or floors	45-60	Walking 2 mi (15 min/mi)	30
Playing touch football	30-45	Swimming laps	20
Wheeling self in wheelchair	30-40	Basketball (playing a game)	15-20
Walking 1 3/4 mi (20 min/mi)	35	Bicycling 4 mi	15
Basketball (shooting baskets)	30	Jumping rope	15
Bicycling 5 mi	30	Running 1 1/2 mi (10 min/mi)	15
Dancing fast (social)	30	Shoveling snow	15

Note: The activities are arranged from less vigorous, and using more time, to more vigorous, and using less time. (1) A "moderate" amount of physical activity uses about 150 calories (kcal), or 1,000 if done daily for a week. (2) Activities can be performed at various intensities; the suggested durations are based on the expected intensity of effort.

Finding Your Target Heart Rate

Source: Carole Casten, EdD, *Aerobics Today;* Peg Jordan, RN, Aerobics and Fitness Assoc. of America

The target heart rate is the heartbeat rate a person should have during aerobic exercise (such as running, fast walking, cycling, or cross-country skiing) to get the full benefit of the exercise for cardiovascular conditioning.

First, determine the intensity level at which one would like to exercise. A sedentary person may want to begin an exercise regimen at the 60% level and work up gradually to the 70% level. Athletes and highly fit individuals must work at the 85-95% level to receive benefits.

Second, calculate the target heart rate. One common way of doing this is by using the American College of Sports Medicine Method.

To obtain cardiovascular fitness benefits from aerobic exercise, it is recommended that an individual participate in an aerobic activity at least 3-5 times a week for 20-30 minutes per session, although cardiac patients and very sedentary individuals can obtain benefits with shorter periods (15-20 minutes). Generally, training changes occur in 4-6 weeks, but they can occur in as little as 2 weeks.

The American College of Sports Medicine Method

Using the American College of Sports Medicine Method to calculate one's target heart rate, an individual should subtract his or her age from 220, then multiply by the desired intensity level of the workout. Then divide the answer by 6 for a 10-second pulse count. (The 10-second pulse count is useful for checking whether the target heart rate is being achieved during the workout. One can easily check one's pulse—at the wrist or side of the neck—counting the number of beats in 10 seconds.)

For example, a 20-year-old wishing to exercise at 70% intensity would employ the following steps:

Maximum Heart Rate	$220 - 20 = 200$
Target Heart Rate	$200 \times .70 = 140$
10-second Pulse Count	$140 \div 6 = 23$

To work at the desired level of intensity, this 20-year-old would strive for a target heart rate of 140 beats per minute, or a 10-second pulse count of 23.

Arthritis

Source: Arthritis Foundation, 1330 West Peachtree Street, Atlanta, GA 30309; phone: (800) 283-7800

The name "arthritis" refers to more than 100 different diseases that cause pain, stiffness, swelling, and restricted movement in joints and connective tissue. The condition is usually chronic. More than 40 million people in the U.S. have some form of arthritis—about 23 million are women and 285,000 are children. The cause for most types of arthritis is unknown; scientists are studying the roles played by genetics, lifestyle, and the environment.

Symptoms of arthritis may develop either slowly or suddenly. A visit to the doctor is indicated when pain, stiffness, or swelling in a joint or difficulty in moving a joint persists for more than two weeks. The doctor analyzes the patient's symptoms, to see if they are consistent with those of arthritis. The doctor examines joint movement, looks for any swelling, and checks for skin rashes. Finally, the doctor may test the blood, urine, or joint fluid, or take X rays of the joints.

Medications to treat arthritis include drugs that relieve pain and swelling, such as analgesics, anti-inflammatory drugs, or glucocorticoids; disease modifiers, which tend to slow the disease process; and sleep medications, which promote deeper sleep and help relax muscles. Most treatment programs call for exercise; use of heat or cold; and joint-protection techniques (such as avoiding excess stress on joints, using assistive devices, and controlling weight). In some cases, surgery can help when other treatments fail.

Of the three most prevalent forms of arthritis, **osteoarthritis** is the most common, affecting more than 20 million Americans; it usually occurs after age 45. In this type, which is also called degenerative arthritis, the cartilage and bones deteriorate, causing pain and stiffness as bones rub against each other. It usually occurs in the fingers, knees, feet, hips, and back.

Fibromyalgia, another common type of arthritis, affects more than 2 million Americans and affects more women than men. In this form, widespread pain and tenderness occur in muscles and their attachments to the bone. Common symptoms include fatigue, disturbed sleep, stiffness, and psychological distress.

Rheumatoid arthritis, which also affects more than 2 million people in the U.S., is one of the most serious and disabling forms of the disease. In this type, which is also more common in women, the joints become inflamed because of an abnormality in the body's immune system. The chronic inflammation may then damage the cartilage and bone. The areas of the body that can be affected are the hands, wrists, feet, knees, ankles, shoulders neck, jaw, and elbows.

Other forms of arthritis and related conditions include lupus, gout, ankylosing spondylitis, and scleroderma; also related are bursitis and tendinitis, which result from injuring or overusing a joint.

Alternative Medicine

Source: National Center for Complementary and Alternative Medicine, National Institutes of Health (NIH)

Alternative medicine comprises a wide variety of healing philosophies, approaches, and therapies. It includes treatments and health care practices not widely taught in medical schools, not generally used in hospitals, and not usually reimbursed by health insurance companies. The NIH cautions people not to seek alternative therapies without the consultation of a licensed health care provider.

Some alternative therapies are described as **holistic**—meaning that the practitioner considers the whole person, including physical, mental, emotional, and spiritual aspects. Some therapies are known as **preventive**, meaning that the practitioner stresses preventing health problems before they arise.

People may use an alternative therapy alone, along with other alternative therapies, or in combination with more standard therapies. Worldwide, only about 10-30% of health care is provided by conventional practitioners; the remaining 70-90% involves alternative practices. An estimated 1 in 3 Americans uses some form of alternative medicine.

An advisory panel to the National Center for Complementary and Alternative Medicine (formerly called the Office of Alternative Medicine) at the National Institutes of Health classified 7 general fields of practice:

Alternative systems of medical practice range from self-care based on folk traditions to care given by practitioners according to established procedures. Included are such therapies as acupuncture, Ayurveda (India's traditional system of natural medicine), environmental medicine (treatment of certain illnesses believed to be caused by exposure to particular foods or chemicals), homeopathic medicine (use of remedies made from naturally occurring plant, animal, or mineral substances), Native American practices, naturopathic medicine (integration of traditional, natural therapeutics with modern scientific medicine), and traditional Oriental medicine.

Bioelectromagnetic applications explore how living things interact with electromagnetic fields. Such therapies include blue light treatment and artificial lighting, electroacupuncture, and electrostimulation.

Diet, nutrition, and lifestyle changes are intended to prevent illness, maintain good health, and reverse the effects of chronic disease. Examples include use of macrobiotics, nutritional supplements, and megavitamins.

Herbal medicine employs plants and plant products for pharmacological use. Some common plants used are echinacea, garlic, ginkgo biloba, ginseng, St. John's wort, and saw palmetto.

Manual healing uses touch and manipulation with the hands therapeutically. Some types are acupressure, chiropractic medicine, massage therapy, osteopathy, and reflexology.

Mind/body control explores the mind's ability to affect the body. Therapies include hypnosis, meditation, psychotherapy, support groups, tai chi, and yoga.

Pharmacological and biological treatments involve drugs and vaccines that are not accepted by mainstream medicine. These include anti-oxidizing agents, metabolic therapy, and oxidizing agents.

Alternative Health Services in the U.S., 1998

Source: Nutrition Business Journal

HEALTH CARE PRACTICE	Licensed practitioners	Lay or other practitioners	Total revenues[1]
Acupuncture	5,000	3,000	$695
Chiropractic	60,000	NA	14,500
Homeopathy	1,600	2,000	560
Massage therapy	30,000	130,000	6,850
Naturopathy	2,000	2,000	460
Osteopathy	39,000	0	NA
Traditional Oriental medicine	10,000	17,000	$3,200
TOTAL	**147,600**	**154,000**	**$26,265**

NA = Not available. (1) In millions of dollars.

Top-Selling Medicinal Herbs in the U.S., 1995-98

Source: Nutrition Business Journal

HERB	Sales in 1995[1]	Sales in 1997[1]	Sales in 1998[1]	% change 1995-98
Echinacea	$170	$270	$300	76
Garlic[2]	150	210	230	53
Ginkgo biloba	170	250	310	82
Ginseng	190	240	250	32
St. John's wort	10	210	290	2,800
Saw palmetto	$40	$100	$120	200
Combinations	900	1,180	1,380	53
All other	840	1,170	1,220	45
TOTAL	**$2,470**	**$3,630**	**$4,100**	**61**

(1) In millions of dollars. (2) Does not include nonmedicinal use.

Where to Get Help

Source: Based on Health & Medical Year Book. Copyright © by Collier Newfield, Inc.; additional data, World Almanac research

Listed here are some of the major U.S. and Canadian organizations providing information about good health practices generally, or about specific conditions and how to deal with them. (Canadian sources are identified as such.) Where a toll-free number is not available, an address is given when possible.

Some entries conclude with an e-mail address for the organization and/or an address for its Internet site, where you can also obtain useful information. In addition to these selected sites, there is a vast array of medical information on the Internet; however, it is very important to be certain that the source of information is reliable and accurate. Always check with a physician before embarking on any new health-related venture.

General Sources

Centers for Disease Control and Prevention Voice Information System
888-232-3228
Recorded information about public health topics, such as AIDS and Lyme disease. Also, you can request to talk with a CDC expert or have information faxed to you.
Website: http://www.cdc.gov

National Health Information Center
800-336-4797; in Maryland, 301-565-4167
Phone numbers for more than 1,000 health-related organizations in the United States. Printed materials offered.
E-mail: nhicinfo@health.org

National Institutes of Health
Bethesda, MD 20892
301-496-4000
Free information, including the latest research findings, on many diseases.
Website: http://www.nih.gov

Tel-Med
Check the phone book for local listings or call Tel-Med at 909-478-0330.
Recorded information on over 600 health topics. Sponsored by local medical societies, health organizations, or hospitals.
E-mail: telmed@ix.netcom.com

Aging
Website: http://www.tel-med.com

National Association of Area Agencies on Aging's Eldercare Locator Line
800-677-1116
Information and assistance on a wide range of services and programs including adult day-care and respite services, consumer fraud, hospital and nursing home information, legal services, elder abuse/protective services, Medicaid/Medigap information, tax assistance, and transportation.
Hours 9 AM-8 PM EST M-F.

National Institute on Aging
800-222-2225
Information and publications about disabling conditions, support groups, and community resources.
E-mail: niainfo@IKACC.com
Website: http://www.nih.gov/nia

AIDS

AIDS Clinical Trials Information Service
800-874-2572
Information on federally and privately sponsored clinical trials for patients with AIDS or HIV.
E-mail: actis@actis.org
Website: http://www.actis.org

Canadian AIDS Society
613-230-3580.
Written materials and referrals.
E-mail: CASinfo@cdnaids.ca

Centers for Disease Control and Prevention National AIDS/HIV Hotline
800-342-AIDS 24 hours;
in Spanish, 800-344-SIDA, M-F, 8 AM-2 AM
for the hearing impaired, 800-AIDS-TTY;
M-F, 10 AM-10 PM
Information on the prevention and spread of AIDS, along with referrals.
Website: http://www.ashastd.org

HIV-AIDS Treatment Information Service
800-HIV-0440
Treatment information to people with AIDS, their families, and health care providers.
E-mail: atis@hivatis.org
Website: http://www.hivatis.org

Alcoholism and Drug Abuse
Alcohol and Drug Helpline
800-821-4357, 24 hours
Referrals to local facilities

Alcoholics Anonymous
212-870-3400
Worldwide support groups for alcoholics. Check phone book for local chapters.
Website: http://www.alcoholics-anonymous.org

American Council on Alcoholism
800-527-5344
Treatment referrals and counseling for recovering alcoholics.
Website: http://www.aca-usa.org

National Clearinghouse for Alcohol and Drug Information
800-729-6686
Provides written materials on alcohol and drug-related subjects.
Website: http://www.health.org

National Council on Alcoholism and Drug Dependence Hopeline
800-622-2255
An answering machine for callers to request information.

National Health Lines
800-262-2463
Answers questions on substance abuse and provides referrals to treatment centers. Operates 24 hours.
Website: http://www.drughelp.org

Alzheimer's Disease
Alzheimer's Association
800-272-3900
Gives referrals to local chapters and support groups; offers information on publications available from the association.
E-mail: info@alz.org
Website: http://www.alz.org

Alzheimer's Society of Canada
20 Eglinton Ave., W., Suite 1200
Toronto, ON M4R 1K8
416-488-8772
Gives phone numbers for local support chapters. Publishes support materials.
E-mail: info@alzheimer.ca
Website: http://www.alzheimer.ca

Amyotrophic Lateral Sclerosis
ALS Association
800-782-4747; in the San Fernando Valley, 818-880-9007
Information about ALS (Lou Gehrig's Disease) and referrals to ALS specialists, local chapters and support groups.
Website: http://www.alsa.org

Arthritis
Arthritis Foundation
800-283-7800
Information, publications, and referrals to local groups.
Website: http://www.arthritis.org

Arthritis Society (Canada)
393 University Ave., Suite 1700
Toronto, ON M5G 1E6
416-979-7228; in Ontario only, 800-321-1433
Phone numbers for local chapters.
E-mail: info@arthritis.ca
Website: http://www.arthritis.ca

National Arthritis and Musculoskeletal and Skin Diseases Information Clearinghouse
301-495-4484
Subject searches and resource referrals.
Website: http://www.nih.gov/niams

Asthma and Allergies
See also *Lung Diseases*
Asthma and Allergy Foundation Information Clearinghouse
800-7-ASTHMA
Written information.

American Academy of Allergy, Asthma, and Immunology Referral Line
800-822-ASMA, 24 hours
Written materials on asthma and allergies.
Website: http://www.aaaai.org

Blindness and Eye Care
Canadian National Institute for the Blind
1929 Bayview Avenue
Toronto, ON M4G 3E8
416-486-2500 or contact your local chapter. National office offers training and library with braille books and audiotapes. Local chapters provide core services: orientation in mobility, sight enhancement, counseling, referrals, career aid, technology services.
Website: http://www.cnib.ca

Foundation Fighting Blindness
800-683-5555; in Maryland, 410-785-1414; for the hearing impaired, 800-683-5551
Answers questions about retinal degenerative diseases; has written materials.
Website: http://www.blindness.org

Library of Congress National Service for the Blind and Physically Handicapped
800-424-9100; in Spanish, 800-345-8901; in Washington, DC, 202-707-5100
Information on libraries that offer talking books and books in braille.
Website: http://lcweb.loc.gov/nls/nls.html

National Association for Parents of the Visually Impaired
800-562-6265
Support and information for parents of individuals who are visually impaired.
Website: http://www.spedex.com/napri

Blood Disorders
Cooley's Anemia Foundation
800-522-7222
Information on patient care and support groups; makes referrals to local chapters.
E-mail: ncaf@aol.com
Website: http://www.thalassemia.org

Sickle Cell Disease Association of America
800-421-8453; in California, 310-216-6363
Genetic counseling and information packet.
E-mail: lascdaa@aol.com
Website: http://www.sicklecelldisease.org

Burns
Phoenix Society
800-888-2876
Counseling for burn survivors and information on self-help services for burn survivors and their families.
E-mail: info@phoenix-society.org
Website: http://www.phoenix-society.org

Cancer
American Cancer Society
800-ACS-2345
Publications and information about cancer and coping with cancer; makes referrals to local chapters for support services.
Website: http://www.cancer.org

Canadian Cancer Information Service
888-939-3333, in Canada only, 9 AM-6PM, Mon.-Fri.
Information on prevention, treatment, drugs, clinical trails, local services.

National Cancer Institute's Cancer Information Service
800-4-CANCER
Information about clinical trials, treatments, symptoms, prevention, referrals to support groups, and screening.
Website: http://www.nci.nih.gov

Y-Me Breast Cancer Support Program
800-221-2141, 24 hours; in Illinois, 312-986-8228
Information and literature on breast cancer, counseling, and referrals.
Website: http://www.y-me.org

Cerebral Palsy

Ontario Federation for Cerebral Palsy
1630 Lawrence Avenue West, Suite 104
Toronto, ON M6L 1C5
416-244-9686
Canada does not have a national
cerebral palsy organization, but the
provincial organizations offer information
on housing, services, and coping with
life, and each one will provide contact
numbers for the others.
E-mail: ofcp@ofcp.on.ca
Website: http://www.ofcp.on.ca

United Cerebral Palsy Associations
800-USA-5UCP; in Washington, DC, 202-
776-0406
Written materials.
Website: http://www.ucpa.org

Child Abuse

See *Domestic Violence*

Children

American Academy of Pediatrics
847-228-5005
Child-care publications and materials;
referrals to pediatricians.
Website: http://www.aap.org

**Childhelp's USA National Child Abuse
Hotline**
800-4-A-CHILD
Crisis intervention, professional
counseling, referrals to local groups
and to shelters for runaways, and
literature. Operates 24 hours.

**National Center for Missing and
Exploited Children**
800-843-5678; for the hearing impaired,
800-826-7653
Hotline for reporting missing children and
sightings of missing children.
Website: www.missingkids.org

Chronic Fatigue Syndrome

CFIDS Association of America
800-442-3437
Literature and a list of support groups.
E-mail: info@cfids.org
Website: http://www.cfids.org

Crisis

National Runaway Switchboard
800-621-4000
Crisis intervention and referrals for
runaways. Runaways can leave messages
for parents, and vice versa.
Operates 24 hours.
Website: http://nrscrisisline.org

Cystic Fibrosis

Canadian Cystic Fibrosis Foundation
416-485-9149; in Canada only,
800-378-2233
Information and brochures; makes referrals
to local chapters.
Website: http://www.ccff.ca

Cystic Fibrosis Foundation
800-FIGHT-CF
Answers questions and offers literature
and referrals to local clinics.
Website: http://www.cff.org

Diabetes

American Diabetes Association
800-342-2383; in Virginia and Washington,
DC, 703-549-1500
Information about diabetes, nutrition,
exercise, and treatment; offers referrals
to diabetes specialists.
Website: http://www.diabetes.org

Canadian Diabetes Association
15 Toronto Street, Suite 800, Toronto, ON
M5C 2E
3416-363-3373; in Ontario only, 800-361-
1306
Information and publications.
Website: http://www.diabetes.ca

Juvenile Diabetes Foundation Hotline
800-223-1138 or 800-533-2873
Answers questions, provides literature
(some in Spanish).
Offers referrals
to local chapters, physicians, and clinics.
Website: http://www.jdfcure.org

Digestive Diseases

**Crohn's and Colitis Foundation of
America**
800-932-2423; in New York, 212-685-3440
Educational materials; offers referrals to
local chapters, which can provide referrals
to support groups and physicians.
Website: http://www.ccfa.org

**Crohn's and Colitis Foundation of
Canada**
21 St. Clair Avenue East, Suite 301,
Toronto, ON M4T 1L9
416-920-5035; in Canada only, 800-387-
1479
Will send out educational materials upon
request.
Website: http://www.ccfc.ca

Domestic Violence

**National Council on Child Abuse and
Family Violence**
800-222-2000;
in Washington, DC, 202-429-6695
A recording provides toll-free numbers to
call for information or referrals.
E-mail: nccafv@aol.com
Website: http://www.nccafv.org

Down Syndrome

National Down Syndrome Congress
800-232-6372; in Georgia, 770-604-9500
Answers questions on all aspects of Down
syndrome. Provides referrals.
E-mail: ndsccenter@aol.com
Website: http://www.ndscenter.org

National Down Syndrome Society
800-221-4602; in New York City, 212-460-
9330
Information; referrals to local programs for
newborns.
Website: http://www.ndss.org

Drug Abuse

See *Alcoholism and Drug Abuse*

Dyslexia

International Dyslexia Association
800-ABCD-123; in Maryland, 410-296-0232
Information on testing, tutoring, and
computers used to aid people with dyslexia
and related disorders.
E-mail: info@interdys.org

Eating Disorders

**National Association of Anorexia
Nervosa and Associated Disorders**
Box 7, Highland Park, IL 60035
847-831-3438
Written materials, referrals to health
professionals treating eating disorders,
telephone counseling, offers 3 self-help
groups and information on how to set up a
self-help group.
E-mail: anad20@aol.com
Website: http://www.anad.org

Endometriosis

Endometriosis Association
800-992-ENDO; in Canada, 800-426-2END
An answering machine for callers to request
information.

Epilepsy

**Epilepsy and Seizure Disorder Service
at the Epilepsy Foundation of America**
800-332-1000
Information and referrals to local chapters.
Website: http://www.efa.org

Food Safety and Nutrition

**Meat and Poultry Hotline of the U.S.
Department of Agriculture's Food,
Safety, and Inspection Service**
800-535-4555
Information on prevention of food-borne
illness and the proper handling, preparation,
storage, labeling, and cooking of meat,
poultry, and eggs.
Website: http://www.fsis.usda.gov

**FDA Center for Food Safety and Applied
Nutrition Outreach & Information Center**
800-FDA-4010; in Washington, DC, 202-
205-4314
Information on how to buy and use food
products and on their proper handling and
storage, women's health, and cosmetics &
colors. Callers may speak to food specialists,
Mon. through Fri., 10 AM to 4 PM (EST).
Website: http://www.cfsan.fda.gov

Headaches

National Headache Foundation
800-843-2256
Literature on headaches and treatment.
Website: http://www.headaches.org

Heart Disease and Stroke

American Heart Association
800-242-8721
Information, publications, and referrals to
organizations.
Website: http://www.americanheart.org

**National Institute of Neurological
Disorders and Stroke**
800-352-9424
Literature and information.
Website: http://www.ninds.nih.gov

National Stroke Association
800-787-6537
Information on support networks for stroke
victims and their families; referrals to local
support groups.
Website: http://www.stroke.org

Hospices

Children's Hospice International
800-242-4453; in Virginia, 703-684-0330
Information, referrals to children's hospices.
E-mail: chiorg@aol.com
Website: http://www.chionline.org

Hospice Education Institute Hospicelink
800-331-1620; in Connecticut, 860-767-
1620
Information, referrals to local programs.
E-mail: hospiceall@aol.com
Website: http://www.hospiceworld.org

Huntington's Disease

Huntington's Disease Society of America
800-345-4372; in New York, 212-242-1968
Information and referrals to physicians and
support groups.
Website: http://www.hdsa.org

Huntington Society of Canada
P.O. Box 1269, 13 Water Street North,
Cambridge, ON N1R 7G6
519-622-1002
Information, including telephone numbers of
local services; publications and referrals.
E-mail: info@hsc-ca.org
Website: http://www.hsc-ca.org

Impotence

Impotence Information Center
800-843-4315
Information on treatment of impotence,
incontinence, and prostate problems.
Website: http://www.visitams.com

Impotence World Institute Hotline
800-669-1603
Written materials, physician referrals, and
telephone numbers of local Impotents
Anonymous chapters.
Website: http://www.impotenceworld.org

Kidney Diseases

Kidney Foundation of Canada
514-369-4806; in Canada only, 800-361-
7494
Educational materials and general
information.
Website: http://www.kidney.ca

**National Kidney and Urologic Diseases
Information Clearinghouse**
3 Information Way
Bethesda, MD 20892-3580
301-654-4415
Information, referrals to organizations.
Website: http://www.niddk.nih.gov

National Kidney Foundation
800-622-9010
Information and referrals.
Website: http://www.kidney.org

Lead Exposure

National Lead Information Center
800-LEAD-FYI
Recommendations (in English and Spanish)
for reducing a child's exposure to lead.
Referrals to state and local agencies.
Website: http://www.epa.gov/lead/nlic

Liver Diseases

American Liver Foundation
800-223-0179; in NJ, 973-256-2550
Information on hepatitis, liver disease, and
gallbladder disease.
Website: http://www.liverfoundation.org

Lung Diseases

See also *Asthma and Allergies*
American Lung Association
Check the phone book for local listings or call the national office at 800-LUNG-USA for automatic connection to the office nearest you. Answers questions about asthma and lung diseases; publications and referrals.
Website: http://www.lungusa.org
Lung Line Information Service at the National Jewish Medical and Research Center
800-222-LUNG; in Denver, 303-355-LUNG
Answers questions on asthma, emphysema, allergies, smoking, and other respiratory and immune system disorders.
Website: http://www.njc.org

Lupus

Lupus Foundation of America
800-558-0121; in Colorado, 301-670-9292
Sends information to those who leave name and address on answering machine.

Lyme Disease

Lyme Disease Foundation
800-886-LYME, 24 hours
Written information; doctor referrals.

Mental Health

National Clearinghouse on Family Support and Children's Mental Health
800-628-1696, 24 hours
Publications, computerized databank, and state-by-state resource file.
National Depressive and Manic Depressive Association
800-826-3632
Support for patients and families, provides publications, and makes referrals to affiliated organizations.
Website: http://www.ndmda.org
National Foundation for Depressive Illness
P.O. Box 2257, NY, NY 10116
800-248-4344, 24 hours
Recorded message describing the symptoms of depression and offering an address for more information and physician referral.
National Institute of Mental Health
6001 Executive Blvd., Room 8184, MSC 9663, Bethesda, MD 20892-9663
301-443-4513
Information on a range of topics, from children's mental disorders to schizophrenia, depression, eating disorders, and others.
Website: http://www.nimh.nih.gov
National Mental Health Association
800-969-6642
Referrals to mental health groups.

Multiple Sclerosis

Multiple Sclerosis Society of Canada
416-922-6065
Counseling, literature, and referrals to local chapters.
Website: http://www.mssoc.ca
National Multiple Sclerosis Society
800-344-4867
Information about local chapters.
Website: http://www.nmss.org

Muscular Dystrophy

Muscular Dystrophy Association
800-572-1717
Written materials on 40 neuromuscular diseases, including muscular dystrophy. Will give information over the phone about such matters as MDA clinics, support groups, summer camps, and wheelchair purchase assistance.
Website: http://www.mdausa.org

Nutrition

See *Food Safety and Nutrition*

Organ Donation

Living Bank
800-528-2971, 24 hours
A registry and referral service for people wanting to commit organs to transplantation or research.
Website: http://www.thelivingbank.org

Osteoporosis

National Osteoporosis Foundation
800-223-9994, in Washington, DC, 202-223-2226

Information packet available on request.
Website: http://www.nof.org

Pain

National Chronic Pain Outreach Association
540-862-9437
Information packet available on request.

Parkinson's Disease

National Parkinson Foundation
800-327-4545; in Florida, 800-433-7022; in Miami, 305-547-6666
Answers questions, makes physician referrals, and provides written information in English and Spanish.
E-mail: mailbox@npf.med.miami.edu
Website: http://www.parkinson.org
Parkinson Foundation of Canada
800-565-3000
Information; referrals to support groups.
Website: http://www.parkinson.ca

Plastic Surgery

Plastic Surgery Information Service
800-635-0635
Referrals to board-certified plastic surgeons in the U.S. and Canada; general information.
Website: http://www.plasticsurgery.org

Polio

International Polio Network
4207 Lindell Blvd., #110
St. Louis, MO 63108
314-534-0475
Information on coping with the late effects of polio; referrals to other organizations.
E-mail: gini_intl@msn.com
Website: http://www.postpolio.org

Prostate Problems

Prostate Information Line
800-543-9632
Advice on treatment.

Rare Disorders

National Organization for Rare Disorders
800-999-6673
Information on diseases and networking programs; referrals to organizations for specific disorders.
Website: http://www.rarediseases.org

Rehabilitation

National Rehabilitation Information Center
800-34-NARIC; in Maryland, 301-588-9284
Research referrals and information on rehabilitation issues.
Website: http://www.naric.com

Scleroderma

United Scleroderma Foundation
800-722-4673
Referrals to local support groups and treatment centers, as well as information on scleroderma and related skin disorders.
E-mail: sfinfo@scleroderma.org
Website: http://www.scleroderma.org

Sexually Transmitted Diseases

See also *AIDS*
National STD Hotline
800-227-8922
Information; confidential referrals.
Website: http://www.ashastd.org

Sjogren's Syndrome

Sjogren's Syndrome Foundation
800-475-6473; in New York, 516-933-6365
Provides an answering machine for callers to request treatment literature.
Website: http://www.sjogren.com

Skin Problems

National Psoriasis Foundation
800-723-9166
Information and referrals.
Website: http://www.psoriasis.org

Speech and Hearing

American Speech-Language-Hearing Association Helpline
800-638-8255 (also TTY); in Maryland, 301-897-8628
Materials on speech and language disorders and hearing impairment; referrals.
Website: http://www.asha.org

Canadian Hard of Hearing Association
2435 Holly Lane, Suite 205
Ottawa, ON K1V 7P2
613-526-1584; TTY 613-526-2692
Publications; answers general questions.
E-mail: chhanational@chha.ca
Website: http://www.chha.ca
Dial a Hearing Screening Test
800-222-EARS
Answers questions on hearing problems. Makes referrals to local telephone numbers for a two-minute hearing test. Also to ear, nose, and throat specialists and to organizations that can provide specialized ear and hearing aid information. 9 AM-5 PM EST
E-mail: dahst@aol.com
Hearing Aid Helpline
800-521-5247
Information and distributes a directory of hearing aid specialists certified by the International Hearing Society.
Website: http://www.hearingihs.org
National Center for Stuttering
800-221-2483; in New York, 212-532-1460
Information on stuttering in all age groups.
Website: http://www.stuttering.com
Stuttering Foundation of America
800-992-9392
Referrals to speech pathologists; resource lists, publications.
E-mail: stutter@vantek.net
Website: http://www.stutterssa.org

Spinal Injuries

National Spinal Cord Injury Association
800-962-9629; in Maryland, 301-588-6959
Peer counseling; referrals to local chapters and other organizations.
Website: http://www.spinalcord.org
National Spinal Cord Injury Hotline
800-526-3456
Written materials on spinal cord injuries; referrals to organizations and support groups.
Website: http://www.scihotline.org

Stroke

See *Heart Disease and Stroke*
Sudden Infant Death Syndrome
American Sudden Infant Death Syndrome Institute
800-232-SIDS; in Georgia, 800-847-7437
Answers questions; literature; referrals to other organizations.
E-mail: prevent@sids.org
Website: http://www.sids.org
National SIDS Foundation
800-221-SIDS; in Maryland, 410-653-8226
Literature on medical information, referrals, and support groups.
Website: http://www.sidsalliance.org

Tourette Syndrome

Tourette Syndrome Association
800-237-0717; in New York, 718-224-2999
Printed information.
E-mail: tourette@ix.netcom.com

Urinary Incontinence

National Association for Continence
800-BLADDER
Information on bladder control, services available for incontinence, and assistive devices.
Website: http://www.nafc.org
Simon Foundation for Continence
800-23-SIMON
Support and literature on incontinence.

Women's Health

National Women's Health Network
514 10th Street NW, Suite 400
Washington, DC 20004
202-347-1140; 202-628-7814 (clearinghouse)
Information and referrals on more than 70 women's health concerns.
Website: http:// www.womenshealthnetwork.org
National Women's Health Resource Center
5255 Loughboro Road
Washington, DC 20016
202-537-4015
A national clearinghouse for women's health information.

POSTAL INFORMATION

(Based on information available as of Oct. 1999)

U.S. Postal Service

The Postal Reorganization Act, creating a government-owned postal service under the executive branch and replacing the old Post Office Department, was signed into law by Pres. Richard Nixon, Aug. 12, 1970. The service officially came into being on July 1, 1971.

The U.S. Postal Service is governed by an 11-person Board of Governors. Nine of the members are appointed by the president with Senate approval. These 9, in turn, choose a postmaster general. The board and the postmaster general choose the 11th member, who serves as deputy postmaster general. An independent Postal Rate Commission of 5 members, appointed by the president, reviews and rules on proposed postal rate increases submitted by the Board of Governors.

In 1998, the U.S. Postal Service, with 792,000 career employees and 38,000 post offices, handled 1.98 billion pieces of mail, up 3.7% from the year before.

U.S. Domestic Rates

Domestic rates apply to the U.S., to its territories and possessions, and to APOs and FPOs. Many changes for domestic postal rates and fees took effect in Jan. 1999.

First Class

First Class includes written matter such as letters, postal cards, and postcards (private mailing cards), plus all other matter wholly or partly in writing, whether sealed or unsealed, except book manuscripts, periodical articles and music, manuscript copy accompanying proofsheets or corrected proofsheets of the same, and the writing authorized by law on matter of other classes. Also included: matter sealed or closed against inspection, bills, and statements of accounts.

Mailing written letters and matter sealed against inspection costs 33¢ for first ounce or fraction, 22¢ for each additional ounce or fraction up to and including 13 oz. U.S. Postal Service cards and private postcards alike cost 20¢ single, 40¢ double. Presort and automation-compatible mail can qualify for lower rates if certain piece minimums, mailing permits, and other requirements are met.

Express Mail

Express Mail Service is available for any mailable article up to 70 lb, and guarantees delivery between major U.S. cities within a specified time frame or your money back. Articles received by the acceptance time authorized by the postmaster at a postal facility offering Express Mail are delivered by 3 PM the next day to some locations or by noon the next day to other destinations. Or, if you prefer, you can pick up the package yourself, as early as 10 AM the next business day.

Second-day service is available to locations that are not on the Next Day Delivery Network. All rates include insurance, shipment receipt, and record of delivery at the destination post office.

The basic rate for Express Mail weighing up to 8 oz is $11.75. Consult postmaster for other Express Mail Services and rates. The Postal Service will refund, upon application to originating office, the postage for any Express Mail shipments not meeting the service standard, except for those delayed by strike or work stoppage, delay or cancellation of flights, or government action beyond the control of the Postal Service.

Periodicals

Periodicals include newspapers and magazines.

For the general public, the applicable Standard Mail or First Class postage is paid for periodicals.

For publishers, rates vary according to (1) whether item is sent to same county, (2) percentage of reading and advertising matter, (3) weight, (4) distance, (5) level of presort, (6) automation compatibility.

Standard Mail (A)

Standard Mail (A) is limited to 16 ounces and bulk mailings (at least 200 pieces or 50 lbs.) of such items as solicitations, newsletters, and advertising materials.

For mailing Standard Mail (A) in bulk (at least 200 pieces or 50 lb of such items as solicitations, newsletters, advertising materials, books, and cassettes, each item of which individually weighs less than 1 lb.), the minimum rate per piece, basic, non-letter, is $0.304 for pieces weighing 3.3087 oz or less. For pieces weighing more than 3.3087 oz, the rate is $0.164 per piece plus $0.677 per pound. Contact your post office for the discounts offered for presorted, letter-shaped, destination entry, and automation-compatible mail.

Separate rates are available for some nonprofit organizations provided with a permit. The permit requires a one-time imprint fee of $100 plus an annual (calendar year) fee of $100.

Parcel Post—Standard Mail (B)

Any matter that weighs 16 oz or more and is not included in First Class or Periodicals goes as Parcel Post, or Standard Mail (B). The post office determines Parcel Post charges according to the weight of the package in pounds and the zone distance it is being shipped. All fractions of a pound are counted as a full pound.

Forwarding Addresses

To obtain a forwarding address, the mailer must write on the envelope or cover the words "Address Correction Requested." The destination post office then will check for a forwarding address on file and provide it for 50¢ per manual correction, 20¢ per automated correction.

Priority Mail Flat Rate

The most expeditious handling and transportation available will be used for fast delivery by "Priority Mail." If the item fits into a special Postal Service flat-rate envelope, the rate is $3.20 regardless of weight.

Pickup service by the Postal Service for Priority Mail is available for an additional $8.25 per stop (not per package).

Priority Mail by Weight

Priority Mail may include packages up to 70 lb and not over 108 in. in length and girth combined, whether sealed or unsealed, including written and other First Class material. Rates are as follows (fractions of a pound are rounded up to the next full pound).

Up to 2 lb	3 lb	4 lb	5 lb
$3.20	$4.30	$5.40	$6.50

For parcels over 5 lb, rates by zone apply. The mileage between the specific geographic locations of 3-digit ZIP codes determines the zone number used. The mileage range by zone number is: Zone 1—up to 50 mi; 2—51 to 150 mi; 3—151 to 300 mi; 4—301 to 600 mi; 5—601 to 1,000 mi; 6—1,001 to 1,400 mi; 7—1,401 to 1,800 mi; 8—over 1,800 mi.

Parcels weighing less than 15 lb and measuring over 84 in. in length or girth, but not exceeding 108 in. in length and girth combined, cost the same as a 15-lb parcel mailed to the same zone.

Special Handling

Parcel Post parcels can be given special, expedited handling upon payment of the following surcharge: up to 10 lb, $5.40; over 10 lb, $7.50. Such parcels must be marked for "Special Handling."

Bound Printed Matter Rates
(single-piece zone rate)

Weight (lbs)	Local	1&2	3	ZONES 4	5	6	7	8
1.5 .	$1.14	$1.54	$1.57	$1.63	$1.72	$1.81	$1.92	$2.02
2 . . .	1.16	1.57	1.61	1.69	1.81	1.93	2.08	2.21
2.5 .	1.18	1.60	1.66	1.76	1.90	2.06	2.24	2.40
3 . . .	1.20	1.63	1.70	1.82	1.99	2.18	2.40	2.60
3.5 .	1.22	1.66	1.74	1.88	2.08	2.30	2.56	2.79
4 . . .	1.24	1.70	1.79	1.94	2.18	2.42	2.72	2.98
4.5 .	1.26	1.73	1.83	2.01	2.27	2.55	2.88	3.17
5 . . .	1.28	1.76	1.88	2.07	2.36	2.67	3.05	3.37
6 . . .	1.31	1.82	1.96	2.20	2.54	2.92	3.37	3.75
7 . . .	1.35	1.89	2.05	2.32	2.73	3.16	3.69	4.14
8 . . .	1.39	1.95	2.14	2.45	2.91	3.41	4.01	4.52
9 . . .	1.43	2.02	2.22	2.57	3.10	3.65	4.33	4.91
10 . . .	1.47	2.08	2.31	2.70	3.28	3.90	4.65	5.29
11 . . .	1.51	2.14	2.40	2.83	3.46	4.15	4.97	5.68
12 . . .	1.55	2.21	2.48	2.95	3.65	4.39	5.29	6.06
13 . . .	1.59	2.27	2.57	3.08	3.83	4.64	5.61	6.45
14 . . .	1.63	2.34	2.66	3.20	4.02	4.88	5.93	6.83
15 . . .	1.67	2.40	2.75	3.33	4.20	5.13	6.26	7.22

(Includes both catalogs and similar bound printed matter.)

(Bound printed matter must weigh at least 1 lb and not more than 15 lb. Bound printed matter includes catalogs, directories, and books.)

Domestic Mail Special Services

Registry—Only matter prepaid with postage at First Class postage rates may be registered. Stamps or meter stamps must be attached. The face of the article must be at least 5" long, 3½" high. The mailer is required to declare the value of mail presented for registration.

Registered Mail

Declared Value	Fee[1]
$0.00	$6.00[2]
$0.01 to $100	6.20
$100.01 to $500	6.75
$500.01 to $1,000	7.30
$1,000.01 to $2,000	7.85
$2,000.01 to $3,000	8.40
$3,000.01 to $4,000	8.95
$4,000.01 to $5,000	9.50
$5,000.01 to $6,000	10.05
$6,000.01 to $7,000	10.60
$7,000.01 to $8,000	11.15
$8,000.01 to $9,000	11.70
$9,000.01 to $10,000	12.25

Consult postmaster for registry fees above $10,000.

(1) Fee for articles with declared value over $0.00 includes insurance; fee is in addition to postage. (2) Without insurance.

C.O.D.: Unregistered: Applicable to First Class, Priority Mail, Standard B, and Express Mail matter. Such mail must be sent as bona fide orders or be in conformity with agreements between senders and addressees. **Registered:** For details, consult postmaster.

Insurance: Applicable to Standard Mail matter. Matter for sale addressed to prospective purchasers who have not ordered it or authorized its sending cannot be insured.

Insured Mail Fees

Declared Value	Fee
$0.01 to $50	$0.85
$50.01 to $100	Add $.95 per $100 or fraction thereof over $100 to $5,000.
$100.01 to $5,000.00	2.75
$200.01 to $300	3.70
$300.01 to $400	4.65
$400.01 to $500	5.60
$500.01 to $600	6.55

(Liability for insured mail is limited to $5,000.) Bulk discount of $.40 per piece is available upon meeting volume requirements. See Postmaster for further details. For Express Mail, insurance is included up to $500. Add $.95 per $100 or fraction thereof over $500 to $5,000.

Certified mail: This service is available for any matter having no intrinsic value on which First Class or Priority Mail postage is paid. A receipt is furnished at the time of mailing, and evidence of delivery is obtained. The basic fee is $1.40 in addition to regular postage. Return receipt and restricted delivery are available upon payment of additional fees. No indemnity.

Special Standard Mail
(limit 70 lbs)

Applies only to: books of at least 8 printed pages consisting wholly of reading matter or scholarly bibliography, or reading matter with incidental blank spaces for notations and containing no advertising matter other than incidental announcements of books; 16-mm or narrower-width films in final form and catalogs of such films of 24 pages or more (at least 22 of which are printed), except films and film catalogs sent to or from commercial theaters; printed music in bound or sheet form; printed objective test materials; sound recordings, playscripts, and manuscripts for books, periodicals, and music; printed educational reference charts; loose-leaf pages and binders consisting of medical information for distribution to doctors, hospitals, medical schools, and medical students; computer-readable media containing prerecorded information and guides for use with such media. Package must be marked "Special Standard."

The rates are: 1st pound or fraction, $1.13; if 500 pieces or more presorted to 5-digit ZIP code, $.64; or $.95 if 500 pieces or more are presorted to Bulk Mail Centers. Through 7 lbs, each additional pound or fraction is $.45; each additional pound after that costs $.28.

Library Mail
(limit 70 lbs)

Applies to books, printed music, bound academic thesis, periodicals, sound recordings, museum materials, and other library materials mailed between schools, colleges, universities, public libraries, museums, veteran and fraternal organizations, and nonprofit religious, educational, scientific, and labor organizations or associations. Also included are slides, transparencies, sound recordings, museum materials and specimens, scientific and mathematical kits, or catalogs of the above-referenced materials mailed to or from schools, universities, public libraries, or museums, and to or from nonprofit religious, educational, scientific, philanthropic, veterans, or fraternal organizations. All packages must be marked "Library Mail." The rates are identical to those for Special Standard Mail.

Delivery Confirmation

Applies to Priority Mail and Standard Mail B. Provides the mailer information about the date and time an article was delivered and, if delivery was attempted, but not successful, the date and time of the delivery attempt. Information is available electronically, for mailers who attach barcodes, and manually on the Internet (http://www.usps.com) or through a toll-free telephone number (800-222-1811) for retail purchasers. The fees are:

Mail Type	Manual	Electronic
Priority	$.35	$.00
Standard B	$.60	$.25

Parcel Post Rate Schedule

(Inter BMC/ASF ZIP codes only, machinable parcels, no discount, no surcharge)

Weight up to but not exceeding—(pounds)	1 and 2	3	4	5	6	7	8
2.	$3.15	$3.15	$3.15	$3.15	$3.15	$3.15	$3.15
3.	3.59	3.90	4.25	4.25	4.25	4.25	4.25
4.	3.73	4.16	4.91	5.35	5.35	5.35	5.35
5.	3.86	4.39	5.33	6.45	6.45	6.45	6.45
6.	3.99	4.62	5.71	7.10	7.40	7.60	8.15
7.	4.11	4.82	6.07	7.72	8.35	8.75	9.85
8.	4.24	5.01	6.38	8.26	9.30	9.90	11.55
9.	4.33	5.19	6.71	8.76	10.25	11.05	13.25
10.	4.45	5.36	6.99	9.23	10.92	12.20	14.95
11.	4.54	5.53	7.27	9.66	11.47	13.30	16.10
12.	4.64	5.68	7.53	10.06	11.97	14.30	17.35
13.	4.73	5.81	7.77	10.44	12.44	15.17	18.65
14.	4.82	5.97	8.01	10.80	12.89	15.74	19.90
15.	4.90	6.10	8.24	11.13	13.31	16.28	21.15
16.	4.98	6.23	8.45	11.45	13.70	16.77	21.85
17.	5.07	6.34	8.66	11.74	14.08	17.25	22.49
18.	5.14	6.46	8.85	12.02	14.42	17.69	23.10
19.	5.23	6.58	9.04	12.29	14.76	18.12	23.67
20.	5.29	6.68	9.20	12.54	15.07	18.52	24.21
21.	5.36	6.80	9.37	12.79	15.38	18.90	24.72
22.	5.43	6.89	9.54	13.02	15.66	19.26	25.21
23.	5.50	7.01	9.71	13.23	15.93	19.60	25.67
24.	5.55	7.10	9.85	13.45	16.19	19.94	26.12
25.	5.62	7.19	10.01	13.64	16.44	20.24	26.54

Consult postmaster for pieces greater than 25 lbs.

Postal Union Mail Special Services

Registration: Available to practically all countries. Fee $4.85. The maximum indemnity payable—generally only in case of complete loss (of both contents and wrapper)—is $42.30. To Canada only, the fee is $4.95, providing indemnity for loss up to $100, $5.40 for loss up to $500, and $5.85 for loss up to $1,000.

Return receipt: Shows to whom and date delivered; $1.25.

Special delivery: As of June 8, 1997, this service was no longer available.

Marking: An article that is intended for special delivery service must have affixed to the cover near the name of the country of destination "EXPRES" (special delivery) label, obtainable at the post office, or the word "EXPRES" (special delivery) may be marked on the cover boldly in red letters.

Air mail: Available daily to practically all countries.

Prepayment of replies from other countries: A mailer who wishes to prepay a reply by letter from another country may do so by sending one or more international reply coupons, available at U.S. post offices. These should be accepted in any country in exchange for stamps to prepay an air mail letter of the first unit of weight to the U.S.

Insurance: Available to many countries for loss of or damage to items paid at parcel post rate. Consult postmaster for indemnity limits for individual countries.

Limit of indemnity Not over	Fees	
	Canada[1]	All other countries[1]
$50	$0.75	$1.60
100	1.60	2.45
200	2.50	3.35
300	3.40	4.25
400	4.30	5.15
500	5.20	6.05
600	6.10	6.95
700		7.40
800		7.85
900		8.30
1,000		8.75
1,100		9.20
1,200		9.65

(1) Not all countries insure items up to the amounts listed in the table. Canada does not insure items for more than $600.

Restricted delivery: Available to many countries for registered mail; some limitations. Fee: $2.75.

Post Office-Authorized 2-Letter State Abbreviations

The abbreviations below are approved by the U.S. Postal Service for use in addresses for the 50 states, the District of Columbia, Puerto Rico, the U.S. Virgin Islands, American Samoa, Guam, and certain other areas in the Pacific.

Alabama	AL	Hawaii	HI	Missouri	MO	Pennsylvania	PA
Alaska	AK	Idaho	ID	Montana	MT	Puerto Rico	PR
American Samoa	AS	Illinois	IL	Nebraska	NE	Rhode Island	RI
Arizona	AZ	Indiana	IN	Nevada	NV	South Carolina	SC
Arkansas	AR	Iowa	IA	New Hampshire	NH	South Dakota	SD
California	CA	Kansas	KS	New Jersey	NJ	Tennessee	TN
Colorado	CO	Kentucky	KY	New Mexico	NM	Texas	TX
Connecticut	CT	Louisiana	LA	New York	NY	Utah	UT
Delaware	DE	Maine	ME	North Carolina	NC	Vermont	VT
Dist. of Col.	DC	Marshall Islands[1]	MH	North Dakota	ND	Virgin Islands	VI
Federated States of Micronesia[1]	FM	Maryland	MD	Northern Mariana Is.	MP	Virginia	VA
		Massachusetts	MA	Ohio	OH	Washington	WA
Florida	FL	Michigan	MI	Oklahoma	OK	West Virginia	WV
Georgia	GA	Minnesota	MN	Oregon	OR	Wisconsin	WI
Guam	GU	Mississippi	MS	Palau [1]	PW	Wyoming	WY

(1) Although an independent nation, this country is currently subject to domestic rates and fees.

Canadian Province and Territory Postal Abbreviations

Source: Canada Post

Alberta	AB	Newfoundland and Labrador	NF	Prince Edward Island	PE
British Columbia	BC	Northwest Territories/Nunavut	NT	Quebec	QC[1]
Manitoba	MB	Nova Scotia	NS	Saskatchewan	SK
New Brunswick	NB	Ontario	ON	Yukon Territory	YT

(1) PQ is also acceptable.

U.S. First Class Stamp Rates Since 1900

At the beginning of the 20th century the cost to send a letter that weighed 1 oz or less was 2 cents. The following table shows the changing rates for first class stamps and the date on which each new rate became effective.

Date	Cost	Date	Cost	Date	Cost	Date	Cost
Nov. 2, 1917	$0.03	Jan. 7, 1968	$0.06	May 29, 1978	$0.15	Apr. 3, 1988	$0.25
July 1, 1919	0.02	May 16, 1971	0.08	Mar. 22, 1981	0.18	Feb. 3, 1991	0.29
July 6, 1932	0.03	Mar. 2, 1974	0.10	Nov. 1, 1981	0.20	Jan. 1, 1995	0.32
Aug. 1, 1958	0.04	Dec. 31, 1975	0.13	Feb. 17, 1985	0.22	Jan. 10, 1999	0.33
Jan. 7, 1963	0.05						

International Air Mail Rates

Aerogrammes — 60¢ from U.S. to all countries.
Air mail postcards (single) — 55¢ to all countries except Canada (45¢ each) and Mexico (35¢ each).
International letters and letter packages: to Canada and Mexico (by air mail; there are no surface rates to these countries)—weight not over 0.5 oz, 46¢ to Canada, 40¢ to Mexico; not over 1.0 oz, 52¢ to Canada, 46¢ to Mexico; not over 2 oz, 72¢ to Canada, 86¢ to Mexico; not over 3 oz, 95¢ to Canada, $1.26 to Mexico.

Air Mail, Letter, and Letter Package Rates to Countries Other Than Canada and Mexico

(weight limit: 64 oz [4 lb])

Weight not over	Rate	Weight not over	Rate	Weight not over	Rate	Weight not over	Rate
0.5 oz	$0.60	12.5 oz	$10.20	24.5 oz	$19.80	41 oz	$29.40
1.0	1.00	13.0	10.60	25.0	20.20	42	29.80
1.5	1.40	13.5	11.00	25.5	20.60	43	30.20
2.0	1.80	14.0	11.40	26.0	21.00	44	30.60
2.5	2.20	14.5	11.80	26.5	21.40	45	31.00
3.0	2.60	15.0	12.20	27.0	21.80	46	31.40
3.5	3.00	15.5	12.60	27.5	22.20	47	31.80
4.0	3.40	16.0	13.00	28.0	22.60	48	32.20
4.5	3.80	16.5	13.40	28.5	23.00	49	32.60
5.0	4.20	17.0	13.80	29.0	23.40	50	33.00
5.5	4.60	17.5	14.20	29.5	23.80	51	33.40
6.0	5.00	18.0	14.60	30.0	24.20	52	33.80
6.5	5.40	18.5	15.00	30.5	24.60	53	34.20
7.0	5.80	19.0	15.40	31.0	25.00	54	34.60
7.5	6.20	19.5	15.80	31.5	25.40	55	35.00
8.0	6.60	20.0	16.20	32.0	25.80	56	35.40
8.5	7.00	20.5	16.60	33.0	26.20	57	35.80
9.0	7.40	21.0	17.00	34.0	26.60	58	36.20
9.5	7.80	21.5	17.40	35.0	27.00	59	36.60
10.0	8.20	22.0	17.80	36.0	27.40	60	37.00
10.5	8.60	22.5	18.20	37.0	27.80	61	37.40
11.0	9.00	23.0	18.60	38.0	28.20	62	37.80
11.5	9.40	23.5	19.00	39.0	28.60	63	38.20
12.0	9.80	24.0	19.40	40.0	29.00	64	38.60

Air Mail Parcel Post Rates

Wt. Not Over (lb.)	Canada[2]	Mexico	A	B	C	D	E	Wt. Not Over (lb.)	Canada[2]	Mexico	A	B	C	D	E
1	$12.61	$9.98	$10.86	$11.76	$13.80	$15.11	$16.69	24	$31.92	$45.93	$63.62	$84.43	$77.30	$109.29	$87.55
2	12.61	12.12	13.71	15.42	17.28	19.83	20.68	25	34.04	48.71	67.12	88.76	81.23	114.53	91.61
3	13.28	13.36	15.57	18.01	19.51	23.19	23.15	26	34.82	50.15	69.28	91.78	83.83	118.44	94.49
4	15.11	15.82	18.87	22.31	23.55	28.76	27.75	27	35.60	51.60	71.45	94.80	86.42	122.36	97.38
5	15.84	17.16	20.88	25.12	25.96	32.40	30.43	28	36.38	53.04	73.61	97.82	89.02	126.27	100.26
6	16.56	18.50	22.89	27.92	28.37	36.03	33.11	29	37.17	54.49	75.78	100.84	91.62	130.19	103.14
7	17.29	19.84	24.90	30.73	30.78	39.67	35.79	30	37.95	55.93	77.94	103.86	94.21	134.11	106.03
8	19.40	22.81	28.99	36.11	35.75	46.63	41.42	31	38.73	57.38	80.11	106.88	96.81	138.02	108.91
9	20.18	24.26	31.15	39.13	38.34	50.55	44.30	32	39.51	58.83	82.27	109.90	99.41	141.94	111.79
10	20.97	25.70	33.31	42.15	40.94	54.47	47.19	33	40.30	60.27	84.44	112.92	102.00	145.85	114.68
11	21.75	27.15	35.48	45.17	43.54	58.38	50.07	34	41.08	61.72	86.60	115.94	104.60	149.77	117.56
12	22.53	28.59	37.64	48.19	46.14	62.30	52.95	35	41.86	63.16	88.77	118.96	107.20	153.68	120.44
13	23.32	30.04	39.81	51.21	48.73	66.21	55.84	36	42.65	64.61	90.93	121.98	109.80	157.60	123.33
14	24.10	31.48	41.97	54.23	51.33	70.13	58.72	37	43.43	66.05	93.10	125.00	112.39	161.52	126.21
15	24.88	32.93	44.14	57.25	53.93	74.04	61.60	38	44.21	67.50	95.26	128.02	114.99	165.43	129.09
16	25.66	34.37	46.30	60.27	56.52	77.96	64.49	39	44.99	68.94	97.43	131.04	117.59	169.35	131.98
17	26.45	35.82	48.47	63.29	59.12	81.88	67.37	40	45.78	70.39	99.59	134.06	120.18	173.26	134.86
18	27.23	37.26	50.63	66.31	61.72	85.79	70.25	41	46.56	71.83	101.76	137.08	122.78	177.18	137.74
19	28.01	38.71	52.80	69.33	64.32	89.71	73.14	41	47.34	73.28	103.92	140.10	125.38	181.09	140.63
20	28.79	40.15	54.96	72.35	66.91	93.62	76.02	43	48.12	74.72	106.09	143.12	127.98	185.01	143.51
21	29.58	41.60	57.13	75.37	69.51	97.54	78.90	44	48.91	76.17	108.25	146.14	130.57	188.95	146.39
22	30.36	43.04	59.29	78.39	72.11	101.45	81.79	Each add'l lb. or fraction:							
23	31.14	44.49	61.46	81.41	74.70	105.37	84.67		0.80	—	1.50	3.00	2.60	4.00	3.00

RATE GROUPS[1]

(1) For countries other than Canada and Mexico, see Country Rate Group table. (2) Canada: Minimum parcel weight is 1 pound; maximum parcel weight is 66 pounds, but only 22 pounds for parcels addressed to members of the Canadian Armed Forces based outside Canada (CFPOs).

Country Rate Groups

(For further information, consult your local post office.)

Country or territory	Rate group	Maximum weight limit (lbs)
Afghanistan[1]	E	44
Albania	D	44
Algeria	D	44
Andorra	B	44
Angola	D	22
Anguilla	A	22
Antigua & Barbuda	A	22
Argentina	E	44
Armenia	E	44
Aruba	B	44
Ascension	no air service	44 (surface)
Australia	D	44
Austria	D	70
Azerbaijan	A	70
Azores	D	66
Bahamas	A	44
Bahrain	C	44
Bangladesh	B	66
Barbados	A	44
Belarus	C	70
Belgium	D	70
Belize	A	44
Benin	A	66
Bermuda	C	44
Bhutan	D	22
Bolivia	C	70
Bosnia & Herzegovina	A	70
Botswana	D	70
Brazil	E	66
British Virgin Islands	A	44
Brunei	A	44
Bulgaria	D	70
Burkina Faso	D	66
Burma	see Myanmar	
Burundi[1]	B	44
Cambodia[2]	B	66 (air only)
Cameroon	D	66
Cape Verde	D	66
Cayman Islands	A	44
Central African Republic	E	44
Chad[3]	D	44 (air only)
Chile	C	44
China (People's Republic of)	D	70
Colombia	B	44
Comoros	B	44
Congo, Dem. Rep. of[1]	B	44
Congo Republic	E	44
Corsica	D	66
Costa Rica	A	66
Côte d'Ivoire	C	70
Croatia	C	70
Cuba[2]	no parcel post	
Cyprus	C	70
Czech Republic	C	33
Denmark	C	70
Djibouti	D	44
Dominica	A	44
Dominican Republic	A	44
East Timor	see Indonesia	
Ecuador	B	70
Egypt	D	66
El Salvador	A	44
Equatorial Guinea	B	44 (surface) 22 (air)
Eritrea	D	44
Estonia	D	66
Ethiopia	B	66
Falkland Islands[2]	no air PP	66 (surface)
Faroe Islands	C	70
Fiji	A	44
Finland	A	70
France	D	66
French Guiana	E	66
French Polynesia	E	66
Gabon	D	44
Gambia, The	A	22
Georgia, Republic of	E	44
Germany	A	70
Ghana	D	70
Gibraltar	B	44
Great Britain N. Ireland	D	66
Greece	A	44
Greenland	C	66
Grenada	A	44
Guadeloupe	E	44
Guatemala	A	44
Guinea	A	70
Guinea-Bissau1	A	22
Guyana	A	44
Haiti	A	55
Honduras	A	44
Hong Kong, China	D	44
Hungary	C	66
Iceland	A	44
India	D	44
Indonesia5	D	44
Iran	E	44
Iraq2	—	44
Ireland	A	66
Israel6	D	44
Italy	A	44
Ivory Coast	see Côte d'Ivoire	
Jamaica	A	22
Japan	D	44
Jordan	B	70
Kazakhstan	B	44
Kenya	D	70
Kiribati	A	44
Korea, Dem. People's Rep. of (North)2	no parcel post	
Korea, Republic of (South)	B	44
Kuwait	D	66
Kyrgyzstan	C	70
Laos	E	44
Latvia	B	70
Lebanon2,3	A	22 (air only)
Lesotho	D	44
Liberia1	—	44
Libya	D	44
Liechtenstein	C	66
Lithuania	D	70
Luxembourg	C	70
Macau	D	44
Macedonia	C	44
Madagascar	E	66
Madeira Islands	D	66
Malawi	D	44
Malaysia	B	44
Maldives	E	70
Mali	D	44
Malta	D	44
Martinique	E	66
Mauritania	A	44
Mauritius	B	44
Moldova	D	66
Mongolia	D	44 (surface) 22 (air)
Montserrat	A	44
Morocco	C	70
Mozambique	E	44
Myanmar	B	44
Namibia	D	44
Nauru	B	44
Nepal	D	44
Netherlands	B	44
Netherlands Antilles	B	44
New Caledonia	E	66
New Zealand	D	66
Nicaragua	A	44
Niger	C	66
Nigeria	D	66
Norway	D	55
Oman	B	44
Pakistan	D	66
Panama	A	70
Papua New Guinea	E	44
Paraguay	C	70
Peru	C	70
Philippines	D	44
Pitcairn Island	D	22
Poland	C	44
Portugal	D	66
Qatar	B	70
Reunion	E	66
Romania	D	70
Russia	E	44
Rwanda	D	66
Saint Helena	C	44
Saint Kitts & Nevis	B	44
Saint Lucia	A	44
Saint Pierre & Miquelon	B	66
Saint Vincent & Grenadines	C	22
Samoa	B	44
San Marino	A	44
São Tomé & Príncipe	B	44
Saudi Arabia	B	44
Senegal	D	44
Serbia-Montenegro (Yugoslavia)	B	33
Seychelles	C	70
Sierra Leone1	D	44
Singapore	D	66
Slovakia	B	66
Slovenia	C	33
Solomon Islands	B	44
Somalia1	—	44
South Africa	D	66
Spain	B	44
Sri Lanka	B	66
Sudan	B	44
Suriname	A	44
Swaziland	B	44
Sweden	E	44
Switzerland	C	66
Syria	C	66
Taiwan	D	44
Tajikistan	C	66
Tanzania	C	44
Thailand	B	44
Togo	D	44
Tonga	A	44
Trinidad & Tobago	A	22
Tristan da Cunha	A	22
Tunisia	B	44
Turkey	B	70
Turkmenistan	D	22
Turks & Caicos Islands	A	22
Tuvalu	A	55
Uganda	D	44
Ukraine	D	22
United Arab Emirates	C	70
United Kingdom	D	66
Uruguay	C	44
Uzbekistan	E	44
Vanuatu	A	44
Vatican City	A	44
Venezuela	A	44
Vietnam	B	44
Wallis & Futuna Islands	E	66
Yemen	C	44
Zambia	D	66
Zimbabwe	E	44

(1) All mail service suspended. (2) Mail service restrictions apply. (3) Surface mail service suspended. (4) Includes Monaco. (5) Includes East Timor. (6) West Bank and Gaza Strip are same rate group as Israel.

CONSUMER INFORMATION

Business Directory

Listed below are major U.S. corporations offering products and services to consumers. Alphabetization is by first key word. Listings generally include examples of products offered. When there is no hyphen or other punctuation at the end of a line in a website address, no spacing or punctuation should be added.

COMPANY NAME; ADDRESS; TELEPHONE NUMBER; WEBSITE; TOP EXECUTIVE; BUSINESS, PRODUCTS, OR SERVICES.

Abbott Laboratories; One Abbott Park Rd., North Chicago, IL 60064; (847) 937-6100; Website: http://www.abbott.com; Duane L. Burnham; health care prods. (Murine, Selsun Blue).

Aetna, Inc.; 151 Farmington Ave., Hartford, CT 06156; (203) 273-0123; Website: http://www.aetna.com; Richard L. Haber; health insurance, financial services.

Alberto Culver; 2525 Armitage Ave., Melrose Park, IL 60160; (708) 450-3000; Website: http://www.alberto.com; Leonard H. Lavin; hair care (VO5), consumer prods. (Mrs. Dash, Sugar Twin), personal care prods. (St. Ives), Sally Beauty Supply stores.

Albertson's, Inc.; 250 Parkcenter Blvd., Boise, ID 83726; (208) 395-6200; Website: http://www.albertsons.com; Gary Michael; supermarkets. (Co. merged with American Stores Co. 6/24/99, making Albertson's, Inc., the largest retail food and drug co. in the U.S.)

Allegheny Teledyne, Inc.; 1000 Six PPG Pl., Pittsburgh, PA 15222-5479; (412) 394-2800; Website: http://www.alleghenyteledyne.com; Richard P. Simmons; electronics, aerospace, industrial, consumer prods. (Water Pik); specialty metals.

AlliedSignal; Morristown, NJ 07960; (973) 455-2000; Web-site: http://www.alliedsignal.com; Lawrence A. Bossidy; aerospace, engineered materials, automotive prods.

Allstate Corp.; Allstate Plaza, Northbrook, IL 60062; (847) 402-5000; Website: http://www.allstate.com; Edward Liddy; property/casualty, life insurance.

Aluminum Co. of America (Alcoa); 425 6th Ave., Pittsburgh, PA 15219; (412) 553-3042; Website: http://www.shareholder.com/Alcoa; Paul O'Neill; world's largest aluminum producer.

Amerada Hess Corp.; 1185 Ave. of the Americas, NY, NY 10036; (212) 997-8500; Website: http://www.hess.com; J. B. Hess; integrated international oil co.

American Express Co.; 200 Vesey St., NY, NY 10285; (212) 640-2000; Website: http://www.americanexpress.com; Harvey Golub; travel, financial, and information services.

American Greetings Corp.; 1 American Rd., Cleveland, OH 44144; (216) 252-7300; Website: http://www.americangreetings.com; Morry Weiss; greeting cards, stationery, party goods, gift items.

American Home Prods. Corp.; 5 Giralda Farms, Madison, NJ 07940; (973) 660-5000; Website: http://www.ahp.com; John R. Stafford; prescription and over-the-counter drugs (Advil, Anacin, Dristan, Robitussin).

American Intl. Group; 70 Pine St., NY, NY 10270; (212) 770-7000; Website: http://www.aig.com; Maurice R. Greenberg; insurance, financial services.

Ameritech; 30 S. Wacker Dr., Chicago, IL 60606; (312) 750-5000; Website: http://www.ameritech.com; Richard C. Notebaert; communications services.

AMR Corp.; PO Box 619616, Dallas/Ft. Worth Airport, TX 75261; (817) 963-1234; Website: http://www.amrcorp.com; Donald J. Carty; air transportation (American Airlines, American Eagle).

Anheuser-Busch Cos., Inc.; 1 Busch Place, St. Louis, MO 63118; (314) 577-2000; Website: http://www.anheuser-busch.com; August A. Busch 3d; world's largest brewer (Budweiser, Michelob, BudLight, Natural Light, Busch, O'Doul's), aluminum can manuf. and recycling, theme parks.

Apple Computer, Inc.; 1 Infinite Loop, Cupertino, CA 95014-2084; (408) 996-1010; Website: http://www.apple.com; Steve Jobs; manuf. of personal computers, software, peripherals.

Aramark Corp.; Aramark Tower, Philadelphia, PA 19107; (215) 238-3000; Website: http://www.aramark.com; Joseph Neubauer; food and support services, uniforms and career apparel, child care and early education.

Archer Daniels Midland Co.; 4666 Faries Pkwy., Decatur, IL 62525; (217) 424-5200; Website: http://www.admworld.com; G. Allen Andreas; agricultural commodities and prods.

Armstrong World Industries, Inc.; 313 W. Liberty St., Lancaster, PA 17604; (717) 397-0611; Website: http://www.armstrong.com; George A. Lorch; interior furnishings, specialty prods.

Arvin Industries, Inc.; 1531 13th St., Columbus, IN 47201; (812) 379-3000; Website: http://www.arvin.com; V. William Hunt; auto emission and ride control systems.

Ashland Inc.; 50 E. River Center, PO Box 391, Covington, KY 41012; (606) 815-3333; Website: http://www.ashland.com; Paul W. Chellgren; petroleum producer and refiner (Valvoline), chemicals, road construction.

Atlantic Richfield Co.; 515 S. Flower St., Los Angeles, CA 90071-2256; (213) 486-3511; Website: http://www.arco.com; M. R. Bowlin; integrated oil and gas producer.

AT&T Corp.; 32 Ave. of the Americas, NY, NY 10013-2412; (212) 387-5400; Website: http://www.att.com; C. Michael Armstrong; communications, global information management. (Co. acquired Tele-Communications, Inc. 3/9/99 for approx. $69.9 bil.)

Avon Prods., Inc.; 1345 Ave. of Americas, NY, NY 10105; (212) 282-5000; Website: http://www.avon.com; Charles Perrin; cosmetics, fragrances, toiletries, fashion jewelry, gift items, casual apparel, lingerie.

Bank of America Corp.; Bank of America Corporate Center, Charlotte, NC 28255; (704) 386-5000; Website: http://www.bankamerica.com; Hugh L. McColl Jr; largest U.S. bank.

Bausch & Lomb Inc.; One Bausch & Lomb Place, Rochester, NY 14604; (716) 338-6000; Website: http://www.bausch.com; William M. Carpenter; vision and health-care prods., accessories.

Baxter International Inc.; 1 Baxter Pkwy., Deerfield, IL 60015; (847) 948-2000; Website: http://www.baxter.com; Vernon R. Loucks Jr; health care prods. & services.

Bear Stearns Cos. Inc.; 245 Park Ave., NY, NY 10167; (212) 272-2000; Website: http://www.bearstearns.com; Alan C. Greenberg; investment banking, securities trading, brokerage.

Becton, Dickinson & Co.; 1 Becton Dr., Franklin Lakes, NJ 07417; (201) 847-6800; Website: http://www.bd.com; C. Castellini; medical, laboratory, diagnostic prods.

Bell Atlantic Corp.; 1717 Arch St., Philadelphia, PA 19103; (215) 963-6000; Website: http://www.bellatlantic.com; Raymond W. Smith; telephone service in mid-Atlantic region, worldwide wireless. (Co. announced 7/28/98 it had agreed to acquire GTE Corp. for $71.3 bil.)

BellSouth Corp.; 1155 Peachtree St. NE, Atlanta, GA 30309; (404) 249-2000; Website: http://www.bellsouth.com; John L. Clendenin; telephone service in southern U.S.

Best Buy Co., Inc.; 7075 Flying Cloud Dr., Eden Prairie, MN 55344; (612) 947-2000; Website: http://www.bestbuy.com; R. M. Schulze; retailer of software, appliances, electronics, cameras, home office equipment.

Bestfoods; International Plaza, Englewood Cliffs, NJ 07632; (201) 894-4000; Website: http://www.bestfoods.com; Charles R. Shoemate; food (Hellmann's, Best Foods mayonnaise, Skippy peanut butter, Knorr soups, Thomas' English muffins, Mueller pasta, Freihofer's, Boboli, Arnold breads, Mazola oils and margarine, Entenmann's cakes).

Bethlehem Steel Corp.; 1170 8th Ave., Bethlehem, PA 18016; (610) 694-2424; Website: http://www.bethsteel.com; Curtis H. Barnette; steel & steel prods.

Black & Decker Corp.; 701 E. Joppa Rd., Towson, MD 21204; (410) 716-3900; Website: http://www.bdk.com; Nolan D. Archibald; manuf. power tools (DeWalt), household prods. (Kwikset, Price Pfister), small appliances (Black & Decker).

H & R Block, Inc.; 4410 Main St., Kansas City, MO 64111; (816) 753-6900; Website: http://www.hrblock.com; Henry Bloch; tax return preparation.

Boeing Co.; 7755 E. Marginal Way, Seattle, WA 98108; (206) 655-2121; Website: http://www.boeing.com; Philip M. Condit; leading manufacturer of commercial, jet aircraft.

Boise Cascade Corp.; 1111 W. Jefferson St., Boise, ID 83728; (208) 384-6161; Website: http://www.bc.com; George J. Harad; timber; paper, wood prods.

Borden, Inc.; 180 E. Broad St., Columbus OH 43215-3707; (614) 225-4000; Website: http://www.bordenfamily.com; C. Robert Kidder; snacks (Wise, Cheez Doodles), adhesives (Elmer's, Krazy Glue), pasta (Prince, Creamette, Goodman's), pasta sauce (Aunt Millie's, Classico), Wyler's bouillon, Soup Starter, Corning Consumer Prods. (Corningware, Corelle, Pyrex, Revere).

Bristol-Myers Squibb Co.; 345 Park Ave., NY, NY 10022; (212) 546-4000; Website: http://www.bms.com; Charles A. Heimhold; toiletries (Ban antiperspirant), haircare (Clairol), drugs (Bufferin, Comtrex, Excedrin), infant formula (Enfamil).

Brown-Forman Corp.; PO Box 1080, Louisville, KY 40201-1080; (502) 585-1100; Website: http://www.brown-forman.com; Owsley Brown 2d; distilled spirits (Jack Daniel's, Southern Comfort), wines (Bolla, Fetzer, Korbel), china and crystal (Dansk, Lenox), Gorham, Kirk Steiff silver prods., Hartmann luggage.

Brown Group, Inc.; 8300 Maryland Ave., P.O. Box 29, St. Louis, MO 63166; (314) 854-4000; Website: http://www.browngroup.com; Ronald A. Fromm; manuf. and retailer (Famous Footwear) of women's, men's, and children's shoes (Buster Brown, Naturalizer).

Brunswick Corp.; 1 N. Field Ct., Lake Forest, IL 60045-4811; (847) 735-4700; Website: http://www.brunswickcorp.com; P. N. Larson; largest U.S. maker of leisure and recreation prods., marine, camping, fitness and fishing equip., bowling centers and equip.

Burlington Northern Santa Fe Inc.; 2650 Lou Menk Dr., Ft. Worth, TX 76131-2830; (817) 333-2000; Website: http://www.bnsf.com; Robert Krebs; one of the largest U.S. rail transportation cos.

Campbell Soup Co.; Campbell Pl., Camden, NJ 08103; (609) 342-4800; Website: http://www.campbellsoup.com; David W. Johnson; soups, Franco-American spaghetti, V-8 vegetable juice, Godiva chocolates, Swanson frozen dinners, Prego spaghetti sauce, Pepperidge Farm.

Carter-Wallace, Inc.; 1345 Ave. of the Americas, NY, NY 10105; (212) 339-5000; H. H. Hoyt; personal care (Arrid, Rise, Pearl Drops, Nair, Trojan condoms).

Caterpillar Inc.; 100 N.E. Adams St., Peoria, IL 61629; (309) 675-1000; Website: http://www.cat.com; Glen A. Barton; world's largest producer of earth moving equip.

Chase Manhattan Corp.; 270 Park Ave., NY, NY 10017; (212) 270-6000; Website: http://www.Chase.com; Walter V. Shipley; 3d largest bank-holding company in U.S.

Chevron Corp.; 475 Market St., San Francisco, CA 94105; (415) 894-7700; Website: http://www.chevron.com; Kenneth T. Derr; integrated oil co.

Chiquita Brands International, Inc.; 250 E. 5th St., Cincinnati, OH 45202; (513) 784-8000; Website: http://www.chiquita.com; Carl H. Lindner; bananas, fruits, vegetables.

Church & Dwight Co., Inc.; 469 N. Harrison St., Princeton, NJ 08543; (609) 683-5900; Website: http://www.armhammer.com; D. C. Minton; world's largest producer of sodium bicarbonate (Arm & Hammer); Brillo.

CIGNA Corp.; 1 Liberty Pl., Philadelphia, PA 19103; (215) 761-1000; Website: http://www.cigna.com; Wilson H. Taylor; insurance holding co.

Circuit City Stores, Inc.; 9950 Mayland Dr., Richmond, VA 23233-1464; (804) 527-4000; Website: http://www.circuitcity.com; Richard L. Sharp; retailer of electronic, audio/video equip., consumer appliances; new and used-car stores (CarMax).

Circus Circus Enterprises, Inc.; 3950 Las Vegas Blvd. S, Las Vegas, NV 89109; (702) 632-6700; Website: http://www.circuscircus.org; Michael Ensign; casino-resort operator (Excalibur, Luxor).

Citigroup; 153 E. 53rd St., NY, NY 10043; (212) 559-1000; Website: http://www.citigroupinfo.com; Sanford A. Weill; diversified financial services.

Clorox Co.; 1221 Broadway, Oakland, CA 94612; (510) 271-7000; Website: http://www.clorox.com; G. Craig Sullivan; retail consumer prods. (Clorox, Formula 409, Pine-Sol, S.O.S., Soft Scrub cleansers; Armor All, STP, Rain Dance automotive prods.; Jonny Cat, Fresh Step cat litters; Kingsford charcoal briquets; StarterLogg; Combat and Black Flag insecticides; Hidden Valley dressing; K.C. Masterpiece barbecue sauce; Brita water systems).

Coastal Corp.; 9 Greenway Plaza, Houston, TX 77046; (713) 877-1400; David A. Arledge; diversified energy company primarily engaged in interstate transmission of natural gas.

Coca-Cola Co.; 1 Coca-Cola Plaza, Atlanta, GA 30313; (404) 676-2121; Website: http://www.cocacola.com; M. Douglas Ivester; world's largest soft drink co. (Coca-Cola, Sprite, Nestea), world's largest dist. of juice prods. (Minute Maid, Five Alive, Hi-C, Fruitopia).

Colgate-Palmolive Co.; 300 Park Ave., NY, NY 10022; (212) 310-2000; Website: http://www.colgate.com; Reuben Mark; soap (Palmolive, Irish Spring), detergent (Fab, Ajax, Fresh Start), toothpaste (Colgate, Ultra Brite), Hill's pet food.

Columbia/HCA Healthcare Corp.; 1 Park Plaza, Nashville, TN 37203; (615) 344-9551; Website: http://www.columbia.net; T. F. Frist Jr; largest hospital mgmt. co. in the U.S.

Compaq Computer Corp.; 20555 SH 249, Houston, TX 77070; (281) 370-0670; Website: http://www.compaq.com; Benjamin M. Rosen; laptop and desktop computers. (Co. acquired Digital Equipment Corp. 6/98.)

CompUSA Inc.; 14951 N. Dallas Pkwy., Dallas, TX 75240; (972) 982-4000; Website: http://www.compusa.com; G. H. Bateman; largest U.S. superstore retailer of microcomputers and peripherals.

Computer Sciences Corp.; 2100 E. Grand Ave., El Segundo, CA 90245; (310) 615-0311; Website: http://www.csc.com; Van B. Honeycutt; technology services.

ConAgra; 1 ConAgra Dr., Omaha, NE 68102; (402) 595-4000; Website: http://www.conagra.com; B. Rohde; 2d largest U.S. food processor.

Continental Airlines, Inc.; 1600 Smith St. HQ511, Houston, TX 77002; (713) 324-5242; Website: http://www.Flycontinental.com; Gordon M. Bethune; air transportation.

Adolph Coors Co.; Golden, CO 80401; (303) 279-6565; Website: http://www.coorsinvestor.com; William K. Coors; brewer (Coors, Killian's, Zima).

Corning Inc.; 1 Riverfront Plaza, Corning, NY 14831; (607) 974-9000; Website: http://www.corning.com; Roger G. Ackerman; specialty materials, optical fiber and cable.

Costco Cos., Inc.; 999 Lake Dr., Issaquah, WA 98027; (425) 313-8100; Website: http://www.costco.com; James D. Sinegal; wholesale-membership warehouses.

Crane Co.; 100 First Stamford Pl., Stamford, CT 06902; (203) 363-7300; Website: http://www.crane.com; R. S. Evans; manuf. fluid control devices, vending machines, fiberglass panels, aircraft brakes.

A. T. Cross Co.; 1 Albion Rd., Lincoln, RI 02865; (401) 333-1200; Website: http://www.cross.com; Bradford R. Boss; writing instruments.

Crown Cork & Seal Co.; 1 Crown Way, Philadelphia, PA 19154-4599; (215) 698-5100; Website: http://www.crowncork.com; William J. Avery; world's leading supplier of packaging prods.

CSX Corp.; 901 E. Cary St., Richmond, VA 23219; (804) 782-1400; Website: http://www.csx.com; John W. Snow; rail, ocean, barge freight transport.

CVS Corp.; 1 CVS Dr., Woonsocket, RI 02895; (401) 765-1500; Website: http://www.CVS.com; Thomas M. Ryan; drugstore chain.

Dana Corp.; 4500 Dorr St., Toledo, OH 43615; (419) 535-4500; Website: http://www.dana.com; Southwood J. Morcott; truck and auto parts, supplies.

Dayton Hudson Corp.; 777 Nicollet Mall, Minneapolis, MN 55402; (612) 370-6948; Website: http:/www.dhc.com; Robert J. Ulrich; department, specialty stores (Marshall Fields, Target, Hudsons).

Deere & Co.; John Deere Rd., Moline, IL 61265; (309) 765-8000; Website: http://www.deere.com; Hans W. Becherer; world's largest manuf. of farm equip.; industrial equip.; lawn and garden tractors.

Dell Computer Corp.; 1 Dell Way, Round Rock, TX 78682; (512) 338-4400; Website: http://www.dell.com; Michael S. Dell; laptop and desktop computers.

Delta Air Lines, Inc.; Hartsfield Atlanta Intl. Airport, Atlanta, GA 30320; (404) 715-2600; Website: http://www.delta-air.com; Gerald Grinstein; air transportation.

Dial Corp.; 15501 N. Dial Blvd., Scottsdale, AZ 85260; (602) 754-3425; Website: http://www.dialcorp.com; Malcolm Jozoff; consumer prods. (Dial, Tone soap, Breck shampoo, Armour Star meats, Renuzit air fresheners).

Diebold, Inc.; PO Box 8230, Canton, OH 44711; (330) 490-4000; Website: http://www.diebold.com; Robert W. Mahoney; manuf. ATMs, security systems and prods.

Dillard's; 1600 Cantrell Rd., Little Rock, AR 72201; (501) 376-5200; Website: http://www.dillards.com; William Dillard; 2d largest dept. store chain in U.S.

Walt Disney Co.; 500 S. Buena Vista St., Burbank, CA 91521-7320; (818) 560-1000; Website: http://www.disney.com; Michael D. Eisner; motion pictures, television (ESPN, ABC, A&E, Lifetime), radio stations, theme parks (Walt Disney World, Disneyland) and resorts, publishing, recordings, retailing (Disney Stores).

Dole Food Co., Inc.; 31365 Oak Crest Dr., Westlake Village, CA 91361; (818) 879-6600; Website: http://www.dole.com; David H. Murdock; food prods., fresh fruits and vegetables.

R. R. Donnelley & Sons Co.; 77 W. Wacker Dr., Chicago, IL 60601-1696; (312) 326-8000; Website: http://www.rrdonnelley.com; William L. Davis; commercial printer.

Dow Chemical Co.; 2030 Dow Center, Midland, MI 48674; (517) 636-1000; Website: http://www.dow.com; W. Stavropoulos; chemicals, plastics.

Dow Jones & Co., Inc.; 200 Liberty St., NY, NY 10281; (212) 416-2000; Website: http://www.dj.com; Peter R. Kann; financial news service, publishing (*Wall Street Journal*, *Barron's*, Ottaway Newspapers).

Dun & Bradstreet Corp.; 1 Diamond Hill Rd., Murray Hill, NJ 07974; (908) 665-5000; Website: http://www.dnbcorp.com; Volney Taylor; business information, publishing (Moody's, "Yellow Pages" phone books).

E. I. du Pont de Nemours & Co.; 1007 Market St., Wilmington, DE 19898; (302) 774-1000; Website: http://www.dupont.com; J. Krol; largest U.S. chemical co.; petroleum, consumer prods.

Eastman Kodak Co.; 343 State St., Rochester, NY 14650; (716) 724-5492; Website: http://www.kodak.com; G. Fisher; world's largest producer of photographic prods.

Eaton Corp.; 1111 Superior Ave., Cleveland, OH 44114; (216) 523-5000; Website: http://www.eaton.com; Steven R. Hardis; manuf. of vehicle powertrain components, controls.

Emerson Electric Co.; 8000 W. Florissant Ave., St. Louis, MO 63136; (314) 553-2000; Website: http://www.emersonelectric.com; C. F. Knight; electrical, electronics prods. & systems.

Exxon Corp.; 5959 Las Colinas Blvd., Irving, TX 75039-2298; (972) 444-1000; Website: http://www.exxon.com; Lee R. Raymond; world's largest publicly owned integrated oil co.

FDX Corp.; Box 727, Memphis, TN 38194; (901) 369-3600; Website: http://www.fedex.com; F. W. Smith; express delivery service.

Fedders Corp.; 505 Martinsville Rd., P.O. Box 813, Liberty Corner, NJ 07938; (908) 604-8686; Website: http://www.fedders.com; Salvatore Giordano Jr; manuf. of room air conditioners (Fedders, Airtemp), dehumidifiers.

Freddie Mac; 8200 Jones Branch Dr., McLean, VA 22102; (703) 903-2000; Website: http://www.freddiemac.com; Leland C. Brendsel; residential mortgage provider.

Fannie Mae; 3900 Wisconsin Ave. NW, Washington, DC 20016; (202) 752-7115; Website: http://www.fanniemae.com; James A. Johnson; largest U.S. provider of residential mortgage funds.

Federated Dept. Stores; 7 W. 7th St., Cincinnati, OH 45202; (513) 579-7000; Website: http://www.Federated-fds.com; James Zimmerman; full-line dept. stores Macy's, Bloomingdale's, Stern's.

First Data Corp.; 5660 New Northside Dr., Atlanta, GA 30328; (800) 735-3362; Website: http://www.firstdatacorp.com; Henry C. Duques; info. retrieval, data processing.

Fleetwood Enterprises, Inc.; 3125 Myers St., Riverside, CA 92503; (909) 351-3500; Website: http://www.fleetwood.com; Glenn F. Kummer; manufactured homes, recreational vehicles.

Fleming Cos. Inc.; 6301 Waterford Blvd., PO Box 26647, Oklahoma City, OK 73126; (405) 840-7200; Website: http://www.fleming.com; Mark S. Hansen; one of largest U.S. wholesale food distrib.

Fluor Corp.; 3333 Michelson Dr., Irvine, CA 92730; (714) 975-6961; Website: http://www.fluor.com; Philip J. Carroll; largest international engineering and construction co. in U.S.

Ford Motor Co.; American Rd., Dearborn, MI 48121; (313) 322-3000; Website: http://www.ford.com; Jacques Nasser; motor vehicle sales (Ford, Lincoln-Mercury, Volvo), rentals (Hertz).

Fortune Brands, Inc.; 1700 E. Putnam Ave., Old Greenwich, CT 06870; (203) 698-5000; Website: http://www.fortunebrands.com; Thomas C. Hays; whiskey (Jim Beam), hardware, office prods. (Swingline), golf and leisure prods. (Titleist, Cobra, Foot-Joy).

Fruit of the Loom, Inc.; 233 So. Wacker Dr., 5000 Sears Tower, Chicago, IL 60606; (312) 876-1724; Website: http://www.fruit.com; William Farley; manuf. of underwear, activewear.

Gannett Co., Inc.; 1100 Wilson Blvd., Arlington, VA 22234; (703) 284-6000; Website: http://www.gannett.com; J. J. Curley; newspaper publishing (*USA Today*), network and cable TV.

The Gap, Inc.; 1 Harrison St., San Francisco, CA 94105; (415) 952-4400; Website: http://www.gap.com; Donald G. Fisher; casual and activewear retailer (Gap, Banana Republic, Old Navy).

General Dynamics; 3190 Fairview Park Dr., Falls Church, VA 22042-4523; (703) 876-3000; Website: http://www.generaldynamics.com; Nicholas D. Chabraja; nuclear submarines (Trident, Seawolf), armored vehicles, combat systems, computing devices, defense systems.

General Electric Co.; 3135 Easton Tpke., Fairfield, CT 06431; (203) 373-2211; Website: http://www.ge.com; John F. Welch; electrical, electronic equip., radio and television broadcasting (NBC), aircraft engines, power generation, appliances.

General Mills, Inc.; PO Box 1113, Minneapolis, MN 55440; (612) 540-2311; Website: http://www.generalmills.com; S. W. Sanger; foods (Total, Wheaties, Cheerios, Chex, Hamburger Helper, Betty Crocker, Bisquick).

General Motors; 3044 W. Grand Blvd., Detroit, MI 48202-3091; (313) 556-5000; Website: http://www.gm.com; John F. Smith Jr; world's largest auto manuf. (Chevrolet, Pontiac, Cadillac, Buick).

Genuine Parts Co.; 2999 Circle 75 Pkwy., Atlanta, GA 30339; (404) 953-1700; Website: http://www.genpt.com; Larry L. Prince; distributes auto replacement parts (NAPA).

Georgia-Pacific Corp.; 133 Peachtree St. NE, Atlanta, GA 30303; (404) 521-5210; Website: http://www.gp.com; A. D. Correll; manuf. of paper and wood prods.

Gillette; Prudential Tower Bldg., Boston, MA 02199; (617) 463-3000; Website: http://www.gillette.com; Michael Hawley; stationery prods. (PaperMate, Parker, Waterman pens), personal care prods. (Sensor, Atra razors, Right Guard, Soft and Dri), appliances (Braun), batteries (Duracell).

The Goodyear Tire & Rubber Co.; 1144 E. Market St., Akron, OH 44316; (330) 796-2121; Website: http://www.goodyear.com; Samir F. Gibara; world's largest rubber manuf.; tires and other auto prods.

W. R. Grace & Co.; 1 Town Center Rd., Boca Raton, FL 33486; (561) 362-2000; Paul J. Norris; chemicals, construction prods.

Great Atlantic & Pacific Tea Co. (A&P); 2 Paragon Dr., Montvale, NJ 07645; (201) 573-9700; Website: http://www.aptea.com; James Wood; supermarkets (A&P, Waldbaum's, Kohl's, Dominion).

GTE Corp.; 1255 Corporate Dr., Irving, TX 75038; (972) 507-2462; Website: http://www.gte.com; Charles R. Lee; large telecommunications co., cellular telephone provider. (Co. announced 7/28/98 it had agreed to be acquired by Bell Atlantic Corp. for $71.3 bil.)

Halliburton Co.; 500 N. Akard St., Dallas, TX 75201; (214) 978-2600; Website: http://www.halliburton.com; Richard Cheney; energy, engineering, and construction services.

Harley-Davidson, Inc.; 3700 W. Juneau Ave., Milwaukee, WI 53208; (414) 343-4680; Website: http://www.harley-davidson.com; Jeffrey Bleustein; manuf. of motorcycles, parts and accessories.

Harrah's Entertainment, Inc.; 1023 Cherry Rd., Memphis, TN 38117; (901) 762-8600; Website: http://www.harrahs.com; Philip G. Satre; casino-hotels and riverboats.

Hartford Life, Inc.; 200 Hopmeadow St., Simsbury, CT 06089; (860) 843-7716; Website: http://www.thehartford.com; Lowndes A. Smith; insurance, finl. svces.

Hartmarx; 101 N. Wacker Dr., Chicago, IL 60606; (312) 372-6300; Elbert O. Hand; apparel manuf. (Hart Schaffner & Marx, Hickey Freeman, Claiborne, Tommy Hilfiger, Pierre Cardin, Perry Ellis).

Hasbro, Inc.; 1027 Newport Ave., Pawtucket, RI 02862; (401) 431-8697; Website: http://www.hasbro.com; Alan G. Hassenfeld; toy and game manuf. (Milton Bradley, Playskool, G. I. Joe, Parker Bros., Tiger Electronics, Play-Doh).

H. J. Heinz Co.; PO Box 57, Pittsburgh, PA 15230; (412) 456-6014; Website: http://www.heinz.com; William R. Johnson; foods (Star-Kist, Ore-Ida, 57 Varieties), pet food (Ken-L Ration, 9 Lives), Weight Watchers.

Hershey Foods Corp.; 100 Crystal A Dr., Hershey, PA 17033; (717) 534-6799; Website: http://www.hersheys.com; Kenneth L. Wolfe; largest U.S. producer of chocolate and confectionery prods. (Reese's, Kit Kat, Mounds, Almond Joy, Cadbury, Jolly Rancher, Twizzler, Milk Duds, Good 'n' Plenty), pasta (San Giorgio, Ronzoni).

Hewlett-Packard Co.; 3000 Hanover St., Palo Alto, CA 94304; (650) 857-1501; Website: http://www.hp.com; Lewis E. Platt; manuf. computers, electronic prods. and systems.

Hillenbrand Industries, Inc.; 700 State Rte. 46, Batesville, IN 47006; (812) 934-8400; Website: http://www.hillenbrand.com; W. A. Hillenbrand; manuf. caskets, adjustable hospital beds, locks (Medeco).

Hilton Hotels Corp.; 9336 Civic Center Dr., Beverly Hills, CA 90210; (310) 278-4321; Website: http://www.Hilton.com; Barron Hilton; hotels, casinos.

Home Depot, Inc.; 2455 Paces Ferry Rd. NW, Atlanta, GA 30339; (770) 433-8211; Website: http://www.homedepot.com; Bernard Marcus; retail building supply, home improvement warehouse stores.

Honeywell Inc.; Honeywell Plaza, Minneapolis, MN 55408; (612) 951-1000; Website: http://www.honeywell.com; Michael R. Bonsignore; industrial and home control systems, aerospace guidance systems.

Hormel Foods Corp.; 1 Hormel Pl., Austin, MN 55912; (507) 437-5611; Website: http://www.hormel.com; Joel W. Johnson; meat processor, pork and beef prods. (SPAM, Dinty Moore, Little Sizzlers).

Houghton Mifflin Co.; 222 Berkeley St., Boston, MA 02116; (617) 351-5000; Website: http://www.hmco.com; Nader F. Dareshori; publisher of textbooks, reference, general interest books.

Huffy Corp.; 225 Byers Rd., Miamisburg, OH 45342; (937) 866-6251; Website: http://www.huffy.com; Don R. Graber; largest U.S. bicycle manuf., sports and hardware equip.

Humana, Inc.; 500 W. Main St., P.O. Box 1438, Louisville, KY 40201-1438; (502) 580-1000; Website: http://www.humana.com; David A. Jones; managed healthcare service provider, financial services.

IBP, Inc.; IBP Ave., PO Box 515, Dakota City, NE 68731; (402) 494-2061; Website: http://www.ibpinc.com; Robert L. Peterson; world's largest processor of fresh beef and pork.

Ingersoll-Rand; Woodcliff Lake, NJ 07675; (201) 573-0123; Website: http://www.ingersoll-rand.com; J. E. Perella; industrial machinery.

Intel Corp.; 2200 Mission College Blvd., Santa Clara, CA 95052-8119; (408) 765-8080; Website: http://www.intc.com; A. S. Grove; manuf. integrated circuits (Pentium).

International Business Machines Corp. (IBM); New Orchard Rd., Armonk, NY 10504; (914) 499-1900; Website: http://www.ibm.com; Louis V. Gerstner Jr; .world's largest supplier of advanced information processing technology equip., services.

International Paper Co.; 2 Manhattanville Rd., Purchase, NY 10577; (914) 397-1500; Website: http://www.internationalpaper.com; John T. Dillon; world's largest paper/forest prods. co., chemicals, minerals.

Interstate Bakeries Corp.; 12 E. Armour Blvd., Kansas City, MO 64111; (816) 502-4000; Website: http://www.irin.com/ibc; Charles A. Sullivan; baked goods wholesaler, distributor (Wonder, Hostess, Dolly Madison, Beefsteak, Home Pride).

Jo-Ann Stores, Inc.; 5555 Darrow Rd., Hudson, OH 44236; (330) 656-2600; Website: http://www.joann.com; Alan Rosskamm; nation's largest specialty fabric and craft stores (Jo-Ann Fabric and Crafts, Jo-Ann etc.).

S.C. Johnson & Son, Inc.; 1525 Howe St., Racine, WI 53403; (414) 260-2000; Website: http://www.scjbrands.com; William Perez; cleaning and other household prods. (Johnson's Wax, Windex, Pledge, fantastik, Raid, Off!, Shout, Glade, Scrubbing Bubbles, Ziploc bags).

Johnson & Johnson; 1 Johnson & Johnson Plaza, New Brunswick, NJ 08933; (732) 524-0400; Website: http://www.jnj.com; Ralph S. Larsen; surgical dressings (Band-Aid), pharmaceuticals (Tylenol), toiletries (Neutrogena).

Johnson Controls; 5757 N. Green Bay Ave., Milwaukee, WI 53201; (414) 228-1200; Website: http://www.jci.com; James H. Keyes; fire protection services, auto seats and batteries.

Jostens Inc.; 5501 Norman Center Dr., Minneapolis, MN 55437; (612) 830-3300; Website: http://www.jostens.com; Robert P. Jensen; school rings, yearbooks, plaques.

Kellogg Co.; 1 Kellogg Sq., Battle Creek, MI 49016; (616) 961-2000; Website: http://www.kelloggs.com; Arnold G. Langbo; world's largest mfgr. of ready-to-eat cereals, other food prods. (Frosted Flakes, Rice Krispies, Froot Loops, Pop-Tarts, Nutri-Grain, Eggo).

Kimberly-Clark Corp.; PO Box 619100, Dallas, TX 75261-9100; (972) 281-1200; Website: http://www.kimberly-clark.com; Wayne R. Sanders; personal care prods. (Kleenex, Scott, Cottonelle, Huggies, Viva, Kotex).

King World Productions, Inc.; 1700 Broadway, NY, NY 10019; (212) 315-4000; Website: http://www.kingworld.com; Roger King; distributor of TV programs (*Oprah Winfrey Show, Wheel of Fortune, Jeopardy!, Inside Edition*).

Kmart Corp.; 3100 W. Big Beaver Rd., Troy, MI 48084; (248) 643-1000; Website: http://www.kmart.com; Floyd Hall; discount stores, home improvement centers (Builders Square).

Knight Ridder, Inc.; 50 W. San Fernando St., San Jose, CA 95113-2413; (408) 938-7700; Website: http://www.Knightridder.com; P. A. Ridder; newspaper publishing.

Kroger Co.; 1014 Vine St., Cincinnati, OH 45202; (513) 762-4000; Website: http://www.Kroger.com/; Joseph A. Pichler; largest U.S. retail grocery chain.

(Estee) Lauder Cos.; 767 5th Ave., NY, NY 10153; (212) 572-4200; Leonard A. Lauder; cosmetics (Clinique), fragrance prods. (Aramis, Aveda, Tommy Hilfiger).

La-Z-Boy Inc.; 1284 N. Telegraph Rd., Monroe, MI 48161; (734) 242-1444; Website: http://www.lazboy.com; Patrick H. Norton; reclining chairs, other furniture.

Leggett & Platt, Inc.; No. 1 Leggett Rd., Carthage, MO 64836; (417) 358-8131; Website: http://www.leggett.com; Harry M. Cornell Jr; furniture and furniture components.

Lehman Bros. Holdings, Inc.; 3 World Financial Ctr., NY, NY 10285; (212) 526-7000; Website: http://www.lehman.com; Richard S. Fuld Jr; investment bank.

Levi Strauss & Co.; 1155 Battery St., San Francisco, CA 94111; (415) 501-6000; Website: http://www.levistrauss.com; Robert D. Haas; blue jeans, casual sportswear.

Eli Lilly and Company; Lilly Corporate Center, Indianapolis, IN 46285; (317) 276-2000; Website: http://www.lilly.com; R. L. Tobias; pharmaceuticals (Axid, Ceclor, Prozac) and animal health prods.

The Limited, Inc.; 3 Limited Pkwy., P.O. Box 16000, Columbus, OH 43216; (614) 479-7000; Website: http://www.limited.com; Leslie H. Wexner; women's apparel stores (Lane Bryant, Lerner, Limited, Express, Structure, Victoria's Secret).

Litton Industries, Inc.; 21240 Burbank Blvd., Woodland Hills, CA 91367; (818) 598-5000; Website: http://www.littoncorp.com; Michael R. Brown; advanced electronic systems, information systems, marine engineering and production electronic components.

Liz Claiborne, Inc.; 1441 Broadway, NY, NY 10018; (212) 354-4900; Website: http://www.lizclaiborne.com; P. Charron; apparel, accessories.

Lockheed Martin Corp.; 6801 Rockledge Dr., Bethesda, MD 20817; (301) 897-6000; Website: http://www.mco.com; Vance Coffman; commercial and military aircraft, electronics, missiles.

Loews Corp.; 667 Madison Ave., NY, NY 10021; (212) 521-2000; James A. Tisch; tobacco prods. (Kent, True, Newport), watches (Bulova), hotels, insurance (CNA Fin'l.), offshore drilling.

Longs Drug Stores, Inc.; 141 N. Civic Dr., P.O. Box 5222, Walnut Creek, CA 94596; (925) 210-6624; Website: http://www.longs.com; R. M. Long; drug store chain.

Lowe's Cos., Inc.; Box 1111, N. Wilkesboro, NC 28656; (336) 658-4000; Website: http://www.lowes.com; Robert L. Tillman; building materials and home improvement superstores.

Luby's Cafeterias, Inc.; 2211 NE Loop 410, P.O. Box 33069, San Antonio, TX 78265; (210) 654-9000; Website: http://www.lubys.com; Barry Parker; operates cafeterias in S and SW.

Lucent Technologies, Inc.; 600 Mountain Ave., Murray Hill, NJ 07974; (908) 582-8500; Website: http://www.lucent.com; Richard A. McGinn; leading developer, designer, and manuf. of telecommunications systems, software, and prods.

Manpower Inc.; 5301 N. Ironwood Rd., Milwaukee, WI 53201; (414) 961-1000; Website: http://www.manpower.com; John Walter; largest non-gov't. employment services co. in the world.

Marriott International, Inc.; 10400 Fernwood Rd., Bethesda, MD 20817; (301) 380-3000; Website: http://www.marriott.com; John Willard Marriott Jr; hotels, retirement communities, food service dist.

Masco Corp.; 21001 Van Born Rd., Taylor, MI 48180; (313) 274-7400; Website: http://www.masco.com; Richard A. Manoogian; manuf. kitchen, bathroom prods. (Delta, Peerless faucets; Fieldstone, Merillat cabinets).

Mattel, Inc.; 333 Continental Blvd., El Segundo, CA 90245; (310) 252-2000; Website: http://www.mattel.com; Jill E. Barad; largest U.S. toymaker (Barbie, Fisher-Price, Hot Wheels, Matchbox, American Girls, Reader Rabbit).

May Department Stores Co.; 611 Olive St., St. Louis, MO 63101; (314) 342-6300; Website: http://www.maycompany.com; Jerome T. Loeb; department stores (Hecht's, Lord & Taylor, Filene's, Foley's).

Maytag Corp.; Newton, IA 50208; (515) 792-8000; Website: http://www.maytagcorp.com; Leonard A. Hadley; major appliance mfgr. (Magic Chef, Admiral, Jenn-Air), Hoover vacuum cleaners, floor care systems.

McDonald's Corp.; 1 McDonald's Plaza, Oak Brook, IL 60523; (630) 623-3000; Website: http://www.mcdonalds.com; Jack Greenberg; fast-food restaurants.

McGraw-Hill Cos.; 1221 Ave. of the Americas, NY, NY 10020; (212) 512-2000; Website: http://www.mcgraw-hill.com; Joseph L. Dionne; book, textbooks, magazine publishing (*Business Week*), information and financial services (Standard and Poor's), TV stations.

MCI WorldCom, Inc.; 515 E. Amite St., Jackson, MS 39201-2702; (601) 360-8600; Website: http://www.wcom.com; Bernard Ebbers; long-distance telephone service. (Co. announced 10/5/99 it had agreed to acquire Sprint Corp. in a stock swap valued at approx. $115 bil, which would be the largest acquisition in corporate history.)

McKesson HBOC Corp.; 1 Post St., San Francisco, CA 94104; (415) 983-8300; http://www.mckesson.com; Alan Seelenfreund; distributor of drugs and toiletries and provides software and services in U.S.; bottled water.

Mead Corporation; Courthouse Plaza NE, Dayton, OH 45463; (937) 495-6323; Website: http://www.mead.com; Jerome F. Tatar; printing and writing paper, paperboard, packaging, shipping containers.

Medtronic, Inc.; 7000 Central Ave. NE, Minneapolis, MN 55432; (612) 514-4000; Website: http://www.medtronic.com; W. W. George; world's largest manuf. of implantable biomedical devices.

Merck & Co., Inc.; PO Box 100, Whitehouse Station, NJ 08889-0100; (908) 423-1000; Raymond V. Gilmartin; pharmaceuticals (Pepcid, Zocor), animal health care prods.

Meredith Corp.; 1716 Locust St., Des Moines, IA 50336; (515) 284-3000; Website: http://www.meredith.com; William T. Kerr; magazine publishing (*Better Homes and Gardens, Ladies Home Journal*), book publishing, broadcasting.

Merrill Lynch & Co., Inc.; World Financial Ctr., N. Tower, NY, NY 10281-1332; (212) 449-1000; Website: http://www.ml.com; David H. Komansky; securities broker, financial services.

Metropolitan Life Ins. Co.; 1 Madison Ave., NY, NY 10010; (212) 578-2211; Website: http://www.metlife.com; Robert H. Benmosche; insurance, financial services.

Microsoft Corp.; 1 Microsoft Way, Redmond, WA 98052-6399; (425) 882-8080; Website: http://www.microsoft.com; William H. Gates; largest independent software maker (Windows, Word, Excel).

Minnesota Mining & Manuf. Co.; 3M Center, St. Paul, MN 55144-1000; (612) 733-1110; Website: http://www.mmm.com; L. D. DeSimone; abrasives, adhesives, electrical, health care, cleaning (Scotch-Brite, O-Cel-O sponges), printing, consumer prods. (Scotch Tape, Post-It).

Mirage Resorts, Inc.; 3400 Las Vegas Blvd. S, Las Vegas, NV 89109; (702) 791-7111; Website: http://www.mirage.com; Stephen A. Wynn; hotel-casino operator (Mirage, Treasure Island, Golden Nugget).

Mobil Corp.; 3225 Gallows Rd., Fairfax, VA 22037; (703) 846-3000; http://www.mobile.com; Lucio A. Noto; integrated international oil and petrochemical co.

Monsanto Company; 800 N. Lindbergh Blvd., St. Louis, MO 63167; (314) 694-1000; Website: http://www.monsanto.com; Robert B. Shapiro; agricultural prods., pharmaceuticals, consumer prods. (Equal, NutraSweet).

J. P. Morgan & Co.; 60 Wall St., NY, NY 10260; (212) 483-2323; Website: http://www.jpmorgan.com; Douglas A. Warner; global financial firm.

Morgan Stanley Dean Witter & Co.; 1585 Broadway, NY, NY 10036; (212) 761-4000; Website: http://www.msdw.com; Phillip J. Purcell; diversified financial services, major U.S. credit-card issuer.

Motorola, Inc.; 1303 E. Algonquin Rd., Schaumburg, IL 60196; (847) 576-5000; Website: http://www.motorola.com; G. L. Tooker; electronic equipment and components.

National Semiconductor Corp.; 2900 Semiconductor Dr., Santa Clara, CA 95052-8090; (408) 721-5000; Website: http://www.national.com; B. Halla; manuf. of semiconductors, integrated circuits.

Navistar Intl. Corp.; 455 N. Cityfront Plaza Dr., Chicago, IL 60611; (312) 836-2000; Website: http://www.navistar.com; John R. Horne; manuf. heavy-duty trucks, parts, school buses.

New York Times Co.; 229 W. 43d St., NY, NY 10036; (212) 556-3660; Website: http://www.nytco.com; A. O. Sulzberger Jr; newspapers (*Boston Globe*), radio and TV stations, magazines (*Golf Digest*).

Newell Rubbermaid Inc.; Newell Center, 29 E. Stephenson St., Freeport, IL 61032; (815) 235-4171; Website: http://www.newellco.com; John McDonough; Calphalon, WearEver cookware; Goody, Ace hair accessories; Anchor Hocking glassware; Levelor, Kirsch, Newell window treatments; Lee Ravan home storage; Eberhard Faber, Sanford writing instruments; Rolodex; Little Tykes, Graco, Century infant and juvenile prods.; Rubbermaid.

Nike, Inc.; 1 Bowerman Dr., Beaverton, OR 97005; (503) 671-6453; Website: http://www.NikeBiz.com; Philip H. Knight; athletic and leisure footwear, apparel.

Nordstrom, Inc.; 1501 5th Ave., Seattle, WA 98101; (206) 628-2111; Website: http://www.nordstrom.com; John J. Whitacre; upscale dept. store chain.

Norfolk Southern Corp.; 3 Commercial Pl., Box 227, Norfolk, VA 23510; (757) 629-2600; Website: http://www.nscorp.com; David R. Goode; operates railway, freight carrier.

Northrop Grumman Corp.; 1840 Century Park East, Los Angeles, CA 90067; (310) 553-6262; Website: http://www.northgrum.com; Kent Kresa; aircraft, electronics, data systems, missiles.

Northwest Airlines Corp.; 2700 Lone Oak Pkwy., Eagan, MN 55121; (612) 726-2111; Website: http://www.nwa.com; John H. Dasburg; air transportation.

Occidental Petroleum Corp.; 10889 Wilshire Blvd., Los Angeles, CA 90024; (310) 208-8800; Website: http://www.oxy.com; Ray R. Irani; oil, natural gas, chemicals, plastics, fertilizers.

Office Depot.; 2200 Old Germantown Rd., Delray Beach, FL 33445; (561) 278-4800; Website: http://www.officedepot.com; David I. Fuente; retail office supply stores.

Owens Corning; Fiberglass Tower, Toledo, OH 43659; (419) 248-8000; Website: http://www.owenscorning.com; Glen H. Hiner; world leader in advanced glass, composite materials.

Owens-Illinois; 1 SeaGate, Toledo, OH 43666; (419) 247-5000; J. H. Lemieux; Website: http://www.o-i.com world's largest producer of glass bottles.

Pacific Gas & Electric Corp. (PG&E); 77 Beale St., San Francisco, CA 94106; (800) 367-7731; Website: http://www.pgecorp.com; Robert D. Glynn Jr.; energy supplier.

PaineWebber Group, Inc.; 1285 Ave. of the Americas, NY, NY 10019-6028; Website: http://www.painewebber.com; Donald B. Marron; controls full-service securities firm.

J.C. Penney Co.; 6501 Legacy Dr., Plano, TX 75024; (972) 431-4757; Website: http://www.jcpenney.com; James E. Oesterreicher; dept. stores, catalog sales, drug stores (Eckerd, Fay's), insurance.

Pennzoil-Quaker State Co.; Pennzoil Pl., P.O. Box 2967, Houston, TX 77252; (713) 546-4000; Website: http://www.pennzoil-quakerstate.com; James L. Pate; automotive consumer products co., franchises Jiffy Lube and Q-Lube service centers.

PepsiCo, Inc.; PepsiCo World Headquarters, Purchase, NY 10577; (914) 253-2000; Website: http://www.pepsico.com; Roger A. Enrico; soft drinks (Pepsi-Cola, Mountain Dew), snacks (Ruffles, Lay's, Fritos, Doritos, Rold Gold). (Co. acquired Tropicana from Seagram Co. 8/25/98 for $3.3 bil cash.)

Pfizer, Inc.; 235 E. 42d St., NY, NY 10017; (212) 573-2323; Website: http://www.pfizer.com; W. C. Steere Jr; pharmaceuticals (Viagra, Zithromax), hospital, agricultural, chemical prods., consumer prods. (Visine eye drops, Ben-Gay pain relief).

Pharmacia & Upjohn, Inc.; 95 Corporate Dr., Bridgewater, NJ 08807; Website: http://www.pnu.com; Fred Hasson; pharmaceuticals (Motrin, Rogaine, Halcion, Xanax), chemicals, agricultural, health-care prods.

Philip Morris Cos. Inc.; 120 Park Ave., NY, NY 10017; (212) 880-5000; Geoffrey C. Bible; cigarettes (Marlboro, Merit, Virginia Slims), beer (Miller, Molson, Red Dog), Kraft Foods products (Jell-O, Maxwell House coffee, Kool-Aid, Oscar Mayer, Tang, Cheez Whiz and Velveeta cheese prods., Post cereals, Lender's Bagels, Tombstone Pizza, and Toblerone chocolate).

Phillips Petroleum Co.; Bartlesville, OK 74004; (918) 661-6600; Website: http://www.phillips66.com; W. W. Allen; integrated oil and petrochemical co.

Pillowtex Corp.; 1 Lake Circle Dr., Kannapolis, NC 28082; (704) 939-2000 Website: http://www.pillowtex.com; Chuck Hansen; household textile prods.

Pitney Bowes, Inc.; Walter H. Wheeler Jr. Dr., Stamford, CT 06926; (203) 356-5000; Website: http://www.pitneybowes.com; Michael J. Critelli; world's largest mfgr. of postage meters, mailing equip.

Polaroid Corp.; Technology Sq., Cambridge, MA 02139; (617) 386-2000; Website: http://www.polaroid.com; Gary T. DiCamillo; photographic equip. and supplies, optical goods.

PPG Industries, Inc.; 1 PPG Place, Pittsburgh, PA 15272; (412) 434-3131; Website: http://www.ppg.com; Raymond W. LeBoeuf; glass prods., fiberglass, chemicals; world's leading supplier of automobile/industrial coatings.

Premark Intl.; 1717 Deerfield Rd., Deerfield, IL 60015; (847) 405-6000; Website: http://www.premarkintl.com; Jim Ringler; food equip. (Hobart), home appliances and cookware (West Bend).

PRIMEDIA Inc.; 745 Fifth Avenue, New York, NY 10151; (212) 745-0100; Website: http://www.primediainc.com; Tom Rogers; consumer magazines (*New York, Seventeen, Modern Bride, American Baby, Soap Opera Digest*), professional magazines, classroom learning, business directories.

Procter & Gamble Co.; 1 Procter & Gamble Plaza, Cincinnati, OH 45202; (513) 983-1100; Website: http://www.pg.com;

John Pepper; soaps and detergents (Ivory, Cheer, Tide, Mr. Clean, Comet, Spic and Span, Zest), toiletries (Crest, Scope, Prell, Head and Shoulders, Noxzema, Oil of Olay, Old Spice), pharmaceuticals (Pepto-Bismol); Vicks cough medicines; Pampers and Luvs disposable diapers, Cover Girl and Max Factor cosmetics, Crisco shortening, Folger's coffee, Pringles, Charmin toilet tissues, Bounty towels, Tampax tampons.

Prudential Ins. Co. of America; 751 Broad St., Newark, NJ 07102-3777; (973) 802-6000; Website: http://www.prudential. com; Arthur F. Ryan; insurance, financial services.

Quaker Oats Co.; PO Box 049001, Chicago, IL 60604; (312) 222-7818; Website: http://www.quakeroats.com; Robert S. Morrison; cereal (Life, Cap'n Crunch), foods (Aunt Jemima, Rice-A-Roni), beverages (Gatorade).

Ralcorp Holdings, Inc.; 800 Market St., St. Louis, MO 63101; (314) 877-7000; Website: http://www.ralcorp.com; Joe R. Micheletto; private-label breakfast cereals, snack foods, baby food (Beech-Nut).

Ralston Purina Group.; Checkerboard Sq., St. Louis, MO 63164; (314) 982-2161; Website: http://www.ralston.com; W. P. Stiritz; world's largest producer of dog and cat food (Purina), and dry-cell batteries (Eveready, Energizer).

Raytheon Co.; 141 Spring St., Lexington, MA 02173; (781) 862-6600; Website: http://www.raytheon.com; Dennis J. Picard; defense systems, electronics.

Reader's Digest Assn., Inc.; Pleasantville, NY 10570; (914) 238-1000; Website: http://www.readersdigest.com; Thomas Ryder; direct-mail marketer of magazines, books, music and video prods.

Reebok Intl., Ltd.; 100 Technology Ctr. Dr., Stoughton, MA 02072; (781) 401-5000; Website: http://www.reebok.com; Paul Fireman; athletic and leisure footwear, apparel.

Revlon, Inc.; 625 Madison Ave., NY, NY 10022; (212) 527-4000; Website: http://www.revlon.com; Ronald O. Perelman; cosmetics, beauty aids, skin care.

Reynolds Metals Co.; 6601 W. Broad St., Richmond, VA 23230; (804) 281-2000; Website: http://www.rmc.com; Jeremiah J. Sheehan; aluminum prods.

Rite Aid Corp.; 30 Hunter Lane, Camp Hill, PA 17011-2404; (717) 761-2633; Website: http://www.riteaid.com; Martin Grass; discount drug stores.

RJR Nabisco Holdings Corp.; 1301 Ave. of the Americas, NY, NY 10019; (212) 258-5600; Website: http://www.rjrnabisco. com; Steven F. Goldstone; cigarettes (Winston, Salem, Camel), largest U.S. mfgr. of cookies and crackers (Oreo, Chips Ahoy!, Newton, SnackWell's, Ritz, Premium, Triscuit); condiments (Grey Poupon, A-1), confections (Life Savers, Breath Savers, Bubble Yum, Carefree), Milk-Bone dog biscuits, Parkay margarine, Planters peanuts.

Rockwell Intl. Corp.; 600 Anton Blvd., Suite 700, Costa Mesa, CA 92628-5090; (714) 424-4200; Website: http://www. rockwell.com; Donald H. Davis; diversified high-tech. co.

Ryder System, Inc.; 3600 NW 82d Ave., Miami, FL 33166; (305) 593-3726; Website: http://www.ryder.com; M. Anthony Burns; truck-leasing service.

Safeway Inc.; 5918 Stoneridge Mall Rd., Pleasanton, CA 94588-3229; (925) 467-3000; Website: http://www.safeway. com; Steven A. Burd; supermarkets.

Salomon Smith Barney; 7 World Trade Ctr., NY, NY 10048; (212) 783-7000; Website: http://www.smithbarney.com; Sandy Weill and John Reed; investment banking, securities and commodities trading.

Sara Lee Corp.; 3 First National Plaza, Chicago, IL 60602; (312) 726-2600; Website: http://www.saralee.com; John H. Bryan Jr; baked goods, fresh and processed meats (Ball Park, Jimmy Dean, Hillshire Farms, Kahn's), hosiery, intimate apparel and knitwear (Hanes, L'eggs, Playtex, Champion), Coach leather goods.

SBC Communications, Inc.; 175 E. Houston, San Antonio, TX 78205; (210) 821-4105; Website: http://www.sbc.com; Edward Whitacre Jr; telephone services (Southwestern Bell, Pacific Bell).

Schering-Plough Corp.; 1 Giralda Farms, Madison, NJ 07940; (973) 822-7000; Website: http://www.sch-plough.com; R.J. Kogan; pharmaceuticals (Claritin, Proventil), consumer prods. (Afrin, Coppertone), animal health prods.

Seagate Technology; 920 Disc Dr., Scotts Valley, CA 95066; (408) 438-6550; Website: http://www.seagate.com; Stephen J. Luczo; manuf. disk drives.

Sears, Roebuck and Co.; 3333 Beverly Rd., Hoffman Estates, IL 60179; (847) 286-2500; Website: http://www.sears.com; Arthur C. Martinez; 2nd largest U.S. retailer, department, specialty stores.

Service Merchandise Co., Inc.; PO Box 24600, Nashville, TN 37202-4600; (615) 660-6000; Website: http://www. servicemerchandise.com; Raymond Zimmerman; discount merchandiser and leading jewelry retailer.

Shaw Industries, Inc.; 616 E. Walnut Ave., Dalton, GA 30720; (706) 278-3812; Website: http://www.shawinc.com; Robert E. Shaw; world's largest carpet mfgr. (Armstrong, Magee, Cabin Craft).

Sherwin-Williams Co.; 101 Prospect Ave. NW, Cleveland, OH 44115-1075; (216) 566-2000; Website: http://www.sherwin. com; John G. Breen; largest North American paint and varnish producer (Dutch Boy, Pratt & Lambert, Minwax).

J. M. Smucker Co.; Strawberry Lane, Orrville, OH 44667; (216) 682-3000; Website: http://www.smuckers.com; T. P. Smucker; preserves, jams, jellies (Dickinson's), toppings (Magic Shell), syrups, juices.

Smurfit Stone Container Corp.; 150 N. Michigan Ave., Chicago, IL 60601; (312) 346-6600; Website: http://www. smurfit-stone.com; R. Curran; industry leader for corrugated containers, paper bags and sacks.

Sprint Corp.; PO Box 11315, Kansas City, MO 64112; (913) 624-3000; Website: http://www.sprint.com; William T. Esrey; long-distance and local telecommunications. (Co. announced 10/5/99 it had agreed to be acquired by MCI WorldCom in a stock swap valued at approx. $115 bil, which would make it the largest acquisition in corporate history.)

Staples, Inc.; 500 Staples Dr., Framingham, MA 01702; (508) 253-5000; Website: http://www.staples.com; Thomas Stemberg; office-supply superstores.

Starwood Hotels and Resorts Worldwide; 777 Westchester Ave., White Plains, NY 10604; (914) 640-8100; Website: http:// www.starwoodlodging.com; Barry S. Sternlicht; hotels and leisure company.

State Farm Mutual Automobile Ins. Co.; 1 State Farm Plaza, Bloomington, IL 61701; (309) 766-2311; Website: http://www. statefarm.com; Edward B. Rust Jr; major insurance co.

Stride Rite Corp.; 191 Spring St., P.O. Box 9191, Lexington, MA 02173; (617) 824-6000; Website: http://www.striderite. com; James Eskridge; high-quality adult's and children's footwear (Keds, Sperry Top-Sider).

Sun Microsystems, Inc.; 2550 Garcia Ave., Mountain View, CA 94043; (415) 960-1300; Website: http://www.sun.com; Scott G. McNealy; supplier of network-based distributed computer systems (Java programming language).

Sunoco, Inc.; 1801 Market St., Philadelphia, PA 19103-1699; (215) 977-3000; R. H. Campbell; energy resources co., markets Sunoco gasoline.

SUPERVALU Inc.; PO Box 990, Minneapolis, MN 55440; (612) 828-4000; Website: http://www.supervalu.com; Michael W. Wright; food wholesaler, retailer.

Sysco Corp.; 1390 Enclave Pkwy, Houston, TX 77077-2099; (281) 584-1390; Website: http://www.sysco.com; John F. Woodhouse; leading U.S. food distributor.

Tandy Corp.; 1800 One Tandy Center, Fort Worth, TX 76102; (817) 390-3700; Website: http://www.tandy.com; Leonard H. Roberts; consumer electronics retailer (Computer City, Radio Shack).

Tenneco, Inc.; 1275 King St., Greenwich, CT 06831; (203) 863-1000; Website: http://www.tenneco.com; Dana G. Mead; packaging materials, (Hefty, Baggies), automotive parts (Monroe, Walker).

Texaco Inc.; 2000 Westchester Ave., White Plains, NY 10650; (914) 253-4000; Website: http://www.texaco.com; Peter I. Bijur; integrated international oil co.

Texas Instruments Inc.; 8505 Forest Lane, P.O. Box 660199,, Dallas, TX 75266-0199; (972) 995-3773; Website: http://www. ti.com; T. J. Engibous; electronics.

Textron, Inc.; 40 Westminster St., Providence, RI 02903; (401) 421-2800; Website: http://www.textron.com; Lewis B. Campbell; aerospace, industrial, automotive prods., financial services.

Times Mirror Publishing Co.; Times Mirror Sq., Los Angeles, CA 90053; (213) 237-3700; Website: http://www.tm.com; M. H. Willes; newspapers, magazines (*Field & Stream, Popular Science*), professional books (Matthew Bender).

Time Warner Inc.; 75 Rockefeller Plaza, NY, NY 10020; (212) 522-1212; Website: http://www.timewarner.com; Gerald M. Levin; magazine publishing (*Time, Sports Illustrated, Fortune, Money, People,* DC Comics), TV and CATV (WB Network, HBO, Cinemax, CNN, TBS, TNT), book publishing (Little, Brown; Warner Books), motion pictures (Warner Bros.), recordings, sports teams (Atlanta Braves, Atlanta Hawks), retailing (Warner Bros. stores).

The TJX Cos., Inc.; 770 Cochituate Rd., Framingham, MA 01701; (508) 390-1000; John Nelson; world's largest off-price apparel retailer (T.J. Maxx, Marshalls).

Tootsie Roll Industries, Inc.; 7401 S. Cicero Ave., Chicago, IL 60629; (773) 838-3400; M. J. Gordon; candy (Tootsie Roll, Mason Dots, Charms, Sugar Daddy, Charleston Chew, Junior Mints).

Toro Co.; 8111 Lyndale Ave. S, Bloomington, MN 55420; (612) 888-8801; Website: http://www.toro.com; Kendrick B. Melrose; lawn and turf maintenance (Lawn-Boy), snow removal equipment, lighting and irrigation systems.

Toys "R" Us; 461 From Rd., Paramus, NJ 07652; (201) 262-7800; Website: http://www.toysrus.com; Michael Goldstein; world's largest children's specialty retailer (Toys "R" Us, Kids "R" Us, Babies "R" Us).

Transamerica Corp.; 600 Montgomery St., San Francisco, CA 94111; (415) 983-4000; Website: http://www.transamerica.com; Frank C. Herringer; insurance, financial services.

Triarc Cos., Inc.; 280 Park Ave., NY, NY 10017; .(212) 451-3000; Website: http://www.triarc.com Nelson Peltz; fast-food restaurants (Arby's), beverages (Royal Crown, Mystic, Nehi, Snapple, Stewart's).

Tribune Co.; 435 N. Michigan Ave., Chicago, IL 60611; (312) 222-9100; Website: http://www.tribune.com; J. W. Madigan; newspaper and book publishing, broadcasting, Chicago Cubs baseball team.

TRICON Global Restaurants, Inc.; 1441 Gardiner Lane, Louisville, KY 40213; (502) 874-8300; Website: http://www.triconglobal.com; Andrall E. Pearson; fast food (Pizza Hut, KFC, Taco Bell).

Trinity Industries, Inc.; P.O. Box 568887, 2525 Stemmons Freeway, Dallas, TX 75207; (214) 631-4420; Website: http://www.trin.net; Timothy R. Wallace; manufactures metal prods., rail and freight prods.

TRW Inc.; 1900 Richmond Rd., Cleveland, OH 44124; (216) 291-7000; Website: http://www.trw.com; J. T. Gorman; car and truck operations, electronics, space and defense systems.

Tyco Intl., Ltd.; 1 Tyco Pk., Exeter, NH 03833; (603) 778-9700; Website: http://www.tycoint.com; L. D. Kozlowski; fire protection systems, pipes, power cables, medical supplies, packaging.

Tyson Foods, Inc.; 2210 W. Oaklawn, Springdale, AR 72764; (501) 290-4000; Website: http://www.tyson.com; Leland Tollett; fresh and processed poultry and seafood prods. (Holly Farms, Weaver, Louis Kemp).

UAL Corp.; 1200 E. Algonquin Rd., Elk Grove Twp., IL 60007; (708) 952-4000; Website: http://www.ual.com; Gerald Greenwald; air transportation (United Airlines).

Union Carbide Corp.; 39 Old Ridgebury Rd., Danbury, CT 06817; (203) 794-6440; Website: http://www.unioncarbide.com; William H. Joyce; chemicals.

Union Pacific Corp.; 1717 Main St., Suite 5900, Dallas, TX 75201; (214) 743-5600; Website: http://www.up.com; Richard Davidson; largest railroad, trucking co. in U.S.

Unisys Corp.; PO Box 500, Blue Bell, PA 19424-0001; (215) 986-5777; Website: http://www.unisys.com; Lawrence A. Weinbach; designs, manuf. computer information systems and related prods..

UnitedHealth Group Corp.; 300 Opus Center, 9900 Bren Rd. East, Minnetonka, MN 55343; (612) 936-1300; Website: http://www.uhc.com; William W. McGuire; owns, manages health maintenance organizations.

United Parcel Service of America, Inc.; 55 Glenlake Pkwy. NE, Atlanta, GA 30328; (404) 828-6000; Website: http://www.ups.com; James Kelley; courier services, truck rentals.

United Technologies Corp.; 1 Financial Plaza, Hartford, CT 06101; (860) 728-7000; Website: http://www.utc.com; George David; aerospace, industrial prods. and services (Otis Elevator, Pratt & Whitney, Sikorsky Aircraft).

Unocal Corp.; 2141 Rosecrans Ave., Ste. 4000, El Segundo, CA 90245; (310) 726-7667; Website: http://www.unocal.com; Roger Beach; integrated oil co.

US Airways Group, Inc.; 2345 Crystal Dr., Arlington, VA 22202; (703) 872-5306; Website: http://www.usairways.com; Stephen M. Wolf; air transportation.

UST Inc.; 100 W. Putnam Ave., Greenwich, CT 06830; (203) 661-1100; Website: http://www.ustshareholder.com; Vincent A. Gierer Jr.; smokeless tobacco (Copenhagen, Skoal), pipe tobacco, wine (Chateau St. Michelle, Conn Creek, Columbia Crest).

USX-Marathon Group; 600 Grant St., Pittsburgh, PA 15230; (412) 433-1121; Website: http://www.marathon.com; Thomas J. Usher; integrated oil co.

Venator Group; 233 Broadway, NY, NY 10279; (212) 553-2000; Website: http://www.venatorgroup.com; Roger Farah; operates retail stores: shoes (Kinney), apparel (Northern group), athletic footwear (Foot Locker), athletic merchandise (Champs), San Francisco Music Box Company.

V.F. Corp.; 628 Green Valley Rd., Suite 500, Greensboro, NC 27408; (336) 547-6000; Website: http://www.vfc.com; M. McDonald; apparel (Lee, Wrangler jeans, Vanity Fair, Healthtex, Jantzen).

Viacom, Inc.; 1515 Broadway, NY, NY 10036; (212) 258-6000; Website: http://www.viacom.com; Sumner M. Redstone; TV broadcast stations and cable systems, channels (Showtime, MTV, VH-1, Nickelodeon); book publishing (Simon & Schuster, Macmillan); produces, distributes movies, TV shows (Paramount); video stores (Blockbuster), theme parks. (Co. announced 9/7/99 it would acquire CBS Corp. for $37.3 bil, which would be the largest media merger ever.)

Walgreen Co.; 200 Wilmot Rd., Deerfield, IL 60015; (847) 940-2500; Website: http://www.walgreens.com; Charles R. Walgreen 3d; nation's largest drugstore chain.

Wal-Mart Stores, Inc.; Box 116, Bentonville, AR 72716; (501) 273-4000; Website: http://www.wal-mart.com; S. Robson Walton; world's largest retailer; discount stores, wholesale clubs.

Warner-Lambert Co.; 201 Tabor Rd., Morris Plains, NJ 07950-2693; (973) 540-2000; Website: http://www.warner-lambert.com; Lodewijk J.R. de Vink; personal health care and consumer prods. (Benadryl, Listerine, Schick, Dentyne, Trident gum, Chiclets, Efferdent dental cleanser, Certs mints, Halls lozenges).

Washington Post Co.; 1150 15th St. NW, Washington, DC 20071; (202) 334-6000; Website: http://www.washpostco.com; D. E. Graham; newspapers, *Newsweek* magazine, TV and CATV stations, Stanley H. Kaplan Educational Centers.

Waste Management; 1001 Fannin, Suite 4000, Houston, TX 77002; (713) 512-6548; Website: http://www.wm.com; John Drury; N. America's largest solid waste collection and disposal co.

Wells Fargo & Co.; 420 Montgomery St., San Francisco, CA 94163; (415) 396-3606; Website: http://www.wellsfargo.com; Paul Hazen; bank holding co.

Wendy's Intl., Inc.; 4288 W. Dublin-Granville Rd., Dublin, OH 43017; (614) 764-3100; Website: http://www.wendys.com; Gordon F. Teter; quick-service restaurants.

Weyerhaeuser Co.; Tacoma, WA 98477; (253) 924-2345; Website: http://www.weyerhaeuser.com; George H. Weyerhaeuser; world's largest private owner of softwood timber, distrib. paper and wood prods.

Whirlpool Corp.; Benton Harbor, MI 49022; (616) 923-5000; Website: http://www.whirlpool.com; David Whitwam; world's largest manuf. of major home appliances (KitchenAid, Kenmore, Roper).

Whitman Corp.; 3501 Algonquin Rd., Rolling Meadows, IL 60008; (708) 818-5000; Website: http://www.whitmancorp.com; Bruce S. Chelberg; beverage bottler and distributor (Pepsi-Cola).

Winn-Dixie Stores, Inc.; 5050 Edgewood Ct., Jacksonville, FL 32205; (904) 783-5000; Website: http://www.winn-dixie.com; A. Dano Davis; supermarkets.

Winnebago Industries, Inc.; PO Box 152, Forest City, IA 50436; (515) 582-3535; Website: http://www.winnebagoind.com; Bruce D. Hertzke; manuf. and financing of motor homes, recreational vehicles.

Wm. Wrigley Jr. Co.; 410 N. Michigan Ave., Chicago, IL 60611; (312) 644-2121; Website: http://www.wrigley.com; William Wrigley; world's largest mfgr. of chewing gum.

Xerox Corp.; PO Box 1600, Stamford, CT 06904; (203) 968-3000; Website: http://www.xerox.com; Paul Allaire; copiers, printers, document publishing equip.

Who Owns What: Familiar Consumer Products

Listed here are consumer brands and their parent companies. For company address and website, see Business Directory.

A-1 steak sauce: RJR Nabisco
ABC broadcasting: Walt Disney
Admiral appliances: Maytag
Advil: American Home Products
Ajax cleanser: Colgate-Palmolive
Almond Joy candy bar: Hershey
American Girl: Mattel
Anacin: American Home Products
Arm & Hammer: Church & Dwight
Arnold breads: Bestfoods
Arrid antiperspirant: Carter-Wallace
Aunt Jemima Pancake mix: Quaker Oats
Aunt Millie's pasta sauce: Borden
Baggies: Tenneco
Ban antiperspirant: Bristol-Myers Squibb
Banana Republic stores: The Gap
Band-Aids: Johnson & Johnson
Barbie dolls: Mattel
Beech-Nut baby food: Ralcorp
Ben-Gay: Pfizer
Betty Crocker prods.: General Mills
Black Flag insecticides: Clorox
Blockbuster video stores: Viacom
Bounty paper towels: Proctor & Gamble
Breck shampoo: Dial
Brillo soap pads: Church & Dwight
Brita water systems: Clorox
Bubble Yum gum: RJR Nabisco
Budweiser beer: Anheuser-Busch
Bufferin: Bristol-Myers Squibb
Bulova watches: Loews
Business Week magazine: McGraw-Hill
Buster Brown shoes: Brown Group
Cadbury: Hershey
Cap'n Crunch cereal: Quaker Oats
Calphalon cookware: Newell Rubbermaid
Certs mints: Warner-Lambert
Charmin toilet tissue: Proctor & Gamble
Cheer detergent: Proctor & Gamble
Cheerios cereal: General Mills
Cheez Whiz: Philip Morris
Cinemax: Time Warner
Clairol hair prods.: Bristol-Myers Squibb
Clorets breath mints: Warner-Lambert
CNN: Time Warner
Coach leather goods: Sara Lee
Combat insecticides: Clorox
Comet cleanser: Proctor & Gamble
Coppertone sun care prods.: Schering-Plough
Crest toothpaste: Proctor & Gamble
Crisco shortening: Proctor & Gamble
Doritos chips: PepsiCo
Dristan: American Home Prods.
Duracell batteries: Gillette
Dutch Boy paints: Sherwin-Williams
Efferdent dental cleanser: Warner-Lambert
Elmer's glue: Borden
ESPN: Walt Disney
Eveready batteries: Ralston Purina
Excedrin: Bristol-Myers Squibb
Fab detergent: Colgate-Palmolive
Fantastik: S.C. Johnson
Fisher Price Toys: Mattel
Foamy shaving cream: Gillette
Folger's coffee: Proctor & Gamble
Formula 409 spray cleaner: Clorox
Franco-American spaghetti: Campbell Soup
Frito-Lays snacks: PepsiCo
Fruitopia drinks: Coca-Cola

Gatorade: Quaker Oats
Godiva chocolate: Campbell Soup
Halcion: Pharmacia & Upjohn
Hamburger Helper: General Mills
Hanes hosiery: Sara Lee
Hawaiian Punch: Procter & Gamble
HBO: Time Warner
Head and Shoulders shampoo: Procter & Gamble
Healthtex: V.F. Corp.
Hellmann's mayonnaise: Bestfoods
Hi-C fruit drinks: Coca-Cola
Hidden Valley prods.: Clorox
Hillshire Farms meats: Sara Lee
Holly Farms: Tyson
Hostess cakes: Interstate Bakeries
Huggies diapers: Kimberly-Clark
Ivory soap: Procter & Gamble
Jack Daniel's Whiskey: Brown-Forman
Java programming language: Sun Microsystems
Jell-O: Philip Morris
Jenn-Air stoves: Maytag
Jif peanut butter: Procter & Gamble
Jim Beam bourbon: Fortune Brands
Keds footwear: Stride Rite
Ken-L-Ration pet foods: H. J. Heinz
Kent cigarettes: Loews
KFC restaurants: TRICON
Kinney shoe stores: Venator Group
KitchenAid appliances: Whirlpool
Kit Kat candy: Hershey's
Kleenex: Kimberly-Clark
Knorr soups: Bestfoods
Kool-Aid: Philip Morris
Krazy Glue: Borden
Kwikset doorknobs: Black & Decker
Ladies Home Journal magazine: Meredith
Lee jeans: V.F. Corp.
L'eggs hosiery: Sara Lee
Lender's bagels: Phillip Morris
Lenox china: Brown-Forman
Lerner stores: The Limited
Life Savers candy: RJR Nabisco
Listerine mouthwash: Warner-Lambert
Lord & Taylor: May Dept. Stores
Marlboro cigarettes: Philip Morris
Maxwell House coffee: Philip Morris
Mazola oils and margarine: Bestfoods
Michelob beer: Anheuser-Busch
Miller beer: Philip Morris
Milton Bradley games: Hasbro
Minute Maid juices: Coca-Cola
MTV: Viacom
Nature Valley granola bars: General Mills
NBC broadcasting: General Electric
Neutrogena soap: Johnson & Johnson
Newsweek magazine: Washington Post
9 Lives cat food: H.J. Heinz
Oil of Olay: Procter & Gamble
Old Navy Clothing: The Gap
Oreo cookies: RJR Nabisco
Oscar Mayer meats: Philip Morris
Pampers: Procter & Gamble
PaperMate pens: Gillette
People magazine: Time Warner
Pepperidge Farm prods.: Campbell Soup
Pepto-Bismol: Procter & Gamble
Pine-Sol cleaner: Clorox
Pizza Hut restaurants: TRICON
Planters nuts: RJR Nabisco

Playskool toys: Hasbro
Playtex apparel: Sara Lee
Post cereals: Philip Morris
Post-It stickers: Minn. Mining & Manuf.
Prego pasta sauce: Campbell Soup
Prell shampoo: Procter & Gamble
Prentice Hall publishing: Viacom
Prozac: Eli Lilly
Radio Shack retail outlets: Tandy
Red Dog beer: Philip Morris
Reese's candy: Hershey
Rice-A-Roni: Quaker Oats
Rice Krispies: Kellogg
Right Guard deodorant: Gillette
Ritz crackers: RJR Nabisco
Robitussin: American Home Products
Rogaine hair growth aide: Pharmacia & Upjohn
Rolaids antacid: Warner-Lambert
Ronzoni pasta: Hershey
Ruffles chips: PepsiCo
San Francisco Music Box Co.: Venator Group
San Giorgio pasta: Hershey
Schick razors: Warner-Lambert
Scope mouthwash: Procter & Gamble
Scotch tape: Minn. Mining & Manuf.
Seventeen magazine: PRIMEDIA
Simon & Schuster publishing: Viacom
Skippy peanut butter: Bestfoods
SnackWell's cookies: RJR Nabisco
Snapple beverages: Triarc
S.O.S. cleanser: Clorox
Southern Comfort liquor: Brown-Forman
SPAM meat: Hormel
Spic and Span: Proctor & Gamble
Sports Illustrated magazine: Time Warner
Sprite soda: Coca-Cola
Star-Kist tuna: H.J. Heinz
Sugar Twin: Alberto Culver
Swanson frozen dinners: Campbell Soup
Taco Bell restaurants: TRICON
Tampax tampons: Procter & Gamble
Thomas' English muffins: Bestfoods
Tide detergent: Procter & Gamble
Titleist: Fortune Brands
Tombstone pizza: Philip Morris
Trident gum: Warner-Lambert
Triscuits: RJR Nabisco
Trojan condoms: Carter-Wallace
Tylenol: Johnson & Johnson
Ultra Brite toothpaste: Colgate-Palmolive
USA Today newspaper: Gannett
V-8 vegetable juice: Campbell Soup
Vanity Fair apparel: V.F. Corp.
Velveeta cheese prods.: Philip Morris
Viagra: Pfizer
Vicks cough medicines: Procter & Gamble
Victoria's Secret stores: The Limited
Visine eye drops: Pfizer
Wall Street Journal: Dow Jones
Waterman pens: Gillette
Weight Watchers: H.J. Heinz
Wheaties cereal: General Mills
Windex: S.C. Johnson
Windows software applications: Microsoft
Wise snacks: Borden
Wonder bread: Interstate Bakeries
Zest soap: Procter & Gamble
Ziploc storage bags: S.C. Johnson

At-Home Shopping—Consumer Tips and Rights

Source: Federal Trade Commission, Consumer Information Center; American Express

TIPS

Deal only with reliable firms. Check with your local consumer protection agency or the Better Business Bureau (BBB) nearest the business.

Review the advertising offer carefully.

Inquire about warranty, refund, and exchange policies.

Never send cash. Pay by money order, check, charge, or credit card so that you have a record of your purchase.

Keep the ad you responded to and a copy of the order form. If there is no order form, record the company's name, address, phone number, date, the item you purchased, amount paid, and the promised delivery date.

Be careful about giving out your credit, debit, charge card, or bank account number.

RIGHTS

Late deliveries. By federal law, a company must ship your order within 30 days, unless the advertisement promises a different shipping time. If the company cannot ship in time, it must give you an "Option Notice." You can either wait longer or cancel and get a prompt refund. If you cancel and your order was charged, the seller has one billing cycle to tell the card issuer to credit your account.

The following are exceptions to this rule:

(1) If a company does not promise a shipping time and if you are applying for credit to pay for your purchase, the company has 50 days after receiving your order to ship.

(2) Other exceptions include spaced deliveries such as magazine subscriptions (except for 1st shipment), items that continue until you cancel (e.g., book or record clubs), COD orders, services, and seeds or growing plants.

Unordered merchandise. If you are shipped a product that you did not order, it's yours. It is illegal for a company to pressure you to pay for it or to return it.

Damaged or spoiled items. If damage is obvious, and if you decide not to accept the package, write "REFUSED" on the wrapper and return it unopened to the seller. No new postage is needed, unless the package came by insured, registered, certified, or COD mail and you signed for it.

Disputes or billing errors. If there is a problem with your order—you were billed the wrong amount, you never got the product, the goods were damaged or merchandise or services were misrepresented—these steps are suggested:

(1) Write immediately to the company, explaining the problem and asking for a specific resolution. Include your name, address, and daytime phone number, order or invoice number, and a copy of the canceled check.

(2) If you charged your purchase or arranged for payment to be withdrawn from a bank account, send a copy of your letter to the card issuer or bank.

You usually have 60 days to dispute charges.

Postal rules allow you to write a check payable to the sender, rather than to the delivery company, on COD orders. If, after examining the merchandise, you believe that there has been misrepresentation or fraud, you can then stop payment on the check and file a complaint with the U.S. Postal Inspector's Office.

ON THE INTERNET

When shopping on the Internet:

Consider using a secured browser, which will encrypt or scramble purchase information that can be intercepted.

If you do not have encryption software, consider shopping by mail, fax, or phone.

If you are unfamiliar with a company, ask for a paper brochure or catalog in the mail.

Be cautious about giving out personal information. It is rarely necessary to give your Social Security number. Never give out your Internet password.

Print out a copy of your order and confirmation number for your records.

For further questions, contact: The Federal Trade Commission, Public Reference, Washington, DC 20580; 202-326-2222; or website at http://www.ftc.gov

CONSUMER INFORMATION CATALOG

The *Consumer Information Catalog* is a listing of more than 200 federal publications, covering at-home shopping and other consumer topics. Many are available free.

Write Consumer Information Catalog, Pueblo, CO 81009, or phone 1-888-8PUEBLO. Publications listed in the catalog are also available online, along with other consumer information, at http://www.pueblo.gsa.gov

Top 10 Shopping Websites

Source: Media Matrix, Inc.

Rank	Site name	Website address	Visitors[1]
1.	Amazon.com	http://www.amazon.com	11,618
2.	AOL Shopping Channel	http://www.aol.com/shopping/home.html	9,509
3.	Bluemountainarts.com	http://www1.bluemountain.com	9,168
4.	Ebay.com	http://www.ebay.com	8,790
5.	Barnesandnoble.com	http://www.barnesandnoble.com	4,819
6.	CNET Software Download Services.	http://www.download.com/?tag=st.cn.1.dir.dl.	4,589
7.	Cdnow.com	http://www.cdnow.com	4,305
8.	Mypoints.com	http://www.mypoints.com	4,103
9.	Valupage.com	http://www.valupage.com	3,543
10.	Freeshop.com	http://www.freeshop.com	3,518

(1) Number of visitors (in thousands) who visited website at least once in Aug. 1999.

The Cost of Raising a Child

Source: Center for Nutrition Policy and Promotion, U.S. Dept. of Agriculture

Estimated annual expenditures in 1998 dollars for a child born in 1998, by income group. Estimates are for the younger child in a 2-parent family with 2 children, for the overall U.S.

Year	Age of child	Income group[1] Low	Income group[1] Middle	Income group[1] High	Year	Age of child	Income group[1] Low	Income group[1] Middle	Income group[1] High
1998	under 1	$5,950	$8,240	$12,260	2008	10	$9,830	$13,410	$19,500
1999	1	6,230	8,630	12,840	2009	11	10,290	14,040	20,420
2000	2	6,520	9,030	13,440	2010	12	12,180	15,960	22,850
2001	3	6,960	9,710	14,380	2011	13	12,750	16,710	23,930
2002	4	7,280	10,170	15,060	2012	14	13,350	17,500	25,050
2003	5	7,620	10,640	15,760	2013	15	13,780	18,600	26,910
2004	6	8,140	11,220	16,390	2014	16	14,430	19,480	28,170
2005	7	8,520	11,750	17,160	2015	17	15,110	20,390	29,500
2006	8	8,920	12,300	17,960					
2007	9	9,390	12,810	18,630	**TOTAL**		**$177,250**	**$240,590**	**$350,210**

(1) In 1998, low annual income is less than $36,000 (average in this range=$22,500); middle income is $36,000–$60,600 (average = $47,900); high income is $60,600 or more (average = $90,700). Projected annual inflation rate is 4.7%.

Interest Laws and Consumer Finance Loan Rates

Source: Revised by Christian T. Jones, Editor, *Consumer Finance Law Bulletin*, Evansville, IN

All states have laws regulating interest rates. These laws fix a legal or conventional rate, which applies when there is no contract for interest. They also fix a general maximum contract rate, but there are so many exceptions that the general contract maximum actually applies to few cases. Also, federal law has preempted state limits on first home mortgages, subject to each state's right to reinstate its own law, and has given depository institutions parity with other state lenders.

Legal rate of interest. The legal or conventional rate of interest applies to money obligations when no interest rate is contracted for, and also to judgments. The rate is usually somewhat below the general contract interest rate.

General maximum contract rates. General interest laws in most states set the maximum contract rate between 8% and 16% per year. Loans to corporations are frequently exempted or subject to a higher maximum. In recent years, it also has been common to provide special rates for home mortgage loans and variable usury rates that are indexed to market rates.

Specific enabling acts. In many states special statutes permit industrial loan companies, second mortgage lenders, and banks to charge 1.5% a month or more. Laws regulating revolving loans, charge accounts, and credit cards generally limit rates to 1.5%-2% per month, plus annual fees for credit cards. Rates for installment sales contracts in most states are somewhat higher. Credit unions generally charge 1%-1.5% a month. Pawnbrokers' rates vary widely. Savings and loan associations and loans insured by federal agencies also are specially regulated. A number of states allow regulated lenders and credit sellers to charge any rate agreed to with the customer for all credit or for credit over a certain amount.

Consumer finance loan statutes. Most consumer finance loan statutes are based on early models drafted by the Russell Sage Foundation (1916-42) to provide small loans to wage earners under license and other protective regulations. Since 1969, the model has frequently been the Uniform Consumer Credit Code, which applies to credit sales and loans for consumer purposes. In general, licensed lenders may charge 3% per month, with reduced rates for higher amounts. An add-on of 17% ($17 per $100) per year amounts to about 2.5% per month if paid in equal monthly installments. Add-on rates are computed on the original principal, not taking into account reduced balances as payments are made. Discount rates are computed on the whole balance of the loan, including interest, to determine how much cash is paid out; thus, for a $1,000 loan at 10% discount rate the borrower would receive only $900.

In the table here, unless otherwise stated, monthly and annual rates are based on reducing principal balances, annual add-on rates are based on the original principal for the full term, and 2 or more rates apply to different portions of the balance or original principal.

Loan Regulations by State

(maximum monthly rates, unless otherwise indicated; as of Sept. 1999)

AL.....Yearly add-on: 15% to $750, 10% to $2,000 (min. 1.5% on unpaid balances). Higher rates to $749. Over $2,000, any agreed rate. Fee: 6% to $2,000; 5% for 2d mortgages.

AK ....3% to $850, 2% to $10,000. Over $10,000, any agreed rate.

AZ.....To $1,000: 3%. Over $1,000: 3% to $500, 2% to $10,000. Over $10,000, any agreed rate. Fee: 4% for 2d mortgages.

CA ....2.5% to $225, 2% to $900, 1.5% to $1,650, 1% to $2,500 (1.6% min.). Over $2,500, any agreed rate. 5% fee (max. $50-$75) to $5,000.

CO ....36% per year to $630, 21% to $2,100, 15% to $25,000 (21% min.).

CT ....Annual add-on: 17% to $600, 11% to $5,000; 11% over $1,800 to $15,000 for certain secured loans. Any agreed rate for 2d mortgages; 8% fee.

DE ....Any agreed rate; 10% fee.

DC ....24% per year.

FL.....30% per year to $2,000, 24% to $3,000, 18% to $25,000; $10 fee.

GA ....10% per year discount to 18 months, add-on to 36 1/2 months; 8% fee to $600, 4% on excess plus $2 per month. Over $3,000, any agreed rate.

HI.....3.5% to $100, 2.5% to $300; 2% on entire balance over $300 or discount rates.

ID .....Any agreed rate.

IL .....Any agreed rate. Fee: 3% for 2d mortgages; $25 for other loans.

IN .....36% per year to $930, 21% to $3,100, 15% to $25,000 (21% min.). Fee: 2% for 2d mortgages; and to $2,000 for other loans.

IA .....3% to $1,000, 2% to $2,800, 1.5% to $10,000; or equivalent flat rate. Over $10,000, 21% per year.

KS ....36% per year to $860, 21% to $2,860, 14.45% to $25,000 (18% min.). Fee: 2% (max. $100); 8% for 2d mortgages.

KY ....3% to $1,000, 2% to $3,000. Over $3,000, 2%.

LA.....36% per year to $1,400, 27% to $4,000, 24% to $7,000, 21% over $7,000, plus $25 fee. Higher rates to $500.

ME ....30% per year to $2,000, 24% to $4,000, 15% to $8,000; 18% on entire balance over $8,000 to $35,000.

MD ....2.75% to $1,000, 2% to $2,000. Over $2,000, 2%.

MA ....23% per year plus $20 annual fee to $6,000; any agreed rate over $6,000.

MI.....25% per year plus 2% fee; 25% for 2d mortgages, plus 5% fee.

MN ....33% per year to $750, 19% over $750 (21.75% min.) plus $25 fee to $4,230.

MS ....36% per year to $1,000, 33% to $1,800, 24% to $5,000, 14% over $5,000. Over $25,000, 18%; 2% fee (max. $50).

MO ....No limits, plus 5% fee (maximum $50 for non-real estate loans). Special rates to $500.

MT ....Any agreed rate.

NE ....24% per year to $1,000. 21% over $1,000, plus fee of 7% to $2,000 and 5% over $2,000 (max. $500). Any agreed rate for real estate loans of $7,500 or more or all loans over $25,000.

NV ....Any agreed rate.

NH ....Any agreed rate.

NJ ....30% per year to $5,000 and for 2d mortgages.

NM ....Any agreed rate.

NY ....25% per year.

NC ....2.5% to $1,000, 1.5% to $7,500; 1.5% on entire amount to $10,000. 1.5% or variable plus 2% fee for 2d mortgages. Higher rates to $3,000.

ND ....2.5% to $250, 2% to $500, 1.75% to $750, 1.5% to $1,000; any agreed rate for amounts over $1,000 up to $35,000.

OH ....28% per year to $1,000, 22% to $5,000; 25% on entire amount over $5,000; plus fee.

OK ....30% per year to $1,120, 21% to $3,400, 15% to $45,000 (21% min.). Higher rates to $680.

OR ....Any agreed rate.

PA ....9.5% per year discount to 48 months, 6% for remaining time plus 2% fee (max. $100); or 2% on unpaid balances. 1.85% for 2d mortgages over $5,000, plus 2% fee.

PR ....21% per year to $2,000.

RI.....3% to $300, 2.5% for loans between $300 and $800; 2% for larger loans to $5,000. 1.75% over $5,000.

SC ....Any agreed and posted rate.

SD   Any agreed rate.

TN ....Over $100, 24% per year or discount rates plus fees.

TX ....Annual add-on: 18% to $1,410, 8% to $11,750 or formula rate (currently 18% per year on unpaid balances). Higher rates to $460.

UT ....Any agreed rate.

VT ....2% to $1,000, 1% to $3,000 (min. 1.5%); 1.5% for 2d mortgages.

VA ....3% to $2,500; any agreed rate to $6,000. Any agreed rate for 2d mortgages, plus 5% fee.

WA ....25% per year plus fees.

WV ....31% per year to $2,000, 27% per year to $10,000, 18% per year to $45,000; fees included in rates.

WI ....Any agreed rate.

WY ....36% per year to $1,000, 21% to $50,000. No limit over $50,000.

How to Check Your Credit File

Any individual can investigate the contents of his or her credit file by directly contacting one or more of the approximately 2,000 credit bureaus, or consumer credit clearinghouses, in the U.S. The nearest ones can be found by calling a local Better Business Bureau or by looking in the telephone Yellow Pages under "Credit Rating or Reporting Agencies."

Although the Fair Credit Reporting Act requires that a bureau give a person no more than an oral or written credit history review, many bureaus will furnish the same computer-generated compilation of facts that they give the banks, retailers, and other companies that subscribe to their service. An individual who has been denied credit on the basis of neg-

ative information from a credit bureau can obtain a review free of charge within 30 days of the denial.

After inspecting this record of past credit behavior, a consumer can question any item believed to be inaccurate, misleading, or vague. The credit bureau must then investigate and remove any item that cannot be substantiated.

When a bureau affirms, rather than removes, a questionable item, an individual can present a 100-word explanation that must be placed in his or her file. Whenever an adverse item is deleted from the file or an explanation added, a consumer may request that the credit bureau inform every credit grantor who received a report within the last 6 months.

Credit Card Rates

Source: Christian T. Jones, Editor, *Consumer Finance Law Bulletin*, Evansville, IN; as of Sept. 1999

Nearly all states have special laws dealing with rates charged for credit cards issued by state banks and other financial institutions. Although some state laws apply only to banks, under federal parity law the same charges can be made by other financial institutions. A bank can charge the highest rates and charges allowed for revolving credit extended by any other creditor for similar types of credit in the state where the bank is located. These rates and charges also may be charged to residents of any other state. Maximum rates and fees are shown below; rates are yearly unless otherwise stated.

AL.... No limit.
AK ... 17% plus fee.
AZ.... No limit.
AR ... 5% over FRB discount rate (max. 17%).
CA ... No limit.
CO ... 21%.
CT ... No limit.
DC ... 24%.
DE ... No limit.
FL.... No limit.
GA ... No limit.
HI .... 24%.
ID .... No limit.
IL .... No limit.
IN .... 36% to $930, 21% to $3,100, then 15%; or 21%.
IA .... No limit.
KS ... No limit.
KY ... 21%; $20 annual fee.

LA ... 18%; 4% cash advance and $12 annual fee.
ME ... No limit; plus annual fee.
MD ... 24%; 2% fee.
MA ... 18% or formula rate.
MI.... No limit.
MN ... 18%; $50 annual fee.
MS ... 21%; or 18% plus $12 annual fee; no limit over $2,000.
MO ... 22% to $1,000, then 10%.
MT ... No limit.
NE ... No limit; plus fees.
NV ... No limit.
NH ... No limit.
NJ ... 30%; $15 annual fee or $50 over $5,000.
NM ... No limit.
NY ... 25% plus annual fee.
NC ... 18%; $24 annual fee.
ND ... No limit.
OH ... 25% plus fee.

OK... 30% to $1,020, 21% to $3,400, then 15%; or 21%.
OR... No limit.
PA ... Variable rate, plus fees.
PR... 26% per year.
RI.... No limit.
SC... No limit.
SD... No limit.
TN ... 24%.
TX ... Set by rule (max. 22%, min. 14%).
UT ... No limit.
VT ... No limit.
VA ... No limit.
WA... 25% loan; no limit for purchases; fees.
WV ... 18%.
WI ... No limit.
WY .. 36% to $1,000, then 21%; no limit over $50,000.

Telephone Area Codes

Source: Lockheed Martin IMS—NANPA
Sorted by number.

Area Code	Location or Service	Area Code	Location or Service	Area Code	Location or Service	Area Code	Location or Service
201	New Jersey	284	British Virgin Island	402	Nebraska	502	Kentucky
202	Dist. of Columbia	301	Maryland	403	Alberta	503	Oregon
203	Connecticut	302	Delaware	404	Georgia	504	Louisiana
204	Manitoba	303	Colorado	405	Oklahoma	505	New Mexico
205	Alabama	304	West Virginia	406	Montana	506	New Brunswick
206	Washington	305	Florida	407	Florida	507	Minnesota
207	Maine	306	Saskatchewan	408	California	508	Massachusetts
208	Idaho	307	Wyoming	409	Texas	509	Washington
209	California	308	Nebraska	410	Maryland	510	California
210	Texas	309	Illinois	411	Local Directory Assistance	512	Texas
212	New York	310	California			513	Ohio
213	California	311	Non-Emergency Access	412	Pennsylvania	514	Quebec
214	Texas			413	Massachusetts	515	Iowa
215	Pennsylvania	312	Illinois	414	Wisconsin	516	New York
216	Ohio	313	Michigan	415	California	517	Michigan
217	Illinois	314	Missouri	416	Ontario	518	New York
218	Minnesota	315	New York	417	Missouri	519	Ontario
219	Indiana	316	Kansas	418	Quebec	520	Arizona
224	Illinois	317	Indiana	419	Ohio	530	California
225	Louisiana	318	Louisiana	423	Tennessee	540	Virginia
228	Mississippi	319	Iowa	424	California	541	Oregon
231	Michigan	320	Minnesota	425	Washington	559	California
240	Maryland	321	Florida	435	Utah	561	Florida
242	Bahamas	323	California	440	Ohio	562	California
246	Barbados	330	Ohio	441	Bermuda	570	Pennsylvania
248	Michigan	331	Illinois	442	California	571	Virginia
250	British Columbia	334	Alabama	443	Maryland	573	Missouri
252	North Carolina	336	North Carolina	450	Quebec	580	Oklahoma
253	Washington	337	Louisiana	456	Inbound International	586	Michigan
254	Texas	340	U.S. Virgin Islands	464	Illinois	600	Canada (Services)
256	Alabama	341	California	469	Texas	601	Mississippi
262	Wisconsin	345	Cayman Islands	473	Grenada	602	Arizona
264	Anguilla	347	New York	480	Arizona	603	New Hampshire
267	Pennsylvania	352	Florida	484	Pennsylvania	604	British Columbia
268	Antigua/Barbuda	360	Washington	500	Personal Comm. Serv.	605	South Dakota
270	Kentucky	361	Texas			606	Kentucky
281	Texas	401	Rhode Island	501	Arkansas	607	New York

Area Code	Location or Service	Area Code	Location or Service	Area Code	Location or Service	Area Code	Location or Service
608...	Wisconsin	707....	California	804...	Virginia	888...	Toll-Free Service
609...	New Jersey	708....	Illinois	805...	California	900...	900 Service
610...	Pennsylvania	709....	Newfoundland	806...	Texas	901...	Tennessee
611...	Repair Service	710....	U.S. Government	807...	Ontario	902...	Nova Scotia
612...	Minnesota	711....	TRS Access	808...	Hawaii	903...	Texas
613...	Ontario	712....	Iowa	809...	Dominican Republic	904...	Florida
614...	Ohio	713....	Texas	810...	Michigan	905...	Ontario
615...	Tennessee	714....	California	811...	Business Office	906...	Michigan
616...	Michigan	715....	Wisconsin	812...	Indiana	907...	Alaska
617...	Massachusetts	716....	New York	813...	Florida	908...	New Jersey
618...	Illinois	717....	Pennsylvania	814...	Pennsylvania	909...	California
619...	California	718....	New York	815...	Illinois	910...	North Carolina
623...	Arizona	719....	Colorado	816...	Missouri	911...	Emergency
626...	California	720....	Colorado	817...	Texas	912...	Georgia
628...	California	724....	Pennsylvania	818...	California	913...	Kansas
630...	Illinois	727....	Florida	819...	Quebec	914...	New York
631...	New York	732....	New Jersey	828...	North Carolina	915...	Texas
636...	Missouri	734....	Michigan	830...	Texas	916...	California
646...	New York	740....	Ohio	831...	California	917...	New York
647...	Ontario	752....	California	832...	Texas	918...	Oklahoma
649...	Turks & Caicos Isl.	757....	Virginia	843...	South Carolina	919...	North Carolina
650...	California	758....	St. Lucia	847...	Illinois	920...	Wisconsin
651...	Minnesota	760....	California	850...	Florida	925...	California
657...	California	763....	Minnesota	856...	New Jersey	931...	Tennessee
660...	Missouri	764....	California	858...	California	931...	California
661...	California	765....	Indiana	859...	Kentucky	937...	Ohio
662...	Mississippi	767....	Dominica	860...	Connecticut	940...	Texas
664...	Montserrat	770....	Georgia	863...	Florida	941...	Florida
669...	California	773....	Illinois	864...	South Carolina	949...	California
670...	N. Mariana Isls.	775....	Nevada	865...	Tennessee	951...	California
671...	Guam	780....	Alberta	867...	Yukon & NW Terr.	952...	Minnesota
678...	Georgia	781....	Massachusetts	868...	Trinidad & Tobago	954...	Florida
686...	Michigan	784....	St. Vincent & Gren.	869...	St. Kitts & Nevis	956...	Texas
700...	IC Services	785....	Kansas	870...	Arkansas	970...	Colorado
701...	North Dakota	786....	Florida	872...	Illinois	971...	Oregon
702...	Nevada	787....	Puerto Rico	876...	Jamaica	972...	Texas
703...	Virginia	800....	Toll-Free Service	877...	Toll-Free Service	973...	New Jersey
704...	North Carolina	801....	Utah	880...	PAID—800 Service	978...	Massachusetts
705...	Ontario	802....	Vermont	881...	PAID—888 Service		
706...	Georgia	803....	South Carolina	882...	PAID—887 Service		

Copyright Law of the United States

Source: Copyright Office, Library of Congress, Oct. 1999

WHAT COPYRIGHT IS

Copyright is a form of protection provided by the laws of the U.S. (title 17, U.S. Code) to "original works of authorship," including literary, dramatic, musical, artistic, and certain other intellectual works. This protection is available to both published and unpublished works. Section 106 of the Copyright Act generally gives the owner of copyright the exclusive right to do and to authorize other parties to do the following:

Reproduce the copyrighted work in copies or phono records;

Prepare derivative works based upon the copyrighted work;

Distribute copies or phono records of the copyrighted work to the public by sale or other transfer of ownership, or by rental, lease, or lending;

Perform the copyrighted work publicly, in the case of literary, musical, dramatic, and choreographic works, pantomimes, and motion pictures and other audiovisual works;

Display the copyrighted work publicly, in the case of literary, musical, dramatic, and choreographic works, pantomimes, and pictorial, graphic, or sculptural works, including individual images of a motion picture or other audiovisual work; and

Perform the work publicly by means of a digital audio transmission, in the case of sound recordings.

It is illegal for anyone to violate any of the rights provided by the act to the owner of copyright. However, sections 107 through 121 of the Copyright Act establish limitations on these rights. In some cases, these limitations are specified exemptions from copyright liability; a major limitation is the doctrine of "fair use," which is given a statutory basis by section 107 of the act. In other instances, the limitation takes the form of a "compulsory license," under which certain limited uses of copyrighted works are permitted upon payment of specified royalties and compliance with statutory conditions.

Copyright protection subsists from the time the work is created in fixed form. The copyright in the work of authorship *immediately* becomes the property of the author who created it. Only the author or those deriving their rights from the author can rightfully claim copyright.

The employer and not the employee is considered the author of any "work made for hire," that is:

(1) a work prepared by an employee within the scope of his or her employment; or (2) a work specially ordered or commissioned for use as a contribution to a collective work, as a part of a motion picture or other audiovisual work, as a translation, as a supplementary work, as a compilation, as an instructional text, as a test, as answer material for a test, or as an atlas, if the parties expressly agree in a written instrument signed by them that the work shall be considered a work made for hire.

The authors of a joint work are co-owners of the copyright, unless there is an agreement to the contrary.

Copyright in each separate contribution to a periodical or other collective work is distinct from copyright in the collective work as a whole and vests initially with the author of the contribution.

WHICH WORKS ARE PROTECTED

Copyright protection is available for all unpublished works, regardless of the author's nationality or domicile.

Published works are eligible for copyright protection in the U.S. if *any* of the following conditions is met:

• On the date of first publication, one or more of the authors is a national or domiciliary of the U.S. or is a national, domiciliary, or sovereign authority of a foreign nation that is a party to a copyright treaty to which the U.S. is also a party, or is a stateless person; or

• The work is first published in the U.S. or in a foreign nation that, on the date of first publication, is a party to the Universal Copyright Convention; or

• The work is a sound recording that was first fixed in a treaty party; or

• The work is a pictorial, graphic, or sculptural work that is incorporated in a building or other structure, or an architectural work that is embodied in a building and the building or structure is located in the U.S. or a treaty party; or

• The work is first published by the United Nations or any of its specialized agencies, or by the Organization of American States; or

• The work is a foreign work that was in the public domain in the U.S. prior to 1996 and its copyright was restored under the

Uruguay Round Agreements Act. Request Circular 38b for further information; or

• The work comes within the scope of a presidential proclamation.

Copyright protects "original works of authorship" that are fixed in a tangible form of expression. The fixation need not be directly perceptible, as long as it may be communicated with the aid of a machine or device. Copyrightable works include the following categories:

(1) literary works; (2) musical works, including any accompanying words; (3) dramatic works, including any accompanying music; (4) pantomimes and choreographic works; (5) pictorial, graphic, and sculptural works; (6) motion pictures and other audiovisual works; (7) sound recordings; and (8) architectural works.

These categories should be viewed quite broadly: for example, computer programs and most "compilations" can be registered as "literary works"; maps and architectural plans are registrable as "pictorial, graphic, and sculptural works."

WHICH WORKS ARE NOT PROTECTED

Several categories of material are generally not eligible for statutory copyright protection. These include among others:

• Works that have not been fixed in a tangible form of expression. For example: choreographic works that have not been notated or recorded, or improvisational speeches or performances that have not been written or recorded.

• Titles, names, short phrases, and slogans; familiar symbols or designs; mere variations of typographic ornamentation, lettering, or coloring; mere listings of ingredients or contents.

• Ideas, procedures, methods, systems, processes, concepts, principles, discoveries, or devices, as distinguished from a description, explanation, or illustration.

• Works consisting entirely of information that is common property and containing no original authorship. For example: standard calendars, height and weight charts, tape measures and rulers, and lists or tables taken from public documents.

NOTICE OF COPYRIGHT

For works first published on or after Mar. 1, 1989, use of the copyright notice is optional. Before Mar. 1, 1989, the use of the notice was mandatory on all published works. Omitting the notice on any work first published before that date could result in the loss of copyright protection if corrective steps are not taken within a certain amount of time.

The Copyright Office does not take a position on whether reprints of works first published with notice before Mar. 1, 1989, which are distributed on or after Mar. 1, 1989, must bear the copyright notice.

Use of the notice is recommended because it informs the public that the work is protected by copyright, identifies the copyright owner, and shows the year of first publication. Furthermore, in the event that a work is infringed, if the work carries a proper notice, the court will not allow a defendant to claim "innocent infringement"—that is, that he or she did not realize that the work is protected. (A successful innocent infringement claim may result in reduced damages.)

The use of the copyright notice is the responsibility of the copyright owner and does not require advance permission from, or registration with, the Copyright Office.

For visually perceptible copies, the notice consists of the copyright symbol © or the word "Copyright" or "Copr.," the year of first publication, and the name of the owner of copyright in the work. Example: © 2000 Judy Smith. The notice must be affixed in such manner and location as to give reasonable notice of the claim of copyright.

The notice of copyright prescribed for all published phono records of sound recordings consists of a symbol (the letter P in a circle), the year of first publication of the sound recording, and the name of the owner of copyright in the sound recording. Example: [the letter P in a circle] 2000 XYZ Records, Inc. The notice on phono records may appear on the surface of the phono record or on the phono record label or container, provided the manner of placement and location give reasonable notice of the claim.

HOW LONG COPYRIGHT PROTECTION ENDURES

For works that are created and fixed in a tangible medium of expression for the first time on and after Jan. 1, 1978, the Copyright Act of 1976 as amended in 1998 establishes a single copyright term and different methods for computing the duration of a copyright. Works of this sort fall into two categories:

Works Created On or After Jan. 1, 1978

For works created after its effective date, the U.S. copyright law adopts the basic "life-plus-seventy" system already in effect in most other countries. A work that is created (fixed in tangible form for the first time) after Jan. 1, 1978, is automatically protected from the moment of its creation and is given a term lasting for the author's life, plus an additional 70 years after the author's death. In the case of "a joint work prepared by two or more authors who did not work for hire," the term lasts for 70 years after the last surviving author's death. For works made for hire, and for anonymous and pseudonymous works (unless the author's identity is revealed in Copyright Office records), the duration of copyright will be 95 years from first publication or 120 years from creation, whichever is shorter.

Works In Existence But Not Published or Copyrighted on Jan. 1, 1978

Works that had been created before the current law came into effect but had neither been published nor registered for copyright before Jan. 1, 1978, automatically are given federal copyright protection. The duration of copyright in these works will generally be computed in the same way as for new works: the life-plus-70 or 95/120-year terms will apply. However, all works in this category are guaranteed at least 25 years of statutory protection. The law specifies that in no case will copyright in a work of this sort expire before Dec. 31, 2002, and if the work is published before that date the term will extend another 45 years, through the end of 2047.

INTERNATIONAL COPYRIGHT PROTECTION

There is no such thing as an "international copyright" that will in itself protect an author's writings throughout the world. Protection against unauthorized use in a particular country basically depends on the laws of that country. However, most countries do offer protection to foreign works under certain conditions which have been greatly simplified by international copyright treaties and conventions. There are 2 principal international copyright conventions, the Berne Union for the Protection of Literary and Artistic Property (Berne Convention) and the Universal Copyright Convention (UCC). The United States became a member of the Berne Convention on Mar. 1, 1989. It has been a member of the UCC since Sept. 16, 1955.

Generally, works of an author who is a national or domiciliary of a country subscribing to these treaties, or works first published in a member country, or works published in a Berne Union country within 30 days of first publication may claim protection. There are no formal requirements under the Berne Convention. Under the UCC, any formality in a national law may be satisfied by the use of a copyright notice in the form and position specified in the UCC. A UCC notice should consist of the copyright symbol accompanied by the year of first publication and the name of the copyright proprietor (example: © 2000 John Doe). This notice must be placed in such a manner and location as to give reasonable notice of the claim to copyright. Since the Berne Convention prohibits formal requirements that affect the "exercise and enjoyment" of the copyright, the U.S. changed its law on Mar. 1, 1989, to make the use of a copyright notice optional. However, U.S. law still provides certain advantages for use of a copyright notice; for example, its use can defeat a defense of "innocent infringement" brought by an alleged copyright violator.

Even if the work cannot be brought under an international convention, protection may be available in other countries by virtue of a bilateral agreement between the U.S. and other countries or under specific provision of a country's laws. (See Circular 38a, *International Copyright Relations of the United States.*)

An author who wishes copyright protection in a particular country should first determine the extent of protection available to works of foreign authors there. If possible, this should be done before the work is published anywhere, because protection may depend on the facts existing at the time of first publication.

There are some countries that offer little or no copyright protection to any foreign works. For current information on the requirements and protection provided by specific countries, it is advisable to consult an expert familiar with foreign copyright laws.

COPYRIGHT REGISTRATION

Copyright registration is a legal formality intended to make a public record of the basic facts of a particular copyright. However, registration is not a condition for protection, but the copyright law provides several inducements or advantages to encourage copyright owners to register. Among these are the following:

• Registration establishes a public record of the copyright claim.

• Before an infringement suit may be filed in court, registration is necessary for works of U.S. origin.

• If made before or within 5 years of publication, registration will establish prima facie evidence in court of the validity of the copyright and of the facts stated in the certificate.

• If registration is made within 3 months after publication of the work or prior to an infringement of the work, statutory damages and attorney's fees will be available to the copyright owner in court actions. Otherwise, only an award of actual damages and profits is available to the copyright owner.

• Registration allows the owner of the copyright to record the registration with the U.S. Customs Service for protection against the importation of infringing copies. For additional information, request Publication No. 563 from Commissioner of Customs, ATTN: IPR Branch, U.S. Customs Service, 1300 Pennsylvania Ave. NW, Washington, DC 20229.

Registration may be made at any time within the life of the copyright. Unlike the law before 1978, when a work has been registered in unpublished form, making another registration when the work becomes published is unnecessary (although the copyright owner may register the published edition, if desired).

The process of registration is simple. Request an appropriate form from the Copyright Office and complete it. Returned it to the Copyright Office along with a $30 nonrefundable filing fee

and the appropriate deposit(s) of the work for which registration is sought. In a common example—a published book—the deposit is 2 copies of the best edition of the book. The Copyright Office sends a certificate of registration when the paperwork is completed, a process that usually takes 12 to 16 weeks because of the large volume of registrations the Office must handle (over 500,000 annually).

Although a copyright registration is not required, the Copyright Act establishes a mandatory deposit requirement for works published in the U.S. In general, the owner of copyright or the owner of the exclusive right of publication in the work has a legal obligation to deposit in the Copyright Office, within 3 months of publication in the U.S., 2 copies (or, in the case of sound recordings, 2 phono records) for the use of the Library of Congress. Failure to deposit these copies can result in fines and other penalties but does not affect copyright protection. Certain categories of works are exempt entirely from the mandatory deposit requirements, and the obligation is reduced for certain other categories.

Information on registration and application forms may be obtained free of charge by writing the Copyright Office, Information Section, LM-401, Library of Congress, Washington, DC 20559. Registration application forms and circulars may be ordered on a 24-hr basis by calling (202) 707-9100. Request Circular 1 for additional general information on copyright, including a list of which application forms to use when registering specific types of works.

For more information, visit the Copyright Office website— http://www.loc.gov/copyright

Median Price of Existing Single-Family Homes

Source: National Association of REALTORS®

CITY[1]	1997	1998	First Quarter 1999	CITY[1]	1997	1998	First Quarter 1999
Akron, OH	$105,900	$106,300	$98,400	Greenville, NC	112,800	113,100	114,500
Albany, NY	105,300	107,000	103,500	Hartford, CT	138,100	142,800	141,900
Albuquerque, NM	126,700	128,200	129,600	Honolulu, HI	307,000	NA	295,000
Amarillo, TX	76,800	79,300	78,700	Houston, TX	90,900	97,500	98,400
Anaheim/Santa Ana, CA[2]	229,800	261,700	272,800	Indianapolis, IN	103,700	108,400	109,100
Appleton/Oshkosh, WI	88,200	92,600	91,600	Jackson, MS	88,300	93,200	92,000
Atlanta, GA	108,400	115,400	N/A	Jacksonville, FL	86,400	95,000	91,900
Atlantic City, NJ	109,700	112,800	109,100	Kalamazoo, MI	97,200	102,300	108,700
Aurora, IL	141,800	146,200	147,200	Kansas City, MO/KS	106,800	114,000	118,000
Austin, TX	NA	NA	NA	Knoxville, TN	99,900	105,000	107,300
Baltimore, MD	118,200	120,600	121,300	Lake County, IL	153,500	159,400	166,300
Baton Rouge, LA	92,300	98,100	99,100	Lansing, MI	89,600	100,200	98,500
Beaumont/Port Arthur, TX	69,000	73,800	80,500	Las Vegas, NV	123,200	128,200	128,900
Biloxi/Gulfport, MS	81,100	86,000	85,500	Lexington/Fayette, KY	100,600	108,300	106,200
Birmingham, AL	118,900	122,700	125,300	Lincoln, NE	92,800	98,600	96,900
Boise City, ID	102,500	109,200	118,900	Little Rock, AR	85,600	91,300	92,700
Boston, MA	196,200	212,600	216,000	Los Angeles, CA[2]	176,500	192,600	192,200
Bradenton, FL	94,900	107,300	108,900	Louisville, KY/IN	96,800	106,100	105,000
Buffalo/Niagara Falls, NY	82,000	84,200	83,100	Madison, WI	126,800	131,800	131,300
Canton, OH	94,300	NA	NA	Melbourne, FL	85,900	86,100	82,400
Cedar Rapids, IA	94,800	102,300	99,400	Memphis, TN/AR/MS	103,700	109,800	109,000
Champaign, IL	84,500	90,100	86,400	Miami, FL	117,700	121,500	133,000
Charleston, SC	103,600	120,000	122,200	Milwaukee, WI	125,300	132,900	130,000
Charleston, WV	87,800	88,000	NA	Minneapolis, MN/WI	118,400	128,000	131,600
Charlotte, NC	124,200	134,000	NA	Mobile, AL	87,200	92,800	90,100
Chattanooga, TN	92,200	97,700	94,600	Montgomery, AL	94,100	98,300	97,400
Chicago, IL	158,900	166,800	166,100	Nashville, TN	115,200	116,700	117,500
Cincinnati, OH/KY/IN	110,500	116,300	117,500	New Haven, CT	134,100	137,800	139,600
Cleveland, OH	116,800	121,800	119,200	New Orleans, LA	93,300	102,100	101,500
Colorado Springs, CO	130,500	138,500	136,900	New York, NY	177,900	188,100	193,800
Columbia, SC	99,100	104,600	108,200	Norfolk/Virginia Bch, VA	NA	109.4	112,200
Columbus, OH	117,600	121,700	123,700	Ocala, FL	64,300	69,900	65,700
Corpus Christi, TX	81,800	84,000	84,600	Oklahoma City, OK	77,000	82,700	80,900
Dallas, TX	112,000	120,400	NA	Omaha, NE	93,600	101,700	96,100
Davenport, IA/IL	72,600	78,600	75,700	Orlando, FL	94,500	98,800	104,500
Dayton/Springfield, OH	96,700	102,800	100,900	Pensacola, FL	88,900	93,300	96,100
Daytona Beach, FL	75,500	79,000	81,700	Peoria, IL	79,700	83,300	82,800
Denver, CO	140,600	152,200	160,700	Philadelphia, PA/NJ	NA	NA	124,800
Des Moines, IA	98,900	106,600	100,900	Phoenix, AZ	113,700	120,200	123,100
Detroit, MI	119,600	132,600	136,400	Pittsburgh, PA	87,000	89,000	88,100
El Paso, TX	75,900	78,100	75,300	Portland, ME	94,500	98,100	127,600
Eugene, OR	119,400	124,400	NA	Portland, OR	152,400	158,100	165,300
Fargo, ND/MN	86,000	91,500	94,500	Providence, RI	119,600	124,400	125,400
Ft. Lauderdale, FL	123,700	128,600	134,900	Raleigh/Durham, NC	152,800	159,800	163,100
Ft. Myers, FL	85,700	88,800	87,400	Reno, NV	143,400	147,200	148,500
Ft. Wayne, IN	85,800	88,000	88,900	Richland, WA	102,600	NA	107,800
Ft. Worth/Arlington, TX	91,800	NA	NA	Richmond, VA	114,200	122,000	124,000
Gainesville, FL	99,600	104,200	97,000	Riverside/San Bern.,CA[2]	114,300	121,500	121,000
Gary/Hammond, IN	97,300	105,600	104,300	Rochester, NY	86,800	89,000	87,100
Grand Rapids, MI	93,600	100,200	105,400	Rockford, IL	88,800	93,000	91,600
Green Bay, WI	100,900	109,000	103,900	Sacramento, CA[2]	116,100	125,600	126,400
Greensboro, NC	117,300	123,500	NA	Saginaw, MI	71,300	78,100	78,200

CITY[1]	1997	1998	First Quarter 1999	CITY[1]	1997	1998	First Quarter 1999
St. Louis, MO/IL	96,900	101,700	99,600	Tallahassee, FL	111,700	114,600	118,100
Salt Lake City, UT	128,600	133,500	135,500	Tampa, FL	83,900	89,300	89,800
San Antonio, TX	86,800	88,700	88,500	Toledo, OH	87,300	94,500	89,800
San Diego, CA[2]	185,200	207,100	216,600	Topeka, KS	77,000	78,300	79,100
San Francisco, CA[2]	286,200	321,700	334,600	Trenton, NJ	137,700	139,500	131,200
Sarasota, FL	114,100	123,100	133,600	Tucson, AZ	106,800	112,600	116,100
Seattle, WA	171,300	175,300	208,700	Tulsa, OK	84,600	89,300	NA
Shreveport, LA	78,200	84,000	84,700	Washington, DC/MD/VA	166,300	172,100	168,100
Sioux Falls, SD	90,200	NA	NA	Waterloo/Cedar Falls, IA	65,200	69,300	70,300
South Bend, IN	78,100	82,600	85,200	W. Palm Beach, FL	133,400	126,600	127,000
Spokane, WA	102,700	102,600	102,700	Wichita, KS	83,200	89,300	89,800
Springfield, IL	83,800	86,100	86,600	Wilmington, DE/NJ/MD	NA	NA	121,000
Springfield, MA	106,500	110,900	110,800	Worcester, MA	135,800	138,700	NA
Springfield, MO	82,200	84,600	83,700	Youngstown, OH	73,900	77,700	75,800
Syracuse, NY	79,000	79,600	78,600	**UNITED STATES**	**$121,800**	**128,400**	**129,300**
Tacoma, WA	NA	NA	NA				

(1) All areas are metropolitan statistical areas (MSAs) as defined by the U.S. Office of Management and Budget. They include the named central city and surrounding areas. (2) Data provided by the California Association of REALTORS®. NA= not available.

Housing Affordability, 1990-99

Source: National Association of REALTORS®

Year	Median-priced existing home	Average mortgage rate[1]	Monthly principal and interest payment	Payment as percentage of median income	Year	Median-priced existing home	Average mortgage rate[1]	Monthly principal and interest payment	Payment as percentage of median income
1990	$92,000	10.04%	$648	22.0%	1995	$110,500	7.85%	$639	18.9%
1991	97,100	9.30	642	21.4	1996	115,800	7.71	661	18.8
1992	99,700	8.11	591	19.3	1997	121,800	7.68	693	18.7
1993	103,100	7.16	558	18.1	1998	128,400	7.10	690	18.1
1994	107,200	7.47	598	18.5	1999[2]	136,900	7.26	748	19.3

(1) The average mortgage rate is based on the effective rate on loans closed on existing homes monitored by the Federal Housing Finance Board. (2) Preliminary figures for June 1999.

Mortgage Loan Calculator

Source: Joyce E. Boulanger, Mortgage Access Corp.

To determine monthly payments, divide loan amount by 1,000 and then multiply the resulting figure by the appropriate factor from this table. To find the appropriate factor use the mortgage term in years and the interest rate percentage. More information on calculating mortgages can be found at http://www.weichert.com/mortgage

EXAMPLE: For a 30-year mortgage at 7.25%, the factor would be 6.82. If the mortgage amount is $220,000, divide by 1,000, which comes to 220. 220 x 6.82 (factor) = $1,500.40 monthly mortgage payment of principal and interest only (there will also be property taxes, home insurance, and other possible costs).

INTEREST RATE	MORTGAGE TERM IN YEARS 5	10	15	20	25	30	35	40
5.00	18.88	10.61	7.91	6.60	5.85	5.37	5.05	4.83
5.25	18.99	10.73	8.04	6.74	6.00	5.53	5.21	4.99
5.50	19.11	10.86	8.18	6.88	6.15	5.68	5.38	5.16
5.75	19.22	10.98	8.31	7.03	6.30	5.84	5.54	5.33
6.00	19.33	11.10	8.44	7.16	6.44	6.00	5.70	5.50
6.25	19.45	11.23	8.57	7.31	6.60	6.16	5.87	5.68
6.50	19.57	11.35	8.71	7.46	6.75	6.32	6.04	5.85
6.75	19.68	11.48	8.85	7.60	6.91	6.49	6.21	6.03
7.00	19.80	11.61	8.99	7.75	7.07	6.65	6.39	6.21
7.25	19.92	11.74	9.13	7.90	7.23	6.82	6.56	6.40
7.50	20.04	11.87	9.27	8.06	7.39	6.99	6.74	6.58
7.75	20.16	12.00	9.41	8.21	7.55	7.16	6.92	6.77
8.00	20.28	12.13	9.56	8.36	7.72	7.34	7.10	6.95
8.25	20.40	12.27	9.70	8.52	7.88	7.51	7.28	7.14
8.50	20.52	12.40	9.85	8.68	8.06	7.69	7.47	7.34
8.75	20.64	12.54	10.00	8.84	8.23	7.87	7.66	7.53
9.00	20.76	12.67	10.15	9.00	8.40	8.05	7.84	7.72
9.25	20.88	12.81	10.30	9.16	8.57	8.23	8.03	7.91
9.50	21.01	12.94	10.45	9.33	8.74	8.41	8.22	8.11
9.75	21.13	13.08	10.60	9.49	8.92	8.60	8.41	8.30
10.00	21.25	13.22	10.75	9.66	9.09	8.78	8.60	8.50
10.25	21.38	13.36	10.90	9.82	9.27	8.97	8.79	8.69
10.50	21.50	13.50	11.06	9.99	9.45	9.15	8.99	8.89
10.75	21.62	13.64	11.21	10.16	9.63	9.34	9.18	9.09
11.00	21.75	13.78	11.37	10.33	9.81	9.53	9.37	9.29
11.25	21.87	13.92	11.53	10.50	9.99	9.72	9.57	9.49
11.50	22.00	14.06	11.69	10.67	10.17	9.91	9.77	9.69
11.75	22.12	14.21	11.85	10.84	10.35	10.10	9.96	9.89
12.00	22.25	14.35	12.01	11.02	10.54	10.29	10.16	10.09
12.25	22.38	14.50	12.17	11.19	10.72	10.48	10.36	10.29
12.50	22.50	14.64	12.33	11.37	10.91	10.68	10.56	10.49
12.75	22.63	14.79	12.49	11.54	11.10	10.87	10.76	10.70
13.00	22.76	14.94	12.66	11.72	11.28	11.07	10.96	10.90
13.25	22.89	15.08	12.82	11.90	11.47	11.26	11.16	11.10
13.50	23.01	15.23	12.99	12.08	11.66	11.46	11.36	11.31
13.75	23.14	15.38	13.15	12.26	11.85	11.66	11.56	11.51
14.00	23.27	15.53	13.32	12.44	12.04	11.85	11.76	11.72

Leasing a Car

Source: Consumer Information Center, U.S. General Services Administration

Leasing a car instead of buying it has become an increasingly common option in recent years. In 1985, only 3.5% of all private vehicles were leased; by 1998, 32% were leased. Under the federal Consumer Leasing Act, consumers have a right to information about the costs and terms of a vehicle lease. The following information can help you compare lease offers and negotiate a lease that best fits your needs, budget, and driving patterns. (The information is mainly for a closed-end lease, the most common type.)

Here are some differences between buying and leasing:

Buying: You own the vehicle and get to keep it at the end of the financing term.

Leasing: You do not own the vehicle. You get to use it but must return it at the end of the lease unless you choose to buy it.

Buying: Up-front costs include the cash price or a down payment, taxes, registration and other fees, and various other charges.

Leasing: Up-front costs may include the first month's payment, a refundable security deposit, a capitalized cost reduction (like a down payment), taxes, registration and other fees, and other charges.

Buying: Monthly loan payments are usually higher than monthly lease payments because you are paying for the entire purchase price of the vehicle, plus interest and other finance charges, taxes, and fees.

Leasing: Monthly lease payments are usually lower because you pay only for depreciation during the lease term, plus rent charges (like interest), taxes, and various fees.

Buying: You are responsible for any pay-off amount if you end the loan early.

Leasing: You may have to pay substantial early termination charges if you end the lease early.

Buying: You may have to sell or trade the vehicle when you decide you want a different vehicle.

Leasing: You may return the vehicle at lease end, pay any end-of-lease costs, and "walk away."

Buying: You have the risk of the vehicle's market value when you trade or sell it.

Leasing: The lessor has the risk of the future market value of the vehicle.

Buying: You may drive as many miles as you want, but higher mileage will lower the vehicle's trade-in or resale value.

Leasing: You may have to pay extra for mileage above a certain limit—12,000-15,000 per year—if you return the vehicle. You can negotiate a higher mileage limit and pay a higher monthly payment.

Buying: There are no limits or charges for excessive wear to the vehicle, but excessive wear will lower the vehicle's trade-in or resale value.

Leasing: Most leases limit wear to the vehicle during the lease term; standards for excess wear, such as for body damage or worn tires, are in your lease agreement. You will likely have to pay extra charges for exceeding those limits if you return the vehicle.

Buying: At the end of the loan term (typically 4-6 years), you have no further loan payments.

Leasing: At the end of the lease (typically 2-4 years), you may have a new payment either to finance the purchase of the existing vehicle or to lease another vehicle.

During the lease, you will have to pay any additional taxes not included in the payment, such as sales, use, and personal property taxes; insurance premiums; ongoing maintenance costs and inspections; and any fees for late payment. At the end of the lease, if you don't buy the vehicle, you may have to pay a disposition fee and charges for excess miles and excess wear.

Among other things to consider when negotiating different lease offers and terms, consider the option to purchase either at lease end or earlier; also check whether your lease includes "gap" coverage, which protects you if the vehicle is stolen or totaled in an accident. Ask for alternatives to advertised specials and other lease offerings.

When you lease a vehicle, you have the right to take advantage of any warranties, recalls, or other services that apply to the vehicle.

For more information on leasing a car, contact your dealer, manufacturer, leasing company, or financial institution. For more information on consumer rights not covered in your lease agreement, contact your state's consumer protection agency or Attorney General's office.

How to Obtain Birth, Marriage, Death Records

The pamphlet "Where to Write for Vital Records: Births, Deaths, Marriages, and Divorces" (Stock # 017-022-01196-4) is available from the Superintendent of Documents, PO Box 371954, Pittsburgh, PA 15250-7954; advance payment of $2.25 is required. Orders can also be placed by calling (202) 512-1800 or via fax, (202) 512-2250, using a credit card.

Wedding Anniversaries

The traditional names for wedding anniversaries go back many years in social usage. As names like "wooden," "crystal," "silver," and "golden" were applied to anniversary years, it was considered proper to present the married couple with gifts made of these products or of something related. Traditional products for gifts are listed here, with a few allowable revisions in parentheses, followed by common modern gifts in each category.

1st	PAPER, clocks	**9th**	POTTERY (CHINA), leather goods	**25th**	SILVER, sterling silver
2d	COTTON, china	**10th**	TIN, ALUMINUM, diamond	**30th**	PEARL, diamond
3d	LEATHER, crystal, glass	**11th**	STEEL, fashion jewelry	**35th**	CORAL (JADE), jade
4th	LINEN (SILK), appliances	**12th**	SILK, pearls, colored gems	**40th**	RUBY, ruby
5th	WOOD, silverware	**13th**	LACE, textiles, furs	**45th**	SAPPHIRE, sapphire
6th	IRON, wood objects	**14th**	IVORY, gold jewelry	**50th**	GOLD, gold
7th	WOOL (COPPER), desk sets	**15th**	CRYSTAL, watches	**55th**	EMERALD, emerald
8th	BRONZE, linens, lace	**20th**	CHINA, platinum	**60th**	DIAMOND, diamond

Birthstones

Source: Jewelry Industry Council

MONTH	Ancient	Modern	MONTH	Ancient	Modern
January	Garnet	Garnet	**July**	Onyx	Ruby
February	Amethyst	Amethyst	**August**	Carnelian	Sardonyx or Peridot
March	Jasper	Bloodstone or Aquamarine	**September**	Chrysolite	Sapphire
April	Sapphire	Diamond	**October**	Aquamarine	Opal or Tourmaline
May	Agate	Emerald	**November**	Topaz	Topaz
June	Emerald	Pearl, Moonstone, or Alexandrite	**December**	Ruby	Turquoise or Zircon

Marriage Laws*

Source: Gary N. Skoloff, Skoloff & Wolfe, Livingston, NJ; as of Oct. 1999

| STATE | Age with parental consent | | Age without consent | | Physical exam & blood test for male and female | | Waiting period | |
	Male	Female	Male	Female	Max. period between exam and license	Scope of medical exam	Before license	After license issuance
Alabama***	14a,b	14a,b	18	18	—	—	—	30 days
Alaska (1917)***	16e	16e	18	18	—	—	3 days, w	—
Arizona**	16e	16e	18	18	—	—	—	1 yr.
Arkansas**	17c,e	16c,e	18	18	—	—	v	—
California**	aa	aa	18	18	30 days h,w	—	—	90 days
Colorado**(x)	16e	16e	18	18	—	—	—	30 days
Connecticut	16e	16e	18	18	—	bb	4 days, w	65 days
Delaware**	18c	16c	18	18	—	—	24 hr, hh	30 days
Florida (1/1/68)***	16a,c	16a,c	18	18	—	—	—	30 days
Georgia (1/1/97)***	16c	16c	16	16	—	bb	—	—
Hawaii**	15e	15e	16	16	—	—	—	—
Idaho (1/1/96)***	16e	16e	18	18	—	s, ee	—	—
Illinois (6/30/05)***	16p	16p	18	18	—	n	1 day	60 days
Indiana (1/1/58)***	18c,e	18c,e	18	18	—	pp	—	60 days
Iowa**	18e	18e	18	18	—	—	3 days	20 days
Kansas**(x)	14	12	18	18	—	—	3 days, w	—
Kentucky**	18e	18e	18	18	—	—	—	—
Louisiana (y)	18e	18e	18	18	10 days	m	—	—
Maine	18e	18e	18	18	—	—	3 days, v, w	90 days
Maryland**	18c,f	18c,f	18	18	—	—	48 hr, w	6 mo
Massachusetts	14j	12j	18	18	3-60 days, jj	—	3 days, v	—
Michigan (1/1/57)***	16	16	18	18	—	gg	3 days, w	—
Minnesota (4/26/41)***	16e	16e	18	18	—	—	5 days, w	—
Mississippi (4/5/56)***	aa,j	aa,j	17	15	30 days	m	3 days, w	—
Missouri (2/31/21)***	15d	15d	18	18	—	—	—	—
Montana**(x)	16e	16e	18	18	—	m	—	180 days
Nebraska (1/1/23)***(x)	17	17	17	17	—	bb	—	1 yr
Nevada	16e	16e	18	18	—	—	—	1 yr
New Hampshire**	18k	18k	18	18	—	gg	3 days, v, ww	90 days
New Jersey (12/1/39)***	16e,c	16e,c	18	18	—	m	72 hr, w	30 days
New Mexico**	16d,c	16d,c	18	18	30 days	m	—	—
New York**	16k	16k	18	18	—	nn	24 hr	60 days
North Carolina**	16c	16c	18	18	—	—	—	—
North Dakota	16	16	18	18	—	—	—	60 days
Ohio (10/10/91)***	18e,c	16e,c	18	18	—	—	5 days, r, ww	60 days
Oklahoma**	16e,c	16e,c	18	18	30 days, w	m	ff	30 days
Oregon**	17t	17t	18	18	—	—	3 days, w	—
Pennsylvania**	16d	16d	18	18	30 days	m	3 days w	60 days
Rhode Island**	18d	16d	18	18	—	pp	—	—
South Carolina**	16c	14c	18	18	—	—	1 day	—
South Dakota (7/1/59)***	16	16	18	18	—	—	—	20 days
Tennessee**	16d	16d	18	18	—	—	3 days cc,w	30 days
Texas**(x)	14j,k	14j,k	18	18	—	—	72 hr w	30 days
Utah**	16a,e	16a,e	18u	18u	—	—	—	30 days
Vermont	16e	16e	18	18	30 days, ww	m	1 day, w	—
Virginia**	16a,c	16a,c	18	18	—	dd	—	60 days
Washington**	17d	17d	18	18	—	gg	3 days	60 days
West Virginia**	18c	18c	18	18	—	m	3 days, ww	—
Wisconsin	16	16	18	18	—	ee	5 days, w	30 days
Wyoming	16d	16d	18	18	—	bb	—	—
Dist. of Columbia**	16a	16a	18	18	30 days	m	3 days, ww	—
Puerto Rico (x)	18c,d	16c,d	21	21c	—	—	—	—

*Most states have other relevant laws or pending legislation, as well as qualifications of the laws shown here. It would be advisable to consult a lawyer in conjunction with use of this chart. **Recognizes common-law marriage. ***Recognizes common-law marriages before listed date. (a) Parental consent not required if minor was previously married. (aa) No age limits. (b) Other statutory requirements apply. (bb) Venereal diseases; rubella (for female). In CT, just rubella. (c) Younger parties may obtain license in case of pregnancy or birth of child. (cc) Unless parties are over 18. (d) Younger parties may obtain license in special circumstances. (dd) Required offer of HIV test and/or must be provided with information on AIDS and tests available. (e) Younger parties may marry with parental consent and/or permission of judge. In CT, judicial approval. (ee) Applicants must receive AIDS information and certify having read it. (f) If parties are at least 16, proof of age and consent of parents in person are required. If a parent is ill, an affidavit by the incapacitated parent and a physician's affidavit required. (ff)If one or both parties are below the age for marriage without parental consent, 3-day waiting period. (g) Unless parties are 18 yr or more, or female is pregnant, or applicants are the parents of a living child born out of wedlock. (gg) No exam required, but parties must file affidavit of non-affliction with contagious venereal disease. In MI, certificate evidencing HIV counseling required. In NH, must sign affidavit affirming Public Health brochure received and discussed. (h) When unmarried man and unmarried woman, not minors, have been living together, they may, without health certificate, be married upon issuance of appropriate authorization. (hh) Residents, before expiration of 24-hr waiting period; non-residents, before expiration of 96-hr waiting period. (j) Parental consent and/or permission of judge required. In MA, under 18 req. court authority. (jj) Doctor's certificate must be filed 30 days prior to notice of intention. (k) Below age of consent parties need parental consent and permission of judge, no younger than 14 for males and 13 for females (14 in NY). (m) Venereal diseases. In WV and OK, Circuit Court judge may waive requirement. (n) Venereal diseases; test for sickle cell anemia given at request of examining physician. (nn) Test for sickle cell anemia may be required. (p) Judicial consent may be given when parents refuse consent. (pp) Physical examination and blood test required; offer of HIV counseling required. (r) Applicants under 18 must state that they have had marriage counseling. (rr) Any unsterilized female under 50 must submit with application for license a medical report stating whether she has immunological response to rubella, or a written record that the rubella vaccine was administered on or after her 1st birthday. Judges in emergencies may dispense with these requirements. (s) Rubella for female; there are certain exceptions, and district judge may waive medical examination on proof that emergency exists. (t)If a party has no parent residing within state, and one party has residence within state for 6 mo, no permission required. (u) Authorizes counties to provide for premarital counseling as a requisite to issuance of license to persons under 19 and persons previously divorced. (v)Parties must file notice of intention to marry with local clerk. (w) Waiting period may be avoided. (ww)Time may be shortened. In WV, a circuit court judge may waive req. (x) Marriages by proxy are valid. In MT, NE under certain conditions. (y) A "covenant marriage" bill provides for an optional, voluntary form of marriage that is more difficult to dissolve.

Divorce Laws

Source: Gary N. Skoloff, Skoloff & Wolfe, Livingston, NJ; as of Oct. 1999

Note: Almost all states also have other laws as well as qualifications of the laws shown here and have proposed divorce-reform laws pending. It would be advisable to consult a lawyer in conjunction with the use of this chart.

Some Grounds for Divorce[1]

	Residence	Adultery	Mental or physical cruelty	Abandonment	Alcoholism	Impotency	Non-support	Insanity	Bigamy	Felony conviction or imprisonment	Drug addiction	Fraud, force, duress
AL.....	6 mo*x	Yes	Yes	1 yr	Yes	Yes	2 yr	5 yr, A	A	2 yr*y	Yes	A
AK	30 days*z	Yes	Yes	1 yr	1 yr	Yes	No	18 mo	A	Yes	Yes	A
AZ.....	90 days	No	No	No	No	No	No	No	No	No	No	No
AR.....	60 days*	Yes	Yes	No	1 yr	Yes	Yes	3 yr	No	Yes	No	A
CA	6 mo*	No	No	No	No	A	No	Yes*	A	No	No	A
CO	90 days	No	No	No	No	A	No	No	A	No	No	A
CT	1 yr*	Yes	Yes	1 yr	Yes	No	Yes	5 yr	A	life*	No	Yes
DE	6 mo	Yes	Yes	Yes	Yes	A	No	6 mo	Yes	Yes	Yes	A
FL.....	6 mo	No	No	No	No	No	No	3 yr	No	No	No	A
GA	6 mo	Yes	Yes	1 yr	Yes	Yes	No	2 yr	A	Yes*	Yes	Yes
HI.....	6 mo	No	No	No	No	No	No	No*	A	No	No	A
ID.....	6 wk	Yes	Yes	Yes	Yes	A	Yes	Perm	A	Yes	No	A
IL.....	90 days	Yes	Yes	1 yr	2 yr	Yes	No	No	A	Yes	2 yr	No
IN.....	6 mo*	No	No	No	No	Yes	No	2 yr	A	Yes	No	A
IA.....	1 yr*	No	No	No	No	A	No	A	A	No	No	No
KS	60 days	No	No	No	No	No	No	2 yr	A	No	No	A
KY	180 days	No	No	No	No	A	No	No	A	No	No	A
LA.....	6 mo*	Yes	No	No	No	No	No	No	A	Yes*	No	A
ME	6 mo*	Yes	Yes	3 yr	Yes	Yes	Yes	7 yr, A	A	No	Yes	No
MD	*	Yes	D	1 yr, D	No	No	No	3 yr	A	1 yr*	No	No
MA	1 yr*	Yes	Yes	1 yr	Yes	Yes	D	A	A	5 yr*	Yes	No
MI.....	180 days*	No	No	No	No	No	No	No	No	No	No	A
MN	180 days	No	No	No	No	No	No	No	No	No	No	A
MS	6 mo	Yes	Yes	1 yr	Yes	Yes, A	No	3 yr, A*	A	Yes	Yes	A
MO	90 days	No	No	No	No	No	No	No	A	No	No	A
MT	90 days	No	No	No	No	A	No	No	A	No	No	A
NE	1 yr*	No	No	No	No	A	No	A	A	No	No	A
NV	6 wk	No	No	No	No	No	No	2 yr	A	No	No	A
NH	1 yr*	Yes	Yes	2 yr	2 yr	Yes	2 yr	No	A	1 yr*	No	No
NJ.....	1 yr*	Yes	Yes	1 yr	1 yr	A	No	2 yr	A	18 mo	1 yr	A
NM	6 mo	Yes	Yes	Yes	No	No	No	No	No	No	No	No
NY	1 yr*	Yes	Yes	1 yr, D	No	No	D	A	A	3 yr, D	No	A
NC	6 mo	No	No	No	No	A	No	3 yr	A	No	No	No
ND	6 mo	Yes	Yes	1 yr	1 yr	A	1 yr	5 yr	A	Yes	No	A
OH	6 mo	Yes	Yes	1 yr, D	Yes	No	Yes	No	Yes	Yes	No	Yes, A
OK	6 mo	Yes	Yes	1 yr	Yes	Yes	Yes	5 yr	Yes	Yes	No	Yes
OR	6 mo*	No	No	No	No	No	No	No	A	No	No	A
PA.....	6 mo	Yes	Yes	1 yr	No	No	No	3 yr	Yes	Yes	No	No
RI.....	1 yr	Yes	Yes	5 yr*	Yes	Yes	1 yr	No	No	Yes	Yes	No
SC	1 yr*	Yes	Yes	1 yr	Yes	No	No	No	No	No	Yes	No
SD	*	Yes	Yes	1 yr, D	1 yr, D	A	1 yr, D	5 yr, A	A	Yes	No	A
TN	6 mo*	Yes	Yes	1 yr	Yes	Yes	Yes	No	Yes	Yes	Yes	A
TX.....	6 mo*	Yes	Yes	1 yr	No	A	No	3 yr	No	1 yr	No	A
UT	3 mo*	Yes	Yes	1 yr	Yes	Yes	Yes	Yes*	A	Yes	No	No
VT.....	6 mo*	Yes	Yes	7 yr	No	No	Yes	5 yr, D	A	3 yr	No	A
VA.....	6 mo bona fide resident	Yes	Yes	1 yr, D	No	A	D	A	A	1 yr	No	No
WA	resident	No	No	No	No	No	No	No	No	No	No	A
WV	1 yr*	Yes	Yes	6 mo	Yes	A	No	3 yr	A	Yes	Yes	No
WI.....	6 mo	No	No	No	No	A	No	No	A	No	No	A
WY	2 mo*	No	No	No	No	A	No	2 yr	A	No	No	A
DC	6 mo	No	No	No	No	A	No	A	A	No	No	A
PR	1 yr	Yes	Yes	1 yr	Yes	Yes	No	7 yr	A	Yes*	Yes	No

(1) Almost all states have "no-fault" divorce laws. Conduct that constitutes "no-fault" divorce may vary from state to state. *indicates qualification; check local statutes. (A) indicates grounds for annulment. (D) indicates grounds for limited divorce or legal separation. (x) grounds for absolute divorce. (y) from a sentence of 7 or more yrs. (z) No residency requirement where plaintiff is a resident and marriage is solemnized in state. If marriage not solemnized in state, suit may be filed regardless of residency.

SOCIAL SECURITY
Social Security Programs

Source: Social Security Administration; World Almanac research; data as of Oct. 15, 1999

Old-Age, Survivors, and Disability Insurance; Medicare; Supplemental Security Income

Social Security Benefits

Social Security benefits are based on a worker's primary insurance amount (PIA), which is related by law to the average indexed monthly earnings (AIME) on which Social Security contributions have been paid. The full PIA is payable to a retired worker who becomes entitled to benefits at age 65 and to an entitled disabled worker at any age. Spouses and children of retired or disabled workers and survivors of deceased workers receive set proportions of the PIA subject to a family maximum amount. The PIA is calculated by applying varying percentages to succeeding parts of the AIME. The formula is adjusted annually to reflect changes in average annual wages.

Automatic increases in Social Security benefits are initiated for Dec. of each year, assuming the Consumer Price Index (CPI) for the 3d calendar quarter of the year increased relative to the base quarter, which is either the 3d calendar quarter of the preceding year or the quarter in which an increase legislated by Congress became effective. The size of the benefit increase is determined by the actual percentage rise of the CPI between the quarters measured.

The average monthly benefit payable to all retired workers amounts to $780 in Dec. 1998. The average benefit for disabled workers in that month amounts to $733.

Minimum and maximum monthly retired-worker benefits payable to individuals who retired at age 65[1]

Year of attainment of age 65	Minimum benefit[2] Payable at retirement	Payable effective Dec. 1998	Maximum benefit[2] Payable at retirement Men	Women[3]	Payable effective Dec. 1998 Men	Women[3]
1970	$64.00	$300.10	$189.80	$196.40	$878.90	$922.00
1980	133.90	300.10	572.00	—	1,267.50	—
1990	(4)	(4)	975.00	—	1,247.50	—
1993	(4)	(4)	1,128.80	—	1,283.10	—
1994	(4)	(4)	1,147.50	—	1,271.30	—
1995	(4)	(4)	1,199.10	—	1,292.30	—
1996	(4)	(4)	1,248.90	—	1,312.00	—
1997	(4)	(4)	1,326.60	—	1,353.80	—
1998	(4)	(4)	1,342.80	—	—	—

(1) Assumes retirement at beginning of year. (2) The final benefit amount payable is rounded to next lower $1 (if not already a multiple of $1). (3) Benefits for women are the same as for men except where shown. (4) Minimum eliminated for workers who reach age 62 after 1981.

Amount of Work Required

To qualify for benefits, the worker generally must have worked a certain length of time in covered employment. Just how long depends on when the worker reaches age 62 or, if earlier, when he or she dies or becomes disabled.

A person is fully insured who has 1 quarter of coverage for every year after 1950 (or year age 21 is reached, if later) up to but not including the year the worker reaches 65, dies, or becomes disabled. In 1999, a person earns 1 quarter of coverage for each $740 of annual earnings in covered employment, up to 4 quarters per year.

The law permits special monthly payments under the Social Security program to certain very old persons who are not eligible for regular benefits since they had little or no opportunity to earn work credits during their working lifetime (so-called special age-72 beneficiaries).

To receive disability benefits, the worker, in addition to being fully insured, must generally have credit for 20 quarters of coverage out of the 40 calendar quarters before he or she became disabled. A disabled blind worker need meet only the fully insured requirement. Persons disabled before age 31 can qualify with a briefer period of coverage. Certain survivor benefits are payable if the deceased worker had 6 quarters of coverage in the 13 quarters preceding death.

Work credit for fully insured status for benefits

Born after 1929; die, become disabled, or reach age 62 in	Years needed	Born after 1929; die, become disabled, or reach age 62 in	Years needed
1983	8	1987	9
1984	8½	1988	9¼
1985	8½	1989	9½
1986	8¾	1990	9¾
		1991 and after	10

Contribution and benefit base

Calendar year	OASDI[1]	HI[2]
1990	$51,300	$51,300
1991	53,400	125,000
1992	55,500	130,200
1993	57,600	135,000
1994	60,600	no limit
1995	61,200	no limit
1996	62,700	no limit
1997	65,400	no limit
1998	68,400	no limit
1999	72,600	no limit
2000	76,200	no limit

(1) Old-Age, Survivors, and Disability Insurance. (2) Hospital Insurance.

Tax-rate schedule
(percentage of covered earnings)

Year	Total	OASDI	HI
	(for employees and employers, each)		
1979-80	6.13	5.08	1.05
1981	6.65	5.35	1.30
1982-83	6.70	5.40	1.30
1984	7.00	5.70	1.30
1985	7.05	5.70	1.35
1986-87	7.15	5.70	1.45
1988-89	7.51	6.06	1.45
1990 and after	7.65	6.20	1.45
For self-employed			
1979-80	8.10	7.05	1.05
1981	9.30	8.00	1.30
1982-83	9.35	8.05	1.30
1984	14.00	11.40	2.60
1985	14.10	11.40	2.70
1986-87	14.30	11.40	2.90
1988-89	15.02	12.12	2.90
1990 and after	15.30	12.40	2.90

What Aged Workers Receive

When a person has enough work in covered employment and reaches retirement age (currently age 65 for full benefit, age 62 for reduced benefit), he or she may retire and receive monthly old-age benefits. The age when unreduced benefits become payable will increase gradually from 65 to 67 over a 21-year period beginning with workers age 62 in the year 2000 (reduced benefits will still be available as early as age 62, but with a larger reduction at that age). If a person age 65-69 has earnings of over $15,500 in 1999, $1 in benefits will be withheld for every $3 above $15,500. For those under 65, the annual exempt amount is $9,600 in 1999, with $1 in benefits withheld for every $2 in earnings above the exempt amount. However, an eligible worker age 70 or over receives the full benefit regardless of earnings. The annual exempt amount has been raised automatically as the general earnings level rises. However, legislation enacted in 1996 provided for bigger increases in the annual exempt amount for persons aged 65-69, rising to $15,500 in 1999 and to $30,000 by 2002. After 2002, the annual exempt amount for those 65-69 will be raised automatically as general earnings levels rise.

For workers who reached age 65 between 1982 and 1989, Social Security benefits are raised by 3% for each year for which the worker between ages 65 and 70 (72 before 1984) failed to receive benefits, whether because of earnings from work or because the worker had not applied for benefits. The delayed retirement credit is 1% per year for workers who reached age 65 before 1982. The delayed retirement credit will gradually rise to 8% per year by 2008. The rate for workers who reached age 65 in 1998-99 is 5.5%. The rate for reaching age 65 in 2001 will be 6.0%.

Effective Dec. 1998, the special benefit for persons aged 72 or over who do not meet the regular coverage requirements became $205.70 a month. Like other monthly benefits, these payments are subject to cost-of-living increases. They are not made to persons on the public assistance or supplemental security income rolls.

Workers retiring before age 65 have their benefits permanently reduced by ⁵⁄₉ of 1% for each month they receive benefits before that age. Thus, workers entitled to benefits in the month they reach age 62 receive 80% of the PIA, while a worker retiring at age 65 receives a benefit equal to 100% of the PIA. The nearer to age 65 the worker is when he or she begins collecting a benefit, the larger the benefit will be.

Benefits for Worker's Spouse

The spouse of a worker who is getting Social Security retirement or disability payments may become entitled to an insurance benefit of one-half of the worker's PIA, when he or she reaches 65. Reduced spouse's benefits are available at age 62 ($^{25}/_{36}$ of 1% reduction for each month of entitlement before age 65). Benefits are also payable to the aged divorced spouse of an insured worker if he or she was married to the worker for at least 10 years.

Benefits for Children of Workers

If a retired or disabled worker has a child under age 18, the child will get a benefit equal to half of the worker's unreduced benefit. So will the worker's spouse, even if under age 62, if he or she is caring for an entitled child of the worker who is under 16 or became disabled before age 22. Total benefits paid on a worker's earnings record are subject to a maximum; if the total that would be paid to a family exceeds that maximum, the dependents' benefits are adjusted downward. (Total monthly benefits paid to the family of a worker who retired in Jan. 1999 at age 65 and always had the maximum earnings creditable under Social Security cannot exceed $2,349.80.)

When entitled children reach age 18, their benefits generally stop, but a child disabled before age 22 may get a benefit as long as the disability meets the definition in the law. Benefits will be paid until age 19 to a child attending elementary or secondary school full-time.

Benefits may also be paid to a grandchild or step-grandchild of a worker or of his or her spouse, in special circumstances.

OASDI	May 1999	May 1998	May 1997
Monthly beneficiaries, total			
(in thousands)[1]	**44,353**	**44,080**	**43,796**
Aged 65 and over, total	31,880	31,806	31,641
Retired workers	25,050	24,873	24,498
Survivors and dependents	6,830	6,933	7,143
Special age-72 beneficiaries . .	(2)	(2)	1
Under age 65, total.	12,473	12,274	12,155
Retired workers	2,505	2,458	2,457
Disabled workers	4,769	4,581	4,407
Survivors and dependents	5,199	5,235	5,291
Total monthly benefits			
(in millions)	**$31,449**	**$30,603**	**$29,542**

(1) Totals may not add because of rounding. (2) Under 500.

What Disabled Workers Receive

A worker who becomes so disabled as to be unable to work may be eligible for a monthly disability benefit. Benefits continue until it is determined that the individual is no longer disabled. When a disabled-worker beneficiary reaches age 65, the disability benefit becomes a retired-worker benefit.

Benefits generally like those for dependents of retired-worker beneficiaries may be paid to dependents of disabled beneficiaries. However, the maximum family benefit in disability cases is generally lower than in retirement cases.

Survivor Benefits

If an insured worker should die, one or more types of benefits may be payable to survivors, again subject to a maximum family benefit as described above.

1. If claiming benefits at age 65, the surviving spouse will receive a benefit equal to 100% of the deceased worker's PIA. The surviving spouse may choose to get the benefit as early as age 60, but it is then reduced by $^{19}/_{40}$ of 1% for each month it is paid before age 65. However, for those whose spouses claimed their benefits before age 65, these are limited to the reduced amount the worker would be getting if alive, but not less than 82½% of the worker's PIA. Marriage after the worker's death ends the surviving spouse's benefit rights. However, if the widow(er) marries and the marriage is ended, he or she regains benefit rights. (A marriage after age 60, age 50 if disabled, is deemed not to have occurred for benefit purposes.) Survivor benefits may also be paid to a divorced spouse if the marriage lasted for at least 10 years.

Disabled widows and widowers may under certain circumstances qualify for benefits after attaining age 50 at the rate of 71.5% of the deceased worker's PIA. The widow or widower

must have become totally disabled before or within 7 years after the spouse's death or the last month in which he or she received mother's or father's insurance benefits.

2. There is a benefit for each child until the child reaches age 18. The monthly benefit for each child of a deceased worker is three-quarters of the amount the worker would have received if he or she had lived and drawn full retirement benefits. A child with a disability that began before age 22 may also receive benefits. Also, a child may receive benefits until reaching age 19 if he or she is in full-time attendance at an elementary or secondary school.

3. There is a mother's or father's benefit for the widow(er) if children of the worker under age 16 are in his or her care. The benefit is 75% of the PIA, and it continues until the youngest child reaches age 16, at which time payments stop even if the child's benefit continues. However, if the widow(er) has a disabled child beneficiary age 16 or over in care, benefits may continue.

4. Dependent parents may be eligible for benefits if they have been receiving at least half their support from the worker before his or her death, have reached age 62, and (except in certain circumstances) have not remarried since the worker's death. Each parent gets 75% of the worker's PIA; if only one parent survives, the benefit is 82½%.

5. A lump sum cash payment of $255 is made when there is a spouse who was living with the worker or a spouse or child who is eligible for immediate monthly survivor benefits.

Self-Employed Workers

A self-employed person who has net earnings of $400 or more in a year must report such earnings for Social Security tax and credit purposes. The person reports net returns from the business. Income from real estate, savings, dividends, loans, pensions, or insurance policies are not included unless it is part of the business.

A self-employed person receives 1 quarter of coverage for each $740 (for 1999), up to a maximum of 4 quarters.

The nonfarm self-employed have the option of reporting their earnings as $2/_3$ of their gross income from self-employment, but not more than $1,600 a year and not less than their actual net earnings. This option can be used only if actual net earnings from self-employment income are less than $1,600, and may be used only 5 times. Also, the self-employed person must have actual net earnings of $400 or more in 2 of the 3 taxable years immediately preceding the year in which he or she uses the option.

When a person has both taxable wages and earnings from self-employment, wages are credited for Social Security purposes first; only as much self-employment income as brings total earnings up to the current taxable maximum becomes subject to the self-employment tax.

Farm Owners and Workers

Self-employed farmers whose gross annual earnings from farming are $2,400 or less may report 2/3 of their gross earnings instead of net earnings for Social Security purposes. Farmers whose gross income is over $2,400 and whose net earnings are less than $1,600 can report $1,600. Cash or crop shares received from a tenant or share farmer count if the owner participated materially in production or management. The self-employed farmer pays contributions at the same rate as other self-employed persons.

Agricultural employees. A worker's earnings from farm work count toward benefits (1) if the employer pays the worker $150 or more in cash during the year; or (2) if the employer spends $2,500 or more in the year for agricultural labor. Under these rules a person gets credit for 1 calendar quarter for each $740 in cash pay in 1999 up to 4 quarters.

Foreign farm workers admitted to the U.S. on a temporary basis are not covered.

Household Workers

Anyone 18 or older employed as maid, cook, laundry worker, nurse, babysitter, chauffeur, gardener, or other worker in the house of another is covered by Social Security if paid $1,100 or more in cash in a calendar year by any one employer. Room and board do not count, but transportation costs count if paid in cash. The job need not be regular or full-time. The employee should get a Social Security card at the Social Security office and show it to the employer.

The employer deducts the amount of the employee's Social Security tax from the worker's pay, adds an identical amount as the employer's Social Security tax, and sends the total amount to the federal government.

Medicare Coverage

The Medicare health insurance program provides acute-care coverage for Social Security and Railroad Retirement beneficiaries age 65 and over, for persons entitled for 24 months to receive Social Security or Railroad Retirement disability benefits, and for certain persons with end-stage kidney disease. What follows is a basic description and may not cover all circumstances.

The basic Medicare plan, available nationwide, is a fee-for-service arrangement, where the beneficiary may use any provider accepting Medicare; some services are not covered and there are some out-of-pocket costs.

Under "Medicare + Choice," persons eligible for Medicare may have the option of getting services through a health maintenance organization (HMO) or other managed care plan. Any such plan must provide at least the same benefits, except for hospice services, and may provide added benefits—such as lower or no deductibles and coverage for some prescription drugs—but is usually subject to restrictions in choice of health care providers. In some plans services by outside providers are still covered for an extra out-of-pocket cost. Also available as options in some areas are Medicare-approved private fee-for-service plans and Medicare medical savings accounts.

Hospital insurance (Part A). The basic hospital insurance program pays covered services for hospital and posthospital care including the following:

- All necessary inpatient hospital care for the first 60 days of each benefit period, except for a deductible ($768 in 1999). For days 61-90, Medicare pays for services over and above a coinsurance amount ($192 per day in 1999). After 90 days, the beneficiary has 60 reserve days for which Medicare helps pay. The coinsurance amount for reserve days was $384 in 1999.
- Up to 100 days' care in a skilled-nursing facility in each benefit period. Hospital insurance pays for all covered services for the first 20 days; for the 21-100th day, the beneficiary pays coinsurance ($96 a day in 1999).
- Part-time home health care provided by nurses or other health workers.
- Limited coverage of hospice care for individuals certified to be terminally ill.

There is a premium for this insurance in certain cases.

Medical insurance (Part B). Elderly persons can receive benefits under this supplementary program only if they sign up for them and agree to a monthly premium ($45.50 if you sign up upon being eligible in 1999). The federal government pays the rest of the cost. The medical insurance program usually pays 80% of the approved amount (after the first $100 in each calendar year) for the following services:

- Covered services received from a doctor in his or her office, in a hospital, in a skilled-nursing facility, at home, or in other locations.
- Medical and surgical services, including anesthesia.
- Diagnostic tests and procedures that are part of the patient's treatment.
- Radiology and pathology services by doctors while the individual is a hospital inpatient or outpatient.
- Other services such as X-rays, services of a doctor's office nurse, drugs and biologicals that cannot be self-administered, transfusions of blood and blood components, medical supplies, physical/occupational therapy and speech pathology services.

In addition to the above, certain other tests or preventive measures are now covered without an additional premium. These include mammograms, bone mass measurement, colorectal cancer screening, and flu shots. Outpatient prescription drugs are generally not covered under the basic plan, nor are routine physical exams, dental care, hearing aids, or routine eye care. There is limited coverage for nonhospital treatment of mental illness.

To get medical insurance protection, persons approaching age 65 may enroll in the 7-month period that includes 3 months before the 65th birthday, the month of the birthday, and 3 months after the birthday, but if they wish coverage to begin in the month they reach age 65, they must enroll in the 3 months before their birthday. Persons not enrolling within their first enrollment period may enroll later, during the first 3 months of each year (coverage begins July 1), but their premium may be 10% higher for each 12-month period elapsed since they first could have enrolled.

The monthly premium is deducted from the cash benefit for persons receiving Social Security, Railroad Retirement, or Civil Service retirement benefits. Income from the medical premiums and the federal matching payments are put in a Supplementary Medical Insurance Trust Fund, from which benefits and administrative expenses are paid.

Further details are available on the Internet at http://www.medicare.gov or by calling 1-800-638-6833.

Medicare card. Persons qualifying for hospital insurance under Social Security receive a health insurance card similar to cards now used by Blue Cross and other health insurers. The card indicates whether the individual has taken out medical insurance protection. It is to be shown to the hospital, skilled-nursing facility, home health agency, doctor, or whoever provides the covered services.

Payments are generally made only in the 50 states, Puerto Rico, Virgin Islands, Guam, and American Samoa.

Social Security Financing

Social Security is paid for by a tax on certain earnings (for 1999, on earnings up to $72,600) for Old Age, Survivors, and Disability Insurance and on all earnings (no upper limit) for Hospital Insurance with the Medicare Program; the taxable earnings base for OASDI has been adjusted annually to reflect increases in average wages. The employed worker and his or her employer share Social Security taxes equally.

Employers remit amounts withheld from employee wages for Social Security and income taxes to the Internal Revenue Service; employer Social Security taxes are also payable at the same time. (Self-employed workers pay Social Security taxes when filing their regular income tax forms.) The Social Security taxes (along with revenues arising from partial taxation of the Social Security benefits of certain high-income people) are transferred to the Social Security Trust Funds—the Federal Old-Age and Survivors Insurance (OASI) Trust Fund, the Federal Disability Insurance (DI) Trust Fund, and the Federal Hospital Insurance (HI) Trust Fund; they can be used only to pay benefits, the cost of rehabilitation services, and administrative expenses. Money not immediately needed for these purposes is by law invested in obligations of the federal government, which must pay interest on the money borrowed and must repay the principal when the obligations are redeemed or mature.

Supplemental Security Income

On Jan. 1, 1974, the Supplemental Security Income (SSI) program established by the 1972 Social Security Act amendments replaced the former federal grants to states for aid to the needy aged, blind, and disabled in the 50 states and the District of Columbia. The program provides both for federal payments, based on uniform national standards and eligibility requirements, and for state supplementary payments varying from state to state. The Social Security Administration administers the federal payments financed from general funds of the Treasury—and the state supplements as well, if the state elects to have its supplementary program federally administered. States may supplement the federal payment for all recipients and must supplement it for persons otherwise adversely affected by the transition from the former public assistance programs. In May 1999, the number of persons receiving federally administered payments was 6,596,748, and the payments totaled $2.6 billion.

The maximum monthly federal SSI payment for individuals with no other countable income, living in their own household, was $500 in 1999. For couples it was $751.

Social Security Statement

On Oct. 1, 1999, the Social Security Administration initiated the mailing of an annual *Social Security Statement* to all workers age 25 and older not already receiving benefits. Workers will automatically receive statements about 3 months before their birth month. The statement provides estimates of potential monthly Social Security retirement, disability, and survivor benefits as well as a record of lifetime earnings. The statement also provides workers an easy way to determine whether their earnings are accurately posted in Social Security records. For further information contact the Social Security Administration toll-free at 1-800-772-1213 or visit its website at http://www.ssa.gov

Examples of Monthly Benefits Available

Description of benefit or beneficiary	For low earnings ($13,380 in 1999)[1]	For avg. earnings ($29,732 in 1999)[2]	For max. earnings ($72,600 in 1999)
Primary insurance amount (worker retiring at 65)	$568.30	$938.00	$1,342.80
Maximum family benefit (worker retiring at 65)	852.60	1,709.30	2,349.80
Maximum family disability benefit (worker disabled at 55; in 1996)*	873.80	1,536.10	2,309.70
Disabled worker (worker disabled at 55)			
Worker alone	621.80	1,024.10	1,539.80
Worker, spouse, and 1 child	873.00	1,536.00	2,307.00
Retired worker claiming benefits at age 62:			
Worker alone[3]	470.00	774.00	1,109.00
Worker with spouse claiming benefits at—			
Age 65 or over	763.00	1,258.00	1,802.00
Age 62[3]	690.00	1,137.00	1,679.00
Widow or widower claiming benefits at—			
Age 65 or over[4]	568.00	938.00	1,342.00
Age 60 (spouse died at 65 without receiving reduced benefits)	406.00	670.00	960.00
Disabled widow or widower claiming benefits at age 50-59[5]	406.00	670.00	960.00
1 surviving child	426.00	703.00	1,007.00
Widow or widower age 65 or over and 1 child[6]	852.00	1,708.00	2,349.00
Widowed mother or father and 1 child[6]	852.00	1,406.00	2,014.00
Widowed mother or father and 2 children[6]	852.00	1,707.00	2,349.00

Effective Jan. 1999, for beneficiaries with first entitlement in 1998. *Assumes work beginning at age 22. (1) 45% of average. (2) Estimate. (3) Assumes maximum reduction. (4) A widow(er)'s benefit amount is limited to the amount the spouse would have been receiving if still living, but not less than 82.5% of the PIA. (5) Effective Jan. 1984, disabled widow(er)s claiming a benefit at ages 50-59 receive a benefit equal to 71.5% of the PIA. (6) Based on worker dying at age 65.

Social Security Trust Funds
Old-Age and Survivors Insurance Trust Fund, 1940-98
(in millions)

Fiscal year[1]	INCOME Total	Net contributions[2]	Income from taxing benefits	Payments from the Treasury fund[3]	Net interest[4]	DISBURSEMENTS Total	Benefit payments[5]	Administrative expenses	Transfers to Railroad Retirement program	Interfund borrowing transfers[6]	Net increase in fund	Fund at end of period
1940	$592	$550	—	—	$42	$28	$16	$12	—	—	$564	$1,745
1950	2,367	2,106	—	$4	257	784	727	57	—	—	1,583	12,893
1960	10,360	9,843	—	—	517	11,073	10,270	202	$600	—	−713	20,829
1970	31,746	29,955	—	442	1,350	27,321	26,268	474	579	—	4,425	32,616
1980	100,051	97,608	—	557	1,886	103,228	100,626	1,160	1,442	—	−3,177	24,566
1990	278,607	261,506	$2,924	34	14,143	223,481	218,948	1,564	2,969	—	55,126	203,445
1995	326,067	289,529	5,114	7	31,417	294,456	288,607	1,797	4,052	—	31,611	447,946
1996	356,843	317,157	5,785	−124	34,026	305,311	299,968	1,788	3,554	—	51,533	499,479
1997	386,465	342,312	6,462	3	37,689	318,548	312,862	1,998	3,688	—	67,916	567,395
1998	415,666	364,871	8,595	2	42,198	329,953	324,256	2,034	3,662	—	85,713	653,108

(1) Fiscal years 1980 and later consist of the 12 months ending on Sept. 30 of each year. Fiscal years prior to 1977 consisted of the 12 months ending on June 30 of each year. (2) Beginning in 1983, includes transfers from general fund of Treasury representing contributions that would have been paid on deemed wage credits for military service in 1957 and later, if such credits were considered covered wages. (3) Includes payments (a) in 1947-52 and in 1967 and later, for costs of noncontributory wage credits for military service performed before 1957;(b) in 1972-83, for costs of deemed wage credits for military service performed after 1956; and (c) in 1969 and later, for costs of benefits to certain uninsured persons who attained age 72 before 1968. (4) Net interest includes net profits or losses on marketable investments. Beginning in 1967, administrative expenses were charged currently to the trust fund on an estimated basis, with a final adjustment, including interest, made in the next fiscal year. The amounts of these interest adjustments are included in net interest. For years prior to 1967, the method of accounting for administrative expenses is described in the 1970 Annual Report. Beginning in Oct. 1973, the figures shown include relatively small amounts of gifts to the fund. During 1983-91, interest paid from the trust fund to the general fund on advance tax transfers is reflected. (5) Beginning in 1967, includes payments for vocational rehabilitation services furnished to disabled persons receiving benefits because of their disabilities. Beginning in 1983, amounts are reduced by amount of reimbursement for unnegotiated benefit checks. (6) Negative figures represent amounts repaid from the OASI Trust Fund to the DI and HI Trust Funds.

Disability Insurance Trust Fund, 1970-98
(in millions)

Fiscal year[1]	INCOME Total	Net contributions[2]	Income from taxation of benefits	Payments from the Treasury fund[3]	Net interest[4]	DISBURSEMENTS Total	Benefit payments[5]	Administrative expenses	Transfers to Railroad Retirement program	Net increase in fund	Fund at end of period
1970	$4,380	$4,141	—	$16	$223	$2,954	$2,795	$149	$10	$1,426	$5,104
1980	17,376	16,805	—	118	453	15,320	14,998	334	−12	2,056	7,680
1990	28,215	27,291	$158	—	766	25,124	24,327	717	80	3,091	11,455
1995	70,209	67,987	335	—	1,888	41,374	40,234	1,072	68	28,835	35,206
1996	59,220	56,571	370	−203	2,482	44,343	43,266	1,074	2	14,877	50,083
1997	60,088	56,162	400	—	3,526	46,689	45,419	1,211	59	13,399	63,483
1998	62,943	57,982	526	—	4,434	49,338	47,619	1,563	157	13,604	77,087

(1) Fiscal years 1977 and later consist of the 12 months ending Sept. 30 of each year. Fiscal years prior to 1977 consisted of the 12 months ending June 30 of each year. (2) Beginning in 1983, includes transfers from general fund of Treasury representing contributions that would have been paid on deemed wage credits for military service in 1957 and later, if such credits were considered to be covered wages. (3) Includes payments (a) for costs of noncontributory wage credits for military service performed before 1957; and (b) in 1972-83, for costs of deemed wage credits for military service performed after 1956. (4) Net interest includes net profits or losses on marketable investments. Administrative expenses are charged currently to the trust fund on an estimated basis, with a final adjustment, including interest, made in the following fiscal year. Figures shown include relatively small amounts of gifts to the fund. During the years 1983-91, interest paid from the trust fund to the general fund on advance tax transfers is reflected. (5) Includes payments for vocational rehabilitation services. Beginning in 1983, amounts are reduced by amount of reimbursement for unnegotiated benefit checks. **NOTE:** Totals may not add because of rounding.

Supplementary Medical Insurance Trust Fund, 1975-98

(in millions)

Fiscal year[1]	INCOME				DISBURSEMENTS			Balance in fund at end of year[4]
	Premium from participants	Government contribu-tions[2]	Interest and other income[3]	Total Income	Benefit payments	Admin-istrative expenses	Total disburse-ments	
1975	$1,887	$2,330	$105	$4,322	$3,765	$405	$4,170	$1,424
1980	2,928	6,932	415	10,275	10,144	593	10,737	4,532
1990	11,494[5]	33,210	1,434[5]	46,138[5]	41,498	1,524[5]	43,022[5]	14,527[5]
1995	19,244	36,988	1,937	58,169	63,491	1,722	65,213	13,874
1996	18,931	61,702	1,392	82,025	67,176	1,771	68,946	26,953
1997	19,141	59,471	2,193	80,806	71,133	1,420	72,553	35,206
1998	19,427	59,919	2,608	81,955	74,837[6]	1,435	76,272	40,889

(1) Fiscal year 1975 consists of the 12 months ending on June 30, 1975; fiscal years 1980 and later consist of the 12 months ending on September 30 of each year. (2) General fund matching payments, plus certain interest-adjustment items. (3) Other income includes recoveries of amounts reimbursed from the trust fund that are not obligations of the trust fund and other miscellaneous income. (4) The financial status of the program depends on both the assets and the liabilities of the program. (5) Includes the impact of the Medicare Catastrophic Coverage Act of 1988 (PL 100-360). (6) Benefit payments less money transferred from the HI trust fund for home health agency costs as provided for by PL 105-33. **NOTE:** Totals do not necessarily equal the sums of rounded components.

Hospital Insurance Trust Fund, 1975-98

(in millions)

Fiscal year[1]	INCOME								DISBURSEMENTS				
	Payroll taxes	Income from taxation of benefits	Transfers from railroad retirement acct.	Reimburse-ment for uninsured persons	Premiums from voluntary enrollees	Pymts. for military wage credits	Interest on invest-ments and other income[2]	Total income	Benefit pymts.[3]	Admin-istrative expense[4]	Total disburse-ments	Net increase in fund	Fund at end of year
1975	$11,291	—	$132	$481	$6	$48	$609	$12,568	$10,353	$259	$10,612	$1,956	$9,870
1980	23,244	—	244	697	17	141	1,072	25,415	23,790	497	24,288	1,127	14,490
1990	70,655	—	367	413	113	107	7,908	79,563	65,912	774	66,687	12,876	95,631
1995	98,053	3,913	396	462	998	61	10,963	114,847	113,583	1,300	114,883	−36	129,520
1996	106,934	4,069	401	419	1,107	−2,293[5]	10,496	121,135	124,088	1,229	125,317	−4,182	125,338
1997	112,725	3,558	419	481	1,279	70	10,017	128,548	136,175	1,661	137,836	−9,287	116,050
1998	121,913	5,067	419	34	1,320	67	9,382	138,203	135,487[6]	1,653	137,140	1,063	117,113

(1) Fiscal year 1975 consists of the 12 months ending on June 30, 1975; fiscal years 1980 and later consist of the 12 months ending Sept. 30 of each year. (2) Other income includes recoveries of amounts reimbursed from the trust fund that are not obligations of the trust fund and a small amount of miscellaneous income, including amounts from the fraud and abuse control system. (3) Includes costs of Peer Review Organizations (beginning with the implementation of the Prospective Payment System on Oct. 1, 1983). (4) Includes costs of experiments and demonstration projects. Beginning in 1997, includes fraud and abuse control expenses, as provided for by PL 104-191. (5) Includes the lump-sum general revenue adjustment of $−2,366 mil, as provided for by PL 98-21. (6) Includes monies transferred to the SMI trust fund for home health agency costs, as provided for by PL 105-33. **NOTE:** Totals do not necessarily equal the sums of rounded components.

Workers and Beneficiaries, 1945–2030

Source: Social Security Administration

In the first years of Social Security, there were a very large number of workers paying payroll taxes to support each person currently receiving benefits—in 1945, an estimated 42 workers per beneficiary. By 1955 there were only 8.6 workers per beneficiary; today there are 3.4. It is estimated that there will only be 2.0 workers per beneficiary by 2030. Given these estimates and current tax and benefit provisions, revenues taken in would fall short of fully covering benefits by around 2012. Trust fund reserves would make up the difference until around 2029, when current revenues would cover only about 75% of benefits.

The bars below show estimated covered workers and beneficiaries for selected years, with projections for the future based on mid-range assumptions. Years before 1958 include the Old Age and Survivors Insurance (OASI) program only; starting in 1958, disabled workers also received benefits and are included as well (OASDI).

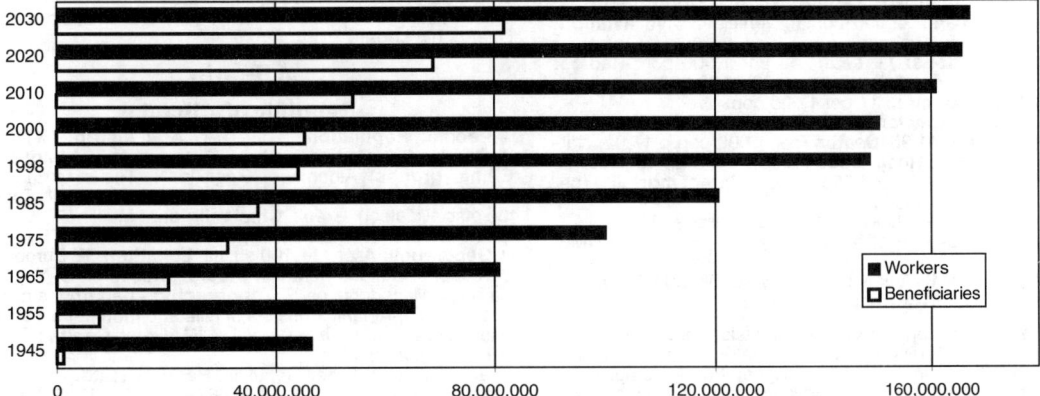

Note: Totals for beneficiaries exclude some uninsured persons, generally having fewer than 3 quarters of coverage, costs for whom are mostly reimbursed by the general fund of the Treasury.

NATIONS OF THE WORLD

Initials used include: AL (Arab League), APEC (Asia-Pacific Economic Cooperation Group), ASEAN (Association of Southeast Asian Nations), CARICOM (Caribbean Community and Common Market), CIS (Commonwealth of Independent States), EU (European Union), FAO (UN Food & Agriculture Org.), ILO (Intl. Labor Org.), IMF (Intl. Monetary Fund), IMO (Intl. Maritime Org.), NATO (North Atlantic Treaty Org.), OAS (Org. of American States), OAU (Org. of African Unity), OECD (Org. for Economic Cooperation and Development), OECS (Org. of Eastern Caribbean States), OSCE (Org. for Security and Cooperation in Europe), UN (United Nations), WHO (World Health Org.), WTrO (World Trade Org., formerly GATT). FY = fiscal year.

Sources: American Automobile Manufacturers Assn.; (U.S.) Census Bureau: Intl. Data Base; (U.S.) Central Intelligence Agency: *The World Factbook;* (U.S.) Dept. of Commerce; (U.S.) Dept. of Energy; Intl. Institute for Strategic Studies: *The Military Balance;* Intl. Monetary Fund; (U.S.) Dept. of State; UN Demographic Yearbook; UN Food and Agriculture Organization; UN Population Division: *World Urbanization Prospects;* UN Statistical Yearbook; Ward's Communications; World Tourism Organization; Encyclopaedia Britannica Book of the Year; The Europa World Year Book; The Statesman's Yearbook. Telephone data (1998 unless otherwise indicated) supplied by the Intl. Telecommunication Union, from the World Telecommunication Indicators database, copyright ITU.

Note: Because of rounding or incomplete enumeration, some percentages may not add to 100%. National population and health figures are mid-1999 estimates, unless otherwise noted. An * after city figures indicates 1995 urban agglomeration. Defense figures are for 1998 unless otherwise noted. Livestock figures are 1998. GDP estimates are based on purchasing power parity calculations, which involve use of intl. dollar price weights applied to quantities of goods and services produced. Tourism figures are 1998 and represent receipts from international tourism. Budget figures are for expenditures, unless otherwise noted. Motor vehicle statistics are for 1996 unless otherwise noted; comm. (commercial) vehicles include trucks and buses. Per-person figures in communications data are post-1994. Literacy rates are 1995 est., unless otherwise noted. Literacy rates given generally measure the percent of population able to read and write on a lower elementary school level, not the (smaller) percent able to read instructions necessary for a job or license. Embassy addresses are Wash., DC, area code (202), unless otherwise noted.

See pages 505–520 for full-color maps and flags.

Afghanistan
Islamic State of Afghanistan

People: Population: 25,824,882. **Age distrib.** (%): <15: 42.8; 65+: 2.8. **Pop. density:** 103 per sq. mi. **Urban:** 20%. **Ethnic groups:** Pashtun 38%, Tajik 25%, Hazara 19%, Uzbek 6%. **Principal languages:** Pashtu 35%, Afghan Persian (Dari) 50% (both official), Turkic (incl. Uzbek, Turkmen) 11%. **Chief religions:** Sunni Muslim 84%, Shi'a Muslim 15%.

Geography: Area: 250,000 sq. mi. **Location:** In SW Asia, NW of the Indian subcontinent. **Neighbors:** Pakistan on E, S; Iran on W; Turkmenistan, Tajikistan, Uzbekistan on N. The NE tip touches China. **Topography:** The country is landlocked and mountainous, much of it over 4,000 ft. above sea level. The Hindu Kush Mts. tower 16,000 ft. above Kabul and reach a height of 25,000 ft. to the E. Trade with Pakistan flows through the 35-mile-long Khyber Pass. The climate is dry, with extreme temperatures, and there are large desert regions, though mountain rivers produce intermittent fertile valleys. **Capital:** Kabul: 2,029,000*.

Government: Type: In transition. **Local divisions:** 32 provinces. **Defense:** 12.5% of GDP. **Active troops:** 429,000.

Economy: Industries: Textiles, soap, furniture, cement. **Chief crops:** Nuts, wheat, fruits. **Minerals:** Gas, oil, copper, coal, zinc, iron. **Other resources:** Wool, karakul pelts, mutton. **Arable land:** 12%. **Livestock** (1997): sheep: 14.30 mil; chickens: 7.20 mil; goats: 2.20 mil; cattle: 1.50 mil. **Electricity prod.** (1997): 485 mil kWh. **Labor force:** 67% agric.

Finance: Monetary unit: Afghani (Sept. 1999: 4,750.00 = $1 U.S.). **GDP** (1997 est.): $19.3 bil. **Per capita GDP:** $800. **Imports** (1996 est.): $150 mil; partners: Japan 14%, EU 11%. **Exports** (1996 est.): $80 mil; partners: EU 10%. **Tourism:** $1 mil.

Transport: Railroad: Length: 16 mi. **Motor vehicles:** 35,000 pass. cars, 32,000 comm. vehicles. **Civil aviation:** 98.3 mil pass.-mi.; 3 airports.

Communications: TV sets: 10 per 1,000 pop. **Radios:** 73.7 per 1,000 pop. **Telephones** (1997): 29,000 main lines. **Daily newspaper circ.:** 11 per 1,000 pop.

Health: Life expectancy: 47.82 male; 46.82 female. **Births** (per 1,000 pop.): 41.93. **Deaths** (per 1,000 pop.): 17.02. **Natural inc.:** 2.491%. **Hosp. beds** (1993): 1 per 2,945 persons. **Physicians** (1993): 1 per 6,690 persons. **Infant mortality** (per 1,000 live births): 140.55.

Education: Compulsory: ages 7-13. **Literacy:** 31.5%.

Major Intl. Organizations: UN (FAO, IBRD, ILO, IMF, WHO).

Embassy: 2341 Wyoming Ave. NW 20008; 234-3770. **Website:** http//www.afghan-web.com

Afghanistan, occupying a favored invasion route since antiquity, has been variously known as Ariana or Bactria (in ancient times) and Khorasan (in the Middle Ages). Foreign empires alternated rule with local emirs and kings until the 18th century, when a unified kingdom was established. In 1973, a military coup ushered in a republic.

Pro-Soviet leftists took power in a bloody 1978 coup and concluded an economic and military treaty with the USSR. In Dec. 1979 the USSR began a massive airlift into Kabul and backed a new coup, leading to installation of a more pro-Soviet leader. Soviet troops fanned out over Afghanistan and waged a protracted guerrilla war with Muslim rebels, in which some 15,000 Soviet troops reportedly died.

A UN-mediated agreement was signed Apr. 14, 1988, providing for withdrawal of Soviet troops, a neutral Afghan state, and repatriation of refugees. Afghan rebels rejected the pact, vowing to continue fighting while "Soviets and their puppets" remained in Afghanistan. The Soviets completed their troop withdrawal Feb. 15, 1989; fighting between Afghan rebels and government forces ensued.

Communist Pres. Najibullah resigned Apr. 16, 1992, as competing guerrilla forces advanced on Kabul. The rebels achieved power Apr. 28, ending 14 years of Soviet-backed regimes. More than 2 million Afghans had been killed and 6 million had left the country since 1979.

Following the rebel victory there were clashes between moderates and Islamic fundamentalist forces. Burhanuddin Rabbani, a guerrilla leader, became president June 28, 1992, but fierce fighting continued around Kabul and elsewhere. The Taliban, an insurgent Islamic fundamentalist faction, gained increasing control and in Sept. 1996 captured Kabul and set up a government. The Taliban executed former President Najibullah and empowered Islamic religious police to enforce codes of dress and behavior that were especially restrictive to women. Rabbani and other ousted leaders fled to the north.

Victories in the northern cities of Mazar-e Sharif, Aug. 8, 1998, and Taloqan, Aug. 11, gave the Taliban control over more than 90% of the country; the killing of several Iranian diplomats during the Mazar-e Sharif takeover further heightened tensions with Iran. On Aug. 20, U.S. cruise missiles struck SE of Kabul, hitting facilities the U.S. alleged were terrorist training camps run by a wealthy Saudi businessman, Osama bin Laden.

Albania
Republic of Albania

People: Population: 3,364,571. **Age distrib.** (%): <15: 32.7; 65+: 6.0. **Pop. density:** 303 per sq. mi. **Urban:** 38%. **Ethnic groups:** Albanians (Gegs in N, Tosks in S) 95%, Greeks 3%. **Principal languages:** Albanian (official; Tosk is the official dialect), Greek. **Chief religions:** Muslim 70%, Albanian Orthodox 20%, Roman Catholic 10%.

Geography: Area: 11,100 sq. mi. **Location:** SE Europe, on SE coast of Adriatic Sea. **Neighbors:** Greece on S, Yugoslavia on N, Macedonia on E. **Topography:** Apart from a narrow coastal plain, Albania consists of hills and mountains covered with scrub forest, cut by small E-W rivers. **Capital:** Tiranë (1995 est.): 270,000.

Government: Type: Republic. **Head of state:** Pres. Rexhep Meidani; b Aug. 17, 1944; in office: July 24, 1997. **Head of gov.:** Prem. Pandeli Majko; b 1967; in office Oct. 2, 1998. **Local divisions:** 36 districts. **Defense:** 6.7% of GDP. **Active troops:** 54,000.

Economy: Industries: Cement, textiles, food processing. **Chief crops:** Corn, wheat, potatoes, watermelon, vegetables. **Minerals:** Chromium, coal, oil, gas. **Crude oil reserves** (1999): 165 mil bbls. **Other resources:** Timber. **Arable land:** 21%. **Livestock** (1997): chickens: 4.60 mil; sheep: 1.89 mil; goats: 1.30 mil; cattle: 780,000; pigs: 100,000;. **Electricity prod.** (1997): 5.425 bil kWh. **Labor force:** 49.5% agric.

Finance: Monetary unit: Lek (Sept. 1999: 135.00 = $1 U.S.). **GDP** (1997 est.): $4.5 bil. **Per capita GDP:** $1,370. **Imports** (1996 est.): $879 mil; partners: Italy 38%, Greece 27%. **Exports** (1996 est.): $228 mil; partners: Italy 52%, Greece 10%. **Tourism:** $27 mil. **Budget** (1997): $996 mil. **Intl. reserves less gold** (Mar. 1999): $345.49 mil. **Gold:** 120,000 oz t. **Consumer prices** (change in 1998): 20.6%.

Transport: Railroad: Length: 419 mi. **Chief ports:** Durres, Sarande, Vlore. **Civil aviation:** 21.9 mil pass.-mi.; 1 airport.

Communications: TV sets: 89 per 1,000 pop. **Radios:** 157 per 1,000 pop. **Telephones:** 115,700 main lines. **Daily newspaper circ.:** 54 per 1,000 pop.

Health: Life expectancy: 65.92 male; 72.33 female. **Births** (per 1,000 pop.): 20.74. **Deaths** (per 1,000 pop): 7.35. **Natural inc.:** 1.339%. **Hosp. beds** (1994): 1 per 333 persons. **Physicians** (1994): 1 per 552 persons. **Infant mortality** (per 1,000 live births): 42.90.

Major Intl. Organizations: UN (IBRD, ILO, IMF, IMO, WHO), OSCE.

Education: Free, compulsory: ages 6-14. **Literacy** (1993): 100%.

Embassy: 2100 S St. NW 20008; 223-4942.

Websites: http://www.undp.tirana.al
http://www.albanian.com

Ancient Illyria was conquered by Romans, Slavs, and Turks (15th century); the latter Islamized the population. Independent Albania was proclaimed in 1912, republic was formed in 1920. King Zog I ruled 1925-39, until Italy invaded.

Communist partisans took over in 1944, allied Albania with USSR, then broke with USSR in 1960 over de-Stalinization. Strong political alliance with China followed, leading to several billion dollars in aid, which was curtailed after 1974. China cut off aid in 1978 when Albania attacked its policies after the death of Chinese ruler Mao Zedong. Large-scale purges of officials occurred during the 1970s.

Enver Hoxha, the nation's ruler for 4 decades, died Apr. 11, 1985. Eventually the new regime introduced some liberalization, including measures in 1990 providing for freedom to travel abroad. Efforts were begun to improve ties with the outside world. Mar. 1991 elections left the former Communists in power, but a general strike and urban opposition led to the formation of a coalition cabinet including non-Communists.

Albania's former Communists were routed in elections Mar. 1992, amid economic collapse and social unrest. Sali Berisha was elected as the first non-Communist president since World War II. Berisha's party claimed a landslide victory in disputed parliamentary elections, May 26 and June 2, 1996. Public protests over the collapse of fraudulent investment schemes in Jan. 1997 led to armed rebellion and anarchy. The UN Security Council, Mar. 28, authorized a 7,000-member force to restore order. Socialists and their allies won parliamentary elections, June 29 and July 6, and international peacekeepers completed their pullout by Aug. 11. During NATO's air war against Yugoslavia, Mar.-June 1999, Albania hosted some 465,000 Kosovar refugees; more than 90% had been repatriated by Sept. 1.

Algeria

Democratic and Popular Republic of Algeria

People: Population: 31,133,486. **Age distrib.** (%): <15: 37.3; 65+: 3.9. **Pop. density:** 34 per sq. mi. **Urban:** 56%. **Ethnic groups:** Arab-Berber 99%. **Principal languages:** Arabic (official), French, Berber dialects. **Chief religion:** Sunni Muslim (state religion) 99%.

Geography: Area: 919,600 sq. mi. **Location:** In NW Africa, from Mediterranean Sea into Sahara Desert. **Neighbors:** Morocco on W; Mauritania, Mali, Niger on S; Libya, Tunisia on E. **Topography:** The Tell, located on the coast, comprises fertile plains 50-100 miles wide, with a moderate climate and adequate rain. Two major chains of the Atlas Mts., running roughly E-W and reaching 7,000 ft., enclose a dry plateau region. Below lies the Sahara, mostly desert with major mineral resources. **Capital:** Algiers (El Djazair): 3,705,000*.

Government: Type: Republic. **Head of state:** Pres. Abdelaziz Bouteflika; b Mar. 2, 1937; in office: Apr. 27, 1999. **Head of gov.:** Prime Min. Ismail Hamdani; b Mar. 11, 1930; in office: Dec. 15, 1998. **Local divisions:** 48 provinces. **Defense:** 4.6% of GDP. **Active troops:** 124,000.

Economy: Industries: Oil, natural gas, light industries, food processing. **Chief crops:** Grains, grapes, citrus, olives. **Minerals:** Iron, oil, gas, phosphates, zinc, lead. **Crude oil reserves** (1999): 9.2 bil bbls. **Arable land:** 3%. **Livestock** (1997): chickens: 132.00 mil; sheep: 16.75 mil; goats: 3.12 mil; cattle: 1.26 mil. **Fish catch** (1996): 99,678 metric tons. **Electricity prod.** (1997): 20.150 bil kWh. **Labor force:** 30% govt.; 27% industry, serv., commerce; 22% agric.

Finance: Monetary unit: Dinar (Sept. 1999: 66.85 = $1 U.S.). **GDP** (1997 est.): $120.4 bil. **Per capita GDP:** $4,000. **Imports** (1997 est.): $10 bil; partners: France 29%, Spain 11%. **Exports** (1997 est.): $13.1 bil; partners: Italy 19%, U.S. 15%. **Tourism:** $20 mil. **Budget** (1996 est.): $13.1 bil. **Intl. reserves less gold** (June 1999): $4.8 bil. **Gold:** 5.58 mil oz t. **Consumer prices** (change in 1997): 3.8%.

Transport: Railroad: Length: 2,965 mi. **Motor vehicles:** 500,000 pass. cars, 420,000 comm. vehicles. **Civil aviation:** 1.95 bil pass.-mi.; 28 airports. **Chief ports:** Algiers, Annaba, Oran.

Communications: TV sets: 71 per 1,000 pop. **Radios:** 122 per 1,000 pop. **Telephone:** 1.6 mil main lines. **Daily newspaper circ.:** 52 per 1,000 pop.

Health: Life expectancy: 68.07 male; 70.46 female. **Births** (per 1,000 pop.): 27.00. **Deaths** (per 1,000 pop.): 5.52. **Natural inc.:** 2.148%. **Hosp. beds** (1994): 1 per 513 persons. **Physicians** (1994): 1 per 1,066 persons. **Infant mortality** (per 1,000 live births): 43.82.

Education: Compulsory: ages 6-15. **Literacy:** 62%.

Major Intl. Organizations: UN (FAO, IBRD, ILO, IMF, IMO, WHO), AL, OAU, OPEC.

Embassy: 2118 Kalorama Rd. NW 20008; 265-2800.

Earliest known inhabitants were ancestors of Berbers, followed by Phoenicians, Romans, Vandals, and, finally, Arabs. Turkey ruled 1518 to 1830, when France took control.

Large-scale European immigration and French cultural inroads did not prevent an Arab nationalist movement from launching guerrilla war. Peace, and French withdrawal, was negotiated with French Pres. Charles de Gaulle. One million Europeans left. Independence came July 5, 1962. Ahmed Ben Bella was the victor of infighting and ruled until 1965, when an army coup installed Col. Houari Boumedienne as leader; Boumedienne led until his death from a blood disorder, 1978.

In 1967, Algeria declared war on Israel, broke ties with U.S., and moved toward eventual military and political ties with the USSR. Some 500 died in riots protesting economic hardship in 1988. In 1989, voters approved a new constitution, which cleared the way for a multiparty system.

The government canceled the Jan. 1992 elections that Islamic fundamentalists were expected to win, and banned all nonreligious activities at Algeria's 10,000 mosques. Pres. Mohammed Boudiaf was assassinated June 29, 1992. There were repeated attacks on high-ranking officials, security forces, foreigners, and others by militant Muslim fundamentalists over the next 7 years; pro-government death squads also were active.

Liamine Zeroual won the presidential election of Nov. 16, 1995. A new constitution banning Islamic political parties and increasing the president's powers passed in a referendum on Nov. 28, 1996. Pro-government parties won the parliamentary election of June 6, 1997. Abdelaziz Bouteflika, who became president after a flawed election on Apr. 15, 1999, made peace with rebels and won approval for an amnesty plan in a referendum on Sept. 16; by then, some 100,000 people had died in the civil war.

Andorra

Principality of Andorra

People: Population: 65,939. **Age distrib.** (%): <15: 14.3; 65+: 12.6. **Pop. density:** 388 per sq. mi. **Urban:** 95%. **Ethnic groups:** Spanish 61%, Andorran 30%, French 6%. **Principal languages:** Catalan (official), French, Castilian. **Chief religion:** Predominently Roman Catholic.

Geography: Area: 170 sq. mi. **Location:** SW Europe, in Pyrenees Mts. **Neighbors:** Spain on S, France on N. **Topography:** High mountains and narrow valleys cover the country. **Capital:** Andorra la Vella (1995 est.): 21,984.

Government: Type: Parliamentary co-principality. **Heads of state:** President of France & Bishop of Urgel (Spain), as co-princes. **Head of gov.:** Marc Forné Molné; b 1947; in office: Dec. 21, 1994. **Local divisions:** 7 parishes. **Defense:** Responsibility of France and Spain.

Economy: Industries: Tourism, sheep, timber, tobacco. **Minerals:** Iron, lead. **Arable land:** 2%.

Finance: Monetary unit: French Franc (Sept. 1999: 6.13 = $1 U.S.). Spanish Peseta (Sept. 1999: 155.44 = $1 U.S.). **GDP** (1995 est.): $1.2 bil. **Per capita GDP:** $18,000. **Imports**

(1995): $1 bil; partners: Spain 41%, France 31%. **Exports** (1995): $47 mil; partners: France 49%, Spain 47%. **Budget** (1993): $177 mil.

Transport: Motor vehicles: 35,358 pass. cars, 4,238 comm. vehicles.

Communications: TV sets: 315 per 1,000 pop. **Radios:** 156 per 1,000 pop. **Telephones:** 33,100 main lines. **Daily newspaper circ.:** 62 per 1,000 pop.

Health: Life expectancy: 80.55 male; 86.55 female. **Births** (per 1,000 pop.): 10.27. **Deaths** (per 1,000 pop.): 5.46. **Natural inc.:** 0.481%. **Hosp. beds** (1993): 1 per 556 persons. **Physicians** (1994): 1 per 491 persons. **Infant mortality** (per 1,000 live births): 4.08.

Education: Free, compulsory: ages 6-16. **Literacy** (1997): 100%.

Major Intl. Organizations: UN.

Embassy: 2 UN Plaza, 25th floor, New York, NY 10017; (212) 750-8064.

Website: http://www.andorra.ad/cniauk.html

Andorra was a co-principality, with joint sovereignty by France and the bishop of Urgel, from 1278 to 1993.

Tourism, especially skiing, is the economic mainstay. A free port, allowing for an active trading center, draws some 13 million tourists annually. Andorran voters chose to end a feudal system that had been in place for 715 years and adopt a parliamentary system of government Mar. 14, 1993.

Angola
Republic of Angola

People: Population: 11,177,537. **Age distrib.** (%): <15: 44.9; 65+: 2.8. **Pop. density:** 23 per sq. mi. **Urban:** 32%. **Ethnic groups:** Ovimbundu 37%, Kimbundu 25%, Bakongo 13%. **Principal languages:** Portuguese (official), varius Bantu and other African languages. **Chief religions:** Indigenous beliefs 47%, Roman Catholic 38%, Protestant 15%.

Geography: Area: 481,400 sq. mi. **Location:** In SW Africa on Atlantic coast. **Neighbors:** Namibia on S, Zambia on E, Congo-Kinshasa (formerly Zaire) on N; Cabinda, an enclave separated from rest of country by short Atlantic coast of Congo-Kinshasa, borders Congo-Brazzaville. **Topography:** Most of Angola consists of a plateau elevated 3,000 to 5,000 feet above sea level, rising from a narrow coastal strip. There is also a temperate highland area in the west-central region, a desert in the S, and a tropical rain forest covering Cabinda. **Capital:** Luanda: 2,081,000*.

Government: Type: Republic. **Head of state:** Pres. José Eduardo dos Santos; b Aug. 28, 1942; in office: Sept. 20, 1979. **Local divisions:** 18 provinces. **Defense:** 8.8% of GDP. **Active troops:** 110,500.

Economy: Industries: Food processing, textiles, mining, brewing, oil. **Chief crops:** Coffee, sugarcane, bananas. **Minerals:** Iron, diamonds (over 1 mil carats a year), gold, phosphates, oil. **Livestock** (1997): chickens: 6.50 mil; cattle: 3.50 mil; goats: 1.45 mil; pigs: 830,000; sheep: 250,000. **Crude oil reserves** (1999): 5.4 bil bbls. **Arable land:** 2%. **Fish catch** (1996): 72,841 metric tons. **Electricity prod.** (1997): 1.865 bil. kWh. **Labor force:** 85% agric., 15% industry & services.

Finance: Monetary unit: Readjusted Kwanza (Sept. 1999: 257,128.00 = $1 U.S.). **GDP** (1996 est.): $8.2 bil. **Per capita GDP:** $800. **Imports** (1995 est.): $1.7 bil; partners: Portugal 30%, U.S. 11%, France 10%. **Exports** (1996 est.): $4 bil; partners: U.S. 70%. **Tourism:** $9 mil. **Budget** (1992 est.): $2.5 bil. **Intl. reserves less gold** (Apr. 1999): $89.16 bil. **Consumer prices** (change in 1997): 111.2%.

Transport: Railroad: Length: 1,739 mi. **Motor vehicles:** 197,000 pass. cars, 26,000 comm. vehicles. **Civil aviation:** 385.3 mil pass.-mi.; 17 airports. **Chief ports:** Cabinda, Lobito, Luanda.

Communications: TV sets: 48 per 1,000 pop. **Radios:** 39 per 1,000 pop. **Telephones:** 72,200 main lines. **Daily newspaper circ.:** 11 per 1,000 pop.

Health: Life expectancy: 46.08 male; 50.82 female. **Births** (per 1,000 pop.): 43.11. **Deaths** (per 1,000 pop.): 16.35. **Natural inc.:** 2.676%. **Infant mortality** (per 1,000 live births): 129.19.

Education: Free, compulsory: ages 7-15. **Literacy** (1992): 40%.

Major Intl. Organizations: UN (FAO, IBRD, ILO, IMF, IMO, WHO, WTrO), OAU.

Embassy: 1050 Connecticut Avenue NW, Suite 760 20036; 785-1156.

Website: http://www.angola.org

From the early centuries AD to 1500, Bantu tribes penetrated most of the region. Portuguese came in 1583, allied with the Bakongo kingdom in the north, and developed the slave trade. Large-scale colonization did not begin until the 20th century, when 400,000 Portuguese immigrated.

A guerrilla war begun in 1961 lasted until 1975, when Portugal granted independence. Fighting then erupted between three rival rebel groups—the National Front, based in Zaire (now Congo), the Soviet-backed Popular Movement for the Liberation of Angola (MPLA), and the National Union for the Total Independence of Angola (UNITA), aided by the U.S. and South Africa. The civil war killed thousands of blacks, drove most whites to emigrate, and completed economic ruin. Cuban troops and Soviet aid helped the MPLA win control of most of the country by 1976, although fighting continued through the 1980s. A peace accord between the MPLA government and UNITA was signed May 1, 1991.

Elections were held in Sept. 1992, but fighting again broke out, as UNITA rejected the results. Large numbers of civilians died from war-related causes, especially starvation. UNITA signed a new peace treaty with the government, Nov. 20, 1994, but the rebels were slow to demobilize. The UN Security Council voted, Aug. 28, 1997, to impose sanctions on UNITA. In Aug. 1998, Angola sent thousands of troops into Congo-Kinshasa (formerly Zaire) to support Laurent Kabila's regime. The UN ended its mission in Angola in Mar. 1999; by July, rebels had reportedly recaptured 70% of the country.

Antigua and Barbuda

People: Population: 64,246. **Age distrib.** (%): <15: 25.8; 65+: 5.5. **Pop. density:** 378 per sq. mi. **Urban:** 36%. **Ethnic groups:** Primarly black. **Principal language:** English (official). **Chief religion:** Predominantly Anglican.

Geography: Area: 170 sq. mi. **Location:** Eastern Caribbean. **Neighbors:** St. Kitts & Nevis to W, Guadeloupe (Fr.) to S. **Capital:** Saint John's (1991): 35,635.

Government: Type: Constitutional monarchy with British-style parliament. **Head of state:** Queen Elizabeth II; represented by Gov.-Gen. James Carlisle; b Aug. 5, 1937; in office: June 10, 1993. **Head of gov.:** Prime Min. Lester Bird; b Feb. 21, 1938; in office: Mar. 9, 1994. **Local divisions:** 6 parishes, 2 dependencies. **Defense:** 0.5% of GDP. **Active troops:** 200.

Economy: Industries: Tourism, manufacturing, construction. **Arable land:** 18%. **Electricity prod.** (1997): 95 mil kWh. **Labor force:** 82% commerce & serv.; 11% agric.; 7% ind.

Finance: Monetary unit: East Caribbean Dollar (Sept. 1999: 2.70 = $1 U.S.). **GDP** (1997 est.): $470 mil. **Per capita GDP:** $7,400. **Imports** (1996 est.): $350.8 mil; partners: U.S. 27%, U.K. 16%. **Exports** (1996 est.): $45 mil; partners: OECS 26%, Barbados 15%. **Tourism:** $260 mil. **Budget** (1995): $132.0 mil. **Intl. reserves less gold** (Apr. 1999): $58.67 mil.

Transport: Motor vehicles: 13,250 pass. cars, 1,423 comm. vehicles. **Civil aviation:** 155.4 mil pass.-mi.; 2 airports.

Communications: TV sets: 435 per 1,000 pop. **Radios:** 776 per 1,000 pop. **Telephones** (1997): 28,000 main lines.

Health: Life expectancy: 69.06 male; 73.98 female. **Births** (per 1,000 pop.): 16.22. **Deaths** (per 1,000 pop.): 5.76. **Natural inc.:** 1.046%. **Hosp. beds** (1992): 1 per 173 persons. **Physicians** (1992): 1 per 1,083 persons. **Infant mortality** (per 1,000 live births): 20.69.

Education: Compulsory: ages 5-16. **Literacy** (1992): 90%.

Major Intl. Organizations: UN (FAO, IBRD, ILO, IMF, IMO, WHO, WTrO), Caricom, the Commonwealth, OAS, OECS.

Embassy: 3216 New Mexico Ave. NW 20016; 362-5211.

Website: http://www.antigua-barbuda.com

Columbus landed on Antigua in 1493. The British colonized it in 1632.

The British associated state of Antigua achieved independence as Antigua and Barbuda on Nov. 1, 1981. The government maintains close relations with the U.S., United Kingdom, and Venezuela. The country was hit hard by Hurricane Luis, Sept. 1995. About 3,000 refugees fleeing a volcanic eruption on Montserrat have settled in Antigua since 1995.

Argentina
Argentine Republic

People: Population: 36,737,664. **Age distrib.** (%): <15: 27.4; 65+: 10.2. **Pop. density:** 34 per sq. mi. **Urban:** 88%. **Ethnic groups:** White 85% (mostly Spanish, Italian); mestizo, Amerindian, other nonwhites 15%. **Principal languages:** Spanish (official), English, Italian. **Chief religion:** Nominally Roman Catholic 90%.

Geography: Area: 1,068,300 sq. mi., second largest country in South America. **Location:** Occupies most of South America. **Neighbors:** Chile on W; Bolivia, Paraguay on N; Brazil, Uruguay on NE. **Topography:** Mountains in the W are: the Andean, Central, Misiones, and Southern ranges. Aconcagua is the highest peak in the western hemisphere, alt. 22,834 ft. E of the Andes are heavily wooded plains, called the Gran Chaco in the N, and the fertile, treeless Pampas in the central region. Patagonia, in the S, is bleak and arid. Rio de la Plata, an estuary in the NE, 170 by 140 mi., is mostly fresh water, from 2,485-mi Parana and 1,000-mi Uruguay rivers. **Capital:** Buenos Aires (the Senate has approved moving the capital to the Patagonia Region). **Cities:** Buenos Aires 11,802,000; Cordoba 1,294,000; Rosario 1,155,000*.

Government: Type: Republic. **Head of state and gov.:** Pres. Carlos Saúl Menem; b July 2, 1930; in office: July 8, 1989. **Local divisions:** 23 provinces, 1 federal district. **Defense:** 1.7% of GDP. **Active troops:** 73,000.

Economy: Industries: Food processing, autos, chemicals, textiles, printing. **Chief crops:** Grains, corn, sugar beets, sorghum, soybeans. **Minerals:** Oil, lead, zinc, iron, copper, tin, uranium. **Crude oil reserves** (1999): 2.6 bil bbls. **Arable land:** 9%. **Livestock** (1997): chickens: 55.00 mil; cattle: 54.60 mil; sheep: 16.43 mil; goats: 3.43 mil; pigs: 3.20 mil. **Fish catch** (1997): 1.35 mil metric tons. **Electricity prod.** (1997): 65.045 bil kWh. **Labor force:** 57% services; 31% ind.; 12% agric.

Finance: Monetary unit: Peso (Sept. 1999: 1.00 = $1 U.S.). **GDP** (1997 est.): $348.2 bil. **Per capita GDP:** $9,700. **Imports** (1997): $30.3 bil; partners: Brazil 21%, Germany 6%, Italy 6%. **Exports** (1997): $25.4 bil; partners: Brazil; 26%, U.S. 9%. **Tourism:** $5.36 bil. **Budget** (1997 est.): $59.0 bil. **Intl. reserves less gold** (June 1999): $22.98 bil. **Gold:** 351,000 oz t. **Consumer prices** (change in 1998): 0.9%.

Transport: Railroad: Length: 21,015 mi. **Motor vehicles:** 4.78 mil pass. cars, 1.29 mil comm. vehicles. **Civil aviation:** 8.9 bil pass.-mi.; 39 airports. **Chief ports:** Buenos Aires, Bahia Blanca, La Plata.

Communications: TV sets: 347 per 1,000 pop. **Radios:** 614 per 1,000 pop. **Telephones:** 7,323,100 main lines. **Daily newspaper circ.:** 137 per 1,000 pop.

Health: Life expectancy: 71.13 male; 78.56 female. **Births** (per 1,000 pop.): 19.91. **Deaths** (per 1,000 pop.): 7.64. **Natural inc.:** 1.227%. **Hosp. beds** (1992): 1 per 227 persons. **Physicians** (1992): 1 per 376 persons. **Infant mortality** (per 1,000 live births): 18.41.

Education: Free, compulsory: ages 6-14. **Literacy:** 96%.

Major Intl. Organizations: UN (FAO, IBRD, ILO, IMF, IMO, WHO, WTrO), OAS.

Embassy: 1600 New Hampshire Ave. NW 20009; 939-6400.

Nomadic Indians roamed the Pampas when Spaniards arrived, 1515-16, led by Juan Diaz de Solis. Nearly all the Indians were killed by the late 19th century. The colonists won independence, 1816, and a long period of disorder ended in a strong centralized government.

Large-scale Italian, German, and Spanish immigration in the decades after 1880 spurred modernization. Social reforms were enacted in the 1920s, but military coups prevailed 1930-46, until the election of Gen. Juan Perón as president.

Perón, with his wife, Eva Duarte (d 1952), effected labor reforms, but also suppressed speech and press freedoms, closed religious schools, and ran the country into debt. A 1955 coup exiled Perón, who was followed by a series of military and civilian regimes. Perón returned to Argentina in 1973, and was once more elected president. He died 10 months later, succeeded by his wife Isabel, who had been elected vice president, and who became the first woman head of state in the western hemisphere.

A military junta ousted Mrs. Perón in 1976 amid charges of corruption. Under a continuing state of siege, the army battled guerrillas and leftists, killed 5,000 people, and jailed and tortured others. On Dec. 9, 1985, after a trial of 5 months and nearly 1,000 witnesses, 5 former junta members were found guilty of murder and human rights abuses.

Argentine troops seized control of the British-held Falkland Islands on Apr. 2, 1982. Both countries had claimed sovereignty over the islands, located 250 miles off the Argentine coast, since 1833. The British dispatched a task force and declared a total air and sea blockade around the Falklands. Fighting began May 1; several hundred lost their lives as the result of the destruction of a British destroyer and the sinking of an Argentine cruiser.

British troops landed on East Falkland Island May 21 and eventually surrounded Stanley, the capital city and Argentine stronghold. The Argentine troops surrendered, June 14; Argentine Pres. Leopoldo Galtieri resigned June 17.

Democratic rule returned to Argentina in 1983 as Raul Alfonsín's Radical Civic Union party gained an absolute majority in the presidential electoral college and Congress. By 1989 the nation was plagued by severe financial and political problems, as hyperinflation sparked looting and rioting in several cities. The government of Perónist Pres. Carlos Saúl Menem, installed 1989, introduced harsh economic measures to curtail inflation, control government spending, and restructure the foreign debt.

About 100 people were killed in the terrorist bombing of a Jewish cultural center in Buenos Aires, July 18, 1994. Following passage of a new constitution in Aug. 1994, Menem was reelected president on May 14, 1995. A pact restoring commercial air links between Argentina and the Falklands was signed July 14, 1999.

Armenia
Republic of Armenia

People: Population: 3,409,234. **Age distrib.** (%): <15: 25.5; 65+: 8.6. **Pop. density:** 296 per sq. mi. **Urban:** 69%. **Ethnic groups:** Armenian 93%, Azeri 3%, Russian 2%, Kurd and others 2%. **Principal language:** Armenian (official). **Chief religion:** Armenian Orthodox 94%.

Geography: Area: 11,500 sq. mi. **Location:** SW Asia. **Neighbors:** Georgia on N, Azerbaijan on E, Iran on S, Turkey on W. **Topography:** Mountainous with many peaks above 10,000 ft. **Capital:** Yerevan: 1,278,000*.

Government: Type: Republic. **Head of state:** Pres. Robert Kocharian; b Aug. 31, 1954; in office: Apr. 9, 1998. **Head of gov.:** Prime Min. Vazgen Sarkissian; b Mar. 5, 1959; in office: June 11, 1999. **Local divisions:** 10 provinces, 1 city. **Defense:** 8.9% of GDP. **Active troops:** 60,000.

Economy: Industries: Manufacturing, machinery, chemicals. **Chief crops:** Vegetables, grapes. **Minerals:** Copper, gold, zinc. **Arable land:** 17%. **Livestock** (1997): chickens: 2.80 mil; sheep: 550,000; cattle; 505,000. **Electricity prod.** (1997): 5.062 bil kWh.

Finance: Monetary unit: Dram (Aug. 1999: 536.00 = $1 U.S.). **GDP** (1997 est.): $9.5 bil. **Per capita GDP:** $2,750. **Imports** (1996): $727 mil; partners: Russia 20%, Turkmenistan 19%. **Exports** (1996): $290 mil; Russia 33%, Turkmenistan 25%. **Tourism:** $14 mil. **Intl. reserves less gold** (June 1999): $304.35 mil. **Gold:** 43,300 oz t. **Consumer prices** (change in 1998): 8.7%.

Transport: Railroad: Length: 515 mi. **Civil aviation:** 476.6 mil pass.-mi.; 1 airport.

Communications: TV sets: 241 per 1,000 pop. **Telephones** (1997): 567,800 main lines. **Daily newspaper circ.:** 23 per 1,000 pop.

Health: Life expectancy: 62.21 male; 71.13 female. **Births** (per 1,000 pop.): 13.53. **Deaths** (per 1,000 pop.): 9.03. **Natural inc.:** 0.450%. **Hosp. beds** (1994): 1 per 125 persons. **Physicians** (1994): 1 per 288 persons. **Infant mortality** (per 1,000 live births): 41.12.

Education: Compulsory: ages 6-17. **Literacy** (1989): 99%.

Major Intl. Organizations: UN (FAO, IBRD, ILO, IMF, WHO), CIS, OSCE.

Embassy: 2225 R St. NW 20008; 319-1976.

Ancient Armenia extended into parts of what are now Turkey and Iran. Present-day Armenia was set up as a Soviet republic Apr. 2, 1921. It joined Georgian and Azerbaijan SSRs Mar. 12, 1922, to form the Transcaucasian SFSR, which became part of the USSR Dec. 30, 1922. Armenia became a constituent republic of the USSR Dec. 5, 1936. An earthquake struck Armenia Dec. 7, 1988; approximately 55,000 were killed and several cities and towns were left in ruins.

Armenia declared independence Sept. 23, 1991, and became an independent state when the USSR disbanded Dec. 26, 1991. Fighting between mostly Christian Armenia and mostly Muslim Azerbaijan escalated in 1992 and continued through 1993. Each country claimed Nagorno-Karabakh, an enclave in Azerbaijan that has a majority population of ethnic Armenians. A temporary cease-fire was announced in May 1994, with Armenian forces in control of the enclave. Voters approved, July 5, 1995, a new constitution strengthening presidential powers. Pres. Levon Ter-Petrosian won reelection on Sept. 22, 1996, amid claims of fraud; he resigned Feb. 3, 1998, in a conflict over Nagorno-Karabakh. Robert Kocharian, a nationalist born in the disputed region, won the presidency on Mar. 30, 1998.

Australia
Commonwealth of Australia

People: Population: 18,783,551. **Age distrib.** (%): <15: 21.0; 65+: 12.5. **Pop. density:** 6 per sq. mi. **Urban:** 85%. **Ethnic groups:** Caucasian 92%, Asian 7%, aboriginal and other 1%. **Principal languages:** English (official), aboriginal languages. **Chief religions:** Anglican 26%, Roman Catholic 26%, other Christian 24%.

Geography: Area: 2,967,900 sq. mi. **Location:** SE of Asia, Indian O. is W and S, Pacific O. (Coral, Tasman seas) is E; they meet N of Australia in Timor and Arafura seas. Tasmania lies 150 mi. S of Victoria state, across Bass Strait. **Neighbors:** Nearest are Indonesia, Papua New Guinea on N; Solomons,

Fiji, and New Zealand on E. **Topography:** An island continent. The Great Dividing Range along the E coast has Mt. Kosciusko, 7,310 ft. The W plateau rises to 2,000 ft., with arid areas in the Great Sandy and Great Victoria deserts. The NW part of Western Australia and Northern Terr. are arid and hot. The NE has heavy rainfall and Cape York Peninsula has jungles. **Capital:** Canberra. **Cities** (1997 est.): Sydney 3.9 mil; Melbourne 3.3 mil; Brisbane 1.5 mil; Perth 1.3 mil; Adelaide 1.1 mil.

Government: Type: Democratic, federal state system. **Head of state:** Queen Elizabeth II, represented by Gov.-Gen. Sir William Patrick Deane; b July 4, 1931; in office: Feb. 15, 1996. **Head of gov.:** Prime Min. John Howard; b July 26, 1939; in office: Mar. 11, 1996. **Local divisions:** 6 states, 2 territories. **Defense:** 2.2% of GDP. **Active troops:** 57,400.

Economy: Industries: Mining, steel, industrial & transportation equip., chemicals, food processing. **Chief crops:** Wheat (a leading export), barley, fruit, sugar. **Minerals:** Bauxite, coal, copper, iron, lead, tin, uranium, zinc. **Crude oil reserves** (1999): 2.9 bil bbls. **Other resources:** Wool (world's leading producer), beef. **Arable land:** 6%. **Livestock** (1997): sheep: 119.60 mil; chickens: 83.00 mil; cattle: 26.71 mil; pigs: 2.68 mil; goats: 230,000. **Fish catch** (1996): 191,533 metric tons. **Electricity prod.** (1997): 173.493 bil kWh. **Labor force:** 73% services; 22% industry; 5% agric.

Finance: Monetary unit: Australian Dollar (Sept. 1999: 1.55 = $1 U.S.). **GDP** (1997 est.): $394.0 bil. **Per capita GDP:** $21,400. **Imports** (1997 est.): $67.0 bil; partners: U.S. 22%, Japan 17%, UK 6%. **Exports** (1997 est.): $68.0 bil; partners: Japan 20%, ASEAN 16%, S. Korea 9%, New Zealand 8%. **Tourism:** $8.58 bil. **Budget** (FY 1997-98 est.): $91.92 bil. **Intl. reserves less gold** (June 1999): $15.97 bil. **Gold:** 2.56 mil oz t. **Consumer prices** (change in 1998): 0.9%.

Transport: Railroad: Length: 22,385 mi. **Motor vehicles:** 8.7 mil pass. cars, 2.05 mil comm. vehicles. **Civil aviation:** 47.2 bil pass.-mi.; 400 airports. **Chief ports:** Sydney, Melbourne, Brisbane, Adelaide, Fremantle, Geelong.

Communications: TV sets: 641 per 1,000 pop. **Radios:** 1,148 per 1,000 pop. **Telephones:** 9,628,000 main lines. **Daily newspaper circ.:** 258 per 1,000 pop.

Health: Life expectancy: 77.22 male; 83.23 female. **Births** (per 1,000 pop.): 13.21. **Deaths** (per 1,000 pop.): 6.90. **Natural inc.:** 0.631%. **Hosp. beds** (1995): 1 per 226 persons. **Physicians** (1996): 1 per 400 persons. **Infant mortality** (per 1,000 live births): 5.11.

Education: Free, compulsory: ages 6-15. **Literacy** (1996): 100%.

Major Intl. Organizations: UN and all of its specialized agencies, APEC, the Commonwealth, OECD.

Embassy: 1601 Massachusetts Ave. NW 20036; 797-3000. **Websites:** http://www.austemb.org
http://www.abs.gov.au

Australia harbors many plant and animal species not found elsewhere, including kangaroos, koalas, platypuses, dingos (wild dogs), Tasmanian devils (raccoon-like marsupials), wombats (bear-like marsupials), and barking and frilled lizards.

Capt. James Cook explored the E coast in 1770, when the continent was inhabited by a variety of different tribes. The first settlers, beginning in 1788, were mostly convicts, soldiers, and government officials. By 1830, Britain had claimed the entire continent, and the immigration of free settlers began to accelerate. The Commonwealth was proclaimed Jan. 1, 1901. Northern Terr. was granted limited self-rule July 1, 1978.

State/Territory, Capital	Area (sq. mi.)	Population (1997)
New South Wales, Sydney.....	309,500	6,274,400
Victoria, Melbourne	87,900	4,605,100
Queensland, Brisbane	666,990	3,401,200
Western Australia, Perth	975,100	1,798,100
South Australia, Adelaide......	379,900	1,479,800
Tasmania, Hobart............	26,200	473,500
Australian Capital Terr., Canberra	900	309,800
Northern Terr., Darwin	519,800	187,100

Racially discriminatory immigration policies were abandoned in 1973, after 3 million Europeans (half British) had entered since 1945. The 50,000 aborigines and 150,000 part-aborigines are mostly detribalized, but there are several preserves in the Northern Territory. They remain economically disadvantaged.

Australia's agricultural success makes the country among the top exporters of beef, lamb, wool, and wheat. Major mineral deposits have been developed, largely for export. Industrialization has been completed. The nation endured a deep recession 1990-93 but has rebounded strongly.

The Labor Party won a majority in Feb. 1983 general elections and was reelected in 1984, 1987, 1990, and 1993. After an election that focused mainly on economic issues, conservatives swept into power in elections Mar. 2, 1996. Constitutional convention delegates voted Feb. 13, 1998, to have a native Australian as head of state, subject to a referendum in Nov. 1999.

Prime Min. John Howard retained power, but with a reduced majority in parliamentary elections Oct. 3, 1998. Australia led an international peacekeeping force into East Timor in Sept. 1999.

Australian External Territories

Norfolk Isl., area 13.3 sq. mi., pop. (1996 est.) 2,209, was taken over, 1914. The soil is very fertile, suitable for citrus, bananas, and coffee. Many of the inhabitants are descendants of the *Bounty* mutineers, moved to Norfolk 1856 from Pitcairn Isl. Australia offered the island limited home rule in 1978.

Coral Sea Isls. Territory, area 1 sq. mi., is administered from Norfolk Isl.

Territory of Ashmore and Cartier Isls., area 2 sq. mi., in the Indian O., came under Australian authority 1934 and are administered as part of Northern Territory. **Heard Isl. and McDonald Isls.,** area 159 sq. mi., are administered by the Dept. of Science.

Cocos (Keeling) Isls., 27 small coral islands in the Indian O. 1,750 mi. NW of Australia. Pop. (1996 est.) 609; area 5.5 sq. mi. The residents voted to become part of Australia, Apr. 1984.

Christmas Isl., area 52 sq. mi., pop. (1996 est.) 813; 230 mi. S of Java, was transferred by Britain in 1958. It has phosphate deposits.

Australian Antarctic Territory was claimed by Australia in 1933, including 2,362,000 sq. mi. of territory S of 60th parallel S Lat. and between 160th-45th meridians E Long. It does not include Adelie Coast.

Austria
Republic of Austria

People: Population: 8,139,299. **Age distrib.** (%): <15: 16.8; 65+: 15.5. **Pop. density:** 251 per sq. mi. **Urban:** 64%. **Ethnic groups:** German 99%, Croatian, Slovene. **Principal language:** German (official). **Chief religions:** Roman Catholic 78%, Protestant 5%.

Geography: Area: 32,378 sq. mi. **Location:** In S Central Europe. **Neighbors:** Switzerland, Liechtenstein on W; Germany, Czech Rep. on N; Slovakia, Hungary on E; Slovenia, Italy on S. **Topography:** Austria is primarily mountainous, with the Alps and foothills covering the western and southern provinces. The eastern provinces and Vienna are located in the Danube River Basin. **Capital:** Vienna (1991 census): 1,540,000*.

Government: Type: Parliamentary democracy. **Head of state:** Pres. Thomas Klestil; b Nov. 4, 1932; in office: July 8, 1992. **Head of gov.:** Chancellor Viktor Klima; b June 4, 1947; in office: Jan. 28, 1997. **Local divisions:** 9 bundeslaender (states), each with a legislature. **Defense:** 0.8% of GDP. **Active troops:** 45,500.

Economy: Industries: Steel, machinery, autos, electrical equip., tourism, mining, paper, textiles, chemicals, food. **Chief crops:** Grains, fruits, potatoes, sugar beets. **Minerals:** Iron ore, oil, magnesite. **Crude oil reserves** (1999): 87 mil bbls. **Other resources:** Forests, hydropower. **Arable land:** 17%. **Livestock** (1997): chickens: 13.95 mil; pigs: 3.68 mil; cattle: 2.20 mil; sheep: 383,655. **Electricity prod.** (1997): 54.226 bil kWh. **Labor force:** 66% services; 30% ind. & crafts.

Finance: Monetary unit: Schilling (Sept. 1999: 12.85 = $1 U.S.). **Euro** (Sept. 1999: 1.07 = $1 U.S.). **GDP** (1997 est.): $174.1 bil. **Per capita GDP:** $21,400. **Imports** (1996): $67.3 bil; partners: Germany 43%. **Exports** (1996): $57.8 bil; partners: Germany 38%. **Tourism:** $12.16 bil. **Budget** (1996 est.): $61.6 bil. **Intl. reserves less gold** (June 1999): $14.84 bil. **Gold:** 13.00 mil oz t. **Consumer prices** (change in 1998): 0.9%.

Transport: Railroad: Length: 3,524 mi. **Motor vehicles in use (1997):** 3.78 mil pass. cars, 324,776 comm. vehicles. **Civil aviation:** 6.3 bil pass.-mi.; 6 airports. **Chief ports:** Linz, Vienna, Enns.

Communications: TV sets: 336 per 1,000 pop. **Radios:** 584 per 1,000 pop. **Telephones:** 3,999,000 main lines. **Daily newspaper circ.:** 465 per 1,000 pop.

Health: Life expectancy: 74.31 male; 80.82 female. **Births** (per 1,000 pop.): 9.62. **Deaths** (per 1,000 pop.): 10.04. **Natural inc.:** −0.042%. **Hosp. beds** (1996): 1 per 117 persons. **Physicians** (1996): 1 per 289 persons. **Infant mortality** (per 1,000 live births): 5.10.

Education: Free, compulsory: ages 6-15. **Literacy** (1994): 100%.

MILLENNIUM FACT BOX

The World in 1900

The political map of the world at the beginning of the 20th century was quite different from what it is at the start of the 21st. Here is a list of the world's countries and territories at that time (with spellings unchanged), as published in *The World Almanac 1900*.

Countries	Population	Sq. Miles	Capitals
China	402,680,000	4,218,401	Peking
British Empire*	385,280,140	11,712,170	London
Russian Empire	128,932,173	8,660,395	St. Petersburg
United States	†78,000,000	3,602,990	Washington
U.S. and Colonies	†87,000,000	3,756,884	Washington
Philippines	8,000,000	143,000	Manila
Porto Rico	900,000	3,600	San Juan
Hawaii	109,029	6,740	Honolulu
Tutuila, Samoa	9,000	500	—
Guam	4,000	54	—
France and Colonies	63,166,967	3,357,856	Paris
France	38,517,975	204,177	Paris
Colonies	21,448,064	2,923,679	—
Algeria	3,870,000	260,000	Algiers
Senegal, etc.	183,237	580,000	St. Louis
Tunis	1,500,000	45,000	Tunis
Cayenne	26,502	46,697	Cayenne
Cambodia	1,500,000	32,254	Saigon
Cochin-China	1,223,000	13,692	—
Tonquin	12,000,000	60,000	Hanoi
New Caledonia	62,752	7,624	Noumea
Tahiti	12,800	462	—
Sahara	1,100,000	1,550,000	—
Madagascar	3,500,000	230,000	Antananarivo
German Empire	52,279,901	211,108	Berlin
Prussia	31,855,123	134,467	Berlin
Bavaria	5,589,382	29,291	Munich
Saxony	3,500,513	5,789	Dresden
Wurtemberg	2,035,443	7,531	Stuttgart
Baden	1,656,817	5,803	Karlsruhe
Alsace-Lorraine	1,603,987	5,602	Strasburg
Hesse	956,170	2,965	Darmstadt
Mecklenburg-Schwerin	575,140	5,137	Schwerin
Hamburg	622,530	158	—
Brunswick	372,580	1,425	Brunswick
Oldenburg	341,250	2,479	Oldenburg
Saxe-Weimar	313,668	1,387	Weimar
Anhalt	247,603	906	Dessau
Saxe-Meiningen	214,697	953	Meiningen
Saxe-Coburg-Gotha	198,717	760	Gotha
Bremen	180,443	99	—
Saxe-Altenburg	161,129	511	Altenburg
Lippe	123,250	472	Detmold
Reuss (Younger line)	112,118	319	Gera
Mecklenburg-Strelitz	98,371	1,131	Neu Strelitz
Schwarzburg-Rudolstadt	83,939	363	Rudolstadt
Schwarzburg-Sond's's'n	73,623	333	S'ndershausen
Lubeck	76,485	115	—
Waldeck	56,565	433	Arolsen
Reuss (Elder line)	53,787	122	Greiz.
Schaumburg-Lippe	37,204	131	Buckeburg
German Africa	5,950,000	822,000	—
Austro-Hungarian Empire	41,827,700	201,591	Vienna
Japan	41,089,940	147,669	Tokio
Netherlands	4,450,870	12,680	The Hague
Netherlands and Colonies	33,042,238	778,187	The Hague
Borneo	1,073,500	203,714	—
Celebes	2,000,000	72,000	—
Java	21,974,161	50,848	Batavia
Moluccas	353,000	42,420	Amboyna
New Guinea	200,000	150,755	—
Sumatra	2,750,000	170,744	—
Surinam	57,141	46,060	Paramaribo

Countries	Population	Sq. Miles	Capitals
Turkish Empire	33,559,787	1,652,533	Constantin'ple
European Turkey	4,790,000	63,850	—
Asiatic Turkey	16,133,900	729,170	—
Tripoli	1,000,000	398,873	Tripoli
Bulgaria	3,154,375	37,860	Sofia
Egypt	9,700,000	400,000	Cairo
Italy	29,699,785	110,665	Rome
Italy and Colonies	34,970,785	425,765	Rome
Abyssinia	4,500,000	189,000	—
Eritrea	660,000	56,100	—
Somal Coast	210,000	70,000	—
Spain	17,550,216	196,173	Madrid
Spanish Africa	437,000	203,767	—
Spanish Islands	127,172	1,957	—
Brazil	18,000,000	3,219,000	Rio Janeiro
Mexico	12,578,861	767,316	City of Mexico
Korea	10,519,000	85,000	Seoul
Congo State	8,000,000	802,000	—
Persia	7,653,600	636,000	Teheran
Portugal	4,708,178	34,038	Lisbon
Portugal and Colonies	11,073,681	951,785	Lisbon
Portuguese Africa	5,416,000	841,025	—
Portuguese Asia	847,503	7,923	—
Sweden and Norway	6,785,898	297,321	—
Sweden	4,784,981	172,876	Stockholm
Norway	2,000,917	124,445	Kristiania
Morocco	6,500,000	314,000	Fez
Belgium	6,030,043	11,373	Brussels
Siam	5,700,000	280,550	Bangkok
Roumania	5,376,000	46,314	Bucharest
Argentine Republic	4,042,990	1,095,013	Buenos Ayres
Colombia	4,600,000	331,420	Bogota
Afghanistan	4,000,000	279,000	Cabul
Chile	3,500,000	256,860	Santiago
Peru	3,000,000	405,040	Lima
Switzerland	2,933,334	15,981	Berne
Bolivia	2,500,000	472,000	La Paz
Greece	2,433,806	24,977	Athens
Denmark	2,172,205	14,780	Copenhagen
Denmark and Colonies	2,288,193	101,403	Copenhagen
Iceland	72,445	39,756	Rejkjavik
Greenland	9,780	46,740	Godthaab
West Indies	33,763	118	—
Venezuela	2,444,816	566,159	Caracas
Servia	2,096,043	18,757	Belgrade
Nepaul	2,000,000	56,800	Khatmandu
Cuba	1,600,000	41,655	Havana
Oman	1,600,000	81,000	Muscat
Guatemala	1,535,632	46,774	N. Guatemala
Ecuador	1,300,000	144,000	Quito
Liberia	1,050,000	14,000	Monrovia
Hayti	1,211,625	29,830	Port au Prince
Transvaal	1,094,156	119,139	Pretoria
Salvador	800,500	7,228	San Salvador
Uruguay	840,725	72,112	Montevideo
Khiva	700,000	22,320	Khiva
Paraguay	600,000	145,000	Asuncion
Honduras	420,000	42,658	Tegucigalpa
Nicaragua	420,000	51,660	Managua
Dominican Republic	600,000	20,596	San Domingo
Montenegro	245,380	3,486	Cettinje
Costa Rica	309,683	19,985	San Jose
Orange Free State	207,503	48,326	Bloemfontein

*These estimates of the population and area of the British Empire include the recently acquired great possessions in Africa.
†Estimated for January 1, 1900.

Major Intl. Organizations: UN and all of its specialized agencies, EU, OECD, OSCE.
Embassy: 3524 International Ct. NW 20008; 895-6700.
Website: http://www.austria.org

Rome conquered Austrian lands from Celtic tribes around 15 BC. In 788 the territory was incorporated into Charlemagne's empire. By 1300, the House of Hapsburg had gained control; they added vast territories in all parts of Europe to their realm in the next few hundred years.

Austrian dominance of Germany was undermined in the 18th century and ended by Prussia by 1866. But the Congress of Vienna, 1815, confirmed Austrian control of a large empire in southeast Europe consisting of Germans, Hungarians, Slavs, Italians, and others. The dual Austro-Hungarian monarchy was established in 1867, giving autonomy to Hungary and almost 50 years of peace.

World War I, started after the June 28, 1914, assassination of Archduke Franz Ferdinand, the Hapsburg heir, by a Serbian nationalist, destroyed the empire. By 1918 Austria was reduced to a small republic, with the borders it has today.

Nazi Germany invaded Austria Mar. 13, 1938. The republic was reestablished in 1945, under Allied occupation. Full independence and neutrality were restored in 1955. Austria joined the European Union Jan. 1, 1995. The rise of the right-wing populist Austrian Freedom Party challenged the dominance of the Austrian Social Democratic Party in the late 1990s.

Azerbaijan
Azerbaijani Republic

People: Population: 7,908,224. **Age distrib.** (%): <15: 32.0; 65+: 6.5. **Pop. density:** 237 per sq. mi. **Urban:** 56%. **Ethnic groups:** Azeri 90%, Dagestani Peoples 3%, Russian 2.5%, Armenian 2%. **Principal languages:** Azeri (official) 89%, Russian 3%, Armenian 2%. **Chief religions:** Muslim 93%, Orthodox 5%.

Geography: Area: 33,400 sq. mi. **Location:** SW Asia. **Neighbors:** Russia, Georgia on N; Iran on S; Armenia on W; Caspian Sea on E. **Capital:** Baku: 1,848,000*.

Government: Type: Republic. **Head of state:** Pres. Haydar A. Aliyev; b May 10, 1923; in office: June 30, 1993. **Head of gov.:** Prime Min. Artur Rasizade; b Feb. 26, 1935; in office: Nov. 26, 1996. **Local division:** 59 rayons, 11 cities, 1 autonomous republic. **Defense:** 4.0% of GDP. **Active troops:** 66,700.

Economy: Industries: Oil refining, chemicals, textiles. **Chief crops:** Grain, rice, cotton, grapes. **Minerals:** Oil, gas, iron. **Crude oil reserves** (1999): 1.2 bil bbls. **Arable land:** 18%. **Livestock** (1997): chickens: 13.00 mil; sheep: 5.87 mil; cattle: 1.84 mil; buffalo: 293,300; goats: 275,000. **Electricity prod.** (1997): 15.865 bil kWh.

Finance: Monetary unit: Manat (Sept. 1999: 3,942.00 = $1 U.S.). **GDP** (1997 est.): $11.9 bil. **Per capita GDP:** $1,460. **Imports** (1996 est.): $1.3 bil; partners: Turkey 21%, Russia 13%. **Exports** (1996 est.): $789 mil; partners: Iran 30%, Russia 18%. **Tourism:** $160 mil. **Budget** (1996 est.): $682 mil. **Intl. reserves less gold** (May 1999): $631.08 mil.

Transport: Railroad: Length: 1,305 mi. **Motor vehicles:** 289,000 pass. cars; 89,000 comm. vehicles. **Civil aviation:** 797.1 mil. pass.-mi.; 1 airport. **Chief port:** Baku.

Communications: TV sets: 212 per 1,000 pop. **Daily newspaper circ.:** 28 per 1,000 pop.

Health: Life expectancy: 58.76 male; 67.63 female. **Births** (per 1,000 pop.): 21.58. **Deaths** (per 1,000 pop.): 9.50. **Natural inc.:** 1.208%. **Hosp. beds** (1994): 1 per 100 persons. **Physicians** (1994): 1 per 256 persons. **Infant mortality** (per 1,000 live births): 82.52.

Education: Compulsory: ages 6-17. **Literacy:** 100%.

Major Intl. Organizations: UN (FAO, IBRD, ILO, IMF, IMO, WHO), CIS, OSCE.
Embassy: 927 15th St. NW 20005; 842-0001.
Website: http://www.president.az/azerbaijan/azerbaijan.htm

Azerbaijan was the home of Scythian tribes and part of the Roman Empire. Overrun by Turks in the 11th century and conquered by Russia in 1806 and 1813, it joined the USSR Dec. 30, 1922, and became a constituent republic in 1936. Azerbaijan declared independence Aug. 30, 1991, and became an independent state when the Soviet Union disbanded Dec. 26, 1991.

Fighting between mostly Muslim Azerbaijan and mostly Christian Armenia escalated in 1992 and continued in 1993 and 1994. Each country claimed Nagorno-Karabakh, an enclave in Azerbaijan with a majority population of ethnic Armenians. A temporary cease-fire was announced in May 1994, with Armenian forces in control of the enclave.

A National Council ousted Communist Pres. Mutaibov and took power May 19, 1992. Abulfez Elchibey became the nation's first democratically elected president June 7, but was ousted from office by Surat Huseynov, commander of a private militia, June 30, 1993. Huseynov became prime minister, and Haydar Aliyev, a pro-Russian former Communist, became president. Huseynov fled the country after his supporters staged an unsuccessful coup attempt Oct. 1994. Voters approved a new constitution expanding presidential powers, Nov. 12, 1995. Pres. Aliyev was reelected Oct. 11, 1998, but international monitors called the election seriously flawed.

The Bahamas
Commonwealth of The Bahamas

People: Population: 283,705. **Age distrib.** (%): <15: 27.5; 65+: 5.7. **Pop. density:** 53 per sq. mi. **Urban:** 87%. **Ethnic groups:** Black 85%, white 15%. **Principal languages:** English (official), Creole. **Chief religions:** Baptist 32%, Anglican 20%, Roman Catholic 19%, other Christian 24%.

Geography: Area: 5,400 sq. mi. **Location:** In Atlantic O., E of Florida. **Neighbors:** Nearest are U.S. on W, Cuba on S. **Topography:** Nearly 700 islands (29 inhabited) and over 2,000 islets in the W Atlantic O. extend 760 mi. NW to SE. **Capital:** Nassau. **Cities** (1990 est.): Nassau 172,196; Grand Bahama 40,898.

Government: Type: Independent commonwealth. **Head of state:** Queen Elizabeth II, represented by Gov.-Gen. Orville A. Turnquest; b July 19, 1929; in office: Jan. 2, 1995. **Head of gov.:** Prime Min. Hubert Ingraham; b Aug. 4, 1947; in office: Aug. 21, 1992. **Local divisions:** 21 districts. **Defense:** 0.6% of GDP. **Active troops:** 900.

Economy: Industries: Tourism (more than 50% of GDP), rum, cement, banking, pharmaceuticals. **Chief crops:** Citrus, vegetables. **Minerals:** Salt, aragonite. **Other resources:** Lobsters. **Arable land:** 1%. **Livestock** (1997): chickens: 3.80 mil. **Electricity prod.** (1997): 1.300 bil kWh. **Labor force:** 40% tourism; 30% govt.; 10% serv.; 5% agric.

Finance: Monetary unit: Dollar (Sept. 1999: 1.00 = $1 U.S.). **GDP** (1997 est.): $5.36 bil. **Per capita GDP:** $19,400. **Imports** (1996): $1.26 bil; partners: U.S. 29%, Finland 10%, Iran 10%. **Exports** (1996): $201.7 mil; partners: U.S. 24%, Spain 14%, UK 7%, Norway 7%. **Tourism:** $1.42 bil. **Budget** (FY 1996-97 est.): $827 mil. **Intl. reserves less gold** (June 1999): $482.6 mil. **Consumer prices** (change in 1998): 1.3%.

Transport: Motor vehicles: 69,000 pass. cars, 14,000 comm. vehicles. **Civil aviation:** 86.9 mil pass.-mi.; 23 airports. **Chief ports:** Nassau, Freeport.

Communications: TV sets: 179 per 1,000 pop. **Radios:** 282 per 1,000 pop. **Telephones** (1997): 96,300 main lines. **Daily newspaper circ.:** 126 per 1,000 pop.

Health: Life expectancy: 70.94 male; 77.64 female. **Births** (per 1,000 pop.): 20.58. **Deaths** (per 1,000 pop.): 5.43. **Natural inc.:** 1.515%. **Hosp. beds** (1993): 1 per 249 persons. **Physicians** (1992): 1 per 709 persons. **Infant mortality** (per 1,000 live births): 18.38.

Education: Free, compulsory: ages 5-14. **Literacy:** 98%.

Major Intl. Organizations: UN (FAO, IBRD, ILO, IMF, IMO, WHO), Caricom, the Commonwealth, OAS.
Embassy: 2220 Massachusetts Ave. NW 20008; 319-2660.
Websites: http://www.bahamas.net.bs/government
        http://www.bahamas.net

Christopher Columbus first set foot in the New World on San Salvador (Watling Isl.) in 1492, when Arawak Indians inhabited the islands. British settlement began in 1647; the islands became a British colony in 1783. Internal self-government was granted in 1964; full independence within the Commonwealth was attained July 10, 1973.

International banking and investment management have become major industries alongside tourism.

Bahrain
State of Bahrain

People: Population: 629,090. **Age distrib.** (%): <15: 30.5; 65+: 2.9. **Pop. density:** 2,621 per sq. mi. **Urban:** 91%. **Ethnic groups:** Bahraini 63%, Asian 13%, other Arab 10%, Iranian 8%. **Principal languages:** Arabic (official), English, Farsi, Urdu. **Chief religions:** Shi'a Muslim 75%, Sunni Muslim 25%.

Geography: Area: 240 sq. mi. **Location:** SW Asia, in Persian Gulf. **Neighbors:** Nearest are Saudi Arabia on W, Qatar on E. **Topography:** Bahrain Island, and several adjacent, smaller islands, are flat, hot, and humid, with little rain. **Capital:** Manama (1995 est.): 148,000.

Government: Type: Traditional monarchy. **Head of state:** Emir Hamad bin Isa al-Khalifa; b Jan. 28, 1950; in office: Mar.

6, 1999. **Head of gov.:** Prime Min. Kahlifa bin Sulman al-Khalifa; b 1935; in office: Jan. 19, 1970. **Local divisions:** 12 municipalities. **Defense:** 6.5% of GDP. **Active troops:** 11,000.

Economy: Industries: Oil products, aluminum smelting. **Chief crops:** Fruits, vegetables. **Minerals:** Oil, gas. **Crude oil reserves** (1999): 160 mil bbls. **Arable land:** 1%. **Livestock** (1997): chickens: 450,000. **Electricity prod.** (1997): 5.000 bil kWh. **Labor force:** 78% ind., commerce, services, 21% govt.

Finance: Monetary unit: Dinar (Sept. 1999: 0.38 = $1 U.S.). **GDP** (1997 est.): $8.2 bil. **Per capita GDP:** $13,700. **Imports** (1996 est.): $3.7 bil; partners: Saudi Arabia 40%, U.S. 13%. **Exports** (1996 est.): $4.6 bil; partners: India 22%, Japan 12%. **Tourism:** $270 mil. **Budget** (1998 est.): $1.9 bil. **Intl. reserves less gold** (June 1999): $1.35 bil. **Gold:** 150,000 oz t. **Consumer prices** (change in 1996): −0.2%.

Transport: Motor vehicles: 141,901 pass. cars, 30,243 comm. vehicles. **Civil aviation:** 1.6 bil pass.-mi.; 1 airport. **Chief ports:** Manama, Sitrah.

Communications: TV sets: 442 per 1,000 pop. **Radios:** 555 per 1,000 pop. **Telephones** (1997): 152,300 main lines. **Daily newspaper circ.:** 128 per 1,000 pop.

Health: Life expectancy: 72.75 male; 77.96 female. **Births** (per 1,000 pop.): 21.86. **Deaths** (per 1,000 pop.): 3.24. **Natural inc.:** 1.862%. **Hosp. beds** (1993): 1 per 352 persons. **Physicians** (1993): 1 per 1,115 persons. **Infant mortality** (per 1,000 live births): 14.81.

Education: Free, compulsory: ages 6-17. **Literacy:** 85%. **Major Intl. Organizations:** UN (FAO, IBRD, ILO, IMF, IMO, WHO, WTrO), AL. **Embassy:** 3502 International Dr. NW 20008; 342-0741.

Long ruled by the Khalifa family, Bahrain was a British protectorate from 1861 to Aug. 15, 1971, when it regained independence.

Pearls, shrimp, fruits, and vegetables were the mainstays of the economy until oil was discovered in 1932. By the 1970s, oil reserves were depleted; international banking thrived.

Bahrain took part in the 1973-74 Arab oil embargo against the U.S. and other nations. The government bought controlling interest in the oil industry in 1975. Shiite dissidents have clashed with the Sunni-led government since 1996.

Bangladesh
People's Republic of Bangladesh

People: Population: 127,117,967. **Age distrib.** (%): <15: 37.7; 65+: 3.3. **Pop. density:** 2,286 per sq. mi. **Urban:** 19%. **Ethnic groups:** Bengali 98%, Bihari, tribals. **Principal languages:** Bangla (official), English. **Chief religions:** Muslim 88%, Hindu 11%.

Geography: Area: 55,600 sq. mi. **Location:** In S Asia, on N bend of Bay of Bengal. **Neighbors:** India nearly surrounds country on W, N, E; Myanmar on SE. **Topography:** The country is mostly a low plain cut by the Ganges and Brahmaputra rivers and their delta. The land is alluvial and marshy along the coast, with hills only in the extreme SE and NE. A tropical monsoon climate prevails, among the rainiest in the world. **Capital:** Dhaka. **Cities:** Dhaka 8,545,000; Chittagong 2,477,000; Khulna 1,071,000*.

Government: Type: Parliamentary democracy. **Head of state:** Pres. Shahabuddin Ahmed; b 1930; in office: Oct. 9, 1996. **Head of gov.:** Prime Min. Hasina Wazed; b Sept. 27, 1947; in office: June 24, 1996. **Local divisions:** 6 divisions. **Defense:** 1.9% of GDP. **Active troops:** 121,000.

Economy: Industries: Food processing, jute, textiles, fertilizers, steel. **Chief crops:** Jute, rice, tea. **Minerals:** Natural gas. **Crude oil reserves** (1999): 5.4 mil bbls. **Arable land:** 73%. **Livestock** (1997): chickens: 152.87 mil; goats: 33.50 mil; cattle: 23.40 mil; sheep: 1.16 mil; buffalo: 854,000. **Fish catch** (1997): 829,992 metric tons. **Electricity prod.** (1997): 12.300 bil kWh. **Labor force:** 63% agric.; 25% services; 10% ind. & mining.

Finance: Monetary unit: Taka (Sept. 1999: 49.50 = $1 U.S.). **GDP** (1997 est.): $167 bil. **Per capita GDP:** $1,330. **Imports** (1996): $6.9 bil; partners: India 21%, China 10%. **Exports** (1996): $3.9 bil; partners: Western Europe 42%, U.S. 30%. **Tourism:** $65 mil. **Budget** (FY 1996-97 est.): $5.3 bil. **Intl. reserves less gold** (June 1999): $1.50 bil. **Gold:** 106,000 oz t. **Consumer prices** (change in 1997): 5.6%.

Transport: Railroad: Length: 1,681 mi. **Motor vehicles** (1997): 134,073 pass. cars, 92,133 comm. vehicles. **Civil aviation:** 2.0 bil pass.-mi.; 8 airports. **Chief ports:** Chittagong, Dhaka, Chalna Port.

Communications: TV sets: 5 per 1,000 pop. **Radios:** 65 per 1,000 pop. **Telephones** (1997): 316,100 main lines. **Daily newspaper circ.:** 0.4 per 1,000 pop.

Health: Life expectancy: 60.73 male; 60.46 female. **Births**

(per 1,000 pop.): 25.20. **Deaths** (per 1,000 pop.): 8.50. **Natural inc.:** 1.670%. **Hosp. beds** (1994): 1 per 3,312 persons. **Physicians** (1994): 1 per 4,759 persons. **Infant mortality** (per 1,000 live births): 69.68.

Education: Free, compulsory: ages 6-11. **Literacy:** 38%. **Major Intl. Organizations:** UN (FAO, IBRD, ILO, IMF, IMO, WHO, WTrO), the Commonwealth. **Embassy:** 2201 Wisconsin Ave. NW 20007; 342-8372. **Website:** http://www.virtualbangladesh.com

Muslim invaders conquered the formerly Hindu area in the 12th century. British rule lasted from the 18th century to 1947, when East Bengal became part of Pakistan.

Charging West Pakistani domination, the Awami League, based in the East, won National Assembly control in 1971. Assembly sessions were postponed; riots broke out. Pakistani troops attacked Mar. 25; Bangladesh independence was proclaimed the next day. In the ensuing civil war, one million died and 10 million fled to India.

War between India and Pakistan broke out Dec. 3, 1971. Pakistan surrendered in the East on Dec. 16. Mujibur Rahman, known as Sheikh Mujib, became prime minister; he was killed in a coup Aug. 15, 1975. During the 1970s the country moved into the Indian and Soviet orbits in response to U.S. support of Pakistan, and much of the economy was nationalized.

On May 30, 1981, Pres. Ziaur Rahman was killed in an unsuccessful coup attempt by army rivals. Vice Pres. Abdus Sattar assumed the presidency but was ousted in a coup led by army chief of staff Gen. H. M. Ershad, Mar. 1982. Ershad declared Bangladesh an Islamic Republic in 1988; a parliamentary system of government was adopted in 1991.

Bangladesh is subject to devastating storms and floods that kill thousands. A cyclone struck Apr. 1991, killing over 131,000 people and causing $2.7 billion in damages. Chronic destitution in the densely crowded population has been worsened by the decline of jute as a world commodity.

Political turmoil led to the resignation, Mar. 30, 1996, of Prime Minister Khaleda Zia, the widow of Ziaur Rahman. Sheikh Mujib's daughter, Hasina Wazed (known as Sheikh Hasina), led the country after the June 12, 1996 election. Bangladesh and India signed a treaty, Dec. 12, resolving their long-standing dispute over the use of water from the Ganges River. A cyclone in May 1997 left an estimated 800,000 people homeless. Floods in July-Sept. 1998 inundated most of the country, killed over 1,400 people (many through disease), and stranded at least 30 million.

Barbados

People: Population: 259,191. **Age distrib.** (%): <15: 23.0; 65+: 10.1. **Pop. density:** 1,525 per sq. mi. **Urban:** 48%. **Ethnic groups:** Black 80%, white 4%, other 16%. **Principal language:** English (official). **Chief religions:** Protestant 67%, Roman Catholic 4%.

Geography: Area: 170 sq. mi. **Location:** In Atlantic O., farthest E of West Indies. **Neighbors:** Nearest are St. Lucia and St. Vincent & the Grenadines to the W. **Topography:** The island lies alone in the Atlantic almost completely surrounded by coral reefs. Highest point is Mt. Hillaby, 1,115 ft. **Capital:** Bridgetown (1990 est.): 6,000.

Government: Type: Parliamentary democracy. **Head of state:** Queen Elizabeth II, represented by Gov.-Gen. Sir Clifford Husbands; b Aug. 5, 1926; in office: June 1, 1996. **Head of gov.:** Prime Min. Owen Arthur; b Oct. 17, 1949; in office: Sept. 7, 1994. **Local divisions:** 11 parishes and Bridgetown. **Defense:** 0.6% of GDP. **Active troops:** 600.

Economy: Industries: Sugar, tourism. **Chief crops:** Sugar, vegetables, cotton. **Minerals:** Oil, gas. **Crude oil reserves** (1999): 2.4 mil bbls. **Other resources:** Fish. **Arable land:** 37%. **Livestock** (1997): chickens: 3.40 mil. **Electricity prod.** (1997): 650 mil kWh. **Labor force:** 75% services; 15% industry; 10% agric.

Finance: Monetary unit: Dollar (Sept. 1999: 2.00 = $1 U.S.). **GDP** (1997 est.): $2.8 bil. **Per capita GDP:** $10,900. **Imports** (1995): $763 mil; partners: U.S. 37%, Trin. & Tob. 11%, UK 10%. **Exports** (1995): $235 mil; partners: U.S. 15%, UK 15%, Trin. & Tob. 9%. **Tourism:** $712 mil. **Budget** (FY 1996-97 est.): $645 mil. **Intl. reserves less gold** (June 1999): $368.44 mil. **Consumer prices** (change in 1998): −1.3%.

Transport: Motor vehicles: 45,000 pass. cars; 3,500 comm. vehicles. **Civil aviation:** 204.9 mil pass.-mi.; 1 airport. **Chief port:** Bridgetown.

Communications: TV sets: 287 per 1,000 pop. **Radios:** 1,134 per 1,000 pop. **Telephones** (1997): 108,500 main lines. **Daily newspaper circ.:** 157 per 1,000 pop.

Health: Life expectancy: 72.22 male; 77.81 female. **Births** (per 1,000 pop.): 14.46. **Deaths** (per 1,000 pop.): 8.16. **Natu-**

ral inc.: 0.630%. **Hosp. beds** (1992): 1 per 134 persons. **Physicians** (1992): 1 per 842 persons. **Infant mortality** (per 1,000 live births): 16.74.

Education: Compulsory: ages 5-16. **Literacy:** 97%.

Major Intl. Organizations: UN (FAO, IBRD, ILO, IMF, IMO, WHO, WTrO), Caricom, the Commonwealth, OAS.

Embassy: 2144 Wyoming Ave. NW 20008; 939-9200.

Barbados was probably named by Portuguese sailors in reference to bearded fig trees. An English ship visited in 1605, and British settlers arrived on the uninhabited island in 1627. Slaves worked the sugar plantations until slavery was abolished in 1834. Self-rule came gradually, with full independence proclaimed Nov. 30, 1966. British traditions have remained.

Belarus
Republic of Belarus

People: Population: 10,401,784. **Age distrib.** (%): <15: 19.4; 65+: 13.5. **Pop. density:** 130 per sq. mi. **Urban:** 72%. **Ethnic groups:** Byelorussian 78%, Russian 13%, Polish 4%. **Principal languages:** Byelorussian (official), Russian. **Chief religions:** Eastern Orthodox 80%, other 20%.

Geography: Area: 80,200 sq. mi. **Location:** E Europe. **Neighbors:** Poland on W; Latvia, Lithuania on N; Russia on E; Ukraine on S. **Capital:** Minsk (1998 met. est.): 1,708,308.

Government: Republic. **Head of state:** Pres. Aleksandr Lukashenko; b Aug. 30, 1954; in office: July 1994. **Head of gov.:** Prime Min. Sergei Ling; b May 7, 1937; in office, Nov. 18, 1996. **Local divisions:** 6 oblasts and 1 municipality. **Defense:** 2.9% of GDP. **Active troops:** 81,800.

Economy: Industries: Manufacturing, chemical fibers, textiles, agricultural & industrial machinery. **Chief crops:** Grain, vegetables, potatoes. **Crude oil reserves** (1999): 198 mil bbls. **Arable land:** 30%. **Livestock** (1997): chickens: 40.00; cattle: 4.80 mil; pigs: 3.68 mil; sheep: 127,200. **Electricity prod.** (1997): 24.304 bil kWh. **Labor force:** 41% services; 40% ind. & const.; 19% agric. & forestry.

Finance: Monetary unit: Ruble (Aug. 1999: 32,000.00 = $1 U.S.). **GDP** (1997 est.): $50.4 bil. **Per capita GDP:** $4,800. **Imports** (1996): $6.7 bil; partners: Russia 46%. **Exports** (1996): $5.4 bil; partners: Russia 47%. **Tourism:** $26 mil. **Budget** (1997 est.): $4.1 bil. **Intl. reserves less gold** (May 1999): $284.73 mil. **Consumer prices** (change in 1998): 72.9%.

Transport: Railroad: Length: 3,480 mi. **Motor vehicles:** 842,500 pass. cars, 10,000 comm. vehicles. **Civil aviation:** 247.9 pass.-mi.; 2 airports. **Chief port:** Mazyr.

Communications: TV sets: 265 per 1,000 pop. **Radios:** 311 per 1,000 pop. **Telephones:** 2,489,900 main lines. **Daily newspaper circ.:** 187 per 1,000 pop.

Health: Life expectancy: 62.04 male; 74.52 female. **Births** (per 1,000 pop.): 9.70. **Deaths** (per 1,000 pop.): 13.71. **Natural inc.:** −0.40%. **Hosp. beds** (1995): 1 per 81 persons. **Physicians** (1995): 1 per 224 persons. **Infant mortality** (per 1,000 live births): 14.39.

Education: Compulsory: ages 6-17. **Literacy** (1994): 98%.

Major Intl. Organizations: UN (IBRD, ILO, IMF, WHO), CIS, OSCE.

Embassy: 1619 New Hampshire Ave. NW 20009; 986-1604.

The region was subject to Lithuanians and Poles in medieval times, and was a prize of war between Russia and Poland beginning in 1503. It became part of the USSR in 1922 although the western part of the region was controlled by Poland. Belarus was overrun by German armies in 1941; recovered by Soviet troops in 1944. Following World War II, Belarus increased in area through Soviet annexation of part of NE Poland. Belarus declared independence Aug. 25, 1991. It became an independent state when the Soviet Union disbanded Dec. 26, 1991.

A new constitution was adopted, Mar. 15, 1994, and a new president was chosen in elections concluding July 1. Russia and Belarus signed a pact Apr. 2, 1996, linking their political and economic systems. An authoritarian constitution enacted in Nov. gave Pres. Aleksandr Lukashenko vast new powers. Lukashenko's insistence on tightening ties with Russia resulted in the signing of new accords in 1997 and 1998.

Belgium
Kingdom of Belgium

People: Population: 10,182,034. **Age distrib.** (%): <15: 17.2; 65+: 16.9. **Pop. density:** 863 per sq. mi. **Urban:** 97%. **Ethnic groups:** Fleming 55%, Walloon 33%. **Principal languages:** Flemish (Dutch) 56%, French 32%, German 1% (all official). **Chief religions:** Roman Catholic 75%, Protestant, other 25%.

Geography: Area: 11,800 sq. mi. **Location:** In W Europe,

on North Sea. **Neighbors:** France on W and S, Luxembourg on SE, Germany on E, Netherlands on N. **Topography:** Mostly flat, the country is trisected by the Scheldt and Meuse, major commercial rivers. The land becomes hilly and forested in the SE (Ardennes) region. **Capital:** Brussels. **Cities** (1995 met. est.): Brussels 948,122; Antwerp 455,852; Ghent 226,464 .

Government: Type: Parliamentary democracy under a constitutional monarch. **Head of state:** King Albert II; b June 6, 1934; in office: Aug. 9, 1993. **Head of gov.:** Premier Guy Verhofstadt; b Apr. 11, 1953; in office: July 12, 1999. **Local divisions:** 10 provinces and Brussels. **Defense:** 1.6% of GDP. **Active troops:** 44,500.

Economy: Industries: Metal products, glassware, autos, textiles, chemicals. **Chief crops:** Grain, fruits, sugar beets, vegetables. **Minerals:** Coal, gas. **Arable land:** 24%. **Livestock** (1997): chickens: 48.00 mil; pigs: 7.44 mil; cattle: 3.18 mil; sheep: 155,000. **Fish catch** (1996): 30,823 metric tons. **Electricity prod.** (1997): 73.742 bil kWh. **Labor force:** 70% services; 27% industry; 3% agric.

Finance: Monetary unit: Franc (Sept. 1999: 37.69 = $1 U.S.). Euro (Sept. 1999: 1.07 = $1 U.S.). **GDP** (1997 est.): $236.3 bil. **Per capita GDP** $23,200. *Note:* Import/Export data include Luxembourg. **Imports** (1997): $158.5 bil; partners: EU 75%, U.S. 5%. **Exports** (1997): $172 bil; partners: EU 67%, U.S. 6%. **Tourism:** $5.36 bil. **Budget** (1997): $69.36 bil. **Intl. reserves less gold** (June 1999): $11.25 bil. **Gold:** 8.00 mil oz t. **Consumer prices** (change in 1998): 1.0%.

Transport: Railroad: Length: 2,093 mi. **Motor vehicles** (1997): 4.42 mil pass. cars, 541,422 comm. vehicles. **Civil aviation:** 7.0 bil pass.-mi.; 2 airports. **Chief ports:** Antwerp (one of the world's busiest), Zeebrugge, Ghent.

Communications: TV sets: 464 per 1,000 pop. **Radios:** 757 per 1,000 pop. **Telephones:** 5,073,000 main lines. **Daily newspaper circ.:** 304 per 1,000 pop.

Health: Life expectancy: 74.31 male; 80.90 female. **Births** (per 1,000 pop.): 9.98. **Deaths** (per 1,000 pop.): 10.43. **Natural inc.:** −0.45%. **Hosp. beds** (1994): 1 per 131 persons. **Physicians** (1996): 1 per 264 persons. **Infant mortality** (per 1,000 live births): 6.17.

Education: Compulsory: ages 6-18. **Literacy:** 99%.

Major Intl. Organizations: UN and all of its specialized agencies, EU, NATO, OECD, OSCE.

Embassy: 3330 Garfield St. NW 20008; 333-6900.

Website: http://www.belgium.fgov.be

Belgium derives its name from the Belgae, the first recorded inhabitants, probably Celts. The land was conquered by Julius Caesar, and was ruled for 1800 years by conquerors, including Rome, the Franks, Burgundy, Spain, Austria, and France. After 1815, Belgium was made a part of the Netherlands, but it became an independent constitutional monarchy in 1830.

Belgian neutrality was violated by Germany in both world wars. King Leopold III surrendered to Germany, May 28, 1940. After the war, he was forced by political pressure to abdicate in favor of his son, King Baudouin. Baudouin was succeeded by his brother, Albert II, Aug. 9, 1993.

The Flemings of northern Belgium speak Dutch, while French is the language of the Walloons in the south. The language difference has been a perennial source of controversy and led to antagonism between the 2 groups. Parliament has passed measures aimed at transferring power from the central government to 3 regions—Wallonia, Flanders, and Brussels. Constitutional changes in 1993 made Belgium a federal state.

Belize

People: Population: 235,789. **Age distrib.** (%): <15: 41.6; 65+: 3.6. **Pop. density:** 26 per sq. mi. **Urban:** 46%. **Ethnic groups:** Mestizo 44%, Creole 30%, Maya 11%, Garifuna 7%. **Principal languages:** English (official), Spanish, Mayan, Garifuna (Carib). **Chief religions:** Roman Catholic 62%, Protestant 30%.

Geography: Area: 8,900 sq. mi. **Location:** Eastern coast of Central America. **Neighbors:** Mexico on N, Guatemala on W and S. **Capital:** Belmopan (1997 est.): 6,785.

Government: Type: Parliamentary democracy. **Head of state:** Queen Elizabeth II, represented by Gov.-Gen. Colville Young; b Nov. 20, 1932; in office: Nov. 17, 1993. **Head of gov.:** Prime Min. Said Musa; b Mar. 19, 1944; in office: Aug. 28, 1998. **Local divisions:** 6 districts. **Defense:** 2.6% of GDP. **Active troops:** 1,100.

Economy: Industries: Garments, food processing, tourism. **Chief crops:** Sugar (main export), citrus, bananas. **Arable land:** 2%. **Livestock** (1997): chickens: 1.50 mil. **Electricity prod.** (1997): 110 mil kWh. **Labor force:** 30% agric.; 16% services; 15% govt.; 11% commerce; 10% manuf.

Finance: Monetary unit: Dollar (Sept. 1999: 2.00 = $1

U.S.). **GDP** (1996 est.): $649 mil. **Per capita GDP:** $2,960. **Imports** (1996): $264 mil; partners: U.S. 53%. **Exports** (1996): $204 mil; partners: U.S. 38%. **Tourism:** $86 mil. **Budget** (FY1996-97): $180 mil. **Intl. reserves less gold** (June 1999): $72.22 mil. **Consumer prices** (change in 1998): –0.9%.

Transport: Motor vehicles: 2,300 pass. cars, 3,100 comm. vehicles. **Chief ports:** Belize City, Big Creek. **Civil aviation:** 9 airports.

Communications: TV sets: 109 per 1,000 pop. **Radios:** 133 per 1,000 pop. **Telephones:** 31,600 main lines.

Health: Life expectancy: 67.23 male; 71.26 female. **Births** (per 1,000 pop.): 30.22. **Deaths** (per 1,000 pop.): 5.39. **Natural inc.:** 2.483%. **Hosp. beds** (1993): 1 per 350 persons. **Physicians** (1995): 1 per 1,546 persons. **Infant mortality** (per 1,000 live births): 31.57.

Education: Compulsory: ages 5-14. **Literacy** (1993): 93%.

Major Intl. Organizations: UN (FAO, IBRD, ILO, IMF, IMO, WHO, WTrO), Caricom, the Commonwealth, OAS.

Embassy: 2535 Massachusetts Ave. NW 20008; 332-9636.

Website: http://www.belizenet.com

Belize (formerly British Honduras) was Britain's last colony on the American mainland; independence was achieved Sept. 21, 1981. Relations with neighboring Guatemala, initially tense, have improved in recent years. Belize has become a center for drug trafficking between Colombia and the U.S.

Benin
Republic of Benin

People: Population: 6,305,567. **Age distrib.** (%): <15: 47.8; 65+: 2.3. **Pop. density:** 145 per sq. mi. **Urban:** 39%. **Ethnic groups:** African (Fon, Adja, Bariba, Yoruba, others) 99%. **Principal languages:** French (official), Fon, Yoruba, tribal languages. **Chief religions:** Indigenous beliefs 70%, Muslim 15%, Christian 15%.

Geography: Area: 43,500 sq. mi. **Location:** In W Africa on Gulf of Guinea. **Neighbors:** Togo on W; Burkina Faso, Niger on N; Nigeria on E. **Topography:** Most of Benin is flat and covered with dense vegetation. The coast is hot, humid, and rainy. **Capital:** Porto-Novo. **Cities** (1994 est.): Cotonou 750,000, Porto-Novo 200,000.

Government: Type: Republic. **Head of state and gov.:** Pres. Mathieu Kerekou; b Sept. 2, 1933; in office: Apr. 4, 1996. **Local divisions:** 6 departments. **Defense:** 1.3% of GDP. **Active troops:** 4,800.

Economy: Industries: textiles, food, beverages, construction materials, petroleum. **Chief crops:** Palm oil, sorghum, cassava, peanuts, cotton, corn, rice. **Minerals:** Oil, limestone, marble. **Crude oil reserves** (1999): 8.2 mil bbls. **Arable land:** 13%. **Livestock** (1997): cattle: 1.40 mil; chickens: 27.00 mil; goats: 1.02 mil; pigs: 580,000; sheep: 605,000. **Fish catch** (1996): 42,000 metric tons. **Electricity prod.** (1997): 6 mil kWh. **Labor force:** 55% agric.; 29% transport, commerce, public services.

Finance: Monetary unit: CFA Franc (Sept. 1999: 612.79 = $1 U.S.). **GDP** (1997 est.): $11.3 bil. **Per capita GDP:** $1,900. **Imports** (1995): $693 mil; partners: France 27%, Thailand 9%. **Exports** (1995): $192 mil; partners: Brazil 18%, Portugal 14%. **Tourism:** $33 mil. **Budget** (1995 est.): $445 mil. **Intl. reserves less gold** (Dec. 1998): $261.5 mil. **Consumer prices** (change in 1998): 5.8%.

Transport: Railroad: Length: 359 mi. **Motor vehicles:** 35,600 pass. cars, 19,300 comm. vehicles. **Civil aviation:** 150.5 mil pass.-mi.; 1 airport. **Chief port:** Cotonou.

Communications: TV sets: 4 per 1,000 pop. **Radios:** 73 per 1,000 pop. **Telephones:** 38,400 main lines. **Daily newspaper circ.:** 2 per 1,000 pop.

Health: Life expectancy: 51.98 male; 56.24 female. **Births** (per 1,000 pop.): 45.37. **Deaths** (per 1,000 pop.): 12.40. **Natural inc.:** 3.297%. **Hosp. beds** (1993): 1 per 4,182 persons. **Physicians** (1993): 1 per 14,216 persons. **Infant mortality** (per 1,000 live births): 97.76.

Education: Free, compulsory: ages 6-12. **Literacy:** 37%.

Major Intl. Organizations: UN (FAO, IBRD, ILO, IMF, IMO, WHO, WTrO), OAU.

Embassy: 2737 Cathedral Ave. NW 20008; 232-6656.

The Kingdom of Abomey, rising to power in wars with neighboring kingdoms in the 17th century, came under French domination in the late 19th century and was incorporated into French West Africa by 1904.

Under the name Dahomey, the country gained independence Aug. 1, 1960; it became Benin in 1975. In the fifth coup since independence Col. Ahmed Kerekou took power in 1972; two years later he declared a socialist state with a "Marxist-Leninist" philosophy. In Dec. 1989, Kerekou announced Marxism-Leninism would no longer be the state ideology.

In Mar. 1991, Kerekou lost to Nicéphore Soglo in Benin's first free presidential election in 30 years. Kerekou defeated Soglo in Mar. 1996 to reclaim the presidency.

Bhutan
Kingdom of Bhutan

People: Population: 1,951,965. **Age distrib.** (%): <15: 40.1; 65+: 4.0. **Pop. density:** 108 per sq. mi. **Urban:** 6%. **Ethnic groups:** Bhote 50%, Nepalese 35%. **Principal languages:** Dzongkha (official), Tibetan, Nepalese dialects. **Chief religions:** Lamaistic Buddhist (state religion) 75%, Hindu 25%.

Geography: Area: 18,000 sq. mi. **Location:** S Asia, in eastern Himalayan Mts. **Neighbors:** India on W (Sikkim) and S, China on N. **Topography:** Bhutan is comprised of very high mountains in the N, fertile valleys in the center, and thick forests in the Duar Plain in the S. **Capital:** Thimphu (1993 est.): 30,300.

Government: Type: Monarchy. **Head of state:** King Jigme Singye Wangchuk; b Nov. 11, 1955; in office: July 21, 1972. **Head of gov.:** Prime Min. Lyonpo Jigme Thinley; in office: July 20, 1998. **Local divisions:** 18 districts.

Economy: Industries: Cement, wood products. **Chief crops:** Rice, corn, citrus. **Other resources:** Timber, hydropower. **Livestock** (1997): cattle: 435,000; chickens: 310,000. **Arable land:** 2%. **Electricity prod.** (1997): 1.757 bil kWh. **Labor force:** 93% agric.; 5% services.

Finance: Monetary unit: Ngultrum (Sept. 1999: 43.49 = $1 U.S.; Indian Rupee also used). **GDP** (1995 est.): $1.3 bil. **Per capita GDP:** $730. **Tourism** (1994): 4.0 mil. **Imports** (1996 est.): $104.1 mil; partners: India 77%. **Exports** (1996 est.): $77.4 mil; partners: India 94%. **Tourism:** $6 mil. **Budget** (FY 1995-96): $152 mil. **Intl. reserves less gold** (Mar. 1999): $270.89 mil. **Consumer prices** (change in 1998): 8.5%.

Transport: Civil aviation: 30.1 mil pass.-mi.; 1 airport.

Communications: Radios: 27 per 1,000 pop. **Telephones** (1997): 6,400 main lines.

Health: Life expectancy: 53.19 male; 52.29 female. **Births** (per 1,000 pop.): 36.76. **Deaths** (per 1,000 pop.): 14.26. **Natural inc.:** 2.250%. **Hosp. beds** (1994): 1 per 825 persons. **Physicians** (1994): 1 per 8,000 persons. **Infant mortality** (per 1,000 live births): 109.33.

Education: Not compulsory. **Literacy:** 42%.

Major Intl. Organizations: UN (FAO, IBRD, IMF, WHO).

Website: http://www.bhutan.org

The region came under Tibetan rule in the 16th century. British influence grew in the 19th century. A monarchy, set up in 1907, became a British protectorate by a 1910 treaty. The country became independent in 1949, with India guiding foreign relations and supplying aid. Most of the population engages in subsistence agriculture.

Bolivia
Republic of Bolivia

People: Population: 7,982,850. **Age distrib.** (%): <15: 39.0; 65+: 4.5. **Pop. density:** 19 per sq. mi. **Urban:** 61%. **Ethnic groups:** Quechua 30%, mestizo 25-30%, Aymara 25%, European 5-15%. **Principal languages:** Spanish, Quechua, Aymara (all official). **Chief religion:** Roman Catholic 95%.

Geography: Area: 424,200 sq. mi. **Location:** In W central South America, in the Andes Mts. (one of 2 landlocked countries in South America). **Neighbors:** Peru and Chile on W, Argentina and Paraguay on S, Brazil on E and N. **Topography:** The great central plateau, at an altitude of 12,000 ft., over 500 mi. long, lies between two great cordilleras having 3 of the highest peaks in South America. Lake Titicaca, on Peruvian border, is highest lake in world on which steamboats ply (12,506 ft.). The E central region has semitropical forests; the llanos, or Amazon-Chaco lowlands are in E. **Capitals:** La Paz (administrative), Sucre (judicial). **Cities 1995 est.:** Santa Cruz 833,307; La Paz 739,453.

Government: Type: Republic. **Head of state and gov.:** Pres. Hugo Banzer Suárez; b May 10, 1926; in office: Aug. 6, 1997. **Local divisions:** 9 departments. **Defense:** 2.0% of GDP. **Active troops:** 33,500.

Economy: Industries: Mining, smelting, tobacco, handicrafts, clothing. **Chief crops:** Coffee, sugarcane, potatoes, cotton, corn, coca. **Minerals:** Antimony, tin, tungsten, silver, zinc, oil, gas, iron. **Crude oil reserves** (1999): 132 mil bbls. **Other resources:** Timber. **Arable land:** 2%. **Livestock** (1997): chickens: 58.80 mil; sheep: 8.41 mil; cattle: 6.39 mil; pigs: 2.64 mil; goats: 1.50 mil. **Electricity prod.** (1997): 3.29 bil kWh.

Finance: Monetary unit: Boliviano (Sept. 1999: 5.87 = $1 U.S.). **GDP** (1997 est.): $23.1 bil. **Per capita GDP:** $3,000. **Imports** (1997): $1.7 bil; partners: U.S. 20% Japan 13%. **Exports** (1997): $1.4 bil; partners: U.S. 22%, Columbia 9%, UK

9%. **Tourism:** $185 mil. **Budget** (1995 est.): $3.75 bil. **Intl. reserves less gold** (June 1999): $838.0 mil. **Gold:** 943,000 oz t. **Consumer prices** (change in 1998): 7.7%.

Transport: Railroad: Length: 2,295 mi. **Motor vehicles:** 199,309 pass. cars, 230,245 comm. vehicles. **Civil aviation:** 1.3 bil pass.-mi.; 14 airports.

Communications: TV sets: 202 per 1,000 pop. **Radios:** 560 per 1,000 pop. **Telephones** (1997): 535,000 main lines. **Daily newspaper circ.:** 69 per 1,000 pop.

Health: Life expectancy: 58.51 male; 64.51 female. **Births** (per 1,000 pop.): 30.72. **Deaths** (per 1,000 pop.): 9.61. **Natural inc.:** 2.111%. **Hosp. beds** (1994): 1 per 1,005 persons. **Physicians** (1994): 1 per 3,663 persons. **Infant mortality** (per 1,000 live births): 62.02.

Education: Free, compulsory: ages 6-14. **Literacy:** 83%.

Major Intl. Organizations: UN (FAO, IBRD, ILO, IMF, IMO, WHO, WTrO), OAS.

Embassy: 3014 Massachusetts Ave. NW 20008; 483-4410.

Website: http:/www.ine.gov.bo

The Incas conquered the region from earlier Indian inhabitants in the 13th century. Spanish rule began in the 1530s and lasted until Aug. 6, 1825. The country is named after Simon Bolivar, independence fighter.

In a series of wars, Bolivia lost its Pacific coast to Chile, the oil-bearing Chaco to Paraguay, and rubber-growing areas to Brazil, 1879-1935.

Economic unrest, especially among the militant mine workers, has contributed to continuing political instability. A reformist government under Victor Paz Estenssoro, 1951-64, nationalized tin mines and attempted to improve conditions for the Indian majority but was overthrown by a military junta. A long series of coups and countercoups continued until constitutional government was restored in 1982.

U.S. pressure on the government to reduce the country's coca output, the raw material for cocaine, has led to clashes between police and coca growers and increased anti-U.S. feeling among Bolivians. Gen. Hugo Banzer Suárez, who ruled as a dictator, 1971-78, became president in Aug. 1997. 105 people died in earthquakes near Aiquile May 22, 1998.

Bosnia and Herzegovina
Republic of Bosnia and Herzegovina

People: Population: 3,482,495. **Age distrib.** (%): <15: 17.4; 65+: 11.9. **Pop. density:** 176 per sq. mi. **Urban:** 42%. **Ethnic groups:** Serb 40%, Muslim 38%, Croat 22%. **Principal language:** Serbo-Croatian (official) 99%. **Chief religions:** Muslim 40%, Orthodox 31%, Catholic 15%.

Geography: Area: 19,781 sq. mi. **Location:** On Balkan Peninsula in SE Europe. **Neighbors:** Yugoslavia on E and SE, Croatia on N and W. **Topography:** Hilly with some mountains. About 36% of the land is forested. **Capital:** Sarajevo (1993 est.): 300,000.

Government: Type: Republic. **Heads of state:** Collective Pres., Alija Izetbegovic (Muslim), b Aug. 8, 1925; Zivko Radisic (Serb); Ante Jelavic (Croat); elected: Sept. 12-13, 1998. **Heads of gov.:** Co-Prime Mins., Haris Silajdzic; b 1945; in office: Jan. 3, 1997; Svetozar Mihajlovic; in office: Feb. 3, 1999. **Local divisions:** Muslim-Croat Federation, divided into 10 cantons; Republika Srpska. **Defense:** 5.0% of GDP. **Active troops:** 40,000.

Economy: Industries: Steel, mining, textiles, timber. **Chief crops:** Corn, wheat, fruits, vegetables. **Minerals:** Bauxite, iron, coal. **Arable land:** 14%. **Livestock** (1997): chickens: 3.87 mil; sheep: 275,600; cattle: 260,000. **Electricity prod.** (1997): 2.3 bil kWh.

Finance: Monetary unit: Conv Mark (Sept. 1999: 1.83 = $1 U.S.). **GDP** (1997 est.): $4.41 bil. **Per capita GDP:** $1,690. **Imports** (1996): $1.88 bil; partners: Croatia 32%. **Exports** (1996): $171 mil; partners: Croatia 34%, Italy 26%. **Tourism:** $15 mil.

Transport: Railroad: Length: 634 mi. **Chief port:** Bosanski Brod. **Civil aviation:** 1 airport.

Communications: TV sets: 94 per 1,000 pop. **Telephones:** 331,900 main lines. **Daily newspaper circ.:** 150 per 1,000 pop.

Health: Life expectancy: 62.55 male; 71.71 female. **Births** (per 1,000 pop.): 9.36. **Deaths** (per 1,000 pop.): 10.81. **Natural inc.:** −0.145%. **Physicians** (1996): 1 per 703 persons. **Infant mortality** (per 1,000 live births): 24.52.

Education: Free, compulsory: ages 7-15. **Literacy** (1991): 86%.

Major Intl. Organizations: UN (FAO, IBRD, ILO, IMF, IMO, WHO), OSCE.

Embassy: 1707 L St. NW, Suite 760, 20036; 833-3612.

Website: http://www.bosnianembassy.org

Bosnia was ruled by Croatian kings c. AD 958, and by Hungary 1000-1200. It became organized c. 1200 and later took control of Herzegovina. The kingdom disintegrated from 1391, with the southern part becoming the independent duchy Herzegovina. It was conquered by Turks in 1463 and made a Turkish province. The area was placed under control of Austria-Hungary in 1878, and made part of the province of **Bosnia and Herzegovina,** which was formally annexed to Austria-Hungary 1908; Bosnia became a province of Yugoslavia in 1918. It was reunited with Herzegovina as a federated republic in the 1946 Yugoslavian constitution.

Bosnia and Herzegovina declared sovereignty Oct. 15, 1991. A referendum for independence was passed Feb. 29, 1992. Ethnic Serbs' opposition to the referendum spurred violent clashes and bombings. The U.S. and EU recognized the republic Apr. 7. Fierce three-way fighting continued between Bosnia's Serbs, Muslims, and Croats. Serb forces massacred thousands of Bosnian Muslims and engaged in "ethnic cleansing" (the expulsion of Muslims and other non-Serbs from areas under Bosnian Serb control). The capital, Sarajevo, was surrounded and besieged by Bosnian Serb forces. Muslims and Croats in Bosnia reached a cease fire Feb. 23, 1994, and signed an accord, Mar. 18, to create a Muslim-Croat confederation in Bosnia. However, by mid-1994, Bosnian Serbs controlled over 70% of the country.

As fighting continued in 1995, the balance of power began to shift toward the Muslim-Croat alliance. Massive NATO air strikes at Bosnian Serb targets beginning Aug. 30 triggered a new round of peace talks, and the siege of Sarajevo was lifted Sept. 15. The new talks produced an agreement in principle to create autonomous regions within Bosnia, with the Serb region (Republika Srpska) constituting 49% of the country. A Croat-Muslim offensive in Sept. recaptured significant territory, leaving Bosnian Serbs in control of approximately that percentage.

A peace agreement initialed in Dayton, Ohio, Nov. 21, 1995, was signed in Paris, Dec. 14, by leaders of Bosnia, Croatia, and Serbia. Some 60,000 NATO troops (about 20,000 from the U.S.) moved in to police the accord. Meanwhile, a UN tribunal began bringing charges against suspected war criminals. Elections were held Sept. 14, 1996, for a 3-person collective presidency, for seats in a federal parliament, and for regional offices. In Dec. a revamped NATO "stabilization force" of some 30,000 members (more than 8,000 from the U.S.) received an 18-month mandate, which was later extended.

A hardline Serb nationalist, Nikola Poplasen, was elected president of the Republika Srpska, Sept. 12-13, 1998. He refused to accept dismissal on Mar. 5, 1999, from Carlos Westendorp, international coordinator for Bosnia, who accused him of obstructing the peace process.

Botswana
Republic of Botswana

People: Population: 1,464,167. **Age distrib.** (%): <15: 41.9; 65+: 3.7. **Pop. density:** 6 per sq. mi. **Urban:** 63%. **Ethnic groups:** Batswana 95%, Kalanga, Basarwa, Kgalagadi. **Principal languages:** English (official), Setswana. **Chief religions:** Indigenous beliefs 50%, Christian 50%.

Geography: Area: 231,800 sq. mi. **Location:** In southern Africa. **Neighbors:** Namibia on N and W, South Africa on S, Zimbabwe on NE; Botswana claims border with Zambia on N. **Topography:** The Kalahari Desert, supporting nomadic Bushmen and wildlife, spreads over SW; there are swamplands and farming areas in N, and rolling plains in E where livestock are grazed. **Capital:** Gaborone (1997 est.): 183,487.

Government: Type: Parliamentary republic. **Head of state and gov.:** Pres. Festus Mogae; b Aug. 21, 1939; in office: Apr. 1, 1998. **Local divisions:** 10 districts, 4 town councils. **Defense:** 6.5% of GDP. **Active troops:** 7,500.

Economy: Industries: Livestock processing, mining. **Chief crops:** Maize, sorghum, millet, pulses, beans. **Minerals:** Copper, coal, nickel, diamonds, salt, silver. **Arable land:** 1%. **Livestock** (1997): chickens: 1.80 mil; cattle: 2.33 mil; goats: 1.82 mil; sheep: 240,000. **Electricity prod.** (1997): 650 mil kWh.

Finance: Monetary unit: Pula (Sept. 1999: 4.56 = $1 U.S.). **GDP** (1997 est.): $5 bil. **Per capita GDP:** $3,300. **Imports** (1996 est.): $1.6 bil. **Exports** (1996 est.): $2.3 bil. **Tourism:** $185 mil. **Budget** (FY 1996-97): $1.8 bil. **Intl. reserves less gold** (May 1999): $5.71 bil. **Consumer prices** (change in 1998): 6.7%.

Transport: Railroad: Length: 603 mi. **Motor vehicles:** 80,000 pass. cars, 19,869 comm. vehicles. **Civil aviation:** 34.2 mil pass.-mi.; 4 airports.

Communications: TV sets: 24 per 1,000 pop. **Radios:** 821 per 1,000 pop. **Telephones** (1997): 85,600 main lines. **Daily newspaper circ.:** 29 per 1,000 pop.

Health: Life expectancy: 39.42 male; 40.37 female. **Births** (1,000 pop.): 31.46. **Deaths** (per 1,000 pop.): 21.00. **Natural inc.:** 1.046%. **Hosp. beds** (1993): 1 per 434 persons. **Physicians** (1994): 1 per 4,395 persons. **Infant mortality** (per 1,000 live births): 59.08.

Education: Not compulsory. **Literacy:** 70%.
Major Intl. Organizations: UN (FAO, IBRD, ILO, IMF, WHO, WTrO), the Commonwealth, OAU.
Embassy: 3400 International Dr. NW, 20008; 244-4990.

First inhabited by bushmen, then Bantus, the region became the British protectorate of Bechuanaland in 1886, halting encroachment by Boers and Germans from the south and southwest. The country became fully independent Sept. 30, 1966, as Botswana. Cattle raising and mining (diamonds, copper, nickel) have contributed to economic growth; economy is closely tied to South Africa.

Brazil
Federative Republic of Brazil

People: Population: 171,853,126. **Age distrib.** (%): <15: 29.8; 65+: 5.2. **Pop. density:** 52 per sq. mi. **Urban:** 79%. **Ethnic groups:** White (incl. Portuguese, German, Italian, Spanish, Polish) 55%, mixed black and white 38%, black 6%. **Principal languages:** Portuguese (official), Spanish, English, French. **Chief religion:** Roman Catholic 70%.
Geography: Area: 3,286,478 sq. mi., largest country in South America. **Location:** Occupies E half of South America. **Neighbors:** French Guiana, Suriname, Guyana, Venezuela on N; Colombia, Peru, Bolivia, Paraguay, Argentina on W; Uruguay on S. **Topography:** Brazil's Atlantic coastline stretches 4,603 miles. In N is the heavily wooded Amazon basin covering half the country. Its network of rivers is navigable for 15,814 mi. The Amazon itself flows 2,093 miles in Brazil, all navigable. The NE region is semiarid scrubland, heavily settled and poor. The S central region, favored by climate and resources, has almost half of the population, produces 75% of farm goods and 80% of industrial output. The narrow coastal belt includes most of the major cities. Almost the entire country has a tropical or semitropical climate. **Capital:** Brasília. **Cities** (1995 est.): São Paulo 10,017,821; Rio de Janeiro 5,606,497; Belo Horizonte 2,097,311, Brasília 1,737,813.
Government: Type: Federal republic. **Head of state and gov.:** Pres. Fernando Henrique Cardoso; b June 18, 1931; in office: Jan. 1, 1995. **Local divisions:** 26 states, 1 federal district (Brasília). **Defense:** 2.3% of GDP. **Active troops:** 314,700.
Economy: Industries: Steel, autos, textiles, shoes, chemicals, machinery. **Chief crops:** Coffee (leading grower), soybeans, sugarcane, cocoa, rice, corn, wheat, citrus. **Minerals:** Iron (largest producer in the world), manganese, phosphates, uranium, gold, nickel, tin, bauxite, oil. **Crude oil reserves** (1999): 7.1 bil bbls. **Arable land:** 5%. **Livestock** (1997): chickens: 900.00 mil; cattle: 161.00 mil; pigs: 31.43 mil; sheep: 18.30 mil; goats: 12.60 mil; buffalo: 1.70 mil. **Fish catch** (1997): 750,000 metric tons. **Electricity prod.** (1997): 303.523 bil kWh. **Labor force:** 26% agric.; 25% services; 13% manuf.
Finance: Monetary unit: Real (Sept. 1999: 1.92 = $1 U.S.). **GDP** (1997 est.): $1.04 tril. **Per capita GDP:** $6,300. **Imports** (1997): $61.4 bil; partners: EU 26%, U.S. 22%. **Exports** (1997): $53 bil; partners: EU 28%, Latin America 23%, U.S. 20%. **Tourism:** $2.78 bil. **Budget** (1996): $96 bil. **Intl. reserves less gold** (Apr. 1999): $42.47 bil. **Gold:** 4.20 mil oz t. **Consumer prices** (change in 1998): 3.2%.
Transport: Railroad: Length: 18,578 mi. **Motor vehicles** (1997): 14.00 mil pass. cars, 4.03 mil comm. vehicles. **Civil aviation:** 26.3 bil pass.-mi.; 139 airports. **Chief ports:** Santos, Rio de Janeiro, Vitoria, Salvador, Rio Grande, Recife.
Communications: TV sets: 193 per 1,000 pop. **Radios:** 348 per 1,000 pop. **Telephones:** 19,986,600 main lines. **Daily newspaper circ.:** 47 per 1,000 pop.
Health: Life expectancy: 59.35 male; 69.01 female. **Births** (per 1,000 pop.): 20.42. **Deaths** (per 1,000 pop.): 8.79. **Natural inc.:** 1.163%. **Hosp. beds** (1993): 1 per 298 persons. **Physicians** (1993): 1 per 681 persons. **Infant mortality** (per 1,000 live births): 35.37.
Education: Free, compulsory: ages 7-14. **Literacy** (1996): 85%.
Major Intl. Organizations: UN and most of its specialized agencies, OAS.
Embassy: 3006 Massachusetts Ave. NW 20008; 238-2700.
Websites: http://www.ibge.gov.br
         http://www.brasil.emb.nw.dc.us/

Pedro Alvares Cabral, a Portuguese navigator, is generally credited as the first European to reach Brazil, in 1500. The country was thinly settled by various Indian tribes. Only a few have survived to the present, mostly in the Amazon basin.
In the next centuries, Portuguese colonists gradually pushed inland, bringing along large numbers of African slaves. (Slavery was not abolished until 1888.)
The King of Portugal, fleeing before Napoleon's army,

moved the seat of government to Brazil in 1808. Brazil thereupon became a kingdom under Dom Joao VI. After his return to Portugal, his son Pedro proclaimed the independence of Brazil, Sept. 7, 1822, and was crowned emperor. The second emperor, Dom Pedro II, was deposed in 1889, and a republic proclaimed, called the United States of Brazil. In 1967 the country was renamed the Federative Republic of Brazil.
A military junta took control in 1930; dictatorial power was assumed by Getulio Vargas, until finally forced out by the military in 1945. A democratic regime prevailed 1945-64, during which time the capital was moved from Rio de Janeiro to Brasília. In 1964, Pres. Joao Belchoir Marques Goulart instituted economic policies that aggravated Brazil's inflation; he was overthrown by an army revolt. The next 5 presidents were all military leaders. Censorship was imposed, and much of the opposition was suppressed amid charges of torture.
Since 1930, successive governments have pursued industrial and agricultural growth and interior area development. Exploiting vast natural resources and a huge labor force, Brazil became the leading industrial power of Latin America by the 1970s, while agricultural output soared. By the 1990s, Brazil had one of the world's largest economies; income was poorly distributed, however, and more than one out of four Brazilians continued to survive on less than $1 a day. Despite protective environmental legislation, development has destroyed much of the Amazon ecosystem. Brazil hosted delegates from 178 countries at the Earth Summit, June 3-14, 1992.
Democratic presidential elections were held in 1985 as the nation returned to civilian rule. Fernando Collor de Mello was elected president in Dec. 1989. In Sept. 1992, Collor was impeached for corruption. He resigned on Dec. 29 as his trial was beginning, and Itamar Franco, who had been acting president, was sworn in as president. In elections held on Oct. 3, 1994, Fernando Henrique Cardoso was elected president. Reelected Oct. 4, 1998, he guided Brazil through a series of financial crises.

Brunei
State of Brunei Darussalam

People: Population: 322,982. **Age distrib.** (%): <15: 32.8; 65+: 4.5. **Pop. density:** 147 per sq. mi. **Urban:** 70%. **Ethnic groups:** Malay 64%, Chinese 20%. **Principal languages:** Malay (official), English, Chinese. **Chief religions:** Muslim (official) 63%, Buddhist 14%, Christian 8%.
Geography: Area: 2,200 sq. mi. **Location:** In SE Asia, on the N coast of the island of Borneo; it is surrounded on its landward side by the Malaysian state of Sarawak. **Capital:** Bandar Seri Begawan (1994 met. est.): 187,000.
Government: Type: Independent sultanate. **Head of state and gov.:** Sultan Sir Muda Hassanal Bolkiah Mu'izzadin Waddaulah; b July 15, 1946; in office: Jan. 1, 1984. **Local divisions:** 4 districts. **Defense:** 6.7% of GDP. **Active troops:** 5,000.
Economy: Industries: Oil & gas (nearly half of GDP is derived from petroleum sector). **Chief crops:** Rice, bananas, cassava. **Crude oil reserves** (1999): 1.35 bil bbls. **Arable land:** 1%. **Livestock** (1997): chickens: 3.00 mil. **Electricity prod.** (1997): 1.5 bil kWh.
Finance: Monetary unit: Dollar (Sept. 1999: 1.69 = $1 U.S.). **GDP** (1997 est.): $5.4 bil. **Per capita GDP:** $18,000. **Imports** (1996 est.): $2.65 bil; partners: Singapore 29%, UK 19%. **Exports** (1996 est.): $2.62 bil; partners: ASEAN 31%, Japan 27%. **Tourism:** $37 mil. **Budget** (1995 est.): $2.6 bil.
Transport: Railroad: Length: 12 mi. **Motor vehicles:** 146,000 pass. cars, 17,780 comm. vehicles. **Civil aviation:** 1.8 bil pass.-mi.; 1 airport.
Communications: TV sets: 308 per 1,000 pop. **Radios:** 417 per 1,000 pop. **Telephones:** 77,700 main lines. **Daily newspaper circ.:** 70 per 1,000 pop.
Health: Life expectancy: 70.35 male; 73.42 female. **Births** (per 1,000 pop.): 24.69. **Deaths** (per 1,000 pop.): 5.21. **Natural increase:** 1.948%. **Hosp. beds** (1993): 1 per 285 persons. **Physicians** (1993): 1 per 1,398 persons. **Infant mortality** (per 1,000 live births): 22.83.
Education: Free, compulsory: ages 5-17. **Literacy:** 88%.
Major Intl. Organizations: UN and some of its specialized agencies, APEC, ASEAN, the Commonwealth.
Embassy: 2600 Virginia Ave. NW, 20037; 342-0159.
Website: http://www.brunet.bn

The Sultanate of Brunei was a powerful state in the early 16th century, with authority over all of the island of Borneo as well as parts of the Sulu Islands and the Philippines. In 1888, a treaty placed the state under the protection of Great Britain.
Brunei became a fully sovereign and independent state on Jan. 1, 1984.

Bulgaria
Republic of Bulgaria

People: Population: 8,194,772. **Age distrib.** (%): <15: 16.1; 65+: 16.3. **Pop. density:** 191 per sq. mi. **Urban:** 69%. **Ethnic groups:** Bulgarian 85%, Turk 9%. **Principal languages:** Bulgarian (official), Turkish. **Chief religions:** Bulgarian Orthodox 85%, Muslim 13%.

Geography: Area: 42,800 sq. mi. **Location:** SE Europe, in E Balkan Peninsula on Black Sea. **Neighbors:** Romania on N; Yugoslavia, Macedonia on W; Greece, Turkey on S. **Topography:** The Stara Planina (Balkan) Mts. stretch E-W across the center of the country, with the Danubian plain on N, the Rhodope Mts. on SW, and Thracian Plain on SE. **Capital:** Sofia. **Cities** (1996 est.): Plovdiv 344,326; Sofia 1,116,823.

Government: Type: Republic. **Head of state:** Pres. Petar Stoyanov; b May 25, 1952; in office: Jan. 19, 1997. **Head of gov.:** Prime Min. Ivan Kostov; b Dec. 23, 1949; in office: May 21, 1997. **Local divisions:** 9 provinces. **Defense:** 3.4% of GDP. **Active troops:** 101,500.

Economy: Industries: Chemicals, machinery, metals, textiles, food processing. **Chief crops:** Grain, fruit, oilseed, vegetables, tobacco. **Minerals:** Bauxite, copper, zinc, lead, coal. **Crude oil reserves** (1999): 15 mil bbls. **Arable land:** 37%. **Livestock** (1997): chickens: 13.77 mil; sheep: 2.85 mil; pigs: 1.48 mil; goats: 966,000; cattle: 612,000. **Fish catch** (1996): 12,723 metric tons. **Electricity prod.** (1997): 39.863 bil kWh. **Labor force:** 41% ind.; 18% agric.

Finance: Monetary unit: Lev (Sept. 1999: 1.82 = $1 U.S.). **GDP** (1997 est.): $35.6 bil. **Per capita GDP:** $4,100. **Imports** (1997 est.): $4.5 bil; partners: OECD 46%, CIS 41%. **Exports** (1997): $4.9 bil; partners: OECD 50%, CIS 32%. **Tourism:** $361 mil. **Budget** (1997 est.): $3.2 bil. **Intl. reserves less gold** (June 1999): $2.58 bil. **Gold:** 1.03 mil oz t. **Consumer prices** (change in 1998): 22.3%.

Transport: Railroad: Length: 4,043 mi. **Motor vehicles:** 1.65 mil pass. cars, 264,196 comm. vehicles. **Civil aviation:** 1.1 bil pass.-mi.; 3 airports. **Chief ports:** Burgas, Varna.

Communications: TV sets: 359 per 1,000 pop. **Telephones:** 2,742,000 main lines. **Daily newspaper circ.:** 141 per 1,000 pop.

Health: Life expectancy: 68.72 male; 76.03 female. **Births** (per 1,000 pop.): 8.71. **Deaths** (per 1,000 pop.): 13.20. **Natural inc.:** −0.449%. **Hosp. beds** (1995): 1 per 94 persons. **Physicians** (1995): 1 per 288 persons. **Infant mortality** (per 1,000 live births): 12.37.

Education: Free, compulsory: ages 7-16. **Literacy:** 98%.

Major Intl. Organizations: UN (FAO, IBRD, ILO, IMF, IMO, WHO, WTrO), OSCE.

Embassy: 1621 22d St. NW 20008; 387-7969.

Bulgaria was settled by Slavs in the 6th century. Turkic Bulgars arrived in the 7th century, merged with the Slavs, became Christians by the 9th century, and set up powerful empires in the 10th and 12th centuries. The Ottomans prevailed in 1396 and remained for 500 years.

A revolt in 1876 led to an independent kingdom in 1908. Bulgaria expanded after the first Balkan War but lost its Aegean coastline in World War I, when it sided with Germany. Bulgaria joined the Axis in World War II but withdrew in 1944. Communists took power with Soviet aid; the monarchy was abolished Sept. 8, 1946.

On Nov. 10, 1989, Communist Party leader and head of state Todor Zhivkov, who had held power for 35 years, resigned. Zhivkov was imprisoned, Jan. 1990, and convicted, Sept. 1992, of corruption and abuse of power. In Jan. 1990, Parliament voted to revoke the constitutionally guaranteed dominant role of the Communist Party. A new constitution took effect July 13, 1991. An economic austerity program was launched in May 1996. Former Prime Min. Andrei Lukanov, a longtime Communist leader, was assassinated Oct. 2 in Sofia. Petar Stoyanov won a presidential runoff election Nov. 3. Bulgaria's deteriorating economy provoked nationwide strikes and demonstrations in Jan. 1997. The Union of Democratic Forces, an anti-Communist group, won parliamentary elections on Apr. 19, 1997.

Burkina Faso
Democratic Republic of Burkina Faso

People: Population: 11,575,898. **Age distrib.** (%): <15: 48.0; 65+ 3.0. **Pop. density:** 109 per sq. mi. **Urban:** 16%. **Ethnic groups:** Mossi (approx. 24%), Gurunsi, Senufo, Lobi, Bobo, Mande, Fulani. **Principal languages:** French (official), Sudanic tribal languages. **Chief religions:** Muslim 50%, indigenous beliefs 40%, Christian (mostly Roman Catholic) 10%.

Geography: Area: 105,900 sq. mi. **Location:** In W Africa, S of the Sahara. **Neighbors:** Mali on NW; Niger on NE; Benin, Togo, Ghana, Côte d'Ivoire on S. **Topography:** Land-locked Burkina Faso is in the savanna region of W Africa. The N is arid, hot, and thinly populated. **Capital:** Ouagadougou 824,000*.

Government: Type: Republic. **Head of state:** Pres. Blaise Compaoré; b 1951; in office: Oct. 15, 1987. **Head of gov.:** Kadré Désiré Ouedraogo; b 1953; in office: Feb. 6, 1996. **Local divisions:** 45 provinces. **Defense:** 2.2% of GDP. **Active troops:** 5,800.

Economy: Industries: Agricultural processing, beverages, soap, textiles. **Chief crops:** Millet, sorghum, rice, peanuts, cotton. **Minerals:** Manganese, limestone, marble. **Arable land:** 13%. **Livestock** (1997): chickens: 20.52 mil; goats: 7.91 mil; sheep: 6.21 mil; cattle: 4.52 mil; pigs: 586,600. **Electricity prod.** (1997): 220 mil kWh. **Labor force:** 92% agric.

Finance: Monetary unit: CFA Franc (Sept. 1999: 612.79 = $1 U.S.). **GDP** (1997 est.): $10.3 bil. **Per capita GDP:** $950. **Imports** (1995 est.): $500 mil; partners: France 25%, Côte d'Ivoire 18%. **Exports** (1995 est.): $298 mil; partners: Côte d' Ivoire 35%, France 21%. **Tourism:** $39 mil. **Budget** (1995 est.): $492 mil. **Intl. reserves less gold** (Dec. 1998): $373.3 mil. **Gold:** 11,000 oz t. **Consumer prices** (change in 1998): 5.1%.

Transport: Railroad: Length: 386 mi. **Motor vehicles:** 35,460 pass. cars, 19,473 comm. vehicles. **Civil aviation:** 154.2 mil pass.-mi.; 2 airports.

Communications: TV sets: 4.4 per 1,000 pop. **Radios:** 48.3 per 1,000 pop. **Telephones:** 41,600 main lines.

Health: Life expectancy: 44.97 male; 46.84 female. **Births** (per 1,000 pop.): 45.84. **Deaths** (per 1,000 pop.): 17.56. **Natural inc.:** 2.828%. **Hosp. beds** (1991): 1 per 1,837 persons. **Physicians** (1991): 1 per 27,158 persons. **Infant mortality** (per 1,000 live births): 107.19.

Education: Free, compulsory: ages 7-14. **Literacy:** 19%.

Major Intl. Organizations: UN and many of its specialized agencies, OAU.

Embassy: 2340 Massachusetts Ave. NW 20008; 332-5577.

The Mossi tribe entered the area in the 11th to 13th centuries. Their kingdoms ruled until they were defeated by the Mali and Songhai empires.

French control came by 1896, but Upper Volta (renamed Burkina Faso on Aug. 4, 1984) was not established as a separate territory until 1947. Full independence came Aug. 5, 1960, and a pro-French government was elected. The military seized power in 1980. A 1987 coup established the current regime, which instituted a multiparty democracy in the early 1990s.

Several hundred thousand farm workers migrate each year to Côte d'Ivoire and Ghana. Burkina Faso is heavily dependent on foreign aid.

Burma
(See Myanmar)

Burundi
Republic of Burundi

People: Population: 5,735,937. **Age distrib.** (%): <15: 47.0; 65+: 2.9. **Pop. density:** 536 per sq. mi. **Urban:** 8%. **Ethnic groups:** Hutu (Bantu) 85%, Tutsi 14%, Twa (Pygmy) 1%. **Principal languages:** Kirundi, French (both official), Swahili. **Chief religions:** Roman Catholic 62%, indigenous beliefs 32%, Protestant 5%.

Geography: Area: 10,700 sq. mi. **Location:** In central Africa. **Neighbors:** Rwanda on N, Dem. Rep. of the Congo (formerly Zaire) on W, Tanzania on E and S. **Topography:** Much of the country is grassy highland, with mountains reaching 8,900 ft. The southernmost source of the White Nile is located in Burundi. Lake Tanganyika is the second deepest lake in the world. **Capital:** Bujumbura (1994 est.): 300,000.

Government: Type: In transition. **Head of state:** Pres. Pierre Buyoya; b Nov. 14, 1949; in office: July 25, 1996. **Local divisions:** 15 provinces. **Defense:** 5.7% of GDP. **Active troops:** 18,500.

Economy: Industries: Light consumer goods, food processing. **Chief crops:** Coffee (80% of exports), cotton, tea. **Minerals:** Nickel, uranium. **Arable land:** 44%. **Livestock** (1997): chickens: 4.70 mil; goats: 900,000; cattle: 346,000; sheep: 320,000. **Electricity prod.** (1997): 122 mil kWh. **Labor force:** 93% agric.

Finance: Monetary unit: Franc (Sept. 1999: 588.31 = $1 U.S.). **GDP** (1997 est.): $4 bil. **Per capita GDP:** $660. **Imports** (1996): $127 mil; partners: EU 47%. **Exports** (1996): $40 mil; partners: EU 60%. **Tourism:** $1 mil. **Budget** (1995 est.): $258 mil. **Intl. reserves less gold** (June 1999): $59.46 mil. **Gold:** 17,000 oz t. **Consumer prices** (change in 1998): 12.5%.

Transport: Motor vehicles: 8,200 pass. cars, 11,800 comm. vehicles. **Civil aviation:** 5.2 mil pass.-mi.; 1 airport. **Chief port:** Bujumbura.

Communications: TV sets: 7 per 1,000 pop. **Radios:** 50 per 1,000 pop. **Telephones** (1997): 15,900 main lines.

Health: Life expectancy: 43.54 male; 47.41 female. **Births** (per 1,000 pop.): 41.27. **Deaths** (per 1,000 pop.): 17.23. **Natural inc.:** 2.404%. **Infant mortality** (per 1,000 live births): 99.36.

Education: Free, compulsory: ages 7-13. **Literacy:** 35%.

Major Intl. Organizations: UN (FAO, IBRD, ILO, IMF, WHO, WTrO), OAU.

Embassy: 2233 Wisconsin Ave. NW 20007; 342-2574.

The pygmy Twa were the first inhabitants, followed by Bantu Hutus, who were conquered in the 16th century by the Tutsi (Watusi), probably from Ethiopia. Under German control in 1899, the area fell to Belgium in 1916, which exercised successively a League of Nations mandate and UN trusteeship over Ruanda-Urundi (now the two countries of Rwanda and Burundi). Burundi became independent July 1, 1962.

An unsuccessful Hutu rebellion in 1972-73 left 10,000 Tutsi and 150,000 Hutu dead. Over 100,000 Hutu fled to Tanzania and Zaire (now Congo). In the 1980s, Burundi's Tutsi-dominated regime pledged itself to ethnic reconciliation and democratic reform. In the nation's first democratic presidential election, in June 1993, a Hutu, Melchior Ndadaye, was elected. He was killed in an attempted coup, Oct. 21, 1993. At least 150,000 Burundians died as a result of ethnic conflict during the next three years. Pres. Cyprien Ntaryamira, elected Jan. 1994, was killed with the president of Rwanda in a mysterious plane crash, Apr. 6. The incident sparked massive carnage in Rwanda; violence in Burundi, initially far more limited, intensified in 1995. Ethnic strife continued after a military coup, July 25, 1996.

Cambodia
Kingdom of Cambodia

People: Population: 11,626,520. **Age distrib.** (%): <15: 45.2; 65+: 3.0. **Pop. density:** 166 per sq. mi. **Urban:** 21%. **Ethnic groups:** Khmer 90%, Vietnamese 5%, Chinese 1%. **Principal languages:** Khmer (official), French. **Chief religion:** Theravada Buddhism 95%.

Geography: Area: 69,900 sq. mi. **Location:** SE Asia, on Indochina Peninsula. **Neighbors:** Thailand on W and N, Laos on NE, Vietnam on E. **Topography:** The central area, formed by the Mekong R. basin and Tonle Sap lake, is level. Hills and mountains are in SE, a long escarpment separates the country from Thailand on NW. 76% of the area is forested. **Capital:** Phnom Penh (1994 est.): 920,000.

Government: Type: Constitutional monarchy. **Head of state:** King Norodom Sihanouk; b Oct. 31, 1922; in office: Sept. 24, 1993. **Head of gov.:** Prime Min. Hun Sen; b Apr. 4, 1952; in office: Nov. 30, 1998. **Local divisions:** 20 provinces and 3 municipalities. **Defense:** 7.3% of GDP. **Active troops:** 140,500.

Economy: Industries: Rice milling, wood & wood products, fishing. **Chief crops:** Rice, corn, rubber, vegetables. **Minerals:** Gemstones, phosphates, manganese. **Other resources:** Timber. **Arable land:** 13%. **Livestock** (1997): chickens: 12.00 mil; cattle: 2.90 mil; pigs: 2.20 mil; buffalo: 770,000. **Fish catch** (1996): 94,710 metric tons. **Electricity prod.** (1997): 200 mil kWh. **Labor force:** 80% agric.

Finance: Monetary unit: Riel (Sept. 1999: 3,800.00 = $1 U.S.). **GDP** (1997 est.): $7.7 bil. **Per capita GDP:** $715. **Imports** (1996 est.): $1 bil; partners: Singapore 36%, Thailand 24%. **Exports** (1996 est.): $615 mil; partners: Thailand 43%. **Tourism:** $143 mil. **Budget** (1995 est.): $496 mil. **Intl. reserves less gold** (June 1999): $365.41 mil. **Consumer prices** (change in 1998): 14.8%.

Transport: Railroad: Length: 380 mi. **Motor vehicles:** 15,000 pass. cars, 15,000 comm. vehicles. **Civil aviation:** 8 airports. **Chief port:** Kompong Som (Sihanoukville).

Communications: TV sets: 8 per 1,000 pop. **Radios:** 124 per 1,000 pop. **Telephones:** 24,300 main lines.

Health: Life expectancy: 46.81 male; 49.75 female. **Births** (per 1,000 pop.): 41.05. **Deaths** (per 1,000 pop.): 16.20. **Natural inc.:** 2.485%. **Hosp. beds** (1994): 1 per 791 persons. **Physicians** (1994): 1 per 7,900 persons. **Infant mortality** (per 1,000 live births): 105.06.

Education: Compulsory: ages 6-12. **Literacy** (1993): 65%.

Major Intl. Organizations: UN (FAO, IBRD, ILO, IMF, IMO, WHO), ASEAN.

Embassy: 4500 16th St. NW 20011; 726-7742.

Website: http://www.cambodia.org

Early kingdoms dating from that of Funan in the 1st century AD culminated in the great Khmer empire that flourished from the 9th century to the 13th, encompassing present-day Thailand, Cambodia, Laos, and southern Vietnam. The peripheral areas were lost to invading Siamese and Vietnamese, and France established a protectorate in 1863. Independence came in 1953.

Prince Norodom Sihanouk, king 1941-1955 and head of state from 1960, tried to maintain neutrality. Relations with the U.S. were broken in 1965, after South Vietnam planes attacked Vietcong forces within Cambodia. Relations were restored in 1969, after Sihanouk charged Viet Communists with arming Cambodian insurgents.

In 1970, pro-U.S. Prem. Lon Nol seized power, demanding removal of 40,000 North Viet troops; the monarchy was abolished. Sihanouk formed a government-in-exile in Beijing, and open war began between the government and Communist Khmer Rouge guerrillas. The U.S. provided heavy military and economic aid.

Khmer Rouge forces captured Phnom Penh Apr. 17, 1975. The new government evacuated all cities and towns, and shuffled the rural population, sending virtually the entire population to clear jungle, forest, and scrub. Over one million people were killed in executions and enforced hardships.

Severe border fighting broke out with Vietnam in 1978 and developed into a full-fledged Vietnamese invasion. Formation of a Vietnamese-backed government was announced, Jan. 8, 1979, one day after the Vietnamese capture of Phnom Penh. Thousands of refugees flowed into Thailand, and widespread starvation was reported.

On Jan. 10, 1983, Vietnam launched an offensive against rebel forces in the west. They overran a refugee camp, Jan. 31, driving 30,000 residents into Thailand. In March, Vietnam launched a major offensive against camps on the Cambodian-Thailand border, engaged Khmer Rouge guerrillas, and crossed the border, instigating clashes with Thai troops. Vietnam withdrew nearly all its troops by Sept. 1989.

Following UN-sponsored elections in Cambodia that ended May 28, 1993, the 2 leading parties agreed to share power in an interim government until a new constitution was adopted. On Sept. 21, a constitution reestablishing a monarchy was adopted by the National Assembly. It took effect Sept. 24, with Sihanouk as king. The Khmer Rouge, which had boycotted the elections, opposed the new government, and armed violence continued in the mid-1990s. Ieng Sary, a Khmer Rouge leader, broke with the guerrillas, formed a rival group, and announced his support for the monarchy in Aug. 1996, as Khmer Rouge strength rapidly diminished.

Co-Prime Min. Hun Sen staged a coup July 5, 1997, ousting his rival, Prince Norodom Ranariddh. Pol Pot, the Khmer Rouge leader who held power during the late 1970s, was denounced by his former comrades at a show trial, July 25, and sentenced to house arrest; he died Apr. 15, 1998. Hun Sen's party won parliamentary elections on July 26. Cambodia was formally admitted to ASEAN on Apr. 30, 1999.

Cameroon
Republic of Cameroon

People: Population: 15,456,092. **Age distrib.** (%): <15: 45.9; 65+: 3.3. **Pop. density:** 84 per sq. mi. **Urban:** 46%. **Ethnic groups:** Cameroon Highlander 31%, Equatorial Bantu 19%, Kirdi 11%, Fulani 10%, NW Bantu 8%. **Principal languages:** English, French (both official), numerous African groups. **Chief religions:** Indigenous beliefs 51%, Christian 33%, Muslim 16%.

Geography: Area: 183,600 sq. mi. **Location:** Between W and central Africa. **Neighbors:** Nigeria on NW; Chad, Central African Republic on E; Congo, Gabon, Equatorial Guinea on S. **Topography:** A low coastal plain with rain forests is in S; plateaus in center lead to forested mountains in W, including Mt. Cameroon, 13,350 ft.; grasslands in N lead to marshes around Lake Chad. **Capital:** Yaoundé. **Cities:** Douala 1,320,000; Yaoundé 1,119,000*.

Government: Type: Republic. **Head of state:** Pres. Paul Biya; b Feb. 13, 1933; in office: Nov. 6, 1982. **Head of gov.:** Prime Min. Peter Mafani Musonge; b Dec. 3, 1942; in office: Sept. 19, 1996. **Local divisions:** 10 provinces. **Defense:** 2.9% of GDP. **Active troops:** 13,100.

Economy: Industries: Oil production and refining, food processing, light consumer goods. **Chief crops:** Cocoa, coffee, cotton. **Crude oil reserves** (1999): 400 mil bbls. **Minerals:** Oil, bauxite, iron ore. **Other resources:** Timber. **Arable land:** 13%. **Livestock** (1997): chickens: 20.00 mil; cattle: 5.90 mil; sheep: 3.80 mil; goats: 3.80 mil; pigs: 1.41 mil. **Fish catch** (1996): 83,984 metric tons. **Electricity prod.** (1997): 2.735 bil kWh.

Finance: Monetary unit: CFA Franc (Sept. 1999: 612.79 = $1 U.S.). **GDP** (1997 est.): $30.9 bil. **Per capita GDP:** $2,100. **Imports** (1996): $1.5 bil; partners: France 40%. **Exports** (1996): $1.9 bil; partners: EU about 60%. **Tourism:** $40 mil. **Budget** (FY 1996-97 est.): $2.23 bil. **Intl. reserves less gold** (Mar. 1999): $2.29 mil. **Gold:** 30,000 oz t. **Consumer prices** (change in 1997): 1.5%.

Transport: Railroad: Length: 686 mi. **Motor vehicles:** 92,200 pass. cars, 60,800 comm. vehicles. **Civil aviation:** 339.9 mil pass.-mi.; 5 airports. **Chief ports:** Douala, Kribi.

Communications: TV sets: 72 per 1,000 pop. **Radios:** 325 per 1,000 pop. **Telephones** (1997): 75,200 main lines.

Health: Life expectancy: 49.75 male; 52.94 female. **Births** (per 1,000 pop.): 41.84. **Deaths** (per 1,000 pop.): 13.95. **Natural inc.:** 2.789%. **Infant mortality** (per 1,000 live births): 75.69.

Education: Free, compulsory: ages 6-12. **Literacy:** 63%.

Major Intl. Organizations: UN (FAO, IBRD, ILO, IMF, IMO, WHO, WTrO), the Commonwealth, OAU.

Embassy: 2349 Massachusetts Ave. NW 20008; 265-8790. **Website:** http://www.camnet.cm

Portuguese sailors were the first Europeans to reach Cameroon, in the 15th century. The European and American slave trade was very active in the area. German control lasted from 1884 to 1916, when France and Britain divided the territory, later receiving League of Nations mandates and UN trusteeships. French Cameroon became independent Jan. 1, 1960; one part of British Cameroon joined Nigeria in 1961, the other part joined Cameroon. Stability has allowed for development of roads, railways, agriculture, and petroleum production.

Pres. Paul Biya retained his office in Oct. 1992 elections, but the results were widely disputed. A new constitution won legislative approval in Dec. 1995. Fraud charges accompanied legislative elections, May 17, 1997, which Biya's party won.

Canada

People: Population: 31,006,347. **Age distrib.** (%): <15: 19.6; 65+: 12.5. **Pop. density:** 8 per sq. mi. **Urban:** 77%. **Ethnic groups:** British Isles 40%, French 27%, other European 20%, Amerindian 1.5%, other (mostly Asian) 11.5%. **Principal languages:** English, French (both official). **Chief religions:** Roman Catholic 45%, United Church 12%, Anglican 8%.

Geography: Area: 3,851,800 sq. mi., the largest country in land size in the western hemisphere. **Topography:** Canada stretches 3,426 miles from east to west and extends southward from the North Pole to the U.S. border. Its seacoast includes 36,356 miles of mainland and 115,133 miles of islands, including the Arctic islands almost from Greenland to near the Alaskan border. **Climate:** While generally temperate, varies from freezing winter cold to blistering summer heat. **Capital:** Ottawa. **Cities** (met. 1996 cen.): Toronto 4.3 mil; Montreal 3.3 mil; Vancouver 1.8 mil; Ottawa-Hull 1.0 mil; Edmonton 862,500; Calgary 821,600; Quebec 671,900; Winnipeg 667,200.

Government: Type: Confederation with parliamentary democracy. **Head of state:** Queen Elizabeth II, represented by Gov.-Gen. Adrienne Clarkson; b Feb. 10, 1939; in office: Oct. 7, 1999. **Head of gov.:** Prime Min. Jean Chrétien; b Jan. 11, 1934; in office: Nov. 4, 1993. **Local divisions:** 10 provinces, 3 territories. **Defense:** 1.3% of GDP. **Active troops:** 61,600.

Economy: Industries: Mining, wood and food prods., transport equip., chemicals, oil, gas. **Chief crops:** Grains, oilseed, tobacco, fruit, vegetables. **Minerals:** Nickel, zinc, copper, gold, lead, molybdenum, potash, silver. **Crude oil reserves** (1999): 4.93 bil barrels. **Arable land:** 5%. **Livestock** (1997): chickens: 139.00 mil; cattle: 13.16 mil; pigs: 11.84 mil; sheep: 634,200. **Fish catch** (1997): 944,562 metric tons. **Electricity prod.** (1996): 547.169 bil kWh. **Labor force:** 75% services, 16% manuf., 3% agric.

Finance: Monetary unit: Dollar (Sept. 1999: 1.50 = $1 U.S.). **GDP** (1997 est.): $658 bil. **Per capita GDP:** $21,700. **Imports** (1997 est.): $194.4 bil; partners: U.S. 67%. **Exports** (1997 est.): $208.6 bil; partners: U.S. 82%. **Tourism:** $9.13 bil. **Budget** (1996): $117.2 bil. **Intl. reserves less gold** (June 1999): $25.23 bil. **Gold:** 2.19 mil oz t. **Consumer prices** (change in 1998): 1.0%.

Transport: Railroad: Length: 44,182 mi. **Motor vehicles:** 13.3 mil pass. cars, 3.52 mil comm. vehicles. **Civil aviation:** 38.4 bil passenger-mi: 301 airports. **Chief ports:** Halifax, Montreal, Quebec, Saint John, Toronto, Vancouver.

Communications: TV sets: 647 per 1,000 pop. **Radios:** 919 per 1,000 pop. **Telephones:** 19,241,000 main lines. **Daily newspaper circ.:** 215 per 1,000 pop.

Health: Life expectancy: 76.12 male; 82.79 female. **Births** (per 1,000 pop.): 11.86. **Deaths** (per 1,000 pop.): 7.26. **Natural inc.:** 0.460%. **Hosp. beds** (1993): 1 per 177 persons. **Physicians** (1994): 1 per 534 persons. **Infant mortality** (per 1,000 live births): 5.47.

Education: Compulsory primary education. **Literacy** (1994): 97%.

Major Intl. Organizations: UN and all of its specialized agencies, APEC, the Commonwealth, NATO, OAS, OECD, OSCE.

Embassy: 501 Pennsylvania Ave. NW 20001; 682-1740. **Websites:** http://www.statcan.ca http://canada.gc.ca/main_e.html

French explorer Jacques Cartier, who reached the Gulf of St. Lawrence in 1534, is generally regarded as Canada's founder. But English seaman John Cabot sighted Newfoundland in 1497, and Vikings are believed to have reached the Atlantic coast centuries before either explorer.

Canadian settlement was pioneered by the French who established Quebec City (1608) and Montreal (1642) and declared New France a colony in 1663.

Britain acquired Acadia (later Nova Scotia) in 1717 and, through military victory over French forces in Canada, captured Quebec (1759) and obtained control of the rest of New France in 1763. The French, through the Quebec Act of 1774, retained the rights to their own language, religion, and civil law. The British presence in Canada increased during the American Revolution when many colonials, proudly calling themselves United Empire Loyalists, moved north to Canada. Fur traders and explorers led Canadians westward across the continent. Sir Alexander Mackenzie reached the Pacific in 1793 and scrawled on a rock by the ocean, "from Canada by land."

In Upper and Lower Canada (later called Ontario and Quebec) and in the Maritimes, legislative assemblies appeared in the 18th century and reformers called for responsible government. But the War of 1812 intervened. The war, a conflict between Great Britain and the United States fought mainly in Upper Canada, ended in a stalemate in 1814.

In 1837 political agitation for more democratic government culminated in rebellions in Upper and Lower Canada. Britain sent Lord Durham to investigate; in a famous report (1839), he recommended union of the 2 parts into one colony called Canada. The union lasted until Confederation, July 1, 1867, when proclamation of the British North America (BNA) Act (now known as the Constitution Act, 1867) launched the Dominion of Canada, consisting of Ontario, Quebec, and the former colonies of Nova Scotia and New Brunswick.

Since 1840 the Canadian colonies had held the right to internal self-government. The BNA Act, which was the basis for the country's written constitution, established a federal system of government on the model of a British parliament and cabinet structure under the crown. Canada was proclaimed a self-governing Dominion within the British Empire in 1931. With the ratification of the Constitution Act, 1982, Canada severed its last formal legislative link with Britain by obtaining the right to amend its constitution.

The so-called Meech Lake Agreement was signed (subject to provincial ratification) June 3, 1987. The accord would have assured constitutional protection for Quebec's efforts to preserve its French language and culture. Critics charged it did not make any provision for other minority groups and it gave Quebec too much power, which might enable Quebec to override the nation's 1982 Charter of Rights and Freedoms (an integral part of the constitution). The accord died June 22, 1990.

Its failure sparked a separatist revival in Quebec, which culminated in Aug. 1992 in the Charlottetown agreement. This called for changes to the constitution, such as recognition of Quebec as a "distinct society" within the Canadian confederation. It was defeated in a national referendum Oct. 26, 1992.

Canada became the first nation to ratify the North American Free Trade Agreement between Canada, Mexico, and the U.S. June 23, 1993. It went into effect Jan. 1, 1994.

On Feb. 24, 1993, Brian Mulroney resigned as prime minister after more than 8 years in office; he was succeeded by Kim Campbell. In elections Oct. 25, 1993, the ruling Conservatives were defeated in a landslide that left them only 2 of the 295 seats in the House of Commons. Jean Chrétien became prime minister. In a Quebec referendum held Oct. 30, 1995, proponents of secession lost by a razor-thin margin. The elections of June 2, 1997, left the Liberals with a slim majority.

On Jan. 7, 1998, the government apologized to native peoples for 150 years of mistreatment and pledged to set up a "healing fund." Canada's highest court ruled, Aug. 20, that Quebec cannot secede unilaterally, even if a majority of the province approves. Nunavut ("Our Land"), carved from Northwest Territories as a homeland for the Inuit, was established Apr. 1, 1999.

Provinces/Territories	Area (sq. mi.)	Population (1996 cen.)
Alberta	255,287	2,696,826
British Columbia.	365,948	3,724,500
Manitoba	250,947	1,113,898
New Brunswick.	28,355	738,133
Newfoundland	156,649	551,792
Nova Scotia	21,425	909,282
Ontario	412,581	10,753,573
Prince Edward Island	2,185	134,557
Quebec.	594,860	7,138,795
Saskatchewan	251,866	990,237
Northwest Territories	503,951	39,672
Yukon Territory	186,661	30,766
Nunavut	818,959	24,730

Prime Ministers of Canada

Canada is a constitutional monarchy with a parliamentary system of government. It is also a federal state. Canada's official head of state, Queen Elizabeth II, is represented by a resident Governor-General. However, in practice the nation is governed by the Prime Minister, leader of the party that commands the support of a majority of the House of Commons, dominant chamber of Canada's bicameral Parliament.

Name	Party	Term
Sir John A. MacDonald	Conservative	1867-1873
Alexander Mackenzie	Liberal	1873-1878
Sir John A. MacDonald	Conservative	1878-1891
Sir John J. C. Abbott	Conservative	1891-1892
Sir John S. D. Thompson	Conservative	1892-1894
Sir Mackenzie Bowell	Conservative	1894-1896
Sir Charles Tupper	Conservative	1896[1]
Sir Wilfrid Laurier	Liberal	1896-1911
Sir Robert Laird Borden	Cons./Union.[2]	1911-1920
Arthur Meighen	Unionist	1920-1921
W. L. Mackenzie King.	Liberal	1921-1926
Arthur Meighen	Conservative	1926[3]
W. L. Mackenzie King.	Liberal	1926-1930
Richard Bedford Bennett	Conservative	1930-1935
W. L. Mackenzie King.	Liberal	1935-1948
Louis St. Laurent	Liberal	1948-1957
John G. Diefenbaker	Prog. Cons.	1957-1963
Lester Bowles Pearson	Liberal	1963-1968
Pierre Elliott Trudeau	Liberal	1968-1979
Joe Clark	Prog. Cons.	1979-1980
Pierre Elliott Trudeau	Liberal	1980-1984
John Napier Turner.	Liberal	1984[4]
Brian Mulroney.	Prog. Cons.	1984-1993
Kim Campbell.	Prog. Cons.	1993[5]
Jean Chrétien.	Liberal	1993-

(1) May-July. (2) Conservative 1911-1917, Unionist 1917-1920. (3) June-Sept. (4) June-Sept. (5) June-Oct.

Cape Verde

Republic of Cape Verde

People: Population: 405,748. **Age distrib.** (%): <15: 45.3; 65+: 6.2. **Pop. density:** 254 per sq. mi. **Urban:** 56%. **Ethnic groups:** Creole (mulatto) 71%, African 28%, **Principal languages:** Portuguese (official), Crioulo. **Chief religion:** Roman Catholic 96%.

Geography: Area: 1,600 sq. mi. **Location:** In Atlantic O., off W tip of Africa. **Neighbors:** Nearest are Mauritania, Senegal to E. **Topography:** Cape Verde Islands are 15 in number, volcanic in origin (active crater on Fogo). The landscape is eroded and stark, with vegetation mostly in interior valleys. **Capital:** Praia (1995 est.): 68,000.

Government: Type: Republic. **Head of state:** Pres. Antonio Mascarenhas Monteiro; b Feb. 16, 1944; in office: Mar. 22, 1991. **Head of gov.:** Prime Min. Carlos Veiga; b 1949; in office: Apr. 4, 1991. **Local divisions:** 16 districts. **Defense:** 1.7% of GDP. **Active troops:** 1,100.

Economy: Industries: Food and beverages, fish processing, shoes and garments. **Chief crops:** Bananas, coffee, sweet potatoes, corn, beans. **Minerals:** Salt. **Other resources:** Fish. **Arable land:** 11%. **Livestock** (1997): chickens: 430,000; pigs: 636,000; goats: 110,000. **Electricity prod.** (1997): 40 mil kWh.

Finance: Monetary unit: Escudo (Sept. 1999: 94.71 = $1 U.S.). **GDP** (1997 est.): $538 mil. **Per capita GDP:** $1,370. **Imports** (1996 est.): $237 mil; partners: Portugal 41%. **Exports** (1996 est.): $12.8 mil; partners: Portugal 50%. **Tourism:** $17 mil. **Budget** (1996): $228 mil. **Intl. reserves less gold** (May 1999): $43.88 mil. **Consumer prices** (change in 1998): 4.4%.

Transport: Motor vehicles: 11,000 pass. cars, 7,000 comm. vehicles. **Civil aviation:** 166.5 mil pass.-mi.; 9 airports. **Chief ports:** Mindelo, Praia.

Communications: TV sets: 2.6 per 1,000 pop. **Radios:** 146 per 1,000 pop. **Telephones:** 40,000 main lines.

Health: Life expectancy: 67.66 male; 74.36 female. **Births** (per 1,000 pop.): 33.49. **Deaths** (per 1,000 pop.): 6.78. **Natural inc.:** 2.671%. **Infant mortality** (per 1,000 live births): 45.50.

Education: Compulsory: ages 7-11. **Literacy:** 72%.

Major Intl. Organizations: UN (FAO, IBRD, ILO, IMF, IMO, WHO), OAU.

Embassy: 3415 Massachusetts Ave. NW 20007; 965-6820.

The uninhabited Cape Verdes were discovered by the Portuguese in 1456 or 1460. The first Portuguese colonists landed in 1462; African slaves were brought soon after, and most Cape Verdeans descend from both groups. Cape Verde independence came July 5, 1975. Antonio Mascarenhas Monteiro won the nation's first free presidential election Feb. 17, 1991; he was reelected without opposition five years later.

Central African Republic

People: Population: 3,444,951. **Age distrib.** (%): <15: 43.7; 65+: 3.7. **Pop. density:** 14 per sq. mi. **Urban:** 40%. **Ethnic groups:** Baya 34%, Banda 27%, Mandjia 21%, Sara 10%. **Principal languages:** French (official), Sangho (national), Arabic, Hunsa, Swahili. **Chief religions:** Protestant 25%, Roman Catholic 25%, indigenous beliefs 24%, Muslim 15%.

Geography: Area: 240,500 sq. mi. **Location:** In central Africa. **Neighbors:** Chad on N, Cameroon on W, Congo-Brazzaville and Congo-Kinshasa (formerly Zaire) on S, Sudan on E. **Topography:** Mostly rolling plateau, average altitude 2,000 ft., with rivers draining S to the Congo and N to Lake Chad. Open, well-watered savanna covers most of the area, with an arid area in NE, and tropical rain forest in SW. **Capital:** Bangui (1995 est.): 553,000.

Government: Type: Republic. **Head of state:** Pres. Ange-Félix Patassé; b Jan. 25, 1937; in office: Oct. 22, 1993. **Head of gov.:** Prime Min. Anicet Georges Dologuele; in office: Feb. 1, 1999. **Local divisions:** 14 prefectures, 2 economic prefectures, 1 commune. **Defense:** 4.0% of GDP. **Active troops:** 2,700.

Economy: Industries: Textiles, breweries, sawmills, diamond mining. **Chief crops:** Cotton, coffee, corn, tobacco, yams. **Minerals:** Diamonds (chief export), uranium. **Other resources:** Timber. **Arable land:** 3%. **Livestock** (1997): chickens: 3.87 mil; cattle: 2.99 mil; goats: 2.34 mil; pigs: 622,100; sheep: 200,600. **Electricity prod.** (1997): 103 mil kWh.

Finance: Monetary unit: CFA Franc (Sept. 1999: 612.79 = $1 U.S.). **GDP** (1997 est.): $3.3 bil. **Per capita GDP:** $1,000. **Imports** (1995): $174 mil; partners: France 37%; Japan 24%. **Exports** (1995): $171 mil; partners: Belg.-Lux. 40%, France 16%. **Tourism:** $6 mil. **Budget** (1994 est.): $1.9 bil. **Intl. reserves less gold** (Mar. 1999): $142.34 mil. **Gold:** 11,000 oz t. **Consumer prices** (change in 1998): −1.9%.

Transport: Motor vehicles: 11,000 pass. cars, 9,000 comm. vehicles. **Civil aviation:** 150.5 mil pass.-mi.; 1 airport. **Chief port:** Bangui.

Communications: TV sets: 5 per 1,000 pop. **Radios:** 75 per 1,000 pop. **Telephones:** 9,600 main lines.

Health: Life expectancy: 45.35 male; 49.09 female. **Births** (per 1,000 pop.): 38.28. **Deaths** (per 1,000 pop.): 16.46. **Natural inc.:** 2.182%. **Hosp. beds** (1991): 1 per 672 persons. **Physicians** (1992): 1 per 18,660 persons. **Infant mortality** (per 1,000 live births): 103.42.

Education: Compulsory: ages 6-14. **Literacy:** 60%.

Major Intl. Organizations: UN (FAO, IBRD, ILO, IMF, WHO, WTrO), OAU.

Embassy: 1618 22d St. NW 20008; 462-2517.

Various Bantu tribes migrated through the region for centuries before French control was asserted in the late 19th century, when the region was named Ubangi-Shari. Complete independence was attained Aug. 13, 1960.

All political parties were dissolved in 1960, and the country became a center for Chinese political influence in Africa. Relations with China were severed after 1965. Pres. Jean-Bedel Bokassa, who seized power in a 1965 military coup, proclaimed himself constitutional emperor of the renamed Central African Empire Dec. 1976.

Bokassa's rule was characterized by ruthless and cruel authoritarianism and human rights violations. He was ousted in a bloodless coup aided by the French government, Sept. 20, 1979. In 1981, Gen. André Kolingba became head of state in another bloodless coup. Multiparty legislative and presidential elections were held in Oct. 1992 but were canceled by the government when Kolingba was losing.

New elections, held in Aug. and Sept. 1993, led to the replacement of Kolingba with a civilian government under Pres. Ange-Félix Patassé. France sent in troops to suppress army mutinies in 1996 and 1997. Patassé loyalists won a narrow majority in legislative elections on Nov. 22 and Dec. 13, 1998, and he was reelected to a 2d 6-year term on Sept. 19, 1999.

Chad
Republic of Chad

People: Population: 7,557,436. **Age distrib.** (%): <15: 44.2; 65+: 3.0. **Pop. density:** 15 per sq. mi. **Urban:** 23%. **Ethnic groups:** Sara 28%, Sudanic Arab 12%, many others. **Principal languages:** French, Arabic (both official), Sara, Sango, more than 100 other languages. **Chief religions:** Muslim 50%, Christian 25%, indigenous beliefs 25%.

Geography: Area: 496,000 sq. mi. **Location:** In central N Africa. **Neighbors:** Libya on N; Niger, Nigeria, Cameroon on W; Central African Republic on S; Sudan on E. **Topography:** Wooded savanna, steppe, and desert in the S; part of the Sahara in the N. Southern rivers flow N to Lake Chad, surrounded by marshland. **Capital:** N'Djamena 826,000*.

Government: Type: Republic. **Head of state:** Pres. Idriss Déby; b 1952; in office: Dec. 4, 1990. **Head of gov.:** Prime Min. Nassour Guelendouksia Ouaido; b 1947; in office: May 17, 1997. **Local divisions:** 14 prefectures. **Defense:** 4.1% of GDP. **Active troops:** 25,400.

Economy: Industries Cotton textiles, meat packing, beer brewing, soap. **Chief crops:** Cotton, sorghum, millet. **Minerals:** Uranium. **Arable land:** 3%. **Livestock** (1997): chickens: 4.70 mil; cattle: 5.58 mil; goats: 4.97 mil; sheep: 2.43 mil. **Fish catch** (1996): 100,000 metric tons. **Electricity prod.** (1997): 90 mil kWh. **Labor force:** 85% agric.

Finance: Monetary unit: CFA Franc (Sept. 1999: 612.79 = $1 U.S.). **GDP** (1997 est.): $4.3 bil. **Per capita GDP:** $600. **Imports** (1996 est.): $301 mil; partners: France 34%, Cameroon 24%. **Exports** (1996 est.): $259 mil; partners: Portugal 30%, Germany 18%. **Tourism:** $10 mil. **Budget** (1998 est.): $218 mil. **Intl. reserves less gold** (Mar. 1999): $99.55 mil. **Gold:** 11,000 oz t. **Consumer prices** (change in 1997): 5.6%.

Transport: Motor vehicles: 9,630 pass. cars, 14,360 comm. vehicles. **Civil aviation:** 153.3 mil pass.-mi.; 4 airports.

Communications: TV sets: 8 per 1,000 pop. **Radios:** 206 per 1,000 pop. **Telephones:** 8,600 main lines.

Health: Life expectancy: 46.13 male; 51.09 female. **Births** (per 1,000 pop.): 43.06. **Deaths** (per 1,000 pop.): 16.57. **Natural inc.:** 2.649%. **Hosp. beds** (1993): 1 per 1,521 persons. **Physicians** (1993): 1 per 27,765 persons. **Infant mortality** (per 1,000 live births): 115.27.

Education: Compulsory: ages 6-14. **Literacy:** 48%.

Major Intl. Organizations: UN (FAO, IBRD, ILO, IMF, WHO, WTrO), OAU.

Embassy: 2002 R St. NW 20009; 462-4009.

Chad was the site of paleolithic and neolithic cultures before the Sahara Desert formed. A succession of kingdoms and Arab slave traders dominated Chad until France took control around 1900. Independence came Aug. 11, 1960.

Northern Muslim rebels have fought animist and Christian southern government and French troops from 1966, despite numerous cease-fires and peace pacts.

Libyan troops entered the country at the request of a pro-Libyan Chad government, Dec. 1980. The troops were withdrawn from Chad in Nov. 1981. Rebel forces, led by Hissène Habré, captured the capital and forced Pres. Goukouni Oueddei to flee the country in June 1982.

In 1983, France sent some 3,000 troops to Chad to assist Pres. Habré in opposing Libyan-backed rebels. France and Libya agreed to a simultaneous withdrawal of troops from Chad in Sept. 1984, but Libyan forces remained in the north until Mar. 1987, when Chad forces drove them from their last major stronghold. In Dec. 1990, Habré was overthrown by a Libyan-supported insurgent group, the Patriotic Salvation Movement.

On Feb. 3, 1994, the World Court dismissed a long-standing territorial claim by Libya to the mineral-rich Aozou Strip, on the Libyan border. Libyan troops reportedly withdrew at the end of May. Following approval of a new constitution in March 1996, Chad's first multiparty presidential election was held in June and July. The U.S. Peace Corps withdrew from Chad in Apr. 1998 because of clashes between rebels and Chad government forces.

Chile
Republic of Chile

People: Population: 14,973,843. **Age distrib.** (%): <15: 27.9; 65+: 7.1. **Pop. density:** 51 per sq. mi. **Urban:** 84%. **Ethnic groups:** White and White-Amerindian 95%, Amerindian 3%. **Principal language:** Spanish (official). **Chief religions:** Roman Catholic 89%, Protestant 11%.

Geography: Area: 292,300 sq. mi. **Location:** Occupies western coast of S South America. **Neighbors:** Peru on N, Bolivia on NE, Argentina on E. **Topography:** Andes Mts. on E border incl. some of the world's highest peaks; on W is 2,650-mile Pacific coast. Width varies between 100 and 250 miles. In N is Atacama Desert, in center are agricultural regions, in S, forests and grazing lands. **Capital:** Santiago (1997 est.) 4,640,635.

Government: Type: Republic. **Head of state and gov.:** Pres. Eduardo Frei Ruiz-Tagle; b June 24, 1942; in office: Mar. 11, 1994. **Local divisions:** 13 regions. **Defense:** 2.8% of GDP. **Active troops:** 94,300.

Economy: Industries: Fish processing, wood products, iron, steel. **Chief crops:** Grain, grapes, fruits, beans, potatoes, sugar beets. **Minerals:** Copper (world's largest producer and exporter), molybdenum, nitrates, iron. **Crude oil reserves** (1999): 150 mil bbls. **Other resources:** Timber. **Arable land:** 5%. **Livestock** (1997): chickens: 70.00 mil; cattle: 3.76 mil; sheep: 3.75 mil; pigs: 1.77 mil; goats: 738,183. **Fish catch** (1997): 5.81 mil metric tons. **Electricity prod.** (1997): 32.331 bil kWh. **Labor force:** 38% serv.; 34% ind. & commerce; 19% agric., forestry, fishing.

Finance: Monetary unit: Peso (Sept. 1999: 489.17 = $1 U.S.). **GDP** (1997 est.): $168.5 bil. **Per capita GDP:** $11,600. **Imports** (1997): $18.2 bil; partners: U.S. 25%, EU 18%, Asia 16%. **Exports** (1997): $16.9 bil; partners: Asia 34%, EU 25%, U.S. 15%. **Tourism:** $991 mil. **Budget** (1996 est.): $17 bil. **Intl. reserves less gold** (May 1999): $16.27 bil. **Gold:** 1.22 mil oz. t. **Consumer prices** (change in 1998): 5.1%.

Transport: Railroad: Length: 4,084 mi. **Motor vehicles:** 900,000 pass. cars, 475,000 comm. vehicles. **Civil aviation:** 5.3 bil pass.-mi.; 23 airports. **Chief ports:** Valparaiso, Arica, Antofagasta.

Communications: TV sets: 280 per 1,000 pop. **Radios:** 305 per 1,000 pop. **Telephones** (1997): 2,630,000 main lines. **Daily newspaper circ.:** 101 per 1,000 pop.

Health: Life expectancy: 72.33 male; 78.75 female. **Births** (per 1,000 pop.): 17.81. **Deaths** (per 1,000 pop.): 5.53. **Natural inc.:** 1.228%. **Hosp. beds** (1994): 1 per 326 persons. **Physicians** (1994): 1 per 875 persons. **Infant mortality** (per 1,000 live births): 10.02.

Education: Free and compulsory, from age 6 or 7, for 8 years. **Literacy:** 95%.

Major Intl. Organizations: UN and all of its specialized agencies, APEC, OAS.

Embassy: 1732 Massachusetts Ave. NW 20036; 785-1746.

Website: http://www.segegob.cl/seg-ingl/index2i.html

Northern Chile was under Inca rule before the Spanish conquest, 1536-40. The southern Araucanian Indians resisted until the late 19th century. Independence was gained 1810-18, under José de San Martin and Bernardo O'Higgins; the latter, as supreme director 1817-23, sought social and economic reforms until deposed. Chile defeated Peru and Bolivia in 1836-39 and 1879-84, gaining mineral-rich northern land.

In 1970, Salvador Allende Gossens, a Marxist, became president with a third of the national vote. His government improved conditions for the poor, but illegal and violent actions by extremist supporters of the government, the regime's failure to attain majority support, and poorly planned socialist economic programs led to political and financial chaos.

A military junta seized power Sept. 11, 1973, and said Allende had killed himself. The junta, headed by Gen. Augusto Pinochet Ugarte, named a mostly military cabinet and announced plans to "exterminate Marxism." Repression continued during the 1980s with little sign of any political liberalization.

In a plebiscite held Oct. 5, 1988, voters rejected the incumbent president, Pinochet. He agreed to presidential elections. In Dec. 1989 voters elected a civilian president, although Pinochet continued to head the army until Mar. 10, 1998. In Mar. 1994 a Chilean human rights group estimated that human rights violations had claimed more than 3,100 lives during Pinochet's rule. At the request of Spanish authorities, Pinochet was arrested by British police in London Oct. 16, 1998; his extradition to Spain was approved by a British magistrate Oct. 8, 1999; an appeal by the ailing former dictator was pending.

Tierra del Fuego is the largest (18,800 sq. mi.) island in the archipelago of the same name at the southern tip of South America, an area of majestic mountains, tortuous channels, and high winds. It was visited 1520 by Magellan and named Land of Fire because of its many Indian bonfires. Part of the island is in Chile, part in Argentina. Punta Arenas, on a mainland peninsula, is a center of sheep raising and the world's southernmost city (pop. about 70,000); Puerto Williams is the southernmost settlement.

China
People's Republic of China
(Statistical data on China do not include Hong Kong.)

People: Population: 1,246,871,951. **Age distrib.** (%): <15: 25.5; 65+: 6.8. **Pop. density:** 337 per sq. mi. **Urban:** 29%. **Ethnic groups:** Han Chinese 92%, Tibetan, Mongol, Korean, Manchu, others. **Principal languages:** Mandarin (official), Yue, Wu, Haka, Xiang, Gan, Minbei, Minnan, others. **Chief religions:** Officially atheist; Buddhism, Taoism; some Muslims, Christians.

Geography: Area: 3,705,400 sq. mi. **Location:** Occupies most of the habitable mainland of E Asia. **Neighbors:** Mongolia on N; Russia on NE and NW; Afghanistan, Pakistan, Tajikistan, Kazakhstan on W; India, Nepal, Bhutan, Myanmar, Laos, Vietnam on S; North Korea on NE. **Topography:** Two-thirds of the vast territory is mountainous or desert; only one-tenth is cultivated. Rolling topography rises to high elevations in the N in the Daxinganlingshanmai separating Manchuria and Mongolia; the Tien Shan in Xinjiang; the Himalayan and Kunlunshanmai in the SW and in Tibet. Length is 1,860 mi. from N to S, width E to W is more than 2,000 mi. The eastern half of China is one of the world's best-watered lands. Three great river systems, the Chang (Yangtze), Huang (Yellow), and Xi, provide water for vast farmlands. **Capital:** Beijing. **Cities:** Shanghai 13,584,000; Beijing 11,299,000; Tianjin 9,415,000; Shenyang 5,116,000; Guangzhou 4,492,000*.

Government: Type: Communist Party-led state. **Head of state:** Pres. Jiang Zemin; b Aug. 17, 1926; in office: Mar. 27, 1993. **Head of gov.:** Premier Zhu Rongji; b Oct. 1, 1928; in office: Mar. 17, 1998. **Local divisions:** 22 provinces (not including Taiwan), 5 autonomous regions, and 4 municipalities, plus the special administrative regions of Hong Kong (as of July 1, 1997) and Macau (as of Dec. 20, 1999). **Defense:** 5.7% of GDP. **Active troops:** 2.840 mil.

Economy: Industries: Iron and steel, textiles and apparel, machine building, armaments, cement (world's leading producer of cotton cloth, cement, steel). **Chief crops:** Grain, rice, cotton, potatoes, tea. **Minerals:** Tungsten, antimony, coal, oil, mercury, iron, lead, manganese, molybdenum, tin. **Crude oil reserves** (1999): 24 bil bbls. **Other resources:** Hydropower. **Arable land:** 10%. **Livestock** (1997): chickens: 3.11 bil; cattle: 96.19 mil; pigs: 485.70 mil; goats: 137.72 mil; sheep: 118.15 mil; buffalo: 20.82 mil. **Fish catch** (1997): 15.72 mil metric tons. **Electricity prod.** (1997): 1,036.350 tril kWh. **Labor force:** 53% agric. & forestry; 26% ind. & commerce; 7% const. & mining.

Finance: Monetary unit: Renminbi (Yuan) (Sept. 1999: 8.28 = $1 U.S.). **GDP** (1997 est.): $4.25 tril. **Per capita GDP:** $3,460. **Imports** (1997): $142.4 bil; partners: Japan 22%, U.S. 12%, Taiwan 11%. **Exports** (1997): $182.7 bil; partners: Hong Kong 24%, Japan 19%, U.S. 17%. **Tourism:** $12.50 bil. **Budget** (1994): $13.7 bil deficit. **Intl. reserves less gold** (May 1999): $150.38 bil. **Gold:** 12.7 mil oz t. **Consumer prices** (change in 1998): −0.8%.

Transport: Railroad: Length: 45,319 mi. **Motor vehicles:** 4.7 mil pass. cars, 6.75 mil comm. vehicles. **Civil aviation:** 45.3 bil passenger-mi, 113 airports. **Chief ports:** Shanghai, Qinhuangdao, Dalian, Guangzhou (Canton).

Communications: TV sets: 189 per 1,000 pop. **Radios:** 177 per 1,000 pop. **Telephones:** 85,020,000 main lines. **Daily newspaper circ.:** 23 per 1,000 pop.

Health: Life expectancy: 68.57 male; 71.48 female. **Births** (per 1,000 pop.): 15.10. **Deaths** (per 1,000 pop.): 6.98. **Natural inc.:** 0.812%. **Hosp. beds** (1995): 1 per 384 persons. **Physicians** (1995): 1 per 628 persons. **Infant mortality** (per 1,000 live births): 43.31.

Education: Compulsory 7-17. **Literacy** (1996): 82%.

Major Intl. Organizations: UN (FAO, IBRD, ILO, IMF, IMO, WHO), APEC.

Embassy: 2300 Conn. Ave. NW 20008; 328-2500.

Website: http://www.china-embassy.org

Remains of various humanlike creatures who lived as early as several hundred thousand years ago have been found in many parts of China. Neolithic agricultural settlements dotted the Huang (Yellow) R. basin from about 5000 BC. Their language, religion, and art were the sources of later Chinese civilization.

Bronze metallurgy reached a peak and Chinese pictographic writing, similar to today's, was in use in the more developed culture of the Shang Dynasty (c. 1500 BC-c. 1000 BC), which ruled much of North China.

A succession of dynasties and interdynastic warring kingdoms ruled China for the next 3,000 years. They expanded Chinese political and cultural domination to the south and west, and developed a brilliant technologically and a culturally advanced society. Rule by foreigners (Mongols in the Yuan Dynasty, 1271-1368, and Manchus in the Ch'ing Dynasty, 1644-1911) did not alter the underlying culture.

A period of relative stagnation left China vulnerable to internal and external pressures in the 19th century. Rebellions left tens of millions dead, and Russia, Japan, Britain, and other powers exercised political and economic control in large parts of the country. China became a republic Jan. 1, 1912, following the Wuchang Uprising inspired by Dr. Sun Yat-sen, founder of the Kuomintang (Nationalist) party. By 1928, the Kuomintang, led by Chiang Kai-shek, succeeded in nominal reunification of China. About the same time, a bloody purge of Communists from the ranks of the Kuomintang fomented hostilities between the two groups that would continue for decades.

For a period of 50 years, 1894-1945, China was involved in conflicts with Japan. In 1895, China ceded Korea, Taiwan, and other areas. On Sept. 18, 1931, Japan seized the Northeastern Provinces (Manchuria) and set up a puppet state called Manchukuo. The border province of Jehol was cut off as a buffer state in 1933. Japan invaded China proper July 7, 1937. After its defeat in World War II, Japan gave up all seized land.

Following World War II, internal conflicts involving the Kuomintang, Communists, and other factions resumed. China came under domination of Communist armies, 1949-1950. The Kuomintang government moved to Taiwan, Dec. 8, 1949.

The Chinese People's Political Consultative Conference convened Sept. 21, 1949; The People's Republic of China was proclaimed in Beijing (Peking) Oct. 1, 1949, under Mao Zedong. China and the USSR signed a 30-year treaty of "friendship, alliance and mutual assistance," Feb. 15, 1950. The U.S. refused recognition of the new regime. On Nov. 26, 1950, the People's Republic sent armies into Korea against U.S. troops and forced a stalemate in the Korean War.

After an initial period of consolidation, 1949-52, industry, agriculture, and social and economic institutions were forcibly molded according to Maoist ideals. However, frequent drastic changes in policy and violent factionalism interfered with economic development. In 1957, Mao admitted an estimated 800,000 people had been executed 1949-54; opponents claimed much higher figures.

The Great Leap Forward, 1958-60, tried to force the pace of economic development through intensive labor on huge new rural communes, and through emphasis on ideological purity. The program caused resistance and was largely abandoned.

By the 1960s, relations with the USSR deteriorated, with disagreements on borders, ideology, and leadership of world Communism. The USSR canceled aid accords, and China, with Albania, launched anti-Soviet propaganda drives.

The Great Proletarian Cultural Revolution, 1965, was an attempt to oppose pragmatism and bureaucratic power and instruct a new generation in revolutionary principles. Massive purges took place. A program of forcibly relocating millions of urban teenagers into the countryside was launched. By 1968 the movement had run its course; many purged officials returned to office in subsequent years, and reforms that had placed ideology above expertise were gradually weakened.

On Oct. 25, 1971, the UN General Assembly ousted the Taiwan government from the UN and seated the People's Republic in its place. The U.S. had supported the mainland's admission but opposed Taiwan's expulsion.

U.S. Pres. Richard Nixon visited China Feb. 21-28, 1972, on invitation from Premier Zhou Enlai, ending years of antipathy between the 2 nations. China and the U.S. opened liaison offices in each other's capitals, May-June 1973. The U.S., Dec. 15, 1978, formally recognized the People's Republic of China as the sole legal government of China; diplomatic relations between the 2 nations were established, Jan. 1, 1979.

Mao died Sept. 9, 1976. By 1978, Vice Premier Deng Xiaoping had consolidated his power, succeeding Mao as "paramount leader" of China. The new ruling group modified Maoist policies in education, culture, and industry, and sought better ties with non-Communist countries. During this "reassessment" of Mao's policies his widow, Jiang Qing, and other "Gang of Four" leftists were convicted of "committing crimes during the 'Cultural Revolution,' " Jan. 25, 1981.

By the mid-1980s, China had enacted far-reaching econom-ic reforms, deemphasizing centralized planning and incorpo-rating market-oriented incentives. Some 100,000 students and workers staged a march in Beijing to demand political reforms, May 4, 1989. The demonstrations continued during a visit to Beijing by Soviet leader Mikhail Gorbachev May 15-18; it was the first Sino-Soviet summit since 1959. As the unrest spread, martial law was imposed, May 20. Troops entered Beijing, June 3-4, and crushed the pro-democracy protests, as tanks and armored personnel carriers rolled through Tiananmen Square. It is estimated that 5,000 died, 10,000 were injured, and hundreds of students and workers were arrested.

China had one of the world's fastest-growing economies in the 1990s. Although human rights violations have persisted, the U.S. has continued to renew China's most-favored-nation trading status. Deng died Feb. 19, 1997, leaving his chosen successor, Jiang Zemin, in firm control as president. Pres. Jiang paid a state visit to the U.S., Oct. 26-Nov. 3, and U.S. Pres. Clinton visited China, June 25-July 3, 1998. Floods in July and Aug. killed at least 3,000 people, left millions home-less, and caused an estimated $20 billion in property dam-age.

NATO bombs hit the Chinese embassy in Belgrade, Yugo-slavia, on May 7, 1999, killing 3 people and wounding 27; the U.S. agreed July 30 to pay $4.5 million to compensate victims and their families. The government banned a popular religious sect, the Falun Gong, July 22, after it staged the largest unau-thorized demonstrations in Beijing since 1989.

By agreement with Great Britain, Hong Kong reverted to Chinese sovereignty July 1, 1997. Portugal agreed to return Macau to China as of Dec. 20, 1999. China, which regardsTai-wan as a renegade province, reacted angrily in July 1999 when Taiwanese leaders used language implying that Taiwan should be treated as a sovereign state.

Manchuria. Home of the Manchus, rulers of China 1644-1911, Manchuria has accommodated millions of Chinese set-tlers in the 20th century. Under Japanese rule 1931-45, the area became industrialized. The region is divided into the 3 NE provinces of Heilongjiang, Jilin, and Liaoning.

Guangxi is in SE China, bounded on N by Guizhou and Hunan provinces, E and S by Guangdong, on SW by Vietnam, and on W by Yunnan. It produces rice in the river valleys and has valuable forest products.

Inner Mongolia was organized by the People's Republic in 1947. Its boundaries have undergone frequent changes, reaching its greatest extent in 1956 (and restored in 1979), with an area of 454,600 sq. mi., allegedly in order to dilute the minority Mongol population. Chinese settlers outnumber the Mongols more than 10 to 1. Pop. (1996 est.): 23.07 mil. Capi-tal: Hohhot.

Xinjiang, in Central Asia, is 635,900 sq. mi., pop. (1996 est.): 16.89 mil (75% Uygurs, a Turkic Muslim group, with a heavy Chinese increase in recent years). Capital: Urumqi. It is China's richest region in strategic minerals.

Tibet, 471,700 sq. mi., is a thinly populated region of high plateaus and massive mountains, the Himalayas on the S, the Kunluns on the N. High passes connect with India and Nepal; roads lead into China proper. Capital: Lhasa. Average altitude is 15,000 ft. Jiachan, 15,870 ft., is believed to be the highest inhabited town on earth. Agriculture is primitive. Pop. (1996 est.): 2.44 mil (of whom about 500,000 are Chinese). Another 4 million Tibetans form the majority of the population of vast adjacent areas that have long been incorporated into China.

China ruled all of Tibet from the 18th century, but indepen-dence came in 1911. China reasserted control in 1951, and a Communist government was installed in 1953, revising the theocratic Lamaist Buddhist rule. Serfdom was abolished, but all land remained collectivized.

A Tibetan uprising within China in 1956 spread to Tibet in 1959. The rebellion was crushed with Chinese troops, and Buddhism was almost totally suppressed. The Dalai Lama and 100,000 Tibetans fled to India.

Hong Kong

Hong Kong (Xianggang), located at the mouth of the Zhu Jiang (Pearl R.) in SE China, 90 mi. S of Canton (Guangzhou), was a British dependency from 1842 until July 1, 1997, when it became a Special Administrative Region of China. Its nucleus is Hong Kong Isl., 31 sq. mi., occupied by the British in 1841 and formally ceded to them in 1842, on which is located the seat of government. Opposite is Kowloon Peninsula, 3 sq. mi., and Stonecutters Isl., added to the territory in 1860. An addi-tional 355 sq. mi. known as the New Territories, a mainland area and islands, were leased from China, 1898, for 99 years. Total area 422 sq. mi.; pop. (1999 est.) 6.85 million, including fewer than 20,000 British.

Hong Kong is a major center for trade and banking. Per cap-ita GDP, $26,500 (1996 est.), is among the highest in the world. Principal industries are textiles and apparel; also tour-ism ($9.24 bil expenditures in 1997); electronics, shipbuilding, iron and steel, fishing, cement, and small manufactures. Hong Kong's spinning mills are among the best in the world.

Hong Kong harbor was long an important British naval sta-tion and one of the world's great transshipment ports. The col-ony was often a place of refuge for exiles from mainland China. It was occupied by Japan during World War II.

From 1949 to 1962 Hong Kong absorbed more than a million refugees fleeing Communist China. Starting in the 1950s, cheap labor led to a boom in light manufacturing, while liberal tax policies attracted foreign investment; Hong Kong became one of the wealthiest, most productive areas in the Far East. Poor living and working conditions and low wages for many led to political unrest in the 1960s, but legislation and public works programs raised the standard of living by the 1970s.

With the end of the 99-year lease on the New Territories drawing near, Britain and China signed an agreement, Dec. 19, 1984, under which all of Hong Kong was to be returned to China in 1997; under this agreement Hong Kong was to be al-lowed to keep its capitalist system for 50 years. In Dec. 1996, an electoral college appointed by China chose a shipping mag-nate, Tung Chee-hwa, to be Hong Kong's chief executive when it reverted to Chinese control.

The July 1 transfer of government was marked by an elabo-rate ceremony. In the immediate wake of the changeover, Hong Kong retained its street names and its currency, the Hong Kong dollar (but without the queen's picture). Official lan-guages remained Chinese (Cantonese dialect) and English. The Legislative Council was disbanded, and an appointed Pro-visional Legislature installed in its place. The new legislature imposed limits on opposition activities and sharply cut back the number of people eligible to vote in legislative elections; de-spite the restrictions, pro-democracy candidates did well in May 24, 1998, balloting.

Macau

Macau, area of 6 sq. mi., is an enclave, a peninsula and 2 small islands, at the mouth of the Xi (Pearl) R. in China. It was established as a Portuguese trading colony in 1557. In 1849, Portugal claimed sovereignty over the territory; this claim was accepted by China in an 1887 treaty. Portugal granted broad autonomy in 1976. In 1987, Portugal and China agreed Macau would revert to China Dec. 20, 1999. Macau, like Hong Kong, was guaranteed 50 years of noninterference in its way of life and capitalist system. Pop. (1999 est.): 437,312.

Colombia

Republic of Colombia

People: Population: 39,309,422. **Age distrib.** (%): <15: 33.0; 65+: 4.6. **Pop. density:** 89 per sq. mi. **Urban:** 73%. **Eth-nic groups:** Mestizo 58%, white 20%, mulatto 14%, black 4%. **Principal language:** Spanish (official). **Chief religion:** Roman Catholic 95%.

Geography: Area: 439,700 sq. mi. **Location:** At the NW corner of South America. **Neighbors:** Panama on NW, Ecua-dor and Peru on S, Brazil and Venezuela on E. **Topography:** Three ranges of Andes—Western, Central, and Eastern Cor-dilleras—run through the country from N to S. The eastern range consists mostly of high tablelands, densely populated. The Magdalena R. rises in the Andes, flows N to Caribbean, through a rich alluvial plain. Sparsely settled plains in E are drained by Orinoco and Amazon systems. **Capital:** Bogotá. (Full name: Santa Fe de Bogotá.) **Cities** (1997 est.): Bogotá 6,004,782; Cali 1,985,906; Medellin 1,970,691.

Government: Type: Republic. **Head of state and gov.:** Pres. Andrés Pastrana Arango; b Aug. 17, 1954; in office: Aug. 7, 1998. **Local divisions:** 32 departments, capital district of Bogota. **Defense:** 4.0% of GDP. **Active troops:** 146,300.

Economy: Industries: Textiles, food processing, clothing, cement, chemicals. **Chief crops:** Coffee, rice, bananas, oil-seed, corn, cotton, sugar, tobacco, coca. **Minerals:** Oil, gas, emeralds, gold, copper, coal, iron, nickel. **Crude oil reserves** (1999): 2.58 bil bbls. **Other resources:** Forest products, cut flowers. **Arable land:** 4%. **Livestock** (1997): chickens: 85.00 mil; cattle: 28.26 mil; pigs: 2.48 mil; sheep: 2.42 mil; goats: 915,000. **Fish catch** (1996): 129,661 metric tons. **Electricity prod.** (1997): 44.345 bil kWh.

Finance: Monetary unit: Peso (Sept. 1999: 1,939.50 = $1 U.S.). **GDP** (1997 est.): $231.1 bil. **Per capita GDP:** $6,200. **Imports** (1997 est.): $13.5 bil; partners: U.S. 36%, EC 18%. **Exports** (1997 est.): $11.4 bil; partners: U.S. 39%, EC 26%. **Tourism:** $955 mil. **Budget** (1996 est.): $30 bil. **Intl. re-**

serves less gold (May 1999): $8.27 bil. Gold: 357,000 oz t. Consumer prices (change in 1998): 20.7%.

Transport: Railroad: Length: 2,007 mi. **Motor vehicles:** 1.15 mil pass. cars, 550,000 comm. vehicles. **Civil aviation:** 4.3 bil pass.-mi.; 43 airports. **Chief ports:** Buenaventura, Barranquilla, Cartagena.

Communications: TV sets: 188 per 1,000 pop. **Radios:** 151 per 1,000 pop. **Telephones:** 6,451,500 main lines. **Daily newspaper circ.:** 55 per 1,000 pop.

Health: Life expectancy: 66.54 male; 74.54 female. **Births** (per 1,000 pop.): 24.45. **Deaths** (per 1,000 pop.): 5.59. **Natural inc.:** 1.886%. **Physicians** (1992): 1 per 1,078 persons. **Infant mortality** (per 1,000 live births): 24.30.

Education: Free and compulsory for 5 years between ages 6-12. **Literacy:** 91%.

Major Intl. Organizations: UN (FAO, IBRD, ILO, IMF, IMO, WHO, WTrO), OAS.

Embassy: 2118 Leroy Pl. NW 20008; 387-8338.

Spain subdued the local Indian kingdoms (Funza, Tunja) by the 1530s and ruled Colombia and neighboring areas as New Granada for 300 years. Independence was won by 1819. Venezuela and Ecuador broke away in 1829-30, and Panama withdrew in 1903.

Colombia is plagued by rural and urban violence. "La Violencia" of 1948-58 claimed 200,000 lives; since 1989, political violence has resulted in more than 35,000 deaths. Attempts at land and social reform and progress in industrialization have not reduced massive social problems.

The government's increased activity against local drug traffickers sparked a series of retaliation killings. On Aug. 18, 1989, Luis Carlos Galán, the ruling party's presidential hopeful for the 1990 election, was assassinated. In 1990, 2 other presidential candidates were assassinated, as drug traffickers carried on a campaign of intimidation.

Charges that Ernesto Samper Pizano's 1994 campaign received money from the Cali drug cartel engulfed his administration in scandal, although the legislature voted, June 12, 1996, not to impeach him. Andrés Pastrana Arango, son of former Pres. Misael Pastrana Borrero (in office 1970-74), won a presidential runoff election, June 21, 1998. An earthquake Jan. 25, 1999, in western Colombia killed at least 1,185 people and left 250,000 homeless.

Comoros
Federal Islamic Republic of the Comoros

People: Population: 562,723. **Age. distrib.** (%): <15: 42.7; 65+: 2.9. **Pop. density:** 703 per sq. mi. **Urban:** 31%. **Ethnic groups:** Antalote, Cafre, Makoa, Oimatsaha, Sakalava. **Principal languages:** Arabic, French, Comorian (all official). **Chief religions:** Sunni Muslim 86%, Roman Catholic 14%.

Geography: Area: 800 sq. mi. **Location:** 3 islands—Grande Comore (Njazidja), Anjouan (Nzwani), and Moheli (Mwali)—in the Mozambique Channel between NW Madagascar and SE Africa. **Neighbors:** Nearest are Mozambique on W, Madagascar on E. **Topography:** The islands are of volcanic origin, with an active volcano on Grande Comore. **Capital:** Moroni (1992 met. est.): 30,000.

Government: Type: In transition. **Head of state:** Pres. Azali Assoumani; in office: May 6, 1999. **Local divisions:** 3 main islands with 4 municipalities.

Economy: Industries: Perfume, textiles. **Chief crops:** Vanilla, copra, perfume essences, cloves. **Arable land:** 35%. **Livestock** (1997): chickens: 440,000; goats: 128,000. **Electricity prod.** (1997): 15 mil kWh. **Labor force:** 80% agric.

Finance: Monetary unit: Franc (Sept. 1999: 459.59 = $1 U.S.). **GDP** (1997 est.): $400 mil. **Per capita GDP:** $685. **Imports** (1996 est.): $70 mil; partners: France 60%. **Exports** (1996 est.): $11.4 mil; partners: France 54%, Germany 18%. **Tourism:** $26 mil. **Budget** (1995 est.): $71 mil. **Intl. reserves less gold** (June 1999): $35.62 mil.

Transport: Civil aviation: 2.1 mil pass.-mi.; 2 airports. **Chief ports:** Fomboni, Moroni, Mutsamudu.

Communications: Radios: 122 per 1,000 pop. **Telephones** (1997): 5,500 main lines.

Health: Life expectancy: 58.39 male; 63.38 female. **Births** (per 1,000 pop.): 40.29. **Deaths** (per 1,000 pop.): 9.23. **Natural inc.:** 3.106%. **Physicians** (1993): 1 per 6,600 persons. **Infant mortality** (per 1,000 live births): 81.63.

Education: Compulsory: ages 7-16. **Literacy:** 57%.

Major Intl. Organizations: UN (FAO, IBRD, ILO, IMF, WHO), AL, OAU.

Embassy: 336 E. 45th St., 2d Fl., New York, NY 10017; (212) 972-8010.

Website: http://www.ksu.edu/sasw/comoros/comoros.html

The islands were controlled by Muslim sultans until the French acquired them 1841-1909. They became a French overseas territory in 1947. A 1974 referendum favored independence, with only the Christian island of Mayotte preferring association with France. The French National Assembly decided to allow each of the islands to decide its own fate. The Comore Chamber of Deputies declared independence July 6, 1975, with Ahmed Abdallah as president. In a referendum in 1976, Mayotte voted to remain French.

A leftist regime that seized power from Abdallah in 1975 was deposed in a pro-French 1978 coup in which he regained the presidency. In Nov. 1989, Pres. Abdallah was assassinated; soon after, a multiparty system was instituted. A Sept. 1995 military coup, assisted by French mercenaries, ousted Pres. Said Mohamed Djohar. French troops invaded, Oct. 4, and forced coup leaders to surrender. Djohar returned from exile in Jan. 1996, and in Mar. a new presidential election was held. A hijacked Ethiopian Airlines Boeing 767 crashed offshore on Nov. 23, killing 123 of the 175 people on board.

Seeking to resume ties with France, Anjouan seceded from the Comoros, Aug. 3, 1997. Comorian troops were unable to put down the rebellion, which was joined by Moheli. Unrest on Grande Comore culminated in a military coup, Apr. 30, 1999.

Congo (formerly Zaire)
Democratic Republic of the Congo

(Congo, officially Democratic Republic of the Congo, is also known as Congo-Kinshasa. It should not be confused with Republic of the Congo, commonly called Congo Republic, and also known as Congo-Brazzaville.)

People: Population: 50,481,305. **Age distrib.** (%): <15: 42.5; 65+: 3.4. **Pop. density:** 56 per sq. mi. **Urban:** 29%. **Ethnic groups:** More than 200 tribes, mostly Bantu. **Principal languages:** French (official), more than 400 dialects. **Chief religions:** Roman Catholic 50%, Protestant 20%, Muslim 10%, Kimbanguist 10%.

Geography: Area: 905,600 sq. mi. **Location:** In central Africa. **Neighbors:** Congo-Brazzaville on W; Central African Republic, Sudan on N; Uganda, Rwanda, Burundi, Tanzania on E; Zambia, Angola on S. **Topography:** Congo includes the bulk of the Congo R. basin. The vast central region is a low-lying plateau covered by rain forest. Mountainous terraces in the W, savannas in the S and SE, grasslands toward the N, and the high Ruwenzori Mts. on the E surround the central region. A short strip of territory borders the Atlantic O. The Congo R. is 2,718 mi. long. **Capital:** Kinshasa. **Cities:** Kinshasa 4,241,000; Lubumbashi 810,000*.

Government: Type: Republic with strong presidential authority (in transition). **Head of state and gov.:** Pres. Laurent Kabila; b Nov. 27, 1939; in office: May 29, 1997. **Local divisions:** 10 provinces, 1 city. **Defense:** 5.3% of GDP. **Active troops:** 40,000.

Economy: Industries: Mining, consumer prods., food processing. **Chief crops:** Coffee, sugar, palm oil, rubber, tea. **Minerals:** Cobalt (65% of world reserves), copper, cadmium, oil, diamonds, gold, silver, tin, germanium, zinc, iron, manganese, uranium, radium. **Crude oil reserves** (1999): 187 mil bbls. **Other resources:** Timber. **Arable land:** 3%. **Livestock** (1997): chickens: 25.00 mil; goats: 4.09 mil; pigs: 1.17 mil; cattle: 1.00 mil; sheep: 1.02 mil. **Fish catch** (1996): 162,261 metric tons. **Electricity prod.** (1997): 6.450 bil kWh. **Labor force:** 65% agric.; 19% services; 16% industry.

Finance: Monetary unit: Congolese Franc (Sept. 1999: 4.50 = $1 U.S.). **GDP** (1996 est.): $18 bil. **Per capita GDP:** $400. **Imports** (1996 est.): $1.1 bil; partners: Belg.-Lux. 15%, U.S. 7%. **Exports** (1996 est.): $1.9 bil; partners: Belg.-Lux. 36%, U.S. 17%. **Tourism:** $2 mil. **Budget** (1996 est.): $244 mil. **Intl. reserves less gold** (Dec. 1996): $82.5 mil **Consumer prices** (change in 1997): 176%.

Transport: Railroad: Length: 3,162 mi. **Motor vehicles:** 330,000 pass. cars, 200,000 comm. vehicles. **Civil aviation:** 189.7 mil pass.-mi.; 12 airports. **Chief ports:** Matadi, Boma, Kinshasa.

Communications: Radios: 79.3 per 1,000 pop. **Telephones** (1997): 21,000 main lines. **Daily newspaper circ.:** 3 per 1,000 pop.

Health: Life expectancy: 47.28 male; 51.67 female. **Births** (per 1,000 pop.): 46.37. **Deaths** (per 1,000 pop.): 14.99. **Natural inc.:** 3.138%. **Infant mortality** (per 1,000 live births): 99.45.

Education: Compulsory: ages 6-12. **Literacy** 77%.

Major Intl. Organizations: UN and most of its specialized agencies, OAU.

Embassy: 1800 New Hampshire Ave. NW 20009; 234-7690.

The earliest inhabitants of Congo may have been the pygmies, followed by Bantus from the E and Nilotic tribes from the N. The large Bantu Bakongo kingdom ruled much of Congo and Angola when Portuguese explorers visited in the 15th century.

Leopold II, king of the Belgians, formed an international group to exploit the Congo region in 1876. In 1877 Henry M. Stanley explored the Congo, and in 1878 the king's group sent him back to organize the region and win over the native chiefs. The Conference of Berlin, 1884-85, organized the Congo Free State with Leopold as king and chief owner. Exploitation of native laborers on the rubber plantations caused international criticism and led to granting of a colonial charter, 1908; the colony became known as the Belgian Congo.

Belgian and Congolese leaders agreed Jan. 27, 1960, the Congo would become independent in June. In the first general elections, May 31, the National Congolese movement of Patrice Lumumba won 35 of 137 seats in the National Assembly. He was appointed premier June 21, and formed a coalition cabinet. The Republic of the Congo was proclaimed June 30.

Widespread violence caused Europeans and others to flee. The UN Security Council, Aug. 9, 1960, called on Belgium to withdraw its troops and sent a UN contingent. Pres. Joseph Kasavubu removed Lumumba as premier in Sept.; Lumumba was murdered in Feb. 1961.

The last UN troops left the Congo June 30, 1964, and Moise Tshombe became president.

On Sept. 7, 1964, leftist rebels set up a "People's Republic" in Stanleyville (now Kisangani). Tshombe hired foreign mercenaries and sought to rebuild the Congolese Army. In Nov. and Dec. 1964 rebels killed scores of white hostages and thousands of Congolese; Belgian paratroopers, dropped from U.S. transport planes, rescued hundreds. By July 1965 the rebels had lost their effectiveness.

In late 1965 Gen. Joseph D. Mobutu was named president. He later changed his name to Mobutu Sese Seko. The country became the Democratic Republic of the Congo (1966) and the Republic of Zaire (1971).

Economic decline and government corruption plagued Zaire in the 1980s and worsened in the 1990s. In 1990, Pres. Mobutu announced an end to a 20-year ban on multiparty politics. He sought to retain power despite mounting international pressure and internal opposition.

During 1994, Zaire was inundated with refugees from the massive ethnic bloodshed in Rwanda. Ethnic violence spread to E Zaire in 1996. In Oct. militant Hutus, who dominated in the refugee camps, fought against rebels (mostly Tutsis) in Zaire, precipitating intervention by government troops. As a result of the fighting, Rwandan refugees abandoned the camps; hundreds of thousands returned to Rwanda, while hundreds of thousands more were dispersed throughout E Zaire. The rebels, led by Gen. Laurent Kabila—a former Marxist and long-time opponent of Mobutu—gained momentum and began to move W across Zaire. As turmoil engulfed his nation, Mobutu stayed in W Europe for most of the last 4 months of 1996, receiving treatment for prostate cancer.

With Mobutu out of the country, the Zairean army put up little resistance; rebels were aided by several of Mobutu's enemies, notably Rwanda and Uganda. Mobutu returned to Zaire in March 1997, but attempts to negotiate with Kabila were ineffectual. On May 17, Kabila's troops entered Kinshasa and Mobutu went into exile. The country again assumed the name Democratic Republic of the Congo. Mobutu died Sept. 7 in Rabat, Morocco.

Kabila, who ruled by decree, alienated UN officials, international aid donors, and former allies. Rebels assisted by Rwanda and Uganda threatened Kinshasa in Aug. 1998, but the assault was turned back with help from Angola, Namibia, and Zimbabwe. Rebel groups agreed to a cease-fire on Aug. 31, 1999.

Congo Republic
Republic of the Congo

(Congo Republic, officially Republic of the Congo, is also known as Congo-Brazzaville. It should not be confused with Democratic Republic of the Congo [formerly Zaire], now commonly called Congo, and also known as Congo-Kinshasa.)

People: Population: 2,716,814. **Age distrib.** (%): <15: 48.2; 65+: 2.6. **Pop. density:** 21 per sq. mi. **Urban:** 59%. **Ethnic groups:** Kongo 48%, Sangha 20%, Teke 17%, M'Bochi 12%. **Principal languages:** French (official); Lingala, Kikongo, other African languages. **Chief religions:** Christian 50%, animist 48%, Muslim 2%.

Geography: Area: 132,000 sq. mi. **Location:** In W central Africa. **Neighbors:** Gabon and Cameroon on W, Central African Republic on N, Congo-Kinshasa (formerly Zaire) on E, Angola on SW. **Topography:** Much of the Congo is covered by thick forests. A coastal plain leads to the fertile Niari Valley. The center is a plateau; the Congo R. basin consists of flood plains in the lower and savanna in the upper portion. **Capital:** Brazzaville: 1,004,000*.

Government: Type: Republic. **Head of state:** Pres. Denis Sassou-Nguesso; b 1943; in office: Oct. 25, 1997. **Local divisions:** 10 regions, 6 communes. **Defense:** 2.5% of GDP. **Active troops:** 10,000.

Economy: Industries: Oil, wood products, brewing, cement. **Chief crops:** Cassava (90% of food output), rice, corn, sugar, cocoa, coffee. **Minerals:** Oil, potash, lead, copper, zinc. **Crude oil reserves** (1999): 1.5 bil bbls. **Livestock** (1997): chickens: 1.80 mil; goats: 280,000; sheep: 114,000. **Fish catch** (1996): 33,785 metric tons. **Electricity prod.** (1997): 453 mil kWh.

Finance: Monetary unit: CFA Franc (Sept. 1999: 612.79 = $1 U.S.). **GDP** (1996 est.): $5.25 bil. **Per capita GDP:** $2,000. **Imports** (1995): $670 mil; partners: France 31%, Netherlands 25%. **Exports** (1995): $1.2 bil; partners: Belg.-Lux. 24%, Taiwan 20%, U.S. 15%. **Tourism:** $3 mil. **Budget** (1997 est.): $970 mil. **Intl. reserves less gold** (Mar. 1999): $760,000 mil.

Transport: Railroad: Length: 494 mi. **Motor vehicles:** 26,000 pass. cars, 21,100 comm. vehicles. **Civil aviation:** 5 airports. **Chief ports:** Pointe-Noire, Brazzaville.

Communications: TV sets: 17 per 1,000 pop. **Radios:** 312 per 1,000 pop. **Telephones** (1997): 22,000 main lines. **Daily newspaper circ.:** 8 per 1,000 pop.

Health: Life expectancy: 45.42 male; 48.92 female. **Births** (per 1,000 pop.): 37.96. **Deaths** (per 1,000 pop.): 16.33. **Natural inc.:** 2.163%. **Infant mortality** (per 1,000 live births): 100.58.

Education: Compulsory: ages 6-16. **Literacy:** 75%.

Major Intl. Organizations: UN (FAO, IBRD, ILO, IMF, IMO, WHO), OAU.

Embassy: 4891 Colorado Ave. NW 20011; 726-5500.

Website: http://www.gksoft.com/govt/en/cg.html

The Loango Kingdom flourished in the 15th century, as did the Anzico Kingdom of the Batekes; by the late 17th century they had become weakened. By 1885, France established control of the region, then called the Middle Congo. Republic of the Congo gained independence Aug. 15, 1960.

After a 1963 coup sparked by trade unions, the country adopted a Marxist-Leninist stance, with the USSR and China vying for influence. France remained a dominant trade partner and source of technical assistance, however, and French-owned private enterprise retained a major economic role. In 1970, the country was renamed People's Republic of the Congo.

In 1990, Marxism was renounced and opposition parties legalized. In 1991 the country's name was changed back to Republic of the Congo, and a new constitution was approved. A democratically elected government came into office in 1992; one of its key problems was a resurgence of ethnic and regional hostilities. Factional fighting broke out in Brazzaville, June 5, 1997, and intensified during the summer, devastating the capital and forcing international aid workers to flee. Troops loyal to former Marxist dictator Denis Sassou-Nguesso took control of the city Oct. 15.

Costa Rica
Republic of Costa Rica

People: Population: 3,674,490. **Age distrib.** (%): <15: 33.1; 65+: 5.1. **Pop. density:** 187 per sq. mi. **Urban:** 50%. **Ethnic groups:** White and mestizo 96%. **Principal language:** Spanish (official). **Chief religion:** Roman Catholic 95%.

Geography: Area: 19,700 sq. mi. **Location:** In Central America. **Neighbors:** Nicaragua on N, Panama on S. **Topography:** Lowlands by the Caribbean are tropical. The interior plateau, with an altitude of about 4,000 ft., is temperate. **Capital:** San José (1996 est.): 324,011.

Government: Type: Republic. **Head of state and gov.:** Pres. Miguel Angel Rodríguez Echeverría; b Jan. 9, 1940; in office: May 8, 1998. **Local divisions:** 7 provinces. **Defense:** 0.7% of GDP. **Active troops:** 7,000 paramilitary (1996).

Economy: Industries: Food processing, textiles, construction materials, fertilizer, plastics. **Chief crops:** Coffee, bananas, sugar, rice, potatoes. **Other resources:** Fish, forests. **Arable land:** 6%. **Livestock** (1997): chickens: 17.00 mil; cattle: 1.53 mil; pigs: 280,000. **Fish catch** (1996): 24,203 metric tons. **Electricity prod.** (1997): 4.715 bil kWh. **Labor force:** 23% serv. & govt.; 21% agric., forestry, fishing; 19% trade; 16% mining & manuf.

Finance: Monetary unit: Colon (Sept. 1999: 290.73 = $1 U.S.). **GDP** (1997 est.): $19.6 bil. **Per capita GDP:** $5,500. **Imports** (1996): $3.4 bil; partners: U.S. 44%. **Exports** (1996): $2.9 bil; partners: U.S. 50%. **Tourism:** $730 mil. **Intl. reserves less gold** (June 1999): $1.44 bil. **Gold:** 2,000 oz t. **Consumer prices** (change in 1998): 11.7%.

Transport: Railroad: Length: 590 mi. **Motor vehicles:** 48,684 pass. cars, 70,308 comm. vehicles. **Civil aviation:** 1.2 bil pass.-mi.; 14 airports. **Chief ports:** Limon, Puntarenas, Golfito.

Communications: TV sets: 102 per 1,000 pop. **Radios:** 224 per 1,000 pop. **Telephones** (1997): 584,500 main lines. **Daily newspaper circ.:** 102 per 1,000 pop.

Health: Life expectancy: 73.60 male; 78.61 female. **Births** (per 1,000 pop.): 22.46. **Deaths** (per 1,000 pop.): 4.16. **Natural inc.:** 1.830%. **Hosp. beds** (1996): 1 per 566 persons. **Physicians** (1996): 1 per 763 persons. **Infant mortality** (per 1,000 live births): 12.89.

Education: Free, compulsory: ages 6-15. **Literacy:** 95%.

Major Intl. Organizations: UN (FAO, IBRD, ILO, IMF, IMO, WHO, WTrO), OAS.

Embassy: 2114 S St. NW 20008; 234-2945.

Guaymi Indians inhabited the area when Spaniards arrived, 1502. Independence came in 1821. Costa Rica seceded from the Central American Federation in 1838. Since the civil war of 1948-49, there has been little violent social conflict, and free political institutions have been preserved. During 1993 there was an unusual wave of kidnappings and hostage-taking, some of it related to the international cocaine trade.

Costa Rica, though still a largely agricultural country, has achieved a relatively high standard of living, and land ownership is widespread. Tourism is growing rapidly.

Côte d'Ivoire
Republic of Ivory Coast

People: Population: 15,818,068. **Age distrib.** (%): <15: 46.6; 65+: 2.2. **Pop. density:** 127 per sq. mi. **Urban:** 44%. **Ethnic groups:** Baoule 23%, Bete 18%, Senoufou 15%, Malinke 11%, Agni, foreign Africans. **Principal languages:** French (official), Dioula and other native dialects. **Chief religions:** Muslim 60%, indigenous beliefs 25%, Christian 12%.

Geography: Area: 124,500 sq. mi. **Location:** On S coast of W Africa. **Neighbors:** Liberia, Guinea on W; Mali, Burkina Faso on N; Ghana on E. **Topography:** Forests cover the W half of the country, and range from a coastal strip to halfway to the N on the E. A sparse inland plain leads to low mountains in NW. **Capital:** Yamoussoukro (official); Abidjan (de facto). **Cities:** Abidjan 2,793,000*.

Government: Type: Republic. **Head of state:** Henri Konan Bédié; b May 5, 1934; in office: Dec. 7, 1993. **Head of gov.:** Prime Min. Daniel Kablan Duncan; b June 30, 1943; in office: Dec. 15, 1993. **Local divisions:** 50 departments. **Defense:** 0.9% of GDP. **Active troops:** 8,400.

Economy: Industries: Food processing, wood products, vehicles, textiles. **Chief crops:** Coffee, cocoa, rubber, palm kernels. **Minerals:** Oil, diamonds, manganese. **Crude oil reserves** (1999): 100 mil bbls. **Other resources:** Timber. **Arable land:** 8%. **Livestock** (1997): chickens: 31.06 mil; sheep: 1.35 mil; cattle: 1.31 mil; goats: 1.05 mil; pigs: 271,130. **Fish catch** (1996): 70,650 metric tons. **Electricity prod.** (1997): 1.865 bil kWh. **Labor force:** 51% agric.; 12% manuf. & mining.

Finance: Monetary unit: CFA Franc (Sept. 1999: 612.79 = $1 U.S.). **GDP** (1997 est.): $25.8 bil. **Per capita GDP:** $1,700. **Imports** (1996): $3.2 bil; partners: France 32%, Nigeria 20%. **Exports** (1996): $4.2 bil; partners: France 18%. **Tourism:** $97 mil. **Budget** (1996 est.): $2.7 bil. **Intl. reserves less gold** (Dec. 1998): $855.5 mil. **Gold:** 45,000 oz t. **Consumer prices** (change in 1998): 4.7%.

Transport: Railroad: Length: 405 mi. **Motor vehicles:** 160,000 pass. cars, 95,000 comm. vehicles. **Civil aviation:** 187.7 mil pass.-mi.; 11 airports. **Chief ports:** Abidjan, Dabou, San-Pédro.

Communications: TV sets: 57 per 1,000 pop. **Radios:** 112 per 1,000 pop. **Telephones:** 170,000 main lines. **Daily newspaper circ.:** 14 per 1,000 pop.

Health: Life expectancy: 44.48 male; 47.67 female. **Births** (per 1,000 pop.): 41.76. **Deaths** (per 1,000 pop.): 16.17. **Natural inc.:** 2.559%. **Hosp. beds** (1993): 1 per 1,698 persons. **Infant mortality** (per 1,000 live births): 94.17.

Education: Free, compulsory: ages 7-13. **Literacy:** 40%.

Major Intl. Organizations: UN and all of its specialized agencies, OAU.

Embassy: 2424 Massachusetts Ave. NW 20008; 797-0300.

Website: http://lcweb2.loc.gov/frd/cs/citoc.html

A French protectorate from 1842, Côte d'Ivoire became independent in 1960. It is the most prosperous of all the tropical African nations, as a result of diversification of agriculture for export, close ties to France, and encouragement of foreign investment. About 20% of the population are workers from neighboring countries. Côte d'Ivoire officially changed its name from Ivory Coast in Oct. 1985.

Students and workers protested, Feb. 1990, demanding the ouster of longtime Pres. Félix Houphouët-Boigny and multiparty democracy. Côte d'Ivoire held its first multiparty presidential election Oct. 1990, and Houphouët-Boigny retained his office. He died Dec. 7, 1993. The National Assembly named a successor, Henri Konan Bédié, who was reelected Oct. 22, 1995.

Croatia
Republic of Croatia

People: Population: 4,676,865. **Age distrib.** (%): <15: 16.8; 65+: 15.1. **Pop. density:** 214 per sq. mi. **Urban:** 56%. **Ethnic groups:** Croat 78%, Serb 12%. **Principal language:** Serbo-Croatian (official) 96%. **Chief religions:** Catholic 77%, Orthodox 11%.

Geography: Area: 21,829 sq. mi. **Location:** SE Europe, on the Balkan Peninsula. **Neighbors:** Slovenia, Hungary on N; Bosnia and Herzegovina, Yugoslavia on E. **Topography:** Flat plains in NE; highlands, low mtns. along Adriatic coast. **Capital:** Zagreb: 981,000*.

Government: Type: Parliamentary democracy. **Head of state:** Pres. Franjo Tudjman; b May 14, 1922; in office: May 30, 1990. **Head of gov.:** Prime Min. Zlatko Matesa; b June 17, 1949; in office: Nov. 4, 1995. **Local divisions:** 21 counties. **Defense:** 5.7% of GDP. **Active troops:** 58,000.

Economy: Industries: Chemicals, plastics, machine tools, aluminum, steel, paper. **Chief crops:** Olives, wheat, corn, sugar beets, fruits. **Minerals:** Oil, bauxite, iron, coal. **Crude oil reserves** (1999): 99.2 mil bbls. **Arable land:** 21%. **Livestock** (1997): chickens: 9.96 mil; pigs: 1.17 mil; sheep: 427,000; cattle: 443,000. **Electricity prod.** (1997): 10.466 bil kWh. **Labor force:** 31% industry & mining.

Finance: Monetary unit: Kuna (Sept. 1999: 7.11 = $1 U.S.). **GDP** (1997 est.): $22.7 bil. **Per capita GDP:** $4,500. **Imports** (1997): $9.1 bil; partners: Germany 21%, Italy 19%. **Exports** (1997): $4.3 bil; partners: Germany 22%, Italy 21%, Slovenia 18%. **Tourism:** $2.74 bil. **Budget** (1997 est.): $6.3 bil. **Intl. reserves less gold** (June 1999): $2.61 bil. **Consumer prices** (change in 1998): 6.4%.

Transport: Railroad: Length: 1,676 mi. **Motor vehicles:** 698,000 pass. cars, 54,000 comm. vehicles. **Civil aviation:** 291.4 mil pass.-mi.; 5 airports. **Chief ports:** Rijeka, Split, Dubrovnik.

Communications: TV sets: 230 per 1,000 pop. **Radios:** 230 per 1,000 pop. **Telephones:** 1,558,000 main lines. **Daily newspaper circ.:** 575 per 1,000 pop.

Health: Life expectancy: 70.69 male; 77.52 female. **Births** (per 1,000 pop.): 10.34. **Deaths** (per 1,000 pop.): 11.14. **Natural inc.:** –0.080%. **Hosp. beds** (1994): 1 per 169 persons. **Physicians** (1994): 1 per 524 persons. **Infant mortality** (per 1,000 live births): 7.84.

Education: Free, compulsory: ages 7-15. **Literacy** (1993): 97%.

Major Intl. Organizations: UN (FAO, IBRD, ILO, IMF, IMO, WHO), OSCE.

Embassy: 2343 Massachusetts Ave. NW 20008; 588-5899.

From the 7th century the area was inhabited by Croats, a south Slavic people. It was formed into a kingdom under Tomislav in 924, and joined with Hungary in 1102. The Croats became westernized and separated from Slavs under Austro-Hungarian influence. The Croats retained autonomy under the Hungarian crown. Slavonia was taken by Turks in the 16th century; the northern part was restored by the Treaty of Karlowitz in 1699. Croatia helped Austria put down the Hungarian revolution 1848-49 and as a result was set up with Slavonia as the separate Austrian crownland of Croatia and Slavonia, which was reunited to Hungary as part of Ausgleich in 1867. It united with other Yugoslav areas to proclaim the Kingdom of Serbs, Croats, and Slovenes in 1918. At the reorganization of Yugoslavia in 1929, Croatia and Slavonia became Savska county, which in 1939 was united with Primorje county to form the county of Croatia. A nominally independent state between 1941 and 1945, it became a constituent republic in the 1946 constitution.

On June 25, 1991, Croatia declared independence from Yugoslavia. Fighting began between ethnic Serbs and Croats, with the former gaining control of about 30% of Croatian territory. A cease-fire was declared in Jan. 1992, but new hostilities broke out in 1993. A cease-fire with Serb rebels forming a self-declared republic of Krajina was agreed to Mar. 30, 1994. Croatian government troops recaptured most of the Serb-held territory Aug. 1995. Pres. Franjo Tudjman signed a peace accord with leaders of Bosnia and Serbia in Paris, Dec. 14. Tudjman won reelection June 15, 1997; international monitors called the vote "free but not fair." The last Serb-held enclave, E Slavonia, returned to Croatian control Jan. 15, 1998.

Cuba
Republic of Cuba

People: Population: 11,096,395. **Age distrib.** (%): <15: 21.7; 65+: 9.6. **Pop. density:** 259 per sq. mi. **Urban:** 76%. **Ethnic groups:** Mulatto 51%, white 37%, black 11%. **Principal language:** Spanish (official). **Chief religion:** Roman Catholic 85% prior to Castro.

Geography: Area: 42,800 sq. mi. **Location:** In the Caribbean, westernmost of West Indies. **Neighbors:** Bahamas and U.S. to N, Mexico to W, Jamaica to S, Haiti to E. **Topography:** The coastline is about 2,500 miles. The N coast is steep and rocky, the S coast low and marshy. Low hills and fertile valleys cover more than half the country. Sierra Maestra, in the E, is the highest of 3 mountain ranges. **Capital:** Havana (1995 est.) 2,184,990.

Government: Type: Communist state. **Head of state and gov.:** Pres. Fidel Castro Ruz; b Aug. 13, 1926; in office: Dec. 3, 1976 (formerly prime min. since Feb. 16, 1959). **Local divisions:** 14 provinces, 1 special municipality. **Defense:** 5.2% of GDP. **Active troops:** 60,000.

Economy: Industries: Oil, food and tobacco processing, sugar. **Chief crops:** Sugar, tobacco, rice, coffee, citrus. **Minerals:** Cobalt, nickel, iron, copper, manganese, salt. **Crude oil reserves** (1999): 283.5 mil bbls. **Other resources:** Timber. **Arable land:** 24%. **Livestock** (1997): chickens: 12.00 mil; cattle: 4.65 mil; goats: 118,000; pigs: 1.50 mil; sheep; 310,000. **Fish catch** (1996): 80,234 metric tons. **Electricity prod.** (1997): 13.158 bil kWh. **Labor force:** 30% services & govt.; 22% ind., 20% agric.; 11% commerce.

Finance: Monetary unit: Peso (Sept. 1999: 1.00 = $1 U.S.). **GDP** (1997 est.): $16.9 bil. **Per capita GDP:** $1,540. **Imports** (1997 est.): $3.2 bil; partners: Russia 12%, Spain 14%, Mexico 9%. **Exports** (1997 est.): $1.9 bil; partners: Russia 18%, Netherlands 14%, Canada 13%. **Tourism:** $1.39 bil.

Transport: Railroad: Length: 3,033 mi. **Motor vehicles:** 16,500 pass. cars, 30,000 comm. vehicles. **Civil aviation:** 2.2 bil pass.-mi.; 14 airports. **Chief ports:** Havana, Matanzas, Cienfuegos, Santiago de Cuba.

Communications: TV sets: 200 per 1,000 pop. **Radios:** 327 per 1,000 pop. **Telephones** (1997): 370,800 main lines. **Daily newspaper circ.:** 122 per 1,000 pop.

Health: Life expectancy: 73.41 male; 78.30 female. **Births** (per 1,000 pop.): 12.90. **Deaths** (per 1,000 pop.): 7.38. **Natural inc.:** 0.552%. **Hosp. beds** (1992): 1 per 134 persons. **Physicians:** 1 per 231 persons. **Infant mortality** (per 1,000 live births): 7.81.

Education: Free, compulsory: ages 6-11. **Literacy:** 96%.

Major Intl. Organizations: UN (FAO, ILO, IMO, WHO, WTrO).

Some 50,000 Indians lived in Cuba when it was reached by Columbus in 1492. Its name derives from the Indian Cubanacan. Except for British occupation of Havana, 1762-63, Cuba remained Spanish until 1898. A slave-based sugar plantation economy developed from the 18th century, aided by early mechanization of milling. Sugar remains the chief product and chief export despite government attempts to diversify.

A ten-year uprising ended in 1878 with guarantees of rights by Spain, which Spain failed to carry out. A full-scale movement under Jose Marti began Feb. 24, 1895.

The U.S. declared war on Spain in Apr. 1898, after the sinking of the USS *Maine* in Havana harbor, and defeated it in the Spanish-American War. Spain gave up all claims to Cuba. U.S. troops withdrew in 1902, but under 1903 and 1934 agreements, the U.S. leases a site at Guantánamo Bay in the SE as a naval base. U.S. and other foreign investments acquired a dominant role in the economy. In 1952, former Pres. Fulgencio Batista seized control and established a dictatorship, which grew increasingly harsh and corrupt. Fidel Castro assembled a rebel band in 1956; guerrilla fighting intensified in 1958. Batista fled Jan. 1, 1959, and in the resulting political vacuum Castro took power, becoming premier Feb. 16.

The government began a program of sweeping economic and social changes, without restoring promised liberties. Opponents were imprisoned, and some were executed. Some 700,000 Cubans emigrated in the first years after the Castro takeover, mostly to the U.S.

Cattle and tobacco lands were nationalized, while a system of cooperatives was instituted. By 1960 all banks and industrial companies had been nationalized, including over $1 billion worth of U.S.-owned properties, mostly without compensation.

Poor sugar crops resulted in farm collectivization, tight labor controls, and rationing, despite continued aid from the USSR and other Communist nations. A U.S.-imposed export embargo in 1962 severely damaged the economy.

In 1961, some 1,400 Cubans, trained and backed by the U.S. Central Intelligence Agency, unsuccessfully tried to invade and overthrow the regime. In the fall of 1962, the U.S. learned the USSR had brought nuclear missiles to Cuba. After an Oct. 22 warning from Pres. John F. Kennedy, the missiles were removed.

In 1977, Cuba and the U.S. signed agreements to exchange diplomats, without restoring full ties, and to regulate offshore fishing. In 1978 and 1980, the U.S. agreed to accept political prisoners released by Cuba, some of whom were criminals and mental patients. A 1987 agreement provided for 20,000 Cubans to emigrate to the U.S. each year; Cuba agreed to take back some 2,500 jailed in the U.S. since 1980.

In 1975-78, Cuba sent troops to aid one faction in the Angola civil war; the last Cuban troops were withdrawn by May 1991. Cuba's involvement in Central America, Africa, and the Caribbean contributed to poor relations with the U.S.

Cuba's economy, dependent on aid from other Communist countries, was severely shaken by the collapse of the Communist bloc in the late 1980s. Stiffer trade sanctions enacted by the U.S. in 1992 made things worse. Antigovernment demonstrations in Aug. 1994 prompted Castro to loosen emigration restrictions. A new U.S.-Cuba accord in Sept. ended the exodus of "boat people" after more than 30,000 had left Cuba. In another policy shift, the U.S. announced May 2, 1995, it would admit 20,000 Cuban refugees held at the Guantánamo base but would send further boat people back to Cuba.

The U.S. imposed additional sanctions after Cuba, Feb. 24, 1996, shot down 2 aircraft operated by an anti-Castro exile group based in Miami. Cuba blamed exile groups for bombings at Havana tourist hotels, July-Sept. 1997. Pope John Paul II visited Cuba, Jan. 21-25, 1998, denouncing U.S. trade sanctions while at the same time pressing Castro to release political prisoners and allow full political and religious freedom. U.S. restrictions on contact with Cuba were eased in 1999.

Cyprus
Republic of Cyprus

(Figures below marked with a # do not include Turkish-held area—Turkish Republic of Northern Cyprus.)

People: Population: 754,064. **Age distrib.** (%): <15: 24.0; 65+: 10.5. **Pop. density:** 209 per sq. mi. **Urban:** 55%. **Ethnic groups:** Greek 78%, Turkish 18%. **Principal languages:** Greek, Turkish (both official), English. **Chief religions:** Greek Orthodox 78%, Muslim 18%.

Geography: Area: 3,600 sq. mi. **Location:** In eastern Mediterranean Sea, off Turkish coast. **Neighbors:** Nearest are Turkey on N, Syria and Lebanon on E. **Topography:** Two mountain ranges run E-W, separated by a wide, fertile plain. **Capital:** Nicosia (1994 est.): 186,400#.

Government: Type: Republic. **Head of state and gov.:** Pres. Glafcos Clerides; b Apr. 24, 1919; in office: Mar. 1, 1993. **Local divisions:** 6 districts. **Defense:** 5.8% of GDP. **Active troop strength#:** 10,000.

Economy: Industries: Food, beverages, textiles. **Chief crops:** Barley, grapes, vegetables, citrus, potatoes, olives. **Minerals:** Copper, pyrites, asbestos. **Arable land:** 12%. **Livestock** (1997): chickens: 3.50 mil; pigs: 414,788; sheep: 262,000; goats: 251,000. **Electricity prod.** (1997): 2.605 bil. kWh. **Labor force#:** 62% serv., 25% ind., 13% agric.

Finance: Monetary unit: Pound (Sept. 1999: 1.81 = $2.13 U.S.). **GDP#** (1997 est.): $9.75 bil. **Per capita GDP#:** $15,000. **Imports#** (1996): $3.6 bil; partners: U.S. 18%, UK 12%, Italy 10%. **Exports#** (1996): $1.3 bil; partners: Russia 19%, Bulgaria 16%, UK 11%. **Tourism:** $1.67 bil. **Budget#** (1997 est.): $3.4 bil. **Intl. reserves less gold** (Mar. 1999): $1.16 bil. **Gold:** 463,000 oz t. **Consumer prices** (change in 1998): 2.2%.

Transport: Motor vehicles (1997)#: 234,976 pass. cars, 108,452 comm. vehicles. **Civil aviation#:** 1.7 bil pass.-mi.; 2 airports. **Chief ports:** Famagusta, Limassol.

Communications: Television sets#: 160 per 1000 pop. **Radios:** 287 per 1,000 pop. **Telephones#:** 404,700 main lines. **Daily newspaper circ.#:** 135 per 1,000 pop.

Health: Life expectancy: 74.91 male; 79.39 female. **Births** (per 1,000 pop.): 13.64. **Deaths** (per 1,000 pop.): 7.42. **Natural inc.:** 0.622%. **Hosp. beds#** (1995): 1 per 201 persons. **Physicians#** (1995): 1 per 667 persons. **Infant mortality** (per 1,000 live births): 7.68.

Education: Free, compulsory: ages $5^1/_2$-15. **Literacy** (1994): 95%.
Major Intl. Organizations: UN (FAO, IBRD, ILO, IMF, IMO, WHO, WTrO), the Commonwealth, OSCE.
Embassy: 2211 R St. NW 20008; 462-5772.

Agitation for enosis (union) with Greece increased after World War II, with the Turkish minority opposed, and broke into violence in 1955-56. In 1959, Britain, Greece, Turkey, and Cypriot leaders approved a plan for an independent republic, with constitutional guarantees for the Turkish minority and permanent division of offices on an ethnic basis. Greek and Turkish Communal Chambers dealt with religion, education, and other matters.
Archbishop Makarios III, formerly the leader of the enosis movement, was elected president, and full independence became final Aug. 16, 1960. Further communal strife led the United Nations to send a peacekeeping force in 1964; its mandate has been repeatedly renewed.
The Cypriot National Guard, led by officers from the army of Greece, seized the government July 15, 1974. On July 20, Turkey invaded the island; Greece mobilized its forces but did not intervene. A cease-fire was arranged but collapsed. By Aug. 16, Turkish forces had occupied the NE 40% of the island, despite the presence of UN peacekeeping forces.
Turkish Cypriots voted overwhelmingly, June 8, 1975, to form a separate Turkish Cypriot federated state. A president and assembly were elected in 1976. Some 200,000 Greeks have been expelled from the Turkish-controlled area, replaced by thousands of Turks, some from the mainland.

Turkish Republic of Northern Cyprus

A declaration of independence was announced by Turkish-Cypriot leader Rauf Denktash, Nov. 15, 1983. The state is not internationally recognized, although it does have trade relations with some countries. Area of TRNC: 1,295 sq mi.; pop. (1995 est.): 134,000, 99% Turkish; capital: Lefkosa (Nicosia).

Czech Republic

People: Population: 10,280,513. **Age distrib.** (%): <15: 16.9; 65+: 13.8. **Pop. density:** 338 per sq. mi. **Urban:** 66%.
Ethnic groups: Czech 94%, Slovak 3%. **Principal languages:** Czech (official), Slovak. **Chief religions:** Atheist 40%, Roman Catholic 39%, Protestant 5%, Orthodox 3%.
Geography: Area: 30,387 sq. mi. **Location:** In E central Europe. **Neighbors:** Poland on N, Germany on N and W, Austria on S, Slovakia on E and SE. **Topography:** Bohemia, in W, is a plateau surrounded by mountains; Moravia is hilly. **Capital:** Prague. **Cities** (1997 est.): Prague 1,200,458; Brno 385,866; Ostrava 323,177.
Government: Type: Republic. **Head of state:** Vaclav Havel; b Oct. 5, 1936; in office: Feb. 15, 1993. **Head of gov.:** Prime Min. Milos Zeman; b Sept. 28, 1944; in office: Jul. 17, 1998.
Local divisions: 8 regions. **Defense:** 2.2% of GDP. **Active troops:** 61,700.
Economy: Industries: Machinery, oil products, iron, glass, motor vehicles. **Chief crops:** Wheat, sugar beets, potatoes, hops, fruit. **Minerals:** Coal, kaolin. **Arable land:** 41%. **Crude oil reserves** (1999): 1.5 mil bbls. **Livestock:** (1997): chickens: 27.85; pigs: 4.0 mil; cattle: 1.69 mil; sheep: 120,900. **Electricity prod.** (1997): 60.775 bil kWh. **Labor force:** 33% ind.; 9% constr.; 7% agric.
Finance: Monetary unit: Koruna (Sept. 1999: 34.29 = $1 U.S.). **GDP** (1997 est.): $111.9 bil. **Per capita GDP:** $10,800. **Imports** (1996): $27.7 bil; partners: EU 61%, Slovakia 12%. **Exports** (1996): $21.7 bil; partners: EU 61%, Slovakia 14%. **Tourism:** $3.51 bil. **Budget** (1997 est.): $14.6 bil. **Intl. reserves less gold** (Mar. 1999): $11.87 bil. **Gold:** 292,000 mil oz t. **Consumer prices** (change in 1998): 10.7%.
Transport: Railroad: Length: 5,860 mi. **Motor vehicles:** 4.41 mil pass. cars, 514,589 comm. vehicles. **Civil aviation:** 1.5 bil pass.-mi.; 2 airports. **Chief ports:** Decin, Prague, Usti nad Labem.
Communications: TV sets: 407 per 1,000 pop. **Telephones:** 3,741,500 main lines. **Daily newspaper circ.:** 219 per 1,000 pop.
Health: Life expectancy: 70.8 male; 77.7 female. **Births** (per 1,000 pop.): 9.84. **Deaths** (per 1,000 pop.): 11. **Natural inc.:** −0.20. **Hosp. beds** (1995): 1 per 138 persons. **Physicians** (1995): 1 per 268 persons. **Infant mortality** (per 1,000 live births): 6.67.
Education: Compulsory: ages 6-15. **Literacy** (1998 est.): 99%.
Major Intl. Organizations: UN (FAO, IBRD, ILO, IMF, IMO, WHO, WTrO), NATO, OECD, OSCE.
Embassy: 3900 Spring of Freedom St. NW 20008; 274-9101.

Bohemia and Moravia were part of the Great Moravian Empire in the 9th century and later became part of the Holy Roman Empire. Under the kings of Bohemia, Prague in the 14th century was the cultural center of Central Europe. Bohemia and Hungary became part of Austria-Hungary.
In 1914-18 Thomas G. Masaryk and Eduard Benes formed a provisional government with the support of Slovak leaders including Milan Stefanik. They proclaimed the Republic of Czechoslovakia Oct. 28, 1918.

Czechoslovakia

By 1938 Nazi Germany had worked up disaffection among German-speaking citizens in Sudetenland and demanded its cession. British Prime Min. Neville Chamberlain, with the acquiescence of France, signed with Hitler at Munich, Sept. 30, 1938, an agreement to the cession, with a guarantee of peace by Hitler and Mussolini. Germany occupied Sudetenland Oct. 1-2.
Hitler on Mar. 15, 1939, dissolved Czechoslovakia, made protectorates of Bohemia and Moravia, and supported the autonomy of Slovakia, proclaimed independent Mar. 14, 1939.
Soviet troops with some Czechoslovak contingents entered eastern Czechoslovakia in 1944 and reached Prague in May 1945; Benes returned as president. In May 1946 elections, the Communist Party won 38% of the votes, and Benes accepted Klement Gottwald, a Communist, as prime minister.
In Feb. 1948, the Communists seized power in advance of scheduled elections. In May 1948 a new constitution was approved. Benes refused to sign it. On May 30 the voters were offered a one-slate ballot and the Communists won full control. Benes resigned June 7 and Gottwald became president. The country was renamed the Czechoslovak Socialist Republic. A harsh Stalinist period followed, with complete and violent suppression of all opposition.
In Jan. 1968 a liberalization movement spread nations explosively through Czechoslovakia. Antonin Novotny, long the Stalinist ruler was deposed as party leader and succeeded by Alexander Dubcek, a Slovak, who supported democratic reforms. On Mar. 22 Novotny resigned as president and was succeeded by Gen. Ludvik Svoboda. On Apr. 6, Prem. Joseph Lenart resigned and was succeeded by Oldrich Cernik, a reformer.
In July 1968 the USSR and 4 Warsaw Pact nations demanded an end to liberalization. On Aug. 20, the Soviet, Polish, East German, Hungarian, and Bulgarian armies invaded Czechoslovakia. Despite demonstrations and riots by students and workers, press censorship was imposed, liberal leaders were ousted from office and promises of loyalty to Soviet policies were made by some old-line Communist Party leaders.
On Apr. 17, 1969, Dubcek resigned as leader of the Communist Party and was succeeded by Gustav Husak. In Jan. 1970, Cernik was ousted. Censorship was tightened, and the Communist Party expelled a third of its members. In 1973, amnesty was offered to some of the 40,000 who fled the country after the 1968 invasion, but repressive policies continued.
More than 700 leading Czechoslovak intellectuals and former party leaders signed a human rights manifesto in 1977, called Charter 77, prompting a renewed crackdown by the regime.
The police crushed the largest antigovernment protests since 1968, when tens of thousands of demonstrators took to the streets of Prague, Nov. 17, 1989. As protesters demanded free elections, the Communist Party leadership resigned Nov. 24; millions went on strike Nov. 27.
On Dec. 10, 1989, the first cabinet in 41 years without a Communist majority took power; Vaclav Havel, playwright and human rights campaigner, was chosen president, Dec. 29. In Mar. 1990 the country was officially renamed the Czech and Slovak Federal Republic. Havel failed to win reelection July 3, 1992; his bid was blocked by a Slovak-led coalition.
Slovakia declared sovereignty, July 17. Czech and Slovak leaders agreed, July 23, on a basic plan for a peaceful division of Czechoslovakia into 2 independent states.

Czech Republic

Czechoslovakia split into 2 separate states—the Czech Republic and Slovakia—on Jan. 1, 1993. Havel was elected president of the Czech Republic on Jan. 26. Record floods in July 1997 caused more than $1.7 billion in damage. The country became a full member of NATO on Mar. 12, 1999.

Denmark
Kingdom of Denmark

People: Population: 5,356,845. **Age distrib.** (%): <15: 18.3; 65+: 14.9. **Pop. density:** 322 per sq. mi. **Urban:** 85%.
Ethnic groups: Scandinavian, Eskimo, Faroese, German. **Principal languages:** Danish (official), Faroese. **Chief religion:** Evangelical Lutheran 91%.

Geography: Area: 16,639 sq. mi. **Location:** In N Europe, separating the North and Baltic seas. **Neighbors:** Germany on S, Norway on NW, Sweden on NE. **Topography:** Denmark consists of the Jutland Peninsula and about 500 islands, 100 inhabited. The land is flat or gently rolling and is almost all in productive use. **Capital:** Copenhagen (1996 est.): 632,246.

Government: Type: Constitutional monarchy. **Head of state:** Queen Margrethe II; b Apr. 16, 1940; in office: Jan. 14, 1972. **Head of gov.:** Prime Min. Poul Nyrup Rasmussen; b June 15, 1943; in office: Jan. 25, 1993. **Local divisions:** 14 counties, 2 kommunes. **Defense:** 1.7% of GDP. **Active troops:** 32,900.

Economy: Industries: Food processing, machinery, textiles, furniture, electronics. **Chief crops:** Grains, potatoes, sugar beets. **Minerals:** Oil, gas, salt. **Crude oil reserves** (1999): 943.5 mil bbls. **Arable land:** 60%. **Livestock** (1997): chickens: 18.16 mil; cattle: 1.97 mil; pigs: 12.00 mil; sheep: 142,000. **Fish catch** (1997): 1.83 mil metric tons. **Electricity prod.** (1997): 39.219 bil kWh. **Labor force:** 70% serv. & govt.; 19% manuf. & mining; 6% constr.

Finance: Monetary unit: Danish Krone (Sept. 1999: 6.95 = $1 U.S.). **GDP** (1997 est.): $122.5 bil. **Per capita GDP:** $23,200. **Imports** (1996): $43.2 bil; partners: Germany 22%, Sweden 12%. **Exports** (1996): $48.8 bil; partners: Germany 23%, Sweden 10%. **Tourism:** $3.63 bil. **Budget** (1996 est.): $66.4 bil. **Intl. reserves less gold** (Dec. 1998): $15.26 bil. **Gold:** 2.0 mil oz t. **Consumer prices** (change in 1998): 1.8%.

Transport: Railroad: Length: 1,780 mi. **Motor vehicles** (1997): 1.79 mil pass. cars, 306,403 comm. vehicles. **Civil aviation:** 3.5 bil pass.-mi.; 13 airports. **Chief ports:** Copenhagen, Alborg, Arhus, Odense.

Communications: TV sets: 516 per 1,000 pop. **Radios:** 988 per 1,000 pop. **Telephones:** 3,465,000 main lines. **Daily newspaper circ.:** 308 per 1,000 pop.

Health: Life expectancy: 73.83 male; 79.33 female. **Births** (per 1,000 pop.): 11.57. **Deaths** (per 1,000 pop.): 10.97. **Natural inc.:** 0.060%. **Hosp. beds** (1994): 1 per 199 persons. **Physicians** (1994): 1 per 358 persons. **Infant mortality** (per 1,000 live births): 5.11.

Education: Compulsory: ages 7-15. **Literacy** (1998): 100%.

Major Intl. Organizations: UN and all of its specialized agencies, EU, NATO, OECD, OSCE.

Embassy: 3200 Whitehaven St. NW 20008; 234-4300.

Website: http://www.denmark.org

The origin of Copenhagen dates back to ancient times, when the fishing and trading place named Havn (port) grew up on a cluster of islets, but Bishop Absalon (1128-1201) is regarded as the actual founder of the city.

Danes formed a large component of the Viking raiders in the early Middle Ages. The Danish kingdom was a major power until the 17th century, when it lost its land in southern Sweden. Norway was separated in 1815, and Schleswig-Holstein in 1864. Northern Schleswig was returned in 1920.

Voters ratified the Maastricht Treaty, the basic document of the European Union, in May 1993, after rejecting it in 1992.

The **Faroe Islands** in the North Atlantic, about 300 mi. NW of the Shetlands, and 850 mi. from Denmark proper, 18 inhabited, have an area of 540 sq. mi. and pop. (1999 est.) of 41,059. They are an administrative division of Denmark, self-governing in most matters. Torshavn is the capital. Fish is a primary export (329, 736 metric tons in 1997).

Greenland (Kalaallit Nunaat)

Greenland, a huge island between the North Atlantic and the Polar Sea, is separated from the North American continent by Davis Strait and Baffin Bay. Its total area is 840,000 sq. mi., 84% of which is ice-capped. Most of the island is a lofty plateau 9,000 to 10,000 ft. in altitude. The average thickness of the cap is 1,000 ft. The population (1999 est.) is 59,827. Under the 1953 Danish constitution the colony became an integral part of the realm with representatives in the Folketing (Danish legislature). The Danish parliament, 1978, approved home rule for Greenland, effective May 1, 1979. With home rule, Greenlandic place names came into official use. The technically correct name for Greenland is now Kalaallit Nunaat; its capital is Nuuk, rather than Godthab. Fish is the principal export (115,940 metric tons in 1996).

Djibouti
Republic of Djibouti

People: Population: 447,439. **Age distrib.** (%): <15: 43.0; 65+: 2.7. **Pop. density:** 53 per sq. mi. **Urban:** 82%. **Ethnic groups:** Somali 60%, Afar 35%. **Principal languages:** French, Arabic (both official); Afar, Somali. **Chief religions:** Muslim 94%, Christian 6%.

Geography: Area: 8,500 sq. mi. **Location:** On E coast of Africa, separated from Arabian Peninsula by the strategically vital strait of Bab el-Mandeb. **Neighbors:** Ethiopia on W and SW, Eritrea on NW, Somalia on SE. **Topography:** The territory, divided into a low coastal plain, mountains behind, and an interior plateau, is arid, sandy, and desolate. The climate is generally hot and dry. **Capital:** Djibouti (1995): 383,000.

Government: Type: Republic. **Head of state:** Pres. Ismail Omar Guelleh; b 1947; in office: May 8, 1999. **Head of gov.:** Prem. Barkat Gourad Hamadou; b 1930; in office: Sept. 30, 1978. **Local divisions:** 5 districts. **Defense:** 5.0% of GDP. **Active troops:** 9,600.

Economy: Based on service activities. **Livestock** (1997): goats: 507,000; sheep: 470,000; cattle: 190,000. **Electricity prod.** (1997): 175 mil kWh.

Finance: Monetary unit: Djibouti Franc (Sept. 1999: 177.72 = $1 U.S.). **GDP** (1997 est.): $520 mil. **Per capita GDP:** $1,200. **Imports** (1996 est.): $200.5 mil; partners: France 15%, Ethiopia 11%. **Exports** (1996 est.): $39.6 mil; partners: Ethiopia 45%, Somalia 38%. **Tourism:** $4 mil. **Budget** (1997 est.): $175 mil. **Intl. reserves less gold** (June 1999): $61.90 mil.

Transport: Railroad: Length: 66 mi. **Motor vehicles:** 13,000 pass. cars, 3,000 comm. vehicles. **Civil avation:** 1 airport. **Chief port:** Djibouti.

Communications: TV sets: 43 per 1,000 pop. **Radios:** 80 per 1,000 pop. **Telephones:** 7,900 main lines.

Health: Life expectancy: 49.48 male; 53.67 female. **Births** (per 1,000 pop.): 41.23. **Deaths** (per 1,000 pop.): 14.41. **Natural inc.:** 2.682%. **Infant mortality** (per 1,000 live births): 100.24.

Education: Literacy: 46%.

Major Intl. Organizations: UN (FAO, IBRD, ILO, IMF, IMO, WHO, WTrO), AL, OAU.

Embassy: Suite 515, 1156 15th St. NW 20005; 331-0270.

France gained control of the territory in stages between 1862 and 1900. As French Somaliland it became an overseas territory of France in 1945; in 1967 it was renamed the French Territory of the Afars and the Issas.

Ethiopia and Somalia have renounced their claims to the area, but each has accused the other of trying to gain control. There were clashes between Afars (ethnically related to Ethiopians) and Issas (related to Somalis) in 1976. Immigrants from both countries continued to enter the country up to independence, which came June 27, 1977.

French aid is the mainstay of the economy, as well as assistance from Arab countries. A peace accord Dec. 1994 ended a 3-year-long uprising by Afar rebels.

Dominica
Commonwealth of Dominica

People: Population: 64,881. **Age distrib.:** (%): <15: 26.5; 65+: 9.7. **Pop. density:** 216 per sq. mi. **Urban:** 70%. **Ethnic groups:** Black, Carib Amerindian. **Principal languages:** English (official), French patois. **Chief religions:** Roman Catholic 77%, Protestant 15%.

Geography: Area: 300 sq. mi. **Location:** In Eastern Caribbean, most northerly Windward Isl. **Neighbors:** Guadeloupe to N, Martinique to S. **Topography:** Mountainous, a central ridge running from N to S, terminating in cliffs; volcanic in origin, with numerous thermal springs; rich deep topsoil on leeward side, red tropical clay on windward coast. **Capital:** Roseau (1991 est.): 15,900.

Government: Type: Parliamentary democracy. **Head of state:** Pres. Vernon Lorden Shaw; in office: Oct. 6, 1998. **Head of gov.:** Prime Min. Edison Chenfil James; b Oct. 18, 1943; in office: June 14, 1995. **Local divisions:** 10 parishes.

Economy: Industries: Soap, tourism. **Chief crops:** Bananas, citrus, mangoes, coconuts. **Other resources:** Forests. **Arable land:** 9%. **Livestock** (1997): chickens: 190,000. **Electricity prod.** (1997): 40 mil kWh. **Labor force:** 31% agric., 28% ind. & commerce; 18% services.

Finance: Monetary unit: East Caribbean Dollar (Sept. 1999: 2.70 = $1 U.S.). **GDP** (1996 est.): $208 mil. **Per capita GDP:** $2,500. **Imports** (1996): $98.1 mil; partners: U.S. 41%. **Exports** (1996): $51.8 mil; partners: UK 36%. **Tourism:** $34 mil. **Budget** (FY 1995-96): $78 mil. **Intl. reserves less gold** (Apr. 1999): $28.28 mil. **Consumer prices** (change in 1998): 1.0%.

Transport: Motor vehicles (1997): 7,560 pass. cars, 3,673 comm. vehicles. **Civil aviation:** 2 airports. **Chief port:** Roseau.

Communications: TV sets: 70 per 1,000 pop. **Radios:** 875 per 1,000 pop. **Telephones** (1997): 18,700 main lines.

Health: Life expectancy: 75.15 male; 81.01 female. **Births** (per 1,000 pop.): 16.92. **Deaths** (per 1,000 pop.): 6.35. **Natural inc.:** 1.057%. **Hosp. beds** (1992): 1 per 298 persons. **Physicians** (1993): 1 per 2,952 persons. **Infant mortality** (per 1,000 live births): 8.75.

Education: Free, compulsory: ages 5-15. **Literacy** (1993): 90%.
Major Intl. Organizations: UN (FAO, IBRD, ILO, IMO, WHO, WTrO), Caricom, the Commonwealth, OAS, OECS.
Embassy: 3216 New Mexico Ave. NW 20016; 364-6781

A British colony since 1805, Dominica was granted self-government in 1967. Independence was achieved Nov. 3, 1978.

Hurricane David struck, Aug. 30, 1979, devastating the island and destroying the banana plantations, Dominica's economic mainstay. Coups were attempted in 1980 and 1981.

Dominica participated in the 1983 U.S.-led invasion of nearby Grenada.

Dominican Republic

People: Population: 8,129,734. **Age distrib.** (%): <15: 34.9; 65+: 4.4. **Pop. density:** 432 per sq. mi. **Urban:** 63%. **Ethnic groups:** Mixed 73%, white 16%, black 11%. **Principal language:** Spanish (official). **Chief religion:** Roman Catholic 95%.
Geography: Area: 18,800 sq. mi. **Location:** In West Indies, sharing isl. of Hispaniola with Haiti. **Neighbors:** Haiti on W, Puerto Rico (U.S.) to E. **Topography:** The Cordillera Central range crosses the center of the country, rising to over 10,000 ft., highest in the Caribbean. The Cibao Valley to the N is major agricultural area. **Capital:** Santo Domingo. **Cities:** Santo Domingo 3,166,000; Santiago de los Caballeros 1,289,000*.
Government: Type: Republic. **Head of state and gov.:** Pres. Leonel Fernández Reyna; b Dec. 26, 1953; in office: Aug. 16, 1996. **Local divisions:** 29 provinces and national district. **Defense:** 1.2% of GDP. **Active troops:** 24,500.
Economy: Industries: Sugar refining, cement, tourism. **Chief crops:** Sugar, cocoa, coffee, cotton, rice. **Minerals:** Nickel, bauxite, gold, silver. **Arable land:** 21%. **Livestock** (1997): chickens: 37.70 mil; cattle: 2.53 mil; pigs: 960,000; goats: 570,000; sheep: 135,000. **Electricity prod.** (1997): 6.989 bil kWh. **Labor force:** 47% serv. & govt.; 13% agric.
Finance: Monetary unit: Peso (Sept. 1999: 15.95 = $1 U.S.). **GDP** (1997 est.): $38.3 bil. **Per capita GDP:** $4,700. **Imports** (1996): $3.7 bil; partners: U.S. 44% EU 16%. **Exports** (1996): $815 mil; partners: U.S. 45%, EU 34%. **Tourism:** $2.15 bil. **Budget** (1996 est.): $2 bil. **Intl. reserves less gold** (June 1999): $536.7 mil. **Gold:** 18,000 oz t. **Consumer prices** (change in 1997): 8.3%.
Transport: Railroad: Length: 1,083 mi. **Motor vehicles:** 113,835 pass. cars, 92,198 comm. vehicles. **Civil aviation:** 9.8 mil pass.-mi.; 7 airports. **Chief ports:** Santo Domingo, San Pedro de Macoris, Puerto Plata.
Communications: TV sets: 97 per 1,000 pop. **Radios:** 154 per 1,000 pop. **Telephones** (1997): 709,200 main lines. **Daily newspaper circ.:** 35 per 1,000 pop.
Health: Life expectancy: 67.86 male; 72.40 female. **Births** (per 1,000 pop.): 25.97. **Deaths** (per 1,000 pop.): 5.66. **Natural inc.:** 2.031%. **Hosp. beds** (1994): 1 per 858 persons. **Physicians** (1994): 1 per 1,076 persons. **Infant mortality** (per 1,000 live births): 42.52.
Education: Compulsory: ages 6-14. **Literacy:** 82%.
Major Intl. Organizations: UN (FAO, IBRD, ILO, IMF, IMO, WHO, WTrO), OAS.
Embassy: 1715 22d St. NW 20008; 332-6280.

Carib and Arawak Indians inhabited the island of Hispaniola when Columbus landed in 1492. The city of Santo Domingo, founded 1496, is the oldest settlement by Europeans in the hemisphere and has the supposed ashes of Columbus in an elaborate tomb in its ancient cathedral.

The western third of the island was ceded to France in 1697. Santo Domingo itself was ceded to France in 1795. Haitian leader Toussaint L'Ouverture seized it, 1801. Spain returned intermittently 1803-21, as several native republics came and went. Haiti ruled again, 1822-44; Spanish occupation occurred 1861-63.

The country was occupied by U.S. Marines from 1916 to 1924, when a constitutionally elected government was installed.

In 1930, Gen. Rafael Leonidas Trujillo Molina was elected president. Trujillo ruled brutally until his assassination in 1961. Pres. Joaquín Balaguer, appointed by Trujillo in 1960, resigned under pressure in 1962.

Juan Bosch, elected president in the first free elections in 38 years, was overthrown in 1963. On Apr. 24, 1965, a revolt was launched by followers of Bosch and others, including a few Communists. Four days later U.S. Marines intervened against pro-Bosch forces. Token units were later sent by 5 South American countries as a peacekeeping force. A provisional government supervised a June 1966 election, in which Balaguer defeated Bosch. Balaguer remained in office for most of the next 28 years, but his May 1994 reelection was widely de-

nounced as fraudulent. He cut short his term and on June 30, 1996, Leonel Fernández Reyna was elected.

Hurricane Georges struck Sept. 22, 1998, causing extensive property damage and claiming more than 200 lives.

Ecuador
Republic of Ecuador

People: Population: 12,562,496. **Age distrib.** (%): <15: 35.2; 65+: 4.5. **Pop. density:** 115 per sq. mi. **Urban:** 60%. **Ethnic groups:** Mestizo 55%, Amerindian 25%, Spanish 10%, black 10%. **Principal languages:** Spanish (official), Quechua, other Amerindian. **Chief religion:** Roman Catholic 95%.
Geography: Area: 109,500 sq. mi. **Location:** In NW South America, on Pacific coast, astride the Equator. **Neighbors:** Colombia on N, Peru on E and S. **Topography:** Two ranges of Andes run N and S, splitting the country into 3 zones: hot, humid lowlands on the coast; temperate highlands between the ranges; and rainy, tropical lowlands to the E. **Capital:** Quito. **Cities:** Guayaquil (1997 est.): 1,973,880; Quito (1996 est.): 1,444,363.
Government: Type: Republic. **Head of state and gov.:** Pres. Jamil Mahuad Witt; b July 29, 1949; in office: Aug. 10, 1998. **Local divisions:** 21 provinces. **Defense:** 3.5% of GDP. **Active troops:** 57,100.
Economy: Industries: Oil, food processing, metalwork, textiles. **Chief crops:** Bananas, cocoa, coffee, rice, sugar, potatoes, plantains. **Minerals:** Oil. **Crude oil reserves** (1999): 2.1 bil bbls. **Other resources:** Forests (leading balsawood producer), seafood (world's 2d lgst. shrimp producer.). **Arable land:** 6%. **Livestock** (1997): chickens: 64.73 mil; cattle: 5.33 mil; pigs: 2.80 mil; sheep: 2.06 mil; goats: 310,000. **Fish catch** (1997): 553,000 metric tons. **Electricity prod.** (1997): 10.061 bil kWh.
Finance: Monetary unit: Sucre (Sept. 1999: 11,150.00 = $1 U.S.). **GDP** (1997 est.): $53.4 bil. **Per capita GDP:** $4,400. **Imports** (1997): $2.9 bil; partners: Latin America 35%, U.S. 32%. **Exports** (1997): $3.4 bil; partners: U.S. 39%, Latin America 25%. **Tourism:** $285 mil. **Budget** (1996 est.): $3.6 bil. **Intl. reserves less gold** (June 1999): $1.58 bil. **Gold:** 414,000 oz t. **Consumer prices** (change in 1998): 36.1%.
Transport: Railroad: Length: 600 mi. **Motor vehicles:** 255,640 pass. cars, 424,120 comm. vehicles. **Civil aviation:** 1.3 bil pass.-mi.; 14 airports. **Chief ports:** Guayaquil, Manta, Esmeraldas, Puerto Bolivar.
Communications: TV sets: 79 per 1,000 pop. **Radios:** 277 per 1,000 pop. **Telephones:** 1,009,800 main lines. **Daily newspaper circ.:** 72 per 1,000 pop.
Health: Life expectancy: 69.54 male; 74.90 female. **Births** (per 1,000 pop.): 22.26. **Deaths** (per 1,000 pop.): 5.06. **Natural inc.:** 1.720%. **Hosp. beds** (1992): 1 per 623 persons. **Physicians** (1993): 1 per 904 persons. **Infant mortality** (per 1,000 live births): 30.69.
Education: Free and compulsory for 6 years between ages 6-14. **Literacy:** 90%.
Major Intl. Organizations: UN (FAO, IBRD, ILO, IMF, IMO, WHO, WTrO), OAS.
Embassy: 2535 15th St. NW 20009; 234-7200.

The region, which was the northern Inca empire, was conquered by Spain in 1533. Liberation forces defeated the Spanish May 24, 1822, near Quito. Ecuador became part of the Great Colombia Republic but seceded, May 13, 1830.

Since 1972, the economy has revolved around petroleum exports; oil revenues have declined since 1982, causing severe economic problems. Ecuador suspended interest payments for 1987 on its estimated $8.2 billion foreign debt following a Mar. 5-6 earthquake that left 20,000 homeless and destroyed a stretch of the country's main oil pipeline.

Ecuadoran Indians staged protests in the 1990s to demand greater rights. A border war with Peru flared from Jan. 26, 1995, until a truce took effect Mar. 1. Vice-Pres. Alberto Dahik resigned and fled Ecuador, Oct. 11, 1995, to avoid arrest on corruption charges. Elected president in a runoff, July 7, 1996, Abdalá Bucaram—a populist known as El Loco, or "The Crazy One"—imposed stiff price increases and other austerity measures. His rising unpopularity and erratic behavior led the National Congress, Feb. 6, 1997, to dismiss him for "mental incapacity." Bucaram went into exile, and Congress, on Feb. 11, confirmed its leader, Fabián Alarcón, as president for 18 months. Voters endorsed the actions in a referendum May 25.

Jamil Mahuad Witt, mayor of Quito, won a presidential runoff election July 12, 1998. In Sept. 1998 and Mar. 1999 he imposed emergency measures to cope with a continuing economic crisis.

The **Galapagos Islands**, pop. (1996 est.) 14,000, about 600 mi. to the W, are the home of huge tortoises and other unusual animals.

Egypt
Arab Republic of Egypt

People: Population: 67,273,906. **Age distrib** (%) <15: 35.6; 65+: 3.7. **Pop. density:** 174 per sq. mi. **Urban:** 45%. **Ethnic groups:** Eastern Hamitic stock (Egyptian, Bedouin, Berber) 99%. **Principal languages:** Arabic (official), English, French. **Chief religions:** Muslim (mostly Sunni) 94%, Coptic Christian and other 6%.

Geography: Area: 386,700 sq. mi. **Location:** Northeast corner of Africa. **Neighbors:** Libya on W, Sudan on S, Israel and Gaza Strip on E. **Topography:** Almost entirely desolate and barren, with hills and mountains in E and along Nile. The Nile Valley, where most of the people live, stretches 550 miles. **Capital:** Cairo. **Cities** (1997 est.): Cairo 6,789,479; Alexandria 3,328,196.

Government: Type: Republic. **Head of state:** Pres. Hosni Mubarak; b May 4, 1928; in office: Oct. 14, 1981. **Head of gov.:** Prime Min. Atef Obeid; b Apr. 14, 1932; in office: Oct. 5, 1999. **Local divisions:** 26 governorates. **Defense:** 4.3% of GDP. **Active troops:** 450,000.

Economy: Industries: Textiles, tourism, chemicals, oil, food processing, cement. **Chief crops:** Cotton, rice, beans, fruits, wheat, vegetables, corn. **Minerals:** Oil, gas, phosphates, gypsum, iron, manganese, limestone. **Crude oil reserves** (1999): 3.5 bil bbls. **Arable land:** 2%. **Livestock** (1997): chickens: 86.00 mil; sheep: 4.30 mil; goats: 3.20 mil; buffalo: 3.15 mil; cattle 3.02 mil. **Fish catch** (1997): 345,240 metric tons. **Electricity prod.** (1997): 54.760 bil kWh. **Labor force:** 44% serv. & gov't; 35% agric.; 22% mining, manuf., const.

Finance: Monetary unit: Pound (Sept. 1999: 3.41 = $1 U.S.). **GDP** (1997 est.): $267.1 bil. **Per capita GDP:** $4,400. **Imports** (FY 1996-97 est.): $15.5 bil; partners: U.S. 19%, Germany 10%. **Exports** (FY 1996-97 est.): $5.1 bil; partners: Italy 19%, U.S. 11%. **Tourism:** $3.84 bil. **Budget** (FY 1996-97): $19.8 bil. **Intl. reserves less gold** (Apr. 1999): $17.29 bil. **Gold:** 2.43 mil oz t. **Consumer prices** (change in 1998): 4.2%.

Transport: Railroad: Length: 2,989 mi. **Motor vehicles:** 1.28 mil pass. cars, 423,300 comm. vehicles. **Civil aviation:** 5.6 bil pass.-mi.; 11 airports. **Chief ports:** Alexandria, Port Said, Suez, Damietta.

Communications: TV sets: 110 per 1,000 pop. **Radios:** 312 per 1,000 pop. **Telephones:** 3,971,500 main lines. **Daily newspaper circ.:** 43 per 1,000 pop.

Health: Life expectancy: 60.39 male; 64.49 female. **Births** (per 1,000 pop.): 26.80. **Deaths** (per 1,000 pop.): 8.27. **Natural inc.:** 1.853%. **Hosp. beds** (1994): 1 per 515 persons. **Physicians** (1996): 1 per 472 persons. **Infant mortality** (per 1,000 live births): 67.46.

Education: Compulsory for 5 years between ages 6-13. **Literacy:** 51%.

Major Intl. Organizations: UN (FAO, IBRD, ILO, IMF, IMO, WHO, WTrO), AL, OAU.

Embassy: 3521 International Ct. NW 20008; 895-5400.

Website: http://www.idsc.gov.eg

Archaeological records of ancient Egyptian civilization date back to 4000 BC. A unified kingdom arose around 3200 BC and extended its way south into Nubia and as far north as Syria. A high culture of rulers and priests was built on an economic base of serfdom, fertile soil, and annual flooding of the Nile.

Imperial decline facilitated conquest by Asian invaders (Hyksos, Assyrians). The last native dynasty fell in 341 BC to the Persians, who were in turn replaced by Greeks (Alexander and the Ptolemies), Romans, Byzantines, and Arabs, who introduced Islam and the Arabic language. The ancient Egyptian language is preserved only in Coptic Christian liturgy.

Egypt was ruled as part of larger Islamic empires for several centuries. The Mamluks, a military caste of Caucasian origin, ruled Egypt from 1250 until defeat by the Ottoman Turks in 1517. Under Turkish sultans the khedive as hereditary viceroy had wide authority. Britain intervened in 1882 and took control of administration, though nominal allegiance to the Ottoman Empire continued until 1914.

The country was a British protectorate from 1914 to 1922. A 1936 treaty strengthened Egyptian autonomy, but Britain retained bases in Egypt and a condominium over the Sudan. Britain fought German and Italian armies from Egypt, 1940-42. In 1951 Egypt abrogated the 1936 treaty; the Sudan became independent in 1956.

The uprising of July 23, 1952 was led by the Society of Free Officers, who named Maj. Gen. Mohammed Naguib commander in chief and forced King Farouk to abdicate. When the republic was proclaimed June 18, 1953, Naguib became its first president and premier. Lt. Col. Gamal Abdel Nasser removed

Naguib and became premier in 1954. In 1956, he was voted president. Nasser died in 1970 and was replaced by Vice Pres. Anwar Sadat.

The Aswan High Dam, completed 1971, provides irrigation for more than a million acres of land. Artesian wells, drilled in the Western Desert, reclaimed 43,000 acres, 1960-66.

When the state of Israel was proclaimed in 1948, Egypt joined other Arab nations invading Israel and was defeated.

After terrorist raids across its border, Israel invaded Egypt's Sinai Peninsula, Oct. 29, 1956. Egypt rejected a cease-fire demand by Britain and France; on Oct. 31 the 2 nations dropped bombs and on Nov. 5-6 landed forces. Egypt and Israel accepted a UN cease-fire; fighting ended Nov. 7.

A UN Emergency Force guarded the 117-mile-long border between Egypt and Israel until May 19, 1967, when it was withdrawn at Nasser's demand. Egyptian troops entered the Gaza Strip and the heights of Sharm el Sheikh and 3 days later closed the Strait of Tiran to all Israeli shipping. Full-scale war broke out June 5; before it ended under a UN cease-fire June 10, Israel had captured Gaza and the Sinai Peninsula, controlled the east bank of the Suez Canal, and reopened the gulf. After sporadic fighting, Israel and Egypt agreed, Aug. 7, 1970, to a new cease-fire.

In a surprise attack Oct. 6, 1973, Egyptian forces crossed the Suez Canal into the Sinai. (At the same time, Syrian forces attacked Israelis on the Golan Heights.) Egypt was supplied by a USSR military airlift; the U.S. responded with an airlift to Israel. Israel counterattacked, crossed the canal, surrounded Suez City. A UN cease-fire took effect Oct. 24.

Under an agreement signed Jan. 18, 1974, Israeli forces withdrew from the canal's W bank; limited numbers of Egyptian forces occupied a strip along the E bank. A second accord was signed in 1975, with Israel yielding Sinai oil fields. Pres. Sadat's surprise visit to Jerusalem, Nov. 1977, opened the prospect of peace with Israel. On Mar. 26, 1979, Egypt and Israel signed a formal peace treaty, ending 30 years of war, and establishing diplomatic relations. Israel returned control of the Sinai to Egypt in Apr. 1982.

Tension between Muslim fundamentalists and Christians in 1981 caused street riots and culminated in a nationwide security crackdown in Sept. Pres Sadat was assassinated on Oct. 6; he was succeeded by Hosni Mubarak.

Egypt was a political and military supporter of the Allied forces in their defeat of Iraq in the Persian Gulf War, 1991.

Egypt saw a rising tide of Islamic fundamentalist violence in the 1990s. Egyptian security forces conducted raids against Islamic militants, some of whom were executed for terrorism. Naguib Mahfouz, winner of the 1988 Nobel Prize for Literature, was stabbed by Islamic militants Oct. 14, 1994. Pres. Mubarak escaped assassination in Ethiopia, June 26, 1995; Egypt blamed Sudan for the attack. On Nov. 17, 1997, near Luxor, Muslim extremists killed 58 foreign tourists and 4 Egyptians. Mubarak, who was grazed by a knife-wielding assailant Sept. 6, 1999, was confirmed by popular vote Sept. 26 for a 4th presidential term.

The **Suez Canal,** 103 mi. long, links the Mediterranean and Red seas. It was built by a French corporation 1859-69, but Britain obtained controlling interest in 1875. The last British troops were removed June 13, 1956. On July 26, Egypt nationalized the canal.

El Salvador
Republic of El Salvador

People: Population: 5,839,079. **Age distrib.** (%): <15: 36.6; 65+: 5.2. **Pop. density:** 721 per sq. mi. **Urban:** 45%. **Ethnic groups:** Mestizo 94%, Amerindian 5%. **Principal language:** Spanish (official). **Chief religions:** Roman Catholic 75%, many Protestant groups.

Geography: Area: 8,100 sq. mi. **Location:** In Central America. **Neighbors:** Guatemala on W, Honduras on N. **Topography:** A hot Pacific coastal plain in the south rises to a cooler plateau and valley region, densely populated. The N is mountainous, including many volcanoes. **Capital:** San Salvador: 1,214,000*.

Government: Type: Republic. **Head of state and gov.:** Pres. Francisco Flores; b Oct. 17, 1959; in office: June 1, 1999. **Local divisions:** 14 departments. **Defense:** 1.9% of GDP. **Active troops:** 28,400.

Economy: Industries: Food and beverages, oil products, chemicals. **Chief crops:** Coffee, corn, sugar, rice. **Other resources:** Hydropower. **Arable land:** 27%. **Livestock** (1997): chickens: 7.98 mil; cattle: 1.16 mil; pigs: 313,500. **Electricity prod.** (1997): 4.000 bil kWh. **Labor force:** 40% agric.; 16% commerce; 15% manuf.; 13% govt.

Finance: Monetary unit: Colon (Sept. 1999: 8.76 = $1 U.S.). **GDP** (1997 est.): $17.8 bil. **Per capita GDP:** $3,000. **Imports** (1997 est.): $3.5 bil; partners: U.S. 42%, Guatemala

11%. **Exports** (1997 est.): $1.96 bil; partners: U.S. 23%, Guatemala 22%. **Tourism:** $125 mil. **Budget** (1997 est.): $1.82 bil. **Intl. reserves less gold** (June 1999): $1.90 bil. **Gold:** 469,000 oz t. **Consumer prices** (change in 1998): 2.5%.

Transport: Railroad: Length: 349 mi. **Motor vehicles:** 35,300 pass. cars, 44,800 comm. vehicles. **Civil aviation:** 1.3 bil pass.-mi.; 1 airport. **Chief ports:** La Union, Acajutla, La Libertad.

Communications: TV sets: 91 per 1,000 pop. **Radios:** 373 per 1,000 pop. **Telephones:** 482,600 main lines. **Daily newspaper circ.:** 53 per 1,000 pop.

Health: Life expectancy: 66.70 male; 73.50 female. **Births** (per 1,000 pop.): 26.19. **Deaths** (per 1,000 pop.): 6.20. **Natural inc.:** 1.999%. **Hosp. beds** (1993): 1 per 588 persons. **Physicians** (1993): 1 per 1,219 persons. **Infant mortality** (per 1,000 live births): 28.38.

Education: Free, compulsory: ages 7-16. **Literacy:** 71%.

Major Intl. Organizations: UN (FAO, IBRD, ILO, IMF, IMO, WHO, WTrO), OAS.

Embassy: 2308 California St. NW 20008; 265-9671.

El Salvador became independent of Spain in 1821, and of the Central American Federation in 1839.

A fight with Honduras in 1969 over the presence of 300,000 Salvadoran workers left 2,000 dead.

A military coup overthrew the government of Pres. Carlos Humberto Romero in 1979, but the ruling military-civilian junta failed to quell a rebellion by leftist insurgents, armed by Cuba and Nicaragua. Extreme right-wing death squads organized to eliminate suspected leftists were blamed for thousands of deaths in the 1980s. The Reagan administration staunchly supported the government with military aid.

The 12-year civil war ended Jan. 16, 1992, as the government and leftist rebels signed a formal peace treaty. The civil war had taken the lives of some 75,000 people. The treaty provided for military and political reforms.

Nine soldiers, including 3 officers, were indicted Jan. 1990 in the Nov. 1989 slaying of 6 Jesuit priests in San Salvador. Two of the officers received maximum 30-year jail sentences. They were released Mar. 20, 1993, when the National Assembly passed a sweeping amnesty.

Francisco Flores, candidate of the right-wing ARENA party, won the presidential election of Mar. 7, 1999.

Equatorial Guinea
Republic of Equatorial Guinea

People: Population: 465,746. **Age distrib.** (%): <15: 43.0; 65+: 3.8. **Pop. density:** 43 per sq. mi. **Urban:** 43%. **Ethnic groups:** Fang 83%, Bubi 10%. **Principal languages:** Spanish, French (both official), Fang, Bubi. **Chief religion:** Predominantly Roman Catholic.

Geography: Area: 10,800 sq. mi. **Location:** Bioko Isl. off W Africa coast in Gulf of Guinea, and Rio Muni, mainland enclave. **Neighbors:** Gabon on S, Cameroon on E and N. **Topography:** Bioko Isl. consists of 2 volcanic mountains and a connecting valley. Rio Muni, with over 90% of the area, has a coastal plain and low hills beyond. **Capital:** Malabo (1991 est.): 58,000.

Government: Type: Republic. **Head of state:** Pres. Teodoro Obiang Nguema Mbasogo; b June 5, 1942; in office: Oct. 10, 1979. **Head of gov.:** Prime Min. Angel Serafin Seriche Dougan; in office: Mar. 29, 1996. **Local divisions:** 7 provinces. **Defense:** 1.3% of GDP. **Active troops:** 1,300.

Economy: Industries: Oil (68% of export earnings), fishing, sawmilling. **Chief crops:** Cocoa, coffee, rice, bananas, yams cassava. **Minerals:** Oil. **Other resources:** Timber. **Crude oil reserves** (1999): 12 mil bbls. **Arable land:** 5%. **Livestock** (1997): chickens: 245,000. **Electricity prod.** (1997): 20 mil kWh. **Labor force:** 72% agric.

Finance: Monetary unit: CFA Franc (Sept. 1999: 612.79 = $1 U.S.). **GDP** (1997 est.): $660 mil. **Per capita GDP:** $1,500. **Imports** (1996 est.): $248 mil; partners: Cameroon 40%, Spain 18%, France 14%. **Exports** (1996 est.): $197 mil; partners: U.S. 34%, Japan 17%, Spain 13%. **Tourism:** $2 mil. **Budget** (1996 est.): $43 mil. **Intl. reserves less gold** (Mar. 1999): $420,000.

Transport: Motor vehicles: 4,000 pass. cars, 3,600 comm. vehicles. **Civil aviation:** 2.8 mil pass.-mi.; 1 airport. **Chief ports:** Malabo, Bata.

Communications: TV sets: 88 per 1,000 pop. **Radios:** 464 per 1,000 pop. **Telephones:** 5,400 main lines.

Health: Life expectancy: 52.03 male; 56.83 female. **Births** (per 1,000 pop.): 38.49. **Deaths** (per 1,000 pop.): 12.98. **Natural inc.:** 2.551%. **Infant mortality** (per 1,000 live births): 91.18.

Education: Free, compulsory: ages 6-11. **Literacy:** 78%.

Major Intl. Organizations: UN (FAO, IBRD, ILO, IMF, IMO, WHO), OAU.

Embassy: Suite 405, 1511 K St. NW 20005; 393-0525.

Fernando Po (now Bioko) Island was reached by Portugal in the late 15th century and ceded to Spain in 1778. Independence came Oct. 12, 1968. Riots occurred in 1969 over disputes between the island and the more backward Rio Muni province on the mainland. Masie Nguema Biyogo, a mainlander, became president for life in 1972.

Masie's reign was one of the most brutal in Africa, resulting in a bankrupted nation. Most of the nation's 7,000 Europeans emigrated. He was ousted in a military coup, Aug. 1979, and Teodoro Mbasogo, leader of the coup, became president. His regime eventually agreed to elections, held Nov. 21, 1993. These were nominally won by the ruling party, but boycotted by opposition parties that maintained the rules were rigged. Elections for president, Feb. 25, 1996, and for the legislature, Mar. 6, 1999, were similarly condemned.

Eritrea
State of Eritrea

People: Population: 3,984,723. **Age distrib.** (%): <15: 43.0; 65+: 3.3. **Pop. density:** 85 per sq. mi. **Urban:** 17%. **Ethnic groups:** Tigrinya 50%, Tigre and Kunama 40%, Afar 4%. **Principal languages:** Tigrinya, Tigre and Kunama, Afar, Amhanc, Arabic. **Chief religions:** Muslim,Coptic Christian, Roman Catholic, Protestant.

Geography: Area: 46,800 sq. mi. **Location:** In E Africa, on SW coast of Red Sea. **Neighbors:** Ethiopia on S, Djibouti on SE, Sudan on W. **Topography:** Includes many islands of the Dahlak Archipelago, low coastal plains in S, mountain range with peaks to 9,000 ft. in N. **Capital:** Asmara (1995 est.): 431,000.

Government: Type: In transition. **Head of state and gov.:** Isaias Afwerki; b Feb. 2, 1946; in office: May 24, 1993. **Local divisions:** 6 provinces. **Defense:** 8.3% of GDP. **Active troops:** 46,000.

Economy: Industries: Food processing, textiles, beverages. **Chief crops:** Cotton, coffee, vegetables, maize, tobacco, lentils, sorghum. **Minerals:** Gold, potash, zinc, copper. **Arable land:** 12%. **Livestock** (1997): chickens: 4.30 mil; sheep: 1.53 mil; goats: 1.40 mil; cattle: 1.32 mil.

Finance: Monetary unit: Birr (Aug. 1999: 6.00 = $1 U.S.); the changeover to a new currency, the nakfa, began Nov. 1997. **GDP** (1996 est.): $2.2 bil. **Per capita GDP:** $600. **Imports** (1996 est.): $499 mil; partners: Saudi Arabia 20%, Italy 18%. **Exports** (1996 est.): $71 mil; partners: Ethiopia 67%. **Tourism:** $75 mil. **Budget** (1996 est.): $453 mil.

Transport: Civil aviation: 2 airports. **Chief ports:** Mitsiwa, Aseb.

Communications: TV sets: 6 per 1,000 pop. **Telephones:** 24,300 main lines.

Health: Life expectancy: 53.61 male; 57.95 female. **Births** (per 1,000 pop.): 42.56. **Deaths** (per 1,000 pop.): 12.32. **Natural inc.:** 3.024%. **Physicians** (1993): 1 per 36,000 persons. **Infant mortality** (per 1,000 live births): 76.84.

Education: Free, compulsory: ages 7-13. **Literacy** (1994): 20%.

Major Intl. Organizations: UN (FAO, IBRD, ILO, IMF, IMO, WHO), OAU.

Embassy: 1708 New Hampshire Ave. NW 20009; 319-1991. **Website:** http://www.NetAfrica.org/eritrea

Eritrea was part of the Ethiopian kingdom of Aksum. It was an Italian colony from 1890 to 1941, when it was captured by the British. Following a period of British and UN supervision, Eritrea was awarded to Ethiopia as part of a federation in 1952. Ethiopia annexed Eritrea as a province in 1962. This led to a 31-year struggle for independence, which ended when Eritrea formally declared itself an independent nation May 24, 1993. A border war with Ethiopia erupted in June 1998.

Estonia
Republic of Estonia

People: Population: 1,408,523. **Age distrib.** (%): <15: 18.2; 65+: 14.5. **Pop. density:** 81 per sq. mi. **Urban:** 73%. **Ethnic groups:** Estonian 65%, Russian 28%. **Principal languages:** Estonian (official), Russian. **Chief religion:** Evangelical Lutheran.

Geography: Area: 17,462 sq. mi. **Location:** E Europe, bordering the Baltic Sea and Gulf of Finland. **Neighbors:** Russia on E, Latvia on S. **Capital:** Tallinn (1996 est.): 423,990.

Government: Type: Republic. **Head of state:** Pres. Lennart Meri; b Mar. 29, 1929; in office: Oct. 5, 1992. **Head of gov.:** Prime Min. Mart Laar; b Apr. 22, 1960; in office: Mar. 25, 1999. **Local divisions:** 15 counties. **Defense:** 2.5% of GDP. **Active troops:** 3,500.

Economy: Industries: Shipbuilding, electric motors, cement. **Chief crops:** Potatoes, fruits, vegetables. **Minerals:** Shale oil, peat, phosphorite. **Other resources:** Dairy prods. **Arable land:** 22%. **Livestock** (1997): chickens: 2.70 mil; pigs: 328,800; cattle: 311,600. **Fish catch** (1996): 107,130 metric tons. **Electricity prod.** (1997): 8.507 bil kWh. **Labor force:** 27% mining, manuf., constr., 12% agric., fishing.

Finance: Monetary unit: Kroon (Sept. 1999: 14.62 = $1 U.S.). **GDP** (1997 est.): $9.34 bil. **Per capita GDP:** $6,450. **Imports** (1996): $3.2 bil; partners: Finland 32%, Russia 13%. **Exports** (1996): $2 bil; partners: Finland 18%, Russia 17%. **Tourism:** $483 mil. **Budget** (1996 est.): $1.8 bil. **Intl. reserves less gold** (June 1999): $765.06 mil. **Gold:** 8,000 oz. t. **Consumer prices** (change in 1998): 10.7%.

Transport: Railroad: Length: 636 mi. **Motor vehicles:** 338,000 pass. cars, 60,000 comm. vehicles. **Civil aviation:** 83.6 mil pass.-mi.; 3 airports. **Chief port:** Tallinn.

Communications: TV sets: 411 per 1,000 pop. **Telephones:** 498,600 main lines. **Daily newspaper circ.:** 242 per 1,000 pop.

Health: Life expectancy: 62.61 male; 75.00 female. **Births** (per 1,000 pop.): 9.05. **Deaths** (per 1,000 pop.): 14.21. **Natural inc.:** –0.516%. **Hosp. beds** (1994): 1 per 119 persons. **Physicians** (1994): 1 per 319 persons. **Infant mortality** (per 1,000 live births): 13.83.

Education: Compulsory: ages 7-16. **Literacy** (1994): 100%. **Major Intl. Organizations:** UN (FAO, IBRD, ILO, IMF, IMO, WHO), OSCE.

Embassy: 2131 Massachusetts Ave. NW 20008; 588-0101. **Websites:** http://www.ciesin.ee/undp/nhdr97/eng/index.html http://www.vm.ee

Estonia was a province of imperial Russia before World War I, was independent between World Wars I and II. It was conquered by the USSR in 1940 and incorporated as the Estonian SSR. Estonia declared itself an "occupied territory," and proclaimed itself a free nation Mar. 1990. During an abortive Soviet coup, Estonia declared immediate full independence, Aug. 20, 1991; the Soviet Union recognized its independence in Sept. 1991. The first free elections in over 50 years were held Sept. 20, 1992. The last occupying Russian troops were withdrawn by Aug. 31, 1994. Center-right parties won the legislative election of Mar. 7, 1999.

Ethiopia
Federal Democratic Republic of Ethiopia

People: Population: 59,680,383. **Age distrib.** (%): <15: 46.1; 65+: 2.8. **Pop. density:** 137 per sq. mi. **Urban:** 16%. **Ethnic groups:** Oromo 40%, Amhara and Tigrean 32%, Sidamo 9%. **Principal languages:** Amharic (official), Tigrinya, Orominga. **Chief religions:** Muslim 45-50%, Ethiopian Orthodox 35-40%, animist 12%.

Geography: Area: 435,185 sq. mi. **Location:** In East Africa. **Neighbors:** Sudan on W, Kenya on S, Somalia and Djibouti on E, Eritrea on N. **Topography:** A high central plateau, between 6,000 and 10,000 ft. high, rises to higher mountains near the Great Rift Valley, cutting in from the SW. The Blue Nile and other rivers cross the plateau, which descends to plains on both W and SE. **Capital:** Addis Ababa: 2,431,000*.

Government: Type: Federal republic. **Head of state:** Pres. Negasso Gidada; b Sept. 8, 1943; in office: Aug. 22, 1995. **Head of gov.:** Prime Min. Meles Zenawi; b May 8, 1955; in office: Aug. 23, 1995. **Local divisions:** 9 administrative regions, 1 federal capital. **Defense:** 2.1% of GDP. **Active troops:** 120,000.

Economy: Industries: Food processing, chemicals, textiles. **Chief crops:** Coffee (62% of export earnings), grains, sugarcane, vegetables. **Minerals:** Platinum, gold, copper. **Arable land:** 12%. **Livestock** (1997): chickens: 55.00 mil; cattle: 29.90 mil; sheep: 21.85 mil; goats: 16.85 mil. **Electricity prod.** (1997): 1.777 bil kWh. **Labor force:** 89% agric.

Finance: Monetary unit: Birr (Sept. 1999: 7.52 = $1 U.S.). **GDP** (1997 est.): $29 bil. **Per capita GDP:** $530. **Imports** (1996 est.): $1.23 bil; partners: Saudi Arabia 15%, U.S.12%, Italy 11%. **Exports** (1996): $418 mil; partners: Germany 32%, Japan 14%. **Tourism:** $40 mil. **Budget** (FY 1996-97): $1.48 bil. **Intl. reserves less gold** (Apr. 1999): $443.1 mil. **Gold:** 3,000 oz t. **Consumer prices** (change in 1997): –3.7%.

Transport: Railroad: Length: 486 mi. **Motor vehicles:** 45,559 pass. cars, 20,462 comm. vehicles. **Civil aviation:** 1.2 bil pass.-mi.; 31 airports.

Communications: TV sets: 4 per 1,000 pop. **Radios:** 153 per 1,000 pop. **Telephones** (1997): 156,500 main lines.

Health: Life expectancy: 39.22 male; 41.73 female. **Births** (per 1,000 pop.): 44.34. **Deaths** (per 1,000 pop.): 21.43. **Natural inc.:** 2.291%. **Infant mortality** (per 1,000 live births): 124.57.

Education: Free, compulsory: ages 7-13. **Literacy:** 35%. **Major Intl. Organizations:** UN (FAO, IBRD, ILO, IMF, IMO, WHO), OAU.

Embassy: 2134 Kalorama Rd. NW 20008; 234-2281.

Ethiopian culture was influenced by Egypt and Greece. The ancient monarchy was invaded by Italy in 1880 but maintained its independence until another Italian invasion in 1936. British forces freed the country in 1941.

The last emperor, Haile Selassie I, established a parliament and judiciary system in 1931 but barred all political parties.

A series of droughts in the 1970s killed hundreds of thousands. An army mutiny, strikes, and student demonstrations led to the dethronement of Selassie in 1974; he died Aug. 1975, while being held by the ruling junta. The junta pledged to form a one-party socialist state and instituted a successful land reform; opposition was violently suppressed. The influence of the Coptic Church, embraced in AD 330, was curbed, and the monarchy was abolished in 1975.

The regime, torn by bloody coups, faced uprisings by tribal and political groups in part aided by Sudan and Somalia. Ties with the U.S., once a major ally, deteriorated, while cooperation accords were signed with the USSR in 1977. In 1978, Soviet advisers and Cuban troops helped defeat Somalian forces. Ethiopia and Somalia signed a peace agreement in 1988.

A worldwide relief effort began in 1984, as an extended drought threatened the country with famine; up to a million people may have died as a result of starvation and disease.

The Ethiopian People's Revolutionary Democratic Front (EPRDF), an umbrella group of 6 rebel armies, launched a major push against government forces, Feb. 1991. In May, Pres. Mengistu Haile Mariam resigned and left the country. The EPRDF took over and set up a transitional government. Ethiopia's first multiparty general elections were held in 1995.

Eritrea, a province on the Red Sea, declared its independence May 24, 1993. Fighting along the border with Eritrea erupted in June 1998.

Fiji
Republic of the Fiji Islands

People: Population: 812,918. **Age distrib.** (%): <15: 33.5; 65+: 3.3. **Pop. density:** 114 per sq. mi. **Urban:** 41%. **Ethnic groups:** Fijian 49%, Indian 46%. **Principal languages:** English (official), Fijian, Hindustani. **Chief religions:** Christian 52%, Hindu 38%, Muslim 8%.

Geography: Area: 7,100 sq. mi. **Location:** In western South Pacific O. **Neighbors:** Nearest are Vanuatu to W, Tonga to E. **Topography:** 322 islands (106 inhabited), many mountainous, with tropical forests and large fertile areas. Viti Levu, the largest island, has over half the total land area. **Capital:** Suva (1996): 167,421.

Government: Type: Republic. **Head of state:** Pres. Ratu Sir Kamisese Mara; b May 13, 1920; in office: Jan. 18, 1994. **Head of gov.:** Prime Min. Mahendra Chaudhry; b Sept 2, 1942; in office: May 19, 1999. **Local divisions:** 4 divisions comprising 14 provinces and 1 dependency. **Defense:** 2.6% of GDP. **Active troops:** 3,600.

Economy: Industries: Sugar refining, light industry, tourism. **Chief crops:** Sugarcane, cassava, coconuts. **Minerals:** Gold, copper. **Other resources:** Timber, fish. **Arable land:** 10%. **Livestock** (1997): chickens: 4.30 mil; cattle: 360,000; goats: 220,000; pigs: 145,000. **Electricity prod.** (1997): 540 mil kWh. **Labor force:** 44% agric; 35% mining, manuf., const.

Finance: Monetary unit: Dollar (Sept. 1999: 1.97 = $1.00 U.S.). **GDP** (1996 est.): $5.1 bil. **Per capita GDP:** $6,500. **Imports** (1996): $947 mil; partners: Australia 30%, N.Z. 17%, Japan 13%. **Exports** (1996): $639 mil; partners: EU 26%, Australia 15%. **Tourism:** $327 mil. **Budget** (1997 est.): $742.65 mil. **Intl. reserves less gold** (May 1999): $403.87 mil. **Gold:** 1,000 oz t. **Consumer prices** (change in 1998): 5.7%.

Transport: Railroad: Length: 370 mi. **Motor vehicles:** 30,000 pass. cars, 29,000 comm. vehicles. **Civil aviation:** 1.2 bil mil pass.-mi.; 13 airports. **Chief ports:** Suva, Lautoka.

Communications: TV sets: 89 per 1,000 pop. **Radios:** 561 per 1,000 pop. **Telephones** (1997): 71,800 main lines. **Daily newspaper circ.:** 68 per 1,000 pop.

Health: Life expectancy: 64.19 male; 69.11 female. **Births** (per 1,000 pop.): 22.76. **Deaths** (per 1,000 pop.): 6.21. **Natural inc.:** 1.655%. **Hosp. beds** (1993): 1 per 438 persons. **Physicians** (1994): 1 per 2,576 persons. **Infant mortality** (per 1,000 live births): 16.30.

Education: Free: ages 6-14. **Literacy** (1996): 91%. **Major Intl. Organizations:** UN (FAO, IBRD, ILO, IMF, IMO, WHO, WTrO), the Commonwealth.

Embassy: 2233 Wisconsin Ave. NW 20007; 337-8320.

A British colony since 1874, Fiji became an independent parliamentary democracy Oct. 10, 1970. Cultural differences between the Indian community (descendants of contract laborers brought to the islands in the 19th century) and indigenous Fijians have led to political polarization.

In 1987, a military coup ousted the government; order was restored May 21 under a compromise granting Lt. Col. Sitiveni Rabuka, the coup's leader, increased power. Rabuka staged a second coup Sept. 25 and declared Fiji a republic. Civilian government was restored in Dec. A new constitution favoring indigenous Fijians was issued July 25, 1990; amendments enacted in July 1997 made the constitution more equitable. Fiji's 1st Indian prime minister took office in May 1999.

Finland
Republic of Finland

People: Population: 5,158,372. **Age distrib.** (%): <15: 18.4; 65+: 14.7. **Pop. density:** 40 per sq. mi. **Urban:** 64%. **Ethnic groups:** Finn 93%, Swede 6%. **Principal languages:** Finnish, Swedish (both official). **Chief religion:** Evangelical Lutheran 89%.

Geography: Area: 130,100 sq. mi. **Location:** In northern Europe. **Neighbors:** Norway on N, Sweden on W, Russia on E. **Topography:** South and central Finland are generally flat areas with low hills and many lakes. The N has mountainous areas, 3,000-4,000 ft. above sea level. **Capital:** Helsinki. **Cities** (1997 est.): Helsinki 532,053; Espoo 196,260; Tampere 186,026.

Government: Type: Constitutional republic. **Head of state:** Pres. Martti Ahtisaari; b June 23, 1937; in office: Mar. 1, 1994. **Head of gov.:** Prime Min. Paavo Lipponen; b Apr. 23, 1941; in office: Apr. 13, 1995. **Local divisions:** 6 laanit (provinces). **Defense:** 1.7% of GDP. **Active troops:** 31,000.

Economy: Industries: Metal prods., shipbuilding, wood processing, chemicals, textiles. **Chief crops:** Grains, sugar beets, potatoes. **Minerals:** Copper, iron, silver, zinc. **Other resources:** Timber, dairy prods. **Arable land:** 8%. **Livestock** (1997): chickens: 5.23 mil; pigs: 1.47 mil; cattle: 1.15 mil; sheep: 102,900. **Fish catch** (1996): 177,984 metric tons. **Electricity prod.** (1997): 62.730 bil kWh. **Labor force:** 46% ind., commerce & finance; 30% public serv.; 9% agric.

Finance: Monetary unit: Markka (Sept. 1999: 5.55 = $1 U.S.). Euro (Sept. 1999: 1.07 = $1 U.S.). **GDP** (1997 est.): $102.1 bil. **Per capita GDP:** $20,000. **Imports** (1996): $29.3 bil; partners: Germany 17%, Sweden 12%. **Exports** (1996): $38.4 bil; partners: Germany 13%, UK 10%, Sweden 10%. **Tourism:** $1.97 bil. **Budget** (1996 est.): $40 bil. **Intl. reserves less gold** (June 1999): $8.00 bil. **Gold:** 1.58 mil oz t. **Consumer prices** (change in 1998): 1.4%.

Transport: Railroad: Length: 3,641 mi. **Motor vehicles** (1997): 1.94 mil pass. cars, 291,235 comm. vehicles. **Civil aviation:** 5.9 bil pass.-mi.; 24 airports. **Chief ports:** Helsinki, Turku, Rauma, Kotka.

Communications: TV sets: 372 per 1,000 pop. **Radios:** 966 per 1,000 pop. **Telephones:** 2,841,000 main lines. **Daily newspaper circ.:** 464 per 1,000 pop.

Health: Life expectancy: 73.6 male; 80 female. **Births** (per 1,000 pop.): 11. **Deaths** (per 1,000 pop.): 10. **Natural inc.:** 0.16%. **Hosp. beds** (1994): 1 per 102 persons. **Physicians** (1995): 1 per 371 persons. **Infant mortality** (per 1,000 live births): 4.

Education: Free, compulsory: ages 7-16. **Literacy** (1997): 100%.

Major Intl. Organizations: UN (FAO, IBRD, ILO, IMF, IMO, WHO, WTrO), EU, OECD, OSCE.

Embassy: 3301 Massachusetts Ave. NW 20008; 298-5800. **Website:** http://www.finland.org

The early Finns probably migrated from the Ural area at about the beginning of the Christian era. Swedish settlers brought the country into Sweden, 1154 to 1809, when Finland became an autonomous grand duchy of the Russian Empire. Russian exactions created a strong national spirit; on Dec. 6, 1917, Finland declared its independence and in 1919 became a republic.

On Nov. 30, 1939, the Soviet Union invaded, and the Finns were forced to cede 16,173 sq. mi. of territory. After World War II, further cessions were exacted. In 1948, Finland signed a treaty of mutual assistance with the USSR; Finland and Russia nullified this treaty with a new pact in Jan. 1992.

Following approval by Finnish voters in an advisory referendum Oct. 16, 1994, Finland joined the European Union effective Jan. 1, 1995.

Aland or **Ahvenanmaa,** constituting an autonomous province, is a group of small islands, 590 sq. mi., in the Gulf of Bothnia, 25 mi. from Sweden, 15 mi. from Finland. Mariehamn is the principal port.

France
French Republic

People: Population: 58,978,172. **Age distrib.** (%): <15: 18.7; 65+: 16.0. **Pop. density:** 279 per sq. mi. **Urban:** 74%. **Ethnic groups:** Celtic and Latin; Teutonic, Slavic, North African, Indochinese, Basque minorities. **Principal language:** French (official). **Chief religion:** Roman Catholic 90%.

Geography: Area: 211,200 sq. mi. **Location:** In western Europe, between Atlantic O. and Mediterranean Sea. **Neighbors:** Spain on S; Italy, Switzerland, Germany on E; Luxembourg, Belgium on N. **Topography:** A wide plain covers more than half of the country, in N and W, drained to W by Seine, Loire, Garonne rivers. The Massif Central is a mountainous plateau in center. In E are Alps (Mt. Blanc is tallest in W Europe, 15,771 ft.), the lower Jura range, and the forested Vosges. The Rhone flows from Lake Geneva to Mediterranean. Pyrenees are in SW, on border with Spain. **Capital:** Paris. **Cities** (1997 est.): Paris 2,152,000; Lyon 1,260,000; Marseilles 1,200,000; Lille 1,000,000.

Government: Type: Republic. **Head of state:** Pres. Jacques Chirac; b Nov. 29, 1932; in office: May 17, 1995. **Head of gov.:** Prime Min. Lionel Jospin; b July 12, 1937; in office: June 3, 1997. **Local divisions:** 22 administrative regions containing 96 departments. **Defense:** 3.0% of GDP. **Active troops:** 380,800.

Economy: Industries: Steel, chemicals, textiles, tourism, wine, perfume, aircraft, machinery, electronics. **Chief crops:** Grains, sugar beets, winegrapes, fruits, potatoes, vegetables. France is largest food producer, exporter, in W Europe. **Minerals:** Bauxite, iron, coal. **Crude oil reserves** (1999): 107 mil bbls. **Other resources:** Timber, dairy. **Arable land:** 33%. **Livestock** (1997): chickens: 238.07 mil; cattle: 20.39 mil; pigs: 15.43 mil; sheep: 10.31 mil; goats: 1.20 mil. **Fish catch** (1997): 542,367 metric tons. **Electricity prod.** (1997): 474.032 bil kWh. **Labor force:** 69% services; 26% ind.; 5% agric.

Finance: Monetary unit: Franc (Sept. 1999: 6.13 = $1 U.S.). Euro (Sept. 1999: 1.07 = $1 U.S.). **GDP** (1997 est.): $1.32 tril. **Per capita GDP:** $22,700. **Imports** (1997): $256 bil; partners: Germany 17%, Italy 10%, U.S. 9%. **Exports** (1997 est.): $275 bil; partners: Germany 17%, Italy 9%, UK 9%. **Tourism:** $29.70 bil. **Budget** (1998 est.): $265 bil. **Intl. reserves less gold** (June 1999): $38.69 bil. **Gold:** 97.24 mil oz t. **Consumer prices** (change in 1998): 0.7%.

Transport: Railroad: Length: 19,847 mi. **Motor vehicles:** in use: 25.50 mil pass. cars, 5.26 mil comm. vehicles. **Civil aviation:** 52.6 bil pass.-mi.; 61 airports. **Chief ports:** Marseille, Le Havre, Bordeaux, Rouen.

Communications: TV sets: 579 per 1,000 pop. **Radios:** 860 per 1,000 pop. **Telephones:** 34.0 mil main lines. **Daily newspaper circ.:** 235 per 1,000 pop.

Health: Life expectancy: 74.76 male; 82.71 female. **Births** (per 1,000 pop.): 11.38. **Deaths** (per 1,000 pop.): 9.17. **Natural inc.:** 0.221%. **Hosp. beds** (1995): 1 per 86 persons. **Physicians** (1994): 1 per 361 persons. **Infant mortality** (per 1,000 live births): 5.62.

Education: Free, compulsory: ages 6-16. **Literacy** (1994): 99%.

Major Intl. Organizations: UN and most of its specialized agencies, EU, NATO, OECD, OSCE.

Embassy: 4101 Reservoir Rd. NW 20007; 944-6000. **Websites:** http://www.info-france-usa.org
http://www.france.org

Celtic Gaul was conquered by Julius Caesar 58-51 BC; Romans ruled for 500 years. Under Charlemagne, Frankish rule extended over much of Europe. After his death France emerged as one of the successor kingdoms.

The monarchy was overthrown by the French Revolution (1789-93) and succeeded by the First Republic; followed by the First Empire under Napoleon (1804-15), a monarchy (1814-48), the Second Republic (1848-52), the Second Empire (1852-70), the Third Republic (1871-1946), the Fourth Republic (1946-58), and the Fifth Republic (1958 to present).

France suffered severe losses in manpower and wealth in the World War I, when it was invaded by Germany. By the Treaty of Versailles, France exacted return of Alsace and Lorraine, provinces seized by Germany in 1871. Germany invaded France again in May 1940, and signed an armistice with a government based in Vichy. After France was liberated by the Allies Sept. 1944, Gen. Charles de Gaulle became head of the provisional government, serving until 1946.

De Gaulle again became premier in 1958, during a crisis over Algeria, and obtained voter approval for a new constitution, ushering in the Fifth Republic. He became president Jan. 1959. Using strong executive powers, he promoted French economic and technological advances in the context of the European Economic Community and guarded French foreign policy independence.

France had withdrawn from Indochina in 1954, and from Morocco and Tunisia in 1956. Most of its remaining African territories were freed 1958-62. In 1966, France withdrew all its troops from the integrated military command of NATO, though 60,000 remained stationed in Germany.

In May 1968 rebellious students in Paris and other centers rioted, battled police, and were joined by workers who launched nationwide strikes. The government awarded pay increases to the strikers May 26. De Gaulle resigned from office in Apr. 1969, after losing a nationwide referendum on constitutional reform. Georges Pompidou, who was elected to succeed him, continued De Gaulle's emphasis on French independence from the two superpowers. After Pompidou's death, in 1974, Valery Giscard d'Estaing was elected president; he continued the basically conservative policies of his predecessors.

On May 10, 1981, France elected François Mitterrand, a Socialist, president. Under Mitterrand the government nationalized 5 major industries and most private banks. After 1986, however, when rightists won a narrow victory in the National Assembly, Mitterrand chose conservative Jacques Chirac as premier. A 2-year period of "cohabitation" ensued, and France began to pursue a privatization program in which many state-owned companies were sold. After Mitterrand was elected to a 2d 7-year term in 1988, he appointed a Socialist as premier. The center-right won a large majority in 1993 legislative elections, ushering in another period of "cohabitation" with a conservative premier.

In 1993, France set tighter rules for entry into the country and made it easier for the government to expel foreigners. In 1994, France sent troops to Rwanda in an effort to help protect civilians there from ongoing massacres. The international terrorist known as Carlos the Jackal (Ilich Ramírez Sánchez) was arrested in Sudan in Aug. 1994 and extradited to France, where he had been sentenced in absentia to life imprisonment.

Former conservative Prime Min. Jacques Chirac won the presidency in a runoff May 7, 1995. A series of terrorist bombings and bombing attempts began in summer 1995; Islamic extremists, opposed to France's support of the Algerian government and its struggle with Islamic fundamentalists, were believed responsible. In Sept. 1995, France stirred widespread protests by resuming nuclear tests in the South Pacific, after a 3-year moratorium; the tests ended Jan. 1996.

Chirac cut government spending to help the French economy meet the budgetary goals set for the introduction of a common European currency. With unemployment at nearly 13%, legislative elections completed June 1, 1997, produced a decisive victory for the leftist parties. The result was a new period of "cohabitation," this time between a conservative president and a Socialist prime minister, Lionel Jospin. France contributed 7,000 troops to the NATO-led security force (KFOR) that entered Kosovo in June 1999.

The island of **Corsica**, in the Mediterranean W of Italy and N of Sardinia, is a territorial collectivity and region of France comprising 2 departments. It elects a total of 2 senators and 3 deputies to the French Parliament. Area: 3,369 sq. mi.; pop. (1996 est.): 258,000. The capital is Ajaccio, birthplace of Napoleon I. Violence by Corsican separatist groups has hurt tourism, a leading industry on the island.

Overseas Departments

French Guiana is on the NE coast of South America with Suriname on the W and Brazil on the E and S. Its area is 33,399 sq. mi.; pop. (1999 est.): 167,982. Guiana sends one senator and 2 deputies to the French Parliament. Guiana is administered by a prefect and has a Council General of 16 elected members; capital is Cayenne.

The famous penal colony, Devil's Island, was phased out between 1938 and 1951. The European Space Agency maintains a satellite-launching center (established by France in 1964) in the city of Kourou.

Immense forests of rich timber cover 88% of the land. Fishing (especially shrimp), forestry, and gold mining are the most important industries.

Guadeloupe, in the West Indies' Leeward Islands, consists of 2 large islands, Basse-Terre and Grande-Terre, separated by the Salt River, plus Marie Galante and the Saintes group to the S and, to the N, Desirade, St. Barthelemy, and over half of St. Martin (the Netherlands' portion is called St. Maarten). A French possession since 1635, the department is represented in the French Parliament by 2 senators and 4 deputies; administration consists of a prefect (governor) as well as an elected general and regional councils.

Area of the islands is 687 sq. mi.; pop. (1999 est.) 420,943, mainly descendants of slaves; capital is Basse-Terre on Basse-Terre Island. The land is fertile; sugar, rum, and bananas are exported. Tourism is an important industry.

Martinique, the northernmost of the Windward Islands, in the West Indies, has been a possession since 1635, and a department since Mar. 1946. It is represented in the French Parliament by 2 senators and 4 deputies. The island was the birthplace of Napoleon's Empress Josephine.

It has an area of 436 sq. mi.; pop. (1999 est.) 411,539, mostly descendants of slaves. The capital is Fort-de-France (pop. 1991: 101,000). It is a popular tourist stop. The chief exports are rum, bananas, and petroleum products.

Réunion is a volcanic island in the Indian O. about 420 mi. E of Madagascar, and has belonged to France since 1665. Area, 970 sq. mi.; pop. (1999 est.) 717,723, 30% of French extraction. Capital: Saint-Denis. The chief export is sugar. It elects 5 deputies, 3 senators to the French Parliament.

Overseas Territorial Collectivities

Mayotte, claimed by Comoros and administered by France, voted in 1976 to become a territorial collectivity of France. An island NW of Madagascar, area is 144 sq. mi., pop. (1999 est.) 149,336. The capital is Mamoutzou.

St. Pierre and Miquelon, formerly an overseas territory (1816-1976) and department (1976-85), made the transition to territorial collectivity in 1985. It consists of 2 groups of rocky islands near the SW coast of Newfoundland, inhabited by fishermen. The exports are chiefly fish products. The St. Pierre group has an area of 10 sq. mi.; Miquelon, 83 sq. mi. Total pop. (1999 est.), 6,966. The capital is St. Pierre.

Both Mayotte and St. Pierre and Miquelon elect a deputy and a senator to the French Parliament.

Overseas Territories

Territory of **French Polynesia** comprises 130 islands widely scattered among 5 archipelagos in the South Pacific; administered by a Council of Ministers (headed by a president). Territorial Assembly and the Council have headquarters at Papeete, on Tahiti, one of the **Society Islands** (which include the **Windward** and **Leeward** islands). Two deputies and a senator are elected to the French Parliament.

Other groups are the **Marquesas Islands,** the **Tuamotu Archipelago,** including the **Gambier Islands,** and the **Austral Islands.**

Total area of the islands administered from Tahiti is 1,544 sq. mi.; pop. (1999 est.), 242,073, more than half on Tahiti. Tahiti is picturesque and mountainous with a productive coastline bearing coconuts, citrus, pineapples, and vanilla. Cultured pearls are also produced.

Tahiti was visited by Capt. James Cook in 1769 and by Capt. Bligh in the *Bounty*, 1788-89. Its beauty impressed Herman Melville, Paul Gauguin, and Charles Darwin. Tahitians angered by French nuclear testing rioted Sept. 1995.

Territory of the **French Southern and Antarctic Lands** comprises **Adelie Land,** on Antarctica, and 4 island groups in the Indian O. Adelie, reached 1840, has a research station, a coastline of 185 mi., and tapers 1,240 mi. inland to the South Pole. The U.S. does not recognize national claims in Antarctica. There are 2 huge glaciers, Ninnis, 22 mi. wide, 99 mi. long, and Mentz, 11 mi. wide, 140 mi. long. The Indian O. groups are:

Kerguelen Archipelago, visited 1772, consists of one large and 300 small islands. The chief is 87 mi. long, 74 mi. wide, and has Mt. Ross, 6,429 ft. tall. Principal research station is Port-aux-Français. Seals often weigh 2 tons; there are blue whales, coal, peat, semiprecious stones. **Crozet Archipelago,** reached 1772, covers 195 sq. mi. Eastern Island rises to 6,560 ft. **Saint Paul,** in southern Indian O., has warm springs with earth at places heating to 120° to 390° F. **Amsterdam** is nearby; both produce cod and rock lobster.

Territory of **New Caledonia** and Dependencies is a group of islands in the Pacific O. about 1,115 mi. E of Australia and approx. the same distance NW of New Zealand. Dependencies are the **Loyalty Islands, Isle of Pines, Belep Archipelago,** and **Huon Islands.**

The largest island, New Caledonia, is 6,530 sq. mi. Total area of the territory is 8,548 sq. mi.; population (1999 est.) 197,361. The group was acquired by France in 1853.

The territory is administered by a High Commissioner. There is a popularly elected Territorial Congress. Two deputies and a senator are elected to the French Parliament. Capital: Noumea.

Mining is the chief industry. New Caledonia is one of the world's largest nickel producers. Other minerals found are chrome, iron, cobalt, manganese, silver, gold, lead, and copper. Agricultural products include yams, sweet potatoes, potatoes, manioc (cassava), corn, and coconuts.

In 1987, New Caledonia voters chose by referendum to remain within the French Republic. There were clashes between French and Melanesians (Kanaks) in 1988. An agreement Apr. 21, 1998, between France and rival New Caledonian factions specified a 15- to 20-year period of "shared sovereignty." The French constitution was amended, July 6, to allow the territory a gradual increase in autonomy, and New Caledonian voters approved the plan Nov. 8, 1998, by a 72% majority.

Territory of the **Wallis and Futuna Islands** comprises 2 island groups in the SW Pacific S of Tuvalu, N of Fiji, and W of Western Samoa; became an overseas territory July 29, 1961. The islands have a total area of 106 sq. mi. and population (1999 est.) of 15,129. **Alofi,** attached to Futuna, is uninhabited. Capital: Mata-Utu. Chief products are copra, yams, taro roots, bananas, and coconuts. A senator and a deputy are elected to the French Parliament.

Gabon
Gabonese Republic

People: Population: 1,225,853. **Age distrib.** (%): <15: 33.5; 65+: 5.6. **Pop. density:** 12 per sq. mi. **Urban:** 51%. **Ethnic groups:** Fang, Eshira, Bapounou, Bateke, other Bantu, other Africans, Europeans. **Principal languages:** French (official), Bantu dialects. **Chief religions:** Christian 55%-75%.
Geography: Area: 103,300 sq. mi. **Location:** On Atlantic coast of W central Africa. **Neighbors:** Equatorial Guinea and Cameroon on N, Congo on E and S. **Topography:** Heavily forested, the country consists of coastal lowlands; plateaus in N, E, and S; mountains in N, SE, and center. The Ogooue R. system covers most of Gabon. **Capital:** Libreville (1993): 362,386.
Government: Type: Republic. **Head of state:** Pres. Omar Bongo; b Dec. 30, 1935; in office: Dec. 2, 1967. **Head of gov.:** Prime Min. Jean-François Ntoutoume-Emane; b Oct. 6, 1939; in office: Jan. 23, 1999. **Local divisions:** 9 provinces. **Defense:** 1.9% of GDP. **Active troops:** 4,700.
Economy: Industries: Oil products, textiles, food and beverages, wood products. **Chief crops:** Cocoa, coffee, palm products. **Minerals:** Oil, manganese, uranium, iron, gold. **Crude oil reserves** (1999): 2.5 bil bbls. **Other resources:** Timber. **Arable land:** 1%. **Livestock** (1997): chickens: 2.70 mil; pigs: 208,000; sheep: 173,000. **Electricity prod.** (1997): 935 mil kWh. **Labor force:** 65% agric.; 30% ind. & commerce.
Finance: Monetary unit: CFA Franc (Sept. 1999: 612.79 = $1 U.S.). **GDP** (1996 est.): $6 bil. **Per capita GDP:** $5,000. **Imports** (1996 est.): $969 mil; partners: France 39%. **Exports** (1996 est.): $3.1 bil; partners: U.S. 50%. **Tourism:** $8 mil. **Budget** (1996 est.): $1.3 bil. **Intl. reserves less gold** (Mar. 1999): $1.49 mil. **Gold:** 13,000 oz t. **Consumer prices** (change in 1997): 4.0%.
Transport: Railroad: Length: 415 mi. **Motor vehicles:** 23,800 pass. cars, 15,700 comm. vehicles. **Civil aviation:** 513.4 mil pass.-mi.; 23 airports. **Chief ports:** Port-Gentil, Owendo, Libreville.
Communications: TV sets: 35 per 1,000 pop. **Radios:** 173 per 1,000 pop. **Telephones** (1997): 37,300 main lines. **Daily newspaper circ.:** 34 per 1,000 pop.
Health: Life expectancy: 53.98 male; 60.08 female. **Births** (per 1,000 pop.): 27.89. **Deaths** (per 1,000 pop.): 13.07. **Natural inc.:** 1.482%. **Infant mortality** (per 1,000 live births): 83.10.
Education: Compulsory: ages 6-16. **Literacy:** 63%.
Major Intl. Organizations: UN (FAO, IBRD, ILO, IMF, IMO, WHO, WTrO), OAU.
Embassy: Suite 200, 2034 20th St. NW 20009; 797-1000.

France established control over the region in the second half of the 19th century. Gabon became independent Aug. 17, 1960. A multiparty political system was introduced in 1990, and a new constitution was enacted Mar. 14, 1991. However, the reelection of longtime Pres. Omar Bongo, on Dec. 5, 1993, prompted rioting and charges of vote fraud; another Bongo victory on Dec. 6, 1998, was likewise allegedly marred by irregularities.

Gabon is one of the most prosperous black African countries, thanks to abundant natural resources, foreign private investment, and government development programs.

The Gambia
Republic of The Gambia

People: Population: 1,336,320. **Age distrib.** (%): <15: 45.7; 65+: 2.7. **Pop. density:** 304 per sq. mi. **Urban:** 30%. **Ethnic groups:** Mandinka 42%, Fula 18%, Wolof 16%, other African. **Principal languages:** English (official), Mandinka, Wolof, Fula. **Chief religions:** Muslim 90%, Christian 9%.
Geography: Area: 4,400 sq. mi. **Location:** On Atlantic coast near W tip of Africa. **Neighbors:** Surrounded on 3 sides

by Senegal. **Topography:** A narrow strip of land on each side of the lower Gambia R. **Capital:** Banjul (1993): 42,407.
Government: Type: Republic. **Head of state and gov.:** Yahya Jammeh; b May 25, 1965; in office: July 23, 1994. **Local divisions:** 5 divisions, 1 city. **Defense:** 3.7% of GDP. **Active troops:** 800.
Economy: Industries: Tourism, peanut processing. **Chief crops:** Peanuts (main export), rice. **Arable land:** 18%. **Livestock** (1997): chickens: 740,000; cattle: 346,295; goats: 250,199; sheep: 181,580. **Fish catch** (1996): 31,521 metric tons. **Electricity prod.** (1997): 75 mil kWh. **Labor force:** 75% agric.; 19% ind., comm., serv.
Finance: Monetary unit: Dalasi (Sept. 1999: 11.44 = $1.00 U.S.). **GDP** (1997 est.): $1.23 bil. **Per capita GDP:** $1,000. **Imports** (1995): $140 mil; partners: China 25%. **Exports** (1995): $160 mil; partners: Belg-Lux 50%, Japan 22%. **Tourism:** $33 mil. **Budget** (FY 1996-97 est.): $98.2 mil. **Intl. reserves less gold** (June 1999): $111.39 mil. **Consumer prices** (change in 1998): 1.1%.
Transport: Motor vehicles: 8,000 pass. cars, 1,000 comm. vehicles. **Civil aviation:** 31.1 mil pass.-mi.; 1 airport. **Chief port:** Banjul.
Communications: Radios: 126 per 1,000 pop. **Telephones:** 25,600 main lines.
Health: Life expectancy: 52.02 male; 56.83 female. **Births** (per 1,000 pop.): 42.76. **Deaths** (per 1,000 pop.): 12.57. **Natural inc.:** 3.019%. **Infant mortality** (per 1,000 live births): 75.33.
Education: Free: ages 7-13. **Literacy:** 39%.
Major Intl. Organizations: UN (FAO, IBRD, ILO, IMF, IMO, WHO, WTrO), the Commonwealth, OAU.
Embassy: Suite 1000, 1155 15th St. NW 20005; 785-1399.
Website: http://www.Gambia.com

The tribes of Gambia were at one time associated with the West African empires of Ghana, Mali, and Songhay. The area became Britain's first African possession in 1588.

Independence came Feb. 18, 1965; republic status within the Commonwealth was achieved in 1970. The country suffered from severe famine in the 1970s. After a coup attempt in 1981, The Gambia formed the confederation of Senegambia with Senegal that lasted until 1989.

On July 23, 1994, after 24 years in power, Pres. Dawda K. Jawara was deposed in a bloodless coup by a military officer, Yahya Jammeh. Jammeh barred political activity, detained potential opponents, and governed by decree. A new constitution was approved by referendum, Aug. 8, 1996. On Sept. 27 Jammeh won the presidential election. Parliamentary balloting on Jan. 2, 1997, completed the nominal return to civilian rule, but Jammeh retained a firm grip on power.

Georgia
Republic of Georgia

People: Population: 5,066,499. **Age distrib.** (%): <15: 21.1; 65+: 12.2. **Pop. density:** 188 per sq. mi. **Urban:** 59%. **Ethnic groups:** Georgian 70%, Armenian 8%, Russian 6%. **Principal languages:** Georgian (official), Russian. **Chief religions:** Georgian Orthodox 65%, Muslim 11%, Russian Orthodox 10%.
Geography: Area: 26,900 sq. mi. **Location:** SW Asia, on E coast of Black Sea. **Neighbors:** Russia on N and NE, Turkey and Armenia on S, Azerbaijan on SE. **Topography:** Separated from Russia on NE by main range of the Caucasus Mts. **Capital:** Tbilisi: 1,342,000*.
Government: Type: Republic. **Head of state:** Pres. Eduard A. Shevardnadze; b Jan. 25, 1928; in office: Mar. 10, 1992. **Local divisions:** 9 districts, 2 autonomous republics, and Tbilisi. **Defense:** 2.9% of GDP. **Active troops:** 33,200.
Economy: Industries: Steel, machinery, trucks, textiles. **Chief crops:** Citrus, potatoes, vegetables, grapes, tea. **Minerals:** Manganese, iron, copper, coal. **Other resources:** Forests. **Crude oil reserves** (1999): 35 mil bbls. **Arable land:** 9%. **Livestock** (1997): chickens: 13.50 mil; cattle: 1.03; sheep: 543,000; pigs: 330,000. **Electricity prod.** (1997): 6.979 bil kWh. **Labor force:** 24% mining, manuf.; 20% services, 9% agric.
Finance: Monetary unit: Lavi (Aug. 1999: 1.84 = $1 U.S.). **GDP** (1997 est.): $8.1 bil. **Per capita GDP:** $1,570. **Imports** (1996 est.): $733 mil; partners: Turkmenistan 71%, Turkey 13%. **Exports** (1996 est.): $400 mil; partners: Russia 46%, Turkey 18%. **Tourism:** $440 mil.
Transport: Railroad: Length: 983 mi. **Motor vehicles:** 442,000 pass. cars, 50,000 comm. vehicles. **Civil aviation:** 128.1 mil pass.-mi.; 1 airport. **Chief ports:** Batumi, Sukhumi.
Communications: TV sets: 220 per 1,000 pop. **Telephones** (1997): 620,400 main lines.
Health: Life expectancy: 61.13 male; 68.32 female. **Births** (per 1,000 pop.): 11.64. **Deaths** (per 1,000 pop.): 14.30. **Nat-

ural increase: -0.266%. **Hosp. beds** (1993): 1 per 95 persons. **Physicians** (1993): 1 per 182 persons. **Infant mortality** (per 1,000 live births): 52.01.
  Education: Compulsory: ages 6-14. **Literacy:** 99%.
  Major Intl. Organizations: UN (FAO, IBRD, ILO, IMF, IMO, WHO), CIS, OSCE.
  Embassy: Suite 424, 1511 K St. NW 20005; 393-5959.
  Website: http://www.parliament.ge

The region, which contained the ancient kingdoms of Colchis and Iberia, was Christianized in the 4th century and conquered by Arabs in the 8th century. It expanded to include an area from the Black Sea to the Caspian and parts of Armenia and Persia before its disintegration under the impact of Mongol and Turkish invasions. Annexation by Russia in 1801 led to the Russian war with Persia, 1804-1813. Georgia entered the USSR in 1922 and became a constituent republic in 1936.

Georgia declared independence Apr. 9, 1991. It became an independent state when the Soviet Union disbanded Dec. 26, 1991. There was fighting during 1991 between rebel forces and loyalists of Pres. Zviad Gamsakhurdia, who fled the capital Jan. 6, 1992. The ruling Military Council picked former Soviet Foreign Minister Eduard A. Shevardnadze to chair a newly created State Council. An attempted coup by forces loyal to Gamsakhurdia was crushed June 24, 1992. Shevardnadze was later elected president. Gamsakhurdia died Jan. 1994, reportedly by suicide.

In Abkhazia, an autonomous republic within Georgia, ethnic Abkhazis, reportedly aided by Russia, launched a bloody military campaign and, by late 1993, had gained control of much of the region. A cease-fire providing for Russian peacekeepers was signed in Moscow May 14, 1994. Intermittent clashes continued into the late 1990s.

On Feb. 3, 1994, Georgia signed agreements with Russia for economic and military cooperation. On Mar. 1, Georgia's Supreme Council ratified membership by Georgia in the Commonwealth of Independent States.

Shevardnadze was wounded by a car bomb Aug. 29, 1995, while on his way to Parliament to sign a new constitution. He was reelected president Nov. 5. Shevardnadze escaped another assassination attempt, Feb. 9, 1998, when gunmen ambushed his motorcade. A mutiny by more than 200 soldiers opposed to Shevardnadze was crushed Oct. 19.

Germany
Federal Republic of Germany

People: Population: 82,087,361. **Age distrib.** (%): <15: 15.4; 65+: 16.1. **Pop. density:** 596 per sq. mi. **Urban:** 87%. **Ethnic groups:** German 92%, Turkish 2%. **Principal language:** German (official). **Chief religions:** Protestant 38%, Roman Catholic 34%.

Geography: Area: 137,800 sq. mi. **Location:** In central Europe. **Neighbors:** Denmark on N; Netherlands, Belgium, Luxembourg, France on W; Switzerland, Austria on S; Czech Rep., Poland on E. **Topography:** Germany is flat in N, hilly in center and W, and mountainous in Bavaria in the S. Chief rivers are Elbe, Weser, Ems, Rhine, and Main, all flowing toward North Sea, and Danube, flowing toward Black Sea. **Capital:** Berlin. **Cities** (1996 est.): Berlin 3,458,763; Munich 1,225,809; Hamburg 1,707,986; Cologne 964,346; Frankfurt 647,304; Essen 611,827.

Government: Type: Federal republic. **Head of state:** Pres. Johannes Rau; b Jan. 16, 1931; in office: July 1, 1999. **Head of gov.:** Chan. Gerhard Schröder; b Apr. 7, 1944; in office: Oct. 27, 1998. **Local divisions:** 16 laender (states). **Defense:** 1.6% of GDP. **Active troops:** 347,100.

Economy: Industries: Steel, ships, vehicles, machinery, electronics, coal, chemicals, iron, cement, food and beverages. **Chief crops:** Grains, potatoes, sugar beets. **Minerals:** Coal, potash, lignite, iron, uranium. **Crude oil reserves** (1999): 388.5 mil bbls. **Other resources:** Timber. **Arable land:** 33%. **Livestock** (1997): chickens: 102.73 mil; goats: 110,000; pigs: 24.80 mil; cattle: 15.23 mil; sheep: 2.30 mil. **Fish catch** (1997): 259,352 metric tons. **Electricity prod.** (1996): 515.777 bil kWh. **Labor force:** 56% services, 41% ind.; 3% agric.

Finance: Monetary unit: Mark (Sept. 1999: 1.83 = $1 U.S.). Euro (Sept. 1999: 1.07 = $1 U.S.). **GDP** (1997 est.): $1.74 tril. **Per capita GDP:** $20,800. **Imports** (1996): $455.7 bil; partners: EU 56%. **Exports** (1996): $521.1 bil; partners: EU 58%. **Tourism:** $16.84 bil. **Budget** (1995): $832.1 bil. **Intl. reserves less gold** (June 1999): $60.75 bil. **Gold:** 111.52 mil oz t. **Consumer prices** (change in 1998): 1.0%.

Transport: Railroad: Length: 54,994 mi. **Motor vehicles** (1997): 41.33 mil pass. cars, 3.17 mil comm. vehicles. **Civil**

aviation: 53.6 bil pass.-mi.; 40 airports. **Chief ports:** Hamburg, Bremen, Bremerhaven, Lubeck, Rostock.
  Communications: TV sets: 551 per 1,000 pop. **Radios:** 1,836 per 1,000 pop. **Telephones:** 46,500,000 main lines. **Daily newspaper circ.:** 375 per 1,000 pop.
  Health: Life expectancy: 74.01 male; 80.50 female. **Births** (per 1,000 pop.): 8.68. **Deaths** (per 1,000 pop.): 10.76. **Natural inc.:** -0.208%. **Hosp. beds** (1996): 1 per 130 persons. **Physicians** (1996): 1 per 293 persons. **Infant mortality** (per 1,000 live births): 5.14.
  Education: Compulsory: ages 6-15. **Literacy** (1993): 100%.
  Major Intl. Organizations: UN and all of its specialized agencies, EU, NATO, OECD, OSCE.
  Embassy: 4645 Reservoir Rd. NW 20007; 298-4000.
  Websites: http://www.undp.org/missions/germany
    http://www.government.de/english/01/newsf.html

Germany is a central European nation originally composed of numerous states, with a common language and traditions, that were united in one country in 1871; Germany was split into 2 countries from the end of World War II until 1990, when it was reunified.

History and government. Germanic tribes were defeated by Julius Caesar, 55 and 53 BC, but Roman expansion N of the Rhine was stopped in AD 9. Charlemagne, ruler of the Franks, consolidated Saxon, Bavarian, Rhenish, Frankish, and other lands; after him the eastern part became the German Empire. The Thirty Years' War, 1618-1648, split Germany into small principalities and kingdoms. After Napoleon, Austria contended with Prussia for dominance, but lost the Seven Weeks' War to Prussia, 1866. Otto von Bismarck, Prussian chancellor, formed the North German Confederation, 1867.

In 1870 Bismarck maneuvered Napoleon III into declaring war. After the quick defeat of France, Bismarck formed the **German Empire** and on Jan. 18, 1871, in Versailles, proclaimed King Wilhelm I of Prussia German emperor (Deutscher kaiser).

The German Empire reached its peak before World War I in 1914, with 208,780 sq. mi., plus a colonial empire. After that war Germany ceded Alsace-Lorraine to France; West Prussia and Posen (Poznan) province to Poland; part of Schleswig to Denmark; lost all colonies and ports of Memel and Danzig.

Republic of Germany, 1919-1933, adopted the Weimar constitution; met reparation payments and elected Friedrich Ebert and Gen. Paul von Hindenburg presidents.

Third Reich, 1933-1945, Adolf Hitler led the National Socialist German Workers' (Nazi) party after World War I. In 1923 he attempted to unseat the Bavarian government and was imprisoned. Pres. von Hindenburg named Hitler chancellor Jan. 30, 1933; on Aug. 3, 1934, the day after Hindenburg's death, the cabinet joined the offices of president and chancellor and made Hitler fuehrer (leader). Hitler abolished freedom of speech and assembly, and began a long series of persecutions climaxed by the murder of millions of Jews and others.

He repudiated the Versailles treaty and reparations agreements, remilitarized the Rhineland (1936), and annexed Austria (Anschluss, 1938). At Munich he made an agreement with Neville Chamberlain, British prime minister, which permitted Germany to annex part of Czechoslovakia. He signed a non-aggression treaty with the USSR, 1939 and declared war on Poland Sept. 1, 1939, precipitating World War II. With total defeat near, Hitler committed suicide in Berlin Apr. 1945. The victorious Allies voided all acts and annexations of Hitler's Reich.

Division of Germany. Germany was sectioned into 4 zones of occupation, administered by the Allied Powers (U.S., USSR, U.K., and France). The USSR took control of many E German states. The territory E of the so-called Oder-Neisse line was assigned to, and later annexed by, Poland. Northern East Prussia (now Kaliningrad) was annexed by the USSR. Administration of the remaining regions, in the W and S (which make up about two-thirds of present-day Germany), was split among the Western Allies.

There was also created the area of Greater Berlin, within but not part of the Soviet zone, administered by the 4 occupying powers under the Allied Command. In 1948 the USSR withdrew, established its single command in East Berlin, and cut off supplies. The Western Allies utilized a gigantic airlift to bring food to West Berlin, 1948-49.

In 1949, 2 separate German states were established; in May the zones administered by the Western Allies became West Germany, capital: Bonn; in Oct. the Soviet sector became East Germany, capital: East Berlin. West Berlin was considered an enclave of West Germany, although its status was disputed by the Soviet bloc.

East Germany. The German Democratic Republic (East Germany) was proclaimed in the Soviet sector of Berlin Oct. 7, 1949. It was proclaimed fully sovereign in 1954, but Soviet troops remained on grounds of security and the 4-power Potsdam agreement.

Coincident with the entrance of West Germany into the European defense community in 1952, the East German government decreed a prohibited zone 3 miles deep along its 600-mile border with West Germany and cut Berlin's telephone system in two. Berlin was further divided by erection of a fortified wall in 1961, after over 3 million East Germans had emigrated West; an exodus of refugees to the West continued, though on a smaller scale.

East Germany suffered severe economic problems at least until the mid-1960s. Then a "new economic system" was introduced, easing central planning controls and allowing factories to make profits provided they were reinvested in operations or redistributed to workers as bonuses. By the early 1970s, the economy was highly industrialized, and the nation was credited with the highest standard of living among Warsaw Pact countries. But growth slowed in the late 1970s, because of shortages of natural resources and labor, and a huge debt to lenders in the West. Comparison with the lifestyle in the West caused many young people to leave the country.

The government firmly resisted following the USSR's policy of *glasnost*, but by Oct. 1989, was faced with nationwide demonstrations demanding reform. Pres. Erich Honecker, in office since 1976, was forced to resign, Oct. 18. On Nov. 4, the border with Czechoslovakia was opened and permission granted for refugees to travel to the West. On Nov. 9, the East German government announced its decision to open the border with the West, signaling the end of the "Berlin Wall," which was the supreme emblem of the cold war. On Aug. 23, 1990, the East German parliament agreed to formal unification with West Germany; this occurred Oct. 3.

West Germany. The Federal Republic of Germany (West Germany) was proclaimed May 23, 1949, in Bonn, after a constitution had been drawn up by a consultative assembly formed by representatives of the 11 laender (states) in the French, British, and American zones. Later reorganized into 9 units, the laender numbered 10 with the addition of the Saar, 1957. Berlin also was granted land (state) status, but the 1945 occupation agreements placed restrictions on it.

The occupying powers, the U.S., Britain, and France, restored civil status, Sept. 21, 1949. The Western Allies ended the state of war with Germany in 1951 (the U.S. resumed diplomatic relations July 2), while the USSR did so in 1955. The powers lifted controls and the republic became fully independent May 5, 1955.

Dr. Konrad Adenauer, Christian Democrat, was made chancellor Sept. 15, 1949, reelected 1953, 1957, 1961. Willy Brandt, heading a coalition of Social Democrats and Free Democrats, became chancellor Oct. 21, 1969. (He resigned May 1974 because of a spy scandal.)

In 1970 Brandt signed friendship treaties with the USSR and Poland. In 1971, the U.S., Britain, France, and the USSR signed an agreement on Western access to West Berlin. In 1972 East and West Germany signed their first formal treaty, implementing the agreement easing access to West Berlin. In 1973 a West Germany-Czechoslovakia pact normalized relations and nullified the 1938 "Munich Agreement."

West Germany experienced strong economic growth starting in the 1950s. The country led Europe in provisions for worker participation in the management of industry.

In 1989 the changes in the East German government and opening of the Berlin Wall sparked talk of reunification of the 2 Germanys. In 1990, under Chancellor Kohl's leadership, West Germany moved rapidly to reunite with East Germany.

A New Era. As Communism was being rejected in East Germany, talks began concerning German reunification. At a meeting in Ottawa, Feb. 1990, the foreign ministers of the World War II "Big Four" Allied nations and of East Germany and West Germany reached agreement on a format for high-level talks on German reunification.

In May, NATO ministers adopted a package of proposals on reunification, including the inclusion of the united Germany as a full member of NATO and the barring of the new Germany from having its own nuclear, chemical, or biological weapons. In July, the USSR agreed to conditions that would allow Germany to become a member of NATO.

The 2 nations agreed to monetary unification under the West German mark beginning in July. The merger of the 2 Germanys took place Oct. 3, and the first all-German elections since 1932 were held Dec. 2. Eastern Germany received more than $1 trillion in public and private funds from western Germany between 1990 and 1995.

In 1991, Berlin again became the capital of Germany; the

legislature, most administrative offices, and most foreign embassies had shifted from Bonn to Berlin by late 1999.

Germany's highest court ruled, July 12, 1994, that German troops could participate in international military missions abroad, when approved by Parliament. Ceremonies were held marking the final withdrawal of Russian troops from Germany, Aug. 31. Ceremonies were held the following week marking the final withdrawal of American, British, and French troops from Berlin. General elections Oct. 16 left Chancellor Helmut Kohl's governing coalition with a slim parliamentary majority. On Oct. 31, 1996, after more than 14 years in office, Kohl surpassed Adenauer as Germany's longest-serving chancellor in the 20th century.

Unemployment hit a postwar high of 12.6% in Jan. 1998. The Kohl era ended with the defeat of the Christian Democrats in parliamentary elections Sept. 27; Gerhard Schröder, of the Social Democratic Party, became chancellor. Germany contributed 8,500 troops to the NATO-led security force (KFOR) that entered Kosovo in June 1999.

Helgoland, an island of 130 acres in the North Sea, was taken from Denmark by a British Naval Force in 1807 and later ceded to Germany to become part of Schleswig-Holstein province in return for rights in East Africa. The heavily fortified island was surrendered to UK, May 23, 1945, demilitarized in 1947, and returned to West Germany, Mar. 1, 1952. It is a free port.

Ghana

Republic of Ghana

People: Population: 18,887,626. **Age distrib.** (%): <15: 42.4; 65+: 3.2. **Pop. density:** 205 per sq. mi. **Urban:** 36%. **Ethnic groups:** Akan 44%, Moshi-Dagomba 16%, Ewe 13%, Ga 8%. **Principal languages:** English (official), Akan, Moshi-Dagomba, Ewe, Ga. **Chief religions:** Indigenous beliefs 38%, Muslim 30%, Christian 24%.

Geography: Area: 92,100 sq. mi. **Location:** On southern coast of W Africa. **Neighbors:** Côte d'Ivoire on W, Burkina Faso on N, Togo on E. **Topography:** Most of Ghana consists of low fertile plains and scrubland, cut by rivers and by the artificial Lake Volta. **Capital:** Accra: 1,673,000*.

Government: Type: Republic. **Head of state and gov.:** Pres. Jerry Rawlings; b June 22, 1947; in office: Dec. 31, 1981. **Local divisions:** 10 regions. **Defense:** 1.5% of GDP. **Active troops:** 7,000.

Economy: Industries: Aluminum, light manufacturing, mining, lumbering, food processing. **Chief crops:** Cocoa, coffee, rice, cassava, peanuts, corn. **Minerals:** Gold, manganese, industrial diamonds, bauxite. **Crude oil reserves** (1999): 16.5 mil bbls. **Other resources:** Timber, rubber. **Arable land:** 12%. **Livestock** (1997): chickens: 13.30 mil; goats: 2.20 mil; sheep: 2.10 mil; cattle: 1.15 mil; pigs: 395,000. **Fish catch** (1997): 446,483 metric tons. **Electricity prod.** (1997): 5.853 bil kWh. **Labor force:** 61% agric. & fishing; 29% services, 10% ind.

Finance: Monetary unit: Cedi (Sept. 1999: 2,652.00 = $1 U.S.). **GDP** (1997 est.): $36.2 bil. **Per capita GDP:** $2,000. **Imports** (1995): $1.84 bil; partners: Germany 14%, UK 12%, U.S. 12%. **Exports** (1996 est.): $1.57 bil; partners: U.K. 16%, Italy 8%. **Tourism:** $274 mil. **Budget** (1996 est.): $1.47 bil. **Intl. reserves less gold** (Mar. 1999): $382.5 mil. **Gold:** 272,000 oz t. **Consumer prices** (change in 1998): 14.6%.

Transport: Railroad: Length: 592 mi. **Motor vehicles:** 90,000 pass. cars, 45,000 comm. vehicles. **Civil aviation:** 436.4 mil pass.-mi.; 1 airport. **Chief ports:** Tema, Takoradi.

Communications: TV sets: 15 per 1,000 pop. **Radios:** 249 per 1,000 pop. **Telephones** (1997): 105,500 main lines. **Daily newspaper circ.:** 64 per 1,000 pop.

Health: Life expectancy: 55.08 male; 59.27 female. **Births** (per 1,000 pop.): 31.79. **Deaths** (per 1,000 pop.): 10.40. **Natural inc.:** 2.139%. **Hosp. beds** (1994): 1 per 638 persons. **Physicians** (1994): 1 per 22,970 persons. **Infant mortality** (per 1,000 live births): 76.15.

Education: Compulsory: ages 6-16. **Literacy:** 64%.

Major Intl. Organizations: UN and all of its specialized agencies, the Commonwealth, OAU.

Embassy: 3512 International Dr. NW 20008; 686-4520.

Named for an African empire along the Niger River, AD 400-1240, Ghana was ruled by Britain for 113 years as the Gold Coast. The UN in 1956 approved merger with the British Togoland trust territory. Independence came Mar. 6, 1957, and republic status within the Commonwealth in 1960.

Pres. Kwame Nkrumah built hospitals and schools, promoted development projects like the Volta R. hydroelectric and aluminum plants but ran the country into debt, jailed opponents, and was accused of corruption. A 1964 referendum gave Nkrumah dictatorial powers and set up a one-party socialist state. Nkrumah was overthrown in 1966 by a police-army coup, which expelled Chinese and East German teachers and technicians. Elections were held in 1969, but 4 further coups occurred in 1972, 1978, 1979, and 1981. The 1979 and 1981 coups, led by Flight Lieut. Jerry Rawlings, were followed by suspension of the constitution and banning of political parties. A new constitution, allowing multiparty politics, was approved in April 1992.

In Feb. 1993 more than 1,000 people were killed in ethnic clashes in northern Ghana. Rawlings won the presidential election of Dec. 7, 1996. Kofi Annan, a career UN diplomat from Ghana, became UN secretary general on Jan. 1, 1997. U.S. Pres. Clinton opened a 12-day African tour with a speech, Mar. 23, 1998, to an Accra audience estimated at over 500,000.

Greece
Hellenic Republic

People: Population: 10,707,135. **Age distrib.** (%): <15: 15.8; 65+: 16.8. **Pop. density:** 210 per sq. mi. **Urban:** 59%. **Ethnic groups:** Greek 98%. (**Note:** Greek govt. states there are no ethnic divisions in Greece.) **Principal languages:** Greek (official), English, French. **Chief religion:** Greek Orthodox 98% (official).

Geography: Area: 50,900 sq. mi. **Location:** Occupies southern end of Balkan Peninsula in SE Europe. **Neighbors:** Albania, Macedonia, Bulgaria on N; Turkey on E. **Topography:** About three-quarters of Greece is nonarable, with mountains in all areas. Pindus Mts. run through the country N to S. The heavily indented coastline is 9,385 mi. long. Of over 2,000 islands, only 169 are inhabited, among them Crete, Rhodes, Milos, Kerkira (Corfu), Chios, Lesbos, Samos, Euboea, Delos, Mykonos. **Capital:** Athens (1995 met. est.): 3,072,922.

Government: Type: Parliamentary republic. **Head of state:** Pres. Konstantinos Stephanopoulos; b 1926; in office: Mar. 8, 1995. **Head of gov.:** Prime Min. Costas Simitis; b June 23, 1936; in office: Jan. 18, 1996. **Local divisions:** 13 regions comprising 51 prefectures. **Defense:** 4.6% of GDP. **Active troops:** 162,300.

Economy: Industries: Tourism, textiles, chemicals, metals, wine, food processing. **Chief crops:** Grains, corn, sugar beets, cotton, tobacco, olives, grapes, citrus and other fruits, tomatoes. **Minerals:** Bauxite, lignite, magnesite, marble, oil. **Crude oil reserves** (1999): 10 mil bbls. **Arable land:** 19%. **Livestock** (1997): chickens: 28.00 mil; sheep: 9.52 mil; goats: 5.88 mil; pigs: 938,000; cattle: 580,000. **Fish catch** (1996): 162,123 metric tons. **Electricity prod.** (1997): 40.746 bil kWh. **Labor force:** 52% services; 25% ind.; 23% agric.

Finance: Monetary unit: Drachma (Sept. 1999: 304.91 = $1 U.S.). **GDP** (1997 est.): $137.4 bil. **Per capita GDP:** $13,000. **Imports** (1997 est.): $27 bil; partners: Italy 18%, Germany 16%. **Exports** (1997 est.): $9.8 bil; partners: Germany 22%, Italy 14%. **Tourism:** $3.93 bil. **Budget** (1998 est.): $45 bil. **Intl. reserves less gold** (June 1999): $20.10 bil. **Gold:** 4.53 mil oz t. **Consumer prices** (change in 1998): 4.8%.

Transport: Railroad: Length: 1,537 mi. **Motor vehicles:** 2.34 mil pass. cars, 939,923 comm. vehicles. **Civil aviation:** 5.8 bil pass.-mi.; 36 airports. **Chief ports:** Piraeus, Thessaloníki, Patrai.

Communications: TV sets: 442 per 1,000 pop. **Radios:** 402 per 1,000 pop. **Telephones:** 5,535,500 main lines. **Daily newspaper circ.:** 135 per 1,000 pop.

Health: Life expectancy: 75.8 male; 81.0 female. **Births** (per 1,000 pop.): 10. **Deaths** (per 1,000 pop.): 9. **Natural inc.:** 0.03%. **Hosp. beds** (1993): 1 per 199 persons. **Physicians** (1994): 1 per 258 persons. **Infant mortality** (per 1,000 live births): 7.

Education: Free, compulsory: ages 6-15. **Literacy** (1993): 95%.

Major Intl. Organizations: UN (FAO, IBRD, ILO, IMF, IMO, WHO, WTrO), EU, NATO, OECD, OSCE.

Embassy: 2221 Massachusetts Ave. NW 20008; 939-5800. **Website:** http://www.hiway.gr/gi

The achievements of ancient Greece in art, architecture, science, mathematics, philosophy, drama, literature, and democracy became legacies for succeeding ages. Greece reached the height of its glory and power, particularly in the Athenian city-state, in the 5th century BC. Greece fell under Roman rule in the 2d and 1st centuries BC. In the 4th century AD it became part of the Byzantine Empire, and after the fall of Constantinople to the Turks in 1453, part of the Ottoman Empire.

Greece won its war of independence from Turkey 1821-1829, and became a kingdom. A republic was established 1924; the monarchy was restored, 1935, and George II, King of the Hellenes, resumed the throne. In Oct. 1940, Greece re-jected an ultimatum from Italy. Nazi support resulted in its defeat and occupation by Germans, Italians, and Bulgarians. By the end of 1944 the invaders withdrew. Communist resistance forces were defeated by Royalist and British troops. A plebiscite again restored the monarchy.

Communists waged guerrilla war 1947-49 against the government but were defeated with the aid of the U.S. A period of reconstruction and rapid development followed, mainly with conservative governments under Premier Constantine Karamanlis. The Center Union, led by George Papandreou, won elections in 1963 and 1964, but King Constantine, who acceded in 1964, forced Papandreou to resign. A period of political maneuvers ended in the military takeover of April 21, 1967, by Col. George Papadopoulos. King Constantine tried to reverse the consolidation of the harsh dictatorship Dec. 13, 1967, but failed and fled to Italy. Papadopoulos was ousted Nov. 25, 1973.

Greek army officers serving in the National Guard of Cyprus staged a coup on the island July 15, 1974. Turkey invaded Cyprus a week later, precipitating the collapse of the Greek junta, which was implicated in the Cyprus coup. Democratic government returned (and in 1975 the monarchy was abolished).

The 1981 electoral victory of the Panhellenic Socialist Movement (Pasok) of Andreas Papandreou brought substantial changes in Greece's internal and external policies. A scandal centered on George Kostokas, a banker and publisher, led to the arrest or investigation of leading Socialists, implicated Papandreou, and contributed to the defeat of the Socialists at the polls in 1989. However, Papandreou, who was narrowly acquitted Jan. 1992 of corruption charges, led the Socialists to a comeback victory in general elections Oct. 10, 1993.

Tensions between Greece and the Former Yugoslav Republic of Macedonia eased when the 2 countries agreed to normalize relations Sept. 13, 1995. The ailing Papandreou was replaced as prime minister by Costas Simitis, Jan. 18, 1996. Simitis led the Socialists to victory in the election of Sept. 22. The International Olympic Committee, Sept. 5, 1997, chose Athens to host the Summer Games in 2004. An earthquake that shook Athens Sept. 7, 1999, killed at least 143 people and left over 60,000 homeless.

Grenada

People: Population: 97,008. **Age distrib.** (%): <15: 42.7; 65+: 4.4. **Pop. density:** 746 per sq. mi. **Urban:** 36%. **Ethnic groups:** Mostly black African. **Principal languages:** English (official), French patois. **Chief religions:** Roman Catholic 53%, Protestant 33%.

Geography: Area: 130 sq. mi. **Location:** In Caribbean, 90 mi. N of Venezuela. **Neighbors:** Venezuela, Trinidid & Tobago to S; St. Vincent & the Grenadines to N. **Topography:** Main island is mountainous; country includes Carriacou and Petit Martinique islands. **Capital:** Saint George's (1991): 4,439.

Government: Type: Parliamentary democracy. **Head of state:** Queen Elizabeth II, represented by Gov.-Gen. Daniel Williams; b Nov. 4, 1935; in office: Aug. 8, 1996. **Head of gov.:** Prime Min. Keith Mitchell; b Nov. 12, 1946; in office: June 22, 1995. **Local divisions:** 6 parishes, 1 dependency.

Economy: Industries: Tourism, textiles, food and beverages. **Chief crops:** Nutmeg, bananas, cocoa, mace. **Resources:** Timber. **Arable land:** 15%. **Livestock** (1997): chickens: 220,000. **Electricity prod.** (1997): 97 mil kWh. **Labor force:** 29% services; 17% agric.; 13% const.

Finance: Monetary unit: East Caribbean Dollar (Sept. 1999: 2.70 = $1 U.S.). **GDP** (1996 est.): $300 mil. **Per capita GDP:** $3,200. **Imports** (1996 est.): $128 mil; partners: U.S. 31%, Caricom 24%, UK 14%. **Exports** (1996 est.): $24 mil; partners: Caricom 32%, UK 20%. **Tourism:** $61 mil. **Budget** (1996 est.): $126.7 mil. **Intl. reserves less gold** (Apr. 1999): $51.21 mil. **Consumer prices** (change in 1997): 1.2%.

Transport: Civil aviation: 2 airports. **Chief ports:** Saint George's, Grenville.

Communications: TV sets: 154 per 1,000 pop. **Radios:** 460 per 1,000 pop. **Telephones:** 27,500 main lines.

Health: Life expectancy: 68.97 male; 74.29 female. **Births** (per 1,000 pop.): 27.62. **Deaths** (per 1,000 pop.): 5.15. **Natural inc.:** 2.247%. **Hosp. beds** (1996): 1 per 223 persons. **Physicians** (1992): 1 per 2,045 persons. **Infant mortality** (per 1,000 live births): 11.13.

Education: Free, compulsory: ages 5-16. **Literacy** (1994): 85%.

Major Intl. Organizations: UN (FAO, IBRD, ILO, IMF, WHO, WTrO), Caricom, the Commonwealth, OAS, OECS.

Embassy: 1701 New Hampshire Ave. NW 20009; 265-2561.

Website: http://www.grenada.org

Columbus sighted Grenada in 1498. First European settlers were French, 1650. The island was held alternately by France and England until final British occupation, 1784. Grenada became fully independent Feb. 7, 1974, during a general strike. It is the smallest independent nation in the western hemisphere.

On Oct. 14, 1983, a military coup ousted Prime Minister Maurice Bishop, who was put under house arrest, later freed by supporters, rearrested, and, finally, on Oct. 19, executed. U.S. forces, with a token force from 6 area nations, invaded Grenada, Oct. 25. Resistance from the Grenadian army and Cuban advisors was quickly overcome as most people welcomed the invading forces. U.S. troops left Grenada in June 1985. Cuban Pres. Castro received an enthusiastic greeting when visiting Grenada Aug. 2-3, 1998.

Guatemala
Republic of Guatemala

People: Population: 12,335,580. **Age distrib.** (%): <15: 42.7; 65+: 3.6. **Pop. density:** 294 per sq. mi. **Urban:** 39%. **Ethnic groups:** Mestizo 56%, Amerindian 44%. **Principal languages:** Spanish (official), Mayan languages. **Religion:** Mostly Roman Catholic, some Protestant, traditional Mayan.

Geography: Area: 42,000 sq. mi. **Location:** In Central America. **Neighbors:** Mexico on N and W, El Salvador on S, Honduras and Belize on E. **Topography:** The central highland and mountain areas are bordered by the narrow Pacific coast and the lowlands and fertile river valleys on the Caribbean. There are numerous volcanoes in S, more than half a dozen over 11,000 ft. **Capital:** Guatemala City: 2,205,000*.

Government: Type: Republic. **Head of state and gov.:** Pres. Alvaro Arzú Irigoyen; b Mar. 14, 1946; in office: Jan. 14, 1996. **Local divisions:** 22 departments. **Defense:** 1.5% of GDP. **Active troops:** 40,700.

Economy: Industries: Furniture, rubber, sugar, chemicals, textiles. **Chief crops:** Coffee, sugar, bananas, corn, cardamom. **Minerals:** Oil, nickel. **Crude oil reserves** (1999): 526 mil bbls. **Other resources:** Rare woods, fish, chicle. **Arable land:** 12%. **Livestock** (1997): chickens: 24.00 mil; cattle: 2.33 mil; pigs: 825,600; sheep: 551,200; goats: 109,200. **Electricity prod.** (1997): 3.362 bil kWh. **Labor force:** 58% agric.; 14% serv.; 14% manuf.

Finance: Monetary unit: Quetzal (Sept. 1999: 7.78 = $1 U.S.). **GDP** (1997 est.): $45.8 bil. **Per capita GDP:** $4,000. **Imports** (1997 est.): $3.3 bil; partners: U.S. 44%. **Exports** (1997 est.): $2.9 bil; partners: U.S. 37%. **Tourism:** $280 mil. **Budget** (1996 est.): $1.88 bil. **Intl. reserves less gold** (June 1999): $1.22 bil. **Gold:** 215,000 oz t. **Consumer prices** (change in 1998): 7.0%.

Transport: Railroad: Length: 549 mi. **Motor vehicles:** 102,000 pass. cars, 97,000 comm. vehicles. **Civil aviation:** 228.9 mil pass.-mi.; 2 airports. **Chief ports:** Puerto Barrios, San Jose.

Communications: TV sets: 45 per 1,000 pop. **Radios:** 52 per 1,000 pop. **Telephones** (1997): 429,700 main lines. **Daily newspaper circ.:** 29 per 1,000 pop.

Health: Life expectancy: 63.78 male; 69.24 female. **Births** (per 1,000 pop.): 35.57. **Deaths** (per 1,000 pop.): 6.80. **Natural inc.:** 2.877%. **Infant mortality** (per 1,000 live births): 46.15.

Education: Free, compulsory: ages 7-14. **Literacy:** 56%.

Major Intl. Organizations: UN (FAO, IBRD, ILO, IMF, IMO, WHO, WTrO), OAS.

Embassy: 2220 R St. NW 20008; 745-4952.

The old Mayan Indian empire flourished in what is today Guatemala for over 1,000 years before the Spanish.

Guatemala was a Spanish colony 1524-1821; briefly a part of Mexico and then of the U.S. of Central America, the republic was established in 1839.

Since 1945 when a liberal government was elected to replace the long-term dictatorship of Jorge Ubico, the country has seen a variety of military and civilian governments and periods of civil war. Dissident army officers seized power Mar. 23, 1982, denouncing a presidential election as fraudulent and pledging to restore "authentic democracy" to the nation. Political violence caused large numbers of Guatemalans to seek refuge in Mexico. Another military coup occurred Oct. 8, 1983. The nation returned to civilian rule in 1986.

The crisis-ridden government of Pres. Jorge Serrano Elías was ousted by the military June 1, 1993. Ramiro de León Carpio was elected president by Congress June 6. A conservative businessman, Alvaro Arzú Irigoyen, won the presidency, Jan. 7, 1996. On Sept. 19 the Guatemalan government and leftist rebels approved a peace accord; the final agreement was signed Dec. 29. During more than 35 years of armed conflict, some 200,000 people were killed or "disappeared"

(and are presumed dead); most of these casualties were attriubutred to the government and its paramilitary allies.

Violent episodes in 1998 included the daylight ambush of a busload of U.S. college students, Jan. 16, resulting in the rape of five young women, and the murder of Bishop Juan José Girardi, a human rights activist, Apr. 26. U.S. Pres. Bill Clinton, on a visit to Guatemala Mar. 10, 1999, apologized for aid the U.S. had given to forces which he said "engaged in violence and widespread repression." On May 16, voters rejected constitutional reforms backed by Arzú that would have granted equal rights to indigenous peoples and curbed the powers of the military.

Guinea
Republic of Guinea

People: Population: 7,538,953. **Age distrib.** (%): <15: 43.7; 65+: 2.7. **Pop. density:** 79 per sq. mi. **Urban:** 30%. **Ethnic groups:** Peuhl 40%, Malinke 30%, Soussou 20%, smaller tribes 10%. **Principal languages:** French (official), tribal languages. **Chief religions:** Muslim 85%, Christian 8%.

Geography: Area: 94,900 sq. mi. **Location:** On Atlantic coast of W Africa. **Neighbors:** Guinea-Bissau, Senegal, Mali on N; Côte d'Ivoire on E; Liberia on S. **Topography:** A narrow coastal belt leads to the mountainous middle region, the source of the Gambia, Senegal, and Niger rivers. Upper Guinea, farther inland, is a cooler upland. The SE is forested. **Capital:** Conakry: 1,558,000*.

Government: Type: Republic. **Head of state:** Pres. Gen. Lansana Conté; b 1934; in office: Apr. 5, 1984. **Head of gov.:** Lamine Sidime; in office: Mar. 8, 1999. **Local divisions:** 33 prefectures, 1 national capital. **Defense:** 1.6% of GDP. **Active troops:** 9,700.

Economy: Industries: Mining, light manufacturing, agricultural processing. **Chief crops:** Bananas, pineapples, rice, palm kernels, coffee, cassava. **Minerals:** Bauxite, iron, diamonds, gold. **Arable land:** 2%. **Livestock** (1997): chickens: 8.70 mil; cattle: 2.34 mil; sheep: 669,000; goats: 820,000. **Fish catch** (1996): 64,580 metric tons. **Electricity prod.** (1997) 525 mil kWh.

Finance: Monetary unit: Franc (Sept. 1999: 1,300.00 = $1 U.S.). **GDP** (1997 est.): $8.3 bil. **Per capita GDP:** $1,100. **Imports** (1995 est.): $809 mil; partners: France 35% Côte d'Ivoire 31%. **Exports** (1995 est.): $748 mil; partners: Belg.-Lux. 21%, U.S. 21%. **Tourism:** $6 mil. **Budget** (1995 est.): $652 mil. **Intl. reserves less gold** (June 1998): $202.16 mil.

Transport: Railroad: Length: 411 mi. **Motor vehicles:** 13,700 pass. cars, 19,300 comm. vehicles. **Civil aviation:** 33.9 mil pass.-mi.; 2 airports. **Chief port:** Conakry.

Communications: TV sets: 10 per 1,000 pop. **Radios:** 34 per 1,000 pop. **Telephones:** 36,800 main lines.

Health: Life expectancy: 44.02 male; 49.06 female. **Births** (per 1,000 pop.): 40.62. **Deaths** (per 1,000 pop.): 17.30. **Natural inc.:** 2.332%. **Infant mortality** (per 1,000 live births): 126.32.

Education: Free, compulsory: ages 7-13. **Literacy:** 36%.

Major Intl. Organizations: UN and most of its specialized agencies, OAU.

Embassy: 2112 Leroy Pl. NW 20008; 483-9420.

Part of the ancient West African empires, Guinea fell under French control 1849-98. Under Sékou Touré, it opted for full independence in 1958, and France withdrew all aid.

Touré turned to Communist nations for support and set up a militant one-party state. Thousands of opponents were jailed in the 1970s, in the aftermath of an unsuccessful Portuguese invasion. Many were tortured and killed.

The military took control in a bloodless coup after the March 1984 death of Touré. A new constitution was approved in 1991, but movement toward democracy was slow. When presidential elections were finally held, in Dec. 1993, the incumbent, Gen. Lansana Conté, was the official winner; outside monitors called the elections flawed. Parliamentary elections June 11, 1995, raised similar complaints. Conté suppressed an army mutiny in Conakry, Feb. 2-3, 1996, and won reelection in Dec. 1998.

Guinea-Bissau
Republic of Guinea-Bissau

People: Population: 1,234,555. **Age distrib.** (%): <15: 42.1; 65+: 2.8. **Pop. density:** 89 per sq. mi. **Urban:** 22%. **Ethnic groups:** Balanta 30%, Fula 20%, Manjaca 14%, Mandinga 13%. **Principal languages:** Portuguese (official), Crioulo, tribal languages. **Chief religions:** Indigenous beliefs 50%, Muslim 45%, Christian 5%.

Geography: Area: 13,900 sq. mi. **Location:** On Atlantic coast of W Africa. **Neighbors:** Senegal on N, Guinea on E and

S. **Topography:** A swampy coastal plain covers most of the country; to the east is a low savanna region. **Capital:** Bissau (1995 est.): 233,000.

Government: Type: In transition. **Head of state:** Pres. Malam Bacai Sanhá; in office: May 14, 1999. **Head of gov.:** Prime Min. Francisco Fadul; in office: Dec. 3, 1998. **Local divisions:** 9 regions. **Defense:** 2.6% of GDP. **Active troops:** 7,300.

Economy: Chief crops: Peanuts, cashews, corn, beans, cotton, rice. **Minerals:** Bauxite, phosphates. **Arable land:** 11%. **Livestock** (1997): chickens: 850,000; cattle: 510,000; pigs: 335,000; goats: 305,000; sheep: 275,000. **Electricity prod.** (1997): 40 mil kWh. **Labor force:** 77% agric.; 18% services.

Finance: Monetary unit: CFA Franc (Sept. 1999: 612.79 = $1 U.S.). **GDP** (1997 est.): $1.15 bil. **Per capita GDP:** $975. **Imports** (1996): $63 mil; partners: Thailand 27%, Portugal 23%. **Exports** (1996): $25.8 mil; partners: Spain 35%. India 30%. **Intl. reserves less gold** (Mar. 1997): $16.41 mil. **Consumer prices** (change in 1997): 49.1%.

Transport: Motor vehicles: 3,500 pass. cars, 2,500 comm. vehicles. **Civil aviation:** 6.2 mil pass.-mi.; 2 airports. **Chief port:** Bissau.

Communications: Radios: 42 per 1,000 pop. **Telephones:** 8,100 main lines.

Health: Life expectancy: 47.91 male; 51.28 female. **Births** (per 1,000 pop.): 38.23. **Deaths** (per 1,000 pop.): 15.13. **Natural inc.:** 2.310%. **Hosp. beds** (1993): 1 per 797 persons. **Infant mortality** (per 1,000 live births): 109.50.

Education: Compulsory: ages 7-13. **Literacy:** 55%.

Major Intl. Organizations: UN (FAO, IBRD, ILO, IMF, IMO, WHO, WTrO), OAU.

Embassy: 1511 K St. NW 20005; 347-3950.

Portuguese mariners explored the area in the mid-15th century; the slave trade flourished in the 17th and 18th centuries, and colonization began in the 19th.

Beginning in the 1960s, an independence movement waged a guerrilla war and formed a government in the interior that had international support. Independence came Sept. 10, 1974, after the Portuguese regime was overthrown.

The November 1980 coup gave Vieira absolute power. Vieira eventually initiated political liberalization; multiparty elections were held July 3, 1994. An army uprising June 7, 1998, triggered a civil war, with Senegal and Guinea aiding the Vieira regime. After a peace accord signed on Nov. 2 broke down, rebel troops ousted Vieira on May 7, 1999.

Guyana
Co-operative Republic of Guyana

People: Population: 705,156. **Age distrib.** (%): <15: 30.4; 65+: 4.6. **Pop. density:** 9 per sq. mi. **Urban:** 36%. **Ethnic groups:** East Indian 49%, black 32%, mixed 12%, Amerindian 6%. **Principal languages:** English (official), Amerindian dialects. **Chief religions:** Christian 57%, Hindu 33%, Muslim 9%.

Geography: Area: 83,000 sq. mi. **Location:** On N coast of South America. **Neighbors:** Venezuela on W, Brazil on S, Suriname on E. **Topography:** Dense tropical forests cover much of the land, although a flat coastal area up to 40 mi. wide, where 90% of the population lives, provides rich alluvial soil for agriculture. A grassy savanna divides the 2 zones. **Capital:** Georgetown (1995 est.): 254,000.

Government: Type: Republic. **Head of state:** Pres. Bharrat Jagdeo; b Jan. 23, 1964; in office: Aug. 11, 1999. **Head of gov.:** Prime Min. Samuel Hinds; b Dec. 27, 1943; in office: Dec. 22, 1997. **Local divisions:** 10 regions. **Defense:** 1.0% of GDP. **Active troops:** 1,600.

Economy: Industries: Mining, textiles. **Chief crops:** Sugar, rice, wheat. **Minerals:** Bauxite, gold, diamonds. **Other resources:** Timber, shrimp, dairy prods. **Arable land:** 2%. **Livestock** (1997): chickens: 11.50 mil; cattle: 220,000; sheep: 130,000. **Electricity prod.** (1997): 325 mil kWh. **Labor force:** 39% agric, forestry, fishing; 24% mining, manuf., const.

Finance: Monetary unit: Dollar (Sept. 1999: 177.30 = $1 U.S.). **GDP** (1997 est.): $1.8 bil. **Per capita GDP:** $2,500. **Imports** (1996 est.): $589 mil; partners: U.S. 29%, Netherlands 17%, Trin. & Tob. 17%. **Exports** (1996): $546 mil; partners: Canada 33%, U.S. 24%. **Tourism:** $40 mil. **Budget** (1996 est): $299 mil. **Intl. reserves less gold** (May 1999): $259.09 mil. **Consumer prices** (change in 1998): 4.6%.

Transport: Motor vehicles: 24,000 pass. cars, 9,000 comm. vehicles. **Civil aviation:** 154.1 mil pass.-mi.; 1 airport. **Chief port:** Georgetown.

Communications: TV sets: 197 per 1,000 pop. **Radios:** 454 per 1,000 pop. **Telephones:** 59,900 main lines. **Daily newspaper circ.:** 585 per 1,000 pop.

Health: Life expectancy: 59.15 male; 64.61 female. **Births**

(per 1,000 pop.): 18.23. **Deaths** (per 1,000 pop.): 9.04. **Natural inc.:** 0.919%. **Physicians** (1993): 1 per 3,148 persons. **Infant mortality** (per 1,000 live births): 48.64.

Education: Free, compulsory: ages 6-14. **Literacy:** 98%.

Major Intl. Organizations: UN (FAO, IBRD, ILO, IMF, IMO, WHO, WTrO), Caricom, the Commonwealth, OAS.

Embassy: 2490 Tracy Pl. NW 20008; 265-6900.

Guyana became a Dutch possession in the 17th century, but sovereignty passed to Britain in 1815. Indentured servants from India soon outnumbered African slaves. Ethnic tension has affected political life.

Guyana became independent May 26, 1966. A Venezuelan claim to the western half of Guyana was suspended in 1970 but renewed in 1982; an agreement was reached in 1989. The Suriname border is disputed. The government has nationalized most of the economy, which has remained severely depressed.

The Port Kaituma ambush of U.S. Rep. Leo J. Ryan and others investigating mistreatment of American followers of the Rev. Jim Jones's People's Temple cult triggered a mass suicide-execution of 911 cultists at Jonestown in the jungle, Nov. 18, 1978.

The People's National Congress, the party in power since Guyana became independent, was voted out of office with the election of Cheddi Jagan in Oct. 1992. When Pres. Jagan died Mar. 6, 1997, Prime Min. Samuel Hinds succeeded him; his widow, Janet Jagan, became prime min. Mar. 17. She won the presidency in a disputed election Dec. 15. She resigned because of ill health Aug. 11, 1999, and was succeeded by Bharrat Jagdeo, who became the youngest head of state in the Americas.

Haiti
Republic of Haiti

People: Population: 6,884,264. **Age distrib.** (%): <15: 41.9; 65+: 4.1. **Pop. density:** 643 per sq. mi. **Urban:** 32%. **Ethnic groups:** Black 95%. **Principal languages:** Haitian Creole, French (both official). **Chief religions:** Roman Catholic 80%, Protestant 16%; Voodoo widely practiced.

Geography: Area: 10,700 sq. mi. **Location:** In Caribbean, occupies western third of Isl. of Hispaniola. **Neighbors:** Dominican Republic on E, Cuba to W. **Topography:** About two-thirds of Haiti is mountainous. Much of the rest is semiarid. Coastal areas are warm and moist. **Capital:** Port-au-Prince (1996 est.): 844,472.

Government: Type: Republic. **Head of state:** Pres. René Préval; b Jan. 17, 1943; in office Feb. 7, 1996. **Head of gov.:** Jacques-Edouard Alexis; b 1947; in office: Mar. 25, 1999. **Local divisions:** 9 departments. **Defense:** 5.2% of GDP. **Active security forces** (1996): 3,000-4,000.

Economy: Industries: Sugar refining, textiles. **Chief crops:** Coffee, sugar, mangoes, corn, rice. **Arable land:** 20%. **Livestock** (1997): chickens: 5.00 mil; goats: 1.62 mil; cattle: 1.30 mil; pigs: 800,000; sheep: 138,000. **Electricity prod.** (1997): 700 mil kWh. **Labor force:** 66% agric.; 25% services; 9% ind.

Finance: Monetary unit: Gourde (Sept. 1999: 16.70 = $1 U.S.). **GDP** (1997 est.): $7.1 bil. **Per capita GDP:** $1,070. **Imports** (1996): $665 mil; partners: U.S. 65%. **Exports** (1996): $90 mil; partners: U.S. 76%. **Tourism:** $96 mil. **Budget** (FY 1996-97 est.): $308 mil. **Intl. reserves less gold** (June 1998): $61.7 mil. **Gold:** 1,000 oz t. **Consumer prices** (change in 1998): 10.6%.

Transport: Motor vehicles: 32,000 pass. cars, 21,000 comm. vehicles. **Civil aviation:** 2 airports. **Chief ports:** Port-au-Prince, Les Cayes, Cap-Haitien.

Communications: TV sets: 4 per 1,000 pop. **Radios:** 41 per 1,000 pop. **Telephones** (1997): 60,000 main lines. **Daily newspaper circ.:** 7 per 1,000 pop.

Health: Life expectancy: 49.53 male; 53.88 female. **Births** (per 1,000 pop.): 32.55. **Deaths** (per 1,000 pop.): 13.97. **Natural inc.:** 1.858%. **Hosp. beds** (1994): 1 per 975 persons. **Physicians** (1994): 1 per 9,846 persons. **Infant mortality rate** (per 1,000 live births): 97.64.

Education: Compulsory: ages 6-12. **Literacy:** 45%.

Major Intl. Organizations: UN and most of its specialized agencies, OAS.

Embassy: 2311 Massachusetts Ave. NW 20008; 332-4090. **Website:** http://www.haiti.org/embassy/

Haiti, visited by Columbus, 1492, and a French colony from 1697, attained its independence, 1804, following the rebellion led by former slave Toussaint L'Ouverture. Following a period of political violence, the U.S. occupied the country 1915-34.

Francois Duvalier was elected president in Sept. 1957; in 1964 he was named president for life. Upon his death in 1971, he was succeeded by his son, Jean Claude. Drought in 1975-77 brought famine, and Hurricane Allen in 1980 destroyed most of the rice, bean, and coffee crops. Following several weeks of unrest, President Jean Claude Duvalier fled Haiti aboard a U.S. Air Force jet Feb. 7, 1986, ending the 28-year dictatorship by the Duvalier family.

A military-civilian council headed by Gen. Henri Namphy assumed control. In 1987, voters approved a new constitution, but the Jan. 1988 elections were marred by violence and boycotted by the opposition. Gen. Namphy seized control, June 20, but was ousted by a military coup in Sept.

Father Jean-Bertrand Aristide was elected president Dec. 1990. In Sept. 1991, Aristide was arrested by the military and expelled from the country. Some 35,000 Haitian refugees were intercepted by the U.S. Coast Guard as they tried to enter the U.S., 1991-92. Most were returned to Haiti. There was a new upsurge of refugees starting in late 1993.

The UN imposed a worldwide oil, arms, and financial embargo on Haiti June 23, 1993. The embargo was suspended when the military agreed to Aristide's return to power on Oct. 30, but the military effectively blocked his return. After renewed sanctions, the UN Security Council authorized, July 31, 1994, an invasion of Haiti by a multinational force. With U.S. troops already en route, an invasion was averted, Sept. 18, by a new agreement for military leaders to step down and Aristide to resume office. As part of the agreement, thousands of U.S. troops began arriving in Haiti, Sept. 19. Aristide returned to Haiti and was restored in office Oct. 15. A UN peacekeeping force exercised responsibility in Haiti from Mar. 31, 1995 to Nov. 30, 1997.

Aristide transferred power to his elected successor, René Préval, on Feb. 7, 1996. Prime Min. Rosny Smarth announced his resignation June 9, 1997, and quit running the government Oct. 20, but Préval and Parliament deadlocked for another 17 months until a successor was appointed by presidential decree. At least 140 people died and more than 160,000 became homeless when Hurricane Georges struck Haiti Sept. 22, 1998.

Honduras
Republic of Honduras

People: Population: 5,997,327. **Age distrib. (%):** <15: 41.4; 65+: 3.5. **Pop. density:** 139 per sq. mi. **Urban:** 44%. **Ethnic groups:** Mestizo 90%, Amerindian 7%. **Principal language:** Spanish (official). **Chief religion:** Roman Catholic 97%.

Geography: Area: 43,300 sq. mi. **Location:** In Central America. **Neighbors:** Guatemala on W, El Salvador and Nicaragua on S. **Topography:** The Caribbean coast is 500 mi. long. Pacific coast, on Gulf of Fonseca, is 40 mi. long. Honduras is mountainous, with wide fertile valleys and rich forests. **Capital:** Tegucigalpa, 995,000*.

Government: Type: Republic. **Head of state:** Pres. Carlos Flores Facusse; b Mar. 1, 1950; in office: Jan. 27, 1998. **Local divisions:** 18 departments. **Defense:** 2.1% of GDP. **Active troops:** 18,800.

Economy: Industries: Textiles, wood prods. **Chief crops:** Bananas, coffee, citrus. **Minerals:** Gold, silver, copper, lead, zinc, iron, antimony, coal. **Other resources:** Timber, fish. **Arable land:** 15%. **Livestock** (1997): chickens: 18.00 mil; cattle: 1.95 mil; pigs: 700,000. **Electricity prod.** (1997): 2.664 bil kWh. **Labor force:** 43% agric.; 22% services; 12% manuf.

Finance: Monetary unit: Lempira (Sept. 1999: 14.42 = $1 U.S.). **GDP** (1997 est.): $12.7 bil. **Per capita GDP:** $2,200. **Imports** (1996): $1.8 bil; partners: U.S. 43%. **Exports** (1996): $1.3 bil; partners: U.S. 54%. **Tourism:** $173 mil. **Budget** (1997 est.): $850 mil. **Intl. reserves less gold** (June 1999): $1.09 bil. **Gold:** 21,000 oz t. **Consumer prices** (change in 1998): 13.7%.

Transport: Railroad: Length: 614 mi. **Motor vehicles:** 80,000 pass. cars, 105,000 comm. vehicles. **Civil aviation:** 189.5 mil pass.-mi.; 8 airports. **Chief ports:** Puerto Cortes, La Ceiba.

Communications: TV sets: 29 per 1,000 pop. **Radios:** 337 per 1,000 pop. **Telephones** (1997): 233,600 main lines. **Daily newspaper circ.:** 45 per 1,000 pop.

Health: Life expectancy: 63.16 male; 66.27 female. **Births** (per 1,000 pop.): 30.98. **Deaths** (per 1,000 pop.): 7.14. **Natural inc.:** 2.384%. **Hosp. beds** (1994): 1 per 1,126 persons. **Infant mortality** (per 1,000 live births): 40.84.

Education: Free, compulsory: ages 7-13. **Literacy:** 73%.

Major Intl. Organizations: UN, (FAO, IBRD, ILO, IMF, IMO, WHO, WTrO), OAS.

Embassy: 3007 Tilden St. NW 20008; 966-7702.
Website: http://www.honduras.com

Mayan civilization flourished in Honduras in the 1st millennium AD. Columbus arrived in 1502. Honduras became independent after freeing itself from Spain, 1821, and from the Fed. of Central America, 1838.

Gen. Oswaldo Lopez Arellano, president for most of the period 1963-75 by virtue of one election and 2 coups, was ousted by the army in 1975 over charges of pervasive bribery by United Brands Co. of the U.S. An elected civilian government took power in 1982. Some 3,200 U.S. troops were sent to Honduras after the Honduran border was violated by Nicaraguan forces, Mar. 1988.

Already one of the poorest countries in the western hemisphere, Honduras was devastated in late Oct. 1998 by Hurricane Mitch, which killed at least 5,600 people and caused more than $850 million in damage to crops and livestock.

Hungary
Republic of Hungary

People: Population: 10,186,372. **Age distrib. (%):** <15: 17.4; 65+: 14.5. **Pop. density:** 284 per sq. mi. **Urban:** 65%. **Ethnic groups:** Hungarian 90%, Gypsy 4%, German 3%. **Principal language:** Hungarian (Magyar; official). **Chief religions:** Roman Catholic 68%, Calvinist 20%, Lutheran 5%.

Geography: Area: 35,900 sq. mi. **Location:** In E central Europe. **Neighbors:** Slovakia, Ukraine on N; Austria on W; Slovenia, Yugoslavia, Croatia on S; Romania on E. **Topography:** The Danube R. forms the Slovak border in the NW, then swings S to bisect the country. The eastern half of Hungary is mainly a great fertile plain, the Alfold; the W and N are hilly. **Capital:** Budapest. **Cities** (1997 est.): Budapest 1,885,000; Debrecen 210,000; Miskolc 178,000.

Government: Type: Parliamentary democracy. **Head of state:** Pres. Arpad Goncz; b Feb. 10, 1922; in office: May 2, 1990. **Head of gov.:** Prime Min. Viktor Orbán; b May 31, 1963; in office: July 8, 1998. **Local divisions:** 19 counties, 1 capital. **Defense:** 1.4% of GDP. **Active troops:** 49,100.

Economy: Industries: Mining, metallurgy, construction materials, processed foods, pharmaceuticals, vehicles. **Chief crops:** Wheat, corn, sunflowers, potatoes, sugar beets. **Minerals:** Bauxite, coal, gas. **Arable land:** 51%. **Livestock** (1997): chickens: 30.98 mil; goats: 108,290; pigs: 4.93 mil; sheep: 858,000; cattle: 871,000. **Crude oil reserves** (1999): 131.4 mil bbls. **Electricity prod.** (1997): 33.437 bil kWh. **Labor force:** 65% services; 27% ind.; 8% agric.

Finance: Monetary unit: Forint (Sept. 1999: 239.80 = $1 U.S.). **GDP** (1997 est.): $73.2 bil. **Per capita GDP:** $7,400. **Imports** (1996): $18.6 bil; partners: Germany 24%. **Exports** (1996): $16 bil; partners: Germany 29%, Austria 11%. **Budget** (1997 est.): $13.8 bil. **Tourism:** $2.57 bil. **Intl. reserves less gold** (Apr. 1999): $8.68 bil. **Gold:** 101,000 oz t. **Consumer prices** (change in 1998): 14.4%.

Transport: Railroad: Length: 8,190 mi. **Motor vehicles:** 2.28 mil pass. cars, 319,424 comm. vehicles. **Civil aviation:** 1.5 bil. pass.-mi.; 1 airport.

Communications: TV sets: 444 per 1,000 pop. **Radios:** 590 per 1,000 pop. **Telephones** (1997): 3,095,300 main lines. **Daily newspaper circ.:** 228 per 1,000 pop.

Health: Life expectancy: 66.85 male; 75.74 female. **Births** (per 1,000 pop.): 10.80. **Deaths** (per 1,000 pop.): 13.29. **Natural inc.:** −0.249%. **Hosp. beds** (1995): 1 per 110 persons. **Physicians** (1995): 1 per 240 persons. **Infant mortality** (per 1,000 live births): 9.46.

Education: Compulsory: ages 6-16. **Literacy** (1993): 99%.

Major Intl. Organizations: UN (FAO, IBRD, ILO, IMF, IMO, WHO, WTrO), NATO, OECD, OSCE.

Embassy: 3910 Shoemaker St. NW 20008; 362-6730.
Website: http://www.hungaryemb.org

Earliest settlers, chiefly Slav and Germanic, were overrun by Magyars from the E. Stephen I (997-1038) was made king by Pope Sylvester II in AD 1000. The country suffered repeated Turkish invasions in the 15th-17th centuries. After the defeats of the Turks, 1686-1697, Austria dominated, but Hungary obtained concessions until it regained internal independence in 1867, with the emperor of Austria as king of Hungary in a dual monarchy with a single diplomatic service. Defeated with the Central Powers in 1918, Hungary lost Transylvania to Romania, Croatia and Bacska to Yugoslavia, Slovakia and Carpatho-Ruthenia to Czechoslovakia, all of which had large Hungarian minorities. A republic under Michael Karolyi and a bolshevist revolt under Bela Kun were followed by a vote for a monarchy in 1920 with Admiral Nicholas Horthy as regent.

Hungary joined Germany in World War II, and was allowed to annex most of its lost territories. Russian troops captured the country, 1944-1945. By terms of an armistice with the Allied powers Hungary agreed to give up territory acquired by the 1938 dismemberment of Czechoslovakia and to return to its borders of 1937.

A republic was declared Feb. 1, 1946; Zoltan Tildy was elected president. In 1947 the Communists forced Tildy out.

Premier Imre Nagy, who had been in office since mid-1953, was ousted for his moderate policy of favoring agriculture and consumer production, April 18, 1955.

In 1956, popular demands to oust Erno Gero, Communist Party secretary, and for formation of a government by Nagy, resulted in the latter's appointment Oct. 23; demonstrations against Communist rule developed into open revolt. On Nov. 4 Soviet forces launched a massive attack against Budapest with 200,000 troops, 2,500 tanks and armored cars.

About 200,000 persons fled the country. Thousands were arrested and executed, including Nagy in June 1958. In spring 1963 the regime freed many captives from the 1956 revolt.

Hungarian troops participated in the 1968 Warsaw Pact invasion of Czechoslovakia. Major economic reforms were launched early in 1968, switching from a central planning system to one based on market forces and profit.

In 1989 Parliament passed legislation legalizing freedom of assembly and association as Hungary shifted away from communism. In Oct. the Communist Party was formally dissolved. The last Soviet troops left Hungary June 19, 1991. Hungary became a full member of NATO on Mar. 12, 1999.

Iceland
Republic of Iceland

People: Population: 272,512. **Age distrib.** (%): <15: 23.4; 65+: 11.8. **Pop. density:** 7 per sq. mi. **Urban:** 92%. **Ethnic groups:** Homogeneous descendants of Norwegians, Celts. **Principal language:** Icelandic (Islenska; official). **Chief religion:** Evangelical Lutheran 96%.

Geography: Area: 40,000 sq. mi. **Location:** Isl. at N end of Atlantic O. **Neighbors:** Nearest is Greenland (Den.), to W. **Topography:** Recent volcanic origin. Three-quarters of the surface is wasteland: glaciers, lakes, a lava desert. There are geysers and hot springs, and the climate is moderated by the Gulf Stream. **Capital:** Reykjavík (1996 est.): 105,487.

Government: Type: Constitutional republic. **Head of state:** Pres. Olafur Ragnar Grímsson; b May 14, 1943; in office: Aug. 1, 1996. **Head of gov.:** Prime Min. David Oddsson; Jan. 17, 1948; in office: Apr. 30, 1991. **Local divisions:** 23 counties, 14 independent towns. **Defense:** Icelandic Defense Force provided by the U.S.

Economy: Industries: Fish products (75% of exports), aluminum. **Chief crops:** Potatoes, turnips. **Livestock** (1997): chickens: 179,000; sheep: 477,306. **Fish catch** (1997): 2.21 mil metric tons. **Electricity prod.** (1997): 5.518 bil kWh. **Labor force:** 60% commerce & services; 13% manuf.; 12% fishing.

Finance: Monetary unit: Krona (Sept. 1999: 72.31 = $1 U.S.). **GDP** (1997 est.): $571 bil. **Per capita GDP:** $21,000. **Imports** (1996): $2 bil; partners: Germany 11%, Norway 10%, UK 10%. **Exports** (1996): $1.8 bil; partners: UK 19%, Germany 14%. **Tourism:** $175 mil. **Budget** (1996 est.): $2.1 bil. **Intl. reserves less gold** (May 1999): $427.0 mil. **Gold:** 56,000 oz t. **Consumer prices** (change in 1998): 1.7%.

Transport: Motor vehicles (1997): 132,468 pass. cars, 17,511 comm. vehicles. **Civil aviation:** 2.0 bil. pass.-mi.; 24 airports. **Chief port:** Reykjavík.

Communications: TV sets: 285 per 1,000 pop. **Radios:** 733 per 1,000 pop. **Telephones** (1997): 167,600 main lines. **Daily newspaper circ.:** 515 per 1,000 pop.

Health: Life expectancy: 76.85 male; 81.19 female. **Births** (per 1,000 pop.): 14.87. **Deaths** (per 1,000 pop.): 7.01. **Natural inc.:** 0.786%. **Hosp. beds** (1993): 1 per 95 persons. **Physicians** (1995): 1 per 335 persons. **Infant mortality** (per 1,000 live births): 5.22.

Education: Free, compulsory: ages 7-15. **Literacy** (1997): 100%.

Major Intl. Organizations: UN (FAO, IBRD, ILO, IMF, IMO, WHO, WTrO), EFTA, NATO, OECD, OSCE.

Embassy: Suite 1200, 1156 15th St. NW 20005; 265-6653. **Website:** http://www.iceland.org

Iceland was an independent republic from 930 to 1262, when it joined with Norway. Its language has maintained its purity for 1,000 years. Danish rule lasted from 1380-1918; the last ties with the Danish crown were severed in 1941. The Althing, or assembly, is the world's oldest surviving parliament.

India
Republic of India

People: Population: 1,000,848,550. **Age distrib.** (%): <15: 34.1; 65+: 4.6. **Pop. density:** 789 per sq. mi. **Urban:** 27%. **Ethnic groups:** Indo-Aryan 72%, Dravidian 25%. **Principal languages:** Hindi (official), English (associate official), 14 regional official languages, others. **Chief religions:** Hindu 80%, Muslim 14%.

Geography: Area: 1,269,300 sq. mi. **Location:** Occupies most of the Indian subcontinent in S Asia. **Neighbors:** Pakistan on W; China, Nepal, Bhutan on N; Myanmar, Bangladesh on E. **Topography:** The Himalaya Mts., highest in world, stretch across India's northern borders. Below, the Ganges Plain is wide, fertile, and among the most densely populated regions of the world. The area below includes the Deccan Peninsula. Close to one quarter of the area is forested. The climate varies from tropical heat in S to near-Arctic cold in N. Rajasthan Desert is in NW; NE Assam Hills get 400 in. of rain a year. **Capital:** New Delhi. **Cities:** Mumbai (Bombay) 15,138,000; Kolkata (Calcutta) 11,923,000; Delhi 9,948,000; Chennai (Madras) 6,002,000; Hyderabad 5,477,000; Bangalore 4,799,000*.

Government: Type: Federal republic. **Head of state:** Pres. Kocheril Raman Narayanan; b Oct. 17, 1920; in office: July 25, 1997. **Head of gov.:** Prime Min. Atal Bihari Vajpayee; b Dec. 25, 1924; in office Mar. 19, 1998. **Local divisions:** 25 states, 6 union territories, 1 national capital territory. **Defense:** 3.3% of GDP. **Active troops:** 1.145 mil.

Economy: Industries: Textiles, steel, processed foods, cement, machinery, chemicals, mining. **Chief crops:** Rice, grains, sugar, spices, tea, cashews, cotton, potatoes, jute, oilseed. **Minerals:** Coal (4th largest reserves in the world), iron, manganese, mica, bauxite, titanium, chromite, diamonds, gas, oil. **Crude oil reserves** (1999): 3.97 bil bbls. **Other resources:** Timber. **Arable land:** 56%. **Livestock** (1997): chickens: 342.5 mil; cattle: 209.49 mil; goats: 120.56 mil; buffalo: 91.78 mil; sheep: 56.47 mil; pigs: 16.01 mil. **Fish catch** (1997): 3.60 mil metric tons. **Electricity prod.** (1997): 425.600 bil kWh. **Labor force:** 67% agric.; 18% services; 15% industry.

Finance: Monetary unit: Rupee (Sept. 1999: 43.49 = $1 U.S.). **GDP** (1997 est.): $1.534 tril. **Per capita GDP:** $1,600. **Imports** (1997): $39.7 bil; partners: U.S. 11%, Germany 9%, Japan 7%. **Exports** (1997): $33.9 bil; partners: U.S. 17%, Japan 7%, UK 6%. **Tourism:** $3.17 bil. **Budget** (FY 1997-98): $61 bil. **Intl. reserves less gold** (June 1999): $31.21 bil. **Gold:** 11.50 mil oz t. **Consumer prices** (change in 1998): 13.2%.

Transport: Railroad: Length: 38,935 mi. **Motor vehicles:** 4.25 mil pass. cars, 2.51 mil comm. vehicles. **Civil aviation:** 15.0 bil pass.-mi.; 66 airports. **Chief ports:** Kolkata (Calcutta), Mumbai (Bombay), Chennai (Madras), Vishakhapatnam, Kandla.

Communications: TV sets: 21 per 1,000 pop. **Radios:** 117 per 1,000 pop. **Telephones** (1997): 17,801,700 main lines. **Daily newspaper circ.:** 21 per 1,000 pop.

Health: Life expectancy: 62.54 male; 64.29 female. **Births** (per 1,000 pop.): 25.39. **Deaths** (per 1,000 pop.): 8.50. **Natural inc.:** 1.689%. **Hosp. beds** (1992): 1 per 1,357 persons. **Physicians** (1992): 1 per 2,173 persons. **Infant mortality** (per 1,000 live births): 60.81.

Education: Theoretically compulsory in 23 states to age 14. **Literacy:** 52%.

Major Intl. Organizations: UN (FAO, IBRD, ILO, IMF, IMO, WHO, WTrO), the Commonwealth.

Embassy: 2107 Massachusetts Ave. NW 20008; 939-7000. **Websites:** http://www.nic.in
    http://www.indianembassy.org
    http://www.tourindia.com/

India has one of the oldest civilizations in the world. Excavations trace the Indus Valley civilization back for at least 5,000 years. Paintings in the mountain caves of Ajanta, richly carved temples, the Taj Mahal in Agra, and the Kutab Minar in Delhi are among relics of the past.

Aryan tribes, speaking Sanskrit, invaded from the NW around 1500 BC, and merged with the earlier inhabitants to create classical Indian civilization.

Asoka ruled most of the Indian subcontinent in the 3d century BC, and established Buddhism. But Hinduism revived and eventually predominated. During the Gupta kingdom, 4th-6th century AD, science, literature, and the arts enjoyed a "golden age."

Arab invaders established a Muslim foothold in the W in the 8th century, and Turkish Muslims gained control of North India by 1200. The Mogul emperors ruled 1526-1857.

Vasco da Gama established Portuguese trading posts 1498-1503. The Dutch followed. The British East India Co. sent Capt. William Hawkins, 1609, to get concessions from the Mogul emperor for spices and textiles. Operating as the East India Co. the British gained control of most of India. The British parliament assumed political direction; under Lord Bentinck, 1828-35, rule by rajahs was curbed. After the Sepoy troops mutinied, 1857-58, the British supported the native rulers.

Nationalism grew rapidly after World War I. The Indian National Congress and the Muslim League demanded constitutional reform. A leader emerged in Mohandas K. Gandhi (called Mahatma, or Great Soul), born Oct. 2, 1869, assassinated Jan. 30, 1948. He advocated self-rule, nonviolence, and removal of the caste system of untouchability. In 1930 he launched a program of civil disobedience, including a boycott of British goods and rejection of taxes without representation.

In 1935 Britain gave India a constitution providing a bicameral federal congress. Muhammad Ali Jinnah, head of the Muslim League, sought creation of a Muslim nation, Pakistan.

The British government partitioned British India into the dominions of India and Pakistan. India became a member of the UN in 1945, a self-governing member of the Commonwealth in 1947, and a democratic republic, Jan. 26, 1950. More than 12 million Hindu and Muslim refugees crossed the India-Pakistan borders in a mass transferral of some of the 2 peoples during 1947; about 200,000 were killed in communal fighting.

After Pakistan troops began attacks on Bengali separatists in East Pakistan, Mar. 25, 1971, some 10 million refugees fled into India. India and Pakistan went to war Dec. 3, 1971, on both the East and West fronts. Pakistan troops in the east surrendered Dec. 16; Pakistan agreed to a cease-fire in the west Dec. 17.

Indira Gandhi, India's prime minister since Jan. 1966, invoked emergency powers in June 1975. Thousands of opponents were arrested and press censorship imposed. These and other actions, including enforcement of coercive birth control measures in some areas, were widely resented. Opposition parties, united in the Janata coalition, turned Gandhi's New Congress Party from power in federal and state parliamentary elections in 1977.

Gandhi became prime minister for the second time, Jan. 14, 1980. She was assassinated by 2 of her Sikh bodyguards Oct. 31, 1984, in response to the government suppression of a Sikh uprising in Punjab in June 1984, which included an assault on the Golden Temple at Amritsar, the holiest Sikh shrine. Widespread rioting followed the assassination. Thousands of Sikhs were killed and some 50,000 left homeless.

Rajiv, Indira Gandhi's son, replaced her as prime minister. He was swept from office in 1989 amid charges of incompetence and corruption, and assassinated May 21, 1991, while campaigning to recapture the prime ministership.

Sikhs ignited several violent clashes during the 1980s. The government's May 1987 decision to bring the state of Punjab under rule of the central government led to violence. Many died during a government siege of the Golden Temple, May 1988. Another trouble spot was Assam in NW India, where thousands were killed in ethnic violence in Feb. 1993; a renewed outburst in July 1994 led to more than 60 deaths.

Nationwide riots followed the destruction of a 16th-century mosque by Hindu militants in Dec. 1992. In the biggest wave of criminal violence in Indian history, a series of bombs jolted Bombay and Calcutta, Mar. 12-19, 1993, killing over 300.

Corruption scandals dominated Indian politics in the mid-1990s. After an inconclusive election, a Hindu nationalist party was unable to form a government, and a center-left coalition took office June 1, 1996. An aircraft collision in midair near New Delhi killed 349 passengers and crew on Nov. 12.

India's 1st lowest-caste pres., K. R. Narayanan, took office July 25, 1997. Mother Teresa of Calcutta, renowned for her work among the poor, died Sept. 5. Parliamentary elections in Feb. 1998 resulted in a Hindu nationalist victory, and Atal Bihari Vajpayee was sworn in as prime minister Mar. 19. India conducted a series of nuclear tests in mid-May, drawing worldwide condemnation and raising tensions with Pakistan. A cyclone June 9 left more than 1,000 people dead in Gujarat state.

An alliance led by Vajpayee won a majority in legislative elections, Sept. 5-Oct. 3, 1999.

Sikkim, bordered by Tibet, Bhutan, and Nepal, formerly British protected, became a protectorate of India in 1950. Area, 2,740 sq. mi; pop., 1994 est., 444,000; capital: Gangtok. In Sept. 1974, India's parliament voted to make Sikkim an associate Indian state, absorbing it into India.

Kashmir, a predominantly Muslim region in the NW, has been in dispute between India and Pakistan since 1947. A cease-fire was negotiated by the UN Jan. 1, 1949; it gave Pakistan control of one-third of the area, in the west and northwest, and India the remaining two-thirds, the Indian state of **Jammu and Kashmir,** which enjoys internal autonomy.

In the 1990s there were repeated clashes between Indian army troops and pro-independence demonstrators triggered by India's decision to impose central government rule. The clashes strained relations between India and Pakistan, which India charged was aiding the Muslim separatists; the heaviest fighting in more than 2 decades took place during May-June 1999.

France, 1952-54, peacefully yielded to India its 5 colonies, former French India, comprising Pondicherry, Karikal, Mahe, Yanaon (which became **Pondicherry Union Territory,** area 190 sq. mi; pop., 1994 est., 894,000) and Chandernagor (which was incorporated into the state of **West Bengal**).

Indonesia
Republic of Indonesia

People: Population: 216,108,345. **Age distrib.** (%): <15: 30.4; 65+: 4.1. **Pop. density:** 292 per sq. mi. **Urban:** 36%. **Ethnic groups:** Javanese 45%, Sundanese 14%, Madurese 8%, Malay 8%. **Principal languages:** Bahasa Indonesian (official), English, Dutch, Javanese. **Chief religions:** Muslim 87%, Protestant 6%.

Geography: Area: 741,100 sq. mi. **Location:** Archipelago SE of Asian mainland along the Equator. **Neighbors:** Malaysia on N, Papua New Guinea on E. **Topography:** Indonesia comprises over 13,500 islands (6,000 inhabited), including Java (one of the most densely populated areas in the world with over 2,000 persons per sq. mi.), Sumatra, Kalimantan (most of Borneo), Sulawesi (Celebes), and West Irian (Irian Jaya, the W half of New Guinea). Also: Bangka, Billiton, Madura, Bali, Timor. The mountains and plateaus on the major islands have a cooler climate than the tropical lowlands. **Capital:** Jakarta. **Cities** (1995 est.): Jakarta 9,112,652; Surabaya 2,663,820; Bandung 2,356,120.

Government: Type: Republic. **Head of state and gov.:** Pres. Abdurrahman Wahid; b Aug. 4, 1940; in office: Oct. 20, 1999. **Local divisions:** 24 provinces, 2 special regions, 1 capital district. **Defense:** 2.2% of GDP. **Active troops:** 284,000.

Economy: Industries: Oil, gas, food processing, textiles, cement, mining. **Chief crops:** Rice, cocoa, peanuts, rubber. **Minerals:** Nickel, tin, oil, bauxite, copper, gas. **Crude oil reserves** (1999): 4.98 bil bbls. **Other resources:** Timber. **Arable land:** 10%. **Livestock** (1997): chickens: 889.15 mil; goats: 15.20 mil; cattle: 12.24 mil; pigs: 10.07 mil; sheep: 8.15 mil; buffalo: 3.15 mil. **Fish catch** (1997): 3.65 mil metric tons. **Electricity prod.** (1997): 72.976 bil kWh. **Labor force:** 44% agric.; 13% manuf.

Finance: Monetary unit: Rupiah (Sept. 1999: 7,835.00 = $1 U.S.). **GDP** (1997 est.): $960 bil. **Per capita GDP:** $4,600. **Imports** (1997): $41.6 bil; partners: Japan 23%, U.S. 12%. **Exports** (1997): $53.4 bil; partners: Japan 27%, U.S. 14%. **Tourism:** $5.14 bil. **Budget** (FY 1997-98 est.): $42.8 bil. **Intl. reserves less gold** (June 1999): $26.32 bil. **Gold:** 3.10 mil oz t. **Consumer prices** (change in 1998): 57.6%.

Transport: Railroad: Length: 4,090 mi. **Motor vehicles** (1997): 2.64 mil pass. cars, 2.16 mil comm. vehicles. **Civil aviation:** 14.6 bil pass.-mi.; 81 airports. **Chief ports:** Jakarta, Surabaya, Palembang, Semarang, Ujungpandang.

Communications: TV sets: 145 per 1,000 pop. **Radios:** 132 per 1,000 pop. **Telephones:** 5,571,600 main lines. **Daily newspaper circ.:** 20 per 1,000 pop.

Health: Life expectancy: 60.67 male; 65.29 female. **Births** (per 1,000 pop.): 22.78. **Deaths** (per 1,000 pop.): 8.14. **Natural inc.:** 1.464%. **Hosp. beds** (1994): 1 per 1,630 persons. **Physicians** (1994): 1 per 6,570 persons. **Infant mortality** (per 1,000 live births): 57.30.

Education: Compulsory: ages 7-16. **Literacy:** 84%.

Major Intl. Organizations: UN and all of its specialized agencies, APEC, ASEAN, OPEC.

Embassy: 2020 Massachusetts Ave. NW 20036; 775-5200.

Hindu and Buddhist civilization from India reached Indonesia nearly 2,000 years ago, taking root especially in Java. Islam spread along the maritime trade routes in the 15th century, and became predominant by the 16th century. The Dutch replaced the Portuguese as the area's most important European trade power in the 17th century, securing territorial control over Java by 1750. The outer islands were not finally subdued until the early 20th century, when the full area of present-day Indonesia was united under one rule for the first time.

Following Japanese occupation, 1942-45, nationalists led by Sukarno and Hatta declared independence. The Netherlands ceded sovereignty Dec. 27, 1949, after 4 years of fighting. A republic was declared, Aug. 17, 1950, with Sukarno as president. West Irian on New Guinea, remained under Dutch control. After the Dutch in 1957 rejected proposals for new negotiations over West Irian, Indonesia stepped up the seizure of Dutch property. In 1963 the UN turned the area over to Indonesia, which promised a plebiscite. In 1969, voting by tribal chiefs favored staying with Indonesia, despite an uprising and widespread opposition.

Sukarno suspended Parliament in 1960, and was named president for life in 1963. He made close alliances with Communist governments. Russian-armed Indonesian troops staged raids in 1964 and 1965 into Malaysia, whose formation Sukarno had opposed. (In 1966 Indonesia and Malaysia signed an agreement ending hostility.)

In 1965 an attempted coup in which several military officers were murdered was successfully put down. The regime blamed the coup on the Communist Party, some of whose members were known to have been involved. In its wake more than 300,000 alleged Communists were killed in army-initiated massacres.

Gen. Suharto, head of the army, was named president in 1968. With military backing he developed a strong government party, restricted the opposition, and allied the country with the West; meanwhile, oil exports spurred economic growth. During Aug.-Nov. 1997, haze from forest fires in Indonesia blanketed large areas of SE Asia. A plane crash near Medan airport, Sept. 26, 1997, killed 234 persons.

Parliament reelected Suharto to a 7th consecutive 5-year term Mar. 10, 1998, as a severe economic downturn focused public anger on nepotism, cronyism, and corruption in the Suharto regime. Price increases in May sparked mass protests and then mob violence in Jakarta and other cities, claiming some 500 lives. Suharto resigned May 21 and was succeeded by his vice-president, Bacharuddin Jusuf Habibie. Abdurrahman Wahid, leader of Indonesia's largest Muslim organization, was elected president Oct. 20, 1999.

In Dec. 1975, Indonesia invaded **East Timor** as Portuguese rule collapsed. Indonesia annexed it in 1976, despite international condemnation; an estimated 200,000 Timorese died as a result of Indonesian oppression and famine. In a referendum held Aug. 30, 1999, under UN auspices, Timorese voted overwhelmingly for independence. Pro-Indonesian militias then went on a rampage, terrorizing the population. In late September an international peacekeeping force led by Australia began efforts to restore order.

Iran

Islamic Republic of Iran

People: Population: 65,179,752. **Age distrib.** (%): <15: 35.9; 65+: 4.5. **Pop. density:** 102 per sq. mi. **Urban:** 60%. **Ethnic groups:** Persian 51%, Azerbaijani 24%, Kurd 7%. **Principal languages:** Persian (Farsi; official), Turkic, Kurdish, Luri. **Chief religions:** Shi'a Muslim 89%, Sunni Muslim 10%.

Geography: Area: 636,000 sq. mi. **Location:** Between the Middle East and S Asia. **Neighbors:** Turkey, Iraq on W; Armenia, Azerbaijan, Turkmenistan on N; Afghanistan, Pakistan on E. **Topography:** Interior highlands and plains surrounded by high mountains, up to 18,000 ft. Large salt deserts cover much of area, but there are many oases and forest areas. Most of the population inhabits the N and NW. **Capital:** Tehran. **Cities** (1994): Tehran 6,750,043; Mashhad 1,964,489; Esfahan 1,220,595.

Government: Type: Islamic republic. **Religious head:** Ayatollah Sayyed Ali Khamenei; b 1940; in office: June 4, 1989. **Head of state and gov.:** Pres. Mohammad Khatami; b 1943; in office: Aug. 3, 1997. **Local divisions:** 28 provinces. **Defense:** 6.6% of GDP. **Active troops:** 518,000.

Economy: Industries: Oil, petrochemicals, cement, sugar refining, carpets. **Chief crops:** Grains, rice, fruits, nuts, sugar beets, cotton. **Minerals:** Chromium, coal, oil, gas. **Crude oil reserves** (1999): 89.7 bil bbls. **Arable land:** 10%. **Livestock** (1997): chickens: 230.00 mil; sheep: 53.00 mil; goats: 27.00 mil; cattle: 8.60 mil; buffalo: 465,000. **Fish catch** (1997): 349,921 metric tons. **Electricity** prod. (1997): 79.600 bil kWh. **Labor Force:** 45% services; 31% mining, manuf., const.; 23% agric.

Finance: Monetary unit: Rial (Sept. 1999: 1,750.00 = $1 U.S.). **GDP** (1997 est.): $371.2 bil. **Per capita GDP:** $5,500. **Imports** (1997 est.): $15.6 bil; partners: Germany 19%, Italy 9%. **Exports** (1997 est.): $19 bil; partners: Japan 15%, U.S. 14%. **Tourism:** $400 mil. **Budget** (FY 1996-97): $34.9 bil. **Consumer prices** (change in 1998): 19.4%.

Transport: Railroad: Length: 4,527 mi. **Motor vehicles:** 1.63 mil pass. cars, 609,000 comm. vehicles. **Civil aviation:** 5.5 bil pass.-mi.; 19 airports. **Chief port:** Bandar-e Abbas.

Communications: TV sets: 117 per 1,000 pop. **Radios:** 213 per 1,000 pop. **Telephones** (1997): 6,513,000 main lines. **Daily newspaper circ.:** 20 per 1,000 pop.

Health: Life expectancy: 68.43 male; 71.16 female. **Births** (per 1,000 pop.): 20.71. **Deaths** (per 1,000 pop.): 5.39. **Natural inc.:** 1.532%. **Hosp. beds** (1995): 1 per 650 persons. **Physicians** (1994): 1 per 1,600 persons. **Infant mortality** (per 1,000 live births): 29.73.

Education: Free, compulsory: ages 6-10. **Literacy** (1997): 79%.

Major Intl. Organizations: UN (FAO, IBRD, ILO, IMF, IMO, WHO), OPEC.

Iran was once called Persia. The Iranians, who supplanted an earlier agricultural civilization, came from the E during the 2d millennium BC; they were an Indo-European group related to the Aryans of India.

In 549 BC Cyrus the Great united the Medes and Persians in the Persian Empire, conquered Babylonia in 538 BC, and restored Jerusalem to the Jews. Alexander the Great conquered Persia in 333 BC, but Persians regained independence in the next century under the Parthians, themselves succeeded by Sassanian Persians in AD 226. Arabs brought Islam to Persia in the 7th century, replacing the indigenous Zoroastrian faith. After Persian political and cultural autonomy was reasserted in the 9th century, arts and sciences flourished.

Turks and Mongols ruled Persia in turn from the 11th century to 1502, when a native dynasty reasserted full independence. The British and Russian empires vied for influence in the 19th century; Afghanistan was severed from Iran by Britain in 1857.

Reza Khan abdicated as shah, 1941; succeeded by his son, Mohammad Reza Pahlavi. He brought economic and social change to Iran, but political opposition was not tolerated.

Conservative Muslim protests led to 1978 violence. Martial law was declared in 12 cities Sept. 8. A military government was appointed Nov. 6 to deal with striking oil workers. The shah, who left Iran Jan. 16, 1979, appointed Prime Min. Shahpur Bakhtiar to head a regency council in his absence.

Exiled religious leader Ayatollah Ruhollah Khomeini named a provisional government council in preparation for his return to Tehran, Feb. 1. Clashes between Khomeini's supporters and government troops culminated in a rout of Iran's elite Imperial Guard Feb. 11, leading to the fall of Bakhtiar's government.

The Iranian revolution was marked by revolts among ethnic minorities and by a continuing struggle between the clerical forces and westernized intellectuals and liberals. The Islamic Constitution established final authority to be vested in a Faghi, the Ayatollah Khomeini.

Iranian militants seized the U.S. embassy, Nov. 4, 1979, and took hostages including 62 Americans. Despite international condemnations and U.S. efforts, including an abortive Apr. 1980 rescue attempt, the crisis continued. The U.S. broke diplomatic relations with Iran, Apr. 7. The shah died in Egypt, July 27. The hostage drama ended Jan. 20, 1981, when an accord, involving the release of frozen Iranian assets, was reached.

A dispute over the Shatt al-Arab waterway that divides the two countries brought Iran and Iraq, Sept. 22, 1980, into open warfare. Iraqi troops occupied Iranian territory, including the port city of Khorramshahr in October. Iranian troops recaptured the city and drove Iraqi troops back across the border, May 1982. Iraq, and later Iran, attacked several oil tankers in the Persian Gulf during 1984.

In Nov. 1986 it became known that senior U.S. officials had secretly visited Iran and that the U.S. had provided arms in exchange for Iran's help in obtaining the release of U.S. hostages held by terrorists in Lebanon. The revelation sparked a major scandal in the Reagan administration.

A U.S. Navy warship shot down an Iranian commercial airliner, July 3, 1988, after mistaking it for an F-14 fighter jet; all 290 aboard the plane died. In Aug. 1988, Iran agreed to accept a UN resolution calling for a cease-fire with Iraq.

An earthquake struck northern Iran June 21, 1990, killing more than 45,000, injuring 100,000, and leaving 400,000 homeless. Some one million Kurdish refugees fled from Iraq to Iran following the Persian Gulf War. To curb Iran's alleged support for international terrorism, the U.S. in 1996 authorized sanctions on foreign companies that invest there.

Mohammad Khatami, a moderate Shiite Muslim cleric, was elected president on May 23, 1997, winning nearly 70% of the vote; his government was repeatedly challenged by religious conservatives. Pro-democracy students and hardline supporters of Islamic rule clashed repeatedly in July 1999.

Iraq

Republic of Iraq

People: Population: 22,427,150. **Age distrib.** (%): <15: 43.7; 65+: 3.1. **Pop. density:** 133 per sq. mi. **Urban:** 75%. **Ethnic groups:** Arab 75-80%, Kurd 15-20%, Turkoman. **Principal languages:** Arabic (official), Kurdish. **Chief religions:** Muslim 97% (Shi'a 60-65%, Sunni 32-37%).

Geography: Area: 168,754 sq. mi. **Location:** In the Middle East, occupying most of historic Mesopotamia. **Neighbors:** Jordan and Syria on W, Turkey on N, Iran on E, Kuwait and Saudi Arabia on S. **Topography:** Mostly an alluvial plain, including the Tigris and Euphrates rivers, descending from mountains in N to desert in SW. Persian Gulf region is marshland. **Capital:** Baghdad. **Cities:** Baghdad 4,336,000; Arbil 1,743,000; Mosul 879,000*.

Government: Type: Republic. **Head of state and gov.:** Pres. Saddam Hussein; b. Apr. 29, 1937; in office: July 16, 1979; also assumed post of prime minister, May 29, 1994. **Local divisions:** 18 governorates (3 in Kurdish Autonomous Region). **Defense:** 7.4% of GDP. **Active troops:** 387,500.

Economy: Industries: Textiles, chemicals, oil refining, cement. **Chief crops:** Grains, dates, cotton. **Minerals:** Oil, gas. **Arable land:** 12%. **Crude oil reserves** (1999): 112.5 bil bbls. **Other resources:** Wool, hides. **Livestock** (1997): chickens: 20.00 mil; sheep: 6.90 mil; cattle: 1.35 mil; goats: 1.50 mil. **Electricity prod.** (1997): 28.600 bil kWh.

Finance: Monetary unit: Dinar (Sept. 1999: 0.31 = $1 U.S.). **GDP** (1997 est.): $42.8 bil. **Per capita GDP:** $2,000. **Imports** (1996 est.): $1.9 bil; partners: Jordan 49%. **Exports** (1994 est.): $450 mil; partners: Jordan 98%. **Tourism:** $13 mil.

Transport: Railroad: Length: 1,263 mi. **Motor vehicles:** 672,000 pass. cars, 368,000 comm. vehicles. **Civil aviation:** 12.4 mil passenger-mi. **Chief port:** Basra.

Communications: TV sets: 48 per 1,000 pop. **Radios:** 167 per 1,000 pop. **Telephones** (1997): 675,000 main lines. **Daily newspaper circ.:** 27 per 1,000 pop.

Health: Life expectancy: 65.54 male; 67.56 female. **Births** (per 1,000 pop.): 38.42. **Deaths** (per 1,000 pop.): 6.56. **Natural inc.:** 3.186%. **Hosp. beds** (1993): 1 per 704 persons. **Physicians** (1993): 1 per 2,181 persons. **Infant mortality** (per 1,000 live births): 62.41.

Education: Free, compulsory: ages 6-12. **Literacy:** 58%.

Major Intl. Organizations: UN (FAO, IBRD, ILO, IMF, IMO, WHO), AL, OPEC.

Website: http://www.Iraqi-mission.org

The Tigris-Euphrates valley, formerly called Mesopotamia, was the site of one of the earliest civilizations in the world. The Sumerian city-states of 3,000 BC originated the culture later developed by the Semitic Akkadians, Babylonians, and Assyrians.

Mesopotamia ceased to be a separate entity after the Persian, Greek, and Arab conquests. The latter founded Baghdad, from where the caliph ruled a vast empire in the 8th and 9th centuries. Mongol and Turkish conquests led to a decline in population, economy, cultural life, and the irrigation system.

Britain secured a League of Nations mandate over Iraq after World War I. Independence under a king came in 1932. A leftist, pan-Arab revolution established a republic in 1958, which oriented foreign policy toward the USSR. Most industry has been nationalized, and large land holdings broken up.

A local faction of the international Baath Arab Socialist party has ruled by decree since 1968. The USSR and Iraq signed an aid pact in 1972, and arms were sent along with several thousand advisers. The 1978 execution of 21 Communists and a shift of trade to the West signalled a more neutral policy, straining relations with the USSR. In the 1973 Arab-Israeli war Iraq sent forces to aid Syria. Within a month of assuming power, Saddam Hussein instituted a bloody purge in the wake of a reported coup attempt against the new regime.

Years of battling with the Kurdish minority resulted in total defeat for the Kurds in 1975, when Iran withdrew support. The fighting led to Iraqi bombing of Kurdish villages in Iran, causing relations with Iran to deteriorate.

After skirmishing intermittently for 10 months over the sovereignty of the disputed Shatt al-Arab waterway that divides the two countries, Iraq and Iran entered into open warfare on Sept. 22, 1980. In the following days, there was heavy ground fighting around Abadan and the port of Khorramshahr, as Iraq launched an attack on Iran's oil-rich province of Khuzistan.

Israeli planes destroyed a nuclear reactor near Baghdad June 7, 1981, claiming it could be used to produce nuclear weapons.

Iraq and Iran expanded their war to the Persian Gulf in Apr. 1984. There were several attacks on oil tankers. An Iraqi warplane launched a missile attack on the USS *Stark*, a U.S. Navy frigate on patrol in the Persian Gulf, May 17, 1987; 37 U.S. sailors died. Iraq apologized for the attack, claiming it was inadvertent. The fierce war ended Aug. 1988, when Iraq accepted a UN resolution for a cease-fire.

Iraq attacked and overran Kuwait Aug. 2, 1990, sparking an international crisis. The UN, Aug. 6, imposed a ban on all trade with Iraq and called on member countries to protect the assets of the legitimate government of Kuwait. Iraq declared Kuwait its 19th province, Aug. 28.

A U.S.-led coalition launched air and missile attacks on Iraq, Jan. 16, 1991, after the expiration of a UN Security Council deadline for Iraq to withdraw from Kuwait. Iraq retaliated by firing scud missiles at Saudi Arabia and Israel. The coalition began a ground attack to retake Kuwait Feb. 23. Iraqi forces showed little resistance and were soundly defeated in 4 days. Some 175,000 Iraqis were taken prisoner, and casualties were estimated at over 85,000. As part of the cease-fire agreement, Iraq agreed to scrap all poison gas and germ weapons and allow UN observers to inspect the sites. UN trade sanctions would remain in effect until Iraq complied with all terms.

In the aftermath of the war, there were revolts against Pres. Saddam Hussein throughout Iraq. In Feb., Iraqi troops drove Kurdish insurgents and civilians to the Iran and Turkey borders, causing a refugee crisis. The U.S. and allies established havens inside Iraq for the Kurds. Iraqi cooperation with UN weapons inspection teams was intermittent.

The U.S. launched a missile attack aimed at Iraq's intelligence headquarters in Baghdad June 26, 1993. The U.S. justified the attack by citing evidence that Iraq had sponsored a plot to kill former Pres. George Bush during his visit to Kuwait in Apr. 1993. In Aug. 1995, two of Saddam Hussein's sons-in-law, who held high positions in the Iraqi government, defected to Jordan; both were killed after returning to Iraq in Feb. 1996. After fighting between two Kurdish factions (one allied with Iraq, the other with Iran) erupted in the protected zone of northern Iraq, the Baghdad government intervened in the conflict by sending troops into Arbil, Aug. 31, 1996. The U.S. retaliated with missile strikes against air defense sites in the south. On Dec. 9 the UN allowed Baghdad to begin selling limited amounts of oil for food and medicine. Saddam Hussein's son Odai was seriously wounded in an assassination attempt in Baghdad Dec. 12.

Iraqi resistance to unrestricted UN access to suspected weapons sites led to diplomatic crises in Nov. 1997, Feb. 1998, and Oct.-Dec. 1998. Threatened with imminent air strikes by the U.S., Iraq on Feb. 22, 1998, embraced peace proposals brought to Baghdad by UN Secretary General Kofi Annan. Renewed disputes over inspections culminated in intensive U.S. and British aerial bombardment of Iraqi military targets, Dec. 16-19, 1998. U.S. and British warplanes continued to strike Iraq on a regular basis, hitting more than 400 targets between Jan. and Aug. 1999.

Ireland
Republic of Ireland

People: Population: 3,632,944. **Age distrib.** (%): <15: 21.4; 65+: 11.3. **Pop. density:** 134 per sq. mi. **Urban:** 58%. **Ethnic groups:** Principally Celtic, English minority. **Principal languages:** English predominates, Irish (Gaelic) spoken by minority (both official). **Chief religions:** Roman Catholic 93%, Anglican 3%.

Geography: Area: 27,100 sq. mi. **Location:** In the Atlantic O. just W of Great Britain. **Neighbors:** United Kingdom (Northern Ireland) on E. **Topography:** Ireland consists of a central plateau surrounded by isolated groups of hills and mountains. The coastline is heavily indented by the Atlantic O. **Capital:** Dublin. **Cities** (1996): Dublin 480,996; Cork 127,092.

Government: Type: Parliamentary republic. **Head of state:** Pres. Mary McAleese; b June 27, 1951; in office: Nov. 11, 1997. **Head of gov.:** Prime Min. Bertie Ahern; b Sept. 12, 1951; in office: June 26, 1997. **Local divisions:** 26 counties. **Defense:** 1.0% of GDP. **Active troops:** 12,700.

Economy: Industries: Food processing, textiles, chemicals, brewing, machinery, crystal. **Chief crops:** Potatoes, grains, sugar beets, turnips. **Minerals:** Zinc, lead, gas, barite, copper, gypsum. **Arable land:** 13%. **Livestock** (1997): chickens: 11.23 mil; cattle: 7.09 mil; sheep: 5.62 mil; pigs: 1.80 mil. **Fish catch** (1997): 292,872 metric tons. **Electricity prod.** (1997): 18.775 bil kWh. **Labor force:** 62% services; 27% manuf. & constr.; 10% agric., forestry & fish.

Finance: Monetary unit: Punt (Sept. 1999: 1.36 = $1 U.S.). Euro (Sept. 1999): 1.07 = $1 U.S.). **GDP** (1996 est.): $59.9 bil. **Per capita GDP:** $16,800. **Imports** (1997): $44.9 bil; partners: UK 29%, U.S. 12%, Germany 10%. **Exports** (1997): $54.8 bil; partners: UK 22%, Germany 13%. **Tourism:** $3.16 bil. **Budget** (1997): $20.3 bil. **Intl. reserves less gold** (June 1999): $5.22 bil. **Gold:** 193,000 oz t. **Consumer prices** (change in 1998): 2.4%.

Transport: Railroad: Length: 1,210 mi. **Motor vehicles:** 1.06 mil pass. cars, 161,355 comm. vehicles. **Civil aviation:** 4.5 bil pass.-mi.; 9 airports. **Chief ports:** Dublin, Cork.

Communications: TV sets: 279 per 1,000 pop. **Radios:** 597 per 1,000 pop. **Telephones:** 1.6 mil main lines. **Daily newspaper circ.:** 151 per 1,000 pop.

Health: Life expectancy: 73.64 male; 79.32 female. **Births** (per 1,000 pop.): 13.58. **Deaths** (per 1,000 pop.): 8.43. **Natural inc.:** 0.515%. **Hosp. beds** (1994): 1 per 301 persons. **Infant mortality** (per 1,000 live births): 5.94.

Education: Compulsory: ages 6-15. **Literacy** (1993): 100%.

Major Intl. Organizations: UN (FAO, IBRD, ILO, IMF, IMO, WHO, WTrO), EU, OECD, OSCE.

Embassy: 2234 Massachusetts Ave. NW 20008; 462-3939.

Websites: http://www.cso.ie/index.html
http://www.genuki.org.uk

Celtic tribes invaded the islands about the 4th century BC; their Gaelic culture and literature flourished and spread to Scotland and elsewhere in the 5th century AD, the same century in which St. Patrick converted the Irish to Christianity. Invasions by Norsemen began in the 8th century, ended with defeat of the Danes by the Irish King Brian Boru in 1014. English invasions started in the 12th century; for over 700 years the Anglo-Irish struggle continued with bitter rebellions and savage repressions.

The Easter Monday Rebellion in 1916 failed but was followed by guerrilla warfare and harsh reprisals by British troops called the "Black and Tans." The Dail Eireann (Irish parliament) reaffirmed independence in Jan. 1919. The British offered dominion status to Ulster (6 counties) and southern Ireland (26 counties) Dec. 1921. The constitution of the Irish Free State, a British dominion, was adopted Dec. 11, 1922. Northern Ireland remained part of the United Kingdom.

A new constitution adopted by plebiscite came into operation Dec. 29, 1937. It declared the name of the state Eire in the Irish language (Ireland in the English) and declared it a sovereign democratic state.On Dec. 21, 1948, an Irish law declared the country a republic rather than a dominion and withdrew it from the Commonwealth. The British Parliament recognized both actions, 1949, but reasserted its claim to incorporate the 6 northeastern counties in the United Kingdom. This claim has not been recognized by Ireland *(see United Kingdom—Northern Ireland)*.

Irish governments have favored peaceful unification of all Ireland and cooperated with Britain against terrorist groups. On Dec. 15, 1993, Irish and British governments agreed on outlines of a peace plan to resolve the Northern Ireland issue. On Aug. 31, 1994, the Irish Republican Army announced a cease-fire; when peace talks lagged, however, the IRA returned to its terror campaign on Feb. 9, 1996. The IRA proclaimed a new cease-fire as of July 20, 1997, and peace talks resumed Sept. 15.

Ireland's first woman president, Mary Robinson, resigned Sept. 12 to become UN high commissioner for human rights. She was succeeded by Mary McAleese, a law professor from Northern Ireland and the first northerner to hold the office. After negotiators in Northern Ireland approved a peace settlement on Good Friday, April 10, 1998, voters in the Irish Republic endorsed the accord on May 22.

Israel
State of Israel

People: Population: 5,749,760. **Age distrib.** (%): <15: 27.9; 65+: 9.9. **Pop. density:** 719 per sq. mi. **Urban:** 91%. **Ethnic groups:** Jewish 82%, non-Jewish (mostly Arab) 18%. **Principal languages:** Hebrew (official), Arabic (used officially for Arab minority), English. **Chief religions:** Judaism 82%, Muslim (mostly Sunni) 14%.

Geography: Area: 8,000 sq. mi. **Location:** Middle East, on E end of Mediterranean Sea. **Neighbors:** Lebanon on N; Syria, West Bank, and Jordan on E; Gaza Strip and Egypt on W. **Topography:** The Mediterranean coastal plain is fertile and well-watered. In the center is the Judean Plateau. A triangular-shaped semi-desert region, the Negev, extends from south of Beersheba to an apex at the head of the Gulf of Aqaba. The E border drops sharply into the Jordan Rift Valley, including Lake Tiberias (Sea of Galilee) and the Dead Sea, which is 1,312 ft. below sea level, lowest point on the earth's surface. **Capital:** Jerusalem (most countries maintain their embassy in Tel Aviv). **Cities** (1997 est.): Jerusalem 591,400; Tel Aviv-Yafo 355,900; Haifa 255,300.

Government: Type: Republic. **Head of state:** Pres. Ezer Weizman; b June 15, 1924; in office: Mar. 24, 1993. **Head of gov.:** Prime Min. Ehud Barak; b Feb 12, 1942; in office: Jul. 6, 1999. **Local divisions:** 6 districts. **Defense:** 11.5% of GDP. **Active troops:** 175,000.

Economy: Industries: Diamond cutting, textiles, electronics, food processing. **Chief crops:** Citrus, fruit, vegetables, cotton. **Minerals:** Copper, phosphates, bromide, potash, clay. **Crude oil reserves** (1999): 3.93 mil bbls. **Arable land:** 17%. **Livestock** (1997): chickens: 22.00 mil; cattle: 370,000; sheep: 360,000; pigs: 168,000. **Electricity prod.** (1997): 31.593 bil kWh. **Labor force:** 31% public services; 20% mfg.; 13% commerce.

Finance: Monetary unit: New Shekel (Sept. 1999: 4.22 = $1 U.S.). **Gross domestic prod.** (1997 est.): $96.7 bil. **Per capita GDP:** $17,500. **Imports** (1997): $28.6 bil; partners: EU 52%, U.S. 20%. **Exports** (1997): $20.7 bil; partners: EU 32%, U.S. 31%. **Tourism:** $2.70 bil. **Budget** (1998 est.): $58 bil. **Intl. reserves less gold** (June 1999): $22.02 bil. **Gold:** 9,000 oz t. **Consumer prices** (change in 1998): 5.4%.

Transport: Railroad: Length: 379 mi. **Motor vehicles**

(1997): 1.24 mil pass. cars, 304,033 comm. vehicles. **Civil aviation:** 7.3 bil pass.-mi.; 7 airports. **Chief ports:** Haifa, Ashdod, Elat.

Communications: TV sets: 290 per 1,000 pop. **Radios:** 489 per 1,000 pop. **Telephones** (1997): 2,656,000 main lines. **Daily newspaper circ.:** 271 per 1,000 pop.

Health: Life expectancy: 76.71 male; 80.61 female. **Births** (per 1,000 pop.): 19.83. **Deaths** (per 1,000 pop.): 6.16. **Natural inc.:** 1.367%. **Hosp. beds** (1997): 1 per 165 persons. **Physicians** (1997): 1 per 206 persons. **Infant mortality** (per 1,000 live births): 7.78.

Education: Free, compulsory: ages 5-15. **Literacy:** 96%.

Major Intl. Organizations: UN (FAO, IBRD, ILO, IMF, IMO, WHO, WTrO).

Embassy: 3514 International Dr. NW 20008; 364-5500.
Website: http://www.israel.org

Occupying the SW corner of the ancient Fertile Crescent, Israel contains some of the oldest known evidence of agriculture and of primitive town life. A more advanced civilization emerged in the 3d millennium BC. The Hebrews probably arrived early in the 2d millennium BC. Under King David and his successors (c.1000 BC-597 BC), Judaism was developed and secured. After conquest by Babylonians, Persians, and Greeks, an independent Jewish kingdom was revived, 168 BC, but Rome took effective control in the next century, suppressed Jewish revolts in AD 70 and AD 135, and renamed Judea Palestine, after the earlier coastal inhabitants, the Philistines.

Arab invaders conquered Palestine in 636. The Arabic language and Islam prevailed within a few centuries, but a Jewish minority remained. The land was ruled from the 11th century as a part of non-Arab empires by Seljuks, Mamluks, and Ottomans (with a crusader interval, 1098-1291).

After 4 centuries of Ottoman rule, during which the population declined to a low of 350,000 (1785), the land was taken in 1917 by Britain, which pledged in the Balfour Declaration to support a Jewish national homeland there. In 1920 a British Palestine Mandate was recognized; in 1922 the land east of the Jordan was detached.

Jewish immigration, begun in the late 19th century, swelled in the 1930s with refugees from the Nazis; heavy Arab immigration from Syria and Lebanon also occurred. Arab opposition to Jewish immigration turned violent in 1920, 1921, 1929, and 1936. The UN General Assembly voted in 1947 to partition Palestine into an Arab and a Jewish state. Britain withdrew in May 1948.

Israel was declared an independent state May 14, 1948; the Arabs rejected partition. Egypt, Jordan, Syria, Lebanon, Iraq, and Saudi Arabia invaded, but failed to destroy the Jewish state, which gained territory. Separate armistices with the Arab nations were signed in 1949; Jordan occupied the West Bank, Egypt occupied Gaza; neither granted Palestinian autonomy.

After persistent terrorist raids, Israel invaded Egypt's Sinai, Oct. 29, 1956, aided briefly by British and French forces. A UN cease-fire was arranged Nov. 6.

An uneasy truce between Israel and the Arab countries, supervised by a UN Emergency Force, prevailed until May 19, 1967, when the UN force withdrew at Egypt's demand. Egyptian forces reoccupied the Gaza Strip and closed the Gulf of Aqaba to Israeli shipping. In a 6-day war that started June 5, the Israelis took the Gaza Strip, occupied the Sinai Peninsula to the Suez Canal, and captured East Jerusalem, Syria's Golan Heights, and Jordan's West Bank. The fighting was halted June 10 by UN-arranged cease-fire agreements.

Egypt and Syria attacked Israel, Oct. 6, 1973 (on Yom Kippur, the most solemn day on the Jewish calendar). Israel counter-attacked, driving the Syrians back, and crossed the Suez Canal. A cease-fire took effect Oct. 24 and a UN peacekeeping force went to the area. Under a disengagement agreement signed Jan. 18, 1974, Israel withdrew from the canal's west bank.

Israeli forces raided Entebbe, Uganda, July 3, 1976, and rescued 103 hostages who had been seized by Arab and German terrorists.

In 1977, the conservative opposition, led by Menachem Begin, was voted into office for the first time. Egypt's Pres. Anwar al-Sadat visited Jerusalem Nov. 1977, and on Mar. 26, 1979, Egypt and Israel signed a formal peace treaty, ending 30 years of war and establishing diplomatic relations. Israel returned the Sinai to Egypt in 1982.

Israel invaded S Lebanon, Mar. 1978, following a Lebanon-based terrorist attack in Israel. Israel withdrew in favor of a 6,000-man UN force, but continued to aid Lebanese Christian militiamen. Israel affirmed the whole of Jerusalem as its capital, July 1980, encompassing the annexed East Jerusalem.

On June 7, 1981, Israeli jets destroyed an Iraqi atomic reactor near Baghdad that, Israel claimed, would have enabled Iraq to manufacture nuclear weapons. Israeli forces invaded

Lebanon, June 6, 1982, to destroy PLO strongholds there. After massive Israeli bombing of West Beirut, the PLO agreed to evacuate the city. Israeli troops entered West Beirut after newly elected Lebanese Pres. Bashir Gemayel was assassinated on Sept. 14. Israel drew widespread condemnation when Lebanese Christian forces, Sept. 16, entered two West Beirut refugee camps and slaughtered hundreds of Palestinian refugees.

In 1989, violence escalated over the Israeli military occupation of the West Bank and Gaza Strip. In a series of uprisings known as the intifada, Palestinian protesters defied Israeli troops, who forcibly retaliated. Israeli police and stone-throwing Palestinians clashed, Oct. 8, 1990, around the al-Aqsa mosque on the Temple Mount in Jerusalem; some 20 Palestinians died.

During the Persian Gulf War in early 1991, Iraq fired a series of scud missiles at Israel. The Labor Party of Yitzhak Rabin won a clear victory in elections held June 23, 1992.

Ongoing peace talks led to historic agreements between Israel and the PLO, Sept. 1993. The PLO recognized Israel's right to exist; Israel recognized the PLO as the Palestinians' representative; the two sides then signed, Sept. 13, an agreement for limited Palestinian self-rule and the West Bank and Gaza.

Israel and Jordan signed, July 25, 1994, in Washington, DC, a declaration ending their 46-year state of war. A formal peace treaty was signed Oct. 26.

Arab and Jewish extremists repeatedly challenged the peace process. A Jewish gunman opened fire on Arab worshippers at a mosque in Hebron, Feb. 25, 1994, killing at least 29 before he himself was killed. On Nov. 4, 1995, an Orthodox Jewish Israeli assassinated Rabin as he left a peace rally in Tel Aviv.

Support for Rabin's successor, Shimon Peres, was shaken by a series of suicide bombings and rocket attacks against Israel by Islamic militants. In Apr. 1996, Israel attacked suspected guerrilla bases in southern Lebanon. Emphasizing security issues, the candidate of the conservative Likud bloc, Benjamin Netanyahu, was elected prime minister on May 29.

On Sept. 24, 1996, Israel opened a tunnel entrance near a sacred Muslim site in Jerusalem, setting off several days of violence between Israeli soldiers and Palestinian demonstrators and police. Pres. Clinton hosted a summit meeting between Netanyahu and PLO leader Yasir Arafat soon after, on Oct. 1-2, and peace talks were resumed.

Two suicide bombings in a Jerusalem market July 30, 1997, left 15 people dead and more than 170 wounded. The parliament (Knesset) reelected Ezer Weizman as president Mar. 4, 1998, despite opposition from Netanyahu.

Under an interim accord brokered by Clinton and signed by Netanyahu and Arafat at the White House, Oct. 23, 1998, Israel yielded more West Bank territory to the Palestinians, in exchange for new security guarantees. Negotiations bogged down, however, and full implementation did not begin until Sept. 1999. In the interim, Netanyahu lost by a landslide to the the Labor party candidate, Ehud Barak, in the general election of May 17.

Gaza Strip

The Gaza Strip, also known as Gaza, extends NE from the Sinai Peninsula for 40 km (25 mi), with the Mediterranean Sea to the W and Israel to the E. The Palestinian Authority is responsible for civil government, but Israel retains control over security. Nearly all the inhabitants are Palestinian Arabs, more than 35% of whom live in refugee camps. Population (1999 est.): 1,112,654. Area: 140 sq. mi.

Israel captured Gaza from Egypt in the 1967 war. It remained under Israeli occupation until May 1994, when the Israel Defense Forces withdrew. Agreements between Israel and the PLO in 1993 and 1994 provided for interim self-rule in Gaza, pending the completion of final status negotiations.

West Bank

Located W of the Jordan R. and Dead Sea, the West Bank is bounded by Jordan on the E and by Israel on the N, W, and S. The Palestinian Authority administers several major cities, but Israel retains control over much land, including Jewish settlements. Population (1998 est.): 1,556,919. Area: 2,270 sq. mi.

Israel captured the West Bank from Jordan in the 1967 war. A 1974 Arab summit conference designated the PLO as sole representative of West Bank Arabs. In 1988 Jordan cut legal and administrative ties with the territory. Jericho was returned to Palestinian control in May 1994. An accord between Israel and the PLO expanding Palestinian self-rule in the West Bank was signed Sept. 28, 1995. Later agreements gave Palestinians full or shared control of 40% of West Bank territory.

Italy
Italian Republic

People: Population: 56,735,130. **Age distrib.** (%): <15: 14.3; 65+: 17.9. **Pop. density:** 488 per sq. mi. **Urban:** 67%. **Ethnic groups:** Italian, small minorities of German, French, Slovene, Albanian. **Principal languages:** Italian (official), German, French, Slovene. **Chief religion:** Roman Catholic 98%.

Geography: Area: 116,300 sq. mi. **Location:** In S Europe, jutting into Mediterranean Sea. **Neighbors:** France on W, Switzerland and Austria on N, Slovenia on E. **Topography:** Occupies a long boot-shaped peninsula, extending SE from the Alps into the Mediterranean, with the islands of Sicily and Sardinia offshore. The alluvial Po Valley drains most of N. The rest of the country is rugged and mountainous, except for intermittent coastal plains, like the Campania, S of Rome. Apennine Mts. run down through center of peninsula. **Capital:** Rome. **Cities:** Rome 2,645,000; Milan 1,304,000; Naples 1,046,000; Turin 920,000.

Government: Type: Republic. **Head of state:** Pres. Carlo Azeglio Ciampi; b Dec. 9, 1920; in office: May 18, 1999. **Head of gov.:** Prime Min. Massimo D'Alema; b Apr. 20, 1949; in office: Oct. 21, 1998. **Local divisions:** 20 regions divided into 94 provinces. **Defense:** 1.9% of GDP. **Active troops:** 325,200.

Economy: Industries: Tourism, steel, machinery, autos, textiles, shoes, clothing, chemicals. **Chief crops:** Grapes, olives, fruits, vegetables, grain. **Minerals:** Mercury, potash, marble, sulphur. **Crude oil reserves** (1999): 821.8 mil bbls. **Arable land:** 31%. **Livestock** (1997): chickens: 138.00 mil; sheep: 10.89 mil; pigs: 8.28 mil; cattle: 7.17 mil; goats: 1.35 mil; buffalo: 162,000. **Fish catch** (1997): 349,677 metric tons. **Electricity prod.** (1997): 234.614 bil kWh. **Labor force:** 61% services; 32% ind.; 7% agric.

Finance: Monetary unit: Lira (Sept. 1999: 1,808.84 = $1 U.S.). Euro (Sept. 1999: 1.07 = $1 U.S.). **GDP** (1997 est.): $1.24 tril. **Per capita GDP:** $21,500. **Imports** (1996): $190 bil; partners: EU 46%. **Exports** (1996): $250.8 bil; partners: EU 53%. **Tourism:** $30.43 bil. **Budget** (1996 est.): $506 bil. **Intl. reserves less gold** (June 1999): $21.28 bil. **Gold:** 78.83 mil oz t. **Consumer prices** (change in 1998): 2.0%.

Transport: Railroad: Length: 9,944 mi. **Motor vehicles** (1997): 31.00 mil pass. cars, 2.99 mil comm. vehicles. **Civil aviation:** 23.6 bil pass.-mi.; 34 airports. **Chief ports:** Genoa, Venice, Trieste, Palermo, Naples, La Spezia.

Communications: TV sets: 436 per 1,000 pop. **Radios:** 790 per 1,000 pop. **Telephones:** 25,986,100 main lines. **Daily newspaper circ.:** 126 per 1,000 pop.

Health: Life expectancy: 75.40 male; 81.82 female. **Births** (per 1,000 pop.): 9.27. **Deaths** (per 1,000 pop.): 10.28. **Natural inc.:** –0.101%. **Hosp. beds** (1993): 1 per 147 persons. **Physicians** (1993): 1 per 193 persons. **Infant mortality** (per 1,000 live births): 6.30.

Education: Free, compulsory: ages 6-13. **Literacy** (1994): 97%.

Major Intl. Organizations: UN and all of its specialized agencies, EU, NATO, OECD, OSCE.

Embassy: 1601 Fuller St. NW 20009; 328-5500. **Website:** http://www.istat.it

Rome emerged as the major power in Italy after 500 BC, dominating the Etruscans to the N and Greeks to the S. Under the Empire, which lasted until the 5th century AD, Rome ruled most of Western Europe, the Balkans, the Middle East, and N Africa. In 1988, archaeologists unearthed evidence showing Rome as a dynamic society in the 6th and 7th centuries BC.

After the Germanic invasions, lasting several centuries, a high civilization arose in the city-states of the N, culminating in the Renaissance. But German, French, Spanish, and Austrian intervention prevented the unification of the country. In 1859 Lombardy came under the crown of King Victor Emmanuel II of Sardinia. By plebiscite in 1860, Parma, Modena, Romagna, and Tuscany joined, followed by Sicily and Naples, and by the Marches and Umbria. The first Italian Parliament declared Victor Emmanuel king of Italy Mar. 17, 1861. Mantua and Venetia were added in 1866 as an outcome of the Austro-Prussian war. The Papal States were taken by Italian troops Sept. 20, 1870, on the withdrawal of the French garrison. The states were annexed to the kingdom by plebiscite. Italy recognized Vatican City as independent Feb. 11, 1929.

Fascism appeared in Italy Mar. 23, 1919, led by Benito Mussolini, who took over the government at the invitation of the king Oct. 28, 1922. Mussolini acquired dictatorial powers. He made war on Ethiopia and proclaimed Victor Emmanuel III emperor, defied the sanctions of the League of Nations, sent troops to fight for Franco against the Republic of Spain, and joined Germany in World War II.

After Fascism was overthrown in 1943, Italy declared war on Germany and Japan and contributed to the Allied victory. It surrendered conquered lands and lost its colonies. Mussolini was killed by partisans Apr. 28, 1945. Victor Emmanuel III abdicated May 9, 1946; his son Humbert II was king until June 10, when Italy became a republic after a referendum, June 2-3.

Since World War II, Italy has enjoyed growth in industrial output and living standards, in part a result of membership in the European Community (now European Union). Political stability has not kept pace with economic prosperity, and organized crime and corruption have been persistent problems.

Christian Democratic leader and former Prime Min. Aldo Moro was abducted and murdered in 1978 by Red Brigade terrorists. The wave of left-wing political violence, including other kidnappings and assassinations, continued into the 1980s.

In the early 1990s, scandals implicated some of Italy's most prominent politicians. In Mar. 1994 voting, under reformed election rules, right-wing parties won a majority, dislodging Italy's long-powerful Christian Democratic Party. After a series of short-lived governments, a coalition of center-left parties won the election of Apr. 21, 1996. Italy led a 7,000-member international peacekeeping force in Albania, Apr.-Aug. 1997. Two earthquakes in central Italy Sept. 26 killed 11 people, left about 12,000 homeless, and damaged priceless frescoes in Assisi.

On Feb. 3, 1998, a low-flying U.S. military aircraft severed a gondola cable at a ski resort in N Italy, killing 20 people. Implementation of a deficit reduction plan enabled Italy to qualify in May to adopt the euro, a common European currency. Italy contributed 2,000 troops to the NATO-led security force (KFOR) that entered Kosovo in June 1999. Turin was chosen June 19 to host the Winter Olympics in 2006.

Sicily, 9,926 sq. mi., pop. (1994 est.) 5,025,000, is an island 180 by 120 mi., seat of a region that embraces the island of **Pantelleria,** 32 sq. mi., and the **Lipari** group, 44 sq. mi., including 2 active volcanoes: **Vulcano,** 1,637 ft., and **Stromboli,** 3,038 ft. From prehistoric times Sicily has been settled by various peoples; a Greek state had its capital at Syracuse. Rome took Sicily from Carthage 215 BC. **Mt. Etna,** an 11,053-ft. active volcano, is its tallest peak.

Sardinia, 9,301 sq. mi., pop. (1994 est.) 1,657,000, lies in the Mediterranean, 115 mi. W of Italy and 7$^{1}/_{2}$ mi. S of Corsica. It is 160 mi. long, 68 mi. wide, and mountainous, with mining of coal, zinc, lead, copper. In 1720 Sardinia was added to the possessions of the Dukes of Savoy in Piedmont and Savoy to form the Kingdom of Sardinia. Giuseppe Garibaldi is buried on the nearby isle of Caprera. **Elba,** 86 sq. mi., lies 6 mi. W of Tuscany. Napoleon I lived in exile on Elba 1814-1815.

Jamaica

People: Population: 2,652,443. **Age distrib.** (%): <15: 31.1; 65+: 6.8. **Pop. density:** 632 per sq. mi. **Urban:** 54%. **Ethnic groups:** Black 90%. **Principal languages:** English (official), Jamaican Creole. **Chief religions:** Protestant 61%, Roman Catholic 4%, spiritual cults and other 35%.

Geography: Area: 4,200 sq. mi. **Location:** In West Indies. **Neighbors:** Nearest are Cuba to N, Haiti to E. **Topography:** Four-fifths of Jamaica is covered by mountains. **Capital:** Kingston (1991 met.): 103,771.

Government: Type: Parliamentary democracy. **Head of state:** Queen Elizabeth II, represented by Gov.-Gen. Sir Howard Cooke; b Nov. 13, 1915; in office: Aug. 1, 1991. **Head of gov.:** Prime Min. Percival J. Patterson; b Apr. 10, 1935; in office: Mar. 30, 1992. **Local divisions:** 14 parishes. **Defense:** 0.6% of GNP. **Active troops:** 3,300.

Economy: Industries: Bauxite mining, tourism. **Chief crops:** Sugar, coffee, bananas, potatoes, citrus. **Minerals:** Bauxite, limestone, gypsum. **Arable land:** 14%. **Livestock** (1997): chickens: 9.50 mil; cattle: 400,000; goats: 440,000; pigs: 180,000. **Fish catch** (1996): 12,843 metric tons. **Electricity prod.** (1997): 5.864 bil kWh. **Labor force:** 26% services; 19% agric.; 18% trade.

Finance: Monetary unit: Dollar (Sept. 1999: 38.18 = $1 U.S.). **GDP** (1996 est.): $9.5 bil. **Per capita GDP:** $3,660. **Imports** (1996 est.): $2.9 bil; partners: U.S. 52%. **Exports** (1996): $1.4 bil; partners: U.S. 37%. **Tourism:** $1.20 bil. **Budget** (FY 1997-98 est.): $3 bil. **Intl. reserves less gold** (Feb. 1999): $702.7 mil. **Consumer prices** (change in 1998): 8.6%.

Transport: Railroad: Length: 129 mi. **Motor vehicles:** 43,500 pass. cars, 15,400 comm. vehicles. **Civil aviation:** 1.7 bil pass.-mi.; 4 airports. **Chief ports:** Kingston, Montego Bay.

Communications: TV sets: 306 per 1,000 pop. **Radios:** 739 per 1,000 pop. **Telephones** (1997): 419,400 main lines. **Daily newspaper circ.:** 65 per 1,000 pop.

Health: Life expectancy: 73.22 male; 78.13 female. **Births** (per 1,000 pop.): 20.22. **Deaths** (per 1,000 pop.): 5.39. **Natural**

inc.: 1.483%. **Hosp. beds** (1993): 1 per 492 persons. **Physicians** (1995): 1 per 6,043 persons. **Infant mortality** (per 1,000 live births): 13.93.

Education: Free, compulsory: ages 6-12. **Literacy:** 85%.

Major Intl. Organizations: UN (FAO, IBRD, ILO, IMF, IMO, WHO, WTrO), Caricom, the Commonwealth, OAS.

Embassy: 1520 New Hampshire Ave. NW 20036; 452-0660. **Website:** http://www.jamaica.com

Jamaica was visited by Columbus, 1494, and ruled by Spain (under whom Arawak Indians died out) until seized by Britain, 1655. Jamaica won independence Aug. 6, 1962.

In 1974 Jamaica sought an increase in taxes paid by U.S. and Canadian bauxite mines. The socialist government acquired 50% ownership of the companies' Jamaican interests in 1976, and was reelected that year. Rudimentary welfare state measures were passed. Relations with the U.S. improved greatly in the 1980s following the election of Edward Seaga, which marked the beginning of a more conservative era.

Japan

People: Population: 126,182,077. **Age distrib.** (%): <15: 15.0; 65+: 16.5. **Pop. density:** 865 per sq. mi. **Urban:** 78%. **Ethnic groups:** Japanese 99.4%. **Principal language:** Japanese (official). **Chief religions:** Buddhism, Shintoism shared by 84%.

Geography: Area: 145,882 sq. mi. **Location:** Archipelago off E coast of Asia. **Neighbors:** Russia to N, South Korea to W. **Topography:** Japan consists of 4 main islands: Honshu ("mainland"), 87,805 sq. mi.; Hokkaido, 30,144 sq. mi.; Kyushu, 14,114 sq. mi.; and Shikoku, 7,049 sq. mi. The coast, deeply indented, measures 16,654 mi. The northern islands are a continuation of the Sakhalin Mts. The Kunlun range of China continues into southern islands, the ranges meeting in the Japanese Alps. In a vast transverse fissure crossing Honshu E-W rises a group of volcanoes, mostly extinct or inactive, including 12,388 ft. Mt. Fuji (Fujiyama) near Tokyo. **Capital:** Tokyo. **Cities** (1996 est.): Tokyo 7,967,614; Osaka 2,599,642; Nagoya 2,151,084; Sapporo 1,774,344; Kyoto 1,463,822.

Government: Type: Parliamentary democracy. **Head of state:** Emp. Akihito; b Dec. 23, 1933; in office: Jan. 7, 1989. **Head of gov.:** Prime Min. Keizo Obuchi; b June 25, 1937; in office: July 30, 1998. **Local divisions:** 47 prefectures. **Defense:** 1.0% of GDP. **Active troops:** 235,600.

Economy: Industries: Electrical & electronic equip., vehicles, machinery, steel, metallurgy, chemicals, fishing. **Chief crops:** Rice, sugar beets, vegetables, fruits. **Crude oil reserves** (1999): 60.2 mil bbls. **Arable land:** 11%. **Livestock** (1997): chickens: 306.00 mil; pigs: 9.80 mil; cattle: 4.70 mil. **Fish catch** (1997): 5.88 mil metric tons. **Electricity prod.** (1997): 972.690 bil kWh. **Labor force:** 50% services & trade; 33% manuf., mining, & constr.; 7% agric. forestry, & fish.

Finance: Monetary unit: Yen (Sept. 1999: 109.86 = $1 U.S.). **GDP** (1997 est.): $3.08 tril. **Per capita GDP:** $24,500. **Imports** (1997): $339 bil; partners: U.S. 22%, SE Asia 15%, EU 14%, China 12%. **Exports** (1997): $421 bil; partners: SE Asia 37%, U.S. 27%. **Tourism:** $4.15 bil. **Budget** (FY 1998-99 est.): $621 bil. **Intl. reserves less gold** (June 1999): $245.25 bil. **Gold:** 24.23 mil oz t. **Consumer prices** (change in 1998): 0.6%.

Transport: Railroad: Length: 12,511 mi. **Motor vehicles** (1997): 46.64 mil pass. cars, 21.39 mil comm. vehicles. **Civil aviation:** 93.9 bil pass.-mi.; 73 airports. **Chief ports:** Tokyo, Kobe, Osaka, Nagoya, Chiba, Kawasaki, Hakodate.

Communications: TV sets: 619 per 1,000 pop. **Radios:** 799 per 1,000 pop. **Telephones** (1997): 60,380,900 main lines. **Daily newspaper circ.:** 578 per 1,000 pop.

Health: Life expectancy: 77.02 male; 83.35 female. **Births** (per 1,000 pop.): 10.48. **Deaths** (per 1,000 pop.): 8.12. **Natural inc.:** 0.236%. **Hosp. beds** (1992): 1 per 74 persons. **Physicians** (1994): 1 per 546 persons. **Infant mortality** (per 1,000 live births): 4.07.

Education: Compulsory: ages 6-15. **Literacy** (1997): 100%.

Major Intl. Organizations: UN and all its specialized agencies, APEC, OECD.

Embassy: 2520 Massachusetts Ave. NW 20008; 939-6700. **Websites:** http://www.embjap.org
http://www.mofa.go.jp

According to Japanese legend, the empire was founded by Emperor Jimmu, 660 BC, but earliest records of a unified Japan date from 1,000 years later. Chinese influence was strong in the formation of Japanese civilization. Buddhism was introduced before the 6th century AD.

A feudal system, with locally powerful noble families and their samurai warrior retainers, dominated from 1192. Central power was held by successive families of shoguns (military dictators), 1192-1867, until recovered by Emperor Meiji, 1868. The Portuguese and Dutch had minor trade with Japan in the 16th and 17th centuries; U.S. Commodore Matthew C. Perry opened the country to U.S. trade in a treaty ratified 1854. Industrialization was begun in the late 19th century. Japan fought China, 1894-95, gaining Taiwan. After war with Russia, 1904-5, Russia ceded S half of Sakhalin and gave concessions in China. Japan annexed Korea 1910.

In World War I Japan ousted Germany from Shandong in China and took over German Pacific islands. Japan took Manchuria in 1931and launched full-scale war in China in 1937. Japan launched war against the U.S. by attacking Pearl Harbor Dec. 7, 1941. The U.S. dropped atomic bombs on Hiroshima, Aug. 6, and Nagasaki, Aug. 9, 1945. Japan surrendered Aug. 14, 1945.

In a new constitution adopted May 3, 1947, Japan renounced the right to wage war; the emperor gave up claims to divinity; the Diet became the sole law-making authority. The U.S. and 48 other non-Communist nations signed a peace treaty and the U.S. a bilateral defense agreement with Japan, in San Francisco Sept. 8, 1951, restoring Japan's sovereignty as of April 28, 1952.

Rebuilding after World War II, Japan emerged as one of the most powerful economies in the world, and as a leader in technology.The U.S. and Western Europe criticized Japan for its restrictive policy on imports, which eventually allowed Japan to accumulate huge trade surpluses.

On June 26, 1968, the U.S. returned to Japanese control the Bonin Isls., Volcano Isls. (including Iwo Jima), and Marcus Isls. On May 15, 1972, Okinawa, the other Ryukyu Isls., and the Daito Isls. were returned by the U.S.; it was agreed the U.S. would continue to maintain military bases on Okinawa.

The Recruit scandal, the nation's worst political scandal since World War II, which involved illegal political donations and stock trading, led to the resignation of Premier Noboru Takeshita in May 1989. Following new political and economic scandals, the ruling Liberal Democratic Party (LDP) was denied a majority in general elections July 18, 1993. On June 29, 1994, Tomiichi Murayama became Japan's first Socialist premier since 1947-48.

An earthquake in the Kobe area in Jan. 1995 claimed more than 5,000 lives, injured nearly 35,000, and caused over $90 billion in property damage. On Mar. 20, a nerve gas attack in the Tokyo subway (blamed on a religious cult) killed 12 and injured thousands. Public anger at the rape of a 12-year-old Okinawa schoolgirl by 3 U.S. servicemen, Sept. 4, led the U.S. to begin reducing its military presence there.

Murayama resigned as prime minister, Jan. 5, 1996, and was replaced by Ryutaro Hashimoto of the LDP. Hashimoto signed a joint security declaration with U.S. Pres. Bill Clinton in Tokyo, Apr. 17, 1996. Nagano hosted the Winter Olympics, Feb. 7-22, 1998. With Japan mired in a lengthy recession, the LDP suffered a sharp rebuke in elections for parliament's upper house, July 12, 1998. Hashimoto resigned office, and on July 24, the LDP chose Keizo Obuchi to become Japan's 23d prime minister since World War II.

An accident Sept. 30, 1999, at a uranium-reprocessing facility in Tokaimura, NE of Tokyo, exposed plant workers and residents of the surrounding area to extremely high radiation levels.

Jordan
Hashemite Kingdom of Jordan

People: Population: 4,561,147. **Age distrib.** (%): <15: 43.0; 65+: 3.0. **Pop. density:** 132 per sq. mi. **Urban:** 72%. **Ethnic groups:** Arab 98%. **Principal language:** Arabic (official), English. **Chief religions:** Sunni Muslim 96%, Christian 4%.

Geography: Area: 34,445 sq. mi. **Location:** In Middle East. **Neighbors:** Israel and West Bank on W, Saudi Arabia on S, Iraq on E, Syria on N. **Topography:** About 88% of Jordan is arid. Fertile areas are in W. Only port is on short Aqaba Gulf coast. Country shares Dead Sea (1,312 ft. below sea level) with Israel. **Capital:** Amman. **Cities:** Amman 483,000*.

Government: Type: Constitutional monarchy. **Head of state:** King Abdullah II; b Jan. 30, 1962; in office: Feb. 7, 1999. **Head of gov.:** Prime Min. Abdul Raouf al-Rawabdeh; b. 1939; in office: Mar. 4, 1999. **Local divisions:** 12 governorates. **Defense:** 6.4% of GDP. **Active troops:** 104,100.

Economy: Industries: Oil refining, cement, light manufacturing. **Chief crops:** Grains, olives, fruits. **Minerals:** Phosphates, potash. **Arable land:** 4%. **Livestock** (1997): chickens: 23.30 mil; sheep: 2.00 mil; goats: 795,000. **Electricity prod.** (1997): 6.025 bil kWh. **Labor force:** 71% services; 21% mining, manuf., const.

Finance: Monetary unit: Dinar (Sept. 1999: 0.71 = $1 U.S.). **GDP** (1997 est.): $20.7 bil. **Per capita GDP:** $4,800. **Im-** ports (1997): $3.7 bil; partners: Iraq 12%, U.S. 10%. **Exports** (1997): $1.53 bil; partners: Saudi Arabia 13%, Iraq 9%. **Tourism:** $790 mil. **Budget** (1997 est.): $2.8 bil. **Intl. reserves less gold** (June 1999): $2.33 bil. **Gold:** 835,000 oz t. **Consumer prices** (change in 1998): 4.4%.

Transport: Railroad: Length: 421 mi. **Motor vehicles:** 175,000 pass. cars, 90,000 comm. vehicles. **Civil aviation:** 3.0 bil pass.-mi.; 2 airports. **Chief port:** Al Aqabah.

Communications: TV sets: 176 per 1,000 pop. **Radios:** 224 per 1,000 pop. **Telephones** (1997): 402,600 main lines. **Daily newspaper circ.:** 62 per 1,000 pop.

Health: Life expectancy: 71.15 male; 75.08 female. **Births** (per 1,000 pop.): 34.31. **Deaths** (per 1,000 pop.): 3.85. **Natural inc.:** 3.046%. **Hosp. beds** (1995): 1 per 567 persons. **Physicians** (1995): 1 per 616 persons. **Infant mortality** (per 1,000 live births): 32.70.

Education: Free, compulsory: ages 6-16. **Literacy:** 87%.

Major Intl. Organizations: UN (FAO, IBRD, ILO, IMF, IMO, WHO), AL.

Embassy: 3504 International Dr. NW 20008; 966-2664.

Website: http://www.nic.gov.jo

From ancient times to 1922 the lands to the E of the Jordan River were culturally and politically united with the lands to the W. Arabs conquered the area in the 7th century; the Ottomans took control in the 16th. Britain's 1920 Palestine Mandate covered both sides of the Jordan. In 1921, Abdullah, son of the ruler of Hejaz in Arabia, was installed by Britain as emir of an autonomous Transjordan, covering two-thirds of Palestine. An independent kingdom was proclaimed, 1946.

During the 1948 Arab-Israeli war the West Bank and East Jerusalem were added to the kingdom, which changed its name to Jordan. All these territories were lost to Israel in the 1967 war, which swelled the number of Arab refugees on the East Bank.

Some 700,000 refugees entered Jordan following Iraq's invasion of Kuwait, Aug. 1990. Jordan was viewed as supporting Iraq during the 1990-1991 Persian Gulf crisis.

Jordan and Israel officially agreed, July 25, 1994, to end their state of war; a formal peace treaty was signed Oct. 26. Following a prolonged bout with cancer, King Hussein died Feb. 7, 1999; his eldest son and designated successor immediately assumed the throne as Abdullah II.

Kazakhstan
Republic of Kazakhstan

People: Population: 16,824,825. **Age distrib.** (%): <15: 28.5; 65+: 7.0. **Pop. density:** 16 per sq. mi. **Urban:** 60%. **Ethnic groups:** Kazakh 46%, Russian 35%, Ukrainian 5%. **Principal languages:** Kazakh, Russian (both official). **Chief religions:** Muslim 47%, Russian Orthodox 44%.

Geography: Area: 1,049,200 sq. mi. **Location:** In Central Asia. **Neighbors:** Russia on N; China on E; Kyrgyzstan, Uzbekistan, Turkmenistan on S; Caspian Sea on W. **Topography:** Extends from the lower reaches of Volga in Europe to the Altay Mts. on the Chinese border. **Capital:** Astana (1997 est.): 270,400.

Government: Type: Republic. **Head of state:** Pres. Nursultan A. Nazarbayev; b July 6, 1940; in office: Apr. 1990. **Head of gov.:** Prime Min. Kasymzhomart Tokayev; in office: Oct. 12, 1999 (acting). **Local divisions:** 14 oblystar, 1 city. **Defense:** 2.3% of GDP. **Active troops:** 35,100.

Economy: Industries: Oil, steel, mining, agricultural machinery. **Chief crops:** Grain, cotton. **Minerals:** Oil, gas, coal, iron, manganese, chrome ore, copper. **Crude oil reserves** (1999): 5.42 bil bbls. **Arable land:** 12%. **Livestock** (1997): chickens: 16.92 mil; sheep: 8.91 mil; cattle: 4.00 mil; pigs: 859,900; goats: 690,900; buffalo: 100,000. **Electricity prod.** (1997): 49.518 bil kWh. **Labor force:** 27% industry; 23% agric., forestry.

Finance: Monetary unit: Tenge (Aug. 1999: 132.80 = $1 U.S.). **GDP** (1997 est.): $50 bil. **Per capita GDP:** $3,000. **Imports** (1996): $6 bil; partners: Russia 65%. **Exports** (1996): $5.6 bil; partners: Russia 64%. **Tourism:** $289 mil. **Budget** (1996 est.): $4.6 bil. **Intl. reserves less gold** (Mar. 1999): $1.10 bil. **Gold:** 1.71 mil oz t. **Consumer prices** (change in 1998): 7.1%.

Transport: Railroad: Length: 13,422 mi. **Motor vehicles:** 1.0 mil pass. cars, 515,000 comm. vehicles. **Civil aviation:** 826.5 mil pass.-mi.; 6 airports. **Chief ports:** Aqtau, Atyrau.

Communications: TV sets: 275 per 1,000 pop. **Telephones** (1997): 1,818,200 main lines.

Health: Life expectancy: 57.92 male; 69.13 female. **Birth rate** (per 1,000 pop.): 17.16. **Death rate** (per 1,000 pop.): 10.34. **Natural inc.:** 0.682%. **Hosp. beds** (1995): 1 per 86 persons. **Physicians** (1995): 1 per 267 persons. **Infant mortality** (per 1,000 live births): 58.82.

Education: Free, compulsory: ages 7-18. **Literacy** (1992): 98%.
Major Intl. Organizations: UN (IBRD, ILO, IMF, IMO, WHO), CIS, OSCE.
Embassy: 3421 Massachusetts Ave. NW 20008; 333-4504.
Website: http://www.undp.org/missions/kazakhstan

The region came under the Mongols' rule in the 13th century and gradually came under Russian rule, 1730-1853. It was admitted to the USSR as a constituent republic 1936. Kazakhstan declared independence Dec. 16, 1991. It became an independent state when the Soviet Union dissolved Dec. 26, 1991. The party chief, Nursultan Nazarbayev, was elected president unopposed. In legislative elections Mar. 7, 1994, criticized by international monitors, his party won a sweeping victory. Kazakhstan agreed, Feb. 14, to dismantle nuclear missiles and adhere to the 1968 Nuclear Nonproliferation Treaty; the U.S. pledged increased aid. Private land ownership was legalized Dec. 26, 1995.

Astana (formerly Akmola) was dedicated as the nation's new capital on June 9, 1998. Pres. Nazarbayev won reelection to a 7-year term Jan. 10, 1999, after his leading opponent, former Prime Min. Akezhan Kazhegeldin, was barred on a technicality.

Kenya
Republic of Kenya

People: Population: 28,808,658. **Age distrib.** (%): <15: 42.9; 65+: 2.7. **Pop. density:** 128 per sq. mi. **Urban:** 30%. **Ethnic groups:** Kikuyu 22%, Luhya 14%, Luo 13%, Kalenjin 12%, Kamba 11%, others including Asian, Arab, European. **Principal languages:** Swahili, English (both official), numerous indigenous languages. **Chief religions:** Protestant 38%, Roman Catholic 28%, indigenous beliefs 26%.
Geography: Area: 225,000 sq. mi. **Location:** E Africa, on coast of Indian O. **Neighbors:** Uganda on W, Tanzania on S, Somalia on E, Ethiopia on N, Sudan on NW. **Topography:** The northern three-fifths of Kenya is arid. To the S, a low coastal area and a plateau varying from 3,000 to 10,000 ft. The Great Rift Valley enters the country N-S, flanked by high mountains. **Capital:** Nairobi. **Cities** (1991 est.): Nairobi 2,000,000; Mombasa 600,000.
Government: Type: Republic. **Head of state and gov.:** Pres. Daniel arap Moi; b Sept. 2, 1924; in office: Aug. 22, 1978. **Local divisions:** Nairobi and 7 provinces. **Defense:** 2.4% of GDP. **Active troops:** 24,200.
Economy: Industries: Tourism, light industry, agricultural processing, oil refining. **Chief crops:** Coffee, corn, tea. **Minerals:** Gold, limestone, salt, rubies, fluorspar, garnets. **Other resources:** Hides, dairy products, cut flowers (world's 4th lgst. exporter). **Arable land:** 7%. **Livestock** (1997): chickens: 29.00 mil; cattle: 14.12 mil; goats: 7.50 mil; sheep: 5.70 mil; pigs: 108,000. **Fish catch** (1996): 178,354 metric tons. **Electricity prod.** (1997): 3.950 bil kWh. **Labor force:** 75-80% agric.
Finance: Monetary unit: Shilling (Sept. 1999: 75.80 = $1 U.S.). **GDP** (1997 est.): $45.3 bil. **Per capita GDP:** $1,600. **Imports** (1996): $2.9 bil; partners: UK 21%, UAE 18%. **Exports** (1996): $2.1 bil; partners: Uganda 23%, UK 20%, Tanzania 19%. **Tourism:** $400 mil. **Budget** (FY 1996-97): $3 bil. **Intl. reserves less gold** (May 1999): $619.1 mil. **Gold:** 80,000 oz t. **Consumer prices** (change in 1998): 5.8%.
Transport: Railroad: Length: 1,885 mi. **Motor vehicles:** 271,000 pass. cars, 75,900 comm. vehicles. **Civil aviation:** 1.1 bil pass.-mi.; 13 airports. **Chief ports:** Mombasa, Kisumu, Lamu.
Communications: TV sets: 18 per 1,000 pop. **Radios:** 103 per 1,000 pop. **Telephones** (1997): 271,800 main lines.
Health: Life expectancy: 46.56 male; 47.49 female. **Births** (per 1,000 pop.): 30.80. **Deaths** (per 1,000 pop.): 14.58. **Natural inc.:** 1.622%. **Hosp. beds** (1994): 1 per 734 persons. **Physicians** (1994): 1 per 5,999 persons. **Infant mortality** (per 1,000 live births): 59.07.
Education: Free, compulsory: ages 6-14. **Literacy:** 78%.
Major Intl. Organizations: UN and all of its specialized agencies, the Commonwealth, OAU.
Embassy: 2249 R St. NW 20008; 387-6101.

Arab colonies exported spices and slaves from the Kenya coast as early as the 8th century. Britain obtained control in the 19th century. Kenya won independence Dec. 12, 1963, 4 years after the end of the violent Mau Mau uprising.
Kenya had steady growth in industry and agriculture under a modified private enterprise system, and enjoyed a relatively free political life. But stability was shaken in 1974-75, with opposition charges of corruption and oppression. Jomo Kenyatta, the country's leader since independence, died Aug. 22, 1978. He was succeeded by his vice president, Daniel arap Moi.

During the first half of the 1990s, Kenya suffered widespread unemployment and high inflation. Tribal clashes in the western provinces claimed thousands of lives and left tens of thousands homeless. Pres. Moi won a third term in Dec. 1992 elections, which were marred by violence and fraud. Clashes in the Mombasa region, Aug. 1997, left more than 40 people dead. Pres. Moi was reelected Dec. 29, in an election again plagued by irregularities.
A truck bomb explosion at the U.S. embassy in Nairobi, Aug. 7, 1998, killed more than 200 people and injured about 5,000. The U.S. blamed the attack on Islamic terrorists associated with a wealthy Saudi businessman, Osama bin Laden.

Kiribati
Republic of Kiribati

People: Population: 85,501. **Pop. density:** 309 per sq. mi. **Urban:** 36%. **Ethnic groups:** Micronesian. **Principal languages:** English (official), Gilbertese. **Chief religions:** Roman Catholic 53%, Protestant 41%.
Geography: Area: 277 sq. mi. **Location:** 33 Micronesian islands (the Gilbert, Line, and Phoenix groups) in the mid-Pacific scattered in a 2-mil sq. mi. chain around the point where the International Date Line formerly cut the Equator. In 1997 the Date Line was moved to follow Kiribati's E border. **Neighbors:** Nearest are Nauru to SW, Tuvalu and Tokelau Isls. to S. **Topography:** Except Banaba (Ocean) Isl., all are low-lying, with soil of coral sand and rock fragments, subject to erratic rainfall. **Capital:** Tarawa (1990): 25,000.
Government: Type: Republic. **Head of state and gov.:** Pres. Teburoro Tito; b Aug. 25, 1953; in office: Oct. 1, 1994. **Local divisions:** 3 units, 6 districts.
Economy: Industries: Fishing, handicrafts. **Chief crops:** Copra, taro, breadfruit, sweet potatoes, vegetables. **Livestock** (1997): chickens: 300,000. **Electricity prod.** (1997): 7 mil kWh.
Finance: Monetary unit: Australian Dollar (Sept. 1999: 1.55 = $1 U.S.). **GDP** (1996 est.): $62 mil. **Per capita GDP:** $800. **Imports** (1996 est.): $37.4 mil; partners: Australia 46%. **Exports** (1996 est.): $6.7 mil; partners: Japan 33%. **Tourism:** $2 mil. **Budget** (1996 est.): $47.7 mil.
Transport: Chief port: Tarawa. **Civil aviation:** 7.0 mil passenger-mi., 17 airports.
Communications: Radios: 75 per 1,000 pop. **Telephones:** 2,800 main lines.
Health: Life expectancy: 61.02 male; 64.98 female. **Births** (per 1,000 pop.): 26.13. **Deaths** (per 1,000 pop.): 7.53. **Natural inc.:** 1.860%. **Physicians** (1993): 1 per 7,687 persons. **Infant mortality** (per 1,000 live births): 48.22.
Education: Free, compulsory: ages 6-14. **Literacy:** 90%.
Major Intl. Organizations: UN (IBRD, IMF, WHO), the Commonwealth.

A British protectorate since 1892, the Gilbert and Ellice Islands colony was completed with the inclusion of the Phoenix Islands, 1937. Tarawa Atoll was the scene of some of the bloodiest fighting in the Pacific during World War II.
Self-rule was granted 1971; the Ellice Islands separated from the colony 1975 and became independent Tuvalu, 1978. Kiribati (pronounced *Kiribass*) independence was attained July 12, 1979. Under a treaty of friendship the U.S. relinquished its claims to several Line and Phoenix islands, including Christmas (Kiritimati), Canton, and Enderbury. Kiribati was admitted to the UN Sept. 14, 1999.

Korea, North
Democratic People's Republic of Korea

People: Population: 21,386,109. **Age distrib.** (%): <15: 25.6; 65+: 6.2. **Pop. density:** 460 per sq. mi. **Urban:** 62%. **Ethnic group:** Korean. **Principal language:** Korean (official). **Chief religions:** Activities almost nonexistent; traditionally Buddhism, Confucianism, Chondogyo.
Geography: Area: 46,500 sq. mi. **Location:** In northern E Asia. **Neighbors:** China and Russia on N, South Korea on S. **Topography:** Mountains and hills cover nearly all the country, with narrow valleys and small plains in between. The N and the E coasts are the most rugged areas. **Capital:** Pyongyang (1993 est.): 2,741,260.
Government: Type: Communist state. **Leader:** Kim Jong Il; b Feb. 16, 1948; officially assumed post Oct. 8, 1997. **Local divisions:** 9 provinces, 3 special cities. **Defense:** 27.0% of GDP. **Active troops:** 1.055 mil.
Economy: Industries: Textiles, chemicals, machinery, food processing. **Chief crops:** Corn, potatoes, soybeans, rice. **Minerals:** Coal, lead, tungsten, zinc, graphite, magnesite, iron, copper, gold, salt. **Arable land:** 14%. **Livestock** (1997): chickens: 8.00 mil; pigs: 1.60 mil; cattle: 500,000; sheep: 150,000; goats:

1.20 mil. **Fish catch** (1997): 236,462 metric tons. **Electricity prod.** (1997): 33.705 bil kWh. **Labor force:** 36% agric.

Finance: Monetary unit: Won (Sept. 1999: 2.20 = $1 U.S.). **GDP** (1997 est.): $21.8 bil. **Per capita GDP:** $900. **Imports** (1996 est.): $1.95 bil; partners: China 30%, Japan 16%. **Exports** (1996 est.): $912 mil; partners: Japan 31%, Austria 17%.

Transport: Railroad: Length: 5,302 mi. **Civil aviation:** 177.5 mil pass.-mi.; 1 airport. **Chief ports:** Chongjin, Hamhung, Nampo.

Communications: TV sets: 85 per 1,000 pop. **Radios:** 200 per 1,000 pop. **Telephones** (1997): 1.1 mil. main lines. **Daily newspaper circ.:** 213 per 1,000 pop.

Health: Life expectancy: 48.9 male; 53.9 female. **Births** (per 1,000 pop.): 15. **Deaths** (per 1,000 pop.): 16. **Natural inc.:** −0.03%. **Infant mortality** (per 1,000 live births): 88.

Education: Free, compulsory: ages 6-17. **Literacy** (1992): 95%.

Major Intl. Organizations: UN (FAO, IMO, WHO).

The Democratic People's Republic of Korea was founded May 1, 1948, in the zone occupied by Russian troops after World War II. Its armies tried to conquer the south, 1950. After 3 years of fighting, with Chinese and U.S. intervention, a cease-fire was proclaimed. For the next four decades, a hardline Communist regime headed by Kim Il Sung kept tight control over the nation's political, economic, and cultural life. The nation used its abundant mineral and hydroelectric resources to develop its military strength and heavy industry.

In Mar. 1993, North Korea became the first nation to formally withdraw from the Nuclear Nonproliferation Treaty, the international pact designed to limit the spread of nuclear weapons. The nation suspended its withdrawal in June in reaction to threats of UN economic sanctions, but was widely believed to be developing nuclear weapons. The U.S. and North Korea reached an interim agreement, Aug. 13, 1994, intended to resolve the nuclear issue, and further negotiations followed.

Kim Il Sung died July 8, 1994. He was succeeded by his son, Kim Jong Il. North Korea suffered from defections by high officials, a deteriorating economy, and severe food shortages in the late 1990s. On Sept. 17, 1999, the U.S. eased travel and trade restrictions on North Korea after Pyongyang agreed to suspend long-range missile testing.

Korea, South
Republic of Korea

People: Population: 46,884,800. **Age distrib.** (%): <15: 22.1; 65+: 6.7. **Pop. density:** 1,234 per sq. mi. **Urban:** 83%. **Ethnic group:** Korean. **Principal language:** Korean (official). **Chief religions:** Christianity 49%, Buddhism 47%.

Geography: Area: 38,000 sq. mi. **Location:** In northern E Asia. **Neighbors:** North Korea on N. **Topography:** The country is mountainous, with a rugged east coast. The western and southern coasts are deeply indented, with many islands and harbors. **Capital:** Seoul. **Cities** (1995 est.): Seoul 10,231,217; Pusan 3,814,325; Taegu 2,449,420.

Government: Type: Republic, with power centralized in a strong executive. **Head of state:** Pres. Kim Dae Jung; b Dec. 3, 1925; in office: Feb. 25, 1998. **Head of gov.:** Prime Min. Kim Jong Pil; b 1926; in office: Aug. 17, 1998 (acting prime min. from Mar. 3). **Local divisions:** 9 provinces, 6 special cities. **Defense:** 3.3% of GDP. **Active troops:** 672,000.

Economy: Industries: Electronics, autos, chemicals, ships, textiles, clothing. **Chief crops:** Rice, barley, vegetables. **Minerals:** Tungsten, coal, graphite. **Arable land:** 19%. **Livestock** (1997): chickens: 88.00 mil; pigs: 6.70 mil; cattle: 3.28 mil; goats: 603,905. **Fish catch:** (1997): 2.20 mil metric tons. **Electricity prod.** (1997): 212.015 bil kWh. **Labor force:** 55% services & other; 22% manuf. & mining; 11% agric.

Finance: Monetary unit: Won (Sept. 1999: 1,190.75 = $1 U.S.). **GDP** (1997 est.): $631.2 bil. **Per capita GDP:** $13,700 **Imports** (1996): $150.2 bil; partners: U.S. 22%, Japan 21%. **Exports** (1996): $129.8 bil; partners: U.S. 17%, Japan 12%. **Tourism** (1994): $5.7 bil. **Budget** (1996 est.): $101 bil. **Intl. reserves less gold** (June 1999): $61.92 bil. **Gold:** 436,000 oz t. **Consumer prices** (change in 1998): 7.5%.

Transport: Railroad: Length: 4,072 mi. **Motor vehicles** (1997): 7.59 mil pass. cars, 2.83 mil comm. vehicles. **Civil aviation:** 34.62 bil pass.-mi.; 14 airports. **Chief ports:** Pusan, Inchon.

Communications: TV sets: 233 per 1,000 pop. **Radios:** 928 per 1,000 pop. **Telephones** (1997): 20,421,900 main lines. **Daily newspaper circ.:** 405 per 1,000 pop.

Health: Life expectancy: 70.4 male; 78.0 female. **Births** (per 1,000 pop.): 16. **Deaths** (per 1,000 pop.): 6. **Natural inc.:** 1.04%. **Hosp. beds** (1995): 1 per 229 persons. **Physicians** (1995): 1 per 784 persons. **Infant mortality** (per 1,000 live births): 8.

Education: Free, compulsory: ages 6-12. **Literacy:** 98%.

Major Intl. Organizations: UN (FAO, IBRD, ILO, IMF, IMO, WHO, WTrO), APEC, OECD.

Embassy: 2450 Massachusetts Ave. NW 20008; 939-5600.

Korea, once called the Hermit Kingdom, has a recorded history since the 1st century BC. It was united in a kingdom under the Silla Dynasty, AD 668. It was at times associated with the Chinese empire; the treaty that concluded the Sino-Japanese war of 1894-95 recognized Korea's complete independence. In 1910 Japan forcibly annexed Korea as Chosun.

At the Potsdam conference, July 1945, the 38th parallel was designated as the line dividing the Soviet and the American occupation. Russian troops entered Korea Aug. 10, 1945; U.S. troops entered Sept. 8, 1945. The Soviet military organized socialists and Communists and blocked efforts to let the Koreans unite their country.

The South Koreans formed the Republic of Korea in May 1948 with Seoul as the capital. Dr. Syngman Rhee was chosen president. A separate, Communist regime was formed in the N; its army attacked the S in June 1950, initiating the Korean War. UN troops, under U.S. command, supported the S in the war, which ended in an armistice (July 1953) leaving Korea divided by a "no-man's land" along the 38th parallel.

Rhee's authoritarian rule became increasingly unpopular, and a movement spearheaded by college students forced his resignation Apr. 26, 1960. In an army coup May 16, 1961, Gen. Park Chung Hee became chairman of a ruling junta. He was elected president, 1963; a 1972 referendum allowed him to be reelected for an unlimited series of 6-year terms. Park was assassinated by the chief of the Korean CIA, Oct. 26, 1979. In May 1980, Gen. Chun Doo Hwan, head of military intelligence, reinstated full martial law and ordered the brutal suppression of pro-democracy demonstrations in Kwangju.

In July 1972 South and North Korea agreed on a common goal of reunifying the 2 nations by peaceful means. But there was no sign of a thaw in relations between the two regimes until 1985, when they agreed to discuss economic issues.

On June 10, 1987, middle-class office workers, shopkeepers, and business executives joined with students in antigovernment protests in Seoul calling for democratic reforms. Following weeks of rioting and violence, Chun, July 1, agreed to permit election of the next president by direct popular vote and other reforms. In Dec., Roh Tae Woo was elected president. In 1990, the nation's 3 largest political parties merged; some 100,000 students protested the merger as undemocratic.

Kim Young Sam took office in 1993 as the first civilian president since 1961. Convicted of mutiny, treason, and corruption, Chun was sentenced to death by a Seoul court, Aug. 26, 1996, for his role in the 1979 coup and 1980 Kwangju massacre; Roh received a 22-1/2 year prison sentence. On Dec. 16, Chun's term was reduced to life in prison, and Roh's to 17 years. The collapse in Jan. 1997 of the Hanbo steel firm triggered a new round of corruption scandals. With currency and stock values plummeting, the nation averted default by agreeing, Dec. 4, on a $57 billion bailout from the IMF. Kim Dae Jung, a longtime dissident, won the presidential election Dec. 18. Chun and Roh were released and pardoned Dec. 22, 1997.

Kuwait
State of Kuwait

People: Population: 1,991,115. **Age distrib.** (%): <15: 31.6; 65+: 2.2. **Pop. density:** 289 per sq. mi. **Urban:** 97%. **Ethnic groups:** Kuwaiti 45%, other Arab 35%. **Principal languages:** Arabic (official), English. **Chief religion:** Muslim 85%.

Geography: Area: 6,900 sq. mi. **Location:** In Middle East, at N end of Persian Gulf. **Neighbors:** Iraq on N, Saudi Arabia on S. **Topography:** The country is flat, very dry, and extremely hot. **Capital:** (1995 est.): Kuwait City 28,859.

Government: Type: Constitutional monarchy. **Head of state:** Emir Sheikh Jabir al-Ahmad al-Jabir as-Sabah; b 1928; in office: Jan. 1, 1978. **Head of gov.:** Prime Min. Sheikh Saad Abdulla as-Salim as-Sabah; b 1930; in office: Feb. 8, 1978. **Local divisions:** 5 governorates. **Defense:** 11.4% of GDP. **Active troops:** 15,300.

Economy: Industries: Oil products. **Minerals:** Oil, gas. **Crude oil reserves** (1999): 94 bil bbls. **Livestock** (1997): chickens: 26.00 mil; goats: 125,000; sheep: 445,000. **Electricity prod.** (1997): 25.000 bil kWh. **Labor force:** 50% gov't. and social services; 40% services; 10% industry and agric.

Finance: Monetary unit: Dinar (Sept. 1999: 0.30 = $1 U.S.). **GDP** (1997 est.): $46.3 bil. **Per capita GDP:** $22,300. **Imports** (1996): $7.7 bil; partners: U.S. 31%, UK 14%, Japan 13%. **Exports** (1996): $14.7 bil; partners: Japan 29%, U.S.

16%. **Tourism:** $188 mil. **Budget** (FY 1997-98 est.): $14.5 bil. **Intl. reserves less gold** (June 1999): $4.14 bil. **Gold:** 2.54 mil oz t. **Consumer prices** (change in 1998): 0.1%.
Transport: Motor vehicles: 538,000 pass. cars, 155,000 comm. vehicles. **Civil aviation:** 3.7 bil pass.-mi.; 1 airport. **Chief port:** Mina al-Ahmadi.
Communications: TV sets: 456 per 1,000 pop. **Radios:** 592 per 1,000 pop. **Telephones** (1997): 411,600 main lines. **Daily newspaper circ.:** 397 per 1,000 pop.
Health: Life expectancy: 75.11 male; 79.30 female. **Births** (per 1,000 pop.): 20.45. **Deaths** (per 1,000 pop.): 2.31. **Natural inc.:** 1.814%. **Hosp. beds** (1995): 1 per 357 persons. **Physicians** (1995): 1 per 464 persons. **Infant mortality** (per 1,000 live births): 10.26.
Education: Free, compulsory: ages 6-14. **Literacy:** 79%.
Major Intl. Organizations: UN (FAO, IBRD, ILO, IMF, IMO, WHO, WTrO), AL, OPEC.
Embassy: 2940 Tilden St. NW 20008; 966-0702.
Website: http://www.moc.kw/

Kuwait is ruled by the Al-Sabah dynasty, founded 1759. Britain ran foreign relations and defense from 1899 until independence in 1961. The majority of the population is non-Kuwaiti, with many Palestinians, and cannot vote.

Oil is the fiscal mainstay, providing most of Kuwait's income. Oil pays for free medical care, education, and social security. There are no taxes, except customs duties.

Kuwaiti oil tankers came under frequent attack by Iran because of Kuwait's support of Iraq in the Iran-Iraq War. In July 1987, U.S. Navy warships began escorting Kuwaiti tankers in the Persian Gulf.

Kuwait was attacked and overrun by Iraqi forces Aug. 2, 1990. The emir and senior members of the ruling family fled to Saudi Arabia to establish a government in exile. On Aug. 28, Iraq announced that Kuwait was its 19th province. Following several weeks of aerial attacks on Iraq and Iraqi forces in Kuwait, a U.S.-led coalition began a ground attack Feb. 23, 1991. By Feb. 27, Iraqi forces were routed and Kuwait liberated. Following liberation, there were reports of abuse of Palestinians and others suspected of collaborating with Iraqi occupiers. Kuwait spent more than $5 billion to repair oil installations damaged during 1990-91.

Former U.S. Pres. George Bush visited Kuwait, Apr. 14-16, 1993, and was honored as the leader of the Persian Gulf War alliance that expelled Iraqi troops. Kuwaiti authorities arrested 14 Iraqis and Kuwaitis for allegedly plotting to assassinate Bush during his visit; 13 were convicted and sentenced to prison or death, June 4, 1994.

Kyrgyzstan
Kyrgyz Republic

People: Population: 4,546,055. **Age distrib. (%):** <15: 35.0; 65+: 6.2. **Pop density:** 59 per sq. mi. **Urban:** 39%. **Ethnic groups:** Kyrgyz 52%, Russian 18%, Uzbek 13%. **Principal languages:** Kyrgyz, Russian (both official). **Chief religions:** Muslim 75%, Russian Orthodox 20%.
Geography: Area: 76,600 sq. mi. **Location:** In Central Asia. **Neighbors:** Kazakhstan on N, China on E, Uzbekistan on W, Tajikistan on S. **Capital:** Bishkek (1997 est.) 589,400.
Government: Type: Republic. **Head of state:** Pres. Askar Akayev; b Nov. 10, 1944; in office: Oct. 28, 1990. **Head of gov.:** Prime Min. Amangeldi Muraliev; b Aug. 7, 1947; in office: Apr. 12, 1999. **Local divisions:** 6 oblasts, 1 city. **Defense:** 2.5% of GNP. **Active troops:** 12,200.
Economy: Industries: Textiles, mining, food processing, cement, small machinery. **Chief crops:** Tobacco, cotton, fruits. **Minerals:** Gold, coal, oil. **Crude oil reserves** (1999): 40 mil bbls. **Arable land:** 7%. **Livestock** (1997): chickens: 2.00 mil; sheep: 3.35 mil; cattle: 830,000; goats: 168,000. **Electricity prod.** (1997): 14.065 bil kWh. **Labor force:** 40% agric. & forestry; 19% ind. & const.
Finance: Monetary unit: Som (Aug. 1999: 42.70 = $1 U.S.). **Gross domestic prod.** (1997 est.): $9.7 bil. **Per capita GDP:** $2,100. **Imports** (1996): $890 mil; partners: Kazakhstan 22%, Russia 22%, Uzbekistan 17%. **Exports** (1996): $506 mil; partners: Russia 26%, China 17%, Uzbekistan 17%, Kazakhstan 16%. **Tourism:** $7 mil. **Intl. reserves less gold** (June 1999): $188.8 mil. **Gold:** 83,100 oz t. **Consumer prices** (change in 1998): 13.6%.
Transport: Railroad: Length: 249 mi. **Motor vehicles:** 164,000 pass. cars. **Civil aviation:** 280.7 mil pass.-mi.; 2 airports. **Chief port:** Ysyk-Kol.
Communications: TV sets: 238 per 1,000 pop. **Telephones:** 358,800 main lines. **Daily newspaper circ.:** 11 per 1,000 pop.
Health: Life expectancy: 59.25 male; 68.10 female. **Birth

rate** (per 1,000 pop.): 21.83. **Death rate** (per 1,000 pop.): 8.74. **Natural inc.:** 1.309%. **Hosp. beds** (1995): 1 per 111 persons. **Physicians** (1995): 1 per 303 persons. **Infant mortality** (per 1,000 live births): 75.92.
Education: Compulsory: ages 6-15. **Literacy** (1993): 97%.
Major Intl. Organizations: UN (FAO, IBRD, ILO, IMF, WHO), CIS, OSCE.
Embassy: 1732 Wisconsin Ave. NW, 20007; 338-5141.

The region was inhabited around the 13th century by the Kyrgyz. It was annexed to Russia 1864. After 1917, it was nominally a Kara-Kyrgyz autonomous area, which was reorganized 1926, and made a constituent republic of the USSR in 1936. Kyrgyzstan declared independence Aug. 31, 1991. It became an independent state when the USSR disbanded Dec. 26, 1991. A constitution was adopted May 5, 1993.

Reelected Dec. 24, 1995, Pres. Askar Akayev gained approval by referendum of a constitutional amendment expanding his presidential powers, Feb. 10, 1996. Amendments restricting the powers of parliament and allowing private ownership of land were ratified by referendum Oct. 17, 1998.

Laos
Lao People's Democratic Republic

People: Population: 5,407,453. **Age distrib. (%):** <15: 45.1; 65+: 3.2. **Pop. density:** 59 per sq. mi. **Urban:** 21%. **Ethnic groups:** Lao Loum 68%, Lao Theung 22%, Lao Soung (includes Hmong and Yao) 9%. **Principal languages:** Lao (official), French, English. **Chief religions:** Buddhism 60%, animist and other 40%.
Geography: Area: 91,400 sq. mi. **Location:** In Indochina Peninsula in SE Asia. **Neighbors:** Myanmar and China on N, Vietnam on E, Cambodia on S, Thailand on W. **Topography:** Landlocked, dominated by jungle. High mountains along eastern border are the source of the E-W rivers slicing across the country to the Mekong R., which defines most of the western border. **Capital:** Vientiane (1996 met. est.): 531,800.
Government: Type: Communist. **Head of state:** Pres. Khamtai Siphandon; b Feb. 8, 1924; in office: Feb. 24, 1998. **Head of gov.:** Prime Min. Sisavat Keobounphan; b 1928; in office: Feb. 24, 1998. **Local divisions:** 16 provinces, 1 municipality, 1 special zone. **Defense:** 3.9% of GDP. **Active troops:** 29,000.
Economy: Industries: Wood products, mining. **Chief crops:** Sweet potatoes, corn, cotton, vegetables, coffee. **Minerals:** Gypsum, tin, gold. **Arable land:** 3%. **Livestock** (1997): chickens: 13.10 mil; pigs: 1.47 mil; buffalo: 1.09 mil; cattle: 1.11 mil; goats: 121,576. **Electricity prod.** (1997): 1.260 bil. kWh. **Labor force:** 80% agric.
Finance: Monetary unit: Kip (Sept. 1999: 7,680.00 = $1 U.S.). **GDP** (1997 est.): $5.9 bil. **Per capita GDP:** $1,150. **Imports** (1996): $678 mil; partners: Thailand 45%. **Exports** (1996): $313.1 mil; partners: Vietnam 49%, Thailand 30%. **Tourism:** $68 mil. **Budget** (1996): $365.9 mil. **Intl. reserves less gold** (Apr. 1999): $112.26 mil. **Gold:** 17,100 oz t. **Consumer prices** (change in 1998): 91.0%.
Transport: Motor vehicles: 9,000 pass. cars, 9,000 comm. vehicles. **Civil aviation:** 29.9 mil passenger-mi. 11 airports.
Communications: TV sets: 17 per 1,000 pop. **Radios:** 116 per 1,000 pop. **Telephones:** 28,500 main lines.
Health: Life expectancy: 52.63 male; 55.87 female. **Births** (per 1,000 pop.): 39.93. **Deaths** (per 1,000 pop.): 12.56. **Natural inc.:** 2.737%. **Infant mortality** (per 1,000 live births): 89.32.
Education: Compulsory for 5 years between ages 6-15. **Literacy:** 57%.
Major Intl. Organizations: UN (FAO, IBRD, ILO, IMF, WHO), ASEAN.
Embassy: 2222 S St. NW 20008; 332-6416.
Website: http://www.laoembassy.com/discover/index.htm

Laos became a French protectorate in 1893, but regained independence as a constitutional monarchy July 19, 1949.

Conflicts among neutralist, Communist, and conservative factions created a chaotic political situation. Armed conflict increased after 1960.

The 3 factions formed a coalition government in June 1962, with neutralist Prince Souvanna Phouma as premier. A 14-nation conference in Geneva signed agreements, 1962, guaranteeing neutrality and independence. By 1964 the Pathet Lao had withdrawn from the coalition, and, with aid from North Vietnamese troops, renewed sporadic attacks. U.S. planes bombed the Ho Chi Minh trail, supply line from North Vietnam to Communist forces in Laos and South Vietnam.

In 1970 the U.S. stepped up air support and military aid. After Pathet Lao military gains, Souvanna Phouma in May 1975

AP/WIDE WORLD PHOTOS

GRAMMY QUEEN

Singer Lauryn Hill won 5 Grammy awards—a record for a female performer—at the Grammy awards ceremony Feb. 24 in Los Angeles. In September she picked up another 4 trophies at the MTV Video Music Awards.

CORBIS/AFP

LA VIDA LOCA

Puerto Rican singing sensation Ricky Martin, known for his breakout hit "Livin' La Vida Loca," earned 5 trophies at the MTV Video Music Awards, Sept. 9 in New York. Sales of his first English-language album soon reached 5 million—an all-time record for a Latin artist.

AP/WIDE WORLD PHOTOS; RICHARD KORNBERG ASSOC.

DEATH OF A SALESMAN

Brian Dennehy, center, as Willie Loman performs with Kevin Anderson, left, and Ted Koch in a Broadway revival of Arthur Miller's powerful tragedy. Dennehy won the 1999 Tony for Best Actor in a Play.

THE EURO

Sign in a Dublin pub shows prices in both Irish punts and euros. Eleven nations of the European Union introduced the euro for some uses—including credit card payments—on Jan. 1, 1999. Bills and coins will be available in 2002.

SCOTTISH LEGISLATURE

With his wife Micheline, Sean Connery arrives in full regalia at the July 1 opening of Scotland's first parliament in nearly 300 years. The actor had campaigned vigorously for Scottish autonomy.

Victim receives treatment

EARTHQUAKE IN TURKEY

Buildings like those shown here collapsed into rubble on Aug. 17, when Turkey's strongest earthquake since 1939 hit the northwest, killing many thousands of people. The government was criticized for alleged lax enforcement of building codes and inefficient efforts to save those trapped.

AP/WIDE WORLD PHOTOS

CORBIS/AFP

AP/WIDE WORLD PHOTOS

Hussein

Abdullah

THE DEATH OF KING HUSSEIN

Family members and Arab leaders at services for King Hussein of Jordan, who died Feb. 7 after a 47-year reign. In his later years, Hussein, 63, was a leading peacemaker. His oldest son was sworn in as King Abdullah II.

NEW ISRAELI PRIME MINISTER

Israeli Prime Min. Ehud Barak, elected in May, sought to speed up the Middle East peace process; he is shown here at a July meeting with Palestinian leader Yasir Arafat.

AP/WIDE WORLD PHOTOS

TOTAL SOLAR ECLIPSE

Sun-gazers in Essen, Germany, protect their eyes as they view the last total solar eclipse of the millennium, visible in Europe and the Middle East on Aug. 11.

AP/WIDE WORLD PHOTOS

JOE DIMAGGIO

NY Yankee center fielder Joe DiMaggio, who died Mar. 8 at 84, is shown at left in 1941, when he hit safely in a record 56 consecutive games. Above, he returns to Yankee Stadium in 1963 for Old Timer's Day.

ROYAL WEDDING

Great Britain's Prince Edward, youngest son of Queen Elizabeth, and his new bride, Sophie Rhys-Jones, wave to the crowd after their wedding on June 19.

UP, UP, AND AWAY

Bertrand Piccard of Switzerland and Brian Jones of Great Britain shake hands outside the *Breitling Orbiter 3*. In March they became the first people ever to circumnavigate the globe nonstop in a balloon.

GEORGE W. BUSH
Texas Gov. George W. Bush, the early front-runner in the race for the Republican presidential nomination in 2000, campaigns in New Hampshire.

AL GORE
Vice President Al Gore courts voters in New Hampshire on a nationwide tour announcing his candidacy for the Democratic presidential nomination.

Our Next
President
Al Gore

HILLARY RODHAM CLINTON
The First Lady shakes hands with children at a middle school during a "listening tour" of New York. Contemplating a bid for the Senate seat of retiring Sen. Daniel P. Moynihan, she established residency in September when she and the president bought a home in Westchester County.

CORBIS/REUTERS NEWMEDIA INC

AP/WIDE WORLD PHOTOS

EDWARD A. THOMAS

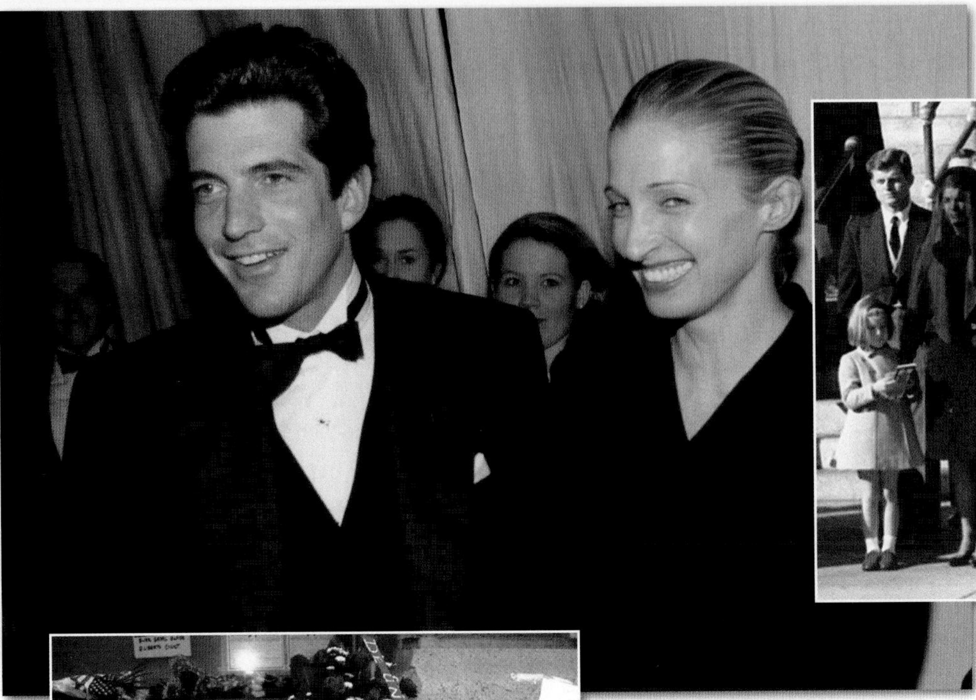

KENNEDY TRAGEDY

John F. Kennedy Jr. and his wife, Carolyn Bessette Kennedy (above), as well as her sister Lauren Bessette, died July 16 when the plane he was piloting plunged into the sea near Martha's Vineyard. Thousands placed floral memorials (left) outside the Kennedys' Manhattan apartment. Above, Kennedy as a 3-year-old salutes the casket of his slain father in November 1963.

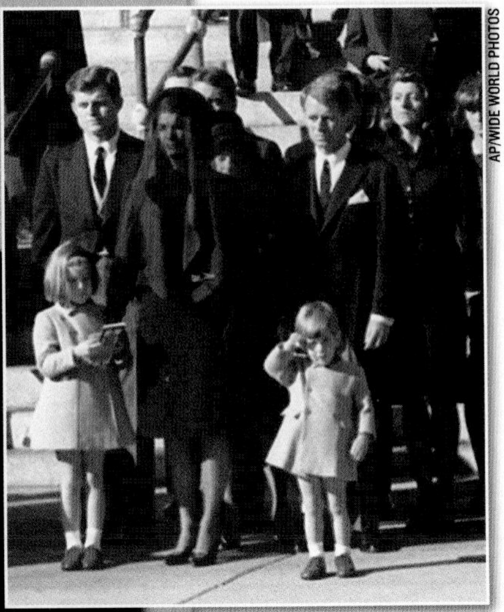

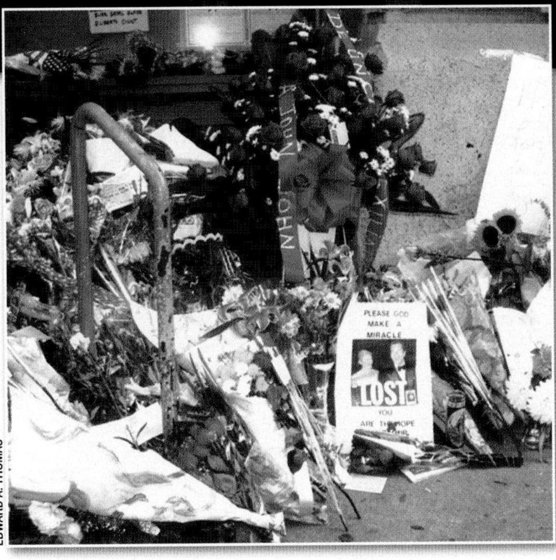

DR. JACK KEVORKIAN

Dr. Jack Kevorkian is handcuffed in Pontiac, MI, Apr. 13, after being sentenced to prison for the "mercy killing" of Thomas Youk, a victim of Lou Gehrig's disease. A tape of Youk's death, by lethal injection, had been played on national television. Kevorkian claimed he had helped 130 people take their own lives before he directly killed Youk.

CORBIS/AFP

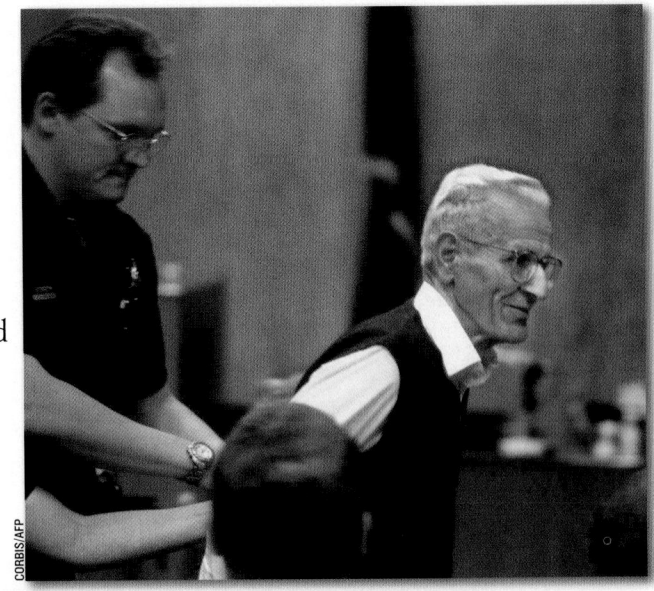

CORBIS/AFP

AP/WIDE WORLD PHOTOS

TORNADOES IN MAY

A woman hugs her granddaughter amid the ruins of a home in Oklahoma, flattened by one of over 70 tornadoes that ripped through the Plains states May 3. About 50 people died and 700 were injured.

AP/WIDE WORLD PHOTOS

DROUGHT AND FLOOD

The eastern United States was parched by a long summer drought that damaged crops and strained water supplies (at left, a dried-out pond in Orefield, PA). In mid-September, however, there was massive flooding and devastation, as Hurricane Floyd battered the east coast (above, evacuees in a boat pass a home swamped by rising waters in Princeville, NC).

Eric Harris

Dylan Klebold

MASSACRE AT LITTLETON

SWAT teams (above) enter Columbine High School, in Littleton, CO, to do a final search, after 2 students, Eric Harris and Dylan Klebold, went on a shooting spree Apr. 20, and killed 13 people, before killing themselves. Several others were seriously injured. At right, the funeral for slain student Cassie Bernall.

ATLANTA KILLINGS

People run from the area where a gunman opened fire at 2 Atlanta securities firms July 29, killing 9 people. Mark Barton, a day-trader who traded at the firms, had murdered his wife and 2 children days earlier. He shot himself to death as police closed in on his van.

CORBIS/AFP

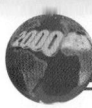

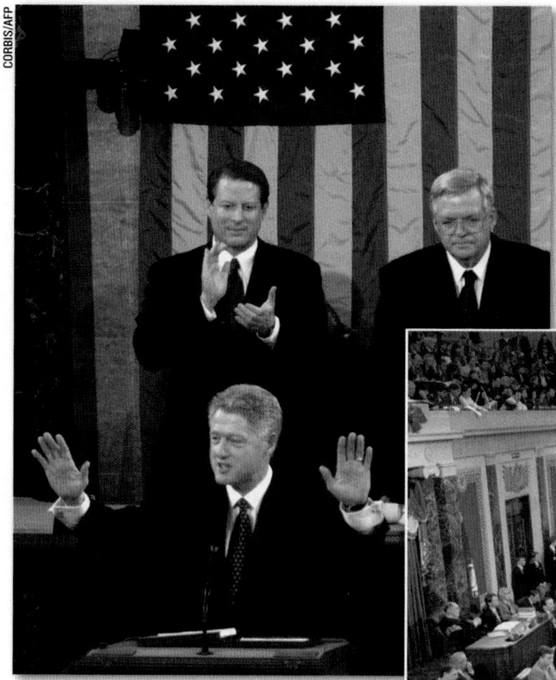

AP/WIDE WORLD PHOTOS; APTN

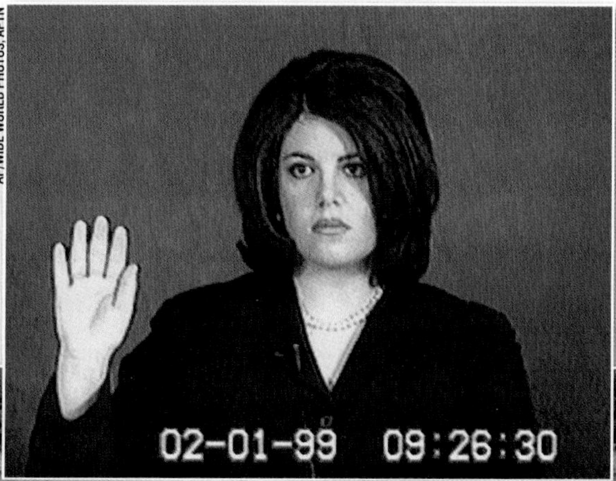

AP/WIDE WORLD PHOTOS; US SENATE

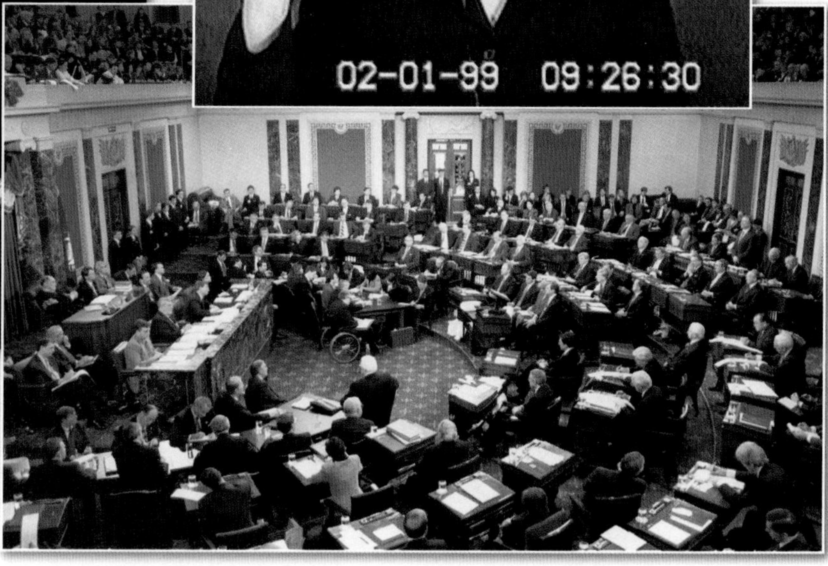

THE PRESIDENTIAL IMPEACHMENT

On trial and facing possible ouster, Pres. Bill Clinton (above) gives his annual State of the Union address Jan. 19. (Polls afterward showed his approval ratings at or near all-time highs.) Above, the impeachment trial is held in the Senate. On Feb. 12 he was found not guilty on charges of perjury and obstruction of justice, brought by the U.S. House in December 1998 in connection with his cover-up of an affair with former White House intern Monica Lewinsky. (She is shown above being sworn in for her Senate deposition.)

CORBIS/AFP

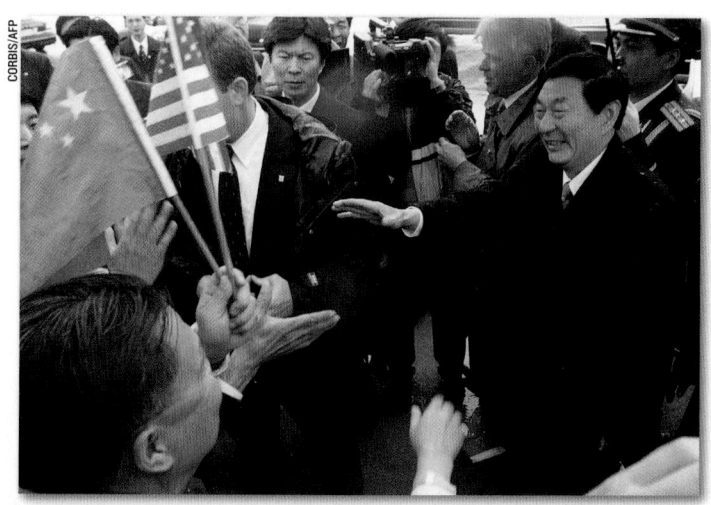

TRADE WITH CHINA

Chinese Prem. Zhu Rongji arrives in Los Angeles Apr. 6, the first stop on a U.S. trip campaigning for a U.S.–China trade agreement. U.S. relations were damaged by China's human rights record and by charges of Chinese involvement in espionage and violation of U.S. campaign finance laws.

Slobodan Milosevic

AIR STRIKES

The Belgrade headquarters of Yugoslav Pres. Slobodan Milosevic's party are destroyed by NATO bombs Apr. 20. Air attacks on Serbian targets were aimed at inducing a Serb withdrawal from Kosovo, which began in early June.

OUTRAGE IN CHINA

Students in China protest the bombing of the Chinese embassy in Belgrade by a U.S. aircraft May 7 during NATO air attacks. They carry pictures of 3 Chinese journalists who were killed; 27 other people were wounded. The U.S. apologized for the incident, which it termed a "tragic mistake."

POWs FREED

The Rev. Jesse Jackson joins Cpl. Steven Gonzalez, Sgt. Christopher Stone, and Sgt. Andrew Ramirez in Belgrade May 2 after they were freed by Yugoslavia. The civil rights leader helped secure release of the 3 Americans, held since their capture near the Serbia-Macedonia border Mar. 31.

REFUGEES

Below, dispirited Kosovars at a refugee camp in Macedonia in April. A Serb offensive against ethnic Albanians in Yugoslavia's Kosovo region, which accelerated after NATO air strikes began in late March, drove hundreds of thousands of ethnic Albanians into refugee camps in neighboring areas. At left, a refugee uses a wheelbarrow to push an elderly woman toward a safe haven in Montenegro.

MASS GRAVES

A man organizes wooden markers, as graves of 2 apparent massacre victims are covered in western Kosovo in July. During the NATO air strikes, reports of massacres of ethnic Albanians by Serb troops proliferated. In April, NATO released aerial photographs showing apparent mass graves; further evidence was found across Kosovo as Serbian troops withdrew and peacekeeping troops arrived.

1999
In PICTURES

02-01-99   09:26:30

THE PHANTOM MENACE

The latest episode in the *Star Wars* saga goes back to the childhood of Anakin Skywalker who was to become Darth Vader. *Episode I: The Phantom Menace* hit theaters in May and grossed over $400 million by summer's end.

LIFE IS BEAUTIFUL

Italian actor-director Roberto Benigni ecstatically accepts two Oscars, Mar. 21 in Los Angeles. He won as Best Actor for the film *Life Is Beautiful*, which also was voted Best Foreign Film.

BLAIR WITCH PROJECT

Actor Joshua Leonard portrays one of 3 amateur filmmakers on a doomed expedition through the forest, in *The Blair Witch Project*. Shot for under $100,000, the movie was a surprise hit that soon topped $100 million at the box office.

THE GREAT ONE

NHL star Wayne Gretzky, skating here in the 1999 All-Star Game, retired in April after 21 seasons. Acclaimed by many as hockey's greatest player, Gretzky holds or shares 61 league records. The Hall of Fame selection committee unanimously voted to let him in without the usual 3-year waiting period.

GOODBYE TO GRAF

Germany's Steffi Graf is shown in action during the 1999 Australian Open. She retired in August, having won a spectacular 22 Grand Slam titles (second only to Margaret Court) and earned more money than any other female athlete.

SUPERBOWL HERO

Denver Bronco quarterback John Elway smiles after scoring a touchdown in what proved to be his final Super Bowl, Jan. 31, 1999, in Miami. The Broncos beat Atlanta, 34–19; Elway, who was named Super Bowl MVP, announced his retirement in May.

SHOOTOUT

U.S. soccer ace Brandi Chastain celebrates after her game-winning overtime penalty kick nailed the Women's World Cup title July 10. The U.S. won the shootout against China 5–4.

TOUR DE FRANCE

Lance Armstrong of the U.S. Postal Service team crosses the finish line of Stage 19 in the Tour de France in July. He went on to become only the second American to win the grueling race. It was an amazing comeback, considering he had been diagnosed with cancer 3 years earlier and given only a 50–50 chance of survival.

TIGER WOODS

Tiger Woods sinks his final putt for par on the 18th hole, for a one-stroke victory in the PGA Championship Aug. 15. At 23 he was the youngest golfer since Seve Ballesteros, in 1980, to have won 2 major tournaments (Woods won the Masters in 1997).

BATTING 3,000

Tampa Bay Devil Rays' Wade Boggs slugs a home run for his 3,000th hit on Aug. 7. He joined 22 other players in the elite 3,000 Club, including Tony Gwynn of the San Diego Padres, who made it one day before.

NEW NBA CHAMPIONS

The Spurs' Tim Duncan puts up a shot over the Knicks' Larry Johnson and Latrell Sprewell in Game 4 of the NBA Finals. When they defeated the NY Knicks in Game 5, 78–77, the San Antonio Spurs won their first NBA championship, 4 games to 1. Duncan was the Finals MVP.

STANLEY CUP FINALS

Joe Nieuwendyk of the Dallas Stars slips a game-winning goal past Buffalo Sabres goalie Dominik Hasek in the 3rd period of Game 3 in the Stanley Cup finals. The Stars won the game, 2–1, and went on to take the series 4 games to 2. Nieuwendyk was named MVP.

ordered government troops to cease fighting; the Pathet Lao took control. The Lao People's Democratic Republic was proclaimed Dec. 3, 1975.

From the mid-1970s through the 1980s, the Laotian government relied on Vietnam for military and financial aid. Since easing its foreign investment laws in 1988, Laos has attracted more than $5 billion from Thailand, the U.S., and other nations. Laos was admitted to ASEAN on July 23, 1997.

Latvia
Republic of Latvia

People: Population: 2,353,874. **Age distrib.** (%): <15: 18.0. 65+: 15.1. **Pop density:** 95 per sq. mi. **Urban:** 73%. **Ethnic groups:** Latvian 57%, Russian 30%. **Principal languages:** Lettish (official), Lithuanian, Russian. **Chief religions:** Lutheran, Roman Catholic, Russian Orthodox.

Geography: Area: 24,700 sq. mi. **Location:** E Europe, on the Baltic Sea. **Neighbors:** Estonia on N, Lithuania and Belarus on S, Russia on E. **Capital:** Riga: (1996 est.): 821,180.

Government: Type: Republic. **Head of state:** Pres. Vaira Vike-Freiberga; b Dec. 1, 1937; in office: July 8, 1999. **Head of gov.:** Prime Min. Andris Skele; b Jan. 16, 1958; in office: July 16, 1999. **Local divisions:** 26 counties, 7 municipalities. **Defense:** 4.6% of GDP. **Active troops:** 4,500.

Economy: Industries: Machinery, vehicles, railway cars. **Chief crops:** Grains, sugar beets, potatoes. **Minerals:** Amber, peat. **Arable land:** 27%. **Livestock** (1997): chickens: 3.00 mil; cattle: 434,396; pigs: 421,113. **Fish catch** (1996): 142,644 metric tons. **Electricity prod.** (1997): 3.648 bil kWh. **Labor force:** 41% ind.; 16% agric. & forestry.

Finance: Monetary unit: Lat (Sept. 1999: 0.59 = $1 U.S.). **GDP** (1997 est.): $10.4 bil. **Per capita GDP:** $4,260. **Imports** (1996): $2.3 bil; partners: Russia 20%, Germany 14%. **Exports** (1996): $1.4 bil; partners: Russia 23%, Germany 14%. **Tourism:** $211 mil. **Intl. reserves less gold** (June 1999): $887.01 mil. **Gold:** 249,300 oz t. **Consumer prices** (change in 1998): 4.6%.

Transport: Railroad: Length: 1,499 mi. **Motor vehicles:** 252,000 pass. cars, 74,000 comm. vehicles. **Civil aviation:** 135.0 mil pass.-mi.; 1 airport. **Chief port:** Riga.

Communications: TV sets: 452 per 1,000 pop. **Radios:** 560 per 1,000 pop. **Telephones:** 941,400 main lines. **Daily newspaper circ.:** 235 per 1,000 pop.

Health: Life expectancy: 61.24 male, 73.66 female. **Births** (per 1,000 pop.): 8.10. **Deaths** (per 1,000 pop.): 15.82. **Natural inc.:** −0.772%. **Hosp. beds** (1995): 1 per 90 persons. **Physicians** (1995): 1 per 298 persons. **Infant mortality rates** (per 1,000 live births): 17.19.

Education: Compulsory: ages 7-16. **Literacy** (1989): 100%.

Major Intl. Organizations: UN (FAO, IBRD, ILO, IMF, IMO, WHO), OSCE.

Embassy: 4325 17th St. NW 20011; 726-8213.
Websites: http://www.latvia-usa.org
http://www.csb.lv

Prior to 1918, Latvia was occupied by the Russians and Germans. It was an independent republic, 1918-39. The Aug. 1939 Soviet-German agreement assigned Latvia to the Soviet sphere of influence. It was officially accepted as part of the USSR on Aug. 5, 1940. It was overrun by the German army in 1941, but retaken in 1945.

During an abortive Soviet coup, Latvia declared independence, Aug. 21, 1991. The Soviet Union recognized Latvia's independence in Sept. 1991. The last Russian troops in Latvia withdrew by Aug. 31, 1994. Responding to international pressure, Latvian voters on Oct. 3, 1998, eased citizenship laws that had discriminated against some 500,000 ethnic Russians. On June 17, 1999, the legislature elected Vaira Vike-Freiberga as Latvia's 1st woman president.

Lebanon
Republic of Lebanon

People: Population: 3,562,699. **Age distrib.** (%): <15: 29.5; 65+: 6.5. **Pop. density:** 891 per sq. mi. **Urban:** 88%. **Ethnic groups:** Arab 95%, Armenian 4%. **Principal languages:** Arabic (official), French, English, Armenian. **Chief religions:** Islam 70%, Christian 30%.

Geography: Area: 4,000 sq. mi. **Location:** In Middle East, on E end of Mediterranean Sea. **Neighbors:** Syria on E, Israel on S. **Topography:** There is a narrow coastal strip, and 2 mountain ranges running N-S enclosing the fertile Beqaa Valley. The Litani R. runs S through the valley, turning W to empty into the Mediterranean. **Capital:** Beirut: 1,826,000*.

Government: Type: Republic. **Head of state:** Pres. Emile Lahoud; b 1936; in office: Nov. 24, 1998. **Head of gov.:** Prime Min. Salim Hoss; b Dec. 20, 1929; in office: Dec. 2, 1998. **Local divisions:** 5 governorates. **Defense:** 4.5% of GDP. **Active troops:** 50,100.

Economy: Industries: Banking, food products, textiles, cement, oil refining. **Chief crops:** Citrus, olives, tobacco, potatoes, vegetables. **Minerals:** Limestone, iron. **Arable land:** 21%. **Livestock** (1997): chickens: 30.00 mil; goats: 450,000; sheep: 350,000. **Electricity prod.** (1997): 5.850 bil kWh. **Labor force:** 62% services, 31% industry; 7% agric.

Finance: Monetary unit: Pound (Sept. 1999: 1,502.00 = $1 U.S.). **GDP** (1997 est.): $15.2 bil. **Per capita GDP:** $4,400. **Imports** (1996): $7.56 bil; partners: Italy 12%, U.S. 11%, Germany 9%, France 8%, **Exports** (1996): $1.02 bil; partners: UAE 23%, Saudi Arabia 14%, **Tourism:** $1.38 bil. **Budget** (1997 est): $5.9 bil. **Intl. reserves less gold** (June 1999): $6.46 bil. **Gold:** 9.22 mil oz t.

Transport: Railroad: Length: 138 mi. **Motor vehicles:** 1.1 mil pass. cars, 83,000 comm. vehicles. **Civil aviation:** 1.3 bil pass.-mi.; 1 airport. **Chief ports:** Beirut, Tripoli, Sidon.

Communications: TV sets: 291 per 1,000 pop. **Radios:** 608 per 1,000 pop. **Telephones:** 620,000 main lines. **Newspaper circ.:** 172 per 1,000 pop.

Health: Life expectancy: 68.34 male; 73.66 female. **Births** (per 1,000 pop.): 22.50. **Deaths** (per 1,000 pop.): 6.45. **Natural inc.:** 1.605%. **Hosp. beds** (1995): 1 per 319 persons. **Physicians** (1995): 1 per 529 persons. **Infant mortality** (per 1,000 live births): 30.53.

Education: Literacy: 92%.

Major Intl. Organizations: UN (FAO, IBRD, ILO, IMF, IMO, WHO), AL.

Embassy: 2560 28th St. NW 20008; 939-6300.
Website: http://www.erols.com/lebanon/stat.htm

Formed from 5 former Turkish Empire districts, Lebanon became an independent state Sept. 1, 1920, administered under French mandate 1920-41. French troops withdrew in 1946.

Under the 1943 National Covenant, all public positions were divided among the various religious communities, with Christians in the majority. By the 1970s, Muslims became the majority and demanded a larger political and economic role.

U.S. Marines intervened, May-Oct. 1958, during a Syrian-aided revolt. Continued raids against Israeli civilians, 1970-75, brought Israeli attacks against guerrilla camps and villages. Israeli troops occupied S Lebanon, Mar. 1978, and again in Apr. 1980.

An estimated 60,000 were killed and billions of dollars in damage inflicted in a 1975-76 civil war. Palestinian units and leftist Muslims fought against the Maronite militia, the Phalange, and other Christians. Several Arab countries provided political and arms support to the various factions, while Israel aided Christian forces. Up to 15,000 Syrian troops intervened in 1976 to fight Palestinian groups. A cease-fire was mainly policed by Syria.

New clashes between Syrian troops and Christian forces erupted, Apr. 1, 1981. By Apr. 22, fighting had also broken out between two Muslim factions. In July, Israeli air raids on Beirut killed or wounded some 800 persons.

Israeli forces invaded Lebanon June 6, 1982, in a coordinated land, sea, and air attack aimed at crushing strongholds of the Palestine Liberation Organization (PLO). Israeli and Syrian forces engaged in the Bekaa Valley. By June 14, Israeli troops had encircled Beirut. On Aug. 21, the PLO evacuated west Beirut after massive Israeli bombings there. Israeli troops entered west Beirut following the Sept. 14 assassination of newly elected Lebanese Pres. Bashir Gemayel. On Sept. 16, Lebanese Christian troops entered 2 refugee camps and massacred hundreds of Palestinian refugees. An agreement May 17, 1983, between Lebanon, Israel, and the U.S. (but not Syria) provided for the withdrawal of Israeli troops; at least 30,000 Syrian troops remained in Lebanon, and Israeli forces continued to occupy a "security zone" in the south.

In 1983, terrorist bombings became a way of life in Beirut as some 50 people were killed in an explosion at the U.S. Embassy, Apr. 18; 241 U.S. servicemen and 58 French soldiers died in separate Muslim suicide attacks, Oct. 23.

Kidnapping of foreign nationals by Islamic militants became common in the 1980s. U.S., British, French, and Soviet citizens were victims. All were released by 1992.

A treaty signed May 22, 1991, between Lebanon and Syria recognized Lebanon as a separate state for the first time since the 2 countries gained independence in 1943.

Israeli forces conducted air raids and artillery strikes against guerrilla bases and villages in S Lebanon, causing over 200,000 to flee their homes July 25-29, 1993. Some 500,000 civilians fled their homes in Apr. 1996 when Israel again struck suspected guerrilla bases in the south. Pope John Paul II visited Lebanon May 10-11, 1997. During May-June 1998 the nation held its 1st municipal elections in 35 years. With Syria's approval, the legislature unanimously elected Lebanese armed forces chief Emile Lahoud as president Oct. 15.

Lesotho
Kingdom of Lesotho

People: Population: 2,128,950. **Age distrib.** (%): <15: 39.8; 65+: 4.6. **Pop. density:** 182 per sq. mi. **Urban:** 25%. **Ethnic groups:** Sotho 99.7%. **Principal languages:** English, Sesotho (both official). **Chief religions:** Christian 80%, indigenous beliefs 20%.

Geography: Area: 11,700 sq. mi. **Location:** In southern Africa. **Neighbors:** Completely surrounded by Republic of South Africa. **Topography:** Landlocked and mountainous, altitudes from 5,000 to 11,000 ft. **Capital:** Maseru (1995 est.): 400,200.

Government: Type: Modified constitutional monarchy. **Head of state:** King Letsie III; b July 17, 1963; in office: Feb. 7, 1996. **Head of gov.:** Pakalitha Mosisili; in office: May 29, 1998. **Local divisions:** 10 districts. **Defense:** 4.6% of GNP. **Active troops:** 2,000.

Economy: Industries: Food processing, textiles. **Chief crops:** Corn, grains, pulses, sorghum. **Other resources:** Diamonds. **Arable land:** 10%. **Livestock** (1997): chickens: 1.60 mil; sheep: 1.10 mil; goats: 730,000; cattle: 580,000. **Labor force:** 86% subsistence agric.

Finance: Monetary unit: Maloti (Sept. 1999: 6.03 = $1 U.S.). **GDP:** (1997 est.): $5.1 bil. **Per capita GDP:** $2,500. **Imports** (1996 est.): $1.1 bil; partners: South Africa 90%. **Exports** (1996 est.): $218 mil; partners: South Africa 52%. **Tourism:** $20 mil. **Budget** (FY 1996-97): $487 mil. **Intl. reserves less gold** (Jan. 1999): $582.57 mil. **Consumer prices** (change in 1996): 9.3%.

Transport: Railroad: Length: 2 mi. **Motor vehicles:** 5,000 pass. cars, 18,000 comm. vehicles. **Civil aviation:** 5.7 mil passenger-mi, 1 airport.

Communications: TV sets: 7 per 1,000 pop. **Radios:** 558 per 1,000 pop. **Telephones:** 20,100 main lines. **Daily newspaper circ.:** 7 per 1,000 pop.

Health: Life expectancy: 51.37 male; 54.65 female. **Births** (per 1,000 pop.): 31.26. **Deaths** (per 1,000 pop.): 13.23. **Natural inc.:** 1.803%. **Hosp. beds** (1992): 1 per 765 persons. **Physicians** (1993): 1 per 14,306 persons. **Infant mortality** 77.58.

Education: Free, compulsory: ages 6-13. **Literacy:** 71%.

Major Intl. Organizations: UN (FOA, IBRD, ILO, IMF, WHO, WTrO), the Commonwealth, OAU.

Embassy: 2511 Massachusetts Ave. NW 20008; 797-5533.

Lesotho (once called Basutoland) became a British protectorate in 1868 when Chief Moshesh sought protection against the Boers. Independence came Oct. 4, 1966. Elections were suspended in 1970. Most of Lesotho's GNP is provided by citizens working in South Africa. Livestock raising is the chief industry; diamonds are the chief export.

South Africa imposed a blockade, Jan. 1, 1986, because Lesotho had given sanctuary to anti-apartheid groups. The blockade sparked a Jan. 20 military coup, and was lifted, Jan. 25, when the new leaders agreed to expel the rebels.

In Mar. 1990, King Moshoeshoe was exiled by the military government. Letsie III became king Nov. 12. In Mar. 1993, Ntsu Mokhehle, a civilian, was elected prime minister, ending 23 years of military rule. After a series of violent disturbances, the king dismissed the Mokhele government Aug. 17, 1994; constitutional rule was restored Sept. 14. Letsie abdicated and Moshoeshoe was reinstated Jan. 25, 1995.

Moshoeshoe died in an automobile accident, Jan. 15, 1996. Letsie was reinstated Feb. 7; his formal coronation was Oct. 31, 1997. South Africa and Botswana sent troops Sept. 22, 1998, to help suppress violent antigovernment protests. Preparations began in Dec. 1998 for new elections within 18 months.

Liberia
Republic of Liberia

People: Population: 2,923,725. **Age distrib.** (%): <15: 44.7; 65+: 3.6. **Pop. density:** 68 per sq. mi. **Urban:** 46%. **Ethnic groups:** Indigenous tribes 95%, Americo-Liberians 2.5%. **Principal languages:** English (official), tribal languages. **Chief religions:** Traditional beliefs 70%, Muslim 20%, Christian 10%.

Geography: Area: 43,000 sq. mi. **Location:** On SW coast of W Africa. **Neighbors:** Sierra Leone on W, Guinea on N, Côte d'Ivoire on E. **Topography:** Marshy Atlantic coastline rises to low mountains and plateaus in the forested interior; 6 major rivers flow in parallel courses to the ocean. **Capital:** Monrovia: 962,000*.

Government: Type: Republic. **Head of state and gov.:** Pres. Charles Taylor; b Jan. 29, 1948; in office: Aug. 2, 1997. **Local divisions:** 13 counties. **Defense:** 3.9% of GDP. **Active troops:** 22,000.

Economy: Industries: Food processing, mining. **Chief crops:** Rice, cassava, coffee, cocoa, sugar. **Minerals:** Iron, di-

amonds, gold. **Other resources:** Rubber, timber. **Arable land:** 1%. **Livestock** (1997): chickens: 3.50 mil; goats: 220,000; sheep: 210,000; pigs: 120,000. **Electricity prod.** (1997): 480 mil kWh. **Labor force:** 70% agric.

Finance: Monetary unit: Dollar (Sept. 1999: 1.00 = $1 U.S.). **GDP:** (1997 est.): $2.6 bil. **Per capita GDP:** $1,000. **Imports** (1995): $5.8 bil; partners: Japan 33%; S. Korea 20%; Italy 9%. **Exports** (1995): $667 mil; partners: Belg.-Lux. 57%; Ukraine 12%.

Transport: Railroad: Length: 306 mi. **Motor vehicles:** 17,400 pass. cars, 10,700 comm. vehicles. **Civil aviation:** 4.3 mil pass.-mi; 1 airport. **Chief ports:** Monrovia, Buchanan, Greenville, Harper.

Communications: TV sets: 20 per 1,000 pop. **Radios:** 263 per 1,000 pop. **Telephones:** (1997): 6,400 main lines. **Daily newspaper circ.:** 15 per 1,000 pop.

Health: Life expectancy: 57.20 male; 62.64 female. **Births** (per 1,000 pop.): 41.49. **Deaths** (per 1,000 pop.): 11.03. **Natural inc.:** 3.046%. **Physicians** (1992): 1 per 8,333 persons. **Infant mortality** (per 1,000 live births): 100.63.

Education: Free, compulsory: ages 7-16. **Literacy:** 38%.

Major Intl. Organizations: UN and most of its specialized agencies, OAU.

Embassy: 5201 16th St. NW 20011; 723-0437.

Liberia was founded in 1822 by U.S. black freedmen who settled at Monrovia with the aid of colonization societies. It became a republic July 26, 1847, with a constitution modeled on that of the U.S. Descendants of freedmen dominated politics.

Charging rampant corruption, an Army Redemption Council of enlisted men staged a bloody predawn coup, April 12, 1980, in which Pres. Tolbert was killed and replaced as head of state by Sgt. Samuel Doe. Doe was chosen president in a disputed election, and survived a subsequent coup, in 1985.

A civil war began Dec. 1989. Rebel forces seeking to depose Pres. Doe made major territorial gains and advanced on the capital, June 1990. In Sept., Doe was captured and put to death. Despite the introduction of peacekeeping forces from several countries, factional fighting intensified, and a series of cease-fires failed. A transitional Council of State was instituted Sept. 1, 1995. Factional fighting flared up again in Apr. 1996, devastating Monrovia.

On Sept. 3, 1996, Ruth Perry became modern Africa's first female head of state, leading another transitional government. By then, the civil war had claimed more than 150,000 lives and uprooted over half the population.

Former rebel leader Charles Taylor was elected president July 19, 1997, in Liberia's 1st national election in 12 years.

Libya
Socialist People's Libyan Arab Jamahiriya

People: Population: 4,992,838. **Age distrib.** (%): <15: 36.5; 65+: 3.8. **Pop. density:** 7 per sq. mi. **Urban:** 86%. **Ethnic groups:** Arab-Berber 97%. **Principal language:** Arabic (official), Italian, English. **Chief religion:** Sunni Muslim 97%.

Geography: Area: 679,400 sq. mi. **Location:** On Mediterranean coast of N Africa. **Neighbors:** Tunisia, Algeria on W; Niger, Chad on S; Sudan, Egypt on E. **Topography:** Desert and semidesert regions cover 92% of the land, with low mountains in N, higher mountains in S, and a narrow coastal zone. **Capital:** Tripoli: 1,681,000*.

Government: Type: Islamic Arabic Socialist "Mass-State." **Head of state and gov.:** Col. Muammar al-Qaddafi; b Sept. 1942; in power: Sept. 1969. **Local divisions:** 13 regions subdivided into about 1,500 communes. **Defense:** 4.7% of GDP. **Active troops:** 65,000.

Economy: Industries: Oil, food processing, textiles. **Chief crops:** Dates, olives, citrus, barley, wheat. **Minerals:** Gypsum, oil, gas. **Crude oil reserves** (1999): 29.5 bil bbls. **Arable land:** 1%. **Livestock** (1997): chickens: 24.00 mil; sheep: 5.70 mil; goats: 1.30 mil.; cattle: 155,000. **Electricity prod.** (1997): 17.5 bil kWh. **Labor force:** 31% ind.; 27% services; 24% govt.; 18% agric.

Finance: Monetary unit: Dinar (Sept. 1999: 0.45 = $1 U.S.). **GDP:** (1997 est.): $38 bil. **Per capita GDP:** $6,700. **Imports** (1995): $6.2 bil; partners: Italy 22%, Germany 14%. **Exports** (1995): $9 bil; partners: Italy 41%, Germany 18%, Spain 10%. **Tourism:** $6 mil. **Budget** (1995 est.): $10.3 bil.

Transport: Motor vehicles: 592,000 pass. cars, 312,000 comm. vehicles. **Civil aviation:** 234.1 mil passenger-mi. **Chief ports:** Tripoli, Banghazi.

Communications: TV sets: 105 per 1,000 pop. **Radios:** 191 per 1,000 pop. **Telephones:** 380,000 main lines. **Daily newspaper circ.:** 15 per 1,000 pop.

Health: Life expectancy: 73.81 male; 77.74 female. **Births** (per 1,000 pop.): 27.33. **Deaths** (per 1,000 pop.): 3.35. **Natural inc.:** 2.398%. **Infant mortality** (per 1,000 live births): 28.15.

Education: Compulsory: ages 6-15. **Literacy:** 76%.
Major Intl. Organizations: UN (FAO, IBRD, ILO, IMF, IMO, WHO), AL, OAU, OPEC.

First settled by Berbers, Libya was ruled in succession by Carthage, Rome, the Vandals, and the Ottomans. Italy ruled from 1912, and Britain and France after WW II. Libya became an independent constitutional monarchy Jan. 2, 1952. In 1969 a junta led by Col. Muammar al-Qaddafi seized power.

Libya and Egypt fought several air and land battles along their border in July 1977. Chad charged Libya with military occupation of its uranium-rich northern region in 1977. Libyan troops were driven from their last major stronghold by Chad forces in 1987, leaving over $1 billion in military equipment behind.

Libya reportedly helped arm violent revolutionary groups in Egypt and Sudan and aided terrorists of various nationalities.

On Jan. 7, 1986, the U.S. imposed economic sanctions against Libya, ordered all Americans to leave that country, and froze all Libyan assets in the U.S. The U.S. commenced flight operations over the Gulf of Sidra, Jan. 27, and a U.S. Navy task force began conducting exercises in the Gulf, Mar. 23. When Libya fired antiaircraft missiles at American warplanes, the U.S. responded by sinking 2 Libyan ships and bombing a missile site in Libya. The U.S. withdrew from the Gulf, Mar. 27.

The U.S. accused Qaddafi of ordering the Apr. 5, 1986, bombing of a West Berlin discotheque, which killed 3, including a U.S. serviceman. In response, the U.S. sent warplanes to attack terrorist-related targets in Tripoli and Banghazi, Libya, Apr. 14.

The UN imposed limited sanctions, Apr. 15, 1992, for Libya's failure to extradite 2 agents linked to the 1988 bombing of Pan American World Airways Flight 103 over Lockerbie, Scotland, and 4 others linked to an airplane bombing over Niger. Sanctions were tightened Dec. 1, 1993. The international embargo ended, although U.S. sanctions remained, after Libya, Apr. 5, 1999, handed over two Lockerbie suspects for trial in the Netherlands under Scottish law.

Liechtenstein
Principality of Liechtenstein

People: Population: 32,057. **Age distrib.** (%): <15: 18.8; 65+: 11.2. **Pop. density:** 534 per sq. mi. **Urban:** 21%. **Ethnic groups:** Alemannic 88%. **Principal languages:** German (official), Alemannic dialect. **Chief religions:** Roman Catholic 80%, Protestant 7%.

Geography: Area: 60 sq. mi. **Location:** Central Europe, in the Alps. **Neighbors:** Switzerland on W, Austria on E. **Topography:** The Rhine Valley occupies one-third of the country, the Alps cover the rest. **Capital:** Vaduz (1997 est.): 5,017.

Government: Type: Hereditary constitutional monarchy. **Head of state:** Prince Hans-Adam II; b Feb 14, 1945; in office: Nov. 13, 1989. **Head of gov.:** Mario Frick; b May 8, 1965; in office: Dec. 15, 1993. **Local divisions:** 11 communes.

Economy: Industries: Precision instruments, electronics, textiles, ceramics. **Chief crops:** Grain, corn, potatoes. **Arable land:** 24%. **Labor force:** 52% services; 46% industry, trade, constr.

Finance: Monetary unit: Franc (Sept. 1999: 1.49 = $1 U.S.). **GDP:** (1996 est.): $713 mil. **Per capita GDP:** $23,000. **Imports** (1996): $917.3 mil. **Exports** (1996): $2.47 bil; partners: Switzerland 16%. **Budget** (1996 est.): $435 mil.

Transport: Railroad: Length: 12 mi.

Communications: TV sets: 371 per 1,000 pop. **Radios:** 384 per 1,000 pop. **Daily newspaper circ.:** 564 per 1,000 pop.

Health: Life expectancy: 75.64 male; 80.69 female. **Births** (per 1,000 pop.): 12.23. **Deaths** (per 1,000 pop.): 7.33. **Natural inc.:** 0.490%. **Physicians** (1995): 1 per 962 persons. **Infant mortality** (per 1,000 live births): 5.23.

Education: Compulsory: ages 7-16. **Literacy** (1997): 100%. **Major Intl. Organizations:** UN (WTrO), EFTA, OSCE.

Liechtenstein became sovereign in 1806. Austria administered Liechtenstein's ports up to 1920; Switzerland has administered its postal services since 1921. Liechtenstein is united with Switzerland by a customs and monetary union. Taxes are low; many international corporations have headquarters there. Foreign workers comprise a third of the population.

Lithuania
Republic of Lithuania

People: Population: 3,584,966. **Age distrib.** (%): <15: 20.0; 65+: 13.2. **Pop. density:** 142 per sq. mi. **Urban:** 73%. **Ethnic groups:** Lithuanian 81%, Russian 9%, Polish 7%. **Principal languages:** Lithuanian (official), Polish, Russian. **Chief religions:** Primarily Roman Catholic.

Geography: Area: 25,200 sq. mi. **Location:** In E Europe, on

SE coast of Baltic. **Neighbors:** Latvia on N, Belarus on E, S, Poland and Russia on W. **Capital:** Vilnius. **Cities** (1997 est.): Vilnius 580,099; Kaunas 418,707.

Government: Type: Republic. **Head of state:** Pres. Valdas Adamkus; b Nov. 3, 1926; in office: Feb. 26, 1998. **Head of gov.:** Prime Min. Rolandas Paksas; b June 10, 1956; in office: May 18, 1999. **Local divisions:** 10 provinces. **Defense:** 4.4% of GDP. **Active troops:** 5,300.

Economy: Industries: Machinery, shipbuilding, textiles. **Chief crops:** Sugar beets, grain, potatoes, vegetables. **Crude oil reserves** (1999): 12 mil bbls. **Arable land:** 35%. **Livestock** (1997): chickens: 7.00 mil; pigs: 1.21 mil; cattle: 1.07 mil. **Electricity prod.** (1997): 12.816 bil kWh. **Labor force:** 42% industry, const.; 20% agric., forestry.

Finance: Monetary unit: Litas (Sept. 1999: 4.00 = $1 U.S.). **GDP:** (1997 est.): $15.4 bil. **Per capita GDP:** $4,230. **Imports** (1996): $4.4 bil; partners: Russia 31%; Germany 15%. **Exports** (1996): $3.3 bil; partners: Russia 20%; Germany 14%. **Tourism:** $418 mil. **Budget** (1997 est.): $1.7 bil. **Intl. reserves less gold** (June 1999): $1.26 bil. **Gold:** 186,100 oz t. **Consumer prices** (change in 1997): 5.1%.

Transport: Railroad: Length: 1,802 mi. **Motor vehicles:** 653,000 pass. cars, 111,000 comm. vehicles. **Civil aviation:** 187.2 mil pass.-mi; 3 airports. **Chief port:** Klaipeda.

Communications: TV sets: 364 per 1,000 pop. **Radios:** 404 per 1,000 pop. **Telephones:** 1,109,800 main lines. **Daily newspaper circ.:** 136 per 1,000 pop.

Health: Life expectancy: 62.91 male; 75.31 female. **Births** (per 1,000 pop.): 10.52. **Deaths** (per 1,000 pop.): 12.93. **Natural inc.:** −0.241%. **Hosp. beds** (1995): 1 per 92 persons. **Physicians** (1995): 1 per 252 persons. **Infant mortality** (per 1,000 live births): 14.71.

Education: Free, compulsory: ages 7-16. **Literacy** (1989): 98%.

Major Intl. Organizations: UN (FAO, IBRD, ILO, IMF, IMO, WHO), OSCE.

Embassy: 2622 16th St. NW 20009; 234-5860.
Website: http://www.std.lt

Lithuania was occupied by the German army, 1914-18. It was annexed by the Soviet Russian army, but the Soviets were overthrown, 1919. Lithuania was a democratic republic until 1926, when the regime was ousted by a coup. In 1939 the Soviet-German treaty assigned most of Lithuania to the Soviet sphere of influence. Lithuania was annexed by the USSR Aug. 3, 1940.

Lithuania formally declared its independence from the Soviet Union Mar. 11, 1990. During an abortive Soviet coup in Aug., the Western nations recognized Lithuania's independence, which was ratified by the Soviet Union in Sept. 1991.

The last Russian troops withdrew on Aug. 31, 1993. Lithuania applied to join the European Union, Dec. 8, 1995. The conservative Homeland Union defeated the former Communists in parliamentary elections Oct. 20 and Nov. 10, 1996. A Lithuanian-American, Valdas Adamkus, won the presidency in a runoff election Jan. 4, 1998.

Luxembourg
Grand Duchy of Luxembourg

People: Population: 429,080. **Age distrib.** (%): <15: 18.1; 65+: 15.0. **Pop. density:** 430 per sq. mi. **Urban:** 90%. **Ethnic groups:** Mixture of French and Germans predominates. **Principal languages:** French, German, Luxembourgian, English. **Chief religion:** Roman Catholic 97%.

Geography: Area: 998 sq. mi. **Location:** In W Europe. **Neighbors:** Belgium on W, France on S, Germany on E. **Topography:** Heavy forests (Ardennes) cover N, S is a low, open plateau. **Capital:** Luxembourg (1996 est.): 77,400.

Government: Type: Constitutional monarchy. **Head of state:** Grand Duke Jean; b Jan. 5, 1921; in office: Nov. 12, 1964. **Head of gov.:** Prime Min. Jean-Claude Juncker; b Dec. 9, 1954; in office: Jan. 19, 1995. **Local divisions:** 3 districts. **Defense:** 0.8% of GDP. **Active troops:** 800.

Economy: Industries: Steel, chemicals, food processing, tires, banking, engineering, metal products. **Chief crops:** Grains, potatoes, wine grapes. **Arable land:** 24%. **Electricity prod.** (1997): 408 mil kWh. **Labor force:** 61% services; 35% ind.

Finance: Monetary unit: Lux. Franc (Sept. 1999: 37.69 = $1 U.S.). Euro (Sept. 1999: 1.07 = $1 U.S.). **GDP:** (1997 est.): $13.48 bil. **Per capita GDP:** $33,700. **Imports** (1996) $9.4 mil; partners: Belgium 38%; Germany 25%. **Exports** (1996): $7.1 mil; partners: Germany 28%; France 18%. **Tourism:** $309 mil. **Budget** (1997 est.): $5.44 bil. **Intl. reserves less gold** (June 1999): $47.7 mil. **Gold:** 76,000 oz t. **Consumer prices** (change in 1997): 1.0%.

Transport: Railroad: Length: 171 mi. **Motor vehicles:** 231,666 pass. cars, 16,665 comm. vehicles. **Civil aviation:** 174.8 mil pass.-mi; 1 airport. **Chief port:** Mertert.

Communications: TV sets: 384 per 1,000 pop. **Radios:** 586 per 1,000 pop. **Telephones:** 293,100 main lines. **Daily newspaper circ.:** 381 per 1,000 pop.

Health: Life expectancy: 74.58 male; 80.83 female. **Births** (per 1,000 pop.): 10.35. **Deaths** (per 1,000 pop.): 9.32. **Natural inc.:** 0.103%. **Hosp. beds** (1994): 1 per 92 persons. **Physicians** (1995): 1 per 454 persons. **Infant mortality** (per 1,000 live births): 4.99.

Education: Compulsory: ages 6-15. **Literacy:** 100%.

Major Intl. Organizations: UN (FAO, IBRD, ILO, IMF, IMO, WHO, WTrO), EU, NATO, OECD, OSCE.

Embassy: 2200 Massachusetts Ave. NW 20008; 265-4171.

Luxembourg, founded about 963, was ruled by Burgundy, Spain, Austria, and France from 1448 to 1815. It left the Germanic Confederation in 1866. Overrun by Germany in 2 world wars, Luxembourg ended its neutrality in 1948, when a customs union with Belgium and Netherlands was adopted.

Macedonia
Former Yugoslav Republic of Macedonia

People: Population: 2,022,604. **Age distrib.** (%): <15: 23.3; 65+: 9.7. **Pop. density:** 207 per sq. mi. **Urban:** 60%. **Ethnic groups:** Macedonian 65%, Albanian 22%. **Principal languages:** Macedonian (official), Albanian, Serbo-Croatian. **Chief religions:** Eastern Orthodox 67%, Muslim 30%.

Geography: Area: 9,781 sq. mi. **Location:** In SE Europe. **Neighbors:** Bulgaria on E, Greece on S, Albania on W, Serbia on N. **Capital:** Skopje (1994): 429,964.

Government: Type: Republic. **Head of state:** Pres. Kiro Gligorov; b May 3, 1917; in office: Jan. 27, 1991. **Head of gov.:** Prime Min. Ljubco Georgievski; b 1966; in office: Nov. 30, 1998. **Local divisions:** 123 municipalities. **Defense:** 10.2% of GDP. **Active troops:** 15,400.

Economy: Industries: Mining, textiles. **Chief crops:** Wheat, rice, cotton, tobacco. **Minerals:** Chromium, lead, zinc. **Arable land:** 24%. **Livestock** (1997): chickens: 3.50 mil; sheep: 1.80 mil; cattle: 297,559; pigs: 193,358. **Electricity prod.** (1997): 6.360 bil kWh. **Labor force:** 40% manuf. & mining.

Finance: Monetary unit: Denar (Sept. 1999: 56.86 = $1 U.S.). **GDP:** (1997 est.): $2.0 bil. **Per capita GDP:** $960. **Imports** (1996): $1.6 bil; partners: Germany 15%. **Exports** (1996): $1.2 bil.; partners: Germany 13%. **Tourism:** $15 mil. **Budget** (1996 est.): $1.0 bil. **Intl. reserves less gold** (June 1999): $322.65 mil. **Gold:** 101,000 oz t. **Consumer prices** (change in 1998): 0.5%.

Transport: Railroad: Length: 573 mi. **Motor vehicles:** 263,000 pass. cars, 23,000 comm. vehicles. **Civil aviation:** 161.1 mil pass.-mi; 2 airports.

Communications: TV sets: 179 per 1,000 pop. **Radios:** 179 per 1,000 pop. **Telephones** (1997): 407,800 main lines. **Daily newspaper circ.:** 21 per 1,000 pop.

Health: Life expectancy: 70.93 male; 75.34 female. **Births** (per 1,000 pop.): 15.21. **Deaths** (per 1,000 pop.): 8.03. **Natural inc.:** 0.718%. **Hosp. beds** (1994): 1 per 195 persons. **Physicians** (1994): 1 per 437 persons. **Infant mortality** (per 1,000 live births): 18.68.

Education: Free, compulsory: ages 7-15. **Literacy** (1996): 89%.

Major Intl. Organizations: UN (FAO, IBRD, ILO, IMF, IMO, WHO).

Embassy: 3050 K St. NW 20007; 337-3063.

Macedonia, as part of a larger region also called Macedonia, was ruled by Muslim Turks from 1389 to 1912, when native Greeks, Bulgarians, and Slavs won independence. Serbia received the largest part of the territory, with the rest going to Greece and Bulgaria. In 1913, the area was incorporated into Serbia, which in 1918 became part of the Kingdom of Serbs, Croats, and Slovenes (later Yugoslavia). In 1946, Macedonia became a constituent republic of Yugoslavia.

Macedonia declared its independence Sept. 8, 1991, and was admitted to the UN under a provisional name in 1993. A UN force, which included several hundred U.S. troops, was deployed there to deter the warring factions in Bosnia from carrying their dispute into other areas of the Balkans.

In Feb. 1994 both Russia and the U.S. recognized Macedonia. Greece, which objected to Macedonia's use of what it considered a Hellenic name and symbols, imposed a trade blockade on the landlocked nation. Apr. 1, the 2 countries agreed to normalize relations Sept. 13, 1995. A car bombing, Oct. 3, seriously injured Pres. Kiro Gligorov. Macedonia and Yugoslavia signed a treaty normalizing relations Apr. 8, 1996. By the end of NATO's air war against Yugoslavia, Mar.-June 1999, Macedonia had a Kosovar refugee population of more than 250,000; more than 90% had reportedly been repatriated by Sept. 1.

Madagascar
Republic of Madagascar

People: Population: 14,873,387. **Age distrib.** (%): <15: 44.6; 65+: 3.3. **Pop. density:** 66 per sq. mi. **Urban:** 27%. **Ethnic groups:** Malayo-Indonesian, Cotiers, French, Indian, Creole, Cornoran. **Principal languages:** Malagasy, French (both official). **Chief religions:** Indigenous beliefs 52%, Christian 41%, Muslim 7%.

Geography: Area: 226,700 sq. mi. **Location:** In the Indian O., off the SE coast of Africa. **Neighbors:** Comoro Isls. to NW, Mozambique to W. **Topography:** Humid coastal strip in the E, fertile valleys in the mountainous center plateau region, and a wider coastal strip on the W. **Capital:** Antananarivo: 876,000*.

Government: Type: Republic. **Head of state:** Pres. Didier Ratsiraka; b Nov. 4, 1936; in office: Jan. 31, 1997. **Head of gov.:** Tantely Andrianarivo; in office: July 23, 1998. **Local divisions:** 6 provinces. **Defense:** 0.8% of GDP. **Active troops:** 21,000.

Economy: Industries: Meat processing, textiles. **Chief crops:** Coffee, cloves, vanilla beans, rice, sugar, cassava, peanuts. **Minerals:** Chromite, graphite, coal, bauxite. **Arable land:** 4%. **Livestock** (1997): chickens: 16.60 mil; cattle: 10.34 mil; pigs: 1.67 mil; goats: 1.34 mil; sheep: 765,000. **Fish catch** (1996): 113,015 metric tons. **Electricity prod.** (1997): 745 mil kWh.

Finance: Monetary unit: Malagasy franc (Sept. 1999: 5,220.00 = $1 U.S.). **GDP:** (1997 est.): $10.3 bil. **Per capita GDP:** $730. **Imports** (1996 est.): $612 mil; partners: France 40%. **Exports** (1996 est.): $493 mil; partners: France 41%. **Tourism:** $74 mil. **Budget** (1996 est.): $706 mil. **Intl. reserves less gold** (Feb. 1999): $181.3 mil. **Consumer prices** (change in 1997): 6.2%.

Transport: Railroad: Length: 640 mi. **Motor vehicles:** 58,100 pass. cars, 15,860 comm. vehicles. **Civil aviation:** 471.0 mil pass.-mi; 44 airports. **Chief ports:** Toamasina, Antsiranana, Mahajanga, Toliara, Antsohimbondrona.

Communications: TV sets: 20 per 1,000 pop. **Radios:** 193 per 1,000 pop. **Telephones** (1997): 43,200 main lines. **Daily newspaper circ.:** 4 per 1,000 pop.

Health: Life expectancy: 52.01 male; 54.51 female. **Births** (per 1,000 pop.): 41.52. **Deaths** (per 1,000 pop.): 13.56. **Natural inc.:** 2.796%. **Infant mortality** (per 1,000 live births): 89.10.

Education: Compulsory for 5 years between ages 6 and 13. **Literacy:** 46%.

Major Intl. Organizations: UN (FAO, IBRD, ILO, IMF, IMO, WHO, WTrO), OAU.

Embassy: 2374 Massachusetts Ave. NW 20008; 265-5525.

Website: http://www3.itu.ch/missions/Madagascar

Madagascar was settled 2,000 years ago by Malayan-Indonesian people, whose descendants still predominate. A unified kingdom ruled the 18th and 19th centuries. The island became a French protectorate, 1885, and a colony 1896. Independence came June 26, 1960.

Discontent with inflation and French domination led to a coup in 1972. The new regime nationalized French-owned financial interests, closed French bases and a U.S. space-tracking station, and obtained Chinese aid. The government conducted a program of arrests, expulsion of foreigners, and repression of strikes, 1979.

In 1990, Madagascar ended a ban on multiparty politics that had been in place since 1975. Albert Zafy was elected president in 1993, ending the 17-year rule of Adm. Didier Ratsiraka. After Zafy was impeached by the legislature, Madagascar's constitutional court removed him from office, Sept. 5, 1996. Prime Min. Norbert Ratsirahonana then became interim president pending national elections, Nov. 3 and Dec. 29, in which Ratsiraka edged Zafy.

Malawi
Republic of Malawi

People: Population: 10,000,416. **Age distrib.** (%): <15: 45.1; 65+: 2.7. **Pop. density:** 219 per sq. mi. **Urban:** 14%. **Ethnic groups:** Chewa, Nyanja, Lomwe, other Bantu tribes. **Principal languages:** English, Chichewa (both official). **Chief religions:** Protestant 55%, Muslim 20%, Roman Catholic 20%.

Geography: Area: 45,700 sq. mi. **Location:** In SE Africa. **Neighbors:** Zambia on W, Mozambique on S and E, Tanzania on N. **Topography:** Malawi stretches 560 mi. N-S along Lake Malawi (Lake Nyasa), most of which belongs to Malawi. High plateaus and mountains line the Rift Valley the length of the nation. **Capital:** Lilongwe. **Cities** (1994 est.): Blantyre 446,800; Lilongwe 395,500.

Government: Type: Multiparty democracy. **Head of state and gov.:** Pres. Bakili Muluzi; b Mar. 17, 1943; in office: May

21, 1994. **Local divisions:** 24 districts. **Defense:** 1.1% of GNP. **Active troops:** 5,000.

Economy: Industries: Agricultural processing, cement. **Chief crops:** Tea, tobacco, sugar, cotton, corn, potatoes. **Arable land:** 18%. **Livestock** (1997): chickens: 14.50 mil; goats: 1.27 mil; cattle: 800,000; pigs: 240,000; sheep: 120,000. **Fish catch** (1996): 63,569 metric tons. **Electricity prod.** (1997): 920 mil kWh. **Labor force:** 86% agric.

Finance: Monetary unit: Kwacha (Sept. 1999: 43.28 = $1 U.S.). **GDP:** (1997 est.): $8.6 bil. **Per capita GDP:** $900. **Imports** (1995): $475 mil; partners: South Africa 44%. **Exports** (1995): $405 mil; partners: South Africa 16%, Germany 15%. **Tourism:** $8 mil. **Budget** (1993): $674 mil. **Intl. reserves less gold** (June 1999): $273.53 mil. **Gold:** 13,000 oz t. **Consumer prices** (change in 1998): 29.7%.

Transport: Railroad: Length: 490 mi. **Motor vehicles:** 25,400 pass. cars, 28,900 comm. vehicles. **Civil aviation:** 208.6 mil pass.-mi; 5 airports.

Communications: Radios: 112 per 1,000 pop. **Telephones:** 37,400 main lines.

Health: Life expectancy: 36.49 male; 36.11 female. **Births** (per 1,000 pop.): 39.54. **Deaths** (per 1,000 pop.): 23.84. **Natural inc.:** 1.570%. **Infant mortality** (per 1,000 live births): 132.14.

Education: Compulsory: ages 6-14. **Literacy:** 56%.

Major Intl. Organizations: UN (FAO, IBRD, ILO, IMF, IMO, WHO, WTrO), the Commonwealth, OAU.

Embassy: 2408 Massachusetts Ave. NW 20008; 797-1007.

Bantus came to the land in the 16th century, Arab slavers in the 19th. The area became the British protectorate Nyasaland in 1891. It became independent July 6, 1964, and a republic in 1966. After 3 decades as a one-party state under Pres. Hastings Kamuzu Banda, Malawi adopted a new constitution and, in multiparty elections held May 17, 1994, chose a new leader, Bakili Muluzi. Banda was acquitted, Dec. 23, 1995, of complicity in the deaths of 4 political opponents in 1983; he died Nov. 25, 1997.

Malaysia

People: Population: 21,376,066. **Age distrib.** (%): <15: 35.4; 65+: 4.0. **Pop. density:** 168 per sq. mi. **Urban:** 54%. **Ethnic groups:** Malay and other indigenous 58%, Chinese 26%, Indian 7%. **Principal languages:** Malay (official), English, Chinese dialects. **Chief religions:** Muslim, Hindu, Buddhist, Christian, tribal.

Geography: Area: 127,300 sq. mi. **Location:** On the SE tip of Asia, plus the N coast of the island of Borneo. **Neighbors:** Thailand on N, Indonesia on S. **Topography:** Most of W Malaysia is covered by tropical jungle, including the central mountain range that runs N-S through the peninsula. The western coast is marshy, the eastern, sandy. E Malaysia has a wide, swampy coastal plain, with interior jungles and mountains. **Capital:** Kuala Lumpur: 1,236,000*.

Government: Type: Federal parliamentary democracy with a constitutional monarch. **Head of state:** Paramount Ruler Sultan Salahuddin Abdul Aziz Shah Alhaj; b Mar. 8, 1926; in office Apr. 26, 1999. **Head of gov.:** Prime Min. Datuk Seri Mahathir bin Mohamad; b Dec. 20, 1925; in office: July 16, 1981. **Local divisions:** 13 states, 2 federal territories. **Defense:** 3.7% of GDP. **Active troops:** 111,500 troops.

Economy: Industries: Rubber goods, logging, electronics, petroleum production. **Chief crops:** Palm oil (world's leading producer), rice, pepper. **Minerals:** Tin (a leading producer), oil, gas, bauxite, copper, iron. **Crude oil reserves** (1999): 3.9 bil bbls. **Other resources:** Rubber, timber. **Arable land:** 3%. **Livestock** (1997): chickens: 110.00 mil; pigs: 3.40 mil; cattle: 725,000; goats: 320,000; sheep: 255,000; buffalo: 150,000. **Fish catch** (1997): 1.17 mil metric tons. **Electricity prod.** (1997): 54.500 bil kWh. **Labor force:** 29% services and trade; 25% manuf.; 21% agric.

Finance: Monetary unit: Ringgit (Sept. 1999: 3.80 = $1 U.S.). **GDP:** (1997 est.): $227 bil. **Per capita GDP:** $11,100. **Imports** (1996): $83.2 bil; partners: Japan 27%, U.S. 16%, Singapore 12%. **Exports** (1996): $78.2 bil; partners: U.S. 21%, Singapore 20%. **Tourism:** $3.37 bil. **Budget** (1996 est.): $22.0 bil. **Intl. reserves less gold** (June 1999): $30.57 bil. **Gold:** 2.34 mil oz t. **Consumer prices** (change in 1997): 5.3%.

Transport: Railroad: Length: 1,113 mi. **Motor vehicles** (1997): 3.33 mil pass. cars, 618,066 comm. vehicles. **Civil aviation:** 17.8 bil pass.-mi; 39 airports. **Chief ports:** Kuantan, Kelang, Kota Kinabalu, Kuching.

Communications: TV sets: 454 per 1,000 pop. **Radios:** 449 per 1,000 pop. **Telephones:** 4,384,000 main lines. **Daily newspaper circ.:** 139 per 1,000 pop.

Health: Life expectancy: 67.62 male; 73.90 female. **Births** (per 1,000 pop.): 26.05. **Deaths** (per 1,000 pop.): 5.29. **Natural inc.:** 2.076%. **Hosp. beds** (1995): 1 per 507 persons. **Physi-**

cians (1995): 1 per 2,153 persons. **Infant mortality** (per 1,000 live births): 21.68.

Education: Free, compulsory: ages 6-16. **Literacy:** 83%.

Major Intl. Organizations: UN (FAO, IBRD, ILO, IMF, IMO, WHO, WTrO), APEC, ASEAN, the Commonwealth.

Embassy: 2401 Massachusetts Ave. NW 20008; 328-2700.

European traders appeared in the 16th century; Britain established control in 1867. Malaysia was created Sept. 16, 1963. It included Malaya (which had become independent in 1957 after the suppression of Communist rebels), plus the formerly British Singapore, Sabah (N Borneo), and Sarawak (NW Borneo). Singapore was separated in 1965, in order to end tensions between Chinese, the majority in Singapore, and Malays in control of the Malaysian government.

A monarch is elected by a council of hereditary rulers of the Malayan states every 5 years.

Abundant natural resources have bolstered prosperity, and foreign investment has aided industrialization. Work on a new federal capital at Putrajaya, south of Kuala Lumpur, began in 1995. However, sagging stock and currency prices forced the postponement of major development projects in Sept. 1997.

As the recession deepened and political unrest grew, Prime Min. Mahathir bin Mohamad imposed new currency controls and fired his popular deputy prime minister, Anwar bin Ibrahim, Sept. 2, 1998. Anwar, who then called for Mahathir's resignation, was arrested Sept. 20; he was convicted of corruption, Apr. 14, 1999, and sentenced to 6 years in prison.

Maldives
Republic of Maldives

People: Population: 300,220. **Age distrib.** (%): <15: 47.0; 65+: 3.1. **Pop. density:** 3,002 per sq. mi. **Urban:** 27%. **Ethnic groups:** Sinhalese, Dravidian, Arab, African. **Principal languages:** Maldivian Divehi (Sinhalese dialect; official), English. **Chief religion:** Sunni Muslim.

Geography: Area: 100 sq. mi. **Location:** In the Indian O., SW of India. **Neighbors:** Nearest is India on N. **Topography:** 19 atolls with 1,190 islands, 198 inhabited. None of the islands are over 5 sq. mi. in area, and all are nearly flat. **Capital:** Male (1995 est.): 62,973.

Government: Type: Republic. **Head of state and gov.:** Pres. Maumoon Abdul Gayoom; b Dec. 29, 1937; in office: Nov. 11, 1978. **Local divisions:** 19 atolls and Male.

Economy: Industries: Fish processing, tourism. **Chief crops:** Coconuts, corn, sweet potatoes. **Arable land:** 10%. **Fish catch** (1996): 105,558 metric tons. **Electricity prod.** (1997): 65 mil kWh. **Labor force:** 25% fishing & agric.; 21% services; 21% manuf. and const.

Finance: Monetary unit: Rufiyaa (Sept. 1999: 11.77 = $1 U.S.). **GDP:** (1997 est.): $500 mil. **Per capita GDP:** $1,800. **Imports** (1996): $302 mil; partners: Singapore 32%. **Exports** (1996): $59 mil; partners: UK 22%, Sri Lanka 18%. **Tourism:** $292 mil. **Budget** (1995 est.): $141 mil. **Intl. reserves less gold** (June 1999): $137.58 mil. **Consumer prices** (change in 1997): –2.2%.

Transport: Civil aviation: 181.2 mil pass.-mi; 5 airports. **Chief ports:** Male, Gan.

Communications: TV sets: 19 per 1,000 pop. **Radios:** 96 per 1,000 pop. **Telephones:** (1997): 18,000 main lines. **Daily newspaper circ.:** 12 per 1,000 pop.

Health: Life expectancy: 66.53 male; 70.15 female. **Births** (per 1,000 pop.): 39.30. **Deaths** (per 1,000 pop.): 5.63. **Natural inc.:** 3.367%. **Hosp. beds** (1995): 1 per 1,192 persons. **Physicians** (1996): 1 per 2,587 persons. **Infant mortality** (per 1,000 live births): 38.14.

Education: Literacy: 93%.

Major Intl. Organizations: UN (FAO, IBRD, IMF, IMO, WHO, WTrO), the Commonwealth.

Websites: http://www.maldives-info.com
http://www.undp.org/missions/maldives/

The islands had been a British protectorate since 1887. The country became independent July 26, 1965. Long a sultanate, the Maldives became a republic in 1968. Natural resources and tourism are being developed; however, the Maldives remains one of the world's poorest countries.

Mali
Republic of Mali

People: Population: 10,429,124. **Age distrib.** (%): <15: 47.4; 65+: 3.2. **Pop. density:** 22 per sq. mi. **Urban:** 28%. **Ethnic groups:** Mande (Bambara, Malinke, Sarakole) 50%, Peul 17%, Voltaic 12%, Tuareg and Moor 10%, Songhai 6%. **Principal languages:** French (official), Bambara, numerous African languages. **Chief religions:** Muslim 90%, indigenous beliefs 9%.

Geography: Area: 479,000 sq. mi. **Location:** In the interior of W Africa. **Neighbors:** Mauritania, Senegal on W; Guinea, Côte d'Ivoire, Burkina Faso on S; Niger on E; Algeria on N. **Topography:** A landlocked grassy plain in the upper basins of the Senegal and Niger rivers, extending N into the Sahara. **Capital** (1996 met. est.): Bamako: 809,552.

Government: Type: Republic. **Head of state:** Pres. Alpha Oumar Konare; b Feb. 2, 1946; in office: June 8, 1992. **Head of gov.:** Prime Min. Ibrahim Boubakar Keita; b Jan. 29, 1945; in office: Feb. 4, 1994. **Local divisions:** 8 regions, 1 capital district. **Defense:** 1.7% of GDP. **Active troops:** 7,400.

Economy: Industries: construction, mining. **Chief crops:** Millet, rice, peanuts, corn, vegetables, cotton. **Minerals:** Gold, phosphates, kaolin. **Arable land:** 2%. **Livestock** (1997): chickens: 24.00 mil; goats: 8.55 mil; cattle: 5.73 mil; sheep: 5.95 mil. **Fish catch** (1996): 111,910 metric tons. **Electricity prod.** (1997): 325 mil kWh. **Labor force:** 80% agric.; 19% services.

Finance: Monetary unit: CFA Franc (Sept. 1999: 612.79 = $1 U.S.). **GDP:** (1997 est.): $6 bil. **Per capita GDP:** $600. **Imports** (1996 est.): $797 mil; partners: Côte d'Ivoire 23%, France 17%. **Exports** (1996 est.): $473 mil; partners: China 13%. **Tourism:** $28 mil. **Budget** (1997 est.): $770 mil. **Intl. reserves less gold** (Dec. 1998): $402.9 mil. **Gold:** 19,000 oz t. **Consumer prices** (change in 1997): 4.0%.

Transport: Railroad: Length: 398 mi. **Motor vehicles:** 24,700 pass. cars, 17,100 comm. vehicles. **Civil aviation:** 150.5 mil pass.-mi; 9 airports. **Chief port:** Koulikoro.

Communications: TV sets: 12 per 1,000 pop. **Radios:** 168 per 1,000 pop. **Telephones:** 26,800 main lines.

Health: Life expectancy: 46.09 male; 48.96 female. **Births** (per 1,000 pop.): 49.50. **Deaths** (per 1,000 pop.): 18.56. **Natural inc.:** 3.094%. **Physicians** (1993): 1 per 18,376 persons. **Infant mortality** (per 1,000 live births): 119.44.

Education: Free, compulsory: ages 7-16. **Literacy:** 31%.

Major Intl. Organizations: UN and most of its specialized agencies, OAU.

Embassy: 2130 R St. NW 20008; 332-2249.

Until the 15th century the area was part of the great Mali Empire. Timbuktu (Tombouctou) was a center of Islamic study. French rule was secured, 1898. The Sudanese Rep. and Senegal became independent as the Mali Federation June 20, 1960, but Senegal withdrew, and the Sudanese Rep. was renamed Mali.

Mali signed economic agreements with France and, in 1963, with Senegal. In 1968, a coup ended the socialist regime. Famine struck in 1973-74, killing as many as 100,000 people. Drought conditions returned in the 1980s.

The military, Mar. 26, 1991, overthrew the government of Pres. Moussa Traoré, who had been in power since 1968. Oumar Konare, a coup leader, was elected president, Apr. 26, 1992. A peace accord between the government and a Tuareg rebel group was signed in June 1994. Konare and his party won a series of flawed elections, Apr.-Aug. 1997. Twice condemned to death for crimes committed in office, Traoré had his sentences commuted to life imprisonment in Dec. 1997 and Sept. 1999.

Malta
Republic of Malta

People: Population: 381,603. **Age distrib.** (%): <15: 20.4; 65+: 11.8. **Pop. density:** 3,180 per sq. mi. **Urban:** 90%. **Ethnic group:** Maltese. **Principal languages:** Maltese, English (both official). **Chief religion:** Roman Catholic 98%.

Geography: Area: 120 sq. mi. **Location:** In center of Mediterranean Sea. **Neighbors:** Nearest is Italy on N. **Topography:** Island of Malta is 95 sq. mi.; other islands in the group: Gozo, 26 sq. mi.; Comino, 1 sq. mi. The coastline is heavily indented. Low hills cover the interior. **Capital:** Valletta (1996 est.): 7,172.

Government: Type: Parliamentary democracy. **Head of state:** Pres. Guido de Marco; b July 22, 1931; in office: Apr. 4, 1999. **Head of gov.:** Prime Min. Edward Fenech-Adami; b Feb. 7, 1934; in office: Sept. 6, 1998. **Local divisions:** 3 regions comprising 67 local councils. **Defense:** 0.9% of GDP. **Active troops:** 2,000.

Economy: Industries: Tourism, electronics, construction, textiles, food & beverages. **Chief crops:** Potatoes, cauliflower, melons, tomatoes. **Minerals:** Salt, limestone. **Arable land:** 38%. **Livestock** (1997): chickens: 820,000. **Electricity prod.** (1997): 1.425 bil kWh. **Labor force:** 34% pub. services; 32% services; 22% manuf. & const.

Finance: Monetary unit: Lira (Sept. 1999: 2.51 = $1 U.S.). **GDP:** (1997 est.): $4.9 bil. **Per capita GDP:** $12,900. **Imports** (1996): $2.8 bil; partners: Italy 27%, Germany 14%, UK 13%. **Exports** (1996): $1.7 bil; partners: Italy 32%, Germany 16%. **Tourism:** $677 mil. **Budget** (1997 est.): $1.5 bil. **Intl. reserves less gold** (Sept. 1997): $1.42 bil. **Gold:** 6,000 oz t. **Consumer prices** (change in 1997): 2.4%.

Transport: Motor vehicles: 122,100 pass. cars, 19,100 comm. vehicles. **Civil aviation:** 1.0 bil pass.-mi; 1 airport. **Chief ports:** Valletta, Marsaxlokk.

Communications: TV sets: 739 per 1,000 pop. **Radios:** 525 per 1,000 pop. **Telephones:** 191,500 main lines. **Daily newspaper circ.:** 145 per 1,000 pop.

Health: Life expectancy: 75.43 male; 80.23 female. **Births** (per 1,000 pop.): 11.02. **Deaths** (per 1,000 pop.): 7.37. **Natural inc.:** 0.365%. **Hosp. beds** (1996): 1 per 174 persons. **Physicians** (1996): 1 per 403 persons. **Infant mortality** (per 1,000 live births): 7.42.

Education: Free, compulsory: ages 5-16. **Literacy:** 91%.

Major Intl. Organizations: UN (FAO, IBRD, ILO, IMF, IMO, WHO, WTrO), the Commonwealth, OSCE.

Embassy: 2017 Connecticut Ave. NW 20008; 462-3611.

Website: http://www.magnet.mt/home/cos

Malta was ruled by Phoenicians, Romans, Arabs, Normans, the Knights of Malta, France, and Britain (since 1814). It became independent Sept. 21, 1964. Malta became a republic in 1974. The withdrawal of the last British sailors, Apr. 1, 1979, ended 179 years of British military presence on the island. From 1971 to 1987 and again from 1996 to 1998, Malta was governed by the socialist Labor Party. The Nationalist Party, which held office 1987-96 and favors Malta's entry into the EU, returned to power after elections Sept. 5, 1998.

Marshall Islands
Republic of the Marshall Islands

People: Population: 65,507. **Age. distrib.** (%): <15: 49.7; 65+: 2.2. **Pop. density:** 936 per sq. mi. **Urban:** 70%. **Ethnic groups:** Micronesian. **Principal languages:** English (official), Marshallese, Japanese. **Chief religion:** Protestant 63%.

Geography: Area: 70 sq. mi. **Location:** In N Pacific Ocean; composed of two 800-mi-long parallel chains of coral atolls. **Neighbors:** Nearest are Micronesia to W, Nauru and Kiribati to S. **Capital:** Majuro (1995 est.) 28,000.

Government: Type: Republic. **Head of state and gov.:** Pres. Imata Kabua; b May 20, 1943; in office: Jan. 22, 1997. **Local divisions:** 24 localities.

Economy: Agriculture and tourism are mainstays.

Finance: Monetary unit: U.S. Dollar. **GDP:** (1996 est.): $98 mil. **Per capita GDP:** $1,680. **Imports** (1996 est.): $70 mil; partners: U.S. 51%. **Exports** (1996 est.): $17.5 mil; partners: U.S. 80%. **Tourism:** $3 mil. **National budget** (FY 1995-96 est.): $77.4 mil.

Transport: Civil aviation: 16.1 mil pass.-mi; 25 airports. **Chief port:** Majuro.

Communications: Telephones: 3,700 main lines.

Health: Life expectancy: 63.21 male; 66.50 female. **Births** (per 1,000 pop.): 45.31. **Deaths** (per 1,000 pop.): 6.73. **Natural inc.:** 3.858%. **Hosp. beds** (1995): 1 per 515 persons. **Physicians** (1995): 1 per 3,269 persons. **Infant mortality** (per 1,000 live births): 43.38.

Education: Compulsory: ages 6-14. **Literacy** (1994): 93%.

Major Intl. Organizations: UN (IBRD, IMF, WHO).

Embassy: 2433 Massachusetts Ave. NW 20008; 234-5414.

The Marshall Islands were a German possession until World War I and were administered by Japan between the World Wars. After WW II, they were administered as part of the UN Trust Territory of the Pacific Islands by the U.S.

The Marshall Islands secured international recognition as an independent nation on Sept. 17, 1991. Amata Kabua, the islands' first and only president since 1979, died Dec. 19, 1996. His cousin Imata Kabua was elected Jan. 13, 1997.

Mauritania
Islamic Republic of Mauritania

People: Population: 2,581,738. **Age distrib.** (%): <15: 46.5; 65+: 2.3. **Pop. density:** 6 per sq. mi. **Urban:** 53%. **Ethnic groups:** Mixed Maur/black 40%, Maur 30%, black 30%. **Principal languages:** Hasaniya Arabic, Wolof (both official), Pular, Soninke. **Chief religion:** Muslim 100%.

Geography: Area: 398,000 sq. mi. **Location:** In NW Africa. **Neighbors:** Morocco on N, Algeria and Mali on E, Senegal on S. **Topography:** The fertile Senegal R. valley in the S gives way to a wide central region of sandy plains and scrub trees. The N is arid and extends into the Sahara. **Capital:** Nouakchott (1995 est.): 735,000.

Government: Type: Islamic republic. **Head of state:** Pres. Maaouya Ould Sidi Ahmed Taya; b 1943; in office: Apr. 18, 1992. **Head of gov.:** Prime Min. Cheikh El Afia Ould Mohamed Khouna; in office: Nov. 16, 1998. **Local divisions:** 12 regions, 1 capital district. **Defense:** 2.2% of GDP. **Active troops:** 15,700.

Economy: Industries: Fish processing, iron mining. **Chief crops:** Dates, grain. **Minerals:** Iron ore, gypsum. **Livestock** (1997): chickens: 3.90 mil; sheep: 6.20 mil; goats: 4.13 mil; cattle: 1.31 mil. **Fish catch** (1996): 85,000 metric tons. **Electricity prod.** (1997): 150 mil kWh. **Labor force:** 47% agric.; 29% services; 14% trade & finance.

Finance: Monetary unit: Ouguiya (Sept. 1999: 205.02 = $1 U.S.). **GDP:** (1996 est.): $4.1 bil. **Per capita GDP:** $1,750. **Imports** (1996): $457 mil; partners: France 30%. **Exports** (1996): $494 mil; partners: Japan 22%. **Tourism:** $11 mil. **Budget** (1996 est.): $265 mil. **Intl. reserves less gold** (June 1999): $181.3 mil. **Gold:** 12,000 oz t. **Consumer prices** (change in 1997): 4.6%.

Transport: Railroad: Length: 437 mi. **Motor vehicles:** 17,300 pass. cars, 9,210 comm. vehicles. **Civil aviation:** 201.1 mil pass.-mi; 9 airports. **Chief ports:** Nouakchott, Nouadhibou.

Communications: Radios: 428 per 1,000 pop. **Telephones:** 13,500 main lines.

Health: Life expectancy: 47.39 male; 53.65 female. **Births** (per 1,000 pop.): 44.10. **Deaths** (per 1,000 pop.): 14.20. **Natural inc.:** 2.990%. **Physicians** (1994): 1 per 11,085 persons. **Infant mortality** (per 1,000 live births 1997): 76.46.

Education: Compulsory: ages 6-12. **Literacy:** 38%.

Major Intl. Organizations: UN (FAO, IBRD, ILO, IMF, IMO, WHO, WTrO), AL, OAU.

Embassy: 2129 Leroy Pl. NW 20008; 232-5700.

Website: http://www.embassy.org/mauritania

Mauritania was a French protectorate from 1903. It became independent Nov. 28, 1960 and annexed the south of former Spanish Sahara (now Western Sahara) in 1976. Saharan guerrillas of the Polisario Front stepped up attacks in 1977; 8,000 Moroccan troops and French bomber raids aided the government. Mauritania signed a peace treaty with the Polisario Front, 1979, resumed diplomatic relations with Algeria while breaking a defense treaty with Morocco, and renounced sovereignty over its share of Western Sahara. Opposition parties were legalized and a new constitution approved in 1991.

Although slavery has been repeatedly abolished, most recently in 1980, an estimated 90,000 Mauritanians continue to live under conditions of servitude.

Mauritius
Republic of Mauritius

People: Population: 1,182,212. **Age distrib.** (%): <15: 26.2; 65+: 6.0. **Pop. density:** 1,689 per sq. mi. **Urban:** 41%. **Ethnic groups:** Indo-Mauritian 68%, Creole 27%. **Principal languages:** English (official), French, Creole, Hindi, Bojpoori. **Chief religions:** Hindu 52%, Christian 28.3%, Muslim 16.6%.

Geography: Area: 700 sq. mi. **Location:** In the Indian O., 500 mi. E of Madagascar. **Neighbors:** Nearest is Madagascar to W. **Topography:** A volcanic island nearly surrounded by coral reefs. A central plateau is encircled by mountain peaks. **Capital:** Port Louis (1996 est.): 145,797.

Government: Type: Republic. **Head of state:** Pres. Cassam Uteem; b Mar. 22, 1941; in office: June 30, 1992. **Head of gov.:** Prime Min. Navin Ramgoolam; b July 14, 1947; in office: Dec. 22, 1995. **Local divisions:** 9 districts, 3 dependencies. **Defense:** 2.1% of GDP. **Active troops** (1996): 1,300.

Economy: Industries: Tourism, textiles, food processing. **Chief crops:** Sugarcane, corn, potatoes, tea. **Arable land:** 49%. **Livestock** (1997): chickens: 3.30 mil. **Electricity prod.** (1997): 1.225 bil kWh. **Labor force:** 36% const. & ind.; 24% services; 14% agric. & fishing.

Finance: Monetary unit: Rupee (Sept. 1999: 25.25 = $1 U.S.). **GDP:** (1996 est.): $11.7 bil. **Per capita GDP:** $10,300. **Imports** (1996 est.): $2.2 bil; partners: France 20%. **Exports** (1996 est.): $1.6 bil; partners: UK 34%, France 21%, U.S. 15%. **Tourism:** $502 mil. **Budget** (FY 1995-96 est.): $1 bil. **Intl. reserves less gold** (June 1999): $613.8 mil. **Gold:** 62,000 oz t. **Consumer prices** (change in 1997): 6.8%.

Transport: Motor vehicles: 69,945 pass. cars, 12,328 comm. vehicles. **Civil aviation:** 2.4 bil pass.-mi; 1 airport. **Chief port:** Port Louis.

Communications: TV sets: 150 per 1,000 pop. **Radios:** 353 per 1,000 pop. **Telephones:** 243,600 main lines. **Daily newspaper circ.:** 49 per 1,000 pop.

Health: Life expectancy: 67.21 male; 74.96 female. **Births** (per 1,000 pop.): 18.49. **Deaths** (per 1,000 pop.): 6.69. **Natural inc.:** 1.180%. **Hosp. beds** (1995): 1 per 351 persons. **Physicians** (1995): 1 per 1,182 persons. **Infant mortality** (per 1,000 live births): 16.20.

Education: Compulsory: ages 5-12. **Literacy:** 83%.

Major Intl. Organizations: UN and all of its specialized agencies, the Commonwealth, OAU.

Embassy: 4301 Connecticut Ave. NW, Suite 441, 20008; 244-1491.

Mauritius was uninhabited when settled in 1638 by the Dutch, who introduced sugarcane. France took over in 1721, bringing African slaves. Britain ruled from 1810 to Mar. 12, 1968, bringing Indian workers for the sugar plantations.

Mauritius formally severed its association with the British crown Mar. 12, 1992.

Mexico
United Mexican States

People: Population: 100,294,036. **Age distrib.** (%): <15: 35.2; 65+: 4.2. **Pop. density:** 132 per sq. mi. **Urban:** 74%. **Ethnic groups:** Mestizo 60%, Amerindian 30%, Caucasian 9%. **Principal languages:** Spanish (official), Mayan dialects. **Chief religions:** Roman Catholic 89%, Protestant 6%.

Geography: Area: 761,600 sq. mi. **Location:** In southern North America. **Neighbors:** U.S. on N, Guatemala and Belize on S. **Topography:** The Sierra Madre Occidental Mts. run NW-SE near the west coast; the Sierra Madre Oriental Mts. run near the Gulf of Mexico. They join S of Mexico City. Between the 2 ranges lies the dry central plateau, 5,000 to 8,000 ft. alt., rising toward the S, with temperate vegetation. Coastal lowlands are tropical. About 45% of land is arid. **Capital:** Mexico City. **Cities** (1995 est.): Mexico City 8,489,007; Guadalajara 1,633,216; Puebla 1,222,569.

Government: Type: Federal republic. **Head of state and gov.:** Pres. Ernesto Zedillo Ponce de León; b Dec. 27, 1951; in office: Dec. 1, 1994. **Local divisions:** 31 states, 1 federal district. **Defense:** 1.0% of GDP. **Active troops:** 175,000.

Economy: Industries: Steel, food & beverages, chemicals, consumer durables, textiles, tourism. **Chief crops:** Cotton, coffee, wheat, rice, beans, soybeans, corn. **Minerals:** Silver, lead, zinc, gold, oil, gas, copper. **Crude oil reserves** (1999): 47.82 bil bbls. **Arable land:** 12%. **Livestock** (1997): chickens: 408.80 mil; cattle: 25.63 mil; pigs: 15.50 mil; goats: 8.61 mil; sheep: 6.50 mil. **Fish catch** (1997): 1.49 mil metric tons. **Electricity prod.** (1997): 164.322 bil kWh. **Labor force:** 28.8% services; 21.8% agric., forestry, hunting, fishing; 17.1% commerce; 16.1% manuf.

Finance: Monetary unit: New Peso (Sept. 1999: 9.38 = $1 U.S.). **GDP:** (1997 est.): $694.3 bil. **Per capita GDP:** $7,700. **Imports** (1997 est.): $109.8 bil; partners: U.S. 75%. **Exports** (1997 est.): $110.4 bil; partners: U.S. 85%. **Tourism:** $7.85 bil. **Budget** (1997 est.): $94 bil. **Intl. reserves less gold** (June 1999): $31.35 bil. **Gold:** 219,000 oz t. **Consumer prices** (change in 1997): 15.9%.

Transport: Railroad: Length: 16,543 mi. **Motor vehicles:** 8.2 mil pass. cars, 4.03 mil comm. vehicles. **Civil aviation:** 14.7 bil pass.-mi; 83 airports. **Chief ports:** Coatzacoalcos, Mazatlan, Tampico, Veracruz.

Communications: TV sets: 192 per 1,000 pop. **Radios:** 227 per 1,000 pop. **Telephones:** 9,926,900 main lines. **Daily newspaper circ.:** 115 per 1,000 pop.

Health: Life expectancy: 68.98 male; 75.17 female. **Births** (per 1,000 pop.): 24.99. **Deaths** (per 1,000 pop.): 4.83. **Natural inc.:** 2.016%. **Hosp. beds** (1994): 1 per 1,196 persons. **Physicians** (1994): 1 per 613 persons. **Infant mortality** (per 1,000 live births): 24.62.

Education: Free, compulsory: ages 6-12. **Literacy:** 90%.

Major Intl. Organizations: UN (FAO, IBRD, ILO, IMF, IMO, WHO, WTrO), APEC, OAS, OECD.

Embassy: 1911 Pennsylvania Ave. NW 20006; 728-1600.

Website: http://www.inegi.gob.mx/homeing/homeinegi/homeing.html

Mexico was the site of advanced Indian civilizations. The Mayas, an agricultural people, moved up from Yucatan, built immense stone pyramids, invented a calendar. The Toltecs were overcome by the Aztecs, who founded Tenochtitlan AD 1325, now Mexico City. Hernando Cortes, Spanish conquistador, destroyed the Aztec empire, 1519-21.

After 3 centuries of Spanish rule the people rose, under Fr. Miguel Hidalgo y Costilla, 1810, Fr. Morelos y Payon, 1812, and Gen. Agustin Iturbide, who made himself emperor as Agustin I, 1821. A republic was declared in 1823.

Mexican territory extended into the present American Southwest and California until Texas revolted and established a republic in 1836; the Mexican legislature refused recognition but was unable to enforce its authority there. After numerous clashes, the U.S.-Mexican War, 1846-48, resulted in the loss by Mexico of the lands north of the Rio Grande.

French arms supported an Austrian archduke on the throne of Mexico as Maximilian I, 1864-67, but pressure from the U.S. forced France to withdraw. Dictatorial rule by Porfirio Diaz, president 1877-80, 1884-1911, led to a period of rebellion and factional fighting. A new constitution, Feb. 5, 1917, brought social reform.

The Institutional Revolutionary Party (PRI) dominated politics from 1929 until the late 1990s. Radical opposition, including some guerrilla activity, was contained by strong measures.

Some gains in agriculture, industry, and social services were achieved, but much of the work force remained jobless or underemployed. Although prospects brightened with the discovery of vast oil reserves, inflation and a drop in world oil prices aggravated Mexico's economic problems in the 1980s.

Mexico reached agreement with the U.S. and Canada on the North American Free Trade Agreement (NAFTA) Aug. 12, 1992; it took effect Jan. 1, 1994.

Guerrillas of the Zapatista National Liberation Army (EZLN) launched an uprising, Jan. 1, 1994, in southern Mexico. A tentative peace accord was reached Mar. 2. The presidential candidate of the governing PRI, Luis Donaldo Colosio Murrieta, was assassinated at a political rally in Tijuana, Mar. 23. The new PRI candidate, Ernesto Zedillo Ponce de León, won election Aug. 21 and was inaugurated Dec. 1, 1994.

An austerity plan and pledges of aid from the U.S. saved Mexico's currency from collapse in early 1995. Popular Revolutionary Army guerrillas launched coordinated attacks on government targets in Aug. 1996. In elections July 6, 1997, the PRI failed to win a congressional majority for the first time since 1929. An armed gang massacred 45 peasants in Chiapas on Dec. 22, 1997.

Weeks of heavy rain in Sept.-Oct. 1999 caused severe flooding, killing at least 350 people and forcing tens of thousands from their homes.

Micronesia
Federated States of Micronesia

People: Population: 131,500. **Pop. density:** 485 per sq. mi. **Urban:** 28%. **Ethnic groups:** 9 ethnic Micronesian and Polynesian groups. **Principal languages:** English (official), Trukese, Pohnpeian, Yapese. **Chief religions:** Roman Catholic 50%, Protestant 47%.

Geography: Area: 271 sq. mi. **Location:** Consists of 607 islands in the W Pacific Ocean. **Capital:** Palikir, on Pohnpei (1994 island pop.) 33,372.

Government: Type: Republic. **Head of state and gov.:** Pres. Leo A. Falcam; b Nov. 20, 1935; in office: May 11, 1999. **Local divisions:** 4 states.

Economy: Industries: Tourism, fish processing. **Chief crops:** Tropical fruits, vegetables, black pepper. **Livestock** (1998): chickens: 185,000.

Finance: Monetary unit: U.S. Dollar. **GDP:** (1996 est.): $220 mil. **Per capita GDP:** $1,760. **Imports** (1996 est.): $168 mil; partners: U.S. 56%, Japan 32%. **Exports** (1996 est.): $73 mil; partners: Japan 80%. **Tourism:** $2.03 bil. **Budget** (FY 1995-96 est.): $52 mil.

Transport: 4 airports. **Chief ports:** Colonia (Yap), Kolonia (Pohnpei), Lele, Moen.

Communications: TV sets: 19 per 1,000 pop. **Radios:** 664 per 1,000 pop. **Telephones** (1997): 8,400 main lines.

Health: Life expectancy: 66.52 male; 70.48 female. **Births** (per 1,000 pop.): 27.32. **Deaths** (per 1,000 pop.): 6.01. **Natural inc.:** 2.131%. **Hosp. beds** (1993): 1 per 318 persons. **Physicians** (1993): 1 per 2,069 persons. **Infant mortality** (per 1,000 live births): 33.99.

Education: Compulsory: ages 6-14. **Literacy** (1991): 90%. **Major Intl. Organizations:** UN (IBRD, IMF, WHO). **Embassy:** 1725 N St. NW 20036; 223-4383.

The Federated States of Micronesia, formerly known as the Caroline Islands, was ruled successively by Spain, Germany, Japan, and the U.S. It was internationally recognized as an independent nation Sept. 17, 1991.

Moldova
Republic of Moldova

People: Population: 4,460,838. **Age distrib. (%):** <15: 24.5; 65+: 9.7. **Pop. density:** 343 per sq. mi. **Urban:** 52%. **Ethnic groups:** Moldovan/Romanian 65%, Ukrainian 14%, Russian 13%. **Principal languages:** Moldovan (official), Russian. **Chief religion:** Eastern Orthodox 99%.

Geography: Area: 13,000 sq. mi. **Location:** In E Europe. **Neighbors:** Romania on W; Ukraine on N, E, and S. **Capital** (1994 est.): Chisinau 655,940.

Government: Type: Republic. **Head of state:** Pres. Petru Lucinschi; b Jan. 27, 1940; in office: Jan. 15, 1997. **Head of gov.:** Prime Min. Ion Sturza; b May 9, 1960; in office: Mar. 12, 1999. **Local divisions:** 21 cities and towns, 48 urban settlements, more than 1,600 villages. **Defense:** 4.4% of GDP. **Active troops:** 11,000.

Economy: Industries: Food processing, machinery, textiles. **Chief crops:** Grain, vegetables, fruits, wine grapes. **Minerals:** Lignite, phosphorites, gypsum. **Arable land:** 53%. **Livestock** (1997): chickens: 12.00 mil; sheep: 1.02 mil; pigs: 761,700; cattle: 485,400; goats: 108,200. **Electricity prod.** (1997): 5.702 bil kWh. **Labor force:** 46.1% agric.; 13.9% industry.

Finance: Monetary unit: Leu (Aug. 1999: 10.92 = $1 U.S.). **GDP:** (1997 est.): $10.8 bil. **Per capita GDP:** $2,400. **Imports** (1997): $1.16 bil; partners: Russia 32%, Ukraine 26%. **Exports** (1997): $816 mil; partners: Russia 47%. **Tourism:** $4 mil. **Budget** (1997 est.): $641 mil. **Intl. reserves less gold** (June 1999): $163.98 mil. **Consumer prices** (change in 1997): 6.6%.

Transport: Railroad: Length: 746 mi. **Motor vehicles:** 169,000 pass. cars, 71,000 comm. vehicles. **Civil aviation:** 37.8 mil pass.-mi; 1 airport.

Communications: TV sets: 30 per 1,000 pop. **Radios:** 209 per 1,000 pop. **Telephones:** 657,500 main lines. **Daily newspaper circ.:** 24 per 1,000 pop.

Health: Life expectancy: 59.76 male; 69.24 female. **Births** (per 1,000 pop.): 14.43. **Deaths** (per 1,000 pop.): 12.50. **Natural inc.:** 0.193%. **Hosp. beds** (1995): 1 per 82 persons. **Physicians** (1995): 1 per 250 persons. **Infant mortality** (per 1,000 live births): 43.52.

Education: Compulsory: ages 7-16. **Literacy:** 96%.

Major Intl. Organizations: UN (FAO, IBRD, ILO, IMF, WHO), CIS, OSCE.

Embassy: 2101 S St. NW 20008; 667-1130.

In 1918, Romania annexed all of Bessarabia that Russia had acquired from Turkey in 1812 by the Treaty of Bucharest. In 1924, the Soviet Union established the Moldavian Autonomous Soviet Socialist Republic on the eastern bank of the Dniester. It was merged with the Romanian-speaking districts of Bessarabia in 1940 to form the Moldavian SSR.

During World War II, Romania, allied with Germany, occupied the area. It was recaptured by the USSR in 1944. Moldova declared independence Aug. 27, 1991. It became an independent state when the USSR disbanded Dec. 26, 1991.

Fighting erupted Mar. 1992 in the Dnestr (Dniester) region between Moldovan security forces and Slavic separatists—ethnic Russians and ethnic Ukrainians—who feared Moldova would merge with neighboring Romania. In a plebiscite on Mar. 6, 1994, voters in Moldova supported independence, without unification with Romania.

Defying the Moldovan government, voters in the breakaway Dnestr region held legislative elections and approved a separatist constitution Dec. 24, 1995. Petru Lucinschi, a former Communist, won a presidential runoff election Dec. 1, 1996. A peace accord with Dnestr separatists was signed in Moscow May 8, 1997. The Communists won the most seats in parliamentary elections Mar. 22, 1998, but a coalition of three center-right parties formed the government.

Monaco
Principality of Monaco

People: Population: 32,149. **Age distrib. (%):** <15: 16.7; 65+: 19.4. **Pop. density:** 42,865 per sq. mi. **Urban:** 100%. **Ethnic groups:** French 47%, Italian 16%, Monegasque 16%. **Principal languages:** French (official), English, Italian, Monegasque. **Chief religion:** Roman Catholic 95%.

Geography: Area: 0.75 sq. mi. **Location:** On the NW Mediterranean coast. **Neighbors:** France to W, N, E. **Topography:** Monaco-Ville sits atop a high promontory, the rest of the principality rises from the port up the hillside. **Capital:** Monaco.

Government: Type: Constitutional monarchy. **Head of state:** Prince Rainier III; b May 31, 1923; in office: May 9, 1949. **Head of gov.:** Min. of State Michel Lévêque; b 1933; in office: Feb. 3, 1997. **Local divisions:** 4 quarters.

Economy: Industries: Tourism, gambling, chemicals, precision instruments.

Finance: Monetary unit: French Franc (Sept. 1999: 6.13 = $1 U.S.) or Monegasque Franc. **GDP:** (1996 est.): $800 mil. **Per capita GDP:** $25,000. **Budget** (1995 est.): $638.7 mil.

Transport: Railroad: Length: 1 mi. **Motor vehicles:** 17,000 pass. cars, 4,000 comm. vehicles. **Civil aviation:** 820,000 pass.-mi; 1 airport. **Chief port:** Monaco.

Communications: TV sets: 690 per 1,000 pop. **Radios:** 941 per 1,000 pop.

Health: Life expectancy: 75.00 male; 82.35 female. **Births** (per 1,000 pop.): 10.70. **Deaths** (per 1,000 pop.): 11.79. **Natural inc.:** −0.109%. **Infant mortality** (per 1,000 live births): 6.47.

Education: Compulsory: ages 6-16.

Major Intl. Organizations: UN (IMO, WHO), OSCE.

An independent principality for over 300 years, Monaco has belonged to the House of Grimaldi since 1297, except during the French Revolution. It was placed under the protectorate of Sardinia in 1815, and under France, 1861. The Prince of Monaco was an absolute ruler until the 1911 constitution. Monaco was admitted to the UN on May 28, 1993.

Monaco's fame as a tourist resort is widespread. It is noted for its mild climate, magnificent scenery, and elegant casinos.

Mongolia

People: Population: 2,617,379. **Age distrib.** (%): <15: 36.1; 65+: 3.7. **Pop. density:** 4 per sq. mi. **Urban:** 61%. **Ethnic groups:** Mongol 90%. **Principal language:** Khalkha Mongol (official). **Chief religion:** Mostly Tibetan Buddhist.

Geography: Area: 604,000 sq. mi. **Location:** In E Central Asia. **Neighbors:** Russia on N, China on E, W, and S. **Topography:** Mostly a high plateau with mountains, salt lakes, and vast grasslands. Arid lands in the S are part of the Gobi Desert. **Capital:** Ulaanbaatar. **Cities** (1997 est.): 627,300.

Government: Type: Republic. **Head of state:** Pres. Natsagiyn Bagabandi; b Apr. 22, 1950; in office: June 20, 1997. **Head of gov.:** Rinchinnyamiyn Amarjargal; b 1961; in office: July 30, 1999. **Local divisions:** 18 provinces, 3 municipalities. **Defense:** 2.0% of GNP. **Active troops:** 9,000.

Economy: Industries: Food processing, mining, construction materials. **Chief crops:** Grain, potatoes. **Minerals:** Coal, oil, tungsten, copper, molybdenum, gold, phosphates, tin. **Arable land:** 1%. **Livestock** (1997): sheep: 14.17 mil; goats: 10.27 mil; cattle: 3.61 mil. **Electricity prod.** (1997): 2.500 bil kWh. **Labor force:** primarily agricultural.

Finance: Monetary unit: Tugrik (Sept. 1999: 1,041.24 = $1 U.S.). **GDP:** (1997 est.): $5.6 bil. **Per capita GDP:** $2,200. **Imports** (1997 est.): $443.4 mil; partners: Russia 34%, China 15%. **Exports** (1997 est.): $418 mil; partners: Russia 21%, China 18%. **Tourism:** $23 mil. **Intl. reserves less gold** (May 1999): $80.48 mil. **Gold:** 4,000 oz t. **Consumer prices** (change in 1997): 9.5%.

Transport: Railroad: Length: 1,294 mi. **Motor vehicles:** 21,000 pass. cars, 27,000 comm. vehicles. **Civil aviation:** 121.0 mil pass.-mi; 1 airport.

Communications: TV sets: 60.7 per 1,000 pop. **Radios:** 74 per 1,000 pop. **Telephones** (1997): 87,000 main lines. **Daily newspaper circ.:** 92 per 1,000 pop.

Health: Life expectancy: 59.71 male; 64.02 female. **Births** (per 1,000 pop.): 22.51. **Deaths** (per 1,000 pop.): 7.97. **Natural inc.:** 1.454%. **Hosp. beds** (1993): 1 per 101 persons. **Physicians** (1993): 1 per 401 persons. **Infant mortality** (per 1,000 live births): 64.63.

Education: Compulsory: ages 6-16. **Literacy** (1991): 83%.

Major Intl. Organizations: UN (FAO, IBRD, ILO, IMF, IMO, WHO, WTrO).

Embassy: 2833 M St. NW 20007; 333-7117.

Website: http://www.MongoliaOnline.mn/english

One of the world's oldest countries, Mongolia reached the zenith of its power in the 13th century when Genghis Khan and his successors conquered all of China and extended their influence as far west as Hungary and Poland. In later centuries, the empire dissolved and Mongolia became a province of China.

With the advent of the 1911 Chinese revolution, Mongolia, with Russian backing, declared its independence. A Communist regime was established July 11, 1921.

In 1990, the Mongolian Communist Party yielded its monopoly on power but won election in July. A new constitution took effect Feb. 12, 1992. A democratic alliance won legislative elections, June 30, 1996. Natsagiyn Bagabandi, a former Communist, won the presidential election of May 18, 1997. A protracted political crisis took a violent turn Oct. 2, 1998, with the murder of Sanjaasuregiyn Zorig, a popular cabinet member seeking to become prime minister.

Morocco
Kingdom of Morocco

People: Population: 29,661,636. **Age distrib.** (%): <15: 35.8; 65+: 4.5. **Pop. density:** 172 per sq. mi. **Urban:** 53%. **Ethnic groups:** Arab-Berber 99%. **Principal languages:** Arabic (official), Berber dialects. **Chief religion:** Sunni Muslim 99%.

Geography: Area: 172,400 sq. mi. **Location:** On NW coast of Africa. **Neighbors:** Western Sahara on S, Algeria on E. **Topography:** Consists of 5 natural regions: mountain ranges (Riff in the N, Middle Atlas, Upper Atlas, and Anti-Atlas); rich plains in the W; alluvial plains in SW; well-cultivated plateaus in the center; a pre-Sahara arid zone extending from SE. **Capital:** Rabat. **Cities** (1993 est. met.): Casablanca 2,943,000; Rabat 1,220,000.

Government: Type: Constitutional monarchy. **Head of state:** King Mohammed VI; b Aug. 21, 1963; in office: Jul 23, 1999. **Head of gov.:** Prime Min. Abderrahmane El Youssoufi; b Mar. 8, 1924; in office: Feb. 4, 1998. **Local divisions:** 16 regions. **Defense:** 4.2% of GDP. **Active troops:** 196,300.

Economy: Industries: Food processing, textiles, leather goods, mining, tourism. **Chief crops:** Grain, citrus, grapes, olives. **Minerals:** Phosphates, iron ore, manganese, lead, zinc. **Crude oil reserves** (1999): 1.97 mil bbls. **Arable land:** 21%. **Livestock** (1997): chickens: 100.00 mil; sheep: 17.60 mil;

goats: 6.20 mil; cattle: 2.60 mil. **Fish catch** (1997): 783,615 metric tons. **Electricity prod.** (1997): 11.800 bil kWh. **Labor force:** 50% agric.; 26% services.

Finance: Monetary unit: Dirham (Sept. 1999: 9.75 = $1 U.S.). **GDP:** (1997 est.): $107 bil. **Per capita GDP:** $3,500. **Imports** (1996): $9.7 bil; partners: France 22%, Spain 9%. **Exports** (1996): $6.9 bil; partners: France 30%, Spain 9%. **Tourism:** $1.60 bil. **Budget** (1996 est.): $10.75 bil. **Intl. reserves less gold** (June 1999): $4.27 bil. **Gold:** 704,000 oz t. **Consumer prices** (change in 1997): 2.9%.

Transport: Railroad: Length: 1,099 mi. **Motor vehicles** (1997): 1.10 mil pass. cars, 333,152 comm. vehicles. **Civil aviation:** 3.3 bil pass.-mi; 11 airports. **Chief ports:** Tangier, Casablanca, Kenitra.

Communications: TV sets: 92.7 per 1,000 pop. **Radios:** 222 per 1,000 pop. **Telephones:** 1,515,000 main lines. **Daily newspaper circ.:** 14.5 per 1,000 pop.

Health: Life expectancy: 66.85 male; 70.99 female. **Births** (per 1,000 pop.): 25.78. **Deaths** (per 1,000 pop.): 6.12. **Natural inc.:** 1.966%. **Hosp. beds** (1994): 1 per 978 persons. **Physicians** (1994): 1 per 2,923 persons. **Infant mortality** (per 1,000 live births): 50.96.

Education: Compulsory: ages 7-13. **Literacy:** 44%.

Major Intl. Organizations: UN (FAO, IBRD, ILO, IMF, IMO, WHO, WTrO), AL.

Embassy: 1601 21st St. NW 20009; 462-7979.

Berbers were the original inhabitants, followed by Carthaginians and Romans. Arabs conquered in 683. In the 11th and 12th centuries, a Berber empire ruled all NW Africa and most of Spain from Morocco.

Part of Morocco came under Spanish rule in the 19th century; France controlled the rest in the early 20th. Tribal uprisings lasted from 1911 to 1933. The country became independent Mar. 2, 1956. Tangier, an internationalized seaport, was turned over to Morocco, 1956. Ifni, a Spanish enclave, was ceded in 1969. Morocco annexed the disputed territory of Western Sahara during the second half of the 1970s.

King Hassan II assumed the throne in 1961, reigning until his death on July 23, 1999; he was immediately succeeded by his eldest son. Political reforms in the 1990s included the establishment of a bicameral legislature in 1997.

Western Sahara

Western Sahara, formerly the protectorate of Spanish Sahara, is bounded the N by Morocco, the NE by Algeria, the E and S by Mauritania, and on the W by the Atlantic Ocean. Phosphates are the major resource. Population (1997 est.): 239,333; capital: Laayoune (El Aaiun). Area: 102,700 sq mi.

Spain withdrew from its protectorate in Feb. 1996. On Apr. 14, 1976, Morocco annexed over 70,000 sq. mi, with the remainder annexed by Mauritania. A guerrilla movement, the Polisario Front, which had proclaimed the region independent Feb. 27, launched attacks with Algerian support. After Mauritania signed a treaty with Polisario on Aug. 5, 1979, Morocco occupied Mauritania's portion of Western Sahara.

After years of bitter fighting, Morocco controlled the main urban areas, but Polisario guerrillas moved freely in the vast, sparsely populated deserts. The 2 sides implemented a ceasefire in 1991, when a UN peacekeeping force was deployed. A UN-sponsored referendum on self-determination for Western Sahara has been repeatedly postponed.

Mozambique
Republic of Mozambique

People: Population: 19,124,335. **Age distrib.** (%): <15: 44.8; 65+: 2.3. **Pop. density:** 62 per sq. mi. **Urban:** 35%. **Ethnic groups:** Indigenous tribal groups. **Principal languages:** Portuguese (official), indigenous dialects. **Chief religions:** Indigenous beliefs 50%, Christian 30%, Muslim 20%.

Geography: Area: 309,500 sq. mi. **Location:** On SE coast of Africa. **Neighbors:** Tanzania on N; Malawi, Zambia, Zimbabwe on W; South Africa, Swaziland on S. **Topography:** Coastal lowlands comprise nearly half the country with plateaus rising in steps to the mountains along the western border. **Capital:** Maputo: 2,212,000*.

Government: Type: Republic. **Head of state:** Pres. Joaquim Chissano; b Oct. 22, 1939; in office: Oct. 19, 1986. **Head of gov.:** Prime Min. Pascoal Mocumbi; b Apr. 10, 1941; in office: Dec. 21, 1994. **Local divisions:** 10 provinces. **Defense:** 3.9% of GDP. **Active troops:** 6,100.

Economy: Industries: Chemicals, petroleum products, textiles. **Chief crops:** Cashews, cotton, sugar, corn, cassava, tea. **Minerals:** Coal, titanium. **Arable land:** 4%. **Livestock** (1997): chickens: 23.50 mil; cattle: 1.30 mil; goats: 388,000; pigs: 176,000; sheep: 123,000. **Fish catch** (1996): 34,915 metric tons. **Electricity prod.** (1997): 550 mil kWh. **Labor force:** 80% agric.

Finance: Monetary unit: Metical (Sept. 1999: 11,495.00 = $1 U.S.). **GDP:** (1997 est.): $14.6 bil. **Per capita GDP:** $800. **Imports** (1996 est.): $802 mil; partners: South Africa 38%. **Exports** (1996 est.): $226 mil; partners: Spain 22%, South Africa 22%, Japan 13%, Portugal 10%. **Budget** (1996 est.): $600 mil. **Intl. reserves less gold** (May 1999): $618.39 mil. **Consumer prices** (change in 1997): 5.5%.

Transport: Railroad: Length: 1,940 mi. **Motor vehicles:** 67,600 pass. cars, 21,200 comm. vehicles. **Civil aviation:** 180.6 mil pass.-mi; 7 airports. **Chief ports:** Maputo, Beira, Nacala, Inhambane.

Communications: TV sets: 3.5 per 1,000 pop. **Radios:** 38 per 1,000 pop. **Telephones:** 75,400 main lines. **Daily newspaper circ.:** 8 per 1,000 pop.

Health: Life expectancy: 44.73 male; 47.09 female. **Births** (per 1,000 pop.): 42.75. **Deaths** (per 1,000 pop.): 17.31. **Natural inc.:** 2.544%. **Hosp. beds** (1993): 1 per 1,133 persons. **Physicians** (1993): 1 per 131,991 persons. **Infant mortality** (per 1,000 live births): 117.56.

Education: Compulsory: ages 7-14. **Literacy:** 40%.

Major Intl. Organizations: UN (FAO, IBRD, ILO, IMF, IMO, WHO, WTrO), the Commonwealth, OAU.

Embassy: 1990 M St. NW, Suite 570, 20036; 293-7146.

Website: http://www.mbendi.co.za/cymzcy.htm

The first Portuguese post on the Mozambique coast was established in 1505, on the trade route to the East. Mozambique became independent June 25, 1975, after a ten-year war against Portuguese colonial domination. The 1974 revolution in Portugal had paved the way for the orderly transfer of power to Frelimo (Front for the Liberation of Mozambique). Frelimo took over local administration Sept. 20, 1974, although opposed, in part violently, by some blacks and whites.

The new government, led by Maoist Pres. Samora Machel, provided for a gradual transition to a Communist system. Economic problems included the emigration of most of the country's whites, a politically untenable economic dependence on white-ruled South Africa, and a large external debt.

In the 1980s, severe drought and civil war caused famine and heavy loss of life. Pres. Machel was killed in a plane crash just inside the South African border, Oct. 19, 1986.

The ruling party formally abandoned Marxist-Leninism in 1989, and a new constitution, effective Nov. 30, 1990, provided for multiparty elections and a free-market economy.

On Oct. 4, 1992, a peace agreement was signed aimed at ending hostilities between the government and the rebel Mozambique National Resistance (MNR). Elections took place Oct. 27-28, 1994. Repatriation of 1.7 million Mozambican refugees officially ended June 1995. In Mar. 1999 the heaviest floods in 4 decades left nearly 200,000 people stranded.

Myanmar (*formerly* Burma)
Union of Myanmar

People: Population: 48,081,302. **Age distrib.** (%): <15: 36.2; 65+: 4.2. **Pop. density:** 184 per sq. mi. **Urban:** 26%. **Ethnic groups:** Burman 68%, Shan 9%, Karen 7%, Rakhine 4%. **Principal language:** Burmese (official). **Chief religions:** Buddhist 89%, Christian 4%, Muslim 4%.

Geography: Area: 262,000 sq. mi. **Location:** Between S and SE Asia, on Bay of Bengal. **Neighbors:** Bangladesh, India on W; China, Laos, Thailand on E. **Topography:** Mountains surround Myanmar on W, N, and E, and dense forests cover much of the nation. N-S rivers provide habitable valleys and communications, especially the Irrawaddy, navigable for 900 miles. The country has a tropical monsoon climate. **Capital:** Yangon (Rangoon) 3,873,000*.

Government: Type: Military. **Head of state and gov.:** Gen. Than Shwe; b 1933; in office: Apr. 24, 1992. **Local divisions:** 7 states, 7 divisions. **Defense:** 7.7% of GDP. **Active troops:** 429,000.

Economy: Industries: Mining, textiles, footwear, wood products, agric. processing. **Chief crops:** Rice, sugarcane, corn, pulses. **Minerals:** Oil, lead, copper, tin, tungsten, precious stones. **Crude oil reserves** (1999): 50 mil bbls. **Arable land:** 15%. **Livestock** (1997): chickens: 33.07 mil; cattle: 10.49 mil; pigs: 3.50 mil; buffalo: 2.34 mil; goats: 1.32 mil; sheep: 369,126. **Fish catch** (1997): 830,346 metric tons. **Electricity prod.** (1997): 4.220 bil kWh. **Labor force:** 65.2% agric.; 14.3% industry, 10% trade.

Finance: Monetary unit: Kyat (Sept. 1999: 6.07 = $1 U.S.). **GDP:** (1997 est.): $55.7 bil. **Per capita GDP:** $1,190. **Imports** (1996): $1.4 bil; partners: Japan 24%, Singapore 15%. **Exports** (1996): $623 mil; partners: Indonesia 16%, Singapore 16%. **Tourism:** $35 mil. **Budget** (FY 1996-97): $12.2 bil. **Intl. reserves less gold** (Apr. 1999): $351.6 mil. **Gold:** 231,000 oz t. **Consumer prices** (change in 1997): 51.5%.

Transport: Railroad: Length: 3,144 mi. **Motor vehicles:** 35,000 pass. cars, 34,000 comm. vehicles. **Civil aviation:** 91.5 mil pass.-mi; 19 airports. **Chief ports:** Bassein, Moulmein.

Communications: TV sets: 22 per 1,000 pop. **Radios:** 72 per 1,000 pop. **Telephones** (1997): 213,500 main lines. **Daily newspaper circ.:** 23 per 1,000 pop.

Health: Life expectancy: 53.24 male; 56.32 female. **Births** (per 1,000 pop.): 28.48. **Deaths** (per 1,000 pop.): 12.39. **Natural inc.:** 1.609%. **Hosp. beds** (1993-94): 1 per 1,586 persons. **Physicians** (1993-94): 1 per 3,554 persons. **Infant mortality** (per 1,000 live births): 76.25.

Education: Free, compulsory: ages 5-10. **Literacy:** 83%.

Major Intl. Organizations: UN (FAO, IBRD, ILO, IMF, IMO, WHO, WTrO), ASEAN.

Embassy: 2300 S St. NW 20008; 332-9044.

Website: http://www.myanmar.com/e-index.html

The Burmese arrived from Tibet before the 9th century, displacing earlier cultures, and a Buddhist monarchy was established by the 11th. Burma was conquered by the Mongol dynasty of China in 1272, then ruled by Shans as a Chinese tributary, until the 16th century.

Britain subjugated Burma in 3 wars, 1824-84, and ruled the country as part of India until 1937, when it became self-governing. Independence outside the Commonwealth was achieved Jan. 4, 1948.

Gen. Ne Win dominated politics from 1962 to 1988, first as military ruler then as constitutional president. His regime drove Indians from the civil service and Chinese from commerce. Economic socialization was advanced, isolation from foreign countries enforced. In 1987 Burma, once the richest nation in SE Asia, was granted less-developed status by the UN.

Ne Win resigned July 1988, following waves of antigovernment riots. Rioting and street violence continued, and in Sept. the military seized power, under Gen. Saw Maung. In 1989 the country's name was changed to Myanmar.

The first free multiparty elections in 30 years took place May 27, 1990, with the main opposition party winning a decisive victory, but the military refused to hand over power. A key opposition leader, Aung San Suu Kyi, awarded the Nobel Peace Prize in 1991, was held under house arrest from July 20, 1989, to July 10, 1995; after her release, the military government continued to restrict her activities and to harass and imprison her supporters. New U.S. economic sanctions took effect on May 21, 1997. Myanmar was admitted to ASEAN July 23, 1997.

Namibia
Republic of Namibia

People: Population: 1,648,270. **Age distrib.** (%): <15: 43.9; 65+: 3.9. **Pop density:** 5 per sq. mi. **Urban:** 37%. **Ethnic groups:** Ovambo 50%, Kavangos 9%, Herero 7%, Damara 7%. **Principal languages:** Afrikaans, English (official), German, indigenous languages. **Chief religions:** Lutheran 50%, other Christian 30%.

Geography: Area: 318,695 sq. mi. **Location:** In S Africa on the coast of the Atlantic Ocean. **Neighbors:** Angola on N, Botswana on E, South Africa on S. **Capital:** Windhoek (1995 est.): 190,000.

Government: Type: Republic. **Head of state:** Pres. Sam Nujoma; b May 12, 1929; in office: Mar. 21, 1990. **Head of gov.:** Prime Min. Hage Geingob; b Aug. 3, 1941; in office: Mar. 21, 1990. **Local divisions:** 13 regions. **Defense:** 3.5% of GDP. **Active troops:** 5,800.

Economy: Mining accounts for 20% of GDP. **Minerals:** Diamonds, copper, gold, tin, lead, uranium. **Arable land:** 1%. **Livestock** (1997): chickens: 2.30 mil; sheep: 2.09 mil; cattle: 2.19 mil; goats: 1.71 mil. **Fish catch** (1997): 291,119 metric tons. **Labor force:** 49% agric., 25% ind. & commerce.

Finance: Monetary unit: Rand (Sept. 1999: 6.03 = $1 U.S.). **GDP:** (1996 est.): $6.2 bil. **Per capital GDP:** $3,700. **Imports** (1996 est.): $1.55 bil; partners: South Africa 85%. **Exports** (1996 est.): $1.45 bil; partners: UK 34%, South Africa 27%. **Tourism:** $339 mil. **Budget** (FY 1996-97 est.): $1.2 bil. **Intl. reserves less gold** (May 1999): $261.31 mil. **Consumer prices** (change in 1997): 8.8%.

Transport: Railroad: Length: 1,480 mi. **Motor vehicles:** 62,500 pass. cars, 66,500 comm. vehicles. **Civil aviation:** 563.0 mil pass.-mi; 11 airports. **Chief ports:** Luderitz, Walvis Bay.

Communications: TV sets: 27.6 per 1,000 pop. **Radios:** 152 per 1,000 pop. **Telephones:** 100,600 main lines. **Daily newspaper circ.:** 27.4 per 1,000 pop.

Health: Life expectancy: 41.64 male; 40.87 female. **Births** (per 1,000 pop.): 35.63. **Deaths** (per 1,000 pop.): 19.92. **Natural inc.:** 1.571%. **Physicians** (1992): 1 per 4,594 persons. **Infant mortality** (per 1,000 live births): 65.94.

Education: Compulsory: ages 6-16. **Literacy** (1993): 76%.

Major Intl. Organizations: UN (FAO, IBRD, ILO, IMF, IMO, WHO, WTrO), the Commonwealth, OAU.
Embassy: 1605 New Hampshire Ave. NW 20009; 986-0540.

Namibia was declared a German protectorate in 1890 and officially called South-West Africa. South Africa seized the territory from Germany in 1915 during World War I; the League of Nations gave South Africa a mandate over the territory in 1920. In 1966, the Marxist South-West Africa People's Organization (SWAPO) launched a guerrilla war for independence. The UN General Assembly named the area Namibia in 1968.

After many years of guerrilla warfare and failed diplomatic efforts, South Africa, Angola, and Cuba signed a U.S.-mediated agreement Dec. 22, 1988, to end South African administration of Namibia and provide for a cease-fire and transition to independence, in accordance with a 1978 UN plan. A separate accord between Cuba and Angola provided for a phased withdrawal of Cuban troops from Namibia. A constitution providing for multiparty government was adopted Feb. 9, 1990, and Namibia gained independence Mar. 21.

Walvis Bay, the principal deepwater port, had been turned over to South African administration in 1922. It remained in South African hands after independence, but South Africa turned control of the port back to Namibia, as of Mar. 1, 1994. Separatist violence flared in the Caprivi Strip region in the late 1990s.

Nauru
Republic of Nauru

People: Population: 10,605. **Pop. density:** 1,326 per sq. mi. **Urban:** 100%. **Ethnic groups:** Nauruan 58%, other Pacific Islander 26%, Chinese 8%, European 8%. **Principal languages:** Nauruan (official), English. **Chief religion:** Predominantly Christian.
Geography: Area: 8 sq. mi. **Location:** In W Pacific O. just S of the Equator. **Neighbors:** Nearest is Kiribati to E. **Topography:** Mostly a plateau bearing high-grade phosphate deposits, surrounded by a sandy shore and coral reef in concentric rings. **Capital:** Govt. offices in Yaren district.
Government: Type: Republic. **Head of state and gov.:** Pres. Rene Harris; in office: Apr. 26, 1999. **Local divisions:** 14 districts.
Economy: Industries: Phosphate mining. **Minerals:** Phosphates. **Electricity prod.** (1997): 32 mil kWh.
Finance: Monetary unit: Australian Dollar (Sept. 1999: 1.55 = $1 U.S.). **GDP:** (1993 est.): $100 mil. **GDP per capita:** $10,000. **Budget** (FY1995-96): $64.8 mil.
Transport: Civil aviation: 151.0 mil passenger-mi. **Chief port:** Nauru.
Communications: Radios: 385 per 1,000 pop.
Health: Life expectancy: 64.30 male; 69.18 female. **Births** (per 1,000 pop.): 18.03. **Deaths** (per 1,000 pop.): 5.10. **Natural inc.:** 1.290%. **Infant mortality** (per 1,000 live births): 40.60.
Education: Free, compulsory: ages 6-16. **Literacy:** 99%.
Major Intl. Organizations: UN (WHO), the Commonwealth.

The island was discovered in 1798 by the British but was formally annexed to the German Empire in 1886. After World War I, Nauru became a League of Nations mandate administered by Australia. During World War II the Japanese occupied the island and shipped 1,200 Nauruans to the fortress island of Truk as slave laborers.

In 1947 Nauru was made a UN trust territory, administered by Australia. It became an independent republic Jan. 31, 1968, and was admitted to the UN Sept. 14, 1999.

Phosphate exports have provided Nauru with per capita revenues that are among the highest in the Third World. Phosphate reserves, however, are expected to be depleted by 2000, and environmental damage from strip-mining has been severe.

Nepal
Kingdom of Nepal

People: Population: 24,302,653. **Age distrib.** (%): <15: 41.4; 65+: 3.4. **Pop. density:** 447 per sq. mi. **Urban:** 11%. **Ethnic groups:** Newars, Indians, Tibetans, Gunings, Sherpas, others. **Principal languages:** Nepali (official), many dialects. **Chief religions:** Hindu (official) 90%, Buddhist 5%, Muslim 3%.
Geography: Area: 54,400 sq. mi. **Location:** Astride the Himalaya Mts. **Neighbors:** China on N, India on S. **Topography:** The Himalayas stretch across the N, the hill country with its fertile valleys extends across the center, while the S border region is part of the flat, subtropical Ganges Plain. **Capital:** Kathmandu. **Cities** (1993 met. est.): Kathmandu 535,000; Lalitpur 190,000; Biratnagar 132,000.
Government: Type: Constitutional monarchy. **Head of state:** King Birendra Bir Bikram Shah Dev; b Dec. 28, 1945; in office: Jan. 31, 1972. **Head of gov.:** Prime Min. Krishna Prasad

Bhattarai; b Dec. 22, 1924; in office: May 31, 1999. **Local divisions:** 5 regions subdivided into 14 zones. **Defense:** 0.9% of GDP. **Active troops:** 46,000.
Economy: Industries: Sugar and jute mills, tourism. **Chief crops:** Sugar, rice, grain. **Minerals:** Quartz. **Other resources:** Forests. **Arable land:** 17%. **Livestock** (1997): chickens: 15.80 mil; cattle: 7.02 mil; goats: 6.00 mil; buffalo: 3.40 mil; sheep: 870,000; pigs: 725,000. **Electricity prod.** (1997): 1.340 bil kWh. **Labor force:** 81% agric., 16% services.
Finance: Monetary unit: Rupee (Sept. 1999: 68.23 = $1 U.S.). **GDP:** (1997 est.): $31.1 bil. **Per capita GDP:** $1,370. **Imports** (1997 est.): $1.6 bil; partners: India 41%, Singapore 32%, Japan 16%. **Exports** (1997 est.): $419 mil; partners: Germany 46%, U.S. 36%. **Tourism:** $124 mil. **Budget** (FY 1996-97 est.): $818 mil. **Intl. reserves less gold** (Mar. 1999): $790.5 mil. **Gold:** 153,000 oz t. **Consumer prices** (change in 1997): 10.0%.
Transport: Railroad: Length: 63 mi. **Civil aviation:** 564.3 mil pass.-mi; 24 airports.
Communications: TV sets: 12 per 1,000 pop. **Radios:** 30 per 1,000 pop. **Telephones:** 194,000 main lines. **Daily newspaper circ.:** 8 per 1,000 pop.
Health: Life expectancy: 58.47 male; 58.36 female. **Births** (per 1,000 pop.): 35.32. **Deaths** (per 1,000 pop.): 10.18. **Natural inc.:** 2.514%. **Hosp. beds** (1995): 1 per 6,387 persons. **Physicians** (1995): 1 per 13,777 persons. **Infant mortality** (per 1,000 live births): 73.58.
Education: Free, compulsory: ages 6-11. **Literacy:** 27%.
Major Intl. Organizations: UN (FAO, IBRD, ILO, IMF, IMO, WHO).
Embassy: 2131 Leroy Pl. NW 20008; 667-4550.
Website: http://www.info-nepal.com

Nepal was originally a group of petty principalities, the inhabitants of one of which, the Gurkhas, became dominant about 1769. In 1951 King Tribhubana Bir Bikram, member of the Shah family, ended the system of rule by hereditary premiers of the Ranas family, who had kept the kings virtual prisoners, and established a cabinet system of government.

Virtually closed to the outside world for centuries, Nepal is now linked to India and Pakistan by roads and air service and to Tibet by road. Polygamy, child marriage, and the caste system were officially abolished in 1963.

The government announced the legalization of political parties in 1990. Elections on Nov. 15, 1994, led to the installation of Nepal's first Communist government, which held power until a no-confidence vote Sept. 10, 1995.

Netherlands
Kingdom of the Netherlands

People: Population: 15,807,641. **Age distrib.** (%): <15: 18.3; 65+: 13.6. **Pop. density:** 986 per sq. mi. **Urban:** 89%. **Ethnic groups:** Dutch 96%. **Principal language:** Dutch (official). **Chief religions:** Roman Catholic 34%, Protestant 25%.
Geography: Area: 16,033 sq. mi. **Location:** In NW Europe on North Sea. **Neighbors:** Germany on E, Belgium on S. **Topography:** The land is flat, an average alt. of 37 ft. above sea level, with much land below sea level reclaimed and protected by some 1,500 miles of dikes. Since 1920 the government has been draining the IJsselmeer, formerly the Zuider Zee. **Capital:** Amsterdam. **Cities** (1996 est.): Amsterdam 716,576; Rotterdam 591,366; The Hague 442,503.
Government: Type: Parliamentary democracy under a constitutional monarch. **Head of state:** Queen Beatrix; b Jan. 31, 1938; in office: Apr. 30, 1980. **Head of gov.:** Prime Min. Wim Kok; b Sept. 29, 1938; in office: Aug. 22, 1994. **Seat of govt.:** The Hague. **Local divisions:** 12 provinces. **Defense:** 1.9% of GDP. **Active troops:** 57,200.
Economy: Industries: Metals, machinery, chemicals, oil refining, diamond cutting, microelectronics, tourism. **Chief crops:** Grains, potatoes, sugar beets, vegetables, fruits, flowers. **Minerals:** Natural gas, oil. **Crude oil reserves** (1999): 125.8 mil bbls. **Arable land:** 27%. **Livestock** (1997): chickens: 98.69; pigs: 11.44 mil; cattle: 4.29 mil; sheep: 1.67 mil. **Fish catch** (1997): 451,799 metric tons. **Electricity prod.** (1997): 81.151 bil kWh. **Labor force:** 75% services; 23% manuf. & constr.; 2% agric.
Finance: Monetary unit: Guilder (Sept. 1999: 2.06 = $1 U.S.). Euro (Sept. 1999: 1.07 = $1 U.S.). **GDP:** (1997 est.): $343.9 bil. **Per capita GDP:** $22,000. **Imports** (1997): $1.791 tril; partners: Germany 22%, Belgium-Lux. 11%, UK 10%. **Exports** (1997): $203.1 bil; partners: Germany 29%, Belgium-Lux. 13%, UK 10%. **Tourism:** $5.75 bil. **Budget** (1998 est.): $112.5 bil. **Intl. reserves less gold** (June 1999): $10.80 bil. **Gold:** 32.53 mil oz t. **Consumer prices** (change in 1997): 2.0%.
Transport: Railroad: Length: 1,702 mi. **Motor vehicles** (1997): 5.81 mil pass. cars, 715,000 comm. vehicles. **Civil avi-

ation: 41.4 bil pass.-mi; 6 airports. **Chief ports:** Rotterdam, Amsterdam, IJmuiden.

Communications: TV sets: 495 per 1,000 pop. **Radios:** 877 per 1,000 pop. **Telephones:** 9,337,000 main lines. **Daily newspaper circ.:** 299 per 1,000 pop.

Health: Life expectancy: 75.28 male; 81.17 female. **Births** (per 1,000 pop.): 11.36. **Deaths** (per 1,000 pop.): 8.69. **Natural inc.:** 0.267%. **Hosp. beds** (1995): 1 per 181 persons. **Physicians** (1995): 1 per 412 persons. **Infant mortality** (per 1,000 live births): 5.11.

Education: Compulsory: ages 5-18. **Literacy:** 100%.

Major Intl. Organizations: UN and all of its specialized agencies, EU, NATO, OECD, OSCE.

Embassy: 4200 Linnean Ave. NW 20008; 244-5300.

Website: http://www.cbs.nl/enindex.htm

Julius Caesar conquered the region in 55 BC, when it was inhabited by Celtic and Germanic tribes.

After the empire of Charlemagne fell apart, the Netherlands (Holland, Belgium, Flanders) split among counts, dukes, and bishops, passed to Burgundy and thence to Charles V of Spain. His son, Philip II, tried to check the Dutch drive toward political freedom and Protestantism (1568-1573). William the Silent, prince of Orange, led a confederation of the northern provinces, called Estates, in the Union of Utrecht, 1579. The Estates retained individual sovereignty, but were represented jointly in the States-General, a body that had control of foreign affairs and defense. In 1581 they repudiated allegiance to Spain. The rise of the Dutch republic to naval, economic, and artistic eminence came in the 17th century.

The United Dutch Republic ended 1795 when the French formed the Batavian Republic. Napoleon made his brother Louis king of Holland, 1806; Louis abdicated 1810 when Napoleon annexed Holland. In 1813 the French were expelled. In 1815 the Congress of Vienna formed a kingdom of the Netherlands, including Belgium, under William I. In 1830, the Belgians seceded and formed a separate kingdom.

The constitution, promulgated 1814, and subsequently revised, provides for a hereditary constitutional monarchy.

The Netherlands maintained its neutrality in World War I, but was invaded and brutally occupied by Germany, 1940-45.

In 1949, after several years of fighting, the Netherlands granted independence to Indonesia. In 1963, West New Guinea (now Irian Jaya) was turned over to Indonesia. Immigration from former Dutch colonies has been substantial.

Although the Netherlands is heavily industrialized, its small farms export large quantities of pork and dairy foods. Rotterdam, located along the principal mouth of the Rhine, is one of the world's leading cargo ports. Canals, extending over 3,400 miles, are important in transportation.

Netherlands Dependencies

The **Netherlands Antilles,** constitutionally on a level of equality with the Netherlands homeland within the kingdom, consist of 2 groups of islands in the West Indies. **Curaçao** and **Bonaire** are near the coast of Venezuela; **St. Eustatius, Saba,** and the southern part of **St. Maarten** are SE of Puerto Rico. The northern two-thirds of St. Maarten belongs to French Guadeloupe; the French call the island St. Martin. Total area of the 2 groups is 309 sq. mi., including Bonaire (111), Curaçao (171), St. Eustatius (8), Saba (5), St. Maarten (Dutch part) (13). St. Maarten suffered extensive damage from Hurricane Luis, Sept. 1995. Total pop. of the Netherlands Antilles (1999 est.) was 207,827. Willemstad, on Curaçao, is the capital. The principal industry is the refining of crude oil from Venezuela. Tourism is also an important industry, as is shipbuilding.

Aruba, about 26 mi. W of Curaçao, was separated from the Netherlands Antilles on Jan. 1, 1986; it is an autonomous member of the Netherlands, the same status as the Netherland Antilles. Area 75 sq. mi.; pop. (1999 est.) 68,675; capital Oranjestad. Chief industries are oil refining and tourism.

New Zealand

People: Population: 3,662,265. **Age distrib.** (%): <15: 22.9; 65+: 11.6. **Pop. density:** 35 per sq. mi. **Urban:** 86%. **Ethnic groups:** European 79%, Maori 15%. **Principal languages:** English (official), Maori. **Chief religions:** Anglican 24%, Presbyterian 18%, Roman Catholic 15%.

Geography: Area: 103,700 sq. mi. **Location:** In SW Pacific O. **Neighbors:** Nearest are Australia on W, Fiji and Tonga on N. **Topography:** Each of the 2 main islands (North and South Isls.) is mainly hilly and mountainous. The east coasts consist of fertile plains, especially the broad Canterbury Plains on South Isl. A volcanic plateau is in center of North Isl. South Isl. has glaciers and 15 peaks over 10,000 ft. **Capital:** Wellington. **Cities** (1996 cen.): Auckland 997,940; Wellington 335,468; Christchurch 331,443.

Government: Type: Parliamentary democracy. **Head of state:** Queen Elizabeth II, represented by Gov.-Gen. Sir Michael Hardie Boys; b Oct. 6, 1931; in office: Mar. 21, 1996. **Head of gov.:** Prime Min. Jenny Shipley; b 1952; in office: Dec. 8, 1997. **Local divisions:** 16 regions subdivided into 15 cities and 59 districts. **Defense:** 1.6% of GDP. **Active troops:** 9,600.

Economy: Industries: Food processing, textiles, machinery, fish, forest prods. **Chief crops:** Grains, potatoes, fruits. **Minerals:** Gold, gas, iron, coal. **Crude oil reserves** (1999): 127 mil bbls. **Other resources:** Wool, timber. **Arable land:** 9%. **Livestock** (1997): chickens: 12.00 mil; sheep: 47.60 mil; cattle: 8.77 mil; pigs: 340,000; goats: 227,000. **Fish catch** (1997): 596,017 metric tons. **Electricity prod.** (1997): 34.840 bil kWh. **Labor force:** 65% services, 25% ind.; 10% agric.

Finance: Monetary unit: N.Z. Dollar (Sept. 1999: 1.93 = $1 U.S.). **GDP:** (1997 est.): $63.4*bil. **Per capita GDP:** $17,700. **Imports** (1997 est.): $19.2 bil; partners: Australia 21%, U.S. 18%, Japan 16%. **Exports** (1997 est.): $18.5 bil; partners: Australia 19%, Japan 15%; U.K. 15%; U.S. 12%. **Tourism:** $1.88 bil. **Budget** (FY 1995-96 est.): $21.8 bil. **Intl. reserves less gold** (May 1999): $4.13 bil. **Consumer prices** (change in 1997): 1.2%.

Transport: Railroad: Length: 2,433 mi. **Motor vehicles** (1997): 1.50 mil pass. cars; 322,889 comm. vehicles. **Civil aviation:** 13.0 bil pass.-mi; 36 airports. **Chief ports:** Auckland, Christchurch, Wellington, Dunedin, Tauranga.

Communications: TV sets: 514 per 1,000 pop. **Radios:** 997 per 1,000 pop. **Telephones:** 1,870,000 main lines. **Daily newspaper circ.:** 239 per 1,000 pop.

Health: Life expectancy: 74.55 male; 81.27 female. **Births** (per 1,000 pop.): 14.42. **Deaths** (per 1,000 pop.): 7.53. **Natural inc.:** 0.689%. **Hosp. beds** (1996): 1 per 164 persons. **Physicians** (1996): 1 per 318 persons. **Infant mortality** (per 1,000 live births): 6.22.

Education: Free, compulsory: ages 6-16. **Literacy** (1997): 100%.

Major Intl. Organizations: UN (FAO, IBRD, ILO, IMF, IMO, WHO, WTrO), APEC, the Commonwealth, OECD.

Embassy: 37 Observatory Cir. NW 20008; 328-4800.

Website: http://www.stats.govt.nz/statsweb.nsf

The Maoris, a Polynesian group from the eastern Pacific, reached New Zealand before and during the 14th century. The first European to sight New Zealand was Dutch navigator Abel Janszoon Tasman, but Maoris refused to allow him to land. British Capt. James Cook explored the coasts, 1769-1770.

British sovereignty was proclaimed in 1840, with organized settlement beginning in the same year. Representative institutions were granted in 1853. Maori Wars ended in 1870 with British victory. The colony became a dominion in 1907, and is an independent member of the Commonwealth.

A progressive tradition in politics dates back to the 19th century, when New Zealand was internationally known for social experimentation; much of the nation's economy has been deregulated in recent years. The National Party, led by Jim Bolger, won general elections in 1990 and 1993. After inconclusive elections, Oct. 12, 1996, Bolger remained as prime minister, heading a National/New Zealand First party coalition. When Bolger lost his party's support, Jenny Shipley became the nation's first female prime minister, Dec. 8, 1997.

The native Maoris number about 550,000. Six of 120 members of the House of Representatives are elected directly by the Maori people.

New Zealand comprises **North Island,** 44,702 sq. mi.; **South Island,** 58,384 sq. mi.; **Stewart Island,** 674 sq. mi.; **Chatham Islands,** 372 sq. mi.; and several groups of smaller islands.

In 1965, the **Cook Islands** (pop., 1998 est., 19,989; area 93 sq. mi.), located halfway between New Zealand and Hawaii, became self-governing although New Zealand retains responsibility for defense and foreign affairs. **Niue** attained the same status in 1974; it lies 400 mi. to W (pop., 1995 est., 1,800; area 100 sq. mi.). **Tokelau** (pop., 1995 est., 1,500; area 4 sq. mi.) comprises 3 atolls 300 mi. N of Samoa.

Ross Dependency, administered by New Zealand since 1923, comprises 160,000 sq. mi. of Antarctic territory.

Nicaragua
Republic of Nicaragua

People: Population: 4,717,132. **Age distrib.** (%): <15: 43.6; 65+: 2.7. **Pop. density:** 94 per sq. mi. **Urban:** 63%. **Ethnic groups:** Mestizo 69%, white 17%, black 9%, Amerindian 5%. **Principal languages:** Spanish (official). **Chief religion:** Roman Catholic 95%.

Geography: Area: 49,998 sq. mi. **Location:** In Central America. **Neighbors:** Honduras on N, Costa Rica on S. **Topography:** Both Caribbean and Pacific coasts are over 200 mi. long. The Cordillera Mts., with many volcanic peaks, run NW-

SE through the middle of the country. Between this and a volcanic range to the E lie Lakes Managua and Nicaragua. **Capital:** Managua 1,124,000*.

Government: Type: Republic. **Head of state and gov.:** Pres. Arnoldo Alemán Lacayo; b Jan. 23, 1946; in office Jan. 10, 1997. **Local divisions:** 15 departments, 2 autonomous regions. **Defense:** 1.4% of GDP. **Active troops:** 17,000.

Economy: Industries: Oil refining, food processing, chemicals, textiles. **Chief crops:** Bananas, cotton, citrus, coffee, sugar, corn, rice. **Minerals:** Gold, silver, copper, tungsten. **Other resources:** Forests, seafood. **Arable land:** 9%. **Livestock** (1997): chickens: 9.00 mil; cattle: 1.69 mil; pigs: 400,000. **Electricity prod.** (1997): 2.120 bil kWh. **Labor force:** 54% services; 31% agric..

Finance: Monetary unit: Gold Cordoba (Sept. 1999: 12.03 = $1 U.S.). **GDP:** (1997 est.): $9.3 bil. **Per capita GDP:** $2,100. **Imports** (1996): $1.19 bil partners: U.S. 31%, Venezuela 12%. **Exports** (1996): $635 mil; partners: U.S. 38%, Germany 10%. **Tourism:** $90 mil. **Budget** (1996 est.): $551 mil. **Intl. reserves less gold** (May 1999): $479.26 mil. **Consumer prices** (change in 1996): 11.6%.

Transport: Motor vehicles: 72,413 pass. cars, 72,227 comm. vehicles. **Civil aviation:** 52.8 mil pass.-mi; 10 airports. **Chief ports:** Corinto, Puerto Sandino, San Juan del Sur.

Communications: TV sets: 48 per 1,000 pop. **Radios:** 206 per 1,000 pop. **Telephones:** 140,000 main lines. **Daily newspaper circ.:** 31 per 1,000 pop.

Health: Life expectancy: 64.70 male; 69.56 female. **Births** (per 1,000 pop.): 35.04. **Deaths** (per 1,000 pop.): 5.60. **Natural inc.:** 2.944%. **Hosp. beds** (1994): 1 per 914 persons. **Physicians** (1994): 1 per 1,566 persons. **Infant mortality** (per 1,000 live births): 40.47.

Education: Free, compulsory: ages 7-13. **Literacy:** 66%.

Major Intl. Organizations: UN and most of its specialized agencies, OAS.

Embassy: 1627 New Hampshire Ave. NW 20009; 939-6570.

Nicaragua, inhabited by various Indian tribes, was conquered by Spain in 1552. After gaining independence from Spain, 1821, Nicaragua was united for a short period with Mexico, then with the United Provinces of Central America, finally becoming an independent republic, 1838.

U.S. Marines occupied the country at times in the early 20th century, the last time from 1926 to 1933.

Gen. Anastasio Somoza Debayle was elected president in 1967. He resigned in 1972, but was re-elected president in 1974. Martial law was imposed in Dec. 1974, after officials were kidnapped by the Marxist Sandinista guerrillas. Violent opposition spread to nearly all classes in 1978; nationwide strikes called against the government touched off a civil war, which ended when Somoza fled Nicaragua and the Sandinistas took control of Managua. In July 1979. Somoza was assassinated in Paraguay, Sept. 17, 1980.

Relations with the U.S. were strained as a result of Nicaragua's aid to leftist guerrillas in El Salvador and U.S. backing of anti-Sandinista contra guerrilla groups. In 1983 the contras launched a major offensive; the Sandinistas imposed rule by decree. In 1985 the U.S. House rejected Pres. Reagan's request for military aid to the contras. The subsequent diversion of funds to the contras from the proceeds of a secret arms sale to Iran caused a major scandal in the U.S.

In a stunning upset, Violeta Barrios de Chamorro defeated Sandinista leader Daniel Ortega Saavedra in national elections, Feb. 25, 1990. Arnoldo Alemán Lacayo, a conservative former mayor of Managua, defeated Ortega in the presidential election of Oct. 20, 1996. Up to 2,000 people died in W Nicaragua Oct. 30, 1998, in a mudslide caused by rains from Hurricane Mitch.

Niger
Republic of Niger

People: Population: 9,962,242. **Age distrib.** (%): <15: 48.1; 65+: 2.3. **Pop. density:** 20 per sq. mi. **Urban:** 19%. **Ethnic groups:** Hausa 56%, Djerma 22%, Fula 9%, Tuareg 8%. **Principal languages:** French (official), Hausa, Djerma. **Chief religion:** Muslim 80%.

Geography: Area: 489,000 sq. mi. **Location:** In the interior of N Africa. **Neighbors:** Libya, Algeria on N; Mali, Burkina Faso on W; Benin, Nigeria on S; Chad on E. **Topography:** Mostly arid desert and mountains. A narrow savanna in the S and the Niger R. basin in the SW contain most of the population. **Capital:** Niamey (1994 est.): 420,000.

Government: Type: In transition. **Head of state:** Daouda Malam Wanké; b 1954; in office: Apr. 11, 1999. **Head of gov.:** Prime Min. Ibrahim Hassane Mayaki; in office: Nov. 27, 1997. **Local divisions:** 7 departments, 1 capital district. **Defense:** 1.4% of GDP. **Active troops:** 5,300.

Economy: Chief crops: Peanuts, cowpeas, cotton. **Miner-**

als: Uranium, coal, iron. **Arable land:** 3%. **Livestock** (1997): chickens: 20.00 mil; goats: 6.15 mil; sheep: 4.10 mil; cattle: 2.10 mil. **Electricity prod.** (1997): 170 mil kWh. **Labor force:** 90% agric.

Finance: Monetary unit: CFA Franc (Sept. 1999: 612.79 = $1 U.S.). **GDP:** (1997 est.): $6.3 bil. **Per capita GDP:** $670. **Imports** (1996): $374 mil; partners: France 24%, Nigeria 19%. **Exports** (1996): $188 mil; partners: France 41%, Nigeria 22%. **Tourism:** $18 mil. **Budget** (1998 est.): $370 mil. **Intl. reserves less gold** (Dec. 1998): $53.1 mil. **Gold:** 11,000 oz t. **Consumer prices** (change in 1997): 4.5%.

Transport: Motor vehicles: 37,500 pass. cars, 14,100 comm. vehicles. **Civil aviation:** 150.5 mil pass.-mi; 6 airports.

Communications: TV sets: 2.8 per 1,000 pop. **Radios:** 48 per 1,000 pop. **Telephones:** 18,100 mains lines.

Health: Life expectancy: 42.22 male; 41.70 female. **Births** (per 1,000 pop.): 41.84. **Deaths** (per 1,000 pop.): 22.78. **Natural inc.:** 2.953%. **Physicians** (1993): 1 per 35,141 persons. **Infant mortality** (per 1,000 live births): 112.79.

Education: Free, compulsory: ages 7-15. **Literacy:** 14%.

Major Intl. Organizations: UN (FAO, IBRD, ILO, IMF, WHO, WTrO), OAU.

Embassy: 2204 R St. NW 20008; 483-4224.

Niger was part of ancient and medieval African empires. European explorers reached the area in the late 18th century. The French colony of Niger was established 1900-22, after the defeat of Tuareg fighters, who had invaded the area from the N a century before. The country became independent Aug. 3, 1960. The next year it signed a bilateral agreement with France.

In 1993, Niger held its first free and open elections since independence; an opposition leader, Mahamane Ousmane, won the presidency. A peace accord Apr. 24, 1995, ended a Tuareg rebellion that began in 1990. A coup, Jan. 27, 1996, followed by a disputed presidential election in July, left the military in control of Niger. On Apr. 9, 1999, Gen. Ibrahim Bare Mainassara, Niger's president since 1996, was assassinated, apparently by members of his security team. Voters approved, July 18, a constitution that would restore civilian rule and grant amnesty to coup leaders.

Nigeria
Federal Republic of Nigeria

People: Population: 113,828,587. **Age distrib.** (%): <15: 44.8; 65+: 2.9. **Pop. density:** 319 per sq. mi. **Urban:** 40%. **Ethnic groups:** Hausa, Yoruba, Ibo, Fulani, others. **Principal languages:** English (official), Hausa, Yoruba, Ibo. **Chief religions:** Muslim (in N) 50%, Christian (in S) 40%.

Geography: Area: 356,700 sq. mi. **Location:** On the S coast of W Africa. **Neighbors:** Benin on W, Niger on N, Chad and Cameroon on E. **Topography:** 4 E-W regions divide Nigeria: a coastal mangrove swamp 10-60 mi. wide, a tropical rain forest 50-100 mi. wide, a plateau of savanna and open woodland, and semidesert in the N. **Capital:** Abuja. **Cities:** Lagos 10,287,000; Ibadan 1,484,000*.

Government: Type: Republic. **Head of state and gov.:** Pres. Olusegun Obasanjo; b Mar. 6, 1935; in office: May 29, 1999. **Local divisions:** 36 states, 1 capital territory. **Defense:** 4.0% of GDP. **Active troops:** 77,000.

Economy: Industries: Crude oil (98% of exports; Africa's leading producer), mining, food processing, textiles. **Chief crops:** Cocoa (main export crop), palm products, corn, rice, yams, cassava. **Minerals:** Oil, gas, lead, zinc, coal, iron, limestone, columbite, tin. **Crude oil reserves** (1999): 22.5 bil bbls. **Other resources:** Timber, rubber, hides. **Arable land:** 33%. **Livestock** (1997): chickens: 126.00 mil; goats: 24.50 mil; cattle: 19.61 mil; sheep: 14.00 mil; pigs: 7.60 mil. **Fish catch** (1997): 365,735 metric tons. **Electricity prod.** (1997): 13.575 bil kWh. **Labor force:** 54% agric.; 19% ind., commerce, serv.; 15% govt.

Finance: Monetary unit: Naira (Sept. 1999: 97.85 = $1 U.S.). **GDP:** (1996 est.): $132.7 bil. **Per capita GDP:** $1,300. **Imports** (1996): $8 bil; partners: EU 50%, U.S. 12%. **Exports** (1996): $15 bil; partners: U.S. 40%, EU 21%. **Tourism:** $124 mil. **Budget** (1998 est.): $13.9 bil. **Intl. reserves less gold** (Dec. 1996): $4.08 bil. **Gold:** 687,000 oz t. **Consumer prices** (change in 1997): 10.3%.

Transport: Railroad: Length: 2,178 mi. **Motor vehicles:** 589,600 pass. cars, 363,900 comm. vehicles. **Civil aviation:** 137.2 mil pass.-mi; 12 airports. **Chief ports:** Port Harcourt, Lagos, Warri, Calabar.

Communications: TV sets: 38 per 1,000 pop. **Radios:** 170 per 1,000 pop. **Telephones** (1997): 412,800 main lines. **Daily newspaper circ.:** 18 per 1,000 pop.

Health: Life expectancy: 52.55 male; 54.06 female. **Births** (per 1,000 pop.): 41.84. **Deaths** (per 1,000 pop.): 12.98. **Natural inc.:** 2.886%. **Hosp. beds** (1994): 1 per 1,070 persons.

Physicians (1994): 1 per 4,496 persons. **Infant mortality** (per 1,000 live births): 69.46.

Education: Free, compulsory: ages 6-15. **Literacy**: 57%.

Major Intl. Organizations: UN (FAO, IBRD, ILO, IMF, IMO, WHO, WTrO), the Commonwealth, OAU, OPEC.

Embassy: 1333 16th St. NW 20036; 986-8400.

Early cultures in Nigeria date back to at least 700 BC. From the 12th to the 14th centuries, more advanced cultures developed in the Yoruba area, at Ife, and in the north, where Muslim influence prevailed.

Portuguese and British slavers appeared from the 15th-16th centuries. Britain seized Lagos, 1861, and gradually extended control inland until 1900. Nigeria became independent Oct. 1, 1960, and a republic Oct. 1, 1963.

On May 30, 1967, the Eastern Region seceded, proclaiming itself the Republic of Biafra, plunging the country into civil war. Casualties in the war were estimated at over 1 million, including many "Biafrans" (mostly Ibos) who died of starvation despite international efforts to provide relief. The secessionists, after steadily losing ground, capitulated Jan. 12, 1970.

Nigeria emerged as one of the world's leading oil exporters in the 1970s, but much of the revenue has been squandered through corruption and mismanagement.

After 13 years of military rule, the nation made a peaceful return to civilian government, Oct. 1979. Military rule resumed, Dec. 31, 1983; a second coup came in 1985.

Headed by Gen. Ibrahim Babangida, the military regime held elections June 12, 1993, but annulled the vote June 23 when it appeared that Moshood Abiola would win. Riots followed and many were killed. Babangida resigned and appointed a civilian to head an interim government, Aug. 26, but that government was ousted Nov. 17 in a coup led by Gen. Sani Abacha. On June 11, 1994, Abiola declared himself president; he was jailed June 23.

Abacha's brutal rule ended June 8, 1998, when he died of an apparent heart attack. Abiola died in prison July 7, as Abacha's successor, Gen. Abdulsalam Abubakar, was reportedly preparing to free him. Abiola's death (also apparently of natural causes) sparked riots in Lagos and other cities; on July 20, Abubakar promised early elections and a return to civilian rule. Olusegun Obasanjo (a former military ruler) won the presidential vote Feb. 27, 1999, to become the head of Nigeria's 1st civilian government in 15 years.

An oil fire caused by a ruptured pipeline in S. Nigeria, Oct. 17, 1998, killed at least 700 people who were scavenging for fuel.

Norway
Kingdom of Norway

People: Population: 4,438,547. **Age distrib.** (%): <15: 19.6; 65+: 15.5. **Pop. density:** 35 per sq. mi. **Urban:** 73%. **Ethnic groups:** Germanic (Nordic, Alpine, Baltic), Lapps (minority). **Principal languages:** Norwegian (official). **Chief religion:** Evangelical Lutheran 88%.

Geography: Area: 125,200 sq. mi. **Location:** W part of Scandinavian peninsula in NW Europe (extends farther north than any European land). **Neighbors:** Sweden, Finland, Russia on E. **Topography:** A highly indented coast is lined with tens of thousands of islands. Mountains and plateaus cover most of the country, which is only 25% forested. **Capital:** Oslo. **Cities** (1996 est.): Oslo 491,726; Bergen 223,773.

Government: Type: Hereditary constitutional monarchy. **Head of state:** King Harald V; b Feb. 21, 1937; in office: Jan. 17, 1991. **Head of gov.:** Prime Min. Kjell Magne Bondevik; b Sept. 3, 1947; in office: Oct. 17, 1997. **Local divisions:** 19 provinces. **Defense:** 2.3% of GDP. **Active troops:** 33,600.

Economy: Industries: Wood & paper prods., shipbuilding, metals, chemicals, food processing, fish, oil, gas. **Chief crops:** Grains, oats. **Minerals:** Oil, gas, copper, pyrites, nickel, iron, zinc, lead. **Crude oil reserves** (1999): 10.91 bil bbls. **Other resources:** Fish, livestock. **Arable land:** 3%. **Livestock** (1997): chickens: 3.24 mil; sheep: 2.45 mil; cattle: 1.02 mil.; pigs: 770,000. **Fish catch** (1997): 2.86 mil metric tons. **Electricity prod.** (1997): 109.339 bil kWh. **Labor force:** 71% services; 23% industry.

Finance: Monetary unit: Krone (Sept. 1999: 7.81 = $1 U.S.). **GDP:** (1997 est.): $120.5 bil. **Per capita GDP:** $27,400. **Imports** (1996): $35.1 bil; partners: EU 71%. **Exports** (1996): $49.3 bil; partners: EU 77%. **Tourism:** $2.21 bil. **Budget** (1994 est.): $53 bil. **Intl. reserves less gold** (June 1999): $16.33 bil. **Gold:** 1.18 mil oz t. **Consumer prices** (change in 1997): 2.3%.

Transport: Railroad: Length: 2,485 mi. **Motor vehicles** (1997): 1.76 mil pass. cars, 412,183 comm. vehicles. **Civil aviation:** 5.7 bil pass.-mi; 50 airports. **Chief ports:** Bergen, Stavanger, Oslo, Kristiansand.

Communications: TV sets: 459 per 1,000 pop. **Radios:** 763 per 1,000 pop. **Telephones** (1997): 2,735,500 main lines. **Daily newspaper circ.:** 498 per 1,000 pop.

Health: Life expectancy: 75.55 male; 81.35 female. **Births** (per 1,000 pop.): 12.54. **Deaths** (per 1,000 pop.): 10.12. **Natural inc.:** 0.242%. **Hosp. beds** (1994): 1 per 197 persons. **Physicians** (1996): 1 per 285 persons. **Infant mortality** (per 1,000 live births): 4.96.

Education: Compulsory: ages 6-16. **Literacy** (1994): 100%.

Major Intl. Organizations: UN and all of its specialized agencies, EFTA, NATO, OECD, OSCE.

Embassy: 2720 34th St. NW 20008; 333-6000.

Website: http://www.ssb.no/www-open/english

The first ruler of Norway was Harald the Fairhaired, who came to power in AD 872. Between 800 and 1000, Norway's Vikings raided and occupied widely dispersed parts of Europe.

The country was united with Denmark 1381-1814, and with Sweden, 1814-1905. In 1905, the country became independent with Prince Charles of Denmark as king.

Norway remained neutral during World War I. Germany attacked Norway Apr. 9, 1940, and held it until liberation May 8, 1945. The country abandoned its neutrality after the war, and joined NATO. In a referendum Nov. 28, 1994, Norwegian voters rejected European Union membership.

Abundant hydroelectric resources provided the base for industrialization, giving Norway one of the highest living standards in the world. The country is a leading producer and exporter of crude oil, with extensive reserves in the North Sea. Norway's merchant marine is one of the world's largest.

Svalbard is a group of mountainous islands in the Arctic O., area 23,957 sq. mi., pop. (1997 est.) 3,231. The largest, Spitsbergen (formerly called West Spitsbergen), 15,060 sq. mi., seat of the governor, is about 370 mi. N of Norway. By a treaty signed in Paris, 1920, major European powers recognized the sovereignty of Norway, which incorporated it in 1925.

Jan Mayen, area 144 sq. mi., is a volcanic island located about 565 mi. WNW of Norway; it was annexed in 1929.

Oman
Sultanate of Oman

People: Population: 2,446,645. **Age distrib.** (%): <15: 40.8; 65+: 2.3. **Pop. density:** 30 per sq. mi. **Urban:** 78%. **Ethnic groups:** Omani Arab 74%, Indian 13%. **Principal languages:** Arabic (official). **Chief religion:** Ibadhi Muslim 75%.

Geography: Area: 82,000 sq. mi. **Location:** On SE coast of Arabian peninsula. **Neighbors:** United Arab Emirates, Saudi Arabia, Yemen on W. **Topography:** Oman has a narrow coastal plain up to 10 mi. wide, a range of barren mountains reaching 9,900 ft., and a wide, stony, mostly waterless plateau, avg. alt. 1,000 ft. Also, an exclave at the tip of the Musandam peninsula controls access to the Persian Gulf. **Capital:** Muscat (1993): 51,969.

Government: Type: Absolute monarchy. **Head of state and gov.:** Sultan Qabus bin Said; b Nov. 18, 1942; in office: July 23, 1970 (also prime min. since Jan. 2, 1972). **Local divisions:** 8 regions subdivided into 59 *wilayats.* **Defense:** 10.9% of GDP. **Active troops:** 43,500.

Economy: Industries: Oil, gas, construction. **Chief crops:** Dates, limes, vegetables, alfalfa, bananas. **Minerals:** Oil (75% of exports). **Livestock** (1997): chickens: 3.00 mil; goats: 725,000; cattle: 146,000; sheep: 155,000. **Crude oil reserves** (1999): 5.28 bil bbls. **Electricity prod.** (1997): 9.0 bil kWh. **Labor force:** 37% agric.

Finance: Monetary unit: Rial Omani (Sept. 1999: 0.39 = $1 U.S.). **GDP:** (1997 est.): $17.2 bil. **Per capita GDP:** $8,000. **Imports** (1997 est.): $4.8 bil; partners: UAE 22%, Japan 15%, UK 15%. **Exports** (1997 est.): $7.6 bil; partners: Japan 29%, South Korea 17%. **Tourism:** $112 mil. **Budget** (1998 est.): $6 bil. **Intl. reserves less gold** (June 1999): $1.50 bil. **Gold:** 291,000 oz t.

Transport: Motor vehicles (1997): 246,097 pass. cars, 101,223 comm. vehicles. **Civil aviation:** 2.0 bil pass.-mi; 6 airports. **Chief ports:** Matrah, Mina' al Fahl.

Communications: TV sets: 711 per 1,000 pop. **Radios:** 426 per 1,000 pop. **Telephones:** 220,000 main lines. **Daily newspaper circ.:** 31 per 1,000 pop.

Health: Life expectancy: 69.31 male; 73.39 female. **Births** (per 1,000 pop.): 37.98. **Deaths** (per 1,000 pop.): 4.29. **Natural inc.:** 3.369%. **Hosp. beds** (1995): 1 per 478 persons. **Physicians** (1995): 1 per 852 persons. **Infant mortality** (per 1,000 live births): 24.71.

Education: Literacy (1993): 59%.

Major Intl. Organizations: UN (FAO, IBRD, ILO, IMF, IMO, WHO), AL.

Embassy: 2535 Belmont Rd. NW 20008; 387-1980.

Oman was originally called Muscat and Oman. A long history of rule by other lands, including Portugal in the 16th century, ended with the ouster of the Persians in 1744. By the early 19th century, Muscat and Oman was one of the most important countries in the region, controlling much of the Persian and Pakistan coasts, and also ruling far-away Zanzibar, which was separated in 1861 under British mediation.

British influence was confirmed in a 1951 treaty, and Britain helped suppress an uprising by traditionally rebellious interior tribes against control by Muscat in the 1950s.

On July 23, 1970, Sultan Said bin Taimur was overthrown by his son, who changed the nation's name to Sultanate of Oman.

Oil is the major source of income.

Oman opened its air bases to Western forces following the Iraqi invasion of Kuwait on Aug. 2, 1990.

Pakistan
Islamic Republic of Pakistan

People: Population: 138,123,359. **Age distrib.** (%): <15: 41.4; 65+: 4.1. **Pop. density:** 445 per sq. mi. **Urban:** 35%. **Ethnic groups:** Punjabi, Sindhi, Pashtun (Pathan), Baloch. **Principal languages:** Urdu, English (both official), Punjabi, Sindhi, Pashtu. **Chief religions:** Sunni Muslim 77%, Shi'a Muslim 20%.

Geography: Area: 310,400 sq. mi. **Location:** In W part of South Asia. **Neighbors:** Iran on W, Afghanistan and China on N, India on E. **Topography:** The Indus R. rises in the Hindu Kush and Himalaya Mts. in the N (highest is K2, or Godwin Austen, 28,250 ft., 2d highest in world), then flows over 1,000 mi. through fertile valley and empties into Arabian Sea. Thar Desert, Eastern Plains flank Indus Valley. **Capital:** Islamabad. **Cities:** Karachi 9,733,000; Lahore 5,012,000; Faisalabad 1,845,000*.

Government: Type: In transition. **Head of state:** Pres. Muhammad Rafiq Tarar; b Nov. 2, 1929; in office: Jan. 1, 1998. **Head of gov.:** Gen. Pervez Musharraf; in office: Oct. 15, 1999 (de facto from Oct. 12). **Local divisions:** 4 provinces and 1 capital territory, plus federally administered tribal areas. **Defense:** 5.8% of GDP. **Active troops:** 587,000.

Economy: Industries: Textiles, food processing, beverages. **Chief crops:** Rice, wheat, cotton. **Minerals:** Natural gas. **Crude oil reserves** (1999): 208 mil bbls. **Arable land:** 23%. **Livestock** (1997): chickens: 200.00 mil; goats: 48.57 mil; sheep: 32.38 mil; buffalo: 21.21 mil; cattle: 18.00 mil. **Fish catch** (1997): 575,137 metric tons. **Electricity prod.** (1997): 57.570 bil kWh. **Labor force:** 47% agric.; 17% mining & manuf.; 17% services.

Finance: Monetary unit: Rupee (Sept. 1999: 51.65 = $1 U.S.). **GDP:** (1997 est.): $344 bil. **Per capita GDP:** $2,600. **Imports** (FY1996-97): $11.4 bil; partners: Japan 9%, Malaysia 9%, U.S. 9%. **Exports** (FY1996-97): $8.2 bil; partners: U.S. 17%. **Tourism:** $111 mil. **Budget** (FY 1996-97): $13.6 bil. **Intl. reserves less gold** (May 1999): $1.71 bil. **Gold:** 2.07 mil oz t. **Consumer prices** (change in 1997): 6.2%.

Transport: Railroad: Length: 5,453 mi. **Motor vehicles:** 800,000 pass. cars, 300,000 comm. vehicles. **Civil aviation:** 7.2 bil pass.-mi; 35 airports. **Chief port:** Karachi.

Communications: TV sets: 16 per 1,000 pop. **Radios:** 76 per 1,000 pop. **Telephones** (1997): 2,557,000 main lines. **Daily newspaper circ.:** 22 per 1,000 pop.

Health: Life expectancy: 58.49 male; 60.30 female. **Births** (per 1,000 pop.): 33.51. **Deaths** (per 1,000 pop.): 10.45. **Natural inc.:** 2.306%. **Hosp. beds** (1995): 1 per 1,517 persons. **Physicians** (1995): 1 per 1,863 persons. **Infant mortality** (per 1,000 live births): 91.86.

Education: Literacy: 38%.

Major Intl. Organizations: UN (FAO, IBRD, ILO, IMF, IMO, WHO, WTrO), the Commonwealth (suspended).

Embassy: 2315 Massachusetts Ave. NW 20008; 939-6200. **Website:** http://www.pak.gov.pk

Present-day Pakistan shares the 5,000-year history of the India-Pakistan subcontinent. At present-day Harappa and Mohenjo Daro, the Indus Valley Civilization, with large cities and elaborate irrigation systems, flourished c. 4,000-2,500 BC.

Aryan invaders from the NW conquered the region around 1,500 BC, forging a Hindu civilization that dominated Pakistan as well as India for 2,000 years.

Beginning with the Persians in the 6th century BC, and continuing with Alexander the Great and with the Sassanians, successive nations to the west ruled or influenced Pakistan. The first Arab invasion, AD 712, introduced Islam. Under the Mogul empire (1526-1857), Muslims ruled most of India, yielding to British encroachment and resurgent Hindus.

After World War I the Muslims of British India began agitation for minority rights in elections. Muhammad Ali Jinnah (1876-1948) was the principal architect of Pakistan. A leader of the Muslim League from 1916, he worked for dominion status for India; from 1940 he advocated a separate Muslim state.

When the British withdrew Aug. 14, 1947, the Islamic majority areas of India acquired self-government as Pakistan, with dominion status in the Commonwealth. Pakistan was divided into 2 sections, West Pakistan and East Pakistan. The 2 areas were nearly 1,000 mi. apart on opposite sides of India. Pakistan became a republic in 1956.

In Oct. 1958, Gen. Mohammad Ayub Khan took power in a coup. He was elected president in 1960, reelected in 1965. He resigned Mar. 25, 1969, after several months of violent rioting and unrest, most of it in East Pakistan, which demanded autonomy. The government was turned over to Gen. Agha Mohammad Yahya Khan and martial law was declared.

The Awami League, which sought regional autonomy for East Pakistan, won a majority in Dec. 1970 elections to a constituent assembly. In March 1971 Yahya postponed the assembly. Rioting and strikes broke out in the East.

On Mar. 25, 1971, government troops launched attacks in the East. The Easterners, aided by India, proclaimed the independent nation of Bangladesh. In months of widespread fighting, countless thousands were killed. Some 10 million Easterners fled into India. Full-scale war between India and Pakistan had spread to both the East and West fronts by Dec. 3. Pakistan troops in the East surrendered Dec. 16; Pakistan agreed to a cease-fire in the West Dec. 17. On July 3, 1972, Pakistan and India signed a pact agreeing to withdraw troops from their borders and seek peaceful solutions to all problems.

Zulfikar Ali Bhutto, leader of the Pakistan People's Party, which had won the most West Pakistan votes in Dec. 1970 elections, became president Dec. 20. Bhutto was overthrown in a military coup July 1977. Convicted of complicity in a 1974 political murder, he was executed Apr. 4, 1979. Over 3 million Afghan refugees flooded into Pakistan after the USSR invaded Afghanistan Dec. 1979; over 1.2 million remained in 1999.

Pres. Mohammad Zia ul-Haq was killed when his plane exploded in Aug. 1988. Following Nov. elections, Benazir Bhutto, daughter of Zulfikar Ali Bhutto, was named prime minister, becoming the first woman leader of a Muslim nation. She was accused of corruption and dismissed by the president, Aug. 1990; her party was soundly defeated in Oct. 1990 elections, and Nawaz Sharif became prime minister. She regained power after elections in Oct. 1993. Opposition to Bhutto centered around Karachi, which was crippled by violent strikes and ethnic clashes during 1995 and 1996. Accusing the Bhutto government of corruption and mismanagement, Pres. Farooq Leghari appointed a caretaker prime minister Nov. 5, 1996. Elections on Feb. 3, 1997, gave Sharif a parliamentary majority.

Responding to nuclear weapons tests by India, Pakistan conducted its own tests, May 28-30, 1998; the U.S. imposed economic sanctions on both countries. Tried in absentia, the exiled Bhutto was convicted Apr. 15, 1999, of receiving kickbacks and sentenced to a 5-year prison term.

In mid-1999 Muslim infiltrators, apparently including Pakistani troops, seized Indian-held positions in the disputed territory of Kashmir, which witnessed its heaviest fighting in over 2 decades. After meeting with Pres. Bill Clinton on July 4, Sharif agreed to a Pakistani pullback. Growing conflict between Sharif and the military climaxed in his firing on Oct. 12 of army chief Gen. Pervez Musharraf, whose supporters staged a bloodless coup later that day. Martial law was imposed and the constitution suspended Oct. 15, and the courts were barred from challenging Musharraf's authority. Because of the military takeover, Pakistan was suspended, Oct. 18, from the Commonwealth.

Palau
Republic of Palau

People: Population: 18,467. **Age distrib.** (%): <15: 27.3; 65+: 4.8. **Pop. density:** 104 per sq. mi. **Urban:** 72%. **Ethnic groups:** Polynesian, Malayan, Melanesian. **Principal languages:** English (official), Palauan, Sonsorolese, Angaur, Japanese, Tobi (all official within certain states). **Chief religions:** Roman Catholic, Protestant, Modekngei.

Geography: Area: 177 sq. mi. **Location:** Archipelago (26 islands, more than 300 islets) in the W Pacific Ocean, about 530 mi SE of the Philippines. **Neighbors:** Micronesia to E, Indonesia to S. **Capital:** Koror (1995) 12,000. (Note: a new capital is being built in Babelthuap.)

Government: Type: Republic. **Head of state and gov.:** Pres. Kuniwo Nakamura; b Nov. 24, 1943; in office: Jan. 1, 1993. **Local divisions:** 16 states.

Economy: Industries: Tourism, fish. **Chief crops:** Coconuts, copra, cassava, sweet potatoes. **Minerals:** Gold.

Finance: Monetary unit: U.S. Dollar. **GDP:** (1997 est.): $160 mil. **Per capita GDP:** $8,800. **Imports** (1996): $72.4 mil; partners: U.S. **Exports** (1996): $14.3 mil; partners: U.S., Japan. **Budget** (1997 est.): $59.9 mil.

Transport: 1 airport.

Communications: TV sets: 98 per 1,000 pop. **Radios:** 550 per 1,000 pop.

Health: Life expectancy: 64.69 male; 70.98 female. **Births** (per 1,000 pop.): 21.55. **Deaths** (per 1,000 pop.): 7.74. **Natural inc.:** 1.381%. **Infant mortality (per 1,000 live births):** 18.50.
Education: Compulsory: ages 6-14. **Literacy** (1990): 98%.
Major Intl. Organizations: UN (WHO).
Embassy: 1150 18th Street NW, Suite 750, 20036; 452-6814.
Website: http://www.visit-palau.com

Spain acquired the Palau Islands in 1886 and sold them to Germany in 1899. Japan seized them in 1914. American forces occupied the islands in 1944; in 1947, they became part of the U.S.-administered UN Trust Territory of the Pacific Islands. In 1981 Palau became an autonomous republic; in 1993 the republic ratified a compact of free association with the U.S., which provides financial aid in return for U.S. use of Palauan military facilities over 15 years. Palau became an independent nation on Oct. 1, 1994.

Panama
Republic of Panama

People: Population: 2,778,526 **Age distrib.** (%): <15: 31.5; 65+: 5.8. **Pop. density:** 92 per sq. mi. **Urban:** 56%. **Ethnic groups:** Mestizo 70%, West Indian 14%, white 10%, Amerindian 6%. **Principal languages:** Spanish (official), English. **Chief religions:** Roman Catholic 85%, Protestant 15%.
Geography: Area: 30,220 sq. mi. **Location:** In Central America. **Neighbors:** Costa Rica on W, Colombia on E. **Topography:** 2 mountain ranges run the length of the isthmus. Tropical rain forests cover the Caribbean coast and eastern Panama. **Capital** (1997 est.): Panama City: 464,928.
Government: Type: Constitutional republic. **Head of state and gov.:** Pres. Mireya Elisa Moscoso; b July 1, 1946; in office: Sept. 1, 1999. **Local divisions:** 9 provinces, 3 territories. **Defense:** 1.3% of GDP. **Active troops** (1996): 11,800 est.
Economy: Industries: Oil refining, cement, construction. **Chief crops:** Bananas, rice, corn, coffee, sugar. **Minerals:** Copper. **Other resources:** Forests (mahogany), shrimp. **Arable land:** 7%. **Livestock** (1997): chickens: 10.00 mil; cattle: 1.44 mil; pigs: 245,000. **Electricity prod.** (1997): 4.672 bil kWh. **Labor force:** 32% govt. & community services; 27% agric. & fishing.
Finance: Monetary unit: Balboa (Sept. 1999: 1.00 = $1 U.S.). **GDP:** (1997 est.): $18 bil. **Per capita GDP:** $6,700. **Imports** (1997 est.): $2.95 bil; partners: U.S. 48%. **Exports** (1997 est.): $592 mil; partners: U.S. 37%. **Tourism:** $376 mil. **Budget** (1997 est.): $2.4 bil. **Intl. reserves less gold** (May 1999): $1.06 bil. **Consumer prices** (change in 1997): 0.6%.
Transport: Railroad: Length: 220 mi. **Motor vehicles:** 144,000 pass. cars, 82,800 comm. vehicles. **Civil aviation:** 679.9 mil pass.-mi; 10 airports. **Chief ports:** Balboa, Cristobal.
Communications: TV sets: 13 per 1,000 pop. **Radios:** 5.1 per 1,000 pop. **Telephones** (1997): 365,700 main lines. **Daily newspaper circ.:** 62 per 1,000 pop.
Health: Life expectancy: 71.91 male; 77.51 female. **Births** (per 1,000 pop.): 21.69. **Deaths** (per 1,000 pop.): 5.14. **Natural inc.:** 1.655%. **Hosp. beds** (1995): 1 per 369 persons. **Physicians** (1995): 1 per 856 persons. **Infant mortality** (per 1,000 live births): 23.35.
Education: Free, compulsory for 6 years between ages 6-15. **Literacy:** 91%.
Major Intl. Organizations: UN (FAO, IBRD, ILO, IMF, IMO, WHO), OAS.
Embassy: 2862 McGill Terrace NW 20008; 483-1407.

The coast of Panama was sighted by Rodrigo de Bastidas, sailing with Columbus for Spain in 1501, and was visited by Columbus in 1502. Vasco Nunez de Balboa crossed the isthmus and "discovered" the Pacific Ocean, Sept. 13, 1513. Spanish colonies were ravaged by Francis Drake, 1572-95, and Henry Morgan, 1668-71. Morgan destroyed the old city of Panama which had been founded in 1519. Freed from Spain, Panama joined Colombia in 1821.
Panama declared its independence from Colombia Nov. 3, 1903, with U.S. recognition. In support of Panama, U.S. naval forces deterred action by Colombia. Panama granted use, occupation, and control of the Canal Zone to the U.S. by treaty, ratified Feb. 26, 1904. In 1978, a new treaty provided for a gradual takeover by Panama of the canal, and withdrawal of U.S. troops, to be completed by Dec. 31, 1999. U.S. payments were substantially increased in the interim.
President Delvalle was ousted by the National Assembly, Feb. 26, 1988, after he tried to fire the head of the Panama Defense Forces, Gen. Manuel Antonio Noriega. Noriega had been indicted by 2 U.S. federal grand juries on drug charges. A general strike followed. Despite U.S.-imposed economic sanctions

Noriega remained in power. Voters went to the polls to elect a new president May 7, 1989. Noriega claimed victory, but foreign observers said that the opposition had won overwhelmingly. The government voided the election May 10, charging foreign interference. A coup against Noriega failed Oct. 3.
U.S. troops invaded Panama Dec. 20, 1989. Noriega took refuge in the Vatican diplomatic mission, but surrendered to U.S. officials Jan. 3, 1990. He was convicted of racketeering and drug trafficking in a U.S. District Court in Miami, FL, Apr. 9, 1992.
On Aug. 30, 1998, voters rejected a constitutional change that would have allowed Pres. Ernesto Pérez Balladares to run for reelection in 1999. Mireya Moscoso, widow of former Pres. Arnulfo Arias, was elected president May 2, 1999, becoming Panama's first female head of state.

Papua New Guinea
Independent State of Papua New Guinea

People: Population: 4,705,126. **Age distrib.** (%): <15: 39.4; 65+: 3.0. **Pop. density:** 26 per sq. mi. **Urban:** 16%. **Ethnic groups:** Papuan, Melanesian. **Principal languages:** English (official), Motu, 700+ indigenous dialects. **Chief religions:** Indigenous beliefs 34%, Roman Catholic 22%, Lutheran 16%.
Geography: Area: 178,700 sq. mi. **Location:** SE Asia, occupying E half of island of New Guinea and about 600 nearby islands. **Neighbors:** Indonesia (West Irian) on W, Australia on S. **Topography:** Thickly forested mts. cover much of the center of the country, with lowlands along the coasts. Included are some islands of Bismarck and Solomon groups, such as the Admiralty Isls., New Ireland, New Britain, and Bougainville. **Capital:** Port Moresby. **Cities** (1991): Port Moresby 192,000.
Government: Type: Parliamentary democracy. **Head of state:** Queen Elizabeth II, represented by Gov-Gen. Silas Atopare; in office: Nov. 1997. **Head of gov.:** Prime Min. Mekere Morauta; in office: July 14, 1999. **Local divisions:** 20 provinces. **Defense:** 1.2% of GDP. **Active troops:** 4,300.
Economy: Industries: Wood products, mining, oil. **Chief crops:** Coffee, coconuts, cocoa. **Minerals:** Gold, copper, silver. **Crude oil reserves** (1999): 333 mil bbls. **Livestock** (1997): chickens: 3.60 mil; pigs: 1.50 mil; cattle: 105,000. **Electricity prod.** (1997): 1.725 bil kWh. **Labor force:** 64% agric.
Finance: Monetary unit: Kina (Sept. 1999: 2.94 = $1 U.S.). **GDP:** (1996 est.): $11.6 bil. **Per capita GDP:** $2,650. **Imports** (1996): $1.7 bil; partners: Australia 52%. **Exports** (1996): $2.5 bil; partners: Australia 30%; Japan 24%. **Tourism:** $75 mil. **Budget** (1997 est.): $1.35 bil. **Intl. reserves less gold** (Apr. 1999): $109.91 mil. **Gold:** 63,000 oz t. **Consumer prices** (change in 1997): 13.6%.
Transport: Motor vehicles: 21,600 pass. cars, 77,700 comm. vehicles. **Civil aviation:** 456.5 mil pass.-mi; 129 airports. **Chief ports:** Port Moresby, Lae.
Communications: TV sets: 23 per 1,000 pop. **Radios:** 68 per 1,000 pop. **Telephones** (1997): 47,000 main lines. **Daily newspaper circ.:** 15 per 1,000 pop.
Health: Life expectancy: 57.58 male; 59.40 female. **Births** (per 1,000 pop.): 32.04. **Deaths** (per 1,000 pop.): 9.47. **Natural inc.:** 2.257%. **Physicians** (1993): 1 per 5,584 persons. **Infant mortality** (per 1,000 live births): 55.58.
Education: Literacy: 72%.
Major Intl. Organizations: UN (FAO, IBRD, ILO, IMF, IMO, WHO, WTrO), the Commonwealth, APEC.
Embassy: 1615 New Hampshire Ave. NW 20009; 745-3680.
Website: http://www.pngembassy.org

Human remains have been found in the interior of New Guinea dating back at least 10,000 years and possibly much earlier. Successive waves of peoples probably entered the country from Asia through Indonesia. Europeans visited in the 15th century, but actual land claims did not begin until the 19th century, when the Dutch took control of the island's western half.
The southern half of eastern New Guinea was first claimed by Britain in 1884, and transferred to Australia in 1905. The northern half was claimed by Germany in 1884, but captured in World War I by Australia, which was first granted a League of Nations mandate and then a UN trusteeship over the area. The 2 territories were administered jointly after 1949, given self-government Dec. 1, 1973, and became independent Sept. 16, 1975.
The indigenous population consists of a huge number of tribes, many living in almost complete isolation with mutually unintelligible languages. Secessionist rebels have clashed with government forces on Bougainville since 1988; a truce signed Oct. 10, 1997, brought a halt to the fighting, which had claimed an estimated 20,000 lives.
The country suffered from a severe drought in 1997. A tsunami killed at least 3,000 people July 17, 1998.

Paraguay
Republic of Paraguay

People: Population: 5,434,095. **Age distrib.** (%): <15: 39.3; 65+: 4.7. **Pop. density:** 35 per sq. mi. **Urban:** 53%. **Ethnic groups:** Mestizo 95%, white & Amerindian 5%. **Principal languages:** Spanish (official), Guarani. **Chief religion:** Roman Catholic 90%.

Geography: Area: 157,000 sq. mi. **Location:** Landlocked country in central South America. **Neighbors:** Bolivia on N, Argentina on S, Brazil on E. **Topography:** Paraguay R. bisects the country. To E are fertile plains, wooded slopes, grasslands. To W is the Gran Chaco plain, with marshes and scrub trees. Extreme W is arid. **Capital** (1994 est.): Asunción 546,637.

Government: Type: Republic. **Head of state and gov.:** Pres. Luis Angel González Macchi; b Dec. 13, 1947; in office: Mar. 28, 1999. **Local divisions:** 17 departments and capital city. **Defense:** 1.5% of GDP. **Active troops:** 20,200.

Economy: Industries: Food processing, textiles, cement. **Chief crops:** Corn, cotton, soybeans, sugarcane. **Minerals:** Iron, manganese, limestone. **Other resources:** Forests. **Arable land:** 6%. **Livestock** (1997): chickens: 14.83 mil; cattle: 9.79 mil; pigs: 2.53 mil; sheep: 387,000; goats: 123,000. **Electricity prod.** (1997): 50.746 bil kWh. **Labor force:** 45% agric.

Finance: Monetary unit: Guarani (Sept. 1999: 3,311.00 = $1 U.S.). **GDP:** (1997 est.): $21.9 bil. **Per capita GDP:** $3,900. **Imports** (1996 est.): $2.5 bil; partners: Brazil 29%, U.S. 22%, Argentina 14%. **Exports** (1997 est.): $1.1 bil; partners: Brazil 48%, Netherlands 22%. **Tourism:** $710 mil. **Budget** (1995 est.): $1.66 bil. **Intl. reserves less gold** (June 1998): $601.09 mil. **Gold:** 35,000 oz t. **Consumer prices** (change in 1997): 11.5%.

Transport: Railroad: Length: 274 mi. **Motor vehicles:** 71,000 pass. cars, 50,000 comm. vehicles. **Civil aviation:** 133.7 mil pass.-mi; 5 airports. **Chief port:** Asunción.

Communications: TV sets: 144 per 1,000 pop. **Radios:** 141 per 1,000 pop. **Telephones** (1997): 218,000 main lines. **Daily newspaper circ.:** 40 per 1,000 pop.

Health: Life expectancy: 70.47 male; 74.49 female. **Births** (per 1,000 pop.): 31.87. **Deaths** (per 1,000 pop.): 5.23. **Natural inc.:** 2.664%. **Hosp. beds** (1993): 1 per 864 persons. **Physicians** (1993): 1 per 1,406 persons. **Infant mortality** (per 1,000 live births): 36.35.

Education: Compulsory: ages 6-12. **Literacy:** 92%.

Major Intl. Organizations: UN (FAO, IBRD, ILO, IMF, IMO, WHO, WTrO), OAS.

Embassy: 2400 Massachusetts Ave. NW 20008; 483-6960.

The Guarani Indians were settled farmers speaking a common language before the arrival of Europeans.

Visited by Sebastian Cabot in 1527 and settled as a Spanish possession in 1535, Paraguay gained its independence from Spain in 1811. It lost much of its territory to Brazil, Uruguay, and Argentina in the War of the Triple Alliance, 1865-1870. Large areas were won from Bolivia in the Chaco War, 1932-35.

Gen. Alfredo Stroessner, who had ruled since 1954, was ousted in a military coup led by Gen. Andrés Rodríguez on Feb. 3, 1989. Rodríguez was elected president May 1. Juan Carlos Wasmosy was elected president May 9, 1993, becoming the nation's first civilian head of state in many years.

A prolonged power struggle involving a popular military leader, Gen. Lino César Oviedo, who was accused of insubordination, culminated in his surrender Dec. 12, 1997. He was freed Aug. 18, 1998, following the inauguration of Pres. Raúl Cubas Grau, Oviedo's successor as Colorado Party nominee. The assassination of Vice Pres. Luis María Argaña, Mar. 23, 1999, by an unidentified gunman, was widely attributed to Cubas and triggered protests and an impeachment vote; Cubas resigned Mar. 28 and was succeeded by Senate leader Luis Angel González Macchi.

Peru
Republic of Peru

People: Population: 26,624,582. **Age distrib.** (%): <15: 35.4; 65+: 4.6. **Pop. density:** 54 per sq. mi. **Urban:** 71%. **Ethnic groups:** Amerindian 45%, mestizo 37%, white 15%. **Principal languages:** Spanish, Quechua (both official), Aymara. **Chief religion:** Predominantly Roman Catholic.

Geography: Area: 496,200 sq. mi. **Location:** On the Pacific coast of South America. **Neighbors:** Ecuador, Colombia on N; Brazil, Bolivia on E; Chile on S. **Topography:** An arid coastal strip, 10 to 100 mi. wide, supports much of the population thanks to widespread irrigation. The Andes cover 27% of land area. The uplands are well-watered, as are the eastern slopes reaching the Amazon basin, which covers half the country with its forests and jungles. **Capital:** Lima. **Cities** (1993 met. est.): Lima 6,742,576; Arequipa 981,272; Callao 684,135.

Government: Type: Republic. **Head of state:** Pres. Alberto Fujimori; b July 28, 1938; in office: July 28, 1990. **Head of gov.:** Prime Min. Alberto Bustamante Belaúnde; nominated Oct. 10, 1999. **Local divisions:** 12 regions, 24 departments, 1 constitutional province. **Defense:** 2.2% of GDP. **Active troops:** 125,000.

Economy: Industries: Fishing, mining, food processing, textiles. **Chief crops:** Cotton, sugar, coffee, rice. **Minerals:** Copper, silver, gold, iron, oil. **Crude oil reserves** (1999): 773.4 mil bbls. **Other resources:** Wool, sardines. **Arable land:** 3%. **Livestock** (1997): chickens: 80.14 mil; sheep: 13.56 mil; cattle: 4.66 mil; pigs: 2.55 mil; goats: 2.02 mil. **Fish catch** (1997): 7.87 mil metric tons. **Electricity prod.** (1997): 17.533 bil kWh.

Finance: Monetary unit: New Sol (Sept. 1999: 3.40 = $1 U.S.). **GDP:** (1997 est.): $110.2 bil. **Per capita GDP:** $4,420. **Imports** (1996): $9.2 bil; partners: U.S. 31%. **Exports** (1996): $5.9 bil; partners: U.S. 20%. **Tourism:** $878 mil. **Budget** (1996 est.): $9.3 bil. **Intl. reserves less gold** (May 1999): $9.18 bil. **Gold:** 1.10 mil oz t. **Consumer prices** (change in 1997): 7.2%.

Transport: Railroad: Length: 1,318 mi. **Motor vehicles:** 500,000 pass. cars, 275,000 comm. vehicles. **Civil aviation:** 1.8 bil pass.-mi; 27 airports. **Chief ports:** Callao, Chimbote, Matarani, Salaverry.

Communications: TV sets: 85 per 1,000 pop. **Radios:** 221 per 1,000 pop. **Telephones** (1997): 1,645,900 main lines. **Daily newspaper circ.:** 87 per 1,000 pop.

Health: Life expectancy: 68.08 male; 72.78 female. **Births** (per 1,000 pop.): 26.09. **Deaths** (per 1,000 pop.): 5.70. **Natural inc.:** 2.039%. **Hosp. beds** (1992): 1 per 509 persons. **Physicians** (1992): 1 per 1,116 persons. **Infant mortality** (per 1,000 live births): 38.97.

Education: Free, compulsory: ages 6-11. **Literacy:** 89%.

Major Intl. Organizations: UN and all of its specialized agencies, APEC, OAS.

Embassy: 1700 Massachusetts Ave. NW 20036; 833-9860.

The powerful Inca empire had its seat at Cuzco in the Andes and covered most of Peru, Bolivia, and Ecuador, as well as parts of Colombia, Chile, and Argentina. Building on the achievements of 800 years of Andean civilization, the Incas had a high level of skill in architecture, engineering, textiles, and social organization.

A civil war had weakened the empire when Francisco Pizarro, Spanish conquistador, began raiding Peru for its wealth, 1532. In 1533 he seized the ruling Inca, Atahualpa, filled a room with gold as a ransom, then executed him and enslaved the natives.

Lima was the seat of Spanish viceroys until the Argentine liberator, José de San Martin, captured it in 1821; Spanish forces were ultimately routed by Simón Bolívar, 1824.

On Oct. 3, 1968, a military coup ousted Pres. Fernando Belaunde Terry. In 1968-74, the military government started socialist programs. Food shortages, escalating foreign debt, and strikes led to another coup, Aug. 29, 1976.

After 12 years of military rule, Peru returned to democratic leadership in 1980 but was plagued by economic problems and by leftist Shining Path (Sendero Luminoso) guerrillas.

Pres. Alberto Fujimori, elected in June 1990, dissolved the National Congress, suspended parts of the constitution, and initiated press censorship, Apr. 5, 1992. The leader of Shining Path was captured Sept. 12.

With the economy booming and signs of significant progress in curtailing guerrilla activity, Fujimori won reelection Apr. 9, 1995. Repressive antiterrorism tactics, however, drew international criticism. On Dec. 17, 1996, leftist Tupac Amaru guerrillas infiltrated a reception at the Japanese ambassador's residence in Lima and took hundreds of hostages, most of whom were later released. Peruvian soldiers stormed the embassy Apr. 22, 1997, rescuing 71 of the remaining hostages; 1 hostage, 2 soldiers, and all 14 guerrillas were killed.

Philippines
Republic of the Philippines

People: Population: 79,345,812. **Age distrib.** (%): <15: 37.3; 65+: 3.6. **Pop. density:** 684 per sq. mi. **Urban:** 55%. **Ethnic groups:** Christian Malay 92%, Muslim Malay 4%. **Principal languages:** Pilipino, English (both official). **Chief religions:** Roman Catholic 83%, Protestant 9%, Muslim 5%.

Geography: Area: 116,000 sq. mi. **Location:** An archipelago off the SE coast of Asia. **Neighbors:** Nearest are Malaysia and Indonesia on S, Taiwan on N. **Topography:** The country consists of some 7,100 islands stretching 1,100 mi. N-S. About 95% of area and population are on 11 largest islands, which are mountainous, except for the heavily indented coastlines and for the central plain on Luzon. **Capital:** Manila. **Cities** (1995): Quezon City 1,989,419; Manila 1,654,761.

Government: Type: Republic. **Head of state and gov.:** Pres. Joseph Ejercito Estrada; b Apr. 19, 1937; in office: June

30, 1998. **Local divisions:** 14 regions and 2 autonomous regions. **Defense:** 1.7% of GNP. **Active troops:** 110,500.

Economy: Industries: Food processing, textiles, chemicals, pharmaceuticals, wood prods. **Chief crops:** Sugar, rice, pineapples, corn, coconuts. **Minerals:** Cobalt, copper, gold, nickel, silver, oil. **Other resources:** Forests (46% of area). **Crude oil reserves** (1999): 228 mil bbls. **Arable land:** 19%. **Livestock** (1997): chickens: 136.89 mil; buffalo: 3.00 mil; pigs: 10.21 mil; goats: 6.50 mil; cattle: 2.40 mil. **Fish catch** (1997): 1.81 mil metric tons. **Electricity prod.** (1997): 33.700 bil kWh. **Labor force:** 43% agric.; 23% services; 18% gov't services; 16% ind. and comm.

Finance: Monetary unit: Peso (Sept. 1999: 39.73 = $1 U.S.). **GDP:** (1997 est.): $244 bil. **Per capita GDP:** $3,200. **Imports** (1997 est.): $34 bil; partners: Japan 21%, U.S. 20%, ASEAN 12%, EU 10%. **Exports** (1997 est.): $25 bil; partners: U.S. 34%, Japan 11%, EU 17%, ASEAN 14%. **Tourism:** $2.42 bil. **Budget** (1996 est.): $16.6 bil. **Intl. reserves less gold** (May 1999): $12.19 bil. **Gold:** 5.76 mil oz t. **Consumer prices** (change in 1997): 9.7%.

Transport: Railroad: Length: 557 mi. **Motor vehicles:** 702,578 pass. cars, 1.35 mil comm. vehicles. **Civil aviation:** 10.1 bil pass.-mi; 21 airports. **Chief ports:** Cebu, Manila, Iloilo, Davao.

Communications: TV sets: 125 per 1,000 pop. **Radios:** 116 per 1,000 pop. **Telephones** (1997): 2,078,000 main lines. **Daily newspaper circ.:** 65 per 1,000 pop.

Health: Life expectancy: 63.79 male; 69.50 female. **Births** (per 1,000 pop.): 27.88. **Deaths** (per 1,000 pop.): 6.45. **Natural inc.:** 2.143%. **Hosp. beds** (1993): 1 per 860 persons. **Physicians** (1993): 1 per 849 persons. **Infant mortality** (per 1,000 live births): 33.89.

Education: Free, compulsory: ages 7-12. **Literacy:** 95%.

Major Intl. Organizations: UN (FAO, IBRD, ILO, IMF, IMO, WHO, WTrO), ASEAN.

Embassy: 1600 Massachusetts Ave. NW 20036; 467-9300. **Website:** http://www.census.gov.ph

The Malay peoples of the Philippine Islands, whose ancestors probably migrated from Southeast Asia, were mostly hunters, fishers, and unsettled cultivators.

The archipelago was visited by Magellan, 1521. The Spanish founded Manila, 1571. The islands, named for King Philip II of Spain, were ceded by Spain to the U.S. for $20 million, 1898, following the Spanish-American War. U.S. troops suppressed a guerrilla uprising in a brutal 6-year war, 1899-1905.

Japan attacked the Philippines Dec. 8, 1941, and occupied the islands during WW II. On July 4, 1946, independence was proclaimed in accordance with an act passed by the U.S. Congress in 1934. A republic was established.

On Sept. 21, 1972, Pres. Ferdinand Marcos declared martial law. Marcos proclaimed a new constitution, Jan. 17, 1973, with himself as president. His wife, Imelda, received wide powers in 1978 to supervise planning and development. Political corruption was widespread. Martial law was lifted Jan. 17, 1981, but Marcos retained broad emergency powers. He was reelected in June to a new 6-year term as president.

The assassination of prominent opposition leader Benigno S. Aquino Jr., Aug. 21, 1983, sparked demonstrations calling for the resignation of Marcos. After a bitter presidential campaign, amid allegations of widespread election fraud, Marcos was declared the victor Feb. 16, 1986, over Corazon Aquino, widow of the slain opposition leader. With his support collapsing, Marcos fled the country Feb. 25.

Recognized as president by the U.S. and other nations, Aquino was plagued by a weak economy, widespread poverty, Communist and Muslim insurgencies, and lukewarm military support. Rebel troops seized military bases and TV stations and bombed the presidential palace, Dec. 1, 1989. Government forces defeated the attempted coup aided by air cover provided by the U.S. F-4s. Aquino endorsed Fidel Ramos in the May 1992 presidential election, which he won.

The U.S. vacated the Subic Bay Naval Station at the end of 1992, ending its long military presence in the Philippines.

The government signed a cease-fire agreement, Jan. 30, 1994, with Muslim separatist guerrillas, but some rebels refused to abide by the accord. A new treaty providing for expansion and development of an autonomous Muslim region on Mindanao was signed Sept. 2, 1996, formally ending a rebellion that had claimed more than 120,000 lives since 1972.

Running as a populist, Joseph (Erap) Estrada, a former movie actor, won the presidential election of May 11, 1998.

Poland
Republic of Poland

People: Population: 38,608,929. **Age distrib.** (%): <15: 19.8; 65+: 12.0. **Pop. density:** 320 per sq. mi. **Urban:** 64%.

Ethnic groups: Polish 98%. **Principal language:** Polish (official). **Chief religion:** Roman Catholic 95%.

Geography: Area: 120,700 sq. mi. **Location:** On the Baltic Sea in E central Europe. **Neighbors:** Germany on W; Czech Rep., Slovakia on S; Lithuania, Belarus, Ukraine on E; Russia on N. **Topography:** Mostly lowlands forming part of the Northern European Plain. The Carpathian Mts. along the S border rise to 8,200 ft. **Capital:** Warsaw. **Cities** (1996 est.): Warsaw 1,632,534; Lodz 820,350; Krakow 747,781.

Government: Type: Republic. **Head of state:** Pres. Aleksander Kwasniewski; b Nov. 15, 1954; in office: Dec. 23, 1995. **Head of gov.:** Prime Min. Jerzy Buzek; b July 3, 1940; in office: Oct. 31, 1997. **Local divisions:** 49 provinces. **Defense:** 2.3% of GDP. **Active troops:** 241,800.

Economy: Industries: Shipbuilding, coal mining, chemicals, metals, machinery, food processing. **Chief crops:** Grains, potatoes, fruits, vegetables. **Minerals:** Coal, copper, silver, lead, sulfur, natural gas. **Crude oil reserves** (1999): 114.88 mil bbls. **Arable land:** 47%. **Livestock** (1997): chickens: 51.12 mil; pigs: 19.17 mil; cattle: 6.96 mil; sheep: 452,913. **Fish catch** (1997): 361,906 metric tons. **Electricity prod.** (1997): 134.263 bil kWh. **Labor force:** 44% services; 29.9% ind. & constr.; 26% agric.

Finance: Monetary unit: Zloty (Sept. 1999: 4.03 = $1 U.S.). **GDP:** (1997 est.): $280.7 bil. **Per capita GDP:** $7,250. **Imports** (1997 est.): $44.5 bil; partners: Germany 26.5%. **Exports** (1997 est.): $26.4 bil; partners: Germany 34.5%. **Tourism:** $8.40 bil. **Budget** (1997 est.): $35.5 bil. **Intl. reserves less gold** (June 1999): $24.99 bil. **Gold:** 3.31 mil oz t. **Consumer prices** (change in 1997): 11.7%.

Transport: Railroad: Length: 14,904 mi. **Motor vehicles:** 7.52 mil pass. cars, 1.55 mil comm. vehicles. **Civil aviation:** 2.6 bil pass.-mi; 8 airports. **Chief ports:** Gdansk, Gdynia, Ustka, Szczecin.

Communications: TV sets: 249.9 per 1,000 pop. **Radios:** 263.3 per 1,000 pop. **Telephones** (1997): 8,812,300 main lines. **Daily newspaper circ.:** 140 per 1,000 pop.

Health: Life expectancy: 68.93 male; 77.41 female. **Births** (per 1,000 pop.): 10.61. **Deaths** (per 1,000 pop.): 9.72. **Natural inc.:** 0.089%. **Hosp. beds** (1996): 1 per 180 persons. **Physicians** (1996): 1 per 436 persons. **Infant mortality** (per 1,000 live births): 12.76.

Education: Free, compulsory: ages 7-14. **Literacy** (1994): 99%.

Major Intl. Organizations: UN (FAO, IBRD, ILO, IMF, IMO, WHO, WTrO), NATO, OECD, OSCE.

Embassy: 2640 16th St. NW 20009; 234-3800. **Website:** http://www.polishworld.com

Slavic tribes in the area were converted to Latin Christianity in the 10th century. Poland was a great power from the 14th to the 17th centuries. In 3 partitions (1772, 1793, 1795) it was apportioned among Prussia, Russia, and Austria. Overrun by the Austro-German armies in World War I, it declared its independence on Nov. 11, 1918, and was recognized as independent by the Treaty of Versailles, June 28, 1919. Large territories to the east were taken in a war with Russia, 1921.

Germany and the USSR invaded Poland Sept. 1-27, 1939, and divided the country. During the war, some 6 million Polish citizens, half of them Jews, were killed by the Nazis. With Germany's defeat, a Polish government-in-exile in London was recognized by the U.S., but the USSR pressed the claims of a rival group. The election of 1947 was completely dominated by the Communists.

In compensation for 69,860 sq. mi. ceded to the USSR, 1945, Poland received approx. 40,000 sq. mi. of German territory E of the Oder-Neisse line comprising Silesia, Pomerania, West Prussia, and part of East Prussia.

In 12 years of rule by Stalinists, large estates were abolished, industries nationalized, schools secularized, and Roman Catholic prelates jailed. Farm production fell off. Harsh working conditions caused a riot in Poznan, June 28-29, 1956. A new Politburo, committed to a more independent Polish Communism, was named Oct. 1956, with Wladyslaw Gomulka as first secretary of the party. Collectivization of farms was ended. Gomulka agreed to permit religious liberty and religious publications, provided the church kept out of politics.

In Dec. 1970 workers in port cities rioted because of price rises and new incentive wage rules. On Dec. 20 Gomulka resigned as party leader; he was succeeded by Edward Gierek. The rules were dropped and price rises revoked.

After 2 months of labor turmoil had crippled the country, the Polish government, Aug. 30, 1980, met the demands of striking workers at the Lenin Shipyard, Gdansk. Among the 21 concessions granted were the right to form independent trade unions and the right to strike. By 1981, 9.5 mil workers had joined the independent trade union (Solidarity). Solidarity leaders proposed, Dec. 12, a nationwide referendum on establishing a non-Communist government if the government failed to agree to a series of demands.

Spurred by fear of Soviet intervention, the government, Dec. 13, imposed martial law. Lech Walesa and other Solidarity leaders were arrested. The U.S. imposed sanctions, which were lifted when martial law was suspended Dec. 1982

On Apr. 5, 1989, an accord was reached between the government and opposition factions on political and economic reforms, including free elections. Candidates endorsed by Solidarity swept the parliamentary elections, June 4. Lech Walesa became president Dec. 22, 1990.

A radical economic program designed to transform the economy into a free-market system led to inflation and unemployment. In Sept. 1993, former Communists and other leftists won a majority in the lower house of Parliament. Walesa lost to a former Communist, Aleksander Kwasniewski, in a presidential runoff election, Nov. 19, 1995.

A new constitution was approved by referendum May 25, 1997. Flooding in July caused more than $1 billion in property damage. Solidarity won parliamentary elections held Sept. 21. Poland became a full member of NATO on Mar. 12, 1999.

Portugal
Portuguese Republic

People: Population: 9,918,040. **Age distrib.** (%): <15: 17.0; 65+: 15.3. **Pop. density:** 278 per sq. mi. **Urban:** 36%. **Ethnic groups:** Homogeneous Mediterranean stock, small African minority. **Principal languages:** Portuguese (official). **Chief religion:** Roman Catholic 97%.

Geography: Area: 35,672 sq. mi., incl. the Azores and Madeira Islands. **Location:** At SW extreme of Europe. **Neighbors:** Spain on N, E. **Topography:** Portugal N of Tajus R., which bisects the country NE-SW, is mountainous, cool and rainy. To the S there are drier, rolling plains, and a warm climate. **Capital:** Lisbon. **Cities** (1996 est.): Lisbon 581,920; Porto 302,467.

Government: Type: Republic. **Head of state:** Pres. Jorge Sampaio; b Sept. 18, 1939; in office: Mar. 9, 1996. **Head of gov.:** Prime Min. Antonio Guterres; b Apr. 30, 1949; in office: Oct. 30, 1995. **Local divisions:** 18 districts, 2 autonomous regions. **Defense:** 2.6% of GDP. **Active troops:** 59,300.

Economy: Industries: Textiles, footwear, cork, wood pulp, chemicals, fish canning, metal working, oil refining, wine, paper. **Chief crops:** Grains, potatoes, grapes, olives. **Minerals:** Tungsten, uranium, iron. **Other resources:** Forests (world leader in cork production). **Arable land:** 26%. **Livestock** (1997): chickens: 27.00 mil; sheep: 6.30 mil; pigs: 2.37 mil; cattle: 1.30 mil; goats: 815,000. **Fish catch** (1997): 221,923 metric tons. **Electricity prod.** (1997): 32.712 bil kWh. **Labor force:** 56% services; 23% manuf.; 11% agric., fish.

Finance: Monetary unit: Escudo (Sept. 1999: 187.29 = $1 U.S.). **Euro** (Sept. 1999: 1.07 = $1 U.S.). **GDP:** (1997 est.): $149.5 bil. **Per capita GDP:** $15,200. **Imports** (1996): $33.9 bil; partners: EU 72%. **Exports** (1996): $23.8 bil; partners: EU 76%. **Tourism:** $4.77 bil. **Budget** (1996 est.): $52 bil. **Intl. reserves less gold** (June 1999): $8.01 bil. **Gold:** 19.50 mil oz t. **Consumer prices** (change in 1997): 2.8%.

Transport: Railroad: Length: 1,909 mi. **Motor vehicles** (1997): 2.95 mil pass. cars, 960,300 comm. vehicles. **Civil aviation:** 5.23 bil pass.-mi; 16 airports. **Chief ports:** Lisbon, Setubal, Leixoes.

Communications: TV sets: 333 per 1,000 pop. **Radios:** 280 per 1,000 pop. **Telephones:** 4,116,900 main lines. **Daily newspaper circ.:** 47 per 1,000 pop.

Health: Life expectancy: 72.51 male; 79.46 female. **Births** (per 1,000 pop.): 10.49. **Deaths** (per 1,000 pop.): 10.25. **Natural inc.:** 0.024%. **Hosp. beds** (1996): 1 per 253 persons. **Physicians** (1996): 1 per 332 persons. **Infant mortality** (per 1,000 live births): 6.73.

Education: Free, compulsory: ages 6-15. **Literacy:** 90%.
Major Intl. Organizations: UN (FAO, IBRD, ILO, IMF, IMO, WHO, WTrO), EU, NATO, OECD, OSCE.
Embassy: 2125 Kalorama Rd. NW 20008; 328-8610.
Website: http://infoline.ine.pt/si/english/port.html

Portugal, an independent state since the 12th century, was a kingdom until a revolution in 1910 drove out King Manoel II and a republic was proclaimed.

From 1932 a strong, repressive government was headed by Premier Antonio de Oliveira Salazar. Illness forced his retirement in Sept. 1968.

On Apr. 25, 1974, the government was seized by a military junta led by Gen. Antonio de Spinola, who became president. The new government reached agreements providing independence for Guinea-Bissau, Mozambique, Cape Verde Islands, Angola, and São Tomé and Príncipe. Banks, insurance companies, and other industries were nationalized.

Parliament approved, June 1, 1989, a package of reforms

that did away with the socialist economy and created a "democratic" economy, denationalizing industries. Portugal was scheduled to return Macau to China on Dec. 20, 1999.

Azores Islands, in the Atlantic, 740 mi. W of Portugal, have an area of 868 sq. mi. and a pop. (1993 est.) of 238,000. A 1951 agreement gave the U.S. rights to use defense facilities in the Azores. The **Madeira Islands,** 350 mi. off the NW coast of Africa, have an area of 306 sq. mi. and a pop. (1993 est.) of 437,312. Both groups were offered partial autonomy in 1976.

Qatar
State of Qatar

People: Population: 723,542. **Age distrib.** (%): <15: 26.9; 65+: 2.2. **Pop. density:** 164 per sq. mi. **Urban:** 92%. **Ethnic groups:** Arab 40%, Pakistani 18%, Indian 18%, Iranian 10%. **Principal languages:** Arabic (official), English. **Chief religion:** Muslim 95%.

Geography: Area: 4,416 sq. mi. **Location:** Middle East, occupying peninsula on W coast of Persian Gulf. **Neighbors:** Saudi Arabia on S. **Topography:** Mostly a flat desert, with some limestone ridges; vegetation of any kind is scarce. **Capital:** Doha (1993 est.): 339,471.

Government: Type: Traditional monarchy. **Head of state:** Emir Hamad bin Khalifa ath-Thani; b 1950; in office: June 27, 1995. **Head of gov.:** Prime Min. Abdullah bin Khalifa ath-Thani; in office: Oct. 29, 1996. **Local divisions:** 9 municipalities. **Defense:** 13.7% of GDP. **Active troops:** 11,800.

Economy: Industries: Oil production and refining, petrochemicals, cement. **Minerals:** Oil, gas. **Crude oil reserves** (1999): 3.7 bil bbls. **Livestock** (1997): chickens: 3.85 mil; sheep: 199,682; goats: 172,071. **Electricity prod.** (1997): 6.300 bil kWh. **Labor force:** 24% agric.; 22% mining, manuf.

Finance: Monetary unit: Riyal (Sept. 1999: 3.64 = $1 U.S.). **GDP:** (1997 est.): $11.2 bil. **Per capita GDP:** $16,700. **Imports** (1997 est.): $5 bil; partners: Italy 14%, UK 12%, France 11%. **Exports** (1997 est.): $5.8 bil; partners: Japan 55%. **Budget** (FY 1997-98 est.): $4.5 bil. **Gold:** 54,000 oz t.

Transport: Motor vehicles: 96,800 pass. cars, 85,600 comm. vehicles. **Civil aviation:** 1.6 bil pass.-mi; 1 airport. **Chief ports:** Doha, Umm Sáid.

Communications: TV sets: 451 per 1,000 pop. **Radios:** 322 per 1,000 pop. **Telephones:** 150,500 main lines. **Daily newspaper circ.:** 143 per 1,000 pop.

Health: Life expectancy: 71.70 male; 76.89 female. **Births** (per 1,000 pop.): 16.75. **Deaths** (per 1,000 pop.): 3.57. **Natural inc.:** 1.318%. **Hosp. beds** (1994): 1 per 509 persons. **Physicians** (1994): 1 per 793 persons. **Infant mortality** (per 1,000 live births): 17.25.

Education: Literacy: 79%.
Major Intl. Organizations: UN (FAO, IBRD, ILO, IMF, IMO, WHO, WTrO), AL, OPEC.
Embassy: 4200 Wisconsin Ave. NW 20016; 274-1600.
Website: http://www.mofa.gov.qa

Qatar was under Bahrain's control until the Ottoman Turks took power, 1872 to 1915. In a treaty signed 1916, Qatar gave Great Britain responsibility for its defense and foreign relations. After Britain announced it would remove its military forces from the Persian Gulf area by the end of 1971, Qatar sought a federation with other British-protected states in the area; this failed and Qatar declared itself independent, Sept. 1, 1971. Crown Prince Hamad bin Khalifa ath-Thani ousted his father, Emir Khalifa bin Hamad ath-Thani, June 27, 1995. In municipal elections held Mar. 8, 1999, women participated for the 1st time as candidates and voters.

Oil and natural gas revenues give Qatar a per capita income among the world's highest.

Romania

People: Population: 22,334,312. **Age distrib.** (%): <15: 18.6; 65+: 13.3. **Pop. density:** 244 per sq. mi. **Urban:** 56%. **Ethnic groups:** Romanian 89%, Hungarian 9%. **Principal languages:** Romanian (official), Hungarian, German. **Chief religions:** Romanian Orthodox 70%, Roman Catholic 6%, Protestant 6%.

Geography: Area: 91,700 sq. mi. **Location:** SE Europe, on the Black Sea. **Neighbors:** Moldova on E, Ukraine on N, Hungary and Yugoslavia on W, Bulgaria on S. **Topography:** The Carpathian Mts. encase the north-central Transylvanian plateau. There are wide plains S and E of the mountains, through which flow the lower reaches of the rivers of the Danube system. **Capital:** Bucharest (1996 est.): 2,037,278.

Government: Type: Republic. **Head of state:** Pres. Emil Constantinescu; b Nov. 19, 1939; in office: Nov. 29, 1996. **Head of gov.:** Prime Min. Radu Vasile; b Oct. 10, 1942; in

office: Apr. 2, 1998. **Local divisions:** 40 counties, 1 municipality. **Defense:** 2.3% of GDP. **Active troops:** 227,000.
Economy: Industries: Mining, timber, construction materials, metals, machinery, oil products, chemicals, food processing. **Chief crops:** Grains, grapes, sunflower seeds, sugar beets, potatoes. **Minerals:** Oil, gas, coal, iron. **Crude oil reserves** (1999): 1.43 bil bbls. **Other resources:** Timber. **Arable land:** 41%. **Livestock** (1997): chickens: 66.62 mil; sheep: 8.94 mil; pigs: 7.10 mil; cattle: 3.24 mil; goats: 609,700. **Fish catch** (1996): 18,259 metric tons. **Electricity prod.** (1997): 58.300 bil kWh. **Labor force:** 34% agric.; 29% industry.
Finance: Monetary unit: Leu (Sept. 1999: 16,249.00 = $1 U.S.). **GDP:** (1997 est.): $114.2 bil. **Per capita GDP:** $5,300. **Imports** (1997 est.): $10.4 bil; partners: Germany 17%, Italy 15.6%, Russia 13%. **Exports** (1997 est.): $8.4 bil; partners: Germany 18%, Italy 16%. **Tourism:** $547 mil. **Budget** (1997 est.): $11.7 bil. **Intl. reserves less gold** (May 1999): $1.94 bil. **Gold:** 3.25 mil oz t. **Consumer prices** (change in 1997): 59.1%.
Transport: Railroad: Length: 7,062 mi. **Motor vehicles:** 2.39 mil pass. cars; 513,312 comm. vehicles. **Civil aviation:** 1.1 bil pass.-mi; 8 airports. **Chief ports:** Constanta, Braila.
Communications: TV sets: 201 per 1,000 pop. **Radios:** 198 per 1,000 pop. **Telephones** (1997): 3,776,900 main lines. **Daily newspaper circ.:** 297 per 1,000 pop.
Health: Life expectancy: 67.05 male; 74.81 female. **Births** (per 1,000 pop.): 10.09. **Deaths** (per 1,000 pop.): 11.55. **Natural inc.:** −0.146%. **Hosp. beds** (1992): 1 per 105 persons. **Physicians** (1993): 1 per 565 persons. **Infant mortality** (per 1,000 live births): 18.12.
Education: Compulsory: ages 6-16. **Literacy** (1992): 97%.
Major Intl. Organizations: UN (FAO, IBRD, ILO, IMF, IMO, WHO, WTrO), OSCE.
Embassy: 1607 23d St. NW 20008; 332-4846.
Website: http://www.embassy.org/romania

Romania's earliest known people merged with invading Proto-Thracians, preceding by centuries the Dacians. The Dacian kingdom was occupied by Rome, AD 106-271; people and language were Romanized. The principalities of Wallachia and Moldavia, dominated by Turkey, were united in 1859, became Romania in 1861. In 1877 Romania proclaimed independence from Turkey, and became an independent state by the Treaty of Berlin, 1878; a kingdom under Carol I, 1881; and a constitutional monarchy with a bicameral legislature, 1886.
Romania helped Russia in its war with Turkey, 1877-78. After World War I it acquired Bessarabia, Bukovina, Transylvania, and Banat. In 1940 it ceded Bessarabia and Northern Bukovina to the USSR, part of southern Dobrudja to Bulgaria, and northern Transylvania to Hungary.
In 1941, Prem. Marshal Ion Antonescu led Romania in support of Germany against the USSR. In 1944 he was overthrown by King Michael and Romania joined the Allies.
After occupation by Soviet troops a People's Republic was proclaimed, Dec. 30, 1947; Michael was forced to abdicate.
On Aug. 22, 1965, a new constitution proclaimed Romania a Socialist Republic. Pres. Nicolae Ceausescu maintained an independent course in foreign affairs, but his domestic policies were repressive. All industry was state-owned, and state farms and cooperatives owned almost all arable land.
On Dec. 16, 1989, security forces opened fire on antigovernment demonstrators in Timisoara; hundreds were buried in mass graves. Ceausescu declared a state of emergency as protests spread to other cities. On Dec. 21, in Bucharest, security forces fired on protesters. Army units joined the rebellion, Dec. 22, and a group known as the Council of National Salvation announced that it had overthrown the government. Fierce fighting took place between the army, which backed the new government, and forces loyal to Ceausescu.
Ceausescu and his wife were captured and, following a trial in which they were found guilty of genocide, were executed Dec. 25, 1989. Former Communists dominated the government in succeeding years. A new constitution providing for a multiparty system took effect Dec. 8, 1991. Many of Romania's state-owned companies were privatized in 1996. The former Communists were swept from power in elections Nov. 3 and 17, 1996. Pope John Paul II visited Romania in May 1999.

Russia
Russian Federation
People: Population: 146,393,569. **Age distrib.** (%): <15: 19.1; 65+: 12.5. **Pop. density:** 22 per sq. mi. **Urban:** 76%. **Ethnic groups:** Russian 82%, Tatar 4%. **Principal languages:** Russian (official), many others. **Chief religions:** Russian Orthodox, Muslim, others.

Geography: Area: 6,592,800 sq. mi., more than 76% of total area of the former USSR and the largest country in the world. **Location:** Stretches from E Europe across N Asia to the Pacific O. **Neighbors:** Finland, Norway, Estonia, Latvia, Belarus, Ukraine on W; Georgia, Azerbaijan, Kazakhstan, China, Mongolia, North Korea on S; Kaliningrad exclave bordered by Poland on the S, Lithuania on the N and E. **Topography:** Russia contains every type of climate except the distinctly tropical, and has a varied topography. The European portion is a low plain, grassy in S, wooded in N, with Ural Mts. on the E, and Caucasus Mts. on the S. Urals stretch N-S for 2,500 mi. The Asiatic portion is also a vast plain, with mountains on the S and in the E; tundra covers extreme N, with forest belt below; plains, marshes are in W, desert in SW. **Capital:** Moscow. **Cities** (1995 est.): Moscow 8,368,449; St. Petersburg 4,232,105; Nizhniy Novgorod 1,375,570; Novosibirsk 1,367,596.
Government: Type: Federal republic. **Head of state:** Pres. Boris Yeltsin; b Feb. 1, 1931; in office: July 10, 1991. **Head of gov.:** Prime Min. Vladimir Putin; b Oct. 7, 1952; in office: Aug. 16, 1999. **Local divisions:** 21 autonomous republics, 68 autonomous territories and regions. **Defense:** 5.8% of GDP. **Active troops:** 1.240 mil.
Economy: Industries: Steel, machinery, machine tools, vehicles, chemicals, mining, footwear, textiles, appliances, paper. **Chief crops:** Grains, sugar beets, vegetables, sunflowers. **Minerals:** Manganese, mercury, potash, bauxite, cobalt, chromium, copper, coal, gold, lead, molybdenum, nickel, phosphates, silver, tin, tungsten, zinc, oil, gas, iron, potassium. **Crude oil reserves** (1999): 48.57 bil bbls. **Other resources:** Forests. **Arable land:** 8%. **Livestock** (1997): chickens: 405.00 mil; cattle: 31.70 mil; sheep: 17.13 mil; pigs: 17.31 mil; goats: 1.63 mil. **Fish catch** (1997): 4.66 mil metric tons. **Electricity prod.** (1997): 788.930 bil kWh. **Labor force:** 24% mining & manuf.; 23% services; 14% agric.
Finance: Monetary unit: Ruble (Aug. 1999: 25.10 = $1 U.S. NOTE: On Jan 1, 1998, Russia eliminated 3 digits from the ruble.) **GDP:** (1997 est.): $692 bil. **Per capita GDP:** $4,700. **Imports** (1997): $66.9 bil; partners: Germany 12%, U.S. 5%. **Exports** (1997): $86.7 bil; partners: Germany 8%, China 6%, U.S. 6%. **Tourism:** $7.11 bil. **Budget** (1997 est.): $70 bil. **Intl. reserves less gold** (June 1999): $8.19 bil. **Gold:** 13.21 mil oz t. **Consumer prices** (change in 1997): 27.8%.
Transport: Railroad: Length: 94,400 mi. **Motor vehicles:** 13.71 mil pass. cars, 9.86 mil comm. vehicles. **Civil aviation:** 30.6 bil pass.-mi; 75 airports. **Chief ports:** St. Petersburg, Murmansk, Arkhangelsk.
Communications: TV sets: 379 per 1,000 pop. **Radios:** 341 per 1,000 pop. **Telephones** (1997): 26,874,600 main lines. **Daily newspaper circ.:** 267 per 1,000 pop.
Health: Life expectancy: 58.83 male; 71.75 female. **Births** (per 1,000 pop.): 9.64. **Deaths** (per 1,000 pop.): 14.96. **Natural inc.:** −0.532%. **Hosp. beds** (1995): 1 per 80 persons. **Physicians** (1995): 1 per 235 persons. **Infant mortality** (per 1,000 live births): 23.
Education: Free, compulsory: ages 7-17. **Literacy:** 99%.
Major Intl. Organizations: UN (IBRD, ILO, IMF, IMO, WHO), APEC, CIS, OSCE.
Embassy: 2650 Wisconsin Ave. NW 20007; 298-5700.
Website: http://www.undp.org/missions/russianfed

History. Slavic tribes began migrating into Russia from the W in the 5th century AD. The first Russian state, founded by Scandinavian chieftains, was established in the 9th century, centering in Novgorod and Kiev. In the 13th century the Mongols overran the country. It recovered under the grand dukes and princes of Muscovy, or Moscow, and by 1480 freed itself from the Mongols. Ivan the Terrible was the first to be formally proclaimed Tsar (1547). Peter the Great (1682-1725) extended the domain and, in 1721, founded the Russian Empire.
Western ideas and the beginnings of modernization spread through the huge Russian empire in the 19th and early 20th centuries. But political evolution failed to keep pace.
Military reverses in the 1905 war with Japan and in World War I led to the breakdown of the Tsarist regime. The 1917 Revolution began in March with a series of sporadic strikes for higher wages by factory workers. A provisional democratic government under Prince Georgi Lvov was established but was quickly followed in May by the second provisional government, led by Alexander Kerensky. The Kerensky government and the freely-elected Constituent Assembly were overthrown in a Communist coup led by Vladimir Ilyich Lenin Nov. 7.

Soviet Union
Lenin's death Jan. 21, 1924, resulted in an internal power struggle from which Joseph Stalin eventually emerged on top. Stalin secured his position at first by exiling opponents, but from

the 1930s to 1953, he resorted to a series of "purge" trials, mass executions, and mass exiles to work camps. These measures resulted in millions of deaths, according to most estimates.

Germany and the Soviet Union signed a non-aggression pact Aug. 1939; Germany launched a massive invasion of the Soviet Union, June 1941. Notable heroic episode was the "900 days" siege of Leningrad (now St. Petersburg), lasting to Jan. 1944, and causing a million deaths; the city was never taken. Russian winter counterthrusts, 1941-42 and 1942-43, stopped the German advance. Turning point was the failure of German troops to take and hold Stalingrad (now Volgograd), Sept. 1942 to Feb. 1943. With British and U.S. Lend-Lease aid and sustaining great casualties, the Russians drove the German forces from eastern Europe and the Balkans in the next 2 years.

After Stalin died, Mar. 5, 1953, Nikita Khrushchev was elected first secretary of the Central Committee. In 1956 he condemned Stalin and "de-Stalinization" began.

Under Khrushchev the open antagonism of Poles and Hungarians toward domination by Moscow was brutally suppressed in 1956. He advocated peaceful co-existence with the capitalist countries, but continued arming the Soviet Union with nuclear weapons. He aided the Cuban revolution under Fidel Castro but withdrew Soviet missiles from Cuba during confrontation by U.S. Pres. Kennedy, Sept.-Oct. 1962. Khrushchev was suddenly deposed, Oct. 1964, and replaced by Leonid I. Brezhnev.

In Aug. 1968 Russian, Polish, East German, Hungarian, and Bulgarian military forces invaded Czechoslovakia to put a curb on liberalization policies of the Czech government.

Massive Soviet military aid to North Vietnam in the late 1960s and early 1970s helped assure Communist victories throughout Indo-China. Soviet arms aid and advisers were sent to several African countries in the 1970s.

In Dec. 1979, Soviet forces entered Afghanistan to support that government against rebels. In Apr. 1988, the Soviets agreed to withdraw their troops, ending a futile 8-year war.

Mikhail Gorbachev was chosen gen. secy. of the Communist Party, Mar. 1985. He held 4 summit meetings with U.S. Pres. Ronald Reagan. In 1987, in Washington, a treaty was signed eliminating intermediate-range nuclear missiles from Europe.

In 1987, Gorbachev initiated a program of reforms, including expanded freedoms and the democratization of the political process, through openness (*glasnost*) and restructuring (*perestroika*). The reforms were opposed by some Eastern bloc countries and many old-line Communists in the USSR. Gorbachev faced economic problems as well as ethnic and nationalist unrest in the republics.

When an apparent coup against Gorbachev became known on Aug. 19, 1991, the pres. of the Russian Republic, Boris Yeltsin, denounced it and called for a general strike. Some 50,000 demonstrated at the Russian Parliament in support of Yeltsin. By Aug. 21, the coup had failed and Gorbachev was restored as president. On Aug. 24, Gorbachev resigned as leader of the Communist Party. Several republics declared their independence, including Russia, Ukraine, and Kazakhstan. On Aug. 29, the Soviet Parliament voted to suspend all activities of the Communist Party.

The Soviet Union officially broke up Dec. 26, 1991, one day after Gorbachev resigned. The Soviet hammer and sickle flying over the Kremlin was lowered and replaced by the flag of Russia, ending the domination of the Communist Party over all areas of national life since 1917.

Russian Federation

In a first major step in radical economic reform, Russia eliminated state subsidies of most goods and services, Jan. 1992. The effect was to allow prices to soar far beyond the means of ordinary workers. In June, Pres. Yeltsin and U.S. Pres. George Bush agreed to massive arms reductions.

Russia launched a drive to privatize thousands of large and medium-sized state-owned enterprises in 1993. Yeltsin narrowly survived an impeachment vote by the Congress of People's Deputies, Mar. 28. He received strong support from voters in a referendum Apr. 25, but he continued to face a legislature dominated by conservatives and former Communists.

On Sept. 21, 1993, Yeltsin called for early elections and dissolved Parliament, which in turn declared him deposed. Anti-Yeltsin legislators then barricaded themselves in the Parliament building. On Oct. 3, anti-Yeltsin forces attacked some facilities in Moscow and broke into the Parliament building. Yeltsin ordered the army to attack and seize the building. About 140 people were killed in the fighting, according to medical authorities. More than 150 were arrested.

In elections Dec. 12, 1993, a Yeltsin-supported constitution was approved, but ultranationalists and Communist hard-liners made strong showings in legislative contests. In Dec. 1994 the Russian government sent troops into the breakaway republic of Chechnya. Grozny, the Chechen capital, fell in Feb. 1995 after heavy fighting, but Chechen rebels continued to resist.

Communists made further gains in parliamentary elections Dec. 17, 1995. Despite poor health, Yeltsin won a presidential runoff election over a Communist opponent, July 3, 1996. On Aug. 14, after rebels embarrassed the Russian military by retaking Grozny, Yeltsin gave his security chief, Alexander Lebed, broad powers to negotiate an end to the Chechnya war. Lebed and Chechen leaders signed a peace accord Aug. 31. On Oct. 17, Yeltsin dismissed Lebed for insubordination. Yeltsin survived quintuple-bypass heart surgery Nov. 5.

Russian troops remaining in Chechnya were pulled out Jan. 1997. A revitalized Yeltsin revamped his cabinet in Mar. to strengthen the hand of reformers. On May 27, he signed a "founding act" increasing cooperation with NATO and paving the way for NATO to admit Eastern European nations.

Russia's economic crisis deepened throughout 1998; in Aug. the ruble plummeted and the country defaulted on its debt. Yeltsin dismissed Prime Min. Viktor Chernomyrdin on Mar. 23 and Chernomyrdin's successor, Sergei Kiriyenko, on Aug. 23. Each move triggered a confrontation with parliament. Yevgeny Primakov became premier Sept. 11, 1998; in subsequent cabinet upheavals, Sergei Stepashin took over on May 19, 1999, followed by Vladimir Putin on Aug. 16. Disagreements over the war in Kosovo strained relations with the U.S. and NATO from Mar. to July. Russia moved forcibly in Aug. to suppress Islamic rebels in Dagestan; the conflict soon spread to neighboring Chechnya.

Rwanda
Republic of Rwanda

People: Population: 8,154,933. **Age distrib. (%):** <15: 44.2; 65+: 2.8. **Pop. density:** 800 per sq. mi. **Urban:** 6%. **Ethnic groups:** Hutu 80%, Tutsi 19%, Twa (Pygmy) 1%. **Principal languages:** French, Kinyarwanda, English (all official). **Chief religions:** Roman Catholic 65%, indigenous beliefs 25%.

Geography: Area: 10,200 sq. mi. **Location:** In E central Africa. **Neighbors:** Uganda on N, Congo (formerly Zaire) on W, Burundi on S, Tanzania on E. **Topography:** Grassy uplands and hills cover most of the country, with a chain of volcanoes in the NW. The source of the Nile R. has been located in the headwaters of the Kagera (Akagera) R., SW of Kigali. **Capital:** Kigali (1993): 234,500.

Government: Type: Republic. **Head of state:** Pres. Pasteur Bizimungu; b 1951; in office: July 19, 1994. **Head of gov.:** Prime Min. Pierre Celestin Rwigema; b 1953; in office: Aug. 31, 1995. **Local divisions:** 12 prefectures subdivided into 155 communes. **Defense:** 5.5% of GDP. **Active troops:** 55,000.

Economy: Industries: Mining, cement. **Chief crops:** Coffee, tea, pyrethrum, bananas. **Minerals:** Tin, gold, wolframite. **Arable land:** 35%. **Livestock** (1997): chickens: 1.40 mil; goats: 980,000; cattle: 500,000; sheep 270,000. **Electricity prod.** (1997): 164 mil kWh. **Labor force:** 93% agric.

Finance: Monetary unit: Franc (Sept. 1999: 338.79 = $1 U.S.). **GDP:** (1996 est.): $3 bil. **Per capita GDP:** $440. **Imports** (1996 est.): $202.4 mil; partners: Belg.-Lux. 17%, Kenya 13%. **Exports** (1996 est.): $62.3 mil; partners: Germany 21%, Netherlands 19%. **Tourism:** $17 mil. **Budget** (1996 est.): $319 mil. **Intl. reserves less gold** (June 1999): $171.12 mil. **Consumer prices** (change in 1998): 9.1%.

Transport: Motor vehicles: 11,900 pass. cars, 15,900 comm. vehicles. **Civil aviation:** 1.2 mil pass.-mi; 2 airports. **Chief ports:** Gisenyi, Cyangugu.

Communications: Radios: 78.4 per 1,000 pop. **Telephones** (1997): 15,000 main lines.

Health: Life expectancy: 40.84 male; 41.80 female. **Births** (per 1,000 pop.): 38.97. **Deaths** (per 1,000 pop.): 19.53. **Natural inc.:** 1.944%. **Physicians** (1992): 1 per 50,000 persons. **Infant mortality** (per 1,000 live births): 112.86.

Education: Compulsory: ages 7-14. **Literacy:** 60%.

Major Intl. Organizations: UN (FAO, IBRD, ILO, IMF, WHO, WTrO), OAU.

Embassy: 1814 New Hampshire Ave. NW 20007; 232-2882.

For centuries, the Tutsi (an extremely tall people) dominated the Hutu (90% of the population). A civil war broke out in 1959 and Tutsi power was ended. Many Tutsi went into exile. A referendum in 1961 abolished the monarchic system. Rwanda, which had been part of the Belgian UN trusteeship of Rwanda-Urundi, became independent July 1, 1962.

In 1963 Tutsi exiles invaded in an unsuccessful coup; a large-scale massacre of Tutsi followed. Rivalries among Hutu led to a bloodless coup July 1973 in which Juvénal Habyarimana took power. After an invasion and coup attempt by Tutsi exiles in 1990, a multiparty democracy was established.

Renewed ethnic strife led to an Aug. 1993 peace accord between the government and rebels of the Tutsi-led Rwandan Patriotic Front (RPF). But after Habyarimana and the president of Burundi were killed Apr. 6, 1994, in a suspicious plane crash, massive violence broke out. At least 500,000 died in massacres, mainly of Tutsi by Hutu militias, and in civil warfare as the RPF sought power. About 2 million Tutsi and Hutu fled to camps in Zaire (now Congo) and other countries, where many died of cholera and other natural causes. French troops under a UN mandate moved into SW Rwanda June 23 to establish a so-called safe zone. The RPF claimed victory, installing a government in July led by a moderate Hutu president. French troops pulled out Aug. 22. A UN peacekeeping mission ended Mar. 8, 1996, but the Rwandan government and a UN-sponsored tribunal in Tanzania continued to gather evidence against those responsible for genocide. More than 1 million refugees (mostly Hutu) flooded back to Rwanda from Tanzania and Zaire in Nov. and Dec. 1996.

Firing squads in Rwanda on Apr. 24, 1998, executed 22 people convicted of genocide. Former Prime Min. Jean Kambanda pleaded guilty May 1 before the UN tribunal and received a life sentence Sept. 4.

Saint Kitts and Nevis
Federation of Saint Kitts and Nevis

People: Population: 42,838. **Age distrib.** (%):<15: 32.7; 65+: 5.9. **Pop. density:** 412 per sq. mi. **Urban:** 34%. **Ethnic groups:** Black African 95%. **Principal language:** English (official). **Chief religion:** Protestant 85%.

Geography: Area: 104 sq. mi. **Location:** In the N part of the Leeward group of the Lesser Antilles in the E Caribbean Sea. **Neighbors:** Antigua and Barbuda to E. **Capital:** Basseterre (1994 est.): 12,600.

Government: Type: Constitutional monarchy. **Head of state:** Queen Elizabeth II, represented by Gov-Gen. Sir Cuthbert M. Sebastian; b Oct. 22, 1921; in office: Jan. 1, 1996. **Head of gov.:** Prime Min. Denzil Llewellyn Douglas; b Jan. 14, 1953; in office: July 7, 1995. **Local divisions:** 14 parishes.

Economy: Industries: Sugar (main industry), tourism. **Arable land:** 22%. **Electricity prod.** (1997): 83 mil kWh. **Labor force:** 69% services; 31% manuf.

Finance: Monetary unit: East Caribbean Dollar (Sept. 1999: 2.70 = $1 U.S.). **GDP:** (1996 est.): $235 mil. **Per capita GDP:** $5,700. **Imports** (1996 est.): $131.5 mil; partners: U.S. 45%, Caricom nations 19%, UK 13%. **Exports** (1996 est.): $39.1 mil; partners: U.S. 47%, UK 26%. **Tourism:** $71 mil. **Budget:** (1996 est.) $100 mil. **Intl. reserves less gold** (Apr. 1999): $49.19 mil. **Consumer prices** (change in 1996): 2.5%.

Transport: Civil aviation: 2 airports. **Chief ports:** Basseterre, Charlestown.

Communications: TV sets: 241 per 1,000 pop. **Radios:** 659 per 1,000 pop. **Telephones** (1997): 17,200 main lines.

Health: Life expectancy: 64.87 male; 71.21 female. **Births** (per 1,000 pop.): 22.60. **Deaths** (per 1,000 pop.): 8.15. **Natural inc.:** 1.445%. **Hosp. beds** (1995): 1 per 142 persons. **Physicians** (1995): 1 per 1,057 persons. **Infant mortality** (per 1,000 live births): 17.39.

Education: Compulsory for 12 years between ages 5-17. **Literacy** (1992): 90%.

Major Intl. Organizations: UN (FAO, IBRD, ILO, IMF, WHO, WTrO), Caricom, the Commonwealth, OAS, OECS.

Embassy: 3216 New Mexico Ave., NW 20016; 686-2636.

St. Kitts (formerly St. Christopher; known by the natives as Liamuiga) and Nevis were reached (and named) by Columbus in 1493. They were settled by Britain in 1623, but ownership was disputed with France until 1713. They were part of the Leeward Islands Federation, 1871-1956, and the Federation of the West Indies, 1958-62. The colony achieved self-government as an Associated State of the UK in 1967, and became fully independent Sept. 19, 1983. A secession referendum on Nevis, Aug. 10, 1998, fell short of the two-thirds majority required.

Saint Lucia

People: Population: 154,020. **Age distrib.** (%): <15: 33.4; 65+: 5.3. **Urban:** 37%. **Pop. density:** 642 per sq. mi. **Ethnic groups:** Black 90%. **Principal languages:** English (official), French patois. **Chief religions:** Roman Catholic 90%, Protestant 7%.

Geography: Area: 240 sq. mi. **Location:** In E Caribbean, 2d largest of the Windward Isls. **Neighbors:** Martinique to N, St. Vincent to S. **Topography:** Mountainous, volcanic in origin; Soufriere, a volcanic crater, in the S. Wooded mountains run N-S to Mt. Gimie, 3,145 ft., with streams through fertile valleys. **Capital:** Castries (1992 est.): 13,615.

Government: Type: Parliamentary democracy. **Head of state:** Queen Elizabeth II, represented by Gov.-Gen. Calliopa Pearlette Louisy; b June 8, 1946; in office: Sept. 17, 1997. **Head of gov.:** Prime Min. Kenny Anthony; b Jan. 8, 1951; in office: May 24, 1997. **Local divisions:** 11 quarters.

Economy: Industries: Clothing, beverages, tourism, manufacturing. **Chief crops:** Bananas, coconuts, vegetables, root crops, cocoa, citrus. **Other resources:** Forests. **Arable land:** 8%. **Livestock** (1997): chickens: 260,000. **Electricity prod.** (1997): 110 mil kWh. **Labor force:** 43% agric.; 39% services; 18% ind. & commerce.

Finance: Monetary unit: East Caribbean Dollar (Sept. 1999: 2.70 = $1 U.S.). **GDP:** (1996 est.): $600 mil. **Per capita GDP:** $3,800. **Imports** (1996 est.): $270.6 mil; partners: U.S. 36%, Caricom countries 22%, UK 11%. **Exports** (1996 est.): $79.5 mil; partners: UK 50%, U.S. 24%, Caricom countries 16%. **Tourism:** $287 mil. **Budget** (FY 1996-97 est.): $169 mil. **Intl. reserves less gold** (Apr. 1999): $74.38 mil.

Transport: Motor vehicles: 10,000 pass. cars, 9,100 comm. vehicles. **Civil aviation:** 2 airports. **Chief ports:** Castries, Vieux Fort.

Communications: TV sets: 172 per 1,000 pop. **Radios:** 619 per 1,000 pop. **Telephones:** 39,500 main lines.

Health: Life expectancy: 68.14 male; 75.74 female. **Births** (per 1,000 pop.): 21.63. **Deaths** (per 1,000 pop.): 5.58. **Natural inc.:** 1.605%. **Hosp. beds** (1995): 1 per 269 persons. **Physicians** (1995): 1 per 2,159 persons. **Infant mortality** (per 1,000 live births): 16.55.

Education: Compulsory: ages 5-15. **Literacy** (1993): 80%.

Major Intl. Organizations: UN (FAO, IBRD, ILO, IMF, IMO, WHO, WTrO), Caricom, the Commonwealth, OAS, OECS.

Embassy: 3216 New Mexico Ave. NW 20016; 364-6792.

St. Lucia was ceded to Britain by France at the Treaty of Paris, 1814. Self-government was granted with the West Indies Act, 1967. Independence was attained Feb. 22, 1979.

Saint Vincent and the Grenadines

People: Population: 120,519. **Age distrib.** (%):<15: 29.6; 65+: 5.6. **Pop. density:** 927 per sq. mi. **Urban:** 50%. **Ethnic groups:** Black 82%, mixed 14%. **Principal languages:** English (official), French patois. **Chief religions:** Anglican, Methodist, Roman Catholic.

Geography: Area: 130 sq. mi. **Location:** In the E Caribbean, St. Vincent (133 sq. mi.) and the northern islets of the Grenadines form a part of the Windward chain. **Neighbors:** St. Lucia to N, Barbados to E, Grenada to S. **Topography:** St. Vincent is volcanic, with a ridge of thickly wooded mountains running its length. **Capital:** Kingstown (1994): 15,924.

Government: Type: Constitutional monarchy. **Head of state:** Queen Elizabeth II, represented by Gov.-Gen. Sir Charles James Antrobus; b May 14, 1933; in office: June 1, 1996. **Head of gov.:** Prime Min. Sir James Fitz-Allen Mitchell; b May 15, 1931; in office: July 30, 1984. **Local divisions:** 6 parishes.

Economy: Industries: Food processing, cement, furniture, clothing. **Chief crops:** Bananas, coconuts, sweet potatoes. **Arable land:** 10%. **Livestock** (1997): chickens: 200,000. **Electricity prod.** (1997): 64 mil kWh. **Labor force:** 57% services; 26% agric; 17% industry.

Finance: Monetary unit: East Caribbean Dollar (Sept. 1999: 2.70 = $1 U.S.). **GDP:** (1996 est.): $259 mil. **Per capita GDP:** $2,200. **Imports** (1996): $127 mil; partners: U.S. 36%, Caricom countries 28%, UK 13%. **Exports** (1996): $46 mil; partners: Caricom countries 49%, UK 16%, U.S. 10%. **Tourism:** $76 mil. **Budget** (1996 est.): $118 mil. **Intl. reserves less gold** (Apr. 1999): $37.84 mil. **Consumer prices** (change in 1997): 2.1%.

Transport: Motor vehicles: 5,000 pass. cars, 3,200 comm. vehicles. **Civil aviation:** 5 airports. **Chief port:** Kingstown.

Communications: TV sets: 161 per 1,000 pop. **Radios:** 591 per 1,000 pop. **Telephones** (1997): 20,500 main lines.

Health: Life expectancy: 72.29 male; 75.36 female. **Births** (per 1,000 pop.): 18.34. **Deaths** (per 1,000 pop.): 5.23. **Natural inc.:** 1.311%. **Hosp. beds** (1995): 1 per 248 persons. **Physicians** (1992): 1 per 2,000 persons. **Infant mortality** (per 1,000 live births): 15.16.

Education: Literacy (1994): 82%.

Major Intl. Organizations: UN (FAO, IBRD, ILO, IMF, IMO, WHO, WTrO), Caricom, the Commonwealth, OAS, OECS.

Embassy: 3216 New Mexico Ave. NW 20016; 364-6730.

Website: http://www.heraldsvg.com

Columbus landed on St. Vincent on Jan. 22, 1498 (St. Vincent's Day). Britain and France both laid claim to the island in the 17th and 18th centuries; the Treaty of Versailles, 1783, finally ceded it to Britain. Associated State status was granted 1969; independence was attained Oct. 27, 1979.

Samoa (*formerly* Western Samoa)
Independent State of Samoa

People: Population: 229,979. **Age distrib.** (%): <15: 39.0; 65+: 4.1. **Pop. density:** 209 per sq. mi. **Urban:** 21%. **Ethnic groups:** Samoan (Polynesian) 93%, Euronesian (mixed) 7%, **Principal languages:** Samoan, English (both official). **Chief religion:** Christian 99.7%.

Geography: Area: 1,100 sq. mi. **Location:** In the S Pacific O. **Neighbors:** Nearest are Fiji to SW, Tonga to S. **Topography:** Main islands, Savaii (659 sq. mi.) and Upolu (432 sq. mi.), both ruggedly mountainous, and small islands Manono and Apolima. **Capital:** Apia (1995 est.): 33,000.

Government: Type: Constitutional monarchy. **Head of state:** Malietoa Tanumafili II; b Jan. 4, 1913; in office: Jan. 1, 1962. **Head of gov.:** Prime Min. Tuilaepa Sailele Malielegaoi; b 1945; in office: Nov. 23, 1988. **Local divisions:** 11 districts.

Economy: Industries: Timber, tourism. **Chief crops:** Coconuts, taro, yams, bananas. **Other resources:** Hardwoods, fish. **Arable land:** 19%. **Livestock** (1997): chickens: 350,000; pigs: 178,800. **Electricity prod.** (1997): 65 mil kWh. **Labor force:** 65% agric.; 30% services; 5% industry.

Finance: Monetary unit: Tala (Sept. 1999: 3.05 = $1 U.S.). **GDP:** (1996 est.): $450 mil. **Per capita GDP:** $2,100. **Imports** (1996): $100 mil; partners: New Zealand 37%, Australia 22%, Fiji 15%, U.S. 13%. **Exports** (1996): $10 mil; partners: New Zealand 48%, Australia 22%, American Samoa 11%. **Budget** (FY1996-97 est.): $99 mil. **Tourism:** $43 mil. **Intl. reserves less gold** (May 1999): $60.28 mil. **Consumer prices** (change in 1997): 2.2%.

Transport: Motor vehicles (1997): 1,200 pass. cars, 1,400 comm. vehicles. **Civil aviation:** 3 airports. **Chief ports:** Apia, Asau.

Communications: TV sets: 30 per 1,000 pop. **Radios:** 448 per 1,000 pop. **Telephones** (1997): 8,500 main lines.

Health: Life expectancy: 67.43 male; 72.33 female. **Births** (per 1,000 pop.): 28.81. **Deaths** (per 1,000 pop.): 5.40. **Natural inc.:** 2.341%. **Hosp. beds** (1991): 1 per 255 persons. **Physicians** (1992): 1 per 2,682 persons. **Infant mortality** (per 1,000 live births): 30.50.

Education: Free, compulsory: ages 6-16. **Literacy** (1989): 100%.

Major Intl. Organizations: UN (FAO, IBRD, IMF, IMO, WHO), the Commonwealth.

Embassy: 820 2nd Ave., Suite 800D, New York, NY 10017; (212) 599-6196.

Samoa (formerly known as Western Samoa to distinguish it from American Samoa, a small U.S. territory) was a German colony, 1899 to 1914, when New Zealand landed troops and took over. It became a New Zealand mandate under the League of Nations and, in 1945, a New Zealand UN Trusteeship.

An elected local government took office in Oct. 1959, and the country became fully independent Jan. 1, 1962.

San Marino
Most Serene Republic of San Marino

People: Population: 25,061. **Age distrib.** (%): <15: 16.1; 65+: 16.8. **Pop. density:** 1,253 per sq. mi. **Urban:** 95%. **Ethnic groups:** Sammarinese 83%, Italian 12%. **Principal language:** Italian. **Chief religion:** Predominantly Roman Catholic.

Geography: Area: 20 sq. mi. **Location:** In N central Italy near Adriatic coast. **Neighbors:** Completely surrounded by Italy. **Topography:** The country lies on the slopes of Mt. Titano. **Capital:** San Marino (1996 est.): 2,316.

Government: Type: Republic. **Heads of state and gov.:** Two co-regents appt. every 6 months. **Local divisions:** 9 castelli. **Defense:** 1% of GDP.

Economy: Industries: Tourism, woolen goods, electronics, wine, cement, ceramics. **Chief crops:** Wheat, grapes, corn. **Arable land:** 17%. **Labor force:** 55% services, 43% industry.

Finance: Monetary unit: Italian Lira (Sept. 1999: 1808.84 = $1 U.S.). **GDP:** (1997 est.): $500 mil. **Per capita GDP:** $20,000. **Budget** (1995 est.): $320 mil.

Transport: Motor vehicles (1997): 24,825 pass. cars, 4,149 comm. vehicles.

Communications: Radios: 514 per 1,000 pop. **Daily newspaper circ.:** 82 per 1,000 pop.

Health: Life expectancy: 77.59 male; 85.35 female. **Births** (per 1,000 pop.): 10.41. **Deaths** (per 1,000 pop.): 8.22. **Natural inc.:** 0.219%. **Infant mortality** (per 1,000 live births): 5.39.

Education: Compulsory: ages 6-13. **Literacy** (1997): 99%.

Major Intl. Organizations: UN (ILO, IMF, WHO), OSCE.

San Marino claims to be the oldest state in Europe and to have been founded in the 4th century. A Communist-led coalition ruled 1947-57; a similar coalition ruled 1978-86. It has had a treaty of friendship with Italy since 1862.

São Tomé and Príncipe
Democratic Republic of São Tomé and Príncipe

People: Population: 154,878. **Age distrib.** (%): <15: 47.6; 65+: 4.1. **Pop. density:** 387 per sq. mi. **Urban:** 44%. **Ethnic groups:** Mestico (Portuguese-African), African minority (Angola, Mozambique immigrants). **Principal language:** Portuguese (official). **Chief religions:** Roman Catholic, Protestant.

Geography: Area: 400 sq. mi. **Location:** In the Gulf of Guinea about 125 miles off W central Africa. **Neighbors:** Gabon, Equatorial Guinea to E. **Topography:** São Tomé and Príncipe islands, part of an extinct volcano chain, are both covered by lush forests and croplands. **Capital:** São Tomé (1993 est.): 43,000.

Government: Type: Republic. **Head of state:** Pres. Miguel Trovoada; b Dec. 27, 1936; in office: Apr. 3, 1991. **Head of gov.:** Prime Min. Guilherme Posser da Costa; b 1945; in office: Jan. 5, 1999. **Local divisions:** 7 districts.

Economy: Industries: light construction, textiles, soap, beer. **Chief crops:** Cocoa, coconuts. **Arable land:** 2%. **Livestock** (1997): chickens: 280,000. **Electricity prod.** (1997): 6.0 mil kWh.

Finance: Monetary unit: Dobra (Sept. 1999: 2,390.00 = $1 U.S.). **GDP:** (1996 est.): $154 mil. **Per capita GDP:** $1,000. **Imports** (1996 est.): $19.6 mil; partners: Portugal 32%; France 17%. **Exports** (1996 est.): $4.9 mil; partners: Netherlands 76%. **Tourism:** $2 mil. **Budget** (1993 est.): $114 mil. **Intl. reserves less gold** (Feb. 1999): $8.64 mil.

Transport: Civil aviation: 5.8 mil pass.-mi; 2 airports. **Chief ports:** São Tomé, Santo Antonio.

Communications: TV sets: 154 per 1,000 pop. **Radios:** 232 per 1,000 pop. **Telephones:** 3,100 main lines.

Health: Life expectancy: 62.9 male; 65.9 female. **Births** (per 1,000 pop.): 43. **Deaths** (per 1,000 pop.): 8. **Natural inc.:** 3.52%. **Infant mortality** (per 1,000 live births): 55.

Education: Compulsory for 4 years between ages 7-14. **Literacy** (1991): 73%.

Major Intl. Organizations: UN (FAO, IBRD, ILO, IMF, IMO, WHO), OAU.

The islands were discovered in 1471 by the Portuguese, who brought the first settlers—convicts and exiled Jews. Sugar planting was replaced by the slave trade as the chief economic activity until coffee and cocoa were introduced in the 19th century.

Portugal agreed, 1974, to turn the colony over to the Gabon-based Movement for the Liberation of São Tomé and Príncipe, which proclaimed as first president its East German-trained leader, Manuel Pinto da Costa. Independence came July 12, 1975. Democratic reforms were instituted in 1987. In 1991 Miguel Trovoada won the first free presidential election following da Costa's withdrawal. A military coup that ousted Trovoada Aug. 15, 1995, was reversed a week later after Angolan mediation. Trovoada defeated da Costa in a presidential runoff election, July 21, 1996.

Saudi Arabia
Kingdom of Saudi Arabia

People: Population: 21,504,613. **Age distrib.** (%): <15: 43.0; 65+: 2.5. **Pop. density:** 28 per sq. mi. **Urban:** 84%. **Ethnic groups:** Arab 90%, Afro-Asian 10%. **Principal language:** Arabic (official). **Chief religion:** Muslim 100%.

Geography: Area: 756,983 sq. mi. **Location:** Occupies most of Arabian Peninsula in Mid-East. **Neighbors:** Kuwait, Iraq, Jordan on N; Yemen, Oman on S; United Arab Emirates, Qatar on E. **Topography:** Bordered by Red Sea on the W. The highlands on W, up to 9,000 ft., slope as an arid, barren desert to the Persian Gulf on the E. **Capital:** Riyadh. **Cities:** Riyadh 2,619,000; Jeddah 1,492,000; Mecca 777,000*.

Government: Type: Monarchy with council of ministers. **Head of state and gov.:** King Fahd ibn Abdul Aziz; b 1923; in office: June 13, 1982 (prime min. since 1982). **Local divisions:** 13 provinces. **Defense:** 12.4% of GDP. **Active troops:** 162,500.

Economy: Industries: Oil, oil products. **Chief crops:** Dates, wheat, barley, tomatoes, melon, citrus. **Minerals:** Oil, gas, gold, copper, iron. **Crude oil reserves** (1999): 259 bil bbls. **Arable land:** 2%. **Livestock** (1997): chickens: 95.00 mil; sheep: 8.04 mil; goats: 4.39 mil; cattle: 200,000. **Electricity prod.** (1997): 100.000 bil kWh. **Labor force:** 40% govt.; 25% industry & oil; 30% services; 5% agric.

Finance: Monetary unit: Riyal (Sept. 1999: 3.75 = $1 U.S.). **GDP:** (1997 est.): $206.5 bil. **Per capita GDP:** $10,300. **Imports** (1996): $25.4 bil; partners: U.S. 22%, UK 12%. **Exports** (1996): $56.7 bil; partners: Japan 17%, U.S. 15%. **Tourism:** $1.46 bil. **Budget** (1998 est.): $52.3 bil. **Intl. reserves less gold** (June 1999): $7.16 bil. **Gold:** 4.60 mil oz t. **Consumer prices** (change in 1997): −0.4%.

Transport: Railroad: Length: 864 mi. **Motor vehicles:** 1.71 mil pass. cars, 1.17 mil comm. vehicles. **Civil aviation:** 11.8 bil pass.-mi; 25 airports. **Chief ports:** Jiddah, Ad Dammam.

Communications: TV sets: 257 per 1,000 pop. **Radios:** 213 per 1,000 pop. **Telephones:** 2,878,100 main lines. **Daily newspaper circ.:** 54 per 1,000 pop.

Health: Life expectancy: 68.67 male; 72.53 female. **Births** (per 1,000 pop.): 37.38. **Deaths** (per 1,000 pop.): 4.86. **Natural inc.:** 3.252%. **Hosp. beds** (1995): 1 per 427 persons. **Physicians** (1995): 1 per 590 persons. **Infant mortality** (per 1,000 live births): 38.80.

Education: Literacy: 63%.

Major Intl. Organizations: UN (FAO, IBRD, ILO, IMF, IMO, WHO), AL, OPEC.

Embassy: 601 New Hampshire Ave. NW 20037; 342-3800.

Before Muhammad, Arabia was divided among numerous warring tribes and small kingdoms and was at times dominated by larger Arabian and non-Arabian kingdoms. It was united for the first time by Muhammad, in the early 7th century AD. His successors conquered the entire Near East and North Africa, bringing Islam and the Arabic language. But Arabia itself soon returned to its former status.

Nejd, in central Arabia, long an independent state and center of the Wahhabi sect, fell under Turkish rule in the 18th century. In 1913 Ibn Saud, founder of the Saudi dynasty, overthrew the Turks and captured the Turkish province of Hasa in E Arabia; he took the Hejaz region in W Arabia in 1925 and most of Asir, in SW Arabia, by 1926. The discovery of oil in the 1930s transformed the new country.

Ibn Saud reigned until his death, Nov. 1953. Subsequent kings have been sons of Ibn Saud. The king exercises authority together with a Council of Ministers. The Islamic religious code is the law of the land. Alcohol and public entertainments are restricted, and women have an inferior legal status. There is no constitution and no parliament, although a Consultative Council was established by the king in 1993.

Saudi Arabia has often allied itself with the U.S. and other Western nations, and billions of dollars of advanced arms have been purchased from Britain, France, and the U.S.; however, Western support for Israel has often strained relations. Saudi units fought against Israel in the 1948 and 1973 Arab-Israeli wars. Beginning with the 1967 Arab-Israeli war, Saudi Arabia provided large annual financial gifts to Egypt; aid was later extended to Syria, Jordan, and Palestinian groups, as well as to other Islamic countries.

King Faisal played a leading role in the 1973-74 Arab oil embargo against the U.S. and other nations. Crown Prince Khalid was proclaimed king on Mar. 25, 1975, after the assassination of Faisal. Fahd became king on June 13, 1982, following Khalid's death.

The Hejaz contains the holy cities of Islam—Medina, where the Mosque of the Prophet enshrines the tomb of Muhammad, and Mecca, his birthplace. More than 2 million Muslims make pilgrimage to Mecca annually. In 1987, Iranians making a pilgrimage to Mecca clashed with anti-Iranian pilgrims and Saudi police; more than 400 were killed. Some 1,426 Muslim pilgrims died July 2, 1990, in a stampede in a pedestrian tunnel leading to Mecca. Nearly 300 pilgrims were killed in a stampede in Mecca, May 26, 1994. More than 340 pilgrims died in a tent fire near Mecca, Apr. 15, 1997.

Following Iraq's attack on Kuwait, Aug. 2, 1990, Saudi Arabia accepted the Kuwaiti royal family and more than 400,000 Kuwaiti refugees. King Fahd invited Western and Arab troops to deploy on its soil in support of Saudi defense forces. During the Persian Gulf War, 28 U.S. soldiers were killed when an Iraqi missile hit their barracks in Dhahran, Feb. 25, 1991. The nation's northern Gulf coastline suffered severe pollution as a result of Iraqi sabotage of Kuwaiti oil fields. Islamic extremists were blamed for truck bombs that killed 7 (5 from the U.S.) at a military training center in Riyadh, Nov. 13, 1995, and 19 Americans at a base in Dhahran, June 25, 1996. U.S. officials repeatedly chided the Saudi government for failing to cooperate fully in the investigation.

With King Fahd ailing, his half-brother, Crown Prince Abdullah, has taken a leading role in recent years.

Senegal
Republic of Senegal

People: Population: 10,051,930. **Age distrib.** (%): <15: 48.0; 65+: 2.8. **Pop. density:** 133 per sq. mi. **Urban:** 44%. **Ethnic groups:** Wolof 36%, Fulani 17%, Serer 17%, Diola 9%, Toucouleur 9%, Mandingo 9%. **Principal languages:** French (official), Wolof, Pulaar, Diola, Mandingo. **Chief religions:** Muslim 92%, indigenous beliefs 6%, Christian 2%.

Geography: Area: 75,700 sq. mi. **Location:** At W extreme of Africa. **Neighbors:** Mauritania on N, Mali on E, Guinea and Guinea-Bissau on S; surrounds Gambia on three sides. **Topography:** Low rolling plains cover most of Senegal, rising somewhat in the SE. Swamp and jungles are in SW. **Capital:** Dakar (1994 est.): 1,641,358.

Government: Type: Republic. **Head of state:** Pres. Abdou Diouf; b Sept. 7, 1935; in office: Jan. 1, 1981. **Head of gov.:** Prime Min. Mamadou Lamine Loum; b Feb. 3, 1952; in office: July 3, 1998. **Local divisions:** 10 regions. **Defense:** 1.6% of GDP. **Active troops:** 13,400.

Economy: Industries: Food processing, fishing, phosphate mining. **Chief crops:** Peanuts, millet, corn, sorghum, rice. **Minerals:** Phosphates, iron. **Arable land:** 12%. **Livestock** (1997): chickens: 44.10 mil; sheep: 4.24 mil; goats: 3.57 mil; cattle: 2.91 mil; pigs: 320,000. **Fish catch** (1997): 506,966 metric tons. **Electricity prod.** (1997): 1.125 bil kWh. **Labor force:** 60% subsistence agric.

Finance: Monetary unit: CFA Franc (Sept. 1999: 612.79 = $1 U.S.). **GDP:** (1997 est.): $15.6 bil. **Per capita GDP:** $1,850. **Imports** (1996): $1.4 bil; partners: France 30%. **Exports** (1996): $968 mil; partners: France 30%. **Tourism:** $165 mil. **Budget** (1996 est.): $885 mil. **Intl. reserves less gold** (Dec.1998): 430.8 mil. **Gold:** 29,000 oz t. **Consumer prices** (change in 1997): 1.2%.

Transport: Railroad: Length: 562 mi. **Motor vehicles:** 110,000 pass. cars, 50,000 comm. vehicles. **Civil aviation:** 164.6 mil pass.-mi; 7 airports. **Chief ports:** Dakar, Saint-Louis.

Communications: TV sets: 6.9 per 1,000 pop. **Radios:** 93 per 1,000 pop. **Telephones:** 130,500 main lines.

Health: Life expectancy: 54.95 male; 60.78 female. **Births** (per 1,000 pop.): 43.88. **Deaths** (per 1,000 pop.): 10.71. **Natural inc.:** 3.317%. **Hosp. beds** (1992): 1 per 1,041 persons. **Physicians** (1992): 1 per 14,825 persons. **Infant mortality** (per 1,000 live births): 59.81.

Education: Compulsory: ages 7-13. **Literacy:** 33%.

Major Intl. Organizations: UN and all of its specialized agencies, OAU.

Embassy: 2112 Wyoming Ave. NW 20008; 234-0540.

Portuguese settlers arrived in the 15th century, but French control grew from the 17th century. The last independent Muslim state was subdued in 1893. Dakar became the capital of French West Africa.

Independence as part, along with the Sudanese Rep., of the Mali Federation, came June 20, 1960. Senegal withdrew Aug. 20. French political and economic influence remained strong.

Senegal, Dec. 17, 1981, signed an agreement with The Gambia for confederation of the 2 countries, without loss of individual sovereignty, under the name of Senegambia. The confederation collapsed in 1989, although in 1991 the 2 nations signed a friendship and cooperation treaty.

Separatists in Casamance Province of S Senegal have clashed with government forces since 1982. Senegal sent troops in June 1998 to help the Guinea-Bissau government suppress an army uprising.

Seychelles
Republic of Seychelles

People: Population: 79,164. **Age distrib.** (%): <15: 29.4; 65+: 6.3. **Pop. density:** 450 per sq. mi. **Urban:** 55%. **Ethnic groups:** Seychellois (mixture of Asians, Africans, Europeans). **Principal languages:** English, French (both official), Creole. **Chief religions:** Roman Catholic 90%, Anglican 8%.

Geography: Area: 176 sq. mi. **Location:** In the Indian O. 700 miles NE of Madagascar. **Neighbors:** Nearest are Madagascar on SW, Somalia on NW. **Topography:** A group of 86 islands, about half of them composed of coral, the other half granite, the latter predominantly mountainous. **Capital:** Victoria (1993 est.): 25,000.

Government: Type: Republic. **Head of state and gov.:** Pres. France-Albert René, b. Nov. 16, 1935; in office: June 5, 1977. **Local divisions:** 23 districts. **Defense:** 2.9% of GDP. **Active troops:** 200.

Economy: Industries: Tourism, food processing, fishing. **Chief crops:** Coconuts, cinnamon, vanilla. **Arable land:** 2%. **Livestock** (1997): chickens: 640,000. **Electricity prod.** (1997): 122 mil kWh. **Labor force:** 57% services; 19% industry & comm.; 14% govt.; 10% agric.

Finance: Monetary unit: Rupee (Sept. 1999: 5.28 = $1 U.S.). **GDP:** (1997 est.): $550 mil. **Per capita GDP:** $7,000. **Imports** (1995): $238 mil; partners: U.S. 27%; UK 11%. **Exports** (1995): $56.1 mil; partners: China 15%, UK 12%, Thailand 12%. **Tourism:** $120 mil. **Budget** (1994 est.): $241 mil. **Intl. re-**

serves less gold (May 1999): $19.31 mil. **Consumer prices** (change in 1997): 0.6%.

Transport: Motor vehicles: 6,620 pass. cars, 1,880 comm. vehicles. **Civil aviation:** 526.5 mil pass.-mi; 2 airports. **Chief port:** Victoria.

Communications: TV sets: 173.4 per 1,000 pop. **Radios:** 667 per 1,000 pop. **Telephones:** 19,000 main lines. **Daily newspaper circ.:** 41 per 1,000 pop.

Health: Life expectancy: 66.61 male; 75.42 female. **Births** (per 1,000 pop.): 19.39. **Deaths** (per 1,000 pop.): 6.56. **Natural inc.:** 1.283%. **Hosp. beds** (1996): 1 per 184 persons. **Physicians** (1996): 1 per 906 persons. **Infant mortality** (per 1,000 live births): 16.65.

Education: Free, compulsory: ages 6-15. **Literacy:** 84%.

Major Intl. Organizations: UN (FAO, IBRD, ILO, IMF, IMO, WHO), the Commonwealth, OAU.

Embassy: 820 2d Ave., Suite 900F, New York, NY 10017; 212-972-1785.

The islands were occupied by France in 1768, and seized by Britain in 1794. Ruled as part of Mauritius from 1814, the Seychelles became a separate colony in 1903. The ruling party had opposed independence as impractical, but pressure from the OAU and the UN became irresistible, and independence was declared June 29, 1976. The first president was ousted in a coup a year later by a socialist leader. A new constitution, approved June 1993, provided for a multiparty state.

Sierra Leone
Republic of Sierra Leone

People: Population: 5,296,651. **Age distrib.** (%): <15: 45.4; 65+: 3.1. **Pop. density:** 191 per sq. mi. **Urban:** 34%. **Ethnic groups:** Temne 30%, Mende 30%, other tribes 30%. **Principal languages:** English (official), Mende, Temne, Krio. **Chief religions:** Muslim 60%, indigenous beliefs 30%, Christian 10%.

Geography: Area: 27,700 sq. mi. **Location:** On W coast of W Africa. **Neighbors:** Guinea on N and E, Liberia on S. **Topography:** The heavily-indented, 210-mi. coastline has mangrove swamps. Behind are wooded hills, rising to a plateau and mountains in the E. **Capital:** Freetown (1990 est.): 669,000.

Government: Type: Republic. **Head of state and gov.:** Ahmad Tejan Kabbah; b Feb. 16, 1932; in office: Mar. 10, 1998. **Local divisions:** 3 provinces, 1 area. **Defense:** 6.9% of GDP. **Active troops:** 15,000.

Economy: Industries: Mining, light manufacturing. **Chief crops:** Cocoa, coffee, palm kernels, rice. **Minerals:** Diamonds, titanium, bauxite. **Arable land:** 7%. **Livestock** (1997): chickens: 6.00 mil; cattle: 400,000; sheep: 350,000; goats: 190,000. **Fish catch** (1996): 61,330 metric tons. **Electricity prod.** (1997): 230 mil kWh. **Labor force:** 65% agric.; 35% ind. & serv.

Finance: Monetary unit: Leone (Sept. 1999: 1,776.20 = $1 U.S.). **GDP:** (1997 est.): $2.65 bil. **Per capita GDP:** $540. **Imports** (1996): $211 mil; partners: U.S. 43%. **Exports** (1996): $47 mil; partners: U.S. 20%, Belgium 20%, Spain 13%. **Tourism:** $57 mil. **Budget** (1996 est.): $150 mil. **Intl. reserves less gold** (June 1999): $25.3 mil. **Consumer prices** (change in 1997): 35.5%.

Transport: Railroad: Length: 52 mi. **Motor vehicles:** 20,860 pass. cars, 21,074 comm. vehicles. **Civil aviation:** 14.9 mil pass.-mi; 1 airport. **Chief ports:** Freetown, Bonthe.

Communications: Radios: 72 per 1,000 pop. **Telephones** (1997): 17,400 main lines.

Health: Life expectancy: 46.07 male; 52.27 female. **Births** (per 1,000 pop.): 45.62. **Deaths** (per 1,000 pop.): 16.77. **Natural inc.:** 2.885%. **Physicians** (1992): 1 per 10,832 persons. **Infant mortality** (per 1,000 live births): 126.23.

Education: Literacy: 31%.

Major Intl. Organizations: UN (FAO, IBRD, ILO, IMF, IMO, WHO, WTrO), the Commonwealth, OAU.

Embassy: 1701 19th St. NW 20009; 939-9261.

Website: http://www.Sierra-Leone.org

Freetown was founded in 1787 by the British government as a haven for freed slaves. Their descendants, known as Creoles, number more than 60,000.

Successive steps toward independence followed the 1951 constitution. Ten years later, full independence arrived Apr. 27, 1961. Sierra Leone declared itself a republic on Apr. 19, 1971. A one-party state approved by referendum in 1978 brought political stability, but mismanagement and corruption plagued the economy.

Mutinous soldiers ousted Pres. Joseph Momoh Apr. 30, 1992. Another coup, Jan. 16, 1996, paved the way for multiparty elections and a return to civilian rule. A peace accord, signed Nov. 30 with the Revolutionary United Front (RUF), brought a temporary halt to a civil war that had claimed over 10,000 lives in 5 years.

A coup on May 25, 1997, was met with widespread international opposition. Armed intervention by Nigeria restored Pres. Ahmad Tejan Kabbah to power on Mar. 10, 1998, but RUF rebels mounted a guerrilla counteroffensive, reportedly killing thousands of civilians and mutilating thousands more. The Kabbah government signed a power-sharing agreement with the RUF on July 7, 1999.

Singapore
Republic of Singapore

People: Population: 3,531,600. **Age distrib.** (%): <15: 21.3; 65+: 7.0. **Pop. density:** 14,126 per sq. mi. **Urban:** 100%. **Ethnic groups:** Chinese 76%, Malay 15%, Indian 6%. **Principal languages:** Chinese, Malay, Tamil, English (all official). **Chief religions:** Buddhist 32%, Taoist 22%, Muslim 15%, Christian 13%, Hindu 3%.

Geography: Area: 250 sq. mi. **Location:** Off tip of Malayan Peninsula in SE Asia. **Neighbors:** Nearest are Malaysia on N, Indonesia on S. **Topography:** Singapore is a flat, formerly swampy island. The nation includes 40 nearby islets. **Capital:** Singapore (1997 est.) 3,737,000.

Government: Type: Republic. **Head of state:** Pres. S. R. Nathan; b July 3, 1924; in office: Sept. 1, 1999. **Head of gov.:** Prime Min. Goh Chok Tong; b May 20, 1941; in office: Nov. 28, 1990. **Defense:** 4.3% of GDP. **Active troops:** 70,000.

Economy: Industries: Oil refining, electronics, banking, food and rubber processing, biotechnology. **Chief crops:** Copra, rubber, fruit, vegetables. **Arable land:** 2%. **Livestock** (1997): chickens: 2.00 mil; pigs: 190,000. **Fish catch** (1996): 9,943 metric tons. **Electricity prod.** (1997): 23.375 bil kWh. **Labor force:** 34% finance, business, other serv.; 26% manuf.; 23% commerce.

Finance: Monetary unit: Dollar (Sept. 1999: 1.69 = $1 U.S.). **GDP:** (1997 est.): $84.6 bil. **Per capita GDP:** $24,600. **Imports** (1997 est.): $133.9 bil; partners: Japan 21%, Malaysia 15%, U.S. 15%. **Exports** (1997 est.): $125.6 bil; partners: Malaysia 19%, U.S. 18%. **Tourism:** $6.50 bil. **Budget** (FY 1997-98 est.): $13.6 bil. **International reserves** (May 1999): $73.20 bil. **Consumer prices** (change in 1997): −0.3%.

Transport: Railroad: Length: 52 mi. **Motor vehicles** (1997): 379,497 pass. cars, 140,827 comm. vehicles. **Civil aviation:** 34.5 bil pass.-mi; 1 airport. **Chief port:** Singapore.

Communications: TV sets: 218 per 1,000 pop. **Radios:** 270 per 1,000 pop. **Telephones:** 1,777,900 main lines. **Daily newspaper circ.:** 340 per 1,000 pop.

Health: Life expectancy: 75.79 male; 82.14 female. **Births** (per 1,000 pop.): 13.38. **Deaths** (per 1,000 pop.): 4.69. **Natural inc.:** 0.869%. **Hosp. beds** (1996): 1 per 285 persons. **Physicians** (1996): 1 per 667 persons. **Infant mortality** (per 1,000 live births): 3.84.

Education: Literacy: 91%.

Major Intl. Organizations: UN (IBRD, ILO, IMF, IMO, WHO, WTrO), the Commonwealth, APEC, ASEAN.

Embassy: 3501 International Pl. NW 20008; 537-3100.

Website: http://www.singstat.gov.sg

Founded in 1819 by Sir Thomas Stamford Raffles, Singapore was a British colony until 1959, when it became autonomous within the Commonwealth. On Sept. 16, 1963, it joined with Malaya, Sarawak, and Sabah to form the Federation of Malaysia. Tensions between Malayans, dominant in the federation, and ethnic Chinese, dominant in Singapore, led to an accord under which Singapore became a separate nation, Aug. 9, 1965.

Singapore is one of the world's largest ports. Standards in health, education, and housing are generally high. International banking has grown rapidly in recent years. The government, dominated by a single party, has taken strong actions to suppress dissent.

Slovakia
Slovak Republic

People: Population: 5,396,193. **Age distrib.** (%): <15: 20.0; 65+: 11.4. **Pop. density:** 286 per sq. mi. **Urban:** 59%. **Ethnic groups:** Slovak 86%, Hungarian 11%. **Principal languages:** Slovak (official), Hungarian. **Chief religions:** Roman Catholic 60%, Protestant 8%.

Geography: Area: 18,859 sq. mi. **Location:** In E central Europe. **Neighbors:** Poland on N, Hungary on S, Austria and Czech Rep. on W, Ukraine on E. **Topography:** Mountains (Carpathians) in N, fertile Danube plane in S. **Capital:** Bratislava. **Cities** (1996 est.): Bratislava 452,278; Kosice 241,163.

Government: Type: Republic. **Head of state:** Rudolf Schuster; b Jan. 4, 1934; in office: June 15, 1999. **Head of gov.:** Prime Min. Mikulás Dzurinda; b Feb. 4, 1955; in office: Oct. 30, 1998. **Local divisions:** 8 departments. **Defense:** 2.1% of GDP. **Active troops:** 41,200.

Economy: Industries: Metal products, food and beverages, oil, chemicals. **Chief crops:** Grains, potatoes, sugar beets, hops, fruit. **Minerals:** Coal, lignite, iron, copper. **Crude oil reserves** (1999): 9.0 mil bbls. **Arable land:** 31%. **Livestock** (1997): chickens: 14.22 mil; pigs: 1.81 mil; cattle: 803,398; sheep: 417,337. **Electricity prod.** (1997): 24.236 bil kWh. **Labor force:** 29% ind.; 9% agric.; 8% constr.

Finance: Monetary unit: Koruna (Sept. 1999: 41.04 = $1 U.S.). **GDP:** (1997 est.): $46.3 bil. **Per capita GDP:** $8,600. **Imports** (1996): $11.1 bil; partners: EU 36.9%, Czech Rep. 24.8%. **Exports** (1996): $8.8 bil; partners: EU 41%, Czech Rep. 30.6%. **Tourism:** $480 mil. **Budget** (1996): $6.4 bil. **Intl. reserves less gold** (May 1999): 2.46 bil. **Gold:** 1.29 mil oz t. **Consumer prices** (change in 1997): 6.7%.

Transport: Railroad: Length: 2,277 mi. **Motor vehicles:** 994,000 pass. cars, 94,000 comm. vehicles. **Civil aviation:** 63.8 mil pass.-mi; 2 airports. **Chief ports:** Bratislava, Komarno.

Communications: TV sets: 216 per 1,000 pop. **Telephones:** 1,539,300 main lines. **Daily newspaper circ.:** 256 per 1,000 pop.

Health: Life expectancy: 69.71 male; 77.40 female. **Births** (per 1,000 pop.): 9.52. **Deaths** (per 1,000 pop.): 9.43. **Natural inc.:** 0.009%. **Hosp. beds** (1995): 1 per 86 persons. **Physicians** (1995): 1 per 381 persons. **Infant mortality** (per 1,000 live births): 9.48.

Education: Compulsory: ages 6-14. **Literacy** (1994): 100%.

Major Intl. Organizations: UN (FAO, IBRD, ILO, IMF, IMO, WHO, WTrO), OSCE.

Embassy: 2201 Wisconsin Ave. NW 20007; 965-5161. **Website:** http://www.slovakemb.com/index.html

Slovakia was originally settled by Illyrian, Celtic, and Germanic tribes and was incorporated into Great Moravia in the 9th century. It became part of Hungary in the 11th century. Overrun by Czech Hussites in the 15th century, it was restored to Hungarian rule in 1526. The Slovaks disassociated themselves from Hungary after World War I and joined the Czechs of Bohemia to form the Republic of Czechoslovakia, Oct. 28, 1918.

Germany invaded Czechoslovakia, 1939, and declared Slovakia independent. Slovakia rejoined Czechoslovakia in 1945.

Czechoslovakia split into 2 separate states—the Czech Republic and Slovakia—on Jan. 1, 1993. Slovakia, with its less developed economy, applied to join the European Union in 1995. A prolonged parliamentary standoff left the country without a president for much of 1998.

Prime Min. Vladimir Meciar, a nationalist, suffered a setback in legislative elections Sept. 25-26, 1998, and was defeated in a presidential runoff vote by Rudolf Schuster, May 29, 1999.

Slovenia
Republic of Slovenia

People: Population: 1,970,570. **Age distrib.** (%): <15: 16.2; 65+: 13.7. **Pop. density:** 252 per sq. mi. **Urban:** 52%. **Ethnic groups:** Slovene 91%, Croat 3%. **Principal languages:** Slovenian (official), Serbo-Croatian. **Chief religion:** Roman Catholic 71%.

Geography: Area: 7,821 sq. mi. **Location:** In SE Europe. **Neighbors:** Italy on W, Austria on N, Hungary on NE, Croatia on SE, S. **Topography:** Mostly hilly; 42% of the land is forested. **Capital:** Ljubljana (1996 est.): 273,000.

Government: Type: Republic. **Head of state:** Pres. Milan Kucan; b Jan. 14, 1941; in office: Apr. 1990. **Head of gov.:** Prime Min. Janez Drnovsek; b May 1950; in office: May 14, 1992. **Local divisions:** 136 municipalities, 11 urban municipalities. **Defense:** 1.7% of GDP. **Active troops:** 9,600.

Economy: Industries: Metallurgy, electronics, vehicles. **Minerals:** Coal, lead, zinc, mercury. **Chief crops:** Potatoes, hops, wheat. **Arable land:** 12%. **Livestock** (1997): chickens: 8.55 mil; pigs: 578,193; cattle: 445,724. **Electricity prod.** (1997): 12.979 bil kWh. **Labor force:** 62% services; 36% ind.

Finance: Monetary unit: Tolar (Sept. 1999: 183.60 = $1 U.S.). **GDP:** (1997 est.): $195 bil. **Per capita GDP:** $10,000. **Imports** (1996): $9.5 bil; partners: Germany 22%, Italy 17%. **Exports** (1996): $8.3 bil; partners: Germany 31%, former Yugoslavia 16.5%, Italy 13%. **Tourism:** $931 mil. **Budget** (1996 est.): $8.53 bil. **Intl. reserves less gold** (May 1999): $3.53 bil. **Gold:** 3,000 oz t. **Consumer prices** (change in 1997): 8.6%.

Transport: Railroad: Length: 746 mi. **Motor vehicles:** 657,000 pass. cars, 37,000 comm. vehicles. **Civil aviation:** 233.0 mil pass.-mi; 1 airport. **Chief ports:** Izola, Koper, Piran.

Communications: TV sets: 374 per 1,000 pop. **Radios:** 317 per 1,000 pop. **Telephones** (1997): 722,500 main lines. **Daily newspaper circ.:** 181 per 1,000 pop.

Health: Life expectancy: 71.71 male; 79.21 female. **Births** (per 1,000 pop.): 8.97. **Deaths** (per 1,000 pop.): 9.62. **Natural inc.:** −0.065%. **Hosp. beds** (1995): 1 per 173 persons. **Physi-**

cians (1995): 1 per 858 persons. **Infant mortality** (per 1,000 live births): 5.28.

Education: Free, compulsory: ages 6-15. **Literacy** (1993): 99%.

Major Intl. Organizations: UN (FAO, IBRD, ILO, IMF, IMO, WHO, WTrO), OSCE.

Embassy: 1525 New Hampshire Ave. NW 20036; 667-5363.

The Slovenes settled in their current territory during the period from the 6th to the 8th century. They fell under German domination as early as the 9th century. Modern Slovenian political history began after 1848 when the Slovenes, who were divided among several Austrian provinces, began their struggle for political and national unification. In 1918 a majority of Slovenes became part of the Kingdom of Serbs, Croats, and Slovenes, later renamed Yugoslavia.

Slovenia declared independence June 25, 1991, and joined the UN May 22, 1992. Linked by trade with the European Union, Slovenia applied for full membership June 10, 1996.

Solomon Islands

People: Population: 455,429. **Age distrib.** (%): <15: 44.8; 65+: 3.0. **Pop. density:** 41 per sq. mi. **Urban:** 18%. **Ethnic groups:** Melanesian 93%, Polynesian 4%. **Principal languages:** English (official); Melanesian, Polynesian languages. **Chief religions:** Anglican 34%, Roman Catholic 19%, Baptist 17%, other Christian 26%.

Geography: Area: 11,000 sq. mi. **Location:** Melanesian Archipelago in the W Pacific O. **Neighbors:** Nearest is Papua New Guinea to W. **Topography:** 10 large volcanic and rugged islands and 4 groups of smaller ones. **Capital:** Honiara (1996 est.): 43,643.

Government: Type: Parliamentary democracy within the Commonwealth of Nations. **Head of state:** Queen Elizabeth II, represented by Gov.-Gen. John Lapli; in office: July 7, 1999. **Head of gov.:** Prime Min. Bartholomew Ulufa'alu; in office: Aug. 27, 1997. **Local divisions:** 9 provinces and Honiara.

Economy: Industries: Copra, tuna. **Chief crops:** Coconuts, rice, cocoa, beans. **Minerals:** Gold, bauxite. **Other resources:** Forests. **Arable land:** 1%. **Livestock** (1997): chickens: 185,000. **Fish catch** (1996): 53,275 metric tons. **Electricity prod.** (1997): 30 mil kWh. **Labor force:** 42% services; 24% agric., forestry, fishing.

Finance: Monetary unit: Dollar (Sept. 1999: 5.01 = $1 U.S.). **GDP:** (1997 est.): $1.27 bil. **Per capita GDP:** $3,000. **Imports** (1995 est.): $152 mil; partners: Australia 34%, Japan 16%. **Exports** (1995): $168 mil; partners: Japan 39%, UK 23%. **Tourism:** $7 mil. **Budget** (1997 est.): $168 mil. **Intl. reserves less gold** (June 1999): $54.35 mil. **Consumer prices:** (change in 1997): 8.1%.

Transport: Civil aviation: 46.0 mil pass.-mi; 21 airports. **Chief port:** Honiara.

Communications: TV sets: 16 per 1,000 pop. **Radios:** 96 per 1,000 pop. **Telephones:** 7,900 main lines.

Health: Life expectancy: 69.55 male; 74.75 female. **Births** (per 1,000 pop.): 35.92. **Deaths** (per 1,000 pop.): 4.11. **Natural inc.:** 3.181%. **Infant mortality** (per 1,000 live births): 23.00.

Education: Literacy (1994): 54%.

Major Intl. Organizations: UN (FAO, IBRD, ILO, IMF, IMO, WHO, WTrO), the Commonwealth.

Embassy: 820 Second Ave., Suite 800, New York, NY 10017; (212) 599-6193.

The Solomon Islands were sighted in 1568 by an expedition from Peru. Britain established a protectorate in the 1890s over most of the group, inhabited by Melanesians. The islands saw major World War II battles. Self-government came Jan. 2, 1976, and independence was formally attained July 7, 1978.

Somalia

People: Population: 7,140,643. **Age distrib.** (%): <15: 44.4; 65+: 2.9. **Pop. density:** 29 per sq. mi. **Urban:** 26%. **Ethnic groups:** Somali 85%, Bantu, Arab. **Principal languages:** Somali (official), Arabic, Italian, English. **Chief religion:** Sunni Muslim.

Geography: Area: 246,200 sq. mi. **Location:** Occupies the eastern horn of Africa. **Neighbors:** Djibouti, Ethiopia, Kenya on W. **Topography:** The coastline extends for 1,700 mi. Hills cover the N; the center and S are flat. **Capital:** Mogadishu 997,000*.

Government: Type: In transition. **Local divisions:** 18 regions. **Defense:** 4.8% of GDP. **Active troops:** 225,000.

Economy: Chief crops: Sugar, bananas, sorghum, corn, mangoes. **Minerals:** Uranium, iron, tin, gypsum, bauxite. **Arable land:** 2%. **Livestock** (1997): chickens: 3.00 mil; sheep: 13.50 mil; goats: 12.50 mil; cattle: 5.30 mil. **Fish catch** (1996):

15,000 metric tons. **Electricity prod.** (1997): 265 mil kWh. **Labor force:** 71% nomadic agric; 29% industry & services.
Finance: Monetary unit: Shilling (Sept. 1999: 2,620.00 = $1 U.S.). **GDP:** (1996 est.): $8 bil. **Per capita GDP:** $600. **Imports** (1994 est.): $269 mil; partners: Kenya 24%, Djibouti 18%. **Exports** (1994 est.): $130 mil; partners: Saudi Arabia 57%, Yemen 14%, Italy 13%.
Transport: Motor vehicles: 10,000 pass. cars, 10,000 comm. vehicles. **Civil aviation:** 86.9 mil pass.-mi; 1 airport. **Chief ports:** Mogadishu, Berbera.
Communications: TV sets: 18 per 1,000 pop. **Radios:** 45 per 1,000 pop. **Telephones** (1997): 15,000 main lines.
Health: Life expectancy: 44.66 male; 47.85 female. **Births** (per 1,000 pop.): 47.98. **Deaths** (per 1,000 pop.): 18.62. **Natural inc.:** 2.936%. **Infant mortality** (per 1,000 live births): 125.77.
Education: Free, compulsory: ages 6-14. **Literacy** (1990): 24%.
Major Intl. Organizations: UN (FAO, IBRD, ILO, IMF, IMO, WHO), AL, OAU.
Website: http://gaia.info.usaid.gov/horn/somalia/somalia.html

British Somaliland (present-day N Somalia) was formed in the 19th century, as was Italian Somaliland (now central and S Somalia). Italy lost its African colonies in World War II. In 1949, the UN approved eventual independence for the former Italian colony (designated the UN Trust Territory of Somalia) after a 10-year period under Italian administration.

British Somaliland gained independence, June 26, 1960, and by prearrangement, merged July 1 with the trust territory of Somalia to create the independent Somali Republic (Somalia).

On Oct. 16, 1969, Pres. Abdi Rashid Ali Shirmarke was assassinated. On Oct. 21, a military group led by Maj. Gen. Muhammad Siad Barre seized power. In 1970, Barre declared the country a socialist state—the Somali Democratic Republic.

Somalia has laid claim to Ogaden, the huge eastern region of Ethiopia, peopled mostly by Somalis. Ethiopia battled Somali rebels in 1977. Some 11,000 Cuban troops with Soviet arms defeated Somali army troops and ethnic Somali rebels in Ethiopia, 1978. As many as 1.5 million refugees entered Somalia. Guerrilla fighting in Ogaden continued until 1988, when a peace agreement was reached with Ethiopia.

The civil war intensified again and Barre was forced to flee the capital, Jan. 1991. Fighting between rival factions caused 40,000 casualties in 1991 and 1992, and by mid-1992 the civil war, drought, and banditry combined to produce a famine that threatened some 1.5 million people with starvation.

In Dec. 1992 the UN accepted a U.S. offer of troops to safeguard food delivery to the starving. The UN took control of the multinational relief effort from the U.S. May 4, 1993. While the operation helped alleviate the famine, efforts to reestablish order foundered, and there were significant U.S. and other casualties. The U.S. withdrew its peacekeeping forces Mar. 25, 1994.

When the last UN troops pulled out Mar. 3, 1995, Mogadishu had no functioning central government, and armed factions controlled different regions. By 1999 a joint police force was operating in the capital, but much of the country, especially in S Somalia, faced continued violence and food shortages.

South Africa
Republic of South Africa

People: Population: 43,426,386. **Age distrib.** (%): <15: 34.4; 65+: 4.6. **Pop. density:** 92 per sq. mi. **Urban:** 50%. **Ethnic groups:** Black 75%, white 14%, colored 9%. **Principal languages:** 11 official languages incl. Afrikaans, English, Ndebele, Pedi, Sotho. **Chief religions:** Christian 68%; traditional, animistic 29%.
Geography: Area: 471,009 sq. mi. **Location:** At the southern extreme of Africa. **Neighbors:** Namibia, Botswana, Zimbabwe on N; Mozambique, Swaziland on E; surrounds Lesotho. **Topography:** The large interior plateau reaches close to the country's 2,700-mi. coastline. There are few major rivers or lakes; rainfall is sparse in W, more plentiful in E. **Capitals:** Cape Town (legislative), Pretoria (administrative), and Bloemfontein (judicial). **Cities:** Cape Town 2,350,157; Johannesburg 1,916,063; Pretoria 1,080,187.
Government: Type: Republic. **Head of state and gov.:** Pres. Thabo Mvuyelwa Mbeki; b: June 18, 1942; in office: June 16 1999. **Local divisions:** 9 provinces. **Defense:** 1.8% of GDP. **Active troops:** 79,400.
Economy: Industries: Mining, steel, chemicals, vehicles, machinery, textiles. **Chief crops:** Corn, wheat, vegetables, sugar, fruit. **Minerals:** Platinum, chromium, antimony, coal, iron, manganese, nickel, phosphates, tin, uranium, gem diamonds, copper, vanadium; world's largest producer of gold (ap-

prox. 30% of total world prod.) **Crude oil reserves** (1999): 29.36 mil bbls. **Other resources:** Wool, dairy products. **Arable land:** 10%. **Livestock** (1997): chickens: 59.00 mil; sheep: 29.98 mil; cattle: 13.80 mil; goats: 7.00 mil; pigs: 1.64 mil. **Fish catch** (1997): 509,390 metric tons. **Electricity prod.** (1997): 195.882 bil kWh. **Labor force:** 35% services; 30% agric.; 20% ind.
Finance: Monetary unit: Rand (Sept. 1999: 6.03 = $1 U.S.). **GDP:** (1997 est.): $270 bil. **Per capita GDP** $6,200. **Imports** (1997): $28 bil; partners: Germany 16%, UK 12%, U.S. 11%. **Exports** (1997): $31.3 bil; partners: Italy 8%, Japan 7%. **Tourism:** $2.37 bil. **Budget** (FY 1994-95 est.): $38 bil. **Intl. reserves less gold** (June 1999): $4.86 bil. **Gold:** 4.05 mil oz t. **Consumer prices** (change in 1997): 6.9%.
Transport: Railroad: Length: 13,418 mi. **Motor vehicles** (1997): 4.35 mil pass. cars, 1.65 mil comm. vehicles. **Civil aviation:** 10.5 bil pass.-mi; 24 airports. **Chief ports:** Durban, Cape Town, East London, Port Elizabeth.
Communications: TV sets: 84 per 1,000 pop. **Radios:** 268 per 1,000 pop. **Telephones** (1997): 4,645,100 main lines. **Daily newspaper circ.:** 29 per 1,000 pop.
Health: Life expectancy: 52.68 male; 56.90 female. **Births** (per 1,000 pop.): 25.94. **Deaths** (per 1,000 pop.): 12.81. **Natural inc.:** 1.313%. **Hosp. beds** (1995): 1 per 239 persons. **Physicians** (1994): 1 per 1,529 persons. **Infant mortality** (per 1,000 live births): 51.99.
Education: Compulsory: ages 7-16. **Literacy:** 82%.
Major Intl. Organizations: UN (FAO, IBRD, ILO, IMF, IMO, WHO, WTrO), the Commonwealth, OAU.
Embassy: 3051 Massachusetts Ave. NW 20008; 232-4400.
Website: http://www.statssa.gov.za

Bushmen and Hottentots were the original inhabitants. Bantus, including Zulu, Xhosa, Swazi, and Sotho, had occupied the area from NE to S South Africa before the 17th century.

The Cape of Good Hope area was settled by Dutch, beginning in the 17th century. Britain seized the Cape in 1806. Many Dutch trekked north and founded 2 republics, Transvaal and Orange Free State. Diamonds were discovered, 1867, and gold, 1886. The Dutch (Boers) resented encroachments by the British and others; the Anglo-Boer War followed, 1899-1902. Britain won and, effective May 31, 1910, created the Union of South Africa, incorporating 2 British colonies (Cape and Natal) with Transvaal and Orange Free State. After a referendum, the Union became the Republic of South Africa, May 31, 1961, and withdrew from the Commonwealth.

With the election victory of Daniel Malan's National Party in 1948, the policy of separate development of the races, or apartheid, already existing unofficially, became official. Under apartheid, blacks were severely restricted to certain occupations, and paid far lower wages than whites for similar work. Only whites could vote or run for public office. Persons of Asian Indian ancestry and those of mixed race (Coloureds) had limited political rights. In 1959 the government passed acts providing for the eventual creation of several Bantu nations, or Bantustans.

Protests against apartheid were brutally suppressed. At Sharpeville on Mar. 21, 1960, 69 black protesters were killed by government troops. At least 600 persons, mostly Bantus, were killed in 1976 riots protesting apartheid. In 1981, South Africa launched military operations in Angola and Mozambique to combat guerrilla groups.

A new constitution was approved by referendum, Nov. 1983, extending the parliamentary franchise to the Coloured and Asian minorities. Laws banning interracial sex and marriage were repealed in 1985.

In 1986, Nobel Peace Prize winner Bishop Desmond Tutu called for Western nations to apply sanctions against South Africa to force an end to apartheid. Pres. P. W. Botha announced in Apr. the end to the nation's system of racial pass laws and offered blacks an advisory role in government. On May 19, South Africa attacked 3 neighboring countries—Zimbabwe, Botswana, Zambia—to strike at guerrilla strongholds of the black nationalist African National Congress (ANC). A nationwide state of emergency was declared June 12, giving almost unlimited power to the security forces.

Some 2 million South African black workers staged a massive strike, June 6-8, 1988. Pres. Botha, head of the government since 1978, resigned Aug. 14, 1989, and was replaced by F. W. de Klerk. In 1990 the government lifted its ban on the ANC. Black nationalist leader Nelson Mandela was freed Feb. 11 after more than 27 years in prison. In Feb. 1991, Pres. de Klerk announced plans to end all apartheid laws.

In 1993 negotiators agreed on basic principles for a new democratic constitution. South Africa's partially self-governing black territories, or "homelands," were dissolved and incorporated into a national system of 9 provinces. In elections Apr. 26-29, 1994, the ANC won 62.7% of the vote, making Mandela president. The National Party won 20.4%. The Inkatha Freedom Party won 10.5% and control of the legislature in a mainly Zulu province. By then, fighting between the ANC and Inkatha (aided, during the apartheid era, by South African defense forces) had killed more than 14,000 people in the Zulu region since the mid-1980s.

In 1995, Mandela appointed a truth commission, led by Desmond Tutu, to document human rights abuses under apartheid. A post-apartheid constitution, modified to meet the objections of the Constitutional Court, became law Dec. 10, 1996, with provisions to take effect over a 3-year period.

The ANC won a landslide victory in elections held June 2, 1999. ANC leader Thabo Mbeki, Mandela's deputy president, thus became South Africa's 2d popularly elected president.

Spain
Kingdom of Spain

People: Population: 39,167,744. **Age distrib.** (%): <15: 14.9; 65+: 16.6. **Pop. density:** 201 per sq. mi. **Urban:** 77%. **Ethnic groups:** Mix of Mediterranean and Nordic types. **Principal languages:** Castilian Spanish (official), Catalan, Galician, Basque. **Chief religion:** Roman Catholic 99%.

Geography: Area: 194,880 sq. mi. **Location:** In SW Europe. **Neighbors:** Portugal on W, France on N. **Topography:** The interior is a high, arid plateau broken by mountain ranges and river valleys. The NW is heavily watered, the S has lowlands and a Mediterranean climate. **Capital:** Madrid. **Cities** (1996 est.): Madrid 2,866,850; Barcelona 1,508,805; Valencia 746,683.

Government: Type: Constitutional monarchy. **Head of state:** King Juan Carlos I de Borbon y Borbon, b Jan. 5, 1938; in office: Nov. 22, 1975. **Head of gov.:** Prime Min. José María Aznar; b Feb. 25, 1953; in office: May 5, 1996. **Local divisions:** 17 automonous communities. **Defense:** 1.4% of GDP. **Active troops:** 197,500.

Economy: Industries: Machinery, metals, textiles, shoes, vehicles, processed foods, tourism. **Chief crops:** Grains, olives, grapes, citrus, vegetables. **Minerals:** Lignite, uranium, iron, mercury, pyrites, fluorspar, gypsum, zinc, lead, coal. **Crude oil reserves** (1997): 14 mil bbls. **Other resources:** Forests. **Arable land:** 30%. **Livestock** (1997): chickens: 127.00 mil; sheep: 24.54 mil; pigs: 19.35 mil; cattle: 5.84 mil; goats: 2.90 mil. **Fish catch** (1997): 1.10 mil metric tons. **Electricity prod.** (1997): 163.976 bil kWh. **Labor force:** 64% serv.; 28% manuf., mining, const.; 8% agric.

Finance: Monetary unit: Peseta (Sept. 1999: 155.44 = $1 U.S.). **Euro** (Sept. 1999: 1.07 = $1 U.S.) **GDP:** (1997 est.): $642.4 bil. **Per capita GDP:** $16,400. **Imports** (1995): $118.3 bil; partners: EU 66%. **Exports** (1995): $94.5 bil; partners: EU 72%. **Tourism:** $29.59 bil. **Budget** (1995): $139 bil. **Intl. reserves less gold** (June 1998): $35.02 bil. **Gold:** 16.80 mil oz t. **Consumer prices** (change in 1997): 1.8%.

Transport: Railroad: Length: 8,252 mi. **Motor vehicles** (1997): 15.30 mil pass. cars, 3.36 mil comm. vehicles. **Civil aviation:** 23.1 bil pass.-mi; 25 airports. **Chief ports:** Barcelona, Bilbao, Valencia, Cartagena.

Communications: TV sets: 490 per 1,000 pop. **Radios:** 306 per 1,000 pop. **Telephones:** 16,288,600 main lines. **Daily newspaper circ.:** 104 per 1,000 pop.

Health: Life expectancy: 73.97 male; 81.71 female. **Births** (per 1,000 pop.): 9.99. **Deaths** (per 1,000 pop.): 9.69. **Natural inc.:** 0.030%. **Hosp. beds** (1994): 1 per 234 persons. **Physicians** (1995): 1 per 241 persons. **Infant mortality** (per 1,000 live births): 6.41.

Education: Free, compulsory: ages 6-16. **Literacy:** 97%.

Major Intl. Organizations: UN and all of its specialized agencies, EU, NATO, OECD, OSCE.

Embassy: 2375 Pennsylvania Ave. NW 20037; 452-0100.

Website: http://www.DocuWeb.ca/SiSpain

Initially settled by Iberians, Basques, and Celts, Spain was successively ruled (wholly or in part) by Carthage, Rome, and the Visigoths. Muslims invaded Iberia from North Africa in 711. Reconquest of the peninsula by Christians from the N laid the foundations of modern Spain. In 1469 the kingdoms of Aragon and Castile were united by the marriage of Ferdinand II and Isabella I. Moorish rule ended with the fall of the kingdom of Granada, 1492. Spain's large Jewish community was expelled the same year.

Spain obtained a colonial empire with the "discovery" of America by Columbus, 1492, the conquest of Mexico by Cortes, and Peru by Pizarro. It also controlled the Netherlands and parts of Italy and Germany. Spain lost its American colonies in the early 19th century. It lost Cuba, the Philippines, and Puerto Rico during the Spanish-American War, 1898.

Primo de Rivera became dictator in 1923. King Alfonso XIII revoked the dictatorship, 1930, but was forced to leave the country in 1931. A republic was proclaimed, which disestablished the church, curtailed its privileges, and secularized education. During 1936-39 a Popular Front composed of socialists, Communists, republicans, and anarchists governed Spain.

Army officers under Francisco Franco revolted against the government, 1936. In a destructive 3-year war, in which some one million died, Franco received massive help and troops from Italy and Germany, while the USSR, France, and Mexico supported the republic. The war ended Mar. 28, 1939. Franco was named caudillo, leader of the nation. Spain was officially neutral in World War II, but its cordial relations with fascist countries caused its exclusion from the UN until 1955.

In July 1969, Franco and the Cortes (Parliament) designated Prince Juan Carlos as the future king and chief of state. After Franco's death, Nov. 20, 1975, Juan Carlos was sworn in as king. In free elections June 1977, moderates and democratic socialists emerged as the largest parties.

In 1981 a coup attempt by right-wing military officers was thwarted by the king. The Socialist Workers' Party, under Felipe González Márquez, won 4 consecutive general elections, from 1982 to 1993, but lost to a coalition of conservative and regional parties in the election of Mar. 3, 1996.

Catalonia and the Basque country were granted autonomy, Jan. 1980, following overwhelming approval in home-rule referendums. Basque extremists, however, have pushed for independence. The militant Basque separatist group ETA proclaimed an indefinite cease-fire as of Sept. 18, 1998.

The **Balearic Islands** in the W Mediterranean, 1,927 sq. mi., are a province of Spain; they include **Majorca** (Mallorca; capital Palma de Mallorca), **Minorca, Cabrera, Ibiza,** and **Formentera.** The **Canary Islands,** 2,807 sq. mi., in the Atlantic W of Morocco, form 2 provinces, and include the islands of **Tenerife, Palma, Gomera, Hierro, Grand Canary, Fuerteventura,** and **Lanzarote;** Las Palmas and Santa Cruz are thriving ports. **Ceuta** and **Melilla,** small Spanish enclaves on Morocco's Mediterranean coast, gained limited autonomy in Sept. 1994.

Spain has sought the return of Gibraltar, in British hands since 1704.

Sri Lanka
Democratic Socialist Republic of Sri Lanka

People: Population: 19,144,875. **Age distrib.** (%): <15: 27.1; 65+: 6.4. **Pop. density:** 757 per sq. mi. **Urban:** 22%. **Ethnic groups:** Sinhalese 74%, Tamil 18%, Moor 7%, others. **Principal languages:** Sinhala (official), Tamil, English. **Chief religions:** Buddhist 69%, Hindu 15%, Christian 8%, Muslim 8%.

Geography: Area: 25,300 sq. mi. **Location:** In Indian O. off SE coast of India. **Neighbors:** India on NW. **Topography:** The coastal area and the northern half are flat; the S-central area is hilly and mountainous. **Capital:** Colombo (1995): 1.3 mil.

Government: Type: Republic. **Head of state:** Pres. Chandrika Bandaranaike Kumaratunga; b June 29, 1945; in office: Nov. 12, 1994. **Head of gov.:** Prime Min. Sirimavo Bandaranaike; b Apr. 17, 1916; in office: Nov. 14, 1994. **Local divisions:** 25 administrative districts. **Defense:** 6.1% of GDP. **Active troops:** 117,000.

Economy: Industries: Clothing, agric. processing, oil refining, textiles. **Chief crops:** Tea, coconuts, rice, sugar. **Minerals:** Graphite, limestone, gems, phosphates. **Other resources:** Forests, rubber. **Arable land:** 14%. **Livestock** (1997): chickens: 9.57 mil; cattle: 1.60 mil; goats: 519,300; buffalo: 720,700. **Fish catch** (1997): 240,000 metric tons. **Electricity prod.** (1997): 4.300 bil kWh. **Labor force:** 46% services; 37% agric.

Finance: Monetary unit: Rupee (Sept. 1999: 71.76 = $1 U.S.). **GDP:** (1997 est.): $72.1 bil. **Per capita GDP:** $3,800. **Imports** (1996): $5.4 bil; partners: Japan 11%, India 10%. **Exports** (1996): $4.1 bil; partners: U.S. 34%, UK 9%. **Tourism:** $233 mil. **Budget** (1997): $4.2 bil. **Intl. reserves less gold** (June 1999): $1.75 bil. **Gold:** 63,000 oz t. **Consumer prices** (change in 1997): 9.4%.

Transport: Railroad: Length: 928 mi. **Motor vehicles:** 220,000 pass. cars, 248,900 comm. vehicles. **Civil aviation:** 2.6 bil pass.-mi; 1 airport. **Chief ports:** Colombo, Trincomalee, Galle.

Communications: TV sets: 39 per 1,000 pop. **Radios:** 182 per 1,000 pop. **Telephones:** 523,500 main lines. **Daily newspaper circ.:** 25 per 1,000 pop.

Health: Life expectancy: 69.89 male; 75.59 female. **Births** (per 1,000 pop.): 18.16. **Deaths** (per 1,000 pop.): 6.02. **Natural inc.:** 1.214%. **Hosp. beds** (1993): 1 per 360 persons. **Physicians** (1993): 1 per 4,745 persons. **Infant mortality** (per 1,000 live births): 16.12.

Education: Free, compulsory: ages 5-12. **Literacy:** 88%.

Major Intl. Organizations: UN (FAO, IBRD, ILO, IMF, IMO, WHO, WTrO), the Commonwealth.
Embassy: 2148 Wyoming Ave. NW 20008; 483-4025.

The island was known to the ancient world as Taprobane (Greek for copper-colored) and later as Serendip (from Arabic). Colonists from N India subdued the indigenous Veddahs about 543 BC; their descendants, the Buddhist Sinhalese, still form most of the population. Hindu descendants of Tamil immigrants from S India account for about one-fifth of the population.

Parts were occupied by the Portuguese in 1505 and the Dutch in 1658. The British seized the island in 1796. As Ceylon it became an independent member of the Commonwealth in 1948, and the Republic of Sri Lanka May 22, 1972.

Prime Min. W. R. D. Bandaranaike was assassinated Sept. 25, 1959. In new elections, the Freedom Party was victorious under Mrs. Sirimavo Bandaranaike, widow of the former prime minister. After May 1970 elections, Mrs. Bandaranaike became prime minister again. In 1971 the nation suffered economic problems and terrorist activities by ultra-leftists, thousands of whom were executed. Massive land reform and nationalization of foreign-owned plantations were undertaken in the mid-1970s. Mrs. Bandaranaike was ousted in 1977 elections. Presidential powers were increased in 1978 in an effort to restore stability.

Tensions between the Sinhalese and Tamil separatists erupted into violence in the early 1980s. More than 55,000 have died in the civil war, which continued in the late 1990s; another 12,000, mostly young Tamils, have "disappeared" after they were taken into custody by government security forces.

Pres. Ranasinghe Premadasa was assassinated May 1, 1993, by a Tamil rebel. Mrs. Bandaranaike's daughter, Chandrika Bandaranaike Kumaratunga, became prime minister after the Aug. 16, 1994, general elections. Elected president Nov. 9, Kumaratunga appointed her mother prime minister.

Sudan

Republic of the Sudan

People: Population: 34,475,690. **Age distrib.** (%): <15: 45.1; 65+: 2.2. **Pop. density:** 36 per sq. mi. **Urban:** 32%. **Ethnic groups:** Black 52%, Arab 39%, Beja 6%. **Principal languages:** Arabic (official), Nubian, Ta Bedawie. **Chief religions:** Sunni Muslim 70%, indigenous beliefs 25%.

Geography: Area: 967,500 sq. mi., the largest country in Africa. **Location:** At the E end of Sahara desert zone. **Neighbors:** Egypt on N; Libya, Chad, Central African Republic on W; Congo (formerly Zaire), Uganda, Kenya on S; Ethiopia, Eritrea on E. **Topography:** The N consists of the Libyan Desert in the W, and the mountainous Nubia Desert in E, with narrow Nile valley between. The center contains large, fertile, rainy areas with fields, pasture, and forest. The S has rich soil, heavy rain. **Capital:** Khartoum. **Cities** (1993): Omdurman 1,271,403; Khartoum 947,483.

Government: Type: Republic with strong military influence. **Head of state and gov.:** Pres. Gen. Omar Hassan Ahmad Al-Bashir; b Jan. 1, 1944; in office: June 30, 1989. **Local divisions:** 26 states. **Defense:** 5.6% of GDP. **Active troops:** 79,700.

Economy: Industries: Cotton ginning, textiles, cement. **Chief crops:** Gum arabic, sorghum, cotton (main export), wheat. **Minerals:** Petroleum, iron, chromium, copper. **Crude oil reserves** (1999): 262.1 mil bbls. **Arable land:** 5%. **Livestock** (1997): chickens: 40.50 mil; sheep: 42.36 mil; cattle: 34.58 mil; goats: 37.35 mil. **Electricity prod.** (1997): 1.320 bil kWh. **Labor force:** 80% agric.; 10% ind. & comm.; 6% govt.

Finance: Monetary unit: Pound (Sept. 1999: 2,540.00 = $1 U.S.), Dinar (Sept. 1999: 254.00 = $1 U.S.). **GDP** (1997 est.): $26.6 bil. **Per capita GDP:** $875. **Imports** 1996): $1.5 bil; partners: Saudi Arabia 10%. **Exports** (1996): $620 mil; partners: Saudi Arabia 20%, UK 14%, China 11%. **Tourism:** $6 mil. **Budget** (1996): $1.5 bil. **Intl. reserves less gold** (Mar. 1999): $121.6 mil. **Consumer prices** (change in 1998): 17.1%.

Transport: Railroad: Length: 2,960 mi. **Motor vehicles:** 35,000 pass. cars, 40,000 comm. vehicles. **Civil aviation:** 292.8 mil pass.-mi; 3 airports. **Chief port:** Port Sudan.

Communications: TV sets: 8.2 per 1,000 pop. **Radios:** 182 per 1,000 pop. **Telephones:** 162,200 main lines. **Daily newspaper circ.:** 21 per 1,000 pop.

Health: Life expectancy: 55.41 male; 57.44 female. **Births** (per 1,000 pop.): 39.34. **Deaths** (per 1,000 pop.):10.60. **Natural inc.:** 2.874%. **Physicians** (1994): 1 per 11,300 persons. **Infant mortality** (per 1,000 live births): 70.94.

Education: Literacy: 46%.

Major Intl. Organizations: UN (FAO, IBRD, ILO, IMF, IMO, WHO), AL, OAU.

Embassy: 2210 Massachusetts Ave. NW 20008; 338-8565.
Website: http://www.sudan.net

Northern Sudan, ancient Nubia, was settled by Egyptians in antiquity. The population was converted to Coptic Christianity in the 6th century. Arab conquests brought Islam to the area in the 15th century.

In the 1820s Egypt took over Sudan, defeating the last of earlier empires, including the Fung. In the 1880s a revolution was led by Muhammad Ahmad, who called himself the Mahdi (leader of the faithful), and his followers, the dervishes.

In 1898 an Anglo-Egyptian force crushed the Mahdi's successors. In 1951 the Egyptian Parliament abrogated its 1899 and 1936 treaties with Great Britain and amended its constitution to provide for a separate Sudanese constitution. Sudan voted for complete independence as a parliamentary government effective Jan. 1, 1956.

In 1969, a Revolutionary Council took power, but a civilian premier and cabinet were appointed; the government announced it would create a socialist state.

Economic problems plagued the nation in the 1980s and 1990s, aggravated by civil war and influxes of refugees from neighboring countries. After 16 years in power, Pres. Jaafar al-Nimeiry was overthrown in a bloodless military coup, Apr. 6, 1985. Sudan held its first democratic parliamentary elections in 18 years in 1986, but the elected government was overthrown in a bloodless coup June 30, 1989.

In the mid-1980s, rebels in the south (populated largely by black Christians and followers of tribal religions) took up arms against government domination by northern Sudan, mostly Arab-Muslim. War and related famine cost an estimated 2 million lives and displaced millions of southerners by the late 1990s. In 1993, Amnesty International accused Sudan of "ethnic cleansing" against the Nuba people in the South.

Egypt publicly blamed Sudan for an attempted assassination of Egyptian Pres. Hosni Mubarak in Ethiopia, June 26, 1995. Opposition groups boycotted elections Mar. 1996.

A new constitution based on Islamic law took effect June 30, 1998. On Aug. 20, in retaliation for bombings in Kenya and Tanzania, U.S. missiles destroyed a Khartoum pharmaceutical plant the U.S. alleged was associated with terrorist activities; independent inquiries later cast some doubt on the U.S. claim.

Suriname

Republic of Suriname

People: Population: 431,156. **Age distrib.** (%): <15: 32.9; 65+: 5.3. **Pop. density:** 7 per sq. mi. **Urban:** 50%. **Ethnic groups:** Hindustani 37%, Creole 31%, Javanese 15%. **Principal languages:** Dutch (official), Sranang Tongo, English, Hindustani. **Chief religions:** Hindu 27%, Protestant 25%, Roman Catholic 23%, Muslim 20%.

Geography: Area: 63,000 sq. mi. **Location:** On N shore of South America. **Neighbors:** Guyana on W, Brazil on S, French Guiana on E. **Topography:** A flat Atlantic coast, where dikes permit agriculture. Inland is a forest belt; to the S, largely unexplored hills cover 75% of the country. **Capital:** Paramaribo (1995 est.): 216,000.

Government: Type: Republic. **Head of state and gov.:** Pres. Jules Wijdenbosch; b May 2, 1941; in office: Sept. 14, 1996. **Local divisions:** 10 districts. **Defense:** 4.4% of GDP. **Active troops:** 1,800.

Economy: Industries: Aluminum, mining, food processing. **Chief crops:** Rice, bananas, palm kernels. **Minerals:** Kaolin, bauxite, gold. **Crude oil reserves** (1999): 74 mil bbls. **Other resources:** Forests, fish, shrimp. **Livestock** (1997): chickens: 2.00 mil. **Electricity prod.** (1997): 1.620 bil kWh. **Labor force:** 39% public admin. & defense; 20% agric. & forestry.

Finance: Monetary unit: Guilder (Sept. 1999: 809.50 = $1 U.S.). **GDP** (1997 est.): $1.4 bil. **Per capita GDP:** $3,400. **Imports** (1997 est.): $490 mil; partners: U.S. 40%, Netherlands 24%. **Exports** (1996 est.): $434 mil; partners: Norway 33%, Netherlands 26%, U.S. 13%. **Tourism:** $17 mil. **Budget** (1997 est.): $333 mil. **Intl. reserves less gold** (May 1999): $70.45 mil. **Gold:** 373,000 oz t. **Consumer prices** (change in 1997): 7.1%.

Transport: Railroad: Length: 187 mi. **Motor vehicles:** 46,408 pass. cars, 19,255 comm. vehicles. **Civil aviation:** 663.5 mil pass.-mi; 3 airports. **Chief ports:** Paramaribo, Nieuw Nickerie, Albina.

Communications: TV sets: 146 per 1,000 pop. **Radios:** 719 per 1,000 pop. **Telephones:** 67,300 main lines. **Daily newspaper circ.:** 107 per 1,000 pop.

Health: Life expectancy: 68.32 male; 73.59 female. **Births** (per 1,000 pop.): 21.75. **Deaths** (per 1,000 pop.): 5.75. **Natural inc.:** 1.600%. **Infant mortality** (per 1,000 live births): 26.52.

Education: Free, compulsory: ages 6-16. **Literacy:** 93%.

Major Intl. Organizations: UN (FAO, IBRD, ILO, IMF, IMO, WHO, WTrO), Caricom, OAS.

Embassy: 4301 Connecticut Ave. NW 20008; 244-7488.

The Netherlands acquired Suriname in 1667 from Britain, in exchange for New Netherlands (New York). The 1954 Dutch constitution raised the colony to a level of equality with the Netherlands and the Netherlands Antilles. Independence was granted Nov. 25, 1975, despite objections from East Indians. Some 40% of the population (mostly East Indians) immigrated to the Netherlands in the months before independence.

The National Military Council took control of the government, Feb. 1982. Civilian rule was restored in 1987, but political turmoil continued until 1992, disrupting the nation's economy.

Swaziland
Kingdom of Swaziland

People: Population: 985,335. **Age distrib.** (%): <15: 46.3; 65+: 2.5. **Pop. density:** 147 per sq. mi. **Urban:** 32%. **Ethnic groups:** African 97%, European 3%. **Principal languages:** siSwati, English (both official). **Chief religions:** Christian 60%, indigenous beliefs 40%.

Geography: Area: 6,700 sq. mi. **Location:** In southern Africa, near Indian O. coast. **Neighbors:** South Africa on N, W, S; Mozambique on E. **Topography:** The country descends from W-E in broad belts, becoming more arid in the low veld region, then rising to a plateau in the E. **Capitals:** Mbabane (administrative), Lobamba (legislative). **Cities:** Mbabane (1990 est.): 47,000.

Government: Type: Constitutional monarchy. **Head of state:** King Mswati III; b 1968; in office: Apr. 25, 1986. **Head of gov.:** Prime Min. Barnabas Sibusiso Dlamini; b 1942; in office: July 26, 1996. **Local divisions:** 4 districts.

Economy: Industries: Wood pulp, mining. **Chief crops:** Sugar, corn, cotton, rice, pineapples, tobacco, citrus, peanuts. **Minerals:** Asbestos, clay, coal. **Other resources:** Forests. **Arable land:** 11%. **Livestock** (1997): chickens: 970,000; cattle: 650,000; goats: 435,000. **Electricity prod.** (1997): 415 mil kWh.

Finance: Monetary unit: Lilangeni (Sept. 1999: 6.03 = $1 U.S.). **GDP:** (1997 est.): $3.9 bil. **Per capita GDP:** $3,800. **Imports** (1996): $1.1 bil; partners: South Africa 88%. **Exports** (1996): $893 mil; partners: South Africa 58%, EU 20%. **Tourism:** $40 mil. **Budget** (FY1996-97): $450 mil. **Intl. reserves less gold** (Apr. 1999): $366.69 mil. **Consumer prices** (change in 1997): 8.1%.

Transport: Railroad: Length: 187 mi. **Motor vehicles:** 28,523 pass. cars, 8,232 comm. vehicles. **Civil aviation:** 26.5 mil pass.-mi; 1 airport.

Communications: TV sets: 96 per 1,000 pop. **Radios:** 129 per 1,000 pop. **Telephones:** 29,000 main lines. **Daily newspaper circ.:** 40 per 1,000 pop.

Health: Life expectancy: 36.86 male; 39.40 female. **Births** (per 1,000 pop.): 40.80. **Deaths** (per 1,000 pop.): 21.72. **Natural inc.:** 1.908%. **Infant mortality rate** (per 1,000 live births): 101.87.

Education: Literacy: 77%.

Major Intl. Organizations: UN (FAO, IBRD, ILO, IMF, WHO, WTrO), the Commonwealth, OAU.

Embassy: 3400 International Dr. NW 20008; 362-6683.

Website: http://www.realnet.co.sz

The royal house of Swaziland traces back 400 years, and is one of Africa's last ruling dynasties. The Swazis, a Bantu people, were driven to Swaziland from lands to the N by the Zulus in 1820. Their autonomy was later guaranteed by Britain and Transvaal (later part of South Africa), with Britain assuming control after 1903. Independence came Sept. 6, 1968. In 1973 the king repealed the constitution and assumed full powers.

A new constitution banning political parties took effect Oct. 13, 1978. As Swaziland slowly moved toward political reform, student and labor unrest grew in the 1990s.

Sweden
Kingdom of Sweden

People: Population: 8,911,296. **Age distrib.** (%): <15: 18.7; 65+: 17.3. **Pop. density:** 51 per sq. mi. **Urban:** 83%. **Ethnic groups:** Swedish 89%, Finnish 2%. **Principal language:** Swedish. **Chief religion:** Evangelical Lutheran 94%.

Geography: Area: 173,732 sq. mi. **Location:** On Scandinavian Peninsula in N Europe. **Neighbors:** Norway on W, Denmark on S (across Kattegat), Finland on E. **Topography:** Mountains along NW border cover 25% of Sweden, flat or rolling terrain covers the central and southern areas, which include several large lakes. **Capital:** Stockholm. **Cities** (1996 est.): Stockholm 718,462; Göteborg 454,016.

Government: Type: Constitutional monarchy. **Head of state:** King Carl XVI Gustaf; b Apr. 30, 1946; in office: Sept. 19, 1973. **Head of gov.:** Prime Min. Goran Persson; b June 20, 1949; in office: Mar. 21, 1996. **Local divisions:** 24 counties. **Defense:** 2.4% of GDP. **Active troops:** 53,400.

Economy: Industries: Steel, precision equipment, vehicles, processed foods, paper. **Chief crops:** Grains, potatoes, sugar beets. **Minerals:** Zinc, iron, lead, copper, silver. **Other resources:** Forests (half the country); yield about 17% of exports. **Arable land:** 7%. **Livestock** (1997): chickens: 7.61 mil; pigs: 2.31 mil; cattle: 1.71 mil; sheep: 407,000. **Fish catch** (1997): 357,406 metric tons. **Electricity prod.** (1997): 143.941 bil kWh. **Labor force:** 38% social & personal services; 21% manuf. & mining.

Finance: Monetary unit: Krona (Sept. 1999: 8.16 = $1 U.S.). **GDP:** (1997 est.): $176.2 bil. **Per capita GDP:** $19,700. **Imports** (1996): $66.6 bil; partners: EU 63%. **Exports** (1996): $84.5 bil; partners: EU 59%. **Tourism:** $3.76 bil. **Budget** (FY 1995-96): $146.1 bil. **Intl. reserves less gold** (June 1999): $13.49 bil. **Gold:** 5.96 mil oz t. **Consumer prices** (change in 1997): −0.1%.

Transport: Railroad: Length: 6,744 mi. **Motor vehicles** (1997): 3.70 mil pass. cars, 336,593 comm. vehicles. **Civil aviation:** 5.54 bil pass.-mi; 48 airports. **Chief ports:** Göteborg, Stockholm, Malmö.

Communications: TV sets: 476 per 1,000 pop. **Radios:** 844 per 1,000 pop. **Telephones** (1997): 6,010,000 main lines. **Daily newspaper circ.:** 515 per 1,000 pop.

Health: Life expectancy: 76.61 male; 82.11 female. **Births** (per 1,000 pop.): 12.00. **Deaths** (per 1,000 pop.): 10.77. **Natural inc.:** 0.123%. **Hosp. beds** (1995): 1 per 194 persons. **Physicians** (1995): 1 per 384 persons. **Infant mortality** (per 1,000 live births): 3.91.

Education: Compulsory: ages 6-15. **Literacy:** 100%.

Major Intl. Organizations: UN and all of its specialized agencies, EU, OECD, OSCE.

Embassy: 1501 M St. NW 20005; 467-2600.

Website: http://www.scb.se/scbeng/keyeng.htm

The Swedes have lived in present-day Sweden for at least 5,000 years, longer than nearly any other European people. Gothic tribes from Sweden played a major role in the disintegration of the Roman Empire. Other Swedes helped create the first Russian state in the 9th century.

The Swedes were Christianized from the 11th century, and a strong centralized monarchy developed. A parliament, the Riksdag, was first called in 1435, the earliest parliament on the European continent, with all classes of society represented.

Swedish independence from rule by Danish kings (dating from 1397) was secured by Gustavus I in a revolt, 1521-23; he built up the government and military and established the Lutheran Church. In the 17th century Sweden was a major European power, gaining most of the Baltic seacoast, but its international position subsequently declined.

The Napoleonic wars, 1799-1815, in which Sweden acquired Norway (it became independent 1905), were the last in which Sweden participated. Armed neutrality was maintained in both world wars.

More than 4 decades of Social Democratic rule ended in the 1976 parliamentary elections; the party returned to power in the 1982 elections. After Prime Min. Olof Palme was shot to death in Stockholm, Feb. 28, 1986, Ingvar Carlsson took office. Carl Bildt, a non-Socialist, became prime minister Oct. 1991, with a mandate to restore Sweden's economic competitiveness. The Social Democrats returned to power following 1994 elections.

Swedish voters approved membership in the European Union Nov. 13, 1994, and Sweden entered the EU as of Jan. 1, 1995. Carlsson retired and was succeeded by Goran Persson in Mar. 1996. Persson forged a coalition with the Left and Green parties after his Social Democrats lost ground in elections Sept. 20, 1998.

Switzerland
Swiss Confederation

People: Population: 7,275,467. **Age distrib.** (%): <15: 17.2; 65+: 15.1. **Pop. density:** 458 per sq. mi. **Urban:** 61%. **Ethnic groups:** German 65%, French 18%, Italian 10%, Romansch 1%. **Principal languages:** German, French, Italian, Romansch (all official). **Chief religions:** Roman Catholic 47%, Protestant 40%.

Geography: Area: 15,900 sq. mi. **Location:** In the Alps Mts. in central Europe. **Neighbors:** France on W, Italy on S, Austria on E, Germany on N. **Topography:** The Alps cover 60% of the land area; the Jura, near France, 10%. Running between, from NE to SW, are midlands, 30%. **Capitals:** Bern (administrative), Lausanne (judicial). **Cities** (1996 est.): Zurich 342,400; Basel 173,396; Geneva 172,885; Bern 128,872.

Government: Type: Federal republic. **Head of state and gov.:** The president is elected by the Federal Assembly to a nonrenewable 1-year term. **Local divisions:** 20 full cantons, 6 half cantons. **Defense:** 1.5% of GDP. **Active troops:** 26,300.

Economy: Industries: Machinery, chemicals, precision instruments, watches, textiles, foodstuffs (cheese, chocolate), banking, tourism. **Chief crops:** Grains, fruits, vegetables. **Minerals:** Salt. **Other resources:** Hydropower potential, timber. **Arable land:** 10%. **Livestock** (1997): chickens: 6.57 mil; cattle: 1.64 mil; pigs: 1.49 mil; sheep: 420,100. **Electricity prod.** (1997): 59.237 bil kWh. **Labor force:** 67% serv.; 29% manuf. & const.

Finance: Monetary unit: Franc (Sept. 1999: 1.49 = 1 U.S.). **GDP:** (1997 est.): $172.4 bil. **Per capita GDP:** $23,800. **Imports** (1997): $16.6 bil; partners: EU 79%. **Exports** (1997): $99.2 bil; partners: EU 61%. **Tourism:** $8.21 bil. **Budget** (1997): $10.8 bil. **Intl. reserves less gold** (June 1999): $35.68 bil. **Gold:** 83.28 mil oz t. **Consumer prices** (change in 1997): 0.1%.

Transport: Railroad: Length: 3,125 mi. **Motor vehicles** (1997): 3.32 mil pass. cars, 302,707 comm. vehicles. **Civil aviation:** 13.83 bil pass.-mi; 5 airports. **Chief port:** Basel.

Communications: TV sets: 370 per 1,000 pop. **Radios:** 791 per 1,000 pop. **Telephones:** 4,803,000 main lines. **Daily newspaper circ.:** 418 per 1,000 pop.

Health: Life expectancy: 75.83 male; 82.32 female. **Births** (per 1,000 pop.): 10.53. **Deaths** (per 1,000 pop.): 9.06. **Natural inc.:** 0.147%. **Hosp. beds** (1994): 1 per 144 persons. **Physicians** (1994): 1 per 592 persons. **Infant mortality** (per 1,000 live births): 4.87.

Education: Compulsory: ages 7-16. **Literacy** (1994): 100%.

Major Intl. Organizations: Many UN specialized agencies (though not a member), EFTA, OECD, OSCE.

Embassy: 2900 Cathedral Ave. NW 20008; 745-7900.

Websites: http://www.swissembassy.org.uk
http://www.admin.ch/bfs/eindex.htm

Switzerland, the former Roman province of Helvetia, traces its modern history to 1291, when 3 cantons created a defensive league. Other cantons were subsequently admitted to the Swiss Confederation, which obtained its independence from the Holy Roman Empire through the Peace of Westphalia (1648). The cantons were joined under a federal constitution in 1848, with large powers of local control retained by each.

Switzerland has maintained an armed neutrality since 1815, and has not been involved in a foreign war since 1515. It is the seat of many UN and other international agencies.

Switzerland is a world banking center. In an effort to crack down on criminal transactions, the nation's strict bank-secrecy rules have been eased since 1990. Stung by charges that assets seized by the Nazis and deposited in Swiss banks in World War II had not been properly returned, the government announced, March 5, 1997, a $4.7 billion fund to compensate victims of the Holocaust and other catastrophies. Swiss banks agreed Aug. 12, 1998, to pay $1.25 billion in reparations.

Syria

Syrian Arab Republic

People: Population: 17,213,871. **Age distrib.** (%): <15: 45.7; 65+: 2.9. **Pop. density:** 241 per sq. mi. **Urban:** 53%. **Ethnic groups:** Arab 90%. **Principal languages:** Arabic (official), Kurdish, Armenian. **Chief religions:** Sunni Muslim 74%, other Muslims 16%, Christian 10%.

Geography: Area: 71,500 sq. mi. **Location:** Middle East, at E end of Mediterranean Sea. **Neighbors:** Lebanon and Israel on W, Jordan on S, Iraq on E, Turkey on N. **Topography:** Syria has a short Mediterranean coastline, then stretches E and S with fertile lowlands and plains, alternating with mountains and large desert areas. **Capital:** Damascus. **Cities** (1994 est.): Damascus 1,549,000; Aleppo 1,542,000.

Government: Type: Republic (under military regime). **Head of state:** Pres. Hafez al-Assad; b Oct. 6, 1930; in office: Feb. 22, 1971. **Head of gov.:** Prime Min. Mahmoud Al-Zoubi; in office: Nov. 1, 1987. **Local divisions:** 14 provinces. **Defense:** 6.3% of GDP. **Active troops:** 320,000.

Economy: Industries: Oil prods., textiles, food processing, tobacco, phosphate mining. **Chief crops:** Cotton, grains, lentils, chickpeas. **Minerals:** Oil, phosphates, chrome, manganese, asphalt, iron. **Crude oil reserves** (1999): 2.5 bil bbls. **Other resources:** Wool, dairy prods. **Arable land:** 28%. **Livestock** (1997): chickens: 20.00 mil; sheep: 14.00 mil; goats: 1.10 mil; cattle: 900,000. **Electricity prod.** (1997): 21.500 bil kWh. **Labor force:** 40% agric.; 40% services; 20% ind.

Finance: Monetary unit: Pound (Sept. 1999: 46.25 = $1 U.S.). **GDP:** (1997 est.): $106.1 bil. **Per capita GDP:** $6,600. **Imports** (1997): $5.7 bil; partners: EU 33%. **Exports** (1997): $4.2 bil; partners: EU 57%. **Tourism:** $1.05 bil. **Budget** (1996 est.): $4.3 bil. **Gold:** 833,000 oz t. **Consumer prices** (change in 1997): –1.2%.

Transport: Railroad: Length: 1,097 mi. **Motor vehicles:** 134,000 pass. cars, 218,900 comm. vehicles. **Civil aviation:** 767.6 mil pass.-mi; 5 airports. **Chief ports:** Latakia, Tartus.

Communications: TV sets: 49 per 1,000 pop. **Radios:** 211 per 1,000 pop. **Telephones** (1997): 1,312,600 main lines. **Daily newspaper circ.:** 19 per 1,000 pop.

Health: Life expectancy: 66.75 male; 69.48 female. **Births** (per 1,000 pop.): 36.95. **Deaths** (per 1,000 pop.): 5.40. **Natural inc.:** 3.155%. **Hosp. beds** (1995): 1 per 832 persons. **Physicians** (1995): 1 per 953 persons. **Infant mortality** (per 1,000 live births): 36.42.

Education: Compulsory: ages 6-12. **Literacy:** 79%.

Major Intl. Organizations: UN (FAO, IBRD, ILO, IMF, IMO, WHO), AL.

Embassy: 2215 Wyoming Ave. NW 20008; 232-6313.

Syria contains some of the most ancient remains of civilization. It was the center of the Seleucid empire, but later became absorbed in the Roman and Arab empires. Ottoman rule prevailed for 4 centuries, until the end of World War I.

The state of Syria was formed from former Turkish districts, separated by the Treaty of Sevres, 1920, and divided into the states of Syria and Greater Lebanon. Both were administered under a French League of Nations mandate 1920-1941.

Syria was proclaimed a republic by the occupying French Sept. 16, 1941, and exercised full independence Apr. 17, 1946. Syria joined the Arab invasion of Israel in 1948.

Syria joined Egypt Feb. 1958 in the United Arab Republic but seceded Sept. 1961. The Socialist Baath party and military leaders seized power Mar. 1963. The Baath, a pan-Arab organization, became the only legal party. The government has been dominated by the Alawite minority.

In the Arab-Israeli war of June 1967, Israel seized and occupied the Golan Heights, from which Syria had shelled Israeli settlements. On Oct. 6, 1973, Syria joined Egypt in an attack on Israel. Arab oil states agreed in 1974 to give Syria $1 billion a year to aid anti-Israel moves. Some 30,000 Syrian troops entered Lebanon in 1976 to mediate in a civil war. They fought Palestinian guerrillas and, later, Christian militiamen. Syrian troops again battled Christian forces in Lebanon, Apr. 1981.

Following Israel's invasion of Lebanon, June 6, 1982, Israeli planes destroyed 17 Syrian antiaircraft missile batteries in the Bekaa Valley, June 9. Some 25 Syrian planes were downed during the engagement. Israel and Syria agreed to a cease-fire June 11. In 1983, Syria backed the PLO rebels who ousted Yasir Arafat's forces from Tripoli.

Syria's role in promoting international terrorism led to the breaking of diplomatic relations with Great Britain and to limited sanctions by the European Community in 1986.

Syria condemned the Aug. 1990 Iraqi invasion of Kuwait and sent troops to help Allied forces in the Gulf War. In 1991, Syria accepted U.S. proposals for the terms of an Arab-Israeli peace conference. Syria subsequently participated in negotiations with Israel, but progress toward peace was slow. Turkey has accused Syria of aiding Kurdish separatists.

Taiwan

Republic of China

People: Population: 22,113,250. **Age distrib.** (%): <15: 22.0; 65+: 8.4. **Pop. density:** 1,591 per sq. mi. **Urban:** 75%. **Ethnic groups:** Taiwanese 84%, mainland Chinese 14%. **Principal languages:** Mandarin Chinese (official), Taiwanese. **Chief religions:** Buddhist, Taoist, and Confucian 93%, Christian 5%.

Geography: Area: 13,900 sq. mi. **Location:** Off SE coast of China, between East and South China seas. **Neighbors:** Nearest is China. **Topography:** A mountain range forms the backbone of the island; the eastern half is very steep and craggy, the western slope is flat, fertile, and well cultivated. **Capital:** Taipei. **Cities** (1997 est.): Taipei 2,595,699; Kaohsiung 1,434,907; Taichung 881,870.

Government: Type: Democracy. **Head of state:** Pres. Lee Teng-hui; b Jan. 15, 1923; in office: Jan. 13, 1988. **Head of gov.:** Prime Min. Vincent Siew; b Jan. 3, 1939, in office: Sept. 1, 1997. **Local divisions:** 16 counties, 5 municipalities, 2 special municipalities (Taipei, Kaohsiung). **Defense:** 4.7% of GDP. **Active troops:** 376,000.

Economy: Industries: Textiles, clothing, electronics, processed foods, chemicals. **Chief crops:** Vegetables, rice, fruit, tea. **Minerals:** Coal, gas, limestone, marble. **Crude oil reserves** (1999): 4 mil bbls. **Arable land:** 24%. **Livestock** (1995): pigs: 10.5 mil. **Fish catch** (1997): 1.04 mil metric tons. **Electricity prod.** (1997): 137.261 bil kWh. **Labor force:** 52% services; 38% ind.; 10% agric.

Finance: Monetary unit: New Taiwan Dollar (Sept. 1999: 31.83 = $1 U.S.). **GDP:** (1997 est.): $308 bil. **Per capita GDP:** $14,200. **Imports** (1997): $114.4 bil; partners: Japan 25.4%, U.S. 20%. **Exports** (1997): $122.1 bil; partners: U.S. 24.2%, Hong Kong 23.5%, EU 15%. **Tourism:** $3.23 bil. **Budget** (1998 est.): $55 bil.

Transport: Railroad: Length: 2,410 mi. **Motor vehicles** (1997): 4.40 mil pass. cars, 833,545 comm. vehicles. **Civil aviation:** 22.8 bil pass.-mi; 13 airports. **Chief ports:** Kaohsiung, Chilung (Keelung), Hualien, Taichung.

Communications: TV sets: 327 per 1,000 pop. **Radios:** 402 per 1,000 pop. **Telephones:** 11,500,400 main lines.

Health: Life expectancy: 74.38 male; 80.85 female. **Births** (per 1,000 pop.): 14.63. **Deaths** (per 1,000 pop.): 5.32. **Natural inc.:** 0.931%. **Hosp. beds** (1995): 1 per 189 persons. **Physicians** (1995): 1 per 867 persons. **Infant mortality** (per 1,000 live births): 6.01.

Education: Free, compulsory: ages 6-15. **Literacy:** 94%.

Major Intl. Organizations: APEC.

Website: http://www.gio.gov.tw

Large-scale Chinese immigration began in the 17th century. The island came under mainland control after an interval of Dutch rule, 1620-62. Taiwan (also called Formosa) was ruled by Japan 1895-1945. Two million Kuomintang supporters fled to the island in 1949, establishing Taiwan as the seat of the Republic of China. The U.S., upon recognizing the People's Republic of China, Dec. 15, 1978, severed diplomatic ties with Taiwan. The U.S. and Taiwan maintain contact via quasi-official agencies.

Land reform, government planning, U.S. aid and investment, and free universal education brought huge advances in industry, agriculture, and living standards. In 1987 martial law was lifted after 38 years, and in 1991 the 43-year period of emergency rule ended. Taiwan held its first direct presidential election Mar. 23, 1996. The ruling Nationalist Party has faced increasing challenge from opposition parties. An earthquake on Sept. 21, 1999, killed more than 2,300 people and injured thousands more.

Both the Taipei and Beijing governments long considered Taiwan an integral part of China, although Taiwanese officials appeared to signal a departure from that policy in July 1999. Taiwan has resisted Beijing's efforts at reunification, including military pressure, but economic ties with the mainland expanded in the 1990s. Taiwan has one of the world's strongest economies and is among the 10 leading capital exporters

The **Penghu Isls.** (Pescadores), 49 sq. mi., pop. (1996 est.) 90,142, lie between Taiwan and the mainland. **Quemoy** and **Matsu,** pop. (1996 est.) 53,286, lie just off the mainland.

Tajikistan
Republic of Tajikistan

People: Population: 6,102,854. **Age distrib.** (%): <15: 40.5; 65+: 4.7. **Pop. density:** 110 per sq. mi. **Urban:** 32%. **Ethnic groups:** Tajik 65%, Uzbek 25%. **Principal languages:** Tajik (official), Russian. **Chief religion:** Sunni Muslim 80%.

Geography: Area: 55,300 sq. mi. **Location:** Central Asia. **Neighbors:** Uzbekistan on N and W, Kyrgyzstan on N, China on E, Afghanistan on S. **Topography:** Mountainous region that contains the Pamirs, Trans-Alai mountain system. **Capital:** Dushanbe (1994 est.): 524,000.

Government: Type: Republic. **Head of state:** Pres. Imomali Rakhmonov; b Oct. 5, 1952; in office: Nov. 19, 1994. **Head of gov.:** Yakhyo Azimov; b Dec. 4, 1947; in office: Feb. 8, 1996. **Local divisions:** 2 viloyats, 1 autonomous viloyat. **Defense:** 12.1% of GDP. **Active troops:** 9,000.

Economy: Industries: Aluminum, cement, mining. **Chief crops:** Cotton, grains, fruits, vegetables. **Minerals:** Oil, uranium, mercury, coal, lead, zinc. **Crude oil reserves** (1999): 12 mil bbls. **Arable land:** 6%. **Livestock** (1997): chickens: 1.00 mil; sheep: 1.60 mil; cattle: 1.04 mil.; goats: 618,000. **Electricity prod.** (1997): 13.853 bil kWh. **Labor force:** 52% agric. & forestry; 31% serv.; 17% manuf., mining, const.

Finance: Monetary unit: Ruble (Aug. 1999: 1,650.00 = $1 U.S.). **GDP:** (1997 est.): $4.1 bil. **Per capita GDP:** $700. **Imports** (1996 est.): $657 mil; partners: FSU 55%. **Exports** (1996 est.): $768 mil; partners: FSU 78%.

Transport: Railroad: Length: 294.5 mi. **Motor vehicles:** 185,000 pass. cars, 3,600 comm. vehicles. **Civil aviation:** 1.1 bil pass.-mi; 1 airport.

Communications: TV sets: 259 per 1,000 pop. **Telephones:** 221,300 main lines. **Daily newspaper circ.:** 14 per 1,000 pop.

Health: Life expectancy: 61.15 male; 67.57 female. **Births** (per 1,000 pop.): 27.46. **Deaths** (per 1,000 pop.): 7.85. **Natural inc.:** 1.961%. **Hosp. beds** (1995): 1 per 115 persons. **Physicians** (1995): 1 per 443 persons. **Infant mortality** (per 1,000 live births): 114.78.

Education: Compulsory for 9 years between ages 7-17. **Literacy:** 100%.

Major International Organizations: UN (FAO, IBRD, ILO, IMF, WHO), CIS, OSCE.

Website: http://www.soros.org/tajkstan.html

There were settled societies in the region from about 3000 BC. Throughout history, it has undergone invasions by Iranians (Arabs who converted the population to Islam), Mongols, Uzbeks, Afghans, and Russians. The USSR gained control of the region 1918-25. In 1924, the Tajik ASSR was created within the Uzbek SSR. The Tajik SSR was proclaimed in 1929.

Tajikistan declared independence Sept. 9, 1991. It became an independent state when the Soviet Union disbanded Dec. 26, 1991. Conservative Communist Pres. Rakhmon Nabiyev was forced to resign, Sept. 1992, by a coalition of Islamic, nationalist, and Western-oriented parties.

Factional fighting led to the installation of a pro-Communist regime, Jan. 1993. A new constitution establishing a presidential system was approved by referendum Nov. 6, 1994. Clashes between Muslim rebels, reportedly armed by Afghanistan, and troops loyal to the government and supported by Russia, claimed at least 30,000 lives by mid-1997, despite a series of peace accords. Constitutional changes including legalization of Islamic political parties were approved by referendum Sept. 26, 1999.

Tanzania
United Republic of Tanzania

People: Population: 31,270,820. **Age distrib.** (%): <15: 44.4; 65+: 2.9. **Pop. density:** 86 per sq. mi. **Urban:** 25%. **Ethnic groups:** African 99%. **Principal languages:** Swahili, English (both official), many others. **Chief religions:** Christian 45%, Muslim 35%, indigenous beliefs 20%; Zanzibar is 99% Muslim.

Geography: Area: 364,900 sq. mi. **Location:** On coast of E Africa. **Neighbors:** Kenya, Uganda on N; Rwanda, Burundi, Congo (formerly Zaire) on W; Zambia, Malawi, Mozambique on S. **Topography:** Hot, arid central plateau, surrounded by the lake region in the W, temperate highlands in N and S, the coastal plains. Mt. Kilimanjaro, 19,340 ft., is highest in Africa. **Capital:** Dar-es-Salaam (capital is being moved to Dodoma). **Cities:** Dar-es-Salaam 1,747,000*.

Government: Type: Republic. **Head of state:** Pres. Benjamin William Mkapa; b Nov. 12, 1938; in office: Nov. 23, 1995. **Head of gov.:** Prime Min. Frederick Tluway Sumaye; May 29, 1950; in office: Nov. 28, 1995. **Local divisions:** 25 regions. **Defense:** 3.4% of GDP. **Active troops:** 34,600.

Economy: Industries: Agricultural processing, mining, textiles. **Chief crops:** Sisal, cotton, coffee, tea, tobacco, corn, spices. **Minerals:** Tin, phospates, iron, coal, gemstones, diamonds, gold. **Other resources:** Pyrethrum (insecticide made from chrysanthemums). **Arable land:** 3%. **Livestock** (1997): chickens: 27.00 mil; cattle: 14.30 mil; goats: 9.69 mil; sheep: 3.96 mil; pigs: 340,000. **Fish catch** (1997): 356,960 metric tons. **Electricity prod.** (1997): 1.700 bil kWh. **Labor force:** 90% agric.; 10% ind. & comm.

Finance: Monetary unit: Shilling (Sept. 1999: 796.13 = $1 U.S.). **GDP:** (1997 est.): $12.1 bil. **Per capita GDP:** $700. **Imports** (1996): $1.4 bil; partners: UK 10%, Kenya 9%. **Exports** (1996): $760 mil; partners: Germany 10%, Japan 9%. **Tourism:** $431 mil. **Budget** (FY 1996-97 est.): $1.1 bil. **Intl. reserves less gold** (May 1999): $602.3 mil. **Consumer prices** (change in 1997): 12.8%.

Transport: Railroad: Length: 2,218 mi. **Motor vehicles:** 55,000 pass. cars; 78,800 comm. vehicles. **Civil aviation:** 143.4 mil pass.-mi; 11 airports. **Chief ports:** Dar-es-Salaam, Mtwara, Tanga.

Communications: TV sets: 2.8 per 1,000 pop. **Radios:** 20 per 1,000 pop. **Telephones:** 121,800 main lines.

Health: Life expectancy: 43.85 male; 48.57 female. **Births** (per 1,000 pop.): 40.37. **Deaths** (per 1,000 pop.): 16.75. **Natural inc.:** 2.362%. **Hosp. beds** (1993): 1 per 1,000 persons. **Physicians** (1993): 1 per 20,511 persons. **Infant mortality** (per 1,000 live births): 95.27.

Education: Free, compulsory: ages 7-14. **Literacy:** 68%.

Major Intl. Organizations: UN and all of its specialized agencies, the Commonwealth, OAU.

Embassy: 2139 R St. NW 20008; 939-6125.

The Republic of Tanganyika in E Africa and the island Republic of Zanzibar, off the coast of Tanganyika, both of which had recently gained independence, joined into a single nation, the United Republic of Tanzania, Apr. 26, 1964. Zanzibar retains internal self-government.

Until resigning as president in 1985, Julius K. Nyerere, a former Tanganyikan independence leader, dominated Tanzania's politics, which emphasized government planning and control of the economy, with single-party rule. In 1992 the constitution was amended to establish a multiparty system. Privatization of the economy was undertaken in the 1990s.

At least 500 people died when an overcrowded Tanzanian

ferry sank in Lake Victoria, May 21, 1996. About 460,000 Rwandan refugees, mostly Hutu, returned from Tanzania to Rwanda in Dec. 1996. A bomb at the U.S. embassy in Dar-es-Salaam, Aug. 7, 1998, killed 11 people and injured at least 70 others; the U.S. blamed the attack on Islamic terrorists associated with a wealthy Saudi businessman, Osama bin Laden.

Former Pres. Nyerere died in London Oct. 14, 1999.

Tanganyika. Arab colonization and slaving began in the 8th century AD; Portuguese sailors explored the coast by about 1500. Other Europeans followed.

In 1885 Germany established German East Africa of which Tanganyika formed the bulk. It became a League of Nations mandate and, after 1946, a UN trust territory, both under Britain. It became independent Dec. 9, 1961, and a republic within the Commonwealth a year later.

Zanzibar, the Isle of Cloves, lies 23 mi. off mainland Tanzania; area 640 sq. mi. and pop. (1995 est.) 456,934. The island of **Pemba,** 25 mi. to the NE, area 380 sq. mi. and pop. (1995 est.) 322,466, is included in the administration.

Chief industry is cloves and clove oil production, of which Zanzibar and Pemba produce most of the world's supply.

Zanzibar was for centuries the center for Arab slave traders. Portugal ruled the region for 2 centuries until ousted by Arabs around 1700. Zanzibar became a British Protectorate in 1890; independence came Dec. 10, 1963. Revolutionary forces overthrew the Sultan Jan. 12, 1964. The new government ousted Western diplomats and newsmen, slaughtered thousands of Arabs, and nationalized farms. Union with Tanganyika followed.

Thailand
Kingdom of Thailand

People: Population: 60,609,046. **Age distrib.** (%): <15: 23.9; 65+: 6.2. **Pop. density:** 305 per sq. mi. **Urban:** 20%. **Ethnic groups:** Thai 75%, Chinese 14%. **Principal languages:** Thai (official), English. **Chief religions:** Buddhist 95%, Muslim 4%.

Geography: Area: 198,500 sq. mi. **Location:** On Indochinese and Malayan peninsulas in SE Asia. **Neighbors:** Myanmar on W and N, Laos on N, Cambodia on E, Malaysia on S. **Topography:** A plateau dominates the NE third of Thailand, dropping to the fertile alluvial valley of the Chao Phraya R. in the center. Forested mountains are in the N, with narrow fertile valleys. The S peninsula region is covered by rain forests. **Capital:** Bangkok 6,547,000*.

Government: Type: Constitutional monarchy. **Head of state:** King Bhumibol Adulyadej; b Dec. 5, 1927; in office: June 9, 1946. **Head of gov.:** Prime Min. Chuan Leekpai; b July 28, 1938; in office: Nov. 9, 1997. **Local divisions:** 76 provinces. **Defense:** 2.1% of GDP. **Active troops:** 266,000.

Economy: Industries: Textiles, agric. processing, tourism. **Chief crops:** Rice (world's largest exporter), corn, cassava, sugarcane. **Minerals:** Tin, tungsten, gas. **Crude oil reserves** (1999): 296.25 mil bbls. **Other resources:** Forests, rubber, seafood (world's largest exporter of farmed shrimp). **Arable land:** 34%. **Livestock** (1997): chickens: 165.00 mil; cattle: 7.00 mil; buffalo: 4.00 mil; pigs: 4.82 mil; goats: 120,000. **Fish catch** (1997): 2.91 mil metric tons. **Electricity prod.** (1997): 94.500 bil kWh. **Labor force:** 54% agric.; 31% serv. & govt.; 15% ind. & comm.

Finance: Monetary unit: Baht (Sept. 1999: 38.76 = $1 U.S.). **GDP:** (1997 est.): $525 bil. **Per capita GDP:** $8,800. **Imports** (1996): $73.5 bil; partners: Japan 25.6%, U.S. 3.9%. **Exports** (1997): $51.6 bil; partners: U.S. 19.6%, Japan 14.9%. **Tourism:** $6.39 bil. **Budget** (FY 1996-97): $25 bil. **Intl. reserves less gold** (June 1999): $30.72 bil. **Gold:** 2.47 mil oz t. **Consumer prices** (change in 1997): 8.1%.

Transport: Railroad: Length: 2,471 mi. **Motor vehicles:** 1.55 mil pass. cars, 4.15 mil comm. vehicles. **Civil aviation:** 19.2 bil pass.-mi; 25 airports. **Chief ports:** Bangkok, Sattahip.

Communication: TV sets: 56 per 1,000 pop. **Radios:** 167 per 1,000 pop. **Telephones** (1997): 4,826,700 main lines. **Daily newspaper circ.:** 47 per 1,000 pop.

Health: Life expectancy: 65.58 male; 73.01 female. **Births** (per 1,000 pop.): 16.46. **Deaths** (per 1,000 pop.): 7.16. **Natural inc.:** 0.930%. **Physicians** (1994): 1 per 4,165 persons. **Infant mortality** (per 1,000 live births): 29.54.

Education: Compulsory: ages 6-15. **Literacy:** 94%.

Major Intl. Organizations: UN (FAO, IBRD, ILO, IMF, IMO, WHO, WTrO), ASEAN, APEC.

Embassy: 1024 Wisconsin Ave. NW 20007; 944-3600.

Website: http://emailhost.ait.ac.th/Asia/info.html

Thais began migrating from southern China during the 11th century. A unified Thai kingdom was established in 1350.

Thailand, known as Siam until 1939, is the only country in SE Asia never taken over by a European power, thanks to King Mongkut and his son King Chulalongkorn. Ruling successively from 1851 to 1910, they modernized the country and signed trade treaties with Britain and France. A bloodless revolution in 1932 limited the monarchy. Thailand was an ally of Japan during World War II and of the U.S. during the postwar period.

The military took over the government in a bloody 1976 coup. Kriangsak Chomanan, prime minister, resigned Feb. 1980 because of soaring inflation, oil price increases, labor unrest, and growing crime. Vietnamese troops crossed the border but were repulsed by Thai forces in the 1980s.

Chatichai Choonhavan was chosen prime minister in a democratic election, Aug. 1988. In Feb. 1991, the military ousted Choonhavan in a bloodless coup. A violent crackdown on street demonstrations in May 1992 led to more than 50 deaths. AIDS reached epidemic proportions in Thailand in the mid-1990s.

A steep downturn in the economy forced Thailand to seek more than $15 billion in emergency international loans in Aug. 1997. A new constitution won legislative approval Sept. 27. As the economic crisis deepened, Chuan Leekpai became prime minister Nov. 9, 1997 and began implementing financial reforms.

Togo
Togolese Republic

People: Population: 5,081,413. **Age distrib.** (%): <15: 48.2; 65+: 2.2. **Pop. density:** 232 per sq. mi. **Urban:** 31%. **Ethnic groups:** Ewe, Mina, Kabre, 34 other tribes. **Principal languages:** French (official), Ewe, Mina, Dagomba, Kabye. **Chief religions:** Indigenous beliefs 70%, Christian 20%, Muslim 10%.

Geography: Area: 21,900 sq. mi. **Location:** On S coast of W Africa. **Neighbors:** Ghana on W, Burkina Faso on N, Benin on E. **Topography:** A range of hills running SW-NE splits Togo into 2 savanna plains regions. **Capital:** Lomé (1990 met. est.): 513,000.

Government: Type: Republic. **Head of state:** Pres. Gnassingbé Eyadéma; b Dec. 26, 1937; in office: Apr. 14, 1967. **Head of gov.:** Prime Min. Koffi Eugene Adoboli; in office: May 21, 1999. **Local divisions:** 5 regions. **Defense:** 2.1% of GDP. **Active troops:** 7,000.

Economy: Industries: Textiles, handicrafts, agric. processing. **Chief crops:** Coffee, cocoa, yams, cotton, millet, rice. **Minerals:** Phosphates, limestone, marble. **Arable land:** 38%. **Livestock** (1997): chickens: 7.50 mil.; goats: 1.11 mil; sheep: 740,000; pigs: 850,000; cattle: 222,800. **Electricity prod.** (1997): 90 mil kWh. **Labor force:** 65% agric., 30% services.

Finance: Monetary unit: CFA Franc (Sept. 1999: 612.79 = $1 U.S.). **GDP:** (1997 est.): $6.2 bil. **Per capita GDP:** $1,300. **Imports** (1996): $404 mil; partners: Ghana 17%, China 13%, France 13%. **Exports** (1996): $196 mil; partners: Canada 9%, U.S. 8%. **Tourism:** $15 mil. **Budget** (1997 est.): $252 mil. **Intl. reserves less gold** (Dec. 1998): $117.7 mil. **Gold:** 13,000 oz t. **Consumer prices** (change in 1997): 1.0%.

Transport: Railroad: Length: 245 mi. **Motor vehicles:** 74,662 pass. cars, 34,605 comm. vehicles. **Civil aviation:** 150.5 mil pass.-mi; 2 airports. **Chief port:** Lomé.

Communications: TV sets: 36 per 1,000 pop. **Radios:** 212 per 1,000 pop. **Telephones:** 31,400 main lines.

Health: Life expectancy: 56.93 male; 61.64 female. **Births** (per 1,000 pop.): 44.78. **Deaths** (per 1,000 pop.): 9.69. **Natural inc.:** 3.509%. **Infant mortality** (per 1,000 live births): 77.55.

Education: Compulsory: ages 6-12. **Literacy:** 52%.

Major Intl. Organizations: UN (FAO, IBRD, ILO, IMF, IMO, WHO, WTrO), OAU.

Embassy: 2208 Massachusetts Ave. NW 20008; 234-4212.

The Ewe arrived in southern Togo several centuries ago. The country later became a major source of slaves. Germany took control in 1884. France and Britain administered Togoland as UN trusteeships. The French sector became the republic of Togo Apr. 27, 1960.

The population is divided between Bantus in the S and Hamitic tribes in the N. Togo has actively promoted regional integration, as a means of stimulating the economy.

In Jan. 1993 police fired on antigovernment demonstrators, killing at least 22. Some 25,000 people fled to Ghana and Benin as a result of civil unrest. In Jan. 1994 at least 40 people were killed when gunmen reportedly attacked an army base. Further violence marred Togo's 1st multiparty legislative elections, held Feb. 1994. In office since 1967, Pres. Gnassingbé Eyadéma was reelected June 21, 1998, in a vote that was disputed as in previous elections.

Tonga
Kingdom of Tonga

People: Population: 109,082. **Pop. density:** 377 per sq. mi. **Urban:** 42%. **Ethnic groups:** Polynesian. **Principal languages:** Tongan, English (both official). **Chief religions:** Free Wesleyan 41%, Roman Catholic 16%, Mormon 14%.

Geography: Area: 289 sq. mi. **Location:** In western South Pacific O. **Neighbors:** Nearest are Fiji to W, Samoa to NE. **Topography:** Tonga comprises 170 volcanic and coral islands, 36 inhabited. **Capital:** Nuku'alofa (1990 est.): 34,000.

Government: Type: Constitutional monarchy. **Head of state:** King Taufa'ahau Tupou IV; b July 4, 1918; in office: Dec. 16, 1965. **Head of gov.:** Prime Min. Baron Vaea; b 1921; in office: Aug. 22, 1991. **Local divisions:** 5 divisions, 23 districts.

Economy: Industries: Tourism, fishing. **Chief crops:** Coconuts, copra, bananas, vanilla beans. **Arable land:** 24%. **Livestock** (1997): chickens: 266,000. **Electricity prod.** (1997): 35 mil kWh. **Labor force:** 65% agric.

Finance: Monetary unit: Pa'anga (Sept. 1999: 1.61 = $1 U.S.). **GDP:** (1996 est.): $239 mil. **Per capita GDP:** $2,250. **Imports** (1996): $82.9 mil; partners: N.Z. 34%, Australia 16%. **Exports** (1996): $15.3 mil; partners: Japan 43%, U.S. 19%, Canada 14%. **Tourism:** $14 mil. **Budget** (FY 1996-97 est.): $120 mil. **Intl. reserves less gold** (May 1999): $27.95 mil. **Consumer prices** (change in 1997): 3.3%.

Transport: Motor vehicles: 3,400 pass. cars, 3,900 comm. vehicles. **Civil aviation:** 6.4 mil pass.-mi; 6 airports. **Chief port:** Nuku'alofa.

Communications: TV sets: 20 per 1,000 pop. **Radios:** 397 per 1,000 pop. **Telephones** (1997): 7,800 main lines. **Daily newspaper circ.:** 70 per 1,000 pop.

Health: Life expectancy: 67.73 male; 72.22 female. **Births** (per 1,000 pop.): 25.92. **Deaths** (per 1,000 pop.): 6.00. **Natural inc.:** 1.992%. **Hosp. beds** (1992): 1 per 320 persons. **Physicians** (1993): 1 per 2,201 persons. **Infant mortality** (per 1,000 live births): 37.93.

Education: Free, compulsory: ages 5-14. **Literacy** (1992): 93%.

Major Intl. Organizations: UN (FAO, IBRD, IMF, WHO), the Commonwealth.

The islands were first visited by the Dutch in the early 17th century. A series of civil wars ended in 1845 with establishment of the Tupou dynasty. In 1900 Tonga became a British protectorate. On June 4, 1970, Tonga became independent and a member of the Commonwealth. It joined the UN on Sept. 14, 1999.

Trinidad and Tobago
Republic of Trinidad and Tobago

People: Population: 1,102,096. **Age distrib.** (%): <15: 26.8; 65+: 7.5. **Pop. density:** 551 per sq. mi. **Urban:** 72%. **Ethnic groups:** Black 43%, East Indian 40%, mixed 14%. **Principal languages:** English (official), Hindi, French, Spanish. **Chief religions:** Roman Catholic 32%, Protestant 28%, Hindu 24%.

Geography: Area: 2,000 sq. mi. **Location:** In Caribbean, off E coast of Venezuela. **Neighbors:** Nearest is Venezuela to SW. **Topography:** Three low mountain ranges cross Trinidad E-W, with a well-watered plain between N and central ranges. Parts of E and W coasts are swamps. Tobago, 116 sq. mi., lies 20 mi. NE. **Capital:** Port-of-Spain (1996 est.): 43,396.

Government: Type: Parliamentary democracy. **Head of state:** Pres. Arthur N. R. Robinson; b Dec. 16, 1926; in office: Mar. 19, 1997. **Head of gov.:** Prime Min. Basdeo Panday; b May 25, 1933; in office: Nov. 9, 1995. **Local divisions:** 14 municipal corporations and Tobago. **Defense:** 1.4% of GDP. **Active troops:** 2,100.

Economy: Industries: Oil products, chemicals, tourism. **Chief crops:** Sugar, cocoa, coffee, citrus, rice. **Minerals:** Asphalt, nat. gas. **Crude oil reserves** (1999): 534 mil bbls. **Arable land:** 15%. **Livestock** (1997): chickens: 9.50 mil. **Electricity prod.** (1997): 4.550 bil kWh. **Labor force:** 62% services; 11% agric.

Finance: Monetary unit: Dollar (Sept. 1999: 6.28 = $1 U.S.). **GDP:** (1996 est.): $13.2 bil. **Per capita GDP:** $10,400. **Imports** (1996): $2.1 bil; partners: U.S. 48%, Caricom countries 15%. **Exports** (1996): $2.5 bil; partners: U.S. 48%. **Tourism:** $108 mil. **Budget** (1997 est.): $1.6 bil. **Intl. reserves less gold** (May 1999): $685.2 mil. **Gold:** 58,000 oz t. **Consumer prices** (change in 1997): 3.6%.

Transport: Motor vehicles: 128,000 pass. cars, 27,000 comm. vehicles. **Civil aviation:** 1.5 bil pass.-mi; 2 airports. **Chief ports:** Port-of-Spain, Scarborough.

Communications: TV sets: 198 per 1,000 pop. **Radios:** 433 per 1,000 pop. **Telephones:** 265,900 main lines. **Daily newspaper circ.:** 139 per 1,000 pop.

Health: Life expectancy: 68.19 male; 73.19 female. **Births** (per 1,000 pop.): 14.46. **Deaths** (per 1,000 pop.): 8.14. **Natural inc.:** 0.632%. **Hosp. beds** (1992): 1 per 340 persons. **Physicians** (1993): 1 per 1,191 persons. **Infant mortality** (per 1,000 live births): 18.56.

Education: Free, compulsory: ages 5-12. **Literacy:** 98%.

Major Intl. Organizations: UN (FAO, IBRD, ILO, IMF, IMO, WHO, WTrO), Caricom, the Commonwealth, OAS.

Embassy: 1708 Massachusetts Ave. NW 20036; 467-6490.

Columbus sighted Trinidad in 1498. A British possession since 1802, Trinidad and Tobago won independence Aug. 31, 1962. It became a republic in 1976.

The nation is one of the most prosperous in the Caribbean. Oil production has increased with offshore finds. Middle Eastern oil is refined and exported, mostly to the U.S.

In July 1990, some 120 Muslim extremists captured the Parliament building and TV station and took about 50 hostages, including Prime Min. Arthur N. R. Robinson, who was beaten, shot in the legs, and tied to explosives. After a 6-day siege, the rebels surrendered.

Basdeo Panday, the country's first prime minister of East Indian ancestry, took office Nov. 9, 1995. Robinson became president on Mar. 19, 1997.

Tunisia
Republic of Tunisia

People: Population: 9,513,603. **Age distrib.** (%) <15: 30.8; 65+: 5.9. **Pop. density:** 151 per sq. mi. **Urban:** 63%. **Ethnic groups:** Arab-Berber 98%. **Principal languages:** Arabic (official), French. **Chief religion:** Muslim 98%.

Geography: Area: 63,200 sq. mi. **Location:** On N coast of Africa. **Neighbors:** Algeria on W, Libya on E. **Topography:** The N is wooded and fertile. The central coastal plains are given to grazing and orchards. The S is arid, approaching Sahara Desert. **Capital:** Tunis (1994): 674,142.

Government: Type: Republic. **Head of state:** Pres. Gen. Zine al-Abidine Ben Ali; b Sept 3, 1936; in office: Nov. 7, 1987. **Head of gov.:** Prime Min. Hamed Karoui; b Dec. 30, 1927; in office: Sept. 27, 1989. **Local divisions:** 23 governorates. **Defense:** 1.8% of GDP. **Active troops:** 35,000.

Economy: Industries: Food processing, textiles, oil products, mining, tourism. **Chief crops:** Grains, dates, olives, sugar beets, grapes. **Minerals:** Phosphates, iron, oil, lead, zinc. **Crude oil reserves** (1999): 307.56 mil bbls. **Arable land:** 19%. **Livestock** (1997): chickens: 35.00 mil; sheep: 6.60 mil; goats: 1.30 mil; cattle: 770,000. **Fish catch** (1996): 83,567 metric tons. **Electricity prod.** (1997): 7.290 bil kWh. **Labor force:** 55% services; 23% industry; 22% agric.

Finance: Monetary unit: Dinar (Sept. 1999: 1.19 = $1 U.S.). **GDP:** (1997 est.): $565 bil. **Per capita GDP:** $6,100. **Imports** (1997 est.): $7.4 bil; partners: EU 80%. **Exports** (1997 est.): $5.6 bil; partners: EU 80%. **Tourism:** $1.55 bil. **Budget** (1997 est.): $6.8 bil. **Intl. reserves less gold** (Feb. 1999): $1.57 bil. **Gold:** 22,000 oz t. **Consumer prices** (change in 1997): 3.1%.

Transport: Railroad: Length: 1,337 mi. **Motor vehicles:** 248,000 pass. cars, 283,000 comm. vehicles. **Civil aviation:** 1.5 bil pass.-mi; 5 airports. **Chief ports:** Tunis, Sfax, Bizerte.

Communications: TV sets: 156 per 1,000 pop. **Radios:** 188 per 1,000 pop. **Telephones** (1997): 654,200 main lines. **Daily newspaper circ.:** 45 per 1,000 pop.

Health: Life expectancy: 71.95 male; 74.86 female. **Births** (per 1,000 pop.): 19.72. **Deaths** (per 1,000 pop.): 5.05. **Natural inc.:** 1.467%. **Hosp. beds** (1994): 1 per 556 persons. **Physicians** (1994): 1 per 1,640 persons. **Infant mortality** (per 1,000 live births): 31.38.

Education: Compulsory: ages 6-16. **Literacy:** 67%.

Major Intl. Organizations: UN (FAO, IBRD, ILO, IMF, IMO, WHO, WTrO), AL, OAU.

Embassy: 1515 Massachusetts Ave. NW 20005; 862-1850. **Website:** http://www.tunisiaonline.com

Site of ancient Carthage and a former Barbary state under the suzerainty of Turkey, Tunisia became a protectorate of France under a treaty signed May 12, 1881. The nation became independent Mar. 20, 1956, and ended the monarchy the following year. Habib Bourguiba, an independence leader, served as president until 1987, when he was deposed by his prime minister, Zine al-Abidine Ben Ali.

Tunisia has actively repressed Islamic fundamentalism.

Turkey
Republic of Turkey

People: Population: 65,599,206. **Age distrib.** (%): <15: 30.4; 65+: 5.9. **Pop. density:** 218 per sq. mi. **Urban:** 71%. **Ethnic groups:** Turk 80%, Kurd 20%. **Principal languages:** Turkish (official), Kurdish, Arabic. **Chief religion:** Muslim 100%.
Geography: Area: 301,400 sq. mi. **Location:** Occupies Asia Minor, stretches into continental Europe; borders on Mediterranean and Black seas. **Neighbors:** Bulgaria, Greece on W; Georgia, Armenia on N; Iran on E; Iraq, Syria on S. **Topography:** Central Turkey has wide plateaus, with hot, dry summers and cold winters. High mountains ring the interior on all but W, with more than 20 peaks over 10,000 ft. Rolling plains are in W; mild, fertile coastal plains are in S, W. **Capital:** Ankara. **Cities** (1997 est.): Istanbul 8,274,921; Ankara 2,937,524; Izmir 2,130,359.
Government: Type: Republic. **Head of state:** Pres. Suleyman Demirel; b Nov. 1, 1924; in office: May 16, 1993. **Head of gov.:** Prime Min. Bülent Ecevit; b 1925; in office: Jan. 11, 1999. **Local divisions:** 80 provinces. **Defense:** 4.2% of GDP. **Active troops:** 639,000.
Economy: Industries: Textiles, steel, mining, processed foods. **Chief crops:** Tobacco, grains, cotton, pulses, citrus, olives, sugar beets. **Minerals:** Antimony, chromium, mercury, copper, coal. **Crude oil reserves** (1999): 317.2 mil bbls. **Other resources:** Wool, forests. **Arable land:** 32%. **Livestock** (1997): chickens: 166.27 mil; sheep: 30.24 mil; cattle: 11.19 mil; goats: 8.38 mil; buffalo: 194,000. **Fish catch** (1997): 454,810 metric tons. **Electricity prod.** (1997): 98.987 bil kWh. **Labor force:** 43% agric.; 30% serv.; 14% ind.
Finance: Monetary unit: Lira (Sept. 1999: 445,869.50 = $1 U.S.). **GDP:** (1997 est.): $388.3 bil. **Per capita GDP:** $6,100. **Imports** (1997): $46.7 bil; partners: Germany 16%. **Exports** (1997): $26 bil; partners: Germany 20%. **Tourism:** $8.30 bil. **Budget** (1997): $52.9 bil. **Intl. reserves less gold** (June 1999): $21.62 bil. **Gold:** 3.75 mil oz t. **Consumer prices** (change in 1997): 84.6%.
Transport: Railroad: Length: 5,252 mi. **Motor vehicles:** 3.27 mil pass. cars, 1.05 mil comm. vehicles. **Civil aviation:** 7.7 bil pass.-mi; 26 airports. **Chief ports:** Istanbul, Izmir, Mersin.
Communications: TV sets: 171 per 1,000 pop. **Radios:** 141 per 1,000 pop. **Telephones:** 16,959,500 main lines. **Daily newspaper circ:** 44 per 1,000 pop.
Health: Life expectancy: 70.81 male; 75.88 female. **Births** (per 1,000 pop.): 20.92. **Deaths** (per 1,000 pop.): 5.27. **Natural inc.:** 1.565%. **Hosp. beds** (1994): 1 per 450 persons. **Physicians** (1995): 1 per 1,200 persons. **Infant mortality** (per 1,000 live births): 35.81.
Education: Free, compulsory: ages 6-14. **Literacy:** 82%.
Major Intl. Organizations: UN (FAO, IBRD, ILO, IMF, IMO, WHO, WTrO), NATO, OECD, OSCE.
Embassy: 1714 Massachusetts Ave. NW 20036; 612-6700.
Website: http://www.turkey.org

Ancient inhabitants of Turkey were among the world's first agriculturalists. Such civilizations as the Hittite, Phrygian, and Lydian flourished in Asiatic Turkey (Asia Minor), as did much of Greek civilization. After the fall of Rome in the 5th century, Constantinople (now Istanbul) was the capital of the Byzantine Empire for 1,000 years. It fell in 1453 to Ottoman Turks, who ruled a vast empire for over 400 years.

Just before World War I, Turkey, or the Ottoman Empire, ruled what is now Syria, Lebanon, Iraq, Jordan, Israel, Saudi Arabia, Yemen, and islands in the Aegean Sea.

Turkey joined Germany and Austria in World War I, and its defeat resulted in the loss of much territory and the fall of the sultanate. A republic was declared Oct. 29, 1923, with Mustafa Kemal (later Kemal Ataturk) as its first president. Ataturk led Turkey until his death in 1938. The Caliphate (spiritual leadership of Islam) was renounced in 1924.

Long embroiled with Greece over Cyprus, off Turkey's south coast, Turkey invaded the island July 20, 1974, after Greek officers seized the Cypriot government as a step toward unification with Greece. Turkey sought a new government for Cyprus, with Greek Cypriot and Turkish Cypriot zones. In reaction to Turkey's moves, the U.S. cut off military aid in 1975. Turkey, in turn, suspended the use of most U.S. bases. Aid was restored in 1978. There was a military takeover, Sept. 12, 1980.

Religious and ethnic tensions and active left and right extremists have caused endemic violence. The military transferred power to an elected Parliament in 1983. Martial law, imposed in 1978, was lifted in 1984.

Turkey was a member of the Allied forces that ousted Iraq from Kuwait, 1991. In the aftermath of the war, millions of Kurdish refugees fled to Turkey's border to escape Iraqi forces. The Turkish government mounted sporadic offensives against separatist Kurds in this border area and in N Iraq, causing heavy casualties among guerrillas and civilians.

Kurdish militants, demanding an independent state for the Kurds, raided Turkish diplomatic missions in some 25 Western European cities June 24, 1993. Tansu Ciller officially became Turkey's first woman prime minister July 5, 1993. The Welfare Party, an Islamic group, gained strength in the 1990s but was unable to form a government until June 1996, when it came to power in coalition with Ciller's True Path Party.

The pro-Islamic government resigned June 18, 1997, under pressure from the military. The European Union rebuffed Turkey's membership bid Dec. 12, 1997. The military stepped up its campaign against Islamic fundamentalism in 1998.

Kurdish rebel leader Abdullah Öcalan was captured Feb. 15, 1999; convicted of terrorism June 29, he was sentenced to death by a Turkish security court. His organization, the Kurdistan Workers' Party, announced Aug. 5 that it would abandon its 14-year-old armed insurgency.

A major earthquake Aug. 17 in NW Turkey killed more than 16,050 people and injured thousands more.

Turkmenistan

People: Population: 4,366,383. **Age distrib.** (%): <15: 38.0; 65+: 4.1. **Pop. density:** 23 per sq. mi. **Urban:** 45%. **Ethnic groups:** Turkmen 77%, Uzbek 9%, Russian 7%. **Principal languages:** Turkmen (official), Russian, Uzbek. **Chief religions:** Muslim 89%, Eastern Orthodox 9%.
Geography: Area: 188,500 sq. mi. **Neighbors:** Kazakhstan on N, Uzbekistan on N and E, Afghanistan and Iran on S. **Topography:** The Kara Kum Desert occupies 80% of the area. Bordered on W by Caspian Sea. **Capital:** Ashgabat (1995 est.): 536,000.
Government: Type: Republic. **Head of state and gov.:** Pres. Saparmurad Niyazov; b Feb. 18, 1940; in office: Oct. 27, 1990. **Local divisions:** 5 regions. **Defense:** 2.7% of GDP. **Active troops:** 18,000.
Economy: Industries: Oil, natural gas, food processing, textiles. **Chief crops:** Grain, cotton. **Minerals:** Coal, sulfur, oils, gas, salt. **Crude oil reserves** (1999): 546 mil bbls. **Arable land:** 3%. **Livestock** (1997): chickens: 2.80 mil; sheep: 5.40 mil; cattle: 900,000; goats: 360,000. **Electricity prod.** (1997): 8.879 bil kWh. **Labor force:** 44% agric. & forestry; 19% ind. & constr.
Finance: Monetary unit: Manat (Aug. 1999: 5,250.00 = $1 U.S.). **GDP:** (1996 est.): $12.5 bil. **Per capita GDP:** $3,000. **Imports** (1996): $1.3 bil, excl. former Soviet Union. **Exports** (1996): $1.7 bil, excl. former Soviet Union. **Tourism:** $119 mil. **Budget** (1996 est.): $548 mil.
Transport: Railroad: Length: 1,317 mi. **Civil aviation:** 679.2 mil pass.-mi; 1 airport. **Chief port:** Turkmenbashi.
Communications: TV sets: 189 per 1,000 pop. **Radios:** 189 per 1,000 pop. **Telephones** (1997): 363,000 main lines.
Health: Life expectancy: 57.48 male; 64.91 female. **Births** (per 1,000 pop.): 25.91. **Deaths** (per 1,000 pop.): 8.77. **Natural inc.:** 1.714%. **Hosp. beds** (1995): 1 per 97 persons. **Physicians** (1995): 1 per 330 persons. **Infant mortality** (per 1,000 live births): 73.10.
Education: Literacy: 100%.
Major Intl. Organizations: UN (FAO, IBRD, ILO, IMF, IMO, WHO), CIS, OSCE.
Embassy: 2207 Massachusetts Ave., NW 20008; 588-1500.
Websites: http://www.soros.org/turkstan.html
http://www.turkmenistan.com

The region has been inhabited by Turkic tribes since the 10th century. It became part of Russian Turkestan in 1881, and a constituent republic of the USSR in 1925. Turkmenistan declared independence Oct. 27, 1991, and became an independent state when the USSR disbanded Dec. 26, 1991.

Extensive oil and gas reserves place Turkmenistan in a more favorable economic position than other former Soviet republics. A new rail line linking Iran and Turkmenistan was inaugurated May 13, 1996. Political power centered around the former Communist Party apparatus, and Pres. Saparmurad Niyazov became the object of a personality cult.

Tuvalu

People: Population: 10,588. **Age distrib.** (%): <15: 34.7; 65+: 4.7. **Pop. density:** 1,059 per sq. mi. **Urban:** 48%. **Ethnic group:** Polynesian 96%. **Principal languages:** Tuvaluan, English. **Chief religion:** Church of Tuvalu (Congregationalist) 97%.
Geography: Area: 10 sq. mi. **Location:** 9 islands forming a NW-SE chain 360 mi. long in the SW Pacific O. **Neighbors:** Nearest are Kiribati to N, Fiji to S. **Topography:** The islands are all low-

lying atolls, nowhere rising more than 15 ft. above sea level, composed of coral reefs. **Capital:** Funafuti Atoll (1995 est.): 4,000.
Government: Head of state: Queen Elizabeth II, represented by Gov.-Gen. Tomasi Puapua; b 1938; in office: 1998. **Head of gov.:** Prime Min. Ionatana Ionatana; in office: Apr. 27, 1999.
Economy: Industries: Copra, fishing, tourism. **Chief crops:** Coconuts.
Finance: Monetary unit: Australian Dollar (Sept. 1999: 1.55 = $1 U.S.). **GDP:** (1995 est.): $7.8 mil. **Per capita GDP:** $800.
Transport: Civil aviation: 1 airport. **Chief port:** Funafuti.
Communications: Radios: 320 per 1,000 pop.
Health: Life expectancy: 63.01 male; 65.34 female. **Births** (per 1,000 pop.): 21.91. **Deaths** (per 1,000 pop.): 8.50. **Natural inc.:** 1.341%. **Physicians** (1993): 1 per 1,152 persons. **Infant mortality** (per 1,000 live births): 25.53.
Education: Compulsory: ages 7-15. **Literacy** (1990): 95%.
Major Intl. Organizations: WHO, the Commonwealth.
Website: http://www.emulateme.com/tuvalu.htm

The Ellice Islands separated from the British Gilbert and Ellice Islands Colony in 1975 and became Tuvalu; independence came Oct. 1, 1978.

Uganda
Republic of Uganda

People: Population: 22,804,973. **Age distrib.** (%): <15: 51.2; 65+: 2.2. **Pop. density:** 250 per sq. mi. **Urban:** 13%. **Ethnic groups:** Baganda 17%, Karamojong 12%, many others. **Principal languages:** English (official), Luganda, Swahili. **Chief religions:** Protestant 33%, Roman Catholic 33%, indigenous beliefs 18%, Muslim 16%.
Geography: Area: 91,100 sq. mi. **Location:** In E Central Africa. **Neighbors:** Sudan on N, Congo (formerly Zaire) on W, Rwanda and Tanzania on S, Kenya on E. **Topography:** Most of Uganda is a high plateau 3,000-6,000 ft. high, with high Ruwenzori range in W (Mt. Margherita 16,750 ft.), volcanoes in SW; NE is arid, W and SW rainy. Lakes Victoria, Edward, Albert form much of borders. **Capital:** Kampala: 954,000*.
Government: Type: Republic. **Head of state:** Pres. Yoweri Kaguta Museveni; b Mar. 1944; in office: Jan. 29, 1986. **Head of gov.:** Prime Min. Apollo Nsibambi; b 1938; in office: Apr. 5, 1999. **Local divisions:** 45 districts. **Defense:** 2.4% of budget. **Active troops:** 55,000.
Economy: Industries: Brewing, textiles, cement. **Chief crops:** Coffee, cotton, tea, corn, tobacco. **Minerals:** Copper, cobalt. **Arable land:** 25%. **Livestock** (1997): chickens: 22.50 mil; cattle: 5.37 mil; goats: 3.60 mil; sheep: 1.96 mil; pigs: 950,000. **Fish catch** (1997): 218,026 metric tons. **Electricity prod.** (1997): 787 mil kWh. **Labor force:** 86% agric.
Finance: Monetary unit: Shilling (Sept. 1999: 1,377.20 = $1 U.S.). **GDP:** (1997 est.): $34.6 bil. **Per capita GDP:** $1,700. **Imports** (1996): $1.2 bil; partners: Kenya 26%, UK 12%. **Exports** (1996): $604 mil; partners: Spain 23%, France 14%, Germany 14%. **Tourism:** $142 mil. **Budget** (1995-96 est.): $985 mil. **Intl. reserves less gold** (Jan. 1999): $681.7 mil. **Consumer prices** (change in 1997): 7.0%.
Transport: Railroad: Length: 771 mi. **Motor vehicles:** 24,400 pass. cars, 26,600 comm. vehicles. **Civil aviation:** 68.4 mil pass.-mi; 1 airport. **Chief ports:** Entebbe, Jinja.
Communications: TV sets: 27 per 1,000 pop. **Radios:** 485 per 1,000 pop. **Telephones:** 54,100 main lines.
Health: Life expectancy: 42.20 male; 43.94 female. **Births** (per 1,000 pop.): 48.54. **Deaths** (per 1,000 pop.): 18.43. **Natural inc.:** 3.011%. **Infant mortality** (per 1,000 live births): 90.68.
Education: Literacy: 62%.
Major Intl. Organizations: UN (FAO, IBRD, ILO, IMF, WHO, WTrO), the Commonwealth, OAU.
Embassy: 5911 16th St. NW 20011; 726-7100.
Website: http://www.nic.ug

Britain obtained a protectorate over Uganda in 1894. The country became independent Oct. 9, 1962, and a republic within the Commonwealth a year later. In 1967, the traditional kingdoms, including the powerful Buganda state, were abolished and the central government strengthened.
Gen. Idi Amin seized power from Prime Min. Milton Obote in 1971. During his eight years of dictatorial rule, he was responsible for the deaths of up to 300,000 of his opponents. In 1972 he expelled nearly all of Uganda's 45,000 Asians. Amin was named president for life in 1976. Tanzanian troops and Ugandan exiles and rebels ousted Amin, Apr. 11, 1979.
Obote held the presidency from Dec. 1980 until his ouster in a military coup July 27, 1985. Guerrilla war and rampant human rights abuses plagued Uganda under Obote's regime.
Conditions improved after Yoweri Museveni took power in Jan. 1986. In 1993 the government authorized restoration of the Buganda and other monarchies, but only for ceremonial pur-

poses. Under a constitution ratified Oct. 1995, nonparty presidential and legislative elections were held in 1996. Uganda helped Laurent Kabila seize power in the Congo (formerly Zaire) in 1997 but sent troops in 1998 to aid insurgents seeking his ouster. Museveni faced several regional insurgencies in the late 1990s.

Ukraine

People: Population: 49,811,174. **Age distrib.** (%): <15: 18.4; 65+: 13.9. **Pop. density:** 214 per sq. mi. **Urban:** 71%. **Ethnic groups:** Ukrainian 73%, Russian 22%. **Principal languages:** Ukrainian (official), Russian. **Chief religions:** Mostly Ukrainian Orthodox, some Ukrainian Catholic.
Geography: Area: 233,100 sq. mi. **Location:** In E Europe. **Neighbors:** Belarus on N; Russia on NE and E; Moldova and Romania on SW; Hungary, Slovakia, and Poland on W. **Topography:** Part of the E European plain. Mountainous areas include the Carpathians in the SW and Crimean chain in the S. Arable black soil constitutes a large part of the country. **Capital:** Kiev. **Cities** (1996 est.): Kiev (Kyiv) 2,630,000; Kharkov 1,555,000; Dnipropetrovsk 1,147,000.
Government: Type: Constitutional republic. **Head of state:** Pres. Leonid Danylovich Kuchma; b Aug. 9, 1938; in office: July 19, 1994. **Head of gov.:** Prime Min. Valery Pustovoitenko; b Feb. 23, 1947; in office: July 16, 1997. **Local divisions:** 24 oblasts, 2 municipalities, 1 autonomous republic. **Defense:** 2.7% of GDP. **Active troops:** 387,400.
Economy: Industries: Chemicals, machinery, food processing. **Chief crops:** Grains, sugar beets, vegetables. **Minerals:** Iron, manganese, coal, gas, oil, sulfur, salt. **Other resources:** Forests. **Crude oil reserves** (1999): 395 mil bbls. **Arable land:** 58%. **Livestock** (1997): chickens: 102.00 mil; cattle: 12.76 mil; pigs: 9.48 mil; sheep: 1.70 mil; goats: 662,000. **Fish catch** (1997): 373,005 metric tons. **Electricity prod.** (1997): 173.435 bil kWh. **Labor force:** 32% ind. & constr.; 24% agric. & forestry; 17% health, edu., culture.
Finance: Monetary unit: Hryvnya (Sept. 1999: 4.47 = $1 U.S.). **GDP:** (1997 est.): $124.9 bil. **Per capita GDP:** $2,500. **Imports** (1997 est.): $20.2 bil; partners: Russia 43%. **Exports** (1997 est.): $15.2 bil; partners: Russia 40%. **Tourism:** $280 mil. **Budget** (1997 est.): $21 bil. **Intl. reserves less gold** (June 1999): $941.3 mil. **Gold:** 127,800 oz t. **Consumer prices** (change in 1997): 15.9%.
Transport: Railroad: Length: 14,100 mi. **Motor vehicles:** 4.5 mil pass. cars. **Civil aviation:** 1.2 bil pass.-mi; 12 airports. **Chief ports:** Odesa, Kiev, Berdiansk.
Communications: TV sets: 233 per 1,000 pop. **Radios:** 346 per 1,000 pop. **Telephones:** 9,410,000 main lines. **Daily newspaper circ.:** 118 per 1,000 pop.
Health: Life expectancy: 60.23 male; 71.87 female. **Births** (per 1,000 pop.): 9.54. **Deaths** (per 1,000 pop.): 16.38. **Natural inc.:** −0.684%. **Hosp. beds** (1995): 1 per 81 persons. **Physicians** (1995): 1 per 224 persons. **Infant mortality** (per 1,000 live births): 21.73.
Education: Compulsory: ages 7-15. **Literacy:** 99%.
Major Intl. Organizations: UN (IBRD, ILO, IMF, IMO, WHO), CIS, OSCE.
Embassy: 3350 M St. NW 20007; 333-0606.
Website: http://www.rada.kiev.ua

Trypilians flourished along the Dnieper River, Ukraine's main artery, from 6000-1000 BC. Ukrainians' Slavic ancestors inhabited modern Ukrainian territory well before the first century AD.
In the 9th century, the princes of Kiev established a strong state called Kievan Rus, which included much of present-day Ukraine. A strong dynasty was established, with ties to virtually all major European royal families. St. Vladimir the Great, ruler of Kievan Rus, accepted Christianity as the national faith in 988. At the crossroads of European trade routes, Kievan Rus reached its zenith under Yaroslav the Wise (1019-1054). Internal conflicts led to the disintegration of the Ukrainian state by the 13th century. Mongol rule was supplanted by Poland and Lithuania in the 14th and 15th centuries. The N Black Sea coast and Crimea came under the control of the Turks in 1478.
Ukrainian Cossacks, starting in the late 16th century, waged numerous wars of liberation against the occupiers of Ukraine: Russia, Poland, and Turkey. By the late 18th century, Ukrainian independence was lost. Ukraine's neighbors once again divided its territory. At the turn of the 19th century, Ukraine was occupied by Russia and Austria-Hungary.
An independent Ukrainian National Republic was proclaimed on January 22, 1918. In 1921, Ukraine's neighbors occupied and divided Ukrainian territory. In 1922, Ukraine became a constituent republic of the USSR as the Ukrainian SSR. In 1932-33, the Soviet government engineered a man-made famine in eastern Ukraine, resulting in the deaths of 7-10 million Ukrainians.
In March 1939, independent Carpatho-Ukraine was the first

European state to wage war against Nazi-led aggression in the region. During World War II the Ukrainian nationalist underground and its Ukrainian Insurgent Army (UPA) fought both Nazi German and Soviet forces. The restoration of Ukrainian independence was declared on June 30, 1941. Over 5 million Ukrainians lost their lives during the war. With the reoccupation of Ukraine by Soviet troops in 1944 came a renewed wave of mass arrests, executions, and deportations of Ukrainians.

The world's worst nuclear power plant disaster occurred in Chernobyl, Ukraine, in April 1986.

Ukrainian independence was restored in Dec. 1991 with the dissolution of the Soviet Union. In the post-Soviet period Ukraine was burdened with a deteriorating economy.

Following a 1994 accord with Russia and the U.S., Ukraine's large nuclear arsenal was transferred to Russia for destruction. A new constitution legalizing private property and establishing Ukrainian as the sole official language was approved by parliament June 29, 1996. In May 1997, Russia and Ukraine resolved disputes over the Black Sea fleet and the future of Sevastopol and signed a long-delayed treaty of friendship.

United Arab Emirates

People: Population: 2,344,402. **Age distrib.** (%): <15: 30.8; 65+: 2.0. **Pop. density:** 73 per sq. mi. **Urban:** 84%. **Ethnic groups:** Arab, Iranian, Pakistani, Indian. **Principal languages:** Arabic (official), Persian, English, Hindi, Urdu. **Chief religions:** Muslim 96%, Christian, Hindu.

Geography: Area: 32,000 sq. mi. **Location:** Middle East, on the S shore of the Persian Gulf. **Neighbors:** Saudi Arabia on W and S, Oman on E. **Topography:** A barren, flat coastal plain gives way to uninhabited sand dunes on the S. Hajar Mts. are on E. **Capital:** Abu Dhabi, 799,000*.

Government: Type: Federation of emirates. **Head of state:** Pres. Zaid ibn Sultan an-Nahayan; b. 1923; in office: Dec. 2, 1971. **Head of gov.:** Prime Min. Sheik Maktum ibn Rashid al-Maktum; b 1946; in office: Nov. 20, 1990. **Local divisions:** 7 autonomous emirates: Abu Dhabi, Ajman, Dubai, Fujaira, Ras al-Khaimah, Sharjah, Umm al-Qaiwain. **Defense:** 5.5% of GDP. **Active troops:** 64,500.

Economy: Industries: Oil, fishing, petrochemicals. **Chief crops:** Vegetables, dates. **Minerals:** Oil, natural gas. **Crude oil reserves** (1999): 97.8 bil bbls. **Livestock** (1997): chickens: 15.00 mil; goats: 1.00; sheep: 385,000. **Electricity prod.** (1997): 19.000 bil kWh. **Labor force:** 65% services; 30% ind. and commerce; 5% agric.

Finance: Monetary unit: Dirham (Sept. 1999: 3.67 = $1 U.S.). **GDP:** (1997 est.): $54.2 bil. **Per capita GDP:** $24,000. **Imports** (1996 est.): $23.5 bil; partners: UK 8%. **Exports** (1996 est.): $33.2 bil; partners: Japan 38%. **Tourism:** $540 mil. **Budget** (1997 est.): $5.4 bil. **Intl. reserves less gold** (May 1999): $9.57 bil. **Gold:** 795,000 oz t.

Transport: Motor vehicles: 320,000 pass. cars, 80,000 comm. vehicles. **Civil aviation:** 8.4 bil pass.-mi; 6 airports. **Chief ports:** Dubai, Abu Dhabi.

Communications: TV sets: 18 per 1,000 pop. **Radios:** 206 per 1,000 pop. **Telephones:** 915,200 main lines. **Daily newspaper circ.:** 135 per 1,000 pop.

Health: Life expectancy: 73.83 male; 76.72 female. **Births** (per 1,000 pop.): 18.86. **Deaths** (per 1,000 pop.): 3.13. **Natural inc.:** 1.573%. **Hosp. beds** (1994): 1 per 360 persons. **Physicians** (1994): 1 per 545 persons. **Infant mortality** (per 1,000 live births): 14.10.

Education: Compulsory: ages 6-12. **Literacy:** 79%.

Major Intl. Organizations: UN (FAO, IBRD, ILO, IMF, IMO, WHO, WTrO), AL, OPEC.

Embassy: Suite 700, 1255 22nd Street NW, 20037, 955-7999. **Websites:** http://www.uae.org.ae
http://www.emirates.org

The 7 "Trucial Sheikdoms" gave Britain control of defense and foreign relations in the 19th century. They merged to become an independent state Dec. 2, 1971.

The Abu Dhabi Petroleum Co. was fully nationalized in 1975. Oil revenues have given the UAE one of the highest per capita GDPs in the world. International banking has grown in recent years.

United Kingdom
United Kingdom of Great Britain and Northern Ireland

People: Population: 59,113,439. **Age distrib.** (%): <15: 19.2; 65+: 15.7. **Pop. density:** 626 per sq. mi. **Urban:** 89%. **Ethnic groups:** English 81.5%, Scottish 9.6%, Irish 2.4%, Welsh 1.9%, Ulster 1.8%; West Indian, Indian, Pakistani, others 2.8%. **Principal languages:** English, Welsh, Scottish,

Gaelic. **Chief religions:** Anglican, Roman Catholic, other Christian, Muslim.

Geography: Area: 94,500 sq. mi. **Location:** Off the NW coast of Europe, across English Channel, Strait of Dover, and North Sea. **Neighbors:** Ireland to W, France to SE. **Topography:** England is mostly rolling land, rising to Uplands of southern Scotland; Lowlands are in center of Scotland, granite Highlands are in N. Coast is heavily indented, especially on W. British Isles have milder climate than N Europe due to the Gulf Stream and ample rainfall. Severn, 220 mi., and Thames, 215 mi., are longest rivers. **Capital:** London. **Cities** (1996 est.): London 7,074,265; Birmingham 1,020,589; Leeds 726,939; Liverpool 467,995; Manchester 430,818.

Government: Type: Constitutional monarchy. **Head of state:** Queen Elizabeth II; b Apr. 21, 1926; in office: Feb. 6, 1952. **Head of gov.:** Prime Min. Tony Blair; b May 6, 1953; in office: May 2, 1997. **Local divisions:** 467 local authorities, including England: 387; Wales: 22; Scotland: 32; Northern Ireland: 26. **Defense:** 2.8% of GDP. **Active troops:** 213,800.

Economy: Industries: Steel, metals, vehicles, shipbuilding, textiles, chemicals, electronics, aircraft, machinery. **Chief crops:** Cereals, oilseeds, potatoes, vegetables. **Minerals:** Coal, tin, oil, gas, limestone, iron, salt, clay. **Crude oil reserves** (1999): 5.19 bil bbls. **Arable land:** 25%. **Livestock** (1997): chickens: 152.89 mil; sheep: 44.47 mil; cattle: 11.52 mil; pigs: 8.15 mil. **Fish catch** (1997): 887,444 metric tons. **Electricity prod.** (1997): 315.116 bil kWh. **Labor force:** 69% services; 18% manuf. & constr.; 11% govt.

Finance: Monetary unit: Pound (Sept. 1999: 1.62 = $1 U.S.). **GDP:** (1997 est.): $1.242 tril. **Per capita GDP:** $21,200. **Imports** (1997): $283.5 bil; partners: EU 50%, U.S. 13.9%. **Exports** (1997): $268 bil; partners: EU 53%, U.S. 11.4%. **Tourism:** $21.30 bil. **Budget** (1996 est.): $470 bil. **Intl. reserves less gold** (Apr. 1999): $29.39 bil. **Gold:** 22.98 mil oz t. **Consumer prices** (change in 1997): 3.4%.

Transport: Railroad: Length: 23,518 mi. **Motor vehicles** (1997): 25.59 mil pass. cars, 3.22 mil comm. vehicles. **Civil aviation:** 98.1 bil pass.-mi; 57 airports. **Chief ports:** London, Liverpool, Cardiff, Belfast.

Communications: TV sets: 612 per 1,000 pop. **Radios:** 1,194 per 1,000 pop. **Telephones** (1997): 31,878,000 main lines. **Daily newspaper circ.:** 383 per 1,000 pop.

Health: Life expectancy: 74.73 male; 80.15 female. **Births:** (per 1,000 pop.): 11.90. **Deaths:** (per 1,000 pop.): 10.64. **Natural inc.:** 0.126%. **Hosp. beds** (1993): 1 per 205 persons. **Physicians** (1993): 1 per 629 persons. **Infant mortality** (per 1,000 live births): 5.78.

Education: Compulsory: ages 5-16. **Literacy** (1993): 100%.

Major Intl. Organizations: UN and all of its specialized agencies, the Commonwealth, EU, NATO, OECD, OSCE.

Embassy: 3100 Massachusetts Ave. NW 20008; 588-6500. **Websites:** http://www.ons.gov.uk/ons_f.htm
http://www.genuki.org.uk

The United Kingdom of Great Britain and Northern Ireland comprises England, Wales, Scotland, and Northern Ireland.

Queen and Royal Family. The ruling sovereign is Elizabeth II of the House of Windsor, b Apr. 21, 1926, elder daughter of King George VI. She succeeded to the throne Feb. 6, 1952, and was crowned June 2, 1953. She was married Nov. 20, 1947, to Lt. Philip Mountbatten, b June 10, 1921, former Prince of Greece. He was created Duke of Edinburgh, and given the title H.R.H., Nov. 19, 1947; he was named Prince of the United Kingdom and Northern Ireland Feb. 22, 1957. Prince Charles Philip Arthur George, b Nov. 14, 1948, is the Prince of Wales and heir apparent. His 1st son, William Philip Arthur Louis, b June 21, 1982, is second in line to the throne.

Parliament is the legislative body for the UK, with certain powers over dependent units. It consists of 2 houses: The **House of Lords** (1999) includes 759 hereditary and 504 life peers and peeresses, 2 archbishops and 24 bishops of the Church of England. Total membership is 1,289. The **House of Commons** has 659 members, elected by direct ballot and divided as follows: England 529; Wales 40; Scotland 72; Northern Ireland 18.

Resources and Industries. Great Britain's major occupations are manufacturing and trade. Metals and metal-using industries contribute more than 50% of exports. Of about 60 million acres of land in England, Wales, and Scotland, 46 million are farmed, of which 17 million are arable, the rest pastures.

Large oil and gas fields have been found in the North Sea. Commercial oil production began in 1975. There are large deposits of coal.

Britain imports all of its cotton, rubber, sulphur, about 80% of its wool, half of its food and iron ore, also certain amounts of paper, tobacco, chemicals. Manufactured goods made from these basic materials have been exported since the industrial age began. Main exports are machinery, chemicals, woolen and synthetic textiles, clothing, autos and trucks, iron and steel, locomotives, ships, jet aircraft, farm machinery, drugs, radio, TV, radar and navigation equipment, scientific instruments, arms, whisky.

Religion and Education. The Church of England is Protestant Episcopal. The queen is its temporal head, with rights of appointments to archbishoprics, bishoprics, and other offices. There are 2 provinces, Canterbury and York, each headed by an archbishop. The most famous church is Westminster Abbey (1050-1760), site of coronations, tombs of Elizabeth I, Mary, Queen of Scots, kings, poets, and of the Unknown Warrior.

The most celebrated British universities are Oxford and Cambridge, each dating to the 13th century. There are about 70 other universities.

History. Britain was part of the continent of Europe until about 6,000 BC, but migration across the English Channel continued long afterward. Celts arrived 2,500 to 3,000 years ago. Their language survives in Welsh, and Gaelic enclaves.

England was added to the Roman Empire in AD 43. After the withdrawal of Roman legions in 410, waves of Jutes, Angles, and Saxons arrived from German lands. They contended with Danish raiders for control from the 8th through 11th centuries. The last successful invasion was by French speaking Normans in 1066, who united the country with their dominions in France.

Opposition by nobles to royal authority forced King John to sign the Magna Carta in 1215, a guarantee of rights and the rule of law. In the ensuing decades, the foundations of the parliamentary system were laid.

English dynastic claims to large parts of France led to the Hundred Years War, 1338-1453, and the defeat of England. A long civil war, the War of the Roses, lasted 1455-85, and ended with the establishment of the powerful Tudor monarchy. A distinct English civilization flourished. The economy prospered over long periods of domestic peace unmatched in continental Europe. Religious independence was secured when the Church of England was separated from the authority of the pope in 1534.

Under Queen Elizabeth I, England became a major naval power, leading to the founding of colonies in the new world and the expansion of trade with Europe and the Orient. Scotland was united with England when James VI of Scotland was crowned James I of England in 1603.

A struggle between Parliament and the Stuart kings led to a bloody civil war, 1642-49, and the establishment of a republic under the Puritan Oliver Cromwell. The monarchy was restored in 1660, but the "Glorious Revolution" of 1688 confirmed the sovereignty of Parliament: a Bill of Rights was granted 1689.

In the 18th century, parliamentary rule was strengthened. Technological and entrepreneurial innovations led to the Industrial Revolution. The 13 North American colonies were lost, but replaced by growing empires in Canada and India. Britain's role in the defeat of Napoleon, 1815, strengthened its position as the leading world power.

The extension of the franchise in 1832 and 1867, the formation of trade unions, and the development of universal public education were among the drastic social changes that accompanied the spread of industrialization and urbanization in the 19th century. Large parts of Africa and Asia were added to the empire during the reign of Queen Victoria, 1837-1901.

Though victorious in World War I, Britain suffered huge casualties and economic dislocation. Ireland became independent in 1921, and independence movements became active in India and other colonies. The country suffered major bombing damage in World War II, but held out against Germany singlehandedly for a year after France fell in 1940.

Industrial growth continued in the postwar period, but Britain lost its leadership position to other powers. Labor governments passed socialist programs nationalizing some basic industries and expanding social security. Prime Min. Margaret Thatcher's Conservative government, however, tried to increase the role of private enterprise. In 1987, Thatcher became the first British leader in 160 years to be elected to a 3d consecutive term as prime minister. Falling on unpopular times, she resigned as prime minister in Nov. 1990. Her successor, John Major, led Conservatives to an upset victory at the polls, Apr. 9, 1992.

The UK supported the UN resolutions against Iraq and sent military forces to the Persian Gulf War.

The Channel Tunnel linking Britain to the Continent was officially inaugurated May 6, 1994. Britain's relations with the European Union were frayed in 1996 when the EU banned British beef because of the threat of "mad cow" disease.

On May 1, 1997, the Labour Party swept into power in a landslide victory, the largest of any party since 1935. Labour Party leader Tony Blair, 43, became Britain's youngest prime

minister since 1812. Diana, Princess of Wales, the divorced wife of Prince Charles and the mother of Prince William, died in a car crash in Paris, Aug. 31. Britain played a leading role in the NATO air war against Yugoslavia, Mar.-June 1999, and contributed 12,000 troops to the multinational security force in Kosovo (KFOR).

Wales

The Principality of Wales in western Britain has an area of 8,019 sq. mi. and a population (1997 est.) of 2,927,000. Cardiff is the capital, pop. (1996 est.) 315,040.

Less than 20% of Wales residents speak English and Welsh; about 32,000 speak Welsh solely. A 1979 referendum rejected, 4-1, the creation of an elected Welsh assembly; a similar proposal passed by a thin margin on Sept. 18, 1997. Elections were held May 6, 1999.

Early Anglo-Saxon invaders drove Celtic peoples into the mountains of Wales, terming them Waelise (Welsh, or foreign). There they developed a distinct nationality. Members of the ruling house of Gwynedd in the 13th century fought England but were crushed, 1283. Edward of Caernarvon, son of Edward I of England, was created Prince of Wales, 1301.

Scotland

Scotland, a kingdom now united with England and Wales in Great Britain, occupies the northern 37% of the main British island, and the Hebrides, Orkney, Shetland, and smaller islands. Length 275 mi., breadth approx. 150 mi., area 30,418 sq. mi., population (1992 est.) 5,111,000.

The Lowlands, a belt of land approximately 60 mi. wide from the Firth of Clyde to the Firth of Forth, divide the farming region of the Southern Uplands from the granite Highlands of the North; they contain 75% of the population and most of the industry. The Highlands, famous for hunting and fishing, have been opened to industry by many hydroelectric power stations.

Edinburgh, pop. (1996 est.) 448,850, is the capital. Glasgow, pop. (1996 est.) 616,430, is Britain's greatest industrial center. It is a shipbuilding complex on the Clyde and an ocean port. Aberdeen, pop. (1996 est.) 227,430, NE of Edinburgh, is a major port, center of granite industry, fish-processing, and North Sea oil exploration. Dundee, pop. (1996 est.) 150,250, NE of Edinburgh, is an industrial and fish-processing center. About 90,000 persons speak Gaelic as well as English.

History. Scotland was called Caledonia by the Romans who battled early Celtic tribes and occupied southern areas from the 1st to the 4th centuries. Missionaries from Britain introduced Christianity in the 4th century; St. Columba, an Irish monk, converted most of Scotland in the 6th century.

The Kingdom of Scotland was founded in 1018. William Wallace and Robert Bruce both defeated English armies 1297 and 1314, respectively.

In 1603 James VI of Scotland, son of Mary, Queen of Scots, succeeded to the throne of England as James I, and effected the Union of the Crowns. In 1707 Scotland received representation in the British Parliament, resulting from the union of former separate Parliaments. Its executive in the British cabinet is the Secretary of State for Scotland. The growing Scottish National Party urges independence. A 1979 referendum on the creation of an elected Scottish assembly was defeated, but a proposal to create a regional legislature with limited taxing authority passed by a landslide Sept. 11, 1997. Elections were held May 6, 1999.

Memorials of Robert Burns, Sir Walter Scott, John Knox, and Mary, Queen of Scots, draw many tourists, as do the beauties of the Trossachs, Loch Katrine, Loch Lomond, and abbey ruins.

Industries. Engineering products are the most important industry, with growing emphasis on office machinery, autos, electronics, and other consumer goods. Oil has been discovered offshore in the North Sea, stimulating on-shore support industries.

Scotland produces fine woolens, worsteds, tweeds, silks, fine linens, and jute. It is known for its special breeds of cattle and sheep. Fisheries have large hauls of herring, cod, whiting. Whisky is the biggest export.

The Hebrides are a group of c. 500 islands, 100 inhabited, off the W coast. The Inner Hebrides include **Skye, Mull,** and **Iona,** the last famous for the arrival of St. Columba, AD 563. The Outer Hebrides include **Lewis** and **Harris.** Industries include sheep raising and weaving. The **Orkney Islands,** c. 90, are to the NE. The capital is Kirkwall, on Pomona Isl. Fish curing, sheep raising, and weaving are occupations. NE of the Orkneys are the 200 **Shetland Islands,** 24 inhabited, home of Shetland pony. The Orkneys and Shetlands are centers for the North Sea oil industry.

MILLENNIUM FACT BOX

Sovereigns of Europe

When the 20th century began, many European countries were still ruled by monarchs. *The World Almanac 1900* published lists of the 40 "sovereigns of Europe" (including the pope) who were on the throne at the time, along with information about each that is given below.

Sovereigns are arranged according to their respective ages.

Sovereigns	Yr of Birth	Age Jan. 1, 1900			Accession
		y.	m.	d.	
Leo XIII, Pope.	1810	89	9	29	1878
Adolphus, Grand Duke of Luxembourg.	1817	82	5	7	1890
Adolphus, Pr. of Schaumburg-Lippe.	1817	82	5	—	1860
Christian IX, King of Denmark	1818	81	8	22	1863
Charles, Grand Duke of Saxe-Weimar	1818	81	6	6	1853
Victoria, Queen of Great Britain	1819	80	7	7	1837
Frederick William, Grand Duke of Mecklenburg-Strelitz	1819	80	2	13	1860
George II, Duke of Saxe-Meiningen	1826	73	8	28	1866
Frederick, Grand Duke of Baden	1826	73	3	21	1852
Ernest, Duke of Saxe-Altenburg.	1826	73	3	14	1853
Peter, Grand Duke of Oldenburg	1827	72	5	23	1853
Albert, King of Saxony	1828	71	8	7	1873
Oscar II, King of Sweden and Norway	1829	70	11	10	1872
Charles, Prince of Schwarzburg-Sondershausen.	1830	69	4	24	1880
Francis Joseph, Emperor of Austria	1830	69	4	13	1848
Frederick, Duke of Anhalt.	1831	68	8	2	1871
Henry XIV, Pr. of Reuss (Younger line).	1832	67	7	3	1867
Leopold II, King of the Belgians	1835	64	8	21	1865
Charles, King of Roumania.	1839	60	8	19	1866
John II, Prince of Liechtenstein	1840	59	2	26	1858
Nicholas, Prince of Montenegro	1841	58	2	23	1860
Abdul Hamid, Sultan of Turkey	1842	57	3	8	1876
Humbert I, King of Italy.	1844	55	9	17	1878
Alfred, Duke of Saxe-Coburg-Gotha.	1844	55	4	26	1893
George I, King of the Greeks	1845	54	0	7	1863
Henry XXII, Pr. of Reuss (Elder line).	1846	53	9	3	1859
William, King of Würtemberg	1848	51	10	3	1891
Otto, King of Bavaria.	1848	51	8	4	1886
Albert, Prince of Monaco	1848	51	1	19	1889
Gunther, Pr. of Schwarzberg-Rudolstadt	1852	47	4	9	1890
William II, German Emperor.	1859	40	11	4	1888
Adolphus, Prince of Lippe-Detmold	1859	40	5	11	1895
Charles I, King of Portugal	1863	36	3	4	1889
Frederick, Prince of Waldeck	1865	34	11	11	1893
Nicholas II, Emperor of Russia.	1868	31	7	14	1894
Ernest Louis, Grand Duke of Hesse.	1868	31	1	6	1892
Alexander I, King of Servia.	1876	23	4	18	1889
Wilhelmina, Queen of Netherlands.	1880	19	4	1	1890
Frederick IV, Grand Duke of Mecklenburg-Schwerin	1882	17	0	1	1897
Alphonso XIII, King of Spain.	1886	13	7	14	1886

Northern Ireland

Northern Ireland was constituted in 1920 from 6 of the 9 counties of Ulster, the NE corner of Ireland. Area 5,452 sq. mi., pop. (1996 est.) 1,663,300, capital and chief industrial center, Belfast, pop. (1996 est.) 297,300.

Industries. Shipbuilding, including large tankers, has long been an important industry, centered in Belfast, the largest port. Linen manufacture is also important, along with apparel, rope, and twine. Growing diversification has added engineering products, synthetic fibers, and electronics. There are large numbers of cattle, hogs, and sheep. Potatoes, poultry, and dairy foods are also produced.

Government. An act of the British Parliament, 1920, divided Northern from Southern Ireland, each with a parliament and government. When Ireland became a dominion, 1921, and later a republic, Northern Ireland chose to remain a part of the United Kingdom. It elects 18 members to the British House of Commons.

During 1968-69, large demonstrations were conducted by Roman Catholics who charged they were discriminated against in voting rights, housing, and employment. The Catholics, a minority comprising about a third of the population, demanded abolition of property qualifications for voting in local elections. Violence and terrorism intensified, involving branches of the Irish Republican Army (outlawed in the Irish Republic), Protestant groups, police, and British troops.

A succession of Northern Ireland prime ministers pressed reform programs but failed to satisfy extremists on both sides. Between 1969 and 1994 more than 3,000 were killed in sectarian violence, many in England itself. Britain suspended the Northern Ireland parliament Mar. 30, 1972, and imposed direct British rule. A coalition government was formed in 1973 when moderates won election to a new one-house Assembly. But a Protestant general strike overthrew the government in 1974 and direct rule was resumed.

The agony of Northern Ireland was dramatized in 1981 by the deaths of 10 Irish nationalist hunger strikers in Maze Prison near Belfast. In 1985 the Hillsborough agreement gave the Rep. of Ireland a voice in the governing of Northern Ireland; the accord was strongly opposed by Ulster loyalists. On Dec. 12,

1993, Britain and Ireland announced a declaration of principles to resolve the Northern Ireland conflict.

On Aug. 31, 1994, the IRA announced a cease-fire, saying it would rely on political means to achieve its objectives; the IRA resumed its terrorist tactics on Feb. 9, 1996. Reinstatement of the IRA cease-fire as of July 20, 1997, led to the resumption of peace talks Sept. 15.

A settlement reached on Good Friday, April 10, 1998, provided for restoration of home rule and election of a 108-member assembly with safeguards for minority rights. Both Ireland and Great Britain agreed to give up their constitutional claims on Northern Ireland. The accord was approved May 22 by voters in Northern Ireland and the Irish Republic, and elections to the assembly were held June 25. IRA dissidents seeking to derail the agreement were responsible for a bomb at Omagh Aug. 15 that killed 29 people and injured over 330. The peace process bogged down in 1999 over the IRA's refusal to disarm.

Education and Religion. Northern Ireland is about 58% Protestant, 42% Roman Catholic. Education is compulsory between the ages of 5 and 16 years.

Channel Islands

The Channel Islands, area 75 sq. mi., pop. (1997 est.) 152,241, off the NW coast of France, the only parts of the one-time Dukedom of Normandy belonging to England, are Jersey, Guernsey and the dependencies of Guernsey—Alderney, Brechou, Great Sark, Little Sark, Herm, Jethou and Lihou. Jersey and Guernsey have separate legal existences and lieutenant governors named by the Crown. The islands were the only British soil occupied by German troops in World War II.

Isle of Man

The Isle of Man, area 227 sq. mi., pop. (1997 est.) 74,504, is in the Irish Sea, 20 mi. from Scotland, 30 mi. from Cumberland. It is rich in lead and iron. The island has its own laws and a lieutenant governor appointed by the Crown. The Tynwald (legislature) consists of the Legislative Council, partly elected, and House of Keys, elected. Capital: Douglas. Farming, tourism, and fishing (kippers, scallops) are chief occupations. Man is famous for the Manx tailless cat.

Gibraltar

Gibraltar, a dependency on the southern coast of Spain, guards the entrance to the Mediterranean. The Rock of Gibraltar has been in British possession since 1704. The Rock is 2.75 mi. long, 3/4 of a mi. wide and 1,396 ft. in height; a narrow isthmus connects it with the mainland. Pop. (1999 est.) 29,165.

Gibraltar has historically been an object of contention between Britain and Spain. Residents voted with near unanimity to remain under British rule, in a 1967 referendum held in pursuance of a UN resolution on decolonization. A new constitution, May 30, 1969, increased Gibraltarian control of domestic affairs (the UK continues to handle defense and internal security matters). Following a 1984 agreement between Britain and Spain, the border, closed by Spain in 1969, was fully reopened in Feb. 1985. A UN General Assembly resolution requested Britain to end Gibraltar's colonial status by Oct. 1, 1996. No settlement has been reached.

British West Indies

Swinging in a vast arc from the coast of Venezuela NE, then N and NW toward Puerto Rico are the Leeward Islands, forming a coral and volcanic barrier sheltering the Caribbean from the open Atlantic. Many of the islands are self-governing British possessions. Universal suffrage was instituted 1951-54; ministerial systems were set up 1956-1960.

The **Leeward Islands** still associated with the UK are **Montserrat**, area 32 sq. mi., pop. (1999 est.) 12,853, capital Plymouth; the **British Virgin Islands**, 59 sq. mi., pop. (1999 est.) 19,156, capital Road Town; and **Anguilla**, the most northerly of the Leeward Islands, 60 sq. mi., pop. (1999 est.) 11,510, capital The Valley. Montserrat has been devastated by the Soufrière Hills volcano, which began erupting July 18, 1995.

The three **Cayman Islands**, a dependency, lie S of Cuba, NW of Jamaica. Pop. (1999 est.) 39,335, most of it on Grand Cayman. It is a free port; in the 1970s Grand Cayman became a tax-free refuge for foreign funds and branches of many Western banks were opened there. Total area 102 sq. mi., capital Georgetown.

The **Turks and Caicos Islands** are a dependency at the SE end of the Bahama Islands. Of about 30 islands, only 6 are inhabited; area 193 sq. mi., pop. (1999 est.) 16,863; capital Grand Turk. Salt, shellfish, and conch shells are the main exports.

Bermuda

Bermuda is a British dependency governed by a royal governor and an assembly, dating from 1620, the oldest legislative body among British dependencies. Capital is Hamilton.

It is a group of about 150 small islands of coral formation, 20 inhabited, comprising 20.0 sq. mi. in the western Atlantic, 580 mi. E of North Carolina. Pop. (1999 est.) 62,472 (about 61% of African descent). Pop. density is high.

The U.S. maintains a NASA tracking facility; a U.S. naval air base was closed in 1995.

Tourism is the major industry; Bermuda boasts many resort hotels. The government raises most revenue from import duties. Exports: petroleum products, medicine. In a referendum Aug. 15, 1995, voters rejected independence by nearly a 3-to-1 majority.

South Atlantic

The **Falkland Islands,** a dependency, lie 300 mi. E of the Strait of Magellan at the southern end of South America.

The Falklands or Islas Malvinas include 2 large islands and about 200 smaller ones, area 4,700 sq. mi., pop. (1995 est.) 2,317, capital Stanley. The licensing of foreign fishing vessels has become the major source of revenue. Sheep-grazing is a main industry; wool is the principal export. There are indications of large oil and gas deposits. The islands are also claimed by Argentina, though 97% of inhabitants are of British origin. Argentina invaded the islands Apr. 2, 1982. The British responded by sending a task force to the area, landing their main force on the Falklands, May 21, and forcing an Argentine surrender at Port Stanley, June 14. A pact resuming commercial air service with Argentina was signed July 14, 1999.

British Antarctic Territory, south of 60° S lat., formerly a dependency of the Falkland Isls., was made a separate colony in 1962 and includes the **South Shetland Islands,** the **South Orkneys,** and the Antarctic Peninsula. A chain of meteorological stations is maintained.

South Georgia and the South Sandwich Islands, formerly administered by the Falklands Isls., became a separate dependency in 1985. South Georgia, 1,450 sq. mi., with no permanent population, is about 800 mi. SE of the Falklands; the South Sandwich Isls., 130 sq. mi., are uninhabited, about 470 mi. SE of South Georgia.

St. Helena, an island 1,200 mi. off the W coast of Africa and 1,800 mi. E of South America, 47 sq. mi. and pop. (1999 est.) 7,145. Flax, lace, and rope-making are the chief industries. After Napoleon Bonaparte was defeated at Waterloo the Allies exiled him to St. Helena, where he lived from Oct. 16, 1815, to his death, May 5, 1821. Capital is Jamestown.

Tristan da Cunha is the principal of a group of islands of volcanic origin, total area 40 sq. mi., halfway between the Cape of Good Hope and South America. A volcanic peak 6,760 ft. high erupted in 1961. The 262 inhabitants were removed to England, but most returned in 1963. The islands are dependencies of St. Helena. Pop. (1993) 300.

Ascension is an island of volcanic origin, 34 sq. mi. in area, 700 mi. NW of St. Helena, through which it is administered. It is a communications relay center for Britain, and has a U.S. satellite tracking center. Pop. (1993) was 1,117, half of them communications workers. The island is noted for sea turtles.

Hong Kong
(See China/Hong Kong)

British Indian Ocean Territory

Formed Nov. 1965, embracing islands formerly dependencies of Mauritius or Seychelles: the Chagos Archipelago (including Diego Garcia), Aldabra, Farquhar, and Des Roches. The latter 3 were transferred to Seychelles, which became independent in 1976. Area 23 sq. mi. No permanent civilian population remains; the U.K. and the U.S. maintain a military presence.

Pacific Ocean

Pitcairn Island is in the Pacific, halfway between South America and Australia. The island was discovered in 1767 by Philip Carteret but was not inhabited until 23 years later when the mutineers of the *Bounty* landed there. The area is 1.7 sq. mi. and 1995 pop. was 54. It is a British dependency and is administered by a British High Commissioner in New Zealand and a local Council. The uninhabited islands of **Henderson, Ducie,** and **Oeno** are in the Pitcairn group.

United States
United States of America

People: Population: 272,639,608 (incl. 50 states & Dist. of Columbia). (Note: U.S. pop. figures may differ elsewhere in *The World Almanac*.) **Age distrib.** (%): <15: 21.6; 65+: 12.6. **Pop. density:** 73 per sq. mi. **Urban:** 76%.

Geography: Area: 3,717,796 sq. mi. (incl. 50 states and DC). **Topography:** Vast central plain, mountains in west, hills and low mountains in east. **Capital:** Washington, D.C.

Government: Federal republic, strong democratic tradition. **Head of state and gov.:** Pres. Bill Clinton; b Aug. 19, 1946; in office: Jan. 20, 1993. **Local divisions:** 50 states and Dist. of Columbia. **Defense:** 3.4% of GDP. **Active troops:** 1.448 mil.

Economy: Minerals: Coal, oil, gas, copper, lead, molybdenum, phosphates, uranium, bauxite, gold, iron, mercury, nickel, potash, silver, tungsten, zinc. **Crude oil reserves** (1999): 22.55 bil bbls. **Other resources:** forests. **Arable land:** 19%. **Livestock** (1997): chickens: 1700.00 bil; cattle: 99.74 mil; pigs: 60.92 mil; sheep: 7.62 mil; goats: 1.40 mil. **Fish catch** (1997): 5.01 mil metric tons. **Electricity prod.** (1997): 3,494.4 bil kWh.

Finance: GDP: (1997 est.): $8.08 tril. **Per capita GDP:** $30,200. **Imports** (1996): $822 bil; partners: Canada 20%, Western Europe 18%, Japan 17%. **Exports** (1996): $625.1 bil; partners: Canada 22%, Western Europe 21%, Japan 11%. **Tourism:** $74.24 bil. **Budget** (1997): $1.601 tril. **Intl. reserves less gold** (June 1999): $60.64 bil. **Gold:** 261.69 mil oz t. **Consumer prices** (change in 1997): 1.6%.

Transport: Railroad: Length: 137,900 mi. **Motor vehicles:** 129.73 mil pass. cars, 76.64 mil comm. vehicles. **Civil aviation:** 599.4 bil pass.-mi; 834 airports.

Communications: TV sets: 776 per 1,000 pop. **Radios:** 2,122 per 1,000 pop. **Telephones** (1997): 172,452,500 main lines. **Daily newspaper circ.:** 238 per 1,000 pop.

Health: Life expectancy: 72.95 male; 79.67 female. **Births** (per 1,000 pop.): 14.30. **Deaths** (per 1,000 pop.): 8.80. **Natural inc.:** 0.550%. **Hosp. beds** (1995): 1 per 243 persons. **Physicians** (1995): 1 per 365 persons. **Infant mortality** (per 1,000 live births): 6.33.

Education: Free, compulsory: ages 7-16. **Literacy** (1994): 97%.

Major Intl. Organizations: UN (FAO, IBRD, ILO, IMF, IMO, WHO, WTrO), APEC, NATO, OAS, OECD, OSCE.

Websites: http://www.census.gov
http://www.whitehouse.gov

Uruguay
Oriental Republic of Uruguay

People: Population: 3,308,523. **Age distrib.** (%): <15: 24.1; 65+: 13.0. **Pop. density:** 49 per sq. mi. **Urban:** 91%. **Ethnic groups:** White 88%, mestizo 8%, black 4%. **Principal language:** Spanish (official). **Chief religion:** Roman Catholic 66%.

Geography: Area: 68,000 sq. mi. **Location:** In southern South America, on the Atlantic O. **Neighbors:** Argentina on W, Brazil on N. **Topography:** Uruguay is composed of rolling, grassy plains and hills, well watered by rivers flowing W to Uruguay R. **Capital:** Montevideo (1996): 1,303,182.

Government: Type: Republic. **Head of state and gov.:** Pres. Julio María Sanguinetti; b Jan. 6, 1936; in office: Mar. 1, 1995. **Local divisions:** 19 departments. **Defense:** 2.3% of GDP. **Active troops:** 25,600.

Economy: Industries: Meat processing, wool and hides, textiles, wine, oil refining. **Chief crops:** Corn, wheat, sorghum, rice. **Arable land:** 7%. **Livestock** (1997): chickens: 12.00 mil; sheep: 17.80 mil; cattle: 10.48 mil; pigs: 270,000. **Fish catch** (1997): 123,361 metric tons. **Electricity prod.** (1997): 6.885 bil kWh. **Labor force** 33% serv.; 25% govt.; 19% manuf.; 12% comm.; 11% agric.

Finance: Monetary unit: Peso (Sept. 1999: 11.48 = $1 U.S.). **GDP:** (1997 est.): $29.1 bil. **Per capita GDP:** $8,900. **Imports** (1997): $3.7 bil; partners: Brazil 22%, Argentina 21%. **Exports** (1999): $2.7 bil; partners: Brazil 35%, Argentina 11%. **Tourism:** $895 mil. **Budget** (1997 est.): $4.3 bil. **Intl. reserves less gold** (Apr. 1999): $1.93 bil. **Gold:** 1.79 mil oz t. **Consumer prices** (change in 1997): 10.8%.

Transport: Railroad: Length: 1,288 mi. **Motor vehicles:** 475,000 pass. cars, 50,000 comm. vehicles. **Civil aviation:** 470.0 mil pass.-mi; 1 airport. **Chief port:** Montevideo.

Communications: TV sets: 191 per 1,000 pop. **Radios:** 586 per 1,000 pop. **Telephones:** 823,500 main lines. **Daily newspaper circ.:** 241 per 1,000 pop.

Health: Life expectancy: 72.69 male; 79.15 female. **Births** (per 1,000 pop.): 16.84. **Deaths** (per 1,000 pop.): 8.81. **Natural inc.:** 0.803%. **Physicians** (1994): 1 per 282 persons. **Infant mortality** (per 1,000 live births): 13.49.

Education: Free, compulsory for 6 years between ages 6-14. **Literacy:** 97%.

Major Intl. Organizations: UN (FAO, IBRD, ILO, IMF, IMO, WHO, WTrO), OAS.

Embassy: 2715 M St. NW 20007; 331-1313.

Website: http://www.embassy.org/uruguay

Spanish settlers began to supplant the indigenous Charrua Indians in 1624. Portuguese from Brazil arrived later, but Uruguay was attached to the Spanish Viceroyalty of Rio de la Plata in the 18th century. Rebels fought against Spain beginning in 1810. An independent republic was declared Aug. 25, 1825.

Terrorist activities led Pres. Juan María Bordaberry to agree to military control of his administration Feb. 1973. In June he abolished Congress and set up a Council of State in its place. Bordaberry was removed by the military in a 1976 coup. Civilian government was restored in 1985.

Socialist measures were adopted in the early 1900s, and the state retains a dominant role in the power, telephone, railroad, cement, oil-refining, and other industries. Uruguay's standard of living remains one of the highest in South America, and political and labor conditions among the freest.

Uzbekistan
Republic of Uzbekistan

People: Population: 24,102,473. **Age distrib.** (%): <15: 37.2; 65+: 4.7. **Pop. density:** 140 per sq. mi. **Urban:** 41%. **Ethnic groups:** Uzbek 80%, Russian 6%, Tajik 5%. **Principal languages:** Uzbek (official), Russian. **Chief religions:** Muslim (mostly Sunni) 88%, Eastern Orthodox 9%.

Geography: Area: 172,700 sq. mi. **Location:** Central Asia. **Neighbors:** Kazakhstan on N and W, Kyrgyzstan and Tajikistan on E, Afghanistan and Turkmenistan on S. **Topography:** Mostly plains and desert. **Capital:** Tashkent 2,282,000*.

Government: Type: Republic. **Head of state:** Pres. Islam A. Karimov; b Jan. 30, 1938; in office: Mar. 24, 1990. **Head of gov.:** Prime Min. Utkir Sultanov; b July 14, 1939; in office: Dec. 21, 1995. **Local divisions:** 12 regions, 1 autonomous republic, 1 city. **Defense:** 3.9% of GDP. **Active troops:** 70,000.

Economy: Industries: Machinery, food processing, natural gas, textiles. **Chief crops:** Vegetables, cotton, fruits, grain. **Minerals:** Gas, oil, coal, gold, uranium, silver, copper. **Crude oil reserves** (1999): 594 mil bbls. **Arable land:** 9%. **Livestock** (1997): chickens: 12.00 mil; sheep: 8.00 mil; cattle: 5.30 mil; goats: 800,000; pigs: 195,000. **Electricity prod.** (1997): 45.183 bil kWh. **Labor force:** 44% agric. & forestry; 20% ind. & constr.

Finance: Monetary unit: Som (Aug. 1999: 168.70 = $1 U.S.). **GDP:** (1997 est.): $60.7 bil. **Per capita GDP:** $2,500. **Imports** (1996): $4.7 bil; partners: Russia 25%, Korea 12%. **Exports** (1996): $3.8 bil; partners: Russia 22%, Italy 9%. **Tourism:** $21 mil.

Transport: Railroad: Length: 2,100 mi. **Motor vehicles:** 865,000 pass. cars, 14,500 comm. vehicles. **Civil aviation:** 2.2 bil pass.-mi; 9 airports. **Chief port:** Termiz.

Communications: TV sets: 176 per 1,000 pop. **Telephones** (1997): 1,490,000 main lines.

Health: Life expectancy: 60.29 male; 67.71 female. **Births** (per 1,000 pop.): 23.43. **Deaths** (per 1,000 pop.): 7.75. **Natural inc.:** 1.568%. **Hosp. beds** (1995): 1 per 120 persons. **Physicians** (1995): 1 per 302 persons. **Infant mortality** (per 1,000 live births): 71.58.

Education: Compulsory: ages 6-14. **Literacy** (1993): 97%.

Major Intl. Organizations: UN (IBRD, ILO, IMF, WHO), CIS, OSCE.

Embassy: 1746 Massachusetts Ave. NW 20036; 887-5300.

Website: http://www.gov.uz

The region was overrun by the Mongols under Genghis Khan in 1220. In the 14th century, Uzbekistan became the center of a native empire—that of the Timurids. In later centuries Muslim feudal states emerged. Russian military conquest began in the 19th century.

The Uzbek SSR became a Soviet Union republic in 1925. Uzbekistan declared independence Aug. 29, 1991. It became an independent republic when the Soviet Union disbanded Dec. 26, 1991. Subsequently, the government of Uzbekistan was dominated by former Communists.

Vanuatu
Republic of Vanuatu

People: Population: 189,036. **Age distrib.** (%): <15: 38.5; 65+: 3.1. **Pop. density:** 33 per sq. mi. **Urban:** 19%. **Ethnic groups:** Melanesian 94%, French 4%. **Principal languages:** French, English, Bislama (all official). **Chief religions:** Presbyterian 37%, Anglican 15%, Catholic 15%, other Christian 10%, indigenous beliefs 8%.

Geography: Area: 5,700 sq. mi. **Location:** SW Pacific, 1,200 mi. NE of Brisbane, Australia. **Neighbors:** Fiji to E, Solomon Isls. to NW. **Topography:** Dense forest with narrow coastal strips of cultivated land. **Capital:** Port-Vila (1996 est.): 31,800.

Government: Type: Republic. **Head of state:** Pres. John Bani; in office: Mar. 24, 1999. **Head of gov.:** Prime Min. Donald Kalpokas; in office: Mar. 30, 1998. **Local divisions:** 6 provinces.

Economy: Industries: Fish-freezing, meat canneries, wood processing. **Chief crops:** Copra, coconuts, cocoa, coffee. **Minerals:** Manganese. **Arable land:** 2%. **Other resources:** Forests, cattle. **Fish catch** (1996): 2,729 metric tons. **Livestock** (1997): chickens: 320,000; cattle: 151,000. **Electricity prod.** (1997): 32 mil kWh.

Finance: Monetary unit: Vatu (Sept. 1999: 129.25 = $1 U.S.). **GDP:** (1996 est.): $231 mil. **Per capita GDP:** $1,300. **Imports** (1996): $97 mil; partners: Japan 47%, Australia 23%. **Exports** (1996): $30 mil; partners: Japan 28%, Spain 21%, Germany 14%. **Tourism:** $52 mil. **Budget** (1996 est.): $99.8 mil. **Intl. reserves less gold** (June 1999): $42.19 mil. **Consumer prices** (change in 1997): 2.8%.

Transport: Motor vehicles: 4,000 pass. cars, 2,500 comm. vehicles. **Civil aviation:** 93.2 mil pass.-mi; 29 airports. **Chief ports:** Foran, Port-Vila.

Communications: Radios: 319 per 1,000 pop. **Telephones:** 5,200 main lines.

Health: Life expectancy: 59.41 male; 63.57 female. **Births** (per 1,000 pop.): 28.49. **Deaths** (per 1,000 pop.): 8.26. **Natural inc.:** 2.023%. **Hosp. beds** (1995): 1 per 450 persons. **Physicians** (1995): 1 per 14,025 persons. **Infant mortality** (per 1,000 live births): 59.58.

Education: Literacy (1997): 36%.

Major Intl. Organizations: UN (FAO, IBRD, IMF, IMO, WHO), the Commonwealth.

The Anglo-French condominium of the New Hebrides, administered jointly by France and Great Britain since 1906, became the independent Republic of Vanuatu on July 30, 1980.

Vatican City (The Holy See)

People: Population: 860. **Urban:** 100%. **Ethnic groups:** Italian, Swiss. **Principal languages:** Italian, Latin. **Chief religion:** Roman Catholic.

Geography: Area: 108.7 acres. **Location:** In Rome, Italy. **Neighbors:** Completely surrounded by Italy.

Monetary unit: Vatican Lira, Italian Lira (equal value) (Sept. 1999: 1,808.84 = $1 U.S.).

Apostolic Nunciature in U.S.: 3339 Massachusetts Ave. NW 20008; 333-7121.

Website: http://www.vatican.va

The popes for many centuries, with brief interruptions, held temporal sovereignty over mid-Italy (the so-called Papal States), comprising an area of some 16,000 sq. mi., with a population in the 19th century of more than 3 million. This territory was incorporated in the new Kingdom of Italy (1861), the sovereignty of the pope being confined to the palaces of the Vatican and the Lateran in Rome and the villa of Castel Gandolfo, by an Italian law, May 13, 1871. This law also guaranteed to the pope and his successors a yearly indemnity of over $620,000. The allowance, however, remained unclaimed.

A Treaty of Conciliation, a concordat, and a financial convention were signed Feb. 11, 1929, by Cardinal Gasparri and Premier Mussolini. The documents established the independent state of Vatican City and gave the Roman Catholic church special status in Italy. The treaty (Lateran Agreement) was made part of the Constitution of Italy (Article 7) in 1947. Italy and the Vatican signed an agreement in 1984 on revisions of the concordat; the accord eliminated Roman Catholicism as the state religion and ended required religious education in Italian schools.

Vatican City includes the Basilica of Saint Peter, the Vatican Palace and Museum covering over 13 acres, the Vatican gardens, and neighboring buildings between Viale Vaticano and the church. Thirteen buildings in Rome, outside the boundaries, enjoy extraterritorial rights; these buildings house congregations or officers necessary for the administration of the Holy See.

The legal system is based on the code of canon law, the apostolic constitutions, and laws especially promulgated for the Vatican City by the pope. The Secretariat of State represents the Holy See in its diplomatic relations. By the Treaty of Conciliation the pope is pledged to a perpetual neutrality unless his mediation is specifically requested. This, however, does not prevent the defense of the Church whenever it is persecuted.

The present sovereign of the State of Vatican City is the Supreme Pontiff John Paul II, born Karol Wojtyla in Wadowice, Po-

land, May 18, 1920, elected Oct. 16, 1978 (the first non-Italian to be elected pope in 456 years).

The U.S. restored formal relations in 1984 after the U.S. Congress repealed an 1867 ban on diplomatic relations with the Vatican. The Vatican and Israel agreed to establish formal relations Dec. 30, 1993.

Venezuela
Republic of Venezuela

People: Population: 23,203,466. **Age distrib.** (%): <15: 33.3; 65+: 4.6. **Pop. density:** 66 per sq. mi. **Urban:** 86%. **Ethnic groups:** Mestizo 67%, white (Spanish, Portuguese, Italian) 21%, black 10%, Amerindian 2%. **Principal language:** Spanish (official). **Chief religion:** Roman Catholic 96%.

Geography: Area: 352,100 sq. mi. **Location:** On Caribbean coast of South America. **Neighbors:** Colombia on W, Brazil on S, Guyana on E. **Topography:** Flat coastal plain and Orinoco Delta are bordered by Andes Mts. and hills. Plains, called llanos, extend between mountains and Orinoco. Guiana Highlands and plains are S of Orinoco, which stretches 1,600 mi. and drains 80% of Venezuela. **Capital:** Caracas. **Cities** (1996 est.): Caracas 3,672,779; Maracaibo 1,220,980; Valencia 910,582.

Government: Type: Federal republic. **Head of state and gov.:** Pres. Hugo Rafael Chávez Frías; b July 28, 1954; in office: Feb. 2, 1999. **Local divisions:** 22 states, 1 federal district (Caracas), 1 federal dependency (72 islands). **Defense:** 1.1% of GDP. **Active troops:** 56,000.

Economy: Industries: Iron mining, steel, oil products, textiles. **Chief crops:** Rice, corn, sorghum, bananas, sugar. **Minerals:** Oil, gas, iron (extensive reserves and production), gold. **Crude oil reserves** (1999): 72.6 bil bbls. **Arable land:** 4%. **Livestock** (1997): chickens: 100.00 mil; cattle: 15.37 mil; goats: 3.20 mil; pigs: 4.76 mil; sheep: 820,000. **Fish catch** (1997): 494,163 metric tons. **Electricity prod.** (1997): 74.598 bil kWh. **Labor force:** 64% services; 23% ind.; 13% agric.

Finance: Monetary unit: Bolivar (Sept. 1999: 622.25 = $1 U.S.). **GDP:** (1997): $185 bil. **Per capita GDP:** $8,300. **Imports** (1996): $10.5 bil; partners: U.S. 40%. **Exports** (1996): $20.8 bil; partners: U.S. & Puerto Rico 55%. **Tourism:** $1.23 bil. **Budget** (1996 est.): $11.48 bil. **Intl. reserves less gold** (June 1999): $11.46 bil. **Gold:** 9.69 mil oz t. **Consumer prices** (change in 1997): 35.8%.

Transport: Railroad: Length: 390 mi. **Motor vehicles:** 1.50 mil pass. cars, 525,000 comm. vehicles. **Civil aviation:** 2.8 bil pass.-mi; 20 airports. **Chief ports:** Maracaibo, La Guaira, Puerto Cabello.

Communications: TV sets: 183 per 1,000 pop. **Radios:** 372 per 1,000 pop. **Telephones:** 2,712,200 main lines. **Daily newspaper circ.:** 215 per 1,000 pop.

Health: Life expectancy: 69.97 male; 76.16 female. **Births** (per 1,000 pop.): 22.25. **Deaths** (per 1,000 pop.): 4.93. **Natural inc.:** 1.732%. **Hosp. beds** (1992): 1 per 382 persons. **Infant mortality** (per 1,000 live births): 26.51.

Education: Free, compulsory: ages 5-15. Literacy: 91%.

Major Intl. Organizations: UN (FAO, IBRD, ILO, IMF, IMO, WHO, WTrO), OAS, OPEC.

Embassy: 1099 30th St. NW 20007; 342-2214.

Website: http://www.embassy.org/embassies/ve.html

Columbus first set foot on the South American continent on the peninsula of Paria, Aug. 1498. Alonso de Ojeda, 1499, was the first European to see Lake Maracaibo. He called the land Venezuela, or Little Venice, because the Indians had houses on stilts. Spain dominated Venezuela until Simón Bolívar's victory near Carabobo in June 1821. The republic was formed after secession from the Colombian Federation in 1830.

Military strongmen ruled Venezuela for most of the 20th century. They promoted the oil industry; some social reforms were implemented. Since 1959, the country has had democratically elected governments.

Venezuela helped found the Organization of Petroleum Exporting Countries (OPEC). The government, Jan. 1, 1976, nationalized the oil industry with compensation. Oil accounts for most of Venezuela's export earnings; the economy suffered a severe cash crisis in the 1980s and 1990s as a result of depressed oil revenues. Government attempts to reduce dependence on oil have met with limited success.

An attempted coup by midlevel military officers was thwarted by loyalist troops Feb. 4, 1992. A second coup attempt was thwarted in Nov. Pres. Carlos Andrés Pérez was removed from office on corruption charges, May 1993; he was convicted, May 1996, of mismanaging a $17 million secret government security fund. Citing an economic crisis, Pres. Rafael Caldera, a populist elected Dec. 5, 1993, suspended many civil liberties June 27, 1994; rights were restored in most regions July 6, 1995.

A 1992 coup leader, Hugo Chávez, who ran as a populist, was elected president Dec. 6, 1998. A constitutional assembly elected July 25, 1999, and controlled by Chávez supporters slashed the powers of Congress and moved to dismiss corrupt judges.

Vietnam
Socialist Republic of Vietnam

People: Population: 77,311,210. **Age distrib.** (%): <15: 33.6; 65+: 5.3. **Pop. density:** 608 per sq. mi. **Urban:** 19%. **Ethnic groups:** Vietnamese 85-90%, Chinese 3%, Muong, Tai, Meo, Khmer, Man, Cham. **Principal languages:** Vietnamese (official), French, Chinese, English. **Chief religions:** Mainly Buddhist and Taoist; also Roman Catholic, indigenous beliefs.

Geography: Area: 127,200 sq. mi. **Location:** SE Asia, on the E coast of the Indochinese Peninsula. **Neighbors:** China on N, Laos and Cambodia on W. **Topography:** Vietnam is long and narrow, with a 1,400-mi. coast. About 22% of country is readily arable, including the densely settled Red R. valley in the N, narrow coastal plains in center, and the wide, often marshy Mekong R. Delta in the S. The rest consists of semi-arid plateaus and barren mountains, with some stretches of tropical rain forest. **Capital:** Hanoi. **Cities:** Ho Chi Minh City 3,521,000; Hanoi 1,236,000*.

Government: Type: Communist. **Head of state:** Pres. Tran Duc Luong; b May 1937; in office: Sept. 24, 1997. **Head of gov.:** Prime Min. Phan Van Khai; b Dec. 1933; in office: Sept. 25, 1997. **Local divisions:** 57 provinces, 3 cities, 1 capital region. **Defense:** 4.1% of GDP. **Active troops:** 492,000.

Economy: Industries: Food processing, textiles, chemical fertilizer. **Chief crops:** Rice, potatoes, soybeans, coffee, tea, corn. **Minerals:** Phosphates, coal, gas, manganese, bauxite, chromate, oil. **Crude oil reserves** (1999): 600 mil bbls. **Other resources:** Forests. **Arable land:** 17%. **Livestock** (1997): chickens: 126.36 mil; pigs: 18.13 mil; cattle: 3.98 mil; buffalo: 2.95 mil; goats: 514,300. **Fish catch** (1997): 1.07 mil metric tons. **Electricity prod.** (1997): 16.800 bil kWh. **Labor force:** 72% agric., forestry, fishing.

Finance: Monetary unit: Dong (Sept. 1999: 13,963.00 = $1 U.S.). **GDP:** (1997 est.): $128 bil. **Per capita GDP:** $1,700. **Imports** (1996 est.): $11.1 bil; partners: Singapore 14%, S. Korea 13%. **Exports** (1996 est.): $7.1 bil. partners: Japan 26%, **Tourism:** $86 mil. **Budget** (1996 est.): $1.7 bil.

Transport: Railroad: Length: 1,619 mi. **Motor vehicles:** 79,079 pass. cars, 97,104 comm. vehicles. **Civil aviation:** 2.4 bil pass.-mi; 12 airports. **Chief ports:** Ho Chi Minh City, Haiphong, Da Nang.

Communications: TV sets: 43 per 1,000 pop. **Radios:** 106 per 1,000 pop. **Telephones:** 2,000,000 main lines. **Daily newspaper circ.:** 8 per 1,000 pop.

Health: Life expectancy: 65.71 male; 70.64 female. **Births** (per 1,000 pop.): 20.78. **Deaths** (per 1,000 pop.): 6.56. **Natural inc.:** 1.422%. **Hosp. beds** (1994): 1 per 380 persons. **Physicians** (1994): 1 per 2,444 persons. **Infant mortality** (per 1,000 live births): 34.84.

Education: Compulsory: ages 6-11. **Literacy:** 94%.

Major Intl. Organizations: UN (FAO, IBRD, ILO, IMF, IMO, WHO), APEC, ASEAN.

Embassy: Suite 501, 1233 20th St. NW 20036; 861-0737. **Website:** http://www.batin.com.vn

Vietnam's recorded history began in Tonkin before the Christian era. Settled by Viets from central China, Vietnam was held by China, 111 BC-AD 939, and was a vassal state during subsequent periods. Vietnam defeated the armies of Kublai Khan, 1288. Conquest by France began in 1858 and ended in 1884 with the protectorates of Tonkin and Annam in the N and the colony of Cochin-China in the S.

Japan occupied Vietnam in 1940; nationalist aims gathered force. A number of groups formed the Vietminh (Independence) League, headed by Ho Chi Minh, Communist guerrilla leader. In Aug. 1945 the Vietminh forced out Bao Dai, former emperor of Annam, head of a Japan-sponsored regime. France, seeking to reestablish colonial control, battled Communist and nationalist forces, 1946-1954, and was defeated at Dienbienphu, May 8, 1954. Meanwhile, on July 1, 1949, Bao Dai had formed a State of Vietnam, with himself as chief of state, with French approval. China backed Ho Chi Minh.

A cease-fire signed in Geneva July 21, 1954, provided for a buffer zone, withdrawal of French troops from the North, and elections to determine the country's future. Under the agreement the Communists gained control of territory north of the 17th parallel, with its capital at Hanoi and Ho Chi Minh as president. South Vietnam came to comprise the 39 southern provinces. Some 900,000 North Vietnamese fled to South Vietnam.

On Oct. 26, 1955, Ngo Dinh Diem, premier of the interim government of South Vietnam, proclaimed the Republic of Vietnam and became its first president.

The North adopted a constitution Dec. 31, 1959, based on Communist principles and calling for reunification of all Vietnam. North Vietnam sought to take over South Vietnam beginning in 1954. Fighting persisted from 1956, with the Communist Vietcong, aided by North Vietnam, pressing war in the South. Northern aid to Vietcong guerrillas was intensified in 1959, and large-scale troop infiltration began in 1964, with Soviet and Chinese arms assistance. Large Northern forces were stationed in border areas of Laos and Cambodia.

A serious political conflict arose in the South in 1963 when Buddhists denounced authoritarianism and brutality. This paved the way for a military coup Nov. 1-2, 1963, which overthrew Diem. Several other military coups followed.

In 1964, the U.S. began air strikes against North Vietnam. Beginning in 1965, the raids were stepped up and U.S. troops became combatants. U.S. troop strength in Vietnam, which reached a high of 543,400 in Apr. 1969, was ordered reduced by President Nixon in a series of withdrawals, beginning in June 1969. U.S. bombings were resumed in 1972-73.

A cease-fire agreement was signed in Paris Jan. 27, 1973 by the U.S., North and South Vietnam, and the Vietcong. It was never implemented.

North Vietnamese forces attacked remaining government outposts in the Central Highlands in the first months of 1975. Government retreats turned into a rout, and the Saigon regime surrendered April 30. North Vietnam assumed control, and began transforming society along Communist lines.

The war's toll included—Combat deaths: U.S. 47,369; South Vietnam more than 200,000; other allied forces 5,225. Total U.S. fatalities numbered more than 58,000. Vietnamese civilian casualties were more than a million. Displaced war refugees in South Vietnam totaled more than 6.5 million.

The country was officially reunited July 2, 1976. The Northern capital, flag, anthem, emblem, and currency were applied to the new state. Nearly all major government posts went to officials of the former Northern government.

Heavy fighting with Cambodia took place, 1977-80, amid mutual charges of aggression and atrocities against civilians. Increasing numbers of Vietnamese civilians, ethnic Chinese, escaped the country, via the sea or the overland route across Cambodia. Vietnam launched an offensive against Cambodian refugee strongholds along the Thai-Cambodian border in 1985; they also engaged Thai troops.

Relations with China soured as 140,000 ethnic Chinese left Vietnam charging discrimination; China cut off economic aid. Reacting to Vietnam's invasion of Cambodia, China attacked 4 Vietnamese border provinces, Feb. 1979.

Vietnam announced reforms aimed at reducing central control of the economy in 1987, as many of the old revolutionary followers of Ho Chi Minh were removed from office.

Citing Vietnamese cooperation in returning remains of U.S. soldiers killed in the Vietnam War, the U.S. announced an end, Feb. 3, 1994, to a 19-year-old U.S. embargo on trade with Vietnam. The U.S. extended full diplomatic recognition to Vietnam July 11, 1995. The Communist Party replaced the country's ill and aging leadership in Sept. 1997.

Western Samoa
See Samoa (formerly Western Samoa)

Yemen
Republic of Yemen

People: Population: 16,942,230. **Age distrib.** (%): <15: 47.8; 65+: 3.2. **Pop. density:** 83 per sq. mi. **Urban:** 34%. **Ethnic groups:** Predominantly Arab, some Afro-Arab, South Asian. **Principal language:** Arabic (official). **Chief religions:** Mostly Muslim (Sha'fi-Sunni, Zaydi-Shi'a).

Geography: Area: 203,800 sq. mi. **Location:** Middle East, on the S coast of the Arabian Peninsula. **Neighbors:** Saudi Arabia on N, Oman on the E. **Topography:** A sandy coastal strip leads to well-watered fertile mountains in interior. **Capital:** Sanaa. **Cities:** (1995 est.) Sanaa 972,000; Aden 562,000.

Government: Type: Republic. **Head of state:** Pres. Ali Abdullah Saleh; b. 1942; in office: July 17, 1978. **Head of gov.:** Prime Min. Abdel Karim al Iriani; b Oct. 12, 1934; in office: May 14, 1998. **Local divisions:** 17 governorates and capital region. **Defense:** 7.0% of GDP. **Active troops:** 66,300.

Economy: Industries: Oil, food processing. **Chief crops:** Grains, fruits, qat, coffee, cotton. **Minerals:** Oil, salt. **Crude oil reserves** (1999): 4 bil bbls. **Arable land:** 3%. **Livestock** (1997): chickens: 25.20 mil; sheep: 4.53 mil; goats: 4.09 mil;

cattle: 1.26 mil. **Fish catch** (1996): 103,743 metric tons. **Electricity prod.** (1997): 2.050 bil kWh.

Finance: Monetary unit: Rial (Sept. 1999: 148.15 = $1 U.S.). **GDP:** (1997 est.): $31.8 bil. **Per capita GDP:** $2,300. **Imports** (1997 est.): $2.3 bil; partners: U.S. 12%, France 11%. **Exports** (1997 est.): $2.3 bil; partners: China 23%, S. Korea 19%. **Tourism:** $69 mil. **Budget** (1998 est.): $2.7 bil. **International reserves less gold** (Feb. 1999): $971.7 mil. **Gold:** 50,000 oz t. **Consumer prices** (change in 1997): 7.9%.

Transport: Motor vehicles: 229,084 pass. cars, 282,615 comm. vehicles. **Civil aviation:** 650.0 mil pass.-mi; 11 airports. **Chief ports:** Al Hudaydah, Al Mukalla, Aden.

Communications: TV sets: 6.5 per 1,000 pop. **Radios:** 43 per 1,000 pop. **Telephones** (1997): 220,300 main lines.

Health: Life expectancy: 58.17 male; 61.88 female. **Births** (per 1,000 pop.): 43.31. **Deaths** (per 1,000 pop.): 9.88. **Natural inc.:** 3.343%. **Hosp. beds** (1995): 1 per 1,582 persons. **Physicians** (1995): 1 per 4,530 persons. **Infant mortality** (per 1,000 live births): 69.82.

Education: Compulsory: ages 6-15. **Literacy** (1994): 43%.

Major Intl. Organizations: UN (FAO, IBRD, ILO, IMF, IMO, WHO), AL.

Embassy: 2600 Virginia Ave. NW 20037; 965-4760.

Website: http://www.nusacc.org/yemen

Yemen's territory once was part of the ancient Kingdom of Sheba, or Saba, a prosperous link in trade between Africa and India. The Bible speaks of its gold, spices, and precious stones as gifts borne by the Queen of Sheba to King Solomon.

Yemen became independent in 1918, after years of Ottoman Turkish rule, but remained politically and economically backward. Imam Ahmed ruled 1948-1962. Army officers headed by Brig. Gen. Abdullah al-Salal declared the country to be the Yemen Arab Republic.

The Imam Ahmed's heir, the Imam Mohamad al-Badr, fled to the mountains where tribesmen joined royalist forces; internal warfare between them and the republican forces continued. About 150,000 people died in the fighting.

There was a bloodless coup Nov. 5, 1967. In April 1970 hostilities ended with an agreement between Yemen and Saudi Arabia. On June 13, 1974, an army group, led by Col. Ibrahim al-Hamidi, seized the government. He was killed in 1977.

Meanwhile, South Yemen won independence from Britain in 1967, formed out of the British colony of Aden and the British protectorate of South Arabia. It became the Arab world's only Marxist state, taking the name People's Democratic Republic of Yemen in 1970 and signing a friendship treaty with the USSR in 1979 that allowed for the stationing of Soviet troops.

More than 300,000 Yemenis fled from the south to the north after independence, contributing to 2 decades of hostility between the 2 states that flared into warfare twice in the 1970s.

An Arab League-sponsored agreement between North and South Yemen on unification of the 2 countries was signed Mar. 29, 1979. An agreement providing for widespread political and economic cooperation was signed in 1988.

The 2 countries were formally united May 21, 1990, but regional clan-based rivalries led to full-scale civil war in 1994. Secessionists declared a breakaway state in S Yemen, May 21, 1994, but northern troops captured the former southern capital of Aden in July.

A new constitution was approved Sept. 28. Parliamentary elections were held Apr. 27, 1997.

A dispute between Yemen and Eritrea over the Hanish Isls. in the Red Sea, which led to armed clashes in 1995, was resolved by arbitration in 1998.

Yugoslavia
Federal Republic of Yugoslavia

People: Population: 11,206,847. **Age distrib.** (%): <15: 20.2; 65+: 13.1. **Pop. density:** 284 per sq. mi. **Urban:** 57%. **Ethnic groups:** Serbian 63%, Albanian 14%, Montenegrin 6%. **Principal languages:** Serbo-Croatian (official) 95%, Albanian 5%. **Chief religions:** Orthodox 65%, Muslim 19%, Roman Catholic 4%.

Geography: Area: 39,500 sq. mi. **Location:** On the Balkan Peninsula in SE Europe. Present-day Yugoslavia consists of the republics of Serbia and Montenegro. **Neighbors:** Croatia, Bosnia and Herzegovina on W; Hungary on N; Romania, Bulgaria on E; Albania, Macedonia on S. **Capital:** Belgrade (Serbia), Podgorica (Montenegro). **Cities:** Belgrade 1,204,000*.

Government: Type: Republic. **Head of state:** Pres. Slobodan Milosevic; b Aug. 29, 1941; in office: July 23, 1997.

Head of gov.: Prime Min. Momir Bulatovic; b Sept. 21, 1956; in office: May 19, 1998. **Local divisions:** 2 republics, 2 autonomous provinces. **Defense:** 7.8% of GDP. **Active troops:** 114,200.

Economy: Industries: Steel, machinery, consumer goods, mining, electronics. **Chief crops:** Cereals, fruits, vegetables. **Minerals:** Oil, gas, coal, antimony, lead, nickel, gold, zinc, pyrite, copper, chrome. **Crude oil reserves** (1999): 77.5 mil bbls. **Livestock** (1997): chickens: 24.02 mil; pigs: 4.15 mil; sheep: 2.40 mil; cattle: 1.89 mil; goats: 312,000. **Electricity prod.** (1997): 38.518 bil kWh. **Labor force:** 41% ind.; 35% services.

Finance: Monetary unit: New Dinar (Sept. 1999: 11.08 = $1 U.S.). **GDP:** (1997 est.): $24.3 bil. **Per capita GDP:** $2,280. **Imports** (1996 est.): $6.2 bil; partners: Germany 13%, Italy 11%. **Exports** (1996 est.): $2.8 bil; partners: Macedonia 12%, Russia 9%. **Tourism:** $39 mil.

Transport: Railroad: Length: 2,505 mi. **Motor vehicles:** 1.00 mil pass. cars, 331,000 comm. vehicles. **Civil aviation:** 93 mil pass.-mi; 4 airports. **Chief ports:** Bar, Novi Sad.

Communications: TV sets: 27 per 1,000 pop. **Radios:** 118 per 1,000 pop. **Telephones:** 2,319,400 main lines. **Daily newspaper circ.:** 256 per 1,000 pop.

Health: Life expectancy: 71.8 male; 77.9 female. **Births** (per 1,000 pop.): 14. **Deaths** (per 1,000 pop.): 10. **Natural inc.:** 0.46%. **Hosp. beds** (1995): 1 per 188 persons. **Physicians** (1995): 1 per 495 persons. **Infant mortality** (per 1,000 live births): 14.

Education: Free, compulsory: ages 7-15. **Literacy:** 98%.

Major Intl. Organizations: Currently suspended from UN and its agencies.

Embassy: 2410 California St. NW 20008; 462-6566.

Website: http://www.gov.yu

Serbia, which had since 1389 been a vassal principality of Turkey, was established as an independent kingdom by the Treaty of Berlin, 1878. Montenegro, independent since 1389, also obtained international recognition in 1878. After the Balkan wars, Serbia's boundaries were enlarged by the annexation of Old Serbia and Macedonia, 1913.

When the Austro-Hungarian empire collapsed after World War I, the Kingdom of Serbs, Croats, and Slovenes was formed from the former provinces of Croatia, Dalmatia, Bosnia, Herzegovina, Slovenia, Vojvodina, and the independent state of Montenegro The name became Yugoslavia in 1929.

Nazi Germany invaded in 1941. Many Yugoslav partisan troops continued to operate. Among these were the Chetniks led by Draja Mikhailovich, who fought other partisans led by Josip Broz, known as Marshal Tito. Tito, backed by the USSR and Britain from 1943, was in control by the time the Germans had been driven from Yugoslavia in 1945. Mikhailovich was executed July 17, 1946, by the Tito regime.

A constituent assembly proclaimed Yugoslavia a republic Nov. 29, 1945. It became a federal republic Jan. 31, 1946, with Tito, a Communist, heading the government. Tito rejected Stalin's policy of dictating to all Communist nations, and he accepted economic and military aid from the West.

Pres. Tito died May 4, 1980. After his death, Yugoslavia was governed by a collective presidency, with a rotating succession. On Jan. 22, 1990, the Communist Party renounced its leading role in society.

Croatia and Slovenia formally declared independence June 25, 1991. In Croatia, fighting began between Croats and ethnic Serbs. Serbia sent arms and medical supplies to the Serb rebels in Croatia. Croatian forces clashed with Yugoslav army units and their Serb supporters.

The republics of Serbia and Montenegro proclaimed a new "Federal Republic of Yugoslavia" Apr. 17, 1992. Serbia, under Pres. Slobodan Milosevic, was the main arms supplier to ethnic Serb fighters in Bosnia and Herzegovina. The UN imposed sanctions May 30 on the newly reconstituted Yugoslavia as a means of ending the bloodshed in Bosnia.

A peace agreement initialed in Dayton, Ohio, Nov. 21, 1995, was signed in Paris, Dec. 14, by Milosevic and leaders of Serbia and Croatia. In May 1996, a UN tribunal in the Netherlands began trying suspected war criminals from the former Yugoslavia. The UN lifted sanctions against Yugoslavia Oct. 1, 1996, after elections were held in Bosnia. Mass protests erupted when Milosevic refused to accept opposition victories in local elections Nov. 17; non-Communist governments took office in Belgrade and other cities in Feb. 1997. Barred from running for a 3d term as Serbian president, Milosevic had himself inaugurated as president of Yugoslavia on July 23, 1997.

Kosovo: A nominally autonomous province in southern Serbia (4,203 sq. mi.), with a population of about 2,000,000, mostly Albanians. The capital is Pristina. Revoking provincial autonomy, Serbia began ruling Kosovo by force in 1989. Albanian secessionists proclaimed an independent Republic of Kosovo in July 1990. Guerrilla attacks by the Kosovo Liberation Army in 1997 brought a ferocious counteroffensive by Serbian authorities.

Fearful that the Serbs were employing "ethnic cleansing" tactics, as they had in Bosnia, the U.S. and its NATO allies sought to pressure the Yugoslav government. When Milosevic refused to comply, NATO launched an air war against Yugoslavia, Mar.-June 1999; the Serbs retaliated by terrorizing the Kosovars and forcing hundreds of thousands to flee, mostly to Albania and Macedonia. A 50,000-member multinational force (KFOR) entered Kosovo in June, and most of the Kosovar refugees had returned by Sept. 1.

Vojvodina: A nominally autonomous province in northern Serbia (8,304 sq. mi.), with a population of about 2,000,000, mostly Serbian. The capital is Novi Sad.

Zaire

See Congo (formerly Zaire)

Zambia

Republic of Zambia

People: Population: 9,663,535. **Age distrib.** (%): <15: 49.0; 65+: 2.5. **Pop. density:** 33 per sq. mi. **Urban:** 43%. **Ethnic groups:** African 99%, European 1%. **Principal languages:** English (official), indigenous. **Chief religions:** Christian 50-75%, Hindu and Muslim 24-49%.

Geography: Area: 290,600 sq. mi. **Location:** In S central Africa. **Neighbors:** Congo (formerly Zaire) on N; Tanzania, Malawi, Mozambique on E; Zimbabwe, Namibia on S; Angola on W. **Topography:** Zambia is mostly high plateau country covered with thick forests, and drained by several important rivers, including the Zambezi. **Capital:** Lusaka 1,317,000*.

Government: Type: Republic. **Head of state and gov.:** Pres. Frederick Chiluba; b Apr. 30, 1943; in office: Nov. 2, 1991. **Local divisions:** 9 provinces. **Defense:** 1.7% of GDP. **Active troops:** 21,600.

Economy: Industries: Mining, construction, foodstuffs, chemicals. **Chief crops:** Corn, cassava, sorghum, sugar. **Minerals:** Cobalt, copper, zinc, emeralds, gold, lead, silver, uranium, coal. **Arable land:** 7%. **Livestock** (1997): chickens: 27.00 mil; cattle: 3.10 mil; goats: 600,000; pigs: 285,000. **Fish catch** (1996): 61,562 metric tons. **Electricity prod.** (1997): 7.740 bil kWh. **Labor force:** 85% agric.

Finance: Monetary unit: Kwacha (Sept. 1999: 2,378.68 = $1 U.S.). **GDP:** (1997 est.): $8.8 bil. **Per capita GDP:** $950. **Imports** (1996 est.): $990 mil; partners: South Africa 28%, UK 11%. **Exports** (1996 est.): $975 mil; partners: Japan 18%, Saudi Arabia 13%. **Tourism:** $90 mil. **Budget** (1995 est.): $835 mil. **Intl. reserves less gold** (May 1999): $67.9 mil. **Consumer prices** (change in 1997): 24.8%.

Transport: Railroad: Length: 791 mi. **Motor vehicles:** 142,000 pass. cars, 73,500 comm. vehicles. **Civil aviation:** 27.7 mil pass.-mi; 4 airports. **Chief port:** Mpulungu.

Communications: TV sets: 32 per 1,000 pop. **Radios:** 99 per 1,000 pop. **Telephones:** 76,700 main lines. **Daily newspaper circ.:** 13 per 1,000 pop.

Health: Life expectancy: 36.72 male; 37.21 female. **Births** (per 1,000 pop.): 44.51. **Deaths** (per 1,000 pop.): 22.56. **Natural inc.:** 2.195%. **Physicians** (1993): 1 per 10,917 persons. **Infant mortality** (per 1,000 live births): 91.85.

Education: Compulsory: ages 7-14. **Literacy:** 78%.

Major Intl. Organizations: UN (FAO, IBRD, ILO, IMF, WHO, WTrO), the Commonwealth, OAU.

Embassy: 2419 Massachusetts Ave. NW 20008; 265-9717. **Website:** http://www.zamnet.zm

As Northern Rhodesia, the country was under the administration of the South Africa Company, 1889 until 1924, when the office of governor was established, and, subsequently, a legislature. The country became an independent republic within the Commonwealth Oct. 24, 1964.

After the white government of Rhodesia (now Zimbabwe) declared its independence from Britain Nov. 11, 1965, relations between Zambia and Rhodesia became strained.

As part of a program of government participation in major industries, a government corporation in 1970 took over 51% of the ownership of 2 foreign-owned copper-mining companies. Privately-held land and other enterprises were nationalized in 1975. In the 1980s and 1990s lowered copper prices hurt the economy and severe drought caused famine.

Food riots erupted in June 1990, as the nation suffered its worst violence since independence. Elections held Oct. 1991 brought an end to one-party rule. The new government sought to sell state enterprises, including the copper industry. Pres. Frederick Chiluba won reelection Nov. 18, 1996, but international observers cited harassment of opposition parties. A coup attempt was suppressed Oct. 28, 1997.

According to UN estmates, the AIDS epidemic had orphaned nearly 500,000 children in Zambia by the late 1990s.

Zimbabwe

Republic of Zimbabwe

People: Population: 11,163,160. **Age distrib.** (%): <15: 43.2; 65+: 2.6. **Pop. density:** 74 per sq. mi. **Urban:** 33%. **Ethnic groups:** Shona 71%, Ndebele 16%. **Principal languages:** English (official), Shona, Sindebele. **Chief religions:** Syncretic (Christian-indigenous mix) 50%, Christian 25%, indigenous beliefs 24%.

Geography: Area: 150,800 sq. mi. **Location:** In southern Africa. **Neighbors:** Zambia on N, Botswana on W, South Africa on S, Mozambique on E. **Topography:** Zimbabwe is high plateau country, rising to mountains on eastern border, sloping down on the other borders. **Capital:** Harare 1,410,000*.

Government: Type: Republic. **Head of state and gov.:** Pres. Robert Mugabe; b Feb. 21, 1924; in office: Dec, 31, 1987. **Local divisions:** 8 provinces, 2 cities. **Defense:** 4.7% of GDP. **Active troops:** 39,000.

Economy: Industries: Clothing, mining, steel, chemicals. **Chief crops:** Tobacco, sugar, cotton, wheat, corn. **Minerals:** Chromium, gold, nickel, asbestos, copper, iron, coal. **Arable land:** 7%. **Livestock** (1997): chickens: 15.50 mil; cattle: 5.45 mil; goats: 2.75 mil; sheep: 520,000; pigs: 270,000. **Electricity prod.** (1997): 7.630 bil kWh. **Labor force:** 46% serv. & transport; 27% agric.; 27% ind.

Finance: Monetary unit: Dollar (Sept. 1999: 38.30 = $1 U.S.). **GDP:** (1996 est.): $24.9 bil. **Per capita GDP:** $2,200. **Imports** (1996 est.): $2.2 bil; partners: South Africa 38%. **Exports** (1996 est.): $2.5 bil; partners: South Africa 12%, UK 12%. **Tourism:** $246 mil. **Budget** (FY 1996-97): $2.9 bil. **Intl. reserves less gold** (June 1999): $163.3 mil. **Gold:** 610,000 oz t. **Consumer prices** (change in 1997): 31.8%.

Transport: Railroad: Length: 1,714 mi. **Motor vehicles:** 250,000 pass. cars, 108,000 comm. vehicles. **Civil aviation:** 582.7 mil pass.-mi; 7 airports. **Chief ports:** Binga, Kariba.

Communications: TV sets: 12 per 1,000 pop. **Radios:** 113 per 1,000 pop. **Telephones** (1997): 212,000 main lines. **Daily newspaper circ.:** 17 per 1,000 pop.

Health: Life expectancy: 38.77 male; 38.94 female. **Births** (per 1,000 pop.): 30.64. **Deaths** (per 1,000 pop.): 20.43. **Natural inc.:** 1.021%. **Hosp. beds** (1996): 1 per 501 persons. **Physicians** (1993): 1 per 6,909 persons. **Infant mortality** (per 1,000 live births): 61.21.

Education: Compulsory: ages 6-13. **Literacy:** 85%.

Major Intl. Organizations: UN (FAO, IBRD, ILO, IMF, WHO, WTrO), the Commonwealth, OAU.

Embassy: 1608 New Hampshire Ave. NW 20009; 332-7100.

Britain took over the area as Southern Rhodesia in 1923 from the British South Africa Co. (which, under Cecil Rhodes, had conquered it by 1897) and granted internal self-government. Under a 1961 constitution, voting was restricted to keep whites in power. On Nov. 11, 1965, Prime Min. Ian D. Smith announced his country's unilateral declaration of independence.

Britain termed the act illegal and demanded that the country (known as Rhodesia until 1980) broaden voting rights to provide for eventual rule by the black African majority. The UN imposed sanctions and, in May 1968, a trade embargo.

Intermittent negotiations between the government and various black nationalist groups failed to prevent increasing guerrilla warfare. An "internal settlement" signed Mar. 1978 in which Smith and 3 popular black leaders would share control of the government until a transfer of power to the black majority was rejected by guerrilla leaders.

In the country's first universal-franchise election, Apr. 21, 1979, Bishop Abel Muzorewa's United African National Council gained a bare majority of the black-dominated Parliament. A cease-fire was accepted by all parties, Dec. 5. Independence as Zimbabwe was finally achieved Apr. 18, 1980.

On Mar. 6, 1992, Pres. Robert Mugabe declared a national disaster because of drought and appealed to foreign donors for food, money, and medicine. An economic adjustment program caused widespread hardship. Mugabe was reelected Mar. 1996 after opposition candidates withdrew.

An estimated 1 mil Zimbabweans have HIV, the virus that causes AIDS.

Area and Population of the World

Source: Bureau of the Census, U.S. Dept. of Commerce; prior to 1950, Rand McNally & Co.

Continent or Region	AREA (1,000 sq. mi.)	% of Earth	POPULATION (est., in thousands)							% World Total, 1999
			1650	1750	1850	1900	1950	1980	1999	
North America	9,400	16.2	5,000	5,000	39,000	106,000	221,000	372,000	476,000	7.9
South America	6,900	11.9	8,000	7,000	20,000	38,000	111,000	242,000	343,000	5.7
Europe	3,800	6.6	100,000	140,000	265,000	400,000	392,000	484,000	727,000	12.1
Asia	17,400	30.1	335,000	476,000	754,000	932,000	1,411,000	2,601,000	3,641,000	60.7
Africa	11,700	20.2	100,000	95,000	95,000	118,000	229,000	470,000	778,000	13.0
Former USSR	—	—	—	—	—	—	180,000	266,000	—	
Oceania, incl. Australia	3,300	5.7	2,000	2,000	2,000	6,000	12,000	23,000	30,000	0.5
Antarctica	5,400	9.3	Uninhabited							
WORLD	**57,900**	**—**	**550,000**	**725,000**	**1,175,000**	**1,600,000**	**2,556,000**	**4,458,000**	**5,996,000**	**—**

Note: Figures may not add to total because of rounding. Figures for 1950 and 1980 count the former U.S.S.R. as a separate area.

World Population Surpasses 6 Billion in 1999

According to estimates by the UN Population Information Network, the world population reached 6 billion on Oct. 12, 1999, having doubled in about 40 years. The U.S. Census Bureau estimated that this historic mark was reached about 3 months earlier in 1999. In either case, it had taken a mere 12 years for a billion people to be added to the population. Although population growth has been slowing down in much of the world, global population is still increasing by 77 million annually; assuming middle-range fertility and mortality trends, it is expected the world population will grow to more than 9 billion by 2050, with most of the growth in countries that are less economically developed.

Leading Countries in Population and Area, 1999

China had the highest population in the world, with an estimated 1.25 billion inhabitants in mid-1999, one-fifth of the world's total population. India, the second-largest country in population, passed the 1-billion mark in 1999. The United States had the third-largest population, with about 273 million in 1999, followed by Indonesia, Brazil, and Russia. Russia is the largest country in land area, with over 6.5 million square miles, followed by Canada, China, the United States, and Brazil.

Population of the World's Largest Cities

Source: United Nations, Dept. for Economic and Social Information and Policy Analysis

The figures given here are United Nations estimates and projections, as revised in 1996, for "urban agglomerations"—that is, contiguous densely populated urban areas, not demarcated by administrative boundaries. These figures may not correspond to figures for cities in other parts of *The World Almanac*.

Rank	City, Country	Pop. (thousands) 1995	Pop. (thousands, projected) 2015	Annual growth rate (percent) 1990-1995	Percentage increase for: 1975-1995	Percentage increase for: 1995-2015	Pop. of city as percentage of: Total pop.[1]	Pop. of city as percentage of: Urban pop.[2]
1.	Tokyo, Japan	26,959	28,887	1.45	36.36	7.15	21.56	27.62
2.	Mexico City, Mexico	16,562	19,180	1.81	47.40	15.81	18.17	24.75
3.	Sao Paulo, Brazil	16,533	20,320	1.84	64.56	22.91	10.40	13.27
4.	New York City, U.S.	16,332	17,602	0.34	2.85	7.78	6.11	8.03
5.	Mumbai (Bombay), India	15,138	26,218	4.24	120.79	73.19	1.63	6.08
6.	Shanghai, China	13,584	17,969	0.36	18.71	32.28	1.11	3.68
7.	Los Angeles, U.S.	12,410	14,217	1.60	39.03	14.56	4.65	6.10
8.	Calcutta, India	11,923	17,305	1.81	51.15	45.14	1.28	4.79
9.	Buenos Aires, Argentina	11,802	13,856	1.15	29.07	17.40	33.95	38.54
10.	Seoul, South Korea	11,609	12,980	1.92	70.52	11.81	25.85	31.81
11.	Beijing, China	11,299	15,572	0.87	32.23	37.82	0.93	3.06
12.	Osaka, Japan	10,609	10,609	0.23	7.77	0.00	8.48	10.87
13.	Lagos, Nigeria	10,287	24,640	5.68	211.73	139.53	9.21	23.28
14.	Rio de Janeiro, Brazil	10,181	11,860	1.00	29.63	16.49	6.40	8.17
15.	Delhi, India	9,948	16,860	3.85	124.76	69.48	1.07	4.0

(1) Denotes percentage of the total population of the country in which the city is located. (2) Denotes percentage of the total urban population of the country in which the city is located.

Current Population and Projections for All Countries: 1999, 2025, and 2050

Source: Bureau of the Census, U.S. Dept. of Commerce

(midyear figures, in thousands)

COUNTRY	1999	2025	2050	COUNTRY	1999	2025	2050
Afghanistan	25,825	48,045	76,231	Belgium	10,182	9,533	7,609
Albania	3,365	4,306	4,609	Belize	236	383	489
Algeria	31,133	47,676	58,880	Benin	6,306	13,541	22,171
Andorra	66	88	69	Bhutan	1,952	3,341	4,935
Angola	11,178	21,598	34,465	Bolivia	7,983	12,007	15,240
Antigua and Barbuda	64	65	51	Bosnia and Herzegovina	3,482	3,471	2,833
Argentina	36,738	48,351	56,258	Botswana	1,464	1,634	2,146
Armenia	3,409	3,434	3,428	Brazil	171,853	209,587	228,145
Australia	18,784	22,191	22,846	Brunei	323	530	704
Austria	8,139	7,822	6,136	Bulgaria	8,195	7,292	5,905
Azerbaijan	7,908	9,429	10,585	Burkina Faso	11,576	21,360	34,956
Bahamas	284	369	404	Burundi	5,736	10,469	17,304
Bahrain	629	923	1,098	Cambodia	11,627	21,434	35,065
Bangladesh	127,118	180,673	211,082	Cameroon	15,456	29,108	48,606
Barbados	259	279	266	Canada	31,006	37,987	40,491
Belarus	10,402	10,248	9,100	Cape Verde	406	532	545

COUNTRY	1999	2025	2050	COUNTRY	1999	2025	2050
Central African Republic	3,445	5,545	7,915	Myanmar	48,081	68,107	87,778
Chad	7,557	14,360	22,504	Namibiaz	1,648	2,310	3,757
Chile	14,974	18,681	19,453	Nauru	11	12	12
China	1,246,872	1,407,739	1,322,435	Nepal	24,303	42,576	60,661
Colombia	39,309	58,287	73,349	Netherlands	15,808	15,852	12,974
Comoros	563	1,160	1,953	New Zealand	3,662	4,445	4,561
Congo (formerly Zaire)	2,717	105,737	184,456	Nicaragua	4,717	8,112	10,817
Congo Republic	50,481	4,246	6,081	Niger	9,962	20,424	33,896
Costa Rica	3,674	5,327	6,321	Nigeria	113,829	203,423	337,591
Côte d'Ivoire	15,818	27,840	44,509	Norway	4,439	4,592	4,012
Croatia	4,677	4,348	3,486	Oman	2,447	5,307	8,453
Cuba	11,096	11,722	10,594	Pakistan	138,123	211,675	260,247
Cyprus	754	870	878	Palau	18	24	26
Czech Republic	10,281	10,128	8,626	Panama	2,779	3,796	4,418
Denmark	5,357	5,334	4,476	Papua New Guinea	4,705	7,597	10,049
Djibouti	447	841	1,329	Paraguay	5,434	9,929	15,001
Dominica	65	67	69	Peru	26,625	39,158	47,855
Dominican Republic	8,130	11,781	14,586	Philippines	79,346	120,519	150,272
Ecuador	12,562	17,800	21,059	Poland	38,609	40,117	36,465
Egypt	67,274	97,431	117,121	Portugal	9,918	9,012	7,256
El Salvador	5,839	8,382	10,814	Qatar	724	1,208	1,348
Equatorial Guinea	466	876	1,394	Romania	22,334	21,417	18,483
Eritrea	3,985	8,438	13,736	Russia	146,394	138,842	121,777
Estonia	1,409	1,237	1,047	Rwanda	8,155	12,159	19,607
Ethiopia	59,680	98,763	159,170	Saint Kitts and Nevis	43	60	69
Fiji	813	1,085	1,285	Saint Lucia	154	203	224
Finland	5,158	5,009	4,170	Saint Vincent and the			
France	58,978	57,806	48,219	Grenadines	121	151	163
Gabon	1,226	1,800	2,518	Samoa (formerly Western			
Gambia, The	1,336	2,678	4,038	Samoa)	230	367	471
Georgia	5,066	4,718	4,365	San Marino	25	27	27
Germany	82,087	75,372	57,429	São Tomé and Príncipe	155	331	518
Ghana	18,888	28,191	34,324	Saudi Arabia	21,505	50,374	97,120
Greece	10,707	10,473	8,362	Senegal	10,052	22,456	39,690
Grenada	97	154	210	Seychelles	79	91	95
Guatemala	12,336	22,344	32,185	Sierra Leone	5,297	11,010	18,369
Guinea	7,539	13,135	20,034	Singapore	3,532	4,231	4,161
Guinea-Bissau	1,235	2,102	2,970	Slovakia	5,396	5,718	5,215
Guyana	705	710	726	Slovenia	1,971	1,864	1,484
Haïti	6,884	10,171	12,746	Solomon Islands	455	840	1,158
Honduras	5,997	8,612	11,001	Somalia	7,141	15,192	26,243
Hungary	10,186	9,374	7,684	South Africa	43,426	49,851	58,972
Iceland	273	298	279	Spain	39,168	36,841	29,405
India	1,000,849	1,415,274	1,706,951	Sri Lanka	19,145	24,088	26,146
Indonesia	216,108	287,985	330,566	Sudan	34,476	64,757	93,625
Iran	65,180	111,891	142,336	Suriname	431	460	380
Iraq	22,427	44,146	65,529	Swaziland	985	1,589	3,059
Ireland	3,633	3,913	3,600	Sweden	8,911	9,158	8,052
Israel	5,750	7,778	8,961	Switzerland	7,275	7,064	5,614
Italy	56,735	50,352	38,290	Syria	17,214	31,684	43,463
Jamaica	2,652	3,355	3,712	Taiwan	22,113	25,897	25,189
Japan	126,182	119,865	101,334	Tajikistan	6,103	9,634	13,261
Jordan	4,561	8,223	11,303	Tanzania	31,271	50,661	76,500
Kazakhstan	16,825	18,565	20,426	Thailand	60,609	70,316	69,741
Kenya	28,809	34,774	43,852	Togo	5,081	11,712	20,725
Kiribati	86	99	100	Tonga	109	133	156
Korea, North	21,386	25,485	25,930	Trinidad and Tobago	1,102	1,083	1,057
Korea, South	46,885	54,256	52,625	Tunisia	9,514	12,760	14,399
Kuwait	1,991	3,559	4,159	Turkey	65,599	89,727	103,649
Kyrgyzstan	4,546	6,066	7,394	Turkmenistan	4,366	6,514	8,422
Laos	5,407	9,805	13,844	Tuvalu	11	15	20
Latvia	2,354	1,965	1,659	Uganda	22,805	49,181	91,398
Lebanon	3,563	4,831	5,598	Ukraine	49,811	45,096	39,096
Lesotho	2,129	2,724	3,533	United Arab Emirates	2,344	3,444	4,057
Liberia	2,924	6,524	10,992	United Kingdom	59,113	59,985	54,116
Libya	4,993	14,185	26,625	United States	272,640	335,360	394,241
Liechtenstein	32	36	31	Uruguay	3,309	3,916	4,256
Lithuania	3,585	3,417	3,063	Uzbekistan	24,102	34,348	42,762
Luxembourg	429	447	360	Vanuatu	189	282	347
Macedonia	2,023	2,171	1,977	Venezuela	23,203	32,474	37,773
Madagascar	14,873	29,306	48,327	Vietnam	77,311	103,909	119,464
Malawi	10,000	12,475	16,884	Yemen	16,942	40,439	76,008
Malaysia	21,376	34,248	47,289	Yugoslavia	11,206	11,244	9,798
Maldives	300	623	949	Zambia	9,664	16,156	26,967
Mali	10,429	22,647	40,433	Zimbabwe	11,163	12,366	16,064
Malta	382	391	325				
Marshall Islands	66	171	348	**REGIONS**			
Mauritania	2,582	5,446	9,329	Asia	3,641,354	4,774,053	5,406,562
Mauritius	1,182	1,488	1,614	Africa	778,434	1,317,493	2,012,567
Mexico	100,294	141,593	167,479	Europe	727,289	696,838	588,240
Micronesia	132	143	143	South America	343,294	451,641	519,878
Moldova	4,461	4,830	4,811	North America	475,815	617,249	728,136
Monaco	32	34	34	Oceania, incl. Australia	30,029	38,512	42,829
Mongolia	2,617	3,555	4,057				
Morocco	29,662	43,228	52,069	**WORLD**[1]	**5,996,215**	**7,895,785**	**9,298,212**
Mozambique	19,124	33,308	47,805				

(1) Figures may not add to total because of rounding and exclusion of certain pseudo-national entities.

Estimated HIV Infection and Reported AIDS Cases

Source: UNAIDS, Joint United Nations Program on HIV/AIDS

Studies, primarily in industrialized nations, have indicated that about 60% of adults infected by the human immunodeficiency virus (HIV) will develop acquired immune deficiency syndrome (AIDS) within 12-13 years of becoming infected; development of the disease might be more rapid in Third World countries. About 75-85% of adult HIV infections worldwide have been transmitted through unprotected sexual intercourse.

The number of people living with HIV/AIDS worldwide as of Dec. 1998 was an estimated 33.4 million, with the largest number in sub-Saharan Africa. The total includes 32.2 million adults (43% are women, up 2% from just a year earlier, and there are no indications this equalizing trend will reverse itself) and about 1.2 million children (under 15 years old; most children are believed to have acquired their HIV infection from their mother before or at birth, or through breastfeeding). The virus continues to spread with great rapidity, causing almost 16,000 new infections daily. Of those, approximately 7,000 are between the ages of 10 and 24, or 5 young persons every minute. UNAIDS estimates that nearly 5.8 million new HIV infections occurred in 1998 (11 men, women, and children per minute) and that 2.5 million people died that year (more than ever before in a single year), including over 500,000 children. Since the start of the global epidemic in the late 1970s, HIV has infected 47.3 million people; an estimated 13.9 million people have died of AIDS, including 3.2 million children.

Estimated Current and Cumulative HIV/AIDS Cases by Region, Dec. 1998

Region	Current cases[1]	Percent of adults[2]	Region	Current cases[1]	Percent of adults[2]
Sub-Saharan Africa	22,500,000	67	Caribbean	330,000	1
South/Southeast Asia	6,700,000	20	North Africa/Middle East	210,000	—
Latin America	1,400,000	4	Eastern Europe/		
North America	890,000	3	Central Asia	270,000	—
Western Europe	500,000	1	Australasia	12,000	—
East Asia/Pacific	560,000	2	**WORLD**	**33,400,000**	**100[3]**

(1) Adults and children living with HIV/AIDS. (2) Percentage of total number of adults worldwide living with HIV. (3) Details do not add to total because of rounding. Dash (—) means less than 1%.

The World's Refugees

Source: *World Refugee Survey 1999*, U.S. Committee for Refugees, a nonprofit corp. These estimates are conservative. The refugees in this table include only those considered in need of protection and/or assistance and generally do not include those who have achieved permanent resettlement.

(as of Dec. 31, 1998; only countries estimated to host 50,000 or more refugees are listed)

Place of asylum	Origin of Most	Number	Place of asylum	Origin of Most	Number
TOTAL AFRICA		**2,922,000**	**TOTAL AMERICAS AND THE CARIBBEAN**		**740,000**
Algeria	Western Sahara, Palestinians	84,000*	United States	El Salvador, Guatemala, Other	651,000**
Congo-Kinshasa[2]	Angola, Sudan, Congo-Brazzaville, Burundi, Uganda	220,000*	**TOTAL EAST ASIA AND THE PACIFIC**		**559,000**
Côte d'Ivoire	Liberia, Sierra Leone, Other	128,000	China	Vietnam, Laos	281,800
Ethiopia	Somalia, Sudan, Kenya, Djibouti	251,000*	Malaysia	Philippines, Burma, Other	50,600
Guinea	Liberia, Sierra Leone, Guinea-Bissau	514,000*	Thailand	Burma, Cambodia, Laos, Other	187,700
Kenya	Somalia, Sudan, Ethiopia, Other	192,000*	**TOTAL MIDDLE EAST**		**5,814,000**
Liberia	Sierra Leone	120,000*	Gaza Strip	Palestinians	773,000
Sudan	Eritrea, Ethiopia, Chad, Congo-Kinshasa[2], Uganda	360,000*	Iran	Afghanistan, Iraq	1,931,000*
Tanzania	Burundi, Congo-Kinshasa[2], Rwanda, Somalia	329,000*	Iraq	Palestinians, Iran, Turkey, Eritrea, Somalia, Sudan	104,000
Uganda	Sudan, Rwanda, Congo-Kinshasa, Somalia	185,000	Jordan	Palestinians, Other	1,463,800
Zambia	Angola, Congo-Kinshasa[2], Other	157,000*	Kuwait	Palestinians, Iraq, Somalia	52,000
			Lebanon	Palestinians, Other	368,300
			Saudi Arabia	Palestinians, Iraq, Afghanistan, Other	128,300
TOTAL EUROPE		**1,728,000**	Syria	Palestinians, Other	369,800
Armenia	Azerbaijan, Georgia, Other	229,000	West Bank	Palestinians	555,000
Azerbaijan	Armenia, Uzbekistan, Georgia, Other	235,300	Yemen	Somalia, Palestinians, Ethiopia, Eritrea, Iraq, Other	68,700
Germany	Bosnia and Herzegovina, Yugoslavia[3], Other	198,000	**TOTAL SOUTH AND CENTRAL ASIA**		**1,706,000**
Russian Federation	Former USSR, Other	161,900*	Bangladesh	Burma, Other	53,100*
United Kingdom		74,000	India	China (Tibet), Sri Lanka, Burma, Afghanistan, Bhutan, Other	292,100
Yugoslavia	Croatia, Bosnia and Herzegovina	480,000	Nepal	Bhutan, China (Tibet)	118,000
			Pakistan	Afghanistan, India, Iraq, Somalia, Other	1,217,400*
			TOTAL REFUGEES		**13,469,000**

(1) Significant variance among sources in number reported. (2) Formerly Zaire. (3) Serbia/Montenegro.

Principal Sources of Refugees

Afghanistan	2,600,000	Liberia	310,000*	Guatemala	151,000**
Iraq	586,000*	Croatia	309,000*	Yugoslavia	145,000
Sierra Leone	480,000*	Angola	302,000	Congo-Kinshasa	136,000
Bosnia and Herzegovina	424,000*	Vietnam	281,000	China (Tibet)	128,000
Somalia	421,000*	El Salvador	250,000*	Sri Lanka	126,000
Sudan	352,000	Burma	238,000*	Burundi	125,000
Eritrea	323,000*	Azerbaijan	218,000	Bhutan	113,000
		Armenia	180,000	Western Sahara	105,000

*Estimates vary widely in number reported. **Includes asylum seekers with cases pending in the United States.

U.S. Immigration Law

Source: Immigration and Naturalization Service, U.S. Dept. of Justice

Most U.S. regulations affecting immigration were modified in the Immigration and Nationality Act of 1952, which has been amended several times since then. Major amendments were made through the Immigration Act of 1990, which was signed by Pres. George Bush on Nov. 29, 1990. New provisions enacted in Sept. 1996 focused extensively on illegal immigration, as did legislation enacted in Nov. 1997. The 1996 legislation also changed rules for the sponsorship of legal immigrants. Provisions enacted in Oct. 1998 increased visa allowances for skilled workers above limits shown below.

The Immigration Act of 1990 raised the number of numerically limited immigrants entering the U.S. annually during fiscal year 1992-94 to 700,000 immigrants (excluding refugees whose admission numbers are announced annually and others not subject to limitation). Beginning in fiscal year 1995, the number dropped from 700,000 to 675,000, subject to adjustment based largely on the number of visas that had been issued in the previous year.

In fiscal year 1997, 226,000 visas were allowed for family immigrants, 140,000 for employment-based immigrants, 55,000 for "diversity immigrants."

Immediate Relatives (Family Immigrants)

Fiscal year 1992-94: 465,000 minus the number of "immediate relatives" admitted the previous fiscal year, plus any numbers unused by the employment-based preference system. During this period, the number of family-sponsored visas could not fall below 226,000. If visa availability dipped below this new floor, the shortfall was made up from the category below.

During this period, 55,000 additional visas were made available to the spouses and children of aliens legalized under the Immigration Reform and Control Act (IRCA) of 1986.

Fiscal year 1995 and beyond: 480,000 minus the number of "immediate relatives" admitted during the previous fiscal year, plus any unused numbers under the employment-based preference system. Family-sponsored visas cannot drop below 226,000.

Family Preference System

First preference—unmarried sons and daughters of U.S. citizens: 23,400 visas in FY 1997 plus unused visas from the 4th preference.

Second preference—spouses and unmarried children of Lawful Permanent Residents (LPRs): 114,200 visas, plus any visas available above the floor of 226,000 family preference visas, plus any unused visas from the previous preference.

The category is subdivided as follows: A minimum of 77% of the visas allocated to the category goes to the spouses and minor children of LPRs; 75% of the visas are issued without regard to per country ceilings, in the order in which the petitions were filed. A maximum of 23% of the category visa allocation goes to the unmarried sons and daughters of LPRs. This group of visas continues to be subject to per country ceilings.

Third preference—married sons and daughters of U.S. citizens; 23,400 visas plus unused visas from all earlier preferences.

Fourth preference—brothers and sisters of U.S. citizens: 65,000 plus unused visas from all earlier preferences.

Employment-Based Immigrants

The law allows a total of 140,000 plus, beginning in fiscal year 1994, any unused numbers under the family-sponsored system. These visas are distributed as follows:

First preference—Priority Workers—28.6% of the employment-based limit plus visas unused by the 4th and 5th employment-based preferences, that is, "investors" and "special immigrants." The category is subdivided as follows: (1) extraordinary ability, demonstrated by sustained national or international acclaim, in the sciences, arts, education, business, and athletics; no U.S. employer required; (2) professors and researchers, seeking to enter in senior positions; U.S. employer required; (3) executives and managers of multinationals—requires one year of prior service with the firm during the preceding 3 years; the terms are extensively defined; U.S. employer required.

Second preference—Professionals with advanced degrees and aliens of exceptional ability—28.6% of the employment-based limit plus any unused "priority worker" visas. A U.S. employer and labor certification are required—although the Attorney General can waive both requirements. Members of the professions with advanced degrees or exceptional ability in the sciences, arts, or business. The possession of a degree, certificate, or license is not by itself considered sufficient evidence of exceptional ability.

Third preference—Skilled workers, professionals, and "other workers"—40,000 visas plus any visas unused by the 2 previous categories. Requires a U.S. employer and labor certification. Skilled workers must be in an occupation that requires at least 2 years training or experience. Professionals need a bachelor's degree. "Other workers" refers to unskilled workers. Their numbers are limited to visas per year.

Fourth preference—Special immigrants—7.1% of the employment-based limit. This category includes ministers of religion, foreign medical graduates, employees of the U.S. government abroad including certain employees of the U.S. mission in Hong Kong who file for admission as special immigrants before Jan. 1, 2002, retired employees of international organizations, etc.

Fifth preference—7.1% of the employment-based limit—7,000 for investors of $1 million in urban areas and 3,000 for investors of no less than $500,000 in rural or high-unemployment areas. The attorney general may increase the required investment amount up to $3 million for high employment areas. Investment must create employment for at least 10 U.S. workers.

Diversity Immigrant (DV) Category

Since FY 1995, the Immigration and Nationality Act has allowed 55,000 immigrant visas each fiscal year, distributed by lottery, to provide immigration opportunities for persons from countries other than the principal sources of immigration. DV visas are divided among six geographic regions. The allotment of FY 2000 visa numbers for each region had yet to be determined as of Oct. 15, 1999. Not more than 3,500 visas may be provided to immigrants from any one country.

The Nicaraguan and Central American Relief Act (NCARA) passed by Congress in Nov. 1997 stipulates that 5,000 of the 55,000 annually allocated diversity visas will be made available under the NCARA program.

The FY 2001 DV registration mail-in was held Oct. 4–Nov. 3, 1999. During this one-month period, the National Visa Center in Portsmouth, NH, expected to receive between 6 to 7 million qualified entries. An estimated 1.5 million entries were expected to be disqualified for not providing the requested information or following published guidelines. Winners of visas were to be notified between Apr. and July 2000.

In order to issue 50,000 visas in FY 2000, the National Visa Center planned to register about 100,000 persons, both principal applicants and their spouses and children. Those selected would receive instructions on how to apply for an immigrant visa. During the visa interview, applicants must provide proof of a high school education or its equivalent, or must show two years of work experience within the past 5 years in an occupation which requires at least 2 years of training or experience. Those selected need to act on their immigrant visa applications quickly. As soon as the 50,000 visas are issued, the program for FY 2001 ends.

Sponsorship of Immigrants

Under the 1996 immigration law, sponsors of an immigrant entering the country as an immediate relative or as an employment-based immigrant who will be employed by either a relative or a relative's company must earn at least 125% of the poverty level. If the sponsor does not earn enough, a cosponsor may be found who will accept joint responsibility for the immigrant. Persons who are active members of the U.S. armed forces need earn only 100% of the poverty level to be accepted as sponsors.

The sponsor must sign a legally binding affidavit of support for an immigrant, which would be enforceable until the immigrant either became a citizen or worked and paid taxes for 40 quarters as determined by the Social Security Administration.

Naturalization: How to Become an American Citizen

Source: Federal Statutes

A person who wishes to be naturalized as a citizen of the United States may obtain the necessary application form as well as detailed information from the nearest office of the Immigration and Naturalization Service.

An applicant must be at least 18 years old and must have been continuously resident in the U.S. for at least 5 years after admission for permanent residence. For husbands and wives of U.S. citizens the period is 3 years in most instances. Special provisions apply to certain veterans of the armed forces.

An applicant must have been physically present in the country for at least half of the required 5 years before filing an application and must:

(1) have been a person of good moral character, attached to the principles of the Constitution, and well disposed to the good order and happiness of the United States for 5 years just before filing the application or for whatever other period of residence is required in the particular case and continue to be such a person;

(2) demonstrate an understanding of the English language, including an ability to read, write, and speak words in ordinary usage in English (persons who are unable to demonstrate this requirement because of physical or developmental disability or mental impairment, are exempt. Persons who, on the date of filing the application, are over 50 years of age and have lived in the U.S. as lawful permanent residents for at least 20 years, or who are over 55 and have been residents for at least 15 years, are exempt); and

(3) demonstrate a knowledge and understanding of the fundamentals of the history, and the principles and form of government, of the United States. This must be done before an INS officer at the interview. Persons who are unable to demonstrate this requirement because of physical or developmental disability or mental impairment, are exempt.

At the interview the applicant may be represented by a lawyer or other representative. If action is favorable, there is a swearing in ceremony. The following oath of allegiance is given:

I hereby declare, on oath, that I absolutely and entirely renounce and abjure all allegiance and fidelity to any foreign prince, potentate, state or sovereignty, to whom or which I have heretofore been a subject or citizen; that I will support and defend the Constitution and laws of the United States of America against all enemies, foreign and domestic; that I will bear true faith and allegiance to the same; that I will bear arms on behalf of the United States when required by the law; that I will perform noncombatant service in the armed forces of the United States when required by the law; that I will perform work of national importance under civilian direction when required by the law; and that I take this obligation freely without any mental reservation or purpose of evasion; so help me God.

Major International Organizations

Asia-Pacific Economic Cooperation Group (APEC), founded Nov. 1989 as a forum to further cooperation on trade and investment between nations of the region and the rest of the world. Members of APEC in 1999 were Australia, Brunei, Canada, Chile, China, Indonesia, Japan, Malaysia, Mexico, New Zealand, Papua New Guinea, Peru, Philippines, Russia, Singapore, South Korea, Taiwan, Thailand, the United States, and Vietnam. Headquarters: Singapore. Website: http://www.apecsec.org.sg

Association of Southeast Asian Nations (ASEAN), formed Aug. 1967 to promote economic, social, and cultural cooperation and development among states of the Southeast Asian region. Members in 1999 were Brunei, Cambodia, Indonesia, Laos, Malaysia, Myanmar, Philippines, Singapore, Thailand, and Vietnam. Annual ministerial meetings set policy; the organization has a central Secretariat and specialized intergovernmental committees. Headquarters: Jakarta. Website: http://www.asean.or.id

Caribbean Community and Common Market (CARICOM), established July 4, 1973. Its aim is to further cooperation in economics, health, education, culture, science and technology, and tax administration, as well as the coordination of foreign policy. Members in 1999 were Antigua and Barbuda, Bahamas (Community only), Barbados, Belize, Dominica, Grenada, Guyana, Haiti (provisional), Jamaica, Montserrat, Saint Kitts and Nevis, Saint Lucia, Saint Vincent and the Grenadines, Suriname, and Trinidad and Tobago. Headquarters: Georgetown, Guyana. Website: http://www.caricom.org

Commonwealth of Independent States (CIS), created Dec. 1991 upon the disbanding of the Soviet Union. An alliance of independent states, it is made up of former Soviet constituent republics. Members in 1999 were 12 of the 15: Armenia, Azerbaijan, Belarus, Georgia, Kazakhstan, Kyrgyzstan, Moldova, Russia, Tajikistan, Turkmenistan, Ukraine, and Uzbekistan. Policy is set through coordinating bodies such as a Council of Heads of State and Council of Heads of Government. Capital of the commonwealth: Minsk, Belarus.

The Commonwealth, originally called the British Commonwealth of Nations, then the Commonwealth of Nations; an association of nations and dependencies that were once parts of the former British Empire. The British monarch is the symbolic head of the Commonwealth.

There are 53 independent nations in the Commonwealth. As of 1999, regular members included the United Kingdom and 14 other nations recognizing the British monarch, represented by a governor-general, as their head of state: Antigua and Barbuda, Australia, Bahamas, Barbados, Belize, Canada, Grenada, Jamaica, New Zealand, Papua New Guinea, Saint Kitts and Nevis, Saint Lucia, Saint Vincent and the Grenadines, and the Solomon Islands. (In addition, Tuvalu, which also recognizes the queen as head of state, was a special member.) Also members in good standing were 37 countries with their own heads of state: Bangladesh, Botswana, Brunei, Cameroon, Cyprus, Dominica, Fiji, The Gambia, Ghana, Guyana, India, Kenya, Kiribati, Lesotho, Malawi, Malaysia, Maldives, Malta, Mauritius, Mozambique, Namibia, Nauru, Pakistan (subject to change), Samoa, Seychelles, Sierra Leone, Singapore, South Africa, Sri Lanka, Swaziland, Tanzania, Tonga, Trinidad and Tobago, Uganda, Vanuatu, Zambia, and Zimbabwe. Nigeria was suspended from membership in 1995. The Commonwealth facilitates consultation among members through meetings of prime ministers and finance ministers and through a permanent Secretariat. Headquarters: London. Website: http://www.thecommonwealth.org/index1.htm

European Free Trade Association (EFTA), created May 3, 1960, to promote expansion of free trade. By Dec. 31, 1966, tariffs and quotas between member nations had been eliminated. Members entered into free trade agreements with the EU in 1972 and 1973. In 1992 the EFTA and EU agreed to create a single market—with free flow of goods, services, capital, and labor—among nations of the 2 organizations. Members in 1999 were Iceland, Liechtenstein, Norway, and Switzerland. Many former EFTA members are now EU members. Headquarters: Geneva. Website: http://www.efta.int/structure/main/index.html

European Union (EU)—known as the European Community (EC) until 1994—the collective designation of 3 organizations with common membership: the European Economic Community (Common Market), the European Coal and Steel Community, and the European Atomic Energy Community (Euratom). The 15 full members in 1999 were Austria, Belgium, Denmark, Finland, France, Germany, Greece, Ireland, Italy, Luxembourg, Netherlands, Portugal, Spain, Sweden, and United Kingdom. Austria, Finland, and Sweden entered the EU on Jan. 1, 1995. Some 70 nations in Africa, the Caribbean, and the Pacific are affiliated under the Lomé Convention. Website: http://europa.eu.int/index.htm

A merger of the 3 communities' executives went into effect July 1, 1967, though the component organizations date back to 1951 and 1958. The Council of Ministers, European Commission, European Parliament, and European Court of Justice comprise the permanent structure. The EU aims to integrate the economies, coordinate social developments, and bring about political union of the member states. Effective Dec. 31, 1992, there are no restrictions on the movement of goods, services, capital, workers, and tourists within the EU. There are also common agricultural, fisheries, and nuclear research policies.

Leaders of member nations (12 at the time) met Dec. 9-11, 1991, in Maastricht, the Netherlands. Treaties and accompanying protocols agreed upon by the leaders committed the organization to launching a common currency (the euro) by 1999; sought to establish common foreign policies; laid the groundwork for a common defense policy; gave the organization a leading role in social policy (Britain was not included in this plan); pledged increased aid for poorer member nations; and slightly increased the powers of the 567-member European Parliament. The treaties went into effect Nov. 1, 1993, following ratification by all 12 members.

In June 1998 the European Central Bank was established. In Jan. 1999, 11 of the 15 EU countries began using the euro for some purposes: Austria, Belgium, Finland, France, Germany, Iceland, Italy, Luxembourg, Netherlands, Portugal, and Spain. This includes all EU countries that wished to participate, except Greece, which did not meet all criteria for inclusion. On July 1, 2002, the 11 countries will change over completely to the euro; at that time current national currencies will no longer be legal tender.

Group of Eight (G-8), established Sept. 22, 1985; organization of 7 major industrial democracies (Canada, France, Germany, Italy, Japan, United Kingdom, and United States) and (later) Russia, meeting periodically to discuss world economic and other issues. At its annual economic summit in May 1998, the name was changed to G-8 from G-7. The original 7 were still free to meet without Russia on some issues, especially those relating to global finance.

International Criminal Police Organization (Interpol), created June 13, 1956, to promote mutual assistance among all police authorities within the limits of the law existing in the different countries. There were 177 members (independent nations), plus 12 subbureaus (dependencies) in 1999.

League of Arab States (Arab League), created Mar. 22, 1945. The League promotes economic, social, political, and military cooperation, mediates disputes, and represents Arab states in certain international negotiations. Members in 1999 were Algeria, Bahrain, Comoros, Djibouti, Egypt, Iraq, Jordan, Kuwait, Lebanon, Libya, Mauritania, Morocco, Oman, Palestine (considered an independent state by the League), Qatar, Saudi Arabia, Somalia, Sudan, Syria, Tunisia, United Arab Emirates, and Yemen. Headquarters: Cairo.

North Atlantic Treaty Organization (NATO), created by treaty (signed Apr. 4, 1949; in effect Aug. 24, 1949). Members in 1999 were Belgium, Canada, Czech Republic, Denmark, France, Germany, Greece, Hungary, Iceland, Italy, Luxembourg, Netherlands, Norway, Poland, Portugal, Spain, Turkey, United Kingdom, and United States. Members agreed to settle disputes by peaceful means, develop their individual and collective capacity to resist armed attack, to regard an attack on one as an attack on all, and take necessary action to repel an attack under Article 51 of the UN Charter. Website: http://www.nato.int

The NATO structure consists of a Council, the Defense Planning Committee, the Military Committee (consisting of 2 commands: Allied Command Europe, Allied Command Atlantic), Nuclear Planning Group, and Canada-U.S. Regional Planning Group. France detached itself from the military command structure in 1966.

With the dissolution of the Soviet Union and the end of the cold war in the early 1990s, members sought to modify the NATO mission, putting greater stress on political action and creating a rapid deployment force to react to local crises. By the mid-1990s, 27 nations, including Russia and other former Soviet republics, had joined with NATO in the so-called Partnership for Peace (PfP; drafted Dec. 1993), which provided for limited joint military exercises, peace-keeping missions, and information exchange. NATO has proceeded gradually toward extending full membership to former Eastern bloc nations. On

Mar. 12, 1999, 3 former Warsaw Pact members, Hungary, Poland, and the Czech Republic, formally became members.

In Dec. 1995, a NATO-led multinational force was deployed to help keep the peace in Bosnia and Herzegovina. Headquarters: Brussels.

Organization of African Unity (OAU), formed May 25, 1963, by 32 African countries (53 members in 1997) to promote peace and security as well as economic and social development. It holds annual conferences of heads of state. Headquarters: Addis Ababa, Ethiopia. Website: http://www.oau-oua.org

Organization of American States (OAS), formed in Bogotá, Colombia, Apr. 30, 1948. It has a Permanent Council, Inter-American Council for Integral Development, Juridical Committee, and Commission on Human Rights. The Permanent Council can call meetings of foreign ministers to deal with urgent security matters. A General Assembly meets annually.

Members in 1999 were: Antigua and Barbuda, Argentina, Bahamas, Barbados, Belize, Bolivia, Brazil, Canada, Chile, Colombia, Costa Rica, Cuba, Dominica, Dominican Republic, Ecuador, El Salvador, Grenada, Guatemala, Guyana, Haiti, Honduras, Jamaica, Mexico, Nicaragua, Panama, Paraguay, Peru, Saint Kitts and Nevis, Saint Lucia, Saint Vincent and the Grenadines, Suriname, Trinidad and Tobago, United States, Uruguay, and Venezuela. In 1962, the OAS suspended Cuba from participation in OAS activities but not from OAS membership. Headquarters: Washington, DC. Website: http://www.oas.org

Organization for Economic Cooperation and Development (OECD), established Sept. 30, 1961, to promote the economic and social welfare of all its member countries and to stimulate efforts on behalf of developing nations. The OECD also collects and disseminates economic and environmental information.

Members in 1999 were Australia, Austria, Belgium, Canada, Czech Republic, Denmark, Finland, France, Germany, Greece, Hungary, Iceland, Ireland, Italy, Japan, Luxembourg, Mexico, Netherlands, New Zealand, Norway, Poland, Portugal, South Korea, Spain, Sweden, Switzerland, Turkey, United Kingdom, and the United States. Headquarters: Paris. Website: http://www.oecd.org

Organization of Petroleum Exporting Countries (OPEC), created Sept. 14, 1960. The group attempts to set world oil prices by controlling oil production. It also pursues members' interests in trade and development dealings with industrialized oil-consuming nations. Members in 1999 were Algeria, Indonesia, Iran, Iraq, Kuwait, Libya, Nigeria, Qatar, Saudi Arabia, United Arab Emirates, and Venezuela. Headquarters: Vienna. Website: http://www.opec.org

Organization for Security and Cooperation in Europe (OSCE), established in 1972 as the Conference on Security and Cooperation in Europe; current name adopted Jan. 1, 1995. The group, formed by NATO and Warsaw Pact members, is interested in furthering East-West relations through a commitment to nonaggression and human rights as well as cooperation in economics, science and technology, cultural exchange, and environmental protection.

There were 55 member states in 1999. Headquarters: Vienna. Website: http://www.osce.org

United Nations

The 54th regular session of United Nations General Assembly opened Sept. 14, 1999.

UN headquarters is in New York, NY, between First Ave. and Roosevelt Drive and E. 42d St. and E. 48th St. The General Assembly Bldg., Secretariat, Conference and Library bldgs. are interconnected.

Some 52,200 people work in the UN system, which includes the Secretariat and 30 other organizations.

The UN has a post office originating its own stamps.

Proposals to establish an organization of nations for maintenance of world peace led to convening of the United Nations Conference on International Organization at San Francisco, Apr. 25-June 26, 1945, where the charter of the United Nations was drawn up.

The charter was signed June 26 by 50 nations, and by Poland, one of the original 51 members of the United Nations, on Oct. 15, 1945. The charter came into effect Oct. 24, 1945,

upon ratification by the permanent members of the Security Council and a majority of other signatories.

Purposes: To maintain international peace and security; to develop friendly relations among nations; to achieve international cooperation in solving economic, social, cultural, and humanitarian problems and in promoting respect for human rights and fundamental freedoms; to be a center for harmonizing the actions of nations in attaining these common ends.

Visitors to the UN: Headquarters is open to the public every day except Thanksgiving, Christmas, and New Year's Day. Guided tours are given approximately every half hour from 9:15 A.M. to 4:45 P.M. daily, except on weekends in January and February.

Groups of 12 or more should write to the Group Program Unit, Public Services Section, Room GA-63, United Nations, New York, NY 10017, or telephone (212) 963-4440. Children under 5 not permitted on tours.

Roster of the United Nations

The 188 members of the United Nations, with the years in which they became members; as of Oct. 1999.

Member	Year	Member	Year	Member	Year	Member	Year
Afghanistan	1946	Djibouti	1977	Lesotho	1966	Saint Lucia	1979
Albania	1955	Dominica	1978	Liberia	1945	Saint Vincent and	
Algeria	1962	Dominican Republic	1945	Libya	1955	the Grenadines	1980
Andorra	1993	Ecuador	1945	Liechtenstein	1990	Samoa (formerly	
Angola	1976	Egypt[3]	1945	Lithuania	1991	Western Samoa)	1976
Antigua and Barbuda	1981	El Salvador	1945	Luxembourg	1945	San Marino	1992
Argentina	1945	Equatorial Guinea	1968	Macedonia[5]	1993	São Tomé and Príncipe	1975
Armenia	1992	Eritrea	1993	Madagascar	1960	Saudi Arabia	1945
Australia	1945	Estonia	1991	Malawi	1964	Senegal	1960
Austria	1955	Ethiopia	1945	Malaysia[6]	1957	Seychelles	1976
Azerbaijan	1992	Fiji	1970	Maldives	1965	Sierra Leone	1961
Bahamas	1973	Finland	1955	Mali	1960	Singapore[6]	1965
Bahrain	1971	France	1945	Malta	1964	Slovakia[2]	1993
Bangladesh	1974	Gabon	1960	Marshall Islands	1991	Slovenia	1992
Barbados	1966	Gambia, The	1965	Mauritania	1961	Solomon Islands	1978
Belarus	1945	Georgia	1992	Mauritius	1968	Somalia	1960
Belgium	1945	Germany	1973	Mexico	1945	South Africa[8]	1945
Belize	1981	Ghana	1957	Micronesia	1991	Spain	1955
Benin	1960	Greece	1945	Moldova	1992	Sri Lanka	1955
Bhutan	1971	Grenada	1974	Monaco	1993	Sudan	1956
Bolivia	1945	Guatemala	1945	Mongolia	1961	Suriname	1975
Bosnia and Herzegovina	1992	Guinea	1958	Morocco	1956	Swaziland	1968
Botswana	1966	Guinea-Bissau	1974	Mozambique	1975	Sweden	1946
Brazil	1945	Guyana	1966	Myanmar (Burma)	1948	Syria[3]	1945
Brunei	1984	Haiti	1945	Namibia	1990	Tajikistan	1992
Bulgaria	1955	Honduras	1945	Nauru	1999	Tanzania[9]	1961
Burkina Faso	1960	Hungary	1955	Nepal	1955	Thailand	1946
Burundi	1962	Iceland	1946	Netherlands	1945	Togo	1960
Cambodia	1955	India	1945	New Zealand	1945	Tonga	1999
Cameroon	1960	Indonesia[4]	1950	Nicaragua	1945	Trinidad and Tobago	1962
Canada	1945	Iran	1945	Niger	1960	Tunisia	1956
Cape Verde	1975	Iraq	1945	Nigeria	1960	Turkey	1945
Central African Republic	1960	Ireland	1955	Norway	1945	Turkmenistan	1992
Chad	1960	Israel	1949	Oman	1971	Uganda	1962
Chile	1945	Italy	1955	Pakistan	1947	Ukraine	1945
China[1]	1945	Jamaica	1962	Palau	1994	United Arab Emirates	1971
Colombia	1945	Japan	1956	Panama	1945	United Kingdom	1945
Comoros	1975	Jordan	1955	Papua New Guinea	1975	United States	1945
Congo, Democratic		Kazakhstan	1992	Paraguay	1945	Uruguay	1945
Republic of the (Zaire)	1960	Kenya	1963	Peru	1945	Uzbekistan	1992
Congo, Republic of the	1960	Kiribati	1999	Philippines	1945	Vanuatu	1981
Costa Rica	1945	Korea, North	1991	Poland	1945	Venezuela	1945
Côte d'Ivoire	1960	Korea, South	1991	Portugal	1955	Vietnam	1977
Croatia	1992	Kuwait	1963	Qatar	1971	Yemen[10]	1947
Cuba	1945	Kyrgyzstan	1992	Romania	1955	Yugoslavia[11]	1945
Cyprus	1960	Laos	1955	Russia[7]	1945	Zambia	1964
Czech Republic[2]	1993	Latvia	1991	Rwanda	1962	Zimbabwe	1980
Denmark	1945	Lebanon	1945	Saint Kitts and Nevis	1983		

(1) The General Assembly voted in 1971 to expel the Chinese government on Taiwan and admit the Beijing government in its place. (2) Czechoslovakia, which split into the separate nations of the Czech Republic and Slovakia on Jan. 1, 1993, was a UN member from 1945 to 1992. (3) Egypt and Syria were original members of the UN. In 1958, the United Arab Republic was established by a union of Egypt and Syria and continued as a single member of the UN. In 1961, Syria resumed its separate membership. (4) Indonesia withdrew from the UN in 1965 and rejoined in 1966. (5) Admitted under the provisional name of The Former Yugoslav Republic of Macedonia. (6) Malaya joined the UN in 1957. In 1963, its name was changed to Malaysia following the accession of Singapore, Sabah, and Sarawak. Singapore became an independent UN member in 1965. (7) The Union of Soviet Socialist Republics was an original member of the UN from 1945. After the USSR's dissolution in 1991, Russia informed the UN it would be continuing the USSR's membership in the Security Council and all other UN organs with the support of the Commonwealth of Independent States (comprised of most of the former Soviet republics). (8) In 1994, the General Assembly accepted the credentials of the South African delegation, which had been rejected for 24 years because of the country's former apartheid policies. (9) Tanganyika was a member of the UN from 1961 and Zanzibar was a member from 1963. Following the ratification in 1964 of Articles of Union between Tanganyika and Zanzibar, the United Republic of Tanganyika and Zanzibar continued as a single member of the UN, later changing its name to United Republic of Tanzania. (10) The Yemen Arab Republic was admitted in 1947; the People's Republic of Yemen, in 1967. The two nations merged in 1990. (11) The Socialist Federal Republic of Yugoslavia became a member in 1945. After four of its six republics (Bosnia and Herzegovina, Croatia, Macedonia, and Slovenia) declared independence in 1991-92, the two remaining republics, Montenegro and Serbia, reconstituted themselves as the Federal Republic of Yugoslavia, which assumed Yugoslavia's UN seat Apr. 8, 1992. In Sept. 1992, the General Assembly decided the Federal Republic of Yugoslavia should apply for membership as it could not automatically take the seat of the former Yugoslavia. **NOTE:** The following sovereign countries are not members of the United Nations: China (Taiwan), Switzerland, Tuvalu, Vatican City (Holy See). Switzerland and Vatican City are, however, permanent observers.

United Nations Secretaries General

Year	Secretary, Nation	Year	Secretary, Nation	Year	Secretary, Nation
1946	Trygve Lie, Norway	1972	Kurt Waldheim, Austria	1992	Boutros Boutros-Ghali, Egypt
1953	Dag Hammarskjold, Sweden	1982	Javier Perez de Cuellar, Peru	1997	Kofi Annan, Ghana
1961	U Thant, Burma				

U.S. Representatives to the United Nations

The U.S. Representative to the United Nations is the Chief of the U.S. Mission to the United Nations in New York and holds the rank and status of Ambassador Extraordinary and Plenipotentiary (A.E.P.).

Year	Representative	Year	Representative	Year	Representative
1946	Edward R. Stettinius, Jr.	1968	James Russell Wiggins	1981	Jeane J. Kirkpatrick
1946	Herschel V. Johnson (act.)	1969	Charles W. Yost	1985	Vernon A. Walters
1947	Warren R. Austin	1971	George Bush	1989	Thomas R. Pickering
1953	Henry Cabot Lodge, Jr.	1973	John A. Scali	1992	Edward J. Perkins
1960	James J. Wadsworth	1975	Daniel P. Moynihan	1993	Madeleine K. Albright
1961	Adlai E. Stevenson	1976	William W. Scranton	1997	Bill Richardson
1965	Arthur J. Goldberg	1977	Andrew Young	1999	Richard C. Holbrooke
1968	George W. Ball	1979	Donald McHenry		

Organization of the United Nations

The text of the UN Charter may be obtained from the Public Inquiries Unit, Department of Public Information, United Nations, New York, NY 10017. (212) 963-4475.

General Assembly. The General Assembly is composed of representatives of all the member nations. Each nation is entitled to one vote.

The General Assembly meets in regular annual sessions and in special session when necessary. Special sessions are convoked by the secretary general at the request of the Security Council or of a majority of the members of the UN.

On important questions a two-thirds majority of members present and voting is required; on other questions a simple majority is sufficient.

The General Assembly must approve the UN budget and apportion expenses among members. A member in arrears can lose its vote if the amount of arrears equals or exceeds the amount of the contributions due for the preceding 2 full years.

Security Council. The Security Council consists of 15 members, 5 with permanent seats. The remaining 10 are elected for 2-year terms by the General Assembly; they are not eligible for immediate reelection.

Permanent members of the Council are: China, France, Russia, United Kingdom, and the United States.

Nonpermanent members are: (with terms expiring Dec. 31, 1999) Bahrain, Brazil, Gabon, Gambia, and Slovenia; (with terms expiring Dec. 31, 2000) Argentina, Canada, Malaysia, Namibia, and the Netherlands.

The Security Council has the primary responsibility within the UN for maintaining international peace and security. The Council may investigate any dispute that threatens international peace and security.

Any member of the UN at UN headquarters may, if invited by the Council, participate in its discussions and a nation not a member of the UN may appear if it is a party to a dispute.

Decisions on procedural questions are made by an affirmative vote of 9 members. On all other matters the affirmative vote of 9 members must include the concurring votes of all permanent members; it is this clause which gives rise to the so-called veto power of permanent members. A party to a dispute must refrain from voting.

The Security Council directs the various peacekeeping forces deployed throughout the world.

Economic and Social Council. The Economic and Social Council consists of 54 members elected by the General Assembly for 3-year terms. The council is responsible for carrying out UN functions with regard to international economic, social, cultural, educational, health, and related matters. It meets once a year.

Trusteeship Council. The administration of trust territories was under UN supervision; however, all 11 Trust Territories have attained their right to self-determination. The work of the Council has, therefore, been suspended.

Secretariat. The Secretary General is the chief administrative officer of the UN. The Secretary General reports to the General Assembly and may bring to the attention of the Security Council any matter that threatens international peace.

Budget: The General Assembly approved a total budget for the biennium 1998-99 of $2.53 billion.

International Court of Justice (World Court). The International Court of Justice is the principal judicial organ of the United Nations. All members are *ipso facto* parties to the statute of the Court. Other states may become parties to the Court's statute.

The Court has jurisdiction over cases which the parties submit to it and matters especially provided for in the charter or in treaties. The Court gives advisory opinions and renders judgments. Its decisions are binding only between parties concerned and in respect to a particular dispute. If any party to a case fails to heed a judgment, the other party may have recourse to the Security Council.

The 15 judges are elected for 9-year terms by the General Assembly and the Security Council. Retiring judges are eligible for reelection. The Court remains permanently in session, except during vacations. All questions are decided by majority. The International Court of Justice sits in The Hague, Netherlands.

Selected Specialized and Related Agencies

These agencies are autonomous, with their own memberships and organs, and have a functional relationship or working agreement with the UN (headquarters), except for UNICEF and UNHCR, which report directly to the Economic and Social Council and to the General Assembly.

Food and Agriculture Organization (FAO), aims to increase production from farms, forests, and fisheries; improve food distribution and marketing, nutrition, and the living conditions of rural people. (Viale delle Terme di Caracalla, 00100 Rome, Italy.)

International Atomic Energy Agency (IAEA), aims to promote the safe, peaceful uses of atomic energy. (Vienna International Centre, PO Box 100, A-1400, Vienna, Austria.)

International Bank for Reconstruction and Development (IBRD) (World Bank), provides loans and technical assistance for projects in developing member countries; encourages cofinancing for projects from other public and private sources. The IBRD has 4 affiliates: (1) The **International Development Association (IDA)** provides funds for development projects on concessionary terms to the poorer developing member countries. (2) The **International Finance Corporation (IFC)** promotes the growth of the private sector in developing member countries; encourages the development of local capital markets; stimulates the international flow of private capital. (3) The **Multilateral Investment Guarantee Agency (MIGA)** promotes private investment in developing countries; guarantees investments to protect investors from noncommercial risks, such as nationalization; advises governments on attracting private investment. (4) The **International Center for Settlement of Investment Disputes (ICSID)** provides conciliation and arbitration services for disputes between foreign investors and host governments which arise out of an investment. (1818 H St., NW, Washington, DC 20433.)

International Civil Aviation Org. (ICAO), promotes international civil aviation standards and regulations. (999 University St., Montreal, Quebec, Canada H3C 5H7.)

International Fund for Agricultural Development (IFAD), aims to mobilize funds for agricultural and rural projects in developing countries. (107 Via del Seratico, 00142 Rome, Italy.)

International Labor Org. (ILO), aims to promote employment; improve labor conditions and living standards. (4 route des Morillons, CH-1211 Geneva 22, Switzerland.)

International Maritime Org. (IMO), aims to promote cooperation on technical matters affecting international shipping. (4 Albert Embankment, London SE1 7SR, England.)

International Monetary Fund (IMF), aims to promote international monetary cooperation and currency stabilization and expansion of international trade. (700 19th St., NW, Washington, DC 20431.)

International Telecommunication Union (ITU), establishes international regulations for radio, telegraph, telephone, and space radio-communications, allocates radio frequencies. (Place des Nations, 1211 Geneva 20, Switzerland.)

United Nations Children's Fund (UNICEF), provides financial aid and development assistance to programs for children and mothers in developing countries. (3 UN Plaza, New York, NY 10017.)

United Nations Educational, Scientific, and Cultural Org. (UNESCO), aims to promote collaboration among nations through education, science, and culture. (7 Place de Fontenoy, 75352 Paris 07SP, France.)

United Nations High Commissioner for Refugees (UNHCR), provides essential assistance for refugees. (Place des Nations, 1211 Geneva 10, Switzerland.)

Universal Postal Union (UPU), aims to perfect postal services and promote international collaboration. (Weltpoststrasse 4, 3000 Berne, 15 Switzerland.)

World Health Org. (WHO), aims to aid the attainment of the highest possible level of health. (1211 Geneva 27, Switzerland.)

World Intellectual Property Org. (WIPO), seeks to protect, through international cooperation, literary, industrial, scientific, and artistic works. (34, Chemin des Colom Bettes, 1211 Geneva, Switzerland.)

World Meteorological Org. (WMO), aims to coordinate and improve world meteorological work. (41, Avenue Giuseppe-Motta, Case Postale 2300, 1211 Geneva 2, Switzerland.)

World Trade Org. (WTrO), replacing the General Agreement on Tariffs and Trade (GATT), is the major body overseeing international trade. The WTrO administers trade agreements and treaties, examines the trade regimes of members, keeps track of various trade measures and statistics, and attempts to settle trade disputes. (Centre William Rappard, 154 rue de Lausanne, 1211 Geneva 21, Switzerland.)

Geneva Conventions

The Geneva Conventions are 4 international treaties governing the protection of civilians in time of war, the treatment of prisoners of war, and the care of the wounded and sick in the armed forces. The first convention, covering the sick and wounded, was concluded in Geneva, Switzerland, in 1864; it was amended and expanded in 1906. A third convention, in 1929, covered prisoners of war. Outrage at the treatment of prisoners and civilians during World War II by some belligerents, notably Germany and Japan, prompted the conclusion, in Aug. 1949, of 4 new conventions. Three of these restated and strengthened the previous conventions, and the fourth codified general principles of international law governing the treatment of civilians in wartime.

The 1949 convention for civilians provided for special safeguards for the following categories of people: wounded persons, children under 15 years of age, pregnant women, and the elderly. Discrimination was forbidden on racial, religious, national, or political grounds. Torture, collective punishment, reprisals, the unwarranted destruction of property, and the forced use of civilians for an occupier's armed forces were also prohibited under the 1949 conventions.

Also included in the new 1949 treaties was a pledge to treat prisoners humanely, feed them adequately, and deliver relief supplies to them. They were not to be forced to disclose more than minimal information.

Most countries have formally accepted all or most of the humanitarian conventions as binding. A nation is not free to withdraw its ratification of the conventions during wartime. However, there is no permanent machinery in place to apprehend, try, or punish violators.

Some Major World's Fairs and Expositions

Source: Bureau of International Expositions, Paris, France; World Almanac reasearch

LOCATION	NAME	DATES	VISITORS
London, England	Crystal Palace Exposition	Apr. 1-Oct. 11, 1851	6,039,195
Paris, France	Paris Universal Exposition	Apr. 1-Nov. 3, 1867	15,000,000
Philadelphia, PA	Centennial Exposition	May 10-Nov. 10, 1876	10,000,000
Paris, France	Paris Universal Exposition	May 20-Nov. 10, 1878	16,156,626
Paris, France	Paris Universal Exposition	May 5-Oct. 31, 1889	32,250,297
Chicago, IL	World's Columbian Exposition	May 1-Oct. 3, 1893	27,500,000
Paris, France	Universal and International Exposition of Paris	Apr. 15-Nov. 12, 1900	50,860,801
Saint Louis, MO	Universal Exposition of Saint Louis	Apr. 3-Dec. 1, 1904	19,694,855
Brussels, Belgium	Universal Expositon of Brussels	Apr. 23-Nov. 7, 1910	13,000,000
San Francisco, CA	Panama-Pacific International Exposition	Feb. 2-Dec. 4, 1915	19,000,000
Chicago, IL	A Century of Progress International Exposition	May 27-Nov. 12, 1933; June 1-Oct. 31, 1934	38,872,000
Brussels, Belgium	Universal Exposition of Brussels	1935	20,000,000
Paris, France	Intl. Exposition of Arts and Techniques in Modern Life	May 25-Nov. 25, 1937	31,040,955
New York, NY	New York World's Fair	Apr. 3-Oct. 31, 1939; May 11-Oct. 27, 1940	44,955,997
Brussels, Belgium	Universal and International Exposition of Brussels	Apr. 17-Oct. 19, 1958	41,454,412
Seattle, WA	Century 21 Exposition	Apr. 21-Oct. 21, 1962	9,609,969
New York, NY	New York World's Fair	Apr. 22-Oct. 18, 1964; Apr. 21-Oct. 17, 1965	51,500,000
Montreal, Canada	Expo 67	Apr. 28-Oct. 27, 1967	50,306,648
Osaka, Japan	Expo 70	Mar. 15-Sept. 13, 1970	64,218,770
Spokane, WA	International Exposition on the Environment	May 1-Nov. 1, 1974	5,600,000
Knoxville, TN	The Knoxville International Energy Exposition	May 1-Oct. 31, 1982	11,127,780
New Orleans, LA	The 1984 Louisiana World Exposition	May 12-Nov. 11, 1984	7,335,000
Tsukuba, Japan	International Exposition	Mar. 17-Sept. 16, 1985	20,334,727
Vancouver, Canada	Expo 86	May 2-Oct. 13, 1986	20,111,578
Brisbane, Australia	International Exposition on Leisure	Apr. 3-Oct. 3, 1988	18,560,447
Seville, Spain	Universal Exposition of Seville	Apr. 2-Oct. 12, 1992	40,000,000
Taejon, S. Korea	The Taejon International Exposition	Aug. 7-Nov. 7, 1993	14,005,808
Lisbon, Portugal	Specialty Exposition: The Oceans	May 22-Sept. 30, 1998	10,128,204
Hannover, Germany	Expo 2000	June 1-Oct. 31, 2000	—

U.S. Aid to Foreign Nations in 1998

Source: Bureau of Economic Analysis, August 1999

Figures are in millions of dollars and may not add because of rounding. (* = Less than $500,000.) Data include military supplies and services furnished under the Foreign Assistance Act and direct Defense Department appropriations, and include credits extended to private entities.

Net grants and credits take into account all known returns to the U.S., including reverse grants, returns of grants, and payments of principal. A minus sign (–) indicates the total of these returns is greater than the total of grants or credits. Nations with net grants or credits under $2 million (+ or –) are included with "Other and unspecified."

Other Assistance represents the transfer of U.S. farm products in exchange for foreign currencies, less the government's disbursements of such currencies as grants, credits, or for purchases.

Amounts do not include investments in the following: Asian Development Bank, $195 mil; Inter-American Development Bank, $83 mil; International Development Assn., $1,087 mil; International Bank for Reconstruction and Development, $49 mil; African Development Fund, $94 mil; IMF-Enhanced Structural Adjustment Facility, $24 mil; European Bank for Reconstruction and Development, $21 mil; Enterprise for the Americas Investment Fund, $23 mil; North American Development Bank, $6 mil.

	Total	Net grants	Net credits	Net other		Total	Net grants	Net credits	Net other
TOTAL	$12,240	$13,165	$–926	(*)	Ethiopia	$116	$117	—	(*)
Western Europe	258	482	–224	—	Gambia, The	5	5	—	—
Bosnia and Herzegovina	220	220	—	—	Ghana	40	39	(*)	—
Croatia	–12	3	–15	—	Guinea	18	18	(*)	—
Ireland	8	8	—	—	Guinea-Bissau	11	11	—	—
Macedonia	6	13	–7	—	Kenya	36	46	–10	(*)
Portugal	–15	(*)	–15	—	Lesotho	3	3	—	—
Slovenia	–16	(*)	–16	—	Liberia	16	16	—	—
Spain	–37		–37	—	Madagascar	44	32	12	—
United Kingdom	–130		–130	—	Malawi	22	22	—	—
Former Yugoslavia –					Mali	36	36	(*)	(*)
Regional	6	9	–3	—	Mauritania	4	4	—	—
Other & unspecified	228	229	–1		Morocco	–49	42	–91	(*)
Eastern Europe	1,785	1,783	2	—	Mozambique	74	74	—	—
Albania	16	16	—	—	Namibia	12	12	—	—
Bulgaria	13	13	—	—	Niger	13	13	(*)	—
Czech Republic	4	4	—	—	Nigeria	–4	5	–9	—
Estonia	5	5	—	—	Rwanda	28	28	—	—
Hungary	12	12	—	—	Senegal	18	18	—	—
Latvia	8	8	—	—	Sierra Leone	13	25	–12	—
Lithuania	14	7	8	—	Somalia	3	3	—	—
Poland	36	52	–16	—	South Africa	89	89	(*)	—
Romania	26	21	5	—	Sudan	23	23	—	—
Slovakia	6	6	—	—	Swaziland	14	14	(*)	—
Independent States of the					Tanzania	26	31	–5	—
Former Soviet Union	—	—			Tunisia	–22	3	–25	(*)
Armenia	38	23	15	—	Uganda	17	17	(*)	—
Azerbaijan	5	5	—	—	Zambia	14	16	–2	—
Belarus	4	4	—	—	Zimbabwe	46	14	31	—
Georgia	27	12	15	—	Other & unspecified	448	453	–4	—
Kazakhstan	67	67	—	—	**East Asia and Pacific**	735	416	319	1
Kyrgyzstan	30	30	—	—	Cambodia	33	35	–2	—
Moldova	5	5	—	—	China	249	(*)	248	—
Russia	435	464	–29	—	Fed. States of Micronesia	64	65	(*)	—
Tajikistan	27	27	—	—	Hong Kong	17		17	—
Turkmenistan	3	3	—	—	Indonesia	19	45	–26	—
Ukraine	140	139	1	—	Korea, Republic of	–52		–52	—
Uzbekistan	6	6	—	—	Laos	3	3	—	—
Former Soviet Union–					Marshall Islands, Rep. of	32	32	—	—
Regional	659	654	5	—	Mongolia	19	19	—	—
Other & unspecified	200	200	—	—	Palau	78	78	—	—
Near East and South Asia	4,979	5,747	–770	2	Philippines	296	67	229	(*)
Bangladesh	27	44	–17	—	Thailand	–68	12	–80	—
Cyprus	14	14	—	—	Vietnam	–10	(*)	–10	—
Egypt	2,014	2,106	–92	—	Other & unspecified	55	59	–5	1
Greece	–240	5	–245	(*)	**Western Hemisphere**	987	1,179	–190	–2
India	160	141	19	—	Argentina	–88	1	–90	—
Israel	2,840	3,034	–194	—	Bolivia	96	88	8	—
Jordan	160	143	17	(*)	Brazil	88	8	80	—
Nepal	20	20	(*)	(*)	Chile	–8	3	–10	—
Oman	20	16	4	—	Colombia	42	59	–17	—
Pakistan	–82	9	–94	2	Costa Rica	–28	7	–35	(*)
Sri Lanka	1	10	–9	(*)	Dominican Republic	–14	15	–30	—
Turkey	–162	2	–164	(*)	Ecuador	2	20	–18	—
Yemen	4	3	2	—	El Salvador	44	48	–4	—
UNRWA	78	78	—	—	Guatemala	35	50	–16	—
West Bank–Gaza	80	79	2	—	Guyana	12	5	8	—
Other & unspecified	45	45	(*)	—	Haiti	87	88	(*)	—
Africa	1,277	1,338	–60	(*)	Honduras	47	50	–3	—
Algeria	45	(*)	44	—	Jamaica	–22	16	–36	–2
Angola	35	25	10	—	Mexico	–126	11	–137	—
Benin	14	14	—	—	Nicaragua	70	61	9	—
Botswana	4	5	–1	—	Panama	–20	11	–31	—
Burkina Faso	17	17	—	(*)	Paraguay	2	5	–3	—
Burundi	6	6	—	—	Peru	88	129	–42	—
Cameroon	9	6	3	—	Trinidad and Tobago	204	(*)	204	—
Cape Verde, Republic of	7	7	—	—	Venezuela	4	1	3	—
Chad	3	3	—	—	Other & unspecified	473	503	–29	—
Côte d'Ivoire	8	17	–9	—	**International Organizations**				
Eritrea	16	6	10	—	**and unspecified**	2,218	2,221	–3	—

Codes for International Direct Dial Calling From the U.S.

Station-to-station: 011 + country code (as shown) + city code (if required) + local number.

Person-to-person (operator-assisted, collect calls, credit card calls, and calls billed to another number): 01 + country code (below) + city code (if required) + local number.

For countries or territories not listed, contact your long distance company.

For area codes within the United States, see U.S. Places of 5,000 or More Population and Telephone Area Codes in the Consumer Information section.

Country/Territory	Code	Country/Territory	Code	Country/Territory	Code	Country/Territory	Code
Afghanistan	93	Cape Verde	238	Italy	39	Portugal	351
Albania	355	Cayman Islands	345*	Jamaica	876*	Puerto Rico	787*
Algeria	213	Central African Rep.	236	Japan	81	Qatar	974
American Samoa	684	Chad	235	Jordan	962	Romania	40
Andorra	33	Chile	56	Kazakhstan	7	Russia	7
Angola	244	China	86	Kenya	254	Rwanda	250
Anguilla	264*	Colombia	57	Kiribati	686	St. Kitts & Nevis	869*
Antarctica	672	Comoros	269	Korea, North	850	St. Lucia	758*
Antigua & Barbuda	268*	Congo (formerly Zaire)	243	Korea, South	82	St. Vincent & the	
Argentina	54	Congo Republic	242	Kuwait	965	Grenadines	784*
Armenia	374	Costa Rica	506	Kyrgyzstan	7	Samoa (formerly	
Aruba	297	Côte d'Ivoire	225	Laos	856	Western Samoa)	685
Ascension Island	247	Croatia	385	Latvia	371	San Marino	378
Australia	61	Cuba	53	Lebanon	961	São Tomé & Príncipe	239
Austria	43	Cyprus	357	Lesotho	266	Saudi Arabia	966
Azerbaijan	994	Czech Republic	42	Liberia	231	Senegal	221
Bahamas	242*	Denmark	45	Libya	218	Seychelles	248
Bahrain	973	Djibouti	253	Liechtenstein	41	Sierra Leone	232
Bangladesh	880	Dominica	767*	Lithuania	370	Singapore	65
Barbados	246*	Dominican Republic	809*	Luxembourg	352	Slovakia	42
Belarus	375	Ecuador	593	Macao	853	Slovenia	386
Belgium	32	Egypt	20	Macedonia	389	Solomon Islands	677
Belize	501	El Salvador	503	Madagascar	261	Somalia	252
Benin	229	Equatorial Guinea	240	Malawi	265	South Africa	27
Bermuda	441*	Eritrea	291	Malaysia	60	Spain	34
Bhutan	975	Estonia	372	Maldives	960	Sri Lanka	94
Bolivia	591	Ethiopia	251	Mali	223	Sudan	249
Bosnia & Herzegovina	387	Falkland Islands	500	Malta	356	Suriname	597
Botswana	267	Fiji	679	Marshall Islands	692	Swaziland	268
Brazil	55	Finland	358	Martinique	596	Sweden	46
Brunei	673	France	33	Mauritania	222	Switzerland	41
Bulgaria	359	French Antilles	596	Mauritius	230	Syria	963
Burkina Faso	226	French Guiana	594	Mexico	52	Taiwan	886
Burundi	257	French Polynesia	689	Micronesia	691	Tajikistan	7
Cambodia	855	Gabon	241	Moldova	373	Tanzania	255
Cameroon	237	Gambia, The	220	Monaco	377	Thailand	66
Canada		Georgia	995	Mongolia	976	Togo	228
Alberta	403*	Germany	49	Montserrat	664*	Tonga	676
British Columbia	250*	Ghana	233	Morocco	212	Trinidad & Tobago	868*
Vancouver	604*	Gibraltar	350	Mozambique	258	Tunisia	216
Manitoba	204*	Greece	30	Myanmar	95	Turkey	90
New Brunswick	506*	Greenland	299	Namibia	264	Turkmenistan	7
Newfoundland	709*	Grenada	473*	Nauru	674	Turks & Caicos Isls.	649*
NW Territories	604*	Guadeloupe	590	Nepal	977	Tuvalu	688
Nova Scotia	902*	Guam	671*	Netherlands	31	Uganda	256
Nunavut	867*	Guantanamo Bay	53	New Caledonia	687	Ukraine	380
Ontario		Guatemala	502	New Zealand	64	United Arab Emirates	971
London	519*	Guinea	224	Nicaragua	505	United Kingdom	44
North Bay	705*	Guinea-Bissau	245	Niger	227	Uruguay	598
Ottawa	613*	Guyana	592	Nigeria	234	Uzbekistan	7
Thunder Bay	807*	Haiti	509	N. Mariana Isls.	670*	Vanuatu	678
Toronto Metro	416*	Honduras	504	Norway	47	Vatican City	379
Toronto Vicinity	905*	Hong Kong	852	Oman	968	Venezuela	58
Prince Edward Isl.	902*	Hungary	36	Pakistan	92	Vietnam	84
Quebec		Iceland	354	Palau	680	Virgin Islands, British	284*
Montreal	514*	India	91	Panama	507	Virgin Islands, U.S.	340*
Quebec City	418*	Indonesia	62	Papua New Guinea	675	Yemen	967
Sherbrooke	819*	Iran	98	Paraguay	595	Yugoslavia	381
Saskatchewan	306*	Iraq	964	Peru	51	Zambia	260
Yukon Territory	403*	Ireland	353	Philippines	63	Zimbabwe	263
		Israel	972	Poland	48		

* These numbers are area codes. Follow Domestic Dialing instructions: dial "1" + area code + number you are calling.

VITAL STATISTICS
Annual Report for the Year 1998
Source: National Center for Health Statistics, U.S. Dept. of Health and Human Services

Highlights
Provisional data for 1998 reported by the National Center for Health Statistics show that the teen birth rate continued a steady decline that began in 1991 (51.1 births per 1,000 women aged 15-19 years in 1998, compared with 62.1 in 1991). Life expectancy reached an all-time high of 76.7 years. Marriage and divorce rates both decreased, and the rate of natural increase rose slightly.

Births
An estimated 3,944,046 babies were born in the U.S. in 1998, an increase of 2% from the 3,880,894 births in 1997. The overall birth rate was higher than the rate for the preceding year (14.6 per 1,000 population as compared to 14.5). The fertility rate (number of live births per 1,000 women aged 15-44 years) for 1998 was 65.6, slightly higher than the rate for 1997 (65.0).

Deaths
The provisional number of deaths during 1998 was 2,338,070, 1% greater than in the previous year (2,314,245). The death rate of 865.0 deaths per 100,000 population was slightly higher than the 1997 death rate of 864.7. The infant mortality rate of 7.2 infant deaths per 1,000 live births was unchanged from 1997.

Natural Increase
As a result of natural increase alone, the excess of births over deaths, an estimated 1,605,976 persons were added to the population in 1998. The rate of 5.90 per 1,000 population was slightly higher than for 1997 (5.85), which was the lowest since 1976 (5.8). The nearly steady rate of natural increase reflected similar slight increases in both birth and death rates.

Marriages
An estimated 2,244,000 marriages were performed in 1998, nearly 6% less than in 1997 (2,384,000). The marriage rate for 1998 (8.3 per 1,000 population), the lowest rate since 1932, was 7% lower than in 1997 (8.9).

Divorces
About 1,135,000 divorces were granted in the U.S. in 1998, 2% fewer than the number for 1997 (1,163,000), and about 7% fewer than the all-time high of 1,215,000 in 1992. The divorce rate per 1,000 population in 1998 (4.2), the lowest divorce rate in over 25 years, was slightly lower than the rate for 1997 (4.3).

Births and Deaths in the U.S.
Source: National Center for Health Statistics, U.S. Dept. of Health and Human Services

Year	BIRTHS Total number	BIRTHS Rate	DEATHS Total number	DEATHS Rate
1960	4,257,850	23.7	1,711,982	9.5
1970	3,731,386	18.4	1,921,031	9.5
1980	3,612,258	15.9	1,989,841	8.7
1990	4,158,212	16.7	2,148,463	8.6
1991	4,110,907	16.3	2,169,518	8.6
1992	4,065,014	15.9	2,175,613	8.5
1993	4,000,240	15.5	2,268,000	8.8
1994	3,952,767	15.2	2,278,994	8.8
1995	3,899,589	14.8	2,312,132	8.8
1996	3,891,494	14.7	2,314,690	8.7
1997	3,880,894	14.5	2,314,245	8.6
1998 (P)	3,944,046	14.6	2,338,070	8.7

(P) = provisional data. **NOTE:** Statistics cover only events occurring within the U.S. and exclude fetal deaths. Rates per 1,000 population; enumerated as of Apr. 1 for 1960 and 1970; estimated as of July 1 for all other years. Beginning 1970 statistics exclude births and deaths occurring to nonresidents of the U.S. Data include revisions.

M I L L E N N I U M   F A C T   B O X

Marriage and Divorce in the 1900s
Source: National Center for Health Statistics, U.S. Dept. of Health and Human Services

The marriage rate dipped during the Depression and peaked just after World War II; in 1998 the rate had fallen to 8.3 per 1,000, the lowest rate since 1932. The divorce rate has generally risen since the 1920s; it peaked at 5.3 per 1,000 in 1981, before declining somewhat.

The graph below shows marriage and divorce rates per 1,000 population since 1920.

Births and Deaths, by States and Regions, 1997-98

Source: National Center for Health Statistics, U.S. Dept. of Health and Human Services

Area	LIVE BIRTHS 1997 Number	1997 Rate	1998 Number	1998 Rate	DEATHS 1997 Number	1997 Rate	1998 Number	1998 Rate
New England.........	**172,126**	**12.9**	**177,435**	**13.2**	**116,718**	**8.7**	**121,957**	**9.1**
Maine	13,524	10.9	13,841	11.1	11,075	8.9	11,496	9.2
New Hampshire	14,456	12.3	13,472	11.4	9,488	8.1	8,911	7.5
Vermont	6,691	11.4	6,286	10.6	5,272	9.0	4,836	8.2
Massachusetts......	82,311	13.5	88,719	14.4	52,101	8.5	58,364	9.5
Rhode Island	12,355	12.5	12,442	12.6	9,784	9.9	9,602	9.7
Connecticut	42,789	13.1	42,675	13.0	28,998	8.9	28,748	8.8
Middle Atlantic	**537,939**	**14.1**	**510,888**	**13.3**	**361,153**	**9.5**	**346,153**	**9.0**
New York	282,389	15.6	249,069	13.7	161,159	8.9	153,175	8.4
New Jersey........	113,235	14.1	116,860	14.4	72,102	9.0	66,021	8.1
Pennsylvania	142,315	11.8	144,959	12.1	127,892	10.6	126,957	10.6
East North Central	**601,719**	**13.7**	**611,931**	**13.8**	**383,605**	**8.7**	**388,330**	**8.8**
Ohio	151,879	13.6	151,289	13.5	105,446	9.4	105,709	9.4
Indiana	69,218	11.8	78,525	13.3	47,043	8.0	46,333	7.9
Illinois	180,621	15.2	181,464	15.1	102,480	8.6	104,153	8.6
Michigan...........	133,627	13.7	133,262	13.6	83,534	8.5	86,292	8.8
Wisconsin..........	66,374	12.8	67,391	12.9	45,102	8.7	45,843	8.8
West North Central....	**255,580**	**13.8**	**255,056**	**13.6**	**170,204**	**9.2**	**171,710**	**9.2**
Minnesota	64,633	13.8	64,998	13.8	37,207	7.9	37,252	7.9
Iowa	36,868	12.9	35,648	12.5	26,179	9.2	27,569	9.6
Missouri	74,538	13.8	75,486	13.9	54,543	10.1	54,703	10.1
North Dakota	8,326	13.0	8,081	12.7	6,027	9.4	5,915	9.3
South Dakota	10,138	13.7	10,299	14.0	6,992	9.5	6,912	9.4
Nebraska	23,243	14.0	23,112	13.9	15,248	9.2	15,207	9.1
Kansas	37,834	14.6	37,432	14.2	24,008	9.3	24,152	9.2
South Atlantic.......	**661,656**	**13.7**	**686,371**	**14.0**	**440,432**	**9.1**	**449,772**	**9.2**
Delaware	10,135	13.9	10,298	13.8	6,424	8.8	6,676	9.0
Maryland	66,253	13.0	70,576	13.7	40,472	7.9	40,792	7.9
District of Columbia ..	8,327	15.7	8,373	16.0	5,984	11.3	5,694	10.9
Virginia	88,669	13.2	95,641	14.1	53,261	7.9	54,274	8.0
West Virginia	20,466	11.3	21,714	12.0	20,870	11.5	20,890	11.5
North Carolina	105,766	14.2	108,166	14.3	66,101	8.9	68,111	9.0
South Carolina	51,252	13.6	53,442	13.9	32,998	8.8	34,208	8.9
Georgia...........	118,365	15.8	122,618	16.0	59,314	7.9	60,788	8.0
Florida	192,423	13.1	195,543	13.1	155,008	10.6	158,339	10.6
East South Central	**230,843**	**14.1**	**232,742**	**14.1**	**160,352**	**9.8**	**164,097**	**10.0**
Kentucky..........	52,696	13.5	54,976	14.0	38,256	9.8	38,224	9.7
Tennessee	74,755	13.9	75,447	13.9	52,113	9.7	54,034	9.9
Alabama	60,921	14.1	62,306	14.3	42,704	9.9	43,989	10.1
Mississippi	42,471	15.6	40,013	14.5	27,279	10.0	27,850	10.1
West South Central ...	**465,914**	**15.7**	**500,649**	**16.7**	**230,942**	**7.8**	**245,603**	**8.2**
Arkansas	36,985	14.7	37,099	14.6	25,545	10.1	26,817	10.6
Louisiana	64,199	14.8	66,172	15.1	37,660	8.7	39,672	9.1
Oklahoma..........	47,985	14.5	50,978	15.2	33,687	10.2	33,750	10.1
Texas	316,745	16.3	346,400	17.5	134,050	6.9	145,364	7.4
Mountain	**270,129**	**16.4**	**272,551**	**16.2**	**120,632**	**7.3**	**125,716**	**7.5**
Montana	10,463	11.9	10,430	11.8	7,697	8.8	7,853	8.9
Idaho	17,979	14.9	19,464	15.8	9,014	7.4	9,269	7.5
Wyoming...........	6,387	13.3	6,363	13.2	3,737	7.8	3,883	8.1
Colorado...........	52,050	13.4	59,789	15.1	25,640	6.6	26,638	6.7
New Mexico	26,571	15.4	27,838	16.0	12,788	7.4	13,410	7.7
Arizona............	86,032	18.9	75,415	16.2	37,334	8.2	38,502	8.2
Utah	43,670	21.2	45,744	21.8	11,437	5.6	11,920	5.7
Nevada	26,977	16.1	27,508	15.7	12,985	7.7	14,241	8.2
Pacific	**674,380**	**15.7**	**695,552**	**16.0**	**309,909**	**7.2**	**317,304**	**7.3**
Washington.........	79,529	14.2	80,612	14.2	42,952	7.7	42,432	7.5
Oregon............	43,394	13.4	44,362	13.5	28,634	8.8	29,529	9.0
California	524,618	16.3	542,476	16.6	228,131	7.1	234,852	7.2
Alaska.............	9,710	15.9	10,898	17.7	2,465	4.0	2,480	4.0
Hawaii.............	17,129	14.4	17,204	14.4	7,727	6.5	8,011	6.7

Note: Data are provisional estimates, reported by state of residence. Figures include revisions, and so may differ from those previously published. Rates for births and deaths are per 1,000 population.

Birth Rates; Fertility Rates by Age of Mother, 1950-98

Source: National Center for Health Statistics, U.S. Dept. of Health and Human Services

	Birth rate[1]	Fertility rate[2]	AGE OF MOTHER 10-14 years	Total 15-19	15-19 years 15-17 years	18-19 years	20-24 years	25-29 years	30-34 years	35-39 years	40-44 years	45-49 years
					Live births per 1,000 women by age group							
1950....	24.1	106.2	1.0	81.6	40.7	132.7	196.6	166.1	103.7	52.9	15.1	1.2
1960....	23.7	118.0	0.8	89.1	43.9	166.7	258.1	197.4	112.7	56.2	15.5	0.9
1970....	18.4	87.9	1.2	68.3	38.8	114.7	167.8	145.1	73.3	31.7	8.1	0.5
1980....	15.9	68.4	1.1	53.0	32.5	82.1	115.1	112.9	61.9	19.8	3.9	0.2
1990....	16.7	70.9	1.4	59.9	37.5	88.6	116.5	120.2	80.8	31.7	5.5	0.2
1991....	16.3	69.6	1.4	62.1	38.7	94.4	115.7	118.2	79.5	32.0	5.5	0.2
1992....	15.9	68.9	1.4	60.7	37.8	94.5	114.6	117.4	80.2	32.5	5.9	0.3
1993....	15.5	67.6	1.4	59.6	37.8	92.1	112.6	115.5	80.8	32.9	6.1	0.3
1994....	15.2	66.7	1.4	58.9	37.6	91.5	111.1	113.9	81.5	33.7	6.4	0.3
1995....	14.8	65.6	1.3	56.8	36.0	89.1	109.8	112.2	82.5	34.3	6.6	0.3
1996....	14.7	65.3	1.2	54.4	33.8	86.0	110.4	113.1	83.9	35.3	6.8	0.3
1997....	14.5	65.0	1.1	52.3	32.1	83.6	110.4	113.8	85.3	36.1	7.1	0.4
1998 P ..	14.6	65.6	1.0	51.1	30.4	82.0	111.2	116.0	87.5	37.4	7.3	0.4

P = preliminary data. (1) Live births per 1,000 population. (2) Live births per 1,000 women 15-44 years of age.

Years of Life Expected at Birth, 1900-98

Source: National Center for Health Statistics, U.S. Dept. of Health and Human Services

Year[1]	ALL RACES Total	Male	Female	WHITE Total	Male	Female	BLACK Total	Male	Female
1900	47.3	46.3	48.3	47.6	46.6	48.7	NA	NA	NA
1910	50.0	48.4	51.8	50.3	48.6	52.0	NA	NA	NA
1920	54.1	53.6	54.6	54.9	54.4	55.6	NA	NA	NA
1930	59.7	58.1	61.6	61.4	59.7	63.5	NA	NA	NA
1940	62.9	60.8	65.2	64.2	62.1	66.6	NA	NA	NA
1950	68.2	65.6	71.1	69.1	66.5	72.2	NA	NA	NA
1960	69.7	66.6	73.1	70.6	67.4	74.1	NA	NA	NA
1970	70.8	67.1	74.7	71.7	68.0	75.6	64.1	60.0	68.3
1975	72.6	68.8	76.6	73.4	69.5	77.3	68.8	62.4	71.3
1980	73.7	70.0	77.5	74.4	70.7	78.1	68.1	63.8	72.5
1981	74.2	70.4	77.8	74.8	71.1	78.4	68.9	64.5	73.2
1982	74.5	70.9	78.1	75.1	71.5	78.7	69.4	65.1	73.6
1983	74.6	71.0	78.1	75.2	71.7	78.7	69.4	65.2	73.5
1984	74.7	71.2	78.2	75.3	71.8	78.7	69.5	65.3	73.6
1985	74.7	71.2	78.2	75.3	71.9	78.7	69.3	65.0	73.4
1986	74.8	71.3	78.3	75.4	72.0	78.8	69.1	64.8	73.4
1987	75.0	71.5	78.4	75.6	72.2	78.9	69.1	64.7	73.4
1988	74.9	71.5	78.3	75.6	72.3	78.9	68.9	64.4	73.2
1989	75.1	71.7	78.5	75.9	72.5	79.2	68.8	64.3	73.3
1990	75.4	71.8	78.8	76.1	72.9	79.4	69.1	64.5	73.6
1991	75.5	72.0	78.9	76.3	72.9	79.2	69.3	64.6	73.8
1992	75.5	72.1	78.9	76.4	73.0	79.5	69.6	65.0	73.9
1993	75.5	72.1	78.9	76.3	73.0	79.5	69.2	64.6	73.7
1994	75.7	72.4	79.0	76.5	73.3	79.6	69.5	64.9	73.9
1995	75.8	72.5	78.9	76.5	73.4	79.6	69.6	65.2	73.9
1996	76.1	73.1	79.1	76.8	73.9	79.7	70.2	66.1	74.2
1997	76.5	73.6	79.4	77.1	74.3	79.9	71.1	67.2	74.7
1998[P]	76.7	73.9	79.4	77.3	74.6	79.9	71.5	67.8	75.0

P = preliminary. NA = Not available. (1) Data prior to 1940 for death-registration states only.

U.S. Infant Mortality Rates, by Race and Sex, 1960-98

Source: National Center for Health Statistics, U.S. Dept. of Health and Human Services

Year	ALL RACES Total	Male	Female	WHITE Total	Male	Female	BLACK Total	Male	Female
1960	26.0	29.3	22.6	22.9	26.0	19.6	44.3	49.1	39.4
1970	20.0	22.4	17.5	17.8	20.0	15.4	32.6	36.2	29.0
1980	12.6	13.9	11.2	11.0	12.3	9.6	21.4	23.3	19.4
1981	11.9	13.1	10.7	10.5	11.7	9.2	20.0	21.7	18.3
1982	11.5	12.8	10.2	10.1	11.2	8.9	19.6	21.5	17.7
1983	11.2	12.3	10.0	9.7	10.8	8.6	19.2	21.1	17.2
1984	10.8	11.9	9.6	9.4	10.5	8.3	18.4	19.8	16.9
1985	10.6	11.9	9.3	9.3	10.6	8.0	18.2	19.9	16.5
1986	10.4	11.5	9.1	8.9	10.0	7.8	18.0	20.0	16.0
1987	10.1	11.2	8.9	8.6	9.6	7.6	17.9	19.6	16.0
1988	10.0	11.0	8.9	8.5	9.5	7.4	17.6	19.0	16.1
1989	9.8	10.8	8.8	8.1	9.0	7.1	18.6	20.0	17.2
1990	9.2	10.3	8.1	7.6	8.5	6.6	18.0	19.6	16.2
1991	8.9	10.0	7.8	7.3	8.3	6.5	17.6	19.4	15.7
1992	8.5	9.4	7.6	6.9	7.7	6.1	16.8	18.4	15.3
1993	8.4	9.3	7.4	6.8	7.6	6.0	16.5	18.3	14.7
1994	8.0	8.8	7.2	6.6	7.2	5.9	15.8	17.5	14.1
1995	7.6	8.3	6.8	6.3	7.0	5.6	15.1	16.3	13.9
1996	7.3	8.0	6.6	6.1	6.7	5.4	14.7	16.0	13.3
1997	7.2	8.0	6.5	6.0	6.7	5.4	14.2	15.5	12.8
1998[1]	7.2	NA	NA	6.0	NA	NA	14.1	NA	NA

Note: Rates per 1,000 live births. NA = Not available. (1) Preliminary data.

U.S. Infant Deaths and Infant Mortality Rates, by Age and Cause, 1997-98

Source: National Center for Health Statistics, U.S. Dept. of Health and Human Services

AGE	1998[1] Number	Rate[2]	1997 Number	Rate[2]	CAUSE	1998[1] Number	Rate[2]	1997 Number	Rate[2]
Under 1 year	**28,486**	**722.3**	**28,045**	**722.6**	Birth trauma	191	4.8	185	4.8
Under 28 days	18,832	477.5	18,524	477.3	Intrauterine hypoxia and birth asphyxia	459	11.6	452	11.6
28 days to 11 months	9,654	244.8	9,521	245.3	Respiratory distress syndrome	1,328	33.7	1,301	33.5
CAUSE					Other conditions originating around the time of birth	7,143	181.1	7,072	182.2
Certain gastrointestinal diseases	305	7.7	276	7.1	Sudden infant death syndrome	2,529	64.1	2,991	77.1
Pneumonia and influenza	400	10.1	421	10.8					
Congenital anomalies	6,266	158.9	6,178	159.2					
Disorders relating to short gestation and unspecified low birthweight	4,011	101.7	3,925	101.1	All other causes	5,856	148.5	5,244	135.1

Note: Because of rounding of estimates, figures may not add to totals. (1) Data are preliminary. (2) Rates per 100,000 live births.

The 10 Leading Causes of Death, 1998[1]

Source: National Center for Health Statistics, U.S. Dept. of Health and Human Services

	Number	Death rate[2]	Percentage of total deaths
ALL CAUSES	**2,338,075**	**865.0**	**100.0**
1. Heart disease	724,269	268.0	31.0
2. Cancer	538,947	199.4	23.1
3. Stroke	158,060	58.5	6.8
4. Chronic obstructive lung diseases and allied conditions	114,381	42.3	4.9
5. Pneumonia and influenza	94,828	35.1	4.1
6. Accidents and adverse effects	93,207	34.5	4.0
7. Diabetes mellitus	64,574	23.9	2.8
8. Suicide	29,264	10.8	1.3
9. Kidney disease	26,295	9.7	1.1
10. Chronic liver disease and cirrhosis	24,936	9.2	1.1

(1) Figures are based on weighted data rounded to the nearest individual. Data are preliminary and may vary somewhat from other sources. (2) Per 100,000 population.

Suicides in the U.S., by Age, Race, and Sex, 1997

Source: National Center for Health Statistics, U.S. Dept. of Health and Human Services

	All ages	1-14 yrs.	15-24 yrs.	25-34 yrs.	35-44 yrs.	45-54 yrs.	55-64 yrs.	65-74 yrs.	75-84 yrs.	85 yrs. & over	Age not stated
All races, both sexes[1]	**30,535**	**307**	**4,186**	**5,672**	**6,730**	**4,948**	**2,946**	**2,663**	**2,260**	**805**	**18**
Male	24,492	233	3,559	4,684	5,223	3,697	2,331	2,183	1,895	671	16
Female	6,043	74	627	988	1,507	1,251	615	480	365	134	2
White, both sexes	**27,513**	**255**	**3,456**	**4,887**	**6,109**	**4,615**	**2,739**	**2,499**	**2,169**	**771**	**13**
Male	22,042	194	2,941	4,026	4,720	3,455	2,166	2,057	1,823	648	12
Female	5,471	61	515	861	1,389	1,160	573	442	346	123	1
Black, both sexes	**2,103**	**40**	**513**	**567**	**453**	**218**	**131**	**107**	**52**	**17**	**5**
Male	1,764	28	447	492	373	164	107	89	45	15	4
Female	339	12	66	75	80	54	24	18	7	2	1

Note: Data are provisional, estimated from a 10% sample of deaths. Because of rounding of estimates, figures may not add to totals. (1) "All races" includes races other than white and black.

U.S. Abortions, by State, 1992-96

Source: Alan Guttmacher Institute, New York, NY

	Number of reported abortions[1]			Rate per 1,000 women[2]			% change
	1992	1995	1996	1992	1995	1996	1992-96
TOTAL U.S.	**1,528,930**	**1,363,690**	**1,365,730**	**25.9**	**22.9**	**22.9**	**−12**
Alabama	17,450	14,580	15,150	18.2	15.0	15.6	−15
Alaska	2,370	1,990	2,040	16.5	14.2	14.6	−11
Arizona	20,600	18,120	19,310	24.1	19.1	19.8	−18
Arkansas	7,130	6,010	6,200	13.5	11.1	11.4	−15
California	304,230	240,240	237,830	42.1	33.4	33.0	−22
Colorado	19,880	15,690	18,310	23.6	18.0	20.9	−12
Connecticut	19,720	16,680	16,230	26.2	23.0	22.5	−14
Delaware	5,730	5,790	4,090	35.2	34.4	24.1	−32
District of Columbia	21,320	21,090	20,790	138.4	151.7	154.5	12
Florida	84,680	87,500	94,050	30.0	30.0	32.0	7
Georgia	39,680	36,940	37,320	24.0	21.2	21.1	−12
Hawaii	12,190	7,510	6,930	46.0	29.3	27.3	−41
Idaho	1,710	1,500	1,600	7.2	5.8	6.1	−15
Illinois	68,420	68,160	69,390	25.4	25.6	26.1	3
Indiana	15,840	14,030	14,850	12.0	10.6	11.2	−7
Iowa	6,970	6,040	5,780	11.4	9.8	9.4	−17
Kansas	12,570	10,310	10,630	22.4	18.3	18.9	−16
Kentucky	10,000	7,770	8,470	11.4	8.8	9.6	−16
Louisiana	13,600	14,820	14,740	13.4	14.7	14.7	10
Maine	4,200	2,690	2,700	14.7	9.6	9.7	−34
Maryland	31,260	30,520	31,310	26.4	25.6	26.3	0
Massachusetts	40,660	41,190	41,160	28.4	29.2	29.3	3
Michigan	55,580	49,370	48,780	25.2	22.6	22.3	−11
Minnesota	16,180	14,910	14,660	15.6	14.2	13.9	−11
Mississippi	7,550	3,420	4,490	12.4	5.5	7.2	−42
Missouri	13,510	10,540	10,810	11.6	8.9	9.1	−21
Montana	3,300	3,010	2,900	18.2	16.2	15.6	−14
Nebraska	5,580	4,360	4,460	15.7	12.1	12.3	−22
Nevada	13,300	15,600	15,450	44.2	46.7	44.6	1
New Hampshire	3,890	3,240	3,470	14.6	12.0	12.7	−13
New Jersey	55,320	61,130	63,100	31.0	34.5	35.8	16
New Mexico	6,410	5,450	5,470	17.7	14.4	14.4	−19
New York	195,390	176,420	167,600	46.2	42.8	41.1	−11
North Carolina	36,180	34,600	33,550	22.4	21.0	20.2	−10
North Dakota	1,490	1,330	1,290	10.7	9.6	9.4	−13
Ohio	49,520	40,940	42,870	19.5	16.2	17.0	−13
Oklahoma	8,940	9,130	8,400	12.5	12.9	11.8	−5
Oregon	16,060	15,590	15,050	23.9	22.6	21.6	−10
Pennsylvania	49,740	40,760	39,520	18.6	15.5	15.2	−18
Rhode Island	6,990	5,720	5,420	30.0	25.5	24.4	−19
South Carolina	12,190	11,020	9,940	14.2	12.9	11.6	−19
South Dakota	1,040	1,040	1,030	6.8	6.6	6.5	−4
Tennessee	19,060	18,240	17,990	16.2	15.2	14.8	−8
Texas	97,400	89,240	91,270	23.1	20.5	20.7	−10
Utah	3,940	3,740	3,700	9.3	8.1	7.8	−16
Vermont	2,900	2,420	2,300	21.2	17.9	17.1	−19
Virginia	35,020	31,480	29,940	22.7	20.0	18.9	−16

	Number of reported abortions[1]			Rate per 1,000 women[2]			% change
	1992	1995	1996	1992	1995	1996	1992-96
Washington	33,190	25,190	26,340	27.7	20.2	20.9	−24
West Virginia	3,140	3,050	2,610	7.7	7.6	6.6	−14
Wisconsin	15,450	13,300	14,160	13.6	11.6	12.3	−9
Wyoming	460	280	280	4.3	2.7	2.7	−37

(1) Rounded to the nearest 10. (2) Only women aged 15-44 years old.

Contraceptive Use in the U.S., 1995

Source: National Center for Health Statistics, U.S. Dept. of Health and Human Services

Age	15-44	15-19	20-24	25-29	30-34	35-39	40-44
				Percent distribution			
Using contraception	64.2	29.8	63.4	69.3	72.7	72.9	71.5
Female sterilization	17.8	0.1	2.5	11.8	21.4	29.8	35.6
Male sterilization	7.0	—	0.7	3.1	7.6	13.6	14.5
Pill	17.3	13.0	33.1	27.0	20.7	8.1	4.2
Implant	0.9	0.8	2.4	1.4	0.5	0.2	0.1
Injectable	1.9	2.9	3.9	2.9	1.3	0.8	0.2
Intrauterine device (IUD)	0.5	—	0.2	0.5	0.6	0.7	0.9
Diaphragm	1.2	0.0	0.4	0.6	1.7	2.2	1.9
Condom	13.1	10.9	16.7	16.8	13.4	12.3	8.8
Female condom	0.0	—	0.1	—	—	—	—
Periodic abstinence	1.5	0.4	0.6	1.2	2.3	2.1	1.8
Natural family planning.....	0.2	—	0.1	0.2	0.3	0.4	0.2
Withdrawal	2.0	1.2	2.1	2.6	2.1	2.3	1.4
Other methods[1]	1.0	0.3	0.9	1.2	1.3	0.9	1.8

(1) Includes morning-after pill, foam, cervical cap, Today sponge, suppository, jelly or cream (without diaphragm), and other methods not shown separately.

U.S. Median Age at First Marriage, 1900-98

Source: Bureau of the Census, U.S. Dept. of Commerce

Year[1]	Men	Women	Year[1]	Men	Women	Year[1]	Men	Women
1998	26.7	25.0	1991	26.3	..... 24.1	1960........	..22.8	20.3
1997	26.8	25.0	1990	26.1	..... 23.9	1950........	..22.8	20.3
1996	27.1	24.8	1985	25.5	..... 23.3	1940........	..24.3	21.5
1995	26.9	24.5	1980	24.7	..... 22.0	1930........	..24.3	21.3
1994	26.7	24.5	1975	23.5	..... 21.1	1920........	..24.6	21.2
1993	26.5	24.5	1970	23.2	..... 20.8	1910........	..25.1	21.6
1992	26.5	24.4	1965	22.8	..... 20.6	1900........	..25.9	21.9

(1) Figures after 1940 based on Current Population Survey data; figures for 1900-40 based on decennial censuses.

Interracial Married Couples in the U.S., 1960-98

Source: Bureau of the Census, U.S. Dept. of Commerce; numbers in thousands

	TOTAL MARRIED COUPLES	INTERRACIAL MARRIED COUPLES				
			Black/White			
Year[1]		Total interracial	Black husband/ White wife	White husband/ Black wife	White/Other race[2]	Black/Other race[2]
1998	55,305	1,348	210	120	975	43
1997	54,666	1,264	201	110	896	57
1996	54,664	1,260	220	117	884	39
1995	54,937	1,392	206	122	988	76
1990	53,256	964	150	61	720	33
1980	49,714	651	122	45	450	34
1970	44,598	310	41	24	233	12
1960	40,491	149	25	26	90	7

(1) Data from Mar. of year, except for 1970 and 1960, which are from decennial census. (2) Any race other than White or Black.

Cigarette Use in the U.S., 1985-98

Source: Substance Abuse and Mental Health Services Administration (SAMHSA), U.S. Dept. of Health and Human Services

(percentage reporting use in the month prior to the survey; figures exclude persons under age 12)

Characteristic	1985	1996	1997	1998	Characteristic	1985	1996	1997	1998
TOTAL	38.7	28.9	29.6	27.7	Age group				
Sex....................					12-17.................	29.4	18.3	19.9	18.2
Male	43.4	31.1	31.2	29.7	18-25.................	47.4	38.3	40.6	41.6
Female	34.5	26.7	28.2	25.7	26-34	45.7	35.0	33.7	32.5
Race/Ethnicity					35 and older	35.5	27.0	27.9	25.1
White	38.9	29.8	30.5	27.9	Education[2]				
Black	38.0	30.4	29.8	29.4	Non-high school graduate .	37.3	36.5	40.0	36.9
Hispanic	40.0	24.7	27.4	25.8	High school graduate	37.0	36.8	36.1	34.3
Other	(1)	17.2	18.8	23.8	Some college..........	32.6	27.5	29.5	29.2
					College graduate	23.0	17.5	17.1	15.2

(1) No estimate reported. (2) Estimates for Education are for persons aged 18 and older.

Drug Use in the General U.S. Population, 1998

Source: Substance Abuse and Mental Health Services Administration (SAMHSA), U.S. Dept. of Health and Human Services

According to the Substance Abuse and Mental Health Services Administration's 1998 National Household Survey on Drug Abuse, an estimated 78 mil Americans 12 years of age and older (36%) had used an illicit drug at least once during their lifetimes, 11% used one during the previous year, and 6% used one in the month before the survey was conducted. Among those 25 years of age and under, an estimated 1.7 mil used cocaine (including crack) and 9.9 mil used marijuana at least once within the previous year. Among those 26 years of age and over, 2.1 mil used cocaine (including crack) and 8.8 mil used marijuana at least once within the previous year.

The Substance Abuse and Mental Health Services Administration's Drug Abuse Warning Network (DAWN) reported 527,058 drug-related episodes in hospital emergency departments nationwide in 1997. The number of these drug-related episodes increased 2% from 514,347 in 1996.

Drug Use: America's Middle and High School Students, 1998

Source: *Monitoring the Future,* Univ. of Michigan Inst. for Social Research and National Inst. on Drug Abuse

Use of illicit drugs by American young people began to decrease in 1998, according to the University of Michigan's 24th annual survey of high school seniors and 8th annual survey of 8th and 10th graders.

The overall decline in drug use followed 6 years of steady increases. It was the 2d year of decline for 8th graders, and the 1st for 10th and 12th graders. Despite the downward trend, the proportion of 8th graders taking illicit drugs in the 12 months prior to the survey (21%) was still nearly double the 1991 level (11%). The proportion of 10th graders using illicit drugs in the prior 12 months has increased by two-thirds (from 21% to 35%) since 1991, and among 12th graders the proportion has increased by nearly half (from 29% to 41%).

Marijuana remained the most commonly used illegal drug among the 3 grade levels. In 1998, the proportion of students that reported using marijuana in the past year declined to just below 17% of 8th graders, 31% of 10th graders, and 38% of 12th graders. Use of marijuana on a daily basis remained level (at 1.1%) among 8th graders and decreased

in the higher grades. About 1 in 18 high school seniors (5.6%) and 1 in 28 10th graders (3.6%) were daily users.

Use of LSD and other hallucinogens decreased in all 3 grades. Use of stimulants declined as well. While the use of inhalants by 8th and 12th graders continued the decline that began in 1996, use by 10th graders leveled off. Although heroin use remained rather low, levels of use in 1998 increased slightly for 8th and 10th graders; use by 12th graders decreased slightly. Although the use of alcohol decreased for all 3 grades in 1998, levels remained high.

Prevalence of cigarette smoking decreased for the 3 grade levels. About 9% of 8th graders, 16% of 10th graders, and 22% of 12th graders reported having smoked daily during the 30 days before they responded to the survey.

In 1998, about 15,800 seniors, 15,400 10th graders, and 18,700 8th graders from 422 public and private secondary schools participated in the survey. It should be noted that the surveys missed the 3-6% of a class group that drops out of school early and about 9-17% who were absentees. These populations tend to have higher rates of drug use overall.

Drug Use: America's High School Seniors, 1975-98

Source: *Monitoring the Future,* Univ. of Michigan Inst. for Social Research and National Inst. on Drug Abuse

PERCENTAGE EVER USED

	Class of 1975	Class of 1980	Class of 1990	Class of 1991	Class of 1992	Class of 1993	Class of 1994	Class of 1995	Class of 1996	Class of 1997	Class of 1998	'97-'98 change
Marijuana/hashish	47.3	60.3	40.7	36.7	32.6	35.3	38.2	41.7	44.9	49.6	49.1	−0.5
Inhalants	NA	11.9	18.0	17.6	16.6	17.4	17.7	17.4	16.6	16.1	15.2	−0.9
Inhalants adjusted[1] . . .	NA	17.3	18.5	18.0	17.0	17.7	18.3	17.8	17.5	16.9	16.5	−0.4
Amyl & butyl nitrites . . .	NA	11.1	2.1	1.6	1.5	1.4	1.7	1.5	1.8	2.0	2.7	+0.7
Hallucinogens.	16.3	13.3	9.4	9.6	9.2	10.9	11.4	12.7	14.0	15.1	14.1	−1.0
Hallucinogens adjusted[2]	NA	15.6	9.7	10.0	9.4	11.3	11.7	13.1	14.5	15.4	14.4	−1.0
LSD	11.3	9.3	8.7	8.8	8.6	10.3	10.5	11.7	12.6	13.6	12.6	−1.0
PCP	NA	9.6	2.8	2.9	2.4	2.9	2.8	2.7	4.0	3.9	3.9	0
Cocaine	9.0	15.7	9.4	7.8	6.1	6.1	5.9	6.0	7.1	8.7	9.3	+0.6
Crack	NA	NA	3.5	3.1	2.6	2.6	3.0	3.0	3.3	3.9	4.4	+0.5
Heroin[3]	2.2	1.1	1.3	0.9	1.2	1.1	1.2	1.6	1.8	2.1	2.0	−0.1
Other opiates[4]	9.0	9.8	8.3	6.6	6.1	6.4	6.6	7.2	8.2	9.7	9.8	+0.1
Stimulants[4,5].	22.3	26.4	17.5	15.4	13.9	15.1	15.7	15.3	15.3	16.5	16.4	−0.1
Sedatives[4]	18.2	14.9	7.5	6.7	6.1	6.4	7.3	7.6	8.2	8.7	9.2	+0.5
Barbiturates[4]	16.9	11.0	6.8	6.2	5.5	6.3	7.0	7.4	7.6	8.1	8.7	+0.6
Methaqualone[4].	8.1	9.5	2.3	1.3	1.6	0.8	1.4	1.2	2.0	1.7	1.6	−0.1
Tranquilizers[4]	17.0	15.2	7.2	7.2	6.0	6.4	6.6	7.1	7.2	7.8	8.5	+0.7
Alcohol[6]	90.4	93.2	89.5	88.0	87.5	87.0	80.4	80.7	79.2	81.7	81.4	−0.3
Cigarettes	73.6	71.0	64.4	63.1	61.8	61.9	62.0	64.2	63.5	65.4	65.3	−0.1

NA=Not available. (1) Adjusted for underreporting of amyl and butyl nitrites. (2) Adjusted for underreporting of PCP. (3) Reflects use with or without injection. (4) Includes only drug use that was not under a doctor's orders. (5) Data for 1990-98 are not directly comparable to prior years. (6) Data for 1994–98 are not directly comparable to prior years.

Alcohol Use by 8th and 12th Graders, 1980-98

Source: *Monitoring the Future,* Univ. of Michigan Inst. for Social Research and National Inst. on Drug Abuse

	1980	1987	1988	1989	1990	1991	1992	1993	1994	1995	1996	1997	1998
ALCOHOL[1]	*Percent using alcohol in the month before the survey*												
All 12th graders	72.0	66.4	63.9	60.0	57.1	54.0	51.3	48.6	50.1	51.3	50.8	52.7	52.0
Male	77.4	69.9	68.0	65.1	61.3	58.4	55.8	54.2	55.5	55.7	54.8	56.2	57.3
Female	66.8	63.1	59.9	54.9	52.3	49.0	46.8	43.4	45.2	47.0	46.9	48.9	46.9
White	75.8	71.8	69.5	65.3	62.2	57.7	56.0	53.4	54.8	54.8	54.7	57.9	57.6
Black	47.7	38.5	40.9	38.1	32.9	34.4	29.5	35.1	33.1	37.4	35.7	33.1	33.6
All 8th graders	—	—	—	—	—	25.1	26.1	24.3	25.5	24.6	26.2	24.5	23.0
Male	—	—	—	—	—	26.3	26.3	25.3	26.5	25.0	26.6	25.2	24.0
Female	—	—	—	—	—	23.8	25.9	28.7	24.7	24.0	25.8	23.9	21.9
White	—	—	—	—	—	26.0	27.3	25.1	25.4	25.4	27.7	25.7	24.0
Black	—	—	—	—	—	17.8	19.2	17.7	20.2	17.3	19.0	16.9	15.4
HEAVY ALCOHOL[2]	*Percent heavily using the 2 weeks before the survey*												
All 12th graders	41.2	37.5	34.7	33.0	32.2	29.8	27.9	27.5	28.2	29.8	30.2	31.3	31.5
Male	52.1	46.1	43.0	41.2	39.1	37.8	35.6	34.6	37.0	36.9	37.0	37.9	39.2
Female	30.5	29.2	26.5	24.9	24.4	21.2	20.3	20.7	20.2	23.0	23.5	24.4	24.0
White	44.6	41.2	38.8	36.9	36.2	32.9	31.3	31.3	31.7	32.9	34.0	36.1	36.6
Black	17.0	15.5	14.9	16.6	11.6	11.8	10.8	14.6	14.2	15.5	15.1	12.0	12.7
All 8th graders	—	—	—	—	—	12.9	13.4	13.5	14.5	14.5	15.6	14.5	13.7
Male	—	—	—	—	—	14.3	13.9	14.8	16.0	15.1	16.5	15.3	14.4
Female	—	—	—	—	—	11.4	12.8	12.3	13.0	13.9	14.5	13.5	12.7
White	—	—	—	—	—	12.6	12.9	12.4	13.4	14.5	15.7	14.6	13.5
Black	—	—	—	—	—	9.9	9.3	11.9	11.8	10.0	10.9	8.8	9.1

(—) = Data not available. **Note:** *Monitoring the Future* study excludes high school dropouts (about 3-6% of the class group, according to a 1996 report) and absentees (about 16-17% of 12th graders and about 9-10% of 8th graders). High school dropouts and absentees have higher alcohol usage than those included in the survey. (1) In 1993 the alcohol question was changed to indicate that a "drink" meant "more than a few sips." (2) Five or more drinks in a row at least once in the prior 2-week period.

Principal Types of Accidental Deaths in the U.S., 1970-98

Source: National Safety Council

Year	Motor vehicle	Falls	Poison (solid, liquid)	Drowning	Fires, burns	Ingestion of food, object	Firearms	Poison (gases)
1970	54,633	16,926	3,679	7,860	6,718	2,753	2,406	1,620
1980	53,172	13,294	3,089	7,257	5,822	3,249	1,955	1,242
1985	45,901	12,001	4,091	5,316	4,938	3,551	1,649	1,079
1990	46,814	12,313	5,055	4,685	4,175	3,303	1,416	748
1991	43,536	12,662	5,698	4,818	4,120	3,240	1,441	736
1992	40,982	12,646	6,449	3,542	3,958	3,182	1,409	633
1993	41,893	13,141	7,877	3,807	3,900	3,160	1,521	660
1994	42,524	13,450	8,309	3,942	3,986	3,065	1,356	685
1995	43,363	13,986	8,461	4,350	3,761	3,185	1,225	611
1996[1]	43,649	14,986	8,872	3,959	3,206	3,206	1,134	638
1997[1]	42,400	15,300	8,400	3,800	4,000	3,100	1,100	600
1998[2]	41,200	16,600	8,400	4,100	3,700	3,200	900	600
Death rates per 100,000 population								
1970	26.8	8.3	1.8	3.9	3.3	1.4	1.2	0.8
1980	23.4	5.9	1.4	3.2	2.6	1.4	0.9	0.5
1985	19.3	5.0	1.7	2.2	2.1	1.5	0.7	0.5
1990	18.8	4.9	2.0	1.9	1.7	1.3	0.6	0.3
1991	17.3	5.0	2.3	1.8	1.6	1.3	0.6	0.3
1992	16.1	5.0	2.5	1.4	1.6	1.2	0.6	0.2
1993	16.3	5.1	3.1	1.5	1.5	1.2	0.6	0.3
1994	16.3	5.2	3.2	1.5	1.5	1.2	0.5	0.3
1995	16.5	5.3	3.2	1.7	1.4	1.2	0.5	0.2
1996[1]	16.5	5.6	3.3	1.5	1.4	1.2	0.4	0.2
1997[1]	15.8	5.7	3.1	1.4	1.5	1.2	0.4	0.2
1998[2]	15.2	6.1	3.1	1.5	1.4	1.2	0.3	0.2

Note: There were 13,500 other accidental deaths in 1998; the most frequently occurring types involved medical and surgical complications, machinery, air transport, water transport (except drownings), mechanical suffocation, and excessive cold. (1) Revised figures. (2) Preliminary figures.

U.S. Motor Vehicle Accidents

Source: National Safety Council

Motor vehicle deaths in the U.S. decreased 3% in 1998 compared to 1997. Of the 185,500,000 licensed drivers in 1998, about 93.7 mil (50.5%) were men and 91.8 mil (49.5%) were women.

Male drivers were involved in more fatal accidents than female drivers in 1998. About 40,800 men and 15,300 women drivers were involved in fatal accidents.

About 12.7 mil male drivers and 8.6 mil female drivers were involved in all types of accidents in 1998. However, since males account for about 63% of the miles driven each year, according to the latest estimates, and females for 37%, women have higher accident involvement rates. At least part of the difference in accident involvement rates between men

and women may be due to differences in the time, place, and circumstance of driving experienced by both groups of drivers. Accident rates were 77 per 10 million miles driven for men and 90 per 10 million miles driven for women.

About 39% of all traffic fatalities in 1997 involved an intoxicated or alcohol-impaired driver or nonoccupant. Of these 16,189 alcohol-related traffic fatalities, an estimated 12,704 occurred in accidents in which a driver or nonoccupant was intoxicated, and the remainder involved a driver or nonmotorist who had been drinking but was not legally intoxicated. Alcohol was also a factor in about 7% of all traffic accidents, both fatal and nonfatal, in 1997. In 1987 alcohol-related fatalities accounted for 51% of all traffic deaths.

	Death total 1998	Percentage change from 1996	Death rate 1998[1]		Death total 1998	Percentage change from 1996	Death rate 1998[1]
All motor vehicle accidents	41,200	−3	15.2	Noncollision accidents	4,100	−2	1.5
Collision between motor vehicles	19,500	−6	7.2	Collision with pedalcycle	700	−13	0.3
				Collision with railroad train	400	0	0.1
Collision with fixed object	10,500	+2	3.9	Other collision (animal, animal-drawn vehicles) . . .	100	0	(2)
Pedestrian accidents	5,900	0	2.2				

(1) Deaths per 100,000 population. (2) Death rate was less than 0.05.

Improper Driving Reported in Accidents, 1997-98

Source: National Safety Council

Type	Percentage of fatal accidents		Percentage of injury accidents		Percentage of all accidents	
	1997	1998	1997	1998	1997	1998
Improper driving	**64.6**	**60.0**	**73.8**	**62.3**	**66.1**	**61.5**
Speed too fast or unsafe	16.8	16.8	13.2	12.8	11.0	13.3
Right of way	15.0	16.0	22.2	21.9	16.8	18.4
Failed to yield	10.1	11.1	15.9	15.9	12.7	13.8
Passed stop sign	3.1	2.6	2.3	4.2	1.5	3.1
Disregarded signal	1.8	2.3	4.0	1.8	2.6	1.5
Drove left of center	7.5	7.3	2.1	1.7	1.5	1.6
Improper overtaking	1.1	1.2	0.8	0.6	1.1	1.0
Made improper turn	3.7	4.1	3.6	4.0	4.0	5.0
Followed too closely	1.2	0.5	11.3	4.3	10.6	5.4
Other improper driving	19.3	14.1	20.6	17.0	21.1	16.8
No improper driving stated	**35.4**	**40.0**	**26.2**	**37.7**	**33.9**	**38.5**

Note: Based on reports from 12 state traffic authorities. When a driver was under the influence of alcohol or drugs, the accident was considered a result of the driver's physical condition—not a driving error. For this reason, accidents in which the driver was reported to be under the influence are classified under "no improper driving."

896   Vital Statistics — Firearm, Home Accident Deaths; Airline Fatalities; Cost of Injuries

Deaths in the U.S. Involving Firearms, by Age, 1996

Source: National Safety Council

	All ages	Under 5	5-14	15-24	25-44	45-64	65-74	75 & over
Total firearms deaths[1] ...	**33,750**	**88**	**604**	**8,697**	**13,292**	**6,341**	**2,345**	**2,383**
Male...............	28,904	49	462	7,795	11,119	5,270	2,056	2,153
Female..............	4,846	39	142	902	2,173	1,071	289	230
Accidents	**1,134**	**17**	**121**	**401**	**332**	**165**	**60**	**38**
Male...............	1,004	11	108	370	288	141	53	33
Female..............	130	6	13	31	44	24	7	5
Suicides	**18,166**	**0**	**162**	**2,724**	**6,537**	**4,549**	**2,034**	**2,160**
Male...............	15,808	0	129	2,424	5,549	3,857	1,831	2,018
Female..............	2,358	0	33	300	988	692	203	142
Homicides...........	**14,037**	**68**	**311**	**5,428**	**6,270**	**1,562**	**235**	**163**
Male...............	11,735	37	215	4,867	5,155	1,220	159	82
Female..............	2,302	31	96	561	1,115	342	76	81
Undetermined[2]	**413**	**3**	**10**	**144**	**153**	**65**	**16**	**22**
Male...............	357	1	10	134	127	52	13	20
Female..............	56	2	0	10	26	13	3	2

(1) Figures exclude firearms deaths by legal intervention. These deaths totaled 290 in 1996. (2) "Undetermined" means that the intention involved (whether accident, suicide, or homicide) could not be determined.

Home Accident Deaths in the U.S., 1950-98

Source: National Safety Council

Year	Total	Falls	Poison (solid, liquid)	Fires, burns[1]	Suffoc., ingesting object	Firearms	Suffoc., mechanical	Poison (gases)	All other
1950.........	29,000	14,800	1,300	5,000	([2])	950	1,600	1,250	4,100
1960.........	28,000	12,300	1,350	6,350	1,850	1,200	1,500	900	2,550
1970.........	27,000	9,700	3,000	5,600	1,800[3]	1,400[3]	1,100[3]	1,100	3,300[3]
1980.........	22,800	7,100	2,500	4,800	2,000	1,100	500	700	4,100[4]
1990.........	21,500	6,700	4,000	3,400	2,300	800	600	500	3,200
1991.........	22,100	6,900	4,500	3,400	2,200	800	700	500	3,100
1992.........	24,000	7,700	4,800	3,700	1,500	1,000	700	400	4,200
1993.........	26,100	7,900	6,000	3,700	1,700	1,100	700	500	4,500
1994.........	26,300	8,100	6,300	3,700	1,600	900	800	500	4,400
1995[5].........	27,200	8,400	6,600	3,500	1,500	900	800	400	5,100
1996[5]........	27,500	9,000	6,800	3,500	1,500	800	800	500	4,500
1997[5]........	27,500	9,800	6,300	3,700	1,400	800	1,000	500	4,000
1998[6]........	28,200	10,700	6,300	3,300	1,500	700	800	500	4,400

(1) Includes deaths resulting from conflagration, regardless of nature of injury. (2) Included under "All other" category. (3) Data for this year and later not comparable with earlier data because of classification changes. (4) Includes about 1,000 deaths attributed to summer heat wave. (5) Revised figures. The National Safety Council adopted the count from the Bureau of Labor Statistics Census of Fatal Occupational Injuries for all work-related unintentional injuries, retroactive to 1992 data. (6) Data for 1998 are preliminary.

Worldwide Airline Fatalities, 1980-98

Source: National Safety Council

Year	Aircraft accidents[1]	Passenger deaths	Death rate[2]	Year	Aircraft accidents[1]	Passenger deaths	Death rate[2]
1980............	22	814	0.14	1990	22	440	0.04
1981............	21	362	0.06	1991	25	510	0.05
1982............	26	764	0.13	1992	25	990	0.09
1983............	20	809	0.13	1993	31	801	0.07
1984............	16	223	0.03	1994	24	732	0.06
1985............	22	1,066	0.15	1995	22	557	0.04
1986............	17	331	0.04	1996	22	1,132	0.08
1987............	24	890	0.10	1997[3]............	27	930	0.06
1988............	25	699	0.08	1998[4]............	22	909	0.06
1989............	27	817	0.08				

(1) Involving 1 or more fatalities only. (2) Passenger deaths per 100 mil passenger mi. (3) Revised. (4) Preliminary.

Cost of Unintentional Injuries in the U.S., 1998

Source: National Safety Council, estimates

The cost of. . .	is equivalent to . . .	
. . .all injuries[1] ($480.5 bil)	*or*	58 cents of every dollar paid in 1998 federal personal income taxes
		59 cents of every dollar spent on food in the U.S. in 1998.
. . . motor vehicle accidents ($191.6 bil)	*or*	purchasing 840 gallons of gasoline per registered vehicle in the U.S.
		more than 19 times greater than the combined profits reported by Exxon, Mobil, and Chevron in 1998.
. . . work injuries ($125.1 bil)	*or*	45 cents of every dollar of 1998 corporate dividends to stockholders
		17 cents of every dollar of 1998 pre-tax corporate profits.
. . . home injuries ($113.5 bil)	*or*	an $89,300 rebate on each new single-family home built in 1998
		52 cents of every dollar of property taxes paid in 1998.
. . . public injuries[2] ($66.3 bil)	*or*	a $6.9 million grant to each public library in the U.S.
		a $79,900 bonus for each police officer and firefighter.

(1) Duplication between motor vehicle accidents and work injuries, which amounted to $16 bil, was eliminated in total of all injuries. (2) Any injuries that occur in public places or places used in a public way and not involving motor vehicles.

U.S. Fires, 1998
Source: National Fire Protection Assn.

Fires
- Public fire departments responded to 1,755,500 fires in 1998, a decrease of 2% from 1997.
- There were 517,500 structure fires in 1998, a decrease of 6% from the 1997 figure.
- 74% of all structure fires, or 381,500 fires, occurred in residential properties.
- There were 381,000 vehicle fires in 1998, a decrease of 4% from the previous year.
- There were 857,000 fires in outside properties, a slight increase of 1% from 1997.
- The South had the highest fire incident rate in the country, with 7.8 fires per 1,000 population.

Civilian deaths
- There were 4,035 civilian fire deaths in 1998, a slight decrease of 0.4% from 1997.
- The number of deaths from fire in the home decreased by 4%, to 3,220.
- About 80% of all fire deaths occurred in the home.
- The South had the highest regional fire death rate, with 18.4 civilian deaths per million population.
- Nationwide, someone died in a fire every 130 minutes.

Civilian injuries
- There were an estimated 23,100 civilian fire injuries in 1998, a decrease of 3% from 1997. This estimate is traditionally low because of underreporting of civilian fire injuries to the fire service.
- Residential properties were the site of 17,175 civilian fire injuries, or 74% of injuries overall; 2,250 injuries, or 10%, occurred in nonresidential structure fires.
- The Northeast had the highest regional injury rate in the U.S., with 115.1 civilian injuries per million population. The next highest rate was in the North Central region, with 98.8 injuries per million.
- Nationwide, a civilian was injured in a fire every 23 minutes.

Property damage
- Property damage resulting from fires slightly decreased in 1998 by 1%, to an estimated $8.629 billion.
- Structure fires resulted in 78% of all property damage, or $6.717 billion.
- 65% of all structure property loss occurred in residential properties, accounting for $4.391 billion.
- The South had the highest property loss rate in the U.S.—about $38.40 per person—followed by the North Central region, with $31.50 per person, and the Northeast, with $31.40 per person.

Incendiary and suspicious fires
- About 15% of all structure fires, or an estimated 76,000 fires, were deliberately set or are suspected of having been deliberately set in 1998. This represents a decrease of 3% from 1997.
- Incendiary or suspicious structure fires resulted in 470 civilian deaths, an increase of 6% from the previous year. Incendiary or suspicious fires caused $1.249 billion in property damage, the lowest figure in 20 years. This represents 19% of all property loss from structure fires.
- The number of vehicle fires of incendiary or suspicious origin in 1998 was 45,000, a decrease of 3% from 1997. They caused an estimated $215 million in property damage, virtually no change from the year before.

Physicians by Age, Sex, and Specialty, 1998
Source: American Medical Assn., as of Dec. 31, 1998

	Total Physicians[1]		Under 35 yrs		35-44 yrs		45-54 yrs		55-64 yrs	
	Male	Female	Male	Female	Male	Female	Male	Female	Male	Female
All Specialties	600,829	177,030	83,744	51,868	150,959	63,461	146,912	36,903	96,036	13,306
Aerospace Medicine	497	35	27	4	123	17	146	11	101	2
Allergy & Immunology	3,056	777	110	95	732	296	971	249	715	84
Anesthesiology	27,130	6,817	3,386	1,196	10,792	2,770	6,913	1,702	3,777	845
Cardiovascular Disease	18,240	1,383	1,492	217	5,924	646	5,888	373	3,204	105
Child Psychiatry	3,479	2,277	222	243	936	851	1,094	703	755	299
Colon/Rectal Surgery	983	66	42	15	297	38	342	11	170	2
Dermatology	6,366	2,853	637	756	1,441	1,217	2,005	629	1,467	188
Diagnostic Radiology	16,517	3,974	2,953	1,094	5,285	1,694	4,870	933	2,623	206
Emergency Medicine	17,394	3,839	3,385	1,235	5,444	1,485	6,139	864	1,642	197
Family Practice	49,375	17,525	7,690	6,163	15,419	6,923	15,060	3,369	5,480	753
Forensic Pathology	388	141	19	6	97	62	111	39	89	22
Gastroenterology	9,109	794	745	161	3,170	400	3,161	189	1,479	39
General Practice	13,975	2,410	111	48	924	431	2,442	776	3,341	606
General Preventive Med.	1,130	558	89	80	316	227	325	163	193	51
General Surgery	36,615	3,833	7,145	1,849	8,219	1,242	8,345	581	7,226	113
Internal Medicine	94,525	33,049	19,222	11,404	27,322	12,668	26,113	6,671	12,609	1,635
Medical Genetics	167	123	13	13	49	45	56	44	34	16
Neurological Surgery	4,739	225	710	74	1,223	98	1,121	45	1,085	6
Neurology	9,628	2,433	1,019	509	2,932	1,027	3,160	653	1,714	181
Nuclear Medicine	1,163	252	69	28	247	76	352	90	311	42
Obstetrics/Gynecology	26,627	12,885	2,647	4,404	6,015	4,844	7,747	2,495	6,065	844
Occupational Medicine	2,544	473	21	11	462	186	733	167	509	70
Ophthalmology	15,559	2,476	1,699	633	4,084	1,078	4,171	551	3,630	152
Orthopedic Surgery	22,387	791	3,571	268	6,214	330	5,797	154	4,574	23
Otolaryngology	8,486	769	1,267	262	2,224	333	2,076	140	1,998	22
Pathology-Anat./Clin.	12,867	5,179	1,139	882	3,084	1,889	3,443	1,415	2,935	691
Pediatric Cardiology	1,041	330	109	70	352	145	282	58	186	37
Pediatrics	30,286	26,752	5,302	8,697	8,152	9,463	8,400	5,662	4,997	2,149
Physical Med./Rehab.	4,103	1,924	714	400	1,635	731	916	482	461	203
Plastic Surgery	5,549	546	360	82	1,672	252	1,737	153	1,287	44
Psychiatry	28,300	11,194	2,077	1,668	5,797	3,775	7,605	3,217	6,554	1,492
Public Health	1,216	425	2	1	132	88	321	124	297	88
Pulmonary Diseases	6,919	891	620	202	2,309	424	2,609	188	997	45
Radiation Oncology	2,888	788	324	130	993	319	775	322	541	86
Radiology	7,230	1,042	393	96	1,280	349	1,370	230	2,498	187
Thoracic Surgery	423	20	140	11	231	9	21	0	19	0
Urological Surgery	9,873	295	1,151	111	2,398	125	2,608	49	2,512	7
Other	5,151	917	107	38	667	242	1,245	262	1,300	183
Unspecified	6,609	3,385	3,796	2,261	1,580	759	659	237	268	77

(1) Includes physicians 65 and older, those living in U.S. possessions, "Not Classified," "Inactive," and "Address Unknown."

U.S. Health Expenditures, 1965-97

Source: *Health, United States, 1999,* National Center for Health Statistics, U.S. Dept. of Health and Human Services

	1965	1970	1975	1980	1985	1990	1994	1995	1996	1997
						Amount in billions				
TOTAL EXPENDITURES	$41.1	$73.2	$130.7	$247.3	$428.7	$699.4	$947.7	$993.7	$1,042.5	$1,092.4
						Percent distribution				
Health services & supplies	91.6	92.7	93.6	95.3	96.2	96.5	96.8	96.9	96.9	96.8
Personal health care	85.5	87.1	87.6	87.8	87.8	87.9	88.0	88.5	88.6	88.7
Hospital care	34.1	38.2	40.2	41.5	39.3	36.7	35.4	34.9	34.6	34.0
Physician services............	19.9	18.5	18.3	18.3	19.5	20.9	20.4	20.3	20.0	19.9
Dentist services..............	6.8	6.4	6.1	5.4	5.0	4.5	4.5	4.5	4.6	4.6
Nursing home care	3.6	5.8	6.6	7.1	7.2	7.3	7.5	7.6	7.6	7.6
Other professional services.....	2.1	1.9	2.1	2.6	3.9	5.0	5.2	5.4	5.5	5.7
Home health care	0.2	0.3	0.5	1.0	1.3	1.9	2.8	2.9	3.0	3.0
Drugs & other medical nondurables	14.3	12.0	10.0	8.7	8.6	8.6	8.6	8.9	9.4	10.0
Vision products & other medical durables............	2.4	2.2	2.0	1.5	1.6	1.5	1.3	1.3	1.3	1.3
Other personal health care	2.0	1.8	1.9	1.6	1.4	1.6	2.3	2.5	2.6	2.7
Program administration & net cost of health insurance	4.7	3.7	3.8	4.8	5.7	5.8	5.8	5.4	5.0	4.6
Government public health activities[1]	1.5	1.8	2.2	2.7	2.7	2.8	3.0	3.1	3.3	3.5
Research & construction..........	8.4	7.3	6.4	4.7	3.8	3.5	3.2	3.1	3.1	3.2
Noncommercial research.........	3.7	2.7	2.5	2.2	1.8	1.7	1.7	1.7	1.6	1.6
Construction.................	4.7	4.6	3.9	2.5	2.0	1.8	1.5	1.4	1.4	1.6
					Average annual % increase from previous year shown					
All expenditures	—	12.2	12.3	13.6	11.6	10.3	7.9	4.9	4.9	4.8
Health services & supplies	—	12.5	12.5	14.0	11.8	10.4	8.0	5.0	4.9	4.6
Personal health care	—	12.7	12.4	13.6	11.6	10.3	7.9	5.4	5.1	4.9
Hospital care	—	14.8	13.4	14.3	10.4	8.8	7.0	3.4	3.9	2.9
Physician services............	—	10.6	12.0	13.6	13.1	11.8	7.2	4.6	3.3	4.4
Dentist services..............	—	10.8	11.2	10.9	10.2	7.8	7.7	6.1	5.6	6.5
Nursing home care	—	23.4	15.5	15.3	11.7	10.7	8.7	6.2	5.2	4.3
Other professional services.....	—	10.2	14.2	18.4	21.2	15.8	9.4	8.1	7.2	7.7
Home health care	—	19.7	23.2	30.7	18.9	18.4	18.9	11.0	7.1	3.7
Drugs & other medical nondurables	—	8.4	8.1	10.7	11.4	10.1	8.0	9.0	10.6	10.7
Vision products & other medical durables............	—	10.2	9.5	8.1	12.4	9.2	4.5	4.9	2.3	3.6
Other personal health care	—	9.5	13.8	10.2	8.8	12.9	18.2	14.5	9.5	9.0
Program administration & net cost of health insurance	—	7.1	12.5	19.3	15.4	10.8	8.0	-3.2	-1.5	-4.8
Government public health activities[1]	—	17.0	16.8	18.1	11.5	11.0	9.5	8.0	11.9	13.1
Research & construction..........	—	9.2	9.4	6.8	7.1	8.4	5.6	0.5	4.3	9.2
Noncommercial research.........	—	5.1	11.2	10.4	7.5	9.3	6.8	5.2	2.6	4.7
Construction.................	—	12.1	8.3	4.1	6.7	7.6	4.4	-4.6	6.3	14.3

Note: Numbers may not add to totals because of rounding. (1) Includes personal care services delivered by government public health agencies.

Ownership of Life Insurance in the U.S. and Assets of U.S. Life Insurance Companies, 1940-98

Source: American Council of Life Insurance

(amounts in millions)

Year	PURCHASES OF LIFE INSURANCE				INSURANCE IN FORCE					Assets
	Ordinary	Group	Industrial	Total	Ordinary	Group	Industrial	Credit	Total	
1940...	$6,689	$691	$3,350	$10,730	$79,346	$14,938	$20,866	$380	$115,530	$30,802
1950...	17,326	6,068	5,402	28,796	149,116	47,793	33,415	3,844	234,168	64,020
1960...	52,883	14,645	6,880	74,408	341,881	175,903	39,563	29,101	586,448	119,576
1970...	122,820	63,690[1]	6,612	193,122[1]	734,730	551,357	38,644	77,392	1,402,123	207,254
1975...	188,003	95,190[1]	6,729	289,922[1]	1,083,421	904,695	39,423	112,032	2,139,571	289,304
1980...	385,575	183,418	3,609	572,602	1,760,474	1,579,355	35,994	165,215	3,541,038	479,210
1985...	910,944	319,503[2]	722	1,231,169[2]	3,247,289	2,561,595	28,250	215,973	6,053,107	825,901
1990...	1,069,660	459,271	220	1,529,151	5,366,982	3,753,506	24,071	248,038	9,392,597	1,408,208
1991...	1,041,508	573,953[1]	198	1,615,659[1]	5,677,777	4,057,606	22,475	228,478	9,986,336	1,551,201
1992...	1,048,135	440,143	222	1,488,500	5,941,810	4,240,919	20,973	202,090	10,405,792	1,664,531
1993...	1,101,327	576,823	149	1,678,299	6,428,434	4,456,338	20,451	199,518	11,104,741	1,839,127
1994...	1,056,976	560,232	257	1,617,465	6,429,811	4,443,179	18,947	189,398	11,081,335	1,942,273
1995...	1,039,102	537,828	156	1,577,086	6,872,252	4,604,856	18,134	201,083	11,696,325	2,143,544
1996...	1,089,137	614,565	130	1,703,832	7,407,682	5,067,804	18,064	210,746	12,704,296	2,327,924
1997...	1,203,552	688,589	128	1,892,269	7,854,570	5,279,042	17,991	212,255	13,363,858	2,579,078
1998...	1,324,565	739,508	106	2,064,179	8,505,894	5,735,273	17,365	212,917	14,471,449	2,819,992

Note: Ordinary purchases, ordinary in force, and group in force numbers were revised for 1994-97. (1) Includes Servicemen's Group Life Insurance, which amounted to $17.1 billion in 1970, $1.7 billion in 1975, and $166.7 billion in 1991. (2) Includes Federal Employees' Group Life Insurance of $10.8 billion.

Health Insurance Coverage,[1] by State, 1997-98

Source: Bureau of the Census, U.S. Dept. of Commerce, Mar. 1999 Current Population Survey

STATE	1998 Total Population[2]	Not covered 1998[2]	% not covered 1998	% not covered 1997	STATE	1998 Total Population[2]	Not covered 1998[2]	% not covered 1998	% not covered 1997
AL........	4,201	714	17.0	15.5	MT........	925	181	19.6	19.5
AK........	647	112	17.3	18.1	NE........	1,716	155	9.0	10.8
AZ........	4,905	1,187	24.2	24.5	NV........	1,862	394	21.2	17.5
AR........	2,563	478	18.7	24.4	NH........	1,224	138	11.3	11.8
CA	33,375	7,373	22.1	21.5	NJ........	8,092	1,329	16.4	16.5
CO	3,971	599	15.1	15.1	NM........	1,829	386	21.1	22.6
CT	3,283	412	12.6	12.0	NY........	18,420	3,177	17.3	17.5
DE	783	115	14.7	13.1	NC........	7,427	1,111	15.0	15.5
DC	512	87	17.0	16.2	ND........	646	92	14.2	15.2
FL........	14,678	2,564	17.5	19.6	OH........	11,225	1,169	10.4	11.5
GA	7,666	1,341	17.5	17.6	OK........	3,269	599	18.3	17.8
HI	1,201	121	10.0	7.5	OR........	3,356	481	14.3	13.3
ID	1,274	225	17.7	17.7	PA........	11,912	1,248	10.5	10.1
IL	12,295	1,842	15.0	12.4	RI........	968	96	10.0	10.2
IN	5,840	839	14.4	11.4	SC........	3,851	594	15.4	16.8
IA	2,837	265	9.3	12.0	SD........	711	102	14.3	11.8
KS	2,616	270	10.3	11.7	TN........	5,572	724	13.0	13.6
KY	3,865	545	14.1	15.0	TX........	19,945	4,880	24.5	24.5
LA........	4,310	817	19.0	14.9	UT........	2,106	293	13.9	13.4
ME	1,266	161	12.7	14.9	VT........	593	58	9.9	9.5
MD	5,046	837	16.6	13.4	VA........	6,688	946	14.1	12.6
MA	6,117	627	10.3	12.6	WA........	5,747	706	12.3	11.4
MI........	10,041	1,328	13.2	11.6	WV........	1,750	302	17.2	17.2
MN	4,833	448	9.3	9.2	WI........	5,129	604	11.8	8.0
MS	2,761	554	20.0	20.1	WY........	486	82	16.9	15.5
MO	5,405	570	10.5	12.6	**TOTAL U.S...**	**271,743**	**44,281**	**16.3**	**16.1**

(1) For population, all ages, including those 65 or over, an age group largely covered by Medicare. (2) In thousands.

Persons Not Covered by Health Insurance, by Selected Characteristics, 1998

Source: Bureau of the Census, U.S. Dept. of Commerce, Mar. 1999 Current Population Survey; in thousands

	Number[1]	Percent		Number[1]	Percent
TOTAL NOT COVERED.............	**44,281**	**16.3**	Black	7,797	22.2
Sex			Asian or Pacific Islander	2,301	21.1
Male..........................	23,014	17.3	Hispanic origin[2]	11,196	35.3
Female	21,266	15.3	**Education[3]**		
Age			No high school diploma.............	9,294	26.7
Under 18 years..................	11,073	15.4	High school graduate, no college	12,094	18.3
18 to 24 years	7,776	30.0	Some college, no degree.............	6,211	15.9
25 to 34 years	9,127	23.7	Associate degree...................	1,730	12.3
35 to 44 years	7,708	17.2	Bachelor's degree or higher	3,880	8.5
45 to 64 years	8,239	14.2	**Work experience[4]**		
65 years and over................	358	1.1	Worked during year	24,655	18.0
Nativity			Worked full-time...................	19,244	16.9
Native........................	35,273	14.4	Worked part-time..................	5,411	23.2
Foreign-born...................	9,008	34.1	Did not work	8,194	27.0
Naturalized citizen	1,891	19.2	**Household income**		
Not a citizen	7,118	42.9	Less than $25,000.................	17,229	25.2
Race and Hispanic origin			$25,000-$49,999..................	14,807	18.8
White.........................	33,588	15.0	$50,000-$74,999..................	6,703	11.7
White, not of Hispanic origin	22,890	11.9	$75,000 or more	5,542	8.3

(1) In thousands. (2) Persons of Hispanic origin may be of any race. (3) Persons aged 18 years and over. (4) Persons aged 18-64.

Health Coverage for Persons Under 65, by Characteristics, 1984, 1995-97

Source: *Health, United States, 1999,* National Center for Health Statistics, U.S. Dept. of Health and Human Services

	PRIVATE INSURANCE				MEDICAID[1]				NOT COVERED[2]			
	1984	1995	1996	1997[3]	1984	1995	1996	1997[3]	1984	1995	1996	1997[3]
Age					Percent of each population group							
Under 18 years............	72.6	65.7	66.4	66.1	11.9	20.6	20.1	18.4	13.9	13.6	13.4	14.0
18-44 years	76.5	71.2	70.6	69.4	5.1	7.4	7.3	6.6	17.1	20.5	21.2	22.4
45-64 years	83.3	80.4	79.5	79.1	3.4	5.3	5.2	4.6	9.6	11.0	11.2	12.4
Race and Hispanic origin[4]												
White, non-Hispanic........	82.3	78.6	78.5	77.9	4.0	7.5	7.5	6.8	11.6	12.7	12.9	13.3
Black, non-Hispanic........	58.5	54.6	55.4	55.1	20.8	26.7	24.2	22.5	19.2	17.8	18.9	19.3
All Hispanic	56.3	47.3	47.5	47.3	13.1	21.2	20.1	17.8	29.0	30.8	31.6	33.2
Percent of poverty level[4]												
Below 100%..............	32.4	21.9	20.0	22.7	32.1	46.9	46.8	41.6	34.0	30.9	32.7	32.8
100-149%................	62.4	47.8	47.1	42.1	7.7	18.4	17.2	19.1	26.4	31.2	32.8	34.8
150-199%................	77.7	66.5	67.9	64.0	3.3	7.7	7.7	8.0	16.7	22.8	22.5	25.1
200% or more.............	91.7	89.3	89.5	87.7	0.6	1.6	1.6	1.9	5.6	7.8	7.4	8.5
Geographic region[4]												
Northeast	80.1	75.1	74.9	74.0	9.4	12.5	12.3	12.4	9.8	12.7	13.2	12.8
Midwest	80.4	77.2	78.4	76.9	7.9	11.0	9.4	9.2	10.9	11.8	11.9	12.6
South	74.0	66.7	65.9	66.8	5.5	11.6	12.0	9.8	17.4	19.1	19.7	20.3
West....................	71.8	67.9	67.1	65.3	7.5	13.2	13.4	12.5	17.6	17.3	18.1	19.8

Note: Data based on household interviews of a sample of the civilian noninstitutionalized population. Percents do not add to 100 because other types of health insurance (e.g., Medicare, military) are not shown and persons with both private insurance and Medicaid appear in both columns. (1) Includes Medicaid or other public assistance. In 1997, the age-adjusted percent of the population under 65 covered by Medicaid was 9.5%; 1.2% were covered by public assistance. (2) Includes persons not covered by private insurance, Medicaid or other public assistance, Medicare, or military plans. (3) Preliminary data. (4) Age adjusted.

Enrollment in Health Maintenance Organizations (HMOs), 1976-98

Source: *Health, United States, 1999*, National Center for Health Statistics, U.S. Dept. of Health and Human Services

	1976	1980	1990	1991	1992	1993	1994	1995	1996	1997	1998
					Number of enrolled in millions						
TOTAL	6.0	9.1	33.0	34.0	36.1	38.4	45.1	50.9	59.1	66.8	76.6
Model type[1]											
Individual practice association[2]	0.4	1.7	13.7	13.6	14.7	15.3	17.8	20.1	26.0	26.7	32.6
Group[3]	5.6	7.4	19.3	17.1	16.5	15.4	13.9	13.3	14.1	11.0	13.8
Mixed	—	—	—	3.3	4.9	7.7	13.4	17.6	19.0	29.0	30.1
Federal program[4]											
Medicaid[5]	—	0.3	1.2	1.4	1.7	1.7	2.6	3.5	4.7	5.6	7.8
Medicare	—	0.4	1.8	2.0	2.2	2.2	2.5	2.9	3.7	4.8	5.7
					Percent of population enrolled in HMOs						
TOTAL	2.8	4.0	13.4	13.6	14.3	15.1	17.3	19.4	22.3	25.2	28.6
Geographic region											
Northeast	2.0	3.1	14.6	15.4	16.1	18.0	20.8	24.4	25.9	32.4	37.8
Midwest	1.5	2.8	12.6	12.7	12.8	13.2	15.2	16.4	18.8	19.5	22.7
South	0.4	0.8	7.1	7.1	7.8	8.4	10.2	12.4	15.2	17.9	21.0
West	9.7	12.2	23.2	23.8	24.7	25.1	27.4	28.6	33.2	36.4	39.1

— = Not available. **Note:** Data as of June 30 in 1976-80, Jan. 1 in 1990-98. Medicaid enrollment in 1990 as of June 30. HMOs in Guam included starting in 1994; Puerto Rico, 1998. Open-ended enrollment in HMO plans, amounting to 11.6 million on Jan. 1, 1998, included from 1994 onwards. (1) In 1976, 11 HMOs with 35,000 enrollment did not report model type. In 1997, 11 HMOs with 153,000 enrollment did not report model type. In 1998, 6 HMOs with 109,000 enrollment did not report model type. (2) This type of HMO contracts with an association of physicians from various settings (a mixture of solo and group practices) to provide health services. (3) Group includes staff, group, and network model types. (4) Enrollment by Medicaid or Medicare beneficiaries, where the Medicaid or Medicare program contracts directly with the HMO to pay the premium. (5) Data for 1990 and later include enrollment in managed-care health insuring organizations.

Physician Contacts Per Person, by Selected Characteristics, 1987-96

Source: *Health, United States, 1999*, National Center for Health Statistics, U.S. Dept. of Health and Human Services

	1987	1988	1989	1990	1991	1992	1993	1994	1995	1996
TOTAL[1,2]	5.4	5.3	5.3	5.5	5.6	5.9	6.0	6.0	5.8	5.8
Age										
Under 15 years	4.5	4.6	4.6	4.5	4.7	4.6	4.9	4.6	4.5	4.4
Under 5 years	6.7	7.0	6.7	6.9	7.1	6.9	7.2	6.8	6.5	6.5
5-14 years	3.3	3.3	3.5	3.2	3.4	3.4	3.6	3.4	3.4	3.3
15-44 years	4.6	4.7	4.6	4.8	4.7	5.0	5.0	5.0	4.8	4.6
45-64 years	6.4	6.1	6.1	6.4	6.6	7.2	7.1	7.3	7.1	7.2
65 years and over	8.9	8.7	8.9	9.2	10.4	10.6	10.9	11.3	11.1	11.7
65-74 years	8.4	8.4	8.2	8.5	9.2	9.7	9.9	10.3	9.8	10.2
75 years and over	9.7	9.2	9.9	10.1	12.3	12.1	12.3	12.7	12.9	13.7
Sex										
Male[1]	4.6	4.6	4.8	4.7	4.9	5.1	5.2	5.2	4.9	5.0
Female[1]	6.0	6.0	5.9	6.1	6.3	6.6	6.7	6.7	6.5	6.5
Race										
White[1]	5.5	5.5	5.5	5.6	5.8	6.0	6.0	6.1	5.9	5.8
Black[1]	5.1	4.8	4.9	5.1	5.2	5.9	6.0	5.7	5.5	5.7
Family income[1,3]										
Less than $16,000	6.8	6.2	6.3	6.3	6.8	7.3	7.3	7.6	7.4	7.5
$16,000-$24,999	5.6	5.3	5.2	5.6	5.6	6.0	5.7	5.9	6.1	5.5
$25,000-$34,999	5.2	5.0	5.5	5.2	5.5	5.7	6.0	5.8	5.3	5.6
$35,000-$49,999	5.2	5.5	5.2	5.7	5.8	5.9	6.0	6.2	5.7	5.9
$50,000 or more	5.4	5.5	6.0	5.6	5.8	5.8	5.8	6.0	5.6	5.3
Geographic region[1]										
Northeast	5.2	5.0	5.3	5.2	5.4	5.9	5.9	5.9	5.6	5.7
Midwest	5.6	5.4	5.4	5.3	5.8	5.9	6.2	6.0	5.8	5.7
South	5.1	5.2	5.3	5.6	5.5	5.8	5.7	5.6	5.8	6.1
West	5.5	5.9	5.5	5.6	5.9	6.1	6.0	6.4	5.8	5.3
Location of residence[1]										
Within MSA	5.5	5.5	5.4	5.6	5.8	6.0	6.1	6.0	5.9	5.8
Outside MSA	4.8	4.9	5.2	4.9	5.1	5.6	5.6	5.7	5.3	5.7

MSA = metropolitan statistical area. **Note:** Data based on household interviews of a sample of the civilian noninstitutionalized population. (1) Age adjusted. (2) Includes all races whether shown separately or not, and unknown family income. (3) The family income categories listed are for 1996. The 5 income categories for 1987 are: less than $10,000; $10,000-$14,999; $15,000-$19,999; $20,000-$34,999; and $35,000 or more. Income categories for 1988 are: less than $13,000; $13,000-$18,999; $19,000-$24,999; $25,000-$44,999; and $45,000 or more. In 1989-94, the 2 lowest income categories are: less than $14,000 and $14,000-$24,999; the 3 higher income categories are as shown. In 1995, the 2 lowest income categories are: less than $15,000 and $15,000-$24,999; the 3 higher income categories are as shown.

Top 20 Reasons Given by Patients for Emergency Room Visits, 1997

Source: National Center for Health Statistics, U.S. Dept. of Health and Human Services

Principal reason for visit	No. of visits (1,000)	Percent of total	Principal reason for visit	No. of visits (1,000)	Percent of total
ALL VISITS	94,936	100	11. Vomiting	1,813	1.9
1. Stomach and abdominal pain, cramps, spasms	5,527	5.8	12. Laceration and cuts—facial area	1,764	1.9
2. Chest pain and related symptoms	5,315	5.6	13. Earache or ear infection	1,683	1.8
3. Fever	4,212	4.4	14. Labored or difficult breathing (dyspnea)	1,603	1.7
4. Headache, pain in head	2,518	2.7	15. Motor vehicle accident, injury unspecified	1,470	1.5
5. Injury—upper extremity	2,383	2.5	16. Injury, other and unspecified type—head, neck, and face	1,383	1.5
6. Shortness of breath	2,242	2.4	17. Vertigo—dizziness	1,289	1.4
7. Cough	2,220	2.3	18. Accident, not otherwise specified	1,286	1.4
8. Back symptoms	2,073	2.2	19. Neck symptoms	1,259	1.3
9. Pain, site not referable to a specific body system	2,040	2.1	20. Hand and finger injury	1,240	1.3
10. Symptoms referable to throat	1,953	2.1	All other reasons	49,664	52.3

Top 20 Reasons Given by Patients for Physicians' Office Visits, 1997

Source: National Center for Health Statistics, U.S. Dept. of Health and Human Services

Principal reason for visit	NUMBER OF VISITS (1,000)	PERCENTAGE DISTRIBUTION Total	PERCENTAGE DISTRIBUTION Female	PERCENTAGE DISTRIBUTION Male
ALL VISITS	787,372	100.0	100.0	100.0
1. General medical examination	59,796	7.6	8.0	7.0
2. Progress visit, not otherwise specified	28,583	3.6	3.2	4.2
3. Cough	25,735	3.3	2.8	4.0
4. Routine prenatal examination	22,979	2.9	4.9	—
5. Postoperative visit	18,861	2.4	2.5	2.2
6. Symptoms referable to throat	17,151	2.2	1.9	2.6
7. Well-baby examination	15,526	2.0	1.6	2.5
8. Vision dysfunctions	13,443	1.7	1.7	1.6
9. Earache or ear infection	13,359	1.7	1.5	1.9
10. Back symptoms	12,863	1.6	1.5	1.9
11. Knee symptoms	12,392	1.6	1.4	1.8
12. Fever	12,374	1.6	1.1	2.2
13. Skin rash	12,316	1.6	1.4	1.9
14. Stomach pain, cramps, and spasms	12,078	1.5	1.7	1.2
15. Hypertension	10,875	1.4	1.3	1.4
16. Nasal congestion	10,564	1.3	1.3	1.4
17. Depression	10,488	1.3	1.5	1.1
18. Headache, pain in head	9,589	1.2	1.3	1.0
19. Medication, other and unspecified kinds	9,056	1.2	1.1	1.2
20. Head cold, upper respiratory infection (coryza)	8,965	1.1	1.2	1.1
All other reasons	450,380	57.2	56.9	57.7

Drugs Most Frequently Prescribed in Physicians' Offices, 1997

Source: National Center for Health Statistics, U.S. Dept. of Health and Human Services; *Physicians' Desk Reference*; in thousands

Rank	Name of drug (principal generic substance)[1]	Times prescribed	Therapeutic use
1.	Amoxicillin	114,148	Antibiotic
2.	Tylenol (acetaminophen)	13,029	Analgesic (for pain relief)
3.	Lasix (furosemide)	12,353	Diuretic, antihypertensive
4.	Claritin (loratadine)	10,962	Antihistamine
5.	Premarin (estrogens)	10,713	Estrogen replacement therapy
6.	Synthroid (levothyroxine)	10,706	Thyroid hormone therapy
7.	Prednisone	10,215	Steroid replacement therapy, anti-inflammatory agent
8.	Amoxil (amoxicillin)	8,924	Antibiotic
9.	Lanoxin (digoxin)	8,492	For congestive heart failure, irregular heartbeat
10.	Coumadin (crystalline warfarin sodium)	8,276	Anticoagulant
11.	Prozac (fluoxetine hydrochloride)	8,225	Antidepressant
12.	Biaxin (clarithromycin)	8,122	Antibiotic
13.	Xanax (alprazolam)	7,846	Anxiety disorders
14.	Prilosec (omeprazole)	7,716	For duodenal or gastric ulcer
15.	Zoloft (sertraline hydrochloride)	7,607	Antidepressant
16.	Augmentin (amoxicillin/clavulanate potassium)	7,501	Antibiotic
17.	A.S.A. (aspirin)	7,492	Analgesic (for pain relief)
18.	Influenza virus vaccine	6,837	Immunization
19.	Paxil (paroxetine hydrochloride)	6,773	Antidepressant
20.	Motrin (ibuprofen)	6,668	Anti-inflammatory agent
	All other	848,292	

(1) The trade or generic name used by the physician on the prescription or other medical records. The use of trade names is for identification only and does not imply endorsement by the Public Health Service or the U.S. Dept. of Health and Human Services.

Hospitals and Nursing Homes in the U.S., 1997

Source: *1999 Hospital Statistics*, Health Forum, L.L.C., An American Hospital Association Company, copyright 1999; *Health, United States, 1999*

For information on choosing a nursing home, go to the website http://www.hcfa.gov/medicare/nurshm1.htm

STATE	Hospitals[1]	% of beds occupied[2]	Nursing homes	% of beds occupied	STATE	Hospitals[1]	% of beds occupied[2]	Nursing homes	% of beds occupied
AL	111	57.8	198	93.0	MT	54	62.0	96	83.1
AK	17	73.0	15	73.6	NE	87	59.5	234	85.4
AZ	63	61.9	118	75.3	NV	19	67.3	42	74.4
AR	82	57.6	256	63.5	NH	28	63.4	78	92.8
CA	414	59.6	1,311	77.1	NJ	85	68.0	314	91.9
CO	67	55.0	222	83.1	NM	36	57.9	77	84.0
CT	34	68.7	252	92.2	NY	225	75.9	561	95.2
DE	6	70.9	43	75.0	NC	118	68.2	384	93.7
DC	12	71.1	22	94.9	ND	41	61.6	80	93.9
FL	206	58.7	638	83.5	OH	170	58.3	931	69.1
GA	158	58.9	346	92.5	OK	111	54.1	400	72.4
HI	19	81.7	42	91.4	OR	61	52.9	152	79.7
ID	42	53.7	78	74.0	PA	217	66.9	794	90.1
IL	202	58.1	841	78.2	RI	11	70.9	93	91.3
IN	112	57.6	549	71.9	SC	65	64.1	176	86.6
IA	115	56.1	424	66.4	SD	49	65.0	100	94.2
KS	130	53.0	403	80.6	TN	124	58.8	341	90.3
KY	105	56.9	266	88.7	TX	407	55.8	1,221	69.8
LA	127	54.2	333	81.1	UT	41	54.4	92	78.3
ME	38	63.6	128	87.9	VT	14	71.2	39	94.0
MD	51	67.6	228	82.1	VA	93	63.6	245	90.8
MA	84	67.3	532	89.3	WA	88	56.9	275	82.3
MI	154	64.7	417	86.4	WV	58	58.6	79	69.5
MN	137	68.3	396	92.6	WI	124	57.0	402	88.3
MS	97	62.6	202	92.6	WY	25	52.4	37	84.0
MO	123	56.6	549	73.3	**U.S.**	**5,057**	**61.8**	**16,052**	**82.0**

(1) Community hospitals. (2) Data exclude hospital units of institutions, facilities for the mentally retarded, and alcoholism and chemical dependency hospitals.

Expected New Cancer Cases and Deaths, by Sex, for Leading Sites, 1999

Source: American Cancer Society

The estimates of expected new cases are offered as a rough guide only. They exclude basal and squamous cell skin cancers and in situ carcinomas except urinary bladder. Carcinoma in situ of the breast accounts for about 39,900 new cases annually, and melanoma carcinoma in situ accounts for about 23,200 new cases annually. About 1,000,000 basal and squamous cell skin cancers occur annually. About 1,900 nonmelanoma skin cancer deaths are included among the deaths in all sites expected in 1999.

EXPECTED NEW CASES

Both sexes		Women		Men	
Prostate	179,300	Breast	175,000	Prostate	179,300
Breast	176,300	Lung	77,600	Lung	94,000
Lung	171,600	Colorectal	67,000	Colorectal	62,400
Colorectal	129,400	Endometrium(uterus)	37,400	Urinary bladder	39,100
Non-Hodgkin's lymphoma	56,800	Ovary	25,200	Non-Hodgkin's lymphoma	32,600
ALL SITES	**1,221,800**	**ALL SITES**	**598,000**	**ALL SITES**	**623,800**

EXPECTED DEATHS

Both sexes		Women		Men	
Lung	158,900	Lung	68,000	Lung	90,900
Colorectal	56,600	Breast	43,300	Prostate	37,000
Breast	43,700	Colorectal	28,800	Colorectal	27,800
Prostate	37,000	Pancreas	14,700	Pancreas	13,900
Pancreas	28,600	Ovary	14,500	Non-Hodgkin's lymphoma	13,400
ALL SITES	**563,100**	**ALL SITES**	**272,000**	**ALL SITES**	**291,100**

U.S. Cancer Incidence for Top 15 Sites, 1990-96

Source: Surveillance, Epidemiology, and End Results (SEER) Program, National Cancer Institute

	Rate[1]	Annual % change		Rate[1]	Annual % change
ALL SITES	399.47	−0.95	Ovary	14.77	−1.39
Prostate	151.75	−2.05	Melanomas of the skin	12.13	2.28
Breast (female)	109.05	0.08	Leukemias	10.35	−1.49
Lung	55.74	−1.58	Oral cavity and pharynx	10.21	−1.89
Colon and rectum	44.27	−2.10	Cervix	9.01	−2.34
Uterus	21.12	−0.25	Kidney and renal pelvis	8.88	0.68
Urinary bladder	16.42	−1.09	Pancreas	8.84	−1.07
Non-Hodgkin's lymphomas	15.48	0.74	Stomach	7.70	−1.62

(1) Per 100,000 population; average for 6-year period.

U.S. Cancer Mortality for Top 15 Sites, 1990-96

Source: Surveillance, Epidemiology, and End Results (SEER) Program, National Cancer Institute

	Rate[1]	Annual % change		Rate[1]	Annual % change
ALL SITES	171.00	−0.61	Stomach	4.36	−2.67
Lung	49.68	−0.41	Brain and other nervous system	4.21	−0.65
Breast (female)	25.91	−1.83	Esophagus	3.56	0.82
Prostate	25.85	−1.57	Kidney and renal pelvis	3.51	0.30
Colon and rectum	17.80	−1.68	Uterus	3.37	−0.56
Pancreas	8.41	−0.35	Liver	3.28	3.87
Ovary	7.69	−0.90	Urinary bladder	3.24	−0.25
Leukemias	6.33	−0.32	Multiple myeloma (bone marrow)	3.08	0.88

(1) Per 100,000 population; average for 6-year period.

Cardiovascular Diseases Statistical Summary, 1997

Source: American Heart Association

Prevalence — 59,700,000[1] Americans had one or more forms of heart and blood vessel disease in 1997.
- high blood pressure — 50,000,000
- coronary heart disease — 12,200,000[1]
- stroke — 4,400,000
- rheumatic heart disease — 1,800,000

Hypertension (high blood pressure) afflicts 50,000,000 Americans age 6 and above, including 1 in 4 adults.

Mortality — 953,110 in 1997 (41.2% of all deaths).
- Someone died from cardiovascular disease every 33 seconds in the U.S. in 1997.

Congenital or inborn heart defects —
- Mortality from such heart defects was 4,698 in 1997.

Coronary heart disease (heart attack and angina pectoris) — caused 466,101 deaths in 1997.
- 12,200,000[1] Americans had a history of heart attack and/or angina pectoris.
- As many as 1,100,000[1] Americans had coronary attacks in 1997.

Stroke — killed about 159,791 Americans in 1997.

Rheumatic heart disease — killed 5,014 in 1997.

(1) Estimate is based on new methodology and therefore not directly comparable to past years.

AIDS Deaths and New AIDS Cases in the U.S., 1985-98

Source: *Health, United States, 1999; HIV/AIDS Surveillance Report,* Vol. 10, No. 2, covering through 1998; National Center for Health Statistics, U.S. Dept. of Health and Human Services

	All years[2]	1985	1990	1993	1994	1995	1996	1997	1998[3]
TOTAL DEATHS[1]	413,576	6,854	31,145	44,108	48,110	47,858	34,557	14,185	NA
				NEW AIDS CASES					
TOTAL NEW CASES	643,350	8,161	41,540	102,082	77,092	70,839	66,398	58,254	24,014
Male									
All males, 13 years and older	536,198	7,510	36,283	85,266	62,811	57,061	52,553	45,291	18,423
White, not Hispanic	271,446	4,755	20,881	43,256	29,497	26,206	23,173	17,557	7,181
Black, not Hispanic.	177,356	1,708	10,267	28,354	22,446	20,945	20,069	18,785	7,595
Hispanic	80,987	991	4,762	12,624	10,083	9,172	8,581	8,248	3,326
American Indian[4]	1,527	7	81	310	207	197	169	165	51
Asian or Pacific Islander[5]	4,184	49	263	662	525	486	481	380	168
13-19 years	1,924	28	107	361	226	228	204	183	81
20-29 years	88,475	1,504	6,943	14,629	9,691	8,426	7,071	5,791	2,223
30-39 years	245,361	3,588	16,718	38,909	28,942	25,842	23,842	20,185	7,992
40-49 years	142,559	1,634	8,854	22,863	17,197	16,273	15,479	13,627	5,702
50-59 years	42,793	597	2,650	6,407	5,060	4,721	4,432	4,124	1,783
60 years and over.	15,086	159	1,011	2,097	1,695	1,571	1,525	1,381	642
Female									
All females, 13 years and older . . .	99,259	523	4,534	15,943	13,310	13,032	13,192	12,515	5,401
White, not Hispanic	23,448	142	1,225	4,043	3,081	3,060	2,867	2,474	1,012
Black, not Hispanic.	58,523	279	2,546	9,105	7,851	7,624	8,104	7,845	3,447
Hispanic	16,299	99	732	2,629	2,284	2,230	2,074	2,040	859
American Indian[4]	294	2	8	61	40	38	43	35	15
Asian or Pacific Islander[5]	537	1	19	97	50	71	80	64	31
13-19 years	1,230	4	66	200	174	156	177	176	84
20-29 years	21,958	178	1,117	3,721	2,944	2,678	2,684	2,427	990
30-39 years	45,260	232	2,079	7,526	6,001	5,966	5,907	5,496	2,297
40-49 years	21,718	45	781	3,217	3,081	3,080	3,265	3,248	1,466
50-59 years	5,936	26	272	848	768	816	833	818	414
60 years and over.	3,157	38	219	431	342	336	326	350	150
Children									
All children, under 13 years	7,893	128	723	873	971	746	653	448	190
White, not Hispanic	1,463	26	158	153	143	117	95	63	36
Black, not Hispanic.	4,810	84	388	535	633	483	428	292	116
Hispanic	1,533	18	168	175	180	135	125	86	37
American Indian[4]	27	0	5	3	2	2	3	2	0
Asian or Pacific Islander[5]	45	0	4	4	11	5	1	3	1
Under 1 year	3,134	63	316	352	350	271	219	131	55
1-12 years	4,759	65	407	521	621	475	434	317	135

NA = Not available. Note: The definition of AIDS cases for reporting purposes was expanded in 1985, 1987, and 1993, as more was learned about the spectrum of human immunodeficiency virus-associated diseases. Data exclude residents of U.S. territories. Figures are updated periodically because of reporting delays, which may affect accuracy of some figures. (1) Based on preliminary figures and subject to revision. (2) Revised figures; includes cases and deaths prior to 1985 and for years not shown. (3) Jan.-June 1998 only, unless otherwise noted. (4) Includes Aleut and Eskimo. (5) Includes Chinese, Japanese, Filipino, Hawaiian and part-Hawaiian, and other Asian or Pacific Islander.

New AIDS Cases in the U.S., 1985-98, by Transmission Category

Source: *Health, United States, 1999,* CDC, National Center for HIV, STD, and TB Prevention, Div. of HIV/AIDS Prevention

TRANSMISSION CATEGORY	All years[1]	1985	1990	1993	1994	1995	1996	1997	1998[2]
All males, 13 years and older	536,198	7,510	36,283	85,266	62,811	57,061	52,553	45,291	18,423
Men who have sex with men	314,241	5,355	23,785	49,514	35,269	30,978	27,538	21,163	8,388
Injecting drug use.	113,705	1,103	6,958	20,066	15,148	13,353	11,838	9,950	3,652
Men who have sex with men and injecting drug use	40,460	656	2,833	7,393	4,593	3,892	3,239	2,357	908
Hemophilia/coagulation disorder .	4,504	68	332	1,052	481	430	309	182	70
Heterosexual contact[3]	19,832	32	715	2,997	2,776	2,826	3,202	2,939	1,147
Sex with injecting drug user . . .	7,235	25	455	1,183	934	882	831	747	327
Transfusion[4]	4,650	103	446	598	367	337	265	214	74
Undetermined[5]	38,806	193	1,214	3,646	4,177	5,245	6,162	8,486	4,184
All females, 13 years and older. . .	99,259	523	4,534	15,943	13,310	13,032	13,192	12,515	5,401
Injecting drug use.	43,259	286	2,331	8,061	5,940	5,323	4,772	4,160	1,541
Hemophilia/coagulation disorder .	214	3	16	32	28	27	23	24	4
Heterosexual contact[3]	37,979	119	1,539	6,056	5,459	5,416	5,747	4,930	1,806
Sex with injecting drug user . . .	16,088	82	1,035	2,782	2,048	1,881	1,894	1,433	525
Transfusion[4]	3,443	63	334	484	306	269	260	179	71
Undetermined[5]	14,364	52	314	1,310	1,577	1,997	2,390	3,222	1,979

Note: The definition of AIDS cases for reporting purposes was expanded in 1985, 1987, and 1993, as more was learned about the spectrum of human immunodeficiency virus-associated diseases. Data exclude residents of U.S. territories. Figures are updated periodically because of reporting delays. (1) Includes cases prior to 1985 and for years not shown. (2) Jan.-June 1998 only. (3) Includes persons who have had heterosexual contact with a person with human immunodeficiency virus (HIV) infection or at risk of HIV infection. (4) Receipt of blood transfusion, blood components, or tissue. (5) Includes persons for whom risk information is incomplete, persons still under investigation, men reported only to have had heterosexual contact with prostitutes, and interviewed persons for whom no specific risk is identified.

CRIME

Crime in U.S. Down Again in 1998

Serious crimes reported to law enforcement agencies in the United States decreased 5.4% in 1998 compared with 1997, according to statistics from *Uniform Crime Reports,* released by the Federal Bureau of Investigation on Oct. 17, 1999. This decrease continued the trend of recent years; reported crime in the U.S. had gone down 2% in 1997, 3% in 1996, 1% in both 1994 and 1995, 2% in 1993, and 3% in 1992.

Serious crime is measured by the Crime Index, which includes four violent crimes and four property crimes. Violent crime dropped 6.4% in 1998, and property crime fell by 5.3%.

All four violent crimes in the Crime Index showed a decrease. Murder fell 7.1%, robbery decreased by 10.4%, aggravated assault declined 4.8%, and forcible rape dropped by 3.2%.

In the property-crime category, motor vehicle theft was down 8.4% in 1998, burglary fell 5.3%, and larceny-theft decreased by 4.8%. Sufficient data were not available to estimate the trend for arson.

Declines in overall Crime Index totals occurred in all four regions of the country: 7% in both the Northeast and West, 5% in the South, and 4% in the Midwest.

Cities with populations between 500,000 and 999,999 showed the largest decline in reported crime between 1997 and 1998—7.5%. Those with 50,000 to 99,999 inhabitants followed closely with a 7.4% decrease, while cities with populations from 250,000 to 499,999 had a 7.2% drop. Crime data for 1997 and 1998 show that suburban counties experienced a 5.9% decrease in their crime level, while rural counties reported a 4.6% decline.

U.S. Crime Index Trends, 1998

Source: FBI, *Uniform Crime Reports,* 1998

(percentage change 1998 over 1997, offenses known to the police)

	No. of agencies[1]	Pop. (thousands)	Crime Index (total)	Violent crime[2]	Property crime[3]	Murder	Forcible rape	Robbery	Aggravated assault	Burglary	Larceny/theft	Motor vehicle theft
TOTAL U.S.		270,296	−5	−6	−5	−7	−3	−10	−5	−5	−5	−8
Cities:												
Over 1,000,000	8	20,149	−6	−6	−6	−11	−4	−12	−2	−10	−4	−11
500,000 to 999,999	19	12,795	−8	−9	−7	−10	−8	−12	−7	−4	−7	−13
250,000 to 499,999	38	13,663	−7	−10	−7	−13	−6	−12	−8	−7	−6	−9
100,000 to 249,999	154	22,169	−6	−9	−6	−10	−3	−11	−8	−6	−6	−8
50,000 to 99,999	327	22,096	−7	−8	−7	−7	−3	−10	−6	−8	−7	−11
25,000 to 49,999	620	21,524	−7	−9	−7	−8	−8	−12	−7	−7	−7	−8
10,000 to 24,999	1,453	22,924	−5	−5	−5	+4	−1	−6	−5	−4	−5	−5
Under 10,000	5,239	17,929	−4	*	−4	−7	+3	−4	+1	−4	−4	−5
Counties:												
Suburban	1,113	51,391	−6	−5	−6	*	−3	−11	−4	−6	−6	−8
Rural[4]	2,112	23,548	−5	−2	−5	−4	−9	−9	*	−5	−5	−2
Areas:												
Suburban area[5]	5,526	93,828	−6	−5	−6	−2	−2	−10	−4	−6	−5	−8

*=Less than 0.5% change. (1) Law-enforcement agencies. (2) Violent crimes are murder, forcible rape, robbery, and aggravated assault. (3) Property crimes are burglary, larceny-theft, and motor vehicle theft. Data for the property crime of arson are not included. (4) Includes state police agencies with no county breakdowns. (5) Includes suburban city and county law enforcement agencies within metropolitan areas, but not central cities. Suburban cities and counties are also included in other groups.

Crime Index Trends by Geographic Region, 1998

Source: FBI, *Uniform Crime Reports,* 1998

(percentage change 1998 over 1997, offenses known to the police)

	Crime Index (total)	Violent crime	Property crime[1]	Murder	Forcible rape	Robbery	Aggravated assault	Burglary	Larceny-theft	Motor vehicle theft
TOTAL U.S.	−5	−6	−5	−7	−3	−10	−5	−5	−5	−8
Northeast	−7	−7	−7	−10	−3	−9	−5	−9	−5	−11
Midwest	−4	−6	−4	−6	−2	−9	−4	−3	−4	−8
South	−5	−6	−5	−6	−4	−11	−4	−4	−5	−6
West	−7	−8	−7	−9	−2	−12	−6	−8	−6	−10

(1) Data for arson not included.

Crime Index Trends, 1992-98

Source: FBI, *Uniform Crime Reports,* 1998

(percentage change over previous year, offenses known to police)

Year	Crime Index (total)	Violent crime	Property crime[1]	Murder	Forcible rape	Robbery	Aggravated assault	Burglary	Larceny-theft	Motor vehicle theft
1992	−3	+1	−4	−4	+2	−2	+3	−6	−3	−3
1993	−2	0	−2	+3	−4	−2	+1	−5	−1	−3
1994	−1	−3	−1	−5	−4	−6	−1	−4	+1	−2
1995	−1	−1	−1	−7	−6	−7	−3	−5	+1	−5
1996	−3	−7	−2	−9	−2	−8	−6	−4	−1	−5
1997	−2	−3	−2	−8	0	−7	−1	−2	−2	−3
1998	−5	−6	−5	−7	−3	−10	−5	−5	−5	−8

(1) Data for arson not included.

Crime in the U.S., 1979-98

Source: FBI, *Uniform Crime Reports*, 1998

Population[1]	Crime Index (total)[2]	Violent crime	Property crime[3]	Murder and non-negligent man-slaughter	Forcible rape	Robbery	Burglary	Larceny-theft
			NUMBER OF REPORTED OFFENSES					
Population by year								
1979–220,099,000	12,249,500	1,208,030	11,041,500	21,460	76,390	480,700	3,327,700	6,601,000
1980–225,349,264	13,408,300	1,344,520	12,063,700	23,040	82,990	565,840	3,795,200	7,136,900
1981–229,146,000	13,423,800	1,361,820	12,061,900	22,520	82,500	592,910	3,779,700	7,194,400
1982–231,534,000	12,974,400	1,322,390	11,652,000	21,010	78,770	553,130	3,447,100	7,142,500
1983–233,981,000	12,108,600	1,258,090	10,850,500	19,310	78,920	506,570	3,129,900	6,712,800
1984–236,158,000	11,881,800	1,273,280	10,608,500	18,690	84,230	485,010	2,984,400	6,591,900
1985–238,740,000	12,431,400	1,328,770	11,102,600	18,980	88,670	497,870	3,073,300	6,926,400
1986–241,077,000	13,211,900	1,489,170	11,722,700	20,610	91,460	542,780	3,241,400	7,257,200
1987–243,400,000	13,508,700	1,484,000	12,024,700	20,100	91,110	517,700	3,236,200	7,499,900
1988–245,807,000	13,923,100	1,566,220	12,356,900	20,680	92,490	542,970	3,218,100	7,705,900
1989–248,239,000	14,251,400	1,646,040	12,605,400	21,500	94,500	578,330	3,168,200	7,872,400
1990–248,709,873	14,475,600	1,820,130	12,655,500	23,440	102,560	639,270	3,073,900	7,945,700
1991–252,177,000	14,872,900	1,911,770	12,961,100	24,700	106,590	687,730	3,157,200	8,142,200
1992–255,082,000	14,438,200	1,932,270	12,505,900	23,760	109,060	672,480	2,979,900	7,915,200
1993–257,908,000	14,144,800	1,926,020	12,218,800	24,530	106,010	659,870	2,834,800	7,820,900
1994–260,341,000	13,989,500	1,857,670	12,131,900	23,330	102,220	618,950	2,712,800	7,879,800
1995–262,755,000	13,862,700	1,798,790	12,063,900	21,610	97,470	580,510	2,593,800	7,997,700
1996–265,284,000	13,493,900	1,688,540	11,805,300	19,650	96,250	535,590	2,506,400	7,904,700
1997–267,637,000[4]	13,194,600	1,636,100	11,558,500	18,210	96,150	498,530	2,460,500	7,743,800
1998–270,296,000	12,475,600	1,531,040	10,944,600	16,910	93,100	446,630	2,330,000	7,373,900
			PERCENT CHANGE: NUMBER OF OFFENSES					
1998/1997	−5.4	−6.4	−5.3	−7.1	−3.2	−10.4	−5.3	−4.8
1998/1994	−10.8	−17.6	−9.8	−27.5	−8.9	−27.8	−14.1	−6.4
1998/1989	−12.5	−7.0	−13.2	−21.3	−1.5	−22.8	−26.5	−6.3
			RATE PER 100,000 INHABITANTS					
Year								
1979	5,565.5	548.9	5,016.6	9.7	34.7	218.4	1,511.9	2,999.1
1980	5,950.0	596.6	5,353.3	10.2	36.8	251.1	1,684.1	3,167.0
1981	5,858.2	594.3	5,263.9	9.8	36.0	258.7	1,649.5	3,139.7
1982	5,603.6	571.1	5,032.5	9.1	34.0	238.9	1,488.8	3,084.8
1983	5,175.0	537.7	4,637.4	8.3	33.7	216.5	1,337.7	2,868.9
1984	5,031.3	539.2	4,492.1	7.9	35.7	205.4	1,263.7	2,791.3
1985	5,207.1	556.6	4,650.5	8.0	37.1	208.5	1,287.3	2,901.2
1986	5,480.4	617.7	4,862.6	8.6	37.9	225.1	1,344.6	3,010.3
1987	5,550.0	609.7	4,940.3	8.3	37.4	212.7	1,329.6	3,081.3
1988	5,664.2	637.2	5,027.1	8.4	37.6	220.9	1,309.2	3,134.9
1989	5,741.0	663.1	5,077.9	8.7	38.1	233.0	1,276.3	3,171.3
1990	5,820.3	731.8	5,088.5	9.4	41.2	257.0	1,235.9	3,194.8
1991	5,897.8	758.1	5,139.7	9.8	42.3	272.7	1,252.0	3,228.8
1992	5,660.2	757.5	4,902.7	9.3	42.8	263.6	1,168.2	3,103.0
1993	5,484.4	746.8	4,737.6	9.5	41.1	255.9	1,099.2	3,032.4
1994	5,373.5	713.6	4,660.0	9.0	39.3	237.7	1,042.0	3,026.7
1995	5,275.9	684.6	4,591.3	8.2	37.1	220.9	987.1	3,043.8
1996	5,086.6	636.5	4,450.1	7.4	36.3	201.9	944.8	2,979.7
1997[4]	4,930.0	611.3	4,318.7	6.8	35.9	186.3	919.4	2,893.4
1998	4,615.5	566.4	4,049.1	6.3	34.4	165.2	862.0	2,728.1
			PERCENT CHANGE: RATE PER 100,000 INHABITANTS					
1998/1997	−6.4	−7.3	−6.2	−7.4	−4.2	−11.3	−6.2	−5.7
1998/1994	−14.1	−20.6	−13.1	−30.0	−12.5	−30.5	−17.3	−9.9
1998/1989	−19.6	−14.6	−20.3	−27.6	−9.7	−29.1	−32.5	−14.0

Note: All rates were calculated on the offenses before rounding. (1) Populations are Bureau of the Census provisional estimates as of July 1, except 1980 and 1990, which are the decennial census counts. (2) Because of rounding, violent and property crime may not add to total. Not all categories of violent and property crime appear separately. (3) Data for arson not included. (4) The 1997 figures have been adjusted.

Law Enforcement Officers, 1998

Source: FBI, *Uniform Crime Reports*, 1998

The U.S. law enforcement community employed an average of 2.5 full-time officers for every 1,000 inhabitants as of Oct. 31, 1998.

Including full-time civilians, the overall law enforcement employee rate was 3.4 per 1,000 inhabitants, according to 13,865 city, county, and state police agencies. These agencies collectively offered law enforcement service covering a population of about 260 million, employing 641,208 officers and 253,327 civilians.

The law enforcement employee average for all cities nationwide was 3.1 per 1,000 inhabitants. The highest city law enforcement employee average was 4.1 per 1,000 inhabitants, in cities with populations of 250,000 or more and in rural and suburban counties.

Regionally, the law enforcement employee rate was 3.6 in the South, 3.5 in the Northeast, 2.8 in the Midwest, and 2.5 in the West. Nationally, males constituted 89 percent of all sworn employees. 92% of the officers in rural counties were males, in suburban counties males accounted for 88 percent.

Civilians made up 28 percent of the total U.S. law enforcement employee force. They represented 22 percent of the police employees in cities, 37 percent of those in rural counties, and 38 percent in suburban counties. Females accounted for 63 percent of all civilian employees.

Sixty-one law enforcement officers were feloniously slain in the line of duty in 1998, 9 fewer than in 1997. Another 78 officers were killed as a result of accidents occurring while performing official duties, 16 higher than in 1997.

Crime Rates by Region, Geographic Division, and State, 1998

Source: FBI, *Uniform Crime Reports*, 1998

(rate per 100,000 population)

	Total rate	Violent crime[1]	Property crime[2]	Murder	Rape	Robbery	Aggra-vated assault	Burglary	Larceny-theft	Motor vehicle theft
U.S. TOTAL	4,615.5	566.4	4,049.1	6.3	34.4	165.2	360.5	862.0	2,728.1	459.0
Northeast	**3,473.5**	**500.5**	**2,973.1**	**4.3**	**23.7**	**184.8**	**287.7**	**589.2**	**2,011.0**	**372.8**
New England	**3,388.1**	**422.5**	**2,965.6**	**2.5**	**26.5**	**86.0**	**307.5**	**608.4**	**2,008.8**	**348.5**
Connecticut	3,786.5	366.3	3,420.2	4.1	22.2	133.8	206.2	665.9	2,366.3	388.1
Maine	3,040.7	125.8	2,914.9	2.0	18.1	21.1	84.6	666.8	2,126.8	121.3
Massachusetts	3,435.9	621.3	2,814.6	2.0	27.4	96.6	495.2	607.3	1,777.7	429.5
New Hampshire	2,419.8	107.2	2,312.7	1.5	33.8	21.5	50.4	325.1	1,863.2	124.4
Rhode Island	3,517.8	312.1	3,205.7	2.4	35.5	66.7	207.5	653.0	2,165.1	387.6
Vermont	3,139.1	106.3	3,032.8	2.2	27.6	9.5	67.0	671.1	2,213.9	147.9
Middle Atlantic	**3,503.5**	**527.8**	**2,975.7**	**4.9**	**22.7**	**219.4**	**280.8**	**582.5**	**2,011.8**	**381.4**
New Jersey	3,654.1	440.1	3,213.9	4.0	20.0	186.2	230.0	671.1	2,109.3	433.6
New York	3,588.5	637.8	2,950.7	5.1	21.1	270.3	341.3	576.7	1,998.9	375.1
Pennsylvania	3,273.0	420.5	2,852.4	5.3	26.9	164.9	223.5	531.4	1,965.4	355.5
Midwest	**4,379.4**	**494.1**	**3,885.3**	**5.7**	**36.8**	**141.4**	**310.2**	**775.7**	**2,713.4**	**396.2**
East North Central	**4,441.2**	**536.9**	**3,904.3**	**6.4**	**37.5**	**161.1**	**331.9**	**789.4**	**2,685.7**	**429.2**
Illinois	4,872.8	807.7	4,065.0	8.4	34.0	248.5	516.9	826.1	2,799.4	439.5
Indiana	4,169.4	431.0	3,738.4	7.7	33.1	111.2	279.0	789.2	2,590.1	359.2
Michigan	4,682.9	620.8	4,062.1	7.3	50.4	155.8	407.3	837.8	2,630.0	594.3
Ohio	4,327.5	362.5	3,965.0	4.0	40.5	133.5	184.5	810.1	2,771.1	383.8
Wisconsin	3,543.1	249.0	3,294.1	3.6	19.9	85.6	139.9	569.3	2,452.8	272.0
West North Central	**4,233.2**	**392.9**	**3,840.3**	**4.3**	**35.1**	**94.8**	**258.7**	**743.4**	**2,778.8**	**318.1**
Iowa	3,500.6	311.5	3,189.1	1.9	25.4	50.9	233.3	673.7	2,306.6	208.7
Kansas	4,858.8	397.0	4,461.7	5.9	42.6	86.8	261.8	892.6	3,341.4	227.8
Minnesota	4,046.5	310.2	3,736.3	2.6	49.9	92.5	165.2	687.5	2,723.6	325.2
Missouri	4,826.4	555.7	4,270.7	7.3	26.9	149.2	372.2	872.5	2,948.4	449.8
Nebraska	4,405.2	451.4	3,953.8	3.1	25.1	77.6	345.7	634.0	2,971.7	348.0
North Dakota	2,681.0	89.3	2,591.7	1.1	33.2	10.2	44.8	356.4	2,058.6	176.6
South Dakota	2,624.1	154.3	2,469.8	1.4	35.0	20.2	97.8	468.6	1,897.8	103.4
South	**5,223.5**	**633.0**	**4,590.5**	**7.8**	**38.1**	**169.1**	**418.0**	**1,045.4**	**3,074.7**	**470.4**
South Atlantic	**5,565.9**	**703.5**	**4,862.4**	**7.8**	**37.5**	**197.8**	**460.5**	**1,088.4**	**3,269.3**	**504.7**
Delaware	5,363.2	762.4	4,600.8	2.8	67.1	194.2	498.3	859.5	3,313.0	428.2
District of Columbia	8,835.6	1,718.5	7,117.0	49.7	36.3	689.5	943.0	1,216.3	4,657.7	1,243.0
Florida	6,886.0	938.7	5,947.4	6.5	49.6	242.7	639.9	1,361.7	3,886.8	698.9
Georgia	5,463.0	572.7	4,890.3	8.1	30.4	187.2	347.0	990.8	3,342.8	556.6
Maryland	5,365.7	796.6	4,569.1	10.0	33.4	298.7	454.5	922.9	3,096.8	549.4
North Carolina	5,322.2	579.4	4,742.8	8.1	30.6	160.8	379.9	1,324.6	3,092.0	326.2
South Carolina	5,777.0	903.2	4,873.8	8.0	45.7	154.9	694.6	1,162.7	3,295.4	415.7
Virginia	3,660.4	325.7	3,334.7	6.2	26.7	105.6	187.2	560.9	2,503.5	270.3
West Virginia	2,547.2	248.6	2,298.6	4.3	18.7	37.3	188.3	613.5	1,497.9	187.2
East South Central	**4,297.6**	**507.6**	**3,790.0**	**8.0**	**37.1**	**131.9**	**330.6**	**953.1**	**2,467.4**	**369.5**
Alabama	4,597.1	512.1	4,085.0	8.1	33.2	130.9	339.9	964.3	2,779.0	341.7
Kentucky	2,889.4	284.0	2,605.3	4.6	29.3	75.4	174.7	637.4	1,750.1	217.8
Mississippi	4,384.0	410.7	3,973.3	11.4	37.3	123.3	238.6	1,144.5	2,490.0	338.7
Tennessee	5,034.4	715.0	4,319.4	8.5	45.8	178.0	482.8	1,075.9	2,726.1	517.4
West South Central	**5,173.2**	**586.8**	**4,586.4**	**7.7**	**39.7**	**142.8**	**396.6**	**1,025.9**	**3,090.5**	**470.0**
Arkansas	4,283.4	490.2	3,793.2	7.9	35.2	96.2	350.9	928.3	2,581.8	283.2
Louisiana	6,098.3	779.5	5,318.8	12.8	36.8	198.0	531.9	1,172.1	3,605.1	541.6
Oklahoma	5,003.9	539.4	4,464.5	6.1	45.2	92.0	396.1	1,143.4	2,915.8	405.3
Texas	5,111.6	564.6	4,547.0	6.8	40.0	145.1	372.6	986.3	3,071.7	489.1
West	**4,879.4**	**593.1**	**4,286.3**	**6.1**	**35.4**	**167.1**	**384.5**	**895.7**	**2,810.0**	**580.6**
Mountain	**5,409.5**	**490.2**	**4,919.4**	**6.4**	**40.2**	**119.8**	**323.8**	**978.8**	**3,383.4**	**557.1**
Arizona	6,575.0	577.9	5,997.0	8.1	31.1	165.2	373.6	1,209.5	3,922.4	865.1
Colorado	4,487.5	377.9	4,109.5	4.6	47.4	81.5	244.4	786.5	2,917.9	405.1
Idaho	3,714.6	282.2	3,432.5	2.9	31.4	21.5	226.4	693.1	2,553.7	185.7
Montana	4,070.7	138.8	3,931.9	4.1	17.8	19.9	96.9	511.5	3,191.6	228.9
Nevada	5,280.5	643.6	4,636.9	9.7	52.1	254.9	326.8	1,137.6	2,711.3	788.0
New Mexico	6,719.1	961.4	5,757.7	10.9	55.1	163.4	732.0	1,394.0	3,743.9	619.9
Utah	5,505.9	314.2	5,191.7	3.1	41.7	66.0	203.5	812.9	4,012.1	366.7
Wyoming	3,807.7	247.6	3,560.1	4.8	27.7	16.2	199.0	560.5	2,860.5	139.1
Pacific	**4,674.3**	**633.0**	**4,041.3**	**6.0**	**33.6**	**185.4**	**408.0**	**863.5**	**2,588.1**	**589.7**
Alaska	4,777.0	653.9	4,123.1	6.7	68.6	86.6	492.0	667.4	3,031.1	424.6
California	4,342.8	703.7	3,639.1	6.6	29.9	210.6	456.6	823.5	2,217.1	598.5
Hawaii	5,333.0	246.9	5,086.1	2.0	29.5	102.7	112.7	936.2	3,681.0	468.9
Oregon	5,646.6	419.8	5,226.8	3.8	39.8	105.2	271.0	927.5	3,773.3	526.0
Washington	5,867.4	428.5	5,438.9	3.9	48.2	115.6	260.8	1,062.5	3,757.7	618.7

Note: Offense totals are based on all reporting agencies and estimates for unreported areas. Totals may not add because of rounding. (1) Violent crimes are murder, forcible rape, robbery, and aggravated assault. (2) Property crimes are burglary, larceny-theft, and motor vehicle theft. Data not included for property crime of arson.

State and Federal Prison Population, Death Penalty, 1997-98[1]

Source: Bureau of Justice Statistics, U.S. Dept. of Justice

The total number of prisoners under the jurisdiction of federal or state adult correctional authorities was at a record high of 1,302,019 at year-end 1998. Overall, the nation's prison population grew 4.8%, which was less than the average annual growth of 6.7% since 1990. In absolute numbers, prison growth during 1998 was equivalent to 1,151 more inmates per week, up from 1,130 per week in 1997. At year-end 1998, state and federal prisons housed two-thirds of the incarcerated population (1,232,538 out of 1,825,000). Jails, which are locally operated and typically hold persons awaiting trial and those with sentences of a year or less, held the other third. Relative to the number of U.S. residents, the rate of incarceration in prisons was 461 sentenced inmates per 100,000 residents, up from 292 in 1990 (1 in every 113 men and 1 in every 1,754 women were sentenced prisoners).

| | SENTENCED TO MORE THAN 1 YEAR | | | DEATH PENALTY, 1997 | | |
	Advance[2] 1998	Final[3] 1997	% change 1997–98	Under sentence of death	Executions	Death penalty
U.S. TOTAL	1,252,830	1,195,498	4.8	3,335	74	—
Federal institutions	103,682	94,987	9.2	15	0	Yes
State institutions	1,149,148	1,100,511	4.4	3,320	74	38
Northeast	169,731	162,744	4.3	232	0	—
Connecticut	12,193	11,920	2.3	4	0	Yes
Maine	1,562	1,542	1.3	—	—	No
Massachusetts	10,739	10,847	-1.0	—	—	No
New Hampshire	2,169	2,168	0.0	0	0	Yes
New Jersey	31,121	28,361	9.7	14	0	Yes
New York	72,289	70,021	3.2	0	0	Yes
Pennsylvania	36,373	34,957	4.1	214	0	Yes
Rhode Island	2,175	2,100	3.6	—	—	No
Vermont	1,110	828	34.1	—	—	No
Midwest	226,788	217,383	4.3	481	10	—
Illinois	43,051	40,788	5.5	159	2	Yes
Indiana	19,016	17,730	7.3	44	1	Yes
Iowa	7,394	6,938	6.6	—	—	No
Kansas	8,183	7,911	3.4	0	0	Yes
Michigan	45,879	44,771	2.5	—	—	No
Minnesota	5,557	5,306	4.7	—	—	No
Missouri	24,949	23,998	4.0	88	6	Yes
Nebraska	3,588	3,329	7.8	11	1	Yes
North Dakota	814	715	13.8	—	—	No
Ohio	48,450	48,016	0.9	177	0	Yes
South Dakota	2,430	2,242	8.4	2	0	Yes
Wisconsin	17,477	15,639	11.8	—	—	No
South	499,184	479,278	4.2	1,838	60	—
Alabama	22,655	21,680	4.5	159	3	Yes
Arkansas	10,561	9,936	6.3	38	4	Yes
Delaware	3,211	3,264	-1.6	15	0	Yes
District of Columbia	9,949	9,353	6.4	—	—	No
Florida	67,193	64,574	4.1	370	1	Yes
Georgia	38,758	35,787	8.3	115	0	Yes
Kentucky	14,987	14,600	2.7	30	1	Yes
Louisiana	32,227	29,265	10.1	70	1	Yes
Maryland	21,540	21,088	2.1	17	1	Yes
Mississippi	15,855	13,676	15.9	64	0	Yes
North Carolina	27,193	27,567	-1.4	176	0	Yes
Oklahoma	20,892	20,542	1.7	137	1	Yes
South Carolina	21,236	20,264	4.8	68	2	Yes
Tennessee	17,738	16,659	6.5	98	0	Yes
Texas	144,510	140,351	3.0	438	37	Yes
Virginia	27,191	27,524	-1.2	43	9	Yes
West Virginia	3,478	3,148	10.5	—	—	No
West	253,445	241,106	5.1	769	4	—
Alaska	2,541	2,571	-1.2	—	—	No
Arizona	23,955	22,353	7.2	120	2	Yes
California	159,109	152,739	4.2	486	0	Yes
Colorado	14,312	13,461	6.3	4	1	Yes
Hawaii	3,670	3,448	6.4	—	—	No
Idaho	4,083	3,911	4.4	19	0	Yes
Montana	2,734	2,517	8.6	7	0	Yes
Nevada	9,651	9,024	6.9	87	0	Yes
New Mexico	4,732	4,450	6.3	4	0	Yes
Oregon	8,596	7,589	13.3	20	1	Yes
Utah	4,337	4,280	1.3	10	0	Yes
Washington	14,154	13,214	7.1	12	0	Yes
Wyoming	1,571	1,549	1.4	0	0	Yes

(1) All information applies to Dec. 31 of the year indicated. (2) The advance count of prisoners is conducted in Jan. and may be revised. (3) Revised from previous tabulations.

Sentences vs. Time Served for Selected Crimes, 1996

Source: Bureau of Justice Statistics, *Truth in Sentencing in State Prisons*, Jan. 1999

The following is a comparison of the average maximum sentence lengths (excluding both life and death sentences) and the actual time served for selected state-court convictions.

Type of offense	Average sentence	Avg. time served[1]	Type of offense	Average sentence	Avg. time served[1]
All violent	7 years, 1 month	3 years, 3 months	Robbery	7 years, 8 months	3 years, 4 months
Homicide	15 years	7 years	Negligent		
Rape	9 years, 8 months	5 years, 1 month	manslaughter	8 years, 1 month	3 years, 5 months
Other sexual			Assault	5 years, 1 month	2 years, 4 months
assault	6 years, 9 months	3 years, 3 months	Other	5 years, 7 months	2 years, 5 months

(1) Includes jail credit and prison time.

Prison Situation Among the States and in the Federal System, 1998

Source: *Prisoners in 1998,* Bureau of Justice Statistics, U.S. Dept. of Justice; Aug. 1999

10 largest prison populations, 1998	Number of inmates	10 highest incarceration rates, 1998	Prisoners per 100,000 residents[1]	10 largest % increases in prison population			
				1997-98	% increase	1990-98	% increase
California	161,904	Louisiana	736	Mississippi	16.7	Texas	12.4
Texas	144,510	Texas	724	North Dakota	14.8	West Virginia	10.5
Federal system	123,041	Oklahoma	622	Wisconsin	13.4	Hawaii	10.0
New York	72,638	Mississippi	574	Vermont	12.3	Idaho	9.6
Florida	67,224	South Carolina	550	Oregon	11.6	Federal system	9.4
Ohio	48,450	Nevada	542	West Virginia	10.5	Mississippi	8.8
Michigan	45,879	Alabama	519	Louisiana	10.1	Montana	8.5
Illinois	43,051	Arizona	507	New Jersey	9.7	North Dakota	8.1
Georgia	39,252	Georgia	502	Federal system	8.9	Colorado	8.1
Pennsylvania	36,377	California	483	South Dakota	8.6	Iowa	8.1

(1) Prisoners with sentences of more than 1 year. The Federal Bureau of Prisons and the District of Columbia are excluded.

Executions, by State and Method, 1977-97

Source: Bureau of Justice Statistics, *Capital Punishment 1997,* Dec. 1998

	No.	METHOD OF EXECUTION						No.	METHOD OF EXECUTION				
		Lethal injection	Electrocution	Lethal gas	Firing squad	Hanging			Lethal injection	Electrocution	Lethal gas	Firing squad	Hanging
TOTAL U.S.	432	284	134	9	2	3	Mississippi	4	0	0	4	0	0
Alabama	16	0	16	0	0	0	Missouri	29	29	0	0	0	0
Arizona	8	7	0	1	0	0	Montana	1	1	0	0	0	0
Arkansas	16	15	1	0	0	0	Nebraska	3	0	3	0	0	0
California	4	2	0	2	0	0	Nevada	6	5	0	1	0	0
Colorado	1	1	0	0	0	0	North Carolina	8	7	0	1	0	0
Delaware	8	7	0	0	0	1	Oklahoma	9	9	0	0	0	0
Florida	39	0	39	0	0	0	Oregon	2	2	0	0	0	0
Georgia	22	0	22	0	0	0	Pennsylvania	2	2	0	0	0	0
Idaho	1	1	0	0	0	0	South Carolina	13	8	5	0	0	0
Illinois	10	10	0	0	0	0	Texas	144	144	0	0	0	0
Indiana	5	2	3	0	0	0	Utah	5	3	0	0	2	0
Kentucky	1	0	1	0	0	0	Virginia	46	22	24	0	0	0
Louisiana	24	4	20	0	0	0	Washington	2	0	0	0	0	2
Maryland	2	2	0	0	0	0	Wyoming	1	1	0	0	0	0

Note: This table shows execution methods used since 1977. Lethal injection was used in 66% of the executions carried out. Eleven states—Arizona, Arkansas, California, Delaware, Indiana, Louisiana, Nevada, North Carolina, South Carolina, Utah, and Virginia— have employed 2 methods.

Total Estimated Arrests,[1] 1998

Source: FBI, *Uniform Crime Reports,* 1998

TOTAL[2]	14,528,300	Stolen property: buying, receiving, possessing	137,900
Murder and nonnegligent manslaughter	17,450	Vandalism	300,200
Forcible rape	31,070	Weapons: carrying, possessing, etc.	190,600
Robbery	120,870	Prostitution and commercialized vice	94,000
Aggravated assault	506,630	Sex offenses (except forcible rape and prostitution)	93,600
Burglary	330,700	Drug abuse violations	1,559,100
Larceny–theft	1,307,100	Gambling	12,800
Motor vehicle theft	150,700	Offenses against family and children	146,400
Arson	17,200	Driving under the influence	1,402,800
Violent crimes[3]	**675,900**	Liquor laws	630,400
Property crime[4]	**1,805,600**	Drunkenness	710,300
Crime Index total[2,5]	**2,481,500**	Disorderly conduct	696,100
Other assaults	1,338,800	Vagrancy	30,400
Forgery and counterfeiting	114,600	All other offenses	3,824,100
Fraud	394,600	Suspicion (not included in totals)	5,200
Embezzlement	17,100	Curfew and loitering law violations	187,800
		Runaways	165,100

(1) Arrest totals are based on all reporting agencies and estimates for unreported areas. (2) Because of rounding, figures may not add to totals. (3) Violent crimes are murder, forcible rape, robbery, and aggravated assault. (4) Property crimes are burglary, larceny-theft, motor vehicle theft, and arson. (5) Includes arson.

Federal Bureau of Investigation

The Federal Bureau of Investigation was created July 26, 1908, and was referred to as Office of Chief Examiner. It became the Bureau of Investigation (Mar. 16, 1909), United States Bureau of Investigation (July 1, 1932), Division of Investigation (Aug. 10, 1933), and Federal Bureau of Investigation (July 1, 1935).

Director	Assumed office	Director	Assumed office
Stanley W. Finch	July 26, 1908	William D. Ruckelshaus, act.	Apr. 27, 1973
A(lexander) Bruce Bielaski	Apr. 30, 1912	Clarence M. Kelley	July 9, 1973
William E. Allen, act.	Feb. 10, 1919	William H. Webster	Feb. 23, 1978
William J. Flynn	July 1, 1919	John E. Otto, act.	May 26, 1987
William J. Burns	Aug. 22, 1921	William S. Sessions	Nov. 2, 1987
J. Edgar Hoover, act.	May 10, 1924	Floyd I. Clarke, act.	July 19, 1993
J. Edgar Hoover	Dec. 10, 1924	Louis J. Freeh	Sept. 1, 1993
L. Patrick Gray, act.	May 3, 1972		

SPORTS
Ten Most Dramatic Sports Events of 1999

The New York Yankees strengthened their claim to "Team of the Century" honors by defeating the Atlanta Braves, 4 games to 0 (Oct. 23-27), for their 25th World Series win and 3d back-to-back Series sweep. In an exciting season, Tony Gwynn (3,067) and Wade Boggs (3,010) joined the 3,000 hit club on Aug. 6 and 7, respectively. Mark McGwire (522) reached the 500 career home-run mark Aug. 5, finishing the season 10th all-time. In addition, McGwire (65) and Sammy Sosa (63) became the first players to hit 60 or more home runs in consecutive seasons.

On Jan. 31 the Denver Broncos defeated the Atlanta Falcons, 34-19, in Super Bowl XXXIII at Pro Player Stadium, Miami, FL. It was Denver's 2d-straight NFL title. Broncos quarterback John Elway threw for 336 yards and one touchdown and was named the game's MVP. Elway improved his Super Bowl record to 2 wins and 3 losses. Denver running back Terrell Davis, the regular season MVP, rushed for 102 yards.

The San Antonio Spurs capped off an incredible 15-2 playoff run by defeating the New York Knicks, 78-77, June 25, in Game 5 of the NBA Finals. San Antonio, led by "twin towers" Tim Duncan and David Robinson, took the series 4 games to 1 for its 1st championship ever. Duncan averaged 27.4 points and 14.0 rebounds per game, and was unanimously chosen as the Finals MVP.

The Dallas Stars defeated the Buffalo Sabres, 2-1, to win the Stanley Cup on June 20 in Buffalo, NY. In the 2d-longest overtime game in Stanley Cup finals history, Brett Hull scored a dramatic, controversial goal to wrap up the series for Dallas, 4 games to 2. The franchise's 1st championship came 6 years after the team moved from Minnesota (North Stars) to Dallas. Dallas center Joe Nieuwendyk was series MVP.

The year 1999 saw the retirement of some of the greatest stars ever to play in their respective sports—including Michael Jordan, who led the Chicago Bulls to 6 championships (Jan. 13), Wayne Gretzky, who held or tied 61 NHL records (Apr. 16), John Elway, who led the Denver Broncos to 2 straight Super Bowl victories (May 2), and Steffi Graf, ranked No. 1 in women's tennis for a record 377 weeks (Aug. 13).

Women's soccer got a huge boost in the U.S. when the U.S. women's soccer team captured its 2d World Cup, July 10, in Pasadena, CA. The title game matched the favored U.S. team against China and drew a crowd of 90,185, a U.S. record for a women's sporting event. Midfielder Brandi Chastain scored on her penalty kick in the 5th round of the shootout to give the U.S. a 5-4 victory (on penalty kicks) after scoreless regulation and overtime play.

In a remarkable turn of events, Lance Armstrong of the U.S. Postal Service team won the 86th Tour de France, July 25. Less than 3 years earlier, he had been diagnosed with cancer and given only a 50-50 chance of survival. The 27-year-old Texan fought back courageously, and became the 1st American cyclist on an American team to win the Tour. He won the prologue and 3 tour stages, and finished 7 minutes and 37 seconds ahead of 2d-place finisher Alex Zülle of Switzerland.

In the 1st year of the Bowl Championship Series (BCS), the top-ranked Tennessee Volunteers took 1st place by defeating the 2d-ranked Florida State Seminoles, 23-16, in the Fiesta Bowl in Tempe, AZ, Jan. 4, 1999. The victory gave Tennessee an undefeated (13-0) season and its 1st national championship since 1951. Tennessee receiver Peerless Price was named MVP.

On Mar. 29 the University of Connecticut Huskies startled the heavily favored Duke Blue Devils, 77-74, to win their 1st national title, at the NCAA men's basketball tournament in St. Petersburg, FL. Connecticut's Richard Hamilton was named tournament MVP. In the women's tournament the top-ranked Purdue Boilermakers also won their 1st national championship, Mar. 28, also by defeating the Duke Blue Devils, 62-45, in San Jose, CA. Ukari Figgs of Purdue was named tournament MVP.

On Sept. 26, Team USA, in a record comeback, defeated Team Europe, 14½-13½, in Brookline, MA, to win the Ryder Cup. Going into the final day, the American golfers trailed, 10-6, but they managed to win 8½ points from the 12 singles matches to clinch their 1st Cup since 1993. They won after Justin Leonard sank a 45-foot putt to take the lead on the 17th hole.

OLYMPIC GAMES
Winter Olympic Games
Sites of Winter Olympic Games

1924 Chamonix, France	**1952** Oslo, Norway	**1972** Sapporo, Japan	**1992** Albertville, France
1928 St. Moritz, Switzerland	**1956** Cortina d'Ampezzo, Italy	**1976** Innsbruck, Austria	**1994** Lillehammer, Norway
1932 Lake Placid, New York	**1960** Squaw Valley, California	**1980** Lake Placid, New York	**1998** Nagano, Japan
1936 Garmisch-Partenkirchen, Germany	**1964** Innsbruck, Austria	**1984** Sarajevo, Yugoslavia	**2002** Salt Lake City, Utah
1948 St. Moritz, Switzerland	**1968** Grenoble, France	**1988** Calgary, Alberta	**2006** Turin, Italy

Winter Olympic Games in 1998

Nagano, Japan, Feb. 7-22, 1998

With a crowd of 50,000 in attendance, a ritual performed by sumo wrestlers to cast off evil spirits, and the lighting of the Olympic torch by skater Midori Ito, the 1998 Winter Olympic Games got underway Feb. 7, 1998, in Nagano, Japan. Over 2,400 athletes from 72 nations participated. Snowboarding and curling made their Olympic debuts.

The 1st gold medal for the host nation came Feb. 10, when Hiroyasu Shimizu won the men's 500-meter speedskating event. He set an Olympic record on his 2d run with a time of 35.59 seconds. The Japanese ski jumping team won 4 medals, including a gold in the team event led by Masashiko "Happy" Harada. Hermann Maier won the gold for Austria in both the giant and the super giant slalom, after recovering from a crash in the downhill; cross-country skier Bjoern Daehlie won 3 golds and 1 silver for Norway, giving him an unprecedented 12 total and 8 gold Winter Olympic medals for his career. Larissa Lazutina, a Russian cross-country skier, won 5 medals, including 3 gold, to become the leading multi-medalist at Nagano. The Czech Republic, led by Dominik Hasek, captured the gold in men's hockey; the U.S. women's hockey team upset Team Canada to win the gold in the event's inaugural year; U.S. skater Tara Lipinski, 15, nailed her long program to become the youngest Olympic figure-skating gold medalist.

Germany won the most medals, 29, and the most gold medals, 12. The U.S. finished 6th in the medal count, with 13 (6 gold), trailing Germany, Norway (25), Russia (18), Austria (17), and Canada (15).

Final Medal Standings

	Gold	Silver	Bronze	Total
Germany	12	9	8	29
Norway	10	10	5	25
Russia	9	6	3	18
Austria	3	5	9	17
Canada	6	5	4	15
U.S.	6	3	4	13
Finland	2	4	6	12
Netherlands	5	4	2	11
Japan	5	1	4	10
Italy	2	6	2	10
France	2	1	5	8
China	0	6	2	8
Switzerland	2	2	3	7

	Gold	Silver	Bronze	Total
Korea	3	1	2	6
Czech Republic	1	1	1	3
Sweden	0	2	1	3
Belarus	0	0	2	2
Kazakhstan	0	0	2	2
Bulgaria	1	0	0	1
Denmark	0	1	0	1
Ukraine	0	1	0	1
Australia	0	0	1	1
Belgium	0	0	1	1
Great Britain	0	0	1	1
TOTAL	**69**	**68**	**68**	**205**

1998 Winter Olympics Medal Winners
(G = Gold, S = Silver, B = Bronze)

Alpine Skiing

MEN

Downhill—G-Jean-Luc Cretier, France; S-Lasse Kjus, Norway; B-Hannes Trinkl, Austria.

Super Giant Slalom—G-Hermann Maier, Austria; S-Didier Cuche, Switzerland; S-Hans Knauss, Austria.

Giant Slalom—G-Hermann Maier, Austria; S-Stefan Eberharter, Austria; B-Michael von Gruenigen, Switzerland.

Slalom—G-Hans-Petter Buraas, Norway; S-Ole Christian Furuseth, Norway; B-Thomas Sykora, Austria.

Combined—G-Mario Reiter, Austria; S-Lasse Kjus, Norway; B-Christian Mayer, Austria.

WOMEN

Downhill—G-Katja Seizinger, Germany; S-Pernilla Wiberg, Sweden; B-Florence Masnada, France.

Super Giant Slalom—G-Picabo Street, U.S.; S-Michaela Dorfmeister, Austria; B-Alexandra Meissnitzer, Austria.

Giant Slalom—G-Deborah Compagnoni, Italy; S-Alexandra Meissnitzer, Austria; B-Katja Seizinger, Germany.

Slalom—G-Hilde Gerg, Germany; S-Deborah Compagnoni, Italy; B-Zali Steggall, Australia.

Combined—G-Katja Seizinger, Germany; S-Martina Ertl, Germany; B-Hilde Gerg, Germany.

Biathlon

MEN

10KM—G-Ole Einar Bjoerndalen, Norway; S-Frode Andresen, Norway; B-Ville Raikkonen, Finland.

20KM—G-Halvard Hanevold, Norway; S-Pier Alberto Carrara, Italy; B-Aleksei Aidarov, Belarus.

30KM Relay—G-Germany; S-Norway; B-Russia.

WOMEN

7.5KM—G-Galina Koukleva, Russia; S-Ursula Disl, Germany; B-Katrin Apel, Germany.

15KM—G-Ekaterina Dafovska, Bulgaria; S-Elena Petrova, Ukraine; B-Ursula Disl, Germany.

30KM Relay—G-Germany; S-Russia; B-Norway.

Bobsledding

4-Man Bob—G-Germany 2; S-Switzerland 1; B-Great Britain 1 and France 1 (tie).

2-Man Bob—G-Canada 1; G-Italy 1; B-Germany 1.

Curling

Men—G-Switzerland; S-Canada; B-Norway.

Women—G-Canada; S-Denmark; B-Sweden.

Figure Skating

Men's Singles—G-Ilia Kulik, Russia; S-Elvis Stojko, Canada; B-Philippe Candeloro, France.

Women's Singles—G-Tara Lipinski, U.S.; S-Michelle Kwan, U.S.; B-Lu Chen, China.

Pairs—G-Oksana Kazakova & Artur Dmitriev, Russia; S-Elena Berezhnaya & Anton Sikharulidze, Russia; B-Mandy Wotzel & Ingo Steuer, Germany.

Ice Dancing—G-Pasha Grishuk & Evgeny Platov, Russia; S-Angelika Krylova & Oleg Ovsyannikov, Russia; B-Marina Anissina & Gwendal Peizerat, France.

Freestyle Skiing

Men's Moguls—G-Jonny Moseley, U.S.; S-Janne Lahtela, Finland; B-Sami Mustonen, Finland.

Men's Aerials—G-Eric Bergoust, U.S.; S-Sebastien Foucras, France; B-Dmitri Dashchinsky, Belarus.

Women's Moguls—G-Tae Satoya, Japan; S-Tatjana Mittermayer, Germany; B-Kari Traa, Norway.

Women's Aerials—G-Nikki Stone, U.S.; S-Nannan Xu, China; B-Colette Brand, Switzerland.

Ice Hockey

Men's—G-Czech Republic; S-Russia; B-Finland.

Women's—G-U.S.; S-Canada; B-Finland.

Luge

Men's Singles—G-Georg Hackl, Germany; S-Armin Zoeggeler, Italy; B-Jens Mueller, Germany.

Men's Doubles—G-Stefan Krausse and Jan Behrendt, Germany; S-Chris Thorpe and Gordy Sheer, U.S.; B-Mark Grimmette and Brian Martin, U.S.

Women's Singles—G-Silke Kraushaar, Germany; S-Barbara Niedernhuber, Germany; B-Angelika Neuner, Austria.

Nordic Skiing

Cross-Country Events

MEN

10KM—G-Bjoern Daehlie, Norway; S-Markus Gandler, Austria; B-Mika Myllylae, Finland.

15KM—G-Thomas Alsgaard, Norway; S-Bjoern Daehlie, Norway; B-Vladimir Smirnov, Kazakhstan.

30KM—G-Mika Myllylae, Finland; S-Erling Jevne, Norway; B-Silvio Fauner, Italy.

50KM—G-Bjoern Daehlie, Norway; S-Niklas Jonsson, Sweden; B-Christian Hoffmann, Austria.

40KM Relay—G-Norway; S-Italy; B-Finland.

WOMEN

5KM—G-Larissa Lazutina, Russia; S-Katerina Neumannova, Czech Republic; B-Benter Martinesen, Norway.

10KM—G-Larissa Lazutina, Russia; S-Olga Danilova, Russia; B-Katerina Neumannova, Czech Republic.

15KM—G-Olga Danilova, Russia; S-Larissa Lazutina, Russia; B-Anita Moen-Guidon, Norway.

30KM—G-Julija Tchepalova, Russia; S-Stefania Belmondo, Italy; B-Larissa Lazutina, Russia.

20KM Relay—G-Russia; S-Norway; B-Italy.

Combined Cross-Country & Jumping Events (Men)

Nordic Combined—G-Bjarte Engen Vik, Norway; S-Samppa Lajunen, Finland; B-Valerij Stoljarov, Russia.

Team Nordic Combined—G-Norway; S-Finland; B-France.

Ski Jumping (Men)

90M (Normal hill)—G-Jani Soininen, Finland; S-Kazuyoshi Funaki, Japan; B-Andreas Widhoelzl, Austria.

120M (Large hill)—G-Kazuyoshi Funaki, Japan; S-Jani Soininen, Finland; B-Masahiko Harada, Japan.

Team 120M—G-Japan; S-Germany; B-Austria.

Snowboarding

Men's Giant Slalom—G-Ross Rebagliati, Canada; S-Thomas Prugger, Italy; B-Ueli Kestenholz, Switzerland.
Men's Halfpipe—G-Gian Simmen, Switzerland; S-Daniel Franck, Norway; B-Ross Powers, U.S.
Women's Giant Slalom—G-Karine Ruby, France; S-Heidi Renoth, Germany; B-Brigitte Koeck, Austria.
Women's Halfpipe—G-Nicola Thost, Germany; S-Stine Brun Kjeldaas, Norway; B-Shannon Dunn, U.S.

Speed Skating

MEN

500M—G-Hiroyasu Shimizu, Japan; S-Jeremy Wotherspoon, Canada; B-Kevin Overland, Canada.
1,000M—G-Ids Postma, Netherlands; S-Jan Bos, Netherlands; B-Hiroyasu Shimizu, Japan.
1,500M—G-Aadne Sondral, Norway; S-Ids Postma, Netherlands; B-Rintje Ritsma, Netherlands.
5,000M—G-Gianni Romme, Netherlands; S-Rintje Ritsma, Netherlands; B-Bart Veldkamp, Belgium.
10,000M—G-Gianni Romme, Netherlands; S-Bob de Jong, Netherlands; B-Rintje Ritsma, Netherlands.

WOMEN

500M—G-Catriona LeMay-Doan, Canada; S-Susan Auch, Canada; B-Tomomi Okazaki, Japan.
1,000M—G-Marianne Timmer, Netherlands; S-Chris Witty, U.S.; B-Catriona LeMay-Doan, Canada.
1,500M—G-Marianne Timmer, Netherlands; S-Gunda Niemann-Stirnemann, Germany; B-Chris Witty, U.S.
3,000M—G-Gunda Niemann-Stirnemann, Germany; S-Claudia Pechstein, Germany; B-Anna Friesinger, Germany.
5,000M—G-Claudia Pechstein, Germany; S-Gunda Niemann-Stirnemann, Germany; B-Lyudmila Prokasheva, Kazakhstan.

Short-Track Speed Skating

Men's 500M—G-Takafumi Nishitani, Japan; S-Yulong An, China; B-Hitoshi Uematsu, Japan.
Men's 1,000M—G-Dong-Sung Kim, Korea; S-Jiajun Li, China; B-Eric Bedard, Canada.
Men's 5,000M Relay—G-Canada; S-Korea; B-China.
Women's 500M—G-Annie Perreault, Canada; S-Yang Yang, China; B-Chun Lee-Kyung, Korea.
Women's 1,000M—G-Chun Lee-Kyung, Korea; S-Yang S. Yang, China; B-Won Hye-Kyung, Korea.
Women's 3,000M Relay—G-Korea; S-China; B-Canada.

Winter Olympic Games Champions, 1924-98

In 1992, the Unified Team represented the former Soviet republics of Russia, Ukraine, Belarus, Kazakhstan, and Uzbekistan.

Alpine Skiing

Men's Downhill

		Time
1948	Henri Oreiller, France	2:55.0
1952	Zeno Colo, Italy	2:30.8
1956	Anton Sailer, Austria	2:52.2
1960	Jean Vuarnet, France	2:06.0
1964	Egon Zimmermann, Austria	2:18.16
1968	Jean-Claude Killy, France	1:59.85
1972	Bernhard Russi, Switzerland	1:51.43
1976	Franz Klammer, Austria	1:45.73
1980	Leonhard Stock, Austria	1:45.50
1984	Bill Johnson, U.S.	1:45.59
1988	Pirmin Zurbriggen, Switzerland	1:59.63
1992	Patrick Ortlieb, Austria	1:50.37
1994	Tommy Moe, U.S.	1:45.75
1998	Jean-Luc Cretier, France	1:50.11

Men's Super Giant Slalom

		Time
1988	Franck Piccard, France	1:39.66
1992	Kjetil-Andre Aamodt, Norway	1:13.04
1994	Markus Wasmeier, Germany	1:32.53
1998	Hermann Maier, Austria	1:34.82

Men's Giant Slalom

		Time
1952	Stein Eriksen, Norway	2:25.0
1956	Anton Sailer, Austria	3:00.1
1960	Roger Staub, Switzerland	1:48.3
1964	Francois Bonlieu, France	1:46.71
1968	Jean-Claude Killy, France	3:29.28
1972	Gustavo Thoeni, Italy	3:09.62
1976	Heini Hemmi, Switzerland	3:26.97
1980	Ingemar Stenmark, Sweden	2:40.74
1984	Max Julen, Switzerland	2:41.18
1988	Alberto Tomba, Italy	2:06.37
1992	Alberto Tomba, Italy	2:06.98
1994	Markus Wasmeier, Germany	2:52.46
1998	Hermann Maier, Austria	2:38.51

Men's Slalom

		Time
1948	Edi Reinalter, Switzerland	2:10.3
1952	Othmar Schneider, Austria	2:00.0
1956	Anton Sailer, Austria	3:14.7
1960	Ernst Hinterseer, Austria	2:08.9
1964	Josef Stiegler, Austria	2:11.13
1968	Jean-Claude Killy, France	1:39.73
1972	Francisco Fernandez Ochoa, Spain	1:49.27
1976	Piero Gros, Italy	2:03.29
1980	Ingemar Stenmark, Sweden	1:44.26
1984	Phil Mahre, U.S.	1:39.41
1988	Alberto Tomba, Italy	1:39.47
1992	Finn Christian Jagge, Norway	1:44.39
1994	Thomas Stangassinger, Austria	2:02.02
1998	Hans-Petter Buraas, Norway	1:49.31

Men's Combined

		Time
1988	Hubert Strolz, Austria	36.55 (pts.)
1992	Josef Polig, Italy	14.58 (pts.)
1994	Lasse Kjus, Norway	3:17.53
1998	Mario Reiter, Austria	3:08.06

Women's Downhill

		Time
1948	Hedi Schlunegger, Switzerland	2:28.3
1952	Trude Jochum-Beiser, Austria	1:47.1
1956	Madeleine Berthod, Switzerland	1:40.7
1960	Heidi Biebl, Germany	1:37.6
1964	Christl Haas, Austria	1:55.39
1968	Olga Pall, Austria	1:40.87
1972	Marie Therese Nadig, Switzerland	1:36.68
1976	Rosi Mittermaier, W. Germany	1:46.16
1980	Annemarie Proell Moser, Austria	1:37.52
1984	Michela Figini, Switzerland	1:13.36
1988	Marina Kiehl, W. Germany	1:25.86
1992	Kerrin Lee-Gartner, Canada	1:52.55
1994	Katja Seizinger, Germany	1:35.93
1998	Katja Seizinger, Germany	1:28.89

Women's Super Giant Slalom

		Time
1988	Sigrid Wolf, Austria	1:19.03
1992	Deborah Compagnoni, Italy	1:21.22
1994	Diann Roffe-Steinrotter, U.S.	1:22.15
1998	Picabo Street, U.S.	1:18.02

Women's Giant Slalom

		Time
1952	Andrea Mead Lawrence, U.S.	2:06.8
1956	Ossi Reichert, Germany	1:56.5
1960	Yvonne Ruegg, Switzerland	1:39.9
1964	Marielle Goitschel, France	1:52.24
1968	Nancy Greene, Canada	1:51.97
1972	Marie Therese Nadig, Switzerland	1:29.90
1976	Kathy Kreiner, Canada	1:29.13
1980	Hanni Wenzel, Liechtenstein (2 runs)	2:41.66
1984	Debbie Armstrong, U.S.	2:20.98
1988	Vreni Schneider, Switzerland	2:06.49
1992	Pernilla Wiberg, Sweden	2:12.74
1994	Deborah Compagnoni, Italy	2:30.97
1998	Deborah Compagnoni, Italy	2:50.59

Women's Slalom

		Time
1948	Gretchen Fraser, U.S.	1:57.2
1952	Andrea Mead Lawrence, U.S.	2:10.6
1956	Renee Colliard, Switzerland	1:52.3
1960	Anne Heggtveigt, Canada	1:49.6
1964	Christine Goitschel, France	1:29.86
1968	Marielle Goitschel, France	1:25.86
1972	Barbara Cochran, U.S.	1:31.24
1976	Rosi Mittermaier, W. Germany	1:30.54
1980	Hanni Wenzel, Liechtenstein	1:25.09
1984	Paoletta Magoni, Italy	1:36.47
1988	Vreni Schneider, Switzerland	1:36.69
1992	Petra Kronberger, Austria	1:32.68
1994	Vreni Schneider, Switzerland	1:56.01
1998	Hilde Gerg, Germany	1:32.40

Women's Combined

		Time
1988	Anita Wachter, Austria	29.25 (pts.)
1992	Petra Kronberger, Austria	2.55 (pts.)
1994	Pernilla Wiberg, Sweden	3:05.16
1998	Katja Seizinger, Germany	2:40.74

Biathlon

Men's 10 Kilometers
		Time
1980	Frank Ullrich, E. Germany	32:10.69
1984	Eirik Kvalfoss, Norway	30:53.80
1988	Frank-Peter Roetsch, E. Germany	25:08.10
1992	Mark Kirchner, Germany	26:02.30
1994	Serguei Tchepikov, Russia	28:07.00
1998	Ole Einar Bjoerndalen, Norway	27:16.20

Men's 20 Kilometers
		Time
1960	Klas Lestander, Sweden	1:33:21.6
1964	Vladimir Melanin, USSR	1:20:26.8
1968	Magnar Solberg, Norway	1:13:45.9
1972	Magnar Solberg, Norway	1:15:55.50
1976	Nikolai Kruglov, USSR	1:14:12.26
1980	Anatoly Aljabiev, USSR	1:08:16.31
1984	Peter Angerer, W. Germany	1:11:52.7
1988	Frank-Peter Roetsch, E. Germany	0:56:33.33
1992	Yevgeny Redkine, Unified Team	0:57:34.4
1994	Serguei Tarasov, Russia	0:57:25.3
1998	Halvard Hanevold, Norway	0:56:16.4

Men's 30-Kilometer Relay
		Time
1968	USSR, Norway, Sweden (40 km)	2:13:02.4
1972	USSR, Finland, E. Germany (40 km)	1:51:44.92
1976	USSR, Finland, E. Germany (40 km)	1:57:55.64
1980	USSR, E. Germany, W. Germany	1:34:03.27
1984	USSR, Norway, W. Germany	1:38:51.70
1988	USSR, W. Germany, Italy	1:22:30.00
1992	Germany, Unified Team, Sweden	1:24:43.50
1994	Germany, Russia, France	1:30:22.1
1998	Germany, Norway, Russia	1:19:43.3

Women's 7.5 Kilometers
		Time
1992	Anfissa Restsova, Unified Team	24:29.20
1994	Myriam Bedard, Canada	26:08.8
1998	Galina Koukleva, Russia	23:08.0

Women's 15 Kilometers
		Time
1992	Antje Misersky, Germany	51:47.2
1994	Myriam Bedard, Canada	52:06.6
1998	Ekaterina Dafovska, Bulgaria	54:52.0

Women's 22.5-Kilometer Relay
		Time
1992	France, Germany, Unified Team	1:15:55.6

Women's 30-Kilometer Relay
		Time
1994	Russia, Germany, France	1:47:19.5
1998	Germany, Russia, Norway	1:40:13.6

Bobsledding
(Driver in parentheses)

4-Man Bob
		Time
1924	Switzerland (Eduard Scherrer)	5:45.54
1928	United States (William Fiske) (5-man)	3:20.50
1932	United States (William Fiske)	7:53.68
1936	Switzerland (Pierre Musy)	5:19.85
1948	United States (Francis Tyler)	5:20.10
1952	Germany (Andreas Ostler)	5:07.84
1956	Switzerland (Franz Kapus)	5:10.44
1964	Canada (Victor Emery)	4:14.46
1968	Italy (Eugenio Monti) (2 races)	2:17.39
1972	Switzerland (Jean Wicki)	4:43.07
1976	E. Germany (Meinhard Nehmer)	3:40.43
1980	E. Germany (Meinhard Nehmer)	3:59.92
1984	E. Germany (Wolfgang Hoppe)	3:20.22
1988	Switzerland (Ekkehard Fasser)	3:47.51
1992	Austria (Ingo Appelt)	3:53.90
1994	Germany (Wolfgang Hoppe)	3:27.28
1998	Germany-2 (Christoph Langen)	2:39.41

2-Man Bob
		Time
1932	United States (Hubert Stevens)	8:14.74
1936	United States (Ivan Brown)	5:29.29
1948	Switzerland (F. Endrich)	5:29.20
1952	Germany (Andreas Ostler)	5:24.54
1956	Italy (Dalla Costa)	5:30.14
1964	Great Britain (Anthony Nash)	4:21.90
1968	Italy (Eugenio Monti)	4:41.54
1972	W. Germany (Wolfgang Zimmerer)	4:57.07
1976	E. Germany (Meinhard Nehmer)	3:44.42
1980	Switzerland (Erich Schaerer)	4:09.36
1984	E. Germany (Wolfgang Hoppe)	3:25.56
1988	USSR (Janis Kipours)	3:54.19
1992	Switzerland (Gustav Weber)	4:03.26
1994	Switzerland (Gustav Weber)	3:30.81
1998	Canada (Pierre Lueders), Italy (Guenther Huber) (tie)	3:37.24

Curling

MEN
1998	Switzerland, Canada, Norway

WOMEN
1998	Canada, Denmark, Sweden

Figure Skating

Men's Singles
1908#	Ulrich Salchow, Sweden
1920#	Gillis Grafstrom, Sweden
1924	Gillis Grafstrom, Sweden
1928	Gillis Grafstrom, Sweden
1932	Karl Schaefer, Austria
1936	Karl Schaefer, Austria
1948	Richard Button, U.S.
1952	Richard Button, U.S.
1956	Hayes Alan Jenkins, U.S.
1960	David W. Jenkins, U.S.
1964	Manfred Schnelldorfer, Germany
1968	Wolfgang Schwartz, Austria
1972	Ondrej Nepela, Czechoslovakia
1976	John Curry, Great Britain
1980	Robin Cousins, Great Britain
1984	Scott Hamilton, U.S.
1988	Brian Boitano, U.S.
1992	Viktor Petrenko, Unified Team
1994	Aleksei Urmanov, Russia
1998	Ilya Kulik, Russia

(#) Event was held at Summer Olympics.

Women's Singles
1908#	Madge Syers, Great Britain
1920#	Magda Julin-Mauroy, Sweden
1924	Herma von Szabo-Planck, Austria
1928	Sonja Henie, Norway
1932	Sonja Henie, Norway
1936	Sonja Henie, Norway
1948	Barbara Ann Scott, Canada
1952	Jeanette Altwegg, Great Britain
1956	Tenley Albright, U.S.
1960	Carol Heiss, U.S.
1964	Sjoukje Dijkstra, Netherlands
1968	Peggy Fleming, U.S.
1972	Beatrix Schuba, Austria
1976	Dorothy Hamill, U.S.
1980	Anett Poetzsch, E. Germany
1984	Katarina Witt, E. Germany
1988	Katarina Witt, E. Germany
1992	Kristi Yamaguchi, U.S.
1994	Oksana Baiul, Ukraine
1998	Tara Lipinski, U.S.

(#) Event was held at Summer Olympics.

Pairs
1908#	Anna Hubler & Heinrich Burger, Germany
1920#	Ludovika & Walter Jakobsson, Finland
1924	Helene Engelman & Alfred Berger, Austria
1928	Andree Joly & Pierre Brunet, France
1932	Andree Joly & Pierre Brunet, France
1936	Maxi Herber & Ernst Baier, Germany
1948	Micheline Lannoy & Pierre Baugniet, Belgium
1952	Ria and Paul Falk, Germany
1956	Elisabeth Schwartz & Kurt Oppelt, Austria
1960	Barbara Wagner & Robert Paul, Canada
1964	Ludmila Beloussova & Oleg Protopopov, USSR
1968	Ludmila Beloussova & Oleg Protopopov, USSR
1972	Irina Rodnina & Alexei Ulanov, USSR
1976	Irina Rodnina & Aleksandr Zaitsev, USSR
1980	Irina Rodnina & Aleksandr Zaitsev, USSR
1984	Elena Valova & Oleg Vassiliev, USSR
1988	Ekaterina Gordeeva & Sergei Grinkov, USSR
1992	Natalia Mishkutienok & Artur Dimitriev, Unified Team
1994	Ekaterina Gordeeva & Sergei Grinkov, Russia
1998	Oksana Kazakova & Artur Dmitriev, Russia

(#) Event was held at Summer Olympics.

Ice Dancing
1976	Ludmila Pakhomova & Aleksandr Gorschkov, USSR
1980	Natalya Linichuk & Gennadi Karponosov, USSR
1984	Jayne Torvill & Christopher Dean, Great Britain
1988	Natalia Bestemianova & Andrei Bukin, USSR
1992	Marina Klimova & Sergei Ponomarenko, Unified Team
1994	Pasha Grishuk & Evgeny Platov, Russia
1998	Pasha Grishuk & Evgeny Platov, Russia

Freestyle Skiing

Men's Moguls
		Points
1992	Edgar Grospiron, France	25.81
1994	Jean-Luc Brassard, Canada	27.24
1998	Jonny Moseley, U.S.	26.93

Men's Aerials
		Points
1994	Andreas Schoenbaechler, Switzerland	234.67
1998	Eric Bergoust, U.S.	255.64

Women's Moguls
		Points
1992	Donna Weinbrecht, U.S.	23.69
1994	Stine Lise Hattestad, Norway	25.97

Women's Moguls	Points
1998	Tae Satoya, Japan	25.06

Women's Aerials	Points
1994	Lina Tcherjazova, Uzbekistan	166.84
1998	Nikki Stone, U.S.	193.00

Ice Hockey

MEN

1920#	Canada, U.S., Czechoslovakia
1924	Canada, U.S., Great Britain
1928	Canada, Sweden, Switzerland
1932	Canada, U.S., Germany
1936	Great Britain, Canada, U.S.
1948	Canada, Czechoslovakia, Switzerland
1952	Canada, U.S., Sweden
1956	USSR, U.S., Canada
1960	U.S., Canada, USSR
1964	USSR, Sweden, Czechoslovakia
1968	USSR, Czechoslovakia, Canada
1972	USSR, U.S., Czechoslovakia
1976	USSR, Czechoslovakia, W. Germany
1980	U.S., USSR, Sweden
1984	USSR, Czechoslovakia, Sweden
1988	USSR, Finland, Sweden
1992	Unified Team, Canada, Czechoslovakia
1994	Sweden, Canada, Finland
1998	Czech Republic, Russia, Finland

(#) Event was held at Summer Olympics.

WOMEN

1998	U.S., Canada, Finland

Luge

Men's Singles

		Time
1964	Thomas Keohler, E. Germany	3:26.77
1968	Manfred Schmid, Austria	2:52.48
1972	Wolfgang Scheidel, E. Germany	3:27.58
1976	Detlef Guenther, E. Germany	3:27.688
1980	Bernhard Glass, E. Germany	2:54.796
1984	Paul Hildgartner, Italy	3:04.258
1988	Jens Mueller, E. Germany	3:05.548
1992	Georg Hackl, Germany	3:02.363
1994	Georg Hackl, Germany	3:21.571
1998	Georg Hackl, Germany	3:18.436

Men's Doubles

		Time
1964	Austria	1:41.62
1968	E. Germany	1:35.85
1972	Italy, E. Germany (tie).	1:28.35
1976	E. Germany	1:25.604
1980	E. Germany	1:19.331
1984	W. Germany	1:23.620
1988	E. Germany	1:31.940
1992	Germany	1:32.053
1994	Italy	1:36.720
1998	Germany	1:41.105

Women's Singles

		Time
1964	Ortun Enderlein, Germany	3:24.67
1968	Erica Lechner, Italy.	2:28.66
1972	Anna M. Muller, E. Germany	2:59.18
1976	Margit Schumann, E. Germany	2:50.621
1980	Vera Zozulya, USSR	2:36.537
1984	Steffi Martin, E. Germany.	2:46.570
1988	Steffi Walter, E. Germany.	3:03.973
1992	Doris Neuner, Austria	3:06.696
1994	Gerda Weissensteiner, Italy	3:15.517
1998	Silke Kraushaar, Germany	3:23.779

Nordic Skiing

Cross-Country Events

Men's 10 Kilometers (6.2 miles)

		Time
1992	Vegard Ulvang, Norway	27:36.0
1994	Bjoern Daehlie, Norway	24:20.1
1998	Bjoern Daehlie, Norway	27:24.5

Men's 15 Kilometers (9.3 miles)

		Time
1924	Thorleif Haug, Norway.	1:14:31
1928	Johan Grottumsbraaten, Norway	1:37:01
1932	Sven Utterstrom, Sweden	1:23:07
1936	Erik-August Larsson, Sweden	1:14:38
1948	Martin Lundstrom, Sweden	1:13:50
1952	Hallgeir Brenden, Norway	1:01:34
1956	Hallgeir Brenden, Norway	0:49:39.0
1960	Haakon Brusveen, Norway	0:51:55.5
1964	Eero Maentyranta, Finland	0:50:54.1
1968	Harald Groenningen, Norway	0:47:54.2
1972	Sven-Ake Lundback, Sweden	0:45:28.24
1976	Nikolai Balukov, USSR.	0:43:58.47
1980	Thomas Wassberg, Sweden	0:41:57.63
1984	Gunde Svan, Sweden	0:41:25.6

Men's 15 Kilometers (9.3 miles)

		Time
1988	Mikhail Deviatiarov, USSR	0:41:18.9
1992	Bjoern Daehlie, Norway	0:38:01.9
1994	Bjoern Daehlie, Norway	0:35:48.8
1998	Thomas Alsgaard, Norway	1:07:01.7

(Note: approx. 18-km course 1924-1952)

Men's 30 Kilometers (18.6 miles)

		Time
1956	Veikko Hakulinen, Finland	1:44:06.0
1956	Veikko Hakulinen, Finland	1:44:06.0
1960	Sixten Jernberg, Sweden	1:51:03.9
1964	Eero Maentyranta, Finland	1:30:50.7
1968	Franco Nones, Italy	1:35:39.2
1972	Vyacheslav Vedenine, USSR	1:36:31.15
1976	Sergei Saveliev, USSR	1:30:29.38
1980	Nikolai Zimyatov, USSR.	1:27:02.80
1984	Nikolai Zimyatov, USSR.	1:28:56.3
1988	Aleksei Prokourorov, USSR.	1:24:26.3
1992	Vegard Ulvang, Norway.	1:22:27.8
1994	Thomas Alsgaard, Norway	1:12:26.4
1998	Mika Myllyla, Finland	1:33:55.8

Men's 50 Kilometers (31.2 miles)

		Time
1924	Thorleif Haug, Norway.	3:44:32.0
1928	Per Erik Hedlund, Sweden	4:52:03.0
1932	Veli Saarinen, Finland	4:28:00.0
1936	Elis Wiklund, Sweden	3:30:11.0
1948	Nils Karlsson, Sweden	3:47:48.0
1952	Veikko Hakulinen, Finland	3:33:33.0
1956	Sixten Jernberg, Sweden	2:50:27.0
1960	Kalevi Hamalainen, Finland	2:59:06.3
1964	Sixten Jernberg, Sweden	2:43:52.6
1968	Ole Ellefsaeter, Norway	2:28:45.8
1972	Paal Tyldum, Norway	2:43:14.75
1976	Ivar Formo, Norway	2:37:30.05
1980	Nikolai Zimyatov, USSR.	2:27:24.60
1984	Thomas Wassberg, Sweden	2:15:55.8
1988	Gunde Svan, Sweden	2:04:30.9
1992	Bjoern Daehlie, Norway	2:03:41.5
1994	Vladimir Smirnov, Kazakhstan	2:07:20.3
1998	Bjoern Daehlie, Norway	2:05:08.2

Men's 40-Kilometer Relay

		Time
1936	Finland, Norway, Sweden	2:41:33.0
1948	Sweden, Finland, Norway	2:32:08.0
1952	Finland, Norway, Sweden	2:20:16.0
1956	USSR, Finland, Sweden	2:15:30.0
1960	Finland, Norway, USSR	2:18:45.6
1964	Sweden, Finland, USSR	2:18:34.6
1968	Norway, Sweden, Finland	2:08:33.5
1972	USSR, Norway, Switzerland	2:04:47.94
1976	Finland, Norway, USSR	2:07:59.72
1980	USSR, Norway, Finland	1:57:03.46
1984	Sweden, USSR, Finland	1:55:06.30
1988	Sweden, USSR, Czechoslovakia	1:43:58.60
1992	Norway, Italy, Finland.	1:39:26.00
1994	Italy, Norway, Finland.	1:41:15.00
1998	Norway, Italy, Finland.	1:40:55.70

Women's 5 Kilometers (approx. 3.1 miles)

		Time
1964	Claudia Boyarskikh, USSR	17:50.5
1968	Toini Gustafsson, Sweden.	16:45.2
1972	Galina Koulacova, USSR	17:00.50
1976	Helena Takalo, Finland	15:48.69
1980	Raisa Smetanina, USSR.	15:06.92
1984	Marja-Liisa Haemaelainen, Finland.	17:04.0
1988	Marjo Matikainen, Finland	15:04.0
1992	Marjut Lukkarinen, Finland	14:13.8
1994	Ljubov Egorova, Russia.	14:08.8
1998	Larissa Lazutina, Russia	17:37.9

Women's 10 Kilometers (6.2 miles)

		Time
1952	Lydia Wideman, Finland	41:40.0
1956	Lyubov Kosyreva, USSR	38:11.0
1960	Maria Gusakova, USSR	39:46.6
1964	Claudia Boyarskikh, USSR	40:24.3
1968	Toini Gustafsson, Sweden.	36:46.5
1972	Galina Koulacova, USSR	34:17.82
1976	Raisa Smetanina, USSR.	30:13.41
1980	Barbara Petzold, E. Germany	30:31.54
1984	Marja-Liisa Haemaelainen, Finland.	31:44.2
1988	Vida Ventsene, USSR	30:08.3
1992	Lyubov Egorova, Unified Team	25:53.7
1994	Lyubov Egorova, Russia	27:30.1
1998	Larissa Lazutina, Russia	46.06.9

Women's 15 Kilometers (9.3 miles)

		Time
1992	Lyubov Egorova, Unified Team	42:20.8
1994	Manuela Di Centa, Italy	39:44.5
1998	Olga Danilova, Russia	46:55.4

Women's 30 Kilometers (18.6 miles)

		Time
1992	Stefania Belmondo, Italy	1:22:30.1
1994	Manuela Di Centa, Italy	1:25:41.6
1998	Julija Tchepalova, Russia	1:22:01.5

Women's 20-Kilometer Relay	Time
1956 Finland, USSR, Sweden (15 km)	1:09:01.0
1960 Sweden, USSR, Finland (15 km)	1:04:21.4
1964 USSR, Sweden, Finland (15 km)	0:59:20.2
1968 Norway, Sweden, USSR (15 km)	0:57:30.0
1972 USSR, Finland, Norway (15 km)	0:48:46.15
1976 USSR, Finland, E. Germany	1:07:49.75
1980 E. Germany, USSR, Norway	1:02:11.1
1984 Norway, Czechoslovakia, Finland	1:06:49.7
1988 USSR, Norway, Finland	0:59:51.1
1992 United Team, Norway, Italy	0:59:34.8
1994 Russia, Norway, Italy	0:57:12.5
1998 Russia, Norway, Italy	0:55:13.5

Combined Cross-Country & Jumping (Men)

Nordic Combined*

1924	Thorleif Haug, Norway
1928	Johan Grottumsbraaten, Norway
1932	Johan Grottumsbraaten, Norway
1936	Oddbjorn Hagen, Norway
1948	Heikki Hasu, Finland
1952	Simon Slattvik, Norway
1956	Sverre Stenersen, Norway
1960	Georg Thoma, W. Germany
1964	Tormod Knutsen, Norway
1968	Franz Keller, W. Germany
1972	Ulrich Wehling, E. Germany
1976	Ulrich Wehling, E. Germany
1980	Ulrich Wehling, E. Germany
1984	Tom Sandberg, Norway
1988	Hippolyt Kempf, Switzerland
1992	Fabrice Guy, France
1994	Fred Barre Lundberg, Norway
1998	Bjarte Engen Vik, Norway

Team Nordic Combined*

1988	W. Germany, Switzerland, Austria
1992	Japan, Norway, Austria
1994	Japan, Norway, Switzerland
1998	Norway, Finland, France

*Medals based on combination of points for jumping events and time for cross-country events.

Ski Jumping (Men)

Normal Hill	Points
1964 Veikko Kankkonen, Finland	229.9
1968 Jiri Raska, Czechoslovakia	216.5
1972 Yukio Kasaya, Japan	244.2
1976 Hans-Georg Aschenbach, E. Germany	252.0
1980 Toni Innauer, Austria	266.3
1984 Jens Weissflog, E. Germany	215.2
1988 Matti Nykanen, Finland	230.5
1992 Ernst Vettori, Austria	222.8
1994 Espen Bredesen, Norway.	282.0
1998 Jani Soininen, Finland	234.5

Large Hill	Points
1924 Jacob Tullin Thams, Norway.	18.960
1928 Alfred Andersen, Norway	19.208
1932 Birger Ruud, Norway	228.1
1936 Birger Ruud, Norway	232.0
1948 Petter Hugsted, Norway	228.1
1952 Arnfinn Bergmann, Norway	226.0
1956 Antti Hyvarinen, Finland.	227.0
1960 Helmut Recknagel, E. Germany.	227.2
1964 Toralf Engan, Norway	230.7
1968 Vladimir Beloussov, USSR.	231.3
1972 Wojciech Fortuna, Poland.	219.9
1976 Karl Schnabl, Austria	234.8
1980 Jouko Tormanen, Finland.	271.0
1984 Matti Nykanen, Finland	231.2
1988 Matti Nykanen, Finland	224.0
1992 Toni Nieminen, Finland.	239.5
1994 Jens Weissflog, Germany	274.5
1998 Kazuyoshi Funaki, Japan	272.3

Team Large Hill	Points
1988 Finland, Yugoslavia, Norway	634.4
1992 Finland, Austria, Czechoslovakia	644.4
1994 Germany, Japan, Austria	970.1
1998 Japan, Germany, Austria	933.0

Snowboarding

Men's Giant Slalom	Time
1998 Ross Rebagliati, Canada	2:03.96

Men's Halfpipe	Points
1998 Gian Simmen, Switzerland.	85.2

Women's Giant Slalom	Time
1998 Karine Ruby, France.	2:17.34

Women's Halfpipe	Points
1998 Nicola Thost, Germany	74.6

Speed Skating

Men's 500 Meters	Time*
1924 Charles Jewtraw, U.S.	0:44.0
1928 Thunberg, Finland & Evensen, Norway (tie) .	0:43.4
1932 John A. Shea, U.S.	0:43.4
1936 Ivar Ballangrud, Norway	0:43.4
1948 Finn Helgesen, Norway	0:43.1
1952 Kenneth Henry, U.S.	0:43.2
1956 Evgeniy Grishin, USSR	0:40.2
1960 Evgeniy Grishin, USSR	0:40.2
1964 Terry McDermott, U.S.	0:40.1
1968 Erhard Keller, W. Germany	0:40.3
1972 Erhard Keller, W. Germany	0:39.44
1976 Evgeny Kulikov, USSR	0:39.17
1980 Eric Heiden, U.S.	0:38.03
1984 Sergei Fokichev, USSR	0:38.19
1988 Uwe-Jens Mey, E. Germany	0:36.45
1992 Uwe-Jens Mey, Germany	0:37.14
1994 Aleksandr Golubev, Russia	0:36.33
1998 Hiroyasu Shimizu, Japan.	0:35.59

*Better time of two runs. Medals based on combined times.

Men's 1,000 Meters	Time
1976 Peter Mueller, U.S	1:19.32
1980 Eric Heiden, U.S.	1:15.18
1984 Gaetan Boucher, Canada	1:15.80
1988 Nikolai Guiliaev, USSR	1:13.03
1992 Olaf Zinke, Germany	1:14.85
1994 Dan Jansen, U.S.	1:12.43
1998 Ids Postma, Netherlands	1:10.64

Men's 1,500 Meters	Time
1924 Clas Thunberg, Finland.	2:20.8
1928 Clas Thunberg, Finland	2:21.1
1932 John A. Shea, U.S.	2:57.5
1936 Charles Mathiesen, Norway	2:19.2
1948 Sverre Farstad, Norway	2:17.6
1952 Hjalmar Andersen, Norway	2:20.4
1956 Grishin, & Mikhailov, both USSR (tie)	2:08.6
1960 Aas, Norway & Grishin, USSR (tie)	2:10.4
1964 Ants Anston, USSR.	2:10.3
1968 Cornelis Verkerk, Netherlands	2:03.4
1972 Ard Schenk, Netherlands	2:02.96
1976 Jan Egil Storholt, Norway	1:59.38
1980 Eric Heiden, U.S.	1:55.44
1984 Gaetan Boucher, Canada	1:58.36
1988 Andre Hoffmann, E. Germany	1:52.06
1992 Johann Koss, Norway	1:54.81
1994 Johann Koss, Norway	1:51.29
1998 Aadne Sondral, Norway.	1:47.87

Men's 5,000 Meters	Time
1924 Clas Thunberg, Finland.	8:39.0
1928 Ivar Ballangrud, Norway	8:50.5
1932 Irving Jaffee, U.S.	9:40.8
1936 Ivar Ballangrud, Norway	8:19.6
1948 Reidar Liaklev, Norway	8:29.4
1952 Hjalmar Andersen, Norway	8:10.6
1956 Boris Shilkov, USSR	7:48.7
1960 Viktor Kosichkin, USSR	7:51.3
1964 Knut Johannesen, Norway	7:38.4
1968 F. Anton Maier, Norway	7:22.4
1972 Ard Schenk, Netherlands	7:23.61
1976 Sten Stensen, Norway.	7:24.48
1980 Eric Heiden, U.S.	7:02.29
1984 Sven Tomas Gustafson, Sweden	7:12.28
1988 Tomas Gustafson, Sweden	6:44.63
1992 Geir Karlstad, Norway	6:59.97
1994 Johann Koss, Norway	6:34.96
1998 Gianni Romme, Netherlands.	6:22.20

Men's 10,000 Meters	Time
1924 Julius Skutnabb, Finland	18:04.8
1928 Event not held because of thawing of ice	
1932 Irving Jaffee, U.S.	19:13.6
1936 Ivar Ballangrud, Norway	17:24.3
1948 Ake Seyffarth, Sweden	17:26.3
1952 Hjalmar Andersen, Norway	16:45.8
1956 Sigvard Ericsson, Sweden	16:35.9
1960 Knut Johannesen, Norway	15:46.6
1964 Jonny Nilsson, Sweden	15:50.1
1968 Jonny Hoeglin, Sweden.	15:23.6
1972 Ard Schenk, Netherlands	15:01.35
1976 Piet Kleine, Netherlands	14:50.59
1980 Eric Heiden, U.S.	14:28.13
1984 Igor Malkov, USSR	14:39.90
1988 Tomas Gustafson, Sweden	13:48.20
1992 Bart Veldkamp, Netherlands	14:12.12

Men's 10,000 Meters		Time
1994	Johann Koss, Norway.................	13:30.55
1998	Gianni Romme, Netherlands............	13:15.33

Women's 500 Meters		Time*
1960	Helga Haase, Germany...............	0:45.9
1964	Lydia Skoblikova, USSR..............	0:45.0
1968	Ludmila Titova, USSR	0:46.1
1972	Anne Henning, U.S.	0:43.33
1976	Sheila Young, U.S.	0:42.76
1980	Karin Enke, E. Germany	0:41.78
1984	Christa Rothenburger, E. Germany	0:41.02
1988	Bonnie Blair, U.S.	0:39.10
1992	Bonnie Blair, U.S.	0:40.33
1994	Bonnie Blair, U.S.	0:39.25
1998	Catriona LeMay-Doan, Canada	0:38.21

* Better time of two runs. Medals based on combined times.

Women's 1,000 Meters		Time
1960	Klara Guseva, USSR	1:34.1
1964	Lydia Skoblikova, USSR..............	1:33.2
1968	Carolina Geijssen, Netherlands	1:32.6
1972	Monika Pflug, W. Germany............	1:31.40
1976	Tatiana Averina, USSR.	1:28.43
1980	Natalya Petruseva, USSR	1:24.10
1984	Karin Enke, E. Germany	1:21.61
1988	Christa Rothenburger, E. Germany	1:17.65
1992	Bonnie Blair, U.S.	1:21.90
1994	Bonnie Blair, U.S.	1:18.74
1998	Marianne Timmer, Netherlands	1:16.51

Women's 1,500 Meters		Time
1960	Lydia Skoblikova, USSR..............	2:52.2
1964	Lydia Skoblikova, USSR..............	2:22.6
1968	Kaija Mustonen, Finland.............	2:22.4
1972	Dianne Holum, U.S.	2:20.85
1976	Galina Stepanskaya, USSR	2:16.58
1980	Anne Borckink, Netherlands...........	2:10.95
1984	Karin Enke, E. Germany	2:03.42
1988	Yvonne van Gennip, Netherlands........	2:00.68
1992	Jacqueline Boerner, Germany	2:05.87
1994	Emese Hunyady, Austria	2:02.19
1998	Marianne Timmer, Netherlands	1:57.58

Women's 3,000 Meters		Time
1960	Lydia Skoblikova, USSR	5:14.3
1964	Lydia Skoblikova, USSR	5:14.9
1968	Johanna Schut, Netherlands............	4:56.2
1972	Christina Baas-Kaiser, Netherlands	4:52.14
1976	Tatiana Averina, USSR	4:45.19
1980	Bjoerg Eva Jensen, Norway	4:32.13
1984	Andrea Schoene, E. Germany	4:24.79
1988	Yvonne van Gennip, Netherlands	4:11.94
1992	Gunda Niemann, Germany	4:19.90
1994	Svetlana Bazhanova, Russia...........	4:17.43
1998	Gunda Niemann-Stirnemann, Germany	4:07.29

Women's 5,000 Meters		Time
1988	Yvonne van Gennip, Netherlands	7:14.13
1992	Gunda Niemann, Germany	7:31.57
1994	Claudia Pechstein, Germany...........	7:14.37
1998	Claudia Pechstein, Germany...........	6:59.61

Short-Track Speed Skating

Men's 500 Meters		Time
1998	Takafumi Nishitani, Japan	42.862

Men's 1,000 Meters		Time
1992	Kim Ki-Hoon, S. Korea	1:30.76
1994	Kim Ki-Hoon, S. Korea	1:34.57
1998	Dong-Sung Kim, S. Korea	1:32.375

Men's 5,000-Meter Relay		Time
1992	S. Korea, Canada, Japan	7:14.02
1994	Italy, U.S., Australia	7:11.74
1998	Canada, S. Korea, China.............	7:06.075
1998	Takafumi Nishitani, Japan	42.862

Women's 500 Meters		Time
1992	Cathy Turner, U.S.	47.04
1994	Cathy Turner, U.S.	45.98
1998	Annie Perreault, Canada	46.568

Women's 1,000 Meters		Time
1998	Chun Lee-Kyung, S. Korea	1:42.776

Women's 3,000 Meter Relay		Time
1992	Canada, U.S., Unified Team	4:36.62
1994	S. Korea, Canada, U.S...............	4:26.64
1998	S. Korea, China, Canada..............	4:16.26

History of the Olympic Games

The modern Olympic Games, first held in Athens, Greece, in 1896, were the result of efforts by Baron Pierre de Coubertin, a French educator, to promote interest in education and culture and to foster better international understanding through love of athletics. His source of inspiration was the ancient Greek Olympic Games, most notable of the 4 Panhellenic celebrations. The games were combined patriotic, religious, and athletic festivals held every 4 years. The first such recorded festival was held in 776 BC, the date from which the Greeks began to keep their calendar by "Olympiads," or 4-year spans between the games. The first Olympiad is said to have consisted merely of a 200-yd foot race near the small city of Olympia, but the games gained in scope and became demonstrations of national pride. Only Greek citizens—amateurs—could participate. Winners received laurel, wild olive, and palm wreaths and were accorded special privileges. Under the Roman emperors, the games deteriorated into professional carnivals and circuses. Emperor Theodosius banned them in AD 394.

Baron de Coubertin enlisted 13 nations to send athletes to the first modern Olympics in 1896; now athletes from nearly 200 nations and territories compete in the Summer Olympics. The Winter Olympic Games were started in 1924.

Olympic Information

Symbol: Five rings or circles, linked together to represent the sporting friendship of all peoples. The rings also symbolize 5 geographic areas—Europe, Asia, Africa, Australia, and America. Each ring is a different color—blue, yellow, black, green, and red.

Flag: The symbol of the 5 rings on a plain white background.

Motto: "Citius, Altius, Fortius." Latin meaning "swifter, higher, stronger."

Creed: "The most important thing in the Olympic Games is not to win but to take part, just as the most important thing in life is not the triumph but the struggle. The essential thing is not to have conquered but to have fought well."

Oath: "In the name of all competitors I promise that we will take part in these Olympic Games, respecting and abiding by the rules which govern them, in the true spirit of sportsmanship for the glory of sport and the honor of our teams."

Flame: Symbolizes the continuity between the ancient and modern Games. The modern version of the flame was adopted in 1936. The torch used to kindle the flame is first lit by the sun's rays at Olympia, Greece, and then carried to the site of the Games by relays of runners. Ships and planes are used when necessary.

Paralympics

The first Olympic games for the disabled were held in Rome after the 1960 Summer Olympic Games; use of the name "paralympic" began with the 1964 games in Tokyo. The Paralympics are held by the Olympic host country in the same year and usually in the same city or venue. A goal of the Paralympics is to provide elite competition to athletes with functional disabilities that prevent their involvement in the Olympics. In 1976 the first Winter Paralympic Games were held in Ornskoldsvik, Sweden.

The VII Paralympic Winter Games, the first to take place outside Europe, were held Mar. 5–Mar. 14, 1998, in Nagano, Japan. The games featured 571 athletes from 32 nations competing for medals in over 30 events in the following sports: Alpine skiing, cross-country skiing, biathlon, ice sledge racing, and ice sledge hockey.

Summer Olympic Games

Sites of Summer Olympic Games

1896	Athens, Greece	**1924**	Paris, France	**1960**	Rome, Italy	**1984**	Los Angeles, U.S.
1900	Paris, France	**1928**	Amsterdam, Netherlands	**1964**	Tokyo, Japan	**1988**	Seoul, South Korea
1904	St. Louis, U.S.	**1932**	Los Angeles, U.S.	**1968**	Mexico City, Mexico	**1992**	Barcelona, Spain
1906*	Athens, Greece	**1936**	Berlin, Germany	**1972**	Munich, W. Germany	**1996**	Atlanta, U.S.
1908	London, England	**1948**	London, England	**1976**	Montreal, Canada	**2000**	Sydney, Australia
1912	Stockholm, Sweden	**1952**	Helsinki, Finland	**1980**	Moscow, USSR	**2004**	Athens, Greece
1920	Antwerp, Belgium	**1956**	Melbourne, Australia				

*Games not recognized by International Olympic Committee. Games 6 (1916), 12 (1940), and 13 (1944) were not celebrated.

Summer Olympics 2000

The 2000 games will be held in Sydney, Australia, Sept. 15-Oct. 1. It is then early spring in Australia; daytime temperatures range from 60 to 68°F. Two new medal sports will be added—taekwondo and the triathlon. Taekwondo will have 4 men's and 4 women's weight classes; both men and women will also compete in the triathlon. A new event, the trampoline, also will be added to the gymnastics program. For the first time, women will be competing in water polo, the modern pentathlon, the pole vault, and the hammer throw. In all, there will be 296 events—166 for men, 118 for women, and 12 mixed. For further information, see the official website for the Sydney games: www.sydney.olympics.org

Summer Olympic Games in 1996

Atlanta, GA, U.S., July 19-Aug. 4, 1996

About 10,750 athletes gathered for 17 days to compete in a record 271 events in the 1996 games. Outstanding athletes included: Kerri Strug, who led the U.S. women's gymnastics team to a gold medal; U.S. runner Michael Johnson, who won the 200 and 400 meters; France's Marie-Jose Perec, winner of the women's 200 and 400 meters; Canada's Donovan Bailey, who set a world record in the 100-meter run; Dan O'Brien of the U.S., who won the decathlon; U.S. swimmer Amy Van Dyken, who won 4 golds; China's Fu Mingxia, who swept both women's diving events; marathon runner Josia Thugwane, South Africa's first black gold medalist; and Carl Lewis of the U.S., who won his 4th consecutive gold in the long jump.

Final Medal Standings

Country	G	S	B	T	Country	G	S	B	T	Country	G	S	B	T
United States	44	32	25	101	Norway	2	2	3	7	Moldova	0	1	1	2
Germany	20	18	27	65	Denmark	4	1	1	6	Uzbekistan	0	1	1	2
Russia	26	21	16	63	Turkey	4	1	1	6	Georgia	0	0	2	2
China	16	22	12	50	New Zealand	3	2	1	6	Morocco	0	0	2	2
Australia	9	9	23	41	Belgium	2	2	2	6	Trinidad & Tobago	0	0	2	2
France	15	7	15	37	Nigeria	2	1	3	6	Burundi	1	0	0	1
Italy	13	10	12	35	Jamaica	1	3	2	6	Costa Rica	1	0	0	1
South Korea	7	15	5	27	South Africa	3	1	1	5	Ecuador	1	0	0	1
Cuba	9	8	8	25	North Korea	2	1	2	5	Hong Kong	1	0	0	1
Ukraine	9	2	12	23	Ireland	3	0	1	4	Syria	1	0	0	1
Canada	3	11	8	22	Finland	1	2	1	4	Azerbaijan	0	1	0	1
Hungary	7	4	10	21	Indonesia	1	1	2	4	Bahamas	0	1	0	1
Romania	4	7	9	20	Yugoslavia	1	1	2	4	Latvia	0	1	0	1
Netherlands	4	5	10	19	Algeria	2	0	1	3	Philippines	0	1	0	1
Poland	7	5	5	17	Ethiopia	2	0	1	3	Taiwan	0	1	0	1
Spain	5	6	6	17	Iran	1	1	1	3	Tonga	0	1	0	1
Bulgaria	3	7	5	15	Slovakia	1	1	1	3	Zambia	0	1	0	1
Brazil	3	3	9	15	Argentina	0	2	1	3	India	0	0	1	1
Great Britain	1	8	6	15	Austria	0	1	2	3	Israel	0	0	1	1
Belarus	1	6	8	15	Armenia	1	1	0	2	Lithuania	0	0	1	1
Japan	3	6	5	14	Croatia	1	1	0	2	Mexico	0	0	1	1
Czech Rep.	4	3	4	11	Portugal	1	0	1	2	Mongolia	0	0	1	1
Kazakhstan	3	4	4	11	Thailand	1	0	1	2	Mozambique	0	0	1	1
Greece	4	4	0	8	Namibia	0	2	0	2	Puerto Rico	0	0	1	1
Sweden	2	4	2	8	Slovenia	0	2	0	2	Tunisia	0	0	1	1
Kenya	1	4	3	8	Malaysia	0	1	1	2	Uganda	0	0	1	1
Switzerland	4	3	0	7										

Summer Olympic Games Champions, 1896-1996

(*indicates Olympic record)

The 1980 games were boycotted by 62 nations, including the U.S. The 1984 games were boycotted by the USSR and by most Eastern bloc nations. East and West Germany competed separately, 1968-88. The 1992 Unified Team consisted of 12 former Soviet republics. The 1992 Independent Olympic Participants (I.O.P.) were athletes from Serbia, Montenegro, and Macedonia.

Track and Field — Men

100-Meter Run

1896	Thomas Burke, United States	12.0s
1900	Francis W. Jarvis, United States	11.0s
1904	Archie Hahn, United States	11.0s
1908	Reginald Walker, South Africa	10.8s
1912	Ralph Craig, United States	10.8s
1920	Charles Paddock, United States	10.8s
1924	Harold Abrahams, Great Britain	10.6s
1928	Percy Williams, Canada	10.8s
1932	Eddie Tolan, United States	10.3s
1936	Jesse Owens, United States	10.3s
1948	Harrison Dillard, United States	10.3s
1952	Lindy Remigino, United States	10.4s
1956	Bobby Morrow, United States	10.5s
1960	Armin Hary, Germany	10.2s
1964	Bob Hayes, United States	10.0s

1968	Jim Hines, United States	9.95s
1972	Valery Borzov, USSR	10.14s
1976	Hasely Crawford, Trinidad	10.06s
1980	Allan Wells, Great Britain	10.25s
1984	Carl Lewis, United States	9.99s
1988	Carl Lewis, United States	9.92s
1992	Linford Christie, Great Britain	9.96s
1996	Donovan Bailey, Canada	9.84s*

200-Meter Run

1900	Walter Tewksbury, United States	22.2s
1904	Archie Hahn, United States	21.6s
1908	Robert Kerr, Canada	22.6s
1912	Ralph Craig, United States	21.7s
1920	Allan Woodring, United States	22.0s
1924	Jackson Scholz, United States	21.6s
1928	Percy Williams, Canada	21.8s

1932	Eddie Tolan, United States	21.2s
1936	Jesse Owens, United States	20.7s
1948	Mel Patton, United States	21.1s
1952	Andrew Stanfield, United States	20.7s
1956	Bobby Morrow, United States	20.6s
1960	Livio Berruti, Italy	20.5s
1964	Henry Carr, United States	20.3s
1968	Tommie Smith, United States	19.83s
1972	Valeri Borzov, USSR	20.00s
1976	Donald Quarrie, Jamaica	20.23s
1980	Pietro Mennea, Italy	20.19s
1984	Carl Lewis, United States	19.80s
1988	Joe DeLoach, United States	19.75s
1992	Mike Marsh, United States	20.01s
1996	Michael Johnson, United States	19.32s*

400-Meter Run

1896	Thomas Burke, United States	54.2s
1900	Maxey Long, United States	49.4s
1904	Harry Hillman, United States	49.2s
1908	Wyndham Halswelle, Great Britain, walkover	50.0s
1912	Charles Reidpath, United States	48.2s
1920	Bevil Rudd, South Africa	49.6s
1924	Eric Liddell, Great Britain	47.6s
1928	Ray Barbuti, United States	47.8s
1932	William Carr, United States	46.2s
1936	Archie Williams, United States	46.5s
1948	Arthur Wint, Jamaica	46.2s
1952	George Rhoden, Jamaica	45.9s
1956	Charles Jenkins, United States	46.7s
1960	Otis Davis, United States	44.9s
1964	Michael Larrabee, United States	45.1s
1968	Lee Evans, United States	43.86s
1972	Vincent Matthews, United States	44.66s
1976	Alberto Juantorena, Cuba	44.26s
1980	Viktor Markin, USSR	44.60s
1984	Alonzo Babers, United States	44.27s
1988	Steven Lewis, United States	43.87s
1992	Quincy Watts, United States	43.50s
1996	Michael Johnson, United States	43.49s*

800-Meter Run

1896	Edwin Flack, Australia	2m. 11s
1900	Alfred Tysoe, Great Britain	2m. 1.2s
1904	James Lightbody, United States	1m. 56s
1908	Mel Sheppard, United States	1m. 52.8s
1912	James Meredith, United States	1m. 51.9s
1920	Albert Hill, Great Britain	1m. 53.4s
1924	Douglas Lowe, Great Britain	1m. 52.4s
1928	Douglas Lowe, Great Britain	1m. 51.8s
1932	Thomas Hampson, Great Britain	1m. 49.8s
1936	John Woodruff, United States	1m. 52.9s
1948	Mal Whitfield, United States	1m. 49.2s
1952	Mal Whitfield, United States	1m. 49.2s
1956	Thomas Courtney, United States	1m. 47.7s
1960	Peter Snell, New Zealand	1m. 46.3s
1964	Peter Snell, New Zealand	1m. 45.1s
1968	Ralph Doubell, Australia	1m. 44.3s
1972	Dave Wottle, United States	1m. 45.9s
1976	Alberto Juantorena, Cuba	1m. 43.50s
1980	Steve Ovett, Great Britain	1m. 45.40s
1984	Joaquim Cruz, Brazil	1m. 43.00s
1988	Paul Ereng, Kenya	1m. 43.45s
1992	William Tanui, Kenya	1m. 43.66s
1996	Vebjoern Rodal, Norway	1m. 42.58s*

1,500-Meter Run

1896	Edwin Flack, Australia	4m. 33.2s
1900	Charles Bennett, Great Britain	4m. 6.2s
1904	James Lightbody, United States	4m. 5.4s
1908	Mel Sheppard, United States	4m. 3.4s
1912	Arnold Jackson, Great Britain	3m. 56.8s
1920	Albert Hill, Great Britain	4m. 1.8s
1924	Paavo Nurmi, Finland	3m. 53.6s
1928	Harry Larva, Finland	3m. 53.2s
1932	Luigi Beccali, Italy	3m. 51.2s
1936	Jack Lovelock, New Zealand	3m. 47.8s
1948	Henri Eriksson, Sweden	3m. 49.8s
1952	Joseph Barthel, Luxembourg	3m. 45.2s
1956	Ron Delany, Ireland	3m. 41.2s
1960	Herb Elliott, Australia	3m. 35.6s
1964	Peter Snell, New Zealand	3m. 38.1s
1968	Kipchoge Keino, Kenya	3m. 34.9s
1972	Pekka Vasala, Finland	3m. 36.3s
1976	John Walker, New Zealand	3m.39.17s
1980	Sebastian Coe, Great Britain	3m.38.4s
1984	Sebastian Coe, Great Britain	3m. 32.53s*
1988	Peter Rono, Kenya	3m. 35.96s
1992	Fermin Cacho Ruiz, Spain	3m. 40.12s
1996	Noureddine Morceli, Algeria	3m. 35.78s

5,000-Meter Run

1912	Hannes Kolehmainen, Finland	14m. 36.6s
1920	Joseph Guillemot, France	14m. 55.6s
1924	Paavo Nurmi, Finland	14m. 31.2s
1928	Willie Ritola, Finland	14m. 38s
1932	Lauri Lehtinen, Finland	14m. 30s
1936	Gunnar Hockert, Finland	14m. 22.2s
1948	Gaston Reiff, Belgium	14m. 17.6s
1952	Emil Zatopek, Czechoslovakia	14m. 6.6s
1956	Vladimir Kuts, USSR	13m. 39.6s
1960	Murray Halberg, New Zealand	13m. 43.4s
1964	Bob Schul, United States	13m. 48.8s
1968	Mohamed Gammoudi, Tunisia	14m. 05.0s
1972	Lasse Viren, Finland	13m. 26.4s
1976	Lasse Viren, Finland	13m. 24.76s
1980	Miruts Yifter, Ethiopia	13m. 21.0s
1984	Said Aouita, Morocco	13m. 05.59s*
1988	John Ngugi, Kenya	13m. 11.70s
1992	Dieter Baumann, Germany	13m. 12.52s
1996	Venuste Niyongabo, Burundi	13m. 07.96s

10,000-Meter Run

1912	Hannes Kolehmainen, Finland	31m. 20.8s
1920	Paavo Nurmi, Finland	31m. 45.8s
1924	Willie Ritola, Finland	30m. 23.2s
1928	Paavo Nurmi, Finland	30m. 18.8s
1932	Janusz Kusocinski, Poland	30m. 11.4s
1936	Ilmari Salminen, Finland	30m. 15.4s
1948	Emil Zatopek, Czechoslovakia	29m. 59.6s
1952	Emil Zatopek, Czechoslovakia	29m. 17.0s
1956	Vladimir Kuts, USSR	28m. 45.6s
1960	Pyotr Bolotnikov, USSR	28m. 32.2s
1964	Billy Mills, United States	28m. 24.4s
1968	Naftali Temu, Kenya	29m. 27.4s
1972	Lasse Viren, Finland	27m. 38.4s
1976	Lasse Viren, Finland	27m. 40.4s
1980	Miruts Yifter, Ethiopia	27m. 42.7s
1984	Alberto Cova, Italy	27m. 47.54s
1988	Brahim Boutaib, Morocco	27m. 21.46s
1992	Khalid Skah, Morocco	27m. 46.70s
1996	Haile Gebrselassie, Ethiopia	27m. 07.34s*

110-Meter Hurdles

1896	Thomas Curtis, United States	17.6s
1900	Alvin Kraenzlein, United States	15.4s
1904	Frederick Schule, United States	16.0s
1908	Forrest Smithson, United States	15.0s
1912	Frederick Kelly, United States	15.1s
1920	Earl Thomson, Canada	14.8s
1924	Daniel Kinsey, United States	15.0s
1928	Sydney Atkinson, South Africa	14.8s
1932	George Saling, United States	14.6s
1936	Forrest Towns, United States	14.2s
1948	William Porter, United States	13.9s
1952	Harrison Dillard, United States	13.7s
1956	Lee Calhoun, United States	13.5s
1960	Lee Calhoun, United States	13.8s
1964	Hayes Jones, United States	13.6s
1968	Willie Davenport, United States	13.3s
1972	Rod Milburn, United States	13.24s
1976	Guy Drut, France	13.30s
1980	Thomas Munkelt, E. Germany	13.39s
1984	Roger Kingdom, United States	13.20s
1988	Roger Kingdom, United States	12.98s
1992	Mark McCoy, Canada	13.12s
1996	Allen Johnson, United States	12.95s*

400-Meter Hurdles

1900	J.W.B. Tewksbury, United States	57.6s
1904	Harry Hillman, United States	53.0s
1908	Charles Bacon, United States	55.0s
1920	Frank Loomis, United States	54.0s
1924	F. Morgan Taylor, United States	52.6s
1928	Lord Burghley, Great Britain	53.4s
1932	Robert Tisdall, Ireland	51.7s
1936	Glenn Hardin, United States	52.4s
1948	Roy Cochran, United States	51.1s
1952	Charles Moore, United States	50.8s
1956	Glenn Davis, United States	50.1s
1960	Glenn Davis, United States	49.3s
1964	Rex Cawley, United States	49.6s
1968	Dave Hemery, Great Britain	48.12s
1972	John Akii-Bua, Uganda	47.82s
1976	Edwin Moses, United States	47.64s
1980	Volker Beck, E. Germany	48.70s
1984	Edwin Moses, United States	47.75s
1988	Andre Phillips, United States	47.19s
1992	Kevin Young, United States	46.78s*
1996	Derrick Adkins, United States	47.54s

400-Meter Relay

1912	Great Britain	42.4s
1920	United States	42.2s
1924	United States	41.0s
1928	United States	41.0s
1932	United States	40.0s
1936	United States	39.8s
1948	United States	40.6s
1952	United States	40.1s
1956	United States	39.5s
1960	Germany (U.S. disqualified)	39.5s
1964	United States	39.0s
1968	United States	38.2s
1972	United States	38.19s
1976	United States	38.33s
1980	USSR	38.26s
1984	United States	37.83s
1988	USSR (U.S. disqualified)	38.19s
1992	United States	37.40s*
1996	Canada	37.69s

1,600-Meter Relay

1908	United States	3m. 29.4s
1912	United States	3m. 16.6s
1920	Great Britain	3m. 22.2s
1924	United States	3m. 16s
1928	United States	3m. 14.2s
1932	United States	3m. 8.2s
1936	Great Britain	3m. 9s
1948	United States	3m. 10.4s
1952	Jamaica	3m. 03.9s
1956	United States	3m. 04.8s
1960	United States	3m. 02.2s
1964	United States	3m. 00.7s
1968	United States	2m. 56.16s
1972	Kenya	2m. 59.8s
1976	United States	2m. 58.65s
1980	USSR	3m. 01.1s
1984	United States	2m. 57.91s
1988	United States	2m. 56.16s
1992	United States	2m. 55.74s*
1996	United States	2m. 55.99s

3,000-Meter Steeplechase

1920	Percy Hodge, Great Britain	10m. 0.4s
1924	Willie Ritola, Finland	9m. 33.6s
1928	Toivo Loukola, Finland	9m. 21.8s
1932	Volmari Iso-Hollo, Finland	10m. 33.4s
	(About 3,450 m; extra lap by error.)	
1936	Volmari Iso-Hollo, Finland	9m. 3.8s
1948	Thore Sjoestrand, Sweden	9m. 4.6s
1952	Horace Ashenfelter, United States	8m. 45.4s
1956	Chris Brasher, Great Britain	8m. 41.2s
1960	Zdzislaw Krzyszkowiak, Poland	8m. 34.2s
1964	Gaston Roelants, Belgium	8m. 30.8s
1968	Amos Biwott, Kenya	8m. 51s
1972	Kipchoge Keino, Kenya	8m. 23.6s
1976	Anders Garderud, Sweden	8m. 08.2s
1980	Bronislaw Malinowski, Poland	8m. 09.7s
1984	Julius Korir, Kenya	8m. 11.8s
1988	Julius Kariuki, Kenya	8m. 05.51s*
1992	Matthew Birir, Kenya	8m. 08.84s
1996	Joseph Keter, Kenya	8m. 07.12s

20-Kilometer Walk

1956	Leonid Spirin, USSR	1h. 31m. 27.4s
1960	Vladimir Golubnichy, USSR	1h. 33m. 7.2s
1964	Kenneth Mathews, Great Britain	1h. 29m. 34.0s
1968	Vladimir Golubnichy, USSR	1h. 33m. 58.4s
1972	Peter Frenkel, E. Germany	1h. 26m. 42.4s
1976	Daniel Bautista, Mexico	1h. 24m. 40.6s
1980	Maurizio Damilano, Italy	1h. 23m. 35.5s
1984	Ernesto Canto, Mexico	1h. 23m. 13.0s
1988	Josef Pribilinec, Czechoslovakia	1h. 19m. 57.0s*
1992	Daniel Plaza Montero, Spain	1h. 21m. 45.0s
1996	Jefferson Perez, Ecuador	1h. 20m.7s

50-Kilometer Walk

1932	Thomas W. Green, Great Britain	4h. 50m. 10s
1936	Harold Whitlock, Great Britain	4h. 30m. 41.4s
1948	John Ljunggren, Sweden	4h. 41m. 52s
1952	Giuseppe Dordoni, Italy	4h. 28m. 07.8s
1956	Norman Read, New Zealand	4h. 30m. 42.8s
1960	Donald Thompson, Great Britain	4h. 25m. 30s
1964	Abdon Pamich, Italy	4h. 11m. 12.4s
1968	Christoph Hohne, E. Germany	4h. 20m. 13.6s
1972	Bern Kannenberg, W. Germany	3h. 56m. 11.6s
1980	Hartwig Gauter, E. Germany	3h. 49m. 24.0s
1984	Raul Gonzalez, Mexico	3h. 47m. 26.0s
1988	Vyacheslav Ivanenko, USSR	3h. 38m. 29.0s*
1992	Andrei Perlov, Unified Team	3h. 50m. 13.0s
1996	Robert Korzeniowski, Poland	3h. 43m. 30s

Marathon

1896	Spiridon Loues, Greece	2h. 58m. 50s
1900	Michel Theato, France	2h. 59m. 45s
1904	Thomas Hicks, United States	3h. 28m. 63s
1908	John J. Hayes, United States	2h. 55m. 18.4s
1912	Kenneth McArthur, South Africa	2h. 36m. 54.8s
1920	Hannes Kolehmainen, Finland	2h. 32m. 35.8s
1924	Albin Stenroos, Finland	2h. 41m. 22.6s
1928	A.B. El Ouafi, France	2h. 32m. 57s
1932	Juan Zabala, Argentina	2h. 31m. 36s
1936	Kijung Son, Japan (Korean)	2h. 29m. 19.2s
1948	Delfo Cabrera, Argentina	2h. 34m. 51.6s
1952	Emil Zatopek, Czechoslovakia	2h. 23m. 03.2s
1956	Alain Mimoun, France.	2h. 25m.
1960	Abebe Bikila, Ethiopia	2h. 15m. 16.2s
1964	Abebe Bikila, Ethiopia	2h. 12m. 11.2s
1968	Mamo Wolde, Ethiopia	2h. 20m. 26.4s
1972	Frank Shorter, United States	2h. 12m. 19.8s
1976	Waldemar Cierpinski, E. Germany	2h. 09m. 55s
1980	Waldemar Cierpinski, E. Germany	2h. 11m. 03s
1984	Carlos Lopes, Portugal	2h. 09m. 21s*
1988	Gelindo Bordin, Italy	2h. 10m. 32s
1992	Hwang Young-Cho, S. Korea	2h. 13m. 23s
1996	Josia Thugwane, South Africa	2h. 12m. 36s

High Jump

1896	Ellery Clark, United States	1.81m.
1900	Irving Baxter, United States	1.90m.
1904	Samuel Jones, United States	1.80m.
1908	Harry Porter, United States	1.905m.
1912	Alma Richards, United States	1.93m.
1920	Richmond Landon, United States	1.935m.
1924	Harold Osborn, United States	1.98m.
1928	Robert W. King, United States	1.94m.
1932	Duncan McNaughton, Canada	1.97m.
1936	Cornelius Johnson, United States	2.03m.
1948	John L. Winter, Australia	1.98m.
1952	Walter Davis, United States	2.04m.
1956	Charles Dumas, United States	2.12m.
1960	Robert Shavlakadze, USSR	2.16m.
1964	Valery Brumel, USSR	2.18m.
1968	Dick Fosbury, United States	2.24m.
1972	Juri Tarmak, USSR	2.23m.
1976	Jacek Wszola, Poland.	2.25m.
1980	Gerd Wessig, E. Germany	2.36m.
1984	Dietmar Mogenburg, W. Germany	2.35m.
1988	Hennady Avdeyenko, USSR	2.38m.
1992	Javier Sotomayor Sanabria, Cuba	2.34m.
1996	Charles Austin, United States	2.39m.*

Long Jump

1896	Ellery Clark, United States	6.35m.
1900	Alvin Kraenzlein, United States	7.18m.
1904	Meyer Prinstein, United States	7.34m.
1908	Frank Irons, United States	7.48m.
1912	Albert Gutterson, United States	7.60m.
1920	William Pettersson (Bjorneman), Sweden	7.15m.
1924	William DeHart Hubbard, United States	7.44m.
1928	Edward B. Hamm, United States	7.73m.
1932	Edward Gordon, United States	7.64m.
1936	Jesse Owens, United States.	8.06m.
1948	Willie Steele, United States	7.82m.
1952	Jerome Biffle, United States	7.57m.
1956	Gregory Bell, United States	7.83m.
1960	Ralph Boston, United States	8.12m.
1964	Lynn Davies, Great Britain	8.07m.
1968	Bob Beamon, United States	8.90m.*
1972	Randy Williams, United States	8.24m.
1976	Arnie Robinson, United States	8.35m.
1980	Lutz Dombrowski, E. Germany	8.54m.
1984	Carl Lewis, United States	8.54m.
1988	Carl Lewis, United States	8.72m.
1992	Carl Lewis, United States	8.67m.
1996	Carl Lewis, United States	8.50m.

Triple Jump

1896	James Connolly, United States	13.71m.
1900	Meyer Prinstein, United States	14.47m.
1904	Meyer Prinstein, United States	14.35m.
1908	Timothy Ahearne, Great Britain, Ireland	14.92m.
1912	Gustaf Lindblom, Sweden	14.76m.
1920	Vilho Tuulos, Finland	14.50m.
1924	Anthony Winter, Australia	15.525m.
1928	Mikio Oda, Japan	15.21m.
1932	Chuhei Nambu, Japan	15.72m.
1936	Naoto Tajima, Japan.	16.00m.
1948	Arne Ahman, Sweden	15.40m.
1952	Adhemar Ferreira da Silva, Brazil	16.22m.
1956	Adhemar Ferreira da Silva, Brazil	16.35m.
1960	Jozef Schmidt, Poland	16.81m.
1964	Jozef Schmidt, Poland	16.85m.
1968	Viktor Saneyev, USSR	17.39m.

1972	Viktor Saneyev, USSR	17.35m.
1976	Viktor Saneyev, USSR	17.29m.
1980	Jaak Uudmae, USSR	17.35m.
1984	Al Joyner, United States	17.26m.
1988	Khristo Markov, Bulgaria	17.61m.
1992	Mike Conley, United States	18.17m.
1996	Kenny Harrison, United States	18.90m.*

Discus Throw

1896	Robert Garrett, United States	29.15m.
1900	Rudolf Bauer, Hungary	36.04m.
1904	Martin Sheridan, United States	39.28m.
1908	Martin Sheridan, United States	40.89m.
1912	Armas Taipale, Finland	45.21m.
1920	Elmer Niklander, Finland	44.685m.
1924	Clarence Houser, United States	46.15m.
1928	Clarence Houser, United States	47.32m.
1932	John Anderson, United States	49.49m.
1936	Ken Carpenter, United States	50.48m.
1948	Adolfo Consolini, Italy	52.78m.
1952	Sim Iness, United States	55.03m.
1956	Al Oerter, United States	56.36m.
1960	Al Oerter, United States	59.18m.
1964	Al Oerter, United States	61.00m.
1968	Al Oerter, United States	64.78m.
1972	Ludvik Danek, Czechoslovakia	64.40m.
1976	Mac Wilkins, United States	67.50m.
1980	Viktor Rashchupkin, USSR	66.64m.
1984	Rolf Dannenberg, W. Germany	66.60m.
1988	Jurgen Schult, E. Germany	68.82m.
1992	Romas Ubartas, Lithuania	65.12m.
1996	Lars Riedel, Germany	69.40m.*

Hammer Throw

1900	John Flanagan, United States	49.73m.
1904	John Flanagan, United States	51.23m.
1908	John Flanagan, United States	51.92m.
1912	Matt McGrath, United States	54.74m.
1920	Pat Ryan, United States	52.875m.
1924	Fred Tootell, United States	53.295m.
1928	Patrick O'Callaghan, Ireland	51.39m.
1932	Patrick O'Callaghan, Ireland	53.92m.
1936	Karl Hein, Germany	56.49m.
1948	Imre Nemeth, Hungary	56.07m.
1952	Jozsef Csermak, Hungary	60.34m.
1956	Harold Connolly, United States	63.19m.
1960	Vasily Rudenkov, USSR	67.10m.
1964	Romuald Klim, USSR	69.74m.
1968	Gyula Zsivotsky, Hungary	73.36m.
1972	Anatoly Bondarchuk, USSR	75.50m.
1976	Yuri Syedykh, USSR	77.52m.
1980	Yuri Syedykh, USSR	81.80m.
1984	Juha Tiainen, Finland	78.08m.
1988	Sergei Litvinov, USSR	84.80m.*
1992	Andrey Abduvaliyev, Unified Team	82.54m.
1996	Balazs Kiss, Hungary	81.24m.

Javelin Throw

1908	Erik Lemming, Sweden	54.825m.
1912	Erik Lemming, Sweden	60.64m.
1920	Jonni Myyra, Finland	64.78m.
1924	Jonni Myyra, Finland	62.96m.
1928	Eric Lundkvist, Sweden	66.60m.
1932	Matti Jarvinen, Finland	72.71m.
1936	Gerhard Stoeck, Germany	71.84m.
1948	Kai Tapio Rautavaara, Finland	69.77m.
1952	Cy Young, United States	73.78m.
1956	Egil Danielson, Norway	85.71m.
1960	Viktor Tsybulenko, USSR	84.64m.
1964	Pauli Nevala, Finland	82.66m.
1968	Janis Lusis, USSR	90.10m.
1972	Klaus Wolfermann, W. Germany	90.48m.
1976	Miklos Nemeth, Hungary	94.58m.
1980	Dainis Kula, USSR	91.20m.
1984	Arto Haerkoenen, Finland	86.76m.
1988	Tapio Korjus, Finland	84.28m.
1992	Jan Zelezny, Czechoslovakia (a)	89.66m.*
1996	Jan Zelezny, Czech Republic	88.16m.

(a) New records were kept after javelin was modified in 1986.

Pole Vault

1896	William Welles Hoyt, United States	3.30m.
1900	Irving Baxter, United States	3.30m.
1904	Charles Dvorak, United States	3.50m.
1908	A. C. Gilbert, United States	
	Edward Cooke Jr., United States	3.71m.
1912	Harry Babcock, United States	3.95m.
1920	Frank Foss, United States	4.09m.
1924	Lee Barnes, United States	3.95m.
1928	Sabin W. Carr, United States	4.20m.
1932	William Miller, United States	4.31m.
1936	Earle Meadows, United States	4.35m.
1948	Guinn Smith, United States	4.30m.
1952	Robert Richards, United States	4.55m.
1956	Robert Richards, United States	4.56m.
1960	Don Bragg, United States	4.70m.
1964	Fred Hansen, United States	5.10m.
1968	Bob Seagren, United States	5.40m.
1972	Wolfgang Nordwig, E. Germany	5.50m.
1976	Tadeusz Slusarski, Poland	5.50m.
1980	Wladyslaw Kozakiewicz, Poland	5.78m.
1984	Pierre Quinon, France	5.75m.
1988	Sergei Bubka, USSR	5.90m.
1992	Maksim Tarassov, Unified Team	5.80m.
1996	Jean Galfione, France	5.92m.*

16-lb. Shot Put

1896	Robert Garrett, United States	11.22m.
1900	Richard Sheldon, United States	14.10m.
1904	Ralph Rose, United States	14.21m.
1908	Ralph Rose, United States	14.21m.
1912	Pat McDonald, United States	15.34m.
1920	Ville Porhola, Finland	14.81m.
1924	L. Clarence Houser, United States	14.99m.
1928	John Kuck, United States	15.87m.
1932	Leo Sexton, United States	16.00m.
1936	Hans Woellke, Germany	16.20m.
1948	Wilbur Thompson, United States	17.12m.
1952	W. Parry O'Brien, United States	17.41m.
1956	W. Parry O'Brien, United States	18.57m.
1960	William Nieder, United States	19.68m.
1964	Dallas Long, United States	20.33m.
1968	Randy Matson, United States	20.54m.
1972	Wladyslaw Komar, Poland	21.18m.
1976	Udo Beyer, E. Germany	21.05m.
1980	Vladimir Kyselyov, USSR	21.35m.
1984	Alessandro Andrei, Italy	21.26m.
1988	Ulf Timmermann, E. Germany	22.47m.*
1992	Michael Stulce, United States	21.70m.
1996	Randy Barnes, United States	21.62m.

Decathlon

1904	Thomas Kiely, Ireland	6,036 pts.
1906-08	not held	
1912	Hugo Wieslander, Sweden (a)	7,724.49 pts.
1920	Helge Lovland, Norway	6,804.35 pts.
1924	Harold Osborn, United States	7,710.77 pts.
1928	Paavo Yrjola, Finland	8,053.29 pts.
1932	James Bausch, United States	8,462.23 pts.
1936	Glenn Morris, United States	7,900 pts.
1948	Robert Mathias, United States	7,139 pts.
1952	Robert Mathias, United States	7,887 pts.
1956	Milton Campbell, United States	7,937 pts.
1960	Rafer Johnson, United States	8,392 pts.
1964	Willi Holdorf, Germany (b)	7,887 pts.
1968	Bill Toomey, United States	8,193 pts.
1972	Nikolai Avilov, USSR	8,454 pts.
1976	Bruce Jenner, United States	8,617 pts.
1980	Daley Thompson, Great Britain	8,495 pts.
1984	Daley Thompson, Great Britain (c)	8,798 pts.*
1988	Christian Schenk, E. Germany	8,488 pts.
1992	Robert Zmelik, Czechoslovakia	8,611 pts.
1996	Dan O'Brien, United States	8,824 pts.

(a) Jim Thorpe of the U.S. won the 1912 Decathlon with 8,413 pts. but was disqualified and had to return his medals because he had played professional baseball prior to the Olympic games. (b) Former point systems used prior to 1964. (c) Scoring change effective Apr. 1985; Thompson's readjusted score is 8,847 pts.

Track and Field—Women

100-Meter Run

1928	Elizabeth Robinson, United States	12.2s
1932	Stella Walsh, Poland (a)	11.9s
1936	Helen Stephens, United States	11.5s
1948	Francina Blankers-Koen, Netherlands	11.9s
1952	Marjorie Jackson, Australia	11.5s
1956	Betty Cuthbert, Australia	11.5s
1960	Wilma Rudolph, United States	11.0s
1964	Wyomia Tyus, United States	11.4s
1968	Wyomia Tyus, United States	11.0s
1972	Renate Stecher, E. Germany	11.07s
1976	Annegret Richter, W. Germany	11.08s
1980	Lyudmila Kondratyeva, USSR	11.6s
1984	Evelyn Ashford, United States	10.97s
1988	Florence Griffith-Joyner, United States	10.54s*
1992	Gail Devers, United States	10.82s
1996	Gail Devers, United States	10.94s

(a) A 1980 autopsy determined that Walsh was a man.

200-Meter Run

1948	Francina Blankers-Koen, Netherlands	24.4s
1952	Marjorie Jackson, Australia	23.7s
1956	Betty Cuthbert, Australia	23.4s
1960	Wilma Rudolph, United States	24.0s
1964	Edith McGuire, United States	23.0s
1968	Irena Szewinska, Poland	22.5s
1972	Renate Stecher, E. Germany	22.40s
1976	Barbel Eckert, E. Germany	22.37s
1980	Barbel Wockel, E. Germany	22.03s
1984	Valerie Brisco-Hooks, United States	21.81s
1988	Florence Griffith-Joyner, United States	21.34s*
1992	Gwen Torrence, United States	21.81s
1996	Marie-Jose Perec, France	22.12s

400-Meter Run

1964	Betty Cuthbert, Australia	52.0s
1968	Colette Besson, France	52.0s
1972	Monika Zehrt, E. Germany	51.08s
1976	Irena Szewinska, Poland	49.29s
1980	Marita Koch, E. Germany	48.88s
1984	Valerie Brisco-Hooks, United States	48.83s
1988	Olga Bryzgina, USSR	48.65s
1992	Marie-Jose Perec, France	48.83s
1996	Marie-Jose Perec, France	48.25s*

800-Meter Run

1928	Lina Radke, Germany	2m. 16.8s
1960	Ludmila Shevtsova, USSR	2m. 4.3s
1964	Ann Packer, Great Britain	2m. 1.1s
1968	Madeline Manning, United States	2m. 0.9s
1972	Hildegard Falck, W. Germany	1m. 58.6s
1976	Tatyana Kazankina, USSR	1m. 54.94s
1980	Nadezhda Olizayrenko, USSR	1m. 53.5s*
1984	Doina Melinte, Romania	1m. 57.6s
1988	Sigrun Wodars, E. Germany	1m. 56.10s
1992	Ellen Van Langen, Netherlands	1m. 55.54s
1996	Svetlana Masterkova, Russia	1m. 57.73s

1,500-Meter Run

1972	Lyudmila Bragina, USSR	4m. 01.4s
1976	Tatyana Kazankina, USSR	4m. 05.48s
1980	Tatyana Kazankina, USSR	3m. 56.6s
1984	Gabriella Dorio, Italy	4m. 03.25s
1988	Paula Ivan, Romania	3m. 53.96s*
1992	Hassiba Boulmerka, Algeria	3m. 55.30s
1996	Svetlana Masterkova, Russia	4m. 00.83s

3,000-Meter Run

1984	Maricica Puica, Romania	8m. 35.96s
1988	Tatyana Samolenko, USSR	8m. 26.53s*
1992	Elena Romanova, Unified Team	8m. 46.04s

5,000-Meter Run

1996	Wang Junxia, China	14m. 59.88s*

10,000-Meter Run

1988	Olga Boldarenko, USSR	31m. 44.69s
1992	Derartu Tulu, Ethiopia	31m. 06.02s
1996	Fernanda Ribeiro, Portugal	31m. 01.63s*

100-Meter Hurdles

1972	Annelie Ehrhardt, E. Germany	12.59s
1976	Johanna Schaller, E. Germany	12.77s
1980	Vera Komisova, USSR	12.56s
1984	Benita Brown-Fitzgerald, United States	12.84s
1988	Jordanka Donkova, Bulgaria	12.38s*
1992	Paraskevi Patoulidou, Greece	12.64s
1996	Ludmila Enquist, Sweden	12.58s

400-Meter Hurdles

1984	Nawal el Moutawakii, Morocco	54.61s
1988	Debra Flintoff-King, Australia	53.17s
1992	Sally Gunnell, Great Britain	53.23s
1996	Deon Hemmings, Jamaica	52.82s*

400-Meter Relay

1928	Canada	48.4s
1932	United States	46.9s
1936	United States	46.9s
1948	Netherlands	47.5s
1952	United States	45.9s
1956	Australia	44.5s
1960	United States	44.5s
1964	Poland	43.6s
1968	United States	42.8s
1972	West Germany	42.81s
1976	East Germany	42.55s
1980	East Germany	41.60s*
1984	United States	41.65s
1988	United States	41.98s
1992	United States	42.11s
1996	United States	41.95s

1,600-Meter Relay

1972	East Germany	3m. 23s
1976	East Germany	3m. 19.23s
1980	USSR	3m. 20.02s
1984	United States	3m. 18.29s
1988	USSR	3m. 15.18s*
1992	Unified Team	3m. 20.20s
1996	United States	3m. 20.91s

10 Kilometer Walk

1992	Chen Yueling, China	44m. 32s
1996	Elena Nikolayeva, Russia	41m. 49s*

Marathon

1984	Joan Benoit, United States	2h. 24m. 52s*
1988	Rosa Mota, Portugal	2h. 25m. 40s
1992	Valentina Yegorova, Unified Team	2h. 32m. 41s
1996	Fatuma Roba, Ethiopia	2h. 26m. 05s

High Jump

1928	Ethel Catherwood, Canada	1.59m.
1932	Jean Shiley, United States	1.657m.
1936	Ibolya Csak, Hungary	1.60m.
1948	Alice Coachman, United States	1.68m.
1952	Esther Brand, South Africa	1.67m.
1956	Mildred L. McDaniel, United States	1.76m.
1960	Iolanda Balas, Romania	1.85m.
1964	Iolanda Balas, Romania	1.90m.
1968	Miloslava Reskova, Czechoslovakia	1.82m.
1972	Ulrike Meyfarth, W. Germany	1.92m.
1976	Rosemarie Ackermann, E. Germany	1.93m.
1980	Sara Simeoni, Italy	1.97m.
1984	Ulrike Meyfarth, W. Germany	2.02m.
1988	Louise Ritter, United States	2.03m.
1992	Heike Henkel, Germany	2.02m.
1996	Stefka Kostadinova, Bulgaria	2.05m.*

Long Jump

1948	Olga Gyarmati, Hungary	5.695m.
1952	Yvette Williams, New Zealand	6.24m.
1956	Elzbieta Krzeskinska, Poland	6.35m.
1960	Vira Krepkina, USSR	6.37m.
1964	Mary Rand, Great Britain	6.76m.
1968	Viorica Viscopoleanu, Romania	6.82m.
1972	Heidemarie Rosendahl, W. Germany	6.78m.
1976	Angela Voigt, E. Germany	6.72m.
1980	Tatyana Kolpakova, USSR	7.06m.
1984	Anisoara Cusmir-Stanciu, Romania	6.96m.
1988	Jackie Joyner-Kersee, United States	7.40m.*
1992	Heike Drechsler, Germany	7.14m.
1996	Chioma Ajunwa, Nigeria	8.50m.

Triple Jump

1996	Inessa Kravets, Ukraine	15.33m.*

Discus Throw

1928	Halina Konopacka, Poland	39.62m.
1932	Lillian Copeland, United States	40.58m.
1936	Gisela Mauermayer, Germany	47.63m.
1948	Micheline Ostermeyer, France	41.92m.
1952	Nina Romaschkova, USSR	51.42m.
1956	Olga Fikotova, Czechoslovakia	53.69m.
1960	Nina Ponomareva, USSR	55.10m.
1964	Tamara Press, USSR	57.27m.
1968	Lia Manoliu, Romania	58.28m.
1972	Faina Melnik, USSR	66.62m.
1976	Evelin Schlaak, E. Germany	69.00m.
1980	Evelin Jahl, E. Germany	69.96m.
1984	Ria Stalman, Netherlands	65.36m.
1988	Martina Hellmann, E. Germany	72.30m.*
1992	Maritza Marten Garcia, Cuba	70.06m.
1996	Ilke Wyludda, Germany	69.66m.

Javelin Throw

1932	"Babe" Didrikson, United States	43.68m.
1936	Tilly Fleischer, Germany	45.18m.
1948	Herma Bauma, Austria	45.57m.
1952	Dana Zatopkova, Czechoslovakia	50.47m.
1956	Inese Jaunzeme, USSR	53.86m.
1960	Elvira Ozolina, USSR	55.98m.
1964	Mihaela Penes, Romania	60.54m.
1968	Angela Nemeth, Hungary	60.36m.
1972	Ruth Fuchs, E. Germany	63.88m.
1976	Ruth Fuchs, E. Germany	65.94m.
1980	Maria Colon Ruenes, Cuba	68.40m.
1984	Tessa Sanderson, Great Britain	69.56m.
1988	Petra Felke, E. Germany	74.68m.*
1992	Silke Renke, Germany	68.34m.
1996	Heli Rantanen, Finland	67.94m.

Shot Put (8 lb., 13 oz.)

1948	Micheline Ostermeyer, France	13.75m.
1952	Galina Zybina, USSR	15.28m.
1956	Tamara Tyshkyevich, USSR	16.59m.
1960	Tamara Press, USSR	17.32m.
1964	Tamara Press, USSR	18.14m.
1968	Margitta Gummel, E. Germany	19.61m.
1972	Nadezhda Chizova, USSR	21.03m.
1976	Ivanka Hristova, Bulgaria	21.16m.
1980	Ilona Slupianek, E. Germany	22.41m.*
1984	Claudia Losch, W. Germany	20.48m.
1988	Natalya Lisovskaya, USSR	22.24m.
1992	Svetlana Krivelyova, Unified Team	21.06m.
1996	Astrid Kumbernuss, Germany	20.56m.

Heptathlon

1984	Glynis Nunn, Australia	6,390 pts.
1988	Jackie Joyner-Kersee, U.S.	7,215 pts.*
1992	Jackie Joyner-Kersee, U.S.	7,044 pts.
1996	Ghada Shouaa, Syria	6,780 pts.

Swimming and Diving—Men

50-Meter Freestyle

1988	Matt Biondi, U.S.	22.14
1992	Aleksandr Popov, Unified Team	21.91*
1996	Aleksandr Popov, Russia	22.13

100-Meter Freestyle

1896	Alfred Hajos, Hungary	1:22.2
1904	Zoltan de Halmay, Hungary (100 yards)	1:02.8
1908	Charles Daniels, U.S.	1:05.6
1912	Duke P. Kahanamoku, U.S.	1:03.4
1920	Duke P. Kahanamoku, U.S.	1:01.4
1924	John Weissmuller, U.S.	59.0
1928	John Weissmuller, U.S.	58.6
1932	Yasuji Miyazaki, Japan	58.2
1936	Ferenc Csik, Hungary	57.6
1948	Wally Ris, U.S.	57.3
1952	Clark Scholes, U.S.	57.4
1956	Jon Henricks, Australia	55.4
1960	John Devitt, Australia	55.2
1964	Don Schollander, U.S.	53.4
1968	Mike Wenden, Australia	52.2
1972	Mark Spitz, U.S.	51.22
1976	Jim Montgomery, U.S.	49.99
1980	Jorg Woithe, E. Germany	50.40
1984	Rowdy Gaines, U.S.	49.80
1988	Matt Biondi, U.S.	48.63*
1992	Aleksandr Popov, Unified Team	49.02
1996	Aleksandr Popov, Russia	48.74

200-Meter Freestyle

1968	Mike Wenden, Australia	1:55.2
1972	Mark Spitz, U.S.	1:52.78
1976	Bruce Furniss, U.S.	1:50.29
1980	Sergei Kopliakov, USSR	1:49.81
1984	Michael Gross, W. Germany	1:47.44
1988	Duncan Armstrong, Australia	1:47.25
1992	Yevgeny Sadovyi, Unified Team	1:46.70*
1996	Danyon Loader, New Zealand	1:47.63

400-Meter Freestyle

1904	C. M. Daniels, U.S. (440 yards)	6:16.2
1908	Henry Taylor, Great Britain	5:36.8
1912	George Hodgson, Canada	5:24.4
1920	Norman Ross, U.S.	5:26.8
1924	John Weissmuller, U.S.	5:04.2
1928	Albert Zorilla, Argentina	5:01.6
1932	Clarence Crabbe, U.S.	4:48.4
1936	Jack Medica, U.S.	4:44.5
1948	William Smith, U.S.	4:41.0
1952	Jean Boiteux, France	4:30.7
1956	Murray Rose, Australia	4:27.3
1960	Murray Rose, Australia	4:18.3
1964	Don Schollander, U.S.	4:12.2
1968	Mike Burton, U.S.	4:09.0
1972	Brad Cooper, Australia	4:00.27
1976	Brian Goodell, U.S.	3:51.93
1980	Vladimir Salnikov, USSR	3:51.31
1984	George DiCarlo, U.S.	3:51.23
1988	Ewe Dassler, E. Germany	3:46.95
1992	Yevgeny Sadovyi, Unified Team	3:45.00*
1996	Danyon Loader, New Zealand	3:47.97

1,500-Meter Freestyle

1908	Henry Taylor, Great Britain	22:48.4
1912	George Hodgson, Canada	22:00.0
1920	Norman Ross, U.S.	22:23.2
1924	Andrew Charlton, Australia	20:06.6
1928	Arne Borg, Sweden	19:51.8
1932	Kusuo Kitamura, Japan	19:12.4
1936	Noboru Terada, Japan	19:13.7
1948	James McLane, U.S.	19:18.5
1952	Ford Konno, U.S.	18:30.3
1956	Murray Rose, Australia	17:58.9
1960	Jon Konrads, Australia	17:19.6
1964	Robert Windle, Australia	17:01.7
1968	Mike Burton, U.S.	16:38.9
1972	Mike Burton, U.S.	15:52.58
1976	Brian Goodell, U.S.	15:02.40
1980	Vladimir Salnikov, USSR	14:58.27
1984	Michael O'Brien, U.S.	15:05.20
1988	Vladimir Salnikov, USSR	15:00.40
1992	Kieren Perkins, Australia	14:43.48*
1996	Kieren Perkins, Australia	14:56.40

100-Meter Backstroke

1904	Walter Brack, Germany (100 yds.)	1:16.8
1908	Arno Bieberstein, Germany	1:24.6
1912	Harry Hebner, U.S.	1:21.2
1920	Warren Kealoha, U.S.	1:15.2
1924	Warren Kealoha, U.S.	1:13.2
1928	George Kojac, U.S.	1:08.2
1932	Masaji Kiyokawa, Japan	1:08.6
1936	Adolph Kiefer, U.S.	1:05.9
1948	Allen Stack, U.S.	1:06.4
1952	Yoshi Oyakawa, U.S.	1:05.4
1956	David Thiele, Australia	1:02.2
1960	David Thiele, Australia	1:01.9
1968	Roland Matthes, E. Germany	58.7
1972	Roland Matthes, E. Germany	56.58
1976	John Naber, U.S.	55.49
1980	Bengt Baron, Sweden	56.33
1984	Rick Carey, U.S.	55.79
1988	Daichi Suzuki, Japan	55.05
1992	Mark Tewksbury, Canada	53.98*
1996	Jeff Rouse, U.S.	54.10

200-Meter Backstroke

1964	Jed Graef, U.S.	2:10.3
1968	Roland Matthes, E. Germany	2:09.6
1972	Roland Matthes, E. Germany	2:02.82
1976	John Naber, U.S.	1:59.19
1980	Sandor Wladar, Hungary	2:01.93
1984	Rick Carey, U.S.	2:00.23
1988	Igor Polianski, USSR	1:59.37
1992	Martin Lopez-Zubero, Spain	1:58.47*
1996	Brad Bridgewater, U.S.	1:58.54

100-Meter Breaststroke

1968	Don McKenzie, U.S.	1:07.7
1972	Nobutaka Taguchi, Japan	1:04.94
1976	John Hencken, U.S.	1:03.11
1980	Duncan Goodhew, Great Britain	1:03.44
1984	Steve Lundquist, U.S.	1:01.65
1988	Adrian Moorhouse, Great Britain	1:02.04
1992	Nelson Diebel, U.S.	1:01.50
1996	Fred Deburghgraeve, Belgium	1:00.60*

200-Meter Breaststroke

1908	Frederick Holman, Great Britain	3:09.2
1912	Walter Bathe, Germany	3:01.8
1920	Haken Malmroth, Sweden	3:04.4
1924	Robert Skelton, U.S.	2:56.6
1928	Yoshiyuki Tsuruta, Japan	2:48.8
1932	Yoshiyuki Tsuruta, Japan	2:45.4
1936	Tetsuo Hamuro, Japan	2:41.5
1948	Joseph Verdeur, U.S.	2:39.3
1952	John Davies, Australia	2:34.4
1956	Masaru Furukawa, Japan	2:34.7
1960	William Mulliken, U.S.	2:37.4
1964	Ian O'Brien, Australia	2:27.8
1968	Felipe Munoz, Mexico	2:28.7
1972	John Hencken, U.S.	2:21.55
1976	David Wilkie, Great Britain	2:15.11
1980	Robertas Zhulpa, USSR	2:15.85
1984	Victor Davis, Canada	2:13.34
1988	Jozsef Szabo, Hungary	2:13.52
1992	Mike Barrowman, U.S.	2:10.16*
1996	Norbert Rozsa, Hungary	2:12.57

100-Meter Butterfly

1968	Doug Russell, U.S.	55.9
1972	Mark Spitz, U.S.	54.27
1976	Matt Vogel, U.S.	54.35
1980	Par Arvidsson, Sweden	54.92
1984	Michael Gross, W. Germany	53.08
1988	Anthony Nesty, Suriname	53.00
1992	Pablo Morales, U.S.	53.32
1996	Denis Pankratov, Russia	52.27*

200-Meter Butterfly

1956	William Yorzyk, U.S.	2:19.3
1960	Michael Troy, U.S.	2:12.8
1964	Kevin J. Berry, Australia	2:06.6
1968	Carl Robie, U.S.	2:08.7
1972	Mark Spitz, U.S.	2:00.70
1976	Mike Bruner, U.S.	1:59.23
1980	Sergei Fesenko, USSR	1:59.76
1984	Jon Sieben, Australia	1:57.04
1988	Michael Gross, W. Germany	1:56.94
1992	Mel Stewart, U.S.	1:56.26*
1996	Denis Pankratov, Russia	1:56.51

200-Meter Individual Medley

1968	Charles Hickcox, U.S.	2:12.0
1972	Gunnar Larsson, Sweden	2:07.17
1984	Alex Baumann, Canada	2:01.42
1988	Tamas Darnyi, Hungary	2:00.17
1992	Tamas Darnyi, Hungary	2:00.76
1996	Attila Czene, Hungary	1:59.91*

400-Meter Individual Medley

1964	Dick Roth, U.S.	4:45.4
1968	Charles Hickcox, U.S.	4:48.4
1972	Gunnar Larsson, Sweden	4:31.98
1976	Rod Strachan, U.S.	4:23.68
1980	Aleksandr Sidorenko, USSR	4:22.89
1984	Alex Baumann, Canada	4:17.41
1988	Tamas Darnyi, Hungary	4:14.75
1992	Tamas Darnyi, Hungary	4:14.23*
1996	Tom Dolan, U.S.	4:14.90

400-Meter Freestyle Relay

1964	United States	3:31.2
1968	United States	3:31.7
1972	United States	3:26.42
1984	United States	3:19.03
1988	United States	3:16.53
1992	United States	3:16.74
1996	United States	3:15.41*

800-Meter Freestyle Relay

1908	Great Britain	10:55.6
1912	Australia	10:11.6
1920	United States	10:04.4
1924	United States	9:53.4
1928	United States	9:36.2
1932	Japan	8:58.4
1936	Japan	8:51.5
1948	United States	8:46.0
1952	United States	8:31.1
1956	Australia	8:23.6
1960	United States	8:10.2
1964	United States	7:52.1
1968	United States	7:52.33
1972	United States	7:35.78
1976	United States	7:23.22
1980	USSR	7:23.50
1984	United States	7:15.69
1988	United States	7:12.51

1992	Unified Team	7:11.95*
1996	United States	7:14.84

400-Meter Medley Relay

1960	United States	4:05.4
1964	United States	3:58.4
1968	United States	3:54.9
1972	United States	3:48.16
1976	United States	3:42.22
1980	Australia	3:45.70
1984	United States	3:39.30
1988	United States	3:36.93
1992	United States	3:36.93
1996	United States	3:34.84*

Springboard Diving — Points

1908	Albert Zurner, Germany	85.5
1912	Paul Guenther, Germany	79.23
1920	Louis Kuehn, U.S	675.40
1924	Albert White, U.S.	97.46
1928	Pete Desjardins, U.S.	185.04
1932	Michael Galitzen, U.S.	161.38
1936	Richard Degener, U.S.	163.57
1948	Bruce Harlan, U.S.	163.64
1952	David Browning, U.S.	205.29
1956	Robert Clotworthy, U.S.	159.56
1960	Gary Tobian, U.S.	170.00
1964	Kenneth Sitzberger, U.S.	159.90
1968	Bernie Wrightson, U.S.	170.15
1972	Vladimir Vasin, USSR	594.09
1976	Phil Boggs, U.S.	619.52
1980	Aleksandr Portnov, USSR	905.02
1984	Greg Louganis, U.S.	754.41
1988	Greg Louganis, U.S.	730.80
1992	Mark Lenzi, U.S.	676.53
1996	Xiong Ni, China	701.46

Platform Diving — Points

1904	Dr. G.E. Sheldon, U.S.	12.75
1908	Hjalmar Johansson, Sweden	83.75
1912	Erik Adlerz, Sweden	73.94
1920	Clarence Pinkston, U.S.	100.67
1924	Albert White, U.S.	97.46
1928	Pete Desjardins, U.S.	98.74
1932	Harold Smith, U.S.	124.80
1936	Marshall Wayne, U.S.	113.58
1948	Sammy Lee, U.S.	130.05
1952	Sammy Lee, U.S.	156.28
1956	Joaquin Capilla, Mexico	152.44
1960	Robert Webster, U.S.	165.56
1964	Robert Webster, U.S.	148.58
1968	Klaus Dibiasi, Italy	164.18
1972	Klaus Dibiasi, Italy	504.12
1976	Klaus Dibiasi, Italy	600.51
1980	Falk Hoffmann, E. Germany	835.65
1984	Greg Louganis, U.S.	710.91
1988	Greg Louganis, U.S.	638.61
1992	Sun Shuwei, China	677.31
1996	Dmitri Sautin, Russia	692.34

Swimming and Diving—Women

50-Meter Freestyle

1988	Kristin Otto, E. Germany	25.49
1992	Yang Wenyi, China	24.76*
1996	Amy Van Dyken, U.S.	24.87

100-Meter Freestyle

1912	Fanny Durack, Australia	1:22.2
1920	Ethelda Bleibtrey, U.S.	1:13.6
1924	Ethel Lackie, U.S.	1:12.4
1928	Albina Osipowich, U.S.	1:11.0
1932	Helene Madison, U.S.	1:06.8
1936	Hendrika Mastenbroek, Holland	1:05.9
1948	Greta Andersen, Denmark	1:06.3
1952	Katalin Szoke, Hungary	1:06.8
1956	Dawn Fraser, Australia	1:02.0
1960	Dawn Fraser, Australia	1:01.2
1964	Dawn Fraser, Australia	59.5
1968	Jan Henne, U.S.	1:00.0
1972	Sandra Neilson, U.S.	58.59
1976	Kornelia Ender, E. Germany	55.65
1980	Barbara Krause, E. Germany	54.79
1984	(tie) Carrie Steinseifer, U.S.	55.92
	Nancy Hogshead, U.S.	55.92
1988	Kristin Otto, E. Germany	54.93
1992	Zhuang Yong, China	54.64
1996	Li Jingyi, China	54.50*

200-Meter Freestyle

1968	Debbie Meyer, U.S.	2:10.5
1972	Shane Gould, Australia	2:03.56
1976	Kornelia Ender, E. Germany	1:59.26
1980	Barbara Krause, E. Germany	1:58.33
1984	Mary Wayte, U.S.	1:59.23
1988	Heike Friedrich, E. Germany	1:57.65*
1992	Nicole Haislett, U.S.	1:57.90
1996	Claudia Poll, Costa Rica	1:58.16

400-Meter Freestyle

1924	Martha Norelius, U.S.	6:02.2
1928	Martha Norelius, U.S.	5:42.8
1932	Helene Madison, U.S.	5:28.5
1936	Hendrika Mastenbroek, Netherlands	5:26.4
1948	Ann Curtis, U.S.	5:17.8
1952	Valerie Gyenge, Hungary	5:12.1
1956	Lorraine Crapp, Australia	4:54.6
1960	Susan Chris von Saltza, U.S.	4:50.6
1964	Virginia Duenkel, U.S.	4:43.3
1968	Debbie Meyer, U.S.	4:31.8
1972	Shane Gould, Australia	4:19.44
1976	Petra Thuemer, E. Germany	4:09.89
1980	Ines Diers, E. Germany	4:08.76
1984	Tiffany Cohen, U.S.	4:07.10
1988	Janet Evans, U.S.	4:03.85*
1992	Dagmar Hase, Germany	4:07.18
1996	Michelle Smith, Ireland	4:07.25

800-Meter Freestyle

1968	Debbie Meyer, U.S.	9:24.0
1972	Keena Rothhammer, U.S.	8:53.68
1976	Petra Thuemer, E. Germany	8:37.14
1980	Michelle Ford, Australia	8:28.90
1984	Tiffany Cohen, U.S.	8:24.95
1988	Janet Evans, U.S.	8:20.20*
1992	Janet Evans, U.S.	8:25.52
1996	Brooke Bennett, U.S.	8:27.89

100-Meter Backstroke

1924	Sybil Bauer, U.S.	1:23.2
1928	Marie Braun, Netherlands	1:22.0
1932	Eleanor Holm, U.S.	1:19.4
1936	Dina Senff, Netherlands	1:18.9
1948	Karen Harup, Denmark	1:14.4
1952	Joan Harrison, South Africa	1:14.3
1956	Judy Grinham, Great Britain	1:12.9
1960	Lynn Burke, U.S.	1:09.3
1964	Cathy Ferguson, U.S.	1:07.7
1968	Kaye Hall, U.S.	1:06.2
1972	Melissa Belote, U.S.	1:05.78
1976	Ulrike Richter, E. Germany	1:01.83
1980	Rica Reinisch, E. Germany	1:00.86
1984	Theresa Andrews, U.S.	1:02.55
1988	Kristin Otto, E. Germany	1:00.89
1992	Krisztina Egerszegi, Hungary	1:00.68*
1996	Beth Botsford, U.S.	1:01.19

200-Meter Backstroke

1968	Pokey Watson, U.S.	2:24.8
1972	Melissa Belote, U.S.	2:19.19
1976	Ulrike Richter, E. Germany	2:13.43
1980	Rica Reinisch, E. Germany	2:11.77
1984	Jolanda De Rover, Netherlands	2:12.38
1988	Krisztina Egerszegi, Hungary	2:09.29
1992	Krisztina Egerszegi, Hungary	2:07.06*
1996	Krisztina Egerszegi, Hungary	2:07.83

100-Meter Breaststroke

1968	Djurdjica Bjedov, Yugoslavia	1:15.8
1972	Cathy Carr, U.S..	1:13.58
1976	Hannelore Anke, E. Germany	1:11.16
1980	Ute Geweniger, E. Germany	1:10.22
1984	Petra Van Staveren, Netherlands	1:09.88
1988	Tania Dangalakova, Bulgaria.	1:07.95
1992	Elena Roudkovskaia, Unified Team.	1:08.00
1996	Penny Heyns, South Africa	1:07.73*

200-Meter Breaststroke

1924	Lucy Morton, Great Britain	3:33.2
1928	Hilde Schrader, Germany	3:12.6
1932	Clare Dennis, Australia	3:06.3
1936	Hideko Maehata, Japan.	3:03.6
1948	Nelly Van Vliet, Netherlands	2:57.2
1952	Eva Szekely, Hungary	2:51.7
1956	Ursula Happe, Germany	2:53.1
1960	Anita Lonsbrough, Great Britain	2:49.5
1964	Galina Prozumenschikova, USSR.	2:46.4
1968	Sharon Wichman, U.S.	2:44.4
1972	Beverly Whitfield, Australia	2:41.71
1976	Marina Koshevaia, USSR	2:33.35
1980	Lina Kachushite, USSR.	2:29.54
1984	Anne Ottenbrite, Canada.	2:30.38
1988	Silke Hoerner, E. Germany	2:26.71
1992	Kyoko Iwasaki, Japan	2:26.65
1996	Penny Heyns, South Africa	2:25.41*

100-Meter Butterfly

1956	Shelley Mann, U.S.	1:11.0
1960	Carolyn Schuler, U.S.	1:09.5
1964	Sharon Stouder, U.S..	1:04.7
1968	Lynn McClements, Australia	1:05.5
1972	Mayumi Aoki, Japan	1:03.34
1976	Kornelia Ender, E. Germany	1:00.13
1980	Caren Metschuck, E. Germany	1:00.42
1984	Mary T. Meagher, U.S.	59.26
1988	Kristin Otto, E. Germany	59.00
1992	Qian Hong, China	58.62*
1996	Amy Van Dyken, U.S.	59.13

200-Meter Butterfly

1968	Ada Kok, Netherlands	2:24.7
1972	Karen Moe, U.S.	2:15.57
1976	Andrea Pollack, E. Germany	2:11.41
1980	Ines Geissler, E. Germany	2:10.44
1984	Mary T. Meagher, U.S..	2:06.90*
1988	Kathleen Nord, E. Germany	2:09.51
1992	Summer Sanders, U.S.	2:08.67
1996	Susan O'Neill, Australia.	2:07.76

200-Meter Individual Medley

1968	Claudia Kolb, U.S.	2:24.7
1972	Shane Gould, Australia	2:23.07

1984	Tracy Caulkins, U.S.	2:12.64
1988	Daniela Hunger, E. Germany	2:12.59
1992	Lin Li, China	2:11.65*
1996	Michelle Smith, Ireland	2:13.93

400-Meter Individual Medley

1964	Donna de Varona, U.S.	5:18.7
1968	Claudia Kolb, U.S.	5:08.5
1972	Gail Neall, Australia	5:02.97
1976	Ulrike Tauber, E. Germany	4:42.77
1980	Petra Schneider, E. Germany	4:36.29*
1984	Tracy Caulkins, U.S.	4:39.24
1988	Janet Evans, U.S.	4:37.76
1992	Krisztina Egerszegi, Hungary	4:36.54
1996	Michelle Smith, Ireland	4:39.18

400-Meter Freestyle Relay

1912	Great Britain	5:52.8
1920	United States	5:11.6
1924	United States	4:58.8
1928	United States	4:47.6
1932	United States	4:38.0
1936	Netherlands	4:36.0
1948	United States	4:29.2
1952	Hungary	4:24.4
1956	Australia	4:17.1
1960	United States	4:08.9
1964	United States	4:03.8
1968	United States	4:02.5
1972	United States	3:55.19
1976	United States	3:44.82
1980	East Germany.	3:42.71
1984	United States	3:43.43
1988	East Germany.	3:40.63
1992	United States	3:39.46
1996	United States	3:39.29*

800-Meter Freestyle Relay

1996	United States	7:59.87*

400-Meter Medley Relay

1960	United States	4:41.1
1964	United States	4:33.9
1968	United States	4:28.3
1972	United States	4:20.75
1976	East Germany.	4:07.95
1980	East Germany.	4:06.67
1984	United States	4:08.34
1988	East Germany.	4:03.74
1992	United States	4:02.54*
1996	United States	4:02.88

Springboard Diving — Points

1920	Aileen Riggin, U.S.	539.90
1924	Elizabeth Becker, U.S.	474.50
1928	Helen Meany, U.S.	78.62
1932	Georgia Coleman U.S.	87.52
1936	Marjorie Gestring, U.S.	89.27
1948	Victoria M. Draves, U.S.	108.74
1952	Patricia McCormick, U.S.	147.30
1956	Patricia McCormick, U.S.	142.36
1960	Ingrid Kramer, Germany	155.81
1964	Ingrid Engel-Kramer, Germany.	145.00
1968	Sue Gossick, U.S..	150.77
1972	Micki King, U.S.	450.03
1976	Jenni Chandler, U.S.	506.19
1980	Irina Kalinina, USSR.	725.91
1984	Sylvie Bernier, Canada.	530.70
1988	Gao Min, China.	580.23
1992	Gao Min, China.	572.40
1996	Fu Mingxia, China.	547.68

Platform Diving — Points

1912	Greta Johansson, Sweden	39.90
1920	Stefani Fryland-Clausen, Denmark.	34.60
1924	Caroline Smith, U.S.	33.20
1928	Elizabeth B. Pinkston, U.S.	31.60
1932	Dorothy Poynton, U.S.	40.26
1936	Dorothy Poynton Hill, U.S.	33.93
1948	Victoria M. Draves, U.S.	68.87
1952	Patricia McCormick, U.S.	79.37
1956	Patricia McCormick, U.S.	84.85
1960	Ingrid Kramer, Germany	91.28
1964	Lesley Bush, U.S.	99.80
1968	Milena Duchkova, Czech.	109.59
1972	Ulrika Knape, Sweden	390.00
1976	Elena Vaytsekhouskaya, USSR	406.59
1980	Martina Jaschke, E. Germany	596.25
1984	Zhou Jihong, China	435.51
1988	Xu Yanmei, China.	445.20
1992	Fu Mingxia, China.	461.43
1996	Fu Mingxia, China.	521.58

Boxing

Light Flyweight (106 lbs)
1968	Francisco Rodriguez, Venezuela
1972	Gyorgy Gedo, Hungary
1976	Jorge Hernandez, Cuba
1980	Shamil Sabyrov, USSR
1984	Paul Gonzalez, U.S.
1988	Ivailo Hristov, Bulgaria
1992	Rogelio Marcelo, Cuba
1996	Daniel Petrov, Bulgaria

Flyweight (112 lbs)
1904	George Finnegan, U.S.
1920	William Di Gennara, U.S.
1924	Fidel LaBarba, U.S.
1928	Antal Kocsis, Hungary
1932	Istvan Enekes, Hungary
1936	Willi Kaiser, Germany
1948	Pascual Perez, Argentina
1952	Nathan Brooks, U.S.
1956	Terence Spinks, Great Britain
1960	GyulaTorok, Hungary
1964	Fernando Atzori, Italy
1968	Ricardo Delgado, Mexico
1972	Georgi Kostadinov, Bulgaria
1976	Leo Randolph, U.S.
1980	Peter Lessov, Bulgaria
1984	Steve McCrory, U.S.
1988	Kim Kwang Sun, S. Korea
1992	Su Choi Choi, N. Korea
1996	Maikro Romero, Cuba

Bantamweight (119 lbs)
1904	Oliver Kirk, U.S.
1908	A. Henry Thomas, Great Britain
1920	Clarence Walker, South Africa
1924	William Smith, South Africa
1928	Vittorio Tamagnini, Italy
1932	Horace Gwynne, Canada
1936	Ulderico Sergo, Italy
1948	Tibor Csik, Hungary
1952	Pentti Hamalainen, Finland
1956	Wolfgang Behrendt, E. Germany
1960	Oleg Grigoryev, USSR
1964	Takao Sakurai, Japan
1968	Valery Sokolov, USSR
1972	Orlando Martinez, Cuba
1976	Yong-Jo Gu, N. Korea
1980	Juan Hernandez, Cuba
1984	Maurizio Stecca, Italy
1988	Kennedy McKinney, U.S.
1992	Joel Casamayor, Cuba
1996	Istvan Kovacs, Hungary

Featherweight (126 lbs)
1904	Oliver Kirk, U.S.
1908	Richard Gunn, Great Britain
1920	Paul Fritsch, France
1924	John Fields, U.S.
1928	Lambertus van Klaveren, Netherlands
1932	Carmelo Robledo, Argentina
1936	Oscar Casanovas, Argentina
1948	Ernesto Formenti, Italy
1952	Jan Zachara, Czechoslovakia
1956	Vladimir Safronov, USSR
1960	Francesco Musso, Italy
1964	Stanislav Stephashkin, USSR
1968	Antonin Roldan, Mexico
1972	Boris Kousnetsov, USSR
1976	Angel Herrera, Cuba
1980	Rudi Fink, E. Germany
1984	Meldrick Taylor, U.S.
1988	Giovanni Parisi, Italy

1992	Andreas Tews, Germany
1996	Somluck Kamsing, Thailand

Lightweight (132 lbs)
1904	Harry Spanger, U.S.
1908	Frederick Grace, Great Britain
1920	Samuel Mosberg, U.S.
1924	Hans Nielsen, Denmark
1928	Carlo Orlandi, Italy
1932	Lawrence Stevens, South Africa
1936	Imre Harangi, Hungary
1948	Gerald Dreyer, South Africa
1952	Aureliano Bolognesi, Italy
1956	Richard McTaggart, Great Britain
1960	Kazimierz Pazdzior, Poland
1964	Jozef Grudzien, Poland
1968	Ronald Harris, U.S.
1972	Jan Szczepanski, Poland
1976	Howard Davis, U.S.
1980	Angel Herrera, Cuba
1984	Pernell Whitaker, U.S.
1988	Andreas Zuelow, E. Germany
1992	Oscar De La Hoya, U.S.
1996	Hocine Soltani, Algeria

Light Welterweight (140 lbs)
1952	Charles Adkins, U.S.
1956	Vladimir Yengibaryan, USSR
1960	Bohumil Nemecek, Czechoslavakia
1964	Jerzy Kulej, Poland
1968	Jerzy Kulej, Poland
1972	Ray Seales, U.S.
1976	Ray Leonard, U.S.
1980	Patrizio Oliva, Italy
1984	Jerry Page, U.S.
1988	Viatcheslav Janovski, USSR
1992	Hector Vinent, Cuba
1996	Hector Vinent, Cuba

Welterweight (147 lbs)
1904	Albert Young, U.S.
1920	Albert Schneider, Canada
1924	Jean Delarge, Belgium
1928	Edward Morgan, New Zealand
1932	Edward Flynn, U.S.
1936	Sten Suvio, Finland
1948	Julius Torma, Czechoslovakia
1952	Zygmunt Chychia, Poland
1956	Nicolae Linca, Romania
1960	Giovanni Benvenuti, Italy
1964	Marian Kasprzyk, Poland
1968	Manfred Wolke, E. Germany
1972	Emilio Correa, Cuba
1976	Jochen Bachfeld, E. Germany
1980	Andres Aldama, Cuba
1984	Mark Breland, U.S.
1988	Robert Wangila, Kenya
1992	Michael Carruth, Ireland
1996	Oleg Saitov, Russia

Light Middleweight (157 lbs)
1952	Laszlo Papp, Hungary
1956	Laszlo Papp, Hungary
1960	Wilbert McClure, U.S.
1964	Boris Lagutin, USSR
1968	Boris Lagutin, USSR
1972	Dieter Kottysch, W. Germany
1976	Jerzy Rybicki, Poland
1980	Armando Martinez, Cuba
1984	Frank Tate, U.S.
1988	Park Si Hun, S. Korea
1992	Juan Lemus, Cuba
1996	David Reid, U.S.

Middleweight (165 lbs)
1904	Charles Mayer, U.S.
1908	John Douglas, Great Britain
1920	Harry Mallin, Great Britain
1924	Harry Mallin, Great Britain
1928	Piero Toscani, Italy
1932	Carmen Barth, U.S.
1936	Jean Despeaux, France
1948	Laszlo Papp, Hungary
1952	Floyd Patterson, U.S.
1956	Gennady Schatkov, USSR
1960	Edward Crook, U.S.
1964	Valery Popenchenko, USSR
1968	Christopher Finnegan, Great Britain
1972	Vyacheslav Lemechev, USSR
1976	Michael Spinks, U.S.
1980	Jose Gomez, Cuba
1984	Joon-Sup Shin, S. Korea
1988	Henry Maske, E. Germany
1992	Ariel Hernandez, Cuba
1996	Ariel Hernandez, Cuba

Light Heavyweight (179 lbs)
1920	Edward Eagan, U.S.
1924	Harry Mitchell, Great Britain
1928	Victor Avendano, Argentina
1932	David Carstens, South Africa
1936	Roger Michelot, France
1948	George Hunter, South Africa
1952	Norvel Lee, U.S.
1956	James Boyd, U.S.
1960	Cassius Clay, U.S.
1964	Cosimo Pinto, Italy
1968	Dan Poznyak, USSR
1972	Mate Parlov, Yugoslavia
1976	Leon Spinks, U.S.
1980	Slobodan Kacar, Yugoslavia
1984	Anton Josipovic, Yugoslavia
1988	Andrew Maynard, U.S.
1992	Torsten May, Germany
1996	Vassili Jirov, Kazakhstan

Heavyweight (201 lbs)
1984	Henry Tillman, U.S.
1988	Ray Mercer, U.S.
1992	Felix Savon, Cuba
1996	Felix Savon, Cuba

Super Heavyweight (Unlimited)
(known as heavyweight, 1904-80)
1904	Samuel Berger, U.S.
1908	Albert Oldham, Great Britain
1920	Ronald Rawson, Great Britain
1924	Otto von Porat, Norway
1928	Arturo Rodriguez Jurado, Argentina
1932	Santiago Lovell, Argentina
1936	Herbert Runge, Germany
1948	Rafael Iglesias, Argentina
1952	H. Edward Sanders, U.S.
1956	T. Peter Rademacher, U.S.
1960	Franco De Piccoli, Italy
1964	Joe Frazier, U.S.
1968	George Foreman, U.S.
1972	Teofilo Stevenson, Cuba
1976	Teofilo Stevenson, Cuba
1980	Teofilo Stevenson, Cuba
1984	Tyrell Biggs, U.S.
1988	Lennox Lewis, Canada
1992	Roberto Balado, Cuba
1996	Vladimir Klitschko, Ukraine

Other Summer Olympics Gold Medal Winners in 1996

Archery
Men's 70-Meter Individual—Justin Huish, U.S.
Men's Team—U.S.
Women's 70-Meter Individual—Kim Kyung Wook, S. Korea
Women's Team—S. Korea

Badminton
Men's Singles—Poul-Erik Hoyer-Larsen, Denmark
Men's Doubles—Rexy Mainaky & Ricky Subagja, Indonesia
Women's Singles—Bang Soo Hyun, S. Korea
Women's Doubles—Ge Fei & Gu Jun, China
Mixed Doubles—Gil Young Ah & Kim Dong Moon, S. Korea

Baseball
G-Cuba; S-Japan; B-U.S.

Basketball
Men—G-U.S.; S-Yugoslavia; B-Lithuania.
Women—G-U.S.; S-Brazil; B-Australia.

Beach Volleyball
Men—Karch Kiraly & Kent Steffes, U.S.
Women—Jackie Silva & Sandra Pires, Brazil

Canoe/Kayak
Men
Kayak Slalom—Oliver Fix, Germany
Kayak 500M Singles—Antonio Rossi, Italy
Kayak 500M Doubles—Kay Bluhm & Torsten Gutsche, Germany
Kayak 1,000M Singles—Knut Holmann, Norway
Kayak 1,000M Doubles—Antonio Rossi & Daniele Scarpa, Italy
Kayak 1,000M Fours—G-Germany; S-Hungary; B-Russia
Canoe Slalom Singles—Michal Martikan, Slovakia
Canoe Slalom Doubles—France
Canoe 500M Singles—Martin Doktor, Czech Rep.
Canoe 500M Doubles—Csaba Horvath & Gyorgy Kolonics, Hungary
Canoe 1,000M Singles—Martin Doktor, Czech Rep.
Canoe 1,000M Doubles—Andreas Dittmer & Gunar Kirchbach, Germany
Women
Kayak Slalom—Stepnka Hilgertova, Czech Rep.
Kayak 500M Singles—Rita Koban, Hungary
Kayak 500M Doubles—Agneta Andersson & Susanne Gunnarsson, Sweden
Kayak 500M Fours—G-Germany; S-Switzerland; B-Sweden

Cycling
Men
Individual Road Race—Pascal Richard, Switzerland
Sprint—Jens Fiedler, Germany
Individual Points Race—Silvio Martinello, Italy
4KM Team Pursuit—G-France; S-Russia; B-Australia
4KM Individual Pursuit—Andrea Collinelli, Italy
1KM Time Trial—Florian Rousseau, France
Individual Time Trial—Miguel Indurain, Spain
Cross-Country—Bart Jan Brentjens, Netherlands
Women
Individual Road Race—Jeannie Longo-Ciprelli, France
Sprint—Felicia Ballanger, France
Individual Points Race—Nathalie Lancien, France
Individual Pursuit—Antonella Bellutti, Italy
Individual Time Trial—Zulfiya Zabirova, Russia
Cross-Country—Paola Pezzo, Italy

Equestrian
Individual Three-Day Event—Blyth Tait, New Zealand
Team Three-Day Event—Australia
Individual Dressage—Isabell Werth, Germany
Team Dressage—Germany
Individual Jumping—Ulrich Kirchoff, Germany
Team Jumping—Germany

Fencing
Men
Individual Foil—Alessandro Puccini, Italy
Team Foil—Russia
Individual Saber—Stanislas Pozdnyakov, Russia
Team Saber—Russia
Individual Épée—Aleksandr Beketov, Russia
Team Épée—Italy
Women
Individual Foil—Laura Badea, Romania
Team Foil—Italy
Individual Épée—Laura Flessel, France
Team Épée—France

Field Hockey
Men—G-Netherlands; S-Spain; B-Australia
Women—G-Australia; S-S. Korea; B-Netherlands

Gymnastics
Men
Floor Exercise—Ioannis Melissanidis, Greece
Horizontal Bar—Andreas Wecker, Germany
Parallel Bars—Rustam Sharipov, Ukraine
Pommel Horse—Li Donghua, Switzerland
Rings—Yuri Chechi, Italy
Vault—Aleksei Nemov, Russia
Individual All-Around—Li Xiaoshuang, China
Team—G-Russia; S-China; B-Ukraine.
Women
Balance Beam—Shannon Miller, U.S.
Floor Exercise—Lilia Podkopayeva, Ukraine
Uneven Bars—Svetlana Chorkina, Russia
Vault—Simona Amanar, Romania
Individual All-Around—Lilia Podkopayeva, Ukraine
Team—G-U.S.; S-Russia; B-Romania.

Rhythmic Gymnastics
Individual All-Around—Yekaterina Serebryanskaya, Ukraine
Team—G-Spain; S-Bulgaria; B-Russia

Judo
Men
132 Pounds—Tadahiro Nomura, Japan
143 Pounds—Udo Quellmalz, Germany
157 Pounds—Kenzo Nakamura, Japan
172 Pounds—Djamel Bouras, France
190 Pounds—Jeon Ki Young, S. Korea
209 Pounds—Pawel Nastula, Poland
Heavyweight—David Douillet, France
Women
106 Pounds—Kye Sun Hi, N. Korea
115 Pounds—Marie-Claire Restoux, France
123 Pounds—Driulis Gonzalez, Cuba
134 Pounds—Yuko Emoto, Japan
146 Pounds—Cho Min Sun, S. Korea
159 Pounds—Ulla Werbrouck, Belgium
Over 159 Pounds—Sun Fuming, China

Modern Pentathlon
Aleksandr Parygin, Kazakhstan

Rowing
Men
Single Sculls—Xeno Müller, Switzerland
Double Sculls—Italy
Lightweight Double Sculls—Switzerland
Quadruple Sculls—Germany
Coxless Pairs—Great Britain
Coxless Fours—Australia
Lightweight Coxless Fours—Denmark
Coxed Eights—Netherlands
Women
Single Sculls—Yekaterina Khodotovich, Belarus
Double Sculls—Canada
Lightweight Double Sculls—Romania
Quadruple Sculls—Germany
Coxless Pairs—Australia
Coxed Eights—Romania

Shooting
Men
Air Pistol—Roberto Di Donna, Italy
Trap—Michael Diamond, Australia
Air Rifle—Artem Khadzhibekov, Russia
Free Pistol—Boris Kokorev, Russia
Double Trap—Russell Mark, Australia
Rapid Fire Pistol—Ralf Schumann, Germany
Rifle Prone—Christian Klees, Germany
Running Game Target—Yang Ling, China
Three-Position Rifle—Jean-Pierre Amat, France
Skeet—Ennio Falco, Italy
Women
Air Pistol—Olga Klochneva, Russia
Three-Position Rifle—Aleksandra Ivosev, Yugoslavia
Double Trap—Kim Rhode, U.S.
Sport Pistol—Li Duihong, China
Air Rifle—Renata Mauer, Poland

Soccer
Men—G-Nigeria; S-Argentina; B-Brazil
Women—G-U.S.; S-China; B-Norway

Softball
G-U.S.; S-China; B-Australia

Synchronized Swimming
G-U.S.; S-Canada; B-Japan

Table Tennis
Men's Singles—Liu Guoliang, China
Men's Doubles—Kong Linghui & Liu Guoliang, China
Women's Singles—Deng Yaping, China
Women's Doubles—Deng Yaping & Qiao Hong, China

Team Handball
Men—G-Croatia; S-Sweden; B-Spain
Women—G-Denmark; S- S. Korea; B-Hungary

Tennis
Men's Singles—Andre Agassi, U.S.
Men's Doubles—Todd Woodbridge & Mark Woodforde, Australia
Women's Singles—Lindsay Davenport, U.S.
Women's Doubles—Gigi Fernandez & Mary Joe Fernandez, U.S.

Volleyball
Men—G-Netherlands; S-Italy; B-Yugoslavia
Women—G-Cuba; S-China; B- Brazil

Water Polo
G-Spain; S-Croatia; B-Italy

Weight Lifting

119 Pounds—Halil Mutlu, Turkey
130 Pounds—Tang Ningsheng, China
141 Pounds—Naim Suleymanoglu, Turkey
154 Pounds—Zhan Xugang, China
161¹/₂ Pounds—Pablo Lara, Cuba
183 Pounds—Pyrros Dimas, Greece
200¹/₂ Pounds—Aleksei Petrov, Russia
218 Pounds—Akakide Kakhiashvilis, Greece
238 Pounds—Timur Taimazov, Ukraine
Over 238 Pounds—Andrei Chemerkin, Russia

Freestyle Wrestling

105¹/₂ Pounds—Kim Il, N. Korea
114¹/₂ Pounds—Valentin Jordanov, Bulgaria
125¹/₂ Pounds—Kendall Cross, U.S.
136¹/₂ Pounds—Tom Brands, U.S.
149¹/₂ Pounds—Vadim Bogiev, Russia
163 Pounds—Bouvaisa Satiev, Russia
180¹/₂ Pounds—Khadzhimurad Magomedov, Russia
198 Pounds—Rasul Khadem, Iran
220 Pounds—Kurt Angle, U.S.

Greco-Roman

286 Pounds—Mahmut Demir, Turkey
105¹/₂ Pounds—Sim Kwan Ho, S. Korea

114¹/₂ Pounds—Armen Nazaryan, Armenia
125¹/₂ Pounds—Yuri Melnichenko, Kazakhstan
136¹/₂ Pounds—Wlodzimierz Zwadzki, Poland
149¹/₂ Pounds—Ryszard Wolny, Poland
163 Pounds—Feliberto Ascuy Aguilera, Cuba
180¹/₂ Pounds—Hamza Yerlikiya, Turkey
198 Pounds—Vyacheslav Oleynyk, Ukraine
220 Pounds—Andrzej Wronski, Poland
286 Pounds—Aleksandr Karelin, Russia

Yachting
Open

Laser—Robert Scheidt, Brazil
Soling—G-Germany; S-Russia; B-U.S.
Star—G-Brazil; S-Sweden; B-Australia
Tornado—G-Spain; S-Australia; B-Brazil

Men

Finn—Mateusz Kusznierewicz, Poland
Mistral—Nikolaos Kaklamanakis, Greece
470—G-Ukraine; S-Great Britain; B-Portugal

Women

Europe—Kristine Rough, Denmark
Mistral—Lee Lai-Shan, Hong Kong
470—G-Spain; S-Japan; B-Ukraine

TRACK AND FIELD
World Track and Field Outdoor Records

As of Oct. 1999

The International Amateur Athletic Federation, the world body of track and field, recognizes only records in metric distances, except for the mile. *Pending ratification.

Men's Records
Running

Event	Record	Holder	Country	Date	Where made
100 meters	9.79 s.	Maurice Greene	U.S.	June 16, 1999	Athens, Greece
200 meters	19.32 s.	Michael Johnson	U.S.	Aug. 1, 1996	Atlanta, GA
400 meters	43.18 s.	Michael Johnson	U.S.	Aug. 26, 1999	Seville, Spain
800 meters	1 m., 41.11 s.	Wilson Kipketer	Denmark	Aug. 24, 1997	Cologne, Germany
1,000 meters	2 m., 11.96 s*.	Noah Ngeny	Kenya	Sept. 5, 1999	Rieti, Italy
1,500 meters	3 m., 26.00 s.	Hicham El Guerrouj	Morocco	July 14, 1998	Rome, Italy
1 mile	3 m., 43.13 s.	Hicham El Guerrouj	Morocco	July 7, 1999	Rome
2,000 meters	4 m., 44.79 s.	Hicham El Guerrouj	Morocco	Sept. 7, 1999	Berlin, Germany
3,000 meters	7 m., 20.67 s.	Daniel Komen	Kenya	Sept. 1, 1996	Rieti, Italy
5,000 meters	12 m., 39.36 s.	Haile Gebreselassie	Ethiopia	June 13, 1998	Helsinki, Finland
10,000 meters	26 m., 22.75 s.	Haile Gebreselassie	Ethiopia	June 1,1998	Hengelo, Netherlands
20,000 meters	56 m., 55.6 s.	Arturo Barrios	Mexico	Mar. 30, 1991	La Fleche, France
25,000 meters	1 hr., 13 m., 55.8 s.	Toshihiko Seko	Japan	Mar. 22, 1981	Christchurch, New Zealand
3,000 meter stpl.	7 m., 55.72 s.	Bernard Barmasai	Kenya	Aug. 24, 1997	Cologne, Germany
Marathon	2 hr., 5m., 42 s.	Khalid Khannouchi	Morocco	Oct. 24, 1999	Chicago

Hurdles

Event	Record	Holder	Country	Date	Where made
110 meters	12.91 s.	Colin Jackson	Gr. Britain	Aug. 20, 1993	Stuttgart, Germany
400 meters	46.78 s.	Kevin Young	U.S.	Aug. 6, 1992	Barcelona

Relay Races

Event	Record	Holder	Country	Date	Where made
400 mtrs. (4x100)	37.40 s.	(Marsh, Burrell, Mitchell, Lewis)	U.S.	Aug. 8, 1992	Barcelona
		(Drummond, Cason, Mitchell, Burrell)	U.S.	Aug. 21, 1993	Stuttgart, Germany
800 mtrs. (4×200)	1 m., 18.68 s.	(Marsh, Burrell, Heard, Lewis)	U.S.	Apr. 17, 1994	Walnut, CA
1,600 mtrs. (4×400)	2 m., 54.20 s.	(Young, Pettigrew, Washington, Johnson)	U.S.	July 22, 1998	Long Island, NY
3,200 mtrs. (4×800)	7 m., 03.89 s.	(Elliott, Cook, Cram, Coe)	Gr. Britain	Aug. 30, 1982	London

Field Events

Event	Record	Holder	Country	Date	Where made
High jump	2.45m	Javier Sotomayor	Cuba	July 27, 1993	Salamanca, Spain
Long jump	8.95m	Mike Powell	U.S.	Aug. 30, 1991	Tokyo
Triple jump	18.29m	Jonathan Edwards	Gr. Britain	Aug. 7, 1995	Göteborg, Sweden
Pole vault	6.14m	Sergei Bubka	Ukraine	July 31, 1994	Sestriere, Italy
16-lb. shot put	23.12m	Randy Barnes	U.S.	May 20, 1990	Los Angeles, CA
Discus	74.08m	Juergen Schult	E. Germany	June 6, 1986	Neubrandenburg, Germany
Javelin	98.48m	Jan Zelezny	Czech Rep.	May 25, 1996	Jena, Germany
16-lb. hammer	86.74m	Yuri Sedykh	USSR	Aug. 30, 1986	Stuttgart, W. Germany
Decathlon	8,994 pts.	Tomás Dvorák	Czech Rep.	July 3-4, 1999	Prague, Czech Rep.

Women's Records
Running

Event	Record	Holder	Country	Date	Where made
100 meters	10.49 s.	Florence Griffith Joyner	U.S.	July 16, 1988	Indianapolis, IN
200 meters	21.34 s.	Florence Griffith Joyner	U.S.	Sept. 29, 1988	Seoul
400 meters	47.60 s.	Marita Koch	E. Germany	Oct. 6, 1985	Canberra, Australia
800 meters	1 m., 53.28 s.	Jarmila Kratochvilova	Czechoslovakia	July 26, 1983	Munich
1,000 meters	2 m., 28.98 s.	Svetlana Masterkova	Russia	Aug. 23, 1996	Brussels
1,500 meters	3 m., 50.46 s.	Qu Yunxia	China	Sept. 11, 1993	Beijing
1 mile	4 m., 12.56 s.	Svetlana Masterkova	Russia	Aug. 14, 1996	Zurich
2,000 meters	5 m., 25.36 s.	Sonia O'Sullivan	Ireland	July 8, 1994	Edinburgh

Event	Record	Holder	Country	Date	Where made
3,000 meters	8 m., 06.11 s.	Wang Junxia	China	Sept. 13, 1993	Beijing
5,000 meters	14 m., 28.09 s.	Jiang Bo	China	Oct. 23, 1997	Shanghai
10,000 meters	29 m., 31.78 s.	Wang Junxia	China	Sept. 8, 1993	Beijing
Marathon	2 h., 21 m., 06 s.	Ingrid Kristiansen	Norway	Apr. 21, 1985	London

Hurdles

Event	Record	Holder	Country	Date	Where made
100 meters	12.21 s.	Yordanka Donkova	Bulgaria	Aug. 20, 1988	Bulgaria
400 meters	52.61 s.	Kim Batten	U.S.	Aug. 11, 1995	Göteborg, Sweden

Field Events

Event	Record	Holder	Country	Date	Where made
High jump	2.09m	Stefka Kostadinova	Bulgaria	Aug. 30, 1987	Rome
Long jump	7.52m	Galina Chistyakova	USSR	June 11, 1988	Leningrad
Triple jump	15.50m	Inessa Kravets	Ukraine	Aug. 10, 1995	Göteborg, Sweden
Pole vault	4.60m	Emma George	Australia	Feb. 20, 1999	Sydney, Australia
	4.60m	Stacy Dragila	U.S.	Aug. 21, 1999	Seville, Spain
Shot put	22.63m	Natalya Lisovskaya	USSR	June 7, 1987	Moscow
Discus	76.80m	Gabriele Reinsch	E. Germany	July 9, 1988	Neubrandenburg, Germany
Hammer	76.07m*	Mihaela Melinte	Romania	Aug. 29, 1999	Rüdlingen, Switz.
Javelin	80.00m	Petra Felke	E. Germany	Sept. 9, 1988	Potsdam, Germany
Heptathlon	7,291 pts.	Jackie Joyner-Kersee	U.S.	Sept. 23-24, 1988	Seoul

Relay Races

Event	Record	Holder	Country	Date	Where made
400 mtrs. (4×100)	41.37 s.	(Gladisch, Rieger, Auerswald, Goehr)	E. Germany	Oct. 6, 1985	Canberra, Australia
800 mtrs. (4×200)	1 m., 28.15 s.	(Goehr, Mueller, Woeckel, Koch)	E. Germany	Aug. 9, 1980	Jena, E. Germany
1,600 mtrs. (4×400)	3 m., 15.17 s.	(Ledovskaya, Nazarova, Pinigina, Bryzgina)	USSR	Oct. 1, 1988	Seoul
3,200 mtrs. (4×800)	7 m., 50.17 s.	(Olizarenko, Gurina, Borisova, Podyalovskaya)	USSR	Aug. 5, 1984	Moscow

World Track and Field Indoor Records

As of mid-Oct. 1999

The International Amateur Athletic Federation began recognizing world indoor track and field records as official on Jan. 1, 1987. World indoor bests set prior to Jan. 1, 1987, are subject to approval as world records providing they meet the IAAF world records criteria, including drug testing. To be accepted as a world indoor record, a performance must meet the same criteria as a world record outdoors, except that a track performance cannot be set on an indoor track larger than 200 meters. *Pending ratification. (a)=altitude.

Men's Records

Event	Record	Holder	Country	Date	Where made
50 meters	5.56 (a)	Donovan Bailey	Canada	Feb. 9, 1996	Reno, NV
	5.56*	Maurice Greene	U.S	Feb. 13, 1999	Los Angeles, CA
60 meters	6.39	Maurice Greene	U.S	Feb. 3, 1998	Madrid
200 meters	19.92	Frankie Fredericks	Namibia	Feb. 18, 1996	Lievin, France
400 meters	44.63	Michael Johnson	U.S.	Mar. 4, 1995	Atlanta, GA
800 meters	1:42.67	Wilson Kipketer	Denmark	Mar. 9, 1997	Paris
1,000 meters	2:15.26	Noureddine Morceli	Algeria	Feb. 22, 1992	Birmingham, England
1,500 meters	3:31.18	Hicham el-Guerrouj	Morocco	Feb. 2, 1997	Stuttgart, Germany
1 mile	3:48.45	Hicham el-Guerrouj	Morocco	Feb. 12, 1997	Ghent, Belgium
3,000 meters	7:24.90	Daniel Komen	Kenya	Feb. 6, 1998	Budapest
5,000 meters	12:50.38	Haile Gebrselassie	Ethiopia	Feb. 14, 1999	Birmingham, England
50-meter hurdles	6.25	Mark McKoy	Canada	Mar. 5, 1986	Kobe, Japan
60-meter hurdles	7.30	Colin Jackson	Gr. Britain	Mar. 6, 1994	Sindelfingen, Germany
High jump	2.43m	Javier Sotomayor	Cuba	Mar. 4, 1989	Budapest
Pole vault	6.15m	Sergei Bubka	Ukraine	Feb. 21, 1993	Donyetsk, Ukraine
Long jump	8.79m	Carl Lewis	U.S.	Jan. 27, 1984	New York, NY
Triple jump	17.83	Eliecer Urrutia	Cuba	Mar. 1, 1997	Sindelfingen, Germany
Shot put	22.66m	Randy Barnes	U.S.	Jan. 20, 1989	Los Angeles, CA

Women's Records

Event	Record	Holder	Country	Date	Where made
50 meters	5.96	Irina Privalova	Russia	Feb. 9, 1995	Madrid
60 meters	6.92	Irina Privalova	Russia	Feb. 9, 1995	Madrid
		Irina Privalova	Russia	Feb. 11, 1993	Madrid
200 meters	21.87	Merlene Ottey	Jamaica	Feb. 13, 1993	Lievin, France
400 meters	49.59	Jarmila Kratochvilova	Czechoslovakia	Mar. 7, 1982	Milan, Italy
800 meters	1:56.36	Maria Mutola	Mozambique	Feb. 22, 1998	Levin, France
1,000 meters	2:30.94	Maria Mutola	Mozambique	Feb. 25, 1999	Stockholm, Sweden
1,500 meters	4:00.27	Doina Melinte	Romania	Feb. 9, 1990	E. Rutherford, NJ
1 mile	4:17.14	Doina Melinte	Romania	Feb. 9, 1990	E. Rutherford, NJ
3,000 meters	8:33.82	Elly van Hulst	Netherlands	Mar. 4, 1989	Budapest
5,000 meters	14:47.35	Gabriela Szabo	Romania	Feb. 13, 1999	Dortmund, Germany
50-meter hurdles	6.58	Cornelia Oschkenat	E. Germany	Feb. 20, 1988	Berlin
60-meter hurdles	7.69	Lyudmila Narozhilenko	USSR	Feb. 4, 1990	Chelyabinsk, USSR
High jump	2.07m	Heike Henkel	Germany	Feb. 8, 1992	Karlsruhe, Germany
Pole vault	4.56m	Nicole Humbert	Germany	Feb. 25, 1999	Stockholm, Sweden
Long jump	7.37m	Heike Drechsler	E. Germany	Feb. 13, 1988	Vienna
Triple jump	15.16m	Ashia Hansen	Gr. Britain	Feb. 28, 1998	Valencia
Shot put	22.50m	Helena Fibingerova	Czechoslovakia	Feb. 19, 1977	Czechoslovakia

GYMNASTICS
World Gymnastics Championships in 1999

On Oct. 14, Maria Olaru of Romania won the women's all-around title at the World Gymnastics Championship, held in Tianjin, China, with a score of 38.774. Nicolay Krukov of Russia won the men's all-around title with a score of 57.485.

NATIONAL FOOTBALL LEAGUE

NFL 1998-99: AFC Wins Again, Elway and Sanders Retire, Instant Replay Is Back

Led by John Elway and Terrell Davis, the Denver Broncos won Super Bowl XXXIII on Jan. 31, 1999, becoming the first AFC team to repeat in nearly 20 years. Davis, the 1998 season MVP, became only the 4th player in history to rush for over 2,000 yards in a season (2,008). Elway, the Super Bowl MVP and winningest quarterback of all time (148-82-1), retired in the off season. Known for spectacular comebacks, Elway engineered an NFL-record 47 game-winning or game-tying drives in the 4th quarter or overtime during his 16-year career. He is the only quarterback ever to start 5 Super Bowls. Detroit Lions running back Barry Sanders, holder of 9 all-time NFL records and the league's 2d-leading career rusher (15,269), announced his retirement after only 10 years. NFL owners voted 28-3 for return of an instant replay system, which offers each team 2 challenges per game and a "replay official" to oversee the final 2 minutes of each half.

Final 1998 Standings

American Football Conference

Eastern Division

	W	L	T	Pct.	Pts.	Opp.
N.Y. Jets	12	4	0	.750	416	266
Miami*	10	6	0	.625	321	265
Buffalo*	10	6	0	.625	400	333
New England*	9	7	0	.563	337	329
Indianapolis	3	13	0	.188	310	444

Central Division

	W	L	T	Pct.	Pts.	Opp.
Jacksonville	11	5	0	.688	392	338
Tennessee	8	8	0	.500	330	320
Pittsburgh	7	9	0	.438	263	303
Baltimore	6	10	0	.375	269	335
Cincinnati	3	13	0	.188	268	452

Western Division

	W	L	T	Pct.	Pts.	Opp.
Denver	14	2	0	.875	501	309
Oakland	8	8	0	.500	288	356
Seattle	8	8	0	.500	372	310
Kansas City	7	9	0	.438	327	363
San Diego	5	11	0	.313	241	342

National Football Conference

Eastern Division

	W	L	T	Pct.	Pts.	Opp.
Dallas	10	6	0	.625	381	275
Arizona*	9	7	0	.563	325	378
N.Y. Giants	8	8	0	.500	287	309
Washington	6	10	0	.375	319	421
Philadelphia	3	13	0	.188	161	344

Central Division

	W	L	T	Pct.	Pts.	Opp.
Minnesota	15	1	0	.938	556	296
Green Bay*	11	5	0	.688	408	319
Tampa Bay	8	8	0	.500	314	295
Detroit	5	11	0	.313	306	378
Chicago	4	12	0	.250	276	368

Western Division

	W	L	T	Pct.	Pts.	Opp.
Atlanta	14	2	0	.875	442	289
San Francisco*	12	4	0	.750	479	328
New Orleans	6	10	0	.375	305	359
Carolina	4	12	0	.250	336	413
St. Louis	4	12	0	.250	285	378

* Wild card team.

AFC Playoffs—Miami 24, Buffalo 17; Jacksonville 25, New England 10; Denver 38, Miami 3; N.Y. Jets 34, Jacksonville 24; Denver 23, N.Y. Jets 10.
NFC Playoffs—Arizona 20, Dallas 7; San Francisco 30, Green Bay 27; Atlanta 20, San Francisco 18; Minnesota 41, Arizona 21; Atlanta 30, Minnesota 27 (OT).
Super Bowl—Denver 34, Atlanta 19.

National Football League Champions

Year	East Winner (W-L-T)	West Winner (W-L-T)	Playoff
1933	New York Giants (11-3-0)	Chicago Bears (10-2-1)	Chicago Bears 23, New York 21
1934	New York Giants (8-5-0)	Chicago Bears (13-0-0)	New York 30, Chicago Bears 13
1935	New York Giants (9-3-0)	Detroit Lions (7-3-2)	Detroit 26, New York 7
1936	Boston Redskins (7-5-0)	Green Bay Packers (10-1-1)	Green Bay 21, Boston 6
1937	Washington Redskins (8-3-0)	Chicago Bears (9-1-1)	Washington 28, Chicago Bears 21
1938	New York Giants (8-2-1)	Green Bay Packers (8-3-0)	New York 23, Green Bay 17
1939	New York Giants (9-1-1)	Green Bay Packers (9-2-0)	Green Bay 27, New York 0
1940	Washington Redskins (9-2-0)	Chicago Bears (8-3-0)	Chicago Bears 73, Washington 0
1941	New York Giants (8-3-0)	Chicago Bears (10-1-1)(a)	Chicago Bears 37, New York 9
1942	Washington Redskins (10-1-1)	Chicago Bears (11-0-0)	Washington 14, Chicago Bears 6
1943	Washington Redskins (6-3-1)(a)	Chicago Bears (8-1-1)	Chicago Bears, 41, Washington 21
1944	New York Giants (8-1-1)	Green Bay Packers (8-2-0)	Green Bay 14, New York 7
1945	Washington Redskins (8-2-0)	Cleveland Rams (9-1-0)	Cleveland 15, Washington 14
1946	New York Giants (7-3-1)	Chicago Bears (8-2-1)	Chicago Bears 24, New York 14
1947	Philadelphia Eagles (8-4-0)(a)	Chicago Cardinals (9-3-0)	Chicago Cardinals 28, Philadelphia 21
1948	Philadelphia Eagles (9-2-1)	Chicago Cardinals (11-1-0)	Philadelphia 7, Chicago Cardinals 0
1949	Philadelphia Eagles (11-1-0)	Los Angeles Rams (8-2-2)	Philadelphia 14, Los Angeles 0
1950	Cleveland Browns (10-2-0)(a)	Los Angeles Rams (9-3-0)(a)	Cleveland 30, Los Angeles 28
1951	Cleveland Browns (11-1-0)	Los Angeles Rams (8-4-0)	Los Angeles 24, Cleveland 17
1952	Cleveland Browns (8-4-0)	Detroit Lions (9-3-0)(a)	Detroit 17, Cleveland 7
1953	Cleveland Browns (11-1-0)	Detroit Lions (10-2-0)	Detroit 17, Cleveland 16
1954	Cleveland Browns (9-3-0)	Detroit Lions (9-2-1)	Cleveland 56, Detroit 10
1955	Cleveland Browns (9-2-1)	Los Angeles Rams (8-3-1)	Cleveland 38, Los Angeles 14
1956	New York Giants (8-3-1)	Chicago Bears (9-2-1)	New York 47, Chicago Bears 7
1957	Cleveland Browns (9-2-1)	Detroit Lions (8-4-0)	Detroit 59, Cleveland 14
1958	New York Giants (9-3-0)(a)	Baltimore Colts (9-3-0)	Baltimore 23, New York 17(b)
1959	New York Giants (10-2-0)	Baltimore Colts (9-3-0)	Baltimore 31, New York 16
1960	Philadelphia Eagles (10-2-0)	Green Bay Packers (8-4-0)	Philadelphia 17, Green Bay 13
1961	New York Giants (10-3-1)	Green Bay Packers (11-3-0)	Green Bay 37, New York 0
1962	New York Giants (12-2-0)	Green Bay Packers (13-1-0)	Green Bay 16, New York 7
1963	New York Giants (11-3-0)	Chicago Bears (11-1-2)	Chicago 14, New York 10
1964	Cleveland Browns (10-3-1)	Baltimore Colts (12-2-0).	Cleveland 27, Baltimore 0
1965	Cleveland Browns (11-3-0)	Green Bay Packers (10-3-1)(a)	Green Bay 23, Cleveland 12
1966	Dallas Cowboys (10-3-1)	Green Bay Packers (12-2-0)	Green Bay 34, Dallas 27

(a) Won divisional playoff. (b) Won at 8:15 of sudden death overtime period.

Year	Conference	Division	Winner (W-L-T)	Playoffs(c)
1967	East	Century	Cleveland Browns (9-5-0)	Dallas 52, Cleveland 14
		Capitol	Dallas Cowboys (9-5-0)	
	West	Central	Green Bay Packers (9-4-1)	Green Bay 28, Los Angeles 7
		Coastal	Los Angeles Rams (11-1-2)(a)	Green Bay 21, Dallas 17
1968	East	Century	Cleveland Browns (10-4-0)	Cleveland 31, Dallas 20
		Capitol	Dallas Cowboys (12-2-0)	
	West	Central	Minnesota Vikings (8-6-0)	Baltimore 24, Minnesota 14
		Coastal	Baltimore Colts (13-1-0)	Baltimore 34, Cleveland 0
1969	East	Century	Cleveland Browns (10-3-1)	Cleveland 38, Dallas 14
		Capitol	Dallas Cowboys (11-2-1)	
	West	Central	Minnesota Vikings (12-2-0)	Minnesota 23, Los Angeles 20
		Coastal	Los Angeles Rams (11-3-0)	Minnesota 27, Cleveland 7
1970	American	Eastern	Baltimore Colts (11-2-1)	Baltimore 17, Cincinnati 0
		Central	Cincinnati Bengals (8-6-0)	Oakland 21, Miami* 14
		Western	Oakland Raiders (8-4-2)	Baltimore 27, Oakland 17
	National	Eastern	Dallas Cowboys (10-4-0)	Dallas 5, Detroit* 0
		Central	Minnesota Vikings (12-2-0)	San Francisco 17, Minnesota 14
		Western	San Francisco 49ers (10-3-1)	Dallas 17, San Francisco 10
1971	American	Eastern	Miami Dolphins (10-3-1)	Miami 27, Kansas City* 24
		Central	Cleveland Browns (9-5-0)	Baltimore 20, Cleveland 3
		Western	Kansas City Chiefs(10-3-1)	Miami 21, Baltimore 0
	National	Eastern	Dallas Cowboys (11-3-0)	Dallas 20, Minnesota 12
		Central	Minnesota Vikings (11-3-0)	San Francisco 24, Washington* 20
		Western	San Francisco 49ers (9-5-0)	Dallas 14, San Francisco 3
1972	American	Eastern	Miami Dolphins (14-0-0)	Miami 20, Cleveland* 14
		Central	Pittsburgh Steelers (11-3-0)	Pittsburgh 13, Oakland 7
		Western	Oakland Raiders (10-3-1)	Miami 21, Pittsburgh 17
	National	Eastern	Washington Redskins (11-3-0)	Washington 16, Green Bay 3
		Central	Green Bay Packers (10-4-0)	Dallas* 30, San Francisco 28
		Western	San Francisco 49ers (8-5-1)	Washington 26, Dallas* 3
1973	American	Eastern	Miami Dolphins (12-2-0)	Miami 34, Cincinnati 16
		Central	Cincinnati Bengals (10-4-0)	Oakland 33, Pittsburgh* 14
		Western	Oakland Raiders (9-4-1)	Miami 27, Oakland 10
	National	Eastern	Dallas Cowboys (10-4-0)	Dallas 27, Los Angeles 16
		Central	Minnesota Vikings (12-2-0)	Minnesota 27, Washington* 20
		Western	Los Angeles Rams (12-2-0)	Minnesota 27, Dallas 10
1974	American	Eastern	Miami Dolphins (11-3-0)	Oakland 28, Miami 26
		Central	Pittsburgh Steelers (10-3-1)	Pittsburgh 32, Buffalo* 14
		Western	Oakland Raiders (12-2-0)	Pittsburgh 24, Oakland 13
	National	Eastern	St. Louis Cardinals (10-4-0)	Minnesota 30, St. Louis 14
		Central	Minnesota Vikings (10-4-0)	Los Angeles 19, Washington* 10
		Western	Los Angeles Rams (10-4-0)	Minnesota 14, Los Angeles 10
1975	American	Eastern	Baltimore Colts (10-4-0)	Pittsburgh 28, Baltimore 10
		Central	Pittsburgh Steelers (12-2-0)	Oakland 31, Cincinnati* 28
		Western	Oakland Raiders (11-3-0)	Pittsburgh 16, Oakland 10
	National	Eastern	St. Louis Cardinals (11-3-0)	Dallas* 17, Minnesota 14
		Central	Minnesota Vikings (12-2-0)	Los Angeles 35, St. Louis 23
		Western	Los Angeles Rams (12-2-0)	Dallas* 37, Los Angeles 7
1976	American	Eastern	Baltimore Colts (11-3-0)	Pittsburgh 40, Baltimore 14
		Central	Pittsburgh Steelers (10-4-0)	Oakland 24, New England* 21
		Western	Oakland Raiders (13-1-0)	Oakland 24, Pittsburgh 7
	National	Eastern	Dallas Cowboys (11-3-0)	Minnesota 35, Washington* 20
		Central	Minnesota Vikings (11-2-1)	Los Angeles 14, Dallas 12
		Western	Los Angeles Rams (10-3-1)	Minnesota 24, Los Angeles 13
1977	American	Eastern	Baltimore Colts (10-4-0)	Oakland* 37, Baltimore 31
		Central	Pittsburgh Steelers (9-5-0)	Denver 34, Pittsburgh 21
		Western	Denver Broncos (12-2-0)	Denver 20, Oakland* 17
	National	Eastern	Dallas Cowboys (12-2-0)	Dallas 37, Chicago* 7
		Central	Minnesota Vikings (9-5-0)	Minnesota 14, Los Angeles 7
		Western	Los Angeles Rams (10-4-0)	Dallas 23, Minnesota 6
1978	American	Eastern	New England Patriots (11-5-0)	Pittsburgh 33, Denver 10
		Central	Pittsburgh Steelers (14-2-0)	Houston* 31, New England 14
		Western	Denver Broncos (10-6-0)	Pittsburgh 34, Houston* 5
	National	Eastern	Dallas Cowboys (12-4-0)	Dallas 27, Atlanta* 20
		Central	Minnesota Vikings (8-7-1)	Los Angeles 34, Minnesota 10
		Western	Los Angeles Rams (12-4-0)	Dallas 28, Los Angeles 0
1979	American	Eastern	Miami Dolphins (10-6-0)	Houston* 17, San Diego 14
		Central	Pittsburgh Steelers (12-4-0)	Pittsburgh 34, Miami 14
		Western	San Diego Chargers (12-4-0)	Pittsburgh 27, Houston* 13
	National	Eastern	Dallas Cowboys (11-5-0)	Tampa Bay 24, Philadelphia* 17
		Central	Tampa Bay Buccaneers (10-6-0)	Los Angeles 21, Dallas 19
		Western	Los Angeles Rams (9-7-0)	Los Angeles 9, Tampa Bay 0
1980	American	Eastern	Buffalo Bills (11-5-0)	San Diego 20, Buffalo 14
		Central	Cleveland Browns (11-5-0)	Oakland* 14, Cleveland 12
		Western	San Diego Chargers (11-5-0)	Oakland* 34, San Diego 27
	National	Eastern	Philadelphia Eagles (12-4-0)	Philadelphia 31, Minnesota 16
		Central	Minnesota Vikings (9-7-0)	Dallas* 30, Atlanta 27
		Western	Atlanta Falcons (12-4-0)	Philadelphia 20, Dallas* 7
1981	American	Eastern	Miami Dolphins (11-4-1)	San Diego 41, Miami 38
		Central	Cincinnati Bengals (12-4-0)	Cincinnati 28, Buffalo* 21
		Western	San Diego Chargers (10-6-0)	Cincinnati 27, San Diego 7
	National	Eastern	Dallas Cowboys (12-4-0)	Dallas 38, Tampa Bay 0
		Central	Tampa Bay Buccaneers (9-7-0)	San Francisco 38, N.Y. Giants* 24
		Western	San Francisco 49ers (13-3-0)	San Francisco 28, Dallas 27
1982(d)	American		Los Angeles Raiders (8-1-0)	Strike-shortened season (see
	National		Washington Redskins (8-1-0)	playoff results after footnote)
1983	American	Eastern	Miami Dolphins (12-4-0)	Seattle* 27, Miami 20
		Central	Pittsburgh Steelers (10-6-0)	L.A. Raiders 38, Pittsburgh 10
		Western	Los Angeles Raiders (12-4-0)	L.A. Raiders 30, Seattle* 14

Year	Conference	Division	Winner (W-L-T)	Playoffs(c)
	National	Eastern	Washington Redskins (14-2-0)	Washington 51, L.A. Rams* 7
		Central	Detroit Lions (9-7-0)	San Francisco 24, Detroit 23
		Western	San Francisco 49ers (10-6-0)	Washington 24, San Francisco 21
1984	American	Eastern	Miami Dolphins (14-2-0)	Miami 31, Seattle* 10
		Central	Pittsburgh Steelers (9-7-0)	Pittsburgh 24, Denver 17
		Western	Denver Broncos (13-3-0)	Miami 45, Pittsburgh 28
	National	Eastern	Washington Redskins (11-5-0)	Chicago 23, Washington 19
		Central	Chicago Bears (10-6-0)	San Francisco 21, N.Y. Giants* 10
		Western	San Francisco 49ers (15-1-0)	San Francisco 23, Chicago 0
1985	American	Eastern	Miami Dolphins (12-4-0)	New England* 27, L.A. Raiders 20
		Central	Cleveland Browns (8-8-0)	Miami 24, Cleveland 21
		Western	Los Angeles Raiders (12-4-0)	New England* 31, Miami 14
	National	Eastern	Dallas Cowboys (10-6-0)	Chicago 21, N.Y. Giants* 0
		Central	Chicago Bears (15-1-0)	L.A. Rams 20, Dallas 0
		Western	Los Angeles Rams (11-5-0)	Chicago 24, L.A. Rams 0
1986	American	Eastern	New England Patriots (11-5-0)	Denver 22, New England 17
		Central	Cleveland Browns (12-4-0)	Cleveland 23, N.Y. Jets* 20
		Western	Denver Broncos (11-5-0)	Denver 23, Cleveland 20
	National	Eastern	New York Giants (14-2-0)	N.Y. Giants 49, San Francisco 3
		Central	Chicago Bears (14-2-0)	Washington* 27, Chicago 13
		Western	San Francisco 49ers (10-5-1)	N.Y. Giants 17, Washington* 0
1987	American	Eastern	Indianapolis Colts (9-6-0)	Cleveland 38, Indianapolis 21
		Central	Cleveland Browns (10-5-0)	Denver 34, Houston* 10
		Western	Denver Broncos (10-4-1)	Denver 38, Cleveland 33
	National	Eastern	Washington Redskins (11-4-0)	Washington 21, Chicago 17
		Central	Chicago Bears (11-4-0)	Minnesota* 36, San Francisco 24
		Western	San Francisco 49ers (13-2-0)	Washington 17, Minnesota* 10
1988	American	Eastern	Buffalo Bills (12-4-0)	Buffalo 17, Houston* 10
		Central	Cincinnati Bengals (12-4-0)	Cincinnati 21, Seattle 13
		Western	Seattle Seahawks (9-7-0)	Cincinnati 21, Buffalo 10
	National	Eastern	Philadelphia Eagles (10-6-0)	Chicago 20, Philadelphia 12
		Central	Chicago Bears (12-4-0)	San Francisco 34, Minnesota* 9
		Western	San Francisco 49ers (10-6-0)	San Francisco 28, Chicago 3
1989	American	Eastern	Buffalo Bills (9-7-0)	Cleveland 34, Buffalo 30
		Central	Cleveland Browns (9-6-1)	Denver 24, Pittsburgh* 23
		Western	Denver Broncos (11-5-0)	Denver 37, Cleveland 21
	National	Eastern	New York Giants (12-4-0)	San Francisco 41, Minnesota 13
		Central	Minnesota Vikings (10-6-0)	L.A. Rams* 19, N.Y. Giants 13
		Western	San Francisco 49ers (14-2-0)	San Francisco 30, L.A. Rams* 3
1990	American	Eastern	Buffalo Bills (13-3-0)	L.A. Raiders 20, Cincinnati 10
		Central	Cincinnati Bengals (9-7-0)	Buffalo 44, Miami* 34
		Western	Los Angeles Raiders (12-4-0)	Buffalo 51, L.A. Raiders 3
	National	Eastern	New York Giants (13-3-0)	San Francisco 28, Washington* 10
		Central	Chicago Bears (11-5-0)	N.Y. Giants 31, Chicago 3
		Western	San Francisco 49ers (14-2-0)	N.Y. Giants 15, San Francisco 13
1991	American	Eastern	Buffalo Bills (13-3-0)	Denver 26, Houston 24
		Central	Houston Oilers (11-5-0)	Buffalo 37, Kansas City* 14
		Western	Denver Broncos (12-4-0)	Buffalo 10, Denver 7
	National	Eastern	Washington Redskins (14-2-0)	Washington 24, Atlanta* 7
		Central	Detroit Lions (12-4-0)	Detroit 38, Dallas* 6
		Western	New Orleans Saints (11-5-0)	Washington 41, Detroit 10
1992	American	Eastern	Miami Dolphins (11-5-0)	Miami 31, San Diego 0
		Central	Pittsburgh Steelers (11-5-0)	Buffalo* 24, Pittsburgh 3
		Western	San Diego Chargers (11-5-0)	Buffalo* 29, Miami 10
	National	Eastern	Dallas Cowboys (13-3-0)	Dallas 34, Philadelphia* 10
		Central	Minnesota Vikings (11-5-0)	San Francisco 20, Washington* 13
		Western	San Francisco 49ers (14-2-0)	Dallas 30, San Francisco 20
1993	American	Eastern	Buffalo Bills (12-4-0)	Buffalo 29, L.A. Raiders* 23
		Central	Houston Oilers (12-4-0)	Kansas City 28, Houston 20
		Western	Kansas City Chiefs (11-5-0)	Buffalo 30, Kansas City 13
	National	Eastern	Dallas Cowboys (12-4-0)	Dallas 27, Green Bay* 17
		Central	Detroit Lions (10-6-0)	San Francisco 44, N.Y. Giants* 3
		Western	San Francisco 49ers (10-6-0)	Dallas 38, San Francisco 21
1994	American	Eastern	Miami Dolphins (10-6-0)	Pittsburgh 29, Cleveland* 9
		Central	Pittsburgh Steelers (12-4-0)	San Diego 22, Miami 21
		Western	San Diego Chargers (11-5-0)	San Diego 17, Pittsburgh 13
	National	Eastern	Dallas Cowboys (12-4-0)	San Francisco 44, Chicago* 15
		Central	Minnesota Vikings (10-6-0)	Dallas 35, Green Bay* 9
		Western	San Francisco 49ers (13-3-0)	San Francisco 38, Dallas 28
1995	American	Eastern	Buffalo Bills (10-6-0)	Indianapolis* 10, Kansas City 7
		Central	Pittsburgh Steelers (11-5-0)	Pittsburgh 40, Buffalo 21
		Western	Kansas City Chiefs (13-3-0)	Pittsburgh 20, Indianapolis* 16
	National	Eastern	Dallas Cowboys (12-4-0)	Dallas 30, Philadelphia* 11
		Central	Green Bay Packers (11-5-0)	Green Bay 27, San Francisco 17
		Western	San Francisco 49ers (11-5-0)	Dallas 38, Green Bay 27
1996	American	Eastern	New England Patriots (11-5-0)	Jacksonville* 30, Denver 27
		Central	Pittsburgh Steelers (10-6-0)	New England 28, Pittsburgh 3
		Western	Denver Broncos (13-3-0)	New England 20, Jacksonville* 6
	National	Eastern	Dallas Cowboys (10-6-0)	Green Bay 35, San Francisco* 14
		Central	Green Bay Packers (13-3-0)	Carolina 26, Dallas 17
		Western	Carolina Panthers (12-4-0)	Green Bay 30, Carolina 13
1997	American	Eastern	New England Patriots (10-6-0)	Pittsburgh 7, New England 6
		Central	Pittsburgh Steelers (11-5-0)	Denver* 14, Kansas City 10
		Western	Kansas City Chiefs (13-3-0)	Denver* 24, Pittsburgh 21
	National	Eastern	New York Giants (10-5-1)	San Francisco 38, Minnesota* 22
		Central	Green Bay Packers (13-3-0)	Green Bay 21, Tampa Bay* 7
		Western	San Francisco 49ers (13-3-0)	Green Bay 23, San Francisco 10

Year	Conference	Division	Winner (W-L-T)	Playoffs(c)
1998	American	Eastern	N.Y. Jets (12-4-0)	Denver 38, Miami* 3
		Central	Jacksonville Jaguars (11-5-0)	N.Y. Jets 34, Jacksonville 24
		Western	Denver Broncos (14-2-0)	Denver 23, N.Y. Jets 10
	National	Eastern	Dallas Cowboys (10-6-0)	Atlanta 20, San Francisco* 18
		Central	Minnesota Vikings (15-1-0)	Minnesota 41, Atlanta* 21
		Western	Atlanta Falcons (14-2-0)	Atlanta 30, Minnesota 27 (OT)

*Wild card team. (c) From 1978 on, only the final 2 conference playoff rounds are shown. (d) A strike shortened the 1982 season from 16 to 9 games. The top 8 teams in each conference played in a tournament to determine the conference champion. See below. **AFC playoffs**—Miami 28, New England 13; L.A. Raiders 27, Cleveland 10; N.Y. Jets 44, Cincinnati 17; San Diego 31, Pittsburgh 28; N.Y. Jets 17, L.A. Raiders 14; Miami 34, San Diego 13; Miami 14, N.Y. Jets 0. **NFC playoffs**—Washington 31, Detroit 7; Green Bay 41, St. Louis 16; Dallas 30, Tampa Bay 17; Minnesota 30, Atlanta 24; Washington 21, Minnesota 7; Dallas 37, Green Bay 26; Washington 31, Dallas 17. **AFC Champion**—Miami Dolphins. **NFC Champion**—Washington Redskins.

Denver Broncos Defeat Atlanta Falcons in Super Bowl XXXIII

The Denver Broncos defeated the Atlanta Falcons, 34-19, in Super Bowl XXXIII, Jan. 31, 1999, in Miami, FL. Denver is the first AFC team since the 1978-79 Pittsburgh Steelers to repeat as champions. Denver quarterback John Elway, in his final Super Bowl, completed 18 of 29 passes for 336 yards and a touchdown and ran in for another score against former Denver coach Dan Reeves, who guided Atlanta to its first Super Bowl in the franchise's 33-year history.

Score by Quarters

Denver	7	10	0	17—34	
Atlanta	3	3	0	13—19	

Scoring

Atlanta—Andersen 32 yd. field goal
Denver—Griffith 1 yd. run (Elam kick)
Denver—Elam 26 yd. field goal
Denver—R. Smith 80 yd. pass from Elway (Elam kick)
Atlanta—Andersen 28 yd. field goal
Denver—Griffith 1 yd. run (Elam kick)
Denver—Elway 3 yd. run (Elam kick)
Atlanta—Dwight 94 yd. kickoff return (Andersen kick)
Denver—Elam 37 yd. field goal
Atlanta—Mathis 3 yd. pass from Chandler (failed 2 pt. conv. pass)

Individual Statistics

Rushing — Denver, Davis 25-102, Griffith 4-9, Elway 3-2, Loville 2-8, R. Smith 1-1, Brister 1-(minus 1). Atlanta, J. Anderson 18-96, Chandler 4-30, Dwight 1-5.

Passing — Denver, Elway 18-29-1-336. Atlanta, Chandler 19-35-3-219.

Receiving — Denver, R. Smith 5-152, McCaffrey 5-72, Chamberlain 3-29, Davis 2-50, Sharpe 2-26, Griffith 1-7. Atlanta, Mathis 7-85, Martin 5-79, J. Anderson 3-16, Harris 2-21, Santiago 1-13, Kozlowski 1-5.

Team Statistics	Denver	Atlanta
First downs	22	21
Total net yards	457	337
Rushes-yards	36-121	23-131
Passing yards, net	336	206
Punt returns-yards	0-0	0-0
Kickoff returns-yards	3-44	7-227
Interception returns-yards	3-136	1-1
Comp.-att.-int.	18-29-1	19-35-3
Field goals made-attempts	2-4	2-3
Sacked-yards lost	0-0	2-13
Punts-average	1-35	1-39
Fumbles-lost	0-0	1-1
Penalties-yards	4-61	0-0
Time of possession	31:23	28:37

Attendance—74,803. Time—3:18.

Super Bowls

	Year	Winner	Loser	Winning coach	Site
I	1967	Green Bay Packers, 35	Kansas City Chiefs, 10	Vince Lombardi	Los Angeles Coliseum, CA
II	1968	Green Bay Packers, 33	Oakland Raiders, 14	Vince Lombardi	Orange Bowl, Miami, FL
III	1969	New York Jets, 16	Baltimore Colts, 7	Weeb Ewbank	Orange Bowl, Miami, FL
IV	1970	Kansas City Chiefs, 23	Minnesota Vikings, 7	Hank Stram	Tulane Stadium, New Orleans, LA
V	1971	Baltimore Colts, 16	Dallas Cowboys, 13	Don McCafferty	Orange Bowl, Miami, FL
VI	1972	Dallas Cowboys, 24	Miami Dolphins, 3	Tom Landry	Tulane Stadium, New Orleans, LA
VII	1973	Miami Dolphins, 14	Washington Redskins, 7	Don Shula	Los Angeles Coliseum, CA
VIII	1974	Miami Dolphins, 24	Minnesota Vikings, 7	Don Shula	Rice Stadium, Houston, TX
IX	1975	Pittsburgh Steelers, 16	Minnesota Vikings, 6	Chuck Noll	Tulane Stadium, New Orleans, LA
X	1976	Pittsburgh Steelers, 21	Dallas Cowboys, 17	Chuck Noll	Orange Bowl, Miami, FL
XI	1977	Oakland Raiders, 32	Minnesota Vikings, 14	John Madden	Rose Bowl, Pasadena, CA
XII	1978	Dallas Cowboys, 27	Denver Broncos, 10	Tom Landry	Superdome, New Orleans, LA
XIII	1979	Pittsburgh Steelers, 35	Dallas Cowboys, 31	Chuck Noll	Orange Bowl, Miami, FL
XIV	1980	Pittsburgh Steelers, 31	Los Angeles Rams, 19	Chuck Noll	Rose Bowl, Pasadena, CA
XV	1981	Oakland Raiders, 27	Philadelphia Eagles, 10	Tom Flores	Superdome, New Orleans, LA
XVI	1982	San Francisco 49ers, 26	Cincinnati Bengals, 21	Bill Walsh	Silverdome, Pontiac, MI
XVII	1983	Washington Redskins, 27	Miami Dolphins, 17	Joe Gibbs	Rose Bowl, Pasadena, CA
XVIII	1984	Los Angeles Raiders, 38	Washington Redskins, 9	Tom Flores	Tampa Stadium, FL
XIX	1985	San Francisco 49ers, 38	Miami Dolphins, 16	Bill Walsh	Stanford Stadium, Palo Alto, CA
XX	1986	Chicago Bears, 46	New England Patriots, 10	Mike Ditka	Superdome, New Orleans, LA
XXI	1987	New York Giants, 39	Denver Broncos, 20	Bill Parcells	Rose Bowl, Pasadena, CA
XXII	1988	Washington Redskins, 42	Denver Broncos, 10	Joe Gibbs	San Diego Stadium, CA
XXIII	1989	San Francisco 49ers, 20	Cincinnati Bengals, 16	Bill Walsh	Joe Robbie Stadium, Miami, FL
XXIV	1990	San Francisco 49ers, 55	Denver Broncos, 10	George Seifert	Superdome, New Orleans, LA
XXV	1991	New York Giants, 20	Buffalo Bills, 19	Bill Parcells	Tampa Stadium, FL
XXVI	1992	Washington Redskins, 37	Buffalo Bills, 24	Joe Gibbs	Metrodome, Minneapolis, MN
XXVII	1993	Dallas Cowboys, 52	Buffalo Bills, 17	Jimmy Johnson	Rose Bowl, Pasadena, CA
XXVIII	1994	Dallas Cowboys, 30	Buffalo Bills, 13	Jimmy Johnson	Georgia Dome, Atlanta, GA
XXIX	1995	San Francisco 49ers, 49	San Diego Chargers, 26	George Seifert	Joe Robbie Stadium, Miami, FL
XXX	1996	Dallas Cowboys, 27	Pittsburgh Steelers, 17	Barry Switzer	Sun Devil Stadium, Tempe, AZ
XXXI	1997	Green Bay Packers, 35	New England Patriots, 21	Mike Holmgren	Superdome, New Orleans, LA
XXXII	1998	Denver Broncos, 31	Green Bay Packers, 24	Mike Shanahan	Qualcomm Stadium, San Diego, CA
XXXIII	1999	Denver Broncos, 34	Atlanta Falcons, 19	Mike Shanahan	Pro Player Stadium, Miami, FL

Super Bowl MVPs

1967	Bart Starr, Green Bay	1978	Randy White, Harvey Martin, Dallas	1989	Jerry Rice, San Francisco
1968	Bart Starr, Green Bay	1979	Terry Bradshaw, Pittsburgh	1990	Joe Montana, San Francisco
1969	Joe Namath, N.Y. Jets	1980	Terry Bradshaw, Pittsburgh	1991	Ottis Anderson, N.Y. Giants
1970	Len Dawson, Kansas City	1981	Jim Plunkett, Oakland	1992	Mark Rypien, Washington
1971	Chuck Howley, Dallas	1982	Joe Montana, San Francisco	1993	Troy Aikman, Dallas
1972	Roger Staubach, Dallas	1983	John Riggins, Washington	1994	Emmitt Smith, Dallas
1973	Jake Scott, Miami	1984	Marcus Allen, L.A. Raiders	1995	Steve Young, San Francisco
1974	Larry Csonka, Miami	1985	Joe Montana, San Francisco	1996	Larry Brown, Dallas
1975	Franco Harris, Pittsburgh	1986	Richard Dent, Chicago	1997	Desmond Howard, Green Bay
1976	Lynn Swann, Pittsburgh	1987	Phil Simms, N.Y. Giants	1998	Terrell Davis, Denver
1977	Fred Biletnikoff, Oakland	1988	Doug Williams, Washington	1999	John Elway, Denver

American Football Conference Leaders

(American Football League, 1960-69)

Passing[1] / Receiving

Player, team (Passing)	Att	Com	YG	TD	Year	Player, team (Receiving)	Rec.	YG	TD
Jack Kemp, L.A. Chargers	406	211	3,018	20	1960	Lionel Taylor, Denver	92	1,235	12
George Blanda, Houston	362	187	3,330	36	1961	Lionel Taylor, Denver	100	1,176	4
Len Dawson, Dallas Texans	310	189	2,759	29	1962	Lionel Taylor, Denver	77	908	4
Tobin Rote, San Diego	286	170	2,510	20	1963	Lionel Taylor, Denver	78	1,101	10
Len Dawson, Kansas City	354	199	2,879	30	1964	Charley Hennigan, Houston	101	1,546	8
John Hadl, San Diego	348	174	2,798	20	1965	Lionel Taylor, Denver	85	1,131	6
Len Dawson, Kansas City	284	159	2,527	26	1966	Lance Alworth, San Diego	73	1,383	13
Daryle Lamonica, Oakland	425	220	3,228	30	1967	George Sauer, N.Y. Jets	75	1,189	6
Len Dawson, Kansas City	224	131	2,109	17	1968	Lance Alworth, San Diego	68	1,312	10
Greg Cook, Cincinnati	197	106	1,854	15	1969	Lance Alworth, San Diego	64	1,003	4
Daryle Lamonica, Oakland	356	179	2,516	22	1970	Marlin Briscoe, Buffalo	57	1,036	8
Bob Griese, Miami	263	145	2,089	19	1971	Fred Biletnikoff, Oakland	61	929	9
Earl Morrall, Miami	150	83	1,360	11	1972	Fred Biletnikoff, Oakland	58	802	7
Ken Stabler, Oakland	260	163	1,997	14	1973	Fred Willis, Houston	57	371	1
Ken Anderson, Cincinnati	328	213	2,667	18	1974	Lydell Mitchell, Baltimore Colts	72	544	2
Ken Anderson, Cincinnati	377	228	3,169	21	1975	Reggie Rucker, Cleveland	60	770	3
						Lydell Mitchell, Baltimore Colts	60	554	4
Ken Stabler, Oakland	291	194	2,737	27	1976	MacArthur Lane, Kansas City	66	686	1
Bob Griese, Miami	307	180	2,252	22	1977	Lydell Mitchell, Baltimore Colts	71	620	4
Terry Bradshaw, Pittsburgh	368	207	2,915	28	1978	Steve Largent, Seattle	71	1,168	8
Dan Fouts, San Diego	530	332	4,082	24	1979	Joe Washington, Baltimore Colts	82	750	3
Brian Sipe, Cleveland	554	337	4,132	30	1980	Kellen Winslow, San Diego	89	1,290	9
Ken Anderson, Cincinnati	479	300	3,754	29	1981	Kellen Winslow, San Diego	88	1,075	10
Ken Anderson, Cincinnati	309	218	2,495	12	1982	Kellen Winslow, San Diego	54	721	6
Dan Marino, Miami	296	173	2,210	20	1983	Todd Christensen, L.A. Raiders	92	1,247	12
Dan Marino, Miami	564	362	5,084	48	1984	Ozzie Newsome, Cleveland	89	1,001	5
Ken O'Brien, N.Y. Jets	488	297	3,888	25	1985	Lionel James, San Diego	86	1,027	6
Dan Marino, Miami	623	378	4,746	44	1986	Todd Christensen, L.A. Raiders	95	1,153	8
Bernie Kosar, Cleveland	389	241	3,033	22	1987	Al Toon, N.Y. Jets	68	976	5
Boomer Esiason, Cincinnati	388	223	3,572	28	1988	Al Toon, N.Y. Jets	93	1,067	5
Boomer Esiason, Cincinnati	455	258	3,525	28	1989	Andre Reed, Buffalo	88	1,312	9
Jim Kelly, Buffalo	346	219	2,829	24	1990	Haywood Jeffires, Houston	74	1,048	8
						Drew Hill, Houston	74	1,019	5
Jim Kelly, Buffalo	474	304	3,844	33	1991	Haywood Jeffires, Houston	100	1,181	7
Warren Moon, Houston	346	224	2,521	18	1992	Haywood Jeffires, Houston	90	913	9
John Elway, Denver	551	348	4,030	25	1993	Reggie Langhorne, Indianapolis	85	1,038	3
Dan Marino, Miami	615	385	4,453	30	1994	Ben Coates, New England	96	1,174	7
Jim Harbaugh, Indianapolis	314	200	2,575	17	1995	Carl Pickens, Cincinnati	99	1,234	17
John Elway, Denver	466	287	3,328	26	1996	Carl Pickens, Cincinnati	100	1,180	12
Mark Brunell, Jacksonville	435	264	3,281	18	1997	Tim Brown, Oakland	104	1,408	5
Vinny Testaverde, N.Y. Jets	421	259	3,256	29	1998	O.J. McDuffie, Miami	90	1,050	7

Scoring / Rushing

Player, team (Scoring)	TD	PAT	FG	Pts	Year	Player, team (Rushing)	Yds	Att	TD
Gene Mingo, Denver	6	33	18	123	1960	Abner Haynes, Dallas Texans	875	156	9
Gino Cappelletti, Boston	8	48	17	147	1961	Billy Cannon, Houston	948	200	6
Gene Mingo, Denver	4	32	27	137	1962	Cookie Gilchrest, Buffalo	1,096	214	13
Gino Cappelletti, Boston	2	35	22	113	1963	Clem Daniels, Oakland	1,099	215	3
Gino Cappelletti, Boston	7	36	25	155	1964	Cookie Gilchrest, Buffalo	981	230	6
Gino Cappelletti, Boston	9	27	17	132	1965	Paul Lowe, San Diego	1,121	222	7
Gino Cappelletti, Boston	6	35	16	119	1966	Jim Nance, Boston	1,458	299	11
George Blanda, Oakland	0	56	20	116	1967	Jim Nance, Boston	1,216	269	7
Jim Turner, N.Y. Jets	0	43	34	145	1968	Paul Robinson, Cincinnati	1,023	238	8
Jim Turner, N.Y. Jets	0	33	32	129	1969	Dick Post, San Diego	873	182	6
Jan Stenerud, Kansas City	0	26	30	116	1970	Floyd Little, Denver	901	209	3
Garo Yepremian, Miami	0	33	28	117	1971	Floyd Little, Denver	1,133	284	6
Bobby Howfield, N.Y. Jets	0	40	27	121	1972	O.J. Simpson, Buffalo	1,251	292	6
Roy Gerela, Pittsburgh	0	36	29	123	1973	O.J. Simpson, Buffalo	2,003	332	12
Roy Gerela, Pittsburgh	0	33	20	93	1974	Otis Armstrong, Denver	1,407	263	9
O.J. Simpson, Buffalo	23	0	0	138	1975	O.J. Simpson, Buffalo	1,817	329	16
Toni Linhart, Baltimore Colts	0	49	20	109	1976	O.J. Simpson, Buffalo	1,503	290	8
Errol Mann, Oakland	0	39	20	99	1977	Mark van Eeghen, Oakland	1,273	324	7
Pat Leahy, N.Y. Jets	0	41	22	107	1978	Earl Campbell, Houston	1,450	302	13
John Smith, New England	0	46	23	115	1979	Earl Campbell, Houston	1,697	368	19
John Smith, New England	0	51	26	129	1980	Earl Campbell, Houston	1,934	373	13
Jim Breech, Cincinnati	0	49	22	115	1981	Earl Campbell, Houston	1,376	361	10
Nick Lowery, Kansas City	0	37	26	115					
Marcus Allen, L.A. Raiders	14	0	0	84	1982	Freeman McNeil, N.Y. Jets	786	151	6
Gary Anderson, Pittsburgh	0	38	27	119	1983	Curt Warner, Seattle	1,446	335	13
Gary Anderson, Pittsburgh	0	45	24	117	1984	Earnest Jackson, San Diego	1,179	296	8
Gary Anderson, Pittsburgh	0	40	33	139	1985	Marcus Allen, L.A. Raiders	1,759	380	11
Tony Franklin, New England	0	44	32	140	1986	Curt Warner, Seattle	1,481	319	13
Jim Breech, Cincinnati	0	25	24	97	1987	Eric Dickerson, L.A. Rams-Ind.	1,288*	283	6
Scott Norwood, Buffalo	0	33	32	129	1988	Eric Dickerson, Indianapolis	1,659	388	14
David Treadwell, Denver	0	39	27	120	1989	Christian Okoye, Kansas City	1,480	370	12
Nick Lowery, Kansas City	0	37	34	139	1990	Thurman Thomas, Buffalo	1,297	271	11
Pete Stoyanovich, Miami	0	28	31	121	1991	Thurman Thomas, Buffalo	1,407	288	7
Pete Stoyanovich, Miami	0	34	30	124	1992	Barry Foster, Pittsburgh	1,690	390	11
Jeff Jaeger, L.A. Raiders	0	27	35	132	1993	Thurman Thomas, Buffalo	1,315	355	6
John Carney, San Diego	0	33	34	135	1994	Chris Warren, Seattle	1,545	333	9
Norm Johnson, Pittsburgh	0	39	34	141	1995	Curtis Martin, New England	1,487	368	14
Cary Blanchard, Indianapolis	0	27	36	135	1996	Terrell Davis, Denver	1,538	345	13
Mike Hollis, Jacksonville	0	41	31	134	1997	Terrell Davis, Denver	1,750	369	15
Steve Christie, Buffalo	0	41	33	140	1998	Terrell Davis, Denver	2,008	392	21

*Includes 277 yards after being traded to NFC; 1,011 yards led AFC. (1) Based on quarterback ranking points.

National Football Conference Leaders
(National Football League, 1960-69)

Passing[1]

Player, team	Att	Com	YG	TD	Year
Milt Plum, Cleveland	250	151	2,297	21	1960
Milt Plum, Cleveland	302	177	2,416	18	1961
Bart Starr, Green Bay	285	178	2,438	12	1962
Y.A. Tittle, N.Y. Giants	367	221	3,145	36	1963
Bart Starr, Green Bay	272	163	2,144	15	1964
Rudy Bukich, Chicago	312	176	2,641	20	1965
Bart Starr, Green Bay	251	156	2,257	14	1966
Sonny Jurgensen, Washington	508	288	3,747	31	1967
Earl Morrall, Baltimore Colts	317	182	2,909	26	1968
Sonny Jurgensen, Washington	442	274	3,102	22	1969
John Brodie, San Francisco	378	223	2,941	24	1970
Roger Staubach, Dallas	211	126	1,882	15	1971
Norm Snead, N.Y. Giants	325	196	2,307	17	1972
Roger Staubach, Dallas	286	179	2,428	23	1973
Sonny Jurgensen, Washington	167	107	1,185	11	1974
Fran Tarkenton, Minnesota	425	273	2,994	25	1975
James Harris, L.A. Rams	158	91	1,460	8	1976
Roger Staubach, Dallas	361	210	2,620	18	1977
Roger Staubach, Dallas	413	231	3,190	25	1978
Roger Staubach, Dallas	461	267	3,586	27	1979
Ron Jaworski, Philadelphia	451	257	3,529	27	1980
Joe Montana, San Francisco	488	311	3,565	19	1981
Joe Thiesmann, Washington	252	161	2,033	13	1982
Steve Bartkowski, Atlanta	432	274	3,167	22	1983
Joe Montana, San Francisco	432	279	3,630	28	1984
Joe Montana, San Francisco	494	303	3,653	27	1985
Tommy Kramer, Minnesota	372	208	3,000	24	1986
Joe Montana, San Francisco	398	266	3,054	31	1987
Wade Wilson, Minnesota	332	204	2,746	15	1988
Joe Montana, San Francisco	386	271	3,521	26	1989
Phil Simms, N.Y. Giants	311	184	2,284	15	1990
Steve Young, San Francisco	279	180	2,517	17	1991
Steve Young, San Francisco	402	268	3,465	25	1992
Steve Young, San Francisco	462	314	4,023	29	1993
Steve Young, San Francisco	461	324	3,969	35	1994
Brett Favre, Green Bay	570	359	4,413	38	1995
Steve Young, San Francisco	316	214	2,410	14	1996
Steve Young, San Francisco	356	241	3,029	19	1997
Randall Cunningham, Minnesota	425	259	3,704	34	1998

Receiving

Year	Player, team	Rec.	YG	TD
1960	Raymond Berry, Baltimore Colts	74	1,298	10
1961	Jim Phillips, L.A. Rams	78	1,092	5
1962	Bobby Mitchell, Washington	72	1,384	11
1963	Bobby Joe Conrad, St. L.	73	967	10
1964	Johnny Morris, Chicago	93	1,200	10
1965	Dave Parks, San Francisco	80	1,344	12
1966	Charley Taylor, Washington	72	1,119	12
1967	Charley Taylor, Washington	70	990	9
1968	Clifton McNeil, San Francisco	71	994	7
1969	Dan Abramowicz, New Orleans	73	1,015	7
1970	Dick Gordon, Chicago	71	1,026	13
1971	Bob Tucker, N.Y. Giants	59	791	4
1972	Harold Jackson, Philadelphia	62	1,048	4
1973	Harold Carmichael, Philadelphia	67	1,116	9
1974	Charles Young, Philadelphia	63	696	3
1975	Chuck Foreman, Minnesota	73	691	9
1976	Drew Pearson, Dallas	58	806	6
1977	Ahmad Rashad, Minnesota	51	681	2
1978	Rickey Young, Minnesota	88	704	5
1979	Ahmad Rashad, Minnesota	80	1,156	9
1980	Earl Cooper, San Francisco	83	567	4
1981	Dwight Clark, San Francisco	85	1,105	4
1982	Dwight Clark, San Francisco	60	913	5
1983	Roy Green, St. Louis Cardinals	78	1,227	14
	Charlie Brown, Washington	78	1,225	8
	Earnest Gray, N.Y. Giants	78	1,139	5
1984	Art Monk, Washington	106	1,372	7
1985	Roger Craig, San Francisco	92	1,016	6
1986	Jerry Rice, San Francisco	86	1,570	15
1987	J.T. Smith, St. Louis Cardinals	91	1,117	8
1988	Henry Ellard, L.A. Rams	86	1,414	10
1989	Sterling Sharpe, Green Bay	90	1,423	12
1990	Jerry Rice, San Francisco	100	1,502	13
1991	Michael Irvin, Dallas	93	1,523	8
1992	Sterling Sharpe, Green Bay	108	1,461	13
1993	Sterling Sharpe, Green Bay	112	1,274	11
1994	Cris Carter, Minnesota	122	1,256	7
1995	Herman Moore, Detroit	123	1,686	14
1996	Jerry Rice, San Francisco	108	1,254	8
1997	Herman Moore, Detroit	104	1,293	8
1998	Frank Sanders, Arizona	89	1,145	3

Scoring

Player, team	TD	PAT	FG	Pts	Year
Paul Hornung, Green Bay	15	41	15	176	1960
Paul Hornung, Green Bay	10	41	15	146	1961
Jim Taylor, Green Bay	19	0	0	114	1962
Don Chandler, N.Y. Giants	0	52	18	106	1963
Lenny Moore, Baltimore Colts	20	0	0	120	1964
Gale Sayers, Chicago	22	0	0	132	1965
Bruce Gossett, L.A. Rams	0	29	28	113	1966
Jim Bakken, St. Louis Cardinals	0	36	27	117	1967
Leroy Kelly, Cleveland	20	0	0	120	1968
Fred Cox, Minnesota	0	43	26	121	1969
Fred Cox, Minnesota	0	35	30	125	1970
Curt Knight, Washington	0	27	29	114	1971
Chester Marcol, Green Bay	0	29	33	128	1972
David Ray, L.A. Rams	0	40	30	130	1973
Chester Marcol, Green Bay	0	19	25	94	1974
Chuck Foreman, Minnesota	22	0	0	132	1975
Mark Moseley, Washington	0	31	22	97	1976
Walter Payton, Chicago	16	0	0	96	1977
Frank Corral, L.A. Rams	0	31	29	118	1978
Mark Moseley, Washington	0	39	25	114	1979
Ed Murray, Detroit	0	35	27	116	1980
Ed Murray, Detroit	0	46	25	121	1981
Rafael Septien, Dallas	0	40	27	121	1982
Wendell Tyler, L.A. Rams	13	0	0	78	1982
Mark Moseley, Washington	0	62	33	161	1983
Ray Wersching, San Francisco	0	56	25	131	1984
Kevin Butler, Chicago	0	51	31	144	1985
Kevin Butler, Chicago	0	36	28	120	1986
Jerry Rice, San Francisco	23	0	0	138	1987
Mike Cofer, San Francisco	0	40	27	121	1988
Mike Cofer, San Francisco	0	49	29	136	1989
Chip Lohmiller, Washington	0	41	30	131	1990
Chip Lohmiller, Washington	0	56	31	149	1991
Morten Andersen, New Orleans	0	33	29	120	1992
Chip Lohmiller, Washington	0	30	30	120	
Jason Hanson, Detroit	0	28	34	130	1993
Fuad Reveiz, Minnesota	0	30	34	132	1994
Emmitt Smith, Dallas	22	0	0	132	
Emmitt Smith, Dallas	25	0	0	150	1995
John Kasay, Carolina	0	34	37	145	1996
Richie Cunningham, Dallas	0	24	34	126	1997
Gary Anderson, Minnesota	0	59	35	164	1998

Rushing

Year	Player, team	Yds	Att	TD
1960	Jim Brown, Cleveland	1,257	215	9
1961	Jim Brown, Cleveland	1,408	305	8
1962	Jim Taylor, Green Bay	1,474	272	19
1963	Jim Brown, Cleveland	1,863	291	12
1964	Jim Brown, Cleveland	1,446	280	7
1965	Jim Brown, Cleveland	1,544	289	17
1966	Gale Sayers, Chicago	1,231	229	8
1967	Leroy Kelly, Cleveland	1,205	235	11
1968	Leroy Kelly, Cleveland	1,239	248	16
1969	Gale Sayers, Chicago	1,032	236	8
1970	Larry Brown, Washington	1,125	237	5
1971	John Brockington, Green Bay	1,105	216	4
1972	Larry Brown, Washington	1,216	285	8
1973	John Brockington, Green Bay	1,144	265	3
1974	Lawrence McCutcheon, L.A. Rams	1,109	236	3
1975	Jim Otis, St. Louis Cardinals	1,076	269	5
1976	Walter Payton, Chicago	1,390	311	13
1977	Walter Payton, Chicago	1,852	339	14
1978	Walter Payton, Chicago	1,395	333	11
1979	Walter Payton, Chicago	1,610	369	14
1980	Walter Payton, Chicago	1,460	317	6
1981	George Rogers, New Orleans	1,674	378	13
1982	Tony Dorsett, Dallas	745	177	5
1983	Eric Dickerson, L.A. Rams	1,808	390	18
1984	Eric Dickerson, L.A. Rams	2,105	379	14
1985	Gerald Riggs, Atlanta	1,719	397	10
1986	Eric Dickerson, L.A. Rams	1,821	404	11
1987	Charles White, L.A. Rams	1,374	324	11
1988	Herschel Walker, Dallas	1,514	361	5
1989	Barry Sanders, Detroit	1,470	280	14
1990	Barry Sanders, Detroit	1,304	255	13
1991	Emmitt Smith, Dallas	1,563	365	12
1992	Emmitt Smith, Dallas	1,713	373	18
1993	Emmitt Smith, Dallas	1,486	283	9
1994	Barry Sanders, Detroit	1,883	331	7
1995	Emmitt Smith, Dallas	1,773	377	25
1996	Barry Sanders, Detroit	1,553	307	11
1997	Barry Sanders, Detroit	2,053	335	11
1998	Jamal Anderson, Atlanta	1,846	410	14

(1) Based on quarterback ranking points.

1998 NFL Individual Leaders
American Football Conference

Passing

	Att	Comp	Pct comp	Yds	Avg gain	Long	TD	Pct TD	Int	Rating points
Vinny Testaverde, N.Y. Jets	421	259	61.5	3,256	7.7	82td	29	6.9	7	101.6
John Elway, Denver	356	210	59.0	2,806	7.9	58	22	6.2	10	93.0
Neil O'Donnell, Cincinnati	343	212	61.8	2,216	6.5	76td	15	4.4	4	90.2
Mark Brunell, Jacksonville	354	208	58.8	2,601	7.4	78td	20	5.6	9	89.9
Doug Flutie, Buffalo	354	202	57.1	2,711	7.7	84td	20	5.6	11	87.4
Drew Bledsoe, New England	481	263	54.7	3,633	7.6	86td	20	4.2	14	80.9
Rich Gannon, Kansas City	354	206	58.2	2,305	6.5	80td	10	2.8	6	80.1
Steve McNair, Tennessee	492	289	58.7	3,228	6.6	47	15	3.0	10	80.1
Dan Marino, Miami	537	310	57.7	3,497	6.5	61td	23	4.3	15	80.0
Warren Moon, Seattle	285	145	56.2	1,632	6.3	45	11	4.3	8	76.6

Rushing

	Att	Yds	Avg	Long	TD
Terrell Davis, Denver	392	2,008	5.1	70	21
Marshall Faulk, Indianapolis	324	1,319	4.1	68td	6
Eddie George, Tennessee	348	1,294	3.7	37td	5
Curtis Martin, N.Y. Jets	369	1,287	3.5	60td	8
Ricky Waters, Seattle	319	1,239	3.9	39td	9
Fred Taylor, Jacksonville	264	1,223	4.6	77td	14
Jerome Bettis, Pittsburgh	316	1,185	3.8	42	3
Corey Dillon, Cincinnati	262	1,130	4.3	66	4
Antowain Smith, Buffalo	300	1,124	3.7	30	8
Robert Edwards, New England	291	1,115	3.8	53	9

Scoring—Non-Kickers

	TD	Rush	Pass	2 Pt	Pts
Terrell Davis, Denver	23	21	2	0	138
Fred Taylor, Jacksonville	17	14	3	0	102
Robert Edwards, New England	12	9	3	0	72
Joey Galloway, Seattle*	12	0	10	0	72
Keyshawn Johnson, N.Y. Jets	11	1	10	0	66

*Total includes 2 punt returns

Receiving

	Rec.	Yds	Avg	Long	TD
O.J. McDuffie, Miami	90	1,050	11.7	61td	7
Marshall Faulk, Indianapolis	86	908	10.6	78td	4
Rod Smith, Denver	86	1,222	14.2	58	6
Keyshawn Johnson, N.Y. Jets	83	1,131	13.6	41td	10
Carl Pickens, Cincinnati	82	1,023	12.5	67td	5
Tim Brown, Oakland	81	1,012	12.5	49	9
Jimmy Smith, Jacksonville	78	1,182	15.2	72td	8
Wayne Chrebet, N.Y. Jets	75	1,083	14.4	63td	8
Frank Wycheck, Tennessee	70	768	11.0	38	2
Ben Coates, New England	67	668	10.0	33	6
Eric Moulds, Buffalo	67	1,368	20.4	84td	9

Scoring—Kickers

	PAT	FG	Long	Pts
Steve Christie, Buffalo	41/41	33/41	52	140
Al Del Greco, Tennessee	28/28	36/39	48	136
Jason Elam, Denver	58/58	23/27	63	127
Adam Vinatieri, New England	32/32	31/39	55	125
John Hall, N.Y. Jets	45/46	25/35	54	120

Interceptions

	No.	Yds	Avg	Long	TD
Ty Law, New England	9	133	14.8	59td	1
Terrell Buckley, Miami	8	157	19.6	61	1
Sam Madison, Miami	8	114	14.3	35	0
Shawn Springs, Seattle	7	142	20.3	56td	2
Aaron Glenn, N.Y. Jets	6	23	3.8	26	0
Greg Jackson, San Diego	6	50	8.3	25	0
Lawyer Milloy, New England	6	54	9.0	32td	1
Kurt Schulz, Buffalo	6	48	8.0	24	0
Rod Woodson, Baltimore	6	108	18.0	60td	2

Kickoff Returns

	No.	Yds	Avg	Long	TD
Chris Harris, Baltimore	35	965	27.6	95td	1
Steve Broussard, Seattle	29	781	26.9	90td	1
Vaughn Hebron, Denver	46	1,216	26.4	95td	1
Tremain Mack, Cincinnati	45	1,165	25.9	97td	1
John Avery, Miami	43	1,085	25.2	55	0

Punt Returns

	No.	Yds	Avg	Long	TD
Reggie Barlow, Jacksonville	43	555	12.9	85td	1
Jermaine Lewis, Baltimore	32	405	12.7	87td	2
Terrell Buckley, Miami	29	354	12.2	35	0
Latario Rachal, San Diego	32	387	12.1	56	0
Desmond Howard, Oakland	45	541	12.0	75td	2

Punting

	No.	Yds	Long	Avg
Craig Hentrich, Tennessee	69	3,258	71	47.2
Tom Rouen, Denver	66	3,097	76	46.9
Chris Gardocki, Indianapolis	79	3,583	62	45.4
Bryan Barker, Jacksonville	85	3,824	65	45.0
Lee Johnson, Cincinnati	69	3,083	69	44.7

Sacks

	No.
Michael Sinclair, Seattle	16.5
Michael McCrary, Baltimore	14.5
Derrick Thomas, Kansas City	12.0
Jason Gildon, Pittsburgh	11.0
Lance Johnstone, Oakland	11.0

National Football Conference

Passing

	Att	Comp	Pct comp	Yds	Avg gain	Long	TD	Pct TD	Int	Rating points
Randall Cunningham, Minnesota	425	259	60.9	3,704	8.7	67td	34	8.0	10	106.0
Steve Young, San Francisco	517	322	62.3	4,170	8.1	81td	36	7.0	12	101.1
Chris Chandler, Atlanta	327	190	58.1	3,154	9.7	78td	25	7.6	12	100.9
Troy Aikman, Dallas	315	187	59.4	2,330	7.4	67td	12	3.8	5	88.5
Steve Beuerlein, Carolina	343	216	63.0	2,613	7.6	68td	17	5.0	12	88.2
Brett Favre, Green Bay	551	347	63.0	4,212	7.6	84td	31	56	23	87.8
Charlie Batch, Detroit	303	173	57.1	2,178	7.2	98td	11	3.6	6	83.5
Erik Kramer, Chicago	250	151	60.4	1,823	7.3	79td	9	3.6	7	83.1
Trent Green, Washington	509	278	54.6	3,441	6.8	75td	23	4.5	11	81.8
Jake Plummer, Arizona	547	324	59.2	3,737	6.8	57	17	3.1	20	75.0

Rushing

	Att	Yds	Avg	Long	TD
Jamal Anderson, Atlanta	410	1,846	4.5	48	14
Garrison Hearst, San Francisco	310	1,570	5.1	96td	7
Barry Sanders, Detroit	343	1,491	4.3	73td	4
Emmitt Smith, Dallas	319	1,332	4.2	32	13
Robert Smith, Minnesota	249	1,187	4.8	74td	6
Duce Staley, Philadelphia	258	1,065	4.1	64td	5
Gary Brown, N.Y. Giants	247	1,063	4.3	45	5
Adrian Murrell, Arizona	274	1,042	3.8	32	8
Warrick Dunn, Tampa Bay	245	1,026	4.2	50	2
Mike Alstott, Tampa Bay	215	846	3.9	37	8

Receiving

	Rec.	Yds	Avg	Long	TD
Frank Sanders, Arizona	89	1,145	12.9	42	3
Antonio Freeman, Green Bay	84	1,424	17.0	84td	14
Herman Moore, Detroit	82	983	12.0	36	5
Jerry Rice, San Francisco	82	1,157	14.1	75td	9
Cris Carter, Minnesota	78	1,011	13.0	54td	12
Michael Irvin, Dallas	74	1,057	14.3	51	1
Larry Centers, Arizona	69	559	8.1	54	2
Raghib Ismail, Carolina	69	1,024	14.8	62	8
Johnnie Morton, Detroit	69	1,028	14.9	98td	2
Randy Moss, Minnesota	69	1,313	19.0	61td	17

Scoring—Non-Kickers

	TD	Rush	Pass	2 Pt	Pts
Randy Moss, Minnesota	17	0	17	2	106
Jamal Anderson, Atlanta	16	14	2	1	98
Terrell Owens, San Francisco	15	1	14	1	92
Emmitt Smith, Dallas	15	13	2	0	90
Antonio Freeman, Green Bay	14	0	14	1	86

Interceptions

	No.	Yds	Avg	Long	TD
Kwamie Lassiter, Arizona	8	80	10.0	29	0
Ray Buchanan, Atlanta	7	102	14.6	34	0
Jimmy Hitchcock, Minnesota	7	242	34.6	79td	3
Sammy Knight, New Orleans	6	171	28.5	91td	2
Eric Davis, Carolina	5	81	16.2	56td	2
Percy Ellsworth, N.Y. Giants	5	92	18.4	43td	2
Robert Griffith, Minnesota	5	25	5.0	17	0
Ronald McKinnon, Arizona	5	25	5.0	17	0
Deion Sanders, Dallas	5	153	30.6	71td	1
Tyrone Williams, Green Bay	5	40	8.0	15	1

Kickoff Returns

	No.	Yds	Avg	Long	TD
Terry Fair, Detroit	51	1,428	28.0	105td	2
Tim Dwight, Atlanta	36	973	27.0	93td	1
Roell Preston, Green Bay	57	1,497	26.3	101td	2

	No.	Yds	Avg	Long	TD
Michael Bates, Carolina	59	1,480	25.1	99td	1
Glyn Milburn, Chicago	62	1,550	25.0	94td	2

Punt Returns

	No.	Yds	Avg	Long	TD
Deion Sanders, Dallas	24	.375	15.6	69td	2
Jacquez Green, Tampa Bay	30	453	15.1	95td	1
Andre Hastings, New Orleans	22	307	14.0	76	0
Glyn Milburn, Chicago	25	291	11.6	93td	1
Brian Mitchell, Washington	44	506	11.5	47	0

Punting

	No.	Yds	Long	Avg
Mark Royals, New Orleans	88	4,017	64	45.6
Brad Maynard, N.Y. Giants	101	4,566	63	45.2
Mitch Berger, Minnesota	55	2,458	67	44.7
Rick Tuten, St. Louis	95	4,202	64	44.2
Matt Turk, Washington	93	4,103	69	44.1

Sacks

	No.
Reggie White, Green Bay	16.0
Chris Doleman, San Francisco	15.0
Kevin Greene, Carolina	15.0
Michael Strahan, N.Y. Giants	15.0
Hugh Douglas, Philadelphia	12.5

First-Round Selections in the 1999 NFL Draft

Team	Player	Pos	College
1. Cleveland	Tim Couch	QB	Kentucky
2. Philadelphia	Donovan McNabb	QB	Syracuse
3. Cincinnati	Akili Smith	QB	Oregon
4. Indianapolis	Edgerrin James	RB	Miami (FL)
5. New Orleans[1]	Ricky Williams	RB	Texas
6. St. Louis	Torry Holt	WR	N. Carolina St.
7. Washington[2]	Roland Bailey	DB	Georgia
8. Arizona[3]	David Boston	WR	Ohio St.
9. Detroit	Chris Claiborne	LB	U.S.C.
10. Baltimore	Chris McAlister	DB	Arizona
11. Minnesota[4]	Daunte Culpepper	QB	Central Florida
12. Chicago[5]	Cade McNown	QB	U.C.L.A.
13. Pittsburgh	Troy Edwards	WR	Louisiana Tech
14. Kansas City	John Tait	OL	Brigham Young
15. Tampa Bay	Anthony McFarland	DL	Louisiana St.
16. Tennessee	Jevon Kearse	LB	Florida
17. New England[6]	Damien Woody	OL	Boston College
18. Oakland	Matt Stinchcomb	OL	Georgia
19. N.Y. Giants	Luke Petitgout	OL	Notre Dame
20. Dallas[7]	Ebenezer Ekuban	DL	North Carolina
21. Arizona	Lonnie Jewel "L.J." Shelton	OL	Eastern Michigan
22. Seattle[8]	Lamar King	DL	Saginaw Valley St. (MI)
23. Buffalo	Antoine Winfield	DB	Ohio St.
24. San Francisco[9]	Reggie McGrew	DL	Florida
25. Green Bay	Antuan Edwards	DB	Clemson
26. Jacksonville	Fernando Bryant	DB	Alabama
27. Detroit[10]	Aaron Gibson	OL	Wisconsin
28. New England[11]	Andy Katzenmoyer	LB	Ohio St.
29. Minnesota	Dimitrius Underwood	DL	Michigan St.
30. Atlanta	Patrick Kerney	DL	Virginia
31. Denver	Al Wilson	LB	Tennessee

(1) From Carolina through Washington. (2) From Chicago. (3) From San Diego. (4) From Washington. (5) From New Orleans through Washington. (6) From Seattle. (7) From New England through Seattle. (8) From Dallas. (9) From Miami. (10) From San Francisco through Washington. (11) From N.Y. Jets.

Number One NFL Draft Choices, 1936-99

Year	Team	Player, Pos., College
1936	Philadelphia	Jay Berwanger, HB, Chicago
1937	Philadelphia	Sam Francis, FB, Nebraska
1938	Cleveland Rams	Corbett Davis, FB, Indiana
1939	Chicago Cards	Ki Aldrich, C, TCU
1940	Chicago Cards	George Cafego, HB, Tennessee
1941	Chicago Bears	Tom Harmon, HB, Michigan
1942	Pittsburgh	Bill Dudley, HB, Virginia
1943	Detroit	Frank Sinkwich, HB, Georgia
1944	Boston Yanks	Angelo Bertelli, QB, Notre Dame
1945	Chicago Cards	Charley Trippi, HB, Georgia
1946	Boston Yanks	Frank Dancewicz, QB, Notre Dame
1947	Chicago Bears	Bob Fenimore, HB, Okla. A&M
1948	Washington	Harry Gilmer, QB, Alabama
1949	Philadelphia	Chuck Bednarik, C, Penn
1950	Detroit	Leon Hart, E, Notre Dame
1951	N.Y. Giants	Kyle Rote, HB, SMU
1952	L.A. Rams	Bill Wade, QB, Vanderbilt
1953	San Francisco	Harry Babcock, E, Georgia
1954	Cleveland	Bobby Garrett, QB, Stanford
1955	Baltimore Colts	George Shaw, QB, Oregon
1956	Pittsburgh	Gary Glick, DB, Col. A&M
1957	Green Bay	Paul Hornung, QB, Notre Dame
1958	Chicago Cards	King Hill, QB, Rice
1959	Green Bay	Randy Duncan, QB, Iowa
1960	L.A. Rams	Billy Cannon, HB, LSU
1961	Minnesota	Tommy Mason, HB, Tulane
1962	Washington	Ernie Davis, HB, Syracuse
1963	L.A. Rams	Terry Baker, QB, Oregon St.
1964	San Francisco	Dave Parks, E, Texas Tech
1965	N.Y. Giants	Tucker Frederickson, HB, Auburn
1966	Atlanta	Tommy Nobis, LB, Texas
1967	Baltimore Colts	Bubba Smith, DT, Michigan St.
1968	Minnesota	Ron Yary, T, USC
1969	Buffalo	O.J. Simpson, RB, USC
1970	Pittsburgh	Terry Bradshaw, QB, La.Tech
1971	New England	Jim Plunkett, QB, Stanford
1972	Buffalo	Walt Patulski, DE, Notre Dame
1973	Houston	John Matuszak, DE, Tampa
1974	Dallas	Ed "Too Tall" Jones, DE, Tenn. St.
1975	Atlanta	Steve Bartkowski, QB, Cal.
1976	Tampa Bay	Lee Roy Selmon, DE, Oklahoma
1977	Tampa Bay	Ricky Bell, RB, USC
1978	Houston	Earl Campbell, RB, Texas
1979	Buffalo	Tom Cousineau, LB, Ohio St.
1980	Detroit	Billy Sims, RB, Oklahoma
1981	New Orleans	George Rogers, RB, S.Carolina
1982	New England	Kenneth Sims, DT, Texas
1983	Baltimore Colts	John Elway, QB, Stanford
1984	New England	Irving Fryar, WR, Nebraska
1985	Buffalo	Bruce Smith, DE, Va.Tech
1986	Tampa Bay	Bo Jackson, RB, Auburn
1987	Tampa Bay	Vinny Testaverde, QB, Miami (FL)
1988	Atlanta	Aundray Bruce, LB, Auburn
1989	Dallas	Troy Aikman, QB, UCLA
1990	Indianapolis	Jeff George, QB, Illinois
1991	Dallas	Russell Maryland, DL, Miami (FL)
1992	Indianapolis	Steve Emtman, DL, Washington
1993	New England	Drew Bledsoe, QB, Washington St.
1994	Cincinnati	Dan Wilkinson, DT, Ohio St.
1995	Cincinnati	Ki-Jana Carter, RB, Penn State
1996	N.Y. Jets	Keyshawn Johnson, WR, USC
1997	St. Louis	Orlando Pace, T, Ohio St.
1998	Indianapolis	Peyton Manning, QB, Tennessee
1999	Cleveland	Tim Couch, QB, Kentucky

NFL MVP, Defensive Player of the Year, and Rookie of the Year

The Most Valuable Player is one of many awards given out annually by the Associated Press. The George Halas Trophy is awarded to the outstanding defensive player of the year, as chosen by a panel of sports experts. Rookie of the Year is one of many awards given out annually by *The Sporting News*. Many other organizations give out annual awards honoring the NFL's best players.

Most Valuable Player

1957 Jim Brown, Cleveland	1971 Alan Page, Minnesota	1986 Lawrence Taylor, N.Y. Giants
1958 Gino Marchetti, Baltimore Colts	1972 Larry Brown, Washington	1987 John Elway, Denver
1959 Charley Conerly, N.Y. Giants	1973 O.J. Simpson, Buffalo	1988 Boomer Esiason, Cincinnati
1960 Norm Van Brocklin, Philadelphia;	1974 Ken Stabler, Oakland	1989 Joe Montana, San Francisco
Joe Schmidt, Detroit	1975 Fran Tarkenton, Minnesota	1990 Joe Montana, San Francisco
1961 Paul Hornung, Green Bay	1976 Bert Jones, Baltimore	1991 Thurman Thomas, Buffalo
1962 Jim Taylor, Green Bay	1977 Walter Payton, Chicago	1992 Steve Young, San Francisco
1963 Y.A. Tittle, N.Y. Giants	1978 Terry Bradshaw, Pittsburgh	1993 Emmitt Smith, Dallas
1964 John Unitas, Baltimore Colts	1979 Earl Campbell, Houston	1994 Steve Young, San Francisco
1965 Jim Brown, Cleveland	1980 Brian Sipe, Cleveland	1995 Brett Favre, Green Bay
1966 Bart Starr, Green Bay	1981 Ken Anderson, Cincinnati	1996 Brett Favre, Green Bay
1967 John Unitas, Baltimore Colts	1982 Mark Moseley, Washington	1997 (tie) Brett Favre, Green Bay
1968 Earl Morrall, Baltimore Colts	1983 Joe Theismann, Washington	Barry Sanders, Detroit
1969 Roman Gabriel, L.A. Rams	1984 Dan Marino, Miami	1998 Terrell Davis, Denver
1970 John Brodie, San Francisco	1985 Marcus Allen, L.A. Raiders	

Defensive Player of the Year

1966 Larry Wilson, St. Louis	1978 Randy Gradishar, Denver	1988 Mike Singletary, Chicago
1967 Deacon Jones, Los Angeles	1979 Lee Roy Selmon, Tampa Bay	1989 Tim Harris, Green Bay
1968 Deacon Jones, Los Angeles	1980 Lester Hayes, Oakland	1990 Bruce Smith, Buffalo
1969 Dick Butkus, Chicago	1981 Joe Klecko, N.Y. Jets	1991 Pat Swilling, New Orleans
1970 Dick Butkus, Chicago	1982 Mark Gastineau, N.Y. Jets	1992 Junior Seau, San Diego
1971 Carl Eller, Minnesota	1983 Jack Lambert, Pittsburgh	1993 Bruce Smith, Buffalo
1972 Joe Greene, Pittsburgh	1984 Mike Haynes, L.A. Raiders	1994 Deion Sanders, San Francisco
1973 Alan Page, Minnesota	1985 Howie Long, L.A. Raiders	1995 Bryce Paup, Buffalo
1974 Joe Greene, Pittsburgh	Andre Tippett, New England	1996 Bruce Smith, Buffalo
1975 Curley Culp, Houston	1986 Lawrence Taylor, N.Y. Giants	1997 Dana Stubblefield, San Francisco
1976 Jerry Sherk, Cleveland	1987 Reggie White, Philadelphia	1998 Reggie White, Green Bay
1977 Harvey Martin, Dallas		

Rookie of the Year

1964 Charley Taylor, Washington	1975 NFC: Steve Bartkowski, Atlanta	1985 Eddie Brown, Cincinnati
1965 Gale Sayers, Chicago	AFC: Robert Brazile, Houston	1986 Rueben Mayes, New Orleans
1966 Tommy Nobis, Atlanta	1976 NFC: Sammy White, Minnesota	1987 Robert Awalt, St. Louis
1967 Mel Farr, Detroit	AFC: Mike Haynes, New England	1988 Keith Jackson, Philadelphia
1968 Earl McCullouch, Detroit	1977 NFC: Tony Dorsett, Dallas	1989 Barry Sanders, Detroit
1969 Calvin Hill, Dallas	AFC: A. J. Duhe, Miami	1990 Richmond Webb, Miami
1970 NFC: Bruce Taylor, San Francisco	1978 NFC: Al Baker, Detroit	1991 Mike Croel, Denver
AFC: Dennis Shaw, Buffalo	AFC: Earl Campbell, Houston	1992 Santana Dotson, Tampa Bay
1971 NFC: John Brockington, Green Bay	1979 NFC: Ottis Anderson, St. Louis	1993 Jerome Bettis, L.A. Rams
AFC: Jim Plunkett, New England	AFC: Jerry Butler, Buffalo	1994 Marshall Faulk, Indianapolis
1972 NFC: Chester Marcol, Green Bay	1980 Billy Sims, Detroit	1995 Curtis Martin, New England
AFC: Franco Harris, Pittsburgh	1981 George Rogers, New Orleans	1996 Eddie George, Houston
1973 NFC: Chuck Foreman, Minnesota	1982 Marcus Allen, L.A. Raiders	1997 Warrick Dunn, Tampa Bay
AFC: Boobie Clark, Cincinnati	1983 Dan Marino, Miami	1998 Randy Moss, Minnesota
1974 NFC: Wilbur Jackson, San Francisco	1984 Louis Lipps, Pittsburgh	
AFC: Don Woods, San Diego		

The Sporting News 1998 NFL All-Pro Team

Offense—Quarterback: Steve Young, San Francisco. Running backs: Jamal Anderson, Atlanta; Terrell Davis, Denver. Wide receivers: Antonio Freeman, Green Bay; Randy Moss, Minnesota. Tight end: Shannon Sharpe, Denver. Tackles: Larry Allen, Dallas; Tony Boselli, Jacksonville. Guards: Bruce Matthews, Tennessee; Randall McDaniel, Minnesota. Center: Dermontti Dawson, Pittsburgh.

Defense—Linebackers: Chad Brown, Seattle; Ray Lewis, Baltimore; Junior Seau, San Diego. Defensive ends: Reggie White, Green Bay; Michael McCrary, Baltimore. Defensive tackles: John Randle, Minnesota; Bryant Young, San Francisco. Cornerbacks: Ty Law, New England; Deion Sanders, Dallas. Safeties: LeRoy Butler, Green Bay; Robert Griffith, Minnesota (tie), Rodney Harrison, San Diego (tie), Darren Woodson, Dallas (tie).

Special Teams—Kicker: Gary Anderson, Minnesota. Punter: Craig Hentrich, Tennessee. Punt returner: Jermaine Lewis, Baltimore. Kick returner: Terry Fair, Detroit.

NFL Head Coaches at Start of 1999 Season

AFC		NFC	
Baltimore—Brian Billick	Miami—Jimmy Johnson	Arizona—Vince Tobin	Minnesota—Dennis Green
Buffalo—Wade Phillips	New England—Pete Carroll	Atlanta—Dan Reeves	New Orleans—Mike Ditka
Cincinnati—Bruce Coslet	N.Y. Jets—Bill Parcells	Carolina—George Seifert	N.Y. Giants—Jim Fassel
Cincinnati—Chris Palmer	Oakland—Jon Gruden	Chicago—Dick Jauron	Philadelphia—Andy Reid
Denver—Mike Shanahan	Pittsburgh—Bill Cowher	Dallas—Chan Gailey	St. Louis—Dick Vermeil
Indianapolis—Jim Mora	San Diego—Mike Riley	Detroit—Bobby Ross	San Francisco—Steve Mariucci
Jacksonville—Tom Coughlin	Seattle—Mike Holmgren	Green Bay—Ray Rhodes	Tampa Bay—Tony Dungy
Kansas City—Gunther Cunningham	Tennessee—Jeff Fisher		Washington—Norv Turner

All-Time NFL Coaching Victories
(at end of 1998 season; *active through 1998)

Coach	Years	Teams	Regular Season W	L	T	Pct	Career W	L	T	Pct
Don Shula	33	Colts, Dolphins	328	156	6	.676	347	173	6	.665
George Halas.	40	Bears	318	148	31	.671	324	151	31	.671
Tom Landry	29	Cowboys	250	162	6	.605	270	178	6	.601
Curly Lambeau.	33	Packers, Cardinals, Redskins	226	132	22	.624	229	134	22	.623
Chuck Noll	23	Steelers	193	148	1	.566	209	156	1	.572
Chuck Knox	22	Rams, Bills, Seahawks	186	147	1	.558	193	158	1	.550
Dan Reeves*	18	Broncos, Giants, Falcons	162	117	1	.580	172	125	1	.579
Paul Brown.	21	Browns, Bengals	166	100	6	.621	170	109	6	.607
Bud Grant.	18	Vikings	158	96	5	.620	168	108	5	.607
Steve Owen	23	Giants	153	100	17	.598	155	108	17	.584
Marv Levy.	17	Chiefs, Bills	143	112	0	.561	154	120	0	.562
M. Schottenheimer*	15	Browns, Chiefs	145	85	1	.630	150	96	1	.609
Bill Parcells*	14	Giants, Patriots, Jets	130	92	1	.585	141	98	1	.590
Joe Gibbs.	12	Redskins	124	60	0	.674	140	65	0	.683
Hank Stram	17	Chiefs, Saints	131	97	10	.571	136	100	10	.573
Weeb Ewbank	20	Colts, Jets	130	129	7	.502	134	130	7	.507
Mike Ditka*	13	Bears, Saints	118	82	0	.590	124	88	0	.585
Sid Gillman.	18	Rams, Chargers, Oilers	122	99	7	.550	123	104	7	.541
George Allen	12	Rams, Redskins	116	47	5	.705	120	54	5	.684
Don Coryell	14	Cardinals, Chargers	111	83	1	.572	114	89	1	.561

All-Time Professional (NFL and AFL) Football Records
(at end of 1998 season; *active through 1998)
Leading Lifetime Scorers

Player	League	Yrs	TD	PAT	FG	Total	Player	League	Yrs	TD	PAT	FG	Total
George Blanda	AFL-NFL	26	9	943	335	2,002	Mark Moseley	NFL	16	0	482	300	1,382
Gary Anderson*	NFL	17	0	585	420	1,845	Jim Bakken	NFL	17	0	534	282	1,380
Morten Andersen*	NFL	17	0	558	401	1,761	Fred Cox	NFL	15	0	519	282	1,365
Nick Lowery	NFL	18	0	562	383	1,711	Lou Groza	NFL	17	1	641	234	1,365
Jan Stenerud	AFL-NFL	19	0	580	373	1,699	Al Del Greco*	NFL	15	0	463	299	1,360
Norm Johnson*	NFL	17	0	613	348	1,657	Jim Breech	NFL	14	0	517	243	1,246
Eddie Murray	NFL	19	0	521	337	1,532	Chris Bahr	NFL	14	0	490	241	1,213
Pat Leahy	NFL	18	0	558	304	1,470	Kevin Butler	NFL	13	0	413	265	1,208
Jim Turner	AFL-NFL	16	1	521	304	1,439	Gino Cappelletti	AFL-NFL	11	42	342	176	1,130
Matt Bahr	NFL	17	0	522	300	1,422	Ray Wersching	NFL	15	0	456	222	1,122

Note: Cappelletti's total includes 4 two-point conversions.

Leading Lifetime Touchdown Scorers

Player	League	Yrs	Rush	Rec	Tot. Ret	Tot. TDs	Player	League	Yrs	Rush	Rec	Tot. Ret	Tot. TDs
Jerry Rice*	NFL	14	10	164	1	175	Steve Largent	NFL	14	1	100	0	101
Marcus Allen	NFL	16	123	21	1	145	Franco Harris	NFL	13	91	9	0	100
Emmit Smith*	NFL	9	125	9	0	134	Eric Dickerson	NFL	11	90	6	0	96
Jim Brown	NFL	9	106	20	0	126	Jim Taylor	NFL	10	83	10	0	93
Walter Payton	NFL	13	110	15	0	125	Tony Dorsett	NFL	12	77	13	1	91
John Riggins	NFL	14	104	12	0	116	Bobby Mitchell	NFL	11	18	65	8	91
Lenny Moore	NFL	12	63	48	2	113	Leroy Kelly	NFL	10	74	13	3	90
Barry Sanders	NFL	10	99	10	0	109	Charley Taylor	NFL	13	11	79	0	90
Don Hutson	NFL	11	3	99	3	105	Don Maynard	AFL-NFL	15	0	88	0	88
Cris Carter*	NFL	12	0	101	1	102	Lance Alworth	AFL-NFL	11	2	85	0	87

Most Points, Season — 176, Paul Hornung, Green Bay Packers, 1960 (15 TDs, 41 PATs, 15 FGs).
Most Points, Game — 40, Ernie Nevers, Chicago Cardinals vs. Chicago Bears, Nov. 28, 1929 (6 TDs, 4 PATs).
Most Touchdowns, Season — 25, Emmitt Smith, Dallas Cowboys, 1995 (25 rushing).
Most Touchdowns, Game — 6, Ernie Nevers, Chicago Cardinals vs. Chicago Bears, Nov. 28, 1929 (6 rushing); Dub Jones, Cleveland Browns vs. Chicago Bears, Nov. 25, 1951 (4 rushing, 2 pass receptions); Gale Sayers, Chicago Bears vs. San Francisco 49ers, Dec. 12, 1965 (4 rushing, 1 pass reception, 1 punt return).
Most Points After Touchdown, Season — 66, Uwe von Schamann, Miami Dolphins, 1984.
Most Consecutive Points After Touchdown — 234, Tommy Davis, San Francisco 49ers, 1959-69.
Most Field Goals, Career — 420, Gary Anderson, Pittsburgh Steelers-Philadelphia Eagles-San Francisco 49ers, 1982-98.
Most Field Goals, Season — 37, John Kasay, Carolina Panthers, 1996.
Most Field Goals, Game — 7, Jim Bakken, St. Louis Cardinals vs. Pittsburgh Steelers, Sept. 24, 1967; Rich Karlis, Minnesota vs. L.A. Rams, Nov. 5, 1989 (OT); Chris Boniol, Dallas vs. Green Bay, Nov. 18, 1996.
Longest Field Goal — 63 yds., Tom Dempsey, New Orleans Saints vs. Detroit Lions, Nov. 8, 1970; Jason Elam, Denver Broncos vs. Jacksonville Jaguars, Oct. 25, 1998.

Defensive Records

Most Interceptions, Career — 81, Paul Krause, Washington Redskins-Minnesota Vikings, 1964-79.
Most Interceptions, Season — 14, Dick "Night Train" Lane, Los Angeles Rams, 1952.
Most Touchdowns, Career — 9, Ken Houston, Houston Oilers-Washington Redskins, 1967-80.
Most Touchdowns, Season — 4, Ken Houston, Houston Oilers, 1971; Jim Kearney, Kansas City Chiefs, 1972; Eric Allen, Philadelphia Eagles, 1993.
Most Sacks, Career (Since 1982) — 192.5, Reggie White, Philadelphia Eagles-Green Bay Packers, 1985-98.
Most Sacks, Season (Since 1982) — 22, Mark Gastineau, New York Jets, 1984.
Most Sacks, Game (Since 1982) — 7, Derrick Thomas, Kansas City Chiefs vs. Seattle Seahawks, Nov. 11, 1990.

Leading Lifetime Rushers
(ranked by rushing yards)

Player	League	Yrs	Att	Yards	Avg	Player	League	Yrs	Att	Yards	Avg
Walter Payton	NFL	13	3,838	16,726	4.4	O.J. Simpson	AFL-NFL	11	2,404	11,236	4.7
Barry Sanders	NFL	10	3,062	15,269	5.0	Ottis Anderson	NFL	14	2,562	10,273	4.0
Eric Dickerson	NFL	11	2,996	13,259	4.4	Earl Campbell	NFL	8	2,187	9,407	4.3
Tony Dorsett	NFL	12	2,936	12,739	4.3	Jim Taylor	NFL	10	1,941	8,597	4.4
Emmitt Smith*	NFL	9	2,914	12,566	4.3	Joe Perry	NFL	14	1,737	8,378	4.8
Jim Brown	NFL	9	2,359	12,312	5.2	Earnest Byner	NFL	14	2,095	8,261	3.9
Marcus Allen	NFL	16	3,022	12,243	4.1	Herschel Walker	NFL	12	1,954	8,225	4.2
Franco Harris	NFL	13	2,949	12,120	4.1	Roger Craig	NFL	11	1,991	8,189	4.1
Thurman Thomas*	NFL	11	2,813	11,786	4.2	Gerald Riggs	NFL	10	1,989	8,188	4.1
John Riggins	NFL	14	2,916	11,352	3.9	Larry Csonka	AFL-NFL	11	1,891	8,081	4.3

Most Yards Gained, Season — 2,105, Eric Dickerson, Los Angeles Rams, 1984.
Most Yards Gained, Game — 275, Walter Payton, Chicago Bears vs. Minnesota Vikings, Nov. 20, 1977.
Most Touchdowns Rushing, Career — 125, Emmitt Smith, Dallas Cowboys, 1990-98.
Most Touchdowns Rushing, Season — 25, Emmitt Smith, Dallas Cowboys, 1995.
Most Touchdowns Rushing, Game — 6, Ernie Nevers, Chicago Cardinals vs. Chicago Bears, Nov. 28, 1929.
Most Rushing Attempts, Game — 45, Jamie Morris, Washington Redskins vs. Cincinnati Bengals, Dec. 17, 1988 (overtime).
Longest Run From Scrimmage — 99 yds., Tony Dorsett, Dallas Cowboys vs. Minnesota Vikings, Jan. 3, 1983 (touchdown).

Leading Lifetime Receivers
(ranked by number of completions)

Player	League	Yrs	No.	Yds	Avg	Player	League	Yrs	No.	Yds	Avg
Jerry Rice*	NFL	14	1,139	17,612	15.5	Gary Clark	NFL	11	699	10,856	15.5
Art Monk	NFL	16	940	12,721	13.5	Andre Rison*	NFL	10	681	9,381	13.8
Andre Reed*	NFL	14	889	12,559	14.1	Tim Brown*	NFL	11	680	9,600	14.1
Cris Carter*	NFL	12	834	10,447	12.5	Ozzie Newsome	NFL	13	662	7,980	12.1
Steve Largent	NFL	14	819	13,089	16.0	Charley Taylor	NFL	13	649	9,110	14.0
Henry Ellard*	NFL	16	814	13,777	16.9	Drew Hill	NFL	14	634	9,831	15.5
Irving Fryar*	NFL	15	784	11,427	15.5	Don Maynard	AFL-NFL	15	633	11,834	18.7
James Lofton	NFL	16	764	14,004	18.3	Raymond Berry	NFL	13	631	9,275	14.7
Charlie Joiner	AFL-NFL	18	750	12,146	16.2	Keith Byars*	NFL	13	610	5,661	9.3
Michael Irvin*	NFL	11	740	11,737	15.9	Herman Moore*	NFL	8	610	8,467	13.9

Most Yards Gained, Career — 17,612, Jerry Rice, San Francisco 49ers, 1985-97.
Most Yards Gained, Season — 1,848, Jerry Rice, San Francisco 49ers, 1995.
Most Yards Gained, Game — 336, Willie "Flipper" Anderson, Los Angeles Rams vs. New Orleans, Nov. 26, 1989 (overtime).
Most Pass Receptions, Season — 123, Herman Moore, Detroit Lions, 1995.
Most Pass Receptions, Game — 18, Tom Fears, Los Angeles Rams vs. Green Bay Packers, Dec. 3, 1950 (189 yards).
Most Touchdown Passes, Career — 164, Jerry Rice, San Francisco 49ers, 1985-98.
Most Touchdown Passes, Season — 22, Jerry Rice, San Francisco 49ers, 1987.
Most Touchdown Passes, Game — 5, Bob Shaw, Chicago Cardinals vs. Baltimore Colts, Oct. 2, 1950; Kellen Winslow, San Diego Chargers vs. Oakland Raiders, Nov. 22, 1981; Jerry Rice, San Francisco 49ers vs. Atlanta Falcons, Oct. 14, 1990.

Leading Lifetime Passers
(minimum 1,500 attempts; ranked by quarterback rating points)

Player	League	Yrs	Att	Comp	Yds	Pts[1]	Player	League	Yrs	Att	Comp	Yds	Pts[1]
Steve Young*	NFL	14	4,065	2,622	32,678	97.6	Len Dawson	NFL-AFL	19	3,741	2,136	28,711	82.6
Joe Montana	NFL	15	5,391	3,409	40,551	92.3	Ken Anderson	NFL	16	4,475	2,654	32,838	81.9
Brett Favre*	NFL	8	3,757	2,318	26,083	89.0	Bernie Kosar	NFL	12	3,365	1,994	23,301	81.8
Dan Marino*	NFL	16	7,989	4,763	58,913	87.3	Danny White	NFL	13	2,950	1,761	21,959	81.7
Mark Brunell*	NFL	6	1,719	1,038	12,512	86.3	Neil O'Donnell*	NFL	9	2,862	1,650	19,026	81.6
Jim Kelly	NFL	11	4,779	2,874	35,467	84.4	R. Cunningham*	NFL	13	3,875	2,177	27,082	81.5
Roger Staubach	NFL	11	2,958	1,685	22,700	83.4	Dave Krieg*	NFL	19	5,311	3,105	38,147	81.5
Troy Aikman*	NFL	10	4,011	2,479	28,346	82.8	Boomer Esiason	NFL	14	5,205	2,969	37,920	81.1
Neil Lomax	NFL	8	3,153	1,817	22,771	82.7	Warren Moon*	NFL	15	6,786	3,972	49,097	81.0
Sonny Jurgensen	NFL	18	4,262	2,433	32,224	82.6	Jeff Hostetler*	NFL	15	2,338	1,357	16,430	80.5

(1) Rating points based on performances in the following categories: Percentage of completions, percentage of touchdown passes, percentage of interceptions, and average gain per pass attempt.

Most Yards Gained, Career — 58,913, Dan Marino, Miami Dolphins, 1983-98.
Most Yards Gained, Season — 5,084, Dan Marino, Miami Dolphins, 1984.
Most Yards Gained, Game — 554, Norm Van Brocklin, Los Angeles Rams vs. New York Yanks, Sept. 18, 1951 (27 completions in 41 attempts).
Most Touchdowns Passing, Career — 408, Dan Marino, Miami Dolphins, 1983-98.
Most Touchdowns Passing, Season — 48, Dan Marino, Miami Dolphins, 1984.
Most Touchdowns Passing, Game — 7, Sid Luckman, Chicago Bears vs. New York Giants, Nov. 14, 1943; Adrian Burk, Philadelphia Eagles vs. Washington Redskins, Oct. 17, 1954; George Blanda, Houston Oilers vs. New York Titans, Nov. 19, 1961; Y.A. Tittle, New York Giants vs. Washington Redskins, Oct. 28, 1962; Joe Kapp, Minnesota Vikings vs. Baltimore Colts, Sept. 28, 1969.
Most Passes Completed, Career — 4,763, Dan Marino, Miami Dolphins, 1983-98.
Most Passes Completed, Season — 404, Warren Moon, Houston Oilers, 1991.
Most Passes Completed, Game — 45, Drew Bledsoe, New England Patriots vs. Minnesota Vikings, Nov. 13, 1994 (overtime).

American Football League Champions

Year	Eastern Division	Western Division	Playoff
1960	Houston Oilers (10-4-0)	Los Angeles Chargers (10-4-0)	Houston 24, Los Angeles 16
1961	Houston Oilers (10-3-1)	San Diego Chargers (12-2-0)	Houston 10, San Diego 3
1962	Houston Oilers (11-3-0)	Dallas Texans (11-3-0)	Dallas 20, Houston 17 (2 overtimes)
1963	Boston Patriots (7-6-1)(a)	San Diego Chargers (11-3-0)	San Diego 51, Boston 10
1964	Buffalo Bills (12-2-0)	San Diego Chargers (8-5-1)	Buffalo 20, San Diego 7
1965	Buffalo Bills (10-3-1)	San Diego Chargers (9-2-3)	Buffalo 23, San Diego 0

Year	Eastern Division	Western Division	Playoff
1966	Buffalo Bills (9-4-1)	Kansas City Chiefs (11-2-1)	Kansas City 31, Buffalo 7
1967	Houston Oilers (9-4-1)	Oakland Raiders (13-1-0)	Oakland 40, Houston 7
1968	New York Jets (11-3-0)	Oakland Raiders (12-2-0)(b)	New York 27, Oakland 23
1969	New York Jets (10-4-0)	Oakland Raiders (12-1-1)	Kansas City 17, Oakland 7(c)

(a) Defeated Buffalo Bills in divisional playoff. (b) Defeated Kansas City Chiefs in divisional playoff. (c) Kansas City Chiefs defeated New York Jets and Oakland Raiders defeated Houston Oilers in divisional playoffs.

Pro Football Hall of Fame, Canton, Ohio

(Asterisks indicate 1999 inductees.)

Herb Adderley	John "Paddy" Driscoll	Sam Huff	Tommy McDonald	Lee Roy Selmon
Lance Alworth	Bill Dudley	Lamar Hunt	Hugh McElhenny	*Billy Shaw
Doug Atkins	Glen "Turk" Edwards	Don Hutson	Johnny "Blood" McNally	Art Shell
Morris "Red" Badgro	Weeb Ewbank	Jimmy Johnson	Mike Michalske	Don Shula
Lem Barney	Tom Fears	John Henry Johnson	Wayne Millner	O.J. Simpson
Cliff Battles	Jim Finks	Charlie Joiner	Bobby Mitchell	Mike Singletary
Sammy Baugh	Ray Flaherty	David "Deacon" Jones	Ron Mix	Jackie Smith
Chuck Bednarik	Len Ford	Stan Jones	Lenny Moore	Bart Starr
Bert Bell	Dr. Daniel Fortmann	Henry Jordan	Marion Motley	Roger Staubach
Bobby Bell	Dan Fouts	Sonny Jurgensen	Anthony Munoz	Ernie Stautner
Raymond Berry	Frank Gatski	Leroy Kelly	George Musso	Jan Stenerud
Charles Bidwill	Bill George	Walt Kiesling	Bronko Nagurski	Dwight Stephenson
Fred Biletnikoff	Joe Gibbs	Frank "Bruiser" Kinard	Joe Namath	Ken Strong
George Blanda	Frank Gifford	Paul Krause	Earle "Greasy" Neale	Joe Stydahar
Mel Blount	Sid Gillman	Earl "Curly" Lambeau	Ernie Nevers	Fran Tarkenton
Terry Bradshaw	Otto Graham	Jack Lambert	*Ozzie Newsome	Charley Taylor
Jim Brown	Red Grange	Tom Landry	Ray Nitschke	Jim Taylor
Paul Brown	Bud Grant	Dick "Night Train" Lane	Chuck Noll	*Lawrence "LT" Taylor
Roosevelt Brown	Joe Greene	Jim Langer	Leo Nomellini	Jim Thorpe
Willie Brown	Forrest Gregg	Willie Lanier	Merlin Olsen	Y.A. Tittle
Buck Buchanan	Bob Griese	Steve Largent	Jim Otto	George Trafton
Dick Butkus	Lou Groza	Yale Lary	Steve Owen	Charley Trippi
Earl Campbell	Joe Guyon	Dante Lavelli	Alan Page	Emlen Tunnell
Tony Canadeo	George Halas	Bobby Layne	Clarence "Ace" Parker	Clyde "Bulldog" Turner
Joe Carr	Jack Ham	Alphonse "Tuffy" Leemans	Jim Parker	Johnny Unitas
Guy Chamberlin	John Hannah	Bob Lilly	Walter Payton	Gene Upshaw
Jack Christiansen	Franco Harris	Larry Little	Joe Perry	Norm Van Brocklin
Earl "Dutch" Clark	Mike Haynes	Vince Lombardi	Pete Pihos	Steve Van Buren
George Connor	Ed Healey	Sid Luckman	Hugh "Shorty" Ray	Doak Walker
Jim Conzelman	Mel Hein	Roy "Link" Lyman	Dan Reeves	Bill Walsh
Lou Creekmur	Ted Hendricks	*Tom Mack	Mel Renfro	Paul Warfield
Larry Csonka	Wilbur "Pete" Henry	John Mackey	John Riggins	Bob Waterfield
Al Davis	Arnold Herber	Tim Mara	Jim Ringo	Mike Webster
Willie Davis	Bill Hewitt	Wellington Mara	Andy Robustelli	Arnie Weinmeister
Len Dawson	Clarke Hinkle	Gino Marchetti	Art Rooney	Randy White
*Eric Dickerson	Elroy "Crazylegs" Hirsch	George Preston Marshall	Pete Rozelle	Bill Willis
Dan Dierdorf	Paul Hornung	Ollie Matson	Bob St. Clair	Larry Wilson
Mike Ditka	Ken Houston	Don Maynard	Gale Sayers	Kellen Winslow
Art Donovan	Cal Hubbard	George McAfee	Joe Schmidt	Alex Wojciechowicz
Tony Dorsett	John "Paddy" Driscoll	Mike McCormack	Tex Schramm	Willie Wood

NFL Stadiums

Team—Stadium, Location, Turf (Year Built)	Capacity
Bears—Soldier Field, Chicago, IL, G (1924)	66,944
Bengals—Cinergy Field[1], Cincinnati, OH, A (1970)	60,389
Bills—Ralph Wilson Stad., Buffalo, NY, A (1973)	80,024
Broncos—Mile High Stad., Denver, CO, G (1948)	76,082
Browns—Cleveland Browns Stad., Cleveland, OH, G (1999)	72,000
Buccaneers—Raymond James Stad., Tampa, FL, G (1998)	66,321
Cardinals—Sun Devil Stad., Tempe, AZ, G (1958)	73,273
Chargers—Qualcomm Stad.[2], San Diego, CA, G (1967)	71,000
Chiefs—Arrowhead Stad., Kansas City, MO, G (1972)	79,409
Colts—RCA Dome, Indianapolis, IN, A (1983)	60,272
Cowboys—Texas Stad., Irving, TX, A (1971)	65,675
Dolphins—Pro Player Stad.[3], Miami, FL, G (1987)	74,916
Eagles—Veterans Stad., Philadelphia, PA, A (1971)	65,352
Falcons—Georgia Dome, Atlanta, GA, A (1992)	71,228
49ers—3Com Park[4], San Francisco, CA, G (1960)	70,140
Giants—Giants Stad.[5], E. Rutherford, NJ, A (1976)	79,469
Jaguars—ALLTEL Stad.[6], Jacksonville, FL, G (1995)	73,000
Jets—Giants Stad.[5], E. Rutherford, NJ, A (1976)	77,803
Lions—Pontiac Silverdome, MI, A (1975)	80,311
Packers—Lambeau Field, Green Bay, WI, G (1957)	60,790
Panthers—Ericsson Stad.[7], Charlotte, NC, G (1996)	75,250
Patriots—Foxboro Stad., MA, G (1971)	60,292
Raiders—Network Associates Coliseum[8], Oakland, CA, G (1966)	63,142
Rams—Trans World Dome, St. Louis, MO, A (1995)	66,000
Ravens—PSINet Stad., Baltimore, MD, SG (1998)	69,354
Redskins—Jack Kent Cooke Stad., Raljon, MD, G (1997)	80,116
Saints—Louisiana Superdome, New Orleans, A (1975)	70,200
Seahawks—Kingdome, Seattle, WA, A (1976)	66,400
Steelers—Three Rivers Stad., Pittsburgh, PA, A (1970)	59,600
Titans—Adelphia Coliseum, Nashville, TN, A (1999)	67,000
Vikings—Hubert H. Humphrey Metrodome, Minneapolis, MN, A (1982)	64,035

G=Grass. A=Artificial turf. SG=SportGrass (hybrid of artificial and natural turf). Stad.=Stadium. (1) Formerly Riverfront Stadium. (2) Formerly San Diego Jack Murphy Stadium. (3) Formerly Joe Robbie Stadium. (4) Formerly Candlestick Park; full name: 3Com Park at Candlestick Point. (5) Although Giants and Jets both play at Giants Stadium, extra seating is made available for Giants games. (6) Formerly Jacksonville Municipal Stadium. (7) Formerly Carolinas Stadium. (8) Formerly Oakland-Alameda County Coliseum.

Future Sites of the Super Bowl

No.	Site	Date	No.	Site	Date
XXXIV	Georgia Dome, Atlanta, GA	Jan. 30, 2000	XXXVI	Louisiana Superdome, New Orleans, LA	Jan. 27, 2002
XXXV	Raymond James Stadium, Tampa, FL	Jan. 28, 2001	XXXVII	Qualcomm Stadium, San Diego, CA	Jan. 26, 2003

CANADIAN FOOTBALL LEAGUE
Grey Cup Championship Game, 1954-98

1954 Edmonton Eskimos 26, Montreal Alouettes 25	1977 Montreal Alouettes 41, Edmonton Eskimos 6
1955 Edmonton Eskimos 34, Montreal Alouettes 19	1978 Edmonton Eskimos 20, Montreal Alouettes 13
1956 Edmonton Eskimos 50, Montreal Alouettes 27	1979 Edmonton Eskimos 17, Montreal Alouettes 9
1957 Hamilton Tiger-Cats 32, Winnipeg Blue Bombers 7	1980 Edmonton Eskimos 48, Hamilton Tiger-Cats 10
1958 Winnipeg Blue Bombers 35, Hamilton Tiger-Cats 28	1981 Edmonton Eskimos 26, Ottawa Rough Riders 23
1959 Winnipeg Blue Bombers 21, Hamilton Tiger-Cats 7	1982 Edmonton Eskimos 32, Toronto Argonauts 16
1960 Ottawa Rough Riders 16, Edmonton Eskimos 6	1983 Toronto Argonauts 18, British Columbia Lions 17
1961 Winnipeg Blue Bombers 21, Hamilton Tiger-Cats 14	1984 Winnipeg Blue Bombers 47, Hamilton Tiger-Cats 17
1962 Winnipeg Blue Bombers 28, Hamilton Tiger-Cats 27	1985 British Columbia Lions 37, Hamilton Tiger-Cats 24
1963 Hamilton Tiger-Cats 21, British Columbia Lions 10	1986 Hamilton Tiger-Cats 39, Edmonton Eskimos 15
1964 British Columbia Lions 34, Hamilton Tiger-Cats 24	1987 Edmonton Eskimos 38, Toronto Argonauts 36
1965 Hamilton Tiger-Cats 22, Winnipeg Blue Bombers 16	1988 Winnipeg Blue Bombers 22, British Columbia Lions 21
1966 Saskatchewan Roughriders 29, Ottawa Rough Riders 14	1989 Saskatchewan Roughriders 43, Hamilton Tiger-Cats 40
1967 Hamilton Tiger-Cats 24, Saskatchewan Roughriders 1	1990 Winnipeg Blue Bombers 50, Edmonton Eskimos 11
1968 Ottawa Rough Riders 24, Calgary Stampeders 21	1991 Toronto Argonauts 36, Calgary Stampeders 21
1969 Ottawa Rough Riders 29, Saskatchewan Roughriders 11	1992 Calgary Stampeders 24, Winnipeg Blue Bombers 10
1970 Montreal Alouettes 23, Calgary Stampeders 10	1993 Edmonton Eskimos 33, Winnipeg Blue Bombers 23
1971 Calgary Stampeders 14, Toronto Argonauts 11	1994 British Columbia Lions 26, Baltimore Football Club* 23
1972 Hamilton Tiger-Cats 13, Saskatchewan Roughriders 10	1995 Baltimore Stallions 37, Calgary Stampeders 20
1973 Ottawa Rough Riders 22, Edmonton Eskimos 18	1996 Toronto Argonauts 43, Edmonton Eskimos 37
1974 Montreal Alouettes 20, Edmonton Eskimos 7	1997 Toronto Argonauts 47, Saskatchewan Roughriders 23
1975 Edmonton Eskimos 9, Montreal Alouettes 8	1998 Calgary Stampeders 26, Hamilton Tiger-Cats 24
1976 Ottawa Rough Riders 23, Saskatchewan Roughriders 20	

* Later Baltimore Stallions.

1998 CFL Review: Calgary Wins Grey Cup, CFL's First 2,000-Yard Man

In the Grey Cup, Mark McLoughlin's 35-yard field goal on the final play of the game gave the Calgary Stampeders a thrilling 26-24 victory over the Hamilton Tiger-Cats. It was the end of a Cinderella season for the Tiger-Cats, who bounced back from a 2-16 record in 1997. For the Stampeders, the win was their 2d in 4 Grey Cup appearances since 1991. Montreal Alouettes running back Mike Pringle became the first CFL player to rush for 2,000 yards in a season (2,065) and was named as the league's Most Outstanding Player.

1998 Final Standings

Western Division	W	L	T	Pct	PF	PA	Eastern Division	W	L	T	Pct	PF	PA
Calgary Stampeders	12	6	0	.667	558	397	Hamilton Tiger-Cats	12	5	1	.706	503	351
Edmonton Eskimos	9	9	0	.500	396	450	Montreal Alouettes	12	5	1	.706	470	435
British Columbia Lions	9	9	0	.500	394	427	Toronto Argonauts	9	9	0	.500	452	410
Saskatchewan Roughriders	5	13	0	.278	411	525	Winnipeg Blue Bombers	3	15	0	.167	399	588

1998 Playoff Results

Divisional Semifinals—Montreal 41, Toronto 28; Edmonton 40, British Columbia 33
Divisional Finals—Hamilton 22, Montreal 20; Calgary 33, Edmonton 10
Grey Cup—(Nov. 22, 1998, Winnipeg, Manitoba) Calgary 26, Hamilton 24

All-Time CFL Records
(through 1998 season; *active during 1998)

Longest Run—The Canadian Football League features 3 downs, 12 players on a side, and a field that is 110 yards long. George Dixon of the Montreal Alouettes made full use of the field with a 109-yard run against Ottawa on Sept. 2, 1963. Willie Fleming of the British Columbia Lions did the same against Edmonton on Oct. 17, 1964.

Leading Lifetime Rushers

	Yrs	No	Yds	Avg	Long	TDs		Yrs	No	Yds	Avg	Long	TDs
George Reed, Sask.	13	3,243	16,116	5.0	71	134	*Damon Allen, Edm.-Ott.-Ham.-Mps.-B.C.	14	1,186	8,394	7.1	51	66
Johnny Bright, Calg.-Edm.	13	1,969	10,909	5.5	90	69	*Tracy Ham, Edm.-Tor.-Balt.-Mtl.	12	1,015	7,832	7.7	80	59
Normie Kwong, Calg.-Edm.	13	1,745	9,022	5.2	60	78	Jim Evenson, B.C.-Ott.	7	1,460	7,060	4.8	68	37
*Mike Pringle, Edm.-Sac.-Balt.-Mtl.	7	1,481	8,923	6.0	86	56	Earl Lunsford, Calg.	6	1,199	6,994	5.8	85	55
Leo Lewis, Wpg.	11	1,351	8,861	6.6	92	48							
Dave Thelen, Ott.-Tor.	9	1,530	8,463	5.5	77	47							

Leading Lifetime Passers

	Yrs	Att	Comp	Yds	Pct	Avg	Long	TDs
Ron Lancaster, Ott.-Sask.	19	6,233	3,384	50,535	54.3	14.9	102	333
Matt Dunigan, Edm.-B.C.-Tor.-Wpg.-Bhm.-Ham.	14	5,476	3,057	43,857	55.8	14.3	89	306
*Damon Allen, Edm.-Ott.-Ham.-Mps.-B.C.	14	5,434	2,949	41,730	54.3	14.1	102	231
Doug Flutie, B.C.-Calg.-Tor.	8	4,854	2,975	41,355	61.3	13.9	106	270
Tom Clements, Ott.-Sask.-Ham.-Wpg.	12	4,657	2,807	39,041	60.3	13.9	105	252
*Tracy Ham, Edm.-Tor.-Balt.-Mtl.	12	4,731	2,542	38,603	53.7	15.1	85	273
Kent Austin, Sask.-B.C.-Tor.-Wpg.	10	4,700	2,709	36,030	57.6	13.3	107	198
Dieter Brock, Wpg.-Ham.	11	4,535	2,602	34,830	57.4	13.4	98	210
Tom Burgess, Ott.-Sask.-Wpg.	10	4,034	2,118	30,308	52.5	14.3	104	190
Sam Etcheverry, Mtl.	7	2,829	1,630	25,582	57.6	15.7	109	183

Leading Lifetime Receivers

	Yrs	No	Yds	Avg	Long	TDs		Yrs	No	Yds	Avg	Long	TDs
Ray Elgaard, Sask.	14	830	13,198	15.9	81	78	Earl Winfield, Ham.	11	573	10,119	17.7	81	75
*Allen Pitts, Calg.	9	792	12,397	15.6	87	101	*Darren Flutie, B.C.-Edm.-Ham.	8	665	9,949	15.0	76	45
*Don Narcisse, Sask.	12	872	11,766	13.5	77	74	Tony Gabriel, Ham.-Ott.	11	614	9,832	16.0	80	69
Brian Kelly, Edm.	9	575	11,169	19.4	97	97	Rocky DiPietro, Ham.	14	706	9,762	13.8	80	45
Tom Scott, Wpg.-Edm.-Calg.	11	649	10,837	16.7	98	88							
Tommy Joe Coffey, Edm.-Ham.-Tor.	14	650	10,320	15.9	83	63							

COLLEGE FOOTBALL

Tennessee Defeats Florida State in the Fiesta Bowl for 1998 National Championship

The #1 ranked University of Tennessee Volunteers held off the #2 ranked Florida State Seminoles, 23-16, in the Fiesta Bowl in Tempe, AZ, Jan. 4, 1999. In accordance with the new Bowl Championship Series format, as a result of the victory Tennessee was ranked first in the USA Today/ESPN coaches' poll. Tennessee also finished first in the Associated Press media poll. The Vols, who hadn't won a national title since 1951, were the only unbeaten team from a major conference. Volunteer receiver Peerless Price, who caught 4 passes for 199 yards and a touchdown, was voted MVP. His 79-yard TD reception was the longest in Fiesta Bowl history.

National College Football Champions, 1936-98

The unofficial champion as selected by the AP poll of writers and USA Today/ESPN (until 1991, UPI; 1991-1996 USA Today/CNN) poll of coaches. When the polls disagree, both teams are listed. The AP poll started in 1936; the UPI poll in 1950.

1936 Minnesota	1952 Michigan St.	1968 Ohio St.	1984 Brigham Young
1937 Pittsburgh	1953 Maryland	1969 Texas	1985 Oklahoma
1938 Texas Christian	1954 Ohio St., UCLA	1970 Nebraska, Texas	1986 Penn St.
1939 Texas A&M	1955 Oklahoma	1971 Nebraska	1987 Miami (FL)
1940 Minnesota	1956 Oklahoma	1972 Southern Cal	1988 Notre Dame
1941 Minnesota	1957 Auburn, Ohio St.	1973 Notre Dame, Alabama	1989 Miami (FL)
1942 Ohio St.	1958 Louisiana St.	1974 Oklahoma, Southern Cal	1990 Colorado, Georgia Tech
1943 Notre Dame	1959 Syracuse	1975 Oklahoma	1991 Miami (FL), Washington
1944 Army	1960 Minnesota	1976 Pittsburgh	1992 Alabama
1945 Army	1961 Alabama	1977 Notre Dame	1993 Florida St.
1946 Notre Dame	1962 Southern Cal	1978 Alabama, Southern Cal	1994 Nebraska
1947 Notre Dame	1963 Texas	1979 Alabama	1995 Nebraska
1948 Michigan	1964 Alabama	1980 Georgia	1996 Florida
1949 Notre Dame	1965 Alabama, Mich. St.	1981 Clemson	1997 Michigan, Nebraska
1950 Oklahoma	1966 Notre Dame	1982 Penn St.	1998 Tennessee
1951 Tennessee	1967 Southern Cal	1983 Miami (FL)	

1998 Final Associated Press and USA Today/ESPN NCAA Football Polls

Associated Press Rankings

1. Tennessee (13-0)	6. Wisconsin (11-1)	11. Texas A&M (11-3)	16. Arkansas (9-3)	21. Missouri (8-4)
2. Ohio St. (11-1)	7. Tulane (12-0)	12. Michigan (10-3)	17. Penn St. (9-3)	22. Notre Dame (9-3)
3. Florida St. (11-2)	8. UCLA (10-2)	13. Air Force (12-1)	18. Virginia (9-3)	23. Virginia Tech (9-3)
4. Arizona (12-1)	9. Georgia Tech (10-2)	14. Georgia (9-3)	19. Nebraska (9-4)	24. Purdue (9-4)
5. Florida (10-2)	10. Kansas St. (11-2)	15. Texas (9-3)	20. Miami (FL) (9-3)	25. Syracuse (8-4)

USA Today/ESPN Rankings

1. Tennessee	6. Florida	11. Georgia Tech	16. Texas	21. Miami (FL)
2. Ohio St.	7. Tulane	12. Michigan	17. Arkansas	22. Notre Dame
3. Florida St.	8. UCLA	13. Texas A&M	18. Virginia	23. Purdue
4. Arizona	9. Kansas St.	14. Georgia	19. Virginia Tech	24. Syracuse
5. Wisconsin	10. Air Force	15. Penn St.	20. Nebraska	25. Missouri

Note: Team records include bowl games. Won-loss records for team in USA Today/ESPN poll are under AP poll.

Annual Results of Major Bowl Games

(Dates indicate year the game was played; bowl games are generally played in late December or early January.)

Rose Bowl, Pasadena, CA

1902	(Jan.) Michigan 49, Stanford 0	1944	Southern Cal 29, Washington 0	1972	Stanford 13, Michigan 12
1916	Washington St. 14, Brown 0	1945	Southern Cal 25, Tennessee 0	1973	Southern Cal 42, Ohio St. 17
1917	Oregon 14, Pennsylvania 0	1946	Alabama 34, Southern Cal 14	1974	Ohio St. 42, Southern Cal 21
1918-19	Service teams	1947	Illinois 45, UCLA 14	1975	Southern Cal 18, Ohio St. 17
1920	Harvard 7, Oregon 6	1948	Michigan 49, Southern Cal 0	1976	UCLA 23, Ohio St. 10
1921	California 28, Ohio St. 0	1949	Northwestern 20, California 14	1977	Southern Cal 14, Michigan 6
1922	Wash. & Jeff. 0, California 0	1950	Ohio St. 17, California 14	1978	Washington 27, Michigan 20
1923	Southern Cal 14, Penn St. 3	1951	Michigan 14, California 6	1979	Southern Cal 17, Michigan 10
1924	Navy 14, Washington 14	1952	Illinois 40, Stanford 7	1980	Southern Cal 17, Ohio St. 16
1925	Notre Dame 27, Stanford 10	1953	Southern Cal 7, Wisconsin 0	1981	Michigan 23, Washington 6
1926	Alabama 20, Washington 19	1954	Mich. St. 28, UCLA 20	1982	Washington 28, Iowa 0
1927	Alabama 7, Stanford 7	1955	Ohio St. 20, Southern Cal 7	1983	UCLA 24, Michigan 14
1928	Stanford 7, Pittsburgh 6	1956	Mich. St. 17, UCLA 14	1984	UCLA 45, Illinois 9
1929	Georgia Tech 8, California 7	1957	Iowa 35, Oregon St. 19	1985	Southern Cal 20, Ohio St. 17
1930	Southern Cal 47, Pittsburgh 14	1958	Ohio St. 10, Oregon 7	1986	UCLA 45, Iowa 28
1931	Alabama 24, Wash. St. 0	1959	Iowa 38, California 12	1987	Arizona St. 22, Michigan 15
1932	Southern Cal 21, Tulane 12	1960	Washington 44, Wisconsin 8	1988	Mich. St. 20, Southern Cal 17
1933	Southern Cal 35, Pittsburgh 0	1961	Washington 17, Minnesota 7	1989	Michigan 22, Southern Cal 14
1934	Columbia 7, Stanford 0	1962	Minnesota 21, UCLA 3	1990	Southern Cal 17, Michigan 10
1935	Alabama 29, Stanford 13	1963	Southern Cal 42, Wisconsin 37	1991	Washington 46, Iowa 34
1936	Stanford 7, SMU 0	1964	Illinois 17, Washington 7	1992	Washington 34, Michigan 14
1937	Pittsburgh 21, Washington 0	1965	Michigan 34, Oregon St. 7	1993	Michigan 38, Washington 31
1938	California 13, Alabama 0	1966	UCLA 14, Mich. St. 12	1994	Wisconsin 21, UCLA 16
1939	Southern Cal 7, Duke 3	1967	Purdue 14, Southern Cal 13	1995	Penn St. 38, Oregon 20
1940	Southern Cal 14, Tennessee 0	1968	Southern Cal 14, Indiana 3	1996	Southern Cal 41, Northwestern 32
1941	Stanford 21, Nebraska 13	1969	Ohio St. 27, Southern Cal 16	1997	Ohio St. 20, Arizona St. 17
1942*	Oregon St. 20, Duke 16	1970	Southern Cal 10, Michigan 3	1998	Michigan 21, Washington St. 16
1943	Georgia 9, UCLA 0	1971	Stanford 27, Ohio St. 17	1999	Wisconsin 38, UCLA 31

*Played at Durham, NC.

Orange Bowl, Miami, FL

1935 (Jan.) Bucknell 26, Miami (FL) 0	1957 Colorado 27, Clemson 21	1979 Oklahoma 31, Nebraska 24
1936 Catholic U. 20, Mississippi 19	1958 Oklahoma 48, Duke 21	1980 Oklahoma 24, Florida St. 7
1937 Duquesne 13, Mississippi St. 12	1959 Oklahoma 21, Syracuse 6	1981 Oklahoma 18, Florida St. 17
1938 Auburn 6, Michigan St. 0	1960 Georgia 14, Missouri 0	1982 Clemson 22, Nebraska 15
1939 Tennessee 17, Oklahoma 0	1961 Missouri 21, Navy 14	1983 Nebraska 21, LSU 20
1940 Georgia Tech 21, Missouri 7	1962 LSU 25, Colorado 7	1984 Miami (FL) 31, Nebraska 30
1941 Mississippi St. 14, Georgetown 7	1963 Alabama 17, Oklahoma 0	1985 Washington 28, Oklahoma 17
1942 Georgia 40, TCU 26	1964 Nebraska 13, Auburn 7	1986 Oklahoma 25, Penn St. 10
1943 Alabama 37, Boston Coll. 21	1965 Texas 21, Alabama 17	1987 Oklahoma 42, Arkansas 8
1944 LSU 19, Texas A&M 14	1966 Alabama 39, Nebraska 28	1988 Miami (FL) 20, Oklahoma 14
1945 Tulsa 26, Georgia Tech 12	1967 Florida 27, Georgia Tech 12	1989 Miami (FL) 23, Nebraska 3
1946 Miami (FL) 13, Holy Cross 6	1968 Oklahoma 26, Tennessee 24	1990 Notre Dame 21, Colorado 6
1947 Rice 8, Tennessee 0	1969 Penn St. 15, Kansas 14	1991 Colorado 10, Notre Dame 9
1948 Georgia Tech 20, Kansas 14	1970 Penn St. 10, Missouri 3	1992 Miami (FL) 22, Nebraska 0
1949 Texas 41, Georgia 28	1971 Nebraska 17, LSU 12	1993 Florida St. 27, Nebraska 14
1950 Santa Clara 21, Kentucky 13	1972 Nebraska 38, Alabama 6	1994 Florida St. 18, Nebraska 16
1951 Clemson 15, Miami (FL) 14	1973 Nebraska 40, Notre Dame 6	1995 Nebraska 24, Miami (FL) 17
1952 Georgia Tech 17, Baylor 14	1974 Penn St. 16, LSU 9	1996 Florida St. 31, Notre Dame 26
1953 Alabama 61, Syracuse 6	1975 Notre Dame 13, Alabama 11	1996 (Dec.) Nebraska 41, Virginia Tech 21
1954 Oklahoma 7, Maryland 0	1976 Oklahoma 14, Michigan 6	1998 (Jan.) Nebraska 42, Tennessee 17
1955 Duke 34, Nebraska 7	1977 Ohio St. 27, Colorado 10	1999 Florida 31, Syracuse 10
1956 Oklahoma 20, Maryland 6	1978 Arkansas 31, Oklahoma 6	

Sugar Bowl, New Orleans, LA

1935 (Jan.) Tulane 20, Temple 14	1957 Baylor 13, Tennessee 7	1979 Alabama 14, Penn St. 7
1936 TCU 3, LSU 2	1958 Mississippi 39, Texas 7	1980 Alabama 24, Arkansas 9
1937 Santa Clara 21, LSU 14	1959 LSU 7, Clemson 0	1981 Georgia 17, Notre Dame 10
1938 Santa Clara 6, LSU 0	1960 Mississippi 21, LSU 0	1982 Pittsburgh 24, Georgia 20
1939 TCU 15, Carnegie Tech 7	1961 Mississippi 14, Rice 6	1983 Penn St. 27, Georgia 23
1940 Texas A&M 14, Tulane 13	1962 Alabama 10, Arkansas 3	1984 Auburn 9, Michigan 7
1941 Boston Col. 19, Tennessee 13	1963 Mississippi 17, Arkansas 13	1985 Nebraska 28, LSU 10
1942 Fordham 2, Missouri 0	1964 Alabama 12, Mississippi 7	1986 Tennessee 35, Miami (FL) 7
1943 Tennessee 14, Tulsa 7	1965 LSU 13, Syracuse 10	1987 Nebraska 30, LSU 15
1944 Georgia Tech 20, Tulsa 18	1966 Missouri 20, Florida 18	1988 Syracuse 16, Auburn 16
1945 Duke 29, Alabama 26	1967 Alabama 34, Nebraska 7	1989 Florida St. 13, Auburn 7
1946 Oklahoma A&M 33, St. Mary's 13	1968 LSU 20, Wyoming 13	1990 Miami (FL) 33, Alabama 25
1947 Georgia 20, N. Carolina 10	1969 Arkansas 16, Georgia 2	1991 Tennessee 23, Virginia 22
1948 Texas 27, Alabama 7	1970 Mississippi 27, Arkansas 22	1992 Notre Dame 39, Florida 28
1949 Oklahoma 14, N. Carolina 6	1971 Tennessee 34, Air Force 13	1993 Alabama 34, Miami (FL) 13
1950 Oklahoma 35, LSU 0	1972 Oklahoma 40, Auburn 22	1994 Florida 41, West Virginia 7
1951 Kentucky 13, Oklahoma 7	1972* (Dec.) Oklahoma 14, Penn St. 0	1995 Florida St. 23, Florida 17
1952 Maryland 28, Tennessee 13	1973 Notre Dame 24, Alabama 23	1995 (Dec.) Virginia Tech 28, Texas 10
1953 Georgia Tech 24, Mississippi 7	1974 Nebraska 13, Florida 10	1997 (Jan.) Florida 52, Florida St. 20
1954 Georgia Tech 42, West Virginia 19	1975 Alabama 13, Penn St. 6	1998 Florida St. 31, Ohio St. 14
1955 Navy 21, Mississippi 0	1977 (Jan.) Pittsburgh 27, Georgia 3	1999 Ohio St. 24, Texas A&M 14
1956 Georgia Tech 7, Pittsburgh 0	1978 Alabama 35, Ohio St. 6	

* Penn St. awarded game by forfeit.

Fiesta Bowl, Tempe, AZ

1971 (Dec.) Arizona St. 45, Florida St. 38	1982 (Jan.) Penn St. 26, USC 10	1991 Louisville 34, Alabama 7
1972 Arizona St. 49, Missouri 35	1983 Arizona St. 32, Oklahoma 21	1992 Penn St. 42, Tennessee 17
1973 Arizona St. 28, Pittsburgh 7	1984 Ohio St. 28, Pittsburgh 23	1993 Syracuse 26, Colorado 22
1974 Okla. St. 16, Brigham Young 6	1985 UCLA 39, Miami (FL) 37	1994 Arizona 29, Miami (FL) 0
1975 Arizona St. 17, Nebraska 14	1986 Michigan 27, Nebraska 23	1995 Colorado 41, Notre Dame 24
1976 Oklahoma 41, Wyoming 7	1987 Penn St. 14, Miami (FL) 10	1996 Nebraska 62, Florida 24
1977 Penn St. 42, Arizona St. 30	1988 Florida St. 31, Nebraska 28	1997 Penn St. 38, Texas 15
1978 UCLA 10, Arkansas 10	1989 Notre Dame 34, W. Virginia 21	1997 (Dec.) Kansas St. 35, Syracuse 18
1979 Pittsburgh 16, Arizona 10	1990 Florida St. 41, Nebraska 17	1999 (Jan.) Tennessee 23, Florida St. 16
1980 Penn St. 31, Ohio St. 19		

Cotton Bowl, Dallas, TX

1937 (Jan.) TCU 16, Marquette 6	1958 Navy 20, Rice 7	1979 Notre Dame 35, Houston 34
1938 Rice 28, Colorado 14	1959 TCU 0, Air Force 0	1980 Houston 17, Nebraska 14
1939 St. Mary's 20, Texas Tech 13	1960 Syracuse 23, Texas 14	1981 Alabama 30, Baylor 2
1940 Clemson 6, Boston Coll. 3	1961 Duke 7, Arkansas 6	1982 Texas 14, Alabama 12
1941 Texas A&M 13, Fordham 12	1962 Texas 12, Mississippi 7	1983 SMU 7, Pittsburgh 3
1942 Alabama 29, Texas A&M 21	1963 LSU 13, Texas 0	1984 Georgia 10, Texas 9
1943 Texas 14, Georgia Tech 7	1964 Texas 28, Navy 6	1985 Boston Coll. 45, Houston 28
1944 Randolph Field 7, Texas 7	1965 Arkansas 10, Nebraska 7	1986 Texas A&M 36, Auburn 16
1945 Oklahoma A&M 34, TCU 0	1966 UCLA 14, Arkansas 7	1987 Ohio St. 28, Texas A&M 12
1946 Texas 40, Missouri 27	1966 (Dec.) Georgia 24, SMU 9	1988 Texas A&M 35, Notre Dame 10
1947 Arkansas 0, LSU 0	1968 (Jan.) Texas A&M 20, Alabama 16	1989 UCLA 17, Arkansas 3
1948 SMU 13, Penn St. 13	1969 Texas 36, Tennessee 13	1990 Tennessee 31, Arkansas 27
1949 SMU 21, Oregon 13	1970 Texas 21, Notre Dame 17	1991 Miami (FL) 46, Texas 3
1950 Rice 27, North Carolina 13	1971 Notre Dame 24, Texas 11	1992 Florida St. 10, Texas A&M 2
1951 Tennessee 20, Texas 14	1972 Penn St. 30, Texas 6	1993 Notre Dame 28, Texas A&M 3
1952 Kentucky 20, TCU 7	1973 Texas 17, Alabama 13	1994 Notre Dame 24, Texas A&M 21
1953 Texas 16, Tennessee 0	1974 Nebraska 19, Texas 3	1995 Southern Cal. 55, Tex. Tech 14
1954 Rice 28, Alabama 6	1975 Penn St. 41, Baylor 20	1996 Colorado 38, Oregon 6
1955 Georgia Tech 14, Arkansas 6	1976 Arkansas 31, Georgia 10	1997 Brigham Young 19, Kansas St. 15
1956 Mississippi 14, TCU 13	1977 Houston 30, Maryland 21	1998 UCLA 29, Texas A&M 23
1957 TCU 28, Syracuse 27	1978 Notre Dame 38, Texas 10	1999 Texas 38, Mississippi St. 11

Sun Bowl, El Paso, TX (John Hancock Bowl, 1989-93)

1936 (Jan.) Hardin-Simmons 14, New Mexico St. 14	1939 Utah 26, New Mexico 0	1943 2d Air Force 13, Hardin-Simmons 7
1937 Hardin-Simmons 34, Texas Mines 6	1940 Catholic U. 0, Arizona St. 0	1944 Southwestern (TX) 7, New Mexico 0
1938 West Virginia 7, Texas Tech 6	1941 Western Reserve 26, Arizona St. 13	1945 Southwestern (TX) 35, U. of Mexico 0
	1942 Tulsa 6, Texas Tech 0	1946 New Mexico 34, Denver 24

1947 Cincinnati 18, Virginia Tech 6	1964 Georgia 7, Texas Tech 0	1982 North Carolina 26, Texas 10
1948 Miami (OH) 13, Texas Tech 12	1965 Texas Western 13, TCU 12	1983 Alabama 28, SMU 7
1949 West Virginia 21, Texas Mines 12	1966 Wyoming 28, Florida St. 20	1984 Maryland 28, Tennessee 27
1950 Texas Western 33, Georgetown 20	1967 UTEP 14, Mississippi 7	1985 Georgia 13, Arizona 13
1951 West Texas St. 14, Cincinnati 13	1968 Auburn 34, Arizona 10	1986 Alabama 28, Washington 6
1952 Texas Tech 25, Pacific (CA) 14	1969 Nebraska 45, Georgia 6	1987 Oklahoma St. 35, West Virginia 33
1953 Pacific (CA) 26, S. Mississippi 7	1970 Georgia Tech. 17, Texas Tech 9	1988 Alabama 29, Army 28
1954 Texas Western 37, S. Miss. 14	1971 LSU 33, Iowa St. 15	1989 Pittsburgh 31, Texas A&M 28
1955 Texas Western 47, Florida St. 20	1972 North Carolina 32, Texas Tech 28	1990 Michigan St. 17, USC 16
1956 Wyoming 21, Texas Tech 14	1973 Missouri 34, Auburn 17	1991 UCLA 6, Illinois 3
1957 Geo. Washington 13, Texas Western 0	1974 Mississippi St. 26, North Carolina 24	1992 Baylor 20, Arizona 15
1958 Louisville 34, Drake 20	1975 Pittsburgh 33, Kansas 19	1993 Oklahoma 41, Texas Tech 10
1958 (Dec.) Wyoming 14, Hardin-Simmons 6	1977 (Jan.) Texas A&M 37, Florida 14	1994 Texas 35, North Carolina 31
1959 New Mexico St. 28, N. Texas St. 8	1977 (Dec.) Stanford 24, LSU 14	1995 Iowa 38, Washington 18
1960 New Mexico St. 20, Utah St. 13	1978 Texas 42, Maryland 0	1996 Stanford 38, Michigan St. 0
1961 Villanova 17, Wichita 9	1979 Washington 14, Texas 7	1997 Arizona St. 17, Iowa 7
1962 West Texas St. 15, Ohio U. 14	1980 Nebraska 31, Mississippi St. 17	1998 Texas Christian 28, USC 19
1963 Oregon 21, SMU 14	1981 Oklahoma 40, Houston 14	

Gator Bowl, Jacksonville, FL

1946 (Jan.) Wake Forest 26, S. Carolina 14	1965 (Jan.) Florida St. 36, Okla.19	1981 N. Carolina 31, Arkansas 27
1947 Oklahoma 34, N. Carolina St. 13	1965 (Dec.) Georgia Tech 31, Texas Tech 21	1982 Florida St. 31, West Virginia 12
1948 Maryland 20, Georgia 20	1966 Tennessee 18, Syracuse 12	1983 Florida 14, Iowa 6
1949 Clemson 24, Missouri 23	1967 Penn St. 17, Florida St. 17	1984 Oklahoma St. 21, S. Carolina 14
1950 Maryland 20, Missouri 7	1968 Missouri 35, Alabama 10	1985 Florida St. 34, Oklahoma St. 23
1951 Wyoming 20, Washington & Lee 7	1969 Florida 14, Tennessee 13	1986 Clemson 27, Stanford 21
1952 Miami (FL) 14, Clemson 0	1971 (Jan.) Auburn 35, Mississippi 28	1987 LSU 30, S. Carolina 13
1953 Florida 14, Tulsa 13	1971 (Dec.) Georgia 7, N. Carolina 3	1989 (Jan.) Georgia 34, Michigan St. 27
1954 Texas Tech 35, Auburn 13	1972 Auburn 24, Colorado 3	1989 (Dec.) Clemson 27, W. Virginia 7
1954 (Dec.) Auburn 33, Baylor 13	1973 Texas Tech 28, Tenn. 19	1991 (Jan.) Michigan 35, Mississippi 3
1955 Vanderbilt 25, Auburn 13	1974 Auburn 27, Texas 3	1991 (Dec.) Oklahoma 48, Virginia 14
1956 Georgia Tech 21, Pittsburgh 14	1975 Maryland 13, Florida 0	1992 Florida 27, N. Carolina St. 10
1957 Tennessee 3, Texas A&M 0	1976 Notre Dame 20, Penn St. 9	1993 Alabama 24, N. Carolina 10
1958 Mississippi 7, Florida 3	1977 Pittsburgh 34, Clemson 3	1994 Tennessee 45, Virginia Tech 23
1960 (Jan.) Arkansas 14, Georgia Tech 7	1978 Clemson 17, Ohio St. 15	1996 (Jan.) Syracuse 41, Clemson 0
1960 (Dec.) Florida 13, Baylor 12	1979 N. Carolina 17, Michigan 15	1997 N. Carolina 20, W. Virginia 13
1961 Penn St. 30, Georgia Tech 15	1980 Pittsburgh 37, S. Carolina 9	1998 N. Carolina 42, Virginia Tech 3
1962 Florida 17, Penn St. 7		1999 Georgia Tech 35, Notre Dame 28
1963 N. Carolina 35, Air Force 0		

Outback Bowl, Tampa, FL (Hall of Fame Bowl Until 1996)

1986 (Dec.) Boston College 27, Georgia 24	1992 Syracuse 24, Ohio St. 17	1996 Penn St. 43, Auburn 14
1988 (Jan.) Michigan 28, Alabama 24	1993 Tennessee 38, Boston College 23	1997 Alabama 17, Michigan 14
1989 Syracuse 23, LSU 10	1994 Michigan 42, N. Carolina St. 7	1998 Georgia 33, Wisconsin 6
1990 Auburn 31, Ohio St. 14	1995 Wisconsin 34, Duke 20	1999 Penn St. 26, Kentucky 14
1991 Clemson 30, Illinois 0		

Liberty Bowl, Memphis, TN

1959 (Dec.) Penn St. 7, Alabama 0	1973 N. Carolina St. 31, Kansas 18	1986 Tennessee 21, Minnesota 14
1960 Penn St. 41, Oregon 12	1974 Tennessee 7, Maryland 3	1987 Georgia 20, Arkansas 17
1961 Syracuse 15, Miami (FL) 14	1975 USC 20, Texas A&M 0	1988 Indiana 34, S. Carolina 10
1962 Oregon St. 6, Villanova 0	1976 Alabama 36, UCLA 6	1989 Mississippi 42, Air Force 29
1963 Mississippi St. 16, N. Carolina St. 12	1977 Nebraska 21, N. Carolina 17	1990 Air Force 23, Ohio St. 11
1964 Utah 32, West Virginia 6	1978 Missouri 20, LSU 15	1991 Air Force 38, Mississippi St. 15
1965 Mississippi 13, Auburn 7	1979 Penn St. 9, Tulane 6	1992 Mississippi 13, Air Force 0
1966 Miami (FL) 14, Virginia Tech 7	1980 Purdue 28, Missouri 25	1993 Louisville 18, Michigan St. 7
1967 N. Carolina St. 14, Georgia 7	1981 Ohio St. 31, Navy 28	1994 Illinois 30, East Carolina 0
1968 Mississippi 34, Virginia Tech 17	1982 Alabama 21, Illinois 15	1995 East Carolina 19, Stanford 13
1969 Colorado 47, Alabama 33	1983 Notre Dame 19, Boston Coll. 18	1996 Syracuse 30, Houston 17
1970 Tulane 17, Colorado 3	1984 Auburn 21, Arkansas 15	1997 So. Mississippi 41, Pittsburgh 7
1971 Tennessee 14, Arkansas 13	1985 Baylor 21, LSU 7	1998 Tulane 41, Brigham Young
1972 Georgia Tech 31, Iowa St. 30		

Insight.com Bowl, Tucson, AZ (Copper Bowl Until 1997)

1989 (Dec.) Arizona 17, N. Carolina St. 10	1993 Kansas St. 52, Wyoming 17	1996 Wisconsin 38, Utah 10
1990 California 17, Wyoming 15	1994 Brigham Young 31, Oklahoma 6	1997 Arizona 20, New Mexico 14
1991 Indiana 24, Baylor 0	1995 Texas Tech 55, Air Force 41	1998 Missouri 34, West Virginia 31
1992 Washington St. 31, Utah 28		

Independence Bowl, Shreveport, LA

1976 (Dec.) McNeese St. 20, Tulsa 16	1984 Air Force 23, Virginia Tech 7	1992 Wake Forest 39, Oregon 35
1977 Louisiana Tech 24, Louisville 14	1985 Minnesota 20, Clemson 13	1993 Virginia Tech 45, Indiana 20
1978 E. Carolina 35, Louisiana Tech 13	1986 Mississippi 20, Texas Tech 17	1994 Virginia 20, Texas Christian 10
1979 Syracuse 31, McNeese St. 7	1987 Washington 24, Tulane 12	1995 LSU 45, Michigan St. 26
1980 So. Mississippi 16, McNeese St. 14	1988 So. Mississippi 38, UTEP 18	1996 Auburn 32, Army 29
1981 Texas A&M 33, Oklahoma St. 16	1989 Oregon 27, Tulsa 24	1997 LSU 27, Notre Dame 9
1982 Wisconsin 14, Kansas St. 3	1990 Louisiana Tech 34, Maryland 34	1998 Mississippi 35, Texas Tech 18
1983 Air Force 9, Mississippi 3	1991 Georgia 24, Arkansas 15	

Florida Citrus Bowl, Orlando, FL (Tangerine Bowl Until 1983)

1947 (Jan.) Catawba 31, Maryville 6	1956 Juniata 6, Missouri Valley 6	1962 Houston 49, Miami (OH) 21
1948 Catawba 7, Marshall 0	1957 West Texas St. 20, So. Miss. 13	1963 Western Ky. 27, Coast Guard 0
1949 Murray St. 21, Sul Ross St. 21	1958 East Texas St. 10, So. Miss. 9	1964 E. Carolina 14, Massachusetts 13
1950 St. Vincent 7, Emory & Henry 6	1958 (Dec.) East Texas St. 26, Missouri Valley 7	1965 E. Carolina 31, Maine 0
1951 Morris Harvey 35, Emory & Henry 14	1960 (Jan.) Middle Tennessee 21, Presbyterian 12	1966 Morgan St. 14, West Chester 6
1952 Stetson 35, Arkansas St. 20		1967 Tenn.-Martin 25, West Chester 8
1953 East Texas St. 33, Tenn. Tech 0	1960 (Dec.) Citadel 27, Tenn. Tech 0	1968 Richmond 49, Ohio U. 42
1954 East Texas St. 7, Arkansas St. 7	1961 Lamar 21, Middle Tennessee 14	1969 Toledo 56, Davidson 33
1955 Neb.-Omaha 7, E. Kentucky 6		1970 Toledo 40, William & Mary 12

1971 Toledo 28, Richmond 3	1981 Missouri 19, So. Mississippi 17	1991 Georgia Tech 45, Nebraska 21
1972 Tampa 21, Kent St. 18	1982 Auburn 33, Boston College 26	1992 California 37, Clemson 13
1973 Miami (OH) 16, Florida 7	1983 Tennessee 30, Maryland 23	1993 Georgia 21, Ohio St. 14
1974 Miami (OH) 21, Georgia 10	1984 Georgia 17, Florida St. 17	1994 Penn St. 31, Tennessee 13
1975 Miami (OH) 20, S. Carolina 7	1985 Ohio St. 10, Brigham Young 7	1995 Alabama 24, Ohio St. 17
1976 Okla. St. 49, Brigham Young 21	1987 (Jan.) Auburn 16, USC 7	1996 Tennessee 20, Ohio St. 14
1977 Florida St. 40, Texas Tech 17	1988 Clemson 35, Penn St. 10	1997 Tennessee 48, Northwestern 28
1978 N. Carolina St. 30, Pittsburgh 17	1989 Clemson 13, Oklahoma 6	1998 Florida 21, Penn St. 6
1979 LSU 34, Wake Forest 10	1990 Illinois 31, Virginia 21	1999 Michigan 45, Arkansas 31
1980 Florida 35, Maryland 20		

Peach Bowl, Atlanta, GA

1968 (Dec.) LSU 31, Florida St. 27	1979 Baylor 24, Clemson 18	1990 Auburn 27, Indiana 23
1969 W. Virginia 14, S. Carolina 3	1981 (Jan.) Miami (FL) 20, Virginia Tech 10	1992 (Jan.) E. Carolina 37, N. Carolina St. 34
1970 Arizona St. 48, N. Carolina 26	1981 (Dec.) W. Virginia 26, Florida 6	
1971 Mississippi 41, Georgia Tech 18	1982 Iowa 28, Tennessee 22	1993 N. Carolina 21, Mississippi St. 17
1972 N. Carolina 49, W. Virginia 13	1983 Florida St. 28, N. Carolina 3	1993 (Dec.) Clemson 14, Kentucky 13
1973 Georgia 17, Maryland 16	1984 Virginia 27, Purdue 22	1995 (Jan.) N. Carolina St. 28, Mississippi St. 24
1974 Vanderbilt 6, Texas Tech 6	1985 Army 31, Illinois 29	
1975 W. Virginia 13, N. Carolina St. 10	1986 Va. Tech 25, N. Carolina St. 24	1995 (Dec.) Virginia 34, Georgia 27
1976 Kentucky 21, N. Carolina 0	1988 (Jan.) Tennessee 28, Indiana 22	1996 LSU 10, Clemson 7
1977 N. Carolina St. 24, Iowa St. 14	1988 (Dec.) N. Carolina St. 28, Iowa 23	1998 (Jan.) Auburn 21, Clemson 17
1978 Purdue 41, Georgia Tech. 21	1989 Syracuse 19, Georgia 18	1998 (Dec.) Georgia 35, Virginia 33

Holiday Bowl, San Diego, CA

1978 (Dec.) Navy 23, Brigham Young 16	1985 Arkansas 18, Arizona St. 17	1992 Hawaii 27, Illinois 17
1979 Indiana 38, Brigham Young 37	1986 Iowa 39, San Diego St. 38	1993 Ohio St. 28, Brigham Young 21
1980 Brigham Young 46, SMU 45	1987 Iowa 20, Wyoming 19 ———	1994 Michigan 24, Colorado St. 14
1981 Brigham Young 38, Washington St. 36	1988 Oklahoma St. 62, Wyoming 14	1995 Kansas St. 54, Colorado St. 21
1982 Ohio St. 47, Brigham Young 17	1989 Penn St. 50, Brigham Young 39	1996 Colorado 33, Washington 21
1983 Brigham Young 21, Missouri 17	1990 Texas A&M 65, Brigham Young 14	1997 Colorado St. 35, Missouri 24
1984 Brigham Young 24, Michigan 17	1991 Iowa 13, Brigham Young 13	1998 Arizona 23, Nebraska 20

Aloha Classics, Honolulu, HI (Aloha Bowl until 1997)

1982 (Dec.) Washington 21, Md. 20	1988 Washington St. 24, Houston 22	1994 Boston Coll. 12, Kansas St. 7
1983 Penn St. 13, Washington 10	1989 Michigan St. 33, Hawaii 13	1995 Kansas 51, UCLA 30
1984 SMU 27, Notre Dame 20	1990 Syracuse 28, Arizona 0	1996 Navy 42, California 38
1985 Alabama 24, USC 3	1991 Georgia Tech 18, Stanford 17	1997 Washington 51, Michigan St. 23
1986 Arizona 30, North Carolina 21	1992 Kansas 23, Brigham Young 20	1998 Colorado 51, Oregon 23 (Aloha)
1987 UCLA 20, Florida 16	1993 Colorado 41, Fresno St. 30	1998 Air Force 45, Washington 25 (Oahu)

Micron PC Bowl, Miami, FL (Blockbuster Bowl, 1990-93; Carquest Bowl 1994-97)

1990 (Dec.) Florida St. 24, Penn St. 17	1994 Boston Coll. 31, Virginia 13	1996 Miami (FL) 31, Virginia 21
1991 Alabama 30, Colorado 25	1995 S. Carolina 24, W. Virginia 21	1997 Georgia Tech 35, W. Virginia 30
1993 (Jan.) Stanford 24, Penn St. 3	1995 (Dec.) N. Carolina 20, Arkansas 10	1998 Miami (FL) 46, North Carolina St. 23

Las Vegas Bowl, Las Vegas, NV

1992 (Dec.) Bowling Green 35, Nevada 34	1994 UNLV 52, Central Michigan 24	1997 Oregon 41, Air Force 13
1993 Utah St. 42, Ball St. 33	1995 Toledo 40, Nevada 37 (OT)	1998 North Carolina 20, San Diego St. 13
	1996 Nevada 18, Ball St. 15	

Alamo Bowl, San Antonio, TX

1993 (Dec.) California 37, Iowa 3	1995 Texas A&M 22, Michigan 20	1997 Purdue 33, Oklahoma St.20
1994 Washington St. 10, Baylor 3	1996 Iowa 27, Texas Tech 0	1998 Purdue 37, Kansas St. 34

Motor City Bowl, Pontiac, MI

1997 (Dec.) Mississippi 34, Marshall 31	1998 Marshall 48, Louisville 29

Humanitarian Bowl, Boise, ID

1997 (Dec.) Cincinnati 35, Utah St. 19	1998 Idaho 42, Southern Mississippi 35

Music City Bowl, Nashville, TN

1998 (Dec.) Virginia Tech 38, Alabama 7

Selected College Division I Football Teams

(1998 record does not include bowl games or Division I-AA playoff games; coaches at the start of 1999 season)

Team	Nickname	Team colors	Conference	Coach	1998 record (W-L)
Air Force	Falcons	Blue & silver	Western Athletic	Fisher DeBerry	11-1
Akron	Zips	Blue & gold	Mid-American	Lee Owens	4-7
Alabama	Crimson Tide	Crimson & white	Southeastern	Mike DuBose	7-4
Arizona	Wildcats	Cardinal & navy	Pacific Ten	Dick Tomey	11-1
Arizona State	Sun Devils	Maroon & gold	Pacific Ten	Bruce Snyder	5-6
Arkansas	Razorbacks	Cardinal & white	Southeastern	Houston Nutt	9-2
Arkansas State	Indians	Scarlet & black	Independent	Joe Hollis	4-8
Army	Cadets, Black Knights	Black, gold, gray	Independent	Bob Sutton	3-8
Auburn	Tigers	Burnt orange & navy	Southeastern	Tommy Tuberville	3-8
Ball State	Cardinals	Cardinal & white	Mid-American	Bill Lynch	1-10
Baylor	Bears	Green & gold	Big Twelve	Kevin Steele	2-9
Boston College	Eagles	Maroon & gold	Big East	Tom O'Brien	4-7
Bowling Green	Falcons	Orange & brown	Mid-American	Gary Blackney	5-6
Brigham Young (BYU)	Cougars	Royal blue & white	Western Athletic	LaVell Edwards	9-4
Brown	Bears	Brown, cardinal, white	Ivy League	Phil Estes	7-3
California	Golden Bears	Blue & gold	Pacific Ten	Tom Holmoe	5-6
Central Michigan	Chippewas	Maroon & gold	Mid-American	Dick Flynn	6-5
Cincinnati	Bearcats	Red & black	Conference USA	Rick Minter	2-9
Citadel	Bulldogs	Blue & white	Southern	Don Powers	5-6

Team	Nickname	Team colors	Conference	Coach	1998 record (W-L)
Clemson	Tigers	Purple & orange	Atlantic Coast	Tommy Bowden	3-8
Colgate	Red Raiders	Maroon, gray, & white	Patriot League	Dick Biddle	8-3
Colorado	Golden Buffaloes	Silver, gold, & black	Big Twelve	Gary Barnett	7-4
Colorado State	Rams	Green & gold	Western Athletic	Sonny Lubick	8-4
Columbia	Lions	Columbia blue & white	Ivy League	Ray Tellier	4-6
Connecticut	Huskies	Blue & white	Atlantic Ten	Randy Edsall	9-2
Cornell	Big Red	Carnelian & white	Ivy League	Peter Mangurian	4-6
Dartmouth	Big Green	Dartmouth green & white	Ivy League	John Lyons	2-8
Delaware	Fightin' Blue Hens	Blue & gold	Atlantic Ten	Harold Raymond	7-4
Delaware State	Hornets	Red & blue	Mid-Eastern Athletic	John McKenzie	0-11
Duke	Blue Devils	Royal blue & white	Atlantic Coast	Carl Franks	4-7
East Carolina	Pirates	Purple & gold	Conference USA	Steve Logan	6-5
East Tennessee State	Buccaneers	Blue & gold	Southern	Paul Hamilton	4-7
Eastern Illinois	Panthers	Blue & gray	Ohio Valley	Bob Spoo	6-5
Eastern Kentucky	Colonels	Maroon & white	Ohio Valley	Roy Kidd	6-5
Eastern Michigan	Eagles	Dark green & white	Mid-American	Rick Rasnick	3-8
Eastern Washington	Eagles	Red & white	Big Sky	Mike Kramer	5-6
Florida	Gators	Orange & blue	Southeastern	Steve Spurrier	9-2
Florida A&M	Rattlers	Orange & green	Mid-Eastern Athletic	Billy Joe	10-1
Florida State	Seminoles	Garnet & gold	Atlantic Coast	Bobby Bowden	11-1
Fresno State	Bulldogs	Cardinal & blue	Western Athletic	Pat Hill	5-6
Furman	Paladins	Purple & white	Southern	Bobby Johnson	5-6
Georgia	Bulldogs	Red & black	Southeastern	Jim Donnan	8-3
Georgia Southern	Eagles	Blue & white	Southern	Paul Johnson	11-0
Georgia Tech	Yellow Jackets	Old gold & white	Atlantic Coast	George O'Leary	9-2
Grambling State	Tigers	Black & gold	Southwestern	Doug Williams	5-6
Harvard	Crimson	Crimson, black, white	Ivy League	Tim Murphy	4-6
Holy Cross	Crusaders	Royal purple	Patriot League	Dan Allen	2-9
Houston	Cougars	Scarlet & white	Conference USA	Kim Helton	3-8
Howard	Bison	Blue, white & red	Mid-Eastern Athletic	Steve Wilson	7-4
Idaho	Vandals	Silver & gold	Big West	Chris Tormey	8-3
Idaho State	Bengals	Orange & black	Big Sky	Larry Lewis	3-8
Illinois	Fighting Illini	Orange & blue	Big Ten	Ron Turner	3-8
Illinois State	Redbirds	Red & white	Gateway	Todd Berry	8-3
Indiana	Hoosiers	Cream & crimson	Big Ten	Cam Cameron	4-7
Indiana State	Sycamores	Blue & white	Gateway	Tim McGuire	5-6
Iowa	Hawkeyes	Old gold & black	Big Ten	Kirk Ferentz	3-8
Iowa State	Cyclones	Cardinal & gold	Big Twelve	Dan McCarney	3-8
Jackson State	Tigers	Blue & white	Southwestern	James Carson	7-4
James Madison	Dukes	Purple & gold	Atlantic Ten	Mickey Matthews	3-8
Kansas	Jayhawks	Crimson & blue	Big Twelve	Terry Allen	4-7
Kansas State	Wildcats	Purple & white	Big Twelve	Bill Snyder	11-1
Kent	Golden Flashes	Navy blue & gold	Mid-American	Dean Pees	0-11
Kentucky	Wildcats	Blue & white	Southeastern	Hal Mumme	7-4
Lafayette	Leopards	Maroon & white	Patriot League	Bill Russo	3-8
Lehigh	Mountain Hawks	Brown & white	Patriot League	Kevin Higgins	11-0
Liberty	Flames	Red, white, blue	Independent	Sam Rutigliano	5-6
Louisiana State (LSU)	Fighting Tigers	Purple & gold	Southeastern	Gerry DiNardo	4-7
Louisiana Tech	Bulldogs	Red & blue	Independent	Jack Bicknell,iii	6-6
Louisville	Cardinals	Red, black, white	Conference USA	John L. Smith	7-4
Maine	Black Bears	Blue & white	Atlantic Ten	Jack Cosgrove	6-5
Marshall	Thundering Herd	Green & white	Mid-American	Bob Pruett	11-1
Maryland	Terrapins	Red, white, black, gold	Atlantic Coast	Ron Vanderlinden	3-8
Massachusetts	Minutemen	Maroon & white	Atlantic Ten	Mark Whipple	8-3
McNeese State	Cowboys	Blue & gold	Southland	Kirby Bruchhaus	9-2
Memphis	Tigers	Blue & gray	Conference USA	Rip Scherer	2-9
Miami (Florida)	Hurricanes	Orange, green, white	Big East	Butch Davis	8-3
Miami (Ohio)	RedHawks	Red & white	Mid-American	Terry Hoeppner	10-1
Michigan	Wolverines	Maize & blue	Big Ten	Lloyd Carr	9-3
Michigan State	Spartans	Green & white	Big Ten	Nick Saban	6-6
Middle Tennessee St.	Blue Raiders	Blue & white	Ohio Valley	Andy McCollum	5-5
Minnesota	Golden Gophers	Maroon & gold	Big Ten	Glen Mason	5-6
Mississippi	Rebels	Cardinal red & navy	Southeastern	David Cutcliffe	6-5
Mississippi State	Bulldogs	Maroon & white	Southeastern	Jackie Sherrill	8-4
Mississippi Valley	Delta Devils	Green & white	Southwestern	LaTraia Jones	1-10
Missouri	Tigers	Old gold & black	Big Twelve	Larry Smith	7-4
Montana	Grizzlies	Copper, silver, gold	Big Sky	Mick Dennehy	8-3
Montana State	Bobcats	Blue & gold	Big Sky	Cliff Hysell	7-4
Morehead State	Eagles	Blue & gold	Independent	Matt Ballard	9-2
Morgan State	Bears	Blue & orange	Mid-Eastern Athletic	Stanley Mitchell	1-10
Murray State	Racers	Blue & gold	Ohio Valley	Denver Johnson	7-4
Navy	Midshipmen	Navy blue & gold	Independent	Charlie Weatherbie	3-8
Nebraska	Cornhuskers	Scarlet & cream	Big Twelve	Frank Solich	9-3
Nevada	Wolf Pack	Silver & blue	Big West	Jeff Tisdel	6-5
Nev.-Las Vegas (UNLV)	Rebels	Scarlet & gray	Western Athletic	John Robinson	0-11
New Hampshire	Wildcats	Blue & white	Atlantic Ten	Sean McDonnell	4-7
New Mexico	Lobos	Cherry & silver	Western Athletic	Rocky Long	3-9
New Mexico State	Aggies	Crimson & white	Big West	Tony Samuel	3-8
Nicholls St.	Colonels	Red & gray	Southland	Daryl Daye	4-7
North Carolina	Tar Heels	Carolina blue & white	Atlantic Coast	Carl Torbush	6-5
North Carolina A & T	Aggies	Blue & gold	Mid-Eastern Athletic	Bill Hayes	8-3
North Carolina State	Wolfpack	Red & white	Atlantic Coast	Mike O'Cain	7-4
North Texas	Mean Green Eagles	Green & white	Big West	Darrell Dickey	3-8
Northeast Louisiana	Indians	Maroon & gold	Independent	Bobby Keasler	5-6
Northeastern	Huskies	Red & black	Atlantic Ten	Barry Gallup	5-6
Northern Arizona	Lumberjacks	Blue & gold	Big Sky	Jerome Souers	6-5
Northern Illinois	Huskies	Cardinal & black	Mid-American	Joe Novak	2-9
Northern Iowa	Panthers	Purple & old gold	Gateway	Mike Dunbar	7-4

Team	Nickname	Team colors	Conference	Coach	1998 record (W-L)
Northwestern	Wildcats	Purple & white	Big Ten	Randy Walker	3-9
Northwestern State	Demons	Purple, white, & burnt orange	Southland	Sam Goodwin	9-2
Notre Dame	Fighting Irish	Gold & blue	Independent	Bob Davie	9-2
Ohio	Bobcats	Ohio green & white	Mid-American	Jim Grobe	5-6
Ohio State	Buckeyes	Scarlet & gray	Big Ten	John Cooper	10-1
Oklahoma	Sooners	Crimson & cream	Big Twelve	Bob Stoops	5-6
Oklahoma State	Cowboys	Orange & black	Big Twelve	Bob Simmons	5-6
Oregon	Ducks	Green & yellow	Pacific Ten	Mike Bellotti	8-3
Oregon State	Beavers	Orange & black	Pacific Ten	Mike Riley	5-6
Penn State	Nittany Lions	Blue & white	Big Ten	Joe Paterno	8-3
Pennsylvania	Quakers	Red & blue	Ivy League	Al Bagnoli	8-2
Pittsburgh	Panthers	Blue & gold	Big East	Walt Harris	2-9
Princeton	Tigers	Orange & black	Ivy League	Steve Tosches	5-5
Purdue	Boilermakers	Old gold & black	Big Ten	Joe Tiller	8-4
Rhode Island	Rams	Light & dark blue, white	Atlantic Ten	Floyd Keith	3-8
Rice	Owls	Blue & gray	Western Athletic	Ken Hatfield	5-6
Richmond	Spiders	Red & blue	Atlantic Ten	Jim Reid	9-2
Rutgers	Scarlet Knights	Scarlet	Big East	Terry Shea	5-6
Sam Houston State	Bearkats	Orange & white	Southland	Ron Randleman	3-8
Samford	Bulldogs	Crimson & blue	Independent	Pete Hurt	6-5
San Diego State	Aztecs	Scarlet & black	Western Athletic	Ted Tollner	7-4
San Jose State	Spartans	Gold, white, blue	Western Athletic	Dave Baldwin	4-8
South Carolina	Fighting Gamecocks	Garnet & black	Southeastern	Lou Holtz	1-10
South Carolina State	Bulldogs	Garnet & blue	Mid-Eastern Athletic	Willie E. Jeffries	5-6
SE Missouri State	Indians	Red & black	Ohio Valley	John Mumford	3-8
Southern California (USC)	Trojans	Cardinal & gold	Pacific Ten	Paul Hackett	8-4
Southern Illinois	Salukis	Maroon & white	Gateway	Jan Quarless	3-8
Southern Methodist (SMU)	Mustangs	Red & blue	Western Athletic	Mike Cavan	5-7
Southern Mississippi	Golden Eagles	Black & gold	Conference USA	Jeff Bower	7-4
SW Missouri State	Bears	Maroon & white	Gateway	Randy Ball	5-6
SW Texas State	Bobcats	Maroon & gold	Southland	Bob DeBesse	4-7
SW Louisiana	Ragin' Cajuns	Vermilion & white	Independent	Jerry Baldwin	2-9
Stanford	Cardinal	Cardinal & white	Pacific Ten	Tyrone Willingham	3-8
Stephen F. Austin State	Lumberjacks	Purple & white	Southland	Mike Santiago	3-8
Syracuse	Orangemen	Orange	Big East	Paul Pasqualoni	8-3
Temple	Owls	Cherry & white	Big East	Bobby Wallace	2-9
Tennessee	Volunteers	Orange & white	Southeastern	Phillip Fulmer	12-0
Tennessee-Chattanooga	Mocs	Navy blue & gold	Southern	Buddy Green	5-6
Tennessee-Martin	Skyhawks	Orange, white, blue	Ohio Valley	Jim Marshall	0-11
Tennessee State	Tigers	Royal blue & white	Ohio Valley	L. C. Cole	9-2
Tennessee Tech	Golden Eagles	Purple & gold	Ohio Valley	Mike Hennigan	4-7
Texas	Longhorns	Burnt orange & white	Big Twelve	Mack Brown	8-3
Texas A & M	Aggies	Maroon & white	Big Twelve	R. C. Slocum	11-2
Texas Christian (TCU)	Horned Frogs	Purple & white	Western Athletic	Dennis Franchione	6-5
Texas Southern	Tigers	Maroon & gray	Southwestern	Bill Thomas	6-5
Texas Tech	Red Raiders	Scarlet & black	Big Twelve	Spike Dykes	7-4
Toledo	Rockets	Blue & gold	Mid-American	Gary Pinkel	7-5
Troy State	Trojans	Cardinal, gray, black	Southland	Larry Blakeney	8-3
Tulane	Green Wave	Olive green & sky blue	Conference USA	Chris Scelfo	11-0
Tulsa	Golden Hurricane	Blue & gold	Western Athletic	David Rader	4-7
UCLA	Bruins	Blue & gold	Pacific Ten	Bob Toledo	10-1
Utah	Utes	Crimson & white	Western Athletic	Ron McBride	7-4
Utah State	Aggies	Navy blue & white	Big West	Dave Arslanian	3-8
UTEP (Texas-El Paso)	Miners	Orange, blue, white	Western Athletic	Charlie Bailey	3-8
Vanderbilt	Commodores	Black & gold	Southeastern	Woody Widenhofer	2-9
Villanova	Wildcats	Blue & white	Atlantic Ten	Andy Talley	6-5
Virginia	Cavaliers	Orange & blue	Atlantic Coast	George Welsh	9-2
Virginia Military Inst. (VMI)	Keydets	Red, white & yellow	Southern	Cal McCombs	1-10
Virginia Tech	Gobblers, Hokies	Orange & maroon	Big East	Frank Beamer	8-3
Wake Forest	Demon Deacons	Old gold & black	Atlantic Coast	Jim Caldwell	3-8
Washington	Huskies	Purple & gold	Pacific Ten	Rick Neuheisel	6-5
Washington State	Cougars	Crimson & gray	Pacific Ten	Mike Price	3-8
Weber State	Wildcats	Royal purple & white	Big Sky	Jerry Graybeal	6-5
West Virginia	Mountaineers	Old gold & blue	Big East	Don Nehlen	8-3
Western Carolina	Catamounts	Purple & gold	Southern	Bill Bleil	6-5
Western Illinois	Leathernecks	Purple & gold	Gateway	Don Patterson	9-2
Western Kentucky	Hilltoppers	Red & white	Independent	Jack Harbaugh	7-4
Western Michigan	Broncos	Brown & gold	Mid-American	Gary Darnell	7-4
William & Mary	Tribe	Green, gold, silver	Atlantic Ten	Jimmye Laycock	7-4
Wisconsin	Badgers	Cardinal & white	Big Ten	Barry Alvarez	10-1
Wyoming	Cowboys	Brown & yellow	Western Athletic	Dana Dimel	8-3
Yale	Bulldogs, Elis	Yale blue & white	Ivy League	Jack Siedlecki	6-4
Youngstown State	Penguins	Red & white	Gateway	Jim Tressel	6-5

Heisman Trophy Winners

Awarded annually to the nation's outstanding college football player by the Downtown Athletic Club.

1935	Jay Berwanger, Chicago, HB	1946	Glenn Davis, Army, HB	1957	John Crow, Texas A & M, HB
1936	Larry Kelley, Yale, E	1947	John Lujack, Notre Dame, QB	1958	Pete Dawkins, Army, HB
1937	Clinton Frank, Yale, HB	1948	Doak Walker, SMU, HB	1959	Billy Cannon, LSU, HB
1938	David O'Brien, Texas Christian, QB	1949	Leon Hart, Notre Dame, E	1960	Joe Bellino, Navy, HB
1939	Nile Kinnick, Iowa, HB	1950	Vic Janowicz, Ohio St., HB	1961	Ernest Davis, Syracuse, HB
1940	Tom Harmon, Michigan, HB	1951	Richard Kazmaier, Princeton, HB	1962	Terry Baker, Oregon St., QB
1941	Bruce Smith, Minnesota, HB	1952	Billy Vessels, Oklahoma, HB	1963	Roger Staubach, Navy, QB
1942	Frank Sinkwich, Georgia, HB	1953	John Lattner, Notre Dame, HB	1964	John Huarte, Notre Dame, QB
1943	Angelo Bertelli, Notre Dame, QB	1954	Alan Ameche, Wisconsin, FB	1965	Mike Garrett, USC, HB
1944	Leslie Horvath, Ohio St., QB	1955	Howard Cassady, Ohio St., HB	1966	Steve Spurrier, Florida, QB
1945	Felix Blanchard, Army, FB	1956	Paul Hornung, Notre Dame, QB	1967	Gary Beban, UCLA, QB

1968	O. J. Simpson, USC, RB	1979	Charles White, USC, RB	1989	Andre Ware, Houston, QB
1969	Steve Owens, Oklahoma, RB	1980	George Rogers, S. Carolina, RB	1990	Ty Detmer, BYU, QB
1970	Jim Plunkett, Stanford, QB	1981	Marcus Allen, USC, RB	1991	Desmond Howard, Michigan, WR
1971	Pat Sullivan, Auburn, QB	1982	Herschel Walker, Georgia, RB	1992	Gino Torretta, Miami, QB
1972	Johnny Rodgers, Nebraska, RB-WR	1983	Mike Rozier, Nebraska, RB	1993	Charlie Ward, Florida St., QB
1973	John Cappelletti, Penn St., RB	1984	Doug Flutie, Boston College, QB	1994	Rashaan Salaam, Colorado, RB
1974	Archie Griffin, Ohio St., RB	1985	Bo Jackson, Auburn, RB	1995	Eddie George, Ohio St., RB
1975	Archie Griffin, Ohio St., RB	1986	Vinny Testaverde, Miami, QB	1996	Danny Wuerffel, Florida, QB
1976	Tony Dorsett, Pittsburgh, RB	1987	Tim Brown, Notre Dame, WR	1997	Charles Woodson, Michigan, CB
1977	Earl Campbell, Texas, RB	1988	Barry Sanders, Oklahoma St., RB	1998	Ricky Williams, Texas, RB
1978	Billy Sims, Oklahoma, RB				

Outland Award Winners

Honoring the outstanding interior lineman selected by the Football Writers Association of America.

1946	George Connor, Notre Dame, T	1964	Steve Delong, Tennessee, T	1982	Dave Rimington, Nebraska, C
1947	Joe Steffy, Army, G	1965	Tommy Nobis, Texas, G	1983	Dean Steinkuhler, Nebraska, G
1948	Bill Fischer, Notre Dame, G	1966	Loyd Phillips, Arkansas, T	1984	Bruce Smith, Virginia Tech, DT
1949	Ed Bagdon, Michigan St., G	1967	Ron Yary, Southern Cal, T	1985	Mike Ruth, Boston College, NG
1950	Bob Gain, Kentucky, T	1968	Bill Stanfill, Georgia, T	1986	Jason Buck, BYU, DT
1951	Jim Weatherall, Oklahoma, T	1969	Mike Reid, Penn St., DT	1987	Chad Hennings, Air Force, DT
1952	Dick Modzelewski, Maryland, T	1970	Jim Stillwagon, Ohio St., MG	1988	Tracy Rocker, Auburn, DT
1953	J. D. Roberts, Oklahoma, G	1971	Larry Jacobson, Nebraska, DT	1989	Mohammed Elewonibi, BYU, G
1954	Bill Brooks, Arkansas, G	1972	Rich Glover, Nebraska, MG	1990	Russell Maryland, Miami (FL), DT
1955	Calvin Jones, Iowa, G	1973	John Hicks, Ohio St., OT	1991	Steve Emtman, Washington, DT
1956	Jim Parker, Ohio St., G	1974	Randy White, Maryland, DE	1992	Will Shields, Nebraska, G
1957	Alex Karras, Iowa, T	1975	Lee Roy Selmon, Oklahoma, DT	1993	Rob Waldrop, Arizona, NG
1958	Zeke Smith, Auburn, G	1976	Ross Browner, Notre Dame, DE	1994	Zach Wiegert, Nebraska, OT
1959	Mike McGee, Duke, T	1977	Brad Shearer, Texas, DT	1995	Jonathan Ogden, UCLA, OT
1960	Tom Brown, Minnesota, G	1978	Greg Roberts, Oklahoma, G	1996	Orlando Pace, Ohio St., OT
1961	Merlin Olsen, Utah St., T	1979	Jim Ritcher, North Carolina St., C	1997	Aaron Taylor, Nebraska, OT
1962	Bobby Bell, Minnesota, T	1980	Mark May, Pittsburgh, OT	1998	Kris Farris, UCLA, OT
1963	Scott Appleton, Texas, T	1981	Dave Rimington, Nebraska, C		

All-Time Division I-A Percentage Leaders

(Classified as Division I-A for the last 10 years; record includes bowl games; ties computed as half won and half lost)

					Bowl Games**									Bowl Games**		
	Years	Won	Lost	Tied	Pct.	W	L	T		Years	Won	Lost	Tied	Pct.	W	L	T
Notre Dame	110	762	231	42	.757	13	10	0	Central Michigan . .	98	506	285	36	.634	0	2	0
Michigan.	119	786	257	36	.745	15	15	0	Georgia	105	625	354	54	.631	17	14	3
Alabama*	104	724	265	43	.722	28	18	3	LSU	105	607	348	47	.629	14	16	1
Ohio St.	109	710	277	53	.708	14	17	0	Arizona St.	86	478	278	24	.628	10	6	1
Texas	106	726	294	33	.705	18	18	2	Army	109	614	356	51	.626	2	2	0
Nebraska	109	731	296	40	.704	18	19	0	Auburn*	106	596	355	47	.621	14	10	2
Oklahoma	104	682	273	53	.703	20	11	1	Colorado	109	601	363	36	.619	10	12	0
Penn St.	112	724	302	41	.698	22	11	2	Miami (FL)	72	452	277	19	.617	12	11	0
Tennessee*	102	690	285	52	.697	22	17	0	Florida	92	545	340	40	.611	13	13	0
USC	106	667	275	54	.697	25	14	0	UCLA	80	474	301	37	.607	11	10	1
Florida St.*	52	369	183	17	.663	16	9	2	Texas A&M	104	594	377	48	.606	12	12	0
Miami (OH) *	110	584	323	44	.637	5	2	0	Syracuse	109	625	401	49	.604	10	8	1
Washington*	109	599	331	50	.637	13	11	1									

*Includes games that were forfeited or changed by action of NCAA Council and/or Committee on Infractions. **Includes major bowl games only; that is, those where team's opponent was classified as a major college team that season or at the time of the bowl game.

College Football Coach of the Year

The Division I-A Coach of the Year has been selected by the American Football Coaches Assn. since 1935 and selected by the Football Writers Assn. of America since 1957. When polls disagree, both winners are indicated.

1935	Lynn Waldorf, Northwestern		Darrell Royal, Texas (FWAA)	1978	Joe Paterno, Penn St.
1936	Dick Harlow, Harvard	1962	John McKay, USC	1979	Earle Bruce, Ohio St.
1937	Edward Mylin, Lafayette	1963	Darrell Royal, Texas	1980	Vince Dooley, Georgia
1938	Bill Kern, Carnegie Tech	1964	Ara Parseghian, Notre Dame, &	1981	Danny Ford, Clemson
1939	Eddie Anderson, Iowa		Frank Broyles, Arkansas (AFCA);	1982	Joe Paterno, Penn St.
1940	Clark Shaughnessy, Stanford		Ara Parseghian (FWAA)	1983	Ken Hatfield, Air Force (AFCA);
1941	Frank Leahy, Notre Dame	1965	Tommy Prothro, UCLA (AFCA);		Howard Schnellenberger, Miami (FL)
1942	Bill Alexander, Georgia Tech		Duffy Daugherty, Mich. St. (FWAA)		(FWAA)
1943	Amos Alonzo Stagg, Pacific	1966	Tom Cahill, Army	1984	LaVell Edwards, Brigham Young
1944	Carroll Widdoes, Ohio St.	1967	John Pont, Indiana	1985	Fisher De Berry, Air Force
1945	Bo McMillin, Indiana	1968	Joe Paterno, Penn St. (AFCA);	1986	Joe Paterno, Penn St.
1946	Earl "Red" Blaik, Army		Woody Hayes, Ohio St. (FWAA)	1987	Dick MacPherson, Syracuse
1947	Fritz Crisler, Michigan	1969	Bo Schembechler, Michigan	1988	Don Nehlen, W. Virginia (AFCA);
1948	Bennie Oosterbaan, Michigan	1970	Charles McClendon, LSU, &		Lou Holtz, Notre Dame (FWAA)
1949	Bud Wilkinson, Oklahoma		Darrell Royal, Texas (AFCA);	1989	Bill McCartney, Colorado
1950	Charlie Caldwell, Princeton		Alex Agase, Northwestern (FWAA)	1990	Bobby Ross, Georgia Tech
1951	Chuck Taylor, Stanford	1971	Paul "Bear" Bryant, Alabama (AFCA);	1991	Don James, Washington
1952	Biggie Munn, Michigan St.		Bob Devaney, Nebraska (FWAA)	1992	Gene Stallings, Alabama
1953	Jim Tatum, Maryland	1972	John McKay, USC	1993	Barry Alvarez, Wisconsin (AFCA);
1954	Henry "Red" Sanders, UCLA	1973	Paul "Bear" Bryant, Alabama (AFCA);		Terry Bowden, Auburn (FWAA)
1955	Duffy Daugherty, Michigan St.		Johnny Majors, Pittsburgh (FWAA)	1994	Tom Osborne, Nebraska (AFCA)
1956	Bowden Wyatt, Tennessee	1974	Grant Teaff, Baylor		Rich Brooks, Oregon (FWAA)
1957	Woody Hayes, Ohio St.	1975	Frank Kush, Arizona St. (AFCA);	1995	Gary Barnett, Northwestern
1958	Paul Dietzel, LSU		Woody Hayes, Ohio St. (FWAA)	1996	Bruce Snyder, Arizona St.
1959	Ben Schwartzwalder, Syracuse	1976	Johnny Majors, Pittsburgh	1997	Mike Price, Washington St.
1960	Murray Warmath, Minnesota	1977	Don James, Washington (AFCA);	1998	Phillip Fulmer, Tennessee
1961	Paul "Bear" Bryant, Ala. (AFCA)		Lou Holtz, Arkansas (FWAA)		

All-Time Division I-A Coaching Victories (Including Bowl Games)

Paul "Bear" Bryant	323	Warren Woodson	203	Johnny Majors	185
Glenn "Pop" Warner	319	Eddie Anderson	201	Darrell Royal	184
Amos Alonzo Stagg	314	Vince Dooley	201	Gil Dobie	180
*Joe Paterno	307	Jim Sweeney	200	Carl Snavely	180
*Bobby Bowden	292	Dana X. Bible	198	Jerry Claiborne	179
Tom Osborne	255	Dan McGugin	197	Ben Schwartzwalder	178
*LaVell Edwards	243	Fielding Yost	196	*John Cooper	178
Woody Hayes	238	Howard Jones	194	Frank Kush	176
Bo Schembechler	234	*Don Nehlen	191	Don James	176
Hayden Fry	232	John Vaught	190	Ralph Jordan	176
Lou Holtz	216	John Heisman	185	*George Welsh	176
Jess Neely	207				

Coaches active in 1998 are denoted by an asterisk (*). Eddie Robinson of Grambling State Univ. (Div. I-AA) holds the record for most college football victories, with 408 at the end of the 1997 season; Robinson retired after the 1997 season.

Selected College Football Conference Champions

Atlantic Coast
1980	North Carolina
1981	Clemson
1982	Clemson
1983	Maryland
1984	Maryland
1985	Maryland
1986	Clemson
1987	Clemson
1988	Clemson
1989	Virginia, Duke
1990	Georgia Tech
1991	Clemson
1992	Florida St.
1993	Florida St.
1994	Florida St.
1995	Virginia, Florida St.
1996	Florida St.
1997	Florida St.
1998	Florida St., Georgia Tech

Ivy Group
1980	Yale
1981	Yale, Dartmouth
1982	Harvard, Dartmouth, Penn
1983	Harvard, Penn
1984	Penn
1985	Penn
1986	Penn
1987	Harvard
1988	Penn, Cornell
1989	Yale, Princeton
1990	Cornell, Dartmouth
1991	Dartmouth
1992	Dartmouth, Princeton
1993	Penn
1994	Penn
1995	Princeton
1996	Dartmouth
1997	Harvard
1998	Penn

Big Eight*
1980	Oklahoma
1981	Nebraska
1982	Nebraska
1983	Nebraska
1984	Nebraska, Oklahoma
1985	Oklahoma
1986	Oklahoma
1987	Oklahoma
1988	Nebraska
1989	Colorado
1990	Colorado
1991	Nebraska, Colorado
1992	Nebraska
1993	Nebraska
1994	Nebraska
1995	Nebraska

Big Ten
1980	Michigan
1981	Iowa, Ohio St.
1982	Michigan
1983	Illinois
1984	Ohio St.
1985	Iowa
1986	Michigan, Ohio St.
1987	Michigan St.
1988	Michigan
1989	Michigan
1990	Iowa, Ill., Mich., Mich. St.
1991	Michigan
1992	Michigan
1993	Ohio St., Wisconsin
1994	Penn St.
1995	Northwestern
1996	Ohio St., Northwestern
1997	Michigan
1998	Ohio St., Wisconsin, Michigan

Mid-American Athletic
1980	Central Michigan
1981	Toledo
1982	Bowling Green
1983	Northern Illinois
1984	Toledo
1985	Bowling Green
1986	Miami (OH)
1987	E. Michigan
1988	W. Michigan
1989	Ball St.
1990	Central Michigan
1991	Bowling Green
1992	Bowling Green
1993	Ball St.
1994	Central Michigan
1995	Toledo
1996	Ball St.
1997	Marshall
1998	Marshall

Southern
1980	Furman
1981	Furman
1982	Furman
1983	Furman
1984	Tenn.-Chattanooga
1985	Furman
1986	Appalachian St.
1987	Appalachian St.
1988	Marshall, Furman
1989	Furman
1990	Furman
1991	Appalachian St.
1992	Citadel
1993	Georgia Southern
1994	Marshall
1995	Appalachian St.
1996	Marshall
1997	Georgia Southern
1998	Georgia Southern

Southeastern
1980	Georgia
1981	Georgia, Alabama
1982	Georgia
1983	Auburn
1984	Florida (title vacated)
1985	Tennessee
1986	LSU
1987	Auburn
1988	Auburn, LSU
1989	Ala., Tenn., Auburn
1990	Tennessee
1991	Florida
1992	Alabama
1993	Florida
1994	Florida
1995	Florida
1996	Florida
1997	Tennessee
1998	Tennessee

Southwest*
1980	Baylor
1981	Texas
1982	SMU
1983	Texas
1984	SMU, Houston
1985	Texas A&M
1986	Texas A&M
1987	Texas A&M
1988	Arkansas
1989	Arkansas
1990	Texas
1991	Texas A&M
1992	Texas A&M
1993	Texas A&M
1994	Baylor, Rice, Texas, Texas Christian, Texas Tech
1995	Texas

Pacific Ten
1980	Washington
1981	Washington
1982	UCLA
1983	UCLA
1984	USC
1985	UCLA
1986	Arizona St.
1987	UCLA, USC
1988	USC
1989	USC
1990	Washington
1991	Washington
1992	Washington, Stanford
1993	UCLA, Arizona, USC
1994	Oregon
1995	USC, Washington
1996	Arizona St.
1997	Washington St., UCLA
1998	UCLA

Big East
1991	Miami (FL), Syracuse
1992	Miami (FL)
1993	West Virginia
1994	Miami (FL)
1995	Virginia Tech, Miami (FL)
1996	Virginia Tech, Miami (FL), Syracuse
1997	Syracuse
1998	Syracuse

Western Athletic
1980	Brigham Young (BYU)
1981	Brigham Young
1982	Brigham Young
1983	Brigham Young
1984	Brigham Young
1985	Brigham Young, Air Force
1986	San Diego St.
1987	Wyoming
1988	Wyoming
1989	Brigham Young
1990	Brigham Young
1991	Brigham Young
1992	Hawaii, Brigham Young, Fresno St.
1993	Wyoming, Fresno St., BYU
1994	Colorado St.
1995	Colorado St., Air Force, Utah, BYU
1996	Brigham Young
1997	Colorado St.
1998	Air Force

Big 12*
1996	Texas
1997	Nebraska
1998	Texas A&M

Big West
1980	Long Beach St.
1981	San Jose St.
1982	Fresno St.
1983	Cal St.-Fullerton
1984	Cal St.-Fullerton
1985	Fresno St.
1986	San Jose St.
1987	San Jose St.
1988	Fresno St.
1989	Fresno St.
1990	San Jose St.
1991	San Jose St., Fresno St.
1992	Nevada
1993	SW Louisiana, Utah St.
1994	Nevada, SW Louisiana, UNLV
1995	Nevada
1996	Nevada, Utah St.
1997	Nevada, Utah St.
1998	Idaho

Conference USA
1996	So. Mississippi, Houston
1997	So. Mississippi
1998	Tulane

(*) After the 1995 season, the Big Eight and Southwest conferences disbanded. In 1996 all former Big Eight Conference teams joined with 4 of the 8 Southwest Conference teams to form the Big 12 Conference.

NATIONAL HOCKEY LEAGUE
1998-99 Review: "Great One" Retires, Dallas Wins Cup, NHL in Atlanta

On Apr. 18, 1999, Wayne Gretzky, regarded as the best-ever player in the NHL, appeared in his last professional game. He was unanimously elected to the Hockey Hall of Fame on June 23, after the usual 3-year wait had been waived.

The Dallas Stars defeated the Buffalo Sabres 2-1, June 19, to take the Stanley Cup. Brett Hull scored a controversial triple overtime goal to win the decisive game. It appeared from instant replay that his foot had been in the crease around Dominik Hasek's goal when he scored. On June 21, the NHL announced that there would no longer be video reviews for crease violations. Another rule change will give both teams in a tie 1 point, regardless of the overtime outcome. A team that wins in overtime gets an additional point. There also will be only 4 skaters (and a goalie) for each team in regular season overtime games.

The Atlanta Thrashers (enfranchised on June 25, 1997) will be the second expansion team to be added in as many years. The Thrashers held their expansion draft on June 25, 1999, and will play in the Southeast Division of the Eastern Conference.

Final Standings 1998-99

(playoff seeding in parentheses; in each conference the three division winners automatically get the number 1, 2, and 3 seeds)

Eastern Conference

Northeast Division

	W	L	T	GF	GA	Pts
Ottawa (2)	44	23	15	239	179	103
Toronto (4)	45	30	7	268	231	97
Boston (6)	39	30	13	214	181	91
Buffalo (7)	37	28	17	207	175	91
Montreal	32	39	11	184	209	75

Atlantic Division

	W	L	T	GF	GA	Pts
New Jersey (1)	47	24	11	248	197	105
Philadelphia (5)	37	26	19	231	196	93
Pittsburgh (8)	38	30	14	242	225	90
N.Y. Rangers	33	38	11	217	227	77
N.Y. Islanders	24	48	10	194	244	58

Southeast Division

	W	L	T	GF	GA	Pts
Carolina (3)	34	30	18	210	202	86
Florida	30	34	18	210	228	78
Washington	31	45	6	200	218	68
Tampa Bay	19	54	9	179	292	47

Western Conference

Central Division

	W	L	T	GF	GA	Pts
Detroit (3)	43	32	7	245	202	93
St. Louis (5)	37	32	13	237	209	87
Chicago	29	41	12	202	248	70
Nashville	28	47	7	190	261	63

Northwest Division

	W	L	T	GF	GA	Pts
Colorado (2)	44	28	10	239	205	98
Edmonton (8)	33	37	12	230	226	78
Calgary	30	40	12	211	234	72
Vancouver	23	47	12	192	258	58

Pacific Division

	W	L	T	GF	GA	Pts
Dallas (1)	51	19	12	236	168	114
Phoenix (4)	39	31	12	205	197	90
Anaheim (6)	35	34	13	215	206	83
San Jose (7)	31	33	18	197	191	80
Los Angeles	32	45	5	189	222	69

1999 Stanley Cup Playoff Results

Eastern Conference

Pittsburgh defeated New Jersey 4 games to 3
Buffalo defeated Ottawa 4 games to 0
Boston defeated Carolina 4 games to 2
Toronto defeated Philadelphia 4 games to 2
Toronto defeated Pittsburgh 4 games to 2
Buffalo defeated Boston 4 games to 2
Buffalo defeated Toronto 4 games to 1

Western Conference

Dallas defeated Edmonton 4 games to 0
Colorado defeated San Jose 4 games to 2
Detroit defeated Anaheim 4 games to 0
St. Louis defeated Phoenix 4 games to 3
Dallas defeated St. Louis 4 games to 2
Colorado defeated Detroit 4 games to 2
Dallas defeated Colorado 4 games to 3

Finals

Dallas defeated Buffalo 4 games to 2 [2-3 (OT), 4-2, 2-1, 1-2, 2-0, 2-1 (3 OT)]

Stanley Cup Champions Since 1927

Year	Champion	Coach	Final opponent	Year	Champion	Coach	Final opponent
1927	Ottawa	Dave Gill	Boston	1964	Toronto	Punch Imlach	Detroit
1928	N.Y. Rangers	Lester Patrick	Montreal	1965	Montreal	Toe Blake	Chicago
1929	Boston	Cy Denneny	N.Y. Rangers	1966	Montreal	Toe Blake	Detroit
1930	Montreal	Cecil Hart	Boston	1967	Toronto	Punch Imlach	Montreal
1931	Montreal	Cecil Hart	Chicago	1968	Montreal	Toe Blake	St. Louis
1932	Toronto	Dick Irvin	N.Y. Rangers	1969	Montreal	Claude Ruel	St. Louis
1933	N.Y. Rangers	Lester Patrick	Toronto	1970	Boston	Harry Sinden	St. Louis
1934	Chicago	Tommy Gorman	Detroit	1971	Montreal	Al MacNeil	Chicago
1935	Montreal Maroons	Tommy Gorman	Toronto	1972	Boston	Tom Johnson	N.Y. Rangers
1936	Detroit	Jack Adams	Toronto	1973	Montreal	Scotty Bowman	Chicago
1937	Detroit	Jack Adams	N.Y. Rangers	1974	Philadelphia	Fred Shero	Boston
1938	Chicago	Bill Stewart	Toronto	1975	Philadelphia	Fred Shero	Buffalo
1939	Boston	Art Ross	Toronto	1976	Montreal	Scotty Bowman	Philadelphia
1940	N.Y. Rangers	Frank Boucher	Toronto	1977	Montreal	Scotty Bowman	Boston
1941	Boston	Cooney Weiland	Detroit	1978	Montreal	Scotty Bowman	Boston
1942	Toronto	Hap Day	Detroit	1979	Montreal	Scotty Bowman	N.Y. Rangers
1943	Detroit	Jack Adams	Boston	1980	N.Y. Islanders	Al Arbour	Philadelphia
1944	Montreal	Dick Irvin	Chicago	1981	N.Y. Islanders	Al Arbour	Minnesota
1945	Toronto	Hap Day	Detroit	1982	N.Y. Islanders	Al Arbour	Vancouver
1946	Montreal	Dick Irvin	Boston	1983	N.Y. Islanders	Al Arbour	Edmonton
1947	Toronto	Hap Day	Montreal	1984	Edmonton	Glen Sather	N.Y. Islanders
1948	Toronto	Hap Day	Detroit	1985	Edmonton	Glen Sather	Philadelphia
1949	Toronto	Hap Day	Detroit	1986	Montreal	Jean Perron	Calgary
1950	Detroit	Tommy Ivan	N.Y. Rangers	1987	Edmonton	Glen Sather	Philadelphia
1951	Toronto	Joe Primeau	Montreal	1988	Edmonton	Glen Sather	Boston
1952	Detroit	Tommy Ivan	Montreal	1989	Calgary	Terry Crisp	Montreal
1953	Montreal	Dick Irvin	Boston	1990	Edmonton	John Muckler	Boston
1954	Detroit	Tommy Ivan	Montreal	1991	Pittsburgh	Bob Johnson	Minnesota
1955	Detroit	Jimmy Skinner	Montreal	1992	Pittsburgh	Scotty Bowman	Chicago
1956	Montreal	Toe Blake	Detroit	1993	Montreal	Jacques Demers	Los Angeles
1957	Montreal	Toe Blake	Boston	1994	N.Y. Rangers	Mike Keenan	Vancouver
1958	Montreal	Toe Blake	Boston	1995	New Jersey	Jacques Lemaire	Detroit
1959	Montreal	Toe Blake	Toronto	1996	Colorado	Marc Crawford	Florida
1960	Montreal	Toe Blake	Toronto	1997	Detroit	Scotty Bowman	Philadelphia
1961	Chicago	Rudy Pilous	Detroit	1998	Detroit	Scotty Bowman	Washington
1962	Toronto	Punch Imlach	Chicago	1999	Dallas	Ken Hitchcock	Buffalo
1963	Toronto	Punch Imlach	Detroit				

Individual Leaders, 1998-99

Points
Jaromir Jagr, Pittsburgh, 127; Teemu Selanne, Anaheim, 107; Paul Kariya, Anaheim, 101; Peter Forsberg, Colorado, 97; Joe Sakic, Colorado, 96.

Goals
Teemu Selanne, Anaheim, 47; Tony Amonte, Chicago, 44; Jaromir Jagr, Pittsburgh, 44; Alexei Yashin, Ottawa, 44; John LeClair, Philadelphia, 43; Joe Sakic, Colorado, 41; Eric Lindros, Philadelphia, 40; Miroslav Satan, Buffalo, 40; Theoren Fleury, Cgy.-Col., 40.

Assists
Jaromir Jagr, Pittsburgh, 83; Peter Forsberg, Colorado, 67; Paul Kariya, Anaheim, 62; Teemu Selanne, Anaheim, 60; Joe Sakic, Colorado, 55.

Power-play goals
Teemu Selanne, Anaheim, 25; Alexei Yashin, Ottawa, 19; Adrian Aucoin, Vancouver, 18; John LeClair, Philadelphia, 16; Brett Hull, Dallas, 15; Markus Nashlund, Vancouver, 15; Petr Sykora, New Jersey, 15.

Shorthanded goals
Scott Pellerin, St. Louis, 5; Joe Sakic, Colorado, 5; Brian Rolston, New Jersey, 5; Magnus Arvedson, Ottawa, 4; Radek Dvorak, Florida, 4; Mike Modano, Dallas, 4; Martin Straka, Pittsburgh, 4.

Shooting percentage
(minimum 82 shots)
Dmitri Khristich, Boston, 20.1; Martin Straka, Pittsburgh, 19.8; Dixon Ward, Buffalo, 19.8; Anson Carter, 19.5; Miroslav Satan, Buffalo, 19.2.

Plus/Minus
Alexander Karpovtsev, N.Y.R.-Tor., 39; John LeClair, Philadelphia, 36; Eric Lindros, Philadelphia, 35; Magnus Arvedson, Ottawa, 33; Al MacInnis, St. Louis, 33.

Penalty minutes
Rob Ray, Buffalo, 261; Jeff Odgers, Colorado, 259; Peter Worrell, Florida, 258; Patrick Cote, Nashville, 242; Krzysztof Oliwa, N.J.-Nash., 240.

Goaltending Leaders
(minimum 25 games)

Goals against average
Ron Tugnutt, Ottawa, 1.79; Dominik Hasek, Buffalo, 1.87; Ed Belfour, Dallas, 1.98; Byron Dafoe, Boston, 1.99; Roman Turek, Dallas, 2.08.

Wins
Martin Brodeur, New Jersey, 39; Ed Belfour, Dallas, 35; Curtis Joseph, Toronto, 35; Chris Osgood, Detroit, 34; Byron Dafoe, Boston, 32; Nikolai Khabibulin, Phoenix, 32; Patrick Roy, Colorado, 32.

Save percentage
Dominik Hasek, Buffalo, .936; Byron Dafoe, Boston, .926; Ron Tugnutt, Ottawa, .925; Arturs Irbe, Carolina, .922; Nikolai Khabibulin, Phoenix, .922; Guy Hebert, Anaheim, .921.

Shutouts
Byron Dafoe, Boston, 10; Dominik Hasek, Buffalo, 9; Nikolai Khabibulin, Phoenix, 8; Guy Hebert, Anaheim, 6; Arturs Irbe, Carolina, 6; Garth Snow, Vancouver, 6; John Vanbiesbrouck, Philadelphia, 6.

Individual Scoring, 1998-99

(40 or more games played; *played for more than one team during 1998-99; g = goalie; E = even)

Mighty Ducks of Anaheim

PLAYER	GP	G	A	PTS	+/-	PIM
Teemu Selanne	75	47	60	107	18	30
Paul Kariya	82	39	62	101	17	40
Steve Rucchin	69	23	39	62	11	22
Fredrik Olausson	74	16	40	56	17	30
Marty McInnis*	81	19	35	54	-15	42
Tomas Sandstrom	58	15	17	32	-5	42
Travis Green	79	13	17	30	-7	81
Matt Cullen	75	11	14	25	-12	47
Ruslan Salei	74	2	14	16	1	65
Ted Drury	75	5	6	11	2	83
Jim McKenzie	73	5	4	9	-18	99
Jeff Nielsen	80	5	4	9	-12	34
Johan Davidsson	64	3	5	8	-9	14
Antti Aalto	73	3	5	8	-12	24
Jason Marshall	72	1	7	8	-5	142
Kevin Haller	82	1	6	7	-1	122
Pascal Trepanier	45	2	4	6	E	48
Pavel Trnka	63	0	4	4	-6	60
Stu Grimson	73	3	0	3	E	158
Jamie Pushor	70	1	2	3	-20	112
Guy Hebert (g)	69	0	1	1	-	0
Coach—Craig Hartsburg						

Boston Bruins

PLAYER	GP	G	A	PTS	+/-	PIM
Jason Allison	82	23	53	76	5	68
Dmitri Khristich	79	29	42	71	11	48
Ray Bourque	81	10	47	57	-7	34
Sergei Samsonov	79	25	26	51	-6	18
Joe Thornton	81	16	25	41	3	69
Anson Carter	55	24	16	40	7	22
Steve Heinze	73	22	18	40	7	30
Kyle McLaren	52	6	18	24	1	48
Rob DiMaio	71	7	14	21	-14	95
Darren Van Impe	60	5	15	20	-5	66
P.J. Axelsson	77	7	10	17	-14	18
Peter Ferraro	46	6	8	14	10	44
Grant Ledyard	47	4	8	12	-8	33
Don Sweeney	81	2	10	12	14	64
Tim Taylor	49	4	7	11	-10	55
Hal Gill	80	3	7	10	-10	63
Ken Belanger*	54	2	5	7	-1	182
Dave Ellett	54	0	6	6	11	25
Ken Baumgartner	69	1	3	4	-6	119
Byron Dafoe (g)	68	0	2	2	-	25
Coach—Pat Burns						

Buffalo Sabres

PLAYER	GP	G	A	PTS	+/-	PIM
Miroslav Satan	81	40	26	66	24	44
Michael Peca	82	27	29	56	7	81
Michal Grosek	76	20	30	50	21	102
Curtis Brown	78	16	31	47	23	56
Dixon Ward	78	20	24	44	10	44
Joe Juneau*	72	15	28	43	-4	22
Jason Woolley	80	10	33	43	16	62
Stu Barnes*	81	20	16	36	-11	30
Brian Holzinger	81	17	17	34	2	45
Alexei Zhitnik	81	7	26	33	-6	96
Vaclav Varada	72	7	24	31	11	61
Geoff Sanderson	75	12	18	30	8	22
Darryl Shannon	71	3	12	15	28	52
Richard Smehlik	72	3	11	14	-9	44
Wayne Primeau	67	5	8	13	-6	38
Erik Rasmussen	42	3	7	10	6	37
James Patrick	45	1	7	8	12	16
Rhett Warrener*	61	1	7	8	2	84
Jay McKee	72	0	6	6	20	75
Rob Ray	76	0	4	4	-2	261
Paul Kruse	43	3	0	3	E	114
Dominik Hasek (g)	64	0	0	0	-	14
Coach—Lindy Ruff						

Calgary Flames

PLAYER	GP	G	A	PTS	+/-	PIM
Cory Stillman	76	27	30	57	7	38
Phil Housley	79	11	43	54	14	52
Valeri Bure	80	26	27	53	E	22
Jarome Iginla	82	28	23	51	1	58
Andrew Cassels	70	12	25	37	-12	18
Derek Morris	71	7	27	34	4	73
Rene Corbet*	73	13	18	31	1	68
Jeff Shantz*	76	13	17	30	14	44
Jason Wiemer	78	8	13	21	-12	177
Clarke Wilm	78	10	8	18	11	53
Andrei Nazarov*	62	7	9	16	-4	73
Steve Smith	69	1	14	15	3	80
Steve Dubinsky*	62	4	10	14	-7	14
Cale Hulse	73	3	9	12	-8	117
Hnat Domenichelli	23	5	5	10	-4	11
Todd Simpson	73	2	8	10	18	151
Ed Ward	68	3	5	8	-4	67
Denis Gauthier	55	3	4	7	3	68
Tommy Albelin	60	1	5	6	-11	8
Bob Bassen	41	1	2	3	-13	35
Coach—Brian Sutter						

Carolina Hurricanes

PLAYER	GP	G	A	PTS	+/-	PIM
Keith Primeau	78	30	32	62	8	75
Sami Kapanen	81	24	35	59	-1	10
Ray Sheppard	74	25	33	58	4	16
Ron Francis	82	21	31	52	-2	34
Gary Roberts	77	14	28	42	2	178
Andrei Kovalenko*	74	19	21	40	-6	32
Jeff O'Neill	75	16	15	31	3	66
Martin Gelinas	76	13	15	28	3	67

PLAYER	GP	G	A	PTS	+/-	PIM
Robert Kron	75	9	16	25	-13	10
Glen Wesley	74	7	17	24	14	44
Paul Ranheim	78	9	10	19	4	39
Kevin Dineen	67	8	10	18	5	97
Jon Battaglia	60	7	11	18	7	22
Kent Manderville	81	5	11	16	9	38
Nolan Pratt*	61	1	14	15	15	95
Paul Coffey*	54	2	12	14	-7	28
Marek Malik	52	2	9	11	-6	36
Sean Hill	54	0	10	10	9	48
Curtis Leschyshyn	65	2	7	9	-1	50
Arturs Irbe (g)	62	0	0	0	-	10

Coach—Paul Maurice

Chicago Blackhawks

PLAYER	GP	G	A	PTS	+/-	PIM
Tony Amonte	82	44	31	75	E	60
Alexei Zhamnov	76	20	41	61	-10	50
Doug Gilmour	72	16	40	56	-16	56
Boris Mironov*	75	11	38	49	13	131
Eric Daze	72	22	20	42	-13	22
Dean McAmmond*	77	10	20	30	8	38
Ed Olczyk	61	10	15	25	-3	29
Dave Manson*	75	6	17	23	1	155
Bob Probert	78	7	14	21	-11	206
Anders Eriksson	72	2	18	20	11	34
Doug Zmolek	62	0	14	14	1	102
Reid Simpson	53	5	4	9	2	145
Jean-Yves Leroux	40	3	5	8	-7	21
Brad Brown*	66	1	7	8	-4	205
Chris Murray*	42	1	6	7	-2	79
Bryan Muir*	54	1	4	5	1	50
Mark Janssens	60	1	0	1	-11	65
Jocelyn Thibault* (g)	62	0	1	1	-	2

Coach—Dirk Graham, Lorne Molleken

Colorado Avalanche

PLAYER	GP	G	A	PTS	+/-	PIM
Peter Forsberg	78	30	67	97	27	108
Joe Sakic	73	41	55	96	23	29
Theoren Fleury*	75	40	53	93	26	86
Claude Lemieux	82	27	24	51	E	102
Adam Deadmarsh	66	22	27	49	-2	99
Milan Hejduk	82	14	34	48	8	26
Chris Drury	79	20	24	44	9	62
Valeri Kamensky	65	14	30	44	1	28
Adam Foote	64	5	16	21	20	92
Sylvain Lefebvre	76	2	18	20	18	48
Shean Donovan	68	7	12	19	4	37
Aaron Miller	76	5	13	18	3	42
Stephane Yelle	72	8	7	15	-8	40
Alexei Gusarov	54	3	10	13	12	24
Dale Hunter*	62	2	9	11	-7	119
Shjon Podein*	55	3	6	9	-5	24
Jeff Odgers	75	2	3	5	-3	259
Greg de Vries*	73	1	3	4	-7	64
Cam Russell*	42	1	2	3	-4	94
Patrick Roy (g)	61	0	2	2	-	28

Coach—Bob Hartley

Dallas Stars

PLAYER	GP	G	A	PTS	+/-	PIM
Mike Modano	77	34	47	81	29	44
Brett Hull	60	32	26	58	19	30
Joe Nieuwendyk	67	28	27	55	11	34
Jere Lehtinen	74	20	32	52	29	18
Sergei Zubov	81	10	41	51	9	20
Darryl Sydor	74	14	34	48	-1	50
Jamie Langenbrunner	75	12	33	45	10	62
Pat Verbeek	78	17	17	34	11	133
Grant Marshall	82	13	18	31	1	85
Derian Hatcher	80	9	21	30	21	102
Benoit Hogue*	74	12	17	29	-10	54
Mike Keane	81	6	23	29	-2	62
Tony Hrkac	69	13	14	27	2	26
Derek Plante*	51	6	14	20	4	16
Dave Reid	73	6	11	17	6	16
Guy Carbonneau	74	4	12	16	-3	31
Richard Matvichuk	64	3	9	12	23	51
Shawn Chambers	61	2	9	11	6	18
Craig Ludwig	80	2	6	8	5	87
Brian Skrudland	40	4	1	5	2	33
Ed Belfour (g)	61	0	0	0	-	26

Coach—Ken Hitchcock

Detroit Red Wings

PLAYER	GP	G	A	PTS	+/-	PIM
Steve Yzerman	80	29	45	74	8	42
Sergei Fedorov	77	26	37	63	9	66
Igor Larionov	75	14	49	63	13	48
Brendan Shanahan	81	31	27	58	2	123

PLAYER	GP	G	A	PTS	+/-	PIM
Vyacheslav Kozlov	79	29	29	58	10	45
Nicklas Lidstrom	81	14	43	57	14	14
Larry Murphy	80	10	42	52	21	42
Wendel Clark*	77	32	16	48	-24	37
Darren McCarty	69	14	26	40	10	108
Chris Chelios*	75	9	27	36	1	93
Tomas Holmstrom	82	13	21	34	-11	69
Martin Lapointe	77	16	13	29	7	141
Doug Brown	80	9	19	28	5	42
Kris Draper	80	4	14	18	2	79
Kirk Maltby	53	8	6	14	-6	34
Mathieu Dandenault	75	4	10	14	17	59
Stacy Roest	59	4	8	12	-7	14
Ulf Samuelsson*	71	4	8	12	5	99
Aaron Ward	60	3	8	11	-5	52
Jamie Macoun	69	1	10	11	-1	36
Todd Gill	51	4	5	9	-10	27
Chris Osgood (g)	63	0	3	3	-	8

Coach—Scotty Bowman

Edmonton Oilers

PLAYER	GP	G	A	PTS	+/-	PIM
Bill Guerin	80	30	34	64	7	133
Josef Beranek	66	19	30	49	6	23
Mike Grier	82	20	24	44	5	54
Pat Falloon	82	17	23	40	-4	20
Rem Murray	78	21	18	39	4	20
Doug Weight	43	6	31	37	-8	12
Todd Marchant	82	14	22	36	3	65
Alexander Selivanov*	72	14	19	33	-8	40
Roman Hamrlik	75	8	24	32	9	70
Ryan Smyth	71	13	18	31	E	62
Janne Niinimaa	81	4	24	28	7	88
Chad Kilger*	77	15	12	27	-4	34
Ethan Moreau*	80	10	11	21	-3	92
Tom Poti	73	5	16	21	10	42
Jason Smith*	72	3	12	15	-9	51
Boyd Devereaux	61	6	8	14	2	23
Christian Laflamme*	73	2	12	14	-3	70
Kelly Buchberger	52	4	4	8	-6	68
Sean Brown*	51	0	7	7	1	188
Marty McSorley	46	2	3	5	-5	101
Tommy Salo* (g)	64	0	0	0	-	0

Coach—Ron Low

Florida Panthers

PLAYER	GP	G	A	PTS	+/-	PIM
Ray Whitney	81	26	38	64	-3	18
Rob Niedermayer	82	18	33	51	-13	50
Viktor Kozlov	65	16	35	51	13	24
Scott Mellanby	67	18	27	45	5	85
Radek Dvorak	82	19	24	43	7	29
Mark Parrish	73	24	13	37	-6	25
Robert Svehla	80	8	29	37	-13	83
Bill Lindsay	75	12	15	27	-1	92
Oleg Kvasha	68	12	13	25	5	45
Bret Hedican*	67	5	18	23	5	51
Johan Garpenlov	64	8	9	17	-9	42
Kirk Muller	82	4	11	15	-11	49
Jaroslav Spacek	63	3	12	15	15	28
Terry Carkner	62	2	9	11	E	54
Paul Laus	75	1	9	10	-1	218
Peter Worrell	62	4	5	9	E	258
Gord Murphy	51	0	7	7	4	16
Alex Hicks*	55	0	7	7	-5	62
Sean Burke (g)	59	0	4	4	-	27

Coach—Terry Murray

Los Angeles Kings

PLAYER	GP	G	A	PTS	+/-	PIM
Luc Robitaille	82	39	35	74	-1	54
Donald Audette	49	18	18	36	7	51
Rob Blake	62	12	23	35	-7	128
Jozef Stumpel	64	13	21	34	-18	10
Glen Murray	61	16	15	31	-14	36
Ray Ferraro	65	13	18	31	E	59
Vladimir Tsyplakov	69	11	12	23	-7	32
Olli Jokinen	66	9	12	21	-10	44
Craig Johnson	69	7	12	19	-12	32
Russ Courtnall	57	6	13	19	-9	19
Garry Galley	60	4	12	16	-9	30
Doug Bodger	65	3	11	14	1	34
Sean O'Donnell	80	1	13	14	1	186
Ian Laperriere	72	3	10	13	-5	138
Dave Babych*	41	2	6	8	-2	22
Philippe Boucher	45	2	6	8	-12	32
Mattias Norstrom	78	2	5	7	-10	36
Sandy Moger	42	3	2	5	-9	26
Matt Johnson	49	2	1	3	-5	131
Stephane Fiset (g)	42	0	0	0	-	2

Coach—Larry Robinson

Montreal Canadiens

PLAYER	GP	G	A	PTS	+/-	PIM
Saku Koivu	65	14	30	44	-7	38
Martin Rucinsky	73	17	17	34	-25	50
Vladimir Malakhov	62	13	21	34	-7	77
Shayne Corson	63	12	20	32	-10	147
Benoit Brunet	60	14	17	31	-1	31
Turner Stevenson	69	10	17	27	6	88
Stephane Quintal	82	8	19	27	-23	84
Brian Savage	54	16	10	26	-14	20
Patrick Poulin	81	8	17	25	6	21
Sergei Zholtok	70	7	15	22	-12	6
Eric Weinrich*	80	7	15	22	-25	89
Jonas Hoglund	74	8	10	18	-5	16
Dainius Zubrus*	80	6	10	16	-8	29
Jason Dawe*	59	6	8	14	E	22
Patrice Brisebois	54	3	9	12	-8	28
Igor Ulanov	47	7	4	11	-2	87
Scott Lachance*	76	2	9	11	-21	41
Craig Rivet	66	2	8	10	-3	66
Brett Clark	61	2	2	4	-3	16
Trent McCleary	46	0	0	0	-1	29
Jeff Hackett* (g)	63	0	1	1	-	12
Coach—Alain Vigneault						

Nashville Predators

PLAYER	GP	G	A	PTS	+/-	PIM
Cliff Ronning*	79	20	40	60	-3	42
Greg Johnson	68	16	34	50	-8	24
Sergei Krivokrasov	70	25	23	48	-5	42
Sebastien Bordeleau	72	16	24	40	-14	26
Scott Walker	71	15	25	40	E	103
Tom Fitzgerald	80	13	19	32	-18	48
Patric Kjellberg	71	11	20	31	-13	24
Andrew Brunette	77	11	20	31	-10	26
Jamie Heward	63	6	12	18	-24	44
Vitali Yachmenev	55	7	10	17	-10	10
Drake Berehowsky	74	2	15	17	-9	140
Denny Lambert	76	5	11	16	-3	218
Joel Bouchard	64	4	11	15	-10	60
John Slaney	46	2	12	14	-12	14
Bob Boughner	79	3	10	13	-6	137
Kimmo Timonen	50	4	8	12	-4	30
Jan Vopat	55	5	6	11	0	28
Darren Turcotte	40	4	5	9	-11	16
Patrick Cote	70	1	2	3	-7	242
Mike Dunham (g)	44	0	0	0	-	4
Coach—Barry Trotz						

New Jersey Devils

PLAYER	GP	G	A	PTS	+/-	PIM
Petr Sykora	80	29	43	72	16	22
Bobby Holik	78	27	37	64	16	119
Brian Rolston	82	24	33	57	11	14
Jason Arnott	74	27	27	54	10	79
Patrik Elias	74	17	33	50	19	34
Brendan Morrison	76	13	33	46	-4	18
Scott Niedermayer	72	11	35	46	16	26
Randy McKay	70	17	20	37	10	143
Lyle Odelein	70	5	26	31	6	114
Dave Andreychuk	52	15	13	28	1	20
Jay Pandolfo	70	14	13	27	3	10
Vadim Sharifijanov	53	11	16	27	11	28
Scott Stevens	75	5	22	27	29	64
Denis Pederson	76	11	12	23	-10	66
Sergei Nemchinov*	77	12	8	20	-13	28
Sergei Brylin	47	5	10	15	8	28
Krzysztof Oliwa	64	5	7	12	4	240
Ken Daneyko	82	2	9	11	27	63
Kevin Dean	62	1	10	11	4	22
Bob Carpenter	56	2	8	10	-3	36
Brad Bombardir	56	1	7	8	-4	16
Sheldon Souray	70	1	7	8	5	110
Martin Brodeur (g)	70	0	4	4	-	4
Coach—Robbie Ftorek						

New York Islanders

PLAYER	GP	G	A	PTS	+/-	PIM
Zigmund Palffy	50	22	28	50	-6	34
Trevor Linden	82	18	29	47	-14	32
Bryan Smolinski	82	16	24	40	-7	49
Mariusz Czerkawski	78	21	17	38	-10	14
Claude Lapointe	82	14	23	37	-19	62
Mark Lawrence*	60	14	16	30	-8	38
Craig Janney*	56	5	22	27	-15	14
Kenny Jonsson	63	8	18	26	-18	34
Mats Lindgren*	60	10	15	25	6	24
Mike Watt	75	8	17	25	-2	12
Barry Richter	72	6	18	24	-4	34
Eric Brewer	63	5	6	11	-14	32
Zdeno Chara	59	2	6	8	-8	83

New York Islanders (continued)

PLAYER	GP	G	A	PTS	+/-	PIM
David Harlock	70	2	6	8	-16	68
Richard Pilon	52	0	4	4	-8	88
Joe Sacco	73	3	0	3	-24	45
Steve Webb	45	0	0	0	-10	32
Coach—Mike Milbury, Bill Stewart						

New York Rangers

PLAYER	GP	G	A	PTS	+/-	PIM
Wayne Gretzky	70	9	53	62	-23	14
John MacLean	82	28	27	55	5	46
Brian Leetch	82	13	42	55	-7	42
Adam Graves	82	38	15	53	-12	47
Petr Nedved	56	20	27	47	-6	50
Marc Savard	70	9	36	45	-7	38
Kevin Stevens	81	23	20	43	-10	64
Niklas Sundstrom	81	13	30	43	-2	20
Mike Knuble	82	15	20	35	-7	26
Mathieu Schneider	75	10	24	34	-19	71
Todd Harvey	37	11	17	28	-1	72
Manny Malhotra	73	8	8	16	-2	13
Mike Maneluk*	45	6	9	15	5	20
Brent Fedyk	67	4	6	10	-11	30
Jeff Beukeboom	45	0	9	9	-2	60
Chris Tamer*	63	1	5	6	-14	124
Peter Popovic	68	1	4	5	-12	40
Eric Lacroix*	64	2	2	4	-12	18
Darren Langdon	44	0	0	0	-3	80
Mike Richter (g)	68	0	0	0	-	0
Coach—John Muckler						

Ottawa Senators

PLAYER	GP	G	A	PTS	+/-	PIM
Alexei Yashin	82	44	50	94	16	54
Shawn McEachern	77	31	25	56	8	46
Andreas Dackell	77	15	35	50	9	30
Magnus Arvedson	80	21	26	47	33	50
Andreas Johansson	69	21	16	37	1	34
Nelson Emerson*	65	13	24	37	8	51
Vaclav Prospal	79	10	26	36	8	58
Jason York	79	4	31	35	17	48
Daniel Alfredsson	58	11	22	33	8	14
Radek Bonk	81	16	16	32	15	48
Marian Hossa	60	15	15	30	18	37
Wade Redden	72	8	21	29	7	54
Ted Donato*	82	11	16	27	-8	41
Igor Kravchuk	79	4	21	25	14	32
Sami Salo	61	7	12	19	20	24
Shaun Van Allen	79	6	11	17	3	30
Bruce Gardiner	59	4	8	12	6	43
Janne Laukkanen	50	1	11	12	18	40
Patrick Traverse	46	1	9	10	12	22
Lance Pitlick	50	3	6	9	7	33
Bill Berg	44	2	2	4	4	28
Damian Rhodes (g)	45	1	1	2	-	4
Ron Tugnutt (g)	43	0	0	0	-	0
Coach—Jacques Martin						

Philadelphia Flyers

PLAYER	GP	G	A	PTS	+/-	PIM
Eric Lindros	71	40	53	93	35	120
John LeClair	76	43	47	90	36	30
Rod Brind'Amour	82	24	50	74	3	47
Keith Jones*	78	20	33	53	23	98
Mark Recchi*	71	16	37	53	-7	34
Eric Desjardins	68	15	36	51	18	38
Daniel McGillis	78	8	37	45	16	61
Mikael Renberg*	66	15	23	38	5	18
Daymond Langkow*	78	14	19	33	-8	39
Steve Duchesne*	71	6	24	30	-6	24
Valeri Zelepukin	74	16	9	25	E	48
Chris Therien	74	3	15	18	16	48
Jody Hull	72	3	11	14	-2	12
Sandy McCarthy*	80	5	8	13	-24	160
Marc Bureau	71	4	6	10	-2	10
Dmitri Tertyshny	62	2	8	10	-1	30
Craig Berube*	77	5	4	9	-10	194
Karl Dykhuis*	78	4	5	9	-23	50
Mikael Andersson*	47	2	4	6	-7	4
Luke Richardson	78	0	6	6	-3	106
Adam Burt*	68	0	4	4	4	60
Roman Vopat*	54	0	3	3	-7	90
John Vanbiesbrouck (g)	62	0	1	1	-	12
Coach—Roger Neilson						

Phoenix Coyotes

PLAYER	GP	G	A	PTS	+/-	PIM
Jeremy Roenick	78	24	48	72	7	130
Robert Reichel*	83	26	43	69	-13	54
Keith Tkachuk	68	36	32	68	22	151
Rick Tocchet	81	26	30	56	5	147
Greg Adams	75	19	24	43	-1	26

PLAYER	GP	G	A	PTS	+/-	PIM
Teppo Numminen	82	10	30	40	3	30
Dallas Drake	53	9	22	31	17	65
Jyrki Lumme	60	7	21	28	5	34
Oleg Tverdovsky	82	7	18	25	11	32
Juha Ylonen	59	6	17	23	18	20
Daniel Briere	64	8	14	22	-3	30
Shane Doan	79	6	16	22	-5	54
Bob Corkum	77	9	10	19	-9	17
Mike Stapleton	76	9	9	18	-6	34
Keith Carney	82	2	14	16	15	62
Deron Quint	60	5	8	13	-10	20
J.J. Daigneault*	70	2	9	11	-12	70
Jim Cummins	55	1	7	8	3	190
Mike Sullivan	63	2	4	6	-11	24
Gerald Diduck	44	0	2	2	9	72
Nikolai Khabibulin (g)	63	0	0	0	-	8
Coach—Jim Schoenfeld						

Pittsburgh Penguins

PLAYER	GP	G	A	PTS	+/-	PIM
Jaromir Jagr	81	44	83	127	17	66
Martin Straka	80	35	48	83	12	26
German Titov	72	11	45	56	18	34
Alexei Kovalev*	77	23	30	53	2	49
Robert Lang	72	21	23	44	-10	24
Kip Miller	77	19	23	42	1	20
Jan Hrdina	82	13	29	42	-2	40
Kevin Hatcher	66	11	27	38	11	24
Rob Brown	58	13	11	24	-15	16
Brad Werenka	81	6	18	24	17	93
Jiri Slegr	63	3	20	23	13	86
Matthew Barnaby*	62	6	16	22	-12	177
Alexei Morozov	67	9	10	19	5	14
Dan Kesa	67	2	8	10	-9	27
Bobby Dollas	70	2	8	10	-3	60
Ian Moran	62	4	5	9	1	37
Maxim Galanov	51	4	3	7	-8	14
Darius Kasparaitis	48	1	4	5	12	70
Martin Sonnenberg	44	1	1	2	-2	19
Tyler Wright	61	0	0	0	-2	90
Tom Barrasso (g)	43	0	3	3	-	20
Coach—Kevin Constantine						

St. Louis Blues

PLAYER	GP	G	A	PTS	+/-	PIM
Pavol Demitra	82	37	52	89	13	16
Pierre Turgeon	67	31	34	65	4	36
Al MacInnis	82	20	42	62	33	70
Scott Young	75	24	28	52	8	27
Chris Pronger	67	13	33	46	3	113
Scott Pellerin	80	20	21	41	1	42
Craig Conroy	69	14	25	39	14	38
Mike Eastwood	82	9	21	30	6	36
Terry Yake	60	9	18	27	-9	34
Pascal Rheaume	60	9	18	27	10	24
Jim Campbell	55	4	21	25	-8	41
Michel Picard	45	11	11	22	5	16
Blair Atcheynum*	65	10	8	18	-8	18
Michal Handzus	66	4	12	16	-9	30
Ricard Persson	54	1	12	13	4	94
Kelly Chase	45	3	7	10	2	143
Tony Twist	63	2	6	8	E	149
Jamie Rivers	76	2	5	7	-3	47
Bryan Helmer*	40	0	4	4	5	42
Jeff Finley	32	1	2	3	11	20
Chris McAlpine	51	1	1	2	-10	50
Marc Bergevin	52	1	1	2	-14	99
Coach—John Quenneville						

San Jose Sharks

PLAYER	GP	G	A	PTS	+/-	PIM
Jeff Friesen	78	22	35	57	3	42
Vincent Damphousse*	77	19	30	49	-4	50
Joe Murphy	76	25	23	48	10	73
Patrick Marleau	81	21	24	45	10	24
Owen Nolan	78	19	26	45	16	129
Mike Ricci	82	13	26	39	1	68
Marco Sturm	78	16	22	38	7	52
Bill Houlder	76	9	23	32	8	40
Alexander Korolyuk	55	12	18	30	3	26
Stephane Matteau	68	8	15	23	2	73
Jeff Norton*	72	4	18	22	2	44
Ronnie Stern	78	7	9	16	-3	158
Dave Lowry	61	6	9	15	-5	24
Mike Rathje	82	5	9	14	15	36
Murray Craven	43	4	10	14	-3	18
Marcus Ragnarsson	74	0	13	13	7	66
Bob Rouse	70	0	11	11	E	44
Ron Sutter	59	3	6	9	-8	40

PLAYER	GP	G	A	PTS	+/-	PIM
Bryan Marchment	59	2	6	8	-7	101
Mike Vernon (g)	49	0	0	0	-	8
Coach—Darryl Sutter						

Tampa Bay Lightning

PLAYER	GP	G	A	PTS	+/-	PIM
Darcy Tucker	82	21	22	43	-34	176
Chris Gratton	78	8	26	34	-28	143
Stephane Richer	64	12	21	33	-10	22
Vincent Lecavalier	82	13	15	28	-19	23
Petr Svoboda*	59	5	18	23	1	81
Pavel Kubina	68	9	12	21	-33	80
Colin Forbes*	80	12	8	20	-5	61
Rob Zamuner	58	8	11	19	-15	24
Cory Cross	67	2	16	18	-25	92
Alexandre Daigle*	63	9	8	17	-13	4
Jassen Cullimore	78	5	12	17	-22	81
Mike Sillinger*	79	8	5	13	-29	36
David Wilkie	46	1	7	8	-19	69
Kjell Samuelsson	46	1	4	5	-6	38
Corey Schwab (g)	40	0	4	4	-	4
Coach—Jacques Demers						

Toronto Maple Leafs

PLAYER	GP	G	A	PTS	+/-	PIM
Mats Sundin	82	31	52	83	22	58
Steve Thomas	78	28	45	73	26	33
Sergei Berezin	76	37	22	59	16	12
Derek King	81	24	28	52	15	20
Igor Korolev	66	13	34	47	11	46
Mike Johnson	79	20	24	44	13	35
Yanic Perreault*	76	17	25	42	7	42
Steve Sullivan	63	20	20	40	12	28
Bryan Berard*	69	9	25	34	1	48
Fredrik Modin	67	16	15	31	14	35
Garry Valk	77	8	21	29	8	53
Sylvain Cote	79	5	24	29	22	28
Dimitri Yushkevich	78	6	22	28	25	88
Alexander Karpovtsev*	58	3	25	28	39	52
Tie Domi	72	8	14	22	5	198
Tomas Kaberle	57	4	18	22	3	12
Todd Warriner	53	9	10	19	-6	28
Daniil Markov	57	4	8	12	5	47
Kris King	67	2	2	4	-16	105
Chris McAllister*	48	1	3	4	-3	102
Curtis Joseph (g)	67	0	5	5	-	6
Coach—Pat Quinn						

Vancouver Canucks

PLAYER	GP	G	A	PTS	+/-	PIM
Markus Naslund	80	36	30	66	-13	74
Mark Messier	59	13	35	48	-12	33
Alexander Mogilny	59	14	31	45	0	58
Bill Muckalt	73	16	20	36	-9	98
Mattias Ohlund	74	9	26	35	-19	83
Adrian Aucoin	82	23	11	34	-14	77
Dave Gagner*	69	6	22	28	-16	63
Ed Jovanovski*	72	5	22	27	-9	126
Dave Scatchard	82	13	13	26	-12	140
Bryan McCabe	69	7	14	21	-11	120
Donald Brashear	82	8	10	18	-25	209
Brad May	66	6	11	17	-14	102
Harry York*	56	7	9	16	-3	24
Peter Zezel	41	6	8	14	5	16
Trent Klatt*	75	4	10	14	-3	12
Darby Hendrickson*	62	4	5	9	-19	52
Murray Baron	81	2	6	8	-23	115
Jason Strudwick	65	0	3	3	-19	114
Steve Staios	57	0	2	2	-12	54
Garth Snow (g)	65	0	1	1	-	34
Coach—Mike Keenan, Marc Crawford						

Washington Capitals

PLAYER	GP	G	A	PTS	+/-	PIM
Peter Bondra	66	31	24	55	-1	56
Adam Oates	59	12	42	54	-1	22
Brian Bellows	76	17	19	36	-12	26
Andrei Nikolishin	73	8	27	35	E	28
Sergei Gonchar	53	21	10	31	1	57
James Black	75	16	14	30	5	14
Calle Johansson	67	8	21	29	10	22
Steve Konowalchuk	45	12	12	24	E	26
Jan Bulis	38	7	16	23	3	6
Ken Klee	78	7	13	20	-9	80
Richard Zednik	49	9	8	17	-6	50
Dmitri Mironov	46	2	14	16	-5	80

PLAYER	GP	G	A	PTS	+/-	PIM	PLAYER	GP	G	A	PTS	+/-	PIM
Joe Reekie	73	0	10	10	11	68	Mark Tinordi	48	0	6	6	-6	108
Brendan Witt	54	2	5	7	-6	87	Enrico Ciccone*	59	3	1	4	-7	127
Kelly Miller	62	2	5	7	-5	29	Olaf Kolzig (g)	64	0	2	2	-	19
Mike Eagles	52	4	2	6	-5	50	**Coach**—Ron Wilson						

Individual Goaltending, 1998-99

(25 or more games played; ranked by goals against average)

Player	GP	GAA	W	L	T	SO	SV%	Player	GP	GAA	W	L	T	SO	SV%
Tugnutt, Ott.	43	1.79	22	10	8	3	.925	Brathwaite, Cgy.	28	2.45	11	9	7	1	.914
Hasek, Buf.	64	1.87	30	18	14	9	.936	Hackett, Mon.	63	2.48	26	26	10	5	.907
Belfour, Dal.	61	1.98	35	15	9	5	.914	Wregget, Cgy.	27	2.52	10	12	4	1	.905
Dafoe, Bos.	68	1.99	32	23	11	10	.926	Barrasso, Pit.	43	2.54	19	16	3	4	.901
Turek, Dal.	26	2.08	16	3	3	1	.914	Joseph, Tor.	67	2.56	35	24	7	3	.910
Khabibulin, Pho.	63	2.13	32	23	7	8	.922	Salo, Edm.	64	2.56	25	28	9	5	.903
Vanbiesbrouck, Phi.	62	2.18	27	18	15	6	.902	Kolzig, Wash.	64	2.57	26	31	3	4	.899
Irbe, Car.	62	2.22	27	20	12	6	.922	Fiset, L.A.	42	2.59	18	21	1	3	.914
Shields, S.J.	37	2.22	15	11	8	4	.920	Shtalenkov, Edm.,Pho.	38	2.61	13	19	4	3	.898
Vernon, S.J.	49	2.26	16	22	10	4	.910	Richter, N.Y.R.	68	2.63	27	30	8	4	.910
Roy, Col.	61	2.28	32	19	8	5	.916	Burke, Fla.	59	2.66	21	24	14	3	.907
Brodeur, N.J.	70	2.29	39	21	10	4	.906	Kidd, Car.	25	2.69	7	10	6	2	.904
McLennan, St.L.	33	2.38	13	14	4	3	.890	Thibault, Mon.,Chi.	62	2.69	24	30	7	5	.905
Storr, L.A.	28	2.40	12	12	2	4	.915	Fitzpatrick, Chi.	27	2.73	6	8	6	0	.906
Hebert, Ana.	69	2.42	31	29	6	6	.921	McLean, Fla.	30	2.74	9	10	4	2	.899
Osgood, Det.	63	2.42	34	25	4	3	.909	Essensa, Edm.	39	2.75	12	14	6	0	.901
Fuhr, St.L.	39	2.43	16	11	8	2	.892	Skudra, Pit.	37	2.78	15	11	5	3	.891
Rhodes, Ott.	45	2.44	22	13	7	3	.904	Snow, Van.	65	2.93	20	31	8	6	.900

Wayne Gretzky Retires

On Apr. 18, 1999, hockey legend Wayne Gretzky, known as "The Great One," played in his last National Hockey League game. Gretzky holds or shares 61 NHL records, including career goals (894), career assists (1,963), career points (2,857), playoff goals (122), playoff assists (260), and playoff points (382). He also holds the season record for goals (92), assists (163), and points (215). From the 1983-84 through the 1986-87 season Gretzky captained the Edmonton Oilers to 4 Stanley Cups.

"The Great One" is an 18-time All-Star and 9-time MVP, and he led the league in scoring 10 times. His stellar NHL career spanned 4 cities, but it was a trade to the Los Angeles Kings in 1988 that shook the hockey world. Gretzky instantly became the game's most influential ambassador ever. His presence on the west coast promoted the NHL in the U.S. and led the way for expansion teams such as the Mighty Ducks of Anaheim and the San Jose Sharks.

Gretzky's Career NHL Scoring Statistics

Year	Team	Regular Season					Playoffs				
		GP	G	A	P	+/-	GP	G	A	P	+/-
1979-80	Edmonton	79	51	86	137	0	3	2	1	3	0
1980-81	Edmonton	80	55	109	164	41	9	7	14	21	0
1981-82	Edmonton	80	92	120	212	81	5	5	7	12	0
1982-83	Edmonton	80	71	125	196	60	16	12	26	38	0
1983-84	Edmonton	74	87	118	205	76	19	13	22	35	0
1984-85	Edmonton	80	73	135	208	98	18	17	30	47	0
1985-86	Edmonton	80	52	163	215	71	10	8	11	19	0
1986-87	Edmonton	79	62	121	183	70	21	5	29	34	0
1987-88	Edmonton	64	40	109	149	39	19	12	31	43	9
1988-89	Los Angeles	78	54	114	168	15	11	5	17	22	-4
1989-90	Los Angeles	73	40	102	142	8	7	3	7	10	-4
1990-91	Los Angeles	78	41	122	163	30	12	4	11	15	0
1991-92	Los Angeles	74	31	90	121	-12	6	2	5	7	-3
1992-93	Los Angeles	45	16	49	65	6	24	15	25	40	6
1993-94	Los Angeles	81	38	92	130	-25	0	0	0	0	0
1994-95	Los Angeles	48	11	37	48	-20	0	0	0	0	0
1995-96	Los Angeles	62	15	66	81	-7	0	0	0	0	0
1995-96	St. Louis	18	8	13	21	-6	13	2	14	16	2
1996-97	N.Y. Rangers	82	25	72	97	12	15	10	10	20	5
1997-98	N.Y. Rangers	82	23	67	90	-11	0	0	0	0	0
1998-99	N.Y. Rangers	70	9	53	62	-23	0	0	0	0	0
TOTAL		**1,487**	**894**	**1,963**	**2,857**	**503**	**208**	**122**	**260**	**382**	**11**

All-Time Leading Scorers

Player	Goals	Assists	Points	Player	Goals	Assists	Points
Wayne Gretzky	894	1,963	2,857	Stan Mikita	541	926	1,467
Gordie Howe	801	1,049	1,850	Bryan Trottier	524	901	1,425
Marcel Dionne	731	1,040	1,771	Dale Hawerchuk	518	891	1,409
Mark Messier*	610	1,050	1,660	Jari Kurri	601	797	1,398
Phil Esposito	717	873	1,590	John Bucyk	556	813	1,369
Mario Lemieux	613	881	1,494	Guy Lafleur	560	793	1,353
Paul Coffey*	385	1,102	1,487	Denis Savard	473	865	1,338
Ron Francis*	449	1,037	1,486	Mike Gartner	708	627	1,335
Steve Yzerman*	592	891	1,483	Gilbert Perreault	512	814	1,326
Ray Bourque*	385	1,083	1,468	Alex Delvecchio	456	825	1,281

Note: Through end of 1998-99 season. * Active at end of 1998-99 season.

Most NHL Goals in a Season

Player	Team	Season	Goals	Player	Team	Season	Goals
Wayne Gretzky	Edmonton	1981-82	92	Teemu Selanne	Winnipeg	1992-93	76
Wayne Gretzky	Edmonton	1983-84	87	Wayne Gretzky	Edmonton	1984-85	73
Brett Hull	St. Louis	1990-91	86	Brett Hull	St. Louis	1989-90	72
Mario Lemieux	Pittsburgh	1988-89	85	Wayne Gretzky	Edmonton	1982-83	71
Phil Esposito	Boston	1970-71	76	Jari Kurri	Edmonton	1984-85	71
Alexander Mogilny	Buffalo	1992-93	76	Bret Hull	St. Louis	1991-92	70

Player	Team	Season	Goals	Player	Team	Season	Goals
Mario Lemieux	Pittsburgh	1987-88	70	Mario Lemieux	Pittsburgh	1995-96	69
Bernie Nicholls	Los Angeles	1988-89	70	Mike Bossy	N.Y. Islanders	1980-81	68
Mike Bossy	N.Y. Islanders	1978-79	69	Phil Esposito	Boston	1973-74	68
Mario Lemieux	Pittsburgh	1992-93	69	Jari Kurri	Edmonton	1985-86	68

Art Ross Trophy (Leading Points Scorer)

1927	Bill Cook, N.Y. Rangers	1952	Gordie Howe, Detroit	1976	Guy Lafleur, Montreal
1928	Howie Morenz, Montreal	1953	Gordie Howe, Detroit	1977	Guy Lafleur, Montreal
1929	Ace Bailey, Toronto	1954	Gordie Howe, Detroit	1978	Guy Lafleur, Montreal
1930	Cooney Weiland, Boston	1955	Bernie Geoffrion, Montreal	1979	Bryan Trottier, N.Y. Islanders
1931	Howie Morenz, Montreal	1956	Jean Beliveau, Montreal	1980	Marcel Dionne, Los Angeles
1932	Harvey Jackson, Toronto	1957	Gordie Howe, Detroit	1981	Wayne Gretzky, Edmonton
1933	Bill Cook, N.Y. Rangers	1958	Dickie Moore, Montreal	1982	Wayne Gretzky, Edmonton
1934	Charlie Conacher, Toronto	1959	Dickie Moore, Montreal	1983	Wayne Gretzky, Edmonton
1935	Charlie Conacher, Toronto	1960	Bobby Hull, Chicago	1984	Wayne Gretzky, Edmonton
1936	Dave Schriner, N.Y. Americans	1961	Bernie Geoffrion, Montreal	1985	Wayne Gretzky, Edmonton
1937	Dave Schriner, N.Y. Americans	1962	Bobby Hull, Chicago	1986	Wayne Gretzky, Edmonton
1938	Gordie Drillon, Toronto	1963	Gordie Howe, Detroit	1987	Wayne Gretzky, Edmonton
1939	Toe Blake, Montreal	1964	Stan Mikita, Chicago	1988	Mario Lemieux, Pittsburgh
1940	Milt Schmidt, Boston	1965	Stan Mikita, Chicago	1989	Mario Lemieux, Pittsburgh
1941	Bill Cowley, Boston	1966	Bobby Hull, Chicago	1990	Wayne Gretzky, Los Angeles
1942	Bryan Hextall, N.Y. Rangers	1967	Stan Mikita, Chicago	1991	Wayne Gretzky, Los Angeles
1943	Doug Bentley, Chicago	1968	Stan Mikita, Chicago	1992	Mario Lemieux, Pittsburgh
1944	Herbie Cain, Boston	1969	Phil Esposito, Boston	1993	Mario Lemieux, Pittsburgh
1945	Elmer Lach, Montreal	1970	Bobby Orr, Boston	1994	Wayne Gretzky, Los Angeles
1946	Max Bentley, Chicago	1971	Phil Esposito, Boston	1995	Jaromir Jagr, Pittsburgh
1947	Max Bentley, Chicago	1972	Phil Esposito, Boston	1996	Mario Lemieux, Pittsburgh
1948	Elmer Lach, Montreal	1973	Phil Esposito, Boston	1997	Mario Lemieux, Pittsburgh
1949	Roy Conacher, Chicago	1974	Phil Esposito, Boston	1998	Jaromir Jagr, Pittsburgh
1950	Ted Lindsay, Detroit	1975	Bobby Orr, Boston	1999	Jaromir Jagr, Pittsburgh
1951	Gordie Howe, Detroit				

James Norris Memorial Trophy (Outstanding Defenseman)

1954	Red Kelly, Detroit	1970	Bobby Orr, Boston	1985	Paul Coffey, Edmonton
1955	Doug Harvey, Montreal	1971	Bobby Orr, Boston	1986	Paul Coffey, Edmonton
1956	Doug Harvey, Montreal	1972	Bobby Orr, Boston	1987	Ray Bourque, Boston
1957	Doug Harvey, Montreal	1973	Bobby Orr, Boston	1988	Ray Bourque, Boston
1958	Doug Harvey, Montreal	1974	Bobby Orr, Boston	1989	Chris Chelios, Montreal
1959	Tom Johnson, Montreal	1975	Bobby Orr, Boston	1990	Ray Bourque, Boston
1960	Doug Harvey, Montreal	1976	Denis Potvin, N.Y. Islanders	1991	Ray Bourque, Boston
1961	Doug Harvey, Montreal	1977	Larry Robinson, Montreal	1992	Brian Leetch, N.Y. Rangers
1962	Doug Harvey, N.Y. Rangers	1978	Denis Potvin, N.Y. Islanders	1993	Chris Chelios, Chicago
1963	Pierre Pilote, Chicago	1979	Denis Potvin, N.Y. Islanders	1994	Ray Bourque, Boston
1964	Pierre Pilote, Chicago	1980	Larry Robinson, Montreal	1995	Paul Coffey, Detroit
1965	Pierre Pilote, Chicago	1981	Randy Carlyle, Pittsburgh	1996	Chris Chelios, Chicago
1966	Jacques Laperriere, Montreal	1982	Doug Wilson, Chicago	1997	Brian Leetch, N.Y. Rangers
1967	Harry Howell, N.Y. Rangers	1983	Rod Langway, Washington	1998	Rob Blake, Los Angeles
1968	Bobby Orr, Boston	1984	Rod Langway, Washington	1999	Al MacInnis, St. Louis
1969	Bobby Orr, Boston				

Vezina Trophy (Outstanding Goalie)*

1927	George Hainsworth, Montreal	1953	Terry Sawchuk, Detroit	1977	Dryden, Larocque, Montreal
1928	George Hainsworth, Montreal	1954	Harry Lumley, Toronto	1978	Dryden, Larocque, Montreal
1929	George Hainsworth, Montreal	1955	Terry Sawchuk, Detroit	1979	Dryden, Larocque, Montreal
1930	Tiny Thompson, Boston	1956	Jacques Plante, Montreal	1980	Sauve, Edwards, Buffalo
1931	Roy Worters, N.Y. Americans	1957	Jacques Plante, Montreal	1981	Sevigny, Larocque, Herron,
1932	Charlie Gardiner, Chicago	1958	Jacques Plante, Montreal		Montreal
1933	Tiny Thompson, Boston	1959	Jacques Plante, Montreal	1982	Bill Smith, N.Y. Islanders
1934	Charlie Gardiner, Chicago	1960	Jacques Plante, Montreal	1983	Pete Peeters, Boston
1935	Lorne Chabot, Chicago	1961	John Bower, Toronto	1984	Tom Barrasso, Buffalo
1936	Tiny Thompson, Boston	1962	Jacques Plante, Montreal	1985	Pelle Lindbergh, Philadelphia
1937	Normie Smith, Detroit	1963	Glenn Hall, Chicago	1986	John Vanbiesbrouck, N.Y.
1938	Tiny Thompson, Boston	1964	Charlie Hodge, Montreal		Rangers
1939	Frank Brimsek, Boston	1965	Sawchuk, Bower, Toronto	1987	Ron Hextall, Philadelphia
1940	Dave Kerr, N.Y. Rangers	1966	Worsley, Hodge, Montreal	1988	Grant Fuhr, Edmonton
1941	Turk Broda, Toronto	1967	Hall, DeJordy, Chicago	1989	Patrick Roy, Montreal
1942	Frank Brimsek, Boston	1968	Worsley, Vachon, Montreal	1990	Patrick Roy, Montreal
1943	Johnny Mowers, Detroit	1969	Hall, Plante, St. Louis	1991	Ed Belfour, Chicago
1944	Bill Durnan, Montreal	1970	Tony Esposito, Chicago	1992	Patrick Roy, Montreal
1945	Bill Durnan, Montreal	1971	Giacomin, Villemure, N.Y.	1993	Ed Belfour, Chicago
1946	Bill Durnan, Montreal		Rangers	1994	Dominik Hasek, Buffalo
1947	Bill Durnan, Montreal	1972	Esposito, Smith, Chicago	1995	Dominik Hasek, Buffalo
1948	Turk Broda, Toronto	1973	Ken Dryden, Montreal	1996	Jim Carey, Washington
1949	Bill Durnan, Montreal	1974	Bernie Parent, Philadelphia;	1997	Dominik Hasek, Buffalo
1950	Bill Durnan, Montreal		Tony Esposito, Chicago	1998	Dominik Hasek, Buffalo
1951	Al Rollins, Toronto	1975	Bernie Parent, Philadelphia	1999	Dominik Hasek, Buffalo
1952	Terry Sawchuk, Detroit	1976	Ken Dryden, Montreal		

*Before 1982, awarded to the goalie or goalies who played a minimum of 25 games for the team that allowed the fewest goals; since 1982, awarded to the outstanding goalie.

Calder Memorial Trophy (Rookie of the Year)

1933	Carl Voss, Detroit	1940	Kilby Macdonald, N.Y. Rangers	1947	Howie Meeker, Toronto
1934	Russ Blinco, Montreal Maroons	1941	John Quilty, Montreal	1948	Jim McFadden, Detroit
1935	Dave Schriner, N.Y. Americans	1942	Grant Warwick, N.Y. Rangers	1949	Pentti Lund, N.Y. Rangers
1936	Mike Karakas, Chicago	1943	Gaye Stewart, Toronto	1950	Jack Gelineau, Boston
1937	Syl Apps, Toronto	1944	Gus Bodnar, Toronto	1951	Terry Sawchuk, Detroit
1938	Cully Dahlstrom, Chicago	1945	Frank McCool, Toronto	1952	Bernie Geoffrion, Montreal
1939	Frank Brimsek, Boston	1946	Edgar Laprade, N.Y. Rangers	1953	Gump Worsley, N.Y. Rangers

1954	Camille Henry, N.Y. Rangers	1970	Tony Esposito, Chicago	1985	Mario Lemieux, Pittsburgh
1955	Ed Litzenberger, Chicago	1971	Gilbert Perreault, Buffalo	1986	Gary Suter, Calgary
1956	Glenn Hall, Detroit	1972	Ken Dryden, Montreal	1987	Luc Robitaille, Los Angeles
1957	Larry Regan, Boston	1973	Steve Vickers, N.Y. Rangers	1988	Joe Nieuwendyk, Calgary
1958	Frank Mahovlich, Toronto	1974	Denis Potvin, N.Y. Islanders	1989	Brian Leetch, N.Y. Rangers
1959	Ralph Backstrom, Montreal	1975	Eric Vail, Atlanta	1990	Sergei Makarov, Calgary
1960	Bill Hay, Chicago	1976	Bryan Trottier, N.Y. Islanders	1991	Ed Belfour, Chicago
1961	Dave Keon, Toronto	1977	Willi Plett, Atlanta	1992	Pavel Bure, Vancouver
1962	Bobby Rousseau, Montreal	1978	Mike Bossy, N.Y. Islanders	1993	Teemu Selanne, Winnipeg
1963	Kent Douglas, Toronto	1979	Bobby Smith, Minnesota	1994	Martin Brodeur, New Jersey
1964	Jacques Laperriere, Montreal	1980	Ray Bourque, Boston	1995	Peter Forsberg, Quebec
1965	Roger Crozier, Detroit	1981	Peter Stastny, Quebec	1996	Daniel Alfredsson, Ottawa
1966	Brit Selby, Toronto	1982	Dale Hawerchuk, Winnipeg	1997	Bryan Berard, N.Y. Islanders
1967	Bobby Orr, Boston	1983	Steve Larmer, Chicago	1998	Sergei Samsonov, Boston
1968	Derek Sanderson, Boston	1984	Tom Barrasso, Buffalo	1999	Chris Drury, Colorado
1969	Danny Grant, Minnesota				

Lady Byng Memorial Trophy (Most Gentlemanly Player)

1925	Frank Nighbor, Ottawa	1950	Edgar Laprade, N.Y. Rangers	1975	Marcel Dionne, Detroit
1926	Frank Nighbor, Ottawa	1951	Red Kelly, Detroit	1976	Jean Ratelle, N.Y.R.-Boston
1927	Billy Burch, N.Y. Americans	1952	Sid Smith, Toronto	1977	Marcel Dionne, Los Angeles
1928	Frank Boucher, N.Y. Rangers	1953	Red Kelly, Detroit	1978	Butch Goring, Los Angeles
1929	Frank Boucher, N.Y. Rangers	1954	Red Kelly, Detroit	1979	Bob MacMillan, Atlanta
1930	Frank Boucher, N.Y. Rangers	1955	Sid Smith, Toronto	1980	Wayne Gretzky, Edmonton
1931	Frank Boucher, N.Y. Rangers	1956	Earl Reibel, Detroit	1981	Rick Kehoe, Pittsburgh
1932	Joe Primeau, Toronto	1957	Andy Hebenton, N.Y. Rangers	1982	Rick Middleton, Boston
1933	Frank Boucher, N.Y. Rangers	1958	Camille Henry, N.Y. Rangers	1983	Mike Bossy, N.Y. Islanders
1934	Frank Boucher, N.Y. Rangers	1959	Alex Delvecchio, Detroit	1984	Mike Bossy, N.Y. Islanders
1935	Frank Boucher, N.Y. Rangers	1960	Don McKenney, Boston	1985	Jari Kurri, Edmonton
1936	Doc Romnes, Chicago	1961	Red Kelly, Toronto	1986	Mike Bossy, N.Y. Islanders
1937	Marty Barry, Detroit	1962	Dave Keon, Toronto	1987	Joe Mullen, Calgary
1938	Gordie Drillon, Toronto	1963	Dave Keon, Toronto	1988	Mats Naslund, Montreal
1939	Clint Smith, N.Y. Rangers	1964	Ken Wharram, Chicago	1989	Joe Mullen, Calgary
1940	Bobby Bauer, Boston	1965	Bobby Hull, Chicago	1990	Brett Hull, St. Louis
1941	Bobby Bauer, Boston	1966	Alex Delvecchio, Detroit	1991	Wayne Gretzky, Los Angeles
1942	Syl Apps, Toronto	1967	Stan Mikita, Chicago	1992	Wayne Gretzky, Los Angeles
1943	Max Bentley, Chicago	1968	Stan Mikita, Chicago	1993	Pierre Turgeon, N.Y. Islanders
1944	Clint Smith, Chicago	1969	Alex Delvecchio, Detroit	1994	Wayne Gretzky, Los Angeles
1945	Bill Mosienko, Chicago	1970	Phil Goyette, St. Louis	1995	Ron Francis, Pittsburgh
1946	Toe Blake, Montreal	1971	John Bucyk, Boston	1996	Paul Kariya, Anaheim
1947	Bobby Bauer, Boston	1972	Jean Ratelle, N.Y. Rangers	1997	Paul Kariya, Anaheim
1948	Buddy O'Connor, N.Y. Rangers	1973	Gil Perreault, Buffalo	1998	Ron Francis, Pittsburgh
1949	Bill Quackenbush, Detroit	1974	John Bucyk, Boston	1999	Wayne Gretzky, N.Y. Rangers

Frank J. Selke Trophy (Best Defensive Forward)

1978	Bob Gainey, Montreal	1986	Troy Murray, Chicago	1993	Doug Gilmour, Toronto
1979	Bob Gainey, Montreal	1987	Dave Poulin, Philadelphia	1994	Sergei Federov, Detroit
1980	Bob Gainey, Montreal	1988	Guy Carbonneau, Montreal	1995	Ron Francis, Pittsburgh
1981	Bob Gainey, Montreal	1989	Guy Carbonneau, Montreal	1996	Sergei Federov, Detroit
1982	Steve Kasper, Boston	1990	Rick Meagher, St. Louis	1997	Michael Peca, Buffalo
1983	Bobby Clarke, Philadelphia	1991	Dirk Graham, Chicago	1998	Jere Lehtinen, Dallas
1984	Doug Jarvis, Washington	1992	Guy Carbonneau, Montreal	1999	Jere Lehtinen, Dallas
1985	Craig Ramsay, Buffalo				

Hart Memorial Trophy (MVP)

1927	Herb Gardiner, Montreal	1952	Gordie Howe, Detroit	1976	Bobby Clarke, Philadelphia
1928	Howie Morenz, Montreal	1953	Gordie Howe, Detroit	1977	Guy Lafleur, Montreal
1929	Roy Worters, N.Y. Americans	1954	Al Rollins, Chicago	1978	Guy Lafleur, Montreal
1930	Nels Stewart, Montreal Maroons	1955	Ted Kennedy, Toronto	1979	Bryan Trottier, N.Y. Islanders
1931	Howie Morenz, Montreal	1956	Jean Beliveau, Montreal	1980	Wayne Gretzky, Edmonton
1932	Howie Morenz, Montreal	1957	Gordie Howe, Detroit	1981	Wayne Gretzky, Edmonton
1933	Eddie Shore, Boston	1958	Gordie Howe, Detroit	1982	Wayne Gretzky, Edmonton
1934	Aurel Joliat, Montreal	1959	Andy Bathgate, N.Y. Rangers	1983	Wayne Gretzky, Edmonton
1935	Eddie Shore, Boston	1960	Gordie Howe, Detroit	1984	Wayne Gretzky, Edmonton
1936	Eddie Shore, Boston	1961	Bernie Geoffrion, Montreal	1985	Wayne Gretzky, Edmonton
1937	Babe Siebert, Montreal	1962	Jacques Plante, Montreal	1986	Wayne Gretzky, Edmonton
1938	Eddie Shore, Boston	1963	Gordie Howe, Detroit	1987	Wayne Gretzky, Edmonton
1939	Toe Blake, Montreal	1964	Jean Beliveau, Montreal	1988	Mario Lemieux, Pittsburgh
1940	Ebbie Goodfellow, Detroit	1965	Bobby Hull, Chicago	1989	Wayne Gretzky, Los Angeles
1941	Bill Cowley, Boston	1966	Bobby Hull, Chicago	1990	Mark Messier, Edmonton
1942	Tom Anderson, N.Y. Americans	1967	Stan Mikita, Chicago	1991	Brett Hull, St. Louis
1943	Bill Cowley, Boston	1968	Stan Mikita, Chicago	1992	Mark Messier, N.Y. Rangers
1944	Babe Pratt, Toronto	1969	Phil Esposito, Boston	1993	Mario Lemieux, Pittsburgh
1945	Elmer Lach, Montreal	1970	Bobby Orr, Boston	1994	Sergei Federov, Detroit
1946	Max Bentley, Chicago	1971	Bobby Orr, Boston	1995	Eric Lindros, Philadelphia
1947	Maurice Richard, Montreal	1972	Bobby Orr, Boston	1996	Mario Lemieux, Pittsburgh
1948	Buddy O'Connor, N.Y. Rangers	1973	Bobby Clarke, Philadelphia	1997	Dominik Hasek, Buffalo
1949	Sid Abel, Detroit	1974	Phil Esposito, Boston	1998	Dominik Hasek, Buffalo
1950	Chuck Rayner, N.Y. Rangers	1975	Bobby Clarke, Philadelphia	1999	Jaromir Jagr, Pittsburgh
1951	Milt Schmidt, Boston				

Conn Smythe Trophy (MVP in Playoffs)

1965	Jean Beliveau, Montreal	1973	Yvan Cournoyer, Montreal	1981	Butch Goring, N.Y. Islanders
1966	Roger Crozier, Detroit	1974	Bernie Parent, Philadelphia	1982	Mike Bossy, N.Y. Islanders
1967	Dave Keon, Toronto	1975	Bernie Parent, Philadelphia	1983	Billy Smith, N.Y. Islanders
1968	Glenn Hall, St. Louis	1976	Reg Leach, Philadelphia	1984	Mark Messier, Edmonton
1969	Serge Savard, Montreal	1977	Guy Lafleur, Montreal	1985	Wayne Gretzky, Edmonton
1970	Bobby Orr, Boston	1978	Larry Robinson, Montreal	1986	Patrick Roy, Montreal
1971	Ken Dryden, Montreal	1979	Bob Gainey, Montreal	1987	Ron Hextall, Philadelphia
1972	Bobby Orr, Boston	1980	Bryan Trottier, N.Y. Islanders	1988	Wayne Gretzky, Edmonton

1989 Al MacInnis, Calgary	1993 Patrick Roy, Montreal	1997 Mike Vernon, Detroit	
1990 Bill Ranford, Edmonton	1994 Brian Leetch, N.Y. Rangers	1998 Steve Yzerman, Detroit	
1991 Mario Lemieux, Pittsburgh	1995 Claude Lemieux, New Jersey	1999 Joe Nieuwendyk, Dallas	
1992 Mario Lemieux, Pittsburgh	1996 Joe Sakic, Colorado		

Maurice "Rocket" Richard Trophy (Most Goals)

1999 Teemu Selanne, Anaheim

Hockey Hall of Fame, Toronto, Ontario

(1999 inductees have an asterisk*)

PLAYERS

Abel, Sid	Foyston, Frank	Maxwell, Fred	Stastny, Peter
Adams, Jack	Fredrickson, Frank	McDonald, Lanny	Stewart, Jack
Apps, Syl	Gadsby, Bill	McGee, Frank	Stewart, Nels
Armstrong, George	Gainey, Bob	McGimsie, Billy	Stuart, Bruce
Bailey, Ace	Gardiner, Chuck	McNamara, George	Stuart, Hod
Bain, Dan	Gardiner, Herb	Mikita, Stan	Taylor, Cyclone
Baker, Hobey	Gardiner, Jimmy	Moore, Dickie	Thompson, Tiny
Barber, Bill	Geoffrion, Bernie	Moran, Paddy	Tretiak, Vladislav
Barry, Marty	Gerard, Eddie	Morenz, Howie	Trihey, Harry
Bathgate, Andy	Giacomin, Eddie	Mosienko, Bill	Trottier, Bryan
Bauer, Bobby	Gilbert, Rod	Nighbor, Frank	Ullman, Norm
Beliveau, Jean	Gilmour, Billy	Noble, Reg	Vezina, Georges
Benedict, Clint	Goheen, Moose	O'Connor, Buddy	Walker, Jack
Bentley, Doug	Goodfellow, Ebbie	Oliver, Harry	Walsh, Marty
Bentley, Max	Goulet, Michel	Olmstead, Bert	Watson, Harry
Blake, Toe	Grant, Mike	Orr, Bobby	(Moose)
Boivin, Leo	Green, Shorty	Parent, Bernie	Watson, Harry Percival
Boon, Dickie	*Gretzky, Wayne	Park, Brad	Weiland, Cooney
Bossy, Mike	Griffis, Si	Patrick, Lester	Westwick, Harry
Bouchard, Butch	Hainsworth, George	Patrick, Lynn	Whitcroft, Fred
Boucher, Frank	Hall, Glenn	Perreault, Gilbert	Wilson, Phat
Boucher, George	Hall, Joe	Phillips, Tom	Worsley, Gump
Bower, Johnny	Harvey, Doug	Pilote, Pierre	Worters, Roy
Bowie, Dubbie	Hay, George	Pitre, Didier	**BUILDERS**
Brimsek, Frank	Hern, Riley	Plante, Jacques	Adams, Charles
Broadbent, Punch	Hextall, Bryan	Potvin, Denis	Adams, Weston
Broda, Turk	Holmes, Hap	Pratt, Babe	Ahearn, Bunny
Bucyk, John	Hooper, Tom	Primeau, Joe	Ahearn, Frank
Burch, Billy	Horner, Red	Pronovost, Marcel	Allan, Sir Montagu
Cameron, Harry	Horton, Tim	Pulford, Bob	Allen, Keith
Cheevers, Gerry	Howe, Gordie	Pulford, Harvey	Arbour, Al
Clancy, King	Howe, Syd	Quackenbush, Bill	Ballard, Harold
Clapper, Dit	Howell, Harry	Rankin, Frank	Bauer, Father David
Clarke, Bobby	Hull, Bobby	Ratelle, Jean	Bickell, J.P.
Cleghorn, Sprague	Hutton, Bouse	Rayner, Chuck	Bowman, Scotty
Colville, Neil	Hyland, Harry	Reardon, Kenny	Brown, George
Conacher, Charlie	Irvin, Dick	Richard, Henri	Brown, Walter
Conacher, Lionel	Jackson, Busher	Richard, Maurice	Buckland, Frank
Conacher, Roy	Johnson, Ching	Richardson, George	Butterfield, Jack
Connell, Alex	Johnson, Ernie	Roberts, Gordie	Calder, Frank
Cook, Bill	Johnson, Tom	Robinson, Larry	Campbell, Angus
Cook, Bun	Joliat, Aurel	Ross, Art	Campbell, Clarence
Coulter, Art	Keats, Duke	Russel, Blair	Cattarinich, Joseph
Cournoyer, Yvan	Kelly, Red	Russell, Ernie	Dandurand, Leo
Cowley, Bill	Kennedy, Ted	Ruttan, Jack	Dilio, Frank
Crawford, Rusty	Keon, Dave	Salming, Borje	Dudley, George
Darragh, Jack	Lach, Elmer	Savard, Serge	Dunn, James
Davidson, Scotty	Lafleur, Guy	Sawchuk, Terry	Francis, Emile
Day, Hap	Lalonde, Newsy	Scanlan, Fred	Gibson, Jack
Delvecchio, Alex	Laperriere, Jacques	Schmidt, Milt	Gorman, Tommy
Denneny, Cy	Lapointe, Guy	Schriner, Sweeney	Griffiths, Frank
Dionne, Marcel	Laprade, Edgar	Seibert, Earl	Hanley, Bill
Drillon, Gordie	Laviolette, Jack	Seibert, Oliver	Hay, Charles
Drinkwater, Graham	LeSueur, Percy	Shore, Eddie	Hendy, Jim
Dryden, Ken	Lehman, Hughie	Shutt, Steve	Hewitt, Foster
Dumart, Woody	Lemaire, Jacques	Siebert, Babe	Hewitt, William
Dunderdale, Tommy	Lemieux, Mario	Simpson, Joe	Hume, Fred
Durnan, Bill	Lewis, Herbie	Sittler, Darryl	Imlach, Punch
Dutton, Red	Lindsay, Ted	Smith, Alf	Ivan, Tommy
Dye, Babe	Lumley, Harry	Smith, Billy	Jennings, William
Esposito, Phil	MacKay, Mickey	Smith, Clint	Johnson, Bob
Esposito, Tony	Mahovlich, Frank	Smith, Hooley	Juckes, Gordon
Farrel, Arthur	Malone, Joe	Smith, Tommy	Kilpatrick, John
Flaman, Fernie	Mantha, Sylvio	Stanley, Allan	Knox, Seymour
	Marshall, Jack	Stanley, Barney	

LeBel, Robert
Leader, Al
Lockhart, Thomas
Loicq, Paul
Mariucci, John
Mathers, Frank
McLaughlin, Frederic
Milford, Jake
Molson, Sen. Hartland
*Morrison, Ian "Scotty"
Murray, Pere Athol
Nelson, Francis
Norris, Bruce
Norris, James
Norris, James Sr.
Northey, William
O'Brien, J. Ambrose
O'Neill, Brian Francis
Page, Frederick
Patrick, Frank
Pickard, Allan
Pilous, Rudy
Poile, Bud
Pollock, Sam
Raymond, Sen. Donat
Robertson, John Ross
Robinson, Claude
Ross, Phillip
Sabetzki, Gunther
Sather, Glen
Selke, Frank
Sinden, Harry
Smith, Frank
Smythe, Conn
Snider, Ed
Stanley, Lord (of Preston)
Sutherland, Capt. James T.
Tarasov, Anatoli
Torrey, Bill
Turner, Lloyd
Tutt, William
Voss, Carl
Waghorne, Fred
Wirtz, Arthur
Wirtz, Bill
Ziegler, John A., Jr.
REFEREES AND LINESMEN
Armstrong, Neil
Ashley, John
Chadwick, Bill
D'Amico, John
Elliott, Chaucer
Hayes, George
Hewiston, Bobby
Ion, Mickey
Pavelich, Matt
Rodden, Mike
Smeaton, Cooper
Storey, Red
Udvari, Frank
*Van Hellemond, Andy

NCAA HOCKEY CHAMPIONS

1948	Michigan	1961	Denver	1974	Minnesota	1987	North Dakota
1949	Boston College	1962	Michigan Tech	1975	Michigan Tech	1988	Lake Superior St.
1950	Colorado College	1963	North Dakota	1976	Minnesota	1989	Harvard
1951	Michigan	1964	Michigan	1977	Wisconsin	1990	Wisconsin
1952	Michigan	1965	Michigan Tech	1978	Boston Univ.	1991	N. Michigan
1953	Michigan	1966	Michigan State	1979	Minnesota	1992	Lake Superior St.
1954	RPI	1967	Cornell	1980	North Dakota	1993	Maine
1955	Michigan	1968	Denver	1981	Wisconsin	1994	Lake Superior St.
1956	Michigan	1969	Denver	1982	North Dakota	1995	Boston Univ.
1957	Colorado College	1970	Cornell	1983	Wisconsin	1996	Michigan
1958	Denver	1971	Boston Univ.	1984	Bowling Green	1997	North Dakota
1959	North Dakota	1972	Boston Univ.	1985	RPI	1998	Michigan
1960	Denver	1973	Wisconsin	1986	Michigan State	1999	Maine

NATIONAL BASKETBALL ASSOCIATION

1998-99 Review: Shortened Season, San Antonio Is Tops, Jordan Retires Again

The 1998-99 NBA season did not start until Feb. 5, 1999, because of a lockout that almost canceled the entire season. On Jan. 6, the NBA and the players' union reached an agreement that was approved by the owners and players the next day, and a season of 50 games per team was set. The San Antonio Spurs, the best team during the ensuing regular season, lost only 2 games in the playoffs on the way to their first-ever NBA Championship. They defeated the surprising New York Knicks, the number-8 seed in the Eastern Conference, in the finals, 4 games to 1. However, the biggest news in the NBA came on Jan. 13, when Michael Jordan, arguably the best player in NBA history, retired. Jordan had retired briefly after the 1992-93 season but returned late in the 1994-95 season.

Final Standings, 1998-99 Season

(playoff seeding in parentheses; in each conference the two division winners automatically get the number 1 and 2 seeds)

Eastern Conference
Atlantic Division

	W	L	Pct	GB
Miami (1)	33	17	.660	—
Orlando (3)...............	33	17	.660	—
Philadelphia (6)	28	22	.560	5
New York (8)...........	27	23	.540	6
Boston	19	31	.380	14
Washington	18	32	.360	15
New Jersey..............	16	34	.320	17

Central Division

	W	L	Pct	GB
Indiana (2)	33	17	.660	—
Atlanta (4)...............	31	19	.620	2
Detroit (5)	29	21	.580	4
Milwaukee (7).............	28	22	.560	5
Charlotte................	26	24	.520	7
Toronto.................	23	27	.460	10
Cleveland	22	28	.440	11
Chicago	13	37	.260	20

Western Conference
Midwest Division

	W	L	Pct	GB
San Antonio (1)	37	13	.740	—
Utah (3)	37	13	.740	—
Houston (5)	31	19	.620	6
Minnesota (8)	25	25	.500	12
Dallas	19	31	.380	18
Denver..................	14	36	.280	23
Vancouver	8	42	.160	29

Pacific Division

	W	L	Pct	GB
Portland (2)	35	15	.700	—
L.A. Lakers (4)............	31	19	.620	4
Sacramento (6)	27	23	.540	8
Phoenix (7)	27	23	.540	8
Seattle	25	25	.500	10
Golden State	21	29	.420	14
L.A. Clippers	9	41	.180	26

NBA Regular Season Individual Highs in 1998-99

Most minutes played, game — 56: Shareef Abdur-Rahim, Vancouver v. Boston, Feb. 17 (OT).

Most points, game — 46: Grant Hill, Detroit v. Washington, Feb. 8; Allen Iverson, Philadelphia v. San Antonio, Feb. 12; Antonio McDyess, Denver v. Vancouver, Feb. 28.

Most field goals made, game — 19: Tim Duncan, San Antonio v. Vancouver, April 1.

Most field goal attempts, game — 36: Allen Iverson, Philadelphia v. L.A. Lakers, Mar. 19.

Most 3-pt. field goals made, game — 9: Dee Brown, Toronto at Milwaukee, April 28.

Most 3-pt. field goal attempts, game — 18: Dee Brown, Toronto at Milwaukee, April 28.

Most free throws made, game — 18: 3 times, most recently by Karl Malone, Utah v. Golden State, April 8.

Most rebounds, game — 25: Lorenzen Wright, L.A. Clippers v. Sacramento, Mar. 11.

Most assists, game — 20: Stephon Marbury, New Jersey v. Indiana, April 25.

Most steals, game — 11: Kendall Gill, New Jersey v. Miami, April 3.

Most blocked shots, game — 9: 5 times, most recently by Alonzo Mourning, Miami at Cleveland, April 19.

Most minutes played, season — 2,060: Jason Kidd, Phoenix.

Most offensive rebounds, season — 210: Danny Fortson, Denver.

Most defensive rebounds, season — 418: Dikembe Mutumbo, Atlanta.

1999 NBA Playoff Results

Eastern Conference
New York defeated Miami 3 games to 2
Indiana defeated Milwaukee 3 games to 0
Philadelphia defeated Orlando 3 games to 1
Atlanta defeated Detroit 3 games to 2
Indiana defeated Philadelphia 4 games to 0
New York defeated Atlanta 4 games to 0
New York defeated Indiana 4 games to 2

Western Conference
San Antonio defeated Minnesota 3 games to 1
Portland defeated Phoenix 3 games to 0
Utah defeated Sacramento 3 games to 2
L.A. Lakers defeated Houston 3 games to 1
San Antonio defeated L.A. Lakers 4 games to 0
Portland defeated Utah 4 games to 2
San Antonio defeated Portland 4 games to 0

Championship
San Antonio defeated New York 4 games to 1 [89-77, 80-67, 81-89, 96-89, 78-77]

The San Antonio Spurs Win Their 1st NBA Championship

In June 1999, the San Antonio Spurs put an end to the reign of the Chicago Bulls when they took the NBA Championship, defeating the New York Knicks 4 games to 1 in the Finals. The Bulls had won the previous 3 championships and 6 of the last 8 before being dismantled after the 1997-98 season. The Spurs, led by "twin towers" David Robinson and Tim Duncan, were the top team in the NBA throughout the season as well as in the playoffs. It was the first title for the team that entered the NBA in 1976, after the American Basketball Association merged with the NBA. Tim Duncan established himself as a star player in only his second year in the league. He averaged 27.4 points per game in the NBA Finals and was named series MVP.

San Antonio Spurs

	FG A-M	FT A-M	Reb O-T	Ast	Avg
Tim Duncan	95-51	44-35	14-70	12	27.4
David Robinson ...	59-25	48-33	14-59	12	16.6
Mario Elie........	38-17	23-20	2-20	13	11.6
Avery Johnson	40-20	10-6	2-13	36	9.2
Sean Elliot	42-14	11-7	4-15	15	8.0
Jaren Jackson	37-12	0-0	1-7	5	6.6
Antonio Daniels ...	5-4	0-0	0-2	4	2.5
Steve Kerr.......	10-4	1-0	2-5	2	1.8
Malik Rose	10-2	4-2	3-12	2	1.2
Jerome Kersey	1-1	0-0	0-0	0	1.0
Gerard King	0-0	0-0	0-0	0	0.0

New York Knicks

	FG A-M	FT A-M	Reb O-T	Ast	Avg
Latrell Sprewell ...	117-48	38-32	6-33	13	26.0
Allan Houston	96-41	26-24	2-16	17	21.6
Marcus Camby ...	42-21	8-6	13-39	1	9.6
Larry Johnson	49-14	13-8	6-24	7	7.6
Charlie Ward	26-12	2-1	6-16	18	5.8
Kurt Thomas	32-11	10-6	13-38	2	5.6
Chris Childs......	22-5	2-1	1-6	11	2.4
Chris Dudley	8-2	6-2	2-19	1	1.2
Herb Williams ...	1-0	0-0	0-0	0	0.0
Rick Brunson	0-0	0-0	0-0	0	0.0

NBA Finals MVP

1969	Jerry West, Los Angeles	1979	Dennis Johnson, Seattle	1989	Joe Dumars, Detroit
1970	Willis Reed, New York	1980	Magic Johnson, Los Angeles	1990	Isiah Thomas, Detroit
1971	Lew Alcindor (Kareem Abdul-Jabbar), Milwaukee	1981	Cedric Maxwell, Boston	1991	Michael Jordan, Chicago
		1982	Magic Johnson, Los Angeles	1992	Michael Jordan, Chicago
1972	Wilt Chamberlain, Los Angeles	1983	Moses Malone, Philadelphia	1993	Michael Jordan, Chicago
1973	Willis Reed, New York	1984	Larry Bird, Boston	1994	Hakeem Olajuwon, Houston
1974	John Havlicek, Boston	1985	Kareem Abdul-Jabbar, L.A. Lakers	1995	Hakeem Olajuwon, Houston
1975	Rick Barry, Golden State			1996	Michael Jordan, Chicago
1976	Jo Jo White, Boston	1986	Larry Bird, Boston	1997	Michael Jordan, Chicago
1977	Bill Walton, Portland	1987	Magic Johnson, L.A. Lakers	1998	Michael Jordan, Chicago
1978	Wes Unseld, Washington	1988	James Worthy, L.A. Lakers	1999	Tim Duncan, San Antonio

NBA Scoring Leaders

Year	Scoring champion	Pts	Avg	Year	Scoring champion	Pts	Avg
1947	Joe Fulks, Philadelphia	1,389	23.2	1973	Nate Archibald, Kans. City-Omaha	2,719	34.0
1948	Max Zaslofsky, Chicago	1,007	21.0	1974	Bob McAdoo, Buffalo	2,261	30.6
1949	George Mikan, Minneapolis	1,698	28.3	1975	Bob McAdoo, Buffalo	2,831	34.5
1950	George Mikan, Minneapolis	1,865	27.4	1976	Bob McAdoo, Buffalo	2,427	31.1
1951	George Mikan, Minneapolis	1,932	28.4	1977	Pete Maravich, New Orleans	2,273	31.1
1952	Paul Arizin, Philadelphia	1,674	25.4	1978	George Gervin, San Antonio	2,232	27.2
1953	Neil Johnston, Philadelphia	1,564	22.3	1979	George Gervin, San Antonio	2,365	29.6
1954	Neil Johnston, Philadelphia	1,759	24.4	1980	George Gervin, San Antonio	2,585	33.1
1955	Neil Johnston, Philadelphia	1,631	22.7	1981	Adrian Dantley, Utah	2,452	30.7
1956	Bob Pettit, St. Louis	1,849	25.7	1982	George Gervin, San Antonio	2,551	32.3
1957	Paul Arizin, Philadelphia	1,817	25.6	1983	Alex English, Denver	2,326	28.4
1958	George Yardley, Detroit	2,001	27.8	1984	Adrian Dantley, Utah	2,418	30.6
1959	Bob Pettit, St. Louis	2,105	29.2	1985	Bernard King, New York	1,809	32.9
1960	Wilt Chamberlain, Philadelphia	2,707	37.9	1986	Dominique Wilkins, Atlanta	2,366	30.3
1961	Wilt Chamberlain, Philadelphia	3,033	38.4	1987	Michael Jordan, Chicago	3,041	37.1
1962	Wilt Chamberlain, Philadelphia	4,029	50.4	1988	Michael Jordan, Chicago	2,868	35.0
1963	Wilt Chamberlain, San Francisco	3,586	44.8	1989	Michael Jordan, Chicago	2,633	32.5
1964	Wilt Chamberlain, San Francisco	2,948	36.5	1990	Michael Jordan, Chicago	2,753	33.6
1965	Wilt Chamberlain, San Francisco, Philadelphia	2,534	34.7	1991	Michael Jordan, Chicago	2,580	31.5
1966	Wilt Chamberlain, Philadelphia	2,649	33.5	1992	Michael Jordan, Chicago	2,404	30.1
1967	Rick Barry, San Francisco	2,775	35.6	1993	Michael Jordan, Chicago	2,541	32.6
1968	Dave Bing, Detroit	2,142	27.1	1994	David Robinson, San Antonio	2,383	29.8
1969	Elvin Hayes, San Diego	2,327	28.4	1995	Shaquille O'Neal, Orlando	2,315	29.3
1970	Jerry West, Los Angeles	2,309	31.2	1996	Michael Jordan, Chicago	2,465	30.4
1971	Lew Alcindor (Kareem Abdul-Jabbar), Milwaukee	2,596	31.7	1997	Michael Jordan, Chicago	2,431	29.6
				1998	Michael Jordan, Chicago	2,357	28.7
1972	Kareem Abdul-Jabbar, Milwaukee	2,822	34.8	1999	Allen Iverson, Philadelphia	1,284	26.8

NBA Most Valuable Player

1956	Bob Pettit, St. Louis	1972	Kareem Abdul-Jabbar, Milwaukee	1984	Larry Bird, Boston
1957	Bob Cousy, Boston			1985	Larry Bird, Boston
1958	Bill Russell, Boston	1973	Dave Cowens, Boston	1986	Larry Bird, Boston
1959	Bob Pettit, St. Louis	1974	Kareem Abdul-Jabbar, Milwaukee	1987	Magic Johnson, L.A. Lakers
1960	Wilt Chamberlain, Philadelphia			1988	Michael Jordan, Chicago
1961	Bill Russell, Boston	1975	Bob McAdoo, Buffalo	1989	Magic Johnson, L.A. Lakers
1962	Bill Russell, Boston	1976	Kareem Abdul-Jabbar, Los Angeles	1990	Magic Johnson, L.A. Lakers
1963	Bill Russell, Boston			1991	Michael Jordan, Chicago
1964	Oscar Robertson, Cincinnati	1977	Kareem Abdul-Jabbar, Los Angeles	1992	Michael Jordan, Chicago
1965	Bill Russell, Boston			1993	Charles Barkley, Phoenix
1966	Wilt Chamberlain, Philadelphia	1978	Bill Walton, Portland	1994	Hakeem Olajuwon, Houston
1967	Wilt Chamberlain, Philadelphia	1979	Moses Malone, Houston	1995	David Robinson, San Antonio
1968	Wilt Chamberlain, Philadelphia	1980	Kareem Abdul-Jabbar, Los Angeles	1996	Michael Jordan, Chicago
1969	Wes Unseld, Baltimore			1997	Karl Malone, Utah
1970	Willis Reed, New York	1981	Julius Erving, Philadelphia	1998	Michael Jordan, Chicago
1971	Lew Alcindor (Kareem Abdul-Jabbar), Milwaukee	1982	Moses Malone, Houston	1999	Karl Malone, Utah
		1983	Moses Malone, Philadelphia		

NBA Champions, 1947-99

	Regular season		Playoffs		
Year	Eastern Conference	Western Conference	Winner	Coach	Runner-up
1947	Washington Capitols	Chicago Stags	Philadelphia	Ed Gottlieb	Chicago
1948	Philadelphia Warriors	St. Louis Bombers	Baltimore	Buddy Jeannette	Philadelphia
1949	Washington Capitols	Rochester	Minneapolis	John Kundla	Washington
1950	Syracuse	Minneapolis	Minneapolis	John Kundla	Syracuse
1951	Philadelphia Warriors	Minneapolis	Rochester	Lester Harrison	New York
1952	Syracuse	Rochester	Minneapolis	John Kundla	New York
1953	New York	Minneapolis	Minneapolis	John Kundla	New York
1954	New York	Minneapolis	Minneapolis	John Kundla	Syracuse
1955	Syracuse	Ft. Wayne	Syracuse	Al Cervi	Ft. Wayne
1956	Philadelphia Warriors	Ft. Wayne	Philadelphia	George Senesky	Ft. Wayne
1957	Boston	St. Louis	Boston	Red Auerbach	St. Louis
1958	Boston	St. Louis	St. Louis	Alex Hannum	Boston
1959	Boston	St. Louis	Boston	Red Auerbach	Minneapolis
1960	Boston	St. Louis	Boston	Red Auerbach	St. Louis
1961	Boston	St. Louis	Boston	Red Auerbach	St. Louis
1962	Boston	Los Angeles	Boston	Red Auerbach	Los Angeles
1963	Boston	Los Angeles	Boston	Red Auerbach	Los Angeles
1964	Boston	San Francisco	Boston	Red Auerbach	San Francisco

	Regular season				Playoffs		
Year	Eastern Conference	Western Conference			Winner	Coach	Runner-up
1965	Boston	Los Angeles			Boston	Red Auerbach	Los Angeles
1966	Philadelphia	Los Angeles			Boston	Red Auerbach	Los Angeles
1967	Philadelphia	San Francisco			Philadelphia	Alex Hannum	San Francisco
1968	Philadelphia	St. Louis			Boston	Bill Russell	Los Angeles
1969	Baltimore	Los Angeles			Boston	Bill Russell	Los Angeles
1970	New York	Atlanta			New York	Red Holzman	Los Angeles
Year	Atlantic	Central	Midwest	Pacific	Winner	Coach	Runner-up
1971	New York	Baltimore	Milwaukee	Los Angeles	Milwaukee	Larry Costello	Baltimore
1972	Boston	Baltimore	Milwaukee	Los Angeles	Los Angeles	Bill Sharman	New York
1973	Boston	Baltimore	Milwaukee	Los Angeles	New York	Red Holzman	Los Angeles
1974	Boston	Capital	Milwaukee	Los Angeles	Boston	Tom Heinsohn	Milwaukee
1975	Boston	Washington	Chicago	Golden State	Golden State	Al Attles	Washington
1976	Boston	Cleveland	Milwaukee	Golden State	Boston	Tom Heinsohn	Phoenix
1977	Philadelphia	Houston	Denver	Los Angeles	Portland	Jack Ramsay	Philadelphia
1978	Philadelphia	San Antonio	Denver	Portland	Washington	Dick Motta	Seattle
1979	Washington	San Antonio	Kansas City	Seattle	Seattle	Len Wilkens	Washington
1980	Boston	Atlanta	Milwaukee	Los Angeles	Los Angeles	Paul Westhead	Philadelphia
1981	Boston	Milwaukee	San Antonio	Phoenix	Boston	Bill Fitch	Houston
1982	Boston	Milwaukee	San Antonio	Los Angeles	Los Angeles	Pat Riley	Philadelphia
1983	Philadelphia	Milwaukee	San Antonio	Los Angeles	Philadelphia	Billy Cunningham	Los Angeles
1984	Boston	Milwaukee	Utah	Los Angeles	Boston	K.C. Jones	Los Angeles
1985	Boston	Milwaukee	Denver	L.A. Lakers	L.A. Lakers	Pat Riley	Boston
1986	Boston	Milwaukee	Houston	L.A. Lakers	Boston	K.C. Jones	Houston
1987	Boston	Atlanta	Dallas	L.A. Lakers	L.A. Lakers	Pat Riley	Boston
1988	Boston	Detroit	Denver	L.A. Lakers	L.A. Lakers	Pat Riley	Detroit
1989	New York	Detroit	Utah	L.A. Lakers	Detroit	Chuck Daly	L.A. Lakers
1990	Philadelphia	Detroit	San Antonio	L.A. Lakers	Detroit	Chuck Daly	Portland
1991	Boston	Chicago	San Antonio	Portland	Chicago	Phil Jackson	L.A. Lakers
1992	Boston	Chicago	Utah	Portland	Chicago	Phil Jackson	Portland
1993	New York	Chicago	Houston	Phoenix	Chicago	Phil Jackson	Phoenix
1994	New York	Atlanta	Houston	Seattle	Houston	Rudy Tomjanovich	New York
1995	Orlando	Indiana	San Antonio	Phoenix	Houston	Rudy Tomjanovich	Orlando
1996	Orlando	Chicago	San Antonio	Seattle	Chicago	Phil Jackson	Seattle
1997	Miami	Chicago	Utah	Seattle	Chicago	Phil Jackson	Utah
1998	Miami	Chicago	Utah	L.A. Lakers	Chicago	Phil Jackson	Utah
1999	Miami	Indiana	San Antonio	Portland	San Antonio	Gregg Popovich	New York

NBA Coach of the Year, 1963-99

1963 Harry Gallatin, St. Louis Hawks
1964 Alex Hannum, San Francisco Warriors
1965 Red Auerbach, Boston Celtics
1966 Dolph Schayes, Philadelphia 76ers
1967 Johnny Kerr, Chicago Bulls
1968 Richie Guerin, St. Louis Hawks
1969 Gene Shue, Baltimore Bullets
1970 Red Holzman, New York Knicks
1971 Dick Motta, Chicago Bulls
1972 Bill Sharman, Los Angeles Lakers
1973 Tom Heinsohn, Boston Celtics
1974 Ray Scott, Detroit Pistons
1975 Phil Johnson, Kansas City-Omaha Kings

1976 Bill Fitch, Cleveland Cavaliers
1977 Tom Nissalke, Houston Rockets
1978 Hubie Brown, Atlanta Hawks
1979 Cotton Fitzsimmons, Kansas City Kings
1980 Bill Fitch, Boston Celtics
1981 Jack McKinney, Indiana Pacers
1982 Gene Shue, Washington Bullets
1983 Don Nelson, Milwaukee Bucks
1984 Frank Layden, Utah Jazz
1985 Don Nelson, Milwaukee Bucks
1986 Mike Fratello, Atlanta Hawks
1987 Mike Schuler, Portland Trail Blazers

1988 Doug Moe, Denver Nuggets
1989 Cotton Fitzsimmons, Phoenix Suns
1990 Pat Riley, Los Angeles Lakers
1991 Don Chaney, Houston Rockets
1992 Don Nelson, Golden State Warriors
1993 Pat Riley, New York Knicks
1994 Lenny Wilkens, Atlanta Hawks
1995 Del Harris, Los Angeles Lakers
1996 Phil Jackson, Chicago Bulls
1997 Pat Riley, Miami Heat
1998 Larry Bird, Indiana Pacers
1999 Mike Dunleavy, Portland Trail Blazers

NBA All-League and All-Defensive Teams, 1998-99

All-League Team			All-Defensive Team	
First team	Second team	Position	First team	Second team
Karl Malone, Utah	Chris Webber, Sacramento	Forward	Tim Duncan, San Antonio	P.J. Brown, Miami
Tim Duncan, San Antonio	Grant Hill, Detroit	Forward	Karl Malone, Utah (tie)	Theo Ratliff, Philadelphia
			Scottie Pippen, Houston (tie)	
Alonzo Mourning, Miami	Shaquille O'Neal, L.A. Lakers	Center	Alonzo Mourning, Miami	Dikembe Mutombo, Atlanta
Allen Iverson, Philadelphia	Gary Payton, Seattle	Guard	Gary Payton, Seattle	Mookie Blaylock, Atlanta
Jason Kidd, Phoenix	Tim Hardaway, Miami	Guard	Jason Kidd, Phoenix	Eddie Jones, Charlotte

NBA Statistical Leaders, 1998-99

Scoring Average
(Minimum 43 games or 854 pts)

	G	FG	FT	Pts	Avg
Iverson, Philadelphia	48	435	356	1,284	26.8
O'Neal, L.A. Lakers	49	510	269	1,289	26.3
Malone, Utah	49	393	378	1,164	23.8
Abdur-Rahim, Vancouver	50	386	369	1,152	23.0
Van Horn, New Jersey	42	322	256	916	21.8
Duncan, San Antonio	50	418	247	1,084	21.7
Payton, Seattle	50	401	199	1,084	21.7
Marbury, New Jersey	49	378	222	1,044	21.3
McDyess, Denver	50	415	230	1,061	21.2
Hill, Detroit	50	384	285	1,053	21.1

Rebounds per Game
(Minimum 43 games or 488 rebounds)

	G	Off	Def	Tot	Avg
Webber, Sacramento	42	149	396	545	13.0
Barkley, Houston	42	167	349	516	12.3
Mutombo, Atlanta	50	192	418	610	12.2
Fortson, Denver	50	210	371	581	11.6
Duncan, San Antonio	50	159	412	571	11.4
Mourning, Miami	46	166	341	507	11.0
McDyess, Denver	50	168	369	537	10.7

	G	Off	Def	Tot	Avg
O'Neal, L.A. Lakers	49	187	338	525	10.7
Garnett, Minnesota	47	166	323	489	10.4
Divac, Sacramento	50	140	361	501	10.0

Field Goal Percentage
(Minimum 183 field goals made)

	FGM	FGA	Pct
O'Neal, L.A. Lakers	510	885	.576
Thorpe, Washington	240	440	.545
Olajuwon, Houston	373	725	.514
Mourning, Miami	324	634	.511
Robinson, San Antonio	268	527	.509
Wallace, Portland	242	476	.508
Dele, Detroit	216	431	.501
Duncan, San Antonio	418	845	.495
Fortson, Denver	191	386	.495
Potapenko, Boston	204	412	.495

Free Throw Percentage
(Minimum 76 free throws made)

	FTM	FTA	Pct
Miller, Indiana	226	247	.915
Billups, Denver	157	172	.913
Armstrong, Orlando	161	178	.904

	FTM	FTA	Pct
Allen, Milwaukee	176	195	.903
Hawkins, Seattle	119	132	.902
Hornacek, Utah	125	140	.893
Mullin, Indiana	80	92	.870
Robinson, Milwaukee	140	161	.870
Elie, San Antonio	103	119	.866
Piatkowski, L.A. Clippers	88	102	.863

3-Point Field Goal Percentage
(Minimum 34 field goals made)

	FG	FGA	Pct
Curry, Milwaukee	69	145	.476
Mullin, Indiana	73	157	.465
Davis, Dallas	65	144	.451
Williams, Portland	63	144	.438
Dickerson, Houston	71	164	.433
Ellis, Seattle	94	217	.433
Hornacek, Utah	34	81	.420
Robinson, Phoenix	58	139	.417
McCloud, Phoenix	69	166	.416
Buechler, Detroit	61	148	.412
Pierce, Boston	84	204	.412

Assists per Game
(Minimum 43 games or 244 assists)

	G	No	Avg
Kidd, Phoenix	50	539	10.8
Strickland, Washington	44	434	9.9
Marbury, New Jersey	49	437	8.9
Payton, Seattle	50	436	8.7
Brandon, Minnesota	36	309	8.6
Jackson, Indiana	49	386	7.9
Knight, Cleveland	39	302	7.7

	G	No	Avg
Stockton, Utah	50	374	7.5
Johnson, San Antonio	50	369	7.4
Van Exel, Denver	50	368	7.4

Steals per Game
(Minimum 43 games or 76 steals)

	G	No	Avg
Gill, New Jersey	50	134	2.68
Jones, Charlotte	50	125	2.50
Iverson, Philadelphia	48	110	2.29
Kidd, Phoenix	50	114	2.28
Christie, Toronto	50	113	2.26
Hardaway, Orlando	50	111	2.22
Payton, Seattle	50	109	2.18
Armstrong, Orlando	50	108	2.16
Snow, Philadelphia	48	100	2.08
Blaylock, Atlanta	48	99	2.06
Ward, New York	50	103	2.06

Blocked Shots per Game
(Minimum 43 games or 61 blocked shots)

	G	Blk	Avg
Mourning, Miami	46	180	3.91
Bradley, Dallas	49	159	3.24
Ratliff, Philadelphia	50	149	2.98
Mutombo, Atlanta	50	147	2.94
Ostertag, Utah	48	131	2.73
Ewing, New York	38	100	2.63
Duncan, San Antonio	50	126	2.52
Olajuwon, Houston	50	123	2.46
Robinson, San Antonio	49	119	2.43
McDyess, Denver	50	115	2.30

NBA Rookie of the Year

Year	Player
1953	Don Meineke, Ft. Wayne
1954	Ray Felix, Baltimore
1955	Bob Pettit, Milwaukee
1956	Maurice Stokes, Rochester
1957	Tom Heinsohn, Boston
1958	Woody Sauldsberry, Philadelphia
1959	Elgin Baylor, Minneapolis
1960	Wilt Chamberlain, Philadelphia
1961	Oscar Robertson, Cincinnati
1962	Walt Bellamy, Chicago
1963	Terry Dischinger, Chicago
1964	Jerry Lucas, Cincinnati
1965	Willis Reed, New York
1966	Rick Barry, San Francisco
1967	Dave Bing, Detroit
1968	Earl Monroe, Baltimore
1969	Wes Unseld, Baltimore

Year	Player
1970	Lew Alcindor, Milwaukee
1971	Dave Cowens, Boston; Geoff Petrie, Portland (tie)
1972	Sidney Wicks, Portland
1973	Bob McAdoo, Buffalo
1974	Ernie DiGregorio, Buffalo
1975	Keith Wilkes, Golden State
1976	Alvan Adams, Phoenix
1977	Adrian Dantley, Buffalo
1978	Walter Davis, Phoenix
1979	Phil Ford, Kansas City
1980	Larry Bird, Boston
1981	Darrell Griffith, Utah
1982	Buck Williams, New Jersey
1983	Terry Cummings, San Diego
1984	Ralph Sampson, Houston

Year	Player
1985	Michael Jordan, Chicago
1986	Patrick Ewing, New York
1987	Chuck Person, Indiana
1988	Mark Jackson, New York
1989	Mitch Richmond, Golden State
1990	David Robinson, San Antonio
1991	Derrick Coleman, New Jersey
1992	Larry Johnson, Charlotte
1993	Shaquille O'Neal, Orlando
1994	Chris Webber, Golden State
1995	Grant Hill, Detroit; Jason Kidd, Dallas (tie)
1996	Damon Stoudamire, Toronto
1997	Allen Iverson, Philadelphia
1998	Tim Duncan, San Antonio
1999	Vince Carter, Toronto

NBA Individual Statistics, 1998-99

(more than 350 minutes played in lockout-shortened season; players ranked by scoring average.)

Atlanta Hawks

	Min	Avg	Reb	Ast	FG%	FT%	Pts
Smith	1,314	18.7	151	118	.402	.849	672
Blaylock	1,763	13.3	224	278	.379	.758	640
Henderson	1,142	12.5	250	27	.442	.671	474
Mutombo	1,829	10.8	610	57	.512	.684	541
L. Ellis	539	10.2	109	18	.421	.705	204
Long	1,380	9.8	296	53	.421	.783	489
Corbin	1,066	7.5	145	43	.392	.650	352
Crawford	784	6.9	89	24	.431	.814	288
Johnson	885	5.0	75	107	.404	.695	244
West	499	1.2	125	13	.373	.356	60

Coach-Lenny Wilkens

Boston Celtics

	Min	Avg	Reb	Ast	FG%	FT%	Pts
Walker	1,549	18.7	359	130	.412	.559	784
Mercer	1,551	17.0	155	104	.431	.790	698
Pierce	1,632	16.5	309	115	.439	.713	791
Anderson	1,010	12.1	103	193	.451	.832	412
Potapenko	1,394	10.0	332	75	.495	.587	499
Barros	1,156	9.3	105	208	.453	.877	464
Battie	1,121	6.7	300	53	.519	.672	335
Mccarty	659	5.7	115	40	.362	.702	181
Minor	765	4.9	117	50	.417	.750	214
Bowen	494	2.3	52	28	.280	.458	70

Coach-Rick Pitino

Charlotte Hornets

	Min	Avg	Reb	Ast	FG%	FT%	Pts
Jones	1,881	15.6	194	186	.437	.782	780
Phills	1,574	14.3	174	149	.433	.685	613
Wesley	1,848	14.1	161	322	.446	.832	706
Coleman	1,178	13.1	328	78	.414	.753	486
Campbell	1,459	12.6	397	69	.477	.639	616
Brown	1,192	8.5	174	57	.472	.678	407
Miller	469	6.3	117	22	.565	.794	238
Person	990	6.1	132	60	.388	.750	303
Recasner	708	5.1	77	91	.446	.872	222
Davis	557	4.5	84	58	.405	.763	209
Shackleford	367	3.3	129	13	.494	.655	107

Coach-Dave Cowens, Paul Silas

Chicago Bulls

	Min	Avg	Reb	Ast	FG%	FT%	Pts
Kukoc	1,654	18.8	310	235	.420	.740	828
Harper	1,107	11.2	180	115	.377	.703	392
B. Barry	1,181	11.1	144	116	.396	.772	412
Simpkins	1,448	9.1	339	65	.463	.642	454
Bryant	1,204	9.0	232	48	.483	.645	407
Brown	1,139	8.8	132	149	.414	.762	344
David	902	6.2	173	40	.449	.811	308
Larue	732	4.7	56	63	.359	1.000	203
C. Carr	624	4.1	49	66	.329	.750	171
Lang	386	3.8	93	13	.323	.696	80
Wennington	451	3.8	79	18	.348	.818	143
Jones	476	3.7	42	41	.317	.500	108
Booth	432	3.1	93	38	.325	.500	120

Coach-Tim Floyd

Cleveland Cavaliers

	Min	Avg	Reb	Ast	FG%	FT%	Pts
Kemp	1,475	20.5	388	101	.482	.789	862
Person	1,342	11.2	142	80	.453	.604	503
Anderson	978	10.8	109	145	.398	.836	409
Knight	1,186	9.6	131	302	.425	.745	373
Henderson	1,517	9.1	197	113	.417	.813	454
Declercq	1,102	7.9	255	31	.500	.674	371
Ferry	1,058	7.0	102	53	.476	.879	349
Newman	949	6.1	75	41	.422	.810	303
Butler	418	5.4	44	22	.482	.719	168
Sura	841	4.3	102	152	.333	.631	214
Blount	530	2.9	151	12	.360	.519	100

Coach-Mike Fratello

Dallas Mavericks

	Min	Avg	Reb	Ast	FG%	FT%	Pts
Finley	205	20.2	263	218	.445	.823	1,011
Trent	1,362	14.4	351	77	.477	.617	719
Ceballos	352	12.5	85	12	.421	.694	163
Davis	1,378	9.1	86	89	.438	.880	457
Pack	468	8.8	36	81	.428	.818	220
Bradley	1,294	8.6	392	40	.480	.748	420
Nowitzki	958	8.2	162	47	.405	.773	385
Nash	1,269	7.9	114	219	.363	.826	315
Strickland	567	7.6	83	64	.403	.815	249
Walker	568	5.9	143	6	.463	.541	229
A.C. Green	924	4.9	228	25	.422	.577	246
Anstey	470	3.3	97	27	.360	.708	134
Williams	403	1.2	83	15	.333	.700	29

Coach-Don Nelson

Denver Nuggets

	Min	Avg	Reb	Ast	FG%	FT%	Pts
Mcdyess	1,937	21.2	537	82	.471	.680	1061
Van Exel	1,802	16.5	113	368	.398	.811	826
Billups	1,488	13.9	96	173	.386	.913	624
Lafrentz	387	13.8	91	8	.457	.750	166
Fortson	1,417	11.0	581	32	.495	.727	550
E. Williams	780	7.3	81	37	.365	.799	277
Alexander	778	7.3	74	119	.373	.841	261
Stith	1,194	7.0	107	82	.393	.859	320
Taylor	724	5.8	101	24	.414	.739	207
Washington	761	5.4	89	30	.397	.688	205
Clark	409	3.3	96	10	.450	.568	93

Coach-Mike D'Antonio

Detroit Pistons

	Min	Avg	Reb	Ast	FG%	FT%	Pts
Hill	1,852	21.1	355	300	.479	.752	1,053
Stackhouse	1,188	14.5	107	118	.371	.850	607
Hunter	1,755	11.9	168	193	.435	.753	582
Dumars	1,116	11.3	68	134	.411	.836	428
Dele	1,177	10.5	273	71	.501	.686	513
J. Williams	1,154	7.1	349	23	.500	.673	355
Buechler	1,056	5.5	133	57	.417	.722	274
Reid	935	5.2	170	33	.557	.608	242
Vaught	481	3.4	146	11	.381	.643	127
Montross	577	2.1	139	14	.525	.344	95

Coach-Alvin Gentry

Golden State Warriors

	Min	Avg	Reb	Ast	FG%	FT%	Pts
Starks	1,686	13.8	163	235	.370	.740	690
Marshall	1,250	11.0	342	66	.421	.727	530
Mills	1,395	10.3	237	103	.411	.823	483
Jamison	1,058	9.6	301	34	.452	.588	449
Coles	1,272	9.5	117	222	.442	.822	455
Cummings	1,011	9.1	255	58	.439	.711	454
Dampier	1,414	8.8	382	54	.389	.588	442
Caffey	876	8.8	205	18	.444	.633	308
Delk	630	6.8	54	95	.364	.648	246
Bogues	714	5.1	73	134	.494	.861	183
Foyle	614	2.9	194	18	.430	.490	129

Coach-P.J. Carlesimo

Houston Rockets

	Min	Avg	Reb	Ast	FG%	FT%	Pts
Olajuwon	1,784	18.9	478	88	.514	.717	945
Barkley	1,526	16.1	516	192	.478	.719	676
Pippen	2,011	14.5	323	293	.432	.721	726
Dickerson	1,558	10.9	83	95	.465	.639	547
Mack	1,083	10.7	95	55	.435	.879	472
Mobley	1,456	9.9	111	121	.425	.818	487
Harrington	902	9.8	246	15	.513	.721	400
Price	807	7.3	78	113	.483	.754	292
Drew	441	3.5	32	52	.364	1.000	118

Coach-Rudy Tomjanovich

Indiana Pacers

	Min	Avg	Reb	Ast	FG%	FT%	Pts
Miller	1,787	18.4	135	112	.438	.915	920
Smits	1,271	14.9	275	52	.490	.818	728
Rose	1,238	11.1	154	93	.403	.791	542
Mullin	1,179	10.1	160	81	.477	.870	507
A. Davis	1,271	9.4	344	33	.471	.703	463
D. Davis	1,374	8.0	416	22	.533	.618	398
Jackson	1,382	7.6	184	386	.419	.823	373
Best	1,043	7.1	80	169	.416	.843	346
Perkins	789	5.0	138	25	.400	.717	238

Coach-Larry Bird

Los Angeles Clippers

	Min	Avg	Reb	Ast	FG%	FT%	Pts
Taylor	1,505	16.8	242	67	.461	.728	773
Murray	1,317	12.2	195	61	.391	.803	612
Piatkowski	1,242	10.5	140	53	.432	.863	513
Nesby	1,288	10.1	175	82	.449	.782	503
Olowokandi	1,279	8.9	357	25	.431	.483	401
Douglas	842	8.2	58	124	.438	.632	247
Martin	941	8.0	48	144	.367	.803	296
Rogers	967	7.4	179	77	.441	.673	348
Hudson	524	6.8	55	92	.400	.895	169
Wright	1,135	6.7	361	33	.458	.692	319

Coach-Chris Ford

Los Angeles Lakers

	Min	Avg	Reb	Ast	FG%	FT%	Pts
O'Neal	1,705	26.3	525	114	.576	.540	1,289
Bryant	1,896	19.9	264	190	.465	.839	996
Rice	985	17.5	99	71	.432	.856	472
J.R. Reid	1,029	9.0	212	48	.477	.766	369
Fox	944	8.9	89	89	.448	.742	394
Harper	1,120	6.9	67	187	.412	.813	309
Fisher	1,131	5.9	91	197	.376	.759	296
Horry	744	5.0	152	56	.459	.739	188
Knight	525	4.2	128	31	.515	.759	156
Rodman	657	2.1	258	30	.348	.436	49

Coach-Del Harris, Kurt Rambis

Miami Heat

	Min	Avg	Reb	Ast	FG%	FT%	Pts
Mourning	1,753	20.1	507	74	.512	.652	924
Hardaway	1,772	17.4	153	352	.400	.812	835
Mashburn	855	14.8	146	75	.451	.721	356
P.J. Brown	1,611	11.4	342	66	.480	.774	571
Porter	1,365	10.5	140	146	.465	.831	525
Weatherspoon	1,040	8.1	243	34	.534	.804	397
Majerle	1,623	7.0	208	150	.396	.717	337
Strickland	357	3.7	78	9	.495	.731	119
Walters	506	3.1	50	58	.365	.826	101
Askin	416	1.6	44	10	.323	.625	53

Coach-Pat Riley

Milwaukee Bucks

	Min	Avg	Reb	Ast	FG%	FT%	Pts
G. Robinson	1,578	18.4	276	100	.459	.870	865
Allen	1,719	17.1	212	178	.450	.903	856
Curry	866	10.1	85	48	.485	.824	423
Gilliam	668	8.3	126	19	.453	.782	281
Thomas	813	7.2	126	46	.473	.652	358
Workman	816	6.9	102	172	.429	.787	200
Del Negro	1,093	5.9	102	174	.422	.800	281
Gatling	776	5.7	179	32	.427	.398	272
Traylor	786	5.3	182	38	.537	.538	259
Johnson	1,028	5.1	320	19	.508	.610	256
Curry	1,141	4.9	108	78	.437	.797	244

Coach-George Karl

Minnesota Timberwolves

	Min	Avg	Reb	Ast	FG%	FT%	Pts
Garnett	1,780	20.8	489	202	.460	.704	977
Brandon	1,217	13.9	134	309	.418	.833	501
Smith	1,418	13.7	354	68	.427	.755	588
Mitchell	1,344	11.2	182	98	.408	.764	561
Peeler	810	9.6	84	78	.379	.732	270
Sealy	731	8.1	92	36	.411	.902	251
Jackson	941	7.1	135	167	.405	.772	353
Scott	738	6.5	58	40	.408	.742	234
J. Robinson	507	5.9	62	56	.362	.683	183
Garrett	1,054	5.5	257	28	.502	.745	270
Hammonds	716	4.3	136	20	.458	.640	212
Curley	372	2.2	51	14	.403	.864	78

Coach-Flip Saunders

New Jersey Nets

	Min	Avg	Reb	Ast	FG%	FT%	Pts
Van Horn	1,576	21.8	358	65	.428	.859	916
Marbury	1,895	21.3	142	437	.428	.799	1,044
Kittles	1,570	12.9	191	116	.370	.772	592
Gill	1,606	11.8	244	123	.398	.683	588
J. Williams	1,020	8.1	360	33	.443	.565	242
Murdock	401	7.9	35	66	.395	.808	119
Burrell	706	6.6	119	45	.361	.810	212

	Min	Avg	Reb	Ast	FG%	FT%	Pts
Feick	852	6.3	288	24	.500	.717	177
Hendrickson	399	5.5	68	13	.443	.840	120
Harris	602	5.4	67	31	.403	.750	193
C. Carr	445	5.3	71	23	.371	.675	207

Coach-John Calipari, Don Casey

New York Knickerbockers

	Min	Avg	Reb	Ast	FG%	FT%	Pts
Ewing	1,300	17.3	377	43	.435	.706	657
Sprewell	1,233	16.4	156	91	.415	.812	606
Houston	1,815	16.3	152	137	.418	.862	813
L. Johnson	1,639	12.0	284	119	.459	.817	587
K. Thomas	1,182	8.1	286	55	.462	.611	406
Ward	1,556	7.6	172	271	.404	.705	378
Camby	945	7.2	253	12	.521	.553	329
Childs	1,297	6.8	133	193	.427	.821	328
Dudley	685	2.5	193	7	.440	.475	115

Coach-Jeff Van Gundy

Orlando Magic

	Min	Avg	Reb	Ast	FG%	FT%	Pts
Hardaway	1,944	15.8	284	266	.420	.706	791
Anderson	1,581	14.9	277	91	.395	.611	701
Armstrong	1,502	13.8	180	335	.441	.904	690
Austin	1,259	9.7	237	89	.408	.669	477
H. Grant	1,660	8.9	351	90	.434	.671	443
Harpring	1,114	8.2	214	45	.463	.713	408
Outlaw	851	6.6	167	56	.545	.432	203
Doleac	780	6.2	147	20	.466	.675	302
Strong	695	5.1	161	17	.422	.717	223
B.J. Armstrong	358	3.3	39	61	.455	.857	105

Coach-Chuck Daly

Philadelphia 76ers

	Min	Avg	Reb	Ast	FG%	FT%	Pts
Iverson	1,989	26.8	236	223	.412	.751	1,284
Geiger	1,540	13.5	362	58	.479	.797	674
Ratliff	1,627	11.2	407	30	.470	.725	560
Hughes	988	9.1	189	77	.411	.709	455
Snow	1,716	8.6	162	301	.428	.733	413
Hill	1,108	8.6	287	35	.455	.540	325
Lynch	1,312	8.3	279	76	.421	.631	356
Mckie	959	4.8	140	100	.401	.710	240
Grant	798	3.1	110	23	.369	.724	146

Coach-Larry Brown

Phoenix Suns

	Min	Avg	Reb	Ast	FG%	FT%	Pts
Gugliotta	1,563	17.0	381	121	.483	.794	729
Kidd	2,060	16.9	340	539	.444	.757	846
C. Robinson	1,740	16.3	226	128	.475	.697	817
Chapman	1,183	12.1	104	109	.359	.835	459
Manning	1,184	9.1	219	113	.486	.696	453
Mccloud	1,245	8.9	162	79	.438	.862	428
Longley	933	8.7	222	45	.485	.776	341
Garrity	538	5.6	75	18	.500	.714	217
Morris	535	4.2	121	23	.430	.870	184
Kleine	374	2.2	67	12	.405	.667	68

Coach-Danny Ainge

Portland Trail Blazers

	Min	Avg	Reb	Ast	FG%	FT%	Pts
Rider	1,385	13.9	196	104	.412	.755	651
Wallace	1,414	12.8	241	60	.508	.732	628
Stoudamire	1,673	12.6	167	312	.396	.730	631
Sabonis	1,349	12.1	393	119	.485	.771	606
B. Grant	1,525	11.5	470	67	.479	.814	550
W. Williams	1,044	9.3	143	80	.424	.832	446
J. Jackson	1,175	8.4	159	128	.411	.842	414
Anthony	806	6.4	63	100	.414	.697	319
Augmon	874	4.3	125	58	.448	.684	208
Cato	546	3.5	150	19	.450	.507	151

Coach-Mike Dunleavy

Sacramento Kings

	Min	Avg	Reb	Ast	FG%	FT%	Pts
Webber	1,719	20.0	545	173	.486	.454	839
Divac	1,761	14.3	501	215	.470	.702	714
Williamson	1,374	13.2	206	66	.485	.638	659
J. Williams	1,805	12.8	153	299	.374	.752	641
Maxwell	1,016	10.5	88	76	.391	.737	494
Abdul-wahad	1,205	9.3	186	50	.435	.691	454
Funderburke	936	8.9	222	30	.559	.708	420
Stojakovic	1,025	8.4	143	72	.378	.851	402

	Min	Avg	Reb	Ast	FG%	FT%	Pts
J. Barry	736	5.0	96	112	.428	.845	213

Coach-Rick Adelman

San Antonio Spurs

	Min	Avg	Reb	Ast	FG%	FT%	Pts
Duncan	1,963	21.7	571	121	.495	.690	1,084
D. Robinson	1,554	15.8	492	103	.509	.658	775
Elliott	1,509	11.2	213	117	.410	.757	561
A. Johnson	1,672	9.7	118	369	.473	.568	487
Elie	1,291	9.7	137	89	.471	.866	455
J. Jackson	861	6.4	99	49	.380	.821	301
Rose	608	6.0	182	29	.463	.671	284
Daniels	614	4.7	54	106	.454	.754	220
Kerr	734	4.4	44	49	.391	.886	192
Kersey	699	3.2	130	41	.340	.429	145
Perdue	445	2.4	138	18	.633	.538	90

Coach-Gregg Popovich

Seattle SuperSonics

	Min	Avg	Reb	Ast	FG%	FT%	Pts
Payton	2,008	21.7	244	436	.434	.721	1,084
Schrempf	1,765	15.0	370	184	.472	.823	752
Baker	1,162	13.8	211	56	.453	.450	468
Maclean	365	10.9	65	16	.396	.625	185
Hawkins	1,644	10.3	201	123	.419	.902	516
D. Ellis	1,232	10.3	115	38	.441	.757	495
Owens	451	7.8	80	38	.394	.800	163
Polynice	1,481	7.7	425	43	.472	.309	368
Crotty	363	6.1	30	58	.405	.851	147
A. Williams	458	4.0	128	22	.423	.730	158

Coach-Paul Westphal

Toronto Raptors

	Min	Avg	Reb	Ast	FG%	FT%	Pts
Carter	1,760	18.3	283	149	.450	.761	913
Christie	1,768	15.2	207	187	.388	.841	760
Willis	1,216	12.0	350	67	.418	.839	504
Brown	1,377	11.2	103	143	.378	.727	549
Mcgrady	1,106	9.4	278	113	.436	.726	458
Wallace	812	8.6	171	46	.432	.700	411
Oakley	1,633	7.0	374	168	.428	.807	348
A. Williams	1,051	5.0	82	130	.401	.846	248
Thomas	593	4.3	134	15	.577	.563	169
Stewart	394	1.5	99	5	.415	.680	61

Coach-Butch Carter

Utah Jazz

	Min	Avg	Reb	Ast	FG%	FT%	Pts
Malone	1,832	23.8	463	201	.493	.788	1,164
Russell	1,770	12.4	266	74	.464	.795	622
Hornacek	1,435	12.2	160	192	.477	.893	587
Stockton	1,410	11.1	146	374	.488	.811	553
Anderson	1,072	8.5	132	56	.446	.712	427
Eisley	1,038	7.4	94	185	.446	.838	368
Ostertag	1,340	5.7	348	23	.476	.620	273
Bailey	543	4.2	94	26	.446	.735	181
Keefe	642	4.0	142	28	.452	.697	174
Fuller	462	3.4	101	6	.452	.600	142
Foster	458	2.8	83	25	.377	.619	118

Coach-Jerry Sloan

Vancouver Grizzlies

	Min	Avg	Reb	Ast	FG%	FT%	Pts
Abdur-rahim	2,021	23.0	374	172	.432	.841	1,152
Bibby	1,758	13.2	136	325	.430	.751	662
Massenburg	1,145	11.2	257	23	.487	.665	481
Reeves	702	10.8	138	37	.406	.578	271
Lopez	1,218	9.3	166	62	.446	.644	437
Parks	1,118	5.5	243	36	.429	.545	266
Smith	1,098	4.8	350	48	.535	.594	230
Wheat	590	4.5	45	102	.378	.727	208
Chilcutt	703	3.7	118	30	.364	.824	168

Coach-Brian Hill

Washington Wizards

	Min	Avg	Reb	Ast	FG%	FT%	Pts
Richmond	1,912	19.7	172	122	.412	.857	983
Howard	1,430	19.0	294	107	.474	.753	684
Strickland	1,632	15.7	212	434	.416	.746	690
Thorpe	1,539	11.3	334	101	.545	.698	554
Cheaney	1,266	7.7	141	73	.414	.493	385
Murray	653	6.5	81	27	.350	.810	233
B. Wallace	1,231	6.0	383	18	.576	.356	275
Whitney	441	4.8	47	69	.410	.871	187
Legler	377	4.0	40	21	.443	.500	119
Mcinnis	427	3.7	21	73	.373	.750	130
T. Davis	578	3.4	139	10	.533	.737	126

Coach-Bernie Bickerstaff, Jim Brovelli

Michael Jordan Ends Unparalleled Career

In January 1999, Michael Jordan, arguably the best player in the history of the National Basketball Association (NBA), announced his retirement from the game. Jordan had retired once before, in October 1993, but returned to the NBA in March 1995. In his 13-year career, all with Chicago, Jordan led the Bulls to 6 NBA Championships (1991-93 and 1996-98) and was named the Finals MVP for all of them. He was a 5-time NBA regular season MVP and was a first-team NBA All-Star 10 times. His 29,277 career points are the 3d highest in league history, he holds the highest career scoring average (31.5 ppg), and he led the league in scoring for a record 10 seasons. Jordan is known for his athleticism, determination, and timely play; fittingly, his last shot in the NBA clinched the Finals against Utah in 1998. Outside the NBA, Jordan is a worldwide icon and one of the most recognized sports figures in the world. He has appeared in numerous television commercials and in the movies.

Michael Jordan's Career Scoring Statistics With the Chicago Bulls

Year	Regular Season					Playoffs				
	Games	FG%	FT%	Points	PPG	Games	FG%	FT%	Points	PPG
1984-85	82	.515	.845	2,313	28.2	4	.436	.828	117	29.3
1985-86	18	.457	.840	408	22.7	3	.505	.872	131	43.7
1986-87	82	.482	.857	3,041	37.1	3	.417	.897	107	35.7
1987-88	82	.535	.841	2,868	35.0	10	.531	.869	363	36.3
1988-89	81	.538	.850	2,633	32.5	17	.510	.799	591	34.8
1989-90	82	.526	.848	2,753	33.6	16	.514	.836	587	36.7
1990-91	82	.539	.851	2,580	31.5	17	.524	.845	529	31.1
1991-92	80	.519	.832	2,404	30.1	22	.499	.857	759	34.5
1992-93	78	.495	.837	2,541	32.6	18	.475	.805	666	35.1
1994-95	17	.411	.801	457	26.9	10	.484	.810	315	31.5
1995-96	82	.495	.834	2,491	30.4	18	.459	.818	552	30.7
1996-97	82	.486	.833	2,431	29.6	19	.456	.831	590	31.1
1997-98	82	.465	.784	2,357	28.7	21	.462	.812	680	32.4
Total	930	.505	.838	29,277	31.5	179	.487	.828	5,987	33.4

1999 NBA Player Draft, First-Round Picks
(held June 30, 1999)

Team	Player, College
1. Chicago	Elton Brand, F, Duke
2. Vancouver	Steve Francis, G, Maryland
3. Charlotte	Baron Davis, G, UCLA
4. L.A. Clippers	Lamar Odom, F, Rhode Island
5. Toronto[1]	Jonathon Bender, C/F, Picayune (High School), Picayune, MS
6. Minnesota[2]	Wally Szczerbiak, F, Miami (OH)
7. Washington	Richard Hamilton, G/F, Connecticut
8. Cleveland[3]	Andre Miller, G, Utah
9. Phoenix[4]	Shawn Marion, F, UNLV
10. Atlanta[5]	Jason Terry, G, Arizona
11. Cleveland	Trajan Langdon, G, Duke
12. Toronto	Aleksandar Radojevic, C, Barton County (KS) Community College
13. Seattle	Corey Maggette[6], F, Duke
14. Minnesota	William Avery, G, Duke
15. New York	Frederic Weis, C, Limoges (France)
16. Chicago[7]	Ron Artest, G/F, St. John's (NY)
17. Atlanta[8]	Cal Bowdler, F, Old Dominion
18. Denver[9]	James Posey, F, Xavier (OH)
19. Utah[10]	Quincy Lewis, F, Minnesota
20. Atlanta[11]	Dion Glover, G, Georgia Tech
21. Golden State[12]	Jeff Foster[13], C/F, SW Texas State
22. Houston	Kenny Thomas, F, New Mexico
23. L.A. Lakers	Devean George, G/F, Augsburg (MN)
24. Utah[14]	Andrei Kirilenko, F, CSKA (Russia)
25. Miami	Tim James, F, Miami (FL)
26. Indiana	Vonteego Cummings[15], G, Pittsburgh
27. Atlanta[16]	Jumaine Jones[17], F, Georgia
28. Utah	Scott Padgett, F, Kentucky
29. San Antonio	Leon Smith[18], C/F, Martin Luther King (High School), Chicago, IL

(1) From Denver. (2) From New Jersey. (3) From Boston. (4) From Dallas. (5) From Golden State. (6) Traded to Orlando. (7) From Phoenix. (8) From Sacramento. (9) From Milwaukee through Phoenix. (10) From Philadelphia. (11) From Detroit. (12) From Atlanta. (13) Traded to Indiana. (14) From Orlando. (15) Traded to Golden State. (16) From Portland through Detroit. (17) Traded to Philadelphia. (18) Traded to Dallas.

Number-One First-Round NBA Draft Picks, 1966-99

Year	Team	Player, college	Year	Team	Player, college
1966	New York	Cazzie Russell, Michigan	1983	Houston	Ralph Sampson, Virginia
1967	Detroit	Jimmy Walker, Providence	1984	Houston	Akeem Olajuwon, Houston
1968	Houston	Elvin Hayes, Houston	1985	New York	Patrick Ewing, Georgetown
1969	Milwaukee	Lew Alcindor[1], UCLA	1986	Cleveland	Brad Daugherty, North Carolina
1970	Detroit	Bob Lanier, St. Bonaventure	1987	San Antonio	David Robinson, Navy
1971	Cleveland	Austin Carr, Notre Dame	1988	L.A. Clippers	Danny Manning, Kansas
1972	Portland	LaRue Martin, Loyola-Chicago	1989	Sacramento	Pervis Ellison, Louisville
1973	Philadelphia	Doug Collins, Illinois St.	1990	New Jersey	Derrick Coleman, Syracuse
1974	Portland	Bill Walton, UCLA	1991	Charlotte	Larry Johnson, UNLV
1975	Atlanta	David Thompson[2], N.C. State	1992	Orlando	Shaquille O'Neal, LSU
1976	Houston	John Lucas, Maryland	1993	Orlando	Chris Webber[3], Michigan
1977	Milwaukee	Kent Benson, Indiana	1994	Milwaukee	Glenn Robinson, Purdue
1978	Portland	Mychal Thompson, Minnesota	1995	Golden State	Joe Smith, Maryland
1979	L.A. Lakers	Magic Johnson, Michigan St.	1996	Philadelphia	Allen Iverson, Georgetown
1980	Golden State	Joe Barry Carroll, Purdue	1997	San Antonio	Tim Duncan, Wake Forest
1981	Dallas	Mark Aguirre, DePaul	1998	L.A. Clippers	Michael Olowokandi, Pacific
1982	L.A. Lakers	James Worthy, North Carolina	1999	Chicago Bulls	Elton Brand, Duke

(1) Later Kareem Abdul-Jabbar. (2) Signed with Denver of the ABA. (3) Traded to Golden State.

All-Time NBA Statistical Leaders
(At the start of the 1999-2000 season. *Player active in 1998-99 season.)

Scoring Average
(Minimum 400 games or 10,000 pts)

	G	Pts.	Avg		G	Pts.	Avg
Michael Jordan	930	29,277	31.5	Jerry West	932	25,192	27.0
Wilt Chamberlain	1,045	31,419	30.1	Bob Pettit	792	20,880	26.4
Elgin Baylor	846	23,149	27.4	George Gervin	791	20,708	26.2
*Shaquille O'Neal	455	12,343	27.1	*Karl Malone	1,110	28,946	26.1
				Oscar Robertson	1,040	26,710	25.7
				*Dominique Wilkins	1,074	26,668	24.8

Field Goal Percentage
(Minimum 2,000 field goals made)

	FGA	FGM	Pct.
Artis Gilmore	9,570	5,732	.599
*Mark West	4,344	2,523	.581
*Shaquille O'Neal	8,545	4,940	.578
Steve Johnson	4,965	2,841	.572
Darryl Dawkins	6,079	3,477	.572
James Donaldson	5,442	3,105	.571
Jeff Ruland	3,734	2,105	.564
Kareem Abdul-Jabbar	28,307	15,837	.559
Kevin McHale	12,334	6,830	.554
Bobby Jones	6,199	3,412	.550
Dale Davis	3,757	2,067	.550

Free Throw Percentage
(Minimum 1,200 free throws made)

	FTA	FTM	Pct.
Mark Price	2,362	2,135	.904
Rick Barry	4,243	3,818	.900
Calvin Murphy	3,864	3,445	.892
Scott Skiles	1,741	1,548	.889
Larry Bird	4,471	3,960	.886
Bill Sharman	3,559	3,143	.883
*Reggie Miller	5,284	4,642	.879
Ricky Pierce	3,871	3,389	.875
*Jeff Hornacek	3,210	2,802	.873
Kiki Vandeweghe	3,997	3,484	.872

Points

Kareem Abdul-Jabbar	38,387
Wilt Chamberlain	31,419
Michael Jordan	29,277
*Karl Malone	28,946
Moses Malone	27,409
Elvin Hayes	27,313
Oscar Robertson	26,710
*Dominique Wilkins	26,668
John Havlicek	26,395
Alex English	25,613

Games Played

Robert Parish	1,611
Kareem Abdul-Jabbar	1,560
Moses Malone	1,329
Buck Williams	1,307
Elvin Hayes	1,303
John Havlicek	1,270
Paul Silas	1,254

*Derek Harper	1,199
*Eddie Johnson	1,199
Alex English	1,193
James Edwards	1,168

Assists

*John Stockton	13,087
Magic Johnson	10,141
Oscar Robertson	9,887
Isiah Thomas	9,061
*Mark Jackson	7,924
Maurice Cheeks	7,392
Lenny Wilkens	7,211
Bob Cousy	6,955
Guy Rodgers	6,917
Kevin Johnson	6,687

Field Goals Made

Kareem Abdul-Jabbar	15,837
Wilt Chamberlain	12,681

Elvin Hayes	10,976
Michael Jordan	10,962
*Karl Malone	10,683
Alex English	10,659
John Havlicek	10,513
*Hakeem Olajuwon	10,079
*Dominique Wilkins	9,963
Robert Parish	9,614

Rebounds

Wilt Chamberlain	23,924
Bill Russell	21,620
Kareem Abdul-Jabbar	17,440
Elvin Hayes	16,279
Moses Malone	16,212
Robert Parish	14,715
Nate Thurmond	14,464
Walt Bellamy	14,241
Wes Unseld	13,769
Buck Williams	13,017

Basketball Hall of Fame, Springfield, MA
(1999 inductees have an asterisk*)

PLAYERS

Abdul-Jabbar, Kareem
Archibald, Nate
Arizin, Paul
Barlow, Thomas
Barry, Rick
Baylor, Elgin
Beckman, John
Bellamy, Walt
Belov, Sergei
Bing, Dave
Bird, Larry
Blazejowski, Carol
Borgmann, Bennie
Bradley, Bill
Brennan, Joseph
Cervi, Al
Chamberlain, Wilt
Cooper, Charles
Cosic, Kresimir
Cousy, Bob
Cowens, Dave
Crawford, Joan
Cunningham, Billy
Curry, Denise
Davies, Bob
DeBernardi, Forrest
DeBusschere, Dave
Denhart, Dutch
Donovan, Anne
Endacott, Paul
English, Alex
Erving, Julius (Dr. J)
Foster, Bud
Frazier, Walt
Friedman, Max
Fulks, Joe
Gale, Lauren
Gallatin, Harry
Gates, Pop
Gervin, George
Gola, Tom
Goodrich, Gail
Greer, Hal
Gruenig, Ace
Hagan, Cliff
Hanson, Victor
Harris-Stewart, Luisa
Havlicek, John

Hawkins, Connie
Hayes, Elvin
Haynes, Marques
Heinsohn, Tom
Holman, Nat
Houbregs, Bob
Howell, Bailey
Hyatt, Chuck
Issel, Dan
Jeannette, Buddy
Johnson, William
Johnston, Neil
Jones, K.C.
Jones, Sam
Krause, Moose
Kurland, Bob
Lanier, Bob
Lapchick, Joe
Lieberman-Cline, Nancy
Lovellette, Clyde
Lucas, Jerry
Luisetti, Hank
Macauley, Ed
Maravich, Pete
Martin, Slater
McCracken, Branch
McCracken, Jack
McDermott, Bobby
McGuire, Dick
*McHale, Kevin
Meyers, Ann
Mikan, George
Mikkelsen, Vern
Miller, Cheryl
Monroe, Earl
Murphy, Calvin
Murphy, Stretch
Page, Pat
Pettit, Bob
Phillip, Andy
Pollard, Jim
Ramsey, Frank
Reed, Willis
Risen, Arnie
Robertson, Oscar
Roosma, John S.
Russell, Bill
Russell, Honey
Schayes, Adolph

Schmidt, Ernest
Schommer, John
Sedran, Barney
Semjonova, Uljana
Sharman, Bill
Steinmetz, Christian
Thompson, Cat
Thompson, David
Thurmond, Nate
Twyman, Jack
Unseld, Wes
Vandivier, Fuzzy
Wachter, Edward
Walton, Bill
Wanzer, Bobby
West, Jerry
White, Nera
Wilkens, Lenny
Wooden, John
Yardley, George

COACHES

Allen, Forrest (Phog)
Anderson, Harold
Auerbach, Red
Barry, Sam
Blood, Ernest
Cann, Howard
Carlson, Dr. H. C.
Carnesecca, Lou
Carnevale, Ben
Carril, Pete
Case, Everett
Conradt, Jody
Crum, Denny
Daly, Chuck
Dean, Everett
Diaz-Miguel, Antonio
Diddle, Edgar
Drake, Bruce
Gaines, Clarence
Gardner, Jack
Gill, Slats
Gomelsky, Aleksandr
Hannum, Alex
Harshman, Marv
Haskins, Don
Hickey, Edgar
Hobson, Howard

Holzman, Red
Iba, Hank
Julian, Alvin
Keaney, Frank
Keogan, George
Knight, Bob
Kundla, John
Lambert, Ward
Litwack, Harry
Loeffler, Kenneth
Lonborg, Dutch
McCutchan, Arad
McGuire, Al
McGuire, Frank
McLendon, John
Meanwell, Dr. W. E.
Meyer, Ray
Miller, Ralph
*Moore, Billie
Newell, Pete
Nikolic, Aleksandar
Ramsay, Jack
Rubini, Cesare
Rupp, Adolph
Sachs, Leonard
Shelton, Everett
Smith, Dean
Taylor, Fred
*Thompson, John
Wade, Margaret
Watts, Stan
Wilkens, Lenny
Wooden, John
Woolpert, Phil

REFEREES

Enright, James
Hepbron, George
Hoyt, George
Kennedy, Matthew
Leith, Lloyd
Mihalik, Red
Nucatola, John
Quigley, Ernest
Shirley, J. Dallas
Strom, Earl
Tobey, David
Walsh, David

CONTRIBUTORS

Abbott, Senda B.
Bee, Clair
Brown, Walter
Bunn, John
Douglas, Bob
Duer, Al O.
*Embry, Wayne
Fagan, Cliff
Fisher, Harry
Fleisher, Larry
Gottlieb, Edward
Gulick, Dr. L. H.
Harrison, Lester
Hepp, Dr. Ferenc
Hickox, Edward
Hinkle, Tony
Irish, Ned
Jones, R. W.
Kennedy, Walter
Liston, Emil
Mokray, Bill
Morgan, Ralph
Morgenweck, Frank
Naismith, Dr. James
O'Brien, John
O'Brien, Larry
Olsen, Harold
Podoloff, Maurice
Porter, H. V.
Reid, William
Ripley, Elmer
St. John, Lynn
Saperstein, Abe
Schabinger, Arthur
Stagg, Amos Alonzo
Stankovich, Boris
Steitz, Edward
Taylor, Chuck
Teague, Bertha
Tower, Oswald
Trester, Arthur
Wells, Clifford
Wilke, Lou
*Zollner, Fred

TEAMS

First Team
Original Celtics
Buffalo Germans
NY Renaissance

All-Time NBA Coaching Victories

(At the start of the 1999-00 season. *Active through 1998-99 season.)

Coach	W-L	Pct.	Coach	W-L	Pct.
*Lenny Wilkens	1,151-927	.554	Alvin Attles	557-518	.518
*Pat Riley	947-404	.701	*Del Harris	556-457	.549
Bill Fitch	944-1,106	.460	Phil Jackson	545-193	.738
Red Auerbach	938-479	.662	*George Karl	531-348	.604
Dick Motta	935-1,017	.479	K.C. Jones	522-252	.674
*Don Nelson	886-710	.555	Kevin Loughery	474-662	.417
Jack Ramsay	864-783	.525	Alex Hannum	471-412	.533
Cotton Fitzsimmons	832-775	.518	Billy Cunningham	454-196	.698
Gene Shue	784-861	.477	Larry Costello	430-300	.589
John MacLeod	707-657	.518	Tom Heinsohn	427-263	.619
Red Holzman	696-604	.535	John Kundla	423-302	.583
*Larry Brown	683-553	.553	*Rick Adelman	384-275	.583
*Jerry Sloan	676-392	.633	*Rudy Tomjanovich	353-219	.617
*Chuck Daly	638-437	.593	Bill Russell	341-290	.540
Doug Moe	628-529	.543	Hubie Brown	341-410	.454
*Mike Fratello	572-465	.552	*Bernie Bickerstaff	338-348	.493

NBA Home Courts

Team	Name (built)	Capacity	Team	Name (built)	Capacity
Atlanta	Philips Arena (1999)	20,000	Milwaukee	Bradley Center (1988)	18,717
	Georgia Tech (1956)	9,300	Minnesota	Target Center (1990)	19,006
Boston	FleetCenter (1995)	18,624	New Jersey	Continental Airlines Arena (1981)	20,049
Charlotte	Charlotte Coliseum (1988)	24,042	New York	Madison Square Garden (1968)	19,763
Chicago	United Center (1994)	21,711	Orlando	Orlando Arena (1989)	17,248
Cleveland	Gund Arena (1994)	20,562	Philadelphia	First Union Center (1996)	20,444
Dallas	Reunion Arena (1980)	18,042	Phoenix	America West Arena (1992)	19,023
Denver	Pepsi Center (1999)	19,309	Portland	The Rose Garden (1995)	21,538
Detroit	Palace of Auburn Hills (1988)	22,076	Sacramento	ARCO Arena (1988)	17,317
Golden State	New Oakland Arena (1997)	19,200	San Antonio	Alamodome (1993)	20,557/34,215
Houston	Compaq Center (1975)	16,285	Seattle	Key Arena (1995)	17,072
Indiana	Conseco Fieldhouse (1999)	18,400	Toronto	Air Canada Centre (1999)	19,800
L.A. Clippers	Staples Center (1999)	20,000	Utah	Delta Center (1991)	19,911
L.A. Lakers	Staples Center (1999)	20,000	Vancouver	GM Place (1995)	19,193
Miami	Miami Arena (1988)	15,200	Washington	MCI Center (1997)	20,674
	American Airlines Arena (as of 12-31-99)	20,000			

WOMEN'S PROFESSIONAL BASKETBALL

WNBA 1999: Houston "Threepeats," Influx of ABL Talent, 4 New Teams in 2000

The Houston Comets defeated the New York Liberty, 59–47, in the decisive Game 3 on Sept. 5, 1999, to win their 3d straight Women's National Basketball Association championship. Spurred by the demise of the competing American Basketball League, which folded on Dec. 22, 1998, the WNBA added 2 new teams in 1999 (Minnesota and Orlando). Franchises were also awarded to Indiana, Miami, Portland, and Seattle for the 2000 season.

WNBA Final Standings, 1999 Season

x-clinched playoff berth; y-clinched top seed

Eastern Conference	W	L	Pct	GB	Western Conference	W	L	Pct	GB
y-New York Liberty	18	14	.563	—	y-Houston Comets	26	6	.813	—
x-Charlotte Sting	15	17	.469	3	x-Los Angeles Sparks	20	12	.625	6
x-Detroit Shock	15	17	.469	3	x-Sacramento Monarchs	19	13	.594	7
Orlando Miracle	15	17	.469	3	Phoenix Mercury	15	17	.469	11
Washington Mystics	12	20	.375	6	Minnesota Lynx	15	17	.469	11
Cleveland Rockers	7	25	.219	11	Utah Starzz	15	17	.469	11

1999 WNBA Playoffs

(Playoff seeding in parentheses; Conference winner automatically gets top seed and 1st round bye)

Eastern Conference	Western Conference
Charlotte (3) defeated Detroit (2) 60-54	Los Angeles (2) defeated Sacramento (3) 71-58
New York (1) defeated Charlotte (3) 2 games to 1	Houston (1) defeated Los Angeles (2) 2 games to 1

WNBA Championship (Best of 3)
Houston defeated New York 2 games to 1 [73-60, 67-68, 59-47]

WNBA Individual Highs and Awards in 1999

Most minutes played — 1,147: Shannon Johnson, Orlando.
Most points — 686: Cynthia Cooper, Houston.
Most points per game — 22.1: Cynthia Cooper, Houston.
Highest field goal percentage — .574: Murriel Page, Washington.
Highest 3-point field goal percentage — .517: Jennifer Azzi, Detroit.
Highest free throw percentage — .984: Eva Nemcova, Cleveland.
Most rebounds — 329: Yolanda Griffith, Sacramento.

Most rebounds per game — 11.3: Yolanda Griffith, Sacramento.
Most assists — 226: Ticha Penicheiro, Sacramento.
Most assists per game — 7.1: Ticha Penicheiro, Sacramento.
Most steals — 78: Teresa Weatherspoon, New York.
Most blocked shots — 77: Malgorzata Dydek, Utah.
MVP — Yolanda Griffith, Sacramento.
MVP Championship Series — Cynthia Cooper, Houston.
Coach of the year — Van Chancellor, Houston.
Rookie of the year — Chamique Holdsclaw, Washington.
Defensive player of the year — Yolanda Griffith, Sacramento.

WNBA Champions

	Regular season		Playoffs		
Year	Eastern Conference	Western Conference	Winner	Coach	Runner-up
1997	Phoenix Mercury	Houston Comets	Houston	Van Chancellor	New York
1998	Cleveland Rockers	Houston Comets	Houston	Van Chancellor	Phoenix
1999	New York Liberty	Houston Comets	Houston	Van Chancellor	New York

COLLEGE BASKETBALL
Final NCAA Division I Conference Standing, 1998-99
(*conference tournament champion)

America East

	Conference W	L	Full Season W	L
Delaware*	15	3	25	6
Drexel	15	3	20	9
Hofstra	14	4	22	10
Maine	13	5	19	9
Hartford	9	9	11	16
Vermont	7	11	11	16
Northeastern	6	12	10	18
Boston U.	5	13	9	18
Towson	4	14	6	22
New Hampshire	2	16	4	23

Atlantic Coast

	Conference W	L	Full Season W	L
Duke*	16	0	37	2
Maryland	13	3	28	6
North Carolina	10	6	24	10
Wake Forest	7	9	17	14
North Carolina St.	6	10	19	14
Georgia Tech	6	10	15	16
Clemson	5	11	20	15
Florida St.	5	11	13	17
Virginia	4	12	14	16

Atlantic 10
Eastern Division

	Conference W	L	Full Season W	L
Temple	13	3	24	11
Rhode Island*	10	6	20	13
Massachusetts	9	7	14	16
St. Bonaventure	8	8	14	15
Fordham	5	11	12	15
St. Joseph's (PA)	5	11	12	18

Western Division

	Conference W	L	Full Season W	L
George Washington	13	3	20	9
Xavier (OH)	12	4	25	11
LaSalle	8	8	13	15
Virginia Tech	7	9	13	15
Dayton	5	11	11	17
Duquesne	1	15	5	23

Big East

	Conference W	L	Full Season W	L
Connecticut*	16	2	34	2
Miami (FL)	15	3	23	7
St. John's (NY)	14	4	28	9
Villanova	10	8	21	11
Syracuse	10	8	21	12
Rutgers	9	9	19	13
Providence	9	9	16	14
Seton Hall	8	10	15	15
Notre Dame	8	10	14	16
Georgetown	6	12	15	16
Pittsburgh	5	13	14	16
West Virginia	4	14	10	19
Boston College	3	15	6	21

Big Sky

	Conference W	L	Full Season W	L
Weber St.*	13	3	25	8
Northern Arizona	12	4	21	8
Portland St.	9	7	17	11
Cal. St. Northridge	9	7	17	12
Montana St.	9	7	16	13
Eastern Wash.	7	9	10	17
Montana	6	10	13	14
Idaho St.	4	12	6	20
Cal. St. Sacramento	3	13	3	23

Big South

	Conference W	L	Full Season W	L
Winthrop*	9	1	21	8
Radford	8	2	20	8
UNC Asheville	5	5	11	18
Charleston Southern	4	6	12	16
Coastal Carolina	4	6	7	20
Liberty	0	10	4	23

Big Ten

	Conference W	L	Full Season W	L
Michigan St.*	15	1	33	5
Ohio St.	12	4	27	9
Indiana	9	7	23	11
Wisconsin	9	7	22	10
Iowa	9	7	20	10
Minnesota	8	8	17	11
Purdue	7	9	21	13
Northwestern	6	10	15	14
Michigan	5	11	12	19
Penn St.	5	11	13	14
Illinois	3	13	14	18

Big 12

	Conference W	L	Full Season W	L
Texas	13	3	19	13
Missouri	11	5	20	9
Kansas*	11	5	23	10
Oklahoma	11	5	22	11
Oklahoma St.	10	6	23	11
Nebraska	10	6	20	13
Kansas St.	7	9	20	13
Colorado	7	9	18	15
Iowa St.	6	10	15	15
Texas A&M	5	11	12	15
Texas Tech	5	11	13	17
Baylor	0	16	6	24

Big West
Eatern Division

	Conference W	L	Full Season W	L
Boise St.	12	4	21	8
New Mexico St.*	12	4	23	10
Idaho	11	5	16	11
Utah St.	8	8	15	13
Nevada	4	12	8	18
North Texas	4	12	4	22

Western Division

	Conference W	L	Full Season W	L
UC Santa Barbara	12	4	15	13
Pacific (CA)	9	7	14	13
Long Beach St.	9	7	13	15
Cal. St. Fullerton	7	9	13	14
Cal. Poly	6	10	11	16
UC Irvine	2	14	6	20

Colonial Athletic Association

	Conference W	L	Full Season W	L
George Mason*	13	3	19	11
Old Dominion	11	5	25	9
Richmond	10	6	15	12
James Madison	9	7	16	11
UNC Wilmington	9	7	11	17
Va. Commonwealth	8	8	15	16
East Carolina	7	9	13	14
William & Mary	3	13	8	19
American	2	14	7	21

Conference USA
American

	Conference W	L	Full Season W	L
Cincinnati	12	4	27	6
Louisville	11	5	19	11
UNC Charlotte*	10	6	23	11
DePaul	10	6	18	13
St. Louis	8	8	15	16
Marquette	6	10	14	15

National

	Conference W	L	Full Season W	L
UAB	10	6	20	12
Southern Mississippi	6	10	14	16
South Florida	6	10	14	14
Memphis	6	10	13	15
Tulane	6	10	12	15
Houston	5	11	10	17

Ivy Group[1]

	Conference W	L	Full Season W	L
Pennsylvania	12	1	21	6
Princeton	11	2	22	8
Dartmouth	10	4	14	12
Harvard	7	7	13	13
Cornell	6	8	11	15
Columbia	5	9	10	16
Brown	2	12	4	22
Yale	2	12	4	22

Metro Atlantic Athletic

	Conference W	L	Full Season W	L
Niagara	13	5	17	12
Siena*	13	5	25	6
Iona	12	6	14	14
Canisius	11	7	15	12
St. Peter's	10	8	14	15
Marist	8	10	16	12
Fairfield	7	11	12	15
Rider	7	11	12	16
Loyola (MD)	6	12	13	15
Manhattan	3	15	5	22

Mid-American
East

	Conference W	L	Full Season W	L
Miami (OH)	15	3	24	8
Kent*	13	5	23	7
Ohio	12	6	18	10
Akron	12	6	18	9
Bowling Green	12	6	18	10
Marshall	11	7	16	11
Buffalo	1	17	5	24

West

	Conference W	L	Full Season W	L
Toledo	11	7	19	9
Ball St.	10	8	16	11
Central Michigan	7	11	10	16
Western Michigan	6	12	11	15
Eastern Michigan	5	13	5	20
Northern Illinois	2	16	6	20

Mid-Continent

	Conference W	L	Full Season W	L
Valparaiso*	10	4	23	9
Oral Roberts	10	4	17	11
Western Illinois	9	5	16	12
Youngstown St.	9	5	14	14
Southern Utah	6	8	13	17
Indiana Purdue	6	8	11	16
Missouri-K.C.	3	11	8	22
Chicago St.	3	11	3	24

Mid-Eastern Athletic

	Conference W	L	Full Season W	L
South Carolina St.	14	4	17	12
Coppin St.	14	4	15	14
Morgan St.	12	6	14	14
Norfolk St.	11	7	15	12
Bethune-Cookman	10	9	11	16
North Carolina A&T	9	9	13	15
Hampton	8	10	8	19
Florida A&M*	8	11	12	19
MD Eastern Shore	7	11	10	17
Delaware St.	5	13	8	19
Howard	2	16	2	25

Midwestern Collegiate

	Conference W	L	Full Season W	L
Detroit*	12	2	25	6
Butler	11	3	22	10
Wisconsin-Green Bay	9	5	20	11
Loyola (IL)	7	7	9	18
Cleveland St.	6	8	14	14
Wisconsin-Milwaukee	5	9	8	19
Wright St.	4	10	9	18
Illinois-Chicago	2	12	7	21

Missouri Valley

	Conference W	L	Full Season W	L
Evansville	13	5	23	10
SW Missouri St.	11	7	22	11
Creighton*	11	7	22	9
Bradley	11	7	17	12
Indiana St.	10	8	15	12
Southern Illinois	10	8	15	12
Illinois St.	7	11	16	15
Wichita St.	6	12	13	17
Northern Iowa	6	12	9	18
Drake	5	13	10	17

Northeast

	Conference W	L	Full Season W	L
Maryland- Baltimore County	17	3	19	9
St. Francis (NY)	16	4	20	8
Robert Morris	12	8	15	12
Cent. Connecticut St.	11	9	19	13
LIU-Brooklyn	10	10	10	17
Mt. St. Mary's (MD)*	10	10	15	15
Fairleigh Dickinson	9	11	12	16
Wagner	7	13	9	18
St. Francis (PA)	7	13	9	17
Quinnipiac	6	14	9	18
Monmouth (NJ)	5	15	5	21

Ohio Valley

	Conference W	L	Full Season W	L
Murray St.*	16	2	27	6
SE Missouri St.	15	3	20	9
Morehead St.	9	9	13	15
Tennessee St.	9	9	12	15
Austin Peay	9	9	11	16
Middle Tennessee St.	9	9	12	19
Eastern Illinois	8	10	13	16
Tennessee Tech.	8	10	12	15
Tennessee-Martin	5	13	8	18
Eastern Kentucky	2	16	3	23

	Conference W L		Full Season W L	
Pacific-10[1]				
Stanford	15	3	26	7
Arizona	13	5	22	7
UCLA	12	6	22	9
Washington	10	8	17	12
California	8	10	22	11
Oregon	8	10	19	13
USC	7	11	15	13
Oregon St.	7	11	13	14
Arizona St.	6	12	14	16
Washington St.	4	14	10	19
Patriot League				
Lafayette*	10	2	22	8
Navy	9	3	20	7
Bucknell	9	3	16	13
Colgate	7	5	14	14
Army	4	8	8	19
Holy Cross	3	9	7	20
Lehigh	0	12	6	22
Southeastern				
Eastern Division				
Tennessee	12	4	21	9
Kentucky*	11	5	28	9
Florida	10	6	22	9
Georgia	6	10	15	15
Vanderbilt	5	11	14	15
South Carolina	3	13	8	21
Western Division				
Auburn	14	2	29	4
Arkansas	9	7	23	11
Mississippi	8	8	20	13
Mississippi St.	8	8	20	13
Alabama	6	10	17	15
LSU	4	12	12	15
Southern				
North				
Appalachian St.	13	3	21	8
Davidson	11	5	16	11
East Tennessee St.	9	7	17	11
VMI	9	7	12	15
U.N.C.-Greensboro	5	11	7	20
Western Carolina	2	14	8	21
South				
Col. of Charleston*	16	0	28	3
Tenn.-Chattanooga	9	7	16	12
Wofford	8	8	11	16
Georgia Southern	6	10	11	17
Furman	5	11	12	16
Citadel	3	13	9	18
Southland				
SW Texas St.	13	5	19	9
Texas-San Antonio*	12	6	18	11
Nicholls St.	12	6	14	15
NE Louisiana	12	6	13	14
Lamar	11	7	17	11
McNeese St.	11	7	13	15
Northwestern St.	8	10	11	15
Texas-Arlington	8	10	10	16
Sam Houston St.	7	11	10	16
SE Louisiana	3	15	6	20
Stephen F. Austin	2	16	4	22
Southwestern Athletic				
Alcorn St.*	14	2	23	7
Southern	13	3	21	7
Jackson St.	11	5	16	12
Mississippi Valley St.	10	6	14	13
Alabama St.	8	8	11	16
Texas Southern	6	10	8	19
Grambling	5	11	6	21
Prairie View	4	12	6	21
Arkansas-Pine Bluff	1	15	3	24
Sun Belt				
Louisiana Tech	10	4	19	9
Arkansas St.*	9	5	18	12
Florida International	7	7	13	16
SW Louisiana	7	7	13	16
Western Kentucky	7	7	13	16
South Alabama	6	8	11	16
New Orleans	5	9	14	16
Arkansas-Little Rock	5	9	12	15
Trans America Athletic				
Samford*	15	1	24	6
Central Florida	13	3	19	10
Georgia St.	11	5	17	13
Stetson	10	6	14	13
Centenary	9	7	14	14
Jacksonville	7	9	12	15
Troy St.	6	10	9	18
Campbell	6	10	9	18
Mercer	5	11	8	18
Jacksonville St.	3	13	8	18
Florida Atlantic	3	13	6	20
West Coast				
Gonzaga*	12	2	28	7
Pepperdine	9	5	19	13
San Diego	9	5	18	9
Santa Clara	8	6	14	15
Loyola Marymount	6	8	11	16
St. Mary's (CA)	5	9	13	18
San Francisco	4	10	12	18
Portland	3	11	9	18
Western Athletic				
Utah*	14	0	28	5
New Mexico	9	5	25	9
Fresno St.	9	5	21	12
Texas-El Paso	8	6	16	12
Brigham Young	6	8	12	16
San Jose St.	5	9	12	16
Hawaii	3	11	6	20
San Diego St.	2	12	4	22
Mountain				
UNLV	9	5	16	13
Tulsa	9	5	23	10
Rice	8	6	18	10
Texas Christian	7	7	21	11
Southern Methodist	7	7	15	15
Colorado St.	7	7	19	11
Wyoming	7	7	18	10
Air Force	2	12	10	16

(1) Conference does not hold a tournament.

All-Time Winningest College Teams by Percentage

School	Years	Won	Lost	Pct.	School	Years	Won	Lost	Pct.
Kentucky	96	1,748	538	.765	Utah	91	1,404	737	.656
North Carolina	89	1,733	609	.740	Indiana	99	1,453	778	.651
UNLV	41	847	320	.726	Louisville	85	1,356	727	.651
Kansas	101	1,688	724	.700	Temple	103	1,520	824	.648
UCLA	80	1,445	621	.699	DePaul	76	1,200	654	.647
St. John's (NY)	92	1,582	715	.689	Weber State	37	679	372	.646
Syracuse	98	1,498	704	.680	Purdue	101	1,404	771	.646
Duke	94	1,585	755	.677	Notre Dame	94	1,441	792	.645
Western Kentucky	80	1,379	685	.668	Illinois	94	1,358	764	.640
Arkansas	76	1,315	682	.658	Pennsylvania	99	1,475	838	.638

Major College Basketball Tournaments

The National Invitation Tournament (NIT), first played in 1938, is the nation's oldest basketball tournament. The first National Collegiate Athletic Association (NCAA) national championship tournament was played one year later. Selections for both tournaments are made in Mar., with the NCAA selecting first from among the top Division I teams.

National Invitation Tournament Champions

Year	Champion	Year	Champion	Year	Champion	Year	Champion
1938	Temple	1954	Holy Cross	1970	Marquette	1985	UCLA
1939	Long Island Univ.	1955	Duquesne	1971	North Carolina	1986	Ohio State
1940	Colorado	1956	Louisville	1972	Maryland	1987	Southern Mississippi
1941	Long Island Univ.	1957	Bradley	1973	Virginia Tech	1988	Connecticut
1942	West Virginia	1958	Xavier (Ohio)	1974	Purdue	1989	St. John's
1943	St. John's	1959	St. John's	1975	Princeton	1990	Vanderbilt
1944	St. John's	1960	Bradley	1976	Kentucky	1991	Stanford
1945	De Paul	1961	Providence	1977	St. Bonaventure	1992	Virginia
1946	Kentucky	1962	Dayton	1978	Texas	1993	Minnesota
1947	Utah	1963	Providence	1979	Indiana	1994	Villanova
1948	St. Louis	1964	Bradley	1980	Virginia	1995	Virginia Tech
1949	San Francisco	1965	St. John's	1981	Tulsa	1996	Nebraska
1950	CCNY	1966	Brigham Young	1982	Bradley	1997	Michigan
1951	Brigham Young	1967	Southern Illinois	1983	Fresno State	1998	Minnesota
1952	LaSalle	1968	Dayton	1984	Michigan	1999	California
1953	Seton Hall	1969	Temple				

1999 NCAA BASKETBALL TOURNAMENT (MEN)

MIDWEST

(1) Michigan St. 76
(16) Mount St. Mary's 53
Michigan St. 74
(8) Villanova 70
(9) Mississippi 72
Mississippi 66
Michigan St. 54
(5) NC Charlotte 81 (OT)
(12) Rhode Island 70
NC Charlotte 72
(4) Arizona 60
(13) Oklahoma 61
Oklahoma 85
Oklahoma 46
Michigan St. 73
(6) Kansas 95
(11) Evansville 74
Kansas 88
(3) Kentucky 82
(14) New Mexico St. 60
Kentucky 92 (OT)
Kentucky 58
(7) Washington 58
(10) Miami (OH) 59
Miami (OH) 66
Kentucky 66
(2) Utah 80
(15) Arkansas St. 58
Utah 58
Miami (OH) 43
Michigan St. 62

EAST

(1) Duke 99
(16) Florida A&M 58
Duke 97
(8) Coll. of Charleston 53
(9) Tulsa 62
Tulsa 56
Duke 78
(5) Wisconsin 32
(12) SW Missouri St. 43
SW Missouri St. 81
(4) Tennessee 62
(13) Delaware 52
Tennessee 51
SW Missouri St. 61
Duke 85
(6) Temple 61
(11) Kent 54
Temple 64
(3) Cincinnati 72
(14) George Mason 48
Cincinnati 54
Temple 77
Duke 68
(7) Texas 54
(10) Purdue 58
Purdue 73
Temple 64
(2) Miami (FL) 75
(15) Lafayette 54
Miami (FL) 63
Purdue 55

SOUTH

(1) Auburn 80
(16) Winthrop 41
Auburn 81
(8) Syracuse 61
(9) Oklahoma St. 69
Oklahoma St. 74
Auburn 64
(5) UCLA 53
(12) Detroit 56
Detroit 44
(4) Ohio St. 72
(13) Murray St. 58
Ohio St. 75
Ohio St. 72
Ohio St. 77
(6) Indiana 108
(11) Geo. Washington 88
Indiana 61
(3) St. John's (NY) 69
(14) Samford 43
St. John's (NY) 86
St. John's (NY) 76
Ohio St. 58
(7) Louisville 58
(10) Creighton 62
Creighton 63
St. John's (NY) 74
(2) Maryland 82
(15) Valparaiso 60
Maryland 75
Maryland 62

WEST

(1) Connecticut 91
(16) Tex.–San Antonio 66
Connecticut 78
(8) Missouri 59
(9) New Mexico 61
New Mexico 56
Connecticut 78
(5) Iowa 77
(12) Ala.–Birmingham 64
Iowa 82
(4) Arkansas 94
(13) Siena 80
Arkansas 72
Iowa 68
Connecticut 67
(6) Florida 75
(11) Pennsylvania 61
Florida 82 (OT)
(3) North Carolina 74
(14) Weber St. 76
Weber St. 74
Florida 72
Connecticut 64
(7) Minnesota 63
(10) Gonzaga 75
Gonzaga 82
Gonzaga 62
(2) Stanford 69
(15) Alcorn St. 57
Stanford 74
Gonzaga 73

Duke 74
Connecticut 77

Connecticut Defeats Duke to Win the 1999 NCAA Men's Basketball Championship

In their first trip to the Final Four, the University of Connecticut Huskies defeated the Duke Blue Devils, 77-74, to capture the NCAA Championship on Mar. 30, in St. Petersburg, FL. The Connecticut win ended number-one-ranked Duke's 32-game winning streak and bid for an NCAA record for wins in a season. Connecticut's Richard Hamilton, who averaged 24.1 points per game, was named the tournament's Most Outstanding Player.

NCAA Division I Champions

Year	Champion	Coach	Final opponent	Score	Outstanding player	Site
1939	Oregon	Howard Hobson	Ohio St.	46-33	None	Evanston, IL
1940	Indiana	Branch McCracken	Kansas	60-42	Marvin Huffman, Indiana	Kansas City, MO
1941	Wisconsin	Harold Foster	Washington St.	39-34	John Kotz, Wisconsin	Kansas City, MO
1942	Stanford	Everett Dean	Dartmouth	53-38	Howard Dallmar, Stanford	Kansas City, MO
1943	Wyoming	Everett Shelton	Georgetown	46-34	Ken Sailors, Wyoming	New York, NY
1944	Utah	Vadal Peterson	Dartmouth	42-40[1]	Arnold Ferrin, Utah	New York, NY
1945	Oklahoma St.[2]	Henry Iba	NYU	49-45	Bob Kurland, Oklahoma St.	New York, NY
1946	Oklahoma St.[2]	Henry Iba	North Carolina	43-40	Bob Kurland, Oklahoma St.	New York, NY
1947	Holy Cross	Alvin Julian	Oklahoma	58-47	George Kaftan, Holy Cross	New York, NY
1948	Kentucky	Adolph Rupp	Baylor	58-42	Alex Groza, Kentucky	New York, NY
1949	Kentucky	Adolph Rupp	Oklahoma St.	46-36	Alex Groza, Kentucky	Seattle, WA
1950	CCNY	Nat Holman	Bradley	71-68	Irwin Dambrot, CCNY	New York, NY
1951	Kentucky	Adolph Rupp	Kansas St.	68-58	None	Minneapolis, MN
1952	Kansas	Forrest Allen	St. John's	80-63	Clyde Lovellette, Kansas	Seattle, WA
1953	Indiana	Branch McCracken	Kansas	69-68	B.H. Born, Kansas	Kansas City, MO
1954	La Salle	Kenneth Loeffler	Bradley	92-76	Tom Gola, La Salle	Kansas City, MO
1955	San Francisco	Phil Woolpert	LaSalle	77-63	Bill Russell, San Francisco	Kansas City, MO
1956	San Francisco	Phil Woolpert	Iowa	83-71	Hal Lear, Temple	Evanston, IL
1957	North Carolina	Frank McGuire	Kansas	54-53[1]	Wilt Chamberlain, Kansas	Kansas City, MO
1958	Kentucky	Adolph Rupp	Seattle	84-72	Elgin Baylor, Seattle	Louisville, KY
1959	California	Pete Newell	West Virginia	71-70	Jerry West, West Virginia	Louisville, KY
1960	Ohio St.	Fred Taylor	California	75-55	Jerry Lucas, Ohio St.	San Francisco, CA
1961	Cincinnati	Edwin Jucker	Ohio St.	70-65[1]	Jerry Lucas, Ohio St.	Kansas City, MO
1962	Cincinnati	Edwin Jucker	Ohio St.	71-59	Paul Hogue, Cincinnati	Louisville, KY
1963	Loyola (IL)	George Ireland	Cincinnati	60-58[1]	Art Heyman, Duke	Louisville, KY
1964	UCLA	John Wooden	Duke	98-83	Walt Hazzard, UCLA	Kansas City, MO
1965	UCLA	John Wooden	Michigan	91-80	Bill Bradley, Princeton	Portland, OR
1966	Texas-El Paso[3]	Don Haskins	Kentucky	72-65	Jerry Chambers, Utah	College Park, MD
1967	UCLA	John Wooden	Dayton	79-64	Lew Alcindor, UCLA	Louisville, KY
1968	UCLA	John Wooden	North Carolina	78-55	Lew Alcindor, UCLA	Los Angeles, CA
1969	UCLA	John Wooden	Purdue	92-72	Lew Alcindor, UCLA	Louisville, KY
1970	UCLA	John Wooden	Jacksonville	80-69	Sidney Wicks, UCLA	College Park, MD
1971	UCLA	John Wooden	Villanova*	68-62	Howard Porter, Villanova*	Houston, TX
1972	UCLA	John Wooden	Florida St.	81-76	Bill Walton, UCLA	Los Angeles, CA
1973	UCLA	John Wooden	Memphis St.	87-66	Bill Walton, UCLA	St. Louis, MO
1974	North Carolina St.	Norm Sloan	Marquette	76-64	David Thompson, N.C. St.	Greensboro, NC
1975	UCLA	John Wooden	Kentucky	92-85	Richard Washington, UCLA	San Diego, CA
1976	Indiana	Bob Knight	Michigan	86-68	Kent Benson, Indiana	Philadelphia, PA
1977	Marquette	Al McGuire	North Carolina	67-59	Butch Lee, Marquette	Atlanta, GA
1978	Kentucky	Joe Hall	Duke	94-88	Jack Givens, Kentucky	St. Louis, MO
1979	Michigan St.	Jud Heathcote	Indiana St.	75-64	Magic Johnson, Michigan St.	Salt Lake City, UT
1980	Louisville	Denny Crum	UCLA*	59-54	Darrell Griffith, Louisville	Indianapolis, IN
1981	Indiana	Bob Knight	North Carolina	63-50	Isiah Thomas, Indiana	Philadelphia, PA
1982	North Carolina	Dean Smith	Georgetown	63-62	James Worthy, N. Carolina	New Orleans, LA
1983	North Carolina St.	Jim Valvano	Houston	54-52	Hakeem Olajuwon, Houston	Albuquerque, NM
1984	Georgetown	John Thompson	Houston	84-75	Patrick Ewing, Georgetown	Seattle, WA
1985	Villanova	Rollie Massimino	Georgetown	66-64	Ed Pinckney, Villanova	Lexington, KY
1986	Louisville	Denny Crum	Duke	72-69	Pervis Ellison, Louisville	Dallas, TX
1987	Indiana	Bob Knight	Syracuse	74-73	Keith Smart, Indiana	New Orleans, LA
1988	Kansas	Larry Brown	Oklahoma	83-79	Danny Manning, Kansas	Kansas City, MO
1989	Michigan	Steve Fisher	Seton Hall	80-79[1]	Glen Rice, Michigan	Seattle, WA
1990	UNLV	Jerry Tarkanian	Duke	103-73	Anderson Hunt, UNLV	Denver, CO
1991	Duke	Mike Krzyzewski	Kansas	72-65	Christian Laettner, Duke	Indianapolis, IN
1992	Duke	Mike Krzyzewski	Michigan	71-51	Bobby Hurley, Duke	Minneapolis, MN
1993	North Carolina	Dean Smith	Michigan	77-71	Donald Williams, N. Carolina	New Orleans, LA
1994	Arkansas	Nolan Richardson	Duke	76-72	Corliss Williamson, Arkansas	Charlotte, NC
1995	UCLA	Jim Harrick	Arkansas	89-78	Ed O'Bannon, UCLA	Seattle, WA
1996	Kentucky	Rick Pitino	Syracuse	76-67	Tony Delk, Kentucky	E. Rutherford, NJ
1997	Arizona	Lute Olson	Kentucky	84-79[1]	Miles Simon, Arizona	Indianapolis, IN
1998	Kentucky	Tubby Smith	Utah	78-69	Jeff Sheppard, Kentucky	San Antonio, TX
1999	Connecticut	Jim Calhoun	Duke	77-74	Richard Hamilton, Connecticut	St. Petersburg, FL

*Declared ineligible after the tournament. (1) Overtime. (2) Then known as Oklahoma A&M. (3) Then known as Texas Western.

Top Career Scorers

Player, school	Years	Points	Avg.	Player, school	Years	Points	Avg.
Pete Maravich, LSU	1968-70	3,667	44.2	Frank Selvy, Furman	1952-54	2,538	32.5
Austin Carr, Notre Dame	1969-71	2,560	34.6	Rick Mount, Purdue	1968-70	2,323	32.3
Oscar Robertson, Cincinnati	1958-60	2,973	33.8	Darrell Floyd, Furman	1954-56	2,281	32.1
Calvin Murphy, Niagara	1968-70	2,548	33.1	Nick Werkman, Seton Hall	1962-64	2,273	32.0
Dwight Lamar, SW Louisiana	1972-73	1,862	32.7	Willie Humes, Idaho State	1970-71	1,510	31.5

John R. Wooden Award

Awarded to the nation's outstanding college basketball player by the Los Angeles Athletic Club.

1977	Marques Johnson, UCLA	1985	Chris Mullin, St. John's	1993	Calbert Cheaney, Indiana
1978	Phil Ford, North Carolina	1986	Walter Berry, St. John's	1994	Glenn Robinson, Purdue
1979	Larry Bird, Indiana State	1987	David Robinson, Navy	1995	Ed O'Bannon, UCLA
1980	Darrell Griffith, Louisville	1988	Danny Manning, Kansas	1996	Marcus Camby, Massachusetts
1981	Danny Ainge, Brigham Young	1989	Sean Elliott, Arizona	1997	Tim Duncan, Wake Forest
1982	Ralph Sampson, Virginia	1990	Lionel Simmons, La Salle	1998	Antawn Jamison, North Carolina
1983	Ralph Sampson, Virginia	1991	Larry Johnson, UNLV	1999	Elton Brand, Duke
1984	Michael Jordan, North Carolina	1992	Christian Laettner, Duke		

Selected Division I Basketball Coaches in 1999[1]

School	Coach	School	Coach	School	Coach
Air Force	Reggie Minton	Illinois	Lon Kruger	Penn St.	Jerry Dunn
Akron	Dan Hipsher	Ill.–Chicago	Jimmy Collins	Pepperdine	Jan van Breda Kolff
Alabama	Mark Gottfried	Illinois St.	Tom Richardson	Pittsburgh	Ben Howland
Alabama A&M	L. Vann Pettaway	Indiana	Bob Knight	Portland	Rob Chavez
Ala.–Birmingham	Murry Bartow	Indiana St.	Royce Waltman	Princeton	Bill Carmody
American	Art Perry	Iowa	Steve Alford	Providence	Tim Welsh
Arizona	Lute Olson	Iowa St.	Larry Eustachy	Purdue	Gene Keady
Arizona St.	Rob Evans	Jacksonville	Hugh Durham	Rhode Island	Jerry DeGregorio
Arkansas	Nolan Richardson	Jacksonville St.	Mark Turgeon	Rice	Willis Wilson
Ark.–Little Rock	Sidney Moncrief	James Madison	Sherman Dillard	Richmond	John Beilein
Arkansas St.	Dickey Nutt	Kansas	Roy Williams	Rutgers	Kevin Bannon
Army	Pat Harris	Kansas St.	Tom Asbury	St. Bonaventure	Jim Baron
Auburn	Cliff Ellis	Kent	Gary Waters	St. John's (NY)	Mike Jarvis
Austin Peay	Dave Loos	Kentucky	Tubby Smith	St. Joseph's	Phil Martelli
Ball St.	Ray McCallum	LSU	John Brady	St. Louis	Lorenzo Romar
Baylor	Dave Bliss	Louisville	Denny Crum	St. Mary's (CA)	Dave Bollwinkel
Boise St.	Rod Jensen	Loyola (IL)	Larry Farmer	San Diego	Brad Holland
Boston College	Al Skinner	Loyola Marymount	Charles Bradley	San Diego St.	Steve Fisher
Boston U.	Dennis Wolff	Manhattan	Bobby Gonzalez	San Francisco	Phil Mathews
Bowling Green	Dan Dakich	Marquette	Tom Crean	San Jose St.	Steve Barnes
Bradley	Jim Molinari	Marshall	Greg White	Santa Clara	Dick Davey
Brigham Young	Steve Cleveland	Maryland	Gary Williams	Seton Hall	Tommy Amaker
Brown	Glen Miller	Massachusetts	Bruiser Flint	South Carolina	Eddie Fogler
Butler	Barry Collier	McNeese St.	Ron Everhart	SE Missouri St.	Gary Garner
California	Ben Braun	Memphis	Tic Price	Southern California	Henry Bibby
Cal Poly	Jeff Schneider	Miami (FL)	Leonard Hamilton	So. Illinois	Bruce Weber
Cal St. Fullerton	Bob Hawking	Miami (OH)	Charlie Coles	Southern Methodist	Mike Dement
Central Mich.	Jay Smith	Michigan	Brian Ellerbe	So. Mississippi	James Green
Cincinnati	Bob Huggins	Michigan St.	Tom Izzo	SW Missouri	Barry Hinson
Citadel	Pat Dennis	Middle Tenn. St.	Randy Wiel	Stanford	Mike Montgomery
Clemson	Larry Shyatt	Minnesota	Dan Monson	Syracuse	Jim Boeheim
Cleveland St.	Rollie Massimino	Mississippi	Roderick Barnes	Temple	John Chaney
Colgate	Emmett Davis	Mississippi St.	Rick Stansbury	Tennessee	Jerry Green
Colorado	Ricardo Patton	Missouri	Quin Snyder	Tennessee St.	Frankie Allen
Colorado St.	Ritchie McKay	Monmouth	Dave Calloway	Tennessee Tech	Jeff Lebo
Columbia	Armond Hill	Montana	Don Holst	Texas	Rick Barnes
Connecticut	Jim Calhoun	Montana St.	Mick Durham	Texas A&M	Melvin Watkins
Cornell	Scott Thompson	Morehead St.	Kyle Macy	Texas Christian	Billy Tubbs
Creighton	Dana Altman	Mt. St. Mary's (MD)	Jim Phelan	Texas Southern	Robert Moreland
Dartmouth	Dave Faucher	Murray St.	Tevester Anderson	Texas Tech	James Dickey
Dayton	Oliver Purnell	Navy	Don DeVoe	Toledo	Stan Joplin
DePaul	Pat Kennedy	Nebraska	Danny Nee	Tulane	Perry Clark
Detroit	Perry Watson	Nevada	Trent Johnson	Tulsa	Bill Self
Drake	Kurt Kanaskie	New Mexico	Fran Fraschilla	UC Irvine	Pat Douglass
Duke	Mike Krzyzewski	New Orleans	Joey Stiebing	UCLA	Steve Lavin
Duquesne	Darelle Porter	N. Arizona	Mike Adras	UC Santa Barbara	Bob Williams
E. Carolina	Bill Herrion	N. Illinois	Brian Hammel	UNLV	Bill Bayno
E. Illinois	Rick Samuels	North Carolina	Bill Guthridge	Utah	Rick Majerus
E. Kentucky	Scott Perry	NC A&T	Curtis Hunter	Utah St.	Stew Morrill
E. Michigan	Milton Barnes	NC–Asheville	Eddie Biedenbach	UTEP	Jason Rabedeaux
E. Washington	Steve Aggers	NC–Charlotte	Bob Lutz	Valparaiso	Homer Drew
E. Tennessee St.	Ed DeChellis	NC–Greensboro	Fran McCaffery	Vanderbilt	Kevin Stallings
Evansville	Jim Crews	North Carolina St.	Herb Sendek	Villanova	Steve Lappas
Fairleigh Dickinson	Tom Green	NC–Wilmington	Jerry Wainwright	Virginia	Pete Gillen
Florida	Billy Donovan	New Mexico St.	Lou Henson	Virginia Tech	Ricky Stokes
Florida Int'l	Shakey Rodriquez	Northwestern	Kevin O'Neill	Wake Forest	Dave Odom
Florida St.	Steve Robinson	Northwestern St.	Mike McConathy	Washington	Bob Bender
Fresno St.	Jerry Tarkanian	Notre Dame	Matt Doherty	Washington St.	Paul Graham
George Mason	Jim Larranaga	Oakland	Greg Kampe	Weber St.	Joe Cravens
Georgetown	Craig Esherick	Ohio	Larry Hunter	W. Kentucky	Dennis Felton
Geo. Washington	Tom Penders	Ohio St.	Jim O'Brien	W. Michigan	Bob Donewald
Georgia	Jim Harrick	Oklahoma	Kelvin Sampson	West Virginia	Gale Catlett
Georgia Tech	Bobby Cremins	Oklahoma St.	Eddie Sutton	Wichita St.	Randy Smithson
Gonzaga	Mark Few	Old Dominion	Jeff Capel	William & Mary	Charlie Woollum
Harvard	Frank Sullivan	Oregon	Ernie Kent	Wisconsin	Dick Bennett
Hawaii	Riley Wallace	Oregon St.	Eddie Payne	Wyoming	Steve McClain
Houston	Clyde Drexler	Pacific (CA)	Bob Thomason	Xavier	Skip Prosser
Idaho	David Farrar	Pennsylvania	Fran Dunphy	Yale	James Jones
Idaho St.	Doug Oliver				

(1) As of mid-Oct.

Most Coaching Victories in the NCAA Tournament Through 1999

Coach, school(s), years	Wins	Tournaments	Coach, school(s), years	Wins	Tournaments
Dean Smith, North Carolina, 1967-97	65	27	Lute Olson, Iowa and Arizona, 1979-99	31	20
Mike Krzyzewski, Duke, 1984-99	48	15	Jerry Tarkanian, Long Beach State		
John Wooden, UCLA, 1950-75	47	16	and UNLV, 1970-91	31	13
Denny Crum, Louisville, 1972-97	42	21	Adolph Rupp, Kentucky, 1942-72	30	20
Bob Knight, Indiana, 1973-99	42	23	Jim Boeheim, Syracuse, 1977-99	29	19
John Thompson, Georgetown, 1975-97	34	20			

WOMEN'S COLLEGE BASKETBALL
Purdue Defeats Duke in the 1999 NCAA Women's Championship

The Purdue Boilermakers defeated the Duke Blue Devils, 62-45, to win their first NCAA women's basketball championship, Mar. 28 in San Jose, CA. After a first half in which Purdue trailed 22-17, setting an NCAA tournament record for fewest points in a half, the Boilermakers rallied behind Ukari Figgs, who scored all of her game-high 18 points in the second half and was named Outstanding Player in the Final Four.

NCAA Division I Women's Champions

Year	Champion	Coach	Final opponent	Score	Outstanding player	Site
1982	Louisiana Tech	Sonja Hogg	Cheyney	76-62	Janice Lawrence, La. Tech	Norfolk, VA
1983	USC	Linda Sharp	Louisiana Tech	69-67	Cheryl Miller, USC	Norfolk, VA
1984	USC	Linda Sharp	Tennessee	72-61	Cheryl Miller, USC	Los Angeles, CA
1985	Old Dominion	Marianne Stanley	Georgia	70-65	Tracy Claxton, Old Dominion	Austin, TX
1986	Texas	Jody Conradt	USC	97-81	Clarissa Davis, Texas	Lexington, KY
1987	Tennessee	Pat Summitt	Louisiana Tech	67-44	Tonya Edwards, Tennessee	Austin, TX
1988	Louisiana Tech	Leon Barmore	Auburn	56-54	Erica Westbrooks, La. Tech	Tacoma, WA
1989	Tennessee	Pat Summitt	Auburn	76-60	Bridgette Gordon, Tennessee	Tacoma, WA
1990	Stanford	Tara VanDerveer	Auburn	88-81	Jennifer Azzi, Stanford	Knoxville, TN
1991	Tennessee	Pat Summitt	Virginia	70-67*	Dawn Staley, Virginia	New Orleans, LA
1992	Stanford	Tara VanDerveer	W. Kentucky	78-62	Molly Goodenbour, Stanford	Los Angeles, CA
1993	Texas Tech	Marsha Sharp	Ohio St.	84-82	Sheryl Swoopes, Texas Tech	Atlanta, GA
1994	North Carolina	Sylvia Hatchell	Louisiana Tech	60-59	Charlotte Smith, North Carolina	Richmond, VA
1995	Connecticut	Geno Auriemma	Tennessee	70-64	Rebecca Lobo, Connecticut	Minneapolis, MN
1996	Tennessee	Pat Summitt	Georgia	83-65	Michelle Marciniak, Tennessee	Charlotte, NC
1997	Tennessee	Pat Summitt	Old Dominion	68-59	Chamique Holdsclaw, Tennessee	Cincinnati, OH
1998	Tennessee	Pat Summitt	Louisiana Tech	93-75	Chamique Holdsclaw, Tennessee	Kansas City, MO
1999	Purdue	Carolyn Peck	Duke	62-45	Ukari Figgs, Purdue	San Jose, CA

* Overtime.

Wade Trophy

Awarded by National Assn. for Girls and Women in Sport for academics, community service, and player performance.

Year	Player, school	Year	Player, school	Year	Player, school
1978	Carol Blazejowski, Montclair St.	1986	Kamie Ethridge, Texas	1993	Karen Jennings, Nebraska
1979	Nancy Lieberman, Old Dominion	1987	Shelly Pennefeather, Villanova	1994	Carol Ann Shudlick, Minnesota
1980	Nancy Lieberman, Old Dominion	1988	Teresa Weatherspoon, Louisiana	1995	Rebecca Lobo, Connecticut
1981	Lynette Woodard, Kansas		Tech	1996	Jennifer Rizzotti, Connecticut
1982	Pam Kelly, Louisiana Tech	1989	Clarissa Davis, Texas	1997	DeLisha Milton, Florida
1983	LaTaunya Pollard, Long Beach St.	1990	Jennifer Azzi, Stanford	1998	Chamique Holdsclaw, Tennessee
1984	Janice Lawrence, Louisiana Tech	1991	Daedra Charles, Tennessee	1999	Stephanie White-McCarty, Purdue
1985	Cheryl Miller, USC	1992	Susan Robinson, Penn St.		

Top Women's Career Scorers
(Minimum 1,500 points; ranked by average)

Player, school	Years	Points	Avg.	Player, school	Years	Points	Avg.
Patricia Hoskins, Mississippi Valley State	1985-89	3,122	28.4	Valorie Whiteside, Appalachian State	1984-88	2,944	25.4
Sandra Hodge, New Orleans	1981-84	2,860	26.7	Joyce Walker, LSU	1981-84	2,906	24.8
Lorri Bauman, Drake	1981-84	3,115	26.0	Tarcha Hollis, Grambling	1988-91	2,058	24.2
Andrea Congreaves, Mercer	1989-93	2,796	25.9	Korie Hlede, Duquesne	1994-98	2,631	24.1
Cindy Blodgett, Maine	1994-98	3,005	25.5	Karen Pelphrey, Marshall	1983-86	2,746	24.1

1999 NCAA BASKETBALL TOURNAMENT (WOMEN)

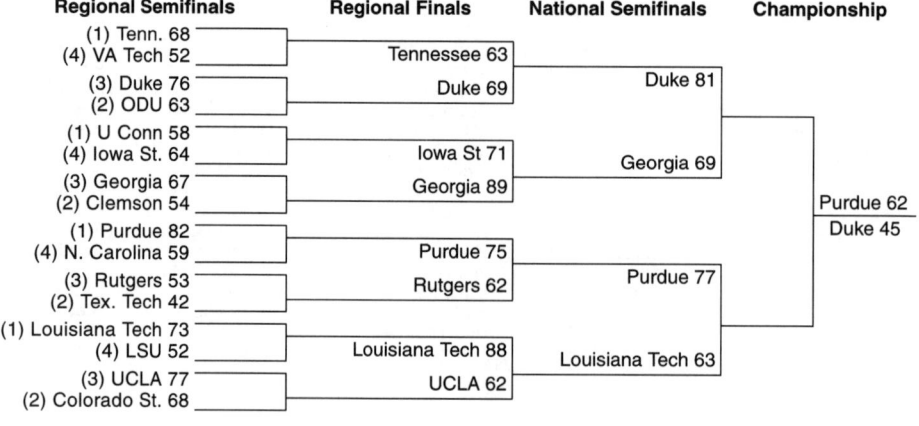

FISHING
Selected IGFA Saltwater & Freshwater All-Tackle World Records

Source: International Game Fish Association; records confirmed to Oct. 15, 1999

Saltwater Fish Records

Species	Weight	Where caught	Date	Angler
Albacore	88 lbs. 2 oz.	Canary Islands, Spain	Nov. 19, 1977	Siegfried Dickemann
Amberjack, greater	155 lbs. 12 oz.	Bermuda	Aug. 16, 1992	Larry Trott
Barracuda, great	85 lbs.	Christmas Island, Kiribati	Apr. 11, 1992	John W. Helfrich
Barracuda, Mexican	21 lbs.	Phantom Isle, Costa Rica	Mar. 27, 1987	E. Greg Kent
Barracuda, Pacific	26 lbs. 8 oz.	Playa Matapalo, Costa Rica	Jan. 3, 1999	Doug Hettinger
Bass, barred sand	13 lbs. 3 oz.	Huntington Beach, CA	Aug. 29, 1988	Robert Halal
Bass, black sea	9 lbs. 8 oz.	Virginia Beach, VA	Jan. 9, 1987	Joe Mizelle Jr.
		Virginia Beach, VA	Dec. 22, 1990	Jack G. Stallings
Bass, giant sea	563 lbs. 8 oz.	Anacapa Island, CA	Aug. 20, 1968	James D. McAdam Jr.
Bass, redeye	8 lbs. 12 oz.	Apalachicola River, FL	Jan. 28, 1995	Carl W. Davis
Bass, striped	78 lbs. 8 oz.	Atlantic City, NJ	Sept. 21, 1982	Albert R. McReynolds
Bluefish	31 lbs. 12 oz.	Hatteras Inlet, NC	Jan. 30, 1972	James M. Hussey
Bonefish	19 lbs.	Zululand, South Africa	May 26, 1962	Brian W. Batchelor
Bonito, Atlantic	18 lbs. 4 oz.	Faial Island, Azores	July 8, 1953	D. Gama Higgs
Bonito, Pacific	21 lbs. 3 oz.	Malibu, CA	July 30, 1978	Gino M. Picciolo
Cabezon	23 lbs.	Juan De Fuca Strait, WA	Aug. 4, 1990	Wesley S. Hunter
Cobia	135 lbs. 9 oz.	Shark Bay, Australia	July 9, 1985	Peter W. Goulding
Cod, Atlantic	98 lbs. 12 oz.	Isle of Shoals, NH	June 8, 1969	Alphonse J. Bielevich
Cod, Pacific	32 lbs.	Unalaska Bay, AK	June 29, 1997	Donald Boston
Conger	133 lbs. 4 oz.	Berry Head, S. Devon, England	June 5, 1995	Vic Evans
Dolphin	88 lbs.	Exuma, Bahamas	May 5, 1998	Richard D. Evans
Drum, black	113 lbs. 1 oz.	Lewes, DE	Sept. 15, 1975	Gerald M. Townsend
Drum, red	94 lbs. 2 oz.	Avon, NC	Nov. 7, 1984	David G. Deuel
Eel, American	9 lbs. 4 oz.	Cape May, NJ	Nov. 9, 1995	Jeff Pennick
Eel, marbled	36 lbs. 1 oz.	Hazelmere Dam, South Africa	June 10, 1984	Ferdie Van Nooten
Flounder, southern	20 lbs. 9 oz.	Nassau Sound, FL	Dec. 23, 1983	Larenza W. Mungin
Flounder, summer	22 lbs. 7 oz.	Montauk, NY	Sept. 15, 1975	Charles Nappi
Grouper, Warsaw	436 lbs. 12 oz.	Gulf of Mexico, Destin, FL	Dec. 22, 1985	Steve Haeusler
Halibut, Atlantic	355 lbs. 6 oz.	Valevag, Norway	Oct. 20, 1997	Odd Arve Gunderstad
Halibut, California	57 lbs. 11 oz.	Santa Rosa Island, CA	June 6, 1989	William Akins
Halibut, Pacific	459 lbs.	Dutch Harbor, AK	June 11, 1996	Jack Tragis
Jack, crevalle	57 lbs. 14 oz.	Southwest Pass, LA	Aug. 15, 1997	Leon D. Richard
Jack, horse-eye	29 lbs. 8 oz.	Ascencion Island, South Atlantic	May 28, 1993	Mike Hanson
Jack, Pacific crevalle	39 lbs.	Playa Zancudo, Costa Rica	Mar. 3, 1997	Ingrid Callaghan
Jewfish	680 lbs.	Fernandina Beach, FL	May 20, 1961	Lynn Joyner
Kawakawa	29 lbs.	Clarion Island, Mexico	Dec. 17, 1986	Ronald Nakamura
Lingcod	69 lbs.	Langara Island, British Columbia	June 16, 1992	Murray M. Rorner
Mackerel, cero	17 lbs. 2 oz.	Islamorada, FL	Apr. 5, 1986	G. Michael Mills
Mackerel, king	93 lbs.	San Juan, PR	Apr. 18, 1999	Steve Perez Graulau
Mackerel, Spanish	13 lbs.	Ocracoke Inlet, NC	Nov. 4, 1987	Robert Cranton
Marlin, Atlantic blue	1,402 lbs. 2 oz.	Vitoria, Brazil	Feb. 29, 1992	Paulo Roberto A. Amorim
Marlin, black	1,560 lbs.	Cabo Blanco, Peru	Aug. 4, 1953	Alfred C. Glassell Jr.
Marlin, Pacific blue	1,376 lbs.	Kaaiwi Pt., Kona, HI	May 31, 1982	Jay W. deBeaubien
Marlin, striped	494 lbs.	Tutukaka, New Zealand	Jan. 16, 1986	Bill Boniface
Marlin, white	181 lbs. 14 oz.	Vitoria, Brazil	Dec. 8, 1979	Evandro Luiz Coser
Permit	56 lbs. 2 oz.	Ft. Lauderdale, FL	June 30, 1997	Thomas Sebestyen
Pollack, European	27 lbs. 6 oz.	Salcombe, Devon, England	Jan. 16, 1986	Robert Samuel Milkins
Pollock	50 lbs.	Salstraumen, Norway	Nov. 30, 1995	Thor-Magnus Lekang
Pompano, African	50 lbs. 8 oz.	Daytona Beach, FL	Apr. 21, 1990	Tom Sargent
Roosterfish	114 lbs.	La Paz, Baja Cal., Mexico	June 1, 1960	Abe Sackheim
Runner, blue	11 lbs. 2 oz.	Dauphin Isl., AL	June 28, 1997	Stacey Michelle Moiren
Runner, rainbow	37 lbs. 9 oz.	Clarion Island, Mexico	Nov. 21, 1991	Tom Pfleger
Sailfish, Atlantic	141 lbs. 1 oz.	Luanda, Angola	Feb. 19, 1994	Alfredo de Sousa Neves
Sailfish, Pacific	221 lbs.	Santa Cruz Island, Ecuador	Feb. 12, 1947	C. W. Stewart
Seabass, white	83 lbs. 12 oz.	San Felipe, Mexico	Mar. 31, 1953	L. C. Baumgardner
Seatrout, spotted	17 lbs. 7 oz.	Ft. Pierce, FL	May 11, 1995	Craig F. Carson
Shark, bigeye thresher	802 lbs.	Tutukaka, New Zealand	Feb. 8, 1981	Dianne North
Shark, bignose	369 lbs. 14 oz.	Markham R., Papua New Guinea	Oct. 23, 1993	Lester J. Rohrlach
Shark, blue	454 lbs.	Martha's Vineyard, MA	July 19, 1996	Pete Bergin
Shark, great hammerhead	991 lbs.	Sarasota, FL	May 30, 1982	Allen Ogle
Shark, Greenland	1,708 lbs. 9 oz.	Trondheimsfjord, Norway	Oct. 18, 1987	Terje Nordtvedt
Shark, porbeagle	507 lbs.	Caithness, Scotland	Mar. 9, 1993	Christopher Bennett
Shark, shortfin mako	1,115 lbs.	Black River, Mauritius	Nov. 16, 1988	Patrick Guillanton
Shark, tiger	1,780 lbs.	Cherry Grove, SC	June 14, 1964	Walter Maxwell
Shark, white	2,664 lbs.	Ceduna, S.A., Australia	Apr. 21, 1959	Alfred Dean
Sheepshead	21 lbs. 4 oz.	New Orleans, LA	Apr. 16, 1982	Wayne Desselle
Skipjack, black	26 lbs.	Thetis Bank, Baja Cal., Mexico	Oct. 23, 1991	Clifford Hamaishi
Snapper, cubera	121 lbs. 8 oz.	Cameron, LA	July 5, 1982	Mike Hebert
Snapper, red	50 lbs. 4 oz.	Gulf of Mexico, LA	June 23, 1996	Capt. Doc Kennedy
Snook, common	53 lbs. 10 oz.	Parismina Ranch, Costa Rica	Oct. 18, 1978	Gilbert Ponzi
Spearfish, Mediterranean	90 lbs. 13 oz.	Madeira Island, Portugal	June 2, 1980	Joseph Larkin
Swordfish	1,182 lbs.	Iquique, Chile	May 7, 1953	L. B. Marron
Tarpon	283 lbs. 4 oz.	Sherbro Island, Sierra Leone	Apr. 16, 1991	Yvon Sebag
Tautog	25 lbs.	Ocean City, NJ	Jan. 20, 1998	Anthony R. Monica
Trevally, bigeye	31 lbs. 8 oz.	Poivre Isl., Seychelles	Apr. 23, 1997	Les Sampson
Trevally, giant	145 lbs. 8 oz.	Makena, Maui, HI	Mar. 28, 1991	Russell Mori
Tuna, Atlantic bigeye	392 lbs. 6 oz.	Canary Islands, Spain	July 15, 1996	Dieter Vogel
Tuna, blackfin	45 lbs. 8 oz.	Key West, FL	May 4, 1996	Sam J. Burnett
Tuna, bluefin	1,496 lbs.	Aulds Cove, Nova Scotia	Oct. 26, 1979	Ken Fraser
Tuna, longtail	79 lbs. 2 oz.	Montague Isl., N.S.W., Australia	Apr. 12, 1982	Tim Simpson
Tuna, Pacific bigeye	435 lbs.	Cabo Blanco, Peru	Apr. 17, 1957	Dr. Russel V. A. Lee

Species	Weight	Where caught	Date	Angler
Tuna, skipjack	45 lbs. 4 oz.	Flathead Bank, Baja Cal., Mexico	Nov. 16, 1996	Brian Evans
Tuna, southern bluefin	348 lbs. 5 oz.	Whakatane, New Zealand	Jan. 16, 1981	Rex Wood
Tuna, yellowfin	388 lbs. 12 oz.	San Benedicto Island, Mexico	Apr. 1, 1977	Curt Wiesenhutter
Tunny, little	35 lbs. 2 oz.	Cap de Garde, Algeria	Dec. 14, 1988	Jean Yves Chatard
Wahoo	158 lbs. 8 oz.	Loreto, Baja Cal., Mexico	June 10, 1996	Keith Winter
Weakfish	19 lbs. 2 oz.	Jones Beach Inlet, NY	Oct. 11, 1984	Dennis Roger Rooney
		Delaware Bay, DE	May 20, 1989	William E. Thomas
Yellowtail, California	80 lbs. 11 oz.	Alijos Rocks, Baja Cal., Mexico	Nov. 12, 1998	Brian Buddell
Yellowtail, southern	114 lbs. 10 oz.	Tauranga, New Zealand	Feb. 5, 1984	Mike Godfrey
		White Island, New Zealand	Jan. 9, 1987	David Lugton

Freshwater Fish Records

Species	Weight	Where caught	Date	Angler
Barramundi	64 lbs. 13 oz.	Lake Tinaroo, N. Queensland, Australia	Dec. 30, 1998	Alf Homewood
Bass, largemouth	22 lbs. 4 oz.	Montgomery Lake, GA	June 2, 1932	George W. Perry
Bass, rock	3 lbs.	York River, Ontario	Aug. 1, 1974	Peter Gulgin
Bass, smallmouth	10 lbs. 14 oz.	Dale Hollow Lake, TN	Apr. 24, 1969	John T. Gorman
Bass, white	6 lbs. 13 oz.	Lake Orange, VA	July 31, 1989	Ronald L. Sprouse
Bass, whiterock	27 lbs. 5 oz.	Greers Ferry Lake, AR	April 24, 1997	Jerald C. Shaum
Bass, yellow	2 lbs. 9 oz.	Waverly, TN	Feb. 27, 1998	John T. Chappell
Bluegill	4 lbs. 12 oz.	Ketona Lake, AL	Apr. 9, 1950	T. S. Hudson
Bowfin	21 lbs. 8 oz.	Florence, SC	Jan. 29, 1980	Robert L. Harmon
Buffalo, bigmouth	70 lbs. 5 oz.	Bastrop, LA	Apr. 21, 1980	Delbert Sisk
Buffalo, black	55 lbs. 8 oz.	Cherokee Lake, TN	May 3, 1984	Edward H. McLain
Buffalo, smallmouth	82 lbs. 3 oz.	Athens Lake, AR	June 6, 1993	Randy Collins
Bullhead, brown	6 lbs. 1 oz.	Waterford, NY	Apr. 26, 1998	Bobby Triplett
Bullhead, yellow	4 lbs. 4 oz.	Mormon Lake, AZ	May 11, 1984	Emily Williams
Burbot	18 lbs. 11 oz.	Angenmanalren, Sweden	Oct. 22, 1996	Margit Agren
Carp, common	82 lbs. 3 oz.	Lake Roduta, Romenia	May 26, 1987	Christian Baldemair
Catfish, blue	111 lbs.	Wheeler Reservoir, Tenn. R.	July 5, 1996	William P. McKinley
Catfish, channel	58 lbs.	Santee-Cooper Res., SC	July 7, 1964	W. B. Whaley
Catfish, flathead	123 lbs. 9 oz.	Independence, KS	May 14, 1998	Ken Paulie
Catfish, white	18 lbs. 14 oz.	Withlacoochee River, FL	Sept. 21, 1991	Jim Miller
Char, Arctic	32 lbs. 9 oz.	Tree River, Canada	July 30, 1981	Jeffrey L. Ward
Crappie, white	5 lbs. 3 oz.	Enid Dam, MS	July 31, 1957	Fred L. Bright
Dolly Varden	19 lbs. 4 oz.	Unnamed river, AK	Sept. 4, 1998	Gary D. Ordway
DoradoYW	51 lbs. 5 oz.	Toledo (Corrientes), Argentina	Sept. 27, 1984	Armando Giudice
Drum, freshwater	54 lbs. 8 oz.	Nickajack Lake, TN	Apr. 20, 1972	Benny E. Hull
Gar, alligator	279 lbs.	Rio Grande, TX	Dec. 2, 1951	Bill Valverde
Gar, Florida	21 lbs. 3 oz.	Boca Raton, FL	June 3, 1981	Jeff Sabol
Gar, longnose	50 lbs. 5 oz.	Trinity River, TX	July 30, 1954	Townsend Miller
Gar, shortnose	5 lbs. 12 oz.	Ren Lake, IL	July 16, 1995	Donna K. Willmert
Gar, spotted	9 lbs. 12 oz.	Lake Mexia, TX	Apr. 7, 1994	Rick Rivard
Grayling, Arctic	5 lbs. 15 oz.	Katseyedie River, N.W.T.	Aug. 16, 1967	Jeanne P. Branson
Inconnu	53 lbs.	Pah River, AK	Aug. 20, 1985	Lawrence E. Hudnall
Kokanee	9 lbs. 6 oz.	Okanagan Lake, Vernon, B.C.	June 18, 1988	Norm Kuhn
Muskellunge	67 lbs. 8 oz.	Lake Court Oreilles, WI	July 24, 1949	Cal Johnson
Muskellunge, tiger	51 lbs. 3 oz.	Lac Vieux-Desert, MI	July 16, 1919	John Knobla
Perch, Nile	213 lbs.	Lake Nasser, Egypt	Dec. 18, 1997	Adrian Brayshaw
Perch, white	4 lbs. 12 oz.	Messalonskee Lake, ME	June 4, 1949	Earl Small
Perch, yellow	4 lbs. 3 oz.	Bordentown, NJ	May, 1865	Dr. C. C. Abbot
Pickerel, chain	9 lbs. 6 oz.	Homerville, GA	Feb. 17, 1961	Baxley McQuaig Jr.
Pike, northern	55 lbs. 1 oz.	Lake of Grefeern, W. Germany	Oct. 16, 1986	Lothar Louis
Redhorse, greater	9 lbs. 3 oz.	Salmon River, Pulaski, NY	May 11, 1985	Jason Wilson
Redhorse, silver	11 lbs. 7 oz.	Plum Creek, WI	May 29, 1985	Neal Long
Salmon, Atlantic	79 lbs. 2 oz.	Tana River, Norway	1928	Henrik Henriksen
Salmon, chinook	97 lbs. 4 oz.	Kenai River, AK	May 17, 1985	Les Anderson
Salmon, chum	35 lbs.	Edye Pass, BC	July 11, 1995	Todd A. Johansson
Salmon, coho	33 lbs. 4 oz.	Salmon River, Pulaski, NY	Sept. 27, 1989	Jerry Lifton
Salmon, pink	13 lbs. 1 oz.	St. Mary's River, Ontario	Sept. 23, 1992	Ray Higaki
Salmon, sockeye	15 lbs. 3 oz.	Kenai River, AK	Aug. 9, 1987	Stan Roach
Sauger	8 lbs. 12 oz.	Lake Sakakawea, ND	Oct. 6, 1971	Mike Fischer
Shad, American	11 lbs. 4 oz.	Connecticut River, MA	May 19, 1986	Bob Thibodo
Sturgeon, beluga	224 lbs. 13 oz.	Guryev, Kazakhstan	May 3, 1993	Merete Lehne
Sturgeon, white	468 lbs.	Benicia, CA	July 9, 1983	Joey Pallotta 3d
Sunfish, green	2 lbs. 2 oz.	Stockton Lake, MO	June 18, 1971	Paul M. Dilley
Sunfish, redbreast	1 lb. 12 oz.	Suwannee River, FL	May 29, 1984	Alvin Buchanan
Sunfish, redear	5 lbs. 7oz.	Diverson Canal, GA	Nov. 6, 1998	Amos M. Gay
Tigerfish, giant	97 lbs.	Zaire River, Kinshasa, Zaire	July 9, 1988	Raymond Houtmans
Tilapia	6 lbs. 5 oz.	Lake Arenal, Costa Rica	Feb. 10, 1995	Marvin C. Smith
Trout, Apache	5 lb. 3 oz.	Apache Res., AZ	May 29, 1991	John Baldwin
Trout, brook	14 lbs. 8 oz.	Nipigon River, Ontario	July, 1916	Dr. W. J. Cook
Trout, bull	32 lbs.	Lake Pend Oreille, ID	Oct. 27, 1949	N. L. Higgins
Trout, cutthroat	41 lbs.	Pyramid Lake, NV	Dec., 1925	John Skimmerhorn
Trout, golden	11 lbs.	Cooks Lake, WY	Aug. 5, 1948	Charles S. Reed
Trout, lake	72 lbs.	Great Bear Lake, N.W.T.	Aug. 9, 1995	Lloyd E. Bull
Trout, rainbow	42 lbs. 2 oz.	Bell Island, AK	June 22, 1970	David Robert White
Trout, tiger	20 lbs. 13 oz.	Lake Michigan, WI	Aug. 12, 1978	Pete M. Friedland
Walleye	25 lbs.	Old Hickory Lake, TN	Aug. 2, 1960	Mabry Harper
Warmouth	2 lbs. 7 oz.	Yellow River, Holt, FL	Oct. 19, 1985	Tony D. Dempsey
Whitefish, lake	14 lbs. 6 oz.	Meaford, Ontario	May 21, 1984	Dennis M. Laycock
Whitefish, mountain	5 lbs. 8 oz.	Elbow River, Calgary, AB	Aug. 1, 1995	Randy G. Woo
Whitefish, round	6 lbs.	Putahow River, Manitoba	June 14, 1984	Allen Ristori
Zander	25 lbs. 2 oz.	Trosa, Sweden	June 12, 1986	Harry Lee Tennison

BASEBALL

1999: Mets-Braves Marathon; Yankee Sweep; Cone Perfect; McGwire at 522

On Oct. 17, 1999, in Game 5 of the National League Championship Series, the New York Mets and Atlanta Braves played the longest postseason game in major league history (5 hrs, 46 min). In the 15-inning contest 126 batters came to the plate; only 7 scored. The Braves eventually lost, 4-3, but went on to win the NLCS 4 games to 2 and face the New York Yankees. What followed was the Yankees' 25th World Series win, in a 4-game sweep. There were 3 no-hitters in 1999, including Yankee David Cone's 6-0 perfect game July 18 against Montreal in interleague play. It was the 2d year in a row that a Yankee pitched a perfect game (David Wells, 1998) and the 14th perfect game in history. On June 25, Jose Jimenez of St. Louis pitched a 1-0 no-hitter against the Arizona Diamondbacks. Minnesota's Eric Milton hurled his 7-0 no-hitter against the Anaheim Angels on Sept. 11. St. Louis's Mark McGwire and the Chicago Cubs' Sammy Sosa continued their assault on the home-run record book, hitting 65 and 63, respectively. McGwire (522) also moved up to 10th on the career home-run list. San Diego's Tony Gwynn (3,067) and Tampa Bay's Wade Boggs (3,010) became the 22d and 23d members of the 3,000-hit club. Tampa Bay's Jose Canseco (431) and Baltimore's Cal Ripken Jr. (402) passed the 400 career home-run mark. Among other milestones: Cleveland's Manny Ramirez batted in 165 runs, the 8th-highest season total ever. Yankee pitcher Roger Clemens, in addition to his 1st World Series victory (Oct. 27, Game 4), set a consecutive win mark of 20 over 2 seasons. Atlanta's Otis Nixon (620) surpassed the 600 career stolen-base mark on his way to 15th on the all-time list.

Major League Pennant Winners, 1901–1968

	National League						American League				
Year	Winner	Won	Lost	Pct	Manager	Year	Winner	Won	Lost	Pct	Manager
1901	Pittsburgh	90	49	.647	Clarke	1901	Chicago	83	53	.610	Griffith
1902	Pittsburgh	103	36	.741	Clarke	1902	Philadelphia	83	53	.610	Mack
1903	Pittsburgh	91	49	.650	Clarke	1903	Boston	91	47	.659	Collins
1904	New York	106	47	.693	McGraw	1904	Boston	95	59	.617	Collins
1905	New York	105	48	.686	McGraw	1905	Philadelphia	92	56	.622	Mack
1906	Chicago	116	36	.763	Chance	1906	Chicago	93	58	.616	Jones
1907	Chicago	107	45	.704	Chance	1907	Detroit	92	58	.613	Jennings
1908	Chicago	99	55	.643	Chance	1908	Detroit	90	63	.588	Jennings
1909	Pittsburgh	110	42	.724	Clarke	1909	Detroit	98	54	.645	Jennings
1910	Chicago	104	50	.675	Chance	1910	Philadelphia	102	48	.680	Mack
1911	New York	99	54	.647	McGraw	1911	Philadelphia	101	50	.669	Mack
1912	New York	103	48	.682	McGraw	1912	Boston	105	47	.691	Stahl
1913	New York	101	51	.664	McGraw	1913	Philadelphia	96	57	.627	Mack
1914	Boston	94	59	.614	Stallings	1914	Philadelphia	99	53	.651	Mack
1915	Philadelphia	90	62	.592	Moran	1915	Boston	101	50	.669	Carrigan
1916	Brooklyn	94	60	.610	Robinson	1916	Boston	91	63	.591	Carrigan
1917	New York	98	56	.636	McGraw	1917	Chicago	100	54	.649	Rowland
1918	Chicago	84	45	.651	Mitchell	1918	Boston	75	51	.595	Barrow
1919	Cincinnati	96	44	.686	Moran	1919	Chicago	88	52	.629	Gleason
1920	Brooklyn	93	60	.604	Robinson	1920	Cleveland	98	56	.636	Speaker
1921	New York	94	56	.614	McGraw	1921	New York	98	55	.641	Huggins
1922	New York	93	61	.604	McGraw	1922	New York	94	60	.610	Huggins
1923	New York	95	58	.621	McGraw	1923	New York	98	54	.645	Huggins
1924	New York	93	60	.608	McGraw	1924	Washington	92	62	.597	Harris
1925	Pittsburgh	95	58	.621	McKechnie	1925	Washington	96	55	.636	Harris
1926	St. Louis	89	65	.578	Hornsby	1926	New York	91	63	.591	Huggins
1927	Pittsburgh	94	60	.610	Bush	1927	New York	110	44	.714	Huggins
1928	St. Louis	95	59	.617	McKechnie	1928	New York	101	53	.656	Huggins
1929	Chicago	98	54	.645	McCarthy	1929	Philadelphia	104	46	.693	Mack
1930	St. Louis	92	62	.597	Street	1930	Philadelphia	102	52	.662	Mack
1931	St. Louis	101	53	.656	Street	1931	Philadelphia	107	45	.704	Mack
1932	Chicago	90	64	.584	Grimm	1932	New York	107	47	.695	McCarthy
1933	New York	91	61	.599	Terry	1933	Washington	99	53	.651	Cronin
1934	St. Louis	95	58	.621	Frisch	1934	Detroit	101	53	.656	Cochrane
1935	Chicago	100	54	.649	Grimm	1935	Detroit	93	58	.616	Cochrane
1936	New York	91	62	.597	Terry	1936	New York	102	51	.667	McCarthy
1937	New York	95	57	.625	Terry	1937	New York	102	52	.662	McCarthy
1938	Chicago	89	63	.586	Hartnett	1938	New York	99	53	.651	McCarthy
1939	Cincinnati	97	57	.630	McKechnie	1939	New York	106	45	.702	McCarthy
1940	Cincinnati	100	53	.654	McKechnie	1940	Detroit	90	64	.584	Baker
1941	Brooklyn	100	54	.649	Durocher	1941	New York	101	53	.656	McCarthy
1942	St. Louis	106	48	.688	Southworth	1942	New York	103	51	.669	McCarthy
1943	St. Louis	105	49	.682	Southworth	1943	New York	98	56	.636	McCarthy
1944	St. Louis	105	49	.682	Southworth	1944	St. Louis	89	65	.578	Sewell
1945	Chicago	98	56	.636	Grimm	1945	Detroit	88	65	.575	O'Neill
1946	St. Louis	98	58	.628	Dyer	1946	Boston	104	50	.675	Cronin
1947	Brooklyn	94	60	.610	Shotton	1947	New York	97	57	.630	Harris
1948	Boston	91	62	.595	Southworth	1948	Cleveland	97	58	.626	Boudreau
1949	Brooklyn	97	57	.630	Shotton	1949	New York	97	57	.630	Stengel
1950	Philadelphia	91	63	.591	Sawyer	1950	New York	98	56	.636	Stengel
1951	New York	98	59	.624	Durocher	1951	New York	98	56	.636	Stengel
1952	Brooklyn	96	57	.627	Dressen	1952	New York	95	59	.617	Stengel
1953	Brooklyn	105	49	.682	Dressen	1953	New York	99	52	.656	Stengel
1954	New York	97	57	.630	Durocher	1954	Cleveland	111	43	.721	Lopez
1955	Brooklyn	98	55	.641	Alston	1955	New York	96	58	.623	Stengel
1956	Brooklyn	93	61	.604	Alston	1956	New York	97	57	.630	Stengel
1957	Milwaukee	95	59	.617	Haney	1957	New York	98	56	.636	Stengel
1958	Milwaukee	92	62	.597	Haney	1958	New York	92	62	.597	Stengel
1959	Los Angeles	88	68	.564	Alston	1959	Chicago	94	60	.610	Lopez
1960	Pittsburgh	95	59	.617	Murtaugh	1960	New York	97	57	.630	Stengel
1961	Cincinnati	93	61	.604	Hutchinson	1961	New York	109	53	.673	Houk
1962	San Francisco	103	62	.624	Dark	1962	New York	96	66	.593	Houk
1963	Los Angeles	99	63	.611	Alston	1963	New York	104	57	.646	Houk
1964	St. Louis	93	69	.574	Keane	1964	New York	99	63	.611	Berra
1965	Los Angeles	97	65	.599	Alston	1965	Minnesota	102	60	.630	Mele
1966	Los Angeles	95	67	.586	Alston	1966	Baltimore	97	63	.606	Bauer
1967	St. Louis	101	60	.627	Schoendienst	1967	Boston	92	70	.568	Williams
1968	St. Louis	97	65	.599	Schoendienst	1968	Detroit	103	59	.636	Smith

Major League Pennant Winners, 1969-1999

National League

Year	East Winner	W	L	Pct	Manager	West Winner	W	L	Pct	Manager	Pennant Winner
1969	N.Y. Mets	100	62	.617	Hodges	Atlanta	93	69	.574	Harris	New York
1970	Pittsburgh	89	73	.549	Murtaugh	Cincinnati	102	60	.630	Anderson	Cincinnati
1971	Pittsburgh	97	65	.599	Murtaugh	San Francisco	90	72	.556	Fox	Pittsburgh
1972	Pittsburgh	96	59	.619	Virdon	Cincinnati	95	59	.617	Anderson	Cincinnati
1973	N.Y. Mets	82	79	.509	Berra	Cincinnati	99	63	.611	Anderson	New York
1974	Pittsburgh	88	74	.543	Murtaugh	Los Angeles	102	60	.630	Alston	Los Angeles
1975	Pittsburgh	92	69	.571	Murtaugh	Cincinnati	108	54	.667	Anderson	Cincinnati
1976	Philadelphia	101	61	.623	Ozark	Cincinnati	102	60	.630	Anderson	Cincinnati
1977	Philadelphia	101	61	.623	Ozark	Los Angeles	98	64	.605	Lasorda	Los Angeles
1978	Philadelphia	90	72	.556	Ozark	Los Angeles	95	67	.586	Lasorda	Los Angeles
1979	Pittsburgh	98	64	.605	Tanner	Cincinnati	90	71	.559	McNamara	Pittsburgh
1980	Philadelphia	91	71	.562	Green	Houston	93	70	.571	Virdon	Philadelphia
1981(a)	Philadelphia	34	21	.618	Green	Los Angeles	36	21	.632	Lasorda	(c)
1981(b)	Montreal	30	23	.566	Williams, Fanning	Houston	33	20	.623	Virdon	Los Angeles
1982	St. Louis	92	70	.568	Herzog	Atlanta	89	73	.549	Torre	St. Louis
1983	Philadelphia	90	72	.556	Corrales, Owens	Los Angeles	91	71	.562	Lasorda	Philadelphia
1984	Chicago	96	65	.596	Frey	San Diego	92	70	.568	Williams	San Diego
1985	St. Louis	101	61	.623	Herzog	Los Angeles	95	67	.586	Lasorda	St. Louis
1986	N.Y. Mets	108	54	.667	Johnson	Houston	96	66	.593	Lanier	New York
1987	St. Louis	95	67	.586	Herzog	San Francisco	90	72	.556	Craig	St. Louis
1988	N.Y. Mets	100	60	.625	Johnson	Los Angeles	94	67	.584	Lasorda	Los Angeles
1989	Chicago	93	69	.571	Zimmer	San Francisco	92	70	.568	Craig	San Francisco
1990	Pittsburgh	95	67	.586	Leyland	Cincinnati	91	71	.562	Piniella	Cincinnati
1991	Pittsburgh	98	64	.605	Leyland	Atlanta	94	68	.580	Cox	Atlanta
1992	Pittsburgh	96	66	.593	Leyland	Atlanta	98	64	.605	Cox	Atlanta
1993	Philadelphia	97	65	.599	Fregosi	Atlanta	104	58	.642	Cox	Philadelphia

Year	Division	Winner	W	L	Pct	Manager	Playoffs	Pennant Winner
1994(d)	East	Montreal	74	40	.649	Alou	—	—
	West	Cincinnati	66	48	.579	Johnson		
	Central	Los Angeles	58	56	.509	Lasorda		
1995	East	Atlanta	90	54	.625	Cox	Atlanta 3, Colorado* 1	Atlanta
	Central	Cincinnati	85	59	.590	Johnson	Cincinnati 3, Los Angeles 0	
	West	Los Angeles	78	66	.542	Lasorda	Atlanta 4, Cincinnati 0	
1996	East	Atlanta	96	66	.593	Cox	Atlanta 3, Los Angeles* 0	Atlanta
	Central	St. Louis	88	74	.543	La Russa	St. Louis 3, San Diego 0	
	West	San Diego	91	71	.562	Bochy	Atlanta 4, St. Louis 3	
1997	East	Atlanta	101	61	.623	Cox	Atlanta 3, Houston 0	Florida* (e)
	Central	Houston	84	78	.519	Dierker	Florida* 3, San Francisco 0	
	West	San Francisco	90	72	.556	Baker	Florida* 4, Atlanta 2	
1998	East	Atlanta	106	56	.654	Cox	Atlanta 3, Chicago* 0	San Diego
	Central	Houston	102	60	.630	Dierker	San Diego 3, Houston 1	
	West	San Diego	97	64	.602	Bochy	San Diego 4, Atlanta 2	
1999	East	Atlanta	103	59	.636	Cox	Atlanta 3, Houston 1	Atlanta
	Central	Houston	97	65	.599	Dierker	New York* 3, Arizona 1	
	West	Arizona	100	62	.617	Showalter	Atlanta 4, New York 2	

American League

Year	East Winner	W	L	Pct	Manager	West Winner	W	L	Pct	Manager	Pennant Winner
1969	Baltimore	109	53	.673	Weaver	Minnesota	97	65	.599	Martin	Baltimore
1970	Baltimore	108	54	.667	Weaver	Minnesota	98	64	.605	Rigney	Baltimore
1971	Baltimore	101	57	.639	Weaver	Oakland	101	60	.627	Williams	Baltimore
1972	Detroit	86	70	.551	Martin	Oakland	93	62	.600	Williams	Oakland
1973	Baltimore	97	65	.599	Weaver	Oakland	94	68	.580	Williams	Oakland
1974	Baltimore	91	71	.562	Weaver	Oakland	90	72	.556	Dark	Oakland
1975	Boston	95	65	.594	Johnson	Oakland	98	64	.605	Dark	Boston
1976	New York	97	62	.610	Martin	Kansas City	90	72	.556	Herzog	New York
1977	New York	100	62	.617	Martin	Kansas City	102	60	.630	Herzog	New York
1978	New York	100	63	.613	Martin, Lemon	Kansas City	92	70	.568	Herzog	New York
1979	Baltimore	102	57	.642	Weaver	California	88	74	.543	Fregosi	Baltimore
1980	New York	103	59	.636	Howser	Kansas City	97	65	.599	Frey	Kansas City
1981(a)	New York	34	22	.607	Michael	Oakland	37	23	.617	Martin	(c)
1981(b)	Milwaukee	31	22	.585	Rodgers	Kansas City	30	23	.566	Frey, Howser	New York
1982	Milwaukee	95	67	.586	Rodgers, Kuenn	California	93	69	.574	Mauch	Milwaukee
1983	Baltimore	98	64	.605	Altobelli	Chicago	99	63	.611	La Russa	Baltimore
1984	Detroit	104	58	.642	Anderson	Kansas City	84	78	.519	Howser	Detroit
1985	Toronto	99	62	.615	Cox	Kansas City	91	71	.562	Howser	Kansas City
1986	Boston	95	66	.590	McNamara	California	92	70	.568	Mauch	Boston
1987	Detroit	98	64	.605	Anderson	Minnesota	85	77	.525	Kelly	Minnesota
1988	Boston	89	73	.549	McNamara, Morgan	Oakland	104	58	.642	La Russa	Oakland
1989	Toronto	89	73	.549	Williams, Gaston	Oakland	99	63	.611	La Russa	Oakland
1990	Boston	88	74	.543	Morgan	Oakland	103	59	.636	La Russa	Oakland
1991	Toronto	91	71	.562	Gaston	Minnesota	95	67	.586	Kelly	Minnesota
1992	Toronto	96	66	.593	Gaston	Oakland	96	66	.593	La Russa	Toronto
1993	Toronto	95	67	.586	Gaston	Chicago	94	68	.580	Lamont	Toronto

Year	Division	Winner	W	L	Pct	Manager	Playoffs	Pennant Winner
1994(d)	East	New York	70	43	.619	Showalter	—	—
	Central	Chicago	67	46	.593	Lamont		
	West	Texas	52	62	.456	Kennedy		
1995	East	Boston	86	58	.597	Kennedy	Cleveland 3, Boston 0	Cleveland
	Central	Cleveland	100	44	.694	Hargrove	Seattle 3, New York* 2	
	West	Seattle	79	66	.545	Piniella	Cleveland 4, Seattle 2	
1996	East	New York	92	70	.568	Torre	Baltimore* 3, Cleveland 1	New York
	Central	Cleveland	99	62	.615	Hargrove	New York 3, Texas 1	
	West	Texas	90	72	.556	Oates	New York 4, Baltimore* 1	

1997	East	Baltimore	98	64	.605	Johnson	Baltimore 3, Seattle 1	Cleveland
	Central	Cleveland	86	75	.534	Hargrove	Cleveland 3, New York* 2	
	West	Seattle	90	72	.556	Piniella	Cleveland 4, Baltimore 2	
1998	East	New York	114	48	.704	Torre	New York 3, Texas 0	New York
	Central	Cleveland	89	73	.549	Hargrove	Cleveland 3, Boston* 1	
	West	Texas	88	74	.543	Oates	New York 4, Cleveland 2	
1999	East	New York	98	64	.605	Torre	New York 3, Texas 0	New York
	Central	Cleveland	97	65	.599	Hargrove	Boston* 3, Cleveland 2	
	West	Texas	95	67	.586	Oates	New York 4, Boston* 1	

*Wild card team. (a) First half. (b) Second half. (c) Montreal, L.A., N.Y. Yankees, and Oakland won the divisional playoffs. (d) In Aug. 1994, a players' strike began that caused the cancellation of the remainder of the season, the playoffs, and the World Series. Teams listed as division "winners" for 1994 were leading their divisions at the time of the strike. (e) Florida manager: Jim Leyland.

The Rawlings Gold Glove Awards in 1998

National League
Greg Maddux, Atlanta, pitcher
Charles Johnson[1], Los Angeles, catcher
J. T. Snow, San Francisco, first base
Bret Boone, Cincinnati, second base
Scott Rolen, Philadelphia, third base
Rey Ordonez, New York, shortstop
Barry Bonds, San Francisco, outfield
Andruw Jones, Atlanta, outfield
Larry Walker, Colorado, outfield

American League
Mike Mussina, Baltimore, pitcher
Ivan Rodriguez, Texas, catcher
Rafael Palmeiro, Baltimore, first base
Roberto Alomar, Baltimore, second base
Robin Ventura, Chicago, third base
Omar Vizquel, Cleveland, shortstop
Jim Edmonds, Anaheim, outfield
Ken Griffey Jr., Seattle, outfield
Bernie Williams, New York, outfield

(1) Also played for Florida.

The following are the players at each position who have won the most Gold Gloves since the award was instituted in 1957.

Pitcher:	Jim Kaat	16	First base:	Keith Hernandez	Shortstop:	Ozzie Smith 13
	Bob Gibson	9		Don Mattingly		Luis Aparicio 9
	Greg Maddux	9	Second base:	Ryne Sandberg	Outfield:	Roberto Clemente 12
Catcher:	Johnny Bench	10		Bill Mazeroski		Willie Mays 12
	Bob Boone	7		Frank White		Al Kaline 10
	Ivan Rodriguez	7	Third base:	Brooks Robinson 16		Ken Griffey Jr. 9
				Mike Schmidt 10		

Home Run Leaders

Note: Asterisk (*) indicates the all-time single-season record for each league.

	National League			**American League**	
Year	**Player, Team**	**HR**	**Year**	**Player, Team**	**HR**
1901	Sam Crawford, Cincinnati	16	1901	Napoleon Lajoie, Philadelphia	13
1902	Thomas Leach, Pittsburgh	6	1902	Socks Seybold, Philadelphia	16
1903	James Sheckard, Brooklyn	9	1903	Buck Freeman, Boston	13
1904	Harry Lumley, Brooklyn	9	1904	Harry Davis, Philadelphia	10
1905	Fred Odwell, Cincinnati	9	1905	Harry Davis, Philadelphia	8
1906	Timothy Jordan, Brooklyn	12	1906	Harry Davis, Philadelphia	12
1907	David Brain, Boston	10	1907	Harry Davis, Philadelphia	8
1908	Timothy Jordan, Brooklyn	12	1908	Sam Crawford, Detroit	7
1909	Red Murray, New York	7	1909	Ty Cobb, Detroit	9
1910	Fred Beck, Boston; Frank Schulte, Chicago	10	1910	Jake Stahl, Boston	10
1911	Frank Schulte, Chicago	21	1911	J. Franklin Baker, Philadelphia	9
1912	Henry Zimmerman, Chicago	14	1912	J. Franklin Baker, Philadelphia; Tris Speaker, Boston	10
1913	Gavvy Cravath, Philadelphia	19	1913	J. Franklin Baker, Philadelphia	13
1914	Gavvy Cravath, Philadelphia	19	1914	J. Franklin Baker, Philadelphia	9
1915	Gavvy Cravath, Philadelphia	24	1915	Robert Roth, Chicago-Cleveland	7
1916	Dave Robertson, N.Y.; Fred (Cy) Williams, Chi.	12	1916	Wally Pipp, New York	12
1917	Dave Robertson, N.Y.; Gavvy Cravath, Phi.	12	1917	Wally Pipp, New York	9
1918	Gavvy Cravath, Philadelphia	8	1918	Babe Ruth, Boston; Tilly Walker, Philadelphia	11
1919	Gavvy Cravath, Philadelphia	12	1919	Babe Ruth, Boston	29
1920	Cy Williams, Philadelphia	15	1920	Babe Ruth, New York	54
1921	George Kelly, New York	23	1921	Babe Ruth, New York	59
1922	Rogers Hornsby, St. Louis	42	1922	Ken Williams, St. Louis	39
1923	Cy Williams, Philadelphia	41	1923	Babe Ruth, New York	41
1924	Jacques Fournier, Brooklyn	27	1924	Babe Ruth, New York	46
1925	Rogers Hornsby, St. Louis	39	1925	Bob Meusel, New York	33
1926	Hack Wilson, Chicago	21	1926	Babe Ruth, New York	47
1927	Hack Wilson, Chicago; Cy Williams, Philadelphia	30	1927	Babe Ruth, New York	60
1928	Hack Wilson, Chicago; Jim Bottomley, St. Louis	31	1928	Babe Ruth, New York	54
1929	Chuck Klein, Philadelphia	43	1929	Babe Ruth, New York	46
1930	Hack Wilson, Chicago	56	1930	Babe Ruth, New York	49
1931	Chuck Klein, Philadelphia	31	1931	Babe Ruth, Lou Gehrig, both New York	46
1932	Chuck Klein, Philadelphia; Mel Ott, New York	38	1932	Jimmie Foxx, Philadelphia	58
1933	Chuck Klein, Philadelphia	28	1933	Jimmie Foxx, Philadelphia	48
1934	Rip Collins, St. Louis; Mel Ott, New York	35	1934	Lou Gehrig, New York	49
1935	Walter Berger, Boston	34	1935	Jimmie Foxx, Philadelphia; Hank Greenberg, Detroit	36
1936	Mel Ott, New York	33	1936	Lou Gehrig, New York	49
1937	Mel Ott, New York; Joe Medwick, St. Louis	31	1937	Joe DiMaggio, New York	46
1938	Mel Ott, New York	36	1938	Hank Greenberg, Detroit	58
1939	John Mize, St. Louis	28	1939	Jimmie Foxx, Boston	35
1940	John Mize, St. Louis	43	1940	Hank Greenberg, Detroit	41
1941	Dolph Camilli, Brooklyn	34	1941	Ted Williams, Boston	37
1942	Mel Ott, New York	30	1942	Ted Williams, Boston	36
1943	Bill Nicholson, Chicago	29	1943	Rudy York, Detroit	34
1944	Bill Nicholson, Chicago	33	1944	Nick Etten, New York	22
1945	Tommy Holmes, Boston	28	1945	Vern Stephens, St. Louis	24
1946	Ralph Kiner, Pittsburgh	23	1946	Hank Greenberg, Detroit	44
1947	Ralph Kiner, Pittsburgh; John Mize, New York	51	1947	Ted Williams, Boston	32

National League			American League		
Year	**Player, Team**	**HR**	**Year**	**Player, Team**	**HR**
1948	Ralph Kiner, Pittsburgh; John Mize, New York	40	1948	Joe DiMaggio, New York	39
1949	Ralph Kiner, Pittsburgh	54	1949	Ted Williams, Boston	43
1950	Ralph Kiner, Pittsburgh	47	1950	Al Rosen, Cleveland	37
1951	Ralph Kiner, Pittsburgh	42	1951	Gus Zernial, Chicago-Philadelphia	33
1952	Ralph Kiner, Pittsburgh; Hank Sauer, Chicago	37	1952	Larry Doby, Cleveland	32
1953	Ed Mathews, Milwaukee	47	1953	Al Rosen, Cleveland	43
1954	Ted Kluszewski, Cincinnati	49	1954	Larry Doby, Cleveland	32
1955	Willie Mays, New York	51	1955	Mickey Mantle, New York	37
1956	Duke Snider, Brooklyn	43	1956	Mickey Mantle, New York	52
1957	Hank Aaron, Milwaukee	44	1957	Roy Sievers, Washington	42
1958	Ernie Banks, Chicago	47	1958	Mickey Mantle, New York	42
1959	Ed Mathews, Milwaukee	46	1959	Rocky Colavito, Cleve.; Harmon Killebrew, Wash.	42
1960	Ernie Banks, Chicago	41	1960	Mickey Mantle, New York	40
1961	Orlando Cepeda, San Francisco	46	1961	Roger Maris, New York	*61
1962	Willie Mays, San Francisco	49	1962	Harmon Killebrew, Minnesota	48
1963	Hank Aaron, Milwaukee; Willie McCovey, S.F.	44	1963	Harmon Killebrew, Minnesota	45
1964	Willie Mays, San Francisco	47	1964	Harmon Killebrew, Minnesota	49
1965	Willie Mays, San Francisco	52	1965	Tony Conigliaro, Boston	32
1966	Hank Aaron, Atlanta	44	1966	Frank Robinson, Baltimore	49
1967	Hank Aaron, Atlanta	39	1967	Carl Yastrzemski, Boston; Harmon Killebrew, Minn.	44
1968	Willie McCovey, San Francisco	36	1968	Frank Howard, Washington	44
1969	Willie McCovey, San Francisco	45	1969	Harmon Killebrew, Minnesota	49
1970	Johnny Bench, Cincinnati	45	1970	Frank Howard, Washington	44
1971	Willie Stargell, Pittsburgh	48	1971	Bill Melton, Chicago	33
1972	Johnny Bench, Cincinnati	40	1972	Dick Allen, Chicago	37
1973	Willie Stargell, Pittsburgh	44	1973	Reggie Jackson, Oakland	32
1974	Mike Schmidt, Philadelphia	36	1974	Dick Allen, Chicago	32
1975	Mike Schmidt, Philadelphia	38	1975	George Scott, Milwaukee; Reggie Jackson, Oakland	36
1976	Mike Schmidt, Philadelphia	38	1976	Graig Nettles, New York	32
1977	George Foster, Cincinnati	52	1977	Jim Rice, Boston	39
1978	George Foster, Cincinnati	40	1978	Jim Rice, Boston	46
1979	Dave Kingman, Chicago	48	1979	Gorman Thomas, Milwaukee	45
1980	Mike Schmidt, Philadelphia	48	1980	Reggie Jackson, New York; Ben Oglivie, Milwaukee	41
1981	Mike Schmidt, Philadelphia	31	1981	Bobby Grich, California; Tony Armas, Oakland; Dwight Evans, Boston; Eddie Murray, Baltimore	22
1982	Dave Kingman, New York	37	1982	Gorman Thomas, Milwaukee; Reggie Jackson, Cal.	39
1983	Mike Schmidt, Philadelphia	40	1983	Jim Rice, Boston	39
1984	Mike Schmidt, Phi.; Dale Murphy, Atlanta	36	1984	Tony Armas, Boston	43
1985	Dale Murphy, Atlanta	37	1985	Darrell Evans, Detroit	40
1986	Mike Schmidt, Philadelphia	37	1986	Jesse Barfield, Toronto	40
1987	Andre Dawson, Chicago	49	1987	Mark McGwire, Oakland	49
1988	Darryl Strawberry, New York	39	1988	Jose Canseco, Oakland	42
1989	Kevin Mitchell, San Francisco	47	1989	Fred McGriff, Toronto	36
1990	Ryne Sandberg, Chicago	40	1990	Cecil Fielder, Detroit	51
1991	Howard Johnson, New York	38	1991	Cecil Fielder, Detroit; Jose Canseco, Oakland	44
1992	Fred McGriff, San Diego	35	1992	Juan Gonzalez, Texas	43
1993	Barry Bonds, San Francisco	46	1993	Juan Gonzalez, Texas	46
1994	Matt Williams, San Francisco	43	1994	Ken Griffey Jr., Seattle	40
1995	Dante Bichette, Colorado	40	1995	Albert Belle, Cleveland	50
1996	Andres Galarraga, Colorado	47	1996	Mark McGwire, Oakland	52
1997[1]	Larry Walker, Colorado	49	1997[1]	Ken Griffey Jr., Seattle	56
1998	Mark McGwire, St. Louis	*70	1998	Ken Griffey Jr., Seattle	56
1999	Mark McGwire, St. Louis	65	1999	Ken Griffey Jr., Seattle	48

(1) In 1997, Mark McGwire hit 58 home runs; 34 with the Oakland Athletics (AL) and 24 with the St. Louis Cardinals (NL).

Runs Batted In Leaders

Note: Asterisk (*) indicates the all-time single-season record for each league.

National League			American League		
Year	**Player, Team**	**RBI**	**Year**	**Player, Team**	**RBI**
1907	Sherwood Magee, Philadelphia	85	1907	Ty Cobb, Detroit	116
1908	Honus Wager, Pittsburgh	109	1908	Ty Cobb, Detroit	108
1909	Honus Wager, Pittsburgh	100	1909	Ty Cobb, Detroit	107
1910	Sherwood Magee, Philadelphia	123	1910	Sam Crawford, Detroit	120
1911	Frank Schulte, Chicago	121	1911	Ty Cobb, Detroit	144
1912	Henry Zimmerman, Chicago	103	1912	J. Franklin Baker, Philadelphia	133
1913	Gavvy Cravath, Philadelphia	128	1913	J. Franklin Baker, Philadelphia	126
1914	Sherwood Magee, Philadelphia	103	1914	Sam Crawford, Detroit	104
1915	Gavvy Cravath, Philadelphia	115	1915	Sam Crawford, Detroit; Robert Veach, Detroit	112
1916	Henry Zimmerman, Chicago-NewYork	83	1916	Del Pratt, St. Louis	103
1917	Henry Zimmerman, New York	102	1917	Robert Veach, Detroit	103
1918	Sherwood Magee, Philadelphia	76	1918	Robert Veach, Detroit	78
1919	Hi Myers, Boston	73	1919	Babe Ruth, Boston	114
1920	George Kelly, N.Y.; Rogers Hornsby, St. Louis	94	1920	Babe Ruth, New York	137
1921	Rogers Hornsby, St. Louis	126	1921	Babe Ruth, New York	171
1922	Rogers Hornsby, St. Louis	152	1922	Ken Williams, St. Louis	155
1923	Emil Meusel, New York	125	1923	Babe Ruth, New York	131
1924	George Kelly, New York	136	1924	Goose Goslin, Washington	129
1925	Rogers Hornsby, St. Louis	143	1925	Bob Meusel, New York	138
1926	Jim Bottomley, St. Louis	120	1926	Babe Ruth, New York	145
1927	Paul Waner, Pittsburgh	131	1927	Lou Gehrig, New York	175
1928	Jim Bottomley, St. Louis	136	1928	Babe Ruth, New York; Lou Gehrig, New York	142
1929	Hack Wilson, Chicago	159	1929	Al Simmons, Philadelphia	157
1930	Hack Wilson, Chicago	*191	1930	Lou Gehrig, New York	174
1931	Chuck Klein, Philadelphia	121	1931	Lou Gehrig, New York	*184
1932	Don Hurst, Philadelphia	143	1932	Jimmie Foxx, Philadelphia	169

Year	National League Player, Team	RBI	Year	American League Player, Team	RBI
1933	Chuck Klein, Philadelphia	120	1933	Jimmie Foxx, Philadelphia	163
1934	Mel Ott, New York	135	1934	Lou Gehrig, New York	165
1935	Walter Berger, Boston	130	1935	Hank Greenberg, Detroit	170
1936	Joe Medwick, St. Louis	138	1936	Hal Trosky, Cleveland	162
1937	Joe Medwick, St. Louis	154	1937	Hank Greenberg, Detroit	183
1938	Joe Medwick, St. Louis	122	1938	Jimmie Foxx, Boston	175
1939	Frank McCormick, Cincinnati	128	1939	Ted Williams, Boston	145
1940	John Mize, St. Louis	137	1940	Hank Greenberg, Detroit	150
1941	Adolph Camilli, Brooklyn	120	1941	Joe DiMaggio, New York	125
1942	John Mize, New York	110	1942	Ted Williams, Boston	137
1943	Bill Nicholson, Chicago	128	1943	Rudy York, Detroit	118
1944	Bill Nicholson, Chicago	122	1944	Vern Stephens, St. Louis	109
1945	Dixie Walker, Brooklyn	124	1945	Nick Etten, New York	111
1946	Enos Slaughter, St. Louis	130	1946	Hank Greenberg, Detroit	127
1947	John Mize, New York	138	1947	Ted Williams, Boston	114
1948	Stan Musial, St. Louis	131	1948	Joe DiMaggio, New York	155
1949	Ralph Kiner, Pittsburgh	127	1949	Ted Williams, Bos.; Vern Stephens, Bos.	159
1950	Del Ennis, Philadelphia	126	1950	Walt Dropo, Bos.; Vern Stephens, Bos.	144
1951	Monte Irvin, New York	121	1951	Gus Zernial, Chicago-Philadelphia	129
1952	Hank Sauer, Chicago	121	1952	Al Rosen, Cleveland	105
1953	Roy Campanella, Brooklyn	142	1953	Al Rosen, Cleveland	145
1954	Ted Kluszewski, Cincinnati	141	1954	Larry Doby, Cleveland	126
1955	Duke Snider, Brooklyn	136	1955	Ray Boone, Detroit; Jackie Jensen, Boston	116
1956	Stan Musial, St. Louis	109	1956	Mickey Mantle, New York	130
1957	Hank Aaron, Milwaukee	132	1957	Roy Sievers, Washington	114
1958	Ernie Banks, Chicago	129	1958	Jackie Jensen, Boston	122
1959	Ernie Banks, Chicago	143	1959	Jackie Jensen, Boston	112
1960	Hank Aaron, Milwaukee	126	1960	Roger Maris, New York	112
1961	Orlando Cepeda, San Francisco	142	1961	Roger Maris, New York	142
1962	Tommy Davis, Los Angeles	153	1962	Harmon Killebrew, Minnesota	126
1963	Hank Aaron, Milwaukee	130	1963	Dick Stuart, Boston	118
1964	Ken Boyer, St. Louis	119	1964	Brooks Robinson, Baltimore	118
1965	Deron Johnson, Cincinnati	130	1965	Rocky Colavito, Cleveland	108
1966	Hank Aaron, Atlanta	127	1966	Frank Robinson, Baltimore	122
1967	Orlando Cepeda, St. Louis	111	1967	Carl Yastrzemski, Boston	121
1968	Willie McCovey, San Francisco	105	1968	Ken Harrelson, Boston	109
1969	Willie McCovey, San Francisco	126	1969	Harmon Killebrew, Minnesota	140
1970	Johnny Bench, Cincinnati	148	1970	Frank Howard, Washington	126
1971	Joe Torre, St. Louis	137	1971	Harmon Killebrew, Minnesota	119
1972	Johnny Bench, Cincinnati	125	1972	Dick Allen, Chicago	113
1973	Willie Stargell, Pittsburgh	119	1973	Reggie Jackson, Oakland	117
1974	Johnny Bench, Cincinnati	129	1974	Jeff Burroughs, Texas	118
1975	Greg Luzinski, Philadelphia	120	1975	George Scott, Milwaukee	109
1976	George Foster, Cincinnati	121	1976	Lee May, Baltimore	109
1977	George Foster, Cincinnati	149	1977	Larry Hisle, Minnesota	119
1978	George Foster, Cincinnati	120	1978	Jim Rice, Boston	139
1979	Dave Winfield, San Diego	118	1979	Don Baylor, California	139
1980	Mike Schmidt, Philadelphia	121	1980	Cecil Cooper, Milwaukee	122
1981	Mike Schmidt, Philadelphia	91	1981	Eddie Murray, Baltimore	78
1982	Dale Murphy, Atlanta; Al Oliver, Montreal	109	1982	Hal McRae, Kansas City	133
1983	Dale Murphy, Atlanta	121	1983	Cecil Cooper, Milwaukee; Jim Rice, Boston	126
1984	Gary Carter, Montreal; Mike Schmidt, Phi.	106	1984	Tony Armas, Boston	123
1985	Dave Parker, Cincinnati	125	1985	Don Mattingly, New York	145
1986	Mike Schmidt, Philadelphia	119	1986	Joe Carter, Cleveland	121
1987	Andre Dawson, Chicago	137	1987	George Bell, Toronto	134
1988	Will Clark, San Francisco	109	1988	Jose Canseco, Oakland	124
1989	Kevin Mitchell, San Francisco	125	1989	Ruben Sierra, Texas	119
1990	Matt Williams, San Francisco	122	1990	Cecil Fielder, Detroit	132
1991	Howard Johnson, New York	117	1991	Cecil Fielder, Detroit	133
1992	Darren Daulton, Philadelphia	109	1992	Cecil Fielder, Detroit	124
1993	Barry Bonds, San Francisco	123	1993	Albert Belle, Cleveland	129
1994	Jeff Bagwell, Houston	116	1994	Kirby Puckett, Minnesota	112
1995	Dante Bichette, Colorado	128	1995	Albert Belle, Cleveland; Mo Vaughn, Boston	126
1996	Andres Galarraga, Colorado	150	1996	Albert Belle, Cleveland	148
1997	Andres Galarraga, Colorado	140	1997	Ken Griffey Jr., Seattle	147
1998	Sammy Sosa, Chicago	158	1998	Juan Gonzalez, Texas	157
1999	Mark McGwire, St. Louis	147	1999	Manny Ramirez, Cleveland	165

Batting Champions

Note: Asterisk (*) indicates the all-time single-season record for each league.

Year	National League Player	Team	Avg.	Year	American League Player	Team	Avg.
1901	Jesse C. Burkett	St. Louis	.382	1901	Napoleon Lajoie	Philadelphia	*.422
1902	Clarence Beaumont	Pittsburgh	.357	1902	Ed Delahanty	Washington	.376
1903	Honus Wagner	Pittsburgh	.355	1903	Napoleon Lajoie	Cleveland	.355
1904	Honus Wagner	Pittsburgh	.349	1904	Napoleon Lajoie	Cleveland	.381
1905	James Seymour	Cincinnati	.377	1905	Elmer Flick	Cleveland	.306
1906	Honus Wagner	Pittsburgh	.339	1906	George Stone	St. Louis	.358
1907	Honus Wagner	Pittsburgh	.350	1907	Ty Cobb	Detroit	.350
1908	Honus Wagner	Pittsburgh	.354	1908	Ty Cobb	Detroit	.324
1909	Honus Wagner	Pittsburgh	.339	1909	Ty Cobb	Detroit	.377
1910	Sherwood Magee	Philadelphia	.331	1910[1]	Ty Cobb	Detroit	.385
1911	Honus Wagner	Pittsburgh	.334	1911	Ty Cobb	Detroit	.420
1912	Henry Zimmerman	Chicago	.372	1912	Ty Cobb	Detroit	.410
1913	Jacob Daubert	Brooklyn	.350	1913	Ty Cobb	Detroit	.390

National League				American League			
Year	Player	Team	Avg.	Year	Player	Team	Avg.
1914	Jacob Daubert	Brooklyn	.329	1914	Ty Cobb	Detroit	.368
1915	Larry Doyle	New York	.320	1915	Ty Cobb	Detroit	.369
1916	Hal Chase	Cincinnati	.339	1916	Tris Speaker	Cleveland	.386
1917	Edd Roush	Cincinnati	.341	1917	Ty Cobb	Detroit	.383
1918	Zach Wheat	Brooklyn	.335	1918	Ty Cobb	Detroit	.382
1919	Edd Roush	Cincinnati	.321	1919	Ty Cobb	Detroit	.384
1920	Rogers Hornsby	St. Louis	.370	1920	George Sisler	St. Louis	.407
1921	Rogers Hornsby	St. Louis	.397	1921	Harry Heilmann	Detroit	.394
1922	Rogers Hornsby	St. Louis	.401	1922	George Sisler	St. Louis	.420
1923	Rogers Hornsby	St. Louis	.384	1923	Harry Heilmann	Detroit	.403
1924	Rogers Hornsby	St. Louis	*.424	1924	Babe Ruth	New York	.378
1925	Rogers Hornsby	St. Louis	.403	1925	Harry Heilmann	Detroit	.393
1926	Eugene Hargrave	Cincinnati	.353	1926	Henry Manush	Detroit	.378
1927	Paul Waner	Pittsburgh	.380	1927	Harry Heilmann	Detroit	.398
1928	Rogers Hornsby	Boston	.387	1928	Goose Goslin	Washington	.379
1929	Lefty O'Doul	Philadelphia	.398	1929	Lew Fonseca	Cleveland	.369
1930	Bill Terry	New York	.401	1930	Al Simmons	Philadelphia	.381
1931	Chick Hafey	St. Louis	.349	1931	Al Simmons	Philadelphia	.390
1932	Lefty O'Doul	Brooklyn	.368	1932	Dale Alexander	Detroit-Boston	.367
1933	Chuck Klein	Philadelphia	.368	1933	Jimmie Foxx	Philadelphia	.356
1934	Paul Waner	Pittsburgh	.362	1934	Lou Gehrig	New York	.363
1935	Arky Vaughan	Pittsburgh	.385	1935	Buddy Myer	Washington	.349
1936	Paul Waner	Pittsburgh	.373	1936	Luke Appling	Chicago	.388
1937	Joe Medwick	St. Louis	.374	1937	Charlie Gehringer	Detroit	.371
1938	Ernie Lombardi	Cincinnati	.342	1938	Jimmie Foxx	Boston	.349
1939	John Mize	St. Louis	.349	1939	Joe DiMaggio	New York	.381
1940	Debs Garms	Pittsburgh	.355	1940	Joe DiMaggio	New York	.352
1941	Pete Reiser	Brooklyn	.343	1941	Ted Williams	Boston	.406
1942	Ernie Lombardi	Boston	.330	1942	Ted Williams	Boston	.356
1943	Stan Musial	St. Louis	.357	1943	Luke Appling	Chicago	.328
1944	Dixie Walker	Brooklyn	.357	1944	Lou Boudreau	Cleveland	.327
1945	Phil Cavarretta	Chicago	.355	1945	George Stirnweiss	New York	.309
1946	Stan Musial	St. Louis	.365	1946	Mickey Vernon	Washington	.353
1947	Harry Walker	St.L.-Phi.	.363	1947	Ted Williams	Boston	.343
1948	Stan Musial	St. Louis	.376	1948	Ted Williams	Boston	.369
1949	Jackie Robinson	Brooklyn	.342	1949	George Kell	Detroit	.343
1950	Stan Musial	St. Louis	.346	1950	Billy Goodman	Boston	.354
1951	Stan Musial	St. Louis	.355	1951	Ferris Fain	Philadelphia	.344
1952	Stan Musial	St. Louis	.336	1952	Ferris Fain	Philadelphia	.327
1953	Carl Furillo	Brooklyn	.344	1953	Mickey Vernon	Washington	.337
1954	Willie Mays	New York	.345	1954	Roberto Avila	Cleveland	.341
1955	Richie Ashburn	Philadelphia	.338	1955	Al Kaline	Detroit	.340
1956	Hank Aaron	Milwaukee	.328	1956	Mickey Mantle	New York	.353
1957	Stan Musial	St. Louis	.351	1957	Ted Williams	Boston	.388
1958	Richie Ashburn	Philadelphia	.350	1958	Ted Williams	Boston	.328
1959	Hank Aaron	Milwaukee	.355	1959	Harvey Kuenn	Detroit	.353
1960	Dick Groat	Pittsburgh	.325	1960	Pete Runnels	Boston	.320
1961	Roberto Clemente	Pittsburgh	.351	1961	Norm Cash	Detroit	.361
1962	Tommy Davis	Los Angeles	.346	1962	Pete Runnels	Boston	.326
1963	Tommy Davis	Los Angeles	.326	1963	Carl Yastrzemski	Boston	.321
1964	Roberto Clemente	Pittsburgh	.339	1964	Tony Oliva	Minnesota	.323
1965	Roberto Clemente	Pittsburgh	.329	1965	Tony Oliva	Minnesota	.321
1966	Matty Alou	Pittsburgh	.342	1966	Frank Robinson	Baltimore	.316
1967	Roberto Clemente	Pittsburgh	.357	1967	Carl Yastrzemski	Boston	.326
1968	Pete Rose	Cincinnati	.335	1968	Carl Yastrzemski	Boston	.301
1969	Pete Rose	Cincinnati	.348	1969	Rod Carew	Minnesota	.332
1970	Rico Carty	Atlanta	.366	1970	Alex Johnson	California	.329
1971	Joe Torre	St. Louis	.363	1971	Tony Oliva	Minnesota	.337
1972	Billy Williams	Chicago	.333	1972	Rod Carew	Minnesota	.318
1973	Pete Rose	Cincinnati	.338	1973	Rod Carew	Minnesota	.350
1974	Ralph Garr	Atlanta	.353	1974	Rod Carew	Minnesota	.364
1975	Bill Madlock	Chicago	.354	1975	Rod Carew	Minnesota	.359
1976	Bill Madlock	Chicago	.339	1976	George Brett	Kansas City	.333
1977	Dave Parker	Pittsburgh	.338	1977	Rod Carew	Minnesota	.388
1978	Dave Parker	Pittsburgh	.334	1978	Rod Carew	Minnesota	.333
1979	Keith Hernandez	St. Louis	.344	1979	Fred Lynn	Boston	.333
1980	Bill Buckner	Chicago	.324	1980	George Brett	Kansas City	.390
1981	Bill Madlock	Pittsburgh	.341	1981	Carney Lansford	Boston	.336
1982	Al Oliver	Montreal	.331	1982	Willie Wilson	Kansas City	.332
1983	Bill Madlock	Pittsburgh	.323	1983	Wade Boggs	Boston	.361
1984	Tony Gwynn	San Diego	.351	1984	Don Mattingly	New York	.343
1985	Willie McGee	St. Louis	.353	1985	Wade Boggs	Boston	.368
1986	Tim Raines	Montreal	.334	1986	Wade Boggs	Boston	.357
1987	Tony Gwynn	San Diego	.370	1987	Wade Boggs	Boston	.363
1988	Tony Gwynn	San Diego	.313	1988	Wade Boggs	Boston	.366
1989	Tony Gwynn	San Diego	.336	1989	Kirby Puckett	Minnesota	.339
1990	Willie McGee	St. Louis	.335	1990	George Brett	Kansas City	.329
1991	Terry Pendleton	Atlanta	.319	1991	Julio Franco	Texas	.341
1992	Gary Sheffield	San Diego	.330	1992	Edgar Martinez	Seattle	.343
1993	Andres Galarraga	Colorado	.370	1993	John Olerud	Toronto	.363
1994	Tony Gwynn	San Diego	.394	1994	Paul O'Neill	New York	.359
1995	Tony Gwynn	San Diego	.368	1995	Edgar Martinez	Seattle	.356
1996	Tony Gwynn	San Diego	.353	1996	Alex Rodriguez	Seattle	.358
1997	Tony Gwynn	San Diego	.372	1997	Frank Thomas	Chicago	.347
1998	Larry Walker	Colorado	.363	1998	Bernie Williams	New York	.339
1999	Larry Walker	Colorado	.379	1999	Nomar Garciapara	Boston	.357

(1) Some baseball researchers have concluded that Ty Cobb actually hit .382 in 1910 while Napoleon Lajoie, Cleveland, hit .383.

Cy Young Award Winners

Year	Player, Team
1956	Don Newcombe, Dodgers
1957	Warren Spahn, Braves
1958	Bob Turley, Yankees
1959	Early Wynn, White Sox
1960	Vernon Law, Pirates
1961	Whitey Ford, Yankees
1962	Don Drysdale, Dodgers
1963	Sandy Koufax, Dodgers
1964	Dean Chance, Angels
1965	Sandy Koufax, Dodgers
1966	Sandy Koufax, Dodgers
1967	(NL) Mike McCormick, Giants
	(AL) Jim Lonborg, Red Sox
1968	(NL) Bob Gibson, Cardinals
	(AL) Dennis McLain, Tigers
1969	(NL) Tom Seaver, Mets
	(AL) (tie) Dennis McLain, Tigers
	Mike Cuellar, Orioles
1970	(NL) Bob Gibson, Cardinals
	(AL) Jim Perry, Twins
1971	(NL) Ferguson Jenkins, Cubs
	(AL) Vida Blue, A's
1972	(NL) Steve Carlton, Phillies
	(AL) Gaylord Perry, Indians
1973	(NL) Tom Seaver, Mets
	(AL) Jim Palmer, Orioles

Year	Player, Team
1974	(NL) Mike Marshall, Dodgers
	(AL) Jim (Catfish) Hunter, A's
1975	(NL) Tom Seaver, Mets
	(AL) Jim Palmer, Orioles
1976	(NL) Randy Jones, Padres
	(AL) Jim Palmer, Orioles
1977	(NL) Steve Carlton, Phillies
	(AL) Sparky Lyle, Yankees
1978	(NL) Gaylord Perry, Padres
	(AL) Ron Guidry, Yankees
1979	(NL) Bruce Sutter, Cubs
	(AL) Mike Flanagan, Orioles
1980	(NL) Steve Carlton, Phillies
	(AL) Steve Stone, Orioles
1981	(NL) Fernando Valenzuela, Dodgers
	(AL) Rollie Fingers, Brewers
1982	(NL) Steve Carlton, Phillies
	(AL) Pete Vuckovich, Brewers
1983	(NL) John Denny, Phillies
	(AL) LaMarr Hoyt, White Sox
1984	(NL) Rick Sutcliffe, Cubs
	(AL) Willie Hernandez, Tigers
1985	(NL) Dwight Gooden, Mets
	(AL) Bret Saberhagen, Royals

Year	Player, Team
1986	(NL) Mike Scott, Astros
	(AL) Roger Clemens, Red Sox
1987	(NL) Steve Bedrosian, Phillies
	(AL) Roger Clemens, Red Sox
1988	(NL) Orel Hershiser, Dodgers
	(AL) Frank Viola, Twins
1989	(NL) Mark Davis, Padres
	(AL) Bret Saberhagen, Royals
1990	(NL) Doug Drabek, Pirates
	(AL) Bob Welch, A's
1991	(NL) Tom Glavine, Braves
	(AL) Roger Clemens, Red Sox
1992	(NL) Greg Maddux, Cubs
	(AL) Dennis Eckersley, A's
1993	(NL) Greg Maddux, Braves
	(AL) Jack McDowell, White Sox
1994	(NL) Greg Maddux, Braves
	(AL) David Cone, Royals
1995	(NL) Greg Maddux, Braves
	(AL) Randy Johnson, Mariners
1996	(NL) John Smoltz, Braves
	(AL) Pat Hentgen, Blue Jays
1997	(NL) Pedro Martinez, Expos
	(AL) Roger Clemens, Blue Jays
1998	(NL) Tom Glavine, Braves
	(AL) Roger Clemens, Blue Jays

Most Valuable Player

National League

Year	Player, team
1931	Frank Frisch, St. Louis
1932	Chuck Klein, Philadelphia
1933	Carl Hubbell, New York
1934	Dizzy Dean, St. Louis
1935	Gabby Hartnett, Chicago
1936	Carl Hubbell, New York
1937	Joe Medwick, St. Louis
1938	Ernie Lombardi, Cincinnati
1939	Bucky Walters, Cincinnati
1940	Frank McCormick, Cincinnati
1941	Dolph Camilli, Brooklyn
1942	Mort Cooper, St. Louis
1943	Stan Musial, St. Louis
1944	Martin Marion, St. Louis
1945	Phil Cavarretta, Chicago
1946	Stan Musial, St. Louis
1947	Bob Elliott, Boston
1948	Stan Musial, St. Louis
1949	Jackie Robinson, Brooklyn
1950	Jim Konstanty, Philadelphia
1951	Roy Campanella, Brooklyn
1952	Hank Sauer, Chicago
1953	Roy Campanella, Brooklyn

Year	Player, team
1954	Willie Mays, New York
1955	Roy Campanella, Brooklyn
1956	Don Newcombe, Brooklyn
1957	Hank Aaron, Milwaukee
1958	Ernie Banks, Chicago
1959	Ernie Banks, Chicago
1960	Dick Groat, Pittsburgh
1961	Frank Robinson, Cincinnati
1962	Maury Wills, Los Angeles
1963	Sandy Koufax, Los Angeles
1964	Ken Boyer, St. Louis
1965	Willie Mays, San Francisco
1966	Roberto Clemente, Pittsburgh
1967	Orlando Cepeda, St. Louis
1968	Bob Gibson, St. Louis
1969	Willie McCovey, San Francisco
1970	Johnny Bench, Cincinnati
1971	Joe Torre, St. Louis
1972	Johnny Bench, Cincinnati
1973	Pete Rose, Cincinnati
1974	Steve Garvey, Los Angeles
1975	Joe Morgan, Cincinnati
1976	Joe Morgan, Cincinnati

Year	Player, team
1977	George Foster, Cincinnati
1978	Dave Parker, Pittsburgh
1979	Willie Stargell, Pittsburgh
(tie)	Keith Hernandez, St. Louis
1980	Mike Schmidt, Philadelphia
1981	Mike Schmidt, Philadelphia
1982	Dale Murphy, Atlanta
1983	Dale Murphy, Atlanta
1984	Ryne Sandberg, Chicago
1985	Willie McGee, St. Louis
1986	Mike Schmidt, Philadelphia
1987	Andre Dawson, Chicago
1988	Kirk Gibson, Los Angeles
1989	Kevin Mitchell, San Francisco
1990	Barry Bonds, Pittsburgh
1991	Terry Pendleton, Atlanta
1992	Barry Bonds, Pittsburgh
1993	Barry Bonds, San Francisco
1994	Jeff Bagwell, Houston
1995	Barry Larkin, Cincinnati
1996	Ken Caminiti, San Diego
1997	Larry Walker, Colorado
1998	Sammy Sosa, Chicago

American League

Year	Player, team
1931	Lefty Grove, Philadelphia
1932	Jimmie Foxx, Philadelphia
1933	Jimmie Foxx, Philadelphia
1934	Mickey Cochrane, Detroit
1935	Hank Greenberg, Detroit
1936	Lou Gehrig, New York
1937	Charley Gehringer, Detroit
1938	Jimmie Foxx, Boston
1939	Joe DiMaggio, New York
1940	Hank Greenberg, Detroit
1941	Joe DiMaggio, New York
1942	Joe Gordon, New York
1943	Spurgeon Chandler, New York
1944	Hal Newhouser, Detroit
1945	Hal Newhouser, Detroit
1946	Ted Williams, Boston
1947	Joe DiMaggio, New York
1948	Lou Boudreau, Cleveland
1949	Ted Williams, Boston
1950	Phil Rizzuto, New York
1951	Yogi Berra, New York
1952	Bobby Shantz, Philadelphia
1953	Al Rosen, Cleveland

Year	Player, team
1954	Yogi Berra, New York
1955	Yogi Berra, New York
1956	Mickey Mantle, New York
1957	Mickey Mantle, New York
1958	Jackie Jensen, Boston
1959	Nellie Fox, Chicago
1960	Roger Maris, New York
1961	Roger Maris, New York
1962	Mickey Mantle, New York
1963	Elston Howard, New York
1964	Brooks Robinson, Baltimore
1965	Zoilo Versalles, Minnesota
1966	Frank Robinson, Baltimore
1967	Carl Yastrzemski, Boston
1968	Denny McLain, Detroit
1969	Harmon Killebrew, Minnesota
1970	John (Boog) Powell, Baltimore
1971	Vida Blue, Oakland
1972	Dick Allen, Chicago
1973	Reggie Jackson, Oakland
1974	Jeff Burroughs, Texas
1975	Fred Lynn, Boston
1976	Thurman Munson, New York

Year	Player, team
1977	Rod Carew, Minnesota
1978	Jim Rice, Boston
1979	Don Baylor, California
1980	George Brett, Kansas City
1981	Rollie Fingers, Milwaukee
1982	Robin Yount, Milwaukee
1983	Cal Ripken, Jr., Baltimore
1984	Willie Hernandez, Detroit
1985	Don Mattingly, New York
1986	Roger Clemens, Boston
1987	George Bell, Toronto
1988	Jose Canseco, Oakland
1989	Robin Yount, Milwaukee
1990	Rickey Henderson, Oakland
1991	Cal Ripken, Jr., Baltimore
1992	Dennis Eckersley, Oakland
1993	Frank Thomas, Chicago
1994	Frank Thomas, Chicago
1995	Mo Vaughn, Boston
1996	Juan Gonzalez, Texas
1997	Ken Griffey Jr., Seattle
1998	Juan Gonzalez, Texas

Rookie of the Year

1947—Combined selection—Jackie Robinson, Brooklyn, 1b; 1948—Combined selection—Alvin Dark, Boston, N.L., ss

National League

Year	Player, team	Year	Player, team	Year	Player, team
1949	Don Newcombe, Brooklyn, p	1966	Tommy Helms, Cincinnati, 2b	1982	Steve Sax, Los Angeles, 2b
1950	Sam Jethroe, Boston, of	1967	Tom Seaver, New York, p	1983	Darryl Strawberry, New York, of
1951	Willie Mays, New York, of	1968	Johnny Bench, Cincinnati, c	1984	Dwight Gooden, New York, p
1952	Joe Black, Brooklyn, p	1969	Ted Sizemore, Los Angeles, 2b	1985	Vince Coleman, St. Louis, of
1953	Jim Gilliam, Brooklyn, 2b	1970	Carl Morton, Montreal, p	1986	Todd Worrell, St. Louis, p
1954	Wally Moon, St. Louis, of	1971	Earl Williams, Atlanta, c	1987	Benito Santiago, San Diego, c
1955	Bill Virdon, St. Louis, of	1972	Jon Matlack, New York, p	1988	Chris Sabo, Cincinnati, 3b
1956	Frank Robinson, Cincinnati, of	1973	Gary Matthews, S.F., of	1989	Jerome Walton, Chicago, of
1957	Jack Sanford, Philadelphia, p	1974	Bake McBride, St. Louis, of	1990	Dave Justice, Atlanta, 1b
1958	Orlando Cepeda, S.F., 1b	1975	John Montefusco, S.F., p	1991	Jeff Bagwell, Houston, 1b
1959	Willie McCovey, S.F., 1b	1976	Butch Metzger, San Diego, p	1992	Eric Karros, Los Angeles, 1b
1960	Frank Howard, Los Angeles, of	(tie)	Pat Zachry, Cincinnati, p	1993	Mike Piazza, Los Angeles, c
1961	Billy Williams, Chicago, of	1977	Andre Dawson, Montreal, of	1994	Raul Mondesi, Los Angeles, of
1962	Ken Hubbs, Chicago, 2b	1978	Bob Horner, Atlanta, 3b	1995	Hideo Nomo, Los Angeles, p
1963	Pete Rose, Cincinnati, 2b	1979	Rick Sutcliffe, Los Angeles, p	1996	Todd Hollandsworth, Los Angeles, of
1964	Richie Allen, Philadelphia, 3b	1980	Steve Howe, Los Angeles, p	1997	Scott Rolen, Philadelphia, 3b
1965	Jim Lefebvre, Los Angeles, 2b	1981	Fernando Valenzuela, Los Angeles, p	1998	Kerry Wood, Chicago, p

American League

Year	Player, team	Year	Player, team	Year	Player, team
1949	Roy Sievers, St. Louis, of	1966	Tommie Agee, Chicago, of	1982	Cal Ripken, Jr., Baltimore, ss
1950	Walt Dropo, Boston, 1b	1967	Rod Carew, Minnesota, 2b	1983	Ron Kittle, Chicago, of
1951	Gil McDougald, New York, 3b	1968	Stan Bahnsen, New York, p	1984	Alvin Davis, Seattle, 1b
1952	Harry Byrd, Philadelphia, p	1969	Lou Piniella, Kansas City, of	1985	Ozzie Guillen, Chicago, ss
1953	Harvey Kuenn, Detroit, ss	1970	Thurman Munson, New York, c	1986	Jose Canseco, Oakland, of
1954	Bob Grim, New York, p	1971	Chris Chambliss, Cleveland, 1b	1987	Mark McGwire, Oakland, 1b
1955	Herb Score, Cleveland, p	1972	Carlton Fisk, Boston, c	1988	Walt Weiss, Oakland, ss
1956	Luis Aparicio, Chicago, ss	1973	Al Bumbry, Baltimore, of	1989	Gregg Olson, Baltimore, p
1957	Tony Kubek, New York, if-of	1974	Mike Hargrove, Texas, 1b	1990	Sandy Alomar, Jr., Cleveland, c
1958	Albie Pearson, Washington, of	1975	Fred Lynn, Boston, of	1991	Chuck Knoblauch, Minnesota, 2b
1959	Bob Allison, Washington, of	1976	Mark Fidrych, Detroit, p	1992	Pat Listach, Milwaukee, ss
1960	Ron Hansen, Baltimore, ss	1977	Eddie Murray, Baltimore, dh	1993	Tim Salmon, California, of
1961	Don Schwall, Boston, p	1978	Lou Whitaker, Detroit, 2b	1994	Bob Hamelin, Kansas City, dh
1962	Tom Tresh, New York, if-of	1979	John Castino, Minnesota, 3b	1995	Marty Cordova, Minnesota, of
1963	Gary Peters, Chicago, p	(tie)	Alfredo Griffin, Toronto, ss	1996	Derek Jeter, New York, ss
1964	Tony Oliva, Minnesota, of	1980	Joe Charboneau, Cleveland, of	1997	Nomar Garciaparra, Boston, ss
1965	Curt Blefary, Baltimore, of	1981	Dave Righetti, New York, p	1998	Ben Grieve, Oakland, of

National League Final Standings, 1999

Eastern Division

	W	L	Pct.	GB	Home	vs. East	vs. Central	vs. West	vs. AL
Atlanta	103	59	.636	—	56-25	35-16	35-13	24-21	9-9
New York*[1]	97	66	.595	6½	49-32	27-23	33-18	25-19	12-6
Philadelphia	77	85	.475	26	41-40	28-22	22-28	16-28	11-7
Montreal	68	94	.420	35	35-46	19-31	22-28	19-25	8-10
Florida	64	98	.395	39	35-45	17-34	17-31	19-26	11-7

Central Division

	W	L	Pct.	GB	Home	vs. East	vs. Central	vs. West	vs. AL
Houston	97	65	.599	—	50-32	25-16	31-31	29-15	12-3
Cincinnati	96	67	.589	1½	45-37	22-20	38-25	29-14	7-8
Pittsburgh	78	83	.484	18½	45-36	19-22	30-32	22-21	7-8
St. Louis	75	86	.466	21½	38-42	16-25	27-34	25-19	7-8
Milwaukee	74	87	.460	22½	32-48	18-23	32-30	16-28	8-6
Chicago	67	95	.414	30	34-47	18-23	28-34	15-29	6-9

Western Division

	W	L	Pct.	GB	Home	vs. East	vs. Central	vs. West	vs. AL
Arizona	100	62	.617	—	52-29	33-12	27-24	33-18	7-8
San Francisco	86	76	.531	14	49-32	23-21	32-21	24-26	7-8
Los Angeles	77	85	.475	23	37-44	21-23	26-26	22-29	8-7
San Diego	74	88	.457	26	46-35	18-26	20-33	25-25	11-4
Colorado	72	90	.444	28	39-42	24-21	21-32	23-29	4-8

*Wild card team. (1) New York beat Cincinnati (5-0) in a one game playoff to win the wild card spot.

National League Playoff Results, 1999

Division Series
Atlanta defeated Houston 3 games to 1 (1-6, 5-1, 5-3, 7-5)
New York defeated Arizona 3 games to 1 (8-4, 1-7, 9-2, 4-3)

Championship Series
Atlanta defeated New York 4 games to 2 (4-2, 4-3, 1-0, 2-3, 3-4, 10-9)

National League Statistics, 1999

(Individual Statistics: Batting—at least 150 at-bats; Pitching—at least 70 innings or 10 saves; *traded within NL during season; entry includes statistics for more than 1 team; # traded to or from AL during season; entry includes only NL stats)

Team Batting

	Avg	AB	R	H	HR	RBI
Colorado	.288	5,717	906	1,644	223	863
New York	.279	5,539	848	1,544	179	809
Arizona	.277	5,658	908	1,566	216	865
Philadelphia	.275	5,598	841	1,539	161	797
Cincinnati	.273	5,621	865	1,534	209	820
Milwaukee	.273	5,582	815	1,524	165	777
San Francisco	.271	5,564	872	1,507	188	828
Houston	.267	5,485	823	1,463	168	784
Atlanta	.266	5,569	840	1,481	197	791
Los Angeles	.266	5,567	793	1,480	187	761
Montreal	.265	5,559	718	1,473	163	680
Florida	.263	5,578	691	1,465	128	655
St. Louis	.262	5,570	809	1,461	194	763
Pittsburgh	.259	5,468	775	1,417	171	735
Chicago	.257	5,482	747	1,411	189	717
San Diego	.252	5,394	710	1,360	153	671

Team Pitching

	ERA	IP	H	BB	SO	Sv
Atlanta	3.65	1,471.0	1,398	507	1,197	45
Arizona	3.77	1,467.1	1,387	543	1,198	42
Houston	3.84	1,459.0	1,485	479	1,205	48
Cincinnati	3.98	1,453.0	1,300	628	1,076	55
New York	4.30	1,447.2	1,370	613	1,165	49
Pittsburgh	4.35	1,433.1	1,444	633	1,083	34
Los Angeles	4.45	1,453.0	1,438	594	1,077	37
San Diego	4.47	1,420.1	1,454	529	1,078	43
Montreal	4.70	1,434.1	1,505	572	1,043	44
San Francisco	4.72	1,456.1	1,486	656	1,076	42
St. Louis	4.76	1,445.1	1,519	667	1,025	38
Florida	4.90	1,435.2	1,560	655	943	33
Philadelphia	4.93	1,438.1	1,494	627	1,030	32
Milwaukee	5.09	1,442.1	1,618	615	986	40
Chicago	5.26	1,429.2	1,617	528	979	32
Colorado	6.03	1,429.0	1,700	737	1,032	33

Arizona Diamondbacks

Batters	AB	R	H	HR	RBI	SB	Avg
Gonzalez	614	112	206	26	111	9	.336
Durazo	155	31	51	11	30	1	.329
*Harris	187	17	58	1	20	2	.310
Williams	627	98	190	35	142	2	.303
Gilkey	204	28	60	8	39	2	.294
Bell	589	132	170	38	112	7	.289
Womack	614	111	170	4	41	72	.277
Frias	150	27	41	1	16	4	.273
Miller	296	35	80	11	47	0	.270
Finley	590	100	156	34	103	8	.264
Fox	274	34	70	6	33	4	.255
Lee	375	57	89	9	50	17	.237
Stinnett	284	36	66	14	38	2	.232

Pitchers	W	L	ERA	IP	H	BB	SO	Sv
Johnson	17	9	2.48	271.2	207	70	364	0
*Mantei	1	3	2.76	65.1	44	44	99	32
Daal	16	9	3.65	214.2	188	79	148	0
Olson	9	4	3.71	60.2	54	25	45	14
Stottlemyre	6	3	4.09	101.1	106	40	74	0
Reynoso	10	6	4.37	167.0	178	67	79	0
Anderson	8	2	4.57	130.0	144	28	75	1
Benes	13	12	4.81	198.1	216	82	141	0

Manager—Buck Showalter

Atlanta Braves

Batters	AB	R	H	HR	RBI	SB	Avg
C. Jones	567	116	181	45	110	25	.319
Lopez	246	34	78	11	45	0	.317
Simon	218	26	69	5	25	2	.317
Klesko	404	55	120	21	80	5	.297
Jordan	576	100	163	23	115	13	.283
A. Jones	592	97	163	26	84	24	.275
Williams	422	76	116	17	68	19	.275
*Hernandez	508	79	135	19	62	11	.266
*Myers	200	19	53	5	24	0	.265
Lockhart	161	20	42	1	21	3	.261
Boone	608	102	153	20	63	14	.252
E. Perez	309	30	77	7	30	0	.249
Hunter	181	28	45	6	30	0	.249
Guillen	232	21	56	1	20	4	.241
Weiss	279	38	63	2	29	7	.226
Nixon	151	31	31	0	8	26	.205
*Fabregas	231	20	46	3	21	0	.199

Pitchers	W	L	ERA	IP	H	BB	SO	Sv
Remlinger	10	1	2.37	83.2	66	35	81	1
Rocker	4	5	2.49	72.1	47	37	104	38
Millwood	18	7	2.68	228.0	168	59	205	0
McGlinchy	7	3	2.82	70.1	66	30	67	0
Smoltz	11	8	3.19	186.1	168	40	156	0
Maddux	19	9	3.57	219.1	258	37	136	0
Glavine	14	11	4.12	234.0	259	83	138	0
*Mulholland	10	8	4.39	170.1	201	45	83	1
*Bergman	5	6	5.21	105.1	135	29	44	0
O. Perez	4	6	6.00	93.0	100	53	82	0

Manager—Bobby Cox

Chicago Cubs

Batters	AB	R	H	HR	RBI	SB	Avg
Grace	593	107	183	16	91	3	.309
Rodriguez	447	72	136	26	87	2	.304
Hill	253	43	76	20	55	5	.300
Sosa	625	114	180	63	141	7	.288
Alexander	177	17	48	0	15	4	.271
Johnson	335	46	87	1	21	13	.260
*Reed	256	29	66	3	28	1	.258
Nieves	181	16	45	2	18	0	.249
Santiago	350	28	87	7	36	1	.249
#Goodwin	157	15	38	0	9	2	.242
Morandini	456	60	110	4	37	6	.241
Blauser	200	41	48	9	26	2	.240
#Houston	249	26	58	9	27	1	.233
Gaetti	280	22	57	9	46	0	.204
*Andrews	348	41	68	16	51	1	.195

Pitchers	W	L	ERA	IP	H	BB	SO	Sv
*Ayala	1	7	3.51	82.0	71	39	79	0
Adams	6	3	4.02	65.0	60	28	57	13
Lieber	10	11	4.07	203.1	226	46	186	0
Tapani	6	12	4.83	136.0	151	33	73	0
Farnsworth	5	9	5.05	130.0	140	52	70	0
Sanders	4	7	5.52	104.1	112	53	89	2
Trachsel	8	18	5.56	205.2	226	64	149	0

Manager—Jim Riggleman

Cincinnati Reds

Batters	AB	R	H	HR	RBI	SB	Avg
Casey	594	103	197	25	99	0	.332
Taubensee	424	58	132	21	87	0	.311
Young	373	63	112	14	56	3	.300
Larkin	583	108	171	12	75	30	.293
Reese	585	85	167	10	52	38	.285
Boone	472	56	132	14	72	17	.280
Hammonds	262	43	73	17	41	3	.279
Cameron	542	93	139	21	66	38	.256
Lewis	173	18	44	6	28	0	.254
Tucker	296	55	75	11	44	11	.253
Vaughn	550	104	135	45	118	15	.245

Pitchers	W	L	ERA	IP	H	BB	SO	Sv
Williamson	12	7	2.41	93.1	54	43	107	19
Sullivan	5	4	3.01	113.2	88	47	78	3
#Guzman	6	3	3.03	77.1	70	21	60	0
Graves	8	7	3.08	111.0	90	49	69	27
Parris	11	4	3.50	128.2	124	52	86	0
Harnisch	16	10	3.68	198.1	190	57	120	0
Villone	9	7	4.23	142.2	114	73	97	2
Neagle	9	5	4.27	111.2	95	40	76	0
Tomko	5	7	4.92	172.0	175	60	132	0
Avery	6	7	5.16	96.0	75	78	51	0

Manager—Jack McKeon

Colorado Rockies

Batters	AB	R	H	HR	RBI	SB	Avg
Walker	438	108	166	37	115	11	.379
Shumpert	262	58	91	10	37	14	.347
Helton	578	114	185	35	113	7	.320
Bichette	593	104	177	34	133	6	.298
Echevarria	191	28	56	11	35	1	.293
Perez	690	108	193	12	70	13	.280
Castilla	615	83	169	33	102	2	.275
Abbott	286	41	78	8	41	3	.273
Barry	168	19	45	5	26	0	.268
Clemente	162	24	41	8	25	0	.253
Blanco	263	30	61	6	28	1	.232
#*McRae	321	36	72	9	37	2	.224

Pitchers	W	L	ERA	IP	H	BB	SO	Sv
Dipoto	4	5	4.26	86.2	91	44	69	1
Wright	4	3	4.87	94.1	110	54	49	0
Astacio	17	11	5.04	232.0	258	75	210	0
Leskanic	6	2	5.08	85.0	87	49	77	0
Veres	4	8	5.14	77.0	88	37	71	31
Bohanon	12	12	6.20	197.1	236	92	120	0
Jones	6	10	6.33	112.1	132	77	74	0
Kile	8	13	6.61	190.2	225	109	116	0

Manager—Jim Leyland

Florida Marlins

Batters	AB	R	H	HR	RBI	SB	Avg
Floyd	251	37	76	11	49	5	.303
Castillo	487	76	147	0	28	50	.302
Redmond	242	22	73	1	27	0	.302
Aven	381	57	110	12	70	3	.289
Bautista	205	32	59	5	24	3	.288
Berg	304	42	87	3	25	2	.286
Millar	351	48	100	9	67	1	.285
Wilson	482	67	135	26	71	11	.280
Gonzalez	560	81	155	14	59	3	.277
Kotsay	495	57	134	8	50	7	.271
Orie	240	26	61	6	29	1	.254
Lowell	308	32	78	12	47	0	.253
Dunwoody	186	20	41	2	20	3	.220
Lee	218	21	45	5	20	2	.206

Pitchers	W	L	ERA	IP	H	BB	SO	Sv
Alfonseca	4	5	3.24	77.2	79	29	46	21
Fernandez	7	8	3.38	141.0	135	41	91	0
Looper	3	3	3.80	83.0	96	31	50	0
*Nunez	7	10	4.06	108.2	95	54	86	1
Dempster	7	8	4.71	147.0	146	93	126	0
Springer	6	16	4.86	196.1	231	64	83	1
Meadows	11	15	5.60	178.1	214	57	72	0
Edmondson	5	8	5.84	94.0	106	44	58	1
Sanchez	5	7	6.01	76.1	84	60	62	0

Manager—John Boles

Houston Astros

Batters	AB	R	H	HR	RBI	SB	Avg
Everett	464	86	151	25	108	27	.325
Bagwell	562	143	171	42	126	30	.304
Biggio	639	123	188	16	73	28	.294
Spiers	393	56	113	4	39	10	.288
Caminiti	273	45	78	13	56	6	.286
*Javier	397	61	113	3	34	16	.285
Johnson	156	24	44	5	23	2	.282
Ward	150	11	41	8	30	0	.273
Eusebio	323	31	88	4	33	0	.272
Gutierrez	268	33	70	1	25	2	.261
Bako	215	16	55	2	17	1	.256
Bogar	309	44	74	4	31	3	.239
Bell	509	61	120	12	66	18	.236
Hidalgo	383	49	87	15	56	8	.227

Pitchers	W	L	ERA	IP	H	BB	SO	Sv
Wagner	4	1	1.57	74.2	35	23	124	39
Hampton	22	4	2.90	239.0	206	101	177	0
Elarton	9	5	3.48	124.0	111	43	121	1
Lima	21	10	3.58	246.1	256	44	187	0
Reynolds	16	14	3.85	231.2	250	37	197	0
Powell	5	4	4.32	75.0	82	40	77	4
Holt	5	13	4.66	164.0	193	57	115	1

Manager—Larry Dierker

Los Angeles Dodgers

Batters	AB	R	H	HR	RBI	SB	Avg
Grudzielanek	488	72	159	7	46	6	.326
Karros	578	74	176	34	112	8	.304
Sheffield	549	103	165	34	101	11	.301
Hollandsworth	261	39	74	9	32	5	.284
Young	456	73	128	2	41	51	.281
Beltre	538	84	148	15	67	18	.275
White	474	60	127	14	68	19	.268
Mondesi	601	98	152	33	99	36	.253
Vizcaino	266	27	67	1	29	2	.252
*Counsell	174	24	38	0	11	1	.218
Hundley	376	49	78	24	55	3	.207

Pitchers	W	L	ERA	IP	H	BB	SO	Sv
Shaw	2	4	2.78	68.0	64	15	43	34
Brown	18	9	3.00	252.1	210	59	221	0
Mills	3	4	3.73	72.1	70	43	49	0
Valdes	9	14	3.98	203.1	213	58	143	0
Dreifort	13	13	4.79	178.2	177	76	140	0
Park	13	11	5.23	194.1	208	100	174	0
Perez	2	10	7.43	89.2	116	39	40	0

Manager—Davey Johnson

Milwaukee Brewers

Batters	AB	R	H	HR	RBI	SB	Avg
Cirillo	607	98	198	15	88	7	.326
Jenkins	447	70	140	21	82	5	.313
Nilsson	343	56	106	21	62	1	.309
Ochoa	277	47	83	8	40	6	.300
Belliard	457	60	135	8	58	4	.295
Loretta	587	93	170	5	67	4	.290
Burnitz	467	87	126	33	103	7	.270
Grissom	603	92	161	20	83	24	.267
Vina	154	17	41	1	16	5	.266
Banks	219	34	53	5	22	6	.242

Batters	AB	R	H	HR	RBI	SB	Avg
Berry	259	26	59	2	23	0	.228
Valentin	256	45	58	10	38	3	.227

Pitchers	W	L	ERA	IP	H	BB	SO	Sv
Wickman	3	8	3.39	74.1	75	38	60	37
Woodard	11	8	4.52	185.0	219	36	119	0
Nomo	12	8	4.54	176.1	173	78	161	0
Peterson	4	7	4.56	77.0	87	25	34	0
Weathers	7	4	4.65	93.0	102	38	74	2
Karl	11	11	4.78	197.2	246	69	74	0
Plunk	4	4	5.02	75.1	71	43	63	0
Roque	1	6	5.34	84.1	96	42	66	1
Pulsipher	5	6	5.98	87.1	100	36	42	0
Abbott	2	8	6.91	82.0	110	42	37	0
Eldred	2	8	7.79	82.0	101	46	60	0

Manager—Phil Garner; Jim Lefebvre

Montreal Expos

Batters	AB	R	H	HR	RBI	SB	Avg
V. Guerrero	610	102	193	42	131	14	.316
White	539	83	168	22	64	10	.312
Vidro	494	67	150	12	59	0	.304
Barrett	433	53	127	8	52	0	.293
W. Guerrero	315	42	92	2	31	7	.292
Fullmer	347	38	96	9	47	2	.277
Merced	194	25	52	8	26	2	.268
Widger	383	42	101	14	56	1	.264
Cabrera	382	48	97	8	39	2	.254
Martinez	331	48	81	2	26	19	.245
Mordecai	226	29	53	5	25	2	.235

Pitchers	W	L	ERA	IP	H	BB	SO	Sv
Urbina	6	6	3.69	75.2	59	36	100	41
Telford	5	4	3.94	96.0	112	38	69	2
Thurman	7	11	4.05	146.2	140	52	85	0
Hermanson	9	14	4.20	216.1	225	69	145	0
Powell	4	8	4.73	97.0	113	44	44	0
Batista	8	7	4.88	134.2	146	58	95	1
Vazquez	9	8	5.00	154.2	154	52	113	0
Pavano	6	8	5.63	104.0	117	35	70	0
Smith	4	9	6.02	89.2	104	39	72	0

Manager—Felipe Alou

New York Mets

Batters	AB	R	H	HR	RBI	SB	Avg
Dunston	243	35	78	5	41	10	.321
Henderson	438	89	138	12	42	37	.315
Hamilton	505	82	159	9	45	6	.315
Cedeno	453	90	142	4	36	66	.313
Alfonzo	628	123	191	27	108	9	.304
Piazza	534	100	162	40	124	2	.303
Ventura	588	88	177	32	120	1	.301
Olerud	581	107	173	19	96	3	.298
Agbayani	276	42	79	14	42	6	.286
Ordonez	520	49	134	1	60	8	.258

Pitchers	W	L	ERA	IP	H	BB	SO	Sv
Benitez	4	3	1.85	78.0	40	41	128	22
J. Franco	0	2	2.88	40.2	40	19	41	19
Wendell	5	4	3.05	85.2	80	37	77	3
#Rogers	5	1	4.03	76.0	71	28	58	0
Leiter	13	12	4.23	213.0	209	93	162	0
Yoshii	12	8	4.40	174.0	168	58	105	0
Hershiser	13	12	4.58	179.0	175	77	89	0
Reed	11	5	4.58	149.1	163	47	104	0
Dotel	8	3	5.38	85.1	69	49	85	0

Manager—Bobby Valentine

Philadelphia Phillies

Batters	AB	R	H	HR	RBI	SB	Avg
Abreu	546	118	183	20	93	27	.335
Glanville	628	101	204	11	73	34	.325
Arias	347	43	105	4	48	2	.303
Lieberthal	510	84	153	31	96	0	.300
Jordan	347	36	99	4	51	0	.285
Brogna	619	90	172	24	102	8	.278
Sefcik	209	28	58	1	11	9	.278
Rolen	421	74	113	26	77	12	.268
Ducey	188	29	49	8	33	2	.261
Gant	516	107	134	17	77	13	.260
Anderson	452	48	114	5	54	13	.252
Relaford	211	31	51	1	26	4	.242

Pitchers	W	L	ERA	IP	H	BB	SO	Sv
Schilling	15	6	3.54	180.1	159	44	152	0
Gomes	5	5	4.26	74.0	70	56	58	19
#Person	10	5	4.27	137.0	130	70	127	0
Byrd	15	11	4.60	199.2	205	70	106	0
Loewer	2	6	5.12	89.2	100	26	48	0
Wolf	6	9	5.55	121.2	126	67	116	0
Ogea	6	12	5.63	168.0	192	61	77	0

Manager—Terry Francona

Pittsburgh Pirates

Batters	AB	R	H	HR	RBI	SB	Avg
Kendall	280	61	93	8	41	22	.332
Giles	521	109	164	39	115	6	.315
Young	584	103	174	26	106	22	.298
Morris	511	65	147	15	73	3	.288
Martin	541	97	150	24	63	20	.277
A. Brown	226	34	61	4	17	5	.270
Sprague	490	71	131	22	81	3	.267
Benjamin	368	42	91	1	37	10	.247
B. Brown	341	49	79	16	58	3	.232
Nunez	259	25	57	0	17	9	.220
Osik	167	12	31	2	13	0	.186

Pitchers	W	L	ERA	IP	H	BB	SO	Sv
Ritchie	15	9	3.49	172.2	169	54	107	0
Benson	11	14	4.07	196.2	184	83	139	0
Schmidt	13	11	4.19	212.2	219	85	148	0
Cordova	8	10	4.43	160.2	166	59	98	0
Williams	3	4	5.09	58.1	63	37	76	23
Schourek	4	7	5.34	113.0	128	49	94	0
Silva	2	8	5.73	97.1	108	39	77	4
Peters	5	4	6.59	71.0	98	27	46	0

Manager—Gene Lamont

St. Louis Cardinals

Batters	AB	R	H	HR	RBI	SB	Avg
Lankford	422	77	129	15	63	14	.306
Tatis	537	104	160	34	107	21	.298
T. Howard	195	16	57	6	28	1	.292
Paquette	157	21	45	10	37	1	.287
McGwire	521	118	145	65	147	0	.278
Polanco	220	24	61	1	19	1	.277
Renteria	585	92	161	11	63	37	.275
McEwing	513	65	141	9	44	7	.275
Castillo	255	21	67	4	31	0	.263
Bragg	273	38	71	6	26	3	.260
Davis	191	27	49	5	30	5	.257
McGee	271	25	68	0	20	7	.251
Drew	368	72	89	13	39	19	.242
Marrero	317	32	61	6	34	11	.192

Pitchers	W	L	ERA	IP	H	BB	SO	Sv
Bottenfield	18	7	3.97	190.1	197	89	124	0
Croushore	3	7	4.14	71.2	68	43	88	3
Stephenson	6	3	4.22	85.1	90	29	59	0
Oliver	9	9	4.26	196.1	197	74	119	0
Bottalico	3	7	4.91	73.1	83	49	66	20
*Mercker	6	5	5.12	103.2	125	51	64	0
Aybar	4	5	5.47	97.0	104	36	74	3
Jimenez	5	14	5.85	163.0	173	71	113	0
Acevedo	6	8	5.89	102.1	115	48	52	4

Manager—Tony La Russa

San Diego Padres

Batters	AB	R	H	HR	RBI	SB	Avg
Gwynn	411	59	139	10	62	7	.338
Sanders	478	92	136	26	72	36	.285
Veras	475	95	133	6	41	30	.280
Magadan	248	20	68	2	30	1	.274
Vander Wal	246	26	67	6	41	2	.272
Nevin	383	52	103	24	85	1	.269
Owens	440	55	117	9	61	33	.266
Gomez	234	20	59	1	15	1	.252
Joyner	323	34	80	5	43	0	.248
Davis	266	29	65	5	30	2	.244
Arias	164	20	40	7	20	0	.244
Jackson	388	56	87	9	39	34	.224
Ru. Rivera	411	65	80	23	48	18	.195

Pitchers	W	L	ERA	IP	H	BB	SO	Sv
Hoffman	2	3	2.14	67.1	48	15	73	40
Wall	7	4	3.07	70.1	58	23	53	0
Boehringer	6	5	3.24	94.1	97	35	64	0
Reyes	2	4	3.72	77.1	76	24	57	1
Ashby	14	10	3.80	206.0	204	54	132	0
Hitchcock	12	14	4.11	205.2	202	76	194	0
Williams	12	12	4.41	208.1	213	73	137	0
Clement	10	12	4.48	180.2	190	86	135	0

Manager—Bruce Bochy

San Francisco Giants

Batters	AB	R	H	HR	RBI	SB	Avg
Rios	150	32	49	7	29	7	.327
Mayne	322	39	97	2	39	2	.301
Benard	562	100	163	16	64	27	.290
Mueller	414	61	120	2	36	4	.290
Kent	511	86	148	23	101	13	.290
Burks	390	73	110	31	96	7	.282
Aurilia	558	68	157	22	80	2	.281
Snow	570	93	156	24	98	0	.274
Servais	198	21	54	5	21	0	.273
Bonds	355	91	93	34	83	15	.262
Santangelo	254	49	66	3	26	12	.260
Hayes	264	33	54	6	48	3	.205

Pitchers	W	L	ERA	IP	H	BB	SO	Sv
Ortiz	18	9	3.81	207.2	189	125	164	0
Nen	3	8	3.98	72.1	79	27	77	37
Nathan	7	4	4.18	90.1	84	46	54	1
*Hernandez	8	12	4.64	199.2	227	76	144	0
Estes	11	11	4.92	203.0	209	112	159	0
Rueter	15	10	5.41	184.2	219	55	94	0
Brock	6	8	5.48	106.2	124	41	76	0
Gardner	5	11	6.47	139.0	142	57	86	0

Manager—Dusty Baker

American League Final Standings, 1999

Eastern Division

	W	L	Pct.	GB	Home	vs. East	vs. Central	vs. West	vs. NL
New York Yankees	98	64	.605	—	48-33	31-18	31-22	27-15	9-9
Boston*	94	68	.580	4	49-32	28-21	36-20	24-15	6-12
Toronto	84	78	.519	14	40-41	24-25	34-20	17-24	9-9
Baltimore	78	84	.481	20	41-40	15-34	27-22	25-21	11-17
Tampa Bay	69	93	.426	29	33-48	25-25	26-22	14-32	4-14

Central Division

	W	L	Pct.	GB	Home	vs. East	vs. Central	vs. West	vs. NL
Cleveland	97	65	.599	—	47-34	26-27	33-16	29-13	9-9
Chicago	75	86	.466	21½	38-42	25-29	24-23	17-25	9-9
Detroit	69	92	.429	27½	38-43	21-34	23-25	17-23	8-10
Kansas City	64	97	.398	32½	33-47	16-33	20-28	22-24	6-12
Minnesota	63	97	.394	33	31-50	18-31	20-28	15-31	10-7

Western Division

	W	L	Pct.	GB	Home	vs. East	vs. Central	vs. West	vs. NL
Texas	95	67	.586	—	51-30	29-26	35-17	21-16	10-8
Oakland	87	75	.537	8	52-29	34-18	26-30	15-21	12-6
Seattle	79	83	.488	16	43-38	24-27	31-25	17-20	7-11
Anaheim	70	92	.432	25	37-44	20-36	24-28	20-16	6-12

*Wild card team.

American League Playoff Results, 1999

Division Series
New York defeated Texas 3 games to 0 (8-0, 3-1, 3-0)
Boston defeated Cleveland 3 games to 2 (2-3, 1-11, 9-3, 23-7, 12-8)

Championship Series
New York defeated Boston 4 games to 1 (4-3, 3-2, 1-13, 9-2, 6-1)

American League Team Statistics, 1999

(Individual Statistics: Batting—at least 150 at-bats; Pitching—at least 70 innings or 10 saves; *traded within AL during season, entry includes statistics for more than one team; # traded to or from NL during season, entry includes only AL stats)

Team Batting

	Avg	AB	R	H	HR	RBI
Texas	.293	5,651	945	1,653	230	897
Cleveland	.289	5,634	1,009	1,629	209	960
Kansas City	.282	5,624	856	1,584	151	800
New York	.282	5,568	900	1,568	193	855
Toronto	.280	5,642	883	1,581	212	856
Baltimore	.279	5,637	851	1,572	203	804
Boston	.278	5,579	836	1,551	176	808
Chicago	.277	5,644	777	1,563	162	742
Tampa Bay	.274	5,586	772	1,531	145	728
Seattle	.269	5,572	859	1,499	244	825
Minnesota	.264	5,495	686	1,450	105	643
Detroit	.261	5,481	747	1,433	212	704
Oakland	.259	5,519	893	1,430	235	845
Anaheim	.256	5,494	711	1,404	158	673

Team Pitching

	ERA	IP	H	BB	SO	Sv
Boston	4.00	1,436.2	1,396	469	1,131	50
New York	4.16	1,439.2	1,402	581	1,111	50
Oakland	4.76	1,438.1	1,537	569	967	48
Baltimore	4.77	1,435.0	1,468	647	982	33
Anaheim	4.79	1,431.1	1,472	624	877	37
Cleveland	4.91	1,450.1	1,503	634	1,120	46
Chicago	4.92	1,438.1	1,608	596	968	39
Toronto	4.93	1,439.0	1,582	575	1,009	39
Minnesota	5.03	1,423.1	1,591	487	927	34
Tampa Bay	5.06	1,433.0	1,606	695	1,055	45
Texas	5.07	1,436.1	1,626	509	979	47
Detroit	5.22	1,421.0	1,528	583	976	33
Seattle	5.25	1,433.2	1,613	684	980	40
Kansas City	5.35	1,420.2	1,608	643	831	29

Anaheim Angels

Batters	AB	R	H	HR	RBI	SB	Avg
Anderson	620	88	188	21	80	3	.303
Vaughn	524	63	147	33	108	0	.281
Palmeiro	317	46	88	1	23	5	.278
Salmon	353	60	94	17	69	4	.266
Huson	225	21	59	0	18	10	.262
Erstad	585	84	148	13	53	13	.253
Edmonds	204	34	51	5	23	5	.250
Greene	321	36	78	14	42	1	.243
Walbeck	288	26	69	3	22	2	.240
Glaus	551	85	132	29	79	5	.240
DiSarcina	271	32	62	1	29	2	.229
Sheets	244	22	48	3	29	1	.197

Pitchers	W	L	ERA	IP	H	BB	SO	Sv
Levine	1	1	3.39	85.0	76	29	37	0
Petkovsek	10	4	3.47	83.0	85	21	43	1
Percival	4	6	3.79	57.0	38	22	58	31
Finley	12	11	4.43	213.1	197	94	200	0
Hill	4	11	4.77	128.1	129	76	76	0
Hasegawa	4	6	4.91	77.0	80	34	44	2
Sparks	5	11	5.42	147.2	165	82	73	0
Belcher	6	8	6.73	132.1	168	46	52	0

Manager—Terry Collins

Baltimore Orioles

Batters	AB	R	H	HR	RBI	SB	Avg
Ripken	332	51	113	18	57	0	.340
Surhoff	673	104	207	28	107	5	.308
Clark	251	40	76	10	29	2	.303
Belle	610	108	181	37	117	17	.297
Conine	444	54	129	13	75	0	.291
Anderson	564	109	159	24	81	36	.282
Bordick	631	93	175	10	77	14	.277
Hairston	175	26	47	4	17	9	.269
DeShields	330	46	87	6	34	11	.264
C. Johnson	426	58	107	16	54	0	.251
Reboulet	154	25	25	0	4	1	.162

Pitchers	W	L	ERA	IP	H	BB	SO	Sv
Mussina	18	7	3.50	203.1	207	52	172	0
Timlin	3	9	3.57	63.0	51	23	50	27
#Guzman	5	9	4.18	122.2	124	65	95	0
Johns	6	4	4.47	86.2	81	25	50	0
Ponson	12	12	4.71	210.0	227	80	112	0
Erickson	15	12	4.81	230.1	244	99	106	0
J. Johnson	8	7	5.46	115.1	120	55	71	0

Manager—Ray Miller

Boston Red Sox

Batters	AB	R	H	HR	RBI	SB	Avg
Garciaparra	532	103	190	27	104	14	.357
Daubach	381	61	112	21	73	0	.294
Offerman	586	107	172	8	69	18	.294
*Huskey	386	62	109	22	77	3	.282
Stanley	427	59	120	19	72	0	.281
O'Leary	596	84	167	28	103	1	.280
Jefferson	206	21	57	5	17	0	.277
Nixon	381	67	103	15	52	3	.270
Varitek	483	70	130	20	76	1	.269
Valentin	450	58	114	12	70	0	.253
Buford	297	39	72	6	38	9	.242
Lewis	470	63	113	2	40	16	.240

Pitchers	W	L	ERA	IP	H	BB	SO	Sv
P. Martinez	23	4	2.07	213.1	160	37	313	0
Lowe	6	3	2.63	109.1	84	25	80	15
Saberhagen	10	6	2.95	119.0	122	11	81	0
Wasdin	8	3	4.12	74.1	66	18	57	2
Rapp	6	7	4.12	146.1	147	69	90	0
*Florie	4	1	4.65	81.1	94	35	65	0
Rose	7	6	4.87	98.0	112	29	51	0
Wakefield	6	11	5.08	140.0	146	72	104	15
Portugal	7	12	5.51	150.1	179	41	79	0
Gordon	0-	2	5.60	17.2	17	12	24	11

Manager—Jimy Williams

Chicago White Sox

Batters	AB	R	H	HR	RBI	SB	Avg
Thomas	486	74	148	15	77	3	.305
Ordonez	624	100	188	30	117	13	.301
Singleton	496	72	149	17	72	20	.300
Fordyce	333	36	99	9	49	2	.297
Durham	612	109	181	13	60	34	.296
Konerko	513	71	151	24	81	1	.294
Lee	492	66	144	16	84	4	.293
Norton	436	62	111	16	50	4	.255
Caruso	529	60	132	2	35	12	.250
Wilson	252	28	60	4	26	1	.238
Johnson	207	27	47	4	16	3	.227

Pitchers	W	L	ERA	IP	H	BB	SO	Sv
Foulke	3	3	2.22	105.1	72	21	123	9
Howry	5	3	3.59	67.2	58	38	80	28
Lowe	4	1	3.67	95.2	90	46	62	0
Simas	6	3	3.75	72.0	73	32	41	2
Sirotka	11	13	4.00	209.0	236	57	125	0
Baldwin	12	13	5.10	199.1	219	81	123	0
Parque	9	15	5.13	173.2	210	79	111	0
Navarro	8	13	6.09	159.2	206	71	74	0
Snyder	9	12	6.68	129.1	167	49	67	0

Manager—Jerry Manuel

Cleveland Indians

Batters	AB	R	H	HR	RBI	SB	Avg
M. Ramirez	522	131	174	44	165	2	.333
Vizquel	574	112	191	5	66	42	.333
R. Alomar	563	138	182	24	120	37	.323
*Baines	430	62	134	25	103	1	.312
Lofton	465	110	140	7	39	25	.301
Cordero	194	35	58	8	32	2	.299
Justice	429	75	123	21	88	1	.287
Diaz	392	43	110	3	32	11	.281
Thome	494	101	137	33	108	0	.277
Wilson	332	41	87	2	24	5	.262
Sexson	479	72	122	31	116	3	.255
Fryman	322	45	82	10	48	2	.255

Pitchers	W	L	ERA	IP	H	BB	SO	Sv
Karsay	10	2	2.97	78.2	71	30	68	1
Shuey	8	5	3.53	81.2	68	40	103	6
Colon	18	5	3.95	205.0	185	76	161	0
Jackson	3	4	4.06	68.2	60	26	55	39
Burba	15	9	4.25	220.0	211	96	174	0
Nagy	17	11	4.95	202.0	238	59	126	0
Wright	8	10	6.06	133.2	144	77	91	0
Gooden	3	4	6.26	115.0	127	67	88	0
*Candiotti	4	6	7.32	71.1	86	30	41	0

Manager—Mike Hargrove

Detroit Tigers

Batters	AB	R	H	HR	RBI	SB	Avg
Polonia	333	46	108	10	32	17	.324
D. Cruz	518	64	147	13	58	1	.284
Clark	536	74	150	31	99	2	.280
Catalanotto	286	41	79	11	35	3	.276
Ausmus	458	62	126	9	54	12	.275
Easley	549	83	146	20	65	11	.266
Palmer	560	92	147	38	100	3	.263
Encarnacion	509	62	130	19	74	33	.255
Kapler	416	60	102	18	49	11	.245
K. Garcia	288	38	69	14	32	2	.240
Higginson	377	51	90	12	46	4	.239
Jefferies	205	22	41	6	18	3	.200

Pitchers	W	L	ERA	IP	H	BB	SO	Sv
Brocail	4	4	2.52	82.0	60	25	78	2
Jones	4	4	3.80	66.1	64	35	64	30
Nitkowski	4	5	4.30	81.2	63	45	66	0
#Mlicki	14	12	4.60	191.2	209	70	119	0
Moehler	10	16	5.04	196.1	229	59	106	0
Thompson	9	11	5.11	142.2	152	59	83	0
Weaver	9	12	5.55	163.2	176	56	114	0
Borkowski	2	6	6.10	76.2	86	40	50	0
Blair	3	11	6.85	134.0	169	44	82	0

Manager—Larry Parrish

Kansas City Royals

Batters	AB	R	H	HR	RBI	SB	Avg
Sweeney	575	101	185	22	102	6	.322
Randa	628	92	197	16	84	5	.314
Damon	583	101	179	14	77	36	.307
Dye	608	96	179	27	119	2	.294
Sanchez	479	66	141	2	56	11	.294
Beltran	663	112	194	22	108	27	.293
Giambi	288	34	82	3	34	0	.285
Febles	453	71	116	10	53	20	.256
Kreuter	324	31	73	5	35	0	.225
Spehr	155	26	32	9	26	1	.206

Pitchers	W	L	ERA	IP	H	BB	SO	Sv
Rosado	10	14	3.85	208.0	197	72	141	0
Suppan	10	12	4.53	208.2	222	62	103	0
*Stein	1	2	4.56	73.0	65	47	47	0
*Rigby	4	6	5.06	83.2	102	31	36	0
Witasick	9	12	5.57	158.1	191	83	102	0
Service	5	5	6.09	75.1	87	42	68	8
*Suzuki	2	5	6.79	110.0	124	64	68	0
Montgomery	1	4	6.84	51.1	72	21	27	12

Manager—Tony Muser

Minnesota Twins

Batters	AB	R	H	HR	RBI	SB	Avg
Koskie	342	42	106	11	58	4	.310
Jones	322	54	93	9	44	3	.289
Cordova	425	62	121	14	70	13	.285
Steinbach	338	35	96	4	42	2	.284
Walker	531	62	148	6	46	18	.279
Allen	481	69	133	10	46	14	.277
Hocking	386	47	103	7	41	11	.267
Coomer	467	53	123	16	65	2	.263
Lawton	406	58	105	7	54	26	.259
Hunter	384	52	98	9	35	10	.255
Gates	306	40	78	3	38	1	.255
Valentin	218	22	54	5	28	0	.248
Mientkiewicz	327	34	75	2	32	1	.229
Guzman	420	47	95	1	39	9	.226

Pitchers	W	L	ERA	IP	H	BB	SO	Sv
Radke	12	14	3.75	218.2	239	44	121	0
Wells	8	3	3.81	87.1	79	28	44	1
Trombley	2	8	4.33	87.1	93	28	82	24
Mays	6	11	4.37	171.0	179	67	115	0
Milton	7	11	4.49	206.1	190	63	163	0
Perkins	1	7	6.54	86.2	117	43	44	0
Hawkins	10	14	6.66	174.1	238	60	105	0
Lincoln	3	10	6.84	76.1	102	26	27	0
Sampson	3	2	8.11	71.0	107	34	56	0

Manager—Tom Kelly

New York Yankees

Batters	AB	R	H	HR	RBI	SB	Avg
Jeter	627	134	219	24	102	19	.349
Williams	591	116	202	25	115	9	.342
Knoblauch	603	120	176	18	68	28	.292
O'Neill	597	70	170	19	110	11	.285
Ledee	250	45	69	9	40	4	.276
Davis	476	59	128	19	78	4	.269
Martinez	589	95	155	28	105	3	.263
Curtis	195	37	51	5	24	8	.262
Brosius	473	64	117	17	71	9	.247
Posada	379	50	93	12	57	1	.245
Girardi	209	23	50	2	27	3	.239
Spencer	205	25	48	8	20	0	.234

Pitchers	W	L	ERA	IP	H	BB	SO	Sv
Rivera	4	3	1.83	69.0	43	18	52	45
Cone	12	9	3.44	193.1	164	90	177	0
Grimsley	7	2	3.60	75.0	66	40	49	1
Hernandez	17	9	4.12	214.1	187	87	157	0
Mendoza	9	9	4.29	123.2	141	27	80	3
Clemens	14	10	4.60	187.2	185	90	163	0
Pettitte	14	11	4.70	191.2	216	89	121	0
Irabu	11	7	4.84	169.1	180	46	133	0

Manager—Joe Torre

Oakland Athletics

Batters	AB	R	H	HR	RBI	SB	Avg
*Velarde	631	105	200	16	76	24	.317
Giambi	575	115	181	33	123	1	.315
Jaha	457	93	126	35	111	2	.276

Batters	AB	R	H	HR	RBI	SB	Avg
Saenz	255	41	70	11	41	1	.275
Grieve	486	80	129	28	86	4	.265
Stairs	531	94	137	38	102	2	.258
Tejada	593	93	149	21	84	8	.251
Chavez	356	47	88	13	50	1	.247
Phillips	406	76	99	15	49	11	.244
Macfarlane	226	24	55	4	31	0	.243
Spiezio	247	31	60	8	33	0	.243
Hinch	205	26	44	7	24	6	.215
Christenson	268	41	56	4	24	7	.209
McDonald	187	26	39	3	8	6	.209

Pitchers	W	L	ERA	IP	H	BB	SO	Sv
Hudson	11	2	3.23	136.1	121	62	132	0
Jones	5	5	3.55	104.0	106	24	63	10
#Taylor	1	5	3.98	43.0	48	14	38	26
*Olivares	15	11	4.16	205.2	217	81	85	0
#Rogers	5	3	4.30	119.1	135	41	68	0
Heredia	13	8	4.81	200.1	228	34	117	0
*Appier	16	14	5.17	209.0	230	84	131	0
Oquist	9	10	5.37	140.2	158	64	89	0
Haynes	7	12	6.34	142.0	158	80	93	0

Manager—Art Howe

Seattle Mariners

Batters	AB	R	H	HR	RBI	SB	Avg
Martinez	502	86	169	24	86	7	.337
Lampkin	206	29	60	9	34	1	.291
Griffey	606	123	173	48	134	24	.285
A. Rodriguez	502	110	143	42	111	21	.285
Bell	597	92	160	21	78	7	.268
Wilson	414	46	110	7	38	5	.266
Ibanez	209	23	54	9	27	5	.258
Davis	432	55	106	21	59	3	.245
Mabry	262	34	64	9	33	2	.244
*Hunter	539	79	125	4	34	44	.232
Buhner	266	37	59	14	38	0	.222

Pitchers	W	L	ERA	IP	H	BB	SO	Sv
Abbott	6	2	3.10	72.2	50	32	68	0
Moyer	14	8	3.87	228.0	235	48	137	0
Paniagua	6	11	4.06	77.2	75	52	74	3
Garcia	17	8	4.07	201.1	205	90	170	0
Halama	11	10	4.22	179.0	193	56	105	0
Meche	8	4	4.73	85.2	73	57	47	0
Mesa	3	6	4.98	68.2	84	40	42	33
F. Rodriguez	2	4	5.65	73.1	94	30	47	3
Cloude	4	4	7.96	72.1	106	46	35	1

Manager—Lou Piniella

Tampa Bay Devil Rays

Batters	AB	R	H	HR	RBI	SB	Avg
McGriff	529	75	164	32	104	1	.310
DiFelice	179	21	55	6	27	0	.307
Boggs	292	40	88	2	29	1	.301
Stocker	254	39	76	1	27	9	.299
Cairo	465	61	137	3	36	22	.295
Trammell	283	49	82	14	39	0	.290
Martinez	514	79	146	6	66	13	.284
Canseco	430	75	120	34	95	3	.279
Flaherty	446	53	124	14	71	0	.278
Winn	303	44	81	2	24	9	.267
Ledesma	294	32	78	0	30	1	.265
Lowery	185	25	48	2	17	0	.259
Perry	209	29	53	6	32	0	.254
#Guillen	168	24	41	2	13	0	.244
Sorrento	294	40	69	11	42	1	.235
Smith	199	18	36	3	19	4	.181

Pitchers	W	L	ERA	IP	H	BB	SO	Sv
Hernandez	2	3	3.07	73.1	68	33	69	43
White	5	3	4.08	132	38	81	0	
Alvarez	9	9	4.22	160.0	159	79	128	0
Rupe	8	9	4.55	142.1	136	57	97	0
Arrojo	7	12	5.18	140.2	162	60	107	0
Eiland	4	8	5.60	80.1	98	27	53	0
Rekar	6	6	5.80	94.2	121	41	55	0
Witt	7	15	5.84	180.1	213	96	123	0

Manager—Larry Rothschild

Texas Rangers

Batters	AB	R	H	HR	RBI	SB	Avg
Rodriguez	600	116	199	35	113	25	.332
Gonzalez	562	114	183	39	128	3	.326
Palmeiro	565	96	183	47	148	2	.324
Greer	556	107	167	20	101	2	.300
Kelly	290	41	87	8	37	6	.300
Zeile	588	80	172	24	98	1	.293
Clayton	465	69	134	14	52	8	.288
Stevens	517	76	146	24	81	2	.282
McLemore	566	105	155	6	45	16	.274
Goodwin	405	63	105	3	33	39	.259
Alicea	164	33	33	3	17	2	.201

Pitchers	W	L	ERA	IP	H	BB	SO	Sv
Zimmerman ..	9	3	2.36	87.2	50	23	67	3
Wetteland....	4	4	3.68	66.0	67	19	60	43
Loaiza.......	9	5	4.56	120.1	128	40	77	0
Sele	18	9	4.79	205.0	244	70	186	0
Helling	13	11	4.84	219.1	228	85	131	0
Burkett	9	8	5.62	147.1	184	46	96	0
Morgan.......	13	10	6.24	140.0	184	48	61	0
*Fassero.....	5	14	7.20	156.1	208	83	114	0
Clark........	3	7	8.60	74.1	103	34	44	0

Manager—Johnny Oates

Toronto Blue Jays

Batters	AB	R	H	HR	RBI	SB	Avg
Fernandez ...	485	73	159	6	75	6	.328
Bush........	485	69	155	5	55	32	.320
Green.......	614	134	190	42	123	20	.309
Stewart.....	608	102	185	11	67	37	.304
*Segui......	440	57	131	14	52	1	.298
Gonzalez	154	22	45	2	12	4	.292

Batters	AB	R	H	HR	RBI	SB	Avg
Fletcher	412	48	120	18	80	0	.291
#Batista	375	61	107	26	79	2	.285
Delgado.....	573	113	156	44	134	1	.272
Cruz........	349	63	84	14	45	14	.241
*Otanez	207	28	49	7	24	0	.237
#Brumfield....	170	25	40	2	19	1	.235
Matheny.....	163	16	35	3	17	0	.215
Greene......	226	22	46	12	41	0	.204

Pitchers	W	L	ERA	IP	H	BB	SO	Sv
Koch	0	5	3.39	63.2	55	30	57	31
Lloyd	5	3	3.63	72.0	68	23	47	3
Halladay.....	8	7	3.92	149.1	156	79	82	1
Carpenter	9	8	4.38	150.0	177	48	106	0
Hentgen......	11	12	4.79	199.0	225	65	118	0
D. Wells.....	17	10	4.82	231.2	246	62	169	0
Escobar	14	11	5.69	174.0	203	81	129	0
Hamilton	7	8	6.52	98.0	118	39	56	0

Manager—Jim Fregosi

National Baseball Hall of Fame and Museum, Cooperstown, NY[1]

#Aaron, Hank
Alexander, Grover Cleveland
Alston, Walt
Anson, Cap
Aparicio, Luis
Appling, Luke
Ashburn, Richie
Averill, Earl
Baker, Home Run
Bancroft, Dave
#Banks, Ernie
Barlick, Al
Barrow, Edward G.
Beckley, Jake
Bell, Cool Papa
#Bench, Johnny
Bender, Chief
Berra, Yogi
Bottomley, Jim
Boudreau, Lou
Bresnahan, Roger
*Brett, George
#Brock, Lou
Brouthers, Dan
Brown, Mordecai (Three Finger)
Bulkeley, Morgan C.
Bunning, Jim
Burkett, Jesse C.
Campanella, Roy
#Carew, Rod
Carey, Max
#Carlton, Steve
Cartwright, Alexander
*Cepeda, Orlando
Chadwick, Henry
Chance, Frank
Chandler, Happy
Charleston, Oscar
Chesbro, John
*Chylak, Nestor
Clarke, Fred
Clarkson, John
Clemente, Roberto
Cobb, Ty[2]
Cochrane, Mickey
Collins, Eddie
Collins, James
Combs, Earle

Comiskey, Charles A.
Conlan, Jocko
Connolly, Thomas H.
Connor, Roger
Coveleski, Stan
Crawford, Sam
Cronin, Joe
Cummings, Candy
Cuyler, Kiki
Dandridge, Ray
Davis, George "Gorgeous"
Day, Leon
Dean, Dizzy
Delahanty, Ed
Dickey, Bill
DiHigo, Martin
DiMaggio, Joe
Doby, Larry
Doerr, Bobby
Drysdale, Don
Duffy, Hugh
Durocher, Leo
Evans, Billy
Evers, John
Ewing, Buck
Faber, Urban
#Feller, Bob
Ferrell, Rick
Fingers, Rollie
Flick, Elmer H.
Ford, Whitey
Foster, Andrew (Rube)
Foster, Bill
Fox, Nellie
Foxx, Jimmie
Frick, Ford
Frisch, Frank
Galvin, Pud
Gehrig, Lou
Gehringer, Charles
#Gibson, Bob
Gibson, Josh
Giles, Warren
Gomez, Lefty
Goslin, Goose
Greenberg, Hank
Griffith, Clark
Grimes, Burleigh
Grove, Lefty

Hafey, Chick
Haines, Jesee
Hamilton, Bill
Hanlon, Ned
Harridge, Will
Harris, Bucky
Hartnett, Gabby
Heilmann, Harry
Herman, Billy
Hooper, Harry
Hornsby, Rogers
Hoyt, Waite
Hubbard, Cal
Hubbell, Carl
Huggins, Miller
Hulbert, William
Hunter, Catfish
Irvin, Monte
#Jackson, Reggie
Jackson, Travis
Jenkins, Ferguson
Jennings, Hugh
Johnson, Byron
Johnson, William (Judy)
Johnson, Walter[2]
Joss, Addie
*Kaline, Al
Keefe, Timothy
Keeler, William
Kell, George
Kelley, Joe
Kelly, George
Kelly, King
Killebrew, Harmon
Kiner, Ralph
Klein, Chuck
Klem, Bill
#Koufax, Sandy
Lajoie, Napoleon
Landis, Kenesaw M.
Lasorda, Tom
Lazzeri, Tony
Lemon, Bob
Leonard, Buck
Lindstrom, Fred
Lloyd, Pop
Lombardi, Ernie
Lopez, Al
Lyons, Ted

Mack, Connie
MacPhail, Larry
MacPhail, Lee
#Mantle, Mickey
Manush, Henry
Maranville, Rabbit
Marichal, Juan
Marquard, Rube
Mathews, Eddie
Mathewson, Christy[2]
#Mays, Willie
McCarthy, Joe
McCarthy, Thomas
#McCovey, Willie
McGinnity, Joe
McGowan, Bill
McGraw, John
McKechnie, Bill
Medwick, Joe
Mize, Johnny
#Morgan, Joe
#Musial, Stan
Newhouser, Hal
Nichols, Kid
Niekro, Phil
O'Rourke, James
Ott, Mel
Paige, Satchel
#Palmer, Jim
Pennock, Herb
Perry, Gaylord
Plank, Ed
Radbourn, Charlie
Reese, Pee Wee
Rice, Sam
Rickey, Branch
Rixey, Eppa
Rizzuto, Phil (Scooter)
Roberts, Robin
#Robinson, Brooks
#Robinson, Frank
#Robinson, Jackie
Robinson, Wilbert
Rogan, Joe "Bullet"
Roush, Edd
Ruffing, Red
Rusie, Amos
Ruth, Babe[2]
*Ryan, Nolan

Schalk, Ray
#Schmidt, Mike
Schoendienst, Red
*Seaver, Tom
*Selee, Frank
Sewell, Joe
Simmons, Al
Sisler, George
Slaughter, Enos
Snider, Duke
*Spahn, Warren
Spalding, Albert
Speaker, Tris
Stargell, Willie
Stengel, Casey
Sutton, Don
Terry, Bill
Thompson, Sam
Tinker, Joe
Traynor, Pie
Vance, Dazzy
Vaughan, Arky
Veeck, Bill
Waddell, Rube
Wagner, Honus[2]
Wallace, Roderick
Walsh, Ed
Waner, Lloyd
Waner, Paul
Ward, John
Weaver, Earl
Weiss, George
Welch, Mickey
Wells, Willie
Wheat, Zach
Wilhelm, Hoyt
Williams, Billy
*Williams, Smokey Joe
#Williams, Ted
Williams, Vic
Wilson, Hack
Wright, George
Wright, Harry
Wynn, Early
#Yastrzemski, Carl
Yawkey, Tom
Young, Cy
Youngs, Ross
*Yount, Robin

(1) Player must generally be retired for five complete seasons before being eligible for induction. (2) Players inducted in 1936 (the year the Hall of Fame began). # Denotes players chosen in first year of Hall of Fame eligibility. * Denotes 1999 inductees. Note: Four players, Babe Ruth (1936), Lou Gehrig (1939), Joe DiMaggio (1955), and Roberto Clemente (1973), were inducted less than five years after retirement or, in Clemente's case, death.

All-Star Baseball Games, 1933-1999

Year	Winner, Score	Host team	Year	Winner, Score	Host team	Year	Winner, Score	Host team
1933*	American, 4-2	Chicago (AL)	1957*	American, 6-5	St. Louis	1977	National, 7-5	New York (AL)
1934*	American, 9-7	New York (NL)	1958*	American, 4-3	Baltimore	1978	National, 7-3	San Diego
1935*	American, 4-1	Cleveland	1959*	National, 5-4	Pittsburgh	1979	National, 7-6	Seattle
1936*	National, 4-3	Boston (NL)	1959*	American, 5-3	Los Angeles	1980	National, 4-2	Los Angeles
1937*	American, 8-3	Washington	1960*	National, 5-3	Kansas City	1981	National, 5-4	Cleveland
1938*	National, 4-1	Cincinnati	1960*	National, 6-0	New York (AL)	1982	National, 4-1	Montreal
1939*	American, 3-1	New York (AL)	1961*	National, 5-4 [3]	San Francisco	1983	American, 13-3	Chicago (AL)

Year	Winner, Score	Host team	Year	Winner, Score	Host team	Year	Winner, Score	Host team
1940*	National, 4-0	St. Louis (NL)	1961*	Called–rain, 1-1	Boston	1984	National, 3-1	San Francisco
1941*	American, 7-5	Detroit	1962*	National, 3-1³	Washington	1985	National, 6-1	Minnesota
1942	American, 3-1	New York (NL)	1962*	American, 9-4	Chicago (NL)	1986	American, 3-2	Houston
1943	American, 5-3	Philadelphia (AL)	1963*	National, 5-3	Cleveland	1987	National, 2-0⁵	Oakland
1944	National, 7-1	Pittsburgh	1964*	National, 7-4	New York (NL)	1988	American, 2-1	Cincinnati
1945	(Not played)		1965*	National, 6-5	Minnesota	1989	American, 5-3	California
1946*	American, 12-0	Boston (AL)	1966*	National, 2-1³	St. Louis	1990	American, 2-0	Chicago (NL)
1947*	American, 2-1	Chicago (NL)	1967*	National, 2-1⁴	California	1991	American, 4-2	Toronto
1948*	American, 5-2	St. Louis (AL)	1968	National, 1-0	Houston	1992	American, 13-6	San Diego
1949*	American, 11-7	Brooklyn	1969*	National, 9-3	Washington	1993	American, 9-3	Baltimore
1950*	National, 4-3¹	Chicago (AL)	1970	National, 5-4²	Cincinnati	1994	National, 8-7³	Pittsburgh
1951*	National, 8-3	Detroit	1971	American, 6-4	Detroit	1995	National, 3-2	Texas
1952*	National, 3-2	Philadelphia (NL)	1972	National, 4-3³	Atlanta	1996	National, 6-0	Philadelphia
1953*	National, 5-1	Cincinnati	1973	National, 7-1	Kansas City	1997	American, 3-1	Cleveland
1954*	American, 11-9	Cleveland	1974	National, 7-2	Pittsburgh	1998	American, 13-8	Colorado
1955*	National, 6-5²	Milwaukee	1975	National, 6-3	Milwaukee	1999	American, 4-1	Boston
1956*	National, 7-3	Washington	1976	National, 7-1	Philadelphia			

*Denotes day game. (1) 14 innings. (2) 12 innings. (3) 10 innings. (4) 15 innings. (5) 13 innings.

Major League Leaders in 1999

American League

Batting
N. Garciaparra, Boston, .357; D. Jeter, New York, .349; B. Williams, New York, .342; E. Martinez, Seattle, .337; M. Ramirez, Cleveland, .333; O. Vizquel, Cleveland, .333.

Runs
R. Alomar, Cleveland, 138; S. Green, Toronto, 134; D. Jeter, New York, 134; M. Ramirez, Cleveland, 131; K. Griffey, Seattle, 123.

Runs Batted In
M. Ramirez, Cleveland, 165; R. Palmeiro, Texas, 148; C. Delgado, Toronto, 134; K. Griffey, Seattle, 134; J. Gonzalez, Texas, 128.

Hits
D. Jeter, New York, 219; B. J. Surhoff, Baltimore, 207; B. Williams, New York, 202; R. Velarde, Anaheim-Oakland, 200; I. Rodriguez, Texas, 199.

Doubles
S. Green, Toronto, 45; J. Dye, Kansas City, 44; M. Sweeney, Kansas City, 44; N. Garciaparra, Boston, 42; J. Fernandez, Toronto, 41; R. Greer, Texas, 41; T. Zeile, Texas, 41.

Triples
J. Offerman, Boston, 11; J. Damon, Kansas City, 9; C. Febles, Kansas City, 9; D. Jeter, New York, 9; R. Durham, Chicago, 8; J. Dye, Kansas City, 8; L. Polonia, Detroit, 8; J. Randa, Kansas City, 8.

Home Runs
K. Griffey, Seattle, 48; R. Palmeiro, Texas, 47; C. Delgado, Toronto, 44; M. Ramirez, Cleveland, 44; S. Green, Toronto, 42; A. Rodriguez, Seattle, 42.

Stolen Bases
B. Hunter, Detroit-Seattle, 44; O. Vizquel, Cleveland, 42; T. Goodwin, Texas, 39; R. Alomar, Cleveland, 37; S. Stewart, Toronto, 37.

Pitching (Most wins: W-L, ERA, Pct.)
P. Martinez, Boston, 23-4, 2.07, .852; B. Colon, Cleveland, 18-5, 3.95, .783; M. Mussina, Baltimore, 18-7, 3.50, .720; A. Sele, Texas, 18-9, 4.79, .667; F Garcia, Seattle, 17-8, 4.07, .680; O. Hernandez, New York, 17.9, 4.12, .654; D. Wells, Toronto, 17-10, 4.82, .630; C. Nagy, Cleveland, 17-11, 4.95, .607.

Strikeouts
P. Martinez, Boston, 313; S. Finley, Anaheim, 200; A. Sele, Texas, 186; D. Cone, New York, 177; D. Burba, Cleveland, 174.

Saves
M. Rivera, New York, 45; R. Hernandez, Tampa Bay, 43; J. Wetteland, Texas, 43; M. Jackson, Cleveland, 39; J. Mesa, Seattle, 33.

National League

Batting
L. Walker, Colorado, .379; L. Gonzalez, Arizona, .336; B. Abreu, Philadelphia, .335; S. Casey, Cincinnatti, .332; J. Cirillo, Milwaukee, .326; M. Grudzielanek, Los Angeles, 326.

Runs
J. Bagwell, Houston, 143; J. Bell, Arizona, 132; E. Alfonzo, New York, 123; C. Biggio, Houston, 123; B. Abreu, Philadelphia, 118; M. McGwire, St. Louis, 118.

Runs Batted In
M. McGwire, St. Louis, 147; M. Williams, Arizona, 142; S. Sosa, Chicago, 141; D. Bichette, Colorado, 133; V. Guerrero, Montreal, 131.

Hits
L. Gonzalez, Arizona, 206; D. Glanville, Philadelphia, 204; J. Cirillo, Milwaukee, 198; S. Casey, Cincinnati, 197; V. Guerrero, Montreal, 193; N. Perez, Colorado, 193.

Doubles
C. Biggio, Houston, 56; L. Gonzalez, Arizona, 45; J. Vidro, Montreal, 45; M. Grace, Chicago, 44; G. Jenkins, Milwaukee, 43.

Triples
B. Abreu, Philadelphia, 11; N. Perez, Colorado, 11; S. Finley, Arizona, 10; T. Womack, Arizona, 10; M. Cameron, Cincinnati, 9; M. Kotsay, Florida, 9.

Home Runs
M. McGwire, St. Louis, 65; S. Sosa, Chicago, 63; C. Jones, Atlanta, 45; G. Vaughn, Cincinnati, 45; J. Bagwell, Houston, 42; V. Guerrero, Montreal, 42.

Stolen Bases
T. Womack, Arizona, 72; R. Cedeno, New York, 66; E. Young, Los Angeles, 51; L. Castillo, Florida, 50; M. Cameron, Cincinnati, 38; P. Reese, Cincinnati, 38.

Pitching (Most wins: W-L, ERA, Pct.)
M. Hampton, Houston, 22-4, 2.90, .846; J. Lima, Houston, 21-10, 3.58, .677; G. Maddux, Atlanta, 19-9, 3.57, .679; K. Bottenfield, St. Louis, 18-7, 3.97, .720; K. Millwood, Atlanta, 18-7, 2.68, .720; K. Brown, Los Angeles, 18-9, 3.00, .667; R. Ortiz, San Francisco, 18-9, 3.81, .667.

Strikeouts
R. Johnson, Arizona, 364; K. Brown, Los Angeles, 221; P. Astacio, Colorado, 210; K. Millwood, Atlanta, 205; S. Reynolds, Houston, 197.

Saves
U. Urbina, Montreal, 41; T. Hoffman, San Diego, 40; B. Wagner, Houston, 39; J. Rocker, Atlanta, 38; R. Nen, San Francisco, 37; B. Wickman, Milwaukee, 37.

Mark McGwire vs. Sammy Sosa, 1999: The Home Run Race II

A year after their 1st record-breaking home run duel, St. Louis Cardinals slugger Mark McGwire and the Chicago Cubs' Slammin' Sammy Sosa resumed their torrid hitting pace and had the baseball world watching in awe. In 1998, McGwire and Sosa hit 70 and 66 home runs, respectively, both shattering the single-season mark of 61 set by New York Yankee Roger Maris in 1961. In 1999 Sosa got off to a fast start in the home run race, leading McGwire 30-23 by the end of June. On July 26, McGwire pulled even with Sosa, at 36. Sosa entered the record books Sept. 18 when he became the first player to hit 60 home runs in consecutive seasons. McGwire followed with his 60th on Sept. 26; 3 days later McGwire hit his 62d and 63d against the San Diego Padres to take the lead for good. McGwire finished with 65 home runs and a consecutive-season record of 135. Sosa finished with 63 homers and a 1998-99 total of 129. McGwire's 128 (1997-98) and Babe Ruth's 114 (1927-28) and 113 (1920-21) are the next highest consecutive season totals in baseball history.

Home Run Comparisons—McGwire, Sosa, Maris, Ruth

Player, year	Month-by-Month Home Run Breakdown Mar.	Apr.	May	Jun.	Jul.	Aug.	Sept.	Oct.	Tot.	GP	AB	AB/HR	Avg.	RBI
Mark McGwire, 1998	1	10	16	10	8	10	15	0	70	155	509	7.27	.299	147
Sammy Sosa, 1998	0	6	7	20*	9	13	11	0	66	159	643	9.74	.308	158
Mark McGwire, 1999	0	5	10	8	16	12	12	2	65	153	521	8.02	.278	147
Sammy Sosa, 1999	0	4	13	13	10	15	7	1	63	162	625	9.92	.288	141
Roger Maris, 1961	0	1	11	15	13	11	9	1	61	161	590	9.67	.269	142
Babe Ruth, 1927	0	4	12	9	9	9	17	0	60	151	540	9.00	.356	164

*Record home runs in any one month. GP=games played. AB=at bats. AB/HR=at bats per home run.

50 Home Run Club

Mark McGwire and Sammy Sosa each hit more than 60 home runs in 2 consecutive seasons—1998 and 1999. They were the only players ever to break Babe Ruth's 60-home-run mark, aside from Roger Maris in 1961, and are among the select group of players who ever hit 50 or more home runs in a season. The following list shows each time a player achieved this mark.

HR	Player, team	Year
70	Mark McGwire, St. Louis Cardinals	1998
66	Sammy Sosa, Chicago Cubs	1998
65	Mark McGwire, St. Louis Cardinals	1999
63	Sammy Sosa, Chicago Cubs	1999
61	Roger Maris, New York Yankees	1961
60	Babe Ruth, New York Yankess	1927
59	Babe Ruth, New York Yankees	1921
58	Jimmie Foxx, Philadelphia Athletics	1932
58	Hank Greenberg, Detroit Tigers	1938
58	Mark McGwire, Oakland Athletics/ St. Louis Cardinals	1997
56	Hack Wilson, Chicago Cubs	1930
56	Ken Griffey Jr., Seattle Mariners	1997
56	Ken Griffey Jr., Seattle Mariners	1998
54	Babe Ruth, New York Yankees	1920
54	Babe Ruth, New York Yankees	1928
54	Ralph Kiner, Pittsburgh Pirates	1949
54	Mickey Mantle, New York Yankees	1961
52	Mickey Mantle, New York Yankees	1956
52	Willie Mays, San Francisco Giants	1965
52	George Foster, Cincinnati Reds	1977
52	Mark McGwire, Oakland Athletics	1996
51	Ralph Kiner, Pittsburgh Pirates	1947
51	Johnny Mize, New York Giants	1947
51	Willie Mays, New York Giants	1955
51	Cecil Fielder, Detroit Tigers	1990
50	Jimmie Foxx, Boston Red Sox	1938
50	Albert Belle, Cleveland Indians	1995
50	Brady Anderson, Baltimore Orioles	1996
50	Greg Vaughn, San Diego Padres	1998

Earned Run Average Leaders

National League

Year	Player, team	G	IP	ERA
1977	John Candelaria, Pittsburgh	33	231	**2.34**
1978	Craig Swan, New York	29	207	**2.43**
1979	J. R. Richard, Houston	38	292	**2.71**
1980	Don Sutton, Los Angeles	32	212	**2.21**
1981	Nolan Ryan, Houston	21	149	**1.69**
1982	Steve Rogers, Montreal	35	277	**2.40**
1983	Atlee Hammaker, San Francisco.	23	172	**2.25**
1984	Alejandro Pena, Los Angeles	28	199	**2.48**
1985	Dwight Gooden, New York	35	276	**1.53**
1986	Mike Scott, Houston	37	275	**2.22**
1987	Nolan Ryan, Houston	34	211	**2.76**
1988	Joe Magrane, St. Louis	24	165	**2.18**
1989	Scott Garrelts, San Francisco	30	193	**2.28**
1990	Danny Darwin, Houston	48	162	**2.21**
1991	Dennis Martinez, Montreal	31	222	**2.39**
1992	Bill Swift, San Francisco	30	164	**2.08**
1993	Greg Maddux, Atlanta	36	267	**2.36**
1994	Greg Maddux, Atlanta	25	202	**1.56**
1995	Greg Maddux, Atlanta	28	209	**1.63**
1996	Kevin Brown, Florida	32	233	**1.89**
1997	Pedro Martinez, Montrea	31	241	**1.90**
1998	Greg Maddux, Atlanta	34	251	**2.22**
1999	Randy Johnson, Arizona	35	271	**2.48**

American League

Year	Player, team	G	IP	ERA
1977	Frank Tanana, California	31	241	**2.54**
1978	Ron Guidry, New York	35	274	**1.74**
1979	Ron Guidry, New York	33	236	**2.78**
1980	Rudy May, New York	41	175	**2.47**
1981	Steve McCatty, Oakland	22	186	**2.32**
1982	Rick Sutcliffe, Cleveland	34	216	**2.96**
1983	Rick Honeycutt, Texas	25	174	**2.42**
1984	Mike Boddicker, Baltimore	34	261	**2.79**
1985	Dave Stieb, Toronto	36	265	**2.48**
1986	Roger Clemens, Boston	33	254	**2.48**
1987	Jimmy Key, Toronto	36	261	**2.76**
1988	Allan Anderson, Minnesota	30	202	**2.45**
1989	Bret Saberhagen, Kansas City	36	262	**2.16**
1990	Roger Clemens, Boston	31	228	**1.93**
1991	Roger Clemens, Boston	35	271	**2.62**
1992	Roger Clemens, Boston	32	246	**2.41**
1993	Kevin Appier, Kansas City	34	238	**2.56**
1994	Steve Ontiveros, Oakland	27	115	**2.65**
1995	Randy Johnson, Seattle	30	214	**2.48**
1996	Juan Guzman, Toronto	27	187	**2.93**
1997	Roger Clemens, Toronto	34	264	**2.05**
1998	Roger Clemens, Toronto	33	234	**2.65**
1999	Pedro Martinez, Boston	31	213	**2.07**

ERA is computed by multiplying earned runs allowed by 9, then dividing by innings pitched.

Yankees Sweep Braves in 4 Games to Win Record 25th World Series Title

In a stellar display, the "Team of the Century" New York Yankees dominated an Atlanta Braves team that many had called the "Team of the '90s." With a 4-1 victory over the Braves on Oct. 27, 1999, the Yankees completed a World Series sweep, winning their 3d World Championship in 4 years and 2d in a row in dominating fashion. Including the final 4 games of the 1996 series against Atlanta and their sweep of the San Diego Padres in 1998, the Yankees won 12 consecutive games in the Fall Classic—a feat equaled only by the legendary Yankee clubs of the "Murderer's Row" era (1927, 1928, 1932). Yankee relief pitcher Mariano Rivera, virtually unhittable in his 3 appearances (2 saves, 1 win), was named Series MVP.

Game One: Yankees 4, Braves 2

New York	ab	r	h	rbi
Knoblauch, 2b	4	1	0	0
Jeter, ss	4	1	2	1
O'Neill, rf	4	0	1	2
B. Williams, cf	2	0	0	0
Martinez, 1b	3	0	0	0
Posada, c	4	0	0	0
Ledee, lf	3	0	0	0
Leyritz, ph	0	0	0	1
Nelson, p	0	0	0	0
Stanton, p	0	0	0	0
Rivera, p	0	0	0	0
Brosius 3b	4	1	3	0
O. Hernandez, p	1	0	0	0
Strawberry, ph	0	0	0	0
Curtis, pr-lf	1	1	0	0
Totals	**30**	**4**	**6**	**4**

Atlanta	ab	r	h	rbi
G. Williams, lf	4	0	0	0
Boone, 2b	4	0	1	0
C. Jones, 3b	2	1	1	1
Jordan, rf	4	0	0	0
Klesko, 1b	3	0	0	0
Hunter, 1	0	0	0	0
Myers, ph	1	0	0	0
A. Jones, cf	2	0	0	0
Perez, c	2	0	0	0
Weiss, ss	2	0	0	0
Guillen, ph	0	0	0	0
J. Hernandez, ph-ss	1	0	0	0
Maddux, p	2	0	0	0
Rocker, p	0	0	0	0
Battle, ph	0	0	0	0
Lockhart, ph	1	0	0	0
Remlinger, p	0	0	0	0
Totals	**28**	**1**	**2**	**1**

New York	0	0	0	0	0	0	0	4	0—4
Atlanta	0	0	0	1	0	0	0	0	0—1

New York	ip	h	r	er	bb	so
O. Hernandez W,1-0	7	1	1	1	2	10
Nelson (H,1)	0.1	0	0	0	1	1
Stanton (H,1)	0.1	0	0	0	0	1
Rivera (S,1)	1.1	1	0	0	1	1

Atlanta

	ip	h	r	er	bb	so
Maddux (L,0-1)..........	7	5	4	2	3	5
Rocker.................	1	1	0	0	2	3
Remlinger..............	1	0	0	0	1	0

(Maddux pitched to 4 batters in the 8th.)

E–Hunter 2 (2). LOB–New York 7, Atlanta 4. HR–C. Jones (1). RBI–Jeter (1), O'Neill 2 (2), Leyritz (1), C. Jones (1). SB–Jeter (1), B. Williams (1). CS–Jeter (1), C. Jones (1). S–O. Hernandez, Knoblauch.

How runs were scored—One in Atlanta fourth: C. Jones solo homer.

Four in New York eighth: Brosius singled. Strawberry ph for O. Hernandez. Strawberry walked, Brosius to second. Curtis pr for Strawberry. Knoblauch bunt sacrifice, Brosius to third, Curtis to second, Knoblauch safe at first on error. Jeter singled, Brosius scored, Curtis to third, Knoblauch to second. O'Neill singled, Curtis and Knoblauch scored, Jeter to second. Jeter to third advancing on throw. O'Neill to second on error. B. Williams intentionally walked. Leyritz ph for Ledee. Leyritz walked, Jeter scored.

Game Two: Yankees 7, Braves 2

New York	ab	r	h	rbi		Atlanta	ab	r	h	rbi
Knoblauch, 2b	4	1	2	1		G. Williams, lf	4	0	0	0
Jeter, ss	5	2	2	0		Guillen, ss	4	0	0	0
O'Neill, rf	4	0	1	1		C. Jones, 3b	3	1	1	0
B. Williams, cf	4	1	3	0		Jordan, rf	3	0	0	0
Martinez, 1b	5	2	2	2		Klesko, 1b	4	0	0	0
Ledee, lf	4	0	2	1		Lockhart, 2b	2	1	0	0
Brosius, 3b	5	1	2	1		Myers, c	3	0	2	1
Girardi, c	4	0	0	0		A. Jones, cf	3	0	0	0
Cone, p	4	0	0	0		McGlinchy, p	0	0	0	0
Mendoza, p	1	0	0	0		Boone, ph	1	0	1	1
Nelson, p	0	0	0	0		Millwood, p	0	0	0	0
Totals	40	7	14	6		Mulholland, p	0	0	0	0
						Fabregas, ph	1	0	0	0
						Springer, p	0	0	0	0
						Nixon, cf	2	0	1	0
						Totals	30	2	5	2

New York ....3 0 2 1 1 0 0 0 0—7
Atlanta ....0 0 0 0 0 0 0 2—2

New York	ip	h	r	er	bb	so
Cone (W,1-0)	7	1	0	0	5	4
Mendoza	1.2	3	2	2	1	0
Nelson	0.1	1	0	0	0	0
Atlanta						
Millwood (L,0-1)	2	8	5	4	2	2
Mulholland	3	3	2	2	1	3
Springer	2	1	0	0	0	1
McGlinchy..............	2	2	0	0	1	2

(Millwood pitched to 3 batters in the 3d.)

E–Cone (1), Guillen (1). LOB–New York 11, Atlanta 7. 2B–Ledee (1), Jeter (1), Brosius (1), Boone (1). RBI–O'Neill (3), Martinez 2 (2), Brosius (1), Ledee (1), Knoblauch (1), Myers (1), Boone (1). S–Girardi. SB–Knoblauch (1).

How runs were scored—Three in New York first: Knoblauch singled. Jeter singled, Knoblauch to second. O'Neill singled, Knoblauch scored, Jeter to second. B. Williams grounded into double play, O'Neill out at second, Jeter to third. Martinez singled, Jeter scored. Ledee walked, Martinez to second. Brosius singled, Martinez scored.

Two in New York third: B. Williams singled. Martinez singled, B. Williams to second. Ledee doubled, B. Williams scored, Martinez to third. Cone safe at first on shortstop fielding error, Martinez scored.

One in New York fourth: Jeter doubled. O'Neill flied out, Jeter to third. B. Williams intentionally walked. Martinez grounded into fielder's choice to shortstop, Jeter scored.

One in New York fifth: Brosius doubled. Girardi sacrificed Brosius to third. Knoblauch singled to left, Brosius scored.

Two in Atlanta ninth: C. Jones singled. Jordan grounded out to third, C. Jones to second. Lockhart walked. Myers singled, C. Jones scored, Lockhart to second. Boone hit for McGlinchy. Boone doubled to left, Lockhart scored.

Game Three: Yankees 6, Braves 5 (10 innings)

Atlanta	ab	r	h	rbi		New York	ab	r	h	rbi
G. Williams, lf	5	2	2	0		Knoblauch, 2b	4	2	2	2
Boone, 2b	5	1	4	1		Jeter, ss	4	0	1	0
Nixon, pr	0	0	0	0		O'Neill, rf	4	0	1	1
Lockhart, 2b	0	0	0	0		B. Williams, cf	4	0	0	0
C. Jones, 3b	4	0	1	1		Davis, dh	4	0	1	0
Jordan, rf	3	1	1	0		Martinez, 1b	4	1	1	1
A. Jones, cf	5	1	1	0		Brosius, 3b	4	0	0	0

Atlanta	ab	r	h	rbi		New York	ab	r	h	rbi
J. Hernandez, dh	4	0	1	2		Curtis, lf	4	2	2	2
Guillen, ph-dh	1	0	0	0		Girardi, c	3	1	2	0
Perez, c	4	0	1	0		Totals	35	6	9	6
Klesko, ph-1b	1	0	1	0						
Hunter, 1b	4	0	1	0						
Myers, ph-c	1	0	0	0						
Weiss, ss	4	0	1	0						
Totals	41	5	14	5						

Atlanta...........1 0 3 1 0 0 0 0 0—5
New York.........1 0 0 0 1 0 1 2 0 1—6

Atlanta	ip	h	r	er	bb	so
Glavine	7	7	5	4	0	3
Rocker	2	1	0	0	0	1
Remlinger (L,0-1)	0	1	1	1	0	0
New York						
Pettitte	3.2	10	5	5	1	1
Grimsley	2.1	2	0	0	2	0
Nelson	2	0	0	0	0	2
Rivera (W,1-0)	2	0	0	0	0	2

(Glavine pitched to 2 batters in the 8th, Remlinger pitched to 1 batter in the 10th.)

E–Jordan (1). LOB–Atlanta 9, New York 2. 2B–Boone 3 (4), J. Hernandez (1), Knoblauch (1). 3B–G. Williams (1). HR–Curtis 2 (2), Martinez (1), Knoblauch (1). RBI–C. Jones (2), Jordan (1), J. Hernandez 2 (2), Boone (2), O'Neill (4), Curtis 2 (2), Martinez (3), Knoblauch 2 (3). SB–J. Hernandez (1). CS–Boone (1), Nixon (1).

How runs were scored—One in Atlanta first: G. Williams singled. Boone doubled, G. Williams to third. C. Jones grounded out, G. Williams scored.

One in New York first: Knoblauch safe at second on right fielder's fielding error. Jeter flied out, Knoblauch to third. O'Neill singled, Knoblauch scored.

Three in Atlanta third: Boone doubled. C. Jones grounded out, Boone to third. Jordan singled, Boone scored. A. Jones singled, Jordan to second. J. Hernandez doubled, Jordan and A. Jones scored.

One in Atlanta fourth: G. Williams tripled. Boone doubled, G. Williams scored.

One in New York fifth: Curtis homered.

One in New York seventh: Martinez homered.

Two in New York eighth: Girardi singled. Knoblauch homered, Girardi scored.

One in New York tenth: Curtis homered.

Game Four: Yankees 4, Braves 1

Atlanta	ab	r	h	rbi		New York	ab	r	h	rbi
G. Williams, lf	4	0	1	0		Knoblauch, 2b	4	1	1	0
Boone, 2b	3	0	1	1		Sojo, 2b	0	0	0	0
C. Jones, 3b	4	0	0	0		Jeter, ss	4	1	1	0
Jordan, rf	3	0	0	0		O'Neill, rf	3	0	0	0
Klesko, 1b	4	0	1	0		B. Williams, cf	3	1	0	0
Lockhart, dh	4	0	1	0		Martinez, 1b	3	0	1	2
Perez, c	2	0	0	0		Strawberry, dh	3	0	1	0
Myers, ph-c	1	0	0	0		Leyritz, ph-dh	1	1	1	1
A. Jones, cf	3	0	0	0		Posada, c	4	0	2	1
Weiss, ss	3	1	1	0		Ledee, lf	3	0	0	0
Totals	31	1	5	1		Curtis, ph-lf	1	0	0	0
						Brosius, 3b	3	0	1	0
						Totals	32	4	8	4

Atlanta...........0 0 0 0 0 0 0 0 1—1
New York.........0 0 3 0 0 0 0 1 X—4

Atlanta	ip	h	r	er	bb	so
Smoltz (L,0-1)	7	6	3	3	3	11
Mulholland................	0.2	2	1	1	0	0
Springer.................	0.1	0	0	0	0	0
New York						
Clemens (W,1-0)	7.2	4	1	1	2	4
Nelson (H,2)	0	1	0	0	0	0
Rivera (S,2)	1.1	0	0	0	0	0

(Nelson pitched to 1 batter in the 8th.)

LOB–Atlanta 5, New York 7. 2B–Posada (1). HR–Leyritz (1). RBI–Boone (3), Martinez 2 (5), Posada (1), Leyritz (2). SB–Jeter 2 (3).

How runs were scored—Three in the New York third: Knoblauch, infield single. Jeter singled, Knoblauch to third. Jeter stole second. B. Williams intentionally walked. Martinez singled to right, Knoblauch and Jeter scored, B. Williams to third. Posada singled to right, B. Williams scored.

One in the Atlanta eighth: Weiss, infield single. G. Williams singled, Weiss to second. Boone singled to center, Weiss scored.

One in New York eighth: Leyritz homered.

World Series Results, 1903-1999

1903	Boston AL 5, Pittsburgh NL 3	1936	New York AL 4, New York NL 2	1968	Detroit AL 4, St. Louis NL 3
1904	No series	1937	New York AL 4, New York NL 1	1969	New York NL 4, Baltimore AL 1
1905	New York NL 4, Philadelphia AL 1	1938	New York AL 4, Chicago NL 0	1970	Baltimore AL 4, Cincinnati NL 1
1906	Chicago AL 4, Chicago NL 2	1939	New York AL 4, Cincinnati NL 0	1971	Pittsburgh NL 4, Baltimore AL 3
1907	Chicago NL 4, Detroit AL 0, 1 tie	1940	Cincinnati NL 4, Detroit AL 3	1972	Oakland AL 4, Cincinnati NL 3
1908	Chicago NL 4, Detroit AL 1	1941	New York AL 4, Brooklyn NL 1	1973	Oakland AL 4, New York NL 3
1909	Pittsburgh NL 4, Detroit AL 3	1942	St. Louis NL 4, New York AL 1	1974	Oakland AL 4, Los Angeles NL 1
1910	Philadelphia AL 4, Chicago NL 1	1943	New York AL 4, St. Louis NL 1	1975	Cincinnati NL 4, Boston AL 3
1911	Philadelphia AL 4, New York NL 2	1944	St. Louis NL 4, St. Louis AL 2	1976	Cincinnati NL 4, New York AL 0
1912	Boston AL 4, New York NL 3, 1 tie	1945	Detroit AL 4, Chicago NL 3	1977	New York AL 4, Los Angeles NL 2
1913	Philadelphia AL 4, New York NL 1	1946	St. Louis NL 4, Boston AL 3	1978	New York AL 4, Los Angeles NL 2
1914	Boston NL 4, Philadelphia AL 0	1947	New York AL 4, Brooklyn NL 3	1979	Pittsburgh NL 4, Baltimore AL 3
1915	Boston AL 4, Philadelphia NL 1	1948	Cleveland AL 4, Boston NL 2	1980	Philadelphia NL 4, Kansas City AL 2
1916	Boston AL 4, Brooklyn NL 1	1949	New York AL 4, Brooklyn NL 1	1981	Los Angeles NL 4, New York AL 2
1917	Chicago AL 4, New York NL 2	1950	New York AL 4, Philadelphia NL 0	1982	St. Louis NL 4, Milwaukee AL 3
1918	Boston AL 4, Chicago NL 2	1951	New York AL 4, New York NL 2	1983	Baltimore AL 4, Philadelphia NL 1
1919	Cincinnati NL 5, Chicago AL 3	1952	New York AL 4, Brooklyn NL 3	1984	Detroit AL 4, San Diego NL 1
1920	Cleveland AL 5, Brooklyn NL 2	1953	New York AL 4, Brooklyn NL 2	1985	Kansas City AL 4, St. Louis NL 3
1921	New York NL 5, New York AL 3	1954	New York NL 4, Cleveland AL 0	1986	New York NL 4, Boston AL 3
1922	New York NL 4, New York AL 0, 1 tie	1955	Brooklyn NL 4, New York AL 3	1987	Minnesota AL 4, St. Louis NL 3
1923	New York AL 4, New York NL 2	1956	New York AL 4, Brooklyn NL 3	1988	Los Angeles NL 4, Oakland AL 1
1924	Washington AL 4, New York NL 3	1957	Milwaukee NL 4, New York AL 3	1989	Oakland AL 4, San Francisco NL 0
1925	Pittsburgh NL 4, Washington AL 3	1958	New York AL 4, Milwaukee NL 3	1990	Cincinnati NL 4, Oakland AL 0
1926	St. Louis NL 4, New York AL 3	1959	Los Angeles NL 4, Chicago AL 2	1991	Minnesota AL 4, Atlanta NL 3
1927	New York AL 4, Pittsburgh NL 0	1960	Pittsburgh NL 4, New York AL 3	1992	Toronto AL 4, Atlanta NL 2
1928	New York AL 4, St. Louis NL 0	1961	New York AL 4, Cincinnati NL 1	1993	Toronto AL 4, Philadelphia NL 2
1929	Philadelphia AL 4, Chicago NL 1	1962	New York AL 4, San Francisco NL 3	1994	No series
1930	Philadelphia AL 4, St. Louis NL 2	1963	Los Angeles NL 4, New York AL 0	1995	Atlanta NL 4, Cleveland AL 2
1931	St. Louis NL 4, Philadelphia AL 3	1964	St. Louis NL 4, New York AL 3	1996	New York AL 4, Atlanta NL 2
1932	New York AL 4, Chicago NL 0	1965	Los Angeles NL 4, Minnesota AL 3	1997	Florida NL 4, Cleveland AL 3
1933	New York NL 4, Washington AL 1	1966	Baltimore AL 4, Los Angeles NL 0	1998	New York AL 4, San Diego NL 0
1934	St. Louis NL 4, Detroit AL 3	1966	Baltimore AL 4, Los Angeles NL 0	1999	New York AL 4, Atlanta NL 0
1935	Detroit AL 4, Chicago NL 2	1967	St. Louis NL 4, Boston AL 3		

World Series MVP

Year	player, position, team	Year	player, position, team	Year	player, position, team
1955	John Podres, p, Brooklyn	1971	Roberto Clemente, of, Pittsburgh	1985	Bret Saberhagen, p, Kansas City
1956	Don Larsen, p, New York, AL	1972	Gene Tenance, c, Oakland	1986	Ray Knight, 3b, NY, NL
1957	Lew Burdette, p, Milwaukee, NL	1973	Reggie Jackson, of, Oakland	1987	Frank Viola, p, Minnesota
1958	Bob Turley, p, NY AL	1974	Rollie Fingers, p, Oakland	1988	Orel Hershiser, p, LA
1959	Larry Sherry, p, LA	1975	Pete Rose, 3b, Cincinnati	1989	Dave Stewart, p, Oakland
1960[1]	Bobby Richardson, 2b, NY, AL	1976	Johnny Bench, c, Cincinnati	1990	Jose Rijo, p, Cincinnati
1961	Whitey Ford, p, NY, AL	1977	Reggie Jackson, of, NY, AL	1991	Jack Morris, p, Minnesota
1962	Ralph Terry, p, NY, AL	1978	Bucky Dent, ss, NY, AL	1992	Pat Borders, c, Toronto
1963	Sandy Koufax, p, LA	1979	Willie Stargell, 1b, Pittsburgh	1993	Paul Molitor, dh, Toronto
1964	Bob Gibson, p, St. Louis	1980	Mike Schmidt, 3b, Philadelphia	1994	no series
1965	Sandy Koufax, p, LA	1981	Ron Cey, 3b, LA	1995	Tom Glavine, p, Atlanta
1966	Frank Robinson, of, Baltimore		Pedro Guerrero, of, LA	1996	John Wetteland, p, NY, AL
1967	Bob Gibson, p, St. Louis		Steve Yeager, c, LA	1997	Livan Hernandez, p, Florida
1968	Mickey Lolich, p, Detroit	1982	Darrell Porter, c, St. Louis	1998	Scott Brosius, 3b, NY, AL
1969	Donn Clendenon, 1b, NY, NL	1983	Rick Dempsey, c, Baltimore	1999	Mariano Rivera, p, NY, AL
1970	Brooks Robinson, 3b, Baltimore	1984	Alan Trammell, ss, Detroit		

(1) Bobby Richardson won the MVP although Pittsburgh beat New York.

World Series Won-Lost Records, by Franchise

Team	Wins	Losses	Team	Wins	Losses
New York Yankees	25	11	Boston/Milwaukee/Atlanta Braves	3	6
Philadelphia/Kansas City/Oakland A's	9	5	Toronto Blue Jays	2	0
St. Louis Cardinals	9	6	New York Mets	2	1
Brooklyn/Los Angeles Dodgers	6	12	Chicago White Sox	2	2
Pittsburgh Pirates	5	2	Cleveland Indians	2	3
Boston Red Sox	5	4	Chicago Cubs	2	8
Cincinnati Reds	5	4	Florida Marlins	1	0
New York/San Francisco Giants	5	11	Kansas City Royals	1	1
Detroit Tigers	4	5	Philadelphia Phillies	1	4
Washington/Minnesota Twins	3	3	Seattle/Milwaukee Brewers	0	1
St. Louis/Baltimore Orioles	3	4	San Diego Padres	0	2

All-Time Major League Leaders

(*player active at end of 1999 season)

Games		At Bats		Runs Batted In		Stolen Bases	
Pete Rose	3,562	Pete Rose	14,053	Hank Aaron	2,297	Rickey Henderson*	1,334
Carl Yastrzemski	3,308	Hank Aaron	12,364	Babe Ruth	2,213	Lou Brock	938
Hank Aaron	3,298	Carl Yastrzemski	11,988	Lou Gehrig	1,995	Billy Hamilton	912
Ty Cobb	3,035	Ty Cobb	11,434	Stan Musial	1,951	Ty Cobb	892
Eddie Murray	3,026	Eddie Murray	11,336	Ty Cobb	1,937	Tim Raines*	807
Stan Musial	3,026	Robin Yount	11,008	Jimmie Foxx	1,922	Vince Coleman	752
Willie Mays	2,992	Dave Winfield	11,003	Eddie Murray	1,917	Eddie Collins	744
Dave Winfield	2,973	Stan Musial	10,972	Willie Mays	1,903	Arlie Latham	739
Rusty Staub	2,951	Willie Mays	10,881	Mel Ott	1,860	Max Carey	738
Brooks Robinson	2,896	Paul Molitor	10,835	Carl Yastrzemski	1,844	Honus Wagner	722

Runs

Ty Cobb	2,246
Hank Aaron	2,174
Babe Ruth	2,174
Pete Rose	2,165
Willie Mays	2,062
Rickey Henderson*	2,103
Stan Musial	1,949
Lou Gehrig	1,888
Tris Speaker	1,882
Mel Ott	1,859

Strikeouts

Nolan Ryan	5,714
Steve Carlton	4,136
Bert Blyleven	3,701
Tom Seaver	3,640
Don Sutton	3,574
Gaylord Perry	3,534
Walter Johnson	3,509
Phil Niekro	3,342
Ferguson Jenkins	3,192
Roger Clemens*	3,153

Shutouts

Walter Johnson	110
Grover Alexander	90
Christy Mathewson	79
Cy Young	76
Eddie Plank	69
Warren Spahn	63
Nolan Ryan	61
Tom Seaver	61
Bert Blyleven	60
Don Sutton	58

Saves

Lee Smith	478
John Franco*	416
Dennis Eckersley	390
Jeff Reardon	367
Randy Myers	347
Rollie Fingers	341
Tom Henke	311
Rich Gossage	310
Jeff Montgomery*	304
Doug Jones*	301

All-Time Home Run Leaders

Player	HR	Player	HR	Player	HR	Player	HR
Hank Aaron	755	Ted Williams	521	Barry Bonds*	445	Ken Griffey Jr.*	398
Babe Ruth	714	Ernie Banks	512	Dave Kingman	442	Joe Carter	396
Willie Mays	660	Ed Mathews	512	Andre Dawson	438	Graig Nettles	390
Frank Robinson	586	Mel Ott	511	Jose Canseco*	431	Fred McGriff*	390
Harmon Killebrew	573	Eddie Murray	504	Billy Williams	426	Johnny Bench	389
Reggie Jackson	563	Lou Gehrig	493	Darrell Evans	414	Dwight Evans	385
Mike Schmidt	548	Stan Musial	475	Duke Snider	407	Frank Howard	382
Mickey Mantle	536	Willie Stargell	475	Cal Ripken Jr.*	402	Jim Rice	382
Jimmy Foxx	534	Dave Winfield	465	Al Kaline	399	Orlando Cepeda	379
Mark McGwire*	522	Carl Yastrzemski	452	Dale Murphy	398	Tony Perez	379
Willie McCovey	521						

Players With 3,000 Major League Hits

Player	Hits	Player	Hits	Player	Hits	Player	Hits
Pete Rose	4,256	Honus Wagner	3,415	George Brett	3,154	Rod Carew	3,053
Ty Cobb	4,189	Paul Molitor	3,319	Paul Waner	3,152	Lou Brock	3,023
Hank Aaron	3,771	Eddie Collins	3,315	Robin Yount	3,142	Wade Boggs*	3,010
Stan Musial	3,630	Willie Mays	3,283	Dave Winfield	3,110	Al Kaline	3,007
Tris Speaker	3,514	Eddie Murray	3,255	Tony Gwynn*	3,067	Roberto Clemente	3,000
Carl Yastrzemski	3,419	Nap Lajoie	3,242				

Pitchers With 300 Major League Wins

Player	Wins	Player	Wins	Player	Wins	Player	Wins
Cy Young	511	Kid Nichols	361	Eddie Plank	326	Tom Seaver	311
Walter Johnson	417	Pud Galvin	360	Nolan Ryan	324	Charley Radbourn	309
Grover Alexander	373	Tim Keefe	342	Don Sutton	324	Mickey Welch	307
Christy Mathewson	373	Steve Carlton	329	Phil Niekro	318	Lefty Grove	300
Warren Spahn	363	John Clarkson	328	Gaylord Perry	314	Early Wynn	300

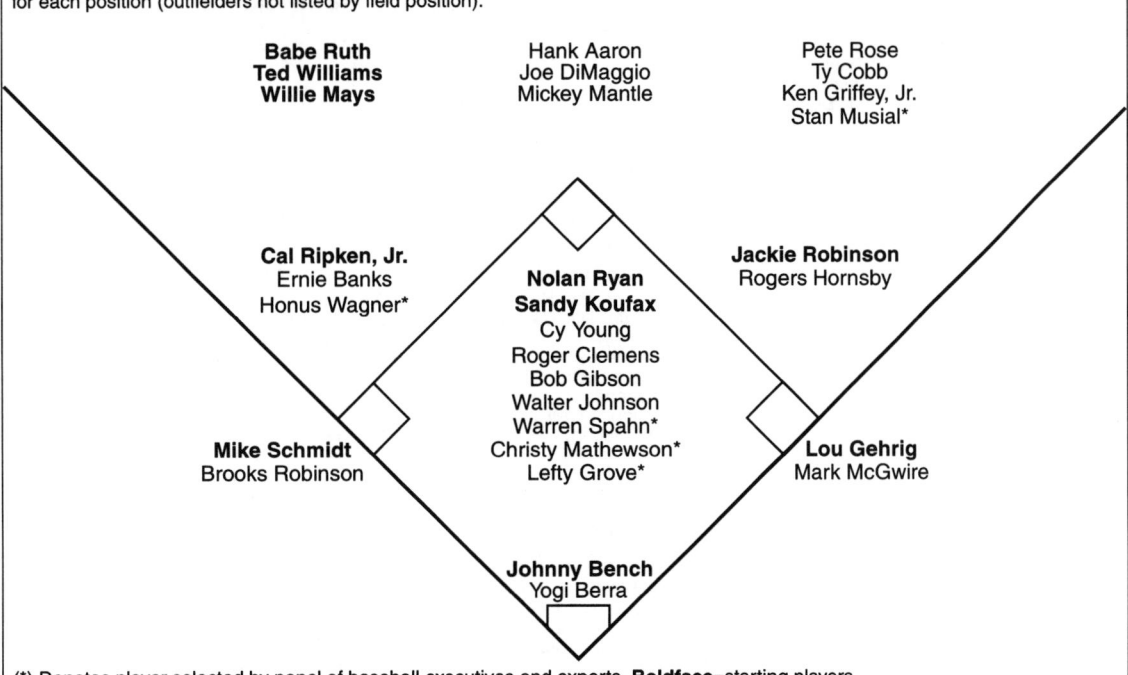

MILLENNIUM FACT BOX

The All-Century Dream Team

Eighteen of the greatest baseball players of all time gathered at Atlanta's Turner Field Oct. 24, 1999, before Game 2 of the World Series; they were the living members selected as part of a 30-player All-Century Team. Announcer Vin Scully read the 30 names to a cheering crowd. Twenty-five were chosen by fans in nationwide balloting; the final 5 were picked by a panel of baseball executives and experts. The diamond below shows the members of this "dream team" in their positions on the field in order of votes received for each position (outfielders not listed by field position).

Babe Ruth
Ted Williams
Willie Mays

Hank Aaron
Joe DiMaggio
Mickey Mantle

Pete Rose
Ty Cobb
Ken Griffey, Jr.
Stan Musial*

Cal Ripken, Jr.
Ernie Banks
Honus Wagner*

Nolan Ryan
Sandy Koufax
Cy Young
Roger Clemens
Bob Gibson
Walter Johnson
Warren Spahn*
Christy Mathewson*
Lefty Grove*

Jackie Robinson
Rogers Hornsby

Mike Schmidt
Brooks Robinson

Lou Gehrig
Mark McGwire

Johnny Bench
Yogi Berra

(*) Denotes player selected by panel of baseball executives and experts. **Boldface**=starting players.

Major League Franchise Shifts and Additions

1953—Boston Braves (NL) became Milwaukee Braves.
1954—St. Louis Browns (AL) became Baltimore Orioles.
1955—Philadelphia Athletics (AL) became Kansas City Athletics.
1958—New York Giants (NL) became San Francisco Giants.
1958—Brooklyn Dodgers (NL) became Los Angeles Dodgers.
1961—Washington Senators (AL) became Minnesota Twins.
1961—Los Angeles Angels (renamed California Angels in 1965 and Anaheim Angels in 1997) enfranchised by the American League.
1961—Washington Senators enfranchised by the American League (a new team, replacing the former Washington club, whose franchise was moved to Minneapolis-St. Paul).
1962—Houston Colt .45's (renamed the Houston Astros in 1965) enfranchised by the National League.
1962—New York Mets enfranchised by the National League.

1966—Milwaukee Braves (NL) became Atlanta Braves.
1968—Kansas City Athletics (AL) became Oakland Athletics.
1969—Kansas City Royals and Seattle Pilots enfranchised by the American League; Montreal Expos and San Diego Padres enfranchised by the National League.
1970—Seattle Pilots became Milwaukee Brewers.
1971—Washington Senators became Texas Rangers (Dallas-Fort Worth area).
1977—Toronto Blue Jays and Seattle Mariners enfranchised by the American League.
1993—Colorado Rockies (Denver) and Florida Marlins (Miami) enfranchised by the National League.
1998—Tampa Bay Devil Rays began play in the American League; Arizona Diamondbacks (Phoenix) began play in the National League (both teams enfranchised in 1995). Milwaukee Brewers moved from the AL to the NL.

Baseball Stadiums[1]

National League

Team	Stadium (year opened)	Surface	Home run distances (ft.) LF	Center	RF	Seating capacity
Arizona Diamondbacks	Bank One Ballpark (1998)	Grass	330	407	334	49,075
Atlanta Braves	Turner Field (1997)	Grass	335	401	330	49,714
Chicago Cubs...........	Wrigley Field (1914)	Grass	355	400	353	38,902
Cincinnati Reds	Cinergy Field (1970)	Artificial	330	404	330	52,953
Colorado Rockies.........	Coors Field (1995)	Grass	347	415	350	50,249
Florida Marlins	Pro Player Stadium (1987)	Grass	325	410	345	41,855
Houston Astros	The Astrodome (1965)	Artificial	325	400	325	54,313
	Enron Field (2000)	Grass	315	435	326	42,000
Los Angeles Dodgers......	Dodger Stadium (1962)	Grass	330	395	330	56,000
Milwaukee Brewers	County Stadium (1953).................	Grass	315	402	315	53,192
Montreal Expos	Olympic Stadium (1976)................	Artificial	325	404	325	46,500
New York Mets	Shea Stadium (1964)	Grass	338	410	338	55,777
Philadelphia Phillies	Veterans Stadium (1971)	Artificial	330	408	330	62,382
Pittsburgh Pirates.........	Three Rivers Stadium (1970)	Artificial	335	400	335	47,687
St. Louis Cardinals........	Busch Stadium (1966)	Grass	330	402	330	49,625
San Diego Padres	Qualcomm Stadium (1967)..............	Grass	327	405	327	59,690
San Francisco Giants......	3Com Park at Candlestick Point (1960)	Grass	335	400	328	63,000
	Pacific Bell Park (2000)	Grass	335	404	307	42,000

American League

Team	Stadium (year opened)	Surface	LF	Center	RF	Seating capacity
Anaheim Angels..........	Edison Intl. Field of Anaheim (1966)	Grass	333	408	333	45,054
Baltimore Orioles	Oriole Park at Camden Yards (1992)	Grass	333	400	318	48,876
Boston Red Sox..........	Fenway Park (1912)	Grass	310	420	302	33,871
Chicago White Sox.......	Comiskey Park (1991)	Grass	347	400	347	44,321
Cleveland Indians.........	Jacobs Field (1994)	Grass	325	405	325	43,368
Detroit Tigers	Tiger Stadium (1912)	Grass	340	440	325	46,945
	Comerica Park (2000).................	Grass	345	422	330	40,000
Kansas City Royals	Kauffman Stadium (1973)...............	Grass	330	400	330	40,625
Minnesota Twins..........	Hubert H. Humphrey Metrodome (1982)	Artificial	343	408	327	48,678
New York Yankees	Yankee Stadium (1923)	Grass	318	408	314	57,545
Oakland A's	Network Associates Coliseum (1968).......	Grass	330	400	330	43,662
Seattle Mariners..........	Safeco Field (1999)	Grass	331	405	327	47,000
Tampa Bay Devil Rays	Tropicana Field (1990)	Artificial	315	407	322	44,207
Texas Rangers	The Ballpark in Arlington (1994)...........	Grass	332	400	325	49,166
Toronto Blue Jays.........	SkyDome (1989)......................	Artificial	328	400	328	50,516

(1) As of 1999 season.

Little League World Series

The Little League World Series is played annually in Williamsport, PA. The team from Osaka, Japan, won the 1999 series by defeating the team from Phenix City, AL, 5-0, on Aug. 28. It was Japan's 1st win since 1976 and its 4th ever. Phenix City had upset the defending champs, the Toms River, NJ, team, 3-2, the previous day to win the U.S. Championship.

Year	Winning / Losing Team	Score	Year	Winning / Losing Team	Score	Year	Winning / Losing Team	Score
1947	Williamsport, PA; Lock Haven, PA	16-7	1962	San Jose, CA; Kankakee, IL	3-0	1981	Taiwan; Tampa, FL	4-2
1948	Lock Haven, PA; St. Petersburg, FL	6-5	1963	Granada Hills, CA; Stratford, CT	2-1	1982	Kirkland, WA; Taiwan	6-0
1949	Hammonton, NJ; Pensacola, FL	5-0	1964	Staten Island, NY; Mexico	4-0	1983	Marietta, GA; Dominican Rep.	3-1
1950	Houston, TX; Bridgeport, CT	2-1	1965	Windsor Locks, CT; Ontario, Canada	3-1	1984	South Korea; Altamonte Springs, FL	6-2
1951	Stamford, CT; Austin, TX	3-0	1966	Houston, TX; W. New York, NJ	8-2	1985	South Korea; Mexico	7-1
1952	Norwalk, CT; Monongahela, PA	4-3	1967	Tokyo, Japan; Chicago, IL	4-1	1986	Taiwan; Tucson, AZ	12-0
1953	Birmingham, AL; Schenectady, NY	1-0	1968	Osaka, Japan; Richmond, VA	1-0	1987	Chinese Taipei; Irvine, CA	21-1
1954	Schenectady, NY; Colton, CA	7-5	1969	Taiwan; Santa Clara, CA	5-0	1988	Chinese Taipei; Pearl City, HI	10-0
1955	Morrisville, PA; Merchantville, NJ	4-3	1970	Wayne, NJ; Campbell, CA	2-0	1989	Trumbull, CT; Chinese Taipei	5-2
1956	Roswell, NM; Delaware, NJ	3-1	1971	Taiwan; Gary, IN	12-3	1990	Chinese Taipei; Shippensburg, PA	9-0
1957	Mexico; La Mesa, CA	4-0	1972	Taiwan; Hammond, IN	6-0	1991	Chinese Taipei; Danville, CA	11-0
1958	Mexico; Kankakee, IL	10-1	1973	Taiwan; Tucson, AZ	12-0	1992	Long Beach, CA; Philippines	6-0
1959	Hamtramck, MI; Auburn, CA	12-0	1974	Taiwan; Red Bluff, CA	12-1	1993	Long Beach, CA; Panama	3-2
1960	Levittown, PA; Ft. Worth, TX	5-0	1975	Lakewood, NJ; Tampa, FL	4-3	1994	Venezuela; Northridge, CA	4-3
1961	El Cajon, CA; El Campo, TX	4-2	1976	Tokyo, Japan; Campbell, CA	10-3	1995	Taiwan; Spring, TX	17-3
			1977	Taiwan; El Cajon, CA	7-2	1996	Taiwan; Cranston, RI	13-3
			1978	Taiwan; Danville, CA	11-1	1997	Mexico; Mission Viejo, CA	5-4
			1979	Taiwan; Campbell, CA	2-1	1998	Toms River, NJ; Japan	12-9
			1980	Taiwan; Tampa, FL	4-3	1999	Japan; Phenix City, AL	5-0

NCAA Baseball Champions

1960	Minnesota	1970	USC	1980	Arizona	1990	Georgia
1961	USC	1971	USC	1981	Arizona St.	1991	LSU
1962	Michigan	1972	USC	1982	Miami (FL)	1992	Pepperdine
1963	USC	1973	USC	1983	Texas	1993	LSU
1964	Minnesota	1974	USC	1984	Cal. St.-Fullerton	1994	Oklahoma
1965	Arizona St.	1975	Texas	1985	Miami (FL)	1995	Cal. St.-Fullerton
1966	Ohio St.	1976	Arizona	1986	Arizona	1996	LSU
1967	Arizona St.	1977	Arizona St.	1987	Stanford	1997	LSU
1968	USC	1978	USC	1988	Stanford	1998	USC
1969	Arizona St.	1979	Cal. St.-Fullerton	1989	Wichita St.	1999	Miami (FL)

SPECIAL OLYMPICS

Special Olympics is an international program of year-round sports training and athletic competition for children and adults with mental retardation. All 50 U.S. states, Washington, DC, Guam, the Virgin Islands, and American Samoa have chapter offices. In addition, there are accredited Special Olympics programs in nearly 150 countries. Persons wishing to volunteer or find out more about Special Olympics can contact Special Olympics, Inc., 1325 G Street NW, Suite 500, Washington, DC 20005, or access the Special Olympics website at http://www.specialolympics.org

1999 Special Olympics World Summer Games/2001 Special Olympics World Winter Games

The 10th Special Olympics World Summer Games were held June 26–July 4, 1999, in Raleigh-Durham/Chapel Hill, NC. Over 7,000 athletes from more than 150 countries participated, along with 2,000 coaches, 45,000 volunteers, and 15,000 family members and friends. It was the largest multi-sport event in the world in 1999. Athletes competed in Athletics, Aquatics, Badminton, Basketball, Bocce, Bowling, Cycling, Equestrian, Golf, Gymnastics, Powerlifting, Roller Skating, Sailing, Soccer, Softball, Table Tennis, Team Handball, Tennis, and Volleyball.

The 7th Special Olympics World Winter Games are scheduled to be held Mar. 4-11, 2001, in and around Anchorage, AK. About 2,000 athletes from over 75 countries are expected to compete in Alpine Skiing, Cross-Country Skiing, Floor Hockey, Figure Skating, Speed Skating, and Snowshoeing; and for the first time, Snowboarding will be held as a demonstration.

CHESS
World Chess Champions
Source: U.S. Chess Federation
Official world champions since the title was first used are as follows:

1866-1894	Wilhelm Steinitz, Austria	**1948-1957**	Mikhail Botvinnik, USSR	**1975-1985**	Anatoly Karpov, USSR
1894-1921	Emanuel Lasker, Germany	**1957-1958**	Vassily Smyslov, USSR	**1985-1993**	Garry Kasparov, USSR/
1921-1927	Jose R. Capablanca, Cuba	**1958-1959**	Mikhail Botvinnik, USSR		Russia (c)
1927-1935	Alexander A. Alekhine,	**1960-1961**	Mikhail Tal, USSR	**1993-**	Garry Kasparov, Russia
	France	**1961-1963**	Mikhail Botvinnik, USSR		(PCA)
1935-1937	Max Euwe, Netherlands	**1963-1969**	Tigran Petrosian, USSR	**1993-1999**	Anatoly Karpov, Russia
1937-1946	Alexander A. Alekhine,	**1969-1972**	Boris Spassky, USSR		(FIDE)
	France (a)	**1972-1975**	Bobby Fischer, U.S. (b)	**1999**	Aleksandr Khalifman,
					Russia (FIDE)

(a) After Alekhine died in 1946, the title was vacant until 1948, when Botvinnik won the 1st championship match sanctioned by the International Chess Federation (FIDE). (b) Defaulted championship after refusal to accept FIDE rules for a championship match, Apr. 1975. (c) Kasparov broke with FIDE, Feb. 26, 1993. FIDE stripped Kasparov of his title Mar. 23. Kasparov defeated Nigel Short of Great Britain in a world championship match played Sept.-Oct. 1993 under the auspices of a new organization the two had founded, the Professional Chess Association (PCA). FIDE held a championship match between Anatoly Karpov (Russia) and Jan Timman (the Netherlands), which Karpov won in Nov. 1993. **Recent matches:** In Feb. 1996, Kasparov defeated Deep Blue (3 wins, 1 loss, 2 draws), a computer designed by IBM, in the 1st multigame regulation match between a world chess champion and a computer. In a May 1997 rematch, however, Kasparov was soundly defeated by the computer; he scored 1 win, 2 losses, 3 draws. Karpov successfully defended the FIDE title in June-July 1996 against 1991 U.S. chess champion Gata Kamsky of New York City, 10½ to 7½. In Aug. 1999, Aleksandr Khalifman (Russia) earned the FIDE title by defeating Vladmir Akopian (Armenia), 3½ to 2½, in Las Vegas. **Further information:** More information on chess and chess champions may be accessed on the U.S. Chess Federation's Internet site: http://www.uschess.org

U.S. Chess Champions
Source: U.S. Chess Federation

	Early champions		**Modern champions**		**Modern champions**
1857-1871	Paul Morphy	1961-1962	Larry Evans	1987	(tie) Joel Benjamin
1871-1890	George Mackenzie	1962-1968	Bobby Fischer		Nick de Firmian
1890-1891	Jackson Showalter	1968-1969	Larry Evans	1988	Michael Wilder
1891-1894	S. Lipschutz	1969-1972	Samuel Reshevsky	1989	(tie) Stuart Rachels
1894	Jackson Showalter	1972-1973	Robert Byrne		Yasser Seirawan
1894-1895	Albert Hodges	1973-1974	Lubosh Kavalek		Roman Dzindzichashvili
1895-1897	Jackson Showalter		John Grefe	1990	Lev Alburt
1897-1906	Harry Pillsbury	1974-1978	Walter Browne	1991	Gata Kamsky
1906-1909	Jackson Showalter	1978-1980	Lubosh Kavalek	1992	Patrick Wolff
1909-1936	Frank Marshall	1980-1981	(tie) Larry Evans	1993	(tie) Alexander Shabaloz
	Modern champions		Larry Christiansen		Alex Yermolinsky
1936-1944	Samuel Reshevsky		Walter Browne	1994	Boris Gulko
1944-1946	Arnold Denker	1981-1983	(tie) Walter Browne	1995	(tie) Patrick Wolff
1946	Samuel Reshevsky		Yasser Seirawan		Nick de Firmian
1946-1948	Arnold Denker	1983	(tie) Walter Browne		Alexander Ivanov
1948-1951	Herman Steiner		Larry Christiansen	1996	Alex Yermolinsky
1951-1954	Larry Evans		Roman Dzindzichashvili	1997	Joel Benjamin
1954-1957	Arthur Bisguier	1984-1985	Lev Alburt	1998	Nick de Firmian
1957-1961	Bobby Fischer	1986	Yasser Seirawan	1999	Boris Gulko

BOSTON MARATHON

Joseph Chebet of Kenya won the 103d Boston Marathon, Apr. 19, 1999, with a time of 2 hrs., 9 mins., 52 secs. He was the 9th Kenyan man in a row to win the race. Fatuma Roba of Ethiopia was the women's winner for the 3d time in 3 years, with a time of 2 hrs., 23 mins., 25 secs. She was only the 2d woman ever to win the race for 3 consecutive years.

SOCCER

World Cup

1999 Women's World Cup

The U.S. team won the women's soccer World Cup by defeating China, 5-4, on penalty kicks, July 10, 1999, in the Rose Bowl in Pasadena, CA, before 90,185 fans, the largest crowd ever at a U.S. women's sporting event. It was the 2d World Cup victory for the U.S.; the U.S. had won the inaugural event, held in China in 1991, by defeating Norway, 2-1.

The next Women's World Cup will be held in 2003; site not selected as of Oct. 1999.

Final Round Results

June 30: San Jose
China 2, Russia 0

June 30: San Jose
Norway 3, Sweden 1

July 4: Boston
China 5, Norway 0

July 1: Washington, D.C.
U.S. 3, Germany 2

July 1: Washington, D.C.
Brazil 4, Nigeria 3 (ot)

July 4: San Francisco
U.S. 2, Brazil 0

July 10: Pasadena
U.S. 0, China 0
   (U.S. 5-4 penalty kicks)

Third Place Final
July 10: Pasedena
Brazil 0, Norway 0
   (Brazil 5-4 penalty kicks)

Women's World Cup, 1991-99

Year	Winner	Final Opponent	Score	Site	Third Place
1991	U.S.	Norway	2-1	China	Germany
1995	Norway	Germany	2-0	Sweden	U.S.
1999	U.S.	China	0-0*	U.S.	Brazil

* U.S. 5-4, penalty kicks

1998 Men's World Cup

In 1998, France became the first host country since 1978 to win the men's soccer World Cup, defeating Brazil, 3-0, on July 12. It was the 2d time the event was held in France; the first time was in 1938. France and Brazil were the only 2 teams in the 1998 World Cup that did not have to qualify for the tournament, because they had automatic bids—France because it was the host country, and Brazil because it was the defending champion. The 2002 World Cup was scheduled to be held jointly in Japan and South Korea.

Final Round Results

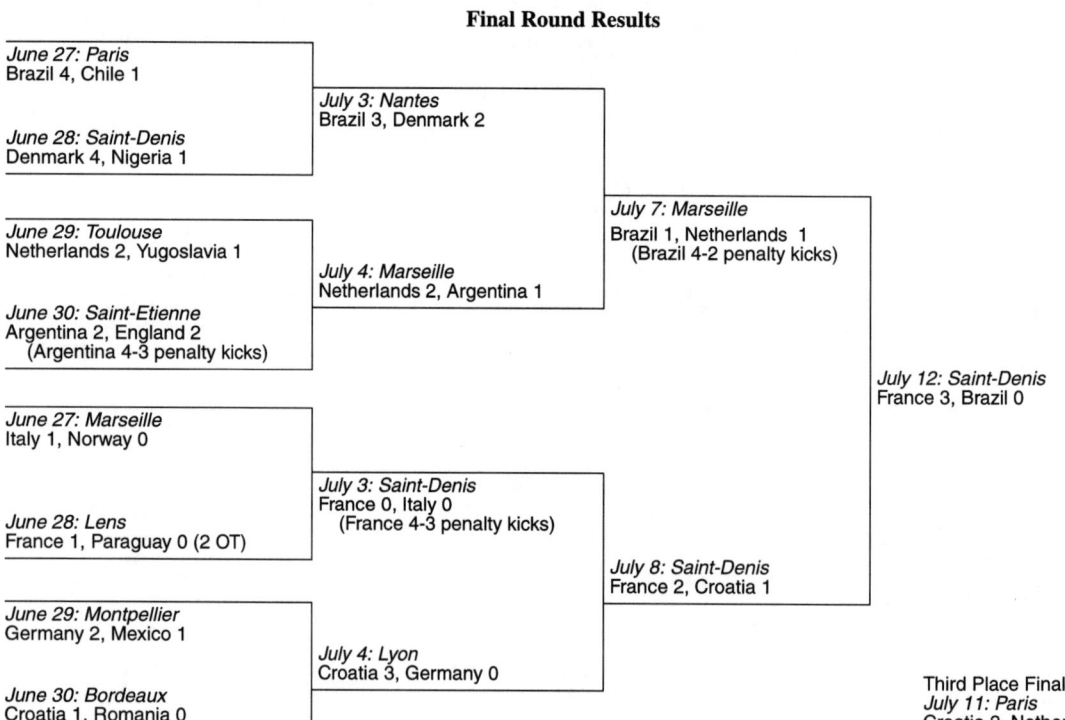

June 27: Paris
Brazil 4, Chile 1

June 28: Saint-Denis
Denmark 4, Nigeria 1

July 3: Nantes
Brazil 3, Denmark 2

June 29: Toulouse
Netherlands 2, Yugoslavia 1

June 30: Saint-Etienne
Argentina 2, England 2
   (Argentina 4-3 penalty kicks)

July 4: Marseille
Netherlands 2, Argentina 1

July 7: Marseille
Brazil 1, Netherlands 1
   (Brazil 4-2 penalty kicks)

July 12: Saint-Denis
France 3, Brazil 0

June 27: Marseille
Italy 1, Norway 0

June 28: Lens
France 1, Paraguay 0 (2 OT)

July 3: Saint-Denis
France 0, Italy 0
   (France 4-3 penalty kicks)

July 8: Saint-Denis
France 2, Croatia 1

June 29: Montpellier
Germany 2, Mexico 1

June 30: Bordeaux
Croatia 1, Romania 0

July 4: Lyon
Croatia 3, Germany 0

Third Place Final
July 11: Paris
Croatia 2, Netherlands 1

Men's World Cup, 1930-98

Year	Winner	Final opponent	Site	Year	Winner	Final opponent	Site
1930	Uruguay	Argentina	Uruguay	1970	Brazil	Italy	Mexico
1934	Italy	Czechoslovakia	Italy	1974	W. Germany	Netherlands	W. Germany
1938	Italy	Hungary	France	1978	Argentina	Netherlands	Argentina
1950	Uruguay	Brazil	Brazil	1982	Italy	W. Germany	Spain
1954	W. Germany	Hungary	Switzerland	1986	Argentina	W. Germany	Mexico
1958	Brazil	Sweden	Sweden	1990	W. Germany	Argentina	Italy
1962	Brazil	Czechoslovakia	Chile	1994	Brazil	Italy	U.S.
1966	England	W. Germany	England	1998	France	Brazil	France

Major League Soccer
1999 Final Standings

Eastern Conference

	W	So	L	GF	GA	Pts
Washington, DC United	23	6	9	65	43	57
Columbus Crew	19	6	13	48	39	45
Tampa Bay Mutiny	14	5	18	51	50	32
Miami Fusion	13	5	19	42	59	29
New England Revolution	12	5	20	38	53	26
NY/NJ MetroStars	7	3	25	32	64	15

Western Conference

	W	So	L	GF	GA	Pts
Los Angeles Galaxy	20	3	12	49	29	54
Dallas Burn	19	3	13	54	35	51
Chicago Fire	18	3	14	51	36	48
Colorado Rapids	20	6	12	38	39	48
San Jose Clash	19	10	13	48	49	37
Kansas City Wizards	8	2	24	33	53	20

Note: 3 points for a regulation-time win, 1 point for a shootout win. So = shootout win.

1999 MLS Individual Statistical Leaders

Leading Scorers (2 points for a goal, 1 point for an assist)

	Name	Team	Games	Goals	Assists	Points
1.	Jason Kreis	Dallas	32	18	15	51
2.	Roy Lassiter	Washington, DC	30	18	11	47
3.	Ronald Cerritos	San Jose	31	15	9	39
4.	Stern John	Columbus	28	18	2	38
	Joe-Max Moore	New England	29	15	8	38
6.	Ante Razov	Chicago	30	14	7	35
7.	Jaime Moreno	Washington, DC	25	10	13	33
	Raul Diaz Arce	Tampa Bay	31	13	7	33
9.	Musa Shannon	Tampa Bay	27	12	5	29
	Jeff Cunningham	Columbus	28	12	5	29

Goalkeeping Leaders (minimum 1,000 minutes)

	Name	Team	Games	Minutes	Shots[1]	Saves	GA	GAA	Wins	Losses
1.	Kevin Hartman	Los Angeles	32	2,870	150	118	29	0.91	20	12
2.	Marcus Hahnemann	Colorado	13	1,170	85	68	14	1.08	10	3
3.	Matt Jordan	Dallas	29	2,584	172	133	31	1.08	17	11
4.	Zach Thornton	Chicago	30	2,633	137	99	32	1.09	17	12
5.	Mark Dougherty	Columbus	31	2,745	152	106	35	1.15	18	12
6.	Ian Feuer	Colorado	19	1,696	99	70	23	1.22	10	9
7.	Scott Garlick	Tampa Bay	28	2,471	193	152	36	1.31	14	13
8.	Joe Cannon	San Jose	24	2,160	129	95	32	1.33	14	10
9.	Tom Presthus	Washington, DC	26	2,227	129	88	34	1.37	16	8
10.	Chris Snitko	Kansas City	16	1,395	95	66	26	1.68	4	10

Note: GA = goals against; GAA = goals against average. (1) Not shots on goal; includes shots over the goal or just past the post.

1999 MLS Playoff Results

Eastern Conference
Washington, DC defeated Miami 2 games to 0
Columbus defeated Tampa Bay 2 games to 0
Columbus vs. Washington, DC (3-game series)

Western Conference
Los Angeles defeated Colorado 2 games to 0
Dallas defeated Chicago 2 games to 1
Dallas vs. Los Angeles (3-game series)

1999 MLS Cup
(scheduled for Nov. 21, 1999, Foxboro, MA)
Eastern Conference champion vs. Western Conference champion

MLS Cup Champions, 1996-98

Year	Winner	Final opponent	Score	Site	MVP
1996	Washington, DC	Los Angeles	3–2 (OT)	Foxboro, MA	Marco Etcheverry
1997	Washington, DC	Colorado	2–1	Washington, DC	Jaime Moreno
1998	Chicago	Washington, DC	2–0	Pasadena, CA	Peter Nowak

NCAA Soccer Champions, 1982-98[1]

Year	Men	Women	Year	Men	Women
1982	Indiana	North Carolina	1990	UCLA	North Carolina
1983	Indiana	North Carolina	1991	Virginia	North Carolina
1984	Clemson	North Carolina	1992	Virginia	North Carolina
1985	UCLA	George Mason	1993	Virginia	North Carolina
1986	Duke	North Carolina	1994	Virginia	North Carolina
1987	Clemson	North Carolina	1995	Wisconsin	Notre Dame
1988	Indiana	North Carolina	1996	St. John's (NY)	North Carolina
1989	Santa Clara (tie, 2 ot)		1997	UCLA	North Carolina
	Virginia	North Carolina	1998	Indiana	Florida

(1) Men's championship dates back to 1959.

GOLF
United States Open Winners

Year[1]	Winner	Year[1]	Winner	Year[1]	Winner	Year[1]	Winner
1903	Willie Anderson	1928	John Farrell	1955	Jack Fleck	1978	Andy North
1904	Willie Anderson	1929	Bobby Jones*	1956	Cary Middlecoff	1979	Hale Irwin
1905	Willie Anderson	1930	Bobby Jones*	1957	Dick Mayer	1980	Jack Nicklaus
1906	Alex Smith	1931	Wm. Burke	1958	Tommy Bolt	1981	David Graham
1907	Alex Ross	1932	Gene Sarazen	1959	Billy Casper	1982	Tom Watson
1908	Fred McLeod	1933	John Goodman*	1960	Arnold Palmer	1983	Larry Nelson
1909	George Sargent	1934	Olin Dutra	1961	Gene Littler	1984	Fuzzy Zoeller
1910	Alex Smith	1935	Sam Parks, Jr.	1962	Jack Nicklaus	1985	Andy North
1911	John McDermott	1936	Tony Manero	1963	Julius Boros	1986	Ray Floyd
1912	John McDermott	1937	Ralph Guldahl	1964	Ken Venturi	1987	Scott Simpson
1913	Francis Ouimet*	1938	Ralph Guldahl	1965	Gary Player	1988	Curtis Strange
1914	Walter Hagen	1939	Byron Nelson	1966	Billy Casper	1989	Curtis Strange
1915	Jerome Travers*	1940	Lawson Little	1967	Jack Nicklaus	1990	Hale Irwin
1916	Chick Evans*	1941	Craig Wood	1968	Lee Trevino	1991	Payne Stewart
1919	Walter Hagen	1946	Lloyd Mangrum	1969	Orville Moody	1992	Tom Kite
1920	Edward Ray	1947	L. Worsham	1970	Tony Jacklin	1993	Lee Janzen
1921	Jim Barnes	1948	Ben Hogan	1971	Lee Trevino	1994	Ernie Els
1922	Gene Sarazen	1949	Cary Middlecoff	1972	Jack Nicklaus	1995	Corey Pavin
1923	Bobby Jones*	1950	Ben Hogan	1973	Johnny Miller	1996	Steve Jones
1924	Cyril Walker	1951	Ben Hogan	1974	Hale Irwin	1997	Ernie Els
1925	Willie MacFarlane	1952	Julius Boros	1975	Lou Graham	1998	Lee Janzen
1926	Bobby Jones*	1953	Ben Hogan	1976	Jerry Pate	1999	Payne Stewart[2]
1927	Tommy Armour	1954	Ed Furgol	1977	Hubert Green		

* Amateur. (1) 1917-18 and 1942-45 not played. (2) Stewart, 42, died Oct. 25, 1999, with 5 others, in a plane crash in South Dakota.

Professional Golfers' Association Championship Winners

Year[1]	Winner	Year[1]	Winner	Year[1]	Winner	Year[1]	Winner
1922	Gene Sarazen	1942	Sam Snead	1962	Gary Player	1981	Larry Nelson
1923	Gene Sarazen	1944	Bob Hamilton	1963	Jack Nicklaus	1982	Ray Floyd
1924	Walter Hagen	1945	Byron Nelson	1964	Bob Nichols	1983	Hal Sutton
1925	Walter Hagen	1946	Ben Hogan	1965	Dave Marr	1984	Lee Trevino
1926	Walter Hagen	1947	Jim Ferrier	1966	Al Geiberger	1985	Hubert Green
1927	Walter Hagen	1948	Ben Hogan	1967	Don January	1986	Bob Tway
1928	Leo Diegel	1949	Sam Snead	1968	Julius Boros	1987	Larry Nelson
1929	Leo Diegel	1950	Chandler Harper	1969	Ray Floyd	1988	Jeff Sluman
1930	Tommy Armour	1951	Sam Snead	1970	Dave Stockton	1989	Payne Stewart
1931	Tom Creavy	1952	James Turnesa	1971	Jack Nicklaus	1990	Wayne Grady
1932	Olin Dutra	1953	Walter Burkemo	1972	Gary Player	1991	John Daly
1933	Gene Sarazen	1954	Melvin Harbert	1973	Jack Nicklaus	1992	Nick Price
1934	Paul Runyan	1955	Doug Ford	1974	Lee Trevino	1993	Paul Azinger
1935	Johnny Revolta	1956	Jack Burke	1975	Jack Nicklaus	1994	Nick Price
1936	Denny Shute	1957	Lionel Hebert	1976	Dave Stockton	1995	Steve Elkington
1937	Denny Shute	1958	Dow Finsterwald	1977	Lanny Wadkins	1996	Mark Brooks
1938	Paul Runyan	1959	Bob Rosburg	1978	John Mahaffey	1997	Davis Love III
1939	Henry Picard	1960	Jay Hebert	1979	David Graham	1998	Vijay Singh
1940	Byron Nelson	1961	Jerry Barber	1980	Jack Nicklaus	1999	Tiger Woods
1941	Victor Ghezzi						

(1) 1943 not played.

Masters Golf Tournament Winners

Year[1]	Winner	Year[1]	Winner	Year[1]	Winner	Year[1]	Winner
1934	Horton Smith	1953	Ben Hogan	1969	George Archer	1985	Bernhard Langer
1935	Gene Sarazen	1954	Sam Snead	1970	Billy Casper	1986	Jack Nicklaus
1936	Horton Smith	1955	Cary Middlecoff	1971	Charles Coody	1987	Larry Mize
1937	Byron Nelson	1956	Jack Burke	1972	Jack Nicklaus	1988	Sandy Lyle
1938	Henry Picard	1957	Doug Ford	1973	Tommy Aaron	1989	Nick Faldo
1939	Ralph Guldahl	1958	Arnold Palmer	1974	Gary Player	1990	Nick Faldo
1940	Jimmy Demaret	1959	Art Wall Jr.	1975	Jack Nicklaus	1991	Ian Woosnam
1941	Craig Wood	1960	Arnold Palmer	1976	Ray Floyd	1992	Fred Couples
1942	Byron Nelson	1961	Gary Player	1977	Tom Watson	1993	Bernhard Langer
1946	Herman Keiser	1962	Arnold Palmer	1978	Gary Player	1994	Jose Maria Olazabal
1947	Jimmy Demaret	1963	Jack Nicklaus	1979	Fuzzy Zoeller	1995	Ben Crenshaw
1948	Claude Harmon	1964	Arnold Palmer	1980	Seve Ballesteros	1996	Nick Faldo
1949	Sam Snead	1965	Jack Nicklaus	1981	Tom Watson	1997	Tiger Woods
1950	Jimmy Demaret	1966	Jack Nicklaus	1982	Craig Stadler	1998	Mark O'Meara
1951	Ben Hogan	1967	Gay Brewer, Jr.	1983	Seve Ballesteros	1999	Jose Maria Olazabal
1952	Sam Snead	1968	Bob Goalby	1984	Ben Crenshaw		

(1) 1943-45 not played.

British Open Winners

Year[1]	Winner	Year[1]	Winner	Year[1]	Winner	Year[1]	Winner
1931	Tommy Armour	1953	Ben Hogan	1969	Tony Jacklin	1985	Sandy Lyle
1932	Gene Sarazen	1954	Peter Thomson	1970	Jack Nicklaus	1986	Greg Norman
1933	Denny Shute	1955	Peter Thomson	1971	Lee Trevino	1987	Nick Faldo
1934	Henry Cotton	1956	Peter Thomson	1972	Lee Trevino	1988	Seve Ballesteros
1935	Alf Perry	1957	Bobby Locke	1973	Tom Weiskopf	1989	Mark Calcavecchia
1936	Alf Padgham	1958	Peter Thomson	1974	Gary Player	1990	Nick Faldo
1937	T.H. Cotton	1959	Gary Player	1975	Tom Watson	1991	Ian Baker-Finch
1938	R.A. Whitcombe	1960	Kel Nagle	1976	Johnny Miller	1992	Nick Faldo
1939	Richard Burton	1961	Arnold Palmer	1977	Tom Watson	1993	Greg Norman
1946	Sam Snead	1962	Arnold Palmer	1978	Jack Nicklaus	1994	Nick Price
1947	Fred Daly	1963	Bob Charles	1979	Seve Ballesteros	1995	John Daly
1948	Henry Cotton	1964	Tony Lema	1980	Tom Watson	1996	Tom Lehman
1949	Bobby Locke	1965	Peter Thomson	1981	Bill Rogers	1997	Justin Leonard
1950	Bobby Locke	1966	Jack Nicklaus	1982	Tom Watson	1998	Mark O'Meara
1951	Max Faulkner	1967	Roberto de Vicenzo	1983	Tom Watson	1999	Paul Lawrie
1952	Bobby Locke	1968	Gary Player	1984	Seve Ballesteros		

(1) 1940-45 not played.

Professional Golf Tournaments in 1999
(official PGA or LPGA tour events only, through mid-Oct.)

Men

Date	Event	Winner	Score	Prize
Jan. 10	Mercedes Championships, Kapalua, HI	David Duval	266	$468,000
Jan. 17	Sony Open, Honolulu, HI	Jeff Sluman	271	468,000
Jan. 24	Bob Hope Chrysler Classic, Palm Springs, CA	David Duval	334	540,000
Jan. 31	Phoenix Open, Scottsdale, AZ	Rocco Mediate	273	540,000
Feb. 7	AT&T Pebble Beach National Pro-Am, Pebble Beach, CA	Payne Stewart	x206	504,000
Feb. 14	Buick Invitational, LaJolla, CA	Tiger Woods	266	486,000
Feb. 21	Nissan Open, Pacific Palisades, CA	Ernie Els	270	504,000
Feb. 28	World Golf Championship Andersen Consult. Match Play, Carlsbad, CA	Jeff Maggert	(38 holes) 1-up	1,000,000
Feb. 28	Tucson Open, Tucson, AZ	Gabriel Hjertstedt	*276	495,000
Mar. 7	Doral-Ryder Open, Miami, FL	Steve Elkington	275	540,000
Mar. 14	Honda Classic, Coral Springs, CA	Vijay Singh	277	468,000
Mar. 21	Bay Hill Invitational, Orlando, FL	Tim Herron	274	450,000
Mar. 28	THE PLAYERS Championship, Ponte Vedra Beach, FL	David Duval	285	900,000
Apr. 4	BellSouth Classic, Duluth, GA	David Duval	270	450,000
Apr. 11	The Masters Tournament, Augusta, GA	Jose Maria Olazabal	280	720,000
Apr. 18	MCI Classic, Hilton Head Island, NC	Glen Day	*274	450,000
Apr. 25	Greater Greensboro Chrysler Classic, Greensboro, NC	Jesper Parnevik	265	468,000
May 2	Shell Houston Open, The Woodlands, TX	Stuart Appleby	279	450,000
May 9	Compaq Classic, New Orleans, LA	Carlos Franco	269	468,000
May 16	GTE Byron Nelson Classic, Irving, TX	Loren Roberts	*262	540,000
May 23	MasterCard Colonial, Ft. Worth, TX	Olin Browne	272	504,000
May 30	Kemper Open, Potomac, MD	Rich Beem	274	450,000
June 6	Memorial Tournament, Dublin, OH	Tiger Woods	273	459,000
June 14y	FedEx St. Jude Classic, Memphis, TN	Ted Tryba	265	450,000
June 20	U.S. Open Championship, Pinehurst, NC	Payne Stewart	279	625,000
June 27	Buick Classic, Rye, NY	Duffy Waldorf	*276	450,000
July 4	Motorola Western Open, Lemont, IL	Tiger Woods	273	450,000
July 11	Greater Milwaukee Open, Milwaukee, WI	Carlos Franco	264	414,000
July 18	British Open, Angus, Scotland	Paul Lawrie	*290	577,500
July 25	John Deere Classic, Coal Valley, IL	J.L. Lewis	*261	360,000
Aug. 1	Canon Greater Hartford Open, Cromwell, CT	Brent Geiberger	262	450,000
Aug. 8	Buick Open, Grand Blanc, MI	Tom Pernice, Jr.	270	432,000
Aug. 15	PGA Championship, Medinah, IL	Tiger Woods	277	630,000
Aug. 22	Sprint International, Castle Rock, CO	David Toms	47 pts.	468,000
Aug. 29	World Golf Championship NEC Invitational, Akron, OH	Tiger Woods	270	1,000,000
Aug. 29	Reno-Tahoe Open, Reno, NV	Notah Begay III	274	495,000
Sept. 5	Air Canada Championship, Surrey, British Columbia	Mike Weir	266	450,000
Sept. 12	Bell Canadian Open, Oakville, Ontario	Hal Sutton	275	450,000
Sept. 19	B.C. Open, Endicott, NY	Brad Faxon	273	288,000
Sept. 26	Westin Texas Open, San Antonio, TX	Duffy Waldorf	270	360,000
Oct. 3	Buick Challenge, Pine Mountain, GA	David Toms	271	324,000
Oct. 10	Michelob Championship, Williamsburg, VA	Notah Begay III	*274	450,000
Oct. 17	Las Vegas Open, Las Vegas, NV	Jim Furyk	331	450,000

Women

Date	Event	Winner	Score	Prize
Jan. 17	HealthSouth Inaugural, Orlando, FL	Kelly Robbins	205	$82,500
Jan. 24	Naples LPGA Memorial, Naples, FL	Meg Mallon	272	112,500
Jan. 30	The Office Depot, West Palm Beach, FL	Karrie Webb	278	101,250
Feb. 14	Valley of the Stars Championship, Glendale, CA	Catrin Nilsmark	*204	97,500
Feb. 20	Sunrise Hawaiian Ladies Open, Kapolei, HI	Alison Nicholas	209	97,500
Feb. 28	Australian Ladies Masters, Ashmore, Australia	Karrie Webb	262	112,500
Mar. 14	Welch's/Circle K Championship, Tucson, AZ	Juli Inkster	273	93,750
Mar. 21	Standard Register PING, Phoenix, AZ	Karrie Webb	274	127,500
Mar. 28	Nabisco Dinah Shore Championship, Rancho Mirage, CA	Dottie Pepper	T269	150,000
Apr. 4	Longs Drugs Challenge, Lincoln, CA	Juli Inkster	280	90,000
Apr. 25	Chick-fil-A Charity Championship, Stockbridge, GA	Rachel Hetherington	*204	120,000
May 2	City of Hope Charity Championship, Murrels Inlet, SC	Rachel Hetherington	x137	101,250
May 10y	Titleholders Championship, Daytona Beach, FL	Karrie Webb	T271	135,000
May 16	Sara Lee Classic, Old Hickory, TN	Meg Mallon	199	112,500
May 23	Philips Invitational Honoring Harvey Penick, Austin, TX	Akiko Fukushima	267	120,000
May 30	LPGA Corning Classic, Corning, NY	Kelli Kuehne	278	112,500
June 6	U.S. Women's Open, West Point, MS	Juli Inkster	T272	315,000
June 13	LPGA Rochester International, Rochester, NY	Karrie Webb	280	150,000
June 20	ShopRite LPGA Classic, Absecon, NJ	Se Ri Pak	198	150,000
June 27	McDonald's LPGA Championship, Wilmington, DE	Juli Inkster	268	210,000
July 4	Jamie Farr Kroger Classic, Sylvania, OH	Se Ri Pak	*276	135,000
July 11	Michelob Light Classic, St. Louis, MO	Annika Sorenstam	*278	120,000
July 18	Japan Airlines Big Apple Classic, New Rochelle, NY	Sherri Steinhauer	*273	127,500
July 25	Giant Eagle LPGA Classic, Warren, OH	Jackie Gallagher-Smith	199	150,000
Aug. 1	du Maurier Classic, Calgary, Alberta, CAN	Karrie Webb	277	180,000
Aug. 8	areaWEB.com Challenge, Sutton, MA	Mardi Lunn	275	120,000
Aug. 15	Weetabix Women's British Open, Milton Keynes, England	Sherri Steinhauer	283	160,000
Aug. 22	Firstar LPGA Classic, Beavercreek, OH	Rosie Jones	*207	97,000
Aug. 29	Oldsmobile Classic, East Lansing, MI	Dottie Pepper	270	105,000
Sept. 6	State Farm Rail Classic, Springfield, IL	Mi Hyun Kim	204	116,250
Sept. 12	Samsung World Championship of Women's Golf, Maple Grove, MN	Se Ri Pak	280	150,000
Sept. 19	SAFECO Classic, Kent, WA	Maria Hjorth	271	97,500
Sept. 26	Safeway LPGA Golf Championship, Portland, OR	Juli Inkster	207	120,000
Oct. 3	New Albany LPGA Classic, New Albany, OH	Annika Sorenstam	269	150,000
Oct. 10	First Union Betsy King Classic, Kutztown, PA	Mi Hyun Kim	280	116,250
Oct. 17	Lifetime's AFLAC Tournament of Champions, Semmes, AL	Akiko Fukushima	279	122,000

* Won playoff. (x) Shortened because of weather. (y) Concluded on Mon. because of inclement weather. (T) Tournament record.

Ryder Cup in 1999
On Sept. 26, Team USA completed the greatest comeback in the history of the Ryder Cup to defeat Team Europe, 14½-13½, in Brookline, MA.

U.S. Women's Open Golf Champions

Year	Winner	Year	Winner	Year	Winner	Year	Winner
1948	"Babe" Zaharias	1961	Mickey Wright	1974	Sandra Haynie	1987	Laura Davies
1949	Louise Suggs	1962	Murle Lindstrom	1975	Sandra Palmer	1988	Liselotte Neumann
1950	"Babe" Zaharias	1963	Mary Mills	1976	JoAnne Carner	1989	Betsy King
1951	Betsy Rawls	1964	Mickey Wright	1977	Hollis Stacy	1990	Betsy King
1952	Louise Suggs	1965	Carol Mann	1978	Hollis Stacy	1991	Meg Mallon
1953	Betsy Rawls	1966	Sandra Spuzich	1979	Jerilyn Britz	1992	Patty Sheehan
1954	"Babe" Zaharias	1967	Catherine Lacoste*	1980	Amy Alcott	1993	Lauri Merten
1955	Fay Crocker	1968	Susie Maxwell Berning	1981	Pat Bradley	1994	Patty Sheehan
1956	Mrs. K. Cornelius	1969	Donna Caponi	1982	Janet Alex	1995	Annika Sorenstam
1957	Betsy Rawls	1970	Donna Caponi	1983	Jan Stephenson	1996	Annika Sorenstam
1958	Mickey Wright	1971	JoAnne Carner	1984	Hollis Stacy	1997	Alison Nicholas
1959	Mickey Wright	1972	Susie Maxwell Berning	1985	Kathy Baker	1998	Se Ri Pak
1960	Betsy Rawls	1973	Susie Maxwell Berning	1986	Jane Geddes	1999	Juli Inkster

*Amateur

PGA Leading Money Winners

Year	Player	Dollars	Year	Player	Dollars	Year	Player	Dollars
1946	Ben Hogan	$42,556	1964	Jack Nicklaus	$113,284	1982	Craig Stadler	$446,462
1947	Jimmy Demaret	27,936	1965	Jack Nicklaus	140,752	1983	Hal Sutton	426,668
1948	Ben Hogan	36,812	1966	Billy Casper	121,944	1984	Tom Watson	476,260
1949	Sam Snead	31,593	1967	Jack Nicklaus	188,988	1985	Curtis Strange	542,321
1950	Sam Snead	35,758	1968	Billy Casper	205,168	1986	Greg Norman	653,296
1951	Lloyd Mangrum	26,088	1969	Frank Beard	175,223	1987	Curtis Strange	925,941
1952	Julius Boros	37,032	1970	Lee Trevino	157,037	1988	Curtis Strange	1,147,644
1953	Lew Worsham	34,002	1971	Jack Nicklaus	244,490	1989	Tom Kite	1,395,278
1954	Bob Toski	65,819	1972	Jack Nicklaus	320,542	1990	Greg Norman	1,165,477
1955	Julius Boros	65,121	1973	Jack Nicklaus	308,362	1991	Corey Pavin	979,430
1956	Ted Kroll	72,835	1974	Johnny Miller	353,201	1992	Fred Couples	1,344,188
1957	Dick Mayer	65,835	1975	Jack Nicklaus	323,149	1993	Nick Price	1,478,557
1958	Arnold Palmer	42,407	1976	Jack Nicklaus	266,438	1994	Nick Price	1,499,927
1959	Art Wall, Jr.	53,167	1977	Tom Watson	310,653	1995	Greg Norman	1,654,959
1960	Arnold Palmer	75,262	1978	Tom Watson	362,429	1996	Tom Lehman	1,780,159
1961	Gary Player	64,540	1979	Tom Watson	462,636	1997	Tiger Woods	2,066,833
1962	Arnold Palmer	81,448	1980	Tom Watson	530,808	1998	David Duvall	2,591,031
1963	Arnold Palmer	128,230	1981	Tom Kite	375,699			

LPGA Leading Money Winners

Year	Player	Dollars	Year	Player	Dollars	Year	Player	Dollars
1954	Patty Berg	$16,011	1969	Carol Mann	$49,152	1984	Betsy King	$266,771
1955	Patty Berg	16,492	1970	Kathy Whitworth	30,235	1985	Nancy Lopez	416,472
1956	Marlene Hagge	20,235	1971	Kathy Whitworth	41,181	1986	Pat Bradley	492,021
1957	Patty Berg	16,272	1972	Kathy Whitworth	65,063	1987	Ayako Okamoto	466,034
1958	Beverly Hanson	12,629	1973	Kathy Whitworth	82,854	1988	Sherri Turner	347,255
1959	Betsy Rawls	26,774	1974	JoAnne Carner	87,094	1989	Betsy King	654,132
1960	Louise Suggs	16,892	1975	Sandra Palmer	94,805	1990	Beth Daniel	863,578
1961	Mickey Wright	22,236	1976	Judy Rankin	150,734	1991	Pat Bradley	763,118
1962	Mickey Wright	21,641	1977	Judy Rankin	122,890	1992	Dottie Mochrie	693,335
1963	Mickey Wright	31,269	1978	Nancy Lopez	189,813	1993	Betsy King	595,992
1964	Mickey Wright	29,800	1979	Nancy Lopez	215,987	1994	Laura Davies	687,201
1965	Kathy Whitworth	28,658	1980	Beth Daniel	231,000	1995	Annika Sorenstam	666,533
1966	Kathy Whitworth	33,517	1981	Beth Daniel	206,977	1996	Karrie Webb	1,002,000
1967	Kathy Whitworth	32,937	1982	JoAnne Carner	310,399	1997	Annika Sorenstam	1,236,789
1968	Kathy Whitworth	48,379	1983	JoAnne Carner	291,404	1998	Annika Sorenstam	1,092,748

RIFLE AND PISTOL INDIVIDUAL CHAMPIONSHIPS

Source: National Rifle Association

National Outdoor Rifle and Pistol Championships in 1999

Pistol—MSG Steve F. Reiter, USA, Sparks, NV, 2638-133X

Civilian Pistol—Darius, R. Young, Malo, WA, 2627-112X

Woman Pistol—SFC Ruby E. Fox, USA, Parker, AZ, 2595-90X

Smallbore Rifle Prone—Lones W. Wigger, Jr., Colorado Springs, CO, 6398-533X

Civilian Smallbore Rifle Prone—Lones W. Wigger, Jr., Colorado Springs, CO, 6398-533X

Woman Smallbore Rifle Prone—Carolyn D. Millard-Sparks, Dunwoody, GA, 6396-531X

Smallbore Rifle NRA 3-Position—SSG Troy A. Bassham, USA, Phenix City, AL, 2301-83X

Civilian Smallbore Rifle NRA 3-Position—Jean A. Foster, Colorado Springs, CO, 2286-90X

Woman Smallbore Rifle NRA 3-Position—Jean A. Foster, Colorado Springs, CO, 2286-90X

High Power Rifle—David Tubb, Canadian, TX, 2377-115X

Civilian High Power Rifle—David Tubb, Canadian, TX, 2377-115X

Woman High Power Rifle—SSGT Julia L. Watson, USMC, Quantico, VA, 2361-110X

High Power Rifle Long Range—Nancy H. Tompkins-Gallagher, Prescott, AZ, 1439-65X

Woman High Power Rifle Long Range—Nancy H. Tompkins-Gallagher, Prescott, AZ, 1439-65X

National Indoor Rifle and Pistol Championships in 1999

Smallbore Rifle 4-Position—Steve Goff, Columbus, GA, 799

Woman Smallbore Rifle 4-Position—Michelle E. Bohren, Taylor, MI, 795

Smallbore Rifle NRA 3-Position—Steve Goff, Columbus, GA, 1191

Woman Smallbore Rifle NRA 3-Position—Nicole Allaire, Kearny, NJ, 1176

International Smallbore Rifle—Robert J. Foth, Colorado Springs, CO, 1185

Woman International Smallbore Rifle—Tammie Forster, Colorado Springs, CO, 1170

Air Rifle—Elizabeth Bourland, Wichita Falls, TX, 594

Woman Air Rifle—Elizabeth Bourland, Wichita Falls, TX, 594

Conventional Pistol—Darius R. Young, Malo, WA, 887

Woman Conventional Pistol—Judy LaVoie, Trumbull, CT, 864

International Free Pistol—Richard L. Eddler, Lower Burrell, PA, 564

Woman International Free Pistol—Elizabeth Lee, Albany, CA, 501

International Standard Pistol—Jimmie R. Dorsey, Harrison, ID, 569

Woman International Standard Pistol—Laura Tyler, Craig, CO, 536

Air Pistol—Richard L. Edder, Lower Burrell, PA, 577

Woman Air Pistol—Susan E. McConnell, Clifton Park, NY, 558

NRA Bianchi Cup National Action Pistol Championships in 1999

Action Pistol—Bruce Piatt, Montvale, NJ, 1920-185X

Woman Action Pistol—Vera Koo, Atherton, CA, 1894-137X

Junior Action Pistol—Mitchell Conrad, Tulsa, OK, 1892-144X

TENNIS
U.S. Open Champions, 1925-99
Men's Singles

Year	Champion	Final opponent	Year	Champion	Final opponent
1925	Bill Tilden	William Johnston	1963	Rafael Osuna	F. A. Froehling 3d
1926	Rene Lacoste	Jean Borotra	1964	Roy Emerson	Fred Stolle
1927	Rene Lacoste	Bill Tilden	1965	Manuel Santana	Cliff Drysdale
1928	Henri Cochet	Francis Hunter	1966	Fred Stolle	John Newcombe
1929	Bill Tilden	Francis Hunter	1967	John Newcombe	Clark Graebner
1930	John Doeg	Francis Shields	1968	Arthur Ashe	Tom Okker
1931	H. Ellsworth Vines	George Lott	1969	Rod Laver	Tony Roche
1932	H. Ellsworth Vines	Henri Cochet	1970	Ken Rosewall	Tony Roche
1933	Fred Perry	John Crawford	1971	Stan Smith	Jan Kodes
1934	Fred Perry	Wilmer Allison	1972	Ilie Nastase	Arthur Ashe
1935	Wilmer Allison	Sidney Wood	1973	John Newcombe	Jan Kodes
1936	Fred Perry	Don Budge	1974	Jimmy Connors	Ken Rosewall
1937	Don Budge	Baron G. von Cramm	1975	Manuel Orantes	Jimmy Connors
1938	Don Budge	C. Gene Mako	1976	Jimmy Connors	Bjorn Borg
1939	Robert Riggs	S. Welby Van Horn	1977	Guillermo Vilas	Jimmy Connors
1940	Don McNeill	Robert Riggs	1978	Jimmy Connors	Bjorn Borg
1941	Robert Riggs	F. L. Kovacs	1979	John McEnroe	Vitas Gerulaitis
1942	F. R. Schroeder Jr.	Frank Parker	1980	John McEnroe	Bjorn Borg
1943	Joseph Hunt	Jack Kramer	1981	John McEnroe	Bjorn Borg
1944	Frank Parker	William Talbert	1982	Jimmy Connors	Ivan Lendl
1945	Frank Parker	William Talbert	1983	Jimmy Connors	Ivan Lendl
1946	Jack Kramer	Thomas Brown Jr.	1984	John McEnroe	Ivan Lendl
1947	Jack Kramer	Frank Parker	1985	Ivan Lendl	John McEnroe
1948	Pancho Gonzales	Eric Sturgess	1986	Ivan Lendl	Miloslav Mecir
1949	Pancho Gonzales	F. R. Schroeder Jr.	1987	Ivan Lendl	Mats Wilander
1950	Arthur Larsen	Herbert Flam	1988	Mats Wilander	Ivan Lendl
1951	Frank Sedgman	E. Victor Seixas Jr.	1989	Boris Becker	Ivan Lendl
1952	Frank Sedgman	Gardnar Mulloy	1990	Pete Sampras	Andre Agassi
1953	Tony Trabert	E. Victor Seixas Jr.	1991	Stefan Edberg	Jim Courier
1954	E. Victor Seixas Jr.	Rex Hartwig	1992	Stefan Edberg	Pete Sampras
1955	Tony Trabert	Ken Rosewall	1993	Pete Sampras	Cedric Pioline
1956	Ken Rosewall	Lewis Hoad	1994	Andre Agassi	Michael Stich
1957	Malcolm Anderson	Ashley Cooper	1995	Pete Sampras	Andre Agassi
1958	Ashley Cooper	Malcolm Anderson	1996	Pete Sampras	Michael Chang
1959	Neale A. Fraser	Alejandro Olmedo	1997	Patrick Rafter	Greg Rusedski
1960	Neale A. Fraser	Rod Laver	1998	Patrick Rafter	Mark Philippoussis
1961	Roy Emerson	Rod Laver	1999	Andre Agassi	Todd Martin
1962	Rod Laver	Roy Emerson			

Women's Singles

Year	Champion	Final opponent	Year	Champion	Final opponent
1925	Helen Willis	Kathleen McKane	1963	Maria Bueno	Margaret Smith
1926	Molla B. Mallory	Elizabeth Ryan	1964	Maria Bueno	Carole Graebner
1927	Helen Wills	Betty Nuthall	1965	Margaret Smith	Billie Jean Moffitt
1928	Helen Wills	Helen Jacobs	1966	Maria Bueno	Nancy Richey
1929	Helen Wills	M. Watson	1967	Billie Jean King	Ann Haydon Jones
1930	Betty Nuthall	L. A. Harper	1968	Virginia Wade	Billie Jean King
1931	Helen Wills Moody	E. B. Whittingstall	1969	Margaret Smith Court	Nancy Richey
1932	Helen Jacobs	Carolin A. Babcock	1970	Margaret Smith Court	Rosemary Casals
1933	Helen Jacobs	Helen Wills Moody	1971	Billie Jean King	Rosemary Casals
1934	Helen Jacobs	Sarah H. Palfrey	1972	Billie Jean King	Kerry Melville
1935	Helen Jacobs	Sarah Palfrey Fabyan	1973	Margaret Smith Court	Evonne Goolagong
1936	Alice Marble	Helen Jacobs	1974	Billie Jean King	Evonne Goolagong
1937	Anita Lizana	Jadwiga Jedrzejowska	1975	Chris Evert	Evonne Goolagong
1938	Alice Marble	Nancye Wynne	1976	Chris Evert	Evonne Goolagong
1939	Alice Marble	Helen Jacobs	1977	Chris Evert	Wendy Turnbull
1940	Alice Marble	Helen Jacobs	1978	Chris Evert	Pam Shriver
1941	Sarah Palfrey Cooke	Pauline Betz	1979	Tracy Austin	Chris Evert Lloyd
1942	Pauline Betz	Louise Brough	1980	Chris Evert Lloyd	Hana Mandlikova
1943	Pauline Betz	Louise Brough	1981	Tracy Austin	Martina Navratilova
1944	Pauline Betz	Margaret Osborne	1982	Chris Evert Lloyd	Hana Mandlikova
1945	Sarah Palfrey Cooke	Pauline Betz	1983	Martina Navratilova	Chris Evert Lloyd
1946	Pauline Betz	Doris Hart	1984	Martina Navratilova	Chris Evert Lloyd
1947	Louise Brough	Margaret Osborne	1985	Hana Mandlikova	Martina Navratilova
1948	Margaret Osborne duPont	Louise Brough	1986	Martina Navratilova	Helena Sukova
1949	Margaret Osborne duPont	Doris Hart	1987	Martina Navratilova	Steffi Graf
1950	Margaret Osborne duPont	Doris Hart	1988	Steffi Graf	Gabriela Sabatini
1951	Maureen Connolly	Shirley Fry	1989	Steffi Graf	Martina Navratilova
1952	Maureen Connolly	Doris Hart	1990	Gabriela Sabatini	Steffi Graf
1953	Maureen Connolly	Doris Hart	1991	Monica Seles	Martina Navratilova
1954	Doris Hart	Louise Brough	1992	Monica Seles	Arantxa Sanchez Vicario
1955	Doris Hart	Patricia Ward	1993	Steffi Graf	Helena Sukova
1956	Shirley Fry	Althea Gibson	1994	Arantxa Sanchez Vicario	Steffi Graf
1957	Althea Gibson	Louise Brough	1995	Steffi Graf	Monica Seles
1958	Althea Gibson	Darlene Hard	1996	Steffi Graf	Monica Seles
1959	Maria Bueno	Christine Truman	1997	Martina Hingis	Venus Williams
1960	Darlene Hard	Maria Bueno	1998	Lindsay Davenport	Martina Hingis
1961	Darlene Hard	Ann Haydon	1999	Serena Williams	Martina Hingis
1962	Margaret Smith	Darlene Hard			

All-England Champions, Wimbledon, 1925-99

Men's Singles

Year	Champion	Final opponent	Year	Champion	Final opponent
1925	Rene Lacoste	Jean Borotra	1965	Roy Emerson	Fred Stolle
1926	Jean Borotra	Howard Kinsey	1966	Manuel Santana	Dennis Ralston
1927	Henri Cochet	Jean Borotra	1967	John Newcombe	Wilhelm Bungert
1928	Rene Lacoste	Henri Cochet	1968	Rod Laver	Tony Roche
1929	Henri Cochet	Jean Borotra	1969	Rod Laver	John Newcombe
1930	Bill Tilden	Wilmer Allison	1970	John Newcombe	Ken Rosewall
1931	Sidney B. Wood	Francis X. Shields	1971	John Newcombe	Stan Smith
1932	Ellsworth Vines	Henry Austin	1972	Stan Smith	Ilie Nastase
1933	Jack Crawford	Ellsworth Vines	1973	Jan Kodes	Alex Metreveli
1934	Fred Perry	Jack Crawford	1974	Jimmy Connors	Ken Rosewall
1935	Fred Perry	Gottfried von Cramm	1975	Arthur Ashe	Jimmy Connors
1936	Fred Perry	Gottfried von Cramm	1976	Bjorn Borg	Ilie Nastase
1937	Donald Budge	Gottfried von Cramm	1977	Bjorn Borg	Jimmy Connors
1938	Donald Budge	Henry Austin	1978	Bjorn Borg	Jimmy Connors
1939	Bobby Riggs	Elwood Cooke	1979	Bjorn Borg	Roscoe Tanner
1940-45	Not held	Not held	1980	Bjorn Borg	John McEnroe
1946	Yvon Petra	Geoff E. Brown	1981	John McEnroe	Bjorn Borg
1947	Jack Kramer	Tom P. Brown	1982	Jimmy Connors	John McEnroe
1948	Bob Falkenburg	John Bromwich	1983	John McEnroe	Chris Lewis
1949	Ted Schroeder	Jaroslav Drobny	1984	John McEnroe	Jimmy Connors
1950	Budge Patty	Frank Sedgman	1985	Boris Becker	Kevin Curren
1951	Dick Savitt	Ken McGregor	1986	Boris Becker	Ivan Lendl
1952	Frank Sedgman	Jaroslav Drobny	1987	Pat Cash	Ivan Lendl
1953	Vic Seixas	Kurt Nielsen	1988	Stefan Edberg	Boris Becker
1954	Jaroslav Drobny	Ken Rosewall	1989	Boris Becker	Stefan Edberg
1955	Tony Trabert	Kurt Nielsen	1990	Stefan Edberg	Boris Becker
1956	Lew Hoad	Ken Rosewall	1991	Michael Stich	Boris Becker
1957	Lew Hoad	Ashley Cooper	1992	Andre Agassi	Goran Ivanisevic
1958	Ashley Cooper	Neale Fraser	1993	Pete Sampras	Jim Courier
1959	Alex Olmedo	Rod Laver	1994	Pete Sampras	Goran Ivanisevic
1960	Neale Fraser	Rod Laver	1995	Pete Sampras	Boris Becker
1961	Rod Laver	Chuck McKinley	1996	Richard Krajicek	MaliVai Washington
1962	Rod Laver	Martin Mulligan	1997	Pete Sampras	Cedric Pioline
1963	Chuck McKinley	Fred Stolle	1998	Pete Sampras	Goran Ivanisevic
1964	Roy Emerson	Fred Stolle	1999	Pete Sampras	Andre Agassi

Women's Singles

Year	Champion	Final Opponent	Year	Champion	Final Opponent
1925	Suzanne Lenglen	Joan Fry	1965	Margaret Smith	Maria Bueno
1926	Kathleen McKane Godfree	Lili de Alvarez	1966	Billie Jean King	Maria Bueno
1927	Helen Wills	Lili de Alvarez	1967	Billie Jean King	Ann Haydon Jones
1928	Helen Wills	Lili de Alvarez	1968	Billie Jean King	Judy Tegart
1929	Helen Wills	Helen Jacobs	1969	Ann Haydon-Jones	Billie Jean King
1930	Helen Wills Moody	Elizabeth Ryan	1970	Margaret Smith Court	Billie Jean King
1931	Cilly Aussem	Hilde Kranwinkel	1971	Evonne Goolagong	Margaret Smith Court
1932	Helen Wills Moody	Helen Jacobs	1972	Billie Jean King	Evonne Goolagong
1933	Helen Wills Moody	Dorothy Round	1973	Billie Jean King	Chris Evert
1934	Dorothy Round	Helen Jacobs	1974	Chris Evert	Olga Morozova
1935	Helen Wills Moody	Helen Jacobs	1975	Billie Jean King	Evonne Goolagong Cawley
1936	Helen Jacobs	Hilde Kranwinkel Sperling	1976	Chris Evert	Evonne Goolagong Cawley
1937	Dorothy Round	Jadwiga Jedrzejowska	1977	Virginia Wade	Betty Stove
1938	Helen Wills Moody	Helen Jacobs	1978	Martina Navratilova	Chris Evert
1939	Alice Marble	Kay Stammers	1979	Martina Navratilova	Chris Evert Lloyd
1940-45	Not held	Not held	1980	Evonne Goolagong	Chris Evert Lloyd
1946	Pauline Betz	Louise Brough	1981	Chris Evert Lloyd	Hana Mandlikova
1947	Margaret Osborne	Doris Hart	1982	Martina Navratilova	Chris Evert Lloyd
1948	Louise Brough	Doris Hart	1983	Martina Navratilova	Andrea Jaeger
1949	Louise Brough	Margaret Osborne duPont	1984	Martina Navratilova	Chris Evert Lloyd
1950	Louise Brough	Margaret Osborne duPont	1985	Martina Navratilova	Chris Evert Lloyd
1951	Doris Hart	Shirley Fry	1986	Martina Navratilova	Hana Mandlikova
1952	Maureen Connolly	Louise Brough	1987	Martina Navratilova	Steffi Graf
1953	Maureen Connolly	Doris Hart	1988	Steffi Graf	Martina Navratilova
1954	Maureen Connolly	Louise Brough	1989	Steffi Graf	Martina Navratilova
1955	Louise Brough	Beverly Fleitz	1990	Martina Navratilova	Zina Garrison
1956	Shirley Fry	Angela Buxton	1991	Steffi Graf	Gabriela Sabatini
1957	Althea Gibson	Darlene Hard	1992	Steffi Graf	Monica Seles
1958	Althea Gibson	Angela Mortimer	1993	Steffi Graf	Jana Novotna
1959	Maria Bueno	Darlene Hard	1994	Conchita Martinez	Martina Navratilova
1960	Maria Bueno	Sandra Reynolds	1995	Steffi Graf	Arantxa Sánchez Vicario
1961	Angela Mortimer	Christine Truman	1996	Steffi Graf	Arantxa Sánchez Vicario
1962	Karen Hantze-Susman	Vera Sukova	1997	Martina Hingis	Jana Novotna
1963	Margaret Smith	Billie Jean Moffitt	1998	Jana Novotna	Nathalie Tauziat
1964	Maria Bueno	Margaret Smith	1999	Lindsay Davenport	Steffi Graf

Davis Cup Challenge Round, 1900-98

Year	Result	Year	Result	Year	Result
1900	United States 3, British Isles 0	1934	Great Britain 4, United States 1	1970	United States 5, W. Germany 0
1901	Not held	1935	Great Britain 5, United States 0	1971	United States 3, Romania 2
1902	United States 3, British Isles 2	1936	Great Britain 3, Australia 2	1972	United States 3, Romania 2
1903	British Isles 4, United States 1	1937	United States 4, Great Britain 1	1973	Australia 5, United States 0
1904	British Isles 5, Belgium 0	1938	United States 3, Australia 2	1974	South Africa (default by India)
1905	British Isles 5, United States 0	1939	Australia 3, United States 2	1975	Sweden 3, Czechoslovakia 2
1906	British Isles 5, United States 0	1940-45	Not held	1976	Italy 4, Chile 1
1907	Australia 3, British Isles 2	1946	United States 5, Australia 0	1977	Australia 3, Italy 1
1908	Australasia 3, United States 2	1947	United States 4, Australia 1	1978	United States 4, Great Britain 1
1909	Australasia 5, United States 0	1948	United States 5, Australia 0	1979	United States 5, Italy 0
1910	Not held	1949	United States 4, Australia 1	1980	Czechoslovakia 4, Italy 1
1911	Australasia 5, United States 0	1950	Australia 4, United States 1	1981	United States 3, Argentina 1
1912	British Isles 3, Australasia 2	1951	Australia 3, United States 2	1982	United States 4, France, 1
1913	United States 3, British Isles 2	1952	Australia 4, United States 1	1983	Australia 3, Sweden 2
1914	Australasia 3, United States 2	1953	Australia 3, United States 2	1984	Sweden 4, United States 1
1915-18	Not held	1954	United States 3, Australia 2	1985	Sweden 3, W. Germany 2
1919	Australasia 4, British Isles 1	1955	Australia 5, United States 0	1986	Australia 3, Sweden 2
1920	United States 5, Australasia 0	1956	Australia 5, United States 0	1987	Sweden 5, India 0
1921	United States 5, Japan 0	1957	Australia 3, United States 2	1988	W. Germany 4, Sweden 1
1922	United States 4, Australasia 1	1958	United States 3, Australia 2	1989	W. Germany 3, Sweden 2
1923	United States 4, Australasia 1	1959	Australia 3, United States 2	1990	United States 3, Australia 2
1924	United States 5, Australasia 0	1960	Australia 4, Italy 1	1991	France 3, United States 1
1925	United States 5, France 0	1961	Australia 5, Italy 0	1992	United States 3, Switzerland 1
1926	United States 4, France 1	1962	Australia 5, Mexico 0	1993	Germany 4, Australia 1
1927	France 3, United States 2	1963	United States 3, Australia 2	1994	Sweden 4, Russia 1
1928	France 4, United States 1	1964	Australia 3, United States 2	1995	United States 3, Russia 2
1929	France 3, United States 2	1965	Australia 4, Spain 1	1996	France 3, Sweden 2
1930	France 4, United States 1	1966	Australia 4, India 1	1997	Sweden 5, United States 0
1931	France 3, Great Britain 2	1967	Australia 4, Spain 1	1998	Sweden 4, Italy 1
1932	France 3, United States 2	1968	United States 4, Australia		
1933	Great Britain 3, France 2	1969	United States 5, Romania 0		

French Open Singles Champions, 1968-99

Men's Singles

Year	Champion	Final Opponent	Year	Champion	Final Opponent
1968	Ken Rosewall	Rod Laver	1984	Ivan Lendl	John McEnroe
1969	Rod Laver	Ken Rosewall	1985	Mats Wilander	Ivan Lendl
1970	Jan Kodes	Zeljko Franulovic	1986	Ivan Lendl	Mikael Pernfors
1971	Jan Kodes	Ilie Nastase	1987	Ivan Lendl	Mats Wilander
1972	Andres Gimeno	Patrick Proisy	1988	Mats Wilander	Henri Leconte
1973	Ilie Nastase	Nikki Pilic	1989	Michael Chang	Stefan Edberg
1974	Bjorn Borg	Manuel Orantes	1990	Andres Gomez	Andre Agassi
1975	Bjorn Borg	Guillermo Vilas	1991	Jim Courier	Andre Agassi
1976	Adriano Panatta	Harold Solomon	1992	Jim Courier	Petr Korda
1977	Guillermo Vilas	Brian Gottfried	1993	Sergi Bruguera	Jim Courier
1978	Bjorn Borg	Guillermo Vilas	1994	Sergi Bruguera	Alberto Berasategui
1979	Bjorn Borg	Victor Pecci	1995	Thomas Muster	Michael Chang
1980	Bjorn Borg	Vitas Gerulaitis	1996	Yevgeny Kafelnikov	Michael Stich
1981	Bjorn Borg	Ivan Lendl	1997	Gustavo Kuerten	Sergei Bruguera
1982	Mats Wilander	Guillermo Vilas	1998	Carlos Moya	Alex Corretja
1983	Yannick Noah	Mats Wilander	1999	Andre Agassi	Andrei Medvedev

Women's Singles

Year	Champion	Final Opponent	Year	Champion	Final Opponent
1968	Nancy Richey	Ann Jones	1984	Martina Navratilova	Chris Evert Lloyd
1969	Margaret Smith Court	Ann Jones	1985	Chris Evert Lloyd	Martina Navratilova
1970	Margaret Smith Court	Helga Niessen	1986	Chris Evert Lloyd	Martina Navratilova
1971	Evonne Goolagong	Helen Gourlay	1987	Steffi Graf	Martina Navratilova
1972	Billie Jean King	Evonne Goolagong	1988	Steffi Graf	Natalia Zvereva
1973	Margaret Smith Court	Chris Evert	1989	Arantxa Sánchez Vicario	Steffi Graf
1974	Chris Evert	Olga Morozova	1990	Monica Seles	Steffi Graf
1975	Chris Evert	Martina Navratilova	1991	Monica Seles	Arantxa Sánchez Vicario
1976	Sue Barker	Renata Tomanova	1992	Monica Seles	Steffi Graf
1977	Mima Jausovec	Florenza Mihai	1993	Steffi Graf	Mary Joe Fernandez
1978	Virginia Ruzici	Mima Jausovec	1994	Arantxa Sánchez Vicario	Mary Pierce
1979	Chris Evert Lloyd	Wendy Turnbull	1995	Steffi Graf	Arantxa Sánchez Vicario
1980	Chris Evert Lloyd	Virginia Ruzici	1996	Steffi Graf	Arantxa Sánchez Vicario
1981	Hana Mandlikova	Sylvia Hanika	1997	Iva Majoli	Martina Hingis
1982	Martina Navratilova	Andrea Jaeger	1998	Arantxa Sánchez Vicario	Monica Seles
1983	Chris Evert Lloyd	Mima Jausovec	1999	Steffi Graf	Martina Hingis

Australian Open Singles Champions, 1969-99

Men's Singles

Year	Champion	Final Opponent	Year	Champion	Final Opponent
1969	Rod Laver	Andres Gimeno	1974	Jimmy Connors	Phil Dent
1970	Arthur Ashe	Dick Crealy	1975	John Newcombe	Jimmy Connors
1971	Ken Rosewall	Arthur Ashe	1976	Mark Edmondson	John Newcombe
1972	Ken Rosewall	Mal Anderson	1977*	Roscoe Tanner	Guillermo Vilas
1973	John Newcombe	Onny Parun		Vitas Gerulaitis	John Lloyd

Year	Champion	Final Opponent	Year	Champion	Final Opponent
1978	Guillermo Vilas	John Marks	1989	Ivan Lendl	Miloslav Mecir
1979	Guillermo Vilas	John Sadri	1990	Ivan Lendl	Stefan Edberg
1980	Brian Teacher	Kim Warwick	1991	Boris Becker	Ivan Lendl
1981	Johan Kriek	Steve Denton	1992	Jim Courier	Stefan Edberg
1982	Johan Kriek	Steve Denton	1993	Jim Courier	Stefan Edberg
1983	Mats Wilander	Ivan Lendl	1994	Pete Sampras	Todd Martin
1984	Mats Wilander	Kevin Curren	1995	Andre Agassi	Pete Sampras
1985	Stefan Edberg	Mats Wilander	1996	Boris Becker	Michael Chang
1986**	—	—	1997	Pete Sampras	Carlos Moya
1987	Stefan Edberg	Pat Cash	1998	Petr Korda	Marcelo Rios
1988	Mats Wilander	Pat Cash	1999	Evgeny Kafelnikov	Thomas Enqvist

Women's Singles

Year	Champion	Final Opponent	Year	Champion	Final Opponent
1969	Margaret Smith Court	Billie Jean King	1984	Chris Evert Lloyd	Helena Sukova
1970	Margaret Smith Court	Kerry Melville Reid	1985	Martina Navratilova	Chris Evert Lloyd
1971	Margaret Smith Court	Evonne Goolagong	1986**	—	—
1972	Virginia Wade	Evonne Goolagong	1987	Hana Mandlikova	Martina Navratilova
1973	Margaret Smith Court	Evonne Goolagong	1988	Steffi Graf	Chris Evert
1974	Evonne Goolagong	Chris Evert	1989	Steffi Graf	Helena Sukova
1975	Evonne Goolagong	Martina Navratilova	1990	Steffi Graf	Mary Joe Fernandez
1976	Evonne Goolagong	Renata Tomanova	1991	Monica Seles	Jana Novotna
1977*	Kerry Reid	Dianne Balestrat	1992	Monica Seles	Mary Joe Fernandez
	Evonne Goolagong	Helen Gourlay	1993	Monica Seles	Steffi Graf
1978	Chris O'Neill	Betsy Nagelsen	1994	Steffi Graf	Arantxa Sánchez Vicario
1979	Barbara Jordan	Sharon Walsh	1995	Mary Pierce	Arantxa Sánchez Vicario
1980	Hana Mandlikova	Wendy Turnbull	1996	Monica Seles	Anke Huber
1981	Martina Navratilova	Chris Evert Lloyd	1997	Martina Hingis	Mary Pierce
1982	Chris Evert Lloyd	Martina Navratilova	1998	Martina Hingis	Conchita Martínez
1983	Martina Navratilova	Kathy Jordan	1999	Martina Hingis	Amelie Mauresmo

* Two tournaments were held in 1977 (Jan. & Dec.). ** Tournament was moved forward to Jan. 1987, so no championship was decided in 1986.

AUTO RACING

Indianapolis 500 Winners

Year	Winner, Car (Chassis-Engine)	MPH[1]	Year	Winner, Car (Chassis-Engine)	MPH[1]
1911	Ray Harroun, Marmon	74.602	1958	Jimmy Bryan, Salih-Offy	133.791
1912	Joe Dawson, National	78.719	1959	Rodger Ward, Watson-Offy	135.857
1913	Jules Goux, Peugeot	75.933	1960	Jim Rathmann, Watson-Offy	138.767
1914	Rene Thomas, Delage	82.474	1961	A.J. Foyt Jr., Trevis-Offy	139.130
1915	Ralph DePalma, Mercedes	89.840	1962	Rodger Ward, Watson-Offy	140.293
1916	Dario Resta, Peugeot	84.001	1963	Parnelli Jones, Watson-Offy	143.137
1917-18	Not held		1964	A.J. Foyt Jr., Watson-Offy	147.350
1919	Howdy Wilcox, Peugeot	88.050	1965	Jim Clark, Lotus-Ford	150.686
1920	Gaston Chevrolet, Frontenac	88.618	1966	Graham Hill, Lola-Ford	144.317
1921	Tommy Milton, Frontenac	89.621	1967	A.J. Foyt Jr., Coyote-Ford	151.207
1922	Jimmy Murphy, Duesenberg-Miller	94.484	1968	Bobby Unser, Eagle-Offy	152.882
1923	Tommy Milton, Miller	90.954	1969	Mario Andretti, Hawk-Ford	156.867
1924	L.L. Corum-Joe Boyer, Duesenberg	98.234	1970	Al Unser, P.J. Colt-Ford	155.749
1925	Peter DePaolo, Duesenberg	101.127	1971	Al Unser, P.J. Colt-Ford	157.735
1926	Frank Lockhart, Miller	95.904	1972	Mark Donohue, McLaren-Offy	162.962
1927	George Souders, Duesenberg	97.545	1973	Gordon Johncock, Eagle-Offy	159.036
1928	Louie Meyer, Miller	99.482	1974	Johnny Rutherford, McLaren-Offy	158.589
1929	Ray Keech, Miller	97.585	1975	Bobby Unser, Eagle-Offy	149.213
1930	Billy Arnold, Summers-Miller	100.448	1976	Johnny Rutherford, McLaren-Offy	148.725
1931	Louis Schneider, Stevens-Miller	96.629	1977	A.J. Foyt Jr., Coyote-Foyt	161.331
1932	Fred Frame, Wetteroth-Miller	104.144	1978	Al Unser, Lola-Cosworth	161.363
1933	Louie Meyer, Miller	104.162	1979	Rick Mears, Penske-Cosworth	158.899
1934	Bill Cummings, Miller	104.863	1980	Johnny Rutherford, Chaparral-Cosworth	142.862
1935	Kelly Petillo, Wetteroth-Offy	106.240	1981	Bobby Unser, Penske-Cosworth	139.084
1936	Louie Meyer, Stevens-Miller	109.069	1982	Gordon Johncock, Wildcat-Cosworth	162.029
1937	Wilbur Shaw, Shaw-Offy	113.580	1983	Tom Sneva, March-Cosworth	162.117
1938	Floyd Roberts, Wetteroth-Miller	117.200	1984	Rick Mears, March-Cosworth	163.612
1939	Wilbur Shaw, Maserati	115.035	1985	Danny Sullivan, March-Cosworth	152.982
1940	Wilbur Shaw, Maserati	114.277	1986	Bobby Rahal, March-Cosworth	170.722
1941	Floyd Davis-Mauri Rose, Wetteroth-Offy	115.117	1987	Al Unser, March-Cosworth	162.175
1942-45	Not held		1988	Rick Mears, Penske-Chevy Indy V8	144.809
1946	George Robson, Adams-Sparks	114.820	1989	Emerson Fittipaldi, Penske-Chevy Indy V8	167.581
1947	Mauri Rose, Deidt-Offy	116.338	1990	Arie Luyendyk, Lola-Chevy Indy V8	185.981*
1948	Mauri Rose, Deidt-Offy	119.814	1991	Rick Mears, Penske-Chevy Indy V8	176.457
1949	Bill Holland, Deidt-Offy	121.327	1992	Al Unser Jr., Galmer-Chevy Indy V8A	134.477
1950	Johnnie Parsons, Kurtis-Offy	124.002	1993	Emerson Fittipaldi, Penske-Chevy Indy V8C	157.207
1951	Lee Wallard, Kurtis-Offy	126.244	1994	Al Unser Jr., Penske-Mercedes Benz	160.872
1952	Troy Ruttman, Kuzma-Offy	128.922	1995	Jacques Villeneuve, Reynard-Ford Cosworth XB	153.616
1953	Bill Vukovich, KK500A-Offy	128.740			
1954	Bill Vukovich, KK500A-Offy	130.840	1996	Buddy Lazier, Reynard-Ford Cosworth	147.956
1955	Bob Sweikert, KK500C-Offy	128.213	1997	Arie Luyendyk, G Force-Aurora	145.827
1956	Pat Flaherty, Watson-Offy	128.490	1998	Eddie Cheever, Dallara-Aurora	145.155
1957	Sam Hanks, Salih-Offy	135.601	1999	Kenny Brack, Dallara-Aurora	153.176

*Race record. **Note:** The race was less than 500 mi in the following years: 1916 (300 mi), 1926 (400 mi), 1950 (345 mi), 1973 (332.5 mi), 1975 (435 mi), 1976 (255 mi). (1) Average speed.

FedEx Championship Series PPG Cup Winners

(U.S. Auto Club Champions prior to 1979; Championship Auto Racing Teams [CART] Champions, 1979-99)

Year	Driver	Year	Driver	Year	Driver	Year	Driver
1959	Roger Ward	1970	Al Unser	1980	Johnny Rutherford	1990	Al Unser Jr.
1960	A. J. Foyt	1971	Joe Leonard	1981	Rick Mears	1991	Michael Andretti
1961	A. J. Foyt	1972	Joe Leonard	1982	Rick Mears	1992	Bobby Rahal
1962	Rodger Ward	1973	Roger McCluskey	1983	Al Unser	1993	Nigel Mansell
1963	A. J. Foyt	1974	Bobby Unser	1984	Mario Andretti	1994	Al Unser Jr.
1964	A. J. Foyt	1975	A. J. Foyt	1985	Al Unser	1995	Jacques Villeneuve
1965	Mario Andretti	1976	Gordon Johncock	1986	Bobby Rahal	1996	Jimmy Vasser
1966	Mario Andretti	1977	Tom Sneva	1987	Bobby Rahal	1997	Alex Zanardi
1967	A. J. Foyt	1978	Tom Sneva	1988	Danny Sullivan	1998	Alex Zanardi
1968	Bobby Unser	1979	Rick Mears	1989	Emerson Fittipaldi	1999	Juan Montoya
1969	Mario Andretti						

Notable One-Mile Land Speed Records

Andy Green, a Royal Air Force pilot, broke the sound barrier and set the first supersonic world speed record on land, Oct. 15, 1997, in Black Rock Desert, NV. Green, driving a car built by Richard Noble, had 2 runs at an average speed of 763.035 mph, as calculated according to the rules of the Federation Internationale Automobiliste (FIA). This record and speed exceeded the speed of sound, calculated at 751.251 mph for this time. On Sept. 25, Green had set a new world mark at 714.144 mph, which eclipsed the old record of 633.468 mph. Both 1997 records were recorded by the United States Auto Club and recognized by the FIA.

Date	Driver	Car	MPH	Date	Driver	Car	MPH
1/26/06	Marriott	Stanley (Steam)	127.659	11/19/37	Eyston	Thunderbolt 1	311.42
3/16/10	Oldfield	Benz	131.724	9/16/38	Eyston	Thunderbolt 1	357.5
4/23/11	Burman	Benz	141.732	8/23/39	Cobb	Railton	368.9
2/12/19	DePalma	Packard	149.875	9/16/47	Cobb	Railton-Mobil	394.2
4/27/20	Milton	Dusenberg	155.046	8/5/63	Breedlove	Spirit of America	407.45
4/28/26	Parry-Thomas	Thomas Spl.	170.624	10/27/64	Arfons	Green Monster	536.71
3/29/27	Seagrave	Sunbeam	203.790	11/15/65	Breedlove	Spirit of America	600.601
4/22/28	Keech	White Triplex	207.552	10/23/70	Gabelich	Blue Flame	622.407
3/11/29	Seagrave	Irving-Napier	231.446	10/9/79	Barrett	Budweiser Rocket	638.637*
2/5/31	Campbell	Napier-Campbell	246.086	10/4/83	Noble	Thrust 2	633.468
2/24/32	Campbell	Napier-Campbell	253.96	9/25/97	Green	Thrust SSC	714.144
2/22/33	Campbell	Napier-Campbell	272.109	10/15/97	Green	Thrust SSC	763.035
9/3/35	Campbell	Bluebird Special	301.13				

*Not recognized as official by sanctioning bodies.

Le Mans 24 Hours Race in 1999

Yannick Dalmas (France), Joachim Winkelhock (Germany), and Pierluigi Martini (Italy) drove their BMW V12 LMR to victory in the 1999 "24 Hours of Le Mans" race, held June 13, 1999. They traveled 4,966 kilometers (3,086 miles) at an average speed of 207 kilometers per hour (128.630 mph). It was the first win for BMW, which was racing for the second time under its name.

World Grand Prix Champions, 1950-99

Year	Driver	Year	Driver	Year	Driver
1950	Nino Farini, Italy	1967	Denis Hulme, New Zealand	1984	Niki Lauda, Austria
1951	Juan Fangio, Argentina	1968	Graham Hill, England	1985	Alain Prost, France
1952	Alberto Ascari, Italy	1969	Jackie Stewart, Scotland	1986	Alain Prost, France
1953	Alberto Ascari, Italy	1970	Jochen Rindt, Austria	1987	Nelson Piquet, Brazil
1954	Juan Fangio, Argentina	1971	Jackie Stewart, Scotland	1988	Ayrton Senna, Brazil
1955	Juan Fangio, Argentina	1972	Emerson Fittipaldi, Brazil	1989	Alain Prost, France
1956	Juan Fangio, Argentina	1973	Jackie Stewart, Scotland	1990	Ayrton Senna, Brazil
1957	Juan Fangio, Argentina	1974	Emerson Fittipaldi, Brazil	1991	Ayrton Senna, Brazil
1958	Mike Hawthorne, England	1975	Niki Lauda, Austria	1992	Nigel Mansell, Britain
1959	Jack Brabham, Australia	1976	James Hunt, England	1993	Alain Prost, France
1960	Jack Brabham, Australia	1977	Niki Lauda, Austria	1994	Michael Schumacher, Germany
1961	Phil Hill, United States	1978	Mario Andretti, United States	1995	Michael Schumacher, Germany
1962	Graham Hill, England	1979	Jody Scheckter, South Africa	1996	Damon Hill, England
1963	Jim Clark, Scotland	1980	Alan Jones, Australia	1997	Jacques Villeneuve, Canada
1964	John Surtees, England	1981	Nelson Piquet, Brazil	1998	Mika Hakkinen, Finland
1965	Jim Clark, Scotland	1982	Keke Rosberg, Finland	1999	Mika Hakkinen, Finland
1966	Jack Brabham, Australia	1983	Nelson Piquet, Brazil		

Grand Prix Races for Formula 1 Cars in 1999

Date	Race	Winner, car	Date	Race	Winner, car
3/7	Australian	Eddie Irvine, Ferrari	7/25	Austrian	Eddie Irvine, Ferrari
4/11	Brazilian	Mika Hakkinen, McLaren-Mercedes	8/1	German	Eddie Irvine, Ferrari
5/2	S. Marino	Michael Schumacher, Ferrari	8/15	Hungary	Mika Hakkinen, McLaren-Mercedes
5/16	Monaco	Michael Schumacher, Ferrari	8/29	Belgian	David Coulthard, McLaren-Mercedes
5/30	Spanish	Mika Hakkinen, McLaren-Mercedes	9/12	Italian	Heinz-Harald Frentzen, Jordan-Mugen, Honda
6/13	Canadian	Mika Hakkinen, McLaren-Mercedes			
6/27	French	Heinz-Harald Frentzen, Jordan-Mugen, Honda	9/26	Europe	Johnny Herbert, Stewart-Ford
			10/17	Malaysia	Eddie Irvine, Ferrari
7/11	British	David Coulthard, McLaren-Mercedes	10/31	Japan	Mika Hakkinen, McLaren-Mercedes

NASCAR Racing

Winston Cup Champions, 1949-98

Year	Driver	Year	Driver	Year	Driver	Year	Driver
1949	Red Byron	1962	Joe Weatherly	1975	Richard Petty	1987	Dale Earnhardt
1950	Bill Rexford	1963	Joe Weatherly	1976	Cale Yarborough	1988	Bill Elliott
1951	Herb Thomas	1964	Richard Petty	1977	Cale Yarborough	1989	Rusty Wallace
1952	Tim Flock	1965	Ned Jarrett	1978	Cale Yarborough	1990	Dale Earnhardt
1953	Herb Thomas	1966	David Pearson	1979	Richard Petty	1991	Dale Earnhardt
1954	Lee Petty	1967	Richard Petty	1980	Dale Earnhardt	1992	Alan Kulwicki
1955	Tim Flock	1968	David Pearson	1981	Darrell Waltrip	1993	Dale Earnhardt
1956	Buck Baker	1969	David Pearson	1982	Darrell Waltrip	1994	Dale Earnhardt
1957	Buck Baker	1970	Bobby Isaac	1983	Bobby Allison	1995	Jeff Gordon
1958	Lee Petty	1971	Richard Petty	1984	Terry Labonte	1996	Terry Labonte
1959	Lee Petty	1972	Richard Petty	1985	Darrell Waltrip	1997	Jeff Gordon
1960	Rex White	1973	Benny Parsons	1986	Dale Earnhardt	1998	Jeff Gordon
1961	Ned Jarrett	1974	Richard Petty				

Daytona 500 Winners, 1959-99

Year	Driver, car	Avg. MPH	Year	Driver, car	Avg. MPH
1959	Lee Petty, Oldsmobile	135.521	1980	Buddy Baker, Oldsmobile	177.602
1960	Junior Johnson, Chevrolet	124.740	1981	Richard Petty, Buick	169.651
1961	Marvin Panch, Pontiac	149.601	1982	Bobby Allison, Buick	153.991
1962	Fireball Roberts, Pontiac	152.529	1983	Cale Yarborough, Pontiac	155.979
1963	Tiny Lund, Ford	151.566	1984	Cale Yarborough, Chevrolet	150.994
1964	Richard Petty, Plymouth	154.334	1985	Bill Elliott, Ford	172.265
1965	Fred Lorenzen, Ford (a)	141.539	1986	Geoff Bodine, Chevrolet	148.124
1966	Richard Petty, Plymouth (b)	160.627	1987	Bill Elliott, Ford	176.263
1967	Mario Andretti, Ford	146.926	1988	Bobby Allison, Buick	137.531
1968	Cale Yarborough, Mercury	143.251	1989	Darrell Waltrip, Chevrolet	148.466
1969	Lee Roy Yarborough, Ford	160.875	1990	Derrike Cope, Chevrolet	165.761
1970	Pete Hamilton, Plymouth	149.601	1991	Ernie Irvan, Chevrolet	148.148
1971	Richard Petty, Plymouth	144.456	1992	Davey Allison, Ford	160.256
1972	A. J. Foyt, Mercury	161.550	1993	Dale Jarrett, Chevrolet	154.972
1973	Richard Petty, Dodge	157.205	1994	Sterling Marlin, Chevrolet	156.931
1974	Richard Petty, Dodge (c)	140.894	1995	Sterling Marlin, Chevrolet	141.710
1975	Benny Parsons, Chevrolet	153.649	1996	Dale Jarrett, Ford	154.308
1976	David Pearson, Mercury	152.181	1997	Jeff Gordon, Chevrolet	148.295
1977	Cale Yarborough, Chevrolet	153.218	1998	Dale Earnhardt, Chevrolet	172.712
1978	Bobby Allison, Ford	159.730	1999	Jeff Gordon, Chevrolet	161.551
1979	Richard Petty, Oldsmobile	143.977			

(a) 322.5 mi. (b) 495 mi. (c) 450 mi.

Winston Cup Series Races in 1999

(through Oct. 24)

Date	Race, site	Winner	Car	Prize
Feb. 14	Daytona 500, Daytona Beach, FL	Jeff Gordon	Chevrolet	$2,194,246
Feb. 21	Dura-Lube/Big Kmart 400, Rockingham, NC	Mark Martin	Ford	104,635
Mar. 7	Las Vegas 400, Las Vegas, NV	Jeff Burton	Ford	336,590
Mar. 14	Cracker Barrell 500, Atlanta, GA	Jeff Gordon	Chevrolet	117,650
Mar. 21	#TranSouth Financial 400, Darlington, SC	Jeff Burton	Ford	161,900
Mar. 28	PRIMESTAR 500, Fort Worth, TX	Terry Labonte	Chevrolet	376,840
Apr. 11	Food City 500, Bristol, TN	Rusty Wallace	Ford	92,435
Apr. 18	Goody's Body Pain 500, Martinsville, VA	John Andretti	Pontiac	113,275
Apr. 25	DieHard 500, Talladega, AL	Dale Earnhardt	Chevrolet	147,795
May 2	California 500 presented by NAPA , Los Angeles, CA	Jeff Gordon	Chevrolet	155,890
May 15	Pontiac Excitement 400, Richmond, VA	Dale Jarrett	Ford	169,715
May 22	*The Winston, Concord, NC	Terry Labonte	Chevrolet	297,500
May 30	Coca-Cola 600, Concord, NC	Jeff Burton	Ford	1,212,500[1]
June 6	MBNA Platinum 400, Dover, DE	Bobby Labonte	Pontiac	144,820
June 13	Kmart 400 pres. by Castrol Super Clean, Brooklyn, MI	Dale Jarrett	Ford	151,240
June 20	Pocono 500, Pocono, PA	Bobby Labonte	Pontiac	151,110
June 27	Save Mart/Kragen 350, Sonoma, CA	Jeff Gordon	Chevrolet	125,040
July 3	Pepsi 400, Daytona Beach, FL	Dale Jarrett	Ford	1,64,965
July 11	Jiffy Lube 300, Loudon, NH	Jeff Burton	Ford	139,490
July 25	Pennsylvania 500, Long Pond, PA	Bobby Labonte	Pontiac	139,385
Aug. 7	Brickyard 400, Indianapolis, IN	Dale Jarrett	Ford	712,240
Aug. 15	Frontier at The Glen, Watkins Glen, NY	Jeff Gordon	Chevrolet	119,860
Aug. 22	Pepsi 400, Brooklyn, MI	Bobby Labonte	Pontiac	121,230
Aug. 28	Goody's Headache Powder 500, Bristol, TN	Dale Earnhardt	Chevrolet	89,880
Sept. 5	#Pepsi Southern 500, Darlington, SC	Jeff Burton	Ford	1,148,170[1]
Sept. 11	Exide NASCAR Select Batteries 400, Richmond, VA	Tony Stewart	Pontiac	136,160
Sept. 19	Dura-Lube/Kmart 300, Loudon, NH	Joe Nemechek	Chevrolet	157,625
Sept. 26	MBNA Gold 400, Dover, DE	Mark Martin	Ford	115,710
Oct. 3	NAPA AutoCare 500, Martinsville, VA	Jeff Gordon	Chevrolet	110,090
Oct. 11	UAW-GM Quality 500, Charlotte, NC	Jeff Gordon	Chevrolet	140,350
Oct. 17	Winston 500, Talladega, AL	Dale Earnhardt	Chevrolet	120,290
Oct. 24	Pop Secret Microwave Popcorn 400, Rockingham, NC	Jeff Burton	Ford	104,715

# Denotes rain-shortened event. *Denotes nonpoint event. (1) Includes $1,000,000 bonus.

BOXING
Champions by Classes

There are many governing bodies in boxing, including the World Boxing Council, World Boxing Assn., International Boxing Federation, World Boxing Org., U.S. Boxing Assn., North American Boxing Federation, and European Boxing Union. Others are recognized by TV networks and the print media. All the governing bodies have their own champions and assorted boxing divisions. The following are the recognized champions—as of mid-Oct. 1999—in the principal divisions of the WBC, WBA, and IBF.

Class, Weight limit	WBC	WBA	IBF
Heavyweight..............	Lennox Lewis, U.K.	Evander Holyfield, U.S.	Evander Holyfield, U.S.
Cruiserweight (190 lb)	Juan Carlos Gomez, Cuba	Fabrice Tiozzo, France	Vassiliy Jirov, U.S./Kazakhstan
Light Heavyweight (175 lb) ...	Roy Jones Jr., U.S.	Roy Jones Jr., U.S.	Roy Jones Jr., U.S.
Super Middleweight (168 lb)...	Richie Woodhall, U.K.	Byron Mitchell, U.S.	Sven Ottke, Germany
Middleweight (160 lb)........	Keith Holmes, U.S.	William Joppy, U.S.	Bernard Hopkins, U.S.
Jr. Middleweight (154 lb)	Javier Castillejo, Spain.	David Reid, U.S.	Fernando Vargas, U.S.
Welterweight (147 lb)	Felix Trinidad, Puerto Rico.	James Page, U.S.	Felix Trinidad, Puerto Rico
Jr. Welterweight (140 lb)......	Kostya Tszyu, Australia	Sharmba Mitchell, U.S.	Terronn Millett, U.S.
Lightweight (135 lb)	Stevie Johnston, U.S.	Stefano Zoff, Italy	Paul Spadafora, U.S.
Jr. Lightweight (130 lb)	Floyd Mayweather Jr., U.S.	Lakva Sim, Mongolia	Roberto Garcia, U.S.
Featherweight (126 lb)	Cesar Soto, Mexico	Freddie Norwood, U.S.	Manuel Medina, Mexico
Jr. Featherweight (122 lb)....	Erik Morales, Mexico	Nestor Garza, Mexico	Benedict Ledwaba, South Africa
Bantamweight (118 lb)	Veeraphol Sahaprom, Thailand	Paulie Ayala, U.S.	Tim Austin, U.S.
Jr. Bantamweight (115 lb)	In-joo Cho, S. Korea	Hideki Todaka, Japan	Mark Johnson, U.S.
Flyweight (112 lb)..........	Medgoen L., Thailand	Sornpichai K., Thailand	Irene Pacheco, Colombia
Jr. Flyweight (108 lb)	Yo-Sam Choi, South Korea	Pichitnoi C. Siriwat, Thailand	Ricardo Lopez, U.S.
Strawweight (105 lb)........	Wande Chareon, Thailand	Noel Arambulet, Venezuela	Zolani Petelo, South Africa

Ring Champions by Years

(*abandoned the title or was stripped of it; IBF champions listed only for heavyweight division)

Heavyweights

1882-1892	John L. Sullivan (a)	1964-1967	Cassius Clay* (Muhammad Ali) (d)	1986-1987	James "Bonecrusher" Smith (WBA)
1892-1897	James J. Corbett (b)	1970-1973	Joe Frazier	1987	Mike Tyson (WBC, WBA)
1897-1899	Robert Fitzsimmons	1973-1974	George Foreman		Tony Tucker (IBF)
1899-1905	James J. Jeffries* (c)	1974-1978	Muhammad Ali	1987-1990	Mike Tyson (WBC, WBA, IBF)
1905-1906	Marvin Hart	1978	Leon Spinks (WBC*, WBA) (e);	1990	James "Buster" Douglas (WBA, WBC, IBF)
1906-1908	Tommy Burns		Ken Norton (WBC)		
1908-1915	Jack Johnson		Larry Holmes (WBC)	1990-1992	Evander Holyfield (WBA, WBC, IBF)
1915-1919	Jess Willard		Muhammad Ali* (WBA)		
1919-1926	Jack Dempsey	1978-1983	Larry Holmes* (WBC) (f)	1992-1993	Riddick Bowe (WBA, IBF, WBC*)
1926-1928	Gene Tunney*	1979-1980	John Tate (WBA)		
1928-1930	Vacant	1980-1982	Mike Weaver (WBA)	1992-1994	Lennox Lewis (WBC)
1930-1932	Max Schmeling	1982-1983	Michael Dokes (WBA)	1993-1994	Evander Holyfield (WBA, IBF)
1932-1933	Jack Sharkey	1983	Gerrie Coetzee (WBA);	1994	Michael Moorer (WBA, IBF)
1933-1934	Primo Carnera		Larry Holmes (IBF) (f)		Oliver McCall (WBC)
1934-1935	Max Baer	1984	Tim Witherspoon (WBC);		George Foreman (WBA*, IBF*)
1935-1937	James J. Braddock		Pinklon Thomas (WBC);	1995	Bruce Seldon (WBA)
1937-1949	Joe Louis*		Greg Page (WBA)		Frank Bruno (WBC)
1949-1951	Ezzard Charles	1985-1986	Tony Tubbs (WBA)		Frans Botha* (IBF)
1951-1952	Joe Walcott	1985-1987	Michael Spinks* (IBF)	1996	Mike Tyson (WBC*, WBA)
1952-1956	Rocky Marciano*	1986	Tim Witherspoon (WBA)		Michael Moorer (IBF)
1956-1959	Floyd Patterson		Trevor Berbick (WBC)		Evander Holyfield (WBA)
1959-1960	Ingemar Johansson		Mike Tyson (WBC)	1997	Lennox Lewis (WBC),
1960-1962	Floyd Patterson		James "Bonecrusher" Smith (WBA)		Evander Holyfield (IBF)
1962-1964	Sonny Liston				

(a) London Prize Ring (bare knuckle champion). (b) First Marquis of Queensberry champion. (c) Jeffries abandoned the title (1905) and designated Marvin Hart and Jack Root as logical contenders. Hart defeated Root in 12 rounds (1905) and in turn was defeated by Tommy Burns (1906), who laid claim to the title. Jack Johnson defeated Burns (1908) and was recognized as champion. Johnson clinched the title by defeating Jeffries in an attempted comeback (1910). (d) Title declared vacant by the WBA and other groups in 1967 after Ali's refusal to fulfill his military obligation. Joe Frazier was recognized as champion by 6 states, Mexico, and South America. Jimmy Ellis was declared champion by the WBA. Frazier KOd Ellis, Feb. 16, 1970. (e) After Spinks defeated Ali, the WBC recognized Ken Norton as champion. Ali defeated Spinks in a 1978 rematch to win the WBA title and then retired in 1979. (f) Holmes relinquished the WBC title in Dec. 1983 and began fighting as champion of the newly formed IBF.

Light Heavyweights

1903	Jack Root, George Gardner	1962-1963	Harold Johnson	1987	Thomas Hearns* (WBC);
1903-1905	Bob Fitzsimmons	1963-1965	Willie Pastrano		Leslie Stewart (WBA);
1905-1912	Philadelphia Jack O'Brien*	1965-1966	Jose Torres		Virgil Hill (WBA);
1912-1916	Jack Dillon	1966-1968	Dick Tiger		Don Lalonde (WBC)
1916-1920	Battling Levinsky	1968-1974	Bob Foster*	1988	Ray Leonard* (WBC)
1920-1922	George Carpentier	1974-1977	John Conteh (WBC);	1989	Dennis Andries (WBC);
1922-1923	Battling Siki		Victor Galindez (WBA)		Jeff Harding (WBC)
1923-1925	Mike McTigue	1977-1978	Miguel Cuello (WBC)	1990	Dennis Andries (WBC)
1925-1926	Paul Berlenbach	1978	Mike Rossman (WBA);	1991	Thomas Hearns (WBA);
1926-1927	Jack Delaney*		Mate Parlov (WBC);		Jeff Harding (WBC)
1927-1929	Tommy Loughran*		Marvin Johnson (WBC)	1992	Iran Barkley* (WBA);
1930-1934	Maxey Rosenbloom	1979	Matthew Saad Muhammad (WBC);		Virgil Hill (WBA)
1934-1935	Bob Olin		Victor Galindez (WBA);	1994-1995	Mike McCallum (WBC)
1935-1939	John Henry Lewis*		Marvin Johnson (WBA)	1995-1996	Fabrice Tiozzo* (WBC)
1939	Melio Bettina	1980	Eddie Mustafa Muhammad (WBA)	1996-1997	Roy Jones Jr. (WBC)
1939-1941	Billy Conn*	1981	Michael Spinks (WBA);	1997	Montell Griffin* (WBC);
1941	Anton Christoforidis (won NBA title)		Dwight Braxton (WBC)		Darius Michalczewski* (WBA);
1941-1948	Gus Lesnevich, Freddie Mills	1983-1985	Michael Spinks*		Roy Jones Jr. (WBC);
1948-1950	Freddie Mills	1985	J. B. Williamson (WBC);		Lou Del Valle (WBA)
1950-1952	Joey Maxim	1986	Marvin Johnson (WBA);	1998	Roy Jones Jr. (WBA)
1952-1960	Archie Moore		Dennis Andries (WBC)	1999	Roy Jones Jr. (IBF)
1961-1962	Vacant				

Middleweights

Years	Champion
1884-1891	Jack "Nonpareil" Dempsey
1891-1897	Bob Fitzsimmons*
1897-1907	Tommy Ryan*
1907-1908	Stanley Ketchel, Billy Papke
1908-1910	Stanley Ketchel
1911-1913	vacant
1913	Frank Klaus; George Chip
1914-1917	Al McCoy
1917-1920	Mike O'Dowd
1920-1923	Johnny Wilson
1923-1926	Harry Greb
1926-1931	Tiger Flowers; Mickey Walker
1931-1932	Gorilla Jones (NBA)
1932-1937	Marcel Thil
1938	Al Hostak (NBA); Solly Krieger (NBA)
1939-1940	Al Hostak (NBA)
1941-1947	Tony Zale
1947-1948	Rocky Graziano
1948	Tony Zale; Marcel Cerdan
1949-1951	Jake LaMotta
1951	Ray Robinson; Randy Turpin; Ray Robinson*
1953-1955	Carl (Bobo) Olson
1955-1957	Ray Robinson
1957	Gene Fullmer; Ray Robinson; Carmen Basilio
1958	Ray Robinson
1959	Gene Fullmer (NBA); Ray Robinson (NY)
1960	Gene Fullmer (NBA); Paul Pender (NY and MA)
1961	Gene Fullmer (NBA); Terry Downes (NY, MA, Europe)
1962	Gene Fullmer; Dick Tiger (NBA); Paul Pender (NY and MA)*
1963	Dick Tiger (universal)
1963-1965	Joey Giardello
1965-1966	Dick Tiger
1966-1967	Emile Griffith
1967	Nino Benvenuti
1967-1968	Emile Griffith
1968-1970	Nino Benvenuti
1970-1977	Carlos Monzon*
1977-1978	Rodrigo Valdez
1978-1979	Hugo Corro
1979-1980	Vito Antuofermo
1980	Alan Minter; Marvin Hagler
1987	Ray Leonard* (WBC); Thomas Hearns (WBC); Sumbu Kalambay (WBA)
1988-1989	Iran Barkley (WBC)
1989	Mike McCallum* (WBA); Roberto Duran (WBC)
1991-1993	Julian Jackson (WBC)
1992-1993	Reggie Johnson (WBA)
1993	Gerald McClellan (WBC); John David Jackson* (WBA)
1994-1995	Jorge Castro (WBA)
1995	Julian Jackson (WBC); Quincy Taylor (WBC); Shinji Takehara (WBA)
1996	Keith Holmes (WBC); William Joppy (WBA)
1997	Julio Cesar Green (WBA)
1998	William Joppy (WBA), Hassine Cherifi (WBC)
1999	Keith Holmes (WBC)

Welterweights

Years	Champion
1892-1894	Mysterious Billy Smith
1894-1896	Tommy Ryan
1896	Kid McCoy*
1900	Rube Ferns; Matty Matthews
1901	Rube Ferns
1901-1904	Joe Walcott
1904-1906	Dixie Kid; Joe Walcott; Honey Mellody
1907-1911	Mike Sullivan
1911-1915	Vacant
1915-1919	Ted Lewis
1919-1922	Jack Britton
1922-1926	Mickey Walker
1926	Pete Latzo
1927-1929	Joe Dundee
1929	Jackie Fields
1930	Jack Thompson; Tommy Freeman
1931	Tommy Freeman; Jack Thompson; Lou Brouillard
1932	Jackie Fields
1933	Young Corbett; Jimmy McLarnin
1934	Barney Ross; Jimmy McLarnin
1935-1938	Barney Ross
1938-1940	Henry Armstrong
1940-1941	Fritzie Zivic
1941-1946	Fred Cochrane
1946	Marty Servo*; Ray Robinson (a)
1946-1950	Ray Robinson*
1951	Johnny Bratton (NBA)
1951-1954	Kid Gavilan
1954-1955	Johnny Saxton
1955	Tony De Marco; Carmen Basilio
1956	Carmen Basilio; Johnny Saxton; Carmen Basilio
1957	Carmen Basilio*
1958-1960	Virgil Akins, Don Jordan
1960	Benny Paret
1961	Emile Griffith; Benny Paret
1962	Emile Griffith
1963	Luis Rodriguez; Emile Griffith
1964-1966	Emile Griffith*
1966-1969	Curtis Cokes
1969-1970	Jose Napoles; Billy Backus
1971-1975	Jose Napoles
1975-1976	John Stracey (WBC); Angel Espada (WBA)
1976-1979	Carlos Palomino (WBC); Jose Cuevas (WBA)
1979	Wilfredo Benitez (WBC); Sugar Ray Leonard (WBC)
1980	Roberto Duran (WBC); Thomas Hearns (WBA); Sugar Ray Leonard (WBC)
1981-1982	Sugar Ray Leonard*
1983-1985	Donald Curry (WBA); Milton McCrory (WBC)
1985-1986	Donald Curry
1986-1987	Lloyd Honeyghan (WBC)
1987	Mark Breland (WBA); Marlon Starling (WBA); Jorge Vaca (WBC)
1988-1989	Tomas Molinares (WBA); Lloyd Honeyghan (WBC)
1989-1990	Marlon Starling (WBC); Mark Breland (WBA)
1990-1991	Maurice Blocker (WBC); Aaron Davis (WBA)
1991	Meldrick Taylor (WBA); Simon Brown (WBC); Buddy McGirt (WBC)
1992-1994	Crisanto Espana (WBA)
1993-1997	Pernell Whitaker (WBC)
1994	Ike Quartey (WBA)
1997	Oscar De La Hoya (WBC)
1998	James Page (WBA)
1999	Felix Trinidad (WBC)

(a) Robinson gained the title by defeating Tommy Bell in an elimination agreed to by the New York Commission and the NBA. Both claimed Robinson waived his title when he won the middleweight crown from LaMotta in 1951.

Lightweights

Years	Champion
1896-1899	Kid Lavigne
1899-1902	Frank Erne
1902-1908	Joe Gans
1908-1910	Battling Nelson
1910-1912	Ad Wolgast
1912-1914	Willie Ritchie
1914-1917	Freddie Welsh
1917-1925	Benny Leonard*
1925	Jimmy Goodrich; Rocky Kansas
1926-1930	Sammy Mandell
1930	Al Singer; Tony Canzoneri
1930-1933	Tony Canzoneri
1933-1935	Barney Ross*
1935-1936	Tony Canzoneri
1936-1938	Lou Ambers
1938	Henry Armstrong
1939	Lou Ambers
1940	Lew Jenkins
1941-1943	Sammy Angott
1944	S. Angott (NBA); J. Zurita (NBA)
1945-1951	Ike Williams (NBA: later universal)
1951-1952	James Carter
1952	Lauro Salas; James Carter
1953-1954	James Carter
1954	Paddy De Marco; James Carter
1955	James Carter; Bud Smith
1956	Bud Smith; Joe Brown
1956-1962	Joe Brown
1962-1965	Carlos Ortiz
1965	Ismael Laguna
1965-1968	Carlos Ortiz
1968-1969	Teo Cruz
1969-1970	Mando Ramos
1970	Ismael Laguna; Ken Buchanan (WBA)
1971	Mando Ramos (WBC); Pedro Carrasco (WBC)
1972-1979	Roberto Duran* (WBA)
1972	Pedro Carrasco; Mando Ramos; Chango Carmona; Rodolfo Gonzalez (all WBC)
1974-1976	Guts Ishimatsu (WBC)
1976-1977	Esteban De Jesus (WBC)
1979	Jim Watt (WBC); Ernesto Espana (WBA)
1980	Hilmer Kenty (WBA)
1981	Alexis Arguello (WBC); Sean O'Grady (WBA); Arturo Frias (WBA)
1982-1984	Ray Mancini (WBA)
1983-1984	Edwin Rosario (WBC)
1984	Livingstone Bramble (WBA); Jose Luis Ramirez (WBC)
1985-1986	Hector (Macho) Camacho (WBC)
1986	Edwin Rosario (WBA); Jose Luis Ramirez (WBC)
1987-1989	Julio Cesar Chavez (WBA)
1989-1990	Edwin Rosario (WBA); Pernell Whitaker (WBC)
1990	Juan Nazario (WBA)
1990-1992	Pernell Whitaker*
1992	Joey Gamache (WBA); Tony Lopez (WBA)
1992-1996	Miguel Angel Gonzalez* (WBC)
1993	Dingaan Thobela (WBA); Orzubek Nazarov (WBA)
1996-1997	Jean-Baptiste Mendy (WBC)
1997	Steve Johnston (WBC)
1998	Jean-Baptiste Mendy (WBA), Cesar Bazan (WBC)
1999	Stefano Zoff (WBA)

Featherweights

1892-1900 George Dixon (disputed)	1950-1957 Sandy Saddler*	1980-1982 Salvador Sanchez (WBC)
1900-1901 Terry McGovern;	1957-1959 Hogan (Kid) Bassey	1982-1984 Juan LaPorte (WBC)
Young Corbett*	1959-1963 Davey Moore	1984 Wilfredo Gomez (WBC);
1901-1912 Abe Attell	1963-1964 Sugar Ramos	Azumah Nelson (WBC)
1912-1923 Johnny Kilbane	1964-1967 Vicente Saldivar*	1985-1986 Barry McGuigan (WBA)
1923 Eugene Criqui;	1968-1971 Paul Rojas (WBA);	1986-1987 Steve Cruz (WBA)
Johnny Dundee	Sho Saijo (WBA)	1987-1991 Antonio Esparragoza (WBA)
1923-1925 Johnny Dundee*	1971-1972 Antonio Gomez (WBA);	1988-1990 Jeff Fenech (WBC)
1925-1927 Kid Kaplan*	Kuniaki Shibada (WBC)	1990-1991 Marcos Villasana (WBC)
1927-1928 Benny Bass; Tony Canzoneri	1972 Ernesto Marcel* (WBA);	1991-1993 Park Yung Kyun (WBA);
1928-1929 Andre Routis	Clemente Sanchez* (WBC);	Paul Hodkinson (WBC)
1929-1932 Battling Battalino*	Jose Legra (WBC)	1993 Goyo Vargas (WBC);
1932-1934 Tommy Paul (NBA)	1973-1974 Eder Jofre (WBC)	Kevin Kelley (WBC);
1933-1936 Freddie Miller	1974 Ruben Olivares (WBA);	Eloy Rojas (WBA)
1936-1937 Petey Sarron	Alexis Arguello (WBA);	Alejandro Gonzalez (WBC);
1937-1938 Henry Armstrong*	Bobby Chacon (WBC)	1995 Manuel Medina (WBC);
1938-1940 Joey Archibald (a)	1975 Ruben Olivares (WBC);	Luisito Espinosa (WBC)
1940-1941 Harry Jeffra	David Kotey (WBC)	1996-1997 Wilfredo Vasquez (WBA)
1942-1948 Willie Pep	1976-1980 Danny Lopez (WBC)	1998 Antonio Ceremeno (WBA);
1948-1949 Sandy Saddler	1977-1978 Rafael Ortega (WBA)	Freddie Norwood (WBA)
1949-1950 Willie Pep	1978 Cecilio Lastra (WBA)	1999 Cesar Soto (WBC)
	Eusebio Pedrosa (WBA)	

(a) After Petey Scalzo knocked out Archibald in an overweight match and was refused a title bout, the NBA named Scalzo champion. NBA title succession: Scalzo, 1938-1941; Richard Lemos, 1941; Jackie Wilson, 1941-1943; Jackie Callura, 1943; Phil Terranova, 1943-1944; Sal Bartolo, 1944-1946.

History of Heavyweight Championship Bouts
(bouts in which title changed hands)

1889—July 8—John L. Sullivan def. Jake Kilrain, 75, Richburg, MS. (Last championship bare knuckles bout.)

1892—Sept. 7—James J. Corbett def. John L. Sullivan, 21, New Orleans. (Big gloves used for first time.)

1897—Bob Fitzsimmons def. James J. Corbett, 14, Carson City, NV.

1899—June 9—James J. Jeffries def. Bob Fitzsimmons, 11, Coney Island, NY. (Jeffries retired as champion in 1905.)

1905—July 3—Marvin Hart KOd Jack Root, 12, Reno, NV. (Jeffries refereed and presented the title to the victor. Jack O'Brien also claimed the title.)

1906—Feb. 23—Tommy Burns def. Marvin Hart, 20, Los Angeles.

1908—Dec. 26—Jack Johnson KOd Tommy Burns, 14, Sydney, Australia. (Police halted contest.)

1915—April 5—Jess Willard KOd Jack Johnson, 26, Havana, Cuba.

1919—July 4—Jack Dempsey KOd Jess Willard, Toledo, OH. (Willard failed to answer bell for 4th round.)

1926—Sept. 23—Gene Tunney def. Jack Dempsey, 10, Philadelphia. (Tunney retired as champion in 1928.)

1930—June 12—Max Schmeling def. Jack Sharkey, 4, New York. (Sharkey fouled Schmeling in a bout generally considered to have resulted in the election of a successor to Tunney.)

1932—June 21—Jack Sharkey def. Max Schmeling, 15, NY.

1933—June 29—Primo Carnera KOd Jack Sharkey, 6, NY.

1934—June 14—Max Baer KOd Primo Carnera, 11, NY.

1935—June 13—James J. Braddock def. Max Baer, 15, NY.

1937—June 22—Joe Louis KOd James J. Braddock, 8, Chicago. (Louis retired as champion in 1949.)

1949—June 22—Ezzard Charles def. Joe Walcott, 15, Chicago; NBA recognition only.

1951—July 18—Joe Walcott KOd Ezzard Charles, 7, Pittsburgh.

1952—Sept. 23—Rocky Marciano KOd Joe Walcott, 13, Philadelphia. (Marciano retired as champion in 1956.)

1956—Nov. 30—Floyd Patterson KOd Archie Moore, 5, Chicago.

1959—June 26—Ingemar Johansson KOd Floyd Patterson, 3, New York.

1960—June 20—Floyd Patterson KOd Ingemar Johansson, 5, New York. (Patterson was 1st heavyweight to regain title.)

1962—Sept. 25—Sonny Liston KOd Floyd Patterson, 1, Chicago.

1964—Feb. 25—Cassius Clay (Muhammad Ali) KOd Sonny Liston, 7, Miami Beach, FL. (In 1967, Ali was stripped of his title by the WBA and others for refusing military service.)

1970—Feb. 16—Joe Frazier KOd Jimmy Ellis, 5, New York. (Frazier def. Ali in 15 rounds, Mar. 8, 1971, in New York.)

1973—Jan. 22—George Foreman KOd Joe Frazier, 2, Kingston, Jamaica.

1974—Oct. 30—Muhammad Ali KOd George Foreman, 8, Zaire.

1978—Feb. 15—Leon Spinks def. Muhammad Ali, 15, Las Vegas. (WBC recognized Ken Norton as champion after Spinks refused to fight him before his rematch with Ali.)

1978—June 9—(WBC) Larry Holmes def. Ken Norton, 15, Las Vegas. (Holmes gave up title in Dec. 1983.)

1978—Sept. 15—(WBA) Muhammad Ali def. Leon Spinks, 15, New Orleans. (Ali retired as champion in 1979.)

1979—Oct. 20—(WBA) John Tate def. Gerrie Coetzee, 15, Pretoria, South Africa.

1980—Mar. 31—(WBA) Mike Weaver KOd John Tate, 15, Knoxville, TN.

1982—Dec. 10—(WBA) Michael Dokes KOd Mike Weaver, 1, Las Vegas.

1983—Sept. 23—(WBA) Gerrie Coetzee KOd Michael Dokes, 10, Richfield, OH.

1983—In Dec., Larry Holmes relinquished the WBC title and was named champion of the newly formed IBF.

1984—Mar. 9—(WBC) Tim Witherspoon def. Greg Page, 12, Las Vegas.

1984—Aug. 31—(WBC) Pinklon Thomas def. Tim Witherspoon, 12, Las Vegas.

1984—Dec. 2—(WBA) Greg Page KOd Gerrie Coetzee, 8, Sun City, Bophuthatswana.

1985—Apr. 29—(WBA) Tony Tubbs def. Greg Page, 15, Buffalo, NY.

1985—Sept. 21—(IBF) Michael Spinks def. Larry Holmes, 15, Las Vegas. (Spinks relinquished title in Feb. 1987.)

1986—Jan. 17—(WBA) Tim Witherspoon def. Tony Tubbs, 15, Atlanta, GA.

1986—Mar. 23—(WBC) Trevor Berbick def. Pinklon Thomas, 12, Miami.

1986—Nov. 22—(WBC) Mike Tyson KOd Trevor Berbick, 2, Las Vegas.

1986—Dec. 12—(WBA) James "Bonecrusher" Smith KOd Tim Witherspoon, 1, New York.

1987—Mar. 7—(WBA, WBC) Mike Tyson def. James "Bonecrusher" Smith, 12, Las Vegas.

1987—May 30—(IBF) Tony Tucker KO'd James "Buster" Douglas, 10, Las Vegas.

1987—Aug. 1—(WBA, WBC, IBF) Mike Tyson def. Tony Tucker, 12, Las Vegas. (Tyson became undisputed champion.)

1990—Feb. 11—(WBA, WBC, IBF) James "Buster" Douglas KOd Mike Tyson, 10, Tokyo.

1990—Oct. 25—(WBA, WBC, IBF) Evander Holyfield KOd James "Buster" Douglas, 3, Las Vegas.

1992—Nov. 13—(WBA, WBC, IBF) Riddick Bowe def. Evander Holyfield, 12, Las Vegas. (Lennox Lewis was later named WBC champion when Bowe refused to fight him.)

1993—Nov. 6—(WBA, IBF) Evander Holyfield def. Riddick Bowe, 12, Las Vegas.

1994—Apr. 22—(WBA, IBF) Michael Moorer def. Evander Holyfield, 12, Las Vegas.

1994—Sept. 24—(WBC) Oliver McCall KOd Lennox Lewis, 2, London.

1994—Nov. 5—(WBA, IBF) George Foreman KOd Michael Moorer, 10, Las Vegas. (In Mar. 1995, Foreman was stripped of the WBA title. In June, Foreman relinquished the IBF title.)

1995—Sept. 2—(WBC) Frank Bruno def. Oliver McCall, 12, London.

1995—Dec. 9—(IBF) Frans Botha def. Axel Schulz, 12, Las Vegas. (Botha was subsequently stripped of title.)

1996—Mar. 16—(WBC) Mike Tyson KOd Frank Bruno, 3, Las Vegas.

1996—June 22—(IBF) Michael Moorer def. Axel Schulz, 12, Dortmund, Germany.

1996—Sept. 7—(WBA, WBC) Mike Tyson KOd Bruce Seldon, 1, Las Vegas. (Tyson was subsequently stripped of the WBC title.)

1996—Nov. 9—(WBA) Evander Holyfield KOd Mike Tyson, 11, Las Vegas.

1997—Feb. 7—(WBC) Lennox Lewis KOd Oliver McCall, 5, Las Vegas.

1997—Nov. 8—(IBF) Evander Holyfield def. Michael Moorer, 8, Las Vegas.

YACHTING
The America's Cup

In the 1995 America's Cup match, the New Zealand yacht *Black Magic 1* defeated the U.S. yacht *Young America* 5-0 in the waters off San Diego, CA. It was only the 2d time since 1851 (the 1st since 1983) that the U.S. lost the Cup. *Black Magic 1* was skippered by Russell Coutts. On Oct. 18, 1999, 11 teams from 7 nations began competing for the right to challenge New Zealand for the America's Cup, in New Zealand from Feb. 14 to March 9, 2000.

Competition for the America's Cup grew out of the first contest to establish a world yachting championship, one of the carnival features of the London Exposition of 1851. The race covered a 60-mile course around the Isle of Wight; the prize was a cup worth about $500, donated by the Royal Yacht Squadron of England, known as the "America's Cup" because it was first won by the U.S. yacht *America*.

Winners of the America's Cup

1851 America
1870 Magic defeated Cambria, England, (1-0)
1871 Columbia (first three races) and Sappho (last two races) defeated Livonia, England, (4-1)
1876 Madeline defeated Countess of Dufferin, Canada, (2-0)
1881 Mischief defeated Atalanta, Canada, (2-0)
1885 Puritan defeated Genesta, England, (2-0)
1886 Mayflower defeated Galatea, England, (2-0)
1887 Volunteer defeated Thistle, Scotland, (2-0)
1893 Vigilant defeated Valkyrie II, England, (3-0)
1895 Defender defeated Valkyrie III, England, (3-0)
1899 Columbia defeated Shamrock, England, (3-0)
1901 Columbia defeated Shamrock II, England, (3-0)
1903 Reliance defeated Shamrock III, England, (3-0)
1920 Resolute defeated Shamrock IV, England, (3-2)
1930 Enterprise defeated Shamrock V, England, (4-0)

1934 Rainbow defeated Endeavour, England, (4-2)
1937 Ranger defeated Endeavour II, England, (4-0)
1958 Columbia defeated Sceptre, England, (4-0)
1962 Weatherly defeated Gretel, Australia, (4-1)
1964 Constellation defeated Sovereign, England, (4-0)
1967 Intrepid defeated Dame Pattie, Australia, (4-0)
1970 Intrepid defeated Gretel II, Australia, (4-1)
1974 Courageous defeated Southern Cross, Australia, (4-0)
1977 Courageous defeated Australia, Australia, (4-0)
1980 Freedom defeated Australia, Australia, (4-1)
1983 Australia II, Australia, defeated Liberty, (4-3)
1987 Stars & Stripes defeated Kookaburra III, Australia, (4-0)
1988 Stars & Stripes defeated New Zealand, New Zealand, (2-0)
1992 America³ defeated Il Moro di Venezia, Italy, (4-1)
1995 Black Magic 1, New Zealand, defeated Young America, (5-0)

POWER BOATING
American Power Boat Assn. Gold Cup Champions

Year	Boat	Driver	Year	Boat	Driver
1978	Atlas Van Lines	Bill Muncey	1989	Miss Budweiser	Tom D'Eath
1979	Atlas Van Lines	Bill Muncey	1990	Miss Budweiser	Tom D'Eath
1980	Miss Budweiser	Dean Chenoweth	1991	Winston Eagle	Mark Tate
1981	Miss Budweiser	Dean Chenoweth	1992	Miss Budweiser	Chip Hanauer
1982	Atlas Van Lines	Chip Hanauer	1993	Miss Budweiser	Chip Hanauer
1983	Atlas Van Lines	Chip Hanauer	1994	Smokin' Joe's	Mark Tate
1984	Atlas Van Lines	Chip Hanauer	1995	Miss Budweiser	Chip Hanauer
1985	Miller American	Chip Hanauer	1996	Pico American Dream	Dave Villwock
1986	Miller American	Chip Hanauer	1997	Miss Budweiser	Dave Villwock
1987	Miller American	Chip Hanauer	1998	Miss Budweiser	Dave Villwock
1988	Circus Circus	Chip Hanauer	1999	Miss PICO	Chip Hanauer

DOGS
Westminster Kennel Club

Year	Best-in-show	Breed	Owner(s)
1989	Ch. Royal Tudor's Wild As The Wind	Doberman	Sue & Art Kemp, Richard & Carolyn Vida, Beth Wilhite
1990	Ch. Wendessa Crown Prince	Pekingese	Ed Jenner
1991	Ch. Whisperwind on a Carousel	Poodle	Joan & Frederick Hartsock
1992	Ch. Registry's Lonesome Dove	Fox Terrier	Marion & Sam Lawrence
1993	Ch. Salilyn's Condor	English Springer Spaniel	Donna & Roger Herzig
1994	Ch. Chidley Willum	Norwich Terrier	Ruth Cooper & Patricia Lussier
1995	Ch. Gaelforce Post Script	Scottish Terrier	Dr. Vandra Huber & Dr. Joe Kinnarney
1996	Ch. Clussexx Country Sunrise	Clumber Spaniel	Judith & Richard Zaleski
1997	Ch. Parsifal Di Casa Netzer	Standard Schnauzer	Rita Holloway & Gabrio Del Torre
1998	Ch. Fairewood Frolic	Norwich Terrier	Sandina Kennels
1999	Ch. Loteki Supernatural Being	Papillon	John Oulton

Iditarod Trail Sled Dog Race in 1999

Doug Swingley of Lincoln, MT, won the 1999 Iditarod Trail Sled Dog Race, Mar. 17, 1999, with a time of 9 days, 14 hours, 31 minutes, 7 seconds. For winning the 1,100-mile race from Anchorage to Nome, AK, Swingley received $54,000 in prize money, as well as a truck and $9,000 in bonuses along the trail. Swingley also won the race in 1995 in record time (9d, 2h, 42m, 19s.)

CYCLING
Tour de France in 1999

On July 25, 1999, Lance Armstrong became only the second American to win the Tour de France, and the first to do so riding for an American team (U.S. Postal Service). Armstrong finished the 21-stage, 2,455-mile (3,950-km) race in 91 hr., 32 min., 16 sec. Alex Zülle of Switzerland finished 2d, 7 min., 37 sec. off the lead. Armstrong's victory was especially dramatic because in late 1996 he learned he had cancer and at that time was given only a 50-50 chance of survival.

THOROUGHBRED RACING
Triple Crown Winners

Since 1920, colts have carried 126 lb. in triple crown events; fillies, 121 lb.
(Kentucky Derby, Preakness, and Belmont Stakes)

Year	Horse	Jockey	Trainer	Year	Horse	Jockey	Trainer
1919	Sir Barton	J. Loftus	H. G. Bedwell	1946	Assault	W. Mehrtens	M. Hirsch
1930	Gallant Fox	E. Sande	J. Fitzsimmons	1948	Citation	E. Arcaro	H. A. Jones
1935	Omaha	W. Sanders	J. Fitzsimmons	1973	Secretariat	R. Turcotte	L. Laurin
1937	War Admiral	C. Kurtsinger	G. Conway	1977	Seattle Slew	J. Cruguet	W. H. Turner Jr.
1941	Whirlaway	E. Arcaro	B. A. Jones	1978	Affirmed	S. Cauthen	L. S. Barrera
1943	Count Fleet	J. Longden	G. D. Cameron				

Kentucky Derby

Churchill Downs, Louisville, KY; inaug. 1875; distance 1-1/4 mi; 1-1/2 mi until 1896. 3-year-olds.
Best time: 1:59.2, by Secretariat, 1973; 1999 time: 2:03.15.

Year	Winner	Jockey	Year	Winner	Jockey	Year	Winner	Jockey
1875	Aristides	O. Lewis	1917	Omr Khayyam	C. Borel	1959	Tomy Lee	W. Shoemaker
1876	Vagrant	R. Swim	1918	Exterminator	W. Knapp	1960	Venetian Way	W. Hartack
1877	Baden Baden	W. Walker	1919	Sir Barton	J. Loftus	1961	Carry Back	J. Sellers
1878	Day Star	Carter	1920	Paul Jones	T. Rice	1962	Decidedly	W. Hartack
1879	Lord Murphy	C. Schauer	1921	Behave Yourself	C. Thompson	1963	Chateaugay	B. Baeza
1880	Fonso	G. Lewis	1922	Morvich	A. Johnson	1964	Northern Dancer	W. Hartack
1881	Hindoo	J. McLaughlin	1923	Zev	E. Sande	1965	Lucky Debonair	W. Shoemaker
1882	Apollo	B. Hurd	1924	Black Gold	J. D. Mooney	1966	Kauai King	D. Brumfield
1883	Leonatus	W. Donohue	1925	Flying Ebony	E. Sande	1967	Proud Clarion	R. Ussery
1884	Buchanan	I. Murphy	1926	Bubbling Over	A. Johnson	1968	Dancer's Image#	R. Ussery
1885	Joe Cotton	E. Henderson	1927	Whiskery	L. McAtee	1969	Majestic Prince	W. Hartack
1886	Ben Ali	P. Duffy	1928	Reigh Count	C. Lang	1970	Dust Commander	M. Manganello
1887	Montrose	I. Lewis	1929	Clyde Van Dusen	L. McAtee	1971	Canonero II	G. Avila
1888	Macbeth II	G. Covington	1930	Gallant Fox	E. Sande	1972	Riva Ridge	R. Turcotte
1889	Spokane	T. Kiley	1931	Twenty Grand	C. Kurtsinger	1973	Secretariat	R. Turcotte
1890	Riley	I. Murphy	1932	Burgoo King	E. James	1974	Cannonade	A. Cordero
1891	Kingman	I. Murphy	1933	Brokers Tip	D. Meade	1975	Foolish Pleasure	J. Vasquez
1892	Azra	A. Clayton	1934	Cavalcade	M. Garner	1976	Bold Forbes	A. Cordero
1893	Lookout	E. Kunze	1935	Omaha	W. Saunders	1977	Seattle Slew	J. Cruguet
1894	Chant	F. Goodale	1936	Bold Venture	I. Hanford	1978	Affirmed	S. Cauthen
1895	Halma	J. Perkins	1937	War Admiral	C. Kurtsinger	1979	Spectacular Bid	R. Franklin
1896	Ben Brush	W. Simms	1938	Lawrin	E. Arcaro	1980	Genuine Risk*	J. Vasquez
1897	Typhoon II	F. Garner	1939	Johnstown	J. Stout	1981	Pleasant Colony	J. Velasquez
1898	Plaudit	W. Simms	1940	Gallahadion	C. Bierman	1982	Gato del Sol	E. Delahoussaye
1899	Manuel	F. Taral	1941	Whirlaway	E. Arcaro	1983	Sunny's Halo	E. Delahoussaye
1900	Lieut. Gibson	J. Boland	1942	Shut Out	W. D. Wright	1984	Swale	L. Pincay
1901	His Eminence	J. Winkfield	1943	Count Fleet	J. Longden	1985	Spend a Buck	A. Cordero
1902	Alan-a-Dale	J. Winkfield	1944	Pensive	C. McCreary	1986	Ferdinand	W. Shoemaker
1903	Judge Himes	H. Booker	1945	Hoop, Jr.	E. Arcaro	1987	Alysheba	C. McCarron
1904	Elwood	F. Prior	1946	Assault	W. Mehrtens	1988	Winning Colors*	G. Stevens
1905	Agile	J. Martin	1947	Jet Pilot	E. Guerin	1989	Sunday Silence	P. Valenzuela
1906	Sir Huon	R. Troxler	1948	Citation	E. Arcaro	1990	Unbridled	C. Perret
1907	Pink Star	A. Minder	1949	Ponder	S. Brooks	1991	Strike the Gold	C. Antley
1908	Stone Street	A. Pickens	1950	Middleground	W. Boland	1992	Lil E. Tee	P. Day
1909	Wintergreen	V. Powers	1951	Count Turf	C. McCreary	1993	Sea Hero	J. Bailey
1910	Donau	F. Herbert	1952	Hill Gail	E. Arcaro	1994	Go for Gin	C. McCarron
1911	Meridian	G. Archibald	1953	Dark Star	H. Moreno	1995	Thunder Gulch	G. Stevens
1912	Worth	C.H. Shilling	1954	Determine	R. York	1996	Grindstone	J. Bailey
1913	Donerail	R. Goose	1955	Swaps	W. Shoemaker	1997	Silver Charm	G. Stevens
1914	Old Rosebud	J. McCabe	1956	Needles	D. Erb	1998	Real Quiet	K. Desormeaux
1915	Regret*	J. Notter	1957	Iron Liege	W. Hartack	1999	Charismatic	C. Antley
1916	George Smith	J. Loftus	1958	Tim Tam	I. Valenzuela			

*Regret, Genuine Risk, and Winning Colors are the only fillies to have won the Derby. # Dancer's Image was disqualified from purse money after tests disclosed that he had run with a pain-killing drug, phenylbutazone, in his system. All wagers were paid on Dancer's Image. Forward Pass was awarded first place money.

The Kentucky Derby has been won 5 times by 2 jockeys: Eddie Arcaro, 1938, 1941, 1945, 1948, and 1952; and Bill Hartack, 1957, 1960, 1962, 1964, and 1969. It was won 4 times by Willie Shoemaker, 1955, 1959, 1965, and 1986; and 3 times by each of 4 jockeys: Isaac Murphy, 1884, 1890, and 1891; Earle Sande, 1923, 1925, and 1930; Angel Cordero, 1974, 1976, and 1985; and Gary Stevens, 1988, 1995, and 1997.

Preakness

Pimlico, Baltimore, MD; inaug. 1873; distance 1-3/16 mi. 3-year-olds.
Best time: 1:53.2, by Tank's Prospect (1985) and Louis Quatorze (1996); 1999 time: 1:55.32.

Year	Winner	Jockey	Year	Winner	Jockey	Year	Winner	Jockey
1873	Survivor	G. Barbee	1888	Refund	F. Littlefield	1906	Whimsical	W. Miller
1874	Culpepper	M. Donohue	1889	Buddhist	G. Anderson	1907	Don Enrique	G. Mountain
1875	Tom Ochiltree	L. Hughes	1890	Montague	W. Martin	1908	Royal Tourist	E. Dugan
1876	Shirley	G. Barbee	1894	Assignee	F. Taral	1909	Effendi	W. Doyle
1877	Cloverbrook	C. Holloway	1895	Belmar	F. Taral	1910	Layminster	R. Estep
1878	Duke of Magenta	C. Holloway	1896	Margrave	H. Griffin	1911	Watervale	E. Dugan
1879	Harold	L. Hughes	1897	Paul Kauvar	C. Thorpe	1912	Colonel Holloway	C. Turner
1880	Grenada	L. Hughes	1898	Sly Fox	W. Simms	1913	Buskin	J. Butwell
1881	Saunterer	W. Costello	1899	Half Time	R. Clawson	1914	Holiday	A. Schuttinger
1882	Vanguard	W. Costello	1900	Hindus	H. Spencer	1915	Rhine Maiden	D. Hoffman
1883	Jacobus	G. Barbee	1901	The Parader	F. Landry	1916	Damrosch	L. McAtee
1884	Knight of Ellerslie	S. H. Fisher	1902	Old England	L. Jackson	1917	Kalitan	E. Haynes
1885	Tecumseh	J. McLaughlin	1903	Flocarline	W. Gannon	1918	War Cloud	J. Loftus
1886	The Bard	S. H. Fisher	1904	Bryn Mawr	E. Hildebrand		Jack Hare, Jr.	C. Peak
1887	Dunboyne	W. Donohue	1905	Cairngorm	W. Davis	1919	Sir Barton	J. Loftus

Year	Winner	Jockey	Year	Winner	Jockey	Year	Winner	Jockey
1920	Man o' War	C. Kummer	1947	Faultless	D. Dodson	1974	Little Current	M. Rivera
1921	Broomspun	F. Coltiletti	1948	Citation	E. Arcaro	1975	Master Derby	D. McHargue
1922	Pillory	L. Morris	1949	Capot	T. Atkinson	1976	Elocutionist	J. Lively
1923	Vigil	B. Marinelli	1950	Hill Prince	E. Arcaro	1977	Seattle Slew	J. Cruguet
1924	Nellie Morse	J. Merimee	1951	Bold	E. Arcaro	1978	Affirmed	S. Cauthen
1925	Coventry	C. Kummer	1952	Blue Man	C. McCreary	1979	Spectacular Bid	R. Franklin
1926	Display	J. Malben	1953	Native Dancer	E. Guerin	1980	Codex	A. Cordero
1927	Bostonian	A. Abel	1954	Hasty Road	J. Adams	1981	Pleasant Colony	J. Velasquez
1928	Victorian	R. Workman	1955	Nashua	E. Arcaro	1982	Aloma's Ruler	J. Kaenel
1929	Dr. Freeland	L. Schaefer	1956	Fabius	W. Hartack	1983	Deputed Testamony	D. Miller
1930	Gallant Fox	E. Sande	1957	Bold Ruler	E. Arcaro	1984	Gate Dancer	A. Cordero
1931	Mate	G. Ellis	1958	Tim Tam	I. Valenzuela	1985	Tank's Prospect	P. Day
1932	Burgoo King	E. James	1959	Royal Orbit	W. Harmatz	1986	Snow Chief	A. Solis
1933	Head Play	C. Kurtsinger	1960	Bally Ache	R. Ussery	1987	Alysheba	C. McCarron
1934	High Quest	R. Jones	1961	Carry Back	J. Sellers	1988	Risen Star	E. Delahoussaye
1935	Omaha	W. Saunders	1962	Greek Money	J.L. Rotz	1989	Sunday Silence	P. Valenzuela
1936	Bold Venture	G. Woolf	1963	Candy Spots	W. Shoemaker	1990	Summer Squall	P. Day
1937	War Admiral	C. Kurtsinger	1964	Northern Dancer	W. Hartack	1991	Hansel	J. Bailey
1938	Dauber	M. Peters	1965	Tom Rolfe	R. Turcotte	1992	Pine Bluff	C. McCarron
1939	Challedon	G. Seabo	1966	Kauai King	D. Brumfield	1993	Prairie Bayou	M. Smith
1940	Bimelech	F.A. Smith	1967	Damascus	W. Shoemaker	1994	Tabasco Cat	P. Day
1941	Whirlaway	E. Arcaro	1968	Forward Pass	I. Valenzuela	1995	Timber Country	P. Day
1942	Alsab	B. James	1969	Majestic Prince	W. Hartack	1996	Louis Quatorze	P. Day
1943	Count Fleet	J. Longden	1970	Personality	E. Belmonte	1997	Silver Charm	G. Stevens
1944	Pensive	C. McCreary	1971	Canonero II	G. Avila	1998	Real Quiet	K. Desormeaux
1945	Polynesian	W.D. Wright	1972	Bee Bee Bee	E. Nelson	1999	Charismatic	C. Antley
1946	Assault	W. Mehrtens	1973	Secretariat	R. Turcotte			

Belmont Stakes

Belmont Park, Elmont, NY; inaug. 1867; distance 1-1/2 mi. 3-year-olds. Best time: 2:24, Secretariat, 1973; 1999 time: 2:27.88.

Year	Winner	Jockey	Year	Winner	Jockey	Year	Winner	Jockey
1867	Ruthless	J. Gilpatrick	1913	Prince Eugene	R. Troxler	1957	Gallant Man	W. Shoemaker
1868	General Duke	R. Swim	1914	Luke McLuke	M. Buxton	1958	Cavan	P. Anderson
1869	Fenian	C. Miller	1915	The Finn	G. Byrne	1959	Sword Dancer	W. Shoemaker
1870	Kingfisher	W. Dick	1916	Friar Rock	E. Haynes	1960	Celtic Ash	W. Hartack
1871	Harry Bassett	W. Miller	1917	Hourless	J. Butwell	1961	Sherluck	B. Baeza
1872	Joe Daniels	J. Rowe	1918	Johren	F. Robinson	1962	Jaipur	W. Shoemaker
1873	Springbok	J. Rowe	1919	Sir Barton	J. Loftus	1963	Chateaugay	B. Baeza
1874	Saxon	G. Barbee	1920	Man o' War	C. Kummer	1964	Quadrangle	M. Ycaza
1875	Calvin	R. Swim	1921	Grey Lag	E. Sande	1965	Hail to All	J. Sellers
1876	Algerine	W. Donohue	1922	Pillory	C. H. Miller	1966	Amberoid	W. Boland
1877	Cloverbrook	C. Holloway	1923	Zev	E. Sande	1967	Damascus	W. Shoemaker
1878	Duke of Magenta	L. Hughes	1924	Mad Play	E. Sande	1968	Stage Door Johnny	H. Gustines
1879	Spendthrift	S. Evans	1925	American Flag	A. Johnson	1969	Arts and Letters	B. Baeza
1880	Grenada	L. Hughes	1926	Crusader	A. Johnson	1970	High Echelon	J. L. Rotz
1881	Saunterer	T. Costello	1927	Chance Shot	E. Sande	1971	Pass Catcher	W. Blum
1882	Forester	J. McLaughlin	1928	Vito	C. Kummer	1972	Riva Ridge	R. Turcotte
1883	George Kinney	J. McLaughlin	1929	Blue Larkspur	M. Garner	1973	Secretariat	R. Turcotte
1884	Panique	J. McLaughlin	1930	Gallant Fox	E. Sande	1974	Little Current	M. Rivera
1885	Tyrant	P. Duffy	1931	Twenty Grand	C. Kurtsinger	1975	Avatar	W. Shoemaker
1886	Inspector B.	J. McLaughlin	1932	Faireno	T. Malley	1976	Bold Forbes	A. Cordero
1887	Hanover	J. McLaughlin	1933	Hurryoff	M. Garner	1977	Seattle Slew	J. Cruguet
1888	Sir Dixon	J. McLaughlin	1934	Peace Chance	W. D. Wright	1978	Affirmed	S. Cauthen
1889	Eric	W. Hayward	1935	Omaha	W. Saunders	1979	Coastal	R. Hernandez
1890	Burlington	S. Barnes	1936	Granville	J. Stout	1980	Temperence Hill	E. Maple
1891	Foxford	E. Garrison	1937	War Admiral	C. Kurtsinger	1981	Summing	G. Martens
1892	Patron	W. Hayward	1938	Pasteurized	J. Stout	1982	Conquistador Cielo	L. Pincay
1893	Comanche	W. Simms	1939	Johnstown	J. Stout	1983	Caveat	L. Pincay
1894	Henry of Navarre	W. Simms	1940	Bimelech	F. A. Smith	1984	Swale	L. Pincay
1895	Belmar	F. Taral	1941	Whirlaway	E. Arcaro	1985	Creme Fraiche	E. Maple
1896	Hastings	H. Griffin	1942	Shut Out	E. Arcaro	1986	Danzig Connection	C. McCarron
1897	Scottish Chieftain	J. Scherrer	1943	Count Fleet	J. Longden	1987	Bet Twice	C. Perret
1898	Bowling Brook	F. Littlefield	1944	Bounding Home	G. L. Smith	1988	Risen Star	E. Delahoussaye
1899	Jean Bereaud	R. R. Clawson	1945	Pavot	E. Arcaro	1989	Easy Goer	P. Day
1900	Ildrim	N. Turner	1946	Assault	W. Mehrtens	1990	Go and Go	M. Kinane
1901	Commando	H. Spencer	1947	Phalanx	R. Donoso	1991	Hansel	J. Bailey
1902	Masterman	J. Bullman	1948	Citation	E. Arcaro	1992	A.P. Indy	E. Delahoussaye
1903	Africander	J. Bullman	1949	Capot	T. Atkinson	1993	Colonial Affair	J. Krone
1904	Delhi	G. Odom	1950	Middleground	W. Boland	1994	Tabasco Cat	P. Day
1905	Tanya	E. Hildebrand	1951	Counterpoint	D. Gorman	1995	Thunder Gulch	G. Stevens
1906	Burgomaster	L. Lyne	1952	One Count	E. Arcaro	1996	Editor's Note	R. Douglas
1907	Peter Pan	G. Mountain	1953	Native Dancer	E. Guerin	1997	Touch Gold	C. McCarron
1908	Colin	J. Notter	1954	High Gun	E. Guerin	1998	Victory Gallop	G. Stevens
1909	Joe Madden	E. Dugan	1955	Nashua	E. Arcaro	1999	Lemon Drop Kid	J. Santos
1910	Sweep	J. Butwell	1956	Needles	D. Erb			

Annual Leading Jockey — Money Won[1]

Year	Jockey	Dollars	Year	Jockey	Dollars	Year	Jockey	Dollars
1957	Bill Hartack	$3,060,501	1965	Braulio Baeza	$2,582,702	1973	Laffit Pincay, Jr.	$4,093,492
1958	Willie Shoemaker	2,961,693	1966	Braulio Baeza	2,951,022	1974	Laffit Pincay, Jr.	4,251,060
1959	Willie Shoemaker	2,843,133	1967	Braulio Baeza	3,088,888	1975	Braulio Baeza	3,695,198
1960	Willie Shoemaker	2,123,961	1968	Braulio Baeza	2,835,108	1976	Angel Cordero, Jr.	4,709,500
1961	Willie Shoemaker	2,690,819	1969	Jorge Velasquez	2,542,315	1977	Steve Cauthen	6,151,750
1962	Willie Shoemaker	2,916,844	1970	Laffit Pincay, Jr.	2,626,526	1978	Darrel McHargue	6,029,885
1963	Willie Shoemaker	2,526,925	1971	Laffit Pincay, Jr.	3,784,377	1979	Laffit Pincay, Jr.	8,193,535
1964	Willie Shoemaker	2,649,553	1972	Laffit Pincay, Jr.	3,225,827	1980	Chris McCarron	7,663,300

Year	Jockey	Dollars	Year	Jockey	Dollars	Year	Jockey	Dollars
1981	Chris McCarron	$8,397,604	1987	Jose Santos.	$12,375,433	1993	Mike Smith	$14,024,815
1982	Angel Cordero, Jr. . .	9,483,590	1988	Jose Santos.	14,877,298	1994	Mike Smith	15,979,820
1983	Angel Cordero, Jr. . .	10,116,697	1989	Jose Santos.	13,838,389	1995	Jerry Bailey.	16,311,876
1984	Chris McCarron	12,045,813	1990	Gary Stevens.	13,881,198	1996	Jerry Bailey.	19,465,376
1985	Laffit Pincay, Jr. . .	13,353,299	1991	Chris McCarron	14,441,083	1997	Jerry Bailey.	18,320,743
1986	Jose Santos	11,329,297	1992	Kent Desormeaux . .	14,193,006	1998	Gary Stevens	19,622,855

(1) Total earnings for all horses that jockey raced in year listed; does not reflect jockey's earnings.

Breeders' Cup

The Breeders' Cup was inaugurated in 1984 and consists of 7 races at one track on one day late in the year to determine Thoroughbred racing's champion contenders. It has been held at the following locations:

1984 Hollywood Park, CA	1989 Gulfstream Park, FL	1994 Churchill Downs, KY
1985 Aqueduct Racetrack, NY	1990 Belmont Park, NY	1995 Belmont Park, NY
1986 Santa Anita Park, CA	1991 Churchill Downs, KY	1996 Woodbine Racetrack, Ontario
1987 Hollywood Park, CA	1992 Gulfstream Park, FL	1997 Hollywood Park, CA
1988 Churchill Downs, KY	1993 Santa Anita Park, CA	1998 Churchill Downs, KY

Juvenile
Distances: 1 mi 1984-85, 1987; 1-1/16 mi 1986 and since 1988

Year		Jockey	Year		Jockey	Year		Jockey
1984	Chief's Crown	D. MacBeth	1989	Rhythm	C. Perret	1994	Timber Country	P. Day
1985	Tasso	L. Pincay, Jr.	1990	Fly So Free	J. Santos	1995	Unbridled's Song	M. Smith
1986	Capote	L. Pincay, Jr.	1991	Arazi	P. Valenzuela	1996	Boston Harbor	J. Bailey
1987	Success Express	J. Santos	1992	Gilded Time	C. McCarron	1997	Favorite Trick	P. Day
1988	Is It True	L. Pincay, Jr.	1993	Brocco	G. Stevens	1998	Answer Lively	J. Bailey

Juvenile Fillies
Distances: 1 mi 1984-85, 1987; 1-1/16 mi 1986 and since 1988

Year		Jockey	Year		Jockey	Year		Jockey
1984	*Outstandingly	W. Guerra	1989	Go for Wand	R. Romero	1994	Flanders	P. Day
1985	Twilight Ridge	J. Velasquez	1990	Meadow Star	J. Santos	1995	My Flag	J. Bailey
1986	Brave Raj	P. Valenzuela	1991	Pleasant Stage	E. Delahoussaye	1996	Storm Song	C. Perret
1987	Epitome	P. Day	1992	Eliza	P. Valenzuela	1997	Countess Diana	S. Sellers
1988	Open Mind	A. Cordero, Jr.	1993	Phone Chatter	L. Pincay, Jr.	1998	Silverbulletday	G. Stevens

*By disqualification.

Sprint
Distance: 6 furlongs

Year		Jockey	Year		Jockey	Year		Jockey
1984	Eillo	C. Perret	1989	Dancing Spree	A. Cordero, Jr.	1994	Cherokee Run	M. Smith
1985	Precisionist	C. McCarron	1990	Safely Kept	C. Perret	1995	Desert Stormer	K. Desormeaux
1986	Smile	J. Vasquez	1991	Sheikh Albadou	P. Eddery	1996	Lit De Justice	C. Nakatani
1987	Very Subtle	P. Valenzuela	1992	Thirty Slews	E. Delahoussaye	1997	Elmhurst	C. Nakatani
1988	Gulch	A. Cordero, Jr.	1993	Cardmania	E. Delahoussaye	1998	Reraise	C. Nakatani

Mile

Year		Jockey	Year		Jockey	Year		Jockey
1984	Royal Heroine	F. Toro	1989	Steinlen	J. Santos	1994	Barathea	L. Dettori
1985	Cozzene	W. Guerra	1990	Royal Academy	L. Piggott	1995	Ridgewood Pearl	J. Murtagh
1986	Last Tycoon	Y. St.-Martin	1991	Opening Verse	P. Valenzuela	1996	Da Hoss	G. Stevens
1987	Miesque	F. Head	1992	Lure	M. Smith	1997	Spinning World	C. Asmussan
1988	Miesque	F. Head	1993	Lure	M. Smith	1998	Da Hoss	J. Velazquez

Distaff
Distances: 1-1/4 mi 1984-87; 1-1/8 mi since 1988

Year		Jockey	Year		Jockey	Year		Jockey
1984	Princess Rooney	E. Delahoussaye	1989	Bayakoa	L. Pincay, Jr.	1994	One Dreamer	G. Stevens
1985	Life's Magic	A. Cordero, Jr.	1990	Bayakoa	L. Pincay, Jr.	1995	Inside Information	M. Smith
1986	Lady's Secret	P. Day	1991	Dance Smartly	P. Day	1996	Jewel Princess	C. Nakatani
1987	Sacahuista	R. Romero	1992	Paseana	C. McCarron	1997	Ajina	M. Smith
1988	Personal Ensign	R. Romero	1993	Hollywood Wildcat	E. Delahoussaye	1998	Escena	G. Stevens

Turf
Distance: 1-1/2 mi

Year		Jockey	Year		Jockey	Year		Jockey
1984	Lashkari	Y. St.-Martin	1989	Prized	E. Delahoussaye	1995	Northern Spur	C. McCarron
1985	Pebbles	P. Eddery	1990	In The Wings	G. Stevens	1996	Pilsudski	W. Swinburn
1986	Manila	J. Santos	1991	Miss Alleged	E. Legrix	1997	Chief Bearhart	J. Santos
1987	Theatrical	P. Day	1992	Fraise	P. Valenzuela	1998	Buck's Boy	S. Sellers
1988	Great Communicator	R. Sibille	1993	Kotashaan	K. Desormeaux			
			1994	Tikkanen	M. Smith			

Classic
Distance: 1-1/4 mi

Year		Jockey	Year		Jockey	Year		Jockey
1984	Wild Again	P. Day	1989	Sunday Silence	C. McCarron	1994	Concern	J. Bailey
1985	Proud Truth	J. Velasquez	1990	Unbridled	P. Day	1995	Cigar	J. Bailey
1986	Skywalker	L. Pincay, Jr.	1991	Black Tie Affair	J. Bailey	1996	Alphabet Soup	C. McCarron
1987	Ferdinand	W. Shoemaker	1992	A.P. Indy	E. Delahoussaye	1997	Skip Away	M. Smith
1988	Alysheba	C. McCarron	1993	Arcangues	J. Bailey	1998	Awesome Again	P. Day

Eclipse Awards

The Eclipse Awards, honoring the Horse of the Year and other champions of the sport, began in 1971 and are sponsored by the *Daily Racing Form,* the Thoroughbred Racing Associations, and the National Turf Writers Assn. Prior to 1971, the DRF (1936-70) and the TRA (1950-70) issued separate selections for Horse of the Year.

Eclipse Awards for 1998

Horse of the Year—Skip Away
2-year-old colt or gelding—Answer Lively
2-year-old filly—Silverbulletday
3-year-old colt or gelding—Real Quiet
3-year-old filly—Banshee Breeze
Older male (4-year-olds & up)—Skip Away
Older female (4-year-olds & up)—Escena
Male turf horse—Buck's Boy

Turf filly or mare—Fiji (GB)
Sprinter—Reraise
Steeplechase horse—Flat Top
Trainer—Bob Baffert
Jockey—Gary Stevens
Apprentice jockey—Shaun Bridgmohan
Breeder—John and Betty Mabee
Owner—Frank Stronach

Horse of the Year

Year	Horse	Year	Horse	Year	Horse	Year	Horse
1936	Granville	1953	Tom Fool	1968	Dr. Fager	1983	All Along
1937	War Admiral	1954	Native Dancer	1969	Arts and Letters	1984	John Henry
1938	Seabiscuit	1955	Nashua	1970	Fort Marcy (DRF)	1985	Spend A Buck
1939	Challedon	1956	Swaps		Personality (TRA)	1986	Lady's Secret
1940	Challedon	1957	Bold Ruler (DRF)	1971	Ack Ack	1987	Ferdinand
1941	Whirlaway		Dedicate (TRA)	1972	Secretariat	1988	Alysheba
1942	Whirlaway	1958	Round Table	1973	Secretariat	1989	Sunday Silence
1943	Count Fleet	1959	Sword Dancer	1974	Forego	1990	Criminal Type
1944	Twilight Tear	1960	Kelso	1975	Forego	1991	Black Tie Affair
1945	Busher	1961	Kelso	1976	Forego	1992	A.P. Indy
1946	Assault	1962	Kelso	1977	Seattle Slew	1993	Kotashaan
1947	Armed	1963	Kelso	1978	Affirmed	1994	Holy Bull
1948	Citation	1964	Kelso	1979	Affirmed	1995	Cigar
1949	Capot	1965	Roman Brother (DRF)	1980	Spectacular Bid	1996	Cigar
1950	Hill Prince		Moccasin (TRA)	1981	John Henry	1997	Favorite Trick
1951	Counterpoint	1966	Buckpasser	1982	Conquistador Cielo	1998	Skip Away
1952	One Count (DRF)	1967	Damascus				
	Native Dancer (TRA)						

HARNESS RACING
Harness Horse of the Year
(Chosen by the U.S. Trotting Assn. and the U.S. Harness Writers Assn.)

Year	Horse	Year	Horse	Year	Horse	Year	Horse
1947	Victory Song	1960	Adios Butler	1973	Sir Dalrae	1986	Forrest Skipper
1948	Rodney	1961	Adios Butler	1974	Delmonica Hanover	1987	Mack Lobell
1949	Good Time	1962	Su Mac Lad	1975	Savoir	1988	Mack Lobell
1950	Proximity	1963	Speedy Scot	1976	Keystone Ore	1989	Matt's Scooter
1951	Pronto Don	1964	Bret Hanover	1977	Green Speed	1990	Beach Towel
1952	Good Time	1965	Bret Hanover	1978	Abercrombie	1991	Precious Bunny
1953	Hi Lo's Forbes	1966	Bret Hanover	1979	Niatross	1992	Artsplace
1954	Stenographer	1967	Nevele Pride	1980	Niatross	1993	Staying Together
1955	Scott Frost	1968	Nevele Pride	1981	Fan Hanover	1994	Cam's Card Shark
1956	Scott Frost	1969	Nevele Pride	1982	Cam Fella	1995	CR Kay Suzie
1957	Torpid	1970	Fresh Yankee	1983	Cam Fella	1996	Continentalvictory
1958	Emily's Pride	1971	Albatross	1984	Fancy Crown	1997	Malabar Man
1959	Bye Bye Byrd	1972	Albatross	1985	Nihilator	1998	Moni Maker

The Hambletonian (3-year-old trotters)

Year	Winner	Driver	Year	Winner	Driver
1965	Egyptian Candor	Del Cameron	1983	Duenna	Stanley Dancer
1966	Kerry Way	Frank Ervin	1984	Historic Freight	Ben Webster
1967	Speedy Streak	Del Cameron	1985	Prakas	Bill O'Donnell
1968	Nevele Pride	Stanley Dancer	1986	Nuclear Kosmos	Ulf Thoresen
1969	Lindy's Pride	Howard Beissinger	1987	Mack Lobell	John Campbell
1970	Timothy T	John Simpson, Sr.	1988	Armbro Goal	John Campbell
1971	Speedy Crown	Howard Beissinger	1989	Park Avenue Joe	Ron Waples
1972	Super Bowl	Stanley Dancer	1990	Harmonious	John Campbell
1973	Flirth	Ralph Baldwin	1991	Giant Victory	Jack Moiseyev
1974	Christopher T	Bill Haughton	1992	Alf Palema	Mickey McNicholl
1975	Bonefish	Stanley Dancer	1993	American Winner	Ron Pierce
1976	Steve Lobell	Bill Haughton	1994	Victory Dream	Michel Lachance
1977	Green Speed	Bill Haughton	1995	Tagliabue	John Campbell
1978	Speedy Somolli	Howard Beissinger	1996	Continentalvictory	Michel Lachance
1979	Legend Hanover	George Sholty	1997	Malabar Man	Malvern Burroughs
1980	Burgomeister	Bill Haughton	1998	Muscles Yankee	John Campbell
1981	Shiaway St. Pat	Ray Remmen	1999	Self Possessed	Mike Lachance
1982	Speed Bowl	Tommy Haughton			

NCAA WRESTLING CHAMPIONS

Year	Champion	Year	Champion	Year	Champion	Year	Champion	Year	Champion
1964	Oklahoma State	1972	Iowa State	1980	Iowa	1988	Arizona State	1996	Iowa
1965	Iowa State	1973	Iowa State	1981	Iowa	1989	Oklahoma State	1997	Iowa
1966	Oklahoma State	1974	Oklahoma	1982	Iowa	1990	Oklahoma State	1998	Iowa
1967	Michigan State	1975	Iowa	1983	Iowa	1991	Iowa	1998	Iowa
1968	Oklahoma State	1976	Iowa	1984	Iowa	1992	Iowa	1999	Iowa
1969	Iowa State	1977	Iowa State	1985	Iowa	1993	Iowa		
1970	Iowa State	1978	Iowa	1986	Iowa	1994	Oklahoma State		
1971	Oklahoma State	1979	Iowa	1987	Iowa State	1995	Iowa		

BOWLING
Professional Bowlers Association
Hall of Fame
(1999 inductees have an asterisk)

PERFORMANCE

Bill Allen	Dave Ferraro	Johnny Petraglia	Walter Ray Williams, Jr.
Glenn Allison	Skee Foremsky	Dick Ritger	Wayne Zahn
Earl Anthony	Jim Godman	Mark Roth	
Barry Asher	Johnny Guenther	Jim St. John	**MERITORIOUS**
Mike Aulby	Billy Hardwick	Carmen Salvino	**SERVICE**
*Tom Baker	Tommy Hudson	Ernie Schlegel	Joe Antenora
*Roy Buckley	Dave Husted	Teata Semiz	John Archibald
Nelson Burton, Jr.	Don Johnson	Bob Strampe	Chuck Clemens
Don Carter	Joe Joseph	Harry Smith	Eddie Elias
Pat Colwell	Larry Laub	Dave Soutar	Frank Esposito
Steve Cook	Mike Limongello	Jim Stefanich	Dick Evans
Dave Davis	Don McCune	Brian Voss	Raymond Firestone
Gary Dickinson	Mike McGrath	Wayne Webb	E. A. "Bud" Fisher
Mike Durbin	Amleto Monacelli	Dick Weber	Lou Frantz
Buzz Fazio	David Ozio	Pete Weber	Harry Golden
	George Pappas	Billy Welu	Ted Hoffman, Jr.

John Jowdy
Joe Kelley
Larry Lichstein
Steve Nagy
*Keijiro Nakano
Chuck Pezzano
Jack Reichert
Joe Richards
Chris Schenkel
Lorraine Stilzlein
Al Thompson
Roger Zeller
Chuck Pezzano

Tournament of Champions

Year	Winner	Year	Winner	Year	Winner	Year	Winner
1965	Billy Hardwick	1974	Earl Anthony	1983	Joe Berardi	1992	Marc McDowell
1966	Wayne Zahn	1975	Dave Davis	1984	Mike Durbin	1993	George Branham, 3d
1967	Jim Stefanich	1976	Marshall Holman	1985	Mark Williams	1994	Norm Duke
1968	Dave Davis	1977	Mike Berlin	1986	Marshall Holman	1996	Dave D'Entremont
1969	Jim Godman	1978	Earl Anthony	1987	Pete Weber	1997	John Gant
1970	Don Johnson	1979	George Pappas	1988	Mark Williams	1998	Bryan Goebel
1971	Johnny Petraglia	1980	Wayne Webb	1989	Del Ballard, Jr.		
1972	Mike Durbin	1981	Steve Cook	1990	Dave Ferraro		
1973	Jim Godman	1982	Mike Durbin	1991	David Ozio		

PBA Leading Money Winners

Total winnings are from PBA, ABC Masters, and BPAA All-Star tournaments only and do not include numerous other tournaments or earnings from special television shows and matches.

Year	Bowler	Amount	Year	Bowler	Amount	Year	Bowler	Amount
1962	Don Carter	$49,972	1975	Earl Anthony	$107,585	1987	Pete Weber	$175,491
1963	Dick Weber	46,333	1976	Earl Anthony	110,833	1988	Brian Voss	225,485
1964	Bob Strampe	33,592	1977	Mark Roth	105,583	1989	Mike Aulby	298,237
1965	Dick Weber	47,674	1978	Mark Roth	134,500	1990	Amleto Monacelli	204,775
1966	Wayne Zahn	54,720	1979	Mark Roth	124,517	1991	David Ozio	225,585
1967	Dave Davis	54,165	1980	Wayne Webb	116,700	1992	Marc McDowell	174,215
1968	Jim Stefanich	67,377	1981	Earl Anthony	164,735	1993	Walter Ray Williams, Jr.	296,370
1969	Billy Hardwick	64,160	1982	Earl Anthony	134,760	1994	Norm Duke	273,753
1970	Mike McGrath	52,049	1983	Earl Anthony	135,605	1995	Mike Aulby	219,792
1971	Johnny Petraglia	85,065	1984	Mark Roth	158,712	1996	Walter Ray Williams, Jr.	241,330
1972	Don Johnson	56,648	1985	Mike Aulby	201,200	1997	Walter Ray Williams, Jr.	240,544
1973	Don McCune	69,000	1986	Walter Ray Williams, Jr.	145,550	1998	Walter Ray Williams, Jr.	238,225
1974	Earl Anthony	99,585						

Leading PBA Averages by Year

Year	Bowler	Average	Year	Bowler	Average	Year	Bowler	Average
1962	Don Carter	212.844	1975	Earl Anthony	219.060	1987	Marshall Holman	216.801
1963	Billy Hardwick	210.346	1976	Mark Roth	215.970	1988	Mark Roth	218.036
1964	Ray Bluth	210.512	1977	Mark Roth	218.174	1989	Pete Weber	215.432
1965	Dick Weber	211.895	1978	Mark Roth	219.834	1990	Amleto Monacelli	218.158
1966	Wayne Zahn	208.663	1979	Mark Roth	221.662	1991	Norm Duke	218.208
1967	Wayne Zahn	212.342	1980	Earl Anthony	218.535	1992	Dave Ferraro	219.702
1968	Jim Stefanich	211.895	1981	Mark Roth	216.699	1993	Walter Ray Williams, Jr.	222.980
1969	Bill Hardwick	212.957	1982	Marshall Holman	214.844	1994	Norm Duke	222.830
1970	Nelson Burton, Jr.	214.908	1983	Earl Anthony	216.645	1995	Mike Aulby	225.490
1971	Don Johnson	213.977	1984	Marshall Holman	213.911	1996	Walter Ray Williams, Jr.	225.370
1972	Don Johnson	215.290	1985	Mark Baker	213.718	1997	Walter Ray Williams, Jr.	222.008
1973	Earl Anthony	215.799	1986	John Gant	214.378	1998	Walter Ray Williams, Jr.	226.130
1974	Earl Anthony	219.394						

American Bowling Congress
ABC Masters Tournament Champions

Year	Winner	Year	Winner	Year	Winner
1980	Neil Burton, St. Louis, MO	1986	Mark Fahy, Chicago, IL	1993	Norm Duke, Oklahoma City, OK
1981	Randy Lightfoot, St. Charles, MO	1987	Rick Steelsmith, Wichita, KS	1994	Steve Fehr, Cincinnati, OH
1982	Joe Berardi, Brooklyn, NY	1988	Del Ballard, Jr., Richardson, TX	1995	Mike Aulby, Indianapolis, IN
1983	Mike Lastowski, Havre de Grace, MD	1989	Mike Aulby, Indianapolis, IN	1996	Ernie Schlegel, Vancouver, WA
1984	Earl Anthony, Dublin, CA	1990	Chris Warren, Dallas, TX	1997	Jason Queen, Decatur, IL
1985	Steve Wunderlich, St. Louis, MO	1991	Doug Kent, Canandaigua, NY	1998	Mike Aulby, Indianapolis, IN
		1992	Ken Johnson, N. Richmond Hills, TX	1999	Brian Boghosian, Middletown, CT

Champions in 1999

Regular Singles Event	Dan Winter, Rockford, IL	**Classified Singles Event**	William Burnett, Greer, SC
Regular Doubles Event	Ryan Lever, Warford, WI / Dale Traber, Cedarburg, WI	**Classified Doubles Event**	Larry Sifton/Thomas Scott, Lafayette, LA
Regular All Events	Thomas Jones, Greenville, SC	**Classified All Event**	William Burnett, Greer, SC
Regular Team	Zawadzki Jewelers, Lackawanna, NY	**Classified Team**	CC Roller #1, Virginia Beach, VA

Most Sanctioned 300 Games

Joe Jimenez, Saginaw, MI 59
Bob Learn Jr., Erie, PA. 56
Jim Johnson Jr., Tampa, FL. 54
Mike Whalin, Cincinnati, OH 53
Jeff Jensen, Wichita, KS 50
Dean Wolf, Reading, PA. 49
Bob Buckery, McAdoo, PA 49
Robert Faragon, Albany, NY 47
Jerry Kessler, Dayton, OH 45
Dave Frascatore Jr., Amsterdam, NY . 43
Ralph Burley Jr., Dayton, OH 42
John Wilcox Jr., Lewisburg, PA 42
John Delp III, West Lawn, PA 42

Jeff Carter, Springfield, IL 42
Ken Hall, Schenectady, NY 40
Bob J. Johnson, Dayton, OH 40
Jason Hurd, Tulare, CA 39
Mike Cowley, Dayton, OH 38
Jim Ewald Jr., Louisville, KY 37
Ron Krippelcz, St. Louis, MO 37
Keith Bruening, St. Charles, MO 36
Steve Gehringer, Reading, PA 36
Eric Roddy, New Orleans, LA 36
Alan Hulsizer, Wernersville, PA 36
Dale Strike, Saginaw, MI 36
Doug Spicer, W. Bloomfield, MI 35

Randy Choat, Granite City, IL 35
Randy Lightfoot, St. Charles MO 35
John Gualtieri, Edison, NJ. 34
Ron Woolet, New Albany, IN 33
Woodrow Crist, Williamsport, PA. . . .33
Richard Vigars, Schenectady, NY . . .33
John Chacko Jr., Larksville, PA 32
Michael Weston, Bay City, MI 32
Kevin Lickers, Wilkes-Barre, PA 31
Stephen Levering, Landisville, PA. . . .31
Hugo McGroty, New York, NY 30
Ron Bohnert, Cincinnati, OH. 30

Women's International Bowling Congress
Champions in 1999

Queens Tournament—Leanne Barrette, Pleasonton, CA
Singles Event—Nikki Gianulis, Vallejo, CA
All Events—Hidemi Mizobuchi, Japan

Doubles Event—Timi McCarvey, Huntsville, AL, and Marianne DiRupoa, Succasunna, NJ
Team—Cascade Beauty College, Renton, Wash

Most Sanctioned 300 Games

Tish Johnson, Panorama City, CA . .27
Aleta Sill, Dearborn, MI 23
Jeanne Naccarato, Tacoma, WA . . .22
Vicki Fischel, Wheat Ridge, CO 21

Leanne Barrette, Yukon, OK 20
Jodi Musto, Schenectady, NY 19
Debbie McMullen, Denver, CO 19

Jodi Hughes, Greenville, SC 17
Cheryl Daniels, Detroit, MI 16
Dede Davidson, Woodland Hills, CA . 16

FIGURE SKATING

| | U.S. Champions | | | World Champions | |
MEN	**WOMEN**	**YEAR**	**MEN**	**WOMEN**
Dick Button	Tenley Albright	1952	Dick Button, U.S.	Jacqueline du Bief, France
Hayes Jenkins	Tenley Albright	1953	Hayes Jenkins, U.S.	Tenley Albright, U.S.
Hayes Jenkins	Tenley Albright	1954	Hayes Jenkins, U.S.	Gundi Busch, W. Germany
Hayes Jenkins	Tenley Albright	1955	Hayes Jenkins, U.S.	Tenley Albright, U.S.
Hayes Jenkins	Tenley Albright	1956	Hayes Jenkins, U.S.	Carol Heiss, U.S.
Dave Jenkins	Carol Heiss	1957	Dave Jenkins, U.S.	Carol Heiss, U.S.
Dave Jenkins	Carol Heiss	1958	Dave Jenkins, U.S.	Carol Heiss, U.S.
Dave Jenkins	Carol Heiss	1959	Dave Jenkins, U.S.	Carol Heiss, U.S.
Dave Jenkins	Carol Heiss	1960	Alain Giletti, France	Carol Heiss, U.S.
Bradley Lord	Laurence Owen	1961	none	none
Monty Hoyt	Barbara Roles Pursley	1962	Don Jackson, Canada	Sjoukje Dijkstra, Netherlands
Tommy Litz	Lorraine Hanlon	1963	Don McPherson, Canada	Sjoukje Dijkstra, Netherlands
Scott Allen	Peggy Fleming	1964	Manfred Schnelldorfer, W. Germany	Sjoukje Dijkstra, Netherlands
Gary Visconti	Peggy Fleming	1965	Alain Calmat, France	Petra Burka, Canada
Scott Allen	Peggy Fleming	1966	Emmerich Danzer, Austria	Peggy Fleming, U.S.
Gary Visconti	Peggy Fleming	1967	Emmerich Danzer, Austria	Peggy Fleming, U.S.
Tim Wood	Peggy Fleming	1968	Emmerich Danzer, Austria	Peggy Fleming, U.S.
Tim Wood	Janet Lynn	1969	Tim Wood, U.S.	Gabriele Seyfert, E. Germany
Tim Wood	Janet Lynn	1970	Tim Wood, U.S.	Gabriele Seyfert, E. Germany
John Misha Petkevich	Janet Lynn	1971	Ondrej Nepela, Czechoslovakia	Beatrix Schuba, Austria
Ken Shelley	Janet Lynn	1972	Ondrej Nepela, Czechoslovakia	Beatrix Schuba, Austria
Gordon McKellen, Jr.	Janet Lynn	1973	Ondrej Nepela, Czechoslovakia	Karen Magnussen, Canada
Gordon McKellen, Jr.	Dorothy Hamill	1974	Jan Hoffmann, E. Germany	Christine Errath, E. Germany
Gordon McKellen, Jr.	Dorothy Hamill	1975	Sergei Volkov, USSR	Dianne de Leeuw, Neth.-U.S.
Terry Kubicka	Dorothy Hamill	1976	John Curry, Gr. Britain	Dorothy Hamill, U.S.
Charles Tickner	Linda Fratianne	1977	Vladimir Kovalev, USSR	Linda Fratianne, U.S.
Charles Tickner	Linda Fratianne	1978	Charles Tickner, U.S.	Anett Poetzsch, E. Germany
Charles Tickner	Linda Fratianne	1979	Vladimir Kovalev, USSR	Linda Fratianne, U.S.
Charles Tickner	Linda Fratianne	1980	Jan Hoffmann, E. Germany	Anett Poetzsch, E. Germany
Scott Hamilton	Elaine Zayak	1981	Scott Hamilton, U.S.	Denise Biellmann, Switzerland
Scott Hamilton	Rosalynn Sumners	1982	Scott Hamilton, U.S.	Elaine Zayak, U.S.
Scott Hamilton	Rosalynn Sumners	1983	Scott Hamilton, U.S.	Rosalynn Sumners, U.S.
Scott Hamilton	Rosalynn Sumners	1984	Scott Hamilton, U.S.	Katarina Witt, E. Germany
Brian Boitano	Tiffany Chin	1985	Aleksandr Fadeev, USSR	Katarina Witt, E. Germany
Brian Boitano	Debi Thomas	1986	Brian Boitano, U.S.	Debi Thomas, U.S.
Brian Boitano	Jill Trenary	1987	Brian Orser, Canada	Katarina Witt, E. Germany
Brian Boitano	Debi Thomas	1988	Brian Boitano, U.S.	Katarina Witt, E. Germany
Christopher Bowman	Jill Trenary	1989	Kurt Browning, Canada	Midori Ito, Japan
Todd Eldredge	Jill Trenary	1990	Kurt Browning, Canada	Jill Trenary, U.S.
Todd Eldredge	Tonya Harding	1991	Kurt Browning, Canada	Kristi Yamaguchi, U.S.
Christopher Bowman	Kristi Yamaguchi	1992	Viktor Petrenko, Ukraine	Kristi Yamaguchi, U.S.
Scott Davis	Nancy Kerrigan	1993	Kurt Browning, Canada	Oksana Baiul, Ukraine
Scott Davis	vacant[1]	1994	Elvis Stojko, Canada	Yuka Sato, Japan
Todd Eldredge	Nicole Bobek	1995	Elvis Stojko, Canada	Chen Lu, China
Rudy Galindo	Michelle Kwan	1996	Todd Eldredge, U.S.	Michelle Kwan, U.S.
Todd Eldredge	Tara Lipinski	1997	Elvis Stojko, Canada	Tara Lipinski, U.S.
Todd Eldredge	Michelle Kwan	1998	Alexei Yagudin, Russia	Michelle Kwan, U.S.
Michael Weiss	Michelle Kwan	1999	Alexei Yagudin, Russia	Maria Butyrskaya, Russia

(1) Tonya Harding was stripped of title.

SKIING
World Cup Alpine Champions

	Men					
1967	Jean Claude Killy, France	1970	Karl Schranz, Austria	1974	Piero Gros, Italy
1968	Jean Claude Killy, France	1971	Gustavo Thoeni, Italy	1975	Gustavo Thoeni, Italy
1969	Karl Schranz, Austria	1972	Gustavo Thoeni, Italy	1976	Ingemar Stenmark, Sweden
		1973	Gustavo Thoeni, Italy	1977	Ingemar Stenmark, Sweden

1978	Ingemar Stenmark, Sweden	1997	Luc Alphand, France	
1979	Peter Luescher, Switzerland	1998	Hermann Maier, Austria	
1980	Andreas Wenzel, Liechtenstein	1999	Lasse Kjus, Norway	
1981	Phil Mahre, U.S.		**Women**	
1982	Phil Mahre, U.S.	1967	Nancy Greene, Canada	
1983	Phil Mahre, U.S.	1968	Nancy Greene, Canada	
1984	Pirmin Zurbriggen, Switzerland	1969	Gertrud Gabl, Austria	
1985	Marc Girardelli, Luxembourg	1970	Michele Jacot, France	
1986	Marc Girardelli, Luxembourg	1971	Annemarie Proell, Austria	
1987	Pirmin Zurbriggen, Switzerland	1972	Annemarie Proell, Austria	
1988	Pirmin Zurbriggen, Switzerland	1973	Annemarie Proell, Austria	
1989	Marc Girardelli, Luxembourg	1974	Annemarie Proell, Austria	
1990	Pirmin Zurbriggen, Switzerland	1975	Annemarie Proell, Austria	
1991	Marc Girardelli, Luxembourg	1976	Rose Mittermaier, W. Germany	
1992	Paul Accola, Switzerland	1977	Lise-Marie Morerod, Switzerland	
1993	Marc Girardelli, Luxembourg	1978	Hanni Wenzel, Liechtenstein	
1994	Kjetil Andre Aamodt, Norway	1979	Annemarie Proell Moser, Austria	
1995	Alberto Tomba, Italy	1980	Hanni Wenzel, Liechtenstein	
1996	Lasse Kjus, Norway			

1981	Marie-Theres Nadig, Switzerland
1982	Erika Hess, Switzerland
1983	Tamara McKinney, U.S.
1984	Erika Hess, Switzerland
1985	Michela Figini, Switzerland
1986	Maria Walliser, Switzerland
1987	Maria Walliser, Switzerland
1988	Michela Figini, Switzerland
1989	Vreni Schneider, Switzerland
1990	Petra Kronberger, Austria
1991	Petra Kronberger, Austria
1992	Petra Kronberger, Austria
1993	Anita Wachter, Austria
1994	Vreni Schneider, Switzerland
1995	Vreni Schneider, Switzerland
1996	Katja Seizinger, Germany
1997	Pernilla Wiberg, Sweden
1998	Katja Seizinger, Germany
1999	Alexandra Meissnitzer, Austria

SWIMMING
World Swimming Records
(As of Oct. 1999)
Men's Records

Freestyle

Distance	Time	Holder	Country	Where made	Date
50 meters	0:21.81	Tom Jager	U.S.	Nashville, TN	Mar. 24, 1990
100 meters	0:48.21	Alexander Popov	Russia	Monte Carlo	June 18, 1994
200 meters	1:46.00	Ian Thorpe	Australia	Sydney, Australia	Aug. 24, 1999
400 meters	3:41.83	Ian Thorpe	Australia	Sydney, Australia	Aug. 22, 1999
800 meters	7:46.00	Kieren Perkins	Australia	Victoria, Canada	Aug. 24, 1994
1,500 meters	14:41.66	Kieren Perkins	Australia	Victoria, Canada	Aug. 24, 1994

Breaststroke

Distance	Time	Holder	Country	Where made	Date
50 meters	0:27.61	Alexander Dzhaburiya	Ukraine	Kharkov	Apr. 27, 1996
100 meters	1:00.60	Fred DeBurghgraeve	Belguim	Atlanta, GA	July 20, 1996
200 meters	2:10.16	Mike Barrowman	U.S.	Barcelona	July 29, 1992

Butterfly

Distance	Time	Holder	Country	Where made	Date
50 meters	0:23.68	Denis Pankratov	Russia	Mulhouse	Aug. 10, 1996
100 meters	0:52.15	Michael Klim	Australia	Brisbane, Australia	Oct. 9, 1997
200 meters	1:55.22	Denis Pankratov	Russia	Canet, France	June 14, 1995

Backstroke

Distance	Time	Holder	Country	Where made	Date
50 meters	0:24.99	Lenny Krayzelburg	U.S.	Sydney, Australia	Aug. 28, 1999
100 meters	0:53.60	Lenny Krayzelburg	U.S.	Sydney, Australia	Aug. 24, 1999
200 meters	1:55.87	Lenny Krayzelburg	U.S.	Sydney, Australia	Aug. 27, 1999

Individual Medley

Distance	Time	Holder	Country	Where made	Date
200 meters	1:58.16	Jani Sievinen	Finland	Rome	Sept. 11, 1994
400 meters	4:12.30	Tom Dolan	U.S.	Rome	Sept. 6, 1994

Medley Relay

Distance	Time	Holder	Country	Where made	Date
400 m. (4×100)	3:34.84	(Rouse, Linn, Henderson, Hall Jr.)	U.S.	Atlanta, GA	July 26, 1996

Freestyle Relays

Distance	Time	Holder	Country	Where made	Date
400 m. (4×100)	3:15.11	(Fox, Hudepohl, Olsen, Hall)	U.S.	Atlanta, GA	Aug. 12, 1995
800 m. (4×200)	7:08.79	(Thorpe, Kirby, Hackett, Klim)	Australia	Sydney, Australia	Aug. 25, 1999

Women's Records

Freestyle

Distance	Time	Holder	Country	Where made	Date
50 meters	0:24.51	Jingyi Le	China	Rome	Sept. 11, 1994
100 meters	0:54.01	Jingyi Le	China	Rome	Sept. 5, 1994
200 meters	1:56.78	Franziska Van Almsick	Germany	Rome	Sept. 6, 1994
400 meters	4:03.85	Janet Evans	U.S.	Seoul	Sept. 22, 1988
800 meters	8:16.22	Janet Evans	U.S.	Tokyo	Aug. 20, 1989
1,500 meters	15:52.10	Janet Evans	U.S.	Orlando, FL	Mar. 26, 1988

Breaststroke

Distance	Time	Holder	Country	Where made	Date
50 meters	0:30.83	Penny Heyns	South Africa	Sydney, Australia	Aug. 28, 1999
100 meters	1:06.52	Penny Heyns	South Africa	Sydney, Australia	Aug. 23, 1999
200 meters	2:23.64	Penny Heyns	South Africa	Sydney, Australia	Aug. 27, 1999

Butterfly

Distance	Time	Holder	Country	Where made	Date
50 meters	0:26.39	Anna Kammerling	Sweden	Halmstad	July 1, 1999
100 meters	0:57.88	Jenny Thompson	U.S.	Sydney, Australia	Aug. 23, 1999
200 meters	2:05.96	Mary T. Meagher	U.S.	Brown Deer, WI	Aug. 13, 1981

Backstroke

Distance	Time	Holder	Country	Where made	Date
50 meters	0:28.78	Sandra Volker	Germany	Monte Carlo	June 12, 1999
100 meters	1:00.16	Cihong He	China	Rome	Sept. 10, 1994
200 meters	2:06.62	Krisztina Egerszegi	Hungary	Athens	Aug. 25, 1991

Individual Medley

Distance	Time	Holder	Country	Where made	Date
200 meters	2:09.72	Yanyan Wu	China	Shanghai	Oct. 17, 1997
400 meters	4:34.79	Yan Chen	China	Shanghai	Oct. 17, 1997

Freestyle Relays

400 m. (4×100). . . **3:37.91**	(Jingyi Le, Shan Ying, Ying Le, Lu Bin).	China	Rome.	Sept. 7, 1994
800 m. (4×200). . . **7:55.47**	(Stellmach, Strauss, Mohring, Friedrich)	E. Germany. . .	Strasbourg, France	Aug. 18, 1987

Medley Relay

400 m. (4×100) **4:01.67**	(Cihong He, Guohong Dai, Limin Liu, Jingyi Le)	China	Rome.	Sept. 10, 1994

LACROSSE
Lacrosse Champions in 1999

World Lacrosse Championship (1998; held every 4 years)—Baltimore, Maryland, USA, July 24: U.S. 15, Canada 14 (OT).

U.S. Club Lacrosse Association Championship—Baltimore, MD, June 13: New York Athletic Club 12, Team Toyota 10.

National Lacrosse League Championship—Toronto, Ontario, Canada, April 23: Toronto defeated Rochester 13-10.

NCAA Division I Championship—College Park, MD, May 31: Virginia 12, Syracuse 10.

NCAA Division II Championship—College Park, MD, May 30: Adelphi 11, C.W. Post 8.

NCAA Division III Championship—College Park, MD, May 29: Salisbury St. 13, Middlebury (VT) College 6.

NCAA Division I All-Star Game—Saratoga Springs, NY, June 15: South 16, North 14.

National Junior College Championship—Arnold, MD, May 9: Nassau C.C. 7, C.C.B.C./Essex 4.

Women's World Cup (1997; held every 4 years)—Edogaua, Japan, May 4: USA 3, Australia 2 (OT).

NCAA Women's Division I Championship—Baltimore, MD, May 14: Maryland 16, Virginia 6.

NCAA Women's Division III Championship—Baltimore, MD, May 16: Middlebury 10, Amherst 9.

1999 NCAA Division I All-America Team

Attack: John Grant, Delaware; Greg McCavera, Georgetown; Tucker Radebaugh, Virginia; Ryan Powell, Syracuse.

Midfield: Mark Frye, Loyola; A.J Haugen, Johns Hopkins; Jay Jalbert, Virginia; Josh Sims, Princeton.

Defense: Ryan Curtis, Virginia; Rob Doerr, Johns Hopkins; Steve Card, Duke.

Goal: Mickey Jarboe, U.S. Naval Academy.

Coach of the Year: Bob Shillinglaw, Delaware.

NCAA Division I Lacrosse Champions

Year	Champion	Year	Champion	Year	Champion	Year	Champion
1971	Cornell	1979	Johns Hopkins	1986	North Carolina	1993	Syracuse
1972	Virginia	1980	Johns Hopkins	1987	Johns Hopkins	1994	Princeton
1973	Maryland	1981	North Carolina	1988	Syracuse	1995	Syracuse
1974	Johns Hopkins	1982	North Carolina	1989	Syracuse	1996	Princeton
1975	Maryland	1983	Syracuse	1990	vacated	1997	Princeton
1976	Cornell	1984	Johns Hopkins	1991	North Carolina	1998	Princeton
1977	Cornell	1985	Johns Hopkins	1992	Princeton	1999	Virginia
1978	Johns Hopkins						

RODEO
Pro Rodeo Cowboy All-Around Champions, 1977-98

Year	Winner	Money won	Year	Winner	Money won
1977	Tom Ferguson, Miami, OK.	$76,730	1988	Dave Appleton, Arlington, TX	$121,546
1978	Tom Ferguson, Miami, OK.	103,734	1989	Ty Murray, Odessa, TX.	134,806
1979	Tom Ferguson, Miami, OK.	96,272	1990	Ty Murray, Stephenville, TX	213,772
1980	Paul Tierney, Rapid City, SD	105,568	1991	Ty Murray, Stephenville, TX	244,230
1981	Jimmie Cooper, Monument, NM	105,862	1992	Ty Murray, Stephenville, TX	225,992
1982	Chris Lybbert, Coyote, CA.	123,709	1993	Ty Murray, Stephenville, TX	297,896
1983	Roy Cooper, Durant, OK	153,391	1994	Ty Murray, Stephenville, TX	246,170
1984	Dee Pickett, Caldwell, ID.	122,618	1995	Joe Beaver, Huntsville, TX	141,753
1985	Lewis Feild, Elk Ridge, UT	130,347	1996	Joe Beaver, Huntsville, TX	166,103
1986	Lewis Feild, Elk Ridge, UT	166,042	1997	Dan Mortensen, Manhattan, MT. . . .	184,559
1987	Lewis Feild, Elk Ridge, UT	144,335	1998	Ty Murray, Stephenville, TX	264,673

JAMES E. SULLIVAN MEMORIAL TROPHY WINNERS

The James E. Sullivan Memorial Trophy, named after the former president of the Amateur Athletic Union (AAU) and inaugurated in 1930, is awarded annually by the AAU to the athlete who "by his or her performance, example and influence as an amateur, has done the most during the year to advance the cause of sportsmanship."

Year	Winner	Sport	Year	Winner	Sport	Year	Winner	Sport
1930	Bobby Jones	Golf	1951	Rev. Robert Richards	Track	1972	Frank Shorter	Track
1931	Barney Berlinger	Track	1952	Horace Ashenfelter	Track	1973	Bill Walton	Basketball
1932	Jim Bausch	Track	1953	Dr. Sammy Lee	Diving	1974	Rick Wohlhutter	Track
1933	Glenn Cunningham	Track	1954	Mal Whitfield	Track	1975	Tim Shaw	Swimming
1934	Bill Bonthron	Track	1955	Harrison Dillard	Track	1976	Bruce Jenner	Track
1935	Lawson Little	Golf	1956	Patricia McCormick	Diving	1977	John Naber	Swimming
1936	Glenn Morris	Track	1957	Bobby Joe Morrow	Track	1978	Tracy Caulkins	Swimming
1937	Don Budge	Tennis	1958	Glenn Davis	Track	1979	Kurt Thomas	Gymnastics
1938	Don Lash	Track	1959	Parry O'Brien	Track	1980	Eric Heiden	Speed Skating
1939	Joe Burk	Rowing	1960	Rafer Johnson	Track	1981	Carl Lewis	Track
1940	Greg Rice	Track	1961	Wilma Rudolph Ward	Track	1982	Mary Decker	Track
1941	Leslie MacMitchell	Track	1962	James Beatty	Track	1983	Edwin Moses	Track
1942	Cornelius Warmerdam	Track	1963	John Pennel	Track	1984	Greg Louganis	Diving
1943	Gilbert Dodds	Track	1964	Don Schollander	Swimming	1985	Joan Benoit Samuelson	Marathon
1944	Ann Curtis	Swimming	1965	Bill Bradley	Basketball	1986	Jackie Joyner-Kersee	Track
1945	Doc Blanchard	Football	1966	Jim Ryun	Track	1987	Jim Abbott	Baseball
1946	Arnold Tucker	Football	1967	Randy Matson	Track	1988	Florence Griffith Joyner	Track
1947	John Kelly, Jr.	Rowing	1968	Debbie Meyer	Swimming	1989	Janet Evans	Swimming
1948	Robert Mathias	Track	1969	Bill Toomey	Track	1990	John Smith	Wrestling
1949	Dick Button	Skating	1970	John Kinsella	Swimming	1991	Mike Powell	Track
1950	Fred Wilt	Track	1971	Mark Spitz	Swimming			

Year	Winner	Sport	Year	Winner	Sport	Year	Winner	Sport
1992	Bonnie Blair	Speed Skating	1994	Dan Jansen	Speed Skating	1997	Peyton Manning	Football
1993	Charlie Ward	Football, Basketball	1995	Bruce Baumgartner	Wrestling	1998	Chamique Holdsclaw	Basketball
			1996	Michael Johnson	Track			

DIRECTORY OF SPORTS ORGANIZATIONS
Major League Baseball
Website: http://www.majorleaguebaseball.com
Note: Teams and leagues as of 1999 season.

Commissioner's Office
245 Park Ave., 31st fl.
New York, NY 10167

American League

American League Office
350 Park Ave.
New York, NY 10022

Anaheim Angels
2000 Gene Autry Way
Anaheim, CA 92806

Baltimore Orioles
333 W. Camden St.
Baltimore, MD 21201

Boston Red Sox
4 Yawkey Way
Boston, MA 02215

Chicago White Sox
333 W. 35th St.
Chicago, IL 60616

Cleveland Indians
2401 Ontario St.
Cleveland, OH 44115

Detroit Tigers
2121 Trumbull Ave.
Detroit, MI 48216

Kansas City Royals
1 Royal Way
Kansas City, MO 64141

Minnesota Twins
34 Kirby Puckett Place
Minneapolis, MN 55415

New York Yankees
Yankee Stadium
Bronx, NY 10451

Oakland Athletics
7677 Oakport, Suite 200
Oakland, CA 94621

Seattle Mariners
83 South King St.
Seattle, WA 98104

Tampa Bay Devil Rays
One Tropicana Dr.
St. Petersburg, FL 33705

Texas Rangers
1000 Ballpark Way
Arlington, TX 76011

Toronto Blue Jays
1 Blue Jays Way, #3200
Toronto, Ont. M5V 1J1

National League

National League Office
350 Park Ave.
New York, NY 10022

Arizona Diamondbacks
401 E. Jefferson St.
Phoenix, AZ 85004

Atlanta Braves
755 Hank Aaron Drive
Atlanta, GA 30315

Chicago Cubs
1060 W. Addison St.
Chicago, IL 60613

Cincinnati Reds
100 Cinergy Field
Cincinnati, OH 45202

Colorado Rockies
2001 Blake St.
Denver, CO 80205

Florida Marlins
2267 NW 199th St.
Miami, FL 33056

Houston Astros
PO Box 288
Houston, TX 77001

Los Angeles Dodgers
1000 Elysian Park Ave.
Los Angeles, CA 90012

Milwaukee Brewers
P.O. Box 3099
Milwaukee, WI 53201

Montreal Expos
4549 Ave. Pierre de Coubertin
Montreal, Que. H1V 3N7

New York Mets
123-01 Roosevelt Ave.
Flushing, NY 11368

Philadelphia Phillies
3501 S. Broad St.
Philadelphia, PA 19148

Pittsburgh Pirates
PO Box 7000
Pittsburgh, PA 15212

St. Louis Cardinals
250 Stadium Plaza
St. Louis, MO 63102

San Diego Padres
8880 Rio San Diego Dr.
San Diego, CA 92112

San Francisco Giants
3Com Park at Candlestick Pt.
San Francisco, CA 94124

National Basketball Association
Website: http://www.nba.com

League Office
Olympic Tower
645 5th Ave.
New York, NY 10022

Atlanta Hawks
One CNN Center, Ste. 405
South Tower
Atlanta, GA 30303

Boston Celtics
151 Merrimac St.
Boston, MA 02114

Charlotte Hornets
100 Hive Dr.
Charlotte, NC 28217

Chicago Bulls
1901 W. Madison St.
Chicago, IL 60612

Cleveland Cavaliers
1 Center Court
Cleveland, OH 44115

Dallas Mavericks
777 Sports St.
Dallas, TX 75207

Denver Nuggets
Pepsi Center
1000 Chopper Pl.
Denver, CO 80204

Detroit Pistons
Two Championship Dr.
Auburn Hills, MI 48326

Golden State Warriors
1011 Broadway
Oakland, CA 94607

Houston Rockets
Two Greenway Plaza, Ste. 400
Houston, TX 77046

Indiana Pacers
125 S. Pennsylvania St.
Indianapolis, IN 46204

Los Angeles Clippers
1111 S. Figueroa St.,
Ste. 1100
Los Angeles, CA 90037

Los Angeles Lakers
555 Nash St.
El Segundo, CA 90245

Miami Heat
SunTrust Intl. Center
One SE 3d Ave., Ste. 2300
Miami, FL 33131

Milwaukee Bucks
1001 N. 4th St.
Milwaukee, WI 53203

Minnesota Timberwolves
600 1st Ave. North
Minneapolis, MN 55403

New Jersey Nets
Nets Champion Center
390 Murray Hill Parkway
E. Rutherford, NJ 07073

New York Knickerbockers
Two Pennsylvania Plaza
New York, NY 10121

Orlando Magic
Two Magic Place
8701 Maitland Summit Blvd.
Orlando, FL 32801

Philadelphia 76ers
First Union Center
3601 S. Broad St.
Philadelphia, PA 19148

Phoenix Suns
201 E. Jefferson
Phoenix, AZ 85004

Portland Trail Blazers
One Center Ct., Ste. 200
Portland, OR 97227

Sacramento Kings
One Sports Parkway
Sacramento, CA 95834

San Antonio Spurs
100 Montana St.
San Antonio, TX 78203

Seattle SuperSonics
190 Queen Anne Ave.
Suite 200
Seattle, WA 98109

Toronto Raptors
40 Bay St., Ste. 400
Toronto, Ont. M5J 2X2

Utah Jazz
301 W. South Temple
Salt Lake City, UT 84101

Vancouver Grizzlies
800 Griffiths Way
Vancouver, B.C. V6B 6G1

Washington Wizards
718 7th St., NW
Washington, DC 20004

National Hockey League
Website: http://www.nhl.com

League Headquarters
1251 Ave. of the Americas
New York, NY 10020

Mighty Ducks of Anaheim
2695 E. Katella Ave.
Anaheim, CA 92803

Atlanta Thrashers
1 CNN Center, Box 105583
Atlanta GA 30348

Boston Bruins
One FleetCenter, Ste. 250
Boston, MA 02114

Buffalo Sabres
Marine Midland Arena
One Seymour H. Knox III Plaza
Buffalo, NY 14203

Calgary Flames
PO Box 1540, Station M
Calgary, Alta. T2P 3B9

Carolina Hurricanes
5000 Aerial Ctr., Ste. 100
Morrisville, NC 27560

Chicago Blackhawks
1901 W. Madison St.
Chicago, IL 60612

Colorado Avalanche
 1635 Clay St.
 Denver, CO 80204

Dallas Stars
 211 Cowboys Parkway
 Irving, TX 75063

Detroit Red Wings
 600 Civic Center Dr.
 Detroit, MI 48226

Edmonton Oilers
 11230 110 St.
 Edmonton, Alta. T5B 4M9

Florida Panthers
 One Panther Parkway
 Sunrise, FL 33323

Los Angeles Kings
 555 N. Nash St.
 El Segundo, CA 90245

Montreal Canadiens
 1260 rue de La Gauchetiere
  Ouest
 Montreal, Que. H3B 5E8

Nashville Predators
 501 Broadway
 Nashville, TN 37203

New Jersey Devils
 50 Rte. 120 N.
 PO Box 504
 E. Rutherford, NJ 07073

New York Islanders
 Nassau Veterans Memorial
  Coliseum
 Uniondale, NY 11553

New York Rangers
 Two Pennsylvania Plaza
 New York, NY 10121

Ottawa Senators
 1000 Palladium Dr.
 Kanata, Ont. K2V 1A5

Philadelphia Flyers
 First Union Center
 3601 South Broad St.
 Philadelphia, PA 19148

Phoenix Coyotes
 Cellular One Ice Den
 9375 E. Bell Rd.
 Scottsdale, AZ 85260

Pittsburgh Penguins
 Civic Arena
 66 Mario Lemieux Place
 Pittsburgh, PA 15219

St. Louis Blues
 1401 Clark
 St. Louis, MO 63103

San Jose Sharks
 525 W. Santa Clara St.
 San Jose, CA 95113

Tampa Bay Lightning
 401 Channelside Dr.
 Tampa, FL 33602

Toronto Maple Leafs
 Air Canada Centre
 40 Bay St.
 Toronto, Ont. M5J 2X2

Vancouver Canucks
 800 Griffiths Way
 Vancouver, B.C. V6B 6G1

Washington Capitals
 601 F St. NW
 Washington, DC 20004

National Football League
Website: http://www.nfl.com

League Office
 280 Park Avenue
 New York, NY 10017

Arizona Cardinals
 PO Box 888
 Phoenix, AZ 85001

Atlanta Falcons
 One Falcon Place
 Suwanee, GA 30024

Baltimore Ravens
 11001 Owings Mills Blvd.
 Owings Mills, MD 2117

Buffalo Bills
 One Bills Drive
 Orchard Park, NY 14127

Carolina Panthers
 800 S. Mint St.
 Charlotte, NC 28202-1502

Chicago Bears
 Halas Hall at Conway Pk.
 1000 Football Dr.
 Lake Forest, IL 60045

Cincinnati Bengals
 One Bengals Dr.
 Cincinnati, OH 45204

Cleveland Browns
 76 Lou Groza Blvd.
 Berea, OH 44017

Dallas Cowboys
 One Cowboys Parkway
 Irving, TX 75063

Denver Broncos
 13655 Broncos Parkway
 Englewood, CO 80112

Detroit Lions
 1200 Featherstone Rd.
 Pontiac, MI 48342

Green Bay Packers
 P.O. Box 10628
 Green Bay, WI 54307

Indianapolis Colts
 PO Box 535000
 Indianapolis, IN 46253

Jacksonville Jaguars
 One ALLTELL Stadium
  Place
 Jacksonville, FL 32202

Kansas City Chiefs
 One Arrowhead Drive
 Kansas City, MO 64129

Miami Dolphins
 7500 SW 30th St.
 Davie, FL 33314

Minnesota Vikings
 9520 Viking Dr.
 Eden Prairie, MN 55344

New England Patriots
 60 Washington St.
 Foxboro, MA 02035

New Orleans Saints
 5800 Airline Drive
 Metairie, LA 70003

New York Giants
 Giants Stadium
 E. Rutherford, NJ 07073

New York Jets
 1000 Fulton Ave.
 Hempstead, NY 11550

Oakland Raiders
 1220 Harbor Bay Parkway
 Alameda, CA 94502

Philadelphia Eagles
 3501 S. Broad St.
 Philadelphia, PA 19148

Pittsburgh Steelers
 300 Stadium Circle
 Pittsburgh, PA 15212

St. Louis Rams
 One Rams Way
 St. Louis County, MO
  63045

San Diego Chargers
 PO Box 609609
 San Diego, CA 92160

San Francisco 49ers
 4949 Centennial Blvd.
 Santa Clara, CA 95054

Seattle Seahawks
 11220 NE 53d St.
 Kirkland, WA 98033

Tampa Bay Buccaneers
 One Buccaneer Place
 Tampa, FL 33607

Tennessee Titans
 460 Great Circle Rd.
 Nashville, TN 37228

Washington Redskins
 PO Box 17247
 Dulles Internat'l Airport
 Washington, DC 20041

Other Sports Organizations

Amateur Athletic Union
 PO Box 10000
 Lake Buena Vista, FL 32830
 http://www.aausports.org

Amateur Softball Assn.
 2801 NE 50th St.
 Oklahoma City, OK 73111
 http://www.softball.org

American Horse Shows Assn.
 220 E. 42d St.
 New York, NY 10017
 http://www.ahsa.org

American Kennel Club
 51 Madison Ave.
 New York, NY 10010
 http://www.akc.org

Canadian Football League
 110 Eglinton Ave. W
 Toronto, Ont. M4R 1A3
 http://www.cfl.ca

CART
 755 W. Big Beaver Rd.
 Troy, MI 48084
 http://www.cart.com

Intl. Game Fish Assn.
 1301 E. Atlantic Blvd.
 Pompano Beach, FL 33060
 http://www.igfa.org

LPGA
 100 International Golf Dr.
 Daytona Beach, FL 32124
 http://www.lpga.com

Little League Baseball
 PO Box 3485
 Williamsport, PA 17701
 http://www.littleleague.org

Major League Soccer
 110 E. 42d St., Ste. 1000
 New York, NY 10017
 http://www.mlsnet.com

NASCAR
 1801 Intl. Speedway Blvd.
 Daytona Beach, FL 32120
 http://www.nascar.com

NCAA
 6201 College Blvd.
 Overland Park, KS 66211
 http://www.ncaa.org

National Rifle Assn.
 11250 Waples Mill Rd.
 Fairfax, VA 22030
 http://www.nra.org

Pro Bowlers Assn.
 PO Box 5118
 Akron, OH 44334
 http://www.pbatour.com

PGA
 100 Ave. of the Champions
 Palm Beach Gardens, FL
  33410
 http://www.pgaonline.com

Pro Rodeo Cowboys Assn.
 101 Pro Rodeo Dr.
 Colorado Springs, CO 80919
 http://www.prorodeo.com

Special Olympics
 1325 G St., NW, Ste. 500
 Washington, DC 20005
 http://www.
  specialolympics.org

Thoroughbred Racing Assns.
 420 Fair Hill Dr.
 Elkton, MD 21921
 http://www.traofna.com

USA Track & Field
 1 RCA Dome, Ste. 140
 Indianapolis, IN 46225
 http://www.usatf.org

U.S. Auto Club
 4910 W. 16th St.
 Speedway, IN 46224

U.S. Figure Skating Assn.
 20 First St.
 Colorado Springs, CO 80906
 http://www.usfsa.org

U.S. Olympic Committee
 One Olympic Plaza
 Colorado Springs, CO 80909
 http://www.olympic-usa.org

U.S. Skiing Assn.
 PO Box 100
 Park City, UT 84060
 http://www.usskiteam.com

U.S. Soccer Federation
 1801 S. Prairie Ave.
 Chicago, IL 60616
 http://www.us-soccer.com

USA Swimming
 One Olympic Plaza
 Colorado Springs, CO 80909
 http://www.usa-swimming.
  org

U.S. Tennis Assn.
 70 W. Red Oak Lane
 White Plains, NY 10604
 http://www.usta.com

U.S. Trotting Assn.
 750 Michigan Ave.
 Columbus, OH 43215
 http://www.ustrotting.com

WNBA
 645 5th Ave.
 New York, NY 10022
 http://www.wnba.com

NOTABLE SPORTS PERSONALITIES

Henry (Hank) Aaron, b. 1934: Milwaukee-Atlanta outfielder; hit record 755 home runs, led NL 4 times; record 2,297 RBIs.

Kareem Abdul-Jabbar, b. 1947: Milwaukee, L.A. Lakers center; MVP 6 times; all-time leading NBA scorer.

Andre Agassi, b. 1970: 1st man since 1938 to win career Grand Slam: Wimbledon, 1992; U.S. Open, 1994; Austral., 1995; French, 1999.

Troy Aikman, b. 1966: quarterback, led Dallas Cowboys to Super Bowl wins in 1993-94, 1996; Super Bowl MVP, 1993.

Grover Cleveland Alexander (1887-1950): pitcher; won 374 NL games; pitched 16 shutouts, 1916.

Muhammad Ali, b. 1942: 3-time heavyweight champion.

Mario Andretti, b. 1940: won Indy 500, 1969; Grand Prix champ, 1978.

Eddie Arcaro, (1916-97): only jockey to win racing's Triple Crown, 1941,1948; rode 4,779 winners in his career.

Henry Armstrong (1912-88): boxer held feather-, welter-, light-weight titles simultaneously, 1937-38.

Arthur Ashe (1943-93): U.S. singles champ, 1968; Wimbledon champ, 1975.

Red Auerbach, b. 1917: coached Boston Celtics to 9 NBA championships.

Donovan Bailey, b. 1967: Canadian Olympic gold medalist in 100 meters (set world record), 1996.

Ernie Banks, b. 1931: Chicago Cubs slugger; hit 512 NL homers; twice MVP.

Roger Bannister, b. 1929: British physician, ran first sub 4-minute mile, May 6, 1954 (3 min. 59.4 sec.).

Rick Barry, b. 1944: NBA scoring leader, 1967; ABA, 1969.

Sammy Baugh, b. 1914: Washington Redskins quarterback; held numerous records upon retirement after 16 pro seasons.

Elgin Baylor, b. 1934: L.A. Lakers forward; 1st team all-star 10 times.

Boris Becker, b. 1967: German tennis star; won U.S. Open 1989; Wimbledon champ 3 times.

Jean Beliveau, b. 1931: Montreal Canadiens center; scored 507 goals; twice MVP.

Johnny Bench, b. 1947: Cincinnati Reds catcher; MVP twice; led league in home runs twice, RBIs 3 times.

Patty Berg, b. 1918: won more than 80 golf tournaments; AP Woman Athlete-of-the-Year 3 times.

Yogi Berra, b. 1925: N.Y. Yankees catcher; MVP 3 times; played in 14 World Series.

Matt Biondi, b. 1965: swimmer, won 5 golds, 1988 Olympics.

Larry Bird, b. 1956: Boston Celtics forward; chosen MVP 1984-86; 1998 coach of the year with Indiana Pacers.

George Blanda, b. 1927: quarterback, kicker; 26 years as active player, scoring record 2,002 points.

Wade Boggs, b. 1958: AL batting champ, 1983, 1985-88; reached 3,000 career hits, 1999 (3,010).

Barry Bonds, b. 1964: outfielder was NL MVP 1990, 1992-93; only player in 400 home runs/400 stolen bases club.

Bjorn Borg, b. 1956: led Sweden to first Davis Cup, 1975; Wimbledon champion 5 times.

Mike Bossy, b. 1957: N.Y. Islanders right wing scored more than 50 goals 8 times.

Ray Bourque, b. 1960: Boston Bruins defenseman, won Norris Trophy 5 times.

Terry Bradshaw, b. 1948: Pittsburgh Steelers quarterback, led team to 4 Super Bowl titles.

George Brett, b. 1953: Kansas City Royals infielder, led AL in batting, 1976, 1980, 1990; MVP, 1980.

Lou Brock, b. 1939: St. Louis Cardinals outfielder, stole NL record 118 bases, 1974; led NL 8 times.

Jim Brown, b. 1936: Cleveland Browns fullback, ran for 12,312 career yards; MVP 3 times.

Paul Brown (1908-91): football owner, coach; led Cleveland Browns to 3 NFL championships.

Paul "Bear" Bryant (1913-83): college football coach with 323 victories.

Sergei Bubka, b. 1963: Ukrainian pole vaulter; first to clear 20 feet; gold medal, 1988 Olympics.

Maria Bueno, b. 1939: U.S. singles champ 4 times; Wimbledon champ 3 times.

Dick Butkus, b. 1942: Chicago Bears linebacker, twice chosen best NFL defensive player.

Dick Button, b. 1929: figure skater; won 1948, 1952 Olympic gold medals; world titlist, 1948-52.

Roy Campanella (1921-93): Brooklyn Dodgers catcher; MVP 3 times.

Earl Campbell, b. 1955: NFL running back; MVP 1978-80.

Rod Carew, b. 1945: AL infielder; won 7 batting titles; MVP, 1977.

Steve Carlton, b. 1944: NL pitcher; won 20 games 5 times, Cy Young award 4 times.

Billy Casper, b. 1931: PGA Player-of-the-Year 3 times; U.S. Open champ twice.

Wilt Chamberlain (1936-99): center; was NBA leading scorer 7 times, MVP 4 times; scored 100 pts. in a game, 1962.

Bobby Clarke, b. 1949: Philadelphia Flyers center; led team to 2 Stanley Cup championships; MVP 3 times.

Roger Clemens, b. 1962: AL pitcher; AL MVP 1986; only pitcher to win Cy Young 5 times, 1986, 1987, 1991, 1997, 1998; twice struck out record 20 batters in a game.

Roberto Clemente (1934-72): Pittsburgh Pirates outfielder; won 4 batting titles; MVP, 1966.

Ty Cobb (1886-1961): Detroit Tigers outfielder; had record .367 lifetime batting average, 12 batting titles.

Sebastian Coe, b. 1956: Briton won, Olympic 1,500-meter run, 1980, 1984.

Nadia Comaneci, b. 1961: Romanian gymnast, won 3 gold medals, achieved 7 perfect scores, 1976 Olympics.

Maureen Connolly (1934-69): won tennis "grand slam," 1953; AP Woman-Athlete-of-the-Year 3 times.

Jimmy Connors, b. 1952: U.S. singles champ 5 times; Wimbledon champ twice.

James J. Corbett (1866-1933): heavyweight champion, 1892-97; credited with being the first "scientific" boxer.

Angel Cordero, b. 1942: leading money winner, 1976, 1982-83; rode 3 Kentucky Derby winners.

Margaret Smith Court, b. 1942: Australian tennis great, won 24 grand slam events.

Bob Cousy, b. 1928: Boston Celtics guard; led team to 6 NBA championships; MVP, 1957.

Bjoern Daehlie, b. 1967: Norwegian cross-country skier; won record 8 Winter Olympic gold medals.

Dizzy Dean (1911-74): colorful pitcher for St. Louis Cardinals "Gashouse Gang" in the 30s; MVP, 1934.

Oscar De La Hoya, b. 1972: won lightweight title, 1995; super lightweight title, 1996; welterweight title, 1997.

Jack Dempsey (1895-1983): heavyweight champ, 1919-26.

Gail Devers, b. 1966: sprinter, won Olympic gold medals in 100-meter run, 1992, 1996.

Eric Dickerson, b. 1960: running back ran for NFL record 2,105 yds., 1984; led NFC 3 times, AFC twice.

Joe DiMaggio (1914-99): N.Y. Yankees outfielder; hit safely in record 56 consecutive games, 1941; AL MVP 3 times.

Dale Earnhardt, b. 1951: auto racer; NASCAR Winston Cup champion 7 times.

Stefan Edberg, b. 1966: U.S. singles champ, 1991, 1992; Wimbledon champ, 1988, 1990.

Gertrude Ederle, b. 1906: first woman to swim English Channel, broke existing men's record, 1926.

John Elway, b. 1960: quarterback; led Denver Broncos to 2 Super Bowl wins, 1998, 1999; regular-season MVP, 1987.

Julius Erving, b. 1950: MVP and leading scorer in ABA 3 times; NBA MVP, 1981.

Phil Esposito, b. 1942: NHL scoring leader 5 times.

Janet Evans, b. 1971: swimmer, won 3 Olympic gold medals, 1988, 1 in 1992.

Chris Evert, b. 1954: U.S. Open tennis champ 6 times, Wimbledon champ 3 times.

Ray Ewry (1873-1937): track-and-field star, won 8 gold medals, 1900, 1904, and 1908 Olympics.

Nick Faldo, b. 1957: won Masters, British Open 3 times each.

Juan Fangio (1911-95): World Grand Prix champ 5 times.

Brett Favre, b. 1969: led Green Bay Packers to Super Bowl win, 1997; NFL regular-season MVP, 1995, 1996; co-MVP, 1997.

Bob Feller, b. 1918: Cleveland Indians pitcher; won 266 games; pitched 3 no-hitters, 12 one-hitters.

Peggy Fleming, b. 1948: world figure skating champion, 1966-68; gold medalist, 1968 Olympics.

Whitey Ford, b. 1928: N.Y. Yankees pitcher, won record 10 World Series games.

George Foreman, b. 1949: heavyweight champion, 1973-74, 1994-95; at 45, the oldest to win a heavyweight title.

Dick Fosbury, b. 1947: high jumper; won 1968 Olympic gold medal; developed the "Fosbury Flop."

Jimmie Foxx (1907-67): Red Sox, Athletics slugger; MVP 3 times; triple crown, 1933.

A. J. Foyt, b. 1935: won Indy 500 4 times; U.S. Auto Club champ 7 times.

Joe Frazier, b. 1944: heavyweight champion, 1970-73.

Lou Gehrig (1903-41): N.Y. Yankees 1st baseman; MVP, 1927, 1936; triple crown, 1934; AL record 184 RBIs, 1931.

George Gervin, b. 1952: top NBA scorer, 1978-80, 1982.

Althea Gibson, b. 1927: twice U.S. and Wimbledon singles champ.

Bob Gibson, b. 1935: St. Louis Cardinals pitcher; won Cy Young award twice; struck out 3,117 batters.

Marc Girardelli, b. 1963: Luxembourg skier, won 5 World Cup titles.

Jeff Gordon, b. 1971: race car driver, youngest to win NASCAR Winston Cup 3 times (1995, 1997-98).

Steffi Graf, b. 1969: German; won tennis "grand slam," 1988; U.S. champ 5 times; Wimbledon champ 7 times.

Otto Graham, b. 1921: Cleveland Browns quarterback; all-pro 4 times.

Red Grange (1903-91): All-America at Univ. of Illinois, 1923-25; played for Chicago Bears, 1925-35.

Joe Greene, b. 1946: Pittsburgh Steelers lineman; twice NFL outstanding defensive player.

Wayne Gretzky, b. 1961: leading scorer in NHL history with record 894 goals, 1,963 assists, 2,857 points; MVP, 1980-87, 1989.

Ken Griffey Jr., b. 1969: Seattle Mariner outfielder; led AL in home runs, 1994, 1997, 1998, 1999; 1997 AL MVP; 9 gold gloves.

Florence Griffith Joyner, (1959-98): sprinter; won 3 gold medals at 1988 Olympics; Olympic record for 100m.

Lefty Grove (1900-75): pitcher; won 300 AL games; 20-game winner 8 times.

Tony Gwynn, b. 1960: 8-time NL batting champ, 1984, 1987-89, 1994-97; reached 3,000 career hits, 1999 (3,067).

Walter Hagen (1892-1969): won PGA championship 5 times; British Open 4 times.

George Halas (1895-1983): founder-coach of Chicago Bears; won 5 NFL championships.

Scott Hamilton, b. 1958: U.S. and world figure skating champion, 1981-84; Olympic gold medalist, 1984.

Bill Hartack, b. 1932: jockey, rode 5 Kentucky Derby winners.

Dominik Hasek, b. 1965: Buffalo Sabres goalie; won Vezina Trophy, 1994-95, 1997-99; NHL MVP, 1997-98.

John Havlicek, b. 1940: Boston Celtics forward scored more than 26,000 NBA points.

Rickey Henderson, b. 1958: outfielder stole record 130 bases, 1982; record lifetime steals; AL MVP, 1990.

Sonja Henie (1912-69): world champion figure skater, 1927-36; Olympic gold medalist, 1928, 1932, 1936.

Martina Hingis, b. 1980: youngest woman to hold No. 1 tennis ranking; won Australian Open 1997-99; Wimbledon, U.S. Open, 1997.

Ben Hogan (1912-97): won 4 U.S. Open championships, 2 PGA, 2 Masters.

Evander Holyfield, b. 1962: 3-time heavyweight champion.

Rogers Hornsby (1896-1963): NL 2d baseman; batted record .424 in 1924; twice won triple crown; batting leader, 1920-25.

Paul Hornung, b. 1935: Green Bay Packers runner-placekicker, scored record 176 points, 1960.

Gordie Howe, b. 1928: hockey forward; NHL MVP 6 times; scored 801 goals in 26 NHL seasons.

Carl Hubbell (1903-88): N.Y. Giants pitcher; 20-game winner 5 consecutive years, 1933-37.

Bobby Hull, b. 1939: NHL all-star 10 times; MVP, 1965-66.

Brett Hull, b. 1964: St. Louis Blues forward; led NHL in goals, 1990-92; MVP, 1991.

Catfish Hunter (1946-99): pitched perfect game, 1968; 20-game winner 5 times.

Don Hutson (1913-97): Green Bay Packers receiver, caught 99 NFL touchdown passes.

Reggie Jackson, b. 1946: slugger; led AL in home runs 4 times; MVP, 1973; hit 5 World Series home runs, 1977.

Earvin (Magic) Johnson, b. 1959: NBA MVP, 1987, 1989, 1990; Playoff MVP, 1980, 1982, 1987; 2d in career assists.

Jack Johnson (1878-1946): heavyweight champion, 1910-15.

Michael Johnson, b. 1967: 1996 Olympic gold medalist and world record-holder in 200-meter and 400-meter runs.

Walter Johnson (1887-1946): Washington Senators pitcher; won 416 games; record 110 shutouts.

Bobby Jones (1902-71): won "grand slam of golf" 1930; U.S. Amateur champ 5 times, U.S. Open champ 4 times.

Michael Jordan, b. 1963: NBA leading scorer, 1987-93, 1996-98; MVP, 1988, 1991-92, 1996, 1998; Playoff MVP, 1991-93, 1996-98.

Jackie Joyner-Kersee, b. 1962: Olympic gold medalist in heptathlon, 1988, 1992.

Duke Kahanamoku (1890-1968): swimmer; won 1912, 1920 Olympic gold medals in 100-meter freestyle; surfing pioneer.

Harmon Killebrew, b. 1936: Minnesota Twins slugger; led AL in home runs 6 times; 573 lifetime.

Jean Claude Killy, b. 1943: French skier; 3 1968 Olympic golds.

Ralph Kiner, b. 1922: Pittsburgh Pirates slugger, led NL in home runs 7 consecutive years, 1946-52.

Billie Jean King, b. 1943: U.S. singles champ 4 times; Wimbledon champ 6 times.

Bob Knight, b. 1940: Indiana U. basketball coach, led team to NCAA championships, 1976, 1981, 1987.

Olga Korbut, b. 1955: Soviet gymnast; 3 1972 Olympic golds.

Sandy Koufax, b. 1935: Dodgers pitcher; won Cy Young award 3 times; lowest ERA in NL, 1962-66; pitched 4 no-hitters, one a perfect game.

Guy Lafleur, b. 1951: forward; led NHL in scoring 3 times; MVP, 1977, 1978.

Tom Landry, b. 1924: Dallas Cowboys head coach, 1960-88.

Rod Laver, b. 1938: Australian; won tennis "grand slam" twice, 1962, 1969; Wimbledon champ 4 times.

Mario Lemieux, b. 1965: 6-time NHL leading scorer; MVP, 1988, 1993, 1996; Playoff MVP, 1991-92.

Ivan Lendl, b. 1960: U.S. Open tennis champ, 1985-87.

Sugar Ray Leonard, b. 1956: boxer, held titles in 5 different weight classes.

Carl Lewis, b. 1961: track-and-field star, won 9 Olympic gold medals in sprinting and the long jump.

Tara Lipinski, b. 1982: youngest figure skater to win U.S. and world championships, 1997, and Winter Olympic gold, 1998.

Vince Lombardi (1913-70): Green Bay Packers coach, led team to 5 NFL championships and 2 Super Bowl victories.

Greg Louganis, b. 1960: won Olympic gold medals in both springboard and platform diving, 1984, 1988.

Joe Louis (1914-81): heavyweight champion, 1937-49.

Sid Luckman (1916-98): Chicago Bears quarterback; led team to 4 NFL championships; MVP, 1943.

Connie Mack (1862-1956): Philadelphia Athletics manager, 1901-50; won 9 pennants, 5 championships.

Greg Maddux, b. 1966: NL pitcher, won 4 consecutive Cy Young awards, 1992-95.

Karl Malone, b. 1963: Utah Jazz forward; was MVP, 1997, 1999; 11-time All-Star; 25,000+ career points.

Moses Malone, b. 1955: NBA center, MVP, 1979, 1982-83.

Mickey Mantle (1931-95): N.Y. Yankees outfielder; triple crown, 1956; 18 World Series home runs; MVP 3 times.

Pete Maravich (1948-88): guard, scored NCAA record 44.2 ppg during collegiate career; led NBA in scoring, 1977.

Rocky Marciano (1923-69): heavyweight champion, 1952-56; retired undefeated.

Dan Marino, b. 1961: Miami Dolphins quarterback; passed for NFL record 5,084 yds and 48 touchdowns, 1984; career records for touchdowns, yds passing, completions.

Roger Maris (1934-85): N.Y. Yankees outfielder; hit AL record 61 home runs, 1961; MVP, 1960 and 1961.

Eddie Mathews, b. 1931: Milwaukee-Atlanta 3d baseman, hit 512 career home runs.

Christy Mathewson (1880-1925): N.Y. Giants pitcher, won 373 games.

Bob Mathias, b. 1930: decathlon gold medalist, 1948, 1952.

Willie Mays, b. 1931: N.Y.-S.F. Giants center fielder; hit 660 home runs, led NL 4 times; had 3,283 hits; twice MVP.

Willie McCovey, b. 1938: S.F. Giants slugger; hit 521 home runs; led NL 3 times; MVP, 1969.

John McEnroe, b. 1959: U.S. Open tennis champ, 1979-81, 1984; Wimbledon champ, 1981, 1983-84.

John McGraw (1873-1934): N.Y. Giants manager, led team to 10 pennants, 3 championships.

Mark McGwire, b. 1963: hit single season record 70 home runs (1998); 500 career home runs in fewest at-bats (5,487); finished 1999 with 522.

Mark Messier, b. 1961: center, chosen NHL MVP, 1990, 1992; Conn Smythe Trophy, 1984.

George Mikan, b. 1924: Minn. Lakers center, considered the best basketball player of the first half of the century.

Stan Mikita, b. 1940: Chicago Black Hawks center, led NHL in scoring 4 times; MVP twice.

Joe Montana, b. 1956: S.F. 49ers quarterback; Super Bowl MVP, 1982, 1985, 1990.

Archie Moore (1913-99): light-heavyweight champ, 1952-62.

Howie Morenz (1902-37): Montreal Canadiens forward, considered best hockey player of first half of the century.

Eddie Murray, b. 1956: durable slugger; 3d player to combine 3,000+ hits with 500+ home runs.

Stan Musial, b. 1920: St. Louis Cardinals star; won 7 NL batting titles; MVP 3 times.

Bronko Nagurski (1908-90): Chicago Bears fullback and tackle; gained more than 4,000 yds. rushing.

Joe Namath, b. 1943: N.Y. Jets quarterback; Super Bowl MVP, 1969.

Martina Navratilova, b. 1956: Wimbledon champ 9 times, U.S. champ 1983-84, 1986-87.

Byron Nelson, b. 1912: won 11 consecutive golf tournaments in 1945; twice Masters and PGA titlist.

Ernie Nevers (1903-76): Stanford star, selected as best college fullback to play between 1919-69.

John Newcombe, b. 1943: Australian; twice U.S. Open tennis champ; Wimbledon titlist 3 times.

Jack Nicklaus, b. 1940: PGA Player-of-the-Year, 1967, 1972; leading money winner 8 times; won 6 Masters.

Chuck Noll, b. 1931: coach, led Pittsburgh Steelers to 4 Super Bowl titles.

Paavo Nurmi (1897-1973): Finnish distance runner, won 6 Olympic gold medals, 1920, 1924, 1928.

Al Oerter, b. 1936: discus thrower, won gold medal at 4 consecutive Olympics, 1956-68.

Hakeem Olajuwon, b. 1963: Houston Rockets center; NBA MVP, 1994, Playoffs MVP, 1994-95; career leader in blocked shots.

Shaquille O'Neal, b. 1972: NBA center; rookie of the year, 1993; scoring leader, 1995.

Bobby Orr, b. 1948: Boston Bruins defenseman; Norris Trophy 8 times; led NHL in scoring twice, assists 5 times.

Mel Ott (1909-1958): N.Y. Giants outfielder hit 511 home runs; led NL 6 times.

Jesse Owens (1913-80): track and field star, won 4 1936 Olympic gold medals.

Satchel Paige (1906-82): pitcher, starred in Negro leagues, 1924-48; entered major leagues at age 42.

Se Ri Pak, b. 1977: youngest winner ever (at 20) of LPGA Championship and of U.S. Women's Open, 1998.

Arnold Palmer, b. 1929: golf's first $1 million winner; won 4 Masters, 2 British Opens.

Jim Palmer, b. 1945: Baltimore Orioles pitcher; Cy Young award 3 times; 20-game winner 8 times.

Joe Paterno, b. 1926: winningest active NCAA football coach; led Penn St. to 2 national titles, 1982, 1986.

Floyd Patterson, b. 1935: twice heavyweight champion.

Walter Payton (1954-1999): Chicago Bears running back; most rushing yards in NFL history; top NFC rusher, 1976-80.

Pele, b. 1940: Brazilian soccer star, scored 1,281 goals during 22-year career.

Bob Pettit, b. 1932: first NBA player to score 20,000 points; twice NBA scoring leader.

Richard Petty, b. 1937: NASCAR national champ 7 times; 7-time Daytona 500 winner.

Laffit Pincay Jr., b. 1946: leading money-winning jockey, 1970-74, 1979, 1985.

Jacques Plante (1929-86): goalie; 7 Vezina trophies; first goalie to wear a mask in a game.

Kirby Puckett, b. 1961: Minn. Twins outfielder; won AL batting title, 1989; led AL in hits, 1987-89, 1992; RBIs, 1994.

Willis Reed, b. 1942: N.Y. Knicks center; MVP, 1970; Playoff MVP, 1970, 1973.

Jerry Rice, b. 1962: S.F. 49ers receiver; Super Bowl MVP, 1989; NFL record for career touchdowns, receptions.

Maurice Richard, b. 1921: Montreal Canadiens forward scored 544 regular season goals, 82 playoff goals.

Branch Rickey (1881-1965): executive; helped break baseball's color barrier, 1947; initiated farm system, 1919.

Pat Riley, b. 1945: coached L.A. Lakers to 4 NBA titles.

Cal Ripken Jr., b. 1960: Baltimore Orioles shortstop; AL MVP 1983, 1991; most consecutive games played, 2,632.

Oscar Robertson, b. 1938: guard; averaged career 25.7 points per game; 3d most career assists; MVP, 1964.

Brooks Robinson, b. 1937: Baltimore Orioles 3d baseman; played in 4 World Series; MVP, 1964; 16 gold gloves.

Frank Robinson, b. 1935: slugger; was MVP in both NL and AL; triple crown, 1966; 586 lifetime home runs; first black manager in majors.

Jackie Robinson (1919-72): broke baseball's color barrier with Brooklyn Dodgers, 1947; MVP, 1949.

Sugar Ray Robinson (1920-89): middleweight champion 5 times, welterweight champion.

Knute Rockne (1888-1931): Notre Dame football coach, 1918-31; revolutionized game by stressing forward pass.

Dennis Rodman, b. 1961: eccentric forward, led NBA in rebounding 1991-98.

Pete Rose, b. 1941: won 3 NL batting titles; hit safely in 44 consecutive games, 1978; has most career hits, 4,256.

Patrick Roy, b. 1965: Montreal-Colorado goalie; was 3-time Vezina Trophy winner, Playoffs MVP, 1993.

Wilma Rudolph (1940-94): sprinter, won 3 1960 Olympic golds.

Adolph Rupp (1901-77): NCAA basketball coach; led Kentucky to 4 national titles, 1948-49, 1951, 1958.

Bill Russell, b. 1934: Boston Celtics center, led team to 11 NBA titles; MVP 5 times; first black coach of major pro sports team.

Babe Ruth (1895-1948): N.Y. Yankees outfielder; hit 60 home runs, 1927; 714 lifetime; led AL 12 times.

Johnny Rutherford, b. 1938: auto racer, won 3 Indy 500s.

Nolan Ryan, b. 1947: pitcher; struck out record 383 batters, 1973; record 5,714 career; pitched record 7 no-hitters; won 324 major league games.

Pete Sampras, b. 1971: tennis star; 1st man in Open era to win 6 Wimbledons; tied for most career Grand Slam wins (12).

Barry Sanders, b. 1968: rushed for 2,053 yards in 1997; led NFL in rushing, 1990, 1994, 1996, 1997.

Gene Sarazen, (1902-99): won PGA championship 3 times, U.S. Open twice; developed the sand wedge.

Gale Sayers, b. 1943: Chicago Bears back, twice led NFL in rushing.

Mike Schmidt, b. 1949: Phillies 3d baseman; led NL in home runs 8 times; 548 lifetime; NL MVP, 1980, 1981, 1986.

Tom Seaver, b. 1944: pitcher; won NL Cy Young award 3 times; won 311 major league games.

Monica Seles, b. 1973: won U.S. Open, 1991-92; Australian Open, 1991-93, 1996; French Open, 1990-92.

Willie Shoemaker, b. 1931: jockey; rode 4 Kentucky Derby and 5 Belmont Stakes winners; leading career money winner.

Eddie Shore (1902-85): Boston Bruins defenseman; MVP 4 times, first-team all-star 7 times.

Don Shula, b. 1930: all-time winningest NFL coach.

Al Simmons (1902-56): AL outfielder batted .334 lifetime.

O. J. Simpson, b. 1947: running back; rushed for 2,003 yds., 1973; AFC leading rusher 4 times.

George Sisler (1893-1973): St. Louis Browns 1st baseman; had record 257 hits, 1920; batted .340 lifetime.

Dean Smith, b. 1931: North Carolina basketball coach; has most career NCAA Division I victories.

Emmitt Smith, b. 1969: Dallas Cowboys running back; led NFL in rushing, 1991-93, 1995; NFL and Super Bowl MVP, 1993; record 25 rushing touchdowns, 1995.

Lee Smith, b. 1957: relief pitcher, all-time saves leader, 478.

Sam Snead, b. 1912: PGA and Masters champ 3 times each.

Sammy Sosa, b. 1968: Chicago Cubs outfielder; 1998 NL MVP; 66 home runs, 1998; 63 home runs, 1999.

Warren Spahn, b. 1921: pitcher; won 363 NL games; 20-game winner 13 times; Cy Young award, 1957.

Tris Speaker (1885-1958): AL outfielder; batted .344 over 22 seasons; hit record 793 career doubles.

Mark Spitz, b. 1950: swimmer, won 7 1972 Olympic golds.

Amos Alonzo Stagg (1862-1965): coached Univ. of Chicago football team for 41 years, including 5 undefeated seasons; introduced huddle, man-in-motion, and end-around play.

Bart Starr, b. 1934: Green Bay Packers quarterback, led team to 5 NFL titles and 2 Super Bowl victories.

Roger Staubach, b. 1942: Dallas Cowboys quarterback; leading NFC passer 5 times.

Casey Stengel (1890-1975): managed Yankees to 10 pennants, 7 championships, 1949-60.

Jackie Stewart, b. 1939: Scot auto racer, retired with 27 Grand Prix victories.

Payne Stewart (1957-99): golfer; won 2 U.S. Opens (1991, 1999), 1 PGA Championship (1989).

John Stockton, b. 1962: Utah Jazz guard; NBA career leader in assists, steals; NBA assists leader, 1988-96.

John L. Sullivan (1858-1918): last bareknuckle heavyweight champion, 1882-1892.

Fran Tarkenton, b. 1940: quarterback, 2d in career touchdown passes.

Lawrence Taylor, b. 1959; linebacker; led N.Y. Giants to 2 Super Bowl titles; played in 10 Pro Bowls.

Frank Thomas, b. 1968: Chicago White Sox 1st baseman; was AL MVP, 1993-94; won AL batting title, 1997.

Jim Thorpe (1888-1953): football All-America, 1911, 1912; won pentathlon and decathlon, 1912 Olympics.

Bill Tilden (1893-1953): U.S. singles champ 7 times; played on 11 Davis Cup teams.

Y. A. Tittle, b. 1926: N.Y. Giants quarterback; MVP, 1961, 1963.

Lee Trevino, b. 1939: golfer, won U.S., British Open twice.

Bryan Trottier, b. 1956: center, played on 6 Stanley Cup championship teams.

Mike Tyson, b. 1966: Undisputed world heavyweight champ, 1987-1990; at 19, youngest to win a heavyweight title (WBC, 1986).

Johnny Unitas, b. 1933: Baltimore Colts quarterback; passed for more than 40,000 yds; MVP, 1957, 1967.

Al Unser, b. 1939: Indy 500 winner 4 times.

Bobby Unser, b. 1934: Indy 500 winner 3 times.

Norm Van Brocklin (1926-83): quarterback; passed for game record 554 yds., 1951; MVP, 1960.

Honus Wagner (1874-1955): Pittsburgh Pirates shortstop, won 8 NL batting titles.

Tom Watson, b. 1949: golfer, won British Open 5 times.

Johnny Weissmuller (1903-84): swimmer; won 52 national championships, 5 Olympic gold medals; set 67 world records.

Jerry West, b. 1938: L.A. Lakers guard; had career average 27 points per game; first team all-star 10 times.

Reggie White, b. 1961: defensive end, all-time NFL sack leader.

Kathy Whitworth, b. 1939: women's golf leading money winner 8 times; first woman to earn more than $300,000.

Lenny Wilkens, b. 1937: winningest coach in NBA history; in Hall of Fame as player and coach.

Ted Williams, b. 1918: Boston Red Sox outfielder; won 6 batting titles, two triple crowns; hit .406 in 1941.

Helen Willis Moody (1905-98): tennis star; won U.S. Open 7 times, Wimbledon 8 times.

Katarina Witt, b. 1965: German figure skater; won Olympic gold medal, 1984, 1988.

John Wooden, b. 1910: coached UCLA basketball team to 10 national championships.

Tiger Woods, b. 1975: only male golfer to win 3 consecutive U.S. Amateur titles, 1994-96; won Masters, 1997; PGA, 1999.

Mickey Wright, b. 1935: won LPGA championship 4 times, Vare Trophy 5 times; twice AP Woman-Athlete-of-the-Year.

Carl Yastrzemski, b. 1939: Boston Red Sox slugger; won 3 batting titles; triple crown, 1967.

Cy Young (1867-1955): pitcher, won record 511 games.

Steve Young, b. 1961: 49ers quarterback; led NFL in passing, 1991-94, 1996, 1997; Super Bowl MVP, 1995.

Babe Didrikson Zaharias (1914-56): track star; won 2 1932 Olympic gold medals; won numerous golf tournaments.

QUICK REFERENCE INDEX

QUICK REFERENCE SPORTS INDEX

For complete Index, see pages 4-32.